The Palgrave Encyclopedia of Urban and Regional Futures

Robert C. Brears
Editor

The Palgrave Encyclopedia of Urban and Regional Futures

Volume 1

A–G

With 418 Figures and 143 Tables

palgrave
macmillan

Editor
Robert C. Brears
Our Future Water
Christchurch, New Zealand

ISBN 978-3-030-87744-6 ISBN 978-3-030-87745-3 (eBook)
https://doi.org/10.1007/978-3-030-87745-3

This Palgrave Macmillan imprint is published by the registered company Springer Nature Switzerland AG.
The registered company address is: Gewerbestrasse 11, 6330 Cham, Switzerland

Preface

The *Palgrave Encyclopedia of Urban and Regional Futures* provides readers (practitioners, academics, researchers, etc.) with expert interdisciplinary knowledge on how urban centers and regions in locations of varying climates, lifestyles, income levels, and stages development are creating synergies and reducing trade-offs in the development of resilient, resource-efficient, environmentally friendly, liveable, socially equitable, integrated, and technology-enabled centers and regions. In particular, the *Palgrave Encyclopedia of Urban and Regional Futures* provides chapters, authored by subject matter experts, on interdisciplinary policies, best practices, lessons learnt, technologies in various stages of development, and case studies of urban centers and regions that aim to decouple economic growth from resource consumption, enhance resilience to climatic extremes, invest in low/zero carbon and smart technologies, lower emissions, reduce economic disparities, improve quality of life, and protect ecosystems and the services they provide for humans and nature.

Christchurch, New Zealand
December 2022

Robert C. Brears
Editor

Acknowledgments

First, I wish to thank Ruth Lefevre and Rachael Ballard for being visionaries who enable Major Reference Works like mine to come to fruition. Second, I wish to thank Anusha Cherian for being an excellent project coordinator. Third, I wish to thank Mum, who has a great interest in the environment and has supported me in this journey.

List of Topics

About the Editor

Robert C. Brears is an international sectoral expert on water for the UN's Green Climate Fund and the World Bank. He is the Editor in Chief of the *Palgrave Handbook of Climate Resilient Societies* and the *Palgrave Encyclopedia of Urban and Regional Futures*. He is the author of 11 books, including the Palgrave Macmillan titles *The Green Economy and the Water-Energy-Food Nexus*, *Blue and Green Cities: The Role of Blue-Green Infrastructure in Managing Urban Water Resources*, *Natural Resource Management and the Circular Economy*, *Developing the Circular Water Economy*, *Developing the Blue Economy*, and *Financing Nature-based Solutions*. He is the founder of Our Future Water, which has knowledge partnerships with World Bank, World Meteorological Organization, and UNEP initiatives.

Contributors

Ayodeji Adeniyi Deception Bay, QLD, Australia

Humera Afaq National University, San Diego, CA, USA

Kristin Agnello Department of Architecture and Planning, Norwegian University of Science and Technology, Trondheim, Norway

Atharv Agrawal University of Toronto, Toronto, ON, Canada

S. Ahilan College of Engineering, Mathematics and Physical Sciences, University of Exeter, Exeter, Devon, UK

Iftekhar Ahmed School of Architecture and Built Environment, University of Newcastle, Callaghan, NSW, Australia

Mubeen Ahmad School of Earth and Environmental Sciences, The University of Queensland, Brisbane, QLD, Australia

Ahmad Ahsan Lahore University of Management Sciences, Lahore, Pakistan

Meredian Alam Sociology and Anthropology Department, Universiti Brunei Darussalam, Gadong, Brunei

Amani Alfarra Land and Water Division, Food and Agriculture Organization of the United Nations, Rome, Italy

Jamal Alibou Department of Civil Engineering, Hydraulic, Environment and Climate, Hassania School of Public Works, Casablanca, Morocco

Sara Alidoust School of Earth and Environmental Sciences, The University of Queensland, Brisbane, QLD, Australia

Angélica Tanus Benatti Alvim Graduate Program in Architecture and Urbanism, Mackenzie Presbyterian University, Sao Paulo, Brazil

P. Ambily Department of Civil Engineering, National Institute of Technology, Calicut, Kerala, India

Grace Andrews Masters Environmental Management, College of Humanities, Arts and Social Sciences, Flinders University, Adelaide, South Australia

Jenna Andrews-Swann School of Liberal Arts, Georgia Gwinnett College, Lawrenceville, GA, USA

T. Angert The Institute for Environmental Security and Well-being Studies, Herzliya, Israel

Shyni Anilkumar National Institute of Technology Calicut, Kozhikode, Kerala, India

Aikaterini Antonopoulou

Hadi Arbabi Department of Civil & Structural Engineering, The University of Sheffield, Sheffield, UK

Md. Arfanuzzaman Food and Agriculture Organization (FAO) of the United Nations, Dhaka, Bangladesh

Felipe Armas Vargas Departamento de Ingeniería de Procesos e Hidráulica, CBI, Universidad Autónoma Metropolitana-Iztapalapa, Ciudad de México, Mexico

S. Arthur Heriot-Watt University, Edinburgh, UK

Hedda Askland University of Newcastle, Callaghan, NSW, Australia

Ditjon Baboci Tirana, Albania

Guy Baeten Urban Studies, Malmö University, Malmö, Sweden

Elham Bahmanteymouri The University of Auckland, Auckland, New Zealand

Nilesh Bakshi School of Architecture, Victoria University of Wellington, Wellington, New Zealand

M. Balasubramanian Centre for Ecological Economics and Natural Resources, Institute for Social and Economic Change, Bangalore, Karnataka, India

Zoran Balukoski School of Geography and Sustainable Communities, The University of Wollongong, Wollongong, NSW, Australia

Jonathan Banfield University of Toronto, Toronto, ON, Canada

Kaya Barry Griffith University, Brisbane, QLD, Australia

Department of Culture and Learning, Aalborg University, Aalborg, Denmark

Matthias Barth Leuphana University, Lüneburg, Germany

Prabal Barua Department of Environmental Sciences, Faculty of Physical and Mathematical Sciences, Jahangirnagar University, Dhaka, Bangladesh

James A. Beckman University of Central Florida, Orlando, FL, USA

Sara Bice Crawford School of Public Policy, The Australian National University, Acton, ACT, Australia

School of Public Policy and Management, Tsinghua University, Beijing, China

S. Birkinshaw University of Newcastle, Newcastle, UK

Stefan Blachfellner The Bertalanffy Center for the Study of Systems Science, Vienna, Austria

Bruno Blanco-Varela Department of Applied Economics, Faculty of Economics, Universidade de Santiago de Compostela, Santiago de Compostela, Galicia, Spain

Tijana Blanusa Royal Horticultural Society, Wisley, UK

University of Reading, Reading, UK

Tinashe Bobo Town Planning Section, Harare City Council, Harare, Zimbabwe

Cherice Bock Portland Seminary of George Fox University, Portland, OR, USA

Antonija Bogadi Department of Urban and Spatial Planning and Research, Technical University Vienna, Vienna, Austria

Simone Borelli Food and Agriculture Organization of the United Nations (FAO), Rome, Italy

Candice Boyd University of Melbourne, Melbourne, VIC, Australia

Christopher T. Boyko Lancaster University, Lancaster, UK

Robert C. Brears Our Future Water, Christchurch, New Zealand

Maria Julieta Brezzo Institutional Relations and Events, Ciudades Globales – CIGLO, Córdoba, Argentina

Katja Brundiers School of Sustainability, Arizona State University, Tempe, AZ, USA

Valerio Alfonso Bruno Università Cattolica del Sacro Cuore, Milan, Italy

Center for European Futures, Naples, Italy

Centre for the Analysis of the Radical Right, Leeds, UK

Felipe Bucci Ancapi Department of Management in the Built Environment, Faculty of Architecture and the Built Environment, Delft University of Technology, Delft, The Netherlands

Felix Bücken Institute of Geography, Osnabrück University, Osnabrück, Germany

Paul Burton Cities Research Institute Griffith University, Gold Coast, QLD, Australia

Alessandro Busà School of Geography, Geology and the Environment, University of Leicester, Leicester, UK

Judy Bush Lecturer in Urban Planning at University of Melbourne, Melbourne, VIC, Australia

Gareth Butler Masters Environmental Management, College of Humanities, Arts and Social Sciences, Flinders University, Adelaide, South Australia

Andrew Butt RMIT University, Melbourne, VIC, Australia

Michael Buxton RMIT University, Melbourne, VIC, Australia

Mohammed Firoz C. Department of Architecture and Planning, National Institute of Technology Calicut, Kozhikode, Kerala, India

Maléne Campbell Department of Urban and Regional Planning, University of the Free State, Bloemfontein, South Africa

Julien Carbonnell Artificial Intelligence on Citizen Engagement, Democracy Studio

M. Cavada School of Architecture, Imagination Lancaster, Lancaster University, Lancaster, UK

Rebecca Cavicchia Department of Urban and Regional Planning, BYREG – Norwegian University of Life Science, Ås, Norway

Lauriane Suyin Chalmin-Pui Royal Horticultural Society, Wisley, UK

The University of Sheffield, Sheffield, UK

Deborah Nabubwaya Chambers Community Health, National University, San Diego, CA, USA

Shenglin E. Chang National Taiwan University, Graduate Institute of Building and Planning, Taipei, Taiwan

Marianna Charitonidou Department of Art Theory and History, Athens School of Fine Arts, Athens, Greece

School of Architecture, National Technocal University of Athens, Athens, Greece

Department of Architecture, ETH Zurich, Zurich, Switzerland

Charles M. Chavunduka Department of Architecture and Real Estate, University of Zimbabwe, Harare, Zimbabwe

Ambika Chawla Urban Climate Innovations, Washington, DC, USA

Fei Chen School of Architecture, University of Liverpool, Liverpool, UK

Andrew Chigudu Department of Demography Settlement and Development, Social & Behavioural Sciences, University of Zimbabwe, Harare, Zimbabwe

Halleluah Chirisa Population Services International Zimbabwe, Harare, Zimbabwe

Innocent Chirisa Department of Demography Settlement and Development, Social & Behavioural Sciences, University of Zimbabwe, Harare, Zimbabwe

Chipo Chitereka Department of Social Work, University of Zimbabwe, Harare, Zimbabwe

N. R. Chithra Department of Civil Engineering, National Institute of Technology, Calicut, Kerala, India

Marcyline Chivenge Department of Demography Settlement and Development, Social & Behavioural Sciences, University of Zimbabwe, Harare, Zimbabwe

Suehyun Cho University of Toronto, Toronto, ON, Canada

Tanya Clark School of Behavioral Sciences, California Southern University, Costa Mesa, CA, USA

M'Lisa Lee Colbert The Nature of Cities, Montreal, QC, Canada

Ramon Fernando Colmenares-Quintero Faculty of Engineering, Universidad Cooperativa de Colombia, Medellín, Colombia

Elif Çolakoğlu Department of Security Sciences, Gendarmerie and Coast Guard Academy, Ankara, Turkey

Michela Conigliaro Food and Agriculture Organization of the United Nations (FAO), Rome, Italy

Sean Connelly University of Otago, Dunedin, New Zealand

A. Contin Politecnico di Milano, Milan, Italy

Rachel Cooper Lancaster University, Lancaster, UK

Samantha Copeland Ethics and Philosophy of Technology, Delft University of Technology, Delft, The Netherlands

João Cortesão Landscape Architecture and Spatial Planning, Wageningen University, Wageningen, The Netherlands

Adriano Cozzolino Center for European Futures, Naples, Italy

Università degli Studi della Campania "Luigi Vanvitelli", Caserta, Italy

Stewart Craine Village Infrastructure Angels, London, UK

Roberta Cucca BYREG – Norwegian University of Life Science, Ås, Norway

Gary Cummisk Central Washington University, Ellensburg, WA, USA

Paul Cureton ImaginationLancaster, Lancaster University, Lancaster, UK

Susan Cyriac Department of Architecture and Planning, National Institute of Technology, Calicut, Kerala, India

Sebastien Darchen School of Earth and Environmental Sciences, The University of Queensland, Brisbane, QLD, Australia

Curt J. Davis University of Delaware, Newark, DE, USA

D. Dawson University of Leeds, Leeds, UK

Evelyne de Leeuw Centre for Health Equity Training, Research and Evaluation (CHETRE), UNSW Australia Research Centre for Primary Health Care & Equity, South Western Sydney Local Health District, Ingham Institute, Sydney, NSW, Australia

Healthy Urban Environments (HUE) Collaboratory, Maridulu Budyari Gumal Sydney Partnership for Health, Education, Research and Enterprise SPHERE, Sydney, NSW, Australia

Ingham Institute for Applied Medical Research, Liverpool, NSW, Australia

Valerio Della Sala Politecnico di Torino (Italy), Interdepartmental Research Centre for Urban Studies (OMERO), Universitat Autonoma de Barcelona (Spain), Turin, Italy

N. Delle-Odeleye Anglia Ruskin University, Chelmsford, UK

Cheryl Desha Cities Research Institute, Griffith University, Brisbane, QLD, Australia

María Mercedes Di Virgilio Instituto de Investigaciones Gino Germani, Universidad de Buenos Aires/ CONICET, Ciudad Autónoma de Buenos Aires, Argentina

Roshini Suparna Diwakar Mahila Housing Trust, New Delhi, India

Timothy J. Dixon School of the Built Environment, University of Reading, Reading, UK

Michelle Duffy University of Newcastle, Callaghan, NSW, Australia

Smart Dumba Department of Demography Settlement and Development, Social & Behavioral Sciences, University Zimbabwe, Harare, Zimbabwe

Nick Dunn Lancaster University, Lancaster, UK

Jenna Dutton Senior Planner – Social Policy, City of Victoria and Research Associate, Center for Civilization, University of Calgary, Calgary, AB, Canada

Vupenyu Dzingirai Department of Community and Social Development, University of Zimbabwe, Harare, Zimbabwe

Charity Edwards Monash University & University of Melbourne, Melbourne, Australia

Huascar Eguino Fiscal Management Division, Inter-American Development Bank (IDB), Washington, DC, USA

Theodore S. Eisenman Department of Landscape Architecture and Regional Planning, University of Massachusetts-Amherst, Amherst, MA, USA

Christina R. Ergler School of Geography, University of Otago, Dunedin, New Zealand

Oscar Escolero Departamento de Dinámica Terrestre y Superficial, Instituto de Geología, Universidad Nacional Autónoma de México, Ciudad de México, Mexico

Javier Esquer Graduate Sustainability Program, Industrial Engineering Department, University of Sonora, Hermosillo, Mexico

G. Everett University of the West of England, Bristol, UK

Caroline Fabianski La Seyne sur Mer, France

Francesco Femia The Center for Climate and Security, an Institute of the Council on Strategic Risks, Washington, DC, USA

Melisha Shavindi Fernando Faculty of Science, Horizon Campus, Malabe, Sri Lanka

Carmen Zuleta Ferrari Food and Agriculture Organization of the United Nations (FAO), Rome, Italy

Carla Sofia Ferreira Research Centre for Natural Resources, Environment and Society (CERNAS), Polytechnic Institute of Coimbra, Coimbra Agrarian Technical School, Coimbra, Portugal

Department of Physical Geography and Bolin Centre for Climate Research, Stockholm University, Stockholm, Sweden

Navarino Environmental Observatory, Messinia, Greece

António Ferreira Research Centre for Natural Resources, Environment and Society (CERNAS), Polytechnic Institute of Coimbra, Coimbra Agrarian Technical School, Coimbra, Portugal

Daniel Fischer School of Sustainability, Arizona State University, Tempe, AZ, USA

Wesley Flannery Urban Planning, School of Natural and Built Environment, David Keir Building, Queen's University Belfast, Belfast, UK

Claudia Fonseca Alfaro Institute for Urban Research, Malmö University, Malmö, Sweden

Mariana Fonseca Braga ImaginationLancaster, Lancaster Institute for the Contemporary Arts (LICA), Lancaster University, Lancaster, Lancashire, UK

Julicn Forbat Univcrsity of Gcncva, Institutc of Global Hcalth, Gcncva, Switzerland

Martin Franz Institute of Geography, Osnabrück University, Osnabrück, Germany

Robert Freestone School of Built Environment, University of New South Wales, Sydney, NSW, Australia

Frances Furio School of Behavioral Sciences, California Southern University, Costa Mesa, CA, USA

Tatiana Gallego Lizon Washington, DC, USA

Emilio Garcia The University of Auckland, Auckland, New Zealand

Birgit Georgi UIA Expert/Strong Cities in a Changing Climate, Egelsbach, Germany

Daniela Getlinger Graduate Program in Architecture and Urbanism, Mackenzie Presbyterian University, Sao Paulo, Brazil

David J. Gilchrist University of Western Australia, Perth, WA, Australia

Brendan Gleeson Monash University & University of Melbourne, Melbourne, Australia

V. Glenis University of Newcastle, Newcastle, UK

Moritz Gold Sustainable Food Processing Laboratory, ETH Zurich, Zurich, Switzerland

Eugenio Gómez Reyes Departamento de Ingeniería de Procesos e Hidráulica, CBI, Universidad Autónoma Metropolitana-Iztapalapa, Ciudad de México, Mexico

Megan Gordon University of Northern British Columbia, Prince George, BC, Canada

Alexa Gower Monash University, Melbourne, VIC, Australia

Sonia Graham School of Geography and Sustainable Communities, The University of Wollongong, Wollongong, NSW, Australia

Institut de Ciència I Tecnologia Ambientals (ICTA), Universitat Autònoma de Barcelona, Barcelona, Spain

Danielle Griego Center for Augmented Computational Design in Architecture, Engineering and Construction, D-BAUG, ETH Zurich, Zurich, Switzerland

Kai Michael Griese Hochschule Osnabrück University of Applied Sciences, Osnabrück, Germany

Carl Grodach Monash University, Melbourne, VIC, Australia

Bern Grush Urban Robotics Foundation, Toronto, Canada

Medhisha Pasan Gunawardena Biodiversity Educational Research Initiative, Colombo, Sri Lanka

Faculty of Science, Horizon Campus, Malabe, Sri Lanka

Hector Manuel Guzman Grijalva Sustainability Graduate Program, University of Sonora, Hermosillo, México

Jochen Hack Technical University of Darmstadt, Section of Ecological Engineering, Institute of Applied Geosciences, Darmstadt, Germany

Perrine Hamel Asian School of the Environment, Nanyang Technological University, Singapore, Singapore

Earth Observatory of Singapore, Nanyang Technological University, Singapore, Singapore

Ben Harris-Roxas School of Population Health, University of New South Wales, Sydney, NSW, Australia

Wolfgang Haupt Leibniz-Insitute for Research on Society and Space, Erkner, Germany

Naomi Hay Australian National University, Canberra, ACT, Australia

Fatime Barbara Hegyi Joint Research Centre – European Commission, Seville, Spain

Hayley Henderson Research Fellow at Crawford School of Public Policy, Australian National University, Canberra, ACT, Australia

Michael Henderson Ramboll Ltd and Oxford Brookes University, London, UK

Cole Hendrigan University of Wollongong and Wollongong City Council, Wollongong, NSW, Australia

Andreas Hernandez Marymount Manhattan College, New York, NY, USA

Victoria Herrmann The Arctic Institute – Center for Circumpolar Security Studies, Washington, DC, USA

Halima Hodzic Food and Agriculture Organization of the United Nations (FAO), Rome, Italy

Karen Horwood The Leeds Planning School, Leeds Beckett University, Leeds, UK

Mette Hotker RMIT University, Melbourne, VIC, Australia

Karin Huber-Heim Circular Economy Forum, Austria, Vienna, Austria

Raisa Binte Huda Department of Geography and Environment, University of Dhaka, Dhaka, Bangladesh

Dan Xuan Thi Huynh School of Economics, Can Tho University, Can Tho, Vietnam

Ligocka Ilona Ministry of Climate and Environment, Warsaw, Poland

Tanya Gottlieb Jacobsen State of Green, Copenhagen, Denmark

Bhanye Johannes Department of Community and Social Development, University of Zimbabwe, Harare, Zimbabwe

Katrina Johnston-Zimmerman THINK.urban, Philadelphia, PA, USA

Kirsty Jones Crawford School of Public Policy, The Australian National University, Acton, ACT, Australia

Alain Jordà Local Development Expert, Manresa, Barcelona, Spain

Gaurav Joshi University of Chinese Academy of Sciences, Beijing, China

Anuja Joy National Institute of Technology Calicut, Kozhikode, Kerala, India

Mahjabin Kabir Adrita Department of Geography and Environment, University of Dhaka, Dhaka, Bangladesh

Zahra Kalantari Department of Physical Geography and Bolin Centre for Climate Research, Stockholm University, Stockholm, Sweden

Navarino Environmental Observatory, Messinia, Greece

Department of Sustainable Development, Environmental Science and Engineering, KTH Royal Institute of Technology, Stockholm, Sweden

Eleni Kalantidou Griffith University, Brisbane, QLD, Australia

Tinashe Natasha Kanonhuhwa Department of Demography Settlement and Development, Social & Behavioral Sciences, University Zimbabwe, Harare, Zimbabwe

L. Kapetas 100 resilient Cities Project, New York, USA

Thomas Karakadzai Department of Demography Settlement and Development, Faculty of Social & Behavioral Sciences, University of Zimbabwe, Harare, Zimbabwe

Abdulrazak Karriem University of the Western Cape, Cape Town, South Africa

Hewa Thanthrige Ashan Randika Karunananda Biodiversity Educational Research Initiative, Colombo, Sri Lanka

Rosemary Kasimba Department of Demography Settlement and Development, University of Zimbabwe, Harare, Zimbabwe

J. O. Kawira County Government of Laikipia, Laikipia, Kenya

Jon Kellett University of Adelaide, Adelaide, SA, Australia

Vlada Kenniff Long Island University, Brookville, NY, USA

Jeffrey Kenworthy Curtin University Sustainability Policy Institute, Curtin University, Perth, WA, Australia

Frankfurt University of Applied Sciences, Frankfurt am Main, Germany

Ganesh Keremane Adelaide, South Australia

Tien Dung Khong School of Economics, Can Tho University, Can Tho, Vietnam

Teng Chye Khoo National University of Singapore, Singapore, Singapore

F. I. Kihara The Nature Conservancy, Nairobi, Kenya

Lorenzo Kihlgren Grandi City Diplomacy Lab, Columbia Global Centers | Paris, Paris, France

C. Kilsby University of Newcastle, Newcastle, UK

Jinhee Kim Centre for Health Equity Training, Research and Evaluation (CHETRE), UNSW Australia Research Centre for Primary Health Care & Equity, South Western Sydney Local Health District, Ingham Institute, Sydney, NSW, Australia

Michael Koh Centre for Liveable Cities, Ministry of National Development, Singapore, Singapore

Victoria Kolankiewicz Faculty of Architecture, Building and Planning, University of Melbourne, Melbourne, VIC, Australia

Weichang Kong The University of Queensland, Brisbane, QLD, Australia

Mrudhula Koshy Norwegian University of Science and Technology, Trondheim, Norway

Maria Kottari School of Transnational Governance, European University Institute, Florence, Italy

Daniel Kozak Universidad de Buenos Aires, Consejo Nacional de Investigaciones CientÃficas y Técnicas (CONICET), Buenos Aires, Argentina

Teresa Kramarz University of Toronto, Toronto, ON, Canada

Tamara Krawchenko University of Victoria, Victoria, BC, Canada

Peleg Kremer Department of Geography and the Environment, Villanova University, Villanova, PA, USA

V. Krivtsov The Royal Botanic Garden, Edinburgh, UK

Arvind Kumar India Water Foundation, New Delhi, India

Gerard Kuperus University of San Francisco, San Francisco, CA, USA

Sigrid Kusch-Brandt Department of Civil, Environmental and Architectural Engineering, University of Padua, Padua, Italy

Faculty of Mathematics, Natural Sciences and Management, University of Applied Sciences Ulm, Ulm, Germany

Ndarova Audrey Kwangwama Department of Architecture and Real Estate, University of Zimbabwe, Harare, Zimbabwe

Oliver Lah Wuppertal Institute for Climate, Environment and Energy, Berlin, Germany

Urban Electric Mobility Initiative (UEMI) a UN-Habitat Action Platform, Berlin, Germany

Khee Poh Lam National University of Singapore, Singapore, Singapore

J. Lamond University of the West of England, Bristol, UK

Martin Larbi Kwame Nkrumah University of Science and Technology, Kumasi, Ghana

Alexander Laszlo The Bertalanffy Center for the Study of Systems Science, Buenos Aires, Argentina

Lucie Laurian School of Planning and Public Affairs, The University of Iowa, Iowa City, IA, USA

Alison Lee Centre for Liveable Cities, Ministry of National Development, Singapore, Singapore

Steffen Lehmann School of Architecture, University of Nevada, Las Vegas, NV, USA

Carlos Leite School of Architecture and Urbanism, Mackenzie Presbyterian University, Sao Paulo, Brazil

Social Urbanism Center, Insper´s Arq.Futuro Cities Lab, Sao Paulo, Brazil

Caitlin Anthea Lewis Architecture Planning and Geomatics, University of Cape Town, Cape Town, South Africa

Nora Libertun de Duren Inter-American Development Bank, Washington, DC, USA

Jade Lindley Law School and Oceans Institute, The University of Western Australia, Crawley, WA, Australia

Yan Liu School of Earth and Environmental Sciences, The University of Queensland, St Lucia, Australia

Adam Loch Centre for Global Food and Resources, School of Economics and Public Policy, Faculty of the Professions, University of Adelaide, Adelaide, SA, Australia

Aynaz Lotfata Department of Geography, Chicago State University, Chicago, IL, USA

Pavel Luksha Global Education Futures, Moscow, Russia

Mengxing Ma Department of Social Work, University of Melbourne, Melbourne, VIC, Australia

Department of Geography, University of Sheffield, Sheffield, UK

Danielle MacCarthy Queen's University Belfast, Belfast, Northern Ireland, UK

Shamiso Hazel Mafuku Department of Architecture and Real Estate, University of Zimbabwe, Harare, Zimbabwe

Kamilia Mahdaoui Hassania School of Public Works, Casablanca, Morocco

Israa H. Mahmoud Laboratorio di Simulazione Urbana Fausto Curti, Department of Architecture and Urban Studies, Politecnico di Milano, Milan, Italy

David Mainenti Palmer iSchool of Library and Information Studies, Long Island University, Brookville, NY, USA

Innocent Maja Faculty of Law, University of Zimbabwe, Harare, Zimbabwe

Soumaya Majdoub Research Group Interface Demography, Department of Sociology, VUB Free University of Brussels, Brussels, Belgium

Brussels Center for Urban Studies (BCUS), Brussels, Belgium

Brussels Interdisciplinary Research Centre for Migration and Minorities (BIRMM), Brussels, Belgium

George Makunde George Makunde Institute, Harare, Zimbabwe

Eleanor Malbon University of New South Wales, Kensington, NSW, Australia

Wendy W. Mandaza-Tsoriyo Department of Rural and Urban Development, Great Zimbabwe University, Harare, Zimbabwe

Manfredo Manfredini School of Architecture and Planning, The University of Auckland, Shanghai University, Auckland, New Zealand

Elton Manjeya Department of Architecture and Real Estate, University of Zimbabwe, Harare, Zimbabwe

Jonathan Manns Rockwell, London, UK

UCL, London, UK

Patrick M. Marchman American Society of Adaptation Professionals/ Climigration Network, Kansas City, MO, USA

Age Mariussen University of Vaasa, Vaasa, Finland

Cecilia Marocchino Food and Agriculture Organization of the United Nations (FAO), Rome, Italy

Andresa Ledo Marques Graduate Program in Architecture and Urbanism, Mackenzie Presbyterian University, Sao Paulo, Brazil

Institute of Urban Design and Planning, Leibniz Universität, Hannover, Germany

Martha Marriner State of Green, Copenhagen, Denmark

Stephen Marshall Bartlett School of Planning, University College London, London, UK

Natalia Martsinovich Department of Chemistry, University of Sheffield, Sheffield, UK

Nesbert Mashingaidze Department of Rural and Urban Development, Great Zimbabwe University, Masvingo, Zimbabwe

Jeofrey Matai Department of Architecture and Real Estate, University of Zimbabwe, Harare, Zimbabwe

Abraham R. Matamanda Department of Urban and Regional Planning, University of the Free State, Bloemfontein, South Africa

Marina Matashova Andorra-LAB, Forward Consulting Group, Barcelona, Spain

Brilliant Mavhima Department of Architecture and Real Estate, University of Zimbabwe, Harare, Zimbabwe

Patience Mazanhi Department of Demography Settlement and Development, Social & Behavioral Sciences, University of Zimbabwe, Harare, Zimbabwe

Chad J. McGuire Department of Public Policy, University of Massachusetts, Dartmouth, MA, USA

Matthew H. McLeskey Department of Sociology, University at Buffalo, State University of New York, Buffalo, NY, USA

Wendy McWilliam School of Landscape Architecture, Faculty of Environment, Society and Design, Lincoln University, Lincoln, New Zealand

Ojilve Ramón Medrano Pérez CONACYT-Centro del Cambio Global y la Sustentabilidad, A.C. (CCGS), Villahermosa, Tabasco, Mexico

Asma Mehan Senior Researcher, CITTA Research Institute, Faculty of Engineering (FEUP), University of Porto, Porto, Portugal

Mahziar Mehan School of Urban Planning, Faculty of Fine Arts, University of Tehran, Tehran, Iran

Prakhar Mehta Digital Transformation: Bits to Energy Lab Nuremberg, School of Business, Economics and Society, Friedrich-Alexander University Erlangen-Nürnberg (FAU), Nuremberg, Germany

Lorena Melgaço Department of Human Geography, Lund University, Lund, Sweden

D. Mendoza Tinoco University of Coahuila, Coahuila, Mexico

Julián Andrés Mera-Paz Faculty of Engineering, Universidad Cooperativa de Colombia, Popayán, Colombia

Magnus Højberg Mernild State of Green, Copenhagen, Denmark

Jessica Ostrow Michel School for Environment and Sustainability, University of Michigan, Ann Arbor, MI, USA

Yoko Mochizuki UNESCO, Paris, France

Itumeleng Mogola C40 Cities, Benoni, South Africa

Mohsen Mohammadzadeh School of Architecture and Planning, Auckland University, Auckland, New Zealand

Abinash Mohanty Council on Energy, Environment and Water (CEEW), New Delhi, India

Mehri Mohebbi Transportation Equity Program, University of Florida (UFTI), Gainesville, FL, USA

Anne Mook University of Georgia, Athens, GA, USA

Eugenio Morello Laboratorio di Simulazione Urbana Fausto Curti, Department of Architecture and Urban Studies, Politecnico di Milano, Milan, Italy

Charlotte Morphet Women and Planning research bursary, Planning, Housing and Human Geography, The Leeds Planning School, Leeds Beckett University, Leeds, UK

Nicky Morrison Western Sydney University, Sydney, NSW, Australia

Sina Mostafavi TU Delft, Delft, The Netherlands

Edmos Mtetwa Department of Social Work, University of Zimbabwe, Harare, Zimbabwe

Tinashe Natasha Mujongonde-Kanonhuwa Department of Rural & Urban Planning, University of Zimbabwe, Harare, Zimbabwe

Manasi R. Mulay Department of Chemistry, University of Sheffield, Sheffield, UK

Grantham Centre for Sustainable Futures, Sheffield, UK

Richard Müller Sustainable Development Institute/Institut udrzatelneho rozvoja, Nitra, Slovakia

Yvonne Munanga Department of Architeture and Real Estate, University of Zimbabwe, Harare, Zimbabwe

Dalia Munenzon College of Architecture, Texas Tech University, Lubbock, TX, USA

Nora Munguia Graduate Sustainability Program, Industrial Engineering Department, University of Sonora, Hermosillo, Mexico

Solomon Muqayi Department of Governance and Public Management, University of Zimbabwe, Harare, Zimbabwe

Cassandra Murphy Department of Psychology, Maynooth University, Maynooth, Ireland

Teagan Murphy University of Maryland, College Park, MD, USA

Brendan Murtagh Urban Planning, School of Natural and Built Environment, David Keir Building, Queen's University Belfast, Belfast, UK

Walter Musakwa Future Earth and Ecosystem Services Research Group, Department of Urban and Regional Planning, University of Johannesburg, Johannesburg, South Africa

Tafadzwa Mutambisi Department of Rural and Urban Planning, University of Zimbabwe, Harare, Zimbabwe

Chipo Mutonhodza Department of Rural and Urban Development, Great Zimbabwe University, Masvingo, Zimbabwe

Valeria Muvavarirwa Department of Demography Settlement and Development, Social & Behavioral Sciences, University of Zimbabwe, Harare, Zimbabwe

Jean Nacishali Nteranya Department of Geology, Faculty of Sciences, Université Officielle de Bukavu (UOB), Bukavu, Democratic Republic of Congo

Anupam Nanda University of Manchester, Manchester, UK

Luzma Fabiola Nava CONACYT-Centro del Cambio Global y la Sustentabilidad, A.C. (CCGS), Villahermosa, Tabasco, Mexico

International Institute for Applied Systems Analysis (IIASA), Laxenburg, Austria

Celeste Nava Jiménez División de Ciencias Económico Administrativas, Campus Guanajuato, Universidad de Guanajuato, Guanajuato, Mexico

Thilini Navaratne Department of Business Economics, Faculty of Management Studies and Commerce, University of Sri Jayewardenepura, Nugegoda, Sri Lanka

Roselin Ncube Women's University in Africa, Harare, Zimbabwe

S. Ncube Heriot-Watt University, Edinburgh, UK

Etienne Nel University of Otago, Dunedin, New Zealand

David Nichols Faculty of Architecture, Building and Planning, University of Melbourne, Melbourne, VIC, Australia

Alejandro Nuñez-Jimenez Sustainability and Technology Group, D-MTEC, ETH Zurich, Zurich, Switzerland

Belfer Center for Science and International Affairs, Harvard University, Cambridge, MA, USA

Gloria Nyaradzo Nyahuma-Mukwashi Department for International Development (DFID), Harare, Zimbabwe

E. O'Donnell University of Nottingham, Nottingham, UK

G. O'Donnell University of Newcastle, Newcastle, UK

Narteh F. Ocansey Water Resources, Freelance, Accra, NA, Ghana

Yukyung Oh King's College London, London, UK

Carolina G. Ojeda Doctorado en Arquitectura y Estudios Urbanos, Pontificia Universidad Católica de Chile, Providencia, Santiago de Chile, Chile

Departamento de Historia, Facultad de Comunicaciones e Historia, Universidad Católica de la Santísima Concepción, Concepción, Chile

Hasan Volkan Oral Faculty of Engineering, Department of Civil Engineering (English), Istanbul Aydın University, Istanbul, Turkey

P. Ortiz International Metropolitan Institute, Madrid, Spain

International Metropolitan Institute, Washington, DC, USA

G. Osei Anglia Ruskin University, Chelmsford, UK

Laura Patricia Otero-Durán Urban Development Institute, Bogotá, Colombia

Maria Pafi Urban Planning, School of Natural and Built Environment, David Keir Building, Queen's University Belfast, Belfast, UK

F. Pascale Anglia Ruskin University, Chelmsford, UK

Maibritt Pedersen Zari School of Architecture, Victoria University of Wellington, Wellington, New Zealand

María Concepción Peñate-Valentín Department of Applied Economics, Faculty of Economics, Universidade de Santiago de Compostela, Santiago de Compostela, Galicia, Spain

Paulo Pereira Environmental Management Laboratory, Mykolas Romeris University, Vilnius, Lithuania

Ben Perks

Shama Perveen Senior Manager (Water), Ceres, Boston, MA, USA

Evi Petersen Institute of Sports, Physical Education and Outdoor Life, University of South-Eastern Norway, Oslo, Norway

Son Phung Department of Civil and Environmental Engineering, Auckland University, Auckland, New Zealand

Francesca Piazzoni

Czarnocki Piotr Ministry of Climate and Environment, Warsaw, Poland

Dorina Pojani The University of Queensland, Brisbane, QLD, Australia

A. Pooley Centre for Alternative Technology, Pantperthog, UK

K. Potter Open University, Milton Keynes, UK

Abdellatif Qamhaieh American University in Dubai, Department of Architecture, Dubai, United Arab Emirates

Md. Anisur Rahman Center for Policy and Economic Research (CPER), Dhaka, Bangladesh

Syed Hafizur Rahman Department of Environmental Sciences, Faculty of Physical and Mathematical Sciences, Jahangirnagar University, Dhaka, Bangladesh

Lakshmi Priya Rajendran The Bartlett School of Architecture, University College London, London, UK

Ritesh Ranjan Department of Architecture & Planning, National Institute of Technology Calicut, Kozhikode, Kerala, India

Andreas Raspotnik High North Center for Business and Governance, Nord University, Bodø, Norway

The Arctic Institute – Center for Circumpolar Security Studies, Washington, DC, USA

Hanna A. Rauf Asian School of the Environment, Nanyang Technological University, Singapore, Singapore

Aaron Redman School of Sustainability, Arizona State University, Tempe, AZ, USA

William E. Rees School of Community and Regional Planning, University of British Columbia, Vancouver, BC, Canada

Christian Reichel University of Applied Sciences for Media, Communication and Management (HMKW), Berlin, Germany

Kimberley Reis Cities Research Institute, Griffith University, Brisbane, QLD, Australia

Catherine E. Richards Center for the Study of Existential Risk (CSER), University of Cambridge, Cambridge, UK

Department of Engineering, University of Cambridge, Cambridge, UK

Lauren Rickards Urban Futures Enabling Capability Platform, RMIT University, Melbourne, Australia

Ritesh Ranjan Department of Architecture and Planning, National Institute of Technology Calicut, Kozhikode, Kerala, India

Alejandra Rivera Vinueza 4CITIES Erasmus Mundus Joint Master Degree (EMJMD) in Urban Studies, Vrije Universitet Brussel (VUB), Brussels, Belgium

Institute for Human Rights and Business (IHRB), Built Environment Global Programme Manager, London, UK

Daniela Rizzi Nature-based Solutions and Biodiversity – Sustainable Resources, Climate and Resilience Team, Freiburg, Germany

Michael Robbins HIPR, New York City, NY, USA

Héctor Rodal Architect and Urban Planner, Barcelona, Spain

Robert Rogerson Institute for Future Cities, University of Strathclyde, Glasgow, UK

Watch Ruparanganda Department of Social Work, University of Zimbabwe, Harare, Zimbabwe

María Carmen Sánchez-Carreira Department of Applied Economics, Faculty of Economics, Universidade de Santiago de Compostela, ICEDE Research Group, CRETUS, Santiago de Compostela, Galicia, Spain

Rami Sabella United Nations Economic and Social Commission for Western Asia, Beirut, Lebanon

Peter Sainsbury School of Medicine, University of Notre Dame, Sydney, NSW, Australia

Samuel Sandoval Solis Department of Land, Air and Water Resources, University of California Davis, Davis, CA, USA

Guido Santini Food and Agriculture Organization of the United Nations (FAO), Rome, Italy

Tom Sanya Architecture Planning and Geomatics, University of Cape Town, Cape Town, South Africa

Hasan Saygın Application, and Research Center for Advanced Studies, Istanbul Aydın University, Istanbul, Turkey

Alice Schmidt Global Health Advisory Service to the European Commission, Mechelen, Belgium

Vienna University of Economics and Business, Vienna, Austria

AS Consulting, Vienna, Austria

Jörg Schröder Institute of Urban Design and Planning, Leibniz Universität, Hannover, Germany

Barbara Schröter Leibniz Centre for Agricultural Landscape Research (ZALF), Working group Governance of Ecosystem Services, Müncheberg, Germany

Lund University Centre for Sustainability Studies (LUCSUS), Lund, Sweden

Kim Philip Schumacher Institute of Geography, Osnabrück University, Osnabrück, Germany

Abel Schumann Organisation for Economic Co-operation and Development, Paris, France

Samad M. E. Sepasgozar School of Built Environment, University of New South Wales, Sydney, NSW, Australia

Alan Shapiro British Columbia Institute of Technology, Vancouver, BC, Canada

Aviram Sharma School of Ecology and Environment Studies, Nalanda University, Rajgir, Bihar, India

Tian Shi Department of Primary Industries and Regions, Adelaide, South Australia

Amna Shoaib Department of City and Regional Planning, Lahore College for Women University (LCWU), Lahore, Pakistan

Yat Shun Kei

Renard Y. J. Siew Climate Change & Sustainability, Centre for Governance & Political Studies (CENT-GPS), Kuala Lumpur, Malaysia

Institute for Globally Distributed Open Research and Education (IGDORE), Bali, Indonesia

David Simon Department of Geography, Royal Holloway, University of London, Egham, UK

Jean Simos Institute of Public Health, Faculty of Medicine, University of Geneva, Geneva, Switzerland

S2D – Health and Sustainable Development, Rennes, France

Neil Sipe School of Earth and Environmental Sciences, The University of Queensland, Brisbane, QLD, Australia

Ben Sonneveld Amsterdam Centre for World Food Studies/Athena Institute, Vrije Universiteit, Amsterdam, The Netherlands

Micol Sonnino The Bertalanffy Center for the Study of Systems Science, Vienna, Austria

Simon Springer Centre for Urban and Regional Studies, Dicipline of Geography and Environmental Studies, University of Newcastle, Australia, Callaghan, NSW, Australia

Janet Stanley Melbourne Sustainable Society Institute, University of Melbourne, Melbourne, Australia

Wendy Steele Centre for Urban Research, RMIT University, Melbourne, VIC, Australia

Justin D. Stewart Department of Ecological Science, Vrije Universiteit Amsterdam, Amsterdam, The Netherlands

Raisa Sultana Department of Geography and Environment, University of Dhaka, Dhaka, Bangladesh

Samantha Suppiah Possible Futures, Manilla, Philippines

Sylvia Szabo Department of Social Welfare and Counselling, University of Seoul, Seoul, South Korea

Gerti Szili College of Humanities, Arts and Social Sciences, Flinders University, Adelaide, South Australia

Bouchra Tafrata Willy Brandt School of Public Policy, University of Erfurt, Erfurt, Germany

Ling Min Tan Department of Civil & Structural Engineering, The University of Sheffield, Sheffield, UK

M. Terdiman The Institute for Environmental Security and Well-being Studies, Jerusalem, Israel

Jacqueline Thomas School of Civil Engineering, The University of Sydney, Sydney, NSW, Australia

M. K. Thomas Rural Focus Limited (RFL), Nanyuki, Kenya

S. Thomas Rural Focus Limited (RFL), Nanyuki, Kenya

C. Thorne University of Nottingham, Nottingham, UK

Karine Tollari Japan Local Government Centre, London, UK

Chiara Tomaselli Consultant – Urban Asset Advisory, Arcadis France, Paris, France

Percy Toriro Municipal Development Partnership for Eastern and Southern Africa, Harare, Zimbabwe

African Centre for Cities, University of Cape Town, Cape Town, South Africa

Isabella Trapani Food and Agriculture Organization of the United Nations (FAO), Rome, Italy

Alejandra Trejo-Nieto Centre for Demographic, Urban and Environmental Studies, El Colegio de Mexico, Mexico City, Mexico

Stella Tsani Department of Economics, University of Ioannina, Ioannina, Greece

Asaf Tzachor Center for the Study of Existential Risk (CSER), University of Cambridge, Cambridge, UK

School of Sustainability, Interdisciplinary Center (IDC) Herzliya, Herzliya, Israel

Zdravka Tzankova Vanderbilt University, Nashville, TN, USA

Kristina Ulm Faculty of Arts, Design and Architecture, University of New South Wales, Sydney, NSW, Australia

Geraldine Usingarawe Department of Architecture and Real Estate, University of Zimbabwe, Harare, Zimbabwe

Luís Valença Pinto Research Centre for Natural Resources, Environment and Society (CERNAS), Polytechnic Institute of Coimbra, Coimbra Agrarian Technical School, Coimbra, Portugal

Environmental Management Laboratory, Mykolas Romeris University, Vilnius, Lithuania

Ellen Van Bueren Department of Management in the Built Environment, Faculty of Architecture and the Built Environment, Delft University of Technology, Delft, The Netherlands

Karel Van den Berghe Department of Management in the Built Environment, Faculty of Architecture and the Built Environment, Delft University of Technology, Delft, The Netherlands

Jeroen van der Heijden School of Government, Victoria University of Wellington, Wellington, New Zealand

School of Regulation and Global Governance, Australian National University, Canberra, ACT, Australia

Wim van Veen Vrije Universiteit, Amsterdam Centre for World Food Studies, Amsterdam, The Netherlands

Lia van Wesenbeeck Vrije Universiteit, Amsterdam Centre for World Food Studies, Amsterdam, The Netherlands

Christopher Vanags Vanderbilt University, Nashville, TN, USA

Kamiya Varshney School of Architecture, Victoria University of Wellington, Wellington, New Zealand

Luis Velazquez Industrial Engineering Department, University of Sonora, Hermosillo, Mexico

Luis Eduardo Velazquez Contreras Sustainability Graduate Program, University of Sonora, Hermosillo, México

T. Vilcan Open University, Milton Keynes, UK

Luiza O. Voinea Urban Planner Certified by The Romanian Register of Urban Planners, Bucharest, Romania

Shreya Wadhawan Council on Energy, Environment and Water (CEEW), New Delhi, India

Sameh N. Wahba The World Bank, Washington, DC, USA

Haiyun Wang School of Earth and Environmental Sciences, The University of Queensland, St Lucia, Australia

Siqin Wang School of Earth and Environmental Sciences, The University of Queensland, St Lucia, Australia

Noelia Wayar National University of Córdoba, Córdoba, Argentina

Oliver Weigel Urban Development Policy Division at the Federal Ministry of the Interior, Building, and Community, Berlin, Germany

Kadmiel H. Wekwete Midlands State University, Gweru, Zimbabwe

Caitlin Werrell The Center for Climate and Security, an Institute of the Council on Strategic Risks, Washington, DC, USA

Andreas Wesener School of Landscape Architecture, Faculty of Environment, Society and Design, Lincoln University, Lincoln, New Zealand

Bettina Wilk ICLEI European Secretariat, Senior Officer for Nature-based Solutions and Biodiversity – Sustainable Resources, Climate and Resilience Team, Freiburg, Germany

Erich Wolff Monash Art, Design and Architecture, Monash University, Melbourne, VIC, Australia

Sam Wong University College Roosevelt, Middelburg, The Netherlands

N. Wright Nottingham Trent University, Nottingham, UK

Junjie Xi

Belinda Young Melbourne Sustainable Society Institute, University of Melbourne, Melbourne, Australia

Asaduz Zaman Centre for Action Research – Barind, Rajshahi, Bangladesh

Asian Development Bank, Dhaka, Bangladesh

Fathima Zehba M. P. Department of Architecture and Planning, Calicut, National Institute of Technology, Calicut, Kerala, India

David Slim Zepeda Quintana Sustainability Graduate Program, University of Sonora, Hermosillo, México

Yuerong Zhang Bartlett School of Planning, University College London, London, UK

Eric Zhao University of Toronto, Toronto, ON, Canada

Metron Ziga University of the Western Cape, Cape Town, South Africa

Monika Zimmermann Urban Sustainability Expert & Former Deputy Secretary General of ICLEI, Freiburg, Germany

Willoughby Zimunya Department of Demography Settlement and Development, University of Zimbabwe, Harare, Zimbabwe

Department of Urban and Regional Planning, University of the Free State, Bloemfontein, South Africa

Michaela Zint School for Environment and Sustainability, University of Michigan, Ann Arbor, MI, USA

Tara Rava Zolnikov School of Behavioral Sciences, California Southern University, Costa Mesa, CA, USA

Department of Community Health, National University, San Diego, CA, USA

A

15-Minute City

▶ Walkable Access and Walking Quality of Built Environment

2030 Agenda for Sustainable Development

▶ Sustainable Development Goals from an Urban Perspective

Accessible Homes

▶ Housing Affordability

Active Living

▶ Hidden Enemy for Healthy Urban Life

Active Transport

▶ Pre-schoolers and Sustainable Urban Transport

Adaptation

▶ From Vulnerability to Urban Resilience to Climate Change
▶ Water Policy in the State of Tabasco

Adapting Cities to Climate Change

The Role of Legal, Financial, and Coordinating Actions

Czarnocki Piotr and Ligocka Ilona
Ministry of Climate and Environment, Warsaw, Poland

Definition

The foundations for adapting cities to climate change in Poland were laid down by the Council of Ministers of the Republic of Poland in 2013 in *Polish National Strategy for Adaptation to Climate Change with the perspective by 2030* (NAS 2020) (Ministry of the Environment 2013). The document has been prepared with a view to ensure the conditions of stable socioeconomic development in the face of risks posed by climate change but also with a view to use the positive impact which adaptation actions may have not only on the state of the Polish environment but also on the

R. C. Brears (ed.), *The Palgrave Encyclopedia of Urban and Regional Futures*,
https://doi.org/10.1007/978-3-030-87745-3

economic growth. NAS 2020 indicates the objectives and directions of adaptation actions to be taken in the most vulnerable sectors and areas, i.e., water management, agriculture, forestry, biodiversity and protected areas, health, energy, building industry, transport, mountain areas, coastal zone, spatial development, and urban areas. On the basis of NAS 2020 and other Polish strategic documents, in the field of climate and environmental protection, Ministry of Climate and Environment (before – Ministry of Climate, Ministry of the Environment) has for many years consistently been implementing actions aimed at adapting society, the environment, and the economy to current and projected climate change.

Introduction

Urban Climate Adaptation Plans Guidelines

One of the first activities of the Ministry of Climate and Environment aimed at adapting cities to climate change was elaboration of *Guidelines for Developing Urban Climate Adaptation Plan* in 2015 (Ministry of the Environment 2015a). It is an example of coordination action at the government level in relation to local level actors and plans, which has provided a methodology and checklist for the process of developing a climate change adaptation plan at the local level. It can be used by any local government to coordinate the development process of climate change adaptation plan. These guidelines may also be used by any interested local government.

Development of Urban Adaptation Plans for Cities with Population over 100,000 in Poland (MPA)

Guidelines for Developing Urban Climate Adaptation Plan were used in an innovative initiative of the Ministry of Climate and Environment – *Development of Urban Adaptation Plans for cities with more than 100,000 inhabitants in Poland* (MPA Project) (Ministry of the Environment 2018). It was an example of Polish government coordination activities at national level as the ministry commissioned the consortium of institutes and a private entity to develop adaptation plans for 44 major Polish cities. The overarching goal of the project was to identify and analyze adaption challenges each city may face, draft plans for local authorities, indicate sources of funding, and raise awareness for the need for adaptation. The MPA Project was cofounded (85%) by the European Union funds (Cohesion Fund within the framework of Operational Program Infrastructure and Environment) and the Polish state budget (15%). All urban adaptation plans were developed in 2017–2019 under one methodology by one contractor. Urban areas have been identified as a priority in the implementation of adaptation to climate change policy in Poland. Over 30% of the Polish population lives in project partner cities.

The following project goals were fulfilled:

- Determination of vulnerability of the largest Polish cities to climate change
- Planning for adaptation actions at the local level
- Raising awareness of the need for adaptation to climate change at the local level

Polish Ministry of Climate and Environment supported local governments from both – organizational and financial side.

The MPA Project is the first step aimed at strengthening resilience and adaptation to changing climate conditions. Due to its scale, this was an innovative and unique project. The result of the project (through the implementation of actions and measures) will enhance cities' resilience to climate change, and consequently the entire country. The project fulfills objectives pointed out in NAS 2020. It is the multispectral approach to decreasing cities' vulnerability to climate change impacts.

Preparation of urban adaptation plans was based on an innovative approach aimed at determining the most effective measures to adapt and protect against the effects of already existing and forecasted threats – extreme temperatures, heat waves, heavy rainfall, storms, and urban floods.

Together with city authorities, strategic solutions have been developed to increase the cities' adaptation potential to the consequences of climate change and to raise the level of citizens' awareness on this issue. Social and economic benefits, an increase in the level of safety, and the improvement of the quality of life in the cities are the objectives which, thanks to the implementation of the adaptation plans, will be achieved by 2030.

Experts and representatives of local authorities stressed that adaptation serves to improve the quality of life in the city, of which the state of the environment is an increasingly important element. Adaptation measures undertaken in cities require, on the one hand, strong organizational and legal support at the national level, and, on the other hand, conscious society and entities responsible for their implementation at the local level. Urban adaptation plans have dynamic and open character, so it will be easy to update or reshape some assumptions.

In addition to developing urban adaptation plans, the project goal was to share knowledge on climate changes and their effects. In parallel to working on the adaptation plans, a vast body of educational materials was prepared for both adults and children (films, brochures, training sessions, newsletters, quizzes, and competitions) and also conferences and debates were organized. All activities in this project were carried out together – by experts, representatives of the local communities, officials, managers of urban utilities and properties, activists, scientists, and business people. Because of such broad participation, the project has a unique value. The awareness of the fact that adaptation to climate change is essential to the protection of our health, and often life, is the foundation of all actions. It also means limiting the costs of mitigating the effects of damage to properties and infrastructure.

Individual adaptation plan is monitored by the Ministry of Environment and Climate, as part of the project sustainability check. Currently, adaptation plans have been adopted by 42 communes. The adopted plans have the status of communal strategic documents, which will be implemented and for which financial resources are guaranteed. The Ministry of Climate and Environment supports cities in the implementation of adaptation plans, including through appropriate programming of the new European Union financial perspective. In the case of investment activities, municipalities will commission the implementation of plans to external contractors.

Climate Risk Assessment Including Environmental Impact Assessment (EIA)/ Strategic Environmental Assessment (SEA) Guide

Another example of coordinating actions at national level is the development of *Guide to Investment Preparation Respecting Climate Change Mitigation and Adaptation as well as Resilience to Natural Disasters* (Ministry of the Environment 2015b). The Guide covers methodologies for integrating climate change adaptation and mitigation into the development of infrastructure projects. These methodologies mostly rely on the rules of risk assessment.

In Poland, projects which are co-financed from European Union funds are obliged to use the methodologies indicated in the Guide. The Guide developed by the Ministry of Climate and Environment indicates the methods that should help investors, including beneficiaries of European Union funds in the period 2014–2020, which are mostly municipalities, in preparation of investment projects and/or in the application of EU funds in the field of issues related to climate change adaptation and mitigation and resilience to natural disasters. The Guide aims to provide methodologies and hints concerning how climate issues should be included in the process of developing investments and projects at the stage of:

- SEA and EIA concerning climate mitigation, climate adaptation, and resilience including ecosystem-based approaches
- Cost-benefit analysis, including calculation of shadow costs and external costs of greenhouse gas emissions, carbon footprint analysis, sensitivity, and vulnerability analysis of projects concerning climate changes and natural disasters
- Risk analysis including climate-related risks
- Climate options analysis and assessment, including climate impact on projects and projects impacts on climate

Climate-Friendly Cities

In March 2020, the Ministry of Climate and Environment launched an initiative *climate-friendly cities* (initially called *Urban Agenda*) (Ministry of Climate and Environment 2020) whose objective is to disseminate modern, effective, and efficient solutions that improve the quality of life of residents and increase cities' resilience to the effects of climate change. Under this initiative, the Ministry of Climate and Environment supports pro-climate transformation of cities, including clean energy and transport, clean air, and increasing resilience to climate change, particularly to drought, floods, and heat waves through the development of green and blue infrastructure.

Key elements of *climate-friendly cities* in 2020 were:

1. **Highlighting Best Practices**

   Ministry of Climate and Environment examined the progress of cities in the context of their environmental policies. Local data concerning waste management, air quality, green areas, sustainable mobility, as well as water, rainwater, and wastewater management were collected and compared. The survey was conducted in two categories – cities up to 100,000 residents and over 100,000 residents. The best practices in adaptation to climate change were also highlighted in the contest with an award for a best project aimed at protecting a city and adapting it to climate change. Another competition envisioned as a part of the initiative was a contest for secondary school students to develop a climate change adaptation plan for the school area and its immediate vicinity, organized by the National Fund for Environmental Protection and Water Management. The winning project received funding for its implementation.
2. **Launching Additional Support Tools**

   Another elements of the implementation of *climate-friendly cities* were additional financial support tools, including a competition for green and blue urban infrastructure. The National Fund for Environmental Protection and Water Management Priority Program was launched under the name "Adaptation to climate change and limiting the effects of environmental hazards," in which a continuous call for applications has been carried out since 2019. The program aims to increase the level of protection against the impacts of climate change, natural hazards, and major accidents, to improve the removal of their impact and to strengthen selected elements of environmental management. The budget for the implementation of the program objective is up to PLN 657 million (ca. USD 170 million) including for grant forms of cofinancing – up to PLN 158 million (ca. USD 41 million). Cofinancing is provided for activities in the field of adaptation to climate change in cities, including green and blue infrastructure, elimination of impermeable surfaces, sustainable rainwater management systems with greenery, and increasing rainwater retention in urban ecosystems. The Ministry of Climate and Environment also provides an opportunity for cities to obtain funding for measures related to sustainable rainwater management in urban areas under the Operational Program Infrastructure and Environment 2014–2020 – the total allocation for this type of projects exceeded PLN 1.1 billion (ca. USD 283 million). Furthermore, calls for proposals for projects on green and blue infrastructure in cities, as well as for awareness-raising activities on climate change adaptation in schools were launched under the European Economic Area and Norway Financial Mechanism 2014–2021. The amount of funds allocated for the indicated types of projects reached PLN 100 million (ca. USD 26 million).

   The Ministry of Climate and Environment has made available from July 2020 a funding budget of PLN 100 million (ca. USD 26 million) for the implementation of the National Fund for Environmental Protection and Water Management Priority Program: "My Water," and this allocation was exhausted in less than 7 months, which demonstrates high interest of beneficiaries to get involved in activities supporting adaptation to climate change. Additional PLN 110 million (ca. USD 28 million) has been allocated in 2021 to the program. The program provides subsidies to natural persons – owners of single-family houses for retaining, retention, and management of rainwater and snowmelt and its aim is to protect water resources by increasing retention on the property and the utilization of collected rainwater and snowmelt, including through the development of green and blue infrastructure. Funds available for the first two editions of the program will allow to retain and use about 2 million m^3 of water per year. Apart from the ecological effect, the success of the program is the increase of social awareness of the need to save water, to collect and use rainwater, and involvement of the inhabitants of Poland in activities contributing to adaptation to climate change.

3. **Developing New Solutions**

   As part of *climate-friendly cities*, a series of workshops was conducted in May–June 2020 in order to develop specific solutions, e.g., legislative initiatives, financial mechanisms, and other mechanisms supporting cities, as well as share good practices for climate-friendly and neutral cities. Workshops gathered stakeholders from variety of institutions (local, regional, and national), experts, representatives of think tanks and nongovernmental organizations, and youth. As a result of these meetings and discussions the Ministry of Climate and Environment has work out certain legislative proposals in order to strengthen the climate dimension of urban policy. The proposed intervention tools consist in introducing the obligatory development of a climate change adaptation plan for cities with more than 20,000 inhabitants and taking into account the aspect of adaptation to climate change in strategic documents, including those relating to urban spatial planning. Another element of the intervention is to strengthen the policy pursued by the voivodship self-government with the context of adaptation to climate change and greening the civic budget in cities.

Future Actions

In 2021 and beyond *climate-friendly cities* will be implemented according to an extended formula, including activities proposed by the units subordinate to and supervised by the Ministry of Climate and Environment (Ministry of Climate and Environment 2021). Based on the application submitted by the cities to the Ministry of Climate and Environment, an analysis of the actions taken by the cities will be carried out to measure the results of the implementation of environmental and climate policies. The analysis will concern several categories, including air quality, urban greenery, transport, energy transition, and rainwater retention. The cities, which win the new edition of the program may receive the title of a *climate-friendly city* in one or several categories. They will be able to boast their distinction on information boards on public buildings, as well as appropriate graphics which they will be able to use in their communication (on the Internet, on letterheads, etc.). Twenty cities which stand out in this regard will also receive specialist strategic advisory services from renowned institutions such as the Institute of Environmental Protection, the Institute for Ecology of Industrial Areas, the Forest Research Institute, and the National Center for Nuclear Research. Cities participating in the analysis will also be offered training courses on the programing and implementation of climate and environmental measures and projects as well as financial support, mainly from the new financial perspective of European Union funds for 2021–2027.

Summary/Conclusions

Measures taken in the years 2013–2021 by the Ministry of Climate and Environment show that a combination of three types of actions is necessary to achieve synergy and foster planning and implementation of climate and environmental policies at the local level:

- Coordinating instruments – which include consultations and cooperation with the widest possible range of stakeholders in the process of

finding the best solutions for pro-climate transformation of cities. Activities in the field of climate change adaptation must be based primarily on building public awareness in this area, information, and promotional activities, including best practices sharing.
- Financial instruments – it is important to channel an adequate funding stream for adaptation measures, including, in particular, funding on sustainable urban rainwater management, small-scale retention in local catchments, as well as developing green and blue infrastructure and promoting nature-based solutions.
- Legal instruments – it is necessary to introduce legal instruments that will oblige relevant actors to take into account specific aspects of climate change in planning and implementing their activities which will support urban adaptation to climate change.

References

Ministry of Climate and Environment, Republic of Poland. (2020). Urban Agenda – Towards sustainable and climate-neutral cities, Warsaw, Poland. https://www.gov.pl/web/climate/urban-agenda%2D%2Dtowards-sustainable-and-climate-neutral-cities

Ministry of Climate and Environment, Republic of Poland. (2021). Climate-friendly cities 2.0 – New edition of the initiative launched, Warsaw, Poland. https://www.gov.pl/web/climate/city-with-climate-20%2D%2Dnew-edition-of-the-initiative-launched

Ministry of the Environment, Republic of Poland. (2013). Polish National Strategy for Adaptation to Climate Change (NAS 2020) with the perspective by 2030, Warsaw, Poland. https://klimada.mos.gov.pl/wp-content/uploads/2014/12/ENG_SPA2020_final.pdf

Ministry of the Environment, Republic of Poland. (2015a). Guidelines for developing urban climate adaptation plan, Warsaw, Poland. https://klimada.mos.gov.pl/wp-content/uploads/2015/09/Podręcznik-adaptacji-dla-miast1.pdf

Ministry of the Environment, Republic of Poland. (2015b). Guide to investment preparation respecting climate change mitigation and adaptation as well as resilience to natural disasters, Warsaw, Poland. https://klimada.mos.gov.pl/wp-content/uploads/2018/02/Poradnik-przygotowania-inwestycji-z-uwzgl%C4%99dnieniem-zmian-klimatu-ich-%C5%82agodzenia-i-przystosowania-do-tych-zmian-oraz-odporno%C5%9Bci-na-kl%C4%99ski_ver_5_2_sierpnia_2017.pdf

Ministry of the Environment, Republic of Poland. (2018). Climate change adaptation plans in 44 Polish cities, Summary report, Warsaw, Poland. http://44mpa.pl/wp-content/uploads/2018/12/MPA_NET-ENG-20-12.pdf

Adapting to a Changing Climate Through Nature-Based Solutions

Abinash Mohanty
Council on Energy, Environment and Water (CEEW), New Delhi, India

Synonyms

CBD – Convention on Biological Diversity; CEEW – Council on Energy, Environment and Water; CEM – Commission on Ecosystem Management; Climate risks; EbA – Ecosystem-based adaptation; Eco-DRR – Ecosystem-led disaster risk reduction; ENSO – El Nino South Oscillation; IMD – India Meteorological Department; INDCs – Intended nationally determined contributions; IPCC – The Intergovernmental Panel on Climate Change; IUCN – International Union for Conservation of Nature; NAPCC – National Action Plan on Climate Change; NbS – Nature-based solutions; NDMA – National Disaster Management Authority of India; OECD – Organization for Economic Cooperation and Development; PIB – Press Information Bureau; SAPCC – State Action Plan on Climate Change; UNDRR – United Nations Office for Disaster Risk Reduction; UNEP – United Nations Environment Programme; UNFCCC – United Nations Framework Convention on Climate Change; WMO – World Meteorological Organization

Definition

Disaster risk reduction (DRR): It is “the concept and practice of reducing disaster risks through systematic efforts to analyse and manage the causal factors of disasters, including through reduced exposure to hazards, lessened vulnerability of people and property, wise management of land and the environment, and improved preparedness for adverse events" (UNISDR Terminology 2009).

Nature-based solutions (NbS): "Nature-based Solutions are actions to protect, sustainably manage and restore natural and modified ecosystems in ways that address societal challenges effectively and adaptively, to provide both human well-being and biodiversity benefits. They are underpinned by benefits that flow from healthy ecosystems and target major challenges like climate change, disaster risk reduction, food and water security, health and are critical to economic development" (IUCN 2016).

Ecosystem-based adaptation (EbA): "EbA is the use of biodiversity and ecosystem services as part of an overall adaptation strategy to help people to adapt to the adverse effects of climate change" (Convention on Biological Diversity 2009).

ENSO: El Niño and the Southern Oscillation, also known as ENSO, is a periodic fluctuation in sea surface temperature (El Niño) and the air pressure of the overlying atmosphere (Southern Oscillation) across the equatorial Pacific Ocean (NOAA).

Adaptation: The process of adjustment to actual or expected climate and its effects. In human systems, adaptation seeks to moderate or avoid harm or exploit beneficial opportunities (WGII, III, IPCC Glossary).

Hydromet hazards: Hydrological and meteorological (or "hydromet") hazards – weather, water, and climate extremes (GFDRR 2018).

Introduction

Climate change impact is imperiling human health, ecosystems, and the consequences are irreversible, as inferred by the Intergovernmental Panel on Climate Change's second tranche of the sixth assessment report. The surge in climate extremities has led to irreversible damages as ecosystems and human beings are pushed beyond thresholds of the ability to adapt (IPCC 2022). The report is a grim reminder that actions on adaptation (and mitigation) are already taken would not be sufficient and there lie gaps with regard to avoiding and reducing risks. The world is en route to face the harsher impacts of climate change through rising climate extremities and slow onset events that will affect the food systems, ecosystems, biodiversity, livelihood patterns, supply chains, among others. South Asia will be one of the worst-hit regions where the likelihood of droughts, floods, heatwaves are supposed to breach the past trends (Ibid.). Hydrometeorological risk is the probability of annual and probable damage caused by hydrometeorological hazards and its interplay with exposure and vulnerability of the impacted populations and ecosystems (Merz et al. 2010). India's vulnerability to climate extremities in all likelihood will impact the developmental trajectories. Research by CEEW found that more than 75% of its districts are extreme events hotspots, of which 40% are showcasing a swapping trend, i.e., flood-prone areas are becoming drought-prone and vice versa (Mohanty 2020). Further, more than 80% of its population is vulnerable to extreme climate risks (Mohanty and Wadhawan 2021). The 1.5 °C global ambition on climate mitigation is tangible: beyond this warming level, impacts and risks will grow increasingly into existential and irreversible. While adaptation actions need to be scaled, maladaptation can increase the impacts beyond projected limits. Restricting climate change is imperative and one of the most thought of, yet less implemented intervention is promoting, implementing, and scaling nature-based solutions (NbS).

IUCN defines NbS as "actions to protect, sustainably manage, and restore natural or modified ecosystems, that address societal challenges effectively and adaptively, simultaneously providing human well-being and biodiversity benefits" (IUCN 2016). The perimeter of NbS goes beyond

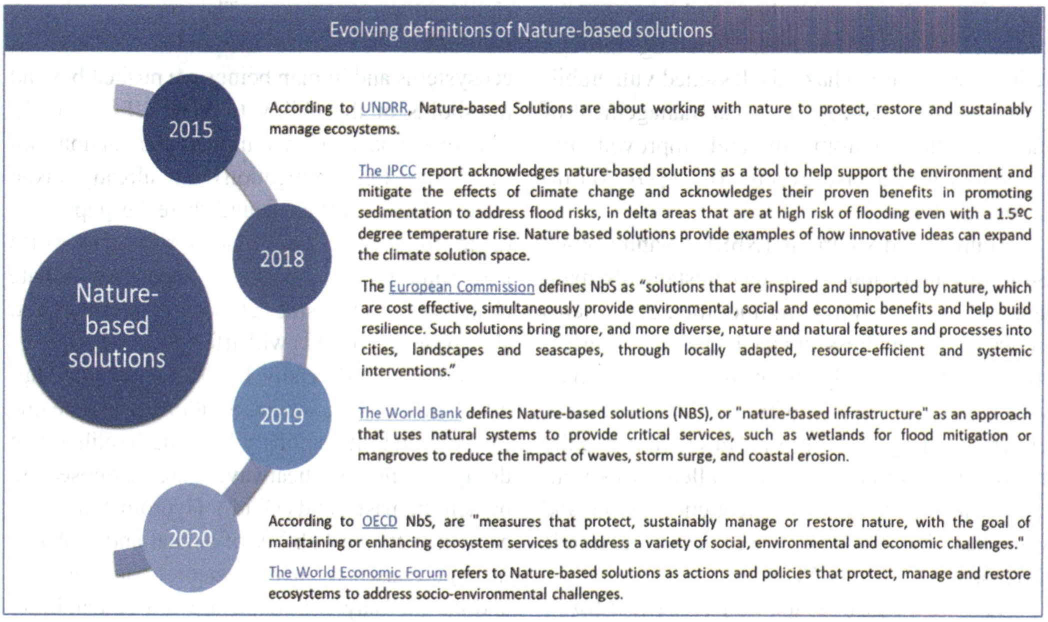

Evolving definitions of Nature-based solutions
Nature-based solutions
2015
According to UNDRR, Nature-based Solutions are about working with nature to protect, restore and sustainably manage ecosystems.
2018
The IPCC report acknowledges nature-based solutions as a tool to help support the environment and mitigate the effects of climate change and acknowledges their proven benefits in promoting sedimentation to address flood risks, in delta areas that are at high risk of flooding even with a 1.5ºC degree temperature rise. Nature based solutions provide examples of how innovative ideas can expand the climate solution space.
The European Commission defines NbS as "solutions that are inspired and supported by nature, which are cost effective, simultaneously provide environmental, social and economic benefits and help build resilience. Such solutions bring more, and more diverse, nature and natural features and processes into cities, landscapes and seascapes, through locally adapted, resource-efficient and systemic interventions."
2019
The World Bank defines Nature-based solutions (NBS), or "nature-based infrastructure" as an approach that uses natural systems to provide critical services, such as wetlands for flood mitigation or mangroves to reduce the impact of waves, storm surge, and coastal erosion.
2020
According to OECD NbS, are "measures that protect, sustainably manage or restore nature, with the goal of maintaining or enhancing ecosystem services to address a variety of social, environmental and economic challenges."
The World Economic Forum refers to Nature-based solutions as actions and policies that protect, manage and restore ecosystems to address socio-environmental challenges.

Adapting to a Changing Climate Through Nature-Based Solutions, Fig. 1 Definitions of NbS by various organizations. (Source: Author's compilation)

biodiversity restoration and a full range of ecosystem services, rather it dwells, advocates, and suggests adaptation and mitigation co-benefits. Figure 1 provides a glimpse of definitions by different organizations.

There is no denial that NbS answers to the multifaceted issues of the adaptation solutions. These solutions are science-based empirical evidence-based approaches aimed to manage landscapes-restore and rehabilitate them, promotes effective resource management, and revitalize the fissured ecosystems. The NbS based on ecosystem-led approaches addresses four major pillars of climate change: (i) climate change adaptation, (ii) climate change mitigation, (iii) environmental management, (iv) disaster risk reduction (UNDRR 2020). Further, promoting NbS would also ensure an equitable and just approach to managing climate change with people and ecosystems at its core. One of the biggest challenges lies in identifying risks and understanding the extent of damage any extreme event will cause and more importantly what are the solutions that are available, feasible, implementable, and scalable. This entry attempts to provide a shortlisted range of nature-based solutions based on their typological benefits interlinked with specific hazards across different climate zones. The enumerated NbS are in the sections below are shortlisted considering the definitional framework of NbS provided by IUCN and its Commission on Ecosystem Management (CEM) (Shacham et al. 2019) (Fig. 2). Furthermore, Convention on Biological Diversity (CBD) also enumerates core guidelines on NbS-based ecosystem-led adaptation (EbA) and ecosystem-led disaster risk reduction (Eco-DRR).

Nature-Based Solutions and Climate Negotiations

Today there is a political consensus that NbS has both adaptation and mitigation co-benefits and will be instrumental for countries to achieve their NDCs. The importance of NbS through EbA and Eco-DRR has gained traction in recent years with the operationalization of Hyogo and UNFCCC negotiations (Seifollahi-Aghmiuni et al. 2019). One hundred and nine of the 189 intended nationally determined contributions (INDCs), countries' commitments under the 2015 Paris Agreement on climate change, and which in most cases cover

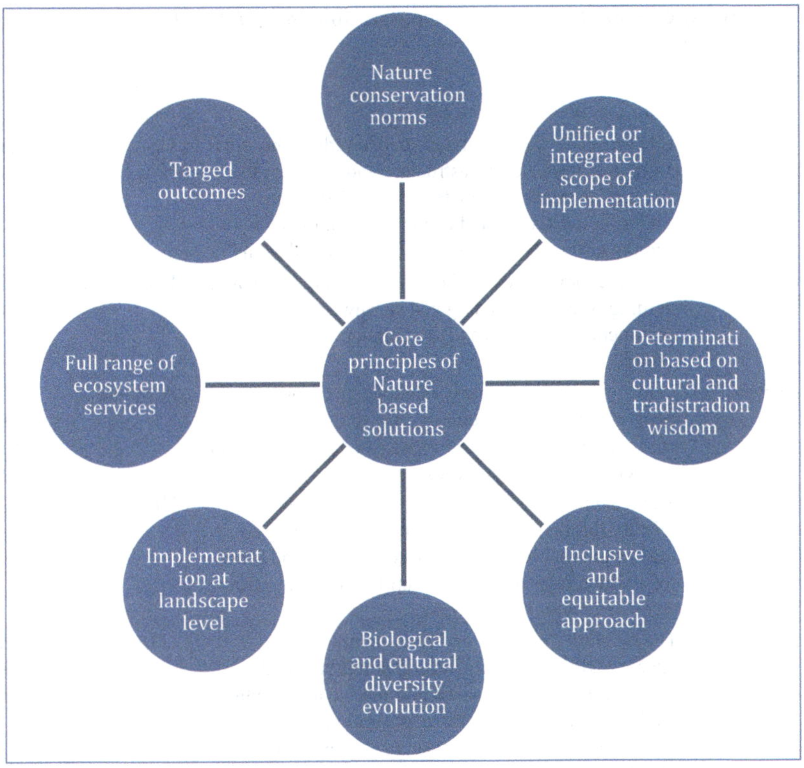

Adapting to a Changing Climate Through Nature-Based Solutions, Fig. 2 Core principles for operationalizing NbS. (Source: Author's compilation)

adaptation as well as mitigation, include ecosystem considerations in their visions for adaptation (UNEP-WCMC and UNEP 2019). However, the term NbS is yet to be actively operationalized climate action plans at regional, national, and subnational levels. Sendai framework does not recognize or identify NbS as a key strategy to mitigate climate risks.

While significant progress has been made in adoption of NbS scaling ecosystem led interventions still remains a challenge. Different ecosystems have different needs, characterizing all the interventions based on typology have hindered the implementation of NbS. This calls for mainstreaming of ecological engineering (Ruangpan et al. 2020). One size fits all does not work in the case of scaling and implementing a particular NbS geography, and topography plays an important role which should be integrated with the hazard profile of the target area (Romnée and De Herde 2015; Zhang and Chui 2018). Climate risk comprises of three major components – hazard, exposure, and vulnerability (IPCC 2014), and empirical evidence from the implementation of NbS suggests that it reduces all three of them and substantially enhances the landscape-based adaptive capacity that has multiple socioeconomic benefits. Given the premise of the challenges, this entry attempts to identify some of the scalable and implementable NbS that are in practice across similar hazard profiles and what are the gaps and challenges with reference to methods and tools that need to be addressed. India is the seventh most vulnerable country globally (Germanwatch 2020) and is already witnessing a glimpse of the large-scale disasters and the resulting socioeconomic upheaval. India suffered an annual loss of USD 87 billion due to extreme climate events (WMO 2020). According to a report by IMD, 24 of the 36 meteorological subdivisions in India witnessed extreme events such as floods, landslides, and heatwaves during the southwest monsoon season (June to September) (IMD 2020), which

Adapting to a Changing Climate Through Nature-Based Solutions, Table 1 Various terminologies used to describe NbS across the globe

Terminology	Definition, objectives, and purpose	Geographical location	Sources
Low-impact development (LIDs)	"LID is used as a retro- fit designed to reduce the stress on urban stormwater infrastructure and/or create the resiliency to adapt to climate changes, LID relies heavily on infiltration and evapotranspiration and attempts to incorporate natural features into design."	USA, New Zealand	Eckart et al. (2017)
Best management practices (BMPs)	"A device, practice or method for removing, reducing, retarding or preventing targeted stormwater run-off constituents, pollutants and contaminants from reaching receiving waters."	USA, Canada	Strecker et al. (2001)
Water-sensitive urban design (WSUD)	"Manage the water balance, maintain and where possible enhance water quality, encourage water conservation and maintain water-related environmental and recreational opportunities."	Australia	Whelans consultants et al. (1994)
Sustainable urban drainage systems (SuDs)	"Replicate the natural drainage processes of an area – typically through the use of vegetation-based interventions such as swales, water gardens and green roofs, which increase localised infiltration, attenuation and/or detention of stormwater."	UK	Ossa-Moreno et al. (2017)
Green infrastructure (GI)	"The network of natural and semi-natural areas, features and green spaces in rural and urban, and terrestrial, freshwater, coastal and marine areas, which together enhance ecosystem health and resilience, contribute to biodiversity conservation and benefit human populations through the maintenance and enhancement of ecosystem services."	USA, UK	Naumann et al. (2011)
Ecosystem-based adaptation (EbA)	"The use of biodiversity and ecosystem services as part of an overall adaptation strategy to help people to adapt to the adverse effects of climate change."	Europe, Canada	CBD (2009)
Ecosystem-based disaster risk reduction (Eco-DRR)	"The sustainable management, conservation, and restoration of ecosystems to reduce disaster risk, with the aim of achieving sustainable and resilient development."	Europe, USA	Estrella and Saalismaa (2013)
Blue–green infrastructure (BGI)	"BGI provides a range of services that include; water supply, climate regulation, pollution control and hazard regulation (blue services/goods), crops, food and timber, wild species diversity, detoxification, cultural services (physical health, aesthetics, spiritual), plus abilities to adapt to and mitigate climate change."	UK	Lawson et al. (2014)
Nature-based solution (NbS)	"NBS aim to help societies address a variety of environmental, social and economic challenges in sustainable ways. They are actions inspired by, supported by or copied from nature, both using and enhancing existing solutions to challenges as well as exploring more novel solutions."	Europe	EC (2015)

Source: Author's compilation based on Ruangpan et al. (2020)

significantly impact the economies, lives, and livelihoods. More than ever, India needs to scale up its fight against climate change by stepping up the implementation of NBS. Table 1 enumerates how NbS has termed cross different regions, and clearly, South Asia and India are lagging far behind in adopting, implementing, and scaling NbS. In this entry, umbrella term of nature-based

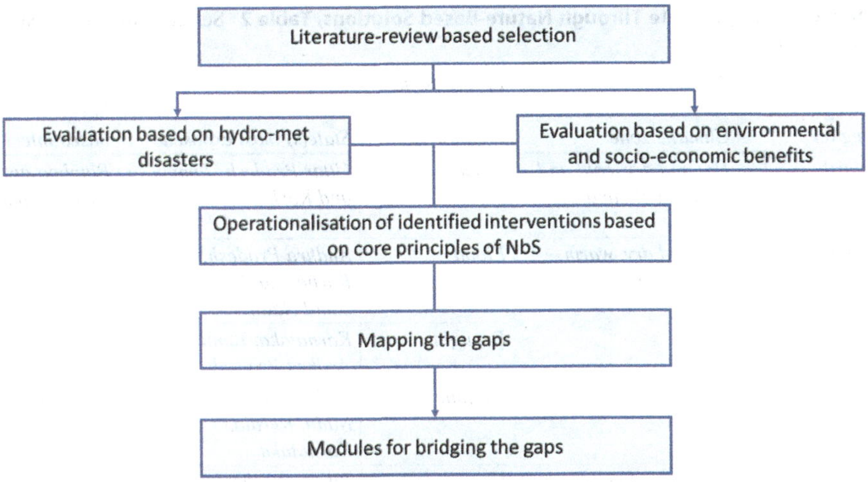

Adapting to a Changing Climate Through Nature-Based Solutions, Fig. 3 A schematic representation of an integrated framework based on core principles of the NbS

solutions (NbS) has been used considering the EbA and Eco-DRR premise which is also acknowledged by IUCN and UNDRR (UNDRR 2020).

The sections below identify some of the scalable and implementable NbS based on a critical review of the literature and provide prospects on the cost to benefits of these solutions through an integrated framework based on core principles of the NbS specified across different hydrometeorological hazards (Fig. 3).

Building India's Resilience Through NbS

The efficacy, paybacks, and recognition of NBSs are reliant on the risk mitigation purpose, hyperlocal context, and socioeconomic setting. Learning from the literature suggest that urban flooding is one of the major challenges that NbS seeks to intervene in compared to other hydrometeorological hazards. Most of the global practices are targeted at large-scale interventions and often overlook the short-term implementation scales. Thus, this entry attempts to identify climate-zone based NbS that can be implemented at scale to mitigate the primary hydro-met disasters and their associated events.

Hazard-Specific Implementable Solutions

Considering the feasibility and the constraints of NbS through the above methodological approach, a hazard-specific tailor-made attempt has been made to characterize and categorize some select solutions. These solutions are shortlisted based on the climate zones, geographical regions, proneness to extreme events, among others. Table 2 illustrates selected interventions based on the above criterion. The progression twitches by opt for potential intervention that resemble to the hyper-local social and landscape attributes.

The abovementioned interventions can be labelled under the EbA and Eco-DRR. However, one of the limitations of the selected hazard and climate zone interlinked NbS does not consider the extent of disease risk mitigation. India is divided into six climate zones: hot and dry, warm and humid, moderate, cold and cloudy, cold and sunny, and composite (when 6 months or more do not fall within any of the above categories) (Bansal and Minke 1988). Within the selected core guidelines mentioned above, ecosystem functions and restoration of the services that have direct influence on the microclimate is prioritized. These interventions have also been categorized to improve the status of ecosystem

Adapting to a Changing Climate Through Nature-Based Solutions, Table 2 Selection of NbS based on climatic zones and geographical regions

S. No.	*Geographic region*	*Climatic zone*	*Proneness to extreme event(s)*	*State(s) most exposed*	*Suggested nature-based intervention*
1.	*North*	*Hot and dry, cold and cloudy, cold and sunny*	*Flood*	*Uttar Pradesh, Jammu and Kashmir*	*Bamboo-grass-based embankments*
2.	*South*	*Hot and dry, warm and humid*	*Flood*	*Andhra Pradesh, Karnataka, Kerala, Tamil Nadu*	*Green corridors*
			Drought	*Karnataka, Tamil Nadu, Andhra Pradesh*	*Anti-salt bunds*
			Cyclone	*Andhra Pradesh, Tamil Nadu, Kerala, Karnataka*	*T-shaped groyne*
3.	*East*	*Warm and humid, composite*	*Flood*	*Bihar, Odisha*	*Bio-dyke*
			Cyclone	*Odisha, Bihar, West Bengal*	*Mangroves as bio-shield*
4.	*West*	*Hot and dry, warm and humid, composite*	*Drought*	*Rajasthan, Maharashtra, Gujarat*	*Agroforestry*
			Cyclone	*Gujarat, Maharashtra*	*Restoration of sand dunes; T-shaped groyne*
5.	*North-East*	*Warm and humid*	*Flood*	*Assam, Manipur, Sikkim, and Arunachal Pradesh*	*Bamboo-grass-based embankments*
6.	*Central*	*Composite, hot and dry*	*Drought*	*Madhya Pradesh*	*Agroforestry*

services. Furthermore, these interventions will have direct impact on the lives and livelihoods of the dependent population and communities. These interventions have both environmental and socioeconomic benefits that enhances the environmental and ecosystem services by interlinking ecosystems and habitats (Reguero et al. 2018). The sector below refers to selected hazard linked interventions that can be piloted and scaled in India to mitigate the extent of loss and damage caused due to hydro-met disasters and will be pivotal in restoring, rehabilitating, and rejuvenating the ecosystems that can hinder the climate change impacts.

Flood-Specific Interventions

An analysis by CEEW suggests that more than 97.51 million people are exposed to extreme flood events with reference to baseline data from Census 2011 (Mohanty 2020). Floods have exposed ecosystems, lives, livelihoods, and estimates suggest that more than 40 million hectares of land are exposed to floods and nearly 8 million hectares of land are impacted by floods annually (Ray et al. 2019). CEEW's analysis suggests that there has been a surge in extreme flood events in recent decades, which is adjunct to the warming of the Indian Ocean (Mohanty and Wadhawan 2021). A warming Indian Ocean supplements to a surge in moisture leading to extreme rainfall events and hence floods in the recent decades (Rao 2012). Given the topography and geography of the Indian subcontinent, the enumerated NbS based on EbA and Eco-DRR framework are suggested (Table 3).

The interventions enumerated are a mix of bioengineering and green infrastructure led intervention targeted at riverine, fluvial, urban flooding, and more importantly can mitigate the compounding flood impacts like soil erosion and

Adapting to a Changing Climate Through Nature-Based Solutions, Table 3 Flood-specific interventions. (Source: Author's compilation)

S. No.	Name of the intervention	Geography or area of implementation	Favorable landscape attributes	Cost and scale of implementation	Co-benefits	Capacity to mitigate risk	Source
1.	Bio-dyke, an environment friendly solution to protect river banks	Bangalipur, a small village in Bardiya, Nepal	Bio-dyke construction is ideally done on clay loam or sandy loam soils. This structure is not suitable at sites where soil is too sandy. This structure is suitable for normal flooding areas but not suitable for heavy flooding areas. The bio-dyke should be built at least 8m away from human settlement areas.	The total cost for constructing a 220-meter-long bio-dyke is 1,153,976 NPR (it includes cost of materials, tools, and labor) USD ~9700	Bio-dyke is a bioengineering measure to manage and control flood water movement such as redirecting flood runoff through the use of flood walls and flood gates. Riverine floods, landslide mitigation, hilly area stabilization, and prevents soil erosion	A 200 m long bio-dyke can prevent 2 Ha of agricultural land from getting eroded during floods	Khadka (2018)
2.	Development of bamboo-grass-based embankment to control river bank erosion (Bangladesh and Nepal)	2018 – ongoing (Bangladesh) and 2015–2020 (Nepal)	Near river banks frequently affected by riverine floods.	The cost of 400-meter long bamboo fence is 400,000 Nepali Rupees (3,300 USD)	Generation of an extra income of 15,000–20,000 Nepali Rupees (NPR) annually (200 USD). Bamboo fencing supplemented with grass plantation reduced the time for bank stabilization. The bamboo fences divert water to the center of river, and over time they trap the silt, leading to extension of the riverbank, rather than its erosion.	Reduce impact of floods, prevent soil erosion	Sinha and Bimson (2021)
3.	Green corridors	Green Infrastructure Chulalongkorn Centenary Park (Bangkok)	Many elevated green spaces can be constructed in urban or semi-urban landscapes facing the issue of urban flooding or water scarcity.	Total monetary benefits, obtained from 697 trees were estimated at USD 101,400 (an average of USD 145.48/ tree/year)	A 3-degree angle model, collects rainwater from its green roof, which can slow down runoff up to 20 times more than regular concrete surfaces.	Reduce the risk of urban floods	UNFCCC (2020)

A

surface runoff. These targeted interventions can mitigate flooding extremities by intervening at a landscape level. The co-benefits of these intervention can support soil quality restoration and build resilience greening the buffer zones.

Globally, India is becoming the world's flood capital, and clearly, it needs to revamp its flood mitigation strategy by implementing and scaling through some of the above mentioned NbS to insulate its communities, lives, livelihoods, and infrastructures.

Cyclone-Specific Interventions

Cyclones are classified under the tropical storm and are defined as "a tropical storm originating over tropical or subtropical waters." "Cyclones are characterised by a warm-core, non-frontal synoptic-scale disturbance with a low-pressure centre, spiral rain bands, and strong winds. Depending on their location, tropical cyclones are referred to as hurricanes (Atlantic, Northeast Pacific), typhoons (Northwest Pacific), or cyclones (South Pacific and the Indian Ocean)" (Em-dat 2022). India has witnessed more than three tropical cyclones yearly in recent decades (Mohanty 2020). IPCC estimates that maximum wind speed in case of a tropical cyclone will increase in the range of 2–11% by the year 2100 (IPCC 2013), while it is estimated that the wind speed is projected to increase by the same range across the north Indian ocean as well (Mohapatra et al. 2012). The impact of a tropical cyclone is sustained for a specific window post-landfall, but its scale of devastation is compounded by the associated events (heavy rainfall, floods, hailstorms, cold waves, tornadoes). Changes in the forest management practices, increase in deforestation, reduction in forest cover, and agricultural practices aggravate the negative impacts of cyclones and even cause the onset of associated hazardous events such as inland flooding and landslides (Srinivas H. 2008). An analysis by CEEW suggests that there has been a fourfold increase in extreme cyclone events across the Indian states, whereas Andhra Pradesh, Maharashtra, and Bihar have registered a twofold increase in extreme cyclone events, and Odisha alone has registered a fourfold surge in extreme cyclone events (Mohanty 2020). Can NbS act as a bio-ecosystemic shield? Table 4 enumerates some of the most feasible NbS that can mitigate the cyclone risks.

Mangroves are the most rampantly used cyclone mitigating NbS that is naturally implemented and scaled across both eastern and western coasts. They can mitigate risks by more than 90% in terms of restricting the tidal flows (Source 1 from the table). Further, mangroves have mitigation co-benefits as well it can sequester carbon fourfold. Further, the mangroves can directly withstand severe cyclonic storms with winds speeds ranging above 178 kmph (Krauss and Osland 2020). Beyond mangroves, T-shaped groyne and restoration of coral reefs and sand dunes are suggested to be pivotal for mitigating the cyclone impacts as enumerated above. Changes in the forest management practices, increase in deforestation, reduction in forest cover, and agricultural practices aggravate the negative impacts of cyclones and even cause the onset of associated hazardous events such as inland flooding and landslides (Srinivas H. 2008). Granular cyclonic risk assessment comprising of inundation levels across the eastern and western coast through these targeted NbS can help in strategizing and deriving hyper-local developmental action plans for preparedness and mitigation.

Drought-Specific Interventions

Droughts have a political, climatological, and economic purview in India. According to Census 2011, more than 52% of India's population is dependent on the agricultural sector (PIB 2020). However, the agriculture sector is the worst hit due to droughts. This self explains that when more than half of the county's earning population is dependent on agriculture, droughts which are most recurrent are devastating livelihoods at an alarming rate. There are primarily three kinds of droughts: (i) meteorological drought (Meteorological drought is defined as the deficiency of precipitation from expected or normal levels over an extended period of time.), (ii) hydrological drought (Hydrological drought is defined as deficiencies in surface and

Adapting to a Changing Climate Through Nature-Based Solutions, Table 4 Cyclone-specific interventions

S. No.	Name of the intervention	Geography or area of implementation	Favorable landscape attributes	Cost and scale of implementation	Co-benefits	Capacity to mitigate risk	Source
1.	Mangroves as bioshield	Various states in India including Odisha and Puducherry	Shorelines with dense vegetation and robust root interlocking system	For 1 ha of mangrove forest in Odisha, the protection value was estimated to be about USD 8700, whereas cleared land fetched only USD 5000 (Benefits = USD 3700)	A mangrove density of 30 trees/100 m^2 in a 100-m wide belt may reduce the maximum giant tidal flow pressure by more than 90%, if the wave height is less than 4–5 m. Sequester carbon four times more efficiently than other ecosystems Act as natural filters of water flowing into the sea, and also retain heavy metals and sediments in their roots and soil substrate Are vital for the existence of many commercially important shrimp, crab, and fish species apart from reptiles, amphibians, birds, and mammals	Mitigating cyclones, storm surges, and tsunamis. They protect tropical shores from erosion by tidal currents.	Sandilya and Kathiresan (2015)
2.	T-shaped groyne	Bang Khun Thian, Bangkok	Near shorelines prone to cyclones or high tidal waves	The construction of a 200 m long T-shaped groyne off the shore would cost 1.5 billion baht (Rs. 35,00,530)	Reduced the rate of soil erosion of the coast from 17.39 m a year from 1979 to 9.5 m a year by 2005. In the future, the inland erosion rate is predicted to be between 1.4 m and 4.5 m a year	Reduce coastal erosion caused by cyclones	Bangkok Metropolitan Administration (BMA). (2014)

(continued)

A

Adapting to a Changing Climate Through Nature-Based Solutions, Table 4 (continued)

S. No.	Name of the intervention	Geography or area of implementation	Favorable landscape attributes	Cost and scale of implementation	Co-benefits	Capacity to mitigate risk	Source
3.	Restoration of oyster reefs, coral reefs, and seagrass beds	Mobile Bay, Alabama, USA	Coastal zones and shorelines prone to cyclones or high tidal waves	The cost of restoring a 5.9 km belt of reef was USD 3.5 million	Reduced average wave heights and energy at the shoreline by 53–91%. Increase in seafood production, reducing nitrogen loads in the coastal waters, and increasing carbon sequestration and storage	Decrease the frequency and intensity of storms and reduce damages from coastal flooding	Kapos, V., et al. (2019)
4.	Restoration of sand dunes	North Norfolk Coastal Restoration (UK)		The cost estimate for creating vegetated dunes is USD 0.3 thousand to USD 5 thousand per linear foot.	Protecting freshwater habitats, enhancing coastal/brackish habitats, contributing to water purification and regulation in coastal aquifers, and creating recreational opportunities	Provides coastal stability by absorbing and dissipating wave energy and prevents stormwater from flooding inland areas	Kapos, V., et al. (2019)

subsurface water supplies, leading to a lack of water for normal and specific needs.), and (iii) agricultural droughts. (Agricultural drought is usually triggered by meteorological and hydrological drought, and occurs when soil moisture and rainfall are inadequate during the crop growing season, causing extreme crop stress and wilting.) Droughts are defined as "an extended period of unusually low precipitation that produces a shortage of water, and operationally, it is defined as the degree of precipitation reduction that constitutes a drought, that varies by locality, climate and environmental sector" (Em-dat 2022). Change in interannual variability of monsoons linked to El Nino South Oscillation (ENSO), warming the Indian Ocean resulting in sea-surface temperature increase is triggering the extreme drought and drought-like conditions across India (Kumar et al. 2013). However, managing extreme droughts through NbS entails primarily watershed management practices, which are prima facie targeted to manage agricultural droughts. Given the purview of EbA and Eco-DRR, the entry suggests the following NbS for mitigating droughts (Table 5). Agroforestry and anti-salt bunds can provide benefits that outlast risks across different regions, including that of coastal.

An analysis by CEEW suggests that there has been a twofold increase in extreme drought events across the Indian states (Mohanty and Wadhawan 2021). Post-2002, there has been substantial improvement in drought monitoring, but gaps still exist in predicting monsoonal drought on seasonal to decadal time scales (Rajeevan et al. 2012). The recent spurt in drought events is triggered by land-use change, urban heat island effect, and changes in precipitation levels. The pentad decadal analysis suggests that the annual average rainfall empirically links the climatological and meteorological linear relationship with drought events until 1990–1999. The drought extremities potentially threaten the economic sectors like agriculture, manufacturing and MSMEs, thereby disrupting the food and water security of the country. Given that only EbA and Eco-DRR led intervention scoped in the entry, water-shed management practices have greater potential to better manage drought and should be supplemented with NbS to fetch better risk-mitigating results.

A

Implementing NbS-at-Scale

The entry explicitly enumerates and generates empirical evidence on the effectiveness, efficacy, and benefits of NbS primarily through the premise of EbA and Eco-DRR. There is no denial that climate is changing-changing fast, and any further delay will etch all developmental trajectories. As the entry pans out, the thrust mainstreaming NbS through piloting and scaling some of the NbS can climate-proof India's communities, economic sectors, and infrastructures. Enumerated are some of the basic recommendations that can implement NbS and bring it from margins to mainstream. First India needs to acknowledge NbS at a system innovation level by integrating it in policies, schemes, and actions. Further, system innovations for NbS can lead to financing NbS thus linking it to social safety nets for its vulnerable populations. Given the financing, India needs to develop an NbS task force balancing the act technological and ecosystem led services.

1. **Implementing NbS-at-Scale Through System Innovation**: System innovation entails policy coherence on NbS that recognized and integrated NbS in India's climate and disaster policies, plans and schemes like NDMA acts, NAPCCs, SAPCCs, so on and so forth. System innovation can bring policy coherence and integrate socioeconomic, developmental environmental governance unification. Furthermore, such innovations can address the sustainable development indices and can enhance regional cooperation at an implementation level. However, it is significant to comprehend that cost of inaction due to lack of system innovation can limit resilience agenda by inequitable natural resource allocation and halting the adaptation modules as committed in NDCs. System innovations will lead to harmonizing fiscal allocations for financing NbS.

Adapting to a Changing Climate Through Nature-Based Solutions, Table 5 Drought-specific interventions

S. No.	Name of the intervention	Geography or area of implementation	Favorable landscape attributes	Cost and scale of implementation	Co-benefits	Capacity to mitigate risk	Source
1.	Agroforestry	Nepal and China	Drought-prone areas with dependence on rain-fed agriculture or fallow lands	Agroforestry interventions can fetch USD 7 worth ecosystem and ecosystem services benefits for every USD 1	Enrichment planting in forests and degraded shrublands Diversifying income earning potential	Reduced risk of water scarcity and rainwater flooding	Mander, M. (2018)
2.	Anti-salt bunds	Coastal and rural areas of Senegal	Landscape prone to soil degradation due to salinization of water and land, attributed to low freshwater inputs during periods of drought, deforestation, and inland fresh water extraction	Approximately USD 440,000 for the construction of 76 anti-salt bunds (includes cost for cumulative production on 232 ha of cultivated lands)	Avoided loss of agricultural yield due to soil degradation (salinization and erosion) Avoided loss of incomes in poultry farming	Reduced soil degradation due to salinization during periods of drought	Raza, A. (2019)

2. **Financing NbS:** India needs to include nature-based infrastructure like wetlands, mangroves, forest ecosystems, and some highlighted and targeted bioengineering interventions like bio-dykes, anti-sand dunes, among others, under the ambit of critical risk-mitigating infrastructure. Built-in infrastructure like buildings, roadways network systems, electric systems, dams, and bridges are currently considered under critical infrastructure's standard definition and practice. Broadening the definition of infrastructure to include natural ecosystems and nature-based solutions offers an opportunity to deploy and enhance nature-based solutions (NbS) to produce sustainable and climate-resilient responses. According to a recent study, wetland and ridge restoration could save USD 7 in avoided damages for every USD 1 invested. Further, more than 45% of the climate risk over 20 years could be averted, saving more than USD 50 billion worth of damages against extreme flood events (Luedke 2019). Restoring, rebuilding, and investing in nature-based solutions can make our cities and villages more climate-resilient and alter the adverse impacts of climate change.
3. **Establishing a NbS Task Force:** Mainstreaming and promoting climate responsive NbS through a constitutionally mandated body can create means and ways to mainstream and implement NbS-at-scale. Through NbS India can adopt a proactive risk mitigation strategy empowered to analyze and identify the cost and benefits of environmental and socioeconomic benefits of NbS interlinked with granular risk assessments. Currently, climate action plans do not consider NbS during the design or implementation phase and hence an NbS task force can be pivotal in bridging these gaps.

These recommendations will prepare India to formulate strategies to climate-proof its population, economies, and infrastructure by integrating, implementing, and scaling NbS. If the impacts of a 1.5 °C warmer future are irreversible, adequate adaptation led actions through NbS can halt the scale and impact of climate extremities. The resilience trajectory of India hinges on embracing NbS as a primary tool for human civilization against climate change.

A

Cross-References

▶ Green Cities: Nature-Based Solutions, Renaturing and Rewilding Cities
▶ Growth, De-growth, and Nature-Based Solutions
▶ Innovation to Bring Nature-Based Solutions to Life: Tales of Two Cities
▶ Policy and Practices of Nature-Based Solutions to Build Resilience in Seoul, Korea

Acknowledgments The author acknowledges the contribution of Ms Shreya Wadhawan, Research Analyst-CEEW for her contribution to the entry.

References

Bangkok Metropolitan Administration (BMA). (2014). https://www.bangkokpost.com/thailand/politics/403518/saving-the-bang-khunthian-coast-will-cost-b1-5bn

Bansal, N K, and Minke, G. (1988). Climatic zones and rural housing in India. Part 1 of the Indo-German project on passive space conditioning. Germany: N. p. https://www.osti.gov/etdeweb/biblio/7784799

Cohen-Shacham, E., et al. (2019). Core principles for successfully implementing and upscaling nature-based solutions. *Environmental Science and Policy, 98*, 20–29. https://doi.org/10.1016/j.envsci.2019.04.014

Eckart, Kyle & Mcphee, Zach & Bolisetti, Tirupati. (2017). Performance and implementation The Science of the total environment, 607–608, 413–432. https://doi.org/10.1016/j.scitotenv.2017.06.254

Eckstein, D., Künzel, V., & Schäfer, L. (2020). Germanwatch. Global climate risk index. Retrieved from https://www.germanwatch.org/sites/default/files/Global%20Climate%20Risk%20Index%202021_2.pdf

EM-DAT. (2022, January). *The CRED/OFDA International Disaster Database*. Retrieved from Available online: http://www.emdat.be/

European Commission. (2015). Nature-based solutions: The EU and nature-based solutions. https://ec.europa.eu/info/research-and-innovation/research-area/environment/nature-based-solutions_en

Ghosh, A., & Raha, S. (2020). *Jobs, growth and sustainability: A new social contract for India's recovery.* New

Delhi: Council on Energy, Environment and Water: CEEW and NIPFP. Retrieved from https://www.ceew.in/publications/jobs-growth-and-sustainability

IMD. (2020). *End of the Season—Southwest Monsoon 2020.* New Delhi: PIB, GoI, Ministry of Earth Science. Retrieved from https://static.pib.gov.in/WriteReadData/userfiles/End%20of%20Season%20Report_2020.pdf

IPCC. (2013). Climate Change 2013: The Physical Science Basis. Contribution of Working Group I to the Fifth Assessment Report of the Intergovernmental Panel on Climate Change [Stocker, T.F., D. Qin, G.-K. Plattner, M. Tignor, S.K. Allen, J. Boschung, A. Nauels, Y. Xia, V. Bex and P.M. Midgley (eds.)]. Cambridge University Press, Cambridge, United Kingdom and New York, NY, USA, 1535 pp. https://www.ipcc.ch/report/ar5/wg1/

IPCC. (2014). Climate Change 2014: Synthesis Report. Contribution of Working Groups I, II and III to the Fifth Assessment Report of the Intergovernmental Panel on Climate Change [Core Writing Team, R.K. Pachauri and L.A. Meyer (eds.)]. IPCC, Geneva, Switzerland, 151 pp. https://www.ipcc.ch/report/ar5/syr/

IUCN. (2016). WCC-2016-Res-069-EN Defining Nature-based Solutions. https://portals.iucn.org/library/sites/library/files/resrecfiles/WCC_2016_RES_069_EN.pdf

Juan Ossa-Moreno, Karl M. Smith, Ana Mijic. (2017). Economic analysis of wider benefits to facilitate SuDS uptake in London, UK. *Sustainable Cities and Society, 28*, 411–419, ISSN 2210-6707, https://doi.org/10.1016/j.scs.2016.10.002. https://www.sciencedirect.com/science/article/pii/S2210670716304541

Kamaljit Ray, P. P. (2019). On the recent floods in India. *Current Science.* https://doi.org/10.18520/cs/v117/i2/204-218

Kapos, V., Wicander, S., Salvaterra, T., Dawkins, K., Hicks, C. (2019). The role of the natural environment in adaptation, background paper for the global commission on adaptation. Rotterdam and Washington, D.C.: Global Commission on Adaptation. https://gca.org/wpcontent/uploads/2020/12/RoleofNaturalEnvironmentinAdaptation

Krauss, K. W., & Osland, M. J. (2020). Tropical cyclones and the organization of mangrove forests: A review. *Annals of Botany, 125*(2), 213–234. https://doi.org/10.1093/aob/mcz161.

Kumar, K., Rajeevan, M., Pai, D., Srivastava, A., & Preethi, B. (2013). On the observed variability of monsoon droughts over India. *Weather and Climate Extremes*, 42–50. https://doi.org/10.1016/j.wace.2013.07.006.

Lawson, E., et al. (2014). Delivering and evaluating the multiple flood risk benefits in blue-green Transactions on Ecology and the Environment 184, 113–124. https://books.google.com/books?hl=en&lr=&id=nYLcAwAAQBAJ&oi=fnd&pg=PA113&ots=QVCR_B9TkA&sig=zTvBkA7Fla_kRgqMZmdfB6-yW-)

Luedke, H. (2019). Nature as Resilient Infrastructure – An Overview of Nature-Based Solutions. October. https://www.eesi.org/papers/view/fact-sheet-nature-asresilient-infrastructure-an-overview-of-nature-based-solutions

Mander, M. (2018). Ecosystem Services Supply, Demand and Values at Petit Barbarons, Seychelles. Ecosystem-based Adaptation through South-South Cooperation (EbA South) Final Report. Nairobi: UNEP. Accessed from: https://gca.org/wpcontent/uploads/2020/12/RoleofNaturalEnvironmentinAdaptation

Merz, B., Kreibich, H., Schwarze, R., & Thieken, A. (2010). Review article "Assessment of economic flood damage". *Natural Hazards and Earth System Sciences, 10,* 1697–1724. https://doi.org/10.5194/nhess-10-1697-2010.

Mohanty, A. (2020). *Preparing India for Extreme Climate Events: Mapping Hotspots and Response Mechanisms.* Council on Energy, Environment and Water. Retrieved from https://www.ceew.in/publications/preparing-india-for-extreme-climate-weather-events

Mohanty, A., & Wadhawan, S. (2021). *Mapping India's Climate vulnerability – A district-level assessment.*

Mohapatra, M., et al. (2012). Classification of cyclone hazard prone districts of India. *Natural Hazards, 63.* https://doi.org/10.1007/s11069-011-9891-8.

Naumann, Sandra & Rayment, Matt. (2011). Design, implementation and cost elements. https://www.researchgate.net/publication/272352149_Design_implementation_and_cost_elements

Nehren, Udo & Sudmeier-Rieux, Karen & Sandholz, Simone & Estrella, Marisol & Lomarda, Mila & Guillén, Tania. (2013). The ecosystem-based disaster risk reduction case study and exercise book. https://www.researchgate.net/publication/disaster_risk_reduction_case_study_and_exercise_book/citation/download

PIB. (2020). Ministry of Agriculture & Farmers Welfare: Agrarian Land. Accessed from: https://pib.gov.in/PressReleasePage.aspx?PRID=1601902

Rajeevan, M., Unnikrishnan, C. K., & Preethi, B. (2012). Evaluation of the ENSEMBLES multi-model seasonal forecasts of Indian summer monsoon variability. *Climate Dynamics, 38*(11–12), 2257–2274. https://doi.org/10.1007/s00382-011-1061-x.

Khadka, R. (2018). Bio-dyke, an environment friendly solution to protect river banks Practical Action: Transforming lives, inspiring change. http://repo.floodalliance.net/jspui/bitstream/44111/2974/1/Bio_Dyke.pdf

Rao, S. D. (2012). Why is Indian Ocean warming consistently? *Climatic Change, 110*, 709–719. https://doi.org/10.1007/s10584-011-0121-x.

Raza, A. (2019). Ecosystems Protecting Infrastructure and Communities Programme (EPIC) Coastal and Rural Areas of Senegal. Accessed from: https://gca.org/wpcontent/uploads/2020/12/RoleofNaturalEnvironmentinAdaptation

Reguero, B. G., Beck, M. W., Bresch, D. N., Calil, J., & Meliane, I. (2018). Comparing the cost effectiveness of nature-based and coastal adaptation: A case study from

the Gulf Coast of the United States. *PLoS One, 13*, 1–24. https://doi.org/10.1371/journal.pone.019213.
Romnée, A., & De Herde, A. (2015). Hydrological efficiency evaluation tool of urban Stormwater best management practices. *International Journal of Sustainable Development and Planning, 10*, 435–452. https://doi.org/10.2495/SDP-V10-N4-435-452.
Ruangpan L., Vojinovic Z., Sabatino S. D., Sandra L., Capobianco L., Oen A.M.P. (2020). McClain Earth Syst. Sci., 20, 243–270. https://doi.org/10.5194/nhess-20-243-2020
Sandilyan, S., & Kathiresan, K. (2015). Mangroves as bioshield: An undisputable fact. *Ocean and Coastal Management, 103*, 94–96. https://doi.org/10.1016/j.ocecoaman.2014.11.011.
Secretariat of the Convention on Biological Diversity (2009). Connecting biodiversity and climate of the second Ad Hoc Technical Expert Group on Biodiversity and Climate Change. Montreal, Technical Series. https://www.cbd.int/doc/publications/cbd-ts-41-en.pdf
Sinha, V. R., & Bimson, K. (Eds.). (2021). *Nature-based Solutions in the Ganges Brahmaputra Meghna (GBM) river basin: Case studies and lessons learned.* Bangkok, Thailand: IUCN ARO. viii + 69pp.
Seifollahi-Aghmiuni, S., Nockrach, M., & Kalantari, Z. (2019). The potential of wetlands in achieving the sustainable Development Goals of the 2030 Agenda. *Water, 11*(3), 609. https://doi.org/10.3390/w11030609.
Srinivas, H., & Nakagawa, Y. (2008). Environmental implications for disaster preparedness: lessons learnt from the Indian Ocean Tsunami. *Journal of environmental management, 89*(1), 4–13. https://www.sciencedirect.com/science/article/pii/S0301479707001430
Strecker, Eric & Quigley, Marcus & Urbonas, Ben & Jones, Jon & Clary, Jane. (2001). Determining Effectiveness. *J Water Resour Plan Man*-9496(2001)127:3 (144)
UNDRR. (2020). *Words into action: Nature-based solutions for disaster risk reduction.* https://www.undrr.org/words-action-nature-based-solutions-disaster-risk-reduction
UNEP-WCMC and UNEP. (2019). *Briefing note 6: Integrating EbA into national planning.* https://wedocs.unep.org/bitstream/handle/20.500.11822/28179/Eba6.pdf?sequence=1&isAllowed=y
UNFCCC. (2020). Nature-Based Solutions to Increase Urban Adaptability, Thailand. Accessed from: https://unfccc.int/climateaction/momentum-forchange/women-for-results/naturebased-solutions
Whelans and Halpern Glick Maunsell. (1994). City of Melbourne WSUD Guidelines. https://www.melbourne.vic.gov.au/SiteCollectionDocuments/wsud-full-guidelines.pdf
WMO. (2020). *The State of the Climate in Asia 2020.* Retrieved from https://library.wmo.int/doc_num.php?explnum_id=10867
Zhang, K., & Chui, T. F. M. (2018). Linking hydrological and bioecological benefits of green infrastructures across spatial scales – A literature review. *Science of the Total Environment, 646*, 1219–1231. https://doi.org/10.1016/j.scitotenv.2018.07.355, 2019.

Adaptive Governance

► Overcoming Barriers in Green Infrastructure Implementation

Adequate Housing

► Housing and Development

Administration

► Closing the Loop on Local Food Access Through Disaster Management

Advanced Environmental Practices

► Voluntary Programs for Urban and Regional Futures

Adversity

► Closing the Loop on Local Food Access Through Disaster Management

Affordable Housing

► Housing Affordability

Affordable Shelter

▶ Housing Affordability

Age-Friendly Community

▶ Age-Friendly Future Cities

Age-Friendly Future Cities

Mengxing Ma[1,2] and Gaurav Joshi[3]
[1]Department of Social Work, University of Melbourne, Melbourne, VIC, Australia
[2]Department of Geography, University of Sheffield, Sheffield, UK
[3]University of Chinese Academy of Sciences, Beijing, China

Synonyms

Age-friendly community; Elder-friendly community; Lifetime neighborhood

Definition

An age-friendly city adapts its structures and services to be inclusive for diverse older adults. By doing so, the enabling environments and communities are built to foster healthy and active aging and to improve the quality of life of older people. An age-friendly future city is a city where older adults' voices and needs are at the heart of the urban agenda, where government commits to making its city age-friendly, and where diverse stakeholders and cross-cutting sectors cooperate to support age-friendly city building. Age-friendly future cities are all-age-friendly cities that foster intergenerational solidarity, embrace social justice lens to promote equal aging, and are evenly developed across the globe.

Introduction

In the twenty-first century, cities across the globe are becoming home to larger and older populations. The age-friendly city approach informed by the active aging paradigm has been seen as an effective response to the two global megatrends of population aging and urbanization. The World Health Organisation (WHO) model of the age-friendly city has widely been adopted by many cities and communities across the world with a modification based on local contexts. Three elements that cities successfully adopt age-friendly initiatives in common are briefly introduced in this chapter, including co-design and co-creation with older adults, co-operation with diverse stakeholders and cross-cutting sectors, top-down political commitment and resources. The chapter also points out three directions that the age-friendly city approach should focus on in future development, namely, all-age-friendly cities, age-friendly cities promoting equal aging, and age-friendly city development in the Global South.

Demographic Drivers

Two major demographic shifts are taking place worldwide in the early twenty-first century: rapid population aging and continued urbanization. In 2019, over 1 billion people were older adults aged 60+ years, making up 13.2% of the world's population. This number is projected to double by 2050, reaching 2.1 billion. At the same time, the process of urbanization is speeding up. According to UN-Habitat, half of the global population are residing in urban areas today, which is expected to be two-thirds by 2030 and increase to 70% by 2050. In other words, there are more older adults living in cities. WHO estimates that 57% of older adults will be living in towns and cities by 2050. The age-friendly city approach has emerged as a key strategy in response to these two challenges on a global scale.

WHO Age-Friendly City Model

The concept of age-friendly cities was first introduced by the director of the WHO's global Ageing and Life Course Programme at the 2005 Congress of the International Association of Geriatrics and Gerontology (IAGG). The WHO age-friendly city model put the idea of active aging into practice. The word active refers to "continuing participation in social, economic, cultural, spiritual and civic affairs, not just the ability to be physically active or to participate in the labor force" (WHO 2002, p. 12). According to WHO (2007), an age-friendly city "encourages active aging by optimizing opportunities for health, participation and security in order to enhance quality of life as people age" and "adapts its structures and services to be accessible to and inclusive of older people with varying needs and capacities" (p. 1). In 2007, the WHO published the *Global Age-friendly Cities: A Guide* (referred to as *the WHO Guide* hereafter) that contains eight domains of urban life to make a city age-friendly. The WHO Guide has become the basis for the majority of age-friendly city strategies across the world. In 2015, WHO (2015) published the *Measuring the Age-friendliness of Cities: A Guide to Using Core Indicators* that provides a set of core and supplementary indicators to monitor and evaluate a city's age-friendliness. In order to support cities and communities to become more age-friendly and to facilitate their mutual learning, WHO established the Global Network for Age-friendly Cities and Communities (GNAFCC) in 2010. By February 2021, over 1000 cities and communities in 44 countries, covering over 262 million people worldwide, have become members of the GNAFCC. Age-friendly city initiative has become a global movement to encounter the challenge of population aging.

The WHO Guide lists 66 core age-friendly features across eight domains (see Fig. 1). Initially, these core features aimed to "provide a universal standard for an age-friendly city" and were claimed to apply to "less developed as well as more developed cities" (WHO 2007, p. 11). Many cities and communities across the world have relied on this guide for baseline assessment and evaluation of their age-friendliness. However, based on the practice of age-friendly city development in many cities, it becomes commonsense that each city is unique and there is no one-size-fits-all recipe for building an age-friendly city. The needs of older adults and the resources of

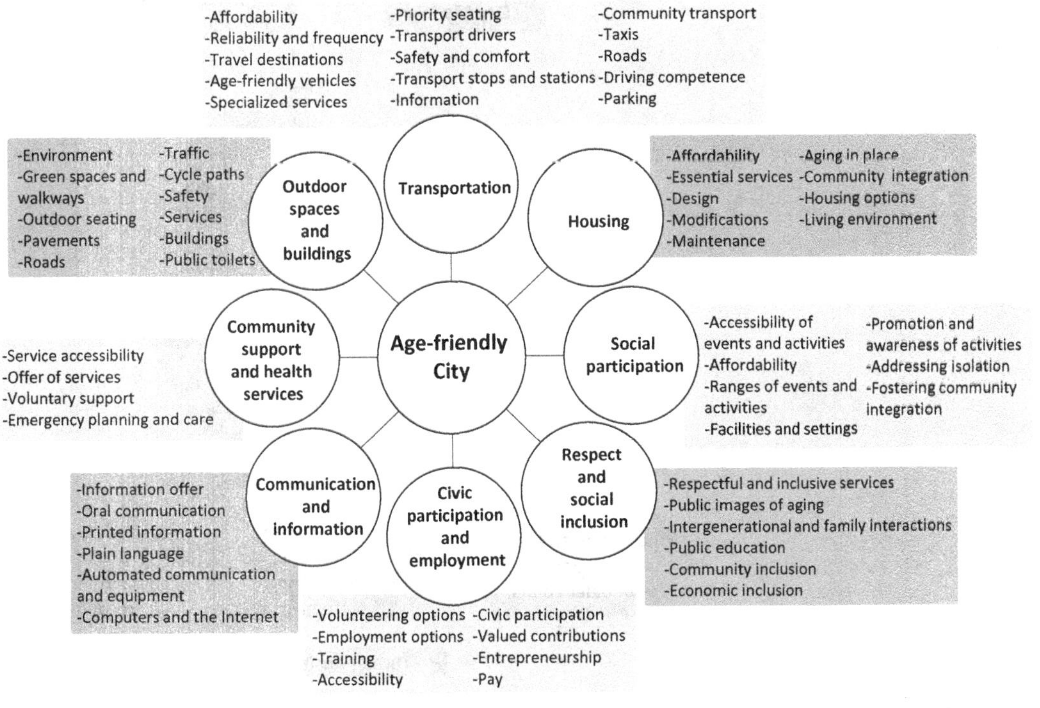

Age-Friendly Future Cities, Fig. 1 The eight domains and 66 core features of an age-friendly city (WHO 2007)

the city are highly contextualized. The older adults' needs listed in the WHO Guide are too broad. It is unrealistic for a city to target all features at one time without considering its local circumstances such as cultural, social, political, or economic settings. Therefore, the WHO Guide and its checklist should not be seen as a prescription for an ideal age-friendly city. Rather, it needs to be reviewed and modified to local models that reflect older adults' needs and sets the priorities within the local contexts of that specific city.

Principles of Age-Friendly Cities Building

As an age-friendly city program is usually a modification of the WHO model based on the city's local context, the strategies of different cities' age-friendly development differ a lot. However, there are some common principles for those who managed to be successful (see Fig. 2).

Co-design and Co-creation with Older Adults

Empowering older adults as place-makers and maximize their participation is a key for the success and sustainability of age-friendly city development. The WHO Guide itself is based on a bottom-up participatory approach. Older adults, caregivers, and service providers from 33 cities from all continents participated in focus group research to identify features of age-friendly cities. However, if a city adopts the WHO checklist as preconceived features to implement age-friendliness, it will become a tick box tokenistic approach. Instead, older adults need to be involved at all stages of age-friendly city development from need assessment, decision-making, service design to project delivery and evaluation. Older adults and their voices should be put at the heart of all the work. They are not only the beneficiaries of age-friendly cities but also advocates, advisors, overseers, and contributors. In the case of Manchester (see Buffel 2018), older adults are also co-researchers in partner with researchers

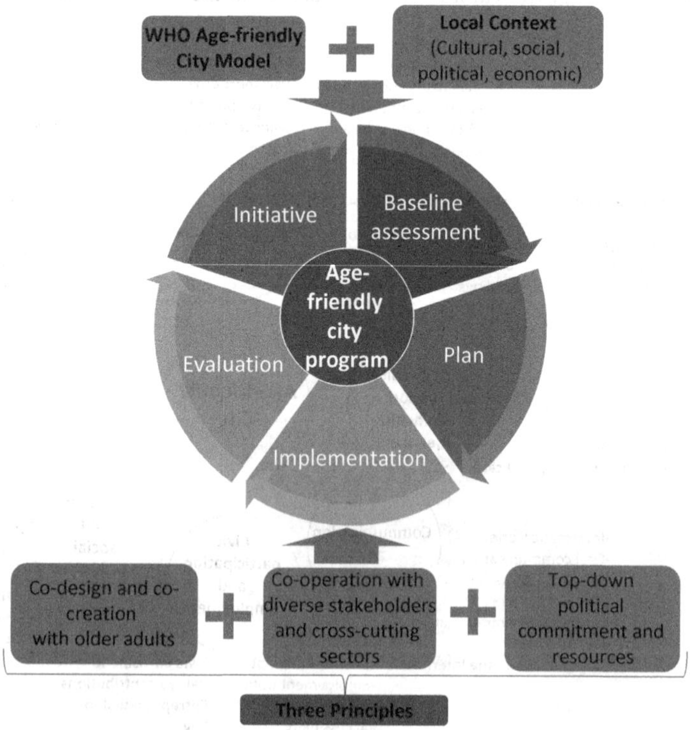

Age-Friendly Future Cities, Fig. 2 Elements for the success of age-friendly city development

from university to investigate how friendly their communities are. Older adults are the experts to pinpoint the specific strength and weak points in their own cities and communities' friendliness. To ensure everyone's inclusion and avoid excluding the marginalized, it is essential to support excluded older groups (e.g., those from minority ethnic communities, LGBTQ communities, and socioeconomic disadvantaged groups) to voice their concerns.

Co-operation with Diverse Stakeholders and Cross-Cutting Sectors

A collaborative partnership with different stakeholders has been recognized as a key element in building age-friendly cities. A variety of stakeholders – policymakers, academics, voluntary and community sectors, and private companies – can build partnerships and use their resources to support communities. Close collaboration at the grassroot and local levels is particularly important in cities and communities where resources are scarce. To actualize the age-friendliness of cities and communities, many sectors – health, transport, housing, police, social protection, education, and others – need to come together in designing and implementing age-friendly city strategies. Steering committees have been demonstrated to be an effective instrument to foster collaborative partnerships (see the example of Quebec at Garon et al. 2014). Partnerships among different stakeholders, cross-cutting sectors, and engagement of older adults are important to establish a sustainable age-friendly city initiative.

Top-Down Political Commitment and Resources

Age-friendly city initiatives need a whole-of-government effort. Therefore, the political commitment of the government is particularly important. The lack of continuing political and resource commitment is one of the causes that make age-friendly city initiatives unrealized and unsustainable. Positive governmental support is indispensable to support the bottom-up initiatives for implementing age-friendly city initiatives otherwise the bottom-up approach may end up with insufficient resources. A mixed approach – adopting a bottom-up participatory approach in combination with top-down commitment and resources – is an effective strategy to actualize age-friendly cities.

Age-Friendly Future Cities

All-Age-Friendly Cities

Although age-friendly city initiatives focus on older adults, making a city age-friendly will make life better for all generations. The goal is to make the city or community a great place to grow older. This embraces a life course perspective, meaning that the status or well-being of an older person is influenced by advantages and disadvantages over his/her life course (Kendig and Nazroo 2016). For instance, age-friendly city initiatives and child-friendly city movements both aim to take account of the needs of specific age groups but they are usually advocated separately. However, children and older adults often co-occupy the same public spaces and have some interests in common. Younger generations also gain optimized opportunities for health, participation, and security that originally for older people. Future cities could be all-age-friendly cities where infrastructure, services, and public spaces are friendly for everyone regardless of their age. By embracing a life-course approach, an age-friendly city can also create greater intergenerational solidarity within its communities.

Age-Friendly Cities Promoting Equal Aging

Unequal aging is one challenge to build age-friendly cities. Older adults are a heterogeneous group. They experience inequality based on their ethnicity, gender, sexuality, disability, health status, socioeconomic status, and other social positions. Implementing age-friendly city initiatives needs to consider the interests of different older groups and establish specific strategies. Some subgroups, such as the oldest-old, older adults with dementia, older migrants, homeless older adults, are largely ignored in mainstream age-friendly city development. A truly age-friendly city or community is the environment that embraces the social justice lens and promotes equal aging.

Age-Friendly City Development in the Global South

One main goal at the initial phase of the WHO age-friendly city initiatives is to help developing countries tackle the challenge of rapid population aging. However, age-friendly cities are developing fast in the Global North but very few in the Global South. For instance, no city or community from Africa has joined the GNAFCC so far. Developing countries, especially low-income countries, have too many challenges to face, and aging fast is just one of them. Consequently, limited resources usually are used to tackle imminent issues. However, cities in the Global South need to be prepared for the challenge of population aging and urbanization rather than passively encounter it. Cross-national collaboration can provide opportunities for cities in the Global South to borrow successful experiences from established age-friendly cities in the Global North. More importantly, it is essential to have a substantial modification of the age-friendly city approach based on lived experiences and age-friendliness perceptions of older adults living in low- and middle-income countries.

Conclusion

Aging in urban settings has become a priority challenge in many countries across the world. The Age-friendly city approach has the potential to effectively respond to this challenge by creating supportive and enabling environments for people to actively grow older within their communities. The success of age-friendly city initiatives will depend on ensuring older adults' major role in shaping and directing the work and seeking alliances with diverse actors and cross-cutting sectors. Making a city age-friendly is to create a future for our older selves as well as meet the needs of current generations. Its potential to benefit all has been well recognized. The age-friendliness of an urban environment is also dynamic as the realities of aging in the future city are changing. Therefore, building an age-friendly city is an ongoing effort. Periodic review and revisions of age-friendly strategies are needed to ensure they are in line with up-to-date thinking, evidence, and practice. Currently, much of the available knowledge and experience are from developed countries. Future developments of age-friendly city frameworks and tools need to consider the application in the developing world. How to implement age-friendly city initiatives in a resource-scarce setting worth exploring? Although this chapter focuses on urban settings, it is important to highlight the importance of building an age-friendly environment in suburban and rural contexts as they are undergoing a similar demographic change of population aging.

Cross-References

► Community Engagement for Urban and Regional Futures
► Future of Urban Governance and Citizen Participation
► Role of Nature for Ageing Populations

References

Buffel, T. (2018). Social research and co-production with older people: Developing age-friendly communities. *Journal of Aging Studies, 44*, 52–60.

Garon, S., Paris, M., Beaulieu, M., Veil, A., & Laliberté, A. (2014). Collaborative partnership in age-friendly cities: Two case studies from Quebec, Canada. *Journal of Aging & Social Policy, 26*(1–2), 73–87.

Kendig, H., & Nazroo, J. (2016). Life course influences on inequalities in later life: Comparative perspectives. *Journal of Population Ageing, 9*(1–2), 1–7.

WHO. (2002). *Active ageing: A policy framework*.

WHO. (2007). *Global age-friendly cities: A guide*.

WHO. (2015). *Measuring the age-friendliness of cities: a guide to using core indicators*.

Further Reading

Moulaert, T., & Garon, S. (Eds.). (2015). *Age-friendly cities and communities in international comparison: Political lessons, scientific avenues, and democratic issues*.

Torku, A., Chan, A. P. C., & Yung, E. H. K. (2020). Age-friendly cities and communities: A review and future directions. *Ageing & Society*, 1–38.
WHO Global Network for Age-friendly Cities and Communities. https://extranet.who.int/agefriendlyworld/network/

Agency

► New Localism: New Regionalism

Agenda Post-2015

► Sustainable Development Goals from an Urban Perspective

Agricultural Production

► Circular Economy and the Water-Food Nexus

Agroforestry

► *Wadi* Sustainable Agriculture Model, The

Allotment

► Urban Food Gardens

Alternative Agriculture

► *Wadi* Sustainable Agriculture Model, The

Alternative Compliance Tools

► Voluntary Programs for Urban and Regional Futures

Alternative Protein

► Future Foods for Urban Food Production

Amsterdam

► Amsterdam's Pathway to Climate Neutrality: Creating an Enabling Environment

Amsterdam's Pathway to Climate Neutrality: Creating an Enabling Environment

Maria Kottari
School of Transnational Governance, European University Institute, Florence, Italy

Synonyms

Amsterdam; Buildings; Citizens; Climate change; Climate neutrality; Decarbonization; Energy transition; Governance; Mobility

Definition

A vibrant European capital and a pioneer in climate-friendly solutions related to citizens services within its urban environment, the municipality of Amsterdam has agreed on a roadmap that aims to decrease the city's CO_2 emissions by 55% in 2030 and by 95% in 2050, compared to 1990, which is used as a reference year. In 2040, Amsterdam would like to be free of the use of natural gas altogether and it aims for all traffic to be emission-free as early as 2030. A package of measures to achieve these goals has been agreed upon and adopted by the Mayor and Cabinet at the beginning of March 2020. The so-called *Amsterdam climate-neutral 2050 Roadmap* includes targets and measures for all relevant sectors: the built environment, mobility, electricity, and industry and harbor.

This chapter aims to describe the climate challenges of Amsterdam's urban environment linked to the goals and targets set in the climate-neutrality roadmap. Further, to break down the goal with a focus on the built environment and mobility sectors for the following reasons. The role the municipality can play in achieving the desired goals and the importance of citizen engagement coupled with the prospects of citizen-driven innovation and profit-sharing in the respective sectors.

Introduction

A densely populated and compact built city, Amsterdam is growing rapidly; by around 2030, the city's population is expected to reach a million residents. The further growth of the city will pose additional challenges associated with urban planning and climate risks.

In more details, the temperature has risen by 1.6 °C since 1950 while in the future, more rainfall is expected with more frequent, heavier downpours. Heatwaves will also be more frequent (indicatively, an increase to 40 summer days is expected by 2085) affecting public health by creating heat stress and sleep disruption. Biodiversity will be threatened and flooding risks will rise due to the sea level rise by 85 cm this century.

The gradual growth of the city (including more residents, homes, jobs, and tourists) after the 1990s has resulted in the amount 5510 kt CO_2 emissions in 2010. Since then, and despite the continuous rapid urban growth, the CO_2 emissions in Amsterdam have fallen since 2010 mainly due to the increasing generation and use of renewable energy and the fall in energy consumption per resident.

As of today, the total greenhouse gas emissions in the Netherlands are already lower than in 1990, but this is not yet the case for Amsterdam. The city's emissions are still falling too slowly.

In the expected growth of the city, we should also consider the BREXIT impact in terms of the relocation of businesses in trading and finance, medicine, and agriculture, as well as logistics and distribution sectors looking to secure their European operations as the UK departed the European Union. This upward trend that started in 2019 already results in population growth and a significant rise in building construction to accommodate the business activities as well as the additional housing needs, creating additional needs for energy supplies.

The climate-neutral 2050 Roadmap is setting Amsterdam's carbon ambitions, to reverse the upwards emissions trend, and the required measures to achieve them.

Setting the Ambition

Amsterdam's emission ambition has been sketched within the interplay of the Dutch climate ambition and a meaningful contribution to meeting the Paris Climate targets.

A collective product of more than 600 agreements between businesses, civil society organizations, and governmental authorities, the Dutch Climate Agreement, enshrined in law in 2019, sets national targets for cutting greenhouse gas emissions: 49% fewer CO_2 emissions in 2030 and 95% fewer in 2050, compared to 1990. Having endorsed the national climate pledges and after having launched an open invitation to all related stakeholders within 2019 – similar to the process that took place at the national level – that led to a first outline of the path to a climate-neutral and gas-free city along with budget estimations, the city council has finalized and endorsed the climate-neutral 2050 Roadmap at the beginning of March 2020

The ambition puts high stakes. A maximum 55% cut in greenhouse emissions by 2030 and a 95% cut by 2050, both compared to 1990 levels. However, it is not certain how far the municipality will succeed in implementing the planned actions and measures as well as what the size of the effects will be. As such the 2030 goal lays in the range between −21% and −55% with the most plausible estimation to be emissions cut of 48% by 2030 which is practically in line with the national reduction target of −49%.

These overarching emissions reduction goals encompass a wider city vision with various components. A climate-neutral and adaptive city able to handle effectively the inevitable consequences of climate change. A city embracing the circular economy principles, where energy is used efficiently and generated sustainably, and where raw and other materials are reused in a never-ending cycle. A municipality that will achieve becoming a climate-neutral organization by 2030.

The emissions reduction ambition will be executed through four transition paths targeting the built environment, mobility sector, electricity, harbor, and industry. As already mentioned, this piece will delve further into the buildings and mobility sectors

Understanding the Challenge

The municipality has put forward a management and governance instrument. The Climate Budget practically consists of the quantitative part of the Roadmap. It provides insights about the emissions targets (delimiting for instance the ranges between the targets) but, also, about the effectiveness of the measures the municipality is taking. It determines which carbon emissions Amsterdam is responsible for and it assesses the measures that do not lead directly to emissions reduction, but that are needed for the transition. More concretely, the Climate Budget helps the municipality to monitor the various aspects of energy transition that vary from purely quantitative indicators, e.g., the number of public electric rapid charging points in Amsterdam to more complex ones such as measuring the citizens' support for the energy transition and the switch to renewable energy, which attained the 75% in 2019. (Data provided by the Research, Information and Statistics (OIS) Department, Municipality of Amsterdam.).

The Climate Budget is not seen as a static mechanism. The purely quantitative part, e.g., the development of CO_2 emissions in Amsterdam or the progress of other standard indicators, will be updated annually and a system monitoring the progress and estimating the impacts of the measures taken will provide the necessary insights to the municipality to adjust its choices and set, if necessary, new priorities.

The Climate Budget is complemented by a municipal financial instrument. The Mayor and Cabinet have deposited a total of 150 million euros in the Climate Fund to support the energy transition in Amsterdam.

Amsterdam Going (Dough)nuts?!

In its pathway towards climate neutrality, Amsterdam will be accompanied by a "fellow traveler." Inspired by the concept of the "Doughnut Economy," developed by the British economist Kate Raworth, according to which an economy can only be sustainable and prosperous if social and economic goals are achieved without exceeding the ecological ceiling, Amsterdam's new vision for urban development will be based on the principles of circular economy, reassessing the impact of the current economic system. Many voices have been risen questioning whether this new model will challenge capitalism in a city that can claim its origins.

It is estimated that 63% of Amsterdam's CO_2 emissions are caused by products and materials. Based on the *Circular Amsterdam 2020–2025* program, the city aims to adopt a smarter approach to the use of raw materials (the aim is to halve the use of new raw materials by 2030 and to achieve a fully circular city by 2050), to avoid waste or use it as fuel, to "close" the production circles for products and buildings. Circularity will target three value chains responsible for the scope-3 emissions: food and organic waste/consumer goods/built environment, and is considered to play a supportive key role in combating climate change.

The municipality has developed a model, the Monitor for Circular Economy, based on input indicators related to the three prioritized value chains (Expressed in terms of weight and will be further broken down in CO_2 emissions and environmental cost.). This model is constantly updated with more accurate data to identify areas

A

that need further intervention, to determine the social and ecological impact of the circular transition, and to define at the end the feasibility of the circularity goals. The Monitor is not socially blind. Amsterdam's ambition is to keep its residents inside the "doughnut circle" safeguarding their "broad prosperity" but without putting more pressure on the planet than is sustainable. Broad prosperity broadens the concept of a good life beyond material wealth to include elements such as efficiency, equitable distribution, and continuity in meeting everyday life needs (housing, food, etc.) while respecting and preserving the natural environment.

Spotlight on Mobility Sector

A city that is known for its biking culture and citizen-friendly transportation system, Amsterdam wants to push environmental-friendly mobility even further, initiating the transition to a system in which people and goods can be transported without emitting CO_2. It is estimated that if all mobility in Amsterdam becomes emissions-free by 2030, this will reduce carbon emissions overall by around 360 kilotons (This includes motorized traffic on Amsterdam's roads, excluding motorways and provincial roads.). Mobility is currently responsible for 18% of CO_2 emissions in Amsterdam, mainly due to fossil fuel consumption, and it is expected to be the transition path that will achieve the greatest reduction in emissions with a −76% in 2030 compared to 2017. It is also considered the transition path that will materialize most rapidly due to the triple-helix innovation model and positive development stemming from private sector initiatives the municipality will subsidize (Set of interactions between academia, industry, and government.).

Amsterdam is already leading the way towards the electrification of urban mobility. Amsterdam offers free public electric vehicles (EV) charging stations, ranking second in Europe (in terms of the number of public chargers per million population) after Oslo and first on a global scale. This is mainly due to the national forward-looking approach to EVs with incentives policies, varying from subsidies, taxes, to charging facilitation, that has been proven so successful that electric car ownership now costs the same as diesel or gasoline car ownership.

Amsterdam itself offers additional subsidies and scrapping schemes, more precisely, since the end of 2019, as part of the Clean Air Action Plan, Amsterdam has provided financial incentives for zero-emission vehicles, offering rebates of up to EUR 3000 (USD 3684) for a taxi and EUR 40,000 (USD 49,129) for a delivery van, truck, or bus. These incentives came along with the municipality's announcement that any cars and motorbikes running on petrol or diesel would be banned from driving in the city center starting in 2030, and nonelectric buses will no longer be able to enter the city center from 2022. The municipality itself is committed to "green" the vehicles including it the municipal fleet.

A critical success factor in Amsterdam's low-carbon transport schemes will be a supportive state policy in the reevaluation of the environmental zones in 2022 mandating the municipality to introduce those zones and impose traffic restrictions.

Focus on the Built Environment

Contrary to the mobility sector, the buildings' transition path will be lengthier and less straightforward. Responsible for 25% of the carbon emissions, primarily caused by the consumption of natural gas and heat consumption by existing and new homes, commercial buildings, and social and civic buildings (such as offices, schools, and hospitals), the majority of the emissions reduction in the built environment is expected to take place after 2030. The planned switch of all existing and new homes from the natural gas grid to sustainable heating by 2040 is expected to deliver a carbon reduction that accounts for around

370 kilotons, almost 30% of current emissions in the buildings sector.

The built environment is the transition path the municipality will play the largest role and where the largest part of the Climate Fund will be allocated. This is mainly because the municipality can have considerable influence on the requirements for new buildings, defining the conditions for the granting of land and challenging the tendering parties to build as sustainably as possible.

However, the situation is different for the existing built environment, which is highly energy-consuming and with a considerable carbon footprint. There the intervention from the part of the municipality will be highly dependent on the collaboration and partnerships with property owners such as housing corporations, owner-occupants, owner associations, and commercial property investors. Additionally, significant investment will be required from property owners, energy companies, and grid operators to make the built environment more energy efficient and improve the living conditions of the resident, phase out the use of natural gas, and provide sustainable heating.

Phasing out natural gas consumption in the built environment will lead to the greatest drop in emissions. This result is highly dependent on national legislation and financial support from the central government. By law, as part of the national Climate Agreement, all seven million homes in the Netherlands should use energy and heating from renewable sources by 2050. However, so far a slow pace of progress is observed and it obvious more stimulus and stronger incentives are required. The financial supporting schemes will have a significant influence on the extent to which residents have a significant financial incentive to disconnect from the natural gas grid. The city of Amsterdam is investing heavily in this domain; however, tougher regulations at the national level accompanied by investment funds will be the real game-changer towards the decarbonization of the built environment.

Installing solar panels on the building roofs, heat pumps, and collective management of heat distribution grids (possibly fueled by hydrogen) are then prioritized and promoted technological alternatives for the gradual replacement of the natural gas grid. The municipality is planning to take a systematic district-by-district approach, outlined in detail in a vision document the "Heat Transition Vision" adopted in September 2020, attempting to identify and implementing the preferred alternative sustainable heating source in each case acknowledging that a one-size-fits-all solution is neither feasible not viable.

The success of this complicated process is highly dependent on the citizens' input and support. The municipality and the private partners proposed a pathway that was open to public consultation until May 2020. After lengthy discussions among residents, companies, and institutions, an agreement at the administrative level has been reached. Citizen-driven and citizen-focused initiatives that support the different groups of residents (tenants, owners, etc.) in implementing the heat transition are supported financially by the municipality, considered an instrumental partner of the process. Those initiatives facilitate two-way communication between the residents and the municipality while providing useful insight to the policy-makers on the barriers the citizens are facing (02025 targeting homeowners, Woon-Thuis in de stad targeting residents in building corporations, Regionaal Energieloket targeting individuals that want to implement retrofitting and energy efficiency solutions by themselves.). However, it is usually observed that the citizens participating actively in those processes have a high educational background and a considerable income. The participation is lower among low-income citizens, usually with an immigrant background. Another crucial element is the rather high percentage of energy poverty (31% in the year 2017) affecting the low-income households not allowing them to invest money in energy efficiency or retrofitting measures.

All in all, despite the municipality's good intentions towards the creation of an inclusive process in the heating transition, obstacles remain and are expected to delay the whole process.

The Many Roles of a (the) Municipality

The scale of ambition described in the Roadmap will challenge the municipality in many different ways. On its part has defined the preconditions needed to meet the intended targets. Those preconditions include activities that may not reduce the CO_2 emissions directly but will play a crucial role in the transition from fossil to renewable energy and achieving a fully circular economy.

The municipality sets "climate justice" as the guiding principle. Acknowledging climate change as a justice problem and the fact that the energy transition will not be equal for everyone, it aims for a fair distribution of the costs and benefits, open access to the decision-making process, and equal opportunities in a changing job market. Seen as a broader social transformation, energy transition and circular economy need to embrace the will of the citizens. Strengthening the positive movement in the city, stimulating knowledge transfer and innovation, managing the urban space and providing the necessary infrastructure are the pillars upon the climate neutrality will be achieved.

As such, the municipality is called upon to undertake different roles in the process. These roles will change over time, depended on the various transition paths progress.

A municipality that performs taking the lead and setting the example. The municipality will work on becoming a sustainable organization. This means making the internal operational management climate-neutral, including the transport fleet and building stock. It also concerns everything the municipality purchases, including the materials, used to organize public space but, also, the behavior of the staff.

A municipality that regulates along with the central government, setting indicators and standards whenever possible and advocating the necessary reforms.

A municipality that cooperates with residents, businesses, and institutions to gradually design shared ambitions and achieve common goals through cooperative platforms and agreements, capacity building, and sharing of resources.

A municipality that supports the assumed self-regulation and self-organization of the different stakeholders, e.g., homeowners associations with instruments including subsidies, advice and information, facilitation of knowledge sharing, the establishment of help desks to support the initiatives, simplifying guidance on procedures and rules, lobbying, and communication.

Quo Vadis Amsterdam? (Conclusions)

With an adopted roadmap, specific emission reduction targets per transition path, and a new economic model underway, one could argue that the city of Amsterdam has already taken important steps towards the right path. Assuming the responsibility as the capital city of the Netherlands, Amsterdam is an urban climate advocate, pioneering especially in the mobility sector. Nevertheless, the local ambition without a robust national regulatory and investment framework is bound to narrower effects.

The renewable energy transition has the support of the majority of the citizens (According to a public survey, 75% of Amsterdamers expressed their support to the renewable energy transition in 2019.). However, some residents or neighborhoods are more vulnerable or will benefit less from the opportunities brought by the energy transition, and this is primarily translated in the slow transition pace we observe in the built environment. Amsterdam's housing rates are already extremely high and the thorny question to answer is how to avoid a rise in the living expenses for households with low or middle incomes due to the costs of the energy transition. Additionally, while the residents may be willing to boost the energy efficiency of their households high costs, uncertainty about how long they will live in the property, and expectations that it will be cheaper to do so in the future, are stopping them from taking action. The rules are still complex, the technological solutions seem promising but not implemented and evaluated to a satisfactory level. Phasing out natural gas in heating will be the biggest challenge to face, but, if accomplished, a one-of-a-kind urban transition and transformation paradigm.

Cross-References

- ▶ Adapting Cities to Climate Change
- ▶ Circular Cities
- ▶ Circular Economy Cities
- ▶ City Visions: Toward Smart and Sustainable Urban Futures
- ▶ Climate Resilience in Informal Settlements: The Role of Natural Infrastructure
- ▶ Collaborative Climate Action
- ▶ Community Engagement for Urban and Regional Futures
- ▶ Future of Urban Governance and Citizen Participation
- ▶ The Governance of Smart Cities
- ▶ Green Belts
- ▶ Green Cities
- ▶ Green Infrastructure
- ▶ Green Infrastructure in Metropolis Dimension: Case Study of Llobregat River, Barcelona Metropolitan Area
- ▶ Housing Affordability
- ▶ How Cities can be Resilient
- ▶ Low-Carbon Transport
- ▶ Making of Smart and Intelligent Cities
- ▶ Overcoming Barriers in Green Infrastructure Implementation
- ▶ Participatory Governance for Adaptable Communities
- ▶ Policies for a Just Transition
- ▶ Public Procurement for Regional and Local Development
- ▶ Public Space
- ▶ Regulation of Urban and Regional Futures
- ▶ Resilient Urban Climates
- ▶ Smart Cities
- ▶ Smart City
- ▶ Sustainable Urban Mobility
- ▶ Systemic Innovation for Thrivable Cities
- ▶ The Sustainable and the Smart City: Distinguishing Two Contemporary Urban Visions
- ▶ Theme Cities Networks
- ▶ Toward a Sustainable City
- ▶ Transport Resilience in Urban Regions
- ▶ Transportation and Mobility
- ▶ Unpacking Cities as Complex Adaptive Systems
- ▶ Urban and Regional Leadership
- ▶ Urban Commons as a Bridge between the Spatial and the Social
- ▶ The Urban Planning-Real Estate Development Nexus
- ▶ Urban Policy and the Future of Urban and Regional Planning in Africa
- ▶ Urban Resilience
- ▶ Urban Resilience: Moving from Idealism to Systems Thinking
- ▶ Urban Structure and its Impact on Mobility Patterns: Reducing Automobile Dependence through Polycentrism

Acknowledgments The author would like to thank the Municipality of Amsterdam, the Amsterdam Institute of Advanced Metropolitan Solutions and 02025 network and platform for sharing their knowledge and insights on the climate-neutrality Roadmap and the various subjects covered in this Chapter.

Further Reading

Amsterdam Circular Strategy 2020–2025. (2020). https://assets.amsterdam.nl/publish/pages/867635/amsterdam-circular-2020-2025_strategy.pdf

Dutch Climate Agreement. (2019). https://www.klimaatakkoord.nl/binaries/klimaatakkoord/documenten/publicaties/2019/06/28/national-climate-agreement-the-netherlands/20190628+National+Climate+Agreement+The+Netherlands.pdf

Raworth, K. (2017). *Doughnut economics. Seven ways to think like a twenty-first-century economist*. Randon House.

Roadmap Amsterdam Climate Neutral 2050. (2020). https://assets.amsterdam.nl/publish/pages/887330/roadmap_climate_neutral.pdf
Roadmap Amsterdam Climate Neutral 2050. (2019). Phase 1: An invitation to the city. http://carbonneutralcities.org/wp-content/uploads/2019/12/Amsterdam-Climate-Neutral-2050-Roadmap_12072019-1.pdf

Anticipatory Cities

► Systemic Innovation for Thrivable Cities

Applying Smart Frameworks to Arctic Cities

Andreas Raspotnik[1,2] and Victoria Herrmann[2]
[1]High North Center for Business and Governance, Nord University, Bodø, Norway
[2]The Arctic Institute – Center for Circumpolar Security Studies, Washington, DC, USA

Definition

This entry considers established metrics for smart city development and evaluates their suitability for implementation in Arctic urban settlements. To do this, the entry first surveys smart city literature and the standardization of "smartness" metrics, with particular interest in the International Organization for Standardization's (ISO) categorization efforts. It then proposes a northern framework of measurement to evaluate smart cities that adjusts smart metrics from current non-Arctic scholarship to the relatively low populations, peripheral development, remote locations, and harsh climate conditions of the circumpolar north. Exploring this is important because these frameworks have implications for how policymakers in northern regions choose to plan and implement their city strategies.

From public discourse to media narrations, the Arctic is most often constructed as an uninhabited, aesthetically spectacular land of ice and snow. And yet, in spite of its popularly imagined identity as one of the last remote and inhospitable frontiers of our shared planet, the Arctic and those who call it home have been a part of, and influenced by, the international economic, political, and cultural developments of the wider world since the fifteenth century (Kunz and Mills 2021).

Since the late 1300s, the global trade routes that crisscrossed the Bering Strait between the North American Arctic and Eurasia have been traveled by Indigenous peoples, bringing beads crafted by glassmakers in Venice to northern Alaska along the Silk Road network. The early forms of cities, with their permanently built environments and concentrated economic activities, came later with the arrival of Euro-American colonialism and capitalism, built as extractive communities rather than places to live (Rasmussen et al. 2015, p. 426). Their vitality was dependent on local resource extraction and their creation as small, remote settlements was based on where the highest concentration of these commercialized sea and land resources could be found.

Today, roughly three-quarters of the Arctic's population live in urban areas, with an increasing trend of urbanization to 2055 as the region "is intensely urbanizing" as "towns grow at a rapid pace and people settle in urban centers, often far away from their home settlements" (Dybbroe et al. 2010, p. 120; Heleniak 2021; Heleniak and Bogoyavlenskiy 2015, pp. 93–94). And though much of the regional economy is still reliant on industrial development, bulk trade from primary industry, and public administration, urban Arctic economies are diversifying to create a distinct knowledge economy based on increasing primary, secondary, and higher education opportunities (Rasmussen et al. 2015, p. 438). Over the past two decades, globalization processes, political centralization, and market volatility from more competitively priced natural resource production southward have created a space to redefine Arctic settlements as livable communities independent of the extraction of a single resource. These shifts in populations, economies, and ultimately cultures that accompany the circumpolar north's urbanization come with novel opportunities and challenges, making it a critical field for future research. This turns our attention to an emerging but underexplored topic in regional studies: *Arctic smart cities*.

Definition and Dimensions of a Smart Sustainable City

The concept of smart cities, developed and most often applied to cities further south than the Arctic Circle, is derived from a straightforward notion: cities are the source of major local and global challenges, and can also be the source of transformative solutions. Cities fuel economic development as centers of capital, workforce, knowledge, information, and technology. However, global cities are also confronted with a multitude of key challenges, including traffic congestion, unplanned development, poor land use regulation, and greenhouse gas emissions (Bansal et al. 2015, p. 551; Garrido-Marijuan et al. 2017, p. 3). To meet these challenges, urban planners have pioneered solutions that are environmentally sustainable and link all areas of a city's economy together in a more efficient way. This often involves more efficient public transportation, energy efficient buildings, and a stronger focus on research, innovation, and knowledge.

Chief among the frameworks to mitigate these challenges and measure success are strategies to develop "sustainable," "smart," "resilient," and "green" cities, all of which are interconnected in their indicators and goals. While there exists a plethora of development frames within which cities can measure their progress, this entry will focus specifically on smart sustainable cities – the intersection of the two interrelated concepts of sustainable cities and smart cities. Sustainable cities achieve "a balance between the development of the urban areas and protection of the environment with an eye to equity in income, employment, shelter, basic services, social infrastructure and transportation in the urban areas" (Hiremath et al. 2013, p. 556). Within this frame, cities must achieve sustainability in the environmental, social, and economic pillars of the concept; however, there has been criticism that sustainable cities elevate environmental indicators at the expense of social and economic performance metrics (Berardi 2013; Robinson and Cole 2015; Tanguay et al. 2010). During the past decade, sustainability has been superseded in popularity by the concept of "smart cities," as a means to achieve urban sustainability (Ahvenniemi et al. 2017, pp. 235–236; Huovila et al. 2019, p. 142). Smart cities build off the previous work of sustainable cities but instead focus development on the implementation of technology, especially information and communication technologies (ICTs). A smart city exists "when investments in human and social capital and traditional (transport) and modern (ICT) communications infrastructure fuel sustainable economic growth and high quality of life, with a wise management of natural resources, through participatory governance" (Caragliu et al. 2011, p. 70). Building on a public-private-people-partnership approach, which is the cooperation between the public sector, companies and individuals, this definition synthesizes the important aspects of ICTs, sustainable growth, and the human component, both in terms of participation and life quality. Set forth in document ISO 37122, a smart city is one that:

> increases the pace at which it provides social, economic and environmental sustainability outcomes and responds to challenges such as climate change, rapid population growth, and political and economic instability by fundamentally improving how it engages society, applies collaborative leadership methods, works across disciplines and city systems, and uses data information and modern technologies to deliver better services and quality of life to those in the city (residents, businesses, visitors), now and for the foreseeable future, without unfair disadvantage of others or degradation of the natural environment. (International Organization for Standardization (ISO) 2019)

Nonetheless, similar to the critique of sustainable cities' environmental focus, the concept of "smart cities" has been widely criticized for its techno-centricity, lacking proper attention to cities' needs and environmental sustainability (Grossi and Pianezzi 2017; Huovila et al. 2019, p. 142; Rosati and Conti 2016). Due to the deficiencies within both sustainable city and smart city frameworks – the former being biased towards environmental components and the latter towards technological ones – there has been a contemporary push towards their integration, founded on the notion that sustainable and technological goals must coexist. In 2015, the United Nations specialized agency for information and communication technologies (ITU) developed a definition combining the two concepts. A "smart sustainable city" is one which

> is an innovative city that uses information and communication technologies (ICTs) and other means to improve quality of life, efficiency of urban operation and services, and competitive- ness, while ensuring that it meets the needs of present and future generations with respect to economic, social, environmental as well as cultural aspects. (International Telecommunication Union (ITU) 2015)

Although the use of modern technology has been recognized as key aspect of a smart city, both academia and policymakers have developed a great number of different definitions with slightly different angles of what smart means in a city (planning) context (Ahvenniemi et al. 2017; Anthopoulos 2017; Meijer and Bolívar 2016; Mora et al. 2017). To adapt this concept to the Arctic region requires reviewing established, standardized metrics for evaluating both the smartness and sustainability of a city; surveying current and anticipated initiatives, policies, and projects in Arctic cities identified as smart through three cases; identifying which smart city metrics are more prominently pursued in Arctic urban development; and considering the challenges that hinder and catalysts such smart pursuits.

Measuring an Arctic Smart City: Metrics and Standardization

In many ways, Arctic cities reflect their counterparts further south; however, both the built environments and the socioeconomic fabrics of cities in far northern latitudes are climatically, geographically, and demographically unique (Raspotnik et al. 2020). These limitations include, but are not limited to, the North's harsh climate; a low density and uneven distribution of population; long distances to global markets; the difficult access to Arctic-based professional skills training and education; depopulation processes, particularly among youth; and a large share of professionals working in the public sector. These problems are similar for many subregions within the Arctic, and are exacerbated by the movement of populations away from rural areas and into the cities of resource-rich regions (Suter et al. 2017, p. 112).

In order to clarify what an Arctic smart city is and how it can be evaluated, a closer examination of the different aspects that make a city smart and the further contextualization of those aspects to an Arctic context is required. ISO catalogues a set of 22 distinct smart dimensions, each with a set of specific measurements of a smart city that can be quantitatively measured and steer the performance of city services and quality of life. These indicators and their measurements are categorized below into six smart city components, *see* Table 1: Smart people, smart governance, smart mobility, smart environment, smart living, and smart energy (Lombardi et al. 2012).

When applied to the Arctic, each of these metrics requires geographic, climatic, and societal modifications to accommodate the uniqueness of the region. While smart economies for southern cities focus on the development of an economy based on the Internet of Things, from a northern point of view, smart economies should also include the development and implementation of new technologies to redefine "remote," so that projects once deemed inaccessible are economically feasible and communication campaigns to combat Arctic-remote bias for investments. Arctic aspects of smart people are similar to southern metrics and focus on access to education, social cohesion, and the robustness of public life. But when measuring smart people for a northern city, more attention should also be allocated to Indigenous education and inclusion programming. Arctic aspects of smart governance do not differ greatly from non-Arctic governance metrics. However, because a number of Arctic cities and regional settlements around urban areas – particularly in North America, Greenland, and Russia – have limited access to high-speed internet, northern local governments must be strategic in pursuing a combination of digital and

Applying Smart Frameworks to Arctic Cities, Table 1 International Organization for Standardization's smart city metrics

People	Governance	Mobility	Environment	Living	Energy
Education	Governance	Transportation	Urban/local agriculture and food security	Housing	Energy
Health	Telecommunications	Finance	Environment and climate change	Population and social conditions	Finance
Economy	Finance	Environment and climate change	Solid waste	Recreation	Telecommunications
Telecommunications		Energy	Wastewater	Safety	Urban/local agriculture and food security
Housing			Water	Sports and culture	
				Urban planning	

non-digital platforms of transparent, accessible government engagement, also inclusive of Indigenous languages. Arctic smart mobility necessitates a widening of both the modes and the geographic reach of urban transportation networks beyond standard multimodal measurements.

Unique modes of Arctic transportation like cross-country skiing, snowmobiles, dog sleds, and mobility infrastructure and vehicles also require cold climate-tested smart technologies that can withstand freezing temperatures and larger vehicles that transport raw materials to markets; for the latter, road and rail systems must be planned at the regional level to connect surrounding economic activities to the urban hub. Arctic aspects of a smart environment adhere to the evaluations of non-Arctic cities with an added urgency and commitment to cold-climate technologies. Energy efficiency of cold-climate housing and integration of renewable energy systems also take on newfound urgency with high costs of fossil fuel-based electricity, heating, and transport fuel in the Arctic. This is particularly important because, though wide-ranging in the measured sectors, scholars note that smart city frameworks, such as the ISO one, lack robust climate change specific indicators (Ahvenniemi et al. 2017). As climate change is warming the circumpolar north at more than twice the rate of the global average, climate mitigation and adaptation are foundational measures for ensuring that Arctic cities meet present needs without compromising future ones. And finally Arctic aspects of smart living are defined by the added dimension of long, dark winters and bright summers with volatile weather. Public health programs, place-based community building, and dedicated cultural spaces need to be smartly designed for inclusive, year-round use.

Thus, a smart and sustainable Arctic city is one that can score highly in each of these measurement through mobilizing and using "available resources to improve its inhabitants' quality of life, significantly improves its resource-use efficiency, reduces its demands on the environment, builds an innovation-driven and green economy, and fosters a well-developed local democracy" (Garau and Pavan 2018, p. 4).

Conclusion

Far from the pristine, uninhabited images of wide expanses of tundra that have captivated the human imagination since the Age of Exploration, the Arctic is very much alive with vibrant communities, economic growth, and inventive urban development. It is anything but the static tabula rasa that Franklin and Scott set forth to conquer and colonize; each Arctic city – both alone and collectively – are sources of innovation and employment; home to culturally diverse identities; integral nodes of the globe's ecological, economic, and social systems; places of pride and democratic empowerment; and, above all, it is a homeland. In the decades to come, it is projected that Arctic residents will continue to depopulate smaller settlements in favor of concentrated metro areas. As the circumpolar north enters the century of the city, it is critical to examine the ways in which we measure the success of livable, smart urban development and investment.

This entry offers a brief overview and framework through which to apply such analyses, upon which future scholars can build and expand. The proposed framework for Arctic urban analysis must integrate metrics of sustainability, and in particular those related to climate change, into the smart city standardization set forth by ISO (Höjer and Wangel 2015). The Arctic smart city survey to follow also considers quantitative indicators for performance measurement of smart cities to meet the need of climate mitigation and adaptation by including particular consideration for greenhouse gas emission targets in smart environment and resilience building in smart environment and smart living.

Cross-References

- ► Big Data for Smart Cities and Inclusive Growth
- ► Circular Economy Cities
- ► City Visions: Toward Smart and Sustainable Urban Futures
- ► Local and Regional Development Strategy
- ► Making of Smart and Intelligent Cities
- ► The Sustainable and the Smart City: Distinguishing Two Contemporary Urban Visions
- ► Toward a Sustainable City

References

Ahvenniemi, H., Huovila, A., Pinto-Seppä, I., & Airaksinen, M. (2017). What are the differences between sustainable and smart cities? *Cities, 60*, 234–245.

Anthopoulos, L. (2017). Smart utopia vs smart reality: Learning by experience from 10 smart city cases. *Cities, 63*, 128–148.

Bansal, N., Shrivastava, V., & Singh, J. (2015). Smart urbanization – Key to sustainable cities. In M. Schrenk, V. V. Popovich, P. Zeile, P. Elisei, & C. Beyer (Eds.), *Real Corp 2015* (pp. 551–560). http://www.corp.at/archive/CORP2015_27.pdf

Berardi, U. (2013). Sustainability assessment of urban communities through rating systems. *Environment, Development and Sustainability, 15*(6), 1573–1591.

Caragliu, A., del Bo, C., & Nijkamp, P. (2011). Smart cities in Europe. *Journal of Urban Technology, 18*(2), 65–82.

Dybbroe, S., Dahl, J., & Müller-Wille, L. (2010). Dynamics of arctic urbanization. *Acta Borealia, 27*(2), 120–124.

Garau, C., & Pavan, V. M. (2018). Evaluating urban quality: Indicators and assessment tools for smart sustainable cities. *Sustainability, 10*(3), 575.

Garrido-Marijuan, A., Pargova, Y., & Wilson, C. (2017). *The making of a smart city: Best practices across Europe*. European Commission. https://smartcities-infosystem.eu/sites/www.smartcities-infosystem.eu/files/document/the_making_of_a_smart_city_-_best_practices_across_europe.pdf

Grossi, G., & Pianezzi, D. (2017). Smart cities: Utopia or neoliberal ideology? *Cities, 69*, 79–85.

Heleniak, T. (2021). The future of the Arctic populations. *Polar Geography, 44*(2), 136–152.

Heleniak, T., & Bogoyavlenskiy, D. (2015). Arctic populations and migration. In J. N. Larsen & G. Fondahl (Eds.), *Arctic human development report: Regional processes and global linkages* (pp. 53–104). Nordic Council of Ministers.

Hiremath, R. B., Balachandra, P., Kumar, B., Bansode, S. S., & Murali, J. (2013). Indicator-based urban sustainability – A review. *Energy for Sustainable Development, 17*(6), 555–563.

Höjer, M., & Wangel, J. (2015). Smart sustainable cities: Definition and challenges. In L. M. Hilty & B. Aebischer (Eds.), *ICT innovations for sustainability. Advances in intelligent systems and computing* (pp. 333–349). Springer.

Huovila, A., Bosch, P., & Airaksinen, M. (2019). Comparative analysis of standardized indicators for Smart sustainable cities: What indicators and standards to use and when? *Cities, 89*, 141–153.

International Organization for Standardization (ISO). (2019). *ISO 37122:2019 Sustainable cities and communities – Indicators for smart cities*. Geneva. https://www.iso.org/standard/69050.html. Accessed on 6 October 2021.

International Telecommunication Union (ITU). (2015). *Focus group on smart sustainable cities*. https://www.itu.int/en/ITU-T/focusgroups/ssc/Pages/default.aspx

Kunz, M. L., & Mills, R. O. (2021). A Precolumbian presence of Venetian glass trade beads in Arctic Alaska. *American Antiquity, 86*(2), 395–412.

Lombardi, P., Giordano, S., Farouh, H., & Yousef, W. (2012). Modelling the smart city performance. *Innovation: The European Journal of Social Science Research, 25*(2), 137–149.
Meijer, A., & Bolívar, M. P. R. (2016). Governing the smart city: A review of the literature on smart urban governance. *International Review of Administrative Sciences, 82*(2), 392–408.
Mora, L., Bolici, R., & Deakin, M. (2017). The first two decades of smart-city research: A bibliometric analysis. *Journal of Urban Technology, 24*(1), 3–27.
Rasmussen, R. O., Hovelsrud, G. K., & Gearheard, S. (2015). Community viability and adaptation. In J. N. Larsen & G. Fondahl (Eds.), *Arctic human development report: Regional processes and global linkages* (pp. 423–473). Nordic Council of Ministers.
Raspotnik, A., Grønning, R., & Herrmann, V. (2020). A tale of three cities: The concept of smart sustainable cities for the Arctic. *Polar Geography, 43*(1), 64–87.
Robinson, J., & Cole, R. J. (2015). Theoretical underpinnings of regenerative sustainability. *Building Research and Information, 43*(2), 133–143.
Rosati, U., & Conti, S. (2016). What is a smart city project? An urban model or a corporate business plan? *Procedia - Social and Behavioral Sciences, 223*, 968–973.
Suter, L., Schaffner, C., Giddings, C., Orttung, R. W., & Streletskiy, D. (2017). Developing metrics to guide sustainable development of arctic cities: Progress & challenges. In L. Heininen, H. Exner-Pirot, & J. Plouffe (Eds.), *Arctic Yearbook 2017* (pp. 112–131). Northern Research Forum.
Tanguay, G. A., Rajaonson, J., Lefebvre, J. F., & Lanoie, P. (2010). Measuring the sustainability of cities: An analysis of the use of local indicators. *Ecological Indicators, 10*(2), 407–418.

Approach

► Public Policies to Increase Urban Green Spaces

Appropriate Technology

► Senegalese Ecovillage Network

Area

► Closing the Loop on Local Food Access Through Disaster Management

Artificial Urban Wetlands

Robert Rogerson
Institute for Future Cities, University of Strathclyde, Glasgow, UK

Synonyms

Constructed urban wetlands; Engineered urban wetlands; Restored urban wetlands; Treatment urban wetlands

Definition

An artificial or constructed urban wetland is a new or restored marsh area within a city designed to manage anthropogenic discharge such as wastewater, stormwater runoff, or sewage treatment and to assist in land reclamation after ecological disturbances associated with mining and urban development. They may also provide habitats for native and migratory wildlife.

Introduction

With climate change bringing about less predictable weather events, altering the patterns of rainfall, and generating more extreme events, renewed attention has been given to how nature-based solutions can be reabsorbed into urban design and planning. One of these responses has been attempts to recreate wetlands through artificial constructions in cities, recognizing their potential natural role in water and land management, biodiversity, hydrology, and human health and reversing in part the loss of natural wetlands that has accompanied urbanization. Despite planning initiatives to reduce urban sprawl and promote more compact cities, the rapid expansion of the world's urban population has put pressure on natural landscapes at an unprecedented rate, and wetland loss has continued to point where it is claimed between one third and one half of all wetlands have been lost over the last past two centuries (Davidson 2014; Hu et al. 2017).

Nontraditional water management approaches such as SuDS (sustainable drainage systems) have helped in meeting the challenges of climate change and urban growth, but there is recognition for a fundamental change in how in the future urban water and flood risk are managed. The constructed or artificial urban wetlands as part of blue-green infrastructure centered on living with and making space for water are increasingly adopted internationally (O'Donnell et al. 2017).

Contributing to Water Management and Habitat Development

While constructed wetlands are acknowledged to have three main applications – wastewater management, flood control, and habitat creation – most scientific research has focused on water management viewing them as engineering systems understood and created through complex processes involving physical, chemical, and biological mechanisms.

Since the early pioneering work in the 1960s and 1970s, ecological engineering technology involved in the construction of artificial wetlands has evolved quickly and now means a diverse set of wastewater can potentially be treated – from municipal wastewater from housing, through industrial waste, to agricultural wastewater as well as stormwater and other runoff (Masi et al. 2018). The core principles remain the same, with the nature of construction for wastewater treatment commonly tanks or waterproofed pods so that treatment systems are self-contained and isolated from the surrounding area. Water flow can be surface or subsurface and depending on site conditions can be naturally flowing or pumped. This focus on artificial wetlands as purification systems, seeking to replicate the various naturally occurring processes under controlled conditions, has dominated the ways in which artificial wetland development has progressed. Scholz (2015) provides an excellent account of the development of scientific knowledge in the use of wetlands to reduce pollution.

In seeking to make gains in water quality through improvement and efficiency of treatment performance, attention has been given to the development of appropriate plants, substrates, and operational parameters. Although debates over the role of plants in water treatment continue (Saggai et al. 2017; Shelef et al. 2013), the selection of macrophytes and plants usually reflects the characteristics of natural treatment systems of wetlands so that hydrophilic and aquatic species are used having the advantage of absorbing pollutants and then utilizing these in their growth. Substrate conditions – essential to the hydraulic connectivity (flow of water through the wetland) and to the growth of plants and microbial community structure – have also been subjected to scrutiny, with the differing benefits and properties of rock types and the use of by-products of other activity (e.g., rubble) selected to form the substrates being shown to influence the efficiency of artificial wetlands. However, as one of the few parameters that can be controlled in individual wetland sites, much research has focused on how to influence the hydrology of selected sites. While, ideally, a natural flow of water through the site makes it easier to manage and be sustainable, this is seldom achievable as more urban-based sites are selected for the construction of artificial wetlands. Emphasis has thus been placed on artificial water flow management, with Gorgoglione and Torretta (2018) providing a useful overview.

Evaluation of artificial wetlands suggests that as water treatment they are less efficient than traditional water treatment plants. As Ingrao et al. (2020) note, their pollutant removal efficiency may be less consistent, as it may vary seasonally in response to changing environmental conditions, including rainfall and drought that require larger spaces. They also required larger spaces making them only economical where land is available and affordable. Offsetting this, constructed wetlands offer environmental quality preservation, landscape conservation, and economic convenience that has assisted to make them a solution in water management and treatment. There also remains a knowledge gap around effective sustainability and long-term management of sites, with most known about its construction and formation.

Rationales for the creation of artificial urban wetlands have thus turned to other contributions they can make to the increasing emphasis in public and urban policy on urban sustainability objectives. City authorities are shifting away from sustainable urban drainage and sewer networks as static means of capturing and diverting water to reduce runoff to more dynamic management practices to add value through the use of captured water to enhance sustainability. As such, artificial wetlands are being repositioned as an opportunity to couple such stormwater functions of flow control, infiltration, detention, and/or retention within landscape-scale ecosystem conservation and/or restoration (Ahn and Schmidt 2019).

Artificial wetlands are increasingly too being viewed as offering opportune sites for the creation of new ecological habitats or for the restoration of a degraded ecosystem, by attracting wildlife species, especially birds, and establishing a green area. As such, wetlands contribute valuable landscape components in urban green corridors while also generating connections between nature and local urban communities. Although some research has mentioned that these systems can also be public recreation and education sites, these are generally viewed as ancillary benefits (Knight et al. 2001). A similar secondary benefit is their value as sources of biomass material helping to offset energy needs or as sources of food and fiber (see Avellan and Gremillion 2019). While this reflects a focus on how constructed urban wetlands may be able to contribute to the urban circular economy, the small scale of operations raises doubts about their efficiency and reliability to be economic.

Positioning in Urban Sustainability Debate

As the Convention for Biological Diversity noted in 2015, wetlands are significant contributors to meeting many of the United Nations' Sustainable Development Goals (SDGs). Although, in their natural state, wetlands sequester some of the largest stores of carbon on the planet with the bulk of sequestered carbon being in the soils rather than in the plant communities, when disturbed or warmed, the reverse action takes place, and they release the three major heat-trapping greenhouse gases (GHGs), carbon dioxide (CO_2), methane (CH_4), and nitrous oxide (N_2O). Understandably, therefore, one of the major priorities for limiting future temperature increase is protecting all types of wetland ecosystems from direct human disturbance (Moomaw et al. 2018).

It is less clear however what contribution artificial urban wetlands can make. Positively, in enhancing the return to economic use of polluted and contaminated land, constructed wetlands are viewed as an efficient and relatively low-cost approach. And in enhancing the provision of accessible drinkable water, life cycle analysis of the environmental impacts of artificial wetlands over traditional wastewater treatment plants indicates their lower GHG emissions, and where specific forms of free water surface artificial wetlands are used, significantly lower levels of CO_2 and CH_4 are emitted (Mander et al. 2014).

Less certain at present is how artificial wetlands can contribute to carbon sequestration as part of urban and regional responses to climate change. Calculating net sequestration on existing evidence is problematic, although studies do suggest that depending on plant settings and operational conditions, there are positive gains in capturing carbon (de Klein and van der Werf 2014). Nevertheless, such contributions are small. Moomaw et al. (2018) offer one estimation of the impact of wetlands on CO_2 suggesting that the area of new wetlands needed to remove 1% of the current annual increase in atmospheric CO_2 is about 2,000,000 km^2, an increase of about 17% of the current wetland coverage of the globe. Despite the fragility of the evidence, these studies do underline the overall assessment of scientific knowledge that (i) priority has to be on the retention of *existing* wetlands as their contribution is much more significant and much more difficult to replace and (ii) that *new or restored* wetlands have to be of significant scale globally to make a marked impact on climate change. Current understanding of the occurrence and variability of carbon storage between wetland types and across regions does represent a major impediment to the ability of nations to include wetlands in greenhouse gas inventories and carbon offset initiatives (Carnell et al. 2018).

In short, the evidence suggests that the **direct** effect of urban wetland development in carbon reduction and thus climate change is limited. It might be that the **indirect** effects are equal or more significant. The Taskforce on Scaling Voluntary Carbon Markets (2020) suggests that these range from increased biodiversity, job creation, and support for local communities, as well as health benefits from avoided pollution. There is the potential that as such projects if located in less developed or socioeconomically poorer areas will result in carbon credits generating flows of private capital into these communities, areas, and nations.

Expanding Future Use, Extending Benefits

Widespread adoption of artificial urban wetlands and indeed blue-green infrastructure (BGI) has been hampered not only by uncertainties regarding the performance and maintenance of the infrastructure itself as noted above but also a lack of confidence that decision-makers and communities will accept, support, and take ownership of such infrastructure (Thorne et al. 2018). While there is considerable research being undertaken to provide the scientific evidence and knowledge to generate insights required to help allay such concerns about functionality, less easily resolved are sociopolitical uncertainties relating, for example, to citizen attitudes and political decision-making. Little research has focused on how to best design wetlands as part of urban infrastructure (Ahn and Schmidt 2019), and there is an absence of guidance on how to integrate successfully community stakeholders into restoration planning.

Natural resource professionals are increasingly faced with the challenges of cultivating community-based support for wetland ecosystem restoration, and this is most acute in the setting of urban projects. Here, often the natural dimensions of wetlands have disappeared, and communities thus struggle to envisage the character and nature of the artificial or restored wetland and its position within their communities.

Based on their study in Newcastle upon Tyne, UK, O'Donnell et al. (2017) advocate that in order to overcome barriers that have constrained blue-green infrastructure projects and the adoption of constructed urban wetland initiatives, there is merit in looking beyond the flood protection and water management benefits derived from them. While these may remain the central rationale for investment in wetland and BGI construction, respondents in their research suggest that promoting the areas as multifunctional space and identifying the multiple benefits of any scheme can be vital to get local support. This means connecting with a wider set of stakeholders, including urban communities and those involved in other areas of local governance and municipal service provision.

Imagining such schemes is likely to be vital for the future expansion of constructed urban wetlands. One recent example is that being proposed in Glasgow ahead of the COP26 summit in 2021 as a pilot project (Fig. 1). Led by a small commercial enterprise company specializing in wetland use, Seawater Solutions, the project envisages how wetland technologies to manage water and flooding, aquaponics, and food production can build local supply chains and enhanced social assets to help in place-making can reconfigure underutilized urban spaces.

Such approaches require changing how wetland developments are planned and delivered, seeking greater collaborative working and co-funding from organizations and departments with a range of different remits and objective while also extending the value of artificial urban wetlands within the local community. Through active community engagement, there is enhanced potential for behavioral and cultural change to embrace artificial wetlands compared with solely

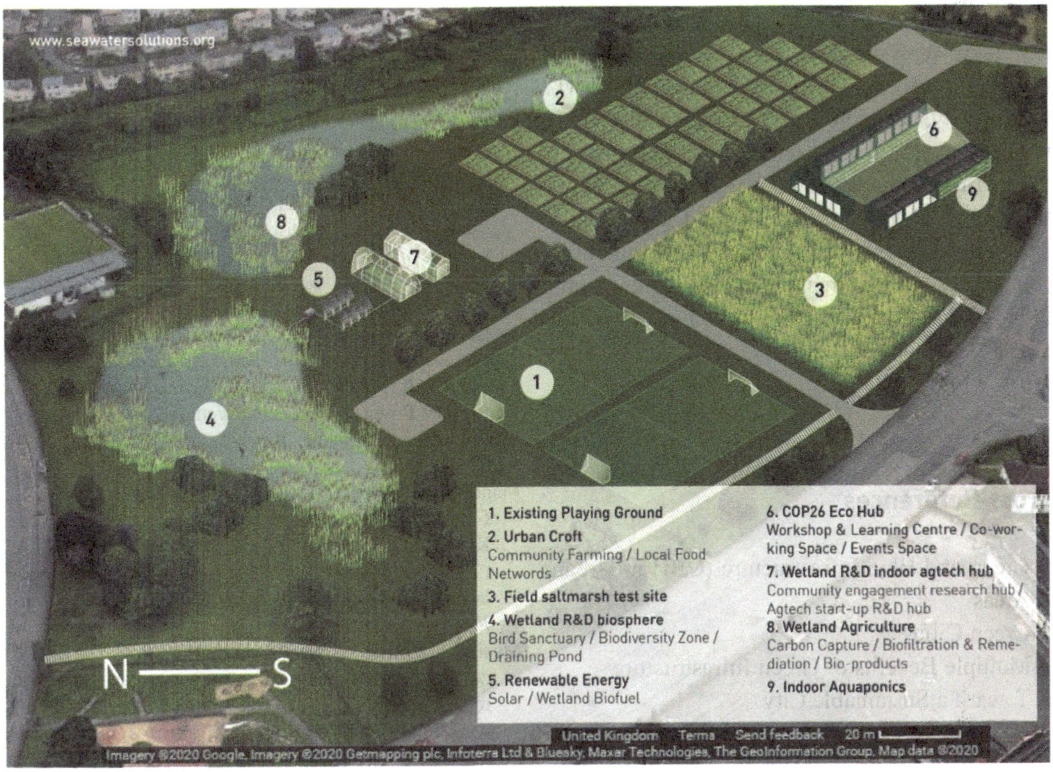

Artificial Urban Wetlands, Fig. 1 Integrating urban wetland technologies into communities

relying on public observation that has to date formed the traditional approach to their development in cities and communities (O'Donnell et al. 2017).

Additional synergies between green technology and urban quality of life can also arise. In areas where wetlands provide important water resources and are a key part of water management approaches, studies in China, USA, and Australia have suggested that both the distance to the nearest wetland and the number of wetlands within close proximity significantly influence house sales price, along with a number of other property-specific and neighborhood attributes (Boyer and Polasky 2004; Du and Huang 2018; Tapsuwan et al. 2009). There is greater recognition of the role wetlands play in improving the quality of human surroundings and providing cultural ecosystem services as aesthetically pleasing places for recreation, education, and spiritual development. Potential benefits include improved physical and psychological health, increased community connection and sense of place, and those derived from community involvement in urban conservation (Carter 2015).

Summary and Conclusion

As a potentially significant contribution to creating more sustainable cities globally, the creation of artificial urban wetlands has been gaining renewed attention. While there is a strong and expanding corpus of scientific knowledge into how artificial wetlands can assist in water management, help rehabilitate contaminated and polluted land, and to a lesser extent assist restore the ecological balance of local areas, encouraging expansion of such wetlands has often proved controversial. With more attention being given as to how wetland technologies can be integrated environmentally, economically, and socially into urban society, there are exciting new opportunities to see artificial wetlands contribute to urban sustainability. Developing local food production, enhancing local social capital, and strengthening community involvement with wetland points offer different and important ways in which urban wetlands will assist globally in shaping urban futures.

Cross-References

- ▶ Green and Blue Infrastructure (GBI) in Urban Areas
- ▶ Green Cities
- ▶ Multiple Benefits of Green Infrastructure
- ▶ Toward a Sustainable City

References

Ahn, C., & Schmidt, S. (2019). Designing wetlands as an essential infrastructural element for urban development in the era of climate change. *Sustainability, 11*(7), 1920.

Avellán, T., & Gremillion, P. (2019). Constructed wetlands for resource recovery in developing countries. *Renewable and Sustainable Energy Reviews, 99*, 42–57.

Boyer, T., & Polasky, S. (2004). Valuing urban wetlands: A review of non-market valuation studies. *Wetlands, 24*(4), 744–755.

Carnell, P. E., Windecker, S. M., Brenker, M., Baldock, J., Masque, P., Brunt, K., & Macreadie, P. I. (2018). Carbon stocks, sequestration, and emissions of wetlands in south eastern Australia. *Global Change Biology, 24*(9), 4173–4184.

Carter, M. (2015). Wetlands and health: How do urban wetlands contribute to community wellbeing? In M. Finlayson, P. Horwitz, & P. Weinstein (Eds.), *Wetlands and human health* (pp. 149–167). Dordrecht: Springer.

Davidson, N. C. (2014). How much wetland has the world lost? Long-term and recent trends in global wetland area. *Marine and Freshwater Research, 65*(10), 934–941.

de Klein, J. J., & van der Werf, A. K. (2014). Balancing carbon sequestration and GHG emissions in a constructed wetland. *Ecological Engineering, 66*, 36–42.

Du, X., & Huang, Z. (2018). Spatial and temporal effects of urban wetlands on housing prices: Evidence from Hangzhou, China. *Land Use Policy, 73*, 290–298.

Gorgoglione, A., & Torretta, V. (2018). Sustainable management and successful application of constructed wetlands: A critical review. *Sustainability, 10*(11), 3910.

Hu, S., Niu, Z., Chen, Y., Li, L., & Zhang, H. (2017). Global wetlands: Potential distribution, wetland loss, and status. *Science of the Total Environment, 586*, 319–327.

Ingrao, C., Failla, S., & Arcidiacono, C. (2020). A comprehensive review of environmental and operational issues of constructed wetland systems. *Current Opinion in Environmental Science & Health, 13*, 35–45.

Knight, R. L., Clarke, R. A., & Bastian, R. K. (2001). Surface flow (SF) treatment wetlands as a habitat for wildlife and humans. *Water Science Technology, 44*, 27–37.

Mander, Ü., Dotro, G., Ebie, Y., Towprayoon, S., Chiemchaisri, C., Nogueira, S. F., Jamsranjav, B., Kasak, K., Truu, J., Tournebize, J., & Mitsch, W. J. (2014). Greenhouse gas emission in constructed wetlands for wastewater treatment: A review. *Ecological Engineering, 66*, 19–35.

Masi, F., Rizzo, A., & Regelsberger, M. (2018). The role of constructed wetlands in a new circular economy, resource oriented, and ecosystem services paradigm. *Journal of Environmental Management, 216*, 275–284.

Moomaw, W. R., Chmura, G. L., Davies, G. T., Finlayson, C. M., Middleton, B. A., Natali, S. M., . . . Sutton-Grier, A. E. (2018). Wetlands in a changing climate: Science, policy and management. *Wetlands, 38*(2), 183–205.

O'Donnell, E. C., Lamond, J. E., & Thorne, C. R. (2017). Recognising barriers to implementation of blue-green infrastructure: A Newcastle case study. *Urban Water Journal, 14*(9), 964–971.

Saggaï, M. M., Aïnouche, A., Nelson, M., Cattin, F., & El Amrani, A. (2017). Long-term investigation of constructed wetland wastewater treatment and reuse: Selection of adapted plant species for metaremediation. *Journal of Environmental Management, 201*, 120–128.

Scholz, M. (2015). *Wetland systems to control urban runoff.* Elsevier.

Shelef, O., Gross, A., & Rachmilevitch, S. (2013). Role of plants in a constructed wetland: Current and new perspectives. *Water, 5*(2), 405–419.

Tapsuwan, S., Ingram, G., Burton, M., & Brennan, D. (2009). Capitalized amenity value of urban wetlands: A hedonic property price approach to urban wetlands in Perth, Western Australia. *Australian Journal of Agricultural and Resource Economics, 53*(4), 527–545.

Taskforce for Scaling Voluntary Carbon Markets. (2020). Consultation paper, November (TSVCM_Consultation_Document.pdf (iif.com)).

Thorne, C. R., Lawson, E. C., Ozawa, C., Hamlin, S. L., & Smith, L. A. (2018). Overcoming uncertainty and barriers to adoption of Blue-Green Infrastructure for urban flood risk management. *Journal of Flood Risk Management, 11*, S960–S972.

At the Intersection and Looking Ahead

COVID-19, Livelihoods, Work, Mobility, Law, and the Urban Resilience

Tafadzwa Mutambisi[1], Tinashe Natasha Kanonhuwa[1], Innocent Maja[2], Roselin Ncube[3] and Innocent Chirisa[1,4]
[1]Department of Rural and Urban Planning, University of Zimbabwe, Harare, Zimbabwe
[2]Department of Demography Settlement and Development, Social & Behavioural Sciences, University of Zimbabwe, Harare, Zimbabwe
[3]Women's University in Africa, Harare, Zimbabwe
[4]Department of Urban and Regional Planning, University of the Free State, Bloemfontein, South Africa

Synonyms

Intersectionality – complex; Law – legal instruments; Resilience – absorption

Definition

Work – employment engagements and activities.

Livelihoods – practices by individuals to enhance a living and get sustenance.

Resilience – ability to withstand shocks – social, economic, emotional, and environmental.

Introduction

The Coronavirus has taken over many urban economies, and its impact has been felt across the globe (African Union 2020). This has threatened urban resilience especially in the Global South as almost half of the global population live in cities, and this share is expected to rise to 55% by 2050. Cities may be better equipped than the rest of their country to respond to the COVID-19 crisis due to their well-developed healthcare facilities. However, cities are densely populated places where people live and gather, thus at risk of spreading the virus due to the close proximity among residents and challenges to implement social distancing. Urban resilience during pandemics has been one of the most unexpected disasters to happen in the twenty-first century (Boldog et al. 2020). This has led to many governments to initiate various laws and policies to flatten the curve as well as to minimize the spread of the virus. Full or partial lockdown was one of the extreme measures that were implemented by various governments across the world as it brought production and consumption almost to a standstill. The virus has led to the closure of many cities; this has seen marginalized, vulnerable, and invisible populations such as those who live in informal settlements and work in the informal economy, being the most affected by the pandemic (Croxford 2020). The different dimensions and nature of this virus shock many small and medium-sized businesses (SMEs) which are among the most exposed to liquidity issues.

The law and urban resilience strategies through governance and various policy measures have had a bearing on the functionality of cities in the pandemic (Sekar et al. 2019). Global governance across the world has seen many urban responses to the pandemic, highlighting the complexity of the virus. COVID-19 is most definitely spreading economic suffering worldwide (Acuto 2020). The virus may in fact be as contagious economically as it is medically, and the most affected region economically is the Global South. This shows that economic policy choices have an important bearing on cushioning the implications of containment measures. They have also seen the speed at which the economy can adjust toward more normal settings and conditions after the virus outbreak. The global community has different and unique laws and regulations concerning various issues (Wekwete 1989). Therefore, government decisions are critical in implementing measures to minimize the spread of the virus. The behavior of the virus is one thing; government reaction is another. The size and persistence of the economic damage are sorely depended on how governments handle situations (Bohoslavsky 2020). It is of grave importance that country laws and regulations work through standardized and universally agreed upon protocols in implementing laws pertaining COVID-19.

The recent outbreak of COVID-19 has affected global mobility in the form of various travel disruptions, restrictions, and blockages. Some of these impacts were magnified by the failure of government to effectively put up disaster mitigation measures (Huiling and Goh 2017). A particular concern has been the outrageous overlooking of warnings to prepare for pandemics and the lack of effective public response from a number of governments to protect the public health. However, many governments in Africa have failed to recognize their informal sector which constitutes more than half of their economic system. This has led to failure of some policies in protecting citizens' health as it put their source of livelihood on the line (LEDRIZ 2015). Through proven effective measures such as social distancing and quarantines to flatten the curve of the pandemic, they have put many livelihoods at stake. Whether lives are protected or some more economic wealth is produced in a given year is a choice that has to be taken from a human rights perspective. The current economic systems of many countries in the Global South especially in Africa are for the most part sustained by gender inequality and discrimination against women in the labor market. Around the world, unpaid and paid care work is too often and mostly performed by women (Gee and Ford 2011).

The coming in of virtual technology has been welcomed by any businesses during the COVID-19 pandemic (Voronkova 2018). With the advent of the digital age and the plethora of Internet of Things (IoT) devices it brings, there has been a substantial rise in the amount of data gathered by these devices in different sectors like transport, environment, entertainment, sport, and health among others (Angelidou 2016). This has assisted many communities to work from home. However, marginalized groups without access to technology as jobs that don't allow them to work at home have been disadvantaged in the pandemic. Therefore, this chapter aims to clearly outline various issues in urban areas that have shifted the way in which cities operate as well as look at the way forward in living life amidst a deadly pandemic called COVID-19 (Coronavirus).

The Conceptual Framework

This section is going to give a brief background on key issues being articulated in this chapter. The main concepts to be discussed are COVID-19, urban livelihoods, work and mobility, the law, and urban resilience.

A cluster of pneumonia cases in Wuhan, China, was reported to the World Health Organization (WHO) on 31 December 2019 (McKibbin and Fernando 2020). The cause of the pneumonia cases was identified as a novel beta Coronavirus, the 2019 novel Coronavirus (2019-nCoV, recently renamed as SARS-CoV-2, the cause of COVID-19). Coronavirus disease (COVID-19) is a very infectious disease which is caused by a newly discovered Coronavirus (FAO 2020). Mild symptoms experienced by most people infected by the virus include mild to moderate respiratory illness, and these symptoms can be managed at home without requiring special treatment. Older people and those with underlying medical problems like diabetes, cardiovascular disease, cancer, and chronic respiratory disease are more likely to develop serious illness (Maqbool 2020). Coronavirus can be spread through various ways such as droplets of saliva or discharge from the nose when an infected person coughs or sneezes. At the moment there is no cure, treatment, or vaccine developed specifically for COVID-19. However, there are many ongoing clinical trials evaluating potential treatments (UNECA 2020).

Urban refers to a city, and livelihood relates to the means of securing basic necessities or a set of economic activities like food, clothing, and shelter (McKinsey & Company 2020). Urban livelihoods therefore refer to the means and ways in which people living in the cities secure basic necessities or the ways in which they earn their income for food, clothing, and shelter. The Coronavirus has seen the disruption of many cities' way of life, therefore threatening people's livelihoods. There could be a huge wave of insolvencies, bankruptcies, or plain inability to pay bills for basic services such as water. A long economic shutdown, left alone, could be a financial catastrophe (Dube 2020). Diminishing of urban livelihoods is also a result of limited mobility to work by urban residents during the COVID-19 pandemic. The current outbreak of COVID-19 has affected global mobility in the form of various travel disruptions, restrictions, and blockages.

Globally, almost half of the global population live in cities; therefore, people in cities are highly mobile than in densely populated areas (van Noorloos and Kloosterboer 2018). This is because cities are the main economic hubs of many regions; thus, people tend to relocate to cities for work (Karmarkar 2017). Since cities are densely populated places where people live and gather, they are therefore at risk of spreading the virus due to the close proximity among residents and challenges to implement social distancing (European Center for Disease and Prevention and Control 2020). This is why many governments have restricted movements of people across cities. This has helped reduce mobility rates and helped to assist in monitoring mobility patterns. Other cities opted for less individualized monitoring options, such as using urban data to observe collective density and mobility patterns. For instance, Mexico City (Mexico) used a partnership with Google maps and Waze to monitor mobility trends, and Budapest (Hungary) is using smart city tools to identify high concentrations of people. Remote working and studying also became the norm for a large part of the population as cities enforced lockdowns or social distancing measures.

Laws are particular set of rules that are engrossed in a system which a particular community or country recognizes as regulating the actions of its members, which may be imposed by the imposition of penalties (Chimusoro et al. 2018). Governments use policies and planning tools to constrain access to and organize urban spaces. These sets of laws in many developing countries have been implemented in a way that puts the urban wealthy at an advantage as well as middle classes ignoring the low-income residents and rural-urban migrants in particular. This has threatened urban residence in the face of a deadly pandemic called the Coronavirus (Harrison et al. 2014). Disease outbreaks are a common occurrence and often result in untoward suffering and loss of life. Delayed response has led to loss of life, economic losses, and disruption of health systems which are already weak especially in low-income countries. Resilience refers to patterns of positive adaptation in the context of significant risk or adversity. Urban resilience is the capacity of urban communities within a city to survive, grow, and adapt no matter what. It refers to the capacity of a dynamic system to adapt successfully to disturbances that threaten its viability, function, and development of that system. Therefore, urban resilience is understood in simpler terms as a continuum of positive responses by urban communities in the face of adverse events (Frantzeskaki 2019). Resilience has been successfully identified as a core component of the sustainability concept. Urban resilience sees urban areas bouncing back, recovering, or stabilizing after facing a crisis (UN-HABITAT 2017). Cities are densely populated places where people who live and gather there are very much at risk of spreading the Coronavirus due to the close proximity among residents and challenges to implement social distancing. Large and secondary cities, in particular, often act as hubs for transnational business and movement. Therefore, laws and urban resilience are important in helping communities survive the pandemic.

Theories Underpinning the Study

There are four major theories that are used in the paper: the theory of travel and mobility, theory of nodality and function, theory of resilience, and theory of survival and succession. These theories are meant to explain the functions of the city before and after the spread of the Coronavirus. They are meant to substantiate the study in explaining the socioeconomic and physical dimensions of a city and the impacts and effects of the Coronavirus.

Travel is the movement of people between different geographic locations, while mobility is the ability to move freely. Theory of travel and mobility thoroughly depended on the laws, context, and transport systems available in an area (Huiling and Goh 2017). Mobility can be divided into five categories: mobility of objects, corporeal mobility, imaginative mobility, virtual mobility, and communicative mobility. In urban theories and concepts, mobility is largely based on population movements between one geographical area to another. Travel and mobility looks at the importance of systematic movements of people for work, school, leisure, and family life (Faus 2020). Mobility is linked across different scales of movement, while traditional transportation tends to focus mainly on particular forms of movement, that is, local traffic studies or household surveys. Mobility has been strongly impacted by the COVID-19 pandemic and provided cities with a momentum to rethink their approach toward urban space and suggest alternative options (Jones and Kassam 2020). While the impact of COVID-19 on public transport systems has been significant, in most countries, transport systems have shown a remarkable capacity to enforce hygiene and distance measures during the lockdown exiting, thus limiting the creation of new transport-related clusters. Moving into more long-term and permanent strategies, cities are now investing in active mobility infrastructure, improved public transport safety and accessibility, and low emission transport options, such as electric vehicles and scooters.

A nodal region is viewed as a case of a functional region that has a single focal point where order and dominance is introduced. A grouping of locational entities may be based on certain criteria where within-group interaction is greater than in-between groups. This is examined without the consideration of roles of each entity in the interaction pattern; a functional region maintains (Appignanesi 2018). On the other hand, if grouping is based on interactions both between local entities and the ranking system of the locational entity to the other, where another single entity is defined as dominating all the other entities, a nodal region maintains. The main distinction that exists between a nodal and functional region is that the inter-locational relationships lack symmetry, particularly if only one type of interaction is considered. This serves consequences of the existence of hierarchy or ordinal ranking among the locational entities. It thus reflects the importance of one locational entity among others. In more recent decades, the inner city in countries such as South Africa has been characterized by the existence of numerous satellite business/administrative nodes within its municipal boundaries (Chirisa et al. 2016). They are designed in such a way that they are serviced by components of a fragmented and hybrid public transport system.

The concept of resilience originated from ecological studies, exploring the varied ability of ecosystems to absorb and adapt to external pressures (DauKuir-Ayius 2016). The ecological perspective views resilience as the capacity of living organisms to survive in a physical environment in relation to each other's functions (UN-HABITAT 2017). The survival of an organism is influenced by the actions of other organisms that are competing for survival in the same environment. The successful survivors are the ones with a greater ability to adapt to disturbance, while the vulnerable, with lesser capacity to withstand the pressures of change, typically vanish; it is likely that many systems encounter disturbances, but the important feature of resilient systems is that they are able to adapt or maintain their usual operations in the face of these disturbances (Sim et al. 2018). Social scientists generally agree with the approach of the natural and physical scientists but tend to apply the idea of resilience to social issues such as public policy. There are criticisms of attempts to integrate and apply the ecological concept of resilience to the social setting of a community, which would be "inadequate to account for the diversity of people's experiences of resilience because individuals' or groups' encounters vary in different contexts." Other critiques on the use of the ecological approach to resilience in the social world point out that this approach focuses more on the individual (organism) (Harrison et al. 2014).

Cities have become crucial zones for the survival of humanity; thus, there is need for resilience strategies in order to help them survive, succeed, and sustain themselves in the middle or after a disaster (Bayang 2006). Development of succession planning policies enables continuity. Continuity is vital in sustaining an organization and local, regional, or national areas. Success and survival ensures continuity, meaning that urban areas can withstand shocks and be present for the future generations to come (Bauer and Scholz 2010). Succession planning also ensures smooth handover of power, creation, sharing, and retention of knowledge. It is also a key strategic tool for business survival and competitive advantage in the knowledge economy. In an organization when succession and survival planning policies are in place, employee's needs are balanced, and suitable replacements are easily identified to fill in senior positions. In urban planning, succession and survival therefore relate to policies that are there to make sure that the socioeconomic state of a region is in place and balanced so as to quickly react to issues that arise. They help build proactive, prescriptive policies as well as manage risks. This is promoting sustainability/survival (Ackankeng 2004).

Research Methodology

Using a desktop and case study approach, this chapter seeks to map urban dynamics as they have been spelt out by COVID-19 in terms of how urban dwellers in different regions – Africa and Asia specifically – are accessing their work and livelihood stations. The paper adopted a critical review of literature anchored in case study analysis and document analysis in assessing impacts of COVID-19 in African and Asian countries. Document analysis enhances the research's validity and reliability as it conveys first-hand information on the topic being researched. Secondary data are data collected earlier by other researchers (Hox and Boeije 2005). Secondary sources such as books, journals, newspapers, and policy documents were perused and analyzed. Literature review helps to determine whether the topic is worth studying, and it provides insight into ways in which the researcher can limit the scope to a needed area of inquiry (Creswell 2014). Case studies are extensive examinations of a single instance of a phenomenon of interest (Sekar et al. 2019). Case studies include Johannesburg, South Africa; Harare, Zimbabwe; Mumbai, India; and Kaesong, North Korea.

The Urban Realities of Africa and Asia Under the New Normal of COVID-19

Urban Africa

The future of the world's urbanization will be in Africa; the continent's urban population will almost triple in the coming 35 years, with more than 1.3 billion Africans living in cities by 2050. New capital cities are more of a phenomenon of the post-independence era. In Africa, urbanization is characterized by unregulated growth, lack of decent and affordable housing, absence of basic social services, pauperization, limited opportunities for gainful employment in the formal economy, failing and neglected infrastructure, severe environmental degradation, criminality, increasing inequalities, and negligent city management. Travelers from China, Korea, Italy, and other affected countries could spread COVID-19 to the African region. This has led to the economic decline of most African countries, and countries such as Zimbabwe saw a re-emergence of ethnic tensions along both regional and political lines (Action Aid Zimbabwe 2014). Most African countries have a pre-colonial history; therefore, they have an inherited development planning and infrastructure system (Bluwstein 2018). This has caused high inequalities, making the poor in most former colonial cities vulnerable to COVID-19. The unevenness in distribution has consequences for human development. Poor people can be found in countries at all levels of development, but poverty rates are higher in low human development countries. There has been a noticeable decline of poverty rates in all regions; however in Sub-Saharan Africa, more than half of the people still live in extreme poverty. Nearly nine out of ten people in Sub-Saharan Africa in 2030 will still be in deep poverty if this trend continues. The poorest people suffer from discriminatory social norms, overlapping deprivations, and lack of political empowerment.

Risk and vulnerabilities are enhancers of fragility in achievements (USAID 2018). Among the developing countries that are off track in Africa, here exist high levels of violence in conflict in more than a third of them. They reflect one of the world's development challenges at a larger scale. Shared characteristics of low health, low tax, and education spending exist in them. Therefore, inequalities put the marginal people at risk to the Coronavirus as they are unable to get access to proper healthcare, sanitization, and protective gear. The COVID-19 pandemic has seen to the closure of industries, travel bans, border closings, and lockdown restrictions (Dube 2020). This has reduced exports of the afflicted sectors in afflicted nations. Evidence clearly shows that lack of adequate work opportunities is a significant driver of migration. Every year, young people migrate from rural areas to cities in search of jobs even in the informal economy of urban areas. Of the ten million more people added to the urban population of Sub-Saharan Africa each year, two-thirds, or seven million, live in informal settlements or slums, and only two million can expect to ever move out of them (Swinkels et al. 2019). The highest inclination to move abroad, at 38%, is found in Sub-Saharan Africa. As of 2015, there were almost 51 million international migrants aged between 15 and 29 years around the globe, accounting for over 21% of the 243 million migrants worldwide. The three largest African cities, Cairo, Kinshasa, and Lagos, have population densities in excess of 12,000 people per km2 but less built-up area per capita at only 54 m2. On the other hand, a small town like Maxixe in Mozambique, whose population is below 100,000, has a population density averaging about 1300 persons per km2 and 528 m2 of built-up area per capital. These are highly mobile; therefore, they have a greater chance of contracting the Coronavirus. The lower densities translate into significantly less congestion and lower COVID-19 exposure risk.

Africa's capacity to effectively contain the pandemic will largely depend on proactive responses and the resilience of its health systems (van Noorloos and Kloosterboer 2018). The New Urban Agenda as reflected in SDG11 put cities at the heart of both national and international developments. Governments use policies and planning tools to constrain access to and organize urban spaces. African cities are persistently growing against their economic status which is weakening continuously, leading to the corrosion in the supply of basic infrastructure and urban services, manifesting through interruptions in water and electricity connections, refuse collection, drainage cleanup, and public road maintenance. These impede urban productivity and negatively affect the well-being of urban residents. It can be argued that the resources available at disposal for African cities cannot feasibly allow the implementation of coping strategies to deal with the growing needs and demand for services by the growing urban population. In addition, Africa has lower ratios of hospital beds and intensive care units (ICUs) relative to other regions (African Union 2020). On average, Africa has 1.8 hospital beds per 1000 people, compared to almost 6.0 in France. Also, 94% of Africa's total stock of pharmaceuticals is imported. With increasing restrictions or outright bans on exports of essential COVID-19 supplies, the outbreak of COVID-19 jeopardizes Africa's access to these life-saving medical supplies. Africa's urban dwellers, especially those living in slums and informal settlements, face challenges in accessing healthcare services and products, notably so in the light of the COVID-19-related loss of incomes.

Johannesburg, South Africa

The city of Johannesburg, a metropolitan city, is among South Africa's six metropolitan municipalities. It is an economic hub for Sub-Saharan Africa, and it attracts both skilled and unskilled workers from within the country and beyond borders (Van Heerden 2016). The city contains a population of an estimated four million people. Of this population, almost 37% are living in poverty. As a city-region, Johannesburg is one of the quickest growing locations globally. Estimates suggest that by 2014, the population of the city-region will exceed 14.5 million. The impact of the COVID-19 pandemic has been felt by many South Africans and will affect many aspects of

South African life (EQUINET 2020). Domestic and international shifts toward social distancing put extreme pressure on all jobs related to face-to-face gatherings. Corporate, government, and private events have already been cancelled under the government's prohibition of gatherings of more than 100 people at a time in South Africa.

COVID-19 will cause job losses as a result of social distancing which is needed to flatten the curve, slowing the virus's spread. But, even more directly, COVID-19 will infect people, making them too ill to work. For those who require ventilation or suffer organ failure, recovery can mean being months away from work or with having limited ability to work. Small, micro, and medium enterprises (SMMEs) typically face some of the greatest liquidity constraints in any economy, since those companies, like many individuals, survive from payment to payment with little reserves. Lockdowns in South Africa have prevented those who live far from the main city to get access to food. Johannesburg is filled with immigrants in its cities who are used to buying and selling to get food for them and their families. Lockdown restrictions and rules have been tough on them as they have been failing to get a source of livelihood for their families. Foreigners such as the Congolese, Zimbabweans, Mozambicans, and Nigerians are excluded from getting aid during this pandemic; therefore, they have no access to any form of income. Therefore, people are resorting to breaking the lockdown rules in order to get food as they cannot survive on empty stomachs.

The urban poor are located within six nodal points and supporting transportation networks, namely, Sandton, Johannesburg (inner city), Midrand, Randburg, Roodepoort, and Lenasia. Inequalities in the housing provision have made the people more vulnerable to COVID-19. Areas such as Alex Park were informal settlements under the threat to COVID-19 as social distancing is a problem. About ten of the city's cases are in Soweto, the city's biggest hotspot. Therefore, poor townships have been hit hardest by the virus. Social distancing is difficult in informal shacks and single-room houses that accommodate whole families. Laws have been implemented; there has been the ban of alcohol as well as curfews. This has been meant to minimize the cases of the virus in Johannesburg as they have been rising. People have been instructed to wear masks all the time. However, mobility is still an issue since people still use mini buses that will be packed to get to their destinations. Conclusively in Johannesburg, inequality is a major setback in combating the virus as there is a need to plan for three societies in one single city, that is, the low, medium, and upper class which is a strain on policy-makers in the city.

Harare, Zimbabwe

The city of Harare was established in 1890, Fort Salisbury by then. Harare is a pre-colonial city that has been vulnerable to obsolescence as the Harare City Council has not been able to replace most of the aging water pumps, pipes, and motors, which have been operated without adequate maintenance even after their lifespan lapsed (Muronda 2008). The consequences of the aging infrastructure and lack of maintenance are apparent in the malfunctioning of the water distribution and sewage systems, as demonstrated by the broken sewage pipe and tank as well as overflow in areas such as Kambuzuma, Budiriro, and Epworth (Gambe and Dube 2015). In the city, there are more of such broken sewage pipes, with overflow of sludge onto streets in residential areas, which cause respiratory complications, especially in densely populated informal settlements. As a result, the city water systems are decrepit and will continuously collapse, and this translates into water shortages. Evidence collected in 2011 from several suburbs and informal settlements of Harare City, including Kambuzuma, Mbare, Epworth (informal settlement), Budiriro, Highfield, and Sunningdale, demonstrate conclusively that the water and sanitation management systems had collapsed. In Chitungwiza, a town of half-a-million people south of Harare, several women are waiting at the water pump with large cans, concern etched on their faces, fearful of the growing threat of the Coronavirus. The spread of the Coronavirus has seen the worsening of situations in the capital.

Urbanization and urban development are skewed toward metropolitan Harare (Moyo 2014).

This trajectory not only is unsustainable but has also led to a major urban crisis in the region. In addition, the concentration of urban development in Harare has suffocated other cities leading to a nation-wide urban crisis. The crisis manifests in failures of services such as water, electricity, transport, and housing provision. Zimbabwe is already trained economically with economic issues such as hyperinflation, unemployment, poor public healthcare services, and political instability (AFP 2020). This has opened Harare to being vulnerable to the spread of the Coronavirus pandemic. Zimbabwe's economy according to the World Bank shrank by 6.5% in 2019, and this is worsening due to the Coronavirus. The inequality gap that exists between the rich and the poor is very wide in Zimbabwe so much that the country is mostly dominated by the informal sectors (which are mostly vendors). Lockdown restrictions ad curfews implemented in Harare have been disastrous to the poor, the vulnerable, and the small and informal businesses. Zimbabwe depends most on imports, and slower imports have caused an increased in inflation where USD$1 is equivalent to $950 Zimbabwean dollars. COVID-119 has affected Zimbabwe's fragile and under-funded health sector as nurses and doctors do not have the necessary equipment to attend to patients. Hospitals have been characterized by a series of strikes since December 2019. This has caused shortage of doctors leading to high rates of deaths especially during child birth. This is because the health sector employs 1.6 physicians and 7.2 nurses for every 1000 people.

Most Zimbabwean cities are characterized by an unsafe and unreliable public transport system. The system uses incidental termini which are not meant for and designed to cater for loading and offloading passengers. As such, there exists immense contestation among public transport users, operators, and local authorities on the management of public transport in cities (Muchadenyika 2015). Therefore, Zimbabwean cities should invest in mass transport systems with defined transport interchanges and loading and offloading bays. The shortage of public transport in Zimbabwe's capital Harare is threatening to derail efforts to contain the COVID-19 pandemic, as commuters are hardly practicing any social distancing while in queues for buses. Overcrowding at bus terminals in the city center and residential areas has become the order of the day as commuters wait for hours on end for the few buses that are operating (Mbara 2015). This has resulted in commuters enduring inordinate delays going to and from work, in addition to exposing themselves to the risk of contracting and spreading the virus. Since the country eased lockdown measures on May 4 to allow formal businesses to resume operations, the volume of workers in the capital who commute to and from work using public transport has increased significantly.

Urban Asia

Asian cities are the drivers of economic growth through manufacturing and services; thus, they attract large numbers of people and activities. In developing countries, the capital city is usually the center of population, commerce, and the nation's most powerful business with some significant industrialization (Dahiya 2012). This has caused cities such as Wuhan and Beijing in China to be one of the main priority cities in the implementation of COVID-19 disaster management strategies (Boldog et al. 2020). 83% of all multidimensionally poor live in South Asia and Sub-Saharan Africa. Sub-national areas stretch from Tajikistan and Kyrgyzstan to most of Afghanistan and, in Southeast Asia, sections of Cambodia and Viet Nam in Central-South Asia. Partial or full lockdowns are extreme measures that have been implemented in countries such as China; they have brought production and consumption almost to a standstill. Shutting down the economy is more of a greater risk than it looks. It is more like shutting down a nuclear reactor.

There is no doubt about the large-scale impact of the latest pandemic to the economy. Equally, from a human rights perspective, potential impacts of the upcoming recession include challenges to access adequate housing, healthcare including mental health, education, water and sanitation, social protection, and work. This period has seen a rise in world hunger, a high increase in evictions, foreclosures, homelessness, and negative impact on affordability of housing. This has been due to limited travel and mobility to work by many people. Many urban public transport systems indeed adapted to this unprecedented crisis, successfully ensuring a minimum level of service and maintenance and rapidly deploying strict hygiene measures to protect the health of employees and transport users, but significant challenges remain. Urban transport agencies around the world have also faced unprecedented low levels of ridership and corresponding losses in fare revenue that are threatening their financial stability and which will continue for months to come in a context where physical distancing may be required in public transport (Yigitcanlar and Dur 2010). Over the past several decades, increasing car traffic in industrialized and developing countries has led to negative side effects, such as air pollution, traffic accidents, greenhouse gas emissions, congestion, and health problems due to lack of physical exercise.

Government in Japan plans to build out a smart city database and get quarantine violators to agree to use tracking bracelets. The database was initially designed to share information between cities. Health authorities plan to leverage that network to reduce the time it takes to find and isolate COVID-19 cases. However, despite the positive impact to contain the epidemic, their use raises privacy concerns. An important lesson from the COVID-19 crisis, largely driven by a combination of the Zoom effect and Greta effect, is that teleworking is compatible with productivity and largely contributes to reducing negative environmental externalities. However, secondly, not everyone can work from home.

Kaesong, North Korea

In the southern part of North Korea (formerly in North Hwanghae Province) and the capital of Korea during the Taebong kingdom and subsequent Goryeo dynasty, the city is near the Kaesong Industrial Region close to the border of South Korea and contains the remains of the Manwoldae palace (Bhagat 2018). Since the 2000s, there has been a change in the thinking of policy-makers about urbanization. The Eleventh Five Year Plan argued that urbanization should be seen as a positive factor in overall development as urban sector contributes to about three-fifths of the GDP. There is also a growing realization that an ambitious goal of 9–10% growth in GDP fundamentally depends upon vibrant urban sector. Urbanization has increased faster than expected as per 2011 Census. This has reversed the declining trend in urban population growth rate observed during th 1980s and 1990s. Also, for the first time since independence, the absolute increase in urban population was higher than rural population.

The urban population has grown from an estimated 286 million people in 2001 to an estimated 377 million people in 2011 – an increment of 91 million compared to rural increment of 90.5 million. However, the urban transition has huge implication for providing urban infrastructure and civic amenities in the urban areas. This chapter presents an assessment of the emerging pattern of urbanization, its spatial pattern, and the components of urban growth, namely, the contribution of natural increase, classification of rural into urban areas, and the contribution of rural to urban migration. The emerging pattern of urbanization indicates that most of the parts of central, eastern, and north-eastern India have very low level of urbanization, and also these areas are characterized by very low level of economic development. This chapter particularly would be helpful to researchers who are interested in the demographic dynamics of urbanization having strong bearing on urban policies and programs (Bhagat 2018). It was called Songdo; while it was the ancient capital of Goryeo, the city prospered as a trade center that produced Korean ginseng. Kaesong now functions as the DPRK's light industry center.

During the Japanese occupation from 1910 to 1945, the city was known. The Kaesong region is a border region and is most vulnerable to COVID-19 due to movement across the border to South Korea. Therefore, North Korean leader Kim Jong Un imposed a total lockdown on the city of Kaesong near the border of South Korea after a person there was found with suspected COVID-19 symptoms. Kaesong, a city with an estimated 200,000 people, is located just north of the heavily fortified land border of South Korea. KCNA said that people who have been in contact with the individual and those who have been to the town in the last 5 days were being "thoroughly investigated, given a medical examination and put under quarantine." Runaway cases receive severe punishments. Due to North Korea's big and fortified military and monarchical state, the city is run by rules and extreme punishments; therefore lockdown rules are imposed by the military.

Lockdown in North Korea has affected Kaesong as it is a major industrial complex. The number of North Koreans who make a living connected in one way or another to markets is greater than the number who subsists through centrally planned agriculture or functioning state industries. Markets account for an increasing proportion of economic transactions, and the monetization of these transactions has allowed for the emergence of a class of elites that trade in the services, financial lending, consumer goods, transportation, and housing sectors. The regime in turn raises considerable revenue from taxation of market transactions, increasing its own dependence on the market system (Ramachandra et al. 2014). Therefore, restrictions have slowed down production as well as deprived citizens of their source of livelihood. The Kaesŏng Industrial Region (KIR) or Kaesŏng Industrial Zone (KIZ) is a special administrative industrial region of North Korea (DPRK). It was formed in 2002 from part of the Kaesŏng Directly-Governed City.

On 10 February 2016, it was temporarily closed by the South Korean government and all staff recalled by the Park Geun-hye administration, although the President of South Korea, Moon Jae-in, has signaled his desire to "reopen and expand" the region. Its most notable feature is the Kaesŏng industrial park, which operated from 2004 to 2016 as a collaborative economic development with South Korea (ROK). The park is located 10 kilometers (6 miles) north of the Korean Demilitarized Zone, an hour's drive from Seoul, with direct road and rail access to South Korea (Clark and Moonen 2014). The park allows South Korean companies to employ cheap labor that is educated, skilled, and fluent in Korean while providing North Korea with an important source of foreign currency. Employees that work at the industrial complex are over 1000. Therefore COVID-19 has left a threat to poverty and hunger. In Daegu (Korea), the epidemiological investigation during the outbreak was able to use the data hub of the smart city to trace patient routes. These research findings also imply that it is not densities alone that make cities vulnerable to COVID-19 but the structural economic and social conditions of cities, making them more or less able to implement effective policy responses. For instance, cities marked with inequalities, inadequate housing conditions, and a high concentration of urban poor are potentially more vulnerable than those that are better resourced, less crowded, and more equal.

Mumbai, India

Mumbai is the capital city of the Indian state of Maharashtra. It is bounded by the Arabian Sea in the west, and it is located at 18 degrees 55'N. It is the sixth most popular city in the world and the most populous city in India. Mumbai is the economic hub of India, and it is also considered to be the commercial capital of the country. In the last three decades, Mumbai witnessed tremendous growth in urban expansion and infrastructure development. Urbanization is the component that makes India vulnerable to the Coronavirus pandemic.

According to 2011 Census, urban population grew to 377 million showing a growth rate of 2.76% per annum during 2001–2011, and the level of urbanization at the country as a whole increased from 27.7% in 2001 to 31.1% in 2011 – an increase of 3.3 percentage points during 2001–2011 (Clark and Moonen 2014). Mumbai has been a constantly evolving, globally engaged city over the past 150 years. In the last 25 years, it has made a rapid economic transition from trade to services and has expanded its national and cross-border roles. In 2014, Mumbai's global significance is visible wherein the extent of housing disparity and unplanned growth clearly affect the city's overall productivity and competitiveness. Unplanned cities have led to the creation of the world's biggest slums in Mumbai (Ramachandra et al. 2014). Slums put the city under enormous threat to the Coronavirus as people living in shacks cannot practice social distancing.

The loss of manufacturing employment left a big gap to fill for new service industries and sectors being transformed by information technology in India. Formal, organized sectors, in banking, finance, oil, ICT, transport, healthcare, communications, and social services, grew quickly. But they were not able to generate enough employment to fill the vacuum. Small businesses in the informal sector — hawkers, taxi drivers, mechanics, vendors, recycling — began to multiply, delinked from trade unions and regulated conditions. The arterial railways and port have been its lifeblood ever since, with huge carrying capacity transporting goods, people, and knowledge along the coast and across Maharashtra. Lockdown restrictions and social distancing have left the informal sector in India struggling to survive. Mumbai has inherited the most extensive rail system in South Asia, which carries over six million passengers daily. Its suburban railway and BEST buses set it apart in India as a city shaped by its mass transport system. Due to overpopulation, people are now at risk of using public transport (Yedla 2003). A high number of vehicles and private cars in the city make it difficult to use private-owned transport as there is already a huge problem of traffic congestion in Mumbai; therefore, public transport is the most common. Connecting people to jobs and opportunity is now a major problem ın Mumbai, despite the new systems that are now operational or in train. This is partly a reflection of governance complexity — the slow progress of large projects and a lack of coordination between agencies that build and maintain rail and road systems plus the new pandemic that has taken shape in the world (Bhagat 2018).

Conclusion and Policy Options

Digitalization has been a crucial lever in cities' response to the pandemic, with tools monitoring contagion risk and ensuring the respect of confinement and social distancing, while also enabling the continuity of certain services and economic activity virtually. These tools and the changes in habits they entailed will remain a permanent component of cities' recovery phase and increased preparedness for potential new waves. This prompted reflections on issues of privacy rights and universality of Internet access. Going forward, it is likely that there will be a "new normal," whereby many employees and companies will leverage the potential of teleworking and adjust their mobility patterns where appropriate and possible. In fact, polls have shown that citizens maintain new work and travel habits after transportation crises. However, people and places are unequal regarding teleworking. While some workers can reduce their exposure to the risk of contagion by teleworking or benefit from preventive measures, many cannot because of the nature of their job, pre-existing inequalities, or the digital divide. One of the weaknesses of developing regions in development and risk management is the lack of a city development framework. As such, the development of cities and in particular metropolitan cities is left uncoordinated.

Compact cities have long been praised for their benefits, which include dense development patterns, better accessibility to local services and jobs, short intra-urban distances, and public transport systems with positive contributions to the efficiency of infrastructure investments. However due to the coming in of the Coronavirus, findings however suggest that urban density is a vulnerability to the spread of the virus; however, it is not automatically correlated with higher infection rates. It is not density alone that makes cities vulnerable to COVID-19 but the structural economic and social conditions of cities making them more or less able to implement effective policy responses. For instance, cities marked with inequalities, inadequate housing conditions, and a high concentration of urban poor are potentially more vulnerable than those that are better resourced, less crowded, and more equal. Globally, almost half of the global population live in cities, and this share is expected to rise to 55% by 2050. Large and secondary cities, in particular, often act as hubs for transnational business and movement. To build inclusive cities that provide opportunities for all, all levels of government should:

- Provide efficient social and community services for disadvantaged groups such as healthcare and home care (i.e., elderly and homeless people), through the design and implementation of ambitious social innovation strategies and a repurposing of empty buildings.
- Ensure that those left behind (i.e., migrants, low-wage workers) are targeted with customized employment and activation programs that are adaptable, relevant, and flexible and respond to the new needs of the local labor market after the crisis.
- Take measures to adjust housing quantity, quality, and affordability to the variety of housing needs, with a view to promote social cohesion and integration with sustainable transport modes.
- Improve accessibility to soft mobility, taking into account the needs of categories of people (i.e., elderly, families with children, disables) when rethinking the future of urban mobility in the post COVID-19 recovery phase (i.e., cycling as transport mode).
- Promote equitable access to quality education and leverage the full potential of online education, especially for low-income youth, and foster collaboration between higher education institutions, businesses, local and regional governments, and civil society.

References

Ackankeng, E. (2004). *Sustainability in Municipal Solid Waste Management in Bamenda and Yaoundé, Cameroon*. Doctor Thesis, The University of Adelaide, Cameroon.

Action Aid Zimbabwe. (2014). The dynamics of devolution in Zimbabwe. A briefing paper on local democracy. https://www.ms.dk/sites/default/files/udgivelser/zimbabwe_report_2014_finale_lav.pdf. Accessed on 15 October 2020.

Acuto, M. (2020). Engaging with global urban governance in the midst of a crisis. *Dialogues in Human Geography, 10*(2), 221–224.

AFP. (2020). Zimbabwe system 'inadequate' to take on coronavirus. *Times live*. https://www.timeslive.co.za/

African Union. (2020). Impact of the Coronavirus (Covid-19) on the African Economy. https://www.tralac.org/news/article/14483-impact-of-the-coronavirus-covid-19-on-the-african-economy.html. Accessed on 28 September 2020.

Angelidou, M. (2016). Four European Smart City strategies. *International Journal of Social Science Studies, 4*(4), 18–30.

Appignanesi, L. (2018). Blurred binary code for the sustainable development of functional systems: Blurred binary code. *Systems Research and Behavioral Science, 35*(4), 386–398.

Bauer, S., & Scholz, I. (2010). Adaptation to climate change in Southern Africa: New boundaries for sustainable development? *Climate and Development, 2*(2), 83–93.

Bayang, A. G. M. (2006). City-region concept body. http://rgdoi.net/10.13140/RG.2.1.2171.8800. 4 August 2020.

Bhagat, R. B. (2018). *Urbanization in India: Trend, pattern and policy issues* (Working Paper No. 17). Mumbai: International Institute for Population Sciences. https://www.researchgate.net/publication/326426013_Urbanization_in_India_Trend_Pattern_and_Policy_Issues?enrichId=rgreq-066da8791c3407c689a5c6e8ac41d8d5-XXX&enrichSource=Y292ZXJQYWdlOzMyNjQyNjAxMztBUzo2NDkwNzYyODI1MTU0NThAMTUzMTc2MzI1MzYwNw%3D%3D&el=1_x_2&_esc=publicationCoverPdf

Bluwstein, J. (2018). From colonial fortresses to neoliberal landscapes in Northern Tanzania: A biopolitical ecology of wildlife conservation. *Journal of Political Ecology, 25*, 145–168. University of Copenhagen, Denmark.

Bohoslavsky, J. P. (2020). *COVID-19: Urgent appeal for a human rights response to the economic recession* (p. 21). United Nations Human Rights Special Procedures, Geneva. Available online: https://reliefweb.int/report/world/covid-19-urgent-appeal-human-rights-response-economic-recession. Accessed on 23 October 2020.

Boldog, P., Tekeli, T., Vizi, Z., Dénes, A., Bartha, F. A., & Röst, G. (2020). Risk assessment of novel coronavirus COVID-19 outbreaks outside China. *Journal of Clinical Medicine, 9*(2), 571.

Chimusoro, A., Maphosa, S., Manangazira, P., Phiri, I., Nhende, T., Danda, S., Tapfumanei, O., Munyaradzi Midzi, S., & Nabyonga-Orem, J. (2018). Responding to Cholera outbreaks in Zimbabwe: Building resilience over time. In D. Claborn (Ed.), *Current issues in global health*. IntechOpen. https://www.intechopen.com/books/current-issues-in-global-health/responding-to-cholera-outbreaks-in-zimbabwe-building-resilience-over-time. 17 May 2020.

Chirisa, I., Bandauko, E., Mazhindu, E., Kwangwama, N. A., & Chikowore, G. (2016). Building resilient infrastructure in the face of climate change in African cities: Scope, potentiality and challenges. *Development Southern Africa, 33*(1), 113–127.

Clark, G., & Moonen, T. (2014). *Mumbai: India's global city. A case study for the Global Cities Initiative: A joint project of Brookings and JPMorgan Chase*. Mumbai: Global Cities Initiative.

Creswell, J. W. (2014). *Research design qualitative, quantitative, and mixed methods approaches* (4th ed.). Thousand Oaks: SAGE. www.sagepublications.com.

Croxford, R. (2020). Coronavirus Ethnic minorities 'are a third' of patients. *BBC News*. https://www.bbc.com. News.

Dahiya, B. (2012). Cities in Asia, 2012: Demographics, economics, poverty, environment and governance. *Cities, 29*, S44–S61.

DauKuir-Ayius, D. (2016). *Building community resilience in mine impacted communities: A study on delivery of health services in Papua New Guinea*. Palmerston North: Massy University. https://mro.massey.ac.nz/bitstream/handle/10179/9882/02_whole.pdf?sequence=2&isAllowed=y. 3 November 2019.

Dube, G. (2020). South Africa declares state of emergency, shuts borders and schools as Coronavirus Grips Nation. https://www.voazimbabwe.com/z/3152

EQUINET. (2020). Equinet Information Sheet 2 on Covid-19. https://www.easp.es/wp-content/uploads/2020/04/EQUINET-COVID-brief2-1April2020.pdf. Accessed on 13 October 2020.

European Center for Disease and Prevention and Control. (2020). Rapid Risk Assessment Outbreak of Novel Coronavirus Disease 2019 (covid-19): Increased Transmission Globally – Fifth Update. https://www.ecdc.europa.eu/sites/default/files/documents/RRA-outbreak-novel-coronavirus-disease-2019-increase-transmission-globally-COVID-19.pdf. 7 April 2020.

FAO. (2020). *Impact of Covid-19 on informal workers* (p. 7). Food and Agriculture Organization of the United Nations, Rome, Italy. Available online: http://www.fao.org/documents/card/en/c/ca8560en/. Accessed on 23 October 2020.

Faus, J. (2020). *This is how coronavirus could affect the travel and tourism industry*. Reuters: World Economic Forum. https://www.weforum.org/agenda/authors/joan-faus

Frantzeskaki, N. (2019). Seven lessons for planning nature-based solutions in cities. *Environmental Science & Policy, 93*, 101–111.

Gambe, T. R., & Dube, K. (2015). Water woes in Harare, Zimbabwe: Rethinking the implications on gender and policy. *International Journal of Innovative Research and Development, 4*(6), 390–397.

Gee, G. C., & Ford, C. L. (2011). Structural racism and health inequities. *Du Bois Review: Social Science Research on Race, 8*(1). https://www.ncbi.nlm.nih.gov/pmc/articles/PMC4306458/. 4 July 2020.

Harrison, P., Bobbins, K., Culwick, C., Humby, T.-L., La Mantia, C., Todes, A. & Weakley, D. (2014). *Urban Resilience Thinking for Municipalities Philip Harrison, Kerry Bobbins, Christina Culwick, Tracy-Lynn Humby, Costanza La Mantia, Alison Todes and Dylan Weakley*. University of the Witwatersrand, Gauteng City-Region Observatory. http://wiredspace.wits.ac.za/bitstream/handle/10539/17082/URreport_1901MR.pdf?sequence=1&isAllowed=y. 6 November 2019.

Hox, J. J., & Boeije, H. R. (2005). Data collection, primary vs. secondary. *Encyclopedia of Social Measurement, 1*, 593–599, Amsterdam, Netherlands: Elsevier.

Huiling, E., & Goh, B. (2017). AI, robotics and mobility as a service: The case of Singapore. *Institut Veolia, 17*, 26–29.

Jones, S., & Kassam, A. (2020). Spain defends response to coronavirus as global cases exceed 500,000. *The Guardian*.

Karmarkar, D. (2017). The Paper presented and published in the proceedings of UGC-Sponsored Two-Day Interdisciplinary International Conference on "Internal and International Migration: Issues and Challenges" to be held on December 19–20, 2014 in Smt. CHM College, Ulhasnagar

LEDRIZ. (2015). Strategies for transitioning the informal economy to formalisation in Zimbabwe. https://library.fes.de/pdf-files/bueros/simbabwe/13714.pdf. Accessed on 24 September 2020.

Maqbool, A. (2020). Coronavirus: Why has the virus hit African Americans so hard. *BBC News*. bbc.com/news/amp/world-us-canada-52245690

Mbara, T. C. (2015). Achieving sustainable urban transport in Harare, Zimbabwe: What are the requirements to reach the milestone? *CODATU, 14*. Available online: http://www.codatu.org/wp-content/uploads/Ttenda-Mbara_pdf. Accessed 20 November 2020.

McKibbin, W. J., & Fernando, R. (2020). The Global Macroeconomic Impacts of COVID-19: Seven Scenarios. *SSRN Electronic Journal*. https://www.ssrn.com/abstract=3547729. 5 April 2020.

McKinsey & Company. (2020). Covid-19: Investing in black lives and livelihoods. McKinsey.com.

Moyo, W. (2014). Urban housing policy and its implications on the low-income earners of a Harare municipality, Zimbabwe. *International Journal of Asian Social Science, 10*.

Muchadenyika, D. (2015). Slum upgrading & inclusive municipal governance in Harare.pdf. *Elsevier, 48*, 1–10.

Muronda, T. (2008). Evolution of Harare as Zimbabwe's Capital City and a Major Central Place in Southern Africa in the Context of, By Byland's Model of Settlement Evolution. *Academic Journals, 1*(12), 034–040.

Ramachandra, T. V., Bharath, H. A., & Sowmyashree, M. V. (2014). Urban footprint of Mumbai – The commercial Capital of India. *Journal of Urban and Regional Analysis, VI*(1), 71–94.

Sekar, S., Lundin, K., Tucker, C., Figueiredo, J., Tordo, S., & Aguilar, J. (2019). *Building resilience, a green growth framework for mobilizing mining investment*. Washington, DC: World Bank Publications. http://documents.worldbank.org/curated/en/6892415566502

41927/pdf/Building-Resilience-A-Green-Growth-Framework-for-Mobilizing-Mining-Investment.pdf. 6 November 2019.

Sim, T., Wang, D., & Han, Z. (2018). Assessing the disaster resilience of megacities: The case of Hong Kong. *Sustainability, 10*(4), 1137.

Swinkels, R., Norman, T., Blankespoor, B., Munditi, N., & Zvirereh, H. (2019). *Analysis of spatial patterns of settlement, internal migration, and welfare inequality in Zimbabwe*. Zimbabwe: World Bank. http://elibrary.worldbank.org/doi/book/10.1596/32190. 10 February 2020.

UNECA. (2020). Potential socio-economic impacts of Coronavirus on West Africa Impediments to Harnessing Demographic Dividend. www.uneca.org

UN-HABITAT. (2017). *Trends in Urban Resilience*. Nairobi: United Nations Human Settlements Program (UN-Habitat). http://urbanresiliencehub.org/wp-content/uploads/2017/11/Trends_in_Urban_Resilience_2017.pdf. 6 November 2019.

USAID. (2018). The intersection of global fragility and climate risks. https://www.climatelinks.org/resources/intersection-global-fragility-and-climate-risks. Accessed on 30 September 2020.

Van Heerden, J. J. (2016). *Sustainable mining communities post mine Closure: Critical Reflection on Roles and Responsibilities of Stakeholders Towards Local Economic Development in the City of Matlosana*. Stellenbosch University. https://scholar.sun.ac.za/. 4 November 2019.

van Noorloos, F., & Kloosterboer, M. (2018). Africa's new cities: The contested future of urbanization. *Urban Studies, 55*(6), 1223–1241.

Voronkova, L. P. (2018). Virtual tourism: On the way to the digital economy. *IOP Conference Series: Materials Science and Engineering, 463*, 042096. https://doi.org/10.1088/1757-899X/463/4/042096.

Wekwete, K. (1989). *Planning laws for urban and regional planning in Zimbabwe – a review* (RUP Occasional Paper No. 20, p. 24). Department of Rural & Urban Planning University of Zimbabw, Harare, University of Zimbabwe.

Yedla, S. (2003). *Urban environmental evolution: The case of Mumbai*. Mumbai: Indira Gandhi Institute of Development Research, p. 26.

Yigitcanlar, T., & Dur, F. (2010). Developing a sustainability assessment model: The sustainable infrastructure, land-use, environment and transport model. *Sustainability, 2*(1), 321–340.

Atmospheric Carbon Dioxide Reduction

► Carbon Sequestration Through Building-Integrated Vegetation

Attributes

► Sustainability Competencies in Higher Education

Augmented Reality: Robotics, Urbanism, and the Digital Turn

Paul Cureton
ImaginationLancaster, Lancaster University, Lancaster, UK

Synonyms

Automated city; Digital turn; Extended reality; Robotics; Smart cities; Urban planning

Definitions

The "Digital Turn" is the embrace of digital tools in everyday life, which in turn shape social behavior. This wide-spanning digital turn has permeated many aspects of society and one such area is Smart Cities from 1994 onwards with Amsterdam's 'digital city' followed by Cisco systems (2005) and IBM (2008) amongst others. Smart Cities are diverse cross-sectoral ICT centric framework developments and innovations for efficiency drives, management and process change in urban settings incorporating a range of technologies, sensors and analytical capabilities. It is important to understand this broad framework and ecosystem to situate the specific use of Extended Reality (XR) which incorporates virtual reality, mixed reality, and augmented reality devices that have been applied for urban simulation, analytics and visualization as explored in this entry for the built environment. XR applied, spans community planning, urban planning, smart city research, human-computer interaction, architectural research, spatial cognition, and wayfinding as well as having a direct relationship with the games industry. In addition to this, the fusion of various hardware, novel techniques and strategies point to future trajectories involving deployments of robotics such as ground and aerial-based systems for urban planning and mobility which have relational aspects or direct relationships with XR in terms of deriving information, simulation and operation in larger smart city ecosystems. This combination of technology is commonly defined as cyber-physical systems or automated cities. Major drivers of XR are formed around immersion and representations of reality in real-time, as well as deeper level semantic information about places. XR has increased the potential for communication and participation in planning at various scales from region to local as well as driving efficiency in design processes and urban management though also faces challenges in terms of skillsets, ethics, security, deployment, interfaces, and community accessibility.

Introduction

The way we live and work will change dramatically with increasingly technological driven urban development agendas in super cities, increasing regional inequalities and changing primary urban areas. These paradigm changes are variegated and uneven as evidenced by the Global Power City Index (GPCI) (See: https://mori-m-foundation.or.jp/english/ius2/gpci2/index.shtml) a mechanism evaluating major cities using six criteria of city function. The criteria are accessibility, economy, research and development, cultural interaction, liveability and environment. This index coupled and juxtaposed with UN studies of urbanization provides further evidence of global super cities (See: https://population.un.org/wup/). For super cities, the form of technological development in city functionality has been defined through "Smart Cities" and the Internet of Things (IoT). Smart City definitions according to Albino using a literature review, identified similar but multifaceted values but also divergent definitions which creates a "fuzziness" and inability for shared universality of the term (Albino et al., 2015). Indeed, the level of smartness beyond super cities is highly diverse. Technological "smartness," and or "intelligence" is more likely to be studied at the regional and or city level due to the specificity of the components, interventions, level of development, implementation and infrastructural capacity. (Batty, 2018, p.178) states that "smartness" definitions have proved difficult because of the process aspect and (Dunn & Cureton, 2019) have critiqued the "frictionless" vision aspect that the Smart City presents. Broad notions of "smartness" and IoT are part of a broader "digital turn" affecting many disciplines and professions. This digital turn is the embrace of digital tools and devices in everyday life which as Wim Westera states in shaping behavior (2012). This entry specifically focuses on the range of applications of extended reality (XR); virtual reality, mixed reality, augmented reality and the use of robotics in a broader paradigm shift for "smartness" and the digital turn in the cultural adoption of these technological tools.

Mario Carpo suggests a "digital turn" in architecture in two historical phases. The first iteration Carpo defines as between 1992-2012 enabled tools for architectural practice for execution and efficiency, of drafting and layout, or crudely the CAD age (2013). A second digital turn in architecture according to Carpo has emerged. The second digital turn involves more "thinking" through digital tools by the shear development of computation in the built environment using artificial intelligence and machine learning (Carpo, 2017). In geography, a digital turn is rooted in socio-technological production and praxis, and the directing and ordering of everyday life through digital mediums has also demonstrated a near-ubiquity (Ash et al., 2018, p.26). These two digital turns discussed in architecture and geography are not limited to the tools of design and production, but also the organization and management of major urban areas and regions through smart city agendas, but also the increasing deployment of automation and robotics in larger metropolitan areas in particular transportation and construction (Yuan & Chao Yan, 2020). The digital turn signals a change in the process beyond disciplines and has had profound effects on the actions and tooling of society as well as a contemporary move for cyber-physical systems (smart systems involving a physical element and digital cyber parts such as network or software systems) in urban nodes. Klaus Schwab defines a fourth industrial revolution as system transformation across countries and industries, with breakthroughs in "artificial intelligence, robotics, the Internet of Things, autonomous vehicles, 3-D printing, nanotechnology, biotechnology, materials science, energy storage, and quantum computing" (Schwab, 2016). This fourth Industrial revolution has also led to the establishment of Construction 4.0 a framework of cyber-physical systems and digital ecosystems for adoption and deployment of these new processes and tools (Sawhney et al., 2020, p.3).

Supporting this digital turn in terms of computing history (Ceruzzi, 2003, 2012) has presented four futurological computing paradigms which are useful to discuss including a digital paradigm, convergence, solid-state electronics, and the human interface. These paradigms have been in large held since the book publications. The Digital age paradigm is defined as coding and a binary logic forming the basis of augmenting human thinking similar to Carpo's claim (Carpo, 2017). The Convergence paradigm is the merging of computing devices and functions as previously mentioned. The Solid-State paradigm is defined as inventions and innovations in hardware such as visualization tables or "smart" glasses and human-computer interaction. The final paradigm the human interface refers to our increasing relationship and communication with computers evidenced by the IoT and personal data abundance which in turn feeds many areas including urban analytics which can analyze crowd presence and behavior among other options (Hazem, 2020). Computer interaction has also risen in importance in urban responses to pandemics. To illustrate these large global paradigms and changes, this entry will briefly backcast as a method to understand future trajectories, discuss specific augmented and virtual reality projects and cases, the use of robotics and cases in the built environment as a basis of evaluating technological development, engaged communities as well as range of solutions and challenges for regional urban futures.

VR from Militarization to the Public

Backcasting is a common method of futuring and the trajectories of XR can be understood by analyzing the early stages of invention. Before a modern understanding of VR/AR applications, there is a history of stereoscopic viewers and Victorian panoramas and a desire for public spectacle. VR and AR link to much longer histories that cannot be discussed in detail here. One of the earliest modern forays into virtual reality was designed by Morton Heilig, who invented the Sensorama a head-mounted display and machine showing a short film on a motorbike, amongst other options in 1962, which also released smells of "on the road" experience. The device was a stereoscopic 3D slide viewer, as Heilig describes the Sensorama was "an object of the present invention to provide an apparatus to simulate the desired experience by developing sensations in a plurality of the senses" (U.S. Patent 3,050,870). VR Development continued with the Sword of Damocles (1968) by Ivan Sutherland and Bob Sproull who presented the first VR head-mounted device interfacing with a computer to display wireframe graphics. In terms of AR, NASA commissioned a system in 1989 which included hand gesture controls and audio headphones. Though within VR/AR development, the work of Tom Furness and head-mounted displays in 1966 for the US air force, for fighter pilots (Furness, 2015), stands out as the rationale was the simulation of environments for real-world solutions. Later the project released an AR system in 1971, and an immersive environment, the Super Cockpit in 1986, which was reliant on topographical mapping and manual dexterity of pilots. Furness work was highly relevant to the emerging games industry and gaming input devices (Furness, 1986). Augmented Reality was later coined by Tom Caudell and David Mizell (1992) which augmented aeroplane schematics (Caudell & Mizell, 1992). Many public and commercial VR /AR applications were derived from military research and VR/AR application remains heavily dependent on the availability of hardware and capability by the games industry. A parallel development in cameras, sensors, components and computing coincided with the first AR toolkit which was developed by Hirokazu Kato and released by the HITLab an open-source computer tracking component library to overlay virtual elements in the real world (1999). The use of AR on smart devices gained ubiquity through Google glasses (2011) and Snapchat (2016) with AR later utilized in Magic Leap from 2010 and Microsoft Hololens from 2016 among other industry actors. In terms of VR, Occulus, and HTC Vive along with Valve index dominate the VR market. This brief backcast demonstrates a fairly embryonic field that has largely focused on the development of physical hardware for VR/AR capability alongside issues of grappling with immersive and accessible experiences. This developmental track runs parallel to the dystopian future set out by Ernest Cline's novel in Ready Player One (2011) and its film adaption (2018). Cline's work sets out a future of a pervasive virtualized world in which users immerse themselves in VR haptic suits. Territory studio executed the visor design for the film and the user interface and this media seems like viable projections concerning the history of XR. From this development history, there are still large technological challenges involving consumer adoption, headsets and interaction, immersive quality, tracking and developer uptake (Rabbi & Ullah, 2016).

Extended Realities (XR)

The communication of the built environment has been a pivotal force in the development of virtual, mixed and augmented reality applications and experiential drivers to mitigate the complexity of the urban and processes of construction. Milgram and Kishino's taxonomy of display is a strong base for understanding the continuum of XR which discusses virtual and mixed display based on knowledge about the world, the fidelity of reproduction, and extent of presence (1995). In essence, virtual interactions are anchored to reality. These mixed reality systems are intended to increase public consultation, deliver efficiency and improve decision-making processes (Sanchez-Sepulveda et al., 2019). XR remains a developmental field for human-computer interaction as well as widespread application in the built environment. The variety of urban management decisions and results using AR/VR vary from conservation and heritage, large scale planning (Zhang et al., 2019) to wayfinding, smart city infrastructure and inspection (Zollmann et al., 2014), architectural simulation and tourism (Yung and Khoo Lattimore 2019) amongst other applications. The degree of the technical application of XR varies from highly mobile discrete commercial units to full-scale industrial urban simulation labs and visualization techniques (Chen et al., 2021). Allied Markets forecast reports that the mixed reality sector is expected to reach $5,362m by 2024 with the Asia-Pacific geographic area exhibiting the highest CAGI with major applications in aerospace, architecture, entertainment, and gaming and education sectors among others (See: https://www.alliedmarketresearch.com/mixed-reality-market). Evans describes this as a "re-emergence" through development by major industries and game cultures drive and desire for immersion (Evans, 2019). In urban settings, VR & AR applications have raised several issues in terms of privacy and security, with AR applications requirements for long-term access to smartphone sensors among other challenges (Roesner et al., 2014). For the "Eyes of the City" 2019 Shenzhen Biennale, the exhibition theme explored the ability of the city to "see" from facial recognition to ubiquitous sensing. A project by Chen et al., (2019) Fig. 1 explored this scopic drive through an AR project titled "Image in Place" in the city of Chongqing, Hualongqiao factory zone, China. Rather than standard AR ground-based experience via headsets or smart devices, the project utilized AR via a drone (Unmanned Aerial Vehicle) using photogrammetry, and 3D modelling as a method of extending the visual experience of users. The degree of extension of reality is a critical debate on the forms of urban assemblage we seek. Moreover, XR is not a passive experience but a tool that actively helps shape decisions and ways of looking at the built environment and raises broader questions of instrumentalization, the forms of social

Augmented Reality: Robotics, Urbanism, and the Digital Turn, Fig. 1 "Eyes of the City" 8th edition of the Bi-City Biennale of Urbanism \Architecture (UABB) "Image in Place" by Chen Hui, Qiao Liu, Zhong Kai, 2019. Copyright Chen Hui, Qiao Liu, Zhong Kai

change XR enacts, XR and ethics, and the relationship between content creator and public (Fisher, 2021).

XR Applications in the Built Environment

Contemporary software workflows of Architecture, Engineering & Construction (AEC) incorporate parameterized objects and libraries (Building Information Modelling) which can be simulated in virtual reality using computer games software engines such as Unity or Unreal, and onsite through AR via smartphone or headset such as the Epson Moverio glasses. However, these workflows can incur issues with polygon counts, file format standards and requirements for data specifications (Graham et al., 2019). These AR directions and challenges have been documented through an extensive literature review by (Wang et al., 2013). Carter and Egliston (2020) have also surveyed the ethical dimensions of XR. In addition, Geographic Information systems (GIS) and the fusion with BIM allow virtual reality applications of large-scale urban centers and regions (GeoBIM). The integration of geo-information is an important supportive value structure for XR and a large-scale study of European application for the EuroSDR GeoBIM project (See: https://3d.bk.tudelft.nl/projects/eurosdr-geobim/) by Noardo et al. (2019) has discussed the challenges in the regulatory environment, data conversion, and diversified user needs. This simulative possibility for multiscale planning information is extended to public consultations and communities more commonly through AR, as Devisch et al. (2016) show "gamification" for urban planning. "Gamification" is game design elements applied to non-game contexts (Deterding et al., 2011). However, urban planning has a much longer history of developing strategies and tactical (gamified) arrangements in analogue and these aspects then became a rich trope in computer games history as seen in Micropolis by Will Wright, the urban planning simulator, later becoming SimCity (1989). For example, Sanchez-Sepulveda et al. (2019) conducted a pilot exploring design options with the public for user feedback for Eixample Esquerra District for Barcelona City Council for participatory VR. The participant sample demographic was narrow as the authors acknowledge, and feedback demanded "real" looking models. However, the process formed a valuable consultation tool. Forms of interaction through VR/AR vary and issues of graphic representation and "realness" have begun to be addressed through reality capture surveys and models and increasing semantic classification of city elements in virtualized models.

A case of applied GeoBIM can be seen in the City of Helsinki and its 3D open data model (See:

Augmented Reality: Robotics, Urbanism, and the Digital Turn, Fig. 2 ZOAN, Virtual Helsinki. 2019. Copyright ZOAN

https://hri.fi/data/en_GB/dataset?q=3D&sort=title+asc). For its "Digital Twin", the company Zoan created a virtual replica "Virtual Helsinki" of its center for tourism and culture (Smart Tourism) Fig. 2. For the 2020 May Day celebrations, a virtual performance attracted 1.4 million virtual tourists (See: https://www.forbes.com/sites/tableau/2021/07/30/a-data-culture-enabled-by-the-cloud-accelerates-time-to-value/?sh=782569435ae3). While this twin enabled civic engagement during a period of Covid travel restrictions, an earlier project in the Kalasatama area explored planning and construction digital interoperability, land and blue infrastructure options and scenarios while also developing engagement through a browser OpenCities application (now Bentley systems) for use on smart devices which visualized development but also used a public participation GIS survey for responses (KIRA-digi pilot project, 2019) (See: https://eu.opencitiesplanner.bentley.com/kymp/kalasatama). Thus, Virtual Helsinki through open data in different projects simultaneously engaged its public through XR in urban planning, but also its cultural events and retail and tourism. Helsinki has one of the most focused VR/AR clusters through its XR Center, a recent project Augmented Urbans (2018–2021) coordinated a series of local actions in the Baltic region (See: https://www.augmentedurbans.eu/). However, as Hemmersam et al. (2015), there are challenges in empowering citizens and providing access to larger planning data moving beyond an institutional setting. Hemmersam et al. inspired by Even Westvang created PlanAR an intuitive app in Oslo for smart devices to provide an accessible structural information layer of the environment specifically to research accessibility issues (See: http://bengler.no/seeplan). Data abundance and XR development have real possibilities in shaping and decision of physical (digitally modelled) elements, future scenarios and semantic information on city structure. These gamified developments, however, are not a panacea for representing urban social aspects and complexity and are reliant on more developed geoinformation and streamlined user interfaces.

Technological Fusion: Eye Tracking, AR, and Machine Learning

With the availability of AR toolkits contemporary applications in urban planning and increasing case uses alongside a wide variety and consumer-friendly eye-tracking (ET) devices a natural cross over between AR and user feedback via ET devices has come to the fore. ET can evaluate spatial preference in urban and landscape settings (Dupont et al., 2014). Reality capture and survey also contributes to a fusion of components and processes with virtual models, being utilized for simulation and evaluation of public spatial behavior. For example, VR simulation of a BIM model has been combined with eye-tracking to good effect in an analysis of Hong Kong road, Fujian by Zhang et al. (2019) to explore perceptions of cityscape and mode to evaluate conservation policy (See also Han et al., 2019). As Kiefer et al. (2017) states real-world studies on navigation and wayfinding are at an early phase, though there are possibilities for fusion limited by mobile eye-tracking computer processing, in-situ tests are prone to anomalies and VR wayfinding may have different visual attentive results. The use of VR/AR and eye tracking can yield deep data on spatial perception, choice zones (Emo, 2018) and cognitive processes (Dalton et al., 2012). The use of machine learning (ML) for urban assemblage and design has been discussed in Carta (2019) and ML analysis for a wide variety of features such as city furniture counts, crowd analysis, and behaviors and identification of defects among others using XR demonstrates the fusion of novel tools and techniques and a future trajectory for deeper level (multitude dimensions of the way data is transformed) semantic web technology immersive experience and display information (Lampropoulos et al., 2020).

The use of mixed reality for simultaneous localization and mapping (SLAM) maps spaces in real-time for measurement and tracking is very much similar to AR in which spatial anchors are required to map models *on* space, and SLAM is a robotic problem in how devices move *through* space. Indeed, urban futures utilizing drones for real-time data in 3D model construction feeds smart city ecosystems and master planning as well as environmental management (Cureton, 2020). The best representation of this is through AI Drones, mapping and navigation through an environment in The Drone Racing League (DRL) and Lockheed Martin and in ground-based systems such as Boston Dynamics SPOT in combination with commands through mixed reality as demonstrated by The Mixed Reality Lab, ETH Zurich and Microsoft. The use of XR for robotic control will also have implications for construction, through the use of industrial robots in architectural fabrication, surveying and inspection and construction site safety. In planning and building control, XR for drone control allows additional real-time data layers as seen in the work of Sysveo, in France part of the European Space Agency (ESA) business incubation (See: https://www.esa.int/Applications/Telecommunications_Integrated_Applications/Business_Incubation/ESA_Business_Incubation_Centres9). These developments also require policy development in terms of operation and regulation as seen in the updates for the use of drones for planning enforcement for the Royal Institute of Town Planners (RTPI) in the United Kingdom (See: https://www.rtpi.org.uk/policy-and-research/practice/planning-enforcement-handbook-for-england/). These examples and cases of technological fusion demonstrate advanced applications some of which require further testing and deployment in urban settings but evidence XR as fundamentally part of these urban futures. These futures have possibilities of emerging from non-traditional tech and innovation invested super cities, as seen in the work of WERobotics, enabling rural community initiatives across Africa, Asia, and Latin America (See: https://werobotics.org/).

Trajectories

Harrison et al. (2021) suggest a paradigm shift in regional planning which they argue has become institutionalized to a mode of planning regional futures and associated problems with heterogeneous sources. This involves working with a consortia of actors, agile working, broadening planner's skills, and the use of novel tools including visualization and VR (ibid p.10). This broad call for changes in regional planning and recognition of the need for agile, and sometime technological solutions to "wicked" urban problems sets the wider issues that augmenting urban environments face, in terms of simulation, community engagement, and mediation as well as the skillsets required and current hardware

Augmented Reality: Robotics, Urbanism, and the Digital Turn, Fig. 3 Block by Block, UN. Habitat, Co-created playground design in Ghana using Minecraft, 2021

limitations of XR. In the Block by Block project, begun in 2012, a collaboration between Mojang, Microsoft, and UN-Habitat, non-traditional demographics, were engaged using a methodology which included the computer game Minecraft, in which users co-construct their world, design elements, and collaborate with professionals and communicating these decisions in mixed reality (See: https://www.blockbyblock.org/resources). Block by Block (Fig. 3) is an example of more meaningful community engagement and side-lines issues of data accessibility, through model creation from scratch in comparison to Helsinki which has a government top-down approach for open data and disseminated projects by entrepreneurial organizations. Both projects have viable and successful approaches and have the same goals. As Block by Block states,

> in order to make urban planning and design processes more participatory, people without design or architectural skills need easy ways to use tools to effectively describe their ideas and desires to professionals. The lack of such tools makes it difficult for non-professionals to engage in dialogue with professionals because they lack the technical skills, confidence and language to adequately communicate their ideas (Block by Block, 2021, p.9).

This trajectory has been central to XR applications in the built environment in many of the cases cited here in the creation, engagement and deeper level understanding of places. In addition, the skillsets required for audiences raises the question of modes of XR education and access to more advanced XR equipment beyond smartphones. The diversity of XR deployments within variegated smart city ecosystems is reliant upon the contemporary development of designed models and proposals and established workflows and interoperability, geoinformation, and GIS science for urban and regional planning. In addition, XR and deployments of robotics in future environments brings forward the possibility of XR control of devices, in-situ data visualization and response and real-time data feed for urban and regional management. These trajectories should all be marshalled carefully, evaluated, and deployed for three dominant rationales of XR the representation of reality, engagement, and added urban intelligence.

Cross-References

- ▶ City Visions: Toward Smart and Sustainable Urban Futures
- ▶ Community Engagement for Urban and Regional Futures
- ▶ Digital Twin and Cities
- ▶ Smart Cities
- ▶ Urban Futures: Pathways to Tomorrow

References

Albino, V., Berardi, U., & Dangelico, R. M. (2015). Smart cities: Definitions, dimensions, performance, and initiatives. *Journal of Urban Technology, 22*(1), 3–21. https://doi.org/10.1080/10630732.2014.942092.

Ash, J., Kitchin, R., & Leszczynski, A. (2018). Digital turn, digital geographies? *Progress in Human Goegraphy 42*(1), 25–43.

Batty, M. (2018). *Inventing future cities*. MIT Press.

Bimber, O., & Raskar, R. (2005). *Spatial augmented reality: Merging real and virtual worlds* (1st ed.). A K Peters/CRC Press.

Block by Block. (2021). *The Block by Block Playbook: Using Minecraft as a participatory design tool in urban design and governance*. UN-Habitat. https://unhabitat.org/the-block-by-block-playbook-using-minecraft-as-a-participatory-design-tool-in-urban-design-and.

Carmigniani, J., Furht, B., Anisetti, M., Ceravolo, P., Damiani, E., & Ivkovic, M. (2011). Augmented reality technologies, systems and applications. *Multimedia Tools and Applications, 51*(1), 341–477.

Carpo, M. (2013). *The digital turn in architecture 1992–2012*. Chichester, West Sussex [England]: Wiley.

Carpo, M. (2017). *The second digital turn: Design beyond intelligence*. MIT Press.

Carta, S. (2019). *Big data, code and the discrete city: Shaping public realms* (1st ed.). Routledge.

Carter, M., & Egliston, B. (2020). Ethical Implications of Emerging Mixed Reality Technologies. Socio-Tech Futures Lab. https://doi.org/10.25910/5ee2f9608ec4d

Caudell, T. P., & Mizell, D. W. (1992). Augmented reality: An application of heads-up display technology to manual manufacturing processes. *Proceedings of the Twenty-Fifth Hawaii International Conference on System Sciences 2*, 659–669.

Ceruzzi, P. E. (2003). *A history of modern computing* (2nd ed.). London, Eng. ; Cambridge, Mass.: MIT Press.

Ceruzzi, P. E. (2012). *Computing*. MIT Press.

Chen, H., Liu, Q., & Zhong, K., (2019). Image In Place, Bi-City Biennale of Urbanism\Architecture (UABB). http://eyesofthecity.net/image-in-place/

Chen, S., Miranda, F., Ferreira, N., Lage, M., Doraiswamy, H., Brenner, C., Defanti, C., Koutsoubis, M., Wilson, L., Perlin, K., & Silva, C. T. (2021). UrbanRama: Navigating cities in virtual reality. *IEEE transactions on visualization and computer graphics*. https://doi.org/10.1109/TVCG.2021.3099012.

Cooper, S. (2018). Civil engineering collaborative digital platforms underpin the creation of "Digital Ecosystems". *Proceedings of the Institution of Civil Engineers – Civil Engineering, 171*(1), 14. https://doi.org/10.1680/jcien.2018.171.1.14.

Cureton, P. (2020). *Drone futures: UAS for landscape & urban design*. London: Routledge. ISBN 9780815380511.

Dalton, R. C., Hölscher, C., & Turner, A. (2012). Understanding space: The nascent synthesis of cognition and the syntax of spatial morphologies. *Environment and Planning B: Planning and Design, 39*(1), 7–11.

Deterding, S. , Dixon, D. , Khaled, R. , & Nacke, L. E. . (2011). From game design elements to gamefulness: Defining "gamification". In *Proceedings of the 15th international academic MindTrek conference: Envisioning future media environments*. ACM. Retrieved from https://doi.org/10.1145/2181037.2181040

Devisch, O., Poplin, A., & Sofronie, S. (2016). The gamification of civic participation: Two experiments in improving the skills of citizens to reflect collectively on spatial issues. *Journal of Urban Technology, 23*(2), 81–102. https://doi.org/10.1080/10630732.2015.1102419.

Dunn, N., & Cureton, P. (2019). Frictionless futures: The vision of smartness and the occlusion of alternatives. In S. M. Figueiredo, S. Krishnamurthy, & T. Schroeder (Eds.), *Architecture and the smart city* (pp. 17–28). (Critiques)). Routledge.

Dunn, N., & Cureton, P. (2021). Digital twins of cities and evasive futures. In A. Aurigi & N. Odendaal (Eds.), *Shaping smart for better cities*. Academic Press.

Dupont, L., Antrop, M., & Eetvelde, V. V. (2014). Eye-tracking analysis in landscape perception research: Influence of photograph properties and landscape characteristics. *Landscape Research, 39*(4), 417–432. https://doi.org/10.1080/01426397.2013.773966.

Egliston, B., & Carter, M. (2020a). *Ethical implications of emerging mixed reality technologies*. Australia: Socio-Tech Futures Lab.

Egliston, B., & Carter, M. (2020b, September). Oculus imaginaries: The promises and perils of facebook's virtual reality. *New Media & Society*. https://doi.org/10.1177/1461444820960411.

Emo, B. (2018). Choice zones: Architecturally relevant areas of interest. *Spatial Cognition & Computation, 18*(3), 173–193. https://doi.org/10.1080/13875868.2017.1412443.

Evans, L. (2019). *The Re-emergence of virtual reality*. New York: Routledge.

Fisher, J. A. (Ed.). (2021). *Augmented and mixed reality for communities* (1st ed.). CRC Press.

Flotynski, J., & Sobocinski, P. (2018). Semantic 4-dimensional modeling of VR content in a heterogeneous collaborative environment. In *Proceedings of the 23rd International ACM Conference on 3D Web Technology*.

Furness, T. A. (1986, September). The super cockpit and its human factors challenges. *Proceedings of the Human Factors Society Annual Meeting, 30*(1), 48–52. https://doi.org/10.1177/154193128603000112.

Furness, T.A. (2015, November 17). '50 years of VR with Tom Furness: The Super Cockpit, Virtual Retinal Display, HIT Lab, & Virtual World Society' *Voices of VR Podcast*. https://voicesofvr.com/245-50-years-of-vr-with-tom-furness-the-super-cockpit-virtual-retinal-display-hit-lab-virtual-world-society/

Graham, K., Pybus, C., Arellano, N., Doherty, J., Chow, L., Fai, S., & Grunt, T. (2019). Defining geometry levels for optimizing BIM for VR: Insights from traditional architectural media. *Technology|Architecture + Design, 3*(2), 234–244. https://doi.org/10.1080/24751448.2019.1640541.

Han, D-I., Weber, J., Bastiaansen, M., Mitas, O., & Lub, X. D. (2019). Virtual and augmented reality technologies to enhance the visitor experience in cultural tourism. In M. Tom Dieck, & T. Jung (Eds.), *Augmented reality and virtual reality: The power of AR and VR for business* (pp. 113–128). (Progress in IS). Springer Nature Switzerland AG. https://doi.org/10.1007/978-3-030-06246-0_9.

Harrison, J., Galland, G., & Tewdwr-Jones, M. (2021). Regional planning is dead: Long live planning regional futures. *Regional Studies, 55*(1), 6–18. https://doi.org/10.1080/00343404.2020.1750580.

Hazem, Z. (2020). The digital crowd. *Architecture and Culture, 8*(3-4), 653–666. https://doi.org/10.1080/20507828.2020.1794419.

Hemmersam, P., Martin, N., Westvang, E., Aspen, J., & Morrison, A. (2015). Exploring urban data visualization and public participation in planning. *Journal of Urban Technology, 22*(4), 45–64. https://doi.org/10.1080/10630732.2015.1073898.

Irizarry, J., Gheisari, M., Williams, G., & Walker, B. N. (2013). InfoSPOT: A mobile augmented reality method for accessing building information through a situation awareness approach. *Automation in Construction, 33*, 11–23., ISSN 0926-5805. https://doi.org/10.1016/j.autcon.2012.09.002.

Kiefer, P., Giannopoulos, I., Raubal, M., & Duchowski, A. (2017). Eye tracking for spatial research: Cognition, computation, challenges. *Spatial Cognition & Computation, 17*(1-2), 1–19. https://doi.org/10.1080/13875868.2016.1254634.

KIRA-digi pilot project. *The Kalasatama Digital Twins Project Ministry of the Environment*, Helsinki. 02/05/2019. https://www.hel.fi/helsinki/en/administration/information/general/3d/potential-uses/

Lampropoulos, G., Keramopoulos, E., & Diamantaras, K. (2020). Enhancing the functionality of augmented reality using deep learning, semantic web and knowledge graphs: A review. *Visual Informatics, 4*(1), 32–42., ISSN 2468-502X. https://doi.org/10.1016/j.visinf.2020.01.001.

Li, X., Yi, W., Chi, H. L., Wang, X., & Chan, A. P. C. (2018). A critical review of virtual and augmented reality (VR/AR) applications in construction safety. *Automation in Construction, 86*, 150–162., ISSN 0926-5805. https://doi.org/10.1016/j.autcon.2017.11.003.

Luisa Caldas, L., & Keshavarzi, M. (2019). Design immersion and virtual presence. *Technology|Architecture + Design, 3*(2), 249–251. https://doi.org/10.1080/24751448.2019.1640544.

Noardo, F., Ellul, C., Harrie, L., Overland, I., Shariat, M., Stoter, J., & Arroyo Ohori, K. (2019). Opportunities and challenges for GeoBIM in Europe: Developing a building permits use-case to raise awareness and examine technical interoperability challenges. *Journal of Spatial Science*. https://doi.org/10.1080/14498596.2019.1627253.

Paul Milgram, P., Takemura, H., Utsumi, A., & Kishino, F. (1995, December 21). Augmented reality: A class of displays on the reality-virtuality continuum, Proceedings of SPIE 2351, *Telemanipulator and Telepresence Technologies*; https://doi.org/10.1117/12.197321

Quattrini, R., Pierdicca, R., Berrocal, A. B., Zamorano, C., Rocha, J., & Varajão, I. (2021). Uncovering the potential of digital technologies to promote railways landscape: Rail to land project. In L. T. De Paolis, P. Arpaia, & P. Bourdot (Eds.), *Augmented reality, virtual reality, and computer graphics* (AVR 2021. Lecture Notes in Computer Science) (Vol. 12980). Cham: Springer. https://doi-org.ezproxy.lancs.ac.uk/10.1007/978-3-030-87595-4_23.

Rabbi, I., & Ullah, S. (2016, December). *A survey on augmented reality challenges and tracking*. Acta Graphica., [S.l.], 24, 1-2, p. 29-46., ISSN 1848-3828. Available at: http://www.actagraphica.hr/index.php/actagraphica/article/view/44. Date Accessed 20 Sept 2021.

Roesner, F., Kohno, T., & Molnar, D. (2014, April) Security and privacy for augmented reality systems. Communications of the ACM 57, 4, 88–96. https://doi-org.ezproxy.lancs.ac.uk/10.1145/2580723.2580730

Sanchez-Sepulveda, M., Fonseca, D., Franquesa, J., & Redondo, E. (2019). Virtual interactive innovations applied for digital urban transformations. *Mixed approach, Future Generation Computer Systems, 91*, 371–381., ISSN 0167-739X. https://doi.org/10.1016/j.future.2018.08.016.

Sawhney, A., Riley, M., & Irizarry, J. (Eds.). (2020). *Construction 4.0 An innovation platform for the built environment*. Routledge.

Schwab, K. (2016). The Fourth Industrial Revolution: what it means, how to respond https://www.weforum.org/agenda/2016/01/the-fourth-industrial-revolution-what-it-means-and-how-to-respond/

Wang, X., Mi Jeong Kim, Peter E.D. Love., & Shih-Chung Kang. (2013). Augmented Reality in built environment: Classification and implications for future research. *Automation in Construction 32*. https://doi.org/10.1016/j.autcon.2012.11.021.

Waterworth, J., & Hoshi, K. (2016). Bridging contextual gaps with blended reality spaces. In *Human-experiential design of presence in everyday blended reality* (Human–Computer Interaction Series). Cham: Springer.

Westera, W. (2012). *The digital turn: How the internet transforms our existence*. Authour House.

Yuan, P. F., & Chao Yan, C. (2020). Collaborative networks of robotic construction. *Architectural Design, 90*(2), 74–81.

Yung, R., & Khoo-Lattimore, C. (2019). New realities: A systematic literature review on virtual reality and augmented reality in tourism research. *Current Issues in Tourism, 22*(17), 2056–2081. https://doi.org/10.1080/13683500.2017.1417359.

Zhang, L. M., Zhang, R. X., Jeng, T. S., & Zeng, Z. Y. (2019). Cityscape protection using VR and eye tracking technology. *Journal of Visual Communication and Image Representation, 64*, 102639., ISSN 1047-3203. https://doi.org/10.1016/j.jvcir.2019.102639.

Zollmann, S., Hoppe, C., Kluckner, S., Poglitsch, C., Bischof, H., & Reitmayr, G. (2014). Augmented reality for construction site monitoring and documentation. *Proceedings of the IEEE, 102*(2), 137–154.

Autarkic City

► Circular Cities

Automated City

► Augmented Reality: Robotics, Urbanism, and the Digital Turn

Automobile Dependence

► Urban Structure and Its Impact on Mobility Patterns: Reducing Automobile Dependence Through Polycentrism

Awareness and Education

► Butterfly Gardening in Colombo, Sri Lanka: Approach to Biodiversity Conservation, Monitoring, Education, and Awareness in Urbanizing Habitats

B

Bag-of-Words Model: Sustainability Lacks Influencers

A Social Network Analysis Using Natural Language Processing (AI) on Tweets in 109 Smart Cities

Julien Carbonnell
Artificial Intelligence on Citizen Engagement, Democracy Studio

At least that's what suggests a lexical analysis of tweets' content, conducted over 109 smart cities worldwide. By using Natural Language Processing techniques, the branch of Artificial Intelligence concerned with the linguistic side of Human-Computer Interaction, the study reveals that the sustainability lexicon is by far the less used between other urban studies topics such as infrastructure, governance, or entrepreneurship.

The book *Democracy Studio* shares a complete analysis of spontaneous communications by Twitter users who mention in a hashtag one of the 109 smart cities listed in the Smart City Index 2020. The original datasets represent 110,862 tweets gathered from the four corners of the globe, totalizing 19,184,388 words associated with one name of a smart city. From the extraction of very basic lexical features such as the number of words in a tweet, the average length of a sentence, to more advanced ones like the polarity of sentiment expressed, NLP techniques point some interesting correlations between the 29 variables collected from tweets. For example, the average number of stop words is highly correlated with all sentiment scores, which means that the more stop words we can found in a tweet, the most probably it expresses an opinion. Similarly, the average number of numerics, hashtags, and punctuation are correlated with the average tweet length and the average number of words, but none of them have a relationship with the intensity of the sentiment expressed.

After having cleaned the tweets from all punctuation, numerics, emojis, and stop words, the datasets result in a very clean lists of meaningful tokens, representing pretty clear semantics used in each of the 109 smart cities worldwide. A sedimentation of the most frequent words associated with *smart grid*, *IoT*, *urban planning*, *urban development*, *innovation*, *gov-tech*, *open-data*, *e-citizenship*, *empowerment*, *transportation*, *mobility*, *environment*, *energy*, *democracy2.0*, *policy*, *economy*, and *business* is proceeded. The resulting lists of words constitute thematic lexicons which are commonly called bag-of-words (BoW) when texts of various lengths are represented as a bag of its own words and used as a reference for document classification or topic modelling of other texts.

R. C. Brears (ed.), *The Palgrave Encyclopedia of Urban and Regional Futures*,
https://doi.org/10.1007/978-3-030-87745-3

After combining these lists of words into six BoWs, each of them hosting the 150 most frequently used words associated to: *smart city, civic tech, infrastructure, sustainability, governance,* and *entrepreneurship,* it is possible to evaluate the predominance of each of them at different scales: city, continent, or global. The overlap between BoWs is avoided by assigning cross-field words to a single category, in order to keep away the possibility to count the same word several times. Therefore, a weighting of each bag-of-words in each city means counting the occurrences of words belonging to each thematic in datasets of different size. This technique consists in filtering the 19 million words collected through different strainers and take the weight of each BoW at the end, to know much this or that urban studies topic has been discussed online. It appears that the sustainability is by far the less discussed topic of the six (see Fig. 1).

For a few years, environmental issues had become the number one priority in global policies, and smart cities are among the few front siders tackling it by communicating massively on green solutions supported by institutional marketing forces. In practice, the sustainability vocabulary including words like *resource, recycling, resilience,* or *biodiversity* represents only 27% of the use of infrastructure words like *supply, system, storage,* or *mobility.* In spontaneous communications of both official media, politics, and random inhabitants on Twitter, online messaging is most of the time motivated by the social reward evaluated by the number clicks received on a publication. This led to the drama of our times: when global knowledge has never been so tangible, dumb content and fake news often collects more success than insightful information. Does this mean that sustainability lacks influencers?

	topic	weight
0	smartcity	31423
1	civictech	27062
2	infrastructure	32193
3	sustainability	8951
4	governance	26538
5	entrepreneurship	25682

Bag-of-Words Model: Sustainability Lacks Influencers, Fig. 1 Weight of each thematic BoW taken on 109 smart cities worldwide

A deeper look at this early observation allows to check different related questions. Is the acknowledgment true to all cities taken individually? Are there some geographical areas in the world where sustainability is more discussed than others?

Taking the average proportion of representation of each topic in each city, by dividing the weight of each BoW by the number of tweets collected in each city, the result confirms that sustainability vocabulary is less used than the others at the city scale taken individually. While the *entrepreneurship* lexicon represents 5.11% of the total of all words used in tweets in the city of Singapore, the *governance* and the *civic technology* ones represent 5.08% and 4.11% of them, respectively, in Hong Kong, and the *infrastructure* and the *smart city* ones represent 4.01% and 3.9% of them, respectively, in Shenzhen. Aside of these, the highest proportion of tweets referring to the sustainability lexicon represents only 1.77% of the tweets in the city of Abu Dhabi, the one city who talks the most about sustainability online. On the map below, we can see the regional distribution of the proportion of sustainability vocabulary used in Twitter (see Fig. 2). It shows clearly the smart cities in the world where the Twitter users communicate the most on the sustainability topic. In decreasing order, the eleventh first are: Abu Dhabi in the United Arab Emirates, Hangzhou, Chongqing, Shanghai, and Nanjing in China, Singapore, Oslo in Norway, Geneva in Switzerland, Madrid in Spain, New Delhi in India, and Gothenburg in Germany.

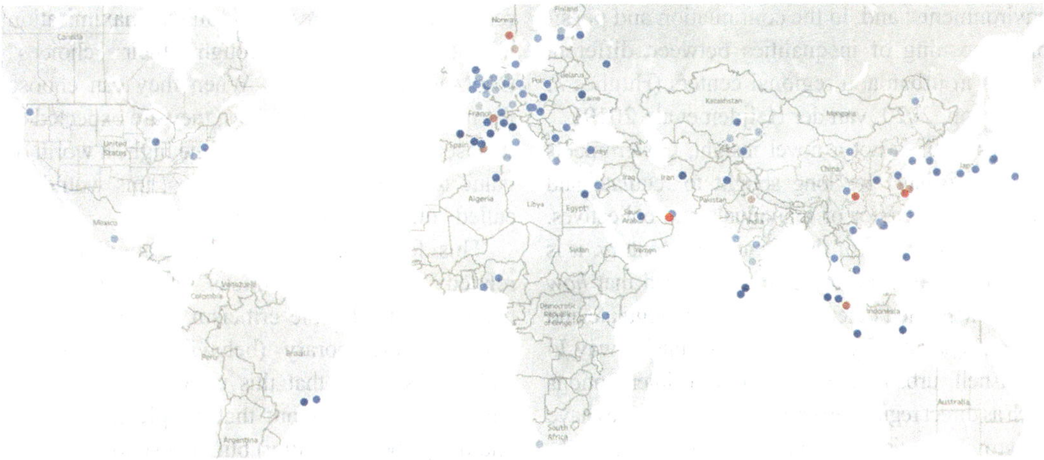

Bag-of-Words Model: Sustainability Lacks Influencers, Fig. 2 Distribution of the use of sustainability lexicon over the world (blue = lowest; red = highest). All the details of the method to duplicate the lexical analysis on tweets is shared on the GitHub repository of *Democracy Studio*. It features ready to implement models in Python to assist all kind of urban projects with such insights

Cross-References

- ▶ Green and Smart Cities in the Developing World
- ▶ Smart City: A Universal Approach in Particular Contexts

Behavioral Science Informed Governance for Urban and Regional Futures

Jeroen van der Heijden
School of Government, Victoria University of Wellington, Wellington, New Zealand
School of Regulation and Global Governance, Australian National University, Canberra, Australia

Synonyms

Nudging; Soft-governance; Soft-law; Soft-regulation

Definition

Insights from the behavioral sciences have rapidly gained prominence to accelerate urban climate action. Effectively, the behavioral sciences challenge the long-held assumption that humans are 'rational', 'self-interested', and 'utility maximizing'. Rather, because of our in-built biases and heuristics, we humans often make decisions that are not in our best interest, let alone that of society at large. Lessons from behavioral science-informed governance for urban climate action that has proven to work may help to tackle other urban challenges too.

It is widely acknowledged that having "the right" technological solutions is only one part of the puzzle for achieving sustainable, resilient, economically viable, and inclusive urban and regional futures (Dodman, 2009; Nisa et al., 2019). Another major part of the puzzle is to make sure that people and collectives behave in "the right" manner. After all, individual and collective behavior plays a key role in the vast amount of emissions and other waste produced at city level; in the resources consumed to develop, operate and maintain urban environments; and, in the continuation and possible deepening of inequalities between different groups in urban and regional centers (Hughes & Hoffmann, 2020; van der Heijden et al., 2019).

This itself is not a novel insight. Governments and others have for long sought to regulate and govern the behavior of individuals and collectives. Yet, since the start of the twenty-first century, it is becoming more and more acknowledged that *how* governments and others have sought to regulate and govern behavior may not be the most suitable way. In a nutshell, urban climate governance interventions such as direct regulation or economic incentives have not provided satisfactory policy solutions to the problems caused by harmful behavior at the individual level (Luque-Ayala et al., 2018; Van der Heijden, 2014). These interventions are typically premised on a rational choice perspective, which assumes that individuals make a cost-benefit analysis when deciding whether to comply with direct regulation (Lehmann Nielsen & Parker, 2012), and make rational self-interested decisions when responding to economic incentives (McMahon, 2015).

However, people do not always behave as predicted by the rational-choice perspective. We often lack the time, information, and mental capacity to make "rational" choices, and instead fall back on heuristics (mental shortcuts) and cognitive biases (including those relating to habits and social norms) when making decisions (Ariely, 2008; Kahneman, 2011). Building on these insights, we now witness an embracing of behavioral science informed interventions and processes for urban and regional futures around the world.

Why Is Behavioral Science Relevant for Governance at all?

Policymaking and implementation have for many years been built on rational choice theory – and that is often still the case. Rational choice theory is an analytical framework in neoclassical economics for understanding and modeling the social and economic behavior of groups of people – for example, the population of a country. A central aspect of this theory is that people are rational beings who have "stable, coherent and well-defined preferences rooted in self-interest and utility maximisation that are revealed through their choices" (McMahon, 2015, 141). When they can choose from a variety of alternatives, they are expected to choose the alternative that has the highest worth or value to them. In technical terms, this would be called "utility maximization."

This focus on utility maximization and the related "Homo Economicus" stereotype have received considerable criticism, however (Read, 2007). Contemporary (behavioral) economists and others claim that this understanding is too narrow. They point out that people may desire one thing (being healthy) but choose to do something else (smoke, eat unhealthy food, and fail to take enough exercise), implicating that human behavior tends to be less "rational" than the predictions of neoclassical economics (Kahneman, 2011). Scholars have also pointed out that, humans are less rational in making choices under uncertainty than is predicted by neoclassical economics. People find it difficult to have a full understanding of many of the problems they are facing. It is often impossible for them to acquire all the relevant information they need to make a rational decision, and, even if they could get all this information, they are very likely to lack the mental capacity or the time to process it. In other words, when making decisions humans possess only "bounded rationality" and must make decisions by "satisficing" – they choose what makes them happy enough (Simon, 1997 [1945]).

A Selected Overview of "Irrational Behaviors"

Recent work from the behavioral sciences has pointed out that people deviate in other predictable ways from the neoclassical assumptions of rationality – we humans rely on cognitive biases ("mental shortcuts") and heuristics when making choices. Sometimes this results in suboptimal outcomes. To name a few dominant heuristics and biases, and their possible suboptimal effects:

- Present bias and hyperbolic discounting: People give stronger weight to a payoff that is received closer to the present time, when they are faced with the choice of getting a payoff at an earlier or a later moment. This is known as *present bias* and is often explained by the psychological desire to have certainty and resolve events immediately (O'Donoghue & Rabin, 2015). The flip side is that people discount the future consequences of their current actions, and postpone losses or dealing with losses until later, a tendency known as *hyperbolic discounting* (Hardisty et al., 2012). The rate of discounting changes with the time horizon faced. That is, people give high discount rates for short time horizons, but low discount rates for long time horizons. Insights into hyperbolic discounting may also explain why people procrastinate about making choices that do not come with immediate and significant gains, such as changing their energy plans, installing solar panels, or switching from travelling by car to travelling by bus (Pollitt & Shaorshadze, 2013).

- Loss aversion: When making decisions, losses loom larger than improvements or gains, and, consequently, people prefer to avoid losses than to acquire gains. It is sometimes argued that, for small or moderate amounts of money, losses loom twice as large as gains (Tversky & Kahneman, 1991). What is relevant to note here is that people define losses and gains relative to a reference point, which is often their status quo at the point of making the decision. Thus, when people are offered solar panels on their house for an installation cost of $1000, and a possible yield of $1500 over the lifetime of the panels, they may choose not to install the panels because the possible loss of the investment looms larger than its gain (i.e., a $2000 yield is required to make up for the $1000 investment; see further, Greene, 2011).
- Anchoring and framing: If people are given a cue or signal (an "anchor") and are then asked to make a choice, they are likely to be heavily influenced by the cue or signal even when it is not related to the object of choice. For example, if people are first asked to recall the last three digits of their social security number and are then asked to estimate the number of cities in the world with a population of over one million inhabitants, those with a low digit-value are likely to underestimate the number of cities, whereas those with a high digit-value are likely to overestimate it (Tversky & Kahneman, 1974). The higher the ambiguity, the lower the familiarity with the problem, or the more trustworthy the source of information, the stronger the anchoring effect (Furnham & Boo, 2011). Also, seemingly inconsequential changes in the formulation of a choice problem ("framing") affect people's preferences. In other words, framing an outcome as a marginal monetary loss or a huge environmental or societal gain may make all the difference in seeking to encourage environmentally sustainable, economic prudent, and socially desirable behavior (Borah, 2011).

Conclusion

In sum, then, for many decades, research has shown that our behavior is less rational than is often assumed by neoclassical economics modeling. This modeling is, however, at the base of many policies and regulations – including much contemporary governance for urban and regional futures. Scholars from the behavioral sciences, therefore, call for policy interventions that are sensitive to the "cognitive failures" of humans (Jolls et al., 1998).

Around the world, we are witnessing an embracing of insights from the behavioral sciences in the development and implementation of governance interventions for urban and regional futures (see among others, Guenard, 2020; Nisa et al., 2019; Sagor, 2020; Sorrell et al., 2020; Van der Heijden, 2020; White et al., 2019; Wi & Chang, 2019). While this is a hopeful development, it remains to be seen if these interventions will deliver on their promises. Ongoing research on the use of behavioral insights in the governance for urban and regional futures will, no doubt, generate important insights.

Cross-References

▶ Financing: Fiscal Tools to Enhance Regional Sustainable Development
▶ Green Cities in Theory and Practice

► Regulation of urban and regional futures
► Unpacking Cities as Complex Adaptive Systems

References

Ariely, D. (2008). *Predictably irrational: The hidden forces that shape our decisions*. New York: HarperCollins.

Borah, P. (2011). Conceptual issues in framing theory: A systematic examination of a decade's literature. *Journal of Communication, 61*(2), 246–263.

Dodman, D. (2009). Blaming cities for climate change? An analysis of urban greenhouse gas emissions inventories. *Environment and Urbanization, 21*(1), 185–201.

Furnham, A., & Boo, H. C. (2011). A literature review of the anchoring effect. *The Journal of Socio-Economics, 40*(1), 35–42.

Greene, D. (2011). Uncertainty, loss aversion, and markets for energy efficiency. *Energy Economics, 33*(4), 608–616.

Guenard, M. (2020). *Key learnings for cities to enable 1.5-degree lifestyles*. Retrieved from https://talkofthecities.iclei.org/key-learnings-for-cities-to-enable-1-5-degree-lifestyles/

Hardisty, D., Appelt, K., & Weber, E. (2012). Good or bad, we want it now: Fixed-cost present bias for gains and losses explains magnitude asymmetries in intertemporal choice. *Journal of Behavioral Decision Making, 21*(4), 348–361.

Hughes, S., & Hoffmann, M. (2020). Just urban transitions: Toward a research agenda. *WIREs Climate Change, 11*(3), 1–11.

Jolls, C., Sunstein, C., & Thaler, R. (1998). A behavioral approach to law and economics. *Stanford Law Review, 50*(5), 1471–1550.

Kahneman, D. (2011). *Thinking fast and slow*. New York: Farrar, Straus and Giroux.

Lehmann Nielsen, V., & Parker, C. (2012). Mixed motives: Economic, social, and normative motivations in business compliance. *Law & Policy, 34*(4), 428–462.

Luque-Ayala, A., Marvin, S., & Bulkeley, H. (Eds.). (2018). *Rethinking urban transitions*. London: Routledge.

McMahon, J. (2015). Behavioral economics as neoliberalism: Producing and governing homo economicus. *Contemporary Political Theory, 14*(2), 137–158. https://doi.org/10.1057/cpt.2014.14.

Nisa, C., Belanger, J., Schumpe, B., & Faller, D. (2019). Meta-analysis of randomised controlled trials testing behavioural interventions to promote household action on climate change. *Nature Communications, 10*(4545), 1–13.

O'Donoghue, T., & Rabin, M. (2015). Present bias: Lessons learned and to be learned. *American Economic Review, 104*(5), 273–279.

Pollitt, M., & Shaorshadze, I. (2013). The role of behavioural economics in energy and climate policy. In R. Fouquet (Ed.), *Handbook on energy and climate change* (pp. 523–546). Cheltenham: Edward Elgar.

Read, D. (2007). Experienced utility: Utility theory from jeremy bentham to Daniel kahneman. *Thinking & Reasoning, 13*(1), 45–61.

Sagor, E. (2020). *Putting behavioural science to work in climate action planning*. Retrieved from https://blogs.lse.ac.uk/progressingplanning/2020/08/17/putting-behavioural-science-to-work-in-climate-action-planning/

Simon, H. A. (1997 [1945]) *Administrative behavior. A study of decision-making processes in administrative organization*. Free Press.

Sorrell, S., Gatersleben, B., & Druckman, A. (2020). The limits of energy sufficiency: A review of the evidence for rebound effects and negative spillovers from behavioural change. *Energy Research & Social Science, 64*(1), 1–17.

Tversky, A., & Kahneman, D. (1974). Judgment under uncertainty: Heuristics and biases. *Science, 185*(4157), 1124–1131.

Tversky, A., & Kahneman, D. (1991). Loss aversion in riskless choice: A reference-dependent model. *The Quarterly Journal of Economics, 106*(4), 1039–1061.

Van der Heijden, J. (2014). *Governance for urban sustainability and resilience: Responding to climate change and the relevance of the built environment*. Cheltenham: Edward Elgar.

Van der Heijden, J. (2020). Urban climate governance informed by behavioural insights: A commentary and research agenda. *Urban Studies*. https://doi.org/10.1177/0042098019864002.

van der Heijden, J., Bulkeley, H., & Certomá, C. (2019). *Urban climate politics: Agency and empowerment*. Cambridge: Cambridge University Press.

White, K., Habib, R., & Hardisty, D. (2019). How to shift consumer behaviors to be more sustainable: A literature review and guiding framework. *Journal of Marketing, 83*(3), 22–49.

Wi, A., & Chang, C.-H. (2019). Promoting pro-environmental behaviour in a community in Singapore – From raising awareness to behavioural change. *Environmental Education Research, 25*(7), 1019–1037.

Benefit

► Furthering the Sustainable Development Agenda by Putting Urban Heritage and Value Extraction at the Center

Benefits

► Peri-urban Regions

Beyond Knowledge: Learning to Cope with Climate Change in Cities

Christian Reichel[1] and Wolfgang Haupt[2]
[1]University of Applied Sciences for Media, Communication and Management (HMKW), Berlin, Germany
[2]Leibniz-Insitute for Research on Society and Space, Erkner, Germany

Synonyms

City-to-city learning; Climate change; Local knowledge; Peer learning; Policy learning

Definition

Climate-related changes in the environment, such as an increase in the magnitude and frequency of extreme weather events, have a particularly drastic impact on cities. To build resilience to climate impacts, cities need to develop flexible adaptation strategies. One way to do this is through the creation and expansion of peer learning networks, as they allow knowledge exchange on adaptation strategies. This chapter describes and discusses the basic conditions that must be in place for cities to build climate adaptive capacity through peer learning. The mere exchange of knowledge, however, does not mean that the people concerned will also change their daily routine. Learning processes are much more complex. An important premise is that the collaborating actors speak the same language. Only then is it possible to classify newly acquired knowledge, to prioritize it, and to adjust one's own behavior according to new attributed values.

Introduction: Cities in Times of Climate Change

Climate change poses major challenges to cities (Bulkeley 2010; Wamsler et al. 2013; Haupt 2021b). Due to higher levels of ground sealing and fewer green areas, cities and urban areas are particularly vulnerable to extreme weather events such as heat waves (Peng et al. 2012; Otto et al. 2021a) or heavy rainfall (Veerbeek 2017; Otto et al. 2021a). Such events pose a threat to citizens' health and lives (Solomon and LaRocque 2019) and can severely damage cities' built environment (Haupt 2021b). Little time is left to adapt to these changing conditions. Therefore, cities cannot rely on their domestic knowledge and learning capacities alone. Instead, city practitioners such as climate, environmental, or resilience specialists also have to learn from the experiences of individuals from other cities and look for solutions that have already been successfully tested elsewhere (Fisher 2014; Haupt 2021a). This contribution explores how learning is theorized, how it depends on demanding and challenging preconditions, how it most commonly takes place, and how it can be accelerated. Several insights presented in this chapter derive from two PhD dissertations that have focused on the production and preservation of local knowledge (Reichel 2020) and processes of policy learning and city-to-city learning (Haupt 2019).

From Knowledge to Learning

Knowledge is the indispensable precondition for learning. Knowing about who knows what is crucial, especially for the development of strategies to cope with climate change. Knowledge can be defined as information "that is meaningful to knowledgeable agents" (Fleck 1997: 384). In other words, information can be knowledge for a certain person that is able to understand its content but remains "just" information for a person that is not a knowledgeable agent. It can be assumed that peers or policymakers working in the same area are more likely to be such knowledgeable agents. With increasing environmental knowledge and people that are able to "digest" such knowledge, environmental changes can – at least theoretically – be perceived in a more nuanced way. Initially, this may contribute to the building of a solid knowledge base and ultimately to a more effective implementation of climate mitigation and adaptation measures. However, the process of moving from climate knowledge to climate action is not self-propelling (at all). Just because people know something does not automatically mean that they will give high priority to this knowledge and change their habitus accordingly. More important is whether a learning process takes place that goes beyond the mere passing on of information or knowledge. Indeed, knowledge sharing should not be mistaken with learning. While learning always consists of processes of knowledge sharing, knowledge sharing alone is no guarantee for learning to take place. Other than information or knowledge sharing, learning is not a singular event but a temporal process (Fleck 1997; Dunlop and Radaelli 2013).

Certainly, there is no shortage of learning definitions. Given the wide range of scholarly definitions and the resulting ambiguity (Bennett and Howlett 1992), the understanding of and perspective on learning needs to be clarified first. The learning processes among public servants (e.g., government officials or city practitioners) can be described and theorized best with a focus on policy learning. A broad definition of policy learning describes it as "the updating of beliefs based on lived or witnessed experiences" (Dunlop and Radaelli, 2013: 599). More specifically, policy learning took place when policymakers "adjust their cognitive understanding of policy development and modify policy in the light of knowledge gained from past policy experience" (Stone 2004: 551). Learning exchanges between individuals holding a similar position within a system are widely known as peer learning (Boud et al. 2014; Andrews and Manning 2016). Many, if not most, policy learning processes can be described as peer learning processes. Learning processes between peers that work in different cities but hold similar positions within their respective city administrations (e.g., climate managers) are also referred to as city-to-city learning (Seymoar et al. 2009; Ilgen et al. 2019; Moodley 2019; Haupt et al. 2020).

Where and How Learning Takes Place

Recent studies have explored how learning between city practitioners happens in practice and how it can be further stimulated (Fisher 2014; Ma 2017; Montero 2017; Haupt 2021a). It appears that learning is most efficient if it takes place rather informally between individuals that have already set up personal relations based on mutual trust (Seymoar et al. 2009; Haupt et al. 2020). Other than formal exchanges, informal exchanges often allow the exchange of "unfiltered information" (Haupt et al. 2020: 151). Nevertheless, formal events or platforms often provide a starting point for city practitioners to establish personal connections, which can result in learning later on (Haupt et al. 2020). International city networks have proven to be helpful platforms to connect city practitioners from different cities (Kern and Bulkeley 2009; Bansard et al. 2017; Busch et al. 2018). Indeed, many of these networks provide its member cities with several tools to establish partnerships. Examples include events such as conferences and workshops but also webinars and the organization of study visits (Haupt and Coppola 2019). Several studies have shown that forerunner cities in governing climate change are usually well-connected with other cities and have joined several networks (Kern 2019; Otto et al. 2021b; Salvia et al. 2021).

Nevertheless, it needs to be highlighted that learning often occurs in a field of tension. This means that people – be it private individuals or public servants – draw on different bodies of knowledge in everyday life to interpret their perceptions and to develop social practices based on these interpretations. While knowledge that has become (supposedly) useless and unimportant is often not passed on and falls into oblivion, newly acquired knowledge is constantly checked for its usefulness and, if necessary, integrated into the existing body of knowledge. If the environment changes due to climate change, that is, the biophysical living environment given by nature, for example, due to the extinction of certain animal species or plant species, or the social environment, for example, as a result of new technical possibilities, knowledge also changes in the field of tension between the empowerment of other forms of knowledge and a creative adaptation or expansion of one's corpus of knowledge

(Linkenbach 2004: 255–256; Reichel 2008: 11–12, Reichel 2020: 193).

Therefore to promote or participate in learning processes for coping with climate change, it is important to have an awareness of the sociocultural structures in which the knowledge to be learned is embedded. Public servants responsible for the development of strategies to cope with climate change should have this awareness. In this context, also power structures play a prominent role. The diffusion of knowledge is the result of "social negotiations that involve contestation, conflict, and negotiation (and are) permeated by aspects of power, authority, and legitimation" (El Berr 2007: 104; cf. also Pottier et al. 2015; Sillitoe 1998; Long and Long 1992; Scoones and Thompson 1994; Mundy and Compton 1995). Michel Foucault also points out that knowledge and power are interdependent: "The criteria of what constitutes knowledge, what is to be excluded and who is designated as qualified to know involves acts of power" (Foucault after Scoones and Thompson 1994: 24). Another example is the heterogeneity of knowledge bearers. Even if the members of a group share common norms and values, their relationship to each other is often characterized less by an egalitarian relationship than by hierarchies and conflicts of interest resulting from economic, ethnic, and social differences. To be able to articulate and assert interests, strategic alliances are formed, which vary greatly depending on the interests at stake (Cornwall 1998: 50ff.; Cleaver et al. 1999: 45 after Reichel 2020: 83).

Summary

This contribution has shown that learning is a complex process that requires a high amount of dedication by the actors involved. Climate knowledge does not automatically lead to climate learning and action. Indeed, the availability or even sharing of information or knowledge is by no means a guarantee for learning to take place. Nevertheless, information and knowledge are the basic preconditions for learning. Eventually, all cities need to adapt to the impacts of global climate change. Cities need to be aware of solutions that have already worked elsewhere and learn from these examples since most of them neither have the time (left) nor do they have the capacities to develop and test solutions on their own. Then again, each city is different and has to cope with specific challenges, which means that solutions cannot "just" be transferred from one place to another. Instead, they have to be translated into the own context. Therefore, it is essential to understand the sociocultural and socioeconomic context of the city or better the individual(s) participating in a learning process. More specifically, there needs to be an understanding of the contextual embeddedness of the knowledge to be obtained. Most commonly, this can only lead to satisfactory learning results if the partners involved in the process are knowledgeable agents that speak a common language, or at least use the same interface to learn a new one.

Cross-References

► The Sustainable and the Smart City: Distinguishing Two Contemporary Urban Visions

References

Andrews, M., & Manning, N. (2016). *A Guide to Peer-to-Peer Learning: How to make peer-to-peer support and learning effective in the public sector?* https://www.effectiveinstitutions.org/files/The_EIP_P_to_P_Learning_Guide.pdf.

Bansard, J. S., Pattberg, P. H., & Widerberg, O. (2017). Cities to the rescue? Assessing the performance of transnational municipal networks in global climate governance. *International Environmental Agreements: Politics, Law and Economics, 17*(2), 229–246. https://doi.org/10.1007/s10784-016-9318-9.

Bennett, C. J., & Howlett, M. (1992). The lessons of learning: Reconciling theories of policy learning and policy change. *Policy Sciences, 25*(3), 275–294. https://doi.org/10.1007/BF00138786.

Boud, D., Cohen, R., & Sampson, J. (2014). *Peer learning in higher education*. London: Routledge. https://doi.org/10.4324/9781315042565.

Bulkeley, H. (2010). Cities and the governing of climate change. *Annual Review of Environment and Resources,*

B

35(1), 229–253. https://doi.org/10.1146/annurev-environ-072809-101747.

Busch, H., Bendlin, L., & Fenton, P. (2018). Shaping local response – The influence of transnational municipal climate networks on urban climate governance. *Urban Climate, 24*, 221–230. https://doi.org/10.1016/j.uclim.2018.03.004.

Cleaver, F. (1999). Paradoxes of participation: Questioning participatory approaches to development. *Journal of International Development, 11*(4), 597–612. https://doi.org/10.1002/(SICI)1099-1328(199906)11:43.0.CO

Cornwall, A. (1998). Gender, participation and the politics of difference. In I. Guijt & M. K. Shah (Eds.), *The myth of community* (Vol. 4, pp. 46–57). Rugby/Warwickshire: Practical Action Publishing. https://doi.org/10.3362/9781780440309.004.

Dunlop, C. A., & Radaelli, C. M. (2013). Systematising policy learning: From monolith to dimensions. *Political Studies, 61*(3), 599–619. https://doi.org/10.1111/j.1467-9248.2012.00982.x.

El Berr, S. (2007). Vom Öko-Heiligen zum Umweltzerstörer und zurück: Indigenes Wissen in der Entwicklungszusammenarbeit. In G. Jilek, S. Kalmring, & S. Müller (Eds.), *Manuskripte / Rosa-Luxemburg-Stiftung: Vol. 73. Von Honig und Hochschulen: Dreizehn gesellschaftskritische Interventionen; zehntes Doktorand Innenseminar der Rosa-Luxemburg-Stiftung* (pp. 94–122). Berlin: Dietz.

Fisher, S. (2014). Exploring nascent climate policies in Indian cities: A role for policy mobilities? *International Journal of Urban Sustainable Development, 6*(2), 154–173. https://doi.org/10.1080/19463138.2014.892006.

Fleck, J. (1997). Contingent knowledge and technology development. *Technology Analysis & Strategic Management, 9*(4), 383–398. https://doi.org/10.1080/09537329708524293.

Haupt, W. (2019). *City-to-city learning in transnational municipal climate networks: An exploratory study.* L'Aquila: Gran Sasso Science Institute. https://iris.gssi.it/handle/20.500.12571/9733#.YPFOsedCQ2w.

Haupt, W. (2021a). How do local policy makers learn about climate change adaptation policies? Examining study visits as an instrument of policy learning in the European Union. *Urban Affairs Review*, 107808742093844. https://doi.org/10.1177/1078087420938443.

Haupt, W. (2021b). The sustainable and the Smart City: Distinguishing two contemporary urban visions. In *The Palgrave encyclopedia of urban and regional futures*. https://doi.org/10.1007/978-3-030-51812-7_177-1.

Haupt, W., & Coppola, A. (2019). Climate governance in transnational municipal networks: Advancing a potential agenda for analysis and typology. *International Journal of Urban Sustainable Development, 11*(2), 123–140. https://doi.org/10.1080/19463138.2019.1583235.

Haupt, W., Chelleri, L., van Herk, S., & Zevenbergen, C. (2020). City-to-city learning within climate city networks: Definition, significance, and challenges from a global perspective. *International Journal of Urban Sustainable Development, 12*(2), 143–159. https://doi.org/10.1080/19463138.2019.1691007.

Ilgen, S., Sengers, F., & Wardekker, A. (2019). City-To-City learning for urban resilience: The case of water squares in Rotterdam and Mexico City. *Water, 11*(5), 983. https://doi.org/10.3390/w11050983.

Kern, K. (2019). Cities as leaders in EU multilevel climate governance: Embedded upscaling of local experiments in Europe. *Environmental Politics, 28*(1), 125–145. https://doi.org/10.1080/09644016.2019.1521979.

Kern, K., & Bulkeley, H. (2009). Cities, Europeanization and multi-level governance: Governing climate change through transnational municipal networks. JCMS. *Journal of Common Market Studies, 47*(2), 309–332. https://doi.org/10.1111/j.1468-5965.2009.00806.x.

Linkenbach, A. (2004). Lokales Wissen im Entwicklungsdiskurs: Abwertung, Aneignung oder Anerkennung des Anderen? In N. Bierschenk & T. Schareika (Eds.), *Mainzer Beiträge zur Afrika-Forschung* (Lokales Wissen: Sozialwissenschaftliche Perspektiven) (Vol. 11, pp. 233–257). Münster: Lit-Verl.

Long, N., & Long, A. (Eds.). (1992). *Battlefields of knowledge: The interlocking of theory and practice in social research and development (1. [publ.]).* London: Routledge.

Ma, L. (2017). Site visits, policy learning, and the diffusion of policy innovation: Evidence from public bicycle programs in China. *Journal of Chinese Political Science, 22*(4), 581–599. https://doi.org/10.1007/s11366-017-9498-3.

Montero, S. (2017). Study tours and inter-city policy learning: Mobilizing Bogotá's transportation policies in Guadalajara. *Environment and Planning a: Economy and Space, 49*(2), 332–350. https://doi.org/10.1177/0308518X16669353.

Moodley, S. (2019). Defining city-to-city learning in southern Africa: Exploring practitioner sensitivities in the knowledge transfer process. *Habitat International, 85*, 34–40. https://doi.org/10.1016/j.habitatint.2019.02.004.

Mundy, P. A., & Compton, J. L. (1995). 7. Indigenous communication and indigenous knowledge. In D. M. Warren, L. J. Slikkerveer, D. Brokensha, & W. H. Dechering (Eds.), *The cultural dimension of development* (pp. 112–123). Rugby/Warwickshire: Practical Action Publishing. https://doi.org/10.3362/9781780444734.007.

Otto, A., Göpfert, C., & Thieken, A. H. (2021a). Are cities prepared for climate change? An analysis of adaptation readiness in 104 German cities. *Mitigation and Adaptation Strategies for Global Change, 26*(8). https://doi.org/10.1007/s11027-021-09971-4.

Otto, A., Kern, K., Haupt, W., Eckersley, P., & Thieken, A. H. (2021b). Ranking local climate policy: Assessing the mitigation and adaptation activities of 104 German cities. *Climatic Change, 167*(1–2). https://doi.org/10.1007/s10584-021-03142-9.
Peng, S., Piao, S., Ciais, P., Friedlingstein, P., Ottle, C., Bréon, F.-M., . . . Myneni, R. B. (2012). Surface urban heat island across 419 global big cities. *Environmental Science & Technology, 46*(2), 696–703. https://doi.org/10.1021/es2030438.
Pottier, J., Bicker, A., & Sillitoe, P. (2015). *Negotiating local knowledge*. Pluto Press. https://doi.org/10.2307/j.ctt18mbd5m.
Reichel, C. (2008). Lokales Wissen als Möglichkeit und Perspektive nachhaltiger Ressourcennutzung und des Schutzes vor Naturkatastrophen am Beispiel der Segara Anakan Lagune (Zentral-Java) und des Taka Bonerate Archipels (Süd- Sulawesi) – Indonesien, Magisterarbeit Freie Universität Berlin. (Master's thesis). Freie Universität Berlin, Berlin.
Reichel, C. (2020). *Mensch – Umwelt – Klimawandel*. Bielefeld: transcript Verlag. https://doi.org/10.14361/9783839446966.
Salvia, M., Reckien, D., Pietrapertosa, F., Eckersley, P., Spyridaki, N.-A., Krook- Riekkola, A., . . . Heidrich, O. (2021). Will climate mitigation ambitions lead to carbon neutrality? An analysis of the local-level plans of 327 cities in the EU. *Renewable and Sustainable Energy Reviews, 135*, 110253. https://doi.org/10.1016/j.rser.2020.110253.
Scoones, I., & Thompson, J. (1994). Introduction; Knowledge, power and agriculture - towards a theoretical understanding. In I. Scoones & J. Thompson (Eds.), *Beyond farmer first* (Vol. 1, pp. 13–31). Rugby/Warwickshire: Practical Action Publishing. https://doi.org/10.3362/9781780442372.002.
Seymoar, N.-K., Mullard, Z., & Winstanley, M. (2009). *City-to-city learning*. Vancouver. Retrieved from https://www.crcresearch.org/files-crcresearch_v2/File/City%20to%20City%20Learning.pdf
Sillitoe, P. (1998). The development of indigenous knowledge. *Current Anthropology, 39*(2), 223–252. https://doi.org/10.1086/204722.
Solomon, C. G., & LaRocque, R. C. (2019). Climate change – A health emergency. *The New England Journal of Medicine, 380*(3), 209–211. https://doi.org/10.1056/NEJMp1817067.
Stone, D. (2004). Transfer agents and global networks in the 'transnationalization' of policy. *Journal of European Public Policy, 11*(3), 545–566. https://doi.org/10.1080/13501760410001694291.
Veerbeek, W. (2017). *Estimating the impacts of urban growth on future flood risk* (Dissertation). IHE – Institute for Water Education.
Wamsler, C., Brink, E., & Rivera, C. (2013). Planning for climate change in urban areas: From theory to practice. *Journal of Cleaner Production, 50*, 68–81. https://doi.org/10.1016/j.jclepro.2012.12.008.

Big Data for Smart Cities and Inclusive Growth

Md. Arfanuzzaman
Food and Agriculture Organization of the United Nations, Dhaka, Bangladesh

Introduction

The UN estimates suggest that 70% of the world's population will live in urban areas by 2050 (UN 2018a). It is also expected that by that time, 90% of population growth, 80% of the increase in wealth, and approximately 60% of energy consumption will occur in the urban areas which will leave sustainability, environmental protection, and natural resource management challenges in the developing countries and emerging economies across the world (UN 2018b). Though the urbanization in developing world is increasing at an unprecedented speed, it throws enormous challenges, such as social vulnerability, pollution, inequality, environmental degradation, food-water-energy insecurity for growing population, and degrades the quality of life to a large extent (Ellis and Roberts 2016). Urban environmental degradation, pollution, poverty, high population growth, inequality, poor water and sanitation system, deforestation, inappropriate land use, natural disaster, and urban heat island effect are the common challenges around the world (Arfanuzzaman and Dahiya 2019; Ellis and Roberts 2016; Jacobs et al. 2019) which can be largely mitigated through big data-driven smart city (Bibri 2019a; UN 2015a, 2016). As increasing population growth and built up causing to debase the urban water sources, tree covers, open spaces, and clean air as well as quality of life of the inhabitants, big data applications can provide low-cost and efficient solutions in solving pressing urban environmental problems (i.e., resource conservation, environmental protection, and natural resource monitoring) (Dubey et al. 2017; Graham and Shelton 2013; Dumbill 2013). Besides, big data can help cities to boost the efficiency of its traditional industries, increase competitiveness and productivity, build socioeconomic and climate resilience, and create decent job opportunities for the youth while helping it to develop newer cutting-edge technologies (Sun and Du 2017; Keeso 2014; Mayer-Schönberger and Cukier 2013).

Current Advancement in Big Data-Driven Smart City and Inclusive Growth

It is now widely accepted that the major driving forces of the Fourth Industrial Revolution are artificial intelligence (AI), big data, and Internet of things which will change the way we lead our life and take our decision. Around 90% of the world's digitized data was captured in past 2 years (Al Nuaimi et al. 2015) through electronic devices, applications, and sensors which provides enormous opportunities to the city government to utilize this data to effectively manage the improve the components of smart cities such as smart governance, smart healthcare, smart transportation, smart education, smart environment, smart safety, and smart energy globally. The world has made some remarkable progress on big data technology in urban and regional development, environmental monitoring, industrial efficiency, early warning, disaster preparedness, natural resource management, and agriculture and food system. Nowadays, big data analytics based on the data generated from sensors can help to monitor water quality variables, track crop and soil health, and predict requirement of irrigation (Illangasekare et al. 2018; Ellis and Roberts 2016; Fu et al. 2016; Mao et al. 2019). Using satellite data to track and monitor urban green space, wetlands, and forests available across the country is now common practice (Bibri 2019b; UN 2015b; Cukier and Mayer-Schoenberger 2013). Though research and innovation efforts on capturing and utilizing are still ongoing in many parts of the world, some scalable approach is available at present to address the urban socioeconomic and environmental challenges. For instance, MIND (Managing Information for Natural Disasters) initiative under UN Global Pulse can support effective logistics planning and information management right after natural disasters and provide stakeholders useful, accurate, and up-to-date data to carry out humanitarian assistance following the disaster by collecting and analyzing traditional and nontraditional datasets. Using diverse spatial data (census data, agrometeorological data, telemetry data, satellite imagery data), Ministry of Environment and Natural Resources of El Salvador established improved early warning systems for natural hazards and disseminated information to people in real time through mobile applications. World Resources Institute (WRI) through its Central Africa Regional Program has identified the most important drivers of forest loss and predicts the likely location of future forest loss in in the Democratic Republic of the Congo using big data approach. Based on machine learning algorithms supported by big data, US Department of Energy models electricity consumption and behavior of distributed energy resources such as rooftop or ground-mount solar and enables real-time optimization and automation of distribution planning and operation decision for utilities. Besides, GSM Association leverages mobile operators' big data capabilities to address humanitarian crises, including epidemics and natural disasters under its Big Data for Social Good program. Integrating big data analytics and remote sensing WRI and Global Cool Cities Alliance identifies urban surface changes to reduce urban heat effects by heat mitigation strategies. Azure FarmBeats of Microsoft Corporation assess farm health using vegetation and water index based on satellite imagery and generate soil moisture map and farm health advisories by fusing the datasets. Using satellite imagery and data from NOAA's Global Forecast System, NOAA Global Hydro Estimator provides global rainfall estimates in 15-min intervals at ~4 km resolution which can be used for urban hydrological planning, monitoring, and building climate resilience. In addition, NOAA generates tens of terabytes of data each day from satellites, radars, ships, weather models, and other sources which can be used for urban environmental protection, ecosystem monitoring, and restoration globally.

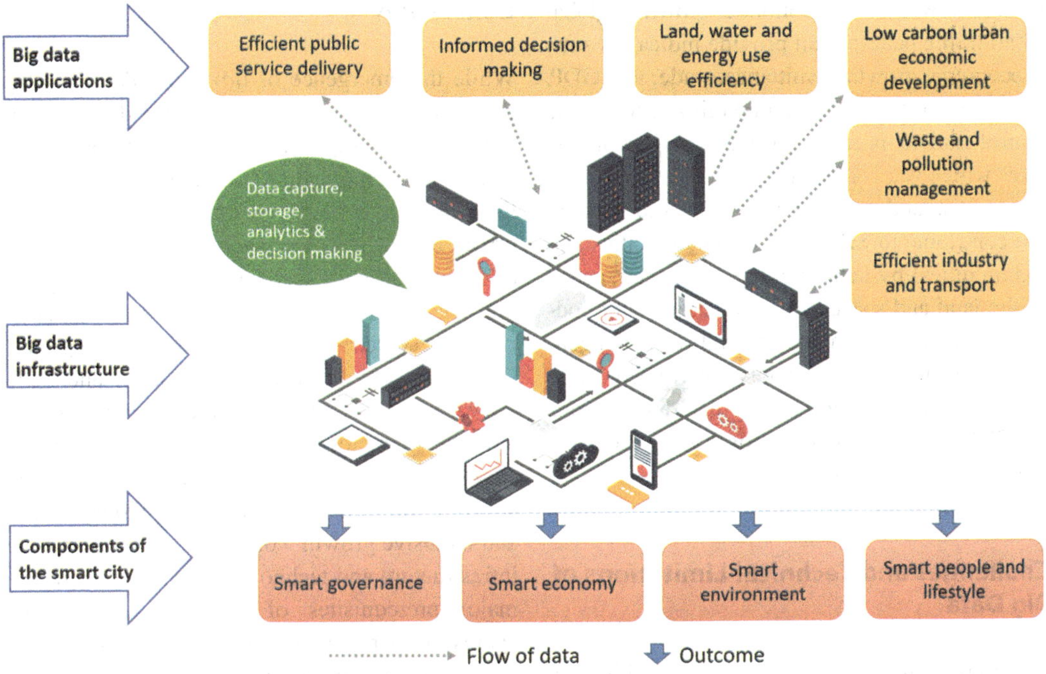

Big Data for Smart Cities and Inclusive Growth, Fig. 1 Interrelationship between big data application, big data infrastructure, and the components of smart city

The Fig. 1 demonstrates the overall concept of big data-driven smart city and interrelationship between big data application, big data infrastructure, and the components of smart city. It illustrates the areas of cities where big data applications can provide robust solution, and necessary data for analytics and decision-making can be gathered from this areas through electronic devices, applications, and sensors. A two-way relationship exists between big data infrastructure and big data application because of their interdependency for data. A well-equipped big data infrastructure can offer cost-effective data collection, processing, storage, analytics, and decision-making services to the cities. The effective utilization of the big data application and big data infrastructure can generate the components of the smart city.

Big Data for SDG Monitoring and Inclusive Growth in Smart Cities

Big data gathers data every second. Present technologies make it possible not just to acquire enormous sets of data but also to solve critical urban issues such as disease outbreak management and monitoring, contribute to the transition toward zero-emission future, reduce urban environmental degradation and unsustainable use of natural resources, promote equity, and ensure aim for public benefits via open data (Estoque 2020; Bibri 2019b; Pan et al. 2016; Kudva and Ye 2017). Big data also has an imperative role in monitoring and attaining several SDGs and advancing inclusive urban economic growth. For instance, spending patterns on mobile phone services can provide proxy indicators of the income level of city dwellers, and this outcome will contribute to SDG 1: no poverty. Mapping the movement of mobile phone users can help predict the spread of infectious diseases and disaster-induced migration that can contribute to SDG 3: good health and well-being. Smart metering in urban areas allows the energy, gas, and water service providers to reduce wastage and help to ensure adequate supply at peak periods that will also support the to the attainment of SDG 7: affordable and clean energy. Correspondingly, in support of SDG

8 (decent work and economic growth), global postal traffic patterns can provide indicators such as economic growth, remittance, trade, and GDP. Data from GPS devices can be utilized for traffic control and to improve public transport in the cities which is a contribution to SDG 9: industry, innovation, and infrastructure. Remote sensing can contribute to SDG 11 (sustainable cities and communities) by help tracking encroachment on public land and spaces such as parks, water bodies, and forests and enable law enforcement agencies to respond immediately. We need more research and innovation on big data applications for SDG monitoring and progressing inclusive growth in smart cities.

Challenges and Technical Limitations of Big Data

There are some technical limitations in big data which impedes its progress in advancing smart city and inclusive growth. Currently the world especially the technologically inefficient nations is facing following constraints in big data utilization.

- Low-cost and sustainable methods for capturing quality data from different sectors
- The limited capacity for storing big data and ensuring its security
- Development of big data and associate IT infrastructure
- Limited searching, sharing, and transferring tools during the data utilization process
- Limited application for visualization and querying of different variety of big data
- Lack of appropriate data processing and analytics platform/application
- Human error in the big data processing and analysis and interpretation
- Inadequate advanced and sophisticated algorithms to deal with big data efficiently
- Limited capacity to design, development, and deployment of big data applications for smart cities

Conclusion

While the emergence of urban big data has the potential to advance targets and indicators for SDGs, studies suggested urban policy-makers to situate the application of these innovations through developing mayoral institutions for urban big data governance, advancing culture and common skill sets for big data, developing big data infrastructure, and investing in specialized research and education programs (Krishnamurthy et al. 2017; Fluckiger and Seth 2016). Scientists and governments countries can work together, not just to respond to the urban and environmental problems of today but work with superior foresight today to advance smart cities and inclusive growth worldwide. As modern data infrastructure and technological readiness are the major prerequisites of big data intervention against pressing urban issues and build smart cities, countries need to work together to seize the optimum benefits of big data. Strapping international cooperation, fiscal incentive, innovative finance, private sector engagement, research and innovation, productivity enhancement, creation of enabling conditions, integration of big data and Internet of things, bridging the technological divide, and social inclusion deserve special focus to build big data-driven smart cities globally. South-South and triangular cooperation need to be enhanced and framed for solving pressing urban issues and advancing smart cities through big data-based innovation, improved IT and data infrastructure, knowledge sharing, technology transfer, and capacity building.

Reference

Al Nuaimi, E., Al Neyadi, H., Mohamed, N., et al. (2015). Applications of big data to smart cities. *J Internet Serv Appl, 6*, 25. https://doi.org/10.1186/s13174-015-0041-5.

Arfanuzzaman, M., & Dahiya, B. (2019). Sustainable urbanization in Southeast Asia and beyond: Challenges of population growth, land use change and environmental health. *Journal of Growth and Change, 50*(2). https://doi.org/10.1111/grow.12297.

Bibri, S. E. (2019a). Data-driven smart sustainable urbanism: The intertwined societal factors underlying its materialization, success, expansion, and evolution. *GeoJournal*. https://doi.org/10.1007/s10708-019-10061-x.

Bibri, S. E. (2019b). The anatomy of the data-driven smart sustainable city: Instrumentation, datafication, computerization and related applications. *J Big Data, 6*, 59. https://doi.org/10.1186/s40537-019-0221-4.

Cukier, K., & Mayer-Schoenberger, V. (2013). The rise of big data. *Foreign Affairs (May/June).*, 28–40.

Dubey, R., Gunasekaran, A., Childe, S. J., Papadopoulos, T., Luo, Z., Wamba, S. F., & Roubaud, D. (2017). Can big data and predictive analytics improve social and environmental sustainability? *Technological Forecasting and Social Change*. https://doi.org/10.1016/j.techfore.2017.06.020.

Dumbill, E. (2013). Making sense of big data. Big data. Vol. 1, No. 1. Mary Ann Liebert, Inc.

Ellis, P., & Roberts, M. (2016). *Leveraging urbanization in South Asia: Managing spatial transformation for prosperity and livability*. Washington, DC: World Bank.

Estoque, R. C. (2020). A review of the sustainability concept and the state of SDG monitoring using remote sensing. *Remote Sensing, 12*, 1770.

Fluckiger, Y., & Seth, N. (2016). Sustainable development goals: SDG indicators need crowdsourcing. *Nature, 531*, 448.

Fu, L., Dallas, P., Sharma, V. K., & Zhang, K. (2016). Sensors for environmental monitoring. *Journal of Sensors, 2016*, 1–1. https://doi.org/10.1155/2016/4108790.

Graham, M., & Shelton, T. (2013). *Geography and the future of big data, big data and the future of geography. Dialogues in human geography* (pp. 255–261). Sage.

Illangasekare, T. H., Han, Q., & Jayasumana, A. P. (2018). Environmental underground sensing and monitoring. *Underground Sensing*, 203–246.

Jacobs, C., Singh, T., Gorti, G., Iftikhar, U., Saeed, S., Syed, A., & Siderius, C. (2019). Patterns of outdoor exposure to heat in three south Asian cities. *Science of The Total Environment*. https://doi.org/10.1016/j.scitotenv.2019.04.087.

Keeso, A. (2014). *Big data and environmental sustainability: A conversation starter, Smith school working paper series: 14–04*. Oxford University.

Krishnamurthy, R., Smith, K. L., & Desouza, C. (2017). Urban Informatics: Critical Data and Technology Considerations. In P. Thakuriah, N. Tilahun, & M. Zellner (Eds.), *Seeing cities through Big Data* (pp. 163–188). Basel: Springer International Publishing.

Kudva, S., & Ye, X. (2017). Smart cities, big data, and sustainability union, Big Data Cogn. *Comput, 1*(1), 4. https://doi.org/10.3390/bdcc1010004.

Mao, F., Khamis, K., Krause, S., Clark, J., & Hannah, D. M. (2019). Low-cost environmental sensor networks: Recent advances and future directions. *Frontiers in Earth Science, 7*, 221. https://doi.org/10.3389/feart.2019.00221.

Mayer-Schönberger, V., & Cukier, K. (2013). *Big data: A revolution that will transform how we live, work and think*. London, UK: John Murray.

Pan, Y., Tian, Y., Liu, X., Gu, D., & Hua, G. (2016). Urban big data and the development of City intelligence. *Engineering, 2016, 2*(2), 171–178.

Sun, Y., & Du, Y. (2017). Big data and sustainable cities: Applications of new and emerging forms of geospatial data in urban studies. *Open Geospatial Data, Softw. Stand., 2*, 24. https://doi.org/10.1186/s40965-017-0037-0.

UN (United Nations). (2015a). Habitat III Issue Papers, 21 – Smart cities (V2.0). In *New York*. https://collaboration.worldbank.org/docs/DOC-20778.

UN (United Nations). (2015b). *Big data and the 2030 agenda for sustainable development*. Prepared by A. Maaroof. http://www.unescap.org/events/call-participants-big-data-and-2030-agendasustainable-development-achieving-development.

UN (United Nations). (2018a). *The World's cities in 2018: Data booklet*. New York: Department of Economic and Social Affairs, United Nations.

UN (United Nations). (2018b). *World urbanization prospects 2018*. New York: Department of Economic and Social Affairs, United Nations.

UN Global Pulse. (2016). *Integrating big data into the monitoring and evaluation of development Programmes*. New York: United Nations.

Biodiversity

▶ Butterfly Gardening in Colombo, Sri Lanka: Approach to Biodiversity Conservation, Monitoring, Education, and Awareness in Urbanizing Habitats

Biodiversity Conservation

▶ Habitat Provisioning

Biophilic City

▶ Green Cities in Theory and Practice

Bioregionalism

Rethinking Place Beyond Human Borders

Gerard Kuperus
University of San Francisco, San Francisco, CA, USA

Synonyms

Ecoregion

Definition

Bioregionalism is an ecological, political, and cultural concept that suggests human activity should be tied to the ecological and geographical region one inhabits. A bioregion is defined in terms of its natural characteristics, and bioregionalism is a movement that attempts to place humans in their bioregion in a positive way. As such, it is tied to the process of "reinhabitation." The latter is the attempt to learn again how to dwell in one's place. Bioregionalism aims to create a positive and constructive relationship between place and its human inhabitants. It provides a challenge to the contemporary global mindset in which it is said that humans do not really live in places, or regions. Bioregionalism is a reaction to our lack of an intricate knowledge of the places in which we live and work. Thus, it hopes to instill wisdom to live in positive relationships with the regions we inhabit.

Introduction

Instead of thinking about places as political regions, in which borders often divide ecosystems, bioregionalism proposes to rethink places in terms of their natural features. Within this proposal, humans are understood as part of nature. The movement encourages practices that aim at making a bioregion thrive. One of the better-known examples of a bioregion is Cascadia, covering parts of Southeast Alaska, Canada, Washington State, the most Northern part of California, Oregon, Idaho, and Western Montana. As the name suggests, Cascadia is defined as water that cascades over land and off mountains toward the Pacific Ocean. In the following, the history of the term, connection to indigenous cultures, and reinhabitation will be discussed.

History of the Term

The term bioregionalism (as well as reinhabitation) was first used by the counterculture movement that emerged in the 1960s and 1970s. The term "bioregion" was introduced by Peter Berg and Raymond Dasmann, while "bioregionalism" was coined by Allen Van Newkirk (see Aberley, in McGinnis (1999) pp. 13–42, for a detailed history of the term).

Indigenous "Bioregional" Practices

While the terminology and political movement appeared in the counterculture movement, some earlier "Western" thinkers with a deep interest in understanding the geographical significance of particular regions were Ralph Waldo Emerson and Wallace Stegner. As already mentioned, Native American and other indigenous cultures often practiced within a place and could be considered as bioregional systems of thinking and living. Limits to natural resources, living with the natural flows of the seasons, and overall knowing one's place were deeply engrained in most, if not all, indigenous cultures. Thus, the practice of bioregionalism – without using that term – is thousands of years old found in indigenous cultures that lived in balance with the ecosystems that constituted their homes.

Native American and other indigenous cultures typically lived (and some still live) with and among non-human animals in ecosystems. An intricate knowledge of the regions in which they lived was developed over many generations. Practices of harvesting, hunting, fishing, and management of the land were benefitting both humans and the natural environment. In *Braiding Sweetgrass*, Robin Wall Kimmerer, a Native American woman trained in Western science, is stunned to learn that many of her college students cannot think of a positive interaction between people and land (Wall Kimmerer (2013), 6). She shares the story of Skywoman, who falls from the sky and lands on Earth which only consists of water. Different animals collaborate and struggle to create some land for her, famously placing some mud on the back of a turtle. Thus, Turtle Island, the name used by many Native American cultures to refer to their continent, was born. While Turtle Island is not a bioregion (it consists of many), the movement to reinstate this name coincides in many ways with bioregionalism as a political movement.

Reinhabitation

Gary Snyder argues for a reinhabitation of Turtle Island. He often refers to indigenous cultures that "lived within territories that conformed to some set of natural criteria" (Snyder (1990), 37). Snyder relates biota, watersheds, landforms, and elevations to "dialects, religions, sorts of arrow release, types of tools, myth motifs, musical scale, art styles" (ibid). Nature and culture are deeply intertwined, something that we lack today and that we need to reinstitute in order to reinhabitate Turtle Island, he argues. For "To know the spirit of a place is to realize that you are a part of a part and that the whole is made of parts, each of which is whole. You start with the part you are whole in" (Snyder (1990), 38). Reinhabitation and bioregionalism thus challenge our contemporary political and cultural boundaries. Political boundaries are often arbitrary, cutting through ecosystems. Cultures, today, often lack sensitivity to the place in which that culture flourishes. Snyder and others argue for the creation of "cultures of nature," i.e., cultures that emerge out of a bioregion and celebrate that place.

Conclusion

B

Bioregionalism is then a political movement as well as a practice of living harmoniously with the land that we inhabit. It suggests that knowing one's place, one's home, means to practice in that place. Instead of understanding a place by its name, state, and country, understanding a place as a bioregion focuses on the natural features. The other animals that humans share a region with have no interest in our political (and cultural) boundaries. They follow and work with the geography of land and rivers, with changes in seasons and tides, and so forth. Bioregionalism suggests that knowing and practicing in a bioregion means that humans, like the other animals, need to work in unison with the land. Instead of changing the land to our practices, bioregionalism suggests that we organize our practices around the natural features of the place we inhabit.

Cross-References

▶ Nature-based Solutions

References

McGinnis, M. (1999). *Bioregionalism*. London: Routledge.

Snyder, G. (1990). *The practice of the wild*. San Francisco: North Point Press.

Wall Kimmerer, R. (2013). *Braiding Sweetgrass*. Minneapolis: Milkweed.

http://cascadia-institute.org/

Further Reading

Barnhill, D. L. (1999). *At home on the earth: Becoming native to our place: A multicultural anthology*. Berkeley: University of California Press.

Berg, P. (2014). *The biosphere and the bioregion: Essential writings of Peter Berg*. New York: Routledge.

Cronon, W. (1996). *Uncommon ground: Rethinking the human place in nature*. New York: W.W. Norton.

Evanoff, R. (2010). *Bioregionalism and global ethics*. New York: Routledge.
Kirkpatrick, S. (2000). *Dwellers in the land: The bioregional vision*. Athens: University of Georgia Press.
Snyder, G. (1995). *A place in space: Ethics, aesthetics, and watersheds*. Washington, DC: Counterpoint.

Blue/Green Infrastructure

► Growth, De-growth, and Nature-Based Solutions

Blue-Green Cities: Achieving Urban Flood Resilience, Water Security, and Biodiversity

V. Krivtsov[1], S. Ahilan[2], S. Arthur[6], S. Birkinshaw[5], D. Dawson[7], G. Everett[8], V. Glenis[5], L. Kapetas[9], C. Kilsby[5], J. Lamond[8], D. Mendoza Tinoco[10], S. Ncube[6], E. O'Donnell[3], G. O'Donnell[5], K. Potter[4], T. Vilcan[4], N. Wright[11] and C. Thorne[3]
[1]The Royal Botanic Garden, Edinburgh, UK
[2]College of Engineering, Mathematics and Physical Sciences, University of Exeter, Exeter, Devon, UK
[3]University of Nottingham, Nottingham, UK
[4]Open University, Milton Keynes, UK
[5]University of Newcastle, Newcastle, UK
[6]Heriot-Watt University, Edinburgh, UK
[7]University of Leeds, Leeds, UK
[8]University of the West of England, Bristol, UK
[9]100 resilient Cities Project, New York, USA
[10]University of Coahuila, Coahuila, Mexico
[11]Nottingham Trent University, Nottingham, UK

Definition

A Blue-Green City aims to recreate a naturally orientated water cycle while contributing to the amenity of the city by bringing together water management and green infrastructure (Hoyer et al. 2011). The UKEPSRC Blue-Green Cities (B-GC, http://www.bluegreencities.ac.uk/) and Urban Flood Resilience (UFR, http://www.urbanfloodresilience.ac.uk/) research consortia further developed the conceptual framework for the B-GC approach to delivering sustainable flood risk management and water security. These consortia emphasized sustainable flood risk management while also addressing the potential of the Blue-Green approach to improve water quality, transform stormwater management, enhance biodiversity, and generate multiple co-benefits that are valued by the beneficiary communities. The B-GC consortium focused on design and implementation of integrated treatment trains comprising the optimal mix of grey and blue-green assets, within the overarching framework of Nature Based Solutions. As well as dealing with water-related issues, the B-GC approach recognizes that environmental and social problems are interlinked, and that the solutions, therefore, also need to be multifaceted: fully considering the interrelations between multiple, urban systems. In this context, the B-GC approach differs from conventional urban water management in that it puts people, planning and communications at the heart of urban flood risk and water management. While the fact that flooding is a social as well as a hydrological problem has long been understood, the integrated approaches and the underpinning science base for their practical implementation in urban spaces have only recently started to be developed, functionally. Notwithstanding this, Blue-Green thinking and practice have become widely accepted, organic components of the move toward urban development that is (a) centered on the real lives of citizens, (b) sustainable and (c) resilient to an uncertain future (O'Donnell and Thorne 2020).

This entry draws heavily on previous research outputs from the UK B-GC and UFR research consortia, plus those of related and follow-up projects and partnerships. It aims to reflect both the resulting consensus and the current state of knowledge, while identifying some significant gaps in knowledge.

Terminology

The overlapping concepts of Green Infrastructure (GI), Blue-Green Infrastructure (BGI), Blue-Green

Cities (B-GC), WSUD, SuDS, LID, BMPs, and "Sponge Cities" are so similar that in practice they are often used interchangeably (O'Donnell and Thorne 2020) . There are, however, significant differences – as outlined below.

Green Infrastructure (GI) is usually characterized as a network of multi-functional green spaces that improve quality of a local resident's life and provide environmental benefits to communities. Some contemporary definitions (e.g., that of the UK Town and Country Planning Association) include sustainable drainage and runoff reduction during flash flooding among its features, although originally those were not among the priorities for the GI concept, which was centered on the enhancement of public amenity (i.e., pleasantness of the environment). Fundamentally, public amenity remains the cornerstone of GI, while other ecosystem services (including flood risk management and water security) are viewed as additional benefits.

The concept of *Blue-Green Infrastructure (BGI)* is basically similar to that of GI, but expands on it by adding the "Blue" component (i.e., water), and the advantages that result when green spaces also fulfil a range of urban water management functions. Development or retrofitting of an integrated BGI network is best achieved when blue, green, and blue-green assets operate synergistically with existing grey assets, these assets being arranged in integrated networks designed and operated to provide a wide range of ecosystem services. In this context, *Grey Infrastructure* refers to conventional, single function, engineered assets designed solely to provide flood control for events up to but not exceeding the magnitude (or return period) of the design event. One of the objectives of BGI is optimization of Blue-Green and Grey systems (B-G+G) to deliver maximum values of multiple co-benefits, while minimizing negative trade-offs. A further advantage of BGI systems is their high capacity for adaptation. For example, a methodology based on 'adaptation pathway tree diagrams' and application of advanced mathematical methods implemented in the SWMM dynamic rainfall-runoff-routing simulation model has been shown to be able to identify optimal adaptation strategies for the next 40 years (Fenner et al. 2019).

Sustainable Drainage Systems (SuDS), also known as or sustainable urban drainage systems, are a collection of water management practices that aim to align modern drainage systems with natural water processes. Together with closely-related principles of *Best Management Practices* (BMPs) and *Low Impact Development* (LID), these terms are often used in a broad sense to refer to systems that slow and/or retain runoff to attenuate peak stormwater flows. Hence, within this broad understanding BGI may be regarded as part of SuDS, BMPs and LID. Many practitioners, however, preferentially associate SuDS predominantly with engineered assets. Consequently, engineered, grey infrastructure (e.g., runoff attenuation tanks, purposefully-designed stormwater retention ponds, water barrels and permeable pavements), can all be regarded as SuDS, but a semi-natural lake or an historic ornamental pond in a city park that could have a flood risk management function would not fall into that category.

Water-Sensitive Urban Design (WSUD) focuses on integrating water cycle management into planning and design, to improve the harmony between people, water and the natural and built environment. Assets falling into the BGI and/or SuDS categories (e.g., green roofs, swales, retention and detention ponds, restored and de-culverted rivers, raingardens, filter drains, water barrels, etc.) are, therefore, instrumental to WSUD implementation.

Urban Flood Resilience (UFR) a city achieves UFR when it has the capacity to maintain future flood risk at acceptable levels by: preventing deaths and injuries, minimizing damage and disruption during floods and recovering quickly afterwards, while ensuring social equity and protecting the city's cultural identity and economic vitality. Implicit to delivering UFR is the requirement to use of Grey and Green/Blue-Green Infrastructure sustainably through maintenance and adaptation of natural and engineered assets.

Blue-Green Cities (B-GC) is a holistic concept incorporating elements of all the concepts mentioned above, which emphasizes the potential to enhance peoples' lived lives through urban development that levers the advantages of working with natural processes. Achieving UFR, combined with protection and enhancement of ecological values of the landscape, follows naturally from broad-scale implementation of sustainable flood risk and water management in a B-GC. The B-GC approach, therefore, not only delivers increased flood resilience, but also provides a range of other, related ecosystem services including improvements in water and air quality, amelioration of climatic abnormalities, increases in local biodiversity and habitats connectivity and, consequently, enhancement of a city's livability, which subsequently translates into improvements to citizens' wellbeing and livelihoods.

China's *"Sponge Cities"* initiative is intended to manage urban flood risk, purify stormwater, and provide water storage opportunities for future usage. Conceptually, this initiative aligns closely with the B-GC vision, though with some notable differences relating to its Chinese context. A detailed description of the Sponge Cities program is beyond the scope of this entry, but the body of literature on this topic is rich and rapidly growing (Qi et al. 2020). In essence, both "B-GC" and "Sponge Cities" aspire to high levels of flood resilience and water security, including sufficient access to adequate spaces and water resources for people, agriculture, industries, and ecosystems. In both approaches, provision of adequate urban water resources are recognized as essential for social stability, economic development, and future prosperity.

Recent Research and Knowledge Gaps

Hydrodynamic Modelling

Application of advanced hydrologic and hydraulic models, such as Shetran (Ewen et al. 2000) and CityCAT (Glenis et al. 2018), provide the basis for simulating urban drainage systems that feature complex combinations of surface water, groundwater, and piped components. These models also have the capability to investigate the physical, chemical, and biological functionality of urban drainage systems, and how they interact with other systems, such as power distribution and transportation. Understanding these interactions lays the foundations for designing them to function interoperably. These and other well-established urban water models provide the basis for the analysis of innovative blue-green+grey plus (B-G+G) solutions that include raingardens, riverine woodlands, ponds and urban public green spaces (Krivtsov et al. 2020a; Krivtsov et al. 2021a; O'Donnell et al. 2020; Krivtsov et al. 2019). An important capability of these models is to simulate *"what if"* scenarios for extreme rain storms that exceed the design event, under a range of antecedent meteorological conditions (Birkinshaw et al. 2021), opening up the possibility of not only analyzing for exceedance, but designing for it as well.

Stormwater as a Resource

Within the Blue-Green Cities (B-GC) conceptual approach, stormwater is regarded as a valuable resource that may be harvested and used beneficially, and outputs from the B-GC and UFR research consortia provide insights into a number of practical applications in this regard (Fenner et al. 2019; O'Donnell et al. 2020). However, converting the hazards associated with stormwater into opportunities requires innovation in the ways that these potentially-valuable water resources are managed. There is a range of options for stormwater use, related to, for example, greywater supply, energy generation, replenishment of groundwater aquifers, and enhancement of ecosystem services through restoration of urban streams and wetlands.

For example, stormwater can be used for micro-hydropower generation. The feasibility of energy recovery from SuDS ponds has been assessed using specially designed screening software that takes into account technical specifications of the installation as well as climatic variability and site characteristics (Costa et al. 2019). Results indicate that the range of suitable sites is limited, but where conditions are right, micro-hydropower could make a valuable contribution to local energy supplies.

Rainwater harvesting (RWH) presents alternative opportunities for stormwater use and can be particularly effective through application of rainwater management systems (RMS), which have been shown to optimize water supply demand, storm water discharge reduction, and energy usage in residential households. This has been demonstrated in case studies of individual houses in Newcastle-upon-Tyne, UK (O'Donnell et al. 2020), while a system dynamics model of Ebbsfleet Garden City, UK projects positive effects if RWH were to be adopted across the urban area (Pluchinotta et al. 2021).

Potentially, the most beneficial uses of stormwater relate to the recharge of natural aquifers and soil moisture (O'Donnell et al. 2020) and maintaining groundwater levels adequate for functioning of ecosystems, especially streams and wetlands. These are particularly important for those areas where dry seasons and droughts lead to water scarcity and associated water security issues. However, negative effects of climate change are projected to increase the occurrence of extreme droughts even in the areas which have traditionally been perceived as water-sufficient (Kirkpatrick et al. 2021). The consequences of such events can be devastating when combined with erosion from flash flooding and/or the occurrence of wildfire (Tallis 1987). These finding highlight the huge potential for using stormwater for managed aquifer recharge (MAR).

Biodiversity and Provision of Ecosystem Services

Closely related to the issues covered in the previous section is management of stormwater to enhance provision of ecosystem services. The B-GC/UFR consortia participated in studies related to: stream restoration that reconnects channels to floodplains and rehydrates wetlands (e.g., Whychus Creek and Johnson Creek, both in Oregon (CIRIA 2019a); BGI ponds in the UK (Krivtsov et al. 2020a, b); pollution reduction by SuDS (CIRIA 2019a); and, enhancement of local biodiversity in floodplain woodlands (Krivtsov et al. 2021a). Knowledge emerging from these and other studies illustrates that blue-green infrastructure networks developed to deliver UFR also provide a range of co-benefits, including improvements in chemical and biological water quality, biodiversity and habitat connectivity, public amenity, and spaces for recreation. Evaluation of these co-benefits (as well as unavoidable trade-offs) is challenging, but can be facilitated by freely available spreadsheet and GIS-based tools, including CIRIA's B£ST, "Multiple Benefit Toolbox," and the Natural Capital Planning Tool (CIRIA 2019b; Fenner 2017; Morgan and Fenner 2019; Hölzinger et al. 2019).

Applications of these tools show that the conventional, pipe and storage-based urban drainage solutions in greenfield developments typically result in a negative net-benefit score. In contrast, blue-green+grey solutions can eliminate, or at least mitigate, the adverse hydrologic and environmental impacts of urban growth (Ncube and Arthur 2021). However, existing evaluation tools do have limitations. For example, many newly created, innovative drainage assets provide opportunities for colonisation by native vegetation (Krivtsov et al. 2021b, c), but the natural capital associated with these novel habitats is underrepresented in existing evaluation software. The point here is that a deep understanding of B-G+G's multi-functionality and capacity to provide a range of ecosystem services is vital. It follows that multi-disciplinary monitoring of B-G+G systems is essential to explaining the complex interrelations between its physical, chemical and biological components. Monitoring and modelling studies combining elements of hydrology, hydraulics, biogeochemistry, water quality, geomorphology, and ecology in Blue-Green systems can provide important insights in this regard. Many of these interactions are specific to particular, local contexts and geographies, and therefore not easily represented in the software tools set up to evaluate generalized situations.

Interoperability

The B-GC concept extends beyond provision of appropriately sequenced treatment trains featuring blue-green and grey infrastructure, to planning and delivering interoperability between multiple urban systems under normal, design storm and extreme hydrological loadings. Interoperability is the capacity of the urban assets, infrastructure and systems (e.g., transportation, power distribution, drainage, water supply, parks and recreational green spaces, etc.) to function synergistically in storing or redirecting surface runoff during rainfall events, without interrupting their primary functions (Vercruysse et al. 2019). Interoperability solutions include co-location of stormwater detention basins with public green spaces, using roads to convey excessive surface runoff without causing unacceptable traffic disruption, and installing elevated crossings over greenways that may be used as temporary watercourses. In practice, achieving urban system interoperability requires both detailed, local knowledge and advanced hydrodynamic simulations using models such as CityCAT and MicroDrainage. It is also vital to understand the urban and physical geographies of the city, so both design and implementation are facilitated using a spatial mapping framework that links runoff sources to areas most at risk of flooding. This helps urban planners and engineers prioritize locations for technical upgrades, design operational protocols, and identify other physical interventions (Dawson et al. 2020).

Water Quality Benefits

The combined BG+G approach also has the potential to achieve remarkable improvements in water quality. Integrated BGI systems and networks have been shown to be effective in trapping suspended sediments and adsorbed chemicals (CIRIA 2019a). Trapped sediments and organic matter (including a range of pollutants) accumulate in bottom of BGI ponds (Ahilan et al. 2019). Monitoring and maintenance (including sediment removal) may be necessary to ensure that the performance of these systems is sustained. Where maintenance is needed, the sediments removed may be of value if, for example, recovery of chemicals such as rare earth elements (REE) is economically feasible (Krivtsov et al. 2020b).

Stakeholder Involvement and the Social Dimension

High levels of constructive collaboration between Local Authorities, urban water practitioners, stakeholders in the commercial, business and industrial sectors, and the public are a prerequisite for any Blue-Green city aspiring to achieve a sustainable and resilient urban water future. Currently, progress toward this goal is hampered by lack of understanding of the language of sustainable urban water management, never mind its technical, social and environmental aspects (Fenner et al. 2019; Arthur et al. 2018). In short, ineffective communications and a lack of appreciation for the views of other stakeholders continue to act as barriers to appreciation and realization of the multiple co-benefits offered by the B-GC concept and implementation of B-G+G innovations (O'Donnell et al. 2020; CIRIA 2019b). That said, there is growing realization that community perceptions and acceptance of BGI is crucial to its success.

Progress in identifying and overcoming barriers to innovation is being made thanks to two types of action research. First, research on stakeholder attitudes and public perceptions is using novel methods such as point of interaction (POI), structured interviews, questionnaires, implicit association tests (IAT) and Likert tests (O'Donnell et al. 2020). Findings to date point to the need for improved stakeholder communications and better ways to manage conflicting views (O'Donnell et al. 2020; Arthur et al. 2018). Second, bringing together multiple stakeholders in local meetings, forums, conferences, and focus groups is proving helpful for increasing awareness, overcoming tunnel vision, facilitating cross-disciplinary exchange, and breaking out of departmental silos (CIRIA 2019b). Experience gained using Learning and Action Alliances (LAAs) is proving particularly fruitful (van Herk et al. 2011; Ashley et al. 2012). LAAs have been found to be conducive to promoting the frank discussion, constructive criticism and mutual understanding needed to delineate and characterize difficult problems and find consensus on feasible solutions.

An LAA established in Newcastle-upon-Tyne, UK by the B-GC consortium has now been running for nearly a decade, while another in Ebbsfleet Garden City has morphed into a useful component of the formal structure for urban water planning and governance. Both forums hold regular meetings at which key stakeholders exchange views on a range of issues related to flood risk management, water security, biodiversity, and public perception. These LAAs have helped make the host city's aspirations to become B-GC achievable in practice. In Newcastle, the LAA led to a joint declaration that commits the signatories to widen implementation of BGI (O'Donnell et al. 2021). In Ebbsfleet, the LAA facilitated stakeholder participation in development of both an urban metabolism model (based on WaterMet), which assesses the sustainability and performance of relevant practices, and evaluates pros and cons of future options for developing the garden city (Fenner et al. 2019), and a system dynamics model (SDM) for analysis of options for future urban water management (Pluchinotta et al. 2021). It is planned to couple the SDM and the urban metabolism models, to train stakeholders in the SDM use, and to design freely accessible teaching materials, thus facilitating participatory modelling by a wide range of stakeholders, and encouraging incorporation of these novel approaches into higher education (Krivtsov et al. 2021c).

Summary and Conclusions

Historically, the vital tasks of making urban environments safer and more amenable have been addressed separately. The concept of using grey infrastructure for stormwater management developed solely to counter the hazards posed by urban flooding, while GI developed to make the urban environments more livable. The holistic concept of the Blue-Green City recognizes both the necessity of integrating and optimizing management of urban water in all its forms, and the multiple co-benefits of so doing. Progress is being made toward making possible the transformative change needed to deliver BGI systems and the B-GC vision, in practice. However, serious barriers to innovation remain. Our assessment of the current situation is that:

- Flood resilience in a B-GC is achievable in concert with enhancing other ecosystem services provided by Blue-Green and Grey infrastructure, but beyond this a Blue-Green City also acquires increased water security, biodiversity and amenity benefits, so that a range of environmental risks are reduced (Krivtsov et al. 2021d).
- Analysis and design of BGI and B-G+G networks are greatly aided by application of advanced urban water models and evaluative tools. However, some aspects of the interactions among system components are context-specific and these are not captured in the currently available, generalized approaches. There is, therefore, a need for expert knowledge for hindcasting past, and forecasting future, system dynamics, evaluating alternative future scenarios and management options, analyzing adaptation pathways, and interpretating the patterns observed.
- It is implicit to the B-GC concept that stormwater should be treated not only as a hazard but also as a resource, important for grey water supply and as natural capital. The potential for expanded RWH is growing thanks to increased environmental awareness, technological innovations and wider acceptance for using recycled grey water to reduce demand for potable water and enhance public green spaces and ecosystems.
- The BGI and B-G+G networks function in conjunction with (and in spaces shared with) other urban systems. Interoperability is therefore crucial to service provision that is reliable, even under extreme rainfall loadings that exceed the "design event". This is best achieved through co-design, monitoring, maintenance and adaptive management of multiple urban systems, and the development of interoperability protocols that are transferrable between cities.

- Bringing together key stakeholders within local forums, discussion and focus groups that provide spaces for frank exchanges of views, concerns and ideas, such as LAAs, helps to identify and characterize the barriers that frustrate those wishing to innovate, while increasing the levels of mutual respect, building trust and promoting co-production of knowledge.
- With respect to urban flooding, local people are not only those who stand to gain or lose depending on the decisions made by urban water managers; they are also the local experts. It is crucial that residents and business owners are fully engaged in UFRM planning and decision-making, and that the designers of urban water systems understand both the concerns and the preferences of the communities they serve.
- Without local support and buy-in by the intended beneficiaries of Blue-Green projects, the transformative changes required to make our cities resilient to future floods and droughts are simply impossible. That is why people, planning and communications must be put at the heart of urban flood risk management.

References

Ahilan, S., Guan, M., Wright, N., Sleigh, A., Allen, D., Arthur, S., et al. (2019). Modelling the long-term suspended sedimentological effects on stormwater pond performance in an urban catchment. *Journal of Hydrology, 571*, 805–818.

Arthur, S., D'Arcy, B. J., Semple, C., Sevilla, A. E., & Krivtsov, V. (2018). *Retrofitting sustainable urban drainage systems to industrial estates*. CREW Report: Scotland, UK.

Ashley, R. M., Blanskby, J., Newman, R., Gersonius, B., Poole, A., Lindley, G., et al. (2012). Learning and action alliances to build capacity for flood resilience. *Journal of Flood Risk Management, 5*(1), 14–22.

Birkinshaw, S. J., O'Donnell, G., Glenis, V., & Kilsby, C. (2021). Improved hydrological modelling of urban catchments using runoff coefficients. *Journal of Hydrology, 594*, 125884.

CIRIA. (2019a). *Blue-green infrastructure – Perspectives on water quality benefits*. London: CIRIA C780b.

CIRIA. (2019b). *Blue-green infrastructure – Perspectives on planning, evaluation and collaboration*. London: CIRIA C780a.

Costa, J., Fenner, R. A., & Kapetas, L. (Eds.). (2019). Assessing the potential for energy recovery from the discharge of storm water run-off. In *Proceedings of the institution of civil engineers-engineering sustainability*. London: Thomas Telford Ltd.

Dawson, D. A., Vercruysse, K., & Wright, N. (2020). A spatial framework to explore needs and opportunities for interoperable urban flood management. *Philosophical Transactions of the Royal Society A., 378*(2168), 20190205.

Ewen, J., Parkin, G., & O'Connell, P. E. (2000). SHETRAN: Distributed river basin flow and transport modeling system. *Journal of Hydrologic Engineering., 5*(3), 250–258.

Fenner, R. (2017). Spatial evaluation of multiple benefits to encourage multi-functional design of sustainable drainage in blue-green cities. *Water., 9*(12), 953.

Fenner, R., O'Donnell, E., Ahilan, S., Dawson, D., Kapetas, L., Krivtsov, V., et al. (2019). Achieving urban flood resilience in an uncertain future. *Water, 11*(5), 1082.

Glenis, V., Kutija, V., & Kilsby, C. G. (2018). A fully hydrodynamic urban flood modelling system representing buildings, green space and interventions. *Environmental Modelling and Software, 109*, 272–292.

Hölzinger, O., Sadler, J., Scott, A., Grayson, N., & Marsh, A. (2019). NCPT–managing environmental gains and losses. *Town and Country Planning, 88*(5), 166–170.

Hoyer, J., Dickhaut, W., Kronawitter, L., & Weber, B. (2011). *Water sensitive urban design*. Hamburg: University of Hamburg: JOVIS Verlag GmBH.

Kirkpatrick, B. F., Stubbs, J. P., & Spray, D. (2021). Anticipating and mitigating projected climate-driven increases in extreme drought in Scotland, 2021–2040. NatureScot research report no. 1228. NatureScot.

Krivtsov, V., Arthur, S., Buckman, J., Bischoff, J., Christie, D., Birkinshaw, S., et al. (2019). Monitoring and modelling SUDS retention ponds: Case studies from Scotland ICONHIC; Chania, Greece. http://www.urbanfloodresilience.ac.uk/documents/krivtsov-et-al.-iconhic-2019b.pdf.

Krivtsov, V., Birkinshaw, S., Arthur, S., Knott, D., Monfries, R., Wilson, K., et al. (2020a). Flood resilience, amenity and biodiversity benefits of an historic urban pond. *Philosophical Transactions of the Royal Society A, 378*(2168), 20190389.

Krivtsov, V., Arthur, S., Buckman, J., Kraiphet, A., Needham, T., Gu, W., et al. (2020b). Characterisation of suspended and sedimented particulate matter in blue-green infrastructure ponds. *Blue-Green Systems, 2*(1), 214–236.

Krivtsov, V., Birkinshaw, S., Forbes, H., Olive, V., Chamberlain, D., Lomax, J., et al. (2020c). Hydrology, ecology and water chemistry of two SuDS ponds: Detailed analysis of ecosystem services provided by blue green infrastructure. *WIT Transactions on The Built Environment, 194*, 167–178.

Krivtsov, V., Buckman, J., Birkinshaw, S., Monteiro, Y., Christie, D., Takezawa, K., et al. (2021a). Ecology,

hydrology and biodiversity of a woodland pond: case study for ecosystem services provided by riverine floodplains. *EMCEI*. in press.
Krivtsov, V., Birkinshaw, S., Yahr, R., & Olive, V. (2021b). Comparative ecosystem analysis of urban ponds: implications for synergistic benefits and potential trade-offs resulting from retrofitting of green roofs in their catchments. *International Journal of Environmental Impacts*. in press.
Krivtsov, V., Pagano, A., Ahilan, S., O'Donnell, E., & Pluchinotta, I. (2021c). Further developments of the Ebbsfleet water management system dynamics model: Adjusting representation of processes and system boundaries, insentivising stakeholder re-engagement, and exploring potential for the use in university teaching. *WIT Transactions on Ecology and the Environment*. in press.
Krivtsov, V., Birkinshaw, S., Lomax, J., Christie, D., & Arthur, S. (2021d). *Multiple benefits of blue-green infrastructure and the reduction of environmental risks: case study of ecosystem services provided by a suds pond. Civil engineering for disaster risk reduction*. New York: Springer.
Morgan, M., & Fenner, R. (2019). Spatial evaluation of the multiple benefits of sustainable drainage systems. *Proceedings of the Institution of Civil Engineers – Water Management, 172*(1), 39–52.
Ncube, S., & Arthur, S. (2021). Influence of blue-green and grey infrastructure combinations on natural and human-derived capital in urban drainage planning. *Sustainability, 13*(5), 2571.
O'Donnell, E., & Thorne, C. (2020). Urban flood risk management: the blue–green advantage. In C. R. Thorne (Ed.), *Blue–green cities: Integrating urban flood risk management with green infrastructure* (pp. 1–13). London: ICE Publishing (Thomas Telford).
O'Donnell, E. C., Netusil, N. R., Chan, F. K. S., Dolman, N. J., & Gosling, S. N. (2021). International perceptions of urban blue-green infrastructure: A comparison across four cities. *Water, 13*(4), 544.
O'Donnell, E. C., & Thorne, C. R. (2020). Drivers of future urban flood risk. *Philosophical Transactions of the Royal Society A., 378*(2168), 20190216.
O'Donnell, E., Thorne, C., Ahilan, S., Arthur, S., Birkinshaw, S., Butler, D., et al. (2020). The blue-green path to urban flood resilience. *Blue-Green Systems, 2*(1), 28–45.
Pluchinotta, I., Pagano, A., Vilcan, T., Ahilan, S., Kapetas, L., Maskrey, S., et al. (2021). A participatory system dynamics model to investigate sustainable urban water management in Ebbsfleet Garden City. *Sustainable Cities and Society, 67*, 102709.
Qi, Y., Chan, F. K. S., Thorne, C., O'Donnell, E., Quagliolo, C., Comino, E., et al. (2020). Addressing challenges of urban water management in chinese sponge cities via nature-based solutions. *Water, 12*(10), 2788.
Tallis, J. H. (1987). Fire and flood at Holme Moss: Erosion processes in an upland blanket mire. *The Journal of Ecology., 75*, 1099–1129.
van Herk, S., Zevenbergen, C., Ashley, R., & Rijke, J. (2011). Learning and action alliances for the integration of flood risk management into urban planning: A new framework from empirical evidence from The Netherlands. *Environmental Science & Policy, 14*(5), 543–554.
Vercruysse, K., Dawson, D. A., & Wright, N. (2019). Interoperability: A conceptual framework to bridge the gap between multifunctional and multisystem urban flood management. *Journal of Flood Risk Management, 12*(S2), e12535.

B

Boundaries

► The Challenges for Wildland-Urban Interfaces (WUI) in Metropolitan Areas: Reducing Fire Risk, Providing Employment Opportunities, and Preserving Natural Habitat

Building Community Resilience

A Design Perspective

Mariana Fonseca Braga
ImaginationLancaster, Lancaster Institute for the Contemporary Arts (LICA), Lancaster University, Lancaster, Lancashire, UK

Introduction

Resilience has attracted attention especially due to the COVID-19 pandemic in these uncertain times. However, many communities live in conditions that require resilience from them because of diverse circumstances, such as exposition to natural and man-made disasters and socio-economic determinants (see United Nations Office for Disaster Risk Reduction [UNDRR] 2017).

Expressions such as "Build back better" (BBB) (The United Nations Office for Disaster Risk Reduction [UNISDR] 2015), "leave no one behind" (UNDRR 2019), "to withstand and bounce back" (United Nations Office for Disaster Risk Reduction [UNDRR] 2017, p. 3) incorporate the sense of resilience and are the new buzzwords that came out of the COVID-19 pandemic. Disadvantaged communities that face the lack of economic facilities and social opportunities are still left behind (see for instance Fonseca Braga et al. 2020, 2021; United Nations [UN] 2020).

Global guidelines on disaster prevention, preparedness, hazards mitigation, and recovery are often approached from a risk management perspective (e.g., UNISDR 2015, 2016), failing to address socio-cultural determinants and livelihood diversity and impacting on the sustainability of proposed policies and solutions to these (see, for instance, Vahanvati and Rafliana 2019).

Community Resilience

Although communities and their organizations play a noteworthy role in dealing with disasters throughout history (Patterson et al. 2010) and community engagement is essential to community resilience and stability, risk assessment and urban planning processes are still often being operated by experts without sufficient community engagement (Meyer et al. 2018).

Successful participatory approaches to recovery favor community empowerment, ownership, commitment to implementation, trust-building between communities, public officials, and key stakeholder groups, contributing to more resilient communities and to sustainable and inclusive actions and solutions. However, "... giving power to communities without support is hardly empowering" (Vahanvati and Rafliana 2019, p. 218). Political backdrop and governance are also critical to community resilience (Choudhury et al. 2019; Hardoy et al. 2019; Wanie and Ndi 2018; Vahanvati and Rafliana 2019).

In the literature on resilience, community resilience is mostly related to natural disasters. It is associated with building or enhancing community capabilities throughout co-development processes of self-organizing, accessing resources, strengthening networks, harnessing collaboration, and creating mechanisms that contribute to holding community-led plans and efforts accountable (Berke et al. 2011) rather than with outcomes themselves (see Berke et al. 2011; Bott and Braun 2019; Schilderman and Lyons 2011; Vahanvati and Beza 2017).

Design principles, approaches, processes, and methods are still under-researched in this context, although the COVID-19 pandemic has recently attracted attention in emergency and recovery contexts in design practices and research.

Design

The idea of design is related to human interventions in the world with a purpose. Even though there is not a commonly agreed definition of design, some of these are helpful to understand the principles that foreground designers' work in different areas, such as:

- Design as a way to transform a current situation into a preferred one (Simon 1996), and
- Design as the "human power of conceiving, planning, and making products that serve human beings in the accomplishment of their individual and collective purposes" (Buchanan 2001, p. 9).

Products in the latter, range from images, words, physical objects to services, activities, processes, experiences, interactions, and their integration into environments and organizations or human systems that influence our ways of "living, working, playing and learning" (Buchanan 2001, p. 12, 2015).

The design field evolved and many specializations are still arising. Design is influenced by an interplay of political and economic factors and influences those as well (Julier 2017). This evolvement required from designers a better interaction with other disciplines and different stakeholder groups to contribute to tackling complex challenges in uncertain environments.

Design approaches and methods, especially participatory design and co-design, have been attracting attention in policymaking, particularly since the austerity period. In this context, communities need to play a more significant role since public resources shrink (see Julier 2017).

Moreover, the need to tackle complex public problems that cannot be solved anymore utilizing traditional policymaking processes – that separate policy plan from implementation (see Junginger 2014), providing preset responses (see Bason 2014; Julier 2017; Mortati et al. 2016) – gives grounds for the value of design for policymaking. Service Design, Social Innovation, Design Policy, Design for Policy or Policy Design are some of the fields that are concerned with issues around this.

Citizens' participation in the decisions that affect their lives is the main principle of democracy (see Sanoff 2007; Sen 1999). In participatory design, participation goes beyond voting and public consultation, involving citizens in the creation and management of their future and environment, and influencing public decision-making (Sanoff 2007). Sanoff (2007) emphasizes the importance to provide citizens "with the information they need to participate in a meaningful way and be informed how their input affects the decision" (p. 59) (see also Vahanvati and Rafliana (2019) for this aspect in recovery situations).

Participation is also influenced by other conditions, requiring "knowledge and basic educational skills" (Sen 1999, p. 32). Diverse factors can hamper effective participation, such as: denied access to civil rights, economic facilities, social opportunities (e.g., education and healthcare), transparency guarantees and protective security (social safety to prevent misery) (Sen 1999).

However, there are also limitations on citizens' engagement in public decision making related to social and political power relations, negligence of implementation aspects, possible and politically desirable outcomes, and impacts.

A Design Lens for Community Resilience

A design perspective for building community resilience brings principles that can support the co-development of situated policies and solutions, considering people's lived experiences, livelihood diversity, social, cultural, and economic determinants, as well as other contextual factors that are often overlooked in global guidelines and policies on hazard mitigation and recovery.

Communities are the experts on their own lives and problems, knowing by experience the barriers that hinder their possibilities to overcome their challenges. Therefore, understanding the perspective and lived experiences of people on the ground earlier is essential to provide appropriate support and policies to tackle complex "glocal" challenges.

Putting global or universal guidelines as "ready-to-use" solutions is misleading as people may not follow those due to specific reasons on the ground that the ones who wrote those are not aware of. Examples from informal settlements during the COVID-19 pandemic include (but are not limited to): measures such as: "to self-isolate" for people who live in overcrowded and intergenerational houses; "to wash your hands with soap and water" for those without access to water and soap.

Design experimental approaches can improve communication and collaboration between different stakeholder groups and communities, bringing diverse perspectives, local knowledge and experience into creativity, innovation, and decision-making processes. Solutions and opportunities can be co-created, prototyped and tested. This ideally generates learning cycles that enable to gradually scale locally meaningful solutions and policies, bridging the gap between policy planning and implementation – as, in conventional policymaking, the ones who make the policy are neither the ones who implement the policies nor those who experience the needs or problems of community members.

A multi-stakeholder approach and collaboration facilitated by design are necessary for enhancing community resilience. However, several factors can hamper the adoption of a design lens in policymaking, including, but not limited to, dysfunctional democracies, the lack of commitment to the common good, and the understanding of design as a synonym of aesthetics.

References

Bason, C. (2014). Introduction: The design for policy nexus. In C. Bason (Ed.), *Design for policy* (pp. 1–8). Abingdon: Grower Publishing.

Berke, P. R., Cooper, J., Salvesen, D., Spurlock, D., & Rausch, C. (2011). Building capacity for disaster

resiliency in six disadvantaged communities. *Sustainability, 3*(1), 1–20. https://doi.org/10.3390/su3010001.
Bott, L.-M., & Braun, B. (2019). How do households respond to coastal hazards? A framework for accommodating strategies using the example of Semarang Bay, Indonesia. *International Journal of Disaster Risk Reduction, 37*, 1–9. https://doi.org/10.1016/j.ijdrr.2019.101177.
Buchanan, R. (2001). Design research and the new learning. *Design Issues, 17*(4), 3–23. http://www.mitpressjournals.org/doi/pdf/10.1162/07479360152681056.
Buchanan, R. (2015). Worlds in the making: Design, management, and the reform of organizational culture. *She Ji: The Journal of Design, Economics, and Innovation, 2*, 5–21. https://doi.org/10.1016/j.sheji.2015.09.003.
Choudhury, M.-U.-I., Uddin, M. S., & Haque, C. E. (2019). Nature brings us extreme events, some people cause us prolonged sufferings: The role of good governance in building community resilience to natural disasters in Bangladesh. *Journal of Environmental Planning and Management, 62*(10), 1761–1781. https://doi.org/10.1080/09640568.2018.1513833.
Fonseca Braga, M., Romeiro Filho, E., Mendonça, R. M. L. O., Oliveira, R. G. L., & Pereira, H. G. G. (2020). Design for resilience: Mapping the needs of Brazilian Communities to Tackle COVID-19 challenges. *Strategic Design Research Journal, 13*(3), 374–386. https://doi.org/10.4013/sdrj.2020.133.07.
Fonseca Braga, M., Romeiro Filho, E., Pereira, H. G. G., Tsekleves, E., & Mendonça, R. M. L. O. (2021). Community-led design capabilities during the COVID-19 pandemic and beyond. In: L. Di Lucchio, L. Imbesi, A. Giambattista, V. Malakuczi (Eds.), *Design Culture(s) Cumulus Conference Proceedings Roma 2021*, pp. 2165–2181. Rome, Italy: Sapienza University of Rome, Cumulus Association, Aalto University.
Hardoy, J., Gencer, E., & Winograd, M. (2019). Participatory planning for climate resilient and inclusive urban development in Dosquebradas, Santa Ana and Santa Tomé. *Environment & Urbanization, 31*(1), 33–52. https://doi.org/10.1177/0956247819825539.
Julier, G. (2017). *Economies of design*. London, UK: Sage Publications.
Junginger, S. (2014). Towards policy-making as designing: Policy-making beyond problem solving and decision-making. In C. Bason (Ed.), *Design for policy* (pp. 57–69). Abingdon: Grower Publishing.
Meyer, M. A., Hendricks, M., Newman, G. D., Masterson, J. H., Cooper, J. T., Sansom, G., Gharaibeh, N., Horney, J., Berke, P., van Zandt, S., & Cousins, T. (2018). Participatory action research: Tools for disaster resilience education. *International Journal of Disaster Resilience in the Built Environment, 9*(4/5), 402–419. https://doi.org/10.1108/IJDRBE-02-2017-0015.
Mortati, M., Villari, B., Maffei, S., & Arquilla, V. (2016). *Le politiche per il design e il design per le politiche. Dal focus sulla soluzione alla centralità della valutazione* [Policies for design and design for policies. From focus on solutions to evaluation centrality]. Santarcangelo di Romagna: Maggioli S.p.A.
Patterson, O., Weil, F., & Patel, K. (2010). The role of community in disaster response: Conceptual models. *Population Research and Policy Review, 29*(2), 127–141. https://doi.org/10.1007/s11113-009-9133-x.
Sanoff, H. (2007). Multiple views of participatory design. *International Journal of Architectural Research (Archnet-IJAR), 2*(1), 57–69.
Schilderman, T., & Lyons, M. (2011). Resilient dwellings or resilient people? Towards people-centred reconstruction. *Environmental Hazards, 10*(3–4), 218–231. https://doi.org/10.1080/17477891.2011.598497.
Sen, A. (1999). *Development as freedom*. Oxford, UK: Oxford University Press.
Simon, H. A. (1996). *The sciences of the artificial* (3rd ed.). Cambridge, MA: MIT Press.
The United Nations Office for Disaster Risk Reduction. (2015). *Sendai framework for disaster risk reduction 2015–2030*. Geneva: UNISDR (United Nations Office for Disaster Risk Reduction). www.unisdr.org/we/coordinate/sendai-framework.
The United Nations Office for Disaster Risk Reduction. (2016). *Terminology on disaster risk reduction*. Geneva: UNISDR (United Nations International Strategy for Disaster Reduction). www.unisdr.org/we/inform/publications/51748.
United Nations. (2020). *UN-Habitat COVID-19 response plan*. United Nations Human Settlements Programme. https://unhabitat.org.
United Nations Office for Disaster Risk Reduction. (2017). Disaster resilience scorecard for cities. Detailed level assessment. https://www.unisdr.org/campaign/resilientcities/toolkit/article/disaster-resiliencescorecard-for-cities
United Nations Office for Disaster Risk Reduction. (2019). *Global assessment report on disaster risk reduction*. Geneva: United Nations Office for Disaster Risk Reduction (UNDRR). https://gar.undrr.org/sites/default/files/reports/2019-05/full_gar_report.pdf.
Vahanvati, M., & Beza, B. (2017). An owner-driven reconstruction in Bihar. *International Journal of Disaster Resilience in the Built Environment, 8*(3), 306–319. https://doi.org/10.1108/IJDRBE-10-2015-0051.
Vahanvati, M., & Rafliana, I. (2019). Reliability of build back better at enhancing resilience of communities. *International Journal of Disaster Resilience in the Built Environment, 10*(4), 208–221. https://doi.org/10.1108/IJDRBE-05-2019-0025.
Wanie, C. M., & Ndi, R. A. (2018). Governance issues constraining the deployment of flood resilience strategies in Maroua, Far North Region of Cameroon. *Disaster Prevention and Management, 27*(2), 175–192. https://doi.org/10.1108/DPM-12-2017-0300.

Building Energy Systems

► Building Energy: How Building Efficiency Can Be Improved in Government Facilities

Building Energy: How Building Efficiency Can Be Improved in Government Facilities

David Slim Zepeda Quintana, Hector Manuel Guzman Grijalva and Luis Eduardo Velazquez Contreras
Sustainability Graduate Program, University of Sonora, Hermosillo, México

Synonyms

Building energy systems; Built environment energy; Energy efficiency in buildings

Definition

Building Energy

Building energy is the energy consumption of all systems and subsystems necessary for the proper functioning of a building, such as lighting, heating, ventilation, and air conditioning (HVAC) systems. Building energy can also be defined as all those energy needs of a building. The improvement of the building energy efficiency is necessary for sustainable development due to its potential to minimize energy consumption and greenhouse gas emissions, as well as the reduction of energy-related costs.

Introduction

Rapid population growth has led to an increase in the built environment. This complex situation brings with it a series of challenges that threaten the possibility of achieving a more sustainable development. One of the most critical challenges is the increase in the global demand for energy from buildings. This situation has exerted pressure on organizations, companies, and governments around the world to redouble efforts in the search for strategies that allow a balance between energy generation and consumption in a more sustainable way.

The efficient use of energy in building construction and operation has become paramount by its association and implications in achieving environmental protection, to the extent it has been incorporated into Goal 7 of the United Nations Sustainable Development Goals. It is, however, a complicated and multifaceted matter involving a wide range of aspects that go from construction material extraction and fabrication and buildings' intended use to government regulations and financial costs related to efficiency improvement.

Energy efficiency plays an important role in the improvement of the efficiency of buildings; through it, the consumption of energy can be controlled, the energy-associated costs can be reduced, and the maintaining of a comfortable environment in buildings can be assured. Governments play a crucial role since they have the responsibility to lead efforts to foster good practices and set an example to society through the implementation and certification of energy efficiency systems in their buildings. In this chapter, a description of the crucial role that energy efficiency plays in the built environment is provided, followed by the identification of a framework of strategies to achieve efficiency in public and government buildings.

Energy Efficiency in the Built Environment

The built environment is responsible for around 40% of global energy consumption, of which the majority comes from residential and non-residential buildings. This energy consumption have a direct and indirect impact on the generation of greenhouse gas emissions, to such a degree that during 2020 alone, the operation of buildings was responsible for around 30% of total greenhouse gas emissions in the planet (United Nations Environment Program 2021). Innovation and the use of current technology can contribute to a more efficient energy consumption and a reduction in emissions; however, the built environment presents inherent challenges in the transition toward a more sustainable performance (Gillott and Spataru 2010).

The energy performance of buildings is complicated since it depends on multiple factors associated with weather conditions, the construction of the building and the thermal properties of the materials used, the occupants and their behavior, the systems necessary for their operation such as lighting, heating, ventilation and air conditioning (HVAC) systems, their performance, hours of use, among others (Fumo 2014; Zhao and Magoulès 2012). It is for this reason that the study and analysis of energy performance, as well as strategies that allow efficiency in buildings has increased. In fact, the search to improve energy efficiency in buildings is included in the Sustainable Development Goals of the 2030 Agenda of the United Nations, specifically, in the goal seven: ensure access to affordable, reliable, sustainable, and modern energy for all; and in goal eleven: make cities and human settlements inclusive, safe, resilient, and sustainable (United Nations 2015).

The government plays a crucial role in this energy transition of buildings. The government is responsible for designing policies that allow more efficient energy management; they can also offer efficiency incentives to builders and building occupants; can set standards (e.g., for appliances and HVAC); can cultivate specialized technical assistance to help state and local governments and private industries identify and implement energy efficiency policies and, probably most importantly, governments can act as leaders in energy efficiency through the promotion of good practices and efficiency certifications in public and government buildings (Doris et al. 2009).

Framework of Building Energy Strategies

One of the most widely used approaches in organizations and in public and government buildings to reduce energy consumption and minimize greenhouse gas emissions is the implementation of Energy Management Systems (EnMSs). The main objective of an EnMSs is to report in detail the energy performance of an organization and, consequently, use this information to increase knowledge about energy efficiency inside and outside organizations (Munguía et al. 2020). On the other hand, numerous national and international green building certification systems have been created and are already in common use (Al-Ghamdi and Bilec 2015).

Currently, LEED (Leadership in Energy and Environmental Design) certification has become the most famous building certification system and is applied worldwide (Amiri et al. 2019). This certification is used by developers, construction companies, and governments to obtain great benefits in their buildings. Some of the advantages of this certification are that it allows to guarantee that the buildings meet the requirements of environmental and economic performance oriented toward the occupants; they need less energy, have lower water consumption, and also significantly reduce operation and maintenance costs; It evaluates the energy behavior of the building, quantifying the amount of energy it needs for its operation and, in turn, the use of renewable energies to improve its efficiency, among others (USGBC 2021). This type of certification is particularly important for government buildings as it establishes leadership in energy efficiency and sustainable construction, it constantly validates achievements through an impartial, external review process, and contributes to the growing knowledge base about sustainable development.

A core component of any EnMSs and/or certification system is auditing the flow of energy throughout a process or system. An energy audit is a tool that helps organizations routinely assess, maintain, and identify opportunities to improve their energy efficiency. Energy audits are not only used in business or industrial settings, but are also relevant to understanding the energy efficiency of various buildings, including government and public buildings. These energy audit reports enable government building managers to understand building energy use, identify sources of energy waste, and explore opportunities to implement energy conservation measures (Ma et al. 2012). Energy audits often reveal several opportunities to improve energy performance despite the unique characteristics of each building.

Summary/Conclusion

Because the built environment is essential to the socioeconomic development of nations, improving its energy efficiency is important. Environmental problems and socioeconomic crises require intense research on energy efficiency and energy savings in buildings to reduce the consumption of fossil fuels and the CO_2 emissions that generate the greenhouse effect. "*Energy is the golden thread that connects economic growth, environmental health, social fairness and opportunity,*" said the former UN General Secretary Ban Ki-Moon

Although improving the efficiency of buildings is a great challenge, there are different strategies that contribute to their energy transition toward more sustainable patterns. EnMSs and green building certifications play a crucial role and have proven to be effective strategies applicable to any environment and context. Through them, government buildings will be able to improve their energy performance and will be able to exercise public leadership in energy efficiency that contributes to the creation of a more sustainable society.

Cross-References

- ▶ Resilient Rural Electrification for the Twenty-First Century
- ▶ Smart Grids to Lower Energy Usage and Carbon Emissions: Case Study Examples from Colombia and Turkey
- ▶ Toward a Sustainable City
- ▶ Toward Smart Public Lighting of Future Cities

References

Al-Ghamdi, S. G., & Bilec, M. M. (2015). Life-cycle thinking and the LEED rating system: Global perspective on building energy use and environmental impacts. *Environmental Science and Technology, 49*(7), 4048–4056. https://doi.org/10.1021/es505938u.

Amiri, A., Ottelin, J., & Sorvari, J. (2019). Are LEED-certified buildings energy-efficient in practice? *Sustainability (Switzerland), 11*(6). https://doi.org/10.3390/su11061672.

Doris, E., Cochran, J., & Vorum, M. (2009). *Energy efficiency policy in the United States: Overview of trends at different levels of government*. National Renewable Energy Laboratory – NREL, December, 63.

Fumo, N. (2014). A review on the basics of building energy estimation. *Renewable and Sustainable Energy Reviews, 31*, 53–60. https://doi.org/10.1016/j.rser.2013.11.040.

Gillott, M., & Spataru, C. (2010). Materials for energy efficiency and thermal comfort in the refurbishment of existing buildings. In *Materials for energy efficiency and thermal comfort in buildings* (pp. 649–680). https://doi.org/10.1533/9781845699277.3.649.

Ma, Z., Cooper, P., Daly, D., & Ledo, L. (2012). Existing building retrofits: Methodology and state-of-the-art. *Energy and Buildings, 55*, 889–902. https://doi.org/10.1016/j.enbuild.2012.08.018.

Munguia, N., Esquer, J., Guzman, H., Herrera, J., Gutierrez-Ruelas, J., & Velazquez, L. (2020). Energy efficiency in public buildings: A step toward the UN 2030 agenda for sustainable development. *Sustainability (Switzerland), 12*(3). https://doi.org/10.3390/su12031212.

United Nations Environment Programme. (2021). *2021 Global Status report for buildings and construction: Towards a Zero-emission, efficient and resilient buildings and construction sector*. Nairobi.

United Nations. (2015). *Transforming our world: The 2030 Agenda for Sustainable Development*. https://sdgs.un.org/publications/transforming-our-world-2030-agenda-sustainable-development-17981

USGBC. (2021). *LEED rating system*. https://www.usgbc.org/leed

Zhao, H. X., & Magoulès, F. (2012). A review on the prediction of building energy consumption. *Renewable and Sustainable Energy Reviews, 16*(6), 3586–3592. https://doi.org/10.1016/j.rser.2012.02.049.

Building Resilient Communities Over Time

Asma Mehan[1,2] and Sina Mostafavi[3]
[1]Leiden University, Leiden, Netherlands
[2]The Netherlands Institute for Advanced Studies, NIAS-KNAW, Amsterdam, Netherlands
[3]TU Delft, Delft, Netherlands

Conceptualizing the Resilient Communities

The term *resilience* (sometimes used interchangeably with *robustness) aims* to describe the ability of an ecological system to continue functioning amid and recover from a disturbance. Based on the 2030 United Nations Agenda for Sustainable Development Goals (SDGs), resiliencies (social, cultural, ecological, environmental, technical, or economical) are defined as the capacity of individuals, societies, communities, institutions, entities, and financial systems within a city to survive, adapt, and recover from the effects of chronic stresses and acute shocks promptly. This mainly reflects the progress toward achieving the Sustainable Development Goals (SDGs), especially SDG 11, which "make cities and human settlements inclusive, safe, resilient and sustainable." The emergency of the theme has been mentioned as part of the Venice Biennale's Italian Pavilion 2021, *Resilient Communities*, which exhibited the seriousness and urgency of the issue of climate change and the significant challenges (urban, productive, and agricultural systems) that architecture is called on to face.

Various academics and policymakers suggest the concept of resilience and environmental activists to empower the communities to respond effectively and positively to the changes, hazards, and risks on various scales. However, as Reyers et al. (2018) put it, "promoting resilience in limited scales or sectors such as households, communities, climate change or recovery from disasters will only suffice for the short term." In better words, resiliency cannot be achieved without a substantive reduction of the potential impact caused by disasters on population, society, culture, and economy. So, the concept of "community resilience" will help provide enough flexibility to cope with a wide range of risks and crises by underpinning the specified resilience where required (Berkes and Ross 2012). To respond to the growing uncertainties, major stresses, and shocks (such as environmental, political, and economic), the increasing complexity, and to enhance the adaptive capacity of the society, the general concept of resilience is continuously evolving to be able to appropriately respond to the societal, political, cultural, and environmental needs of the people and society (Bazazzadeh et al. 2021; Mehan 2017). In another definition, community resilience has been defined as the "capacity of a distinct community to absorb disturbance and reorganize while changing to retain key elements of structure and identity that preserve its distinctness" (Fleming and Ledogar 2008). So, in this definition, the maintenance of structure and identity is critical in resilient systems.

The concept of resilience has a long history in the local communities, which is embedded in their culture around shared values based on a strong and dedicated collaboration among diverse groups of the community and the various actors from different backgrounds. However, providing the holistic and multi-benefit response from this wide variety of actors with varying interests to a diverse, complex, and sometimes unpredictable range of changes is a significant challenge to overcome, but opportunities do co-exist (Kozlowski et al. 2020; Rahdari et al. 2019; Mehan and Soflaei 2017; Repellino et al. 2016). Norris et al. (2008) present a thorough review of "community resilience" definitions before proposing a comprehensive conceptualization that describes it as a "process linking a set of networked adaptive capacities to a positive trajectory of functioning and adaptation in constituent populations after a disturbance." Norris et al. (2008) make one key point: community resilience implies "networked adaptive capacities" from which the "community wellnesses" emerges. Therefore, "community resilience" is a collective endeavor and participatory process that will increase societies' adaptive capacities. Among the sets of adaptive capabilities highlighted by Norris et al. (2008), the role of "communal/collective narratives" deserves particular attention for further research. To foster the sense of belonging and identity that directly affects the "community resilience" (both in terms of response and recovery), "communal/collective narratives" provide a shared meaning of experience and social memory.

As Crane (2010) clarifies, the plurality and diversity of cultural behaviors within any system are essential to its resilience. Here, the aim is to deal with (and support) the diversity not just based on individual characteristics alone (such as culture, cultural values, language, customs, norms, etc.), but also from the support of more significant sociocultural factors (such as the national values and organizations) (Clauss-Ehlers 2010). Building upon this definition, it is important to take into consideration the elements of the resilient communities that remain stable despite adaptation or even transformation of other elements of that system which are inherently related to the resilience of communities to respond to change in an adaptive and transformative manner, without traversing the social thresholds that secure their cohesiveness and identity (Rotarangi and Stephenson 2014). In resilient communities, the collective narratives provide the system's stability to cope with unpredictable change. In contrast, the plurality and diversity of individual records provide the transformative agent that is critical for adaptation and innovation. Thus, it is critical to focus on the communities and the socio-cultural context to see their effects on resilient outcomes by considering the larger variables to help individuals overcome the obstacles they face (Clauss-Ehlers 2004). However, due to the lack of empirical methods and innovative assessment tools to capture the collective narratives of change, measuring the "community resilience remains a challenge" (Steiner and Markantoni 2014).

Building Resilient Communities Over Time

The resiliency of a community in a built environment needs to be analyzed, assessed, and refined when a radical change threatens the stability and sustainability of the context. Such radical shifts might range from economic crises, natural or man-made disasters, environmental hazards, pandemics, and changes to societal turmoil. While there is a consensus on the definition of "community resilience," there is less clarity on the building process of community resilience over time. These domains have been rather broad and lack the specificity required for implementation in a timely manner.

In the community setting, shared memory and oral history play a crucial role in facilitating communications, sharing information, learning from experiences, and transmitting and articulating the built environment. Because the new social means and digital media have changed the world, it is essential to focus on the latest digital technologies, smart apps, automation, and data-driven methods for urban research in a selected community to encourage self-resilience. To gather and analyze a large pool of data, crowdsourcing techniques, data journalism, and data visualization, are recommended to promote the dialogue between different parts of the society, which can make research more democratic by increasing the more significant number of public participation (Del Savio et al. 2016). These new digital methods combine the competence of journalists, planners, citizens, data analysts, and graphic designers to work collaboratively on the online tools: interactive charts, maps, timelines, and visualizations.

The other urban technique is "cultural mapping," also called "community-based mapping," "counter mapping," or "participatory mapping," to develop the intercultural dialogue. In practical terms, cultural mapping is defined as the "process of collecting, recording, analyzing and synthesizing information to describe the cultural resources, networks, links and patterns of usage of a given community or group" (Stewart 2007). As Lydon (2003) puts it, "ordinary people and communities can make maps to express the stories about their lives and home places. Community mapping is both the recovery and discovery of the connections and common ground that all communities share to enhance participatory learning, community empowerment, and sustainable development." In addition, the use of "artistic interventions," "participatory art" or "socially engaged art," "augmented reality (AR)," and "urban living lab" as an approach, not an output, can merge the terms both in theory and practice (Hoskyns 1999). So, organizing the experimental workshops and living labs, artistic representations, training schools, museum installations, cultural and recreational events, conferences, seminars, and festivals can enhance the knowledge sharing and building the community resilience.

It is important to note the role of time in creating and sustaining resilient communities which tie the past to the future visions. This includes a holistic and systematic approach of the community on how it uses material and energy resources or how a society educates the members over time to learn from the past and adapt to present and future opportunities and threads (Bears 2021; Rauf et al. 2021). To open up a conversation on community resilience, it is essential to use interdisciplinary and cross-disciplinarily methods, new media, civic engagement, and digital techniques to contribute to the co-creation of diverse and inclusive narratives of change.

Cross-References

- ▶ Circular Economy and the Water-Food Nexus
- ▶ Climate Resilience in Informal Settlements: The Role of Natural Infrastructure

References

Bazazzadeh, H., Nadolny, A., Mehan, A., & Safaei, S. S. H. (2021). the importance of flexibility in adaptive reuse of industrial heritage: Learning from Iranian cases. *International Journal of Conservation Science, 12*(1).

Brears R. C. (2021). Circular economy and the water-food nexus. In: R. Brears (Eds.), *The palgrave encyclopedia of urban and regional futures*. Cham.: Palgrave Macmillan. https://doi.org/10.1007/978-3-030-51812-7_98-1.

Berkes, F., & Ross, H. (2012). Community resilience: Toward an integrated approach. *Society & Natural Resources, 26*(1), 5–20. https://doi.org/10.1080/08941920.2012.736605.

Clauss-Ehlers, C. S. (2004). Re-inventing resilience: A model of "culturally-focused resilient adaptation". In C. S. Clauss-Ehlers & M. D. Weist (Eds.), *Community planning to foster resilience in children* (pp. 27–41). New York: Kluwer Academic.

Clauss-Ehlers, C. S. (2010). Cultural resilience. In C. S. Clauss-Ehlers (Ed.), *Encyclopedia of cross-cultural school psychology*. Boston: Springer.

Crane, T. A. (2010). Of models and meanings: Cultural resilience in social-ecological. *Ecology and Society, 15*(4).

Del Savio, L., Prainsack, B., & Buyx, A. (2016). Crowdsourcing the human gut. Is crowdsourcing also "Citizen science"? *Journal of Science Communication, 15*(03), A03.

Fleming, J., & Ledogar, R. J. (2008). Resilience, an evolving concept: A review of literature relevant to aboriginal research. *Pimatisiwin, 6*(2), 7–23.

Hoskyns, M. W. (1999). Operating the door for people's participation. In S. White (Ed.), *The art of facilitating participation: Releasing the power of grassroots communication* (p. 287). New Delhi: SAGE.

Kozlowski, M., Mehan, A., & Nawratek, K. (2020). *Kuala Lumpur: Community, infrastructure and urban inclusivity* (1st ed.). Routledge. https://doi.org/10.4324/9781315462417.

Lydon, M. (2003). Community mapping: The recovery (and discovery) of our common ground. *Geomatica, 57*(2), 131–143.

Mehan, A. (2017). An integrated model of achieving social sustainability in urban context through theory of affordance. *Procedia Engineering, 198*, 17–25. https://doi.org/10.1016/j.proeng.2017.07.070.

Mehan, A., & Soflaei, F. (2017). Social sustainability in urban context: Concepts, definitions, and principles. In *Architectural research addressing societal challenges* (pp. 293–300). CRC Press.

Norris, F. H., Stevens, S. P., Pfefferbaum, B., Wyche, K. F., & Pfefferbaum, R. L. (2008). Community resilience as a metaphor, theory, set of capacities, and strategy for disaster readiness. *American Journal of Community Psychology, 41*(1–2), 127–150. https://doi.org/10.1007/s10464-007-9156-6.

Rahdari, A., Mehan, A., & Malekpourasl, B. (2019). Sustainable real estate in the Middle East: Challenges and future trends. In T. Walker, C. Krosinsky, L. Hasan, & S. Kibsey (Eds.), *Sustainable real estate. Palgrave studies in sustainable business in association with future earth*. Cham: Palgrave Macmillan. https://doi.org/10.1007/978-3-319-94565-1_16.

Rauf, H. A., Wolff, E., & Hamel, P. (2021). Climate resilience in informal settlements: The role of natural infrastructure. In: R. Brears (Eds.), *The palgrave encyclopedia of urban and regional futures*. Cham.: Palgrave Macmillan. https://doi.org/10.1007/978-3-030-51812-7_39-1.

Repellino, M. P., Martini, L., & Mehan, A. (2016). Growing environment culture through urban design processes 城市设计促进环境文化. *NANFANG JIANZHU, 2*(2), 67–73.

Reyers, B., Folke, C., Moore, M.-L., Biggs, R., & Galaz, V. (2018). Social-ecological systems insights for navigating the dynamics of the anthropocene. *Annual Review of Environment and Resources, 43*(1), 267–289. https://doi.org/10.1146/annurev-environ-110615-085349.

Rotarangi, S. J., & Stephenson, J. (2014). Resilience pivots: Stability and identity in a social-ecological-cultural system. *Ecology and Society, 19*(1). https://doi.org/10.5751/ES-06262-190128.

Steiner, A., & Markantoni, M. (2014). Unpacking community resilience through capacity for change. *Community Development Journal, 49*(3), 407–425.

Stewart, S. (2007). *Cultural mapping toolkit*. Vancouver: Creative City Network of Canada and 2010 Legacies Now. http://www.creativecity.ca/database/files/6Editorial/City,CultureandSociety7(2016)1e7library/cultural_mapping_toolkit.pdf. p. 8.
United Nations Office for Disaster Risk Reduction – New York UNHQ Liaison Office. https://sdgs.un.org/goals/goal11.

Building-Integrated Agriculture

► Future Foods for Urban Food Production

Buildings

► Amsterdam's Pathway to Climate Neutrality: Creating an Enabling Environment

Built Environment

► Toward a Sustainable City

Built Environment Energy

► Building Energy: How Building Efficiency Can Be Improved in Government Facilities

Bushfire

► Managing the Risk of Wildfire Where Urban Meets the Natural Environment

Busing

► Pre-schoolers and Sustainable Urban Transport

Butterfly Gardening

► Butterfly Gardening in Colombo, Sri Lanka: Approach to Biodiversity Conservation, Monitoring, Education, and Awareness in Urbanizing Habitats

Butterfly Gardening in Colombo, Sri Lanka: Approach to Biodiversity Conservation, Monitoring, Education, and Awareness in Urbanizing Habitats

Hewa Thanthrige Ashan Randika Karunananda[1] and Medhisha Pasan Gunawardena[1,2]
[1]Biodiversity Educational Research Initiative, Colombo, Sri Lanka
[2]Faculty of Science, Horizon Campus, Malabe, Sri Lanka

Synonyms

Butterfly gardening, Urban parks, Urbanization, Conservation, Biodiversity, Capacity-building, Awareness and education

Definition

The rate of urbanization has increased over the past few decades causing a huge threat to biodiversity. It has caused a major threat to urban wildlife including butterflies. The presence of urban oases in the form of gardens is identified as a vital resource for butterflies to survive in an anthropogenic environment. Therefore, butterfly gardening is considered as a vital biodiversity conservation approach to protect the decreasing butterfly populations. The butterfly gardening concept is a simple way to attract butterflies to a garden, regardless the size of the land space available. By building a properly planned butterfly garden, a variety species could be attracted if the needs of adult butterflies as well as the larvae are provided. Colombo is considered as a city containing an extensive system of wetlands within the South Asian Region, many urban parks were developed close to these wetlands over the past decade. Few of these parks contain butterfly gardens, hence scientific studies have proved these spaces provide an ideal habitat to these species. Even though only a research had been conducted in a home garden in Colombo, a properly managed home garden provides a suitable habitat for urban butterflies. Organizations in the government sector, private, and other environment societies have taken measures to protect the declining butterfly fauna by promoting the establishment of butterfly gardens in public areas and by creating awareness among the public which has boosted the engagement in butterfly conservation in Sri Lanka. Further, concepts such as "butterfly houses" are becoming popular worldwide, these concepts could be implemented to further support the conservation of butterflies.

Introduction

The rate of urbanization has increased over the past few decades causing severe problems to nature. It creates a huge threat to the biodiversity (Savard et al. 2000; Bergerot et al. 2010; Fontaine et al. 2016). However, according to Knapp et al. 2008, cities are hotspots of species diversity especially in temperate countries like Central Europe and North America, providing a home for more species than in rural areas. Nevertheless, for species such as butterflies, birds, lichens, and mosses urbanization has created a negative impact reducing their numbers over the past decade. In tropical countries such as Sri Lanka common species could be observed in green spaces in cities in large numbers. However, rare and endemic species are more concentrated in forest areas and are lower in numbers. Due to the presence of both high species diversity as well as increased human impacts, conservation approaches must focus not only on natural habitats but also on urban areas (Cincotta et al. 2000; Liu et al. 2003). Butterflies are identified as a valuable model for such studies because of their importance to the ecosystem as pollinators, as a food source for the predator, and due to the short life cycle (Strong et al. 2000). Furthermore, past studies have proved that the butterfly diversity and abundance is impacted by the landscape size and the vegetation type (Rundlöf and Smith 2006). According to Toms et al. 2010, the presence of urban oases in the form of gardens is a vital resource for butterflies to survive in an anthropogenic environment. Therefore, butterfly gardening is a vital biodiversity conservation approach to protect the decreasing butterfly populations.

Butterfly Gardening

The butterfly gardening concept is a simple way to attract butterflies to a garden, regardless of the size of the land space available. This concept could be developed by simply identifying the specific requirements needed by the butterfly which changes throughout the life cycle. By establishing a properly planned butterfly garden, a variety of species could be attracted if the needs of adult butterflies as well as the larvae are provided (Daniels et al. 1990). For example, most adult butterflies feeds mainly on nectar. While, most of the larvae feed on leaves of plants (Poorten and Poorten 2016). Therefore, proper awareness and implementation of the butterfly gardening concept with the above-mentioned fundamental knowledge is vital in protecting the decreasing butterfly populations in urban areas.

Urban Parks and Butterfly Gardening in Colombo, Sri Lanka

The development of urban parks has become popular worldwide. These spaces provide several benefits in social, environmental, and economic aspects (Oshani and Wijethissa 2015). Colombo is considered as a city containing an extensive system of wetlands within the South Asian Region

(Hettiarachchi et al. 2014). Colombo was designated as a Ramsar wetland site during the 13th Ramsar Convention on Wetlands (COP 13), the only wetland city in the South Asian Region. Wetlands provide a range of services: management of floods, reduce peak air temperature, improve water quality, and provide an area for education and recreation (McInnes and Everard 2017). Several urban parks were established close to wetland ecosystems in Colombo, Sri Lanka, over the past decade. These include Diyatha Uyana, Beddagana Wetland Park, Diyasaru Wetland Park, and Weli Park. These parks provide a habitat for wild plants such as *Heliotropium indicum* and *Stachytarpheta jamaicensis*. These plants attract butterflies to these urban parks (Jayasinghe 2015; Jayasinghe et al. 2015; Poorten and Poorten 2016). More importantly, a butterfly garden was established in one of these parks: "Diyasaru Wetland Park," a 60 acre land space located in Thalawathugoda, opened in 2017. According to Oshani and Wijethissa (2015), with the increase in urbanization there is a tendency of individuals becoming closer to the natural environment. Hence, by building butterfly gardens in these parks it helps to provide awareness and education to the visitors. Furthermore, butterfly gardens have been developed in government institutions within Colombo: Lady Ridgeway Hospital, The National Museum (Wings of Beauty at Lyceum 2021), the Survey Department, and many other places. These places are visited by locals frequently, hence creates awareness on the importance of utilizing urban space for conservation. A nonstate organization in Sri Lanka plays a pivotal role in the butterfly conservation. They established the first butterfly garden in Colombo and supports awareness creation by publishing field guides such as Common Butterflies of Sri Lanka (Sri Lanka Butterfly Garden|Butterfly Conservation by Dilmah 2021). The Department of National Zoological Gardens has established a "closed" butterfly garden to educate the visitors (Butterfly Garden – Department of National Zoological Gardens 2021).

Figure 1 indicates the "open" and "closed" butterfly gardens located in Colombo, found in urban parks, private home gardens, and in government and private organizations. As depicted on the map there is a lack of studies conducted in butterfly gardens in private and government organizations. Furthermore, all the gardens are clustered in a small area, by conducting proper awareness, scientific research, and capacity building programs this problem could be minmized.

The Importance of Butterfly Gardens, Urban Parks, and Urban Home Gardens for Butterfly Conservation

Several scientific studies conducted in parks and gardens in Colombo, has demonstrated the importance of these patches for butterfly conservation (Table 1). A study conducted at "Diyasuru" Wetland Park has recorded 35 species of butterflies (Karunananda and Gunawardena 2019). According to Hewavithana et al. (2017), 42 butterfly species have been recorded at the Beddagana Wetland Park. A total of 75 species of butterflies had been recorded at the Bellanwilla Attidiya Wetland (Karunarathna et al. 2010). A relatively handful of studies had been conducted on the importance of properly managed home gardens in the conservation of butterflies. A study conducted at a home garden in Colombo has recorded 25 species during 4 months and 32 species in 8 months (Karunanada et al. 2018; Karunananda 2019). These butterfly species were recorded in very small land space within the highly urbanizing Colombo City. These scientific studies prove that butterfly gardens, urban parks, and urban home gardens play a vital role in the conservation of urban butterflies. More importantly, butterflies are considered a good indicator species (Poorten and Poorten 2016); they are very sensitive to the changes in the environment. Therefore, by establishing more butterfly gardens in urban areas and conducting continuous research, the impacts of urbanization can be monitored by the changes in the butterfly populations in these areas. Hence, proper awareness must be created to the public on the importance of the implementation and maintaining of such concepts.

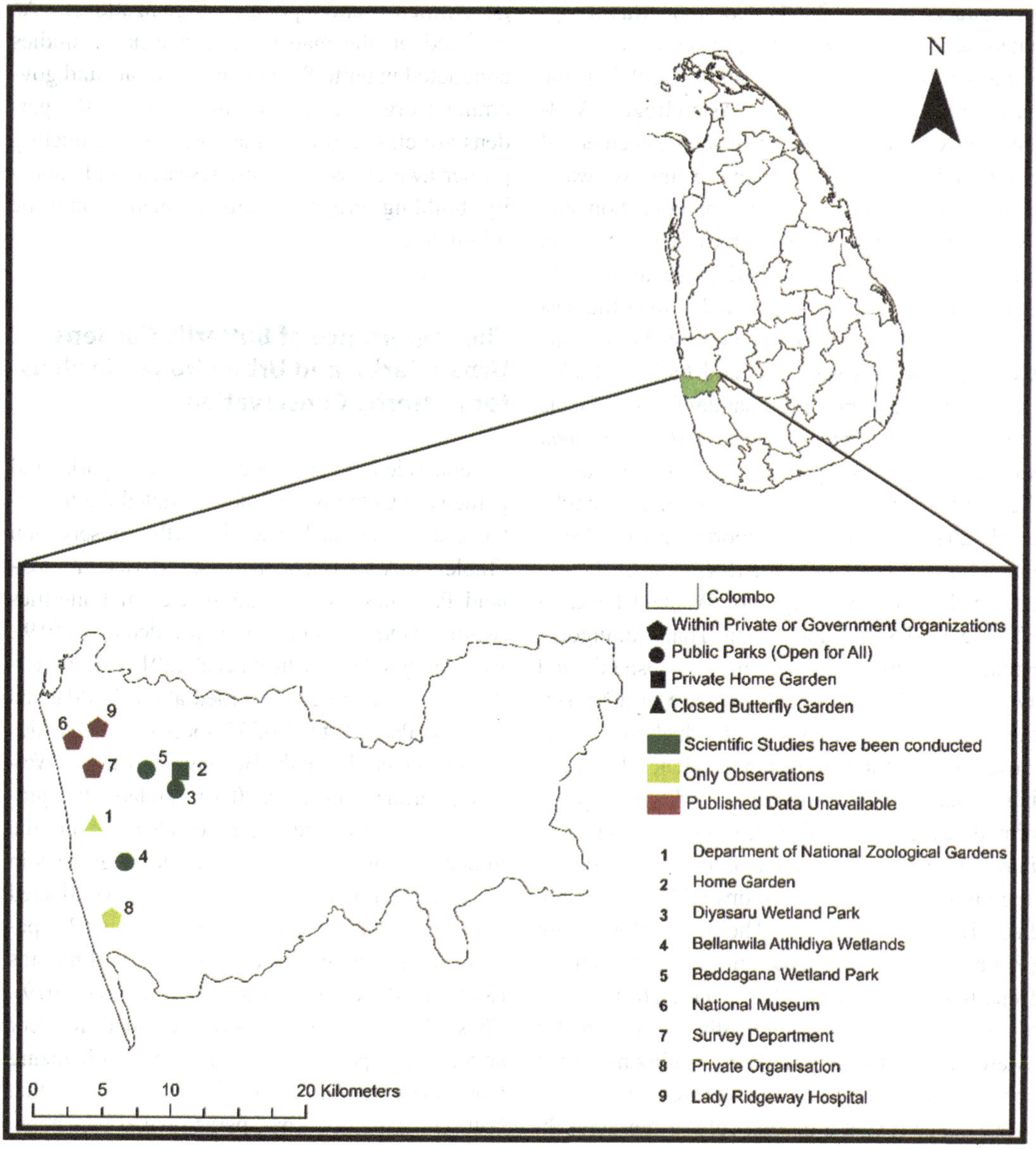

Butterfly Gardening in Colombo, Sri Lanka: Approach to Biodiversity Conservation, Monitoring, Education, and Awareness in Urbanizing Habitats, Fig. 1 Map of butterfly gardens in Colombo, Sri Lanka

Awareness Creation on Butterflies in Sri Lanka

The awareness creation on butterflies and butterfly gardening has increased over the past years in Sri Lanka. The Butterfly Conservation Society of Sri Lanka is the pioneer organization established in Sri Lanka for the conservation of butterflies. There are several initiatives taken to create awareness on butterflies. The annual Butterfly and Dragonfly Race is an event where several teams get together and record butterflies and dragonflies during a specific time interval. The event is mostly conducted in an urban park within Colombo. The winners and participants are provided with certificates that encourages them to engage more on these activities in the urban landscapes. The newsletter “Krumithuru”

Butterfly Gardening in Colombo, Sri Lanka: Approach to Biodiversity Conservation, Monitoring, Education, and Awareness in Urbanizing Habitats,

Table 1 Summary of research findings on butterfly species and endemics in Colombo and sub-urban areas in relation to the area of the island.

Location	No. of families	No. of species (% compared to Sri Lanka	No. of endemics	Area of land
Sri Lanka	6	248	31	65, 610 km^2
"Diyasaru" Wetland Park	5	35 (14%)	0	0.24 km^2
Beddagana Wetland Park	5	42 (17%)	1	0.18 km^2
Bellanwilla Attidiya Wetland Park	5	75 (30%)	1	3.72 km^2
Home garden	5	32 (13%)	0	0.003 km^2

(The Meaning – Friends of Insects) is a bi-annual publication which shares interesting articles and images on insects mainly on butterflies. These articles educate readers on butterfly gardening concepts and life cycles of butterflies. More importantly, the Butterfly Conservation Society maintains a webpage which includes photographs of all butterfly species and larval food plants which helps in developing butterfly gardens. This creates awareness to the public especially the youth, which support in butterfly conservation (BCSSL 2019). Due to these awareness creations (Fig. 2) youth entities have begun to collaborate with environmental organizations to take actions to conserve butterflies. One such example is "Heal the World of Butterflies," a conservation project by enhancing butterfly gardening initiative done by a youth organization in Sri Lanka. The butterfly gardening concept is becoming popular among urban schools and universities in Colombo as well. An International School within the Colombo urban limits has taken the effort to develop a butterfly garden within the premises to educate students (Wings of Beauty at Lyceum 2021). Many government and private universities has established environment societies to educate students on utilizing urban cityscapes for wildlife and nature conservation via butterfly gardening. These programs have also supported in building awareness on butterfly conservation. Figs. 3, 4, and 5 include some species recorded in Colombo; butterflies are colorful and attractive, these features can be utilized efficiently when building more awareness to the public. Currently, the butterfly gardens in Colombo are fragmented, these are not resilient and will be destroyed by small environmental changes. Proper awareness and capacity building activities must be conducted in government and private institutions and to the general public and provide government incentives to develop home gardens in the available green spaces, in the Colombo urban area as well as other cities. The process will develop a butterfly-friendly ecosystem which connects the fragmented patches and makes the system more resilient and further enhance the butterfly populations in the urban areas.

Status of Butterfly Gardening/Awareness in the World and Initiatives That Could Be Taken to Improve Butterfly Gardening in Colombo, Sri Lanka

Many initiatives had been taken worldwide for the protection of butterfly species. The development of urban parks has increased worldwide especially in India to support the conservation of butterflies (Kasambe 2016). The Changi Airport in Singapore is the first airport to have a "closed" butterfly garden designed where it provides a habitat for over 40 species of butterflies. The passengers visiting the airport are able to observe the breeding and feeding habits of these species (Changi Airport 2021). In Sri Lanka more butterfly gardens must be developed in regions where people visit more frequently to share more knowledge

Butterfly Gardening in Colombo, Sri Lanka: Approach to Biodiversity Conservation, Monitoring, Education, and Awareness in Urbanizing Habitats,

Fig. 2 Programs conducted by universities to build awareness on urban butterflies among students

efficiently. Over the past decade new concepts have developed to conserve butterflies out of which the establishment of butterfly houses are becoming popular in countries like Britain (Boppre and Vane-Wright 2012; National History Museum 2021). The temperature in Britain is not ideal for butterflies, however these spaces are maintained to suite their natural habitats. Usually, only 2–3% of eggs laid transform into adult butterflies in the wild, however in the National History Museum it is 70–80% indicating a higher conversion rate (National History Museum 2021). A government institution can encourage butterfly breeding in Sri Lanka, the emerging butterflies can then be released to the wild which helps in boosting populations. The Sensational Butterflies, an exhibition conducted by the National History Museum where over 500 species of butterflies from several countries could be seen. The pupae are collected from farms across Central America, South America, Africa, and Asia. Similar exhibitions can be conducted in Sri Lanka

A. Glassy Tiger (*Parantica aglea aglea)* **B.** Common Tiger (*Danaus genutia*) **C.** Tawny Coster (*Acraea terpsicore*) **D.** Great Eggfly (*Hypolimnas bolina bolina*) **E.** Blue Glassy Tiger (*Tirumala limniace exoticus*) **F.** Common Crow (*Euploea core asela*) **G.** Common Leopard (*Phalanta palantha*) **H.** Common Sailor (*Neptis hylas varmona*) **I.** Gray Pansy (*Junonia atlites atlites*) **J.** Baron (*Euthalia aconthea vasanta*) **K.** Common Palmfly (*Elymnias hypermnestra fraternal*) **L.** White Four-ring (*Ypthima ceylonica*) **M.** Chocolate Soldier (*Junonia iphita*) **N.** Blue Tiger (*Tirumala limniace exoticus*) **O.** Lime Blue (*Chilades lajus lajus)*

Butterfly Gardening in Colombo, Sri Lanka: Approach to Biodiversity Conservation, Monitoring, Education, and Awareness in Urbanizing Habitats, **Fig. 3** Butterflies found in Colombo. (Photographed by: D. Eranda Nipunika Mandawala, Medhisha Pasan Gunawardena & Ashan Karunananda)

which would attract more people, hence more knowledge could be shared with them. Even though the butterfly houses are upcoming there are many questions to be answered on whether it is ethical, has it produced new knowledge, and whether it will cause problems in gene transfer

A. Common Mormon (*Papilio polytes*) **B.** Common Mime (*Papilio clytia*) **C.** Tailed Jay (*Graphium agamemnon*) **D.** Lesser Grass Blue (Zizina Otis) **E.** Common Jay (*Graphium doson*) **F.** Crimson Rose (*Pachliopta hector*) **G.** Monkey Puzzle (Rathinda amor) **H.** Small Branded Swift (*Pelopidas mathias mathias*) **I.** Tropic Dart (*Potanthus confuscius*) **J.** Common Redeye (*Matapa aria*) **K.** Chesnut Bob (*Lambrix salsala luteipalpus*) **L.** Common Cerulean (*Jamides celeno*) **M.** Dark Cerulean (*Jamides bochus bochus*) **N.** Red Pierrot (*Talicada nyseus nyseus*) **O.** Tiny Grass Blue (*Zizula hylax hylax*)

Butterfly Gardening in Colombo, Sri Lanka: Approach to Biodiversity Conservation, Monitoring, Education, and Awareness in Urbanizing Habitats,

Fig. 4 Butterflies found in Colombo. (Photographed by: D. Eranda Nipunika Mandawala, Medhisha Pasan Gunawardena & Ashan Karunananda)

A. Common Jezebel (*Delias eucharis*) **B.** Common Grass Yellow (*Eurema hecabe simulate*) **C.** Lemon Emigrant (*Catopsilia pomona pomona)* **D.** Psyche (*Leptosia nina nina*) **E.** Mottled Emigrant (*Catopsilia pyranthe minna*)

Butterfly Gardening in Colombo, Sri Lanka: Approach to Biodiversity Conservation, Monitoring, Education, and Awareness in Urbanizing Habitats, **Fig. 5** Butterflies found in Colombo. (Photographed by: D. Eranda Nipunika Mandawala, Medhisha Pasan Gunawardena & Ashan Karunananda)

(Boppre and Vane-Wright 2012). The Kuala Lumpur Butterfly Park located in Malaysia is a home for over 120 butterflies species (Kuala Lumpur Butterfly Park Malaysia 2021). Hence, it has become a popular tourist attraction. In Sri Lanka, the development of such parks will encourage more locals as well as tourists to visit these areas which will help promote butterfly conservation measures.

Conclusion

The concept of urban parks and urban gardening has supported the biodiversity in Sri Lanka. These patches provide a feeding, residing, and breeding habitats for urban wildlife including butterflies. The butterfly gardening concept is gradually becoming popular among the public with awareness programs conducted by the private and government sector. Several organizations have taken the lead to create awareness on urban butterflies and butterfly gardening concept. However, butterfly gardens must be encouraged to be built in small scale in home gardens which will further support the conservation of butterflies. These actions will be important to support the declining butterfly fauna.

Cross-References

▶ Ecological Restoration
▶ Green Cities
▶ Nature-Based Solutions
▶ Wildlife Corridors

Acknowledgments The authors would like to thank Prof. S.J.B.A. Jayasekara – Vice Chancellor, Horizon Campus, Sri Lanka, Dr. K. Ruwan Perera – Deputy Vice Chancellor, Horizon Campus, Sri Lanka, Dr. Lekha Wanasekara – Former Dean, Faculty of Science, Horizon Campus, Sri Lanka, Mr. J.A.T.P. Gunawardena – Former Director, Agriculture Department of Sri Lanka, Dr. K.V.S.N. Bandara – Lecturer, Faculty of Zoology, University of Ruhuna, Mr. Himesh Dilruwan Jayasinghe – Founder, Butterfly Conservation Society of Sri Lanka, Mr. Amila Prasanna Sumanapala – Former President, Butterfly Conservation Society of Sri Lanka, Mr. D. Eranda Nipunika Mandawala – Senior Lecturer, Faculty of Science, Horizon Campus, Sri Lanka., Mr. Ravindu Anjana – Former Environment Management &

Sustainability Executive, Theme Resorts & Spas, Ms. Melisha Fernando – Former Secretary, Nature Beyond the Horizon – The Environment Society of Horizon Campus, Chairman – Sri Lanka Land Reclamation & Development Corporation, Ms. Sadhini Lokuliyana – University of the Visual and Performing Arts, Sri Lanka, and Staff Members of Sri Lanka Land Reclamation & Development Corporation and "Diyasaru" Wetland Park.

References

BCSSL. (2019). Butterfly conservation. Retrieved January 17, 2021, from Butterfly Conservation Society of Sri Lanka. http://bcssl.lk/.

Bergerot, B., Fontaine, B., Julliard, R., & Baguette, M. (2010). Landscape variables impact the structure and composition of butterfly assemblages along an urbanization gradient. *Landscape Ecology, 26*(1), 83–94. https://doi.org/10.1007/s10980-010-9537-3.

Boppre, M., & Vane-Wright, R. I. (2012). The butterfly house industry: Conservation risks and education opportunities. *Conservation and Society, 10*(3), 258–303.

Butterfly Garden – Department of National Zoological Gardens. (2021). Retrieved 17 January 2021, from http://nationalzoo.gov.lk/animal/butterfly-garden/

Changi Airport. (2021). Retrieved 30 January 2021, from https://www.changiairport.com/en/discover/attractions/butterfly-garden.html

Cincotta, R., Wisnewski, J., & Engelman, R. (2000). Human population in the biodiversity hotspots. *Nature, 404*(6781), 990–992. https://doi.org/10.1038/35010105.

Daniels, J. C., Schaefer, J., Huegel, C.N., & Mazzotti, F. J. (1990). *Butterfly gardening in Florida*. IFAS Extension.

Fontaine, B., Bergerot, B., Le Viol, I., & Julliard, R. (2016). Impact of urbanization and gardening practices on common butterfly communities in France. *Ecology and Evolution, 6*(22), 8174–8180. https://doi.org/10.1002/ece3.2526.

Hettiarachchi, M., Morrison, T. H., Wickramasinghe, D., Mapa, R., De Alwis, A., & McAlpine, C. (2014). The eco-social transformation of urban wetlands: A case study of Colombo, Sri Lanka. *Land and Urban Planning, 135*, 55–68.

Hewavithana, D., Peries, N., Weerakoon, D., & Wijesinghe, M. R. (2017). Establishing baseline information for dragonflies and butterflies of the newly established Beddagana Wetland Park. *The Sri Lanka Forestor, 38*(New Series), 47–60.

Jayasinghe, H. (2015). *Common butterflies of Sri Lanka*. Peliyagoda: Ceylon Tea Services PLC.

Jayasinghe, H. D., Rajapakse, S. S., & De Alwis, C. (2015). A pocket guide for butterflies of Sri Lanka. The Butterfly Conservation Society of Sri Lanka.

Karunananda, H. T. A. R. (2019). *Urban butterfly diversity, habitat relationships and threats in Colombo, Sri Lanka*. Unpublished Manuscript.

Karunananda, H. T. A. R., & Gunawardena, M. P. (2019). *Butterfly diversity, conservation, awareness and action at urban wetland park- Colombo, Sri Lanka*. In 6th Asian Lepidoptera conservation symposium. Zoological Survey of India, Kolkata, p. 19.

Karunananda, H. T. A. R., Gunawardena, M. P., & Prasadini, P. M. (2018). *Urban parks and urban gardening for sustainable development and conservation: A case study from Sri Lanka*. In Twenty-Third International Forestry and Environment Symposium. Waskaduwa: Department of Forestry and Environmental Science University of Sri Jayewardenepura, Sri Lanka, p. 97.

Karunarathna, D. M., Amarasinghe, A. A., Gabadage, D. E., Bahir, M. M., & Harding, L. E. (2010). Current status of faunal diversity in Bellanwila-Attidiya sanctuary, Colombo District – Sri Lanka. *Taprobanica, 2*(1), 48–63.

Kasambe, R. (2016). Butterfly gardening in India. https://doi.org/10.13140/RG.2.1.4934.6164.

Knapp, S., Kuhn, I., Mosbrugger, V., & Klotz, S. (2008). Do protected areas in urban and rural landscapes differ in species diversity? *Biodiversity and Conservation*, 1595–1612.

Kuala Lumpur Butterfly Park Malaysia. (2021). Retrieved 04 February 2021, from http://klbutterflypark.com/about.html

Liu, J., Daily, G., Ehrlich, P., & Luck, G. (2003). Effects of household dynamics on resource consumption and biodiversity. *Nature, 421*(6922), 530–533. https://doi.org/10.1038/nature01359.

McInnes, R., & Everard, M. (2017). Rapid assessment of wetland ecosystem services (RAWES): An example from Colombo, Sri Lanka. *Ecosystem Services, 25*, 89–105. https://doi.org/10.1016/j.ecoser.2017.03.024.

Oshani, P., & Wijethissa, K. (2015). Visitors' effective problems in urban park in Sri Lanka: special reference within Diyatha Uyana urban park Colombo. 3rd Research Symposium for Temporary Academic Staff-RSTAS 2015 (p. 31). The Research Center for Social Sciences, Faculty of Social Sciences, University of Kelaniya, Sri Lanka.

Rundlof, M., & Smith, H. (2006). The effect of organic farming on butterfly diversity depends on landscape context. *Journal of Applied Ecology, 43*(6), 1121–1127. https://doi.org/10.1111/j.1365-2664.2006.01233.x.

Savard, J., Clergeau, P., & Mennechez, G. (2000). Biodiversity concepts and urban ecosystems. *Landscape and Urban Planning, 48*(3–4), 131–142. https://doi.org/10.1016/s0169-2046(00)00037-2.

Sensational Butterflies: bringing the jungle to London| National History Museum. (2021). Retrieved 23 January 2021, from https://www.nhm.ac.uk/discover/sensational-butterflies-bringing-the-jungle-to-london.html

Sri Lanka Butterfly Garden|Butterfly Conservation by Dilmah. (2021). Retrieved 17 January 2021, from https://www.dilmahconservation.org/initiatives/biodiversity/butterfly-garden.html

Strong, A., Sherry, T., & Holmes, R. (2000). Bird predation on herbivorous insects: Indirect effects on sugar maple saplings. *Oecologia, 125*(3), 370–379. https://doi.org/10.1007/s004420000467.

Toms, M. P., Humphreys, L., & Kirkland, P. (2010). Monitoring butterflies within an urbanised landscape: The role of garden butterfly populations.

van der Poorten, G. M., & van der Poorten, N. E. (2016). *The butterfly Fauna of Sri Lanka.* Toronto: Lepodon Books.

Wings of Beauty at Lyceum. (2021). Retrieved 17 January 2021, from http://edu.dailymirror.lk/articles/news/829/Wings-of-Beauty-at-Lyceum

C

Capabilities

▶ Sustainability Competencies in Higher Education

Capacities

▶ Sustainability Competencies in Higher Education

Capacity-Building

▶ Butterfly Gardening in Colombo, Sri Lanka: Approach to Biodiversity Conservation, Monitoring, Education, and Awareness in Urbanizing Habitats

Carbon Capture and Storage

▶ Carbon Sequestration Through Building-Integrated Vegetation

Carbon Dioxide Removal

▶ Carbon Sequestration Through Building-Integrated Vegetation

Carbon Emissions

▶ Circular Economy Cities
▶ Resource Effectiveness in and Across Urban Systems

Carbon Neutral Adelaide

Roadmap or Roadblock to Sustainability?

Gerti Szili[1], Gareth Butler[2] and Grace Andrews[2]
[1]College of Humanities, Arts and Social Sciences, Flinders University, Adelaide, South Australia
[2]Masters Environmental Management, College of Humanities, Arts and Social Sciences, Flinders University, Adelaide, South Australia

Synonyms

Carbon neutrality; Urban climate governance; Urban decarbonization; Zero emissions

Definition

The carbon neutral city has many iterations but can be broadly perceived as a transformative roadmap to deeply decarbonize urban centers through the reformation of energy, transportation, built environment, and solid waste sectors while

R. C. Brears (ed.), *The Palgrave Encyclopedia of Urban and Regional Futures*,
https://doi.org/10.1007/978-3-030-87745-3

offsetting any remaining emissions through carbon sequestration projects and trading schemes. Carbon Neutral Adelaide is one example of an Australian Central Business District that has formulated and adopted a Carbon Neutral Action Plan in a bid to become a global leader in urban climate governance.

Introduction

Today, more than half of the world's population live in cities, and by 2050, this number is estimated to increase to 70% (United Nations 2019). In terms of their footprint, cities occupy only 2% of global land area; however, they contribute to more than 70% of global greenhouse gas emissions and consume 78% of global energy supplies (Davidson and Gleeson 2018; Hutyra et al. 2014; While and Whitehead 2013). In Australia, the percentage of the population living in low-to-medium housing density urban areas is more than 85% and growing, with per capita carbon emissions second among OECD countries reaching 16.8 t of CO_2 in 2013 (Newton et al. 2018; Trading Economics 2020). As one of the world's driest and warmest countries, Australia is particularly vulnerable to the effects of climate change (Iping et al. 2019), including declining annual rainfall, a rise in extreme weather events (i.e., tropical storms, droughts, fires, and flooding), extreme heat events, and ocean acidification (Newton 2009). While rural communities have already experienced a recent spate of such unprecedented weather events (see, for example, the 2019/2020 "Black Summer" bushfire season (Davey and Sarre 2020) and the 2021 "East Coast" floods (Deacon 2021)), Australian cites are also acutely affected by climate change due to their high-density populations and proximity to the coastline, increasing their vulnerability to extreme weather occurrences (Newton 2009; Newton et al. 2018). Moreover, as urban landscapes are disproportionately covered by concrete and other hard surfaces, and analogously reducing greenspace, they are more likely to experience localized warming due to the materials and structures used in urban developments that trap in heat from the sun (Iping et al. 2019). Thus, urban climate change impacts will only intensify as urban centers continue to expand and compete for open space (UN Habitat 2016). Considering these vulnerabilities and the number of Australian urban dwellers, urban climate mitigation and adaptation strategies must be prioritized by all tiers of Australian government, with the support from local communities and the business sector.

Urban Climate Governance

In the past, global climate governance has been highly centralized, relying on international conventions held by the United Nations Framework Convention for Climate Change (UNFCCC) and national commitments to global targets, such as the Paris Agreement, for reducing carbon emissions to a safe level (Gordon 2013). However, these attempts have often been impeded by politics and largely ineffective in solidarity (Gordon 2013; van der Heijden 2019). In Australia, for example, there has been a lag in uniform commitment to meet carbon reduction targets, especially at the federal level (Newton and Rogers 2020). Consequently, municipalities have been gaining recognition for the crucial role they can play in climate change responses and their potential to advance climate governance from the bottom-up (Bulkeley 2010; Bulkeley et al. 2012; While and Whitehead 2013). Indeed, research on municipal strategies, policies, and implementation measures for climate change governance has been growing since the 1990s (Broto and Bulkeley 2013). Bernstein and Hoffmann (2018) recognized that over this period, approaches to climate governance had shifted away from multilateral treaty-making to experimental governance where multiple actors are engaged in processes of trial and error for decarbonizing the urban environment. In the last decade, one such urban climate governance experiment that has been popularized as a roadmap to urban sustainability is the *carbon neutral city*.

The Carbon Neutral City: A Roadmap Towards Sustainability?

The carbon neutral city model has become one of the most aggressive strategies to reduce urban carbon emissions and assist in transitioning to a low-carbon economy (Laine et al. 2020). Rather than taking an incremental approach as often extolled by national governments and international agreements, its premise is to take a swifter transformative pathway to deeply decarbonize the city through the reformation of energy, transportation, built environment, and solid waste sectors while offsetting any remaining emissions through carbon sequestration projects and trading schemes (C.N.C. Alliance 2015; Newton and Rogers 2020). However, despite these laudable goals, questions remain as to how these can be realistically achieved, especially considering the multiple imaginings and pathways that cities are currently taking (Laine et al. 2020; Tozer and Klenk 2018). To collectivize these disparate visions and actions, the global Carbon Neutral Cities Alliance (CNCA) was formed to assist those cities at the vanguard of carbon neutral pledges reach their ambitious goals. Nevertheless, despite such global alliances, cities are still ultimately beholden to partnerships with and between their local constituents, business community and governments, and thus often face competing priorities in urban decision-making (Evans et al. 2013; Kenis and Lievens 2017; McGuirk et al. 2015). Indeed, the complexities of these partnerships have led to barriers in implementing carbon neutral initiatives that are truly transformative (Tozer and Klenk 2018). One of the newest members of the CNCA, the Central Business District (CBD) of Adelaide in South Australia appears to be experiencing some of these roadblocks toward a low-carbon urban transition.

Carbon Neutral Adelaide

Adelaide is the fifth largest capital city of South Australia, and the greater metropolitan area is home to 1.3 million people, accounting for 75% of the State's population (City of Adelaide 2016). Its Central Business District (CBD) is a one square mile area surrounded by parklands on all sides and is located just over 6 miles from the Gulf of St Vincent (Government of South Australia 2020). Given its greenbelt, the Adelaide CBD and its suburbs has long been cited as one of the cleanest and greenest cities in the world and was recently voted Australia's most livable and the world's third most livable city (Government of South Australia 2020; The Economist Intelligence Unit 2021). To maintain this reputation, in 2015, the State Premier at the time, Jay Weatherill, and the Lord Mayor of the City of Adelaide (Martin Haese) signed two international network agreements (i.e., the Compact of Mayors by Council and the Compact of States and Regions) and a Sector Agreement to herald Adelaide as the first carbon neutral city in the world (City of Adelaide 2015). To help guide this process, the Carbon Neutral Strategy 2015–2025 was developed by the local Council to provide the context of carbon neutrality and the outcomes it would provide to Adelaide (City of Adelaide 2015). Shortly after, the first Action Plan (i.e., Carbon Neutral Adelaide Action Plan 2016–2021) was released and included a total of 104 actions that would place the Adelaide CBD on a trajectory to carbon neutrality by 2025.

The Action Plan provided a roadmap that identified both the quantitative measures and political mechanisms needed for Adelaide to become carbon neutral. Quantitatively, emissions would need to decrease by 65% from 2007 levels across the energy, transport, and waste and water industries (City of Adelaide 2016). This would leave a remaining 421,174 kilotons CO_2e that would need to be offset to attain net-zero emissions (City of Adelaide 2016). To do this, the Action Plan allocated a set of strategies and actions to five pathways:

1. Energy-efficient built form
2. Zero-emission transport
3. Towards 100% renewable energy
4. Reduce emissions from waste and water
5. Offset carbon emissions (City of Adelaide 2016)

In terms of the political mechanisms, a set of actions were also established for how these pathways would be implemented and measured through good governance and partnerships with international networks, the community, and the private sector.

Carbon Neutral Adelaide's Roadblocks to Sustainability

Despite the clear set of strategies and actions towards carbon neutrality, and state government support, the ambitious goal to become the world's first carbon neutral city has been stymied. The roadblocks toward CBD sustainability have been multifarious, but include conflicting political views and agendas, mercurial engagements with some stakeholders, and an entrenched dependence on fossil fuels. Indeed, just a year after the adoption of the Action Plan, the local government Council quietly repealed its "world's first" ambition, in favor of becoming "*one of the world's first* carbon neutral cities by 2025" (Siebert 2017). Responding to the diminished target, then Lord Mayor, Martin Haese, claimed that while the end goal towards carbon neutrality remained, there was nothing stopping any other city from becoming carbon neutral tomorrow through carbon offset purchasing, adding that it was more important to be "authentically" carbon neutral than the "world's first" (Siebert 2017). Interestingly, the discreet amendment of its own target came 7 months after the Council voted against buying any carbon offsets until "after all cost-effective and reasonable measures to reduce city emissions had been exhausted" (Siebert 2016). The notions of "authenticity" and the process of achieving carbon neutrality through offsets provides an interesting juncture in imagining a carbon neutral CBD. To date, no contemporary city can realistically function without emitting carbon (Newton and Rogers 2020; Siebert 2017), especially given how deeply carbon is embedded in its built environment and transportation networks (Newton et al. 2019). Thus, to claim genuine carbon neutral status, a city must resort to carbon offset purchases through carbon sequestration projects or forest conversions "over and above their own emissions reduction efforts, to reduce net emissions to zero" (Siebert 2017; see also Laine et al. 2020; Ravetz et al. 2021). The Council's decision to vote against purchasing carbon offsets, which correspondingly flouted their own strategy towards carbon neutrality (City of Adelaide 2016), clearly highlights the contentious sociotechnical imaginings of carbon neutral cities (Kenis and Lievens 2017; Tozer and Klenk 2018) and calls in to question whether carbon neutral cities are indeed at all possible under the dominant neoliberal, pro-economic growth matrix in which cities remain embedded (Cugurullo 2018).

Nevertheless, despite the change in semantics, the Council remained committed to reduce carbon emissions, reporting a 15% reduction of greenhouse gases (GHGs) from the 2007 baseline in 2018 (City of Adelaide 2019). However, by 2019/2020, the Council had again questioned their target of global climate leadership with its Carbon Neutral Adelaide progress report (2019) and 2020–2024 Strategic Plan revealing that only 18 of the 104 actions identified to achieve carbon neutrality in the city had been completed. Eighteen actions were listed as "ongoing/ annual," 59 were "in progress," and 9 were "not started" (City of Adelaide 2019; Richards 2020). Perhaps even more compelling is that less than 22% of zero emissions transport targets had been met, 8% of energy efficient built form targets achieved, and none of the 8 actions for partnerships to achieve carbon neutrality had been completed to date (City of Adelaide 2019). Certainly, without significant progress in the two most energy intensive, resource-consuming, and GHG emitting urban sectors – transport and the built environment (Newton and Rogers 2020) – the Adelaide CBD's carbon neutral ambitions seem like a pipedream at best. In response to this poor progress, the current Lord Mayor Sandy Verschoor acknowledged that the City's biggest challenge is the rise in transport emissions, which increased by 27% from 2017 to 2018 (City of Adelaide 2019; Richards 2020). However, Verschoor was quick to defend the Council, emphasizing that the "Action Plan is in place" (Richards 2020), and strong public and private sector support remained. To Verschoor, and several other councilors, the primary roadblock to carbon neutrality was attributed to an absence of federal government level policies, which facilitate transitions to low emissions and zero emission vehicles (Richards 2020). While it would be convenient to consider this as political blame shifting, Verschoor and the Adelaide city councilors make a compelling point. In an assessment of the aspiring carbon neutral Finnish city, Vantaa, Laine et al. (2020, p. 8) noted similar roadblocks citing that most of the carbon neutral actions are in fact "outside of the City's jurisdiction, limiting its capability to ensure the achievement of a carbon-neutral city." Thus, the responsibility of a truly carbon-neutral city must consider the energy and waste flows outside of its borders and be supported by robust central/federal government policies (Dowling et al. 2018; Laine et al. 2020; Newton and Rogers 2020).

Moreover, partnerships to navigate the complex roadmap towards carbon neutrality are an equally essential waypoint. Indeed, they provide a vital platform to produce and share normative beliefs, encourage community action, knowledge sharing, and capacity building (Glasbergen et al. 2007; Wolfram et al. 2019). The Carbon Neutral Adelaide Action Plan explicitly recognizes the role of partnerships as they form a key part of the Plan's framework (Andrews 2020). However, as previously mentioned, none of these partnership actions had yet materialized. According to a series of interviews held with public and private stakeholders in 2020 (Andrews 2020), the roadblocks to partnership materialization were multifaceted. First, the relationship between the local council and state government remained fractured and both levels of government remained locked-in to carbon. For example, they both supported a "drive into the city month" initiative and continue to champion fossil fuel companies such as Santos (Andrews 2020). Second, while both the council and partners in the business sector and civil society were committed to carbon neutrality, these groups remained mostly siloed. Furthermore, according to several respondents, the Adelaide community outside of these networks remain unaware of the carbon neutral ambitions of the CBD, highlighting a lack of network building which is key to enabling long-term, effective partnerships for climate governance (Bernstein and Hoffmann 2018). Thus, according to both public and private stakeholders, the partnership roadblocks to carbon neutrality has been a result of a lack of collaboration, dialogue, and network building.

Conclusion: Carbon Neutral Adelaide – Roadworks in Urban Sustainability

Given the increasing urbanization of global populations, and the attendant role that cities play in facilitating anthropogenically induced climate change, urban centers must reimagine their identities and functioning to pivot towards more sustainable futures. Australian cities should especially heed this call considering they are among the highest GHG emitters in the world. Indeed, the potential for cities to act as catalysts for genuine climate action has gained traction over the last few decades (Bulkeley 2013), recently amplified by an explicitly urban-centered United Nation's Sustainable Development Goal, SDG 11 to "make cities inclusive, safe, resilient and sustainable." In response to this call, cities across the world have started to experiment with climate governance where multiple actors are involved in decarbonizing the environment and at various scales. One such experiment has been the carbon neutral city. While disparate understandings of how a genuinely carbon neutral city may look and operate remain (Laine et al. 2020; Newton and Rogers 2020), their popularity as a viable roadmap to urban sustainability continues to grow. Australia's fifth most populous city, Adelaide, is one example where the laudable goal to become carbon neutral has been adopted by the CBD local government. Initially set out to achieve all 104 targets by 2025, a mere 10 years since the Action Plan's inception, this case study revealed several roadblocks in the local government's attempts to become global climate leaders. Conflicting political views and agendas, sporadic engagement with various stakeholders, and a deeply rooted dependence on fossil fuels added great complexity to governing their transition to carbon neutrality, to the point where climate leadership was almost abandoned. This case study investigated the City of Adelaide's minimal progress towards a carbon neutral goal not to criticize its failures but to reignite its initial enthusiasm for reducing carbon emissions and provide lessons for other cities striving for similar goals. In addition to more radical changes to the built infrastructure and attendant sociocultural imaginings, a departure from "siloed thinking" (Newton and Rogers 2020) with a greater focus on partnerships and network governance may provide a more straightforward roadmap to successfully decarbonize the Adelaide CBD's pathways to carbon neutrality. If successful, Adelaide may well become a city that inspires other municipal governments to take similar deep decarbonization reform, ultimately mitigating the impacts of anthropogenically induced climate change.

Cross-References

- ▶ Amsterdam's Pathway to Climate Neutrality: Creating an Enabling Environment
- ▶ Circular Economy Cities
- ▶ Networking Collaborative Communities for Climate-Resilient Cities
- ▶ Sustainability Transition and Climate Change Adaption of Logistics
- ▶ Sustainable Development Goals from an Urban Perspective
- ▶ Urban Climate Resilience

References

Alliance, C. N. C. (2015). *Framework for long-term deep carbon reduction planning*. CNCA.

Andrews, G. (2020). *Governing pathways to decarbonization: A case study of carbon neutral Adelaide partnerships*. Master of environmental management thesis, College of Humanities, Arts and Social Sciences, Flinders University.

Bernstein, S., & Hoffmann, M. (2018). The politics of decarbonization and the catalytic impact of subnational climate experiments. *Policy Sciences, 51*(2), 189–211.

Broto, V. C., & Bulkeley, H. (2013). A survey of urban climate change experiments in 100 cities. *Global Environmental Change, 23*(1), 92–102.

Bulkeley, H. (2010). Cities and the governing of climate change. *Annual Review of Environment and Resources, 35*, 229–253.

Bulkeley, H. (2013). *Cities and climate change*. Routledge.

Bulkeley, H., Broto, V. C., & Edwards, G. (2012). Bringing climate change to the city: Towards low carbon urbanism? *Local Environment, 17*(5), 545–551.

City of Adelaide. (2015). *Carbon neutral strategy 2015–2025*. Adelaide City Council.

City of Adelaide. (2016). *Carbon neutral Adelaide action plan 2016–2021*. Adelaide City Council & Government of South Australia.

City of Adelaide. (2019). *Carbon neutral adelaide status report July 2019*. City of Adelaide.

Cugurullo, F. (2018). Exposing smart cities and eco-cities: Frankenstein urbanism and the sustainability challenges of the experimental city. *Environment and Planning A: Economy and Space, 50*(1), 73–92.

Davey, S. M., & Sarre, A. (2020). Editorial: The 2019/20 Black Summer bushfires. *Australian Forestry, 83*(2), 47–51.

Davidson, K., & Gleeson, B. (2018). New socio-ecological imperatives for cities: Possibilities and dilemmas for Australian metropolitan governance. *Urban Policy and Research, 36*, 230–241.

Deacon, B. (2021). NSW floods break 120-year-old rain records during march rain event, BOM says. *ABC News*. Available online: https://www.abc.net.au/news/2021-04-20/nsw-floods-break-120-year-old-rain-records/100079400. Accessed 18 June 2021.

Dowling, R., McGuirk, P., & Maalsen, S. (2018). Multiscalar governance of urban energy transitions in Australia: The cases of Sydney and Melbourne. *Energy Research & Social Science, 44*, 260–267.

Economist Intelligence Unit. (2021). Global Liveability index, 2021. Available online: https://www.eiu.com/n/campaigns/global-liveability-index-2021/. Accessed 20 July 2021.

Evans, B., Joas, M., Sundback, S., & Theobald, K. (2013). *Governing sustainable cities*. Routledge.

Glasbergen, P., Biermann, F., & Mol, A. P. (Eds.). (2007). *Partnerships, governance and sustainable development:*

Reflections on theory and practice. Edward Elgar Publishing.
Gordon, D. J. (2013). Between local innovation and global impact: Cities, networks, and the governance of climate change. *Canadian Foreign Policy Journal, 19*(3), 288–307.
Government of South Australia. (2020). *Living in South Australia*. Available online: https://www.sa.gov.au/topics/about-sa/living-in-sa. Accessed 22 July 2021.
Hutyra, L. R., Duren, R., Gurney, K. R., Grimm, N., Kort, E. A., Larson, E., & Shrestha, G. (2014). Urbanization and the carbon cycle: Current capabilities and research outlook from the natural sciences perspective. *Earth's Future, 2*(10), 473–495.
Iping, A., Kidston-Lattari, J., Simpson-Young, A., Duncan, E., & McManus, P. (2019). (Re) presenting urban heat islands in Australian cities: A study of media reporting and implications for urban heat and climate change debates. *Urban Climate, 27*, 420–429.
Kenis, A., & Lievens, M. (2017). Imagining the carbon neutral city: The (post) politics of time and space. *Environment and Planning A: Economy and Space, 49*(8), 1762–1778.
Laine, J., Heinonen, J., & Junnila, S. (2020). Pathways to carbon-neutral cities prior to a national policy. *Sustainability, 12*(6), 2445.
McGuirk, P., Dowling, R., Brennan, C., & Bulkeley, H. (2015). Urban carbon governance experiments: The role of Australian local governments. *Geographical Research, 53*(1), 39–52.
Newton, G. (2009). Australia's environmental climate change challenge: Overview with reference to water resources. *Australasian Journal of Environmental Management, 16*(3), 130–139.
Newton, P. W., & Rogers, B. C. (2020). Transforming built environments: Towards carbon neutral and blue-green cities. *Sustainability, 12*(11), 4745.
Newton, P., Bertram, N., Handmer, J., Tapper, N., Thornton, R., & Whetton, P. (2018). *Australian cities and the governance of climate change*. Australia's Metropolitan Imperative: An Agenda for Governance Reform, CSIRO Publishing, Clayton, pp 193–209.
Newton, P., Prasad, D., Sproul, A., & White, S. (Eds.). (2019). *Decarbonizing the built environment: Charting the transition*. Springer.
Ravetz, J., Neuvonen, A., & Mäntysalo, R. (2021). The new normative: Synergistic scenario planning for carbon-neutral cities and regions. *Regional Studies, 55*(1), 150–163.
Richards, S. (2020). City council gives up carbon neutral leadership goal. *In daily*. Available online: https://indaily.com.au/news/local/2020/03/06/city-council-gives-up-carbon-neutral-leadership-goal/. Accessed 20 June 2021.
Siebert, B. (2016). City council kicks carbon neutral own goal. *In daily*. Available online: https://indaily.com.au/news/local/2016/09/28/city-council-kicks-carbon-neutral-own-goal/. Accessed 20 June 2021.
Siebert, B. (2017). Council quietly dumps "world's first" carbon neutral target. *In daily*. Available online: https://indaily.com.au/news/local/2017/04/27/council-quietly-dumps-worlds-first-carbon-neutral-target/. Accessed on 21 Apr 2020.
Tozer, L., & Klenk, N. (2018). Discourses of carbon neutrality and imaginaries of urban futures. *Energy Research & Social Science, 35*, 174–181.
Trading Economics. (2020). *Australia- urban population (% of total)*. Available online: https://tradingeconomics.com/australia/urban-population-percent-of-total-wb-data.html. Accessed 15 Apr 2020.
UN HABITAT. (2016). *Urbanization and development: Emerging futures*. World Cities Report 2016.
United Nations, Department of Economic and Social Affairs, Population Division. (2019). *World urbanization prospects: The 2018 revision (ST/ESA/SER.A/420)*. United Nations.
Van der Heijden, J. (2019). Studying urban climate governance: Where to begin, what to look for, and how to make a meaningful contribution to scholarship and practice. *Earth System Governance, 1*, 100005.
While, A., & Whitehead, M. (2013). Cities, urbanisation and climate change. *Urban Studies, 50*(7), 1325–1331.
Wolfram, M., Borgström, S., & Farrelly, M. (2019). Urban transformative capacity: From concept to practice. *Ambio, 48*(5), 437–448.

Carbon Neutrality

► Carbon Neutral Adelaide
► Policy and Practices of Nature-Based Solutions to Build Resilience in Seoul, Korea

Carbon Sequestration Through Building-Integrated Vegetation

Kamiya Varshney, Maibritt Pedersen Zari and Nilesh Bakshi
School of Architecture, Victoria University of Wellington, Wellington, New Zealand

Synonyms

Atmospheric carbon dioxide reduction; Carbon capture and storage; Carbon dioxide removal; Carbon sink; Carbon uptake

C

Definition

Urban vegetation absorbs carbon dioxide from the atmosphere through photosynthesis and stores it in biomass of trunk, branches, foliage, stem, roots, and soils (Lorenz and Lal 2009). Out of which, a fraction of carbon is returned to the atmosphere through respiration, and some amount of carbon is stored in vegetation.

Carbon sequestration is the process of actively removing carbon dioxide from the atmosphere, often but not exclusively through the growth of vegetation, that is, when carbon absorption from the atmosphere (input) is greater than carbon released (output) to the atmosphere over a specified time period and within a site or a region (Fig. 1). Carbon storage is the process of storing the sequestered carbon. It is one of the approaches for reducing the amount of carbon dioxide in the atmosphere and thus a potential strategy to address global climate change.

This chapter specifically defines how the built environment can augment the natural process of carbon sequestration by integrating vegetation into architecture. Associated technologies or strategies are often described as achieving *net-positive carbon* in the building industry, where carbon emissions are offset by carbon sequestration or storage (e.g., in timber construction) and/or the use of energy derived from renewable sources such as wind, solar, or hydro (Renger et al. 2015).

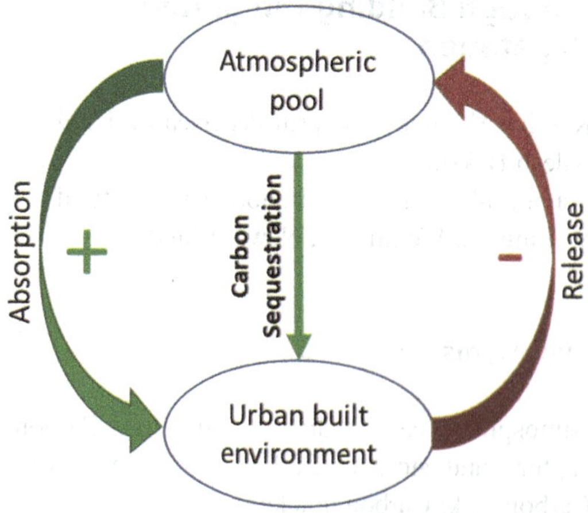

Carbon Sequestration Through Building-Integrated Vegetation, Fig. 1 Carbon sequestration process

Introduction

Urban Ecosystems and Climate Change

An inherent link has been established between climate change and human settlements across the globe (Churkina 2012; IPCC 2014; Revi et al. 2014). This link has been observed by various measurable criteria and phenomena, such as greenhouse gas (GHG) emissions and adverse effects on various landmasses and habitats due to increased erosion, landslips, flooding, and coastal inundation (IPCC 2018). As a consequence, this climate change has adversely affected the built environments within various regions, often increasing air conditioning loads due to overheating.

There is an increasing awareness amongst designers to consider the implications of design interventions on local climates and ecosystems. The Intergovernmental Panel on Climate Change (IPCC) has identified four strategies to address climate change. These include mitigation, adaptation, remedial measures, and carbon dioxide removal (IPCC 2018). This chapter focuses on carbon dioxide removal. According to IPCC (2018), carbon dioxide removal strategies aim to reduce existing carbon dioxide concentrations in the atmosphere.

The building industry has explored or adopted many mitigations, adaptation, and remedial strategies to address climate change. This has been achieved through built environment design which focuses on reducing emissions and encouraging optimized energy usage. However, little attention has been given to carbon dioxide removal from the atmosphere (sequestration) or carbon storage through building design (Kuittinen et al. 2016; Lal 2012). Despite considerable policies and guidelines highlighting economic, biodiversity, and human well-being benefits of providing vegetation in urban areas, uptake is still relatively low (Kabisch et al. 2016; Kondo et al. 2018; Lee and Maheswaran 2010; van den Bosch and Sang 2017). Lal (2012) identified that this lack of adoption is partially due to limited socioeconomic and scientific knowledge about the acceptance and utilization of carbon storage and sequestration. Moreover, region-specific climate change impact pathways through carbon dioxide removal or sequestration are not well understood (Honegger et al. 2018). This chapter focuses on how building design can facilitate the natural process of carbon sequestration by translating ecological knowledge into an architectural context.

Background

The building industry is observing a significant change in its design targets, moving from sustainable to regenerative goals, that is, from net-zero to net-positive and from carbon emission reduction to carbon-positive buildings (Birkeland 2008; Cole 2012; Du Plessis 2012; Pedersen Zari 2018). In a regenerative development, human aspirations are reconnected with the evolution of natural systems for co-evolution (Mang 2016). Regenerative developments have the potential to significantly contribute to ecological health and human well-being rather than just reducing negative impacts on the surrounding ecosystem.

Humans are dependent on nature for survival and well-being (Bolund and Hunhammar 1999; Gómez-Baggethun et al. 2013). The benefits humans derive from nature, either directly or indirectly, are known as ecosystem services (Costanza et al. 1997). Carbon sequestration is one of the ecosystem services that help human beings and ecosystems by reducing atmospheric carbon naturally, and as a consequence, regulating the climate.

The natural process of long-term sequestration and storage of carbon from the atmospheric pools is predominantly through oceans, vegetation, soil, and geologic formations (Lal 2008). Rapidly increasing land-based vegetation (forests) is perhaps the most viable and effective way to increase carbon sequestration. Despite this, the potential for the built environment to make a valuable contribution cannot be ignored. There are two main categories of anthropogenically driven carbon sequestration process: (1) biotic sequestration (enhancing existing natural processes, such as increasing uptake of carbon through biomass of stems, branches, roots, and soil, that remove the carbon from the atmosphere), or (2) abiotic sequestration (engineering process to capture carbon directly from the air and store it elsewhere; such as deep injection in the ocean, geologic injection, and mineral carbonation) (IPCC 2018; Lal 2008).

The goal is to create urban centers that can also be a medium of carbon sequestration and sinking through increased biomass. Moreover, monitoring and analyzing carbon emissions and sequestration at design, construction, maintenance, and deconstruction phases could contribute to carbon-positive buildings and regenerative goals.

Carbon Sequestration in the Urban Built Environment

There are many ways that buildings and urban design can mitigate carbon emissions. However, the built environment has the potential to sequester carbon more effectively or expand the storage of carbon in urban settings. This can be achieved by increasing the area and carbon density of vegetation. For example:

- By strategically increasing and retaining existing vegetation in the built environment (Churkina 2012; Lal 2012; Reichle et al. 1999; Renger et al. 2015)
- By retrofitting existing buildings to provide a substrate or surface for vegetation (Birkeland 2012; Girardet 2015)
- By strategically improving carbon sequestration and storage through, for example, improving soil quality, increasing the depth and density of carbon in the soil, and increasing the longevity of biomass carbon, either around or on buildings (Reichle et al. 1999).

Typically, in a building or a site, carbon is sequestered and/or stored in four ways: introducing urban trees and shrubs; lawns and grassed areas; soil; and building materials. Each strategy is discussed in detail below.

1. **Urban trees and shrubs**

Carbon is stored in the aboveground biomass of stems, branches, and leaves; and belowground biomass of roots and soil. Trees and shrubs capture, store, and release carbon as a part of their biophysical processes. Leaves absorb carbon dioxide during the photosynthesis process and, utilizing the light energy and water, convert it into oxygen, carbohydrates, and water (Aguaron and McPherson 2012). Trees and shrubs obtain energy from carbohydrates to grow, maintain, and respire. Plants return some stored carbon to the atmosphere through respiration. Stored carbon is released from trees (or timber, if they have been cut down) during the decomposition processes. Therefore, the net storage of carbon is a result of the balance between carbon capture by photosynthesis and carbon release by respiration, deforestation, and decomposition (Lorenz and Lal 2009) (Fig. 2).

2. **Lawns and grassed areas**

Carbon is stored in the lawn and/or grassed areas through the development of roots, thatch, and shoots. Lawns and/or grassed urban landscapes capture and store carbon through photosynthesis and release through respiration and decomposition processes. This stored carbon is also released during lawn management processes such as clipping/mowing, irrigation, or N-fertilizers (Guertal 2012). Therefore, the net carbon sequestered in a lawn or grassed landscape is evaluated based on the carbon capture by photosynthesis and carbon release by respiration, decomposition, mowing, irrigation, and fertilization (Fig. 2). The amount of energy released by different lawn management practices, from manufacturing to lawn care, is known as hidden carbon cost (HCC) (Zirkle et al. 2011). Lawn management practices involving a higher level of management (and, therefore, a higher HCC) can negate the net carbon sequestration rates (Zirkle et al. 2011). In contrast, a lawn requiring less HCC to maintain has a higher rate of carbon

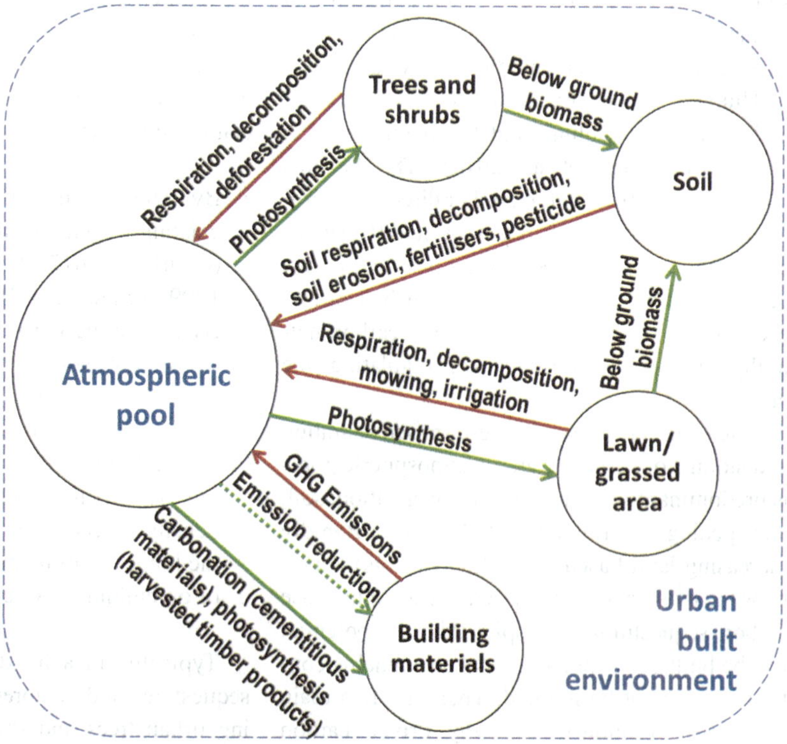

Carbon Sequestration Through Building-Integrated Vegetation, Fig. 2 Interchange of carbon from atmospheric pools and the urban built environment

sequestration (Kuittinen et al. 2016; Selhorst and Lal 2013; Zirkle et al. 2011, 2012).

3. **Soil**

Carbon captured through trees, shrubs, or lawns is stored in the soil as soil organic matter (Fig. 2). Soil organic carbon (contained in soil organic matter) is formed by plants, animals, and microorganisms in all stages of growth, adding carbon to the soil (Lal 2004). Soil organic carbon is released into the atmosphere through respiration and decomposition by soil organisms (Chapin et al. 2006). Moreover, this stored carbon can be lost due to soil management practices (e.g., use of fertilizers, pesticides).

4. **Building materials***

Carbon can be stored in certain building materials through:

- Photosynthesis (e.g., bio-based building materials, such as harvested timber products, bamboo, and straw (Vallas and Courard 2017))
- Carbonation mechanisms: a process in which carbon dioxide reacts with calcium oxide of cement, resulting in the formation of calcium carbonates (Kuittinen et al. 2016) (e.g., through cementitious materials, such as concrete and mortar (Pacheco-Torgal et al. 2018).
- Both carbonation and photosynthesis (e.g., using hempcrete (Arehart et al. 2021).
- Bio-mineralization: a process by which living organisms produce minerals that biologically convert the sequestered carbon into carbonates (Heveran et al. 2020; Pedersen Zari 2017) (e.g., employing cyanobacteria to produce building materials) (Reddy and Joshi 2018)

However, carbon emissions during harvesting, processing, manufacturing, and/or transportation must be considered while accounting for sequestered carbon.

Carbon storage through building materials is important and should be encouraged, but it is not the same as carbon sequestration, which in this case is using the built environment to actively pull carbon out of the atmosphere.

*[*This sub-section is included here to show the potential of building materials for carbon storage through buildings. However, it is not in the scope of this chapter.]*

C

Carbon Sequestration Strategies Through Building Design

Vegetation can be integrated into a building or a site for sequestering carbon through strategies, such as green roofs, green walls, internal courtyards/atriums, in-ground landscaping, urban agriculture, and living machines (Table 1). Moreover, strategies are required to protect, restore, and increase the carbon sequestration in above- and belowground biomass and in the soil pool. There are significant benefits of increasing the storage of carbon in vegetation and soils, such as improved soil and water quality, decreased nutrient loss, reduced soil erosion, better wildlife habitats, increased water conservation, and more biomass products (Smit 2014). Therefore, an integrated system approach is required to execute strategies to sequester carbon through urban vegetation and adopt management processes to accelerate and increase current carbon sequestration capacity in the built environment.

Other techniques, such as the use of carbon-based materials (for carbon storage), do exist, but their efficacy is in question.

Carbon Sequestration Dynamics

Carbon sequestration through building-integrated vegetation can be improved by increasing the amount of carbon stored in living plant matter, roots, and soil. The phenological stage of the plant and the growing conditions (light availability, temperature, moisture, soil type, and fertility levels) largely affect the net amount of carbon sequestration (Zirkle et al. 2012). Other parameters that influence the rate of carbon sequestration in the urban built environment are:

Carbon Sequestration Through Building-Integrated Vegetation, Table 1 Strategies for carbon sequestration through building-integrated vegetation

Green roofs  Source: Newtben33, https://commons.wikimedia.org/wiki/File:Green_Roof_Shed.JPG, CC BY-SA 3.0	**Green walls**  Source: Thomas Claveirole, https://www.flickr.com/photos/80318369@N00/277493405, CC BY-SA 2.0
Internal courtyards/atriums  Source: jann_on, https://www.flickr.com/photos/48018466@N00/4240227903, CC BY-NC-SA 2.0	**In-ground landscaping** 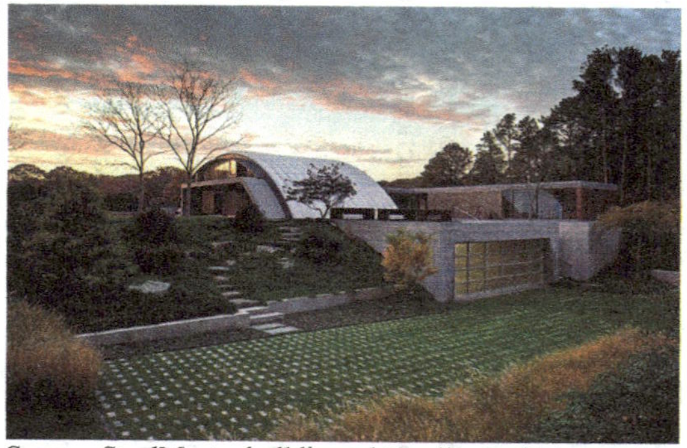 Source: SteelMaster buildings, https://www.flickr.com/photos/35055618@N07/17298528912, CC BY-SA 2.0

Urban agriculture	**Living machines**
 Source: Sustainable sanitation Alliance (SuSanA), https://www.flickr.com/photos/gtzecosan/3009777019/, CC BY 2.0	 Source: Living Machine Systems, L3C, https://commons.wikimedia.org/wiki/File:Living_Machine_at_Port_of_Portland-interior.jpg, CC BY-SA 3.0

- The more extensive the vegetation area coverage, the higher the carbon sequestration rate typically (Shafique et al. 2020).
- Higher substrate depth sequesters more carbon compared to lower substrate depth (Shafique et al. 2020).
- The more the density of leaves or vegetation, the higher the carbon capture (Dymond 2013).
- Perennial evergreens have a permanent leaf canopy and thus capture more carbon compared to annuals (Renger et al. 2015).
- Mature trees store more carbon (but sequester less carbon) compared to young trees (Lorenz and Lal 2009).
- Shaded lawn sequesters less carbon compared to the lawn with high sunlight availability (Zirkle et al. 2012).

Conclusions

Urban centers are a major source of carbon emissions. However, if designed for carbon sequestration, they have the potential to contribute to carbon sequestration and storage needs. The carbon emissions for building construction and maintenance could potentially be offset by sequestering carbon within the green spaces if integrated and strategic design occurs. Carbon emissions reduction is still vital for achieving net-zero or regenerative targets, even if carbon sequestration strategies are adopted. Therefore, it is important to identify how the built environment could bring positive changes to human and environmental health concurrently by considering carbon sequestration right from the pre-design stage.

Net removal of CO_2 from the atmosphere by building-integrated vegetation occurs when plant photosynthesis exceeds all processes of consumption, respiration, and decomposition resulting in aboveground plant growth and increases in root and microbial biomass in the soil (Chapin et al. 2006). The main objective of anthropogenically driven carbon sequestration process is to balance carbon emissions from the building industry in a way that future development results in carbon-positive and no net gain in the atmospheric carbon pool. The other objective is to challenge the notion that the construction of buildings must necessarily lead to the destruction of preexisting ecosystems. Furthermore, carbon sequestration through building-integrated vegetation has multiple co-benefits that can synergistically work toward climate change mitigation and adaptation agendas, increased biodiversity, and increased human well-being.

Cross-References

▶ Ecological Restoration
▶ Green Cities
▶ Nature-Based Solutions

References

Aguaron, E., & McPherson, E. G. (2012). Comparison of methods for estimating carbon dioxide storage by Sacramento's urban forest. In *Carbon sequestration in urban ecosystems* (pp. 43–71). Springer.

Arehart, J. H., Hart, J., Pomponi, F., & D'Amico, B. (2021). Carbon sequestration and storage in the built environment. *Sustainable Production and Consumption, 27*, 1047–1063.

Birkeland, J. (2008). *Positive development: From vicious circles to virtuous cycles through built environment design*. Earthscan.

Birkeland, J. (2012). *Positive development: From vicious circles to virtuous cycles through built environment design*. Routledge.

Bolund, P., & Hunhammar, S. (1999). Ecosystem services in urban areas. *Ecological Economics, 29*(2), 293–301.

Chapin, F. S., Woodwell, G. M., Randerson, J. T., Rastetter, E. B., Lovett, G. M., Baldocchi, D. D., Clark, D. A., Harmon, M. E., Schimel, D. S., & Valentini, R. (2006). Reconciling carbon-cycle concepts, terminology, and methods. *Ecosystems, 9*(7), 1041–1050.

Churkina, G. (2012). Carbon cycle of urban ecosystem. In R. A. B. Lal (Ed.), *Carbon sequestration in urban ecosystems*. Dordrecht: Springer.

Cole, R. J. (2012). Transitioning from green to regenerative design. *Building Research & Information, 40*(1), 39–53.

Costanza, R., d'Arge, R., De Groot, R., Farber, S., Grasso, M., Hannon, B., Limburg, K., Naeem, S., O'Neill,

R. V., & Paruelo, J. (1997). The value of the world's ecosystem services and natural capital. *Nature, 387*(6630), 253–260.

Du Plessis, C. (2012). Towards a regenerative paradigm for the built environment. *Building Research & Information, 40*(1), 7–22.

Dymond, J. R. (2013). *Ecosystem services in New Zealand: Conditions and trends*. Manaaki Whenua Press, Landcare Research.

Girardet, H. (2015). *Creating regenerative cities*. Routledge.

Gómez-Baggethun, E., Gren, Å., Barton, D. N., Langemeyer, J., McPhearson, T., O'Farrell, P., Andersson, E., Hamstead, Z., & Kremer, P. (2013). Urban ecosystem services. In *Urbanization, biodiversity and ecosystem services: Challenges and opportunities* (pp. 175–251). Dordrecht: Springer.

Guertal, E. A. (2012). Carbon sequestration in turfed landscapes: A review. In R. Lal, & B. Augustin (Eds.), *Carbon sequestration in urban ecosystems* (pp. 197–213). Dordrecht: Springer. https://doi.org/10.1007/978-94-007-2366-5_10.

Heveran, C. M., Williams, S. L., Qiu, J., Artier, J., Hubler, M. H., Cook, S. M., Cameron, J. C., & Srubar, W. V. (2020). Biomineralization and successive regeneration of engineered living building materials. *Matter, 2*(2), 481–494. https://doi.org/10.1016/j.matt.2019.11.016.

Honegger, M., Derwent, H., Harrison, N., Michaelowa, A., & Schäfer, S. (2018). *Carbon removal and solar geoengineering: Potential implications for delivery of the sustainable development goals*. Climate Strategies.

IPCC. (2014). Buildings. In O. Edenhofer, R. Pichs-Madruga, Y. Sokona, E. Farahani, S. Kadner, K. Seyboth, A. Adler, I. Baum, S. Brunner, P. Eickemeier, B. Kriemann, J. Savolainen, S. Schlömer, C. von Stechow, T. Zwickel, & J. C. Minx (Eds.), *Climate change 2014: Mitigation of climate change. Contribution of Working Group III to the Fifth Assessment Report of the Intergovernmental Panel on Climate Change*. IPCC.

IPCC. (2018). *Global warming of 1.5 C An IPCC special report on the impacts of global warming of 1.5 C above pre-industrial levels and related global greenhouse gas emission pathways, in the context of strengthening the global response to the threat of climate change, sustainable development, and efforts to eradicate poverty*. Geneva: Intergovernmental Panel on Climate Change.

Kabisch, N., Frantzeskaki, N., Pauleit, S., Naumann, S., Davis, M., Artmann, M., Haase, D., Knapp, S., Korn, H., & Stadler, J. (2016). Nature-based solutions to climate change mitigation and adaptation in urban areas: Perspectives on indicators, knowledge gaps, barriers, and opportunities for action. *Ecology and Society, 21*(2), 39.

Kondo, M. C., Fluehr, J. M., McKeon, T., & Branas, C. C. (2018). Urban green space and its impact on human health. *International Journal of Environmental Research and Public Health, 15*(3), 445.

Kuittinen, M., Moinel, C., & Adalgeirsdottir, K. (2016). Carbon sequestration through urban ecosystem services: A case study from Finland. *Science of the Total Environment, 563*, 623–632.

Lal, R. (2004). Soil carbon sequestration to mitigate climate change. *Geoderma, 123*(1–2), 1–22.

Lal, R. (2008). Carbon sequestration. *Philosophical Transactions of the Royal Society B: Biological Sciences, 363*(1492), 815–830.

Lal, R. (2012). Towards greening of urban landscape. In *Carbon sequestration in urban ecosystems* (pp. 373–383). Springer.

Lee, A. C. K., & Maheswaran, R. (2010). The health benefits of urban green spaces: A review of the evidence. *Journal of Public Health (Oxford, England), 33*(2), 212–222. https://doi.org/10.1093/pubmed/fdq068.

Lorenz, K., & Lal, R. (2009). *Carbon sequestration in forest ecosystems*. Springer.

Mang, P. (2016). *Regenerative development and design: A framework for evolving sustainability*. Wiley.

Pacheco-Torgal, F., Shi, C., & Palomo, A. (2018). *Carbon dioxide sequestration in cementitious construction materials*. Woodhead Publishing.

Pedersen Zari, M. (2017). Utilizing relationships between ecosystem services, built environments, and building materials. In *Materials for a healthy, ecological and sustainable built environment: Principles for evaluation* (pp. 1–28). Duxford: Woodhead Publishing.

Pedersen Zari, M. (2018). *Regenerative urban design and ecosystem biomimicry*. Routledge.

Reddy, M. S., & Joshi, S. (2018). Chapter 10: Carbon dioxide sequestration on biocement-based composites. In F. Pacheco-Torgal, C. Shi, & A. P. Sanchez (Eds.), *Carbon dioxide sequestration in cementitious construction materials* (pp. 225–243). Woodhead Publishing. https://doi.org/10.1016/B978-0-08-102444-7.00010-1.

Reichle, D., Houghton, J., Kane, B., & Ekmann, J. (1999). *Carbon sequestration research and development*. Washington, DC: U.S. Department of Energy's Office of Fossil Energy.

Renger, B. C., Birkeland, J. L., & Midmore, D. J. (2015). Net-positive building carbon sequestration. *Building Research and Information, 43*(1), 11–24. https://doi.org/10.1080/09613218.2015.961001.

Revi, A., Satterthwaite, D. E., Aragón-Durand, F., Corfee-Morlot, J., Kiunsi, R. B. R., Pelling, M., Roberts, D. C., & Solecki, W. (2014). Urban areas. In C. B. Field, V. R. Barros, D. J. Dokken, K. J. Mach, M. D. Mastrandrea, T. E. Bilir, M. Chatterjee, K. L. Ebi, Y. O. Estrada, R. C. Genova, B. Girma, E. S. Kissel, A. N. Levy, S. MacCracken, P. R. Mastrandrea, & L. L. White (Eds.), *Climate change 2014: Impacts,*

adaptation, and vulnerability. Part A: global and sectoral aspects. contribution of working group II to the fifth assessment report of the intergovernmental panel on climate change. New York: Cambridge University Press.
Selhorst, A., & Lal, R. (2013). Net carbon sequestration potential and emissions in home lawn turfgrasses of the United States. *Environmental Management, 51*(1), 198–208.
Shafique, M., Xue, X., & Luo, X. (2020). An overview of carbon sequestration of green roofs in urban areas. *Urban Forestry and Urban Greening, 47*, 126515.
Smit, B. (2014). *Introduction to carbon capture and sequestration*. Imperial College Press.
Vallas, T., & Courard, L. (2017). Using nature in architecture: Building a living house with mycelium and trees. *Frontiers of Architectural Research, 6*(3), 318–328.
van den Bosch, M., & Sang, Å. O. (2017). Urban natural environments as nature-based solutions for improved public health–A systematic review of reviews. *Environmental Research, 158*, 373–384.
Zirkle, G., Lal, R., & Augustin, B. (2011). Modeling carbon sequestration in home lawns. *HortScience, 46*(5), 808–814.
Zirkle, G., Lal, R., Augustin, B., & Follett, R. (2012). Modeling carbon sequestration in the US residential landscape. In *Carbon sequestration in urban ecosystems* (pp. 265–276). Springer.

Carbon Sink

► Carbon Sequestration Through Building-Integrated Vegetation

Carbon Uptake

► Carbon Sequestration Through Building-Integrated Vegetation

Care Ethics

► Moving Towards Sustainable, Liveable, and Care-Full Urban Environments: Pre-schoolers' Rights and Visions for Planning Just, Socially, and Ecologically Integrated Cities

Care-full Cities

► Moving Towards Sustainable, Liveable, and Care-Full Urban Environments: Pre-schoolers' Rights and Visions for Planning Just, Socially, and Ecologically Integrated Cities

CBD – Convention on Biological Diversity

► Adapting to a Changing Climate Through Nature-Based Solutions

CEEW – Council on Energy, Environment and Water

► Adapting to a Changing Climate Through Nature-Based Solutions

Cellular Agriculture

► Future Foods for Urban Food Production

CEM – Commission on Ecosystem Management

► Adapting to a Changing Climate Through Nature-Based Solutions

The Centrality of Ellensburg

Gary Cummisk
Central Washington University, Ellensburg, WA, USA

Introduction

Ellensburg is a resilient community of approximately 21,000, in Kittitas County (population

49,100). The town is located at the geographical center of the state of Washington, east of the Cascade Mountains and at the western edge of the Columbia Plateau. Due to its unique geography, even with transportation innovations over the centuries, such as wagon trails, railroads, and eventually interstate highways, Ellensburg and Kittitas County have remained vital. Proximity to the lowest pass in the Cascades north of the Columbia River is key to economic vitality and urban growth in Ellensburg and other county towns such as Kittitas, Cle Elum, Thorp, and Roslyn. Recent history has seen increases of populations of retirees and telecommuters, as suburbs expand and exurbs develop. For example, the west side of the valley has attracted wealthy occupants to communities like Suncadia (adjacent to Roslyn), replete with golf courses and country-club dining and recreation. The burgeoning of population in Seattle and the Puget Sound region, starting one hundred miles to the west, has no doubt spurred some of central Washington's recent growth, as people are drawn to the Kittitas Valley by relatively lower real estate costs, a sunnier climate, and a more relaxed lifestyle. But the presence of the towering Cascade Mountains between these regions still serves as a formidable barrier, especially in winter, helping the Kittitas Valley to essentially retain its rural character. During the past half century, the valley has become a major producer of timothy hay and it still produces the traditional cattle, apples, and potatoes; it was long known for, and like many rural areas in the state, is home to a nascent wine industry. The nearby Yakima Valley is the major producer of fine wines in Washington. Ellensburg is also home to Central Washington University, one of six institutions in the state University system offering 4-year and graduate degrees, serving approximately 12,000 students on its main campus. Founded in 1891, CWU serves as a major employer and a vital cultural center of town, affording numerous artistic, scientific, and sports forums to the greater community. The town and the university have long worked as partners coordinating infrastructural development. It is a favorable combination of factors that has contributed to centuries of viability in the valley and over a century of prosperity to Ellensburg itself.

Native Occupancy of the Valley

Native Americans occupied the Kittitas Valley for millennia. They had numerous villages upon the arrival of the first Euro-American explorers to the valley. The local Kittitas Indians, close relatives of the larger Yakama Nation, had an ongoing trade with the Snoqualmie Indians west of the Cascades, despite the fact that they are from two distinct language families—Salish and Sahaptin. These peoples even intermarried and stayed at each other's villages. Rivers and lake outlets were frequented by Native peoples harvesting the bountiful seasonal anadromous fish runs of salmon and steelhead. They hunted game in the valley and gathered roots and berries in the mountains.

One early trapper named Alexander Ross recorded the presence of over 3000 men in the valley in his 1814 account of visiting Che-Ho-Lan, a village that occupied a site close to present-day Ellensburg (Prater 6, 10: Spaulding 23). Some extrapolations from his account have put the population of the valley at over 10,000 persons. And Ross indicated the presence of three-fold as many horses. Che-Ho-Lan was an active trading center and was the site of periodic large gatherings of Native peoples from as far away as the Rockies, and even the Great Plains, as evidenced by the presence of trade goods, including buffalo robes (Prater 10). Beyond trade, they engaged in horse racing, gaming, and celebrations. Indians east of the Cascades had been trading widely and possessed goods from the Pacific Coast and even herded cattle bound for markets as far away as California before Pioneer cattlemen had even got started in the trade.

Ross's account of so many horses in the valley, was probably indicative of active trading, and points to the excellent grazing conditions that existed then, as they do now. Horses were introduced into the Americas by Spaniards, and according to Prater, arrived on the Columbia Plateau by the 1740s. But, in spite of the presence of numerous people and horses, it would be a miscalculation to assume large permanent settlements. Geographer Morris Uebelacker noted in *Time Ball,* his comprehensive study of the

Yakama people and their territorial usage, that settlement patterns among the Yakama were seasonal, revolving around fishing, hunting, gathering, and trade. In a personal refection at a conference honoring her life, Ida Nason, the last full-blooded Kittitas to live in the valley, recalled her girlhood and youth in the late nineteenth and early twentieth centuries as "all travel all the time." She spent time processing fish from the Columbia River and traveled seasonally into the Kittitas Valley with her kinship group to collect camas root and other herbs, and to gather with other groups. Thus, the Kittitas Valley, rich in resources, and along a key trade route between the Pacific (Puget Sound) and the interior of what would become Washington, long supported populations of people. Even after pioneer settlement, Native Americans continued to use the resources that long supported their transhumance. The Native Americans in the region continue to contribute to the greater history of the region. They have contributed numerous place names, and bring insights into the deep history of the land.

Pioneer Days, the Coming of the Railroads, and Road Improvement

Before permanent pioneering settlement, the United States sent surveyors to plot routes through the mountain passes. Among these was George McClellan, [later to be noted as a commanding general of the Union Forces, thus achieving Civil War fame], representing the Army Corps of Engineers. He camped near a Kittitas village in 1853 and explored the possibilities of potential railroad routes through Snoqualmie and Yakima passes. Failing to explore the passes sufficiently under winter conditions as his party started their survey late in the season, he traveled the southern Yakima Pass missing the more northerly but lower Snoqualmie Pass. The Snoqualmie Pass was longer but the grade was more suitable as a railroad passing. McClellan's incomplete recognizance disappointed then Washington Territorial Governor Isaac Stevens. The following year, in 1854, a young engineer named Abiel Tinkham was able to traverse the Snoqualmie with the help of Native guides. This work later led to development of a railroad pass through the mountains.

In 1855 hostilities between whites and Indians led to the Yakima War, and the resultant Treaty of 1855, which saw the Yakama people relegated to living on a reservation that comprised only a portion of their original territory. The Kittitas Valley was not part of those lands. This, in part, opened the Kittitas Valley to Euro-American settlement. Cattlemen, such as Ben Snipes, were using the valley as early as 1859 (Kittitas County Museum).

Sixteen years after the Treaty of 1855, Ellensburg was founded in 1871 by John Shoudy, and named for his wife Mary Ellen. Ellensburg [originally Ellensburgh] grew around the sight of an older trading post called "Robber's Roost" dating to 1855, and purchased by Shoudy from A.J. Splawn. A plaque on 3rd Street in central Ellensburg still commemorates the site.

In 1886 the first train arrived in Ellensburg from Yakima. The following year the railroad reached the Puget Sound, thus realizing the dreams of McClellan, Tinkham, and Stevens. Both the Northern Pacific and Milwaukee Railroads eventually plied their routes through the Kittitas Valley. The building of improved roads and the arrival of railways fueled population growth due to both encouraging agricultural trade and immigration. In the days of steam, water stops were necessary, and numerous towns grew at these railroad stops. Thus the Kittitas Valley of the late 19th and early 20th centuries contained numerous towns that have since disappeared. Others that got their starts during the Railroad era have persisted because they have served as nodes of the farming community. But the trend has been toward fewer urban hubs. Ellensburg and Cle Elum emerged as the most viable connection points in the Kittitas's so-called Lower and Upper valleys.

Official incorporation was bestowed on Ellensburg in 1883. Because of Ellensburg's central location in the state, it was a contender as a sight of the state capital. Among other considerations, a major fire in 1889 served to defeat that campaign. Nevertheless, Ellensburg has remained a viable community throughout its history.

Offering hardware, and livery, grocery stores, bakeries, and essential necessities, Ellensburg was, and still is, the major regional hub of the Kittitas Valley. It also offered luxuries and services such as furniture (example: Fitterer's Furniture founded in 1896 and still in business), later automobiles (Kelleher's, founded 1911), repair services, tractors, medical services, etc.

The Sunset Highway followed an old wagon road and Native American trail as it passed through the Cascades on its way to Ellensburg and beyond. The route had served the region long before Euro-American settlement. In 1926 this officially became part of the national highway system as Route 10 and served as the major highway travel route across the state. Hotels and restaurants sprung up to serve the clientele highway commerce generated. In the 1970s and 80s, as many U.S. towns saw their businesses close as big box stores gravitated to serve communities at corridors serving interstate access, Ellensburg consciously resisted this pattern as I-90 and I-82 linked western Washington to the greater hubs in the central and eastern United States. Determined to keep the core of Ellensburg as a viable center, the community organized to resist big box efforts. As a result, Ellensburg retains to this day am active commercial center. In more recent years, Ellensburg has founded a first-Friday community Art Walk with agencies such as Gallery One, The Clymer Museum, and 420 Building, and The Gard Winery participating as central drivers. In good weather a Saturday Farmer's Market brings the community together in the downtown. Special events help viability as well. For the last two decades Ellensburg has hosted *Jazz in the Valley* bringing regional and national acts to the community. Central Washington University's music program has played a major collaborative effort in this endeavor as well. And since 1923, Ellensburg has hosted a major rodeo which annually brings commerce and visitation to town. Thus urban and rural characters merge in a collective identity. Ellensburg remains a "farm town" and a "university town," but it stands above these designations as an entity in its own right. Distinctive due to its unique geographical position, it successfully blends characteristics of eastern and western Washington, to the chagrin of some and the delight of others, no doubt. It blends liberal and conservative ideas and politics, literally having its wellbeing tied to both sides of the state.

Expansion of the Highway System

The relationship between the railroads and the highway system remain closely connected as the both forms of transport utilized the same mountain passes. Prater noted that "Unofficial records indicate[d] that 105 cars crossed Snoqualmie Pass in 1909"(47), but the process was certainly tough-going and uncertain. Breakdowns were frequent, for wagons and cars. The roadway was little more than an improved path, and frequently suffered the deprivations of winter, and spring erosion. Efforts from motorist clubs on both sides of the Cascades led to political pressures to improve the highway. In 1915 Governor Ernest Listed dedicated the route known as Sunset Highway. By 1926, the Milwaukee Railroad had built a tunnel and ceded part of its right-of-way over the summit to highway department. The Pass was on its way to significant improvement. From 1926, and into the 1970s, the main route between Seattle and Ellensburg (and beyond to central Washington) became Highway 10. Motels and services sprang up along the route to service trucks and cars. Highway 10 ran directly through Ellensburg. By the late 1970s, Interstate 90, the only part of the interstate highway system to travel directly east and west in Washington, became the predominant corridor linking Seattle to the interior. Interstate 82, linking south to north, intersects with I-90 a few miles east of Ellensburg. Thus, Ellensburg now lies close to the juncture of two major interstate highways.

The University Legacy

Washington State Normal School began in 1891. Historic Barge Hall was dedicated in 1893 and houses the University administration today. In 1937 the school name was changed to Central

Washington College of Education, and in 1961 reflecting an expanded curriculum and mission, the institution became Central Washington State College. Sixteen years later in 1977, it became Central Washington University. Beyond expanding cultural opportunities, training teachers, medical professionals, and contributing to scientific understanding, the university is a major employer in the region. During the last decade, the university has generated tens of millions of dollars in tax revenue (Schactler). Offering undergraduate degrees and competitive graduate programs in the sciences, the arts, social sciences, and education, CWU has carved out a significant niche in the state of Washington and beyond.

Contemporary Trends

Agriculture: Timothy Hay is the most valuable agricultural commodity in the Kittitas Valley and ships to Pacific Rim countries (Chamber of Commerce). Cattlemen continue to produce beef cattle, although they are not as prominent as in the early days of settlement. Wine and spirit production are increasingly a component of the local agricultural scene. Local farmers contribute to a lively Farmer's Market offered in downtown Ellensburg in the late summer and early fall.

Rodeo Town: Perhaps reflecting to the days of the Native gatherings at Che-Ho-Lan, the Kittitas County Fair was recorded as early as 1885. In 1922 over 500 local men volunteered to build the Rodeo Fairgrounds where the fair and the annual rodeo are still conducted. The Ellensburg Rodeo is the second largest (behind Calgary) rodeo in the Pacific Northwest and draws participants and spectators from throughout the West, generating millions of dollars in annual revenue for the local economy.

Culture and the Arts: Today Ellensburg has a lively arts and culture scene. Noted for its monthly First Friday Art Walks (started in 2000), numerous galleries and businesses participate showing art works by local and regional artists, offering musical venues, and dining opportunities. The 2021 "Central Washington Visitor's Guide" lists Gallery One, the 420 Loft Gallery, the Clymer Gallery, and other establishments as providing a

central core to festivities and viewing. Many of these institutions work cooperatively in establishing an Arts and Culture Plan including the Arts Commission, the Ellensburg Downtown Association, the Chamber of Commerce and the University Chief of Staff. Events like the annual Jazz in the Valley, with an annual budget exceeding $100,000 draw acts of regional stature and generate significant income for local hotels and restaurants. Gallery One exemplifies how arts institutions benefit the community. With an annual budget exceeding 500 K and six employees, the gallery offers arts programs to children and adults, hosts monthly exhibits, and musical venues in cooperation with Jazz in the Valley. Additionally, it rents studio space to artists and craftsmen at subsidized rates. In attesting that the arts have been so successful in supercharging the local economy of Ellensburg, the city contributes 50 K annually to community arts development. All the major arts associations in Ellensburg are active in procuring grants to support further activities. Additionally, the University has an active calendar of musical and theatrical performances and art exhibitions at the Sarah Spurgeon Gallery.

Internet Highway

As pathway gave way to wagon road, then the railroads, State Route 10, then the Interstate Highway systems, each had an impact on population dynamics and settlement patterns. The coming of the internet has freed many from the necessity of geographical proximity to their workplaces. Telecommuting has come to the Kittitas Valley, Ellensburg, and surrounding communities. The result is an influx of population seeking greater solitude, less traffic, and a more relaxed lifestyle. Early retirements and flex-scheduling are becoming commonplace. The amenities offered by the Kittitas Valley have not gone unnoticed as tech-savvy workers seek recreation, cultural amenities, and value without traditional constraints. Each transportation and conceptual revolution brings changes. The unfolding matrix of settlement is conditioned by rich historic past and a dynamic and ever-changing present.

Summary

Ellensburg and the Kittitas valley have remained central geographically, not only due to the political map, but to the essential characteristics of the land. It is these features—climate, rivers, mountains, proximity to passes—that cultures must adapt to. In a state with a high population growth rate, Ellensburg and the Kittitas Valley will continue to see change. So far, they have remained resilient and culturally adaptive, while retaining their distinctive identities.

Cross-References

- ▶ Growth, Expansion, and Future of Small Rural Towns
- ▶ Shrinking Towns and Cities
- ▶ Small Towns in Asia and Urban Sustainability

Further Reading

Allen, M. (2021, January 10). *Interview at Ellensburg Rodeo Hall of Fame.*
Ellensburg: Daily Record. (2021). *Central Washington visitors guide.*
"Ellensburg History and Information". (2022). Ellensburg: Chamber of commerce.
Kinnick, J., & Chery. (2007). *Snoqualmie Pass.* Charleston: Arcadia.
Miller, M. (2021, January 10). Interview at Gallery One.
Prater, Y. (1981). *Snoqualmie pass.* The Mountaineers.
Schactler, L. 2022, January 22). Phone interview at Central Washington University.
Spaulding, K. (Ed.). (1956). *The fur hunters of the far west.* Norman: University of Oklahoma Press.
Uebelacker, M. (1984). *Time Ball.* Yakima: Shields Bag.
Wikipedia. (2022a, January 27). Central Washington University. https://en.wikipedia.org/wiki/Central_Washington_University
Wikipedia. (2022b, January 23). George B. McClellan. https://en.wikipedia.org/wiki/George_B._McClellan
Wikipedia. (2022c, January 27). Yakama. https://en.wikipedia.org/wiki/Yakama
Williams, C. (1980). *Bridge of the gods, mountains of fire: A return to the Columbia Gorge.* Seattle: Friends of the Earth.

Centralization

▶ Urban Structure and Its Impact on Mobility Patterns: Reducing Automobile Dependence Through Polycentrism

The Challenges for Wildland-Urban Interfaces (WUI) in Metropolitan Areas: Reducing Fire Risk, Providing Employment Opportunities, and Preserving Natural Habitat

Carolina G. Ojeda
Doctorado en Arquitectura y Estudios Urbanos, Pontificia Universidad Católica de Chile, Providencia, Santiago de Chile, Chile
Departamento de Historia, Facultad de Comunicaciones e Historia, Universidad Católica de la Santísima Concepción, Concepción, Chile

Synonyms

Boundaries; Ecological corridor; Landscape fragmentation; Peri-urban spaces; Urban-rural interfaces

Definition

Wildland-Urban Interfaces (WUI): They are boundaries between consolidated urban areas (cities, conurbations, and metropolises) and wilderness/productive areas (forestry, agriculture, etc.). Similarly, these urban and natural interfaces are constantly pressured by their interaction results in the fragmentation of non-human habitats, and is clearly different from planned greenbelts and peri-urban areas. Moreover, in some metropolitan areas, the housing development often creates chaotic intermixed-like interfaces, unlike the more orderly WUI. The probability of occurrence of forest fires has increased considerably as human/

non-human elements are mixed in the territory incrementing the overall vulnerability in these areas due to the highly concerning tendency of wildfire mismanagement. In that sense, it could be observed that the agroforestry or agroindustry leaves flammable material, there is not sufficient equipment for first responders (eg., access roads for fire trucks, availability of water, etc.), the anthropic infrastructure associated with transportation-highways, roads, ports, and railways-, and electricity distribution (high-voltage towers, powerlines, and transformers). Lastly, the main challenges to maintaining a good quality of WUI to reducing fire risk are: improving fire prevention, providing work for their human inhabitants, avoiding urban sprawl – illegal resource extraction, and preserving natural habitats for non-humans.

Introduction: The Problem with the Boundaries Between Nature and Humankind

In 1931, the historian Marc Bloch established the factors that shaped the physical and social characteristics of the French countryside between the desegregation of the Carolingian Empire and the eighteen-century agrarian revolution in Europe (Bloch 1966). The most important change that Bloch considered in his work was the slow but steady increment of the arable land on expenses of European forest. At that time, Europeans considered those forests as the hardest obstacle to conquer, and during centuries in medieval France – as well as in the rest of the continent – the trees delayed substantially the progress of the plow. According to David Arnold (1996), for the first time in contemporary historiography, Bloch explained that the people were adapted to the landscape but most importantly, the same people were simultaneously submitted to its influence and productive control, leaving behind an exclusively anthropocentric view of history.

Consequently, the relationship of people and nature is the center of the seminal works from environmental historians such as Rachel Carson's (1962) "*Silent Spring*," William Cronon's (2011) "*Changes in the land: Indians, colonists, and the ecology of New England*," David Arnold's (1996) "*The Problem of Nature. Environment, Culture, and European Expansion*," Stephen Pyne's (1997) "*World of fire: the culture of fire on Earth*," Aldo Leopold's "*Sand County Almanac*" (1986), and Mike Davis's (1998) "*Ecology of fear: Los Angeles and the imagination of disaster.*" All of them narrate the conflict about the prevalence of humans and non-humans' activities intensified at the boundaries of those interfaces (ESRI 2021). Regrettably, for nonhumans, this predominance of anthropic activities has decanted in an enclosing phenomenon first described in the "*Theory of the Commons*" suggested by Elinor Ostrom in the 1990s (Hess and Ostrom 2007). Due to anthropic affectation, the commons are confined to a determined space and they cannot connect with each other because of the transference to property rights, destroying its essence as "*lands, waters, and resources that are not legally owned and controlled by a single private entity, such as ocean and coastal areas, the atmosphere, public lands, freshwater aquifers, and migratory species*" (Burger et al. 2001).

Recently, the conflict for predominance in the Earth's surface (lands, seas, mountains, forests, deserts, etc.) has been studied by scientists from all over the world with the help of technology: from rustic printed maps to multitemporal satellite images. This last element is relevant for Earth sciences since the launch of LANDSAT technologies in the 1970's and the uprising evolution of other geospatial technologies such as drones, mobile GPS, tablets, LIDAR, and geoprocessing software like ArcGIS© (ESRI 2016).

There are few wild areas without human intrusions, such as urban sprawl or extractive activities such as forestry, livestock, farming, fishing, and mining. According to J. Duncombe (2019), satellite data showed that 56% of the earth's land (excluding areas covered by ice and snow) has a relatively low human impact is being parsed as a shrinking part: there are approximately 990,000 fragments larger than 1km^2 on land with less human impact, and the same area will naturally be broken into 73,000 pieces. These phenomena, which make the world less amicable for non-humans, are well known as landscape fragmentation and Land Use and Land Cover Change (LULCC). As a result, these pressures have been creating multiple types of zones between urban and rural territories that will be briefly defined next: peri-urban spaces, greenbelts, intermixed areas, and wildland-urban interfaces (WUI) (Fig. 1).

Peri-urban spaces (see chapter ▶ "Peri-Urbanization") could be defined as the spatial zones between the artificial distinctions of "rural" and "urban," which are characterized in environmental planning and management processes by three key features: "*the specific ecological nature of peri-urban systems; the heterogeneity and vulnerability of peri-urban communities; and the difficulty in identifying the boundaries of a system subject to rapid change by overlapping institutions*" (Allen 2003).

Greenbelts (see chapter ▶ "Green Belts") are an implementation in urban planning and landscape architecture introduced in the early twentieth century to control the urban growth by separating the "urban" and "rural" areas through a vast space with distinctive vegetal elements (Amati 2016). Those green spaces were not meant for extractive activities; instead, they were planned for recreation, incineration, small agriculture, salvage yards, parks, pollution mitigation, and small-scale forestry.

The intermixed areas are those WUI that have a "*threshold of 6.17 housing units/km^2 dominated by wildland vegetation (coniferous, deciduous, and mixed forest; shrubland, grasslands/*

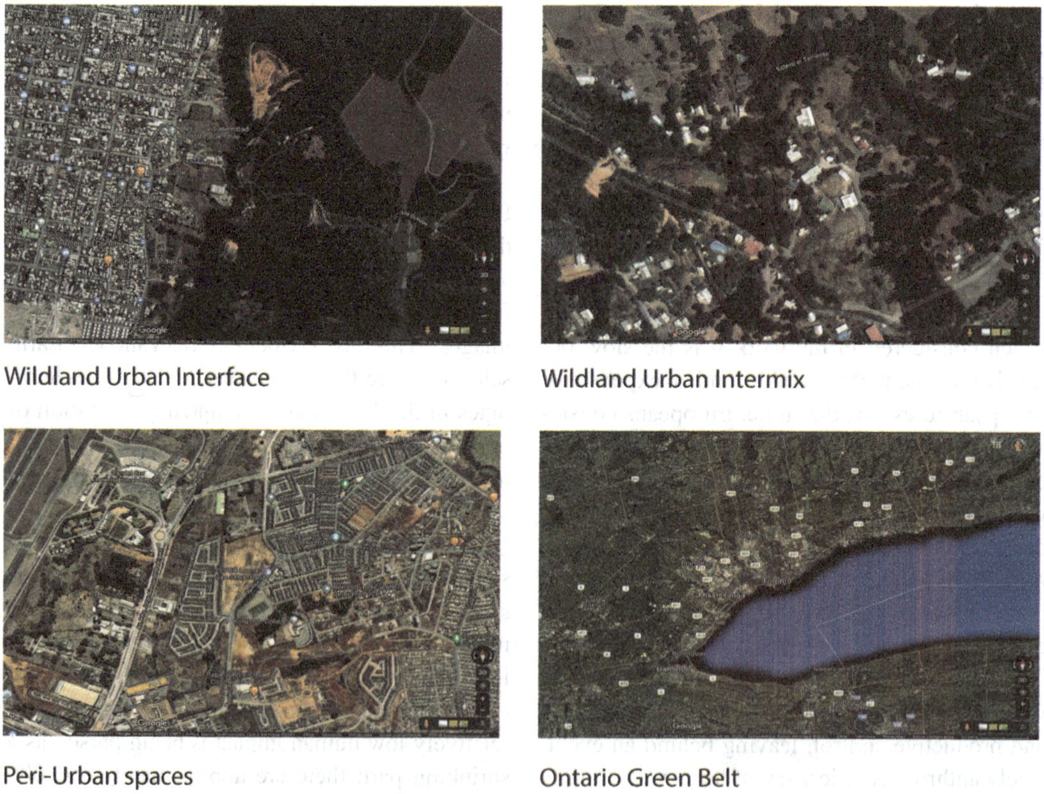

The Challenges for Wildland-Urban Interfaces (WUI) in Metropolitan Areas: Reducing Fire Risk, Providing Employment Opportunities, and Preserving Natural Habitat, Fig. 1 Spatial differences among peri-urban spaces, greenbelts, intermixed areas, and wildland-urban interfaces (WUI). (Source: Google Maps © 2021 and author's personal archive)

herbaceous; transitional; woody and emergent herbaceous wetlands) in at least 50% of the terrestrial area of a given census block" (Radeloff et al. 2005). These areas are difficult to address in urban planning and hazard management and represent a constant danger for their inhabitants to suffer wildfires, landslides, floods, and tsunamis.

WUI is defined as "*where houses meet or intermingle with underdeveloped wildland vegetation*" (USDA and USDI 2001), which means that the "rural" and "urban" interfaces are in contact following productive or planning patterns. They have three key features: contains unorganized urban features (houses, condos, apartments, etc.), large areas of agriculture, forestry, or industry – which could not be categorized as entirely rural – and heavy urban infrastructure associated with transport (highways, roads, ports, and railways) and energy supply (high-voltage towers, etc.). Particularly, all the abovementioned areas are at risk of wildfires due to the proximity of human activities, high amounts of vegetal fuel load, and geographical accidents such as mountains and ravines (see chapter ▶ "Managing the Risk of Wildfire Where Urban Meets the Natural Environment").

According to the United Nations, the WUI and metropolitan areas will attract more population in the future and this land expansion will pose a threat to rural lifestyles in the small towns (see chapter ▶ "Growth, Expansion, and Future of Small Rural Towns"), which could be managed through urban planning zonification (Fig. 2) and addressing three key challenges: reducing fire risk, create jobs and preserve natural habitats.

The First WUI Challenge in Metropolitan Areas: Reducing Fire Risk

Fire risk can be defined as the probability of ignition in a determined area considering three main variables: low humidity (>20% of relative air humidity), high temperatures above 30°C or 86 °F, and high wind speed above 10 km/hr. or 6.21 miles/hr.). These key features are well known

Habitat, Fig. 2 Proposal to integrate the WUI in urban planning. (Source: California Wildland Urban Interface Code Information, Redwood City)

The Challenges for Wildland-Urban Interfaces (WUI) in Metropolitan Areas: Reducing Fire Risk, Providing Employment Opportunities, and Preserving Natural

in literature as the "triangle of fire" because they represent the perfect scenery to create hotspots regardless of the place. If the fire risk occurs in WUI or rural areas, it will be called "wildfire" (in this chapter both terms will be used indistinctly).

Fire ignition itself does not construct the risk, however, it is complemented by a vulnerability that represents the anthropic characteristics that augment or diminish that risk. There are several categories of vulnerability, but in general, they all refer to the ability of people's preparedness and the capability of infrastructures to manage the wildfire cycle: risk assessment, ignition suppression, spreading control, and landscapes/housing recovery.

In that sense, the first challenge to be addressed by the WUI is reducing the fire risk by improving fire prevention policies for the community, the private sector, and the government. This actor-collaborative perspective has been slowly adopted since the 1980s when the necessity to involve academics and stakeholders in the development of risk assessment and disaster mitigation, not just fire suppression, became evident. Undoubtedly, a participatory model of disaster assessment was more than justified with the development of that perspective (Paveglio et al. 2018). In this regard, it will be relevant to adopt several recommendations to tackle vulnerability and reduce fire risk:

- **Managing the WUI according to the context and not leaving the territorial planning aside**: It is necessary to increase participatory processes in regional and metropolitan planning. Public-private partnerships are a good starting point to settle proper planning management to create a healthy urban and industrial development, especially considering that the WUI are more prone to be affected by a wildfire and be a spreader of wildfires (Radeloff et al. 2018). In addition, the cities that would like to embrace a smart city development should consider the evolution of these areas into their planning through remote sensing.
- **Fire education is a keystone to prevent wildfires**: Fire risk reduction cannot be possible without proper fire education for citizens and private sector stakeholders. According to Prestemon et al. (2010) "*wildfire prevention education efforts involve a variety of methods, including airing public service announcements, distributing brochures, and making presentations*." In this context, the best practices are those that encourage homeowners to reduce flammable material nearby their houses (eg., garbage, old leaves, branches, bushes, etc.), to promote social bonding among neighbors, and to educate through demonstrations of good fire prevention practices such aschecklists, fire drills, homeowners' certification, videos, workshops, between others (Sturtevant and McCaffrey 2006).
- **Infrastructure is a game-changer into fire risk mitigation**: A key difference between proper and improper management of WUI is the presence of infrastructure that helps fire suppression and adequate management that guarantees the access to water infrastructure (e.g., fire hydrants, pools, artificial lagoons, and fog-catching equipment) could help to quickly reduce fire at the beginning without major intervention. Moreover, an acceptable organization of transportation network and width road design could improve the access of emergency vehicles such as fire trucks or firefighter's brigades reducing significantly the fire spreading.
- **Do not be afraid of engaging in citizen science**: Sources of open access data such as apps and social networks may improve the stakeholder's ability to understand what is happening in the WUI in real-time. Implementation of citizen science initiatives may well reduce fire vulnerability within a reasonable budget: complaint apps for fire-related issues (eg., presence of garbage, old leaves, branches, lack of fire hydrants, etc.), online forms to book a fire evacuation drill, homeowner's fire preparedness surveys, real-time monitoring through

satellite images, or having drones equipped with thermal cameras.

The Second WUI Challenge in Metropolitan Areas: Providing Work in Agriculture, Small-Scale Forestry, and Tourism

Managing the WUI in metropolitan areas is tougher than it sounds, especially when the planning instruments or zoning laws are not well developed to ensure that goal. In paper, the division between "urban" and "rural" areas is well defined. However, some urban phenomena are being developed faster than the planning process itself: urban sprawl, peri-urban poverty belts, rural housing dispersion, industrial sites expansion, and forestry plantations without proper firewalls. Furthermore, the mismanagement of WUI's could decrease the healthy benefits for cities obtained from the vicinity of small and medium towns (see chapter ► "Growth, Expansion, and Future of Small Rural Towns").

Through the implementation of progressive updates to zoning divisions in the WUI, based on citizen participation, it is possible to provide a steady income for their inhabitants through the preservation of some natural characteristics of the WUI to ensure meaningful jobs such as urban agriculture, small-scale forestry, and special interest tourism.

First, **the creation of community gardens may strengthen social bonds among neighbors** around the seasonality of crops and harvests. The maintainance of these community gardens might create meaningful occupations for younger and elder people, such as repairing fences, preparing the seeds, giving tours to school kids. Also, it could create socially valued oriented jobs such as community garden manager or community organizer. Besides, not only public spaces and sidewalks are useful for this purpose but the neighbors' patios could be used for growing vegetables to secure each household's food and nutrition. In that sense, it has been studied that having a backyard kitchen garden is relevant in vulnerable neighborhoods because it guarantees access to a variety of nutritious foods, increases the money flow by selling vegetables, and creates food provision during an economic crisis, massive layoffs, or after socio-natural disasters (FAO 2010).

Second, regulating the wildland side of the WUI could be beneficial for traditionally disadvantaged groups – mainly youth, immigrants, and women – through job creation in fire hazard management, youth groups' activities, landscaping, and fuel material cleaning. Likewise, if they exist in the area, the management of the commons would enhance sustainable small forestry that balances the productive plantations (eg., pines, eucalyptus, fruit trees, etc.) and native forests, guaranteeing access to local wood products with better prices and reducing the global production chain's carbon footprint. Additionally, it may generate more wood-related jobs for men and women, like wood cutting, lumber extraction, and carpentry. However, it is necessary to supplement that forestry production with regulatory laws: condemning overexploitation of trees, subsidizing sustainable wood production, developing an incentive scheme to protect its soil from erosion, and overwatch planted species to safeguard the right amount of native forest preservation by monitoring through citizen science projects.

Third, special interest tourism is a growing branch of tourism around the world (Trauer 2006). It is characterized by the selection or creation of routes grounded on people's particular curiosities such as cave exploring, birdwatching tours, astronomical expeditions, sea diving, safaris, or vineyard-growing experiences. In general, it responds to more curated content tourism based on field experiences rather than just quick sightseeings in the usual tourist traps. Special interest's tourism activities could change the general view about WUI which are often perceived as wasted landscapes – i.e., underused, polluted, or abandoned sites (Rigillo et al. 2018) – making them more appealing to travelers. Similarly, the path from being illegal dumpsters to desired places might generate tourism-related jobs like farmers' markets, food trucks, insect or bird reserves, florists, and small astronomical observatories. Additionally, it could maintain a flow of seasonal works by guiding field trips for tourists to watch endangered species' habitats in spring–summer, and enjoy the snowy places in winter–autumn.

The Third WUI Challenge in Metropolitan Areas: Preserving Natural Habitats Through Technology

Some urban planning phenomena that could mismanage WUI's are urban sprawl, peri-urban poverty belts, rural housing dispersion, industrial sites expansion, and forestry plantations without proper firewalls. This section will add other urban processes that might damage the relationship between humans and non-humans who live in the WUI: lack of habitat's connectivity, the significantly increase of illegal resource's extraction, and the destruction of non-human's natural habitats (see chapter ► "Peri-Urbanization").

As its name establishes, the WUI has important components of wildland like mountains, water bodies, wetlands, or forests. Indeed, those components are cataloged as non-humans, which means that they could be vegetal (eg., trees or shrubs), mineral (eg., rocks, soil, etc.), or animal (eg., birds, fishes, etc.). Traditionally, all these non-humans have been considered as part of a big umbrella term known as "Nature" and have been contemplated as part of humankind's possessions. That perspective has slowly changed with the arrival of a noteworthy change in research framework's in social sciences called "animal turn" (Ritvo 2007), which has given more relevance to the non-human part of cities in its functioning and leading to a recognition of their agency in areas like urban planning or infrastructure development.

If the WUI is not properly managed, it can damage the habitat connectivity for non-humans, followed by indirect destruction of the human economy when losing the necessary interactions between the species to thrive: agriculture needs pollinators, fisheries need rivers without sedimentation, and livestock needs grass for pasture. A possible solution might be applying mandatory regulations to include land bridges and resting places for key species once some heavy urban infrastructure is developed (eg., towers, industries, ports, etc.), creating an opportunity to embrace greenbelts and greener cities (see chapters ► "Green Belts"; ► "Green Cities: Nature-Based Solutions, Renaturing and Rewilding Cities"). Another solution is to promote citizen science using monitoring apps focused on distributing knowledge about the non-humans that inhabit the WUI for conservation purposes or live stocking.

In that sense, the illegal extraction of natural resources is a threat that affects non-humans directly as it destroys their habitat. This phenomenon is more visible in the WUI due to the direct pressure from urban sprawl and industrial expansion through the land. Satellite observation, which started with the launch of LANDSAT in the 1970s, is the most relevant technological tool that has been helping scientists and governments to track changes from decades past. Moreover, with the advances in AI and drone technology, it is possible to evaluate LULCC in real-time by establishing control points in conflictive areas prone to natural resources' illegal extraction, like forests, rivers, abandoned mines, mountains, lakes, and wetlands.

For the WUI, the destruction of the wildland is particularly devastating because it means that it will be absorbed for urban or industrial uses. The technology can help in that matter to evaluate the already-done damages to the environment. In that case, it is necessary to raise awareness about preventing the illegal extraction of natural resources, especially in countries where the informal economy forces their inhabitants to exploit the resources to maintain a household. For example, some people have been facilitating the destruction of tree forests to supply the firewood market.

Summary/Conclusion

Due to its dynamic composition, the WUI is one of the most fast-changing places in metropolitan areas. They are spatially affected by the pressures of housing development and agroforestry/industrial expansion, which can be observed in the reduction of wildland. Adequate planning management of the WUI is necessary to preserve this interface and to avoid hazards like wildfires and landslides. Likewise, proper management of the WUI may improve the benefits for its inhabitants by offering more job opportunities in agriculture, small-scale forestry, and special interest tourism (eg., birdwatching, etc.). Lastly, it might help to preserve biodiversity hotspots, key ecosystems for endangered species, and improve the overall relationship between humans and non-humans.

Cross-References

- ▶ Green Belts
- ▶ Green Cities: Nature-Based Solutions, Renaturing and Rewilding Cities
- ▶ Growth, Expansion, and Future of Small Rural Towns
- ▶ Managing the Risk of Wildfire Where Urban Meets the Natural Environment
- ▶ Peri-Urbanization

Acknowledgments This work was partially funded by ANID – Chile Beca de Doctorado Nacional folio 21200455.

References

Allen, A. (2003). Environmental planning and management of the peri-urban interface: Perspectives on an emerging field. *Environment and Urbanization, 15*(1), 135–148. https://doi.org/10.1177/095624780301500103.

Amati, M. (Ed.). (2016). *Urban green belts in the twenty-first century.* London: Routledge.

Arnold, D. (1996). *The problem of nature. Environment, culture and European expansion.* London: Wiley-Blackwell.

Bloch, M. (1966). *French rural history: An essay on its basic characteristics.* London.

Burger, J., Ostrom, E., Norgaard, R., Policansky, D., Goldstein, B. D., & (Eds.). (2001). *Protecting the commons: A framework for resource management in the Americas.* Island Press.

Carson, R., & Lear, L. (1962/2002). *Silent spring.* Anniversary edition. New York: Houghton Mifflin Company.

Cronon, W. (2011). *Changes in the land: Indians, colonists, and the ecology of New England.* New York: Hill and Wang.

Davis, M. (1998). *Ecology of fear: Los Angeles and the imagination of disaster.* New York: Macmillan.

Duncombe, J. (2019). We have broken nature into more than 990,000 little pieces. *Eos, 100.* https://doi.org/10.1029/2019EO136354.

ESRI. (2016). The ArcGIS Imagery Book. New view. New Vision. Redlands: ESRI Press. http://downloads.esri.com/LearnArcGIS/pdf/The-ArcGIS-Imagery-Book.pdf?utm_source=PR&utm_medium=KR&utm_term=June&utm_content=pdf&utm_campaign=imagery-book

ESRI. (2021). The living Atlas sentinel-2 2020 land cover map. https://www.arcgis.com/apps/mapviewer/index.html?layers=d6642f8a4f6d4685a24ae2dc0c73d4ac

FAO. (2010). Improving nutrition through home gardening. http://www.fao.org/ag/agn/nutrition/household_gardens_en.stm

Hess, C., & Ostrom, E. (Eds.). (2007). Understanding knowledge as a commons: From theory to practice (p. 24). Cambridge, MA: MIT Press.

Leopold, A. (1986). *Sand County Almanac (Outdoor essays & reflections).* New York: Ballantine Books.

Paveglio, T. B., Carroll, M. S., Stasiewicz, A. M., Williams, D. R., & Becker, D. R. (2018). Incorporating social diversity into wildfire management: Proposing "pathways" for fire adaptation. *Forest Science, 64*(5), 515–532.

Pyne, S. J. (1997). *World fire: The culture of fire on Earth.* Washington: University of Washington press.

Radeloff, V. C., Hammer, R. B., Stewart, S. I., Fried, J. S., Holcomb, S. S., & McKeefry, J. F. (2005). The Wildland–Urban Interface in the United States. *Ecological Applications, 15,* 799–805. https://doi.org/10.1890/04-1413.

Radeloff, V. C., Helmers, D. P., Kramer, H. A., Mockrin, M. H., Alexandre, P. M., Bar-Massada, A., . . . Stewart, S. I. (2018). Rapid growth of the US wildland-urban interface raises wildfire risk. *Proceedings of the National Academy of Sciences, 115*(13), 3314–3319.

Rigillo, M., Amenta, L., Attademo, A., Boccia, L., Formato, E., & Russo, M. (2018). Eco-innovative solutions for wasted landscapes. *Ri-Vista. Research for Landscape Architecture, 16*(1), 146–159.

Ritvo, H. (2007). On the animal turn. *Daedalus, 136*(4), 118–122.

Sturtevant, V., & McCaffrey, S. (2006). Encouraging wildland fire preparedness: Lessons learned from three wildfire education programs. In: McCaffrey, S. M. (tech. ed.), *The public and wildland fire management: Social science findings for managers.* Gen. Tech. Rep. NRS-1. Newtown Square, PA. US Department of Agriculture, Forest Service, Northern Research Station: 125–136. https://www.fs.usda.gov/treesearch/pubs/18690

Trauer, B. (2006). Conceptualizing special interest tourism – Frameworks for analysis. *Tourism Management, 27*(2), 183–200.

USDA & USDI. (2001). Urban wildland interface communities within vicinity of Federal Lands that are at high risk from wildfire. *Federal Register, 66,* 751–777.

Challenges of Delivering Regional and Remote Human Services and Supports

David J. Gilchrist and Ben Perks
University of Western Australia, Perth, WA, Australia

Synonyms

Community support; Countryside; Rural; Social services; Welfare services

Definitions

The inherent difficulties in delivering human services and supports to remote areas falling outside of metropolitan districts generally stem from physical isolation related to the absence of major transport, communications, and other infrastructure and restrictions in relation to the allocation of scarce resources within a political framework where value for money is often considered a priority. That said, there is no universal criteria nor measurement for regionality or remoteness since what is considered regional or remote changes depending on the social, geographical and cultural context, as well as the nature of the service or support required. Human services and supports are often programs delivered consistent with social welfare objectives and can include the provision of health care and supports in the form of primary or allied services. These services can include such supports as those provided for people with disability and the elderly, in addition to employment and education programs.

Introduction

Despite the radical increase in the urbanization of many regions of the world, and the subsequent concentration of employment opportunities within these metropolises, a significant proportion of the world's population still resides in regional and remote areas. Estimates derived from the United Nations Population Division's World Urbanization Prospects report estimates that approximately 44% of the world's population live in such areas (United Nations 2019). Furthermore, deficiencies demonstrated by key welfare indicators have been identified repeatedly in regional and remote communities – regardless of the relative wealth of the nation in question.

Delivering human services is a distinctly different activity to that of delivering private or public goods. Unlike access to public pools or commercial shopfronts, human services and supports are often not facilitated through traditional supply and demand mechanisms as acute needs for supports, the cost of their provision, and the populations required to reach economies of scale are often incompatible, especially in the context of remoteness (Rolfe et al. 2017), while affordability is always a challenge for those in need (Gilchrist and Charlton 2014). As such, these services are often provided directly or funded by the state.

Some definitive features of human services delivery include prescribed and mutually obligatory funding timelines, a high level of human resource requirement with concomitant cost as these services are labour intensive, unique infrastructure and equipment, engagement with vulnerable groups, and prolonged, repeated contact with recipients. The relevance in outlining and evaluating the challenges experienced in the delivery of human services and potential solutions in these regions is a palpable concern for many governments, for the communities themselves, and for the largely nonprofit and charitable service providers.

Geography and Infrastructure

The most obvious challenge to service delivery in regional and remote (R&R) areas is the physical travel distance from metropolitan centers coupled with the added difficulty of access. The physical sparsity of these areas generates several notable difficulties for both providers and clients, the first being the added costs to transport outlays (Gilchrist and Emery 2020). Survey evidence shows the need for reliable and flexible cars to not only travel long distances but to be sufficiently robust to navigate under-developed roads and potential climatic extremes. These vehicles are more expensive to purchase and maintain and must be properly insured and equipped at increased cost (Gallego et al. 2017).

Additional costs can arise from the lack of accommodation and equipment available in R&R areas which necessitates day-trip service models (Fly-In-Fly-Out, Drive-In-Drive-Out). These factors further increase travel costs for providers which are often reimbursed in one-size-fits-all government funding models (O'Sullivan et al. 2016). While these service delivery models do increase access to otherwise non-serviced areas, there are concerns that they potentiate an under-investment in local service infrastructure and support coordination (Hussain et al. 2015).

However, service hubs used as an alternative have been observed to hamper accessibility by shifting the costs in foregone wages, accommodation, and travel expenses, as well as in-kind loses (i.e., informal support networks) to the recipient or their carer/family (Barr et al. 2018).

When seeking to localize private investment in service infrastructure, low population density is the most cited impediment. For example, Zhao and Malyon (2010) estimated the average highest per capita expenditure in remote clinics in the Northern Territory of Australia occurs in communities with populations of less than 200. The largest non-staffing costs were incurred in maintaining the necessary infrastructure, while the necessary employee incentives kept staffing costs higher than in urban clinics, as increased remuneration and the provision of accommodation, amongst other things, drove up costs.

The underlying issue here is that, in high population areas such as cities, service delivery costs are contained as a result of the density of the population – the reaching of a critical mass allows for economies of scale to develop whereby average cost per iteration of service declines as output increases (Rolfe et al. 2017). Therefore, dedicated equipment and infrastructure are increasingly difficult to justify in low-population areas, both in terms of sustained cost and aggregate demand regardless of the severity of R&R needs (OECD 2010).

Workforce Shortages and Retention

The heterogeneity of the workforce that provides human services means that many of the challenges in recruitment and retention are service specific but varying in their degree of severity. However, there are two interconnected challenges – the shortage of trained individuals with appropriate experience who are able and willing to service R&R areas and the difficulties encountered when trying to retain these people either as sole traders or by service providers as employees. The international literature details an array of disincentives facing the workforce in R&R areas. The increased costs of living touched on above is a pertinent example.

Limited compensation for time traveled to or between services or the equipment required to access and execute services in rural and remote areas dissuades professionals who could otherwise operate in the metropolitan centres (Schubert et al. 2018).

Studies have identified the unrecognized costs of delivery dampening incentive schemes designed to entice additional workers to remote workplaces. Other prominent factors include poor lifestyle due to excessive travel and long working hours, practicing personal liability concerns, limited supporting workforce availability, as well as a lack of opportunity for professional development and networking (Johnsson et al. 2017; Schubert et al. 2018).

That said, survey evidence from New South Wales in Australia shows high job satisfaction among fixed-location aged care and primary health workers (Gallego et al. 2016). Although, the average age of these workers was 55, which in many countries is nearing retirement; this is also indicative of the challenges impacting service delivery in the context of demographic change. Trends suggest that fewer people would be available to replace future retirees which would further limit the availability and quality of R&R services (World Health Organization 2009).

Unfortunately, there is currently a notable dearth of research examining the effectiveness of trialed incentive schemes.

Deficiencies in the comprehensiveness and continuities of services presuppose staff shortages and high staff turnovers. Many support services rely on an early intervention or preventative approach to produce compounding benefits (Schoo et al. 2016). When disjointed, these supports fail to produce the desired outcomes, often at the cost of additional scarce resources and the development of more acute needs that could have been prevented. In health care provision in remote areas, continuity is further interrupted by the necessary prioritization of acute problems and the displacement of those with less urgent issues (Dossetor et al. 2019).

Cultural Security

It is critical, considering the often diverse cultural priorities of First Nations People in regional and remote areas, that human service delivery is implemented in a way that is culturally secure. For instance, an international review of Indigenous health care models conducted by Harfield et al. (2018), unanimously found that the centralization of culture in service delivery practice was an essential component of Indigenous cultural security. Further research suggests a consistent barrier to accessibility is the hesitance of Indigenous people in accessing services due to previous experiences of discrimination and the stigma associated with Western-centric medical labels (Stephens et al. 2014). The training of staff, engagement of the broader community, and the lateral aspects of delivery of care must be consistent with, and reflective of, the relevant cultural cues. However, this necessity also adds unavoidable cost to the delivery of services.

In Australia, the undertraining, shortage, or absence of Indigenous health workers in services coalesces preferences for non-medical community support over Allied health services (Parikh et al. 2021). A concomitant barrier is underinvestment in relationship-building with community Elders prior to service delivery. Elder approval is often essential for access to the community and to facilitate service uptake yet it is often not resourced in funding plans (Gibson et al. 2015). These challenges can be extended to other rural communities, as insufficient priority is given to building the levels of trust and reciprocity that would facilitate greater service efficacy (OECD 2010).

Importantly, studies in regional United States show an aversion to social services grounded in the ideological preference for individualism and self-sufficiency. Many communities were seen to be distrusting of service staff and the government in general. Inadequate education and employment opportunities further constrains people's willingness to pay for, or demand, services delivered in these areas (Carson and Mattingly 2018).

Appropriateness of Timelines and Outcomes

The development of policy for implementing rural and remote human services and supports varies considerably between countries and service types. A consensus generated in extant literature is that policy criteria often fail to accurately reflect the essential components of service delivery, as well as the unique human, geographic, and other characteristics of remote regions. Policy document analyses conducted in Australia, the USA, and the UK (see – Atterton 2008; Carson and Mattingly 2018; Dew et al. 2014) evidence this disconnect between the broader policy objectives and metropolitan-centric operational content. Often, these policy and practice documents fail to acknowledge the barriers to the provision of services impacting workers and participants, partly due to assuming costings and circumstances consistent with urban areas.

A detrimental result of this policy misalignment is a neglect of the necessary funding required to cover extra costs that constrain service delivery in R&R areas, along with the forwarding of unsuitable and seldom realistic funding timelines. For instance, Dew et al. (2016) note uncompensated travel and context-specific equipment as key demotivators to workforce growth and retention. Further, funding timelines reflect serviced hours rather than considering the resources impacts of the precursory relational and cultural needs present in many R&R communities (Massey et al. 2013). Scholars lay much of this tension on the funding trade-off between impact and demand as determined by population size (Carson and Mattingly 2018; Weinhold and Gurtner 2014). As such, standard cost/benefit analyses disadvantage rural areas as demand is considered in the context of population density not its severity. As governments continue to retreat from the direct provision of service, it will be vital that the true costs incurred by private providers – whether nonprofit or commercial – are fully compensated so as to minimize market barriers (O'Toole et al. 2010).

Evidenced and Emerging Alternatives

With the advancement of technology and our understanding of the key challenges, notable solutions have been identified aimed at lowering costs, creating context-specific frameworks, and establishing spaces whereby communities can engage meaningfully with service providers.

For example, Telehealth, the delivery of services via communication technology, is one of the more studied responses to high service costs, professional development concerns, and challenges to achieving economies of scale. The review of Telehealth services in Australia by Bradford et al. (2016) substantiated the capacity for its scalability and for its financial sustainability among providers. Likewise, as a training and professional development tool for organizations, Telehealth has been successful in increasing job satisfaction and improving participants' outcomes. However, there are still many R&R areas that lack internet connectivity to facilitate the delivery of these virtual services and little is known about their effectiveness in mediating relational and cultural barriers.

Suggestions for incentivizing the greater uptake of regional employment opportunities include flexible employment and career development options to ensure staff have meaningful career progression opportunities. These institutional solutions can be coupled with tailored pathways for local high school students to enter relevant fields (Johnsson et al. 2017). That said, youth migration to metropolitan centers is prevalent in many areas, thus remuneration would need to better represent the inherent challenges faced by rural workers, including in relation to career opportunities if long-term retention of the necessary workforce is to be realized. Similar examinations have been undertaken regarding integrated service approaches whereby providers coordinate resources between them (Schoo et al. 2016) and in the dispersal of appropriate information to afford communities the ability to better advocate for their needs (Massey et al. 2013). Underpinning all of these solutions is the need for greater stewardship by the government to ensure that policy solutions and resourcing encompass the unique challenges considered.

Summary

With a sizable proportion of people still living in remote and regional areas globally, understanding and overcoming the significant challenges restricting the delivery of human services and supports is vital. While there is an expanding body of research identifying and enumerating these challenges which, alongside practitioner experiences, has gone some way to producing potential solutions, these issues are complex and enduring. They are also impacted by democratic processes that often prioritize the allocation of resources toward metropolitan-based issues.

To advance and ensure that those living in these areas are properly serviced, it is necessary for governments to ensure that the true costs of rural service delivery are reflected accurately in policy and funding criteria. This includes the added costs related to private sector sustainability, as well as the appropriate workforce incentive structures and cultural training. Lastly, it is evident that efficacious human services outcomes for people in these areas will require mixed and localized approaches that reflect the contextual diversity inherent in remote and regional areas.

References

Atterton, J. (2008). *Rural proofing in England: A formal commitment in need of review.*

Barr, M., Duncan, J., & Dally, K. (2018). A systematic review of services to DHH children in rural and remote regions. *Journal of Deaf Studies and Deaf Education, 23*(2), 118–130. https://doi.org/10.1093/deafed/enx059.

Bradford, N. K., Caffery, L. J., & Smith, A. C. (2016). Telehealth services in rural and remote Australia: A systematic review of models of care and factors influencing success and sustainability. *Rural and Remote Health, 16*(4), 1–23.

Carson, J. A., & Mattingly, M. J. (2018). "We're all sitting at the same table": Challenges and strengths in service delivery in two rural New England counties. *Social Service Review, 92*(3), 401–431. https://doi.org/10.1086/699212.

Dew, A., Gallego, G., Bulkeley, K., Veitch, C., Brentnall, J., Lincoln, M., Bundy, A., & Griffiths, S. (2014). Policy development and implementation for disability services in rural New South Wales, Australia. *Journal of Policy and Practice in Intellectual Disabilities, 11*(3), 200–209. https://doi.org/10.1111/jppi.12088.

Dew, A., Barton, R., Ragen, J., Bulkeley, K., Iljadica, A., Chedid, R., Brentnall, J., Bundy, A., Lincoln, M., Gallego, G., & Veitch, C. (2016). The development of a framework for high-quality, sustainable and accessible rural private therapy under the Australian National Disability Insurance Scheme. *Disability and Rehabilitation, 38*(25), 2491–2503. https://doi.org/10.3109/09638288.2015.1129452.

Dossetor, P. J., Thorburn, K., Oscar, J., Carter, M., Fitzpatrick, J., Bower, C., Boulton, J., Fitzpatrick, E., Latimer, J., Elliott, E. J., & Martiniuk, A. L. C. (2019). Review of aboriginal child health services in remote Western Australia identifies challenges and informs solutions. *BMC Health Services Research, 19*(1), 1–15. https://doi.org/10.1186/s12913-019-4605-0.

Gallego, G., Chedid, R., Hons, B. O. T., Dew, A., Hons, M. A., Brentnall, J., Ot, B., Veitch, C., & Ther, D. R. (2016). Private practice disability therapy workforce in rural New South Wales, Australia. *Journal of Allied Health, 45*(3), 225–229.

Gallego, G., Dew, A., Lincoln, M., Bundy, A., Chedid, R. J., Bulkeley, K., Brentnall, J., & Veitch, C. (2017). Access to therapy services for people with disability in rural Australia: A carers' perspective. *Health and Social Care in the Community, 25*(3), 1000–1010. https://doi.org/10.1111/hsc.12399.

Gibson, O., Lisy, K., Davy, C., Aromataris, E., Kite, E., Lockwood, C., Riitano, D., McBride, K., & Brown, A. (2015). Enablers and barriers to the implementation of primary health care interventions for Indigenous people with chronic diseases: A systematic review. *Implementation Science, 10*(1), 1–11. https://doi.org/10.1186/s13012-015-0261-x.

Gilchrist, D. J., & Charlton, A. (2014). *Home ownership and affordability for people living with disability in Western Australia.*

Gilchrist, D., & Emery, T. (2020). *Value of the not-for-profit sector 2020: The second examination of the economic contribution of the not-for-profit human services sector in the Northern Territory.* Northern Territory: NTCOSS.

Harfield, S. G., Davy, C., McArthur, A., Munn, Z., Brown, A., & Brown, N. (2018). Characteristics of Indigenous primary health care service delivery models: A systematic scoping review. *Globalization and Health, 14*(1), 1–11. https://doi.org/10.1186/s12992-018-0332-2.

Hussain, R., Maple, M., Hunter, S. V., Mapedzahama, V., & Reddy, P. (2015). The fly-in fly-out and drive-in drive-out model of health care service provision for rural and remote Australia: Benefits and disadvantages. *Rural and Remote Health, 15*(3), 1–7. https://doi.org/10.22605/rrh3068.

Johnsson, G., Kerslake, R., Crook, S., & Cribb, C. (2017). Investigation of training and support needs in rural and remote disability and mainstream service providers: Implications for an online training model. *Australian Health Review, 41*(6), 693–697. https://doi.org/10.1071/AH16132.

Massey, L., Jane, A., Lindop, N., & Christian, E. (2013). *Disability audit* (issue June).

O'Sullivan, B. G., McGrail, M. R., Joyce, C. M., & Stoelwinder, J. (2016). Service distribution and models of rural outreach by specialist doctors in Australia: A national cross-sectional study. *Australian Health Review, 40*(3), 330–336. https://doi.org/10.1071/AH15100.

O'Toole, K., Schoo, A., & Hernan, A. (2010). Why did they leave and what can they tell us? Allied health professionals leaving rural settings. *Australian Health Review, 34*(1), 66–72. https://doi.org/10.1071/AH09711.

OECD. (2010). The service delivery challenge in rural areas. In *OECD rural policy reviews: Strategies to improve rural service delivery* (pp. 13–54). https://doi.org/10.1787/9789264083967-2-en.

Parikh, D. R., Diaz, A., Bernardes, C., de Ieso, P. B., Thachil, T., Kar, G., Stevens, M., & Garvey, G. (2021). The utilization of allied and community health services by cancer patients living in regional and remote geographical areas in Australia. *Supportive Care in Cancer, 29*(6), 3209–3217. https://doi.org/10.1007/s00520-020-05839-6.

Rolfe, M. I., Donoghue, D. A., Longman, J. M., Pilcher, J., Kildea, S., Kruske, S., Kornelsen, J., Grzybowski, S., Barclay, L., & Morgan, G. G. (2017). The distribution of maternity services across rural and remote Australia: Does it reflect population need? *BMC Health Services Research, 17*(1), 1–13. https://doi.org/10.1186/s12913-017-2084-8.

Schoo, A., Lawn, S., & Carson, D. (2016). Towards equity and sustainability of rural and remote health services access: Supporting social capital and integrated organisational and professional development. *BMC*

Health Services Research, 16(1), 1–5. https://doi.org/10.1186/s12913-016-1359-9.

Schubert, N., Evans, R., Battye, K., Gupta, T., Larkins, S., & McIver, L. (2018). International approaches to rural generalist medicine: A scoping review. *Human Resources for Health, 16*(1), 1–27. https://doi.org/10.1186/s12960-018-0332-6.

Stephens, A., Cullen, J., Massey, L., & Bohanna, I. (2014). Will the National Disability Insurance Scheme improve the lives of those most in need? Effective service delivery for people with acquired brain injury and other disabilities in remote Aboriginal and Torres Strait Islander communities. *Australian Journal of Public Administration, 73*(2), 260–270. https://doi.org/10.1111/1467-8500.12073.

United Nations. (2019). World urbanization prospects. In *Demographic research* (vol. 12). https://population.un.org/wup/Publications/Files/WUP2018-Report.pdf

Weinhold, I., & Gurtner, S. (2014). Understanding shortages of sufficient health care in rural areas. *Health Policy, 118*(2), 201–214. https://doi.org/10.1016/j.healthpol.2014.07.018.

World Health Organization. (2009, February 2–4). *Increasing access to health workers in remote and rural areas through improved retention*. World Health Organization. http://www.who.int/hrh/migration/background_paper.pdf

Zhao, Y., & Malyon, R. (2010). Cost drivers of remote clinics: Remoteness and population size. *Australian Health Review, 34*(1), 101–105. https://doi.org/10.1071/AH09685.

Changing Paradigms in Urban Planning 2000–2020

The Case Study of the 22 @ District in Barcelona

Héctor Rodal
Architect and Urban Planner, Barcelona, Spain

Introduction

Urbanism is the physical materialization of the concepts and social, cultural, and technological paradigms of each era and each place. This fact is reflected in the historical development of our cities, in which the different layers that have been shaping them throughout history as we know them overlap. These changes usually occurred over long periods of time, sometimes hundreds of years, such as the transformation of ancient Roman squared cities to more irregular medieval cities after the fall of the Roman Empire.

Nonetheless, in all areas of human life, there is an increasing acceleration of events and changes. From 2000 to 2020, there have been numerous relevant events that have made us see and design cities in a very different way in each of those moments. Concepts, theoretical and practical bases that were considered good 20 years ago are being questioned and revised. Relevant facts and trends, already taken as scientific evidence – climate change, environmental deterioration, scarcity of resources and raw materials, increase in inequalities, etc. – as well as disruptive events and technologies – expansion and penetration of Information and Communication Technologies, industrial relocation, global terrorism, the Covid-19 pandemic, etc. – are part of the challenges posed.

This entry takes the case of the 22 @ Innovation District in Barcelona, which has served as a model and inspiration for various cities around the world to undertake similar urban programs. Started in 2000, it has been modified and rethought under new insights and criteria since 2017. Indeed, the concretion of urban regulations and guidelines is being finalized at the same time this entry is being written. Thus, this case allows us to reflect on the changes and differences between both moments and the urban planning criteria and practice associated with each of them.

The World Between 2000 and 2020: General Context

Before getting into the specific case of the Barcelona 22 @ Innovation District, we will try to make an introductory sketch of the situation and general trends and differences between 2000 and 2020 at a global level and specifically in the Spanish and Barcelona's context.

Technological progress has modified the economic and social system worldwide.

If we went back 20 years, we would realize that we currently have totally internalized behaviors that were unthinkable at that time. In 2000, the majority of the world's population still did not have a mobile device. Internet spread occurs from 2010, with an improvement in speed and connections for mobile phones. It has only been 20 years since the use of mobile phones (not yet "smartphone") spread.

As we entered the new millennium, Wi-Fi also began to spread, an element that today has become essential in our day-to-day life. Now with a messaging application, our messages arrive practically instantly with all the multimedia files that you want to add: audio, video, image, gif, etc. or even a video call in real time.

All this has also been integrated with social networks. Let us bear in mind that before 2000 there were none of the most popular social media platforms today. These, in addition, have greatly expanded their applications and utilities. From instant communication to geolocation and online shopping.

Technology has allowed us the possibility of establishing contact in just a few seconds, but it has also allowed us to avoid crowds and waiting. And all this just with a mobile device. Time has been saved, and production and logistics processes have been synchronized in an unthinkable way just two decades ago, which generates efficiency and economic value and this has had an impact on the economic system and social relations that will be even greater in the future.

In 2012, a new disruptive technology emerged, the blockchain. This system of transactions and data storage in a decentralized and distributed way has had its best-known application so far in cryptocurrencies, but the development and traceability possibilities that it allows will make it have an important impact on a large multitude of sectors in the next years. Other phenomena that are in the process of fast development and practical implementation are Artificial Intelligence and the Internet of Things, in which blockchain technology can also be applied.

The space tourism trips are becoming normal and beyond the medical advances in genetic engineering, proposals and projects are now on the horizon advocating a closer integration between human beings and machines, and quantum computing is making its first advances as well.

Although there is still no clear set of how to address these issues, the configuration of cities and urban planning cannot be indifferent to all these phenomena. The answers will probably be initially hesitant and sometimes misguided, which forces us to already propose much more flexible and adaptive urban planning processes.

Indeed, it is not longer possible to understand the current world without the progress and diffusion of the Internet, which is the key piece of the current technological take-off based on the growth and immediate exchange of information. The network of networks that has brought devices from all over the world into contact. That is why the level of use and penetration of the Internet in the population is taken as a marker of all this technological and social change that we have briefly outlined here.

In Spain, the General Media Study collects data on the population's daily Internet use. In 2000, the percentage of respondents who had connected to the network was 5.6% compared to 80.2% in the last report. In other words, the daily use of the Internet in Spain has multiplied by 14 in these years, as has happened in the rest of the world with similar intensity.

Finally, the latest disruptive event that has had an impact on our cities is the Covid-19 pandemic. In addition to the paralysis and interruption for several months of some of the world's supply chains, there are other effects on economic, logistics and social trends and behaviours that may have longer-term consequences still difficult to define.

The need for more open spaces both in homes and in public realm has become evident, especially in urban areas with high population density. It is necessary to evaluate what impact the forced displacement towards teleworking modalities will have in the medium and long term on the office market. The concept of the traditional office has entered into crisis and will be redefined in the coming years, probably towards a more flexible concept that combines the modality of face-to-face and remote work.

Both questions – the need for more open spaces and flexibility in the concept and designs of offices and corporate headquarters – will also impact on the design of numerous urban areas within cities in next years.

Political, Economic, and Urban Context in Barcelona in 2000

In order to present the situation in Barcelona in 2000, it is necessary to make a previous reference to the 1992 Olympic Games held in the city, which were a great success in updating infrastructure and urban deficits accumulated over decades and a success in promoting the Barcelona internationally.

The Olympic Games placed the City and its brand or model – the so called Barcelona model– on the world map of cities. The Barcelona model promotes a model of city that has been copied, sometimes not without criticism, with more or less success in other cities around the world.

The event involved the urban redevelopment of different areas of the city, with four main focal points (Fig. 1): 1. Montjuic mountain, where the so-called Olympic Ring was located, with several sports facilities, among which the Palau Sant Jordi. 2. Another area of action was Vall d'Hebron neighborhood, which combined green areas with sports facilities. 3. In the western area of the city, at Avenida Diagonal, there was also an area for sports activities.

4. Finally, the Olympic Village was built in the Sant Martí district between 1985 and 1992 to accommodate athletes. Built on the remains of an old industrial area of Poblenou, a new residential maritime neighborhood was projected to open the city center to the Mediterranean Sea and to transform the coastline together with the Olympic Port. The project was conceived as a new development from scratch. The entire area covered an area of 720,000 m^2, of which a third was intended for residential areas. The planning and development process was complex as different crucial issues had to be addressed: with the aim of improving urban connectivity in the area, the cut and cover method was applied to tunnel the coastal railway, discharges were channeled and treatment plants were set up, the Olympic Port was built, new beaches were established and regenerated, and new road axes were drawn.

The Olympic Village is located in the southeast of the area of what is now the 22 @ arroba innovation district and stands between 22@ neighborhood and the sea.

Regarding the organization and financing of the Olympic Games, the participation of the private sector was approximately one third of the total investment. Private investment was

Changing Paradigms in Urban Planning 2000–2020, Fig 1 The four Olympic areas in the city of Barcelona

concentrated in the construction of housing, hotels, and road network. Despite the fact that the financing was mainly carried out by the public sector, the contribution of real estate developers, hotel entrepreneurs, and construction companies became essential for the achievement of the infrastructures and facilities needed by the city.

It is in this way that with the celebration of the Olympic Games in Barcelona in 1992, one of the projects with the greatest transformative impact of the twentieth century in the city was concluded and a way of development and urban management was established that would continue in the following years. The moment also coincided with the end of a period of economic expansion and the irruption in the municipality of the economic crisis that had already begun in other European countries at the beginning of the nineties.

Thus, after the celebration of the Olympic Games, the following stages took place:

- 1992–1997. This period covers the post-Olympic crisis period. The result of the Olympic Games is highly appreciated by the local Administration and numerous social agents, both in terms of its financing, the infrastructure created and the international projection achieved by Barcelona. Despite this, the city is facing the post-Olympic hangover with a significant slowdown in economic activity, which especially affects sectors linked to tourism.

At the urban policy level, and in the absence of a political or administrative structure at the metropolitan level, the Municipality recovered its vocation as a central power in the metropolitan region, creating agencies and technical proposals on a metropolitan scale. For instance, BarcelonaRegional was created in 1993 as a public limited company with the function of providing technical assistance to public entities and companies in the metropolitan area.

During this period, economic competitiveness and efficiency became essential themes of local politics. The second strategic plan approved in 1994 focused on economic development issues, emphasizing the need to create technological infrastructure and improve the efficiency and performance, both from the public and private sectors.

- 1997–2006. This period starts with the approval of the Barcelona Municipal Charter to establish a special legal regime for the city. This is the period of the consolidation of the "Barcelona model," especially with regard to the participation of private investment in the financing of urban transformation and the realization of urban projects. The drafting of the city's first strategic plan in 1990 is part of this effort to create a scheme of social consensus and public-private cooperation.

At an urban level, the "Barcelona model" advocates an urban model that is based on density, compactness and diversity and mixed uses, which are taken as defining characteristics of the Mediterranean city.

On the other hand, the urban remodeling linked to the celebration of the Forum of Cultures in 2004 stands out, which is part of a large-scale transformation process in the border area of the municipality next to the Besòs River. Two other important macro-projects were focused on improving transport infrastructure, with the creation of a new high-speed train network with a terminal in the Sagrera area, and the expansion of the airport at the other end of the city, along the Llobregat River.

Thus, the evolution of local politics shows how, since the early 1980s, the characteristics of Barcelona's governance and its urban model have been transforming a local political agenda initially motivated by redistributive issues and improvement of urban quality of life at the neighborhood scale, towards a model more focused on economic growth and large-scale urban transformation.

Principles and Objectives of Urban Planning in the Year 2000 in Barcelona

As previously stated, the first Barcelona 2000 Economic and Social Plan was approved in March 1990. It is the result of a long process that included a broad executive committee, the Barcelona Strategic Plan Association, an organization promoted by the City Council of Barcelona that had the participation of the most relevant institutions and economic and social agents in the city.

The learning and recognition of the mutual dependence between the City Council and the business sector initiated in the context of the Games was consolidated in the post-Olympic period. Following the model of North American cities, the reasons for creating formulas for public-private collaboration were multiple. The key element is the attempt to combine the resources of the Administration with the management and financing capacity of the private sector. Thus, the model of public-private collaboration initiated in the framework of the Olympic Games is repeated in the different urban projects of the city.

Begun in 2000, the urban renewal of the Poblenou district, an intensive industrial area of the city during the nineteenth and part of the twentieth century, is also carried out through public-private collaboration. The objective of the project was the creation of a new productive district focused on knowledge-intensive activities. At the beginning of the project the company 22 @ bcn S.A. was in charge of carrying out the project. 22 @ bcn S.A. was a private municipal company whose objectives exemplify the characteristics of the public-private collaboration model that has been consolidated in the process of urban transformation of the city.

The first objective of the agency was the development and execution of all urban actions related to the 22 @ project, both in the area of planning and management, and execution. The second objective focused on the planning, promotion, design, construction and management of infrastructures, urban services, facilities and public spaces. The third and last objective of the company referred to the national and international promotion of the area to attract or generate new companies in its area of influence. To meet these objectives, 22 @ bcn S.A. held powers that combined the capabilities of public agencies and private companies. These capabilities included having the initiative in the drafting, processing and approval of the planning instruments, and acting in the name of the City Council. Likewise, it was in charge of the Infrastructure Plan, as well as the acquisition and transfer and alienation of municipal land.

The company participated in several initiatives with the private sector under different cooperation modes, from the negotiation of the transfer of municipally owned land to private companies, to participation as a shareholder in public-private agencies created ad hoc for specific projects. The agency operated in a highly effective framework of cooperation between the public sector, which provided the Administration's resources, and the private sector, which provided investment and financial resources.

In short, in the case of Barcelona, urban policies at the beginning of the twentieth century were the result of a process of intense public-private cooperation.

Principles and Objectives of the 22 @ Innovation District in 2000

Context and Urban Background

In the 1990s, it became clear for the Municipality that the industrial land of Poblenou showed symptoms of stagnation leading to a process of urban degradation, to which must be added problems of coexistence between industrial use and existing housing, and the lack of urban recognition of the latter.

That situation occurred in a context of industrial relocation to other areas of the metropolitan region (Fig. 2) and offshoring, along with the lack of identification by the urban planning framework of the changes that were taking place in industrial processes. These facts highlighted the need for a regulatory change to regenerate this large area of land with significant urban centrality.

The reflection on this modification took the form of two planning documents, which began the urban planning process that led to the Modification of the Master Plan of the area in 2000.

The first of these was based on an exhaustive analysis of the territory and was committed to maintaining the productive nature of the area, already outlining the idea of orienting it towards activities related to new technologies,

Changing Paradigms in Urban Planning 2000–2020, Fig 2 Aerial View

culture, research, and knowledge in general (Fig. 3).

The second document supposed a greater concretion of the proposal, and established the main criteria for 22@ Innovation District, such as the progressive transformation of the area, a new land use regulations to attract new activities, the presence of a certain percentage of housing, the provision of a plurality of instruments for intervention, the implementation of new structuring elements to provide a coherent urban structure, and the preservation of industrial heritage. On June 9, 1998, the document of Criteria, objectives and general planning solutions for the renovation of the industrial areas of Poblenou was presented to the public, which was approved on March 25, 1999.

The justification for the development of District 22 @ can be summarized in the following four sections:

1. Continue the process of urban regeneration started with the construction of the Olimpic Village, which had led to a profound transformation of the city's coastal skyline, thus completing the recovery of the area in the historic center of the Poblenou neighborhood.
2. Boost the modernization of local economy, at a time when many traditional industrial activities in the City were declining.
3. Promote the social development of the Poblenou area, both in relation to the residents of the 22 @ district, as well as the professionals who work in it.
4. Strengthen the project based on a solid strategy of public-private collaboration, given the good results that had been obtained through this formula in other key development initiatives of the city in previous years.

On the one hand, and from an urban perspective, the need to complete the recovery of Poblenou was raised. After the Olimpic Games a new road system for the city had been built, which had provided this area with excellent connectivity with the metropolitan region, the port

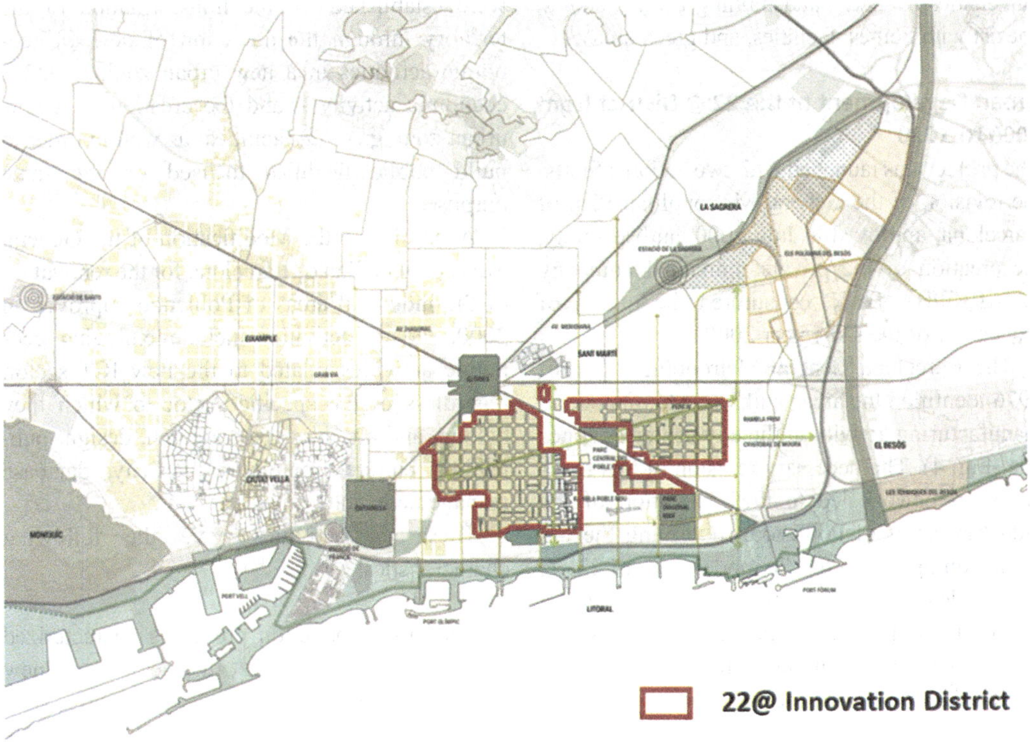

Changing Paradigms in Urban Planning 2000–2020, Fig 3 Location

and the airport, the beaches had been recovered for urban use, and the Olimpic Village had been built. Subsequently, in February 1999, the last section of Avenida Diagonal, one of the main streets of the City, was also opened, thus connecting the Plaza de las Glorias with the Universal Forum of Cultures. The opening of this avenue improved communication between the Poblenou neighborhood and the city's business center.

On the other hand, from an economic point of view the Poblenou area was in a deep crisis situation. This area was known as the Catalan Manchester, since it had been one hotspot of the industrial revolution in Catalonia. In the middle of the eighteenth century, textile activity began in the neighborhood, and during the nineteeth century, the process of industrial activity expanded in the area. From 1940 to 1964, there was a new impulse in Poblenou with the establishment of companies in the metallurgy and automobile sectors. However, in the mid-sixties a stage of progressive decline in economic activity in the neighborhood began. From then on, companies related to logistics and transport services were mainly established in the area. In this scenario of decadence, the need for a reconversion of the economic activity was raised, consistent with the challenges that the city of Barcelona faced globally.

Third, intervention in the neighborhood was proposed with the aim of facilitating the mix of uses and giving priority to quality of life, in contrast to the model of the diffuse city resulting from the industrial city. Thus, the promotion of housing in Poblenou was planned, including the promotion of social housing, which would allow people to live close to their job and favoring the development of local businesses. Finally, the provision of quality public space in the district was contemplated, through the creation of new green areas and public spaces, and reducing acoustic and environmental pollution in Poblenou. In short, the territory was conceived as a space where companies, universities, research and training centers should coexist with homes, facilities, and green areas.

Urban Development of the 22 @ District from 2000 to 2020

The project was launched with two actions: firstly, the revision of the General Metropolitan Plan of Barcelona, approved in July 2000, and secondly, the creation of the private municipal company 22 ARROBA BCN, constituted the month of November of the same year 2000.

The Barcelona General Metropolitan Plan of 1976 identified the areas with a predominance of manufacturing activity in the city center as zones 22a (Fig. 4). The necessary revision of this regulatory framework led to the renovation of the industrial areas of Poblenou, modifying the old urban zoning of 22a for the modern one of 22 @, which clearly indicated the orientation of the change to be promoted: the economy linked to Information and Communication Technologies (ICTs). Specifically, in the 22 @ area, incentives were established for the transformation of the territory through the attraction of new @ economic activities in a new urban zoning –22@ economic activity – and the creation of a new urban zoning –7 @ facilities- as well in order to build public facilities focused on the same purposes.

According to the Modification of the General Metropolitan Plan of Barcelona for the renovation of the industrial areas of Poblenou approved in 2000, the @ activities include emerging economic activities related to the new ICT sector, regardless of the specific sector to which they belong, and are related to research, design, publishing, culture, multimedia activity, database management, and knowledge.

These @ activities have the following characteristics:

- They use production processes characterized by the intensive use of media and new technologies.

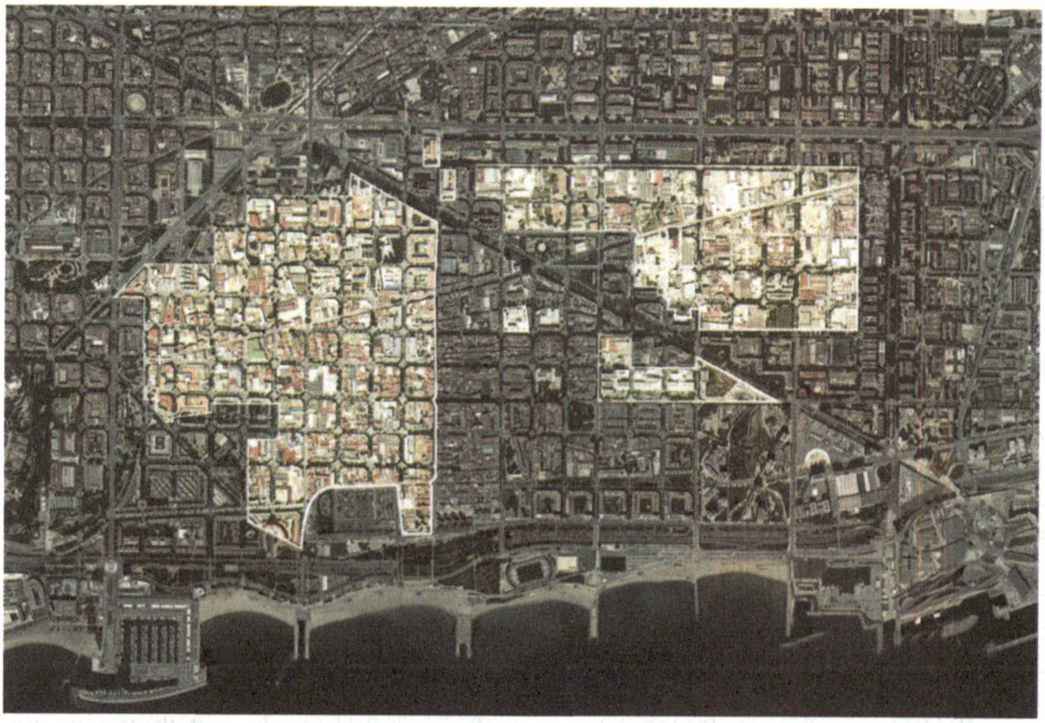

Changing Paradigms in Urban Planning 2000–2020, Fig 4 Area of 22@ Innovation District

- They are labour-intensive (number of workers or users / surface).
- They generate high added value.
- They are directly related to the generation, processing and transmission of information and knowledge.
- They are not polluting or annoying and can be developed in central urban environments.

On the other hand, the 7 @ facilities are based on the following uses: permanent training activities, dissemination of new technologies, and private productive activities related to training in the field of ICT. Specifically, the aforementioned plan foresees that approximately 10% of the transformed land will become public property and will be developed as 7 @ facilities.

The second decisive action in the launch of 22 @ was the creation of the private municipal company 22 ARROBA BCN, an entity with its own legal personality and with powers to promote the plan with the following objectives:

- The development and execution of any type of urban actions related to the economic areas of the city with the urban qualification 22 @ (planning, design, management, elaboration, and execution of projects; design, construction and management of infrastructures, urban services, and public spaces).
- National and international promotion missions of the aforementioned industrial and productive areas, as well as actions for the creation and attraction of companies and activities related to information and communication technologies.
- Unique and whole management of the plan, which has allowed for a highly professional and independent organizational structure, which has ensured comprehensive management of its two most characteristic dimensions: urban planning and knowledge-intensive activities.

The 22 @ BCN company was dissolved in 2011, and its powers passed from then on to the general services of the City Council.

Current Situation

In its first 20 years, the 22 @ Barcelona project was developed in two major phases:

1. Urban renewal actions: they had a fundamental and priority character from the beginning of the project in 2000 to 2005. The ambitious process of reconversion of an area in decline to create a diverse and balanced urban environment required focusing several urbanprograms and actions of the City in this area during the first years.
2. Promotion of economic, social and cultural activities.

Consequently, once the initial impulse of urban interventions had been consolidated, 22 @ opened a new phase of vigorous development in the economic, social and cultural spheres. To this end, projects were developed aimed at creating areas of European excellence in various sectors where it was considered that Barcelona could achieve international leadership: audiovisual, ICT, medical technology, energy, or design.

The strategy followed consisted in the promotion of clusters in the territory to favor the concentration of companies, public organizations, and scientific and technological centers of reference in these strategic areas of knowledge. Specifically, the following four clusters were initially configured.

Media, Medical Bio-Technologies, ICT, and Energy.

In 2008 a new line of activity began in a fifth strategic sector: design.

The results achieved at the urban level by the 22 @ project in its 10 years of existence may be summarized in the following terms:

- Renovation of approximately 65% of the industrial areas of Poblenou.
- Construction of 85,000 m2 for facilities
- Management of 70% of the planned social housing.
- Eight public green areas totaling more than 22,000 m2
- 12 kilometers of reurbanized streets

In the field of economic activity:

- Location in the district of 10 university centers with more than 25,000 students, and 12 R&D and technology transfer centers.
- The 22 @ innovation district has 7064 companies and some 4400 self-employed workers. The total number of existing companies is 105.5% more than in 2000. The increase is much higher than that registered in the province of Barcelona (57.3%) and in the whole of Catalonia (60%).
- The estimated number of workers is 90,000 people, 62.5% more than in 2000, which represents an increase of 56,200 workers. On the other hand, the global business volume, not only of the @ activity companies, amounts to about 8900 million euros per year.

As a consequence of the results obtained, the 22 @ Innovation District of Barcelona has become a reference model for urban, economic and social transformation on an international scale. Likewise, it has received recognition from the International Association of Science and Technology Parks (IASP), and also from the World Network of Clusters (TCI).

However, one of the most critical points of the project has been the need to establish greater protection of the industrial and historic heritage, the maintenance of traditional economic activities and the improvement of communication with social agents. Thus, in 2006 the modification of the heritage catalog of the city of Barcelona was approved, with the aim of recognizing the importance of Poblenou in the industrial past of the city.

Beside this and after the change of government in the municipal administration, a process of citizen participation began in 2017 to rethink the development of the area that led to the signing of the "Pact for a more inclusive and sustainable 22 @." This document, which had the participation of social agents, neighborhood entities, companies, and the academic sector, gathered the claims and complaints of the population residing in the neighborhood.

Principles and Objectives of Urban Planning in Barcelona in 2020

Current Urban Context: Environmental, Social and Economic Crises

Cities are presently hyper-connected places of agglomeration that, on the one hand, concentrate economic activity, opportunities, talent, culture, and diversity; but, on the other hand, they are in the first line managing the different urban crises: environmental, social, and economic (Fig. 5).

In the context of these current crises, Barcelona is presently deploying an Urban Regeneration Strategy that seeks to revitalize the city in all the dimensions, scales, and functions: from the global and metropolitan area, to the neighborhood, street, housing, and citizens scale with specific measures in each dimension of the crisis.

Principles and Objectives of the 22 @ Innovation District in 2022

22 @ has proven to be a project of great importance for the innovative economic development not only of the Sant Martí district, but of the whole of Barcelona and its metropolitan area. In the last 18 years, the knowledge economy sector driven by 22 @ has adapted to different economic, political and technological changes, with stages of great expansion and also less accelerated stages. The city project proposed in 2000, to promote an industrial model adapted to the technological revolution based on the knowledge economy, made it possible to promote a new productive base.

However, after 17 years of development, the need to respond to accumulated deficits and new urban, social and economic challenges became evident. Thus, in 2017 the City Council began a participatory reflection with all the local agents with the aim of drawing up a shared roadmap that could guide the future transformation of Poblenou. More than 1000 people from different entities and sectors participated in the Rethink 22 @ process (Fig. 6).

As a result of this reflection, in 2018 the "Towards a Poblenou with a more inclusive and

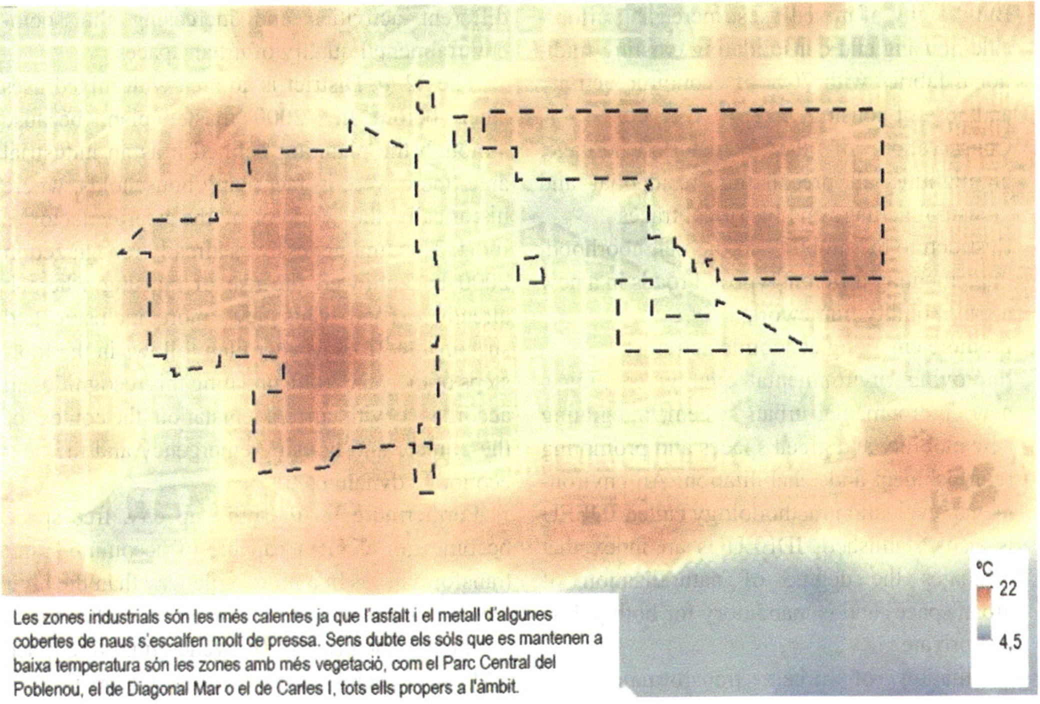
Les zones industrials són les més calentes ja que l'asfalt i el metall d'algunes cobertes de naus s'escalfen molt de pressa. Sens dubte els sòls que es mantenen a baixa temperatura són les zones amb més vegetació, com el Parc Central del Poblenou, el de Diagonal Mar o el de Carles I, tots ells propers a l'àmbit.
°C
22
4,5

Changing Paradigms in Urban Planning 2000–2020, Fig 5 Urban surface temperature in the area

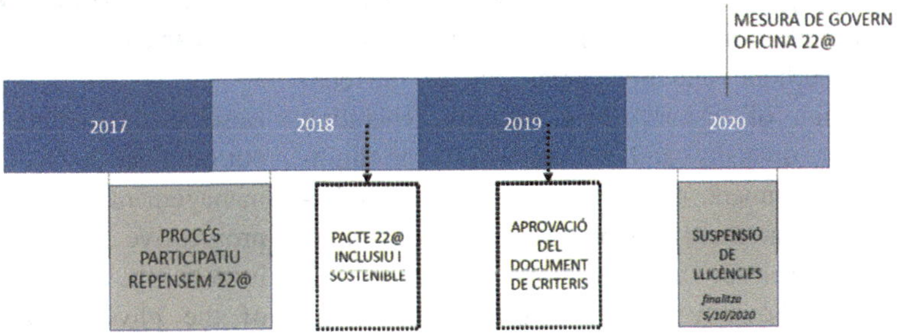

Changing Paradigms in Urban Planning 2000–2020, Fig 6 Participation process

sustainable 22 @" document was signed with the participation of citizens, the economic sector, universities, and public administrations. A year later, that agreement allowed the development of the criteria document that established the bases of the new Masterplan for a more inclusive and sustainable 22 @, which is currently in the initial approval phase.

The urban proposal for the 22 @ area is the development of an Urban Regeneration strategy that aims to consolidate existing urban fabrics and uses with great social, identity, and business value, but also the transformation of some areas to achieve a city of mixed uses and environmental quality.

Objectives

Based on the 22 @ 2000 urban model, the main objective of the new Master Plan is to plan an area of urban regeneration in order to develop a mixed city of environmental quality, which enhances daily life and productive activities. To address this, a number of specific objectives have been defined in order to speed up the transformation and facilitate the development and consolidation of the area:

- Build a city of mixed uses, increasing affordable housing and consolidating existing traditional fabrics with 70% of economic activity and 30% of housing
- Conservation of the historic heritage, highlighting all preexisting, residential and industrial heritage and historical traces
- Restructure the mobility in the neighborhood, implementing different types of roads in a new urban mobility framework
- Promote emerging economic activities
- Improving environmental quality: creating a new environmental urban system, integrating new mobility and green spaces and promoting refurbishment and rehabilitation. An environmental evaluation methodology called IDEEU is also established. IDEEU is an index that evaluates the degree of naturalization of urban spaces and is mandatory for both public and private.
- Facilitation of urban transformation of undeveloped areas

In this sense, the proposal articulates an urban area of mixed uses through a process of urban regeneration that allows the renewal of obsolete neighborhoods and urban areas, but also the rehabilitation of historic urban fabrics generating greater diversity and versatility of the built-up environment, maintenance of existing activities in the area and also for the consideration of the urban landscape and the improvement of the environmental quality. Beyond the preservation of the architectural and historical value of some buildings, pre-existences become elements that shape places of identity and identification, and that characterize the different areas of a diverse city.

This urban regeneration is a process that requires an exhaustive knowledge of the built reality, but also of the tendencies and requirements of the population and the economic sectors of the city in order to generate spaces of opportunity while mitigating the social and environmental impact.

Urban areas that host mixed uses facilitate daily life, generating greater intensity of use of public space, improve the economic vitality of the area and promote active mobility between different activities and increasing the socio-environmental quality of urban spaces.

The 22 @ District is an area with mixed uses since before the 2000 master plan, because although the planning defined it as an industrial area, there were in fact many households, where historically the workers of the nearby factories lived. This uniqueness was already recognized in 2000 and allowed the use of housing where it already existed. This new, more inclusive and sustainable 22 @ Master Plan follows in the footsteps of the year 2000, updating this recognition in accordance with current legislation, the context of the climate and housing emergency and the new economic dynamics.

Furthermore, in the built up city, free space becomes an element capable of accommodating transformations in a more agile way than the built fabric. That is why, in the face of an urban environment in the process of regeneration, free space has the capacity to provide the city with new socio-environmental requirements and needs.

Urban Proposals of the 22 @ 2022 Master Plan

Based on the fact that the transformation is carried out on a previously occupied land, planning and management mechanisms are designed to allow a progressive intervention as well as the balance between the transformation and the maintenance of the physical and economic pre-existences (Fig. 7).

Regarding economic objectives, it favors the reorientation of the productive character of the area to more information and knowledge-based economic activities, which have needs of density, employment, centrality, and new infrastructures different from those of the traditional industry.

On the social side, the new Masterplan sought the recognition of existing values in the territory with a transformation that consolidates the existing residential fabric, allowing the permanence of the population as well as existing activities. At the same time, the area will be developed with greater density, and a mix of uses with an increase in public space as well.

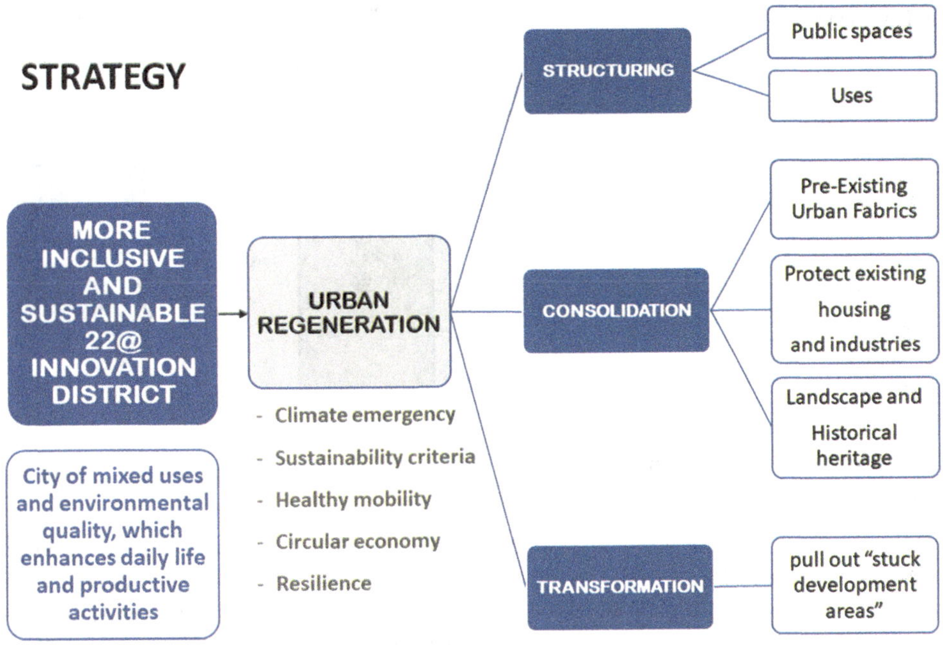

Changing Paradigms in Urban Planning 2000–2020, Fig 7 Urban regeneration strategy

The new Land use regulations, the Buildability, and the IDEEU Index

The old 22 @ Master Plan developed a new urban framework by establishing common rights and duties for land owners and defining urban planning instruments of different scales and characteristics.

This common framework is very briefly summarized in some rights: the possibility of increasing the buildability depending on the uses implemented and compliance with some requirements: the transfer to the city of land to maintain pre-existing or install new uses (Fig. 8).

Thus, with the aim of achieving the heterogeneity of uses, and favoring the implementation of new productive activities, it is possible to move from an exclusively industrial productive soil to a range of uses that includes, in addition to industrial and housing, office, residential, commercial, health, religious, cultural, recreational, sports, technical, and environmental services. Among the new productive uses are the activities @, defined as those emerging activities related to the new sector of information and communication technologies and those related to research, design, publishing, culture, multimedia activity, database, and knowledge management.

Thus, the increase in buildability is linked to the transformation of uses.

On the other hand, with regard to existing residential land, the use is recognized, although in two different situations: as consolidated and non-consolidated housing which, in the event of physical remodeling of the area in which it is located, entitles the pre-existing dwellers to rehousing.

Thus, in the scope of the Master Plan 22 @ we have as a situation of origin a land that, from the urban legal regulation is defined as an industrial area, but from the point of view of its reality is diverse and heterogeneous. In this situation it is possible to achieve a more diversified, complex and rich urban environment taking advantage of any of the new urban and legal instruments and regulations provided by the Master Plan for this purpose.

Economic Activity

Currently, considering the evolution of the project and the new opportunities for innovation, it is

Buildability and Density

	2000		2020	
Floor Area Ratio	3.0	100%	3.2	100%
Activity	2.7	90%	2.2	69%
Social Housing	0.3	10%	1.0	31%

It is planned to increase density from 47 units / ha → 80 units / ha.

Density reference data:

City center: 174 units / ha

Average density mixed city: 100-120 units / ha

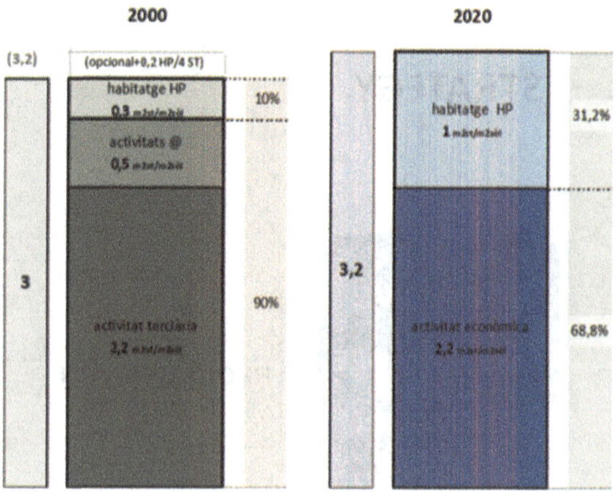

Changing Paradigms in Urban Planning 2000–2020, Fig 8 Urban parameters

necessary to readjust some aspects to respond both to the needs of the residents of the neighborhoods involved and to the needs of the business sector and in general to the needs of the city, reinforcing its innovative capacity and global competitiveness.

It should be noted that, although one of the objectives of the previous Masterplan was to promote a diverse urban environment, today there is still a lack of different kind of urban spaces, capable of hosting different activities (Figs. 9 and 10).

Diversity, on the other hand, can be found in the fact that some industrial activities have remained in unprocessed areas. Activities that last over time and, therefore, can work in coexistence with other urban uses. The mixture of transformed and non-transformed areas has made 22 @ an area that has offered different types of spaces for activities with different needs.

22 @ is a neighborhood of industrial origin and marked by its productive capacity, but also by innovation and transformation. As we have seen in previous sections, history has made the 22 @ realm a territory of continuous reinvention, housing the different activities over time, which were transformed with the requirements of each period of time.

In the year 2020 and 2021, at the time of writing the Masterplan for a more inclusive and sustainable 22 @, Barcelona and its citizens are facing a global pandemic that has made rethink many production systems in order to become less fragile and dependent and more resilient and socially and environmentally responsible. The importance of cities remaining productive, and therefore ready to host various economic activities, rather than relying mainly on tourism as a major economic sector, has also been further emphasized.

For all this, it is necessary to analyze current economic and productive trends, and understand their spatial requirements, in order to accommodate the 22 @ area, as long as it is a benefit for the city and the citizens of Barcelona and its metropolitan area, such as the green and circular economy and the social and solidarity economy, industry 4.0, the maker movement, and other creative and productive activities.

Green and Circular Economy and the IDEEU index

The green and circular economy has become increasingly important at the global and municipal levels in recent years. An example of this is the study on the circular economy carried out by Barcelona City Council in 2018, where the principles of the green and circular economy are translated into the promotion of environmental quality, the naturalization of the city by applying the environmental assessment methodology based on the IDEEU index, the regeneration and reuse of water, local loops, responsible consumption and environmental culture. Furthermore, the environmental assessment methodology based on the IDEEU

C

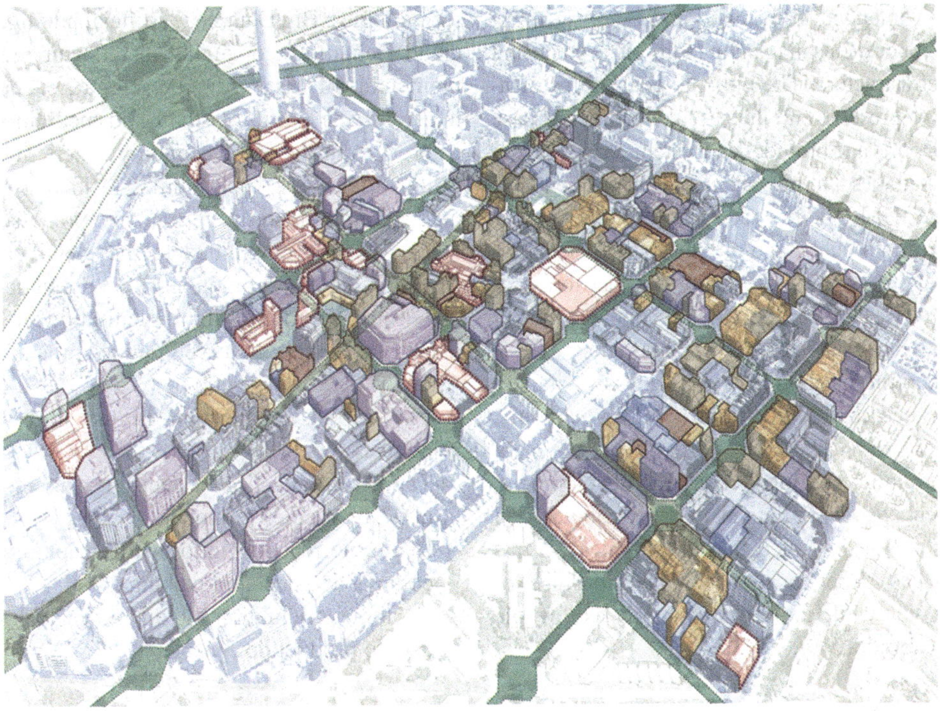

Changing Paradigms in Urban Planning 2000–2020, Fig 9 General Structure

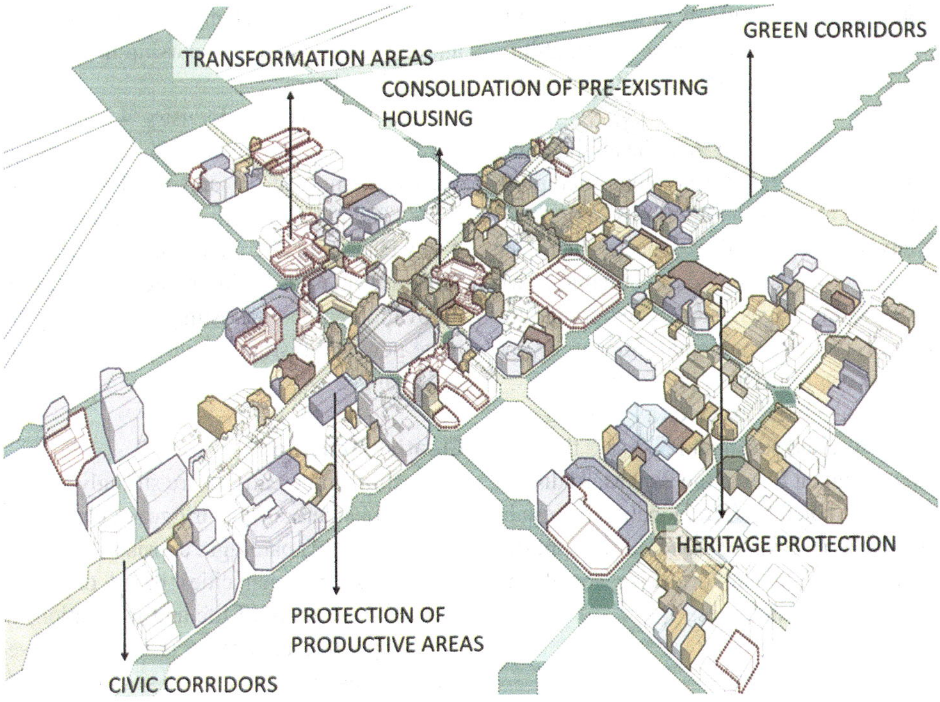

Changing Paradigms in Urban Planning 2000–2020, Fig 10 Urban Systems

index is intended to evaluate and establish minimum requeriments in relation to the city's green and blue infrastructure. The ultimate goal of this measure is to preserve and increase the ecological base that supports and provides ecosystem services to the city and makes it more sustainable and resilient. For this, the IDEEU index takes into account the surfaces, of both public and private spaces, their coverage, biomass and biodiversity in order to assess the services they can provide: evapotranspiration, infiltration of rainwater, shaded areas, etc.

Economic promotion and business revitalization related to the green and circular economy, together with the execution of employment plans, training and local economic development, represent specific initiatives that are being carried out. It should be noted that employment in this sector has experienced an annual increase of around 4% while the city's economy as a whole is only about 2.5%.

Social and Solidarity Economy

This area represents an opportunity for 22 @. Its relevance is not only economic, but also promotes comprehensive sustainability focused on economic development based on values that unite the city, such as equity, solidarity, sustainability, participation, inclusion, and commitment to the community.

This sector represents 2.8% of the total number of companies and 8% of Barcelona's employment, although in order to identify its importance for the city's economy other aspects should be considered, such as the degree of equitable distribution of labor and wages, internal democracy, the reduction of the environmental impact or the value of their contribution to the commons.

In line with the Pact "Towards a Poblenou with a more inclusive and sustainable 22 @," the Economic Development Plan of the Sant Martí district 2018–2022 promoted by Barcelona Activa establishes the urban revision of 22 @ as an opportunity to accommodate and prioritize social and solidarity economy initiatives.

Industry 4.0 and Makers

The fourth industrial revolution is based on the availability of relevant information about a product, obtained through a network of different technologies: Big Data, IoT, artificial intelligence, cloud technology, or digital manufacturing. The momentum of this sector represents a great opportunity for the fabric of small and medium-sized enterprises in the city's economy, where digitalization offers new opportunities to increase productivity and modernize production processes.

The City is committed to the identification, development and promotion of good practices in this sector, promoting the digitalisation of local industry following four well-defined pillars:

- The vertical connection in the form of a network
- The decentralization of decision-making, executed by predictive models
- Real-time reaction: information capture, processing, and decision-making
- Customer orientation and direct feedback from user to product designer

It is a sector with a large increase in employment and the number of companies, with a growth of 36% and 42%, respectively, between 2011 and 2016. This increase is reflected in the international positioning of Barcelona in reference to digital and technology entrepreneurship in various international rankings.

Another aspect of Industry 4.0 is the maker movement. It is a trend that promotes imagination, creativity, innovation, and the learning of new skills. The main objective is to make Poblenou a "New District of Art and Creativity in Barcelona." This movement is highly related to creativity and innovation and is interesting for its ability to promote new and disruptive initiatives, technologies and changes.

Creative Activities

Innovation and creativity are strongly related and can be considered as the trigger of change. At the urban level, this paradigm consists of a new competitiveness between cities based on creativity, knowledge, quality of life and innovation, where the competitive advantage is not only based on economic aspects, but also on social, cultural, and environmental aspects.

Creative activities represent a well-established reality in the economic reality of Barcelona that must continue to be promoted, with an employment of close to 130,000 people and an increase of 13% compared to 2008. The dynamic behavior of non-traditional creative industries is relevant such as research and creative development, advertising, software, video games and electronic publishing, etc., which grew by 45.7% in the same period.

Inclusive Urban Planning

Inclusive urban planning focuses on the knowledge of the life processes of the different social groups, and seeks ways to allow and provide spatial and temporal support to daily needs and, at the same time, be respectful of ecosystems.

Life support activities such as buying food, visiting a medical center, caring for children or the sick, playing, socializing, or participating in the community have a spatial and temporal translation. Depending on the forms that cities take and how they organize activities, it will be more or less easy to carry them out and share and socialize them.

The 22 @ in 2000 did not consider this perspective, and as a result the district has not developed by paying attention to these aspects. That is why in the participatory process spaces and activities are vindicated to coexist and to carry out tasks of daily life.

The data that have been analyzed and shared for this purpose are: daily facilities, urban green spaces, shops and metro network stations. Population density per block has also been added. With the superposition of all this information spaces and services associated with the development of daily life can be identified.

In this sense, it should be noted that in the 22 @ area, in general, there is a lack of space for daily tasks.

The 22 @ area has a large surface area of free space, which is approximately 40%.

However, most of this space is for vehicles and mobility. In addition, the low density of housing in coexistence with economic activities, both industrial and tertiary, also generates very different uses of public space throughout the day. These uses in environments with a greater presence of spaces for daily life are more constant during the different hours of the day.

In relation to the built space, the façades and their ground floors define the character of the public spaces, which is why their continuity allows for proximity and daily life. The current state of the 22 @ district, with many areas still undergoing transformation, mean urban discontinuities that make it difficult for daily activities or to form urban axes capable of hosting urban and social activities.

Therefore, the lack of spaces for daily life in the 22 @ area is a consequence of the difference in population density compared to other areas of the surrounding neighborhoods, which absorb and concentrate more activities and more continuously throughout the day.

Conclusions

Twenty years does not constitute a long period of time in the development of a city. Traditionally, any paradigm shift and urban transformation involved much longer periods. For this reason, the causes of this change in such a short time must be sought in the increasingly changing conditions of the world in which cities must develop and evolve. The scientific, citizen and political awareness of the effects of different emerging phenomena, such as climate change, environmental deterioration, the lack of energy and resources, etc., are triggers for such a quick change in premises and goals. In this sense, the new Master Plan for 22@ Innovation District can be included in what is known generically as Green Urbanism.

Green urbanism, also known as ecological urbanism, is an urban model whose design and use of resources reduces environmental impact. Green urbanism is interdisciplinary, since there are several different elements, theories and skills needed to be implemented. It has a sustainable vision and is based on the circular economy.

Although the proposals for a more conscious and respectful urbanism with the environment and the use of resources are not new, it can be said that now there is a decided bet to implement its postulates in practice.This implementation encompasses different scales and measures: from the urban codification itself, with norms and ordinances that impose the inclusion of measures and elements focused on greater urban sustainability and resilience, to a greater awareness of technicians, companies, and citizens and a proliferation of methodologies for evaluating and verifying the results obtained.

Perhaps it is in the determined will to implement concrete measures and to evaluate the effects resulting from these measures that this mind shift can be perceived to a greater degree, for example with the establishment and mandatory enforcement for both public and private of the environmental assessment methodology of urban spaces based on the IDEEU index. Many cities now have sustainable action plans which are roadmaps towards sustainability. Green Urbanism has gone from textbooks to real action plans.

Thus, if 20 years ago the main concern was the achievement of an economically attractive city, capable of hosting and attracting economic activity linked to the incipient information and communication technologies to overcome the post-Olympic economic crisis and continue to position Barcelona as a City of reference at the international level – objectives that were met to a large extent – we now perceive a broadening of the focus of interest towards the effective inclusion of environmental and climatic variables in urban policy and planning, at the same time that it seeks to redirect the application of ICTs towards more decentralized and distributed operating models, which allow a more egalitarian distribution and the closing of matter and energy loops in a more circular economy.

A shift from highly technological solutions to more environmentally sensitive solutions for the 22@ Innovation District can also be perceived.

Finally, it is foreseeable that emerging trends and disruptive changes will accelerate more and more. This leads us to also consider the need for new implementation and urban planning systems and procedures that, without losing consistency and coherence, should be able to react much faster and in a more flexible way.

Child-Friendly Cities

► Moving Towards Sustainable, Liveable, and Care-Full Urban Environments: Pre-schoolers' Rights and Visions for Planning Just, Socially, and Ecologically Integrated Cities

Children – Minors, Infants

► Children, Urban Vulnerability, and Resilience

Children, Urban Vulnerability, and Resilience

Options for the Future

Gloria Nyaradzo Nyahuma-Mukwashi[1], Marcyline Chivenge[2] and Innocent Chirisa[2]
[1]Department for International Development (DFID), Harare, Zimbabwe
[2]Department of Demography Settlement and Development, Social & Behavioural Sciences, University of Zimbabwe, Harare, Zimbabwe

Synonyms

Children – minors, infants; Resilience – absorptive capacity; Urban vulnerability – poverty

Definition

Resilience – the ability to bounce back to form after a shock

Vulnerability – proneness to danger

Introduction

Cities are both drivers of prosperity and inequity. Children vulnerability has a spatial dimension of inequity (Ryder 2017). Spatial inequity is evident in numerous ways and reveals the importance of land value, land tenure, land use, and the planning and management of the spatial characteristics of the built environment (Stephens 2012). Children especially those from disadvantaged families are confronted with spatial inequity in several ways. These include the high cost of living and access to urban services, the unequal geospatial distribution to urban services, the poor characteristics of the built environment, and the inequitable spatial distribution of land and urban space.

Methodologically the chapter draws upon evidence from recent research and literature to gain insights, outlining the most important areas of interest for child vulnerability and future policy measures that would improve the resilient of children. The chapter draws upon evidence from recent research and literature to highlight the importance of children vulnerabilities and to recommend the need for increased support for meaningful participation of children in urban planning and other decision-making processes concerning their lives in urban contexts. Children are most vulnerable to both natural disasters and effects of bad governance. It is therefore imperative for local planning authorities to reduce vulnerability and increase resilience of children living in cities. Children's voices need to be heard and acted upon, and inclusive participatory governance processes and mechanisms need to be strengthened. When children are vulnerable, it means that the views and concerns of the most disadvantaged young people are likely to lead to the extension of protection risks as well as rights violations and will result in failure to develop effective, responsive, and accountable programming.

Putting into Perspective the Issues and Context

Urban areas occupy approximately one billion children. The rapid rate at which urbanization is occurring has noteworthy consequences for the realization of children's rights to survival, protection, development, and participation. This calls for an increased focus on urbanization and deprivation on the lives of children. Urbanization has been identified as the major contributor of vulnerability as it affects government efforts to provide adequate services to its residents. Rapid urbanization is associated with urban informality and poverty. The rate and complex nature of urbanization in developing countries cripple effects of local authorities to provide infrastructure (Boyden et al. 2019). Protecting children means ensuring that the increased urban attention reflects children's views and experiences and supports their meaningful participation in urban governance so that children's realities can be better addressed to createl inclusive, safe, child-friendly cities (Eide 2006; Mota Borges 2016).

Children living in low-income regions are more susceptible to hazards and risks. Poverty is concentrated in areas and neighborhoods that are economically, physically, socially, and environmentally vulnerable (Robinson et al. 2018). The poor characteristics of the built environment, the social insecurity, and the environmental vulnerability disclose the socioeconomic status of a community. Unequal spatial distribution of land use and urban space (the living environment) – spaces and built environment programs do not prioritize the needs of the urban poor or vulnerable groups such as large families, youth without parental support, children with disabilities, women, and elderly (Douglas et al. 2008). When space and programs are not designed with or for the poor and vulnerable, there is more likely to be a lack of adequate housing, space to walk, and bike in the streets and public space for children.

The urban paradox reviews that not everyone benefits from living in cities.

Children are often placed in the most disadvantaged positions, as the built environment of a city is constructed by adults for their own use, to respond to their daily needs. Cities have been built by the adults and mainly cater for their uses neglecting needs of children. The benefits of urban life bypass children, and the negative aspects can impact them hard. Children are more susceptible to these impacts also because they are unaware of the risks they are exposed to. Children's vulnerability is experienced in a number of ways. Children are vulnerable to environmental risks. The environment in which fetal and childhood development occurs is critically important. The occurrence of an environmental hazard has an impact on the social and economic well-being of children. Several factors influence the risk of health effects from exposure to hazards, including the degree of susceptibility or vulnerability. Children are more likely to suffer from adverse reactions and long-term health effects than adults, since they are subject to greater exposure and are more vulnerable to environmental hazards, especially to contaminants (Landrigan et al. 2004;

Salvi 2007). Children need more food, water, and air per kilogram of body weight compared to adults. Also, child's immune system and other key physical defense mechanisms are not fully developed, leading to a greater risk from exposure to contaminants in the outdoor and indoor air, water, and soil. Children's immature immune systems may increase their susceptibility to infections and allergic reactions. Also, children are unaware of risks they may experience and therefore might lack defense mechanisms.

Urban poor children are highly prone to the effects of climatic disasters, extremes, and environmental emergencies (Dodman and Satterthwaite 2008; Douglas et al. 2008). Climatic hazards affecting urban poor children commonly include flooding, water scarcity, and urban heat. Complex emergencies resulting from reaction of climatic incidences with non-climatic stressors such as chemicals, hazardous and infectious wastes, drainage and sewerage failure, pollution, accidents, fire, and dilapidating and unsafe buildings are a growing concern in relation to children's vulnerability. Urban children particularly those living in poverty face multiple deprivations rendering them vulnerable in urbanizing cities (Christian et al. 2017). Young people are extremely restricted independent movement and opportunities for recreation as well as play while increasing their susceptibility to hazards, violence, and accidental injuries (Rudner 2012). The increasing consequence of such risks rigorously challenges the adaptive capacities of young boys and girls to climate change. Appreciating these risks is significant, as policies intended to reduce pressures on resources, manage environmental threats, as well as upsurge the welfare of the deprived community members may concurrently advance to sustainable development goals through improved adaptive capacity and lessening of vulnerability to climate change as well as non-climatic risks (Stringer et al. 2009).

Water, sanitation, wastes, climatic hazards, and environmental emergencies coupled with ecological decline and poor governance are known to compromise not only children's health, safety, and protection but also multiply and complicate their future vulnerability to such or other risks (Blaikie et al. 2014; Gupta et al. 2017). Several slums are located along the edges of sanitation corridors on untenable lands. Servicing slums on ecologically fragile lands, besides being additionally complex, is also fraught with legal problems. Children in informal settlements are susceptible to child labor and health and safety problems. Slums also lack social services such as clinics and schools which limits ability of children to access these resources.

Children, Urban Vulnerability, and Resilience: A Review

Vulnerabilities refer to individual or household exposure to risks as well as the capacity to respond to them. This brings the importance of social protection which is defined by UNICEF (2012:14) as "the set of public and private policies and programs aimed at preventing, reducing and eliminating economic and social vulnerabilities to poverty and deprivation." The "social risk management" approach developed by the World Bank focuses on risks, differentiating between idiosyncratic risks (affecting individuals or households) and covariant risks (affecting the entire communities or countries) (Holzmann et al. 2003). Drivers of vulnerability include gender, ethnicity, economic disadvantage, disability, displacement, and invisibility. These drivers determine the extent of vulnerability faced by different children in different settings.

Children's vulnerabilities in urban settings are divided into different categories. The correlation between the built environment and the unavailability of urban services presents potential vulnerabilities in urban areas. Unavailability in this case relates to the gap between supply and demand. Unavailability of urban services includes lack of health, protection, and participation. Majority of children living in slums are exposed to numerous health and safety risks without a voice or legal status. Children are exposed to ambient air pollution, physical inactivity, and stress as well as unsafe roads and transportation options.

The reality of young boys and girls in city lifestyles differs depending on childhood diversity based on age, gender, disability, ethnicity, family income, sociocultural context, as well as external factors such as climate change, disaster, conflict, as well as displacement (Heckenberg and Johnston 2012; Haynes et al. 2010). For children living in urban poor communities where there are varied sociocultural contexts, their reality is often characterized by poverty, inequality, and discrimination; challenges in accessing education; poor health, nutrition, and sanitation; protection risks (lack of family care and attention, child labor and exploitation, physical and sexual violence, gang violence, and police harassment); negative impacts of drugs, alcohol, and gambling; unsafe play; and pollution (Aubrey 2017; Boyden et al. 2019; Grant 2010). Exposure to the use of advanced modern technology has both positive and negative effects for children.

Children have the greatest needs; unavailability of these needs makes them vulnerable. They are susceptible to highest abuses of children's rights. The most underprivileged and susceptible are mainly frequently excluded from development and inaccessible. They need specific consideration both to secure their entitlements and as a matter of safeguarding the recognition of everyone's rights. Young boys and girls "...living in urban poverty have the full range of civil, political, social, cultural and economic rights [acknowledged] by international human rights instruments" (Freeman and Nkomo 2006: 25). The greatest fast and extensively approved of these is the Convention on the Rights of the Child. Every child has the right to "...survival; development to the fullest; protection from abuse, exploitation and discrimination; and full participation in family, cultural and social life" (Eide 2006: 78). These rights are often violated because of absence of effective governance. There is a gap in protection of rights with respect to health care, education, and legal, civil, and social protection.

Analysis and Emerging Issues

Examination of urbanization impacts on city developments shows that it does not induce sustainable urban environments for children. Informal settlements are increasing despite upgrading efforts. A larger population of people in low-income areas occupy slum settlements. Of the number of people in slums, children suffer from multiple deprivations, as they live without a voice and have no access to land, housing, and services (Grasham et al. 2019). Also, with absence of investment in planning, urban expansion mostly occurs in a fragmented way, with limited centrality, a lack of public space, and no compactness in urban form (Gbadegesin and Olayide 2019). That is urbanization and creation of new suburbs without supporting infrastructure and social services. For children, it means unhealthy and unsafe environments, limited options for walking and playing, and limited connectivity to social networks, services, and local economy. Existing urban areas are responsible for proportionally higher energy consumption and carbon dioxide (CO2) emissions, thereby putting stress on the environment and the cities themselves. Improved use of urban resource systems necessitates innovation in terms of energy efficiency and in forging sustainable lifestyles. As children's behavior is molded by their ongoing interaction with the urban environment, children's participation in shaping sustainable cities will be a determinant for the future of our cities and for our planet (Haynes et al. 2010).

It is crucial to involve children in urban planning policies and programs so as to reduce vulnerability. In several countries, urban development, management, and governance systems continue to be underdeveloped and have led to increased urban problems, rather than allowing procedures which might generate more inclusive and safer urban areas with sufficient services for children and families (Heckenberg and Johnston 2012; Rudner 2012). To respond to the complications of urbanization and the multiplicity of local wants, there is growing realization among government and other stakeholders that developments are required in urban planning to effectively budget and strengthen participatory governance processes that include the urban poor as active citizens.

Efficient urban planning policies and good implementation reduce vulnerability and increase resilient among children. Children's participation in disaster risk reduction, climate change adaptation, and emergency preparedness helps to identify and reduce risks affecting girls and boys, reducing vulnerability and enhancing resilience (Shang et al. 2005). When integrated efforts are made to support children's participation in urban contexts, it contributes to more resilient children, families, and communities. There is growing evidence concerning the importance of children and young people's participation in climate change adaptation and disaster reduction efforts. Participation is crucial for the acceptance and relevance of any policy or program. Young people should be part of implementation of policies which are crucial to them. Their involvement ensures that their wants and needs are incorporated.

Urban child poverty is increasing among developing countries. The numbers are rising rapidly due to population growth and urbanization. Moreover, multidimensional poverty is higher than monetary poverty: almost half of all children under 5 living in large cities in less economically developed countries have unsatisfied basic needs relating to housing conditions, sanitary facilities, economic dependency, and household crowding (Robinson et al. 2018). An exclusive focus on income poverty could lead to inadequate policy focus on the multiple deprivations that are experienced by almost half of urban children in developing countries. Rural poverty rates are much higher than those in urban areas. One in four rural children under 18 experience extreme monetary poverty (UNICEF 2012), and eight out of ten rural children under 5 do not have their basic needs met.

Overall, urban children and rural children account for 19% and 81% of all extremely poor children, respectively. But the character of poverty and the nature of the deprivations that children face are quite different in urban and rural contexts, which is why disaggregated needs assessments and tailored policy responses are called for. The lower poverty levels and share in poverty for children living in urban areas have led to a strong focus on rural child poverty and limited acknowledgment of the plight of children in cities (Rudner 2012), and such acknowledgment has often focused very narrowly on street children (Landrigan et al. 2004). However, there are strong reasons for paying more attention to child poverty in urban areas and for considering its specific characteristics.

Poverty estimates and measures are premised on indicators and thresholds that are primarily applicable to rural settings. Indicators reflecting housing conditions based on materials used for walls or roofing, for example, fail to reflect issues that are core to the predicament of urban children and their families, such as land ownership and informal settlement (Salvi 2007) or deprivations in areas of crowding, smoky fuels, and lack of electricity (Sheffield and Landrigan 2011). Similarly, water sources and sanitation facilities that can be considered "improved" in rural settings prove inadequate in densely populated urban settings, including hard-to-maintain pit latrines and water points with long lines and irregular supply (Kendall and O'Gara 2007). Costs of living are higher in urban areas, as reflected in expenditure profiles. Urban residents generally pay more for housing, transport, and utilities (e.g., electricity or water). Being more market-dependent for their food and basic needs, they are also more exposed to inflation and price spikes, but these differentiated cost and spending patterns are not captured in aggregated national poverty lines.

High levels of informality mean that marginalized urban populations are often invisible in official data and are therefore underrepresented in poverty estimates (World Vision 2014) 0.2 Informal settlements that house many of the poorest urban dwellers tend to be excluded from sampling frames or under-sampled, leading to underestimates of urban (child) poverty (Gupta et al. 2017). A simple rural-urban dichotomy masks vast disparities between different groups and types of urban settlements. Dimensions along which experiences of children in urban areas will differ include size and location of the city/town, governance, and quality of infrastructure. Urban child poverty is often associated with – but not limited to – living in slums.

Conclusion and Options for the Future

Urbanization impacts on young people by their vulnerability to disease and death threats as pollution levels are usually higher in urban centres than rural regions. Urbanization therefore needs to be associated with infrastructure and policies limiting the vulnerability of young people. Urbanization processes are a matter of credibility, of principle, and as a means of making programs more effective and accountable, and it is crucial that the organization makes more systematic efforts to support children and young people's participation in urban planning and other relevant decision-making processes. Supporting children's participation in urban planning will help the organization to achieve change as it supports efforts to be the voice ensuring children's voices are heard (particularly those children who are most deprived living in urban poverty) and to build partnership and to be an innovator ensuring space and value for children's role as active citizens in urban governance processes together with other key stakeholders, which will enable us to achieve results at scale. Children's participation is also a critical approach to achieving our breakthroughs, including breakthrough of violence against children that is no longer tolerated.

Children's participation should not be seen in isolation but should be supported as part of more integrated efforts to empower and engage families and communities. Children are participating and contributing in numerous ways in their daily lives, their families, schools, and communities, and they are the experts when it comes to understanding their own lives, difficulties, and aspirations. Children and young people are a huge untapped resource of creativity, competency, energy, and agency and can make a real contribution to the development of their towns, cities, and broader society. Thus, it is crucial that increasing strategic and practical efforts with government, business, academic, civil society, child and youth organizations, and community members to proactively engage with girls and boys as active citizens so that their voices are heard and acted upon in governance processes affecting them. Furthermore, to ensure quality participation processes, nine basic requirements for meaningful participation should be applied when planning and monitoring participation, namely, that participation is transparent and informative, voluntary, respectful, relevant, child-friendly, inclusive, supported by training, safe and sensitive to risk, and accountable.

Urban living and livelihoods pose some specific challenges for the design and implementation of social protection, as compared to rural areas (Grant 2010; Kendall and O'Gara 2007). A primary reason for this relates to the specific vulnerabilities of urban residents. Urban poverty is characterized by volatility in income, reliance on fully monetized means of exchange, insecurity of employment and income, insecure housing tenure, population mobility (including influxes of refugee populations), diverse population groups with diverse needs, and weaker social networks to rely on in times of distress or shock. All these factors mean that designing social protection for the urban poor and vulnerable requires that programs are tailored to these needs.

The unsustainable built environment has diminishing returns on service delivery for children and increases children vulnerability to certain risks. Cities have strong correlation between the vulnerability of the most disadvantaged children and the built environment (Christian et al. 2017). Children's access to physical urban services is affected by unsustainable practices on the built environment through ineffective planning as well as lack of quality in design and construction. This leads to urban-specific environmental health problems that health support systems cannot address alone and shifts the focus from communicable to non-communicable diseases.

Urban environments increase the vulnerability of children as they present threats when children fail to evaluate, prepare for, prevent, and cope from risks (Sheffield and Landrigan 2011; World Health Organization 2017). Lack of children's participation increases their vulnerability as their concerns would not be heard. Children are often neglected in terms of public space where they can assemble and other infrastructure that allows physical, social, and digital connectivity. Urbanization, climate change, and urban poverty are among the major driving issues of the twenty-first century. These issues present challenges on urban policymakers whereby they are unable to manage changes in the urban environment. Urban local authorities fail to manage urban poverty and inequity as well as to identify the most vulnerable urban children. As a result of lack of sufficient services in urban communities, young children are tasked with certain responsibilities which may not necessarily be for children.

The chapter concludes that future policies need to focus on children's participation in strategies dealing with vulnerability. Minimizing child vulnerability means planning and budgeting for children's participation in urban programming. Children living in low-income regions are most vulnerable to economic, social, and environmental conditions. This chapter has considered the higher vulnerability of children to sanitation, to water- and health-related illness, as well as to environmental and poor economic structure. These affect children's mental and social development. It is recommended that global, regional, and country programs increasing budgets and capacity building support for child led action as well as advocacy initiatives in urban settings. These encompass supporting organization, research, as well as media initiatives led by children.

Cross-References

▶ From Vulnerability to Urban Resilience to Climate Change
▶ The Future of Reducing Urban Vulnerability with Perspectives of Child Development in Zimbabwe
▶ Vulnerability to Food Insecurity Among the Urban Poor in Sri Lanka: Implications for Policy and Practice

References

Aubrey, C. (2017). Sources of inequality in south African early child development services. *South African Journal of Childhood Education, 7*(1), 1–9.

Blaikie, P., Cannon, T., Davis, I., & Wisner, B. (2014). *At risk: Natural hazards, people's vulnerability and disasters*. Routledge.

Boyden, J., Dawes, A., Dornan, P., & Tredoux, C. (2019). *Tracing the consequences of child poverty: Evidence from the Young Lives study in Ethiopia, India, Peru and Vietnam*. Bristol: Policy Press.

Christian, H., Ball, S. J., Zubrick, S. R., Brinkman, S., Turrell, G., Boruff, B., & Foster, S. (2017). Relationship between the neighborhood built environment and early child development. *Health & Place, 48*, 90–101.

Dodman, D., & Satterthwaite, D. (2008). Institutional capacity, climate change adaptation and the urban poor. *IDS Bulletin, 39*, 67.

Douglas, I., Alam, K., Maghenda, M., Mcdonnell, Y., McLean, L., & Campbell, J. (2008). Unjust waters: Climate change, flooding and the urban poor in Africa. *Environment and Urbanization, 20*(1), 187–205.

Eide, A. (2006). *Article 27: The right to an adequate standard of living* (Vol. 27). Martinus Nijhoff Publishers.

Freeman, M., & Nkomo, N. (2006). Guardianship of orphans and vulnerable children. A survey of current and prospective south African caregivers. *AIDS Care, 18*(4), 302–310.

Gbadegesin, T. K., & Olayide, O. E. (2019). Assessment of women and children vulnerability to water use in low-income urban area of Agbowo community, Ibadan, Nigeria. *International Journal of Sustainability Management and Information Technologies, 5*(2), 29.

Grant, U. (2010). *Spatial inequality and urban poverty traps*. London: Overseas Development Institute.

Grasham, C. F., Korzenevica, M., & Charles, K. J. (2019). On considering climate resilience in urban water security: A review of the vulnerability of the urban poor in sub-Saharan Africa. *Wiley Interdisciplinary Reviews: Water, 6*(3), e1344.

Gupta, A.K., Wajih, S.A. and Mani, N., 2017. Vulnerabilities of urban poor children and urban/Peri-urban ecosystem based resilience.

Haynes, K., Lassa, J., & Towers, B. (2010). *Child-centred disaster risk reduction and climate change adaptation: Roles of gender and culture in Indonesia. Children in a Changing Climate Working Paper*. Brighton: Institute of Development Studies.

Heckenberg, D., & Johnston, I. (2012). Climate change, gender and natural disasters: Social differences and environment-related victimisation. In *Climate change from a criminological perspective* (pp. 149–171). New York: Springer.

Holzmann, R., Sherburne-Benz, L., & Telsuic, E. (2003). Social risk management. In *The World*

Bank's approach to social protection in a globalized world. Washington, DC: Social Protection Department, World Bank.

Kendall, N., & O'Gara, C. (2007). Vulnerable children, communities and schools: Lessons from three HIV/AIDS affected areas. *Compare: A Journal of Comparative and International Education, 37*(1), 5–21.

Landrigan, P. J., Kimmel, C. A., Correa, A., & Eskenazi, B. (2004). Children's health and the environment: Public health issues and challenges for risk assessment. *Environmental Health Perspectives, 112*(2), 257–265.

Mota Borges, I. (2016). The responsibility of transnational corporations in the realization of Children's rights. *University of Baltimore Journal of International Law, 5*(1), 2–8.

Robinson, C., Bouzarovski, S., & Lindley, S. (2018). Underrepresenting neighborhood vulnerabilities? The measurement of fuel poverty in England. *Environment and Planning A: Economy and Space, 50*(5), 1109–1127.

Rudner, J. (2012). Public knowing of risk and children's independent mobility. *Progress in Planning, 78*(1), 1–53.

Ryder, S.S., 2017. A bridge to challenging environmental inequality: Intersectionality, environmental justice, and disaster vulnerability.

Salvi, S. (2007). Health effects of ambient air pollution in children. *Paediatric Respiratory Reviews, 8*(4), 275–280.

Shang, X., Wu, X., & Wu, Y. (2005). Welfare provision for vulnerable children: The missing role of the state. *The China Quarterly*, 122–136.

Sheffield, P. E., & Landrigan, P. J. (2011). Global climate change and children's health: Threats and strategies for prevention. *Environmental Health Perspectives, 119* (3), 291–298.

Stephens, C. (2012). Urban inequities; urban rights: A conceptual analysis and review of impacts on children, and policies to address them. *Journal of Urban Health, 89*(3), 464–485.

Stringer, L. C., Dyer, J. C., Reed, M. S., Dougill, A. J., Twyman, C., & Mkwambisi, D. (2009). Adaptations to climate change, drought and desertification: Local insights to enhance policy in southern Africa. *Environmental Science & Policy, 12*(7), 748–765.

UNICEF. (2012). *UNICEF social protection strategic framework*. New York: UNICEF.

World Health Organization. (2017). *Inheriting a sustainable world?* Atlas on children's health and the environment: World Health Organization.

Church Gardens

► Faith Communities as Hubs for Climate Resilience

Circular Cities

Felipe Bucci Ancapi, Ellen Van Bueren and Karel Van den Berghe
Department of Management in the Built Environment, Faculty of Architecture and the Built Environment, Delft University of Technology, Delft, The Netherlands

Synonyms

Autarkic city; Climate-neutral city; Metabolic city; Regenerative city; Resource-efficient city; Self-sustainable city

Definition

Circular city is a concept inspired by biological metabolic systems that seeks to apply the principles and strategies of the circular economy at the different scales of urban functioning. By doing so, a circular city is meant to reduce the intake of primary resources and energy and resulting environmental impacts, such as waste and emissions. Its functioning is (re)defined by efforts aiming to close, narrow, and/or slow material and energy flows. A circular city is a normative concept, implying thus there is an ambition to switch the current – linear – consumption-production system into one that works and develops circularly, in closed loops. It is also normative as it proposes the urban scale as the main spatial level of implementing circularity. As cities in the twenty-first century deal with their historical ecological impacts, circular cities also embrace ecological regeneration and adaptation measures to maintain their development within the carrying capacity of Earth.

This definition is a compendium of the perspectives contained in this chapter. Thus, although not exhaustively, this definition seeks to provide a common frame of reference for the study of circular cities.

Introduction

Cities are complex systems of production and consumption. Their ecological impacts have grown significantly in the last decades. Currently, cities consume 60–80% of natural resources globally, while producing around 50% of global waste and 75% of greenhouse gas emissions (UN 2019). The urban population is expected to increase in the coming decades, reaching 6.5 billion by 2050, the equivalent of two thirds of the future global population (UN 2017b).

While the attention for the environmental impact of cities and material flows is not something new (cf. Wolman 1965), arguably, it has recently become more popular within sustainable urban development, along with the increasing popularity of the concept of circular cities (Williams 2019b). A circular city aims to close material and energy flows that are used by and within its boundaries and thus reducing its overall environmental externalities, such as ecosystem degradation, greenhouse gas emissions, and waste generation. In some cases, a circular city also includes social and economic goals, but in general, the focus of circular cities and the circular economy (CE) is on material and energy flows (Korhonen et al. 2018b). Following this, circular cities received a fair amount of critique. The most heard critique is that the knowledge development and implementation of circularity and/or the CE is too technical and fails to include other dimensions such as the economy, culture, social affairs, politics, governance, design, or spatial planning (Korhonen et al. 2018b; Pomponi and Moncaster 2017; Williams 2019b).

This chapter explores the concept of circular cities. In the first section, a broader concept of CE is provided. Secondly, the challenge of scale and responsibility in the CE are explained. Thirdly, the chapter continues by tracing the origin of the concept of circular cities. Fourthly, different contemporary definitions of circular cities are covered, as well as their recent increase in publishing. Finally, this chapter ends with an outlook of challenges for circular cities in their implementation.

The Circular Economy

The CE gained momentum from 2010 onwards in the western world when the Ellen MacArthur Foundation (EMF) developed the "butterfly diagram" depicting closing loops of biological and technical resources (EMF 2012, 2016). However, one must not forget that it was the Chinese government that first clearly introduced the concept in its 1996 Five Year Plan (Su et al. 2013). In the years following, the CE has been put to the forefront, among others by the UN (2017a), the OECD (2019), and the European Union (EC 2019) as a focus strategy. In a nutshell, "the objective of a CE is to reduce the societal production-consumption systems' linear material and energy throughput flows by applying materials cycles, renewable and cascade-type energy flows to the linear system" (Korhonen et al. 2018b, p. 547). Often, the CE is linked to the so-called hierarchical ladder of R-strategies to prevent and to Rethink, Reduce, Reuse, Remanufacture, Recycle, Recover the use of materials (Reike et al. 2018), as it builds upon the waste management hierarchy developed by Lansink, a Dutch Member of Parliament in 1979, and later introduced in the EU legislation with the 2008 Waste Framework Directive. The Directive distinguishes prevention, preparing for reuse, recycling, recovery, and landfill on a preferential scale (EC 2008). Simply said, the rule of thumb is the higher on the R-ladder, or earlier in the production-consumption system (Korhonen et al. 2018a), the less resources and energy are needed. During the last decades, the focus of waste management has changed. While first the challenge was to avoid landfilling and incineration, the main attention changed to increasing reusing and recycling of primary and secondary materials (Van den Berghe et al. 2020). However, by now it is known that there are not enough secondary materials that can substitute the use of primary materials (PBL 2021). To achieve CE-ambitions, it will be pivotal to move up the R-ladder, beyond recycling (PBL 2019). As the CE finds its ways within urban development, different aspects of a city's daily operations require adaptation at different scales of urban aggregation – i.e., at the household, neighborhood, city, or regional level. To illustrate the question of scales, in the next section we examine an elemental aspect of (circular) cities: its built environment.

Applying Circularity at Multiple Levels of the Built Environment

Analytically, different layers or levels of spatial scales can be identified in the built environment. These can range from fine-grained scales such as materials and components to more coarse scales such as neighborhoods, districts, cities, countries, and the global. When circularity is understood as closing material and energy loops while minimizing input and output with minimized impact on the human and natural environment, it can more easily be applied to the lower scales than to the higher ones. Up to the scale of the building level, the concept of closing loops has an inherent logic, pleading for the reuse of materials, building components, and buildings. While circularity can be best understood to the lower scales, the other way around, the circular *economy* can be better understood in line with higher scales, such as nations or the global level. Conceptually, the lower scales deal more with the circularity of products and the design of those, but only to a minor level consider the material and immaterial flows, institutions, and agency, better known as the economy, that enable these to be produced and consumed. From the global level downwards, it is better to imagine what a CE implies, but it becomes more difficult if it is translated to the exact locations where these consumption-production networks take place. Conceptually, they confluence at the area level, city level, or regional level. Arguably, this scale is where the circular produced components and built environment come together with the CE consumption and production system. Otherwise said, a circular area/city/region cannot exist without a circular built environment and circular products, and a circular consumption-production system where that area and those assets, people, institutions, and materials are part of (Fig. 1).

This is not really something new. If one imagines how an area (should) function(s), logically one (implicitly) connects how the built environment and exact locations of assets, people, and institutions, interact in networked systems, crossing borders in many different aspects. However, as the next paragraph explains, this reasoning is not at all something that is followed within contemporary CE literature.

From Metabolic to Circular Cities

To start off with, it is important to underline that a circular city is a normative concept, for it states that a city is a reality, one that consumes and produces materials, which – apparently – to date tends to be mostly linear (take, make, use, waste) and should become circular. The latter is influenced by the increasing attention for environmental issues since the 1970s, and by now as an idea arguably easy to understand. Yet, taking the city as a given is not straightforward. Brenner and Schmid (2014) question the abundant non-critical

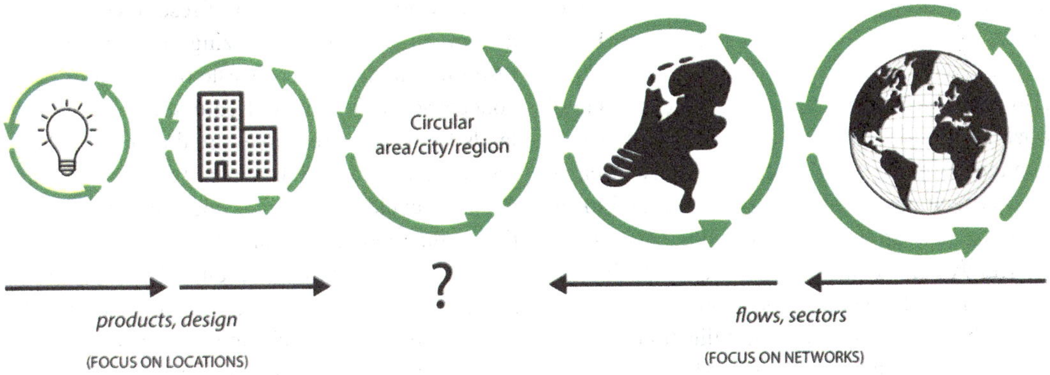

Circular Cities, Fig. 1 How conceptually circularity (cf. design and locations) and CE (cf. flows and networks) come together at a circular area/city/region. (Adapted from Van den Berghe (2021))

use of the city as a given object and argue that the city is at highest a statistical artifact that shall always remain a subject of reconsideration and, consequently, a circular city is all but a clear concept. To understand why circular cities are nonetheless so prominent, even though it is not very clear what it is, we first need to understand the epistemological history of a city, and how it confluences with (material) flows, that are normatively expected to be circular instead of linear.

Although always arbitrary, following Wachsmuth (2012), through several stages in time the two conceptualizations – city and (circular) material flows – became intertwined and increasingly the city became seen as both the problem and the solution to environmental problems, the latter thus illustrated by a circular city. Firstly, in the era of industrialization, the idea of an industrially provisioned city started to emerge. Industrial capitalism, the factory system of production, was significantly changing the relation between human and nature, resulting in a society-nature divide (Foster 2000; Polanyi 1944; van Driel 2016). During the industrial revolution, manufacturing concentrated increasingly in and around urban areas. This in turn created a new working class with new political ideas, new organizational forms, and collaborative infrastructures. As such, cities emerged as political, social, and economic bodies as opposite to the non-city, or countryside. The city became seen as the human social optimum, in contrast to the non-human countryside, that was primarily there in support of the city. This idea can be found back within the urban studies and sociology works at the late nineteenth and early twentieth century. Ebenezer Howard described the opposition of town and country (Howard 1989), but foremost the Chicago School established the widespread idea of city versus non-city. Among their words, the city became seen as a self-contained system of people and social relations that grows along with the increase of interactions. Here we encounter a contradiction: How can a self-contained system grow?

By the 1960s and 1970s, best illustrated by the work of the "Club of Rome" (Meadows et al. 1972), it became rather clear that the social growth of "self-contained" systems was impacting the non-city, or nature, in a very negative way. The observed environmental problems caused by human actions triggered researchers to increase their understanding. Here lies the birth of industrial ecology (IE), examining how materials and energy flow through industrial systems of consumption and production, in analogy with ecosystems (Erkman 1997). Subsequently, IE and the perspective of the city as a system started to intertwine. The city became seen as a system that converts natural resources, also known as urban metabolism. Especially the work of Wolman (1965) was pivotal in conceptualizing the city as a metabolic system. By carefully graphically analyzing the metabolism of the city of Brussels, Wolman showed how the city is an open system.

Wolman's understanding of the city also demarcated an epistemological shift: while before the city was primarily seen as an isolated social system, without the inclusion of natural sources, thereafter the city was seen as a system fueled by natural resources, but without the inclusion of the human (cf. Newell and Cousins 2015). The city was understood as a sort of machine, without a reference by whom, why, and how natural resources are converted. This urban metabolic perspective eventually became normative once it linked to circularity (Stahel 1982). Figure 2 shows the current linear urban metabolism of cities, and how this metabolism could be improved by reducing resource inputs into the urban system (Rogers and Gumuchdjian 1997). In a circular urban metabolism, resources are used and reused as much as possible once they are in the system, while avoiding degradation of resources as much as possible, and minimizing the output of resources, in the form of waste or emissions. The concentric circles in the figure show the hinterlands from where resources are drawn and where resources are collected, remanufactured, and up- or down-cycled, adding a geographical perspective to the modeled resource flows.

The taking up of circular cities by the IE communities is no coincidence, as sustainable resource use and the closing of loops is key to IE. This has also brought a strong emphasis on resource flows. Within this community, the circular city is mainly examined as a technical artifact

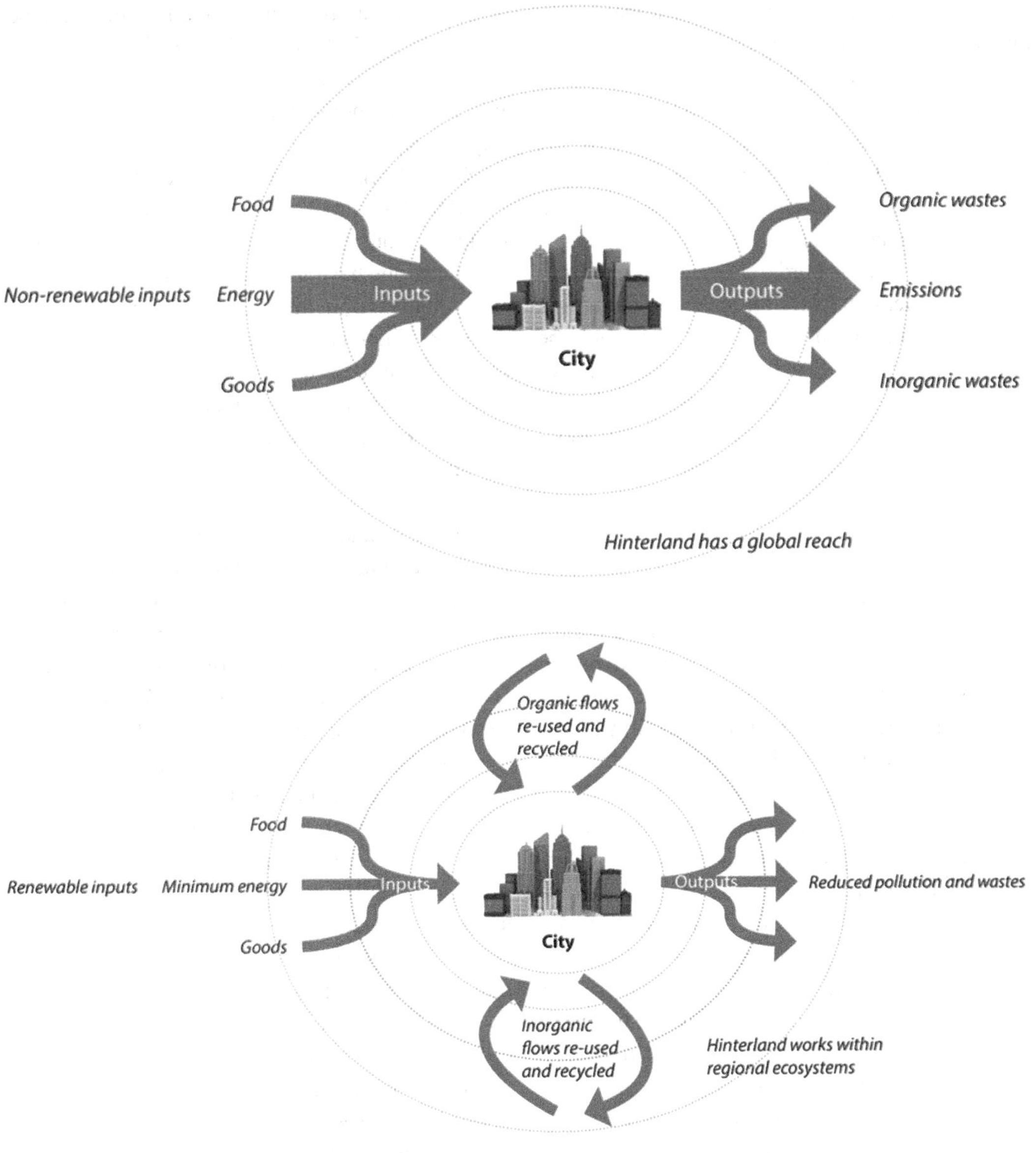

Circular Cities, Fig. 2 Linear and circular metabolisms of cities (van Bueren 2015). (Adapted from Rogers and Gumuchdjian (1997))

that converts natural resources in a way that needs changing. Despite the fact that circular cities are only one application of the concept of circularity and many others can also be linked to it (e.g., an economic sector that should become circular, a specific circular region), circular cities are the most popular. The main explanation is that we now live in the so-called urban age (Brenner and Schmid 2014), as more than half of the human population lives in urban areas. Next, increasingly cities have gained agency as centers of value creation in economic and cultural perspective (e.g., Florida 2005), which in turn also led cities to organize and empower themselves in fora such as the Global Parliament of Mayors (See https://globalparliamentofmayors.org/) or the Resilient Cities Network (See https://resilientcitiesnetwork.org/).

Nonetheless, authors have claimed that focusing on the circularity of material flows without a reference to social and economic processes is problematic (Corvellec et al. 2021; Korhonen et al. 2018b). One could also question if the focus on circular cities, without a reference to the "outer city" or hinterland is the way forward. Many environmental problems are transboundary: they do not stop at borders, hence the environmental problem and solution of cities, it being a circular city in this case, will most likely only be achieved if we are able to improve our perspective of the city, beyond the late nineteenth and beginning twentieth century perspective. Whether more recent academic proposals to define a circular city elaborate about these conceptual issues is examined in the following section.

Circular Cities: Existing Perspectives

Despite the lack of a commonly accepted definition, circular cities receive political and public attention and of policy makers. Different international organizations, governments, firms, and scholars have come up with perspectives regarding what a circular city entails.

Among international organizations, the EMF is arguably the most prominent private actor fostering a circular approach to cities. The EMF understands a circular city as one that thrives in the long-term, bringing prosperity to its citizens while respecting the planetary boundaries. For the EMF, cities provide a workable system boundary for action. Special attention is put on buildings, mobility, products and services, and food systems (EMF 2017). The C40 network of megacities has also put attention on circular transition at the city level. They provide a concept based on long-lasting resource use, maximum value extraction, recovery, and regeneration of product and materials at the end of their lifecycles (C40 2018). Similarly, but less straightforward, the United Nations' (UN) "Waste Wise Cities" initiative is arguably its closest attempt to a circular city perspective. In this initiative, among 12 principles aimed for coping with the ever-increasing global waste management crisis, a call for designing incentives to promote a CE in cities is included. Although a circular perspective to cities may be linked to UN's Sustainable Development Goal 11 on sustainable cities and communities, the international body does not provide a circular city concept.

Some firms have also started to work with circularity in cities. Prominent work has been done by young Dutch consultancy firms. Both organizations have guided dozens of cities, especially in the EU, in their intention to become more circular. Their approaches focus on identifying crucial resource flows in city areas to be used to create visions and agreements among local stakeholders to close resource loops and minimize resource use (Circle-Economy 2017; Metabolic 2021).

National and local governments have also established their own strategies for circular cities, regions, and countries. For instance, in Europe, several countries and cities have already launched their visions and strategies towards a more circular economy. The European CE Stakeholder Platform by the European Union (EU) offers a policy repository wherein more than 40 city and national level strategies can be found, most of them including construction, buildings, infrastructure, and/or city's daily operations as part of their scope (See https://circulareconomy.europa.eu/platform/en/strategies?populate=).

The last 5 years are characterized by a growing number of academic attempts towards a circular city conceptualization. We now discuss the conceptual directions towards circular cities as provided by eight recent and already well-cited publications. Together, they give an impression of the width and depth of the conceptual development of circular cities. Petit-Boix and Leipold (2018) reviewed CE initiatives in cities and grouped them according to four urban targets: infrastructure, social consumption, industries and businesses, and urban planning. They looked at the number of city initiatives in place, leaving out the degree of effectiveness of their implementation, and found that while cities themselves focus mainly on urban infrastructure, circular city research is mainly concerned with industrial and commercial practices. They concluded that more attention should be put on social consumption, urban planning, and how to define the environmental impact of adopted circular strategies at the city level. Gravagnuolo et al. (2019) identified sectors for circular city implementation. These are the built environment, energy and mobility, waste, water, industrial production, agro-food, and citizens and communities. Their idea of a circular city is that of self-sustainable systems that require not only technical and business innovation, but a cultural paradigm shift characterized by changes in governmental organization and educational structures by which the city works cooperatively to create niches of circular innovation. Specifically on governmental aspects, Bolger and Doyon (2019) analyzed the role of strategic planning and resource management at the local scale to promote CE strategies. After identifying ways by which two different municipalities are integrating strategies in their planning instruments, they pointed out the difficulties posed by the absence of a clear circular city framework, as well as the need for introducing circular thinking in urban planning and to understand the role of different levels of government in sustainability urban transitions. Although the above-mentioned characterizations provide insights or directions towards a circular city concept, the authors agree on the fact that such a shared concept is still lacking.

Some authors do provide more concrete concepts. Girard and Nocca (2019) claim that the circular city is a metaphor to illustrate the functioning of a city as that of natural systems (cf. Wachsmuth 2012). More particularly, a circular city is the territorialization of the CE, a human-centered system wherein resources are recycled, and the use of primary resources is minimized. The built environment of a circular city is or should be therefore constructed in a flexible and modular way. Kębłowski et al. (2020) see circular cities as a promise of fundamental change towards the re-territorialization of production, distribution, consumption, waste management, and innovation although such promise is restricted by major capitalistic ways of production. Paiho et al. (2020) focus on actions tending to either close, slow, or narrow resource loops in the urban space. Yet, these actions are only applicable "after the potential for conservation, efficiency improvements, resource sharing, servitization and virtualization has been exhausted" (p. 6). Localization of production and productive processes powered by renewable energy are also inherent to their understanding. A more comprehensive conceptualization is given by Williams (2019a, p. 10), who defines the city as a "complex, heterotrophic artificial ecosystem in which resources are produced and consumed by a variety of activities, initiated by inter-dependent actors, across multiple sectors and scales." Hence, whatever the changes a circular approach to cities intends to accomplish, they must be understood in a context of ever-changing demands, patterns of consumption, and systems of provision. The basics for circular city functioning will be determined by three circular actions – looping, regenerating, and adapting – and four supporting actions – optimization, sharing, substitution, and localization (Williams 2021).

The publications discussed in this section show similarities, differences, and research directions of the circular city concept. A central common ground is provided by the need for identifying relevant systems of provision, production and consumption, and scales of circular intervention. This resonates with the historically developed systemic view of cities, not only as urban systems fueled by their hinterlands, but as a process governed by the interplay of different stakeholders, space, institutions, and resources. A marked difference among authors is the tendency to either encapsulate the concept of circular city as the implementation of R-strategies in urban areas or expand it to embrace aspects such as territorial planning, ecological regeneration, and multiple levels of governance. The quest for a circular city concept is another difference. While some authors approach circularity in cities through the identification of circular initiatives, others attempt to provide circular city definitions to test its operationalization. As this conceptual examination is not exhaustive, bibliometric analysis may offer broader perspectives on circular city research and understanding.

Bibliometrics

The evolution of circular cities research can be traced by looking at major research databases, such as Scopus. This section shows the results after searching for ["circular economy" AND ("city" OR "cities" OR "urban")] in titles, abstracts and keywords. This search string resulted in 1059 documents between 2000 and 2020, as for March 26, 2021. Firstly, publications increased from less than 30 in 2015 to more than 350 in 2020. Secondly, about 60% of results correspond to articles, 21% to conference papers and 7% to reviews. Thirdly, most contributing countries are China, Italy, the United Kingdom, Spain, The Netherlands, the United States, and Germany. Fourthly, when it comes to affiliation, most documents are linked to Delft University of Technology, Chinese Academy of Science, and Università degli Study di Napoli Federico II, respectively. Fifthly, in terms of subject areas, environmental science (26%), social science (14%), engineering (14%), and energy (12%) are most predominant. Finally, funding sponsors have mainly been the Chinese government and the European Union. Figure 3 shows four resulting graphs of our search in Scopus.

The findings provided by this bibliometric analysis help the reader to situate circular city research by pointing out who is contributing to research, which institutions lead its scientific progress, and what governments have invested in major research funding. There is a clear link between the governmental bodies that have included the circular economy in their political agendas, the geographical location of the institutions where research takes place, and the authors that produce research output. Williams (2021) claims that circular cities are a European phenomenon, yet the bibliometric findings show that China is among the major contributor to circular city research. These findings may also be seen as a reminder of the essential role of governments in fostering and incentivizing more circular systems of production and consumption. However, this bibliometric analysis is blind to the contributions of the private sector globally. This is worth noting as the circular economy is characterized by ever-increasing reports from consulting firms (Kirchherr et al. 2017).

An Outlook of Circular Cities: Barriers and Challenges

Translated into policy, in general, circular cities tend to focus on the waste of the construction sector, organic waste, and consumer goods (Williams 2021) – all three accounting significantly for the ecological impact of cities. In accordance with this, several existing and newly developed frameworks have been proposed to understand and facilitate the journey in improving cities' ecological impact. These frameworks are the R-frameworks, focusing on the waste (prevention) hierarchy, or more recently, the ReSOLVE framework by EMF et al. (2015), which builds around six actions to businesses shifting towards circularity, namely: regenerate, share, optimize, loop, virtualize, and exchange. Consequently, within the policy documents of circular cities, these frameworks are often mentioned and operationalized.

The challenge of implementing more circular cities is quite significant. As shown in Fig. 1, a circular city lies on the confluence of the local and global. It involves a systemic change of the current consumption and production system at global level, and at local level, a change of the locations where production and consumption take place, changing the material and immaterial design of those places. In this article, the reviewed academic and policy sources on circular city concepts show that technical, ecological/environmental, and social aspects are all addressed. The main challenge, however, is a political one and deals with scales and responsibilities. A circular city is a normative concept, implying that there is an ambition to change the current (linear) consumption-production system. As explained, this involves the conceptual – and eventually operational – confluence of scales. Questions to address are: What should be organized on what scale and when? Who is responsible?

There is by no means an easy answer to these questions. It is a utopia that all relevant circular functions can be located within a particular city to match the consumption and production – cf. an

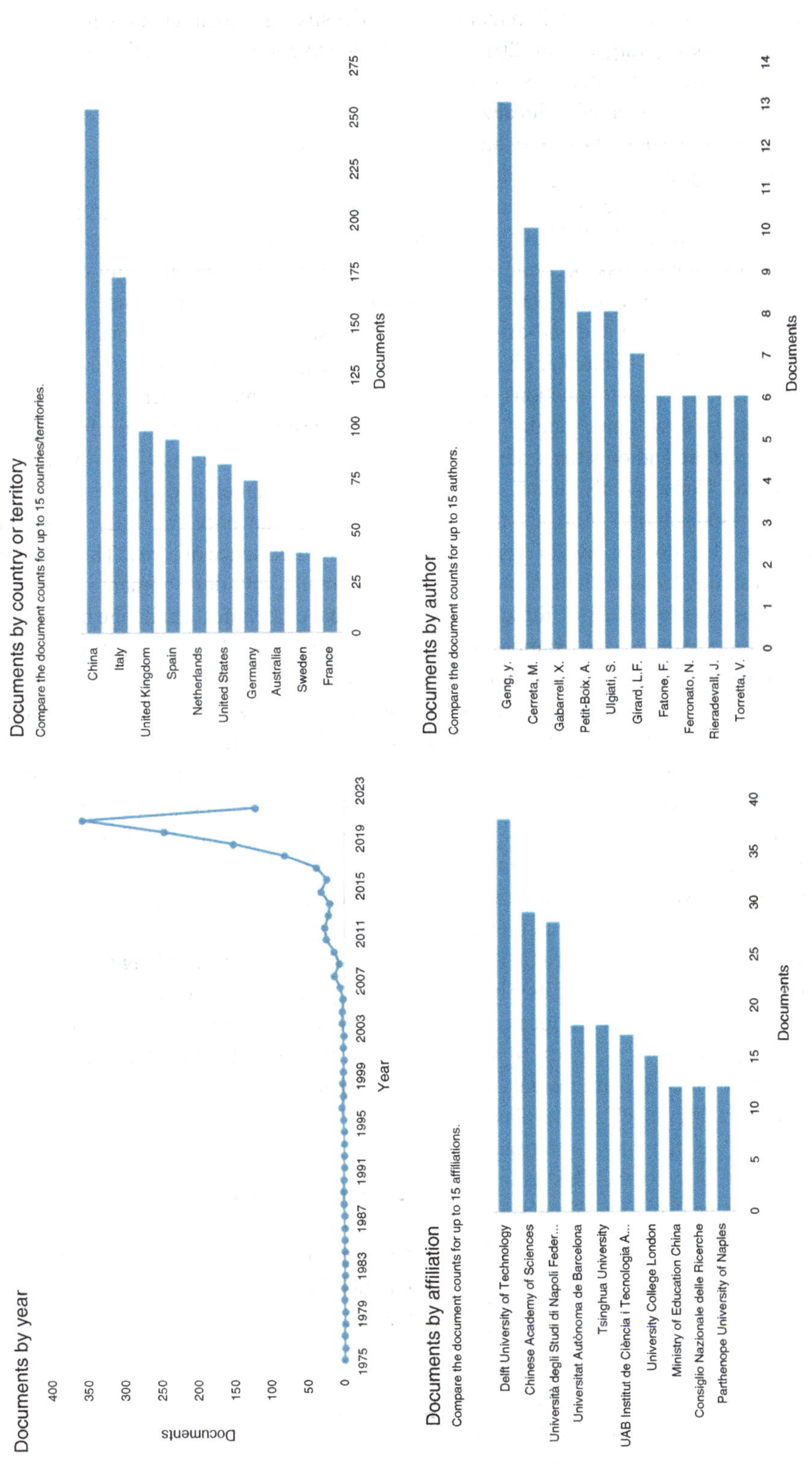

Circular Cities, Fig. 3 Circular city bibliometrics sorted by year (upper-left), country or territory (upper-right), affiliation (bottom-left), and author (bottom-right). (Made by the authors based on Scopus search, March 26, 2021)

autarkic system. Even a circular world will remain a globally connected world (Burger et al. 2019), though most likely differently organized than today. For a circular city it is essential to localized functions conditional and in support of a CE, such as remanufacturing, logistics, and agriculture; functions that are essential to process and supply the demand of (circular) materials. Without such functions, circular city policies risk becoming no more than marketing talk – or a "circular washing" of traditional good housekeeping and end-of-pipe waste reduction strategies. Key to circular cities is thus the question of what circular functions and what kind of (im)material flows cities should "(re)-capture" or (re)manufacture and on what scale this should be organized? Consequently, what scale comes with which responsibility? And are there scales and locations that do not and/or cannot take up this responsibility? Again, it is a utopia that all materials and the processing of these can remain within a certain region – for example, to create a circular built environment – as well as it is a utopia that loops can be closed without leakages and without negative environmental effects. Consequently, the extent to which a circular city can become a reality will to a large extent depend on what other – institutional – places decide to do. The plastics case can serve as an example. At the time of writing, 2021, many Western-European cities, regions, and countries have optimized the collection and separation of plastic, with the idea that this would improve the reuse of those materials. However, the plastics processing plants are located in other non-Western places, places – as it was revealed – with less strict environmental and labor regulations. In reality, much of the plastics arrived at landfills (see for more information Ananthalakshmi and Chow 2019). This example shows that the policy goal of one place should consider the whole (re)supply production chains of products.

Eventually, the insights provided in research and policies for circular cities add up to the argument of Williams (2021) that for circular cities, not materials but space is the key concern. Space to accommodate – extra – functions that enable matching the consumption and production within a circular (urban) economy; space that is scarce in these densifying urban areas with rising land prices due to continuing urbanization.

Conclusion

Circular cities are increasingly a popular concept and policy goal. This chapter has given a brief overview of the conceptual origin of the concept and explained why in some cases it is difficult to match consumption and production on an urban scale. We explained that a circular city is where different scales come together – cf. the location of circularly designed products or buildings, and circular economic systems. The former cannot exist without the latter, and vice versa. It is, however, a utopia that both can fully be accommodated in a limited space of a city. The way forward towards circular city development is not so much a conceptual or technical challenge but primarily a political one. A circular city, a city with a normative goal to become more circular, must find out for itself what is essential to move towards this policy goal (Van den Berghe and Vos 2019). Exchange of the experiences with circular city development among cities, practitioners, and academics will contribute to conceptual clarity, which in turn will provide guidance in the fragmented governance setting in which circular city policies are formulated and implemented. Summarized, the main challenge is how circular cities can go beyond the marketing of the circular city concept and effectively take up their responsibilities that come with the scale they are operating on.

Cross-References

- ▶ Circular Economy and the Water-Food Nexus
- ▶ Circular Economy Cities
- ▶ Circular Water Economy

References

Ananthalakshmi, A., & Chow, E. (2019). *Malaysia, flooded with plastic waste, to send back some scrap to source*. Retrieved from https://www.reuters.com/article/us-malaysia-waste-plastic-idINKCN1SR1KA

Bolger, K., & Doyon, A. (2019). Circular cities: Exploring local government strategies to facilitate a circular economy. *European Planning Studies, 27*(11), 2184–2205. https://doi.org/10.1080/09654313.2019.1642854.

Brenner, N., & Schmid, C. (2014). The 'urban age' in question. *International Journal of Urban and Regional Research, 38*(3), 731–755. https://doi.org/10.1111/1468-2427.12115.

Burger, M., Stavropoulos, S., Ramkumar, S., Dufourmont, J., & van Oort, F. (2019). The heterogeneous skill-base of circular economy employment. *Research Policy, 48*(1), 248–261. https://doi.org/10.1016/j.respol.2018.08.015.

C40. (2018). *Municipality-led circular economy case studies*. Retrieved from https://www.c40.org/researches/municipality-led-circular-economy

Circle-Economy. (2017). *A future-proof built environment*. Retrieved from Amsterdam: https://www.circle-economy.com/resources/a-future-proof-built-environment

Corvellec, H., Stowell, A. F., & Johansson, N. (2021). Critiques of the circular economy. *Journal of Industrial Ecology*. https://doi.org/10.1111/jiec.13187.

EC. (2008). *Directive 2008/98/EC on waste and repealing certain Directives*. Brussels Retrieved from https://eur-lex.europa.eu/legal-content/EN/TXT/PDF/?uri=CELEX:32008L0098&from=EN

EC. (2019). *Communication on The European Green Deal*. Brussels Retrieved from https://ec.europa.eu/info/sites/default/files/european-green-deal-communication_en.pdf

EMF. (2012). *Towards the circular economy. Economic and business rationale for an accelerated transition. Vol. 1*. Retrieved from https://emf.thirdlight.com/link/x8ay372a3r11-k6775n/@/preview/1?o

EMF. (2016). *What is a circular economy?*. Retrieved from https://ellenmacarthurfoundation.org/topics/circular-economy-introduction/overview

EMF. (2017). *Cities in the circular economy: An initial exploration*. Retrieved from https://ellenmacarthurfoundation.org/cities-in-the-circular-economy-an-initial-exploration

EMF, SUN, & Environment, M. C. f. B. a. (2015). *Growth within: A circular economy vision for a competitive Europe*. Retrieved from https://emf.thirdlight.com/link/8izw1qhml4ga-404tsz/@/preview/1?o

Erkman, S. (1997). Industrial ecology: An historical view. *Journal of Cleaner Production, 5*(1), 1–10. https://doi.org/10.1016/S0959-6526(97)00003-6.

Florida, R. (2005). *Cities and the creative class*. Routledge.

Foster, J. B. (2000). *Marx's ecology: Materialism and nature*. New York: Monthly Review Press.

Girard, L. F., & Nocca, F. (2019). Moving towards the circular economy/city model: Which tools for operationalizing this model? *Sustainability (Switzerland), 11*(22). https://doi.org/10.3390/su11226253.

Gravagnuolo, A., Angrisano, M., & Fusco Girard, L. (2019). Circular Economy strategies in eight historic port cities: Criteria and indicators towards a Circular City assessment framework. *Sustainability, 11*(13), 3512. Retrieved from https://www.mdpi.com/2071-1050/11/13/3512.

Howard, E. (1989). *To-morrow: A peaceful path to real reform*. London: Swan Sonnenschein &.

Kębłowski, W., Lambert, D., & Bassens, D. (2020). Circular economy and the city: An urban political economy agenda. *Culture and Organization, 26*(2), 142–158. https://doi.org/10.1080/14759551.2020.1718148.

Kirchherr, J., Reike, D., & Hekkert, M. (2017). Conceptualizing the circular economy: An analysis of 114 definitions. *Resources, Conservation and Recycling, 127*, 221–232. https://doi.org/10.1016/j.resconrec.2017.09.005.

Korhonen, J., Honkasalo, A., & Seppälä, J. (2018a). Circular economy: The concept and its limitations. *Ecological Economics, 143*, 37–46. https://doi.org/10.1016/j.ecolecon.2017.06.041.

Korhonen, J., Nuur, C., Feldmann, A., & Birkie, S. (2018b). Circular economy as an essentially contested concept. *Journal of Cleaner Production, 175*, 544–552. https://doi.org/10.1016/j.jclepro.2017.12.111.

Meadows, D. H., Meadows, D. L., Randers, J., & Bejrems, W. W. (1972). *The limits to growth: A report for the Club of Rome's project on the predicament of mankind*. New York: Universe Books.

Metabolic. (2021). *Framework voor circulaire bestaande gebouwen*. Retrieved from Amsterdam: https://www.metabolic.nl/publications/framework-voor-circulaire-bestaande-gebouwen/

Newell, J. P., & Cousins, J. J. (2015). The boundaries of urban metabolism: Towards a political–industrial ecology. *Progress in Human Geography, 39*(6), 702–728. https://doi.org/10.1177/0309132514558442.

OECD. (2019). Business models for the circular Economy. Retrieved from: https://www.oecd.org/environment/business-models-for-thecircular-economy-g2g9dd62-en.htm

Paiho, S., Mäki, E., Wessberg, N., Paavola, M., Tuominen, P., Antikainen, M., . . . Jung, N. (2020). Towards circular cities—Conceptualizing core aspects. *Sustainable Cities and Society, 59*. https://doi.org/10.1016/j.scs.2020.102143.

PBL. (2019). *Circulaire economie in kaart*. Netherlands. Retrieved from https://www.pbl.nl/sites/default/files/downloads/pbl-2019-circulaire-economie-in-kaart-3401_0.pdf

PBL. (2021). *Integrale Circulaire Economie Rapportage*. Den Haag Retrieved from https://www.pbl.nl/sites/default/files/downloads/pbl-2021-integrale-circulaire-economie-rapportage-2021-4124.pdf

Petit-Boix, A., & Leipold, S. (2018). Circular economy in cities: Reviewing how environmental research aligns with local practices. *Journal of Cleaner Production, 195*, 1270–1281. https://doi.org/10.1016/j.jclepro.2018.05.281.

Polanyi, K. (1944). The great transformation. In *New York*. Toronto: Farrar & Rinehart.

Pomponi, F., & Moncaster, A. (2017). Circular economy for the built environment: A research framework. *Journal of Cleaner Production, 143*, 710–718. https://doi.org/10.1016/j.jclepro.2016.12.055.
Reike, D., Vermeulen, W. J. V., & Witjes, S. (2018). The circular economy: New or refurbished as CE 3.0? — Exploring controversies in the conceptualization of the circular economy through a focus on history and resource value retention options. *Resources Conservation and Recycling, 135*, 246–264. https://doi.org/10.1016/j.resconrec.2017.08.027.
Rogers, R. G., & Gumuchdjian, P. (1997). *Cities for a small planet*. London: Faber and Faber.
Stahel, W. (1982). The product life factor. In *An inquiry into the nature of sustainable societies: The role of the private sector* (pp. 72–105). Houston Area Research Center.
Su, B., Heshmati, A., Geng, Y., & Yu, X. (2013). A review of the circular economy in China: Moving from rhetoric to implementation. *Journal of Cleaner Production, 42*, 215–227. https://doi.org/10.1016/j.jclepro.2012.11.020.
UN. (2017a). *Circular economy: the new normal?.* Retrieved from https://unctad.org/webflyer/circular-economy-new-normal
UN. (2017b). *New Urban Agenda*: Habitat III Secretariat.
UN. (2019). *2019 Global Status Report for Buildings and Construction: Towards a zero-emission, efficient and resilient building and construction sector.* United Nations Environment Programme.
van Bueren, E. (2015). The great urban bake-off. Inaugural lecture, Delft University of Technology.
Van den Berghe, K. (2021). De paradox van circulariteit. *Ruimte, 49*, 4.
Van den Berghe, K., & Vos, M. (2019). Circular area design or circular area functioning? A discourse-institutional analysis of circular area developments in Amsterdam and Utrecht, The Netherlands. *Sustainability, 11*(18), 4875. Retrieved from https://www.mdpi.com/2071-1050/11/18/4875.
Van den Berghe, K., Bucci Ancapi, F., & van Bueren, E. (2020). When a fire starts to burn. The relation between an (inter)nationally oriented incinerator capacity and the port cities' local circular ambitions. *Sustainability, 12*(12). https://doi.org/10.3390/su12124889.
van Driel, J. (2016). *The filthy and the fat: Oeconomy, chemistry and resource management in the age of revolutions, 1700–1850.* (PhD Thesis – Research UT, graduation UT). Universiteit Twente,
Wachsmuth, D. (2012). Three ecologies: Urban metabolism and the society-nature opposition. *The Sociological Quarterly, 53*(4), 506–523. https://doi.org/10.1111/j.1533-8525.2012.01247.x.
Williams, J. (2019a). Circular cities. *Urban Studies, 56*(13), 2746–2762. https://doi.org/10.1177/0042098018806133.
Williams, J. (2019b). Circular cities: Challenges to implementing looping actions. *Sustainability (Switzerland), 11*(2). https://doi.org/10.3390/su11020423.
Williams, J. (2021). *Circular cities: A revolution in urban sustainability.* London: Routledge.
Wolman, A. (1965). The metabolism of cities. *Scientific American, 213*, 11. https://doi.org/10.1038/scientificamerican0965-178.

Circular Economy

► Circular Economy and the Water-Food Nexus
► Circular Economy Cities
► Resource Effectiveness in and Across Urban Systems
► Systemic Innovation for Thrivable Cities

Circular Economy and the Water-Food Nexus

Robert C. Brears
Our Future Water, Christchurch, New Zealand

Synonyms

Agricultural production; Circular economy; Water conservation; Water-food nexus; Water recycling

Introduction

In our current economic model, manufactured capital, human capital, and natural capital all contribute to human welfare by supporting the production of goods and services in the economic process. Natural capital – the world's stock of natural resources (provided by nature before their extraction or processing by humans) – is typically used for material and energy inputs into production and acts as a "sink" for waste from the economic process. This economic model can be best described as "linear" which typically involves economic actors – who are people or organizations engaged in any of the four economic activities of production, distribution, consumption, and resource maintenance – harvesting and extracting natural resources, using them to manufacture a product, and selling a product to other economic actors, who then discard it when it no longer serves its purpose (Brears 2018). A key challenge to the linear economy is rising water-food nexus pressures. This chapter will first provide an overview of water-food nexus trends before introducing the concept of the circular economy. The chapter will then discuss how water managers, following the circular economy principles of reduce, reuse and recycle, recover, and restore, are implementing policy innovations to reduce water-food nexus pressures.

Water-Food Nexus Pressures

Agriculture accounts for 70% of all water withdrawn. Annual global agricultural water consumption includes crop water consumption for food, fiber, and seed production plus evaporation losses from the soil and open water associated with agriculture, for example, rice fields, irrigation canals, and reservoirs. By 2050, the world will require at the minimum 60% more food production to maintain current consumption patterns. The result is that water demand is set to increase with irrigated food production likely to rise by more than 50% (FAO 2017). The increase in agricultural production will impact the quality of water resources due to non-point source pollution. Fundamental problems include sediment runoff that causes siltation problems; nutrient runoff with nitrogen and phosphorus being key pollutants found in agricultural runoff having been applied to farmland in several ways including as a fertilizer, animal manure, and municipal wastewater; microbial runoff from livestock or use of excreta as fertilizer; and chemical runoff from pesticides, herbicides, and other agrichemicals contaminating surface water and groundwater. Around 220 million tonnes of phosphate rock is mined worldwide every year, with phosphate rock geographically concentrated in five countries (Morocco, China, Algeria, Syria, and Jordan), which combined control 85–90% of the world's remaining reserves, making other regions of the word vulnerable to phosphate scarcity, volatile pricing, monopolization, and geopolitical tensions (Brears 2020b).

Circular Economy

The circular economy, in contrast to the linear "take-make-consume-dispose" economy, aims to keep resources in use for as long as possible, extract value from them while in use, and recover and regenerate products and materials at the end of each service life. As such, the circular economy focuses on recycling, limiting, and reusing the physical inputs of the economy and using waste as a resource, leading to reduced primary resource consumption, with the ultimate aim being the decoupling of economic growth from resource use. The notion of decoupling is that economic output shall continue to increase at the same time as rates of increasing resource use and environmental impact are slowed and in time brought into decline. There are two modes of decoupling that can be distinguished:

- Resource decoupling or "dematerialization" involves reducing the rate at which natural resources are used per unit of economic output.
- Impact decoupling seeks to increase economic activity while decreasing negative environmental impacts from pressures such as pollution, carbon emissions, or destruction of biodiversity (Brears 2020b).

Circular Economy Reducing Water-Food Nexus Pressures

Water managers can reduce water-food nexus pressures by encouraging the development of a circular economy that follows the principles of reduce, reuse and recycle, restore, and recover (Brears 2018, 2020a; Kirchherr et al. 2017; Morseletto 2020), in particular:

Reduce

In the circular economy, the concept of reduce is achieved through water conservation and water efficiency measures: using less and, where we do need to use water, making sure water is used efficiently. Water conservation is essentially doing fewer things that use water and usually involves people changing their behavior. Water

C

conservation is essential when water supplies become unusually low, such as during severely dry summers or after natural events that disrupt water supplies. Water efficiency is when new hardware or management techniques are used to get the same level of benefits from using water, with water users needing less water to receive those benefits (Brears 2020b).

Case Study of Reduce: New Jersey's Soil and Water Conservation Grants

New Jersey's State Agriculture Development Committee (SADC) provides grants to landowners to fund up to 50% of the costs of approved soil and water conservation projects. Landowners apply to their local soil conservation districts, who will assist them in developing farm conservation plans and ensure projects are necessary and feasible. Soil and water conservation projects eligible include projects designed to impound, store, and manage water for agricultural purposes, improve management of land and soils to achieve maximum agricultural productivity, and control pollution of farmland. Farms must be permanently preserved or enrolled in a term (8-year or 16-year) farmland preservation program to be eligible for grant funding, with permanently preserved farms receiving priority for grant funding. Projects must be completed within 3 years of SADC funding approval, and grants may be renewed for 1 year under certain circumstances, such as seasonal constraints (New Jersey State Agriculture Development Committee 2019).

Reuse and Recycle

Regarding water conservation hierarchy terminology, reuse is defined as the reuse of water within a single process or the use of harvested water for another purpose without treatment. The increased competition for water supplies and the increased availability of sources of water for reuse provide an opportunity to develop this resource for agricultural production, particularly during times of drought when regular water supplies are limited or nonexistent. Water recycling is defined as the use of harvested water for another purpose, after treatment. Treatment can be tailored to meet the water quality requirements of planned use. Specifically, water can be treated to three different levels: primary (removes some of the suspended solids and organic matter from wastewater), secondary (biological processes involving microorganisms removing organic matter and suspended material), and tertiary (further removes suspended and dissolved materials through chemical disinfection and filtration of the wastewater). Water recycling for agricultural production provides numerous benefits, including a year-round supply of water, together with nutrients and organic matter, that supports crop production (Brears 2020b; FAO 2012; Pacific Institute 2016).

Case Study of Recycle: South Australia's Northern Adelaide Irrigation Scheme

The $155.6 million Northern Adelaide Irrigation Scheme (NAIS) will make available up to 12 gigaliters per annum of affordable, secure, recycled water to a new irrigation area. The recycled water is climate- and season-independent water. It is not subject to drought-related restrictions or events that occur in river-based irrigation systems, such as low flows, algal blooms, or black water. The scheme will offer long-term tradeable NAIS Water Contracts for the life of the scheme (45 years+). The long-term contracts, fully tradeable within the scheme, can be sold by private treaty or made available to other scheme participants as temporary annual or multi-annual entitlements. Agribusiness operators and investors will make a capital contribution toward the scheme, with an indicative cost of $3.09/kL, as well as pay an availability charge (an annual charge based on contracted water volumes), with an indicative cost of $0.26 kL (SA Water 2020).

Recover

With further treatment, sewage sludge can yield biosolids, which are defined by the US EPA as "nutrient-rich organic materials resulting from the treatment of domestic sewage in a treatment facility. . .that can be recycled and applied as fertiliser to improve and maintain productive soils and stimulate plant growth." Biosolids that are land applied have been treated to minimize odors and to reduce or eliminate pathogens. The use of biosolids increases crop yields and maintains nutrients in the root zone. Unlike chemical fertilizers, biosolids provide nitrogen that is slowly released over the growing season as the nutrient is mineralized and made available for plant uptake (Brears 2020b; US EPA 2020).

Case Study of Recover: Welsh Water's Biosolids for Agriculture

Welsh Water provides cost-effective and sustainable biosolids to farmers within 40 miles of the utility's advanced anaerobic digestion facilities – Port Talbot, Cardiff, Hereford, and Wrexham. To help agricultural customers fully utilize the potential for biosolids on their farms, the utility provides a biosolid service that involves soil testing, guidance on utilizing biosolids in the farm system, and ongoing support to make sure farmers get the best out of their biosolids. In addition to biosolids enhancing the water holding capacity of the soil, the recycling of biosolids uses less energy than intensive mineral fertilizer production, providing an alternative that reduces carbon footprint (Welsh Water 2020).

Restore

Environmental flows are defined as the quantity, quality, and timing of water flows required to sustain freshwater ecosystems and the human livelihoods and well-being that depend on these ecosystems. The circular economy aims to maximize environmental flows by reducing consumptive and non-consumptive uses of water, preserving and enhancing natural capital (e.g., pollution prevention, quality of effluent, etc.), and ensuring minimum disruption to natural water systems from human interaction and use. Also, water reuse and recycling can meet water demand, free up considerable amounts of water for the environment, and increase flows to vital ecosystems (Brears 2020a; WMO 2019).

Case Study of Restore: South Australia's Irrigation Industry Improvement Program

South Australia's $240 million Irrigation Industry Improvement Program (3IP) is a competitive grant program created by industry to support the restoration of a healthy Murray-Darling Basin environment. 3IP is a component of the South Australian River Murray Sustainability (SARMS) program that supports the growth of strong and sustainable irrigation communities. The program aims to recover 40 gigaliters of water access entitlements from participating irrigators. 3IP has to date offered funding to 186 projects throughout the River Murray regions and helped secure around 35 gigaliters of water ready to be returned to the river. Projects that have been funded include high-tech hydroponic and irrigation systems as well as initiatives to reposition businesses toward higher-value crops (Department of Primary Industries and Regions 2020).

Conclusion

Our current linear economy involves economic actors harvesting and extracting natural resources, using them to manufacture a product, and selling a product to other economic actors, who then discard it when it no longer serves its purpose. A key challenge to the linear economy is rising water-food nexus pressures. In contrast, the circular economy focuses on recycling, limiting, and reusing the physical inputs of the economy and using waste as a resource, leading to reduced primary resource consumption, with the ultimate aim being the decoupling of economic growth from resource use. Following the circular economy principles of reduce, reuse and recycle, recover, and restore, water managers are implementing policy innovations to reduce water-food nexus pressures, including providing grants to agricultural producers to conserve water resources as well as control pollution from farmland, developing recycled water infrastructure for farmers to provide climate- and season-independent irrigation water, recovering nutrients from wastewater treatment facilities for the production of sustainable biosolids for farmers, and providing grants to improve irrigation efficiency to restore natural flows in river basins.

Cross-References

- ▶ Circular Economy Cities
- ▶ Circular Water Economy

References

Brears, R. C. (2018). *Natural resource management and the circular economy*. Cham: Springer International Publishing.

Brears, R. C. (2020a). Building circular economy cities. In *The Palgrave handbook of climate resilient societies* (pp. 1–23). Cham: Springer International Publishing.

Brears, R. C. (2020b). *Developing the circular water economy*. Cham: Palgrave Macmillan.

Department of Primary Industries and Regions. (2020). Irrigation industry improvement program. Retrieved from https://www.pir.sa.gov.au/regions/sarms/3ip

FAO. (2012). Water reuse: Agriculture and urban water management in a recycling society. Retrieved from http://www.fao.org/3/a-i2616e.pdf

FAO. (2017). Water for Sustainable Food and Agriculture: A report produced for the G20 Presidency of Germany. Retrieved from http://www.fao.org/3/a-i7959e.pdf

Kirchherr, J., Reike, D., & Hekkert, M. (2017). Conceptualising the circular economy: An analysis of 114 definitions. *Resources, Conservation and Recycling, 127*, 221–232. https://doi.org/10.1016/j.resconrec.2017.09.005.

Morseletto, P. (2020). Restorative and regenerative: Exploring the concepts in the circular economy. *Journal of Industrial Ecology, 24*(4), 763–773. https://doi.org/10.1111/jiec.12987.

New Jersey State Agriculture Development Committee. (2019). Soil & water conservation grants. Retrieved from https://www.nj.gov/agriculture/sadc/publications/soil&watergrants.pdf

Pacific Institute. (2016). Using recycled water on agriculture: Sea Mist Farms and Sonoma County. Retrieved from http://agwaterstewards.org/wp-content/uploads/2016/08/recycled_water_and_agriculture3.pdf

SA Water. (2020). Invest in NAIS. Retrieved from https://www.sawater.com.au/nais/invest-in-nais

US EPA. (2020). Agriculture nutrient management and fertilizer. Retrieved from https://www.epa.gov/agriculture/agriculture-nutrient-management-and-fertilizer

Welsh Water. (2020). Biosolids for agriculture. Retrieved from https://www.dwrcymru.com/en/our-services/wastewater/biosolids-for-agriculture

WMO. (2019). Guidance on environmental flows: Integrating E-flow science with fluvial geomorphology to maintain ecosystem services. Retrieved from https://library.wmo.int/doc_num.php?explnum_id=9808

Circular Economy Cities

Robert C. Brears
Our Future Water, Christchurch, Canterbury, New Zealand

Synonyms

Carbon emissions; Circular economy; Environment; Linear economy; Resource efficiency

Introduction

Our current industrialized economy is essentially a linear model in which resource consumption follows a “take-make-consume-dispose” pattern where natural resources are harvested for the manufacturing of products, which are then disposed of after consumption. In terms of volume, global material use is projected to more than double from 79 gigatonnes (Gt) in 2011 to 167 Gt in 2060. Total emissions are projected to reach 75 Gt CO_2 equivalent by 2060, of which material management would constitute approximately 50 Gt CO_2 equivalent (OECD 2018). With rapid urbanization, the proportion of the global economy living in urban areas projected to increase from 54% in 2015 to 66% by 2050. Following a business-as-usual path, annual resource requirements of urban areas are estimated to increase from 40 billion tonnes in 2010 to nearly 90 billion tonnes by mid-century (International Resource Panel 2018).

As such, it has become increasingly clear that the current economic model is untenable due to a growing shortage of materials, increased levels of pollution, increased material demand, and growing demand for responsible products by consumers. In contrast, the circular economy aims to decouple economic growth from resource use and associated environmental impacts (Brears 2018). It is estimated that if cities become more circular in the areas of transport, commercial buildings, and building heating/cooling, they could achieve reductions of between 36% and 54% in energy use, GHG emissions, and metal, land, and waste use (International Resource Panel 2018).

The Linear Economy

In our current economic model, manufactured capital, human capital, and natural capital all contribute to human welfare by supporting the production of goods and services in the economic process. Natural capital – the world's stock of natural resources (provided by nature before their extraction or processing by humans) – is typically used for material and energy inputs into production and acts as a "sink" for waste from the economic process. This economic model can be best described as "linear" which typically involves economic actors – who are people or organizations engaged in any of the four economic activities of production, distribution, consumption, and resource maintenance – harvesting and extracting natural resources, using them to manufacture a product, and selling a product to other economic actors, who then discard it when it no longer serves its purpose. As such, natural resources in the linear economic model:

- *Become inputs*: Material resources used in the economy come from raw materials that are extracted from domestic natural resource stocks or extracted from natural resource stocks abroad and imported in the form of raw materials, semi-finished materials, or materials embedded in manufactured goods.
- *Become outputs*: After use in production and consumption activities, materials leave the economy as an output either to the environment in the form of residuals (pollution, waste) or in the form of raw materials, semi-finished materials, and materials embedded in manufactured goods.
- *Accumulate in man-made stocks*: Some materials accumulate in the economy where they are stored in the form of buildings, transport infrastructure, or durable and semi-durable goods such as cars, industrial machinery, and household appliances. If these materials are not recovered from waste, they flow back into the environment.
- *Create indirect flows*: When materials or goods are imported for use in an economy, their upstream production is associated with unused materials that remain abroad including raw materials needed to produce the goods and the generation of residuals. These indirect flows of materials consider the life cycle dimension of the production chain but are not physically imported. As such, the environmental consequences occur in countries from which the imports originate (Brears 2018).

The Circular Economy

The circular economy, in contrast to the linear "take-make-consume-dispose" economy aims to keep resources in use for as long as possible, extract value from them while in use, and recover and regenerate products and materials at the end of each service life. As such, the circular economy focuses on recycling, limiting, and reusing the physical inputs of the economy and using waste as a resource, leading to reduced primary resource consumption, with the ultimate aim being the decoupling of economic growth from resource use. The notion of decoupling is that economic output shall continue to increase at the same time as rates of increasing resource use and environmental impact are slowed and in time brought into decline. There are two modes of decoupling that can be distinguished:

- Resource decoupling or "dematerialization" involves reducing the rate at which natural resources are used per unit of economic output.

- Impact decoupling seeks to increase economic activity while decreasing negative environmental impacts from pressures such as pollution, carbon emissions, or destruction of biodiversity.

The Role of Cities in Developing a Circular Economy

Cities have an essential role in developing the circular economy with some of the critical aspects cities need to consider in encouraging the development of a circular economy, including:

- *Encouraging better product design*: Better design can make products more durable or easier to repair, upgrade, or remanufacture. It can help recyclers disassemble products to recover valuable materials or components, which overall reduces resource use.
- *Facilitating better consumption choices*: The choices consumers make can support or hamper the development of the circular economy. These choices are shaped by the information consumers have access to, the range and prices of existing products, and the availability of economic instruments to ensure that product prices better reflect environmental costs.
- *Improving waste management*: Waste management plays a central role in the circular economy. The way that waste is collected and managed determines whether there are high levels of recycling or not and whether valuable materials find their way back into the economy or to an inefficient system where most recyclable waste ends up in landfills or is incinerated, with potentially harmful environmental costs and significant economic losses.
- *Creating a market for waste to resources*: In the circular economy, materials can be recycled back into the economy as new raw materials, thereby increasing the security of supply. Secondary raw materials can be used just like primary raw materials from traditional extractive resources (Brears 2020b).

The 5Rs

Cities can guide the development of a circular economy by encouraging all sectors of the economy to follow the 5R approach of reduce, reuse, recycle, restore, and recover (Brears 2018, 2020a; Kirchherr et al. 2017; Morseletto 2020), in particular:

- *Reduce*: Increase efficiency in product manufacture or use by consuming few natural resources and materials.
- *Reuse*: Reuse by another consumer of discarded product which is still in good condition and fulfils its original function.
- *Recycle*: Process materials to obtain the same (high-grade) or lower (low-grade) quality.
- *Restore*: Restoring of natural capital.
- *Recover*: Recovery of resources from waste.

Case Study of Reduce: Singapore's Building Retrofit Energy Efficiency Financing Scheme

Singapore's Building Retrofit Energy Efficiency Financing (BREEF) Scheme for Existing Buildings aims to encourage building owners to undertake energy-efficient retrofits by providing financing for building owners, management corporations, energy performance contracting firms, and energy services companies, among others. The scheme, facilitated by the Building and Construction Authority and participating financial institutions, offers financing to pay for the upfront costs of energy retrofits of existing buildings, through an energy performance contract arrangement. BREEF can cover the cost of equipment, installation, and professional fees. The financing can only be used for energy efficiency retrofits of existing buildings and lead to achieving a minimum Green Mark Certification standard, which is maintained through the loan period. The scheme's

Circular Economy Cities, Table 1 Building Retrofit Energy Efficiency Financing (BREEF) Scheme details

Detail	Description
Maximum loan quantum	Up to $4 million or 90% of costs, whichever is lower
Interest rate	Determined by financial institution
Maximum loan tenure	Five years
Scheme availability	Until 31 March 2023

details are listed in Table 1 (Building and Construction Authority 2020).

Case Study of Reuse: San Francisco's Onsite Water Reuse Grant Program

The San Francisco Public Utilities Commission's (SFPUC) Onsite Water Reuse Grant Program provides grant funding to encourage commercial water users to collect, treat, and reuse alternative water sources including rainwater, stormwater, gray water, foundation drainage, air conditioning condensate, and blackwater for non-potable uses, such as toilet flushing, irrigation, and cooling tower makeup. Grant funding is available for three types of projects:

- Projects that are installing onsite water systems voluntarily (voluntary projects)
- Projects that are installing onsite water systems on a mandatory basis in compliance with the Nonpotable Water Ordinance that go above and beyond baseline compliance (above and beyond projects)
- Projects that are installing onsite treatment and reuse of brewery process water

For projects to be funded, they must meet one of the following criteria listed in Table 2. The types of activities considered for funding include the installation of collection systems for onsite alternate water sources, installation of treatment systems to improve the water quality of onsite alternate water sources for beneficial reuse, and/or storage of the treated water (San Francisco Public Utilities Commission 2019).

Circular Economy Cities, Table 2 San Francisco's Onsite Water Reuse Grant Program funding

Criteria	Grant funding available
Projects that replace at least 450,000 gallons of SFPUC water per year	Up to $200,000
Projects that replace at least 1,000,000 gallons of SFPUC water per year	Up to $500,000
Projects that replace at least 3,000,000 gallons of SFPUC water per year	Up to $1,000,000

Case Study of Recycle: City of Toronto's Electronic Waste Programme

The City of Toronto collects unwanted electronics for free to ensure they are disposed of safely, recycled, and kept out of the landfill. Electronic items can be put out on garbage day for pickup, brought to a drop-off depot or Community Environment Day, or donated for reuse. Items that are accepted as electronic waste are:

- Cell phones and home phones
- Computer cables
- Laptop computers and accessories
- VCR/DVD players
- Video recorders
- TVs
- Desktop computers and monitors
- Printer and fax machines
- Accessories
- Cameras
- Receivers and speakers
- Stereos, tuners, and turntables

The City of Toronto also provides the TOwaste App to help residents never miss their garbage or recycling day again. The App lets users know what waste items go where, what bins to put out when, and where to donate used items or find a city drop-off depot from their smartphone or tablet. The key features of the App are:

- Waste Wizard – the City's quick and easy search tool with information on how to properly sort over 2,000 items
- Collection schedules and the ability to set reminders and opt-in for alerts about service changes
- A map of drop-off depot locations where residents can properly dispose of items such as hazardous waste
- A map of locations where residents can donate or buy used items (City of Toronto 2020)

Case Study of Restore: City West Water's Stormwater Harvesting Partnering Fund

City West Water, a Victorian government-owned retail water corporation in Melbourne, provides the Stormwater Harvesting Partnering Fund to work with the community to develop projects that promote sustainable water management. The Fund aims to develop collaborative partnerships on projects that promote water sustainability and develop thriving green spaces. One such project is the Laverton Stormwater Harvesting Project, a partnership project between City West Water and Hobsons Bay City Council. The stormwater harvesting scheme is designed to treat around 92 million liters per annum of harvested stormwater from Laverton Creek, located near the Princes Freeway. The stormwater harvesting scheme supplies 80 percent of the Laverton Recreation Reserve's water demand. The scheme's flows represent around 2% of the total runoff that could be harvested from an estimated 4,000 hectares of the Laverton Creek catchment. The stormwater treatment facility involves a diversion that removes pollutants. Then the stormwater is treated in a series of wetlands which remove sediment and reduce nutrient levels. The stormwater is then pumped into a 4.5 million liter open storage basin used to irrigate the Laverton Recreational Reserve, which is becoming a vital biodiversity asset for the area, with vegetation providing habitat for waterbirds. At the same time, a conservation team conducts targeted weed control and revegetation of local indigenous species (City West Water 2020).

Case Study of Recover: Berliner Wasserbetriebe Recovering Resources from Wastewater

Berliner Wasserbetriebe's Schönerlinde sewage treatment plant is turning sewage sludge into sewage gas to generate power and heat. Additionally, the utility has constructed three wind turbines, with a capacity of 2 MW each, as well as two micro-gas turbines to complement the plant's combined heat and power unit. Overall, around 84 percent of the energy required by the plant is produced internally, saving up to 13,000 tonnes of carbon emissions per year. Berliner Wasserbetriebe has also developed a patented process for recovering phosphorus from its sewage treatment plants. The recovered phosphorus is sold under the brand name "Berliner Pflanze" (Berlin Plant) to horticulture and agriculture producers in the surrounding areas of the city. Several years ago, Berliner Pflanze won the GreenTec Award for environmentally friendly recycling products. Berliner Plant is now available for purchase from the utility's customer center:

- 0.5 kg for the price of 2.50 EUR
- 2 kg for the price of 4.00 EUR
- 5 kg for the price of 10.00 EUR
- 16 kg box (8 × 2 kg bag) for 28.00 EUR (Berliner Wasserbetriebe 2020)

Conclusion

Our current industrialized economy is essentially a linear model in which resource consumption follows a "take-make-consume-dispose" pattern where natural resources are harvested for the manufacturing of products, which are then disposed of after consumption. As such, it has become increasingly clear that the current economic model is untenable due to a growing shortage of materials, increased levels of pollution, increased material demand, and growing demand for responsible products by consumers. In contrast, the circular economy aims to decouple economic growth from resource use and associated environmental impacts. Cities can guide the development of the circular economy by implementing policy innovations that encourage all sectors of the economy to follow the 5R approach of reduce, reuse, recycle, restore, and recover. From the case studies of cities implementing various aspects of the 5Rs, cities can provide financing for existing buildings to become energy efficient; provide grants for commercial buildings to reuse alternative water sources for non-potable uses; provide free electronic waste collection services to ensure they are disposed of safely, recycled, and kept out of the landfill; utilize stormwater runoff for the irrigation of public land as well as the restoration of the natural environment; and recover heat and energy from wastewater plants for renewable energy generation as well as recover phosphorus for agricultural use.

Cross-References

- ▶ Circular Cities
- ▶ Circular Economy and the Water-food Nexus
- ▶ Circular Water Economy

References

Berliner Wasserbetriebe. (2020). We recycle phosphorus to fertilizer. Retrieved from https://www.bwb.de/de/4951.php

Brears, R. C. (2018). *Natural resource management and the circular economy*. Cham: Springer International Publishing.

Brears, R. C. (2020a). Building circular economy cities. In *The Palgrave handbook of climate resilient societies* (pp. 1–23). Cham: Springer International Publishing.

Brears, R. C. (2020b). *Developing the circular water economy*. Cham: Palgrave Macmillan.

Building and Construction Authority. (2020). Building Retrofit Energy Efficiency Financing (BREEF) Scheme. Retrieved from https://www1.bca.gov.sg/buildsg/sustainability/green-mark-incentive-schemes/building-retrofit-energy-efficiency-financing-breef-scheme#:~:text=Contact%20Details&text=The%20BREEF%20scheme%2C%20facilitated%20by,equipment%2C%20installation%20and%20professional%20fees

City of Toronto. (2020). Electronic waste. Retrieved from https://www.toronto.ca/services-payments/recycling-organics-garbage/electronic-waste/

City West Water. (2020). Stormwater Harvesting Partnering Fund. Retrieved from https://www.citywestwater.com.au/saving_water/partnering_fund

International Resource Panel. (2018). The weight of cities: Resource requirements of future urbanization. Retrieved from https://www.resourcepanel.org/reports/weight-cities

Kirchherr, J., Reike, D., & Hekkert, M. (2017). Conceptualizing the circular economy: An analysis of 114 definitions. *Resources, Conservation and Recycling, 127*, 221–232. https://doi.org/10.1016/j.resconrec.2017.09.005.

Morseletto, P. (2020). Restorative and regenerative: Exploring the concepts in the circular economy. *Journal of Industrial Ecology, 24*(4), 763–773. https://doi.org/10.1111/jiec.12987.

OECD. (2018). Global Material Resources Outlook to 2060. Economic drivers and environmental consequences: Highlights. Retrieved from https://www.oecd.org/environment/waste/highlights-global-material-resources-outlook-to-2060.pdf

San Francisco Public Utilities Commission. (2019). Onsite water reuse grant program rules. Retrieved from https://sfwater.org/Modules/ShowDocument.aspx?documentID=5445

Circular Lifestyle

▶ Systemic Innovation for Thrivable Cities

Circular Water Economy

Robert C. Brears
Our Future Water, Christchurch, New Zealand

Synonyms

Renewable energy; Water conservation; Water efficiency; Water recycling

Introduction

In our current economic model, manufactured capital, human capital, and natural capital all contribute to human welfare by supporting the production of goods and services in the economic process, where natural capital – the world's stock of natural resources (provided by nature before their extraction or processing by humans) – is typically used for material and energy inputs into production and acts as a "sink" for waste from the economic process. This economic model can be best described as "linear" which typically involves economic actors – who are people or organizations engaged in any of the four economic activities of production, distribution, consumption, and resource maintenance – harvesting and extracting natural resources, using them to manufacture a product, and selling a product to other economic actors, who then discard it when it no longer serves its purpose.

While this model has generated unprecedented levels of growth, the model has led to resource scarcity, the generation of waste, and environmental degradation from a variety of climatic and non-climatic challenges. Climate change is impacting the availability of good-quality water of sufficient quantities necessary for both humans and nature. Rapid population growth is increasing demand for food, which is placing immense stress on water resources, while agricultural production is impacting water quality. At the same time, economic growth and rising income levels are increasing demand for water resources, resulting in environmental degradation and biodiversity loss (Brears 2016, 2018b, 2020b). This chapter discusses the circular economy in general and in the context of water resources management. The chapter then discusses how the circular water economy can be enacted.

Circular Economy

The circular economy, in contrast to the linear "take-make-consume-dispose" economy, aims to decouple economic growth from resource use and associated environmental impacts. The notion of decoupling is that economic output shall continue to increase at the same time as rates of increasing resource use and environmental impact are slowed and in time brought into decline. In the context of water, the circular water economy aims to design out externalities and keep resources in use, all the while regenerating natural capital. Specifically:

- *Designing out externalities*: The circular water economy optimizes the amount of energy, minerals, and chemicals used in the operation of water systems in concert with other systems, optimizes consumptive use of water, and uses measures or solutions which deliver the same outcome without using water.
- *Keeping resources in use*: The circular water economy aims to optimize resource yields (water use and reuse, energy, minerals, and chemicals) within water systems, optimize energy or resource extraction from the water system and maximize their reuse, and optimize value generated in the interfaces of water systems with other systems.
- *Regenerating natural capital*: The circular water economy aims to maximize environmental flows by reducing consumptive and non-consumptive uses of water, preserve and enhance natural capital (e.g., pollution prevention, quality of effluent, etc.), and ensure minimum disruption to natural water systems from human interaction and use (Brears 2018c, 2020a)

Circular Water Economy

To action the circular water economy, water authorities can follow the circular economy 5R approach of reduce, reuse, recycle, recover, and restore:

- *Reduce*: Water conservation and water-use efficiency best management practices (drip irrigation, water metering, pricing of water etc.) can reduce demand for scarce water resources.
- *Reuse*: Water users can implement reuse systems (without treatment) that harvest rainwater and gray water for various non-potable uses.
- *Recycle*: Municipal wastewater treatment plants and housing developments can recycle

water (treated) for a variety of uses including irrigation of crops.

- *Recover*: Water authorities can recover nutrients from wastewater and process them to healthy levels for use as fertilizer. This provides an additional revenue stream for wastewater treatment plants while reducing demand for carbon-intensive fertilizer.
- *Restore*: Water authorities can utilize nature-based solutions to restore environmental flows, including recharging aquifers and ensuring minimum flow levels in waterways for environmental preservation and restoration.

Reduce

Demand management promotes water conservation and water efficiency during both normal and abnormal conditions, through changes in practices, culture, and people's attitudes toward water resources. Demand management seeks to reduce the loss and misuse of water, optimize the use of water, and facilitate significant financial and infrastructural savings by minimizing the need to meet increasing demand with new water supplies. The benefits of demand management include reduced water and electricity bills, reduced carbon emissions from pumping and heating water, reduced leakage, and more water for a healthier environment (Brears 2018b). There are a variety of demand management tools available to water utilities to promote water conservation and water efficiency including the following:

Pricing of Water

The main pricing structures used by water utilities to promote the conservation of water resources are:

- *Volumetric rates*: A volumetric rate is a charge based on the volume used at a constant rate. Therefore, the amount users pay for water is strictly based on the amount of water consumed.
- *Increasing block tariffs*: An increasing block tariff contains different prices for two or more pre-specified quantities (blocks) of water, with the price increasing with each successive block.
- *Two-part tariff systems*: A two-part tariff system contains a fixed and variable component. In the fixed component, water users pay one amount independently of consumption, and this covers infrastructural and administrative costs of supplying water. Meanwhile, the variable amount is based on the quantity of water consumed and covers the costs of providing water as well as encouraging conservation (Brears 2016).

Water Meters

Before water users can be charged for the amount of water consumed, the dwelling or building must have water meters to measure the volume of water consumed. Two types of water meters can be used:

- *Automated meter readers (AMRs)*: One-way readers that send usage data back to the utility.
- *Advanced metering infrastructure (AMI)*: AMIs is a "two-way" solution that creates a network between the meters and the utility's information system. Data flows both ways facilitating not only remote meter reading but the ability to send alerts to customers, for example, when a leak is detected (Brears 2020a).

Regulations

Temporary regulations are used to restrict water usage during specified times, such as during times of water shortages. Permanent regulations are used to mandate the installation of water-saving devices in new developments or retrofits of existing buildings (Brears 2016).

Consumer Education and Awareness

Water authorities can promote water conservation in schools to increase young people's knowledge of the water cycle and encourage the sustainable use of scarce water resources. Meanwhile, water utilities can raise public awareness of the need to conserve water resources through a variety of formats, including public information and public events (Brears 2014).

C

Reuse

There are a variety of onsite non-potable water systems that are available to reduce demand for potable water for non-potable uses, such as cooling buildings, irrigating landscapes, and flushing toilets and urinals.

Gray Water

Gray water is reusable wastewater from residential, commercial, and industrial bathroom sinks, bathtub shower drains, and clothes washing equipment drains. Gray water is reused onsite, usually for toilet flushing and irrigation. Gray water systems vary significantly in their complexity and size from small systems with simple treatment to large systems with complex treatment processes. Nevertheless, most have standard features including a tank for storing the treated water, a pump, a distribution system for transporting the treated water to where it is needed, and some treatment (Brears 2020a).

Blackwater

Blackwater, or sewage, is the wastewater from toilets. In blackwater recycling systems, all the blackwater is routed to an initial tank via gravity, from which it settles, and a primary colony of bacteria eats at the waste. The blackwater then goes through an aeration stage, a sludge settling stage, before it is chlorinated and used as irrigation water (watering lawns or nonfood gardens) (Brears 2020a).

Rainwater Harvesting

Rainwater harvesting systems collect and store rainfall for later use. When designed appropriately, they slow down and reduce runoff and provide a source of water. There are two main types of rainwater harvesting systems: passive and active.

- *Passive harvesting systems*: These systems are generally small and utilize rain barrels to capture rooftop runoff for garden and landscape irrigation or car washing.
- *Active harvesting systems*: These systems are usually large and utilize cisterns to capture large volumes of water for non-potable water replacement, for example, toilet flushing, clothes washing, and evaporative cooling, etc. (Brears 2020a).

Stormwater Harvesting

Stormwater harvesting involves collecting, storing, and treating stormwater from urban areas, which can then be used to water public parks, gardens, sports fields, and golf courses. Stormwater harvesting systems conserve potable water for essential uses, provide an alternative to potable water during time of peak demand, reduce or limit withdrawals from ground or surface water supply, and maintain reliable water supply in the event of municipal service disruption (Brears 2018a).

Recycle

Water recycling is the use of harvested water for the same or a different function, after treatment, where treatment can be tailored to meet the water quality requirements of planned use. Recycled water can be used for a variety of non-potable and potable uses.

Non-potable Use

Water recycling systems can be developed for non-potable projects that treat wastewater for specific purposes other than drinking. The water is usually of a lower quality than potable systems, and the level of treatment varies depending on the end use. Industrial water can be recycled within a business itself or between several businesses for a variety of uses. Recycled water can be used for agricultural production, particularly during times of drought when regular water supplies are limited or nonexistent. In urban areas, water recycling systems can be used for irrigating lawns and parks and other non-potable uses (Brears 2020a).

Potable Use

Potable water reuse involves the use of a community's wastewater as a source of drinking water. Two forms of planned potable reuse exist as follows:

- *Indirect potable reuse*: Recycled wastewater can be returned into the current/natural water cycle well upstream of a drinking water treatment plant.
  - Planned indirect potable reuse means there is an intent to reuse the water for potable use. The point of return could be either into a major water supply reservoir, a stream feeding a reservoir, or a water supply aquifer.
  - Unplanned indirect potable reuse is where wastewater entering the natural water (creeks, rivers, lakes, aquifers) is eventually extracted for drinking water: usually with no awareness that the natural system contains wastewater.
- *Direct potable reuse*: Direct potable reuse is where recycled water is directly injected into the potable water supply distribution system downstream of the water treatment plant or into the raw water supply immediately upstream of the water treatment plant. This means water used by consumers could contain either undiluted or slightly diluted recycled water (Brears 2020a).

Recover

Wastewater treatment plants in the circular water economy provide multiple opportunities, including generating renewable energy and recovering nutrients.

Biogas from Anaerobic Digestion

Anaerobic digestion is a series of biological processes in which microorganisms break down biodegradable material in the absence of oxygen. One of the end products is biogas, which is combusted to generate electricity and heat or can be processed into renewable natural gas and transportation fuels. Meanwhile, anaerobic digestion of energy-rich organic waste materials, including restaurant grease and food waste along with wastewater treatment sludge, is defined as "co-digestion." In addition to diverting food waste and fats, oils, and grease (FOG) away from landfills and collection, it reduces greenhouse gases and energy bills of utilities (Brears 2020a).

Biogas for Combined Heat and Power

Combined heat and power (CHP) systems generate electricity and capture the heat that would otherwise be lost to provide useful thermal energy. They can be installed at wastewater treatment plants in different forms including anaerobic digester gas-fuelled CHP, non-biogas-fuelled (CHP), for example, natural gas, heat recovery from a sludge incinerator, and combined heat and mechanical power system, for example, an engine-driven pump or blower with heat recovery (US EPA 2011).

Thermal Conversion of Wastewater and Biosolids

The process of converting biosolids to energy through thermal oxidation (incineration), which is the complete oxidation of organics (biomass) to carbon dioxide and water in the presence of excess air, is a well-established technology. The benefits of thermal conversion include a reduction in biosolids mass, generation of heat for use in heating or electricity generation, reduction in the facility's overall carbon footprint, lowering the reliance on fossil fuels, generation of ash for use in building materials, and generation of additional revenue to utilities (Brears 2020a).

Recovery of Nutrients from Wastewater

In the circular water economy, numerous resources can be recovered from wastewater, including:

- *Nitrogen and phosphorus*: Nitrogen and phosphorous can be recovered from wastewater and turned into highly pure, slow-release, environmentally friendly fertilizer.
- *Bioplastic*: One of the most nontraditional technologies under development is the production of biodegradable plastic using polymers isolated from biosolids. The plastic can be used as a substitute for conventional petroleum-based plastics.
- *Bricks*: Sewage sludge ash is the by-product produced during the combustion of dewatered

sewage sludge in an incinerator. Sewage sludge ash is primarily a silty material with some sand-sized particles. The ash can be used in the brick and tile industry.
- *Metals*: Metals can be potentially mined from wastewater, for instance, silver and cadmium are increasingly being found in wastewater and are expensive enough to potentially warrant recovery (Brears 2020a).

Restore

The circular water economy aims to maximize environmental flows by reducing consumptive and nonconsumptive uses of water, preserve and enhance natural capital (e.g., pollution prevention, quality of effluent, etc.), and ensure minimum disruption to natural water systems from human interaction and use.

Riparian Buffers

Riparian buffers, for example, wetlands and forests, protect waterways from potential pollutants from the surrounding land area such as those from agricultural land and industrial activities. Riparian buffers are usually naturally present but are often under pressure due to human activities, including urbanization. Degraded riparian buffers can be re-established through restoration and revitalization of riparian vegetation. The projects should be carefully planned and the correct plant species selected. It is also essential to recreate conditions for animal habitats. Also, as riparian restoration projects can affect existing community uses of waterways, including fishing, water collection, and recreation, projects should take these uses into account and try to preserve them with community consultation. When applied throughout a watershed, riparian buffers provide multiple environmental benefits including contributing to stream base flows as well as providing an interconnected network of habitats. By preserving interconnected networks of habitats, riparian buffers can increase wildlife diversity in urban areas while providing recreational opportunities, including cycling and walking trails (Brears 2020b).

Wetlands

Wetland restoration is the process of supporting the wetland to regain its health and function, therefore restoring its ability to provide a variety of ecosystem services. Restoring a wetland involves re-establishing both its extent and ecological functions by removing the non-wetland features and recreating portions of the wetlands that were lost. This may include excavation, removal of hydraulic structures, clearing of invasive and non-wetland vegetation, diversion of polluted runoff away from the wetland, and flooding of some sections. Protecting, restoring, or constructing wetlands can help provide clean water for ecosystems, drinking water needs, and wildlife habitats. Wetlands retain large volumes of water, which they release slowly, making them essential for drought mitigation and flood control. Wetlands provide the necessary conditions for microorganisms to live there, which transform and remove pollutants from the water, for example, nutrients, including nitrogen and phosphorous, are deposited in wetland soils from stormwater runoff, with excess nutrients absorbed by wetland soils and taken up by plants and organisms. Because wetlands are multifunctional ecosystems, their restoration can also lead to multiple social benefits including recreation, fishing, and the production of reeds, fodder, and fruits that can be harvested (Brears 2020b).

Cross-References

▶ Circular Cities
▶ Circular Economy and the Water-Food Nexus
▶ Circular Economy Cities

References

Brears, R. (2014). Transitioning towards urban water security in Asia Pacific. Retrieved from http://www.diss.fu-berlin.de/docs/receive/FUDOCS_document_000000021440?lang=de

Brears, R. C. (2016). *Urban water security*. Chichester/Hoboken: Wiley.

Brears, R. C. (2018a). *Blue and green cities: The role of blue-green infrastructure in managing urban water resources*. London: Palgrave Macmillan.
Brears, R. C. (2018b). *Climate resilient water resources management*. Cham: Palgrave Macmillan.
Brears, R. C. (2018c). *Natural resource management and the circular economy*. Cham: Springer International Publishing.
Brears, R. C. (2020a). *Developing the circular water economy*. Cham: Palgrave Macmillan.
Brears, R. C. (2020b). *Nature-based solutions to 21st century challenges*. Oxfordshire: Routledge.
US EPA. (2011). Opportunities for combined heat and power at wastewater treatment facilities: Market analysis and lessons from the field. Retrieved from https://www.eesi.org/files/wwtf_opportunities.pdf

Cities

▶ Sustainable Development Goals and Urban Policy Innovation

Cities in Low-and-Middle Income Countries

▶ Multi-stakeholder Partnerships to Support Climate Migrants in Fragile Cities

Cities in Nature

Khee Poh Lam and Teng Chye Khoo
National University of Singapore, Singapore, Singapore

Evolution of City Concepts

The Industrial Revolution (IR) in the eighteenth century transformed the way cities grew. Mass production in urban factories led to large numbers of workers relocating into the cities to work. As the cities grew, homes and other commercial and community facilities like schools and hospitals had to be planned and built. Urban transport and communication also transformed with the building of suburban railway networks, metros, automobiles, and bicycles. However, industrialization and rapid growth also resulted in environmental pollution and poor living conditions.

City planning concepts evolved in response to these technological trends in the last two centuries. But as cities grew denser, more polluted, and congested with industrialization and with the proliferation of cars, the urban planners' challenge has primarily been to keep cities in harmony with nature and to retain the human scale even with higher density and high-rise buildings. These ideas have therefore influenced the shape of cities especially in western cities as policymakers, planners, architects, and engineers adopted them in their policies, plans, and designs. These well-known city concepts are summarized in Table 1.

Many Asian countries became independent from their western colonialists in the mid-twentieth century. Rapid economic and population growth resulted also in rapid urbanization, very often with serious sprawl, pollution, and congestion, with haphazard high-rise buildings and poor infrastructures, much like many western industrial cities in the nineteenth century.

Singapore became an island city state in 1965 with very little land and basic resources like water, food, and energy. The challenge was to plan for, build, and run a city that was able to accommodate people and density and yet be liveable, sustainable, and resilient with these constraints. Its founding Prime Minister, Lee Kuan Yew, having been a student in England, was familiar with the failure of planned new towns there. He was not keen to see Singapore develop as a high-rise concrete jungle like some other cities in Asia. Instead, he wanted a Singapore that is clean and green, like a garden city.

How Singapore has since evolved, from a Garden City towards being a City in Nature even with increasing density, is an approach to city planning and development that has been documented by the Centre for Liveable Cities (CLC) as the Liveability Framework (LF). This approach draws on many of the good city concepts mentioned earlier but has combined and adapted them in a way that is more suited to many developing cities in Asia. The next section elaborates on this urban systems

Cities in Nature, Table 1 How city concepts have evolved

City concepts
1. Garden City (Encyclopedia Britannica, inc, 1998) The Garden City concept was developed and promoted in "Tomorrow: A Peaceful Path to Social Reform" published in 1898, by Ebenezer Howard, an English Town Planner, to address issues such as overcrowding and congestion, caused by the Industrial Revolution. Howard advocated for quality urban life by planning a series of small cities and residential community for 30,000 inhabitants that provide urban amenities and access to both nature and the rural environment. The integral elements of a successful Garden City include establishing the city's greenbelt areas and population densities up front.
2. Modern City Movement (O'Donnell 2019) French-Swiss architect Le Corbusier, inspired by the machine age as an opportunity to refine the society and the quality of life, was the champion of the modernist movement, in the 1900's. Corbusier's Radiant City concept advocated an ultimate form of perfect and ordered environment by destructing the existing city structure. Inspired by the industrial revolution, s architecture and cities were expected to be mass produced in the most economical way and served as a functional machine to fulfill its purpose. The functionalist ideology emphasized on raw geometry and separated the architecture from preexisting culture. The Contemporary City (Ville Contemporaine), with a population of three million, is an example of Corbusier's urban planning work. The main features included a central business district consisting of twenty four identical glass skyscrapers located on a grid with generous public park spaces in between. The skyscraper was integrated with a complex transportation system, while the public open spaces comprised of various squares, restaurants, and theatres. The geometric low-rise housing and sporting grounds were designed around the center to enable the density to increase and congestion to decrease.
3. Automobile City This city planning concept, coined by Jeffrey Kenworthy, prioritized the usage of cars and trucks, allocated vast amount of land for multilane highways and parking and had a lower density in population and jobs compared to older and more compact cities (Newman 1945– (1989)). Such a city planning concept was extensively implemented, from the 1940's, in cities such as Houston and Phoenix in America, and Perth and Brisbane in Australia. The central city high-rise office block was the primary focal point as people were expected to commute from the suburbs to the inner city. The public transport and non-motorized modes of transportation received limited attention. This city planning concept resulted in urban sprawl, air pollution, increased travelling time from place to place, and reduced sense of place due to the ease of long-distance car travel (Marshall 2010).
4. Humanistic City Planning An urbanist and activist, Jane Jacobs wrote the book *Death and Life of Great American Cities*, published in 1961, to critique the 1950's urban planning policies, responsible for the decline of city neighborhoods in the United States (Dreier 2006). Jacobs opined that the modernist urban planning ignored the intricacy of human lives. She opposed large-scale urban renewal programs, especially freeways through inner cities. Instead, she advocated for walkable streets with dense mixed-use developments and preached that the "eyes on the street" of passers-by would enhance public order. These ideas eventually led to a humanistic planning approach for the neighborhood level. Jan Gehl, a Danish architect and urban planner, has been inspired by the works of Jane Jacobs. He focused on the re-orientation of city design towards the pedestrian and cyclist to improve the quality of urban life. Gehl argued that the modernist ideas separating functions as work, living, recreation, and communication had resulted in the focus on objects instead of spaces. When human scale and left-over spaces are ignored, the automobiles would be the main focus of city planning. Such planning paradigms, with limited considerations on humans, led to the decline of cities. The main concept of Humanistic City Planning involves the understanding of how people interact at the microlevel with their urban environments as a reaction to the fast-paced city developments. Gehl's book *Life between buildings*, published in 1971, shared how spaces and places influenced people and focused on making people visible in architecture and city planning.
5. Compact Cities The term "Compact City" was coined in 1973, by mathematicians, George Dantzig and Thomas L. Saaty, who advocated the need to use resources efficiently. In contrast to the past conventional planning and transport planning, they advocated for a compact city with high residential density, mixed land uses, efficient public transport system, and an urban layout to encourage low energy consumption and nonpolluting transportation. When compared to urban sprawl, a compact city will be a more sustainable urban settlement type due to less dependency on automobiles and requiring less infrastructure provision (Dempsey 2010).

(continued)

Cities in Nature, Table 1 (continued)

6. Biophilia and Biophilic City Edward O. Wilson, an American biologist, popularized the word "biophilia" in 1984, by emphasizing the idea that humans have an innate love for nature and living things. Since 2010's, there is increased interest in incorporating nature into urban environments, by improving people's interaction with parks and green spaces, building eco-bridges to enhance habitat connectivity, and seeking quantifiable impacts of nature on human (National Parks Board 2015). Dr Timothy Beatley, a city researcher, focused on inventive strategies for cities to become more liveable by easing their ecological footprints. Beatley advocated the use of the biophilia hypothesis to implement the essential elements of a biophilic city and shared examples of cities with positive integration of biophilic elements (Beatley 2011). Professor Peter Newman, an environmental scientist, is recognized for his research and contributions to improve Perth's transport system, by introducing sustainable public transport practice to replace the city's dependence on automobiles. Together with Isabella Jennings, Newman authored the book *Cities as Sustainable Ecosystems: Principles and Practices* to illustrate the integration of city residents into their bioregional environment, and the planning of cities with ecological sustainability in mind (Newman and Jennings 2012). Newman advocated for urban redevelopments to consider sustainable elements such as greening roofs, rainwater harvesting, and renewable energy production, supported the biodiversity in parks, community gardens for food productions, and pedestrian-friendly spaces suitable for walking and cycling.

approach which has shaped modern Singapore for more than five decades. The subsequent section documents Singapore's transformation from a Garden City to a City in Nature, even as a very high-density city.

Singapore's Urban Transformation: From Mudflats to Metropolis

The archaeological excavations from the Singapore River and Fort Canning Hill have suggested that Singapore as a city probably dates back 700 years (National Heritage Board 2020). A wide variety of trade wares were unearthed and shed light on Singapore's history as a harbor and trade settlement. The city of Singapore came about upon the signing of an official treaty, on 6 February 1819, between Raffles and both Sultan Hussein and Temenggong Abdul Rahman. Planning for the growth of the early settlement was based on Raffles' vision of Singapore as a port city and strategic British trading post for South East Asia (Makepeace et al. 1991). Raffles' land-use plan for Singapore, published in 1828, was known as the Jackson Plan. It focused on the downtown area, around the Singapore River, spanning from Telok Ayer to the Kallang River. The key features of the town plan include a grid layout for the road network, which accentuate the clear segregation of residential communities by ethnic group, and the reserved areas designated for government functions and commercial activities (Fig. 1).

The area around Fort Canning (formerly known as Government Hill), Singapore River, and the Padang were reserved strictly for government use. Fort Canning was identified as Raffles' place of residence and also became an important communication center with a lighthouse and a telegraph office. Fort Canning was converted into a fort in 1860 and was further upgraded into an artillery fort within a few years. Commercial activities were planned to take place at the "Commercial Square," which was later renamed as Raffles Place. Inspired by the greenery in Calcutta, Raffles advocated for greenery in Singapore, to give the impression of prosperity. Greenery was provided in parklands and gardens, especially at the Botanic Gardens and the Padang.

When conceiving the town plan, Raffles might have been inspired by the Renaissance ideals of urban planning, where radial streets outstretched from a strategic point of governmental power. Such a model was first implemented in Florence and was widely imitated in European cities. Influenced by early established cities, such as Barcelona and Paris, the paved streets were laid out in right angles to create a grid pattern to celebrate the hierarchy of streets and to improve drainage, urban sanitation, hygiene, fire safety and facilitate policing and troop movement. The Padang, Raffles Place, Fort Canning, and the

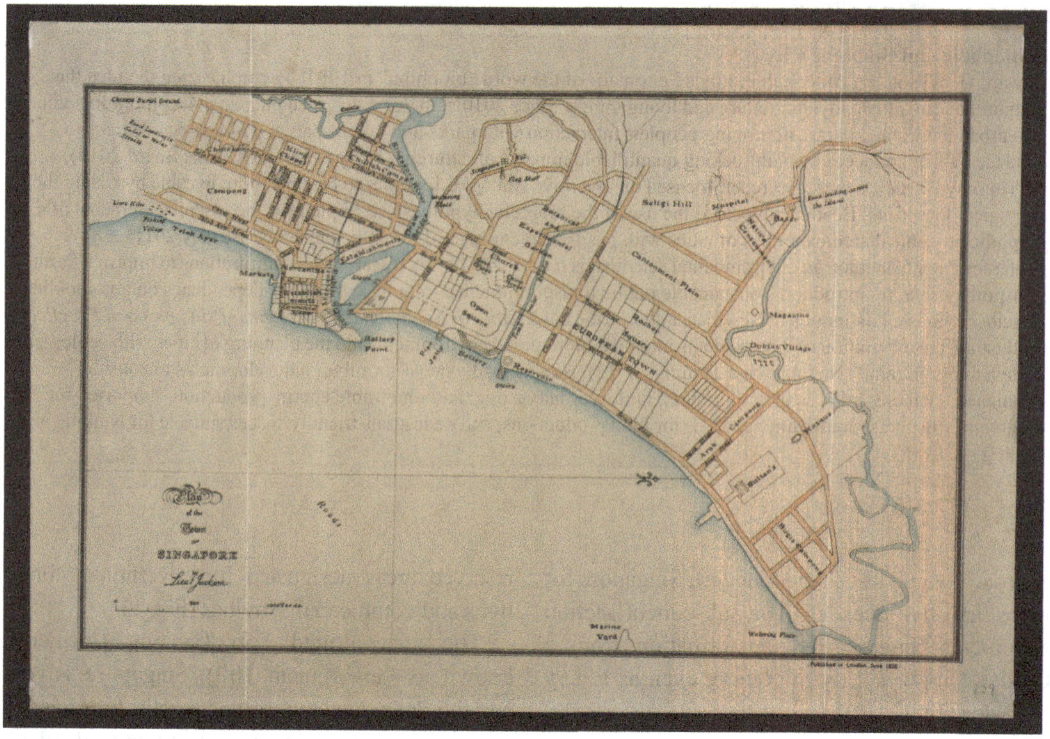

Cities in Nature, Fig. 1 The 1828 Jackson Plan (2020). (Source: National Heritage Board)

street grid pattern, found in the city center, are reminders of Singapore's colonial legacy and are today conserved as the Civic District.

Under British administration, planning was largely ad hoc and mainly confined to the control of private developments before the Second World War. This resulted in haphazard growth, overcrowding, and acute housing shortage in the central city. By early 1900's, slums had infused into the older sectors of the city and the roads were unable to handle the automobiles. Rapiddeforestation continued as the town grew and cash crops became more popular (Corlett 1992). With limited regulations over forest clearance activities, most of Singapore's forests were felled to set up gambier plantations.

Sir Nathaniel Cantley was appointed Singapore's first Superintendent of the Forest Department in the 1880s, when only 7% of Singapore's primary forests remained. He managed to establish 14 forest reserves formally, which occupied around 11% of Singapore's land area. By salvaging what was left regardless of the reserves' locations, Cantley's initial demarcation of forest reserves was not spatially systematic.

The competing land use needs resulted in the revocation of all the forest reserves in 1936. Although most reserves had been depleted due to over-harvesting of timber for economic development by 1937, Bukit Timah Forest Reserve was retained as it was under the administration of the Singapore Botanic Gardens. Over the next three decades, the Kranji and Pandan mangrove reserves were intermittently re-gazetted, but were eventually still released for development upon Singapore's independence. The protection of Bukit Timah Forest Reserve was strengthened finally, with the enactment of a Nature Reserves Ordinance and the establishment of a Nature Reserves Board in 1951. Today, Singapore's four remaining nature reserv es are Bukit Timah, Sungei Buloh, Labrador, and the Central Catchment.

To address the poor urban living conditions as the city center was overcrowded with slums, the authorities established the Singapore Improvement Trust (SIT), in 1927. The enactment of the Singapore Improvement Ordinance and the creation of a Master Plan Committee facilitated the detailed study of the urban issues to draw up a new master plan. Despite the completion of 23,000 housing units and limited road improvement works, before dissolving SIT in 1959, the housing supply was still insufficient due to rapid population increase. SIT's limited accomplishments was because it had no mandate to undertake the overall physical planning to regulate development until 1951. However, the Master Plan, adopted in 1958, did not achieve the desired outcomes as it was based on traditional ideas of town planning that were more suited to United Kingdom, rather than that of a rapidly growing Asian city. By the time the Planning Ordinance (now known as the Planning Act) was implemented on 1 February 1960, to lay down the basic legal framework controlling the use and development of land set out by the 1958 Master Plan, it was already too late and had limited impact on the city's development (Fig. 2).

In 1959, Mr Lee Kuan Yew set out a strong political mandate to remake Singapore into a "First World oasis in a Third World region," with his vision of a clean, green, and modern city that was not a concrete jungle and a home where people could live in well-integrated, high-rise communities, and not in ghettoes (Lee 2000). This manifested into an era of continuous official involvement on comprehensive and systematic programmed exercise of urban development for Singapore with a multi-ethnic society but without resources. In a 2012 interview by the CLC, he said, "I am pleased that we redeveloped the city when there was a chance to do it... That was a chance of a lifetime (Liu 2012)."

Mr Lee was mindful to avoid the duplication of residential high-rise towers, built in Great Britain

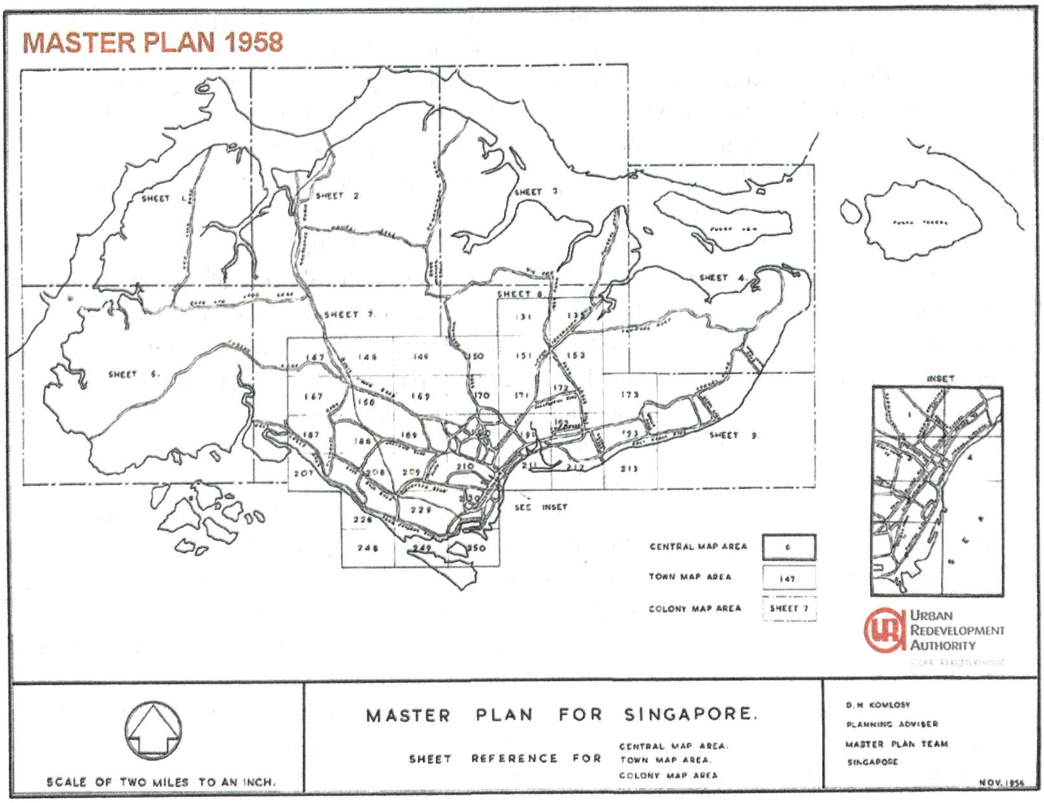

Cities in Nature, Fig. 2 1958 Master Plan (Source: Urban Redevelopment Authority)

after the Second World War, which eventually became high-rise slums. The high-rise towers were meant to resolve the crumbling and unsanitary nineteenth-century dwellings and replace buildings destroyed during the war. Influenced by Corbusier's high-rise architecture, the towers were surrounded by public open space to promote social interactions and were deemed cheaper to build, as they shared similar population density as the terraced housings they replaced. However, when the towers deteriorated over the years, they were perceived as undesirable low-cost housings, and the rising crime levels within the vicinity increased their unpopularity.

Hence, the HDB apartments (or flats, as they are commonly referred to locally) evolved in quite a different way, by fostering a greater mix of ethnic groups well as income groups. Mr Lee was very clear that each town must have a mix of flats that catered to different income groups. Hence, different flat types from rental flats, 2 to 5 rooms flats, and executive condominiums are built for catering different housing needs. Within the HDB Towns, land has been safeguarded in the planning for private condominiums to provide additional housing options to complement the HDB's home ownership scheme. The provision of planned community facilities like schools, libraries, community centers, and social facilities (i.e., childcare and senior care centers) enhance the conveniences of the residents. The HDB Towns are managed by Town Councils, which maintain and improve the common property of the residential and commercial properties to keep them in a state of good and serviceable condition (MND 2020). The service and conservancy charges (S&CC) collected from residents and commercial operators fund the Town Councils' operational costs. Learning from the ethnically-charged violence in the 1960s, such town management arrangements help to build social cohesion and a sense of community to ensure harmony within a diverse society. To avoid ethnic enclaves from re-forming in certain housing estates, the introduction of Ethnic Integration Policy (EIP) in 1989 ensures a mix of ethnic groups at both the apartment block and town level. The mixing of the races also contributed to the even spread of the less well-off or under privilege families in all estates. The social mixing and the preventing of enclave formation in HDB estates have remained an important principle in Singapore. These safeguards prevented the fragmentation of communities and crime within the poor enclaves. This physical landscape, integrated with parks and open spaces, connected seamlessly by roads lined with trees, and the demographic mix has created a distinctive character of Singapore society. Singapore became more liveable even as it urbanized.

With HDB breaking the backbone of the severe housing crunch with its delivery of 55,000 flats in the first five years, and the acquisition of land for urban renewal in the city center, the Government was in a position to plan more ambitiously for the growth of the city. It sought technical assistance and advice from the United Nations (UN) in the early 1960s. This led to the establishment of a State and City Planning Department (SCPD) in 1967, to prepare a comprehensive, long-term urban development plan for Singapore. Working with an UN appointed team, the SCPD group comprised a diverse mix of agencies, with multidisciplinary expertise, such as the Public Works Department (PWD) and the Urban Renewal Department (URD), set up under HDB in 1964 to "rejuvenate the old core of the city by making better use of land... by rebuilding the city completely." This laid the foundations for an integrated approach to urban planning that would eventually be realized as the Concept Plan of 1971.

The 1971 Concept Plan: A Radical Plan for a Modern City (Garden City)

In contrast to the 1958 Masterplan, the Concept Plan 1971 envisioned the comprehensive development of residential estates with varied densities, commercial centers, and industrial areas, in a ring formation around the central water catchment area (Fig. 3). A network of expressways and a mass rapid transit (MRT) system were planned, to

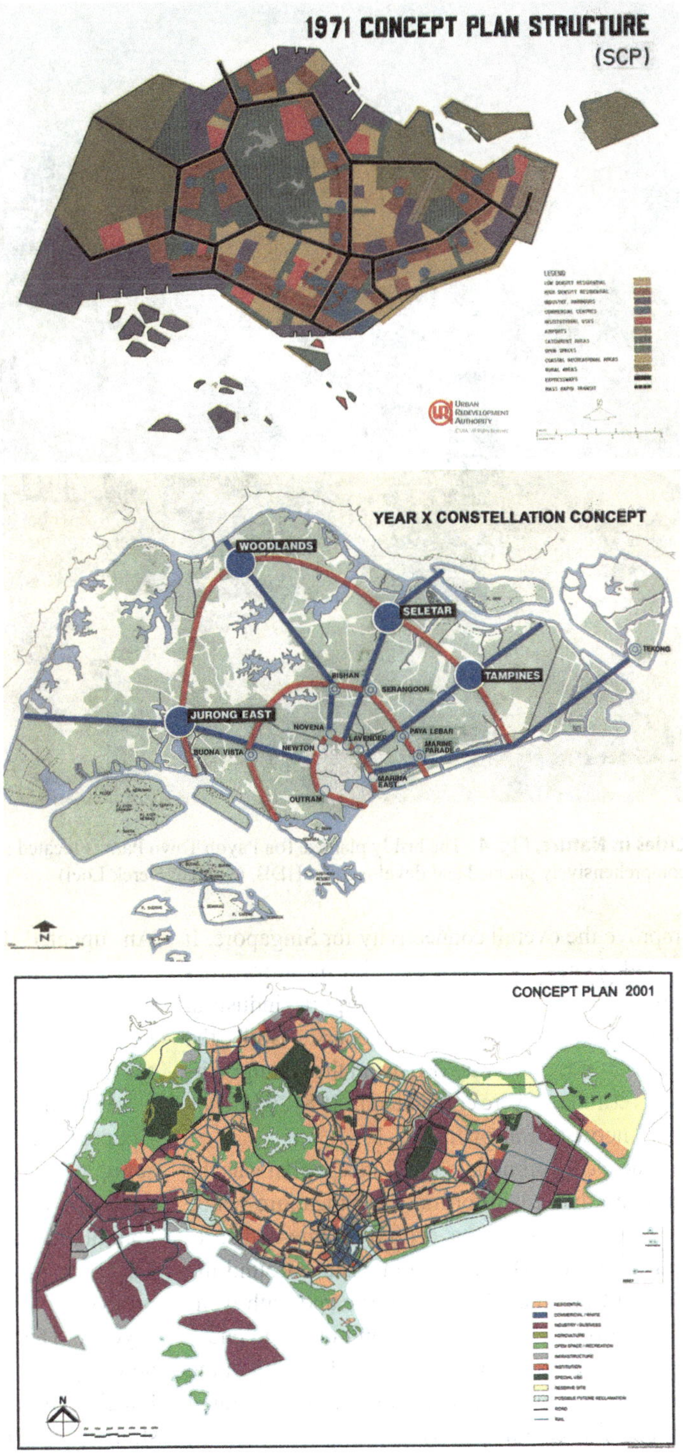

Cities in Nature, Fig. 3 The Concept Plans of 1971, 1991, and 2001. (Source: Urban Redevelopment Authority)

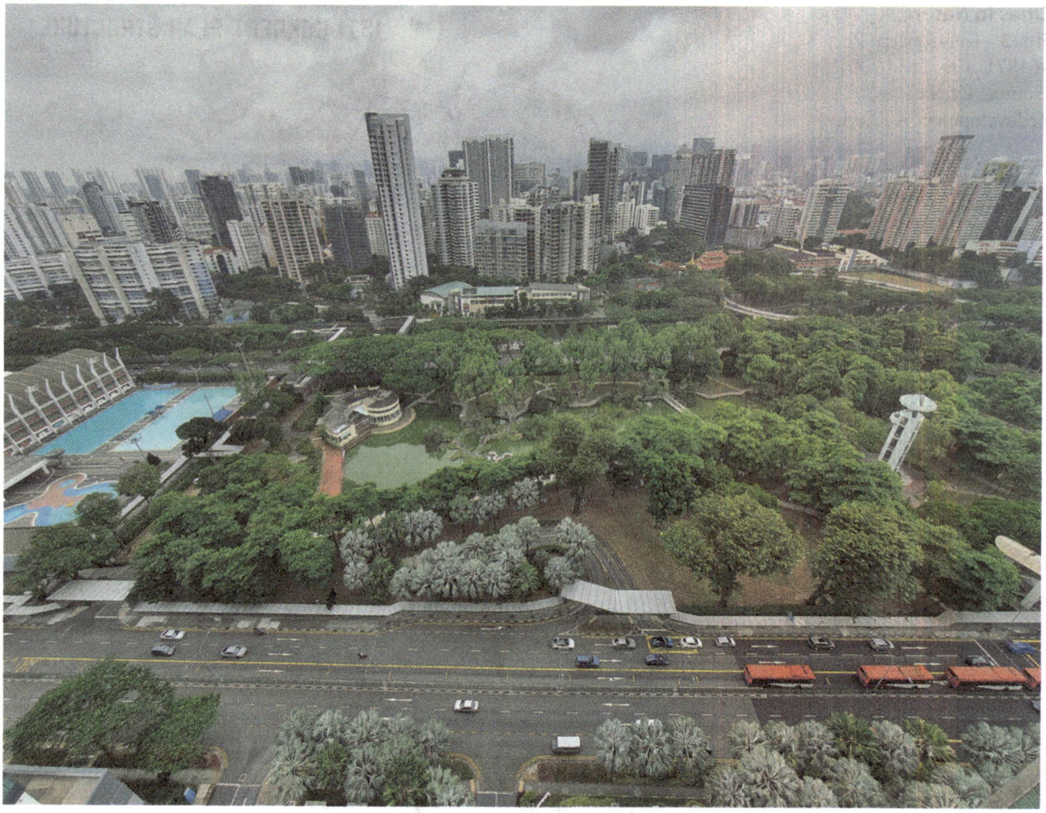

Cities in Nature, Fig. 4 The lushly planted Toa Payoh Town Park is located in Toa Payoh Town, the first satellite town comprehensively planned and developed by HDB. (Source: Derek Loei)

improve the overall connectivity for Singapore. It provided the framework to build the infrastructures such as utilities, port, airport, industrial parks, new towns, and greenery. Both the Pan-Island Expressway (PIE) and the Changi Airport Terminal 1 were completed in 1981, while the MRT network was opened in 1987.

The clear hallmark of the Concept plan was the structure that it laid for modern Singapore which remains to this day: a central business district ringed by 26 HDB towns and estates around the Central Catchment Nature reserve and with well-planned regional, town parks and other greenery to achieve a Garden City despite a doubling of the population. How this was achieved is elaborated in the next section (Fig. 4).

The 1991 Concept Plan and the Reformed Urban Planning System (City in a Garden)

Key policy changes in the late 1980s laid the foundations of today's urban planning system. An updated Concept Plan was completed in 1991 and several major changes were made to create a more formal and transparent urban planning system, which remains essentially the system till today.

With the completion of most proposals from Concept Plan 1971 by 1989, a formal process was set up to involve all ministries and agencies, professional institutes, trade associations, and the community, to prepare for Concept Plan 1991. The process includes the Concept Plan review on every decade, with a mid-term review every 5 years. The Concept Plan 1991 crystalized the new aspirations to become a "Tropical City of Excellence" and focused on a "decentralization strategy" to bring jobs closer to homes. The development of the four regional commercial centers in Woodlands, Tampines, Jurong East, and Seletar, had progressively eased the congestions in the city center. By emphasizing the quality-of-life aspirations, Concept Plan 1991 envisioned a Singapore that balanced work and play, culture and commerce. Furthermore, the nature, waterbodies, and urban development within Singapore are interfaced and integrated seamlessly. The amalgamation of Planning Department and Research and Statistics Unit as part of URA in 1989 allowed the consolidation of urban planning functions. This results in the formation of the present-day URA, a national planning, and conservation authority for Singapore.

The new masterplan, which was created from an intensive exercise involving the drawing up of 55 Development Guide Plans, was forward looking. It gave clear guides on every plot of land on land use, plot ratios, and building heights and became the statutory plan which had to be reviewed every 5 years. The 1958 masterplan remained as the base for calculation of development charge, Singapore's very successful mechanism to capture the land value enhancement from private landowners, as a result of the land use, plot ratio, and building height changes which increase land value substantially.

With planned increases in population to be accommodated on just the same island, the focus could not just be on increasing heights and density, but to also for the city to become even greener as it densified. During this period, the green cover of the city in fact increased from about one-third of the island to about half in the City in a Garden strategy.

The 2001 Concept Plan: A City with Identity (City in a Garden)

10 years later, Concept Plan 2001 focused on high quality live work play to become a "thriving world class city in the twenty-first century," with character, diversity, rich heritage, and identity. More housing options in the city were provided to inject vibrancy into the central area. As part of identity enhancement and retaining social memories, the charm and endearing ambience of Balestier, Holland Village, Jalan Kayu, Joo Chiat, and Serangoon Gardens were safeguarded and upgraded, to preserve the scale and well-loved activities. Commitment to the City in a Garden vision was demonstrated with the setting aside of more than 100 ha of prime land in the city center to be a new people's gardens, the Gardens by the Bay, which was designed and implemented through an international design competition.

The 2011 Concept Plan (City of Gardens and Waters)

With the PUB launching the ABC Waters program to turn Singapore's 17 reservoirs, 32 rivers, and 8000 km of stormwater drains and canals into beautifully landscaped streams, rivers, and lakes, Singapore now aimed to be a city of gardens and waters (section 3).

2021 and Beyond (City in Nature)

Singapore is currently undergoing its next Concept Plan (now termed Long Term Plan) review. While population growth is likely to slow, the challenge now is for sustainable urbanization. This has been articulated in the Singapore Green Plan 2030, of which a key concept is City in Nature.

Liveable and Sustainable Cities Framework by the Centre for Liveable Cities (CLC)

Singapore's evolving approach to urban planning in the past decades has been documented by the CLC (Fig. 5). Back then, Singapore's pioneering leaders did not have a particular framework in mind or a written set of principles and guidelines when they started to build Singapore (Khoo 2012). CLC facilitated in-depth interviews with the key individuals to capture their tacit knowledge and document the roles of key pioneers and the enabling processes and innovative policies that contributed positively to Singapore's key urban systems and urban development. The Urban Systems Studies distilled the principles that were applied in solving urban challenges like water, transport, housing, the environment, and industrial infrastructure. These principles have been consolidated in the CLC Liveability Framework.

The urban systems approach that Singapore has adopted builds on the urban planning theories mentioned in section 1 (Table 1). It is an approach probably more relevant to rapidly growing high-density cities than any particular theory of city planning. The three liveability outcomes, which are easily measurable policy goals, are supported by two elements: an integrated approach to planning and development and dynamic urban

C

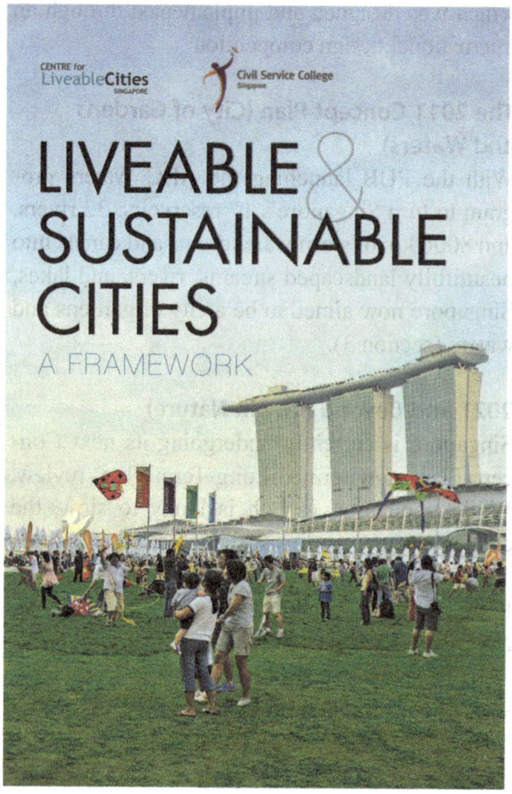

Cities in Nature, Fig. 5 Liveable and sustainable cities. (Source: Centre for Liveable Cities)

governance. This approach is further elaborated in the next section, that is, the way Singapore transformed from a Garden City to an aspiring City in Nature (Fig. 6).

Liveability and Density: The Journey from Garden City to a City in Nature

This section elaborates on the green and blue characteristics of Singapore's urban transformation; which is a key element of achieving liveability even as the city densified. Figure 7 showed the progression as Singapore developed from Garden City to City in Nature in terms of higher density and liveability. This apparent contradiction of becoming more green and in harmony in nature even with higher density has been possible with the urban systems approach depicted in the Liveability Framework.

Garden City

The Garden City phase of urban development was characterized by a systemic approach not just to plant trees, but a whole of society effort to green the city, along with legislation to set up and empower a Parks and Recreation Department and an enabling Act. The Parks and Trees Act was a key enabler to protect nature reserves and mature trees and mandate road codes that guarantee planting verges for trees along all roads and requiring public and private developers to set aside green buffers and peripheral planting verges for landscaping within all developments, effectively extending the streetscape beyond the road reserves (Attorney-General's Chambers 2001). The first nationwide tree-planting campaign to signify the start of the greening movement was also launched in 1963.

The formation of the Garden City Action Committee (GCAC) in 1970, one of the first formal Whole-of-Government initiatives, was directly overseen by the Prime Minister to ensure coordinated greening efforts across governmental agencies. The GCAC ensured that any conflicting agendas could be resolved within the different government stakeholders at an early stage. Functioning as a coordination mechanism and leadership guidance provider, the GCAC had demonstrated an innovative process within the bureaucratic level.

Beyond tree planting, the GCAC focused significantly on the greening of the harsh concrete urban infrastructures that were being built. Standards and guidelines were established and executed by agencies to plant creepers on retaining walls, concrete pillars, and even lamp posts and provide planter boxes alongside pedestrian and road bridges.

A hierarchy of parks (precinct, neighborhood, town, regional, and national parks) and open space planning standards were established and implemented not just by the Parks and Recreation Department but other regulatory and implementing agencies overseen by the GCAC, which resides in the Ministry of National Development that has both strong regulatory and execution authority on housing, public infrastructure, urban land use, and urban development.

C

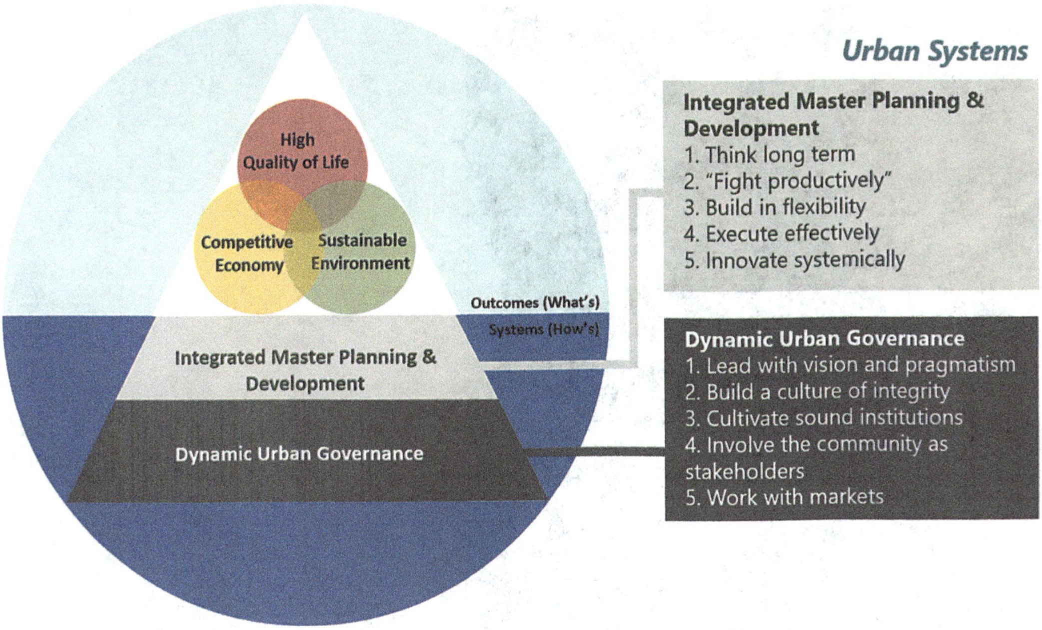

Cities in Nature, Fig. 6 Singapore's urban systems approach. (Source: Centre for Liveable Cities)

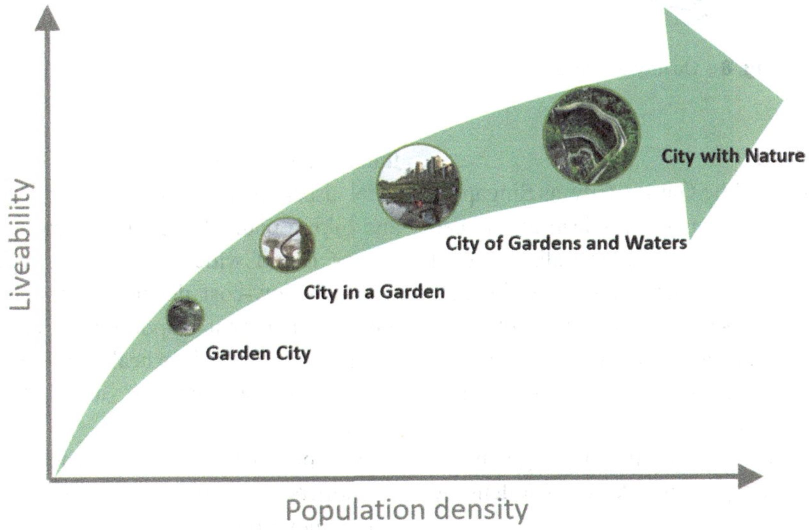

Cities in Nature, Fig. 7 Nature provision as the means to achieving high liveability even as urban density increases. (Source: Centre for Liveable Cities)

While the land was being greened, the government took a systematic, multi-agency approach to clean up the polluted Singapore River. The goal set in 1977 to clean up the river in 10 years was achieved, in line with the vision of a Clean and Green Singapore (Fig. 8).

Cities in Nature, Fig. 8 Quality greening for the harsh concrete urban infrastructures. (Source: Derek Loei)

City in a Garden

With the success of the Garden City, as Singapore became renowned for being a Clean and Green city, the "City in a Garden" plan to realize the next vision was officially announced in 1998. The fundamental change was to plan for the city not just to be green but urbanization to be seen as taking place within a matrix of an ever-greening island and to bring people closer to greenery. This was achieved with imaginative planning, given the land scarcity, with projects like the Gardens by the Bay.

In the Master Plan 2003, released with the Parks and Waterbodies Plan, the matrix of greenery was achieved with a network of parks carefully planned within the dense HDB towns and other centers. The Park Connector Network (PCN) is a brilliant concept of carving out strips of land along drainage reserves or space under transport viaducts in land-scarce Singapore, for PCN users to travel through tree-lined walking and biking paths that connect parks and housing estates island wide, without interruption. As a relatively low-cost development, the PCN provides recreational opportunities and promotes diverse benefits such as healthier lifestyles, biodiversity enhancements, and "carbon-free" transportations.

New policies were also launched to incentivize developers to increase Singapore's green cover. The Landscaping for Urban Spaces and High-Rises (LUSH) program and the Skyrise Greenery Incentive Scheme (SGIS) were launched by URA and NParks, respectively, in 2009. The LUSH program is a "landscape replacement" policy to ensure developers replace the greenery lost from the site due to development, by providing greenery at suitable locations (i.e., the ground, rooftops

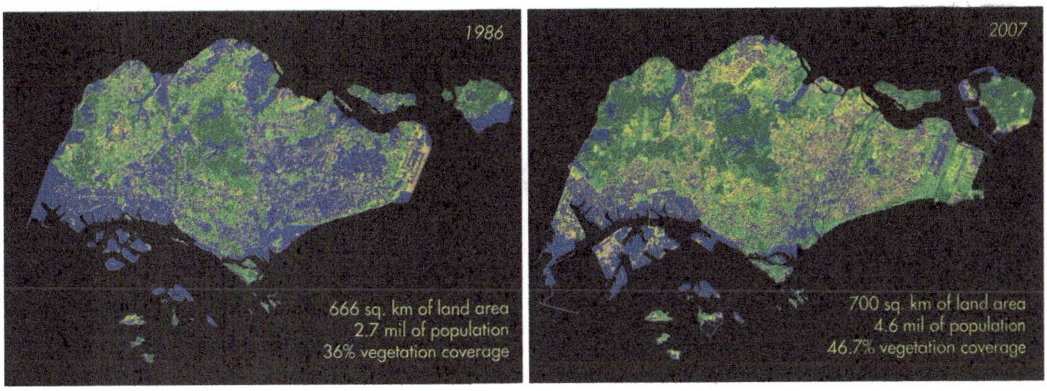

Cities in Nature, Fig. 9 Green cover comparison between Singapore in 1986 and 2007. (Source: National Parks Board)

or mid-level sky terraces) within the development. Since 2019, the LUSH program has added more than 250 ha of greenery, consisting of urban farms, communal gardens, and green walls, to reduce the environmental ambient temperature and provide visual relief to the community. Complimenting LUSH is the SGIS, which provides funding to developers, to encourage the increase of greenery provision on built structures. Despite Singapore's intense urbanization, its green cover increased from 36% in 1986 to 47% in 2007. With holistic master planning, governance and careful design, both the LUSH and SGIS, have contributed to the mainstreaming of pervasive greenery in Singapore (Fig. 9).

Beside enhancing the hardware infrastructures, the Community in Bloom (CIB) initiative and programs from the Centre for Urban Greenery and Ecology (CUGE) were launched to provide the community and industry partners with more awareness, knowledge, and hands-on experience. The Garden City Fund (GCF) was set up as a registered charity and Institution of a Public Character (IPC) in 2002, to partner and facilitate organizations and individuals to contribute to Singapore's greening (National Parks Board 2002). The donations were used to promote conservation, research, outreach, and education initiatives. There was also increasing pressure from the public to preserve Singapore's rich history and natural environment. These interactions between the local Nature Groups and the public agencies led to increased awareness of Singapore's natural assets, leading to both Labrador Park (Tan and Cornelius-Takahama 1997) and Sungei Buloh Nature Park (Loo 2015) being gazetted as nature reserves in 2002.

City of Garden and Waters

Singapore has an annual rainfall of 2.4 m across the island. However, Singapore's limited land could not collect and store all that rain water for the population. Building more reservoirs is not feasible since land is required for other more pressing needs, such as housing. The scarcity of land and water in Singapore was overcome by the efficient planning and use of land resources. Singapore's National Water Agency (PUB) worked with many agencies to manage land use and environmental pollutions, again using a systems approach. This made it possible to harvest rain from the most urbanized water catchments and built 17 urban reservoirs by damming river estuaries, something which was earlier thought to be impossible because of the extensive land and water pollutions. Two-thirds of Singapore's land area is now a water catchment. The Marina Barrage, built at a cost of S$226 million (Tortajada 2018) and completed only in 4 years in 2008, formed Singapore's 15th reservoir and first reservoir in the world to harvest rainwater in the central business district, with a catchment area of about

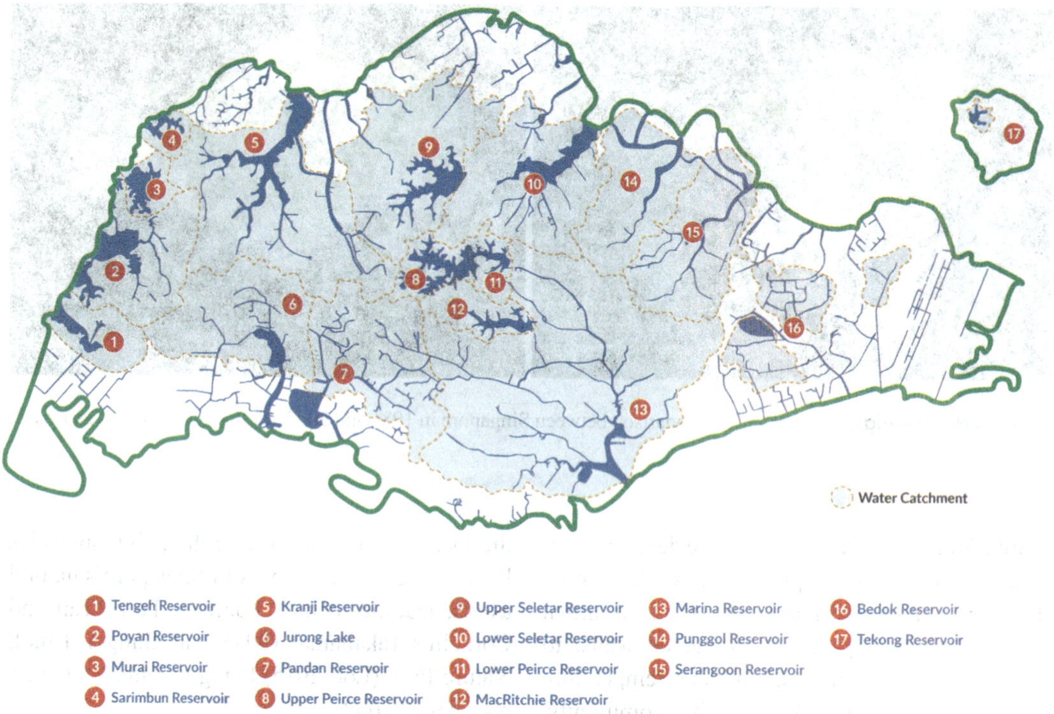

Cities in Nature, Fig. 10 Location of Singapore's water catchment areas and reservoirs. (Source: PUB, Singapore's National Water Agency)

one-sixth the size of Singapore, or 10,000 ha (Figs. 10 and 11).

Beyond recognizing water as a resource, Singapore unlocked the potential of water as an urban asset by visioning the re-naturalization of Singapore's engineered waterways, such as drains, canals, and stormwater collection ponds. This built upon the efforts by the Waterbodies Design Panel (WDP), an advisory body within the government in 1989, chaired by Dr Liu Thai Ker, to advise on the design and aesthetics of major waterways (Tan et al. 2009). By involving the Ministry of Environment, Urban Redevelopment Authority, Housing and Development Board, the WDP provided positive benchmarks by creating outstanding projects, such as Sungei Api Api, located in Pasir Ris Town. Instead of a typical monsoon canal, the Sungei Api Api was designed as a scenic river lined with lush mangroves. The success of the WDP facilitated the URA to draw up a Parks and Water Bodies Plan to drive the implementation of re-naturalizing Singapore's engineered waterways.

The ABC Waters Journey

In 2006, the PUB launched the Active, Beautiful, Clean Waters (ABC Waters) Programme (PUB 2006) to improve water quality, enhance liveability, and transform Singapore's canals, rivers, and reservoirs into beautiful recreational spaces where people can be close to water and naturally become stewards of the waterways (Centre for Liveable Cities 2019). Prime Minister Lee Hsien Loong coined the inspirational phrase "a City of Gardens and Water" when he launched the ABC Waters Exhibition in 2007. He advocated that more should be done and commented, "To realise the exciting potential of ABC Waters, we should mainstream it in our vision for the future."

Besides launching demonstration projects in Bedok, Kallang River, and MacRitchie reservoir, PUB partnered agencies to conduct proof-of concept pilot projects to test specific ideas. One such example was the implementation of rain gardens at Balam Estate. The project was possible due to the planned infrastructure works of Pelton Canal

Cities in Nature, Fig. 11 Marina Reservoir forms Singapore's 15th reservoir. (Source: Derek Loei)

Cities in Nature, Fig. 12 Schematic section of Pelton Canal and the KPE Tunnel. (Source: Housing & Development Board)

diversion to make way for the Kallang-Paya Lebar Expressway (KPE) tunnel in 2001. Upon the completion of the KPE tunnel, upgrading of Pelton Canal, and realignment of adjacent vehicular roads in 2008, the new open spaces at Balam Estate were re-landscaped. The reinstatement works involved HDB, PUB, LTA, NParks, and Marine Parade Town Council, to incorporate the rain gardens into the new landscapes (Fig. 12).

Extensive community consultations were carried out during the design process to improve the designs and to show the viability of ABC Waters to other agencies and the general public. Such consultations enhanced the understanding of the functions and benefits of ABC Waters and avoided the misperceptions of the rain gardens as possible mosquito breeding ground.

When PUB launched the ABC Waters Certification Scheme in 2010, both private and public developers are encouraged to adopt ABC Waters design features in their projects. To cultivate the industry and mainstream ABC Waters, the developers are provided with recognition and allowed to promote their "certified" ABC Waters projects. The PUB also shared the design guidelines and engineering procedures including technical standards, in the public domain, and partnered the Institution of Engineers Singapore (IES) to conduct seminars and professional programs to train and certify professionals. Since then, the local institutes of higher learning (IHLs) comprising universities and polytechnics, have introduced ABC Waters design into their curriculums.

In the flagship Kallang River at Bishan-Ang Mo Kio Park, PUB and NParks introduced a combination of softscapes, natural materials, and civil engineering solutions to create a softer and more natural appearance for the waterway. Designed as a flood plain, the once utilitarian concrete canal is now a useable park space. During dry days, the flow of water is confined to a narrow stream that park users can access and explore, improving the park user experience. During a storm event, the adjacent park area functions as a conveyance channel to carry the run-off downstream steadily. The stormwater surge in January 2017 validated its capacity to hold and slow down runoff. Additionally, a hedonic pricing analysis carried out by the Ministry of Environment and Water Resources assessed that this project had a positive and statistically significant impact on property prices there. The estimated total impact was at $157 million, more than offsetting its implementation cost of $76 million (Ministry of Environment and Water Resources 2016). A 2018 study by Future Cities Laboratory at Singapore-ETH Centre and National University of Singapore estimated that the presence of green spaces could account for $179 million in value of all homes sold during 2013 to 2014, due to the positive effect coming almost entirely from managed vegetation surrounding the estate (Au-Yong 2018). Beyond the monetary value, re-naturalized riverscape now provides opportunities for the community to have a closer interaction with water. It is common to see families walking towards the stream to catch fishes and spot wildlife such as egrets and otters. This type of living environment, with outdoor learning opportunities, and urban lifestyle is rarely experienced by most other city dwellers.

ABC Waters' Contributions to City in Nature

The ABC Waters Programme systematically build on the good work of the Waterbodies Design Panel (WDP), by institutionalizing the Blue in the city in the same way that the Green has been achieved through the Garden City and City in a Garden strategies. It was guided by an overarching master plan and was being implemented on a larger scale. By developing small-scale pilot projects and demonstration sites, the ABC Waters Programme is able to innovate systemically to ensure that all new ideas could keep stormwater flowing efficiently and at the same time, gauge the level of involvement by communities. Guided by research, pilot projects, and experts, the ABC Waters Design Guidelines were published and regularly updated as new projects were completed.

Recognizing the need to provide quality outdoor environment to create pleasing high-density living, HDB worked with PUB and stakeholders to push the creative boundaries of landscape design to bring many firsts to the HDB public housing estates (Table 2) (Fig. 13).

The mainstreaming of ABC Waters in HDB estates has certainly contributed to cleaner runoff, improved esthetic quality of the outdoors, biodiversity enhancement, build up resilience to climate change, and increased education opportunities. By designing sensitively, there is minimal land take and no compromise of functional spaces. The ABC Waters design features, especially those with elaborated designs, generate more interests among the community. Separately, there are challenges on the ground as well, for example, the various upgrading and renovation works by third parties may damage the ABC Waters design features and affect the hydraulic connectivity.

Cities in Nature, Table 2 Examples of ABC Waters projects located in the heartlands

No	Projects	Completion Date
1	1st Bioretention System in SG: Balam Estate Raingarden	2008
2	1st Neighbourhood Park-scale ABC Waters: Greenwood Sanctuary @ Admiralty	2009
3	1st Precinct-scale ABC Waters: Waterway Ridges @ Punggol	2015
4	1st Public Integrated Development with Skyrise ABC Waters: Kampung Admiralty	2018
5	1st Estate-scale ABC Waters: Bidadari Estate	Estimated to be completed in 2025

Public education and community engagements facilitate a better understanding of the integration of blue layer and the green matrix. Through dynamic urban governance, initiatives like the ABC Waters Learning Trails have encouraged the adoption of various water sites by schools, to keep the waterways clean and bring people closer to water.

City in Nature

The Meteorological Service Singapore Centre for Climate Research Singapore indicated that Singapore is almost 1° hotter today than in the 1950s (Ng 2019). This is an increase of about 0.25 °C per decade and by year 2100, the maximum daily temperatures could reach 35–37 °C. Climate scientists at the Crowther Lab (based at ETH Zurich) warned that 22% of the cities will experience unprecedented climate conditions, such as more intense dry and wet monsoon seasons, heightening the risks of drought and flooding by 2050 (TODAY 2019a). With the

Cities in Nature, Fig. 13 Wide range of ABC Waters design implemented in HDB. (Source: Housing & Development Board)

Cities in Nature, Fig. 14 Bishan Ang Mo Kio Park is an early example of implementing nature-based solution in the urban landscape. Source: Derek Loei]

warmer temperature regime, the coral reefs have bleached and died in the warmer oceans and affected the global food supply chain. The breeding of more dengue mosquitoes during the warmer months of June to October has led to a higher transmission of dengue in Singapore (Tan 2015). There is an urgent need to build resilience and mitigate the impacts of climate change, an existential challenge for Singapore.

Singapore had laid out mitigation and adaptation strategies in the Singapore Climate Action Plan (2020) and the Singapore Green Plan 2030. It has formulated "City in Nature" plans to provide citizens with a better quality of life, while co-existing with the biodiversity (Heng 2020). Nature-based solutions are applied to restore nature into the city for liveability, sustainability, and well-being, seeking to attain climate, ecological, and social resilience. (Fig. 14).

The Implementation Strategy

With the integration of the green and blue strategies, Singapore continues to build upon its strong foundation, to restore nature into Singapore's urban fabric, with an urban systems approach. Four key strategies have been identified to guide Singapore towards fulfilling the vision of City in Nature:

- Extend Singapore's natural capital
- Intensify nature in our gardens and parks
- Restore nature into the urban landscape
- Strengthen connectivity between our green spaces

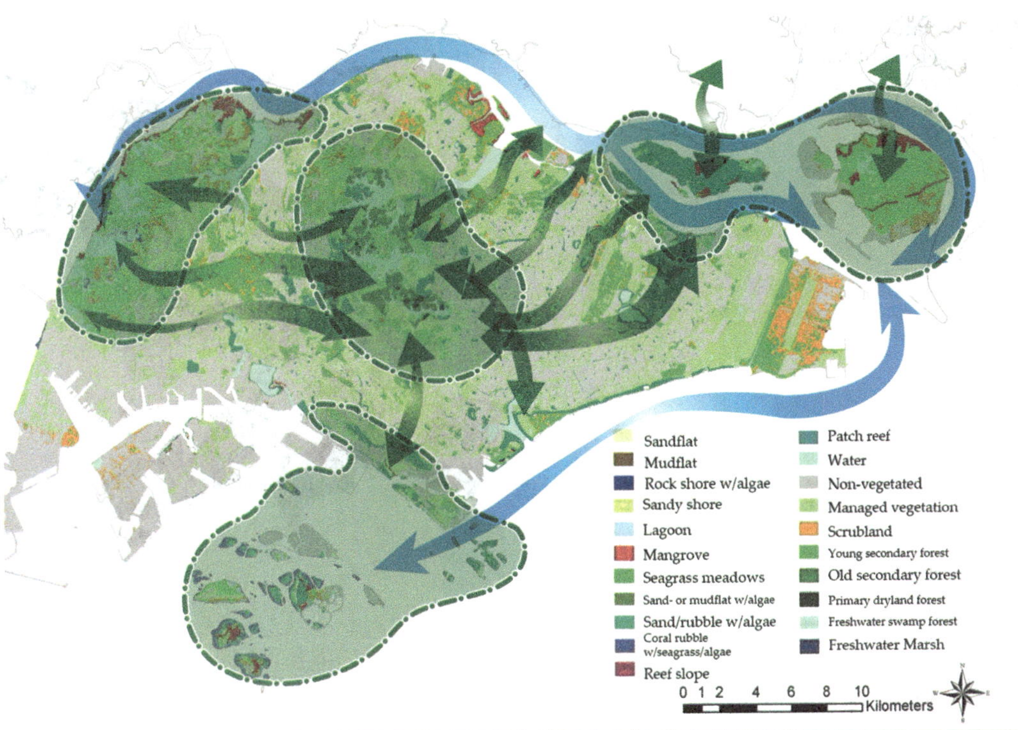

Cities in Nature, Fig. 15 The four biodiversity core areas of Singapore. (Source: National Parks Board)

Strategy 1: Extend Singapore's Natural Capital

Singapore has consciously left four zones of high biodiversity relatively untouched, each characterized by different ecosystems. These four Biodiversity Core Areas house the majority of the native biodiversity, as well as Singapore's four Nature Reserves. The nature reserves occupy a land area of 3347 ha, which is about 4.6% of Singapore's total area. Singapore's coastal waters are home to over 250 species of hard corals, which is roughly one third of the world's total number. Various coastal and marine habitats ranging from the rocky shore of Labrador Nature Reserve to the mangroves and mudflats of Sungei Buloh Wetland Reserve, and Sisters' Islands Marine Park, contribute to biodiversity and recreation options in Singapore (Fig. 15).

Growing the Nature Park Network As the most valuable parts of the natural ecosystems reside in the core biodiversity areas, it is important to extend such natural capitals beyond the boundary of the nature reserves, to other areas of Singapore. The Nature Park Network seeks to create rustic and forested nature parks, around the nature reserves, to act as buffer to mitigate the impact of urbanization.

The new Nature Parks provide complementary habitats for the flora and fauna from Nature Reserves and serve as compatible nature-based recreation for the population. With more than 350 ha of Nature Parks currently, Singapore plans to provide an additional 200 ha of nature parks by 2030. These new nature parks allow the community to enjoy a myriad of nature-based recreational activities such as bird watching and trekking. The mainstreaming of Nature Ways along roads and park connectors, and ABC Waters along all major canals, will enhance the ecological connectivity across Singapore. By softening the urban matrix sufficiently, nature and biodiversity can permeate our everyday experience, increasing the city's liveability (Figs. 16 and 17).

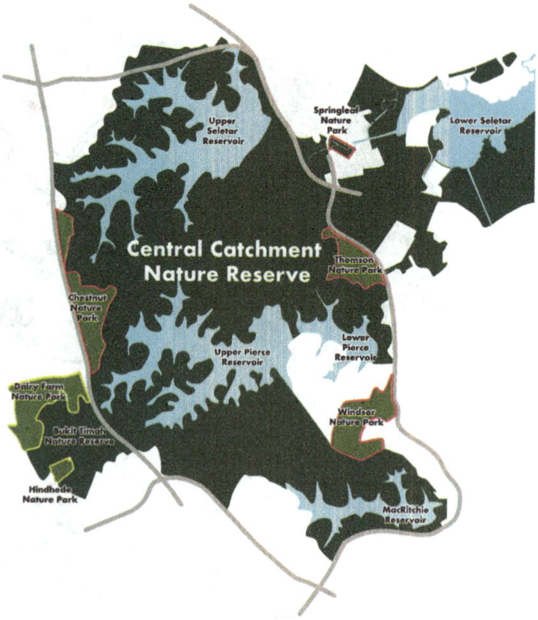

Cities in Nature, Fig. 16 (Left) Examples of Nature Park Network (NPN). (Source: National Parks Board)

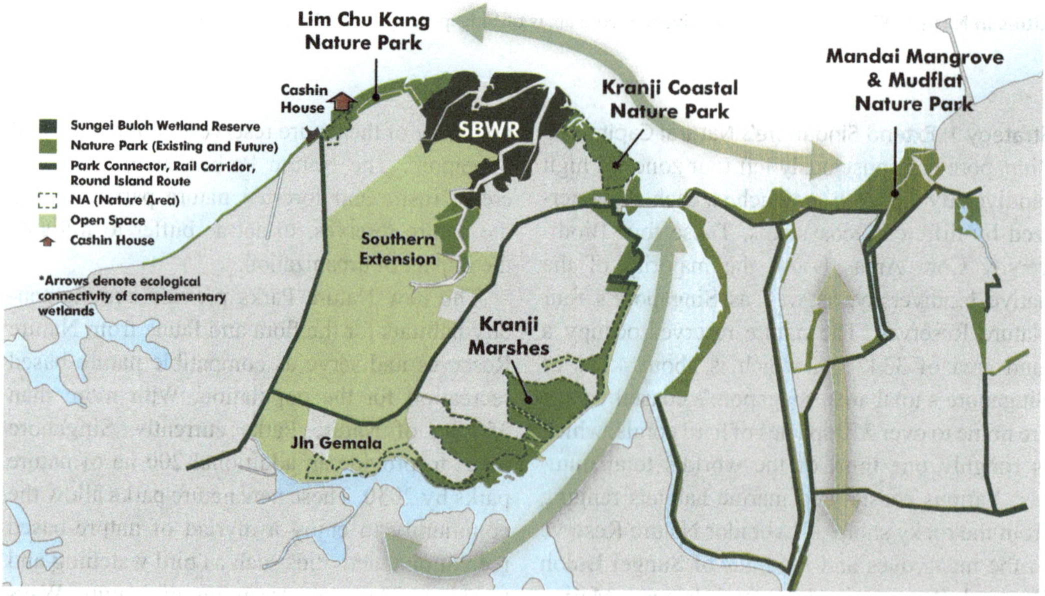

Cities in Nature, Fig. 17 (Right) Details of Sungei Buloh Nature Park Network (NPN). (Source: National Parks Board)

Strategy 2: Intensify Nature in Our Gardens and Parks

Singapore is intensifying nature in the public parks and gardens by incorporating biophilic design, nature inspired planting designs, and conserving more native flora to create positive impact on health and well-being. More waterways and water bodies in existing parks will be naturalized

Cities in Nature, Fig. 18 A therapeutic garden located at Bishan-Ang Mo Kio Park. (Source: Derek Loei)

and function as floodplains to protect homes and properties from flooding while supporting biodiversity. The plans include the implementation of therapeutic landscapes, nature playgardens for children, naturalized waterways, and conservation of mangrove areas and species recovery and habitat restoration.

Implementation of More Therapeutic Landscapes With Singapore's aging population, therapeutic landscapes have been introduced in gardens and parks, as therapeutic horticulture has shown to improve wellbeing of elderly. The therapeutic gardens are able to meet the physical, psychological, and social needs of the users. Besides spaces for relaxation, strolling, and resting, these gardens also comprise activity spaces where programs for seniors, such as gardening and exercise, may be organized. The customised design and programming for therapeutic gardens are aimed at catering to a range of users including those with dementia, autism and attention deficit hyperactivity disorder (ADHD). In a government-wide effort to mainstream therapeutic gardens, NParks targets to implement 30 therapeutic gardens by 2030 (Fig. 18).

Implementation of Nature Playgardens The nature playgardens seek to reconnect children with nature, to improve children's mental and physical well-being, cognitive functions, and self-esteem and confidence. Taking reference from the concept of "biophilia," research has shown that interacting with the outdoors helps children develop creativity and playing freely with nature, reduces stress (TODAY 2019b). These positive experiences strengthen the love and stewardship of nature from young.

The 0.35-hectare nature playgarden built at HortPark in 2019 (National Parks Board 2020)

Cities in Nature, Fig. 19 Nature play opportunities are also available at Jacob Ballas Children's Garden. (Source: Derek Loei)

Cities in Nature, Fig. 20 Concrete canal between Bishan Road and Braddell Road in 2015 (Source: PUB, Singapore's National Water Agency)

serves as a testing ground for NParks to develop design guidelines, to help developers and education providers in building more of such sites for children to experience and learn about nature. In March 2021, the Como Adventure Grove nature playgarden, located within the 8-hectare Gallop extension of the Singapore Botanic Gardens was completed. Based on the nature play principles, the design familiarizes children with nature through play, by recreating the experience of climbing and playing in trees (Amin 2021) (Fig. 19).

Naturalize Waterways and Conservation of Mangrove Areas Flood resilient designs have been introduced to Riverine and Coastal Parks and mangrove areas, to mitigate flooding risk and sea level rise. Such designs include the naturalizing of waterways and waterbodies. Strategic mangrove areas have been identified for conservation and restoration, so as to avert coastline erosion and protect seagrass beds and coral reefs from siltation. Mangrove forests serve as a source of food and nursery ground for numerous marine organisms and also contribute as an excellent carbon sink, storing up to five times as much carbon per hectare compared to tropical forests (Figs. 20 and 21).

Species Recovery and Habitat Restoration The ongoing biodiversity conservation efforts have enabled Singaporeans to encounter rare plant and animal species, such as the

Cities in Nature, Fig. 21 The naturalized waterway between Bishan Road and Braddell Road was completed in 2019. (Source: Derek Loei)

Hairless Rambutan (Nephelium maingayi), Straw headed Bulbul (Pycnonotus zeylanicus), and Tiger Tail Seahorse (Hippocampus comes). Technology such as video traps has been utilized in the nature areas for biodiversity monitoring to track the species recovery of Porcupines, Pangolin, Lesser Mouse Deer, Leopard Cat, Sunda Colugo, Slow Loris, Raffles Banded Langur, etc.

The 10-year Forest Restoration Action Plan was launched by NParks in 2019, to strengthen the resilience of native rainforests in Singapore. The action plan charts the restoration progress over the next 10 years to regenerate the secondary forests in the nature parks buffering the Bukit Timah Nature Reserve and Central Catchment Nature Reserve, including the disturbed patches within the reserves (Juraimi 2019). The community have been involved in the planting of more than 250,000 plants, including fruit-bearing trees like the Kumpang (*Horsfieldia polyspherula*) and nitrogen-fixing plants like the Petai (*Parkia speciosa*), across the nature parks and open areas in the nature reserves. Such initiatives allow the community to learn about the science and the broader strategic considerations behind the planting program and cultivate the community stewardship over the forests.

Strategy 3: Restore Nature into the Urban Landscape

Industrial Estates: Bringing Nature Back Built urban hard surfaces such as roads typically absorb and retain a lot of heat from solar radiation. Hence, 170,000 more new trees will be planted in industrial estates, in a "multi-tiered manner," made to resemble the "look and feel of natural forests," as part of the "One Million Trees movement", over the next 10 years (Co 2020a). The tree planting efforts seek to cool the industrial areas by up to 6°C, improve air quality, and infuse the therapeutic effects of greenery closer to people's workplaces.

Residential Estates: Incorporating Nature into HDB New and Existing Towns The HDB towns and estates are steered by the HDB Biophilic Town Framework, launched in 2013, to guide the incorporation of greenery into the functional landscapes, to enhance the natural environment and the well-being of residents (Wong 2018). The framework encourages the intertwine of nature with the built environment of all HDB Towns, to enhance the sense of place and liveable spaces (Fig. 22).

More Parks in the Heartland There are plans to implement more parks in the heartlands. As part of the Jurong Lake District project, the Jurong Lake Gardens (JLG) is Singapore's first national garden in the heartlands, which provides a distinctive recreational node in a lakeside setting. The multi-agency Steering Committee for the Jurong Lake District oversaw the planning process, which consisted of extensive consultations with the

Cities in Nature, Fig. 22 A lushly planted precinct garden located in Dawson Estate, where a row of matured existing trees is retained during development. (Source: Derek Loei)

community and resulted in strong emphasis on nature, community, and play, in an effort to create a "people's garden." The nature-inspired design exemplifies Singapore's efforts to create a City in Nature through the restoration of nature in the urban environment by recreating freshwater swamp forest habitats and naturalized storm-resilient waterways.

Strategy 4: Strengthening Connectivity Between Green Spaces

Growing the Network of Nature Ways Nature Ways are routes planted with specific trees and shrubs to facilitate the movement of animals like birds and butterflies between two green spaces (National Parks Board 2021b). Besides enhancing the quality of our living environment, Nature Ways bring birds and butterflies from nature areas and parks to urban spaces so that Singapore residents can develop a greater appreciation of Singapore's rich biodiversity. The plant species are curated and planted strategically along the identified corridors progressively to replicate the natural structure of forests.

There are 39 Nature Ways in Singapore, stretching 150 km in total (as of 2020), connecting high biodiversity areas such as Bukit Timah Nature Reserve, Central Catchment Nature Reserve, and Western Catchment. With an

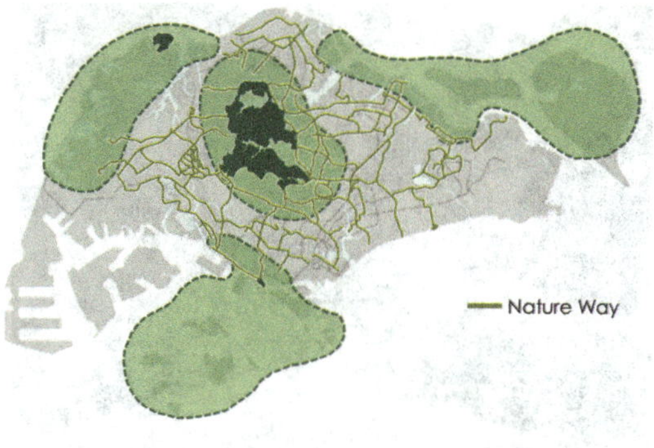

Cities in Nature, Fig. 23 Using Nature Ways to facilitate movements of birds. (Source: National Parks Board)

Cities in Nature, Fig. 24 (Right) A Nature Way located in Clementi Town. (Source: Derek Loei)

aspirational vision of "Every Road a Nature Way," NParks is committed to increase the Nature Ways to 300 km by 2030 (National Parks Board 2021c) (Figs. 23 and 24).

Expanding the Park Connector Network (PCN) The interlinking network of park connectors which improve connectivity and expand green spaces by integrating the PCN with the surrounding heartlands amenities, park spaces and allowing users to engage in recreational activities such as cycling, jogging, and walking had proven successful (Poon 2013).

To date, Singapore has 340 km of park connectors island wide, although the PCN network was initially planned to be 300 km. The 36 km Coast-to-Coast Trail (Choo 2019), which stretches from Coney Island Park to the Jurong Lake Gardens,

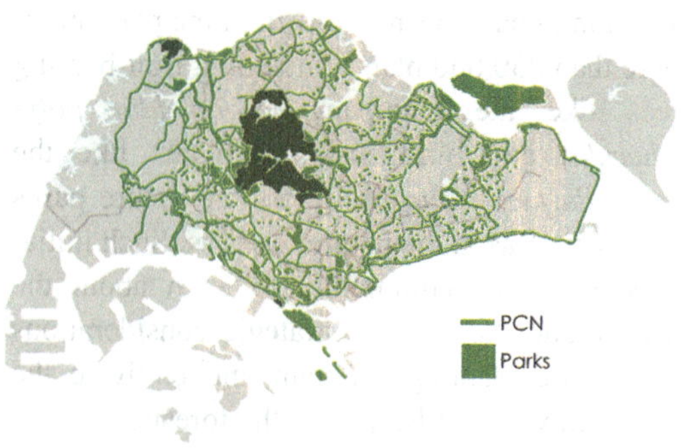

Cities in Nature, Fig. 25 (Left) Park Connector Network (PCN). (Source: National Parks Board)

was completed in 2019, and the 150 km Round Island Route (RIR), that loops around Singapore, will be completed from 2021, respectively. By the year 2030, NParks aims to increase the PCN to 500 km, in the form of green links and recreational corridors, so that 100% of Singapore's households will be within a 10-min walk from a park (Figs. 25 and 26).

Building Expertise, an Industry, and Living with Nature

The population's general health and well-being are interconnected with the health and welfare of animals, found in the community. While the community animals play a vital role as human's companions, the diversity of animals and birds in the wild, such as oriental pied hornbill, long-tailed macaque, smooth-coated otters, and wild boar, enrich the Singapore's urban ecosystem. There is a need to inculcate a shared responsibility to conserve Singapore's existing biodiversity and safeguard Singapore's public health, animal health, and welfare.

Coupled with more development projects in the works, and the increased greening of Singapore's urban areas, there will be closer contacts between humans and animals (Chong 2021). The Animal & Veterinary Service (AVS) was created under NParks in April 2019, to take on the role as first responder for animal-related feedback, as well as maintaining standards in animal welfare and health. This allows NParks to provide a one-stop service on animal management and welfare issues, manage "human-animal interactions," and improve the detection and response to animal

Cities in Nature, Fig. 26 Right) A PCN located in Sembawang Town. (Source: Derek Loei)

diseases which can be transmitted to humans (Mohan 2018). The development of a science-based management approach, with holistic strategies, comprising urban animal population management, bio-surveillance, public education, and community stewardship, is facilitating in-depth assessments to balance different viewpoints and mitigate the conflicts between Singapore's diverse population, nature, and wildlife.

Build Science and Technology and Industry Capacity Higher-order skills are necessary for greenery and landscape management, as Singapore transforms into a City in Nature. The deepening of Science and Technology capabilities is essential for Singapore's strategic and operational decision-making. Since 2007, the Centre for Urban Greenery and Ecology (CUGE) has up-skilled many local workers through the Workforce Skills Qualifications (WSQ) programs and helped industry organisations to restructure operations to increase productivity. CUGE is also working with the Animal & Veterinary Service (AVS) to jointly develop relevant trainings, to enhance the skills and knowledge on animal welfare, health, ecology, and behavior.

Inspire Communities to Cocreate and Be Stewards of Nature The success of a City in Nature is to forge closer bonds within the community, through active participations and stewardship of the greenery and biodiversity. With the building of more parks, communities can partake in the design, building, management, and programming of parks within their estates through the "Friends of the Parks initiative," which aims to promote greater stewardship in the management of the green spaces. This ground-led initiative spearheaded by the community will complement the "OneMillionTrees movement" (Co 2020b). The Community in Bloom program and the

Cities in Nature, Fig. 27 An example of allotment garden found in Bishan Ang Mo Kio Park. Source Derek Loei

allotment gardening scheme have been expanded, as part of the Growing with Edibles initiative. The distribution of complimentary packets of seeds to households and the wealth of online resources on gardening encourage the growing of produce at home. The Community in Nature initiative, launched in 2011, is continuing to involve more educational institutes and partners to participate in citizen science projects and reforestation efforts. Furthermore, NParks will take the lead to introduce options for up to 25,000 youths to participate in biodiversity conservation yearly, through the Youth@SGNature initiative (Ng 2020) (Fig. 27).

What Does the Future Holds?

As Singapore strives to recover from the current pandemic to be a City in Nature, these are key issues to address:

(a) Planning a Postpandemic City

When COVID-19 first emerged in Singapore in January 2020, it was the start of a series of unprecedented closures and restrictions (Goh 2020). Beyond Singapore, the world faced serious challenges and drastic changes both socially and economically. The cities with high population densities and concentration of activities were more vulnerable due to COVID-19's ability to spread rapidly within confined spaces and there was no known vaccine for the new virus then. These cities grappled with issues such as access to quality healthcare, trust in government, and ensuring the population complied with mask-wearing and safe management measures.

With the implementation of Work-From-Home and Home-Based-Study, there are fewer people on the streets, transport nodes, public spaces, commercial centers, and the

Central Business Districts (CBDs). With the mainstreaming of remote working, there is a need to review the design of CBDs, homes, and public transport networks to consider the human experience and the innate need to come together creatively and socially in shared spaces.

The URA has already in its latest Concept Plan or Long Term Plan as it is now called, planned for an increasingly polycentric city with new centers like Jurong Lake District, subregional centers like Paya Lebar as well as the Greater Southern Waterfront, with 30 km of coastline which extends from Pasir Panjang to Marina East (Co 2019). The urgency to develop these new centers will probably be greater postpandemic to increase the city's resilience.

(b) Care for the People: General Wellness and Mental Health

How can people can live, work, and play well in cities during lockdowns (or circuit breakers) in a pandemic? With people living in high-rise, small flats in HDB Towns, the accessibility of nearby amenities like parks and park connectors, community clubs, sports facilities, schools, and child care centers became a key asset of HDB towns. However, this is an area that needs more research and experimentation, especially how the urban environment affects the mental health and general wellness of people living and working in high rise, high density cities.

(c) Pragmatic Solutions for Climate Resilience

Singapore is a low-lying island at the equator and is hence vulnerable to the impacts of climate change: rising temperature, sea level, and extreme weather resulting in flooding and drought. The challenge is the uncertainty of these impacts, and how collectively as a city and society, it can build up its climate defense. With established governance and strong financial position, Singapore will explore both engineering and technological solutions and nature-based solutions. Singapore continues to undertake intensive research and experimentation and is learning and sharing with other countries and experts. For example, as part of strengthening Singapore's coastal defenses, empoldering, a Dutch reclamation method involving the building of a seawall in the water, and pumping the water out behind the seawall to create dry land, is being tried out on an island in the north, Pulau Tekong (Mohan 2021; Tan and Fogarty 2019).

Cities in Nature, Fig. 28 Curious children interacting with nature. (Source: Derek Loei)

(d) Living with Nature

Singapore's sustainable urbanization journey to be a "City in Nature" is not just about technical challenges related to urban planning and building to incorporate nature. While people generally love nature, inevitable human-wildlife conflicts pose challenges to this relationship. Learning how to live with nature is a complex and ongoing process that involves education, starting from a young age. (Fig. 28).

Epilogue

Singapore has evolved from a basket case of urbanization gone wrong in the early 1960s with a population of less than 2 million to a liveable city of 5.7 million. It has done so by adopting ideas from many city planning theories but uniquely used an urban system approach, the Liveability Framework as it transformed from a Garden City, to a City in a Garden, a City of Gardens, and Water to a City in Nature even as its urban density tripled. The principles of this systems approach will continue to apply even as Singapore face a more challenging future with climate change, an aging society, and other yet unknown challenges while remaining constrained as an island without much land and other natural resources. It is an approach that could well be adopted or adapted by other rapidly growing cities.

Acknowledgments The authors acknowledge the research assistance provided by Derek Loei in the preparation of this Chapter

References

'Plan of the Town of Singapore', also known as the "Jackson Plan", Singapore, 1828, lithograph. ROOTS. (2020, October 27). https://www.roots.gov.sg/stories-landing/stories/the-singapore-story-through-60-objects/colonial/jackson-plan/story

Amin, H. (2021). First look: New features open at Singapore Botanic Gardens' Gallop Extension. CNA. http://www.channelnewsasia.com/news/singapore/botanic-gardens-gallop-extension-playground-first-look-14392014

Attorney-General's Chambers. (2001). Search within legislation: Parks and Trees Act. Singapore Statutes Online: Parks and Trees Act. https://sso.agc.gov.sg/Act/PTA2005.

Au-Yong, R. (2018). Green the colour of money where property is concerned? *The Straits Times*. https://www.straitstimes.com/singapore/green-the-colour-of-money-where-property-is-concerned

Beatley, T. (2011). *Biophilic cities integrating nature into urban design and planning*. Island Press.

Centre for Liveable Cities. (2019). Active, Beautiful, Clean Waters (ABC Waters) Programme. Infopedia. https://eresources.nlb.gov.sg/infopedia/articles/SIP_2019-08-21_203240.html

Chong, C. (2021). Feeding wildlife a major reason for spike in wild animal numbers: NParks. *The Straits Times*. http://www.straitstimes.com/singapore/feeding-wildlife-a-major-reason-for-spike-in-wild-animal-numbers-nparks

Choo, F. (2019, April 13). Coast-to-Coast trail launched. *The Straits Times*. https://www.straitstimes.com/singapore/environment/coast-to-coast-trail-launched

Co, C. (2019). NDR 2019: New attractions, housing and office spaces to be developed in Greater Southern Waterfront. CNA. https://www.channelnewsasia.com/news/singapore/ndr-2019-greater-southern-waterfront-pulau-brani-sentosa-keppel-11819376

Co, C. (2020a). Singapore to plant 170,000 more trees in industrial estates over the next 10 years. CNA. http://www.channelnewsasia.com/news/singapore/170000-more-trees-industrial-estates-cooling-therapeutic-13374114

Co, C. (2020b). Singapore to plant 1 million trees, develop more gardens and parks by 2030. CNA. https://www.channelnewsasia.com/news/singapore/singapore-to-plant-1-million-trees-develop-more-gardens-and-12500858

Corlett, R. (1992). The ecological transformation of Singapore, 1819-1990. *Journal of Biogeography, 19*(4), 411–420. https://doi.org/10.2307/2845569

Dempsey, N. (2010). Revisiting the compact city? *Built Environment (1978-), 36*(1), 4-8. Retrieved July 19, 2021, from http://www.jstor.org/stable/23289980

Dreier, P. (2006). Jane Jacobs' radical legacy. Jane Jacobs' radical legacy. https://web.archive.org/web/20060928205849/http://www.nhi.org/online/issues/146/janejacobslegacy.html

Encyclopedia Britannica, inc. (1998, July 20). Garden city. Encyclopedia Britannica. https://www.britannica.com/topic/garden-city-urban-planning

Goh, T. (2020). Six months of Covid-19 in Singapore: A timeline. *The Straits Times*. http://www.straitstimes.com/singapore/six-months-of-covid-19-in-singapore-a-timeline

Heng, M. (2020). Parliament: More nature parks, park connectors as Singapore transforms into city in nature in next decade. *The Straits Times*. https://www.straitstimes.com/politics/parliament-more-nature-parks-park-connectors-as-singapore-transforms-into-city-in-nature-in

Hermesauto. (2015). Stretch of Kallang River to be transformed from concrete canal to cascading stream. *The Straits Times*. https://www.straitstimes.com/singapore/environment/stretch-of-kallang-river-to-be-transformed-from-concrete-canal-to-cascading

Juraimi, F. (2019). NParks unveils 10-year action plan to make Singapore's rainforests more resilient. CNA. http://www.channelnewsasia.com/news/singapore/nparks-unveils-10-year-action-plan-to-make-singapore-s-11167608

Khoo, T. C. (2012). The CLC framework for liveable and sustainable cities. Urban Solution Issue 1. https://www.clc.gov.sg/docs/default-source/urban-solutions/urb-sol-iss-1-pdfs/researchandreports-the-clc-framework-for-liveable-and-sustainable-cities.pdf

Lee, K. Y. (2000). *From Third World to first: The Singapore story, 1965–2000: Singapore and the Asian economic boom*. HarperCollins Publishers.

Liu, T. K. (2012). Lee Kuan Yew: The chance of a lifetime. Urban Solutions Issue 2. https://www.clc.gov.sg/docs/default-source/urban-solutions/urb-sol-iss-2-pdfs/interview-lee-kuan-yew.pdf

Loo, J. (2015). Sungei Buloh Wetland Reserve. Infopedia. https://eresources.nlb.gov.sg/infopedia/articles/SIP_566_2005-02-01.html

Makepeace, W., Brooke, G. E., & Braddell, R. St. J. (Eds.). (1991). *One hundred years of Singapore* (pp. 6–7). Singapore: Oxford University Press. (Call no.: RSING 959.57 ONE-[HIS]).

Marshall, A. (2010). How cities work: Suburbs, sprawl, and the roads not taken (pp. 44–45). University of Texas Press. ISBN 0292792433.

Master Plan 1958. Republic of Singapore – Master Plan 1958. (n.d.). https://www.ura.gov.sg/dc/mp58/mp58map_index.htm.

Ministry of Environment and Water Resources. (2016). A policymaker's toolkit for policy design and evaluation. Cost-benefit analysis: Hedonic Pricing Study for ABC Waters. Extracted from CLC Insights City-in-Nature

MND. (2020, July 30). About Town Councils. MND. https://www.mnd.gov.sg/our-work/regulating-town-councils/about-town-councils

Mohan, M. (2018). New stat board to oversee food safety and security; AVA to be disbanded. CNA. http://www.channelnewsasia.com/news/singapore/new-statutory-board-singapore-food-agency-ava-disbanded-10563864

Mohan, M. (2021). Engineering solutions to tackle rising sea levels important but more research vital: Experts. CNA. http://www.channelnewsasia.com/news/singapore/climate-change-rising-sea-levels-research-poldering-ndr2019-11822204

National Heritage Board. (2020). Archaeological excavation site at Fort Canning Park. ROOTS. https://www.roots.gov.sg/places/places-landing/Places/surveyed-sites/Archaeological-Excavation-Site-at-Fort-Canning-Park

National Parks Board. (2002). About GCF. Garden City Fund. https://www.gardencityfund.org/who-we-are/about-gcf

National Parks Board. (2015, September 23). Biophilia Symposium. National Parks Board. https://www.nparks.gov.sg/biodiversity/urban-biodiversity/biophilia-symposium

National Parks Board. (2020). Nature playgarden at HortPark. National Parks Board. https://www.nparks.gov.sg/gardens-parks-and-nature/parks-and-nature-reserves/hortpark/nature-playgarden

National Parks Board. (2021a). City in nature. Extracted from NParks slides for the "Liveable and Sustainable Cities: Lessons from Singapore and Other Cities" module

National Parks Board. (2021b). Nature ways. National Parks Board. https://www.nparks.gov.sg/gardens-parks-and-nature/nature-ways

National Parks Board. (2021c). NParks to work with the community to transform Singapore into a City in Nature. National Parks Board. https://www.nparks.gov.sg/news/2020/3/nparks-to-work-with-the-community-to-transform-singapore-into-a-city-in-nature

Newman, P., 1945– (1989). Cities and automobile dependence: a sourcebook. Kenworthy, Jeffrey R., 1955–. Aldershot, Hants: Gower Technical. ISBN 0-566-07040-5. OCLC 20132931.

Newman, P., & Jennings, I. (2012). *Cities as sustainable ecosystems principles and practices*. Island Press.

Ng, D. (2019). Why Singapore is heating up twice as fast as the rest of the world. CNA. http://www.channelnewsasia.com/news/cnainsider/singapore-hot-weather-urban-heat-effect-temperature-humidity-11115384

Ng, M. (2020). NParks to nurture more young S'poreans to be stewards of nature. *The Straits Times*. http://www.straitstimes.com/singapore/nparks-to-nurture-more-young-sporeans-to-be-stewards-of-nature

Ng, M. (2021). Over 2,000 BTO flats in Bidadari estate completed. *The Straits Times*. https://www.straitstimes.com/singapore/housing/over-2000-bto-flats-in-bidadari-estate-completed

O'Donnell, T. (2019, June 1). Le Corbusier: From the Contemporary City to the Radiant City. Urban Utopias. https://urbanutopias.net/2019/06/01/le-corbusier/

Poon, H. Y. (2013). Park connectors, living large in small spaces. Urban Solutions Issue 2: Park Connectors, Living Large in Small Spaces. https://www.clc.gov.sg/docs/default-source/urban-solutions/urb-sol-iss-2-pdfs/case-study-singapore-park-connectors.pdf

PUB. (2006). Active, beautiful, clean waters programme. PUB, Singapore's National Water Agency. http://www.pub.gov.sg/abcwaters/about

Tan, A. (2015). More people, urbanisation 'behind rise in dengue cases'. *The Straits Times*. https://www.straitstimes.com/singapore/health/more-people-urbanisation-behind-rise-in-dengue-cases#:~:text=The%20study%2C%20which%20looked%20at,the%20increase%20in%20dengue%20cases

Tan, B., & Cornelius-Takahama, V. (1997, December 3). Labrador Park/Fort Pasir Panjang. Infopedia. https://eresources.nlb.gov.sg/infopedia/articles/SIP_14_2005-01-25.html

Tan, A., & Fogarty, D. (2019). Singapore will use nature-based solutions to deal with sea level rise: Masagos. *The Straits Times*. http://www.straitstimes.com/singapore/environment/singapore-will-incorporate-nature-based-solutions-that-go-beyond-coastal

Tan, Y. S., Lee, T. J., & Tan, K. (2009). Clean, green and blue: Singapore's journey towards environmental and water sustainability (p. 221). Singapore: ISEAS Publishing. (Call no.: RSING 363.70095957 TAN)

The Straits Times. (1965). Island's newest satellite town. https://eresources.nlb.gov.sg/newspapers/Digitised/Article/straitstimes19650823.2.31.1

TODAY. (2019a). S'pore among world's major cities to face 'unprecedented' climate conditions by 2050. TODAY online. https://www.todayonline.com/world/spore-among-worlds-major-cities-face-unknown-climate-conditions-2050

TODAY. (2019b). NParks opens nature playground at HortPark as testbed for future 'biophilic' sites. TODAY online. https://www.todayonline.com/singapore/nparks-opens-nature-playground-hortpark-testbed-future-biophilic-sites

Tortajada, C. (2018). Commentary: The Marina Barrage, a dream 20 years in the making. CNA. https://www.channelnewsasia.com/news/commentary/singapore-water-marina-barrage-lee-kuan-yew-challenge-visitors-10898664

Wong, D. (2018). Nature to play bigger role in HDB estates. *The Straits Times*. http://www.straitstimes.

C

com/singapore/housing/nature-to-play-bigger-role-in-hdb-estates
Zhang, L. M. (2019). Stretch of Kallang River gets $86 million upgrade, with flood protection and water features. *The Straits Times*. https://www.straitstimes.com/singapore/stretch-of-kallang-river-gets-86-million-upgrade-to-protect-against-flooding-and-with

Citizens

► Amsterdam's Pathway to Climate Neutrality: Creating an Enabling Environment

City

► Pre-schoolers and Sustainable Urban Transport

City Digital Twin

► Digital Twin and Cities

City Financing and Social Urbanism in Latin America: The Importance of Good Fiscal Management

Carlos Leite[1,2] and Huascar Eguino[3]
[1]School of Architecture and Urbanism, Mackenzie Presbyterian University, Sao Paulo, Brazil
[2]Social Urbanism Center, Insper´s Arq.Futuro Cities Lab, Sao Paulo, Brazil
[3]Fiscal Management Division, Inter-American Development Bank (IDB), Washington, DC, USA

Each territory and culture within Latin American cities has unique characteristics which sometimes make it difficult to define a common ground between their urban dynamics. However, through more specialized reading, some dysfunctional symptoms can be perceived as recurrent patterns among South American urban systems. This includes the urgency for decent affordable housing and basic infrastructure and the need for public facilities and transportation systems in the most peripheral and vulnerable regions of larger cities.

In developing countries, the highest population densities are found precisely in highly socially vulnerable informal territories on the cities' peripheries; at the same time, urban centers do not fully utilize their capacity in terms of existing infrastructure. It is a land use unbalance that reflects the incongruity of urban economic development.

An analysis of these territories under the light of social urbanism shows that the priority of the public sector must be focused on social-spatial inclusion, directing investments towards the development of the most vulnerable areas. However, to promote greater social balance, it is necessary to address the scarcity of public resources. How to finance urban transformation? What are the legal, economic, and urban instruments available for local governments to promote such transformation?

The frequent limitation of the public budget requires municipal administrations to seek diversification of funding sources in order to complement the available revenue and guarantee the economic sustainability of urban investments. In terms of fiscal management, three of the biggest challenges are: (i) insufficient funds available to provide basic infrastructure and public facilities and services, (ii) inadequate institutional capacity to generate financial resources, and (iii) limited access to external financing (Leite et al. 2020; Rojas 2018).

Challenges of Municipal Governments

Latin America and the Caribbean countries (LAC) require an annual investment of around 5% of GDP to close their significant infrastructure gap. However, figures indicate that investment levels have barely reached an average of 2.7% of GDP in the last decade. The investment deficit highlights the lack of financial resources at all levels of government, but mainly in Municipal Governments (MGs) which are at the forefront of citizen's demands (Eguino and Leite 2018).

To understand the reason for the urgency: there are 113 million people in LAC living in slums (approximately 25% of the total urban population), without adequate access to basic facilities, public services, transportation, sanitation, and drainage infrastructure, in addition to a huge housing deficit. In Brazil alone, such a deficit is estimated at 7 million of housing units. High vulnerability is also influenced by the annual growth rate of 6.5 million inhabitants in LAC cities, with an urban population expected to reach 40% by 2050 and an increase in urban land consumption two to four times faster than population growth. It is estimated that by 2025, 100 million people will reside in just six megacities (Mexico City, São Paulo, Buenos Aires, Rio de Janeiro, Lima, and Bogotá), which will generate a great demand for services and infrastructure. In addition, the emergence of new challenges associated with climate change and changes in the labor market must be considered. These trends set a complex picture for MGs and, in particular, test their ability to generate more resources to finance their investments.

LAC's MGs are highly dependent on intergovernmental transfers, increasing their average expenditure in relation to GDP by more than two percentage points between 2000 and 2015. To reduce this high subordination to the national level, MGs need to improve their monetary independence, which in turn depends on efforts to improve their tax base systems as well as collection of fees for services and other sources of revenue.

Currently, fundraising is below its potential level and can be observed in the underutilization of property taxes, which barely reached 0.4% of GDP on average in 2015. This number is just over a third in comparison to the collection of the OECD (Organization for Economic Cooperation and Development). On the other hand, there is timidity in the implementation of Land Value Capture (LVC) instruments, as well as a low recovery of service costs and poor management of public assets and properties (Eguino and Leite 2018).

The under-utilization of local resources is caused both by regulatory restrictions imposed by national governments and by factors inherent to each municipal administration. Secondary causes include: outdated databases and taxpayer registration, limited capacity for automated collection of revenues, low capacity for tax adjustments, and limited capacity of enforcement. Local fiscal dynamics is also affected by the political calendar, especially during electoral years, which makes it necessary to strengthen the autonomy of tax administration at the municipal level. With regard to Property Tax, there is not always the necessary technical staff to carry out basic tasks such as periodic evaluation of property values and keeping taxpayer databases updated, among other necessities.

The result is that MGs are unable to fully employ their tax potential, which in turn restricts their potential to obtain social benefits through public investment in urban development. In this context, it is essential to strengthen the payment culture, improve database systems, modernize management, and carry out necessary institutional reforms to enforce and increase revenues through tax collection.

Another challenge is limited access to external financing. There are MGs which, even with access to debt within the framework of fiscal responsibility, are prevented from doing so. The reasons include lack of technical aptitude to generate a portfolio of "finance-ready" projects, little knowledge about financing instruments such as Land Value Capture (LVC) or public-private partnerships (PPP), or the existence of regulatory limitations, such as restrictions from repayment timeline of loans linked to the extension of the administration's term. In some cases, MGs accumulate old debts with other public entities (e.g., social security), which prevents them from accessing the market. The low levels of revenue generation derived from own resources also limits the amounts of indebtedness, or they are not even sufficient to be considered credit counterparts, although there are instruments that allow the pooling of resources from many entities in a single operation (as in the case of joint financing operations) (Eguino and Leite 2018).

National level technical assistance to MGs, aiming at their financial recovery and subsequent graduation into the market, is generally insufficient. This is due in large part to the lack of information and monitoring by Finance Ministries, which limits MGs to be considered according to their payment credibility. In this sense, the effective implementation of fiscal responsibility frameworks (as in the case of Brazil and Colombia) is the greatest tool that the central government has to monitor the situation of subnational public finances, in particular of the larger MGs, and may authorize debt operations with low risk of default.

Fiscal Management Measures to Increase Municipal Resources

Taking into account the challenges mentioned above, a set of measures for fiscal management was selected that can positively impact the increase in MG resources. The proposed lines of action emphasize aspects of public management, distinguishing sources of internal and external resources (UN-Habitat 2016) (Fig. 1).

Among actions to stimulate local level revenue, special attention is given to two resources from internal sources: IPTU (Property Tax) and LVC instruments. In general, management measures that can optimize the performance of property taxes imply improving registrations, updating property values, and modernizing municipal tax administration systems (De Cesare 2012).

As for the need to update the real estate value of properties, traditional methods of direct inspection are disappearing in favor of self-assessments (based on Bogotá's famous success story) and bulk appraisals. Regarding the first, it is necessary that there is specific legislation to define the limits of the reported values (e.g., it cannot be lower than those previously registered for each property). This type of approach works best when the values calculated by the tax authority are used as a reference. In addition, in the case of smaller municipalities, the formation of consortia is recommended to reduce the costs associated with contracting value maps. In this area, there is a

Resources from internal sources		
Land-related income	**Taxes for basic and administrative services**	**Other income sources**
• property taxes • transfer fees • land value capture instruments	• taxes for services: water, sewerage, parking • administrative quotas: construction permit, registration of new businesses, commercial licenses	• vehicle taxes • income related to asset management (rentals, investments, others)

Resources from external sources		
Intergovernmental transfers	**Private financing**	**Public funding**
• unconditional transfers • conditional transfers	• loans from the banking system • bond issues • public-private partnerships	• state bank loans • development funds • development programs of international organizations

City Financing and Social Urbanism in Latin America: The Importance of Good Fiscal Management, Fig. 1 Resources from internal and external sources (developed by the authors)

growing relevance of the Real Estate Market Observatories, which capture the market dynamics and facilitate the application of modern methodologies for real estate appraisals.

The collection of property taxes in LAC is generally insufficient, and its local tax administration systems could be modernized.

The Importance of Land Policy and LVC Tools

Municipal resources may also come from LVC instruments, that is, from the increase in the value of the land that such governments have produced. If, on the one hand, the public budget is very limited for investments, on the other, urban development generates a lot of value in its main input – land – and the real estate sector absorbs enormous benefits from this land value appreciation. The intensification of land use and the increase in urban density are desirable in order to optimize the use of the territory, but are only possible in view of existing or development of new basic infrastructure. The investment in new infrastructure generates, as an external factor, the appreciation of the land value in that territory, normally higher than the cost of the original investment. If the increase in the price of land occurs as a result of actions promoted by the public sector, the corresponding capture of that value can later be redistributed in favor of the community through new municipal projects (Leite et al. 2020).

Several studies show that changes in land use categories from rural to urban can multiply its value up to four times on average, and the installation of the necessary infrastructure for urbanization tends to produce a value higher than its cost. For example, in a sample of 10 LAC cities, a basic package of urban services with an average cost of US $25/m^2 increases the land value by more than US $70/m^2. This appreciation can be captured by MGs through taxes, contributions, and regulations, generating new sources of local financing (Eguino and Leite 2018).

LVC instruments can be classified into: (i) taxes, both as differentiated rates of property tax or tax increase financing; (ii) tariffs, such as contributions for improvement or appreciation; and (iii) regulations, such as construction rights or land readjustment fees (Eguino and Leite 2018).

In Colombia and Brazil, in particular, there are already some legal-urban instruments of reasonable technical aptitude, which still suffer from unjustified and enormous resistance to their effective utilization in our cities. The following stand out:

- Onerous Grant to the Right to Build (OODC): sale of additional construction rights (Floor Area Ratio) of the land, in order to raise funds for urban development.
- Consortium Urban Operations: sets of interventions in specific territories, financed by the sale of additional construction rights.
- Transfer of Building Rights (TDC): an instrument that allows the transfer of unused construction rights ratio (e.g., property protected by historical interest) to another property.
- Special Social Interest Zones (ZEIS): territories specially destined for the provision of affordable housing, land tenure regularization, and urbanization of socially vulnerable areas.

Land Value Capture stems from public investments in basic infrastructure. Therefore, since the public sector is responsible for the development of urban and regulatory instruments that allow specific conditions for capture of land use-based values, it is also its duty to establish actions and policies for subsequent redistribution of revenues in favor of the community. The capture of this appreciation, finally, must finance the basic infrastructure, especially in the most vulnerable regions (Fig. 2).

With regard to external sources of financing, there are certain actions that MGs can take in order to improve their chances of access to finance, either via banks, through the issuance of bonds, or through different alternatives regarding public-private partnerships.

Although almost all MGs have some kind of territorial planning instruments, they do not

The presence of Land Value Capture instruments in Brazil	
• **OODC** Charges for Additional Building Rights	regulated by law in 1,946 municipalities
• **CEPACs** Certificates of Additional Potential Construction Bonds	regulated by law in 1,401 municipalities

Results from Instruments of Land Value Capture in Sao Paulo (OODC and CEPACs)
• The resources represent approximately 20% of the total annual urban investments in the city, reaching more than 30% of the investment capacity of the municipality
• Resources represent less than 5% of the so-called global GSV (general sales value), the total annual volume of the city's real estate market (ie there is potential for growth)
• US$ 890 million was collected with OODC in the period 2002 to 2017, representing only approximately 1% of the total global GSV for the period
• US$ 3 billion was raised with CEPACs for 2017

City Financing and Social Urbanism in Latin America: The Importance of Good Fiscal Management, Fig. 2 LVC instruments and its enormous potential as has been applied in LAC cities like Sao Paulo, Brazil (developed by the authors)

always have investment plans that reflect their priorities, establish financing capacities based on the available fiscal space, and develop a strategy to mobilize resources from external sources. The actions for formulating a capital investment plan are:

- Identify infrastructure needs and establish priorities

  MGs decide, through a participatory process, what are the priority investments and how to finance them. During the preparation of the term's budget, each department assesses urban infrastructure needs. This long initial list encompasses many competitive proposals; therefore it is necessary to establish a selection criteria based on their economic feasibility. To rank the projects, there are tools such as Cost-Benefit Analysis (CBA), the Internal Rate of Return (IRR), or the Net Present Value (NPV). To complement the process, other techniques may be used, such as analyzing the degree of coherence with the medium and long-term planning instruments, the project's contribution to closing the infrastructure and municipal services gaps, or the use of multiple variables in the project selection process (Eguino and Leite 2018).
- Assess of credit solvency levels

  Before determining the terms and conditions of any financial transaction (loans or bonds), investors assess credit solvency levels, which measure the ability to borrow and pay off debts. This is a delicate process that considers a detailed analysis of the MG's financial situation, an assessment of the local economy and an overview of the national macroeconomic environment. In this sense, it is recommended that city governments carry out a self-assessment of their credit capabilities before turning to a financial institution, in order to stipulate how much money they can borrow without jeopardizing their stability.
- Choosing the best combination of instruments
  - Once the MG knows the amount of financing it is entitled to obtain, it can choose the

best combination of external resources to match the timeline of the investment with its medium-term fiscal space. Some of the alternatives are bank loans, capital market financing (municipal bonds), and public-private partnerships.

- Bank credits: access to bank credit can be either locally based or provided by multilateral credit organizations such as IDB or World Bank. These credits are normally amortized by funds such as service fees, local taxes, or intergovernmental transfers. It is recommended that this type of financing be used to cover capital expenditures prioritized in the Investment Plan and that the amount contracted is compatible with the sustainability of the municipal debt.
- Municipal bonds: access to capital market financing through the issuance of municipal bonds and securities is still timid in LAC. The experience in countries with consolidated capital markets indicate that the securities have a number of advantages, such as: lower financing costs, greater transparency in the management of financial information, and the creation of a new market for alternative instruments for investors. However, it presents an advantage only for operations of significant value, since there are costs inherent to this process that are only justified if there is a minimum issue value. Once the sale is made, the funds obtained should be used only for the execution of the projects included in the investment program that will be financed with the bonds.
- Public-private partnerships: in the face of high infrastructure demands in an environment of budget constraints, many municipal governments have made increasing use of public-private partnerships (PPPs). These are contracts for the construction, operation, and maintenance of infrastructure. The asset may be owned by the municipal government from the beginning, or by the contractor until the end of the contract. In addition to complementing restricted public budgets, they reduce excess costs and increase transparency, contributing to greater efficiency. However, they also present challenges in their implementation, especially in relation to project preparation, high-cost of contracts, and difficulties in specifying appropriate operating standards. Therefore, it is very important that governments that carry out PPPs have well-developed institutional skills.

It is not always possible to apply all the modalities of resource mobilization discussed above in LAC countries. The improvement of the regulatory framework is a central aspect, mainly with regard to its "fiscal pact." Through it, agreements are defined that regulate the process of decentralization and performance of the municipal financing system. The creation of new taxes is one of the alternatives to be considered in the "fiscal pact"; however, before creating additional taxes, it is necessary to study the impact on the incentives of economic agents and devise strategies to control the possible political cost that this can generate (Eguino and Leite 2018).

Hopefully, the presentation of the challenges regarding the financing of infrastructure construction and, subsequently, the likely measures that can be taken to address them, has elucidated concrete actions for municipal access to alternative sources of revenues. Optimizing local level resources and shaping the internal organization to obtain credit conditions is key for public administrations that aim to overcome limitations in the existing budget. Social urbanism and its induction tools must always prioritize the balancing of land use through social-spatial inclusion.

It is necessary to take advantage of the economic arrangements available to provide a socially just and sustainable urban development process, bringing the private sector also as an active agent in the city's process of transformation. After all, the cities from the Global South, "often struggle to maintain a physical, social and economic infrastructure for their populations, raising concerns among global organizations like the United Nations and the World Health Organization" (Arefi and Kickert 2019, p. 1).

References

Arefi, M., & Kickert, C. (Eds.). (2019). *The Palgrave handbook of bottom-up urbanism*. https://doi.org/10.1007/978-3-319-90131-2.

De Cesare, C. (2012). *Improving the performance of the property tax in Latin America*. Policy Focus Report. Cambridge: Lincoln Institute of Land Policy.

Eguino, H., & Leite, C (2018). ¿Como mejorar los ingresos municipales y prepararse para acceder a financiamiento?: la importancia de una buena gestión fiscal. In R. de Ciudades (ed.), *Reunión Anual de Alcades y Seminario: "Ciudades Incluyentes: Aprendiendo de Medellín"* (pp. 161–174). Medellin. Technical document from Inter-American Development Bank (IADB). https://issuu.com/ciudadesemergentesysostenibles/docs/reuni_n_anual_de_alcaldes__medell_n/2?fbclid=IwAR2mt3X8xHDFAOw-oH9EcLklXDd8tqw9xU5ND1lWYZELDz5DkEPxfBAIXyQ. Accessed 20 Apr 2019.

Leite, C., et al. (2020). *Social urbanism in Latin America: Cases and instruments of planning, land policy and financing the city transformation with social inclusion* (1st ed.). Cham: Springer. https://doi.org/10.1007/978-3-030-16012-8.

Rojas, E. (2018). No time to waste: Applying the lessons from Latin America's 50 years of housing policies to rapidly urbanizing countries. *Environment and Urbanization*. https://doi.org/10.1177/0956247818781499.

UN-Habitat. (2016). *Finance for city leaders*. Nairobi: United Nations Human Settlements Programme.

City Vision

► Green and Smart Cities in the Developing World
► The Governance of Smart Cities

City Visions: Toward Smart and Sustainable Urban Futures

Timothy J. Dixon
School of the Built Environment, University of Reading, Reading, UK

Definition

Today the world is heavily urbanized, and this is set to grow by 2050. The climate crisis and the recent COVID pandemic are providing opportunities and threats to urban living. This has meant that decision-makers need to develop long-term visions for cities. Urban futures thinking (based on city foresight methods) offers us the opportunity to imagine what cities and urban areas will be like in the long term, how they will operate, what infrastructure and governance systems will underpin and coordinate them, and how they can be best shaped and influenced by their primary stakeholders. This chapter therefore begins by examining urbanization and the main urban challenges that cities face today. A discussion of what is meant by "urban futures" then follows, before reviewing the emergence of "smart" and "sustainable" thinking in cities. The chapter also examines city visioning as a futures-based technique and the emergence of city visions. An example of a UK city vision (Reading 2050) is then reviewed, before the chapter examines what future lies beyond COVID-19 for cities. Finally, a summary and conclusions are presented to help the reader see the wider implications of urban futures thinking for cities.

Introduction

The recent COVID-19 crisis has reminded us all about the vital role that cities play in our local, regional, national, and global economies. Without fully functioning city ecosystems, it is clear that reduced economic growth, financial hardship, social unrest, and socioeconomic disruptions are major risks in our urban areas. Yet the COVID crisis has also taught us some important lessons about how we could change the way in which we live, work, and play in our cities in order to tackle climate change, improve the urban environment, and benefit the health and quality of life of people in our cities. After all, there is strong evidence to suggest that in many cities across the world, carbon emissions fell, and air quality improved, at least in the short term, as people travelled less and workplaces closed because of the pandemic crisis (OECD 2020a). Today, as city leaders begin to consider how best to emerge from the crisis, it is crucial to think about how we might do things differently beyond the short term, into a long-term future (beyond 20 years), and reimagine our urban futures in the context of climate change and resource depletion and environmental impact.

Although cities present us with huge environmental challenges and are at the heart of the COVID crisis simply because the majority of people live in cities, there are also many inherent opportunities for transformation related to a city's unique characteristics: for example, it is not only the close proximity of people that provides economies of scale and capacity for social learning and could transform the way in which we work and live in our cities, but cities are also the main source of innovation, R & D, and experimentation which could potentially tackle urban environment issues. This duality of problems and solutions is often referred to as the "urban paradox" (Iossifova et al. 2018).

To think about the long-term future, however, requires us to go beyond short-term political perspectives and to also overcome the disconnection which is inbuilt into many urban planning systems and their separation for the longer-term environmental challenges. In other words, we need an analytical framework of structured thinking to get us beyond the "here and now" and to think explicitly about the long-term future of our cities. This is where "urban futures" thinking and "city visioning" come into play.

This chapter therefore begins by examining urbanization and the main urban challenges that cities face today. A discussion of what is meant by "urban futures" then follows, before reviewing the emergence of "smart" and "sustainable" thinking in cities. The chapter also examines city visioning as a futures-based technique and the emergence of city visions. An example of a city vision (Reading 2050) is then reviewed, before the chapter examines what lies beyond COVID-19 for cities. Finally, a summary and conclusions are presented to help the reader see the wider implications of urban futures thinking for cities.

Urbanization and Urban Challenges

Cities are not a recent invention of humankind. The world's first great cities are known to have been built 4000 years ago, and they brought together people to make markets and create trading opportunities (Knox 2014). Foundational cities such as Athens and Rome followed later, before the emergence of more "modern" cities from medieval times through to the industrial revolution and later to the present day (Clark 2016). The unique feature of the twentieth- and twenty-first-century city has been its rapid growth however, and hence the level of global urbanization has increased commensurately. Today, according to UN statistics, some 55% of the world's population lives in cities, and this is set to grow to 68% by 2050 (UN 2018). All of the world's population growth between 2016 and 2050 was expected to be in urban areas, as a result of natural increase, migration, and some degree of reclassification as to what is really meant by the term "city" (UN 2018). This is expected to result not only in the growth of smaller medium-sized cities (of fewer than 1 million people) but also the number of megacities (cities of more than 10 million people) to 43 by 2030.

Historically this surge in urbanization has been caused and is likely to continue to be caused, by economic development, because cities attract people who seek out education and employment opportunities (i.e., the "pull factor"). Yet the "urban paradox" remains: although cities are hubs of economic growth and innovation, they face a wide range of challenges ranging from climate change through to environmental degradation, traffic congestion, health risks from poor air quality, and socioeconomic inequalities (EU 2016). To put this in context, if global warming is to be limited to 1.5 °C, then emissions from global urban consumption must halve by 2030, and all cities will need to be net zero by 2050 at the very latest (C40 Cities 2019).

Urban challenges are examples of "wicked" problems or ones that are complex and interrelated (Rittel and Weber 1973). For example, many of the global sustainability challenges that we face, such as biodiversity decline, climate change, energy supply, and environmental justice, are persistent, complex, and "wicked," and they are also "urban scale" problems (Wolfram et al. 2019). The COVID-19 pandemic, which has had substantial impacts in our cities, is another example of a wicked problem. Tackling, managing, and resolving such problems therefore require not only an integrated understanding of their interrelationships but also urban planning responses that recognize their mutual and interconnected complexity.

Urban Futures, City Foresight and City Visioning

It has been argued by some authors that the inherent complexity and unpredictability of cities means that although we can develop models of cities as complex systems (which can help us understand how cities have evolved and how they behave in what is termed a "science of cities"), we cannot predict their future with any degree of certainty because we, as inhabitants of a city, are all part of that future (see, e.g., Batty 2018). On the other hand, it can be argued that although the future may not be "predictable," it is crucial to find other ways of developing desirable and shared visions for our future cities in the light of the many complex and "wicked" problems that we face (Dixon and Tewdwr-Jones 2021).

Therefore, to overcome the disconnection between relatively short-term planning horizons of 5–10 years and longer-term environmental changes (20 years or more), it is vital for cities to develop specific longer-term "visions" that open up a possibility space to explore multiple futures and also provide a roadmap of how to achieve a shared and desirable future. This does not negate the importance of recognizing the inherent complexity of cities, the continued desire for immediate and short-term political decision-making, or the important role that the "science of cities" plays in our understanding of cities. But it does require us to develop new ways of seeing and planning for a transition to a sustainable urban future.

This is what can be termed "urban futures," which is a term used to "imagine what cities and urban areas will be like in the long-term, how they will operate, what infrastructure and governance systems will underpin and co-ordinate them, and how they are best shaped and influenced by their primary stakeholders (civil society, governments, businesses and investors, academia and others)" (Dixon and Tewdwr-Jones 2021).

Urban futures thinking requires city stakeholders to work together in terms of co-creating a city vision in a highly participatory way. This means that four main groups need to work together to build and develop city visions: namely, civil society, local government, academia, and business in what is known a "quadruple helix" partnership (Goddard and Tewdwr-Jones 2016). As part of "urban futures" thinking, city visioning is the formal process of creating a "city vision," or a shared and desirable future for a particular city or urban area. However, in practice the city vision either can relate to a single preferred urban future or can explore a variety of different and alternative urban futures. City foresight, which includes city, is therefore the "science of thinking about the future of cities" (GOfS 2016) and includes a range of futures-based methods and tools to help build and develop a city vision: for example, "backcasting" which starts with defining a desirable future and then works backward to identify policies and programs and pathways that will connect the present with the specified future, and "three horizons" (3H) thinking, which is designed to help visioning participants think about three overlapping waves (e.g., short (now)-, medium (near future)-, and long-term (far future)) into the future.

City Visions and City Visioning

Visionary thinking has been part of human culture, religion, and politics for many thousands of years. Visions are fundamental to thinking about the future and often related to preferred or desirable futures and to a shared sense of change and transformation. Early examples of what might be termed humanistic visionary thinking emerge in the writings of Plato (fourth century BC) and, later on, Thomas More's city-based Utopia (sixteenth century). This sense of "futurism" is also seen in the writings of Patrick Geddes and Ebenezer Howard, two of the early visionary planners in the late nineteenth/early twentieth centuries, who developed particular generic visions of what an ideal city should be.

In the context of urban planning, the idea of "city visioning" (or having a clear and formal sense of where a particular city wants to be in the long-term future) emerged during the 1980s and 1990s, particularly in the USA, not only as a way of understanding the future but also to plan for a desirable, or preferred, set of sustainable outcomes (see, e.g., Atlanta and Portland) (Dixon et al. 2018). Newman and Jennings (2008) also highlight "successful" examples of city visions in Perth, Vancouver, and Chicago during this period. This emergence of thinking about the future of cities also reflected a growing body of literature focusing on "visioning sustainability" in a range of other contexts, such as energy futures (Wiek and Iwaniec 2014). Since the early 2000s, we have also seen the development of more "formal" visioning processes (or what might be termed "city foresight" methods) in many cities and urban areas which have been used to develop city visions (see, e.g., Phoenix, Johannesburg, and Vancouver or, in the UK, Reading (Dixon et al. 2018) and Newcastle) (Tewdwr-Jones et al. 2015; Dixon and Tewdwr-Jones 2021).

The UK Government Office for Science (GOfS) Future Cities Programme (2013–2016) also highlighted the importance of "city foresight," which was founded on the science of thinking about the future of cities and which can be used to enable city stakeholders to explore urban futures not only in a local and regional context but as part of a wider connected network of cities (GOfS 2016). A number of UK city visions were created as part of this program, resulting from partnerships based on the "quadruple helix" model of innovation (Arnkil et al. 2011; Goddard and Tewdwr-Jones 2016). Some of these visions have also linked with and underpinned the existing statutory local plans in cities (see, e.g., Dixon et al. 2018).

Discourses about the Future: Smart Cities and Sustainable Cities

Throughout the history of urban studies, we have seen shifts and changes in the way in which the city is viewed. This has also paralleled thinking about makes an "ideal city," which has been typified by visions of the future which revolve around how new cities could be built or how cities might be redesigned or reconfigured to represent new or reimagined futures. Two dominant city futures discourses have been (i) "the sustainable city" and (ii) the "smart city." The origins of the term "sustainable city" (or "eco city") can be found in previous "organic" city visions such as Patrick Geddes' biopolis and Ebenezer Howard's garden city. It was not until the 1960s and 1970s, however, that the concept of what a "sustainable city" might be started to permeate the world of urban studies. Whitehead (2003, 2011) suggests that this increasing focus was the result of the interweaving of an "ecological crisis" and the "urban crisis," and Richard Register (1987) is credited with first using the term "eco city" in which he outlined the eco city as one built according to the principles of living within environmental limits (set within the ecological capacity of the city's bioregion).

Although the sustainable city concept continues to run strongly through policy and practice discourses, over the last decade, the "smart city" leitmotif has gained traction as a major "signifier" and "global discourse network" in urban development (Joss et al. 2019). Essentially, the smart city discourse relates to a normative view of the future founded on a technology-led ecological modernization (Trencher and Karvonen 2017). There are a very large number of definitions for smart city which not only reflect the differing origins of the term but also the varying disciplinary and institutional lenses through which a city can be viewed (Kitchin 2015). For example, some highlight the smart city as an urban environment that is idealistic, alluring, and more liveable than the complex, messy environments that we inhabit today. For others, the smart city provides a new market for urban management systems and an opportunity to sell technology-led solutions to city authorities facing environmental, economic, and social challenges (Dixon and Tewdwr-Jones 2021). This lack of consensus, as in the case of sustainable cities, has led to a growing critical literature on smart cities, particularly as issues over the role of citizens, privacy and security are raised.

However, from the mid-2010s onward, we have also seen the emergence of a new term, the "smart and sustainable city," as a result of growing sustainability awareness, continued urban growth, and the development of new technologies (Bibri 2018; Dixon 2018). This rebranding is intended to highlight the fact that not every smart city is necessarily a sustainable city – for example, smart transport technologies may continue to promote car use at the expense of more sustainable modes of transport such as bus, walking and cycling (Dixon 2018).

Case Study Example: Smart and Sustainable Reading 2050 City Vision

One example of a city vision which combines smart and sustainable thinking is the Reading 2050 vision in the UK (Dixon et al. 2018). Although Reading is not yet officially a "city," it forms part of one of the most economically vibrant and connected urban areas in the UK: Reading, as part of a wider Reading/Wokingham urban area (including Arborfield, Woodley, Theale (West Berkshire), Crowthorne, Earley), has a population of 318,000 (based on 2011 ONS data), and this is set to grow to 362,000 by 2037 (Dixon and Cohen 2015; Dixon and Farrelly 2020). This presents big challenges in maintaining its competitive edge and dealing with the important environmental and socioeconomic issues arising from its continued economic growth. Developing a Reading 2050 vision which was both "smart" (making the best use of technology) and "sustainable" (creating a truly sustainable city) was seen an important step in supporting longer-term planning and development in Reading. The starting point for this vision was provided through a formal definition of a smart and sustainable city as one (ITU 2014, pp. 12–13):

> that leverages the ICT infrastructure to:
>
> - Improve the quality of life of its citizens.
> - Ensure tangible economic growth for its citizens.
> - Improve the well-being of its citizens.
> - Establish an environmentally responsible and sustainable approach to development.
> - Streamline and improve physical infrastructure.
> - Reinforce resilience to natural and man-made disasters.
> - Underpin effective and well-balanced regulatory, compliance and governance mechanisms.

In 2013 the Reading 2050 project brought together the University of Reading (School of the Built Environment), Barton Willmore (a major planning and design consultancy), and Reading UK (the economic development unit for Reading) to lead the development of the vision. Drawing on previous research which had scoped out retrofit visions for Cardiff and Manchester (Dixon et al. 2014), the Reading 2050 project combined elements of a smart city with those of a sustainable city. This was because Reading already has a long-term aspiration to be "low-carbon" by 2050 but also has a strong technology and green technology focus in its existing economy. Moreover, a 2050 time horizon provides space to think beyond today's immediate problems and facilitates a greater sense of strategic thinking by identifying desirable as well as undesirable outcomes.

The visioning process which ran from 2013 to 2017 (and is ongoing) adopted a "quadruple helix" approach which brought together business, local government, civil society, and higher education (Arnkil et al. 2011) and was based on workshops and the adoption of a backcasting approach. This is where a desirable future is co-created with stakeholders through a participatory-based foresight approach, and then look stakeholders work together to look backward from that future to the present in order to strategize and to plan how it could be achieved. During the course of its work, to date, the Reading 2050 program has engaged with 21,000 people and more than 400 businesses with some 15 linked events (Dixon and Farrelly 2020).

As a result, three interrelated urban futures were developed for the Reading 2050 vision (Fig. 1):

- *Green Tech City*: A city that builds upon its established technology focus. It celebrates

City Visions: Toward Smart and Sustainable Urban Futures, Fig. 1 Three main elements from the Reading 2050 vision (top to bottom: "green tech city"; "city of rivers and parks"; "city of diversity and culture"). (Source: Reading 2050 website (www.reading2050.co.uk). Image courtesy of Reading 2050 – a collaborative initiative, jointly led by Barton Willmore, Reading UK and the University of Reading)

and encourages diversity through business incubation units, "idea factories" and a city center university campus through which to exhibit and test cutting edge ideas and approaches, no matter what discipline they are emerging from.

- *City of Diversity and Culture:* A city that builds on the success of the iconic Reading Festival to deliver arts and culture to people of all ages and ethnicities. Reading would facilitate community interaction and opportunity. The city would integrate, enhance, and celebrate our heritage, bringing it to life through modern interpretations and uses of space as well as preservation.
- *City of Rivers and Parks*: A city that recognizes how water has shaped much of Reading would celebrate its waterways, opening them up to offer recreational spaces such as animated parks, a lido, food production opportunities, and city center waterside living.

The vision is strongly linked with the development of the new Reading Borough Council Local Plan (which looks ahead to 2036) and is directly referenced within it as an important longer-term framework for Reading. A similar synergy is highlighted in the corporate plan where the council describes its endorsement of the vision and its commitment to integrating the 2050 ambitions into its priorities. Finally, the vision also links with the Reading Climate Emergency Strategy (2025–2030) which targets net zero emissions by 2030.

Futures Thinking for Cities: What Lies Beyond COVID-19?

Like many other cities in the UK and around the world, city leaders in Reading are currently developing plans and strategies that look to boost economic recovery in the aftermath of COVID-19 (or what is still currently *life with COVID*). The COVID crisis has been very much an urban crisis which has particularly affected the urban poor – for example, over 95% of total cases are in our urban areas (UN Habitat 2020), and it is clear that city economies which are less diversified have been harder hit. During the pandemic we saw that many cities in the UK and elsewhere took steps to increase active mobility (walking and cycling) through the provision of additional pedestrianized areas and cycleways. In Paris, for example, the equivalent of 30 miles of roads were made available for cycling, and the city's mayor decided to formally promote the concept of the 15-minute city (developed by Sorbonne Professor Carlos Moreno). This means developing and promoting services and everything a neighborhood needs within 15 min travel time (OECD 2020a).

As people returned to work, however, we saw things returning to "normal," so many city authorities are trying to look longer term to see how the hard won environmental gains for cities under COVID-19 could be integrated with an economic recovery based on green jobs and clean growth (UN 2020). We have also seen how new technologies have been used to help people work from home more easily and so travel to work less but how smart technologies can manage social distancing and monitor the spread of the virus in cities (OECD 2020a).

Finally, besides the continued importance of "smart" and "sustainable" thinking, we are also seeing an increasing emphasis on the "resilience" of cities which focuses on their ability to bounce back from environmental, socioeconomic shocks, and natural disasters (Wray 2020). Quite how cities will change in the future, however, is open to debate: will there be de-urbanization, re-urbanization or the development of enclaves? (OECD 2020b). Much will depend on if or when a vaccine is found, but what is clear, however, is that city stakeholders need to think clearly about the long-term futures of our cities.

Summary/Conclusions

Creating a coherent vision for a city is a challenging process. It requires resources, a clear plan, and leadership. Thinking at city scale also requires thinking across boundaries and across interest groups and using imaginative and innovative ways of engaging with communities (Dixon and Cohen 2015). The experiences of cities (including Reading) which have developed long-term visions also have important lessons for interdisciplinary research and the way in which city visions are co-created through a city foresight approach.

These include (Dixon et al. 2018; Dixon and Farrelly 2020):

- Framings of the problem for transformation: how is the problem framed from the outset? What is the overall ambition or goal of the vision?
- Urban foresight activities – how can these be best developed to include a truly participatory element and a balance between structured activities and "blue sky" thinking?
- Ownership and leadership – who is responsible for the leadership of the vision? Who "owns" the city vision?
- Vision and implementation – how does the city vision link with existing local city plans and the aspirations of the city authorities, the public, and other stakeholders? To what extent do the city authorities support the vision and its implementation?
- Contrasting partnership ambitions – related to leadership, can the differing ambitions of those creating and leading the vision be reconciled and balanced?
- Structural change and reform (vis-à-vis environment and design) – what are the wider implications of the vision, for example, in relation to governance structures and city status?
- Interdisciplinary challenges – how can different disciplines and different professionals work with each other, other stakeholders, and the public to help develop the vision? Can built environment professionals really think "longer term" beyond the constraints of the present?

Ultimately, city foresight techniques (which underpin urban futures thinking) can provide a powerful addition to longer-term planning and the more detailed master plan approach adopted in many cities in continental Europe. If we are to develop the longer-term, unconstrained thinking that is required to move to a more sustainable future, futures-based studies offer us a potentially powerful set of tools to help achieve this and mobilize resources in the best possible way (Dixon and Tewdwr-Jones 2021). Cities will almost certainly survive just as they have done before, but in living with COVID and the climate crisis, we need to fast forward the development long-term visions for our cities so that we can plan for smart, sustainable (and resilient) futures.

Cross-References

▶ Age-friendly Future Cities
▶ Behavioral Science Informed Governance for Urban and Regional Futures
▶ Future of the City-Region Concept and Reality
▶ Regulation of Urban and Regional Futures
▶ Smart Cities
▶ Smart City
▶ Smart(er) Cities
▶ Spatial Justice and the Design of Future Cities in the Developing World

References

Arnkil, R., Jarvesivu, A., Koski, P., & Piirainen, T. (2011). *Exploring quadruple helix: Outlining user-oriented innovation models* (Working paper). Tampere: University of Tampere.

Batty, M. (2018). *Inventing future cities*. Cambridge, MA/ Cambridge, UK: MIT Press.

Bibri, S. (2018). Backcasting in futures studies: A synthesized scholarly and planning approach to strategic smart sustainable city development. *European Journal of Futures Research, 6*(13), 1–27.

C40 Cities. (2019). *The future of urban consumption in a 1.5 °C world.* London: C40 Cites and Arup. https://c40-production-images.s3.amazonaws.com/other_uploads/images/2236_WITH_Forewords_-_Main_report__20190612.original.pdf?1560421525. Accessed Nov 2020

Clark, G. (2016). *Global cities*. Washington, DC: Brookings Institution Press.

Dixon, T. (2018). Smart and sustainable?: The future of 'future cities'. In T. Dixon, J. Connaughton, & S. Green (Eds.), *Sustainable futures in the built environment to 2050: A foresight approach to construction and development* (pp. 94–116). Oxford: Wiley-Blackwell.

Dixon, T., & Cohen, K. (2015). Towards a smart and sustainable Reading 2050 vision. *Town and Country Planning*, January, pp. 20–27.

Dixon, T., & Farrelly, L. (Eds.). (2020). *Reading 2050: A smart and sustainable city?* School of the Built Environment, University of Reading. https://livingreading.co.uk/public/downloads/AhO8o/Reading%202050%20Lecture%20Series.pdf. Accessed Nov 2020.

Dixon, T. J., & Tewdwr-Jones, M. (2021). *Urban futures: Planning for city foresight and city visions.* Bristol: Policy Press/Bristol University Press.

Dixon, T., Eames, M., Hunt, M., & Lannon, S. (Eds.). (2014). *Urban retrofitting for sustainability: Mapping the transition to 2050*. London: Routledge.

Dixon, T., Montgomery, J., Horton-Baker, N., & Farrelly, L. (2018). Using urban foresight techniques in city visioning: Lessons from the Reading 2050 vision. *Local Economy, 33*(8), 777–799.

European Union. (2016). *Urban Europe: Statistics on cities, towns and suburbs*. Brussels: EU. https://ec.europa.eu/eurostat/statistics-explained/index.php/Urban_Europe_%E2%80%94_statistics_on_cities,_towns_and_suburbs. Accessed Nov 2020.

Goddard, M., & Tewdwr-Jones, M. (2016). *City futures and the civic university*. Newcastle: Newcastle University.

Government Office for Science (GOfS). (2016). *Future of cities: Foresight for cities*. London: GOfS. https://www.gov.uk/government/publications/future-of-cities-foresight-for-cities. Accessed Nov 2020.

Iossifova, D., Doll, C., & Gasparatos, A. (2018). Defining the urban-why do we need definitions? In D. Iossifova, C. Doll, & A. Gasparatos (Eds.), *Defining the urban: Interdisciplinary and professional perspectives*. London: Routledge.

ITU. (2014). *Smart sustainable cities – An analysis of definitions*. Geneva: International Telecommunication Union.

Joss, S., Sengers, F., Schraven, D., Caprotti, F., & Dayot, Y. (2019). The smart city as global discourse: Storylines and critical junctures across 27 cities. *Journal of Urban Technology, 26*(1), 3–34.

Kitchin, R. (2015). Making sense of smart cities: Addressing present shortcomings. *Cambridge Journal of Regions, Economy and Society, 8*, 131–136.

Knox, P. (Ed.). (2014). *Atlas of cities*. Princeton: Princeton University Press.

Newman, P., & Jennings, I. (2008). *Cities as sustainable ecosystems: Principles and practices*. Washington, DC: Island Press.

OECD. (2020a). *Cities policy responses*, July. https://www.oecd.org/coronavirus/policy-responses/cities-policy-responses-fd1053ff/. Accessed Nov 2020

OECD. (2020b). *Strategic foresight for the COVID-19 crisis and beyond: Using futures thinking to design better public policies*. http://www.oecd.org/coronavirus/policy-responses/strategic-foresight-for-the-covid-19-crisis-and-beyond-using-futures-thinking-to-design-better-public-policies-c3448fa5/. Accessed Nov 2020

Register, R. (1987). *Ecocity Berkeley*. Berkeley: North Atlantic Books.

Rittel, H., & Weber, M. (1973). Dilemmas in a general theory of planning. *Policy Sciences, 4*(2), 155–169.

Tewdwr-Jones, M., Goddard, J., & Cowie, P. (2015). *Newcastle city futures 2065: Anchoring universities in cities through urban foresight*. Newcastle: Newcastle Institute for Social Renewal, Newcastle University.

Trencher, G., & Karvonen, A. (2017). Stretching "smart": Advancing health and well-being through the smart city agenda. *Local Environment, 24*(7), 610–627.

UN. (2018). *World urbanization prospects: The 2018 revision-full report*. New York: UN. https://population.un.org/wup/Publications/Files/WUP2018-Report.pdf. Accessed Nov 2020.

UN (2020). Policy Brief: COVID-19 in an Urban World UN, NY Available: https://www.un.org/en/coronavirus/covid-19-urban-world

UN Habitat. (2020). *UN Habitat COVID-19 response plan*. https://unhabitat.org/un-habitat-covid-19-response-plan. Accessed Nov 2020.

Whitehead, M. (2003). (Re)analysing the sustainable city: Nature, urbanisation and the regulation of socio-environmental relations in the UK. *Urban Studies, 40* (7), 1183–1206.

Whitehead, M. (2011). The sustainable city: An obituary? On the future form and prospects of sustainable urbanism. In J. Flint & M. Raco (Eds.), *The future of sustainable cities* (pp. 29–46). Bristol: Policy Press.

Wiek, A., & Iwaniec, D. (2014). Quality criteria for visions and visioning in sustainability science. *Sustainability Science, 9*, 497–512.

Wolfram, M., Borgstrom, S., & Farrelly, M. (2019). Urban transformative capacity: From concept to practice. *Ambio, 48*, 437–448.

Wray, S. (2020, October 26). COVID-19 is shifting the focus from smart cities to resilient cities. *Cities Today*. https://cities-today.com/covid-19-shifts-the-focus-from-smart-cities-to-resilient-cities/#:~:text=COVID%2D19%20is%20shifting%20the%20focus%20from%20smart%20cities%20to%20resilient%20cities,-26th%20October%202020&text=This%20perfect%20storm%20looks%20set,whether%20economic%2C%20social%20or%20environmental. Accessed Nov 2020.

Further Reading

Reading 2050 website: www.reading2050.co.uk

City-to-City Learning

► Beyond Knowledge: Learning to Cope with Climate Change in Cities

Civic Society

► Philanthropy in Sustainable Urban Development: A Systems Perspective

Clean Water

► Meeting SDG6: Ensuring Safe Drinking Water for All in Rural India

Climate Adaptation

► Urban Climate Resilience

Climate Change

► Adapting to a Changing Climate Through Nature-Based Solutions
► Amsterdam's Pathway to Climate Neutrality: Creating an Enabling Environment
► Beyond Knowledge: Learning to Cope with Climate Change in Cities
► Responsibility to Prepare and Prevent (R2P2): Applying Unprecedented Foresight to Addressing Unprecedented Climate Risks
► Role of Disaster Relief Policy in Building Resilient Coastal Regions in the United States
► The State of Extreme Events in India
► Urban Management in Bangladesh
► Water-Smart Cities

Climate Change Adaptation

► Stewarding Street Trees for a Global Urban Future

Climate Change Adjustment

► Sustainability Transition and Climate Change Adaption of Logistics

Climate Change and Surface Water Resources in Sri Lanka

Medhisha Pasan Gunawardena[1,2] and
Melisha Shavindi Fernando[1]
[1]Faculty of Science, Horizon Campus, Malabe, Sri Lanka
[2]Biodiversity Educational Research Initiative, Colombo, Sri Lanka

Synonyms

Climate deviations and surface water resources in Sri Lanka; Shift in atmospheric conditions and surface water resources in Sri Lanka; Climatic variations and surface water resources in Sri Lanka; Meteorological conditions' distortion and surface water resources in Sri Lanka; Weather pattern turnover and surface water resources in Sri Lanka

Definition

Sri Lanka is a tropical island that is extremely vulnerable to climate change in terms of rainfall, temperature, and sea level. Climate change is currently influencing many components in the ecosystem. These impacts were evident in the past, can be observed at present, and are expected to cause complications in the future as well. The main focus of this chapter is to highlight the effects of climate change to surface water resources in Sri Lanka. Population growth, overall expansion in economic activities, and rapid urbanization have significantly increased the regional and urban demand for freshwater in Sri Lanka over the past decade. Flooding, water quality, runoff changes, water temperatures, nutrient and oxygen concentration, water level, evapotranspiration, and soil moisture deficit are the most evident effects of climate change to surface water resources. Studies have confirmed that varying rainfall and temperature trends have directly begun to affect water resources leading to major problems in agriculture, industry, and community. However, among the negative impacts, there are several positive consequences as well. Evaluation of the various impacts of climate change on water resources in Sri Lanka has clearly depicted that the entire country is under threat. Exposure to droughts and floods can result in complications in human settlements, transport infrastructure, tourism, agriculture, fisheries, and supply of water for both drinking and irrigation needs. Rainwater harvesting, renovating the existing tanks, shifting to alternative energy sources, practicing water-saving methods are several actions that can be taken to reduce the negative impacts of climate change to water resources. Although the exact climate change impacts on these major surface water bodies are still unknown, they are clearly under stress. Actions must be taken in order to preserve them as they play a significant role in the urban lifestyle of Sri Lankans.

Introduction

Sri Lanka has multiple rainfall origins and varied temperature patterns across the country (DOM 2019). Sri Lanka is extremely vulnerable to climate change. Climate change affects many areas such as water resources, agriculture, health, coastal zones, and industry. The main focus of this chapter is to highlight the effects of climate change to surface water resources in Sri Lanka. Surface water resources such as rivers, tanks, and groundwater resources such as aquifers are the major water resources in Sri Lanka. Eighty percentage of rural domestic water requirement is from groundwater, while surface water supports majority of the urban water supply requirements. Impacts of climate change in Sri Lanka were evident in the past, can be observed at present, and are expected to cause complications in the future as well. Contradicting projections of climatic conditions makes it difficult to reach a conclusion of how climate changes affect water resources (Eriyagama et al. 2010). Rainfall and temperature variations are the major climatic factors that affect water resources. The most evident effects of climate change to water resources are related to water quality (nutrient and oxygen concentration, DO, BOD levels), runoff changes (flooding), water temperatures, moisture level, evapotranspiration, and soil moisture deficit. Surface water is any body of water above ground, including streams, rivers, lakes, wetlands, reservoirs, and creeks. The variation of the permanent and seasonal surface water area is divergent in different climatic zones within the country. Climate change imposing negative impacts on water resources gives rise to many different issues in agriculture and infrastructure. The efficacy of forecasting the climate for the next 50 years is a concern, as is Sri Lanka's adaptation to such changes. Ongoing studies aim to predict the effects of future climate change. However, the precise effects of climate change on major resources such as surface water bodies are still uncertain (Gunawardena and Najim 2017).

Sri Lanka, a tropical island located in the Indian ocean, has multiple rainfall origins and varied temperature patterns across the country. A major share of the annual rainfall is accounted for by monsoonal, convectional, and depressional rain. The mean annual rainfall in the dry southeastern and northwestern parts and wet western slopes of the central highland is under 900 mm and over 5000 mm, respectively. The mean monthly temperature depends on the movement of the sun and rainfall. The mean annual temperature ranges from 27 °C in the coastal lowlands to 16 °C in the central highlands (1900 m above mean sea level). These rainfall and temperature variations results in the unique feature of experiencing warm and cold climatic conditions within a few hours' journey in the island (DOM 2019).

Surface water resources such as rivers, tanks, and groundwater resources such as aquifers are the major water resources in Sri Lanka. The country consists of 103 district river basins with a total length of 4500 km, covering 90% of the island (UNESCO and MoAIMD 2006). In the southwest part of the island, there are seven major basins with catchment areas ranging from 620 to 2700 km^2. Deep confined aquifers, the shallow karstic aquifer of the Jaffna Peninsula, alluvial aquifers, coastal sand aquifers, the shallow regolith aquifer of the Hard Rock Zone, and the southwestern lateritic (cabook) aquifer are the six types of aquifers in Sri Lanka (Water Resources Board (WRB) 2005). Storage of water is done in many river basins in the *yala* season (Amarasinghe 2009). A region of 169,941 hectares is covered by ancient irrigation reservoirs and recently constructed multipurpose reservoirs (MENR & UNEP 2009). The total capacity of irrigation dams is approximately 3.37 km^3 (Illangasinghe 2012). Groundwater resources in the country is close to 7800 million m^3 per year. About 72% of the rural population depend on groundwater for domestic use (Nandalal 2010). Accordingly, the total renewable water resources are approximately 52.8 km^3/year (Illangasinghe 2012). According to the per capita average based on the 2001 population census records, groundwater utilization was 420 m^3/per person and surface water utilization was 1850 m^3/per person. This clearly depicts that surface water is utilized at a greater extent than groundwater.

Several projections have been made about the possible shift in temperature, rainfall, and sea level in Sri Lanka. A temperature rise of 0.8–0.2 °C is expected by the year 2060. Precipitation patterns are changing. In 2017, the floods in Sri Lanka were triggered by a strong southwestern monsoon, intensified by the presence of Cyclone Mora. Fifteen districts were impacted by the flood. About 700,000 people were affected and 208 people were killed (Lacombe et al. 2019). The sea level would rise by 20–58 cm by 2060. The frequency and severity of extreme weather events increases. Therefore, it is clear that climatic patterns are changing in various aspects.

These conditions affect many areas such as agriculture, water resources, health, coastal zones, and industry (Climate Links 2018). Effects related to agriculture are salinization, increased drought frequency, decreased food security, and decreased yield of crops. Climate-related risks to water resources are less hydropower generation, reduced water availability for agriculture, and decreased drinking water availability. Health can be affected in aspects such as decreased food security and nutrition, decreased water quality and availability, and shifts in infectious disease patterns. Groundwater salinization, loss of ocean biodiversity, and damage to coastal infrastructure are few climate-related risks in the coastal zones. Industries will face increased energy costs, damages, and thereby decreased economic output due to drastic climate changes. However, the main focus of this chapter would be the effects of climate change to surface water resources in Sri Lanka.

Climatic Change Effects Impacting Water Resources

Climate change is predicted to have varying effects on rainfall and temperature fluctuations across areas, as well as on the spatial and temporal distributions of various water resource components (Abbaspour et al. 2009).

Sri Lanka receives rainfall by the influence of monsoonal winds of the Indian Ocean and Bay of Bengal. The rainfall pattern consists of two monsoonal (southwest and northeast monsoons) and two inter-monsoonal seasons. The northeast monsoon is predicted to decrease while the southwest monsoon is predicted to increase across the country. This would increase the surface water area in lowland areas receiving rainfall from the southwest monsoon and lead to flooding problems due to inadequate drainage systems, which would result in serious traffic and road accidents. Also, drought problems will be created in areas receiving rainfall from the northeast monsoon and increase irrigation requirement for paddy and other field crop cultivation.

Impacts of climate change in Sri Lanka were evident in the past, can be observed at present, and are expected to cause complications in the future as well. In the past, Sri Lanka has experienced many severe rainfall incidents (Deheragoda and Karunanayake 2003). Landslides in the hill country occurred in 1986 as a result of daily rainfall varying from 90 mm to 299 mm in a 24-h period (De Silva 2006a). From 1869 to 2007, the temperature has increased; the mean monthly minimum and maximum temperatures were 1.7 °C per 100 years and 2.6 °C per 100 years, respectively.

The number of consecutive dry days in the country as a whole has increased, whereas the number of consecutive wet days has decreased (Ratnayake and Herath 2005; Premalal 2009). Both the daily rainfall intensity (amount of rain per rainy day) and the average rainfall per spell have risen, resulting in an increase in landslides (Ratnayake and Herath 2005). Rainfall projections for Sri Lanka over the next century tend to be perplexing and often inconsistent. Although the majority of them forecast higher mean annual rainfall, a few forecast lower mean annual rainfall. During December–February (northeast) and during May–September (southwest) monsoon seasons in Sri Lanka, the mean temperature will be increased from 2.9 °C and 2.5 °C over the baseline by the year of 2100 and cause for changes in the quantity and spatial distribution of rainfall (Eriyagama and Smakhtin 2009). These past, present, and possible future climate conditions show that Sri Lanka's climate change-related weather aberrations are becoming increasingly common. It is evident that this condition will remain the same in the future or increase.

Varied temperatures among regions are mainly due to the change in altitude. Temperature rises are most evident in the country's north, northeastern, and northwestern areas. Predicted temperature rises along with reduced rainfall in dry zone areas will intensify drought conditions. Water availability and surface water dynamics displayed moderate positive and negative relationships with temperature in the dry and wet zones, with respect to precipitation and temperature. In the intermediate region, there were moderate and weak associations with precipitation and temperature, respectively (Somasundaram et al. 2020).

Effects of Climate Change to Surface Water Resources

Contradicting projections of climatic conditions makes it difficult to reach a conclusion of how climate changes affect water resources (Eriyagama et al. 2010). Sixty percentage of agricultural water uses in South Asia rely on surface water and 40% on groundwater.

Climate change imposing negative impacts on water resources gives rise to many different issues. Impacts to agriculture are damage to crops due to soil erosion and waterlogging, lower yields due to degradation, increased livestock death, less water availability for agriculture, changes in crop productivity and growing season, and changes in freshwater species composition and biodiversity. Impacts to industry and community are disruption of settlements and infrastructure, migration, water shortages for settlements and industry, and degradation of freshwater quality.

A decrease in rainfall, particularly in the dry zone, in addition to the increase in temperature, evapotranspiration, and soil moisture deficit, would have serious consequences for the country's food production, livelihoods, and economy. It would have significant implications for the country's water supplies and will jeopardize poverty-reduction efforts. By 2040, Sri Lanka's population is expected to rise by 15%. This clearly indicates that there will be an increase in the demand for water.

In order to maintain the irrigation efficiency of the country, it has to be increased up to 45% from the currently assumed level of 35%. Thereby, the irrigation demand shall decrease by 22%. This increase of efficiency will lead to a 78% reduction in demand in the major irrigated areas. If irrigation efficiency is improved to 55%, irrigation demand will be decreased by 35% (Fig. 1). In such a situation, irrigation demand falls by more than 3.9 km^3, which equates to approximately 32% of overall water demand (Amarasinghe 2009). This means that even though the existing water source is adequately handled, only a portion of the water savings would be adequate to satisfy future irrigation demand.

Reduced rainfall and increased possible evapotranspiration during January and February may have an effect on paddy production, leading to early planting or short-term varieties (De Silva et al. 2007). Agricultural GDP have shown positive effects on surface water area, which is the area covered by a body of water above ground, including rivers, streams, lakes, reservoirs, wetlands, and creeks.

In contrast to the negative impacts, there are several positive consequences as well. Along with

Climate Change and Surface Water Resources in Sri Lanka, Fig. 1 Irrigation efficiency versus demand

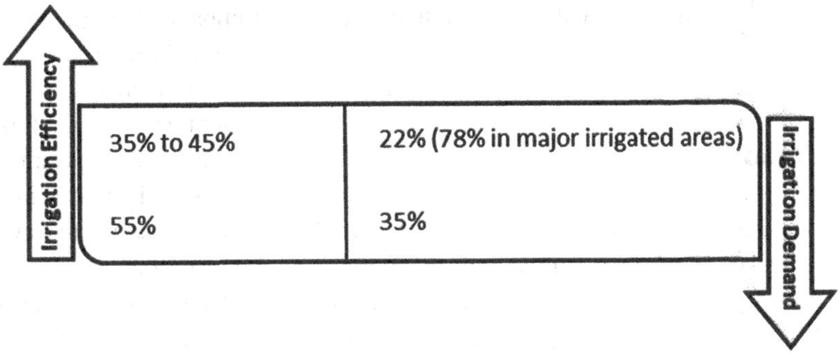

C

the increase in rainfall, temperature, and evapotranspiration, potential crop growth will rise due to elevated CO_2 levels. Climate changes can result in prolonged growing season and a higher yield (Droogers 2004). In Sri Lanka, the southern tip of the island is expected to have beneficial effects due to climate change (Droogers 2004; De Silva et al. 2007). For example, the impact of climate change on Walawe river basin will have more positive than negative aspects (Droogers 2004).

Heavy precipitation events lead to flooding and adverse effects on water quality. Changes in runoff and water quality are caused by poor dissolution of sediments, nutrients, dissolved organic carbon, bacteria, toxins, and salt due to high fluctuations in precipitation and increased droughts. Increased temperature leads to increased water temperature, evaporation, and decrease in nutrient and oxygen concentration in water resources. Other effects of climate change to water resources are related to moisture level, evapotranspiration, and potential soil moisture deficit.

Water Level

Although water availability is expected to increase due to climate change, the increased uncertainty and unequal spatial distribution would adversely affect agriculture and food security. Wet areas are expected to get wetter while dry areas get drier (Basnayake 2008; Basnayake et al. 2004; Basnayake and Vithanage 2004). The water resources in the northeastern and eastern dry zone are expected to be drier by the 2050s (De Silva 2006b).

Evapotranspiration

Evapotranspiration is the sum of water evaporation and transpiration from a surface area to the atmosphere. Evaporative loss of water could have negative impacts on surface water resources.

High evapotranspiration could increase the salinity of soils in irrigated areas and surface waters, especially in lakes and reservoirs with long residence times. This leads to the accumulation of contaminants on land surfaces during prolonged droughts. During runoff due to rainfall, these contaminants runoff leading to problems related to water quality. Therefore, these impacts may magnify into rather complex situations.

The average reference evapotranspiration is expected to increase mainly in the north and northeastern regions of Sri Lanka due to temperature, rainfall, and various other factors (De silva 2006b). This suggests that the influence of climate change is greatest in the northeastern regions with the highest agriculture activity. For example, all subbasins of Kalu Ganga possesses an evapotranspiration rate of 525 mm/year (De costa 1995). If there is sufficient water in the atmosphere, real evapotranspiration (or green water flow) is predicted to increase as the temperature rises (Abbaspour et al. 2009).

The effects of various land uses on water demand were studied in order to explore the feasibility of manipulating evapotranspiration loss by land use management (Gunawardena 1998). Warmer and drier temperatures improve field evapotranspiration and necessitate more irrigation water. Irrigated crops' water usage is calculated using their evapotranspiration rate, and withdrawals are calculated by assigning irrigation water-use efficiency.

Soil Moisture Deficit

Critical climatic conditions will significantly increase the maximum annual soil moisture deficit in dry zone, necessitating further irrigation. The reference evapotranspiration (ETo) is used to estimate possible soil moisture deficit (Smith 2000). Potential soil moisture deficit (PSMD) is measured in any month of the year by adding the difference between potential evapotranspiration (of short grass) and rainfall in the current month to the PSMD from the previous month. PSMD calculation begins in January, when the soil moisture deficit in Sri Lanka is normally negligible due to the onset of the wet season rainfall in October. The maximum PSMD for the whole year is the maximum annual PSMD (PSMDmax). By the 2050s, the maximum annual potential soil moisture deficit (PSMDmax) may be increasing across the country. A high potential soil moisture deficit occurs in areas of the country where agricultural activities are intense and the stability and supply of water supplies are under extreme stress. Areas with a low potential soil moisture deficiency have a low need for irrigation. Because of the slightly higher PSMDmax, the northern part of the country becomes is predicted to become drier than it is today (De Silva 2006b).

Zonal Water Dynamics

The variation of the permanent and seasonal water area is divergent in different climatic zones within the country. Permanent water in Sri Lanka expanded due to the construction of several reservoirs in the last decade. The intermediate zone reservoirs are Deduru Oya (46 km^2 in 2014) and Moragahakanda (47 km^2 during 2017–2018), the wet zone reservoir is Upper Kotmale (6 km^2 during 2013–2014), and the dry zone reservoir is Rambaken Oya (22 km^2 during 2011–2013). For the past three decades, the dry zone seasonal water area has increased significantly every year. Permanent water, in contrast, increased every year in the wet zone. In the intermediate zone, there were significant increases in both their permanent and seasonal water area every year (Somasundaram et al. 2020).

Annual water occurrence frequency provides information on changes of water resources based on the period of existence. Accordingly, the proportions of permanent and seasonal water remained almost the same throughout 1988–2019. Furthermore, since 2015, there has been a nearly 20% reduction in the permanent water area (WOF > 80%). Seasonal water area with WOF 40%, in contrast, expanded rapidly. Over the last three decades, these factors suggested a rise in competing demand and shortage of water, with a recent escalation.

Current Status of Sri Lankan Water Resources

Studies have been conducted to evaluate the various impacts of climate change on water resources in Sri Lanka (Table 1).

Studies have observed a small rise in annual total rainfall due to an increase in rainfall during the southwestern monsoon. However, the northeast monsoon rains are decreasing. In the meantime, global annual temperature has increased. These increases in rainfall and temperature, along with other climatic conditions, will greatly increase the maximum annual soil moisture deficit in dry areas, requiring higher irrigation needs (De Silva 2006b).

Other studies indicate that climate change would have major impacts on soil moisture deficits. Hence, irrigation needs are not only projected to rise in Sri Lanka's dry zone but also in the intermediate zones during the Yala season (dry season) (De Silva 2006b).

Across most of Sri Lanka, the impacts of climate change would raise the requirements for wet

Climate Change and Surface Water Resources in Sri Lanka, Table 1 Studies conducted related to water resources and climate change

Study	Location	References
Analysis		
District level water supply and demand analysis	Sri Lanka	Amarasinghe et al. (1998, 1999)
Rainwater harvesting capacity computation	Moneragala District	Gunawardana et al. (2005)
Assessment of streamflow	Kotmale Oya	Wijesekera (1999)
Annual water accounting	Walawe river basin	Nandalal and Sakthivadivel (2003)
Long-term hydrologic study	Nilwala basin	Mungai et al. (2004)
Investigation of storage characteristics	Kalu Ganga	De Costa (1995)
Preliminary assessment of surface water resources	Deduru Oya basin	Wickramaarachchi (2004,2010)
Monograph on water resources	Sri Lanka	Manchanayaka and Madduma Bandara (1999)
Development		
Flow duration development for long-term forecast of daily streamflow	Talawakelle, Caledonia and Hulu -Ganga basins	Hunukumbura et al. (2004)
Water and related development works carried out in the past and present	Kalu Ganga	Wickramasuriya (1996)
Recommendations and suggestions		
Recommendations to strengthen the water resources assessment	Walawe river	Smakhtin and Weragala (2005)
Issues associated with the development of water resources for irrigated agriculture	Sri Lanka	Imbulana (2000)

season (Maha). Paddy cultivation is going to get more water stressful. However, some beneficial impacts are visible in the extreme south of Sri Lanka (De Silva et al. 2007).

Vulnerability of Various Sectors

Sri Lanka is also prone to several natural disasters, especially flooding, landslides, and droughts. The occurrence and severity of these threats are predicted to escalate as a result of climate change (Ministry of Environment 2011). Sea level rises, tidal floods, coastal erosions, changes in climate patterns, and an increase in ambient temperature are all predicted. These challenges would be intensified by numerous anthropogenic causes that have already threatened the island's freshwater supplies and created several socioeconomic and environmental problems.

Exposure to droughts and floods can result in complications in human settlements, transport infrastructure, tourism, agriculture, fisheries, and supply of water for both drinking and irrigation needs.

Drought vulnerability of human settlements is common, but more localized in the north-central and southern areas. Highly drought-sensitive settlements are located in 19 DS Divisions (DSDs). The three most vulnerable DSDs are Siyambalanduwa (Moneragala District), Embilipitiya (Ratnapura District), and Kalpitiya (Puttalam District). However, the populations of these areas mainly depend on groundwater, therefore it is clear that droughts have not highly affected human settlements that depend on surface water resources. The vulnerability of communities to the anticipated rise in flooding tends to be high in the country's western region.

The vulnerability of the paddy sector to the projected rise in droughts due to climate change is widespread across the region, with a concentration in the dry and intermediate zones. In the 16 DSDs that emerge as highly vulnerable, 3153 tanks covering a total area of 88,395 ac are located. Farmers in the most vulnerable DSDs earn 63% of their revenue from agriculture on average. In the 6 DSDs that emerge as highly vulnerable to floods, 283 tanks covering an area of 16,717 ac are located. Therefore, it can be concluded that climate change effects on surface water resources has an impact on paddy cultivation.

According to the climate change vulnerability data book's drinking water sector vulnerability to drought exposure, 9.2% of households in the most vulnerable DSDs rely on rivers, streams, and tanks as their primary source of drinking water. 8.3% of relatively poor households receive their water from rivers, streams, and other sources. According to the vulnerability to flood exposure in the drinking water sector, 11% of households in the most vulnerable DSDs depend on rivers, streams, and tanks as their primary source of water. 6.8% of households in moderately vulnerable DSDs use water from rivers, streams, and other sources for drinking.

Irrigation is vulnerable to predicted changes due to drought conditions around the island, although it is more concentrated in the dry zone, where cultivation is heavily reliant on irrigation. In this respect, 9 DS Divisions (DSDs) tend to be extremely vulnerable. These DSDs have a total of 2375 tanks occupying an area of 240 km^2. Thanamalwila (Moneragala District), Anamaduwa (Puttalam District), and Horowpothana are the three most vulnerable DSDs (Anuradhapura District). Thanamalwila alone has 464 tanks (the second most in the country, according to DSD) covering 27.6 km^2. Another 18 DSDs are classified as moderately vulnerable. They employ 145,880 workers in agriculture-related employment.

In addition to these, transport and tourism also have been indirectly affected. The vulnerability of transportation networks to the predicted rise in frequency and severity of flooding caused by climate change is widespread and prevalent throughout the country. Gampaha is the most vulnerable district in this regard. The tourism industry is especially vulnerable to the projected rise in floods triggered by climate change in the country's western region. Colombo is the most vulnerable district in this regard.

Current Status of Demand for Water

Over the last decade, urban and regional demand for freshwater in Sri Lanka has risen significantly (Hussain et al. 2002). Residential, private, and industrial water use grows as a result of population growth, rapid urbanization, and overall economic expansion. Water intake in the domestic, private, and industrial sectors rose by more than 34%, 27%, and 37%, respectively, from 1994 to 1998 (Hussain et al. 2002). By 2025, the urban population is projected to rise from 5.6 million to 15 million (Sri Lanka National Water Partnership (SLNWP) 2000).

Required Management Responses

There are several actions that can be taken to reduce the negative impacts of climate change to water resources. Rainwater harvesting, renovating the existing tanks, shifting to alternative energy sources, practicing water-saving methods are few among those.

De Silva (2006a) introduced a National Rainwater Harvesting Program to include rainwater harvesting systems to all households in drought-prone regions. This will improve water quality in these areas while also providing relief to the population. Renovating existing tanks in the dry and intermediate zones would aid in the storage of excess rain during the southwest monsoon. The collected water can then be transferred and used for a number of purposes (De Silva 2006a). In the energy sector, a greater transition away from hydropower and fossil fuels is proposed (Shantha and Jayasundera 2005), while the Coast Conservation Department (CCD) is preparing a Climate Change Action Plan to respond to sea level rise (Wickramarachchi n.d.). Drip irrigation and other water-saving measures would be more useful in sustaining agricultural operations throughout the water-shortage season (De Silva 2006a).

Conclusion

In conclusion, climate change is impacting Sri Lanka's water resources in a greater scale than highlighted by various studies. Although the exact climate change impacts on these major surface water bodies are still unknown, they are clearly under stress. Varying rainfall and temperature trends have directly begun to affect water resources leading to major problems in agriculture, industry, and community. Necessary actions must be taken in this regard as they play a major role in the urban and regional lifestyles of Sri Lankans.

Cross-References

- ▶ Circular Water Economy
- ▶ Water Security and the Green Economy
- ▶ Water-Smart Cities

References

Abbaspour, K. C., Faramarzi, M., Ghasemi, S. S., & Yang, H. (2009). Assessing the impact of climate change on water resources in Iran. *Water Resources Research, 45*(10), 1–16. https://doi.org/10.1029/2008WR007615.

Amarasinghe, U. A. (2009). Spatial variation of water supply and demand in Sri Lanka. In Jinapala et al. (Eds.), *National conference on water, food security and climate change in Sri Lanka. Vol. 3: Policies, institutions and data needs for water management*. International Water Management Institute. Available at: http://www.environmentportal.in/files/SLWC_Vol_3_final-low.pdf

Amarasinghe, U. A., Mutuwatte, L., & Sakthivadivel, R. (1998). Water supply and demand in Sri Lanka, processing and management: Sri Lanka. In M. Samad, N. T. S. Wijesekera, & S. Birch (Eds.), *National water conference on status and future direction of water research in Sri Lanka*. Colombo: International Water Management Institute.

Amarasinghe, U. A., Mutuwatta, L., & Sakthivadivel, R. (1999). *Water scarcity variations within a country: A case study of Sri Lanka*. Research Report 32. Colombo, Sri Lanka: IWMI.

Basnayake, B. R. S. B. (2008). *Climate change: Present and future perspective of Sri Lanka*. Meteorological Department of Sri Lanka. http://www.meteo.gov.lk/Non_%20Up_Date/pages/ccinsl_1.html

Basnayake, B. R. S. B., & Vithanage, J. C. (2004). Rainfall change scenarios for Sri Lanka under the anticipated climate change. In *Proceedings of the international conference on sustainable water resources management in the changing environment of the monsoon region*. Colombo, Sri Lanka.

Basnayake, B. R. S. B., Rathnasiri, J., & Vithanage, J. C. (2004). Rainfall and temperature scenarios for Sri Lanka under the anticipated climate change. In *2nd AIACC regional workshop for Asia and the Pacific*, Manila, Philippines

Climate Risk Profile: Sri Lanka. (2018). Climatelinks.org. https://www.climatelinks.org/resources/climate-risk-profile-sri-lanka

De Costa, S. (1995). Response characteristics of the Kalu Ganga catchment and lumped system model for flood characteristics. *Engineer Journal of the Institution of Engineers, Sri Lanka*. December 1995.

De Silva, C. S. (2006a). Impacts of climate change on water resources in Sri Lanka. In: Fisher, J. (ed.), *Sustainable development of water resources, water supply and environmental sanitation* (pp. 289–295). Colombo, Sri Lanka: Proceedings of the 32nd WEDC International Conference.

De Silva, C. S. (2006b). Impacts of climate change on potential soil moisture deficit and its use as a climate indicator to forecast irrigation need in Sri Lanka. In: *Water resources research in Sri Lanka* (pp. 79–90). Post graduate institute of Agriculture. University of Peradeniya. ISBN-955-1308-05-0.

De Silva, C. S., Weatherhead, E. K., Knox, J. W., & Rodriguez-Diaz, J. A. (2007). Predicting the impacts of climate change – A case study of paddy irrigation water requirements in Sri Lanka. *Agricultural Water Management, 93*(1–2), 19–29. https://doi.org/10.1016/j.agwat.2007.06.003.

Deheragoda, C. K. M., & Karunanayake M. M. (2003). *Landslide Disaster May 2003: Research Report on Kotapola Divisional Secretariat Division, Matara District of Sri Lanka*. Department of Geography. University of Sri Jayewardenepura, Sponsored by National Building Research Organization, Ministry of Housing & Plantation Infrastructure, Colombo.

Department of Meteorology (DOM). (2019). *Climate of Sri Lanka*. http://www.meteo.gov.lk/index.php?option=com_content&view=article&id=94&Itemid=310&lang=en

Droogers, P. (2004). Adaptation to climate change to enhance food security and preserve environmental quality: Example for southern Sri Lanka. *Agriculture Water Management, 66*, 15–33.

Eriyagama, N., & Smakhtin, V. (2009). How prepared are water and agricultural sectors in Sri Lanka for climate change? In *A review* (pp. 1–26). Water for Food Conference, International Water Management Institute (IWMI).

Eriyagama, N., Smakhtin, V., Chandrapala, L., & Fernando, K. (2010). *Impacts of climate change on water resources and agriculture in Sri Lanka: A review and preliminary vulnerability mapping*. IWMI Research Report 135. https://doi.org/10.5337/2010.211.

Gunawardana, I. P. P., De Silva, R. P., & Dayawansa, N. D. K. (2005). *Identification of appropriate technology for rainwater harvesting in Moneragala District*. Water Professionals' Symposium. http://www.gissl.lk

Gunawardena, E. R. N. (1998). Assessing the impact of land use conservation on water demand: A prerequisite for policy formulation for water allocation. In M. Samad, N. T. S. Wijesekera, & S. Birch (Eds.), *National water conference on status and future direction of water research in Sri Lanka*. Colombo: International Water Management Institute.

Gunawardena, M. P., & Najim, M. M. M. (2017). *Adapting Sri Lanka to climate change: Approaches to water modelling in the upper Mahaweli catchment area* (pp. 95–115). Climate Change Research at Universities. https://doi.org/10.1007/978-3-319-58214-6.

Hunukumbura, P. B., Weerakoon, S. B., & Herath, S. (2004). Estimation of flow duration curves for mini hydropower plant design in ungauged streams using measurements during a short period. In *International conference on sustainable water resources management in the changing environment of the monsoon region*. Colombo: National Water Resources Secretariat.

Hussain, I., Thrikawala, S., & Barker, R. (2002). Economic analysis of residential, commercial, and industrial uses of water in Sri Lanka. *Water International, 27*(2), 183–193. https://doi.org/10.1080/02508060208686991.

Illangasinghe, K. (2012). *State of water Sri Lanka*. Water Environment Partnership in Asia. https://doi.org/10.1111/j.1758-6623.2012.00160.x.

Imbulana, K. A. U. S. (2000). Water resources development for irrigated agriculture in Sri Lanka: Present issues and future challenges. *Journal of the Institutional of Engineers Sri Lanka, 32*(2), 40–52.

Lacombe, G., Chinnasamy, P., & Nicol, A. (2019). *Review of climate change science, knowledge and impacts on water resources in South Asia* (Background Paper 1). International Water Management Institute. https://doi.org/10.5337/2019.202.

Manchanayaka, P., & Madduma Bandara, C. M. (1999). *Water resources of Sri Lanka*. Colombo: National Science Foundation. ISBN 955-590-008-6, ISSN 1391-2488.

Ministry of Energy and Natural Resources (MENR) & United Nations Environment Programme (UNEP). (2009). *Sri Lanka environment outlook 2009*. Battaramulla: MENR & UNEP.

Ministry of Environment (MOE). (2011). *Climate change vulnerability data book*. MOE.

Mungai, D. N., Ong, C. K., Kiteme, B., Elkaduwa, W., & Sakthivadivel, R. (2004). Lessons from two long-term hydrological studies in Kenya and Sri Lanka. *Agriculture, Ecosystems and Environment, 104*, 135–143.

Nandalal, K. D. W. (2010). Groundwater resources. In *National Forum on water research identification of gaps and priorities*, 16–17 September, Colombo, Sri Lanka.

Nandalal, K. D. W., & Sakthivadivel, R. (2003). Availability, use and productivity of water in Walawe River Basin. *Engineer, 36*, 30–38.

Premalal, K. H. M. S. (2009). *Weather and climate trends, climate controls & risks in Sri Lanka*. Presentation made at the Sri Lanka Monsoon Forum, April 2009. Department of Meteorology, Sri Lanka.

Ratnayake, U., & Herath, G. (2005). Changes in water cycle: Effect on natural disasters and ecosystems. In N. T. S. Wijesekera, K. A. U. S. Imbulana, & B. Neupane (Eds.), *Workshop on Sri Lanka National Water Development Report*. Paris: World Water Assessment Programme.

Shantha, W. W. A., & Jayasundara J. M. S. B. (2005). Study on changes of rainfall in the mahaweli upper watershed in Sri lanka due to climatic changes and develop a correction model for global warming. In *International symposium on the stabilisation of greenhouse gas concentrations*. Hadley Centre, Met Office, Exeter, UK 64.

Smakhtin, V., & Weragala, N. (2005). *An assessment of hydrology and environmental flows in the Walawe river basin, Sri Lanka* (Working Paper 103). Colombo: International Water Management Institute (IWMI). ISBN 92-9090-619-7, 2005.

Smith, M. (2000). The application of climatic data for planning and management of sustainable rain-fed and irrigated crop production. *Agricultural and Forest Meteorology, 103*, 99–108.

Somasundaram, D., Zhang, F., Ediriweera, S., Wang, S., Li, J., & Zhang, B. (2020). Spatial and temporal changes in surface water area of Sri Lanka over a 30-year period. *Remote Sensing, 12*(22), 1–23. https://doi.org/10.3390/rs12223701.

Sri Lanka National Water Partnership (SLNWP). (2000). *Water vision 2025, Sri Lanka*. Colombo: SLNWP.

United Nations Educational, Scientific and Cultural Organization (UNESCO) & Ministry of Agriculture, Irrigation and Mahaweli Development (MoAIMD). (2006). *Sri Lanka water development report*. http://unesdoc.unesco.org/images/0014/001476/147683E.pdf

Water Resources Board (WRB). (2005). *Groundwater resources of Sri Lanka*. Available at: http://tsunami.obeysekera.net/documents/Panabokke_Perera_2005_Sri_Lanka.pdf

Wickramaarachchi, T. N. (2004). An assessment of surface water resources in Deduru Oya Basin of Sri Lanka – A preliminary approach. In *International conference on sustainable water resources management in the changing environment of the monsoon region*. Colombo: National Water Resources Secretariat.

Wickramaarachchi, T. N. (2010). *Preliminary assessment of surface water resources – A study from Deduru Oya Basin of Sri Lanka*. http://www.sljol.info/index.php/JAS/article/viewPDFInterstitial/1641/1397

Wickramarachchi, B. (n.d.). *Risk profiling for sea level rise in South-West Coast, Sri Lanka*. Coast Conservation Department.
Wickramasuriya, A. G. T. A. (1996). *Challenges of a changing environment in the development of Sri Lanka's water resources: Past, present and future*. Transactions Annual Sessions of the Institution of Engineers Sri Lanka.
Wijesekera, N. T. S. (1999). *An evaluation of the upper Kotmale hydropower project alternatives*. Transactions Annual Sessions of the Institution of Engineers Sri Lanka.

Climate Change Mitigation

▶ Sustainability Transition and Climate Change Adaption of Logistics

Climate Change Resilience

▶ Faith Communities as Hubs for Climate Resilience

Climate deviations and surface water resources in Sri Lanka

▶ Climate Change and Surface Water Resources in Sri Lanka

Climate Disaster Preparedness

▶ Faith Communities as Hubs for Climate Resilience

Climate Displaced Persons

▶ Multi-stakeholder Partnerships to Support Climate Migrants in Fragile Cities

Climate Gentrification

An Emerging Phenomenon in Coastal Cities in the Era of Climate Change

Haiyun Wang, Siqin Wang and Yan Liu
School of Earth and Environmental Sciences, The University of Queensland, St Lucia, Australia

Introduction

Global climate change, which has its origins in the nineteenth and twentieth century, poses significant risks to cities around the world. Since many major cities are located in low-lying coastal areas near oceans, natural harbors, and other types of waterbodies, they are highly susceptible to the projected rise in sea levels, increasingly occurring storm surges, flooding, and other associated hazards (Intergovernmental Panel on Climate Change 2019). Such impacts subsequently influence the urban built environment, property markets, the desirability of residential locations, as well as the demographic and socioeconomic profile of neighborhoods, thereby affecting the social justice, fairness, stability, and harmony of neighborhoods (Bryson 2013; Graham et al. 2018; Uribe-Toril et al. 2018).

This type of climate-triggered human resettlement is termed "climate gentrification" (Keenan et al. 2018; Wiggins 2018), which is embedded in the conventional concept of gentrification as the displacement of a lower socioeconomic group by a higher one, resulting in a change of social classes within a neighborhood (Shaw 2008). As a newly emerging concept across the social, geographic, and environmental sciences, climate gentrification addresses a primary concern regarding the way in which humans' residential adaptation within the urban built environment responds to the security of the natural environment, namely, the threat of rising sea levels and other natural hazards along the coasts (Ahsan 2019; Keenan et al. 2018; Pearman 2019; Wiggins 2018). Climate gentrification also leads to the redistribution of social resources and is a matter concerning social justice, equality, and fairness, given that climate-response measures (in the form, for example, of adaptation plans and hazard prevention strategies undertaken by governments) and the discourse surrounding them have stratified outcomes for vulnerable populations (Featherstone 2013; Graham et al. 2018).

With the phenomenon of climate gentrification gained increasing attention from the mainstream media in recent years, there is a pressing need to conceptualize climate gentrification and explore its research paradigm and empirical evidence potentially related to climate gentrification. In order to reveal the nature of the heterogeneous impacts of climate change on human societies, and to identify the vulnerable population and places subject to climate gentrification in different geographic contexts, this chapter aims to achieve three objectives: (1) to conceptualize climate gentrification through a narrative review of existing literature in relation to the various forms of gentrification in its evolutionary history and periodization; (2) to propose a tri-disciplinary research paradigm for further research on climate gentrification; and (3) to compile a global survey of cities and areas with potential sign on the occurrence of climate gentrification. This, then, becomes an opening basis for productive debate around the environmental politics of climate change and shedding new light on how global environmental change affects human settlements and human adaptation behaviors (Swyngedouw 2013).

The Evolving Concept of Gentrification

The term "gentrification" was first introduced in 1964 by British sociologist Ruth Glass to describe the phenomenon of residential displacement whereby lower-income residents relocate – voluntarily or involuntarily – from their current homes to a different neighborhood, resulting in changes in the demographics and socioeconomic status of the neighborhoods (Atkinson et al. 2011; Hackworth and Smith 2001). This displacement further resulted in changes in the racial and ethnic composition of a neighborhood (Lees 2008). This concept of gentrification originates from the demographic and socioeconomic perspective, but over time, it has changed to become subjective to different causes and consequences, resulting in the so-called "wave theory" (Lees et al. 2016). Table 1 summarizes the triggers, gentrifiers, and empirical studies of the five waves that are commonly accepted in the scholarly literature. The fifth wave, concerning the impact of environmental and ecological factors, relates more closely to climate gentrification and is the subject of this chapter.

The first wave of gentrification, known as "sporadic gentrification" (Hackworth and Smith 2001), occurred from the end of the 1960s to 1973 during a period where urban decline and the downturn in property values was the dominant discourse in cities. This enabled middle-class homeowners, small-scale developers, and investors to occupy and renovate the cheaper inner-city suburbs close to their jobs, commercial or services centers – suburbs that were previously occupied by the working classes (Davidson and Lees 2005; Lees 2008). The early studies in this period rested on examining class-based population change and was heavily focused on small neighborhoods in large European and US cities (Freeman and Braconi 2004). This wave of gentrification is commonly considered as a bottom-up and piecemeal process (Davidson and Lees 2005).

The second wave of gentrification, known as "anchoring gentrification" (Hackworth and Smith 2001), occurred in the late 1970s to early 1990s and took place more commonly in urban centers but also expanded to other parts of small- and medium-sized cities (Aalbers 2019; Cheshire et al. 2019). Political initiatives and planning strategies were implemented so as to facilitate the flow of capital into underinvested neighborhoods: residents with lesser means were replaced (Lees 2008) by the middle classes or the so-called "creative class" by attracting the latter to these new socially mixed neighborhoods in centrally located areas of cities (Florida 2002). This wave shifted the definition of gentrification from simply a compositional change of residents in a neighborhood to also consider a change in cultural and commercial spheres (Lees et al. 2016). However, with the stock market crash of 1987 and the start of a

Climate Gentrification, Table 1 Positioning climate gentrification in the wave theory and periodization of gentrification

Period	Wave	Drivers/triggers	Gentrifiers	Gentrified areas	Empirical studies
Prior to 1973	First Wave: sporadic gentrification	Bottom-up and piecemeal; based on investment by individual homeowners and renovators who are attracted by the symbolic value of old working-class areas.	Individual homeowners, small-scale developers, investors, and renovators.	Small and sporadic neighborhoods in inner city and central areas.	Large cities in the USA and Western Europe (e.g., Betancur 2011; De Verteuil 2011; Goetz 2011; Timberlake and Johns-Wolfe 2017)
1973 to early 1990s	Second Wave: anchoring gentrification	Largely bottom-up; driven by private developers through the construction of large residential and consumption spaces.	Developers, creative laborers, and middle-class families.	Inner city suburbs but also expanding to other parts of cities.	Small- to medium-sized cities in the USA and Europe (e.g., Kalantaridis 2010; Kovács et al. 2013; Solana-Solana 2010)
Early 1990s to early 2000s	Third Wave: state-led gentrification	Large-scale capital investment by developers who work with local and central government through state-sponsored urban renewal programs.	The middle-class, with support by local and state/federal government.	Brown-field development from city center to more remote suburbs.	Most cities in both developed and developing countries (e.g., Betancur 2014; Cheshire et al., 2019; Ghertner 2014; Janoschka et al. 2014)
Early 2000s to 2007	Fourth Wave: third-wave plus "financialization of the home"	Lower mortgage rates due to the recession; state support slowed down but with intensified financialization of housing and urban consolidation policies.	The middle-class (led by local and state/nation)	Diverse neighborhoods and more remote suburbs in large parts of cities.	Cities in the USA since 2007 and expanding to European and other developing countries (e.g., Betancur 2014; Dong 2017; Greene and Wang 2010; Janoschka et al. 2014; Lin and Chung 2017; Shin 2016)
After global financial crisis in 2007–2010 and onwards	Fifth Wave: environmental/ecological gentrification, including climate gentrification	Variegated and characterized by austerity urbanism; state continuously plays a leading role but its prominent role is displaced by accelerating globalization, diverse industries, sharing economy, international tourism, and global environmental change	Residents resilient to changes in physical and environmental conditions, or induced industries (e.g., tourism).	Areas with particular types of physical or built environment conditions or encountered by global environmental change.	Urban and rural areas in both developed and developing countries; areas ripe for tourism; coastal cities (e.g., Anguelovski et al. 2018; Atkinson et al. 2011; Huang-Lachmann and Lovett 2016; Liang and Bao 2015; Nelson and Nelson 2011; Solana-Solana 2010)

recession, the process of gentrification paused or slowed down, even prompting the claim, in the early 1990s, of "degentrification" (Lees and Bondi 1995), a reverse process of gentrification where a residential area previously only affordable to affluent people becomes affordable to those who are poor.

The third wave started after the 1987 recession and is known as "post-recession," "government-sponsored," or "state-led" gentrification (Hackworth and Smith 2001). Distinct from the previous waves, gentrification in the third wave was more linked to large-scale capital investment: the restructuring and globalization of the real estate industry attracted corporate developers and local and federal authorities to use their regulatory and financial powers to optimize neighborhood environments and improve the prospect of profit to be made from the real estate market (Aalbers 2019; Cheshire et al. 2019). Gentrified neighborhoods moved further away from the inner city to more remote suburbs, with some case studies broaching the context of developing countries (Bryson 2013; Pearman 2019).

The fourth wave of gentrification, which was arguably the continuation or intensification of the third wave, occurred from the mid-1990s until the global financial crisis in 2007–2008 (Aalbers 2019). This process was temporarily affected by the recession: state-led support indeed slowed down, though this was met with an intensified financialization of the housing market and urban consolidation policies (August and Walks 2018). Studies in this period expanded to the developing world in which the entire neighborhoods were renewed – often with state support – and were physicalized as an upward spatial transformation in many newly industrialized areas (He 2007, 2012).

The fifth wave of gentrification coincided with the breakout of the 2007 global financial crisis in the United States, and sequentially, the European sovereign debt crisis in 2009 and the Chinese stock market bubble in 2015 (Aalbers 2019). Jointly supported by large developers, local, and global investors, the fifth wave of gentrification was targeted primarily at the commercial real estate market, though residential real estate for low- and middle-income households also appeared in both new and old gentrifying areas. This fifth-wave gentrification, according to Aalbers (2019), was classified more from a socio-economic and political perspective. However, an earlier review on gentrification by Bryson (2013) criticized that such wave thinking ignores geographic and environmental heterogeneities and the differences between places. Since gentrification has expanded to cities with different sizes, this gentrification phenomenon becomes more complex in different countries, coupled by the changing economic, demographic, social, cultural, physical, and environmental landscapes (Lees et al. 2016; Pearman 2019).

Thus, rather than following Aalbers' philosophy, we argue that the fifth wave of gentrification is variegated and can be induced by factors such as accelerating globalization, diverse industries, the sharing economy, international tourism, and global environmental change (Forrest 2016). These gentrification-related phenomena have been under-explored and described in the past decade using various new terms such as rural gentrification (López-Morales 2018), new-build gentrification (Gentile et al. 2015), tourism gentrification (González-Pérez 2019), and education-led gentrification (Wu et al. 2016). Among these new terms, environmental or ecological gentrification (Aalbers 2019; Gould and Lewis 2016; Pearsall and Anguelovski 2016; Rice et al. 2019) has become a promising direction for research, with increasing literature in the field of environmental sciences (Uribe-Toril et al. 2018). Indeed, many aspects of neighborhood transformation can be placed under the aegis of environmental gentrification; these are the social outcomes of an environmental rationality. Driven by both the urban built environment and the natural environment, such social transformations are often expressed through the implementation of an environmental plan (Pearsall and Anguelovski 2016; Pearman 2019). Most of the existing literature defines environmental gentrification as the upgrading process of urban neighborhoods via the provision of green space, transit accessibility, outdoor or high-class amenities (e.g., Starbucks) (Haase et al. 2017; Wolch et al. 2014). In this sense, a wide range of new forms of gentrification can be categorized as a synonym or subclass of environmental gentrification, such as high-amenity gentrification (Greene and Wang 2010), transit-induced gentrification (e.g., metro, railway) (Dong 2017; Lin and Chung 2017), tourism and commercial gentrification (Cocola-Gant 2018), new-build gentrification (He 2010), green gentrification (Anguelovski et al. 2018; Gould and Lewis 2016), low-carbon gentrification (Bouzarovski et al. 2018), and climate gentrification (Ahsan 2019; Keenan et al. 2018; Pearman 2019; Wiggins 2018).

The limited body of existing literature defines climate gentrification as a dynamic process whereby the displacement of urban inhabitants is primarily focused on the underlying environmental risks driven by climate change, such as sea level rise, storm surge, desertification and erosion, as well as extreme weather events (Ahsan 2019; Wiggins 2018). This definition is based on a simple proposition that climate change arguably affects property markets and prices, as well as associated infrastructure and the built environment more broadly, which in turn alters residential mobility and human settlements by virtue of their capacity to cope with the threat of climate change (Florida 2018). In this regard, the proposition of climate gentrification is consistent with the conventional framing of gentrification: the price variability associated with the real estate market, rental costs, speculative investment, or superior purchasing act as drivers which lead to the displacement of populations with lesser means and potentially cause social injustice, inequality, and unfairness in the longer term (Graham et al. 2018; Lees et al. 2013).

A Tri-disciplinary Research Paradigm

Although the impact of climate change has drawn worldwide attention, research on climate gentrification is still in its infancy and needs enrichment, both in terms of its theoretical foundation as well as its empirical validation (Florida 2018). A recent empirical study by Keenan et al. (2018) presents a descriptive theory depicting three pathways along which climate change – or more specifically, sea level rise – affects residential mobility, property prices, and ultimately, the transformation of gentrifying neighborhoods. The first pathway, the "superior investment pathway," is a simple proposition that investors start to shift capital to more elevated properties, resulting in gentrified neighborhoods being associated with a higher elevation. The second pathway, the "cost-burden pathway," occurs when climate change causes a rise in the cost of living (i.e., escalating costs of insurance, property taxes, and repairs) so that only wealthier households can afford to stay in place; and so, lower-income households are forced to relocate. The third pathway, the "resilience investment pathway," occurs when the environment is reengineered to be more resilient, and thereby making it more attractive. Although this descriptive theory provides a solid framework to guide the empirical study of climate gentrification in Miami, Florida in the United States, it is primarily economically orientated, thereby limiting the adaptation behaviors to post-hazard conditions and excludes adaptive conditions prior to hazards. The theory is also less testable by empirical studies due to deficits in the analytical design, for instance, by capturing gentrification merely through changes in property prices and neglecting the effects of the urban built environment. Thus, this chapter forges a more comprehensive research paradigm for studies on climate gentrification from a tri-disciplinary perspective within the social, built, and natural environments (Fig. 1). These three environmental systems collectively determine the contextual vulnerability of a community and forms the basis for analyzing a community's coping strategies in the face of hazards (Intergovernmental Panel on Climate Change 2002). When a large-scale hazard poses a threat to humans and their welfare, a community with higher vulnerability has greater exposure to the hazard and is more susceptible to losses, thus, may encounter higher risks. Thus, a consistent assessment across these three environmental perspectives would inform more suitable climate adaptation strategies linking to the ramifications of post-hazard residential resettlement and the formation of climate gentrification.

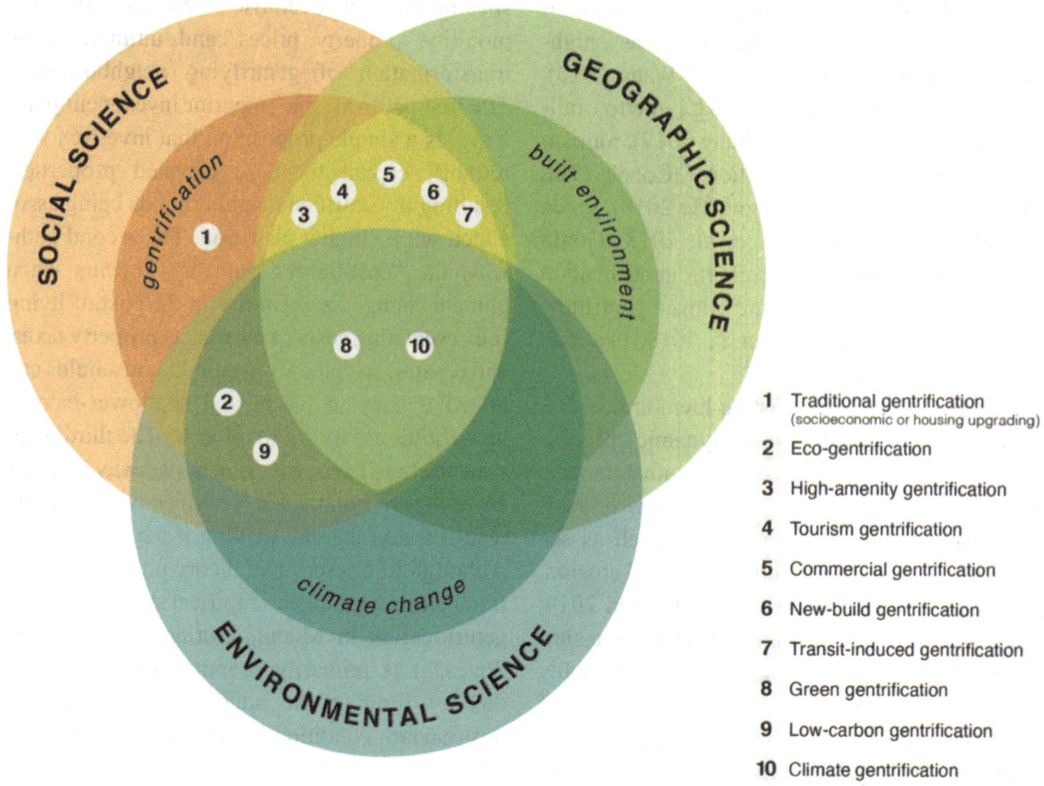

Climate Gentrification, Fig. 1 Climate gentrification as a tri-disciplinary concept integrating social, geographic, and environmental sciences

The social environment refers to a set of characteristics at the neighborhood or community level in terms of demographics, socioeconomic levels, and the political/institutional situation. Earlier studies have found that the gentrification of a neighborhood is subject to significant increases in residents with one or more of the following characteristics: white, young, well-educated, middle- or high-income, having small families, and having professional or management occupations (Freeman et al. 2016; Hwang 2016; Landis 2016; Skaburskis and Nelson 2014; Walks and Maaranen 2008). Thus, age structure, ethnic composition, educational level, personal/household income, family composition, and occupation are considered important indicators from the socioeconomic and demographic perspectives (O'Neill et al. 2014). With regard to political and institutional situations, neighborhoods favored by policies, regulations, and plans for hazard prevention (such as hazard reduction and mitigation plans, and emergency response and services) have stronger resilience to natural disasters, which may trigger the increase of property values and further lead to climate gentrification (Cutter et al. 2008; Eriksen et al. 2015; Warner 2010). Other triggers include state-led investments and large financial capital investments to areas which offer risk-adjusted returns for real estate, infrastructure, engineered facilities and utilities (e.g., open assembly space, recreational parks, elevated embankment, flooding discharge area) which have the aim of preventing damage caused by natural disasters (Cerè et al. 2017; Hollnagel 2014; Keenan et al. 2018).

The built environment refers to the constructed or human-made environment surrounding neighborhoods that may potentially be gentrified (Cerè et al. 2017; Hollnagel 2014), and is closely associated with state-led developments and large-scale financial capital inputs to a geographic area (which offers risk-adjusted returns for investment in real estate and infrastructure) (Keenan et al. 2018). Neighborhoods with better quality housing construction, higher housing values, older but upgraded or renovated buildings, increased rent and home values, a shift in property tenure from rental to ownership, high share of small apartments, and more stories of buildings are more likely to have a stronger resilience to climate-related hazards and are therefore positively associated with the occurrence of gentrification (Immergluck 2009; Kahn 2007; Skaburskis and Nelson 2014). In addition, it has been observed in conventional gentrification studies that neighborhoods closer to the city center tend to be more susceptible to gentrification (Bengtsson and Kopsch 2019), as are neighborhoods in close proximity and easy access to high amenities, transit facilities, educational institutes, or green space (Anguelovski et al. 2018; Dong 2017; Gould and Lewis 2016; Haase et al. 2017; Lin and Chung 2017; Wolch et al. 2014).

The natural environment system refers to hazards or disaster characteristics, including frequency, duration, intensity, magnitude, and type of hazards, all of which vary depending on the geography of the region (Ahsan 2019; Cutter et al. 2003; Wiggins 2018). Coastal cities are more likely to be susceptible to the threat of sea level rise compared to inland cities; so are coastal areas within a city as compared to inland areas (Ahsan 2019; Wiggins 2018). In this regard, climatic and typological factors such as elevation should be considered in measuring climate gentrification in the natural environment perspective. In the Miami study by Keenan et al. (2018), the results show that properties at a higher elevation would have a larger increase in value than those at lower ground, and thus higher-elevated areas are more likely to encounter climate gentrification – though this study only considers the threat of sea level rise. When considering more diverse impacts of climate change, other factors such as land slope of a property, exposure to wind and heat circulation, and proximity to water bodies also need to be considered (Burkart et al. 2015; Forzieri et al. 2016; Mudd et al. 2014; Wolf et al. 2010).

A Global Survey of Potential Case Studies

We conducted a global survey of cities and areas potentially subject to climate gentrification by continent. A complete list of neighborhoods with potential signs of gentrification in coastal cities and/or areas threatened by climate change at a global scale is provided in the Appendix. Overall, we note that existing empirical studies of climate gentrification and human adaptations that consider sea level rise as the primary threat have dominantly focused on the coastal cities distributed in developed countries (Araos et al. 2016; Lees 2012). On the other hand, the process of urban renewal and gentrification in the developing countries is generally described as the accumulation of residential displacement or urban eviction because of the complex process involving not only class-based displacement but also dispossession along demographic and socioeconomic dimensions (Wu 2016); it is also related to a formal and powerful nexus of politicians, planners, and developers are profiting from land use policies and planning controls (Ghertner 2015; Harris 2008; Lees et al. 2016). Thus, the same magnitude in sea level rise may have heterogeneous impacts on human societies in different geographic contexts, as well as differing responses from population groups and communities that have various capacities of adaptation, vulnerability, and resilience (Arnall 2014; Arnall and Kothari 2015; Kelman et al. 2017).

Northern America

In Northern America, coastal cities in the United States and Canada have been the main point of discussion: studies concerning climate change adaption and environmental change have focused on coastal cities such as Miami, Houston, Oakland, Tampa, New York City, Seattle, Los Angeles, New Orleans, Toronto, and Vancouver (Anguelovski et al. 2017; Bouzarovski et al. 2018; Florida 2018; Kashem et al. 2016; Rice et al.

2019; Shkuda 2016), where an ongoing process of gentrification is still being observed. However, there is no empirically tested interplay between gentrification and climate change to prove the existence of climate gentrification – except for the Miami case study (Keenan et al. 2018). Beyond coastal and beachfront communities, and apart from sea level rise as the only threat causing climate gentrification, it has been reported that an influx of mostly wealthier people has tried to escape from Phoenix to Flagstaff (a cooler city in the north of Arizona) due to the extreme heatwaves and potential smoke hazards from wildfires in deserts around Phoenix (Yip et al. 2008). Similarly, a pressing concern has been expressed in the eastern and southern coasts of the United States, which have been pummeled by ever stronger hurricanes (Mudd et al. 2014). For example, New Orleans has observed increasing number of gentrified neighborhoods after the Hurricane Katrina in 2005 in St. Bernard and Plaquemines parishes, and Mid-City (Van Holm and Wyczalkowski 2018). Although the terms "heat-induced," "cyclone-induced," and "hurricane-induced gentrification" have not yet officially being defined in any academic literature, the phenomenon itself is generating significant public awareness and is predicted to precipitate crises of poverty, fresh water supplies, shelter, food, and energy; this could further lead to greater socioeconomic disparity and inequality, and imbalance of social structures in neighborhoods, and thereby potentially leading to climate gentrification.

Europe

European and British cities such as Copenhagen, Venice, Rotterdam, Amsterdam, Stockholm, Hamburg, and London have also attracted the attention of public media for climate gentrification-related reasons (Anguelovski et al. 2017; Florida 2018; Hedin et al. 2012; Huang-Lachmann and Lovett 2016; Uitermark and Bosker 2014; Wiggins 2018). Among them, Copenhagen and Venice have been empirically studied by Keenan et al. (2018), who report the occurrence of climate gentrification caused by sea level rise, displacing the population from St Kjeld to Norrebro (mainland Copenhagen) in the case of Denmark, and from Venice to Mestre (mainland Veneto) in the case of Italy. In addition, Rotterdam in The Netherland and Hamburg in Germany have been discussed by Huang-Lachmann and Lovett (2016) as having a high potential to experience gentrification caused by carbon dioxide emissions and flooding. London, the original city where the concept of gentrification was formulated, is mentioned by Florida (2018) as one of the top 20 risk cities in terms of total projected economic loss due to exposure to water bodies, and is thus potentially highly vulnerable to climate gentrification.

Oceania

In Oceania, larger cities are usually located along the coast, and the threat of sea level rise has long been discussed with regards to cities such as Sydney, Melbourne, Brisbane, Adelaide in Australia, and Auckland in New Zealand (Araos et al. 2016; Atkinson et al. 2011; Darchen and Ladouceur 2013; Murphy 2008; Wiesel et al. 2013). According to Araos et al. (2016), most Oceanian cities have implemented moderate to extensive adaptation policies to cope with various impacts of climate change. Meanwhile, these cities have also experienced the process of gentrification, especially for neighborhoods in and around the inner city and along the coasts: Arncliffe in Sydney, St Kilda in Melbourne, and Wai O Taiki Bay in Auckland (Atkinson et al. 2011; Murphy 2008; Wiesel et al. 2013). Changes in social structures and population displacement in these areas are considered to be associated with economic uplifting and an upgrading of the built environment, which may be articulated to relate to sea level rise, as a future case study for climate gentrification.

Asia

In Asia, population-dense megacities such as Shanghai, Ho Chi Minh City, Mumbai, Hong Kong, Tokyo, and Singapore are also potential high-risk cities that may be vulnerable to climate gentrification, especially in neighborhoods close to water bodies, as indicated by Florida (2018). In addition, these cities are seen to be vulnerable to the heat-island effect (Phi 2007; Thomas and Warner 2019; Tran et al. 2006), which is also closely related to climate gentrification. Bangladesh is another typical example of climate-affected countries as reported widely in the media, though population resettlement in Bangladesh usually occurs as large-scale mass migration from the Delta region to Dhaka and other inland cities; this type of migration is different from the short-distance mobility across neighborhoods mostly featured in climate gentrification (Ahsan 2019; Black et al. 2011; Reuveny 2007). Such mass migration would affect the change of social structures and the development of new residential areas in Dhaka, forcing the influx of low-income residents toward urban slums and congested inner city areas, for easier access to jobs, or toward urban fringes with lower rents (Parvin and Shaw 2011). Not only in Bangladesh, urban slums in Mumbai have become the first destination for such displaced populations who may face limited access to employment, shelter, water, sanitation, and other basic amenities (Harris 2008; Mukhija 2017). Thus, these Asian cities are set up as good case studies for climate gentrification beyond the American and European contexts.

Latin America

A number of coastal megacities in Latin America, such as Buenos Aires in Argentina, Santiago and Lima in Chile, Rio de Janeiro and Sao Paulo in Brazil, have also been discussed in terms of gentrification and climate change (Betancur 2014; Janoschka and Sequera 2016; Reguero et al. 2015), though the forms of gentrification in these cities may vary from one another. For example, neoliberal gentrification triggered by creative classes is claimed to be occurring intensively in Buenos Aires (Herzer et al. 2015); gentrification in Santiago takes place in urban peripheral developments (López-Morales 2016); while gentrification driven by large-scale, high-rise urban renewal is observed in central areas of Sao Paulo (Sotto 2018). Nevertheless, these cities share a common experience in "urban renaissance" based on new public transit networks, restoration of historic districts, and upgrading programs involving local residents (Betancur 2014; Janoschka and Sequera 2016; Lees et al. 2015). Meanwhile, these megacities are also at risk of water scarcity, intensifying urban flooding, and rising urban heat-island effects, all of which bring vulnerability to socioeconomically disadvantaged neighborhoods threatened by the consequences of such changes in climate (Reguero et al. 2015). The intracity migration may affect the spatial inequality of different social groups, further triggering CG across urban space.

Africa

In contrast to the aforementioned cases in developed countries and megacities in Asia, all of which have been relatively large economies, only two coastal major cities in South Africa – Cape Town and Durban – have been discussed in relation to climate change and gentrification (Bond and Browder 2019; Mösgen et al. 2018). Araos et al. (2016) claims that these cities, which have limited financial resources in low- and middle-income countries (as defined by the World Bank), have indeed implemented moderate adaption policies to cope with climate change. Public media has reports of gentrification occurring in neighborhoods close to the coastline, such as Woodstock and areas around Signal Hill in Cape Town, and Rivertown and Point Precinct in Durban. The gentrifying process here not only results in an influx of moderate- or high-level income White people, but also exacerbates the gap between these neighborhoods and the surrounding urban slums, where most local Black people are concentrated (Mösgen et al. 2018). Herein, the key issues relating to climate gentrification are about how the needs and interests of local Africans are identified and accommodated, and how measures are put in place to ensure such efforts will reduce their vulnerability and not exacerbate existing inequalities or create new inequalities (Holland 2017). Thus, future exploration of climate gentrification in these and other coastal cities of Africa would bring new insights to test and validate the existence and extent of climate gentrification in this context.

Conclusion

This chapter conceptualizes climate gentrification through a review of the evolution and periodization of gentrification and propose a tri-disciplinary paradigm for research on climate gentrification from the perspective of three "environmental" disciplines – the social environment, the built environment, and the natural environment – so as to extend the possibilities of future empirical studies on climate gentrification. It is followed by a systematic scan of the literature and the mainstream media at a global scale to investigate various signs of gentrification relating to climate change and human adaptive responses. A new paradigm for research on climate gentrification is just commencing and will burgeon in response to the increasingly critical issues of climate change globally. In this regard, we encourage scholars in the related fields of sociology, psychology, politics, economy, urban planning, human and physical geography, as well as in the spatial and environmental sciences to contribute empirical studies and experiences that enrich the theorization of climate gentrification across a broader interdisciplinary scope, with the aim of exploring how global environmental changes will affect human settlement and human adaptation behaviors.

Appendix: (See Table 2)

Climate Gentrification, Table 2 A global scan of cities that are (potentially) being affected by climate gentrification, as reported in the literature and pubic media

Country	City	(Potentially) climate gentrification-affected Areas	Reference
North America			
USA	Houston	Harris County	Kashem et al. 2016
	Miami*	Miami Beach, Little Haiti (mainland Miami)	Keenan et al. 2018; Wiggins 2018
	Oakland	Tassafaronga Village	Alkon et al. 2019; Ramírez 2019
	Tampa	Hillsborough County	Kashem et al. 2016
	New York City	Lower Manhattan, Brooklyn, Queens and Bronx	Araos et al. 2016; Florida 2018; Gould and Lewis 2016; Shkuda 2016; Rice et al. 2019
	Seattle*	Capitol Hill, First Hill, South Lake Union, Central District	Rice et al. 2019
	Los Angeles*	Jefferson Park and West Adams, City Terrace, Cypress Park, Inglewood, Wrigley	Rice et al. 2019; Scott 2019
	New Orleans*	St. Bernard and Plaquemines parishes; Mid-City	Anguelovski et al. 2017; Kashem et al. 2016; Van Holm and Wyczalkowski 2018
Canada	Vancouver*	Chinatown, Grandview Woodlands and Northeast False Creek, Downtown Eastside, Little Mountain,	Anguelovski et al. 2018; Bouzarovski et al. 2018; Ley and Dobson 2008; Mösgen et al. 2018; Quastel et al. 2012
	Toronto*	Parkdale, High Park, Cabbagetown, Clinton, South Riverdale, Leslieville, Trinity Bellwoods, the Junction, Dovercourt-Wallace Emerson, Little Portugal North, James Town, Danforth, Annex	Araos et al. 2016; August and Walks 2018; Keatinge and Martin 2016; Rice et al. 2019
Europe			
Denmark	Copenhagen*	St Kjeld to Norrebro	Baron and Petersen 2016; Keenan et al. 2018

(continued)

Climate Gentrification, Table 2 (continued)

Country	City	(Potentially) climate gentrification-affected Areas	Reference
Italy	Venice*	Venice, Mestre (mainland Venice)	Carbognin et al. 2010; Keenan et al. 2018
Netherland	Amsterdam	De Jordaan	Van Gent 2013; Uitermark and Bosker 2014
	Rotterdam*	Oude Noorden, Nieuwe Westen, Middelland, het Liskwatier, Nieuw Crooswijk, KralingenWest, Lloydkwatier, Katendrecht, the Kop van Zuid-Entrepot	Huang-Lachmann and Lovett 2016
Germany	Hamburg*	Hafen City	Huang-Lachmann and Lovett 2016
Sweden	Stockholm	Bromma, Danderyd, Lidingö, Nacka, Saltsjöbaden Norrmalm, Östermalm Långedrag, Askim, Hovås, Gothenburg	Franzén 2005; Hedin et al. 2012
Spain	Barcelona	Sant Antoni, Poblenou, El Raval	Anguelovski et al. 2017
UK	London	Outer London	Araos et al. 2016; Florida 2018; Harris 2008; Paccoud and Mace 2018
	Lyng	East Lyng	Wiggins 2018
Asia			
China	Shanghai*	Downtown, eastern coast areas	Florida 2018; Weinstein and Ren 2009
	Hong Kong	Repulse Bay, Tuen Mun, Lantau Island, Tai O, Cheung Chau, Discovery Bay, Peng Chau, Tuen Mun, Sham Tseng, SaiKung, Lei Yue Mun, Northern New Territory	Florida 2018; La Grange and Pretorius 2016
Bangladesh	Khulna, Dhaka	Khulna	Ahsan 2019; Black et al. 2011; Reuveny 2007; Wiggins 2018
Vietnam	Ho Chi Minh City*	Thu Duc, Phu My Hung	Florida 2018; Huynh 2015; Phi 2007
India	Mumbai	Dharavi, Mariman Point, Port, New Bombay, Prabhadevi	Chatterjee and Parthasarathy 2018; Florida 2018; Harris 2008; Mukhija 2017; Saryal 2018; Weinstein and Ren 2009
Japan	Tokyo	Shibuya, Central Tokyo, Chuo Ward	Florida 2018; Lützeler 2008; Radovic 2010
Singapore	Singapore	Bukit Timah, Orchard Road and CBD (Mackenzie Road), Central and North Singapore	Chang 2016; Wong 2006; Yuen 2008
South America			
Caribbean	Barbuda	Codrington	Gould and Lewis 2016
Argentina	Buenos Aires	Palermo, Puerto Madero, La Boca, San Telmo, Barracas	Betancur 2014; Herzer et al. 2015; Janoschka and Sequera 2016; Lees et al. 2015; Reguero et al. 2015
Chile	Santiago	Inner comunas, Santiago central, Santa Isabel	Anguelovski et al. 2016; Betancur 2014; Janoschka and Sequera 2016; Lees et al. 2015; López-Morales 2016; Reguero et al. 2015
	Lima	Historic centre, Barranco, Miraflores, Parque de la Reserva	Betancur 2014; Reguero et al. 2015

(continued)

C

Climate Gentrification, Table 2 (continued)

Country	City	(Potentially) climate gentrification-affected Areas	Reference
Brazil	Rio de Janeiro	Santa Marta, Vidigal, Charles Heck, Babilônia Hill, Leme beach, Babilônia	Betancur 2014; Janoschka and Sequera 2016; Reguero et al. 2015
	Sao Paulo	Downtown, Luz, Santa Efigênia	Betancur 2014; Leite 2015; Reguero et al. 2015; Sotto 2018
Oceania			
Australia	Adelaide	Thebarton, West Croydon and Hindmarsh	Araos et al. 2016
	Brisbane	Annerley, Lutwyche and Woolloongabba, Fortitude Valley	Darchen and Ladouceur 2013
	Sydney	Arncliffe, St. Peters and Tempe	Atkinson et al. 2011; Wiesel et al. 2013
	Melbourne	St Kilda, Carlton Fitzroy, Richmond, Braybrook, Footscray, West Footscray, Yarraville, Sunshine, Preston, Reservior, Thomastown, Lalor, South Morang	Atkinson et al. 2011
New Zealand	Auckland	West Auckland, Glen Eden, Wai o Taiki Bay, Onehunga, Ponsonby, Glen Innes	Murphy 2008
Africa			
South Africa	Durban	Rivertown, Glenwood, Florida, Point Precinct	Araos et al. 2016; Bond and Browder 2019; Mösgen et al. 2018
	Cape Town	Woodstock, Inner City, Bo-Kaap, Signal Hill	Araos et al. 2016; Mösgen et al. 2018; Teppo and Millstein 2015
Cote d'lvoire	Abidjan	Ivory Coast	Wiggins 2018

Note: Cities labelled with an asterisk have been discussed in the current climate gentrification-related studies. Non-asterisked cities were frequently studied in the broader research on gentrification or climate change adaptation, which also attracted significant attention in public media and raised concerns from the general public

References

Aalbers, M. B. (2019). Introduction to the forum: from third to fifth-wave gentrification. *Tijdschrift voor Economische en Sociale Geografie, 110*(1), 1–11.

Ahsan, R. (2019). Climate-induced migration: Impacts on social structures and justice in Bangladesh. *South Asia Research, 39*(2), 184–201.

Alkon, A. H., Cadji, Y. J., & Moore, F. (2019). Subverting the new narrative: Food, gentrification and resistance in Oakland, California. *Agriculture and Human Values, 36*(4), 793–804.

Anguelovski, I., Brand, A. L., Chu, E., & Goh, K. (2017). Urban planning, community (re) development and environmental gentrification: Emerging challenges for green and equitable neighbourhoods. In R. Holifield, J. Chakraborty, & G. Walker (Eds.), *The Routledge handbook of environmental justice* (pp. 449–462). Routledge: Oxfordshire.

Anguelovski, I., Connolly, J. J., Masip, L., & Pearsall, H. (2018). Assessing green gentrification in historically disenfranchised neighborhoods: A longitudinal and spatial analysis of Barcelona. *Urban Geography, 39* (3), 458–491.

Anguelovski, I., Shi, L., Chu, E., Gallagher, D., Goh, K., Lamb, Z., Reeve, K., & Teicher, H. (2016). Equity impacts of urban land use planning for climate adaptation: Critical perspectives from the global north and south. *Journal of Planning Education and Research, 36*(3), 333–348.

Araos, M., Berrang-Ford, L., Ford, J. D., Austin, S. E., Biesbroek, R., & Lesnikowski, A. (2016). Climate change adaptation planning in large cities: A systematic global assessment. *Environmental Science and Policy, 66*, 375–382.

Arnall, A. (2014). A climate of control: Flooding, displacement and planned resettlement in the Lower Zambezi River valley, Mozambique. *The Geographical Journal, 180*(2), 141–150.

Arnall, & Kothari, U. (2015). Challenging climate change and migration discourse: Different understandings of timescale and temporality in the Maldives. *Global Environmental Change, 31*, 199–206.

Atkinson, R., Wulff, M., Reynolds, M., & Spinney, A. (2011). Gentrification and displacement: The household impacts of neighbourhood change. *AHURI Final Report, 160*, 1–89.

August, M., & Walks, A. (2018). Gentrification, suburban decline, and the financialization of multi-family rental housing: The case of Toronto. *Geoforum, 89*, 124–136.

Baron, N., & Petersen, L. K. (2016). Understanding controversies in urban climate change adaptation. A case study of the role of homeowners in the process of climate change adaptation in Copenhagen. *Nordic Journal of Science and Technology Studies, 3*(2), 4–13.

Bengtsson, I., & Kopsch, F. (2019). Indicators of candidates for gentrification: A spatial framework. *International Journal of Housing Markets and Analysis, 12*(4), 736–745.

Betancur, J. J. (2011). Gentrification and community fabric in Chicago. *Urban Studies, 48*(2), 383–406.

Betancur, J. J. (2014). Gentrification in Latin America: Overview and critical analysis. *Urban Studies Research, 2014*, 1–14.

Black, R., Bennett, S. R., Thomas, S. M., & Beddington, J. R. (2011). Climate change: Migration as adaptation. *Nature, 478*(7370), 447–449.

Bond, P., & Browder, L. (2019). Deracialized Nostalgia, reracialized community, and truncated gentrification: Capital and cultural flows in Richmond, Virginia and Durban, South Africa. *Journal of Cultural Geography, 36*(2), 211–245.

Bouzarovski, S., Frankowski, J., & Tirado Herrero, S. (2018). Low-carbon gentrification: When climate change encounters residential displacement. *International Journal of Urban and Regional Research, 42* (5), 845–863.

Bryson, J. (2013). The nature of gentrification. *Geography Compass, 7*(8), 578–587.

Burkart, K., Meier, F., Schneider, A., Breitner, S., Canário, P., Alcoforado, M. J., Scherer, D., & Endlicher, W. (2015). Modification of heat-related mortality in an elderly urban population by vegetation (urban green) and proximity to water (urban blue): Evidence from Lisbon, Portugal. *Environmental Health Perspectives, 124*(7), 927–934.

Carbognin, Teatini, P., Tomasin, A., & Tosi, L. (2010). Global change and relative sea level rise at Venice: What impact in term of flooding. *Climate Dynamics, 35*(6), 1039–1047.

Cerè, G., Rezgui, Y., & Zhao, W. (2017). Critical review of existing built environment resilience frameworks: Directions for future research. *International Journal of Disaster Risk Reduction, 25*, 173–189.

Chang, T. C. (2016). 'New uses need old buildings': Gentrification aesthetics and the arts in Singapore. *Urban Studies, 53*(3), 524–539.

Chatterjee, D., & Parthasarathy, D. (2018). Gentrification and rising urban aspirations in the inner city: Redefining urbanism in Mumbai. In J. Mukherjee (Ed.), *Sustainable urbanization in India* (pp. 239–255). Springer: Singapore.

Cheshire, L., Fitzgerald, R., & Liu, Y. (2019). Neighbourhood change and neighbour complaints: How gentrification and densification influence the prevalence of problems between neighbours. *Urban Studies, 56*(6), 1093–1112.

Cocola-Gant, A. (2018). Tourism gentrification. In L. Lees & M. Phillips (Eds.), *Handbook of gentrification studies*. Edward Elgar Publishing.

Cutter, S. L., Barnes, L., Berry, M., Burton, C., Evans, E., Tate, E., & Webb, J. (2008). A place-based model for understanding community resilience to natural disasters. *Global Environmental Change, 18*(4), 598–606.

Cutter, S. L., Boruff, B. J., & Shirley, W. L. (2003). Social vulnerability to environmental hazards. *Social Science Quarterly, 84*(2), 242–261.

Darchen, S., & Ladouceur, E. (2013). Social sustainability in urban regeneration practice: A case study of the Fortitude Valley Renewal Plan in Brisbane. *Australian Planner, 50*(4), 340–350.

Davidson, M., & Lees, L. (2005). New-build 'gentrification' and London's riverside renaissance. *Environment and Planning A, 37*(7), 1165–1190.

De Verteuil, G. (2011). Evidence of gentrification-induced displacement among social services in London and Los Angeles. *Urban Studies, 48*(8), 1563–1580.

Dong, H. (2017). Rail-transit-induced gentrification and the affordability paradox of TOD. *Journal of Transport Geography, 63*, 1–10.

Eriksen, S. H., Nightingale, A. J., & Eakin, H. (2015). Reframing adaptation: The political nature of climate change adaptation. *Global Environmental Change, 35*, 523–533.

Featherstone, D. J. (2013). The contested politics of climate change and the crisis of neo-liberalism. *ACME: An International E-Journal for Critical Geographies, 12*(1), 44–64.

Florida, R. (2002). *The rise of the creative class.* New York: Basic Books.

Florida, R. (2018). Climate gentrification will deepen urban inequality. City Lab. https://www.citylab.com/equity/2018/07/the-reality-of-climate-gentrification/564152/

Forrest, R. (2016). Commentary: Variegated gentrification? *Urban Studies, 53*(3), 609–614.

Forzieri, G., Feyen, L., Russo, S., Vousdoukas, M., Alfieri, L., Outten, S., Migliavacca, M., Bianchi, A., Rojas, R., & Cid, A. (2016). Multi-hazard assessment in Europe under climate change. *Climatic Change, 137*(1–2), 105–119.

Franzén, M. (2005). New social movements and gentrification in Hamburg and Stockholm: A comparative study. *Journal of Housing and the Built Environment, 20*(1), 51–77.

Freeman, L., & Braconi, F. (2004). Gentrification and displacement New York City in the 1990s. *Journal of the American Planning Association, 70*(1), 39–52.

Freeman, L., Cassola, A., & Cai, T. (2016). Displacement and gentrification in England and Wales: A quasi-experimental approach. *Urban Studies, 53*(13), 2797–2814.

Gentile, M., Salukvadze, J., & Gogishvili, D. (2015). New build gentrification, teleurbanization and urban growth: Placing the cities of the post-Communist South in the gentrification debate. *Geografie, 120*(2), 134–163.

Ghertner, D. A. (2014). India's urban revolution: geographies of displacement beyond gentrification. *Environment and Planning A, 46*(7), 1554–1571.

Ghertner, D. A. (2015). Why gentrification theory fails in 'much of the world'. *City, 19*(4), 552–563.

Goetz, E. (2011). Gentrification in black and white: The racial impact of public housing demolition in American cities. *Urban Studies, 48*(8), 1581–1604.

González-Pérez, J. M. (2019). The dispute over tourist cities. Tourism gentrification in the historic Centre of Palma (Majorca, Spain). *Tourism Geographies, 22*(1), 171–191.

Gould, K. A., & Lewis, T. L. (2016). *Green gentrification: Urban sustainability and the struggle for environmental justice*. Oxfordshire: Routledge.

Graham, S., Barnett, J., Mortreux, C., Hurlimann, A., & Fincher, R. (2018). Local values and fairness in climate change adaptation: insights from marginal rural Australian communities. *World Development, 108*, 332–343.

Greene, R. P., & Wang, S. (2010). *Identifying high amenity zones in the US and China with advanced GIS techniques.* 2010 18th International Conference on Geoinformatics. Beijing, China. https://doi.org/10.1109/GEOINFORMATICS.2010.5567587.

Haase, D., Kabisch, S., Haase, A., Andersson, E., Banzhaf, E., Baró, F., Brenck, M., Fischer, L. K., Frantzeskaki, N., Kabisch, N., Krellenberg, K., Kremer, P., Kronenberg, J., Larondelle, N., Mathey, J., Pauleit, S., Ring, I., Rink, D., Schwarz, N., & Wolff, M. (2017). Greening cities –To be socially inclusive? About the alleged paradox of society and ecology in cities. *Habitat International, 64*, 41–48.

Hackworth, J., & Smith, N. (2001). The changing state of gentrification. *Tijdschrift voor Economische en Sociale Geografie, 92*(4), 464–477.

Harris, A. (2008). From London to Mumbai and back again: Gentrification and public policy in comparative perspective. *Urban Studies, 45*(12), 2407–2428.

He, S. (2007). State-sponsored gentrification under market transition the case of Shanghai. *Urban Affairs Review, 43*(2), 171–198.

He, S. (2010). New-build gentrification in central Shanghai: Demographic changes and socioeconomic implications. *Population, Space and Place, 16*(5), 345–361.

He, S. (2012). Two waves of gentrification and emerging rights issues in Guangzhou, China. *Environment and Planning A, 44*(12), 2817–2833.

Hedin, K., Clark, E., Lundholm, E., & Malmberg, G. (2012). Neoliberalization of housing in Sweden: Gentrification, filtering, and social polarization. *Annals of the Association of American Geographers, 102*(2), 443–463.

Herzer, H., Di Virgilio, M. M., & Rodriguez, C. (2015). Gentrification in Buenos Aires: Global trends and local features. In L. Lees, H.B. Shin & E. López-Morales (Eds.), *Global gentrifications: Uneven development and displacement* (pp. 199–222). Bristol, UK; Chicago, IL: Bristol University Press. https://doi.org/10.2307/j.ctt1t894bt.

Holland, B. (2017). Procedural justice in local climate adaptation: political capabilities and transformational change. *Environmental Politics, 26*(3), 391–412.

Hollnagel, E. (2014). Resilience engineering and the built environment. *Building Research and Information, 42*(2), 221–228.

Huang-Lachmann, J. T., & Lovett, J. C. (2016). How cities prepare for climate change: Comparing Hamburg and Rotterdam. *Cities, 54*, 36–44.

Huynh, D. (2015). Phu my hung new urban development in Ho Chi Minh City: Only a partial success of a broader landscape. *International Journal of Sustainable Built Environment, 4*(1), 125–135.

Hwang, J. (2016). The social construction of a gentrifying neighborhood: Reifying and redefining identity and boundaries in inequality. *Urban Affairs Review, 52*(1), 98–128.

Immergluck, D. (2009). Large redevelopment initiatives, housing values and gentrification: The case of the Atlanta Beltline. *Urban Studies, 46*(8), 1723–1745.

Intergovernmental Panel on Climate Change (2002). *Vulnerability assessment for climate adaptation*. https://www.ipcc.ch/apps/njlite/ar5wg2/njlite_download2.php?id=10996

Intergovernmental Panel on Climate Change (2019). *Global warming of 1.5°C.* https://www.ipcc.ch/sr15/

Janoschka, M., & Sequera, J. (2016). Gentrification in Latin America: Addressing the politics and geographies of displacement. *Urban Geography, 37*(8), 1175–1194.

Janoschka, M., Sequera, J., & Salinas, L. (2014). Gentrification in Spain and Latin America – A critical dialogue. *International Journal of Urban and Regional Research, 38*(4), 1234–1265.

Kahn, M. E. (2007). Gentrification trends in new transit-oriented communities: evidence from 14 cities that expanded and built rail transit systems. *Real Estate Economics, 35*(2), 155–182.

Kalantaridis, C. (2010). In-migration, entrepreneurship and rural–urban interdependencies: The case of East Cleveland, North East England. *Journal of Rural Studies, 26*(4), 418–427.

Kashem, S. B., Wilson, B., & Van Zandt, S. (2016). Planning for climate adaptation: Evaluating the changing patterns of social vulnerability and adaptation challenges in three coastal cities. *Journal of Planning Education and Research, 36*(3), 304–318.

Keatinge, B., & Martin, D. G. (2016). A 'Bedford Falls' kind of place: Neighbourhood branding and commercial revitalisation in processes of gentrification in Toronto, Ontario. *Urban Studies, 53*(5), 867–883.

Keenan, J. M., Hill, T., & Gumber, A. (2018). Climate gentrification: From theory to empiricism in Miami-Dade County, Florida. *Environmental Research Letters, 13*(5), 054001.

Kelman, I., Upadhyay, H., Simonelli, A. C., Arnall, A., Mohan, D., Lingaraj, G. J., Shadananan, N., &

Webersik, C. (2017). Here and now: Perceptions of Indian Ocean islanders on the climate change and migration nexus. *Geografiska Annaler: Series B, Human Geography, 99*(3), 284–303.

Kovács, Z., Wiessner, R., & Zischner, R. (2013). Urban renewal in the inner city of Budapest: Gentrification from a post-socialist perspective. *Urban Studies, 50*(1), 22–38.

La Grange, A., & Pretorius, F. (2016). State-led gentrification in Hong Kong. *Urban Studies, 53*(3), 506–523.

Landis, J. D. (2016). Tracking and explaining neighborhood socioeconomic change in US metropolitan areas between 1990 and 2010. *Housing Policy Debate, 26*(1), 2–52.

Lees, L. (2008). Gentrification and social mixing: Towards an inclusive urban renaissance? *Urban Studies, 45*(12), 2449–2470.

Lees, L. (2012). The geography of gentrification: Thinking through comparative urbanism. *Progress in Human Geography, 36*(2), 155–171.

Lees, L., & Bondi, L. (1995). De-gentrification and economic recession: The case of New York City. *Urban Geography, 16*(3), 234–253.

Lees, L., Shin, H. B., & López-Morales, E. (2016). *Planetary gentrification*. Cambridge, UK: John Wiley and Sons.

Lees, L., Shin, H. B., & Morales, E. L. (Eds.). (2015). *Global gentrifications: Uneven development and displacement. Bristol, UK.* Chicago, IL: Bristol University Press. https://doi.org/10.2307/j.ctt1t894bt.

Lees, L., Slater, T., & Wyly, E. (2013). *Gentrification*. New York: Routledge.

Leite, R. P. (2015). Cities and gentrification in contemporary Brazil. *Current Urban Studies, 3*(03), 175–186.

Ley, D., & Dobson, C. (2008). Are there limits to gentrification? The contexts of impeded gentrification in Vancouver. *Urban Studies, 45*(12), 2471–2498.

Liang, Z. X., & Bao, J. G. (2015). Tourism gentrification in Shenzhen, China: Causes and socio-spatial consequences. *Tourism Geographies, 17*(3), 461–481.

Lin, J. J., & Chung, J. C. (2017). Metro-induced gentrification: A 17-year experience in Taipei. *Cities, 67*, 53–62.

López-Morales, E. (2016). Gentrification in Santiago, Chile: A property-led process of dispossession and exclusion. *Urban Geography, 37*(8), 1109–1131.

López-Morales, E. (2018). A rural gentrification theory debate for the Global South? *Dialogues in Human Geography, 8*(1), 47–50.

Lützeler, R. (2008). Population increase and "New-Build Gentrification" in Central Tōkyō. *Erdkunde, 62*(4), 287–299.

Mösgen, A., Rosol, M., & Schipper, S. (2018). State-led gentrification in previously 'un-gentrifiable'areas: Examples from Vancouver/Canada and Frankfurt/Germany. *European Urban and Regional Studies, 26*(4), 419–433.

Mudd, L., Wang, Y., Letchford, C., & Rosowsky, D. (2014). Assessing climate change impact on the US East Coast hurricane hazard: Temperature, frequency, and track. *Natural Hazards Review, 15*(3), 04014001. https://doi.org/10.1061/(ASCE)NH.1527-6996.0000128.

Mukhija, V. (2017). *Squatters as developers? Slum redevelopment in Mumbai*. Routledge: Oxfordshire.

Murphy, L. (2008). Third-wave gentrification in New Zealand: The case of Auckland. *Urban Studies, 45*(12), 2521–2540.

Nelson, L., & Nelson, P. B. (2011). The global rural: Gentrification and linked migration in the rural USA. *Progress in Human Geography, 35*(4), 441–459.

O'Neill, B. C., Kriegler, E., Riahi, K., Ebi, K. L., Hallegatte, S., Carter, T. R., Mathur, R., & van Vuuren, D. P. (2014). A new scenario framework for climate change research: The concept of shared socioeconomic pathways. *Climatic Change, 122*(3), 387–400.

Paccoud, A., & Mace, A. (2018). Tenure change in London's suburbs: Spreading gentrification or suburban upscaling? *Urban Studies, 55*(6), 1313–1328.

Parvin, G. A., & Shaw, R. (2011). Climate disaster resilience of Dhaka city corporation: An empirical assessment at zone level. *Risk, Hazards and Crisis in Public Policy, 2*(2), 1–30.

Pearman, F. A. (2019). Gentrification and academic achievement: A review of recent research. *Review of Educational Research, 89*(1), 125–165.

Pearsall, H., & Anguelovski, I. (2016). Contesting and resisting environmental gentrification: Responses to new paradoxes and challenges for urban environmental justice. *Sociological Research Online, 21* (3), 1–7.

Phi, H. L. (2007). Climate change and urban flooding in Ho Chi Minh City. In C. Jayasena (Eds.), *Proceedings of third international conference on climate and water* (pp. 194–199).

Quastel, N., Moos, M., & Lynch, N. (2012). Sustainability-as-density and the return of the social: The case of Vancouver, British Columbia. *Urban Geography, 33* (7), 1055–1084.

Radovic, D. (2010). The roles of gentrification in creation of diverse urbanities of Tokyo. *Open House International, 35*(4), 20–28.

Ramírez, M. M. (2019). City as borderland: Gentrification and the policing of Black and Latinx geographies in Oakland. *Environment and Planning D: Society and Space, 38*(1), 147–166. https://doi.org/10.1177/0263775819843924.

Reguero, B. G., Losada, I. J., Diaz-Simal, P., Mendez, F. J., & Beck, M. W. (2015). Effects of climate change on exposure to coastal flooding in Latin America and the Caribbean. *PLoS One, 10*(7), e0133409.

Reuveny, R. (2007). Climate change-induced migration and violent conflict. *Political Geography, 26*(6), 656–673.

Rice, J. L., Cohen, D. A., Long, J., & Jurjevich, J. R. (2019). Contradictions of the climate-friendly city: New perspectives on eco-gentrification and housing justice. *International Journal of Urban and Regional*

Research, 44(1), 145–165. https://doi.org/10.1111/1468-2427.12740.
Saryal, R. (2018). Climate change policy of India: Modifying the environment. *South Asia Research, 38*(1), 1–19.
Scott, A. J. (2019). Residential adjustment and gentrification in Los Angeles, 2000–2015: theoretical arguments and empirical evidence. *Urban Geography, 40*(4), 506–528.
Shaw, K. (2008). Gentrification: What it is, why it is, and what can be done about it. *Geography Compass, 2*(5), 1697–1728.
Shin, H. B. (2016). Economic transition and speculative urbanisation in China: Gentrification versus dispossession. *Urban Studies, 53*(3), 471–489.
Shkuda, A. (2016). *The Lofts of SoHo: Gentrification, art, and industry in New York, 1950–1980*. Chicago: The University of Chicago Press.
Skaburskis, A., & Nelson, K. (2014). Filtering and gentrifying in Toronto: neighbourhood transitions in and out from the lowest income decile between 1981 and 2006. *Environment and Planning A, 46*(4), 885–900.
Solana-Solana, M. (2010). Rural gentrification in Catalonia, Spain: A case study of migration, social change and conflicts in the Empordanet area. *Geoforum, 41*(3), 508–517.
Sotto, D. (2018). Parque Minhocão, São Paulo–Brazil: a case study on urban rehabilitation, place-making and gentrification. *Revista de Direito da Cidade, 10*(3), 1895–1910.
Swyngedouw, E. (2013). The non-political politics of climate change. *ACME: An International Journal for Critical Geographies, 12*(1), 1–8.
Teppo, A., & Millstein, M. (2015). The place of gentrification in Cape Town. In L. Lees, H.B. Shin, & E. López-Morales (Eds.), *Global gentrifications: Uneven development and displacement* (pp. 419–440). Bristol, UK; Chicago, IL: Bristol University Press. https://doi.org/10.2307/j.ctt1t894bt.
Thomas, K. A., & Warner, B. P. (2019). Weaponizing vulnerability to climate change. *Global Environmental Change, 57*, 101928.
Timberlake, J. M., & Johns-Wolfe, E. (2017). Neighborhood ethnoracial composition and gentrification in Chicago and New York, 1980 to 2010. *Urban Affairs Review, 53*(2), 236–272.
Tran, H., Uchihama, D., Ochi, S., & Yasuoka, Y. (2006). Assessment with satellite data of the urban heat island effects in Asian mega cities. *International Journal of Applied Earth Observation and Geoinformation, 8*(1), 34–48.
Uitermark, J., & Bosker, T. (2014). Wither the 'Undivided City'? An assessment of state-sponsored gentrification in Amsterdam. *Tijdschrift voor Economische en Sociale Geografie, 105*(2), 221–230.
Uribe-Toril, J., Ruiz-Real, J., & de Pablo Valenciano, J. (2018). Gentrification as an emerging source of environmental research. *Sustainability, 10*(12), 4847. (1–14).
Van Gent, W. P. (2013). Neoliberalization, housing institutions and variegated gentrification: How the 'Third Wave' broke in Amsterdam. *International Journal of Urban and Regional Research, 37*(2), 503–522.
Van Holm, E. J., & Wyczalkowski, C. K. (2018). Gentrification in the wake of a hurricane: New Orleans after Katrina. *Urban Studies, 56*(13), 2763–2778.
Walks, R. A., & Maaranen, R. (2008). Gentrification, social mix, and social polarization: Testing the linkages in large Canadian cities. *Urban Geography, 29*(4), 293–326.
Warner, K. (2010). Global environmental change and migration: Governance challenges. *Global Environmental Change, 20*(3), 402–413.
Weinstein, L., & Ren, X. (2009). The changing right to the city: Urban renewal and housing rights in globalizing Shanghai and Mumbai. *City and Community, 8*(4), 407–432.
Wiesel, I., Freestone, R., & Randolph, B. (2013). Owner-driven suburban renewal: Motivations, risks and strategies in 'knockdown and rebuild' processes in Sydney, Australia. *Housing Studies, 28*(5), 701–719.
Wiggins, M. (2018). Eroding paradigms: Heritage in an age of climate gentrification. *Change Over Time, 8*(1), 122–130.
Wolch, J. R., Byrne, J., & Newell, J. P. (2014). Urban green space, public health, and environmental justice: The challenge of making cities 'just green enough'. *Landscape and Urban Planning, 125*, 234–244.
Wolf, J., Adger, W. N., Lorenzoni, I., Abrahamson, V., & Raine, R. (2010). Social capital, individual responses to heat waves and climate change adaptation: An empirical study of two UK cities. *Global Environmental Change, 20*(1), 44–52.
Wong, T. C. (2006). Revitalising Singapore's central city through gentrification: The role of waterfront housing. *Urban Policy and Research, 24*(2), 181–199.
Wu, F. (2016). State dominance in urban redevelopment: Beyond gentrification in urban China. *Urban Affairs Review, 52*(5), 631–658.
Wu, Q., Zhang, X., & Waley, P. (2016). Jiaoyufication: When gentrification goes to school in the Chinese inner city. *Urban Studies, 53*(16), 3510–3526.
Yip, F. Y., Flanders, W. D., Wolkin, A., Engelthaler, D., Humble, W., Neri, A., Lewis, L., Backer, L., & Rubin, C. (2008). The impact of excess heat events in Maricopa County, Arizona: 2000–2005. *International Journal of Biometeorology, 52*(8), 765–772.
Yuen, B. (2008). Conclusion: Beyond sustainable development? In T. Wong, B. Yuen, & C. Goldblum (Eds.), *Spatial planning for a sustainable Singapore* (pp. 205–209). Springer: Dordrecht.

Climate Migration

► Climate-Induced Relocation

Climate Neutrality

► Amsterdam's Pathway to Climate Neutrality: Creating an Enabling Environment

Climate Refugees

► Climate-Induced Relocation

Climate Resilience in Informal Settlements: The Role of Natural Infrastructure

A Focus on Climate Adaptation

Hanna A. Rauf[1], Erich Wolff[2] and Perrine Hamel[1,3]
[1]Asian School of the Environment, Nanyang Technological University, Singapore, Singapore
[2]Monash Art, Design and Architecture, Monash University, Melbourne, VIC, Australia
[3]Earth Observatory of Singapore, Nanyang Technological University, Singapore, Singapore

Introduction

According to the United Nations (2018), over one billion inhabitants reside in informal dwellings – a number inflated by rapid urbanization and lack of affordable housing. Informal settlements are often off-the-grid, disconnected from basic infrastructure services, and constructed within proximity of hazardous locations, for example, dumpsites and flood-prone areas. Despite the high exposure to climate risks, e.g., flooding and heat waves, informal dwellers often opt for unsafe locations due to lack of options, in the absence of legal citizenship and formal tenure status (Satterthwaite et al. 2020). Affordability has been the main deterrence for informal dwellers in accessing decent infrastructure due to low household income (McGranahan 2015). The dysfunctional role of the state in providing inclusive public services for improving standards of living of informal dwellers raises a question of whether this is "a technical crisis or governance crisis" (Allen et al. 2006: p. 20; Satterthwaite et al. 2020).

Compounding the issue of informality, climate change affects informal dwellers all around the world. Direct impacts of climate change include inland and coastal flooding, sea -level rise, and heat waves, and cities increasingly consider climate resilience as a major objective of urban development (Revi et al. 2014). Recognizing the interlinkage between many urban challenges, the Sustainable Development Goal (SDG) 11 and the New Urban Agenda have called for a more sustainable "way-of-doing." An important dimension of such approach is the reliance and protection of natural ecosystems that is central not only for protecting livelihoods but also for mitigating climate risks. This approach is implemented in practice under the umbrella terms "ecosystem-based adaptation" or "nature-based solutions" and relies on the use of natural infrastructure – "natural and semi-natural elements capable of providing multiple functions and ecosystem services" (Bartesaghi Koc et al. 2017; Benedict and McMahon 2006).

The importance of informal settlements in the Global South cities and calls to improve climate resilience raise the question of the role of natural infrastructure for building climate resilience in informal settlements. "Climate resilience" is here defined using a situated perspective of socio-ecological systems in the context of informality. The following sections define these terms and explain challenges and opportunities for using natural infrastructure in informal settlements, illustrated by case studies in Colombia, Indonesia, and Fiji.

Defining Climate Resilience and Informality

Resilience in Informal Settlements

The mushrooming of informal dwellings in the Global South in recent years suggests that it has become "a way of life," inseparable from the city-making process (Al-Sayyad 2004). According to UN Habitat ("Habitat III Issue Paper", 2015), a particular residential area is considered informal when it is characterized by: (i) uncertain land tenure, (ii) limited access to public services and infrastructure, and (iii) nonconformity in relation to regulation and planning frameworks. The term is sometimes used interchangeably with slums, although not all slums are informal (Jason 2018). The notion of urban informality is often positioned on a binary model of legal and illegal, or temporariness and permanence, yet it exists as "gray spaces" that negotiate conflicting interests and "the politics of unrecognition" of the urban poor as part of city dwellers (Yiftachel 2009). Therefore, understanding informality requires the recognition of "formality" as a continuous process that jeopardizes the poor groups, and informality is where the residents grapple with everyday hardship in response to the dysfunctional of formal system (Banks et al. 2020).

Resilience, in a general sense, refers to the ability of a system, entity, or individual to withstand and cope with shocks from unexpected crises including climate risk (Leichenko 2011; Satterthwaite et al. 2020). Although the modern definition is often traced back to ecological theory, multiple perspectives coexist in the fields of engineering, ecological, socio-ecological, and psychological. Socio-ecological resilience, which is influential in the field of urban transformation, recognizes the intricacy of socio-ecological systems that continue to transform and adapt without having to return to equilibrium in response to shock or pressure. Following Elmqvist et al. (2019), resilience can be defined as "the capacity of an urban system to absorb disturbance, reorganize, maintain essentially the same functions and feedbacks over time and continue to develop along a particular trajectory. This capacity stems from the character, diversity, redundancies, and interactions among and between the components involved in generating different functions." (p. 269).

The concept of "continued development," or transformation, in the definition of socio-ecological lens resilience is important to distinguish it from social or psychological resilience. The latter can have a negative connotation in some communities – its use being perceived as glorifying their struggles or romanticizing their ability to survive with limited capacity. For instance, reflecting on the resilient labeling on the victims of Hurricane Katrina and BP Oil Spills, the President of Louisiana Justice Institute pointed out:

> every time you say, "Oh, they're resilient, [it actually] means you can do something else, [something] new to [my community]. . . . We were not born to be resilient; we are *conditioned* to be resilient. I don't want to be resilient . . . . [I want to] fix the things that [create the need for us to] be resilient [in the first place] [emphasis added]" (Josh 2015 quoted in Kaika 2017: p. 95).

In the context of informal settlements, shocks and pressure are not in the future but experienced in the everyday risks faced by inhabitants (Ziervogel et al. 2017). Everyday risks range broadly – mosquito-borne diseases, premature death, and workplace injuries (Satterthwaite and Bartlett 2017), and can occur at home, workplace, or public spaces. In the uncertainty of urban conditions searching for a better life (Simone 2010), the ubiquity of informal settlements demonstrates their resilience to everyday risks and challenges, i.e., physical displacement and urban poverty (Satterthwaite and Bartlett 2017). Repositioning resilience alongside everyday risks and environmental hazards in building climate adaptation in informal settlements thus provides room for more "just and negotiated resilience" under socio-ecological approach (Roberts et al. 2020).

Upgrading Informal Settlements for Climate Resilience

Effects of climate change include rising temperatures, leading to urban heat stress, increased flood risk (coastal, stormwater, and riverine flooding) and, often, increased stress to the water supply system by increasing the intensity of precipitation events. To address such risks and adapt to a changing climate, informal housing upgrading is one of the most viable and cost-effective options (Satterthwaite et al. 2020). While most settlement upgrading projects do not have an explicit climate adaption objective, improving built infrastructure could reduce hazard risks and withstand disasters (Dodman et al. 2019). Figure 1 indicates varying degrees of negotiation of power relations between communities and local government through different types of upgrading. Implementing this, however, can be challenging when life conditions in informal settlements are permanently changing due to risk of eviction and lack of recognition from the state. Participatory governance is the key to reduce exclusion when framing resilience for climate adaptation, through a more negotiated coproduction process (Roberts et al. 2020).

Natural Infrastructure as Climate Adaptation Strategy in Informal Settlements

Natural Infrastructure for Climate Adaptation

To mitigate the effects of climate change, investments in natural infrastructure is increasingly considered by cities and funders all around the world. Natural infrastructure encompasses a broad range of vegetated systems that are either engineered – designed to manage stormwater (e.g., bioswales, bioretention systems, constructed wetlands, etc.) – or naturally occurring (e.g., natural wetlands, trees, lakes, etc., Depietri and McPhearson 2017). Because of its potential for urban water management, natural infrastructure has been studied for a long time in the fields of engineering and stormwater management. It is often termed "green infrastructure" and is an important component of sustainable urban drainage systems (or water-sensitive urban design, low impact development, Fletcher et al. 2015). Essential hydrologic services provided by natural infrastructure, such as water retention, peak flow attenuation, or water quality improvement, are well recognized. In addition, a key feature of natural infrastructure is its multifunctionality – not only providing multiple hydrologic services but also reducing heat stress, providing aesthetic amenities, and, to a lesser extent, attenuating noise and air pollution.

Compared to conventional gray infrastructure, natural infrastructure offers additional key advantages, summarized in Table 1. Natural infrastructure can be cheaper in the long term (e.g., protecting forests requires minimal investment that provide a flood mitigation service) and is often a low-regret strategy (leaving options for future development). These characteristics make natural infrastructure an important component of what has been called a "water sensitive" approach to urban management. Such an approach relies on three pillars: "Cities as Water Supply Catchments," whereby cities use a portfolio of water sources and treatment options; "Cities Providing Ecosystem Services," whereby vegetated systems within cities provide benefits for people; and "Cities Comprising Water-Sensitive Communities," reflecting the cities' social and institutional capital to promote, implement, and improve ecologically sensitive solutions (Wong et al. 2020). The last two principles often translate in practice into inclusive and adaptive management approaches for natural infrastructure, codesigned and coproduced by a range of stakeholders and involving communities.

Challenges and Opportunities for Implementation in Informal Settlements

Opportunities

Given the potential of natural infrastructure for climate adaptation, there have been recent calls for implementing natural infrastructure in informal settlements (French et al. 2020; Satterthwaite et al. 2020). As noted above, an important feature of natural infrastructure is its **multifunctionality**: they provide multiple potential benefits such as heat stress or flood risk reduction, water quality improvement, and additional ecosystem services. Williams et al. (2019), for example, highlights the multiple benefits of floodplain protection in Durban, South Africa, highlighting the role of the river for regulating and cultural services.

Relatedly, natural infrastructure can be used in a **leapfrogging** strategy, enabling cities in developing countries to adopt more advanced water management approaches, possibly avoiding unnecessary

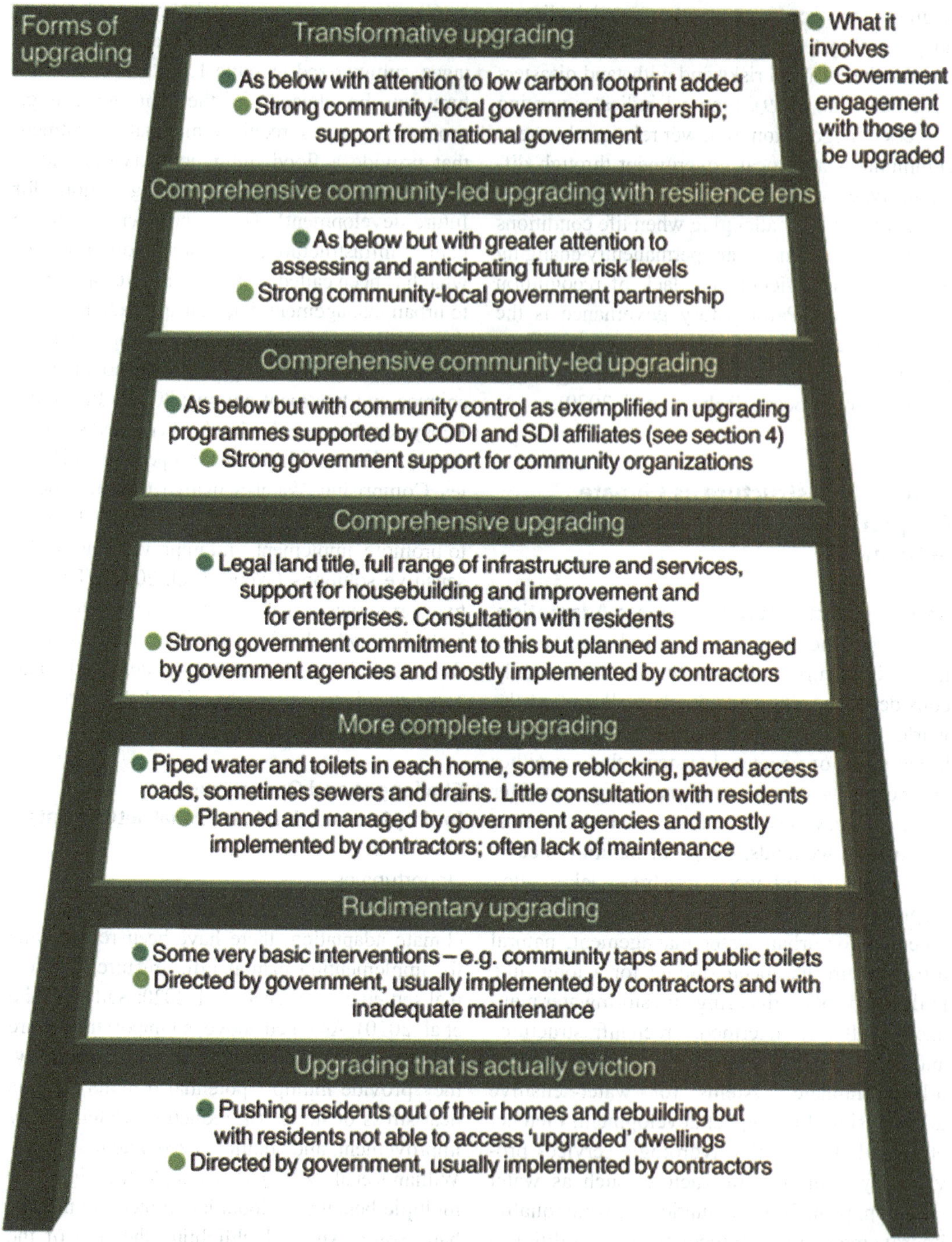

Forms of upgrading
What it involves
Government engagement with those to be upgraded
Transformative upgrading
As below with attention to low carbon footprint added
Strong community-local government partnership; support from national government
Comprehensive community-led upgrading with resilience lens
As below but with greater attention to assessing and anticipating future risk levels
Strong community-local government partnership
Comprehensive community-led upgrading
As below but with community control as exemplified in upgrading programmes supported by CODI and SDI affiliates (see section 4)
Strong government support for community organizations
Comprehensive upgrading
Legal land title, full range of infrastructure and services, support for housebuilding and improvement and for enterprises. Consultation with residents
Strong government commitment to this but planned and managed by government agencies and mostly implemented by contractors
More complete upgrading
Piped water and toilets in each home, some reblocking, paved access roads, sometimes sewers and drains. Little consultation with residents
Planned and managed by government agencies and mostly implemented by contractors; often lack of maintenance
Rudimentary upgrading
Some very basic interventions – e.g. community taps and public toilets
Directed by government, usually implemented by contractors and with inadequate maintenance
Upgrading that is actually eviction
Pushing residents out of their homes and rebuilding but with residents not able to access 'upgraded' dwellings
Directed by government, usually implemented by contractors

Climate Resilience in Informal Settlements: The Role of Natural Infrastructure, Fig. 1 Different forms of upgrading according to the level of engagement of government. (Source: Satterthwaite et al. 2020)

tradeoffs experienced by wealthier economies (Rogers et al. 2019). This includes, for example, avoiding centralized drainage systems in resettlement projects, but rather leveraging the benefits of decentralized water management (for stormwater, graywater, or wastewater management,

Climate Resilience in Informal Settlements: The Role of Natural Infrastructure, Table 1 Key characteristics of natural infrastructure compared to gray infrastructure: e.g., stormwater and wastewater piped network

Aspect	Gray infrastructure	Hybrid approaches using natural infrastructure
Feasibility in urban context	High (occupies a reduced area)	Medium to high (some systems are space intensive)
Reliability	Medium to high (does not completely eliminate risk)	Medium to high (limited capacity for extreme events)
Reversibility and flexibility	Low (no leapfrogging)	Medium (option value of protecting existing greenspaces)
Cost effectiveness	Low (high building costs, depreciates over time)	Medium (protection may be less expensive, especially accounting for co-benefits)
Biodiversity and co-benefits	None to low	Medium to high

Adapted from Depietri and McPhearson (2017)

thereby reducing flood risk and water quality issues).

Finally, the paradigm behind water-sensitive cities and natural infrastructure is one of **inclusive governance**. This aspect makes it suitable for implementation in informal settlements where participatory and inclusive approaches are critical (Reid et al. 2015). For example, Mulligan et al. (2020) highlights that "co-development of small-scale green infrastructure changed people's valuation, perception, and stewardship of nature-based systems and ecosystem services," emphasizing the importance of participatory approaches to natural infrastructure." Given the low resources and demands for basic needs it is even more critical to embed a climate adaptation strategy in local knowledge and responding to community values and needs (French et al. 2020).

Challenges

Key challenges for natural infrastructure implementation in informal settlements reflect to some extent those in formal areas, albeit at a different scale. Constraints of **space** are commonly cited (Sinharoy et al. 2019; Mulligan et al. 2020), highlighting the need for place-based design. The work of the Community Organizations Development Institute illustrates redevelopment projects that create new open spaces, providing heat mitigation and stormwater management benefits in addition to recreational and social benefits. Importantly, designing natural infrastructure requires **documentation** of the existing infrastructure, as well as environmental data. Multiple organizations undertake such mapping, highlighting the long term, trust-building process it involves (Patel and Baptist 2012).

Another important constraint concerns **financial resources**. Access to infrastructure funding is often limited by the perception of informal settlements as temporary, suggesting that more funding should be directed to infrastructure development rather than post-disaster. Hybrid financing models are increasingly seen in informal settlements, including funds from community savings, grant and loan assistance from governments and international agencies, and state subsidies. Although the development of such local funds requires much cooperation, they have the added benefit of involving stakeholders early on in the codevelopment projects.

Finally, an important dimension is that of governance in a context of **contested land tenure**, and low **implementation capacity** of local governments. Informal settlers may be unwilling to participate in codevelopment processes due to fear of eviction. Conversely, local governments may generate distrust by making unrealistic requests from informal settlers, e.g., to pay for expensive infrastructure upgrades or bring buildings to official standards, or directly by relocating communities. Investment in natural infrastructure in that context is no different from investments in other basic services, in that it should be promoted as part of supporting community-led initiatives that aim to improve residents' health and safety.

Case Studies of Natural Infrastructure Implementation in Informal Settlements

Moravia: Natural Infrastructure in the Management of Landslides and Contamination

The urban transformation of Medellin's informal settlements has been widely documented in media over the last decades (Duque Franco and Ortiz 2020). Among the many projects that have gained notoriety in the urban upgrading literature, the city hosts one of the first remarkable projects of natural infrastructure in informal settlements: the closure of the city's former landfill and revitalization of the surrounding settlement of Moravia.

At the time of the closure of the landfill in 1984, the neighborhood of Moravia was characterized by precarious dwellings encroaching on an unstable mountain of waste known as "El Morro" (Ramirez et al. 2018). In 2004, the neighborhood was considered the most densely populated in Colombia, and faced frequent floods, landslides, and the direct exposure to highly contaminated soil (Ahlert et al. 2018). Over the last few decades, however, the authorities have gradually involved the community in a series of actions including the implementation of natural infrastructure to treat the former landfill and the creation of a cultural center. The former "El Morro" has been used as a testing ground for nature-based contamination remediation (Gomez et al. 2011) and is now a vibrant green mountain dotted with flowers and visited by tourists.

The project has completely transformed the neighborhood, with multiple cultural and social impacts (Daza Vargas et al. 2017). It also significantly reduced landslide and flood risks by implementing a successful participatory approach. Located in an area that has been historically linked to disasters (Departamento de Geologia 2010), the interventions are expected to reduce runoff from more frequent and intense storms and improve the climate resilience of Moravia.

Exemplifying the challenges of this kind of approach, the authorities had to relocate several families that had settled on top of the waste dump and in the surrounding flood-prone areas. While most of the surrounding houses were maintained, a part of the population was relocated to a distant complex of social housing in Ciudadela Nuevo Occidente, losing important social and economic connections with the community (Ramirez et al. 2018).

Now housing more than 45,000 people, the case of Moravia is a project still in progress. While it created economic opportunities and increased local resilience, some problems persist such as the ongoing challenge of land tenure faced by many dislocated families. The case of Moravia, however, offers important insights into how the tensions of risk management and infrastructure provision can be negotiated in contexts of informality (Gouverneur 2014). It shows that natural infrastructure can be implemented using a community-oriented approach and, if well designed, create new economic opportunities, promote stewardship, and community participation (Vilar and Cartes 2016).

RISE Program: Citizen Science to Support the Design of Natural Infrastructure in Flood-Prone Areas

The Revitalizing Informal Settlements and their Environments (RISE) program is an action-research investigation on the effects of water-sensitive urban design, including natural infrastructure, in the context of informal settlements. Working with 12 settlements in the city of Makassar (Indonesia) and 12 settlements in Suva (Fiji), the program is conducting a randomized control trial to investigate the results of the implementation of NI in the sites (Brown et al. 2018; French et al. 2021).

The main natural infrastructure implemented in settlements consists of a series of wetland systems for wastewater treatment. From its inception in 2017, the program has been designed to implement natural infrastructure according to a community-oriented approach. This kind of interaction with the community allowed for community-members to familiarize themselves with the program and participate actively in RISE's decision-making processes through a codesign framework (Ramirez-Lovering et al. 2018).

One example of the codesign process is an extensive flood monitoring project, which engaged local communities. Developed using a citizen science framework (Haklay 2015), the project involved community volunteers from six settlements in Makassar and seven settlements in Suva, who were asked to take pictures of local water bodies. In total, more than 5000 photos were taken between the years 2018 and 2020 (Wolff 2021). While insufficient to fully characterize the local hydrology, the project generated a systematic database of flood documentation that supported the calibration of flood models and localized understanding of water level variations in the region. This knowledge was critical to inform the location of the constructed wetlands and is expected to significantly improve resilience to floods over the coming years.

The RISE project exemplifies common data challenges of implementing natural infrastructure in informal settlements, where collecting and storing data to support the infrastructure design requires creative solutions (Prescott et al. 2020). An additional challenge was the need to respond to a constantly changing built environment, uncertain land tenure, and significant hazards, in particular floods (French et al. 2021; Josey and Ramirez-Lovering 2020). Addressing this challenge was critical to design systems natural infrastructure that can operate efficiently over long timeframes.

Beyond data collection, evidence suggests that the community-based flood monitoring was successful at creating a virtual community in which community members were proud to contribute to the mapping of floods in their settlements. This shows that participatory methods, such as citizen science, can not only inform the design of natural infrastructure in the context of informal settlements, but also plays an important role in terms of improving risk communication and awareness among community members. The extensive research agenda of the RISE program provides important evidence to inform the implementation, operation, and maintenance of innovative natural infrastructure systems in informal settlements by involving the communities in all stages of the process.

C

References

Ahlert, M., Becker, M., Kreisel, A., Misselwitz, P., Pawlicki, N., & Schrammek, T. (2018). *Moravia Manifesto: Coding strategies for informal neighborhoods (Estrategias de Codificacion Para Barrios Populares).* Berlin: Jovis. www.jovis.de/de/buecher/product/moravia-manifesto.html.

Allen, A., Dávila, J. D., Hofman, P., & Aref, M. F. (2006). *Governance of water and sanitation services for the peri-urban poor: A framework for understanding and action in Metropolitan regions.*

AlSayyad, N. (2004). Urbanism as a "new" way of life. In A. Roy & N. AlSayyad (Eds.), *Urban informality: Transnational perspectives from the Middle East, South Asia and Latin America* (pp. 7–30). Lanham: Lexington Books.

Banks, N., Lombard, M., & Mitlin, D. (2020). Urban informality as a site of critical analysis. *The Journal of Development Studies, 56*(2), 223–238. https://doi.org/10.1080/00220388.2019.1577384.

Bartesaghi Koc, C., Osmond, P., & Peters, A. (2017). Towards a comprehensive green infrastructure typology: a systematic review of approaches, methods and typologies. *Urban Ecosyst 20*, 15–35. https://doi.org/10.1007/s11252-016-0578-5.

Benedict, M., & McMahon, E. (2006). Green Infrastructure: Linking Landscapes and Communities. Washington, DC: Island Press.

Brown, R., Leder, K., Wong, T., French, M., Ramirez-Lovering, D., Chown, S. L., Luby, S., et al. (2018). Improving human and environmental health in urban informal settlements: The revitalising informal settlements and their environments (RISE) programme. *The Lancet Planetary Health, 2*(May), S29. https://doi.org/10.1016/S2542-5196(18)30114-1.

Daza Vargas, Y., Rodríguez Murcia, S., Florez Yepes, G., & Montoya Restrepo, J. (2017). Análisis de los cambios socioambientales en el Morro de Moravia en Medellín (Antioquia-Colombia). *Anales De Geografía De La Universidad Complutense, 37*(2), 325–348. https://doi.org/10.5209/aguc.57728.

Departamento de Geologia. (2010). Amenazas y Riesgos En El Valle de Aburra. In M. Hermelin, A. Echeverri, & J. Giraldo (Eds.), *Medellín Medio-Ambiente Urbanismo Sociedad*. Medellín: Fondo Editorial Eafit.

Depietri, Y., & McPhearson, T. (2017). Integrating the grey, green, and blue in cities: Nature-based solutions for climate change adaptation and risk reduction. In N. Kabisch, H. Korn, J. Stadler, & A. Bonn (Eds.), *Nature-based solutions to climate change adaptation in urban areas: Linkages between science, policy and practice* (pp. 91–109). Cham: Springer International Publishing. https://doi.org/10.1007/978-3-319-56091-5_6.

Dodman, D., Archer, D., & Satterthwaite, D. (2019). Editorial: Responding to climate change in contexts of urban poverty and informality. *Environment and Urbanization, 31*(1), 3–12. https://doi.org/10.1177/0956247819830004.

Duque Franco, I., & Ortiz, C. (2020). Medellín in the headlines: The role of the media in the dissemination of urban models. *Cities, 96*. https://doi.org/10.1016/j.cities.2019.102431.

Elmqvist, T., Andersson, E., Frantzeskaki, N., McPhearson, T., Olsson, P., Gaffney, O., Takeuchi, K., & Folke, C. (2019). Sustainability and resilience for transformation in the urban century. *Nature Sustainability, 2*(4), 267–273. https://doi.org/10.1038/s41893-019-0250-1.

Fletcher,TD., Shuster, W., Hunt, WF., Ashley, R., Butler, D., Arthur, S., Trowsdale, S., Barraud, S., Semadeni-Davies, A., BertrandKrajewskim J-L., Mikkelsen, PS., Rivard, G., Uhl, M., Dagenais, D., & Viklander, M. (2015). SUDS, LID, BMPs, WSUD and more—the evolution and application of terminology surrounding urban drainage. *Urban Water J 12*, 525–542. https://doi.org/10.1080/1573062X.2014.916314.

French, M., Ramirez-Lovering, D., Sinharoy, S.S., Turgabeci, A., Latif, I., Leder, K., & Brown, R. (2020). Informal settlements in a COVID-19 world: moving beyond upgrading and envisioning revitalisation, Cities & Health. https://doi.org/10.1080/23748834.2020.1812331

French, M. A., S., Barker F., Taruc R. R., Ansariadi A., Duffy G. A., Saifuddaolah M., Agussalim A. Z., et al. (2021). A Planetary Health Model for Reducing Exposure to Faecal Contamination in Urban Informal Settlements: Baseline Findings from Makassar, Indonesia. Environment International 155 (October): 106679. https://doi.org/10.1016/j.envint.2021.106679.

Gomez, A. M., Yannarell, A. C., Sims, G. K., Cadavid-Restrepo, G., & Moreno Herrera, C. X. (2011). Characterization of bacterial diversity at different depths in the Moravia Hill landfill site at Medellín, Colombia. *Soil Biology and Biochemistry, 43*(6), 1275–1284. https://doi.org/10.1016/j.soilbio.2011.02.018.

Gouverneur, D. (2014). Dealing with informal settlements of the developing world: Lessons from Venezuela and Colombia. In *Planning and design for future informal settlements: Shaping the self-constructed city*. Routledge.

Haklay, M. (2015). *Citizen science and policy: A European perspective*. Washington, DC: Woodrow Wilson International Center for Scholars. https://www.wilsoncenter.org/publication/citizen-science-and-policy-european-perspective.

Jason, W. (2018). *Slums, informal settlements, and the role of land policy*. https://www.lincolninst.edu/publications/articles/sustainable-development

Josey, B. C., & Ramirez-Lovering, D. (2020). (Temporary) appropriation (of Space), Makassar, and Urban Kampung. In A. Melis, J. A. Lara-Hernandez, & J. Thompson (Eds.), *Temporary appropriation in cities: Human spatialisation in public spaces and community resilience* (pp. 171–193). Springer. https://doi.org/10.1007/978-3-030-32120-8_11.

Kaika, M. (2017). 'Don't call me resilient again!': The New Urban Agenda as immunology ... or ... what happens when communities refuse to be vaccinated with 'smart cities' and indicators. *Environment and Urbanization, 29*(1), 89–102. https://doi.org/10.1177/0956247816684763.

Leichenko, R. (2011). Climate change and urban resilience. *Current Opinion in Environmental Sustainability, 3*(3), 164–168. https://doi.org/10.1016/j.cosust.2010.12.014.

McGranahan, G. (2015). Realizing the right to sanitation in deprived urban communities: Meeting the challenges of collective action, coproduction, affordability, and housing tenure. *World Development, 68*, 242–253. https://doi.org/10.1016/j.worlddev.2014.12.008.

Mulligan, J., Bukachi,V., Clause, J.C., Jewell, R., Kirimi, F., Odbert, C. (2020). Hybrid infrastructures, hybrid governance: New evidence from Nairobi (Kenya) on green-blue-grey infrastructure in informal settlements. *Anthropocene 29*,100227. https://doi.org/10.1016/j.ancene.2019.100227

Patel, S., & Baptist, C. (2012). Editorial: Documenting by the undocumented. *Environment and Urbanization, 24* (1), 3–12. https://doi.org/10.1177/0956247812438364.

Prescott, M. F., Ramirez-Lovering, D., & Hamacher, A. (2020). RISE planetary health data platform: Applied challenges in the development of an interdisciplinary data visualisation platform. *Journal of Digital Landscape Architecture, 5*, 567–574.

Ramirez, H., Maria, C., & Monsalve, D. (2018). Fenix de La Basura: Cuento de Moravia/Phoenix from the Waste: The story of Moravia. In M. Ahlert, K. A. Becker, P. Misselwitz, N. Pawlicki, & T. Schrammek (Eds.), *Moravia Manifesto: Coding strategies for informal neighborhoods (Estrategias de Codificacion Para Barrios Populares)*. Berlin: Jovis Publishers. www.jovis.de/de/buecher/product/moravia-manifesto.html.

Ramirez-Lovering, D., Prescott, M., & Kamalipour, H. (2018). RISE: A case study for design research in informal settlement revitalisation interdisciplinary design research in informal settlements. In D. W. Maxwell (Ed.), *Proceedings of the 1st annual design research conference (ADR18)* (pp. 461–478). University of Sydney. https://sydney.edu.au/content/dam/corporate/documents/sydney-school-of-architecture-design-and-planning/research/ADR18-Proceedings-Final.pdf.

Reid, H., Swiderska, K., King-Okumu, C., & Archer, D. (2015). *(Rep.). International Institute for Environment and Development*. Retrieved January 20, 2021, from http://www.jstor.org/stable/resrep17971

Revi, A., Satterthwaite, D. E., AragÃn-Durand, F., Corfee-Morlot, J., Kiunsi, R. B. R., Pelling, M., Roberts, D. C., & Solecki, W. (2014). Urban areas. In C. B. Field, V. R. Barros, D. J. Dokken, K. J. Mach, M. D. Mastrandrea, T. E. Bilir, M. Chatterjee, K. L. Ebi, Y. O. Estrada, R. C. Genova, B. Girma, E. S. Kissel, A. N. Levy, S. MacCracken, P. R. Mastrandrea, & L. L. White (Eds.), *Climate change 2014: Impacts, adaptation, and vulnerability. Part A: Global and sectoral aspects. contribution of working group II to the fifth assessment report of the intergovernmental panel on climate change* (pp. 535–612). Cambridge/New York: Cambridge University Press.

Roberts, D., Douwes, J., Sutherland, C., & Sim, V. (2020). Durban's 100 resilient cities journey: Governing resilience from within. *Environment and Urbanization, 32* (2), 547–568. https://doi.org/10.1177/0956247820946555.

Rogers, B., Ramirez, D., Marthanty, DR., Arifin, HS., Farrelly, M., Fowdar, H., Gunn, A., Holden, J., Kaswanto, RL., Marino, Zamudio R., McCarthy, D., Novalia, W., Payne, E., Suwarso, R., Syaukat, Y., Urich, C., Wright, A., & Yuliantoro, D. (2019). Leapfrogging pathways for a water sensitive Bogor. Australia-Indonesia Centre (AIC). https://research.monash.edu/en/publications/leapfrogging-pathways-for-a-water-sensitivebogor

Satterthwaite, D., & Bartlett, S. (2017). Editorial: The full spectrum of risk in urban centres: Changing perceptions, changing priorities. *Environment and Urbanization, 29*(1), 3–14. https://doi.org/10.1177/0956247817691921.

Satterthwaite, D., Archer, D., Colenbrander, S., Dodman, D., Hardoy, J., Mitlin, D., & Patel, S. (2020). Building resilience to climate change in informal settlements. *One Earth, 2*(2), 143–156. https://doi.org/10.1016/j.oneear.2020.02.002.

Simone, A. (2010). *City life from Jakarta to Dakar* (pp. 1–60). New York: Routledge.

Sinharoy, S. S., Pittluck, R., & Clasen, T. (2019). Review of drivers and barriers of water and sanitation policies for urban informal settlements in low-income and middle-income countries. *Utilities Policy, 60*, 100957. https://doi.org/10.1016/j.jup.2019.100957.

UN-Habitat. (2015). *Habitat III issue paper 22 – Informal settlements.* Available at https://unhabitat.org/habitat-iii-issue-papers-22-informal-settlements/

Vilar, K., & Cartes, I. (2016). Urban design and social capital in slums. Case study: Moravia's neighborhood, Medellin, 2004–2014. *Procedia – Social and Behavioral Sciences, 216*, 56–67. https://doi.org/10.1016/j.sbspro.2015.12.008.

Williams, D. S., Máñez Costa, M., Sutherland, C., Celliers, L., & Scheffran, J. (2019). Vulnerability of informal settlements in the context of rapid urbanization and climate change. *Environment and Urbanization, 31* (1), 157–176. https://doi.org/10.1177/0956247818819694.

Wolff, E. (2021). The Promise of a "People-Centred" Approach to Floods: Types of Participation in the Global Literature of Citizen Science and Community-Based Flood Risk Reduction in the Context of the Sendai Framework. *Progress in Disaster Science 10*, 100171. https://doi.org/10.1016/j.pdisas.2021.100171.

Wong, T., Rogers, B., & Brown, R. (2020). Transforming Cities through Water-Sensitive Principles and Practices. *One Earth, 3*(4), 436–447. https://doi.org/10.1016/j.oneear.2020.09.012.

Yiftachel, O. (2009). Theoretical notes on 'Gray Cities': The coming of urban apartheid? *Planning Theory, 8* (1), 88–100. https://doi.org/10.1177/1473095208099300.

Ziervogel, G., Pelling, M., Cartwright, A., Chu, E., Deshpande, T., Harris, L., Hyams, K., Kaunda, J., Klaus, B., Michael, K., Pasquini, L., Pharoah, R., Rodina, L., Scott, D., & Zweig, P. (2017). Inserting rights and justice into urban resilience: A focus on everyday risk. *Environment and Urbanization, 29*(1), 123–138. https://doi.org/10.1177/0956247816686905.

C

Climate Risk Atlas (CRA)

► The State of Extreme Events in India

Climate Risks

► Adapting to a Changing Climate Through Nature-Based Solutions
► The State of Extreme Events in India

Climate Security

► Responsibility to Prepare and Prevent (R2P2): Applying Unprecedented Foresight to Addressing Unprecedented Climate Risks

Climate Smart Agriculture

► Smart Agriculture and ICT

Climate Vulnerability

▶ The State of Extreme Events in India

Climate-Adaptive Urban Areas

▶ Urban Climate Resilience

Climate-Induced Relocation

climate migration to managed retreat

Patrick M. Marchman
American Society of Adaptation Professionals/
Climigration Network, Kansas City, MO, USA

Synonyms

Climate migration; Climate refugees; Managed retreat

Definition

The relocation or movement of structures, land uses, and/or humans themselves as a result of direct or indirect impacts of climate change. The study of climate-induced relocation attempts to answer the question "where will people go?", and, increasingly, "where are people going now?". Climate-induced relocation is an umbrella term for numerous other ways of describing the general idea, but two of the most frequent are climate migration and managed retreat.

Introduction

The movement of people and structures has been a defining characteristic of humanity since before the beginning of Homo sapiens. Early modern humans, as well as other hominids, moved to other places for a variety of reasons including climatic. New conditions produced by ice ages pushed early humans into new habitats and opened up new places for habitation. Events such as the Toba supervolcano around 70,000 BCE had severe impacts upon human populations, leaving its mark in sharply constrained genetic diversity.

The onset of the Holocene in 12,000 BCE heralded an era of unprecedented (for modern humans) climatic stability. Even so, further changes in regional climate such as the drying of parts of the Middle East and associated deforestation created incentives for successful hunter-gatherer and gardening societies to adopt more intensive agriculture, paving the way for the emergence of cities and state societies.

Changes in climate continued to prompt movements of people during recorded history as well as increased strain on large state societies, in some cases contributing to their reorganization or even collapse. But migration as a result of anthropogenic climate change brought on by industrial society has begun emerging as a definable stream of migration only in the last few decades.

Climate-induced relocation generally refers to the movement of individuals or populations as a result of changes in the climate. There have been several terms that have been introduced to describe this, but two of the most prominent have been climate migration and managed retreat.

Climate Migration

Climate migration is the movement of individuals, groups, or populations due to changes in climatic conditions. The term generally refers to the movement of people themselves and also tends to refer to the movement of larger groups.

With the exception of small island nations and low-lying continental areas and the impacts of sea-level rise, climate change is rarely thought of as the immediate or sole cause of migration. Instead, it is often a secondary or tertiary cause – for example, changes in climate and weather patterns in parts of Central America have led to greater difficulties for traditional farmers in maintaining income or even sustenance, inducing large numbers to migrate to either larger cities elsewhere in Central America, Mexico, or the United States. This makes it difficult to gain a clear perspective on just how many migrants are "climate migrants." The term "climate refugees" is an even more difficult one. The perceived pejorative nature of the word "refugee" leads to a situation in which refugees by any other name – middle-class Americans or Australians forced from their homes due to wildfires or hurricanes – rarely have the term applied to them. Refugee also can have a specific legal meaning, implying the duty of states and other entities to care for and resettle them, which can lead to states fighting against the claims of people displaced from plainly climate-amplified adversity.

Due to the difficulty of clearly determining the immediate cause of migration and political sensitivities, broadly accepted examples of climate migration can be hard to identify. Some situations currently proposed for this are African and Syrian migrants to Europe and the migration of Pacific Islanders from some small island states such as the Marshall Islands and Kiribati.

As climate change continues to accelerate, some studies have placed the numbers of "climate refugees" over the next several decades in the hundreds of millions or higher. One study estimated that nearly 3 billion people would live in areas where temperatures made many human activities impossible by 2070. Climate migration in this case could include mass immigration or migration to countries in latitudes closer to the poles such as Russia, Canada, the Scandinavian countries, Chile, or Argentina. This would inevitably have huge social and geopolitical impacts as well as ecological that are very hard to predict. Analogies to this kind of mass migration and the potential for dramatic changes in social structures can be found in the latter period of the Western Roman Empire as well as the invasion and conquest of the Americas and Australia by Europeans.

Managed Retreat

If climate migration generally, but not always, refers to larger-scale movements of peoples, managed retreat generally, but not always, refers to movements of individual homeowners, neighborhoods, or communities as a result of intensifying natural hazards. This has mostly been in response to sea-level rise and flooding but could also be in response to increased risks from fire, drought, changes in permafrost, or other environmental conditions.

Instances of managed retreat are occurring in many countries, from the household level up to Indonesia's recent announcement of plans to construct a new capital in central Borneo in response to the combined impact of sea-level rise and subsidence in its current capital Jakarta. The term is thought to have originated in the United States, where managed retreat often takes the form of property buyouts financed ultimately through the Federal Emergency Management Agency (FEMA). One of the first relocations of an entire community took place in Valmeyer, Illinois, where the town was relocated 2 miles to the east following massive floods in 1993. Over the past decades as buyouts have become increasingly recognized as a viable tool, they have also increased in price and complexity. Today, some of the more recent examples of large-scale community relocation include the village of Newtok, Alaska, the Native American community inhabiting Isle de Jean Charles offshore of Louisiana, and buyouts in New York and New Jersey as a result of Hurricane Sandy. Due to conceptions of property rights in the United States, the vast majority of buyouts have been voluntary. Increasing pressures may lead government agencies in the future to invoke eminent domain to speed the process, as is already happening in scattered cases.

As sea-level rise continues, the conversation around managed retreat is expanding to include more and more cities and communities worldwide. Even without full submergence or destruction of individual properties, continued flooding of roads and damage to other infrastructure from repeated inundation and saltwater intrusion can impact property values, contributing in countries in which local jurisdictions are highly dependent on property tax revenues such as the United States experiencing a "vicious spiral" of lower property values leading to reduced property tax revenues and less money for effective climate adaptation measures, which leads in turn to further reductions in property values, reductions in municipal and regional services, and a continued erosion of the tax base necessary for adaptation measures. Managed retreat, in this situation, is an alternative to the kind of unmanaged retreat that takes place through the actions of individuals acting in the housing market. Unmanaged retreat can be difficult to identify, but signals that it is taking place can be found in land use patterns in Louisiana following Hurricane Katrina in 2005 as well as the reduction in the rate of property values for many coastal properties relative to properties not directly exposed to coastal risk seen in the United States since the 2008 housing market crash.

Summary/Conclusion

Researchers around the world are examining various aspects of climate-induced relocation. The Hugo Observatory at the University of **Liège is currently leading the HABITABLE consortium of 20 universities in Europe, Asia, and Africa examining questions related to the environment and migration. In North America, several researchers in conjunction with various levels of government in the United States and Canada as well as professional networks such as the Climigration Network and the American Society of Adaptation Professionals are leading efforts to understand and prepare for future climate-induced relocation.**

Climate migration and managed retreat are states on a continuum, and, as such, the boundary between the two is often unclear and shifting. As climate change accelerates, climate-induced relocation will intensify, affecting the ever-increasing number of people and, in some cases, entire communities or even nations. The scale and speed of such migrations necessitate revisiting many of the metaphors that the idea of migration generates in receiving societies. The views of migrants as outsiders inherently hostile to a society's deepest values or parasites taking advantage of a nation's generosity interfere with the process of developing new concepts to more effectively allow societies to coexist with migrants. And the increasing visibility of managed retreat within a society may prompt governments to develop policies that are less harsh than if the only migrants were outsiders.

Climate-induced relocation will affect every society at some point. Interdisciplinary collaboration and popular organizing hold the potential of developing those new metaphors and new ideas needed to equitably address the challenges of both climate migration and managed retreat.

Climate-Neutral City

▶ Circular Cities

Climate-Resilient City

▶ Urban Climate Resilience

Climate-Resilient Technologies and Innovations for Sustainable Agriculture, Improved Landscape, and Food Security

Md. Arfanuzzaman[1] and Md. Anisur Rahman[2]
[1]Food and Agriculture Organization (FAO) of the United Nations, Dhaka, Bangladesh
[2]Center for Policy and Economic Research (CPER), Dhaka, Bangladesh

Definition

This paper has presented and assessed the promising technologies and innovations to promote sustainable land management and climate-resilient agriculture in climate-vulnerable landscapes of Bangladesh (i.e., High Barind Tract (HBT), Chittagong Hill Tract (CHT), waterlogging, and salinity-prone areas). In the view of increasing intensity and frequency of extreme weather events combined with low socio-economic situation, limited human and ecological coping mechanisms, and inadequate enabling conditions, climate-resilient technologies and innovations can reduce the vulnerability of the agriculture sector, enhance food security, and build landscape resilience in Bangladesh. The paper recommends that, climate-resilient agricultural technologies of Bangladesh need to be designed and promoted considering the geographical situation, hydroclimatic condition, natural resource availability, and ecological vulnerability. No technology, innovation or management practice will sustain unless these ecological challenges are solved.

Introduction

In the view of increasing climate stress, land degradation, livelihood vulnerabilities, and limited resilient technologies, the countries need to sustainably manage agriculture, landscape, food systems, natural resources, and ecosystems. Feeding everyone, every day and everywhere, remains a key challenge in most of the developing countries of the world which can be mitigated largely by reaching sustainable agriculture with improved land management, resilient food system, and technological advancement in inputs, cultivation, postharvest, and processing. Bangladesh is a deltaic country with a total area of 147,570 km^2 covering 133,910 km^2 land and 10,090 km^2 water. The country has one of the highest population densities in the world (1,253 people per km^2), with an estimated 163 million people (UN 2019). Due in part to its low-lying geographical location, Bangladesh has been historically vulnerable to river flooding, storm surges, tropical cyclones, sea-level rise, hydro-morphological changes, coastal erosion, and salinity intrusion, and the magnitude and frequency of sudden and slow-onset events have grown in recent years as climate change manifests (MoFA 2018; MoEF 2009). According to the Global Climate Risk Index 2019, Bangladesh was the seventh most affected country by extreme weather events in 2017, reflective of exposure and vulnerability to climate-related risks (Eckstein et al. 2019). To reduce climate-induced crop loss and damage, overextraction of natural resources, and land degradation and attain sustainable agricultural and rural development, locally suitable, cost-effective, and climate-resilient agricultural technologies and innovations are obligatory (CIAT and World Bank 2017).

On 30 October 2019, a technical consultation on technologies and innovations to promote sustainable land management and climate-resilient agriculture was held in Dhaka, Bangladesh, which was organized by the Food and Agriculture Organization (FAO) of the United Nations with support from the Bangladesh Agricultural Research Institute (BARI) and Bangladesh Agricultural Research Council (BARC). Renowned scientists from four institutions, i.e., BARI, Bangladesh Institute for Nuclear Agriculture (BINA), and Soil Resources Development Institute (SRDI), and officials from FAO participated in the technical consultation and discussed the critical ecological, climatic, and socioeconomic issues in the four climate-vulnerable landscapes of Bangladesh (i.e., High Barind Tract (HBT), Chittagong Hill Tract (CHT), waterlogging, and salinity-pronc arcas), and a priority list of agricultural technologies and innovations which are suitable and can potentially address the issues was identified. This was followed by some discussion on critical constraints, in the enabling environment and along agricultural value chains, that function as a barrier to farmers adopting and adapting these technologies and innovations. This paper synthesizes expert opinion on technologies, management practices, and innovations – to promote sustainable agriculture, improved land management, food security, and resilient livelihoods of communities in ecologically degraded and climate-vulnerable landscape and identify the ground-level constraints which need to be simultaneously addressed.

Vulnerabilities of Degraded Landscapes in Bangladesh

The HBT is located in the northwest region of Bangladesh covering the part of Rajshahi, Rangpur, and Bogra districts. HBT differs from other parts of Bangladesh due to its topography and distinct physiography of terraced lands at 20–30 m above sea level. Dry climate, low rainfall, high evapotranspiration, heat and cold wave, and drought are the major threats to crops, livestock, and fishery sectors in HBT (Aziz et al. 2015). About 55% of the land remains fallow in the Rabi season (Rabi crops are winter crops and are sown in October/November and harvested in April), reducing the income-generation opportunities for smallholders. The lands of HBT are degraded, and hence the biological activity of soils and their moisture retention capacity needs to be enhanced (Nasim et al. 2017). Overexploitation of groundwater, siltation, shortage of labor, low soil fertility and organic matter, and soil moisture retention are the major challenges in HBT.

The southwestern coastal region of Bangladesh is highly saline prone which has a clear impact on crop agriculture, sweet water fish, land, and water resources. There are 19 coastal districts in this region, covering 32% of the country and accommodating more than 35 million people where primary production system, coastal biodiversity, ecosystem, livelihoods, and human health are at high risk due to salinity and extreme weather events such as flood, storm, cyclone, erratic rainfall, etc. Out of about 1,689 million hectares of coastal land 1,056 million ha are already affected by the salinity of different degrees which are projected to be increased further in the mid and long term. Both saline groundwater and seawater intrusion intensifies the soil salinity, posing a significant risk to agriculture and aquaculture. Gher-based (saline water stored in the agricultural field with raised mud wall) shrimp farming practices have also contributed to increased salinity.

Increasing waterlogged areas is one of the major areas of concern in the southwest part of Bangladesh, particularly in the coastal districts of Jessore, Khulna, and Satkhira. In 2015, a total of 68,197 ha of land were affected by waterlogging in Khulna, Satkhira, and Jessore districts which adversely affected the food production, landscape, human lives, and livelihoods. Silted up rivers, unplanned and poor maintenance of infrastructure projects, and drainage blockage are the major causes of waterlogging. Standing water persisting for any period up to 6 months after monsoons is not favorable to agriculture. Soils of this region are mostly alkaline, and soil organic matter is low except in the peat soils. The remoteness of the waterlogged area makes transport and communication very challenging.

The CHT comprises Khagrachari, Rangamati, and Bandarban districts, and two-thirds of the area is characterized by steep slopes, and the remaining area encompasses undulating topography. The CHT covers 10% of the country's total landmass. A severe drought during Rabi and Kharif-I and robust weed growth during Kharif-I, along with the slope' plantation, high insect and pest infestation/diseases, and low soil fertility, are the major challenges in the CHT. Soils in the elevated areas of CHT are well drained but classified as moderately to strongly acidic, have low natural fertility, and are highly leached. The combination of topsoil and vegetation degradation and intensive rainfall has increased the risks of landslides in CHT. Besides, like HBT, low rainfall in the dry season is an emerging threat in CHT. The people usually depend on the hill stream for crop production, and household activities are now at tremendous risk for changing climate (Fig. 1).

Promising Technologies and Innovations for Sustainable Agriculture, Improved Landscape, and Food Security

HBT of Northwestern Bangladesh

Technologies and innovations which can promote sustainability and climate resilience were prioritized for dissemination and scaling up, by the On-Farm Research Division (OFRD), BARI (Rajshahi), including:

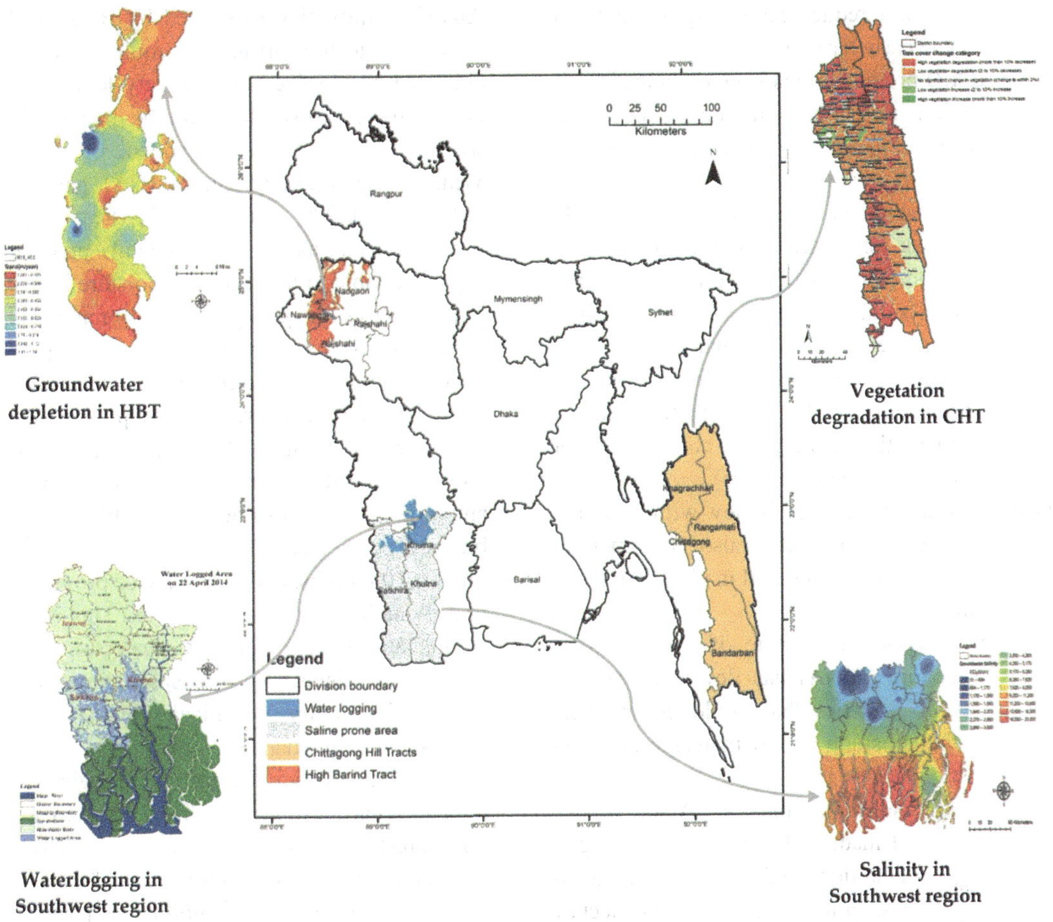

Climate-Resilient Technologies and Innovations for Sustainable Agriculture, Improved Landscape, and Food Security, Fig. 1 Major ecological risks in HBT, CHT, and the Southwest region of Bangladesh. (Source: FAO)

(i) Conservation agriculture for wheat and rice to address soil fertility and moisture issues.
(ii) Mustard, Boro rice, and T. Aman (transplanted Aman) rice (short-duration varieties) cropping pattern to effectively increase cropping intensity/productivity.
(iii) Chickpea or lentils after T. Aman to use residual soil moisture and to increase cropping intensity/productivity.
(iv) Rice cultivation in unpuddled (zero tillage) transplanting system.
(v) Four crop (potato-Boro-T. Aus-T. Aman)-based cropping pattern is also proved to be useful for HBT.
(vi) Mixed cropping of mustard with lentil and maize cultivation in Kharif season.
(vii) Alternate furrow irrigation for raised bed crops (potato, maize, tomato, brinjal, etc.) to reduce water loss and enhance water conservation.
(viii) Rainwater harvesting for supplementary irrigation (16X16X3 meters pond will cost US$ 1446 (BDT 120,000), and this cost can be reduced further by using a digging machine).
(ix) Direct seeding of Aus rice (traditionally, Aus rice is grown from May to August under rainfed conditions) to help conserve

water and reduce labor requirement and enable early harvest.

(x) BARI production package (balanced fertilization, irrigation, spraying of urea/pesticides, etc.) can increase the quality of mango production.

(xi) Establishment of a fruit orchard with BARI varieties for better adaptation to drought and increased incomes. Pulse or legume crops could be established inside these orchards, e.g., mango orchards.

(xii) Homestead vegetable and fruit production can increase employment opportunities for women (this is pertinent considering the small size of landholdings).

(xiii) Mixed cropping of barley with chickpea and cropping of onion is also recommended bearing dry weather and market demand in mind.

- Bangladesh Institute of Nuclear Agriculture (BINA) has developed several crop varieties that could complement the transition to new management practices. These are:
  (i) Short-duration Aus rice (Binadhan-19) to escape drought: Duration is 95–100 days; the average grain yield is 4.5 tons/hectare (t/ha).
  (ii) Short-duration Aus rice (Binadhan-21) to escape drought: Duration is 100–105 days; the average grain yield is 4.5 t/ha.
  (iii) Short-duration Aman rice (Binadhan-7): Duration is 115–120 days; the average yield is 5.0 t/ha.

Major Discussion Points

- Participants suggested that the residual moisture of mustard has the potentiality to reduce the water requirement of Boro rice. Herbicides need to be applied to control weeds in the unpuddled (zero tillage) rice cultivation system. As HBT is a drought-prone and water-scarce region, most of the participants recommended to stop promoting Boro and Aus rice cultivation suggested by BARI and BINA due to high irrigation requirement. As chickpea and maize are relatively less water consumptive crops, they can be planted in water stress conditions, they added.
- While conservation agriculture (CA), particularly for 80,000 hectares of wheat, was recommended as a management practice highly suitable to manage soil moisture issues in HBT, access to quality machinery remains a constraint. Scientists noted that locally produced machines are fragile and unable to handle HBT soils, have trouble dropping seeds, and increase drudgery for farmers (because they are forced to walk behind the machine and work it). It is also not sustainable from a business perspective because machines remain idle for 10 months, and local service providers find it challenging to make profits. This is all complicated by the lack of a dealership network, constrained access to spare parts, and limited in-country technical capacity for repairs and maintenance.
- Poor quality of fertilizers in Bangladesh remains a barrier to soil fertility enhancement.
- Agroforestry, including grass systems (fodder), could be another area of focus in HBT. Studies reveal that *Acacia nilotica* is particularly suited to this ecology, is preferred by birds as a nesting site, grows quickly, and is a protection against winds. Sal forests used to cover HBT pre-colonial times and are an indirect indication that greening initiatives in HBT are feasible. It was noted that the banks of ponds – about 70,000 such ponds in HBT – can be utilized for trees, fodder crops, vegetables, or fruit production.
- Though BARI suggested direct seeding for Aus rice to conserve water and reduce labor, there is a possibility of low yield. Hence, the room is there to improve the Aus variety.
- BARI and BINA prioritized short-duration T. Aman rice, fruit orchard, mustard, maize, and wheat cultivation through conservation agriculture and agroforestry for HBT in the view of climate stress and groundwater depletion.

- Popularizing of rainwater harvest techniques for T. Aman rice and Rabi crops is required to reduce the pressure from groundwater. However, the digging machine can remarkably reduce the cost of establishing a water harvesting pond, but there is a regulatory issue as it requires permission from the appropriate authority. This barrier needs to be resolved.

Saline-Prone Area of Southwest Bangladesh
Innovations prioritized by OFRD, BARI (Khulna), for saline-prone areas include:

(i) Dyke and ditch (modified sorjan) for vegetables, fruits, and fish cultivation to increase cultivation intensity/productivity
(ii) Raised beds for vegetable production, including potato with mulch to avoid inundation and water stagnation issues during monsoons
(iii) Mung bean and cowpea production through zero/minimum tillage with crop residue retention to use residual soil moisture and save production costs
(iv) Mulching and drip irrigation for vegetable production
(v) Alternate furrow irrigation for tomato/sunflower/ brinjal, etc. to save irrigation water and other co-benefits
(vi) Sweet gourd and T. Aman rice relay cropping to enable soil moisture retention, higher productivity, and early harvest and reduce irrigation frequency
(vii) Sunflower cultivation – short/dwarf varieties – that could be more resilient to high winds/cyclonic weather during harvest
(viii) Zero tillage cultivation of potato, maize, or wheat
(ix) Intercropping of short-duration vegetables with elephant foot yam
(x) Homestead/gher boundary vegetable production
(xi) Pit-based crops (watermelon and other fruits/vegetables)
(xii) Cultivation of *Chui Jhal* (spice) in saline land

- Climate-resilient technologies and innovations prioritized for sustainable agriculture with improved land management by SRDI include:

(i) Farm pond technology (FPT).
(ii) Pitcher irrigation.
(iii) Double layer mulching.
(iv) Shallow furrow and ridge system.
(v) Flying bed agriculture (in this system, a crop bed is created within a big plastic drum and kept above the bamboo/concrete structure).
(vi) Dibbling and transplanting of maize.

- Innovations for sustainable land management and climate-resilient agriculture by the Bangladesh Institute of Nuclear Agriculture (BINA) include:

(i) Short-duration Aman rice Binadhan-7; duration, 115–120 days; yield, 5.0 t/ha
(ii) Salinity-tolerant Boro rice Binadhan-8; crop duration, 140–145 days; salinity tolerance – seedling stage, 10–12 dS/m, and mature stage, 8–10 dS/m; yield, 5.0–5.5 t/ha in saline condition and 7.0–8.0 tons/ha in normal condition
(iii) Salinity-tolerant Boro rice Binadhan-10; salinity tolerance – seedling stage, 12–14 dS/m, and mature stage, 10–12 dS/m; yield, 5.5–6.0 tons /ha in saline condition and 8.0–9.0 tons/ha in normal condition
(iv) Dual tolerance (salinity and submergence) Aman rice Binadhan-23; crop duration, 125 days; salinity tolerance (8 dS/m); submergence tolerance (14 days); yield, 5.3 t/ha
(v) Salinity-tolerant wheat Binagom-1; crop duration, 100–120 days; salinity tolerance, 12 dS/m; yield, 3.5 t/ha
(vi) Salt-tolerant rapeseed Binasarisha-4; crop duration, 80 days; salinity-tolerant level, 4dS/m; average yield, 1.9 t/ha
(vii) Salt-tolerant groundnut Binachinabadam-6; salinity tolerance, 8 dS/m; larger seed size; high yielding; duration 140 days; average yield, 2.4 t/ha

(viii) Boro rice cultivation in the salinity-affected coastal area through transplanting rearrangement and gypsum application

Major Discussion Points

- BARI suggested raised bed and dyke and ditch farming have very good potential in both saline and waterlogged areas, but it requires substantial initial investment which is a barrier at the farmer's level. Availability and access to soft credit can help to scale up this practice.
- Though BARI recommended mung bean and cowpea for saline areas, they may not grow well without salinity management.
- In the view of scientists, there is an enormous market opportunity for sunflower seeds to produce oil, and entrepreneurs have the machines to process seeds, but the supply of enough quantity of seeds is very limited. Besides, currently available sunflower seeds are not much high yielding, and storm damages the plant most of the time. Participants urged for high yielding and dwarf-type sunflower variety as there is good demand for edible sunflower oil.
- There is a huge scope for fruit production in this region, particularly the ones that do not have high freshwater requirements and tolerate salinity: watermelon seeds are imported from Japan, and dragon fruit and blood orange are other possibilities.
- Though *Chui Jhal* is a high-value spice suggested by BARI, its saline tolerance level is still unknown.
- BARI has ranked crops by saline tolerance, and dragon fruit, sugar beet, watermelon, sunflower, and barley rank the highest for tolerance. Scientists of BARI, SRDI, and BINA opined for raised bed farming, zero tillage cropping system, farm pond technology, pit-based vegetable, and fruit cultivation technologies in the view of salinity.
- BINA has an adaptive research division; they also collaborate with DAE and BARI (especially in Khulna) to disseminate new varieties of seeds.
- BINA has also developed a management practice – silicon application and transplanting arrangement – to reduce salinity in Boro rice fields.

Waterlogging-Prone Areas of Southwest Bangladesh

- Vegetable production on water hyacinth-based floating bed, fruits and vegetable production in sorjon method, year-round fruit and vegetable cultivation on gher and dyke, and pyramid cultivation are the major technologies in waterlogging areas developed by BARI.
- "Floating bed farming" is a traditional innovation of the farmers, and BARI research helped improve the practice and introduce new ideas/techniques to farmers.
- Promising innovations prioritized by BARI (Satkhira) include:
  - (i) Spices (turmeric) on floating water hyacinth beds result in high yields.
  - (ii) Napier grass on floating beds to address fodder shortage issues in Bangladesh.
  - (iii) Pyramid cropping in areas where land is always under a few cms of water.
  - (iv) Cultivation of water chestnuts on floating beds.
  - (v) Off-farm activities (making handicrafts) using grass buss and the leaves of date palms to create earning opportunities for poor women.
  - (vi) Flying bed agriculture in submerged land.
  - (vii) Vegetable cultivation through vertical gardening in marginal land.
- Innovations for waterlogged areas prioritized by the Bangladesh Institute of Nuclear Agriculture (BINA) include:
  - (i) Submergence tolerant Aman rice Binadhan-11; crop duration, 110–115 days; yield, 5.5 t/ha in normal condition and 4.5/ha in submerged condition.
  - (ii) Submergence-tolerant Aman rice Binadhan-12, crop duration 125–130 days, can tolerate up to 20 days of submergence – yield 4.0–4.5 t/ha in normal condition and 3.5–4.0 t/ha in submerged condition.

(iii) Short-duration Aman rice Binadhan-16; crop duration, 100–105 days; yield, 5.0 t/ha.

(iv) Green super Aman rice Binadhan-17 requires 30% less urea fertilizer and 40% less water – crop duration, 112–118 days; yield, 6.8 t/ha.

Major Discussion Points

- Since water hyacinth is a naturally good source of organic matter, vegetables grown on water hyacinth beds tend to be organic. Developing linkages to urban markets and highlighting the heritage value of this practice could improve farmer remuneration.
- Pyramid cropping practice enables farmers to grow Rabi and Kharif-I (pre-monsoon cropping season that starts in April and ends in June) vegetables in relative lowlands that remain under a few centimeters of floodwater at the time of harvest of the transplanted Aman rice. Scientists urged to ensure that soils to be used for developing pyramids are free from salinity.
- Though water chestnut cultivation on floating beds is suggested by BARI, participants recommended developing the marketing channel to ensure the farmer's income.
- The suggested off-farm activities are promising, but the capacity of rural women needs to be developed, and micro SME needs to be established to promote such goods.
- As the vertical garden is a new technology, farmers need special training on these technologies.
- Though BINA proposed some waterlog-tolerant varieties, their yield performance requires to be examined widely before promoting.
- Late draining condition, poor polder management/defective sluice gate, low soil organic matter, late transplantation and harvesting of Aman rice, lack of waterlogging-tolerant crops, and poor communication and marketing facilities are impeding livelihoods in the waterlogging-prone areas.

CHT of Southeastern Bangladesh

- While there is a high potential for root crops and tubers in the hills, scientists had an equivocal opinion on whether this contributes to soil erosion.
- Both SRDI and BARI had several recommendations for hilly areas. These include:
  (i) Establishment of cowpea and country bean/yard-long bean using residual soil moisture just after harvest of Jhum crops
  (ii) Establishment of fruit orchards with BARI varieties (BARI Aam-1, 2, 3, 4, and 8, BARI Malta-1 and 2, BARI Orange-2, BARI Pumello-3 and 4, BARI Litchu-5, BARI Dragon fruit-1, etc.)
  (iii) Mixed fruit orchard and integrated mango and banana production
  (iv) Bagging of mangoes to improve the quality of fruit and prevent fruit flyer
  (v) Sugarcane, which can withstand flash floods and can be intercropped with vegetables like radish, soybean, and French beans
  (vi) Rainwater harvesting technology for supplementary irrigation in the 5 months of the dry season is necessary.
  (vii) Cultivation of high-value early winter vegetables just after rainfed Aman rice in the valley areas
  (viii) Homestead vegetables and fruit production following the modified Khagrachari model (a model for year-round homestead vegetable production in the hills was developed by Hill Agricultural Research Station, BARI, Khagrachari, called "Khagrachari model")
  (ix) Production of aroids (BARI Kachu-2) in the wet valley lands
  (x) Introduction of crops that require less water like high-value cucurbits (gourd family) and legumes (e.g., BARI Lau-4; BARI Teasel gourd-1; BARI Country bean-6, 9, and 10; BARI Bush bean-2 and 3)
  (xi) Summer tomato production using BARI Hybrid Summer Tomato-4

(xii) Hedge species in farmer fields (*Indigofera* – dyes, vetiver – personal care products) to reduce soil erosion
(xiii) Ornamental and medicinal plants also have good potential in CHT

- Innovations for sustainable land management and climate-resilient agriculture by the Bangladesh Institute of Nuclear Agriculture (BINA) include:
  (i) Short-duration Aus rice Binadhan-19 suitable for CHT, crop duration 95–100 days, average grain yield 4.5 cultivation t/ha
  (ii) Short-duration Aus rice Binadhan-21 suitable for CHT, crop duration 100–105 days, average grain yield 4.5 t/ha

Major discussion points

(i) Lakes, rivers, creeks, streams, or charas (small waterfalls) are the major source of water. Conservation and effective management are required for future water security in CHT.
(ii) CHT is a flood-free area. So, there is a good opportunity to cultivate monsoon crops.
(iii) The minimum soil moisture requires to be present before planting of cowpea and country bean seeds using residual soil moisture after the harvest of Jhum crops.
(iv) Participants agreed that CHT is suitable for growing high-value horticultural crops and vegetables but many of the landowners are absentee and water availability is a challenge in CHT. Unless water harvest technology is available on the hill, farmers will not go for the orchard and vegetable cultivation.
(v) As bagging of mangoes is suggested by BARI to improve the quality of fruit and prevent fruit flyer, farmers may not afford the high cost of the imported bag. The locally produced low-cost bag may solve this problem.
(vi) The modified Khagrachari model may not work in CHT if the soil fertility problem and irrigation situation are not improved.
(vii) Participants opined that communication and marketing are the major problems in CHT. Farmers will not get the return properly unless or until the communication and marketing channel are improved in this region.
(viii) The establishment of agro-processing industry and value-added product development based on mango, banana, radish, soybean, French bean, and sugarcane is required for the overall socioeconomic development and climate resilience.

Conclusion

Climate-resilient technologies, innovation, and sustainable land management practice can help to contribute to livelihood protection, vulnerability reduction, productivity enhancement, emission reduction, and food security identified in the Bangladesh Climate Change Strategy and Action Plan (BCCSAP), National Adaptation Programme of Action (NAPA), seventh and eighth 5-year plan, and new agricultural policy 2018 (GED 2018; Béné et al. 2015; Arfanuzzman et al. 2020). Scaling up of promising agricultural technologies and innovations can enhance inclusive growth, resilient food system, rural development, environmental protection, and natural resource sustainability (FAO 2019; CIAT and World Bank 2017; Hossain et al. 2015; Carrijo et al. 2017). Nevertheless, the agricultural technologies of Bangladesh need to be designed and promoted considering the geographical situation, hydroclimatic condition, natural resource availability, and ecological vulnerability (i.e., drought and groundwater depletion in HBT, land degradation and irrigation water availability in CHT, increasing salinity and waterlogging in the southwest region). No crop, technology, or management practice will sustain unless these ecological challenges are solved.

Acknowledgment This work was funded by the Global Environment Facility (GEF) through the Project Preparation Grant for Building climate-resilient livelihoods in vulnerable landscapes in Bangladesh (GCP/BGD/069/LDF).

Conflicts of Interest The authors declare no conflict of interest. The funders had no role in the design of the study; in the collection, analyses, or interpretation of data; in the writing of the manuscript; or in the decision to publish the results.

Disclaimer The views expressed in this entry are those of the authors alone and do not necessarily reflect the views of the Food and Agriculture Organization (FAO) of the United Nations, Center for Policy and Economic Research (CPER), and the Global Environment Facility (GEF).

References

Arfanuzzaman, M., Hasan, T., & Syed, A. (2020). *Cost-benefit of promising adaptation for resilient development in climate hotspots: Evidence from Teesta river basin, water and climate change.* IWA Publishing. https://doi.org/10.2166/wcc.2020.130.

Aziz, M. A., et al. (2015). Groundwater depletion with expansion of irrigation in barind tract: A case study of Rajshahi district of Bangladesh. *International of Journal of Geology Agriculture Environment Science, 3*, 32–38.

Béné, C., Waid, J., Jackson-deGraffenried, M., Begum, A., Chowdhury, M., Skarin, V., Rahman, A., Islam, N., Mamnun, N., Mainuddin, K., & Amin, S. M. A. (2015). *Impact of climate related shocks and stresses on nutrition and food security in selected areas of rural Bangladesh.* World Food Programme (WFP): Dhaka.

Carrijo, D. R., et al. (2017). Rice yields and water use under alternate wetting and drying irrigation: A meta-analysis. *Field Crops Research, 203*, 173–180.

CIAT, & World Bank. (2017). *Climate-smart agriculture in Bangladesh: CSA country profiles for Asia series.* International Center for Tropical Agriculture (CIAT). Washington, DC: World Bank, 28 pp.

Eckstein, D., Hutfils, M., & Winges, M. (2019). *Global climate risk index 2019. Who suffers the Most from extreme weather events? Weather-related loss events in 2017 and 1998 to 2017* (Briefing paper). Bonn: Germanwatch.

FAO. (2019). *Agriculture and climate change challenges and opportunities at the global and local level collaboration on climate-smart agriculture.* Rome: Food and Agriculture Organization of the United Nations.

GED. (2018). *Bangladesh delta plan 2100.* Dhaka: General Economics Division, Bangladesh Planning Commission, Ministry of Planning.

Hossain, M. I., et al. (2015). Status of conservation agriculture based tillage technology for crop production in Bangladesh. *Bangladesh Journal of Agricultural Research, 40*(2), 235–248.

MoEF. (2009). *National adaptation program of action (NAPA), updated version of 2005.* Ministry of Environment and Forests (MoEF): Dhaka.

MoFA. (2018). *Climate change profile Bangladesh.* The Hague: Ministry of Foreign Affairs of the Netherlands.

Nasim, M., et al. (2017). Distribution of crops and cropping patterns in Bangladesh. *Bangladesh Rice Journal, 21*(2), 294.

UN (United Nations). (2019). Department of Economic and Social Affairs, Population Division. 2019. World Population Prospects 2019: Data Booklet (ST/ESA/SER.A/424).

C

Climate-Responsive Urban Areas

► Urban Climate Resilience

Climatic Variations and Surface Water Resources in Sri Lanka

► Climate Change and Surface Water Resources in Sri Lanka

Closing the Loop on Local Food Access Through Disaster Management

Kimberley Reis and Cheryl Desha
Cities Research Institute, Griffith University, Brisbane, QLD, Australia

Synonyms

Administration; Adversity; Area; Contribute; Coordinate; County; Distribute; District; Involvement; Municipal; Neighbourhood; Participate; Precedent; Provincial; Provision; Public; Regional; Resource; Secure; Share; Stock; Sustenance

Definition

Local and regional food resilience is enhanced when governments, businesses, and community groups work together for developing and maintaining short food supply chains. Local food contingency plans aim to formalize arrangements, so we can all share in responsibility for our food-related disaster resilience.

Introduction

When we consider the principles of "circular economy" (i.e., sustainable flows of materials and energy through our economies around the world), it is evident that our current food systems of production and supply require excessive energy, water, nutrients, and other materials. Over the last 200 years since the Industrial Revolution, food has transitioned from being a locally produced and locally consumed commodity to being globally sourced, stored, and traded virtually anywhere on the planet. While our grandparents ate fruit and vegetables from the local area on the same day or week that it was picked, our children are eating fruit and vegetables that are often from a different country and which may have been stored for more than one season. There is an urgent need to "close the loop" on our demand for inputs, encouraging local food systems.

There are many environmental impacts of food traveling long distances from "paddock to plate." Within five generations, our relationship with food has fundamentally shifted, with environmental and social consequences. For example, global supply chains mean earlier harvest of unripened fruit, the use of chemicals to reduce bacterial or fungal problems, long-term cold storage, and a mix of trucking, shipping, air freight, and rail transport to reach the end consumer. While yoghurt and oranges literally pass each other in the air or by ship, for consumers, (demanding different brands for breakfast), our generation is consuming food that has unprecedented embodied food miles. There are also immediate cultural implications of such a significant shift in food patterns, wherein our generation has very limited appreciation of food seasonality – i.e., when food would naturally ripen and become available locally. There is also a very limited appreciation of the natural variety of various foods such as carrots, when we almost always see the standard orange straight carrots on the supermarket shelf.

In the face of significant climate change disruptions and other disruptions such as natural disasters, public health, and cyber security, there is also the urgent need to consider alternative food options (i.e., food contingencies). Long food supply chains have numerous points in the system that are fragile in the face of disruption. This type of vulnerability is substantially different to the locality-based vulnerabilities of our previous "local food" systems last century. Whereas local weather, soil type, and access to labor were the previous major constraining factors, today's supply chain vulnerabilities are embedded in remote production through to large-scale harvesting, en masse processing, cold storage, transport, purchase, and finally consumption. The extent of vulnerabilities is particularly evident during and immediately following severe weather events and other disasters such as pandemics, when storage facilities and transportation routes may be disrupted. The Covid-19 pandemic has highlighted the fragility of food supply systems when consumption behaviors are disrupted – in this case physical distancing disrupted the stocking rates for supermarkets and how food was transported from supermarket to the home.

Our "closing the loop" metaphor is derived from our "call to action" within the Australian disaster management community to enable local food contingency planning (see Fig. 1) (Reis et al. 2019).

In the spirit of the principles of "circular economy," these five actions aim for greater degrees of sustainable flows of local food materials and energy through our local economies around the world. Within this context, this chapter presents the context for "local food contingency" as a way forward in addressing circular economy principles within food production and supply and resilience in the face of food security challenges. Using a disaster management lens

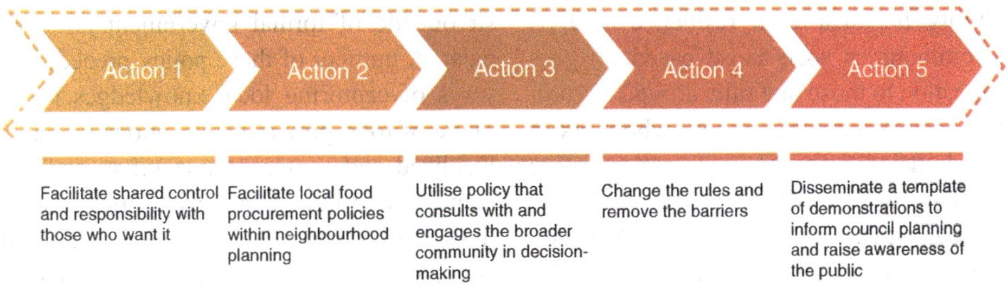

Closing the Loop on Local Food Access Through Disaster Management, Fig. 1 Actions to close the loop for enabling local food contingency planning

C

(i.e., considering prevention, preparedness, response, and recovery), the authors draw on their research experiences in three Australian examples, to highlight perspectives and opportunities arising for urban and regional food futures. The following sections overview the concept of "shared responsibility" in disaster management and the potential for using this concept to build community-level food resilience. Regulatory and policy precedents for such contingency planning are presented, which are helping the transition from global long supply chains to shorter regional and urban supply chains that are good for people and planet. The chapter concludes with a call to use disaster management initiatives and thinking to inform improved decision-making for food resilience and, in turn, resilient communities.

Food Insecurity

Each year, Australian records are broken for higher temperatures, with the 2019–2020 bushfires ushering the highest "Forest Fire Index" since records began (Australian Government 2020). Severe drought conditions are punctuated by flood events that are more intense and occur more often. The impacts to farming communities and the road and communication infrastructures to distribute and access our food are ongoing and relentless. Compounding this is the global Covid-19 pandemic, revealing a stressed food system with emptied shelves (Reis et al. 2020).

The Food and Agriculture Organization of the United Nations (FAO) states that: "Food security exists when all people, at all times, have physical, social and economic access to sufficient, safe and nutritious food which meets their dietary needs and food preferences for an active and healthy life" (2018, p. 1). In spite of stockpiling in anticipation of food scarcity, food insecurity is experienced across a spectrum from hunger and insufficient food to worrying about relying on food handouts and anxiety about running out of food (Reis and Ferreira 2015).

As a wealthy nation, many do not think of Australia as a country that can experience food insecurity. The Australian Government Institute of Family Studies identifies groups that experience food insecurity at a higher rate than the general population, including indigenous people, unemployed people, single parent households, low-income earners, rental households, and young people. Additionally, those who are more vulnerable to food insecurity include some culturally and linguistically diverse groups; refugees; those without access to transport options to acquire food; those who experience alcohol and substance misuse; and those who are disabled, unwell, or frail. These people are on the front line, relying upon food aid when times get tougher than normal (Rosier 2011).

Australian Disaster Risk Management and Shared Responsibility

The Australian Government, through its *National Strategy for Disaster Resilience,* intends that the capacities and confidence of individuals and communities should be improved so they may share responsibility for their own resilience (COAG 2011). Communities span those based on geographic location to those that rally around circumstances or areas of interest, to those who work in particular domains – communities are therefore rich and contextualized (COAG 2011).

To this end, Australian Disaster Risk Management (DRM) has incorporated engagement with communities as core business for building resilience (Australian Government 2013a). There is a rich diversity with and among urban and rural contexts. Acknowledging that "one-size" solutions do not work for all communities, there is an emerging movement within DRM toward grassroots and place-based approaches, already mainstreamed in the social services sector, that give voice to the local knowledges and the varied risk perceptions that are essential for building resiliencies (Australian Government 2013a; QCOSS 2020). These resiliencies, depending on the circumstance, may be focused upon withstanding events, bouncing back to "normality," or morphing to new ways of being (Reis 2019). There is much diversity to draw upon and consider here in terms of what will be the "right" mix of actions for each circumstance.

Local Food Procurement

Local food procurement is generally about shortening food supply lines in a way that minimizes actors who handle and trade food between the producer and consumer (Reis 2019). More common examples can include farmer's markets; community-supported agriculture; community and school gardens; farming and distribution cooperatives; online delivery systems; a variety of urban farms including city farms, mini farms, and pop-up farms; and charities that repurpose food "waste" for food aid. These innovations tend to be driven by grassroots efforts and reflect the values of the places and spaces under which the efforts are undertaken.

In the absence of a national food policy, and in the absence of a food policy that meaningfully considers grassroots and place-based approaches to the food supply chain – a range of policies that sit outside of formal government arrangements exist. A number of these policy precedents show ways of capturing local knowledges to build a range of food procurement options that are more localized and centered upon urban and regional contexts that have relatively shorter food supply lines. For example, by demonstrating alternative and complementary models, the "bounce-back" mentality to the "business-as-usual" model of reliance on longer and globalized supply chains is questioned (Australian Food Sovereignty Alliance 2013). Furthermore, propositions on modifying the governance and hence leadership of farming and food practices are offered (Australian Conservation Foundation 2009) including reimagining urban planning and design for improving human health outcomes (Heart Foundation 2011). These kinds of policy considerations sit at the heart of improved public health outcomes (Public Health Association Australia 2012). Networks that build upon the values of betterment for their groups and communities and act upon those values in a coordinated way are known to offer a greater variety of food options than circumstances where this dynamic is absent (Crowe and Smith 2012).

Contingency Planning and Local Food: An Emerging Alliance to Enhance Food Access

In a time of uncertainty around food supply with impacts from severe weather events and pandemic conditions, contingency planning plays a key role in clarifying what we will do when things go wrong. When we ask the "what if" questions ahead of the events, we can undertake timely and meaningful action to be prepared ahead of the next event and hence reduce the severity of impacts and recovery (IFRC 2012; EMQ 2012). Having a Plan B and a Plan C is prudent, practical, and necessary (Reis et al. 2020; Chakraborty et al. 2011).

The community engagement model for Australian DRM shows a clear scope to gauge the existing knowledge, experiences, and strengths of the local food procurement networks.

Through the explicit acknowledgement that these varied communities represent varied perceptions of risk to the existing food supply chain, more formal arrangements may be made to partnering with these networks and mobilize their strategies to empower local action – where communities want it (Australian Government 2013a). This invites a broader, networked governance approach (Howes et al. 2014) that lessens the pressure of the traditional industry-government-based services that form the longer supply chains, to step forward with solutions for all circumstances (Reis 2013).

Our project entitled "Enabling action for local food resilience and contingency" is a call to action to forward the emerging alliance between contingency planning and local food initiatives. See the project website https://www.griffith.edu.au/cities-research-institute/research/digital-earth-and-resilient-infrastructure/food-contingency located within the Cities Research Institute (Griffith University). The main objective of this project is to formalize local food resilience and contingency arrangements for urban and regional centers. Food provides an important enabler of community resilience to disruptions such as weather extremes, pandemic, and other disaster-related response and recovery. Without food-system knowledge, we are missing opportunities to nurture and build upon local leadership in food-related community endeavors and organize needed contingency arrangements.

Australian Example

The Australian Government has set the national agenda for shared responsibility to build resilience to disasters. Furthermore, it calls for a "lessons management" approach where projects may identify and share the learnings from previous experiences. This is seen as an essential pathway for raising the awareness of the public and showing the possibilities of what can be achieved (Australian Government 2013b). Governments at the state level are responsible for guiding and funding statewide initiatives. In partnership with the Queensland Government's Office of the Inspector-General Emergency Management (IGEM), the longer-term goal of our project is to establish an online "community of practice" to enable nationwide action among local councils and community groups to access shorter and more locally based food supply chains and showcase the achievements of their demonstration sites. However, it is the local governments that are responsible for leading community resilience efforts (Queensland Government 2017).

The following paragraphs summarize three examples of work undertaken in local council districts in the State of Queensland, from the far north of the state to its southern border. As examples of innovators and early adopters (Balfour and Alter 2016) in local food procurement for contingency plans, these are demonstration sites/examples of what may be replicated to suit the circumstances and aspirations of other local councils.

Cairns: Regional Perspective

The Far North Queensland (FNQ) region is a World Heritage area in the state of Queensland, spanning the Daintree rainforest to the Great Barrier Reef. At approximately 380,000 km^2 (the size of Japan or Zimbabwe), it contains the expansive Cape York Peninsula and three local government authorities (Aurukun Shire, Burke Shire and the Cairns Region). The Cairns region sits within the state's cyclone corridor and for many decades has been working on documentation and community awareness about local-level empowerment to prevent, prepare, respond, and recover. In 2019–2020, the authors have been working with the Disaster Management Unit of the Cairns Regional Council to consider how the disaster management processes and protocols for cyclones could be enhanced with a documented approach to local food that rewards local initiative in the production and supply of local food all year round.

In Stage 1 of the pilot trial project (during February 2020–2021), the Council's existing "Disaster Resilience Strategy" was reviewed for places to integrate plans for accessing local food options and formalizing contingency arrangements for accessing those options. The particular focus in Stage 1 was to facilitate access to local food options to the organizations that service the care and advocacy of the most vulnerable members in the Cairns Region. This includes those experiencing homelessness, disability, and aging. It includes indigenous, multicultural groups and refugees where English may not be the first spoken language. Disconnected youth and those experiencing multiple complex issues are also included (Reis et al. 2021).

Stage 2 (commencing March 2021 to January 2022) of the pilot trial will create an online "Local Food Resilience Hub" on the Cairns Regional Council web pages that aims to provide an online presence of local food options available in the region. This undertakes a holistic approach to food that considers groups and enterprises that may strengthen social and local economic benefits. Prospects within the existing Disaster Dashboard (Cairns Regional Council 2020) will be considered to formalizing local food access options in anticipation of and during disaster events.

Logan City: Growth Corridor Perspective

Logan City is in the South East Queensland (SEQ) region and experiences the highest growth rate in the nation. At 958 km^2 (half of the size of New Caledonia), it is bordered by the City of Brisbane (north), the City of Gold Coast (south), and three other smaller local government authorities. Logan City is multicultural and home to more than 310,000 people from more than 215 different cultures where more than 86,000 were born overseas (Logan Together 2020). It is also one of the fastest growing areas in Australia, with an estimated 200,000 people arriving over the next two decades.

This project builds on and opens a social innovation and entrepreneurship angle to the work led by the Cities Research Institute and involving researchers from across the Griffith Business School, the Logan Campus, and a number of key partner organizations in the Logan region, including Logan City Council. It aims to building a demonstration of what can be achieved for a resilient food ecosystem in a growth corridor context. The research team is creating an evidence base in the form of an interactive map, report, and short videos, to support local and regional decision-makers regarding food initiatives and food production in Logan growth corridor.

The focus of the project is to map the existing food ecosystem in the Logan region including the environment- and climate-related vulnerabilities that may impact food access. This interactive GIS map also identifies the pockets of socioeconomic disadvantage that will benefit from access to the local food sites. It details key aspects of food resilience in the Logan region, which can be overlaid with SEIFA data (socioeconomic disadvantage), data associated with the location of food enterprises, land use data, and future development data. The map will enable identification of clusters of relative vulnerability and strengths. The aim is to enable emergency food relief to tap into the possibilities and over the longer term to enable broader entrepreneurship and social innovation around local food. Please access the map here: https://regionalinnovationdatalab.shinyapps.io/Logan_Food_Mapping/. As the project links to other food ecosystem initiatives undertaken by the Cities Research Institute, the goal is to work with the Policy Innovation Hub to undertake a meta-analysis of this work that could build and inform a Food Resilience Policy framework for local and state governments.

Lockyer Valley: Rural Perspective

The Lockyer Valley is a rural food-producing area located in the southeast corner of Queensland, between the cities of Ipswich and Toowoomba. At 2200 km^2 (slightly larger than Mauritius), it is rated among the top 10 most fertile farming areas in the world (Lockyer Valley Regional Council 2011), providing Queensland and Australia with a "national food bowl" of produce throughout the year.

The Lockyer Valley has also featured over the last decade for its exposure to the extremes of drought, flood, and fire. In recent years, the Council has been piloting opportunities for community-based contingency arrangements. The author's project involves working with the Council and local community groups to explore how business continuity planning (BCP) can be enhanced with local food information. In particular, the team is exploring BCPs as an avenue to create contingency arrangements for accessing locally produced food to support family farms (rural producers).

Conclusion

This chapter explored the opportunity for "local food contingency" to address the twenty-first-century food resilience in the face of food security challenges. There are significant opportunities for using a disaster management lens to address food security challenges, through planning for prevention, preparedness, response and recovery procedures and protocols that could embed access to local food as a consideration. The three Australian examples highlighted next step opportunities for urban and regional areas, including the importance of "shared responsibility" in building community-level food resilience. We invite readers to consider harnessing the existing energy around disaster management initiatives and thinking, to inform and integrate improved decision-making for food resilient communities.

Cross-References

- ▶ Community Engagement for Urban and Regional Futures
- ▶ Future of Urban Governance and Citizen Participation
- ▶ Participatory Governance for Adaptable Communities
- ▶ Planning for Peri-urban Futures
- ▶ Public Procurement for Regional and Local Development
- ▶ The Practice of Resilience Building in Urban and Regional Communities
- ▶ Urban and Regional Leadership

References

Australian Conservation Foundation. (2009). *Paddock to plate: Policy propositions for sustaining food and farming systems. The future food and farm project propositions paper.* Prepared by A, Campbell, Melbourne. Available: https://apo.org.au/sites/default/files/resource-files/2009/10/apo-nid19512-1149991.pdf

Australian Food Sovereignty Alliance. (2013). *People's food plan: A common-sense approach to a fair, sustainable and resilient food system. Revised edition following community input.* Available: https://afsa.org.au/wp-content/uploads/2012/11/AFSA_PFP_WorkingPaper-FINAL-15-Feb-2013.pdf

Australian Government. (2013a). *Australian disaster resilience handbook 6: National Strategy for disaster resilience: Community engagement framework.* Australian Institute for Disaster Resilience. Commonwealth of Australia. Available: https://knowledge.aidr.org.au/resources/handbook-6-community-engagement-framework/

Australian Government. (2013b). *Australian disaster resilience handbook 8: Lessons management.* Australian Institute for Disaster Resilience. Commonwealth of Australia. Available: https://knowledge.aidr.org.au/media/1760/handbook-8-lessons-management-kh-final.pdf

Australian Government. (2020). *Royal Commission into national natural disaster arrangements report, October 28.* Available: https://naturaldisaster.royalcommission.gov.au/system/files/2020-11/Royal%20Commission%20into%20National%20Natural%20Disaster%20Arrangements%20-%20Report%20%20%5Baccessible%5D.pdf

Balfour, B., & Alter, T. (2016). Mapping community innovation: Using social network analysis to map the interactional field, identify facilitators, and foster community development. *Community Development, 47*(4), 431–448.

Cairns Regional Council. (2020). *Cairns disaster dashboard.* Available: http://disaster.cairns.qld.gov.au/

Chakraborty, A., Kaza, N., Knaap, G., & Deal, B. (2011). Robust plans and contingent plans. *Journal of the American Planning Association, 77*(3), 251–266.

Council of Australian Governments (COAG). (2011). *National strategy for disaster resilience: Building our nation's resilience to disasters.* Barton: NEMC Working Group tasked by the COAG. Available: https://www.ag.gov.au/EmergencyManagement/Documents/NationalStrategyforDisasterResilience.PDF

Crowe, J., & Smith, J. (2012). The influence of community capital toward a community's capacity to respond to food insecurity. *Community Development, 43*(2), 169–186.

Emergency Management Queensland (EMQ). (2012). *Queensland re-supply guidelines.* Brisbane: State of Queensland. Available: http://www.disaster.qld.gov.au/dmp/Archive/Documents/Queensland%20Resupply%20Guidelines.pdf

Food and Agriculture Organisation of the United Nations (FAO). (2018). *Food security statistics.* Available: http://www.fao.org/economic/ess/ess-fs/en/

Heart Foundation. (2011). *Food-sensitive planning and urban design: A conceptual framework for achieving a sustainable and healthy food system*. Melbourne: VicHealth and VEIL. Available: http://www.healthyplaces.org.au/userfiles/file/Design%20elements/foodsensitive_planning.pdf

Howes, M., Tangney, P., Reis, K., Grant-Smith, D., Heazle, M., Bosomworth, K., & Burton, P. (2014). Towards networked governance: Improving interagency communication and collaboration for disaster risk management and climate change adaptation in Australia. *Journal of Environmental Planning and Management*, 1–20. https://doi.org/10.1080/09640568.2014.891974.

International Federation of Red Cross and Red Crescent Societies (IFRC). (2012). *Contingency planning guide*. Geneva. Available: http://www.ifrc.org/PageFiles/40825/1220900-CPG%202012-EN-LR.pdf

Lockyer Valley Regional Council. (2011). *Lockyer Valley community recovery plan* (p. 13). Available: http://www.floodcommission.qld.gov.au/__data/assets/pdf_file/0005/8528/QFCI_Exhibit_173_-_Lockyer_Valley_Community_Recovery_Plan.pdf

Logan Together. (2020). About Logan. Available: https://logantogether.org.au/why/about-logan/

Public Health Association Australia. (2012). A future for food 2: Healthy. Sustainable, Fair. PHAA Food and Nutrition Special Interest Group. Available: http://www.phaa.net.au/documents/item/562

Queensland Council of Social Services (QCOSS). (2020). Place-based approaches. Available: https://www.qcoss.org.au/our-work/place-based-approaches/

Queensland Government. (2017). *Queensland strategy for disaster resilience 2017: Making Queensland the Most disaster resilience state in Australia*. Brisbane: State of Queensland. Available: http://qldreconstruction.org.au/u/lib/cms2/QLD-Strategy-for-Disaster-Resilience.pdf

Reis, K. (2013). *Food for thought: The governance of garden networks for building local food security and community-based disaster resilience*. PhD dissertation, Griffith University. Available: https://research-repository.griffith.edu.au/handle/10072/366226

Reis, K. (2019). Five things government can do to facilitate local food contingency plans. *Journal of Environmental Planning and Management, 63*(13), 2295–2312. https://doi.org/10.1080/09640568.2018.1540772.

Reis, K., & J. Ferreira. (2015). Community and school gardens as spaces for learning social resilience. *Canadian Journal of Environmental Education, 20*, 63–77. Available: https://cjee.lakeheadu.ca/article/view/1341/843

Reis, K., Desha, C., & Rifai, A. (2019). Planning for food contingencies: A call to action. *Australian Journal of Emergency Management*, 14–15. Available: https://knowledge.aidr.org.au/resources/ajem-october-2019-planning-for-food-contingencies-a-call-to-action/

Reis, K., Desha, C., & Burton, P. (2020). We've had a taste of disrupted food supplies: Here are 5 ways we can avoid a repeat. *The Conversation*, 4 May. Available: https://theconversation.com/weve-had-a-taste-of-disrupted-food-supplies-here-are-5-ways-we-can-avoid-a-repeat-135822

Reis, K., Desha, C., Bailey, M., Liddy, P., & Campbell, S. (2021). Towards local food resilience. Key considerations for building local food resilience and contingency plans: A focus on the Cairns region. Final Report, 26 March. Cities Research Institute in consultation with the Cairns Regional Council. Available: https://www.griffith.edu.au/__data/assets/pdf_file/0029/1334297/CRI-Towards-local-food-resilience-research-report.pdf

Rosier, K. (2011). *Food insecurity in Australia: What is it, who experiences it and how can child and family services support families experiencing it?* Prepared for the Australian Government, Institute of Family Studies. Available: https://aifs.gov.au/cfca/publications/food-insecurity-australia-what-it-who-experiences-it-and-how-can-child

Coastal Management

► Role of Disaster Relief Policy in Building Resilient Coastal Regions in the United States

Co-creation Processes

► Social Urbanism: Transforming the Built and Social Environment

Collaborative

► Participatory Planning: A Useful Tool for the Development of Sustainable Mega-City Regions

Collaborative Adaptation

► Collectively Adapting to Sea-Level Rise Through Disaster Response, Commons Management, and Social Mobilization

Collaborative Climate Action

Monika Zimmermann
Urban Sustainability Expert & Former Deputy Secretary General of ICLEI, Freiburg, Germany

Synonyms

Multi-level governance; Vertical integration

Definition

Collaborative Climate Action (CCA) is defined as the politically intended, well-organized cooperation across different levels of government to achieve – ideally jointly defined – climate goals. Effective and more ambitious climate protection can only be achieved with and through CCA.

Summary

Collaborative Climate Action as a term has been coined a few years ago. One could understand it as a general notion for cooperation between different types of actors.

Here, Collaborative Climate Action (CCA) is defined as the politically intended, well-organized cooperation across different levels of government to achieve – ideally jointly defined – climate goals. Effective and more ambitious climate protection can only be achieved with and through CCA.

CCA contributes to coherence and consistency in climate policy, from the international to the local level. Many refer to this coherence as the "architecture of climate policy" or climate governance. CCA is a part of it, normally within a country. Looking to the European Union, however, CCA in European countries also includes this regional government level. On the international level, cities and regions argue for many years that their contribution is key for fulfilling climate targets.

Climate Collaboration among Government Levels

This (or: the following) approach to CCA has been designed within the project "Climate Policy Meets Urban Development (CPMUD) (https://www.giz.de/en/worldwide/75947.html)," which also supports the Partnership for Collaborative Climate Action (https://collaborative-climate-action.org/), launched by the German Ministry for the Environment (BMU) in 2019 on the occasion of the International Conference on Climate Action (ICCA2019) in Heidelberg, Germany (https://www.icca2019.org/). This partnership invites other governments and key actors to join efforts for promoting Collaborative Climate Action and has 61 members at the beginning of 2021, among these eight national governments (https://collaborative-climate-action.org/partner/) (Graph 1).

CCA has two main and interrelated messages (Graph 2):

1. When all levels of government utilize their competences, tasks, budgets, experts, contacts, and cooperation partners in a targeted and coordinated manner, they can achieve more together than each actor alone. At the same time, joint and coordinated action prevents the creation of contradictory incentives and can save significant resources, whether it be money, personnel, or time – as those involved support one another.
2. The subnational, particularly the local, level is a key actor and must gain the necessary recognition and support. The reason for the importance of the subnational level is obvious and has been described in many publications (here, a link within the Palgrave Encyclopedia of Urban and Regional Futures would be great, rather than to write a particular text about this hat would disturb the text flow).

National and regional governments have primary authority over around 35% of urban abatement potential, and municipalities are responsible for roughly 28%. However, the highest level of

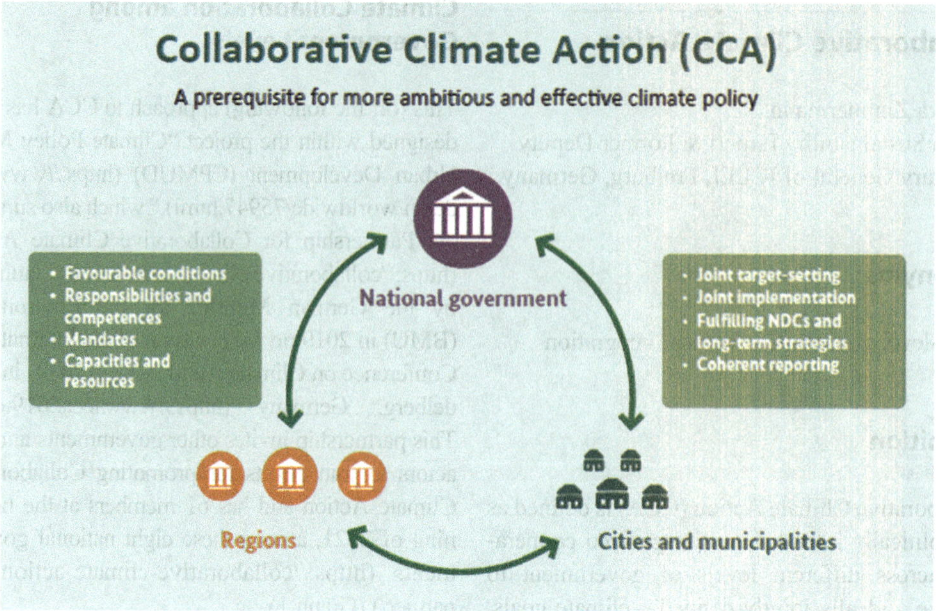

Collaborative Climate Action, Graph 1 Collaborative Climate Action refers to the cooperation of (all) government levels (Zimmermann et al. 2021)

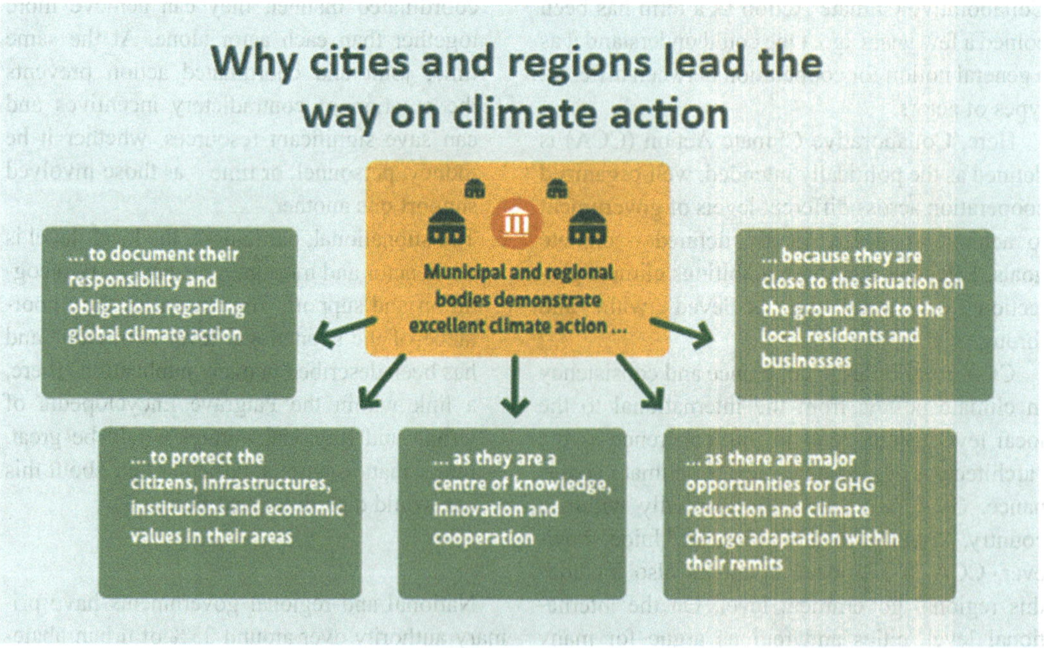

Collaborative Climate Action, Graph 2 Cities and regions are key for successful climate action and can unfold their potential when being strong and accepted partners of national governments (Zimmermann et al. 2021)

abatement potential (approx. 37%) can only be tapped through joint action by all government levels, according to the Climate Emergency – Urban Opportunity report (not taking into account decarbonization of the power sector) (https://coalitionforurbantransitions.org/urban-opportunity/) (Graph 3).

Levels of Government …

… are the public actors (parliaments, governments, administrations) within a country between which tasks, competences, and resources are split.

C

(continued)

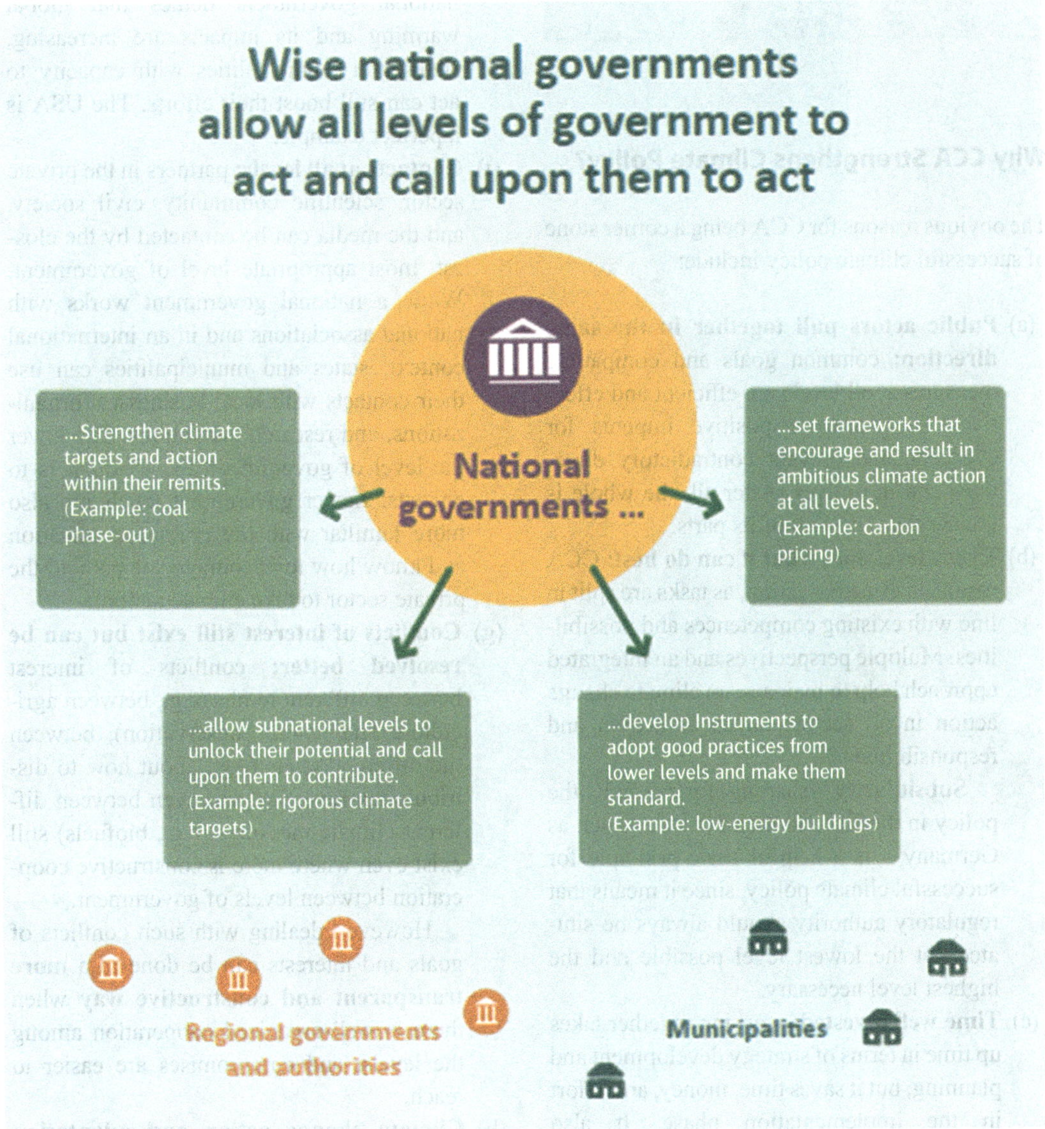
Wise national governments allow all levels of government to act and call upon them to act
National governments …
…Strengthen climate targets and action within their remits. (Example: coal phase-out)
…set frameworks that encourage and result in ambitious climate action at all levels. (Example: carbon pricing)
…allow subnational levels to unlock their potential and call upon them to contribute. (Example: rigorous climate targets)
…develop instruments to adopt good practices from lower levels and make them standard. (Example: low-energy buildings)
Regional governments and authorities
Municipalities

Collaborative Climate Action, Graph 3 National governments have many reasons to strengthen the subnational level (Zimmermann et al. 2021)

All countries have a national government and municipalities; many have other levels in between, such as territories, states, provinces, and districts.

Federations assign special competences to the state level, which may result in disparities between states, for instance, when they pass different legislations for their municipalities.

Why CCA Strengthens Climate Policy?

The obvious reasons for CCA being a corner stone of successful climate policy include:

(a) **Public actors pull together in the same direction:** common goals and compatible measures at all levels are efficient and effective. They provide positive impetus for everyone and prevent contradictory directives and incentives. After all, the whole is greater than the sum of its parts.

(b) **Every level does what it can do best:** CCA results in effective action, as tasks are split in line with existing competences and possibilities. Multiple perspectives and an integrated approach help to mainstream climate change action in all sectors, areas of action, and responsibilities.

Subsidiarity – shaping, for example, the policy in the EU and member states such as Germany – is a helpful basic principle for successful climate policy, since it means that regulatory authority should always be situated "at the lowest level possible and the highest level necessary."

(c) **Time well invested:** working together takes up time in terms of strategy development and planning, but it saves time, money, and effort in the implementation phase. It also increases the acceptance of measures and thus reduces subsequent (time-consuming) conflicts.

(d) **Ownership:** the more an actor is involved in developing and planning goals and activities (and this also applies to government levels), the more willing it is to get involved in implementation and overcoming hurdles.

(e) **CCA increases the chances of implementation:** even if one level of government is "absent" (e.g., political reluctance or financial constraints), other levels can continue or even step up their activities. For instance, if a national government denies that global warming and its impacts are increasing, states and municipalities with capacity to act can still boost their efforts. The USA is a perfect example.

(f) **Contacts at all levels:** partners in the private sector, scientific community, civil society, and the media can be contacted by the closest, most appropriate level of government. While a national government works with national associations and in an international context, states and municipalities can use their contacts with local businesses, organizations, and research institutions. The lower the level of government, the closer it is to citizens. Lower government levels are also more familiar with the grassroots situation and know how to encourage society and the private sector to take climate action.

(g) **Conflicts of interest still exist but can be resolved better:** conflicts of interest between different remits (e.g., between agriculture and nature conservation), between the different levels (e.g., about how to distribute tax revenue), and even between different climate measures (e.g., biofuels) still exist even where there is constructive cooperation between levels of government.

However, dealing with such conflicts of goals and interests can be done in a **more transparent and constructive way** when there is well-organized cooperation among the levels, and compromises are easier to reach.

(h) **Climate change action and adaptation always considered:** the better levels of government work together, the greater the likelihood that climate issues will be incorporated into all decisions. This integrated approach, also known as mainstreaming, is also the objective of the 17 UN Sustainable Development Goals.

(i) **CCA represents a modern governing style:** many governments across the globe work together to pool their strengths and boost each government level's possibilities for action.

(j) **CCA makes everyone stronger:** climate cooperation makes everyone stronger; there are no losers. Climate cooperation among all levels of government helps to achieve joint climate goals and achieve success at each individual level. Each level wins because all those involved combine their potential. Instead of a feared loss of power (particularly national governments often act according to such a fear), actually each level gains recognition. Rather than worrying about having to share funding, there is a joint responsibility for future-proof, competitive, and socially responsible investment in the future. Instead of limiting international contacts, additional sources of funding for municipalities and regions can be unlocked.

How to Achieve "Better" CCA?

Cooperation between governmental levels can be sporadic, voluntary, and temporary. Its real strength as a politically intended, well-organized, and long-term approach comes to the fore when constitutions and laws define competences and rules for cooperation and when predictable procedures determine and define how joint targets can be reached. Stronger, institutionalized cooperation across levels of government places demands on all actors: old routines must be adjusted, and new behavior must be adopted.

When the national government and provinces, states and municipalities, or national and subnational government units agree that they can only achieve success by working together, a whole host of opportunities to act opens up, including (Graph 4):

- **From exchanging ideas to institutionalized cooperation**: Meeting and exchanging information across different levels of government can evolve from one-off meetings and temporary processes to safeguarded participatory rights and duties. Direct exchange is always at the start of real cooperation. Participatory rights enshrined in law are the most reliable form of cooperation. Those involved know their rights and duties with regard to decision-making and implementation. In this case, all levels of government must contribute to ambitious climate policy, and citizens can demand action. An excellent example is the co-legislative rights of parliament chambers representing states and regions, thus as the senate in the USA or the Bundesrat in Germany. Here, this second chamber refused to agree to the government's proposal for a national climate law at the end of 2019 and pushed through a change, resulting in higher CO2 taxes, for example.
- **Joint target setting and planning:** Different levels of government can set their own climate goals, derive them (e.g., municipalities adopt their government's climate targets), aggregate them (e.g., governments present the sum total of their municipalities' climate targets), or, ideally, develop them together and coordinate with one another. All national climate plans should always incorporate municipalities as a key level of action, urge them to take action, and make improvements to the framework conditions and the necessary support available.
- **Mandatory incorporation of climate policy at all levels**: Mainstreaming climate change mitigation and adaptation means creating predictable mechanisms that are effective in the long term, rather than making isolated decisions. Examples might include laws, municipal statutes, basic resolutions, taxes, incentives, subsidies, income opportunities for the different levels, defined competences and human resources, and recurring processes, for example, in sustainability management and links with budgetary law.
- **Take-up and scaling up:** National governments have a special role to play in

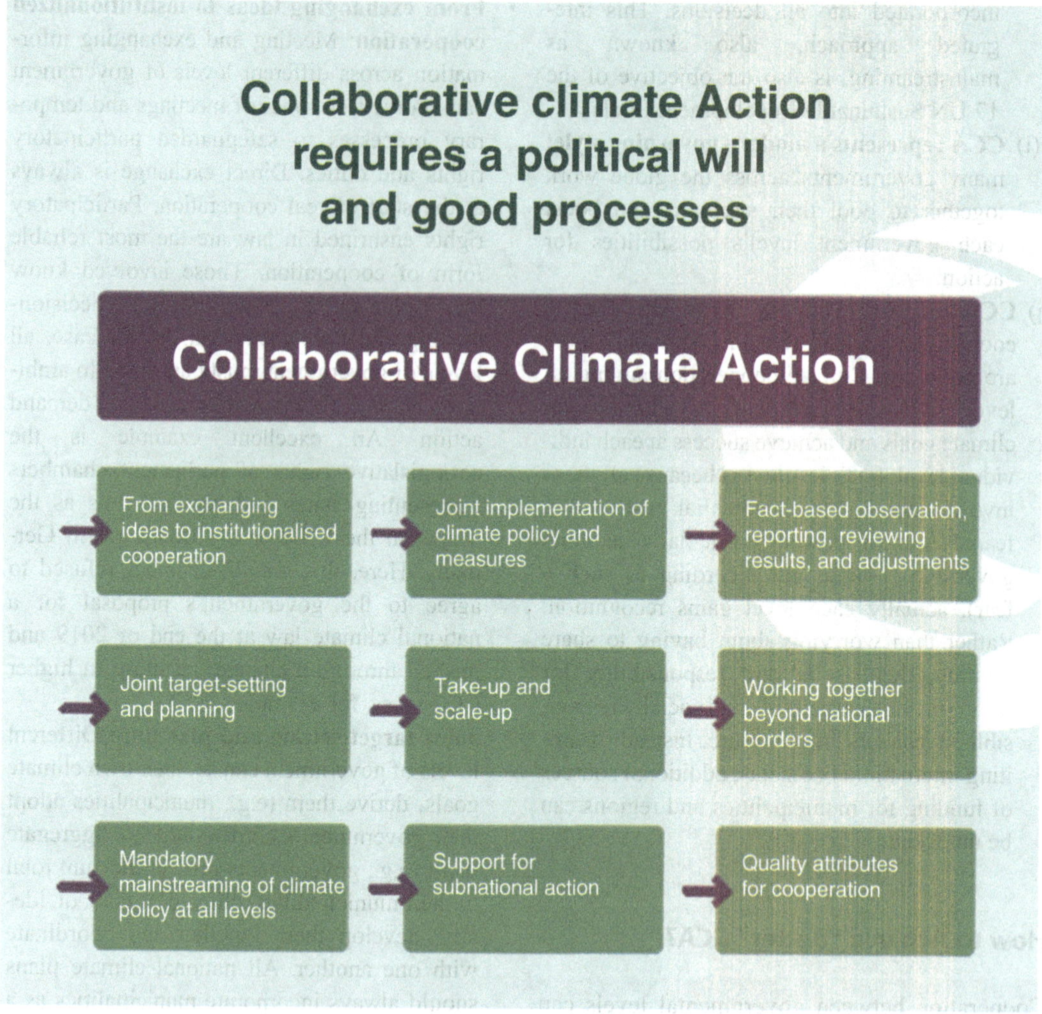

Collaborative Climate Action, Graph 4 Key components of successful CCA (Zimmermann et al. 2021)

disseminating effective examples. They can highlight specific projects and pioneering approaches by cities (lighthouse projects) to encourage broad take-up or make them the general standard by enacting rules and procedures. This scaling-up approach is also a good starting point for stronger cooperation between the municipal, regional, and national levels. In addition to individual projects, such as a zero-energy city hall, drawing on experience gained at the subnational level of government can foster or even launch long-term transformation processes.

- **Fact-based observation, reporting, verifying, and adjusting:** To make both success and potential for improvement visible, target implementation and achievement need to be underpinned by fact-based monitoring and evaluation. It is very useful to install and run climate inventories to which entities of all governmental levels make their entries.

- **Learning and cooperation beyond national borders:** Being inspired by colleagues is the best way to start acting yourself. CCA allows this learning to take place across levels of government and, indeed, requires it in order to make continuous improvements to complex processes.
- **Support for subnational action:** Higher levels of government have a special responsibility to delegate, share, and build capacities on lower levels. They must encourage, enable, empower, and resource these lower levels to allow them to take bold action.

It is important that parliaments, ministries, subordinate authorities, departments, offices, and other actors at different levels of government adhere to the principle of cooperation. As a continuous process, cooperation is easier if everybody focuses on specific factors that contribute to success and quality. These factors include:

- **A common understanding of basic principles:** Good cooperation requires a common understanding of the need to take action, with climate policy as a central task for the future and derived principles for action. Guidance can be found in the principles of the Partnership for Collaborative Climate Action described in Chapter 3.1, which were defined jointly at the national, regional, and local levels.
- **Long-term and well-organized collaboration between two or more levels of government:** The higher level ensures climate-friendly, reliable, predictable, and action-guiding frameworks for all sectors and supports the lower levels in their actions. The subnational level has mechanisms at its disposal to influence climate policy and to help shape both its own and the superordinate climate strategy. Regular and long-term cooperation mechanisms (e.g., participation and coordination committees, support programs) are preferred to selective instruments.
- **Managing and steering CCA in accordance with rules:** Steering systems that specify competences for success clearly define roles and establish mechanisms for areas where their remits overlap ensuring that there is interplay between the national and subnational level. It is also important to consider the role of CCA in order to document and continuously improve the impact of cooperation.
- **Process-oriented approach and process management:** CCA's success is rooted in clearly defined processes for decision-making, planning, implementation, review, and intervention. Due to the complex nature of these processes, agile approaches that seek steady and gradual improvement are more promising than static implementation model
- **Fact-based decision-making**: Facts and (scientific) insights are the starting point for decision-making. Their results should be made visible and verifiable in order to demonstrate successes and, if necessary, make specific adjustments.
- **Available resources for independent climate action and cooperation between levels:** The levels of government have resources (data, financial resources, appropriately qualified personnel) at their disposal that enable independent climate action as well as cooperation between the levels. Resources can be made available in the form of advisory or training services, research funding, or financing initiatives.
- **Opportunities for commitment and participation:** Politicians and administrators become experts on climate issues and on the organization of cooperation processes. Forums and formats for the exchange of experience and the joint shaping of climate policies have been created at various levels, in which actors are proactively involved.
- **Opportunities for cooperation with non-governmental partners:** Appropriate partners from business, science, civil society, and citizens' groups are involved in order to use their knowledge and potential for climate protection

and adaptation. The broad acceptance that can be achieved in this way means that decisions and activities are supported by a broader base and implementation is ensured.

- **Capacity to learn:** CCA requires and facilitates experimentation and the opportunity to learn from experience. Trying things out constantly improving results is crucial to successful climate policy, as is building trust and courage through cooperation.
- **Dissemination and exchange with actors in other countries:** The exchange enables mutual learning. It encourages the dissemination and use of good mechanisms and practices while avoiding.

Beyond these ingredients, CCA needs trust, courage, and patience of individuals and institutions, as well as joint products and successes. Successful cooperation across levels of government is not contingent on money, external support, or internationally set rules. A political will to engage in cooperation is crucial.

Starting Points and Initiatives for CCA

CCA does exist in various countries, in some more as a goal, in others more as anchored policies and realities. There are many role models, drivers, initiatives, and starting points that provide motivation and offer concrete examples for strengthening CCA. Many of these act internationally.

The current UN climate policy processes, the NDCs and long-term strategies (LTS), are creating a special dynamic. Various UN processes are recognizing the subnational level more and more and involving it in different ways.

The Paris Climate Agreement (2015) addresses all levels of government and incorporates the importance of cities and regions. The preamble (Sect. 15) states: "… recognising the importance of the engagements of all levels of government and various actors, in accordance with respective national legislations of Parties, in addressing climate change…." This is a helpful point of reference for CCA all over the world.

The NDC partnership, for example, has described a host of ways in which national governments can involve their subnational governments and offers support for practicing it (https://ndcpartnership.org).

The UN Climate Conference in 2019 (COP25) also recognized cities and regions as actors. National governments and subnational units participate equally in the new Climate Ambition Alliance (https://cop25.mma.gob.cl/wp-content/uploads/2020/02/Annex-Alliance-ENGLISH.pdf), a group of actors especially committed to taking action. By mid-2020, 120 cities and 12 regions had pledged to achieve climate neutrality (the term has varying definitions) by 2050.

Networks of cities, international organizations, international development cooperation work, research institutions, and foundations are also identifying ways to improve climate policy through CCA.

The Local Governments and Municipal Authorities (LGMA) Constituency, managed by ICLEI on behalf of the international city networks, coordinates cities and regions in the international climate process and provides new impetus for cooperation between the international, national, and municipal level.

Together with Chile, which held the presidency of COP25 (2019), organized cities and regions have put forward the theme of "Multilevel Action COP" in order to make the 2021 UN Climate Conference (COP26) in Glasgow a success and described a roadmap to a COP for *all levels of government* is supported by the Chilean COP presidency and a growing number of organizations (https://www.cities-and-regions.org/press-release-lgma-calls-for-multilevel-action-cop26/]).

Some governments came together on an informal basis as the Friends of Cities to signal their support for boosting municipalities in the international climate process. Founded in 2013, pioneers of the group include Germany, France, Mexico, the Netherlands, Indonesia, Peru, Poland, Senegal, and South Africa. Since 2019, municipalities have proposed developing this group into Friends of Multilevel Action (http://old.iclei.org/index.php?id=2202).

Germany has not only taken the lead for the Partnership for Collaborative Climate Action but also for the so-called New Leipzig Charter. The Leipzig Charter (2007) laid the groundwork for a shared urban policy in Europe. EU member states adopted a new version of the Leipzig Charter on November 30, 2020. The New Leipzig Charter is intended to create a legally nonbinding framework for the implementation of new European and global agendas, including the implementation of the Paris Agreement and the EU Green Deal. Its principles include close cooperation across different levels of government (https://www.bmi.bund.de/SharedDocs/downloads/EN/themen/building-housing/city-housing/new-leipzig-charta.html; https://www.weforum.org/events/the-davos-agenda-2021/?stream=day-two-1&stream-item=coming-up-special-address-by-emmanuel-macron-president-of-france#stream-header).

CCA is not the single silver bullet for effective climate policy, nor does it mean that responsibility for action can be pushed from one level to another. CCA is a precondition for success of climate policies and goes well together with other such conditions.

Collaborative Climate Action (CCA)

- ... Is a principle for policy-making and an instrument for achieving and strengthening climate goals
- ... Aims to prevent contradictory measures and thus supports coherence between policy and municipal action
- ... Allows planning processes to be shaped together and measures implemented jointly at the most expedient level
- ... Requires and promotes dialogue and cooperation among levels of government and with many other actors in society and the private sector
- ... Provides an opportunity to involve the subnational level more and therefore allows cities and regions to develop their ability to innovate
- ... Dictates that climate issues are taken into account in all political and administrative decisions
- ... Should through continuous learning improve over time

Cross-References

- ▶ Adapting Cities to Climate Change
- ▶ Amsterdam's Pathway to Climate Neutrality: Creating an Enabling Environment
- ▶ Building Community Resilience
- ▶ Connecting Urban and Regional Innovation Ecosystems to Enhance Competitiveness
- ▶ From Vulnerability to Urban Resilience to Climate Change
- ▶ Future of the City-region Concept and Reality
- ▶ Future of Urban Governance and Citizen Participation
- ▶ How Cities can be Resilient
- ▶ How Cities Cooperate to Address Transnational Challenges
- ▶ New Localism: New Regionalism
- ▶ Resilient Urban Climates
- ▶ Strategies for Liveable and Sustainable Cities: The Singapore Experience
- ▶ The Governance of Smart Cities
- ▶ The New Leipzig Charter: From Strategy to Implementation

References

Zimmermann, M., Lindemann, J., Marmon, T., Meinert, G., & Ziegler, M. (2021). *Collaborative climate action – A prerequisite for more ambitious climate policy*. Berlin: Deutsche Gesellschaft für Internationale Zusammenarbeit (GIZ) GmbH. https://collaborative-climate-action.org/cca-report/

Collective Climate Adaptation

▶ Collectively Adapting to Sea-Level Rise Through Disaster Response, Commons Management, and Social Mobilization

Collective Emotions and Resilient Regional Communities

Michelle Duffy[1], Candice Boyd[2], Kaya Barry[3,4] and Hedda Askland[1]
[1]University of Newcastle, Callaghan, NSW, Australia
[2]University of Melbourne, Melbourne, VIC, Australia
[3]Griffith University, Brisbane, QLD, Australia
[4]Associate Professor in Tourism, Department of Culture and Learning, Aalborg University, Aalborg, Denmark

Synonyms

Disaster; Emergency; Response; Rural

Definition

Collective emotions: Refers to the various ways that emotions are both individually experienced and collectively "felt."

Resilience: Originally used in physical and natural sciences to define the capacity of a system to return to an equilibrium or steady state after a disturbance. In social-ecological systems resilience is broadly defined as the capacity of an individual or community to cope with stress, overcome adversity, or adapt positively to change.

Introduction

The Anthropocene heralds unparalleled changes that impact communities across a range of scales and locations. Supporting regional communities and places is increasingly essential given these are where the destructive impacts of climate change – including prolonged drought, catastrophic bushfires and flooding – are keenly experienced, and will ripple out and greatly affect the lives of those in metropolitan regions. In addition to environmental loss, shifting weather patterns and the flow-on effects that impact industry, agriculture, energy consumption, jobs and education, as well as COVID-19, will continue to disrupt recovery strategies. These are strategies based on current development approaches which rely on industries such as tourism, sport and agriculture (Greenville et al. 2020; Shipway and Miles 2020) and the vulnerable groups and communities dependent on these industries (Berg and Farbenblum 2020; O'Kane 2020).

This chapter presents the argument that people formulate their own "lived" versions of community resilience through everyday *emotional* practices. Such practices may differ from institutionally sanctioned "resilient behaviors" promoted by governmental and other organizations but they deserve to be taken seriously as a crucial (perhaps *the* crucial) point of contact between the discourse of resilience and the reality of human (in)action. It is essential to understand how individuals feel connected to the place in which they live, how these feelings contribute to collective emotions, and how such feelings of connection (and disconnection) shape notions of what resilience means in their everyday lives.

A particular impact of prolonged or chronic disasters is that "people can be left feeling 'in limbo' when danger, risk and health effects are being considered" (Teague et al. 2014, p. 386). The impacts of disasters range from immediate physical danger, income loss, economic and productivity losses, infrastructure damage, housing loss, health impacts including psychological trauma, loss of social connectedness, a loss of a sense of belonging, and a decline in community well-being (Yell et al. 2019). The burden, then, of such disasters is significant, not only in terms of hectares burned, property destroyed, or animal and human lives lost but also the associated impacts on health, services, and the economy. Usual practice focuses on emergency management for a specified period of time after a disaster, followed by recovery strategies, and then longer-term planning for resilience in communities (Gallopin 2006; Porter and Davoudi 2012; Tehan 2017). While these forms of support from government and nongovernment sectors are necessary, how this support is conceptualized and delivered in terms of strategy and planning often overlooks the ways that people, plants, animals, water, weather, and light are embedded within complex sets of emotional, social, physical, and cultural relationships and how these then shape localized practices of resilience.

Significance of Collective Emotions

Emotions are important to comprehension of the world and how sense is made of it (Boyd and Hughes 2020; Wright 2012). Emotions are a powerful force in drawing individuals into communities and shaping social identities (Ahmed 2004). Without acknowledging the role of emotions, there is failure to understand how people "both know, and intervene in, the world" (Anderson and Smith 2001, p. 7). Therefore, understanding the role of the emotions in responding to change is fundamental to understanding the motivations and mobilization of people in their everyday places (Duffy et al. 2019).

Collaborations between the neurosciences and the humanities have broadened our understanding of emotions beyond private, internal experiences (Gross 2017). This is not to deny that emotions are produced by the body and the brain, but it does challenge the notion that emotions are "basic" and universal. As neuroscientist Lisa Feldman Barrett argues (2017), emotions are not "responses" but "predictions" – *concepts* based on past experiences that are constructed in the brain. Similarly, neuroscientist Stephanie Preston, in collaboration with rhetorical studies scholar Daniel Gross, asserts that "the situation in which an emotion concept is experienced shapes how the emotion is instantiated in the brain" (2017, p. 141). Put yet another way, emotions are *always* situated. As such, accounting for the social and cultural dimensions of emotional experience is integral to a new science of emotion (Callard and Fitzgerald 2015).

Barbara Rosenwein, a scholar of the history of emotion, coined the term "emotional communities" to reflect the way in which emotions are both individually experienced and *collectively* "felt." For her, emotional communities are social, but

> the researcher looking at them seeks above all to uncover systems of feeling, to establish what these communities (and the individuals within them) define and assess as valuable or harmful to them (for it is about such things that people express emotions); the emotions that they value, devalue, or ignore; the nature of the affective bonds between people that they recognize; and the modes of emotional expression that they expect, encourage, tolerate, and deplore. (2010, p. 11)

In this way, it is possible to understand how emotions are not just spatial but also temporal. Collective emotions, therefore, are not only a reflection of the demands of particular environments but also the historical time period in which people and their communities find themselves (Martín-Moruno and Beatriz 2019).

In the context of the Anthropocene there is a "need to urgently find ways for society to bear, and bear witness to, the painful emotions around climate change" (Head 2016, p. 6). This urgent need is outlined in detail by environmental philosopher Glenn Albrecht (2019) who declares that the Anthropocene is marked by "an open emotional war between the forces of creation and the forces of destruction" (2019, p. xi). Emotions are both the cause of the environmental decline and destruction marking the age of the Anthropocene and the source for environmental revival. The beginning of the third decade of the twenty-first century is a critical point that requires urgent action. Emotions are critical, because it is through relations with a declining environment, human and nonhuman relations, that climate change is experienced and interpreted. The affective dimension of these relations makes climate change real; it translates the abstract models and political discourse into everyday lived experiences. As Myers et al. (2012, n.p.) explain, "[e]motions serve as affective prompts for engagement with an issue and lead to forming predispositions for action when a relevant situation arises" (cf. Baumeister et al. 2007) and they "can serve as frames by which future information is interpreted."

Current Framings of Community Resilience

Resilience is most often defined in strategy and planning in response to disaster in order for communities to recover (Bach et al. 2015; Hunt et al. 2011). In devising strategies and policies, resilience is commonly referred to as a means to address a diverse range of challenges; these include disaster risk management, sustainable development, and climate change adaptation, which often invoke an underlying assumption of a capacity to return to some sort of "business as usual" (O'Hare and White 2013). In Australia, planning policy tends to be based on a strategy of governance focused on hazard management and emergency response, one supported by a broader neoliberal framework where responsibilities for risk governance and management have shifted from the state to the private sector and communities (Porter and Davoudi 2012). Some argue that resilience has become a tool of governance in an effort to exert control over challenges faced by contemporary society (O'Hare and White 2013). Others contend that resilience is now embedded within practices of "integrated place-making" (Coaffe 2013, p. 333) in which responsibility for resilience is dispersed through a range of locally based professionals and communities.

The appropriation of resilience into a socio-ecological framework – originally used within, first, the physical and, second, natural sciences (Holling 1973; Porter and Davoudi 2012) – has led to tensions between stakeholders involved in shaping our environments, as well as within the communities that policy and practice seek to support (O'Hare and White 2013). In addition, a tendency for decision-makers to view practices of resilience in conservative terms, often because of the complexity of the workings of government and related techno-institutional facilities, means existing structures can bias (in)action, and often promote maintenance of the *status quo* (Rickards et al. 2014). Critiques of intergovernmental responses for building resilience point to a lack of engagement with resources from "places understood as marginal to environmental preservation; Indigenous communities, gardens, suburbs, farms, domestic homes" *and* the opportunities these offer "to revisit empirical evidence [and] consider capacity and vulnerability in new ways" (Head 2016, p. 3). These vernacular understandings and practices are significant to understanding how resilience is achieved in an era of accelerated change, because they offer deep knowledge that arises out of ongoing connection and emotional attachment with and in place (Askland and Bunn 2018; Barnett et al. 2016; Duffy et al. 2019). The lack of attention to the importance of these collective emotions in the emergence of resilience is a missed opportunity for better understanding collective and individual choices and preferences (Rohse et al. 2020). Without acknowledging the power of collective emotions in the lived experience of profound environmental and economic change, strategies and policy attempting to ensure the resilience of communities will be flawed.

Nonetheless, and increasingly following significant recent disasters in regional Australia, government and agencies acknowledge the necessity of working with communities to foster community resilience. As Neil Comrie (cited in Tehan 2017, p. 2), former Australian Victorian police and emergency services leader, stated:

> We need to facilitate the education of communities so we can empower them. We need to work in conjunction with them, rather than directing them when disasters strike. We need to encourage communities to discover their own learning about resilience in preparation for natural disasters.

For many agencies, participation by community members in decision-making has been found to be crucial for effective recovery from a disaster and enhancing resilience for future events. Community involvement that recognizes active community engagement is also advocated by the communities themselves. This is borne out in disaster studies, such as that of the impacts of Canterbury (Aotearoa–New Zealand) earthquakes in 2010 and 2011, which found that community members wanted official agencies to have "a greater understanding of, and links to, the communities they serve" and that disaster preparedness and planning needs to be characterized by "transparency, good communication, partnership and respect for local knowledge, skills and priorities" (Thornley et al. 2015, p. 29). Community-based organizations are recognized as an essential interface between government agencies and communities because the highly localized knowledge they hold and their services "play a key role bridging government policy to hyper-local implementation" (Drennan and Morrissey 2019, p. 329). In addition, facilitating greater community participation in decision-making is viewed as an integral part of recovery. Many studies have argued this needs to be more than a community-engaged process, rather it is a process led and defined by the communities themselves (Drennan and Morrissey 2019; Koliou et al. 2020; Thornley et al. 2015). Such communities of practice are essential to developing appropriate "location-specific, group-specific and time-specific coping pathways and structures within a vulnerable environment" (Cradock-Henry et al. 2018, p. 6).

While an extensive overview of resilience literature is available elsewhere (e.g., Koliou et al. 2020; Tiernan et al. 2019), it is worth noting that this body of research has tended to focus on urban resilience in minority world locations, with studies on rural resilience and vulnerability more associated with that of the majority world (Cutler et al. 2016). Regional areas often face environmental, resource, and infrastructure vulnerabilities that differ to that of urban locations, including a lack of human and financial resources as well as limited communications infrastructure (Cutler et al. 2016), yet they may also thrive. Regional communities are often more self-reliant, arising out of what scholars have determined as a framework of assets in terms of capital, including human (knowledge, skills, and health status of the population), produced (financial resources, equipment, and infrastructure driving the local economy), natural (the state of the natural biophysical environment) and, importantly, social (Hunt et al. 2011). It is social capital – the relationships and social groupings within the community – that builds a strong sense of community and the strong social networks that contribute to resilience (Poortinga 2012).

In Australia, the regional communities most likely to suffer adversely from climate change have recognized the importance of harnessing their social capital to build resilience locally. One example is the *Farmers for Climate Action* group in the Shepparton-Goulburn Valley region – an area that has already experienced increased riverine flooding and its effects on farming. The climate action group combines the grass roots knowledge of farmers with the expertise of climate scientists and community psychologists. Another example is the now wound-up community organization *Groundswell Gloucester Inc* on the New South Wales mid-north coast, which brought together people and organizations to fight to protect their local community from the disastrous short- and long-term implications of gas and coal exploration in their region (Bailey and Osborne 2020). Working with local community members from all brackets of life, as well as activists and experts from the field of environmental planning and health, *Groundswell Gloucester* fought an emotional campaign resting on affective notions of place and place attachment to win a landmark court case against a coal mine on the grounds of social impacts and climate change (Lawrence and Askland 2019). Both *Farmers for Climate Action* and *Groundswell Gloucester* exemplify an approach that is fully cognizant of the importance of emotional practices to the resilience of regional communities (Farmers for Climate Change 2020). Nevertheless, the multifaceted nature of resilience means there is no single, consistently applicable definition, and "for the policy practitioner, this both simplifies policy design and contains hidden pitfalls" (Tiernan et al. 2019, p. 66).

This lack of definitional rigor is also problematic as it can frame communities as lacking agency and knowledge in decision-making processes. Complicating these practices of resilience is the larger shadow of climate change that unfolds complex sets of relations between slow-onset disasters (such as drought) and fast-onset disasters (exemplified in the Australian bushfire emergencies across most of the continent in the 2019–2020 summer season; see Readfearn 2020). An important aspect that defines and differentiates disasters is timing, both in the speed of onset and the duration of the disaster. A definitive understanding of resilience remains elusive and the implications are that what is lacking is a clear understanding as to how communities of resilience emerge in some locations and situations yet dissipate in others. The importance of how meaning and interpretation attributed to practices of resilience has been underestimated (Cline et al. 2010). Responses to the impacts of climate change point to the significance of emotion and affect in everyday vernacular understandings and practices of resilience.

Returning to the example of Gloucester on the Australian mid-north coast, the combined spatial and temporal role of emotions in shaping community resilience is evident. Preparing to act as an expert witness in the aforementioned court case in the NSW Land and Environment Court, Askland conducted interviews with close to 29 residents in the small township of about 2300 people. The court case concerned the proposal to mine for thermal and cooking coal in the valley just a few kilometres south of the center of the township. The valley is known for its unique topographic features, with the mountain ranges to the west and east covered in forests and foothills marked by talus and slopes, broken gullies and creeks. Generations have made the valley home, and newer residents speak of Gloucester as a place that upon first encounter generated a sense of solace, of home. Although the vast majority of residents were against the mine and spoke strongly for the protection of the valley from extractive industries and the need to reduce carbon emissions to act on climate change – a real threat to the rural community, which has lived with the severe impact of drought for many years and in 2020 ran out of water – some spoke strongly in favor of the mine. Regardless of their position on the mine, they fought for the preservation of their community and place; their emotional bond with Gloucester as a distinct place lived and experienced both spatially and temporally generated an intense emotional response towards proposals that were seen to threaten the future of the place. The threat of physical destruction of a loved environment, the risks associated with climate change, and the challenges associated with depopulation and loss of industry, all spoke to the affective dimension of resilience, pulling people in to protect their community, thereby building its resilience (Askland and Bunn 2018; see also Askland 2020).

Alongside consideration as to how emotions form part of relationships with and in places, how these are assembled, and how they then shape action (Duffy et al. 2019), what is also needed is careful consideration of the ways in which emotional experiences are captured and communicated. Particular tropes are commonly used in news and social media to convey feelings of shock, fear, loss, grief, or resignation when describing the impacts of trauma (Duffy and Yell 2014). This lexicon of emotional response creates publics that are interpellated into communities through the emotional address of disaster narratives (Duffy and Yell 2014). However, critical media and social analysts argue that, while emotional expression may enable us to reflect and help manage our responses to traumatic events, these discourses of emotion also shape political and ethical judgements, orienting us towards "normative and moral orders, to responsibility and blame, intentionality, and social evaluation" (Edwards 2001, p. 242). Such narratives generate a politics of emotion (Ahmed 2004) that may lead to people and communities feeling overwhelmed and immobilized (Shaw and Bonnett 2016).

Conclusion

The threat of climate change is no longer something of the future. For many people, the indicators of climate change have become nested into their everyday and measures for proactive mitigation is beyond reach. While climate change is a global phenomenon, it is, as all measures of globalization, lived and felt locally. Conservative science and political framings of resilience ignore the importance of individual and community-driven practices that arise out of the firsthand, lived experiences of being in a place. Without acknowledging the emotional registers of how people endure and prepare for the unknown futures of climate change that lie ahead, there is failure to understand why people make decisions to the detriment of our environment and to themselves.

Albrecht argues that the world has entered the "age of solastalgia," marked by affective dimensions of loss and mourning. Negative earth emotions (terraphthoric forces) channel emotional disturbances through their tension with positive earth emotions, expressed as love, care, and responsibility. To redress uncertain environmental futures, what must be resurrected are the terranascient emotions – the emotions that seek to protect and that are the premise of survival. Only after this can people enter "a new era that is compatible with human flourishing, and a whole Earth that is rich, bountiful, and beautiful" (Albrecht 2019, p. 11). The centrality of emotions in shaping people's experiences and understandings of environmental change, and their response and reaction to such change, presents an immanent challenge to social scientists and humanities scholars. There is an urgent need to understand the affective dimensions of climate change as both an outcome of environmental change and a driver for social and political action to address such change. Emotions, then, are key in building resilience capacity, drawing a map for the future, and, not least, fostering and shaping social action.

Cross-References

- ▶ From Vulnerability to Urban Resilience to Climate change
- ▶ How Cities Can Be Resilient
- ▶ Policy and Practices of Nature-Based Solutions to Build Resilience in Seoul, Korea
- ▶ Resilient Urban Climates
- ▶ Strategies for Liveable and Sustainable Cities: The Singapore Experience
- ▶ Urban Resilience

References

Ahmed, S. (2004). *The cultural politics of emotion*. New York: Routledge.

Albrecht, G. A. (2019). *Earth emotions. New words for a new world*. Ithaca/London: Cornell University Press.

Anderson, K., Smith, J. (2001). Editorial: Emotional geographies. *Transactions of the Institute of British Geographers, 26*(1), 7–10.

Askland, H. H. (2020). *Mining voids*. Extraction and emotion at the Australian coal frontier. *Polar Record, 56*, E5. https://doi.org/10.1017/S0032247420000078.

Askland, H. H., & Bunn, M. (2018). Lived experiences of environmental change: Solastalgia, power and place. *Emotion, Space and Society, 27*, 16–22.

Bach, R., Kaufman, D., & Dahns, F. (2015). What works to support community resilience? In R. Bach (Ed.), *Strategies for supporting community resilience: Multinational experiences* (pp. 309–340). Stockholm: Crismart: The Swedish Defence University, Multinational Resilience Policy Group.

Bailey, M.-S., & Osborne, N. (2020). Extractive resources and emotional geographies: The battle for treasured places in the Gloucester Valley. *Geoforum, 116*, 153–162.

Barnett, J., Tschakert, P., Head, L., & Adger, N. (2016). A science of loss. *Nature Climate Change, 6*, 976–978.

Berg, L., & Farbenblum, B. (2020). *As if we weren't humans: The abandonment of temporary migrants in Australia during COVID-19*. Migrant Worker Justice Initiative. Available at https://papers.ssrn.com/sol3/papers.cfm?abstract_id=3709527

Boyd, C. P., & Hughes, R. (2020). *Emotion and the contemporary museum: Development of a geographically-informed method of visitor evaluation*. Singapore: Palgrave.

Callard, F., & Fitzgerald, D. (2015). *Rethinking interdisciplinarity across the social sciences and neurosciences*. New York: Palgrave Macmillan.

Cline, R., Orom, H., Berry-Bobovski, L., Hernandez, T., Black, C., Schwartz, A., & Ruckdeschel, J. (2010). Community-level social support responses in a slow-motion technological disaster: The case of Libby, Montana. *American Journal Community Psychology, 46*, 1–18.

Coaffe, J. (2013). Towards next-generation urban resilience in planning practice. *Planning, Research & Practice, 28*(3), 323–339.

Cradock-Henry, N., Fountain, J., & Buelow, F. (2018). Transformations for resilient rural futures: The case of Kaikōura, Aotearoa-New Zealand. *Sustainability, 10*. https://doi.org/10.3390/su10061952.

Cutler, S., Ash, K., & Emrich, C. (2016). Urban-rural differences in disaster resilience. *Annals of the American Association of Geographers, 106*(6), 1236–1252.

Drennan, L., & Morrissey, L. (2019). Resilience policy in practice: Surveying the role of community based organisations in local disaster management. *Local Government Studies, 45*(3), 328–349.

Duffy, M., & Yell, S. (2014). Mediated public emotion: Collective grief and Australian natural disasters. In D. Lemmings & A. Brooks (Eds.), *Emotions and social change: Historical and sociological perspectives* (pp. 113–130). London: Routledge.

Duffy, M., Gallagher, M., & Waitt, G. (2019). Emotional and affective geographies of sustainable community leadership: A visceral approach. *Geoforum, 106*, 378–384.

Edwards, D. (2001). Emotion. In M. Wetherell, S. Taylor, & S. J. Yates (Eds.), *Discourse theory and practice: A reader* (pp. 236–246). London: Sage.

Farmers for Climate Change. (2020). https://farmersforclimateaction.org.au/media-release/new-forum-to-build-shepparton-and-goulbourn-valley-community-resilience/. Accessed 22 Jan 2021.

Feldman Barrett, L. (2017). *How emotions are made: The secret life of the brain*. New York: Palgrave.

Gallopin, G. C. (2006). Linkages between vulnerability, resilience, and adaptive capacity. *Global Environmental Change, 16*(3), 293–303.

Greenville, J., McGilvray, H., Cao, L.Y., & Fell, J. (2020). *Impact of COVID-19 on Australian agriculture, forestry and fisheries trade*. ABARES Research report. Canberra. https://doi.org/10.25814/5e9539b7cb004

Gross, D. M. (2017). *Uncomfortable situations: Emotions between science and the humanities*. Chicago: The University of Chicago Press.

Head, L. (2016). *Hope and grief in the Anthropocene: Reconceptualising human-nature relations*. Abingdon/Oxon/New York: Routledge.

Holling, C. S. (1973). Resilience and stability of ecological systems. *Annual Review of Ecology and Systematics, 4*, 1–23.

Hunt, W., Vanclay, F., Birch, C., Coutts, J., Flittner, N., & Williams, B. (2011). Agricultural extension: Building capacity and resilience in rural industries and communities. *Rural Society, 20*, 112–127.

Koliou, M., van de Lindt, J. W., McAllister, T. P., Ellingwood, B. R., Dillard, M., & Cutler, H. (2020). State of the research in community resilience: Progress and challenges. *Sustainable and Resilient Infrastructure., 5*(3), 131–151.

Lawrence, R., & Askland, H. H. (2019). *The Rocky Hill Mine has been refused – So what will happen to the land?* Sydney Environment Institute. Available online at https://sei.sydney.edu.au/opinion/rocky-hill-mine-refused-will-happen-land/

Martín-Moruno, D., & Beatriz, P. (2019). *Emotional bodies: The historical performativity of emotions*. Urbana: University of Illinois Press.

Myers, T. A., Nisbet, M. C., Maibach, E. W., & Leiserowitz, A. A. (2012). A public health frame arouses hopeful emotions about climate change. *Climatic Change, 113*, 1105–1112.

O'Hare, P., & White, I. (2013). Deconstructing resilience: Lessons from planning practice. *Planning, Research & Practice, 28*(3), 275–279.

O'Kane, G. (2020). COVID-19 puts the spotlight on food insecurity in rural and remote Australia. *Australian Journal of Rural Health, 28*(3), 319–320.

Poortinga, W. (2012). Community resilience and health: The role of bonding, bridging, and linking aspects of social capital. *Health & Place, 18*(2), 286–295.

Porter, L., & Davoudi, S. (2012). The politics of resilience for planning: A cautionary note. *Planning, Research & Practice, 13*(2), 329–333.

Readfearn, G. (2020). Australia's hottest year on record. *The Guardian*. 2 January.

Rickards, L., Wiseman, J., & Kashima, Y. (2014). Barriers to effective climate change mitigation. *WIREs Climate Change., 5*(6), 753–773.

Rohse, M., Day, R., & Llewellyn, D. (2020). Towards and emotional energy geography: Attending to emotions and affects in a former coal mining community in South Wales, UK. *Geoforum, 110*, 136–146.

Rosenwein, B. H. (2010). Problems and methods in the history of emotions. *Passions in Context: Journal of the History and Philosophy of the Emotions, 1*, 1–32.

Shaw, W., & Bonnett, A. (2016). Environmental crisis, narcissism and the work of grief. *Cultural Geographies, 23*(4), 565–579.

Shipway, R., & Miles, L. (2020). Bouncing back and jumping forward: Scoping the resilience landscape of international sports events and implications for events and festivals. *Event Management, 24*, 185–196.

Teague, B., Catford, J., & Petering, S. (2014). *Hazelwood mine fire inquiry report* (Vol. 1). Melbourne: Vic Government Printers. Available online at http://report.hazelwoodinquiry.vic.gov.au/.

Tehan, B. (2017). *Building resilient communities: Working with the community sector to enhance emergency management*. Report prepared for VCOSS (Victorian Council of Social Service). Available at: https://apo.org.au/node/89656

Thornley, L., Ball, J., Signal, L., Lawson-Te Aho, K., & Rawson, E. (2015). Building community resilience: Learning from the Canterbury earthquakes. *Kōtuitui: New Zealand Journal of Social Sciences., 10*(1), 23–35.

Tiernan, A., Drennan, L., Nalau, J., Onyango, E., Morrissey, L., & Mackey, B. (2019). A review of themes in disaster resilience literature and international practice since 2012. *Policy Design and Practice, 2*(1), 53–74.

Wright, S. (2012). Emotional geographies of development. *Third World Quarterly., 33*(6), 1113–1127.

Yell, S., Duffy, M., Whyte, S., Walker, L., Carroll, M., & Walker, J. (2019). *Hazelwood health study community wellbeing stream report volume 1: Community perceptions of the impact of the smoke event on community wellbeing and of the effectiveness of communication during and after the smoke event*. Report for the Department of Health and Human Services, Victorian State Government. Available at https://hazelwoodhealthstudy.org.au/__data/assets/pdf_file/0018/2052540/CW-Report-Volume-1_v2.0.pdf

Collectively Adapting to Sea-Level Rise Through Disaster Response, Commons Management, and Social Mobilization

Sonia Graham[1,2], Eleanor Malbon[3] and Zoran Balukoski[4]
[1]School of Geography and Sustainable Communities, The University of Wollongong, Wollongong, NSW, Australia
[2]Institut de Ciència I Tecnologia Ambientals (ICTA), Universitat Autònoma de Barcelona, Barcelona, Spain
[3]University of New South Wales, Kensington, NSW, Australia
[4]School of Geography and Sustainable Communities, The University of Wollongong, Wollongong, NSW, Australia

Synonyms

Collaborative adaptation; Collective climate adaptation; Commons-based adaptation; Coordinated adaptation

Definition

Collective adaptation refers to actions taken together by communities, government organizations, and/or nongovernment organizations (NGOs) to prepare for the impacts of climate change. There are at least three types of community-led collective adaptation. Collective disaster response incorporates diverse actions that communities take to prepare for, respond to, and recover from extreme events associated with climate change. Collective adaptation planning involves communities working together to identify how they would like to prepare for the changes that climate change will have on their local built, natural and social environments, and lobbying governments to bring those plans to fruition. This form of collective adaptation can also be considered to be a form of social mobilization. Collective implementation of climate adaptation involves communities providing public goods or maintaining common resources to minimize the impacts of climate change.

Introduction

Communities around the world are already experiencing the effects of climate change, from more frequent and severe storm events along the coast (Horton et al. 2015), coral bleaching events (Hughes et al. 2017), to longer and more frequent droughts (Dai 2010), among others. Recognition that these impacts will continue to become more frequent and intense has led to a growing body of research on how we can best prepare for, and adapt to, environmental changes. While the focus of such research is on diverse adaptations at multiple scales, there is one consistent message: collective action is needed to efficiently and effectively mobilize physical and social resources to reduce vulnerability (Baldwin and Chandler 2010; Campos et al. 2016; Keys et al. 2016; Cinner et al. 2018).

While repeated calls are made for collective climate adaptation, there are few studies that define what it is, provide examples of when and where it is needed, or under what circumstances it evolves. For example, Adger's (2003) seminal article on "Social capital, collective action, and climate adaptation" provides no definition of collective adaptation. The limited examples provided include: "voluntary wardens, lobbying for improvements in the disposal of sewerage, and other regulatory tasks" for managing a marine protected area and agricultural cooperatives "repairing and maintaining the sea-dyke system" (p. 398). More recently, Cinner et al. (2018) have argued that "the ability to organize and act collectively" (p. 117) is one of five key domains needed to build adaptive capacity. Yet, they also fail to define "collective action" or similar concepts such as "cooperation," "knowledge sharing," and "coordinated action." The examples provided are also cursory: "(1) establishment and strengthening of networks across scales… (2) community currency, or time banking systems, where individuals are incentivized to volunteer... and; (3) creation of interaction arenas where people can work together towards shared goals, build trust, and develop social cohesion" (p. 119). Such conceptual ambiguity makes it difficult to know what collective climate adaptation is, whether it reduces vulnerability to climate change, and if so, how further collective adaptation can be encouraged.

One way to improve the conceptual clarity of collective climate adaptation research is to consider how existing theories of collective action apply. For example, the descriptions of collective adaptation provided by Adger (2003) reflect community-based management of commons (as described by Ostrom 1990). Similarly, Cinner et al.'s (2018) call for social organization does not consider political social mobilization or social movements, despite the rich theory of such collective action provided by sociologists, political scientists, social psychologists, anthropologists, and historians (Roggeband and Klandermans 2010). Analyzing collective climate adaptation with these theoretical traditions in mind can provide insights into how collective adaptation is similar to, or different from, adaptation in other contexts. Thus, this entry synthesizes existing empirical research to describe examples of community-led collective action in the context of climate adaptation; identify contextual factors that enable such forms of collective adaptation to emerge; and relate examples of collective adaptation to existing theoretical understandings from commons and social movement scholarship on collective action. Given that adaptation options vary depending on the nature of the climate change impacts being prepared for, this entry focuses on adaptation to sea-level rise.

Types of Community-Led Collective Adaptation to Sea-Level Rise

Empirical research has documented examples of community-led collective adaptation to sea-level rise on every continent. The examples can be grouped into short-term responses to extreme weather events or longer-term actions aimed at preparing for, and responding to, climatic changes, such as coastal erosion and coastal flooding. Social mobilization forms of collective action are evident during the planning phase, while commons management occurs in the implementation phase.

Disaster Response

Short-term collective adaptation involves sharing disaster warnings and information about extreme events (Ahamed 2013; Galappaththi et al. 2020). This includes the use of tidal calendars (Petzold 2018), social media (Dwirahmadi et al. 2019), or bells and drums to communicate impending danger to residents (Massuanganhe et al. 2015). Communities also work together to collect and share information during a disaster, such as where utilities have been restored, which is particularly important when such information is not available through official government announcements or traditional media (Schmeltz et al. 2013).

Other forms of collective adaptation are also evident during emergency events (Ahamed 2013; Süsser 2018), such as by rescuing people (Massuanganhe et al. 2015; Petzold 2016), filling and distributing sand bags (Petzold 2016, 2018; Brink and Wamsler 2019), organizing impromptu first response medical teams (Petzold 2016; Schmeltz et al. 2013), and collecting donations and distributing them to those most in need (Schmeltz et al. 2013). Conversely, Ahamed (2013) noted that during disasters in four coastal communities in Bangladesh, no one undertook collective action that would improve river embankments or levees because they considered that to be the responsibility of government.

Following extreme weather events, communities work together to repair communal and private property (Petzold 2016) and undertake beach cleans (Petzold 2016; Ratter et al. 2016). In some communities, collective action post-disaster involves the pooling of finances to help people cope. For example, in coastal Bangladesh, households borrowed money from relatives, as well as from governments, NGOs, and banks (Ahamed 2013). In one case, experience of a severe cyclone acted as a catalyst for a village to relocate inland to adapt to climate change (Bertana 2020).

Sea-Level Rise Adaptation Planning

Longer-term collective adaptations focus on planning or implementations phases. For communities involved in adaptation planning, collective action is evident in planning processes that build consensus, reconcile differences and identify common ground (Baldwin and Chandler 2010), develop shared goals and a shared vision (Campos et al. 2016; Keys et al. 2016; Austin et al. 2019; Dwirahmadi et al. 2019), achieve community ownership (Lieske et al. 2015; Volk et al. 2015), and reach agreement on a particular management plan (Remling and Veitayaki 2016) or set of adaptation pathways (Campos et al. 2016; Basel et al. 2020).

Collective action that enables the collecting, sharing, and generation of information and knowledge is pivotal in the planning phase. Such information sharing is key to developing a mutual understanding of the problem and risks (Lieske et al. 2015; Campos et al. 2016; Basel et al. 2020), within and beyond the communities (Zografos 2017), to enable participatory learning and action (Remling and Veitayaki 2016), so that communities can share their knowledge and ideas for adaptation (Lieske et al. 2015). Such information can help governments inspire action (Bertana 2020; Boda and Jerneck 2019) and lead to evidence-based decisions (Austin et al. 2019). Yet, collective action is also used to challenge external expertise and knowledge. For example, Tebboth (2014) explained how an advocacy group developed a coherent set of logics that shaped the way the public interpreted adaptation planning and influenced public policy debate.

Local advocacy groups are often established in the planning phase to share information and campaign for particular adaptation solutions (Tebboth 2014; Karlsson and Hovelsrud 2015; Petzold 2016; Blunkell 2017; Zografos 2017). Although, in some cases, the advocacy groups advocate for solutions that would exacerbate coastal impacts, or are co-opted by the causes they are seeking to protest (Ratter et al. 2016). Often, advocacy-based collective action not only focuses on what is happening within communities and small islands (Petzold 2016, 2018; Remling and Veitayaki 2016), but also on how to leverage networks and resources beyond the community (Karlsson and Hovelsrud 2015; Keys et al. 2016; Petzold 2016; Blunkell 2017).

Sea-Level Rise Adaptation Implementation

In the implementation phase, commons-management collective action is evident. Communities work together to provide public goods, such as hard protection. For example, farmers, fishers, and community members build defensive structures (Ratter et al. 2016; Süsser 2018), and donate or distribute material for structures (Karlsson and Hovelsrud 2015; Petzold 2018). Some communities collectively clean and maintain drainage systems to minimize flooding. This works well where dike and sluice associations organize and coordinate drainage maintenance (Süsser 2018; Dwirahmadi et al. 2019) or when the community agrees on rules for operating sluice gates (Barua and Rahman 2017). Where there are no formal local institutions to manage drainage systems, public goods may not be provided because only a few people contribute to the cleaning, maintenance, and management of the drainage systems or only focus on infrastructure surrounding their property (Middelbeek et al. 2014; Mycoo 2014; Barua and Rahman 2017).

Communities also work together to maintain common pool resources that provide soft protection. For example, communities restore and protect mangrove forests (Massuanganhe et al. 2015; Ratter et al. 2016; Remling and Veitayaki 2016) and (re)plant coastal vegetation to stabilize shorelines (Basel et al. 2020) and dunes (Petzold 2016, 2018). Such collective action requires engaging community members with diverse interests (Ratter et al. 2016), achieving a common understanding of the importance of these common pool resources for minimizing coastal erosion (Massuanganhe et al. 2015) and developing conflict resolution mechanisms (Ratter et al. 2016). However, as with many collective action problems, one challenge involves conflict between individual and collective interests. For example, Liski et al. (2019) found that some farmers preferred hard over soft protection because coastal nature-based solutions reduce the amount of land available for farming, with associated reductions on yields and incomes.

Factors That Enable Collective Adaptation

There are three cross-cutting factors that shape collective adaptation through disaster response, adaptation planning, and adaptation implementation: the presence of social capital; commitment to sharing information; and the role of the State in adaptation.

Social Capital

Of the three key components of social capital – networks, trust, and norms – networks receive the greatest attention in empirical studies of collective adaptation. Local and regional networks are important for commons-based adaptation because they facilitate local actions (Mycoo 2014) as well as regional monitoring of the resource condition (Remling and Veitayaki 2016). For social mobilizations, bonding, bridging, and linking networks facilitate communication about adaptation preferences within and beyond communities. The networks held by individual community leaders are particularly important for ensuring community perspectives are heard in political settings (e.g., Blunkell 2017; Karlsson and Hovelsrud 2015; Keys et al. 2016). In studies that document multiple types of collective adaptation, there is some evidence that the underlying social networks are the same. For example, the Belizean coastal erosion advocacy group who gathered information and built awareness of the coastal erosion problem, also lobbied the government and contributed to the provision of labor and resources for the coastal protection public good (Karlsson and Hovelsrud 2015).

Trust is also important for collective adaptation. This includes trust in neighbors and relatives (Ahamed 2013), stakeholders, the community (Lieske et al. 2015; Petzold 2016), community leaders (Tebboth 2014), and local institutions responsible for managing the coastline (Tebboth 2014; Petzold 2016). Beyond the community, trust is required in scientific reports (Volk et al. 2015), risk information, and that engineering structures will decrease the impacts of climate change (Hung et al. 2019). Trust is more commonly mentioned as important for social mobilization than commons-based adaptation. Consistent with the social movement literature, a lack of trust in institutions was seen to drive the establishment of social mobilizations (e.g., Tebboth 2014; Petzold 2016). Trust was also important in the adaptation planning processes to achieve a shared vision for the future (Lieske et al. 2015). The role of norms is notably absent in empirical research on collective adaptation. Both commons-based and social movement theories of collective action acknowledge the role of norms. For example, Ostrom (2010) describes the importance of shared reciprocity norms and the social movement literature recognizes the importance of challenging prevailing norms and creating new norms (Jasper 2010). Thus, future research into the role of norms in collective adaptation may shed new light on how communities work together and are influenced by one another in adapting to climate change.

Sharing Information

A lack of knowledge or awareness is frequently identified as a barrier to collective adaptation. This includes awareness of climate change and its impacts (Petzold 2018; Boda and Jerneck 2019; Basel et al. 2020), the advantages and disadvantages of various adaptation options (Brink and Wamsler 2019; Bertana 2020), actions that community members can take (Archie et al. 2018; Boda and Jerneck 2019; Basel et al. 2020), and even who is being given an opportunity to participate (Bertana 2020). High community awareness about flood risks increases the likelihood of community preparedness (Dwirahmadi et al. 2019).

To address concerns about awareness, sharing information is pivotal to all three types of collective adaptation. In disaster response, sharing information enables individuals and communities to respond to an emerging crisis and allows resources to be allocated to those most at risk (Schmeltz et al. 2013). In managing coastal commons, the collection and sharing of information about a resource and the associated climate risks facilitates a common understanding about the state and dynamics of the socio-ecological system, allowing collective adaptations to be designed and implemented based on this knowledge (Massuanganhe et al. 2015; Remling and

Veitayaki 2016). In social mobilizations, sharing information is crucial to activism and outreach. For example, scientific information is sometimes used to support local causes (e.g., Karlsson and Hovelsrud 2015; Volk et al. 2015; Zografos 2017) and the development of adaptation plans (Paolisso et al. 2012), while in other cases the challenging of external expertise is a form of collective action (e.g., Tebboth 2014).

Collective Adaptation and the Role of the State

Across multiple studies, collective action emerged as a response to an absence of support from government. The clearest example is provided by Petzold (2016), where the highest levels of community effort to provide and maintain public goods, such as sea walls, occurs in the most remote islands of the Isles of Scilly where the government rarely maintains or repairs them. Similarly, Mycoo (2014) found that a perceived lack of government capacity led respondents to clean drains adjacent to their property. Conversely, when residents expected that governments would provide support, community-led collective action was absent. Ahamed (2013) argued that there was no collective action to improve embankments or levees during disasters because residents believed this to be the responsibility of the local government. Similarly, Middelbeek et al. (2014) found that where residents expected the government to maintain the drainage system, few residents cleaned and maintained it. These examples highlight the importance of political context to all three forms of collective climate adaptation.

Conclusion

It is repeatedly argued that collective action is important for successful climate adaptation. Synthesis of empirical research reveals that collective action enables community-led climate adaptation through: (1) building consensus and mobilizing communities to advocate for the adaptations they want; (2) enabling communities to work together to provide public goods and manage common pool resources that protect them from rising sea levels; and (3) responding cooperatively to coastal disasters in real time. All three types of collective action depend on local and regional social networks and involve information sharing. The relationship between communities and weak governance needs to be further explored. Given that most empirical studies are cross-sectional, it is unclear whether communities will continue to engage in collective action over the long timeframes required for adaptation. Research is needed to explore how collective adaptation evolves and persists over time.

Cross-References

- ▶ Climate Resilience in Informal Settlements: The Role of Natural Infrastructure
- ▶ The Interplay of Intersectionality and Vulnerability Towards Equitable Resilience

References

Adger, W. N. (2003). Social capital, collective action, and adaptation to climate change. *Economic Geography, 79*(4), 387–404.

Ahamed, M. (2013). Community based Approach for Reducing Vulnerability to Natural Hazards (Cyclone, Storm Surges) in Coastal Belt of Bangladesh. *Procedia Environmental Sciences, 17*(The 3rd International Conference on Sustainable Future for Human Security, SUSTAIN 2012, 3–5 November 2012, Clock Tower Centennial Hall, Kyoto University, JAPAN), 361–371.

Archie, K. M., Chapman, R., & Flood, S. (2018). Climate change response in New Zealand communities: Local scale adaptation and mitigation planning. *Environmental Development, 28*, 19–31.

Austin, S. E., Ford, J. D., Berrang-Ford, L., Biesbroek, R., & Ross, N. A. (2019). Enabling local public health adaptation to climate change. *Social Science & Medicine, 1982*(220), 236–244.

Baldwin, C., & Chandler, L. (2010). "At the water's edge": Community voices on climate change. *Local Environment, 15*(7), 637–649.

Barua, P., & Rahman, S. H. (2017). Indigenous knowledge practices for climate change adaptation in the Southern Coast of Bangladesh. *IUP Journal of Knowledge Management, 15*(1), 44–62.

Basel, B., Goby, G., & Johnson, J. (2020). Community-based adaptation to climate change in villages of Western Province, Solomon Islands. *Marine Pollution Bulletin, 156*(111), 266.

Bertana, A. (2020). The role of power in community participation: Relocation as climate change adaptation in

Fiji. *Environment and Planning. C, Politics and Space, 38*(5), 902–919.

Blunkell, C. T. (2017). Local participation in coastal adaptation decisions in the UK: Between promise and reality. *Local Environment, 22*(4), 492–507.

Boda, C. S., & Jerneck, A. (2019). Enabling local adaptation to climate change: Towards collective action in Flagler Beach, Florida, USA. *Climatic Change, 157*(3–4), 631–649.

Brink, E., & Wamsler, C. (2019). Citizen engagement in climate adaptation surveyed: The role of values, worldviews, gender and place. *Journal of Cleaner Production, 209*, 1342–1353.

Campos, I., Vizinho, A., Coelho, C., Alves, F., Truninger, M., Pereira, C., … Penha Lopes, G. (2016). Participation, scenarios and pathways in long-term planning for climate change adaptation. *Planning Theory and Practice, 17*(4), 537–556.

Cinner, J. E., Adger, W. N., Allison, E. H., Barnes, M. L., Brown, K., Cohen, P. J., … Morrison, T. H. (2018). Building adaptive capacity to climate change in tropical coastal communities. *Nature Climate Change, 8*(2), 117–123.

Dai, A. (2010). Drought under global warming: A review. *Wiley Interdisciplinary Reviews: Climate Change, 2*(1), 45–65.

Dwirahmadi, F., Rutherford, S., Phung, D., & Chu, C. (2019). Understanding the operational concept of a flood-resilient urban community in Jakarta, Indonesia, from the perspectives of disaster risk reduction, climate change adaptation and development agencies. *International Journal of Environmental Research and Public Health, 16*(20), 3993.

Galappaththi, E. K., Ford, J. D., & Bennett, E. M. (2020). Climate change and adaptation to social-ecological change: The case of indigenous people and culture-based fisheries in Sri Lanka. *Climatic Change, 162*(2), 279–300.

Horton, R., Little, C., Gornitz, V., Bader, D., & Oppenheimer, M. (2015). New York City Panel on Climate Change 2015 Report Chapter 2: Sea Level Rise and Coastal Storms. *Annals of the New York Academy of Sciences, 1336*(1), 36–44.

Hughes, T. P., Kerry, J. T., Álvarez-Noriega, M., Álvarez-Romero, J. G., Anderson, K. D., Baird, A. H., Babcock, R. C., Beger, M., Bellwood, D. R., Berkelmans, R., Bridge, T. C., Butler, I. R., Byrne, M., Cantin, N. E., Comeau, S., Connolly, S. R., Cumming, G. S., Dalton, S. J., Diaz-Pulido, G., … Wilson, S. K. (2017). Global warming and recurrent mass bleaching of corals. *Nature, 543*, 373.

Hung, H. C., Lu, Y. T., & Hung, C. H. (2019). The determinants of integrating policy-based and community-based adaptation into coastal hazard risk management: A resilience approach. *Journal of Risk Research, 22*(10), 1205–1223.

Jasper, J. M. (2010). Cultural approaches in the sociology of social movements. In *Handbook of social movements across disciplines* (pp. 59–109). Boston: Springer.

Karlsson, M., & Hovelsrud, G. K. (2015). Local collective action: Adaptation to coastal erosion in the Monkey River Village, Belize. *Global Environmental Change, 32*, 96–107.

Keys, N., Thomsen, D. C., & Smith, T. F. (2016). Adaptive capacity and climate change: The role of community opinion leaders. *Local Environment, 21*(4), 432–450.

Lieske, D. J., Roness, L. A., Phillips, E. A., & Fox, M. J. (2015). Climate change adaptation challenges facing New Brunswick coastal communities: A review of the problems and a synthesis of solutions suggested by regional adaptation research. *Journal of New Brunswick Studies/Revue Detudes Sur Le Nouveau-Brunswick (JNBS/RENB), 6*(1), 32.

Liski, A. H., Ambros, P., Metzger, M. J., Nicholas, K. A., Wilson, A. M. W., & Krause, T. (2019). Governance and stakeholder perspectives of managed re-alignment: Adapting to sea level rise in the Inner Forth estuary, Scotland. *Regional Environmental Change, 19*(8), 2231–2243.

Massuanganhe, E. A., Macamo, C., Westerberg, L.-O., Bandeira, S., Mavume, A., & Ribeiro, E. (2015). Deltaic coasts under climate-related catastrophic events – Insights from the Save River delta, Mozambique. *Ocean and Coastal Management, 116*, 331–340.

Middelbeek, L., Kolle, K., & Verrest, H. (2014). Built to last? Local climate change adaptation and governance in the Caribbean – The case of an informal urban settlement in Trinidad and Tobago. *Urban Climate, 8*, 138–154.

Mycoo, M. A. (2014). Autonomous household responses and urban governance capacity building for climate change adaptation: Georgetown, Guyana. *Urban Climate, 9*, 134–154.

Ostrom, E. (1990). *Governing the commons: The evolution of institutions for collective action*. New York: Cambridge University Press.

Ostrom, E. (2010). Analyzing collective action. *Agricultural Economics, 41*(s1), 155–166.

Paolisso, M., Douglas, E., Enrici, A., Kirshen, P., Watson, C., & Ruth, M. (2012). Climate change, justice, and adaptation among African American Communities in the Chesapeake Bay Region. *Weather, Climate, and Society, 4*(1), 34.

Petzold, J. (2016). Limitations and opportunities of social capital for adaptation to climate change: A case study on the Isles of Scilly. *Geographical Journal, 182*(2), 123–134.

Petzold, J. (2018). Social adaptability in ecotones: Sea-level rise and climate change adaptation in Flushing and the Isles of Scilly, UK. *Island Studies Journal, 13*(1), 101–118.

Ratter, B. M. W., Petzold, J., & Sinane, K. (2016). Considering the locals: Coastal construction and destruction in times of climate change on Anjouan, Comoros. *Natural Resources Forum, 40*(3), 112–126.

Remling, R., & Veitayaki, J. (2016). Community-based action in Fiji's Gau Island: A model for the Pacific?

International Journal of Climate Change Strategies and Management, 8(3), 375–398.
Roggeband, C., & Klandermans, B. (2010). Introduction. In *Handbook of social movements across disciplines* (pp. 1–12). New York: Springer.
Schmeltz, M. T., González, S. K., Fuentes, L., Kwan, A., Ortega-Williams, A., & Cowan, L. P. (2013). Lessons from Hurricane Sandy: A community response in Brooklyn, New York. *Journal of Urban Health: Bulletin of the New York Academy of Medicine, 90*(5), 799–809.
Süsser, D. (2018). Coastal dwellers-power against climate change: A place-based perspective on individual and collective engagement in North Frisia. *Journal of Coastal Conservation, 22*(1), 169–182.
Tebboth, M. (2014). Understanding intractable environmental policy conflicts: The case of the village that would not fall quietly into the sea. *Geographical Journal, 180*(3), 224–235.
Volk, M., Frank, K., & Nettles, B. B. (2015). Managing coastal change in the cultural landscape: A case study in Yankeetown and Inglis. *Change Over Time,* Florida, 5(2), 226–246.
Zografos, C. (2017). Flows of sediment, flows of insecurity: Climate change adaptation and the social contract in the Ebro Delta, Catalonia. *Geoforum, 80*, 49–60.

Colombia Energy Transformation

► Smart Grids to Lower Energy Usage and Carbon Emissions: Case Study Examples from Colombia and Turkey

Combined Green and Grey Infrastructure

► Integrated Urban Green and Grey Infrastructure

Commons-Based Adaptation

► Collectively Adapting to Sea-Level Rise Through Disaster Response, Commons Management, and Social Mobilization

Communicable Diseases (Infectious Diseases, Contagious Diseases)

► Epidemiological Shifts in Urban Bangladesh

C

Community

► Master Planned Estates and the Promises of Suburbia
► Participatory Planning: A Useful Tool for the Development of Sustainable Mega-City Regions
► The Social and Solidarity Economy

Community Development

► Senegalese Ecovillage Network

Community Engagement and Climate Change: The Value of Social Networks and Community-Based Organizations

Nicky Morrison
Western Sydney University, Sydney, NSW, Australia

Synonyms

Climate change; Community engagement; Community-based organizations; Impact knowledge; Normative; Procedural; Trust; Values

Definition

Encouraging individuals to adopt energy-efficient and other sustainable ways of living has predominantly focused on trying to change individual behavior. Yet this approach has faced criticism from those seeking a more meaningful engagement process that communicates a "bigger than self" responsibility to act. This chapter contributes to this discussion by suggesting the use of social networks and role of community-based organizations within such networks, to encourage individuals to collectively move to low-carbon living. Drawing on the work of one such organization, Cambridge Carbon Footprint in the UK, it considers more broadly the ability of these "trusted messengers" in improving access to procedural knowledge (i.e., instructions on what to do) but also normative and impact knowledge (respectively, an individual's beliefs about the consequences of their actions and an understanding of the behaviors of others). Approaching the climate challenge through sustained community engagement, with knowledge, trust, and reciprocity developed through community-based organizations, offers the potential of bringing more people into a constructive process of engagement.

Introduction

The global challenge of climate change, together with rising energy prices and concerns over future energy sources, has served to solidify an international policy agenda towards increasing each nation's energy efficiency in conjunction also with national targets to reduce energy-related carbon emissions (Popovski 2019; IPCC, 2021; UN, 2021). The Paris UN Climate Agreement recognizes:

> That sustainable lifestyles and sustainable patterns of consumption and production, with development country parties taking the lead, play an important role in addressing climate change (UN, 2015)

While technological fixes and financial instruments exist to assist, they are, in effect, insufficient without broader public engagement in climate change issues and an overall move towards more sustainable forms of living (Corner, Markowitz, & Pidgeon, 2014). Here, responsibility is considered to lie with the individual, whose behavioral choices can make a difference (Brick, Bosshard, & Whitmarsh, 2021). Framing the problem this way was, for instance, central to the UN Environment Programme project to encourage individuals to "Kick the CO2 habit" (UNEP, 2008). However, motivating people to act has traditionally been, and remains, a significant challenge.

The dominant approach to encouraging individuals to adopt energy-efficient and other sustainable ways of living has primarily drawn upon theories of behavioral change. In the United Kingdom (UK), understanding individual behavior and ways to influence the adoption of sustainable behaviors into the future has been a core theme within central government policy agendas. The UK Department for Environment, Food and Rural Affairs, for instance, has adopted an "Engage, enable, encourage and exemplify" (4Es) model of public intervention (DEFRA, 2011) and the UK Government's Behavioural Insights Team's work also underpins current government strategy (2011, 2020, 2021). This approach is centered around a need to "nudge" individuals into specific behavioral change and draws upon work by Thaler R and Sunstein (2008) in the United States.

However, this approach has been criticized by climate change campaigners and academics alike, who seek a more meaningful engagement process (Whitmarsh, O'Neill & Lorenzoni, 2011; Bouman & Steg, 2019; Brick et al., 2021; Corner et al., 2014; Vesely, Klockner, & Brick, 2020). Campaigners object to the role of the citizen being restricted to that of a consumer, encouraged to make better choices (Howarth & Parsons, 2021; Randall, 2011). Rather than "nudging" the message or "marketing" it to segments of society, Crompton (2010), in the World Wildlife Fund report "Common Cause," argues that such an approach does not create sufficient effect or impact within the necessary timescale. Instead, he advocates a more effective engagement with people at a deeper level, focusing on personal values, social identities, and practices, particularly around the way people live and consume energy. By emphasizing a reshaping value, or guiding principles, across all sectors of society, the adoption of more energy-efficient, sustainable ways of living then becomes a collective responsibility.

Academics, like Shove, a social scientist, are equally critical of the dominant model of behavioral change, and comprising, as she encapsulates it, a linear process of attitude-behavior-choice (or "ABC"). Framing the problem of climate change as one that is primarily affected by individual human behavior overlooks the way that actions of individuals are the outcome of shared conventions and practices (Shove, 2010). Moreover, the ABC model itself does not contain the relevant terms and concepts required to debate significant societal transformations in daily lives and marginalizes engagement with other possible analyses grounded in social theories. However, Whitmarsh, Poortinga, and Capstick (2021), while agreeing that the overall theoretical orientation of conventional behavioral theorists tends towards (a limiting) overly individualistic and decontextualized view, also suggest that social theorists lean too much towards the overly structural and undifferentiated. Commentators therefore endorse constructive attempts to extract insights from both sides of the disciplinary line.

This chapter contributes to this discussion by exploring the way in which the climate change agenda could move from a predominant focus on the micro level (i.e., trying to change individual behavior) to looking at the meso level (concerned with collective behavior). Here, the chapter draws on an approach used in the social sciences, namely, social network analysis. This strand of social analysis is potentially useful and influential in that it not only seeks to understand the social world, but also considers how it might change. As Granovetter (Granovetters, 1973, 1983) argues, seeing the world in terms of social networks can offer considerable analytic leverage. Rather than treating individuals as discrete units of analysis, the connections between individuals warrants similar attention and exploration. Moreover, this chapter suggests that community-based organizations (CBOs), with a sustainability remit and working at the local level, may be best positioned within a social network to mediate the way individuals understand, interpret, engage with, and respond to such fundamental issues as climate change. Individuals are deeply embedded in social situations, and as Shove (2010) notes, "reference to contextual factors and social norms represents an attempt to handle some of the problems of the (individualistic) ABC (attitude-behaviour-choice) model" (p.1276). Rather than individuals perceiving themselves solely as consumers of energy, it is important that they also see themselves as members of a wider, interconnected community (McNamara & Buggy, 2016).

Using CBOs embedded in social networks, it may be possible to not only improve access to *procedural* knowledge (i.e., instructions on what to do) but also the equally necessary and influential *normative* and *impact* knowledges (respectively, an individual's beliefs about the consequences of their actions and an understanding of the behaviors of others) that guide and influence individuals' choice decisions. In doing so, individuals may be encouraged to follow through on making energy and carbon savings in their daily lives, knowing also that their actions are being collectively reinforced to help curb climate change.

The chapter first provides a critique of the core tenets of behavioral economics and associated psychology perspectives on the way individuals are considered to act. Second, it outlines the premise of social network theory and the role of CBOs as trusted messengers within these networks. Here, the chapter draws on the work of one such organization, Cambridge Carbon Footprint, to consider more broadly the ability of CBOs to encourage community engagement with climate change, to then promote increased individual action towards low-carbon living.

From Economic Theory to the Psychology of the Individual: Why We Act

The conventional theoretical orientation of behavioral change discourse concentrates on concepts of individual choice (central to economics) and driving factors (important to psychology). Modes of human behavior, posited in classical economic theory, center on the concept of maximizing expected utility. Humans are viewed as single-minded utility maximizers who display objective rational behavior; possess the ability to consider all relevant options available to them; know the probability of each option; and thereby make informed choices in a predictable and consistent manner (Akerlof & Shiller, 2009).

Such economic "rationality" implies that household investment decisions are based on independent rational calculations of costs and benefits of all evidence when making a choice, with no difficulties in problem recognition, information search, alternative evaluations, and decisions about future consequences. Rational economic agents, or "homo economicus," have perfect knowledge and perfect foresight. Further, decisions are driven by personal rather than societal interests (List, 2004).

More recent insights by behavioral economists as well as psychologists studying human cognition, beliefs, and values have now led to a fundamental questioning of this line of economic reasoning (Brick et al., 2021; Steg, Lindenberg, & Keizer, 2016; Whitmarsh et al., 2021). In reality, individuals deviate from the standard rational choice model. It is inherently difficult to accurately recognize the extent of a problem and to then assess which option will provide the highest utility. As Smith, Searle, and Cook (2008) suggest, individuals are not aware of what to invest in, how to make comparisons, and therefore prioritize accurately to achieve maximized economic benefits.

Further, daily behaviors around consumption, such as the use of heating and lighting, are largely habitual and reliant on automatic processes; individuals may therefore be resistant to change (Marechal, 2010), whereas decisions to make structural changes to a home, such as improved roof and wall insulation, airtightness, and heating systems, entail inter-temporal choices, with individuals having to make investment decisions with future consequences. Individuals have a greater aversion to financial losses than achieving expected gains and tend to discount long-term gains. Investments in energy efficiency measures within a home, with upfront costs and rewards in the long term, are therefore likely to be discounted by individual households.

A consequent focus on motivators and barriers to encourage the installation of energy-saving measures is in keeping with the dominant model of behavioral change. Behavioral (or internal) drivers, including belief, experience, attitudes, habits, knowledge, lifestyle, and personality, are cited in evidence-based policy research as factors that influence an individual's decision to change domestic energy (and other) consumption patterns. Situational (or external) factors, such as culture, institutional frameworks, and access to capital and information, then also affect an individual's ability to implement measures (Moser & Kleinhuckelkotten, 2018). While useful to examine these common motivators, often there is no method of establishing the dynamic qualities, history, and interdependence of these drivers, or their precise role in promoting or preventing different behaviors. Instead, it is often assumed in government behavioral change programs that by giving households "better information and more appropriate incentives, individuals will choose to act more responsibly and adopt pro environmental behaviour" (Shove, 2010).

Providing a person with information, however, does not necessarily encourage them to act (Bouman & Steg, 2019). Policy incentives (such as reducing purchasing taxes on roof insulation) can be rendered ineffective and misaligned if the type of pro-environmental behavior being sought (whether to buy insulation or not) is not on a household's investment decision-making radar in the first place (Moser & Kleinhuckelkotten, 2018; Vesely et al., 2020). Whatever technological solutions or financial incentives exist, they will not gain popularity and durability if an individual does not act upon the information provided (Blake, 1999).

For Constanzo, Archer, Aronson, and Pettigrew (1986), there is a "difficult path from information to action" (p.1). Information campaigns, often instigated by governments and centering on procedural knowledge, are likely to be ineffective without any deeper level of personal engagement. Here, impact and normative knowledge also necessarily come into play. Both are required to create a consensus to act in the belief that others share their views (Cooke, Fielding, & Louis, 2016).

Uzzell (2010) equally argues that the "information deficit model" (that increasing procedural knowledge will translate into a change in behavior), and which government programs often fixate upon, diverts attention from larger and more significant problems concerning the ways in which people think they need to live and consume energy. For Bouman and Steg (2019), individuals need to understand the climate change issue, care about it, know what to do, and know that doing something will help to solve the problem. Also, individuals are heavily influenced by what others around them are doing; in effect, "what we see around us provides a more effective avenue for changing behaviour than information campaigns" (Jackson, 2005, p.1). Moreover, as Corner et al. (2014) suggest, people also want others to live up to what they think they and others should be doing. Adding both impact and normative knowledge into this dynamic then starts to reframe community values and norms towards making socially beneficial choices about energy matters for the collective good.

Nevertheless, Crompton (2010) argues that the associated language around such approaches can still fail to inspire people, particularly when faced with counter narratives, as promoted by the climate change denial lobby, for instance, and which speaks more effectively to people's current values and aspirations about how they want to live. As Rees (2010) notes in his paper, "What is blocking sustainability? Human nature, cognition and denial', "the modern world remains mired in a swamp of cognitive dissonance and collective denial seemingly dedicated to maintaining the status quo" (p.1). It remains that the consequences of individual actions, and an understanding that the scale of any domestic change *can* make a difference, is hard to grasp without a deeper level of public engagement with the climate change issue.

From the Individual to Social Networks: A Potential to Change the Way We Act

Societal values act as powerful guiding principles that individuals then use to judge situations and determine their course of action (Crompton, 2010). Values are in effect foundational to humans' motivational systems and correlate strongly with patterns of behavior. Critical here is that individuals tend to engage in more pro-environmental behaviors if they display strong "self-transcendent" values. They will, in turn, sustain that engagement over the longer term. As Crompton (2010) also notes, if campaigners in support of climate change action want to widen or extend public engagement, they need to appeal not to individual self-interest, but rather communicate – and promote – intrinsic values of "bigger than self" universalism. Such "framings," and which he defines as language and knowledge constructs able to activate and strengthen particular cultural values, then also offer a way for shared conventions and contemporary standards of life to be re-examined. Frames are not static, and deep frames of understanding, rather than mere surface ones, will result in individuals becoming more willing to adopt new practices. In turn, complex practices constituting daily lives may then be reconfigured on a larger societal scale.

Leading authors, such as Putnam (Putman, 2000), conjecture that a resurgence of civic engagement should follow from a palpable national or even global crisis. Halpern (2010) equally notes that for any wider social action, such as the need to curb climate change, to be successful, there needs to be an inherent movement from the micro (individual) to the meso (community) to the macro (world). The aggregate significance of individual actions then become mutually reinforcing to achieve a "bigger than self" goal, helping to close what is referred to as the "value-action" gap: the fact that what people think is generally not matched by what they do and how they act (Randall, 2011).

However, others point to an opposite, and substantial, dynamic also in play. Quimby and Angelique (2011), for instance, note that, perhaps paradoxically, crises-level calls for action can backfire. Increased involvement with a complex problem like climate change can magnify its scope and make solutions appear less attainable, leading people to believe that their individual actions are in vain. Closing the "values-action" gap can be particularly hard in relation to the private sphere of one's home (Moser & Kleinhuckelkotten, 2018). The home is closely linked with household identity and control (over this domestic situation). For Randall (2011), "behaviour is a surface phenomenon, beneath it lie more complex motivations and meanings…even those who are willing (to change) in principle find themselves surprisingly resistant in practice…mitigating climate change has the capacity to upset, disturb and raise people's defences" (p.2).

Social network analysis provides a means to understand how all this might work, and then be used to achieve positive change; how, for instance, knowledge, practices, trust, and reciprocity transcend through human social relations, assisting individuals to act (Putman, 2000). A social network is a social structure made up of individuals (who can be thought of as nodes) connected through ties. Granovetter's (Granovetters, 1973, 1983) work demonstrates how new ideas and practices are more likely to be diffused in networks comprising numerous loose connections (or "weak ties") than in networks comprising stronger ties and with a more closed and inward-looking demeanour. By having access to other "social worlds", the range of knowledge is effectively widened; and opportunities open up for existing and new social networks to act as "social vehicles" to promote messages of social change.

Ultimately, policy makers are not the drivers of change alone (UN, 2021). Rather, the creation of trust and momentum for change often comes from outside government, with bottom-up networks playing pivotal roles in not only disseminating procedural knowledge but in also shaping constructive community norms and values (Steg et al., 2016). People gain understanding and knowledge from each other, with tacit knowledge transmitted laterally through non-hierarchical, "trusted messengers" or intermediaries rather than through top-down communications from governments and which can engender sceptics and suspicion (Keys, Thomsen, & Smith, 2013).

Using the Role of Community-Based Organizations as Trusted Messengers

For Granovetters (1973, 1983), trusted intermediators effectively "spread influence," operating as brokers by strategically introducing new ideas and practices, as well as connecting individuals to further knowledge pools and social networks to explore. Community-based organizations operate, either implicitly or explicitly, in this way, acting as "hub" or "bridge" between individuals and other organizations within the community in which they are situated. In turn, such characteristics mean that the CBO "structure" is well-positioned to provide the critical networks needed to foster a broad community engagement in issues around climate change and the need to foster sustainable patterns of consumption and production (McNamara & Buggy, 2016; Stokes, Mandarano, & Dilworth, 2014).

Difficulties in leading sustainable lifestyles, however, are often compounded by individuals lacking basic procedural knowledge. They remain unclear, for instance, of their energy usage or the amounts they spent on energy supplies in their homes. Infrequent energy bills mean that individuals have little way of knowing which of their everyday behaviors contributes most to their energy bills or what the simplest domestic changes are to make to bring their bills down and improve their ability to act. Uncertainty about what an individual could achieve in relation to savings on electricity bills and potential returns on a housing investment can therefore result in little action (Moser & Kleinhuckelkotten, 2018). Mistrust in information disseminated by the government, media, and sales' people and even scientists can be cited as reasons not to act, reaffirming the difficulty in top-down information campaigns reaching individuals. "What to do and the benefits that accrue if we do" from locally based, trusted intermediaries can therefore offer one solution to breaking down these barriers to act (Sloot & Steg, 2018).

Trusted advice and support through CBOs can effectively assist households in negotiating options in the marketplace. Poor procedural knowledge, which leads to sub-optimal choices, can be addressed, for instance, through providing professional advice, through recommending relevant tradespeople, and through loaning out energy monitors to track household progress in making energy savings. Yet it is normative and impact knowledge that invariably makes an individual follow through on their intention to act (Shultz, 1999). CBOs need to not only diffuse procedural knowledge but also impact and normative knowledge, particularly among those who seek their advice. For Crompton (2010), intrinsic values embodied in the desire to address environmental issues is more likely to result in an individual's continued adoption of energy and carbon saving measures. CBOs can actively encourage and facilitate individuals' deeper levels of engagement with climate change concerns.

Cambridge Carbon Footprint

Cambridge Carbon Footprint (CCF) provides an exemplary case in point of a CBO approach to addressing climate change. Established in 2005, in Cambridge (UK), the CCF "mission" is to raise awareness of climate change issues and directly support individuals in their own personal efforts to reduce their carbon footprint and move to low-carbon living. The understanding behind its approach is that the challenge of carbon reduction is equally psychological and technical, with this reflecting the backgrounds of the two co-founders, being, respectively, in psychotherapy and low-energy buildings. Since its inception, CCF has run a varied program of events and activities providing both practical advice and psychological support for personal action.

In 2006, CCF established a 6-week facilitated course, Carbon Conversations, with the aim "to connect, explore and then act on climate change." The course provides a supportive and non-judgmental "safe space," taking groups of people through trust-building exercises, discussions, and explorations of their carbon footprints and lifestyles, particularly in relation to decarbonizing their homes. What was found when conducting the conversations was that those involved started to appreciate the range of unconscious dynamics and complex emotions and defenses that invariably develop around change, thereby also enabling individuals to be effectively supported into accepting new practices (Randall, 2011). This depth of engagement, the openness of experiences recounted, and the encouragement to share and explore the emotional as well as the rational responses to the climate challenge reflect the intrinsic "self-transcendent" values which Crompton (2010) embodied. As Randall, one of the co-founders of CCF, concluded from the experience,

> I am not proposing that the nation be invited to take part in "climate psychotherapy", simply that more imaginative and personal responses are required in working with people to achieve change in their individual lives.

This form of "psycho-education" effectively allows the individual to be understood at a more complex level than that of the traditional linear – and now entirely inadequate – model of behavioral change embodied in attitude-behavior-choice. It also provides an avenue to further explore the complex and dynamic qualities, histories, and interdependencies of behavioral and situational drivers. During the discussions, the trusted CBO facilitators can continually "frame" and "re-frame" the "climate" problem in ways that can more effectively explain the rationale behind global and national carbon reduction targets and the need to collectively act in meeting them, in response to the attitudes and values of the other participants.

Defenses, however, can still be raised, with individuals reacting to the arguments given, reticent to make the link between their own individual lifestyles, domestic energy consumption, and the environment (Randall, 2011). Defensiveness occurs not just at the level of the individual. Facilitated discussions not only provide insights into individual thought processes and reasonings behind their actions (or inaction), the group discussion itself also produces insights that would be less accessible without the interactions found in a group setting. However, such interactions can in turn also contribute to a kind of collective avoidance towards changing domestic behaviors. The language of barriers can easily be adopted, such as and in relation to the matter of energy consumption, "I fly all the time with work, so what's the point in trying to make carbon savings in my home?". Group discussion can also risk reinforcing perceived disruption and financial barriers, such as in relation to installing high-technology measures, like photovoltaic solar panels. Fixating on the more expensive measures can engender a "can't do" and "wait and see how somebody else gets on with it" mindset. Here, the CBO's professional experts can – more positively – heighten the group's awareness of alternative and cheaper investment possibilities.

Moreover, individuals attending such facilitated courses invariably have weak ties with each other, which can thereby allow the infiltration of new ideas and practices. Drawing on the work by Stock Whitaker (1975), Randall (2011) suggests that the act of listening to others (strangers) verbalizing their experiences stimulates those listeners to act. New group norms can be formed, comprising informal rules about what should be done in relation to appropriate practices. Like norms, group values (being the goals or ideas that come to serve as guiding principles) can then also arise.

Interestingly, when participants are guided to explore what would positively motivate them to adopt energy efficiency (and other) measures, intrinsic values embodied in a desire to address wider environmental issues often start to emerge. And then, while many decisions remain driven by personal economic benefits and aversion to cost, at the same time attempts are likely to be made to adhere to constructive social norms about what others are doing. As Randall (2011) notes, when pursuing behavioral change, underlying defenses and denial need to be brought into a consciousness where "they can be talked about and worked through" (p.2). And, as Stock Whitaker (1975) also notes, this can be especially facilitated through exchanges within a group.

Further, even if the underlying motivation of individual personal interest may not alter during the length of a course, like Carbon Conversations, it may well be that, through the social connections of the group (and CBO), their engagement with the broader climate change issue and commitment to act may progressively develop and strengthen over time. The potential for knowledge to be transmitted in this way is though dependent on the acceptance and commitment of that group to addressing the issue (here climate change). It also depends on how the message is communicated and the nature of the connections made within the social network. Here, an CBO provides an active "hub of learning" within this (network) structure. Moreover, it can also serve as a "directional influence," encouraging participants to deliberate further on ways to decarbonize their homes and lives in general. The trained facilitators can provide not only a forum for such discussions to unfold, but also for knowledge, trust, and reciprocity to develop.

Since establishing Carbon Conversations, CCF has gone on to more political work, prompted by the UN Climate Talks in Copenhagen in 2009, including rallying with other groups as a member of the Stop Climate Chaos national coalition (now The Climate Coalition). They have also established Open Eco Homes, first held in 2010 and held annually ever since and where households across Cambridgeshire open their low-energy homes to inspire visitors to make their own home energy savings. They have launched other initiatives aimed at reducing consumption footprints, including the Second-Hand style (2013) clothes swap, and the Circular Cambridge network of repair cafes. This latter project is now working with other CBOs to build one of the largest networks of local repair cafes, globally (CCF, 2019, 2021).

To capitalize on this sort of momentum, it is important that these (community-based) organizations maintain on-going connections (weak ties) with their participants, such as offering feedback on progress in making energy and carbon savings over an extended period. Further, the services and engagement activities that these local intermediaries provide need to be promoted across their localities or regions' different social networks, to widen their take-up. Moreover, by forging partnerships with other organizations, like local government authorities, and pooling knowledge and resources, the capacity to collectively act can strengthen, relative to being left to individual CBO endeavors. Through such initiatives, CCF has been acclaimed nationally and replicated by community groups across other cities (COIN, 2021).

The cascading effects of a CBO's activities however rely upon the organization's coordination of volunteers to run the forums and on those community members who have participated in events, projects, and courses to spread the procedural, normative, and impact knowledge they have gained into their wider social networks. Although this type of endeavor can reach certain groups of individuals committed to tackling climate change, it can remain difficult for a CBO to gain leverage and trigger widespread engagement without a concurrent greater community commitment to responding to climate change issues. Here, understanding the motivations behind "early adopters" in moving to low-carbon living is important, and the gleaning of knowledge on the ways in which they act (Quimby & Angelique, 2011). But not just this, observing and engaging the "laggards" in dissonance and denial of imminent climate change is equally important, and a crucial next step.

Conclusion

Universal awareness and understanding of the global challenge of climate change is crucial (IPCC, 2021). Yet securing behavioral change to address the issue is not a single event; it is a long-term and complex process. Further, neither government nor individual action will suffice; it requires a sustained commitment from everyone, and at all levels. A sense of personal ownership of an issue and a belief that the scale of any domestic change can make a difference is a critical aspect of efficacy. But neither do calls for deeper levels of community engagement with the climate agenda simply shift the allocation of responsibility from the state.

Moreover, social network analysis offers an opportunity to also explore the meso-level of how knowledge, trust, and reciprocity developed through community-based organizations, as trusted messengers, encouraging individuals to act, can similarly contribute and in doing so support both levels of micro household and macro government.

Approaching the climate challenge at – and with – this community level is likely to deliver change at a much greater scale, with new and existing social networks offering the potential to bring more people into a constructive process of engagement. Cambridge Carbon Footprint provides a useful example of how community-based organizations embedded in these networks can encourage individuals to reduce their carbon footprint and empower them to join to build a low-carbon future. By doing so, ground-level action will play its crucial part in meeting macro level targets already set for 2050 and beyond.

Cross-References

▶ Adapting Cities to Climate Change
▶ Amsterdam's Pathway to Climate Neutrality: Creating an Enabling Environment
▶ An Overview of the Relationship of the Sustainable Development Goals and Urban and Regional Development
▶ Building Community Resilience
▶ Carbon Neutral Adelaide
▶ City Visions: Toward Smart and Sustainable Urban Futures
▶ Collaborative Climate Action
▶ Community Engagement for Urban and Regional Futures
▶ Environmental Education and Non-governmental Organizations
▶ Faith Communities as Hubs for Climate Resilience
▶ Increasing Young People's Environmental Awareness

- Multi-stakeholder Partnerships to Support Climate Migrants in Fragile Cities
- Networking Collaborative Communities for Climate-Resilient Cities

References

Akerlof, G., & Shiller, R. (2009). *Animal spirits: How human psychology drives the economy, and why it matters for global capitalism*. Princeton: Princeton University Press.

Blake, J. (1999). Overcoming the `value-action gap' in environmental policy: Tensions between national policy and local experience. *Local Environment, 4*(3), 257–278.

Bouman, T., & Steg, L. (2019). Motivating society-wide pro-environmental change. *One. Earth, 1*(1), 27–30.

Bouman, T., Verschoor, M., Albers, C., Bohm, G., Fisher, S., Poortinga, W., Whitmarsh, L., & Steg, L. (2020). When worry about climate change leads to climate action: How values, worry and personal responsibility relate to various climate actions. *Global Environmental Change, 62*, 102061.

Brick, C., Bosshard, A., & Whitmarsh, L. (2021). Motivation and climate change: A review. *Current Opinion in Psychology, 42*, 82–77.

Cambridge Carbon Footprint. (2019). *Annual Report* https://cambridgecarbonfootprint.org/wp-content/uploads/2020/09/2019-Annual-Report_compressed.pdf

Cambridge Carbon Footprint. (2021). https://cambridgecarbonfootprint.org/

COIN. (2021). Climate outreach and information network https://doit.life/organisation/707289

Constanzo, M., Archer, D., Aronson, E., & Pettigrew, T. (1986). Energy conservation and behaviour: The difficult path from information to action. *American Psychologist, 41*, 521–528.

Cooke, A., Fielding, K. S., & Louis, W. (2016). Environmentally active people: The role of autonomy, relatedness, competence and self-determined motivation. *Environmental Education Research, 22*(5), 631–657.

Corner, A., Markowitz, E., & Pidgeon, N. (2014). Public engagement with climate change: The role of human values. *Wiley Interdisciplinary Reviews: Climate Change, 5*(3), 411–422.

Crompton, T. (2010). *Common cause: The case for working with our cultural values*. Oxford: World Wildlife Fund.

DEFRA. (2011). *A framework for pro-environmental behaviours*. London: The Stationary Office.

Granovetters, M. (1973). The strength of weak ties. *American Journal of Sociology, 78*(6), 1360–1380.

Granovetters, M. (1983). The strength of weak ties revisited. *Sociological Theory, 1*, 201–233.

Halpern, D. (2010). *The hidden wealth of nations*. Cambridge: Polity Press.

Howarth, C., & Parsons, L. (2021). Assembling a coalition of climate change narratives on UK climate action: A focus on the city, countryside, community and home. *Climatic Change, 164*, 8. https://doi.org/10.1007/s10584-021-02959-8.

Intergovernmental Panel on Climate Change (IPCC). (2021). *Climate change 2021: the physical science basis.*

Jackson, T. (2005). *Motivating sustainable consumption: A review of evidence on consumer behaviour and behavioural change*. London: Policy Studies Institute. https://www.ipcc.ch/report/ar6/wg1/downloads/report/IPCC_AR6_WGI_Full_Report.pdf.

Keys, N., Thomsen, D., & Smith, T. (2013). Adaptive capacity and climate change: The role of community opinion leaders. *The International Journal of Justice and Sustainability, 21*(4), 432–450.

List, J. (2004). Neoclassical theory versus prospect theory: Evidence from the market place. *Econometrica, 72*, 615–625.

Marechal, K. (2010). Not irrational but habitual: The importance of `behavioural lock-in' in energy consumption. *Ecological Economics, 69*, 1104–1114.

McNamara, K., & Buggy, L. (2016). Community-based climate change adaptation: A review of academic literature. *The International Journal of Justice and Sustainability, 22*(4), 443–460.

Moser, S., & Kleinhuckelkotten. (2018). Good intents, but low impacts: Diverging importance of motivational and socioeconomic determinants explaining pro-environmental behaviour, energy use and carbon footprint. *Environment and Behaviour, 50*, 626–656.

Popovski, V. (2019). *The implementation of the Paris agreement on climate change*. Abingdon: Routledge.

Putman. (2000). *Bowling alone: The collapse and revival of American community*. New York: Simon & Schuster Audio.

Quimby, C., & Angelique, H. (2011). Identifying barriers and catalysts to fostering pro-environmental behaviour: Opportunities and challenges for community psychology. *American Journal Community Psychology, 47*, 388–396.

Randall, R. (2011). *Is it time to stop talking about behaviour change? Conference proceedings `Future climate change'*. London: The Institute of Mechanical Engineering.

Rees, W. (2010). What's blocking sustainability? Human nature, cognition, and denial. *Sustainability: Science, Practice and Policy, 6*(2), 13–25.

Shove, E. (2010). Beyond the ABC: Climate change policy and theories of social change. *Environment and Planning A, 42*, 1273–1285.

Shultz, P. (1999). Changing behaviour with normative feedback interventions: A field experiment in curbside recycling. *Basic and Applied Social Psychology, 21*(1), 25–36.

Sloot, D., & Steg, L. (2018). Can community energy initiatives motivate sustainable energy behaviours? The role of initiative involvement and personal

pro-environmental motivation. *Journal of Environmental Psychology, 57*, 99–106.
Smith, S., Searle, B., & Cook, N. (2008). Rethinking the risks of home ownership. *Journal of Social Policy, 38*(1), 83–102.
Steg, L., Lindenberg, S., & Keizer, K. (2016). Intrinsic motivation, norms and environmental behaviour: The dynamics of overarching goals. *International Review Environmental and Resource Economics, 9*(1–2), 179–207.
Stock Whitaker, D. (1975). Some conditions for effective work with groups. *British Journal of Social Work, 5*(4), 423–439.
Stokes, R., Mandarano, L., & Dilworth, R. (2014). Community-based organisations in city environment policy regimes: Lessons from Philadelphia Local Environment. *The International Journal of Justice and Sustainability, 19*(4).
Thaler R & Sunstein C. (2008) Nudge: Improving decisions about health, health and happiness.
The Behavioural Insights Team. (2011). Behaviour change and energy use https://www.bi.team/wp-content/uploads/2015/07/behaviour-change-and-energy-use.pdf
The Behavioural Insights Team. (2020). The behavioural economy https://www.bi.team/wp-content/uploads/2020/11/The-Behavioural-Economy-1.pdf
The Behavioural Insights Team & WPI Economics. (2021). Applying behavioural insights to support flood resilience https://www.bi.team/wp-content/uploads/2021/08/210621-EA-Flood-resilience-report_final-draft.pdf
United Nations. (2015). Paris Agreement to the United Nations Framework Convention on Climate Change https://unfccc.int/sites/default/files/english_paris_agreement.pdf
United Nations Environment Programme (UNEP) (2008). "Twelve steps to help you kick the CO2 habit" United Nations Environment Programme http://www.unep.org/wed/2008/english/information_material/factsheet.asp
United Nations. (2021). Climate Action https://www.un.org/en/climatechange/net-zero-coalit
Uzzell D. (2010). Psychology and climate change: Collective solutions to a global problem *Joint British Academy/British Psychological Society Annual Lecture*
Vesely, S., Klockner, C., & Brick, C. (2020). Pro-environmental behaviour as a signal of cooperativeness: Evidence from a social dilemma experience. *Journal of Environmental Psychology, 67*, 101362.
Whitmarsh, L., O'Neill, S., & Lorenzoni, I. (2011). Climate change or social change? Debate within, amongst and beyond disciplines. *Environment and Planning A, 43*, 258–261.
Whitmarsh, L., Poortinga, W., & Capstick, S. (2021). Behaviour change to address climate change. *Current Opinion in Psychology, 42*, 76–81.

Community Engagement for Urban and Regional Futures

Sara Bice[1,2] and Kirsty Jones[1]
[1]Crawford School of Public Policy, The Australian National University, Acton, ACT, Australia
[2]School of Public Policy and Management, Tsinghua University, Beijing, China

C

Synonyms

Community impact; Consultation; Social impact; Socio-environmental impact; Stakeholder consultation; Stakeholder engagement

Definitions

Community engagement	The active information exchange and discussion between affected community members and governments or project/policy proponents to define and solve shared problems, design or implement policies or services, and make better informed and more representative decisions.
Social risks	The direct and indirect economic, societal, and environmental hazards related to projects or policies that generate perceived or actual threats to individuals' or communities' livelihoods and sustainability.
Social license to operate	A policy or project's level of acceptance within a defined community.

Introduction

Requirements for new and upgraded infrastructure are increasing throughout the world, fueled by population growth, climate change mitigation, and the ascension of a new middle class in the Global South (Bice et al. 2018). For both urban and regional futures, housing, transport, energy, water, telecommunications, social and health infrastructure needs must be met. Significant investment is required to meet these demands, in addition to that needed to support socioeconomic recovery from the COVID-19 pandemic (Akhtar et al. 2021). The Global Infrastructure Hub (GI Hub) estimates that over the next two decades US$79 trillion will be invested in infrastructure. Total estimated investment needs are closer to US $94 trillion, however, leaving a $US15 trillion gap.

The unprecedented investment levels required to meet contemporary challenges are reflected in record-breaking government commitments being made in countries including the USA's US1.2 trillion "Build Back Better" infrastructure bill, Australia's $AU110 billion Commonwealth government investment to support infrastructure's role in pandemic recovery, and the China State Council's recent commitment to accelerate its already enormous infrastructure investments in a push to stabilize its pandemic-disrupted economy. While these investments hold tremendous potential for economic stimulus and societal improvement, infrastructure selection, planning, and delivery is not without externalities. Project delivery necessitates considerable disruptions, from noise and dust to traffic congestion and the influx of construction teams from outside local communities. Longer-term impacts involve both voluntary and involuntary land acquisition and resettlement; cultural heritage and land access impacts; loss of or substantial change to local amenity and community identity; environmental degradation, including but not limited to effects on biodiversity, water, and soil; and the politicization of projects that may be decades in the making. Moreover, as leading economists Larry Summers and Ed Glaeser note, infrastructure may not be the best lever for economic stimulus in regional areas, at least in the USA (Summers and Glaeser 2021). Infrastructure costs, therefore, come in forms other than financial. Impacts on individuals' and communities' livelihoods, reputational impacts on the governments and corporations that deliver and operate infrastructure, and a "social licence to operate" for these projects all comprise social risks that must be managed.

Social risks present clear, immediate, and costly considerations for infrastructure projects. Community opposition and protest is increasing across all areas of public policy concerns (Carothers and Youngs 2015). Protests around mining or gas fields projects reveal incurred closure costs of up to US$20 million per week (Davis and Franks 2014). In Australia, as an example, community opposition to major infrastructure projects is estimated to have contributed to at least AU$30 billion in costs due to delays or cancellations (Bice et al. 2019). And these are only the costs borne by proponents or government investors. Costs to communities in opposition are also high but difficult to quantify. Many of those who oppose projects do so in their own time as volunteers representing their communities (Mati et al. 2016). Studies in engagement demonstrate that such situations impact communities across a variety of indicators, including well-being (Cox et al. 2010). Oppositional relationships take a toll on individuals, contributing to increased stress levels, possible time away from paid work, reduced resilience, and a reduced sense of control over one's life (Vella-Brodrick 2017). Major projects also bring socio-environmental costs, including loss of visual amenity and decreased land access or substantial changes to an environment that communities call home. The latter may lead to "solistalgia" – a sense of distress induced by environmental change and a longing for the pre-existing landscape – an abstract but heartbreaking situation in which communities feel a collective homesickness for the place that once was (Albrecht 2005).

In order to realize the urban and regional futures possible through sustainable infrastructure investment, social risks – encompassing economic, societal, and environmental costs to projects, livelihoods, and communities – must be addressed. Community engagement, as a principle and a practice, is core to this redress and response. Community engagement is a widely recognized discipline in the infrastructure sector.

While not formally acknowledged as a profession, community engagement incorporates a number of traits common to professions. This includes an identifiable cohort of experts (Mayo et al. 2007), discipline-focused member associations, including the International Association for Public Participation (IAP2) and the International Association for Impact Assessment (IAIA), and practice-specific guidance to support best practice (e.g., IAP2 2015; Vanclay et al. 2015). Training in community engagement practice is readily available, and community engagement roles are increasingly disaggregated from communications or marketing departments into distinct business divisions.

Contemporary infrastructure delivery environments regularly involve escalating community opposition, protests, and related project costs. These circumstances signal a breakdown in project-community relations. Yet the global infrastructure boom is occurring within the most advanced community engagement environment in history. Corporate social responsibility (CSR) and related sustainable development practices are widely institutionalized, especially by multinational corporations of the type responsible for infrastructure delivery (Bice 2017). In principle, widespread acceptance of the need for community engagement should result in smoother project delivery, reduced opposition or protest, and consequent cost savings. This chapter explores the principles and practices of community engagement that can support the urban and regional futures possible through major infrastructure delivery, within a challenging contemporary environment.

This chapter proceeds by first defining community engagement theory and practice, as it exists beyond the infrastructure sector. This broader survey of the field equips readers with an understanding foundational to the chapter's subsequent concern: how community engagement can contribute to more successful infrastructure delivery and optimal community outcomes. It outlines suggestions for community engagement's potential contributions to sustainable infrastructure delivery by exploring a major case study from the Australian infrastructure sector. This case example introduces the Infrastructure Engagement Excellence Standards (IEE Standards). This world-first set of quality assurance standards for major infrastructure projects clearly sets out the ten standards, related attributes, and indicators that support best practice engagement. Wide uptake of standards such as these is critical to facilitating infrastructure's potential to support regional and urban futures and to ensuring that the major investments being made in infrastructure are worthwhile, financially and societally.

What Is Community Engagement?

Before discussing the role of high-quality community engagement in successful infrastructure project delivery, it is helpful to define it, beyond the infrastructure sector. Community engagement is the active information exchange and discussion between affected community members and governments or project/policy proponents to define and solve shared problems, design or implement policies or services, and make better informed and more representative decisions. In scholarly terms community engagement involves the use of active "coordinated network" relationships and solutions that prioritize joint planning and programming of autonomous members (Head 2007), focused on project delivery with reduced or avoided negative impacts and optimized community benefits.

Community engagement can be practiced in many legitimate ways and is recognized as a core element to successful project planning and delivery in a variety of sectors and policy areas, including cultural heritage, development, disability services, education, energy, finance, health, infrastructure, indigenous people's concerns, social welfare, and technology. It supports policymaking and implementation to shore up relationships between citizens and governments, supporting democratic processes. Community engagement supports improved understanding of policies or projects, encourages public participation in decision-making, and takes into consideration public perceptions about issues including financing models, transparency, and accountability. For instance, relative to infrastructure, this might include public consultation about public-private partnerships (PPPs), especially selection and accountability of government partners. The outcomes of successful community engagement are often tightly interlinked with concerns about a project or policy's "social licence to operate" – a policy or project's level of acceptance within a defined community (Bice 2014). Taken together, these issues offer a helicopter view of the concourse of concerns covered by community engagement.

Community engagement practice is supported and understood through a considerable foundation of scholarly literature that advances both the theory and practice of community engagement. Dating at least to Arnstein's 1969 ladder of citizen participation, community engagement theory articulates the sociopsychological basis for engagement, the importance of trust in relationships, and the fundamental role of social capital in project and policy success. An empirically based subset of literature focuses on practice, providing advice on the wide range of methods effective for involving community members in policy and project planning, delivery, or evaluation processes. Such involvement might range from informing communities through direct communication (e.g., newsletters or websites) to fielding questions at town hall forums, from establishment of community consultative committees to deliberative democratic processes. Aligned literature highlights the proactive involvement of communities in public service delivery through nascent practices of co-production, smoothing power, and influence between "the engagers" and "the engaged."

In practice, community engagement has often historically been conflated with marketing and communications roles or public relations. This represents an outmoded and narrow understanding of community engagement as both a philosophy and a discipline. Today, community engagement is far more sophisticated and challenging than marketing or PR. It extends beyond fire-fighting efforts to douse community protests or outrage. It focuses instead on the establishment of mutually respectful and meaningful relationships that assist policymakers or project developers to better understand and work with community members, even in spite of clear differences. Contemporary engagement practice commonly espouses bottom-up approaches and public participation. This is not to say that community engagement is necessarily consensus-committed, nor is it inherently deliberatively democratic in nature. Instead, it is about the active incorporation of representative community perspectives, concerns, and priorities into project or policy considerations.

It is equally important to consider what is meant by "community" when considering community engagement. From a sociological perspective, "community" is socially constructed, acknowledging that its boundaries and meaning arise from a series of social interactions, values, and behaviors that result in a largely shared, collective understanding. In practice, the idea of community is regularly deployed as a political football or wedging device and will inevitably be defined differently depending upon the position and perspective of a particular stakeholder. Disparate approaches to how community is defined and deployed are evident in recent research that looks into national and international efforts to integrate community engagement into policy concerning infrastructure planning or project delivery (Cowell and Devine-Wright 2018; Nabatchi and Jo 2018). Transport, urban planning and housing scholars contribute to an entire subgenre of research concerning community responses to particular projects (De Martinis and Moyan 2017), planning politicization (Legacy et al. 2018), and the role of new planning paradigms in what certain scholars argue is a "post-political" era (Legacy 2016). Taken together, such studies reinforce the challenges of defining community and of consequent difficulties defining clear boundaries of any particular community. The challenge of defining community boundaries is particularly pertinent to infrastructure delivery where a locally affected community may bear the brunt of construction disruptions but where a much wider "community" will realize a project's benefits.

Regulating Community Engagement

Regulation has a central role to play in establishing and clarifying the function of community engagement for infrastructure project delivery. Scholarly literature also explores the relationship between policy mandates and effective community engagement, with a further subset of studies focused on evidence-based community engagement (Day 1997; Innes and Booher 2004). In Australia, the focus of this chapter's case study, many states and territories require stakeholder engagement as part of project approvals processes at various points in a project's lifecycle. Very recently, several states have begun investigating ways to improve the regulation shaping these engagements. Recent regulatory improvements included an inquiry into stakeholder engagement practices to extend guidelines beyond mining and extractives into state significant infrastructure projects and recommendations for whole of government frameworks to guide public servants in public participation. These regulatory concerns extend to "governance and oversight, [engagement] capability development, access to expertise and monitoring, and evaluation mechanisms" (Victorian Auditor General's Office 2017). Regulatory advances, at least in Australia, reflect concerns in critical literature that point out the lack of accountability mechanisms in community engagement practice (Innes and Booher 2004).

While advances in regulation are improving community engagement practice, the academic literature clearly articulates inadequacies in the effectiveness and appropriateness of certain engagement efforts (Abelson et al. 2003; Cowell and Devine-Wright 2018). Indigenous and vulnerable communities remain disadvantaged in community engagement processes and rarely achieve equal participation (Nish and Bice 2012). Tokenistic community engagement remains unfortunately common, with critics pointing out a general lack of meaningful, truly deliberative engagement that is also influential (Dryzek 2002). The literature serves as an important reminder that, despite advances in theory and practice in recent decades, the policy and regulation to guide best practice community engagement remains wanting in many places. However, recent government-led initiatives, including those summarized above, suggest that in places such as Australia, governments are demonstrating an interest in improving the situation.

Digital Era Community Engagement

Establishing the boundaries that define a particular community is an especially delicate task, involving a great deal of poetic license. Today, such boundary setting is palpably more perplexing due to the geographic boundary dissolution and increased stakeholder interconnectedness possible through online means, especially social media. These virtual communities are genuine and meaningful to their members. Recent studies, for instance, suggest that online "communities" fuel new forms of activism and protest (Einfeld et al. 2018). A small but emerging field of research into digital community engagement introduces previously impossible modes of interaction, made possible primarily through real-time monitoring of satellite-based geographic information systems (GIS) (Brown and Kyttä 2014), the use of "big data" or government data sharing (Moon 2002; Desouza and Jacob 2017), and online networks (Mandarano et al. 2010).

The evolution of online communities as a concern for community engagement reflects broader considerations of "digital era governance" (DEG) – an approach that acknowledges information and communication technology's centrality and its commensurate changes to management systems, citizen interactions, cultures, behaviors, and organizations (Dunleavy et al. 2006). DEG's rise is especially pertinent for the public-private partnerships (PPPs) that will play a major role in future infrastructure delivery. This is because research in other sectors suggests that private corporations are more active social media participants and consequently more likely to come under community fire via social media (Einfeld et al. 2018). The COVID-19 pandemic has further accelerated shifts to digital interaction, resulting in widespread adoption of online tools to facilitate individuals' engagement. It is anticipated, therefore, that as digital community engagement grows and as more PPPs come on stream, the regulation, policies, and practices needed to promote and govern community engagement will be further and rapidly complicated by a need to adapt to social media, digital engagement, and online communities.

Key Concepts that Relate to and Influence Community Engagement

When considering the role of community engagement for urban and regional futures, it is helpful to mention two closely related, distinct concepts: social risk and social license to operate. These concepts encapsulate the primary considerations for community engagement for infrastructure projects. As noted, a social license to operate represents the level of acceptance of a policy or project by an affected community (Moffat and Zhang 2014; Bice and Moffat 2014). Governments investing in major infrastructure projects often reference the need for a project to earn or maintain a social license and certain policy instruments in Australia prioritize a social license as an important component to project implementation (Bice et al. 2017). Other studies reveal the costs of lost social licenses (Davis and Franks 2014), but also caution against adoption of tokenistic community engagement or superficial development practices as a means to secure a social license (Harvey and Bice 2014). Social risk management also plays a role here, as the community engagement work necessary to earning and maintaining a social license is frequently understood in terms of managing project risk. While social risk remains relatively poorly defined and managed, and certainly lacks the systematic attention or legitimacy of financial, actuarial, or environmental risks (Haines 2011), it is emerging as a key consideration for community engagement on major projects. At a very practical level, the work to earn and maintain a social license, and related social risk management, commonly rests with community engagement practitioners (Kemp and Owen 2013).

Community Engagement as Practice

It is also useful here to consider what community engagement practice looks like, in a contemporary project environment. This includes education, training, position descriptions, and the aforementioned professionalization of community engagement. While community engagement roles are increasingly common, it remains an emerging profession still in a transition period in which it is refining its identity and separating from historical attachments to marketing and communications disciplines. Perhaps it is this history or a lack of formal professionalization, or a combination of both, that results in community engagement's struggle for legitimacy. This is especially the case in the infrastructure sector, where community engagement must be performed alongside more technical (and therefore more "valuable") roles necessary to major infrastructure projects, such as project management, engineering, risk management, and finance. In many instances, discussion of role legitimacy is linked back to difficulties measuring the indirect or intangible impacts and benefits of community engagement efforts, leading some to describe the work of community engagement practitioners as "a black box" (Bice 2013).

While tertiary degrees specific to community engagement are limited, leading membership organizations, including IAP2 and IAIA, offer professional development, certifications, and training. Organizations like these also provide their considerable membership bases (IAIA has approximately 7,000 members) with regularly updated, research-based guidelines, principles, and ethical standards (IAP2 2015; Vanclay et al. 2015). These documents aim both to encourage and improve best practice but also to provide legitimacy through shared methods, via organizational reputation, and through creation of a global community of practitioners.

The preceding sections introduced community engagement, its theoretical and practical foundations, connections to regulation and policy, relationship to social license and social risk, and its development as a profession. Ongoing work seeks to raise the profile and legitimacy of community engagement roles and work, in a variety of sectors. For major infrastructure, community engagement practice is being supported through the development of quality assurance standards that work to define the qualities of best practice engagement while simultaneously delivering a framework for quality assessment. The chapter now turns to a case example of these standards from Australia as a means of demonstrating the qualities of community engagement necessary to realizing urban and regional futures via sustainable infrastructure delivery.

Case Study: Community Engagement in Australia

Background to Australia's Infrastructure Sector

The Australian infrastructure sector is thriving. Pre-pandemic investments were historically high, and further investment is now being prioritized as a major lever for post-pandemic economic recovery. In June 2020 Australian Prime Minister Scott Morrison announced the fast-tracking of $1.5 billion in national infrastructure projects as part of the Government's COVID-19 stimulus package, with a further $9.7 billion in additional funding announced in the October 2020 budget. States hard-hit by the pandemic are similarly fast-tracking major infrastructure. Victoria approved more than $7.5 billion in projects since only March under its Recovery Taskforce, while NSW is fast-tracking $4.3 billion in projects and has streamlined planning and approvals processes under state emergency measures.

Population growth – and now economic recovery – is driving Australia's urgent requirements for new and upgraded infrastructure, including energy, water, transport, housing, and social needs. Infrastructure Australia (IA) – the nation's independent infrastructure advisor – estimates that there is a $600 billion infrastructure need nationally to 2035 (Infrastructure Australia 2019). IA recently released a three-pronged strategy for infrastructure in COVID-19 recovery, involving "enabling the long-term sustainability of the sector, build[ing] community resilience and the transformation of the Australian economy" (Infrastructure Australia 2020). Responsibility to deliver these resources stretches across government and these diverse industries, also involving disciplines of architecture, construction, urban planning, engineering, and social welfare. The investments are massive, the stakeholders varied. Project fast-tracking, economic recovery, and the substantial expenditure it requires, however, do not mean that all projects are acceptable to Australian communities.

Costing hundreds of millions (and often billions) major infrastructure projects are time- and resource-intensive, demand a multitude of expertise, and are highly visible and disruptive for neighboring residents, workers, businesses, and commuters. Infrastructure delivery is, undoubtedly, one of the most conspicuous areas of policy implementation. Institutional investors globally rank public policy uncertainty as the number one barrier to investment in infrastructure (Blanc-Brude et al. 2016). Eighty-three percent of investors responding to a recent Infrastructure Partnerships Australia survey reported that "uncertainty in Australia's policy and regulatory settings is limiting their willingness to invest" (Infrastructure Partnerships Australia 2019). Very public and well-studied examples, including Melbourne's East-West Link, Sydney's Westconnex and Perth's Freight Link, demonstrate the close interconnections between community opposition to major projects, politics, and policy changes (Legacy et al. 2017).

Additionally, the PPPs used to deliver a considerable portion of major infrastructure projects have come under scrutiny (Hodge et al. 2017), raising accountability issues (Stafford and Stapleton 2017) and spurring socioeconomic critique (Zwalf et al. 2017). Recent examples from around Australia show that many communities are unhappy with the way certain projects have been proposed or delivered (De Martinis and Moyan 2017), regardless of funding model. From the industry perspective, social opposition contributes to increased costs and barriers to infrastructure delivery, including substantial delays and cancelations (Bice et al. 2019; Harris et al. 2003). This is not to mention the stresses and difficulties placed upon infrastructure project staff. From communities' perspectives, there may be dissatisfaction with the ways in which they are being engaged in the planning and delivery of major projects, or simply with the projects themselves. Meanwhile, the combined stresses of

C

the pandemic mean that many communities are feeling compounded pressures which may result in compassion fatigue (Montemurro 2020; Luo et al. 2020), resulting in reduced resilience and diminished willingness to accommodate the disruptions inherent to infrastructure delivery.

Community engagement will be core to the success of Australia's "Big Build" and to the achievement of societal benefits from infrastructure delivery. In recent years in Australia, community opposition to major infrastructure projects contributed to at least $30 billion in lost investment, with industry professionals consistently ranking "stakeholder and community pressure" as one of the top three reasons for project delays or cancellations (State of Infrastructure and Engagement 2019–2020) (Bice et al. 2021). Social risks, including land acquisition, cultural heritage concerns, and controversial projects, also played a major role in project delays or cancellations. Yet this chapter reveals community engagement as a professional practice remains secondary to and often misunderstood by major infrastructure disciplines of engineering, project management, and finance.

The Australian context is representative of current infrastructure climates in many developed economies. The Australian case demonstrates the need for improved frameworks to enhance the ability to assess, model, plan for, or evaluate the characteristics of community engagement quality to support optimal project outcomes and societal benefits. Such planning and quality assessments are essential to supporting the capacity of infrastructure projects to deliver maximum benefit for communities. By pursuing community engagement within a quality assurance framework, connections between responsible industry behaviors and sound community engagement, and between poor engagement, lost social license and community opposition is made clear. Standards can also play a role in reducing the disparity between community engagement and other major disciplines critical to infrastructure delivery. They can equip community engagement professionals with the types of assessment and reporting instruments readily available to their colleagues in other disciplines. Throughout the world major infrastructure project delivery has long been criticized for poor project selection that is misaligned with community needs, failure to deliver touted social benefits, cost overruns, and high-risk levels, especially from unexpected events, including project opposition (Flyvbjerg 2009). Standards offer an important and proven means of supporting community engagement for infrastructure.

The Infrastructure Engagement Excellence Standards

While community engagement standards do exist within Australia, many of these standards and frameworks are vague, lacking the detail required to address the challenges of infrastructure planning and delivery. The Infrastructure Engagement Excellence (IEE) Standards are a series of ten research-derived standards with assessment indicators that detail the fundamentals, standards, and indicators of community engagement to achieve optimal project benefits for investors, developers, and communities. The IEE Standards advance knowledge for industry by offering an innovative, reliable, and comparable quality assessment framework for engagement for major infrastructure projects. The Standards are supporting government and industry to benchmark, compare, and evaluate community engagement within organizations, between projects and between sectors (e.g., water, energy, and transport) for the first time in Australia. This information is critical to improving the ability of the Australian infrastructure industry to generate an accurate understanding of the quality of community engagement being delivered and the implications of engagement quality for project delays, community opposition, and benefit delivery.

The IEE Standards are research-derived and outcomes-focused. They seek to deliver engagement outcomes that are community-centered, coordinated, and cost-effective. The Standards are comprised of three fundamentals which outline the essentials that must be present for Infrastructure Engagement Excellence to be achieved (See, Fig. 1). Each fundamental is comprised of standards, with a total of ten standards outlining the characteristics of Infrastructure Engagement

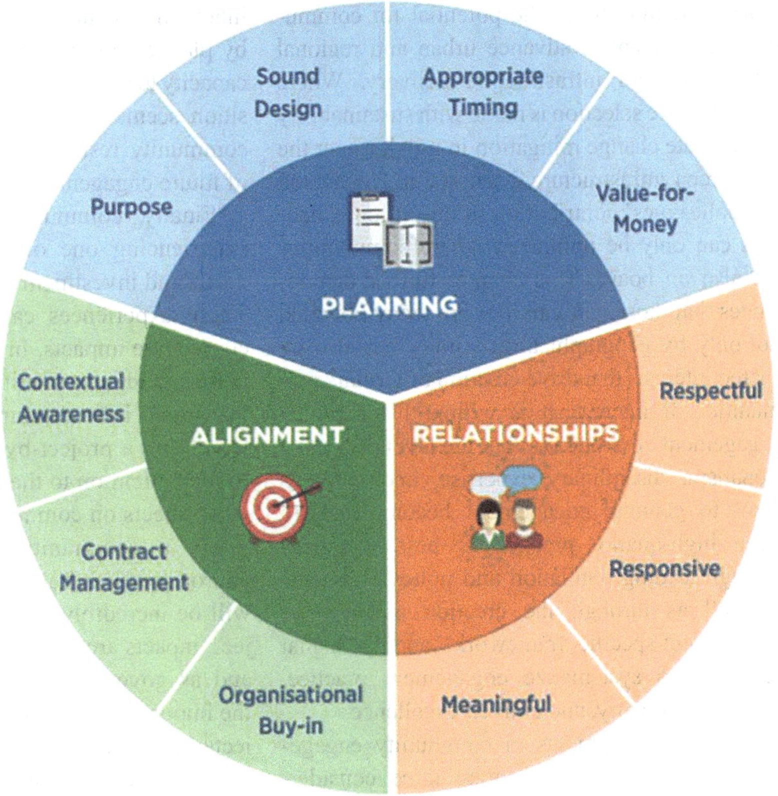

Community Engagement for Urban and Regional Futures, Fig. 1 Infrastructure Engagement Excellence Standards. (Source: Authors)

Excellence and the indicators that allow assessment of their presence and quality in practice.

By defining community engagement excellence for infrastructure planning and delivery, the IEE Standards support the sector in several crucial management processes. Firstly, the Standards can provide guidance during the planning phases of a project or new engagements. Secondly, the Standards can support the ongoing monitoring of engagements throughout the life course of project delivery. A third use for the Standards is to support the evaluation of engagement performance upon project completion through post-project assessments. Finally, the Standards can be used to compare and contrast engagement within or between government departments, organizations, or sectors. This type of analysis can allow community engagement performance quality to be benchmarked, in addition to capturing lessons learnt and supporting a culture of continuous improvement.

Beyond Australia, a framework for assessing and understanding community engagement quality and its implications will enable governments to make better informed investment decisions. Community engagement professionals will be better positioned to articulate the value of the work they do for major project delivery. Project developers will be better able to judge the resources required to deliver quality engagement, and therefore be better equipped to make appropriate resource and budgeting decisions. Communities should therefore benefit from higher-quality community engagement to meet their needs and deliver project benefits.

Summary and Conclusion: Future Considerations for Community Engagement in Urban and Regional Futures

The broad but thorough understanding of community engagement theory and practice offered in this chapter, combined with a recommended use of quality assurance standards to encourage best practice, demonstrates the potential for community engagement to advance urban and regional futures through infrastructure delivery. Where infrastructure selection is made with sustainability and climate change mitigation in mind, as per the blue-green infrastructure discussed in Henderson and colleagues' contribution to this volume, benefits can only be optimized where communities are also on board. This chapter further demonstrates that those communities are today bounded not only by geographical proximity but also by shared interests that drive creation of virtual communities of interest. It is within this complex engagement environment that the level of professionalism, disciplinary expertise, and commitment to genuine engagement become critical. Such high-quality practice is best supported through leading regulation and policy guidance, as well as through the creation and use of engagement-specific frameworks and tools that can help to systematize engagement practice, benchmark quality, and enliven excellence.

Future developments in community engagement will need to incorporate three considerations: digital engagement, community resilience, and cumulative impacts. Digital engagement will require a unique skill set in which community engagement practitioners remain human while going digital. This means careful consideration of selection of online tools and concerted effort to build trust and generate warmth through an emotional register that can be difficult to achieve online. Yes, digital engagement may enhance the speed, safety and spread of community engagement but if it does so at the loss of the more emotional element of human interaction, it compromises the trust building that is vital to successful engagement.

Resilience – the capacity of a community to "bounce back" from change or adverse events – will also play a major role in future community engagement. As more communities face climate change events from severe weather, as the pandemic wanes but fallout continues, and as new crises emerge, communities around the globe are suffering and exhausted. While infrastructure may often be seen as a positive intervention it is also an inherently disruptive one. Communities fatigued by plague and disaster have less willingness and capacity to accept disruptions and increased opposition seems likely. Attending to and nurturing community resilience will be an important part of future engagements.

Finally, communities across the globe are experiencing one of the largest infrastructure needs and investment periods in modern history. These experiences can best be understood as cumulative impacts. In many locations the regulations and community engagement practices that guide infrastructure project delivery do so largely on a project-by-project basis with no or limited attention to the effects of contemporaneous projects on communities. This does not represent a community-centered or place-based approach to infrastructure. Such approaches will be incredibly important as cumulative project impacts are felt, consultation fatigue sets in, and as governments and developers recognize the importance of context to their individual projects. Together, digital engagement, resilience, and cumulative impacts define the terrain of community engagement in future infrastructure delivery.

Cross-References

▶ Building Community Resilience
▶ Community in a Changing Climate: Shaping Urban and Regional Futures

References

Abelson, J., Forest, P.-G., Eyles, J., Smith, P., Martin, E., & Gauvin, F.-P. (2003). Deliberations about deliberative methods: Issues in the design and evaluation of public participation processes. *Social Science & Medicine, 57*, 239–251.

Akhtar, S., Bhattacharya, A., Buchoud, N. J. A., Hendriyetty, N. S., & Yoshino, N. (2021). How quality infrastructure can bring private sector finance into infrastructure investment to recover from the Covid-19 crisis. www.G20-insights.org: G20 Global Infrastructure Hub.

Albrecht, G. (2005). 'Solastalgia'. A new concept in health and identity. *PAN: Philosophy Activism Nature*, (3), 41.

Bice, S. (2013). No more sun shades please: Experiences of corporate social responsibility in remote Australian mining communities. *Rural Society Journal, 22*, 138–152.

Bice, S. (2014). What gives you a social licence? An exploration of the social licence to operate in the Australian Mining Industry. *Resources, 3*, 62.

Bice, S. (2017). Corporate social responsibility as institution: A social mechanisms framework. *Journal of Business Ethics: Springer, 143*, 17.

Bice, S., & Moffat, K. (2014). Social licence to operate and impact assessment. *Impact Assessment and Project Appraisal, 32*, 257–263.

Bice, S., Brueckner, M., & Pforr, C. (2017). Putting social license to operate on the map: A social, actuarial and political risk and licensing model (SAP Model). *Resources Policy, 53*, 46–55.

Bice, S., Poole, A., & Sullivan, H. (2018). Public policy in the 'Asian century'. In *Concepts, cases and futures* (1st ed.). London: Palgrave Macmillan.

Bice, S., Neely, K., & Einfeld, C. (2019). Next generation engagement: Setting a research agenda for community engagement in Australia's infrastructure sector. *Australian Journal of Public Administration, 78*, 290–310.

Bice, S., Emerson, S., & O'Connell, K. (2021). State of infrastructure and engagement report: 2019–2020. Canberra.

Blanc-Brude, F., Chen, G., & Whittaker, T. (2016). *Towards better infrastructure investment products? A survey of investor's perceptions and expectations from investing in infrastructure*. Singapore: The EDHEC Infrastructure Institute-Singapore and Global Infrastructure Hub.

Brown, G., & Kyttä, M. (2014). Key issues and research priorities for public participation GIS (PPGIS): A synthesis based on empirical research. *Applied Geography, 46*, 122–136.

Carothers, T., & Youngs, R. (2015). *The complexities of global protests*. Washington, DC: Carnegie Endowment for International Peace, Publications Department.

Cowell, R., & Devine-Wright, P. (2018). A 'delivery-democracy dilemma'? Mapping and explaining policy change for public engagement with energy infrastructure. *Journal of Environmental Policy & Planning, 20*, 499–517.

Cox, D., Frere, M., West, S., & Wiseman, J. (2010). Developing and using local community wellbeing indicators: Learning from the experience of Community Indicators Victoria. *Australian Journal of Social Issues, 45*, 71–88.

Davis, R., & Franks, D. M. (2014). Costs of company-community.

Day, D. (1997). Citizen participation in the planning process: An essentially contested concept? *Journal of Planning Literature, 11*, 421–434.

De Martinis, M., & Moyan, L. (2017). The East West link PPP Project's failure to launch: When one crash-through approach is not enough. *Australian Journal of Public Administration, 76*, 352–377.

Desouza, K. C., & Jacob, B. (2017). Big data in the public sector: Lessons for practitioners and scholars. *Administration & Society, 49*, 1043–1064.

Dryzek, J. S. (2002). *Deliberative democracy and beyond: Liberals, critics, contestations*. Oxford University Press on Demand.

Dunleavy, P., Margetts, H., Bastow, S., & Tinkler, J. (2006). New public management is dead – Long live digital-era governance. *Journal of Public Administration Research and Theory, 16*, 467–494.

Einfeld, C., Bice, S., & Li, C. (2018). *Social media and community relations: Five key challenges and opportunities for future practice*. Extracting Innovations: Mining, Energy, and Technological Change in the Digital Age, 287–312.

Flyvbjerg, B. (2009). Survival of the unfittest: Why the worst infrastructure gets built – And what we can do about it. *Oxford Review of Economic Policy, 25*, 344–367.

Haines, F. (2011). *The paradox of regulation: What regulation can achieve and what it cannot*. Edward Elgar.

Harris, C., Hodges, J., & Schur, M. (2003). *Infrastructure projects: A review of canceled private projects*. World Bank.

Harvey, B., & Bice, S. (2014). Social impact assessment, social development programmes and social licence to operate: Tensions and contradictions in intent and practice in the extractive sector. *Impact Assessment and Project Appraisal, 32*, 327–335.

Head, B. W. (2007). Community engagement: Participation on whose terms? *Australian Journal of Political Science, 42*, 441–454.

Hodge, G., Greve, C., & Boardman, A. (2017). Public-private partnerships: The way they were and what they can become. *Australian Journal of Public Administration, 76*, 273–282.

IAP2. (2015). *Quality assurance standard in community and stakeholder engagement*. Brisbane: IAP2.

Infrastructure Australia. (2019). *Australian infrastructure audit 2019*. Sydney: Infrastructure Australia.

Infrastructure Australia. (2020). *Common principles of infrastructure recovery (COVID19)* [Online]. Canberra: Infrastructure Australia. Available: https://www.infrastructureaustralia.gov.au/common-principles-infrastructure-recovery-covid-19. Accessed 12 Dec 2020.

Infrastructure Partnerships Australia. (2019). *Australian infrastructure investment report 2019*. Sydney: Infrastructure Partnerships Australia.

Innes, J. E., & Booher, D. E. (2004). Reframing public participation: Strategies for the 21st century. *Planning Theory & Practice, 5*, 419–436.

Kemp, D., & Owen, J. R. (2013). Community relations and mining: Core to business but not "core business". *Resources Policy, 38*, 523–531.

Legacy, C. (2016). Transforming transport planning in the postpolitical era. *Urban Studies, 53*, 3108–3124.
Legacy, C., Curtis, C., & Scheurer, J. (2017). Planning transport infrastructure: Examining the politics of transport planning in Melbourne, Sydney and Perth. *Urban Policy and Research, 35*, 44–60.
Legacy, C., Cook, N., Rogers, D., & Ruming, K. (2018). Planning the post-political city: Exploring public participation in the contemporary Australian city. *Geographical Research, 56*, 176.
Luo, M., Guo, L., Yu, M., Jiang, W., & Wang, H. (2020). The psychological and mental impact of coronavirus disease 2019 (COVID-19) on medical staff and general public – A systematic review and meta-analysis. *Psychiatry Research, 291*, 113190.
Mandarano, L., Meenar, M., & Steins, C. (2010). Building social capital in the digital age of civic engagement. *Journal of Planning Literature, 25*, 123–135.
Mati, J. M., Wu, F., Edwards, B., El Taraboulsi, S. N., & Smith, D. H. (2016). *Social movements and activist-protest volunteering*. Palgrave Macmillan.
Mayo, M. C., Hoggett, P., & Miller, C. (2007). Navigating the contradictions of public service modernisation; the case of community engagement professionals. *Policy & Politics, 35*, 667.
Moffat, K., & Zhang, A. (2014). The paths to social licence to operate: An integrative model explaining community acceptance of mining. *Resources Policy, 39*, 61–70.
Montemurro, N. (2020). The emotional impact of COVID-19: From medical staff to common people. *Brain, Behavior, and Immunity, 87*, 23–24.
Moon, M. J. (2002). The evolution of e-government among municipalities: Rhetoric or reality? *Public Administration Review, 62*, 424–433.
Nabatchi, T., & Jo, S. (2018). The future of public participation. In *Conflict and collaboration: for better or worse* (p. 75). Routledge.
Nish, S., & Bice, S. (2012). Participatory planning and monitoring in the extractive industries. In *New directions in social impact assessment: conceptual and method* (pp. 59–77). Chelthenham: Edward Elgar.
Stafford, A., & Stapleton, P. (2017). Examining the use of corporate governance mechanisms in public–private partnerships: Why do they not deliver public accountability? *Australian Journal of Public Administration, 76*, 378–391.
Summers, L., & Glaeser, E. (2021). *Debate: Funding infrastructure investment in a post COVID-19 economy* [Online]. https://www.mckinsey.com/business-functions/operations/our-insights/debate-funding-infrastructure-investment-in-a-post-covid-19-economy. Mckinsey Insights. Accessed 27 Nov 2021.
Vanclay, F., Esteves, A. M., Aucamp, I., & Franks, D. (2015). *Social Impact Assessment: Guidance for assessing and managing the social impacts of projects*. International Association for Impact Assessment.
Vella-Brodrick, D. (2017). *Community engagement in infrastructure: Fostering well-being and resilience*. [Online]. www.nextgenengagement.org/wp-content/uploads/2017/07/Next_Gen_Expert_Commentary_Series_Vella-Brodrick-1wldapv.pdf. Accessed 17 Dec 2018.
Victorian Auditor General's Office. (2017). *Public participation in government decision-making*. Melbourne: VAGO.
Zwalf, S., Hodge, G., & Alam, Q. (2017). Choose your own adventure: Finding a suitable discount rate for evaluating value for money in public–private partnership proposals. *Australian Journal of Public Administration, 76*, 301–315.

Community Foundations

► Philanthropy in Sustainable Urban Development: A Systems Perspective

Community Garden

► Urban Food Gardens

Community Impact

► Community Engagement for Urban and Regional Futures

Community in a Changing Climate: Shaping Urban and Regional Futures

Eleni Kalantidou[1] and Naomi Hay[2]
[1]Griffith University, Brisbane, QLD, Australia
[2]Australian National University, Canberra, ACT, Australia

Definition

Community, in its current context, is a construct based upon uniformity, commonality rooted in nationality/ethnicity, and shared cultural, socio-economic, or ideological conditions. When confronted with climate change, issues of vulnerability, exclusion, and social injustice are exacerbated, creating both visible and invisible divisions. To proactively shape urban and regional futures, a reconceptualization of community is required that acknowledges humanity's shared exposure to multiple planetary crises. This entails a collective effort in order to prepare for and adapt to present and forthcoming challenges.

Introduction

The present relationship between the natural environment and current modes of habitation is increasingly fragile and inherently unsustainable. Population growth, globalization, and hyperconsumption are placing unprecedented strain upon the planet. In the face of global ecological, geopolitical, and socioeconomic vulnerability, climate change is expected to further escalate conditions of vulnerability for millions of people around the world. Countries at the frontline of climate change face an increasingly uncertain future as coastlines and deltas become uninhabitable due to sea-level rise. Against this backdrop, resources including fresh water become scarce while agricultural land becomes arid. Cities, where most of the global population resides, are challenged by heat waves and heat islanding thereby reducing livability. Simultaneously, global health concerns arising from climate change escalate. These vulnerabilities accelerate uneven global development and socioeconomic inequality, with millions of global citizens currently experiencing homelessness and/or a lack of access to basic needs including water, food, clothing, and health care. In order to address these issues and create conditions of adaptation, communities are confronted by the need to reevaluate the values and premises upon which they exist. Difference and plural identities have been ignored by contemporary communities, exposing populations to elevated levels of risk. To overcome environmental and social inequality, the concept of a homogenized and unified community imposed by neoliberal globalization needs to be redefined (Rose 1999). Recomposing communities is a crucial task related to shaping urban and regional futures so as to effectively respond to present and future challenges.

Settlements in a Changing Climate

Climate change (both warming and cooling of the earth) predates human existence; however, the rapid acceleration of warming in recent years reflects the impact of unsustainable human activities on the planet. This point is supported by the Intergovernmental Panel on Climate Change (IPCC 2021, p. 4), which argued that "it is unequivocal that human influence has warmed the atmosphere, ocean and land," with each of the last four decades being successively warmer than any proceeding since 1850. At current rates of warming, global surface temperature rises will continue to increase under all emissions' scenarios considered by the IPCC, and unless significant reductions in CO_2 and other greenhouse gas emissions occur in the coming decades, global warming of 1.5 °C and 2 °C will likely be exceeded in the twenty-first century (IPCC 2021). The repercussions of this continued warming are of critical concern to the future of humanity and global settlements, with the IPCC contending that the scale of recent changes across the global system are unprecedented over centuries to thousands of years. As global warming increases, so does the frequency and intensity of heat waves, tropical cyclones, reductions in Arctic sea ice, and droughts in some regions. Many of these changes, especially in regard to ocean changes, ice sheets, and sea-level rises, are expected to be irreversible for centuries to millennia (IPCC 2021). As a response to this crisis, 195 nations at the Paris 2015 United Nations Climate Change Conference in December 2015 made a historic, binding agreement to keep global temperature rise to below 2 °C and drive efforts to limit temperature further to 1.5 °C above preindustrial levels (UNEP, 2015). However, the damage already caused despite current and future responses to minimize anthropogenic climate change, has resulted in many human settlements facing diminishing viability moving forward.

Environmental degradation, as a result of inappropriately sited and rapidly expanding cities and towns, exacerbates this. The effects of flooding and storm surges are escalated through human intervention in natural watercourses, with the destruction of natural vegetation exposing sand dunes and clearing mangroves. Other human settlements are subject to rising temperatures, lack of available fresh water, and extreme heat islanding due to replacing vegetation with hard surfaces, buildings, and infrastructure.

Despite humanity's inherent capacity for adaptation, as can be seen in rising to multiple occasions of climate change over many thousands of years, humans have become increasingly vulnerable as communities no longer live a flexible, nomadic life. Present-day settlement in urban centers with ever-growing populations has made adaptation to climate change prohibitively expensive and extremely difficult to respond to (Fagan 2005). The failure of humanity to better understand the effects of climate change in shaping society, and to comprehend human reliance upon the environment, severely compromises the capacity of preparing for current and upcoming challenges. In neoliberal conditions of uneven development and a culture of profit making, goals such as zero emissions, rewilding, and reducing the overconsumption of material resources are compromised by a lack of long-term planning, funding, and policy implementation. Coastal nations and islands are experiencing the side effects of sea-level rises and saltwater contamination to agricultural land, increasing the risk of food insecurity for vulnerable populations. As climate change leads to increased levels of migration from unviable land, existing geopolitical tensions will inevitably escalate, resulting in conflict over access to fresh water, food, and habitable environments. As stated by Lewis Mumford, "civilizations have risen and fallen without apparently perceiving the full import of their relations with the earth" (Mumford 1940, p. 316), a quote that makes evident the ineptness of humanity in recognizing its dependence on nature's volatility. More than 80 years later, in the face of escalating conditions of unparalleled ecological vulnerability, it is evident that little has changed.

Examining and Redefining the Construct of Community

Considering the present dangers and forthcoming uncertainties of climate change, communities must be prepared for, and develop adaptive capacity to, both fast and slow disasters. In this context, "community" as a concept needs to shift from conventional definitions, which exclude equally as they include populations according to race, religion, or ethnic background. Perceptions of community vary significantly across social contexts: these might encompass a group of people living in the same neighborhood; a social unit sharing common values, religion, and culture; or a population being governed by the same rules. In this sense, community means different things to different people depending on their sociopolitical context and is therefore open to constant change and diverse interpretations. Community as a connection of separate individuals operating with varying agendas, or conversely, as a single body of being with constraints upon the political, the individual, and social identity, has been disputed. Jean Luc-Nancy (1991) argued that underpinning the core of Western social and political thought is a nostalgic longing for an intimate community – a residue of the misconstrued idea that we once lived in a harmonious and connected community that has deteriorated over history. Nancy (1991) considered the concept of idealized community as a mythology, an imagined fusion of individuals translated into a "being-in" attempting to function as a single identity. In view of this, a community designed to fulfil a set of nostalgic ideals through its "work" would inevitably self-destruct. Contrary to the "being-in" construct, Nancy's (1991) "inoperative community" was conceived as a "being-with" – existing as unplanned and spontaneous without fixed identity or collective goals. Georgio Agamben (1993, p. 25) similarly asserted that most people preconceive community as a being-in, which constitutes "an absolutely unrepresentable community." Community's unrepresentability translates into an abstraction based upon tradition, myth, history, and sentimentality that generates preconceived divisions of class, race, and religious affiliation. To deconstruct it, it is necessary to question notions of singular identity and community, and transverse these imaginary boundaries (Agamben 1993). Maurice Blanchot (1988, p. 111) further contested that the relationships that a community creates – a drawing together yet opposing tension between individual beings – where death (finitude), disaster, and absence lie at the core, presents "an impossible, absent community." Further interpretations define community as an imposition disguised in individual choice, a romanticized condition yet to be achieved (Bauman 2000, 2001), and an effort to fill a void, which is falsely represented as a bond by modern politics (Esposito 2000).

Community Within Urban and Regional Environments

There have been multiple attempts to constitute community beyond the conceptual and into the realm of the physical, including small-scaled communities envisioned in Marxist ideals, a myriad of planning movements (the City Beautiful, the Garden City, the Garden Suburb, the New Town, and New Urbanism), and movements supporting multiple occupancy and co-living communities globally. Early proponents of grassroots movements in community engagement and empowerment, including Murray Bookchin, advocated for new relationships between the city, nature, and people in the form of eco-communities. Bookchin's vision of the urban was small scaled, self-administering, and controllable, bringing the garden to the city while targeting balanced living based on high ecological, political, and intellectual standards, with active participation of citizens in urban transformation (White 2008). Additionally, Colin Ward (Ward et al. 2011) proposed self-organization and administration, championing community cooperation, and self-managed social practices across all areas of urban renewal, including housing, urban policy, and planning. This would include everything from architecture, urban design, and community gardens and spaces, through to self-managed credit-unions and finance sectors. While approaches such as these play an important role in supporting alternate models of living in communities, they do not provide a blueprint for scaling up, or frameworks on how to undo existing structures that perpetuate social and environmental injustice. For human settlements to be prepared for major climatic disasters, food shortage, and territorial conflicts, pragmatic models of implementation are required, which would address the broken policies and uncontrolled sprawl of urban environments (Knuth et al. 2020). Moreover, idealized visions of community do not fit well with today's uneven relations in power, exclusion, and disenfranchisement of millions of people around the planet. Community may reject outsiders, discriminate within itself, and is ever-changing and shifting over generations. In this respect, David Harvey (2002) identified the dangers entailed in the idealization of bottom-up social movements on account of their vulnerability to gradual disempowerment and corruption by external political forces.

Within a city, region, or town, community is never equal, and therefore it is problematic to present its construct as unifying. Coming together due to necessity does not imply shared values, needs, or principles. As a case in point, millions of people around the planet are forced to move to towns and cities for work, while confronted by uneven global development, displacement, growth of informal settlements, and increasing socioeconomic disparities. In many parts of the world, there is a rapid rise of informal settlements, often inappropriately sited and with minimal or no basic infrastructure and services, facing increased risk from natural hazards and the impacts of climate change. As a result, impromptu communities are born in order for individuals to respond to their immediate situation. These communities manage to design innovative approaches regarding housing, disaster management, and resources without state support and external intervention. Their improvised configurations contrast with a growing number of Western communities that are becoming increasingly technological, controlled, and surveilled, making evident the disparity between communal modes of adaptation.

Moreover, technology has brought with it a new form of nomadism. Communities created online with the support of fast internet and social media, bring to fore new communal structures yet to be untangled. To quote Ulrich Beck (2000, p. 74), "to live in one place no longer means to live together and living together no longer means living in the same place." This translates to a fragmentation of traditional, fixed communities, due to the emergence of transnational communities being no longer dependent upon geography and physical connectedness.

All the above demonstrate the complex and dynamic social ecology that can be found in human settlements. Considering the current and upcoming challenges humanity is facing, outdated concepts of community as a reflection of uniformity and sameness need to be abandoned. Inclusiveness could adopt different shapes and sizes outside the scope of "being-in," in order to adapt to a changing climate and respond to growing social and environmental inequalities. In this regard, how community is defined needs to move beyond traditional notions of togetherness, and instead be explored in conditions of interdependence. Furthermore, community as a construct requires repositioning in contemporary circumstances of fluid borders, online communication, and population mobility, where it is formed in nontraditional ways. In brief, community is shifting to new configurations and structures led by climate change, the technologization and globalization of everyday life, and increasing precariousness for billions of people around the planet. The latter necessitates a new understanding of how humans position themselves in community, beyond locality and homogeneity.

Reframing Community in Conditions of Climate Change

Community in the context of a changing climate does not only come together in times of adversity but also acts together to minimize future hardships and risk. In doing so, community moves beyond relationships between individuals, existing only within the confines of their own finitude. As demonstrated in the previous sections of this entry, there is a growing need for communities to adapt to multiple levels of crisis as the face of human settlements change. On the one hand, grassroots movements, despite addressing the benefits of bottom-up governance and citizen participation in decision-making, lack access to financial and infrastructural resources and therefore have limited power to replace the economic models they are fighting against with new ones. On the other hand, traditional governmental approaches often overlook the complex diversity and differing needs of communities, treating them as a cohesive whole, which exacerbates inequality and exclusion to basic services including housing, education, and health care. The question that arises is how communities can become actively involved in local governance while acknowledging the role global conditions play, become inclusive and diverse, and evolve beyond outdated notions of nationalism and tribalism. To tackle issues of complexity at a structural level, Ezio Manzini (2015) proposed co-design processes, which he outlined as a vast social conversation between networks of individuals and groups. Manzini (2015) claimed that incremental change is insufficient, and instead, systematic change is needed that puts in place a shift from hierarchical systems to larger distributed systems – an interweaving of small scale, locally situated social organizations and systems. This would allow for decision-making processes to remain in the local community while, at the same time, digital and social media would enable a flow of communication, ideas, money, and goods, with communities becoming nodes in short and long-distance networks thereby reducing geographic isolation (Manzini 2015).

From another perspective, Arturo Escobar (2018), drawing from Humberto Maturana and Francisco Varela's concepts of autopoiesis and biological autonomy, suggested a process of "autonomous design." This is grounded in an ontological process, with communities adopting design practices enabling transformational change, which respectively puts in place conditions that strengthen community moving forward. Autonomous design is described as a design praxis, whereby every design activity starts with the presupposition that people are practitioners of their own knowledge. Escobar (2018) perceived this process starting with the community designing a learning system about itself and developing a statement of problems and possibilities. Bringing problems and possibilities to the table puts in place conditions for an adaptive framework to plan for, mitigate, and adapt to a changing climate. This requires understanding environmental, spatial, socioeconomic, and political vulnerabilities at a structural level through a process of risk mapping. The latter, undertaken within each and every community, encompasses natural, built, and social environments. Considerations to be addressed, while specific to each community, involve possibilities for economic diversification and transition to new economies, rethinking current agricultural practices, improving land use, and land buy-back schemes. Likewise, challenges that will need to be tackled include upgrading flood levees, the construction of sea walls for future sea-level rises, relocating essential services and communication systems at risk, and ensuring robust emergency, health, aged care, social, and volunteer services. Among these tasks should be a thorough analysis of available local materials and resources, facilities, industries, existing skillsets, and specialized knowledge within each community. To ensure the long-term application of adaptation plans, potential sources of future public and private funding should also be in place.

Activating these networks, and making available this information within communities and by communities, supports community building beyond the gestural or given constructs that currently exist. In this respect, the condition of climate change serves to bring community together, not only in times of adversity, but as future-proofing too. What needs to be contested is the perpetuation of the notion that human settlements are permanent. Such thinking leads to reactive, rather than proactive design and urban methodologies. From this perspective, communities that are united despite differences in the face of climate change progress beyond the individual and pursue collective identification of the steps necessary to scaffold viable futures. Designing speculative scenarios that envisage what communities could look like in the future has been employed to help people immerse into situations of change, express individual thoughts and ideas, and identify how change could affect their circumstances moving forward. This approach might sound similar to examples such as urban labs that are usually connected to conventional understandings of adaptation and community and intertwined with socio-technical systems. Nonetheless, speculative scenarios produced with user participation while acknowledging different epistemologies and ontologies, the problematic nature of definitions of community, and the dependence on state governance, aim to go beyond the limited framework provided by existing knowledge/practice schemes. Overall, new ways of thinking will need to be employed by communities in order to better prepare and protect them from catastrophes.

Conclusion

Community, as a concept and a construct, has undergone multiple iterations across multidimensional contexts. In its current imagining, the notion of community creates an illusion of unity, making invisible exclusion, injustice, socioeconomic inequity, and unequal development. The recent example of the COVID-19 pandemic illustrates the vulnerability of disenchanted populations, despite many of these living in, and being citizens of affluent countries with extensive resources. To overcome the illusion of solidarity, inherent differences within communities, whether based upon gender, race, identity, culture, religion, or privilege, need to be addressed. So too does the assumption that uniformity is required for communities to exist and function. Within this framework, the level of privilege and unique positioning of each community within a globalized world (geographically, economically, politically, and socially) must be considered in a shared condition of climate change. Escobar (2018) nominated a model of self-governance where communities identify their knowledge systems, map their natural and built resources, and detect how to innovatively and sustainably fund a remodeling of everyday life. Another suggestion in shaping urban and regional futures is the design of speculative scenarios with the active participation of communities so as to capture how future change would look and be experienced. These scenarios would be based on different knowledge schemata and move beyond socio-technical systems that are rooted in existing models of governance. To conclude, making urban and regional futures possible requires a reconfiguration of what community means and how people interconnect; it necessitates a level of interdependence, reliant upon active involvement in decision-making within community, so as to put in place viable plans for transformation and social change.

Cross-References

▶ The Practice of Resilience Building in Urban and Regional Communities

References

Agamben, G. (1993). *The coming community. Theory out of bounds*. Minneapolis: University of Minnesota Press.

Bauman, Z. (2000). *Liquid modernity*. Cambridge: Polity Press.

Bauman, Z. (2001). *Community: Seeking safety in an insecure world*. Malden: Polity Press.

Beck, U. (2000). *What is globalization?* (Trans. Patrick Camiller). Cambridge/Malden: Polity Press.

Blanchot, M. (1988). *The Unavowable community*. Barrytown: Station Hill Press.

Escobar, A. (2018). *Designs for the Pluriverse: Radical Interdependance, autonomy, and the making of worlds*. Durham/London: Duke University Press.

Esposito, R. (2000). *Communitas*. Stanford University Press.

Fagan, B. M. (2005). *The long summer: How climate changed civilization*. New York: Basic Books.

Harvey, D. (2002). *Spaces of capital: Towards a critical geography*. New York: Routledge.

IPCC. (2021). Summary for policymakers. In V. Masson-Delmotte, P. Zhai, A. Pirani, S. L. Connors, C. Péan, S. Berger, N. Caud, Y. Chen, L. Goldfarb, M. I. Gomis, M. Huang, K. Leitzell, E. Lonnoy, J. B. R. Matthews, T. Maycock, T. Waterfield, O. Yelekçi, R. Yu, & B. Zhou (Eds.), *Climate Change 2021: The physical science basis. Contribution of Working Group I to the Sixth Assessment Report of the Intergovernmental Panel on Climate Change*. Cambridge, U. K. and New York, NY, USA. Cambridge University Press.

Knuth, S., Stehlin, J., & Millington, N. (2020). Rethinking climate futures through urban fabrics:(De) growth, densification, and the politics of scale. *Urban Geography, 41*(10), 1335–1343.

Manzini, E. (2015). *Design when everybody designs: An introduction to Design for Social Innovation*, (Trans. Rachel Coad). Cambridge, MA: The MIT Press.

Mumford, L. (1940). *The culture of cities*. London: Seeker & Warburg.

Nancy, J. (1991). *The inoperative community*. Minneapolis/Oxford: University of Minnesota Press.

Rose, N. (1999). *Powers of freedo Reframing political thought*. Cambridge: Cambridge University Press.

Ward, C., White, D. F., & Wilbert, C. (2011). *Autonomy, solidarity, possibility: The Colin Ward reader*. Edinburgh: AK.

White, D. F. (2008). *Bookchin: A critical appraisal*. London: Pluto Press.

Further Reading

https://link.springer.com/referenceworkentry/10.1007/978-3-030-51812-7_296-1

Community Leadership

▶ Participatory Governance for Adaptable Communities

Community Resilience

▶ Faith Communities as Hubs for Climate Resilience

Community Resources

▶ Urban Commons as a Bridge Between the Spatial and the Social

Community Support

▶ Challenges of Delivering Regional and Remote Human Services and Supports

Community Vulnerability to Extractive Industry Disasters

Atharv Agrawal, Jonathan Banfield, Suehyun Cho, Teresa Kramarz and Eric Zhao
University of Toronto, Toronto, ON, Canada

Introduction

Environmental disasters have been increasingly characterized by the interaction of human activities and natural hazards. For example, the extraction and burning of fossil fuels increases carbon emissions that alter the Earth's climate and provokes extreme weather events such as wildfires, droughts, flooding of low-lying areas, and melting of permafrost (Park 2021). These environmental disasters overwhelmingly affect communities that are already marginalized because of their economic, gender, and racial status (Ribot 2011). These communities are vulnerable to external shocks because they live at the edge of their capacities to respond. Their vulnerabilities are a product of social and place-based inequities (Cutter et al. 2003). The specific focus here is on social vulnerabilities which subsumes the more traditional and narrower focus of the disaster literature on biophysical hazards such as infrastructure. The concept of social vulnerability highlights that the state of being susceptible to shocks is a "socially-constructed phenomenon influenced by institutional and economic dynamics" (Adger and Kelly 1999, p. 253). This entry focuses on disasters caused by oil and mining, the reproduction of social vulnerabilities, and the factors that curtail or enable resilient responses.

Shifting Definition of Disasters

Disasters have been defined in various ways by states, international organizations, nongovernmental organizations, and academia (UNDRR 2016; WHO 2002; Farber 2011; Perry 2017). The definition has also shifted across historical periods (Kim and Sohn 2018). For example, in the context of the Cold War, Fritz (1961) referred to disasters as events concentrated in time and space, which incur losses to a society by disrupting its structures and essential functions. Importantly, he suggested that disasters originate outside of a focal society to raise awareness of external attacks (Perry 2017). Quarantelli (1985) considered disasters shocks that exceeded a community's knowledge of, or capacity for, effective mitigation. This definition demonstrates the gradual shift of emphasis to communities' inherent vulnerabilities. Richardson (1994) explained that disasters go beyond the physical destruction of structures and outlined the diversity of human-induced organizational disruptions, recognizing the onset of industrial disasters throughout the 1980s and 1990s.

Natural and Human-Induced Disasters

Natural disasters involve natural phenomena that create devastation (Shaluf 2007). Natural hazards can stem from beneath the Earth's surface (earthquakes, tsunamis, volcanic eruptions) or from topographical, meteorological, hydrological, and biological phenomena (Shaluf 2007; Ranke 2016). Human-induced disasters refer to the short- and long-term devastation of communities that directly occur due to human activity (The International Federation of Red Cross and Red Crescent Societies 2003). Among these are war, pollution, nuclear explosions, fires, exposures to hazardous materials, and transportation accidents (Zibulewsky 2001; Shaluf 2007).

Extractive Disasters

Extractive industry hazards produce technological disasters, which are considered a subset of human-induced disasters. Extractive industries develop oil, mining, and natural gas projects that generate profit by unearthing valuable resources (World Bank 2021). These projects are operated by private, state, or hybrid corporations with mixed profit or development goals and varying degrees of accountability to shareholders and citizens for the health and safety of affected ecosystems and communities. Examples of extractive disasters include oil spills or mining waste leakages (Sigam and Garcia 2012). These disasters are very toxic to the environment and human health. They generate adverse impacts on ecosystems and resource dependent economic activities, such as fishing or tourism (Sigam and Garcia 2012). They also cause direct or indirect negative health effects through workplace exposure or environmental contamination (Schrecker et al. 2018).

Hybrid Disasters

Hybrid disasters are those which result from both natural and human-induced hazards, where a natural phenomenon merges with human agency (Shaluf 2007). The positive feedback between climate change and extractive activities exacerbates the frequency of hybrid disasters: increased emissions of greenhouse gases following greater demands for extractives heighten climate change, which in turn destroys industry infrastructure (Tansel 1995; Fenton and Hanfling 2019; Mulvihill 2020; Gibson et al. 2021). An example of this is the Norilsk oil spill of 2020, where the abnormal freeze-thaw cycles of permafrost induced by climate change collapsed the structural support for industrial oil tanks leading to a major hybrid environmental disaster (Rajendran et al. 2021). As of 2021, it is the largest oil spill observed in the Arctic, but the frequency and geographical range of hybrid environmental hazards similar in nature are projected to expand with climate change (Sakirko 2021; Melvin et al. 2016).

Communities and Social Vulnerability

Community systems produce the conditions necessary for survival, health, and needs attainment (Streeter 1991). However, disasters cause community-wide disruption of the social, economic, and environmental conditions needed for well-being (Streeter 1991). Moreover, disasters are regarded as a severe type of stressor capable of having significant impacts on the psychosocial functioning of community members (Greene and Greene 2009). The collective stress of disaster events and their disruptive impacts over time can lead to social dysfunction and experiences of dislocation among affected communities.

While there is no singularly agreed-upon definition, conceptualizations of community in the social sciences tend to be underpinned by groups of households existing within the same locality. This geographical demarcation depends on the linkage of households with others by functional interdependencies (Bell and Newby 1974). As such, implicit within community is an interactional foundation which may include a sense of belonging and commitment, social ties, the sharing of perspectives, and joint action anchored by a shared physical environment (Deeming et al. 2019). Over time, a high frequency of interaction among individuals forges a density of acquaintanceship characterized by depth of knowing, shared memories, and intergenerational attachments (Freudenburg 1986).

In this context, the social vulnerability of a community has been defined as the socioeconomic and demographic characteristics that influence both its sensitivity to natural hazards and ability to prepare for, respond to, and recover from their impacts (Cutter and Finch 2008). The impact of disaster events on communities is determined by a complex set of interacting conditions (Morrow 1999). These tend to reflect lesser degrees of social inclusion and the presence of material inequities within society, often drawn along lines of gender, ethnicity, and occupation. In social theory, the growth of social vulnerability has been attributed to the disembedding of the traditionally embedded community by the homogenizing forces of modernity (Barrett 2015). Specifically, globalization and the proliferation of market forces and urbanization have led to the breakdown of communal bonds and limited collective action to prevent and respond to disasters (Barrett 2015).

What Makes Communities Vulnerable to Disasters?

Social vulnerability to extractive industry-based disasters results from the interaction of multiple factors and community characteristics. The hazards-of-place model characterizes risk as "an objective measure of the likelihood of a hazard event" and mitigation as "measures to lessen risks or reduce their impact" (Cutter et al. 2003, p. 243). The interaction of risk and mitigation creates the "hazard potential" of a community. Risk can be reduced through effective and timely mitigation or accentuated due to poor, inadequate, or delayed mitigation (Cutter 1996). This hazard potential is shaped by the biophysical, social, economic, political, and legal characteristics of a community. The aggregate effect of factors that constitute a hazard potential determines the vulnerability of a community (Cutter et al. 2003; Cutter 1996). It is worth examining each of these factors in greater detail.

Biophysical

Some communities are more susceptible to environmental disasters because of the natural and geological features of its physical space. A community's dependence on its local environment for food, water, and sources of income for livelihood increases the extent of potential disruption caused by extractive disasters that disturb or pollute the environment, thus increasing community vulnerability (Lindon et al. 2014). Geographic proximity to sites of extraction greatly increases the susceptibility to disasters (Kroll-Smith and Couch 1990; Stephens 2015; Gulson et al. 2009; Auyero et al. 2019; Fry et al. 2015; Pearce et al. 2011). The geographic location of a community can also increase its vulnerability if it plays a role in the resource supply chain, despite not being proximate to the site of extraction. Coastal proximity often serves as a general risk factor for communities located along international shipping routes (Gainey 2018; Lewis 2020). The climate experienced by a community can also inadvertently contribute to its vulnerability. Particularly strong winds, currents, and other such extreme weather conditions can help facilitate the spread of harmful industry effluents to regions that are situated far from the site of extraction itself (Gulson et al. 2009; Grigalunas et al. 1986; Moldan et al. 1985; Aiken 2018).

Biophysical characteristics can also help shape the attitude of communities to disasters in general (Kroll-Smith and Couch 1990). When loss of life and property due to natural disasters are assimilated as part of normal life, it can have desensitizing effects to calamity from extractive disasters and their ensuing loss (Kroll-Smith and Couch 1990, p. 22).

Social

Social capital plays a crucial role in the ability of communities to mobilize against extractive industry-based technological disasters. A decent socioeconomic status and relatively uniform distribution of wealth within a community tends to enable the formation of social capital. This leads to more elaborate, inclusive, and collaborative community responses, with dialogue and deliberation being the preferred methods of mitigation in the event of disputes (Hames 2007; Boerchers et al. 2016). Some economically marginalized communities lack significant social capital, or strong networks and bonds between community members, in order to generate collective responses to hazards from extraction (Conde and Le Billon 2017). Social capital can be especially hard to establish in sparsely populated settings among households that are overwhelmingly preoccupied with economic survival. This can limit the formation of strong community groups. That said, even communities with greater socioeconomic privileges can have a weak social fabric due to a lack of shared identity or internal rifts that inhibit collective action, rendering them vulnerable to extractive disasters (Kroll-Smith and Couch 1990).

Economic

Economic dependence on extractive industries substantially increases community vulnerability to disasters. Economic reliance on oil and mining companies affects communities at multiple levels. From a general public and individual standpoint, industries are a major source of employment, providing jobs to locals (Davidson 2017). This kind of dependency generates a form of "economic blackmail," when communities are willing to endure gradual pollution if it avoids short-term unemployment (Gould 1991). During periods of economic recession, the incentives of extractive industries are diminished, and layoffs are common. At these times, communities become more receptive to environmental and human health concerns as they do not see much reason to bear the costs of extractive activities, further underscoring the influence of economic blackmail in times of profit (Gould 1991).

Extractive industries can also generate significant government revenue. Extractive companies often craft revenue-sharing or royalty programs with local governments, cementing themselves in communities through the promise of new revenue streams in addition to wages. At a national level, federal governments can become dependent on revenue from extractive companies, and thus welcome their increased presence into local communities over time (Davidson 2017; Filer and Macintyre 2006). Excessive dependence on industry also empowers the industry with great political influence, which leads to bureaucratic and judicial negligence when it comes to regulation of extractive activities, making communities increasingly vulnerable in the name of sustaining resource extraction (Hames 2007; Kuhlberg and Miller 2018; Fry et al. 2015). Both of these circumstances are further amplified in undiversified economies with inadequate alternative employment and revenue generation prospects (Davidson 2017; Filer and Macintyre 2006). This economic dependence ultimately hinders states and communities from taking action to change their baseline vulnerabilities (Couto 1989, p. 322; Turner 1976).

The commodification of land and the proliferation of corporate activity in rural areas has also produced community vulnerability (Kilborn 2000). Mining and oil companies aggressively buy land from people for extractive activities with the promise of employment and wages (Kilborn 2000). In many cases, this process can provide communities with short-term economic benefits. However, it changes the relationship people have with land from one of nourishment and sustenance in a sustainable manner to unencumbered extraction, while increasing their proximity to sites of extraction (Lewis et al. 2017). This is particularly true in cases of vulnerable populations like Indigenous communities (Lewis et al. 2017).

Political

The level of political involvement and engagement of a community impacts its vulnerability. Limited engagement, in the form of poor electoral turnout and participation in local political activity, reduces the amount of influence the community wields over lawmakers (O'Faircheallaigh 2013). This is particularly true in rural and peripheral areas with limited avenues for such engagement (O'Faircheallaigh 2013). The effective functioning of state institutions affords community members the means seek responses to political claims. Democratic accountability mechanisms such as elections and petitions allow them to exert political influence and impact legislation (Hames 2007; Fry et al. 2015; Auyero et al. 2019). The socioeconomic status of a community plays a role in determining its political influence. In contrast to residents of affluent communities, economically marginalized ones tend to be less likely to have the time or resources to become involved in political advocacy surrounding extractive projects – if they have a general awareness of the projects at all (Conde and Le Billon 2017). Access to information is sparse or isolated communities, be it geographic isolation or socioeconomic marginalization, further hinders their ability to organize and mobilize against sources of risk (Conde and Le Billon 2017; Kroll-Smith and Couch 1990).

Legal

Communities may face added risks of extractive disasters based on legal factors such as labor or land rights. Mining and oil companies have often exploited labor laws in particular states to deprive workers, largely composed of community members themselves, of fair compensation (Hart 2016). Lower income levels often lead to increased debt and bankruptcy, further weakening the community's economic status (Hart 2016). This also compounds the community's inability to relocate from the site of extraction to safer locales (Hart 2016).

Land rights enable communities to have a say in the decision-making process, making them less vulnerable to extractive disasters. Thus, communities that do not have land ownership are excluded from consultation processes and are inherently more vulnerable (Germond-Duret 2014). This is particularly true for Indigenous communities in settler nations like Canada, Australia, and the United States. Measures like the International Labor Organization Convention 169 (ILO 169) and Free, Prior, and Informed Consent (FPIC) (specifically in the case of Indigenous communities), which require states and industries to obtain consent from involved communities and engage in sustained consultation with them before decisions are taken help mitigate this dimension of community vulnerability.

Access to land rights on its own is an insufficient legal protection to decrease community vulnerability. Effective access to channels of accountability is an equally important measure. In many cases, mining and oil companies lobby to limit communities' abilities to pursue litigation (Kuhlberg and Miller 2018; Kroll-Smith and Couch 1990; Smith 2010). Many countries also divide property rights into two domains, surface rights and mineral rights (Fry et al. 2015). When communities lack access to mineral rights, it inhibits them from undertaking any legal action even if they have land ownership (Kuhlberg and Miller 2018; Fry et al. 2015). Communities with even partial access to mineral rights have been found to be better positioned because they have legal authority in consultations, further underscoring the importance of land rights in determining community vulnerability to extractive industry disasters (Fry et al. 2015).

Community Responses to Extractive Disasters

Defining Community Resilience

Within the environmental disaster literature, resilience refers to the capacity of systems to absorb shocks and reorganize themselves into fully functional states. As articulated by Cutter et al. (2008), resilience encompasses not only a system's ability to return to an initial state, but also its capacity to evolve into a new, more resistant state through learning and adaptation. At the community level, resilience refers to the ability to direct available resources towards absorbing disturbances, organize local responses to disasters, and prevent further harm. Opportunities for community resilience can be influenced by a variety of factors relating to community cohesion, disaster-induced stress, and local knowledge (Cutter et al. 2008)). Particularly resilient communities are often those that can effectively mobilize resources, networks, and social capital towards various response strategies (Adger et al. 2005). Across these communities, common mobilization strategies may include protests, transnational alliances, shareholder activism, and making legal claims.

Factors Inhibiting Community Resilience

Community Polarization

Disasters can evoke a wide range of reactions among community members, ranging from unified responses to highly divisive reactions. One potential framework accounts for these diverging responses by delineating between "therapeutic" and "corrosive" communities, which may emerge depending on the type of disaster (PWSRCAC 2019). Natural disasters that typically occur without a specific party perceived as responsible, such as hurricanes and earthquakes, tend to create therapeutic communities where response efforts are focused, coordinated, and quick to mobilize. These communities tend to be fairly unified and solidaristic, seeking to return to a pre-disaster state through collaborative recovery efforts (Picou et al. 2004; PWSRCAC 2019). However, extractive industry disasters such as oil spills tend to generate much less unanimity and cooperation, often producing corrosive communities characterized by tension and conflict. Crucially, these events involve a clearly identifiable entity at fault, which can lead to polarization as corporate actors staunchly deny culpability and community members become distrustful of one another and divided over whether to pursue reparative actions (Cope et al. 2020; Miller 2006; PWSRCAC 2019). These tensions can generate a corrosive atmosphere that delays collective responses and undermines community resilience.

Psychological Impacts

In addition to the impact from corrosive community dynamics, individuals facing disaster events can experience heightened psychological distress that hinders collective mobilization and resilience. The urgent need for disaster victims and their families to escape from or adapt to disasters can impose an added cognitive load on individuals that diverts attention from other priorities. Psychologists studying cognition under financial stress argue that there is a finite mental bandwidth, or a fixed reserve of mental resources, available for cognitive processes (Schilbach et al. 2016). When individuals face added stressors in their physical environment, such as poverty or food scarcity, more cognitive resources must be devoted towards meeting immediate needs and fewer can be dedicated to complex mental processes (Schilbach et al. 2016). Sudden shocks and material losses associated with disasters may create distress and capture individual attention that might otherwise be given to joining protests or lobbying public officials for relief efforts. Disasters effectively add new sets of problems to individuals' lives, overburdening cognitive capacities that can hinder mobilization efforts among communities.

Psychological distress to extractive industry disasters can also generate resignation and apathy among individuals. Tacit acceptance among community members that extractive activities are polluting environments and raising surrounding toxicity levels can contribute to a profound sense of powerlessness. Particularly in communities with limited political representation, individuals may feel betrayed by public representatives and begin to view resistance to ecological destruction as futile (Hirsch et al. 2018). In studies of individual attitudes among those living near fracking "sacrifice zones," locals often express cynicism and disillusionment with the viability of achieving change through conventional political processes (Dunlop et al. 2020). Exacerbated by the perceived lobbying power of extractive corporations over local officials, individuals can view the current state of extraction as unchangeable (Dunlop et al. 2020). Consequently, some community members may disengage from visible activism and become politically resigned.

Community Knowledge

To effectively adapt to extractive disasters, local communities first require several forms of knowledge: that a disaster is genuinely occurring, that it poses substantial threats to human health or natural environments, and that specific extractive activities are responsible. Limited knowledge of extractive disasters, especially surrounding their health and environmental impacts, presents a significant barrier to community responses and resilience. Low knowledge is often rooted in interactions between high marginalization and trust in institutions. Marginalized and low-income groups face greater difficulties in acquiring vital technical information, due to low educational attainment, political exclusion, and limited funds and expertise to conduct investigations on local toxicity levels (Burningham et al. 2008; Omanga et al. 2014). In the absence of available knowledge, locals may default to deceptive messaging spread by political and industry officials, which act as trusted sources of information (Conde 2017; Conde and Le Billon 2017; Dougherty and Olsen 2014). Researchers on community resistance to mining describe this tendency as high institutional trust, characterized by faith in authority figures and institutions to preserve community interests (Dougherty and Olsen 2014). In contrast to communities with lower institutional trust and higher relational trust, characterized by strong bonds with neighbors and fellow community members, communities with strong ties to institutions may be more susceptible to corporate messaging and reluctant to mobilize following extractive industry disasters. Both institutional trust and marginalization can therefore contribute to low levels of community knowledge, hampering resilience to disaster events.

Mobilization Strategies Among Resilient Communities

Local Protests

Protests and civil disobedience are among the most common community responses following extractive disasters, typically to voice discontent with extractive projects and to demand policies that enable communities to adapt to changing circumstances. Protests encompass not only large-scale demonstrations but also a broad set of actions that can generate disturbances and threaten the profits or public images of responsible companies and governments. Communities facing ongoing and unaddressed extractive disasters have been known to coordinate very public disturbances, such as collective walkouts and rallies, as well as more targeted acts of civil disobedience against officials including vandalizing regulators' offices (Hanna et al. 2016; McLean and Johnes 1999; Coumans 2002). A potential advantage of protests, as opposed to pursuing legal avenues of recourse, involves their relatively low costs and barriers to entry for marginalized participants (Lober 1995; van Rooij 2010). However, local acts of civil disobedience often encounter hurdles in drawing sufficient attention and generating a credible threat against their targets. Marginalized or politically disenfranchised communities can especially struggle to prompt a response from negligent industries and governments (Kirsch 2007). Resilient communities that employ protests effectively are typically those capable of targeting actions towards specific officials and galvanizing support from external allies, such as journalists and international advocacy groups (Jackson 2016; Kirsch 2007; Lea et al. 2018).

Case: Aberfan Disaster, Wales

The Aberfan Disaster of 1966 involved the collapse of a coal mining spoil tip, resulting in deadly floods of coal slurry transported by heavy rainfall. After public officials denied requests by local community members to remove Aberfan's remaining spoil tips, protestors responded by dumping bags of coal slurry in the lobby of the Welsh Office. The government subsequently agreed, drawing 150,000 euros from disaster funds to remove the remaining tips (Jackson 2016).

C

Transnational Connections

Connections with transnational allies, external actors, and global media networks can potentially serve three functions for local communities responding to extractive disasters. First, transnational activists can place additional pressure on companies and governments to address community concerns, effectively magnifying local struggles to a global audience (Evans et al. 2020; Kirsch 2007). International visibility can then allow for further forms of protest, such as wider boycotts and shareholder activism. Second, external civil society organizations can supplement relief efforts and offer monetary support in the absence of formal compensation mechanisms, enhancing communities' adaptive capacities. Finally, in certain rare cases, international allies may even command additional authority or resources to confer tangible environmental protections. For instance, at the request of local communities, the UNESCO World Heritage Committee has declared certain mining sites as ecologically significant and lobbied for their restoration (GAC 2011; Lea et al. 2018). In each of these situations, the presence of external actors can contribute to community resilience following extractive disasters.

Case: Uranium Mining in Kakadu National Park, Australia

Since the 1970s imposition of mining at the Jabiluka, Koongarra, and Ranger uranium deposits in Australia, over 200 spills and license breaches have occurred (GAC 2011). In 1998, local activists initiated an 8-month blockade at the Jabiluka deposit, inviting global media companies and UNESCO officials to attend (GAC 2011; Lea et al. 2018). Activists further campaigned to add uranium deposits as officially listed World Heritage Sites, eventually securing protected status for the Koongarra deposit in 2013 (Meskell 2013).

Shareholder Activism

This is a strategy deployed by shareholders to pressure changes in corporate behavior, using their stakes and status as partial owners as leverage. For instance, shareholders can take actions such as filing and voting on shareholder resolutions, negotiating privately with managers, initiating publicity campaigns, and threatening lawsuits or sell-offs. Social activists and nongovernmental organizations can also become shareholders, which entitles them to attend annual meetings and file resolutions (Rehbein et al. 2004). Supporters of shareholder activism cite its ability to prompt internal dialogues and policy changes within corporations, which may supplement and increase the efficacy of protests, boycotts, or divestment campaigns (Flammer et al. 2021; Vasi and King 2012). However, scholars critical of shareholder activism argue that its effects on target firms are often negligible, particularly in many cases where activists own only a small proportion of overall shares (Denes et al. 2017; Karpoff 2001).

Case: Mariana Dam Disaster, Brazil

After the sudden failure of a tailings dam operated by Vale, a Brazilian mining corporation, in 2015, one method that local activists employed was acquiring shares to influence internal corporate policy. Activists coordinated purchases of Vale stocks, entitling them to join shareholder meetings, voice concerns within the company, and initiate resolutions (Cezne 2019).

Legal Tools

Legal claims act as formal avenues for communities to express grievances and seek monetary damages, potentially facilitating local resilience and adaptation to extractive disasters. Filing lawsuits against extractive industries can offer certain advantages, such as necessitating a response from defending parties and offering the possibility of direct compensation, in contrast to protests making informal demands that may go largely ignored. However, preparing legal claims can involve substantial financial and time costs, as well as considerable personal risks, without a guarantee of success (Coggins 1972). While some lawsuits successfully push defendants into settlements, others can evoke highly volatile responses from corporations. Many companies view legal battles as a matter of attrition, and opt for heavily funded legal teams to extend litigation processes or file countersuits until plaintiffs become unwilling to continue (Kass 2011; UNHRC 2015). This may be strategic for giant extractive conglomerates with high budgets and market shares, while mid-sized companies that fear reputational damage have fewer incentives to respond through aggressive and prolonged litigation (Hanson and Stuart 2001). Depending on factors such as the strength of legal cases to reactions of corporate actors, legal tools can either enhance community resilience or have a limited effect.

Case: Ok Tedi Mine Disaster, Papua New Guinea

Following BHP's decision to dump mining waste directly into the Ok Tedi River in Papua New Guinea in 1984, local community members sought legal action in the Supreme Court of Victoria, where BHP was incorporated. BHP ultimately admitted culpability for the disaster, settled out of court in 1996, and fully withdrew from the mine in 2002 (Kirsch 2007).

Conclusion

Disasters have been historically understood and framed in varying ways. Initially they were thought to originate from the supernatural and were defined as acts of God. Following the Enlightenment, disasters were seen as acts of nature. More recently, the causes for disasters are associated with human agency (Quarantelli 2000). Toxic and technological disasters are primary examples of the role that humans play in the construction and reproduction of hazardous environments. Extractive disasters are a common example of human-induced catastrophes that devastate communities and ecosystems.

The shift in focus to the role humans play in creating toxic and hazardous environments has generated greater scholarly and policy interest in what some call the vulnerability paradigm (Furedi 2007). Focusing on communities as primary sites of social vulnerability and resilience we examined the antecedent conditions that shape the risk profiles of populations that are disproportionately affected by extractive disasters, and the factors that condition community responses. The interaction between economic dependencies on extractive activities, political exclusions, demographic features, and access to legal rights shape the hazard potential of a given community and opportunities for resilience.

Capacities to respond to a disaster are constrained or enabled by different factors. Community polarization, the psychological impacts of a disaster, and knowledge of hazards inhibit community resilience. On the other hand, communities that can organize through protest, connect with transnational activists and other allies to raise the visibility of a toxic hazard, promote shareholder activism, and deploy legal tools to contest extractive activities that pose significant risk, are better positioned to limit future disasters.

References

Adger, W. N., & Kelly, P. M. (1999). Social vulnerability to climate change and the architecture of entitlements. *Mitigation and Adaptation Strategies for Global Change, 4*(3), 253–266. https://doi.org/10.1023/A:1009601904210

Adger, W. N., Hughes, T. P., Folke, C., Carpenter, S. R., & Rockström, J. (2005). Social-ecological resilience to coastal disasters. *Science, 309*(5737), 1036–1039. https://doi.org/10.1126/science.1112122

Aiken, G. (2018). Timeline: 25 years on from Braer oil spill – News for the energy sector. *Energy Voice*. https://www.energyvoice.com/other-news/160250/timeline-25-years-braer-oil-spill/

Auyero, J., Hernandez, M., & Stitt, M. E. (2019). Grassroots activism in the belly of the beast: A relational account of the campaign against urban fracking in Texas. *Social Problems, 66*(1), 28–50. https://doi.org/10.1093/socpro/spx035

Barrett, G. (2015). Deconstructing community: Deconstructing community. *Sociologia Ruralis, 55*(2), 182–204. https://doi.org/10.1111/soru.12057

Bell, C., & Newby, H. (Eds.). (1974). *The sociology of community: A selection of readings*. London: Cass.

Boerchers, M., Fitzpatrick, P., Storie, C., & Hostetler, G. (2016). Reinvention through regreening: Examining environmental change in Sudbury, Ontario. *The Extractive Industries and Society, 3*(3), 793–801. https://doi.org/10.1016/j.exis.2016.03.005

Burningham, K., Fielding, J., & Thrush, D. (2008). 'It'll never happen to me': Understanding public awareness of local flood risk. *Disasters, 32*(2), 216–238. https://doi.org/10.1111/j.1467-7717.2007.01036.x

Cezne, E. (2019). Forging transnational ties from below: Challenging the Brazilian mining giant Vale S.A. across the South Atlantic. *The Extractive Industries and Society, 6*(4), 1174–1183. https://doi.org/10.1016/j.exis.2019.10.007

Coggins, G. C. (1972). Preparing an environmental lawsuit, part II: Doctrinal barriers and pre-trial preparation. *Iowa Law Review, 58*(3), 487–530.

Conde, M. (2017). Resistance to mining. A review. *Ecological Economics, 132*, 80–90. https://doi.org/10.1016/j.ecolecon.2016.08.025

Conde, M., & Le Billon, P. (2017). Why do some communities resist mining projects while others do not? *The Extractive Industries and Society, 4*(3), 681–697. https://doi.org/10.1016/j.exis.2017.04.009

Cope, M. R., Slack, T., Jackson, J. E., & Parks, V. (2020). Community sentiment following the deepwater horizon oil spill disaster: A test of time, systemic community, and corrosive community models. *Journal of Rural Studies, 74*, 124–132. https://doi.org/10.1016/j.jrurstud.2019.12.019

Coumans, C. (2002). *The successful struggle against submarine tailings disposal in marinduque, Philippines*. MiningWatch Canada. https://miningwatch.ca/sites/default/files/marinduque_std_struggle_0.pdf

Couto, R. A. (1989). Economics, experts, and risk: Lessons from the catastrophe at Aberfan. *Political Psychology, 10*(2), 309–324. https://doi.org/10.2307/3791650

Cutter, S. (1996). Vulnerability to environmental hazards. *Progress in Human Geography, 20*, 529–539. https://doi.org/10.1177/030913259602000407

Cutter, S. L., & Finch, C. (2008). Temporal and spatial changes in social vulnerability to natural hazards. *Proceedings of the National Academy of Sciences, 105*(7), 2301–2306. https://doi.org/10.1073/pnas.0710375105

Cutter, S. L., Barnes, L., Berry, M., Burton, C., Evans, E., Tate, E., & Webb, J. (2008). A place-based model for understanding community resilience to natural disasters. *Global Environmental Change, 18*(4), 598–606. https://doi.org/10.1016/j.gloenvcha.2008.07.013

Cutter, S. L., Boruff, B. J., & Shirley, W. L. (2003). Social vulnerability to environmental hazards. *Social Science Quarterly, 84*(2), 242–261. https://doi.org/10.1111/1540-6237.8402002

Davidson, H. (2017). "They'll get rich and go": Glencore's McArthur River mine could take 300 years to clean up. *The Guardian*. https://www.theguardian.com/australia-news/2017/may/15/theyll-get-rich-and-go-glencores-mcarthur-river-mine-could-take-300-years-to-clean-up

Deeming, H., Fordham, M., Kuhlicke, C., Pedoth, L., Schneiderbauer, S., & Shreve, C. (2019). *Framing community disaster resilience: Resources, capacities, learning, and action*. https://books-scholarsportal-info.myaccess.library.utoronto.ca/en/read?id=/ebooks/ebooks3/wiley/2018-11-23/1/9781119166047

Denes, M. R., Karpoff, J. M., & McWilliams, V. B. (2017). Thirty years of shareholder activism: A survey of empirical research. *Journal of Corporate Finance, 44*, 405–424. https://doi.org/10.1016/j.jcorpfin.2016.03.005

Dougherty, M. L., & Olsen, T. D. (2014). "They have good devices": Trust, mining, and the microsociology of environmental decision-making. *Journal of Cleaner Production, 84*, 183–192. https://doi.org/10.1016/j.jclepro.2014.04.052

Dunlop, L., Atkinson, L., & Diepen, M. T. (2020). Perspectives on fracking from the sacrifice zone: Young people's knowledge, beliefs and attitudes. *Chemistry Education Research and Practice, 21*(3), 714–729. https://doi.org/10.1039/D0RP00022A

Evans, E. M., Schofer, E., & Hironaka, A. (2020). Globally visible environmental protest: A cross-national analysis, 1970–2010. *Sociological Perspectives, 63*(5), 786–808. https://doi.org/10.1177/0731121420908899

Farber, D. A. (2011). Environmental disasters: An introduction. *SSRN Electronic Journal*. https://doi.org/10.2139/ssrn.1898401

Fenton, E., & Hanfling, D. (2019). Natural and industrial disaster events, public health, and ethics. In *The Oxford handbook of public health ethics* (pp. 785–796). https://doi.org/10.1093/oxfordhb/9780190245191.013.68

Filer, C., & Macintyre, M. (2006). Grass roots and deep holes: Community responses to mining in Melanesia. *The Contemporary Pacific, 18*(2), 215–231. https://doi.org/10.1353/cp.2006.0012

Flammer, C., Toffel, M. W., & Viswanathan, K. (2021). *Shareholder activism and firms' voluntary disclosure of climate change risks* (SSRN scholarly paper ID 3468896; issue ID 3468896). Social Science Research Network. https://doi.org/10.2139/ssrn.3468896

Freudenburg, W. R. (1986). The density of acquaintanceship: An overlooked variable in community research? *American Journal of Sociology, 92*(1), 27–63. https://doi.org/10.1086/228462

Fritz, C. E. (1961). *Disaster*. Institute for Defense Analyses, Weapons Systems Evaluation Division.

Fry, M., Briggle, A., & Kincaid, J. (2015). Fracking and environmental (in)justice in a Texas city. *Ecological Economics, 117*, 97–107. https://doi.org/10.1016/j.ecolecon.2015.06.012

Furedi, F. (2007). The changing meaning of disaster. *Area, 39*(4), 482–489. https://doi.org/10.1111/j.1475-4762.2007.00764.x

GAC. (2011). *Uranium Mining – Uranium mining has long caused controversy in the Kakadu region*. Gundjeihmi Aboriginal Corporation. https://www.mirarr.net/uranium-mining

Gainey, T. (2018). *Torrey canyon disaster – Cornwall remembers 51 years on*. CornwallLive. https://www.cornwalllive.com/news/history/torrey-canyon-disaster-cornwall-remembers-1340394

Germond-Duret, C. (2014). Extractive industries and the social dimension of sustainable development: Reflection on the Chad–Cameroon pipeline. *Sustainable Development, 22*(4), 231–242. https://doi.org/10.1002/sd.1527

Gibson, C. M., Brinkman, T., Cold, H., Brown, D., & Turetsky, M. (2021). Identifying increasing risks of hazards for northern land-users caused by permafrost thaw: Integrating scientific and community-based research approaches. *Environmental Research Letters, 16*(6), 064047. https://doi.org/10.1088/1748-9326/abfc79

Gould, K. A. (1991). The sweet smell of money: Economic dependency and local environmental political mobilization. *Society & Natural Resources, 4*(2), 133–150. https://doi.org/10.1080/08941929109380749

Greene, R., & Greene, D. (2009). Resilience in the face of disasters: Bridging micro- and macro-perspectives. *Journal of Human Behavior in the Social Environment, 19*(8), 1010–1024. https://doi.org/10.1080/10911350903126957

Grigalunas, T. A., Anderson, R. C., Brown, G. M., Congar, R., Meade, N. F., & Sorensen, P. E. (1986). Estimating the cost of oil spills: Lessons from the Amoco Cadiz incident. *Marine Resource Economics, 2*(3), 239–262. https://doi.org/10.1086/mre.2.3.42628902

Gulson, B., Korsch, M., Matisons, M., Douglas, C., Gillam, L., & McLaughlin, V. (2009). Windblown lead carbonate as the main source of lead in blood of children from a seaside community: an example of local birds as "canaries in the mine". *Environmental Health Perspectives, 117*(1), 148–154. https://doi.org/10.1289/ehp.11577

Hames, K. D. (2007). *Inquiry into the cause and extent of lead pollution in the esperance area* (No. 8). Perth, Australia: Parliament of Western Australia. https://parliament.wa.gov.au/Parliament/commit.nsf/(InqByName)/Inquiry+into+the+Cause+and+Extent+of+Lead+Pollution+in+the+Esperance+Area#Report

Hanna, P., Vanclay, F., Langdon, E. J., & Arts, J. (2016). Conceptualizing social protest and the significance of protest actions to large projects. *The Extractive Industries and Society, 3*(1), 217–239. https://doi.org/10.1016/j.exis.2015.10.006

Hanson, D., & Stuart, H. (2001). Failing the reputation management test: The case of BHP, the big Australian. *Corporate Reputation Review, 4*(2), 128–143. https://doi.org/10.1057/palgrave.crr.1540138

Hart, P. (2016). *Financial struggles and (rare) successes of miners in general and at Te aroha in particular* (Working paper). Historical Research Unit, University of Waikato. https://researchcommons.waikato.ac.nz/handle/10289/10361

Hirsch, J. K., Bryant Smalley, K., Selby-Nelson, E. M., Hamel-Lambert, J. M., Rosmann, M. R., Barnes, T. A., Abrahamson, D., Meit, S. S., GreyWolf, I., Beckmann, S., & LaFromboise, T. (2018). Psychosocial impact of fracking: A review of the literature on the mental health consequences of hydraulic fracturing. *International Journal of Mental Health and Addiction, 16*(1), 1–15. https://doi.org/10.1007/s11469-017-9792-5

Jackson, C. (2016). Aberfan: The mistake that cost a village its children. *BBC News*. https://www.bbc.co.uk/news/resources/idt-150d11df-c541-44a9-9332-560a19828c47

Karpoff, J. M. (2001). *The impact of shareholder activism on target companies: A survey of empirical findings* (SSRN scholarly paper ID 885365; issue ID 885365). Social Science Research Network. https://doi.org/10.2139/ssrn.885365

Kass, S. L. (2011). *Lessons from Lago Agrio environmental pollution case | New York Law Journal*. Carter Ledyard & Milburn LLP. https://www.clm.com/publication.cfm?ID=344

Kilborn, P. T. (2000). A torrent of sludge muddies a town's future. *The New York Times*. https://www.nytimes.com/2000/12/25/us/a-torrent-of-sludge-muddies-a-town-s-future.html

Kim, Y.-K., & Sohn, H.-G. (2018). Disaster risk management in the Republic of Korea. *Disaster Risk Reduction*. https://doi.org/10.1007/978-981-10-4789-3

Kirsch, S. (2007). Indigenous movements and the risks of counterglobalization: Tracking the campaign against Papua New Guinea's Ok Tedi mine. *American Ethnologist, 34*(2), 303–321. https://doi.org/10.1525/ae.2007.34.2.303

Kroll-Smith, J. S., & Couch, S. R. (1990). *The real disaster is above ground: a mine fire and social conflict*. Lexington, KY: University Press of Kentucky.

Kuhlberg, M., & Miller, S. (2018). "Protection to the sulphur-smoke tort-feasors": The tragedy of pollution in sudbury, ontario, the World's Nickel Capital, 1884–1927. *Canadian Historical Review, 99*(2), 225–257. https://doi.org/10.3138/chr.99.2.03

Lea, T., Howey, K., & O'Brien, J. (2018). Waging paperfare: Subverting the damage of extractive capitalism in Kakadu. *Oceania, 88*(3), 305–319. https://doi.org/10.1002/ocea.5203

Lewis, D. (2020). How Mauritius is cleaning up after major oil spill in biodiversity hotspot. *Nature, 585*(7824), 172–172. https://doi.org/10.1038/d41586-020-02446-7

Lewis, J., Hoover, J., & MacKenzie, D. (2017). Mining and environmental health disparities in native American Communities. *Current Environmental Health Reports, 4*(2), 130–141. https://doi.org/10.1007/s40572-017-0140-5

Lindon, J. G., Canare, T. A., & Mendoza, R. U. (2014). Corporate and public governance in mining: Lessons from the Marcopper mine disaster in Marinduque, Philippines. *Asian Journal of Business Ethics, 3*(2), 171–193. https://doi.org/10.1007/s13520-014-0038-3

Lober, D. J. (1995). Why protest? *Policy Studies Journal, 23*(3), 499–518. https://doi.org/10.1111/j.1541-0072.1995.tb00526.x

McLean, I., & Johnes, M. (1999). Regulating gifts of generosity: The Aberfan disaster fund and the Charity Commission*. *Legal Studies, 19*(3), 380–396. https://doi.org/10.1111/j.1748-121X.1999.tb00101.x

Melvin, A. M., Larsen, P., Boehlert, B., Neumann, J. E., Chinowsky, P., Espinet, X., Martinich, J., Baumann, M. S., Rennels, L., Bothner, A., Nicolsky, D. J., & Marchenko, S. S. (2016). Climate change damages to Alaska public infrastructure and the economics of proactive adaptation. *Proceedings of the National Academy of Sciences, 114*(2). https://doi.org/10.1073/pnas.1611056113

Meskell, L. (2013). UNESCO and the fate of the World Heritage indigenous peoples Council of Experts (WHIPCOE). *International Journal of Cultural Property, 20*(2), 155–174.

Miller, D. S. (2006). Visualizing the corrosive community: looting in the aftermath of hurricane katrina. *Space and Culture, 9*(1), 71–73. https://doi.org/10.1177/1206331205283762

Moldan, A. G. S., Jackson, L. F., McGibbon, S., & Van Der Westhuizen, J. (1985). Some aspects of the Castillo de Bellver oilspill. *Marine Pollution Bulletin, 16*(3), 97–102. https://doi.org/10.1016/0025-326X(85)90530-2

Morrow, B. H. (1999). Identifying and mapping community vulnerability. *Disasters, 23*(1), 1–18. https://doi.org/10.1111/1467-7717.00102

Mulvihill, P. R. (2020). The ambiguity of environmental disasters. *Journal of Environmental Studies and Sciences, 11*(1), 1–5. https://doi.org/10.1007/s13412-020-00646-1

O'Faircheallaigh, C. (2013). Extractive industries and indigenous peoples: A changing dynamic? *Journal of Rural Studies, 30*, 20–30. https://doi.org/10.1016/j.jrurstud.2012.11.003

Omanga, E., Ulmer, L., Berhane, Z., & Gatari, M. (2014). Industrial air pollution in rural Kenya: Community awareness, risk perception and associations between risk variables. *BMC Public Health, 14*(1), 377. https://doi.org/10.1186/1471-2458-14-377

Park, S. (2021). The role of the Sovereign state in 21st century environmental disasters. *Environmental Politics, 0*(0), 1–20. https://doi.org/10.1080/09644016.2021.1892983

Pearce, T. D., Ford, J. D., Prno, J., Duerden, F., Pittman, J., Beaumier, M., Berrang-Ford, L., & Smit, B. (2011). Climate change and mining in Canada. *Mitigation and Adaptation Strategies for Global Change, 16*(3), 347–368. https://doi.org/10.1007/s11027-010-9269-3

Perry, R. W. (2017). Defining disaster: An evolving concept. In *Handbook of disaster research* (pp. 3–22). https://doi.org/10.1007/978-3-319-63254-4_1

Picou, J. S., Marshall, B. K., & Gill, D. A. (2004). Disaster, litigation, and the corrosive community*. *Social Forces, 82*(4), 1493–1522. https://doi.org/10.1353/sof.2004.0091

PWSRCAC. (2019). *Coping with technological disasters: A user friendly guidebook* (version 3). Prince William Sound Regional Citizens' Advisory Council (PWSRCAC).

Quarantelli, E. L. (1985). *What is disaster? The need for clarification in definition and conceptualization in research*. UDSpace Home. http://udspace.udel.edu/handle/19716/1119

Quarantelli, E. L. (2000). *Disaster planning, emergency management and civil protection: The historical development of organized efforts to plan for and to respond to disasters*. Disaster Research Center, University of Delaware. https://udspace.udel.edu/handle/19716/673

Rajendran, S., Sadooni, F. N., Al-Kuwari, H. A.-S., Oleg, A., Govil, H., Nasir, S., & Vethamony, P. (2021). Monitoring oil spill in Norilsk, Russia using satellite data. *Scientific Reports, 11*(1). https://doi.org/10.1038/s41598-021-83260-7

Ranke, U. (2016). *Natural disaster risk management: Geosciences and social responsibility*. Cham: Springer International Publishing Switzerland.

Rehbein, K., Waddock, S., & Graves, S. B. (2004). Understanding shareholder activism: Which corporations are targeted? *Business & Society, 43*(3), 239–267. https://doi.org/10.1177/0007650304266869

Ribot, J. (2011). Vulnerability before adaptation: Toward transformative climate action. *Global Environmental Change, 21*, 1160–1162. https://doi.org/10.1016/j.gloenvcha.2011.07.008

Richardson, B. (1994). Socio-technical disasters: Profile and prevalence. *Disaster Prevention and Management: An International Journal, 3*(4), 41–69. https://doi.org/10.1108/09653569410076766

Sakirko, E. (2021). *Remember the Norilsk oil spill? Well, the polluters will pay*. Greenpeace Aotearoa. https://www.greenpeace.org/aotearoa/story/remember-the-norilsk-oil-spill-well-the-polluters-will-pay/.

Schilbach, F., Schofield, H., & Mullainathan, S. (2016). The psychological lives of the poor. *American Economic Review, 106*(5), 435–440. https://doi.org/10.1257/aer.p20161101

Schrecker, T., Birn, A.-E., & Aguilera, M. (2018). How extractive industries affect health: Political economy underpinnings and pathways. *Health & Place, 52*, 135–147. https://doi.org/10.1016/j.healthplace.2018.05.005

Shaluf, I. M. (2007). Disaster types. *Disaster Prevention and Management: An International Journal, 16*(5), 704–717. https://doi.org/10.1108/09653560710837019

Sigam, C., & Garcia, L. (2012). *Extractive industries: Optimizing value reteintion in host countries*. United Nations. https://digitallibrary.un.org/record/725984?ln=en.

Stephens, D. B. (2015). Analysis of the groundwater monitoring controversy at the Pavillion, Wyoming natural gas field. *Groundwater, 53*(1), 29–37. https://doi.org/10.1111/gwat.12272

Streeter, C. L. (1991). Disasters and development: Disaster preparedness and mitigation as an essential component of development planning. *Social Development Issues, 13*(3), 100–110.

Tansel, B. (1995). Natural and manmade disasters: Accepting and managing risks. *Safety Science, 20*(1), 91–99. https://doi.org/10.1016/0925-7535(94)00070-j

The World Bank. (2021). *Extractive industries: Overview*. The World Bank. https://www.worldbank.org/en/topic/extractiveindustries/overview

Turner, B. A. (1976). The organizational and interorganizational development of disasters. *Administrative Science Quarterly, 21*(3), 378–397. https://doi.org/10.2307/2391850

UNDRR. (2016). *Report of the open-ended intergovernmental expert working group on indicators and terminology relating to disaster risk reduction*. https://www.undrr.org/publication/report-open-ended-intergovernmental-expert-working-group-indicators-and-terminology.

UNHRC. (2015). *Chevron's activities impair freedom of expression of victims, academics, students and activists*. United Nations Office of the High Commissioner for Human Rights. https://ap.ohchr.org/documents/dpage_e.aspx?si=A/HRC/29/NGO/23

van Rooij, B. (2010). The people vs. pollution: Understanding citizen action against pollution in China. *Journal of Contemporary China, 19*(63), 55–77. https://doi.org/10.1080/10670560903335777

Vasi, I. B., & King, B. G. (2012). Social movements, risk perceptions, and economic outcomes: The effect of primary and secondary stakeholder activism on firms' perceived environmental risk and financial performance. *American Sociological Review, 77*(4), 573–596. https://doi.org/10.1177/0003122412448796

WHO. (2002). Disasters & emergencies: Definitions. https://apps.who.int/disasters/repo/7656.pdf.

Zibulewsky, J. (2001). Defining disaster: The emergency department perspective. *Baylor University Medical Center Proceedings, 14*(2), 144–149. https://doi.org/10.1080/08998280.2001.11927751

Community-Led Interventions

► Social Urbanism: Transforming the Built and Social Environment

Community-Supported Agriculture

► *Wadi* Sustainable Agriculture Model, The

Company Town

► New Cities

Competitiveness of Cities

▶ Connecting Urban and Regional Innovation Ecosystems to Enhance Competitiveness

Competitiveness of Regions

▶ Connecting Urban and Regional Innovation Ecosystems to Enhance Competitiveness

Composing an Urban Sphere Computation

▶ Computational Urban Planning

Computational Urban Planning

Gaps, Challenges, and Strides in Africa

Innocent Chirisa[1], Patience Mazanhi[1] and Abraham R. Matamanda[2]
[1]Department of Demography Settlement and Development, Social & Behavioral Sciences, University of Zimbabwe, Harare, Zimbabwe
[2]Department of Urban and Regional Planning, University of the Free State, Bloemfontein, South Africa

Synonyms

Urban design; Composing an urban sphere computation; Computer applications

Definitions

Urban planning – technical and political process that determine decisions for land use in towns and cities.

Urban design – technical process that determines form and size of spatial units in towns and cities.

Computation urban design – the application of computing and statistical tools to design places and spaces.

Introduction

The volatile development arena has not spared urban planning from the impacts it has wrought. With many changes toward modernity, sustainability, and smart growth, urban planning has seen more changes resulting in rising complexity that has made it difficult to proffer long-lasting solutions to development challenges (Healey 2007). Whereas urban planning is complex, involving various aspects and disciplines which require asynchronous approaches (Batty 2008), much difficulty has resulted in trying to tackle facet by facet especially in Africa (Wilson et al. 2019). It is the purpose of this chapter to shed light on a more better approach to urban planning. The chapter aims at providing an understanding of computational urban planning and explores its importance in helping planners provide better design solutions. It explores how the computational approach has been appreciated in various continents and how Africa fairs concerning the use of computational urban planning techniques in planning. It unravels the efforts of African cities, the challenges they face in trying to use the quantitative models in planning, and the opportunities they have. The desktop study is foundational to this chapter as it is used to capture the center of the argument. Case studies are used in the chapter to assess how computational urban planning has been successful in cities of other continents.

Computational urban planning is complementary to the usual planning methods of using CAD software as it allows for quantitative analysis and data assimilation (Wilson et al. 2019). The urban planning environment, being a complex dynamic system, requires the alignment of planning strategies in a way that is compatible with the modern urban environment (Chadwick 1971; Dalberg

2016). It has become imperative for the players in the urban design field to adapt to the changes and become knowledgeable of better techniques that can effectively be used to achieve optimal solutions for urban problems. The process of formulating master plans has been a time-consuming and complicated one as more tasks like data collection and even data analysis are done in a piecemeal way that does not fully solve the planning challenges (Lane 2005; Taylor 2009; Chigara et al. 2013). Thus, applying such traditional planning approaches becomes a slow process which may even encounter hiccups along the way, and by the time the master plans are implemented, a lot more and unanticipated changes would have taken place. Therefore, the adoption of more effective planning methods such as computational urban planning models becomes very necessary and hence should be used early and consistently in the planning process (Wilson et al. 2019). This chapter, therefore, explains the relevance of applying computational urban planning techniques in African cities as an effective means toward designing for a more robust, resilient, and sustainable city environment.

Background and Context

Urban planning plays a big role in influencing development in cities and as such needs to prioritize innovation, modernization, and technological improvement if ever sustainable development is to be achieved (Watson 2009; Healey 2010; Lennon 2015). To appreciate the relevance of change toward modernized planning methods, it is important to understand how complex the planning environment has become and the necessity of adjusting to meet up with the demands for sustainable and resilient development. Urban planning has not been spared from the ever-changing modern environment's impacts both negative and positive (Campbell 2013). There is an infinite continuity in the changes taking place in the modern world, and this has seen the urban planning environment becoming more complex. The urban planning arena is in itself a complex system composed of various interrelated disciplines and elements both being physical and human elements (Archibugi 2004). Complexity is explained by Erem (2003), as a state that is hard or very complicated to understand involving various pieces which are related. Complexity is just heterogeneity in nature (Moor 2001). Thus, a complex urban environment becomes very unpredictable and ambiguous and poses irreversible impacts on the urban environment such that the old approaches to planning cannot alone counter all the challenges that may be resultant. The emergence of the planning field arose with a mandate to address difficulties and impacts arising from complex activities. Thus, it does not solely focus on managing the existing environment but does work toward positive and futuristic potentialities (Healey 2007). The current approaches used in planning are no longer, alone, effective in harnessing and solving all the rising urban challenges emanating from the complex nature of the planning environment.

With the complex nature of cities, there is need for a constant adaptation to the changing environment, and this means that the traditional and conventional methods or approaches in planning are no longer adequate to curtail development challenges (Bettencourt 2019). Therefore, the aura of computational urban planning and design techniques becomes more relevant in complementing the old planning methods to achieve sustainable and resilient urban environments. The modern world is speedily advancing in technology and innovation such that planners need to pace up and move along with the transformational environment. A transformational environment requires transformational players to have design solutions that are compatible with the prevalent design problems. Computational urban planning empowers the planners and city decision-makers to manage and make proper choices concerning the future of urban environments (Geertman et al. 2019). However, the rate at which cities mostly in the African continent are progressing technologically is very slow.

Many local planning authorities recently adopted the use of computers and computer-aided design (CAD) techniques, and some are still in the pencil-and-paper era. This, therefore, explains why sometimes the planning challenges overhaul the piecemeal planning solutions given by planning authorities to development problems. Pisano, De Luca, and Dastgerdi (2019) state that the use of such old methods in the management of cities has slowly lost validity. Therefore, there is an urgent need to adopt smart urban design techniques to manage cities, and computational urban planning proves its relevance, Africa not excluded. Computational urban planning models are effective toward accommodating rapid urbanization, easing master plan formulation, and ensuring sustainable and livable city environments among many other benefits. When adopted, there would be an evident improvement and reduction of city chaos in developments as proper urban management would be engaged.

Computational Urban Planning: A Review

The relevance of applying computational planning techniques in Africa needs to be assessed based on the knowledge of the current planning practices in Africa. The introduction of a new approach in planning can be successful when a background analysis is done to capture the gaps, challenges, and potentials currently existing in Africa, hence exploring the urban planning practices in Africa. Africa, among other continents, is a harbor of many countries that are still in their developing stages. Urbanization and rapid population increases are mostly the common and defining features of many African cities especially those in their developing stages. As urbanization is scaling up at a fast pace, many opportunities and possibilities are looming up which include labor productivity and economic opportunities (Freire 2006). However, urbanization is not without its challenges. It has caused so much pressure over resources such as water, energy, and sanitation (DESA 2013).

Many social, economic, and environmental challenges have been and are still experienced in the African continent due to the influx of people in the urban cities and generally due to the natural population increases. The rapid growth in the urban population in Africa is not at par with productivity as it is far below the rate of population increase (UNICEF 2019). Lall et al. (2017) allude that many cities in the African continent have three common features which affect urban planning as these create challenges for city management. These features include being overcrowded, spatially disconnected and dispersed, as well as being costly to live in. This has exacerbated the environmental, economic, and social challenges in most African cities, and the traditional design methods are failing to solve adequately these ever-increasing urban problems.

Urban planning and development in Africa have been leaning toward rural regions despite the high rate of urbanization in the continent (Matamanda 2020). Some of the development strategies that have been applied in the rural regions include capacity building and education that enhances the livelihoods of the rural people. However, the cities have not been neglected in the process, and their relevance to the future of Africa's development is appreciated and recognized in the principles that hold the African Union's (AU) Agenda 2063 (UNECA 2017).

The AU Agenda applauds the high development potential the cities have for the future and the opportunities it brings. However, less is said toward bettering the means of achieving better and sustainable cities that are user friendly. The rate of urbanization in African cities requires immediate action on the part of planners and development officers as well as policymakers. This has had an irreversible impact and thus calls for a more considerate and effective approach to enhance the effectiveness of urban planning (UNECA 2017). If ever cities are to become adaptive and sustainable, more and better means of planning ought to be applied. However, the planning field in many African cities is lax, and as such the urban challenges are increasing at a fast pace than is anticipated. Such has been evidenced by increasing informal settlements, transportation challenges, and high cost of living. Therefore, urgent action is needed to harness the opportunities that globalization and technology have brought to enhance the planning approaches.

Chigara et al. (2013) state that local development authorities in many African countries lack the capacity and resources to formulate master plans and even local plans. A good example of that is Zimbabwe. Zimbabwe has most of its local authorities relying on the Department of Physical Planning (DPP) for most of their planning needs. Such local authorities lack the capacity and expertise needed to draw up development plans. When they source out for planning services from the DPP, they get plans that do not meet the exact needs on the ground. This is the case with many urban areas in Africa which face similar challenges (Watson 2014). Such challenges include insufficient resources such as capital and technical expertise and poor commitment by the higher planning boards (Chigara et al. 2013). The existence of such inefficiencies renders development planning strategies unachievable and difficult. There is a failure to coordinate social, environmental, and economic requirements necessary to the society (Chigara et al. 2013). Therefore, along with the adoption of the computational planning methods, it is also imperative that decentralization of planning activities be done to make local authorities independent to plan for their problems.

With many daunting challenges like overcrowding, poor transportation, and environmental degradation among many chaotic issues in African cities, not so much has changed toward upgrading the means to solve the problems. The cities fail to meet up with the modern planning approaches because of lack of skills, education, and the weak hand of the public sector (Freire 2006). As highlighted earlier, urban design is now characterized by rising complexities, dynamics, and changes. However, many African cities continue to rely on traditional methods of urban design whose approaches are sectoral and static, thus failing to meet the requirements of the present dynamic and complex planning environment (Miao et al. 2017). Modernization is the trend which requires the planning professionals and city managers to embrace technology through the use of computers and models that improve efficiency in the planning process as almost everything has become automated (Miao et al. 2017). Therefore, Africa has to embrace better planning methods that enhance the planning process and makes it easier to forecast the future.

Even though heavy reliance on the colonial planning standards has not yielded tangible results, many countries in Africa are still reluctant to accommodate the modern approaches that effectively tackle planning problems. Urban planning was born to bring harmony, efficiency, beauty, and economy for the cities (Murphy and Fox-Rogers 2015; Lennon 2020). However, the current state in development and planning may indicate a failure of its original mandate. On a positive note, urban planning in Africa is not yet incorrigible as it is never too late to attain an orderly cityscape and produce layout designs that meet development needs. Commitment, passion, and hard work can be applied to redeem the intended beauty of the built environment. When such concern is worked on, the adoption of technological means to planning such as computational models becomes inevitable. Computational urban planning enables the design of effective and efficient master and local plans that can counter urban sprawling and unorganized and inharmonious developments (Chigara et al. 2013). When other continents have managed to achieve sustainable environments and work with modern technology successfully, there is a huge possibility that Africa can do it and improve its cities.

Application of Computational Planning: International Cases

Learning from the modernized world helps inform the African cities on how to implement better planning approaches. Most of the first-world countries have transitioned to the modern planning models and as such are in a more capable state to accomplish sustainable smart city environments.

Redesigning of Rochor District, Singapore

One of the continents to also have applied the use of computational methods in planning is Asia. The computational planning tools were used in the research design project for Rochor in Singapore. According to Klein and Koeinig

(2016), the Rochor region required urgent redesigning more comprehensively. Existing data for the streets and built structures in the region was made available to facilitate the planning. Computational planning works on already available data, hence the necessity of collecting data from old and existing designs. Thus, the new approach needs to be used to complement the already existing approaches to planning. The software used in the computational design are very interactive and allow for planner creativity and consistency in the design process. Evidence-based computational planning was used for Rochor which gives flexibility in improving the results through adding informative restrictions and values that are objective. According to Klein and Koeinig (2016), this type of computational planning used for Rochor enables designers to fulfill well-formulated design demands, thus eliminating potential solutions that are problematic. Reviewing the efforts and successes from other continents gives a glimpse of the efficiency and effectiveness of using computational urban planning models to achieve livable urban cities.

Creation of a New Mixed-Use District: Hangzhou, China

Computational design techniques were applied in Hangzhou, China, on a 620-acre district master plan. The government laws in the region regulate at least 2-h duration for direct sunlight on the residential units on the winter solstice. Generally, such regulations made it difficult to produce smaller-sized blocks as intended by urban designers. Therefore, in trying to solve the challenge, the computational design tools were applied which produced approximately 7400 design options which were later evaluated to check their compliance with the regulations for direct sunlight in Hangzhou. The use of these models gave a variety of solutions which met the optimal urban environment that was desirable, at the same time complying with the federal regulations (Wilson et al. 2019). Had not the computational methods been used in the process, it would have been difficult to come up with an optimal solution that met the region's regulations, at the same time satisfying the expectations of the urban designers. Technology has made planning easier by widening the choices for better planning. Unlike the old approaches, which in this case if it had been used would not have resulted in at least 7400 design options as humans tend to be limited, computational planning gives the edge of variety.

Master Planning of the Toronto Sidewalk Project, Canada

Computational urban planning, when used complementarily with the principle of public engagement or participation in decisionmaking, yields good results that are favorable and satisfy all stakeholders. For the sidewalk project in Toronto, various stakeholders were involved in the process which includes the public and Sidewalk Labs. Various stakeholders were allowed to explore various density combinations, streets, and public spaces by inputting various design inputs. Such engagement of the public in planning allowed for the development of a neighborhood that they all appreciated. It is imperative therefore that there be user engagement in planning processes and implementations which allow for feedback and reciprocal communication of the planners, the design, and the users (Wilson et al. 2019). The use of computational models in the Toronto sidewalk master plan yielded optimal results that saw the formulation of a user-friendly master plan as the users were engaged in the process.

Whereas computational urban planning promises a more iterative and quantitative way to planning, its position in the process of master planning is still questionable in many planning authorities and even cities (Wilson et al. 2019). Despite computational planning being the trending planning approach in this modern world, cities have become more reluctant and unnoticing. It promises a good future in planning through generating insightful projects that can influence development within a complex, dynamic, and multi-stakeholder urban environment (Wilson et al. 2019). Therefore, it is becoming an urgent call for all urban planners to be capacitated and be skilled to use the computational models. However, this does not dismiss the traditional process of making master plans; computational urban planning is complementary to the old ideology in master planning. With the rising complexity in the urban system, it has become of more importance that the process of master planning adopts the computational planning methods to enhance urban design (Wilson et al. 2019).

Computational Urban Planning Attempts in Africa

There have been attempts at applying computational urban planning in Africa. Exploring these attempts helps to assess the success, challenges, and potential opportunities that can be gained from such planning approaches. This informs the future attempts, thus enabling a more successful application of modernized planning methods.

Empower Shack Project: Cape Town South Africa (Case Study)

Among many African countries, South Africa has attempted to apply computational planning methods in the city of Cape Town. Cape Town is one of the cities considered as prosperous among other cities in the African continent. However, it is not without its problems. The city has an estimated population of about 7.5 million that lives in the city's informal settlements, and nearly 2.5 million units of housing are needed according to Miao et al. (2017). As a way to plan for housing, the city came up with the "Empower Shack Project" that aimed to formulate strategies for upgrading these informal settlements. To facilitate the project, computational prototyping tools were developed as a way to enhance fast urban design synthesis (Miao et al. 2017). The use of computational urban planning in the Empower Shack Project in Cape Town yielded fruitful results. Miao et al. (2017) denote that it was time-effective as it saved on the time required by the urban planners to try different possible outcomes for "mock-up urban layouts"; it was cost-effective as little manpower was needed and enabled quality design solutions as it allowed various outcomes to be tried (Miao et al. 2017).

The study by Miao et al. (2017) for the shack project in Cape Town points out that computational planning tools are very effective in reducing the workload of urban designers and in producing quality solutions to planning challenges. However, a few limitations were encountered in the Empower Shack Project. One of the limitations was that the presentation of data relating to the parcels could not fully meet the expectations or needs of the designers (Miao et al. 2017). However, this required manipulation to yield the desired results and cannot erase the effectiveness of applying computational methods in urban design. No positive design attempt is without its challenges during implementation and such is with the computational design methods. When weighed against the old planning approaches, it still promises a good future to plan, and thus it is recommendable.

Although urban planning issues such as climate change, population growth, and transportation are diverse and challenging for urban management, many African countries still lag in effectively addressing them. There is a need for a more effective and data-driven approach to proffer lasting solutions. However, despite the greater need for such design and planning tools, the urban designers are only exposed to the traditional ones which make it difficult integrating the various city needs (Wilson et al. 2019). Much difficulty has been experienced in the use of old manual design methods in urban planning, especially in many African cities. They involve time-consuming processes which make it difficult to integrate various and conflicting ideas from stakeholders involved in the formulation of master plans. Therefore, it becomes very imperative that cities adopt computational tools in every step of planning; urban planners need to find a way to capacitate themselves (Wilson et al. 2019).

Discussion

The twenty-first-century cities have experienced very complex challenges emanating from rapid urbanization, population growth, and climate changes that are unpredictable. These challenges have rendered the traditional master planning approaches not fully effective, thus the search for a more efficient alternative. One of the defining characteristics of urban designers is to proffer solutions to urban challenges and to plan for the future sustainably. Such planning that provides lasting solutions and makes the urban environments better places no longer needs to depend solely on the traditional approach to planning. A more modern approach is needful, and thus computational urban planning proves to be the required solution. Computational urban design methods improve the planning process as they bring about a measurable methodology with the principle to ensure a "design process with new digital planning tools and evaluation framework to ensure scalability and relocability" (Miao et al. 2017, p. 408). Using computational planning methods is recommendable to planners as it makes it possible to automatically formulate layouts using simple and easy parameters like the parcel length and width as stated or specified by the user. It is also easy to revise interactively the layout through moving street sections and thus obtain satisfying results (Miao et al. 2017). If computational urban planning thus proves to be effective in helping planners, no hindrances such as lack of finance can completely make it futile. It is advisable, therefore, for planners to assess the possible limitations to the implementation of computational planning. For example, if the limitation to the adoption is financing, funding options for planning authorities need to be worked on to facilitate and enable the adoption of computational planning methods.

The software prototypes for computational urban planning as explained by Klein and Koeinig (2016) allow for mapping that gives the advantage of getting similar and workable solutions which enable quality decisionmaking in the selection of a more effective solution. It allows for interaction with the planner. In using this planning tool, urban planners are guided efficiently in a search space that is ever-changing, thus generating evidence-based solutions. According to Wilson et al. (2019), the computational design models have successfully been applied at the building level to examine various designs to quantify their level of performance. However, much difficulty has been experienced in trying to apply them at an urban scale as there are much stakeholder involvement, high computational costs, and more difficulties in trying to limit the inputs (Wilson et al. 2019). One of the limitations in using this method as stated by Klein and Koeinig (2016) is that it may be difficult to precisely specify the parcel length. But this alone cannot scrape away the usefulness of applying computational methods in planning. When such few limitations are noticed, it becomes easier to manipulate to design requirements and specifications, thus achieving the targeted result.

It is high time urban planners modify their planning toolset and adopt the modern-day smart design means that are effective. This serves in making better futuristic plans through an enhanced understanding and assessment of the current conditions. There is need for a paradigm shift from managing cities and their performance through old ideologies that advocate for the use of predetermined master or local plans (Pisano et al. 2019) to a more effective one. Compared to the traditional design means that are used in the design, which refines solutions manually, computational methods make use of "parametric CAD tools to explore larger design spaces" (Wilson et al. 2019, p. 2). The old way of city management was effective in the past when urban challenges were manageable from paper as probably there was low urban population and urban complexities. However, with the rising complexity in urban planning emanating from population, environmental, and social problems, the urban planners ought to adjust accordingly if ever-resilient and livable cities are to achieved.

Urban planning in Africa has been very reluctant to embrace the potential that comes with computational planning models as the cities heavily rely on the old planning that has been passed on from the colonial era (Chigara et al. 2013). Using such traditional methods has numbed the innovative and creative capacities of the cities such that urban challenges are becoming more and more difficult to solve using the traditional means. If urban planning would be recognized as a top priority in most African countries, there would be improved planning as resources and support from governments will be readily available. However, the issue of capacity building has been left to individual practitioners who may not be able to afford the costs associated with technology in planning and gaining new skills. It is now very imperative that urban planning authorities support planning practitioners through on-the-job training schemes that bring innovative technology, thus promoting computational urban planning. This also motivates planners and helps in improving city management.

The processes involved in urban planning of data collection, data analysis, and information dissemination that help to improve decisionmaking by planning practitioners make urban planning an essential part of every development (Klosterman 1997). A more accurate collection of data is in these modern times enhanced by the use of computers and the Internet of things. The grand problems of overpopulation, depletion of resources, and degradation of the natural environment demand a computer-based assessment, evaluation, and forecasting. Thus, academic institutions ought to adjust quickly to technological changes and even innovate new designing prototypes (Klosterman 1997). Almost all facets of planning now require the use of computer technology, be it from research to design and implementation. Thus, computational urban planning has become a local necessity than a foreign ideology.

The technological revolution taking place in the modern world plays an important part in addressing economic and environmental challenges that are facing the world at the present moment. Laffta and Alrawi (2018) explain green technology as a better means to solve urban challenges sustainably at all planning levels. The concept of sustainability was introduced as an approach to address such problems, and the most effective way to aid sustainable development is the adoption of better planning models (Klosterman 1997). More work and determination are needed on the part of practitioners and policy influencers to prioritize urban planning. National budgets formulated by governments need to consider the planning and development sector and channel sufficient resources to enable the planning authorities and the practitioners to embrace new technology.

Every planning and development practitioner wishes for a perfect, safe, and livable urban environment, though many from the developing world are incapacitated to achieve that dream. Klosterman (1997) states the various obstacles that have slowed down the adoption of computerized planning methods include the reluctance of the urban practitioners to actively consider technology, and the small market for the design models, which limits the developers from further improving their design tools. Likewise, the modern-day planners spend most of their time on administrative work such as the processing of permits, policy formulation, and enforcement of development regulations. As a result, it becomes difficult to meet the demands of the communities sustainably. Therefore, the chapter makes an effort to realign urban planning toward meeting design needs by way of adopting quantitative tools that enhance effective, harmonious, and sustainable development. We are now living in an era of exciting technological innovations that require earnest advances and adjustments by urban planning practitioners to rejuvenate urban design (Klosterman 1997). Thus, to be successful in this development era, it is important to use the right tools.

Conclusion and Recommendations

The study observed that, despite the modernization that has brought efficient design tools for planning, Africa still lags in terms of urban management. The mismanagement of funds in most African nations has heightened capital and technical challenges in the management of cities. The chapter has highlighted the rising challenges in African cities, many of which are results of rapid urbanization, population growth, and climate changes. Such challenges no longer need the traditional approaches to planning alone as such planning has to a greater extent failed to give lasting solutions. It has become very important for urban practitioners to gear up to the changes by adopting new technology and learning new skills that promote the development of sustainable urban environments. Capacity building needs to begin as early as at the learning institutions so that planners get into practice already equipped. It is time for African cities to reconsider the planning field and prioritize it. Computational urban planning has become the modern means to counter urban challenges. It provides planners with a variety of planning solutions which could not have been realized had they stuck to the old planning regime. Thus, where there are various options or solutions, better planning decisions that satisfy all stakeholders are brought up. It is recommendable also that there be public engagement at every stage of the planning process as this would build up smart cities with smart citizens that know city management. Such can be best achieved through using computational planning models complementing the traditional planning ways. Urban planning is not an isolated field; African cities need to learn from technologically advanced countries and other continents and adopt the smart planning techniques if ever smart cities are to be realized. Though it may not be necessary to copy all the strategies used by other nations, as such may not be compatible to the African city environment, it is best to appreciate, adopt, and implement that which is suitable to the needs of a particular city.

Cross-References

▶ Spatial Justice and the Design of Future Cities in the Developing World

References

Archibugi, F. (2004). Planning theory: Reconstruction or requiem for planning? *European Planning Studies, 12* (3), 425–444.

Batty M. (2008). *Cities as complex systems: Scaling, interactions, networks, dynamics and urban morphologies* (UCL working papers series 131 – Feb 08). London: Centre for Advanced Spatial Analysis (CASA), University of London.

Bettencourt, L. M. (2019). Designing for complexity: The challenge to spatial design from sustainable human development in cities. *Technology Architecture Design, 3*(1), 24–32.

Campbell, S. (2013). Sustainable development and social justice: Conflicting urgencies and the search for common ground in urban and regional planning. *Michigan Journal of Sustainability, 1*, 75–91.

Chadwick, G. F. (1971). *A systems view of planning: Towards a theory of the urban and regional planning process*. Oxford: Pergamon.

Chigara, B., Magwaro-ndiweni, L., Mudzengerere, F., & Ncube, A. (2013). An analysis of the effects of piecemeal planning on development of small urban centres in Zimbabwe: Case study of Plumtree. *Journal of Sustainable Development in Africa, 15*(2), 27–40.

Dalberg. (2016). *100 resilient cities. Opportunities for building a more resilient Durban: Summary of systems analysis*. Durban: Dalberg Global Development Advisors. Available at http://www.durban.gov.za/City_Services/development_planning_management/environmental_planning_climate_protection/About%20Durban%E2%80%99s%20Resilience%20Programm/Phase%20Two/Documents/DurbanSystemsAnalysis_DalbergFinalReport.pdf. Accessed 12 Nov 2018.

Department of Economic and Social Affairs (DESA). (2013). *Sustainable development challenges*. New York: United Nations Publishing.

Erem, Ö. (2003). *An approach to the evaluation of the legibility for holiday villages*. Unpublished doctoral thesis, Institute of Science and Technology, Istanbul Technical University.

Freire, M. (2006). *Urban planning: Challenges in developing countries*. Madrid: World Bank International Congress in Human Development.

Geertman, S., Allan, A., Zhan, Q., & Pettit, C. (2019). Computational urban planning and management for smart cities: An introduction. In S. Geertman, A. Allan, Q. Zhan, & C. Pettit (Eds.), *Computational urban planning and management for smart cities* (pp. 1–14). Cham: Springer.

Healey, P. (2007). *Urban complexity and spatial strategies: Towards a relational planning for our times*. London: Routledge and Royal Town Planning Institute.

Healey, P. (2010). *Making better places: The planning project in the twenty-first century*. Hampshire: Palgrave Macmillan.

Klein, B., & Koeinig, R. (2016). Computational urban planning: Using the value lab as control center. https://www.researchgate.net/publication/304265608_Computational_Urban_Planning_Using_the_Value_Lab_as_Control_Center/link/5783b2a408ae3f355b4a28ef/download

Klosterman, R. E. (1997). Planning support systems: A new perspective on computer-aided planning. *Journal of Planning Education and Research, 17*(1), 45–54.

Laffta, S. J., & Alrawi, A. (2018). *Green technologies in sustainable urban planning*. Baghdad: University of Baghdad.

Lall, S. V., Henderson, J. V., & Venables, A. J. (2017). *Africa's cities: Opening doors to the world.* Washington, DC: World Bank Group.
Lane, M. B. (2005). Public participation: An intellectual history. *Australian Geographer, 36*(3), 283–299.
Lennon, M. (2015). Finding purpose in planning. *Journal of Planning Education and Research, 35*(1), 63–75.
Lennon, M. (2020). Planning as justification. *Planning Theory & Practice*. https://doi.org/10.1080/14649357.2020.1769918.
Matamanda, A. R. (2020). Mugabe's urban legacy: A postcolonial perspective on urban development in Harare, Zimbabwe. *Journal of Asian and African Studies*. https://doi.org/10.1177/0021909620943620.
Miao, Y., Koenig, R., Buš, P., Chang, M.-C., Chirkin, A., & Treyer, L. (2017). Empowering urban design prototyping: A case study in Cape Town with interactive computational synthesis methods. In P. Janssen, P. Loh, A. Raonic, & M.A. Schnabel (Eds.), *Protocols, flows and glitches – Proceedings of the 22nd CAADRIA conference* (pp. 407–416).
Moor, J. (2001). Cities at risk. *Habitat Debate, 7*(4), 1–6.
Murphy, E., & Fox-Rogers, L. (2015). Perceptions of the common good in planning. *Cities, 42*, 231–241.
Pisano, C., De Luca, G., & Dastgerdi, A. S. (2019). Smart techniques in urban planning: An insight to the rule-based design. *Sustainability, 12*(1), 1–11. MDPI Center.
Taylor, N. (2009). Anglo-American town planning theory since 1945: Three significant developments but no paradigm shifts. In E. L. Birch (Ed.), *The urban and regional planning reader* (pp. 96–105). London: Routledge.
UNICEF. (2019). *Shaping urbanization for children: A handbook on child-responsive urban planning*. New York: United Nations.
United Nations Economic Commission for Africa (UNECA). (2017). *Urbanization and national development planning in Africa*. Addis Ababa: UNECA.
Watson, V. (2009). Urban challenges and the need to revisit urban planning. In B. C. Arimah, I. Jensen, N. D. Mutizwa-Mangiza, & E. A. Yemeru (Eds.), *Planning sustainable cities: Global report on human settlements 2009*. London: Earthscan.
Watson, V. (2014). African urban fantasies: Dreams or nightmares? *Environment and Urbanization, 26*(1), 215–231.
Wilson, L., Danforth, J., Davila, C. C., & Harvey, D. (2019). *How to generate a thousand master plans: A framework for computational urban design*. New York: KPF.

Computer Applications

► Computational Urban Planning

Concepts, Approaches, and Methodologies for Ecological Flood Resilience Assessment: A Review

P. Ambily[1], N. R. Chithra[1] and C. Mohammed Firoz[2]
[1]Department of Civil Engineering, National Institute of Technology, Calicut, Kerala, India
[2]Department of Architecture and Planning, National Institute of Technology, Calicut, Kerala, India

Synonyms

Ecosystem based adaptation; Green blue infrastructure; Socio-ecological resilience

Definitions

Resilience

The origin of the concept of resilience is often credited to C S Holling, who initially used the term in a descriptive ecological perspective (Holling 1973). Since then, a variety of interdisciplinary definitions have been introduced. Currently, there are about 70 definitions of resilience proposed by the research community. According to Brand (2007), the definitions of resilience can be categorized under three main heads, namely the descriptive concept, hybrid concept, and normative concept. To the purpose of this entry, the most widely used definition of resilience by Folke (2016) has been adopted as:

> The capacity of a system to absorb disturbance and reorganize while undergoing change so as to still retain essentially the same function, structure, feedbacks, and therefore identity that is, the capacity to change in order to sustain identity; resilience is a dynamic concept focusing on how to persist with change how to evolve with change.

Adaptive Capacity

Adaptability, adaptation, and adaptive capacity are closely used with the resilience regime, and the same is defined by Steve Carpenter B. W. (2001) and Gunderson (2000) as:

> Adaptive capacity is a component of resilience that reflects the learning aspect of system behaviour in response to disturbance.

Transformability

One important character of transformability or transformation in the resilience perspective is the discontinuous and "not-so-smooth" cycles of reorganization. According to Folke (2016), transformability can be defined as:

> The resilience approach allows the new identity of the social-ecological system to emerge through interactions of individuals, communities, and societies, and through their interplay with the biosphere within and across scales.

Ecological Flood Resilience

According to Osvaldo M. Rezende and Miranda (2019):

> Flood resilience is the ability of a city to resist flooding over time, being able to adapt itself and continue functioning, even under stress conditions, and recover rapidly from material losses.

Introduction

The term disaster resilience has gained wide significance in the last few decades due to its ability to explain long-term temporal and spatial variations of a system subjected to external disturbances. Clarity of the resilience concept and an established method of assessment is of utmost importance in the context of disaster research, especially in the case of urban floods. The hydrometeorological hazards constitute 80–90% of the total documented disasters in recent decades (WHO 2022). Such hazards have a high impact on the governance, planning, and management of human settlements. There are several concepts, assessments, and methods to analyze flood resilience and the present entry aims to summarize such researches already done in different works of literature. Accordingly, the present entry is structured into four parts. The first part discusses resilience as a concept followed by the discussions on ecological resilience in the second part. This ecological resilience concept is further assessed from the flood perspective in the third part. Here, assessing flood resilience based on complex adaptive system thinking and the relevant methodologies and approaches for assessment are also elaborated. The final part of the entry discusses relevant case studies of urban resilience projects from the Netherlands, the Room for the River Project in Nijmegen and Sustainable Urban Development Master Plan, Leidsche Rijn, Utrecht District.

Concept of Resilience

The concept of resilience is broadly divided into three categories, namely engineering, ecological, and complex adaptive system resilience. According to Kerry McClymont, (Kerri McClymont 2019), engineering resilience considers "withstand, resist and bounce back" as its key components. Ecological resilience considers "cope, resist and bounce forth" as the main characteristics. Thus, the major difference between ecological resilience and engineering resilience relies on the assumption that dynamic and variable stability domains and multiple states of equilibrium exist. The engineering resilience of a system measures the time required to reach back to its single or global equilibrium state, which is often denoted as the "return time." Ecological resilience focus on the unsteady system conditions that can trigger transformation or adaptation to a new behavior regime (Gunderson 2000). A new terminology evolved from these two predecessors termed complex adaptive system resilience or socioecological resilience, which consider "adapt, learn and transform" as its major strategy.

Assessing Resilience

Methodologies for assessing resilience have been further classified as qualitative and quantitative methods. These methodologies can be further grouped based on the scale of assessment. As per Kerri McClymont (2019), 39% of the total evaluation methods are based on quantitative assessment. The most popular quantitative methods for resilience assessment are index-based and simulation-based approaches. Descriptive methods based on field-based data collection, development of a theoretical framework, etc. are part of qualitative approaches.

The scale of assessment ranges over temporal and spatial scales, in which spatial scale varies over household, community, city, and national scale. However, many of the assessments consider multiscale methods to analyze the interrelations and network loops across the scales.

Concept of Ecological Resilience

The beginnings of the concept of resilience can be tracked down to 1973 from the works of C.S Holling, on ecological resilience and adaptive cycle. F. S Brand and K. Jax consider three characteristics of ecological resilience: (1) the ability of a system to withstand a certain amount of change while remaining within its stability domain; (2) the ability of a system to reorganize; and (3) the capability of a system to learn and adapt in the process of shifting from its stability domain (Brand 2007). These three characteristics form the basis of assessing the ecological resilience for a system.

As the concepts of ecological resilience are highly related to the theories of the adaptive cycle and adaptive capacity, it is important to understand the phases of the adaptive cycle. A loss of ecological resilience happens while a system changes from one phase to another which is discussed in further sections.

Ecological Resilience and Adaptive Cycle

Ecological systems undergo a highly complex and gradual process of self-organization when it is subjected to a series of sudden disturbances. This process is often referred to as a regime shift. According to C.S. Holling and Folke (2004), *"Regime* Shifts are large and constant, often irreversible shifts in the structure and function of an ecosystem." The process of passing through each threshold of the adaptive cycle to transform and reorganize itself by an ecosystem forms the basis of ecological resilience. The four phases of the adaptive cycle proposed by C.S. Holling and Folke (2004) are:

- The **exploitation phase** which is marked by the colonization of recently disturbed areas
- The **conservation phase** where the system accumulates and store materials and energy
- The **creative destruction phase** in which the structure that is accumulated in the previous phases is influenced by disturbances
- The **reorganization phase** where the system passes through self-organization to enter another cycle of explosive phase

The above four phases of the adaptive cycle can be illustrated as given in Fig. 1.

Assessing Ecological Resilience

The biggest challenge in measuring ecological resilience is answering questions on *"resilience of what to what"* (Rocha et al. 2022). According to Brian Walker (2004), the four critical aspects of assessing ecosystem resilience are (1) latitude, (2) resistance, (3) precariousness, and (4) cross-scale relations.

Latitude refers to the width of the stability domain. In other words, it is the extreme extent to which a change is accommodated by the system before losing its capability for reorganization. Resistance refers to the comparative difficulty in shifting the system; large-scale disruptions are necessary to alter the current state of the system with a deep basin of attraction. Precariousness is the proximity of the current direction of the system shift to the threshold of destruction. If that threshold is broken, a return to the original state and restructuring will be hard or even impossible for the system. Also termed as panarchy, cross-scale relations refer to the influence on the three abovementioned aspects by the state and dynamics of the (sub)systems.

The measures of resilience should answer characteristics of the system, as given in Table 1.

The notion of each phase of an adaptive cycle as a measure of resilience is further elaborated by Berkes. F, who devised a two-tier set of variables which are categorized around four major dimensions: (1) N – representing the learning phase to

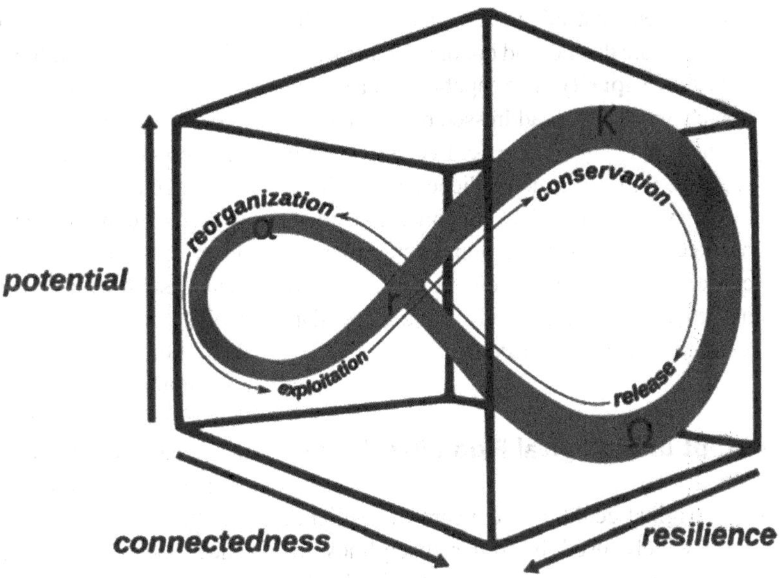

Concepts, Approaches, and Methodologies for Ecological Flood Resilience Assessment: A Review, Fig. 1 Hollings' adaptive cycle. (Source: Wolfgang zu Castell 2020)

Concepts, Approaches, and Methodologies for Ecological Flood Resilience Assessment: A Review, Table 1 Resilience measures. (Source: authors' interpretation of resilience characteristics by Steve Carpenter (2001))

Sl No	Characteristic of the measurement	Description
1.	Resilience *of* what	The system that is under consideration
2.	Resilience *to* what	Perturbations of interest
3.	Measure in model	Based on the size of the basin, attractor size in quasi-steady-state analysis
4.	Biophysical field measures	Slowly changing variables that affect the ability to maintain the physical equilibrium of the system
5.	Interpretation of biophysical field measures	The intervention of local solutions to the problem triggered by the flexibility of agents, and the presence of incentives to increase resilience
6.	Socioeconomic field measures	Pollution costs, impact on the market, effects on income, etc.
7.	Interpretation of socioeconomic field measures	An incentive to stabilize the equilibrium

adapt with uncertainties and instability; (2) D – representing the diversity nurturing phase where the system undergoes restructuring and renewal; (3) E – combination of knowledge and learning by the system; and (4) S – representing prospects for self-organization (Berkes 2003).

To put it concisely, ecological resilience measures the capacity of a socioecological system under three broad principles (Steve Carpenter B. W. 2001). They are:

- **Biophysical measurements** decide the boundaries of the socioecological system. The extent of the disruption is related to the likelihood of the system staying within the stability domain. Some examples are soil characteristics, frequency of flood, sedimentation, land tenure systems, cultural practices in land management, etc.
- **The capacity of force**: The ability to reorganize should be indicated by the capacity of force by the management system, rather than self-organization within the management system. An example is in forest fire management, rigorous repression of fire leads to monotonous vegetation and high-energy loads. But fire

management that allows a variety to develop leads to more distinct and continual ecosystems.
- **Adaptive capacity**: The capability of a system to manage change is addressed by the adaptive capacity indicators. Genetic diversity and heterogeneity indicate the resilience capability of biotic systems. Flexible networks and power balance among interest groups decide adaptive capacity in a social system. It is a matter of interest that, the ability to learn, decides adaptive capacity in human systems.

Concept of Ecological Flood Resilience

As the original ecological concept of resilience became widely used by other disciplines, its descriptive meaning and concept of adaptive capacity got partially diluted. K. McClymont indicates that most of the flood resilience measurements do not consider resilience as an adaptive and iterative process (Kerri McClymont 2019). A complex mechanism of absorbing disturbances before the system adjusts itself to a new state of equilibrium is often termed as socio-ecological flood resilience. Since this entry focuses mainly on ecological resilience, flood resilience measurements based on complex adaptive system thinking have been discussed further in detail.

Assessing Flood Resilience Based on Complex Adaptive System Thinking

The Concept of Complex Adaptive Systems

According to Y. Shi et al., there are three components that form an urban system, namely the system environment, system elements, and the system structure (Yijun Shi 2021). The mutual connections, combinations, and influences in the networks constitute a complex urban system. When this system is hit by a disaster or an external shock, the entire network of the urban system will undergo varying levels of impact. This complex urban system then goes through the different phases of the adaptive cycle, which was discussed earlier. As the system passes through the release phase denoted as Ω, some of the components and attributes of the system networks may be lost which will be followed by a reorganization phase denoted as α, during which a novel stable system can arise (Steve Carpenter B. W. 2001). This process forms the basis for measuring the resilience of complex urban adaptive systems.

Three components constitute resilience of urban environments, namely the urban environment system which considers the sensitivity of the ecological systems to natural and anthropogenic disruptions; the disaster risk environment which constitutes the natural and man-made disasters that affect the normal function of the system; and the socioeconomic environment which considered the socioeconomic stability of the system. As the environment sensitivity increases, resilience decreases. Similarly, resilience also decreases with high disaster risk. A stable or well-established socioeconomic component makes the system more resilient toward sudden shocks. Thus, the interdependencies of these three components decide the resilience of a complex urban adaptive system.

In the following sections, some of the index categories, subcategories, and variables to assess mutual relation between system environments, system elements, and system structures are discussed.

Assessing Complex Adaptive System Resilience: Ecological Systems

In the process of developing complex adaptive system resilience, the social, ecological, and other physical systems should evolve together. Thus, the assessment of ecological flood resilience focuses on the biophysical and social characteristics of the system. The assessment methods also consider the relationship between frequency of the event, magnitude, and the scale of impact. Some of the critical aspects of assessment according to Brujin (2003) are:

- **Reaction threshold** – the maximum discharge or intensity of rainfall that does not trigger floods
- **The amplitude of reaction** – the anticipated damage that can result from peak discharge or extreme flood event

- **Graduality** – the extent of damages rises with the increase in the intensity of flood waves
- **The recovery rate or return time** indicates how fast the system can recover
- **Robustness**, which is essentially a response curve, is used in synonym with socioecological resilience

The robustness curve is like a risk curve in which the peak discharge and impact of system response are represented by probabilities. The sum of resistance range and resilience range can be used to calculate socioecological resilience. Resistance range represents discharges causing no flooding impacts while resilience range represents the discharges causing limited impacts to the system from which the system can recover (Klijn 2001).

Assessing Complex Adaptive System Resilience: Sociopolitical Systems

As discussed in the earlier section, the urban system environment is associated with system elements and system structures. The trajectory of flood resilience is governed by policy decisions, action plans, and social learning in cities. Therefore, it is of utmost importance to assess the direction of resilience thinking by sociopolitical systems in addressing flood resilience to assess various degrees of impacts in the ecology. Accordingly, the set of variables devised to assess the resilience thinking trajectory in cities by governance mechanism and society are discussed in the subsequent sections.

According to Heejun Chang (2021), the social, ecological, or technological orientation of flood resilience measures can be assessed using an integrated framework called SETS – social-ecological-technological systems. Based on the characteristics of flood resilience management, systems can be classified as resistant, resilient, or transformative. The variables for assessment under the social category include disaster planning and management, laws and regulations, participatory approaches in flood management, economic initiatives, etc. The significance is given to ecology restoration and preservation; adopting green and blue infrastructure for flood management, etc. constitutes the variables under the ecological framework. Construction, operation, and maintenance of engineered solutions, design codes, data-driven disaster management, etc. contribute to technological variables.

C

Assessing Complex Adaptive System Resilience: Urban Infrastructure Systems

Urban infrastructure systems, especially stormwater drainage systems, play an important role in deciding the flood resilience of an urban area. Thus, most of the studies on urban flood resilience are directed toward assessing the resilience capacity of urban drainage systems. In this section, some of the index methods devised to assess the resilience capacity of urban infrastructure systems and their impacts on related networks are discussed.

Several multicriteria index-based methods have been evolved to measure urban flood resilience. Some of the most cited examples are Flood Resilience Index (FResI), adapted Flood Resilience Index (aFResI), Urban Flood Resilience Index (UFRI), etc. Most of these indices are based on hydrological simulation models. The UFRI, proposed by Osvaldo M. Rezende and Miranda (2019) combines three characteristics of ecological resilience, namely absorptive capacity represented as the risk to resistance capacity; adaptive capacity represented as the risk to system functional capacity; and restorative capacity is denoted as the risk to material recovery capacity. The variable of measuring absorptive capacity includes structural exposure, urban infrastructure exposure, and flood depth. Adaptive capacity is represented by variables of structural susceptibility, the vulnerability of the population, and the velocity factor. Material recovery mechanisms and permanence factors denote restorative capacity.

Assessing Complex Adaptive System Resilience: Community Resilience

Agreeing to R. Jacinto and Reis (2020), there are six dimensions of flood resilience, namely individuals, society, governance, built environment, environment, and disaster. In that, four deal with social aspects. Individual adaptive capacity is measured based on the capacity to learn. Variables of evaluation include psychological characteristics like knowledge, social skills, confidence level, health, age, and migratory aspects. A society's resilience capacity is dependent on the ability to associate, networking, livelihood status, and status of governance. Governance is one of the most important aspects of community resilience. This factor is highly dependent on policy decisions, strategies, participatory approach, research and development, etc. The built environment resilience is discussed in the earlier section on urban infrastructure.

Methodologies and Approaches for Assessment

According to Tong (2021), some of the widely used methods in urban resilience assessment are:

1. **Conceptual/Theoretical framework** – which is a theoretical approach to resilience. However, this model is not widely recognized as it is highly subjective and depends on the perspective, experience, and expertise of concerned entities only.
2. **Mapping method**: The data related to geographical evidence are collected by this method. The quality of a single system's elements and relationships are often quantified as resilience in this method.
3. **Delphi method**: It is a theoretical method to select, define, and weigh variables based on subjective information from experts.
4. **Interview method**: The advantage of the interview method is that it can facilitate the transfer of respondents' understanding by evaluating evidence. This method also enables the exchange of knowledge and experience. However, this is a qualitative method that may lack standardization and representativeness.
5. **Indicators/Index method**: The application of diverse metrics to calculate attributes and dimensions of resilience is the character of this method.
6. **Numerical method**: In this method, resilience is dependent on the reliability and stability of input data. It quantifies resilience as the quality and efficacy of network systems. The major features of the numerical method are "performance curve, simulated scenarios, and multi-criteria analysis" (Tong 2021). Thus, the numerical method is considered one of the most reliable methods in resilience assessment.
7. **Survey method**: This method is used to gather a wide array of data from a specific population; however, there is a chance of rigidity and potential bias.

Approaches Toward Ecological Flood Resilience Measurement

A variety of approaches have been employed by various researchers in assessing flood resilience, including, but not limited to, input- and output-based resilience capacity and response-based, wellbeing-based, and stress-based measurements. However, to the purpose of this entry, two major approaches toward urban flood resilience, i.e., bottom-up and top-down approaches to flood resilience measurements, are only discussed.

According to Serra-Llobet et al. (2016) and Kerri McClymont (2019), the top-down approach to flood resilience refers to the goals and approaches set by a top authority and its operationalization. In the top-down approach of measurement, predetermined indicators set by the researcher are employed for measurement. The bottom-up approach promotes the participation of the affected community in decision-making and enhanced capacity building. A balanced composition of both approaches is necessary to evolve flood resilience indicators.

Case Studies

The case studies discussed in this entry are based on a promising and popular concept on urban flood resilience, the sponge city. Sponge projects refer to a set of low-impact developments for integrated water management that replicate the natural water cycles (Scott Hawken 2021; Chan 2017). It is adopted and operationalized by many countries like Australia, the Netherlands, China, etc. In the Netherlands, the flagship project – Making Room for the River, is an example of a sponge project. This entry discusses two major examples of sponge projects, Nijmegen and

Utrecht, both from the Netherlands, and their assessment methodologies.

Making Room for the River: Nijmegen, the Netherlands

The city of Nijmegen is the oldest in the Netherlands in the banks of River Waal, a tributary of the River Rhine. The city got affected by severe floods in the year 1995 due to extreme rainfall in Europe and snow melting in the upper regions. A major policy shift happened following the repeated floods, from "fighting with water" to "making room for the river" (Rădulescu 2021).

The River Waal sharply bends toward the east of the city, thereby increasing risk of water currents and flooding. Thus, the bottleneck of River Waal was selected for the pilot project in the "Making Room for the River" approach. Relocation of the existing dike was proposed, which was initially opposed by the municipality and residents. However, the project was successfully implemented with the creation of a new ancillary channel 350 m apart from the initial location.

A reduction of 35 cm flood water height is achieved by the project. The case of Nijmegen is an example of a series of engineered and ecosystem-based top-down solutions and represents socioecological resilience by absorbing, adapting, and transforming according to external disturbances. The transformation of River Waal is shown in Figs. 2 and 3.

Adaption Measures

Restoring rivers and flood plains and their rehabilitation is an important aspect of nature-based solutions in flood mitigation. Flood plain and river restoration helps in flood control by a delayed release of floodwater and increased infiltration. A buffer is formed by the flood plain between the river and catchment (EEA 2020). The buffer performs multiple functions including water quality improvement, temporary storage of water and

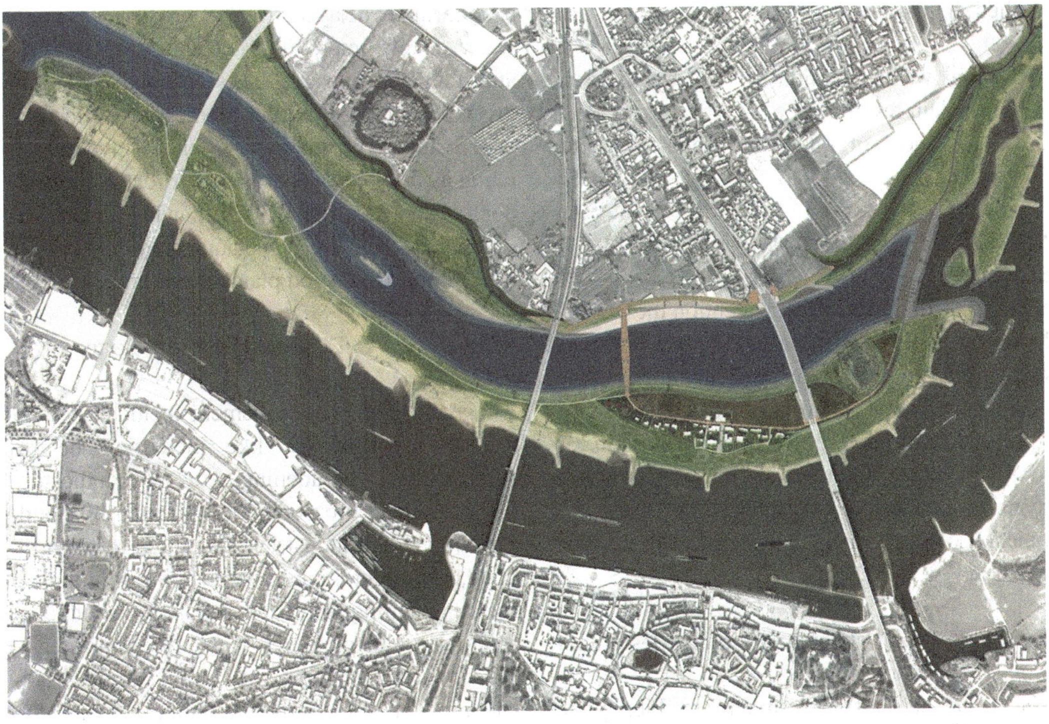

Concepts, Approaches, and Methodologies for Ecological Flood Resilience Assessment: A Review, Fig. 2 River Waal before implementation of Room for the River Project. (Source: Rădulescu 2021, https://medium.com/)

C

Concepts, Approaches, and Methodologies for Ecological Flood Resilience Assessment: A Review, Fig. 3 River Waal after implementation of "Room for the River Project." (Source: https://www.environmentandsociety.org/)

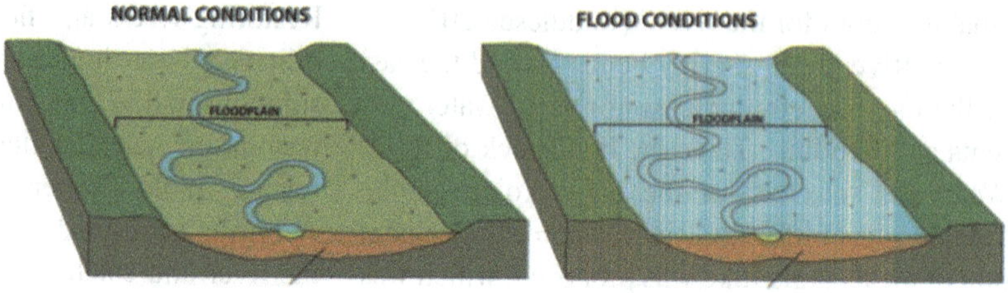

Concepts, Approaches, and Methodologies for Ecological Flood Resilience Assessment: A Review, Fig. 4 Function of a flood plain during an event. (Source: NWRM 2013)

its delayed release, sustaining aquatic habitats, etc. Two significant adaptation measures are applied in this case:

1. Natural water retention (NWR) by river and flood plain restoration
2. Improvement of dikes and dams for better adaptation

Natural Water Retention (NWR) in River Ecosystems

Some of the important water retention measures in hydromorphological systems are wetland and flood plain restoration and management, natural bank stabilization, restoring lakes and streams, increased infiltration, etc. In that flood plain restoration is one of the most important measures in retaining water through natural measures. The function of flood plains during an event is illustrated in Fig. 4.

Assessing Natural Water Retention by Restoration and Rehabilitation

The parameters of assessment are biophysical impacts, ecosystem service benefits, policy objectives, design guidance, cost, governance, and incentives (NWRM 2013). Each of these parameters is subcategorized into various subindicators. Some examples of biophysical impacts are temporary storage and delayed release, reducing runoff, water quality improvement, habitat preservation, and climatic impacts. These subindicators are given a rating based on their comparative effectiveness. In this case, "store runoff" can be given a high rating as flood plains play a major role in the temporary storage of direct runoff, while "increased infiltration" is given a medium rating due to the limited impact of flood plains on infiltration.

Improvement of Dikes and Dams for Better Adaptation

Dikes and dams' improvement included several engineering solutions like increasing height, strengthening, broadening, etc. The indicators for assessment are population affected by flood, agriculture, and industrial production affected by flood, etc. In addition to engineering solutions, a spatial planning approach can be adapted, i.e., creating protected zones (e.g., dike ring areas), by compartmentalizing the region. Some disadvantages of such structural solutions are the increased cost factor, effect on cultural and historic structures in the river basin, need for long-term maintenance, etc. Thus, engineered and structural measures have to be adopted in combination with ecological solutions like flood plain and wetland restoration.

Nature-Based Solutions for Urban Resilience: Case of Sustainable Urban Development in Utrecht, the Netherlands

Leidsche Rijn is an expansion of the Utrecht district in the Netherlands. The participatory master planning of Leidsche Rijn for integrated urban water management and sustainability illustrates one of the most promising examples of urban water resilience. The major objectives of the project are water quality improvement and flood mitigation. A significant part of the master planning project deals with natural water retention through spatial planning. Some of the important measures are creating permeable surfaces, bioswales, retention and infiltration basis, discharge ponds, etc. The project reports retention of 2,200,000 m^3/year of water and shows increased storage of 1000 m^3/ha (nwrm 2015).

An illustration of the various concepts of master planning of Leids che Rijn shows various aspects of integrated water management in Fig. 5.

Parking lots and low traffic roads were provided with permeable pavements to capture runoff from low permeable areas. Bioswales, typically located near roads, replaced the traditional drainage systems with vegetated channels to improve water quality and reduce runoff. An example of the bioswale provided in Leids che Rijn is shown in Fig. 6.

Another design solution provided for increased infiltration is filter strips, which is a linear strip of uniform vegetation for slow conveyance and infiltration. Soak ways or underground chambers proposed in this project are highly effective infiltration devices that can also help in water filtration and groundwater recharge. The principle of a soak way is shown in Fig. 7. Vegetated depressions called detention ponds also allow temporary storage, increased infiltration, and delayed discharge of runoff. Another temporary water storage measure is the retention pond, which consists of a pond area surrounded by an additional vegetated area to capture additional rainwater during rainfall. Thus, several water retention measures are combined in Leids che Rijn for integrated water management.

Assessment of Natural Water Retention Through Spatial Planning

The indicators of assessment are physical infrastructure, application of technologies, networks and organizations, values and principles, policy framework, knowledge systems, funding, and economic systems. One of the most important aspects of Leids che Rijn is the role played by the municipality in negotiating with developers for water management (Hade Dorst and van der Jagt 2021). The existing ecological system of waterways and soil conditions proved favorable to the implementation of nature-based solutions. In addition, understanding and involvement of the

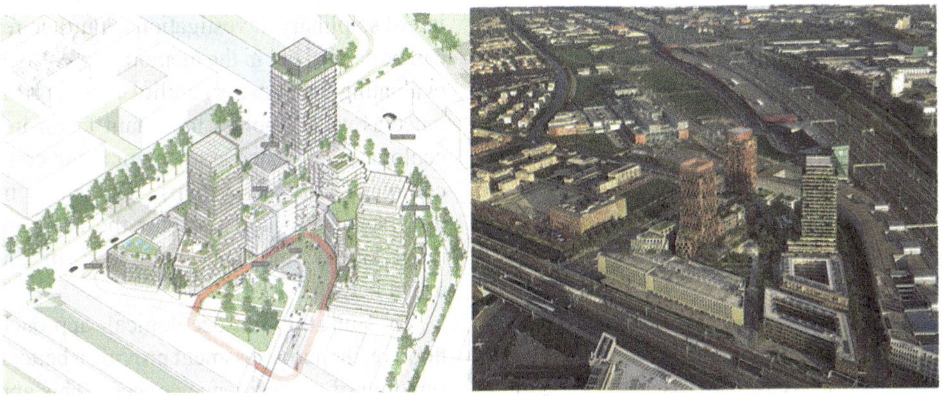

Concepts, Approaches, and Methodologies for Ecological Flood Resilience Assessment: A Review, Fig. 5 Conceptual overview, Leids che Rijn. (Source: mecanoo 2022)

Concepts, Approaches, and Methodologies for Ecological Flood Resilience Assessment: A Review, Fig. 6 Bioswale in Leids che Rijn. (Source: Hade Dorst and van der Jagt 2021)

Concepts, Approaches, and Methodologies for Ecological Flood Resilience Assessment: A Review, Fig. 7 Principle of a soak way. (Source: nwrm 2015)

stakeholders is an important factor in the success of the project. This case study is an example of the dependency of urban resilience projects on social, ecological, technical, and political scenarios.

Conclusion

In this entry, ecological flood resilience is considered as an integrated process of the adaptive cycle, when a system moves between stability cycles under external disturbances. Evaluation of resilience measures should consider this continuing process of exploitation, conservation, release, and reorganization phases. Thus, in addition to adaptive and absorptive capacity, "transformability" is also an indicator of significance. Though many studies undertake resilience assessment and evaluation, conceptual clarity on the adaptive cycle is missing in many of the established research. This entry therefore attempts to address the lack of theoretical cohesion among interdisciplinary investigations. Further research can be carried out in the methodological aspects of evaluating the transformability of ecological resilience as well. Another important factor in socioecological resilience assessment is the concept of complex adaptive systems. The entry tries to understand the indicators for the assessment of interrelations and mutual impacts of complex urban networks.

To sum up, socioecological approaches to flood resilience assessment provide a better understanding of the resilience concept. This approach will allow the research community and policymakers in the decision-making process of equipping cities of tomorrow for climate change adaptation and disaster resilience.

Cross-References

- ▶ Adapting Cities to Climate Change
- ▶ Blue-Green Cities: Achieving Urban Flood Resilience, Water Security, and Biodiversity
- ▶ How Cities Can Be Resilient
- ▶ Urban Climate Resilience

References

Anna Serra-Llobet, E. C. (2016). Governing for integrated water and flood risk management: Comparing top-down and bottom-up approaches in Spain and California. *Water, 8*, 445.

Berkes, F. J. (2003). Navigating social-ecological systems: Building resilience for complexity and change. *Ecology and Society*. https://www.ecologyandsociety.org/vol9/iss1/art1/

Brand, F. S. (2007). Focusing the meaning(s) of resilience: Resilience as a descriptive concept and a boundary object. *Ecology and Society*. https://doi.org/10.1146/annurev-environ-051211-123836.

Brian Walker, C. S. (2004). Resilience, adaptability and transformability in social–ecological systems. *Ecology and Society*. http://www.ecologyandsociety.org/vol9/iss2/art5/

Bruijn, K. M. (2003). Resilience strategies for flood risk management under uncertainities. Retrieved from www.iwra.org. https://iwra.org/member/congress/resource/MADRID2003_KARIN_DE_BRUIJN_EN.pdf

Chan, F. K. (2017). "Sponge City" in China – A breakthrough of planning and flood risk management in the urban context. *Land Use Policy*. https://doi.org/10.3389/fenvs.2021.748231.

Dorst, H., & van der Jagt, A. (2021). Structural conditions for the wider uptake of urban nature-based solutions – A conceptual framework. *Cities*. https://doi.org/10.2166/bgs.2019.199.

EEA, E. E. (2020). *Flood plains- a natural system to preserve and restore*. Luxomberg: European Environment Agency. https://www.eea.europa.eu/publications/floodplains-a-natural-system-to-preserve-and-restore

Folke, C. (2016). Resilience (republished). *Ecology and Society*. https://doi.org/10.5751/ES-09088-210444.

Gunderson, L. H. (2000). Ecological resilience – In theory and application. *Annual Review of Ecological Systems*. https://doi.org/10.1146/annurev.ecolsys.31.1.425.

Hawken, S. (2021). What makes a successful Sponge City project? Expert perceptions of critical factors in integrated urban water management in the Asia-Pacific. *Sustainable Cities and Society*. https://doi.org/10.1016/j.scs.2021.103317.

Heejun Chang, D. J. (2021). Understanding urban flood resilience in the Anthropocene: A social–ecological–technological systems (SETS) learning framework. *Annals of the American Association of Geographers*. https://doi.org/10.1080/24694452.2020.1850230.

Holling, C. S. (1973). Resilience and stability of ecological systems. *Annual Review of Ecology and Systematics*. https://doi.org/10.1146/annurev.es.04.110173.000245.

Holling, C. S., & Folke, C. (2004). Regime shifts, resilience and biodiversity in ecosystem management. *Annual Review of Ecology Evolution and Systematics*. https://doi.org/10.1146/annurev.es.04.110173.000245.

Jacinto, R., & Reis, E. (2020). Indicators for the assessment of social resilience in flood-affected communities – A text mining-based methodology. *Science of the Total Environment*. https://doi.org/10.1016/j.scitotenv.2020.140973.

Kerri McClymont, D. M. (2019). Flood resilience: A systematic review. *Journal of Environmental Planning and Management*. https://doi.org/10.1080/09640568.2019.1641474.

Klijn, V. M. (2001). *Living with floods: Resilience strategies for flood risk management and multiple land use in the lower Rhine River Basin*. Delft: NCR. https://doi.org/10.1080/15715124.2003.9635190.

mecanoo. (2022). www.mecanoo.nl/Projects. Retrieved from www.mecanoo.nl: https://www.mecanoo.nl/Projects/project/245/AIR-Leidsche-Rijn

nwrm. (2015, Jan 12). www.nwrm.eu/casestudy. Retrieved from www.nwrm.eu: http://nwrm.eu/case-study/leidsche-rijn-sustainable-urban-development-netherlands#

NWRM, N. W. (2013). *Flood plain restoration and management*.

Rădulescu, M. A. (2021). Metamorphosis of a waterway: The City of Nijmegen embraces the river Waal. *Environment & Society Portal*. Retrieved from www.environmentandsociety.org

Rezende, O. M., & Miranda, F. M. (2019). A framework to evaluate urban flood resilience of design alternatives for flood defence considering future adverse scenarios. *Water*. https://doi.org/10.3390/w11071485.

Rocha, J., Lanyon, C., & Peterson, G. (2022). Upscaling the resilience assessment through comparative analysis. *Global Environmental Change*. https://doi.org/10.1016/j.gloenvcha.2021.102419.

Steve Carpenter, B. W. (2001). From metaphor to measurement: Resilience of what to what? *Ecosystems*. https://doi.org/10.1007/s10021-001-0045-9.

Tong, P. (2021). Characteristics, dimensions and methods of current assessment for urban resilience to climate-related disasters: A systematic review of the literature. *International Journal of Disaster Risk Reduction*. https://doi.org/10.2478/rtuect-2020-0101.

WHO. (2022, January 29). www.who.int/health-topics/floods. Retrieved from www.who.int: https://www.who.int/health-topics/floods#tab=tab_1

C

Wolfgang zu Castell, H. S. (2020). Computing the adaptive cycle. *Nature*. https://doi.org/10.1038/s41598-020-74888-y.

Yijun Shi, G. Z. (2021). Assessment methods of urban system resilience: From the perspective of complex adaptive system theory. *Cities*. https://doi.org/10.1016/j.cities.2021.103141.

Conceptualizing the Urban Commons

Asma Mehan[1] and Mahziar Mehan[2]
[1]Senior Researcher, CITTA Research Institute, Faculty of Engineering (FEUP), University of Porto, Porto, Portugal
[2]School of Urban Planning, Faculty of Fine Arts, University of Tehran, Tehran, Iran

The concept of the commons was made widely known by the research of economist Elinor Ostrom (1990), allowing the "commoners" of that community the right to sustain themselves by grazing animals and collecting wood and wild food (Bingham-Hall 2016:2). This concept denotes the public land and natural resources – such as water and air – accessible to all members of society for development and survival, around which, historically, commoners organized themselves as self-governing collectives (Brears 2021). Referring to Lessig (2001) and the Oxford English Dictionary (Simpson and Weiner 1989), the commons is any collectively owned resource held in common use or possession to which anyone has access without obtaining permission of anyone else. Urban Commons "suggests a community of commoners that actively utilize and upkeep whatever is being commoned. In the new social definition, the term has taken on through grassroots projects and scholarly rethinking (...) common access has the potential to offer a richer form of interaction with the city than public ownership" (Bingham-hall, p.2). Huron defines urban commons as a way of "experiencing collective work, among strangers, to govern non-commodified resources in spaces saturated with people, conflicting uses, and capitalist investment" (2015: 977). This means that the urban commons conceptualization is a representation of resistance against the capitalist order and spatial commodification. The urban commons exist "as a dynamic and collective resource – a variegated form of social wealth – governed by emergent custom and constant negotiating, rebuffing, and evading the fixity of law" (Gidwani and Baviskar 2011: 42).

Urban Governance: Definition, Introduction, Theoretical Framework

The literature on State Theory and the transformation of state spatiality theory under contemporary capitalism has grown rapidly during the past three decades. In general, power and politics are central to the urban governance literature. The core of these analyses focuses upon the state policy and governance alignment that has occurred in recent decades. During the 1970s, the term "entrepreneurial urban governance" was introduced to promote economic development from below to respond to the contemporary challenges of urban industrial decline, inclusion and integration policies, and globalization (Harvey 1989). As of the early 1980s, during the early stages of the institutionalization of neoliberal ideology intertwined with the fiscal crisis of the Keynesian welfare in the Global North, national states began promoting the economic rejuvenation within localized territorial competitiveness (Brenner and Theodore 2012). Such urban locational policies entailed a fundamental redefinition of the national state's roles as an institutional mediator of uneven geographical development (Amin and Malmberg 1994).

The term "governance" (Rhodes 1996; Stoker 1998) appeared more than 20 years ago as a quite obscure and undefined concept, becoming a "buzz-word" in recent decades, particularly when referring to positive ways of governing. Governance brought a different set of actors to the scenario of social development implementation such as private initiatives, institutions, and people. Thus, society as a whole is engaged in the process. Following this perspective, the government is understood to refer to the formal and institutional methods which operate at the level of the nation-state to maintain public order and facilitate collective action. However, "there is a baseline agreement that governance refers to the development of governing styles in which boundaries between and within public and private sectors have become blurred" (Stoker 1998, p. 15). Therefore, the government is linked with globalization and state territoriality (Brenner 1999). As (Folke et al. 2010) put it, governance refers to "the structures and processes by which people in societies make decisions and share power" (Folke et al. 2010, p. 444), which brings more complexity when it comes to solving societal problems (Munaretto et al. 2014).

Although the term has been applied in a wide range of issues and different contexts (such as urban, environment, political, etc.) not always with a clear-cut definition, there is a consensus regarding its core meaning. However, its consistency fades away when applied in those specific contexts. In general, there are three main assets related to governance: governance as a broader concept than government, governance comprising a set of rules and processes, and finally, governance as an analytical framework (Obeng-Odoom 2012). However, the key insights have been developed based on Henri Lefebvre's "Space and the State," Bob Jessop's "State Theory" (1990), Eric Swyngedouw and Neil Brenner's "New State Spaces" as the major benchmarks of this intellectual journey (Lefebvre 2003; Jessop 2003; Brenner 1999). For Swyngedouw (2005), "Governance is an arrangement of governing beyond-the-state (but often with the explicit inclusion of parts of the state apparatus) organized as [apparently] horizontal associational networks of the private market, civil society (usually NGO) and state actors. In addition, governance often promises greater democracy and grassroots empowerment but also exhibits a series of contradictory tendencies" (Swyngedouw 2005, p. 1992).

In this way, governance brought a different set of actors to the scenario of social development implementation. The process involves collaboration between government institutions at all levels, NGOs, individual and private organizations, and the society as a whole (Florini and Pauli 2018) which brings more complexity when it comes to solving societal problems (Munaretto et al. 2014; Folke et al. 2010). In this way, urban governance needs to enrich the "soft infrastructures" which connects governance activity to its milieu and which relates a fine-grain understanding of the range and complexity of evolutions forming this milieu to a strategic understanding of the dynamics within the broader worlds in which the relations of urban area exist (Cars et al. 2002, p. 225).

In general, urban governance studies aim to unpack the strategies to frame the potential future urbanisms we might produce. As McCann (2017) puts it, "studies of urban governance have addressed the actors and interests that make urban policy decisions to define and enact what it means to be a citizen, and address existential challenges, including environmental crises" (McCann 2017, p. 314). Focusing on Neoliberal governmentality, Davoudi and Madanipour (2015) state that "the tension between the perceived moral and responsible individuals, communities and localities, and their identification as rational economic actors, whose decisions are solely motivated by the cost-benefit analysis of their self-interests, remains high; and finding ways of bridging the two remains a critical challenge" (Davoudi and Madanipour 2015, p. 98).

Concluding Notes: Governing the Urban Commons

Many scholars and activists see strengthening "the urban commons" as a crucial means of achieving more sustainable use of environmental resources and a more equitable future for humans and more than human habitats and settlements (Mehan 2021; Rahdari et al. 2019). Governing commons in an urban context could be about finding the right path to regulate something as dynamic, spontaneous, and agile as a community (URBACT 2021). Identifying commoning as the creation of formal rules and management systems or as social relations and existing informal norms (Bollier 2010:3), governance could be identified as a tool to identify and justify those relations

(Mehan and Mostafavi 2022). Polycentric urban governance involves resource pooling and cooperation between five possible categories of actors – social innovators or the unorganized public, public authorities, businesses, civil society organizations, and knowledge institutions – the so-called "quintuple helix governance" approach (Iaione and Cannavo 2015). These co-governance arrangements have three main aims including the social innovation enhancement in urban welfare provision, prompting collaborative participatory economies as a driver of local sustainable development plans, and promoting the inclusive urban regeneration of run-sown residential neighborhoods. Public authorities play an essential enabling role in creating and sustaining the co-city. The ultimate goal is making a more just and democratic city, consistent with the Lefebvrian approach of the right to the town (Foster and Iaione 2016; Iaione 2017).

Cross-References

▶ Circular Economy and the Water-Food Nexus

References

Amin, A., & Malmberg, A. (1994). Competing structural and institutional influences on the geography of production in Europe. In A. Amin (Ed.), *Post-fordism: A reader.* Oxford: Blackwell Publishers Ltd.

Bingham-Hall, J. (2016). Future of cities: Commoning and collective approaches to urban space. *Future of cities*, UK Government Office for Science.

Bollier, D. (2010). The commons: a neglected sector of wealth-creation. *Heinrich-Boell Stiftung*. North America.

Brears, R. C. (2021). Circular economy and the water-food nexus. In R. Brears (Ed.), *The palgrave encyclopedia of urban and regional futures*. Cham: Palgrave Macmillan. https://doi.org/10.1007/978-3-030-51812-7_98-1.

Brenner, N. (1999). Globalisation as reterritorialisation: The re-scaling of urban governance in the European Union. *Urban Studies, 36*(3), 431–451.

Brenner, N., & Theodore, N. (2012). *Spaces of neoliberalism: Urban restructuring in North America and Western Europe*. Blackwell Publishers Ltd..

Cars, G., Healey, P., Madanipour, A., & De Magalhaes, C. (2002). *Urban governance, institutional capacity and social milieux*. London/New York: Routledge.

Davoudi, S., & Madanipour, A. (2015). Localism and the 'post-social' governmentality. In S. Davoudi & A. Madanipour (Eds.), *Reconsidering localism* (pp. 77–103). London: Taylor & Francis.

Florini, A., & Pauli, M. (2018). Collaborative governance for the sustainable development goals. *Asia & the Pacific Policy Studies, 5*, 583–598.

Folke, C., Carpenter, S. R., Walker, B., Scheffer, M., Chapin, T., & Rockstrom, J. (2010). Resillience thinking: Integrating ressilience, adaptability and transformability. *Ecology and Society, 15*(4), 20.

Foster, S., & Iaione, C. (2016). The City as a Commons. *Yale Law & Policy Review, 34*, 281.

Gidwani, V., & Baviskar, A. (2011). Urban commons. *Economic and Political Weekly, 46*(50), 42–43.

Harvey, D. (1989). From managerialism to entrepreneurialism: the transformation in urban governance in late capitalism. *Geografiska Annaler B, 71*(1), 3–18.

Huron, A. (2015). Working with strangers in saturated space: Reclaiming and maintaining the urban commons. *Antipode, 47(4)*, 963–979.

Iaione, C. (2017). The right to the co-city. *Italian Journal of Public Law, 9, 80.*

Iaione, C., & Cannavo, P. (2015). The Collaborative and polycentric governance of the urban and local commons. *5 Urb. Pamphleteer, 5*, 29.

Jessop, B. (2003). *State theory: Putting the capitalist state in its place*. Oxford: Polity Press.

Lefebvre, H. (2003). Space and the state. In N. Brenner, B. Jessop, M. Jones, & G. Macleod (Eds.), *State/Space: A reader*. Boston: Blackwell Publishing.

Lessig, L. (2001), The Internet under Siege. *Foreign Policy, no. 127*, 56–65.

McCann, E. (2017). Governing urbanism: Urban governance studies 1.0, 2.0 and beyond. *Urban Studies, 54*(2), 312–326.

Mehan, A. (2021). EUKN webinar "Port cities and megatrends: Glocal approaches to sustainable transitions". *The Port City Futures (PCF) Blog*, Leiden. Delft. Erasmus (LDE) Initiative.

Mehan, A., & Mostafavi, S. (2022). Building resilient communities over time. In R. Brears (Ed.), Living Edition ed *The Palgrave Encyclopedia of urban and regional futures* (p. 4). Cham: Palgrave Macmillan. https://doi.org/10.1007/978-3-030-51812-7_322-1.

Munaretto, S., Siciliana, G., & Turwani, M. E. (2014). Integrating adaptive governance and participatory multicriteria methods: A framework for climate adaptation governance. *Ecology and Society, 19*(2), 74.

Obeng-Odoom, F. (2012). On the origin, meaning, and evaluation of urban governance. *Norwegian Journal of Geography, 66*(4), 204–212.

Ostrom, E. (1990). *Governing the commons: The evolution of institutions for collective action*. New York: Cambridge University Press.

Rahdari, A., Mehan, A., & Malekpourasl, B. (2019). Sustainable real estate in the middle east: Challenges and future trends. In *Sustainable Real Estate* (pp. 403–426). Cham: Palgrave Macmillan.

Rhodes, R. (1996). The new governance: Governing without government. *Political Studies*, 652–667.
Simpson, E. S. C., & Weiner, J. A. (Eds.). (1989). *The Oxford Encyclopedic English Dictionary*. Oxford: Clarendon Press.
Stoker, G. (1998). Governance as theory: five propositions. *International Social Science Journal, 50*, 17–28.
Swyngedouw, E. (2005). Governance innovation and the citizen: The Janus Face of governance-beyond-the-state. *Urban Studies, 42*(11), 1991–2006.
URBACT. (2021, September 14). *Governing commons, is it even possible?*. https://urbact.eu/governing-the-commons-is-it-even-possible

Conflict

▶ Responsibility to Prepare and Prevent (R2P2): Applying Unprecedented Foresight to Addressing Unprecedented Climate Risks

Connecting Urban and Regional Innovation Ecosystems to Enhance Competitiveness

Fatime Barbara Hegyi[1] and Age Mariussen[2]
[1]Joint Research Centre – European Commission, Seville, Spain
[2]University of Vaasa, Vaasa, Finland

Synonyms

Competitiveness of cities; Competitiveness of regions; Regional eco-systems of innovation; Territorial competitiveness; Territorial development; Urban eco-systems of innovation

Definition

This paper explores why an increasing number of European regions and cities in close cooperation with quadruple helix partners, such as clusters, firms, universities, nongovernmental organizations, and member states, combine place-based innovation strategies with cross-border networks of innovation. Such collaborations may result in a rich variety of organizational solutions and approaches, which allow actors and stakeholders to overcome different barriers and concerns of innovation. The chapter outlines a conceptual framework of how cooperation between urban and regional innovation ecosystems may strengthen regional place-based development strategies and improve regional innovation capabilities. Key analytical concepts are proximity, knowledge complexity, entrepreneurial discovery processes, and stakeholder analysis.

Introduction

Linking interregional innovation ecosystems may improve regional innovation capabilities and drive institutional change. It may even contribute to entrepreneurial discovery processes (EDP). Combining spatial/geographic proximity inside regions with complementary forms of transnational proximity, such as cognitive, temporal, and organizational proximity, enables transnational synergies across different regions with related knowledge domains. These synergies may create knowledge complexity, new knowledge combinations, which open up for new locus of innovation, where different forms of proximity are combined in different phases of the entrepreneurial discovery process. This process results in new emerging clusters. Micro-level cluster emergence may have the power to remove institutional barriers of innovation and improve place-based innovation capabilities of regions.

Regions and cities with shared research and innovation priorities collaborate to exploit complementing research and innovation capabilities while building up necessary capacities and overcoming interregional fragmentation and lack of critical mass across urban and regional ecosystems of innovation. Furthermore, such collaborations lead to improved business environment by identification of barriers to innovation, new investment, or skills.

Conceptual Framework for Connected Urban and Regional Innovation Ecosystems

The concept of "open regions" refers to proactive policy measures aiming at "redesigning the dialectic interplay between territorial openness and closure" (Schmidt et al. 2018, pp. 187). In parallel, opening up and connecting urban and regional innovation ecosystems has been identified as a challenge as regards smart vertical regional strategies (Mariussen et al. 2016) that shape opportunities for innovation within the sphere of influence of policy makers (Schmidt et al. 2018, pp. 193) (see chapter on ► "Making of Smart and Intelligent Cities").

The motivation to link innovation ecosystems is influenced by structural and institutional factors. As Rutten puts it, geographical distance is more accurately seen as a dynamic trade-off between effort, preference, and dependency (Rutten 2017, pp. 159–177). Such preferences and dependencies facilitate the creation of cross-border/transregional networks resulting in diverse forms of proximity.

Distance and Proximity

While spatial and temporary proximities refer to interactions within a place and to social interactions that connect people from different places, organizational and cognitive proximities are such that result in organized networks connecting hubs of shared knowledge and expertise of different places (Boschma 2005). These proximities can be linked to connecting urban and regional ecosystems of innovation, presented on Table 1.

As presented on Table 1, connecting innovation eco-ecosystems builds on urban and regional development strategies relating to spatial proximity. Such strategies open for the discovery of shared domains that lead to cognitive proximities. Within innovation ecosystems, shared domains are built on complementary competencies, expertise, and skills that require an organizational setup that is based on agreed methods and principles guided by shared vision toward common objectives. Collaboration across borders requires tailor-made governance structures that allow urban and regional spaces to work toward their shared objectives, while temporary proximities are created by regular interaction between stakeholders. Consequently, connecting innovation ecosystems across urban and regional borders requires combinations of different forms of proximity that lead to access – among others – to new knowledge.

New knowledge is to be transformed and translated through diverse processes from search through problem-solving to industrial upscaling. Diverse forms of knowledge, such as tacit, codified, industrial engineering, and science-based knowledge, have to be combined, which is often based on trial and error and on continuous dialogue among actors, who decide to share trust and cooperate for sustained periods of time

Connecting Urban and Regional Innovation Ecosystems to Enhance Competitiveness, Table 1 Distance and proximity: connecting urban and regional ecosystems of innovation by typologies of proximities.

Typology of proximities		Urban and regional ecosystems of innovation
Spatial	Interaction within a place	Urban or regional development strategy; a strategy aiming to develop a space
Temporary	Interaction in conferences, workshops, meetings, or other ways of connecting people from different places	Regular meetings and exchanges between regions, clusters, and other stakeholders
Organizational	Interaction within an organization or an organized network located in several places	Set up governance mechanisms for the partnerships, ensuring regular dialogue
Cognitive	Interaction between specialists, who share the same knowledge	Learn and connect regions within shared domains of innovation with the objective of developing complementary strength and capacities of innovation and realizing joint investment projects

Source: Own adaptation based on Boschma (2005, pp. 61–74)

(Mariussen and Hegyi 2020). While connecting innovation ecosystems, a combination of several sources of knowledge can enhance the innovation capacity of place-based development strategies, leading to a "living knowledge" (see chapter on ► "Local and Regional Development Strategy").

Living knowledge refers to practical knowledge that is shared and communicated resulting in the development of an ecosystem of innovation leading to entrepreneurial discoveries of new (business) opportunities. Complexity theory relates to knowledge, innovation, and biological ecosystems when sharing some of the same properties (Byrne and Callaghan 2014, pp. 17–38). There is a crucial difference between complicated and complex systems.

Complex systems in nature are not designed top down, but they are the result of self-organization by many autonomous interrelated components. Complex system theory emphasizes that sophisticated entrepreneurial ecosystems have *emergent properties* in the sense that they have the capacity to combine different forms of knowledge and create new products, value chains, and clusters. Complex systems, such as entrepreneurial ecosystems, are able to create something new by increasing the system scale. These systems start a process, which may go at different pace, but then as they grow and develop, they are able to attract more and more human or financial resources, making complex systems inherently unstable. At the same time, in order to be able to mobilize more stakeholders, dynamic ecosystems should be open without rigid borders. While openness is an essential feature of complex systems, these systems also tend to be dissipative on how they interact with their environments as they are likely to experience a continual inflow and outflow of resources including information. By opening the borders between the knowledge domains of urban and regional territories, connecting urban and regional ecosystems of innovation increases knowledge complexity, by involving different stakeholders (Mariussen and Hegyi 2020). The advantages of complex knowledge domains, as compared to more simple, noncomplex structures, may be illustrated with the discussion of the advantages of scale (critical mass) and scope in corporate organization. It is well known that large companies with a wide variety of knowledge domains have an ability to diversify and adapt to changes better than small, narrowly specialized companies.

The theory of stakeholder involvement can be applied to connecting innovation ecosystems (Mitchell et al. 1997, pp. 853–886) viewing the role of stakeholders along the following dimensions (Fig. 1).

Stakeholders with power indicate a type of relationship among social actor that influences one's actions, while legitimacy can be understood as a "a generalized perception that the actions of an entity are desirable, proper, or appropriate within the socially constructed system of norms, values, beliefs and definitions" (Mitchell et al. 1997, p. 866). In the case of cross-border innovation actions, urgency can be caused by challenging/shocks of a value chain dynamic. Diverse stakeholders of the same value chain are exploring new common opportunities that impact the dynamics of their relations. Through such exploration, actors grow unique forms of knowledge and create shared domains that are more competitive together. They may be able to grow more power and diversify their markets. These three main dimensions make it possible to define seven types of stakeholders as shown in Fig. 2.

The dormant, discretionary, and demanding stakeholders are latent stakeholders with low salience. Dominant, dangerous, and dependent stakeholders are expectant stakeholders representing two attributes according to the classification and might show a high level of engagement. Definite stakeholders are the ones with all three attributes, representing high salience; therefore, there is an immediate priority of involving them (Mitchell et al. 1997). Definitive stakeholders are the initiators and leaders of the cross-border collaborations of different institutions, such as regional authorities, universities, or clusters. The composition of stakeholders, their engagement, and their agility vary considerably (see chapter on ► "Collaborative Climate Action"). According to Morgan, barriers to interregional collaboration lead to lack of access to knowledge, lack of political support, and/or lack of synergies between policy sectors (Morgan 2018).

Correspondingly, motivations of regions differ depending on their level of

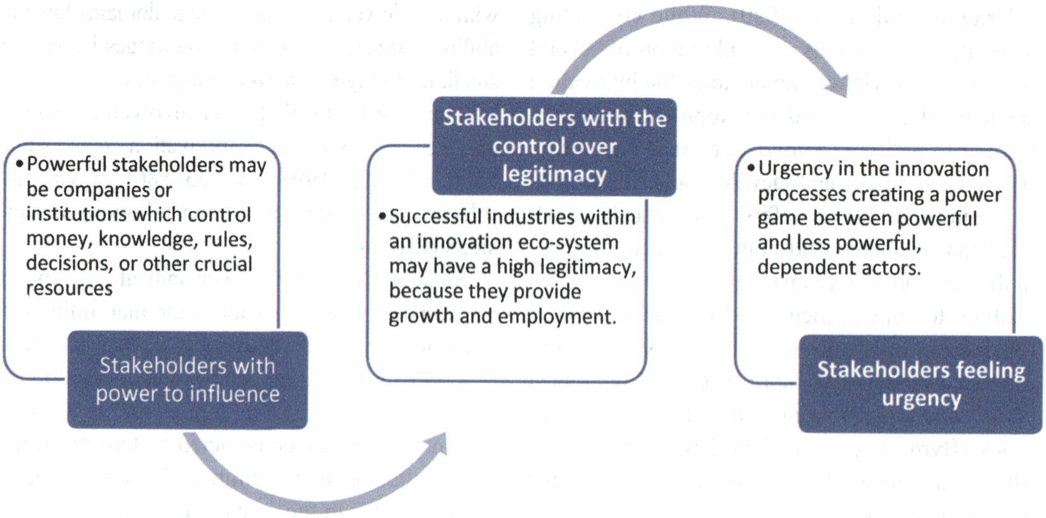

Connecting Urban and Regional Innovation Ecosystems to Enhance Competitiveness, Fig. 1 Distance and proximity: stakeholder analysis in connecting ecosystems of innovation. (Source: Own adaptation based on Mitchell et al. (1997, pp. 853–886))

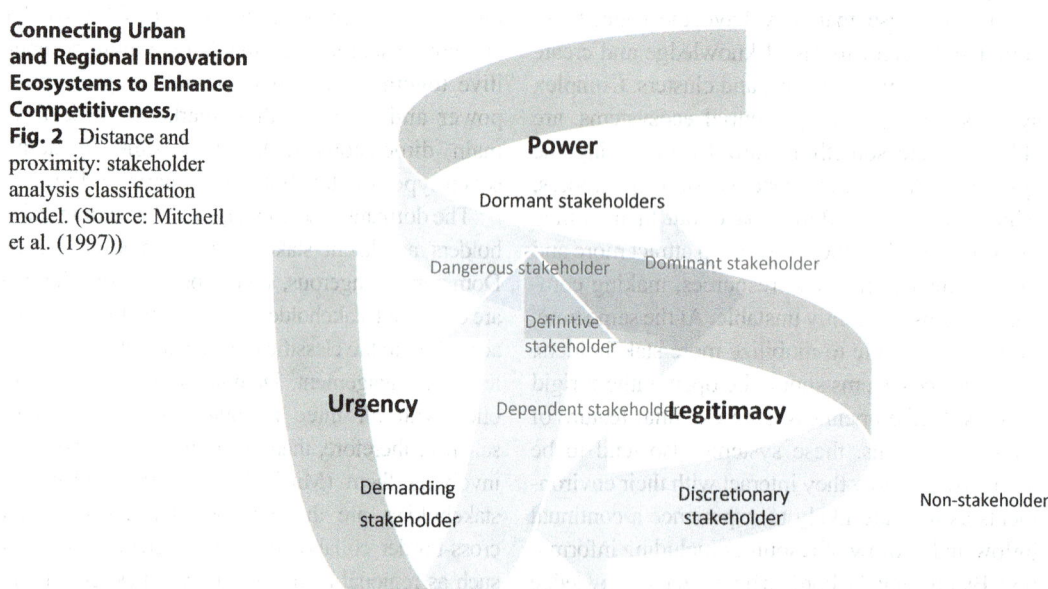

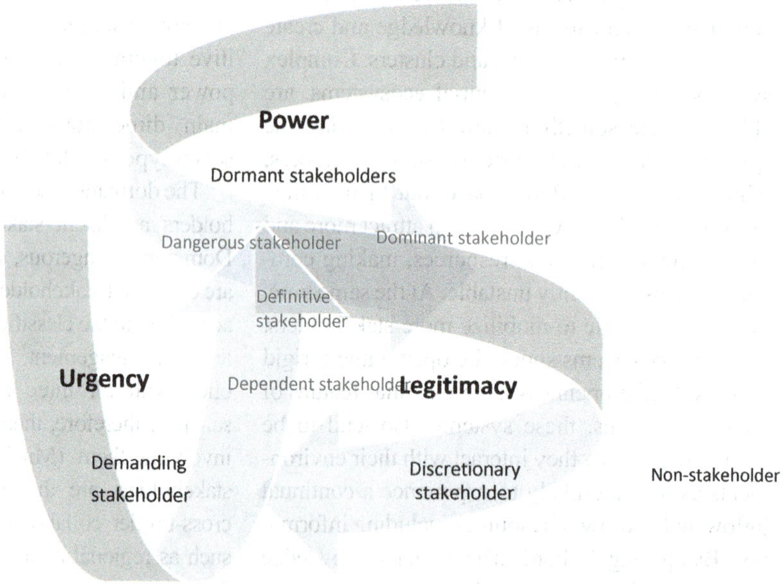

Connecting Urban and Regional Innovation Ecosystems to Enhance Competitiveness, Fig. 2 Distance and proximity: stakeholder analysis classification model. (Source: Mitchell et al. (1997))

innovativeness. For example, moderate innovator regions may be more motivated to actively participate in cross-border collaboration to get access to existing technologies and knowledge, which may be exploited in a shorter-term perspective.

Building new European value chains and clusters and to close the gap between innovation leaders and followers allow urban and regional ecosystems of innovation to become more competitive. When looking at European regions, the competitive forces are the United States and

China, which may or may not be able to get access to European research and take advantage of growth possibilities of industrial upscaling. In this respect, the European Commission is applying a long-term perspective in order to overcome market failures, critical mass, or parallel investments across regional borders, which furthermore enables growth and regional convergence between innovation leaders and lagging regions.

Cluster Emergence Transforming Regions

In a static comparison, it might seem obvious that differences between innovation leader regions and innovation followers are both structural and institutional. If we shift perspective and look at how successful clusters develop, we can see how successful innovation and cluster development co-evolve and drive institutional change. As a successful growth experience, this may help shaping regional institutions.

The cluster life cycle literature explains how clusters are emerging from small micro-level entrepreneurial discoveries and start to grow and get bigger (Menzel and Fornahl 2010; Isaksen 2011; Fornahl and Hassink 2007). At a certain point in the process of growth, the small firm and the embryonic network of a new value chain are to be noticed at the macro-level of the region as a new export sector and as an addition to the regional labor market. This growing cluster may require spatial planning, regulations, and improved educational frameworks (see chapter on ▶ "Metropolitan Discipline: Management and Planning"). The new sector will be copied by followers, whether they will collaborate or compete. An emerging cluster will ask for – and sometimes even get – more innovation-friendly regional institution, which signals the co-evolution between economic change and institutional transformation (Virkkala and Mariussen 2019).

On a longer term, something deeper might happen inside the region. The new institutional arrangements initiated by the cluster may become generalized at regional level. The experience of new path creation may be repeated, which is likely to provide a new framework for other stakeholders with novel ideas. What started as a movement from the bottom and up and has created a more innovative region becomes a process, which goes from the top to the bottom. The stimulation of further innovation creates a self-reinforcing process of co-evolution of economic growth of the new clusters driving institutional modernization and transformation, which in turn creates new clusters in the region.

Combining Industrial and Science-Based Innovation to Create New Paths of Development

Less innovative regions in Europe are often squeezed by markets pushing for lower costs and higher productivity, combined with a weak regional and/or national support from science. These regions face competition from low-cost producers in Asia, in Latin America, and increasingly also in Africa. This lock-in effect significantly restricts growth. The impact of such lock-in is illustrated in Fig. 3, which shows GDP per capita in OECD countries, seen in relation to private industrial investments in R&D per capita.

As Fig. 3 suggests, in European countries, lower level of private investments on science-based innovation is associated with the lower levels of GDP/capita. The upshot is that firms, which are not investing in R&D, tend to focus on low-cost competition and the parts of value chains, where value creation is somewhat lower. Some countries are richer than they "should have been," given their level of R&D investments. On the other hand, countries with income from natural resource extraction, like Norway, have a higher GDP/capita than we should expect, given the private investments in R&D. Some countries with relatively low GDP/capita, such as Israel and South Korea, have high private investments in R&D. South Korea has a strong domestic industry with large corporate actors that invest heavily in R&D, while Israel has a national labor market, which is attractive to large ICT companies from the United States. Consequently, a combination of industrial innovation and science-based innovation in regions and countries provides a mix that

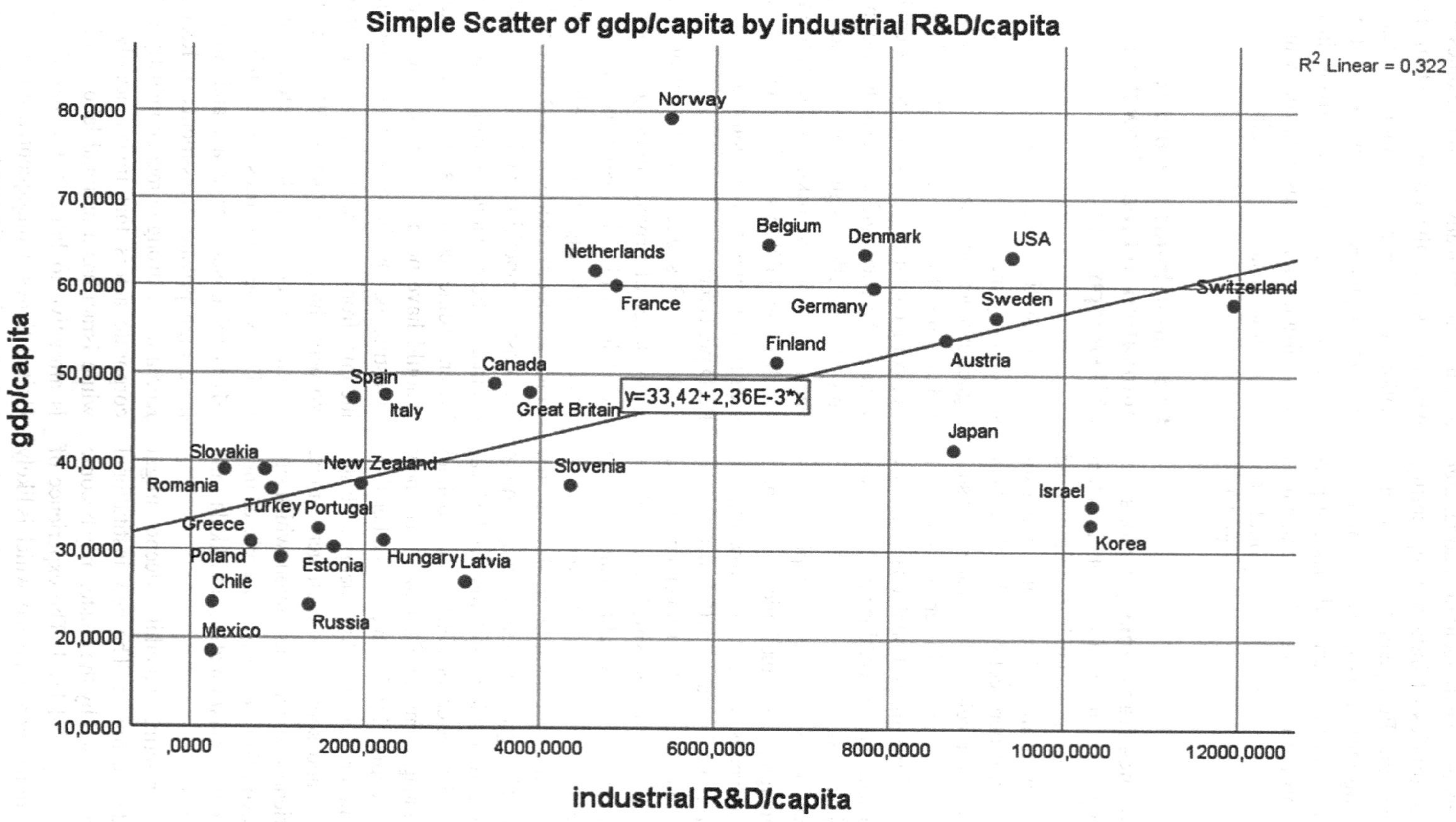

Connecting Urban and Regional Innovation Ecosystems to Enhance Competitiveness, Fig. 3 Combining industrial and science-based innovation to create new paths of development: GDP/capita and industrial investments in R&D/capita in OECD countries. (Source: Own compilation, based on OECD data)

is promoting innovation, industrialization, and creation of new paths of development, leading to competitiveness and economic growth. The innovation ecologies of these regions reach a higher level of complexity with a mix of science-based knowledge and industrial skills and knowledge.

This combination of science-based knowledge and industrial knowledge depends on the degree of spatial proximity that indicates a place-based dynamic within a region combined with "global pipelines" providing knowledge from abroad. The absorption and application of science-based knowledge are rarely straightforward. Examples of how this is achieved through proactive policy interventions can be found in China. The Chinese strategy of enabling growth in additive manufacturing value chains (Xu et al. 2018) provides a definition of complex innovation ecosystems consisting of *three sub-systems*:

- Business ecosystems, making up value networks
- Science ecosystems consisting of universities and research institutions
- Technological ecosystems, which include firms, government agencies, industrial players, universities, and research institutions

This poses the question of whether economic complexity might be enough and if European regions really need science and technology to create growth.

The complexity model of Hausmann also shows expected annual growth, which relies on complexity inside the economy (Hausmann et al. 2013; The Growth Lab at Harvard University 2019). The winners with fairly high economic complexity have positions in the global economy having had moved up on the value chain. Moving away from complete dependence on raw material extraction, there are still possibilities for improvement on the positioning of the value chain, where innovation through related varieties without R&D inputs is efficient (Nguyen and Mariussen 2019). Table 2 shows countries with high growth expectancies based on economic complexity worldwide and in Europe.

Connecting Urban and Regional Innovation Ecosystems to Enhance Competitiveness, Table 2 Combining industrial and science-based innovation to create new paths of development: estimated annual growth among economic complexity winners

Worldwide		Europe	
India	+7.89	Ukraine	+5.05
Uganda	+7.48	Serbia	+4.88

Source: Atlas of Economic Complexity (2018)

Connecting Urban and Regional Innovation Ecosystems to Enhance Competitiveness, Table 3 Combining industrial and science-based innovation to create new paths of development: estimated annual growth among economic complexity losers

Worldwide		Europe	
Qatar	+1.55	Germany	+2.38
Venezuela	+1.88	Norway	+2.54

Source: Atlas of Economic Complexity (2018)

Based on the same method, the European countries with the highest growth expectation are Ukraine and Serbia, which have fairly complex industries that can easily be diversified into production of products with low costs, which are not new for the world, but new for them. Some losers have a low economic complexity, because they are locked into resource dependencies (Qatar, Venezuela, or Norway). Other countries have exhausted their potential for growth through related varieties based on economic complexity (Germany). Table 3 shows the annual growth of countries with low growth expectations.

The implication of this is that countries with a high level of economic complexity should realize that their potential for growth based on relatedness is restricted (Xiao et al. 2018), also confirmed by other methods:

> We find that relatedness is a more important driver of diversification in regions with a weaker innovation capacity. The effect of relatedness appears to decrease monotonically as the innovation capacity of a regional economy increases. This is consistent with the argument that high innovation capacity allows an economy to break from its past and to develop, for the economy, truly new industry specializations. We infer from this that innovation

capacity is a critical factor for economic resilience and diversification. (Jing et al. 2016).

Some European countries have fairly high levels of economic complexity but low levels of firm investments in research driven innovations, which shows a challenging market failure. In the race to innovate without R&D investments, they are competing on costs with several successful countries in Africa and Asia. Some of these challenges are related to institutional and system failures.

Institutional Challenges and Their Solutions

As extensively discussed by Foray, there are market failures resulting in firms investing in new business areas related to their existing strengths, which in addition are often copied or imitated by followers (Foray 2015). In many European regions, dominating firms may be sleeping giants. These firms might have foreign owners, and they mainly compete by increasing productivity and decreasing their operational costs. In such situations, these firms might have no incentives or only weak or short-term incentives to develop new products or explore new business opportunities, effectively resulting in a significant underinvestment in innovation. To achieve change, it would require a proactive regional governance strategy identifying new and challenging priorities and directions (Mäenpää and Lundström 2019).

Regional planning is generally expected to satisfy many different considerations for diverse consumer groups; furthermore, regional planning needs to respond grand challenges that require the "reinvention of established economic, social and political conventions" (Hegyi 2020). This move signals a shift from a supply-driven to a demand-driven approach in planning and implementation, also serving to set – besides others – economic, sustainability, and environmental goals that incentivize innovation activities (and mindset) of firms (Hegyi 2007). Along the instructional factors, the impact of leadership has been recognized as an important factor in achieving growth and development at regional or local levels (OECD 2010) and effective leadership contributing to the success of places (Beer and Clower 2014; Hegyi 2020). As regards industrial development, there are different sectors with different interests and potentially powerful stakeholder organizations protecting vested interests, presenting a demanding problem to public sector planners. According to Mäenpää and Lundström, regional planning may reach a state of paralysis called wicked problems. Wicked problems are problems with no good solutions (Mäenpää and Lundström 2019).

In several regions, it proved hard to make what was assumed to be potential enabler in a regional triple helix model. According to Blažek and Morgan, "cooperation among business, researchers and the public sector have to start literally from scratch, underlining the fact that the triple helix model of regional development is a triumph of rhetoric over reality in the vast majority of less developed regions" (Blažek and Morgan 2018). A core problem seems to be the gap between universities, regional planners, and the industry (Mieszkowski et al. 2015). Research is carried out at universities and institutes, not inside firms, resulting in a shift of focus on copying best practices (Muscio et al. 2015). Therefore, real synergies need to be achieved between existing industrial skills and capacities found in industrial ecosystems and science-based knowledge. These are coming from across borders that may start with dialogues and developments of common platforms of shared knowledge resources, expertise, and methods and continue through learning and through joint investments of diverse resources in pilots.

A pilot is an experimental approach to create a new economic activity, which may require combinations of different forms of knowledge. Some of this knowledge may be available locally, such as industrial skills, and some may come from outside, such as industrial applications of scientific knowledge. Building on place-based development and opening up to transnational learning, innovation ecosystems may become more dynamic and may be able to move in new directions. If a region wants to build a complex open knowledge space, where actors from different places, sectors, and helices cooperate, it is crucial to follow and nurture some of the best ideas, innovations, and projects in the direction of success. Successes should be made visible and used as building blocks for new and even more advanced projects, which could become the foundation of new and institutionalized ways of cooperation that generate self-reinforcing loops. Thus, it is important to look for preconditions for initiating a process of exploration that leads to discovery that may result in growth. An important starting point of such a process is a network with many potential partners. The process of exploration needs linking to other innovation ecosystems, networks, which may result in successful pilots. To explore, to discover, and to initiate pilots, scale and complexity are important. Scale means to increase the number of potential relations and innovation ideas, which can be used to discover and initiate pilots. Accordingly, an important precondition for successful processes of exploration is the formation of complex knowledge spaces. A knowledge space is defined by a context, where knowledge is shared, which may be an innovation platform that defines concepts of a technological paradigm or a combination of scientific disciplines or skills in understanding the dynamics of markets. Knowledge spaces may be separated, like in epistemic communities and communities of expertise protecting their skills from others, or they may be overlapping, as cross-sector and cross-disciplinary forms of knowledge.

Conclusion

This publication looks at how connecting urban and regional innovation ecosystems strengthens place-based development and improves regional innovation capabilities by combining spatial proximity within regions with complementary forms of transnational proximity, including cognitive, temporal, and organizational. Doing so, regional innovation ecosystems allow cross-border synergies with shared domains of knowledge and expertise. The process leads to a strengthened and more dynamic knowledge base of the regional innovation ecosystem, leading to new competitive advantages within regions and to improved positioning of regional actors in global value chains.

Accordingly, through aligning innovation agendas across regions and borders, cities and regions can combine complementary strengths in research and innovation, can exploit research and innovation competencies, and may acquire necessary research capacities while overcoming lack of critical mass and fragmentation. Furthermore, learning via the institutionalized network of knowledge and expertise regions overcomes challenges of transnational collaboration. Innovative process has been showing a shift from in-house policy development to networked learning efforts involving peers along structured frameworks (Hegyi and Rakhmatullin 2020). Peer learning can boost advancement, which then contributes to enhanced ecosystem dynamics at regional and urban levels. Supported by literature on network analysis, networks provide access to information, resources, and markets that offer gains in terms of learning, effectiveness, innovation, legitimacy, or internationalization (Human and Provan 2000; Provan and Sydow 2008; Porter and Powell 2006).

Stakeholder analysis helps to understand the motivating factors that diverse stakeholders bring to the table and how synergies may be found through multi-level governance strategies. Likewise, it shows how mobilizing stakeholders with different perspectives and timescales can be enhanced in promoting European innovation ecosystems and value chains.

Cross-References

▶ Collaborative Climate Action
▶ Local and Regional Development Strategy
▶ Making of Smart and Intelligent Cities
▶ Metropolitan Discipline: Management and Planning

References

Atlas of Economic Complexity. (2018). Harvard University, Press Release, May 3.

Beer, A., & Clower, T. (2014). Mobilising leadership in cities and regions. *Regional Studies, Regional Science, 1*(1), 10–34.

Blažek, J., & Morgan, K. (2018). The institutional worlds of entrepreneurial discovery: Finding a place for less developed regions. In Å. Mariussen, S. Virkkala, H. Finne, & T. M. Aasen (Eds.), *The entrepreneurial discovery process and regional development: New knowledge emergence, conversion and exploitation*. Abingdon: Routledge.

Boschma, R. (2005). Proximity and innovation: A critical assessment. *Regional Studies, 39*(1), 61–74. https://doi.org/10.1080/0034340052000320887.

Byrne, D., & Callaghan, G. (2014). *Complexity theory and the social sciences: The state of the art*. Abingdon: Routledge.

Foray, D. (2015). *Smart specialisation. Opportunities and challenges for regional innovation policy*. London: Routledge.

Fornahl, D., & Hassink, R. (Eds.). (2007). *The life cycle of clusters: A policy perspective*. Cheltenham: Edward Elgar.

Hausmann, R., Hidalgo, C., Bustos, S., Coscia, M., Chung, S., Jimenez, J., Simoes, A., & Yildirim, M. (2013). *The Atlas of economic complexity*. Cambridge, MA: MIT Press.

Hegyi, F. B. (2007). *Market oriented city development model*. Gyor: Szechenyi Istvan University.

Hegyi, F. B. (2020). Leadership to address urban environmental challenges. *Mark and Focus Magazine*, 2(3).

Hegyi, F. B., & Rakhmatullin, R. (2020). *Evaluation framework integrating results of thematic Smart Specialisation approach*. Seville: European Commission.

Human, S. E., & Provan, K. G. (2000). Legitimacy building in the evolution of small-firm multilateral networks: A comparative study of success and demise. *Administrative Science Quarterly, 45*(2), 327–365.

Isaksen, A. (2011). Cluster evolution. In P. Cooke, B. Asheim, R. Boschma, R. Martin, D. Schwartz, & F. Tödtling (Eds.), *Handbook of regional innovation and growth*. Cheltenham: Edward Elgar.

Jing X., Boschma, R., & Andersson, M. (2016). Industrial diversification in Europe: The differentiated role of relatedness. *Evolutionary Economic Geography* (PEEG) 1627, Utrecht University, Department of Human Geography and Spatial Planning, Group Economic Geography.

Mäenpää, A., & Lundström, N. (2019). Entrepreneurial discovery processes through a wicked game approach: Civil society engagement as a possibility for exploration. In Å. Mariussen, S. Virkkala, H. Finne, & T. M. Aasen (Eds.), *The entrepreneurial discovery process and regional development. New knowledge emergence, conversion and exploitation*. Abingdon: Routledge.

Mariussen, A., & Hegyi, F. B. (2020). *Creating growth by connecting smart place-based development strategies*. EUR 30417 EN, Publications Office of the European Union, Luxembourg, ISBN 978-92-76-23862-1, https://doi.org/10.2760/013479.

Mariussen, Å., Rakhmatullin, R., & Stanionyte, L. (2016). *Smart Specialisation: Creating Growth through Transnational cooperation and Value Chains. Thematic Work on the Understanding of Transnational cooperation and Value Chains in the context of Smart Specialisation*. Luxembourg: Publications Office of the European Union.

Menzel, M. P., & Fornahl, D. (2010). Cluster life cycles – Dimensions and rationales of cluster evolution. *Industrial and Corporate Change, 19*, 205–238.

Mitchell, R. K., Agle, B. R., & Wood, D. J. (1997). Toward a theory of stakeholder identification and salience: Defining the principle of who and what really counts. *Academy of Management Review, 22*(4), 853–886.

Mieszkowski, K., Kardas, M. (2015). Facilitating an entrepreneurial discovery process for smart specialisation. The case of Poland. *Journal of the knowledge economy, 6*(2), 357–384

Morgan, K. (Presenter) (2018, November 27). Steering Committee meeting, Opening Session, Thematic S3 Platform for Industrial Modernisation. Available at http://s3platform.jrc.ec.europa.eu/-/s3-thematic-platforms-conference-and-meetings-bilbao-27-28-november-2018

Muscio, A., Reid, A., & Rivera, L. (2015). An empirical test of the regional innovation paradox: Can smart specialisation overcome the paradox in Central and Eastern Europe? *Journal of Economic Policy Reform, 18*(2), 153–171.

Nguyen, N., & Mariussen, Å. (2019). Moving beyond related variety, creating firm level ambidexterity for economic growth via entrepreneurial discovery process. In Å. Mariussen, S. Virkkala, H. Finne, & T. M. Aasen (Eds.), *The entrepreneurial discovery process and regional development. New knowledge emergence, conversion and exploitation*. Abingdon: Routledge.

Organisation for Economic Co-operation and Development. (2010). *Regions matter*. Paris: OECD.

Organisation for Economic Co-operation and Development. Data from https://stats.oecd.org/

Porter, K. A., & Powell, W. W. (2006). Networks and organisations. In S. Clegg, C. Hardy, T. B. Lawrence, & W. R. Nord (Eds.), *The SAGE handbook of organisational studies* (pp. 776–799). London: Sage.

Provan, K. G., & Sydow, J. (2008). Evaluating Interorganizational relationships. In S. Copper, M. Ebers, C. Huxham, & P. S. Ring (Eds.), *Handbook of Interorganizational relations*. Oxford: Oxford University Press.

Rutten, R. (2017). Beyond proximities: The socio-spatial dynamics of knowledge creation. *Progress in Human Geography, 41*(2), 159–177.
Schmidt, S., Ibert, O., & Brinks, V. (2018). Open region: Creating and exploiting opportunities for innovation at the regional scale. *European Urban and Regional Studies, 25*(2), 187–205.
The Growth Lab at Harvard University. (2019). Growth Projections and Complexity Rankings, V2 [Data set]. https://doi.org/10.7910/dvn/xtaqmc
Virkkala, S., & Mariussen, Å. (2019). Emergence of new business areas in regional economies through entrepreneurial discovery processes. In Å. Mariussen, S. Virkkala, H. Finne, & T. M. Aasen (Eds.), *The entrepreneurial discovery process and regional development. New knowledge emergence, conversion and exploitation*. Abingdon: Routledge.
Xiao, J., Ron B. & Martin, A. (2018). *Industrial diversification in europe: the differentiated role of relatedness, economic geography, 94*(5), 514–549. https://doi.org/10.1080/00130095.2018.1444989
Xu, G., Wu, Y., & Minshall, T. (2018). Exploring innovation ecosystems across science, technology, and business: A case of 3D printing in China Technological Forecasting and Social Change. *Technological Forecasting and Social Change, 136*, 208–221. https://doi.org/10.1016/j.techfore.2017.06.030

Connectivity

▶ Transport Resilience in Urban Regions

Conservation

▶ Butterfly Gardening in Colombo, Sri Lanka: Approach to Biodiversity Conservation, Monitoring, Education, and Awareness in Urbanizing Habitats

Constructed Urban Wetlands

▶ Artificial Urban Wetlands

Consultation

▶ Community Engagement for Urban and Regional Futures

Contribute

▶ Closing the Loop on Local Food Access Through Disaster Management

Coordinate

▶ Closing the Loop on Local Food Access Through Disaster Management

Coordinated Adaptation

▶ Collectively Adapting to Sea-Level Rise Through Disaster Response, Commons Management, and Social Mobilization

Countryside

▶ Challenges of Delivering Regional and Remote Human Services and Supports

County

▶ Closing the Loop on Local Food Access Through Disaster Management

Coupled Green and Grey Infrastructure Systems

▶ Integrated Urban Green and Grey Infrastructure

Crops and Panels: A Farm Model with Trade-offs in the Water-Energy-Food Nexus

Wim van Veen[1], Rami Sabella[2], Lia van Wesenbeeck[1], Amani Alfarra[3] and Ben Sonneveld[4]
[1]Vrije Universiteit, Amsterdam Centre for World Food Studies, Amsterdam, The Netherlands
[2]United Nations Economic and Social Commission for Western Asia, Beirut, Lebanon
[3]Land and Water Division, Food and Agriculture Organization of the United Nations, Rome, Italy
[4]Amsterdam Centre for World Food Studies/ Athena Institute, Vrije Universiteit, Amsterdam, The Netherlands

Abbreviations

FAO	Food and Agriculture Organization of the United Nations
JRB	Jordan River Basin
GAMS	General Algebraic Model Solver
MENA	Middle East and North Africa
SPIS	Solar Powered Irrigated System
WEF	Water-Energy-Food

Definitions

Annual crops	Crops that occupy the land for less than one year (as opposed to tree crops)
Crop yield	Output of the main crop product in, say, ton per hectare
Endogenous variable	Variable that can be adjusted by the farmer in the farm model in order to maximize net revenue
Exogenous variable	Variable that cannot be adjusted by the farmer in the farm model (such as market prices or the volume of available resources)
Farm model	Stylized mathematical description of farmer decision-making, in particular regarding crop area allocation and optimal use of labor, water, fertilizer, and other commodity inputs
Farmgate price	Price received by the farmer when selling a crop
First-order conditions	Mathematical conditions that should be satisfied by the optimal solution of the farm model
Irrigated cultivation	Crop cultivation in which water comes largely from man-made intervention (irrigation infrastructure) and only partly from direct rainfall
Market price	Price paid by the farmer for input commodities (such as fertilizer, pesticides, or packaging material)
Model constraint	Limitation (such as resource availability or technical input-output relation) that should be respected in the optimal solution of the farm model
Model objective	Target variable (e.g., net revenue of the farmer) that should be optimized in the farm model
Model solver	Algorithm in modeling software that calculates the solution of a mathematical optimization model
Net revenue	Value of crop sales minus the market costs of purchased inputs (equals gross value added, in terms of National Accounting)
Shadow price	Mathematical concept that indicates the increase in net optimal revenue if one additional unit of a resource would be available
Technical coefficients	Parameters in the farm model that describe the relation between crop output and required inputs (land, labor, water, fertilizer, other commodities)

Rainfed cultivation	Crop cultivation in which all water comes from direct rainfall
Resource	Total volume of a production factor (such as land, labor, or water) that is available to the farmer
Water-Energy-Food nexus	Interlinkages between water, , energy and food supply

Introduction

Ever since the introduction of the Water-Energy-Food Nexus in 2011 by the World Economic Forum Water Initiative (Waughray, 2011), interlinkages between water, energy, and food supply have gained increasing attention in scholarly articles. This growth in attention was strongly fueled by the lively policy debates in this "age of sustainable development" (Sachs, 2015), debates in which water, energy, and food are crucial elements. By 2050, the world will have to feed another 2 billion people, with on average higher incomes and more diversified food preferences. As such, this is already a challenging task, but it has to be realized in a period of global energy transition away from fossil fuels while, furthermore, climate change is leading to shifts in rainfall patterns and temperature throughout the world. Therefore, the research focus on the interlinkages between water, energy, and food is fully justified. However, the outcomes of the Water-Energy-Food Nexus research are, to our knowledge, not concrete enough to feed into in actual policy making.

A clear example of a specific policy area is crop farming. Climate change and the energy transition have a huge impact on crop farming, as climate change will lead to higher temperatures and affect rainfall patterns. Reduced availability of fossil fuels is likely to lead to higher energy prices for cooling, heating, and machinery use, to increased fertilizer prices, and to higher irrigation costs (pumping, desalination). Furthermore, it will affect transportation costs, possibly with a negative effect on farm gate prices if farmers' bargaining power is weak compared to that of traders and transporters. On the other hand, the energy transition provides also new options to farmers to generate own electricity (solar panels, wind mills) and even to sell surplus electricity to the grid, while it may also improve the prospects of growing biofuel crops.

Assessment of these new threats and options is important not only for individual farmers but also for regional policy makers. Quantitative insight in changing farming conditions and in the reactions of farmers is important, as it highlights potential infrastructural bottlenecks and/or worsening socioeconomic conditions.

To this end, we present in this entry the structure of a farm optimization model as a tool of analysis for assessing the reaction of a "representative" farmer in a specific region to changing conditions of water availability, energy supply, and food markets. The model can be used to predict the impact of different scenarios about future climate change and the speed of the energy transition, as input for policy discussions. The presentation of the model focuses on the Middle East (more specifically, the Jordan River Basin), but the approach is applicable also to other regions, with their own characteristics.

The Water-Energy-Food Nexus in the Middle East

With the prevailing water scarcity as a daily concern and rainfall patterns becoming more erratic under climate change, the sustainability of food production in the Middle East is balancing at the verge of despair. Middle East countries face a raft of challenges to meet the water demand of the agricultural sector and are confronted with the uncertainties of fuel costs for various components of the food production process. On the one hand, irrigation and cooling require substantial amounts of energy while, on the other hand, conventional energy generation consumes large quantities of already scarce water, reducing further the availability for food production and other uses (households, municipality, industry, and tourism). The tension is likely to mount in the near future under persistent demographic pressure, with a growing and more affluent urban population that demands not only its water rights but also food security at reasonable prices. Allocation of transboundary water resources from water-rich countries like Lebanon and Turkey is currently no option as it would aggravate already serious tensions. Hence, solutions should be found within the countries themselves, in particular by focusing on high-intensity farming systems with increased water and energy efficiency and, particularly, by using renewable energy sources. The challenge is to meet, simultaneously, multiple potentially conflicting objectives of the involved stakeholders without compromising the natural resource base.

When the World Economic Forum introduced the Water-Energy-Food Nexus (Initially called the Water-Energy-Food-Climate Nexus.) in 2011, it proposed an integrated platform to manage interlinked resources to enhance water, energy, and food security and reduce greenhouse gas emissions (Waughray, 2011). The concept (commonly abbreviated as the WEF nexus) should increase water and energy efficiency, reduce negative trade-offs, build synergies, and improve governance across water, energy, and food sectors while mitigating climate change effects. For the Middle East, it could provide opportunities for reducing the tensions due to transboundary water conflicts.

Yet, reality is hard to change. The assessment in (Zolfaghari & Farzaneh, 2020) is that the WEF concept does not give meaningful guidelines for water, energy, and food policies for the Middle East and North Africa (MENA) region. According to (Al-Zubari, 2021), the deficiency of widespread WEF implementation is due to the fragmented policies in most Arab countries that impair an effective coordination and implementation of the concept. Whatever the reason, few if any countries in the MENA region have made progress in implementing a WEF approach (Hoff et al., 2019). Hence, despite the fact that water issues in the Middle East region are intrinsically linked to other development challenges, the decisions on energy, land management, and water resources are commonly still taken in clearly demarcated silos within governments. Needless to say that these shortcomings in inter-sectorial coordination also have transboundary effects that easily increase frictions between riparian countries, especially given the unprecedented water stress in the region (UNICEF, 2021).

Implementation of the Nexus Concept

There is also more fundamental criticism on the WEF concept. Some articles, e.g. (Albrecht et al., 2018), refer to the relatively immature character of the WEF nexus that, as yet, does not have accepted definitions, methods, and frameworks. Other articles express their doubts on the novelty of the concept (Benson et al., 2015), or argue that interlinkages between water, energy, and food are difficult to identify as each region has different needs (Al-Muqdadi et al., 2021). The critical review of (Purwanto et al., 2021) finds that published WEF frameworks lack guidelines for resource management and governance and are bound by gaps and omissions which impede a practical implementation. Due to these deficiencies, even the increasing attention for studies claiming to implement a nexus approach is questioned (Wichelns, 2017).

Given the foregoing, the limited success of implementing a WEF approach in the MENA region is not a surprise. Even well-intended policy briefs such as (PSI and AUB, 2017) to use the WEF-nexus for enhancing regional cooperation seem too optimistic unless a more practical interpretation is given to the concept.

However, empirical research on elements of the WEF nexus is available that points at bottlenecks. For the MENA region, there is only a limited dependence of energy systems on fresh water, but water abstraction is highly dependent on energy (Siddiqi & Anadon, 2011). For example, in Saudi Arabia 9% of total annual electrical energy is used for groundwater pumping and desalination while Arabian Gulf countries use 5–12% of total electricity for desalination. This finding emphasizes the importance of further development of renewable energy sources, which will release water for other purposes than power generation (Ferroukhi et al., 2015). WEF scenario analyses for the MENA region indeed show that novel techniques can reduce water scarcity while transitioning to renewable energies, thereby also significantly reducing greenhouse gas emissions (Borgomeo et al., 2018).

At a more practical level, the WEF concept is implemented by solar-powered irrigation systems (SPIS). However, at present most countries lack the level of policy and institutional coordination required to sustainably scale up solar irrigation. Therefore, based on concrete observations for several cases, a framework has been proposed to support policy, regulation, and monitoring for environmentally sustainable and socioeconomically inclusive solar irrigation investments (Lefore et al., 2021). Indeed, designing such practical guidelines for low-cost, off-grid drip irrigation systems could significantly contribute to adoption of SPIS in the MENA region (Grant et al., 2020). Yet, investments in SPIS have to be based on a full analysis of costs and benefits, as shown by a case study in Tunisia (Keskes et al., 2019) where off-grid pumping only became beneficial when lower government subsidies increased the price of diesel.

Summarizing, in order to arrive at concrete policy contributions based on the WEF concept, the task ahead is twofold. Institutional improvements are necessary to facilitate and guide local initiatives on the ground, while policy design must be supported via quantitative analysis of WEF-inspired policy options. To this end, nexus-specific models are necessary to elucidate the underlying mechanisms (Zhang et al., 2018), preferably based on integrated software tools for systematic analysis of the nexus (Liu et al., 2017).

Farm Modeling As a Contribution to Analyze the Nexus

In response to the observed lack of concrete tools for turning the WEF concept into concrete policy options, this entry presents a formal mathematical model that incorporates all water, energy, and food considerations relevant for the planting decisions of crop farmers, including the impact of changes in water availability and quality. In addition, the model explores the possible profitability of installing solar panels on farm land in the Jordan River Basin (JRB). Although the outcomes presented here are specific for the JRB, the model structure is generic, with the aim to contribute to the practical development of the WEF-nexus concept by presenting a concrete empirical case in a fully formalized format.

From a purely technological point of view, farmers in the JRB have the possibility to install solar panels at affordable prices, while the irradiation maps of the JRB show a great potential to produce a high volume of solar energy in a very efficient way. Yet, so far, empirical research shows negligible initiatives by farmers to do so. The model at hand explicitly takes incentives and motivation of farmers to engage in solar energy production as point of departure, in its relation with food production and water use. The research addresses the following questions: Relative to growing crops, is producing solar energy profitable? If yes, to which extent will farmers produce it? If farmers invest in solar energy production, how does this affect crop area? What happens to farm value-added? What is the effect on employment? What happens to the output of the different crops when water quality and availability change? Will reduction in water availability under climate change provide a boost to solar energy production?

Theoretically, the model stands in the tradition of profit maximization under technological and resource constraints, the standard approach in microeconomics (Varian, 2020). This approach is a proven method with the possibility to incorporate knowledge from several disciplines (agronomy, energy technology) as well as to account for the seasonal patterns and the variety in quality of local resources (land, water, and labor). The empirical underpinning is based on (i) statistical data of land resources, land quality, and land use, (ii) technical information on crop calendars, crop yields, and input requirements, principally labor and water, (iii) economic data on selling price and cost structure of crops, (iv) assessment of availability and quality of rainwater and irrigation water, and (v) literature-based data of cost structure and profitability of solar panels on land surfaces.

Plan of the Remainder of the Entry

The remainder of the entry is structured as follows. Section "General Structure Farm Model" presents the general outline for a crop farm model in the microeconomic tradition. Section "Reflecting the Nexus in a Farm Model" discusses vital processes, activities, and resources that are included in an applied model, in order to reflect the main aspects of the water-energy-food nexus. Section "Mathematical Representation of the Model" gives the corresponding mathematical structure of the applied farm model while section "Model Application" outlines its use for policy purposes.

General Structure Farm Model

Premises and Formulation

As point of departure, we consider a representative regional farmer who maximizes net revenue from cropping subject to the prevailing production technology and available productive resources, and at given selling and buying prices for commodities. The following premises are postulated:

(a) The farmer grows crops c, buys commodity inputs k (fertilizer, services), and has three resources:
- Land with availability $\overline{A}$ (in ha)
- Labor with availability $\overline{L}$ (in 1000 person-days)
- Water with availability $\overline{W}$ (in 1000 m^3).

(b) Crop-specific production functions f_c determine how much output q_c (in tons) can be produced at each combination of land a_c, labor ℓ_c, water w_c, and commodity inputs v_{kc}.

(c) Water salinity $\overline{s}$ (in dS/m) is a parameter of each production function, where dS stands for deci-Siemens, a measure for electrical conductivity.

(d) Crop output prices p_c and commodity input prices $\widehat{p}_k$ are given (for the farmer).

In these premises, one may note the different treatments of resources and commodities. For resources (land, labor, and water), the concept of unlimited buying or renting at a single price is less plausible than for commodities due to heterogeneity, limited tradability, and fragmented supply. Therefore, the regional availability of the resource is postulated as constraint whereas its value is derived as endogenous shadow price (see below).

With these premises, the farmer's optimal cropping decisions follow from maximizing net revenue (defined as crop output value minus commodity input costs) subject to the production functions and resource constraints:

$$\max_{q_c, v_{kc}, \ell_c, w_c, a_c \geq 0} \sum_c \left(p_c q_c - \sum_k \widehat{p}_k v_{kc} \right)$$

$$\text{subject to } \sum_c a_c \leq \overline{A} \qquad (\lambda)$$

$$\sum_c w_c \leq \overline{W} \qquad (\mu)$$

$$\sum_c \ell_c \leq \overline{L} \qquad (\rho)$$

$$q_c \leq f_c(a_c, \ell_c, w_c, v_c; \overline{s}) \qquad (\sigma_c)$$

where the shadow prices of the constraints are indicated between brackets. Crop yields (in ton/ha) can be calculated ex post as: $y_c = q_c/a_c$.

Optimal Allocation

As mentioned already in the introduction, this farm model structure is rooted in microeconomic theory. It covers simultaneously physical limitations (land and water resources), technology (knowledge from agronomy), and economic considerations (farmer decisions). Standard assumptions are that the production functions f_c are nonnegative, nondecreasing, and concave in the inputs (Varian, 2020). The latter property signifies decreasing marginal returns of the inputs. Under these conditions, the maximum exists and the optimal allocation satisfies the mathematical first-order conditions (Karush-Kuhn-Tucker conditions (Lancaster, 1968)).

The optimal levels of the variables and the shadow prices of the constraints are calculated from the first-order conditions. The shadow prices (indicated between brackets in the formulation of the model) reflect the *increase in net optimal revenue if one additional unit of the resource would be available.*

For each input (land, labor, water, and commodities), the value of marginal productivity is equal across all crops in the optimal solution. In particular, according to the first-order conditions for water:

$$p_c \frac{\partial f_c}{\partial w_c} = \mu \quad \text{if} \quad w_c > 0$$

Hence, in the optimal solution the value of the marginal productivity of water is *for each crop the same* and equal to shadow price μ of the water constraint. If the marginal revenue of water for a certain crop c^* is too low compared to the shadow price (hence, if $p_c \frac{\partial f_c}{\partial w_c} < \mu$), crop c^* will not be produced. Similar conditions apply to land and labor. For the commodities, the first-order conditions are slightly different and imply that the commodities are applied to a crop until the marginal productivity equals the buying price:

$$p_c \frac{\partial f_c}{\partial v_{kc}} = \widehat{p}_k \quad \text{if} \quad v_{kc} > 0$$

Targeted Use

The general model structure presented here has been widely used in applied agro-economic modeling since the 1970s, both in stand-alone mode and as part of a full-economy model with endogenous market prices (Ginsburgh & Keyzer, 2002). However, throughout time, the emphasis has changed from staple crop output and farm income to crop diversification and environmental impact. Major ongoing international research efforts in this spirit are, e.g., the global modeling activities of the International Food Policy Research Institute (Dietrich et al., 2019) and the MAgPIE group (Robinson et al., 2015).

C

Reflecting the Nexus in a Farm Model

Against the background of the general model structure of the previous section, we discuss here the WEF-requirements of an applied farm model for the JRB. Two districts in particular are discussed, where farmers face the traditional water-food tensions but have also new options due to the possibility of installing solar panels on their land. Figure 1 divides the JRB geopolitically in 48 districts, which are part of five different countries, and shows the location of the two

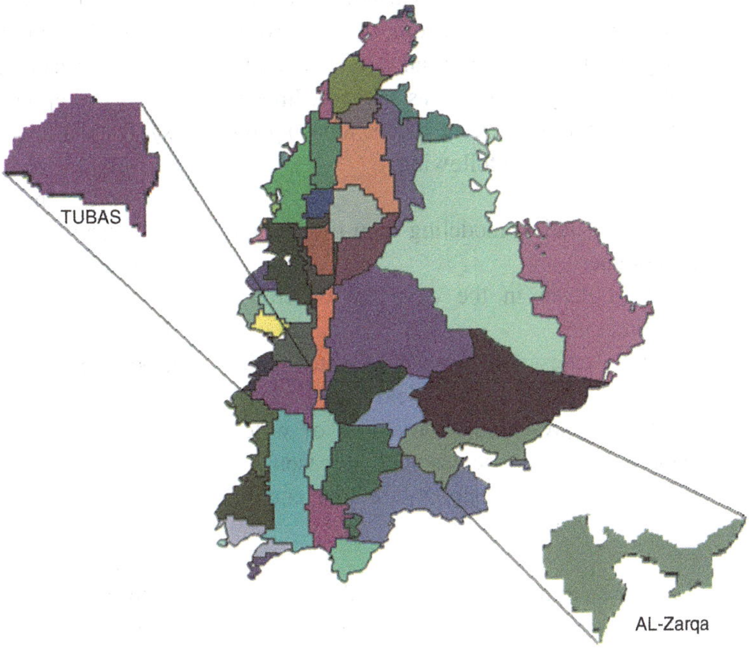

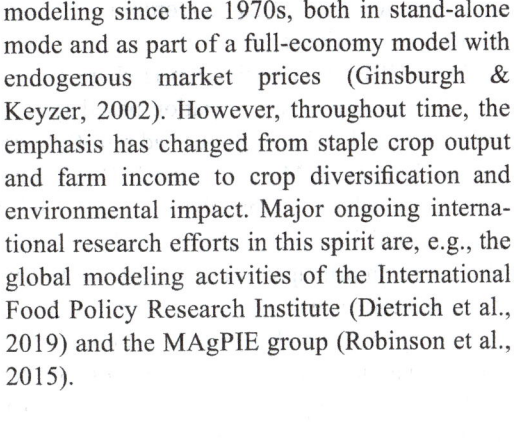

Crops and Panels: A Farm Model with Trade-offs in the Water-Energy-Food Nexus, Fig. 1 JRB with selected districts as case studies

study districts, viz. Al-Zarqa (Jordan) and Tubas (Palestinian Territories).

In the JRB, crop cultivation is characterized by a dual system, where irrigation-based agriculture (protected and open-field) and rain-fed agriculture both exist. Moreover, the JRB is known to have good irradiation characteristics and a great potential for solar generation of electricity. Therefore, in principle, it is plausible to assume that each farmer will try to maximize profits by allocating scarce land and other resources in the most rewarding way to cultivation of rain-fed and irrigated crops and production of solar energy or to any combination of those activities. However, in doing so the farmer has to respect several constraints, primarily not only arising from the limited availability of resources (in particular, land, labor, and water) and the input-output relations in production, but also arising from precautionary considerations such as producing minimal amounts of food for the own household and spreading output risk by keeping sufficient variation in the cropping pattern. This section explains how the farm model takes these principles into account.

WEF Modeling Needs

The farm model covers one calendar year. Its baseline scenario is taken to refer to a relatively normal year in terms of rainfall and irradiation. For the JRB, the year 2015 was selected. Important requirements for a farm model focusing on the WEF-nexus are the following:

- Sufficient detail in modeling crop production technologies
- Sufficient detail in the coverage of crops, including trees
- Ability to represent the impact of water availability and salinity on crop yields
- Ability to represent the variability of water quantity and salinity across periods/seasons
- Ability to represent energy use and energy production by the farmer

With these requirements, the premises in formulating the model are more specific than for the general model structure of the previous section:

(i) There is one representative farm and farmer by district; in other words, the farm sector of each district is modeled as a representative farm.
(ii) For activities and constraints, the year is exhaustively subdivided into periods (seasons or, more refined, monthly or two-weekly periods).
(iii) Farmers optimally allocate land to field crops, garden crops, trees, and solar panels over time and across technologies.
(iv) Land is classified by cultivation type (open-field irrigation, protected irrigation, and rain-fed) and suitability (for cropping and/or solar panel installation).
(v) For trees, a distinction is made between existing and newly planted trees.
(vi) Energy can be bought from the grid or sold to the grid, depending on the farmer's net energy needs.
(vii) Farmers are responsive to prices of inputs, outputs, taxes, and subsidies.
(viii) Three types of constraints are taken into account:
    (a) Availability of resources (in particular land, labor, and water)
    (b) Input-output relations in growing crops and producing solar energy
    (c) Precautionary considerations (like spreading risks and assuring home consumption of own produce).
(ix) Installation of solar panels only refers to private decisions of farmers; in other words, the model does not account for solar projects initiated by the government or other actors.

Below, the implementation of these principles will be discussed, covering resources, activities, constraints, and farmer decisions.

Resources

Land

The model captures land in three dimensions: The first dimension classifies land according to the cultivation technology used, the second classifies it according to its quality in terms of fertility, and the third classifies it according to topography for energy production purposes.

Concerning cultivation technology, the following different subcategories are distinguished (We note that in the JRB, flooding is not an issue, and hence, this possibility is not explicitly represented in the model):

- Open-field irrigated cultivation: where farmers grow crops in open fields requiring water for irrigation
- Protected irrigated cultivation: where farmers grow crops in greenhouses or plastic tunnels requiring water for irrigation
- Rain-fed cultivation: where land is without irrigation infrastructure and crop growth is fully dependent on local rainfall

The total land available for cropping is empirically determined and kept exogenous in the crop selection part of the model; other land that would be suitable for cropping but is actually not cultivated for political, social, or economic reasons is excluded from the land resources in the farm model. For this "underutilized" land, private owners apparently lack the willingness and/or ability to invest in it for reasons of low fertility, limited access to irrigation water, high cost of reclamation, or uncertainty as to land rights. The choice of how much of the total available crop land to irrigate is included in the farm model as an ex ante ("beginning of the year") decision of the farmer, in response to expected water availability. This decision precedes the allocation of the resulting areas of irrigated and rain-fed land to crops and solar panels.

The second dimension focuses on land suitability in terms of fertility, based on determinants such as soil attributes, slope, stoniness, and depth of land. Using these determinants, land (both irrigated and rain-fed) is classified into suitability classes. The higher the quality of land, the higher the yield per hectare for given input use.

The third dimension classifies land according to its suitability for solar energy production. On the basis of national irradiation maps, the level of irradiation for each district was determined. Combined with an assessment of exposure to the sun (flatness of surface, areas facing south or north), suitability classes for solar panels are defined. For the JRB-districts, three classes are distinguished.

The representative farmer is assumed to possess full ownership and control over the irrigated and rain-fed land. If farmers were to install solar panels, they would choose to do so on sites that can be protected relatively well and on land that is relatively less suitable for cropping. Since irrigated land is assumed to have higher yields than rain-fed land and already has immobile infrastructure (water pipes for irrigation and possibly greenhouses), the farm model allows solar energy production only on rain-fed land. Thus, solar energy production competes with rain-fed crops, although the subdivision in crop and solar panel suitability classes could make the competition less fierce, depending on the district.

Labor

The treatment of labor is kept relatively simple in the model. Only one type of labor is considered, hence there is no explicit distinction of male, female, and child labor, and no separation into skill classes. Labor volumes are expressed in person-days. Labor availability is exogenous in the farm model and specified by period of the year.

On the demand side, the model specification starts from average labor requirements by crop, cultivation type, and period, expressed in person-days per hectare. Empirically, these requirements are based on regional farm input studies, after cross-checking with statistical data on total district employment in agriculture. For each crop and cultivation type, the crop calendar allows further allocation of the labor requirements to the different periods in the growing season, with the largest values in the peak periods (planting, harvesting).

However, these average requirements are not directly translated into linear coefficients in the model. They are used as benchmark data for a nonlinear specification which reflects a narrative on changes in the average quality of labor when the total volume changes: Labor use is modeled with diminishing marginal returns. This is shown in Fig. 2. When the cultivated area a_c of a certain crop c expands, more specialized labor is needed, but well-trained laborers may be harder to find

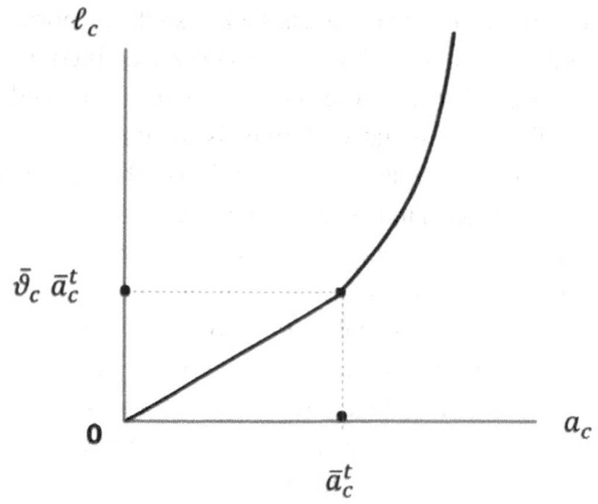

Crops and Panels: A Farm Model with Trade-offs in the Water-Energy-Food Nexus, Fig. 2 Labor demand in cropping

every time farms expand the production of this crop. Therefore, the model defines a threshold $\bar{a}_c^t$ (in hectares) for each crop, by period t. Once the threshold is reached, farms need to hire increasingly more labor per hectare because of decreasing efficiency of labor. This implies that labor requirements ℓ_c increase more than proportionally when additional land is used in producing a specific crop.

Labor requirements for solar energy are specified as fixed coefficients per hectare of solar panels, reflecting mainly time for management, surveillance, and maintenance.

The exogenous labor availability data is defined on the basis of observed empirical total labor use in cropping and on an assessment of available labor surplus in the region throughout the year. The labor constraint in the farm model is period specific.

Water

Both the quantity and quality of available water for cultivation are exogenous to the farm model. For rain-fed and irrigated land, rain water that directly percolates to the root zones represents one source, the volume of which differs considerably across periods. For irrigated cultivation, farmers are assumed to receive user rights to an additional exogenous annual amount of irrigation water of given quality, accumulated in a reservoir. They then use it during the different periods to optimally irrigate the crop mix they choose to produce.

Whenever farmers irrigate their land, the irrigation volumes applied are assumed to be sufficient for an optimal yield of the crop under cultivation. This assumption is consistent with the ex ante decision on the irrigated area, mentioned earlier. However, irrigation water may contain (a high level of) salt that has a negative impact on yields. For rain-fed areas, the situation is different. Here, yields will respond to changes in water availability while salinity poses no problem.

Data on water availability and water quality are not always readily available, especially not by period of the year. Usually, a special empirical substudy is required that takes into account the regional climate conditions, irrigation possibilities, and water quality assessments in the region, as well as the actual distribution of water between households, industry, and agriculture. In the case of the JRB, use could be made of a regional water economy model constructed in an earlier project (ACWFS, 2016).

We note that the representation of irrigation water availability does not only reflect natural and infrastructural conditions but depends implicitly also on assumptions about policy regulations for groundwater pumping. For instance, in the JRB both Jordanian and Palestinian farmers face restrictions on the extractable amount of water, either for political reasons (Palestinian territories) or for protection of the aquifers (Jordan).

The model imposes water constraints in each period for rain-fed and irrigated areas separately. Analogue to the restrictions on labor, these water constraints lead to endogenous shadow prices for water (by period and cultivation type) that can be used for the valuation of water inputs. Hence, the model disregards (regulated) prices that may prevail in the region for irrigation water. However, if regulated prices exist, ex post calculations can be made of the implicit subsidies or taxes in the provision of irrigation water to the farmer by comparing the shadow prices reflecting scarcity with monetary prices charged.

Cropping Activities

Classification and Planting Options

For the JRB-districts, we distinguish a range of 55 typical crops, although not all are grown in each district. Table 1 provides the list, as well as their aggregation into main categories. Field and garden crops are aggregated into 10 main categories, while tree crops are classified into three main categories. Hence, the farm model operates in terms of these 13 crop categories.

Crops and Panels: A Farm Model with Trade-offs in the Water-Energy-Food Nexus, Table 1 Annual crops and tree crops

Annual crops
Cereals
Wheat, Food maize, Barley, Sorghum, Other grain
Dry pulses
Lentils, Chick peas, Other dry pulses,
Oilseed crops
Groundnuts, Soybeans, Sunflowers, Sesame seed, Other oil crops
Spices (field)
Anise, Other field spices
Industrial crops
Sugar beets, Cotton, Tobacco, dry, Other industrial crops
Leafy vegetables
Green beans and peas, Cabbages, Lettuce and chicory, Green peppers, Other leafy vegetables
Forage crops
Fodder maize, Vetches (leguminous), Luzerne (alfalfa), Other forage crop
Root crops
Potatoes, Dry onions, Other root crops
Fruit vegetables
Tomatoes, Cucumbers, Watermelons, Sweet melons, Strawberries, Pumpkins and gourds, Eggplants
Spices (garden)
Hot peppers, Other garden spices
Trees
Citrus fruit
Oranges, Tangerines, Other citrus fruit
Nuts
Walnuts, Almonds, Pistachios, Other nuts
Other trees
Olives, Bananas, Avocados, Apples and pears, Peaches, plums and apricot, Dates, Figs, Grapes and table grapes, Other tree fruit

All crop categories can be cultivated both on irrigated and rain-fed land. However, protected cultivation (greenhouse or tunnel) is relevant only for vegetables, root crops, and spices. The growing seasons of the annual crops are detailed in a crop calendar, specified by cultivation type (protected, open-field irrigation, and rain-fed), that shows when the land is occupied by a certain crop. Evidently, trees occupy land throughout the year, but the model accounts for the fact that land used for annual crops can potentially be cultivated more than once in a year time, and, where relevant, the model allows the farmer to choose between several possible "planting options" for a given crop.

Crop Yields

For each crop and cultivation type, an exogenous reference yield is specified (in kg per hectare), based on local empirical observations. This reference yield is assumed to be the maximum yield for the current state of farming practices in the region, at optimal water use and quality. The actual yield will be different when water volume and/or water quality do not meet these reference values. The reaction functions for changes in water volume and changes in water quality are based on the information in technical studies of the Food and Agricultural Organization, in particular the original studies (Doorenbos & Kassam, 1979; Ayers & Westcot, 1985), with updates in (Steduto et al., 2012). Water quality is captured via its salinity.

Figure 3 shows, for crop c, the yield reduction factor β_c as function of water volume per hectare ω_c. This function is piecewise linear, with the slope depending on the crop-specific parameter κ_c^y and an upper bound of one when the reference water requirement ω_c^{ref} is reached. Segment (a) applies if $\kappa_c^y < 1$, segment (b) if $\kappa_c^y > 1$.

Figure 4 shows, for crop c, the yield reduction factor γ_c as function of water salinity $\bar{s}$, measured in dS/m. As long as salinity is below the crop-specific threshold level s_c^*, the reduction factor is one. However, for higher salinity levels the factor starts falling until the yield is zero at salinity level s_c^{**}.

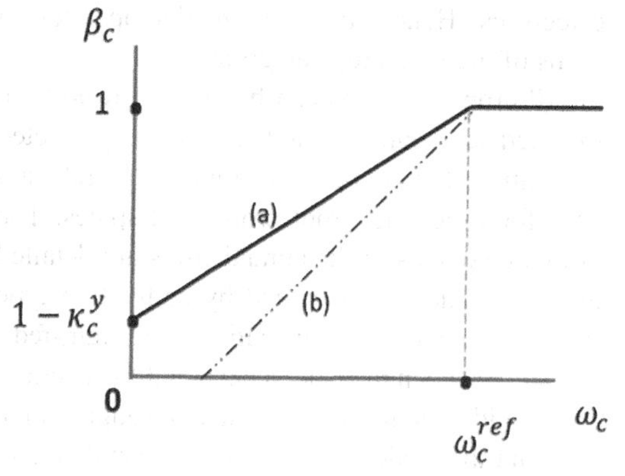

Crops and Panels: A Farm Model with Trade-offs in the Water-Energy-Food Nexus, Fig. 3 The impact of water volume on crop yield

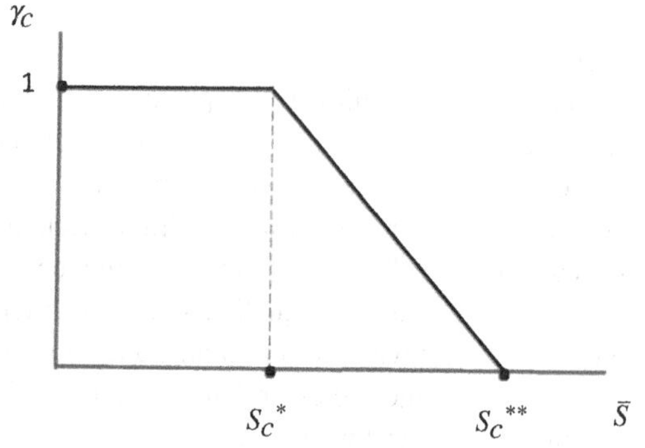

Crops and Panels: A Farm Model with Trade-offs in the Water-Energy-Food Nexus, Fig. 4 The impact of salinity on crop yield

In the model, these reduction factors are calculated and applied by period. Therefore, also the growth curve of the crop must be included in the model, building up to the reference yield at harvest time.

To obtain this curve, information is collected from crop-specific agronomic research.

As mentioned already, for irrigated land (whether open-field or protected), the water volumes are assumed to be sufficient to realize the reference yield, following the ex ante decision of the farmer on the size of irrigated land. Therefore, the yields on irrigated land are determined by the salinity equation only. For rain-fed land, the opposite applies as salinity is not an issue, but the actual yield depends on the availability of percolated rainwater in each period.

The reference yields (and, hence, also the actual yields) of a specific crop differ across cultivation types and across land suitability classes. On irrigated land, the yields are higher than on rain-fed land, while within each of these cultivation types the yields gradually decline when moving from the most suitable to the least suitable land.

New and Existing Trees

The model pays special attention to cultivation of trees as these are very important for the agricultural sector in the JRB. A distinction is made between existing and new trees. Existing trees bear fruits until they become too old. Hence, their area is exogenous and follows from earlier decisions by the farmer or his ancestors. For new trees, the planting decision is based on the discounted flow of expected revenue and costs. The model for the JRB assumes an average life span of 25 years, of which the first 5 years are nonproductive implying that farmers incur costs only. The remaining 20 years, however, are productive and generate revenues. For the assessment of future costs and earnings, projections are made of future output prices and production costs.

Energy

Electricity Supply and Demand

For the JRB region, energy use consists of fuel (for machinery, vehicles, and pumping) and electricity. In the model, fuel is part of the intermediate commodities purchased at market prices by the farmer. For electricity, three forms of demand are considered: (i) for cooling and/or ventilating greenhouses, (ii) for cold storage of perishable fruits and vegetables, and (iii) for lighting and use of small devices.

The farmers can buy the electricity from the grid or produce the electricity themselves via the installation of solar panels. The buying and selling prices are exogenous, and the model assumes that famers can sell surplus electricity to the grid without constraints. The selling price is set at 60–80% of the buying price, depending on the district. We note that these prices are scenario assumptions, due to lack of empirical information.

The assumption that electricity can be sold by private individuals to the grid without constraints is mainly for the purpose of exploring possible future pathways. Current regulations in the JRB do not allow it, at least not with a direct financial compensation. Furthermore, a decentralized system of net-metering is required for a proper registration of the volumes bought or sold to the grid.

Crops and Panels: A Farm Model with Trade-offs in the Water-Energy-Food Nexus, Table 2 Irradiation levels in the JRB, in Mwh/ha per week

	Irradiation class of the district		
Suitability class of the land	High	Medium	Low
Optimal	34.0	32.4	30.6
Suboptimal	17.0	16.2	15.3
Unsuitable	0	0	0

Yield of Solar Panels

When deciding if and where to produce solar electricity, the farmer will consider the irradiation levels on the various parts of the farm land. Apart from the overall amount of sunshine in the region, these levels depend on site-specific conditions. Therefore, the farm model has a special dimension of land classification that indicates the suitability for installing solar panels. This suitability is based on both slope and orientation, together accounting for differences in exposure level to the sun. Three categories are distinguished: (i) land facing south with optimal exposure to the sun, (ii) land facing south with suboptimal exposure to the sun, and (iii) land facing north, which is unsuitable for solar energy production. Furthermore, there are differences in average irradiation levels across the JRB districts due to their geographical characteristics. Therefore, the districts are classified as high, medium, and low in terms of irradiation volume. Table 2 gives an overview of the irradiation levels by land suitability class and district irradiation class, measured in Mwh (megawatt-hour) per ha per week.

The values in the table are based on information from technical studies for the JRB region such as (Rabi & Ghanem, 2016; Alrwashdeh et al., 2018), with additional assumptions for the differences across classes. When moving from a district with high irradiation levels to a district in a lower category, a reduction of 5 percent is assumed. When moving from the optimal class to the suboptimal class, the yield is assumed to drop by 50 per cent, while it becomes zero in unsuitable areas.

The irradiation levels of Table 2 are also used as yield per hectare of solar panels in the farm model. These yields are exogenous. Since the input costs (capital, labor) are equal across suitability classes, the farmer who decides to install solar panels will do so first on land that is optimal in terms of irradiation and the least fertile for cropping.

Farmer Optimization

Two-Stage Decisions

To conclude this section, we summarize how the representative regional farmer maximizes net revenue subject to the prevailing production technology and available productive resources, and at given selling and buying prices for commodities. This decision is divided in two stages.

First, the farmer is assumed to foresee in the beginning of the year how much water will be available, allowing him, within bounds, to change the size of irrigated land, leaving the remainder of the land as rain-fed area. Due to this assumption, the water volumes on irrigation land are sufficient for an optimal yield of the crop under cultivation, apart from the impact of salinity.

In the second stage, when the irrigated and rain-fed areas are known as well as the yields per hectare, the farmer decides on the optimal allocation, in each period, of the available land, labor and water to crops, and solar panels, subject to the technological constraints and at given prices for crops and input commodities.

Precautionary Constraints

Apart from resource constraints and production technology (input-output relations), farm behavior is influenced by other constraints such as the need to produce minimal amounts of food for the own household or the desire to spread output risk by keeping sufficient variation in the cropping pattern. These considerations can be indicated as precautionary constraints. In the farm model, they are implemented via exogenous lower bounds on

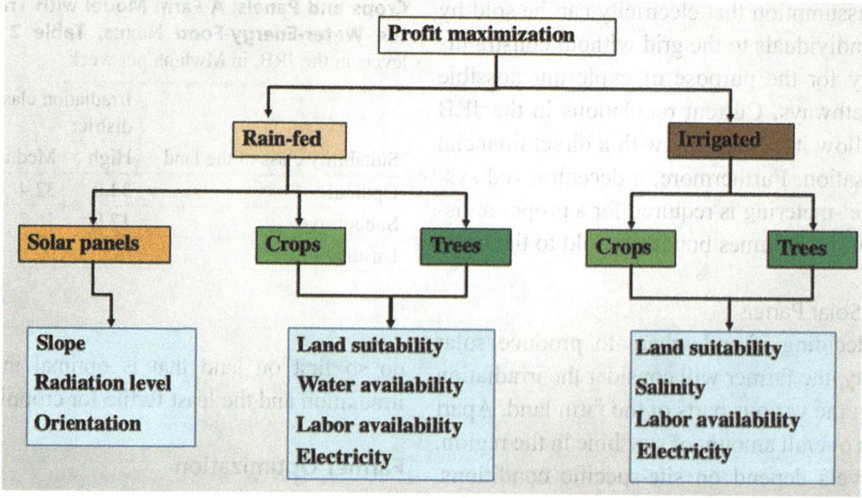

Crops and Panels: A Farm Model with Trade-offs in the Water-Energy-Food Nexus, Fig. 5 Farmer optimization and yield determinants (simplified)

the amount of each crop produced. Quantitatively, these bounds may be determined as a certain percentage of the volumes produced in the reference year of the model. In the JRB farm model, for instance, they were set at 75% of the observed crop output.

Schematic Illustration

The optimal allocation (based on profit maximization) by the farmer is schematically illustrated in Fig. 5. Rain-fed land can be allocated to annual crops, trees, and solar panels, while irrigated land can only be allocated to annual crops and trees. The overall objective of the farmer is to maximize the net revenue (output value minus input costs) of all activities together. In fact, the figure simplifies the use of irrigated land since the model actually has two types, viz. open-field and protected irrigated land. The blue rectangles indicate the main factors that determine the yields per hectare of crops, trees, and solar panels. These factors have been addressed in the discussion above.

Figure 5 not only simplifies the farmer's decisions problem by taking open-field and protected irrigated land together, but also hides the different qualities of rain-fed, respectively, irrigated land, each with their own maximal yield levels. Moreover, it suppresses the timing of planting of crops, which is based on crop calendars specifying the growing season of each crop for the various districts of the JRB. The next section provides the full specification, in mathematical terms.

Mathematical Representation of the Model

In this section, the mathematical formulation of the farm model is given, as follow-up to the verbal description in the previous section. Subsection "Exogenous Resources and Ex Ante Calculations" addresses the resources and the first stage of the farmer decisions. The optimization program itself is specified in subsections "Endogenous Activities", "Resource Constraints", "Yield and Output", "Input Needs", "Electricity Balance", "Precautionary Constraints", and "Objective", divided into the list of endogenous activities, resource constraints, input needs, yield functions and output, the electricity balance, precautionary constraints, and the objective. Finally, subsection "Summarizing the Model" summarizes. For clarity, this section redefines the variables and indices used in the general model formulation introduced earlier.

The model has the following indices:

c crop
g planting date option

k input commodity
m cultivation type
t time period (within the year)
z land suitability (for crops and solar panels combined)

The set of cultivation technologies, indicated by subscript m, has three elements: irrigated cultivation in protected areas m^p, irrigated cultivation in open fields m^o, and rain-fed cultivation m^r. Crops c are divided into two subsets, viz the set of annual crops C^a and the set of tree crops C^τ.

In the description of the model, the index g will be suppressed, in order to keep the presentation as simple as possible. However, in fact, the model has more than one planting date option g for most of the annual crops $c \in C^a$.

Exogenous Resources and Ex Ante Calculations

This subsection presents the exogenous resources land, water, and labor, partly determined by the ex ante decisions of the farmer, as well as the crop yield and water requirement coefficients that are related directly to these ex ante decisions.

Resources

Land, water, and labor are the exogenous resources in the optimal allocation decisions of the farmer, with the following dimensions:

$\overline{A}_{mz}$ land area by cultivation type m and suitability class z, in ha
$\overline{W}_{mt}$ water volume available for cultivation type m in period t, in 1000 m^3
$\overline{L}_t$ labor volume available in period t, in 1000 person-days

Empirical labor volumes $\overline{L}_t$ are directly determined in the definition above, but water volumes and land areas are calculated from more detailed exogenous variables and coefficients. In this way, scenario assumptions on water and land availability can be built up in a more refined way. These calculations are done "ex-ante," i.e., before actually solving the model's optimal endogenous variables.

Water resources. For rain-fed land, available water volumes $\overline{W}_{mt}$ are determined as total rainfall percolation to the root zone (by period) multiplied by the share that ends up in the root zone of rain-fed areas. For dry periods, the resulting volume is increased with an estimate of the amount of water that trees can extract from deeper layers via their root system. For irrigated land (open field or protected), the available water volumes $\overline{W}_{mt}$ are calculated as the sum of percolated rain water that ends up in the root zone of irrigated land and the amount of irrigation water that is available.

Land resources. For land resources, the ex ante calculation consists first of allocating the empirically determined exogenous farm land $\overline{A}^{tot}$ to the three cultivation types, i.e., open-field irrigated, protected irrigated, and rain-fed land. To this end, farmers are assumed to plan at the beginning of the year how many hectares to cover with irrigation infrastructure. This plan is based on their expectation of the quantity of water they will receive for irrigated land in the coming year and their estimate of the annual average water requirement per hectare. For the expectation of water availability, they follow the value of $\overline{W}_{mt}$ as described above, while their estimate of the average water requirement is based on the observed crop mix of the reference year of the model. The essential steps of this calculation can be described as follows.

Let, for $m = m^p$ and $m = m^o$, area $\overline{a}^0_{cm}$ be the area (hectare) assigned to irrigated crop c in the reference year, and let ω^{ref}_{cmt} be the water required in m^3 per hectare of irrigated land for an optimal yield increase of crop c in period t. Then, the average annual water requirement in m^3 per hectare ω^{av}_m can be determined as:

$$\omega^{av}_m = \frac{\sum_c \overline{a}^0_{cm}\left(\sum_t \omega^{ref}_{cmt}\right)}{\sum_c \overline{a}^0_{cm}} \quad (1)$$

With these requirements, the planned areas of irrigated cultivation A^{plan}_m are, in hectare:

$$A^{\text{plan}}_m = \frac{\sum_t \overline{W}_{mt}}{\omega^{av}_m} \quad \text{for} \quad m = m^p, m^o \quad (2)$$

Planned land with rain-fed cultivation is then calculated residually:

$$A_{m^r}^{\text{plan}} = \overline{A}^{tot} - A_{m^p}^{\text{plan}} - A_{m^o}^{\text{plan}} \tag{3}$$

Equations (1)–(3) show the essential steps of the derivation of the planned crop areas. In fact, the actual calculation has two refinements. It applies a correction for multiple cropping (when relevant) and imposes upper bounds on the irrigated areas to avoid too large increases of irrigated land compared to the reference year.

Finally, the model resources $\tilde{A}_{mz}$ are derived by allocating A_m^{plan} to land suitability classes z, following for each cultivation type the same relative proportions θ_z of the suitability classes of the reference year:

$$\tilde{A}_{mz} = \theta_z \, A_m^{\text{plan}} \quad \forall \quad m,z \tag{4}$$

From here onward, the area with irrigation infrastructure will also be referred to as the "prepared" area.

Ex Ante Crop Yields and Water Requirements

Due to the planning of irrigated land by the farmer, the water volumes on irrigation land are sufficient for an optimal yield of the crops, apart from the impact of salinity, as discussed earlier in section "Reflecting the Nexus in a Farm Model". For rain-fed land, the expected water volumes may not be sufficient to reach the optimal yield, while salinity can be neglected. In both cases, the crop yields can be determined directly via ex ante calculations on the basis of the yield relations in Figs. 3 and 4, applied by period.

Let y_{cmtz}^{ref} be the optimal yield increase of crop c with cultivation type m in period t on land of suitability class z, measured in ton/ha, and ω_{cmt}^{ref} (already introduced) the corresponding water requirements in m^3/ha. Furthermore, let β_{cmt} and γ_{cmt} be the yield reduction factors due to water shortage and salinity, respectively. Then, the ex ante calculation of yield $\overline{y}_{cmz}$ of crop c with cultivation type m on land of suitability z, in ton/ha, is as follows:

$$\overline{y}_{cmz} = \sum\nolimits_t \beta_{cmt} \, \gamma_{cmt} \, y_{cmtz}^{ref} \tag{5}$$

where, as in Fig. 3:

$$\beta_{cmt} = \min\left(1 - \kappa_c^y\left(1 - \frac{\omega_{cmt}}{\omega_{cmt}^{ref}}\right), 1\right) \tag{6}$$

and, as in Fig. 4:

$$\gamma_{cmt} = \min\left(\frac{\tau_c \, \overline{s}_{mt} + \nu_c}{100}, 1\right) \tag{7}$$

Exogenous variable ω_{cmt} in Eq. (6) represents available water in m^3/ha for crop c with cultivation type m in period t. For irrigation cultivation, it equals the reference requirement, hence $\beta_{cmt} = 1$ for $m = m^p$ and $m = m^o$. For rain-fed cultivation, it is derived from total available water $\overline{W}_{mt}$ by applying the water allocation shares χ_{cm}^0 and the crop areas $\overline{a}_{cm}^0$ of the reference year:

$$\omega_{cmt} = \chi_{cm}^0 \, \overline{W}_{mt} / \overline{a}_{cm}^0 \tag{8}$$

Exogenous variable $\overline{s}_{mt}$ in Eq. (7) represents the salinity of the water available for cultivation type m in period t, expressed in dS/m. Coefficient τ_c is negative, and coefficient ν_c larger than 100. For rain-fed cultivation, salinity is zero and, hence, $\gamma_{cmt} = 1$.

As mentioned already in the explanation in section "Reflecting the Nexus in a Farm Model", Eqs. (6) and (7) rely heavily on the Irrigation and Drainage Papers of the Food and Agricultural Organization (Doorenbos & Kassam, 1979; Ayers & Westcot, 1985; Steduto et al., 2012). The reference water requirements ω_{cmt}^{ref} (in m^3/ha) are directly derived from the location-specific reference evapotranspiration rate ET_0 (in mm) and the crop-specific water demand factors κ_{cmt} in these reports, with appropriate unit conversion:

$$\omega_{cmt}^{ref} = 10 \, \kappa_{cmt} \, ET_0 \tag{9}$$

As a final step in the ex ante calculations, we determine the water requirement coefficients $\overline{\omega}_{cmt}$ that will be used in the optimal allocation decision of the farmer model, expressed in m^3/ha. They are set equal to the available water volumes ω_{cmt}, to maintain consistency with the crop yields:

$$\overline{\omega}_{cmt} = \omega_{cmt} \tag{10}$$

For irrigated cultivation, these coefficients coincide with the reference requirements. For rain-fed cultivation, they may be higher or lower than the reference requirement.

Endogenous Activities

Before presenting the structure of the optimal allocation model, we list its most essential endogenous variables. They are, for land of suitability z and cultivation type m:

- a_{cmtz} area planted with crop c in period t, in ha
- a^s_{mtz} area occupied by solar panels in period t, in ha
- w_{cmtz} water use by crop c in period t, in 1000 m^3
- $w^{\tau x}_{cmtz}$ water use of existing tree crop $c \in C^\tau$ in period t, in 1000 m^3
- ℓ_{cmtz} labor use by crop c in period t, in 1000 man-days
- $\ell^{\tau x}_{cmtz}$ labor use of existing tree crop $c \in C^\tau$ in period t, in 1000 man-days
- ℓ^s_{mtz} labor use for solar panels in period t, in 1000 man-days
- v_{kcmz} input use of commodity k for growing crop c, in kg
- $v^{\tau x}_{kcmz}$ input use of commodity k for growing existing tree crop $c \in C^\tau$, in kg
- v^s_{kmz} input use of commodity k for solar panels, in kg
- q_{cmz} output of (annual) crop $c \in C^a$, in ton
- $q^{\tau x}_{cmz}$ output of existing tree crop $c \in C^\tau$, in ton
- $q^{\tau n}_{fcmz}$ expected output of newly planted tree crop $c \in C^\tau$ in year f, in ton
- q^s_{mtz} electricity output of solar panels in period t, in megawatt-hour

The optimal values of these endogenous variables follow from solving the farmer's profit maximization program of which the resource constraints, output functions, input needs, electricity balance, precautionary constraints, and objective are described successively in the subsections below.

Resource Constraints

In this subsection, we discuss successively farm land, water availability, and labor constraints.

Farm Land Constraints

For irrigated land with protected cultivation:

$$\sum_c a_{cmtz} \le \overline{A}_{mz} \quad m = m^p \tag{11}$$

This constraint implies that in any period t and for each land suitability class z, the sum of all areas allocated to producing crops c must be less than or equal to the prepared land.

For irrigated land with open-field cultivation:

$$\sum_c a_{cmtz} \le \overline{A}_{mz} - \sum_{c \in C^\tau} \overline{a}^{\tau x}_{cmz} \quad m = m^o \tag{12}$$

This constraint is similar to the one for protected cultivation above, with one exception: Part of open-field land is occupied by existing tree crops. These areas are exogenous, denoted by $\overline{a}^{\tau x}_{cmz}$, and subtracted in the constraint from the available resources. Please note in this regard that, for tree crops c, the endogenous variables a_{cmtz} refer only to newly planted areas.

For land with rain-fed cultivation:

$$\sum_c a_{cmtz} + a^s_{mtz} \le \overline{A}_{mz} - \sum_{c \in C^\tau} \overline{a}^{\tau x}_{cmz} \quad m = m^n \tag{13}$$

This constraint is similar as the one for open-field irrigated cultivation, but with the additional solar panel areas a^s_{mtz} since solar panels compete with crops on rain-fed land.

To make sure that farmers utilize land following the crop calendars, the following additional constraint must be imposed:

$$a_{cmtz} = \tilde{a}_{cmz} \quad \text{for} \quad t \in T_{cm} \tag{14}$$

In this equation, set T_{cm} represents all periods of the growth season of crop c with cultivation type m. The additional variable $\tilde{a}_{cmz}$ is time independent. Therefore, the constraint implies that the areas occupied by crops respect the crop calendars.

Similarly, for the area under solar panels the following additional constraint applies:

$$a^s_{mtz} = \tilde{a}^s_{mz} \quad m = m^n \quad \text{for all} \quad t \tag{15}$$

Again, the additional variable $\tilde{a}^s_{mz}$ is time independent. Therefore, the equation implies that solar panels occupy the land throughout the year (and that farmers are assumed to install the solar panels at the beginning of the year).

C

At the end of a growing season, farmers have to decide – given the planting options and corresponding crop calendars – whether it is possible to start another crop cycle, or whether the land has to remain idle until next year. These decisions are also part of the profit maximization program and guided by the constraints above.

Water Availability Constraints

The water constraints, in 1000 m^3, have to apply in each period t:

$$\sum\nolimits_{c,z} w_{cmtz} + \sum\nolimits_{c\in c^{\tau},z} w^{\tau x}_{cmtz} \leq \overline{W}_{mt} \quad (16)$$

These constraints imply that for each cultivation type m water use by newly planted crops (w_{cmtz}) and existing trees ($w^{\tau x}_{cmtz}$) should not exceed water availability, in each period.

However, if farmers have access to irrigation water, they can use it optimally over time and the water constraint is relaxed for irrigated areas. In the most flexible case, the constraint would be an annual one and apply to protected and open-field irrigation together.

Labor Constraints

The constraint on the quantity of labor employed in agriculture applies to each period, expressed in 1000 person-year:

$$\sum\nolimits_{c,m,z} \ell_{cmtz} + \sum\nolimits_{c\in c^{\tau},m,z} \ell^{\tau x}_{cmtz} + \sum_{m,z} \ell^{s}_{mtz} \leq \overline{L}_t \quad (17)$$

The constraint implies that for each period t the sum of labor use for seasonal crops and new trees, labor use for existing tree crops, and labor use for solar panels cannot exceed the available labor volume $\overline{L}_t$. The specification of these demand volumes is discussed below in the subsection on input needs.

Yield and Output

Here, we specify yield and output of annual crops (irrigated and rain-fed crop cultivation), tree crops (existing and newly planted), and solar panels.

Annual Crops (Irrigated and Rain-Fed)

Crop output equals yield per hectare multiplied by the cultivated area. The yields $\overline{y}_{cmz}$ have been determined already, as part of the ex ante calculations. Hence, output of annual crop $c \in C^a$ with cultivation type m on land suitability class z can be calculated directly from the endogenous area:

$$q_{cmz} = \overline{y}_{cmz}\, \tilde{a}_{cmz} \quad \text{for} \quad c \in C^a \quad (18)$$

Since the yield is expressed in ton/ha and the area in ha, output q_{cmz} follows in ton.

Existing Irrigated and Rain-Fed Trees

For land with currently existing trees, a similar calculation applies since their yields are also known from the ex ante calculations. The area is exogenous, and denoted by $\overline{a}^{\tau z}_{cmz}$, expressed in ha.

$$q^{\tau x}_{cmz} = \overline{y}_{cmz}\, \overline{a}^{\tau x}_{cmz} \quad \text{for} \quad c \in C^{\tau} \quad (19)$$

New Irrigated and Rain-Fed Trees

For new trees, whether on irrigated or on rain-fed land, there is no immediate output. The trees become productive only in the future, say from year F onward. Then, in year $f \geq F$, yield $y^{\tau n}_{fcmz}$ equals the annual reference yield (which is the sum over the periods t of the reference yield increases y^{ref}_{cmtz} defined in the ex ante calculations) and an annual growth rate μ representing technical progress. There is no impact of water availability and water quality since future changes in these conditions are hard to predict. Hence, the expected yields are, in ton/ha:

$$y^{\tau n}_{fcmz} = (1+\mu)^f \sum_t y^{ref}_{cmtz} \quad \text{for} \quad f \geq F \quad \text{and} \quad c \in C^{\tau} \quad (20)$$

Thus, the output of the newly planted tree crop c grown on land suitability class z with cultivation type m becomes in future year f, expressed in ton:

$$q^{\tau n}_{fcmz} = y^{\tau n}_{fcmz}\, \tilde{a}_{cmz} \quad \text{for} \quad f \geq F \quad \text{and} \quad c \in C^{\tau} \quad (21)$$

Solar Energy

The output of solar energy is straightforwardly given by multiplying for each period the yield, i.e., solar energy produced in megawatt-hour per hectare, by the area covered with solar panels:

$$q^{s}_{mtz} = \overline{y}^{s}_{mtz}\tilde{a}^{s}_{mz} \quad \text{for} \quad m = m^{r} \tag{22}$$

The resulting output is expressed in megawatt-hours. As mentioned earlier, it applies only to rain-fed land. Yield per hectare $\overline{y}^{s}_{mtz}$ is exogenous and varies with land suitability z while it depends in principle also on period t.

Input Needs

Here, we formulate the input use of water, labor, electricity, and other intermediate inputs.

Water Use

The volumes of water use, in 1000 m^3, can directly be calculated from the crop areas and the water requirements per hectare $\overline{\omega}_{cmt}$ determined in the ex ante calculations. For annual crops and newly planted trees, the water use by crop c on land of suitability class z and with cultivation type m is in period t:

$$w_{cmtz} = \overline{\omega}_{cmt}\, a_{cmtz} \tag{23}$$

For existing tree crops, the same water use coefficients are used as for newly planted crops but now applied to the exogenous tree crop area $\overline{a}^{\tau x}_{cmz}$, leading to the following water use volume on land of suitability class z with cultivation type m, in period t:

$$w^{\tau x}_{cmtz} = \overline{\omega}_{cmt}\, \overline{a}^{\tau x}_{cmz} \quad \text{for} \quad c \in C^{\tau} \tag{24}$$

Labor Input

Labor use is specified relative to crop area, with as coefficient the product of a fixed reference requirement in man-days per hectare ($\overline{\theta}_{cmt}$) and an endogenous labor-augmenting factor (η_{cmtz}) that reflects declining labor efficiency (see Fig. 2).

Thus, labor use in growing crop c on land of suitability z with cultivation type m in period t can be calculated as:

$$\ell_{cmtz} = \eta_{cmtz}\, \overline{\theta}_{cmt} a_{cmtz}/1000 \tag{25}$$

In this equation, dividing by 1000 ensures that the resulting labor volume is expressed in 1000 person-years. It is assumed that efficiency of labor in crop cultivation starts decreasing when farmers reach a certain threshold of cultivated area, denoted as $\overline{a}^{T}_{cmtz}$. The more land above the threshold is used, the more labor is needed per hectare, in particular since the additional labor is assumed to have lower average skills. Therefore, the labor-augmenting factor η_{cmtz} is defined as follows:

$$\eta_{cmtz} = 1 + \frac{\max\left(0, a_{cmtz} - \overline{a}^{T}_{cmtz}\right)}{\overline{a}^{D}_{cmtz}} \tag{26}$$

Hence, $\eta_{cmtz} \geq 1$. Denominator $\overline{a}^{D}_{cmtz}$, which is expressed in hectares also, neutralizes the impact of the unit of measurement of the areas on the outcome. One may note that these equations together introduce a nonlinearity in the farm program.

Labor use for existing tree crops follows merely from the reference requirements per hectare:

$$\ell^{\tau x}_{cmtz} = \overline{\theta}_{cmt} a^{\tau x}_{cmtz}/1000 \tag{27}$$

Labor use for solar panels is based on fixed requirements $\overline{\theta}^{s}_{mt}$ in person-days per hectare:

$$\ell^{s}_{mtz} = \overline{\theta}^{s}_{mt}\, a^{s}_{mtz}/1000 \tag{28}$$

Electricity Demand

Here, we specify the annual total volume of on-farm electricity demand D^{e}, expressed in megawatt-hour. It is directly specified for all crop activities, cultivation types, and land suitability classes together, as follows:

$$\begin{aligned} D^{e} = &\sum\nolimits_{c \in C^{a}} \left(\hat{e}_{c} \sum\nolimits_{m,z} q_{cmz}\right) \\ &+ \sum\nolimits_{c \in C^{\tau}} \left(\hat{e}_{c} \sum\nolimits_{m,z} q^{\tau x}_{cmz}\right) \\ &+ \sum\nolimits_{c \in C^{a}} \left(\breve{e}_{c} \sum\nolimits_{z} \tilde{a}_{cm^{p}z}\right) \end{aligned} \tag{29}$$

The coefficients $\hat{e}_{c}$ are the electricity requirements for cooling and cold storage of perishable products, expressed in megawatt-hour per ton of crop output, and $\breve{e}_c$ the electricity requirements for ventilation of greenhouses, expressed in megawatt-hour per hectare of crop land. The first expression, on the right side of the equation, determines the quantity demanded of electricity for cooling and storing harvested field crops, namely leafy vegetables and fruit vegetables. The second expression determines the quantity demanded of electricity for cooling and storing harvested tree crops, namely citrus and other fruits. The third expression determines the quantity demanded of electricity for ventilation and cooling of greenhouses. The coefficients include a general allowance for electricity used by small devices and lamps.

Other Intermediate Inputs

All intermediate inputs in crop growing other than water and electricity are treated as inputs of commodities k, including seedlings, fertilizers, pesticides, package material, and banking services. They also cover energy inputs other than electricity, such as fuel for pumping and machinery. Furthermore, although not fully according to standard accounting principles, they are assumed to cover capital depreciation costs, as capital constraints themselves are not active in the model.

These inputs have no direct impact on the yield per hectare since their usage is assumed to conform to the needs of the current technology, hence compatible with the reference yield that is applied in the ex ante calculations in subsection "Exogenous Resources and Ex Ante Calculations". Since there is no direct link with output, they can be expressed relative to the areas under cultivation.

The intermediate inputs are modeled as fixed coefficients $\widehat{v}_{kcm}$, expressed in kg per hectare. Then, the input volumes of commodity k in growing crop c with cultivation type m on land of suitability class z can be calculated as follows for annual crops and existing tree crops, respectively:

$$v_{kcmz} = \widehat{v}_{kcm}\, \tilde{a}_{cmz} \quad \text{for} \quad c \in C^a \tag{30}$$

$$v^{\tau x}_{kcmz} = \widehat{v}_{kcm}\, \overline{a}^{\tau x}_{cmz} \quad \text{for} \quad c \in C^\tau \tag{31}$$

The resulting input volumes are expressed in kg. For newly planted tree crops, the volumes have an additional dimension f since they recur in all future years:

$$v^{\tau n}_{fkcmz} = \widehat{v}_{kcm}\, \tilde{a}_{cmz} \quad \text{for} \quad c \in C^\tau \tag{32}$$

Intermediate input costs of producing solar energy are mainly due to maintenance, repair, and insurance. They are represented by fixed coefficients $\widehat{v}^s_k$ per hectare. The values are based on a literature review for the region.

$$v^s_{kmz} = \widehat{v}^s_k\, \tilde{a}^s_{mz} \tag{33}$$

Electricity Balance

The electricity balance in the farm model is annual. The following new symbols are used:

B^e annual volume of electricity bought by the farmer from the grid, in megawatt-hours

S^e annual volume of electricity sold by the farmer to the grid, in megawatt-hours

Then, the balance can be specified as follows, in megawatt-hours:

$$D^e + S^e = \sum\nolimits_{m,t,z} \left(q^s_{mtz}\right) + B^e \tag{34}$$

This equation specifies that electricity demand by the farmer (D^e) plus sales of electricity to the grid (S^e) must be equal to the solar energy produced by the farmer (sum of q^s_{mtz} over m, t and z) and the electricity bought from the grid (B^e). In an optimal solution, the farmer will buy from the grid or sell to the grid or be purely self-sufficient, but he will not buy and sell at the same time (since the buying price is higher than the selling price, as will be specified below).

Precautionary Constraints

The precautionary considerations mentioned in the discussion of the model structure in section "Reflecting the Nexus in a Farm Model" are implemented by imposing lower bounds $\overline{B}_c$ on annual crop output, expressed in ton:

$$\sum\nolimits_{m,z} q_{cmz} + \sum\nolimits_{m,z} q^{\tau x}_{cmz} \geq \overline{B}_c \tag{35}$$

In these constraints, the parameters $\overline{B}_c$ are set such that the model forces production of each crop to be at least a substantial share of actual production in the reference year. In the case of the JRB farm model, production of each crop should not be below 75% of this actual production volume, as mentioned earlier in section "Reflecting the Nexus in a Farm Model".

Objective

The farmer's objective is to select the combination of activities that maximizes net revenues subject to the resource constraints, the technical (input, output) relations, the electricity balance, and the precautionary constraints presented in the previous subsections. All crop output is valued at market price (even the part that is consumed by the farm household itself). Hence, for practical purposes the model assumes that all crops are sold to the market. Electricity can be bought or sold, depending on the farmer's optimal strategy. Future crop output (of newly planted tree crops) is included in the objective at discounted projected prices. Production costs subtracted in the objective are the costs of electricity and the costs of the intermediate input bundle of all goods other than electricity and water. The costs of water and labor are not subtracted from the objective but calculated ex post from the shadow prices of their constraints in the optimal solution.

These considerations lead to the following objective, in which current crop output prices are denoted by p_c, commodity input prices by $\widehat{p}_k$, future crop output prices by $p_{fc}^{\tau n}$, future commodity input prices $\widehat{p}_{fk}^{\tau n}$, sales prices of solar energy by p^s, and buying prices of electricity by p^b ($>p^s$), while sales of annual crops, existing tree crops, and future crops of newly planted trees are indicated by S_c^a, $S_c^{\tau x}$, and $S_{fc}^{\tau n}$, respectively:

Maximize net revenue : (36)

Annual crops $\sum_{c \in C^a} \left[p_c S_c^a - \sum_{k,m,z} \widehat{p}_k v_{kcmz} \right] +$

Existing trees $\sum_{c \in C^\tau} \left[p_c S_c^{\tau x} - \sum_{k,m,z} \widehat{p}_k v_{kcmz}^{\tau x} \right] +$

New trees $\sum_{c \in C^\tau} \left[\sum_{f \geq F} \left(\frac{p_{fc}^{\tau n} S_{fc}^{\tau n}}{(1+r)^f} \right) - \sum_f \left(\frac{\sum_{k,m,z} \widehat{p}_{fk}^{\tau n} v_{fkcmz}^{\tau n}}{(1+r)^f} \right) \right] +$

Solar energy $p^s S^e - \sum_{k,m,z} \widehat{p}_k v_{kmz}^s \quad -$

Elec.purchases $p^b B^e$

Since all crop quantities produced are assumed to be sold, sales of seasonal crops, existing tree crops, and future crops of newly planted trees are given by, respectively:

$$S_c^a = \sum\nolimits_{m,z} q_{cmz} \quad c \in C^a \tag{37}$$

$$S_c^{\tau x} = \sum\nolimits_{m,z} q_{cmz}^{\tau x} \quad c \in C^\tau \tag{38}$$

$$S_{fc}^{\tau n} = \sum\nolimits_{m,z} q_{fcmz}^{\tau n} \quad c \in C^\tau \tag{39}$$

Thus, the objective equals the output values of all current and future crops and solar energy minus their intermediate input costs (including electricity but excluding water). (For ease of presentation, we discard here in the objective the expected future electricity costs for cooling, but in applications it should be added to the intermediate costs of new trees.) Newly planted trees become productive in year F, while revenues and costs are discounted by annual rate *r*.

Summarizing the Model

This section has described the two-stage decision model of the farmer. In the first stage, the farmer plans the division of his land resources into cultivation types and makes ex ante calculations of the yields per hectare and the corresponding water requirements. These decisions are given in Eqs. (1)–(4) and Eqs. (5)–(10), respectively. In the second stage, given these resources and ex ante parameters, the farmer optimally allocates land to crops and solar panels by maximizing objective (36) subject to the resource constraints (11)–(17), output calculations (18)–(22), input use specifications (23)–(33), electricity balance (34), precautionary constraint (35), and definitions (37)–(39). Furthermore, the endogenous variables should be nonnegative. The core endogenous variables by cultivation type and land suitability class are listed in subsection "Endogenous Activities".

In the equations, the index *g*, representing different planting date options for a crop, is suppressed for the sake of transparency. However, applied models may add the index without changing the structure of the model. Indeed, in the application for the JRB most annual crops have alternative cropping calendars, each with a different starting date.

The optimal allocation model has one nonlinearity, viz. the labor demand equation, as can be seen in Fig. 2. This nonlinearity is relevant for the model simulations since it helps to avoid the intrinsic tendency of fully linear programs to specialize in the most profitable activities. At the same time, the feasible set of the model remains convex, which makes it possible to use a standard nonlinear optimization procedure to solve the model.

As in the general model of section "General Structure Farm Model", the optimal allocation satisfies the mathematical first-order conditions as established by Lagrange and Kuhn-Tucker, while the shadow prices of water and labor can be calculated simultaneously with the optimal endogenous variables.

The model is written in GAMS-software and solved using its nonlinear solver MINOS (GAMS, 2020). Dedicated facilities are added that ensure that the model outcomes for each scenario are saved with well-identified names in both tabulation format and Excel format.

Model Application

As mentioned already in the introduction, the empirical underpinning of a model of this type is based on a variety of data sources:

(i) Statistical data of land resources, land quality, and land use
(ii) Technical information on crop calendars, crop yields, and input requirements, principally labor and water
(iii) Economic data on selling price and cost structure of crops
(iv) Assessment of availability and quality of rainwater and irrigation water
(v) Literature-based data of cost structure and profitability of solar panels on land surfaces

Once all this information has been collected and processed, the second empirical task is to ensure that for the observed values of the exogenous variables, the model can "reasonably closely" reproduce the realized values of the endogenous variables. To this end, a careful calibration process is necessary. Exact replication is for a model of this type not possible, as opposed to farm models that operate mainly with nonlinear specifications. However, the current structure has the advantage that it is well-suited to incorporate agronomic knowledge and process-based information.

When the model has been sufficiently calibrated, alternative scenarios for the exogenous variables and parameters can be simulated that show the impacts of WEF-related changes in government policies, climate conditions, or infrastructural works on the farm sector. In this vein, model simulations for the districts Tubas and Al-Zarqa in the JRB (Sabella et al., 2020) show that farmers will indeed install solar panels on their land if the options of delivery to the grid are adequate, but the areas with panels remain very limited. Furthermore, the selling price must be high enough, otherwise the farmers will produce electricity for own use only. Concerning the link between water and food, model simulations for the Jericho district (also in the JRB) explore the possibility of water transfers from the Al-Faksha springs (Sonneveld et al., 2020). The outcomes show that these transfers will substantially improve the cropping options for the Jericho farmers, although the water is relatively saline. Farmers will use the water to expand the area of both annual and perennial crops, but the largest income gains will come from new date palms, due to their high salinity tolerance.

The model's generic structure can be widely applied in decision-making processes that concern the interlinkages of the WEF nexus. The model is particularly suitable to make informed choices between the use of solar panels, electricity from the grid, or direct use of fossil fuels for the energy component, on the wider selection of crop composition that is made possible due to guaranteed water deliveries and, finally, on the planning of water volumes while accounting for water qualities. Moreover, including the suitability of land and soil characteristics makes a comprehensive assessment of cropping possibilities feasible.

In water-scarce and remote dryland areas, installation of solar panels may significantly increase the stability of energy supply compared to full dependence of the farmer on the, often uncertain, fossil-based electricity supply from the grid. In this case, application of the farm model may show whether the benefits of the investment in solar panels are sufficient for the farmer to recover the costs and whether the improved irrigation possibilities will lead to adjustment of the cropping pattern, possibly by storing surplus water to assure continuous supply of irrigation during dry spells.

Extensions to allow for other types of renewable energy generation (specifically wind turbines) can be included in analogy to the present representation of solar panels. In general, the model can support Environmental Impact Assessments in ecologically fragile conditions to evaluate the effect of replacing polluting fossil-fuel energy sources by solar or wind energy. One promising area seems to be the use of the model in the fragile and interconnected eco-systems on Small Island Developing States that could benefit immensely by replacing fossil fuels by solar cells to pump fresh water that can be used to cultivate much demanded fresh and nutritious food.

Cross-References

▶ Circular Economy and the Water-Food Nexus
▶ Climate-Resilient Technologies and Innovations for Sustainable Agriculture, Improved Landscape, and Food Security
▶ Smart Agriculture and ICT

References

ACWFS. (2016). *Towards concerted sharing: Development of a regional water economy model in the Jordan River Basin, Final report of the SIDA-sponsored Concerted Sharing Project*. Amsterdam: Amsterdam Centre for World Food Studies, VU University.

Albrecht, T., Crootof, A., & Scott, C. (2018). The water-energy-food nexus: A systematic review of methods for nexus assessment. *Environmental Research Letters, 13*(4), 1–26.

Al-Muqdadi, S., Khalaifawi, A., Abdulrahman, B., Kittana, F., Alwadi, K., Abdulkhaleq, M., Al-Saffar, S., Al-Taie, S., Merz, S., & Al-Dahmani, R. (2021). Exploring the challenges and opportunities in the water, energy, food nexus for arid regions. *Journal of Sustainable Development of Energy, Water and Environment Systems, 9*(4), 1–30.

Alrwashdeh, S., Al-saraireh, F., & Saraireh, M. (2018). Solar radiation map of Jordan governorates. *International Journal of Engineering & Technology, 7*(3), 1664–1667.

Al-Zubari, W. (2021). The water–energy–food nexus in the Arab region: Governance and role of institutions. In C. Carmona-Moreno et al. (Eds.), *Implementing the water–energy–food–ecosystems nexus and achieving the sustainable development goals* (pp. 139–150). UNESCO, European Union and IWA Publishing.

Ayers, R., & Westcot, D. (1985). *Water quality for agriculture, FAO Irrigation and Drainage Paper 29 (revision 1)*. Rome: Food and Agricultural Organization of the United Nations.

Benson, D., Gain, A., & Rouillard, J. (2015). Water governance in a comparative perspective: From IWRM to a 'nexus' approach? *Water Alternatives, 8*(1), 756–773.

Borgomeo, E., Jagerskog, A., Talbi, A., Wijnen, M., Hejazi, M., & Miralles-Wilhelm, F. (2018). *The water-energy-food nexus in the Middle East and North Africa: Scenarios for a sustainable future*. Washington, DC: World Bank Report.

Dietrich, J., Bodirsky, B., & F. (2019). Humpenoder and seventeen, "MAgPIE 4: A modular open-source framework for modeling global land systems". *Geoscientific Model Development, 12*, 1299–1317.

Doorenbos, J., & Kassam, A. (1979). *Yield response to water, FAO irrigation and drainage paper 33*. Rome: Food and Agriculture Organization of the United Nations.

Ferroukhi, R., Nagpal, D., Lopez-Peña, A., Hodges, T., Mohtar, R., Daher, B., Mohtar, S., & Keulertz, M. (2015). *Renewable energy in the water, Energy & Food Nexus*. Report of the International Renewable Energy Agency (IRENA), United Arab Emirates.

GAMS. (2020). *General algebaric model solver*, https://www.gams.com.

Ginsburgh, V., & Keyzer, M. (2002). *The structure of applied general equilibrium models*. Cambridge, MA: MIT Press.

Grant, F., Sheline, C., Amrose, S., Brownell, E., Nangia, V., Talozi, S., & Winter, A. V. (2020). Validation of an analytical model to lower the cost of solar-powered drip irrigation systems for smallholder farmers in the Mena region. In *Proceedings of the ASME 2020 international design engineering technical conferences and computers and information in engineering conference*.

Hoff, H., Alrahaife, S., El-Hajj, R., Lohr, K., Mengoub, F., Farajalla, N., Fritzsche, K., Jobbins, G., Özerol, G., Schultz, R., & Ulrich, A. (2019). A nexus approach for the MENA region: From concept to knowledge to action. *Frontiers in Environmental Science, 7*, 48.

Keskes, T., Zahar, H., Ghezal, A., & Bedoui, K. (2019). Impact of solar-powered irrigation systems in Tunisia. In *Nexus regional dialogue in the MENA region, organized by league of Arab states, European Union and BMZ (Germany)*.

Lancaster, K. (1968). *Mathematical economics*. New York: MacMillan Company.

Lefore, N., Closas, A., & Schmitter, P. (2021). Solar for all: A framework to deliver inclusive and environmentally sustainable solar irrigation for smallholder agriculture. *Energy Policy, 154*, 112313.

Liu, J., Yang, H., Cudennec, C., Gain, A., Hoff, H., Lawford, R., Qi, J., DeStrasser, L., Yillia, P., & Zheng, C. (2017). Challenges in operationalizing the water–energy–food nexus. *Hydrological Sciences Journal, 62*(11), 1714–1720.

PSI and AUB. (2017). *Enhancing regional cooperation in the Middle East and North Africa through the WaterEnergy-food security nexus*. Planetary Security Initiative (The Hague) and the American University of Beirut.

Purwanto, A., Sušnik, J., Suryadi, F., & de Fraiture, C. (2021). Water-energy-food nexus: Critical review, practical applications, and prospects for future research. *Sustainability, 13*(4), 1–18.

Rabi, A., & Ghanem, I. (2016). *Pre master plan solar energy production in palestine*. Palestinian Environmental NGOs Network.

Robinson, S., Mason-D'Croz, D., Islam, S., Sulser, T., Robertson, R., Zhu, T., Gueneau, A., Pitois, G., & Rosegrant, M. (2015). *The international model for policy analysis of agricultural commodities and trade (IMPACT), model description for version 3*. Washington, DC: International Food Policy Research Institute.

Sabella, R., van Veen, W., van Wesenbeeck, L., & Sonneveld, B. (2020). *The water-energy-food nexus in a farm model for the Jordan River basin: Introducing solar panels on farmland*. Amsterdam: SunShine Project Report, Amsterdam Centre for World Food Studies.

Sachs, J. (2015). *The age of sustainable development*. New York: Columbia University Press.

Siddiqi, A., & Anadon, L. (2011). The water–energy nexus in Middle East and North Africa. *Energy Policy, 39*(8), 4529–4540.

Sonneveld, B., van Veen, W., & Marei, A. (2020). *Improving agricultural conditions in the Jericho district, Lower Jordan Valley: The prospects of large-scale water transfers from the Al-Faksha springs*. Amsterdam: Paduco Project Report, Amsterdam Centre for World Food Studies.

Steduto, P., Hsiao, T., Fereres, E., & Raes, D. (2012). *Crop yield response to water, FAO Irrigation and Drainage Paper 66*. Rome: Food and Agricultural Organization of the United Nations.

UNICEF. (2021). *Running dry: The impact of water scarcity on children in the Middle East and North Africa*. Amman: UNICEF Regional Office.

Varian, H. (2020). *Intermediate micro-economics* (9th ed.). New York and London: Norton and Company.

Waughray, D. (Ed.). (2011). *Water security: The water-food-energy-climate nexus*. Washington, DC: Island Press, published for the World Economic Forum Water Initiative.
Wichelns, D. (2017). The water-energy-food nexus: Is the increasing attention warranted, from either a research or policy perspective? *Environmental Science & Policy, 69*, 113–123.
Zhang, C., Chen, X., Li, Y., Ding, W., & Fu, G. (2018). Water-energy-food nexus: Concepts, questions and methodologies. *Journal of Cleaner Production, 195*, 625–639.
Zolfaghari, M., & Farzaneh, J. (2020). *Water-energy-food nexus in the middle east and north african countries (MENA)*, MPRA paper no. 104583, Munich Personal RePEc Archive. Available online at https://mpra.ub.uni-muenchen.de/104583/

Cross-cultural Adaptation

▶ Residential Crowding in Urban Environments

Crowding

▶ Residential Crowding in Urban Environments

Cultural Ecosystem Services

▶ Faith Communities as Hubs for Climate Resilience

Curated Emergence

▶ Systemic Innovation for Thrivable Cities

D

Decarbonization

▶ Amsterdam's Pathway to Climate Neutrality: Creating an Enabling Environment
▶ European Green Deal and Development Perspectives for the Mediterranean Region

Decentralization

▶ Urban Structure and Its Impact on Mobility Patterns: Reducing Automobile Dependence Through Polycentrism

Decentralized Concentration

▶ Urban Structure and Its Impact on Mobility Patterns: Reducing Automobile Dependence Through Polycentrism

Decentralized Cooperation and Political Strategy

▶ Internationalization of Cities

Decline

▶ Shrinking Towns and Cities

Delivery Bots

▶ Personal Delivery Robots: How Will Cities Manage Multiple, Automated, Logistics Fleets in Pedestrian Spaces?

Delivery Robots

▶ Personal Delivery Robots: How Will Cities Manage Multiple, Automated, Logistics Fleets in Pedestrian Spaces?

Demand Management

▶ Water-Smart Cities

R. C. Brears (ed.), *The Palgrave Encyclopedia of Urban and Regional Futures*,
https://doi.org/10.1007/978-3-030-87745-3

Democratic Processes

► Participatory Governance for Adaptable Communities

Demography

► Spatial Demography as the Shaper of Urban and Regional Planning Under the Impact of Rapid Urbanization

Density

► Urban Structure and Its Impact on Mobility Patterns: Reducing Automobile Dependence Through Polycentrism

Design

► Participatory Planning: A Useful Tool for the Development of Sustainable Mega-City Regions

Design Thinking

► Systemic Innovation for Thrivable Cities

Developing Countries

► Smart Agriculture and ICT

Developing World Cities – Cities of the Global South

► Urbanization, Planning Law, and the Future of Developing World Cities

Development

► Sustainable Development Goals and Urban Policy Innovation

Development – Expansion/ Growth

► Future of Urban Land-use Planning in the Quest for Local Economic Development

Development Goals

► An Overview of the Relationship of the Sustainable Development Goals and Urban and Regional Development

Digital City Modeling and Emerging Directions in Public Participation in Planning

Alexa Gower[1], Mette Hotker[2] and Carl Grodach[1]
[1]Monash University, Melbourne, VIC, Australia
[2]RMIT University, Melbourne, VIC, Australia

Introduction/Definition

City planning represents a key intersectional point for environmental, equity, and economic considerations. These critical issues are the pillars of sustainable development, and are reflected in The United Nations 17 Sustainable Development Goals (SDGs) (2015) as a transformative cross-sectoral agenda, whereof planning plays a key role. Whilst such a broad agenda offers direction for city planning it also presents challenges due to the complexities of information to be considered by planning professionals and community members alike. There is growing recognition that collaborative planning approaches are the best way to address planning decisions relating to trade-offs, uncertainty, and equity inherent in a sustainable development agenda and future-orientated decisions (Somarakis & Stratigea, 2019; Soomro, Khan and Ludlow, no date; Schulz & Newig, 2015). The complexities of this place further demands on effective and meaningful collaboration and public participation. In particular, improved public understanding of the trade-offs involved in complex planning decisions has been highlighted as necessary to produce the sophisticated discussions needed between all stakeholders for better decision-making (International association for public participation, 2021; Baker et al., 2007).

Digital participation models (DPMs) are increasingly utilized as a planning tool to grapple with the complexities of sustainable development. DPMs are scenario-based, 3D representations of cities that enable three key functions: DPMs are a means of communicating complex information in a visual format, gathering public feedback on social priorities and values (Anttiroiko, 2021) and may encourage interactive engagement around the intricacies and conflicts embedded in a particular planning context. The models can visualize and facilitate greater understanding of the decision trade-offs required by these intricacies and conflicts to encourage public consideration and input into these difficult choices and thereby, providing greater legitimacy in the final decision made (Schulz & Newig, 2015). While traditional planning and participation approaches are challenged by the layers of complexities inherent in these issues, DPM present an opportunity for complex engagement via more accessible communication, collaboration, engagement, creativity, consensus building, and integration of the UN Sustainable Development Goals (SDGs) into the planning system (Anttiroiko, 2021; Wilson et al., 2019). In this way, the models also offer a possible way in which to address criticisms of the "smart city" movement and the exclusive focus on rational efficiency and functionality that ignore social and community considerations (Cardullo & Kitchin, 2019).

This chapter discusses the need for DPM in public participation in complex issues, the potential role of the models in fostering comprehension of trade-offs in these issues and the significance of this improved understanding for sustainable development. It will then outline the implications of design in DPM and the future challenges existing for this planning tool.

Digital Participation Models Driven by the Need for Improved Public Participation in Complex Planning Issues

Historically, the uptake of digital technology in planning has been sporadic due to data and computational limitations (Kitchin et al., 2021). The most common focus has been on collecting and analyzing information via Smart City Data initiatives. Now, digital technology is being recognized for its potential as a collaborative tool and is the focus of DPM (Staffans et al., 2020). This shift is not only attributed to technological advancements but also the increase in community familiarity with technology (Houghton et al., 2014) and the need for greater consultation on complex issues (Schulz & Newig, 2015). Complex issues, such as environmental sustainability, social justice, and economic development, include various interrelated and conflicting trade-offs that are seen to benefit from the legitimacy and broader understanding offered by inclusive and wide-reaching participation. Traditional modes of engagement have struggled to capture sufficient breadth of access, whereas DPM have the potential to reach a greater diversity of voices including different citizens and their interests, stakeholders, and professional sectors, particularly those not typically included. Via the model, perspectives can be heard from younger demographics that typically attend face-to-face consultation workshops, as well as people who are employed full-time or otherwise occupied, making it difficult to attend day-time workshops (Bouzguenda et al., 2021; Kahila-Tani et al., 2016). Ease of access has been noted as an important factor for reaching a diversity of voices either in terms of the simplicity of the model (Dembski et al., 2020) or accessibility via readily available devices such as smartphones and watches (Wilson et al., 2019).

The notion that digital devices automatically broaden the diversity of voices is not unanimously agreed upon with technical and process limitations highlighted. Glaas and colleagues et al. (2020), for example, showed that whilst planners were impressed by the improved reach facilitated by digital participation models, participants were still critical of the extent of voices missed in the process; Jelokhani-Niaraki et al. (2019) showed that elderly people found it difficult to grasp the tool; and Akbar et al. (2020) found that fewer women were involved in the digital consultation than typical; and for those women that did attend, they experienced more conflict and power struggles with male participants than typical.

Digital Participation Models Can Improve Public Understanding of the Trade-Offs Involved in Complex Planning Decisions

Digital models are increasingly seen as an important tool to assist participants in understanding the trade-offs involved in complex planning. They have the ability to foster sophisticated and complex discussions between all stakeholders. DPM may improve communication and planning literacy by stepping participants through a proposal and quickly facilitating understanding of likely outcomes and unintended consequences that should be considered in the decision (Aspen & Amundsen, 2021; Newell et al., 2020; Dembski et al., 2020; Wilson et al., 2019). The models can also showcase the interrelationship of factors occurring in a given planning scenario (Newell et al., 2021) and in turn how this informs/guides decision-making and planning processes (Videira et al., 2017; Totin et al., 2018).

The real-time feedback through 3D scenario modeling also encourages participants to reflect more critically on alternative scenarios and the implications of their priorities compared to others (Wilson et al., 2019; Lieven et al., 2021). Functionalities such as galleries of preferences and suggestions by other participants with voting and commenting capabilities can further develop critical reflection and foster important discussion of options between participants (Mueller et al., 2018).

The 3D visualization invites place-based discussion for participants, which can be very useful to people not familiar with planning (Wilson et al., 2019), as it helps to locate themselves within a larger scale and encourage thinking at multiple scales concurrently (Aguilar et al., 2020; Akbar et al., 2020). Responses from participants were also more likely to be received spatially, which assisted planners to use this information gathered effectively (Kahila-Tani et al., 2016). Finally, the more comprehensive understanding of the issue and proposal facilitated by digital models also has been found to better manage community expectations in consultation by moving away from delivering a "perfect world" or "shopping wish-list" (Videira et al., 2017).

The Significance of Improved Public Understanding of Trade-Offs Via Digital Participation Models

Assisting the public to better understand the complexities of sustainable planning issues and the trade-offs involved can improve the sophistication of public contributions and insight. While this does not reduce or remove the trade-offs occurring in the issue, a more detailed understanding enables the public to make better-informed choices (Khan et al., 2014) and gain greater authority of their perspective by articulating holistic solutions (Khan et al., 2017; Glaas et al., 2020). This has been found to empower the public in planning issues (Webster & Leleux, 2018), encourage critical reflection on the built environment (Wilson et al., 2019), and foster a mutually beneficial planning-public relationship (Newell et al., 2020).

Specifically, a digital participation model has been found to improve public understanding of trade-offs and dialogue by communicating data in a visually objective manner. By being able to visualize a scenario and all of the consequences of this option when compared to alternative options, digital models are able to challenge misinformed assumptions held by participants on the topic (Schmitt Olabisi et al., 2016). An example of this is the preconceived ideas held around policies which support increases in residential density, with DPM able to illustrate the associated amenity and service provision gains that underpin this support as opposed to ideas that the development is arbitrary or without consideration of community needs. This understanding can broaden participants' outlook and thereby facilitate a more open discussion on the issue.

Additionally, this visual understanding creates a boundary object or common language between stakeholders to better communicate and discuss these trade-offs between participants (Schmitt Olabisi et al., 2016). It can break down planner jargon such as policy buzzwords like SDGs, or ways of thinking and analyzing such as assessing policy interventions from an overarching, bird's-eye view rather than the human street scale that a participant may more easily relate to (Aspen & Amundsen, 2021). It can also help contribute place-based knowledge, potentially previously undocumented (Falco et al., 2019) that can serve to inform planners of the public's social choices (Thoneick, 2021). DCM have been identified that by developing a common language, DPM have opened up previously, one-way planning processes of knowledge production to a two-way, more inclusive power dynamic (Staffans et al., 2020). Kahila-Tani et al. (2019) note that the potential for common language development and two-way communication can be further encouraged in digital models by providing all data freely available and open source to the public.

In addition to the planner/ public communication, a digital model can improve dialogue between the participants. Examples include fostering dialogue and understanding between the public and professionals (Bouzguenda et al., 2021; Lin & Benneker, 2021), the public and government policy-makers (Newell et al., 2021), and differently motivated members of the public (Akbar et al., 2020). Importantly, Lin and Benneker (2021) found that engagement via digital models between the public and development professionals provided information to the public more readily and avoided the establishment and spread of misinformed assumptions. For government policy-makers, early collaboration with the public through digital models has encouraged the co-discovery of shared values (Newell et al., 2021). This was found to be more effective for nurturing buy-in to a project than retroactive methods.

The Role for Design in Effective Digital Participation Models

For digital models to successfully improve public understanding of planning issues and the trade-offs present and assist public participation, a series of necessary design and technical features have been identified. Firstly, it has been stressed that the planning issue, the purpose of the digital participation, and model interface must both be clearly and directly presented to the user (Mueller et al., 2020; Sabri et al., 2016). Although Staffans et al. (2020) note that some traditional participation practices were also unclear in purpose and presentation, Newell et al. (2021) highlight how digital models also have the additional challenge of differing levels of technical literacy. This stresses the importance of clear design in the models as the model interface can, therefore, either assist understanding or become a source of further confusion and misunderstanding (Mueller et al., 2020). Visually prioritizing information and the narrative is commonly cited as necessary to effectively communicate the issue (Mueller et al., 2020), avoid information overload (Houghton et al., 2014), and focus feedback into areas that are pragmatic and useful (Bouzguenda et al., 2021). It is important to note that in a study by Glaas et al. (2020), planners lauded the communicative ability of a 3D model in a municipal-level planning participation around climate change issues, but public participants requested greater contextual information on, for example, the policy and the tool assumptions, alongside this 3D visualization to help their understanding of the issue.

Additional design issues include accuracy, with Dembski et al. (2020) finding that the omission of model aspects like traffic flow or inaccuracy in data decreased the trust held of the model as being a reliable representation of a scenario and trade-offs. Other studies have suggested improving creative input ability for the public, either increasing the editability of objects/data in the model, a pinned comment function or inclusion of social media (Lock et al., 2020; Mueller et al., 2018). Wilson et al. (2019) note that while technical advancements in these design problems have occurred, progress is still slow.

Future Challenges for Digital Participation Models

While DPM has greatly improved in capability and the specific design problems noted above over the past decade, many researchers still comment on the lack of uptake by planning for participation (Staffans et al., 2020; Kahila-Tani et al., 2016; Schulz & Newig, 2015). The research outlined below shows both practical reasons for this lack of connection between digital models and participation and broader, more fundamental issues of planning democracy.

Practically, planners are still grappling with how to integrate DPM and the broad-based information that they can produce into existing traditional policy systems (Aspen & Amundsen, 2021; Staffans et al., 2020) and siloed municipal departments (Glaas et al., 2020; Schulz & Newig, 2015). Furthermore, the initial and ongoing high cost of model development, data collection, and staff training are barriers many local planning departments have to digitally uptake (Khan et al., 2017). As models are often linked to specific project time and budget constraints, development is also hindered by sporadic spurts of activity, sudden stops, and uneven project implementation even across phases of the same project (Glaas et al., 2020; Gebetsroither-Geringer et al., 2018). This sporadic development phase impacts the prevalence of digital uptake by planners, further discouraging use. Data scarcity and sufficient accuracy is another problem, particularly in developing countries (Akbar et al., 2020).

More fundamentally, DPM is yet to satisfactorily address issues of transparency, dialogue capabilities and obscurity of power relations between professionals and different community stakeholders. Dembski et al. (2020) stress the importance of development for DPM in this area. There is a clear need to ensure that smart-planning processes and tools also are democratic tools. Currently, there is limited transparency and/or accountability regarding the incorporation or consideration of public contributions gathered from the models towards final decisions (Staffans et al., 2020; Thoneick, 2021) nor feedback to the public on how their input was used (Glaas et al., 2020). Cardullo and Kitchin (2019) also caution that the functionality for DPM must continue to develop around active participation goals and two-way dialogue despite being more difficult. The models mustn't devolve into simply one-way communication tools that inform but do not empower the public in planning issues. Lock et al. (2020) notes the importance of the public being involved not only in data/information provision but also in the evaluation criteria and supporting values that underpin the model. Lastly, the digital divide that the model encourages between participants is noted (Aspen & Amundsen, 2021). Self-selection towards projects that develop and progress the models may shape development further towards the already digitally literate, away from participants disempowered to take up the technology, and further exacerbate this digital divide (Kahila-Tani et al., 2019).

Conclusion

DPM holds promises for a technological advancement for planning participation, facilitating engagement, and reaching a broader audience than traditional consultation methods. The visual representation of the spatial issue can assist public understanding of the complexities and specifically the trade-offs involved in planning, thereby leading to more sophisticated discussions between stakeholders and decision-makers. While the need for clear and purposeful design in digital models is recognized and being improved, it is important that this technology development addresses fundamental issues of sustainability and planning democracy to continue to assist effective and meaningful public participation in planning.

Cross-References

- ▶ Community Engagement and Climate Change: The Value of Social Networks and Community-Based Organizations
- ▶ Community Engagement for Urban and Regional Futures

References

Aguilar, R., Flacke, J., & Pfeffer, K. (2020). Towards supporting collaborative spatial planning: Conceptualization of a Maptable tool through user stories. *ISPRS International Journal of Geo-Information, 9*(1), 29. https://doi.org/10.3390/ijgi9010029.

Akbar, A., et al. (2020). Knowing my village from the sky: A collaborative spatial learning framework to integrate spatial knowledge of stakeholders in achieving sustainable development goals. *ISPRS International Journal of Geo-Information, 9*(9), 515. https://doi.org/10.3390/ijgi9090515.

Anttiroiko, A.-V. (2021). Digital urban planning platforms: The interplay of digital and local embeddedness in urban planning. *International Journal of E-Planning Research, 10*(3), 35–49. https://doi.org/10.4018/IJEPR.20210701.oa3.

Aspen, D. M., & Amundsen, A. (2021). Developing a participatory planning support system for sustainable regional planning—A problem structuring case study. *Sustainability, 13*(10), 5723. https://doi.org/10.3390/su13105723.

Baker, M., Coaffee, J., & Sherriff, G. (2007). Achieving successful participation in the new UK spatial planning system. *Planning, Practice & Research, 22*(1), 79–93.

Bouzguenda, I., Fava, N., & Alalouch, C. (2021). Would 3D digital participatory planning improve social sustainability in smart cities? An empirical evaluation study in less-advantaged areas. *Journal of Urban Technology*, 1–31. https://doi.org/10.1080/10630732.2021.1900772.

Cardullo, P., & Kitchin, R. (2019). Being a "citizen" in the smart city: Up and down the scaffold of smart citizen participation in Dublin, Ireland. *GeoJournal, 84*(1), 1–13. https://doi.org/10.1007/s10708-018-9845-8.

Dembski, F., et al. (2020). Urban digital twins for smart cities and citizens: The case study of Herrenberg, Germany. *Sustainability, 12*(6), 2307. https://doi.org/10.3390/su12062307.

Falco, E., Zambrano-Verratti, J., & Kleinhans, R. (2019). Web-based participatory mapping in informal settlements: The slums of Caracas, Venezuela. *Habitat International, 94*, 102038. https://doi.org/10.1016/j.habitatint.2019.102038.

Gebetsroither-Geringer, E., Stollnberger, R., & Peters-Anders, J. (2018). 'Interactive spatial web-applications as new means of support for urban decision-making processes', *ISPRS annals of the photogrammetry, remote sensing and spatial. Information Sciences, IV-4/W7*, 59–66. https://doi.org/10.5194/isprs-annals-IV-4-W7-59-2018.

Glaas, E., et al. (2020). Visualization for citizen participation: User perceptions on a mainstreamed online participatory tool and its usefulness for climate change planning. *Sustainability, 12*(2), 705. https://doi.org/10.3390/su12020705.

Houghton, K., Miller, E., & Foth, M. (2014). Integrating ICT into the planning process: Impacts, opportunities and challenges. *Australian Planner, 51*(1), 24–33. https://doi.org/10.1080/07293682.2013.770771.

International association for public participation (2021) *Public participation pillars*. Available at: https://cdn.ymaws.com/www.iap2.org/resource/resmgr/Communications/A3_P2_Pillars_brochure.pdf. Accessed 17 Aug 2021.

Jelokhani-Niaraki, M., Hajiloo, F., & Samany, N. N. (2019). A web-based public participation GIS for assessing the age-friendliness of cities: A case study in Tehran, Iran. *Cities, 95*, 102471. https://doi.org/10.1016/j.cities.2019.102471.

Kahila-Tani, M., et al. (2016). Let the citizens map—Public participation GIS as a planning support system in the Helsinki master plan process. *Planning Practice & Research, 31*(2), 195–214. https://doi.org/10.1080/02697459.2015.1104203.

Kahila-Tani, M., Kytta, M., & Geertman, S. (2019). Does mapping improve public participation? Exploring the pros and cons of using public participation GIS in urban planning practices. *Landscape and Urban Planning, 186*, 45–55. https://doi.org/10.1016/j.landurbplan.2019.02.019.

Khan, Z., et al. (2014). ICT enabled participatory urban planning and policy development: The urban API project. *Transforming Government: People, Process and Policy, 8*(2), 205–229. https://doi.org/10.1108/TG-09-2013-0030.

Khan, Z., et al. (2017). Developing knowledge-based citizen participation platform to support Smart City decision making: The Smarticipate case study. *Information, 8*(2), 47. https://doi.org/10.3390/info8020047.

Kitchin, R., Young, G. W., & Dawkins, O. (2021). Planning and 3D spatial media: Progress, prospects, and the knowledge and experiences of local government planners in Ireland. *Planning Theory & Practice*, 1–19. https://doi.org/10.1080/14649357.2021.1921832.

Lieven, C., et al. (2021). Enabling digital co-creation in urban planning and development. In A. Zimmermann, R. J. Howlett, & L. C. Jain (Eds.), *Human Centred Intelligent Systems* (pp. 415–430). Singapore: Springer Singapore (Smart Innovation, Systems and Technologies). https://doi.org/10.1007/978-981-15-5784-2_34.

Lin, Y., & Benneker, K. (2021). Assessing collaborative planning and the added value of planning support apps in the Netherlands. *Environment and Planning B: Urban Analytics and City Science*, 239980832110092. https://doi.org/10.1177/23998083211009239.

Lock, O., et al. (2020). A review and reframing of participatory urban dashboards. *City, Culture and Society, 20*, 100294. https://doi.org/10.1016/j.ccs.2019.100294.

Mueller, J., et al. (2018). Citizen design science: A strategy for crowd-creative urban design. *Cities, 72*, 181–188. https://doi.org/10.1016/j.cities.2017.08.018.

Mueller, J., Asada, S., & Tomarchio, L. (2020). Engaging the crowd: Lessons for outreach and tool design from a creative online participatory study. *International Journal of E-Planning Research, 9*(2), 66–79. https://doi.org/10.4018/IJEPR.2020040101.oa.

Newell, R., Picketts, I., & Dale, A. (2020). Community systems models and development scenarios for integrated planning: Lessons learned from a participatory approach. *Community Development, 51*(3), 261–282. https://doi.org/10.1080/15575330.2020.1772334.

Newell, R., et al. (2021). Communicating complexity: Interactive model explorers and immersive visualizations as tools for local planning and community engagement. *FACETS. Edited by V. Metcalf, 6*(1), 287–316. https://doi.org/10.1139/facets-2020-0045.

Sabri, S., et al. (2016). Leveraging VGI integrated with 3D spatial technology to support urban intensification in Melbourne, Australia. *Urban Planning, 1*(2), 32–48. https://doi.org/10.17645/up.v1i2.623.

Schmitt Olabisi, L., et al. (2016). Participatory, dynamic models: A tool for dialogue. In A. S. Parris et al. (Eds.), *Climate in Context* (pp. 99–116). Chichester: Wiley. https://doi.org/10.1002/9781118474785.ch5.

Schulz, D., & Newig, J. (2015). Assessing online consultation in participatory governance: Conceptual framework and a case study of a national sustainability-related consultation platform in Germany: Online consultation in participatory governance. *Environmental Policy and Governance, 25*(1), 55–69. https://doi.org/10.1002/eet.1655.

Somarakis, G., & Stratigea, A. (2019). Guiding informed choices on participation tools in spatial planning: An E-decision support system. *International Journal of E-Planning Research, 8*(3), 38–61. https://doi.org/10.4018/IJEPR.2019070103.

Soomro, K., Khan, Z. & Ludlow, D. (n.d.) *Participatory Governance in Smart Cities: the urbanAPI case study.* p. 28.

Staffans, A., et al. (2020). Communication-oriented and process-sensitive planning support. *International Journal of E-Planning Research, 9*(2), 1–20. https://doi.org/10.4018/IJEPR.2020040101.

Thoneick, R. (2021). Integrating online and onsite participation in urban planning: Assessment of a digital participation system. *International Journal of E-Planning Research, 10*(1), 1–20. https://doi.org/10.4018/IJEPR.2021010101.

Totin, E., et al. (2018). Can scenario planning catalyse transformational change? Evaluating a climate change policy case study in Mali. *Futures, 96*, 44–56. https://doi.org/10.1016/j.futures.2017.11.005.

United Nations (2015) *Transforming our world: the 2030 Agenda for Sustainable Development*. Available at: https://sdgs.un.org/2030agenda. Accessed 7 Sept 2021.

Videira, N., Antunes, P., & Santos, R. (2017). Engaging stakeholders in environmental and sustainability decisions with participatory system dynamics modelling. In S. Gray et al. (Eds.), *Environmental modelling with stakeholders* (pp. 241–265). Cham: Springer International Publishing. https://doi.org/10.1007/978-3-319-25053-3_12.

Webster, C. W. R., & Leleux, C. (2018). Smart governance: Opportunities for technologically-mediated citizen co-production. *Information Polity, 23*(1), 95–110. https://doi.org/10.3233/IP-170065.

Wilson, A., Tewdwr-Jones, M., & Comber, R. (2019). Urban planning, public participation and digital technology: App development as a method of generating citizen involvement in local planning processes. *Environment and Planning B: Urban Analytics and City Science, 46*(2), 286–302. https://doi.org/10.1177/2399808317712515.

Digital Nature

► Emerging Concepts Exploring the Role of Nature for Health and Well-Being

Digital Turn

► Augmented Reality: Robotics, Urbanism, and the Digital Turn

Digital Twin and Cities

Samad M. E. Sepasgozar
School of Built Environment, University of New South Wales, Sydney, NSW, Australia

Synonyms

City digital twin; Spatial digital twin; Urban digital twin; Virtual city

Definition

The digital twin, by definition, is tied up with a range of concepts such as digital representation, integration, connection, real-time 3D visualization, computation, prediction, and decision-making. It is defined as a digital replica of a physical entity with bidirectional data flow that means both physical and virtual twins are synchronized, and simulations, optimizations, or visualizations are real-time.

Introduction

Digital twin is an emerging technology and has been well-defined in the aerospace, automotive, and manufacturing contexts along with the demonstration of various prototypes or practices. However, when it comes to spatial planning and social contexts or any individual less-industrialized sectors, it can be confused by other information systems and modeling technologies that are currently in use. This compels the clarification of the concepts and definitions in the city context and spatial planning. The purpose of this chapter is to discuss the concept of city digital twin and review the collocation of terminologies and technologies in the city context and review the applications. While initial concepts of a digital twin have been voiced in the early 1990s, clearly specified definitions along with successful prototypes have been recently developed. The following presents how the concept of the digital twin has been evolved and revised over previous decades.

1990s: indicators of a mirror word were introduced, such as a deep and live picture of the world, including historical data gathered previously from an organization. The mirror world gave an image of how an organization, including its processes and data, can be represented by a computer. Mirror world is defined as software-based models representing a part of reality which was originally referred to as business management and organization environment (Gelernter 1993).

2000s: it referred to the virtual replica or counterpart of a physical object. In this decade, information mirroring or connections became an important element of digital twins. Digital twins were mainly used for manufacturing and product lifecycle management when a new generation of lean thinking was introduced by Githens (2007).

2010s: refers to physical objects, digital models, and the connections between these two through sensing technologies and network platforms for predicting the performance of the physical system. This was applied mainly to assets such as aircraft, locomotives, industrial equipment, and large plants. In this period, the integration of multi-physics at different scales and more complex scenarios were considered. The synchronization of the data, consistency of data format, and utilization of algorithms to process data in real-time was the core demand based on the evolved definition. NASA developed a road map focusing on a flying twin with a plan for using quantum computing, developing flight computation and information processing systems with lower energy consumption and real-time vision processing mechanisms (Shafto et al. 2012). The awareness of the digital twin has been significantly increased in this decade due to the fourth industrial revolution and the popularity of the Industry 4.0 concept.

2020s: digital twins were applied in different contexts, and the definitions tend to be revised and adjusted to each context. The integration of digital twins with a range of disciplinary tools and technologies is investigated. For example, integration of building information modeling (BIM) and digital twins in construction or the interoperability of geographic information systems (GIS) and digital twins in city planning are perceived as industry demand.

Digital Twin Versions and Variations

The DT applications are being increasingly developed, and its multi-billion market will be exponentially expanded in the near future. The significance of the DT is not in the nature of the technology and its advancements; rather, it lies in the impact of its utilization on the current processes, decision-making processes, and operation mechanisms. While developing a DT for a product or asset involves the integration of many tools and technologies, it will be much more complex when the digital twin must cover a city fully or partially, considering citizens, their activities, and the physical body of the city. The physical body such as buildings, streets, and other objects can be less complex to be twinned compared to dynamic objects such as human activities and their engagement. This means the exercise of developing a city digital twin may take years to be fully developed and become useful to policymakers and users. In practice, city twining faces some challenging risks and barriers such as cybersecurity, privacy, and the expenses of development and maintenance, which negatively affect the rate and speed of the digital twin diffusion within sectors and societies.

The following concepts go hand-in-hand with digital twins that could be one of the required components of the digital twin in various sectors. Sometimes, these concepts or technologies are used without developing an actual digital twin. To understand what an actual digital twin is, relevant concepts or components should be defined in various contexts. This also will help to define the digital twin for cities.

3D Data

Three-dimensional (3D) data incorporates an additional dimension to the traditional x and y coordinates. This is important in geospatial analysis at the city scale. Laser scanners and Light Detection and Ranging (lidar) tools are key sources of producing point clouds, including 3D data, digital elevation models (DEM), or digital surface models (DSM). These datasets and models are mainly used for building and city analysis. See examples: Spatiotemporal assessment of 3D urban areas (Shirowzhan and Trinder 2017) and applying autocorrelation-based computation over urban areas (Shirowzhan et al. 2016).

3D Models and Simulations

A digital 3D model is the visualization of data and mathematical representation of a physical entity that is generated by a computer. A simulation mainly represents a particular process and helps to predict what could happen to a physical object under different situations. Simulations may improve processes and decision-making processes within organizations. However, simulations are not necessarily linked to real-time data. Voxel automata are suggested as an important simulation technique for spatial planning at the city scale that is mainly based on key concepts of location automata systems. A voxel represents the smallest size of a 3D spatial object, different than a pixel that is a small 2D of an image (Shirowzhan et al. 2018; Gorte et al. 2019). Figure 1 shows some examples of voxelization (Shirowzhan et al. 2018). Voxelization techniques and principles are suggested for investigating the 3D spatial behavior of cities and large buildings; details are discussed by Shirowzhan et al. (2018) and Shirowzhan et al. (2021).

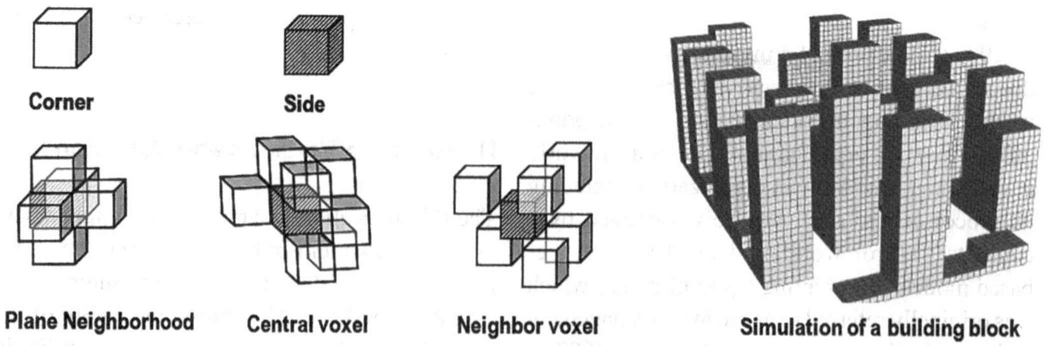

Digital Twin and Cities, Fig. 1 Voxels (left) and voxelated representation of the simulated 3D building block (right)

Digital Thread

Digital thread (DTh) refers to the development of 3D models created in a computer-aided design (CAD) environment to drive a computer numerically controlled (CNC) machine. This is possible if the digital data flow is unbroken during the entire process of manufacturing the physical object.

Physical Internet

Physical internet (PI) refers to digital supply chain and transportation networks as open global systems where the system components are hyper-connected, and the relevant networks are interconnected (Sternberg and Norman 2017). The PI is used for increasing the sustainability and efficiency of transportation and supply chain management (Nguyen et al. 2022).

Digital Shadow

The digital representation of a physical object with one-directional data flow between the physical entity and the 3D model is considered a digital shadow. Similar to other 3D models, the data can be captured by laser scanners and be updated by a range of sensing technologies. However, there will not be bidirectional data flow between both virtual and physical entities, and prediction algorithms are not used for creating a digital shadow. Digital as-builts (Li et al. 2020) and part-built (Sepasgozar et al. 2018) are categorized as a digital shadow. The difference between a digital shadow and a digital twin is discussed in detail by Sepasgozar (2021).

Digital Twin for City and Spatial Planning

The abovementioned technologies can be combined for developing a set of digital twins for a city of part of a city. Kevin Lynch describes an image of a city by considering people and their activities along with the physical elements of the city (Lynch 2020). A digital twin for a city should be able to consider three parameters: people, their activities, and the physical environment to be able to provide useful information predicting citizens' behavior, engagement, and needs. Thus, city digital twins can be defined as:

> A virtual live representation of a spatial unit of a city, including people and/or their activities, offering a probabilistic simulation of changes, spatio-temporal visualization, and public participation. Changes and time-varying behavior may refer to people, their activities, local mobilities, movement, urban growth, carbonization, vegetation, and population dynamics. The simulations and visualizations can be multi-physics and multi-scales.

In this definition, the city digital twin is not only a digital twin at the urban "scale" rather, it refers to the analysis of dynamics in a city or part of it. The key consideration is that city is not a product or a simple asset. Digital twins can be limited to a component or part called part twins, or it can cover the whole asset, a system, or a process. Figure 2 shows an overall view of a city digital twin. Examples of digital twins at the urban scale are presented by Charitonidou (2022).

Ecosystem Data Management

Digital twin ecosystem data management (DT-EDM) refers to applications and the infrastructure established in an organization for data acquisition, cloud-based storage, cloud-based computation, sharing, governance and stewardship, leverage, curation, and standards. Different organizations may have different data ecosystems.

The key considerations of EDM for developing a city digital twin are data variety, size or volume of data, data mining, efficient data query, and complexity of algorithms. The requirement of an efficient DT-EDM is to deploy high-performance computing (HPC) for processing the relevant real-time data and performing advanced computing algorithms in a speedy manner.

Governance, Standards, and Regulations

Digital twin at the city scale processes big datasets which can be created and shared by a wide range of stakeholders. More than any industrial digital twins, a city digital twin requires the participation and collaboration of as many stakeholders as possible, and this means the need for a clear set of governance and regulations become very important. Otherwise, there will be many legal disputes which may affect the usefulness of the city digital

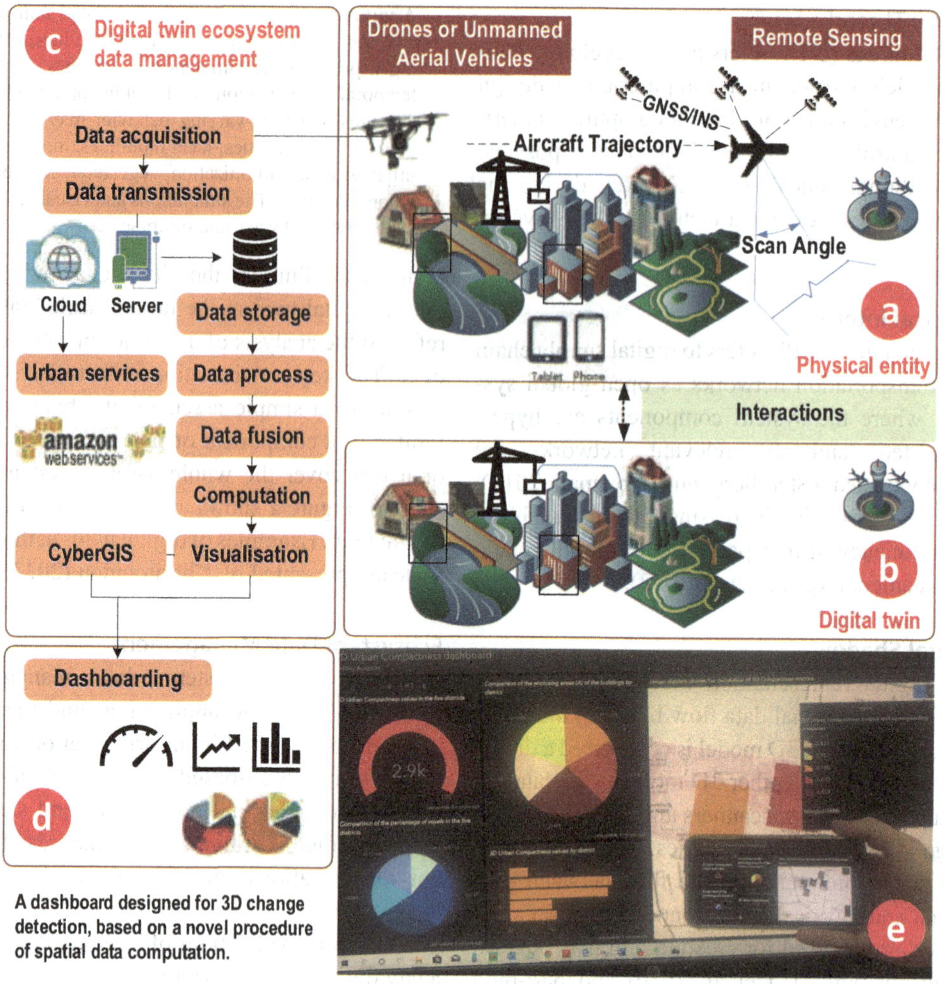

Digital Twin and Cities, Fig. 2 Illustration of digital twin of a city including dashboards for visualization of data

twin negatively. There are some organizations such as ANZLIC – the Spatial Information Council, DT Consortium, DT cities center, and center for DT Britain as pioneers of digital twin standardization to support national digital twin development schemes in Australia, the USA, Sweden, and the UK, respectively. There are needs in each country to establish DT-EDM and connect various businesses and institutions at the city and regional levels. Key considerations of the governance and accountability for city digital twin are security and privacy, curation with clear responsibilities and ownerships, standards and interoperability, and a federated model for sharing data custodians.

Summary and Conclusions

City digital twins provide real-time datasets which may transform the spatial planning process and city-level operations. A wide range of technologies and techniques such as artificial intelligence augmented reality, drones, the internet of things, blockchain, and digital 3D modeling can be combined for developing a city digital twin. A city digital twin is supposed to consider people and their activities along with physical elements of the city, which means it is not only covering physical entities or a product. This consideration will make the twin be useful and help in predicting citizens' behavior, engagement, and daily needs. Digital twins do not necessarily cover the entire city, and they can be limited to a component or part called part twins addressing a specific need of the city. These needs can be monitoring and controlling air quality, water, and energy consumption, Co_2 emissions, and noise at building and city level, or providing live transport movements, connecting all smart systems, and improve citizens' experience by offering them real-time services.

Cross-References

► Big Data for Smart Cities and Inclusive Growth
► Smart Cities

References

Charitonidou, M. (2022). Urban scale digital twins in data-driven society: Challenging digital universalism in urban planning decision-making. *International Journal of Architectural Computing*, 14780771211070005.

Gelernter, D. (1993). *Mirror worlds: Or the day software puts the universe in a shoebox... How it will happen and what it will mean*. Oxford University Press.

Githens, G. (2007). *Product lifecycle management: Driving the next generation of lean thinking by Michael grieves*. Wiley Online Library.

Gorte, B., Zlatanova, S., & Fadli, F. (2019). Navigation in indoor voxel models. *ISPRS Annals of the Photogrammetry, Remote Sensing and Spatial Information Sciences, IV-2/W5*, 279.

Li, Y., Li, W., Tang, S., Darwish, W., Hu, Y., & Chen, W. (2020). Automatic indoor as-built building information models generation by using low-cost RGB-D sensors. *Sensors, 20*(1), 293.

Lynch, K. (2020). *"The City image and its elements": From the image of the City (1960)* (pp. 570–580). The City Reader: Routledge.

Nguyen, T., Duong, Q. H., Nguyen, T. V., Zhu, Y., & Zhou, L. (2022). Knowledge mapping of digital twin and physical internet in supply chain management: A systematic literature review. *International Journal of Production Economics, 244*, 108381.

Sepasgozar, S. M. (2021). Differentiating digital twin from digital shadow: Elucidating a paradigm shift to expedite a smart, sustainable built environment. *Buildings, 11*(4), 151.

Sepasgozar, S. M., Forsythe, P., & Shirowzhan, S. (2018). Evaluation of terrestrial and mobile scanner technologies for part-built information modeling. *Journal of Construction Engineering and Management, 144*(12), 04018110.

Shafto, M., Conroy, M., Doyle, R., Glaessgen, E., Kemp, C., LeMoigne, J., & Wang, L. (2012). *Modeling, simulation, information technology & processing roadmap*. National Aeronautics and Space Administration.

Shirowzhan, S., & Trinder, J. (2017). Building classification from lidar data for spatio-temporal assessment of 3D urban developments. *Procedia Engineering, 180*, 1453–1461.

Shirowzhan, S., Lim, S., & Trinder, J. (2016). Enhanced autocorrelation-based algorithms for filtering airborne lidar data over urban areas. *Journal of Surveying Engineering, 142*(2), 04015008.

Shirowzhan, S., Sepasgozar, S. M. E., Li, H., & Trinder, J. (2018). Spatial compactness metrics and constrained voxel automata development for analyzing 3D densification and applying to point clouds: A synthetic review. *Automation in Construction, 96*, 236–249.

Shirowzhan, S., Sepasgozar, S. M., & Trinder, J. (2021). Developing metrics for quantifying buildings' 3D compactness and visualizing point cloud data on a web-based app and dashboard. *Journal of Construction Engineering and Management, 147*(3), 04020178.

Sternberg, H., & Norrman, A. (2017). The physical internet–review, analysis and future research agenda. *International Journal of Physical Distribution & Logistics Management, 47*, 736.

Further Reading

Glaessgen, E., & Stargel, D. (2012). April. The digital twin paradigm for future NASA and US Air Force vehicles. In *53rd AIAA/ASME/ASCE/AHS/ASC structures, structural dynamics and materials conference 20th AIAA/ASME/AHS adaptive structures conference 14th AIAA* (p. 1818). https://doi.org/10.2514/6.2012-1818.

Sepasgozar, S. M., Ghobadi, M., Shirowzhan, S., Edwards, D. J., & Delzendeh, E. (2021). Metrics development and modelling the mixed reality and digital twin adoption in the context of industry 4.0. *Engineering, Construction and Architectural Management*. https://doi.org/10.1108/ECAM-10-2020-0880.

Sepasgozar, S., Karimi, R., Farahzadi, L., Moezzi, F., Shirowzhan, S., Ebrahimzadeh, S. M., Hui, F., & Aye, L. (2020). A systematic content review of artificial intelligence and the internet of things applications in smart home. *Applied Sciences, 10*(9), 3074. https://doi.org/10.3390/app10093074.

Shirowzhan, S., Sepasgozar, S. M., & Trinder, J. (2021). Developing metrics for quantifying buildings' 3D compactness and visualizing point cloud data on a web-based app and dashboard. *Journal of Construction Engineering and Management, 147*(3), 04020178. https://orcid.org/0000-0003-2568-3111.

Shirowzhan, S., Tan, W. and Sepasgozar, S.M., 2020. Digital twin and CyberGIS for improving connectivity and measuring the impact of infrastructure construction planning in smart cities. ISPRS International Journal of Geo-Information, 9(4), p. 240. https://doi.org/10.3390/ijgi9040240.

D

Sepasgozar, S. M. (2020). Digital twin and web-based virtual gaming technologies for online education: A case of construction management and engineering. *Applied Sciences, 10*(13), 4678. https://doi.org/10.3390/app10134678.

Sepasgozar, S. M., & Shirowzhan, S. (2016). Challenges and opportunities for implementation of laser scanners in building construction. In *ISARC. Proceedings of the international symposium on automation and robotics in construction* (Vol. 33, p. 1). IAARC Publications.

Shirowzhan, S., Lim, S., Trinder, J., Li, H., & Sepasgozar, S. M. (2020). Data mining for recognition of spatial distribution patterns of building heights using airborne lidar data. *Advanced Engineering Informatics, 43*, 101033. https://doi.org/10.1016/j.aei.2020.101033.

Munawar, H. S., Qayyum, S., Ullah, F., & Sepasgozar, S. (2020). Big data and its applications in smart real estate and the disaster management life cycle: A systematic analysis. *Big Data and Cognitive Computing, 4*(2), 4. https://doi.org/10.3390/bdcc4020004.

Teisserenc, B., & Sepasgozar, S. (2021). Adoption of Blockchain technology through digital twins in the construction industry 4.0: A PESTELS approach. *Buildings, 11*(12), 670. https://doi.org/10.3390/buildings11120670.

Teisserenc, B., & Sepasgozar, S. (2021). Project data categorization, adoption factors, and non-functional requirements for Blockchain based digital twins in the construction industry 4.0. *Buildings, 11*(12), 626. https://doi.org/10.3390/buildings11120626.

Digitalization

► The Sustainable and the Smart City: Distinguishing Two Contemporary Urban Visions

Digitalization, Urbanization, and Urban-Rural Divide

Anupam Nanda
University of Manchester, Manchester, UK

Introduction

Throughout the history of human civilization, location choice has been an area of constant consideration and exploration. In the early days, locational decisions were primarily driven by climatic conditions and food sources. In recent times, technological progress, location of manufacturing industries and income prospect have played important roles in locations of human settlement and these factors have also fueled a shift from agriculture-based economies to industry-based economies. Such shift has naturally caused rapid urbanization in recent years. As we look into the future, digitalization and digital technologies appear to be the most significant driver that would shape how and where we live. With the widespread use of the internet and mobile technologies, we are experiencing massive transformations across all sectors and aspects of human life. As it stands and it is often noted by many commentators, this is a cross-cutting trend along with other mega-trends with profound economic and social implications for the future (see Nanda 2019, pp. 212 for a note on these trends, and also Ch. 5 for a discussion). These trends are somewhat interrelated as noted below:

- ***Rapid urbanization*** – According to the United Nations report (https://www.un.org/ga/Istanbul+5/bg10.htm), the extent of urbanization rose from only about 2% of the world's population in 1800 to almost 30% in 1950, 47% in 2000 and 54% in 2016. The projections indicate that it would reach 60% by 2030 and as high as 70% by 2050. An important implication for high level of urbanization is increasing concentration of socio-economic activities and massive demand for public amenities and services.
- ***Deepening climate crisis*** – Rise in average temperature has widespread implications for all aspects of human life across all urban and rural areas (see the recent report from IPCC – https://www.ipcc.ch/2021/08/09/ar6-wg1-20210809-pr/). There is an urgent need to address those implications in terms of optimal resource use, development of alternative sources of resource, disaster management, building resiliencies along with robust adaptation and mitigation strategies.
- ***Ageing population*** – Population ageing is evident across many parts of the world. Innovations in medical science (such as vaccinations, drugs and medical procedures) along with easy

and affordable access to medical products and services and more health awareness are contributing to rising life expectancy. This has significant implications for demand for healthcare, social care, and other amenities in urban and rural areas, as communities start to have a higher proportion of elderly cohort.
- ***Large-scale migration*** – We are experiencing intensification of a long-standing trend – migration – across many parts of the world. Since urban areas are the centers of economic activities and offer job and income potential, migrants flock to those areas. In future, we may also see climate-driven migration.

However, our world is unequal and none of the above-mentioned trends is being experienced equally across the world. Some regions and their populations are better prepared, more equipped and resourced, compared to others. Those areas enjoy advantages in terms of resources, access to technology and pace of growth and development. Such inequalities impede progress toward meeting the United Nations' millennium development goals (MDGs). This seems to be the most important area of policy-making across all levels – from subnational to international. Therefore, in this review, I examine urbanization and digitalization by focusing on the implications for existing and new inequalities, and shed light on some of the policy concerns.

Extent of Urbanization

The pace of urbanization has been especially high since the early 1980s, with reforms across major economies fueling the process. Urban industrial hubs have boosted job creation and income generation. Therefore, arguably, urbanization has improved GDP growth for many countries and lifted many out of poverty. Rural-to-urban migration has provided labor resource to fuel the industrial activities. However, this has also inflicted a loss of population in rural hinterlands. These phenomena have come with their problems, the most prominent being the unequal impacts across the geographies – across and within countries. Some regions have benefitted, and some have lagged, thus widening the economic inequalities. The role of inequality in the growth–poverty nexus has been studied extensively by several authors (see Bourguignon 2003; Epaulard 2003; Fosu 2009; Kalwij and Verschoor 2007; Ravallion 1997). Fosu (2017) noted that countries with the same level of growth rates experienced different realities in attaining goal 1 of the MDGs (MDG1) of halving poverty levels by 2015. The current sustainable development goal 1 (SDG1) of eradicating poverty by 2030 may have similar realities and require country-level growth targets.

Figure 1 shows the extent of urbanization across the world as of 2017. Urbanization patterns reveal different trajectories and show varied characteristics across the countries. Comparing the rate of urbanization across developed and developing countries, Jedwab et al. (2017) point out a faster urban expansion in the developing world, compared to that across the developed countries.

The inequalities within the countries remained significant as much of the growth stories was concentrated in cities and urban areas. Both the factors of production as well as the aspects of consumption are heavily concentrated within cities and urban areas. Chauvin et al. (2017) studied four major developed and developing economies (Brazil, China, India, and the USA) and examined whether the well-known facts about urbanization in the USA had also held true for other three countries. Their analysis suggests that both Gibrat's Law and Zipf's Law seem to hold as well in Brazil as in the US, but China and India look quite different. In Brazil and China, the implications of the spatial equilibrium hypothesis as the central organizing idea of urban economics are not rejected. (Gibrat's Law is a rule defined by Robert Gibrat (1904–1980) in 1931, stating that the proportional rate of growth of a firm is independent of its absolute size. Gibrat's Law can be applied to city size and growth rate, where the proportionate growth process is independent of city size and may give rise to a distribution of city sizes that is log-normal, as predicted by Gibrat's Law (source: Eeckhout (2004) and Wikipedia). "Zipf's Law claims that the number of people in a city is inversely proportional to the

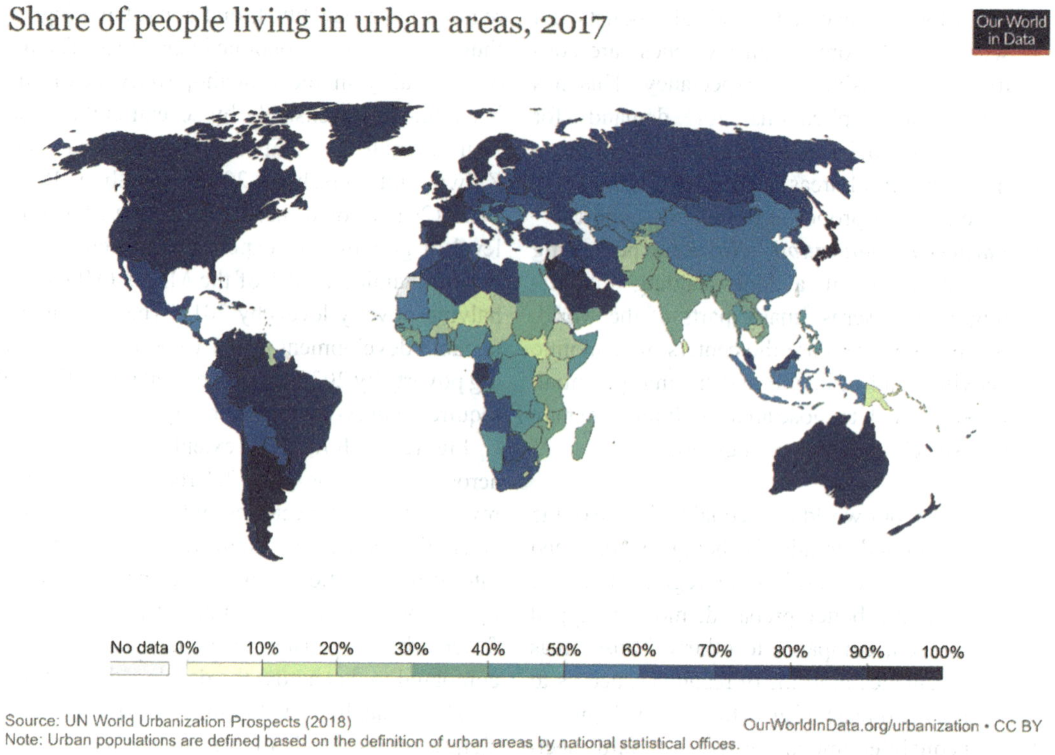

Digitalization, Urbanization, and Urban-Rural Divide, Fig. 1 Extent of Urbanization. (Sources: World Bank, Our World in Data https://ourworldindata.org/grapher/share-of-population-urban)

city's rank among all cities. In other words, the biggest city is about twice the size of the second biggest city, three times the size of the third biggest city, and so forth. Zipf's Law is named after the linguist George Kingsley Zipf, who discovered the Law when studying the distribution of words: the second most common word in a text typically shows up one-half as often as the most commonly used word" (source: Glaeser, E., https://economix.blogs.nytimes.com/2010/04/20/a-tale-of-many-cities/). Spatial equilibrium models solve equilibrium conditions in multiple regional markets simultaneously, assuming that transportation costs exist between two regions.) The India data, however, repeatedly rejects tests for the spatial equilibrium assumption. One hypothesis is that the spatial equilibrium only emerges with economic development, as markets replace social relationships and as human capital spreads more widely. In all four countries, there is strong evidence of agglomeration economies and human capital externalities. The correlation between density and earnings is stronger in both India and China than in the US and strongest in China. In India, the gap between urban and rural wages is huge, but the correlation between the city size and earnings is more modest. The cross-sectional relationships between area-level skills, earnings and area-level growth are also stronger in the developing world than in the US. The forces that determine urban successes appear to be similar in the rich and poor world, even if limited migration, restrictive migration policies and unaffordable housing markets make it harder for a spatial equilibrium to emerge. However, the idea of spatial equilibrium also faces significant implications as digitalization intensifies and enables delivery of an increasing number of economic goods through digital platforms and processes.

Extent of Digitalization

Technological innovations since 1990s, fueled by widespread access and use of internet, have led to significant restructuring across most economies in the world. Especially, the recent decade has experienced a high level of internet-based economic activities, a prominent sector to face that being the retail industry (see Nanda et al. 2021). Looking into the future, it can only be expected to intensify more as Internet penetration improves across the world. A key implication of the digitalization is blurring of distance and greater ability to reach remote areas. Digital technologies can create significant economic opportunities with new types of jobs, more streamlined processes, reduced transaction costs, less information asymmetry, less demand-supply uncertainties, etc. If we were to eradicate poverty and inequalities, these opportunities need to be harnessed better and delivered to the section of the population with targeted policies and robust governance structure.

Figure 2 shows the share of the population using Internet in 2017. It shows a significant amount of disparities across the countries. While the North America and Europe show very high level of share of population using Internet, some of the largest economies in the world such as India and China are still far away from reaching very high level. A large part of the African continent shows very low level of Internet users.

The Inclusive Internet Index, commissioned by Facebook and conducted by The Economist Intelligence Unit (EIU), measures Internet inclusion in 120 countries across four categories: Availability, Affordability, Relevance, and Readiness. Table 1 shows ranks based on the III Index and also shows the extent of urbanization across a set of selected countries – a mixture of developed, less developed, large and small countries around the world.

A crude correlation measure between the Inclusive Internet Index, 2021 (column 3 in Table 1) and percent of urban population in 2017 (column 4 in Table 1) stands at almost 65% within this sample of countries. While this correlation calculation is far from being a robust measure of the complex relationship between urbanization and digitalization, it can perhaps indicate that economic dynamics of urbanization can be fueled

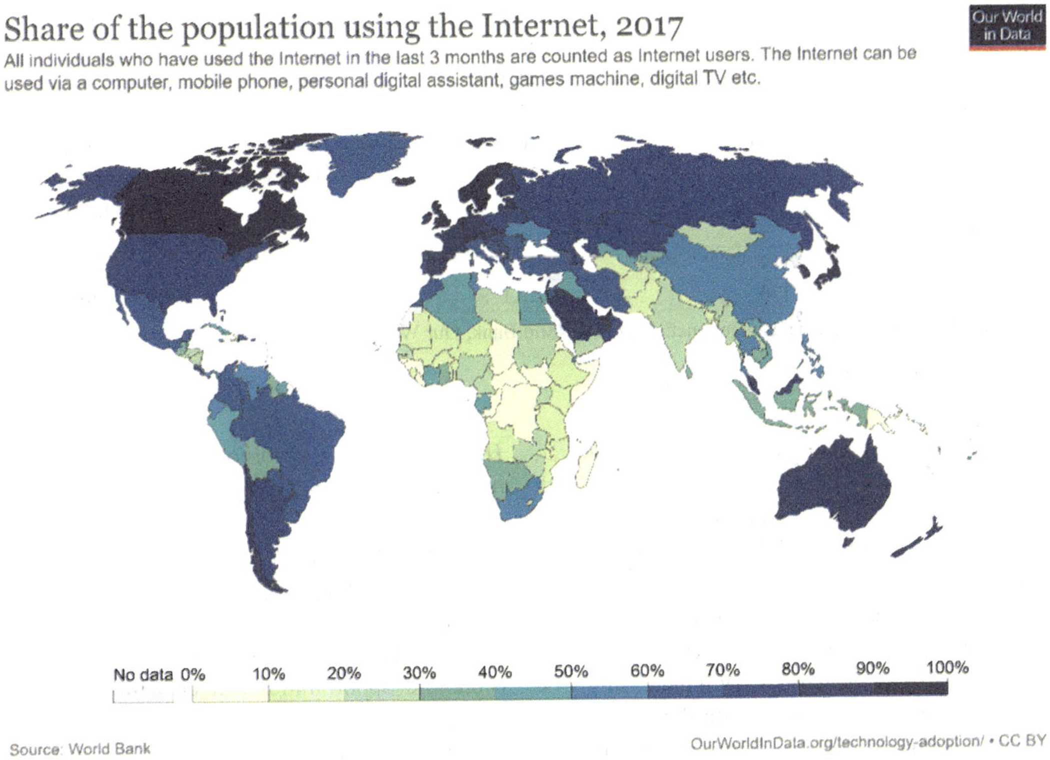
Share of the population using the Internet, 2017
All individuals who have used the Internet in the last 3 months are counted as Internet users. The Internet can be used via a computer, mobile phone, personal digital assistant, games machine, digital TV etc.
Our World in Data
No data 0% 10% 20% 30% 40% 50% 60% 70% 80% 90% 100%
Source: World Bank
OurWorldInData.org/technology-adoption/ • CC BY

Digitalization, Urbanization, and Urban-Rural Divide, Fig. 2 Extent of Internet Use. (Sources: World Bank, Our World in Data https://ourworldindata.org/grapher/share-of-individuals-using-the-internet)

Digitalization, Urbanization, and Urban-Rural Divide, Table 1 Internet Use and Urbanization across a set of selected countries

Country	Rank (The Inclusive Internet Index, 2021)	The Inclusive Internet Index 2021	Urban population (% of total) 2017	Absolute change in urban population (% of total) 1960–2017
	Source: https://theinclusiveinternet.eiu.com/explore/countries/performance	Source: https://theinclusiveinternet.eiu.com/explore/countries/performance	Source: https://ourworldindata.org/urbanization	Source: https://ourworldindata.org/urbanization
Sweden	1	87.4	87.2%	+14.66 pp
United States	2	86.8	82.1%	+12.06 pp
United Kingdom	=9	84.7	83.1%	+4.70 pp
Brazil	36	77.9	86.3%	+40.17 pp
China	39	76.3	58.0%	+41.76 pp
Argentina	43	75.6	91.8%	+18.14 pp
Mexico	46	73.8	79.9%	+29.11 pp
South Africa	47	73.7	65.9%	+19.23 pp
India	=49	73.4	33.6%	+15.68 pp
Thailand	=49	73.4	49.2%	+29.53 pp
Indonesia	66	67.8	54.7%	+40.07 pp
Philippines	=68	67.4	46.7%	+16.39 pp
Kenya	71	66.4	26.6%	+19.20 pp
Bangladesh	82	59.9	35.9%	+30.72 pp
Uganda	93	55.1	23.2%	+18.78 pp
Zimbabwe	105	46.6	32.2%	+19.63 pp
Sudan	106	45.7	34.4%	+23.62 pp
Mali	107	43.8	41.6%	+30.51 pp
Ethiopia	108	43.4	20.3%	+13.88 pp
Mozambique	109	41.5	35.5%	+28.59 pp
Angola	=110	40.7	64.8%	+54.40 pp
Niger	117	33.5	16.4%	+10.56 pp
Congo (DRC)	118	33.4	66.5%	+34.86 pp
Liberia	119	31.6	50.7%	+32.06 pp
Burundi	120	29.1	12.7%	+10.63 pp

Sources: EIU – https://theinclusiveinternet.eiu.com/, World Bank, Our World in Data

by digitalization, among other drivers. However, there are also several instances of countries with low level of digitalization and per capita income but accompanied by a high rate of urbanization (see Glaeser 2014 for a discussion).

There are clear indications that in the foreseeable future, the extent of digitalization will only grow and economic activities will increasingly be based on digital platforms. With the above context, I next highlight three areas of policy concerns regarding urban-rural inequalities.

Hypothesis 1 *Accessibility is a key determinant of unequal economic growth.*

Transport and accessibility to places of economic interest play a significant role in determining unequal economic outcomes. Weiss et al. (2018) suggested that the lack of easy access to economic opportunities can be a major barrier to improved livelihoods and overall development. It is important to stress that the equity agenda of "leaving no one behind" established by the SDGs of the United Nations depends on the alleviation of accessibility barriers. Using a novel approach, Weiss et al. (2018) developed and validated a map that quantifies travel time to cities for 2015 at a spatial resolution of approximately 1 × 1 kilometer. This is done by integrating ten global-scale surfaces that characterize factors affecting human movement rates and 13,840 high-density urban centers within an established geospatial-modelling framework. More importantly, the results showed significant disparities in accessibility relative to wealth. The key result is that 50.9% of individuals living in low-income settings (concentrated in sub-Saharan Africa) reside within an hour of a city compared to 90.7% of individuals in high-income settings. There is little doubt that this gap can be addressed with an appropriate level of infrastructure spending and improvements in technology.

New technologies have been developed and commercialized over the last few years. Technologies such as driverless cars and autonomous public transit systems are experiencing huge growth, gaining public interest, and a growing level of acceptance. The take-up is likely to be high once the costs are reduced from prohibitive to more affordable for the city administrations. Bajpai (2016) suggested that these technologies have substantial potential to generate positive externalities by improving road safety, lowering fuel consumption and emissions of vehicles, and providing much-needed mobility options for the vulnerable population, including young, old, and persons with disabilities. Although such technologies show superior performances against the negative impacts on transport systems in the controlled environment, it is not yet clear how in the scaled applications, road congestion, and low-density expansion of cities can be addressed by these technologies. Many cities in LMICs (lower middle income countries) face severe forms of these twin problems. Moreover, these solutions need to be affordable and within the reach of poorer states, else the inequality gap may not close and it may even widen further.

***Hypothesis** 2 The urban-rural divide is not uniform.*

While much focus has been put on the urban areas, the rural areas across the world have also undergone significant changes. The scale of issues is staggering, and, as noted by Anríquez and Stamoulis (2007), the global challenges of poverty, inequality, and food security are centered on the rural areas. Almost 1.2 billion poor people (living on $1.25 or less each day) live in rural areas (World Development Indicators 2013). However, it is noted in the literature that development policies tend to favor industrial, urban, and service sectors more than agricultural and other rural sectors (Anríquez and Stamoulis 2007). It is possible that urban-focused development policies had to be prioritized due to the sheer scale of urbanization, and therefore, the local government policies had to be concentrated on the urban areas to meet the burgeoning demand for public amenities (such as housing, education, public transport). However, it also creates funding gaps, leaving efforts to close inequalities within and across rural areas less resourced.

Rodriguez-Pose and Hardy (2015) analyzed the approaches to resolving poverty and inequality in rural areas. They found that there has been limited progress on alleviating rural poverty, inequalities, and food insecurity, with the traditional place-neutral approaches largely failing to deliver the desired outcomes. In a very useful chart (Fig. 2 page 12 of the paper), the authors showed a clear association between poverty and rurality across various countries. In terms of the regional variation, the incidence of rural poverty (defined at the $1.25 [PPP] level) is most severe in Latin America and the Caribbean, followed closely by sub-Saharan Africa and South Asia. The authors also showed a significant variation in the rural poverty headcount ratio. According to their analysis, since the 1970s, many countries have seen regional disparities falling, including Brazil, Peru, China, Pakistan, and South Africa. However, in Colombia, Mexico and Malaysia, there has been a rising trend.

The urban-rural migration can be a significant contributor to the urban-rural divide. Although south-north migration has important implications for development and poverty reduction across developing countries, migration is not primarily a south-north phenomenon. Mendola (2012) discussed the relationship between rural out-migration and the economic development at the origin. As noted by the author, most migration, predominantly driven by the labor mobility of the poor, takes place within and between the developing countries, with many implications for those regions experiencing inflows and outflows. Mendola (2012) outlines the direct impacts on the migrant-sending households due to the spillover effects and highlights migration impact on the labor market at the origin as a major knowledge gap and an area of policy concerns.

Hypothesis 3 *Agriculture-based rural economies may not always benefit from technological changes.*

The complex relationships between rural labor mobility and the economic behavior of people left behind require policies and approaches to target the specific areas for rural development. The urban-rural divide can be closed with technological interventions in the rural areas that would support the agrarian communities facing resource constraints. It can address the supply-chain issues and may reduce the possibilities of misappropriation of economic profits by intermediaries. Digital platforms can provide many benefits including education, best practice sharing, information sharing, better communications related to natural disaster and weather events. However, the success of such interventions cannot be guaranteed across the board due to lack of good quality, stable Internet connections.

A growing population and rising income levels (especially rising purchasing power of the middle class) often translate into an increasing demand for food, creating food insecurity and other basic goods. There may be a case favoring technologies suitable for higher yield and opting for more capital-intensive (less labor-intensive) agricultural processes. However, as pointed out by Rodriguez-Pose and Hardy (2015), this can also pose a significant challenge for the rural regions (especially in the LMICs) with small, family-run farms and subsistence agriculture. Therefore, technological innovations need to be weighed appropriately for suitability in specific regions to protect the rural economy and citizens' livelihoods.

Since the rural economies are driven by agriculture and food production, it is crucial to consider rural land management and focus on governance issues to create robust institutions around resource allocation. Li et al. (2014) examined the accelerated rural "hollowing" driven by vast and increasing out-migration of rural labors under an urban-rural dual-track system, which appeared to have posed significant barriers to improving land-use efficiency and coordinating urban-rural development in China. The authors report two community-based practices showing positive effects on improving local living conditions, increasing farmland area and developing the rural industries. Their experiences, including self-organized rural planning, democratic decision-making, and endogenous institutional innovation, may benefit future land consolidation programs. As often seen, institutional constraints can be alleviated by the use of digital processes. Therefore, it is vital to recognize the needs of the rural areas and evaluate the effectiveness of the technologies that can break down the information asymmetries and barriers.

Concluding Remarks

The last few decades have seen significant technological advancements with widespread applications across all aspects of modern living. However, all sections of human society must experience the benefits of these technological innovations. Otherwise, we risk rifts in social cohesion as well as widening of the inequality gap among communities and within and across places. The technology applications can alleviate or even eradicate the massive challenges that we face today, such as climate change, food shortage, potable water shortage, inadequate supply of shelter for all, lack of access to quality healthcare for all, lack of sanitation, inadequate waste management systems, etc. However, the technology solutions should also need to consider the issues related to ethics, privacy, trust, and human rights. There is little doubt that the impacts of technological innovations on economic growth have been unequal across the countries, and urbanization has intensified with unequal impacts over the last few decades. More recently, the Covid-19 pandemic has certainly accentuated much greater use of Internet and digital technologies. At the same time, a large section of the population with poor or no access to Internet and more likely to be concentrated in rural areas seems to have suffered more compared to those with good Internet access who tend to be located in urban areas. Such unequal trends and challenges are likely to persist unless appropriate policy interventions are devised. Digitalization, if channelized and governed carefully, can be a great leveler.

Cross-References

- ▶ Adapting Cities to Climate Change
- ▶ Age-Friendly Future Cities
- ▶ Big Data for Smart Cities and Inclusive Growth
- ▶ Building Community Resilience
- ▶ Climate Gentrification
- ▶ Climate-Induced Relocation
- ▶ Growth, Expansion, and Future of Small Rural Towns
- ▶ Meeting SDG6: Ensuring Safe Drinking Water for All in Rural India
- ▶ Neither Rural Nor Urban: A Critical Review of the Fringe Dynamics of Settlements

References

Anríquez, G., & Stamoulis, K. (2007). *Rural development and poverty reduction: Is agriculture still the key?* (ESA working paper no. 07–02). Rome: The Food and Agriculture Organization of the United Nations.

Bajpai, J. N. (2016). Emerging vehicle technologies & the search for urban mobility solutions. *Urban, Planning and Transport Research, 4*(1), 83–100.

Bourguignon F. (2003). The growth elasticity of poverty reduction: Explaining heterogeneity across countries and time periods. In Eicher T., Turnovsky S. (eds) Inequality and growth. Theory and policy implications. Cambridge, MIT Press, pp 3–26.

Chauvin, J. P., Glaeser, E., Ma, Y., & Tobio, K. (2017). What is different about urbanisation in rich and poor countries? Cities in Brazil, China, India and the United States. *Journal of Urban Economics, 98*, 17–49.

Eeckhout, J. (2004). Gibrat's Law for (all) cities. *American Economic Review, 94*(5), 1429–1451. https://pubs.aeaweb.org/doi/pdfplus/10.1257/0002828043052303

Epaulard, M. A. (2003). *Macroeconomic performance and poverty reduction* (No. 3/72). International Monetary Fund.

Fosu, A. K. (2009). Inequality and the impact of growth on poverty: Comparative evidence for sub-Saharan Africa. *Journal of Development Studies, 45*(5), 726–745.

Fosu, A. (2017). Growth, inequality, and poverty reduction in developing countries: Recent global evidence. *Research in Economics, 71*(2), 306–336.

Glaeser, E. L. (2014). A world of cities: The causes and consequences of urbanisation in poorer countries. *Journal of the European Economic Association, 12*(5), 1154–1199.

Jedwab, R., Christiaensen, L., & Gindelsky, M. (2017). Demography, urbanisation and development: Rural push, urban pull and. . .urban push? *Journal of Urban Economics, 98*, 6–16.

Kalwij, A., & Verschoor, A. (2007). Not by growth alone: The role of the distribution of income in regional diversity in poverty reduction. *European Economic Review, 51*(4), 805–829.

Li, Y., Liu, Y., Long, H., & Cui, W. (2014). Community-based rural residential land consolidation and allocation can help to revitalise hollowed villages in traditional agricultural areas of China: Evidence from Dancheng County, Henan Province. *Land Use Policy, 39*, 188–198.

Mendola, M. (2012). Rural out-migration and economic development at origin: A review of the evidence. *Journal of International Development, 24*(1), 102–122.

Nanda, A. (2019). *Residential real estate: Urban & regional economic analysis*. Routledge.

Nanda, A., Xu, Y., & Zhang, F. (2021). How would the COVID-19 pandemic reshape retail real estate and high streets through acceleration of E-commerce and digitalization? *Journal of Urban Management, 10*(2), 110–124.

Ravallion, M. (1997). Can high-inequality developing countries escape absolute poverty? *Economics Letters, 56*(1), 51–57.

Rodriguez-Pose, A., & Hardy, D. (2015). Addressing poverty and inequality in the rural economy from a global perspective. *Applied Geography, 61*, 11–23.

WDI. (2013). *World development indicators*. Washington, DC: World Bank.

Weiss, D. J., Nelson, A., Gibson, H. S., Temperley, W., Peedell, S., Lieber, A., & Mappin, B. (2018). A global map of travel time to cities to assess inequalities in accessibility in 2015. *Nature, 553*(7688), 333.

D

Dimension/Size

► Metropolitan Discipline: Management and Planning

Direction

► Spatial Demography as the Shaper of Urban and Regional Planning Under the Impact of Rapid Urbanization

Directive

► Regulation of Urban and Regional Futures

Disaster

► Collective Emotions and Resilient Regional Communities

Disaster Management

► Urban Management in Bangladesh

Disaster Preparedness

► Urban Resilience

Disaster Relief

► Role of Disaster Relief Policy in Building Resilient Coastal Regions in the United States

Disaster Risk in Informal Settlements and Opportunities for Resilience

Iftekhar Ahmed
School of Architecture and Built Environment, University of Newcastle, Callaghan, NSW, Australia

Synonyms

Shantytowns; Slums; Squatter settlements

Definition

Informal settlements are mostly prevalent in Global South countries, although there are also limited examples in the Global North. Here the focus is on the Global South where the vast majority of informal settlements are located. Informal settlements are primarily located in urban areas and characterized by poor living conditions, poverty, and significantly, uncertain land tenure status which has contributed to use of the term "informal." The term "informal" implies that these settlements have been established outside formal, institutional processes associated with planning and building approvals from governmental authorities, that is, they are unplanned settlements. There are also widespread examples in the Global South of buildings built and neighborhoods established informally, which do not have a legal status, and where local land-use planning and building regulations are not followed or bypassed through corrupt means, but these are differentiated from the informal settlements discussed here because of their relatively higher level of wealth. Typically, informal settlements are distinguished by conditions of poverty and are usually the poorest urban neighborhoods. The words "slums" and "shantytowns" are also used widely and synonymously, but because they can sometimes have negative connotations, the term "informal settlements" is considered more neutral and an objective description of such settlements. The term "squatter settlements" is also used, however, that essentially implies a temporary and transitory settlement, whereas informal settlements may grow and consolidate over a long period of time. A large number of people in the Global South, and indeed, increasingly in the Global North, are homeless and live on the streets or in other makeshift arrangements, but since they do not comprise a settlement in the true sense of the word, they are not included in the discussion here. Such "floating" populations can in some instances be considered an earlier stage of formation of an informal settlement.

Introduction

The widespread and rapid urbanization process in the Global South over the last few decades has been accompanied by the growth of informal settlements characterized by impoverished and difficult living conditions, and vulnerability to natural and human-induced hazards. In response to the widespread absence of affordable urban land, housing, and services in the formal sector, informal settlements are typically established without governmental sanction and are beyond formal planning regulations and building codes. Informal settlements present a critical challenge to governments and urban management agencies, an issue widely explored with diverse approaches implemented around the world. Yet, the challenge persists, pointing to an intractable issue facing the rapidly growing cities of the Global South.

More than a billion people around the world live in informal settlements (UN-Habitat 2016), with around 80% concentrated in Asia and Africa (United Nations Statistics Division 2019). While the vast agglomerations of the population in these continents draw institutional and media attention, it is less recognized that small island-developing states are also urbanizing rapidly, accompanied by the growth of informal settlements; in many of the Pacific island countries, urbanization is occurring at more than three times the global rate (Kuruppu 2016). The Solomon Islands is a case in point – although still predominantly rural, a high urbanization rate of close to 5% has resulted in 25% urbanization in 2020, jumping from 20% a decade ago (Plecher 2020; UN-Habitat 2012); 35–40% of the growing population in the capital city, Honiara, live in informal settlements (Kiddle and Hay 2017; UN-Habitat 2012).

The worldwide growing trend of disasters has significant impacts on informal settlements, widely documented, most evident in rapid-onset disasters such as earthquakes. In Haiti where almost 90% of the population was living in informally constructed housing, the 2010 earthquake killed around 220,000 people and devastated nearly 300,000 houses (Blaranova and Christiaens 2012; United Nations 2012). Eleven years later, many people had still not recovered and were living in displacement camps and informal settlements and another earthquake caused severe devastation (Council on Foundations 2021). Another such example is the 2015 earthquakes in Nepal when more than 600,000 buildings were destroyed in a context where 80% of settlements were informal (von Meding et al. 2017) (Fig. 1).

Alongside the prevalence of poor construction methods, the location of informal settlements is also often a risk factor. Such settlements are often built on marginal land, exposing them to additional risks such as floods which can have a high impact on already vulnerable housing construction. While the loss of lives and damage to property and assets are usually less in floods than rapid-onset disasters such as earthquakes, floods occur on a regular basis and over time have a cumulative impact and economic cost, which the urban poor can ill afford. Furthermore, spurred by climate change, floods are increasing in magnitude and becoming more widespread (Braganza 2012; Power et al. 2017), amplifying the risk for informal settlements.

Institutional Approaches

Two key global frameworks led by the United Nations – the Sustainable Development Goals (SDGs) and Sendai Framework for Disaster Risk Reduction (SFDRR) – acknowledge the importance of addressing the problems experienced in informal settlements and to reduce their disaster

Disaster Risk in Informal Settlements and Opportunities for Resilience, Fig. 1 A large informal settlement in Dhaka, Bangladesh, which is exposed to regular flooding

risks through institutional action. The SDGs have a specific focus on informal settlements in its Goal 11, target 11.1, where the importance of upgrading informal settlements to improve housing and services is specified (United Nations n.d.). In the declaration of the UN General Assembly on the SFDRR, the assessment of disaster risk in informal settlements and consequent action is stated in terms of the development of contextualized building codes and standards, and rehabilitation and reconstruction practices (United Nations 2015). This is a broad prescription for translating to national and local levels with a specific suite of actions. The diversity of institutional approaches toward informal settlements has been succinctly captured by Bah et al. (2018, p. 222–223) as ranging from "… benign neglect, laissez-faire, forced eviction and demolition, resettlement or relocation, slum upgrading programs, and the adoption of enabling strategies." A common approach followed by many governments in the Global South is to demolish informal settlements and evict residents. Even approaches termed as "slum rehabilitation" or "slum improvement" that attempt to convey well-intended policies can in reality be eviction programs. There have been advocates against such an approach since the 1970s (for example, Turner and Fichter 1972), and there are organizations around the world that seek to prevent forced evictions and advocate the rights of informal settlement residents (for example, the Asian Coalition for Housing Rights). These approaches stem from a human rights perspective, and they point out that if informal settlers are provided tenure security and the fear of eviction is removed, they make significant investments over time to improve their housing and settlement.

An alternative to eviction, the concept of upgrading is promoted by prominent international agencies such as the World Bank (2011) and United Nations (UN-Habitat 2014). Typically upgrading projects include the provision of basic infrastructure and services in informal settlements. Insecure tenure is a key aspect of informal settlements, and increasingly agencies address the improvement of tenure security in upgrading programs. Most upgrading projects focus on infrastructure and services, and very few deal with housing improvement, although housing is such an essential and valuable asset.

Relocation and Disaster Risk Reduction

A consequence of the widely practiced clearance of informal settlements in the Global South is involuntary relocation, which may sometimes be accompanied by resettlement on alternative sites often on the pretext of providing safety from natural hazards. Such actions often result in a range of negative social, economic, and psychological outcomes for the displaced and resettled communities ranging from disruption of social and community networks to difficulty of accessing livelihood opportunities to trauma, isolation, and marginalization. In many cases, the new settlements are uninhabitable for the people made to settle there and are thus abandoned or preempted by groups with higher incomes.

Relocation and resettlement are widely practiced in a postdisaster context with the primary aim of reducing future disaster risk; again, such projects have similar mixed outcomes. The importance of relocation from areas with high-disaster risk is often recognized by governments and nongovernmental organizations (NGOs), but forced eviction is not a long-term solution whereas in situ upgrading for disaster risk reduction offers a better opportunity for effectiveness because existing employment and community networks can be maintained. Relocation is also a highly expensive option compared to upgrading. Only in situations of extreme risk should relocation be considered, where the existing location is much too hazardous and precarious for safe habitation and faces a certain likelihood of being obliterated by future disasters.

Opportunities for Resilience Through In Situ Upgrading

There have been various approaches to addressing the challenges faced by informal settlements as discussed above, and despite the many constraints, there are some effective cases that point to potential opportunities for building resilience in informal settlements. Principally, in situ upgrading offers the best opportunities, demonstrated by a range of projects around the world (for example, see The World Bank Group 2001). While the focus of most of these examples has been on economic and physical improvements, it allows gaining lessons for the additional and important purpose of building resilience to disasters. Significantly, the prerequisite for the projects was first to ensure tenure security, without which upgrading might end up in being counterproductive. Also important is to identify funding sources; many large donor agencies such as the World Bank and governmental bilateral international aid agencies support upgrading programs, where a case of upgrading with a view to imparting disaster resilience can be made. An upgrading project along this line in Bangladesh funded by the Australian Government, which was unique in its specific focus on resilience, where the author was involved, offers useful lessons (Ahmed 2016). The following sequential progression of activities in the project offers a framework for wider applicability (see Fig. 2):

- Staff training: The capacity of staff members of the implementing organization and its partner organizations first needs to be strengthened through training on conceptual and hands-on aspects of the project framework.
- Community-based participatory risk assessment: The trained staff then carry out hazard and risk assessments at the informal settlement community level through participatory engagement with community members. They are provided with a customized toolkit for the

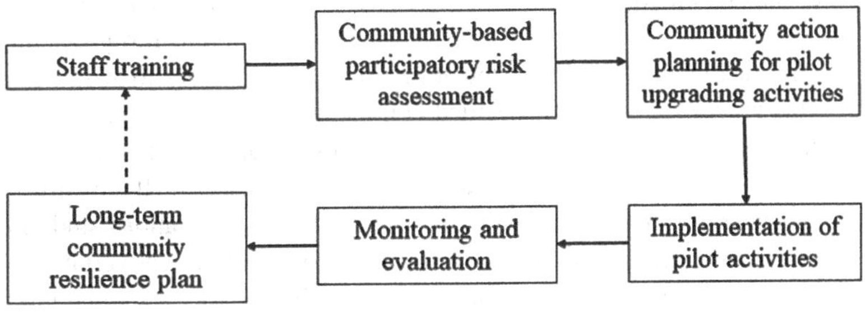

Disaster Risk in Informal Settlements and Opportunities for Resilience, Fig. 2 Sequential activity flow in an informal settlement upgrading project for community resilience assessments. Respecting the local culture, necessary separate assessment sessions should be conducted for different groups such as women, men, and youth.

- Community action planning for pilot upgrading activities: Based on the risk assessment, again through participatory consultations with the community, an action plan for pilot-upgrading activities is developed, placing priority to the hazards that most impact the community' resilience.
- Implementation of pilot activities: Following the community action plan, a set of pilot-upgrading activities are undertaken together with the community.
- Monitoring and evaluation: A rigorous process of monitoring the pilot activities and an evaluation after completion of the pilot activities allow gaining lessons on the efficacy of the intervention and indicate future directions. If possible, the evaluation should be conducted after a disaster cycle (such as seasonal flooding event) to understand the achievements and gaps of the intervention.
- Long-term community resilience plan: The pilot activities and evaluation are the foundations for developing a long-term plan for resilience of the community. At this stage, community-based committees or groups need to be established to sustain the initiative over the long term and foster community ownership and independence. The plan can also allow exploring prospects for upscaling the pilot initiative and replicating in other areas of the settlement or elsewhere.

Some of the key strategic considerations in upgrading informal settlement for disaster resilience include the following:

- Prioritization according to the level of risk: This requires developing detailed community profiles, conducting localized risk assessments, and building and land audits for detailed documentation.
- Establishing tenure security: Tenure security in informal settlements is a complex process with diverse arrangements, and upgrading initiatives need to work with this reality.
- A participatory community-based approach is essential for in situ upgrading. If there are any existing community-based organizations in the informal settlement, they should be the focal points, where capacity building for resilient housing and infrastructure by enhancing existing skills through a training-of-trainers approach can lead to positive long-term outcomes.
- In many cases, local building codes and planning regulations are based on Western standards, which might not be relevant in the context of informal settlements. At a minimum, if only the disaster resilient aspects of the codes and regulations can be implemented, it would save lives and property.
- In addition to developing new housing and infrastructure, retrofitting of existing stock offers an opportunity for incorporating resilient features. There might already be some efforts in place, which can be supported while encouragement through incentives might prove worthwhile where such efforts are lacking.
- To back such upgrading initiatives, there is a need for financial incentives beyond project funding. Microcredit programs offer such possibilities, and loans, if such programs can be combined with capacity building and provision of technical support.
- There are a range of hazards that can occur in conjunction and can have a compounded impact. Therefore, a multihazard approach is required, informed by localized risk assessments.

Conclusion

It should be understood that a one-size-fits-all approach to upgrading of informal settlements is unlikely to work, any broad framework such as the one presented above requires adaptation and contextualization to fit specific circumstances. It requires detailed community-based consultations and multistakeholder engagement between governmental, nongovernmental, private sector, and community organizations to develop a comprehensive suite of physical planning guidelines applicable for specific informal settlements. These guidelines require testing through pilot initiatives to enable mainstreaming, sustainability, and long-term implementation.

Cross-References

- ▶ Building Community Resilience
- ▶ Building Resilient Communities Over Time
- ▶ Climate Resilience in Informal Settlements: The Role of Natural Infrastructure
- ▶ Community Engagement and Climate Change: The Value of Social Networks and Community-Based Organizations
- ▶ From Vulnerability to Urban Resilience to Climate Change
- ▶ Global Homelessness: Neoliberalism, Violence, and Precarious Urban Futures
- ▶ Housing and Development
- ▶ The Practice of Resilience Building in Urban and Regional Communities

References

Ahmed, I. (2016). Building resilience of urban slums in Dhaka, Bangladesh. *Procedia – Social and Behavioural Sciences, 218*, 202–213.

Bah, E.-h. M., Faye, I., & Geh, Z. F. (2018). Slum upgrading and housing alternatives for the poor. In E.-h. M. Bah, I. Faye, & Z. F. Geh (Eds.), *Housing market dynamics in Africa* (pp. 215–253). London: Palgrave Macmillan.

Blaranova, L., & Christiaens, B. (2012). *Project proposal: Community development Delmas 30*. Port-au-Prince: International Federation of Red Cross and Red Crescent Societies.

Braganza, K. (2012). A land of (more extreme) droughts and flooding rains? *The Conversation*. Retrieved on September 9, 2021, from https://theconversation.com/a-land-of-more-extreme-droughts-and-flooding-rains-5184

Council on Foundations. (2021). *Disaster in a crisis zone: Understanding the impact of Haiti's earthquake*. Retrieved on September 8, 2021, from https://www.cof.org/event/center-disaster-philanthropy-disaster-crisis-zone-understanding-impact-haitis-earthquake

Kiddle, G. L., & Hay, I. (2017). Informal settlement upgrading: Lessons from Suva and Honiara. *Development Bulletin, 78*, 25–29.

Kuruppu, N. (2016, September 28). *Turning the tide on urbanisation policy in the Pacific islands*. Retrieved on September 9, 2021, from https://unu.edu/publications/articles/urbanisation-in-pacific-islands.html#:~:text=But%20despite%20their%20comparatively%20lower,three%20times%20the%20global%20average

Plecher, H. (2020). *Urbanization in the Solomon Islands 2009–2019*. Retrieved on July 13, 2020, from https://www.statista.com/statistics/728723/urbanization-in-the-solomon-islands/

Power, S. B., Murphy, B., Chung, C., Delage F., & Ye, H. (2017). *Droughts and flooding rains already more likely as climate change plays havoc with Pacific weather*. Retrieved on September 9, 2021 from https://theconversation.com/droughts-and-flooding-rains-already-more-likely-as-climate-change-plays-havoc-with-pacific-weather-71614

The World Bank Group. (2001). *Upgrading urban communities – Resource framework: Case examples*. Retrieved on September 8, 2021, from http://web.mit.edu/urbanupgrading/upgrading/case-examples/index.html

Turner, J. F. C., & Fichter, R. (1972). *Freedom to build: Dweller control of the housing process*. New York: Macmillan.

UN-Habitat. (2012). *Solomon Islands: Honiara urban profile*. Nairobi: UN-Habitat.

UN-Habitat. (2014). *Participatory slum upgrading programme (PSUP): Halving the number of slum dwellers by 2020*. Nairobi: UN-Habitat.

UN-Habitat. (2016). *Slum almanac 2015/2016: Tracking improvement in the lives of slum dwellers*. Nairobi: UN-Habitat.

United Nations. (2012). *Key statistics: Haiti relief*. New York: UN Office of the Secretary-General's Special Adviser.

United Nations. (2015). Resolution adopted by the General Assembly on 3 June 2015: 69/283. Sendai Framework for Disaster Risk Reduction 2015–2030. Retrieved on September 8, 2021, from https://www.un.org/en/development/desa/population/migration/generalassembly/docs/globalcompact/A_RES_69_283.pdf

United Nations. (n.d.). *A short guide to human settlement indicators Goal 11*. New York: United Nations.

United Nations Statistics Division. (2019). *Make cities and human settlements inclusive, safe, resilient and sustainable*. Retrieved on September 9, 2021, from https://unstats.un.org/sdgs/report/2019/goal-11/

von Meding, J., Shrestha, H. D., Kabir, H., & Ahmed, I. (2017). Nepal earthquake reconstruction won't succeed until the vulnerability of survivors is addressed. *The Conversation*. Retrieved on September 9, 2021, from https://theconversation.com/nepal-earthquake-reconstruction-wont-succeed-until-the-vulnerability-of-survivors-is-addressed-87335

World Bank. (2011). *Urban poverty and slum upgrading*. Retrieved on 24 July, 2015, from http://web.worldbank.org/WBSITE/EXTERNAL/TOPICS/EXTURBANDEVELOPMENT/EXTURBANPOVERTY/0,,menuPK:341331~pagePK:149018~piPK:149093~theSitePK:341325,00.html

Discipline/Practice

► Metropolitan Discipline: Management and Planning

Discrimination

► Hidden Enemy for Healthy Urban Life

Displaced

► Global Homelessness: Neoliberalism, Violence, and Precarious Urban Futures

Dispossessed

► Global Homelessness: Neoliberalism, Violence, and Precarious Urban Futures

Disruptive Innovation

► Systemic Innovation for Thrivable Cities

Disruptive Mobility: Sharing Electric Autonomous Vehicles (SEAVs) Reshape Our Future Cities

Mohsen Mohammadzadeh
School of Architecture and Planning, Auckland University, Auckland, New Zealand

Introduction

Emerging Industry 4.0 is fundamentally changing transport technologies and consequently transforming cities around the world (Martínez-Gutiérrez et al. 2021). This new stage of industrialization encompasses two trends: (1) the advancement of pre-existing technologies such as information communication technologies (ICTs) and electrification and (2) the evolvement of emerging technologies such as artificial intelligence (AI) and the Internet of Things (IoT). These trends are radically altering transportation by generating new opportunities including, but not limited to, electrification, automation, and sharing platforms. Mohammadzadeh (2021, p. 1) argues that "emerging mobility technologies, including connected shared platforms and automation, are disrupting urban transportation." We have reached a critical juncture in determining how these new emerging transport technologies will shape our transport systems, cities, and the built environment in the foreseeable future (Legacy et al. 2019).

Urban history shows that the previous stages of the industrial revolution have transformed mobility technologies. The evolution of mobility technologies has consequently changed cities around the world. In the first stage of the industrial revolution, the invention of the steam engine and railroad transport in the late eighteenth century expanded cities by generating opportunities for fast, affordable, and reliable access to locations beyond the city (Mohammadzadeh 2018). Hall (2014) points out that planned suburbs, as suburbanization, were developed around the new railways that radically reshaped urban forms. The second stage of the industrial revolution incorporated mass production including, but not limited to, the automobile industry, particularly after Fordism. Mass car production made car ownership affordable for middle-class households. Sprei (2018, p. 238) argues that "interestingly enough the transport sector has already experienced a disruptive change from horses to motorized transport dominated by cars. This transition went fairly rapidly. Data for the US shows that the car replaced the horse in around 12 years, even if it took 30–50 to completely phase out." The car has since generated a "new scale of local distance" because it shortens residents' commuting time. "When we talk about urban sprawl, we talk about cars" (Fox 2017, p. 2). The third stage of industrialization reinforced car dependency by improving the quality of vehicles and particularly by reducing fuel consumption and costs. There is an expectation that the fourth stage of industrialization will radically transform urban forms through emerging disruptive mobility technologies such as smart sharing platforms, electrification, and automation (Mohammadzadeh 2018, 2019, 2021; Golbabaei et al. 2021; Loeb et al. 2018; Sprei 2018; Stocker and Shaheen 2017; Yigitcanlar et al. 2019; Zhang and Guhathakurta 2017). To investigate the potential impacts of emerging disruptive mobility on cities, it is important to define what "disruptive" means.

What is the Disruptive Technology?

The term "disruption" has often been used with a negative connotation. However, disruption also has a positive meaning. In *The Innovator's Dilemma* (2000), Clayton Christensen defines "disruption" by distinguishing "sustaining technologies" from those that are "disruptive." He (2000, xv) argues that:

> Some sustaining technologies can be discontinuous or radical in character, while others are of an incremental nature. What all sustaining technologies have in common is that they improve the performance of established products, along the dimensions of performance that mainstream customers in major markets have historically valued.... Disruptive technologies bring to a market a very different value proposition than had been available previously. Generally, disruptive technologies underperform established products in mainstream markets.... Products based on disruptive technologies are typically cheaper, simpler, smaller, and, frequently, more convenient to use.

Christensen (2000) argues that sustaining technologies seldom precipitate the failure of major companies. In contrast, disruptive technology can challenge leading firms, resulting in their potential failure and even elimination from the market. Sprei (2018) divides disruptive technologies into two categories. The first category includes technologies that have a worse performance but a lower price than mainstream technologies; however, given their convenience and the possibility of their technological improvement, they could take over the market. The second group of disruptive technologies are superior in performance to the mainstream but are more expensive. The positive aspects of disruptive technologies, such as improving the quality of service, consequently generate new demand for these disruptive technologies in the market. "The challenge with potentially disruptive technology is, however, that how this demand is expressed in behavioural choices today does not predict how this demand will be expressed in the future" (Ronald et al. 2017, p. 276). Bahmanteymouri and Farzaneh (2020) argue that the demand for disruptive technologies can be ephemeral, and they will be constantly replaced by new disruptive technologies. However, disruptive technologies often provide new alternative solutions for complex issues such as pollutions.

Disruptive mobility is mostly perceived as a transportation solution that challenges the existing transport system (Millard-Ball 2019; Sprei 2018; Yigitcanlar et al. 2019; Zakharenko 2016). Sprei (2018, p. 239) argues that, "Given the role of transport in a larger socio-technical system it is interesting to look at disruption from a broader perspective and its systemic effects. It might be so that innovation has a disruptive effect on the automotive industry but not on, e.g., the transport system." Mohammadzadeh (2021) suggests that disruptive mobility should be considered as a technological transformation rather than a paradigm shift in transportation. Yet, the mobility technological transformation will dramatically change cities and the built environment. The next section will go through three of the major disruptive mobilities that shape future transportation systems: smart shared mobility platforms, electrification, and autonomous vehicles. For each of these, the section will investigate how these mobility innovations might disrupt the transport system. There are many synergies between these mobility innovations that simultaneously transform transportation and cities.

Three Disruptive Mobility Technologies

This section investigates three transport technologies: vehicle automation, vehicle electrification, and online sharing platforms. Some researchers have used the term "disruptive mobility" when conceptualizing and studying the impacts of vehicle automation, vehicle electrification, and online sharing platforms because of their disruptive impacts on transportation, cities, and the built environment (Golbabaei et al. 2021; Loeb et al. 2018; Mohammadzadeh 2018; Sprei 2018; Stocker and Shaheen 2017; Zhang and Guhathakurta 2017). Sperling (2018, p. xi) conceptualizes these transport technologies as three mobility revolutions, regarding them as "a fundamental change in the way of thinking. . .and a changeover in use or preference, especially in technology." Smart sharing mobility and electric and autonomous vehicles are perceived as the most disruptive technological advances of the century (Mohammadzadeh 2021). These are revolutionary and disruptive because their impacts are not limited to improving the performance of established transport technologies. These disruptive mobility technologies offer a different value proposition and a larger number of alternative options than those previously available in transportation.

The electrification of vehicles and the automation of vehicles and smart sharing platforms are often perceived as distinct mobility technologies developing concurrently. Several researchers have recognised the notable synergies that exist between these emerging mobility technologies (Chen et al. 2016; Sperling 2018; Mohammadzadeh 2018). Shared Electric Autonomous Vehicles (SEAVs) will be the next generation of mobilities (Shaheen et al. 2018; Golbabaei et al. 2021). Large companies, such as Google, Ford, and Baidu, claim that their AVs will be mass-produced and commercialised in a few years (Etherington 2017; Holland 2018; Muoio 2016). However, it is crucial to understand how the disruptive impacts of these mobility technologies will influence the transformation of future cities. It is therefore necessary to define these disruptive mobility technologies.

Electric Vehicles (EVs)

The history of electric vehicles began with the invention of the first EVs in the late nineteenth century. Wakefield (1993) named the period 1895–1905 as the "golden age of electric vehicles" with the early development of EVs in the UK, France, and the USA. The main reason for the popularity of EVs at this time was their functionality. "EVs were also preferable to steam-powered vehicles because they were capable of longer ranges on a single charge, and were more convenient during colder weather. Under similar weather conditions, steam-powered vehicles suffered from start-up times of up to 45 minutes" (Khajepour et al. 2014, p. 2). Later, in the twentieth century, combustion vehicles became popular because they were more powerful and cheaper and offered higher flexibility than the initial EVs. In 1909, the production of the Ford Model T was a turning point in the car industry, leading to the market supremacy of combustion engine vehicles over EVs (Khajepour et al. 2014). Henry Ford's use of assembly lines to produce the Ford Model T ensured it was cheaper than EVs and made car ownership affordable for middle-class households. The first EVs could cover a maximum of 100 km without recharging. However, at the beginning of the twentieth century, the electricity power network was not developed outside major cities, even less in remote areas and regions. Thus, the use of EVs was limited to cities. In contrast, petrol was a flexible and accessible energy source that could be stored and used cheaply and easily in remote areas. Consequently, the market for EVs gradually disappeared by WW1 (Beretta 2013).

For decades, the electrification of vehicles has been perceived as the foreseeable future of mobility, and several factors have recently reignited the enthusiasm for EVs. Current technological advancements have made EVs more affordable and reduced travel costs and charging time, thus ensuring EVs are responsive to the needs of people in the twenty-first century (Khajepour et al. 2014). Previously, the major impediments to EV adoption included the relatively high cost of EVs, battery costs, limited driving range, and potentially long charging time requirements. These limitations have mostly been overcome in the last decade through technological improvements (Udaeta et al. 2015). The Intergovernmental Panel on Climate Change (2019) considers the increase in greenhouse gases (GHG) as a major challenge in dealing with global warming, climate change, and air quality. Road transport generates

17% of worldwide carbon dioxide (CO_2) emissions, which are an important ingredient of GHGs (Olivier and Peters 2019). The transition from combustion to electric vehicles (EVs) is often perceived as a solution to the issue of greenhouse gas emissions and the need to improve air quality, particularly in major cities (Degirmenci and Breitner 2017). Environmental concerns over air pollution, global warming, and climate change have convinced countries around the world to set EV adoption targets in order to reduce greenhouse gas emissions. Many cities have encouraged the use of EVs as one of their climate change mitigation strategies (Heidrich et al. 2017).

EVs are categorized as disruptive technologies because they offer a very different value proposition and a larger number of alternative options than had previously been available. EVs are fundamentally disrupting the car and oil industry, transportation and infrastructure system, and cities.

Smart Shared Mobility Platforms

Sharing mobility is a component of emerging, innovative, digital-based economic systems that are conceptualized as the on-demand economy, the online platform economy, and the sharing economy, among others. The on-demand economy focuses on "on-demand" needs by providing immediate access-based goods and services when required. The online platform economy provides direct access to customers according to their needs (Kenney and Zysman 2016) based on the exchange of goods and the supply of services (Sundararajan 2016). Although the concept of "sharing" is not new, the concept of a "sharing economy" is a new paradigm that enables access to goods and services beyond ownership (Botsman and Rogers 2010). The sharing economy persuades people to share their underutilized assets and resources including their vehicles with others for financial gains. Over the last decade, the smart economy has grown rapidly based on digital online platforms.

Urban mobility and transportation experts pervasively believe that we are moving in "an era of shared mobility due to the advent of alternative transportation services which offer the possibility of major changes" in the transport system (Machado et al. 2018, p. 2). Shared mobility refers to the shared use of a mode of transport (Ronald et al. 2017). Sharing vehicles, scooters, and bicycles offers users short-term access to transportation modes on an "as-needed" basis (Stocker and Shaheen 2017). The advancements in information communication technologies (ICTs) have enabled a growing number of mobility innovations and services. Over the last decade, several online sharing platforms have been developed around the world to facilitate shared mobility. The pervasive usage of smartphones and gadgets has provided new opportunities for shared mobility options that have existed for a longer time, such as shared bicycles and shared cars. Smart shared mobility involves new mobility technologies that "depend entirely on the real-time connectivity between drivers and passengers to agree on a trip" (Tirachini et al. 2020, p. 2).

Smart shared mobility is radically transforming the existing transportation system, which largely centers on car ownership. The popularity of smart sharing mobility platforms will lessen the importance of private car ownership in future cities (Mohammadzadeh 2021). Researchers have conceptualized "mobility on demand" (MOD) and "mobility as a service" (MaaS), among others, to elucidate the transition from car ownership to a sharing service. Lucken and Shaheen (2021, p. 410) describe mobility on demand as a transport concept whereby "consumers can access mobility, goods, and services on demand by dispatching or using shared mobility, courier services, unmanned aerial vehicles and public transportation solutions." Mobility as a service (MaaS) functions on a digital platform and integrates various forms of transport into a united place, "deploying the most appropriate mode for each journey, recognising customer preferences and real-time conditions of the transport network when the trip is requested" (Ho et al. 2020, p. 70). Smart sharing mobility should be considered as the next generation of mobility technologies and is a component of Industry 4.0 and its ICT advancement.

There are different models of shared mobility services such as car-sharing, bike-sharing, scooter-sharing, ride-sharing, car-renting, and on-demand mobility services (Stocker and Shaheen 2017). Smart shared mobility options are gradually changing people's travel mode choices, and subsequently, reducing the existing level of car ownership in cities. The transition from car ownership to a sharing mobility service is a nascent phenomenon. Newman and Kenworthy (2015) observe that the percentage of private car ownership has gone down for the first time in history. Shared mobility services have been growing rapidly around the world. In 2014, there were over 4.8 million car-sharing members worldwide and over 100,000 vehicles, a 65% and 55% increase, respectively, over 2 consecutive years (Shaheen 2016). On-demand mobility services, like Lyft and Uber, are also growing at a rapid pace. In 2016, Uber claimed that more than 50 million riders worldwide had taken more than 2 billion rides since its founding in 2009 (Meyer and Beiker 2018).

Smart sharing mobility is a disruptive phenomenon. Major car manufacturers, such as Ford, GM, Mercedes, Fiat Chrysler, BMW, Peugeot, PSA, and Volvo, intend to evolve beyond manufacturing driver-centric cars to become mobility companies that provide shared mobility services (Sperling 2018). Novikova (2017) argues that governments, nonprofit organizations, and communities facilitate smart sharing mobility services with the intention of reducing car dependency in cities. Smart sharing mobility potentially improves vehicle usage efficiency, lowering the number of vehicles produced, decreasing single-occupant vehicles on the roads, and possibly reducing the total number of vehicles on the streets and roads.

Automated Vehicles (AVs)

The story of AVs and their potential benefits begins nearly a century ago (Mohammadzadeh 2018). Weber (2014) discusses that the first attempt to create an AV dates back to the 1920s. Perhaps the first AV was a radio-controlled car that required a second car behind it to send out radio signals to its transmitting antennae. This radio-controlled AV was generally known as a "phantom auto" in the 1920s and 1930s (Sentinel 1926). The development of automated highway systems in the 1980s generated a new opportunity for "semiautonomous and autonomous vehicles to be connected to the highway infrastructure" (Bagloee et al. 2016, p. 287). However, AVs are more recognized as a technological achievement of the fourth stage of the industrial revolution – Industry 4.0. AVs are alternatively signified as self-driving vehicles or driverless vehicles. "AVs operate on a three-phase design known as 'sense-plan-act' which is the premise of many robotic systems" (Bagloee et al. 2016, p. 287). Using various sensors, transmitters, computing technologies, the Internet of Things (IoTs), and artificial intelligence (AI), a new generation of vehicles has been created that can drive themselves (SAE 2016). While the terms "automated vehicles" and "autonomous vehicles" are used interchangeably, these terms are not the same:

1) Automated vehicles can drive themselves, yet they depend extensively on artificial hints in the environment. These external inputs are often referred to as vehicle-to-infrastructure (V2I) (Maitipe et al. 2012).
2) Autonomous vehicles can handle uncertainty and compensate for system failure without external intervention (Antsaklis et al. 1989).

AVs can be categorized according to their different characteristics. However, some of these characteristics, such as their features, sizes, and types, are inherited from current non-AVs. Some other characteristics of AVs are embedded in smart technology such as the level of automation and smart economy. Autonomous vehicles also vary in size, with autonomous technologies applied to cars, light vans, trucks, and heavy freight (Skinner and Bidwell 2016).

Providing a clear, consistent definition and terminology of vehicle automation is crucial to advancing the discussion around its impact on cities. However, there is no concrete and widely accepted definition of vehicle automation. Academics and professionals have used a variety of terms such as self-driving, autonomous, driverless, and highly automated, among others, to describe various forms and levels of vehicle automation. There is no (in)correct terminology and this paper chooses to use the term "autonomous vehicles" (AVs). "The automotive industry has been continuously incorporating parts of the algorithms and devices required by AVs into regular vehicles, including the four performance categories: (1) steering, acceleration, and deceleration, (2) monitoring of driving environment, (3) fall-back performance of dynamic driving task, and (4) system capability (driving modes)" (Duarte and Ratti 2018, p. 5). There are different levels of car automation resulting in different types of AVs on the road. The Society of Automotive Engineers (SAE) (2016) categorizes the automation of vehicles into five stages of automation based on the capabilities of autonomous technologies that decrease the responsibility of the driver:

- No Automation (Level 0): The full-time performance by the human driver of all aspects of the dynamic driving task, even when enhanced by warning or intervention systems.
- Driver Assistance (Level 1): The human driver performs all aspects of the dynamic driving task excluding the mode-specific execution by a driver assistance system of either steering or acceleration/deceleration using information about the driving environment.
- Partial Automation (Level 2): The human driver performs most aspects of the dynamic driving task excluding driving mode-specific execution by one or more driver assistance systems of both steering and acceleration/deceleration using information about the driving environment.
- Conditional Automation (Level 3): The human driver responds appropriately to a request to intervene in driving. The mode-specific performance is based on an automated driving system of all aspects of the dynamic driving task.
- High Automation (Level 4): An automated driving system controls all aspects of the dynamic driving task, even if a human driver does not respond appropriately to a request to intervene.
- Full Automation (Level 5): The full-time performance by an automated driving system of all aspects of the dynamic driving task under all roadway and environmental conditions.

AV often refers to a high (Level 4) or full (Level 5) automated vehicle. The different levels of vehicle automation are gradually being developed, tested, and marketed, providing enough time for cities to embrace this new technology (Anderson et al. 2014).

AV cars are receiving a great deal of media, policymaker, academic, and professional attention. High and full AVs are currently being tested in many countries around the world, and the expected timeline for commercial sales is shortening. Several major companies have announced their plans to offer full AVs in the global market over the next decade. Mercedes-Benz, Google, Tesla, and others have already developed and tested AV prototypes in cities. These companies have made huge investments to make AVs more viable, affordable, and safer, thus making AVs accessible to a large number of people around the world (Talebian and Mishra 2018). Navigant Consulting predicts that 75% of new light-duty vehicles will be automated by 2035 (Shepard and Jerram 2015) and the AV market is predicted to reach $77 billion by the same year (Noyman et al. 2016). IHS Automotive (2014) further predicts that every car on the road will be autonomous by 2050. Many governments and their transport agencies consider the widespread use of AVs in the future as an opportunity to address current urban transportation problems such as reducing traffic congestion, crashes, energy consumption, and pollution while at the same time increasing transport accessibility (Bagloee et al. 2016; Simoudis 2017).

AVs can provide a number of advantages to their users including, but not limited to, the ability to work, sleep, or entertain during a trip while reducing travel time. AVs can also provide access to mobility for people without a driving license such as children. In addition, it is expected that AVs will have a positive influence on traffic flow stability and capacity (Talebpour and Mahmassani

2016). Further advantages arise with people giving up their cars and switching to shared autonomous vehicles. Overall. AVs will disrupt both the transportation system and cities.

There have been recent developments in electric and shared mobility in combination with partial or conditional AVs. Some major mobility companies intend to expand their AV ride-sourcing services based on owning or leasing a portion of their vehicle fleet instead of relying on personal vehicles owned by the drivers themselves (Mohammadzadeh 2018). For example, Lyft rides will utilize fully automated vehicles in their fleet (Zimmer 2016). Tesla Motors announced that the company intends to develop its shared fleet named "Tesla Network" in the future (Shaheen and Cohen 2019). It is pervasively expected that the technological achievements and massive investments in shared electric autonomous vehicles (SEAVs) will result in their popularity and use in cities around the world.

Disruptive Mobility and Future Cities

Around 1 billion cars were produced over the twentieth century, approximately 1.2 billion cars are currently used around the world, and it is estimated that the number of cars will increase to 2 billion by 2035 (Mohammadzadeh 2018). Since the late nineteenth century, our cities have significantly changed due to the private car becoming the main mode of transportation. The twentieth century is named the "automobile century" (Wells 2013). "The development of automobile dependence in cities is a complex process, enacted over decades of land-use and infrastructure development linked to the dominant economic waves of innovation" (Newman and Kenworthy 2015, p. 2). Urban scholars have conceptualized various terms such as "car culture," "car-dependent urban planning," and "car architecture" to describe the role of cars in the transformation of urban form, the design of cities and neighborhoods, and even the architecture of buildings.

If SEAVs were erstwhile imaginary technologies in science-fiction movies and books, they are now the realities on roads around the world. There is an emerging expectation that SEAVs will significantly disrupt our cities and their functions. In this context, it is generally expected that SEAVs will replace existing mobility technologies because of their efficiencies and effectiveness. The pervasive use of SEAVs and their impact on cities are not fully apparent because this disruptive technology is in the process of technological development and improvements, as well as in the process of utilization and adoption by people in cities (Mohammadzadeh 2021). However, it is important to study and predict the potential impact of SEAVs on cities to avoid, or at least to mitigate, potential undesirable side effects on cities in the future.

A growing number of studies have been carried out on the possible impacts of SEAVs on cities (Soteropoulos et al. 2019; Stead and Vaddadi 2019; Yigitcanlar et al. 2019). Two contrasting perspectives are dominant in (non)academia. The utilization of SEAVs may provide new opportunities to mitigate urban issues such as reducing air and noise pollution. However, this disruptive mobility may exacerbate urban problems such as traffic congestion and urban sprawl (Duarte and Ratti 2018). Some adherents of SEAVs primarily emphasize their positive consequences such as decreasing the number of vehicles on the roads, providing cleaner/greener alternative transportation, and decreasing fatal accidents (Faisal et al. 2019; Sperling 2018). Papa and Ferreira (2018, p. 5) argue that utilizing SEAVs "will be associated with radical changes for the better in the built environment." However, others argue that SEAVs will have negative effects on urban sustainability by increasing the number of trips, lengthening distances, increasing pollution, and reinforcing further urban sprawl (Cugurullo et al. 2020; Winter 2020). The following paragraphs review some of the potential impacts of SEAVs on cities.

SEAVs could significantly reshape cities spatially. Stead and Vaddadi (2019) argue that the widespread utilization of disruptive mobility technologies will improve the quality of the built environment through recentralization or regeneration of inner areas, re-densification, and land-use changes to new green public areas and residential locations, among other positive impacts. Disruptive mobility technologies will also generate new opportunities for higher population intensification in the metropolitan centers and near transit-oriented development (TOD) stations and corridors by supporting public transport as an alternative travel mode option. SEAVs could function as a complement to the public transport system by bringing passengers from the surrounding neighborhoods to TOD stations and corridors (Mohammadzadeh 2019). In contrast, some researchers believe that the pervasive use of SEAVs will increase suburbanization and urban sprawl by providing more convenient and feasible travel options that make peripheral areas more attractive for potential residents (Stead and Vaddadi 2019). SEAVs could change the perceptions of people by reducing the stress related to commuting and turning the wasted time of driving into a productive or pleasant time. The cost of using SEAVs based on vehicle kilometers traveled (VKT) will be significantly lower than it is for existing petrol cars. The distance between home and work could lose its priority as a factor when deciding where to live (Duarte and Ratti 2018). Therefore, people could decide to move even farther from cities. Mohammadzadeh (2019) argues that good urban plans, policies, and development strategies are crucial to regulate the deployment of SEAVs and their impacts on future urban development.

Roads and streets should be redesigned to align with SEAVs' capacities. Pervasive use of SEAVs may dramatically reduce the number of cars on the streets. Fagnant and Kockelman (2014) argue that replacing 10% of cars with AVs will reduce traffic up to 15%, and a 90% car fleet could mitigate 60% of traffic congestion. To prepare for SEAVs, streets should be redesigned to prioritize pedestrians, cyclists, and transit riders. That means the existing design codes and standards should be fundamentally revised to address the new requirements of the autonomous era. NACTO (2017) suggests that the future street should provide smaller and fewer lanes for SEAVs to mitigate conflicts and provide pedestrians crossing and space for the expansion of cycle networks. The speed of SEAVs could also be controlled based on the automated speed restriction system to 20 km per hour to improve safety and liveability in residential areas (Riggs et al. 2020). "The utilization of SEAVs will assist us to revitalize our streets as places for socio-economic and political activities rather than car movements" (Mohammadzadeh 2018, p. 40). However, König and Grippenkoven (2017, p. 295) state that SEAVs "may not lead to a reduction of traffic but rather increase it. Future mobility is challenged to bundle up traffic demands to handle an increasing mobility demand caused by spatial sprawl, economic growth and flexible working hours."

"The contemporary car is not a driving machine but a parking machine" (Hawken 2017, p. 185). Parking spaces occupy a large area of cities. Ben-Joseph (2012) points out that 4400 square km, approximately 75 times the area of Manhattan, is used for parking space in the USA. Parking spots in Melbourne cover an area equivalent to 76% of its downtown (Lipson and Kurman 2016). In Los Angeles, 110,000 on-street parking spots cover an area of 331 hectares, equivalent to 81% of its downtown area. "Cars are idle 96 percent of their life span, and AVs could have a utilization rate higher than 75 percent" (Duarte and Ratti 2018, p. 6). It is generally believed that the replacement of private cars by SEAVs will substantially reduce the number of vehicles on streets, subsequently mitigating parking needs in cities. Zhang and Guhathakurta (2017, p. 80) maintain that the SEAV "system can reduce parking land by 4.5% in Atlanta at a 5% market penetration level." Most of the land now devoted to urban parking can be reclaimed because SEAVs will be used more frequently based on shared platforms; therefore, fewer parking spaces will be required in cities. Reclamation of parking land will generate

D

new opportunities for cities to allocate land for required urban amenities and housing. Based on GIS analysis, Mohammadzadeh (2019, pp. 25–27) points out that "around 223 hectares of land is allocated for parking in Auckland's CBD and its other metropolitan centres and the total value of allocated land is NZ$8,905 million." He argues that the pervasive utilization of SEAVs will significantly contribute to economic growth through the transformation of parking space as an unproductive use of land into a productive use such as mixed land use. The transition requires new urban design codes and planning regulations to embrace SEAVs in future cities.

Conclusion

SEAVs are an important disruptive mobility technology that will fundamentally transform cities, their transportation systems, and the built environment. SEAVs can bring propositions to the market that have previously been unavailable in transportation. This technology challenges existing established mobility options by offering a cheaper, reliable, more convenient, and sustainable alternative mobility service. The implementation of adequate regulations, plans, and policies is crucial in maximizing the benefits of SEAVs and mitigating their potential adverse impacts such as urban sprawl.

Disruptive mobility technologies and their impacts on cities are not limited to what has been covered in this article. Online sharing platforms, automation, and the electrification of vehicles are in different stages of development. The level of SEAV adoption and consequently the impact of this technology on cities vary. Overall, the impact of SEAVs should be considered as an ephemeral phenomenon as Industry 4.0 is continuously developing new opportunities for radical disruptive mobilities such as autonomous aerial vehicles, delivery drones, and autonomous delivery robots. Urban transport scholars and professionals must constantly study the new technological achievements of the mobility industry to predict their impacts on cities. These investigations are crucial in mitigating the adverse impacts of disruptive mobility technologies while also maximizing their benefits for cities and their residents.

References

Anderson, J. M., Nidhi, K., Stanley, K. D., Sorensen, P., Samaras, C., & Oluwatola, O. A. (2014). *Autonomous vehicle technology: A guide for policymakers*. Rand Corporation.

Antsaklis, P. J., Passino, K. M., & Wang, S. (1989). Towards intelligent autonomous control systems: Architecture and fundamental issues. *Journal of Intelligent and Robotic Systems, 1*(4), 315–342.

Bagloee, S. A., Tavana, M., Asadi, M., & Oliver, T. (2016). Autonomous vehicles: Challenges, opportunities, and future implications for transportation policies. *Journal of Modern Transportation, 24*(4), 284–303.

Bahmanteymouri, E., & Farzaneh, H. (2020). Airbnb as an ephemeral space: Towards an analysis of a digital heterotopia. In S. Ferdinand, I. Souch, & D. Wesselman (Eds.), *Heterotopia and globalisation in the twenty-first century* (pp. 131–146). Routledge.

Ben-Joseph, E. (2012). Rethinking a Lot. Cambridge, MA: MIT Press.

Beretta, J. (2013). *Automotive electricity: Electric drives*. Wiley.

Bertolini, L. (2020). From "streets for traffic" to "streets for people": Can street experiments transform urban mobility? *Transport Reviews, 40*(6), 734–753.

Botsman, R., & Rogers, R. (2010). *What's mine is yours: How collaborative consumption is changing the way we live*. London: Collins.

Chen, T. D., Kockelman, K. M., & Hanna, J. P. (2016). Operations of a shared, autonomous, electric vehicle fleet: Implications of vehicle & charging infrastructure decisions. *Transportation Research Part A: Policy and Practice, 94*(December), 243–254.

Christensen Clayton, M. (2000). The Innovator's Dilemma: the revolutionary book that will change the way you do business/Clayton M. Christensen.

Cugurullo, F., Acheampong, R. A., Gueriau, M., & Dusparic, I. (2020). The transition to autonomous cars, the redesign of cities and the future of urban sustainability. *Urban Geography, 41*(1), 1–27.

Degirmenci, K., & Breitner, M. H. (2017). Consumer purchase intentions for electric vehicles: Is green more important than price and range? *Transportation Research Part D: Transport and Environment, 51*, 250–260.

Duarte, F., & Ratti, C. (2018). The impact of autonomous vehicles on cities: A review. *Journal of Urban Technology, 25*(4), 3–18.

Etherington, D. (2017). Baidu Plans to Mass Produce Level 4 Self-Driving Cars with BAIC by 2021. Retrieved from TechCrunch.

Fagnant, D. J., & Kockelman, K. (2014). The travel and environmental implications of shared autonomous vehicles, using agent-based model scenarios. *Transportation Research Part C: Emerging Technologies, 40*-(Supplement C), 1–13.

Faisal, A., Kamruzzaman, M., Yigitcanlar, T., & Currie, G. (2019). Understanding autonomous vehicles. *Journal of Transport and Land Use, 12*(1), 45–72.

Fox, S. J. (2017). Planning for density in a driverless world. *NEULJ, 9*(1), 151–202.

Golbabaei, F., Yigitcanlar, T., & Bunker, J. (2021). The role of shared autonomous vehicle systems in delivering smart urban mobility: A systematic review of the literature. *International Journal of Sustainable Transportation, 15*(10), 731–748.

Hall, P. (2014). *Cities of tomorrow: An intellectual history of urban planning and design since 1880*. Wiley.

Hawken, P. (Ed.). (2017). *Drawdown: The most comprehensive plan ever proposed to reverse global warming*. Penguin.

Heidrich, O., Hill, G. A., Neaimeh, M., Huebner, Y., Blythe, P. T., & Dawson, R. J. (2017). How do cities support electric vehicles and what difference does it make? *Technological Forecasting and Social Change, 123*(October), 17–23.

Ho, C. Q., Mulley, C., & Hensher, D. A. (2020). Public preferences for mobility as a service: Insights from stated preference surveys. *Transportation Research Part A: Policy and Practice, 131*(January), 70–90.

Holland, F. (2018). Here's How Ford's Autonomous Vehicles Will Shake up Ride Hailing and Delivery Services. Retrieved from CNBC.

IHS Automotive. (2014). *Emerging technologies: Autonomous cars – Not If, But When*. IHS Automotive Study, http://press.ihs.com/press-release/automotive/selfdriving-cars-moving-industrys-drivers-seat. Accessed 22 Sept 2021.

Intergovernmental Panel on Climate Change. (2019). *Technical summary: Global warming of 1.5°C*, M. Allen, P. Antwi-Agyei, F. Aragon-Durand, M. Babiker, P. Bertoldi, M. Bind, ... & K. Zickfeld (Eds.). World Meteorological Organization. https://www.ipcc.ch/site/assets/uploads/sites/2/2019/06/SR15_Full_Report_Low_Res.pdf. Accessed 10 Sept 2021.

Kenney, M., & Zysman, J. (2016). The rise of the platform economy. *Issues in Science and Technology, 32*(3), 61.

Khajepour, A., Fallah, M. S., & Goodarzi, A. (2014). *Electric and hybrid vehicles: Technologies, modeling and control-a mechatronic approach*. Wiley.

König, A., & Grippenkoven, J. (2017). From public mobility on demand to autonomous public mobility on demand–learning from dial-a-ride services in Germany. In E. Sucky, R. Kolke, N. Biethahn, J. Werner, & G. Koch (Eds.), *Mobility in a globalised world 2016* (pp. 295–305). Bamberg: University of Bamberg Press.

Legacy, C., Ashmore, D., Scheurer, J., Stone, J., & Curtis, C. (2019). Planning the driverless city. *Transport Reviews, 39*(1), 84–102.

Lipson & M. Kurman, (2016). Driverless: Intelligent Cars and the Road Ahead (Cambridge MA: MIT Press).

Loeb, Benjamin, Kockelman, Kara M., Liu, Jun, (2018). Shared autonomous electric vehicle (SAEV) operations across the Austin, Texas network with charging infrastructure decisions. *Transportat. Res. Part C: Emerg. Technol., 89*, 222–233. https://doi.org/10.1016/j.trc.2018.01.019.

Lucken, E., & Shaheen, S. (2021). Incorporating Mobility-on-Demand (MOD) and Mobility-as-a-Service (MaaS) automotive services into public transportation. In G. Currie (Ed.), *Handbook of public transport research* (pp. 410–433). Edward Elgar Publishing.

Machado, C. A. S., de Salles Hue, N. P. M., Berssaneti, F. T., & Quintanilha, J. A. (2018). An overview of shared mobility. *Sustainability, 10*(12), 4342–4363.

Maitipe, B. R., Ibrahim, U., Hayee, M. I., & Kwon, E. (2012). Vehicle-to-infrastructure and vehicle-to-vehicle information system in work zones: Dedicated short-range communications. *Transportation Research Record, 2324*(1), 125–132.

Martínez-Gutiérrez, A., Díez-González, J., Ferrero-Guillén, R., Verde, P., Álvarez, R., & Perez, H. (2021). Digital twin for automatic transportation in industry 4.0. *Sensors, 21*(10), 3344–3367.

Meyer, G., & Beiker, S. (Eds.). (2018). *Road vehicle automation 4*. Springer.

Millard-Ball, A. (2019). The autonomous vehicle parking problem. *Transport Policy, 75*, 99–108.

Mohammadzadeh, M. (2018). The disruptive mobility and the future of our neighbourhoods. *Journal of Consumer Research, 39*(4), 881–898.

Mohammadzadeh, M. (2019). The disruptive mobility and the potential for land reclamation: The case. *Development, 131*(4), 233–245.

Mohammadzadeh, M. (2021). Sharing or owning autonomous vehicles? Comprehending the role of ideology in the adoption of autonomous vehicles in the society of automobility. *Transportation Research Interdisciplinary Perspectives, 9*(March), 100294–100304.

Muoio, D. (2016). These 20 Companies Are Racing to Build Self-Driving Cars in the next 5 Years. Retrieved from Business Insider.

NACTO. (2017). *Blueprint for autonomous urbanism*. New York. Retrieved from https://nacto.org/publication/bau/blueprint-for-autonomous-urbanism/.

Newman, P., & Kenworthy, J. (2015). *The end of automobile dependence*. Washington, DC: Island Press.

Novikova, O. (2017). The sharing economy and the future of personal mobility: New models based on car sharing. *Technology Innovation Management Review, 7*(8), 27–31.

Noyman, A., Stibe, A., & Larson, K. (2016). *Autonomous cities and the urbanism of the 4th machine age: Should AV industry design future cities*? Changing Places Research Group, MIT Media Lab. https://www.semanticscholar.org/paper/Autonomous-Cities-and-the-Urbanism-of-the-4-th-Age-Noyman-Stibe/fa09f350f79e2a76fcf68415ca803619b5587015. Accessed 13 Sept 2021.

D

Olivier, J. G. J., & Peters, J. A. H. W. (2019). *Trends in global CO2 and total greenhouse gas emissions: 2019 report*. PBL Netherlands Environmental Assessment Agency, The Hague. Retrieved from https://www.pbl.nl/sites/default/files/downloads/pbl-2020-trends-in-global-co2-and-total-greenhouse-gas-emissions-2019-report_4068.pdf

Papa, E., & Ferreira, A. (2018). Sustainable accessibility and the implementation of automated vehicles: Identifying critical decisions. *Urban Science, 2*(1), 5. 1–14.

Riggs, W., Appleyard, B., & Johnson, M. (2020). A design framework for livable streets in the era of autonomous vehicles. *Urban, Planning and Transport Research, 8*(1), 125–137.

Ronald, N., Navidi, Z., Wang, Y., Rigby, M., Jain, S., Kutadinata, R., Thompson, R., & Winter, S. (2017). Mobility patterns in shared, autonomous, and connected urban transport. In G. Meyer & S. Shaheen (Eds.), *Disrupting mobility* (pp. 275–290). Cham: Springer.

SAE. (2016). *Taxonomy and definitions for terms related to on-road motor vehicle automated driving systems*. SAE International.

Sentinel, M. (1926). Phantom Auto will tour city. *The Milwaukee Sentinel, 4*.

Shaheen, S. A. (2016). Mobility and the sharing economy. *Transport Policy, 51*(Supplement C), 141–142.

Shaheen, S., & Cohen, A. (2019). Shared ride services in North America: Definitions, impacts, and the future of pooling. *Transport Reviews, 39*(4), 427–442.

Shaheen, S., Totte, H., & Stocker, A. (2018). *Future of mobility white paper*. UC Berkeley: Institute of Transportation Studies at UC Berkeley. Retrieved from https://escholarship.org/uc/item/68g2h1qv

Shepard, S., & Jerram, L. (2015). Transportation forecast: Light duty vehicles. Boulder, CO: Navigant Consulting, Inc.

Simoudis, E. (2017). *The big data opportunity in our driverless future*. Corporateinovation.

Skinner, R., & Bidwell, N. (2016). Making Better Places: Autonomous vehicles and future opportunities. WSP Parsons Brinckerhoff in association with Farrells.

Soteropoulos, A., Berger, M., & Ciari, F. (2019). Impacts of automated vehicles on travel behaviour and land use: An international review of modelling studies. *Transport Reviews, 39*(1), 29–49.

Sperling, D. (2018). Three revolutions: Steering automated, shared, and electric vehicles to a better future. Island Press.

Sprei, F. (2018). Disrupting mobility. *Energy Research & Social Science, 37*(March), 238–242.

Stead, D., & Vaddadi, B. (2019). Automated vehicles and how they may affect urban form: A review of recent scenario studies. *Cities, 92*(September), 125–133.

Stocker, A., & Shaheen, S. (2017). Shared automated vehicles: Review of business models. International Transport Forum Discussion Paper.

Sundararajan, A. (2016). *The sharing economy: The end of employment and the rise of crowd-based capitalism*. Boston: MIT Press.

Talebian, A., & Mishra, S. (2018). Predicting the adoption of connected autonomous vehicles: A new approach based on the theory of diffusion of innovations. *Transportation Research Part C: Emerging Technologies, 95*(October), 363–380.

Talebpour, A. & Mahmassani, H.S. (2016). Influence of connected and autonomous vehicles on traffic flow stability and throughput. *Transportation Research Part C: Emerging Technologies, 71*, 143–163. https://doi.org/10.1016/j.trc.2016.07.007

Tirachini, A., Chaniotakis, E., Abouelela, M., & Antoniou, C. (2020). The sustainability of shared mobility: Can a platform for shared rides reduce motorized traffic in cities? *Transportation Research Part C: Emerging Technologies, 117*(August), 102707–102722.

Udaeta, M. E. M., Chaud, C. A., Gimenes, A. L. V., & Galvao, L. C. R. (2015). Electric vehicles analysis inside electric mobility looking for energy efficient and sustainable metropolis. *Open Journal of Energy Efficiency, 4*(1), 1–14.

Wakefield, E. H. (1993). History of the Electric Automobile: Battery-Only Powered Cars. Society of Automotive Engineers Inc., Warrendale, PA, USA.

Weber, M. (2014). Where to? A History of Autonomous Vehicles, Computer History Museum at http://www.computerhistory.org/atchm/where-toa-history-of-autonomous-vehicles/

Wells, C. W. (2013). Car country: An environmental history. University of Washington Press.

Winter, S. (2020). Wayfinding and navigation research for sustainable transport. *Journal of Spatial Information Science, 2020*(20), 103–107.

Yigitcanlar, T., Wilson, M., & Kamruzzaman, M. (2019). Disruptive impacts of automated driving systems on the built environment and land use: An urban planner's perspective. *Journal of Open Innovation: Technology, Market, and Complexity, 5*(2), 24.

Zakharenko, R. (2016). Self-driving cars will change cities. *Regional Science and Urban Economics, 61*, 26–37.

Zhang, W., & Guhathakurta, S. (2017). Parking spaces in the age of shared autonomous vehicles: How much parking will we need and where?. *Transportation Research Record, 2651*(1), 80–91.

Zimmer, J. (2016). The Third Transportation Revolution. Medium, Sept.

Distribute

► Closing the Loop on Local Food Access Through Disaster Management

Distributed Energy

▶ Faith Communities as Hubs for Climate Resilience

District

▶ Closing the Loop on Local Food Access Through Disaster Management

Drinking Water Standards

▶ Meeting SDG6: Ensuring Safe Drinking Water for All in Rural India

Driver Policy

▶ Spatial Demography as the Shaper of Urban and Regional Planning Under the Impact of Rapid Urbanization

E

EbA – Ecosystem-Based Adaptation

▶ Adapting to a Changing Climate Through Nature-Based Solutions

Eco-city

▶ Green Cities
▶ Green Cities in Theory and Practice

Eco-DRR – Ecosystem-Led Disaster Risk Reduction

▶ Adapting to a Changing Climate Through Nature-Based Solutions

Eco-efficient City

▶ Green Cities in Theory and Practice

Eco-industrial Network

▶ Industrial Symbiosis

Ecological Corridor

▶ The Challenges for Wildland-Urban Interfaces (WUI) in Metropolitan Areas: Reducing Fire Risk, Providing Employment Opportunities, and Preserving Natural Habitat

Ecological Engineering

▶ Growth, De-growth, and Nature-Based Solutions
▶ Nature-Based Solutions for River Restoration in Metropolitan Areas

Ecological Restoration

▶ Nature-Based Solutions for River Restoration in Metropolitan Areas

Ecological Security

▶ Water Security and Its Role in Achieving SDG 6

Economic Sustainability

▶ Sustainable Development Goals

R. C. Brears (ed.), *The Palgrave Encyclopedia of Urban and Regional Futures*,
https://doi.org/10.1007/978-3-030-87745-3

Economical Security

► Water Security and Its Role in Achieving SDG 6

Ecoregion

► Bioregionalism

Ecosystem Based Adaptation

► Concepts, Approaches, and Methodologies for Ecological Flood Resilience Assessment: A Review

Ecosystem Services

► Growth, De-growth, and Nature-Based Solutions

Ecosystem-Based Adaptation

► Growth, De-growth, and Nature-Based Solutions

► Nature-Based Solutions for River Restoration in Metropolitan Areas

► Overcoming Barriers in Green Infrastructure Implementation

Ecosystem-Based Approach

► Growth, De-growth, and Nature-Based Solutions

Ecosystem-Based Disaster Risk Reduction or Building with Nature

► Nature-Based Solutions for River Restoration in Metropolitan Areas

Ecosystem-Based Management

► Nature-Based Solutions for River Restoration in Metropolitan Areas

Ecosystems Restoration and Habitats Enhancement

A Cast Study of Weir Pool Manipulation Along the South Australian River Murray

Tian Shi
Department of Primary Industries and Regions, Adelaide, South Australia

Introduction

In South Australia, flow in the River Murray was naturally highly variable. Flows increased through late winter, peaked in spring and reduced over summer. Lowest flows commonly occurred over autumn and early winter. The river experienced large seasonal and annual variations in flow and water level. Flows above 150,000 megaliters per day (ML/day) and below 2,000 ML/day were common. River levels changed as a result of the changes in flow, with variations of $\geq$5 m in the majority of years.

The wide variation in flow and water level was fundamental to the healthy functioning of the river, wetland, and floodplain ecosystems. In many years, spring and early summer high flows were sufficient to inundate over half of the floodplain providing the river and wetlands with nutrients and organic matter. Floods are particularly important for the breeding cycles of fish and water birds, regeneration and maintenance of floodplain vegetation such as river red gums and black box, and flushing salt from the landscape. The periods of low flow between floods were also important. During these periods the wetlands and backwaters dried out, oxygenating the sediments, aiding nutrient exchange, and decomposition of organic matter, as well as providing habitat for terrestrial plants and animals. These periods of low flow also reduced groundwater levels beneath the floodplain, reducing floodplain salinization and its impact on vegetation health (DEWNR 2012; Bice et al. 2016).

During the first half of the twentieth century, the River Murray was used extensively for commercial navigation, carrying agricultural products to markets, and at that time river waters began to be developed for consumptive uses (e.g., irrigation, industry, and urban supply). The variable nature of flows in the River Murray provided difficulties for navigation and water supply and consequently a series of weir and locks were constructed along the River Murray to regulate the river. Weirs are structures built to maintain water levels and locks enable boats to pass through weirs. Construction commenced in 1922 at Blanchetown, South Australia, and was completed at Euston, Victoria in 1937. Today there are 14 weirs on the River Murray, weirs 1 to 6 are located in South Australia (see Fig. 1).

The natural variation in water levels in the lower River Murray has been dramatically reduced as a result of river regulation and extraction for urban and irrigation supplies. River regulation has reduced hydrological variability, resulting in reduced connectivity and, in places (e.g., downstream of weirs), the complete disconnection of floodplain areas from the main channel of the River Murray. Loss of connectivity has significantly impacted the health and resilience of the river, anabranch, backwater, and floodplain ecosystems. Almost all native plants and animals along the River Murray rely on regular changing water levels to be healthy, diverse, and abundant. Locks and weirs have changed the river and connected or disconnected wetlands to a series of stable pools, which experience very little water level variation (see Fig. 2). As a result, 70% of wetlands that were once seasonally inundated are now permanently connected to the river at pool level. The system needs variations in water level to remain robust.

As DEWNR (2014) summarized the impacts of current river operations include:

- Unseasonal and prolonged wetting in low-level wetlands and floodplains
- Storage of large volumes of salt on the floodplain and maintenance of high saline groundwater levels under the floodplain
- Reduced frequency and duration of small to medium floods
- Stress and death of floodplain vegetation
- Increased sedimentation and higher risk of algal blooms
- Reduced opportunities and cues that trigger breeding cycles of birds, fish and invertebrates
- Conditions that favor exotic species such as European carp and willows.

Under this background, this entry reviewed the history of weir pool manipulation (WPM) trials along the South Australian River Murray, and through a case study to illustrate it as an instrument of delivering environmental water and optimizing river operations to help improving the health of wetlands, floodplains, and the river and achieving the long-term benefits for ecosystems restoration and habitats enhancement.

Weir Pool Manipulation in South Australia

A Brief History

WPM includes weir pool raising (WPR) and weir pool lowering (WPL). WPR for environmental outcomes was first undertaken in South Australia in October 2000 at Lock 5. A flow of approximately 32,000 ML/day was increased with discharge from Lake Victoria to provide a peak flow of about 42,000 ML/day. The Lock 5 weir pool water level was raised by 50 centimeters (cm) above normal pool level (NPL) for a period of about 4 weeks and the resulting inundation of hundreds of hectares (ha) of floodplain and wetlands was similar to that normally experienced during a flow of 70,000 ML/day.

From 15 March through to early April 2005, the Lock 6 water level was slowly raised to 15 cm above NPL for a period of about 4 weeks. The Lock 5 weir pool was also raised a small amount at this time to reduce the hydraulic pressure on Lock 6. These levels were maintained until early May when the level was slowly lowered back to normal. Raising the levels increased the amount

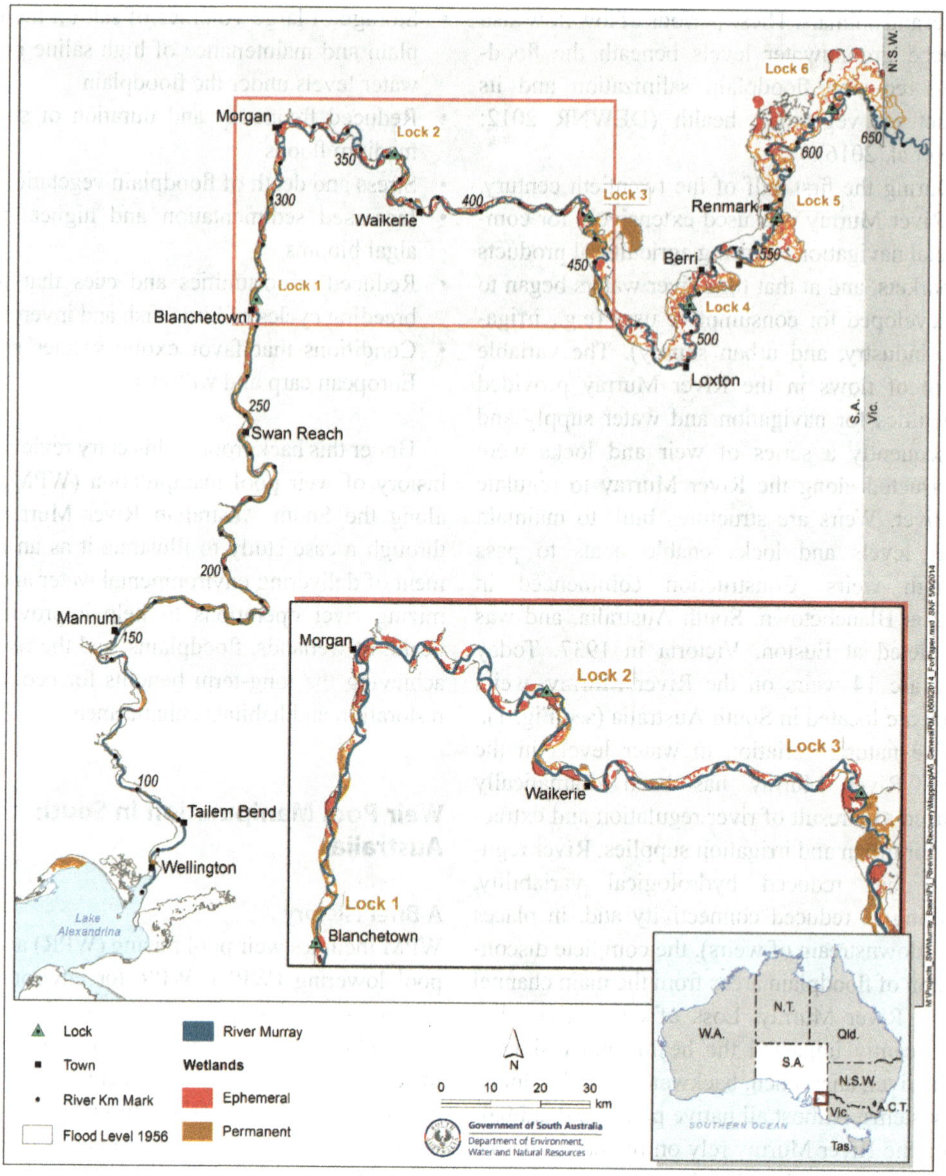

Ecosystems Restoration and Habitats Enhancement, Fig. 1 River Murray in South Australia showing locks 1 to 6 and associated weir pools (Source: Bonifacio et al. 2016)

of water flowing through the Chowilla anabranch, pushing water into creeks and inundating low-lying areas of the floodplain and wetlands. About 50 ha of floodplain and wetlands were watered in addition to many kilometers of creeks.

Further WPRs were undertaken at Lock 6 (+15 cm), Lock 5 (+50 cm), Lock 4 (+30 cm), and Lock 1 (+10 cm) between October 2005 and January 2006. These WPRs inundated areas along the river channel; pushed water into anabranch creek systems; and allowed the flooding of a number of wetlands. This resulted in a range of benefits primarily by increasing vegetation health. Weirs can also be used to increase the area of the floodplain inundated for a particular volume of flow and thus provide a valuable method of increasing the effects of environmental flow.

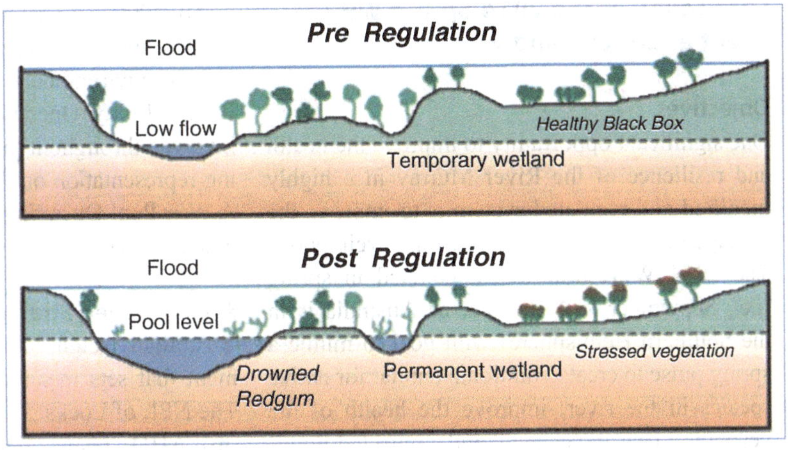

Ecosystems Restoration and Habitats Enhancement, Fig. 2 River Murray water levels pre and post regulation

While WPL has occurred in association with river management and maintenance activities, lowering for environmental benefit has not been undertaken in South Australia. Initial WPLs would be relatively conservative (e.g., less than 30 cm below NPL). Physically, weir pools can be lowered from between two and three meters. The exact amount depends on which weir is being managed.

The extent to which any weir pool can be raised or lowered is constrained by a number of factors. These include limitations on raising due to the capacity of the lock and weir infrastructure. Lowering limitations include location and capacity of pumps on the river and backwaters to tolerate changes in water level, ability to navigate the river at different levels, salinity impacts and structural stability of the weir and associated infrastructure. Information on these factors is being collected and analyzed, and will be carefully considered prior to any changes in weir management to ensure minimum impacts on river users. When there is confidence that any impacts on river users can be effectively managed, the extent of lowering will likely be increased.

Rationale and Expected Ecological Benefits

Water level variability is a key driver of the ecosystem health of the River Murray. Weirs can be used to raise and lower water levels in a weir pool to mimic more natural water level variability. This means that lowering will occur when water levels in the river would naturally be low – during autumn and winter, which is also the time when diversions for irrigation are at their lowest. Raising weir pool levels will occur when flooding would have naturally occurred during spring and early summer. WPM affect the length of a weir pool through raising or lowering water levels. WPM is, along with wetland regulation, one of the means to provide targeted delivery of environmental water to help restore ecological function to the river, wetland, and floodplain.

Fully restoring the pre-regulation ecology via water level manipulation is not a realistic goal. However, improving aspects of the ecology is possible, especially in localized areas of the River Murray. Changing how weirs are managed has been shown to deliver real benefits to the river channel, ephemeral wetlands and anabranch creeks and low-lying parts of the floodplain.

As DEWNR (2014) summarized the benefits from weir pool raising and lowering include:

- Improved vegetation coverage and recruitment
- Reduced water stress in floodplain vegetation
- Improved water level variability to benefit waterbirds, fish and invertebrates
- Improved biofilm (algae, micro-organisms) value as a food source
- Drying of inundated sediments to consolidate them
- Providing positive impacts in the river channel including flowing water habitats.

Case Study of the 2014 Spring WPR Event at Locks 1 and 2

Objectives

One significant opportunity to improve the health and resilience of the River Murray in a highly modified and regulated system is to improve the management of water levels within weir pools. The 2014 WPR trial was conducted in spring (i.e., September to November as Australia is in the Southern Hemisphere) with goal to mimic a spring pulse to create additional habitat for native species in the river; improve the health of the vegetation near the channel and connected floodplains and wetlands; and increase the numbers of frogs, birds, and fish in the system.

As Shi (2014a) summarized the objectives of trial event were to:

- Test state government processes and instruments for environmental water delivery
- Return some of the natural hydrological variation in water levels
- Provide benefits to in channel and low-lying wetland and floodplain areas
- Learn more about the operational requirements and ecological effects of such event
- Raise public awareness of the benefits of water level variations
- Document the process and learnings to inform future weir pool planning and operation.

A Coordinated Multi-agency Trial

Department for Environment, Water and Natural Resources (DEWNR) led the trial in conjunction with Murray-Darling Basin Authority (MDBA), Commonwealth Environmental Water Office (CEWO), SA Water and Nature Foundation South Australia (NFSA). Each has representation on the Weir Pool Manipulation Steering Committee (WPMSC). A Weir Pool Manipulation Advisory Committee (WPMAC), comprised members from DEWNR, MDBA, SA Water, University of Adelaide and NFSA, continued to provide technical input into the event. Both committee members were consulted since the commencement of event planning and were supportive of the scope and intent of the trial.

DEWNR's Sustainable Water Resources Branch included weir pool water requirements into the environmental water planning. DEWNR's River Murray Operations Branch was integrally involved throughout the life of the project, including representation on the WPMSC and WPMAC. A Weir Pool Operating Protocol was developed to guide the management of such actions in the future.

A Two-Staged Strategy in Raising

Australian Height Data (AHD) is the measurement that sets mean sea level at zero elevation. The NPL of Locks 1 and 2 are 3.2 and 6.1 meters (m) AHD, respectively. The normal operating range is defined as the expected water level variations under routine River Murray operations above or below NPL.

The 2014 spring trial was conducted in a two-staged strategy to mimic the natural variability of water levels that occurred before the weirs and locks were built (see Figs. 3 and 4). The water levels were first raised within normal operating range for Lock 1 from 3.2 to 3.5 m AHD, and for Lock 2 from 6.1 to 6.4 m AHD by 10 cm increments, and assessed the water flow across the landscape at each increment. Once this was achieved, weir pool water levels were raised beyond the normal operating range by a further 20 cm at Locks 1 and 2 to 3.7 m AHD and 6.6 m AHD, respectively, pending ongoing risk assessment during operation and ensuring all potentially affected landholders and lessees are appropriately notified. The water levels remained at the maximum achieved level until late spring and then was drawn down over a few weeks, subject to flow conditions in the River Murray.

Risk Assessment

The aim of WPM is to achieve the greatest ecological benefits by manipulating water levels in order to mimic preregulation conditions to which native species are adapted. At the same time, a range of operational constraints and risks need to be identified and effectively managed. The area inundated will be highly dependent on flow conditions. Estimates of areas inundated under different flow conditions are shown in Table 1.

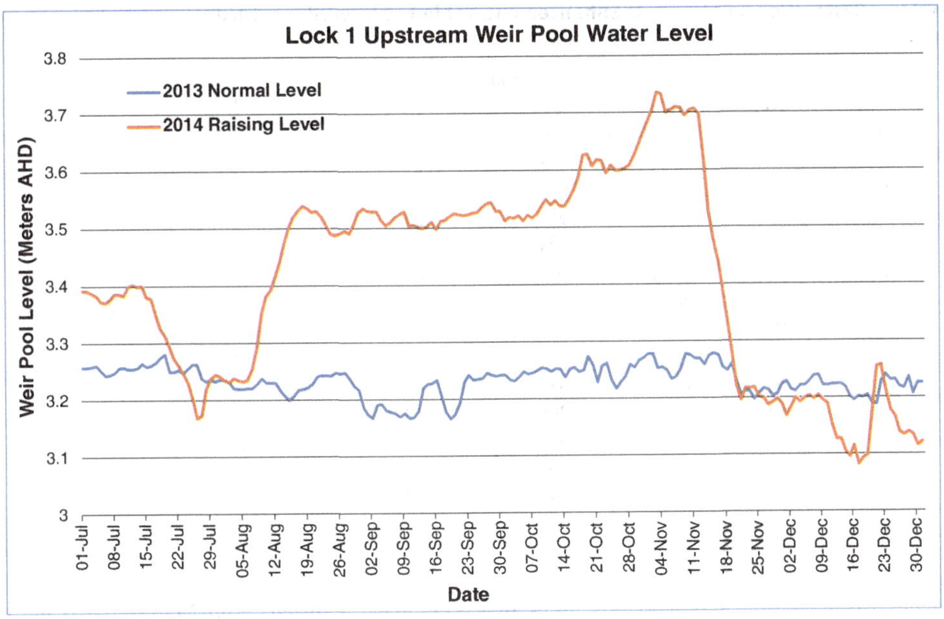

Ecosystems Restoration and Habitats Enhancement, Fig. 3 2014 raised Lock 1 weir pool water levels from relatively constant in 2013 (Data Source: MDBA)

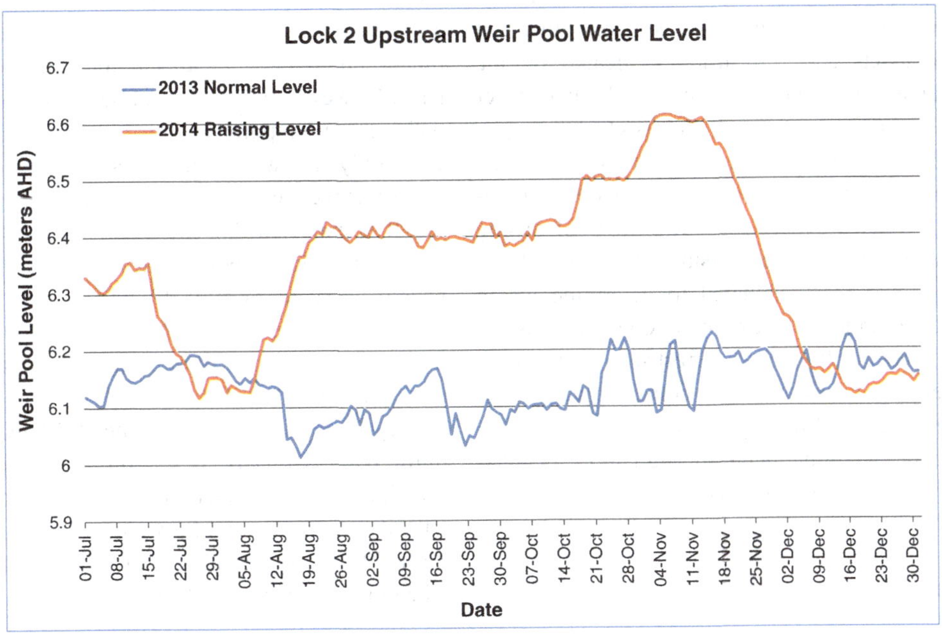

Ecosystems Restoration and Habitats Enhancement, Fig. 4 2014 raised Lock 2 weir pool water levels from relatively constant in 2013 (Data Source: MDBA)

Ecosystems Restoration and Habitats Enhancement, Table 1 Modelled inundated areas and volumes for weir pool raising scenarios

Lock	Flow (GL/day)	Weir raising (cm)	Area inundated (ha)	Additional area inundated (ha)	Relative increase in inundated area (%)	Total volume in pool (GL)	Additional volume required for raising (GL)
1	10	0	2275	0	0	58	0
		15	2333	58	3	61	3
		50	2712	437	19	70	12
	20	0	2361	0	0	63	0
		15	2418	56	2	66	3
		50	2859	498	21	75	12
2	10	0	1557	0	0	41	0
		20	1620	63	4	44	3
		50	1766	209	13	49	7
	20	0	1681	0	0	46	0
		20	1771	89	5	49	3
		50	1922	241	14	53	7

Source: Shi 2014b

Estimates of vegetation and wetlands areas that may be inundated in the process and addition areas (i.e., areas above the normal level inundation for the given flow) to receive ecological benefits are shown in Table 2.

Generally, the higher the raising, the greater the ecological benefits as this inundated a greater area of floodplain. By raising water levels above the normal operating range, there is a possibility that some assets and infrastructure located on the floodplain may be impacted by higher water levels. Such infrastructure and assets can be privately, local or state government owned. The numbers of roads, assets, and structures potentially affected by each of the weir pool raising scenarios were estimated using desktop GIS techniques and then evaluated to determine where modelled inundation extents overlapped existing data layers of floodplain infrastructure. The estimates were considered to be conservative to align with the gradual raising of weir pool and benefit from further refinements through more detailed analysis and on-ground verification.

Discussion

The monitoring investigations conducted during (September and November 2014) and after (January, February and March 2015) the raising found discernible ecological benefits associated with the 2014 spring WPR event (e.g., an increase in the abundance of littoral understorey vegetation at the lower elevations of the riverbank of the weir pools and in wetlands, a limited number of shifts in biofilm community composition). This assisted nutrient cycling and benefited food web pathways, which are essential for supporting a healthy functioning river, wetland, and floodplain ecosystems. However, the ecological benefits were spatially and/or temporally limited and overall, there were only a few statistically significant differences that were considered ecologically meaningful. It is expected that the magnitude, diversity, and spatial and temporal scales of the ecological benefits derived from WPR will become more apparent with regular WPR events, and by refining WPR operations through adaptive management practices (Bonifacio et al. 2016).

Key risks associated with the WPR included third party impacts (e.g., property damage, flooding of roads, stock loss, inundation of private jetties) and negative public responses to the event. Shi (2014a) outlined mitigation activities undertaken to minimize impacts including:

- Maintained WPR at moderate levels (i.e., raise the water levels of Locks 1 and 2 weir pools by

Ecosystems Restoration and Habitats Enhancement, Table 2 Modelled areas of vegetation and wetlands inundated during the weir pool raising

Lock	Scenario	Vegetation				Wetlands			
		Area inundated (ha)	Additional area inundated (ha)	% of total vegetation area not inundated at normal level, now inundated	% increase in inundated area when compared to normal level for same flow	Area inundated (ha)	Additional area inundated (ha)	% of total vegetation area not inundated at normal level, now inundated	% increase in inundated area when compared to normal level for same flow
1	10GL/d normal	42	0	0	0	970	0	0	0
	10GL/d 15 cm	52	9	0.2	22	1001	31	1.7	3
	10GL/d 50 cm	213	171	3.1	406	1162	192	10.6	20
	20GL/d normal	68	0	0	0	1001	0	0	0
	20GL/d 15 cm	92	24	0.4	35	1019	19	1	2
	20GL/d 50 cm	285	217	3.9	318	1220	220	12.1	22
2	10GL/d normal	46	0	0	0	463	0	0	0
	10GL/d 20 cm	61	14	0.3	31	499	36	3.8	8
	10GL/d 50 cm	114	68	1.4	146	569	106	11.1	23
	20GL/d normal	80	0	0	0	530	0	0	0
	20GL/d 20 cm	115	35	0.7	44	572	42	4.4	8
	20GL/d 50 cm	193	114	2.3	142	624	94	9.8	18

Source: Shi 2014b

E

up to 50 cm above NPL), pending ongoing risk assessment during operation
- Commenced discussions with community at least 2 months prior to the event
- Engaged with potentially affected landholders and lessees prior to and during the event
- Identified potential hazards prior to the event and undertook corrective actions, as required
- Closely monitored infrastructure during the event.

An extensive communications program must start at least 2 months before the event so the wider community are informed of the proposed trial event and can implement any necessary actions (e.g., raising irrigation infrastructure, see Fig. 5). This is also an essential part of mitigation activities to manage potential third-party impacts (e.g., property damage, flooding of roads, stock loss, inundation of private jetties) and potential negative public responses to the trial.

Concluding Remarks

Before regulation, flows in the South Australian River Murray were highly variable. Major floods caused large-scale inundation of wetlands and floodplains. Conversely, during periods of low-flow, the river level dropped and the floodplains dried. The construction of weirs and locks after 1900 stabilized water levels in river primarily to benefit navigation and irrigation. The stable pool levels have limited exchange of water, nutrients, and biota between the river and floodplain, significantly impacting ecosystem health and resilience of the River Murray.

WPM has now been demonstrated as an effective mechanism to provide targeted delivery of environmental water to improving the health of wetlands, floodplains, and the river and giving security to regional communities. With increased community understanding and support of changing river operations for ecological and environmental benefits, weir pool water level manipulation will become an increasingly

Ecosystems Restoration and Habitats Enhancement, Fig. 5 Pump and pipes on adjustable slide to accommodate changes in water level

common feature of routine river operations and greater hydrological variability can be expected in the future to achieve the long-term benefits for ecosystems restoration and habitats enhancement along the South Australian River Murray.

References

Bice, C. M., Zampatti, B. P., & James, C. (2016). The influence of weir pool raising on main channel hydraulics in the lower River Murray, South Australia, in 2015. South Australian Research and Development Institute (SARDI), SARDI Research Report Series No. 904, 40 pp.

Bonifacio, R. S., Hanisch, D. R., & Hillyard, K. A. (2016). Riverine recovery: Synthesis of the ecological response to weir pool raising at locks 1 and 2 of the River Murray, spring 2014. Department of Environment, Water and Natural Resources, Adelaide, South Australia, 58 pp.

DEWNR. (2012). Riverine recovery monitoring and evaluation program: Conceptual understanding of the ecological response to water level manipulation. Department of Environment, Water and Natural Resources (DEWNR), Adelaide, South Australia, 76 pp.

DEWNR. (2014). Riverine recovery: Overview of monitoring to assess the ecological response to the locks 1 and 2 weir pool raising, spring 2014. Department of Environment, Water and Natural Resources (DEWNR), Adelaide, South Australia.

Shi, T. (2014a). Riverine recovery 2014 weir Pool raising: Weir 1 event plan. Department for Environment, Water and Natural Resources, Adelaide, South Australia, 26 pp.

Shi, T. (2014b). Weir Pool raising: Benefits, Risks and Management. Information Paper prepared for Chair of the Murray-Darling Basin Coordinating Committee. Department of Environment, Water and Natural Resources, Adelaide, South Australia, 13 pp.

Ecotourism is *responsible tourism*

► Sustainable Development and Responsible Tourism: The Grijalva-Usumacinta Lower River Basin

Ecovillages

► Senegalese Ecovillage Network

Education for Inclusive and Transformative Urban Development

Putting the Future into Perspective

Chipo Chitereka[1], Edmos Mtetwa[1], Watch Ruparanganda[1] and Innocent Chirisa[2]
[1]Department of Social Work, University of Zimbabwe, Harare, Zimbabwe
[2]Department of Demography Settlement and Development, Social & Behavioural Sciences, University of Zimbabwe, Harare, Zimbabwe

E

Introduction

Education plays a fundamental role in enhancing inclusive and transformative urban development. Inclusive and transformative urban development should foster a holistic improvement of all aspects of life of the residents. Development by the people and for the people encourages and empowers people to participate in their own development. This chapter seeks to explore the significance of education toward inclusive and transformative urban development evidenced by growth, positive change, or the addition of physical, economic, environmental, social, and demographic components. Inclusive and transformative development ought to be demonstrated by the quality of life of citizens through the creation or expansion of income and employment opportunities while maintaining the resources of the environment. This type of urban development should also be visible and useful, including an aspect of quality of change and the establishment of conditions for continuation of change. It is also argued that inclusive and transformative urban development cannot be achieved where there is continual existence of educational inequalities. Thus, more

"Education is the premise of progress, in every society, in every family. It is a human right with immense power to transform. On its foundation rest the cornerstones of freedom, democracy and sustainable human development."
Kofi Annan

emphasis should be given to equitable access to quality education. The chapter explores factual materials on the meaning of development, the concept of education, the relationship between education and development, the emerging issues restraining education from enhancing inclusive and transformative urban development, and future directions.

The Meaning of Development

The term development has a multitude of meanings to different people and can be explained in different perspectives, which makes it complex, contested, ambiguous, and elusive. By and large, development is an event constituting a new stage in a changing situation or the processes of change per se (Atkinson 1970). According to Rabie (2016), development is a comprehensive societal process to move the underdeveloped from their state of economic backwardness and slow sociocultural change to a dynamic state characterized by sustained economic growth and sociocultural and political transformation that improves quality of life of all members of society. When denoting to cities and towns, development means improvement of physical, social, cultural, spiritual, and economic aspects of the populations which is maintainable.

Todaro (1999, p. 101) asserts that "development must be conceived of as a multidimensional process involving changes in structures, attitudes and institutions as well as the acceleration of economic growth, the reduction of inequalities and eradication of absolute poverty." Bellue (2011, p. 2) has also noted that "development is a multidimensional concept in its nature because any improvement of complex systems, as indeed actual socio-economic systems are, can occur in different parts or ways, at different speeds and driven by different forces." The multidimensionality of the concept of development also requires an intrinsic multidimensional exercise to determine whether and to what extent has development taken place. In essence, development must represent an array of changes by which an entire social system, turned to the diverse basic needs and desires of individuals and social groups within that system, moves away from a condition of life widely perceived as unsatisfactory and toward a situation and condition of life regarded as materially and spiritually better (Todaro 1999).

Development should as well have aspects of continuity to ensure that improvements occurring should not be detrimental to the future status of generation to come (i.e., sustainable development). The concept of sustainable development was initially introduced by Brundtland (1987) who envisions development as sustainable if it meets the needs of the present without compromising the ability of future generations to meet their own needs. This entails that sustainable development is both a quantitative increase in economic production and a qualitative improvement in life conditions while protecting the environment (Rabie 2016). As maintained by Bellue (2011), sustainable development should minimize the use of exhaustible resources or at least ensure that revenues obtained from them are used to create constant flow of income across generations and making appropriate use of renewable resources. Sustainable development is a holistic concept addressing economic development, social inclusion, environmental sustainability, and good governance (Sustainable Development Solution Network (SDSN 2013)).

Mahadevia (2001) has echoed the same sentiments emphasizing that a holistic approach incorporates all dimensions of development including the interests of the poor and the disempowered. The SDSN proposed a post-2015 Sustainable Development Goal for cities to "empower inclusive and resilient cities" that aims at making all cities socially inclusive, economically productive, environmentally sustainable, and resilient to climatic change and other risks. Additionally, the goal proposes that cities should develop participative, accountable, and effective city governance to support rapid and equitable urban transformation. Therefore, for development to be sustainable, people should be the major agents of development and the main beneficiaries of transformation produced by it.

It can be concluded that development is both a physical reality and a state of mind in which society has, through some combination of social, economic, and institutional processes, secured the means for obtaining a better life (Todaro 1999). Development as asserted by Todaro (1999, p. 102) must have the components of the following three objectives:

- To increase the availability and widen the distribution of basic life-sustaining goods such as food, shelter, health, and protection to all members of society.
- To raise levels of living, including, in addition to higher incomes, the provision of more jobs, better education, and more attention to cultural and humanistic values. These all serve not only to enhance material well-being but also to generate greater individual and national self-esteem.
- To expand the range of economic and social choice to individuals and nations by freeing them from servitude and dependence not only in relation to other people and nation-states but also to the forces of ignorance and human misery.

To achieve inclusive and transformative urban development, the following three targets proposed and outlined by the SDSN (2013) should be taken into cognizance:

- Eliminate extreme poverty, expand employment and productivity, and raise living standards, especially in slums and informal settlements.
- Ensure universal access to secure and affordable built environment and basic urban services: housing, water, sanitation, and waste management, low carbon energy and transportation, and communication.
- Ensure safe air quality and water quality for all, and integrate reduction in greenhouse gas emissions, efficient land and resource use, and climate and disaster resilience into investments and standards.

The aforementioned three targets provide a framework of the sustained elevation of inhabitants in towns and cities and the social system toward a better or more humane life. Thus, more emphasis should be given to the person-centered development, with elements of a broad base and bottom up, redistributive and just, and empowering and environmentally sustainable, seeking to meet the needs of both the present and future generations.

The Menu of Inclusive and Transformative Urban Development

Inclusive development is the seductive idea that the more dynamic and productive economy can go hand in hand with reduced inequality and exclusion (Turok and Visagie 2018). Inclusive urban development is development which involves every citizen regardless of their economic means, gender, race, ethnicity, or religion is enabled and empowered to fully participate in the social, economic, and political opportunities cities and towns have to offer (Asian Development Bank 2017). Urban spaces and the flows between these spaces can be exclusive, where economic activities are preserved for the few and the poor are trapped at the periphery, or inclusive, where people are able to access economic opportunities across the city (South African Cities Network 2016). Thus, inclusive urban development should emphasize the placement of the vision of the poor and the marginalized urban sectors at the center of urban policy-making.

Transformation involves fundamental change, which, in the context of sustainability, requires radical, systematic shift in values, and beliefs, patterns of social behavior, and multi-level governance and management regimes (Olsson et al. 2014). The transformation and the growth of a city must be governed by harmony between its new needs and the preservation of buildings and symbols of its past and current existence; thus, city planning must consider the enormous impact of the urban environment on the development of all individuals on the integration of their personal and social aspirations and resist the segregation of generations and the segregation of people from different cultures, who have much to learn from each other (Rodrigues and Rodrigues 2014). Therefore, transformative urban development should encompass every aspect of change intended for improving citizens' welfare.

Education is a necessity without which man can live a life. It is the most cherished goal to be achieved by a human being that ever existed or is yet to come in this world. Philosophers, educationists, and great thinkers have tried their best in defining education, though there is no universally agreed definition. Etymologically the word education was derived from the Latin words "educare," meaning to train or to mold, and "educere," meaning to lead out (Bass and Good 2004; Kumar and Ahmad 2008; Dagar 2017). According to Bass and Good (2004), though the two words are quite different, they are both represented in the English word "education." On one hand, education means the preservation and passing down of knowledge and the shaping of youths in the image of their parents, while on the other hand, education is seen as preparing a new generation for changes that are to come – readying them to create solutions to problems yet unknown (Bass and Good 2004).

One calls for rote learning memorization and becomes good workers, and the other requires questioning, thinking, and creating. In the same token, Paulo Freire (2005) in his book *Pedagogy of the Oppressed* talks of the banking education and the problem-posing education. However, he argues that the banking education inhibits creativity and domesticates the intentionality of consciousness by isolating consciousness from the world, thereby denying people the ontological and historical vocation of becoming more fully human. Freire (2005, pp. 83–84) advocates for problem-posing education that fosters creativity and stimulates true reflection and action upon reality, thereby responding to the vocation of persons as beings who are authentic only when engaged in inquiry and creative transformation.

Various educationists, philosophers, and great thinkers have defined education in innumerable ways. For instance, the following definitions of education were coined (Table 1).

From the definitions above, it can be said that education provides an individual with a holistic development which enables him to be a full functional member of society, enabling an individual to cope with the continuous rapidly changing world. It is a purposive, conscious, or unconscious psychological, sociological, scientific, and philosophical process which brings about the development of an individual to the fullest extent and also the maximum development of society in such a way that both enjoy maximum happiness and prosperity (Kumar and Ahmad 2008). Therefore, it can be noted that education is a continuous process; it is knowledge or experience; it is the development of certain aspects of human personality; it is a discipline or a vocational course; it is a

Education for Inclusive and Transformative Urban Development, Table 1 Some Definitions of Education

Proponent	Definition	Source of material
Aristotle	Education is the process of training man to fulfil his aim by exercising all the faculties to the fullest extent as a member of society.	ExamPlanning (2020)
Mahmud Yunus	Education efforts are deliberately chosen to influence and assist children with the aim of improving knowledge, physical and morals that can gradually deliver the child to the highest goal.	ExamPlanning (2020)
Socrates	Education means the bringing out of the ideas of universal validity which are latent in the mind of every man.	ExamPlanning (2020)
John Dewey	Education is all one with growing; it has no end beyond itself. (Education is everything along with growth; education itself has no final destination behind him).	ExamPlanning (2020)
H. H Horne	In the broadest sense, education is the device by which a social group continued existence renew yourself and defend his ideals.	ExamPlanning (2020)
Froebel	Education is enfoldment of what is enfolded in the germ. It is the process through which the child makes the internal-external.	Kumar and Ahmad (2008)
Rousseau	Education is the child's development from within.	Kumar and Ahmad (2008)
Mahatma Gandhi	By education I mean an all-round drawing out in child and man's body, mind and spirit.	Kumar and Ahmad (2008)

Source: Authors

measure of social control, a preserver of culture, and instrument of social change and social reconstruction; and it is conducive for the well-being of an individual and that of the society.

For education to promote inclusive and transformative urban development, it has to be accessible, acceptable, available, and adaptable. According to Batista (2015), Socrates' reflections on education are of paramount importance to improving individuals and communities. Socrates argues that knowledge belongs to mankind, not only to a privileged portion of it; so, depriving a human being of knowledge means to infringe it a capital right. In addition, Socrates made mention of the fact that knowledge is free; consequently, it is not to be bargained or commercialized; being, thus, illicit to be enriched due to it, there is here his condemnation to sophists. As asserted by Freire (2005), education liberates the oppressed: liberation from ignorance which shrouds the mind, liberation from superstition which blinds the vision of truth, and liberation from an attitude that hinders progress. Education cultivates knowledge and understanding by transfiguring the human personality into a pattern of perfection through an ongoing process of development of the body, the enrichment of the mind, sublimation of the emotions, and illumination of the spirit. Basically, education is preparation for a living and life, for the present and the future.

Good quality education is essential for sustainable development. The United Nations 2030 Agenda of Sustainable Development Goals (SDGs) envisages a future of inclusive equity, justice, and prosperity within environmental limits and places an important emphasis on education as stated in Goal 4 (Kioup and Voulvoulis 2019). Education is a means or an instrument to support and strengthen the process of sustainable development. To enable inclusive and transformative urban development to take place, education ought to be sustainable. Education for sustainable development promotes the development of knowledge, skill, understanding, values, and actions required to create sustainable cities and towns that ensure environmental protection and conservation, promotes social equity, and encourages economic sustainability (Nevin 2018). The aim of education for sustainable development is to enable people to make decisions and carry out actions to improve their lives without compromising the future of the planet as well as to integrate the values inherent in sustainable development in all aspects and levels of learning (Nevin 2018). Sustainability competences should not only include cognitive components, such as knowledge and understanding of environmental, social, economic, and political systems, and higher-order thinking abilities such as reasoning and synthesizing but also social skills, values, and emotions collectively referred to as the affective domain of learning (Kioup and Voulvoulis 2019).

The Relationship Between Education and Development

Education has always been a significant part of humanity development since time immemorial. The proposition that education promotes or determines the rate of the overall GNP growth remains unquestionable. Middle- or high-level skilled manpower, which can only be created by the formal education system, leads to the development of both the public and private sectors (Todaro 1999). Absence of knowledgeable and skilled people needed to plan, manage, and run economic growth will be retarded if not totally curtailed. Numerous quantitative studies have shown that sources of economic growth have repeatedly demonstrated that it was not only the physical capital that generated economic progress in the developed world but human capital (Todaro 1999). Educational opportunities at all levels contribute in creating a more productive labor force by endowing it with increased knowledge and skills; providing widespread employment and income-earning opportunities (e.g., teachers, school ancillary workers, builders, textbook and paper printers, school uniform manufacturers, etc.); creating a class of educated leaders in governmental services, public corporations, private businesses, and professions; and presumably providing the kind of training that would promote literacy, numeracy, and basic skills while encouraging modern attitudes on the part of diverse segments of the population. Durkheim argues that education enables people to have mutual responsibility and the value of collective good as well as acquire skills needed to perform roles in increasingly specialized occupations (Giddens 2009). Accordingly, an educated and skilled labor force is a necessary requirement for sustained economic growth, resulting in the overall improvement of society's well-being.

All cities and towns possess educational potentials. People who live in urban areas have opportunities to obtain great knowledge in various areas. The Charter of Educating Cities based upon the Declaration of Human Rights (1948); the International Covenant on Economic, Social and Cultural Rights (1966); the Convention on the Rights of the Child (1989); the World Declaration on Education for All (1990); and the Universal Declaration on Cultural Diversity (2001) established principles for working toward and promoting educational potentials within the cities (Rodrigues and Rodrigues 2014). Municipalities and other institutions of the local sphere and local government have the responsibility of educating cities. The tasks assigned to the local government by the charter of educating cities are oriented to generally allow the city to become an agent of education and change for every person without distinction irrespective of cultural, generational, gender, ethnic, or other differences (Rodrigues and Rodrigues 2014). The Charter of Educating Cities, in general, aims to build a complex project, addressing the need to make social life more democratic, to produce a more active citizenship which works to promote parity (Messina and Valdes-Cotera 2013). Cities that embrace lifelong learning for all have seen significance improvements in terms of public health, economic growth, reduced criminality, and democratic participation (Rodrigues and Rodrigues 2014).

According to the conflict perspective, education threatens to put some groups at a continuing disadvantage in the equity and quality of education (Andersen and Taylor 2011). Thus, the highest performing education systems are those that combine quality with equity. In these education systems, the vast majority of students have the opportunity to attain high level of skills regardless of their own personal and socioeconomic circumstances (Organization for Economic Cooperation and Development (OECD 2012)). Equity in education means that personal or social circumstances such as gender, ethnic origin, or family background are not obstacles for achieving educational potential (definition of fairness) and that all individuals reach at least a basic minimum level of skills (definition inclusion) (OECD 2012). However, the major factors affecting education systems are the resources and money needed to support those systems in different nations to promote equal access and provision of quality education.

Mostly, the urban poor send their children to public schools which are poorly resourced. Urban public schools face a plethora of challenges that impede the capacity to effectively educate youths. Due to inadequate resources in these schools, children from low-income households are at a greater disadvantage (overcrowded classrooms, unqualified teachers, inadequate school facilities, etc.), and this usually impacts negatively on their educational outcomes. Consequently, educational systems of many nations rather than being a general force for equality are increasing income inequalities (Todaro 1999).

Poverty and unequal educational opportunities are inextricably linked (Reay 2019). Concentrated poverty in urban areas exacerbates the challenges of being poor which are also closely linked to achieving an education. Children coming from households living in poverty whose parents have low educational qualifications and no or low-status job who live in inadequate housing and disadvantaged neighborhoods are less likely to gain good qualifications themselves at school. Poor urban children have also less chances of completing their education compared to relatively rich children. The costs of educating a child, especially the opportunity cost of a child's labor for poor families, are higher for the poor student than the rich student; thus, a student of low socioeconomic status might prefer working to schooling and drop out of school (Todaro 1999).

Compared to their counterparts, the poor are more likely to drop out of school during early years of schooling, while higher-income individuals are more likely to complete secondary and tertiary education than their poorer counterparts. The result is inequalities in social status as well as economic status between the rich and the poor. The large income inequalities will continue to be reinforced, and the magnitude of poverty perpetuated by these dissimilarities in accessing education is represented disproportionately in secondary and tertiary school enrolments. If the poor are denied secondary and higher educational opportunities, then the educational system results in perpetuating inequalities among urban populations.

The educational divide between the rich and the poor is significant in terms of quantity (years of education) and quality (educational outcomes) (Omic 2017). Those in the upper income bracket tend to complete higher levels of education and take away more from the years spent in education. Lower-income students obtain much lower learning outcomes than rich students, meaning that they gain less in the education system. The inequalities are evident even before children start school from the socioeconomic groups they belong. Research has shown that poverty decreases a child's readiness for school through aspects of health, home life, schooling, and neighborhoods. A child's home has predominantly a strong impact on school reediness since children from low-income families often do not receive the stimulation and do not learn the social skills required to prepare them for school (Ferguson et al. 2007). It is therefore not surprising that the performance of most children from low socioeconomic backgrounds in school becomes very poor.

Empirical studies also show that income inequality, which is closely connected to educational inequality, limits growth (Blanden and McNally 2015). In general, investments in human capital yield a widespread distribution of earnings, since there are higher rates of return in the labor market associated with higher schooling, meaning that educational inequality positively impacts on income inequality (Petcu 2014). Todaro (1999) also noted that a positive correlation between a person's education and his lifetime earnings is particularly true for those who are able to complete secondary and tertiary education, where income differentials over workers who have only completed part or all of their primary education can be of the order of 300–800 percent, since levels of earned income are so clearly dependent on years of schooling (Todaro 1999). It can be noted that educational achievement strongly impacts on urban development and individuals' general well-being.

Conclusion and Future Direction

Education has remarkable contributions in promoting inclusive and transformative urban development. Development is both a physical reality and a state of mind in which society has, through some combination of social, economic, and institutional processes, secured the means for obtaining a better life (Atkinson 1970; Todaro 1999; Mahadevia 2001; Bellue 2011; Rabie 2016).

As well, development is perceived as a multidimensional process involving changes in structures, attitudes, and institutions as well as the acceleration of economic growth, reduction of inequalities, and eradication of absolute poverty. Development should have aspects of continuity to ensure that improvements occurring should not be detrimental to the future status of generations to come (i.e., sustainable development). Education is a means or an instrument to support and strengthen the process of sustainable development. The aim of education for sustainable development is to enable people to make decisions and carry out actions to improve their lives without compromising the future of the planet as well as to integrate the values inherent in sustainable development in all aspects and levels of learning. For education to promote inclusive and transformative urban development, it has to be accessible, acceptable, available, and adaptable. Good quality education is essential for sustainable development.

Additionally, for inclusive and transformative urban development to take place, education ought to be sustainable. Empirical evidence revealed that education for sustainable development promotes the development of knowledge, skill, understanding, values, and actions required to create sustainable cities and towns that ensure environmental protection and conservation, promotes social equity, and encourages economic sustainability. Middle- or high-level skilled manpower, which can only be created by the formal education system, leads to the development of both the public and private sectors. Educational opportunities at all levels contribute in creating a more productive labor force by endowing it with increased knowledge and skills and provide widespread employment and income-earning opportunities.

Education enables people to have mutual responsibility and the value of collective good as well as acquire skills needed to perform roles in increasingly specialized occupations. However, though education plays a pivotal role in fostering inclusive and transformative urban development, urban communities still experience educational inequalities that have a negative impact on equal participation and improvement of individuals, families, and communities' social and economic growth. In order to achieve inclusive and transformative urban development, the following actions are necessary:

- Children should be able to achieve their educational potential by being able to access educational opportunities regardless of their individual outcomes. Ensuring equality in education can further accelerate the achievement of the other SDGs.
- Advocating for an educational system that ensures a minimum standard for all children to achieve upon completion of the formal education system (e.g. literacy, mathematics), thus ensuring that investment in education is effective in achieving this goal. This can help in reducing educational disparities in urban areas.
- Investment in compulsory primary and secondary school.
- Empowering the urban poor with civic values and critical thinking needed to contribute actively in their communities. It is a lever to promote greater social cohesion and shared prosperity in a longer term.
- For those who earn little or who are unable to find paid jobs, social safety nets must be established to address challenges faced by the urban poor in accessing education.
- Governments to improve quality of education offered in public schools by increasing resources that ensure provision of quality education. Increase of resources that are effective for the disadvantaged groups leads to higher average achievements as well as more equitable outcomes.
- The use of effective educational policies that focus on directly improving the achievements of the disadvantaged children.
- Developing policies with the most needs of the disadvantaged in mind that aim to reduce economic inequality and promote economic growth.
- Advocate for support intervention programs that provide academic, social, and community support to raise the success of disadvantaged children and youth.
- Advocate for and support schools which strive to achieve equity of outcomes. Socioeconomic background is a predictor of educational attainment; if the economically disadvantaged acquire education and skills to equal the more advantaged peers, they would at least be rewarded for their productivity and contribute to economic growth.

Cross-References

▶ Education for Inclusive and Transformative Urban Development

References

Andersen, N., & Taylor, H. F. (2011). *Sociology. The essential*. Belmont: Wadsworth.

Asian Development Bank. (2017). *Enabling inclusive cities. Tool kit for inclusive development*. Mandalayong: Asian Development Bank.

Atkinson, A. (1970). On the measurement of inequality. *Journal of Economic Theory, 2*(3), 244–263.

Bass, R. V., & Good, J. W. (2004). Educare Educere. Is a Balance Possible in the Educational System. *The Educational Forum, 68*, 161–168.

Batista, G. A. (2015). Socrates: Philosophy applied to education. Search for virtue. *Athens Journal of Education, 2*(2), 149–156.

Bellue, L. G. (2011). *Development and development paradigms. A (Reasoned) review of prevailing visions*. Rome: FAO United Nations.
Blanden, J., & McNally, S. (2015). Reducing inequality in education and skills: Implications for economic growth. European Network Expert on Economics of Education Analytical Report Number 21. Prepared for the European Commission. London: EENEE.
Dagar, B. S. (2017). Concept and meaning of education. Available online: http://www.egyankosh.ac.in/bitstream/123456789/82661l/unit-1.pdf. Accessed 10/08/2020.
ExamPlanning. (2020). Definition of education by different authors. Available from examplanning.com/definition-of-education-by-different-authors/?. Accessed 10/08/2020.
Ferguson, H. B., Bovaird, S., & Mueller, M. P. (2007). The impact of poverty on educational outcomes for children. *Paediatric Child Health, 12*(8), 701–706.
Freire, P. (2005). *Pedagogy of the oppressed*. 30th Anniversary Edition. London: Continuum.
Giddens, A. 2009 Sociology. (6th ed). Cambridge: Polity.
Kioup, V., & Voulvoulis, N. (2019). Education for Sustainable Development: A systematic framework for connecting the SDGs to educational outcomes. *Sustainability, 11*, 1–18.
Kumar, S., & Ahmad, S. (2008) Meaning, aim and process of education. Available online: http://www.sol.du.ac.in/solsite/courses/English/sm-1.pdf. Accessed 10/08/2020.
Mahadevia, D. (2001). Sustainable Development in India: An inclusive perspective. *Development Practice, 11*(2), 242–254.
Messina, G., & Valdes-Cotera. (2013). Educating cities in America Latina. *International Review of Education. Journal of Life Long Learning, 59*(4), 425–441.
Nevin, E. (2018). Education for Sustainable Development. Available online: http://www.sustainablesummitspeakerelainnevin-national-director-eco-unesco/. Accessed 10/08/2020.
Olsson, P., Galaz, V., & Boonstra, W. J. (2014). Sustainable transformation s: A resilient perspective. *Ecology and Society, 19*(14), 1–14.
Omic, E. (2017). Educational inequality in Europe. In *Tackling inequalities in Europe: The social investment*. London: Council of Europe Development Bank.
Organisation for Economic Cooperation and Development. (2012). *Equity in education. Supporting disadvantaged students and schools*. Paris: OECD.
Petcu, C. (2014). Does educational inequality explain income inequality across countries. Honours Projects Paper 125. Available online: http://digitalcommons.iwueducation.econ_honproj/125. Accessed 10/08/2020.
Rabie, M. (2016). *A theory of sociocultural and economic development*. London: Palgrave Macmillan.
Reay, D. (2019). Poverty and education. Available online: http://www.researchgate.net/publication/334398213_Poverty_and_Education. Accessed 10/08/2020.
Rodrigues, P., & Rodrigues, A. (2014). Education and development post-2015. The City for the Education. Available online: http://www.unesco.org/santiago. Accessed 10/08/2020.
South African Cities Network. (2016). *State of South African Cities Report*. Johannesburg: SACN.
Sustainable Development Solution Network. (2013). *The urban opportunity: Enabling transformative and sustainable development, background paper for the high level panel of eminent persons on post-2015 development agender*. New York: United Nations.
Todaro, M. P. (1999). *Economics for a developing world. An introduction to principles, problems and policies for development* (7th ed.). New York: Longman.
Turok, I., & Visagie, J. (2018). *Inclusive urban development in South Africa. What does it mean and how can it be measured?* South Africa: Available online: researchgate.net/publication/326410270-Inclusive-Urban-Development-In-South-Africa-What-Does-It-Mean-and-How-Can-It-Be-Measured. Accessed 10/08/2020

E

El Nino South Oscillation (ENSO)

▶ The State of Extreme Events in India

Elder-Friendly Community

▶ Age-Friendly Future Cities

The Elusive Quest for Affordable Housing: Five Principles of a Comprehensive Approach

Sameh N. Wahba
The World Bank, Washington, DC, USA

Introduction: The Affordable Housing Challenge

The world is urbanizing at an unprecedented scale and speed. Today, about 55 percent of the world's population lives in cities. By 2050, the figure is expected to increase to over two-thirds of the world's population (UN-DESA, 2019). This means cities will need to accommodate an additional 2.5 billion inhabitants. This is equivalent to adding to cities each year the entire population of Turkey or ten cities the size of New York. This requires not only proactive urban planning to accommodate such growth but also significant investment by national and local governments for the delivery of housing, infrastructure, and basic municipal services.

The most critical element to facilitating such rapid urbanization is enabling access to affordable housing, but it is also arguably the most elusive dimension. Overall, an estimated 3 billion people will require access to affordable housing by 2030 (UN, 2019). This figure encompasses both future urban population growth and the large share of the urban population that currently lives in inadequate living conditions.

Indeed, today some one billion urban inhabitants live in slums and informal settlements, where they mostly lack security of tenure, live in housing of sub-standard quality and in overcrowding situations, and lack access to public spaces, infrastructure, and basic services (Ibid). This is mainly the result of dysfunctional land and housing markets, where poor people lack access to affordable housing with adequate quality and within reasonable commuting distance from job opportunities. As such, poor people often find themselves with no option but to live in overcrowded slums and informal settlements which have formed on marginal land in at risk areas.

The delivery of affordable housing is a particularly complicated challenge. Unlike access to health, education, and water and sanitation infrastructure, which are largely considered basic rights and public goods that warrant public intervention and subsidies, opinions about access to (affordable) housing tend to be divided. Some view access to shelter as a human right while others view it as a private good, and more importantly opinions about housing standards – especially what constitutes minimally acceptable housing quality or affordability – vary greatly within and across countries (see, e.g., Hartman, 1998; Bratt et al., 2013; Leckie, 2001; Mikkola, 2008; Whitehead, 1984; Jaffe, 1989).

The Need for a Comprehensive Approach

Tackling the challenge of affordable housing requires a comprehensive approach. The World Bank's housing strategy of 1993 – entitled Enabling Housing Markets to Work – highlighted the importance of three supply-side interventions (access to land, infrastructure and services, building materials and the construction industry), three demand-side interventions (property rights, access to finance, subsidies), as well as institutions (World Bank, 1993). Its most important tenet was to call for a shift in the public sector's role from a direct housing provider to an enabler of the market, with the bulk of housing supply to be delivered by private initiative (whether private developers, non-for-profit institutions, and/or households). Since then, the World Bank's comprehensive approach to housing was further refined (World Bank and IFC, 2015). Such comprehensive approach comprises of five key dimensions:

Tackling Supply-Side Issues, Demand-Side Issues, and Subsidies in a Sequenced Manner

The first and most critical step is tackling housing supply-side challenges in order to lower the cost of formal housing supply to the extent possible and enhancing the efficiency of private supply. The key elements are: (a) improving access to land, through improved land and property rights, and efficient public land asset management; (b) provision of off-site infrastructure and basic services; and (c) tackling market distortions affecting building materials and the construction and development industry (such as import duties and other supply bottlenecks facing building materials, tax disincentives).

The most difficult impediment to lowering the cost of housing supply is invariably access to land. This challenge manifests itself in countries such as Kenya where developers report that land reaches as much as 60 percent of total development cost in affordable housing projects (World Bank, 2016). The release of public land reserves for affordable housing can often serve as a regulator of land markets. Similarly, the introduction of vacant land tax can serve to dampen speculation and the retention of undeveloped land for long periods (Haas & Kopanyi, 2017).

In parallel with interventions to lower the cost of housing supply, housing demand interventions are essential to boost households' ability to afford housing. In the absence of housing finance, households have no option but to build their own housing incrementally in line with their gradual savings (which is the predominant way through which informal development operates, but which is not without its share of inefficiency) or to combine sale of assets and savings for extended periods to be able to afford to buy housing in cash. Beyond economic growth and a stable macro-economic environment which can boost purchase power and yield stable interest rates respectively, key demand side interventions include: (a) the development of mortgage finance markets, which in turn rests on strengthening property rights and facilitating foreclosure procedures in case of defaults; (b) the introduction of liquidity or refinancing facilities to boost the activities of primary lenders; and (c) the development of housing micro-finance and/or guarantee funds to provide solutions to enhance the affordability of poorer, less creditworthy borrowers.

Lastly, after addressing both supply side issues to reduce the cost of formal housing supply and demand side issues to facilitate households' ability to acquire housing, housing subsidies can be introduced to enable low and moderate-income households to access housing, either through homeownership or rental, by plugging the gap between what the market provides and what they can afford. Introducing housing subsidies before tackling supply or demand side issues will only serve to subsidize market inefficiencies. Similarly, with both supply and demand-side interventions, it is important to note that tackling one without the other will not lead to optimal outcomes. Having a functional housing mortgage finance market without tackling supply-side distortions will translate into households purchasing unnecessarily expensive housing units. Usually, the "house price to income" ratio, which compares median housing price to median household income, would be in the range of 3–6 in reasonably well-functioning markets, whereas in markets suffering from supply-side constraints would have a ratio of over 10, which means far fewer households can afford to buy housing, even in the presence of mortgage finance and liquidity (see Florida, 2018).

Morocco is a notable example of a country that has taken such a comprehensive approach. On the supply side, the Royal Court released nearly 10,000 hectares of public land for affordable housing development (IEG, 2009), while on the demand side the Government set up guarantee funds to scale up access to mortgage lending by civil servants and informal sector workers, with the latter – called FOGARIM – having benefited 191,651 informal income earning households by May 2021 (Ouchagour, 2021).

Combining the Curative and the Preventive

With one billion people living in slums and informal settlements in cities in developing countries and with such ratio reaching as much as 75 percent of Sub-Saharan Africa's urban population (World Bank, 2015), a key tenet of housing policy will require focus on improving living conditions in poor, marginalized neighborhoods. This includes the regularization of land and property tenure for squatters and informal settlers – which typically unleashes significant private investment in housing consolidation. Equally important are the improvement of infrastructure and basic services and the introduction of home improvement subsidies that aim to improve livability and strengthen the resilience of housing against natural hazards and extreme weather events.

However, investing in curative interventions through slum upgrading alone is insufficient to prevent future slum formation. In fact, it may become a moral hazard if households perceive that squatting is systematically rewarded with official recognition. As such, a parallel pillar of housing policy will need to invest in improving access to land for housing developers and households as well as boosting the supply of affordable housing, which are key measures to accommodate future demand and prevent future slum formation or the densification of existing slums.

Housing policy in most countries tend to favor new housing construction, even when there are cheaper, and greener, alternatives. By contrast, slum upgrading and home improvement subsidies tend to cost much less per unit, and thus have the ability to reach a much larger number of households. Colombia is a case in point. Closing the qualitative housing deficit in Colombia's cities with new housing construction is estimated to cost US$24.6 billion, more than three times the subsidy amount needed for improving existing homes (World Bank, 2021a). Not to mention, home improvement interventions produce a fraction of the greenhouse gas emissions generated by new housing.

Mauritania is one country that carried a national urban development program, financed by the World Bank, which included a massive slum upgrading program together with a large-scale land subdivision program which allocated plots to households being resettled from overcrowded slums as well as newly formed households (Roquet et al., 2017). In total, this $156 million program benefited 377,000 urban poor in Mauritanian cities as direct beneficiaries, with 181,035 slum dwellers in the capital, Nouakchott, having access to improved services, and with 100,042 land titles regularized and registered. One of the outcomes has been to halve the cost of water in project-supported slums from $\$2.00/m^3$ to $\$1.00/m^3$, which resulted in a tripling of daily potable water consumption from 15 to 41 liters/person/day (IEG, 2014).

Similarly, Indonesia has been investing significantly in improving living conditions in underserviced neighborhoods through the National Slum Upgrading Program (NSUP), and its predecessor the Kampung Improvement Program. NSUP has upgraded infrastructure to benefit over 7 million urban poor beneficiaries and improve 7345 hectares of slum areas (World Bank, 2021c). At the same time, the Government introduced a new program for affordable housing subsidies that combined upfront grants for households to acquire new housing units (coupled with access to finance) that benefited 7328 low-income families, as well as a home improvement subsidy program to enable households to improve existing housing that benefited more than 140,000 poor families across the country (World Bank, 2021d). NSUP also creates collaborative platforms, via city-level Slum Improvement Actions Plans and Community Settlement Plans, to attract broader investments to slum areas. To date, the program has leveraged USD 1 billion in financing from other public and private sources.

Working Across the Entire Housing Value Chain

For housing sector interventions to be effective, it is important to tackle the obstacles to affordable housing provision and acquisition along the entire value chain. This includes: (a) facilitating access to land; (b) streamlining land development regulations; (c) improving access to infrastructure and basic services; (d) designing well-targeted subsidies; (e) improving access to mortgage finance for home buyers; and (f) improving access to finance for developers. The value chain, which spans supply and demand side and subsidies, will be as weak as its weakest link.

While access to land is often the most intractable problem, the reform of land development regulations tends to present a low-hanging fruit. Limitations on land development densities whether through low floor area ratio (FAR), limited building heights or large setbacks limiting plot coverage all contribute to reducing the amount of affordable housing development per unit of land and increasing the share of land to total development cost. Similarly, the introduction of unreasonably high minimum plot size requirements in planning legislation – e.g., 300 m^2 plots in Dar Essalam, Tanzania, or half-acre plots in some US cities – either forces people into informality or hampers the ability of delivering affordable housing, thus amounting to exclusionary zoning (see Whittemore, 2021 and World Bank, 2021b).

Where such limitations are imposed on centrally located land, the opportunity cost becomes higher. The city of São Paulo, Brazil, took effective steps to remedy its low FAR in central locations. The city introduced a regulation called *Outorga Onerosa* allowing for the vertical expansion of existing buildings beyond the permissible heights in return for a payment to a citywide fund for infrastructure improvement. On the other hand, São Paulo designated special urban intervention zones in central locations called *Operações Urbanas*, wherein the city auctioned additional development rights (called *certificados de potencial adicional de construção* or CEPACs) to increase the FAR from a base of 1.5–2.5 up to 4 (Sandroni, 2010). The proceeds from the sale of these additional development rights were used to upgrade the infrastructure in the area to support the increased density. In fact, the sale of CEPACs in the two centrally zones of Faria Lima and Agua Espraiadas had generated over US$800 million in a five-year period, equivalent to 58 percent of total property tax collected in the city over the same period (Ibid).

Developing Segmented Solutions Tailored to Different Income Groups

Housing affordability is usually measured by the share of a household's income that is devoted to housing cost – whether rent or mortgage loan repayment – and is often set at 25–33 percent of income (Herbert et al., 2018). Some indicators add the cost of utilities or the cost of transportation to reflect the trade-off between more expensive housing located near job opportunities versus cheaper housing that requires longer and more expensive commutes (see Dewita et al., 2018). Irrespective of the metric used, the reality is that housing affordability varies by income level. As such, government policy should support housing solutions that are segmented by income group.

In developing countries, the poorest households – especially in the bottom 2–3 deciles of the income distribution, are much harder to reach, especially for access to home ownership. In such cases, rental subsidies and/or slum upgrading would enable access to affordable housing solutions. Where subsidies were used to enable access to homeownership to the poorest households, such as Brazil's *Minha Casa Minha Vida* (my house, my life) program that targeted households earning below 3 minimum wages, the level of subsidies reached a whopping 90 percent of total development cost, which makes it hard to scale up such programs (Acolin et al., 2019; Biderman et al., 2018).

The next group – those with moderate or lower-middle incomes, and who would represent the 3rd to the 5th, 6th, or 7th decile – would benefit from Government subsidies to afford home ownership. Countries employ a large range of subsidies including supply-side (developers benefiting from free or subsidized access to land, per unit grants, subsidized financing, and/or tax benefits) and demand-side (households benefiting from grants, access to subsidized finance, and/or tax benefits). Demand-side subsidies are considered more efficient because they reach directly the beneficiary households, whereas developer subsidies will invariably include mark-ups. And within the range of demand-side subsidies, capital grants toward households' equity contribution and which are linked to access to credit or interest rate buy-down are preferred to interest rate subsidies because they are one-off and don't have multi-year budgetary implications. In Brazil's *Minha Casa Minha Vida* program, households earning between 3 and 6 minimum wages received on average upfront grants linked with access to credit and averaging about 25 percent of the unit's cost. A guarantee fund to de-risk informal sector workers as FOGARIM in Morocco or to provide insurance against temporary unemployment as the one used in Brazil are examples of targeted interventions (see Nadal & Linka, 2018 and Tagma, 2016).

Finally, for higher income groups who do not need subsidies, interventions to enable access to land and to facilitate for private developers and lenders to go further down market will benefit everyone and improve the overall efficiency of the housing market.

Offering Different Housing Solutions Across a Range of Tenure Options

Most countries with housing policies tend to favor access to homeownership through the dedication of the bulk of subsidies. However, not all households can afford to buy a home nor are ready or interest to buy a home. Younger households and

E

those at the beginning of their professional careers may want to rent until their incomes grow or stabilize and they are in a measure to acquire a housing unit that is in line with their aspirations, and which could accommodate a growing family. This is especially the case in housing markets which have high transaction costs to home buying, which limits residential mobility. As such, housing policy should aim to offer differentiated solutions to a range of tenure options including home ownership, rental, as well as rent-to-own arrangements.

In particular, the rental sector tends to be underdeveloped in most countries. Housing developers rarely find it advantageous to develop rental properties, which entail a burden of property management and rent collection. As such, incentives targeted at rental property developers would be useful to offer a range of solutions that are more affordable and more suited to some households' preferences.

Conclusions

In summary, facilitating households' access to affordable housing requires a comprehensive approach with five different pillars – tackling the supply and demand side, with subsidies in a sequenced manner; combining the curative (slum upgrading) with the preventive (new affordable housing construction); working across the entire value chain; developing differentiated solutions by income group; and offering housing solutions across multiple tenure options. A comprehensive approach will also require action from multiple stakeholders. This includes governments – primarily as enablers of markers to work, but also as market makers where conditions warrant it; the private sector – housing developers and finance institutions, including non-profits and microfinance institutions; and communities and households, given the prevalence of auto-construction in developing countries and the need for community participation in slum upgrading.

What is clear is that there is no one-size-fits-all housing policy. The mix of ingredients will differ by country and in accordance with the key challenges, but a housing policy with these ingredients should improve the functioning of the overall housing market and enable urban households to access more housing options that are better suited to their needs, priorities, and affordability. Combining slum upgrading, home improvement and rental support with homeownership subsidies will also enable Governments to offer differentiated solutions and scale up.

Cross-References

▶ Housing Affordability

Acknowledgement The author acknowledges Loic Chiquier, Britt Gwinner, and Marja Hoek-Smit for their contributions to the development of the World Bank Group's updated housing strategy.

Disclaimer The findings, interpretations, and conclusions expressed in this [article/chapter] are entirely those of the author. They do not necessarily represent the views of the International Bank for Reconstruction and Development/ The World Bank and its affiliated organizations, or those of the Executive Directors of the World Bank or the governments they represent.

References

Acolin, A., Hoek-Smit, M. C., & Eloy, C. M. (2019). High delinquency rates in Brazil's Minha Casa Minha Vida housing program: Possible causes and necessary reforms. *Habitat International, 83*(2019), 99–110. https://doi.org/10.1016/j.habitatint.2018.11.007.

Biderman, C., Hiromoto, M. H., & Ramos, F. (2018). *The Brazilian Housing Program Minha Casa Minha Vida: Effect on Housing Sprawl. Working Paper WP18CB2.* Lincoln Institute for Land Policy. Available at: https://www.lincolninst.edu/sites/default/files/pubfiles/biderman_wp18cb2_0.pdf.

Bratt, R. G., Stone, M. E., & Hartman, C. (2013). Why a right to housing is needed and makes sense: Editors' introduction. In J. R. Tighe & E. J. Mueller (Eds.), *The affordable housing reader* (pp. 53–72). London and New York: Routledge.

Dewita, Y., Yen, B. T. H., & Burke, M. (2018). The effect of transport cost on housing affordability: Experiences from the Bandung Metropolitan Area, Indonesia. *Land Use Policy, 79*(2018), 507–519. https://doi.org/10.1016/j.landusepol.2018.08.043.

Florida, R. (2018). *Where the house-price-to-income-ratio is most out of whack.* Bloomberg City Lab.

Available at: https://www.bloomberg.com/news/articles/2018-05-29/how-many-years-of-income-does-a-home-in-your-city-cost (last accessed November 4, 2011).

Haas, A., & Kopanyi, M. (2017). *Taxation of vacant urban land: From theory to practice. Policy note*. International Growth Centre. Available at: https://www.theigc.org/wp-content/uploads/2017/07/201707TaxationVacantLandPolicyNote_Final.pdf.

Hartman, C. (1998). The case for a right to housing. *Housing Policy Debate, 9*(2), 223–246. https://doi.org/10.1080/10511482.1998.9521292.

Herbert, C., Hermann, A., & McCue, D. (2018). *Measuring housing affordability: Assessing the 30 percent of income standard*. Joint Center for Housing Studies, Harvard University. Available at: https://www.jchs.harvard.edu/sites/default/files/Harvard_JCHS_Herbert_Hermann_McCue_measuring_housing_affordability.pdf.

Independent Evaluation Group. (2009). *Morocco housing sector development policy loan ICR review*. Available at: https://documents1.worldbank.org/curated/en/760751475112018557/pdf/000020051-20140619110847.pdf (last accessed November 7, 2021).

Independent Evaluation Group. (2014). *Mauritania urban development program (P069095) ICR review*. Report Number: ICRR 14209. Available at: https://documents1.worldbank.org/curated/en/505771475111749926/text/000020051-20140626105416.txt (last accessed November 7, 2021).

Jaffe, A. J. (1989). Concepts of property, theories of housing, and the choice of housing policy. *The Netherlands Journal of Housing and Environmental Research, 4*(4), 311–320. http://www.jstor.org/stable/43928426.

Leckie, S. (2001). The human right to adequate housing. In A. Eide et al. (Eds.), *Economic, social and cultural rights: A textbook* (2nd revised ed., pp. 149–168). The Netherlands: Kluwer Law International. https://doi.org/10.1163/9789047433866_013.

Mikkola, M. (2008). Housing as a human right in Europe. *European Journal of Social Security., 10*(3), 249–294. https://doi.org/10.1177/138826270801000303.

Nadal, L., & Linka, C. (2018). *Minha Casa Minha Vida (MCMV), access and mobility: A case for transit-oriented, low-income housing in Rio de Janeiro. Working paper WP18LN1*. Lincoln Institute for Land Policy. Available at: https://www.lincolninst.edu/sites/default/files/pubfiles/nadal_wp18ln1.pdf (last accessed November 7, 2021).

Ouchagour, L. (2021). *Damane Assakane: Près de 47 MMDH de Prêts Accordés et Plus de 250.000 Familles Bénéficiaires*. Aujourd'hui Le Maroc. Available at: https://aujourdhui.ma/economie/immobilier/damane-assakane-pres-de-47-mmdh-de-prets-accordes-et-plus-de-250-000-familles-beneficiaires (last accessed November 7, 2021).

Roquet, V., Bornholdt, L., Sirker, K., & Lukic, J. (2017). *Urban land acquisition and involuntary resettlement: Linking innovations and local benefits. Directions in development: Environment and sustainable development*. Washington, DC: The World Bank. Available at: https://openknowledge.worldbank.org/bitstream/handle/10986/26070/9781464809804.pdf?sequence=2&isAllowed=y (last accessed November 7, 2021).

Sandroni, P. (2010). A new financial instrument of value capture in Sao Paulo: Certificates of additional construction potential. In G. Ingram & Y.-H. Hong (Eds.), *Municipal revenues and land policies. Proceedings of the 2009 land policy conference*. Cambridge, MA: Lincoln Institute of Land Policy.

Tagma, M. (2016). *The FOGARIM: A housing loan guarantee for the informally employed. Case study series 1*. Centre for Affordable Housing Finance. Available at: https://housingfinanceafrica.org/app/uploads/CAHF-Case-Study-1_FOGARIM.pdf (last accessed November 7, 2021).

United Nations. (2019). *The sustainable development goals report*. New York: United Nations. Available at: https://unstats.un.org/sdgs/report/2019/The-Sustainable-Development-Goals-Report-2019.pdf (last accessed November 7, 2021).

United Nations, Department of Economic and Social Affairs, Population Division. (2019). *World urbanization prospects: The 2018 revision (ST/ESA/SER.A/420)*. New York: United Nations.

Whitehead, C. M. E. (1984). Privatisation and housing. In J. le Grand & R. Robinson (Eds.), *Privatisation and the welfare state*. London: Routledge.

Whittemore, A. H. (2021). Exclusionary zoning. *Journal of the American Planning Association, 87*(2), 167–180. https://doi.org/10.1080/01944363.2020.1828146.

World Bank. (1993). *Housing: Enabling markets to work. World Bank policy paper*. Washington, DC: World Bank. https://doi.org/10.1596/0-8213-2434-9.

World Bank. (2015). *Stocktaking of the Housing Sector in Sub-Saharan Africa: Challenges and opportunities*. Washington, DC. World Bank. Available at: https://openknowledge.worldbank.org/handle/10986/23358 License: CC BY 3.0 IGO.

World Bank. (2016). *Kenya urbanization review*. Washington, DC: World Bank. Available at: https://openknowledge.worldbank.org/handle/10986/23753 License: CC BY 3.0 IGO.

World Bank. (2021a). *Striking a balance: Toward a comprehensive housing policy for a post-COVID Colombia. Global program for resilient housing*. Washington, DC: World Bank. Available at: https://documents.worldbank.org/en/publication/documents-reports/documentdetail/481511633672773921/striking-a-balance-toward-a-comprehensive-housing-policy-for-a-post-covid-colombia.

World Bank. (2021b). *Transforming Tanzania's cities: Harnessing urbanization for competitiveness, resilience and livability*. Washington, DC: The World Bank. Available at: https://openknowledge.worldbank.org/bitstream/handle/10986/35930/Transforming-

Tanzania-s-Cities-Harnessing-Urbanization-for-Competitiveness-Resilience-and-Livability.pdf?sequence=1&isAllowed=y (last accessed November 7, 2021).

World Bank. (2021c). *Indonesia national slum upgrading program (P154782) implementation status and results (ISR) report*. Available at: https://documents1.worldbank.org/curated/en/372751636385629855/pdf/Disclosable-Version-of-the-ISR-Indonesia-National-Slum-Upgrading-Project-P154782-Sequence-No-11.pdf (last accessed November 7, 2021).

World Bank. (2021d). *Indonesia national affordable housing program (P154948) implementation status and results (ISR) report*. Available at: https://documents1.worldbank.org/curated/en/993771636385294966/pdf/Disclosable-Version-of-the-ISR-National-Affordable-Housing-Program-P154948-Sequence-No-10.pdf (last accessed November 7, 2021).

World Bank and International Finance Corporation. (2015). *Global housing: An integrated World Bank Group approach*. Washington, DC: Unpublished PowerPoint Presentation.

Embedding Justice in Resilient Climate Change Action

Asma Mehan[1] and Bouchra Tafrata[2]
[1]CITTA Research Institute, University of Porto, Porto, Portugal
[2]Willy Brandt School of Public Policy, University of Erfurt, Erfurt, Germany

Definitions, Conceptualisation and Synonyms

Based on the 2030 United Nations Agenda for Sustainable Development Goals (SDGs), it is urgent to effectively address the climate change's urgency linked to all other 16 SDGs. This issue mainly reflects the progress made toward achieving the United Nation's Sustainable Development Goals (SDGs), especially SDG 13 binding targets including improving education and public awareness-raising mechanisms for raising capacities of management, participation, mitigation, and adaptation strategies especially focusing on marginalized communities. The 2021 United Nations Climate Change Conference, also known as the COP26 summit (UNFCCC), highlighted this importance by bringing 25,000 delegates from 200 countries together in order to enhance international ambition toward mitigating climate change as outlined in the Paris agreement.

From the Kyoto Protocol to the recent 2021 COP26 International Summit and Paris Agreement, decision-makers continue to grapple with achieving an effective climate change mitigation. The shifts of institutionalization from top-down to bottom-up engendered hope and discontent among governments in various countries. The Paris Agreement paved the way for a new form of climate change governance that allows different decision-makers, such as non-state actors and local governments or municipal networks, to decarbonize and transition toward a fossil fuel-free planet. However, the fragmentation of climate change governance led to critical debates on the employed mechanisms and hindered political action. Additionally, the latest report by the Intergovernmental Panel on Climate Change (IPCC) has struck more fear internationally (IPCC 2021).

Climate hazards are more present in the everyday environment. They continue to highlight the interconnectedness between the local and the global, as Jon (2021) underlined that what humans are enduring locally is "fundamentally and inevitably related to the planetary environmental degradation which requires us to behave as a part of 'the whole'" (p. 12). Climate change has affected various places and brought inequalities to many populations, particularly those who have emitted the least CO2 throughout history (Davis and Todd 2017). Settler colonialism and the capitalism of extraction and exploitation constructed the injustice weathered by post-colonies (Whyte 2016). While the Western narrative and institutions of power continue to conceal the implications of colonialism on ecologies and confining climate action between national borders and green policy agendas, climate justice is bringing forward critical discussions that dissect marginalization, environmentalism, and the production of systemic inequalities, through the same prism (Nawratek and Mehan 2020; Schlosberg and Collins 2014; Pettit 2004). Climate justice forages for a reassessment of systems of governance, including fossil fuel dependency, fighting green gentrification, withstanding racial capitalism, and mobility injustice and displacements (Sheller 2018; Vergès 2017; Jon 2021; Moore and Patel 2017). Also, feminist scholarship sheds light on the intersection between gender issues and climate justice, promoting more equity and inclusion, which centers on race, class, and gender (Agostino and Lizarde 2012; Arora-Jonsson 2011). Climate change exacerbates women's unpaid labor as it disturbs accessibility to resources and land (Dankelman and Naidu 2020). This ecological crisis amplifies patriarchal oppression and exploitation systems and extends socio-economic inequalities, notably displacing women from marginalized areas and communities (Güiza et al. 2017).

Toward a Resilient Climate Action

The COVID-19 pandemic and economic crisis have shown the urgent need for collective, multi-scaler, and multi-lateral resilient responses to urgencies to reduce climate change, inequality, and poverty (Mehan 2021). With ongoing and increasing uncertainties in place and accelerated by the COVID-19 pandemic, as well as the climate risks, different governments must be able to respond to how social-ecological systems can adapt to these situations, especially when dealing with high levels of uncertainty and the non-linearity of ecological systems (Wyborn 2015; Karpouzoglou et al. 2016; Rijke et al. 2012). Moreover, a transdisciplinary collaboration among the different fields is critical since dealing with climate uncertainties and the impacts requires multi-scalar cooperation between other sectors and various actors of society.

Implementing environmental assessment measures adapted locally but with a globally cooperative ethic will help slow the pace of ecological crisis. New dynamics and collaborative glocalization are required, predicated on economic growth and environmental awareness, economic equity, and spatial justice (Goffman 2020, p. 48). The current global plan on climate change is focused mainly on the technical and ecological perspectives. There is a knowledge gap rooted in the global south's local communities' social, historical, cultural, and economic circumstances (Kozlowski et al. 2020; Rahdari et al. 2019). It means an urgent need to analyze the power and politics in conjunction with the socio-economic factors that often determine how vulnerable local communities in developing countries respond to climate change (Mikulewicz 2018). The emergency of the theme has been mentioned as part of the Venice Biennale's Italian Pavilion 2021, Resilient Communities, which exhibited the seriousness and urgency of the issue of climate change during the pandemic outbreak and the significant challenges (urban, productive, and agricultural systems) that architecture is called on to face (Mehan and Mostafavi 2022).

Achieving climate justice in governance requires a practice of acknowledging the past, being attentive to the communities affected by the deterioration of the ecosystem, and building policies and measures on principles of solidarity and aid. In others words, governments must aim to repair through accountability and to refrain from propagating narratives on vulnerability and resilience (Arora-Jonsson 2011). Thus, climate justice can be met through planning within the complexity of social entities and their heterogeneity. As Mehan and Mostafavi (2022) articulated, "plurality and diversity of individual records provide the transformative agent that is critical for adaptation and innovation" (p. 2). The pandemic highlighted the diversity of actors such as urban authorities, activists, policy-makers, public citizens, NGOs, and various private and non-private stakeholders and the variety of territorial scales for impacts of imminent threats and hazards on the long-term development in terms of enhancing the resiliency and sustainability of cities (Brears 2021). This diverse makeup will allow the critical and multiscale analysis of the different aspects and actors of climate change challenges and the resilient urban future transitions.

Cross-References

- ▶ Adapting Cities to Climate Change
- ▶ An Overview of the Relationship of the Sustainable Development Goals and Urban and Regional Development

References

Agostino, A., & Lizarde, R. (2012). Gender and climate justice. *Development, 55*(1), 90–95. https://doi.org/10.1057/dev.2011.99.

Arora-Jonsson, S. (2011). Virtue and vulnerability: Discourses on women, gender and climate change. *Global Environmental Change, 21*(2), 744–751. https://doi.org/10.1016/j.gloenvcha.2011.01.005.

Brears, R. C. (2021). Circular economy and the water-food nexus. In R. Brears (Ed.), *The palgrave encyclopedia of urban and regional futures*. Cham: Palgrave Macmillan. https://doi.org/10.1007/978-3-030-51812-7_98-1.

Dankelman, I., & Naidu, K. (2020). Introduction: Gender, development, and the climate crisis. *Gender and Development, 28*(3), 447–457. https://doi.org/10.1080/13552074.2020.1843830.

Davis, H., & Todd, Z. (2017). On the importance of a date, or, decolonizing the anthropocene. *ACME: An International Journal for Critical Geographies, 16*(4), 761–780. Retrieved from https://www.acme-journal.org/index.php/acme/article/view/1539.

Goffman, E. (2020). In the wake of COVID-19, is glocalization our sustainability future? *Sustainability: Science, Practice and Policy, 16*(1), 48–52.

Güiza, F., Méndez-Lemus, Y., & McCall, M. K. (2017). Urbanscapes of disaster: The sociopolitical and spatial processes underpinning vulnerability within a slum in Mexico. *City & Community, 16*(2), 209–227. https://doi.org/10.1111/cico.12230.

IPCC. (2021). *Climate change 2021: The physical science basis. Contribution of working group I to the sixth assessment report of the intergovernmental panel on climate change.* https://www.ipcc.ch/report/ar6/wg1/downloads/report/IPCC_AR6_WGI_Full_Report.pdf

Jon, I. (2021). *Cities in the Anthropocene: New ecology and urban politics*. London: Pluto Press.

Karpouzoglou, T., Dewulf, A., & Clark, J. (2016). Advancing adaptive governance of social-ecological systems through theoretical multiplicity. *Environmental Science & Policy, 57*(C), 1–9. https://doi.org/10.1016/j.envsci.2015.11.011.

Kozlowski, M., Mehan, A., & Nawratek, K. (2020). *Kuala Lumpur: Community, infrastructure and urban inclusivity* (1st ed.). Routledge. https://doi.org/10.4324/9781315462417.

Mehan, A. (2021). EUKN webinar "Port Cities and Mega-Trends: Glocal Approaches to Sustainable Transitions". *The Port City Futures (PCF) Blog*, Leiden. Delft.Erasmus (LDE) Initiative.

Mehan, A., & Mostafavi, S. (2022). Building resilient communities over time. In R. Brears (Ed.), *The Palgrave encyclopedia of urban and regional futures* (Living ed., p. 4). Cham: Palgrave Macmillan. https://doi.org/10.1007/978-3-030-51812-7_322-1.

Mikulewicz, M. (2018). Politicizing vulnerability and adaptation: On the need to democratize local responses to climate impacts in developing countries. *Climate and Development, 10*(1), 18–34. https://doi.org/10.1080/17565529.2017.1304887.

Moore, J., & Patel, R. (2017). Unearthing the capitalocene: Towards a reparation's ecology. *Roar Magazine*, 18–26. https://roarm ag.org/magazine/moore-patel-seven-cheap-things-capitalocene/.

Nawratek, A., & Mehan, A. (2020). De-colonising public spaces in Malaysia. Dating in Kuala Lumpur. *Cultural Geographies, 27*(4), 615–629. https://doi.org/10.1177/1474474020909457.

Pettit, J. (2004). Climate justice: A new social movement for atmospheric rights. *IDS Bulletin, 35*(3), 102–106. Available from: https://doi.org/10.1111/j.1759-5436.2004.tb00142.x.

Rahdari, A., Mehan, A., & Malekpourasl, A. (2019). Sustainable real estate in the Middle East: Challenges and future trends. In I. T. Walker, C. Krosinsky, L. N. Hasan, & S. D. (Eds.), *Kibsey, sustainable real estate: Multidisciplinary approaches to an evolving system* (pp. 403–426). Palgrave Macmillan. https://doi.org/10.1007/978-3-319-94565-1_16.

Rijke, J., Herk, S., Zevebergen, C., & Richard, A. (2012). Room for the river: Delivering integrated river basin management in the Netherlands. *International Journal of River Basin Management, 10*(4), 369–382. https://doi.org/10.1080/15715124.2012.739173.

Schlosberg, D., & Collins, L. B. (2014). From environmental to climate justice: Climate change and the discourse of environmental justice. *Wiley Interdisciplinary Reviews: Climate Change, 5*(3), 359–374. Available from: https://doi.org/10.1002/wcc.275.

Sheller, M. (2018). *Mobility justice: The politics of movement in an age of extremes*. London; Brooklyn: Verso.

Vergès, F. (2017). Racial capitalocene. In G. Johnson & A. Lubin (Eds.), *Futures of black radicalism.* London, UK: Verso.

Whyte, K. (2016). Indigenous experience, environmental justice and settler colonialism.

Wyborn, A. C. (2015). Connecting knowledge with action through co-productive capacities: Adaptive governance and connectivity conservation. *Ecology and Society, 20*(1), 11. https://doi.org/10.5751/ES-06510-20011.

Emergency

▶ Collective Emotions and Resilient Regional Communities

Emerging Concepts Exploring the Role of Nature for Health and Well-Being

Cassandra Murphy[1], Danielle MacCarthy[2] and Evi Petersen[3]
[1]Department of Psychology, Maynooth University, Maynooth, Ireland
[2]Queen's University, Belfast, UK
[3]Institute of Sports, Physical Education and Outdoor Life, University of South-Eastern Norway, Oslo, Norway

Synonyms

Digital nature; Health; Nature connectedness; Nature relatedness; Physical activity; Virtual reality; Well-being

Definition

As society changes so too does the environment around us. Urbanization has become a threat for natural environments. As our understanding of the benefits of nature engagement continues to evolve, it is important to understand what our relationship with nature looks like and how to foster it. With the looming challenge of urbanization it is time to consider alternatives to the outside nature that we are used to. Developments in digital nature and virtual reality may hold the key to helping individuals maintain a connection to nature.

Introduction

Spending time in the mountains, at the beach, or simply in an urban green space – experiencing the outdoors seems to receive growing popularity worldwide, most notably across Western societies. While personal motives for increased interest in being in nature often vary and are diverse, public discourses highlight the positive impact of nature on health and well-being. Awareness that natural environments can play a preventative role in illness at a fraction of the costs of treatment approaches has influenced government policy and health professionals to be open to a more ecological approach to public health (Ward Thompson 2011). As a result, society and policymakers are turning to landscape planners, wilderness therapists, outdoor educators, and increasingly to innovators in virtual reality for answers to a myriad of questions about how to optimize our experience with diverse forms of nature.

It is vital to understand how people's relationships with different forms of nature are shaped, how they influence personal values and can cultivate conservation activities. As we shall discuss later, the role of nature experiences in contributing to health and well-being-related goals is especially pertinent for neglected societal groups.

Establishing What Is Meant by Nature

Asking the question "why is nature important?" has led to novel and evolving developing paradigms. One of the most common observations articulated is the health and well-being benefits from nature interactions (Donnelly and Macintyre 2019). However, this necessitates a prior question before we can truly understand its importance – "what is nature?". Furthermore, the paradigms employed to understand human–nature interactions vary according to the discipline, organization, or individual. Recent work on indigenous local knowledge (ILK) systems draws attention to the profound differences that exist in terms of questions such as, "what is 'nature'," which have their own processes of validation, verification, implementation within ILK production (Díaz et al. 2015). By contrast, Western scientific approaches have adopted frameworks, which frequently delineate boundaries between where we encounter nature; between everyday and non-everyday nature exposure; the development of distinct typologies of nature; and lastly, distinguishing between the type of engagement with nature.

Early research on nature engagement often considers "nature" as a place to visit. The related terms "nature tourism," "soft-tourism," and "slow-adventure" have been coined to explain the popular concept of nature as a destination. It is associated with an imagination of nature being wild, inaccessible, and pristine. Definitions of nature in this sense are conceived of as an "escape" from social, civic life, for example, nature has been commonly studied in national parks and harder to access nature spots. To this extent, nature engagement is a novel and non-daily event. It involves traveling outside of the home and neighborhood to reach and consequently return from. Criticisms of these definitions exist in that the populations and activities studied in relation to this nature engagement commonly exclude populations and tend to concentrate on able-bodied engagement. Definitions may also depict a "white-gendered" Western landscape concept of nature. However, more recently, in line with the growing recognition of multiple ecosystem benefits and critical role of nature in cities, work in this area has begun to increasingly highlight the role of "nature nearby" in urban environments and multiple communities, which interact with it (Beatley 2011; Beatley and Newman 2013). Attention has been directed to the nature that surrounds us and a nature that has benefits to humans across the lifecycle. Research agendas have come to emphasize the importance of everyday exposures to nature across a range of mental health, physical health, and social benefits across the spectrum of communities and populations. Nature in this sense can embody all forms of flora and fauna in the environment. As a result, typologies of nature have been created and are categorized, i.e., blue/ green nature (Pouso et al. 2021), studied as singular and constituent parts of the urban environment, i.e., vegetation that are commonly found as we move around urban environments, i.e., street trees (Mullaney et al. 2015), parks (Sturm and Cohen 2014; Wood et al. 2017), and trails, which can then be examined with their respective value and impact on health and well-being measures. NBS infrastructure *Nature as a concept can also be determined by the type of behavioral engagement associated with it. The Japanese practice of "shinrin yoku" or "forest-bathing" has recently gained in popularity globally, emphasizing moving beyond a purely physical exposure in nature to a practice encouraging the use of all senses and immersion. In this respect, nature, more specifically forests, represent large, treed areas and require an ability to move through them. By contrast, other forms of engagement with nature such as gardening can represent smaller "natural" areas in residential areas or may involve designated allotments in otherwise highly urbanized areas. Engagement with nature may not always distinguish between indoors and outdoors as increasing work exploring house plants (Dzhambov et al. 2021) and their impact upon well-being has been explored. The discipline or field of inquiry may also shape how nature is conceived. Urban planning and city design fields have traditionally emphasized "greenness" as a concept of nature; however this definition is being challenged to include greater levels of biodiversity (Marselle et al. 2021), which can be explored from perceptions of biodiversity to ecological definitions of nature and microbiome.

Recent work has redefined and expanded our concept of nature prompting new definitions, which take into account where nature can be found, how we engage with it, and how often we interact. These new definitions promote a greater examination and awareness of the multiple benefits nature holds for us. These will be explored in the following section.

A Relationship with Nature

As mentioned above there are many benefits to nature. This section will focus on outlining these benefits and how an individual can attain them. To fully understand the advantages of nature we must first look at how a relationship with nature is formed. Researchers are familiar with Wilson's (1984) biophilia hypothesis in which he describes the innate human urge to seek a connection with other living things such as nature. It is believed this urge is a result of our evolution in specific types of natural environments. Wilson's claim, albeit not a specifically testable hypothesis, is supported by research in the area of nature engagement and well-being, alongside the emerging construct of nature connectedness, in which an individual feels a sense of closeness to nature (Nisbet et al. 2009). Nature connectedness is a relatively recent construct used to explain a connection that goes beyond a sense of affinity for nature and focuses more on the concept that one's identity is grounded in nature. When a person feels connected to nature, it is believed that a sense of community is created between nature and the self (Mayer et al. 2009). This connection is not formed or measured in the amount of time a person spends in nature but rather how you engage with the nature around you. The activities used to engage with nature are not limited to the outdoors (Passmore and Holder 2017; Richardson et al. 2021). Activities such as indoor gardening and digital nature (more on this in the next section) are becoming more accessible to those who cannot readily access nature outdoor settings. Thus, Wilson's original concept, the connection to natural systems, facilitates what are termed the restorative benefits of nature, which we will explain in more detail below. This can also help us understand how the hypothesis can be expanded to include nature in urban environments. Research suggests that citizens report significantly higher levels of overall health when perceiving higher amounts of "greenness" within their living environment, independently of urban or rural settings (Callaghan et al. 2021; Nisbet and Zelenski 2011).

Health and Well-Being Benefits

Different research perspectives provide a variety of explanations of the well-being benefits of nature engagement. Some include, but are not limited to, social well-being, psychological well-being, and subjective well-being. Our social well-being can refer to our ability to create connections with other living things while psychological well-being is linked to finding the meaning of life or your sense of purpose in the world. Subjective well-being is associated with the construct of happiness and the idea of life satisfaction, which is often dictated by the different emotions we experience throughout life (Diener et al. 1999; Larson 1993; Ryff and Keyes 1995).

A recent qualitative study by Iqbal and Mansell (2021) outlined the different ways in which engaging with nature contributes to these three facets of well-being. Many interesting discussion topics emerged in this contemporary thematic analysis, which clearly resonate with prior research findings. The authors linked the concept of freedom that participants spoke of to the sense of autonomy that plays a role in understanding of psychological well-being. Autonomy, integral to theoretical explanations of motivation (e.g., SDT; Ryan & Deci 2000) is perhaps a central concept in human–nature interactions, as unlike sport- or gym-based activities, natural settings afford us more opportunities for choice and decision-making. Preserving autonomy can also give us a sense of control, reducing stressors, which is a consistent finding from human–nature interactions (Nisbet and Zelenski 2011). This calmness can also act as a break for individuals from everyday tasks, allowing for nature to provide a restorative function in that it gives people a chance to become refreshed and feel revitalized (Ulrich et al. 1991). Again, Iqbal and Mansell (2021) reported that restoration was linked to reflection and decision making in that participants felt that having taken their break they now were able to return with a clearer mind to make decisions and enhance their cognitive function. Finally, they reported a key link between nature and social connection. This theme focused on how engaging with nature offered everyone the opportunity to feel a part of a community often by taking part in nature activities with local social groups. This idea connects back to the construct of nature connectedness we discussed earlier and the belief that you are connected to something larger. Cartwright et al. (2018) conducted research that suggests nature can act as a "buffer" for social connection in situations where an individual may be isolated, such as those recently experienced with the mitigating behavioral restrictions (e.g., lockdowns) implemented as a result of the coronavirus pandemic. These lockdown periods saw many millions of people isolated for long periods of time with their only escape being a short walk

E

outdoors in their local area (i.e., often restricted to 200 m or 2 km from home). Having access to nearby nature offers a short-term solution to what can be a long-term problem for some.

One area we cannot ignore when speaking about health and well-being is our physical health (de Vries et al. 2003; Maas et al. 2006; Mitchell and Popham 2007). While residential greenness might indirectly influence people's mental health by modulating heat and air quality (Larkin et al. 2017), many explanations draw on the positive impact of contact with nature on people's stress level and mental well-being. Moreover, natural environments such as urban parks and coasts provide attractive locations for walking, jogging, playing, and other activities. Even visiting natural environments for the enjoyment of nature's quietness most often requires some degree of physical activity in the form of active transportation (e.g., walking, hiking, biking, climbing, kayaking, etc.). Although regular physical activity has an impact on health per se, physical activity in nature can potentially provide health benefits above and beyond those provided by physical activity taking place in other environments such as indoors or in urban settings (Calogiuri and Chroni 2014; Rogerson and Giddings 2020).

Having focused on the diverse outcomes that nature engagement and connectedness has for human health and well-being, we must not overlook the consequences for the health of the environment. When engaging with the many news stories on climate change it is understandable that one may believe that this relationship is a one-sided transaction; however, it is known to be a mutually beneficial exchange. Those who experience a sense of nature connectedness are shown to experience environmental concern (Nisbet et al. 2009) in addition to engaging in pro-environmental behavior and consumption (Alcock et al. 2020; Mackay and Schmitt 2019; Mayer et al. 2009). These behaviors are seen in the form of recycling habits, buying greener or energy saving products. This strengthened relationship may even encourage some to take on an advocacy role, engaging in citizen-science initiatives or joining in political activism movements.

While the benefits of connecting to nature are clear to see, in reality access to natural settings is arguably decreasing due to urbanization, degradation of natural environments, and a general reduction in time spent outdoors. A WHO report (2017) recommended that all individuals have access to at least two hectares of green space no further than 300 m (a 5 min walk) from their home. However, this is not a reality for many, specifically those living in urban areas. This is why we need to think beyond the realm of what we understand nature to be and take a look at innovative forms such as digital nature, primarily provided by virtual reality (VR) technology, that are emerging. The following paragraph discusses how emerging forms of digital nature link to aspects of health and well-being.

The Emerging Role of Digital Nature

Entering the so-called digital age, new digital forms of technologies have begun to shape people's daily living routines. This phenomenon is particularly observable in the context of how people consume and process new information as well as in their communication patterns with others. Consequently, digital technologies have a profound impact on societal interactions in a variety of cultural settings. Simultaneously, new digital technologies also have begun to affect humans' encounters with the natural world in many ways. For instance, on a daily user scale, people are using smartphone apps to get directions to natural environments, to check the local weather conditions, to identify a flower or a tree, or to engage in GPS-based games (e.g., geocaching), where one is encouraged to find specific outdoor spots or to take a specific walking route in nature.

Moreover, digital inventions are also reconfiguring the way we interact with nature when we are not physically outdoors. In this scope, immersive technologies are increasingly promoted, implemented, and tested as virtual alternatives or supplements to provide engagement opportunities with nature (Depledge et al. 2011). Immersive technologies, often called extended reality technologies or XR, refer to virtual reality, mixed reality, as well as to augmented reality technologies such as VR headsets, HoloLens, and augmented reality applications in smartphones and tablets. These digital technology applications have in common that they enable computer-generated simulation of a three-dimensional image or environment that allows a certain degree of interaction, creating the illusion of reality. For a successful illusion of reality to be in place, both immersion and presence are essential elements of immersive virtual environments (Litleskare et al. 2020). While commercial interest in virtual reality applications already appeared in the 1980s, it diminished across the 1990s. This development was mainly linked to the fact that technologies were too costly or ineffective due to hardware limitations, which caused, for instance, cyber sickness to be an issue. However, new immersive technologies are subject to a new global wave of interest due to intensive technological improvements.

As mentioned earlier, limited or less frequent access to natural outdoor environments is becoming a global public health concern. With a decrease in accessibility of physical natural areas and an increase of accessibility of digital forms of nature, the question emerges, how these new ways of engaging with nature may differ regarding their impact on health and well-being.

The Current State of Knowledge

On the one hand, early critical voices have noticed that digital (and particularly immersive technology) could cause alienation from physical outdoor nature and thus could provoke the opposite of the intended positive health and well-being impact (Levi and Kocher 1999). Further, previous research has shown that sedentary behavior is linked to increased screen time, which has several negative implications for well-being (Larson et al. 2019). Inevitably, physical contact to nature offers well-being benefits, which virtual nature cannot provide, such as enhanced immunity from exposure to microbiomes and phytoncides from trees or increased vitamin D synthesis induced by sunlight.

On the other hand, the potential benefits to psychological well-being from immersive virtual nature engagement should not be underestimated. For instance, research suggests that these interactions are more beneficial for health and well-being than an absence of human–nature interactions, even though they might not have an equally psychological benefit as physical nature exposure (Plante et al. 2006). Similar reports are found in studies comparing virtual to physical green exercise (Calogiuri et al. 2018). Furthermore, another study with healthy undergraduate students by Browning et al. (2020) could show that interactions with virtually simulated nature can have similar positive impacts upon well-being as interactions with outdoor nature. The authors found that both types of nature exposure increased physiological arousal and perceived restoration; however, mood levels increased for the outdoor nature exposure but not for the virtual nature exposure.

Researchers have also begun to stress the potential of immersive technology on health and well-being in clinical settings (White et al. 2018). For instance, a review of virtual nature in hospital inpatients identified the use of virtual reality technology across different medical settings to be safe and found high patient satisfaction (Dascal et al. 2017). Additionally, a 360-virtual reality study with cancer patients undergoing chemotherapy in Florida (USA) found that feelings of relaxation were elicited by watching virtual local nature sceneries (Scates et al. 2020), and exposure to a tropical island virtual reality scenery correlated significantly with a general reduction in anxiety within individuals with Generalized Anxiety Disorder (Gorini et al. 2010). Moreover, nature-based virtual reality has been successfully implemented for short-term distraction from pain, anxiety, or distress during surgical procedures (Furman et al. 2009; Mosso et al. 2009; Tanja-Dijkstra et al. 2018).

Future Questions, Concerns, and Developments

Current immersive applications still face technical challenges in incorporating other sensory inputs beyond visual and auditive sense activation, such as the smell of nature or the kinesthetic experience of walking on a natural ground, to create more immersive and potentially restorative experiences

E

(Hedblom et al. 2019; Litleskare et al. 2020). Thus, instead of expecting immersive technologies to replicate the multisensory experience of the physical nature outdoors, better questions to ask may be how and for whom immersive technologies can be a valuable alternative and supplemental tool when balancing people's need to experience nature in times of decreasing access to physical nature environments. Although researchers seem particularly interested in the dosages of virtual nature needed to elicit well-being effects (Browning et al. 2020), more studies are needed to investigate who and in which setting users can benefit from immersive virtual nature. For example, potential health and well-being effects could be elicited for populations with limited access to physical nature. These populations may include vulnerable groups such as older adults in healthcare settings, people with mobility restrictions, as well as people with access to nature, who do not always feel comfortable going outside or having insufficient time to do so. Additionally, with the recent example of a pandemic, people living in large urban agglomerations that suffer from a lack of nature contact might benefit from supplemental virtual reality opportunities.

Conclusion

In past decades, a growing body of literature has proposed salutogenic effects of exposure to nature (for a comprehensive scoping review, see Dzhambov et al. 2021). In this entry we have looked at emerging concepts of nature, which are considered within this spectrum ranging from traditional and physical forms of nature to digital forms of nature. For work that considers physical nature, emerging concepts are challenging the geographies of nature, taxonomies, and scales of nature, which make the idea of nature a unified concept, a complex and multifaceted notion in the literature.

The benefits of spending time in creating a relationship with nature are clear. However, the trend of urbanization and reduced access to nature are not entirely unrelated. Consequently, we must innovate in our thinking and research alternatives to nature engagement. Research in the scope of digital technology is still in its infancy, suggesting an untapped potential for enhancing health and well-being. Immersive technologies may be a valuable alternative or supplemental tool when balancing people's need to experience nature in times of decreasing access to physical nature environment, especially for vulnerable groups and within clinical settings. However, up to date, the long-term effects of immersive virtual nature are still unknown, inviting future research to explore the potential and limitations of digital nature interactions.

Cross-References

- ▶ Augmented Reality: Robotics, Urbanism, and the Digital Turn
- ▶ Cities in Nature
- ▶ Digitalization, Urbanization, and Urban-Rural Divide
- ▶ Health and the City: How Cities Impact on Health, Happiness, and Well-Being
- ▶ Health and the Role of Nature in Enhancing Mental Health

Acknowledgments This project (GoGreenRoutes; www.gogreenroutes.eu) has received funding from the European Union's Horizon 2020 research and innovation programme under grant agreement No 869764

References

Alcock, I., White, M. P., Pahl, S., Duarte-Davidson, R., & Fleming, L. E. (2020). Associations between pro-environmental behaviour and neighbourhood nature, nature visit frequency and nature appreciation: Evidence from a nationally representative survey in England. *Environment International, 136*, 105441. https://doi.org/10.1016/j.envint.2019.105441.

Beatley, T. (2011). *Biophilic cities: Integrating nature into Urban Design and Planning*. Island Press/Center for Resource Economics.

Beatley, T., & Newman, P. (2013). Biophilic cities are sustainable, resilient cities. *Sustainability, 5*(8), 3328–3345.

Browning, M. H. E. M., Mimnaugh, K. J., van Riper, C. J., Laurent, H. K., & LaValle, S. M. (2020). Can simulated nature support mental health? Comparing short, single-doses of 360-degree nature videos in virtual reality with the outdoors [Original Research]. *Frontiers in Psychology, 10*(2667). https://doi.org/10.3389/fpsyg.2019.02667.

Callaghan, A., McCombe, G., Harrold, A., McMeel, C., Mills, G., Moore-Cherry, N., & Cullen, W. (2021). The impact of green spaces on mental health in urban settings: A scoping review. *Journal of Mental Health, 30*(2), 179–193. https://doi.org/10.1080/09638237.2020.1755027.

Calogiuri, G., & Chroni, S. (2014). The impact of the natural environment on the promotion of active living: An integrative systematic review. *BMC Public Health, 14*, 873. https://doi.org/10.1186/1471-2458-14-873.

Calogiuri, G., Litleskare, S., Fagerheim, K. A., Rydgren, T. L., Brambilla, E., & Thurston, M. (2018). Experiencing nature through immersive virtual environments: Environmental perceptions, physical engagement, and affective responses during a simulated nature walk [Original Research]. *Frontiers in Psychology, 8*(2321). https://doi.org/10.3389/fpsyg.2017.02321.

Cartwright, B. D. S., White, M. P., & Clitherow, T. J. (2018). Nearby nature 'Buffers' the effect of low social connectedness on adult subjective wellbeing over the last 7 days. *International Journal of Environmental Research and Public Health, 15*(6). https://doi.org/10.3390/ijerph15061238.

Dascal, J., Reid, M., Ishak, W. W., Spiegel, B., Recacho, J., Rosen, B., & Danovitch, I. (2017). Virtual reality and medical inpatients: A systematic review of randomized, controlled trials. *Innovations in Clinical Neuroscience, 14*(1–2), 14–21.

de Vries, S., Verheij, R. A., Groenewegen, P. P., & Spreeuwenberg, P. (2003). Natural environments – Healthy environments? An exploratory analysis of the relationship between greenspace and health. *Environment and Planning A: Economy and Space, 35*(10), 1717–1731. https://doi.org/10.1068/a35111.

Depledge, M. H., Stone, R. J., & Bird, W. J. (2011). Can natural and virtual environments be used to promote improved human health and wellbeing? *Environmental Science & Technology, 45*(11), 4660–4665. https://doi.org/10.1021/es103907m.

Díaz, S., Demissew, S., Carabias, J., Joly, C., Lonsdale, M., Ash, N., . . . Zlatanova, D. (2015). The IPBES conceptual framework – Connecting nature and people. *Current Opinion in Environmental Sustainability, 14*, 1–16. https://doi.org/10.1016/j.cosust.2014.11.002.

Diener, E., Suh, E. M., Lucas, R. E., & Smith, H. L. (1999). Subjective well-being: Three decades of progress. *Psychological Bulletin, 125*(2), 276–302. https://doi.org/10.1037/0033-2909.125.2.276.

Donnelly, A., & Macintyre, T. (2019). *Physical activity in natural settings: Green and blue exercise*. https://doi.org/10.4324/9781315180144.

Dzhambov, A. M., Lercher, P., Browning, M., Stoyanov, D., Petrova, N., Novakov, S., & Dimitrova, D. D. (2021). Does greenery experienced indoors and outdoors provide an escape and support mental health during the COVID-19 quarantine? *Environmental Research, 196*, 110420. https://doi.org/10.1016/j.envres.2020.110420.

Furman, E., Jasinevicius, T. R., Bissada, N. F., Victoroff, K. Z., Skillicorn, R., & Buchner, M. (2009). Virtual reality distraction for pain control during periodontal scaling and root planing procedures. *J Am Dent Assoc, 140*(12), 1508–1516. https://doi.org/10.14219/jada.archive.2009.0102.

Gorini, A., Pallavicini, F., Algeri, D., Repetto, C., Gaggioli, A., & Riva, G. (2010). Virtual reality in the treatment of generalized anxiety disorders. *Studies in Health Technology and Informatics, 154*, 39–43.

Hedblom, M., Gunnarsson, B., Iravani, B., Knez, I., Schaefer, M., Thorsson, P., & Lundström, J. N. (2019). Reduction of physiological stress by urban green space in a multisensory virtual experiment. *Scientific Reports, 9*(1), 10113. https://doi.org/10.1038/s41598-019-46099-7.

Iqbal, A., & Mansell, W. (2021). A thematic analysis of multiple pathways between nature engagement activities and well-being. *Frontiers in Psychology, 12*, 580992. https://doi.org/10.3389/fpsyg.2021.580992.

Larkin, A., Geddes, J. A., Martin, R. V., Xiao, Q., Liu, Y., Marshall, J. D., . . . Hystad, P. (2017). Global land use regression model for nitrogen dioxide air pollution. *Environmental Science & Technology, 51*(12), 6957–6964. https://doi.org/10.1021/acs.est.7b01148.

Larson, J. M. (1993). Exploring reconciliation. *Mediation Quarterly, 11*(1), 95–106. https://doi.org/10.1002/crq.3900110110.

Larson, L. R., Szczytko, R., Bowers, E. P., Stephens, L. E., Stevenson, K. T., & Floyd, M. F. (2019). Outdoor time, screen time, and connection to nature: Troubling trends among rural youth?. *Environment and Behavior, 51*(8), 966–991.

Levi, D., & Kocher, S. (1999). Virtual nature:The future effects of information technology on our relationship to nature. *Environment and Behavior, 31*(2), 203–226. https://doi.org/10.1177/00139169921972065.

Litleskare, S., MacIntyre, T. E., & Calogiuri, G. (2020). Enable, reconnect and augment: A new ERA of virtual nature research and application. *International Journal of Environmental Research and Public Health, 17*(5). https://doi.org/10.3390/ijerph17051738.

Maas, J., Verheij, R. A., Groenewegen, P. P., de Vries, S., & Spreeuwenberg, P. (2006). Green space, urbanity, and health: How strong is the relation? *Journal of Epidemiology and Community Health, 60*(7), 587–592. https://doi.org/10.1136/jech.2005.043125.

Mackay, C. M. L., & Schmitt, M. T. (2019). Do people who feel connected to nature do more to protect it? A meta-analysis. *Journal of Environmental Psychology, 65*, 101323. https://doi.org/10.1016/j.jenvp.2019.101323.

Marselle, M. R., Lindley, S. J., Cook, P. A., & Bonn, A. (2021). Biodiversity and health in the urban environment. *Current Environmental Health Reports, 8*(2), 146–156. https://doi.org/10.1007/s40572-021-00313-9.

Mayer, F. S., Frantz, C. M., Bruehlman-Senecal, E., & Dolliver, K. (2009). Why is nature beneficial?: The role of connectedness to nature. *Environment and Behavior, 41*(5), 607–643. https://doi.org/10.1177/0013916508319745.

Mitchell, R., & Popham, F. (2007). Greenspace, urbanity and health: Relationships in England. *Journal of Epidemiology and Community Health, 61*(8), 681. https://doi.org/10.1136/jech.2006.053553.

Mosso, J. L., Gorini, A., De La Cerda, G., Obrador, T., Almazan, A., Mosso, D., ... Riva, G. (2009). Virtual reality on mobile phones to reduce anxiety in outpatient surgery. *Studies in Health Technology and Informatics, 142*, 195–200.

Mullaney, J., Lucke, T., & Trueman, S. J. (2015). A review of benefits and challenges in growing street trees in paved urban environments. *Landscape and Urban Planning, 134*, 157–166. https://doi.org/10.1016/j.landurbplan.2014.10.013.

Nisbet, E. K., & Zelenski, J. M. (2011). Underestimating nearby nature: Affective forecasting errors obscure the happy path to sustainability. *Psychological Science, 22*(9), 1101–1106. https://doi.org/10.1177/0956797611418527.

Nisbet, E. K., Zelenski, J. M., & Murphy, S. A. (2009). The nature relatedness scale: Linking individuals' connection with nature to environmental concern and behavior. *Environment and Behavior, 41*(5), 715–740. https://doi.org/10.1177/0013916508318748.

Passmore, H.-A., & Holder, M. D. (2017). Noticing nature: Individual and social benefits of a two-week intervention. *The Journal of Positive Psychology, 12*(6), 537–546. https://doi.org/10.1080/17439760.2016.1221126.

Plante, T. G., Cage, C., Clements, S., & Stover, A. (2006). Psychological benefits of exercise paired with virtual reality: Outdoor exercise energizes whereas indoor virtual exercise relaxes. *International Journal of Stress Management, 13*(1), 108–117. https://doi.org/10.1037/1072-5245.13.1.108.

Pouso, S., Borja, Á., Fleming, L. E., Gómez-Baggethun, E., White, M. P., & Uyarra, M. C. (2021). Contact with blue-green spaces during the COVID-19 pandemic lockdown beneficial for mental health. *Science of the Total Environment, 756*, 143984. https://doi.org/10.1016/j.scitotenv.2020.143984.

Richardson, M., Passmore, H.-A., Lumber, R., Thomas, R., & Hunt, A. (2021). Moments, not minutes: The nature-wellbeing relationship. *International Journal of Wellbeing, 11*(1), 8–33. https://doi.org/10.5502/ijw.v11i1.1267.

Rogerson, R. J., & Giddings, B. (2020). The future of the city centre: Urbanisation, transformation and resilience – A tale of two Newcastle cities. *Urban Studies, 58*(10), 1967–1982. https://doi.org/10.1177/0042098020936498.

Ryff, C. D., & Keyes, C. L. M. (1995). The structure of psychological well-being revisited. *Journal of Personality and Social Psychology, 69*(4), 719–727. https://doi.org/10.1037/0022-3514.69.4.719.

Scates, D., Dickinson, J. I., Sullivan, K., Cline, H., & Balaraman, R. (2020). Using nature-inspired virtual reality as a distraction to reduce stress and pain among cancer patients. *Environment and Behavior, 52*(8), 895–918. https://doi.org/10.1177/0013916520916259.

Sturm, R., & Cohen, D. (2014). Proximity to urban parks and mental health. *The Journal of Mental Health Policy and Economics, 17*(1), 19–24.

Tanja-Dijkstra, K., Pahl, S., White, M. P., Auvray, M., Stone, R. J., Andrade, J., ... Moles, D. R. (2018). The Soothing Sea: A virtual coastal walk can reduce experienced and recollected pain. *Environment and Behavior, 50*(6), 599–625. https://doi.org/10.1177/0013916517710077.

Ulrich, R. S., Simons, R. F., Losito, B. D., Fiorito, E., Miles, M. A., & Zelson, M. (1991). Stress recovery during exposure to natural and urban environments. *Journal of Environmental Psychology, 11*(3), 201–230. https://doi.org/10.1016/S0272-4944(05)80184-7.

White, M. P., Yeo, N. L., Vassiljev, P., Lundstedt, R., Wallergård, M., Albin, M., & Lõhmus, M. (2018). A prescription for "nature" – The potential of using virtual nature in therapeutics. *Neuropsychiatric Disease and Treatment, 14*, 3001–3013. https://doi.org/10.2147/NDT.S179038.

WHO. (2017). *Urban green spaces: A brief for action*. Copenhagen: Regional Office for Europe.

Wilson, E. O. (1984). *Biophilia*. Harvard University Press.

Wood, L., Hooper, P., Foster, S., & Bull, F. (2017). Public green spaces and positive mental health – investigating the relationship between access, quantity and types of parks and mental wellbeing. *Health & Place, 48*, 63–71. https://doi.org/10.1016/j.healthplace.2017.09.002.

Ryan, R. M., & Deci, E. L. (2000). Self-determination theory and the facilitation of intrinsic motivation, social development, and well-being. American Psychologist, 55 (1), 68–78. https://doi.org/10.1037/0003-066X.55.1.68

Ward Thompson, C. (2011). Linking landscape and health: The recurring theme. *Landscape and Urban Planning, 99*(3–4), 187–195. https://doi.org/10.1016/j.landurbplan.2010.10.006

Endowments

► Furthering the Sustainable Development Agenda by Putting Urban Heritage and Value Extraction at the Center

Energy Efficiency

► Smart Grids to Lower Energy Usage and Carbon Emissions: Case Study Examples from Colombia and Turkey

Energy Efficiency in Buildings

▶ Building Energy: How Building Efficiency Can Be Improved in Government Facilities

Energy Transition

▶ Amsterdam's Pathway to Climate Neutrality: Creating an Enabling Environment

Engineered Urban Wetlands

▶ Artificial Urban Wetlands

ENSO – El Nino South Oscillation

▶ Adapting to a Changing Climate Through Nature-Based Solutions

Environment

▶ Circular Economy Cities
▶ Resource Effectiveness in and Across Urban Systems

Environmental Education and Non-governmental Organizations

Meredian Alam
Sociology and Anthropology Department, Universiti Brunei Darussalam, Gadong, Brunei

Environmental Education

Environmental education is a body of knowledge and pedagogical practices that has developed in popularity across the globe in recent decades with the aim of understanding and solving a set of problems that emerge from social relationships, especially between education and the environment. Environmental education has rapidly extended its influence beyond the boundaries of schools (Jickling and Wals 2008). Although not formally conducted by social organizations, it is portrayed as a new teaching method aimed at changing habits, attitudes, and social practices, thereby leading to a solution to the current world's socio-environmental degradation (Smaldone and Dey 2010; Sureda and Calvo 2001). Considering the current context of risks, uncertainties, and challenges in the world, investing in education pertaining to a behavioral model that creates a new link between people and nature has a strategic significance for environmental education. Through environmental education, individuals learn to make informed choices and take responsible action by increasing their understanding of the environment and its problems; developing the skills and competence necessary to solve challenges; and developing attitudes towards, motivations to, and commitments to the preserving the environment (Graci and Dodds 2008; Hays et al. 1996; Neumayer 2002).

When properly understood, environmental education is viewed as a full lifelong education that is responsive to changes in a continually changing world. It prepares individuals for life by providing knowledge about the world's major problems and revealing the skills and qualities needed to play a positive, ethical role in improving life and protecting the environment.

Non-governmental Organization

A non-governmental organization (NGO) is a non-profit organization that operates outside the government. Sometimes known as civil societies, NGOs are established on a local, national, and worldwide level to serve a social or political objective such as humanitarian issues or the environment (Detenber et al. 2018). These self-created social organizations unite people in a shared goal. In general, NGOs take direct action, which may be political, religious, social, or environmental. These actions are instructive.

Experience may be a fantastic teacher. An approach termed "natural learning" addresses the learning of very young children and follows the path of experiential enquiry (Gamel et al. 2017; Prévot et al. 2018). Thus, involvement in an NGO activity may transmit informal environmental knowledge and therefore contribute towards environmental education. Based on UNESCO's definition, non-formal environmental education includes any structured educational activity that occurs outside the current formal educational system and particularly that which is produced in groups that assemble themselves for reasons other than learning.

Non-governmental organizations (NGOs) at the local, national, and regional levels have played an increasingly significant role in the region's economic development and conservation efforts (Haigh 2006). Indeed, in terms of empowering people, they are on the front lines. They also assist in identifying problems, raising awareness, and providing data to grassroots communities. In other words, they articulate the issues and needs of communities and bring this information to the attention of those who have the power to effect change.

With regard to national issues, NGOs are primarily concerned with policy, and they play a critical role in identifying flaws and inadequacies in current frameworks for policy and legislative development. Furthermore, they often gather information and educate people, as well as members of the business sector and representatives of the government.

Roles of NGOs in Promoting Environmental Education

Importantly, environmental NGOs have raised awareness about ecological issues and have promoted sustainable development. Legislation promoting public participation in environmental management has given NGOs and major groups a boost in recent years. Thailand, for example, recognizes people's right to participate in environmental protection under Article 56 of the 1997 Constitution (Kang 2019).

Unfortunately, the Asian financial crisis of the late 1990s wreaked havoc on the finances of many large public interest organizations, especially in Southeast Asian nations like Indonesia, Thailand, and Malaysia, forcing them to restrict much of their activity. For example, within the Indonesian NGO Wahana Lingkungan Hidup (WALHI), over 600 groups are active in a variety of activities, including campaigning against the burning of the country's tropical forest (Nugroho 2019; Yani et al. 2019). The roles of non-governmental organizations in promoting environmental education are elucidated below.

Environmental Awareness and Campaigning

Across the diverse field of NGOs, organizations strive to promote environmental awareness and advocate for policy and development program reforms. These organizations raise environmental awareness and campaign locally, nationally, and globally, with some campaigns taking place at all three levels at the same time. The Kerala Sastra Sahitya Parishad (KSSP) of India, for example, has received worldwide acclaim for its efforts in mobilizing public opinion among people's groups in Kerala (Prabhash 2004). With over 20,000 members, the KSSP is considered one of India's most well-informed and well-organized grassroots movements (Prabhash 2004). This organization has received worldwide acclaim for its efforts in mobilizing public opinion among people's groups in Kerala (Heller 1996) United Nations 1995. With over 20,000 members, the KSSP is considered one of India's most well-informed and well-organized grassroots movements (Heller 2001).

According to the Pakistani Environmental Society (SCOPE), established in 1988 with an emphasis on establishing relationships with local NGOs, research institutes, universities, and government agencies, environmental efforts in Pakistan have been quite successful (Imran et al. 2021). As well as inspiring grassroots organizations, SCOPE is actively involved in public interest litigation and lobbying on behalf of the public interest.

Non-governmental organizations (NGOs) working in the fields of science and technology aid in bridging the knowledge gap that exists between scientists, policymakers, and the general population (Farooqi and Fatimah 2010). They are proven to be a useful resource during the decision-making process. For instance, the Citizen's Report on the Environment in India focuses on specific environmental issues such as pollution in cities and flood management (Dillon and Teamey 2002). Their studies, which are written in layman's terms, aid in the public's understanding of the challenges being encountered.

Numerous long-standing NGOs in the field of environmental education are involved in major national projects that include a range of promotional activities, from grassroots awareness-raising to lobbying and media campaigns. Such efforts include fields like research, education, awareness-raising, and advocacy. The Species 2000 Campaign, run by the WWF in Malaysia, aims to organize effective national action to protect Malaysia's biodiversity (Osman and Pudin 2009). Thus, WWF Malaysia has forged partnerships with a wide range of conservation organizations, from federal and state government agencies to universities and non-governmental organizations and to local community groups. India, Malaysia, and the Philippines have established similar coalitions of environmental organizations to increase awareness of biodiversity loss and its consequences for governments and the public.

Global campaigns frequently include regional NGOs, such as Greenpeace, China's well-publicized campaign against possibly hazardous phthalate-containing PVC toys and children's goods. As a result of that campaign, Toys R Us announced a global product withdrawal of phthalate-containing teethers, rattles, and pacifiers in November 1998 (Wilkinson and Lamb IV 1999). NGOs throughout the area collaborate with regional and international organizations to commemorate important days like World Water Day (March 22), World Environment Day (June 5), and World Habitat Day (October 1).

Several fields related to water resources, solid and hazardous waste management, and biodiversity (genetic engineering) have gained popularity in recent years. New NGOs are emerging in Southern Asia to promote awareness and disseminate knowledge about these issues, while older, established NGOs are redirecting their research efforts in response to the new obstacles. The Centre for Science and Environment (CSE) in India, for example, was essential in convincing 28 pulp and paper mills to provide data such as the US EPA's "toxic discharges" database (Fraas and Egorenkov 2015). Only two companies were determined to be complying, but the research also showed that a mill with excellent environmental practices is more likely to maintain consistent profits than other facilities. Nine of the companies upgraded their pollution controls because of the program. The CSE aims to give green ratings to two sectors each year; the automotive industry in India is being targeted next (Roy 2018).

Educating the public on the dangers of environmental destruction is a task of environmental NGO. The Pasig River Movement, a collection of NGOs working to clean Metro Manila's major river, gathers data on companies that dump garbage into the river on a regular basis in the Philippines (Segovia and Galang 2002). The Lason (Poison) Award is given to the ten worst polluters of the year; this award receives significant media attention and has pushed companies to build effective wastewater treatment systems (Van Damme and Neluvhalani 2004). Several "poison" winners have later redeemed themselves, receiving "environment" honors in the following year.

Animal Products

In the fight against trading in endangered animal products, non-governmental organizations have achieved worldwide success. Even though most Asian and Pacific nations have signed the CITES Convention, there is still an active trade in endangered species (Kopnina and Cherniak 2015). Non-governmental organizations (NGOs) are not only pushing governments to efficiently implement current laws to preserve endangered species, but are also increasing public awareness about the significance of conservation. For example, the Asian Conservation Awareness Programme (ACAP) has successfully utilized the media, presenting horrific pictures of animals being killed for their tusks, fur, and gall bladders in a series of hard-hitting TV, movie, and poster advertisements

(Gigliotti 1990). NGOs in Asia have used international and national campaigns to convince the public to alter their attitudes towards unusual foods and clothes, as well as to persuade the government to implement the CITES rules.

Religion

One of the most difficult tasks for environmental NGOs involves delivering the message to important religious groups (Swan 1978). The Alliance of Religions and Conservation (ARC) has been collaborating with religious leaders from across the world to develop new, realistic forms of religious engagement in environmental problems (Seiffert 1991). According to the Ohito Declaration on Religions, Land, and Conservation, issued in 1995, for people of faith, preserving and sustaining natural living systems is a religious duty (Sherkat and Ellison 2007). All of the world's main faiths have rediscovered "sacred ground" and the idea of the necessity for Man to conserve and protect the environment because of the Ohito Declaration and the work of groups like ARC (Schlosberg and Carruthers 2010). The Taoists were the eighth religion to join ARC in 1995, demonstrating the breadth of the organization's network operations (El Jurdi et al. 2017). They requested that ARC join them in starting a campaign to preserve their hallowed holy mountains in China, which were endangered by changes in forestry, agriculture, urban development, and, more recently, tourism, after talks with the WWF/ARC personnel.

Reporting

Many non-governmental organizations in Bangladesh work together to produce "State of the Environment" reports on a quarterly basis, whereas the Centre for Science and Environment in India publishes the Citizen's State of Environment Report (Dutt et al. 2013). Many of these broader analyses are accompanied by more in-depth research that contributes to a better understanding of environmental issues and increases public awareness of the need of environmental preservation efforts.

Investigative reports on specific environmental issues have grown in number alongside State of the Environment reports. An environmental reporting handbook for journalists was created by the Asia-Pacific Forum of Environmental Journalists (APFEJ) with support from the Economic and Social Commission for Asia and the Pacific (ESCAP). The APFEJ has affiliates working in Nepal, India, Sri Lanka, and Malaysia, as well as the Philippines and Thailand (Hanaoka and Regmi 2011). They have published Citizen's Reports on the Environment and newspaper articles related to these nations' environmental issues. Seminars on environmental reporting have been supported by the Asian Institute of Development Communication and the Asian Media Information and Communication Centre, both of which are APFEJ affiliates. As a result of the Pacific Islands News Association (PINA) and the Pacific Forum of Environmental Journalists, the PINA Pacific Journalism Centre in Suva, Fiji, now offers a regular environmental news service to the 22 countries and territories of the South Pacific (Richstad et al. 2021).

Monitoring and Geographic Information System (GIS)

NGOs have made a significant contribution to environmental conservation in Southeast Asia by monitoring and reporting on the region's environmental conditions. Such a project on the Philippines Pangasinan Island resulted in better coral ecosystem protection and, eventually, higher catches for local fishermen (Panapasa and Singh 2018). In addition, Bantay Dagat ("Sea Watch") is an organization established by locals in Cabacongan (a province of Bohol) to track down and arrest illegal fishing vessels (Catedrilla et al. 2012).

For the purposes of environmental monitoring, many NGOs have used Geographic Information System (GIS) technology. The International Mountain Development (ICIMOD) has, for example, developed a decentralized network for collecting, storing, and disseminating important biophysical and socioeconomic data (Catedrilla et al. 2012). The University of the South Pacific in Fiji has improved its GIS training, as has the United Nations (Rakai and Williamson 1995).

Monitoring at the national or regional level can allow the NGOs to keep tabs on legislation's effectiveness and investigate issues like hazardous waste movement; species migration; the trade in endangered or restricted animals or plants; or research into the health of rivers, forests, or other ecosystems. Campaign tactics and policy proposals are developed by NGOs via this kind of activity. With regard to Greenpeace China's toxic waste trade monitoring, the Special Administrative Region of Hong Kong was able to prohibit hazardous waste imports from being moved into or through the region (Lin et al. 2020).

A fundamental requirement for effective environmental decision-making is the availability of reliable information on the environmental impact of development and economic policies. Several NGOs and other groups are attempting to offer information to influence government decision-making. For example, the Economy and Environment Programme for Southeast Asia, or EEPSEA, is based in Singapore and operates across the region, coordinating and supporting a network of academics that research policies' environmental impacts (Yusuf 2008). It works closely with a variety of institutions, including the Vietnamese University of Agriculture and Forestry, the University of the Philippines at Los Baños, and the China Centre for Economic Research in Beijing (Zamora 2009). Research stemming from EEPSEA and its associated network has affected the outcome of a range of issues, including pollution in the People's Republic of China, the water supply in Manila, the effect of international law on farmers in Sri Lanka, and policy formulation for Thailand's National Parks (EEPSEA 1998; Danh 2008).

Several NGO initiatives have been created through the UN Development Programme's regional offices to monitor the urban environment. Asia-Pacific 2000, an initiative to help NGOs tackle the urban environment problem, and the Urban Governance Initiative are two examples (Smedby and Neij 2013). Both initiatives have worked extensively with regional and national NGOs to organize events to increase awareness and knowledge about urban environmental issues. Several books have also been produced, including *Our Cities, Our Homes, an A-to-Z Guidebook on Human Settlement Issues* and *Water Watch*, a community action guide issued in collaboration with the Asia-Pacific People's Environmental Network (APPEN) (Tolkach et al. 2016).

Training and Capacity Building

An increasing number of NGOs are using education to encourage individuals to join in conservation initiatives. The International Union for the Conservation of Nature and Natural Resources (IUCN) is one of the most active international NGOs engaged in environmental education in the area (Turner 2009; Tilbury and Wortman 2004). The IUCN fulfils this function through its Commission on Education and Communication (CEC), which focuses on providing information, training, capacity building, and networking to improve environmental awareness among teachers in the region (Murphy 2012).

Other NGOs have worked extensively with governments to develop and implement national environmental education initiatives. In Nepal, for example, NGOs collaborated on the implementation of the country's National Conservation Strategy's environmental education component and provided technical assistance to government agencies (Aryal et al. 2012; Dressler et al. 2010).

Government Partnership

Partnerships between the government and non-governmental groups are encouraged in a number of ways. In some countries, there are regulations in place to ensure that community and big groups are both involved in the formulation of regional and national policies and plans. In Thailand, for example, the 1992 Environment Act devolved environmental management to provincial and municipal governments and promoted citizen participation through environmental non-governmental groups (Kuasirikun 2005; Tongcumpou and Harvey 1994). The influence and potential political power of NGOs increases when countries adopt explicit laws requiring public participation in decision-making. In the Philippines, for example, the EIA standards and implementing rules linked to the Mining Act and the Indigenous Peoples Rights Act require indigenous people's freely informed consent before projects may proceed (Dombrowski 2010). In many cases, people have little knowledge or access to information regarding the issues at hand,

and NGOs are relied upon to offer advice on hearings and consultations, as well as negotiating and resolution techniques.

However, elsewhere in the region, some public interest groups have been focused on their own political agenda and have not been held accountable. This has created concern, and in some cases, their legitimacy and/or sanction has been lost. Governments around the globe have made efforts to legitimize the role of independent NGOs in decision-making bodies by giving them representation (Alger 2003). Furthermore, many governments in the Asian region provide or allow funding for environmental education and natural resource conservation NGO activities (Alger 2003; Spiro 1995). Governments, philanthropists, and non-governmental groups have also joined together to establish endowment funds to support environmental projects. Bhutan (Bhutan Trust Fund for Environmental Conservation), Indonesia (Indonesian Biodiversity Foundation or Yayasan KEHATI), and the Philippines (Foundation for the Philippine Environment) are examples (Gaioni 1989; Permana et al. 2011). In addition, the Japan Fund for Global Environment was established using contributions from the government and the corporate sector to assist NGOs in their worldwide environmental conservation activities.

The Global Environmental Facility (GEF) recognizes that civil society has emerged as a major participant in Agenda 21 implementation (Handl 1998; James et al. 1999; Wonham 1998). NGOs are engaged in the design, planning, and/or implementation of approximately 20% of the money invested by the GEF (McCrea-Strub et al. 2011). The GEF also supports the UNDP's Small Grants Program (SGP) and has established a medium-sized grants program under which NGOs may receive up to $1 million for biodiversity conservation projects (Carugi and Bryant 2020).

Conclusion

We have pointed out different ways of working with and providing public environmental education established by global and national environmental organizations. This article illustrates that environmental education is not confined to formal schooling institutions, but can be implemented by non-governmental organizations. By recognizing the multiple commitments and responsibilities that NGOs can undertake, as noted above, people should realize that non-government organizations can create more flexible environmental education events that can have wider national and international impacts.

References

Alger, C. F. (2003). Evolving roles of NGOs in member state decision-making in the UN system. *Journal of Human Rights, 2*(3), 407–424.

Aryal, A., Raubenheimer, D., Sathyakumar, S., Poudel, B. S., Ji, W., Kunwar, K. J., … Brunton, D. (2012). Conservation strategy for brown bear and its habitat in Nepal. *Diversity, 4*(3), 301–317.

Carugi, C., & Bryant, H. (2020). A joint evaluation with lessons for the sustainable development goals era: The joint GEF-UNDP evaluation of the small grants programme. *American Journal of Evaluation, 41*(2), 182–200.

Catedrilla, L. C., Espectato, L. N., Serofia, G. D., & Jimenez, C. N. (2012). Fisheries law enforcement and compliance in District 1, Iloilo Province, Philippines. *Ocean & Coastal Management, 60*, 31–37.

Danh, V. T. (2008). *Household switching behavior in the use of ground water in the Mekong Delta, Vietnam*. Singapore: EEPSEA, IDRC Regional Office for Southeast and East Asia.

Detenber, B. H., Ho, S. S., Ong, A. H., & Lim, N. W. (2018). Complementary versus competitive framing effects in the context of pro-environmental attitudes and behaviors. *Science Communication, 40*(2), 173–198.

Dillon, J., & Teamey, K. (2002). Reconceptualizing environmental education: Taking account of reality. *Canadian Journal of Science, Mathematics and Technology Education, 2*(4), 467–483.

Dombrowski, K. (2010). Filling the gap? An analysis of non-governmental organizations responses to participation and representation deficits in global climate governance. *International Environmental Agreements: Politics, Law and Economics, 10*(4), 397–416.

Dressler, W., Büscher, B., Schoon, M., Brockington, D. A. N., Hayes, T., Kull, C. A., … Shrestha, K. (2010). From hope to crisis and back again? A critical history of the global CBNRM narrative. *Environmental Conservation, 37*(1), 5–15.

Dutt, B., Garg, K. C., & Bhatta, A. (2013). A quantitative assessment of the articles on environmental issues published in English-language Indian dailies. *Annals of Library and Information Studies (ALIS), 60*(3), 219–226.
Economy and Environmental Program for Southest Asia (EEPSEA). (1998). *The economic valuation of tropical forest land use options: A manual for researcher*. Singapore: EEPSEA.
El Jurdi, H. A., Batat, W., & Jafari, A. (2017). Harnessing the power of religion: Broadening sustainability research and practice in the advancement of ecology. *Journal of Macromarketing, 37*(1), 7–24.
Farooqi, A., & Fatimah, H. (2010). Historical perspective of environment education and its objectives in Pakistan. *Science Technology and Development, 21*(3), 1–7.
Fraas, A. G., & Egorenkov, A. (2015). A Retrospective Study of EPA's Rules Setting Best Available Technology Limits for Toxic Discharges to Water Under the Clean Water Act. *Resources for the Future Discussion Paper*, 15–41.
Gaioni, D. T. (1989). Ecophilia and ecocide: The struggle for the Philippine environment. *Philippine Studies*, 345–356.
Gamel, J., Menrad, K., & Decker, T. (2017). Which factors influence retail investors' attitudes towards investments in renewable energies? *Sustainable Production and Consumption, 12*, 90–103.
Gigliotti, L. M. (1990). Environmental education: What went wrong? What can be done? *The Journal of Environmental Education, 22*(1), 9–12.
Graci, S., & Dodds, R. (2008). Why go green? The business case for environmental commitment in the Canadian hotel industry. *Anatolia, 19*(2), 251–270.
Haigh, M. J. (2006). Promoting environmental education for sustainable development: The value of links between higher education and non-governmental organizations (NGOs). *Journal of Geography in Higher Education, 30*(2), 327–349.
Hanaoka, S., & Regmi, M. B. (2011). Promoting intermodal freight transport through the development of dry ports in Asia: An environmental perspective. *Iatss Research, 35*(1), 16–23.
Handl, G. (1998). The legal mandate of multilateral development banks as agents for change toward sustainable development. *American Journal of International Law, 92*(4), 642–665.
Hays, S. P., Esler, M., & Hays, C. E. (1996). Environmental commitment among the states: Integrating alternative approaches to state environmental policy. *Publius: The Journal of Federalism, 26*(2), 41–58.
Heller, P. (1996). Social capital as a product of class mobilization and state intervention: Industrial workers in Kerala, India. *World Development, 24*(6), 1055–1071.
Heller, P. (2001). Moving the state: The politics of democratic decentralization in Kerala, South Africa, and Porto Alegre. *Politics and Society, 29*(1), 131–163.
Imran, M., Akhtar, S., Chen, Y., & Ahmad, S. (2021). Environmental education and women: Voices from Pakistan. *SAGE Open, 11*(2), 21582440211009469.
James, A. N., Gaston, K. J., & Balmford, A. (1999). Balancing the Earth's accounts. *Nature, 401*(6751), 323–324.
Jickling, B., & Wals, A. E. (2008). Globalization and environmental education: Looking beyond sustainable development. *Journal of Curriculum Studies, 40*(1), 1–21.
Kang, K. (2019). On the problem of the justification of river rights. *Water International, 44*(6–7), 667–683.
Kopnina, H., & Cherniak, B. (2015). Cultivating a value for non-human interests through the convergence of animal welfare, animal rights, and deep ecology in environmental education. *Education Sciences, 5*(4), 363–379.
Kuasirikun, N. (2005). Attitudes to the development and implementation of social and environmental accounting in Thailand. *Critical Perspectives on Accounting, 16*(8), 1035–1057.
Lin, S., Man, Y. B., Chow, K. L., Zheng, C., & Wong, M. H. (2020). Impacts of the influx of e-waste into Hong Kong after China has tightened up entry regulations. *Critical Reviews in Environmental Science and Technology, 50*(2), 105–134.
McCrea-Strub, A., Zeller, D., Sumaila, U. R., Nelson, J., Balmford, A., & Pauly, D. (2011). Understanding the cost of establishing marine protected areas. *Marine Policy, 35*(1), 1–9.
Murphy, K. (2012). The social pillar of sustainable development: A literature review and framework for policy analysis. *Sustainability: Science, Practice and Policy, 8*(1), 15–29.
Neumayer, E. (2002). Do democracies exhibit stronger international environmental commitment? A cross-country analysis. *Journal of Peace Research, 39*(2), 139–164.
Nugroho, A. A. (2019). Analisis Putusan PTUN NO. 7/G/LH/2019/PTUN. DNA Antara Walhi Melawan Gubernur Aceh Atas Penerbitan Izin Pinjam Pakai Kawasan Hutan untuk Pembangunan PLTA Tampur [A Jurisdictive Analysis of PTUN NO. 7/G/LH/2019/PTUN. BNA between WALHI (Friends of the Earth Indonesia) against Aceh Governor over Issue of Permit Land Use of Forested Area for Tampur Water-Generated Electric Powerhouse]. *Jurnal Hukum Lingkungan Indonesia, 6*(1), 126–144.
Osman, K., & Pudin, S. (2009). The adults non-formal environmental education (EE): A scenario in Sabah, Malaysia. *Procedia-Social and Behavioral Sciences, 1*(1), 2306–2311.
Panapasa, G., & Singh, S. (2018). Pacific media under siege: A review of the PINA 2018 Summit. *Pacific Journalism Review, 24*(2), 135–145.
Permana, R. C. E., Nasution, I. P., & Gunawijaya, J. (2011). Kearifan lokal tentang mitigasi bencana pada masyarakat Baduy [Local Wisdom for Disaster

Mitigation amongst Baduy Community]. *Makara Human Behavior Studies in Asia, 15*(1), 67–76.
Prabhash, J. (2004). State and public policy in Kerala: An overview. *The Indian Journal of Political Science*, 403–418.
Prévot, A. C., Clayton, S., & Mathevet, R. (2018). The relationship of childhood upbringing and university degree program to environmental identity: Experience in nature matters. *Environmental Education Research, 24*(2), 263–279.
Rakai, M. E., & Williamson, I. P. (1995). Implementing LIS/GIS from a customary land tenure perspective: The Fiji experience. *Australian Surveyor, 40*(2), 112–121.
Richstad, J., McMillan, M., & Bowen, J. (2021). *Mass communication and journalism in the Pacific Islands*. Hawai'i: University of Hawaii Press.
Roy, D. (2018). Review of Centre for Science and Environment. 2017. Annual State of India's Environment—SOE 2017. *Ecology, Economy and Society-the INSEE Journal, 1*(2354-2020-1076), 105–107.
Schlosberg, D., & Carruthers, D. (2010). Indigenous struggles, environmental justice, and community capabilities. *Global Environmental Politics, 10*(4), 12–35.
Segovia, V. M., & Galang, A. P. (2002). Sustainable development in higher education in the Philippines: The case of Miriam College. *Higher Education Policy, 15*(2), 187–195.
Seiffert, M. (1991). Environmental education, religion and moral education. *Journal of Christian Education, 1*, 41–52.
Sherkat, D. E., & Ellison, C. G. (2007). Structuring the religion-environment connection: Identifying religious influences on environmental concern and activism. *Journal for the Scientific Study of Religion, 46*(1), 71–85.
Smaldone, D., & Dey, S. E. (2010). Developing a successful state-level environmental education organization: A nationwide assessment. *Applied Environmental Education and Communication, 9*(3), 159–172.
Smedby, N., & Neij, L. (2013). Experiences in urban governance for sustainability: The constructive dialogue in Swedish municipalities. *Journal of Cleaner Production, 50*, 148–158.
Spiro, P. J. (1995). New global communities: Nongovernmental organizations in international decision making institutions. *Washington Quarterly, 18*(1), 45–56.
Sureda, J., & Calvo, A. M. (2001). Organization of school centres and environmental education: In search of action models for the greening of school. *The Environmentalist, 21*(4), 287–296.
Swan, J. A. (1978). Environmental education: A new religion? *The Journal of Environmental Education, 10*(1), 44–48.
Tilbury, D., & Wortman, D. (2004). *Engaging people in sustainability*. IUCN.
Tolkach, D., Chon, K. K., & Xiao, H. (2016). Asia Pacific tourism trends: Is the future ours to see? *Asia Pacific Journal of Tourism Research, 21*(10), 1071–1084.
Tongcumpou, C., & Harvey, N. (1994). Implications of recent EIA changes in Thailand. *Environmental Impact Assessment Review, 14*(4), 271–294.
Turner, S. D. (2009). *Strategic review of the IUCN programme on protected areas*. Gland: IUCN.
Van Damme, L. S. M., & Neluvhalani, E. F. (2004). Indigenous knowledge in environmental education processes: Perspectives on a growing research arena. *Environmental Education Research, 10*(3), 353–370.
Wilkinson, C. F., & Lamb, J. C., IV. (1999). The potential health effects of phthalate esters in children's toys: A review and risk assessment. *Regulatory Toxicology and Pharmacology, 30*(2), 140–155.
Wonham, J. (1998). Agenda 21 and sea-based pollution: Apportunity or apathy? *Marine Policy, 22*(4–5), 375–391.
Yani, R. F., Asrinaldi, A., & Rahmadi, D. (2019). Peran WALHI Sumatra Barat Dalam Investigasi Tambang Emas Ilegal di Kota Padang. *Jurnal Demokrasi dan Politik Lokal, 1*(1), 88–100.
Yusuf, A. A. (2008). *The distributional impact of environmental policy: The case of carbon tax and energy pricing reform in Indonesia*. Singapore: Environment and Economy Program for Southeast Asia, Research Report, (2008-RR1).
Zamora, O. B. (2009). Sustainable Agriculture Education and Research at the University of the Philippines Los Banos: Status, challenges, and needs. *Journal of Developments in Sustainable Agriculture, 4*(1), 41–49.

Environmental Literacy

▶ Pre-schoolers and Sustainable Urban Transport

Environmental Movement

▶ Senegalese Ecovillage Network

Environmental Psychology

▶ Residential Crowding in Urban Environments

Environmental Sustainability

▶ Sustainable Development Goals

Epidemiological Shifts in Urban Bangladesh

Mahjabin Kabir Adrita and Raisa Sultana
Department of Geography and Environment, University of Dhaka, Dhaka, Bangladesh

Synonyms

Communicable diseases (infectious diseases, contagious diseases); Non-communicable diseases (non-infectious diseases)

Definitions

Epidemiology: The term epidemiology has been used as the study of determinants, distribution, and occurrence of incidences related to diseases in a specific population group (Brachman 2011) which encompasses both infectious and non-infectious diseases.

Epidemiological trends: This can refer to various aspects of epidemiology, but in this review this term will only be confined to secular trends of epidemiology which indicates the occurrence of diseases over a period of time, usually years; seasonal trends indicating disease incidence shift due to environmental changes and epidemic trends (Brachman 2011).

Communicable or infectious diseases (CDs): These are diseases that are contagious, spreading via human contact (touch, fecal, oral, food, sexual activity, etc.), water, air, insect bites, animal contact, sexual activities, etc.

Non-communicable diseases (NCDs): Also referred as chronic diseases, these are not contagious and are of long duration, caused by a combination of environmental, behavioral, psychological, and genetic factors.

Risk factors: These are characteristics associated with lifestyle and living conditions that promote disease incidence.

Diarrheal diseases: These are caused by various microorganisms and share the common symptom of diarrhea.

Vaccine-preventable diseases (VPDs): These are diseases that can be prevented via immunization.

Vector-borne diseases: These are diseases that spread through humans via blood-feeding arthropods.

Immerging and re-emerging diseases: These are diseases that are newly discovered and/or were not observed before in the study area.

Other diseases: These refer to diseases that spread via blood transfusion and/or venereal activity.

E

Introduction

Urban Bangladesh and Increasing Health Risks

Bangladesh has seen rapid growth in economy in the past decade. This economic growth was mostly due to the contribution of rapid urbanization and industrialization. More and more people are drawn to urban areas for better opportunities. But this prosperity came at a price. As the population of the cities increased, problems that were unique to the urban areas, such as high population densities, poverty and slum creation, congestion, environmental pollution, etc., became evident. These problems gave rise to health issues that were unique as well. This was mostly because the unplanned cities could not ensure proper nutrition sanitation, proper housing facilities, and inclusive healthcare.

Bangladesh has been plagued by numerous emerging and re-emerging diseases throughout the years. Being at the center of human traffic, the urban population has always been a hub for these maladies. With the increase of this densely packed bustling population group, there have been concomitant alterations in urban epidemiological determinants and frequencies. While the urban population is beset with various infectious diseases, affliction rates for noninfectious diseases have also gained momentum due to prolonged exposure to the urban lifestyle and living conditions. The assessment of these public health problems and outcomes is of significance in predicting future patterns to improve healthcare surveillance and control. This literature review seeks to understand the epidemiological trend in the urban areas and address the changes in the conditions throughout the years.

Communicable and Non-communicable Diseases: Risk Factors

The urban population of Bangladesh has seen a fair share of communicable and non-communicable diseases throughout the years. There are many risk factors behind diseases. While some risk factors are more associated with NCDs and others with CDs, they aren't confined within this classification. Any risk factor (both for NCD and CD) could jeopardize life in any kind of diseases. In the case of NCDs, it is more associated with risk factors such as modifiable behavioral risk factors, socioeconomic risk factors, and metabolic risk factors which relate to water, sanitation, hygiene, nutrition, access to healthcare, etc.

Risk factors such as malnutrition or high blood pressure can lead to deaths much like CDs and NCDs but are not a disease itself, for example, in cardiovascular diseases high blood pressure is a metabolic risk factor and a health condition, but it is not considered an NCD (Fig. 1).

Methodology

This review seeks to know what the current disease trends are and whether they have altered from the past tense and if so how. To answer these questions, many literature were sourced by using keywords such as epidemiology, urban, Bangladesh infectious diseases, non-infectious disease, communicable diseases, non-communicable diseases trends, etc. on Google scholar, ScienceDirect, and ResearchGate. There was a secondary search using the disease name such as keywords like dengue, cardiovascular disease, pneumonia, etc. along with the initial search engines; in this search, government and non-government disease databases, i.e., DGHS, BBS, IEDCR, UN IGME, WHO, etc. were also searched for relevant data.

Along with these searches, sub-terms arose relating to the study. These sub-terms were then collected from the titles and abstracts of the articles. These expressed details about the diseases such as specific trend, patterns, determinants, frequencies, public perceptions, etc.

The documents were chosen as such that they were relevant to the research question and were from the public health point of view, relating to distribution, determinants, and frequency of diseases. For this purpose the titles and abstracts were screened. The documents relating medical science were excluded. Since this review aims to understand trends of the past and predict the future, the documents were not limited to any time frame. However, some parts of the data

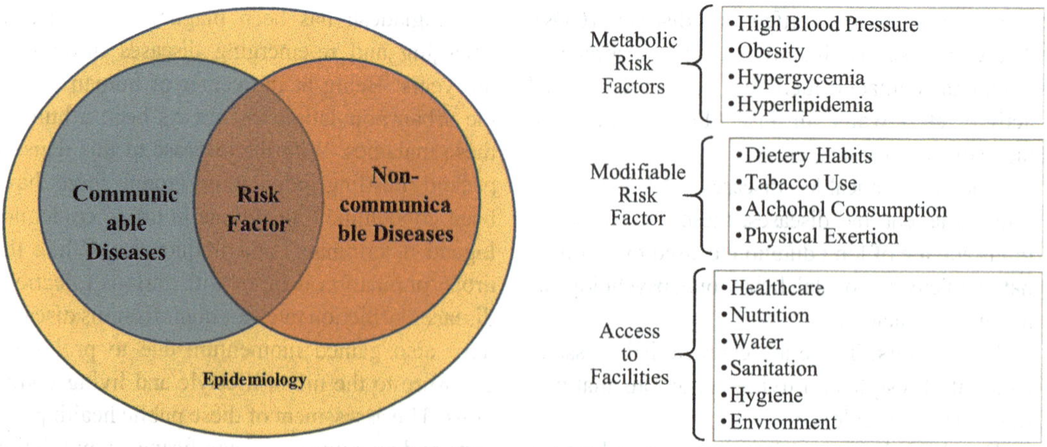

Epidemiological Shifts in Urban Bangladesh, Fig. 1 The relationship between epidemiology, types of diseases, and associated risk factors

were analyzed keeping the time frame of 8 years because the last decade has witnessed some major changes in county's health sector, also depending on the availability of the data sources. By following this method, 35 research articles and 15 reports were identified that solely related to the disease trends that transpired in Bangladesh, if not the urban part of Bangladesh (Fig. 2).

Concurrent Disease Trends

Through analyzing data and research work, it is evident that some maladies are unique to the urban area, and often times these centers are the source of disease proliferation. This is due to the structure and nature of the area and the busting lifestyle and demographic characteristics of the residents.

Communicable Diseases in Urban Bangladesh

There are many communicable diseases found throughout the world but not all of them are prominent in Bangladesh. And among these diseases, not all are found in urban areas. Some of these are endemic such as malaria which is prominent in the southeast and northeast of Bangladesh; some of these are prominent in rural areas such as kala-azar or rabies. The urban areas in particular have seen high affliction rates of a variety of diseases shown in Table 1.

VPDs and Diarrheal Diseases

In the 1970s and 1980s, a vast majority of child death were comprised of vaccine-preventable diseases (VPDs). To contain such high death rates, the government took on national immunization programs among 12–23-month-old children, from 1974 (Jamil et al. 1999) for VPDs. On the other hand, there were government and non-government organization efforts for combating diarrheal diseases through implementing sanitary practices, promoting of oral rehydration solutions, and improving female perception on the matter at scale (Billah et al. 2019). Through consistent implementation of these initiatives throughout the years, the rates have gone down significantly. The mortality rates have gone down from 260.2 per thousand live births to 30.8 per thousand live births; however the deaths amounted to 0.1 million in 2016, but the affliction rates of these diseases remained stable (UN IGME 2017a, b; 2020).

While assessing why the deaths due to communicable diseases had dropped, it was seen that Bangladesh achieved above 90% vaccine coverage for VPDs throughout the span of 1980–2018 (Fig. 3). It displays the coverage of BCG vaccine for tuberculosis; DTP vaccine for diphtheria, pertussis, and tetanus; OPV vaccine for polio; and MCV vaccine for meningococcal diseases.

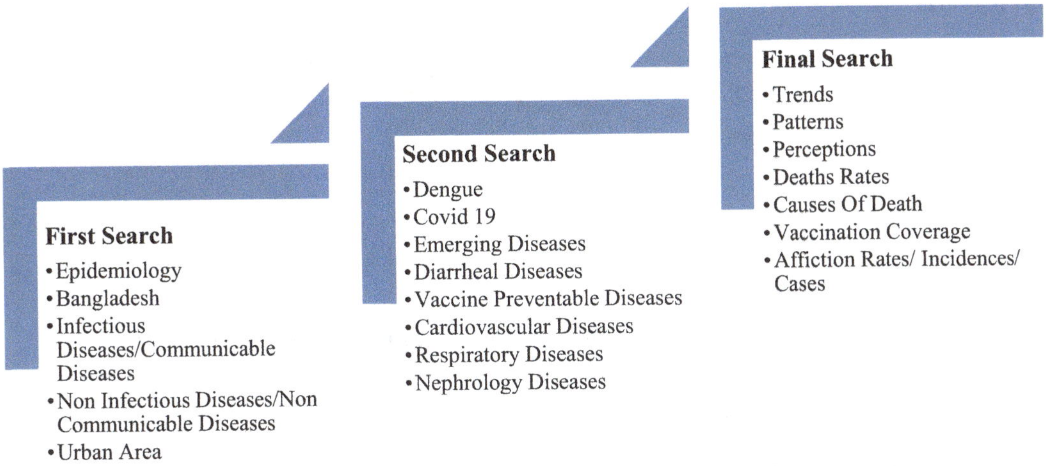

First Search
• Epidemiology
• Bangladesh
• Infectious Diseases/Communicable Diseases
• Non Infectious Diseases/Non Communicable Diseases
• Urban Area
Second Search
• Dengue
• Covid 19
• Emerging Diseases
• Diarrheal Diseases
• Vaccine Preventable Diseases
• Cardiovascular Diseases
• Respiratory Diseases
• Nephrology Diseases
Final Search
• Trends
• Patterns
• Perceptions
• Deaths Rates
• Causes Of Death
• Vaccination Coverage
• Affiction Rates/ Incidences/ Cases

Epidemiological Shifts in Urban Bangladesh, Fig. 2 Search strings

Vector-Borne Diseases

While the most troublesome diseases for Bangladesh were contained through vaccination programs and awareness campaigns, some infections remained uncontrollable. The most prominent of them are dengue incidences, with the first major outbreak in the year 2000 (Hossain et al. 2003). Since then there were some fluctuations in hospitalized confirmed cases, for example, 6232 in 2002, 3934 in 2004, 3162 in 2015, 6060 in 2016, and 10,148 in 2018 (Mamun et al. 2019).

Epidemiological Shifts in Urban Bangladesh, Table 1 Prominent communicable diseases in urban Bangladesh

Category	Disease name
Diarrheal diseases	Diarrhea, cholera, typhoid, dysentery
Vaccine-preventable diseases (VPDs)	Tuberculosis, measles, chicken pox, whooping cough (pertussis), tetanus, diphtheria, polio, meningococcal diseases, pneumonia, hepatitis
Vector-borne diseases	Dengue, chikungunya, malaria
Emerging and re-emerging diseases	Nipah virus, SARS, chikungunya virus (CHIKV), H1N1 virus (swine flu), and COVID-19
Others	HIV-AIDs, STDs

Source: Developed by authors, 2021

Dhaka had the highest number of dengue cases between 2012 and 2019 (Banu et al. 2012; Sharmin et al. 2018). In 2019, there were 100,201 hospitalized patients, a ninefold increase rate from 2018 (Hsan et al. 2019; Wilder-Smith et al. 2019). One study showed that in 2019, up until July prior to Eid (major religious festival for Muslims), 81% of all dengue cases were localized in Dhaka city. But the dengue incidence in Dhaka came down to 57% 1 week prior to Eid when people began to travel to their hometowns from the capital to different parts of the country, and dengue incidences increased to 43% across the country. This trend continued in the upcoming weeks as well, with the dengue cases outside Dhaka rising to almost 55% (Hossain et al. 2020a, b).

From this scenario it is evident that such occurrences coincide with mass mobility events which escalate infection rates. The simultaneous conglomeration and long-distance movement of people stimulate affliction (Hermes 2018; Eritja et al. 2017). The 2017 chikungunya outbreak also conforms to such grounds (Eritja et al. 2017), because they both are *Aedes*-borne and their geographical hotspots overlap (Dzul-Manzanilla et al. 2021). However, in case of malaria, the transmission confines only in a clustered forms or in hotspots unlike chikungunya and dengue, though the

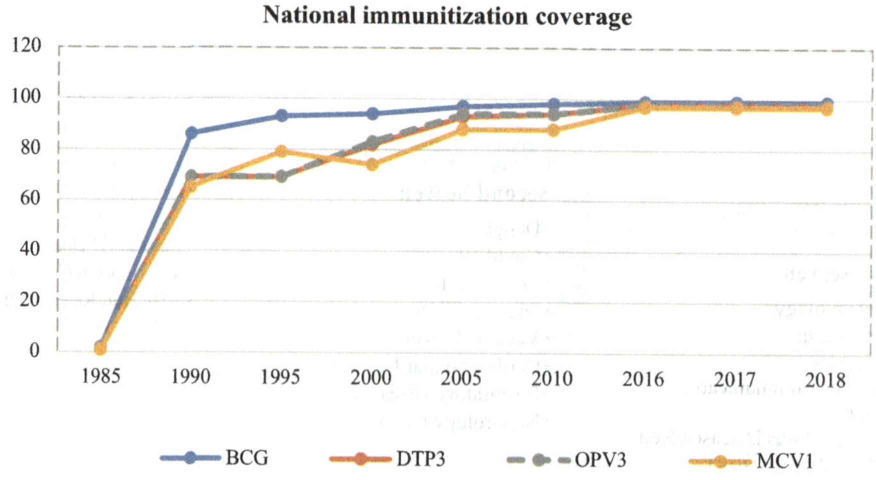

Epidemiological Shifts in Urban Bangladesh, Fig. 3 National immunization coverage during 1985–2018. (Data source: WHO 2019)

disease also takes devastating forms as seasonal disease and claims thousands of deaths (especially children) every year (Noé et al. 2018).

Emerging Diseases and COVID-19

The urban population has also been the primary subject of emerging disease outbreaks. From the beginning of the twenty-first century, the country saw new disease strands and variants such as the Nipah virus in 2001, SARS in 2003, chikungunya virus (CHIKV) in 2008, H1N1 virus (swine flu) in 2009, and COVID-19 in 2020. Although affliction rates have been low for most of them, COVID-19 infections have been a cause for concern. It has been perceived that urban centers had been the major risk zones during the initial breakout, especially Dhaka and Chittagong (IEDCR 2020). The government promptly implemented strict lockdown and initiated awareness campaigns to handle the situation and then moved to a partial lockdown in the later part of 2020. The incidence rate of COVID-19 follows the pattern of dengue affliction patterns. While lax restrictions were one of the causes of the spike in COVID-19 cases, combining it with mass mobility events produced a peak COVID-19 affliction rates (Ghosh and Mollah 2020; Shammi et al. 2020).

Non-communicable Diseases (NCDs) in Urban Areas

As many contagious diseases keep rendering heavy death tolls for the younger generation, NCDs are often forgotten because they do not concur out of the blue and drastically. These diseases are caused by specific and prolonged habits, lifestyles, and environments which include high air pollution, limited scope of physical exertion, dietary habits, smoking, alcohol consumption, and overall environmental degradation (Saquib et al. 2012; Chowdhury et al. 2018; WHO 2010). Low-income and middle-income countries are seen to the urban dwellers to be especially susceptible to these illnesses because these characteristics are distinctive to the urban areas.

NCDs are the cause of approximately 71% deaths globally, and among these, deaths in lower middle-income and middle-income countries account for 77%. Bangladesh is no exception from this; NCDs have increased significantly in the twenty-first century (Saquib et al. 2012; Hussain and Sullivan 2013). The situation of NCD in Bangladesh is quite alarming as 75% of the population are exposed to two or more modifiable NCD risk factors, 5% of the adult population are diabetic, and 21% are hypertensive (M. M. Zaman et al. 2015). Deaths are more prevalent as well. In 2012, it was estimated that over two thirds of total deaths were due to NCDs which amounted to over 886,000 people (WHO 2014). Between 2009 and 2019, deaths due to different cardiovascular diseases have increased 20–50% and respiratory diseases have increased about 20% (GBD 2019 Diseases and Injuries Collaborators 2020). Specifically for the urban dwellers in Bangladesh, the death tolls for NCDs are higher than rural areas as seen from 2013 to 2020. And the predominant of these diseases are cardiovascular diseases (heart diseases, heart attack, stroke, etc.) which were over 37% on average in the past 8 years (Fig. 4). During this 8-year span, 55.8% of the deaths were caused by NCDs and related risk factors, and 6.86% deaths were caused by CDs. The breakdown of the NCDs is provided in Fig. 4. It shows the causes of deaths for all in average including the associated risk factors that can lead to it.

Analysis of Changing Trends

Upon glancing at published works, it seems that aside from dengue and the COVID-19 situation, most communicable disease incidence rates have dropped over the years. This is mostly due to government awareness campaigns and immunization programs. Nongovernment effort has also been substantial in this case. Vaccination programs and other interventions have been the key in controlling diseases in Bangladesh. Due to higher education levels and access to mass media, the people in urban areas have been the first to respond to these initiatives.

It is evident from Fig. 5 that there has truly been a shift in epidemiological trends in the past decade just by glancing at the graph. Deaths due to CDs have come well below 10% in the last decade. While deaths due to NCDs are above 60% and on the rise, in this analysis the associated

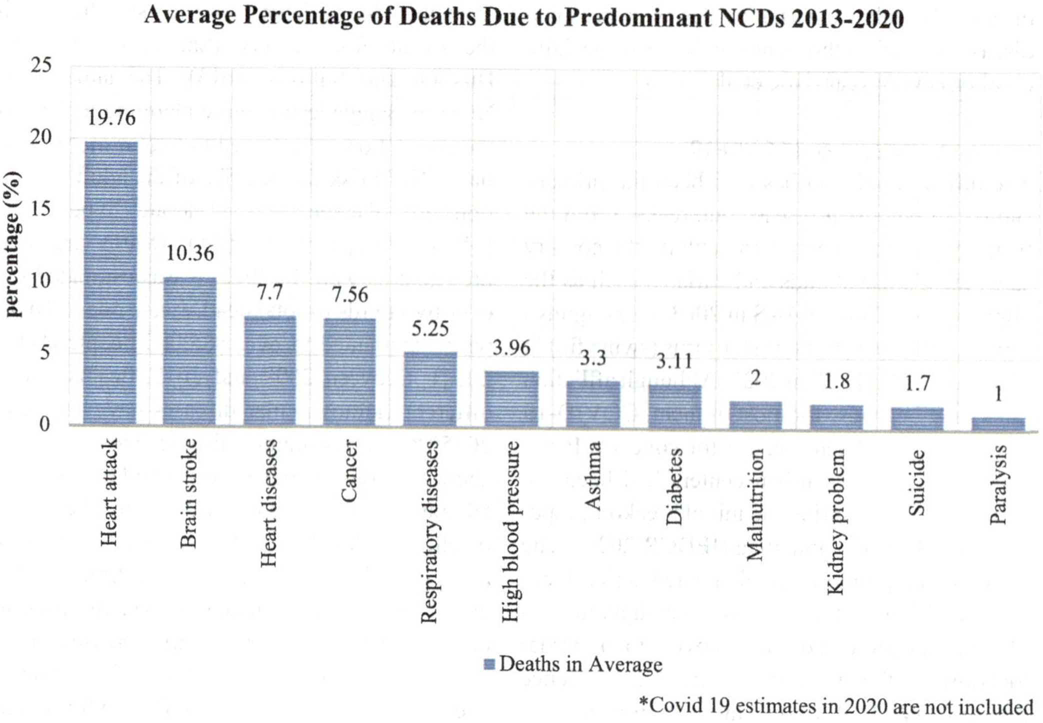

Epidemiological Shifts in Urban Bangladesh, Fig. 4 Average percentage of deaths due to predominant NCDs in urban areas of Bangladesh during 2013–2020. (Data source: BBS 2021, 2020, 2019, 2018, 2017, 2016, 2015, 2014)

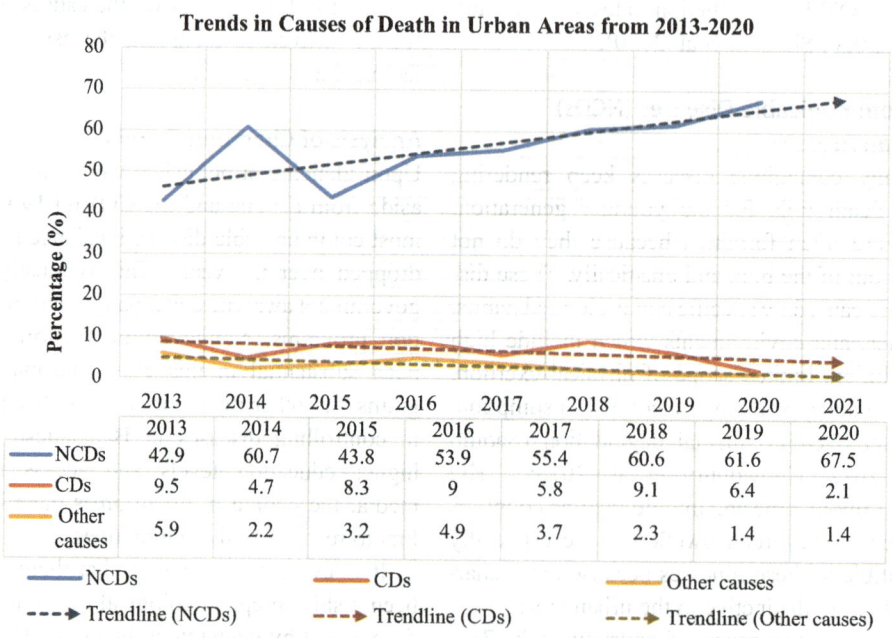

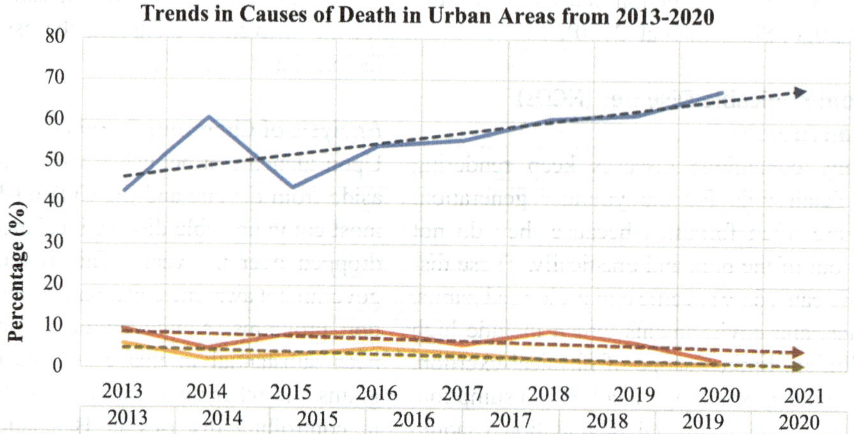

	2013	2014	2015	2016	2017	2018	2019	2020
NCDs	42.9	60.7	43.8	53.9	55.4	60.6	61.6	67.5
CDs	9.5	4.7	8.3	9	5.8	9.1	6.4	2.1
Other causes	5.9	2.2	3.2	4.9	3.7	2.3	1.4	1.4

NCDs CDs Other causes
Trendline (NCDs) Trendline (CDs) Trendline (Other causes)

*Covid 19 estimeates in 2020 are not included

Epidemiological Shifts in Urban Bangladesh, Fig. 5 Causes of death in urban areas of Bangladesh from 2013 to 2020. (Data source: BBS 2021, 2020, 2019, 2018, 2017, 2016, 2015, 2014) risk factors and other causes of death are included for a clearer comparison of causes of deaths in urban areas (Fig. 5).

Another aspect of these disease incidences is that the trend is not universal for all age groups. The data is represented in Fig. 5 in a generalization for all age groups. But if the trends for different age groups were considered, they would turn out quite different from each other. For example, in the case of children under 5 during the span of 2013–2020 (Fig. 6), they had high mortality rates due to pneumonia (above 31%). In the second and third position, there are malnutrition and respiratory diseases, but in comparison to pneumonia, the values are five times lower. Chickenpox (0.375%), cancer (0.3%), complex dysentery (0.23%), cholera (0.16%), whooping cough/pertussis (0.11%), epilepsy (0.16%), leprosy (0.18%), etc. are listed as other causes of death for children under 5 years. When compared to the data of elderly aged 60 and above, the major cause of death is cardiovascular diseases which in total range above 35%, including associated health risk factors. Not accounting for geriatric causes, the second and third causes were respiratory illnesses including asthma (over 10%) and different types of cancer predominantly stomach, liver, and blood cancer.

Discussions

The available data, literature, and resources strongly suggest that infectious diseases are mostly under control and there are no prominent outbreak of them in Bangladesh as urban areas had lower infection rates than rural areas (Zaman et al. 2012). This could be due to the fact that urban areas have higher immunization coverage. In a study it was seen that 88.5% of the respondents from urban areas had the full coverage of the vaccine doses which was higher than that of rural areas (Sarker et al. 2019). However this is not the entire picture; according to many sources, despite the containment by such programs the child mortality due to VPDs and diarrheal diseases is still not negligible. This is especially true for the urban poor. They are still seen suffering from diseases that are now almost absent in other economic classes. The statistics found from different sources

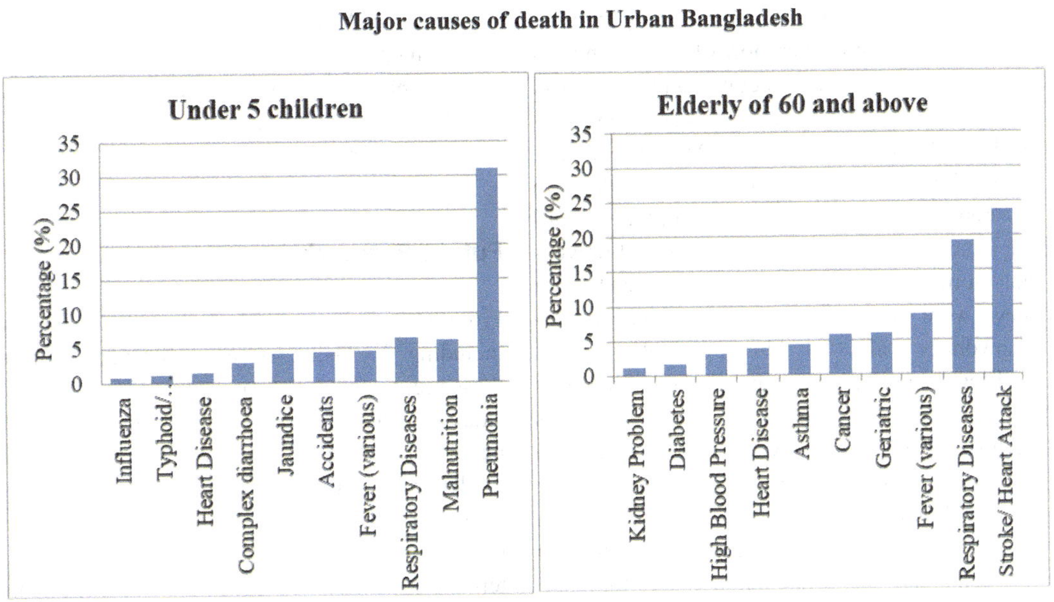

Epidemiological Shifts in Urban Bangladesh, Fig. 6 Top 10 causes of deaths among children under 5 and above 60 populations on average in urban areas of Bangladesh during 2013–2020. (Data source: BBS 2021, 2020, 2019, 2018, 2017, 2016, 2015, 2014)

do not conform to the statistics from studies done strictly on the urban poor. Diarrheal diseases and VPDs are still a major cause for concern among them. For example, one study showed high prevalence of tuberculosis in urban slums (Banu et al. 2013). They also have the tendency to not take recurring symptoms seriously and misattribute them as common ailments (i.e., common cold, coughs, or headaches) (Tune et al. 2020). A study in Dhaka showed that about 80% people were unlikely to seek professional help regarding these matters initially (Chowdhury et al. 2015).

This case of negligence was also noticeable in the case of the COVID-19 pandemic, only it wasn't limited to any economic class. Studies showed that despite urban residents being highly fearful and having more awareness regarding preventative measures such as social distancing, sanitation, and hygiene practices, they had disregard for these practices (Hossain et al. 2020a, b). These practices were not ubiquitously spread throughout areas or social classes or even among genders but are more common among districts neighboring Dhaka and Chittagong Hill Tracts (Hosen et al. 2021).

The opposite scenario is seen for dengue incidences. The general perception of people on these types of diseases tends to differ vastly. Studies showed that nearly half or more of the respondents were unaware of the breeding and ecological aspects of the mosquitoes (Dhar-Chowdhury et al. 2014; Bashar et al. 2020; Abir et al. 2021). Approximately, two thirds of the respondents in different studies had an acceptable attitude toward the issue and engage in practices toward prevention. Some didn't implement practice even though they were aware of complications (Dhar-Chowdhury et al. 2014; Abir et al. 2021).

There is a prominent trend on the entire statistics analysis that in the case of all age groups deaths due to respiratory problems have been prominent and the mortality rates are rising steadily. In many studies it is seen that this pattern is linked to the increased amount of air pollution in urban areas. One of the major causes for respiratory problems is PM 2.5 which is very high in urban areas (Rouf et al. 1970). Even the case of pneumonia incidences is linked with indoor air pollution and long-term exposure to ambient air pollution (Adaji et al. 2019; Ji et al. 2017). Therefore air pollution is a risk factor for not only communicable diseases but also non-communicable diseases as well. Along with the high incidences of cardiovascular diseases and the rising rates of respiratory problems, the urban population is headed toward a transition in epidemiology, for which the healthcare system needs to prepare because these diseases are long lasting, covert, and expensive.

The trends have started leaning toward NCDs from the beginning of twenty-first century, and analyzing the data from the last 8 years, NCDs have already become the major cause of deaths in urban areas in the older generation and are dominant among young adults as well. These diseases are gradually becoming evident among infants and children as well. It could be surmised that while the trends do indicate a shift in epidemiology of Bangladesh, the infectious diseases cannot simply be overlooked. The measures and practices that have contained these maladies have to be more extensive and thorough and constantly kept under high supervision because any hindsight could result in a prominent outbreak given the practices and outlook of the urban dwellers in Bangladesh. At the same time, it needs to be environment friendly and healthy (relating to diet and physical activity). These practices combined allow healthy lifestyle which lessen the possibility of NCDs.

Future Research

From studies it is evident that the maladies, including NCDs, have never originated from these regions but have spread there. There are no studies regarding the pattern of this spread or people's responses. Therefore it becomes difficult to determine the actual trend of epidemiology from just morbidity statistics. The studies related to epidemiology in Bangladesh are mostly centered on dengue, VPDs, and NCDs, literature found on other diseases is few and far apart in time. For example, researches on STDs (sexually transmitted diseases), HIV, measles, chicken pox, or rubella are still scarce. Little research has been done on the environmental impacts on disease incidence rates. In the case of NCDs, there are nationwide studies done on the practices that lead to higher risk of these complexities, but similar to CDs there are not many specific studies done linking environmental changes to NCDs. Furthermore, there are many resources available on trends of disease occurrences in Dhaka, but the same cannot be said about other urban areas of the country. Dhaka has characteristics which are unique to other urban areas of the country. Therefore the trends in this region may not be commensurate with others. The data regarding other regions are generalizations done from nationwide studies. Hence, there is room for further research which will make the literature pool even more substantial.

Conclusion

Communicable diseases were a huge burden for urban Bangladesh during the 1970–1980s. While these maladies where successfully combated through government and nongovernment efforts, non-communicable diseases became another cause for concern when they started slowly crippling up since the 2000s. From analyzing the data from government cross-sectional survey reports and other associated resources, it is evident that in 2020, NCDs have become a major cause of death in urban areas. 67.5% of the deaths in urban Bangladesh is caused by NCDs. The most prominent of these are cardiovascular diseases (over 37%). Though death is relatively low among children, among adults it is highly prevalent and life threatening. Chronic respiratory diseases are on the rise and is deadly for both children and adults. This sudden increase in deaths is due to poor lifestyle, environmental degradation, improper living conditions, and indulgence in associated risk factors.

Cross-References

▶ Urban Health

References

Abir, T., Ekwudu, O., Kalimullah, N. A., Nur-A Yazdani, D. M., Al Mamun, A., Basak, P., & Agho, K. E. (2021). Dengue in Dhaka, Bangladesh: Hospital-based cross-sectional KAP assessment at Dhaka North and Dhaka South City Corporation area. *PLoS One, 16*(3), e0249135.

Adaji, E. E., Ekezie, W., Clifford, M., & Phalkey, R. (2019). Understanding the effect of indoor air pollution on pneumonia in children under 5 in low- and middle-income countries: A systematic review of evidence. *Environmental Science and Pollution Research International, 26*(4), 3208–3225.

Banu, S., Hu, W., Hurst, C., Guo, Y., Islam, M. Z., & Tong, S. (2012). Space-time clusters of dengue fever in Bangladesh: Space-time clusters of dengue fever. *Tropical Medicine & International Health: TM & IH, 17*(9), 1086–1091.

Banu, S., Rahman, M. T., Uddin, M. K. M., Khatun, R., Ahmed, T., Rahman, M. M., & Van Leth, F. (2013). Epidemiology of tuberculosis in an urban slum of Dhaka city, Bangladesh. *PLoS One, 8*(10), e77721.

Bashar, K., Mahmud, S., Asaduzzaman, T., & E. A., & Zaman, A. B. (2020). Knowledge and beliefs of the city dwellers regarding dengue transmission and their relationship with prevention practices in Dhaka city, Bangladesh. *Public Health in Practice, 1*(100051), 100051.

BBS. (2014). Sample Vital Registration System (SVRS) 2013, Bangladesh Bureau of Statistics. Statistics and Informatics Division (SID), Ministry of Planning, 54–56.

BBS. (2015). Bangladesh Sample Vital Statistics 2014. Bangladesh Bureau of Statistics. Statistics and Informatics Division (SID), Ministry of Planning, 52–54.

BBS. (2016). Bangladesh Sample Vital Statistics 2015. Bangladesh Bureau of Statistics. Statistics and Informatics Division (SID), Ministry of Planning, 53–55.

BBS. (2017). Bangladesh Sample Vital Statistics 2016. Bangladesh Bureau of Statistics. Statistics and Informatics Division (SID), Ministry of Planning, 59–61.

BBS. (2018). Bangladesh Sample Vital Statistics 2017. Bangladesh Bureau of Statistics. Statistics and Informatics Division (SID), Ministry of Planning, 58–60.

BBS. (2019). Bangladesh Sample Vital Statistics 2018. Bangladesh Bureau of Statistics. Statistics and Informatics Division (SID). Ministry of Planning, 65–67.

BBS. (2020). Report on Sample Vital Registration System (SVRS) 2019. Bangladesh Bureau of Statistics. Statistics and Informatics Division (SID). Ministry of Planning, 67–69.

BBS. (2021). Report on Sample Vital Registration System (SVRS) 2020. Bangladesh Bureau of Statistics. Statistics and Informatics Division (SID). Ministry of Planning, 71–73.

Billah, S. M., Raihana, S., Ali, N. B., Iqbal, A., Rahman, M. M., Khan, A. N. S., & El Arifeen, S. (2019).

Bangladesh: A success case in combating childhood diarrhoea. *Journal of Global Health, 9*(2), 020803.

Brachman, P. S. (2011). Epidemiology. In S. Baron (Ed.), *Medical microbiology*. Galveston (TX): University of Texas Medical Branch at Galveston.

Chowdhury, F., Khan, I. A., Patel, S., Siddiq, A. U., Saha, N. C., Khan, A. I., & Ali, M. (2015). Diarrheal illness and healthcare seeking behavior among a population at high risk for diarrhea in Dhaka, Bangladesh. *PLoS One, 10*(6), e0130105.

Chowdhury, M. Z. I., Haque, M. A., Farhana, Z., Anik, A. M., Chowdhury, A. H., Haque, S. M., & Turin, T. C. (2018). Prevalence of cardiovascular disease among Bangladeshi adult population: A systematic review and meta-analysis of the studies. *Vascular Health and Risk Management, 14*, 165–181.

Dhar-Chowdhury, P., EmdadHaque, C., Michelle Driedger, S., & Hossain, S. (2014). Community perspectives on dengue transmission in the city of Dhaka, Bangladesh. *International Health, 6*(4), 306–316.

Dzul-Manzanilla, F., Correa-Morales, F., Che-Mendoza, A., Palacio-Vargas, J., Sánchez-Tejeda, G., González-Roldan, J. F., & Vazquez-Prokopec, G. M. (2021). Identifying urban hotspots of dengue, chikungunya, and Zika transmission in Mexico to support risk stratification efforts: A spatial analysis. *The Lancet Planetary Health, 5*(5), e277–e285.

Eritja, R., Palmer, J. R. B., Roiz, D., Sanpera-Calbet, I., & Bartumeus, F. (2017). Direct evidence of adult aedesalbopictus dispersal by car. *Scientific Reports, 7*(1), 14399.

GBD 2019 Diseases and Injuries Collaborators. (2020). Global burden of 369 diseases and injuries in 204 countries and territories, 1990–2019: A systematic analysis for the Global Burden of Disease Study 2019. *Lancet, 396*(10258), 1204–1222.

Ghosh, P., & Mollah, M. M. (2020). The risk of public mobility from hotspots of COVID-19 during travel restriction in Bangladesh. *Journal of Infection in Developing Countries, 14*(7), 732–736.

Hermes. (2018, June 15). *Eid exodus: Asia's Muslims on the move at the end of Ramadan, Asia*. Retrieved August 1, 2021, from Straitstimes.com website: https://www.straitstimes.com/asia/eid-exodus-asias-muslims-on-the-move-at-the-end-of-ramadan

Hosen, I., Pakpour, A. H., Sakib, N., Hussain, N., Al Mamun, F., & Mamun, M. A. (2021). Knowledge and preventive behaviors regarding COVID-19 in Bangladesh: A nationwide distribution. *PLoS One, 16*(5), e0251151.

Hossain, M. A., Khatun, M., Arjumand, F., Nisaluk, A., & Breiman, R. F. (2003). Serologic evidence of dengue infection before onset of epidemic, Bangladesh. *Emerging Infectious Diseases, 9*(11), 1411–1414.

Hossain, M. S., Siddiqee, M. H., Siddiqi, U. R., Raheem, E., Akter, R., & Hu, W. (2020a). Dengue in a crowded megacity: Lessons learnt from 2019 outbreak in Dhaka, Bangladesh. *PLoS Neglected Tropical Diseases, 14*(8), e0008349.

Hossain, M. A., Jahid, M. I. K., Hossain, K. M. A., Walton, L. M., Uddin, Z., Haque, M. O., & Hossain, Z. (2020b). Knowledge, attitudes, and fear of COVID-19 during the Rapid Rise Period in Bangladesh. *PLoS One, 15*(9), e0239646.

Hsan, K., Hossain, M. M., Sarwar, M. S., Wilder-Smith, A., & Gozal, D. (2019). Unprecedented rise in dengue outbreaks in Bangladesh. *The Lancet Infectious Diseases, 19*(12), 1287.

Hussain, S. A., & Sullivan, R. (2013). Cancer control in Bangladesh. *Japanese Journal of Clinical Oncology, 43*(12), 1159–1169.

IEDCR. (2020). *Bangladesh COVID-19 update*. Retrieved May 29, 2020, from https://www.iedcr.gov.bd/

Jamil, K., Bhuiya, A., Streatfield, K., & Chakrabarty, N. (1999). The immunization programme in Bangladesh: Impressive gains in coverage, but gaps remain. *Health Policy and Planning, 14*(1), 49–58.

Ji, W., Park, Y. R., Kim, H. R., Kim, H.-C., & Choi, C.-M. (2017). *Prolonged effect of air pollution on pneumonia: a nationwide cohort study*. Respiratory Infections. European Respiratory Society.

Mamun, M. A., Misti, J. M., Griffiths, M. D., & Gozal, D. (2019). The dengue epidemic in Bangladesh: Risk factors and actionable items. *Lancet, 394*(10215), 2149–2150.

Noé, A., Zaman, S. I., Rahman, M., Saha, A. K., Aktaruzzaman, M. M., & Maude, R. J. (2018). Mapping the stability of malaria hotspots in Bangladesh from 2013 to 2016. *Malaria Journal, 17*(1). https://doi.org/10.1186/s12936-018-2405-3.

Rouf, M. A., Nasiruddin, M., Hossain, A. M. S., & Islam, M. S. (1970). Trend of particulate matter PM 2.5 and PM 10 in Dhaka City. Bangladesh. *Journal of Scientific and Industrial Research, 46*(3), 389–398.

Saquib, N., Saquib, J., Ahmed, T., Khanam, M. A., & Cullen, M. R. (2012). Cardiovascular diseases and type 2 diabetes in Bangladesh: A systematic review and meta-analysis of studies between 1995 and 2010. *BMC Public Health, 12*(1), 434.

Sarker, A. R., Akram, R., Ali, N., & Sultana, M. (2019). Coverage and factors associated with full immunisation among children aged 12–59 months in Bangladesh: Insights from the nationwide cross-sectional demographic and health survey. *BMJ Open, 9*(7), e028020.

Shammi, M., Bodrud-Doza, M., Islam, A. R. M. T., & Rahman, M. M. (2020). Strategic assessment of COVID-19 pandemic in Bangladesh: Comparative lockdown scenario analysis, public perception, and management for sustainability. *Environment Development and Sustainability, 23*(4), 1–44.

Sharmin, S., Glass, K., Viennet, E., & Harley, D. (2018). Geostatistical mapping of the seasonal spread of under-reported dengue cases in Bangladesh. *PLoS Neglected Tropical Diseases, 12*(11), e0006947.

Tune, S. N. B. K., Hoque, R., Naher, N., Islam, N., Mazedul Islam, M., & Ahmed, S. M. (2020). Health, illness and healthcare-seeking behaviour of the street

dwellers of Dhaka City, Bangladesh: Qualitative exploratory study. *BMJ Open, 10*(10), e035663.
UN IGME. (2017a). *United Nations Inter-agency Group for Child Mortality Estimation*. Levels & Trends in Child Mortality: Report 2017.
UN IGME. (2017b). *United Nations Inter-agency Group for Child Mortality Estimation*. Under-Five Mortality Rate – Total.
UN IGME. (2020). *Under-five mortality rate – Total*. Available at: https://childmortality.org/data/Bangladesh. Accessed 30 Aug 2021.
WHO. (2010). *Non-communicable disease risk factor survey Bangladesh*. Retrieved from https://www.who.int/docs/default-source/searo/bangladesh/pdf-reports/year-2007-2012/non-communicable-disease-risk-factor-survey-bangladesh-2010.pdf?sfvrsn=37e45e81_2
WHO. (2014). *Non communicable diseases country profile, 2014*. Geneva: WHO.
WHO. (2019). *Expanded programme on immunization (EPI) factsheet 2019*. South-East Asia Region.
Wilder-Smith, A., Ooi, E.-E., Horstick, O., & Wills, B. (2019). Dengue. *Lancet, 393*(10169), 350–363.
Zaman, K., Hossain, S., Banu, S., Quaiyum, M. A., Barua, P. C., Salim, M. A. H., … Van Leth, F. (2012). Prevalence of smear-positive tuberculosis in persons aged ≥ 15 years in Bangladesh: Results from a national survey, 2007–2009. *Epidemiology and Infection, 140*(6), 1018–1027.
Zaman, M. M., Bhuiyan, M. R., Karim, M. N., MoniruzZaman, R., Rahman, M. M., Akanda, A. W., & Fernando, T. (2015). Clustering of non-communicable diseases risk factors in Bangladeshi adults: An analysis of STEPS survey 2013. *BMC Public Health, 15*(1), 659.

Equal Cities

▶ Women in Urbanism, Perpetuating the Bias?

Equitable Transitions

▶ Policies for a Just Transition

Equity

▶ Feminist Planning and Urbanism: Understanding the Past for an Inclusive Future
▶ Hidden Enemy for Healthy Urban Life

ESG in Real Estate Investment

Issues for the Future

Anupam Nanda
University of Manchester, Manchester, UK

Introduction

Sustainability is not only the most talked about subject area now, but also a critical area of research across all fields. Real estate is part of it, with several sustainability concerns. The carbon footprint of real estate is significant, and reducing the environmental impact of real estate can go a long way towards meeting the net zero targets. However, the environmental aspects are only one side of sustainability concerns. Societal and governance issues are also critical in understanding and addressing the concerns related to the wider concept of sustainability. All three components are brought together to define ESG (environmental, social, and governance). All these three components are critical and variously interlinked. Therefore, sustainability strategies for the future should encompass all three dimensions. In this chapter, the concept, measures and implications of ESG for future real estate investment strategies are discussed.

Definition of ESG and Relevance to Real Estate

The scope and definition of ESG have expanded in recent times to recognize the multidimensional aspects of sustainability. MSCI, which provides ESG ratings for companies, defines ESG across a range of attributes. The MSCI definition of ESG is based on three pillars of the E (environment), the S (social), and the G (governance). Those three pillars comprise ten themes, which are then further tracked across 35 key issues (see MSCI, 2022 for a detailed discussion and their methodology of measuring ESG).

While the E pillar is well understood with various climate-related metrics and evidence base, the S and G are still evolving as our understanding develops and there is a growing recognition of the concern areas. Moreover, metrics under these two pillars are more difficult to measure as the data items are somewhat scattered. However, with new ways of capturing data based on non-traditional methods and the application of big data principles and data science tools, a clearer picture has started to emerge and further improvements are expected in the future.

Specifically, under the **Environment** pillar, MSCI (2022) considers four themes: climate change, natural capital, pollution and waste, and environmental opportunities. The climate change theme addresses carbon emissions, product carbon footprints, climate change vulnerabilities, and the extent of financing available to mitigate environmental impact. The natural capital theme deals with issues that are related to the use and extraction of natural resources, e.g., stress on water as a resource, sources of raw materials (e.g., minerals), impact on and preservation of biodiversity, and land use patterns. The theme of pollution and waste mainly deals with three types of pollution and waste: toxic, electronic, and packaging materials. Under the theme of environmental opportunities, three areas are covered, namely, clean tech, renewable energy, and green building.

Land use, building materials (e.g., more environmentally friendly materials for new buildings and retrofits), impact of real estate development on biodiversity, water stress (e.g., potable water sources, waste water disposal), energy sources and the use of energy-efficient technologies in building operations, and climate change vulnerabilities (e.g., extreme weather events such as flooding, drought, wide fires, and sea level rise) are of particular relevance in real estate.

Climate change vulnerabilities are of real concerns for real estate as those can threaten the building stock through physical risks to the structure, the resident population's lives and livelihoods, and may pose as catastrophic risks to investment capital that is linked to the buildings. The world's key real estate markets are around the coastlines, rivers, or major water bodies, and those are now under existential threats due to the potential sea level rise from rising average temperatures. Such concerns and risks are relevant to building insurance companies who need to evaluate long-term risks to the properties under consideration.

MSCI (2022) defines the **Social** pillar with four themes. First, the human capital theme tracks labor issues both within the organizations and in the supply chain; health and safety at workplaces; and human capital development. Second, product liability evaluates the safety and quality of products; chemical safety; data ethics; and financial product safety. It also assesses health and demographic risks, as well as the extent of responsible investment. The pillar also evaluates relations with the community and other stakeholders. An important theme under the pillar is the evaluation of opportunities in healthcare, finance, and communication.

From the real estate perspective, the social pillar includes a variety of issues that need better understanding and data collection. There are several areas of indirect and secondary linkages, which are important but require a deeper level of information gathering. For example, labor issues in the supply chain of building materials lead to a better understanding of the suppliers' business operations. The site selection for real estate development should consider health and demographic risks to future occupants as well as the local community. Sources of finance for the development projects and issues of money laundering in real estate transactions are also relevant. Issues such as citizen engagement, public consultations for development projects, clear communication to all stakeholders and local residents on development and regeneration projects are other relevant examples for real estate. As noted before, the social (or "S") pillar is more difficult to understand and measure compared to the "E," as it often requires information from third parties and deals with non-real estate aspects. Nonetheless, these are important for achieving long-term sustainability.

Finally, MSCI (2022) includes two themes under the **Governance** pillar: corporate governance and corporate behaviour, which track ownership and control of the company; board representation; fair pay; adherence to accounting standards; various aspects of business ethics; and compliance with tax regulations. From the real estate perspective, these are all relevant. For example, publicly listed real estate companies like Real Estate Investment Trusts (REITs) are typically regulated by a range of mandates and special tax status on specific aspects of income that require careful application of appropriate accounting standards. Moreover, representation of minorities and women, and the makeup of a REIT's board can also be areas of concern and opportunity, which have been the subjects of academic research in recent years.

Notably, the ESG factors as defined by MSCI (2022) are increasingly being factored into the decision-making processes of real estate investors. The increasing importance that is now being placed on ESG factors is, in many ways, driven by a range of emerging legal mandates and changes in customer taste and preferences. The legal mandates are the direct outcomes of the sovereign governments committing to means and timescales for achieving net zero targets. Moreover, the relationship between ESG and investment performance, in particular, is complex with various considerations for financial, risk management, and public relations. With an increasing level of awareness of ESG issues across both the customer base and the investment community, this can have significant implications for an organization's reputational risk and the ability to raise capital. The capital market has put strong emphasis on socially responsible investments (SRI). The lessons from the recently held COP26 indicate that more importance needs to be placed and a sense of urgency should be attached to climate change matters in particular.

The current endeavor is very much focused on understanding all relevant issues across the three pillars of ESG and on finding tools and information to measure the extent of impacts as precisely as possible. To be able to evaluate how these various issues may increase risks and influence returns on investment, robust metrics and benchmarks are needed. Major providers (e.g., MSCI, S&P DJ, Global Reporting Initiative, and Carbon Disclosure Project) as well as specialized groups have developed ESG rating systems and benchmarks. While these measures and benchmarks show an increasing level of coverage of ESG issues and more emphasis on the social and governance aspects of ESG, they still have several biases, inaccuracies, and an inadequate information base that restrict ability to draw robust inferences.

Newell et al. (2020) provide an evaluation of current benchmarks and categorize them into four types, i.e., benchmarks at the fund level, the listed property level, the delivery level, and the reporting level. They describe various benchmarks that can be used to address both property-level and broader investor-level issues in order to track both asset and portfolio-level property performance, set improvement goals, identify best practices, examine asset-level enhancement, and communicate the ESG message to investors and other stakeholders. However, they also point out that, while some companies have developed their own internal benchmarks to suit their needs, external benchmarks may work better as those allow cross-industry and cross-competitor comparisons (see Newell et al. 2020 for details).

In recent years, the investment community has also become quite proactive in channelizing capital into businesses that are serious about ESG issues as the long-run benefits of investing in ESG matters have become clearer over the time period. Several key international and intergovernmental organizations have also made concerted efforts to increase awareness at all levels and recognize how ESG issues can be embedded into the United Nations' Sustainable Development Goals (SDGs). For example, the United Nations Finance Initiative's Principles for Responsible Investment (PRI, https://www.unpri.org/download?ac=10948), the Institutional Investors Group on Climate Change (IIGCC, https://www.iigcc.org/) and the Task Force on Climate-Related Financial Disclosures (TCFD, https://www.fsb-tcfd.org/about/) have been promoting how ESG participation benefits investors and businesses. Nowadays, aligning with SDGs is an important part of corporate strategies for companies. This can facilitate a strong and action-based linkage between investment motives and societal objectives.

Areas of Extant Research

ESG as a concept is not new, although emergence and recognition of several factors as part of the ESG definition are developments that are more recent. In many ways, ESG is linked to the concept of corporate social responsibility (CSR), which has been an area of active academic enquiry over the last few decades. From the business perspective, several ESG issues have been analyzed for many years under the CSR concept. Decades of research have investigated various business challenges. However, research is still quite active due to emerging concerns and challenges – from the role of CSR activities to the financial return of investments into CSR activities. As businesses use various resources such as natural resources and human capital, the role of business in tackling environmental and social issues has been continuously questioned. This has been especially important for companies that typically make heavy use of natural resources to produce their goods and services. The prime examples include oil, gas, and mining companies. In many ways, real estate assets are similar as buildings use various natural resources in both production and consumption phases during their life cycle, and therefore, several aspects of real estate fall under the ESG scanner.

At the asset level, a building's operational aspects come directly under the scope. A major focus is on the energy efficiency aspects of a building's operation. It is not only the monitoring of energy consumption and the efficiency standards of the appliances, but also the sources of energy are the key areas of concerns. Academic and industry research has variously investigated heating, air-conditioning, water preservation, and conservation of energy through improvement in a building's physical features. Green building standards (e.g., LEED, BREEAM, Energy Star, and NABERS), as well as numerous Green Building Councils and the International Initiative for a Sustainable Built Environment, have all contributed to raising awareness among all stakeholders and providing significant impetus for product innovation.

Several papers have investigated the importance of ESG issues for real estate investors (see, for example, Falkenbach et al., 2010; Cajias et al., 2014, Hirsch et al., 2019). While initial focus was on the US market, over the past 7–8 years, studies focusing on other markets, such as Europe, Australia, and Asia, have emerged. However, the findings of all these studies are somewhat mixed and not strongly conclusive, and they depend on modelling biases and data quality issues. Due to the complex nature of correlations, it is, however, quite difficult to separate out the effects across all relevant variables. In many cases, joint determination does not allow a clear demarcation of the effects. Statistically, it is difficult to establish individual effects without extensive, granular data collection. Take, for example, the studies on whether energy performance ratings affect market prices. The results have been somewhat mixed due to the fact that some of the energy efficiency factors also offer significant benefits in terms of security, noise cancellation, and aesthetics. Several studies have also established a link between green office buildings and real estate performance (e.g., sales price, values, rent, vacancy, and yield), supporting the E dimension of the ESG agenda. Overall, while there are several methodological challenges, studies broadly confirm some evidence of a price effect in both residential and non-residential real estate.

Issues for the Future

As ESG issues become prominent, it becomes more important to record, analyze, and quantify key ESG attributes at the asset, portfolio, and firm levels. There are still significant gaps in our current understanding of environmental, social, and governance (ESG) issues. Much research effort is needed to understand the efficacy of the measures, mandates, and standards. As mentioned earlier, while the "E" elements are clearer, the "S" and the "G" elements are a lot less definitive.

The behaviors of various stakeholders toward ESG concerns are an area of growing research emphasis. Occupants, managers, and investors of real estate may approach the ESG issues differently based on their varied goals and motivations. While the long-run objectives should ideally be in alignment across all stakeholders, the short-run economic and financial realities may drive behaviors that may not always be in line with the long-run objectives. However, a growing public outcry about climate risks, as well as an increasing amount of evidence base identifying the risks that they pose to all stakeholders, may aid in aligning interests.

A key area of research is policy evaluation around ESG issues. As ESG concerns and opportunities have become more established, the governments around the world have now enacted a range of regulations to drive behavior towards the desired levels. However, as with any regulation, effectiveness crucially depends on the nature and structure of the laws and their implementation strategies. A misguided roll-out of a legislation or any ambiguity in legal texts and the process of rulemaking can easily lead to loopholes, become political and make the mandates ineffective or less effective, which not only makes those redundant but also leads to waste of valuable public funding and resources. Therefore, it is important to undertake a detailed impact analysis before enacting regulations around ESG issues.

Long-run projection of climate impacts utilizing non-real estate contextual data such as regional and local climate information, elevation data, and micro-area weather pattern data is an area of active research. Studies should utilize non-real estate data more to be able to evaluate and measure specific risks to the property markets. Extensive use of data science tools should enable a more thorough and rigorous examination of the issues. This will also require investments in education and training activities to raise further awareness, allow ready take-up, and encourage further innovations. New training modules at various levels will need to be developed.

Finally, while numerous studies have focused on the developed world, much effort is now needed to understand the ESG issues across the developing economies around the world. With a massive concentration of people facing a range of risks and uncertainties, the developing economies are at the brink of several ESG crises. Research effort and subsequent development of tools and solutions should be focused on those markets, without which the world may see a widening inequality gap between the rich and poor nations across the ESG space. However, it also requires significant resource commitments as the developing countries face severe resource constraints. It would require intergovernmental collaborations devoid of any political interferences and the global conglomerates joining forces to develop tools and deliver the solutions. There is no doubt that ESG actions require massive amount of resources that only the rich nations and big companies can undertake successfully. However, that would not lead to sector-wide improvements and the achievement of net-zero targets at a global scale. Best practices around ESG issues should be shared without restrictions to enable adoption and adaptation at a minimal cost. Active knowledge transfer and support among countries and organizations would enable faster and more efficient approaches to tackle the issues effectively.

References

Cajias, M., Fuerst, F., McAllister, P., & Nanda, A. (2014). Do responsible real estate companies outperform their peers? *International Journal of Strategic Property Management, 18*(1), 11–27.

Falkenbach, H., Lindholm, A., & Schleich, H. (2010). Environmental sustainability: Drivers for the real estate investor. *Journal of Real Estate Literature, 18*(2), 203–223.

Hirsch, J., Spanner, M., & Bienert, S. (2019). The carbon risk real estate monitor- developing a framework for science-based decarbonising and reducing stranding risks within the commercial real estate sector. *Journal of Sustainable Real Estate, 11*(1), 174–190.

MSCI (2022). MSCI ESG Ratings Methodology. https://www.msci.com/documents/1296102/21901542/ESG-Ratings-Methodology-Exec-Summary.pdf (accessed on June 10, 2022).

Newell, G., Moss, A., & Nanda, A. (2020). Benchmarking Real Estate Investment Performance: The Role of ESG Factors (July 2020) Summary Report. https://www.ipf.org.uk/resourceLibrary/benchmarking-real-estate-investment-performance%2D%2Dthe-role-of-esg-factors%2D%2Djuly-2020%2D%2Dsummary-report.html

United Nations (2021) Principles for Responsible Investment. Available at https://www.unpri.org/download?ac=10948 (accessed June 10, 2022).

Ethical Finance

► Financing: Fiscal Tools to Enhance Regional Sustainable Development

Ethics

► Sustainable Development Goals and Urban Policy Innovation

European Green Deal and Development Perspectives for the Mediterranean Region

Stella Tsani
Department of Economics, University of Ioannina, Ioannina, Greece

Synonyms

Decarbonization; Just energy transition; Local content; Public policies; Sustainable development

Definitions

European Green Deal: Europe aims to become the first climate-neutral continent by 2050. The aim is to become a modern resource-saving economy. The European Green Deal sets the blueprint for this transformational change. This will impact profoundly the neighbor countries of the European Union in the South Mediterranean that export energy and other core resources to the European Union, that are recipients of significant European investment capital, and that are subject to EU policy priorities with regard to regional cross-border cooperation and integration.

Just energy transition: Just energy transition refers to the transformation of the energy system in a fair, equitable, and resilient manner. Just energy transition aims at the greening of the energy sector and to the phasing out of fossil fuels in the energy mix, without leaving behind any regions, countries, and/or local communities and by considering the creation of decent work, quality employment, and the development priorities of the nation(s) (Tsani 2020). Just energy transition necessitates the design and the implementation of cooperation programs, between countries, regions, and sectors of production that facilitate technology, know-how, and skills transfer.

Local content policies: Local content policies are policies that target the utilization and the development of the local industrial and human capital base through the creation of economic interactions between the energy sector and the rest of the productive sectors of the economy. Local content policies aim at increasing the value added beyond that directly derived from the energy production activities as well as creating opportunities for employment, innovation, and transfer of know-how. Local content policies are usually formulated through input thresholds for local goods and services in foreign investments set by legislation and regulatory requirements (Tsani 2020).

Introduction

The European Commission revealed in 2019 the European Green Deal, the European Union (EU) plan to make its economy sustainable (European Commission 2021). EU energy policies are expected to impact the Southern Mediterranean (South-Med) countries, that is, Algeria, Egypt, Israel, Jordan, Lebanon, Libya, Morocco, Palestine, and Tunisia, as the two regions have strong trade links. Following the United Nations Agenda to 2030, countries worldwide agreed in 2015 on 17 Sustainable Development Goals (SDGs) to guide national, regional, and global decisions until 2030 (United Nations 2015). The Paris Agreement and the Nationally Determined Contributions (NDCs) adopted by 196 Parties at

COP 21 in 2015 aim at limiting global warming to below 2 °C, compared to the preindustrial levels (UNFCCC 2021). The South-Med countries have committed to the UN Agenda to 2030 (Palestine is a nonmember observer state of the UN) and are signatory parties of the Paris Agreement. Thus, the commitments deriving from these international agreements should guide the national policies of the South-Med countries with horizon to 2030. COVID-19 further intensifies the urgency for sustainability transition (World Bank 2020; IMF 2020; WHO 2020; European Commission 2020).

The European Green Deal, the Paris Agreement, and the UN Agenda 2030 make a priority the increase of renewable energy sources in the energy mix and the decarbonization of the economies. This energy transition will have a significant impact on the South-Med countries, especially on hydrocarbon producers: Egypt, Libya, and Algeria. Challenges are similar to those faced by other oil-dependent countries in the Middle East, in Central Asia, and in the Sub-Saharan Africa. They may be faced with abrupt shocks in their primary sectors of production and in their labor markets if no timely action is taken. Oil-producing countries must decarbonize and at the same time diversify their economies and no longer rely on oil to provide their main source of revenue and employment generation. The challenge of exploiting conventional energy sources while progressing with the energy transition is also of relevance for the new-comers in the exploration of hydrocarbons in the Mediterranean, that is, Israel, Lebanon, Cyprus, and Greece.

For a resilient future, energy transition should go hand in hand with just transition in the labor markets, that is, the inclusive and fair labor market transformation. This is important for the labor market segments that are directly and/or indirectly linked to the energy sector (e.g., extraction, construction, transport, maintenance services). Energy transition makes sense only if expensive carbon lock-in (Unruh 2000), silos between industries, gaps between developed and developing countries, the global North and South are overcome (Pauw et al. 2019; Eicke et al. 2019). This requires the uptake of new skills, time, and flexibility from employees, businesses, and policy makers (Tsani 2020; Facchinetti et al. 2016). If transition comes as an uneven shock to the labor force, it may lead to increased social unrest, supply bottlenecks, and market distortions (Matsuo and Schmidt 2019).

Scientific analysis suggests taking a closer look at the policies that can foster the backward and forward links of the energy sector with other sectors in the economy, if just energy transition remains a priority target. The design and the implementation of such policies in the South-Med countries can benefit from the cooperation with countries that face similar challenges (i.e., Middle East, Central Asia, sub-Saharan countries), with regions and countries that develop policies to turn energy projects into an engine of industrial growth and employment creation (e.g., EU, Norway), and with institutions that bring significant knowledge on this matter (like the World Bank).

This entry aims at contributing to the policy debate related to energy transition. The end goal is to shed more light on the policies which can support just transition in the labor markets in the face of a clean energy future in the Mediterranean. These may relate to the COVID-19 fiscal responses, and to the cooperation opportunities that can capitalize on the South-Med links with the EU and other countries and regions worldwide. The discussion builds on the review of the state-of-the-art scientific literature on the economic, social, and welfare implications of energy transition and labor markets transformations. Recommendations include: (i) the well-informed policy design, (ii) promotion of cooperation through clusters and regional schemes, (iii) emphasis on know-how, technology, and skills transferability beyond the energy sector, (iv) timely intervention in the education system, (v) priority allocation of public funding and international assistance to education, skills, and technological upgrade of the labor force, which can improve job prospects, resilience, and long-term sustainability.

The remainder of the entry is structured as follows: Section "The Mediterranean Region in

Transition: European Green Deal, Sustainability, and Energy Perspectives" presents the regional context with focus on energy and labor markets. Section "Labor Policies for Just and Inclusive Energy Transition" discusses the opportunities and challenges related to energy transition employment-focused policies and interventions. The last section concludes by offering some policy recommendations.

The Mediterranean Region in Transition: European Green Deal, Sustainability, and Energy Perspectives

According to the 2021 Sustainable Development Report (Sachs et al. 2021) South-Med countries score a total SDG Index of above 65 (Table 1). This means that they are approximately two-thirds of the way away from achieving the SDGs.

European Green Deal and Development Perspectives for the Mediterranean Region, Table 1 2021 SDG Dashboard for South-Med countries and selected country groups

	2021 SDG Index		Susainable Development Goals																
Country	Score	Rank	SDG1	SDG2	SDG3	SDG4	SDG5	SDG6	SDG7	SDG8	SDG9	SDG10	SDG11	SDG12	SDG13	SDG14	SDG15	SDG16	SDG17
Algeria	70,9	66	Yellow	Red	Red	Yellow	Orange	Orange	Yellow	Red	Orange	Yellow	Orange	Yellow	Orange	Red	Red	Orange	Orange
Egypt	68,6	82	Orange	Red	Red	Yellow	Red	Orange	Yellow	Red	Orange	Orange	Orange	Yellow	Yellow	Red	Red	Red	Orange
Israel	75,0	38	Orange	Red	Green	Red	Orange	Orange	Orange	Yellow	Yellow	Red	Orange	Red	Red	Red	Red	Orange	Red
Jordan	70,1	72	Yellow	Orange	Orange	Orange	Red	Orange	Yellow	Red	Orange	Red	Orange	Yellow	Yellow	Orange	Red	Orange	Orange
Lebanon	66,8	93	Green	Red	Orange	Grey	Red	Orange	Yellow	Red	Orange	Orange	Red	Orange	Orange	Red	Red	Orange	Orange
Libya			Grey	Red	Red	Grey	Red	Red	Orange	Orange	Red	Grey	Orange	Orange	Orange	Red	Orange	Red	Orange
Morocco	70,5	69	Yellow	Red	Red	Yellow	Red	Orange	Yellow	Red	Orange	Orange	Orange	Green	Green	Red	Orange	Orange	Orange
Tunisia	71,4	60	Yellow	Red	Red	Yellow	Orange	Orange	Yellow	Red	Orange	Orange	Orange	Yellow	Yellow	Red	Red	Orange	Yellow
Eastern Europe and Central Asia	71,4		Yellow	Orange	Orange	Yellow	Orange	Orange	Yellow	Orange	Orange	Orange	Orange	Orange	Orange	Grey	Red	Red	Yellow
Middle East and North Africa	67,1		Yellow	Red	Red	Yellow	Red	Orange	Orange	Red	Orange	Red	Orange	Yellow	Orange	Red	Red	Red	Orange
OECD members	77,2		Green	Orange	Yellow	Green	Yellow	Yellow	Yellow	Yellow	Green	Red	Yellow	Orange	Red	Red	Red	Orange	Yellow
Sub-Saharan Africa	51,9		Red	Red	Red	Red	Red	Red	Red	Red	Red	Red	Red	Yellow	Yellow	Red	Red	Red	Orange
Low-income Countries	51,0		Red	Red	Red	Red	Red	Red	Red	Red	Red	Red	Red	Yellow	Green	Red	Red	Red	Red
Lower-middle-income Countries	60,1		Orange	Red	Red	Orange	Red	Red	Red	Orange	Red	Red	Red	Green	Green	Red	Red	Red	Red
Upper-middle-income Countries	70,8		Yellow	Orange	Red	Yellow	Yellow	Orange	Orange	Orange	Orange	Red	Orange	Yellow	Orange	Red	Red	Red	Orange
High-income Countries	78,1		Green	Orange	Yellow	Green	Yellow	Orange	Yellow	Yellow	Green	Orange	Yellow	Red	Red	Red	Red	Yellow	Yellow

Major challenges | Significant challenges | Challenges remain | SDG achieved | Information unavailable

Source: Sachs et al. (2021). The SDG dashboard shows the overall performance assessment in relation to the 17 SDGs. Performance is indicated with the use of a specific color: Green color indicates SDG achievement (i.e., all indicators under the goal have been rated green). Yellow, orange, and red indicate increasing distance from SDG achievement, with red indicating the largest distance away from the target achievement of SDGs. Data for Palestine are not available. 2021 SDG index score and rank for Libya are not available

Nevertheless, the detailed review of the SDG7 (clean and affordable energy) and of the 2019 Arab region SDG indicators (Luomi et al. 2019) shows the distance that South-Med countries must cover with regard to the uptake of renewable energy (solar, wind, etc.) in the energy mix. Oil-dependent countries in the Middle East, in sub-Saharan Africa, and in Central Asia also face this challenge. South-Med (excluding Israel), Middle East, sub-Saharan, and Central Asian oil-producing countries are found at the lower end of the 2021 Energy transition Index (WEF 2021), recording low system performance and transition readiness (Fig. 1). Robust action is also required from the oil-dependent countries and regions to address unemployment, particularly among the youth (Table 2).

In the face of global and national commitments to energy transition South-Med countries, like the Middle East, Central Asia, and sub-Saharan countries, are faced with the challenges of timely transition in the labor markets. Energy transition is expected to lower labor demand in the energy intensive sectors and in the extractive (coal, oil, gas) industries. On the other hand, investments in renewable energy should have a direct positive employment effect and an indirect positive effect on sectors providing inputs to these investments (Pestel 2019).

Moving to a low-carbon economy increases the demand for middle-skill employment (electricians, mechanics, technicians) and expertise with technical education (Cedefop 2010). The uptake of new technologies triggers demand for specific new skills of the workforce. These relate to new design and construction methods, development and use of novel materials and technologies, new energy-efficient technical solutions, development

E

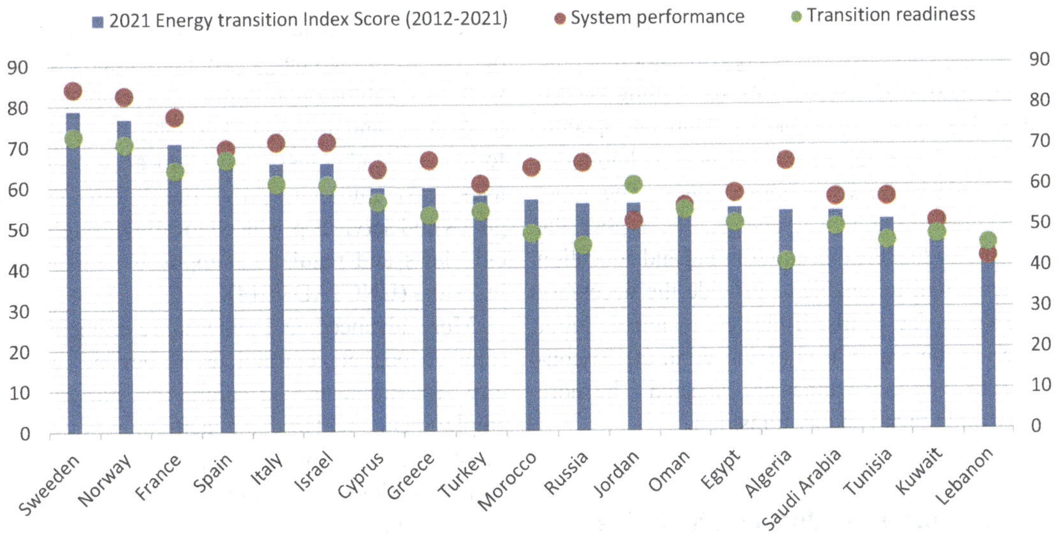

European Green Deal and Development Perspectives for the Mediterranean Region, Fig. 1 Energy transition in the South-Med countries and in selected countries. (Source: WEF (2021)). The Energy Transition Index benchmarks countries on the performance of their energy system, and on their readiness for transition to a secure, sustainable, affordable, and reliable energy future. ETI 2021 score on a scale from 0 to 100%. The ETI 2021 ranks 115 countries. Sweden ranks first (1) and Zimbabwe ranks last (115) in the ETI 2021. Ranking shows the order of the countries in the ETI 2021. The ETI is the average of the two sub-indices, System performance and Transition readiness. System performance provides an assessment of countries' energy system related to their delivery in three key priorities: the ability to support economic development and growth, universal access to a secure and reliable energy supply, and environmental sustainability across the energy value. Transition readiness assesses the presence of an enabling environment for energy system's ability to deliver on the transition imperatives. Energy transition readiness is captured by the stability of the policy environment and the level of political commitment, the investment climate and access to capital, the level of consumer engagement, the development and adoption of new technologies, etc. (WEF 2021)

European Green Deal and Development Perspectives for the Mediterranean Region, Table 2 Renewable electricity output and unemployment in the South-Med and in selected Arab countries

	Algeria	Egypt, Arab Rep.	Jordan	Lebanon	Libya	Morocco	West Bank and Gaza	Tunisia	Qatar	United Arab Emirates	Oman	Saudi Arabia	Bahrain	Kuwait
Renewable electricity output (% of total electricity output, average 2010-2015)*	0.6	9.0	0.5	4.7	0.0	13.0	0.0	2.0	0.0	0.1	0.0	0.0	0.0	0.0
SDG7 indicator: Renewable electricity output (% of total electricity output, reference year 2015)**														
Unemployment (% of total labor force, average 2010-2019)*	10.8	11.6	13.9	6.3	18.8	9.2	22.3	15.7	0.3	2.1	3.0	5.7	1.3	2.3
SDG 8 indicator: Unemployment rate (% of total labor force, reference year 2018)**														
Youth unemployment (% of total labor force ages 15-24, average 2010-2019)*	26.6	30.9	32.3	17.4	49.6	20.4	37.4	35.9	0.9	6.7	12.4	28.4	5.9	13.5
SDG 8 indicator: Unemployment, youth total (% of total labor force ages 15-24, reference year 2018)**														

Major challenges Significant challenges Challenges remain SDG achieved

Source:

[a]Luomi et al. (2019). The dashboard shows the performance assessment of the selected indicator. Performance is indicated with the use of a specific color: Green color indicates target indicator achievement. Yellow, amber, and red indicate increasing distance from target achievement, with red indicating the largest distance away from the target achievement
[b]Author's estimations. Time series data are extracted from the World Bank databank, World Development Indicators. 2015 latest year for data on renewable electricity output. 2019 latest year for data on total and youth unemployment

and use of energy efficiency, and carbon emission measurements. Overall, energy transition should be coupled with demand for cross-cutting knowledge. The impact on demand for labor and specific skills relates to the speed that new technologies will replace the conventional ones. The more rapid the technological change, the higher the possibility that the existing training and education systems will be unable to provide the necessary and up to date skills. If barriers to female labor force in the South-Med (Tsani et al. 2013) continue to exist, then women are expected to be more vulnerable to the energy transition.

Labor Policies for Just and Inclusive Energy Transition

Labor market implications of energy transition highlight the need for a careful consideration of "local content" policies. Local content policies target the utilization and the development of the local industrial and human capital base through the creation of economic interactions between the energy sector and the rest of the productive sectors of the economy (Tordo et al. 2013; Tsani 2020; Kalyuzhnova 2021). Local content policies aim at increasing the value added beyond that directly derived from the energy production activities as well as creating opportunities for employment, innovation, and transfer of know-how. In a simple format, and rather less efficient, local content policies make minimum provisions for the use of goods and services produced locally through specific laws and regulations applicable to foreign investors (UNCTAD 2014).

More advanced, and rather more efficient versions of local content policies go beyond regulatory requirements on domestic goods and employment, and target the facilitation of domestic innovation, knowledge transfer, and integration of the domestic suppliers in the international supply chain.

While local content policies have been so far relevant for hydrocarbon producing countries, scientific analysis shows that they are also relevant for renewable energy projects (Binz et al. 2017; Rennkamp et al. 2017). Similar to hydrocarbons, investments in renewables (solar, wind etc.) are often undertaken by large international companies, which come with established international supply chains. Renewable energy projects are also characterized by high-capital intensity and expert knowledge requirements. If no appropriate measures are taken in the countries where renewable energy projects materialize, then local suppliers and labor may not benefit from the investments made and from the resulting activities. Beyond the potential positive impact on the local supply chain and labor markets, local content policies may: (i) have an accelerating impact on the political momentum for ambitious renewable energy policies, (ii) support the move of the economic base away from consumption and rent-seeking closer to the creation of domestic value added, (iii) "leap-frog" existing barriers to technological transfer, (iv) bring more new mature players to the global market, which increase competition and innovation (UNCTAD 2014).

Several countries include local content policies in their green strategy agendas. Examples include both developed (e.g., Canada, USA, Spain, Italy, France, Greece) and developing (e.g., China, Brazil, Argentina, India, South Africa, Tunisia, Ecuador) countries. Indicative examples include: (i) Ontario, Canada, where feed in tariffs require that certain components for wind and solar energy come from local manufacturing, (ii) Spain: local content policies have been related to the remarkable Gamesa's growth as a global wind manufacturer (Kuntze and Morenhoute 2013). In Navara, local content policies are linked to the creation of 4000 jobs (Hao et al. 2010), (iii) Brazil's local content experience with oil and gas has been used for the development of local content requirements (60%) for wind energy. Access to subsidized loans from Brazil's national Development Bank has been conditional on local content requirements, (iv) in Saudi Arabia the "In-Kingdom Total Value Added" program encourages the purchase of goods and services from the local suppliers' base aiming to double Saudi Aramco's percentage of locally manufactured energy-related goods and services to 70% by 2021 (see: Saudi Arabia Country Commercial Guide). Local content targets can also be identified in the Egypt Vision 2030, in Jordan's National Employment Strategy 2011–2020, in Morocco's National Energy Strategy launched in 2009, and in the 2020 Vision of the National Agency for the Promotion of Employment and skills (Cote 2019).

If properly designed and implemented, local content policies can support employment. On the other hand, if local content policies are too restrictive or not fully grasping the local potential to address the sectoral needs, they may hurt competitiveness, increase projects' costs, and hinder international interest in energy projects with large development potential for the host countries. Additional challenges related to local content policies in the energy sector relate to the: (i) technological complexity, (ii) existing domestic capacity, and (iii) time needed to build links between the energy sector and the local economy (European Committee of the Regions 2017; Levett and Chandler 2012). (Small) developing economies often have a thin industrial base and thus may find it difficult to provide fast competitive inputs (goods and services) to the green energy projects (Tsani 2020). Foreign investors usually show preference to tap their established technology and service supply networks, thus importing the value chains in the host countries rather than putting effort in integrating local suppliers into the value chain (Matsuo and Schmidt 2019). Lack of skills is often recorded in the early stages of an energy project. Skill shortages are related to the education system, quality, and requirements of the existing industrial base or to the progress of the investments in the energy projects, for example, if projects do not allow for the early development of skills in the domestic market (Tsani 2021).

Implications and Policy Recommendations for the South-Med Region

COVID-19 pandemic is likely to have a significant and persistent impact on the global economy and the energy system. While the economic activity is expected to partially recover over the next years, some effects may persist (for instance people may choose to travel less, or work from home). The impact of COVID-19 on the oil-dependent economies is indicative of the impact that global energy transition will have on hydrocarbon producers if no timely action is taken. If labor transition policies are not implemented, oil-based economies run the risk of being left with outdated/low skills for their human capital, low diversity in the labor markets, poor job prospects, and increasing need for public support.

For a resilient future, energy transition should be coupled with just labor transition policies. Into this direction several considerations and recommendations are relevant to the South-Med countries. Hydrocarbon producers in the South-Med share similar socioeconomic structure, just energy, and labor market transition challenges with other oil-dependent countries in the Middle East, in the sub-Saharan Africa and in Central Asia. In the quest for (just) energy transition oil-dependent countries should join forces and capitalize on cooperation in different fronts. These may include the decarbonization of the energy system, innovation in clean hydrocarbon exploration, and the identification of optimal sectoral and labor market policies tailored to the needs of the oil-dependent countries. In this regard a core recommendation calls for the exploration of collaboration and solution-driven partnerships with other countries and regions (Middle East, Central Asia, Sub-Sahara) faced with similar structural issues (e.g., oil-dependency) in line with SDG17. Focus should be on (i) knowledge sharing and cooperation for access to science, technology, and innovation and (ii) encouragement of effective public, public-private, and civil society partnerships.

Localities where conventional energy production, manufacturing, and use take place, but also where green energy projects materialize, pose a challenge for employment and human capital development in energy transition. The stance of public policies toward these localities (e.g., conventional fuel substitution, delay in regional education support, upgrade or creation of technology and skill upgrade clusters) can impact on the extent and the speed of energy and labor markets transition. In this context policies should target the design and the implementation of well-informed industrial, fiscal, and employment policies which consider the present and the future capacity of the domestic economy. These should target the creation of long-term backward and forward links of the energy sector with the domestic economy. Exchange of best practices through partnerships with Middle East, Central Asia, and with Sub-Saharan countries can be supportive to these efforts.

Policy interventions in the labor markets and policies aiming at local content in the energy sector should not impose excess management costs and bureaucratic burdens. Policy design can benefit from the accumulated knowledge on this matter in the countries implementing similar policies and/or in the international institutions (e.g., the World Bank). Into this direction policy design and implementation should follow the undertaking of a detailed socioeconomic, financial, and technology impact assessment and cost-benefit analysis of local content policies related to the energy sector across time and space.

Energy transition calls for overcoming systemic weaknesses of human capital such as lack of science, technology, engineering knowledge, and cross-cutting skills. Any legal and regulatory local content requirements may impact on the industrial competitiveness, on the emergence of bottleneck problems arising from the skills upgrade requirements, on the speed and on the direction of the transition. It is thus important local content and regional development policies to identify and timely address skill gaps, gender inequality, and industry needs for specific cross-cutting skills. Policies can benefit from the experience of other countries and regions that have developed similar agendas (like the EU or Norway).

COVID-19 has indicated the existence of large technology, human capital, gender, and economic inequalities and the need to address them (Tsani et al. 2021a). Technology, know-how, and skills transfers should be at the core of local content and just energy transition policies. In order to meet these targets policy makers should aim at the design and the implementation of cooperation programs, between countries, regions (EU, Middle East), and sectors (e.g., education, public administration, energy) that facilitate technology, know-how, and skills transfer (e.g., exchange programs, internships, close collaboration between education institutions and foreign investors). The renewed partnerships and neighbor policies brought forward with the EU Green Deal and the strategy agenda of the EU for the period 2019–2024 promote and support cooperation vehicles and schemes that can also support just transition in the South-Med beyond promoting specific regional cooperation with social and geopolitical focus.

Creating clusters is especially important for the energy sector that is dominated by large companies which invest heavily in technological upgrade and innovation. Clusters can serve as a platform of integration and exchange of best practices for small and medium size enterprises, which can be an important actor of sustainable development in the South-Med. The creation of clusters can be geographically and/or sectorally distributed, within or outside national borders (e.g., formations can be expanded at transnational or regional level to exploit economies of scale and scope). The EU Green Deal and the renewed neighbor policies brought forward in the European Union pose a prime opportunity for the Mediterranean countries to establish and to develop industrial and/or technology clusters across the knowledge triangle (research, education, and innovation) in support of business development, access to innovation, better coordination, effective use of public goods, and dissemination of best practices.

Oil and gas producers (Egypt, Algeria, Libya) should consider the role that hydrocarbons are foreseen to play in the global energy mix and in the European energy imports in the years to come. Global energy scenarios to 2050 foresee a peak in demand in the period 2020–2030 (depending on policy assumptions of each scenario). After this period of plateauing demand, hydrocarbons are expected to account for a smaller share in the global energy mix as compared to renewable energy sources. The European Green Deal signals a clear moving away of the European Union from conventional energy sources and a firm intention to lower dependence on conventional energy imports. These developments call for the need to prioritize policies that can support the local supply chain, employment, and innovation ecosystem in a way that extends beyond the hydrocarbon sector. This structural change, that the hydrocarbon sector is faced with, is understood by the major international oil companies (Tsani 2021). Major hydrocarbon players such as ExxonMobil, Saudi Aramco, Equinor, BP, and Shell, many of which engage in exploration and production activities in the Mediterranean, have joined forces under the Oil and Gas Climate Initiative (OGCI). The OGCI aims at accelerating the industry response to climate change and to support the Paris Agreement. The members of the OGCI have collectively invested over $7B each year in low carbon solutions with the intention to lead industry response to climate change. The shift away from oil and gas is also explored by large oil producing countries like Norway. Taking the above into consideration, policy efforts should look closer to the support of education and cooperation in favor of development of skills and technology beyond hydrocarbons (e.g., offshore safety, environmental monitoring, IT tools for large data management, remote monitoring). Priority can be given to actions that align to OGCI priorities in cooperation with national research institutions and suppliers.

COVID-19 responses worldwide and in the Mediterranean have mobilized public funds for the maintenance of the current consumption and production, income, and employment levels (Tsani et al. 2021b). Given the long-lasting implications of the pandemic and of the public fund mobilization patterns, COVID-19 responses pose a prime opportunity for incentive provision, skills, and technology upgrade that are needed for the just transition in the labor markets. Policy makers should consider the importance of digitalization, IT, and green infrastructure so as to speed-up the greening of the economies and employment generation in the emerging sectors. Infrastructure investments may pay-off in terms of employment and long-term productivity. Fiscal measures should look beyond providing direct financial support to the most vulnerable. Instead, they should actively seek for a combination of financial, training, and education support packages that can ensure resilience of the most vulnerable under the current conditions and readiness for future uncertainties and skill requirements. It emerges thus as a core policy recommendation for the mobilization and the use of the public funding and of the international assistance response to COVID-19 toward education, skills, and technological upgrade of the labor force, that can improve job prospects during the sustainability transition and in the future.

Acknowledgments Earlier version of the present work has been published under the FEMISE Network/Center for Mediterranean Integration (CMI) COVID-19 MED Policy Briefs, Medbrief 17. COVID-19 Implications in the Mediterranean (April 2021). The author wishes to thank Blanca Moreno Dodson, Maryse Louis, and Constantin Tsakas for their useful comments. Any errors and omissions remain with the author. The work presented therein has received, in part, financial support from the EEA Grants 2014–2021, Bilateral Fund, under the project "Energy Governance for Sustainable Development".

References

Binz, C., Gosens, J., Hansen, T., & Hansen, U. E. (2017). Toward technology-sensitive catching-up policies: Insights from renewable energy in China. *World Development, 96*, 418–437.

Cedefop. (2010). *Skills for green jobs: European synthesis report*. Luxembourg: Publications Office of the European Union.

Cote, S. (2019). *Renewable energy and employment: The experience of Egypt, Jordan and Morocco*. KAPSARC. https://www.kapsarc.org/file-download.php?i=42077&m=o

Eicke, L., Weko, S., & Goldthau, A. (2019). Countering the risk of an uneven low-carbon energy transition, IASS Policy Brief, November 2019.

European Commission. (2020). European Parliament resolution of 10 July 2020 on the EU's public health strategy post-COVID-19, 2020/2691. https://www.europarl.europa.eu/doceo/document/B-9-2020-0217_EN.html

European Commission. (2021). A European green Deal. Striving to be the first climate-neutral continent. https://ec.europa.eu/info/strategy/priorities-2019-2024/european-green-deal_en

European Committee of the Regions. (2017). *Financing climate action: Opportunities and challenges for local and regional authorities*. European Union. https://doi.org/10.2863/329600.

Facchinetti, E., Eid, C., Bollinger, A., & Sulzer, S. (2016). Business model innovation for local energy management: A perspective from Swiss utilities. *Frontiers in Energy Research*. https://doi.org/10.3389/fenrg.2016.00031.

Hao, M., Mackenzi, M., Pomerant, A., & Strachran, K. (2010). *Local content requirements in British Columbia's wind power industry*. Pacific Institute for Climate Solutions. https://pics.uvic.ca/sites/default/files/uploads/publications/WP_Local_Content_Requirements_December2010.pdf

IMF. (2020). Special series on COVID-19. https://www.imf.org/en/Publications/SPROLLs/covid19-special-notes

Kalyuzhnova, Y. (2021). Local content policies and institutional capacity for sustainable resource management. In S. Tsani & I. Overland (Eds.), *The sustainable politics and economics of natural resources*. Edward Elgar.

Kuntze, J. C., & Morenhoute, T. (2013). *Local content requirements and the renewable energy industry – A good match?*. International Institute for Sustainable Development, September 2012.

Levett, M., & Chandler, A. E. (2012). *Maximizing development of local content across industry sectors in emerging markets*. Washington, DC: Center for Strategic and International Studies.

Luomi, M., Fuller, G., Dahan, L., Lisboa Båsund, K., de la Mothe Karoubi, E., & Lafortune, G. (2019). *Arab region SDG index and dashboards report 2019*. Abu Dhabi/New York: SDG Centre of Excellence for the Arab Region/Emirates Diplomatic Academy and Sustainable Development Solutions Network.

Matsuo, T., & Schmidt, T. S. (2019). Managing tradeoffs in green industrial policies: The role of renewable energy policy design. *World Development, 122*, 11–26.

Pauw, W. P., Castro, P., Pickering, J., & Bhasin, S. (2019). Conditional nationally determined contributions in the Paris agreement: Foothold for equity or Achilles heel? *Climate Policy, 1*(17).

Pestel, N. (2019). Employment effects of green energy policies. *IZA World of Labor, 76*(2). https://doi.org/10.15185/izawol.76.v2.

Rennkamp, B., Haunss, S., Wongsa, K., Ortega, A., & Casamadrid, E. (2017). Competing coalitions: The politics of renewable energy and fossil fuels in Mexico, South Africa and Thailand. *Energy Research and Social Science, 34*(2016), 214–223.

Sachs, et al. (2021). *The sustainable development goals. Sustainable development report 2021*. Cambridge: Cambridge University Press.

Tordo, S., Warner, M., Manzano, E. M., & Anouti, Y. (2013). *Local content policies in the oil and gas sector*. Washington, DC: The World Bank.

Tsani, S. (2020). Public policies for just transition: local content, employment, and human capital. In W. Leal Filho, A. Azul, L. Brandli, A. Lange Salvia, & T. Wall (Eds.), *Decent work and economic growth. Encyclopedia of the UN sustainable development goals*. Cham: Springer. ISBN: 978-3-319-71058-7.

Tsani, S. (2021). Energy sector developments, local content policies and economic growth. In V. In Vlachos, A. Bitzenis, & B. Sergi (Eds.), *Modeling economic growth in contemporary Greece: The role of changing economic and industrial contexts*. Emerald Publishing.

Tsani, S., Paroussos, L., Fragkiadakis, K., Charalambidis, I., & Capros, P. (2013). Female labour force participation and economic growth in the South Mediterranean countries. *Economics Letters, 120*(2), 323–328.

Tsani, S., Riza, E., Tsiamagka, P., & Nassi, M. (2021a). Public policies, one-health and global inequalities under the Covid-19 lens. In F. W. Leal, A. Azul, L. Brandli, P. Özuyar, & T. Wall (Eds.), *Reduced inequalities. Encyclopedia of the UN sustainable development goals*. Cham: Springer.

Tsani, S., Riza, E., Tsiamagka, P., & Nassi, M. (2021b). Public financing and management for a sustainable healthcare sector: Some lessons from the Covid-19 pandemic. In W. Leal Filho (Ed.), *COVID-19: Paving the way for more sustainable world*. Cham: Springer.

UNCTAD. (2014). *Local content requirements and the green economy*. United Nations Conference on Trade and Development. UNCTAD/DITC/TED/2013/7. United Nations publication.

UNFCCC. (2021). Nationally Determined Contributions (NDCs). The Paris agreement and NDCs. https://unfccc.int/process-and-meetings/the-paris-agreement/nationally-determined-contributions-ndcs/nationally-determined-contributions-ndcs

United Nations. (2015). *Transforming our world: The 2030 agenda for sustainable development*. United Nations. https://sdgs.un.org/publications/transforming-our-world-2030-agenda-sustainable-development-17981

Unruh, G. C. (2000). Understanding carbon lock-in. *Energy Policy, 28*(12), 817–830.

WEF. (2021). *WEF fostering effective energy transition 2021 edition*. Geneva: World Economic Forum.

WHO. (2020). COVID-19 and sustainable development goals. *Bulletin of the World Health Organization, 98*, 646. https://doi.org/10.2471/BLT.20.263533.

World Bank. (2020). COVID-19 intensifies the urgency to expand sustainable energy solutions worldwide. Press release 28 May 2020. https://www.worldbank.org/en/news/press-release/2020/05/28/covid-19-intensifies-the-urgency-to-expand-sustainable-energy-solutions-worldwide

Evidence-Based Research

▶ Smart Agriculture and ICT

Extended Reality

▶ Augmented Reality: Robotics, Urbanism, and the Digital Turn

Extended Urbanization

▶ New Orbital Urbanization

F

Faith Communities

▶ Faith Communities as Hubs for Climate Resilience

Faith Communities as Hubs for Climate Resilience

Cherice Bock
Portland Seminary of George Fox University, Portland, OR, USA

Synonyms

Church gardens; Climate change resilience; Climate disaster preparedness; Community resilience; Cultural ecosystem services; Distributed energy; Faith communities; Food justice; Microgrids; Places of worship; Sacred sites; Social-ecological systems; Spiritual services

Definition

As impacts of climate change are felt with increasing regularity in everyday life and in climate-related natural disasters, it becomes imperative to building resilient communities that can mitigate and reverse climate change impacts, adapt to local changes, and recover quickly and equitably from climate disasters. This chapter considers faith communities as potential hubs for community resilience to respond to vulnerabilities due to climate change. By first situating faith communities as integral members of social-ecological systems offering cultural as well as provisioning and regulating ecosystem services, the chapter then discusses the ways faith communities contribute to community resilience.

Specific examples from faith communities in the State of Oregon, USA, are offered in regard to the ways faith community buildings and grounds contribute to spiritual and recreational services, food justice, resource cycling, and climate disaster preparedness and response. This chapter shows that faith communities can be valuable partners in working toward social and ecological resilience as the climate changes, helping decrease inequitable outcomes, implementing more efficient and sustainable energy and landscaping options, and providing accessible green spaces within urban and suburban settings. Places of worship can also be used as community spaces in the case of disasters, particularly if equipped with solar power and backup batteries, contributing to a distributed energy grid. Faith communities, embedded as they are in their local communities and networked with local partners as well as larger national and international groups, are well situated to respond quickly to climate disasters in context-specific ways.

R. C. Brears (ed.), *The Palgrave Encyclopedia of Urban and Regional Futures*,
https://doi.org/10.1007/978-3-030-87745-3

Introduction

As impacts of climate change are felt with increasing regularity in everyday life and in climate-related natural disasters, it becomes imperative to building resilient communities that can mitigate and reverse climate change impacts, adapt to local changes, and recover quickly and equitably from climate disasters. Faith communities can be valuable partners in creating community resilience for the purpose of more holistic and functional social-ecological systems, mitigating and adapting to climate change, and preparing for and responding to climate-related and other natural disasters.

Places of worship and the communities who worship in them can be found in most cities and towns around the world. These groups often hold a significant amount of land in their care, in addition to their buildings. Members of each faith community work in a variety of occupational sectors and hold a network of social connections and relationships within a town or city. Many faith communities hold ties with other groups from their religious tradition to create national and international webs of relationship and support across the globe. Additionally, many faith communities maintain partnerships with social service agencies, educational institutions, government agencies, and community groups at the local level. With land, social capital, and large buildings that can serve as community centers for education and in the case of disasters, faith communities are well situated to serve as community hubs within local climate resilience strategies.

Resilience, in the context of sustaining healthy social-ecological systems and responding to the challenges of climate change, can be defined as: "having the ability to live with change, and develop with it" (Biggs et al. 2015, xx). It is preferable to mitigate and reverse the causes of climate change, but communities the world over are already experiencing the impacts of unhealthy levels of greenhouse gas emissions, habitat fragmentation, environmental pollution, biodiversity loss, and unjust distribution of resources. Therefore, it is imperative to reorganize human communities in ways that can respond to climate disturbances with flexibility and equity and in ways that contribute to the overall health of social-ecological systems.

The following six capacities have been identified as important to consider when preparing rural and urban contexts to respond to climate disasters with community resilience: social, economic, physical, human, institutional, and environmental (Haase et al. 2021). Strengthening these six capacities can help a community respond to disturbances in ways that maintain essential services while experiencing novel situations. This does not necessarily mean returning to the status quo prior to the disturbance but will include creative and inclusive innovation so the community can transform in ways that meet current needs (Magis 2010; Young 2014). Additionally, three types of ecosystem services have been identified as essential in maintaining a resilient social-ecological system: provisioning services (food and other resources), regulating services (cycling clean water and air), and cultural services (recreational and spiritual) (Biggs et al. 2015; de Groot et al. 2005).

Although each faith community and each location will be different, this chapter offers examples of ways faith communities can be helpful in developing community resilience and responding to climate crises. Faith communities can offer important anchor points in the three nodes of ecosystem services identified above and hold meaningful stakes and leverage points in each of the six capacities for community resilience. Examples are drawn from the experiences of the researcher in academic work and faith-based community organizing around climate justice in the State of Oregon, USA. This chapter will highlight two main ways in which faith communities can contribute to community resilience and climate disaster preparedness: use of faith community lands to maintain and improve resilient social-ecological systems and use of faith community facilities in the case of climate-related and other natural disasters.

Faith Community Lands

Cultural Ecosystem Services: Spiritual and Recreational

Green spaces, including trees and plants, provide measurable spiritual services to individuals who experience increased well-being after time spent in natural spaces (Clayton and Myers 2015; De Lacy and Shackleton 2017; Kaczyńska 2020; Rall et al. 2017; Laband 2013). Faith community lands can offer this form of spiritual services, in addition to the cultural and spiritual services individuals experience from participation in religious rituals and communities. Those who spend time in church gardens experience the spiritual service of connection to nature as well as connection to their God.

The benefits of these faith community green spaces do not have to be limited to members of the faith community, however: green spaces within an urban environment can decrease stress and improve well-being for everyone who sees or visits them. For those faith communities offering outdoor recreational spaces to the community, such as trails, ball fields, or recreational access to waterways, faith community lands can also offer recreational services (Baghwat 2009; Laband 2013; Radford and James 2013; Rall et al. 2017).

Some faith communities intentionally open up their land to the broader community to offer spiritual and recreational services. For example, North Valley Friends Church in Newberg, Oregon, USA, partnered with the local parks and recreation district to build a 0.75-mile paved walking trail around a 20-acre property on the edge of a suburban town. Community members are welcome to walk, run, and exercise dogs on the trail. To fulfill his Eagle Scout project, a local teenager built circuit training exercise elements spaced out along the trail to encourage community fitness.

In addition to these recreational services, the trail also offers spiritual services: it contains forested portions of the trail as well as open prairie with a diversity of species (Laband 2013). The faith community also installed a dozen "peace poles" at intervals along the trail. These poles are engraved with the words, "May peace prevail on Earth," in a number of languages, and each pole includes an inspirational quote from famous civil rights and peace activists. Additionally, a 60-foot diameter prayer labyrinth is accessible to community members along the trail. This painted cement prayer labyrinth offers an opportunity for meditative reflection and prayer as one journeys along serpentine paths into the center and back (Baghwat 2009; Rall et al. 2017).

North Valley Friends also hosts a small community garden, a free used clothing space, and a tiny home village with three units and plans for more. Community garden produce is shared at times when the clothing space is open, with members of the congregation when they come to worship, and with the local food bank. The tiny homes help people who are experiencing homelessness, offering a safe space for people to get back on their feet. The tiny homes are designed to be energy efficient and with sustainable features. Students from the local high school build a tiny home each year to contribute to this village, so students learn about sustainable building practices.

These actions by North Valley Friends contribute to cultural ecosystem services including spiritual and recreational services, and they also contribute to community resilience through wise use of community resources, helping develop students with awareness and experience in sustainability practices, contributing to equity through sharing of food, clothing, and housing and through engaging community resources as active agents via partnerships with many local agencies and community groups.

Food Justice

An important aspect of functional and equitable social-ecological systems is the food system, and as the climate changes and natural disasters increase the incidence of both crop failure and supply chain disruptions, producing more food at the local level and in distributed sites rather than large-scale monocrops will increase community resilience (Biggs et al. 2015). Those with the least economic means are often those who are impacted most profoundly by food scarcity and price increases, and therefore, local food production that is accessible to those with lower socioeconomic statuses is a justice issue (Ryan-Simkins 2021; Sta Maria 2020).

Many faith communities have begun hosting community or church gardens on their lands (Byassee 2018; Grenfell-Lee 2017; Oliver et al. 2017). There is a range of models for these gardens, depending on the particular needs and assets of a given garden. Some church gardens provide spaces for members of the faith community to grow food for themselves and their families, while other such gardens grow food to contribute to a local food bank or to use in community meals they host. Other faith communities offer space for a community garden with plots available to neighbors for free or for a nominal fee. Other community gardens offer education about planting, growing, harvesting, cooking, and storing the produce grown.

McMinnville Cooperative Ministries in McMinnville, Oregon, USA, offers an example of a community garden on church lands. This community garden exists "to provide fresh produce to folks who are at nutritional risk" (OSU Extension Service 2021). The garden is organized by volunteers from the church and a number of community groups, including master gardeners, social service agencies, local businesses, a university extension service, and community groups connected to international organizations such as Rotary and Kiwanis. The garden includes a variety of ways to be involved: through a personal raised bed for a small fee, a personal raised bed for free with volunteer hours contributed, or volunteering to work in the garden to produce food for a local food bank. Children are able to garden their own small raised bed with instruction from master gardeners. In 2016, the garden produced over 14,000 pounds of food for the food bank (McMinnville Community Garden 2017).

Produce is distributed by the local food bank, and it is also used in weekly meals hosted by McMinnville Cooperative Ministries (before the COVID-19 pandemic). Church members and other neighbors volunteer to cook and clean up a meal each Saturday morning for the community, free of charge.

In addition to offering the spiritual and psychological benefits of a neighborhood green space, this community garden offers healthy, local, organic food to low-income neighbors free of charge in food boxes and in a prepared meal each week. This garden also offers the opportunity for neighbors to grow their own food, contributing to food justice in cases where individuals do not have their own access to land. There are some formal educational opportunities for children to learn how to grow food, creating a more resilient community in the future as these children grow up with skills and knowledge of growing their own food, and the garden also affords the opportunity for informal education as individuals work their garden plots alongside master gardeners, learning tips and tricks and developing relationships that strengthen community ties.

Regulating Services

One other aspect of faith community lands that contributes to climate and community resilience is that of regulating services: faith community lands can help cycle water and air to ensure a healthier environment for everyone. This can occur more or less naturally in the case of sacred groves, ancient trees growing in aged cemeteries, and forested areas of large acreages such as monasteries (De Lacy and Shackleton 2017; Kaczyńska 2020).

Even in urban and suburban contexts, faith community lands can help provide regulating services through implementation of sustainable landscaping, planting native species, and restoring wetlands and riparian zones.

St. Andrew Lutheran Church in Beaverton, Oregon, USA, provides an example of a faith community that is restoring a wetland on their land, planting native species. In partnership with Clean Water Services for their county, St. Andrew Lutheran has planted thousands of native trees, shrubs, and other wetland plants to restore acres of forest and wetlands. This helps cycle water and air and offers habitat for native insects and animals.

A suburban congregation, St. Andrew Lutheran also has a small community garden and a solar array. They have instituted a Carbon Gardens project, helping community members learn about the importance of healthy soil in capturing carbon dioxide and working with individuals and other faith communities to plan and implement their own carbon gardens, building soil and growing plants in their yards or planter boxes (Carbon Garden 2021; St. Andrew Lutheran Church). These projects have the direct intention of offering regulating services through their faith community lands and through educating other community members about carbon dioxide capture, helping them install projects that will contribute to healthier air quality and climate change mitigation.

An urban faith community implementing regulating services with a focus on water is Common Ground, a Presbyterian church in Portland, Oregon, USA. In partnership with the Portland Bureau of Environmental Services, Common Ground created a plan for more sustainable water cycling on their property through creating rain gardens. They worked with city agencies to hold public meetings in their building for neighbors disgruntled about the city's plan for a "green streets" effort, offering space for dialogue and communication to facilitate stakeholder buy-in. Common Ground received a grant to put in rain gardens, and they also received help from local master gardeners. This allowed them to put in two rain gardens and plant nine new trees. Rainwater is diverted from the roof and parking area into the rain gardens in order to stay in the neighborhood rather than running immediately into storm drains. Common Ground also removed invasive species on their land, replacing them with native species and receiving a Backyard Habitat certification through the Audubon Society, as well as becoming an Earth Care Congregation through the Presbyterian Church (USA).

In addition to these actions that contribute to regulating services, particularly in relation to water cycling, Common Ground also has a community garden, a rest and meditation walk, and a summer day camp incorporating environmental themes for neighborhood children. These programs offer cultural and spiritual services to members as well as the broader community. The actions they have undertaken on their grounds provide regulating services that contribute to practical aspects of community resilience.

Faith Communities and Disaster Preparedness

As climate-related disasters become more and more prevalent, community resilience plans must include preparation for disasters in addition to maintaining or fortifying existing social-ecological systems. Community resilience in the face of the climate crisis must include the ability to adapt while maintaining healthy capacity in all six areas of climate resilience: social, economic, physical, human, institutional, and environmental (Haase et al. 2021).

Faith communities can be important partners in preparing for and responding to climate-related disasters. They can be helpful in providing material aid in the form of money and goods, they can provide space for those who are displaced, and they can help with rebuilding after a disaster. Faith communities have long been involved in disaster assistance, with disaster response teams available to help after hurricanes, tsunamis, earthquakes, floods, and fires all over the world. Through systems of voluntary donations, supported by regular giving and special calls for donations for particular incidents, faith-based nonprofit organizations or groups within particular religious traditions can be ready to provide a range of aid from medical to material goods, with groups of volunteers spending time cleaning up and rebuilding. These groups include examples such as Mercy Corps, Church World Service, Habitat for Humanity, and a range of disaster response teams from various religious traditions and denominations such as Jewish Federations of North America, Buddhist Tzu Chi Foundation, Lutheran Disaster Response, Mennonite Disaster Service, Church of the Brethren Disaster Ministries, and Presbyterian Disaster Assistance. Because of faith communities' global networks, they are able to raise funds and distribute aid through local connections supported by outside donations and volunteer aid.

While these disaster response teams have previously been focused on sending teams from afar to contribute to response and relief after a disaster, some are turning to local preparedness for more frequent climate disasters (e.g., Iqbal 2021; Rumahuru and Kakiay 2020). This is the case for Lutheran Disaster Response, Region 1, which is based in the Pacific Northwest, USA. This organization is working to equip local congregations for preparedness, rather than simply giving funds to a national or international aid organization. They identify "systemic disaster issues," naming racism and climate change as the two most important concerns. In 2020, they held a webinar series entitled, "Revealing Relationships," in which they discussed connections between disaster preparedness and racism, climate change, and migration due to displacement (Region 1 Be Prepared 2020). In addition to this educational offering, Lutheran Disaster Response, Region 1, is working with congregations to prepare their buildings and have supplies on hand to assist their neighbors in the case of local climate and other disasters.

Places of worship can be very useful in the case of disasters since many of them are large enough to hold hundreds of people. Many are already equipped with large kitchens and are used regularly to feed groups of people. They often have gyms and sanctuaries spacious enough for many people to gather, and they have large parking areas where those displaced can park or where food and goods can be distributed.

Additionally, congregations can install solar projects with backup battery storage. In the case of a natural disaster where the electrical grid is compromised, buildings with their own backup storage will be able to continue to provide electricity. Individuals may also purchase solar and backup batteries for their own homes, but these do not aid the community very much in the case of a disaster. Congregational spaces equipped with solar and backup storage can contribute greatly to a distributed energy microgrid (Mishra et al. 2020). As large central locations, these buildings can offer help for people with medications that need to be refrigerated and offer charging stations so people can keep cell phone power and access important emergency updates, and continue to have power to store and cook food. In the case of natural disasters that include extreme heat or cold, these spaces can stay at a livable temperature, offering space to hundreds of others who would otherwise be at high risk of exposure.

In 2020–2021, in addition to the global pandemic, Oregon experienced two of the most extreme wildfire seasons on record, an ice storm that took down the electrical grid in some areas for over a week, and two periods of extreme heat, one of which repeatedly broke Oregon's all-time high temperature records on subsequent days. During the 2020 fire season, fires came dangerously close to densely populated regions, and a majority of the state was blanketed in unhealthy levels of smoke. Through a statewide group of faith communities called Ecumenical Ministries of Oregon, congregations willing to provide space for evacuees were identified, and a network of food, clothing, masks, and other aid was organized. Monetary donations from around the country and the world were received by faith-based disaster assistance groups to make this aid possible. In the winter and summer of 2021, many congregations opened up their spaces as warming and cooling shelters on the most extreme days. These faith communities did not do everything required to deal with these climate emergencies: a collaborative effort emerged, including local, state, and federal government support, firefighters and first responders, mutual aid groups, businesses offering their services, and volunteers from all walks of life offering their time and energy to make sure their neighbors could make it through these challenging times. Faith communities did, however, offer an important piece of the overall effort.

Conclusion

In times and places that are more or less normal and without disturbance, faith communities can help maintain sustainable and equitable social-ecological systems, contributing to community resilience. When disasters occur, faith communities can be valuable partners in maintaining essential services and helping meet immediate needs until the emergency has passed. Faith communities provide spiritual services not found elsewhere, and their grounds often offer green spaces within urban and suburban environments that not only can provide the social benefits to mental health and well-being of natural areas but can also help cycle air and water and create native habitats. Many faith communities contribute to more local and sustainable food systems through hosting community gardens on their grounds and contribute to food justice by offering food and the means of producing it to low-income neighbors. In times of disaster, faith communities are local partners with national and international networks of support, and in many cases, local congregations are ready and willing to help their neighbors with safe space, food, and other aid in the first hours and days after a disaster strikes. Therefore, faith communities are an invaluable partner to state and federal agencies and to community partners wishing to prepare in advance for disaster relief.

Not every faith community will be ready and willing to help, and there are also negative examples of congregations unwilling to share their space. Many others downplay or deny the reality of anthropogenic climate change.

Policymakers and officials tasked with climate resilience and disaster preparedness would do well, however, to partner with the many willing faith communities who care about environmental and climate justice and want to do their part to help their community weather the coming storms. Likewise, faith leaders and faith-based disaster response teams are encouraged to implement sustainable and just solutions to mitigate human-caused ecological system breakdowns and to reach out to partner organizations and government agencies in order to prepare the community well for disaster response and recovery.

Cross-References

- ▶ City Visions: Toward Smart and Sustainable Urban Futures
- ▶ Closing the Loop on Local Food Access Through Disaster Management
- ▶ Collective Emotions and Resilient Regional Communities
- ▶ From Vulnerability to Urban Resilience to Climate Change
- ▶ Green Cities
- ▶ Green Cities in Theory and Practice
- ▶ Multiple Benefits of Green Infrastructure
- ▶ Need for Greenspace in an Urban Setting for Child Development
- ▶ Policies for a Just Transition
- ▶ Role of Disaster Relief Policy in Building Resilient Coastal Regions in the United States
- ▶ Sustainable Cities via Urban Ecosystem Restoration
- ▶ Toward a Sustainable City
- ▶ Urban Food Gardens

F

References

Baghwat, S. A. (2009). Ecosystem services and sacred natural sites: Reconciling material and non-material values in nature conservation. *Environmental Values, 18*(4), 417–427. https://doi.org/10.3197/096327109X12532653285731.

Biggs, R., Schlüter, M., & Schoon, M. L. (Eds.). (2015). *Principles for building resilience: Sustaining ecosystem services in social-ecological systems*. New York: Cambridge University Press.

Byassee, J. (2018). God's gardeners: In British Columbia, a community tends the watershed. *The Christian Century, 135*(9), 24–26.

Carbon Garden. (2021). Retrieved 9 Oct 2021 from Carbon Garden. http://www.carbongarden.org/

Clayton, S., & Myers, G. (2015). *Conservation psychology: Understanding and promoting human care for nature*. Hoboken: Wiley.

de Groot, R., Ramakrishnan, P. S., et al. (2005). Cultural and amenity services. In R. Hassan, R. Scholes, & N. Ash (Eds.), *Ecosystems and human well-being: Current state and trends, vol. 1. Millennium ecosystem assessment* (pp. 455–476). Washington, DC: Island Press.

De Lacy, P., & Shackleton, C. (2017). Aesthetic and spiritual ecosystem services provided by urban sacred sites. *Sustainability, 9*(9), 1628–1642. https://doi.org/10.3390/su9091628.

Grenfell-Lee, T. Z. (2017). Earth as community garden: The bounty, healing, and justice of holy permaculture. In J. Hart (Ed.), *The Wiley Blackwell companion to religion and ecology* (pp. 410–426). Oxford, UK: Wiley.

Haase, T. W., Wang, W., & Ross, A. D. (2021). The six capacities of community resilience: Evidence from three small Texas communities impacted by hurricane Harvey. *Natural Hazards, 109*(1), 1097–1118. https://doi.org/10.1007/s11069-021-04870-y.

Iqbal, Z. (2021). When Texas froze, Muslim volunteers mobilized to help. *The Christian Century, 138*(6), 17–18.

Kaczyńska, M. (2020). The church garden as an element improving the quality of city life – A case study in Warsaw. Urban Forestry & Urban Greening, 54, 126765. https://doi.org/10.1016/j.ufug.2020.126765.

Laband, D. N. (2013). The neglected stepchildren of forest-based ecosystem services: Cultural, spiritual, and aesthetic values. *Forest Policy and Economics, 35*, 39–44. https://doi.org/10.1016/j.forpol.2013.06.006.

Magis, K. (2010). Community resilience: An indicator of social sustainability. *Society and Natural Resources, 23*(5), 401–416.

McMinnville Community Garden. (2017). Community participation. Retrieved 9 Oct 2021 from McMinnville Community Garden. https://www.mcminnvillecg.org/about

Mishra, S., Anderson, K., Miller, B., Boyer, K., & Warren, A. (2020). Microgrid resilience: A holistic approach for assessing threats, identifying vulnerabilities, and designing corresponding mitigation strategies. *Applied Energy, 264*. https://doi.org/10.1016/j.apenergy.2020.114726.

Oliver, E., Tshabele, V., Baartman, F., Masooa, A., & Laister, L. (2017). Can Christians really make a difference?: A response to the call for change to make the world a better place. *HTS Theological Studies, 73*(3), 1–11. https://doi.org/10.4102/hts.v73i3.4351.

OSU Extension Service. (2021). Demonstration gardens across Oregon: Yamhill county – McMinnville Community Garden. Retrieved 9 Oct 2021 from Oregon State University. https://extension.oregonstate.edu/mg/demonstration-gardens-across-oregon

Radford, K. G., & James, P. (2013). Changes in the value of ecosystem services along a rural–urban gradient: A case study of Greater Manchester, UK. *Landscape and Urban Planning, 109*(1), 117–127. https://doi.org/10.1016/j.landurbplan.2012.10.007.

Rall, E., Bieling, C., Zytynska, S., & Haase, D. (2017). Exploring city-wide patterns of cultural ecosystem service perceptions and use. *Ecological Indicators, 77*, 80–95. https://doi.org/10.1016/j.ecolind.2017.02.001.

Region 1 Be Prepared. (2020). Region 1 disaster preparedness and response, Evangelical Lutheran Church in America. Revealing relationships. Retrieved 11 Oct 2021 from Region 1 Be Prepared. https://www.region1beprepared.org/webinars/

Rumahuru, Y. Z., & Kakiay, A. C. (2020). Rethinking disaster theology: Combining Protestant theology with local knowledge and modern science in disaster response. *Open Theology, 6*, 623–635. https://doi.org/10.1515/opth-2020-0136.

Ryan-Simkins, K. (2021). The intersection of food justice and religious values in secular spaces: Insights from a nonprofit urban farm in Columbus, Ohio. *Agriculture and Human Values, 38*(3), 767–781. https://doi.org/10.1007/s10460-020-10188-5.

St. Andrew Lutheran Church (2021). Earth care. Retrieved 9 Oct 2021 from St. Andrew Lutheran Church. https://standrewlutheran.com/earth-care/

Sta Maria, F. P. (2020). Fitting food to circumstances: Potential contributions of Philippine culinary heritage to disaster risk reduction. *Budhi, 24*(2), 37–84.

Young, R. (2014). Some behavioral aspects of energy descent: How a biophysical psychology might help people transition through the lean times ahead. *Frontiers in Psychology, 5*, 1–16.

Feminism

► Feminist Planning and Urbanism: Understanding the Past for an Inclusive Future

Feminist Planning and Urbanism: Understanding the Past for an Inclusive Future

Jenna Dutton[1], Chiara Tomaselli[2], Mrudhula Koshy[3], Kristin Agnello[4], Katrina Johnston-Zimmerman[5], Charlotte Morphet[6] and Karen Horwood[7]

[1]Senior Planner – Social Policy, City of Victoria and Research Associate, Center for Civilization, University of Calgary, Calgary, AB, Canada

[2]Consultant – Urban Asset Advisory, Arcadis France, Paris, France

[3]Norwegian University of Science and Technology, Trondheim, Norway

[4]Department of Architecture and Planning, Norwegian University of Science and Technology, Trondheim, Norway

[5]THINK.urban, Philadelphia, PA, USA

[6]Women and Planning research bursary, Planning, Housing and Human Geography, The Leeds Planning School, Leeds Beckett University, Leeds, UK

[7]The Leeds Planning School, Leeds Beckett University, Leeds, UK

Synonyms

Equity; Feminism; Gender equality; Gender mainstreaming; Intersectional feminism; Planning; Urbanism

Definitions

Feminist Planning: The process driving the equitable involvement of all persons in the planning profession, which was originally initiated with consideration for women and later expanded to include all persons regardless of age, gender, place of origin, and sexual identity.

Feminist Urbanism: Various facets of contemporary urban design and planning practices that developed as a reaction to traditional planning processes that were shaped by patriarchal values, with an intention to facilitate equitable access to the benefits and opportunities that can be offered in multilevel agendas.

Introduction

Women have always held pivotal roles in society; however, the vast majority of history has shed light on the contributions of a select few and has tended to focus disproportionately on the achievements of privileged men, particularly those of white cis males (Hendler et al. 2017, p. 25; Allick 2020). These patriarchal norms and the structures that they have enabled have encompassed all facets and all corners of the world and the ensuing top-down exclusionary processes have in many ways impacted how cities have been planned, designed, and developed for centuries (Eichler 1995; Kern 2020a). This inequity has thereby impacted women's daily experiences, comfort, and sense of place regardless of the geographical context.

A woman's experience in the urban environment is entirely distinct from a man's: from perceptions of fear and safety in public spaces to the unique biometrics of a woman's stride when carrying a package or pushing a stroller (Perez 2019; Kern 2020a). Women are more likely to experience fear and intimidation in the built environment, are more likely to be limited in terms of access to and movement through urban spaces, are more often observed by others, and are more likely to be accompanied by children, seniors, or other people in their care. Despite these tangible impacts, these considerations have and continue to be absent from the majority of local, national, and global priorities and policymaking at the detriment to half of the world's population.

As part of the Sustainable Development Goals (SDGs) adopted at the United Nations in 2016, Goal 5 strives to achieve gender equality and empower all women and girls. This widespread inequality has been further exposed during the 2020 global pandemic as lockdowns have increased domestic violence toward women and girls, highlighted that 70% of health and social workers are women, and shown that women take on the majority of unpaid domestic work (UN 2020). To improve these trends during and after the global economic recovery with community and city building practices, there is a call for an increase in the presence and consideration of women in politics and leadership positions (Didi 2020), in foreign policy (Centre for Feminist Foreign Policy 2020), an improvement in women's active participation outside government (McFarlane 1995), a reduced or eliminated gender pay gap (Auspurg et al. 2017; Rubery et al. 2005), and a reduction of the gender data gap to include sex disaggregated data (Klein and D'Ignazio 2020). It is clear that inequity is rampant in all segments of society and thereby intersects to structure traditional urbanist practices.

Though previous chapters have explored many facets of urbanism, feminist urbanism involves different considerations that include the systematic impacts of development. Though there are some general definitions of feminist urbanism, it has been generally described to explore how the built environment excludes the consideration of gender. Though urban planning and urban design have intended to reflect the human experience, both practices have been conducted by and for a specific segment of society, at the detriment of everyone else (Kern 2020b; Perez 2019; Moghadam and Rafieian 2019). Feminist Urbanism acknowledges that cities and city building practices have been created and evolved through predominantly patriarchal structures that have reenforced sexism in both public and private contexts.

Choosing to focus on urbanism for this chapter is not at the exclusion of planning, rather the acknowledgement that the profession of planning and its conventional history has been limited to a few male planning visionaries (Hendler et al. 2017, p. 24). Women's efforts were not recognized as "legitimate" planning and due to the "narrative of professionalization" many scholars have acknowledged that community building movements outside of token city building have played an equally important role (2017, p. 25). As will become further apparent in the following sections, evidenced by the diversity and breadth of examples from across the world, the definition of feminist urbanism is always evolving.

Nevertheless, to apply a true feminist lens to urbanism is to acknowledge that equity is equally as essential in informal settlements in India, major cities in north America, small rural towns in Europe, favelas in Brazil, islands in the Caribbean, and for all groups of people in those contexts.

The chapter is divided in the following sections. Firstly, it traces the development of feminist urbanism by exploring the links between various theoretical traditions such as feminism, intersectional feminism and feminist geography, and their influence on the planning field. It then discusses the advances made in global and national policymaking in countries around the world in efforts to mainstream gender in planning cities. The next section highlights examples of women-led planning networks and organizations on a global, national, city, and local level. This is followed by a section on best practices on physical and spatial interventions in places around the world and the barriers to achieve them. Finally, the chapter concludes with reflections on the contextual limitations of gender-based interventions and the implications of mainstreaming feminist urbanism.

Theoretical Foundations

The theoretical foundation of this chapter is based on various strands of feminist theories and traces the arc of feminist urbanism by drawing links between the following theoretical traditions, namely, feminism, intersectional feminism, feminist geography, and ecofeminism. Given the extensive and complex history and multiple streams of feminism and its corresponding waves, we have notably not included an entirely fulsome summary. The selected focus intends to connect most directly with the current policy and practices that are described in the following sections.

The presence of women in the built environment has long been used as an indicator of the safety, prosperity, and values of a community (Day 1999; Hayden 1980; Kern 2020a). Still today, women – or, more specifically, white, heteronormative women – are often used as markers for the gentrification and revitalization of urban neighborhoods (Kern 2020a). However, women are not a homogenous population; therefore, it is critical to consider intersecting identities in urban responses, including racial and ethnic diversity, sexuality, age, ability, income, and life stage (Brown-Saracino 2015; Crenshaw 1989; De Madariaga and Neuman 2016; Irazábal and Huerta 2016; Kern 2020a). Without due consideration of how these identities overlap and impact socio-spatial experiences, then the needs of women and other marginalized populations can be understood only to the extent that they intersect with the dominant population.

Feminism and Intersectional Feminism

Feminism has a myriad of definitions attributed to its evolving waves in history. Although traditionally defined as "the belief and aim that women should have the same rights and opportunities as men; the struggle to achieve this aim" (Oxford University Press 2020). The feminist activist Bell Hooks stated that "feminism is a movement to end sexism, sexist exploitation, and oppression" (Hooks 1984). The feminist movement has been widely criticized for its predominance of white, privileged feminists (Biana 2020) and thereby exclusion of other groups. Additionally, it has been acknowledged that feminist discourses must evolve and be informed by diverse identities and sexualities of queer-identified people in urban planning, geography, and public safety (Angeles and Roberton 2020). As Leslie Kern notes in Feminist City, "feminism must be intersectional if it seeks to address the challenges of the present moment" (2020a). Intersectionality looks beyond a single-axis framework to acknowledge that intersectional experience is more than the sum of racism and sexism (Crenshaw 1989). Therefore, our definition of feminist urbanism intends to involve intersectional feminism to accurately reflect the zeitgeist.

As coined by Kimberlé Crenshaw in 1989, intersectionality is a "method and a disposition, a heuristic and analytic tool" introduced to address the marginalization of Black women in antidiscrimination law, feminist and antiracist theory, and politics (Carbado et al. 2013, p. 303). In 1991, Crenshaw further elaborated the framework "to highlight the ways in which social movement and organization and advocacy around violence against women elided the vulnerabilities of women of color, particularly those from immigrant and socially disadvantaged communities." (2013, p. 304).

This was however not the first intervention of its kind as there were previous attempts by Marxist-feminist theorists, interventions from lesbian feminism perspectives, and connections made between gender and disability (Supik et al. 2012). The concept has since been critiqued for only reflecting the intersectional experience of black American women and not women of color in Europe (Kerner 2017, p. 848). Nonetheless, this does not negate the global interdisciplinary engagement that the intellectual history of intersectionality has created (Carbado et al., p. 303). Kerner (2017) posits that feminist theory could benefit from both intersectional feminism and postcolonial feminism. Acknowledging inequality among subgroups of women, power relations among women and interactions among feminists could assist with framing the understanding and discourse (2017, p. 847).

Feminist Geography

The discipline of feminist geography is a subset of human geography that deals more explicitly with the aspects the female experience as it relates to space. This can be public or private space, as is the case of the seminal text by Dolores Hayden "What Would a Non-Sexist City Be Like? Speculations on Housing, Urban Design, and Human Work" (Hayden 1980). In this article Hayden deconstructs the notion of "home" and a woman's place within it by suggesting alternative models for cohabitation in order to break down gender norms of work (and by extension, society) through reverting isolation and confinement of women to a section of the standard dwelling unit. This is considered one of the earliest references to overtly calling the city "sexist" and was a part of the launching point of feminist geographical discussion beginning in the 1970s.

Ecofeminism

Through its joint name, ecofeminism recognizes that the domination of women and nature cannot be understood without recognizing the conceptual ties between the two (Mallory 2018). Ecofeminism began in the 1970s as a political movement by Francis D'Eaubonne who argued that "the destruction of the planet is due to the profit motive inherent in male power" (Nhanenge 2011, p. 11). There are many perspectives on ecofeminism; however, it does generally note that women's position in society is due to prevailing social and economic structures that expose them to particular environmental incivilities (Buckingham 2004, p. 147). Further, that as women share from being disadvantaged by environmental degradation, they are therefore in the best position to argue on nature's behalf (2004, p. 147). Importantly, ecofeminists caution the value in pursuing unlimited economic growth at the expense of the environment (2004, p. 149). This has recently been echoed by Jacinda Ardern in New Zealand's national budget by focusing on well-being rather than traditional economic growth (Peat 2019).

These considerations are inherently tied to climate change, especially with disaster management in informal settlements. A glaring example of this was demonstrated in Gujarat, a state in Western India in post-disaster management after an earthquake. As the majority of housing had been destroyed new homes were needed but as women were not included or consulted in the process, homes were built entirely without kitchens, thereby impacting essential daily tasks (Perez 2019, p. 575). Fundamentally, the climate crisis is a feminist issue as it disproportionally impacts women and amplifies existing gender inequalities across the world (Baker 2019).

Influence of Feminist Theoretical Foundations on Planning Theory

Planning theory has been isolated from the field in which it operates (Fainstein 2005, p. 127) and speaks to what planners do with "little reference either to the socio-spatial constraints under which they do it or the object they seek to affect" (2005, p. 121). Planning theorists delved into an abstract process that was isolated from social conditions, planning practice, and the physical city (Bureagard in Fainstein 2005, p. 211). This detachment between theory and practice resulted in a disconnect between the real need for community planning, people in a community wishing to improve their environment (Hodge and Gordon 2008, p. 3). Historically plans have not reflected the "public good" (2008, p. 6) despite the intent to be in the public interest, because it is not "monolithic and neutral" (2008, p. 7) and has resulted in inequity in community planning. "Progressive" ideas of acknowledging equity in planning are difficult to implement given long established policies that concentrate low-income and minority households and exclude them from other neighborhoods through zoning and building codes (Carmon and Fainstein 2013, p. 126). Despite the "trickle-down" policy in cities that has produced "few benefits for increasingly destitute residents" (2013, p. 124) planners have a substantial amount of power and ability to see comprehensively that could be leveraged in the future. However, the lack of gender diversity in planning affects not only the way we design and plan but also who we design and plan for (RTPI 2020).

Policymaking

Although there have been attempts at integrating feminist concepts in urbanist practices over the past 20 years, a full integration into policy has been lacking. While some contexts in the global north have achieved more success in this regard, it is especially glaring in many developmental contexts in the broadly defined "Global South" where initiatives often remain on paper with many barriers for implementation.

Organizations in many countries in the Global South are often not explicitly focused on women-centric planning issues but have in recent decades, mandated various policies ranging from women's rights and empowerment, safety issues in public spaces, and bottom-up, community-oriented planning. Urban planning in the broadest sense in these contexts is often top-down and bureaucratic, functioning through a combination of zonal and master plans and tends to leave out nuances in spatial interventions like walkable street profiles and vital public spaces that can enable safer cities. It thus contributes to fragmented and therefore unsafe cities reinforced by huge gaps in income and livelihoods and unequal access to safe public transport in different urban areas. Together with weak governance systems, these planning models tend to perpetuate inequality in public spaces and affect the safety of women.

In these contexts "unequal gender relations" (Falú 2017) also mean that leadership positions in the planning sector are increasingly occupied by male professionals. These factors collectively contribute to a lack of focus toward addressing the dearth of women planning professionals in public and private domains and negate the possible spheres of influence for gender mainstreaming in city planning, leaving women planners out of key decision-making processes. Recent changes strive to be more strategic and inclusive through communicative and participative forms of planning with a mandated gender focus However, these are not explicitly often women-led and tend to get lesser priority due to a bureaucratic way of functioning and local political dynamics. We discuss these through concrete examples in the next section. In including these examples from across the globe, the intent is not to provide copy-paste approaches to feminist urbanism practices. Rather, the intention is to acknowledge that cities evolve at different rates and through entirely different contexts, and these examples can be referenced to assist in progression. Despite the examples that are typically associated with feminist urbanism, it does not solely concern women's safety; rather it entails an equal opportunity to interact, move around, and enjoy the vitality offered by the urban environment regardless of the season or time of day.

The following section looks at examples of feminist policy and gender inclusion. Despite its inclusion and mandate by some governments, it is important to note that "women are underrepresented as social actors in policy making" (Mósesdóttir and Erlingsdóttir 2005), and this results in difficulties actioning and implementing these policies.

Global Policymaking

The structural connections between gender equality and sustainable urban development are already well acknowledged at the global level in the Sustainable Development Goals (SDGs). Indeed, SDG 11 explicitly commits to gender equality by making cities and urban settlements inclusive, safe, resilient, and sustainable and providing universal access to safe, inclusive, and accessible green public spaces. Following this path, the New Urban Agenda, adopted in 2016 at the United Nations Conference on Housing and Sustainable Urban Development, advocates for gender-inclusive and gender-responsive cities. UN Women, in partnership with other UN agencies and partners, participates in the implementation of integrated and evidence-based approach in cities though the *Safe Cities and Safe Public Spaces* Global Flagship Programme (UN Women 2017) and the *Safe Cities Free of Violence against Women and Girls* Global Programme (Bhatla et al. 2013).

UN Women advocates for ensuring women's voices and needs count, achieving equal participation in decision-making and including gender perspective in policies (UN Women 2020). In Albania, for example, UN Women worked with the southern cities of Fier, Berat, and Përmet on gender-responsive budgeting through open discussion and identification of community needs (UN Women Albania n.d.). The agency also works on gender mainstreaming in institutional development, such as the initiative in Tbilisi, Georgia, to conduct a participatory gender audit with the objective of serving as an entry point for mainstreaming gender into urban development and planning (Government of Georgia 2017).

Organizations such as the Centre for Feminist Foreign Policy have acknowledged the restrictive and exclusionary lens of policy at a global scale. Looking beyond the existing foreign policy discourse toward "a multidimensional policy framework that aims to elevate women's and marginalised groups' experiences and agency to scrutinise the destructive forces of patriarchy, colonisation, heteronormativity, capitalism, racism, imperialism, and militarism" (Centre for Feminist Foreign Policy 2020).

Canada

In Canada, the federal government initially committed to using Gender-Based Analysis plus (GBA+) in 1995 as an analytical process to assess how diverse groups of women, men, and non-binary people may experience policies, programs, and initiatives (Government of Canada 2018). The commitment was renewed in 2015 to support the full implementation across federal departments and mandated the Minister of Status of Women to ensure that any government policy and legislation is sensitive to the impacts on diverse groups of people (2018). The Fall 2015 Report of the Auditor General of Canada entitled "Implementing Gender-Based Analysis," provided a 2016–2020 Action Plan to address full implementation of GBA+ across government by identifying and addressing barriers and reporting on progress (Government of Canada 2015).

Vienna, Austria

The municipality of Vienna, in Austria, is known to be a pioneer in enforcing gender mainstreaming into urban planning by introducing different user profiles, with the objective of improving and ensuring "fair shares in the city" for all. The city took a stand to assume a change in its planning culture through the adoption of gender-sensitive planning in 1991, putting respect for the everyday life of women and men of all ages and at different life phases at the very center of the urban planning process (Foran 2013). By referring to different user profiles, the municipality all refers to further variables of discrimination such as age, sociocultural background, religion, or physical or psychological abilities. To support the daily life of these different users, Vienna came up with strategic visioning to guide cross-sector actions. The models go from polycentric urban structure to improve urban infrastructure and services accessibility to the idea of supporting a "city of short distance" allowing combination of paid work, family chores, and services users but also focusing on developing adequate public spaces (Bauer 2009). After two decades of pilot projects and processes, gender mainstreaming is at the center of Vienna's structuring urban documents (Strategy Plan for

F

Vienna, urban plan and sectoral programs, master plans, and urban design concepts etc.).

Iceland

Iceland has been the world's leader in gender equality since 2009 (WEF 2019, p. 9). Issues of gender equality are not addressed by a national employment policy, rather dealt with in special equality action programs (Mósesdóttir and Erlingsdóttir 2005). These programs are connected to the Iceland Gender Equality Fund and Gender Equality Implementation Fund and tied to the 2008 Gender Equality Act (Centre for Gender Equality Iceland 2012). The Gender Equality Act aims to "establish and maintain equal status and equal opportunities for women and men and promote gender equality in all spheres of the society" (European Parliament 2010).

Women-Led Planning Networks and Organizations

This section aims to highlight women-led planning networks and organizations from different parts of the world and functioning at a global, national, city, and local scale. Many of these organizations were borne out of a need to bring women planning professionals together for various initiatives, to create a sense of solidarity and belonging, and to address together the urgent need for a feminist lens on urban planning.

Global

Commonwealth Women in Planning Network

The Commonwealth Women in Planning Network is a subgroup of the Commonwealth Association of Planners (CAP), which represents over 40,000 professional planners from across 28 Commonwealth countries. Formed in 2018, the Commonwealth Women in Planning Network recognizes the potential of intersectional, gender-inclusive planning, policy, and design to positively contribute to global economic, social, cultural, and environmental objectives. The Network released the unprecedented *Women in Planning Manifesto* (Commonwealth Women in Planning Network 2018), an international call to action for advancing the role of women in the planning profession and highlighting the impact of planning and design on women's safety, prosperity, and empowerment. The Manifesto has since been adopted by CAP and a number of global planning organizations, supporting the Network's objective of fostering inclusiveness and equality, as advocated for by the 2030 Agenda, the New Urban Agenda, and the Paris Agreement.

Slum Dwellers International

It is also important to highlight the role played by women in initiating transformative initiatives in informal settlements. The network Slum/Shack Dwellers International (SDI) formed by the urban poor across 15 countries of the Global South originally initiated by federations founded by women of savings groups in local communities (Patel and Mitlin 2007). The roles of these federations have grown from managing savings and credits to taking actions aiming at improving access to basic infrastructure and services in local neighborhoods, securing land, upgrading homes, and more generally improving the quality of life in informal settlements. The SDI approach means to and foster participation to address women's needs and create and protect local space where women can engage. SDI also adopted a methodology based on a "culture to aspire" to empower women and support them to operate in a dominant culture to bring change.

Urbanistas Global

Urbanistas is a global network founded in 2012 in the UK. There are chapters across the UK, Rotterdam, and in New York and Sydney (Urbanistas 2020). The collaborative women-led network aims to amplify women's voice and ideas to improve cities (Urbanistas 2020). The aim of the network is to support women to start and deliver their own projects by crowd sourcing support from women within the network. Projects have included the Office of Displaced Designers (ODD). ODD is a project based in Greece which seeks to provide refugees with professional development opportunities in architecture and design (Urbanistas 2020).

Urbanists Rotterdam, Netherlands

Urbanistas Rotterdam, based in the city of Rotterdam, is a community of women urbanists working in and around the Randstad region in The Netherlands (comprising the urban agglomeration between Amsterdam, The Hague, Rotterdam and Utrecht). While The Netherlands is considered a progressive country in many aspects, there persists a glass ceiling wherein the ratio of women occupying leadership positions is skewed (ref xx). This is also reflected in the urban planning and design sector. Urbanistas Rotterdam strives to overcome this gap through bottom-up and community-oriented initiatives where young women urbanists can showcase their work and ideas to each other and gain peer support.

National

Women in Urbanism Aotearoa, New Zealand

Women in Urbanism Aotearoa is a society located in New Zealand. Aotearoa is the Maori name for New Zealand, and the organization has a mission to "transform our towns and cities into more beautiful, inspiring and inclusive places for everyone" (Women in Urbanism Aotearoa 2020a). They are membership-based for those who have an interest in cities, sustainability, climate change, and good design outcomes for women and live in or have a strong connection to New Zealand (Women in Urbanism Aotearoa Incorporated 2020). The organization advocates for efforts such as Equitable Light Rail for Auckland (Women in Urbanism Aotearoa 2020b) and a Street Harassment Campaign (Women in Urbanism Aotearoa 2020c).

Women in Planning UK

Women in Planning UK is an independent network for women working in planning and related disciplines. It was started in 2012 and is separate from the Royal Town Planning Institute. It has 14 branches across the UK including outposts in the devolved nations. Women in Planning aims to promote equality, inclusion, and diversity within the planning profession; with a focus on gender. The network hosts events, publishes blogs such as the "Day in the Life of" series and in 2020 has set up a new initiative called Thought Exchange. Thought Exchange aims to provide a platform for women in the profession to share ideas through short videos, hosted on YouTube channels, and social media posts. In addition, the network has published research on gender equality within the UK planning profession (Women in Planning 2019). The research focused on how many women held the position of director and above at private planning consultancies in the UK. The research found that women held 17% of these positions with men filling 83% (Women in Planning 2019).

The Association of Women in Property, UK

Women in Property is a UK-wide network founded in the 1980s (FIND REF). The network is focused on gender equality in the property and construction industry. It runs programs including the Mid-Career Taskforce, mentoring program, and annual National Student Awards. The awards have been running for 14 years with over 124 students nominated by their lecturers across 60 different universities (Women in Property 2020).

Safetipin, India

Safetipin, based in Delhi and founded in 2013, is a social developmental organization that aims to tackle safety for women in public spaces through crowdsourced data on mobility patterns via three apps, My Safetipin, Safetipin Nite, and Safetipin Site and safety audits. Through the apps, the users can express their observation on issues such as lack of street lighting, broken footpaths, open electrical wiring, etc. which enables the apps to plan safer routes for women to walk to their destinations. They also work closely with public service providers by acting as a mediator through the crowdsourced information, with NGOs, and companies that strive to provide safe access to their women employees.

Tanzania Women Architects for Humanity, Tanzania

Tanzania Women Architects for Humanity (TAWAH) is a multidisciplinary consortium of women architects, engineers, scientists, and quantity surveyors who undertake collective initiatives through participatory methods by mobilizing women from low-income groups to build adequate housing for marginalized communities. This includes those living in poverty, indigenous communities, minorities, and refugees who are traditionally excluded from an equitable access to basic services and appropriate living conditions. By imparting practical building skills and know-how for women from low-income groups, this NGO strives for social justice by enabling the women to generate income and economic opportunities for themselves.

TAO-Pilipinas, The Philippines

TAO Pilipinas is a women-led NGO in The Philippines that started in 1986 following the end of 20 years of dictatorship through a People Power revolution and subsequent socio-civic student activism bolstered through community participation. Currently the organization assists the urban and rural poor through various capacity building and community and research projects to enable sustainable and inclusive forms of development.

Women in Real Estate, Kenya

Women in Real Estate (WIRE) functions as a networking organization that strives for capacity building and increasing the presence of women professionals in the increasingly male dominated architecture and building industry in Kenya (Maichuhie 2020). Through various initiatives pertaining to business development, industry research, career outreach, advocacy, and mentoring programs, WIRE serves as a platform that enables women to broaden their knowledge and leadership skills and grow their network and business opportunities.

Women in Planning South Africa (WiPSA)

Women in Planning South Africa (Appendix A) is a network seeking to promote gender mainstreaming in the town planning and development sector, still predominantly dominated by males. Their activities revolve around space planning but also economic and social community development. The association aims at transforming the profession by building a platform to foster support and collaboration among women. Through a set of seminars, workshops, and online courses and by developing mentorship programs, the network also aspires to ensure that young women find guidance and opportunities to enter the job market.

Local

Women Transforming Cities, Vancouver, Canada

Women Transforming Cities (WTC) was formed in 2009 from the City of Vancouver's first Women's Advisory Committee with the vision to live in cities where all self-identified women and girls, in all their diversity, have real social, economic, and political power (Women Transforming Cities 2020). Since inception they have led initiatives and challenges to connect to the community including the Women-Led Cities Initiative, an initiative to action on systemic barriers on women's participation in local government and the Hot Pink Paper Campaign to incentivize women to vote. Most recently they were a main contributor to the Federation of Canadian Municipalities (FCM) Toward Parity Framework (FCM 2020).

Women in Cities International, Montréal, Canada

Women in Cities International (WICI) was founded in 2002 in Montréal, Canada, following nearly a decade of work on women's safety in Canadian cities. WICI champions the participatory potential of women's safety audits, maintaining that women and girls are experts in their own safety. The nonprofit organization works alongside local and national governments, urban planners, transit authorities, international organizations, and community groups to build safe, inclusive, and accessible cities. WICI was instrumental in the development of the First International Conference on the Safety of Women in 2002 and continues to generate and exchange knowledge on women's and girls' experiences in urban environments. WICI has maintained a commitment to using participatory approaches to support dialogue and change, while expanding their organizational focus to incorporate conversations around diversity, disability, gender policy, and women's access to water and sanitation.

The Women-Led Cities Initiative, Philadelphia, USA

In 2018 Women-Led Cities (WLC) began with a pilot working conference for women in the Greater Philadelphia region working in various urbanism fields. Attention was specifically placed on recruiting women from a wide range of backgrounds and levels of experience, including students, urban planners, artists, policymakers, and grassroots organizers. An intersectional cohort of mostly Women of Color was assembled to join three working conferences over the following year thanks to a grant from the Knight Foundation. Though the project is no longer active following the funding period, it was featured at SXSW, Placemaking Week in Amsterdam, and various other planning conferences where workshops helped shed light on the gender inequity of urban design and leadership.

Women in Urbanism YYC, Calgary, Canada

Women in Urbanism YYC is based in Calgary, Canada and was founded in March, 2019 as a way to connect women working in urbanism-related fields such as planning, architecture, landscape design, community associations, civil society, and other realms. Inspired by Women in Planning UK and Women in Urbanism Aotearoa and similar efforts, the network aims at promoting equity, diversity, and inclusion in urbanism (Women in Urbanism YYC 2020). Since adoption the network has served as a point of connection and resource sharing for local women, has held numerous discussion and brainstorming sessions, and held a popular cross-disciplinary mentorship and networking event and webinar discussion on feminism during COVID-19.

Promising Practices and the Barriers to Achieve Them

This section showcases examples of best practices from different parts of the world. While examples from countries in Europe and North America focus on policy and spatial interventions that were a result of feminist planning practices, those from developmental contexts in Africa, Asia, and South and Central America highlight the ongoing efforts to mainstream gender conscious planning practices as well as the systemic issues that act as barriers to achieving them. While this is not an exhaustive list, it serves as examples of how different contexts have responded to the need to put gender-based planning on the agenda.

Europe

Sweden Based on initial efforts from the town of Karlskoga, many towns in Sweden changed their seemingly gender-neutral approach to snow-clearing to shift the prioritization from main roads to local streets. This was based on a 2011 gender equality initiative that resulted from data collection that demonstrated how male and female mobility patterns are quite different (99pi 2019). When the town councilors changed the snow-clearing order, there was no obvious cost associated; however, they ended up saving money by reduced hospital admissions from injuries caused from slippery or icy pedestrian conditions (Perez, p. 4). Caroline Criado-Perez also looked at data from other Swedish cities showing that "the cost of pedestrian accidents in icy conditions was about twice the cost of winter road maintenance" (p. 6).

Barcelona, Spain Barcelona's first female mayor Ada Colau Ballano was elected in 2015 under the municipalist political platform Barcelona En Comu. Along with Paris mayor Anne Hidalgo, she was recognized for her advocacy of the Right to the City and its inclusion in the UN's New Urban Agenda ratified at Habitat III in Quito in 2016 (Johnston-Zimmerman 2017). As a part of this reform, she stated very clearly her personal, feminist, incentive for the ratification which was otherwise opposed:

The way in which this prioritization manifests in the public realm comes in the form of Ada's radical rethink on streets as public spaces, or Superblocks, as she calls them. The Superblock program (or Superilles) transforms neighborhood streets within the block system of the urban form into car-free or low-car streets in order to prioritize broader arteries for car traffic and provide more public spaces for community use. The program is a part of the Right to the City ethos, as well as the feminist perspective that Ada espouses, in that it prioritizes children, the elderly, and women and girls' health and well-being in the public realm.

According to research conducted 1 year after pilot activations, it is predicted that the Superblocks, if implemented citywide, could save hundreds of lives each year and cut air pollution by one quarter (Burgen 2019). Additional positive impacts have come from decreased noise pollution as well as heat island effects, and an increase in walking/cycling and green spaces which boost physical and mental health. Ada has connected car-dominated cities with patriarchal design in part because of these benefits, especially with regard to climate change and a focus on the collective good. "We urgently need a paradigm shift away from the car-centred urban planning model and toward a people-centred approach," she said in a Guardian article detailing the program's impact (Colau 2016).

Women Design Service, UK WDS was set up in 1983 (Berglund and Wallace 2013) and ran until February 2012 (WDS 2020). Set up by a group of women architects, designers, and planners who wanted to ensure that cities across the UK took account of women's needs, it produced research and training for women's groups (WDS 2020). WDS focused on all women, including women from ethnic minorities and women with disabilities (WDS 2020). Themes occurred through WDS, the need for toilets, nappy changing, creches, housing design, parks, pavements, safety, and transport; and the WDS undertook research on these themes (WDS 2020). The research led to a series of publications and guides (WDS 2020).

When the WDS was established in the 1980s, it was an era of radical policymaking in London (Berglund and Wallace 2013). The Greater London Council provided political support for women's issues (Berglund and Wallace 2013). At this time, there was a London-wide Women and Planning group, which was made up of local authority officers across London (Greed 2006) and the Women Design Service was the secretary for this group.

In the 1990s and onward, the political support for women's issues wavered, and WDS was operating in a very different context, and it became harder to secure funding and support (Berglund and Wallace 2013). During the UK coalition government and the global recession, the WDS was unable to find funding or resources to continue its work (WDS 2020).

The Royal Town Planning Institute's Gender Mainstreaming Toolkit Gender mainstreaming is the process of embedding the different needs of men and women into decision-making. The EU Amsterdam Treaty of 1997 makes "the elimination of inequalities and the promotion of equality between women and men" a key consideration for all activities at a local authority level in the UK. Research suggests that the impact of gender mainstreaming on planning practice in the UK was limited in scope (Greed and Reeves 2005; Greed 2006).

In this context in 2003 the Royal Town Planning Institute published their Gender Mainstreaming toolkit developed by academics and practitioners (Greed et al. 2003). Aimed at all those engaged in the planning process, the toolkit sought to give practical guidance for those working in planning to incorporate gender issues into their work. The toolkit provides users with a series of questions to consider in order to better understand the impact of planning policy or other activities on gender equality.

Women and Planning Conference In 2019 The Leeds Planning School at Leeds Beckett University hosted the Women and Planning Conference. The conference sought to further develop the conversation around the need for diversity in the planning profession. It brought together academics and practitioners (and those in between) to discuss the histories of women and planning, issues facing gender diversity in contemporary planning, and ambitions for the future (Leeds Beckett University 2020). Following the conference the Women and Planning Research Group was formed to take this research forward.

Women-Friendly Leeds Women-Friendly Cities is a UN Joint Programme which seeks to further equality in the financial, social, and political opportunities a city can offer. A Women-Friendly City is described as one where women have full access to health, social, employment, and urban services and where decision-makers include women's issues and needs in their planning. In 2019 Women's Lives Leeds, a consortium of 11 women's organizations in Leeds commenced their Women-Friendly Leeds movement which aims to be the first UK Women-Friendly City. Key initiatives to date include listening to women and girls to better understand their needs and engaging with strategic partners and decision-makers. Work is ongoing to seek to ensure women's voices are heard in planning the post-COVID19 city.

North America

Canada

Vancouver, British Columbia: A City for All Women In 2016, Vancouver City Council passed a motion to undertake the development of a comprehensive, intersectional gender equity strategy. The resulting document, A City for All Women (2018), is a 10-year strategy that recognizes Vancouver's evolving political and social landscape. The Strategy identifies five key priority areas that contribute to women's equity in the built environment, including intersectionality, safety, childcare, housing, and representation. By providing a number of measurable short and long-term goals for each municipal department, the strategy recognizes the importance of taking a coordinated approach to gender equity. Notably, the City of Vancouver aligned this strategy with a number of cross-departmental policies and initiatives, including Vancouver's Healthy City Strategy, Childcare Strategy, and Housing Strategy, and has committed to ongoing progress measurement and reporting throughout the project lifespan.

Montréal, Conseil des Montréalaises, Quebec Created in 2004, the Conseil des Montréalaises (CM) is a women's advisory panel serving Montréal's City Council and Executive Committee. The Conseil consists of 15 women, nominated on the recommendation of female-elected officials, which acts as a consultative body to City Council and municipal administration matters relating to women's status, equity, or quality of life in Montréal. CM works alongside stakeholders and women's community groups to develop relationships with elected officials and municipal employees in order to raise awareness of issues related to the status and equity of women.

The USA

Feminist Economic Recovery Plan, the USA While continuing to recover from COVID-19, the state of Hawai'i adopted a Feminist Economic Recovery Plan. The plan was produced by the Hawai'i State Commission on the Status of Women and is designed to incorporate "the unique needs of indigenous and immigrant women, caregivers, elderly women, femme-identifying and nonbinary people, incarcerated women, unsheltered women, domestic abuse and sex trafficking survivors, and women with disabilities" (Nguyen 2020). The comprehensive document and call to action importantly notes that "this is our moment to build a system that is capable of delivering gender equality" (Hawai'i State Commission on the Status of Women 2020).

Stalled, the USA While it has been widely acknowledged that the needs of transgender, non-binary, and intersex people should be reflected in the built environment and public realm, there is limited urbanist practices reflecting this necessity. An important example of this valuable consideration is the design of public washrooms. In the USA, the debate over sex segregated washrooms has been lengthy and contentious (99pi 2020). Stalled was formed in 2018 by a cross-disciplinary research team to address an urgent need to create "safe, sustainable and inclusive public restrooms for everyone regardless of age, gender, race, religion and disability" (Stalled 2020).

Asia

South Asia

India In India, for example, the sustained coexistence of patriarchal systems in multiple home and work environments, traditionally followed norms in social hierarchies, and enduring unequal power relationships between men and women have meant that historically, women have been prohibited from equitable use of public spaces (Phadke et al. 2011). This has roots in tendencies to censor the behavior of women and protectionism, and, in turn, limits the mobility of women and their opportunities.

However, recent feminist developments within the planning field around the world and collective activism of various women's groups within India has meant that a paternalistic way of planning that used to dictate urban design standards has gradually made way for awareness of gender-sensitive planning and the need to include this explicitly in planning agendas. For example, as a result of sustained campaigns by women's groups and citizens, the Mumbai development plan 2034 has established an "Advisory committee on gender," a first in the country to develop infrastructure for working women. The eight-member committee consists of a gender activist, urban planner, architect, academic, communication specialist, lawyer, and a representative of the municipal corporation. Some of the suggestions put forth by the committee include care centers, skill centers, working women's hostels, shelters for the homeless, and reservation for women vendors (Dhupkar 2019). While many aspects of the development plan such as reduction of green spaces, lack of heritage preservation and the intensity of allocated Floor Space Index (FSI) has been critiqued (Koppikar 2015), the inclusion of gender in a development plan is considered a positive step.

Bangladesh In recent decades, while gender dimensions are increasingly taken into account in governance, economic, and developmental issues, as well as in empowering low-income women entrepreneurs in Bangladesh, these have not made notable inroads into the planning field. While there is an increase in women planning professionals in various sectors, their formal and informal collaboration currently exists mainly through efforts on social media platforms, for example, "Women Architects Engineers Planners Association (WAEPA Bangladesh)." Buoyed by the growing economy and a rising middle class, Dhaka, the capital city has witnessed in recent decades, a surge in contemporary architecture practices. These however remain overwhelmingly dominated by male professionals. Interventions are restricted to "individual plot based on standalone practices" (Morshed 2018), while the city faces multiple planning issues typical of other Asian mega cities.

Nepal In efforts to mainstream gender in the public sphere by increasing equitable access to women, urban poor, and marginalized communities, the Ministry of Urban Development (MoUD) in Nepal published Gender Equality and Social Inclusion (GESI) operational guidelines which was adopted by various sectors such as urban development, building construction, housing, water supply, and sanitation. However, the GESI has attracted critique because of a lack of genuine participation by women in decision-making processes. To address this gap, UNOPS Nepal Office and Cities Alliance have constituted "The City for Women Laboratory" to explore potentials for empowerment and inclusion of women in issues pertaining to urban development and governance (Cities Alliance 2020).

Southeast Asia

Singapore Singapore is regarded as one of the safest cities in the world. This is largely due to the presence of the state in the surveillance of the public spaces and the meting of fines and harsh punishments in case the laws and regulations are not adhered to. The founding Prime Minister of Singapore Lee Kuan Yew prioritized safety as one of the bulwarks of modern Singapore where "a woman can run at three o' clock in the morning" (Tan 2019; p. 1). Perhaps because of this top-down approach to safety, gender mainstreaming is seen in collusion to safe public spaces and does not receive an explicit mention in the urban documents published by the "Centre for Livable Cities" which is jointly established by the Ministry of National Development and the Ministry of the Environment and Water Resources.

The Philippines In the Philippines, major gains have been made in the recent decades regarding women's participation in development work and governance, and various policies have been initiated to ensure gender equality and to prevent discrimination. The Philippine Commission on Women bears main responsibility for institutionalizing gender mainstreaming in the national development plans and further coordinates, prepares, assesses, and updates the National Plan for Women as well as ensures its implementation. It also monitors the performance of government agencies regarding the implementation of the plan at all levels (Phillipine Commission on Women 2020).

South and Central America

In countries in South and Central America, gender issues and violence against women are exacerbated because of its coexistence with class differences, race prejudice, political ideologies, climate justice, and income gaps subsequently impacting women's access to the benefits of urbanization (Hiramatsu 2018). To counter the absence of women from urban planning policies, several women have emerged as thought leaders, activists, and public intellectuals from an individual capacity as well as in collaboration with NGOs, local communities, and municipalities to bring the complex nature of these issues to the forefront. For example, The Latin America Women and Habitat Network and CISCSA Ciudades Feministas in Argentina strive to develop a feminist way of urban planning (Appendix A).

Brazil In Brazil, many activists and organizations have held discussions at various levels to shed light on the deeply complex societal fissure and inequalities brought about by gender bias and race and its subsequent impact on the rights to the city, safety, walkability, and lack of viable public transport options, for example, "Cidadeapé – Associação pela Mobilidade a Pé em São Paulo', 'Instituto Pólis', 'Mobilize: Mobilidade Urbana Sustentável" among other organizations that are functioning at the intersection of architecture and urban planning (Appendix A). Leadership positions in most favelas and housing movements are led by women; however, since most of these women are poor and less educated, they have very limited or no access to ongoing discussions about gender mainstreaming in the planning field.

Mexico In the city of Monterrey in Mexico, as a reaction against the traditional gender norms in the society, after several years of lobbying by women's advocates, the city instituted "the Gender Equality Program of the Secretariat of Sustainable Development and the State Program for Urban Development 2030" to enable women to have a greater say in how the cities are planned and designed (García 2019). Both in 2019 and 2020, Mexico City witnessed marches by women to protest the violence against women and the rising femicides, numbering up to ten registered cases a day. This is attributed in a large measure to the unsafe public spaces in Mexican cities (Arellano 2020).

Africa

East Africa Several countries in East Africa comprising Burundi, Comoros, Djibouti, Eritrea, Ethiopia, Kenya, Rwanda, Seychelles, Somalia, Tanzania, and Uganda are witnessing a construction boom. However, planning is an emerging field in these contexts, and allied fields such as architecture, real estate, contracting, and quantity surveying are considered as male-oriented disciplines. In addition, more focus is devoted to addressing urgent developmental issues of poverty, infant mortality, access to basic services, health, and education, and this could explain partly why gender issues in the planning sector are afforded lesser priority.

A few notable examples include a woman-led social movement for ethical and sustainable development in Tanzania which pushes for legislative reform in built environment practices

(Wood 2019). In Kenya, there are acknowledgements that women are underrepresented in professions which concern the built environment, and this is also reflected in the universities where fewer women enroll for these courses (see Maichuhie 2020 for statistics).

Reflections

Through collaboration on this chapter, we have had the opportunity to explore intersections between feminist planning and urbanism. Looking at a range of networks and organizations across the globe, we have observed that a majority of the initiatives are at a local grassroots scale rather than within professional networks. This suggests that the feminist urbanism discourse is not necessarily accepted and mainstreamed within existing professional structures. Regardless of the scale, the existence of a policy or network does not mean that the corresponding built environment is equitable. This demonstrates the importance of monitoring, implementation, and widespread education to ensure that policies, goals, and targets are continuously reflected in city building in the immediate, medium, and long term.

There is not a singular effective approach to feminist urbanism as it requires engagement, continuous and intentional conversations with communities, and an understanding of the context to properly reflect the intersectionality within the built environment. Given that planning is a relatively new profession in many parts of the world and the practice of planning continues to involve mostly white cis males, there is a great deal of work that still needs to be done. Female representation at leadership levels typically lessens the higher you move up in hierarchical organizations. This demonstrates an inherent conflict where in leadership decisions and the systems that they reside within continue to be exclusionary. To ensure planning and urbanist practices are equitable, there should be a diversity of representation at all levels to reflect the communities that are being planned and designed in collaboration with. Ultimately, what is considered a "good" city remains entirely subjective. Nonetheless, the built environment and its harbingers should be far more equitable so that everyone has the opportunity to feel safe and welcome and is able to prosper irrespective of their gender.

Reflexivity Statement

The authors of this chapter come from a diversity of backgrounds and expertise. While we were assisted by a range of colleagues, practitioners, and authors from different contexts and included best practices from across the world, we were not able to include all female identifying community builders and visionaries. Inherently, our personal and professional experience provides us with a research bias that we acknowledge may have limited our scope and focus. With this acknowledgement we understand that this online publication can evolve to include additional examples that may contribute toward achieving a more inclusive feminist urbanism in every corner of the world.

Cross-References

► Gender Inequalities in Cities: Inclusive Cities

Acknowledgments The authors would like to thank the following planning, architecture, and urbanism professionals for their insights and perspectives from their cities and home countries: Ankit Bhargava, Brenda Kamande, Carolina Monteiro de Carvalho, Geomilie Tumamao-Guittap, Flavia Gwiza, Hema Kabali, Huraera Jabeen, Marcia Trento, Mariana Cunhaos, Moniek Driesse, Riya Rahiman, Ruchi Varma, Samrawit Yohannes Yoseph, Sunita Shreshta, Tanya Chandra, Tasfin Aziz, Wahyu Pratomo, and Yamini Jain.

Appendix

List of feminist planning and urbanism-related organizations

Organization	Location	Website
American Planning Association (APA) Women and Planning Division	The USA	https://women.planning.org/

(continued)

Organization	Location	Website
Arquitetas em Rede	Brazil	https://www.instagram.com/arquitetasemrede/
BAME in Property	The UK	https://www.bameinproperty.com/
BAME Planners Network	The UK	https://twitter.com/bameplanners
Black Females in Architecture	The UK	https://www.blackfemarc.com/
Cidadeapé – Associação pela Mobilidade a Pé em São Paulo	Sao Paulo, Brazil	https://cidadeape.org/
CISCSA Ciudades Feministas	Argentina	https://www.ciscsa.org.ar/quienes-somos
Commonwealth Women in Planning Network (CWIP)	Global (Commonwealth countries)	https://tinyurl.com/yxh262qj
Instituto Pólis	Brazil	https://polis.org.br/
Mahila Housing SEWA Trust	India	https://www.mahilahousingtrust.org/
Mobilize: Mobilidade Urbana Sustentável	Brazil	https://www.mobilize.org.br/
Network for Feminist Urban Planning	Global (based in Sweden)	https://tryggaresverige.org/tankesmedja/eng-nfu
Safecity	Global (based in India)	https://safecity.in/
Safetipin	India	https://safetipin.com/
Tanzania Women Architects for Humanity (TAWAH)	Tanzania	http://outbox.co.tz/test/tawah/
TAO-Pilipinas	The Philippines	https://tao-pilipinas.org/
The Association of Women in Property	The UK	https://www.womeninproperty.org.uk
The Latin America Women and Habitat Network	Latin America	https://www.redmujer.org.ar/the-network
Urbanistas Rotterdam		https://www.urbanistasrdam.nl/
Urbanistas UK	The UK	https://www.urbanistas.org.uk/
WEDO (Women's Environmental and Development Organization)	Global	https://wedo.org/
WIEGO (Women in Informal Employment Globalizing and Organizing)	Global	https://www.wiego.org/
Women in Cities International	Montreal, Canada	https://femmesetvilles.org/
Women in Planning, South Africa (WIPSA)	South Africa	http://wipsa.org.za/
Women in Real Estate (WIRE)	Kenya, Africa	http://www.wire.or.ke/
Women Transforming Cities	Vancouver, Canada	https://www.womentransformingcities.org/
Women in Urbanism Aotearoa	New Zealand	https://www.womeninurban.org.nz/
Women in Urbanism YYC	Calgary, Canada	https://twitter.com/Womeninyyc

(continued)

References

99pi. (2019, July 23). *Invisible women*. https://99percentinvisible.org/episode/invisible-women/

99pi. (2020, September 8). *Where do we go from here?* https://99percentinvisible.org/episode/where-do-we-go-from-here/

Allick, C. (20 April 2020). When cities are built for white men. https://thewalrus.ca/when-cities-are-built-for-white-men/

Angeles, L. C., & Roberton, J. (2020). Empathy and inclusive public safety in the city: Examining LGBTQ2+ voices and experiences of intersectional discrimination. *Women's Studies International Forum, 78*, 102313.

Arellano, M. Mujeres restauradoras se pronuncian ante las pintas de los monumentos en la Ciudad de México. https://www.archdaily.mx/mx/924586/mujeres-restauradoras-se-pronuncian-ante-las-pintas-del-angel-de-la-independencia-en-la-ciudad-de-mexico. Last accessed 5 Oct 2020).

Auspurg, K., Hinz, T., & Sauer, C. (2017). Why should women get less? Evidence on the gender pay gap from multifactorial survey experiments. *American Sociological Review, 82*(1), 179–210.

Baker, C. (2019, September 26). *The climate crisis is a feminist issue*. https://msmagazine.com/2019/09/26/the-climate-crisis-is-a-feminist-issue/

Bauer, U. (2009). Gender mainstreaming in Vienna: How the gender perspective can raise the quality of life in a big city. *Kvinder, Køn & Forskning*, 3–4. https://tidsskrift.dk/KKF/article/download/27973/24602

Berglund, E., & Wallace, B. (2013). Women's design service as counter-expertise. In: I. Sanchez de Madariaga, & M. Roberts (Eds.), *Fair shared cities. The impact of gender planning in Europe* (pp. 249–264). Farnham: Ashgate.

Bhatla, N., Achyut, P., Ghosh, S., Gautam, A., & Verma, R. (2013). *Safe cities free from violence against women and girls*. https://www.icrw.org/publications/safe-cities-free-from-violence-against-women-and-girls-baseline-finding-from-the-safe-cities-delhi-programme/

F

Biana, H. T. (2020). Extending bell hooks' feminist theory. *Journal of International Women's Studies, 21*(1), 13–29. https://vc.bridgew.edu/jiws/vol21/iss1/3.

Brown-Saracino, J. (2015). How places shape identity: The origins of distinctive LBQ identities in four small U.S. Cities. *American Journal of Sociology, 121*(1), 1–63. https://doi.org/10.1086/682066.

Buckingham, S. (2004). Ecofeminism in the twenty-first century. *The Geographical Journal, 170*(2), 146–154. https://doi.org/10.1111/j.0016-7398.2004.00116.x.

Burgen, S. (2019, September 10). Barcelona's car-free 'superblocks' could save hundreds of lives. *The Guardian.* https://www.theguardian.com/cities/2019/sep/10/barcelonas-car-free-superblocks-could-save-hundreds-of-lives

Carbado, D., Crenshaw, K., Mays, V., & Tomlinson, B. (2013). Intersectionality: Mapping the movements of a theory. *Du Bois Review: Social Science Research on Race, 10*(2), 303–312. https://doi.org/10.1017/S1742058X13000349.

Carmon, N., & Fainstein, S. S. (2013). *Policy, planning, and people: Promoting justice in urban development* (1st ed.). Philadelphia: University of Pennsylvania Press.

Centre for Feminist Foreign Policy. (2020). *Our story.* https://centreforfeministforeignpolicy.org/our-story

Centre for Gender Equality Iceland. (2012). *Gender equality in Iceland.* https://www.stjornarradid.is/media/utanrikisraduneyti-media/media/mannrettindi/Gender-Equality-in-Iceland.pdf

Cities Alliance. (2020). *City for women laboratory in Nepal.* https://www.citiesalliance.org/newsroom/news/cities-alliance-news/city-for-women-laboratory-nepal

City of Vancouver. (2018). Vancouver: *A city for all women.* https://vancouver.ca/docs/council/Women%27sEquityStrategy.pdf

Colau, A. (2016, October 20). *After Habitat III: A stronger urban future must be based on the right to the city.* https://www.theguardian.com/cities/2016/oct/20/habitat-3-right-city-concrete-policies-ada-colau

Commonwealth Women in Planning Network. (2018). *Women in planning manifesto.* https://www.commonwealth-planners.org/cwip-network

Crenshaw, K. (1989). *Demarginalizing the intersection of race and sex: A Black feminist critique of antidiscrimination doctrine, feminist theory and antiracist politics.* https://chicagounbound.uchicago.edu/cgi/viewcontent.cgi?article=1052&context=uclf

Day, K. (1999). Introducing gender to the critique of privatized public space. *Journal of Urban Design, 4*(2), 155–178. https://doi.org/10.1080/13574809908724444.

De Madariaga, I. S., & Neuman, M. (2016). Mainstreaming gender in the city. *Town Planning Review, 87*(5), 493–504. https://doi.org/10.3828/tpr.2016.33.

Dhupkar, A. (2019). Development plan takes a gender sensitive step. https://www.hindustantimes.com/columns/make-room-for-gender-in-mumbai-of-the-future/story-0BPZEFqspGr30KVtsQHroJ.html. Last accessed 7 Sept 2020.

Didi. (2020, May 4). Why we need more women in politics. https://www.kcl.ac.uk/news/why-we-need-more-women-in-politics

Eichler, M. (1995). Change of plans. In *Change of plans.* Broadview Press. https://doi.org/10.3138/j.ctt2ttv77.

European Parliament. (2010). *The policy on gender equality in Iceland.* https://www.europarl.europa.eu/document/activities/cont/201107/20110725ATT24624/20110725ATT24624EN.pdf

Fainstein, S. S. (2005). Planning theory and the city. *Journal of Planning Education and Research, 25*(2), 122–130.

Falú. (2017). Women's right to the city: Reflections on inclusive urban planning. https://www.urbanet.info/womens-right-to-the-city/. Last accessed 31 Aug 2020.

Federation of Canadian Municipalities. (2020). Retrieved from https://fcm.ca/en/resources/wilg/run-win-lead-toward-parity-framework

Foran, C. (2013, September 16). How to design a city for women. https://www.bloomberg.com/news/articles/2013-09-16/how-to-design-a-city-for-women

García, A. N. (2019). Designing a woman-friendly city: Gender mainstreaming and urban planning in Monterrey, Mexico. https://wafmag.org/2019/03/designing-a-woman-friendly-city-gender-mainstreaming-and-urban-planning-in-monterrey-mexico/. Last accessed 5 Oct 2020.

Government of Canada. (2015). *Status of Women Canada, Privy Council Office and Treasury Board of Canada Secretariat Action Plan (2016–2020) Audit of Gender-based Analysis Fall 2015 Report of the Auditor General of Canada.* https://cfc-swc.gc.ca/gba-acs/plan-action-2016-en.PDF

Government of Canada. (2018). *What is GBA+?* https://cfc-swc.gc.ca/gba-acs/index-en.html

Government of Georgia. (2017, July). *Implementation of gender equality policy in Georgia.* http://www.parliament.ge/en/ajax/downloadFile/72000/Gender_Equality_NAP_report_2016_ENG_Edited_Final_July_2017

Greed, C., Davies, L., Brown, C., & Dühr, S. (2003). *Gender equality and plan making: The gendermainstreaming toolkit.* https://www.rtpi.org.uk/media/3518/genderequality-planmaking.pdf

Greed, C., & Reeves, D. (2005). Mainstreaming equality into strategic spatial policy making: Are town planners losing sight of gender? *Construction Management and Economics, 23*(10), 1059–1070.

Greed, C. (2006). Institutional and conceptual barriers to the adoption of gender mainstreaming within spatial planning departments in England. *Planning Theory & Practice, 7*(2), 179–197.

Hawai'i State Commission on the Status of Women. (2020, April 14). Building bridges not walking on backs: A feminist economic recovery plan for COVID-19. https://humanservices.hawaii.gov/wp-content/uploads/2020/04/4.13.20-Final-Cover-D2-Feminist-Economic-Recovery-D1.pdf

Hayden, D. (1980). What would a nonsexist city be like? Speculations on housing, urban design and human work. *Ekistics, 52*(310), 99–107. https://doi.org/10.1086/495718.

Hiramatsu, A. (2018). Inclusive cities: 4 examples of urban productivity through gender equality. https://blogs.iadb.org/ciudades-sostenibles/en/inclusive-cities-urban-productivity-gender-equality/. Last accessed 5 Oct 2020.

Hendler, S. A., Hendler, S., & Markovich, J. (2017). I was the Only Woman. In: *Women and planning in Canada*. UBC Press.

Hodge, G., & Gordon, D. L. A. (2008). *Planning Canadian communities* (5th ed.). Toronto: Thomson Nelson.

Hooks, B. (1984). *Feminist theory: From margin to center*. Cambridge, MA: South End Press.

Irazábal, C., & Huerta, C. (2016). Intersectionality and planning at the margins: LGBTQ youth of color in New York. *Gender, Place & Culture, 23*(5), 714–732. https://doi.org/10.1080/0966369X.2015.1058755.

Johnston-Zimmerman, K. (2017). Urban planning has a sexism problem. Retrieved from https://nextcity.org/features/view/urban-planning-sexism-problem

Kern, L. (2020a). *Feminist city: Claiming space in a man-made World*. Verso Books.

Kern, L. (2020b). *Cities are even worse for women than you might imagine*. https://www.curbed.com/2020/7/7/21315882/feminist-cities-leslie-kern-book

Kerner, I. (2017). Relations of difference: Power and inequality in intersectional and postcolonial feminist theories. *Current Sociology. La Sociologie Contemporaine, 65*(6), 846–866.

Klein, L., & D'Ignazio, C. (2020). Introduction: Why data science needs feminism. (2020). In *Data feminism*. Retrieved from https://data-feminism.mitpress.mit.edu/pub/frfa9szd

Koppikar, S. (2015). Make room for gender in Mumbai of the future. https://www.hindustantimes.com/columns/make-room-for-gender-in-mumbai-of-the-future/story-0BPZEFqspGr30KVtsQHroJ.html. Last accessed 7 July 2020.

Leeds Beckett University. (2020, March 10). RTPI's women of influence 2020. https://www.leedsbeckett.ac.uk/blogs/school-of-bee/2020/03/rtpiwmnofinfluence/

Maichuhie, K. (2020). Where are the women in built industry. https://nation.africa/kenya/gender/where-are-the-women-in-built-industry-259086. Last accessed 20 Sept 2020.

Mallory, C. (2018). What's in a name? In defense of eco-feminism (not ecological feminisms, feminist ecology, or gender and the environment): Or "why ecofeminism need not be ecofeminine–but so what if it is?" *Ethics & the Environment, 23*(2), 11+. https://link-gale-com.ezproxy.lib.ucalgary.ca/apps/doc/A562973800/EAIM?u=ucalgary&sid=EAIM&xid=5f19c212

McFarlane, L. (1995, August–September). Women and urban planning. *OECD Observer*, (195), 37+. https://link-gale-com.ezproxy.lib.ucalgary.ca/apps/doc/A17350233/EAIM?u=ucalgary&sid=EAIM&xid=e6feaa0d

Moghadam, S. N. M., & Rafieian, M. (2019). What did urban studies do for women? A systematic review of 40 years of research. *Habitat International, 92*, 1–21. https://doi.org/10.1016/j.habitatint.2019.102047.

Morshed, A. Z. (2018). Does architecture define a "new" Bangladesh? https://www.thedailystar.net/supplements/rethinking-urban-spaces/does-architecture-define-new-bangladesh-1538080. Last accessed 25 Sept 2020.

Mósesdóttir, L., & Erlingsdóttir, R. G. (2005). Spreading the word across Europe. *International Feminist Journal of Politics, 7*(4), 513–531.

Nguyen. F. (2020). *This state says it has a 'feminist economic recovery plan.' Here's what that look like*. https://www.thelily.com/this-state-says-they-have-a-feminist-economic-recovery-plan-heres-what-that-looks-like/

Nhanenge, J. (2011). *Ecofeminism towards integrating the concerns of women, poor people, and nature into development*. Lanham: University Press of America.

Oxford University Press. (2020). Feminism noun. https://www.oxfordlearnersdictionaries.com/definition/english/feminism

Patel, S., & Mitlin, D. (2007). *Gender and urban federations, making gender and generation matter for sustainable development*. IIED: https://pubs.iied.org/pdfs/G03089.pdf

Perez, C. C. (2019). *Invisible women: Exposing data bias in a world designed for men*. Random House.

Peat, J. (2019, May 31). Economic growth is an unnecessary evil. Jacinda Arden is right to deprioritize it. https://www.thelondoneconomic.com/opinion/economic-growth-is-an-unnecessary-evil-jacinda-ardern-is-right-to-deprioritise-it/31/05/

Phadke, S., Khan, S., & Ranade, S. (2011). *Why loiter?: Women and risk on Mumbai streets*. Penguin Books India.

Phillipine Commission on Women. (2020). *Herstory*. https://pcw.gov.ph/herstory/

RTPI. (2020). *Women and planning: An analysis of gender related barriers to professional advancement*. https://www.rtpi.org.uk/media/4325/women-and-planning.pdf

Rubery, J., Grimshaw, D., & Figueiredo, H. (2005). How to close the gender pay gap in Europe: Towards the gender mainstreaming of pay policy. *Industrial Relations Journal, 36*, 184–213. https://doi.org/10.1111/j.1468-2338.2005.00353.x.

Stalled. (2020). *Stalled*. https://www.stalled.online/

Supik, L. M., Herrera, V. M. T. M., & Lutz, H. P. (Eds.). (2012). *Framing intersectionality: Debates on a multi-faceted concept in gender studies*. ProQuest Ebook Central. https://ebookcentral-proquest-com.ezproxy.lib.ucalgary.ca

Tan, S. (2019). *Planning for a secure city*, Urban Systems Studies. Centre for Livable Cities (CLC) Singapore, First Edition, CLC Publications. https://www.clc.gov.sg/docs/default-source/urban-systems-studies/uss-planning-for-a-secure-city.pdf

UN Women. (2017). *Safe cities and safe public spaces – Global results report*. https://www.unwomen.org/-/media/headquarters/attachments/sections/library/

F

publications/2017/safe-cities-and-safe-public-spaces-global-results-report-en.pdf?la=en&vs=45
UN Women. (2020). *About UN women*. https://www.unwomen.org/en/about-us/about-un-women
UN Women Albania. (n.d.). *National planning and budgeting*. https://albania.unwomen.org/en/what-we-do/national-planning-and-budgeting
United Nations. (2020a). Retrieved from https://www.un.org/sustainabledevelopment/development-agenda-retired/#:~:text=On%201%20January%202016%2C%20the,Summit%20%E2%80%94%20officially%20came%20into%20force
United Nations. (2020b). *Achieve gender equality and empower all women and girls*. https://sdgs.un.org/goals/goal5
Urbanistas. (2020). *Urbanistas*. https://www.urbanistas.org.uk/
Women in Planning. (2019, March). *Who's leading planning? How many women are working in leadership role in private sector planning consultancy?* https://docs.wixstatic.com/ugd/9848d7_86c89186387242ea981ccb427740de51.pdf
Women in Property. (2020). *National student awards*. https://www.womeninproperty.org.uk/initiatives/national-student-awards/
Women in Urbanism Aotearoa. (2020a). *Women in urbanism*. https://www.womeninurban.org.nz/
Women in Urbanism Aotearoa. (2020b). *Equitable light rail*. https://www.womeninurban.org.nz/equitable-light-rail
Women in Urbanism Aotearoa. (2020c). *Street harassment*. https://www.womeninurban.org.nz/street-harassment
Women in Urbanism Aotearoa Incorporated. (2020). *Rules of Women in Urbanism Aotearoa Incorporated*. https://84228a0a-ebd0-44b7-95b7-aaa038be47f6.filesusr.com/ugd/a523fc_d2e9aab1609b44d6a0cb6bf60614b8f6.pdf
Women in Urbanism YYC. (2020). Twitter page. https://twitter.com/Womeninyyc
Women Transforming Cities. (2020). *Women transforming cities*. https://www.womentransformingcities.org/about-us
Women's Design Service. (2020). *Women's design service*. https://www.wds.org.uk/
Wood, H. (2019). The emerging female architects of East Africa. https://archinect.com/features/article/150164665/the-emerging-female-architects-of-east-africa. Last accessed 5 Oct 2020.
World Economic Forum. (2019). *Mind the 100 year gap*. https://www.weforum.org/reports/gender-gap-2020-report-100-years-pay-equality

Finance

► Financing: Fiscal Tools to Enhance Regional Sustainable Development

Financing: Fiscal Tools to Enhance Regional Sustainable Development

Valerio Alfonso Bruno[1,2,3] and Adriano Cozzolino[2,4]
[1]Università Cattolica del Sacro Cuore, Milan, Italy
[2]Center for European Futures, Naples, Italy
[3]Centre for the Analysis of the Radical Right, Leeds, UK
[4]Università degli Studi della Campania "Luigi Vanvitelli", Caserta, Italy

Synonyms

Ethical finance; Finance; Future; Globalization and financialization; Green finance; Regional development; Sustainability

Definition

The role played by finance and the relevance of fiscal tools in order to enhance regional sustainable development are here analytically considered. Regional development is increasingly related to sustainability and refers to a wide set of dynamics: among others, the wise use of natural resources, the transition to green/circular economy, job creation and tackling poverty, growth of clean technologies, production, and well-being. The debate on the sustainability of finance and financial instruments has seen, in the last decade, an important acceleration, including the analysis of best practices from specific countries and regions, or evidences from developing countries. In this respect, ethical and green finance, in contrast to predatory forms of financialization (a key feature of globalization), are emerging and currently regarded as necessary to accompany the sustainable development of regions. In particular, green finance and ethical finance play a role in sustainable regional development by encouraging policies, both public and private, in the direction of environmental and green initiatives. They also aim at building a typology of finance based upon the ethics of prudent risk-taking, while recognizing financial dilemmas as an area of possible consequences of the action.

Introduction

Regions can be understood as portions of the Earth that share certain factors (language, culture, geography), and/or are functionally interdependent under several respects (trade and the economy, or political relations), and/or share forms of political and administrative authority and institutions. Regions can be sub-national (i.e., parts of the nation-state) or supranational territories, as in the case of the European Union. Specifically to regional development, this is a broad topic concerning the policy strategies and tools of socio-economic change – comprising inequalities, poverty, demography, and occupation – at regional level. Regional development can be also understood beyond socioeconomic factors, thus involving, among other things, the development of innovation, environmental resources, cultural heritage, entrepreneurship, technology, and education (Bærenholdt 2009; Pike et al. 2010). Regional development is also distinguished in (i) exogenous, thus driven by the external action of, for example, international financial institutions and other actors, or (ii) endogenous, accordingly, driven by bottom-up processes and tied to a wide range of regional and local factors (from local resources to technology, from social forces to political relations, from labor to entrepreneurship) (Tödtling 2020).

The question of regional development is increasingly tied to sustainability, a concept that – as used for example by actors such as the European Union and the United Nation – refers to the wise use of natural resources, the transition to green/circular economy, job creation and tackling poverty, growth of clean technologies and production, and well-being. In this respect, ethical and green finance (more below), in contrast to predatory forms of financialization, is regarded as necessary to accompany the sustainable development of regions. In particular, green finance and ethical finance may play a role in the frame of regional sustainable development (a) by encouraging sustainable policies, both public and private, in the direction of environmental and green initiatives, and (b) by building an ethics of prudent risk-taking, embedding and recognizing financial dilemmas as the area of possible consequences of the action.

Globalization, Finance, and Regional Development

Especially since the crisis of Keynesian approaches to macro-economic policy (late 1970s), finance has taken on a predominant role in the strategies of economic growth. In broad terms, finance can be framed as: (1) money or other liquid resources of a government, business, group, or individual; (2) the system that includes the circulation of money, the granting of credit, the making of investments, and the provision of banking facilities; (3) the science or study of the management of funds; and (4) the obtaining of funds or capital. Thus, the relevance of finance worldwide has growth exponentially over the last decades, in the wake of that wide and complex phenomenon known as globalization, namely a set of uneven and contradictory processes of expansion and restructuring of trade and finance spanning different spatial scales – from local to global (Hirst et al. 2015), especially under the influence of neoliberal ideology (Harvey 2007). In the first decades of the twenty-first century, a number of authoritative studies and researches pointed out the key role finance plays in shaping globalization, coining the notion of "financialization" (Aalbers 2016; Van der Zwan 2014). Financialization is defined by Epstein (2005) as "the increasing importance of financial motives, financial institutions and financial elites in the functioning of the economy," while Nölke (2017) adds that it is "an especially aggressive form of economic globalisation." On other hand, other authors also noted that, within economic geography, the role of finance in local and regional economic development is still overlooked, in turn reflecting a tendency to focus on the flow of money rather than analyzing the specific places and processes that give rise to such flows.

Nonetheless, global financial integration and globalization have strengthened, over time, the attention to the regional dimension of development in both macro-regions (as the European Union) and micro-regions. The regional dimension of development can be conceived as the second face of transnational processes of socio-economic and financial integration. In fact, regions are central actors in shaping global integration (and the organization of capitalism), namely "places" where specific policy and political choices variously intersected with other regions and transnational dynamics, constituting the structure of the global political economy. Thus, globalization and financialization do not imply a reduction of the relevance of regional and local levels. Rather, at stake are processes of complex interconnection between different spatial dimensions, whereby "through these local, micro-regional and macro-regional processes, 'regions' are now seen as playing a crucial role in constituting economic globalization" (ÓRiain 2010, 17; Coe et al. 2004).

The notion of "development" is matter of longstanding and widespread debates among scholars, and greatly varies according to the geo-social and historical context, and the theoretical and disciplinary standpoint (Mohan 2010). Actually, "changing and contested definitions of development seek to encompass and reflect geographical variation and uneven economic, social, political, cultural and environmental conditions and legacies in different places across the world" (Pike et al. 2010, 2). However, if definitions of regional development are inherently contested due to a wide range of context-dependent and historical factors, "perceptions of local and regional development across the world share numerous characteristics" (ibid). Following Pike et al. (2010), three elements in particular stand up. First, regions are interconnected with global political economy processes. Interdependency, rapid changes, uncertainty are inherently tied to the global context in which regional development occurs. In this regard, regions continuously adapt to the changing international context, for example as in the case of unfolding of financial crises (e.g., the Asian crisis of 1997, or the global financial crisis of 2008). Second, as already noted, the current notion of regional development is broadening its meaning beyond a merely quantitative focus of "growth" – thus embracing social, cultural, political, and environmental concerns. Finally, (1) the set of processes linked to the changing international environment and (2) the broadening of the notion of development to a wide array of factors have increasingly opened the study of regional development to interdisciplinary perspectives. This "means that any single discipline – regardless of its predicament or status – is ill-equipped and perhaps ultimately unable to capture the evolving whole" (Pike et al. (2010, 3).

In specific relation to macroeconomic patterns of regional development, it is important to note that since the late 1970s onwards, capital flows and the transnationalization of production (through the operations of transnational corporations) have significantly reshaped the internal dynamics of regional economies. Actually, while specific regional varieties clearly remain in place, on the other hand – within the pressures exerted by transnational capitalist competition – regions increasingly tried to boost policy competitiveness to attract foreign capitals and/or create a capital-friendly context (Gordon 2010). Thus, "even as regions become more central to capitalist accumulation the range of policy strategies available is narrowed to 'entrepreneurial' efforts to enhance 'competitiveness'" (ÓRiain 2010, 19). On the other hand, financial integration, coupled with the neo-liberal model of capital accumulation through increased competition, led to the accumulation of widespread and persistent inequalities, within and between regions, since the 1970s. It is in this framework that the debate on finance and sustainable development looms up.

Finance and Regional Sustainable Development: Ethics and Green Finance

The debate on the sustainability of finance and financial instruments had in the last decade an important acceleration (Fatemi and Fooladi 2013; Jeucken 2010; Lagoarde-Segot 2019; La Torre et al. 2019; Lehner 2016; Ramiah and Gregoriou 2015; Schoenmaker and Schramade 2018), including the analysis of best practices from specific countries (Soundarrajan and Vivek 2016 on India, Volz et al. 2015 on Indonesia, Falcone and Sica 2019 on Italy, Dörry and Schulz 2018 on Luxembourg, Popescu and Popescu 2019 on Romania, Wray et al. 2010 on United Kingdom), specific contexts (Ahmed et al. 2015 and Nugroho et al. 2019 on sustainable finance implementation in Islamic bank) but also entire regions (Liyanage et al. 2021 on Africa and Asia) or evidences specifically from developing countries (Banga 2019).

In the debate covering financial sustainability, the role and the function of "green finance" deserves a special mention. Even if green finance lacks a precise and commonly accepted definition (Lindenberg 2014), it has become of the utmost importance on the debate on sustainable finance in the last decade (Berensmann and Lindenberg 2016; Falcone 2020; Migliorelli and Dessertine 2019; Sachs et al. 2019a; Sachs et al. 2019b; Taghizadeh-Hesary and Yoshino 2019; Wang and Zhi 2016). After considering the reasons behind the lack of a common definition of green finance, Lindenberg (2014: 2) propose what green finance does comprise: (1) the financing of public and private green investments (including preparatory and capital costs) in the following areas: (a) environmental goods and services (such as water management or protection of biodiversity and landscapes), (b) prevention, minimization, and compensation of damages to the environment and to the climate (such as energy efficiency or dams); (2) the financing of public policies (including operational costs) that encourage the implementation of environmental and environmental-damage mitigation or adaptation projects and initiatives (e.g., feed-in-tariffs for renewable energies); (3) components of the financial system that deal specifically with green investments, such as the Green Climate Fund or financial instruments for green investments (e.g., green bonds and structured green funds), including their specific legal, economic, and institutional framework conditions.

Along with the question "green finance," a second element that deserves a specific mention in the complex and vast debate over sustainable finance for development is the issue of "ethical finance" (see for instance Baranes 2009 on sustainable and ethical finance) and the relationship between ethics and finance (Boatright 2010; Chelawat and Trivedi 2013; Gangi and Trotta 2015; Grasso 2010; Omarova 2017; Sahut and Switzer 2015; Wilson 1997). As for green finance, it is not possible to provide a precise and widely accepted definition of ethical finance. However it can be useful to report what Omarova (2017: 797), writing about the United States in post global financial crisis, asserts: "[. . .] the current resurgence of financial regulators' *interest in the role of ethical and cultural norms* in shaping financial institutions' and professionals' behavior is hardly a surprising development. Thus, beginning approximately in late 2013, U.S. financial regulators became particularly and increasingly vocal in their *calls for ethically sound behavior and a culture of prudent risk-taking* within financial firms (Emphasis ours)."

The interest in the role of ethical and cultural norms parallels the attention to ethical behavior and culture of prudent risk-taking. To this regard, contemporary authors have articulated several analyses on this particular point. Dembinski (2015), among others, has framed the problem not only in the terms of "ethics *and* finance," or of "ethics *of* finance" – but as ethics *in* finance. Introducing and framing ethics as the place where dilemmas are expressed, Dembinski systematically analyzes the ethical dilemmas of financial actors. In particular, (a) the owner of the funds, from the saver to the rentier (2015: 51–68), faces different ethical dilemmas, exactly as (b) the user of the funds, may it belong to the public or the private (2015: 69–88) and (c) the financial intermediator (2015: 89–106). As the notion of responsibility, put aside the legal dimension, leads the ethical analysis to design the area of possible consequences of the action, the freedom of choice is what characterized the ethical domain (2015: 43). Complementarily, Marc Chesney (2018) has heavily criticized the toxic role of specific "financial oligarchies" in support an unsustainable *modus operandi*, defining this form of "casino finance" as not only completely uninterested to real economy, but also as practices that seek and relie on government support through public money bailouts in moment of crisis. In synthesis, (a) while green finance encourages policies, both public and private, directed toward environmental and green initiatives (b) ethical finance focuses on building an ethics of prudent risk-taking, embedding, and recognizing financial dilemmas as the area of possible consequences of the action.

Conclusion: The Futures of Regions and Regional Development

The conditions and futures of local and regional development were severely hit by the global financial crisis of 2008, which interrupted a prolonged phase of expansion of economic growth – though, an expansion marked by a parallel rise of inequalities and imbalances since the 1970s. The great crisis of 2008 showed, and at the same time increased, the economic and social imbalances manifested in the global economy, especially in the financial sector. In addition, the prolonged application of austerity therapy in the years following the great crisis inhibited a possible and structural demand-led economic recovery, triggering a phase of "permanent stagnation" (Benigno and Fornaro 2018). Thus, regions suffered not only a period of persistent stagnation but also trade imbalances, unemployment, and inequalities in both wealth and living standards – inequalities within and between regions. Worth of noting is also that, since the global financial crisis, the international order (and signally the G20) followed new patterns and relations. On the one hand, the rise of new actors, particularly China, India and Brazil, while generally reshaping trade and economic relations at global level, embodied also an alternative *state-led* model of economic growth – dwarfing the poor performance and growing imbalances marking the neoliberal market-led model of the West, and marking also the poor global projection of the European Union. Thus, the balance of power between different regions of the world increasingly changed.

In addition, the Covid-19 crisis – a unique historically event due to the joint action of health and economic crisis – severely affected the regional dimension of development, throwing this into an unprecedented uncertainty concerning the future (s) of world regions. First of all, it is important to stress that, as occurred after the great crisis of 2008, states intervened with bold fiscal action to sustain household and firms, avoiding the collapse of entire market economies. In fact, international financial institutions (IFIs) – prominently the International Monetary Fund, among others – called for expansive fiscal stimuli in all world regions and economies (advanced, developing and poor). Interestingly, and partly differently from the crisis of 2008, for the post-Covid-19 phase IFIs are also envisioning a development model based on a more incisive role for the state – for example, through public investments and economic planning, and with the purpose of creating jobs (this latter especially being a process usually left, according to neoliberal orthodoxy, to market forces).

Thus, in light of the (i) aftermath of the great crisis, permanent stagnation, and accumulation of imbalances and inequalities (especially in the West and other advanced economies), (ii) the rise of new economic and political actors, China prominently, and (iii) the widespread and unique uncertainty posed by Covid-19 crisis, it is not an easy task to envision the possible futures of regions and regional development. Furthermore, climate change, migration movements, and demographic dynamics (with both increasing rates of population growth in some regions, and de-growth in other as Europe) constitute other important factors that will shape regional development. For all this reason, and considering both the global financial crisis and its aftermath, and the current Covid-19 crisis, the debate concerning sustainability through responsible finance is key to the futures of regions and regional development.

Cross-References

▶ An Overview of the Relationship of the Sustainable Development Goals and Urban and Regional Development

- ▶ Big Data for Smart Cities and Inclusive Growth
- ▶ Circular Economy Cities
- ▶ City Financing and Social Urbanism in Latin America: the Importance of Good Fiscal Management
- ▶ Global Homelessness: Neoliberalism, Violence, and Precarious Urban Futures
- ▶ Green Economy Policies to Achieve Water Security

References

Aalbers, M. B. (2016). Financialization. In *International encyclopedia of geography: People, the Earth, Environment and Technology* (pp. 1–12). Hoboken: Wiley.

Ahmed, H., Mohieldin, M., Verbeek, J., & Aboulmagd, F. (2015). *On the sustainable development goals and the role of Islamic finance*. The World Bank.

Banga, J. (2019). The green bond market: A potential source of climate finance for developing countries. *Journal of Sustainable Finance & Investment, 9*(1), 17–32.

Baranes, A. (2009). Towards sustainable and ethical finance. *Development, 52*(3), 416–420.

Bærenholdt, J. O. (2009). Regional development and non-economic factors. In *International Encyclopedia of Human Geography* (pp. 181–186). Pergamon Press.

Benigno, G., & Fornaro, L. (2018). Stagnation traps. *The Review of Economic Studies, 85*(3), 1425–1470.

Berensmann, K., & Lindenberg, N. (2016). Green finance: Actors, challenges and policy recommendations. *German Development Institute/Deutsches Institut für Entwicklungspolitik (DIE) Briefing Paper, 23*.

Boatright, J. R. (2010). Ethical implications of finance. *Finance Ethics, 23*.

Chelawat, C. A., & Trivedi, I. V. (2013). Ethical finance: Trends and emerging issues for research. *International Journal of Business Ethics in Developing Economies, 2*, 2.

Chesney, M. (2018). *A permanent crisis: The financial Oligarchy's seizing of power and the failure of democracy*. Springer.

Coe, N. M., Hess, M., Yeung, H. W.-C., Dicken, P., & Henderson, J. (2004). "Globalising" regional development: A global production networks perspective. *Transactions of Institute of British Geographers, NS, 29*, 468–484.

Dembinski, P. H. (2015). *Ethique et responsabilité en finance. Quo vadis?* Paris: Revue Banque Edition. English edition (2017). *Ethics and responsibility in finance* (Vol. 2), Routledge.

Dörry, S., & Schulz, C. (2018). Green financing, interrupted. Potential directions for sustainable finance in Luxembourg. *Local Environment, 23*(7), 717–733.

Epstein, G. A. (2005). *Financialization and the world economy*. Edward Elgar Publishing.

Falcone, P. M. (2020). Environmental regulation and green investments: The role of green finance. *International Journal of Green Economics, 14*(2), 159–173.

Falcone, P. M., & Sica, E. (2019). Assessing the opportunities and challenges of green finance in Italy: An analysis of the biomass production sector. *Sustainability, 11*(2), 517.

Fatemi, A. M., & Fooladi, I. J. (2013). Sustainable finance: A new paradigm. *Global Finance Journal, 24*(2), 101–113.

Gangi, F., & Trotta, C. (2015). The ethical finance as a response to the financial crises: An empirical survey of European SRFs performance. *Journal of Management & Governance, 19*(2), 371–394.

Gordon, I. (2010). Territorial competition. In *Handbook of local and regional development* (pp. 52–64). Routledge.

Grasso, M. (2010). An ethical approach to climate adaptation finance. *Global Environmental Change, 20*(1), 74–81.

Harvey, D. (2007). *A brief history of neoliberalism*. Oxford University Press.

Hirst, P., Thompson, G., & Bromley, S. (2015). *Globalization in question* (Third Edition). John Wiley & Sons.

Jeucken, M. (2010). *Sustainable finance and banking: The financial sector and the future of the planet*. Earthscan.

La Torre, M., Trotta, A., Chiappini, H., & Rizzello, A. (2019). Business models for sustainable finance: The case study of social impact bonds. *Sustainability, 11*(7), 1887.

Lagoarde-Segot, T. (2019). Sustainable finance. A critical realist perspective. *Research in International Business and Finance, 47*, 1–9.

Lehner, O. M. (Ed.). (2016). *Routledge handbook of social and sustainable finance*. Routledge.

Lindenberg, N., Definition of Green Finance (2014). DIE mimeo, 2014, Available at SSRN: https://ssrn.com/abstract=2446496

Liyanage, S. I. H., Netswera, F. G., & Motsumi, A. (2021). Insights from EU policy framework in aligning sustainable Finance for sustainable development in Africa and Asia. *International Journal of Energy Economics and Policy, 11*(1), 459–470.

Migliorelli, M., & Dessertine, P. (2019). The rise of green finance in Europe. In *Opportunities and challenges for issuers, investors and marketplaces*. Palgrave Macmillan.

Mohan, G. (2010). Local and regional 'development studies'. In *Handbook of local and regional development* (pp. 65–78). Routledge.

Nölke, A. (2017). Financialisation as the core problem for "social Europe". *Revista de Economia Mondial, 46*(2017), 27–47.

Nugroho, L., Badawi, A., & Hidayah, N. (2019). Discourses of sustainable finance implementation in Islamic bank (Cases studies in Bank Mandiri Syariah 2018). *International Journal of Financial Research, 10*(6), 108–117.

Omarova, S. T. (2017). Ethical Finance as a systemic challenge: Risk, culture, and structure. *Cornell JL & Public Policy, 27*, 797.

F

Pike, A., Rodríguez-Pose, A., & Tomaney, J. (2010). Introduction: A handbook of local and regional development. In *Handbook of local and regional development* (pp. 23–36). Routledge.
Popescu, C. R. G., & Popescu, G. N. (2019). An exploratory study based on a questionnaire concerning green and sustainable finance, corporate social responsibility, and performance: Evidence from the Romanian business environment. *Journal of Risk and Financial Management, 12*(4), 162.
Ramiah, V., & Gregoriou, G. N. (Eds.). (2015). *Handbook of environmental and sustainable finance*. Academic.
Riain, S. Ó. (2010). Globalization and regional development. In *Handbook of local and regional development* (pp. 39–51). London: Routledge.
Sachs, J., Woo, W. T., Yoshino, N., & Taghizadeh-Hesary, F. (2019a). Importance of green finance for achieving sustainable development goals and energy security. In *Handbook of green finance: Energy security and sustainable development* (pp. 3–12). Singapore: Springer.
Sachs, J., Woo, W. T., Yoshino, N., & Taghizadeh-Hesary, F. (Eds.). (2019b). *Handbook of green Finance: Energy security and sustainable development*. Springer.
Sahut, J. M., & Switzer, L. N. (2015). Ethical finance and governance. Introduction. *Journal of Management & Governance, 19*(2), 255–257.
Schoenmaker, D., & Schramade, W. (2018). *Principles of sustainable finance*. Oxford University Press.
Soundarrajan, P., & Vivek, N. (2016). Green finance for sustainable green economic growth in India. *Agricultural Economics, 62*(1), 35–44.
Taghizadeh-Hesary, F., & Yoshino, N. (2019). The way to induce private participation in green finance and investment. *Finance Research Letters, 31*, 98–103.
Tödtling, F. (2020). Regional Development, Endogenous. In *International Encyclopedia of Human Geography* (Second Edition) (pp. 303–308). Elsevier.
Van der Zwan, N. (2014). Making sense of financialization. *Socio-Economic Review, 12*(1), 99–129.
Volz, U., Böhnke, J., Knierim, L., Richert, K., Roeber, G. M., & Eidt, V. (2015). *Financing the green transformation: How to make green finance work in Indonesia*. Springer.
Wang, Y., & Zhi, Q. (2016). The role of green finance in environmental protection: Two aspects of market mechanism and policies. *Energy Procedia, 104*, 311–316.
Wilson, R. (1997). Islamic finance and ethical investment. *International Journal of Social Economics, 24*(11), 1325–1342.
Wray, F., Marshall, N., & Pollard, J. (2010). Finance and local and regional economic development. In A. Pike, A. Rodriguez-Pose, & J. Tomaney (Eds.), *Handbook of Local and Regional Development*. Routledge.

Further Readings

Bruno, V. A., & Cozzolino A. (2019). *Radical-right populists and financialisation*. Social Europe, 26 February. Available at the link: https://www.socialeurope.eu/radical-right-populists-finance.
Center for European Futures. (2020). Projects, activities and publications. Available at the link: https://www.instituteforthefuture.it/center-for-european-futures/?lang=en
Dembinski, P. H. (2009). Finance: Servant or deceiver. In *Financialization at the crossroad (Observatoire de la Finance)*. Palgrave Macmillan.
Dembinski, P. H., Bonvin, J. M., Dommen, E., & Monnet, F. M. (2003). The ethical foundations of responsible investment. *Journal of Business Ethics, 48*(2), 203–213.
Observatoire de la Finance (2020), Activities and publications. Available at the link: http://www.obsfin.ch/activities/journal-of-ethics-in-economics-and-finance.
Paetzold, F., Busch, T., & Chesney, M. (2015). More than money: Exploring the role of investment advisors for sustainable investing. *Annals in Social Responsibility, 1*(1), 195–223.
Seele, P., & Chesney, M. (2017). Toxic sustainable companies? A critique on the shortcomings of current corporate sustainability ratings and a definition of 'financial toxicity. *Journal of Sustainable Finance, 7*(2), 139–146.

Fiscal Tools

▶ Green Economy Policies to Achieve Water Security

Fisheries Crime and Ocean Resilience

Jade Lindley
Law School and Oceans Institute, The University of Western Australia, Crawley, WA, Australia

The ocean is one of the most important features of the Earth. It produces more than half the oxygen we breathe, regulates the climate, supports economies through trade and tourism, and provides sources of food, medicine, and recreation to not only littoral, but also landlocked states (National Oceanic and Atmospheric Administration, 2021). The balance of the ocean ecosystem must be maintained to ensure it thrives for generations to come. As such, the health of our world ocean must be prioritized at the international, national, and local levels, requiring collective buy-in.

Resilient oceans are necessary for a sustainable future. To achieve ocean resilience requires a collective effort to prevent illegitimate and unsustainable impacts, such as through the disruption of the natural fish stock replenishment. It is now well understood that the natural balance of ocean health and life within it can be irreparably upset by human forces. The world over, there is increasing reliance on fish for food security, meanwhile the Food and Agriculture Organization (FAO) estimates around 58 percent of fish stocks globally are already decimated due to unsustainable practices, pollution and habitat loss, illegal fishing, and is now further threatened by climate change (Food and Agriculture Organization of the United Nations, 2016). The continuation of ocean-based crimes threatens ocean resilience, particularly crimes in the fishing industry.

Fisheries crimes and more specifically, illegal fishing are indeed worrying. Estimates suggest 26 million tons of fish are illegally caught each year amounting to approximately 15 percent of global capture fisheries, valued at US$23.5 billion annually (Food and Agriculture Organization of the United Nations, 2016). In developing nations alone, illegal fishing costs up to $15 billion annually (Liddick, 2014: 290), and due to the reliance on the ocean to support the livelihoods of local fishers, the problem is magnified (Lymer et al., 2010).

Continuation of harmful fishing practices must be addressed before the ocean's health is irreversible. This contribution explores fisheries crimes in the context of ocean resilience. Specifically, it provides an overview of fisheries crime, including the internationally accepted legal definition of illegal fishing, and includes examples of how it may present. Then, to enable ocean resilience, it discusses innovative measures that could be coupled with traditional law enforcement to control fisheries crimes at the global and local levels.

Definition

Fisheries crimes involve a vast array of activities. Crimes that occur within the fishing industry may relate to fishing themselves (commonly referred to as illegal, unreported, and unregulated (IUU) fishing) or indeed extend to other crimes that may enable it in some way, such as corruption of border and fishing license officials, transhipment of (legal and illegal) catches in international waters, or trafficking of humans to supply cheap or free labor aboard vessels (Lindley et al., 2018; Martini, 2013; Steele & Adomeit, 2019; United Nations Office on Drugs and Crime, 2011). It may also involve fraud, whereby mislabeled fish and seafood are on sold to the consumer unknowingly (Lindley, 2021b). Evidence also exists of strong links to transnational organized crime (Lindley, 2018; Martini, 2013; United Nations Office on Drugs and Crime, 2011; Witbooi et al., 2020).

International Perspective

Sustainable Development Goals (SDGs) 14.4 (effectively regulate harvesting and end overfishing, IUU fishing and destructive fishing practices) and 14.6 (prohibit certain forms of fisheries subsidies which contribute to overcapacity and overfishing) optimistically outlined plans to end illegal fishing through greater regulatory control by 2020 (United Nations, 2015). These SDGs are more nationally focused, encouraging greater buy-in to support the global challenge to undo previous generations of ocean and marine life devastation. National support via SDGs is useful, given that fishing beyond any nation's territorial waters is well regulated via a range of international hard and soft laws (Lindley & Techera, 2017). Several international instruments exist to address IUU fishing in some way.

IUU fishing is understood to defy ocean resilience. IUU fishing is a broad catch-all for (mostly) illicit fishing activities. It fails international, regional, and local fishing laws and regulations in favor of profit with little consideration of maximum sustainable yield (United Nations, 1982: Articles 61 and 119). While some IUU fishing may involve artisanal fishers low-scale catches, the majority are organized, syndicated criminals. Regardless, "lack of awareness of relevant rules is inadequate justification for violation" (Lindley, 2018). IUU fishing is defined in the United Nations Food and Agricultural Organization (FAO) International Plan of Action to Prevent, Deter and Eliminate Illegal, Unreported and Unregulated Fishing (IPOA-IUU) (Food and Agriculture Organization of the United Nations, 2001). The definition involves three separate but intertwined elements: *illegal* fishing contravenes existing domestic and regional fishery management organization (RFMO) regulations; *unreported* fishing extends to nonreporting, misreporting, or underreporting catches, failing to comply with laws and RFMO measures; and *unregulated* fishing relates to activities yet to be regulated but elsewhere defy state and RFMO regulation (Food and Agriculture Organization of the United Nations, 2001: paragraphs 3.1, 3.2 and 3.3). While RFMO regulations are soft law and thus recommendary only, it serves as one of the most important of many international instruments in response to illegal fishing.

There is no shortage of international laws and regulations focused on supporting sustainable fisheries (Lindley, 2021a; Lindley & Techera, 2017), in fact, given the extent of legal and regulatory responses already available, illegal fishing should be controlled; however, this is not the case (Telesetsky, 2014). Rather, treaty congestion may overwhelm responses to illegal fishing (Lindley & Techera, 2017). The inability for some states to meet obligations set out within international instruments due to political, economic, and legal barriers, for example, may be among the reasons preventing greater buy-in and by extension, lacking universal adoption of these important fisheries controls. As such, illegal fishing is the result of failing political will to prevent it. Innovatively addressing the failure can be achieved in two ways, responding to: (1) corruption in fisheries and (2) consumer demand for legal and sustainable fish.

Corruption in Fisheries

Fisheries crime involves several crimes, and regulatory and enforcement failures, such as enabling illegal fishing through corruption, money laundering, document fraud, trafficking (namely people), and lax regulatory enforcement at several points along the supply chain (Organisation for Economic Co-operation and Development, 2013; Steele & Adomeit, 2019; United Nations Office on Drugs and Crime, 2011). Most commonly, illegal fishing is enabled by vessels flagged to open registries (commonly referred to as flags of convenience), using transhipment (mixing illegal with legal catches at sea to hide the illegal catches), entering ports of convenience (whereby officials are paid off to legitimize entry of illegal catches), and hide profits in tax havens (Martini, 2013: 3). Each of these methods intends to disregard or in some way evade controls in place to protect marine life, promote fair market share, and provide the relevant state with its due revenues. Given that fisheries are government-administered resources, for the most part, overcoming corruption of officials must be central to fisheries crime responses to secure ocean resilience. The following two case studies provide a glimpse of the failures to protect these important common resources.

Case Study: Indonesian Corruption Scandal High-demand, high-cost species are increasingly vulnerable to crime and requires strong regulations with harsh penalties to ensure compliance. However, the failure within governments to adhere to available scientific evidence to protect marine life challenges ocean resilience. In 2020, the Indonesian Fisheries Minister was charged of accepting bribes relating to *lobster larvae* export permits. Due to declining lobster populations and to combat illegal smuggling of the vulnerable species, the Indonesian government introduced a ban in 2016 to prevent lobster larvae exports. When the Minister came into power, the ban was overturned increasing trade of the lobster larvae due to the potential opportunity for earnings. Given the limited available export permits, the Minister accepted bribes from a number of private companies to secure permits to export the lobster larvae (Dao, 2021; Walden & Souisa, 2020). Reversing the ban put in place to protect the species was against the scientific advice but worsened by the Minister and other high level officials financially benefiting personally from corruptly issuing permits to

31 companies out of the 100 that sought permits (Dao, 2021). The Indonesian ban of lobster larvae exports was reinstated in 2021 to protect local marine resources.

Case Study: "Fishrot" Corruption Scandal Competition among international fishing corporations to secure coveted fishing licenses may increase potential for corruption among government officials, disregarding good fisheries management (Lindley, 2021a). Icelandic fishing company Samherji was central to *the Fishrot scandal* investigation whereby bribes were made to Namibian government officials and businessmen in return for access to Namibia's *horse mackerel* fishing quota (Mogotsi, 2019). An involved whistleblower exposed the scandal, revealing several other Samherji executives bribed high-level Namibian government officials, hiding kickbacks of at least N$150 million over four years in tax havens (Mogotsi, 2019). In February 2021, Namibian prosecutors charged 26 people, including three current Samherji employees, with racketeering, money laundering, and tax evasion (Oirere, 2021). The scandal highlights inadequate domestic antimoney laundering oversight and regulations that parties to the relevant international instruments are obliged to adopt – and that both Iceland and Namibia are party to (United Nations, 2004; United Nations Office on Drugs and Crime, 2020).

These case studies exemplify how important regulations and systems put in place to preserve ocean resilience can be disregarded. These examples relating to corruption, accepting bribes, money laundering, falsifying documents, and lax law enforcement show the damaging impact at the local level by crippling the fishing industry and threatening livelihoods and potentially expose many to poverty, but more broadly, a fish stock imbalance can devastate the ability to sustainably manage fishing industries globally. All but universally, corruption is regarded as a problem and political buy-in to respond exists (United Nations, 2004). Responding to fisheries crimes via the corruption backchannel may appear arbitrary but would be a suitable foundation to build on (Lindley, 2020).

Realigning Consumer Demand for Legal and Sustainable Fish

The absence of transparent labeling fails the consumer by disempowering an informed decision. If consumers were aware of the potential criminal activities, including document fraud to create fake fishing licenses, forced labor of fishers, offshore transhipment to mix legal and illegal fish intending to hoodwink authorities, and corruption at the border to dodge inspection, they may opt for fish with clearly labeled provenance. While standards and regulations exist to protect the consumer, it is not enough. Emerging technology is proving to be a sound investment to increase traceability of catches (see for example Borit & Olsen, 2020; Tarmizia & Zainuddinb, 2020; Tsolakisa et al., 2020; Widowati et al., 2019). Political will to enhance fisheries control and drive down demand for unsustainable fish is part of the solution.

Fisheries crimes directly or indirectly affect everyone as seafood is among the most traded food. Legal controls and strong enforcement over fisheries are critical to protecting ocean resilience. However, despite the existence of adequate regulatory controls over fisheries crimes, the threat remains. Given the transnational nature of supply chains, there are countless points along whereby criminals may infiltrate. Indeed, the transnational nature of the supply chains may aid rather than encumber criminals; therefore, in reality, no market is secure. Minor legal, testing, and definitional differences may enable criminals to operate in some markets rather than others and consumer demand also plays an important role. The following case study of seafood fraud provides insight into how criminals may operate between borders.

Case Study: Seafood Fraud Once fished, whether legally or illegally, further crimes relating to the catch may occur, such as fish and seafood fraud. Fish and seafood fraud (herein collectively called seafood fraud) occurs when intentional criminal label tampering (or mislabeling) occurs to dupe consumers and ultimately, increase profit potential. Seafood fraud may involve mislabeling country of origin, species, and/or gear and equipment used to fish. Seafood fraud supports illegal and unsustainable fishing practices and when those fraudulent products enter the legitimate market, fails to protect the consumer from making informed decisions about seafood being purchased (Lindley, 2019).

While the correlation between seafood fraud and ocean resilience may appear intangible, the continuation of mislabeled seafood has the potential to devastate entire stocks targeted by these criminals. For example, the strictest regulated seafood labeling practices exist in the European Union (EU). In addition to international standards that all foods must adhere to, such as sell and use by dates, weight, ingredients, and product name, the EU demands seafood labels to include details such as scientific and commercial fish name, catch area and country of production, production method, fishing gear used, and defrosted and date of freezing (European Commission, 2014). Given its large seafood import footprint, the EU's tough stance on seafood import controls is positive; however, there is a void between what is legally expected by states exporting to the EU and what those same exporting states expect for their own imported seafood supply. Indeed, Australia meets the EU's labeling requirements for exports but lowers its labeling standards for its own consumers – despite having the same available information (Lindley, 2021b). These additional labeling requirements position the consumer to make informed and (likely) sustainable choices; thus, the regulations enable consumers to make sustainable choices when purchasing seafood.

Additionally, there exists a disconnect between international and domestic standards and demand for exotic food may reduce the vigilance of some consumers in order to obtain the unobtainable. Despite the existence of the International Union for Conservation of Nature's (IUCN) Red List – a list that raises awareness of endangered species – some governments still allow imports of endangered and threatened species, such as caviar from the critically endangered Beluga sturgeons (Gesner et al., 2010; Lindley, 2021b). These threatened species farmed to harvest their caviar should be harmoniously banned around the world (Gesner et al., 2010; International Union for Conservation of Nature, 2020). Alignment of fish import policies with the IUCN Red List enables governments to support stock recovery management and reduce opportunity for illegal fishing and food fraud.

Understanding where the loopholes exist that may enable fisheries crimes is critical to ensuring ocean resilience. Without such, sustainable fishing practices may not be prioritized, preventing consumers from making sustainable choices about the seafood they purchase. Seafood fraud provides an example of how variations in legal responses may have a devastating impact on sustainable fishing and available fish stocks in the future.

Innovative Control of Fisheries Crimes to Build Ocean Resilience

Crime within the fishing industry is not new. Fisheries crimes date back to the commencement of the industry, though in modern times fisheries crimes are growing increasingly sophisticated, akin to modernization of the fishing industry. Thus, the increasingly sophisticated crimes must be matched with commensurate formal and informal regulatory responses to ensure the ocean remains resilient from threats.

International waters are free from national sovereignty, and with that means they are unpoliceable in the traditional sense (United Nations, 1982: Article 87). As such, innovative methods of monitoring the high seas are critical to protecting ocean resilience. To address and combat illegal fishing, a network of global, local, and informal innovative measures is emerging beyond formal, traditional responses.

Public Data Sharing

As an example of a global initiative, launched in 2015, the Global Fishing Watch website draws on available technology, supported by the Leonardo DiCaprio Foundation (Nugent, 2019). The purpose of the site was to "create and publicly share knowledge about human activity at sea to enable fair and sustainable use of our ocean" (Global Fishing Watch, 2021) enabling overwise unavailable oversight over activities such as legitimate and illegal fishing. A collaborative effort by Oceana, SkyTruth and Google, Global Fishing Watch draws on Automatic Identification System (AIS) to monitor vessels movements in near real time.

Triangulating datasets between data holdings geo-maps vessels, aptly tracking movements to understand regularities and anomalies from which algorithms can be generated of particular fisheries within particular locations as a baseline (Merten et al., 2016). Data deviating from the algorithms enables further enquiry as to illegitimate activities and results can be shared with relevant authorities and states and potentially lead to vessel interception and inspection.

In practice, visibility of fisheries activities resulting from the Global Fishing Watch data can also increase legitimate participation. For example, Indonesian fisheries have long been suspected in illegal fishing and with the launch of the Global Fishing Watch, opted to share vessel movement data to show its commitment towards legitimate and sustainable fishing (Ramadhan & Dugis, 2018). The pressure of increased oversight and potential for governmental failure in legitimizing its fisheries may reduce implicit support for illegal fishing activities and resultantly reduce fisheries crime.

Sharing data enables greater transparency, legitimacy, and benefit to and beyond the participants. While websites and other platforms exist that hold similar data, these may only be available through paid subscription meaning it reduces availability to wealthier nations or private companies. Global Fishing Watch provides freely available data to any users – including policing organizations, fishing stakeholders, and the public – to monitor, prevent, and intercept illegal fishing vessels in territorial waters and on the high seas (Merten et al., 2016). Collectively, data sharing increases knowledge and awareness of a region that is otherwise out of sight and unpoliceable, and therefore by extension, builds ocean resilience.

Innovative Market Regulation

In the local domain, an example of a response to illegal fishing is the EU's tough regulations restricting seafood imports from states with irresponsible fishing practices (European Commission, 2008). Collectively as one of the largest importers of seafood globally, the EU's regulations pressure states to improve practices to incentivize the continued seafood trade with the EU. States noncompliant with practices to prevent illegal fishing states and their illegal fishing vessels may be banned or blacklisted, reversible by improving standards (European Commission, 2008: Articles 25 and 26).

While having tough and clear regulations to prevent unsustainable practices is necessary, the EU introduced an innovative system whereby states in breach of the EU's zero-tolerance regulations may initially be issued a "yellow card" as a warning to improve practices or face harsher penalties, escalating to a "red card" banning trade with continued failed compliance (European Commission, 2021). The system essentially shames states into compliance, by incentivizing sustainable practices. Financial penalties may also discourage participation in irresponsible fishing; however, often these are incommensurate with the harm caused, or meaningless compared to the potential profits available (Martini, 2013), and thus, the innovative incentivization system may be more effective. Nonetheless, the EU approach sends a clear message to the fishing industry and states to encourage a higher standard in responsible fishing, deterring those implicitly or explicitly supporting illegal fishing (Lindley & Techera, 2020). These regulations are effective due to the EU state's collectively large seafood purchasing power; feasibility for individual states to adopt and implement such strong regulations for scaled-down imports would be a challenge.

Smaller scale measures exist to boost ocean resilience and prevent and protect against illegal fishing. For example, Eastern African states bordering the Western Indian Ocean – Comoros, Kenya, Madagascar, Mauritius, Mozambique, Seychelles, Somalia, and Tanzania – combined to respond collaboratively to illegal fishing. *FISH-i Africa* is a multilateral effort aimed at protecting the vast marine resources vital for food, jobs, and local economies (Stop Illegal Fishing, 2020). Limited resources within these states to work alone, through information sharing, and financial and physical resource collaboration enable greater effort in responding to illegal fishing in the region. Localized responses to build regional capacity and share responsibility to improve resilience, prevent illegal fishing, and protect ocean commons are becoming increasingly prevalent and necessary (Kubiak, 2020).

Emerging Technology to Close Policing Gaps and Build Resilience

Blockchain is one of the most promising technological advancements in supply chain resilience. It digitizes secure transactions or *blocks* at every point along the supply chain, viewable only by those with access to that blockchain ledger, and eliminates the need for intermediaries along the supply chain (Yadav et al., 2020). Each block is encrypted with a unique, nonmanipulable identifier and therefore completely decentralized, transparent, and traceable, starting with the source – assuming the producer is operating legitimately. Blockchain has been successfully applied outside the food industry and increasingly tested within it.

Blockchain offers the ability to track in real time and removes time-consuming (potentially fraudulent) document processing, while realizing cost saving increases in efficiency (Keogh et al., 2020). Though supply chain vulnerabilities are evident in transnational transportation, blockchain can close the loop on potential criminal infiltration along the supply chain and thus reduce the likelihood of fisheries crimes after the catch is landed. The uses of blockchain are endless and have the very real potential to control crime in the fishing industry by essentially making the previously opaque fishing industry transparent and therefore increasing consumer unwillingness to accept seafood without clear provenance (Lindley & Graycar, 2020).

Fisheries crimes prevent ocean resilience and as such, this contribution provides greater understanding as to how it presents and can be innovatively responded to. Sustainably protecting fisheries and oceans through innovative measures requires a collective effort for the benefit of everyone, now and into the future to enable and enhance ocean resilience.

Cross-References

- ► Climate-Resilient Technologies and Innovations for Sustainable Agriculture, Improved Landscape, and Food Security
- ► Planning for Food Security in the New Urban Agenda
- ► Role of Urban Agriculture Policy in Promoting Food Security in Bulawayo, Zimbabwe
- ► Sustainable Development and Responsible Tourism: The Grijalva-Usumacinta Lower River Basin
- ► Sustainable Development Goals
- ► Sustainable Development Goals from an Urban Perspective
- ► Sustainable Development Goals in Relation to Urban and Regional Development in Japan

References

Borit, M., & Olsen, P. (2020). *Beyond regulatory compliance seafood traceability benefits and success cases*. FAO Retrieved from http://www.fao.org/3/ca9550en/CA9550EN.pdf

Dao, T. (2021). *Former fisheries minister of Indonesia charged in bribery case*. https://www.seafoodsource.com/news/supply-trade/former-fisheries-minister-of-indonesia-charged-in-bribery-case

European Commission. (2008). COUNCIL REGULATION No 1005/2008, 'Establishing a Community system to prevent, deter and eliminate illegal, unreported and unregulated fishing, amending Regulations (EEC) No 2847/93, (EC) No 1936/2001 and (EC) No 601/2004 and repealing Regulations (EC) No 1093/94 and (EC) No 1447/1999'. In.

European Commission. (2014). *A pocket guide to the EU's new fish and aquaculture consumer labels*. European Commission Retrieved from https://ec.europa.eu/fisheries/sites/fisheries/files/docs/body/eu-new-fish-and-aquaculture-consumer-labels-pocket-guide_en.pdf

European Commission. (2021). *EU rules to combat IUU fishing*. European Commission. https://ec.europa.eu/oceans-and-fisheries/fisheries/rules/illegal-fishing_en

Food and Agriculture Organization of the United Nations. (2001). *Food and agriculture organization of the United Nations international plan of action to prevent, deter and eliminate illegal, unreported and unregulated fishing*. Rome, Italy. Retrieved from http://www.fao.org/3/y1224e/Y1224E.pdf

Food and Agriculture Organization of the United Nations. (2016). *The state of world fisheries and aquaculture: Contributing to food security and nutrition for all*. FAO Retrieved from http://www.fao.org/3/a-i5555e.pdf

Gesner, J., Chebanov, M., & Freyhof, J. (2010). *Huso huso (Beluga): The IUCN red list of threatened species*.

ICUN. Retrieved from https://www.iucnredlist.org/species/10269/3187455

Global Fishing Watch. (2021). *About us: Our purpose*. https://globalfishingwatch.org/about-us/

International Union for Conservation of Nature. (2020). *Sturgeons*. ICUN. Retrieved 7 August from https://www.iucnredlist.org/search/stats?query=Sturgeons&searchType=species

Keogh, J. G., Rejeb, A., Khan, N., Dean, K., & Hand, K. J. (2020). Optimizing global food supply chains: The case for blockchain and GSI standards. *Building the Future of Food Safety Technology*, 171–204. https://doi.org/10.1016/B978-0-12-818956-6.00017-8

Kubiak, L. (2020). *Protecting our global ocean commons*. https://www.iucn.org/crossroads-blog/202008/protecting-our-global-ocean-commons

Liddick, D. (2014). The dimensions of a transnational crime problem: The case of IUU fishing. *Trends in Organized Crime, 17*(4), 290–312.

Lindley, J. (2018). Nexus between illegal, unreported and unregulated fishing and other organised maritime crimes. In R. G. Smith (Ed.), *Organised crime research in Australia 2018* (Vol. 10). Australian Institute of Criminology. https://www.aic.gov.au/publications/rr/rr10

Lindley, J. (2019). Sustainably sourced seafood: A criminological approach to reduce demand for illegal seafood supply. In *Sustainability and the humanities*. Springer. https://doi.org/10.1007/978-3-319-95336-6_23

Lindley, J. (2020). Criminal threats undermining indo-Pacific maritime security: Can international law build resilience? *Journal of Asian Economic Integration, 2*(2), 206–220. https://doi.org/10.1177/2631684620940477

Lindley, J. (2021a). Crime and the environment. In E. Techera, J. Lindley, K. Scott, & A. Telesetsky (Eds.), *Routledge handbook of international environmental law* (2nd ed.). Routledge.

Lindley, J. (2021b). Food security amidst crime: Harm of illegal fishing and fish fraud on sustainable oceans. In R. C. Brears (Ed.), *The Palgrave handbook of climate resilient societies*. Springer.

Lindley, J., & Graycar, A. (2020). *Enhancing food supply chain regulation through blockchain*. University of Pennsylvania. https://www.theregreview.org/2020/12/28/lindley-graycar-regulating-food-supply-chain-blockchain/

Lindley, J., & Techera, E. (2017). Overcoming complexity in illegal, unregulated and unreported fishing to achieve effective regulatory pluralism. *Marine Policy, 81*.

Lindley, J., & Techera, E. (2020). Using routine activity theory to explain illegal fishing in the indo-Pacific. In S. Hufnagel & A. Moiseienko (Eds.), *Criminal networks and law enforcement: Global perspectives on illegal Enterprise*. Routledge.

Lindley, J., Percy, S., & Techera, E. (2018). Illegal fishing and Australian security. *Australian Journal of International Affairs, 73*(1). https://doi.org/10.1080/10357718.2018.1548561

Lymer, D., Funge-Smith, S., Clausen, J., & Miao, W. (2010). *Status and potential of fisheries and aquaculture in Asia and the Pacific 2008*. FAO. Retrieved from http://www.fao.org/docrep/011/i0433e/I0433E04.htm

Martini, M. (2013). *Illegal, unreported and unregulated fishing and corruption*. Transparency International. Retrieved from https://www.u4.no/publications/illegal-unreported-and-unregulated-fishing-and-corruption.pdf

Merten, W., Reyer, A., Savitz, J., Amos, J., Woods, P., & Sullivan, B. (2016). *Global fishing watch: Bringing transparency to global commercial fisheries*. Cornell University Computers and Society. https://doi.org/https://arxiv.org/abs/1609.08756

Mogotsi, K. (2019). LPM weighs in on Fishrot scandal. *The Namibian*. https://www.namibian.com.na/85626/read/LPM-weighs-in-on-Fishrot-scandal

National Oceanic and Atmospheric Administration. (2021). *Why should we care about the ocean? Our ocean provides countless benefits to our planet and all the creatures that live here*. NOAA. https://oceanservice.noaa.gov/facts/why-care-about-ocean.html

Nugent, J. (2019). Global fishing watch. *Science Scope, 42* (5). https://www.jstor.org/stable/26898882

Oirere, S. (2021). *US acts on Fishrot scandal by banning two Namibian ministers*. https://www.seafoodsource.com/news/business-finance/us-acts-on-fishrot-scandal-by-banning-two-namibian-ministers

Organisation for Economic Co-operation and Development. (2013). *Evading the net: Tax crime in the fisheries sector*. Retrieved from http://www.oecd.org/ctp/crime/evading-the-net-tax-crime-fisheries-sector.pdf

Ramadhan, P., & Dugis, V. (2018). *Indonesia's decision to share data of vessel monitoring system with global fishing watch* Airlangga conference on international relations. https://www.researchgate.net/profile/Vinsensio-Dugis/publication/349918045_Indonesia's_Decision_to_Share_Data_of_Vessel_Monitoring_System_with_Global_Fishing_Watch/links/60abd60f299bf1031fc83389/Indonesias-Decision-to-Share-Data-of-Vessel-Monitoring-System-with-Global-Fishing-Watch.pdf

Steele, T., & Adomcit, M. (2019). *Rotten fish: A guide on addressing corruption in the fisheries sector*. UN. Retrieved from https://www.unodc.org/documents/Rotten_Fish.pdf

Stop Illegal Fishing. (2020). *FISH-i Africa*. Retrieved 22 June, from https://stopillegalfishing.com/initiatives/fish-i-africa/

Tarmizia, D., & Zainuddinb, Z. (2020). DNA barcoding: A tool to Deect fraud in seafood products. *Big Data in Agriculture, 2*(1), 20–22. https://doi.org/10.26480/bda.01.2020.20.22.

Telesetsky, A. (2014). Laundering fish in the global undercurrents: Illegal, unreported, and unregulated fishing and transnational organized crime. *Ecology Law Quarterly, 41*(4), 939–998. https://heinonline.org/HOL/Page?handle=hein.journals/eclawq41&div=36&g_sent=1&casa_token=3nRQNxuIj6IAAAAA:fExR4nWYHubDZNkrWxtrhDaFuEiscE2FLTZPRE_HXyJkjNBVeozS1gRr0RbWa4GZGfqp67dipw&collection=journals

Tsolakisa, N., Niedenzub, D., Simonettob, M., Dorac, M., & Kumarb, M. (2020). Supply network design to

F

address United Nations sustainable development goals: A case study of blockchain implementation in Thai fish industry. *Journal of Business Research*. https://doi.org/10.1016/j.jbusres.2020.08.003

United Nations. (1982). *Convention on the law of the sea*. UN.

United Nations. (2004). *United Nations convention against corruption*. Retrieved from https://www.unodc.org/documents/treaties/UNCAC/Publications/Convention/08-50026_E.pdf

United Nations. (2015). *Transforming our world: The 2030 agenda for sustainable development A/RES/70/1*. UN. https://www.un.org/ga/search/view_doc.asp?symbol=A/RES/70/1&Lang=E

United Nations Office on Drugs and Crime. (2011). *Fisheries crime*. UNODC Retrieved from https://www.unodc.org/documents/about-unodc/Campaigns/Fisheries/focus_sheet_PRINT.pdf

United Nations Office on Drugs and Crime. (2020). *UN convention against corruption: Signature and ratification status*. Retrieved 12 May, from https://www.unodc.org/unodc/en/corruption/ratification-status.html

Walden, M., & Souisa, H. (2020). *Indonesian fisheries Minister Edhy Prabowo arrested over alleged lobster corruption*. https://www.abc.net.au/news/2020-11-27/rolex-watches-seized-as-indonesian-minister-arrested-corruption/12922796

Widowati, T. A., Andayani, N., & Maryanto, A. E. (2019). Application of DNA barcoding to detect mislabeling of fish fillet products from Jabodetabek's market. *IOP Conference Series: Earth and Environmental Science, 538*(012021). https://doi.org/10.1088/1755-1315/538/1/012021

Witbooi, E., Ali, K.-D., Achmad Santosa, M., Hurley, G., Husein, Y., Maharaj, S., Okafor-Yarwood, I., Quiroz, I. A., & Salas, O. (2020). *Organised crime in the fisheries sector*. World Resources Institute. Retrieved from https://bluejustice.org/wp-content/uploads/2020/08/Organised-Crime-Fisheries-Sector.pdf

Yadav, J., Misra, M., & Goundar, S. (2020). An overview of food supply chain virtualisation and granular traceability using blockchain technology. *International Journal of Blockchains and Crytocurrencies, 1*(2), 154–178. https://www.researchgate.net/profile/Jitendra_Yadav25/publication/338447663_An_overview_of_food_supply_chain_virtualization_and_granular_traceability_using_blockchain_technology/links/5f1a89e845851515ef44d04d/An-overview-of-food-supply-chain-virtualization-and-granular-traceability-using-blockchain-technology.pdf

Food Governance

▶ Governing for Food Security: A Cultural Perspective

Food Justice

▶ Faith Communities as Hubs for Climate Resilience

Food Loss

▶ Global Survey of Food Waste Policies

Food Losses and Waste

▶ Global Survey of Food Waste Policies

Food Security

▶ Governing for Food Security: A Cultural Perspective

Food Security – Food Governance

▶ Planning for Food Security in the New Urban Agenda

Forest

▶ The Source Waters of Tanga

Forest Fire

▶ Managing the Risk of Wildfire Where Urban Meets the Natural Environment

Formulating Sustainable Foodways for the Future: Tradition and Innovation

Jenna Andrews-Swann
School of Liberal Arts, Georgia Gwinnett College, Lawrenceville, GA, USA

Synonyms

Heritage cuisines; Local food traditions; Natural foodways; Sustainable food systems

Definition

Foodways represent the complex interaction of factors that result in food access and experiences, from the global to the hyper-local. Sustainable foodways include considerations of food production and distribution in the context of sociocultural norms and economies, along with patterns of food preparation and consumption, which, by virtue of their resilience and flexibility, might exist indefinitely.

Introduction

Sustainability goals and metrics often focus on the future, an observation not surprising owing to the current state of the planet. But discounting the past – traditions, histories, and heritage – commits a disservice to the communities that have established long-standing, unique, and creative means of interacting with their environment. *Sustainable foodways* are one example representing a combined approach to sustainability: one that is rooted in tradition and custom, which also acknowledges the importance of forward-looking resilience and innovation.

Foodways is a term that appears most commonly in works by historians, cultural scholars, and food specialists (Freedman 2007; Klein and Watson 2019). Foodways are the result of generations of decisions, norms, and habits around food production and consumption, and thus they are informed by myriad forces ranging from culture to ecology to economy (Katz and Weaver 2003). By their very nature, foodways are simultaneously taken for granted, constantly reinvented, and often contested parts of cultural identity (Brown and Mussell 1984; Chaudhury and Albinsson 2015; Timothy and Ron 2013). The *sustainable* foodways concept is a bit more recent, making early appearances in the mid-to-late 1990s. It clearly adds an environmental sustainability lens to considerations of food traditions, but scholars also use this model to demonstrate social justice elements of (continued) control over cultural heritage and for food sovereignty (Alkon and Vang 2016; Valley et al. 2020; Mares 2012).

This entry briefly outlines some of the scholarly work on the topic of sustainable foodways and considers the perspectives of food production, distribution, and consumption in turn. The goal of this overview is to highlight the utility of the combined (past + future, culture + environment) approach that sustainable foodways provides to sustainability studies more generally so that it may continue to provide a useful framework for promoting sustainable human-environment interactions.

Sustainable Foodways, Old and New

Food, of course, is basic to human survival. But its production and consumption are complex and often impactful processes that need attention if the species is to endure. Lessons from around the world, past and present, can provide guidance for policymakers and provide a pathway toward innovative and more sustainable foodways.

Production

Humans have been a foraging species for the majority of their existence, and it is only in the last 10,000 years or so that this subsistence strategy has changed much; in many cases, subsistence has morphed into a resource-intensive industrial agriculture that dominates much of the world today (Clunies-Ross and Hildyard 2013). Concentrated animal feed operations (CAFOs) are one of the more egregious examples of modern industrialized food production strategies; as the name suggests, they concentrate large numbers of domesticated animals into small areas, generating immense amounts of (also concentrated) waste and pollution (Hribar 2010). CAFOs have developed in a context of meat industry subsidies, meat-heavy foodways, and cheap animal feed produced with petrochemicals (Bunton et al. 2007; Hribar 2010).

Small-scale food production tends to be more flexible and sustainable (Horrigan et al. 2002), though economies of scale often mean that small farmers and makers must charge more for their products and so may find it difficult to compete with big agribusiness (Bray 2019). Counter to this, some small producers have found success in marketing their foods as artisanal, luxury, or local thus differentiating them from their uber-industrial counterparts (West 2019). The flexibility and resilience of small-scale food production is evident in a host of historical examples, from North American and British Victory Gardens during World Wars I and II to home gardening in Special Period Cuba (Endres and Endres 2009).

Distribution

The global economy impacts foodways, most notably through the distribution of foods from one part of the world to another. Modern shipping patterns are situated in a history of unbalanced economic ties across nation-states (Mintz 1985) and often exacerbate extant inequalities, leading to food injustice and other social ills (D'Odorico et al. 2019). (Re)localizing food and shrinking the network of global food distribution can ameliorate these injustices by empowering locals to make more decisions about their food, plus the extra inputs for packaging and shipping are minimized (Van Passel 2013).

For instance, the oft-cited Slow Food Movement embraces seasonal local foods and traditional preparations, in contrast to highly processed, globally-sourced "fast" food (see Chrzan 2004, for a good overview of this movement). Attention to *terroir*, the interplay of soil, ecology, and atmosphere, centers the source of food (and drink) as a mark of authenticity and quality (Lucini et al. 2020). Culinary tourism presents opportunities for travelers to taste another culture, and may provide economic and pride-in-heritage-based motivation to retain or revitalize local sustainable foodways – though, of course, travel presents its own set of sustainability issues (Wondirad et al. 2021). And the global popularity of farmers markets and of community supported agriculture (CSA) in North America of late is another nod to small and local (Alkon 2008; Chiffoleau and Dourian 2020).

Efforts to regain or retain food sovereignty, the ownership and control over foodway elements like cuisine and farmland, are a way to understand some of the socioecological impacts of globalized food networks. Examples of successful, sustained foodway revitalization efforts that emphasize local production and consumption, landraces, and sustainable production strategies in contrast to industrial agriculture include Afro-diasporic and indigenous communities in North America (Garth and Reese 2020; Joseph and Turner 2020; Spigelski and Erasmus 2013; Twitty 2017); Peruvian indigenous and traditional growers (Matta 2019; Nazarea 2014); subsistence farmers in the Philippines (Nazarea 2006); and smallholder agriculture for food security in South Africa (Baiphethi and Jacobs 2009) among many others.

Consumption

Both overconsumption and food insecurity are issues that plague human populations, and each can have negative impacts on the sustainability of local foodways. For instance, overconsumption is not only closely tied to physiological problems like heart disease and obesity, but it exacts a significant toll on landscapes and the environment more generally owing to the huge amount of inputs, making sustained access to preferred foods more problematic (Lang and Barling 2012). Food insecurity, or lack of access to fresh, healthy foods, typically means higher reliance on packaged and highly processed food items, rather than the whole foods that often comprise more desirable and sustainable foodways (Berry et al. 2015).

Plant-based diets have gained visibility (if not always popularity) in recent years, and are purported to be more sustainable and healthful as well as being friendlier to animals. But these diets are not novel; indeed, they echo long traditions of eating locally, seasonally, and economically, as well as eating in accordance with religious guidelines that limit animal products (Contento 2011; Sobal and Bisogni 2009; Weibel et al. 2019). In fact, high levels of meat and dairy consumption have only become the norm (first in the West, eventually in much of the rest of the world) in the last century so this practice is actually quite new (Sabate and Soret 2014). A complimentary approach to the plant-based "trend" encourages entomophagy as a means of increasing the sustainability of many Western foodways and appreciation for the diversity of food traditions in a given culture and place (Looy et al. 2014).

Summary: Foodways for the Future

The future of foodways rests on their sustainability and propensity for reinvention. Extant sustainable foodways tend to have some elements in common: equitable access to land and economic supports; preference for (agro)biodiversity; lower levels of pesticide/herbicide usage; centering local stakeholders; food seasonality; smaller-scale production; and a higher proportion of plant-based ingredients in traditional or authentic preparations.

Empowering locals, be it via sustainable tourism, cultural revitalization, economic development, or other community-led means, may help to promote those foodways that already have sustainability at their core. And these traditions can serve as models, in actuality and in spirit, for populations – and policymakers – in need of inspiration for the future.

References

Alkon, A. H. (2008). *Black, white and green: A study of urban farmers markets*. Davis: University of California.

Alkon, A. H., & Vang, D. (2016). The Stockton farmers' market: Racialization and sustainable food systems. *Food, Culture & Society, 19*(2), 389–411.

Baiphethi, M. N., & Jacobs, P. T. (2009). The contribution of subsistence farming to food security in South Africa. *Agrekon, 48*(4), 459–482.

Berry, E. M., Dernini, S., Burlingame, B., Meybeck, A., & Conforti, P. (2015). Food security and sustainability: Can one exist without the other? *Public health nutrition, 18*(13), 2293–2302.

Bray, F. (2019). Feeding farmers and feeding the nation in modern Malaysia: The political economy of food and taste. In J. A. Klein & J. L. Watson (Eds.), *The handbook of food and anthropology* (2nd ed., pp. 173–199). Bloomsbury.

Brown, L. K., & Mussell, K. (1984). *Ethnic and regional foodways in the United States: The performance of group identity*. Knoxville: University of Tennessee Press.

Bunton, B., O'Shaughnessy, P., Fitzsimmons, S., Gering, J., Hoff, S., Lyngbye, M., . . . Werner, M. (2007). Monitoring and modeling of emissions from concentrated animal feeding operations: Overview of methods. *Environmental Health Perspectives, 115*(2), 303–307.

Chaudhury, S. R., & Albinsson, P. A. (2015). Citizen-consumer oriented practices in naturalistic foodways: The case of the slow food movement. *Journal of Macromarketing, 35*(1), 36–52. https://doi.org/10.1177/0276146714534264. Publisher version of record available at: https://journals.sagepub.com/doi/full/10.1177/0276146714534264.

Chiffoleau, Y., & Dourian, T. (2020). Sustainable food supply chains: Is shortening the answer? A literature review for a research and innovation agenda. *Sustainability, 12*(23), 9831.

Chrzan, J. (2004). Slow food: What, why, and to where? *Food, Culture & Society, 7*(2), 117–132.

Clunies-Ross, T., & Hildyard, N. (2013). *The politics of industrial agriculture*. Routledge.

Contento, I. (2011). Overview of determinants of food choice and dietary change: Implications for nutrition education. In *Nutrition Education Linking Research Theory Practice* (2nd ed.). Jones & Bartlett Learning.

D'Odorico, P., Carr, J. A., Davis, K. F., Dell'Angelo, J., & Seekell, D. A. (2019). Food inequality, injustice, and rights. *Bioscience, 69*(3), 180–190.

Endres, A. B., & Endres, J. M. (2009). Homeland security planning: What victory gardens and Fidel Castro can teach us in preparing for food crises in the United States. *Food & Drug LJ, 64*, 405.

Freedman, P. (2007). *Food: The history of taste*. University of California Press.

Horrigan, L., Lawrence, R. S., & Walker, P. (2002). How sustainable agriculture can address the environmental and human health harms of industrial agriculture. *Environmental health perspectives, 110*(5), 445–456.

Hribar, C. (2010). *Understanding concentrated animal feeding operations and their impact on communities*. National Association of Local Boards of Health.

Joseph, L., & Turner, N. J. (2020). "The Old Foods Are the New Foods!": Erosion and revitalization of indigenous food Systems in Northwestern North America. *Frontiers in Sustainable Food Systems, 23*. https://doi.org/10.3389/fsufs.2020.596237.

Katz, S. H., & Weaver, W. W. (2003). *Encyclopedia of food and culture*. New York: Scribner.

Klein, J. A., & Watson, J. L. (2019). *The handbook of food and anthropology* (2nd ed.). Bloomsbury.

Lang, T., & Barling, D. (2012). Food security and food sustainability: Reformulating the debate. *The Geographical Journal, 178*(4), 313–326.

Looy, H., Dunkel, F. V., & Wood, J. R. (2014). How then shall we eat? Insect-eating attitudes and sustainable foodways. *Agriculture and human values, 31*(1), 131–141.

Lucini, L., Gabriele Rocchetti, G., & Trevisan, M. (2020). Extending the concept of terroir from grapes to other agricultural commodities: An overview. *Current Opinion in Food Science, 31*, 88–95. https://doi.org/10.1016/j.cofs.2020.03.007.

Mares, T. M. (2012). Tracing immigrant identity through the plate and the palate. *Latino Studies, 10*(3), 334–354.

Matta, R. (2019). Heritage foodways as matrix for cultural resurgence: Evidence from rural Peru. *International Journal of Cultural Property, 26*(1), 49–74. https://doi.org/10.1017/S094073911900002X.

Mintz, S. W. (1985). *Sweetness and power: The place of sugar in modern history*. Viking-Penguin.

Nazarea, V. D. (2006). *Cultural memory and biodiversity*. University of Arizona Press.

Nazarea, V. D. (2014). Potato eyes: Positivism meets poetry in food systems research. *CAFÉ, 36*, 4–7. https://doi.org/10.1111/cuag.12024.

Sabate, J., & Soret, S. (2014). Sustainability of plant-based diets: Back to the future. *The American Journal of Clinical Nutrition, 100*(suppl_1), 476S–482S.

Sobal, J., & Bisogni, C. A. (2009). Constructing food choice decisions. *Annals of Behavioral Medicine, 38*, 37–46.

Spigelski, D., & Erasmus, B. (2013). *Indigenous peoples' food systems & well-being: Interventions & policies for healthy communities*. Rome: Food and Agriculture Organization of the United Nations.

Timothy, D. J., & Ron, A. S. (2013). Heritage cuisines, regional identity and sustainable tourism. In *Sustainable culinary systems: Local foods, innovation, and tourism & hospitality* (pp. 275–283). London: Routledge.

Twitty, M. W. (2017). *The cooking gene: A journey through African American culinary history in the Old South*. New York: Harper Collins.

Valley, W., Anderson, M., Tichenor Blackstone, N., Sterling, E., Betley, E., Akabas, S., Koch, P., Dring, C., Burke, J., & Spiller, K. (2020). Towards an equity competency model for sustainable food systems education programs. *Elementa: Science of the Anthropocene, 8*, 33. https://doi.org/10.1525/elementa.428.

Van Passel, S. (2013). Food miles to assess sustainability: A revision. *Sustainable Development, 21*(1), 1–17.

Weibel, C., Ohnmacht, T., Schaffner, D., & Kossmann, K. (2019). Reducing individual meat consumption: An integrated phase model approach. *Food quality and preference, 73*, 8–18.

West, H. G. (2019). Artisanal foods and the cultural economy: Perspectives on craft, heritage, authenticity and reconnection. In J. A. Klein & J. L. Watson (Eds.), *The handbook of food and anthropology* (2nd ed., pp. 406–434). Bloomsbury.

Wondirad, A., Kebete, Y., & Li, Y. (2021). Culinary tourism as a driver of regional economic development and socio-cultural revitalization: Evidence from Amhara National Regional State, Ethiopia. *Journal of Destination Marketing & Management, 19*, 100482.

Further Reading

Timothy, D. J. (Ed.). (2015). *Heritage cuisines: Traditions, identities and tourism*. Routledge.

Foundations

► Philanthropy in Sustainable Urban Development: A Systems Perspective

From Vulnerability to Urban Resilience to Climate Change

Kamilia Mahdaoui[1] and Jamal Alibou[2]
[1]Hassania School of Public Works, Casablanca, Morocco
[2]Department of Civil Engineering, Hydraulic, Environment and Climate, Hassania School of Public Works, Casablanca, Morocco

Synonyms

Adaptation; Mitigation; Resilience

Definition

Today, cities tend to reduce their vulnerability to the effects of climate change, a direct consequence of human activities that consume a lot of energy and aggressively pollute nature through the emission of greenhouse gases. The transition from this vulnerability to urban resilience requires a good reflection on how to act. We must first define and understand these two notions of vulnerability/resilience in order to move for one to the other.

Introduction

The urban environment, considered as one of the most complex systems and a place where the laws of economics, ecology and society confront each other, is today exposed to the effects of climate change, a global, irreversible and anthropogenic phenomenon which it contributes to reinforce or generate through the various anthropogenic activities. In fact, the increase of climate risks in urban areas has a "*twofold anthropogenic dimension: it is linked to the anthropization of hazards such as flooding by urban runoff and to the increase of the vulnerability of societies*" (Quenault et al. 2011). The term of vulnerability is generally used in relation to an inability to cope with external changes including avoiding harm due to hazards.

Nowadays, stakeholders are focusing more on mitigating the impacts and vulnerabilities generated by the hazard than on mitigating the phenomena themselves by trying to link urban development with adequate adaptation strategies in order to have an urban resilient to climate risks. So, Particular attention must be paid to the notions of vulnerability and resilience for a better understanding of the impacts of climate change on the different components of urban society and for a better identification of the factors that determine the capacity to respond to risks.

Context of Climate Change

First of all, it is necessary to clarify the difference between "weather" and "climate" and also between "climate change" and "global warming". Indeed, weather was the state of the atmosphere that existed at any given place and time, and it could be observed or measured with suitable instruments. Climate was the long-term average of the weather at a place, as shaped by various factors like its latitude, elevation or distance from the ocean. Regarding global warming, it is measured as the average increase in Earth's global surface temperature. It is responsible to climate change. As the planet's temperature rises more than it would naturally, the climate varies.

Greenhouse gases (CO_2, CH_4, N_2O, Fluorinated gases) retains some of the Sun's heat. Thanks to them, the temperature of the earth is on average (15 °C) allowing it to maintain the necessary conditions to host life. Without them, it will be unbearably cold (−18 °C). The problem is that daily human activities maximize the greenhouse effect: the Industrial Revolution was the turning point when emissions of greenhouse effect gases entering the atmosphere began to soar. In addition, the population growth cause an exploding resource use, increasing energy demand and production, mainly from fossil fuels. As a result, the global average temperature of our planet is rising (Global warming). Our planet has already warmed by an average of 0.85 °C. It could reach 1.1 °C to 6.4 °C by 2100 according to scientists.

The climate change due to the global warming brings significations perturbations: the rise in temperature, the evolution of precipitation, the occurrence and severity of the risk of flooding, the expected rise in sea level, the risk of erosion, etc. Consequently, many key economic sectors are affected by these perturbations. For example, energy is used to keep buildings warm in winter and cool in summer: Changes in temperature would thus affect energy demand. Climate change also affects energy supply through the cooling of thermal plants, through wind, solar and water resources for power, and through transport and transmission infrastructure. Water supply depends on precipitation patterns and temperature, and water infrastructure is vulnerable to extreme weather. Heath care systems are also affected. Agriculture is arguably the most climate-sensitive sector: A warming climate has a negative effect on crop production and generally reduces yields of staple cereals such as wheat, rice, and maize, which differ between regions and latitudes.

The different regions of the world are very unequally threatened by catastrophic events and climate change is expected to accentuate these inequalities.

Concepts and Terminology

It would be desirable to get acquainted with some of the terminologies and concepts that are commonly encountered in the discussions about climate change.

The city appears both vulnerable and best positioned to deal with climate risk. The concept of the city has been defined differently by many researchers. For some, the city is an object constructed by our representations: "We grasp the city through our representations that means the mental images that emerge from our personality or culture, our desire to understand, our willingness to act. Thus, we see the city through what we are, but also through our way of thinking, through what we want. So, the representations of the city are multiple, as is the man who thinks it is multiple" (Lhomme 2013). For others, the city is seen as a system that is "a set of elements in dynamic interaction organized according to a purpose" (Donnadieu et al. 2003) more precisely, it will be seen as a socio-technical system where critical infrastructures occupy a central place. "The city, which today is home to more than half of the world's inhabitants, is not only a form of densely populated area, which brings together a certain number of political, administrative and economic functions, but also a form of management of drinking water, wastewater, energy and waste issues" (Petitet 2011). Without repeating all the existing literature on the city, *the city can be defined as a sociotechnical system in which critical infrastructures play a primordial role in its functioning.*

As for the concept of **vulnerability**, there are a variety of definitions in the available literature. They can be grouped into three groups: The first one is related to physical sciences, it is a technical vulnerability, and refers to the whole physical damage of a hazard on issues (infrastructures, buildings, populations, etc.). The second is part of the humanities and social sciences, it is a social vulnerability, it refers to the absence of society's capacity to face a crisis or change, the difficulty of a person, a group of people, an organization or a territory to anticipate a destructive phenomenon, to face it, to resist it and to recover after its occurrence (Quenault 2015). This type of vulnerability denounces, in a way, social fragility and social inequalities (income, age, etc.). For the third group, it is *the global* vulnerability which brings together the two previous vulnerabilities. So, *the city's global vulnerability to climate change expresses at what point materials and humans issues can resist the hazard (According to their degree of sensitivity and exposure) and what a point the society is capable of coping and rebuilding itself (According to their intrinsic characteristics).*

The notion of vulnerability is associated with the notion of **risk** which the Intergovernmental Panel on Climate Change (IPCC) has defined as "The possible and uncertain consequences [= impact or effect] of an event on something of value (...). Risk results from the interaction of vulnerability, exposure and hazards (...)" (Groupe d'experts intergouvemememental sur l'évolution du climat et al. 2015).*Climate risk then refers to climate events that have a particular impact on the territory, its inhabitants and its assets.* They can take different forms: floods, cold waves and snowfall, heat waves, storms, rising sea levels, risk of erosion and loss of space...

Like vulnerability, **resilience** is a polysemic concept, subject to multiple definitions and specifications. In physical science, resilience is the ability of a system to return to its initial state after a shock or continuous pressure, it is physical resilience (Dauphiné and Provitolo 2007). For ecological resilience, it is defined as the capacity of an ecosystem to adapt to new situations (Quenault et al. 2011). In psychology, resilience is the ability to live, succeed and develop in spite of adversity. In social science, Adger has defined social resilience as the capacity of human communities to support and recover from external shocks or disruptions. Most recently, he contributed to evolve the terminology in order to use the notion of eco-social resilience in the context of climate change. It concerns humans and nature as interdependent systems (Barroca et al. 2013). In dynamic systems science, according to Dovers and Handmer (1992, 1996), systemic resilience takes two forms: the first is reactive resilience which presents the maximum of disturbance a system can undergo while remaining in the same state. It is directly linked to self-organization. The second is proactive resilience which is the system's ability to build and increase its adaptive capacity and learning capacity, which means its ability to recover from damage suffered by transforming, reorganizing or renewing the structures and functions of a system (Quenault et al. 2011). So, urban resilience corresponds to *the city's ability to prepare and plan for the occurrence of hazards by anticipating response measures and its ability to function despite the disruption of some of its components (to operate in degraded mode), then to put them back into service so that they can return to optimal functioning as quickly as possible while taking into account the impacts endured with a view to minimize them during future disruptions.*

The use of the term **adaptation** is recent in the context of climate change. The IPCC has given a universal definition of adaptation to climate change: "The process of adjusting to actual or expected climate and its effects. For human systems, it is about mitigating adverse effects and exploiting beneficial effects" (Groupe d'experts intergouvememental sur l'évolution du climat et al. 2015). *Adaptation then consists in making systems less vulnerable to climate change, through actions that reduce the actual impacts of climate change or improve the response capacities of societies and the environment.*

How to Improve Urban Resilience?

Two complementary and inseparable components to respond to climate change and its impacts: a "mitigation" component that aims to act directly on the causes of climate change, in other words on responsibilities, and an "adaptation" component that aims to limit the harmful impacts of climate risk by reducing vulnerability to the present and future impacts of climate change. All components of the city are exposed to the various impacts of climate risks (floods, drought, etc.), long-term adaptation options depend on the nature of the exposed element and the type of climatic hazard. The first key step is then to determine the impact chain, a diagram that describes the cause-effect relationships between its different elements (Fig. 1).

The elements of the impact chain must be defined:

Hazard is defined as "[...] the potential occurrence of a natural or human-induced physical event or trend, or physical impact that may cause loss of life, injury, or other health impacts, as well as damage and loss to property, infrastructure, livelihoods, service provision, and environmental resources." (Groupe d'experts intergouvememental sur l'évolution du climat et al. 2015). The concept of effective exposure is the portion of the exposed assets that is actually affected by a specific hazard occurrence.

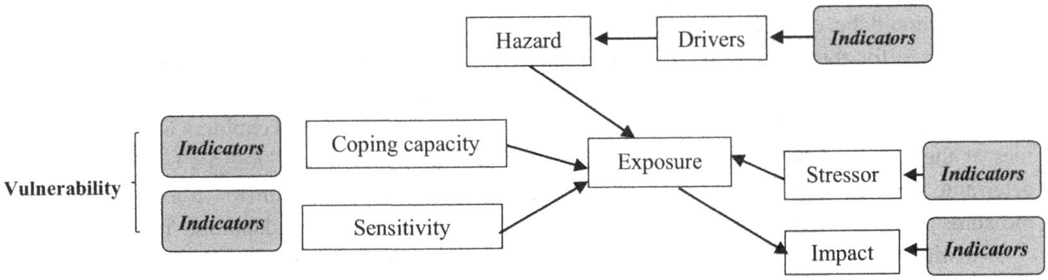

From Vulnerability to Urban Resilience to Climate Change, Fig. 1 Impact chain

Drivers are climatic factors that causes those hazards.

Stressors are Non-climatic events that may have a major effect on the system exposed. For example, Population growth or percentage of sealed surface will increase the risk of inundation.

Sensitivity is defined by the degree to which a system could be affected by the considered hazard. Different systems are more or less sensitive to a hazard.

Coping capacity is the ability of people, institutions, organizations and systems to deal with, manage and overcome pejorative situations in the short to medium term using available skills, values, resources and opportunities.

Impact refers to undesired effects on natural, human and material systems like, livelihoods, health, services, infrastructure, ecosystems, economies, etc. of extreme weather and events due to climate change within a specific time period. Impacts are also called consequences and outcomes.

Indicators are employed to quantify the intensifying or mitigating elements of an exposed system with regard to selected hazard(s), as well as the potential impacts hazards may have on the exposed system. For instance, indicator socio-economic like population density, environmental indicator like change in average temperature or infrastructure indicator as water supply measured in number day, etc. The values of the selected indicators will later be aggregated to risk components and provide the basis for the calculation of the composite risk score (Rome et al. 2019).

Once the elements of the impact chain have been identified, the indicators assigned to each attribute should be quantified by collecting the data required for this purpose. Then, operations of standardization, weighting and aggregation of the indicators are inevitable to assess vulnerabilities at the end, which will make it possible to determine the effective and optimal adaptation measures to improve the resilience to climate risk.

A distinction must be made between reactive and proactive adaptation. The first corresponds to post-crisis responses, i.e. actions taken to overcome the crisis after its onset. The second corresponds to proactive responses, i.e. actions taken to reduce vulnerability to a hazard before it results in undesirable impacts.

To sum up, series of adaptation actions must be taken, from the most anticipated to the most reactive: First prevent to reduce the probability of a climate impact occurring, for example building flood defenses, then prepare to improve knowledge of risks to reduce vulnerability and improve resilience such as raising awareness among inhabitants, after respond to a climatic event to limit its consequences like restricting water consumption in times of drought and finally repair to allow a rapid and cost-effective return to normal or a more sustainable state after a climatic event as improving resilience in the event of flooding by undertaking repairs of damage.

After drafting a list of options, it is important to decide which ones are the most important. Different criteria is used to evaluate and prioritize adaptation options: Risk or type of loss, efficacy, urgency, cost, alignment with development planning, social acceptance and political support, environmental integrity, etc.

Conclusion

An effective response at the city level must combine mitigation measures "to avoid the unmanageable" and adaptation measures "to manage the unavoidable" (World Bank 2007). Given the urgency of protecting against the adverse effects of climate change, adaptation is less and less perceived as a simple option, but as one of the key elements of sustainable development trajectories.

"A successful shift to an inclusive, resilient and low-emission society will require a distinct way of decision-making to negotiate disagreements, build trust among stakeholders and foster a long-term common vision." *Rebeca Grynspan, Associate Administrator, United Nations Development Programme.* Every adaptation process, whether planning for actions at the national level or designing a local adaptation project, is more sustainable when it is built on a consultative process.

In other words, effective adaptation action brings together in a meaningful way all the relevant stakeholders, beyond the operational team.

The question that comes to mind after reading this chapter is the following: Can resilience make the sustainable city a reality? The implication between resilience and sustainability is not obvious and requires a good understanding of both concepts.

Cross-References

- ▶ Climate Resilience in Informal Settlements: The Role of Natural Infrastructure
- ▶ Green Cities
- ▶ Urban Resilience

References

Barroca, B., DiNardo, M., & Mboumoua, I. (2013). De la vulnérabilité à la résilience: Mutation ou bouleversement? *EchoGéo, 24*, Art 24.

Dauphiné, A., & Provitolo, D. (2007). La résilience: Un concept pour la gestion des risques. *Annales de géographie, 654*(2), 115.

Donnadieu, G., Durand, D., Neel, D., Nunez, E., & Saint-Paul, L. (2003). L'approche systémique: De quoi s'agit-il? Synthèse des travaux du groupe AFSCET "Diffusion de la pensée systémique." *L'Approche Systémique* 11.

Groupe d'experts intergouvememental sur l'évolution du climat, Pachauri, R. K., & Meyer, L. A. (2015). *Changements climatiques 2014: Rapport de synthèse: contribution des Groupes de travail I, II et III au cinquième Rapport d'évaluation du Groupe d'experts intergouvememental sur l'évolution du climat*. GIEC.

Lhomme, S. (2013). *Les réseaux techniques comme vecteur de propagation des risques en milieu urbain—Une contribution théorique et pratique à l'analyse de la résilience urbaine*. 376.

Petitet, S. (2011). Eau, assainissement, énergie, déchets: Vers une ville sans réseau? *Métropolitiques, 5*. http://www.metropolitiques.eu/Eau-assainissement-energie-dechets.html

Quenault, B. (2015). *La vulnérabilité, un concept central de l'analyse des risques urbains en lien avec le changement climatique—Persée*. France: Renne II University.

Quenault, B., Pigeon, P., Bertrand, F., & Blond, N. (2011). *Vulnérabilités et résilience au changement climatique en milieu urbain: Vers de nouvelles stratégies de développement urbain durable ?* 204. France: Renne II University.

Rome, E., Ullrich, O., Lückerath, D., Worst, R., Xie, J., & Bogen, M. (2019). IVAVIA: Impact and vulnerability analysis of vital infrastructures and built-up areas. In E. Luiijf, I. Zutautaitè, & B. M. Hâmmerli (Éds.), *Critical information infrastructures security* (11260, pp. 84–97). Germany: Springer International Publishing, Fraunhofer-institut for Intelligent Analyse and Information System.

Furthering the Sustainable Development Agenda by Putting Urban Heritage and Value Extraction at the Center

Chipo Mutonhodza[1], Nesbert Mashingaidze[1] and Innocent Chirisa[2]

[1]Department of Rural and Urban Development, Great Zimbabwe University, Masvingo, Zimbabwe

[2]Department of Demography Settlement and Development, Social & Behavioural Sciences, University of Zimbabwe, Harare, Zimbabwe

Synonyms

Heritage – Inheritance; Endowments; Sustainability – Maintainability; Value – Benefit; Development – Growth; Progression

Definitions

Sustainable development: "Development that meets the needs of the present without compromising the ability of future generations to meet their own needs" (World Commission on Environment and Development 1987).

Heritage: Property that is or may be inherited; an inheritance; factor endowments; assets achieved over time.

Urban Heritage: "The list of heritage elements located in urban arcas: archaeological vestiges, historical buildings, vernacular architecture, historical gardens, social practices, rituals, and festive events, among others" (García-Hernández and de la Calle-Vaquero 2019).

F

Value extraction: This refers to the capacity to capture social, economic, and technological returns from a property, factor endowment, or asset.

Introduction

This chapter seeks to make a contribution that interrogates how heritage resources are of importance in urban planning and the urban spaces and consider how such may be critical in future. As the discussion on sustainability of urban settlements gathers momentum, the discussion on urban heritage and planning has not been spurred. Placing heritage in the planning of urban areas remains critical in balancing between physical, historical, environmental, social, and economic fabrics of urban areas. It is of paramount significance to note that heritage and culture remains essential, both as an asset and as a tool, both for urban sustainability and sustainable development of urban areas, urban reconstruction and urban recovery. Why? This is because much emphasis is given on the importance of culture and urban heritage, once urban heritage and culture is lost, it means some cultural and historical attributes of cities cannot be reversed.

Janssen et al. (2014) argue that preservation and conservation of heritage has become a topical issue under growing and expanding cities. Cultural heritage buildings can be former places of religious worship, aristocratic or royal residence, community meeting places, industrial production sites, early modern office buildings, or military objects. Therefore, these objects are not only for sustainable significance in urban areas, but something of value can be extracted from these objects. With today's rapidly urbanizing cities that are characterized with informal expansion, uncontrolled growth, and management deficiencies, the irreplaceable urban heritage is at risk of extinction (Swain 2018).

However, it is not only about preservation and conservation of heritage that is placed by urban heritage and culture, but there is more of value that can be extracted from urban culture and heritage. Value extraction from urban heritage resources entails capturing of value from the endowed resources which are heritage in nature, either outside or inside the urban areas by manipulating or incorporation of the comprehensive advantage. Therefore, this means that areas endowed with some heritagical resources such as provinces and towns are not only significant to urban planning in terms of its presence and culture resemblance, but also generates multiple benefits from the endowments. Foster (2020) argues that cultural heritage buildings hold a unique niche in the urban landscape.

It is further argued that besides the value of shelter, heritage provides, they also embody the local cultural and historical characteristics that define communities (Foster and Jones 2019). Therefore, extending their useful lifespan has multiple benefits that extend beyond the project itself to the surrounding area, contributing to economic and social development. Despite the fact that sustainable solutions for those buildings in urban development is required, it is also of importance to think of value extraction from these resources. Additionally, resourceful and innovative approaches for the built environment in general and existing buildings in particular are key to accomplishing future sustainability (Foster and Jones 2019). Urban culture heritage buildings are of particular interest because they may be underutilized or abandoned, nevertheless are important for the heritage of local, and possibly international communities. More so, given the existing increasing urban infrastructure services demands, municipalities are putting low priority on investments and developments that foster heritage preservation and conservation (Lynne and Yung 2006). Additionally, it is interesting to note that value has always been the common reason underlying the preservation and conservation of heritage. Therefore, it is clearly evident that no any society makes an effort to conserve and preserve what it does not consider of value to them.

Conceptual Framework

The concept of value suggests usefulness and benefits (Getty Conservation Institute 2002). In relation to heritage, heritage is valued not as an intellectual enterprise but because it plays an instrumental, symbolic, and other functions in society. In the sphere of material heritage, such as a given heritage place or site, a given building structure or any object considered of heritage significance has multiple different values attached to it, and this why it is argued that heritage is multivalent (Getty Conservation Institute 2002). For instance, historical buildings form part and parcel of urban heritage and thus provide a chance for a city to become a creative city as they are used for cultural tourism, thus creating employment for the people while preserving the history of the city. Additionally, a church can be given as an hypothetical example, it has spiritual values as a place of worship, it has historical values because of the events that have transpired there, or is simply because it is too old, or it has aesthetic value because it is beautiful and fine work of architecture, it has economic value as a piece of real estate, and it has political value or a symbolic representation.

Sharp et al. (2005) argue that heritage is widely acknowledged as a development catalyst. Therefore, adapting current spatial, economic, and cultural systems of cities to ease the integration with the new economy is important and leads to creativity of cities. Historical buildings display a history of the past, an identity for the present, and a story for the future to compare and appreciate (Wallace 2007). They tell stories of the people who lived or live in a place, type of technology they used, and how it has changed as well as the building materials used in building their buildings which have withstood the weather and time. Poulios (2010) argue that historical buildings are a storytelling tool which tell a history and the re-meaning of urban sites. This view is supported by Rojas-Sola (2018) who argues that building blocks of a city are not just simply buildings but a collective of memories and narratives of a city. The memories and narratives brought by historical buildings make them important structures in every city or town hence they are used for different uses in modern-day cities. The concept of cultural heritage preservation has undergone a major metamorphosis during the modern era (Markeviciene 2012). It is further argued that volume, typology, and spectrum of cultural heritage but first values and reasons for preservation of historic environments are subject to change (Markeviciene 2012).

Getty Conservation Institute (2002) suggests that cultural significance refers to the importance of a site. This importance attached to a site is determined by the aggregate of values ascribed to it. The values considered in the process should include those held by a range of experts, to include art historians, archaeologists, architects, and other values such as social and economic which are brought forth by other stakeholders. However, from the anthropological perspective, culture is often verified not only with human activities, but also with an environment as well. Culture is man-made environment, continuum of things and events in a cause-and-effect relationship (Markeviciene 2012).

Scholarship and Experiences in Urban Heritage and Value Extraction

While acknowledging built environmental heritages resources emphasized since the nineteenth century in developed countries such as France and England, it is recently when management of built environmental heritage resources have become a global movement (Mangara 2016). Despite the fact that heritage has become a movement, the notion of value extraction form urban heritage has not been much emphasized. This is why Getty Conservation Institute (2002) argues that little much about how practically heritage values can be assessed in different context of planning and decision-making. Additionally, assessments of the values attributed to heritage is a very important activity in a conservation effort, since value is widely understood to be critical in urban planning. Though it is argued that heritage management in urban areas has become a critical global issue as evidenced by global and international commitments on the management of natural heritage, but the notion of value extraction remains isolated (Janssen 2016).

Major world heritage preservation and conservation advisory bodies include those from International Council on Monuments and Sites (ICOMOS) and United Nations Educational, Scientific, and Cultural Organization (UNESCO) (UNESCO 2015). In 1976, UNESCO adopted the *Recommendation Concerning the Safeguarding and Contemporary Role of Historic Areas* as the roadmap in safeguarding existing natural heritage sites (ibid). Likewise, the ICOMOS later on in 1987 adopted the *Washington Charter for the Conservation of Historic Towns and Urban Areas* (*ibid.*). In 2011, ICOMOS and UNESCO adopted the Valletta Principles and the Historic Urban Landscapes (HUL) approach. The Valletta Principle emphasizes more on the safeguarding and management of historical cities, towns, and urban areas. The HUL approach is a heritage management tool that provide guidelines for urban development for all cities with heritage.

However, it is observed that these principles did not much emphasize the significance of value extraction from these resources. By simply managing and safeguarding resources means something of value can be extracted from the management of heritage.

In 2015, as the Millennium Development Goals reached their deadline, the UN proposed 17 new Sustainable Development Goals for 2030. Specifically, SDG 11.4 provides a basis for a New Urban Agenda that targets cultural heritage as part of the emerging urban challenges. In order to transcend the interdependence between sustainable development and cultural heritage conservation from an official global acknowledgment to the implementation of local sustainable practices, cities need to address the challenging task of setting clear sustainable priorities to redirect community-level actions (Budd et al. 2008; Chamberlain 2008; Swain 2018). However, this was supposed to be guided by the value of heritage and culture so that its sustainability will be much emphasized. Cities like Amsterdam, in the Netherlands, included the HUL approach in its World Heritage management (Veldpaus et al. 2013). Moreover, it is interesting to note that on yearly basis, the World Heritage Center and the advisory bodies such as ICCROM, ICOMOS, and IUCN give a report on conservation issues of selected natural and cultural heritage worldwide (Guzmán et al. 2018). However, this is mainly done on individual basis. To date, there are over 3000 reports documenting and detailing heritage at a global scale (Guzmán et al. 2018).

Despite major international and global commitments on heritage preservation and conservation, heritage in cities in most parts of the world is experiencing biggest threat from management deficiency and aggressive development (Pendlebury 2015). Therefore, this means that in terms of value extraction, little much can be extracted from these heritagical resources. Evidence from literature indicates that integration between tangible and intangible heritage management and sustainable urban development is far from being a common practice (Wang 2015; Vecco 2010; UNESCO 2009). Nonetheless, the general goal of UNESCO in natural heritage is to identify cultural heritage of outstanding universal value and ensure its protection, conservation, and presentation in the spirit of sustainable development and transmission to future generations (UNESCO WHC 2015). The sites considered to be of "Outstanding Universal Value" are those which meet at least one of the ten selection criteria of the World Heritage List, as well as conditions of authenticity and integrity and the requirement for the existence of adequate protection and management (UNESCO WHC 2015).

Value of World Heritage also contributes to and promotes the activities within them that can bring benefits both to the heritage and to the local community (UNESCO 2009). Vecco (2010) argues that from 1990s onwards, the discourse on heritage definition across Western European states has expanded progressively. There is a growing recognition that the historic environment is an integral part of our cities and landscapes, rather than a world set apart (Fairclough 2008). However, because of the dynamic nature of cultural landscape, preservation is no longer the sole objective. Instead, a more suitable approach in the form of "management of change" seems to provide a more suitable definition for current conservation activity (Fairclough and Rippon 2002). In this scenario, there is a growing demand to link preservation activities with spatial planning process. It is argued that the proactive role of spatial planning is best suited for combining the past with contemporary use to ensure the continued existence of heritage assets (Denhez 1997). In this case, spatial planning is now viewed as a catalyst which enables value extraction of these asserts, specifically heritage resources.

A cultural turn is happening to world's cities and towns especially in developed countries by refurbishing and rebranding cities as cultural havens in an attempt to revitalize their economies (Mercer 2006). Therefore, this is clearly evident heritage is of more value as they provide benefits and are useful to people and the environment. The UN Habitat State of the World's Cities (2004) notes that the growing trend of rebranding cities as cultural havens has proved to be a blessing for many urban officials and planners. For example, in Pakistan the presence of historical buildings has made the city attractive for local and international tourists (Ahmad and Sharif 2011). In Warsaw tourism has increased due to existence of historical buildings, thereby promoting trade in the city, social cohesion, and the regional economy (Petkovšek et al. 2021; Erokhin 2020).

Europe has long been endowed with a developed urban system; this developed urban system is a result of a layering process, whereby pre-existing structures were continuously transformed or reused (Porfyriou and Sepe 2017). Despite the fact that global policy concepts promoting the integration of planning and the historical built environment have been developing since the beginning of the twentieth century, a gap can be observed between these concepts and their adoption in planning frameworks and practice (Mangara 2016). Indeed, whereas the work of heritage professionals and planners is infused by current ideas on heritage, they at the same time have to conform to the policies and regulations that are in force (Kalman 2014).

During the 1980s and 1990s, however, major changes took place, which transformed thinking about the conservation of historic buildings and landscapes into a much more dynamic concept of heritage (Janssen et al. 2014). This development reflected international trends in heritage definition, discursively widening the scale, scope, and ambition of heritage conservation: from monumental objects (including townscapes) to a more holistic idea of heritage landscape, which also depicts immaterial aspects, and from expert-led authoritarian procedures toward more inclusive and participative community-led practices (Vecco 2010).

The African continent is one of the richest continents in the world with a plurality of cultural heritage components such as monuments and heritage sites. However, one common challenge in the African region is that most heritage is not documented and its value has not been extensively researched. Nonetheless, for the documented heritage, it is observed that more than 40% of the documented sites are in dangers of damage and extinction. The situation is most prevalent in conflict and post-conflict environments (UNESCO 2014). The region has legislations which guide development of heritage sites, as well as protect and preserve heritage in various ways (Un-Habitat 2008). However, scholars argue that the heritage legal frameworks in Africa are often inadequate and outdated, leaving loopholes for heritage destruction and damage (UN-Habitat 2008; Tanyanyiwa and Chikwanha 2011; Mangara 2016). More so, most African governments generally are grappling with implementation and enforcement of legislations (Mukwende et al. 2018).

Additionally, Africa is experiencing significant economic growth and associated demographic changes, including rising urbanization without the requisite infrastructure, as well as spatial and settlement planning. The proportion of urban residents living in informal settlements is higher in Africa, as a region, than any other part of the world (UNDP 2012). It is also interesting to note that sub-Saharan Africa still have an important form of tradition of urban settlements, this form of settlement is dated back as early as the eighth century, which is gradually being rediscovered (Chirikure et al. 2010). The uniqueness of African cities of having a traditional background which relates to their precolonial, colonial, and postcolonial urban heritage should be acknowledged, emphasizing the significance of value extraction.

It is argued that vagaries of urbanization and unregulated land use patterns threaten urban heritage (Chirikure et al. 2010). This resulted in loss of traditional community values and practices. Additionally, the present development challenges, such as access to education, basic urban services, and infrastructure amenities, calls for inclusive and meaningful urban conservation (UN-Habitat 2008). Moreover, African cities are experiencing urbanized populations and widening inequalities which are deeply rooted in the colonial era, patterns of segregation which were experienced during the colonial era (UN-Habitat 2008; Muchemwa 2010). This therefore means that culture-based approaches are particularly relevant in fostering a sense of belonging. It is also interesting to note that as a new generation of cultural entrepreneurs is emerging, this therefore means that culture can increasingly gain relevance and importance as it provides a stage for community participation and renewed links among stakeholders such as the government. However, it is sad to realize that some urban development strategies have often disregarded the social and cultural realities of African cities of which these social and cultural realities provides some different forms of values if scholars and researchers believes that value can be extracted from heritage objects. Culture can now become a strategic tool to regenerate marginalized and informal areas (UNESCO 2016).

It is suggested that culture has taken on a growing role in urban regeneration strategies (Hove and Muchemwa 2013). Therefore, strategies such as revitalization of public spaces and the rehabilitation of declining industrial areas can make use of culture as a tool for revitalizing dilapidated areas. However, the need and practice of urban conservation has unlocked new potential approaches and instruments for achieving different forms of urban sustainability which emphasizes local knowledge, creativity, and well-being of societies (UN-Habitat 2008).

Zimbabwe has got many historical sites with World Heritage status which are guided by the National Museums and Monuments Act. Historic parts of Zimbabwe's cities and towns form an important part of the national collective memory (Jackson 2004). Town planning in Zimbabwe is guided by the Regional, Town, and Country Planning Act, Chapter 29:12 (1996). The Act provides for the planning of all areas but there is no specific mention of the planning of heritage sites. The Act does provide for the preservation of historic buildings in Section 30. Spatial planning in heritage sites follows the normal procedure as prescribed by the Act without any special consideration being taken into preserving or inculcating the unique importance inherent in heritage sites.

The major goal of urban planning in heritage sites areas should be to preserve and build upon the unique community character through the preservation of existing heritage structures and an insistence on design quality, sensitivity, and compatibility for new development (Roders and Oers 2011). This means that the value of urban heritage in urban areas can be recognized and given its due recognition. All aspects of design should reflect the community's special significance and international profile. Spatial patterns and architecture have evolved over time and these have important implications on the image of the city and its identity as well as urban regeneration and local economic development (Janssen et al. 2014). Katrinka (2009) observes the imperative nature of the planning of settlements that have heritage value be responsive to this need by developing successful strategies for managing both the challenges and opportunities posed by heritage on city planning. There must be consideration of historical linkages when planning for conservation of historical settlements. This is in line with the Sustainable Development Goal (SDG) Number 11 on sustainable cities and communities which call for efforts of culture preservation and conservation. Spatial planning is an important tool for the management, preservation, and enhancement of heritage sites (Katrinka 2009).

Some towns in Zimbabwe such as Masvingo are rich in cultural and natural heritage as it houses the Great Zimbabwe Ruins which has world heritage status (Ndoro 1994). The materials, colors, scale, and building styles found in the regions provide the basis for a cohesive set of design guidelines

(Tanyanyiwa and Chikwanha 2011). The indigenous architecture found in this city perfectly reflects the local natural environment, building technologies, materials, and social values important to the community (Chirikure et al. 2016). Therefore, this means that the notion of value has been recognized but not extensively emphasized.

Given that this settlement relies on its heritage values to attract tourism, it seems appropriate to draw upon the heritage wealth of the area to promote its future. The Great Zimbabwe Ruins are the flagship of the Zimbabwe tourist industry and therefore have a national importance (ibid). This settlement developed on a natural landscape. This makes it a unique cityscape based on its heritage status and hence its planning must be based on heritage and environmental enhancement than common planning and design practice (Macheka 2016). It provides a framework for thinking about and understanding the concept of heritage and environmental conservation in spatial planning. While quality of development is important in all resort communities, its location in a World Heritage Site makes it absolutely essential that the quality of the built environment strives to match the quality of its spectacular heritage and natural environment (Macheka 2016).

Emerging Issues and Synthesis

Urban heritage conservation has emerged as a fundamental attribute of the Sustainable Development Agenda (Guzmán et al. 2018). Comprehending how places alter knowing the importance of their history paves the way for successful and sustainable urban regeneration. Nowadays, it is widely acknowledged by the community that conservation of heritage buildings plays a critical role for obtaining significant economic, cultural, and social benefits. Promoting sustainability in the built environment is a part of a wider revitalization strategy. The consequences of heritage-led urban regeneration can be generally regarded as economic, social, cultural, and environmental. These consequences can be either positive or negative and may be both quantitative and qualitative.

There is need to consider historical linkages when planning for conservation of historical settlements. This is also important in fulfilling the demands of SDGs with much respect to goal number 11. Within the new Sustainable Development Goals and the New Urban Agenda, the achievement of sustainable urban development (SUD) calls for different forms in which cities preserve and conserve urban heritage. This not only demands the integration of cultural heritage conservation in wider urban policies and planning, but also that fundamental principles of sustainability must be also applied when managing urban heritage. A culturally vibrant city is likely to improve its economic health, this is because businesses are mainly attracted to those locations with strong cultural amenities (Vecco 2010). During the period of capitalism, a distinction was made between cities of industry and commerce and those cities of art and culture.

Urban heritage endowments are increasingly becoming recognized as invaluable economic resources both in developing and developed countries. Urban areas have significant concentrations of heritage sites, private sector activity, infrastructure services, and human resources (Talakudar et al. 2015). As such, urban areas are the most focal points for development that is based on these natural heritage resources (Janssen et al. 2014). Bearing that in mind, improvements on management and conservation of urban heritage have to shift from mere preservation of heritage but to utilize the potential of urban heritage to increase urban livability, urban competitiveness, and increased income-generating opportunities. It is noted that by preserving city heritage, cities can create a unique sense of place and singular urban landscapes, developing strong branding and conditions which attract investors (Katrinka 2009). This is mostly true for investments in urban tourism which is emerging as one of the industries in the world.

It is acknowledged that success urban heritage preservation and conservation is hinged on their relationship with spatial planning. Guided by the "Belvedere Memorandum," Amsterdam has a longstanding tradition of monument protection and urban planning, both anchored in local and national policy and both now forming key parts of the management plan for the canal district as WH site (Veldpaus et al. 2013, p. 7). The existence and strictness of urban regulatory frameworks in Amsterdam creates a specific and controlled process through which developments have to be approved. The regulatory frameworks in Amsterdam integrates historic heritage conservations into the city's urban planning regulations and building codes. These historical buildings thus are a physical record of what our country used to be and how it has changed through new buildings which are compared to the old existing historical buildings. Historical buildings also tell a story of how urban planning has evolved by comparing new designs and planning systems to the old ways of planning.

Conclusion and Future Direction

At the future of heritage management lies in all-inclusive approaches that embrace change and transition of urban local contexts (UN-Habitat 2016). The human factor remains a key consideration in the management of urban heritage (Chirikure et al. 2016). This is so because heritage encompasses objects and processes that are used and/or valued by people. Management of such objects and processes implies that humans have to be at the center stage of such initiatives. Historic urban environments sustain social structures of human societies and ensure their continuity to some extent. Every culture ensures its continuity through a mechanism of tradition. Historic towns are viewed as the monuments of the past. From this perspective, urban heritage exposes itself mainly as a collection of monuments and landmarks, and honorable relicts of the past, symbolizing our history, giving meaning to our cultural identity, and embellishing our lives (Markeviciene 2012).

Cross-References

► An Overview of the Relationship of the Sustainable Development Goals and Urban and Regional Development
► Building Community Resilience

References

Ahmad, I., & Sharif, B. (2011). *Role of government agencies in the preservation of cultural heritage*. 47th ISOCARP Congress.

Budd, W., Lovrich Jr, N., Pierce, J. C., & Chamberlain, B. (2008). Cultural sources of variations in US urban sustainability attributes. *Cities, 25*(5), 257–267.

Chamberlain, K. (2008). The recognition and enforcement of foreign cultural heritage laws: Iran v. Barakat. *Art Antiquity & L., 13*, 161.

Chirikure, S., Manyanga, M., Ndoro, W., & Pwiti, G. (2010). Unfulfilled promises? Heritage management and community participation at some of Africa's cultural heritage sites. *International Journal of Heritage Studies, 16*(1–2), 30–44.

Chirikure, S., Mukwende, T., & Taruvinga, P. (2016). Postcolonial heritage conservation in Africa: Perspectives from drystone wall restorations at Khami World Heritage site, Zimbabwe. *International Journal of Heritage Studies, 22*(2), 165–178. https://doi.org/10.1080/13527258.2015.1103300.

Denhez, M. C. (1997). *The heritage strategy planning handbook*. Oxford: Dundrum Press.

Erokhin, V. (2020). Produce internationally, consume locally: Changing paradigm of China's food security policy. In *Handbook of research on agricultural policy, rural development, and entrepreneurship in contemporary economies* (pp. 273–295). IGI Global.

Fairclough, G. (2008). A new landscape for cultural heritage management: Characterisation as a management tool. In *Landscapes under pressure* (pp. 55–74). Boston, MA: Springer.

Fairclough, G., & Rippon, S. (2002). *Europe's landscape: Archeologists and the management of change (EAC, Occasional Paper 2)*. Brussels: EAC.

Foster, G. (2020). Circular economy strategies for adaptive reuse of cultural heritage buildings to reduce environmental impacts. *Resources, Conservation and Recycling, 152*, 104507.

Foster, S. M., & Jones, S. (2019). The untold heritage value and significance of replicas. *Conservation and Management of Archaeological Sites, 21*(1), 1–24.

García-Hernández, M., & de la Calle-Vaquero, M. (2019). Urban heritage. https://www.oxfordbibliographies.com/view/document/obo-9780199874002/obo-9780199874002-0208.xml

Getty Conservation Institute. (2002). *Management planning for archaelogical sites. An international workshop*. Corinth: Loyola Marymount University.

Guzmán, P., Pereira Roders, A., & Colenbrander, B. J. F. (2018). Measuring links between cultural heritage management and sustainable urban development: An overview of global monitoring tools. *Cities, 60*, 192–201. https://doi.org/10.1016/j.cities.2016.09.005.

Habitat III Conference. (2016). *Quito declaration on sustainable cities and human settlements for all*. New York. http://www.eukn.eu/news/detail/agreedfinal-draft-of-the-new-urban-agenda-is-now-available/

Hove, N., & Muchemwa. (2013). The urban crisis in Sub Saharan Africa: A threat to human security and sustainable development. *Stability, 2*(1), *7*, 1–14. https://doi.org/10.5334/sta.op.

Jackson. (2004). *Urban sprawl and public health: Designing, planning and building healthy communities*. Washington, DC: Island Press. Corelo. London.

Janssen, J. (2016). Religiously inspired urbanism: Catholicism and the planning of the southern Dutch provincial cities Eindhoven and Roermond, c. 1900 to 1960. *Urban History, 43*(1), 135–157.

Janssen, J., Luiten, E., Renes, H., & Rouwendal, J. (2014). Heritage planning and spatial development in the Netherlands: Changing policies and perspectives. *International Journal of Heritage Studies, 20*(1), 1–21. https://doi.org/10.1080/13527258.2012.710852.

Kalman, H. (2014). *Heritage planning: Principles and process*. London: Routledge.

Katrinka, E. (2009). *Infrastructure and heritage conservation: Opportunities for urban revitilization and economic development. directions in urban development*. Washington, DC: World Bank. © World Bank. https://openknowledge.worldbank.org/handle/10986/10260 License: CC BY 3.0 IGO.

Lynne, A., & Yung, Y. (2006). *Heritage protection in the built environment in Hong Kong and Queensland: Across cultural comparison*. Hong Kong: 12th annual conference of the Pacific Real Estate.

Macheka, M. (2016). Great Zimbabwe heritage site and sustainable development. *Journal of Cultural Heritage Management and Sustainable Development, 6*(3), 226.

Mangara, F. (2016). *Towards integrating urban conservation and urban: Klerksdorp as a case study*. North West University.

Markeviciene, J. (2012). The spirit of the place-the problem of (re)creating. *Journal of Architecture and Urbanism, 36*(1), 73–81.

Mercer, C. (2006). Cultural planning for urban development and creative cities. http://www.kulturplanoresund.dk/pdf/Shanghai_cultural_planning_paper.pdf.

Muchemwa, K. Z. (2010). Galas, biras, state funerals and the necropolitan imagination in reGconstructions of the Zimbabwean nation, 1980–2008. *Social Dynamics: A Journal of African Studies, 36*(3), 504–514.

Mukwende, T., Bandama, F., Chirikure, S., & Nyamushosho, R. T. (2018). The chronology, craft production and economy of the Butua capital of Khami, southwestern Zimbabwe. *Azania: Archaeological Research in Africa, 53*(4), 477–506.

Ndoro, W. (1994). *The preservation and presentation of Great Zimbabwe*. Cambridge University Press.

Pendlebury, J. (2015). Heritage and policy. In *The Palgrave handbook of contemporary heritage research* (pp. 426–441). London: Palgrave Macmillan.

Petkovšek, V., Hrovatin, N., & Pevcin, P. (2021). Local public services delivery mechanisms: A literature review. *Lex Localis, 19*(1), 39–64.

Poulios, I. (2010). Moving beyond a values-based approach to heritage conservation. *Conservation and Management of Archaeological Sites, 12*(2), 170–185.

Regional, Town and Country Planning Act, Chapter 29:12. (1996). *Government printers*. Zimbabwe.

Rojas-Sola, J. I. (2018). Digital 3D reconstruction of Betancourt's historical heritage: The dredging machine in the Port of Kronstadt. *Virtual Archaeology Review, 9* (18), 44–56.

Roders, A., & Oers, R. (2011). Bridging cultural heritage and sustainable development. *Journal of Cultural Heritage Management and Sustainable Development, 1*(1), 5.

Sharp, J., Pollock, V., & Paddison, R. (2005). Just art for a just city: Public art and social inclusion in urban regeneration. *Urban Studies, 42*(5–6), 1001–1023.

Swain, S. K. (2018). *Heritage and media in India: Built haritage of Delhi and the role of print media*. India: Doctoral dissertation, Maharaja Sayajirao University of Baroda.

Talakudar, et al. (2015). Development of decadal (1985–1995–20050 Land use and Land cover database for India.

Tanyanyiwa, V., & Chikwanha, M. (2011). The role of indigenous knowledge systems in the management of forest resources in Mugabe Area, Masvingo.

UN Habitat. (2004). *The state of the world's cities 2004/ 2005 globalization and urban culture*. London. www.unhabitat.org

UNDP. (2012). *St, Lucia – Sustainable development national synthesis report*. Government of St: Lucia.

UNESCO. (2009). *World heritage and buffer zones, World heritage papers 25*. UNESCO.

UNESCO. (2014). *Rice terraces of the Philippine cordilleras. Electronic document*. http://whc.unesco.org/en/list/722. Accessed August 23.

UNESCO. (2015). *Operational guidelines for the implementation of the world heritage convention*. Paris: UNESCO.

UNESCO. (2016). *Interactive map*. http://whc.unesco.org/en/interactive-map

UNESCO World Heritage Centre. (2015). Operational guidelines for the implementation of the World Heritage Convention, July, p. 167.

Vecco, M. (2010). A definition of cultural heritage: From the tangible to the intangible. *Journal of Cultural Heritage, 11*(3), 321–324. https://doi.org/10.1016/j.culher.2010.01.006.

Veldpaus, L., Roders, A. R., & Colenbrander, B. F. (2013). *Urban heritage: Putting the past into the future*. Maney.

Wallace, T. (2007). Went the day well: Scripts, glamour and performance in war-weekends. *International Journal of Heritage Studies, 13*(3), 200–223.

Wang, M. (2015). *Historical layering and historic preservation in relation to urban planning and protecting local identity: City study of Nanjing*. Philadelphia: University of Pennsylvania.

World Commission on Environment and Development. (1987). *Our common future*. Oxford: Oxford University Press.

Future

► Financing: Fiscal Tools to Enhance Regional Sustainable Development

Future – Prospects

► Urbanization, Planning Law, and the Future of Developing World Cities

Future – The Prospective

► Spatial Justice and the Design of Future Cities in the Developing World

Future Cities

► Urban Futures: Pathways to Tomorrow

Future Foods for Urban Food Production

Asaf Tzachor[1,2] and Catherine E. Richards[1,3]
[1]Center for the Study of Existential Risk (CSER), University of Cambridge, Cambridge, UK
[2]School of Sustainability, Interdisciplinary Center (IDC) Herzliya, Herzliya, Israel
[3]Department of Engineering, University of Cambridge, Cambridge, UK

Synonyms

Alternative protein; Plant-based protein; Cellular agriculture; Vertical farming; Urban agriculture; Building-integrated agriculture; Organic engineering systems

Definition

Future foods refer to a list of edible items that are considered non-customary and nutritious, including species of micro-algae, macro-algae, bivalve mollusks, insects, and in-vitro meat. Future food systems, or future foods farming systems, encompass the technological configurations that are required to produce these edible items at a large scale in controlled-environment agriculture conditions, such as advanced photo-bioreactors for microalgae or breeding systems for insect larvae.

In the context of global environmental change, and biotic and abiotic risks that undermine conventional agriculture, future food systems ensure a risk-resilient supply of safe and nutritious foods, through a modular design of discrete, stand-alone food production units that are deployed in decentralized, distributed food networks.

Furthermore, future food systems may mitigate climate change and environmental degradation. For example, by using municipal organic waste as feed stock for insects, or by utilizing carbon dioxide for the cultivation of microalgae.

Urban production of future foods may be enhanced by sensor and data technologies (e.g., internet of things) and automation technologies (e.g., artificial intelligence), if these are integrated into enclosed production units, thereby improving resources efficiency, environmental sustainability, and nutritional quality of urban food systems.

Evolution of Food Production, Food Provision, and Urbanization

The contemporary city, as a specific spatial, economically productive, prosperous, and often-praised form of social organization (Glaeser 2008; Glaeser et al. 1992), is a consequence of agricultural revolutions and advancements in food security.

From small-scale trial cultivation of "proto-weeds" in pre-Neolithic sedentary settlements some 23,000 years ago (Snir et al. 2015) to widespread social transition from hunting and foraging for wild foods to farming domesticated plant and animal species during the Neolithic Revolution some 12,000 years ago (Weisdorf 2005), improvements in agronomic, horticulture, and husbandry practices enabled calorie-rich grain surpluses, stockpiling and trade, specialization in non-subsistence activities, economic diversification (Childe 1950; Simmons 2011), and permanent human settlements. Subsequently, the increase of population in early urban centers of the Fertile Crescent, of the Indus Valley, of Northern China, of Mesoamerica, and of Great Zimbabwe, instigated human expansions and migration afield (Bocquet-Appel 2011; Clark 2013).

The second agricultural revolution, approximately between the seventeenth and nineteenth centuries AD – its exact period is contested (Jones 2016) – exploited fertilizers, fodder crops, agrarian reforms, and foreign trade to see a growth in land and labor productivity, crop yields, and production of meat, milk, and eggs from livestock (Overton 1996; Thompson 1968).

Concomitantly, improvements in food processing, including tinning and canning to preserve foods, and pasteurization to reduce microbial activity, gave rise to global supply chains.

With agricultural mechanization, productivity gains in food production, and new possibilities to process, preserve, and provide foodstuffs, cities grew by natural population increase as well as rural to urban migration (Keyfitz 1980). Over the twentieth century, urbanization accelerated. In 1950, approximately 30% of the world's population, some 751 million individuals, lived in urban areas. By 2018, this figure reached 55%, corresponding to some 4.2 billion people. By 2030 it is expected to reach 60% (UN DESA 2018).

Continued innovations, for instance in telecommunications and computers, realized further benefits of supply chains, such as efficient just-in-time provisioning of food.

Culturally, the emergence of agri-food supply chains induced a process of "distancing." That is, in becoming city dwellers, humans have removed themselves from the farming process of the plant-source foods (PSF), which include staple crops of wheat, maize, rice, and soybean, and animal-source foods (ASF), which include beef cattle, dairy cattle, pigs, and poultry, on which they rely for sustenance (Parodi et al. 2018).

With these trends, *global food security* – defined by the United Nations as a state in which all people, at all times, have physical, social, and economic access to sufficient, safe, and nutritious food that meets their food preferences and dietary needs for an active and healthy life (IFPRI 2021) – is increasingly becoming *urban food security*.

Furthermore, *urban food security* relies not only on available, accessible, and nutritious food but also on efficient, sustainable, and risk-resilient agri-food supply chains (Garnett 2014; Grote 2014; Krejci and Beamon 2010). Yet, global environmental change poses manifold risks to the interconnected food system (Richards et al. 2021; Beard et al. 2021; Rosegrant and Cline 2003), supply chains, and consequently to safe and prosperous urban futures.

Risks to Food Supply Chains and Urban Food Security

Supply chains are generally conceptualized in terms of distinct units of analysis (Peck 2006). Food, and feed-stock, supply chains typically include four phases, described using various terms (Bourlakis and Weightman 2004; Jaffee et al. 2010; Manzini and Accorsi 2013). Herein, these phases are defined as: primary agricultural production of plant and animal products; processing of primary products into consumable goods; distribution of foodstuffs, including temporary storage and transportation in land and maritime vessels; and retailing and provision of food for consumption.

While retailing and provision of food occurs increasingly in the built environment of urban regions, where a growing number of humans reside, the former phases mainly occur in rural regions, or in transit to urban centers, where they are exposed to risks of three types: biotic, abiotic, and institutional (Tzachor et al. 2021).

As it stands, the relatively open environment of conventional farming, such as traditional open-field, soil-based, rain-fed cultivation of PSF, and concentrated animal feeding operations (CAFO) of ASF, renders urban food security particularly vulnerable.

Biotic risks to primary production of PSF and ASF include pests, parasites, and pathogens, such as viruses and bacteria (Wrigley et al. 2015). For example, an outbreak of the fungi rice blast (*Magnaporthe oryzae*) in 2013 reduced global yield by 30%, which is equivalent to food supply for 60 million people (Nalley et al. 2016).

Abiotic risks, including deviations and alternations in physical conditions such as sunlight, precipitation, and soil nutrient content, also threaten primary production (Minhas et al. 2017). For instance, rice yields have declined in relation to global increase in temperatures (Peng et al. 2004). Additionally, direct and indirect effects of global warming are expected to exacerbate biotic risks (Harvell et al. 2002).

Furthermore, biotic and abiotic factors may spoil produce during processing and handling. For example, unregulated wheat cargoes are susceptible to contamination of grains with mycotoxins (Goswami and Kistler 2004). For ASF processing, microbial contamination of animal carcasses and meat cuts is a risk in slaughterhouse environments (Koutsoumanis and Sofos 2004).

Moreover, abiotic risk factors, such as abnormal or extreme weather events, may disrupt storage and transportation operations, as was the case in USA flour supply chain disruptions by flooding of the Mississippi river (Bailey and Wellesley 2017).

In response to disturbances in production, processing, and distribution phases, institutions may interfere unilaterally to halt food trade with implications for food provision and retailing. Such scenarios were observed during the 2007–2008 global food crises and recent coronavirus pandemic, when an estimated 44 million people (Headey and Fan 2010) and 265 million people (FSIN 2020), respectively, faced food insecurity related to export restrictions and trade sanctions.

Considering the increasingly chronic and acute nature of these risks, recent articles (Parodi et al. 2018; Tzachor et al. 2021) proposed future foods and their unconventional production systems as a means to achieve provision of nutritious and sustainable food through risk-resilient supply chains.

Future Foods and Their Production Systems

"Future foods" herein refers to foodstuffs that are appropriate alternatives to conventional PSF and ASF, as the bulk of the human diet, made possible through technological development. Currently, nine future foods – four aquatic and five terrestrial – have been identified as most promising for at-scale production. These include: mussels (*Mytilus* spp.), a bivalve mollusk; sugar kelp (*Saccharina latissima*), a macroalgae; chlorella (*Chlorella vulgaris*) and spirulina (*Arthrospira platensis*), both microalgae; larvae of the black soldier fly (*Hermetia illucens*), housefly (*Musca domestica*), and mealworm beetle (*Tenebrio molitor*), all insects; mycoprotein from the fungi *Fusarium venenatum*; and cultured meat.

In terms of nutritional value, these future foods have similar or higher content of essential micro- and macronutrients compared to PSF and ASF. For example, mealworm larvae have similar protein content to chicken, sugar kelp has higher calcium content than milk, and spirulina is rich in vitamin B12 lacking in PSF and ASF. The novel production processes of future foods also make them more sustainable alternatives. Future foods require significantly less land compared to ASF, and similar or less land compared to PSF, for equivalent production of essential nutrients. While they presently show similar GHG intensities to ASF and higher GHG intensities to PSF, futures foods' GHG emissions are dissimilarly restricted to process energy input currently dependent on fossil fuels, which can be proportionally reduced by transitions to renewable energy (Parodi et al. 2018).

Recent advancements in biotechnology indicate that the cultivation of several future foods at scale, in controlled-environment agriculture (CEA) conditions, is now a viable option.

CEA designs may reduce exposure to open-environment abiotic and biotic hazards. Modular architecture of CEA configurations achieves operational redundancy, including containment of biochemical contaminations, flexible response to failures, and ease of offline maintenance, as well as ease of production scaling in response to oscillating demand. The risk-resilient nature of these production systems, reviewed in turn below, provides for more consistent and efficient supply of future foods relative to PSF and ASF (Tzachor et al. 2021), and supports sustainable urban futures.

Macroalgae and Mussels in Integrated Multi-Trophic Aquaculture

Advances in aquaculture have seen development of industrial-scale farms in coastal and offshore locations, selected for optimal environmental conditions, outside native growth zones, to cultivate bivalve mollusks such as mussels and macroalgae such as sugar kelp.

A typical monotrophic production system consists of three stages and employs automated equipment. In the first stage, fertile organisms are collected, the reproductive process is induced, and produced seedlings are grown in onshore nursery tanks under controlled conditions on substrate strings. Once of appropriate size for outplanting, the seed-strings are spooled around carrier rope for deployment. In the second stage, the seed-spools are transferred from the nursery to the open-water farm, where they are grown to maturity. The farm consists of moored submerged longlines, from which seed spools are suspended in either a vertical or parallel configuration. In the third stage, the mollusks or macroalgae are harvested and processed onshore.

Further progress in the form of integrated multi-trophic aquaculture systems (IMTA), such as coupled fish-mussel-sugar kelp production units, offers economies of scale and environmental benefits, by combination of these monotrophic cultivation processes (Buck et al. 2018).

Microalgae in Photobioreactors

As unicellular organisms, microalgae, such as chlorella and spirulina, are considered promising, not only as food but also as feed, fuel, and fertilizer, due to their fast growth rates. Recent developments have seen microalgae, previously cultivated in open ponds, produced in photobioreactors (PBRs), including tubular, vertical column and flat-plate PBRs (Tredici 2004). The microalgae culture is grown in a liquid medium, irradiated by light-emitting diodes (LED) to achieve high photosynthetic photon flux at the wavelengths of photosynthetic interest, improving the efficiency of photosynthesis (Tzachor 2019). Mechanical or air lift pump systems provide circulation, enabling homogenization of the culture, and aeration targeted at maximizing gas-liquid mass transfer efficiency (Khan et al. 2018).

Mycoprotein in Fermenters

Mycoprotein, a single-cell protein derived from filamentous fungi, has been produced at scale for commercially available alternative meat products for several decades. Fungal biomass is cultivated in continuous-flow aerobic bioreactors. Substrate growth media is inoculated with fungal spores and fermentation proceeds under temperature- and pH-controlled conditions with oxygen and nutrients provided to support growth. The biomass is heat-treated to reduce RNA content and centrifuged to obtain a concentrated paste, while the liquid stream is extracted for use in flavoring. The mycoprotein is mixed with binding agent and flavoring, and undergoes heating, cooling, and freezing processes to obtain a meat-like structure (Hashempour-Baltork et al. 2020).

Insect Larvae in Breeding Systems

Larvae breeding systems consist of processes contained in stackable multicompartment units assisted by automated equipment that leverage the high reproduction yields and natural growth stage migratory habits of insects. The cultivation processes are similar but nuanced for different species, such as black soldier fly, house fly, and mealworm beetle. A colony of adult insects are maintained and mated in a breeding compartment. Eggs produced by the colony are grown to larvae form on organic substrate in a nursery compartment. Once of appropriate size, most larvae are harvested for final processing, while a small percentage remain in the unit to replenish the adult colony. The larvae remaining continue to grow to adult form in a pupal compartment. The larvae harvested for processing undergo screening, heating, cooling, pressing, drying, or centrifuging treatments to kill the larvae, remove potential contaminants, and manipulate physical characteristics. To form final products, larvae may be freeze-dried and packed whole, pulverized into a powder or flour, or further processed with other ingredients to form food items (Makkar et al. 2014).

Cultured Meat in Bioreactors

The production process for cultured meat, also referred to as lab-grown meat, is a form of cellular agriculture involving in vitro cell cultures and tissue engineering to grow skeletal muscle mimicking conventional meat. Typically, starter cells are first obtained. These are either primary cells derived from an animal biopsy, typically embryonic stem cells, adult stem cells, or skeletal muscle myoblasts, or secondary cell lines, including induced pluripotent stem cells, derived from master primary cells in the lab. The cells are combined with growth medium in a bioreactor where they proliferate and differentiate under controlled conditions. To obtain three-dimensional tissue, a scaffold is added to structure and order the cultured cells as skeletal muscle fibers. Finally, standard food processing technology is used to shape the tissue into final products (Post et al. 2020).

Future Foods Production Systems in Urban Landscapes

From the perspective of urban futures, the opportunity to decentralize food production and localize food supply chains, while leveraging smart technology, is a salient benefit of the compact, discrete, and risk-resilient design of future foods production systems.

By nature, future food production systems may be incorporated into both new and retrofit developments, with ease of connection to the urban water and electricity grid. Urban design consolidating the full "farm-to-table" supply chain of future foods into mixed-use developments, as depicted in Fig. 1, would facilitate the rise of "compact cities" (Bibri et al. 2020; Vorontsova et al. 2016).

These arrangements could contribute to an urban circular economy by generating co-benefits.

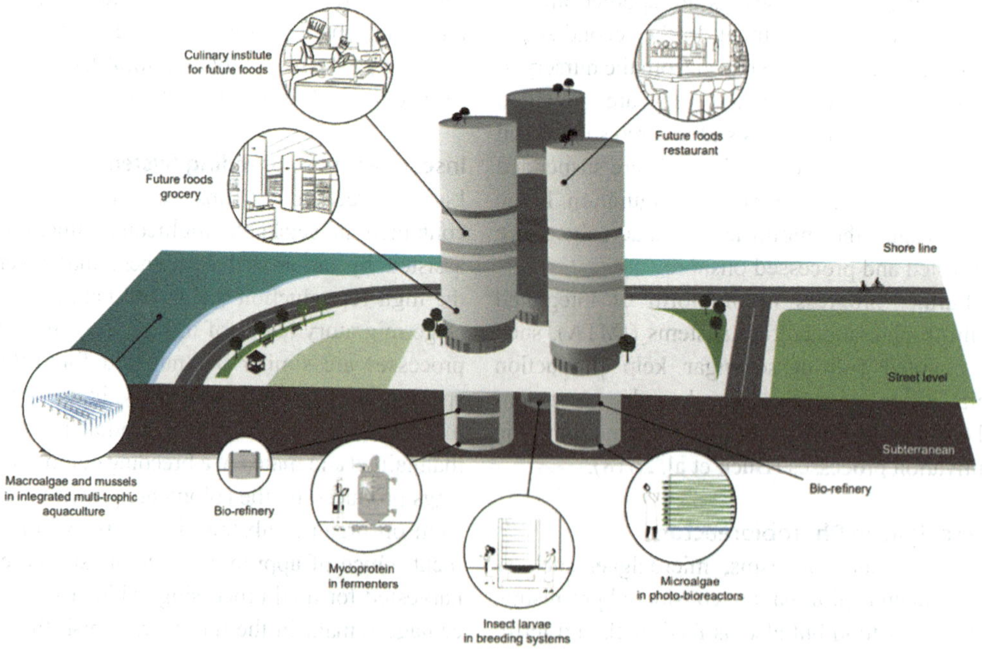

Future Foods for Urban Food Production, Fig. 1 Urban future foods mixed-uses complex

For instance, future foods production units coupled with biorefineries and heat pumps could transform waste biomass and residual heat into building energy (Cherubini 2010).

Additionally, some processes, such as in IMTA and insect larvae cultivation, can recycle organic matter, including agricultural by-products and human food scraps, as feedstock, thereby reducing waste.

Shorter supply chains may also prompt democratization of the food system, with culinary artists, culinary institutes, and consumers (see Fig. 1) becoming prosumers, thereby enabling innovative contributions the evolution of future foods cultivation practices and various end use products (Nature Food 2020).

Furthermore, enclosed and modular future foods production systems are fundamentally CEA facilities, making them well suited for applications leveraging the computer technologies of Agriculture 4.0 and smart cities (Batty et al. 2012). Sensor and automation technology, such as that used for controlled-environment greenhouses of conventional crops, can enhance control of internal physical, chemical, and biological parameters for optimized cultivation (Jiménez et al. 2012). Wireless sensor networks (WSN), for instance, may be used to regulate microclimate in PBRs, insect farms, and mycoprotein fermenters (e.g., temperature and humidity), to monitor culture development and maintain safety standards, to adjust cultivation conditions (e.g., nutrient support), as well as to automize farming with robotic assistance (Abdullah et al. 2016).

Real-time dynamic data infrastructure, such as Internet of Things (IoT) and cellular networks, participatory sensing, cloud computing, and data mining and analysis, could facilitate municipal oversight and coordination of urban food production, and connection of production systems with local processing plants, urban retailing, and consumers. Greater connectivity should reduce the response time to fluctuations in demand and mitigate market volatility (De Clercq et al. 2018). The engineering of future foods production systems, and of advanced computer technologies, enables efficient, resilient, and participatory urban landscapes.

Realizing the Potential of Future Foods for Urban Food Production

Currently, several future foods may be deployed for at-scale decentralized production, but are not yet prevalent, in large metropolitan areas. Similar potential is demonstrated in other frontiers, such as cultured meat, with successful development of prototypes and pilot plants under construction. Prospective advancements in food technologies and urban development, in parallel, promise to yield synergies in addressing technical and institutional viability of urban food production that is more sustainable and resilient.

In a few urban agglomerations, shifts in consumer preferences toward healthy and sustainable lifestyles may drive efforts on regulatory approvals and normalize adoption of future foods as alternatives in human diets (Nyberg et al. 2020). Additionally, positive trends in smart city technologies strengthen the sustainability premise of future foods, such as where smart grids facilitate greater penetration of variable renewable energy sources (Hossain et al. 2016). Increased focus on building urban resilience, and awareness of "black sky hazards" (Schnurr 2018), may also propel deployment of future foods systems given their propensity to mitigate abiotic, biotic, and institutions risks and build polycentric food supply networks (Ribeiro and Gonçalves 2019; Tzachor et al. 2021).

Developing and deploying closed-environment, modular, and distributed future foods production systems in cities is expected to be increasingly important in achieving a sustainable and risk-resilient alternative to conventional globalized food supply chains, necessary to ensure food security within the context of a growing human population and expanding urbanization.

Cross-References

► An Overview of the Relationship of the Sustainable Development Goals and Urban and Regional Development
► Big Data for Smart Cities and Inclusive Growth
► Circular Cities

- ► Circular Economy Cities
- ► Climate-Resilient Technologies and Innovations for Sustainable Agriculture, Improved Landscape, and Food Security
- ► Future of Urban Governance and Citizen Participation
- ► How Cities can be Resilient

References

Abdullah, A., Al Enazi, S., & Damaj, I. (2016, March). AgriSys: A smart and ubiquitous controlled-environment agriculture system. In *2016 3rd MEC International Conference on Big Data and Smart City (ICBDSC)* (pp. 1–6). IEEE.

Bailey, R., & Wellesley, L. (2017). *Chokepoints and vulnerabilities in global food trade* (p. 111). London: Chatham House.

Batty, M., Axhausen, K. W., Giannotti, F., Pozdnoukhov, A., Bazzani, A., Wachowicz, M., … & Portugali, Y. (2012). Smart cities of the future. *The European Physical Journal Special Topics, 214*(1), 481–518.

Beard, S., Holt, L., Tzachor, A., Kemp, L., Avin, S., Torres, P., & Belfield, H. (2021). Assessing climate Change's contribution to global catastrophic risk. *Futures, 127*, 102673.

Bibri, S. E., Krogstie, J., & Kärrholm, M. (2020). Compact city planning and development: Emerging practices and strategies for achieving the goals of sustainability. *Developments in the Built Environment, 4*, 100021.

Bocquet-Appel, J. P. (2011). The agricultural demographic transition during and after the agriculture inventions. *Current Anthropology, 52*(S4), S497–S510.

Bourlakis, M. A., & Weightman, P. W. (Eds.). (2004). *Food supply chain management*. Blackwell.

Buck, B. H., Troell, M. F., Krause, G., Angel, D. L., Grote, B., & Chopin, T. (2018). State of the art and challenges for offshore integrated multi-trophic aquaculture (IMTA). *Frontiers in Marine Science, 5*, 165.

Cherubini, F. (2010). The biorefinery concept: Using biomass instead of oil for producing energy and chemicals. *Energy Conversion and Management, 51*(7), 1412–1421.

Childe, V. G. (1950). The urban revolution. *The Town Planning Review, 21*(1), 3–17.

Clark, P. (Ed.). (2013). *The Oxford handbook of cities in world history*. Oxford: Oxford University Press.

De Clercq, M., Vats, A., & Biel, A. (2018). *Agriculture 4.0: The future of farming technology*. World Government Summit.

Democratizing food systems. (2020). *Nature Food, 1*, 383. https://doi.org/10.1038/s43016-020-0126-6.

FSIN. (2020). *Global report on food crisis*. Food Security Information Network & Global Network Against Food Crises.

Garnett, T. (2014). Three perspectives on sustainable food security: Efficiency, demand restraint, food system transformation. What role for life cycle assessment? *Journal of Cleaner Production, 73*, 10–18.

Glaeser, E. L. (2008). *Cities, agglomeration, and spatial equilibrium*. Oxford University Press.

Glaeser, E. L., Kallal, H. D., Scheinkman, J. A., & Shleifer, A. (1992). Growth in cities. *Journal of Political Economy, 100*(6), 1126–1152.

Goswami, R. S., & Kistler, H. C. (2004). Heading for disaster: *Fusarium graminearum* on cereal crops. *Molecular Plant Pathology, 5*(6), 515–525.

Grote, U. (2014). Can we improve global food security? A socio-economic and political perspective. *Food Security, 6*(2), 187–200.

Harvell, C. D., Mitchell, C. E., Ward, J. R., Altizer, S., Dobson, A. P., Ostfeld, R. S., & Samuel, M. D. (2002). Climate warming and disease risks for terrestrial and marine biota. *Science, 296*(5576), 2158–2162.

Hashempour-Baltork, F., Khosravi-Darani, K., Hosseini, H., Farshi, P., & Reihani, S. F. S. (2020). Mycoproteins as safe meat substitutes. *Journal of Cleaner Production, 253*, 119958.

Headey, D., & Fan, S. (2010). *Reflections on the global food crisis: How did it happen? How has it hurt? And how can we prevent the next one?* (Vol. 165). International Food Policy Research Institute.

Hossain, M. S., Madlool, N. A., Rahim, N. A., Selvaraj, J., Pandey, A. K., & Khan, A. F. (2016). Role of smart grid in renewable energy: An overview. *Renewable and Sustainable Energy Reviews, 60*, 1168–1184.

IFPRI. (2021). *Food security*. International Food Policy Research Institute.

Jaffee, S., Siegel, P., & Andrews, C. (2010). Rapid agricultural supply chain risk assessment: A conceptual framework. *Agriculture and Rural Development Discussion Paper, 47*(1), 1–64.

Jiménez, A., Jiménez, S., Lozada, P., & Jiménez, C. (2012, April). Wireless sensors network in the efficient management of greenhouse crops. In *2012 Ninth International Conference on Information Technology-New Generations* (pp. 680–685). IEEE.

Jones, P. M. (2016). *Agricultural enlightenment: Knowledge, technology, and nature, 1750–1840*. Oxford University Press.

Keyfitz, N. (1980). Do cities grow by natural increase or by migration. *Geographical Analysis, 12*(2), 142.

Khan, M. I., Shin, J. H., & Kim, J. D. (2018). The promising future of microalgae: Current status, challenges, and optimization of a sustainable and renewable industry for biofuels, feed, and other products. *Microbial Cell Factories, 17*(1), 1–21.

Koutsoumanis, K., & Sofos, J. N. (2004). Microbial contamination of carcasses and cuts. In *Encyclopedia of meat sciences* (Vol. 67(1), pp. 1624–1629). Elsevier.

Krejci, C. C., & Beamon, B. M. (2010). Environmentally-conscious supply chain design in support of food security. *Operations and Supply Chain Management, 3*(1), 14–29.

Makkar, H. P., Tran, G., Heuzé, V., & Ankers, P. (2014). State-of-the-art on use of insects as animal feed. *Animal Feed Science and Technology, 197*, 1–33.

Manzini, R., & Accorsi, R. (2013). The new conceptual framework for food supply chain assessment. *Journal of Food Engineering, 115*(2), 251–263.
Minhas, P. S., Rane, J., & Pasala, R. K. (2017). Abiotic stresses in agriculture: An overview. In *Abiotic stress management for resilient agriculture* (pp. 3–8). Springer.
Nalley, L., Tsiboe, F., Durand-Morat, A., Shew, A., & Thoma, G. (2016). Economic and environmental impact of rice blast pathogen (*Magnaporthe oryzae*) alleviation in the United States. *PLoS One, 11*(12), e0167295.
Nyberg, M., Olsson, V., & Wendin, K. (2020). Reasons for eating insects? Responses and reflections among Swedish consumers. *International Journal of Gastronomy and Food Science, 22*, 100268.
Overton, M. (1996). *Agricultural revolution in England: The transformation of the agrarian economy 1500–1850 (No. 23)*. Cambridge University Press.
Parodi, A., Leip, A., De Boer, I. J. M., et al. (2018). The potential of future foods for sustainable and healthy diets. *Nat Sustain, 1*, 782–789.
Peck, H. (2006). Reconciling supply chain vulnerability, risk and supply chain management. *International Journal of Logistics: Research and Applications, 9*(2), 127–142.
Peng, S., Huang, J., Sheehy, J. E., Laza, R. C., Visperas, R. M., Zhong, X., . . . & Cassman, K. G. (2004). Rice yields decline with higher night temperature from global warming. *Proceedings of the National Academy of Sciences, 101*(27), 9971–9975.
Post, M. J., Levenberg, S., Kaplan, D. L., Genovese, N., Fu, J., Bryant, C. J., . . . & Moutsatsou, P. (2020). Scientific, sustainability and regulatory challenges of cultured meat. *Nature Food, 1*(7), 403–415.
Ribeiro, P. J. G., & Gonçalves, L. A. P. J. (2019). Urban resilience: A conceptual framework. *Sustainable Cities and Society, 50*, 101625.
Richards, C. E., Lupton, R. C., & Allwood, J. M. (2021). Re-framing the threat of global warming: An empirical causal loop diagram of climate change, food insecurity and societal collapse. *Climatic Change, 164*(3–4), 1. https://doi.org/10.1007/s10584-021-02957-w.
Rosegrant, M. W., & Cline, S. A. (2003). Global food security: Challenges and policies. *Science, 302*(5652), 1917–1919.
Schnurr, A. (2018). Black sky hazards: Systems engineering as a unique tool to prevent National Catastrophe. In *Disciplinary convergence in systems engineering research* (pp. 987–1004). Cham: Springer.
Simmons, A. H. (2011). *The Neolithic revolution in the near east: Transforming the human landscape*. University of Arizona Press.
Snir, A., Nadel, D., Groman-Yaroslavski, I., Melamed, Y., Sternberg, M., Bar-Yosef, O., & Weiss, E. (2015). The origin of cultivation and proto-weeds, long before Neolithic farming. *PLoS One, 10*(7), e0131422.
Thompson, F. M. L. (1968). The second agricultural revolution, 1815–1880. *The Economic History Review, 21*(1), 62–77.
Tredici, M. R. (2004). Mass production of microalgae: Photobioreactors. In *Handbook of microalgal culture: Biotechnology and applied phycology* (Vol. 1, pp. 178–214). Wiley.
Tzachor, A. (2019). The future of feed: Integrating technologies to decouple feed production from environmental impacts. *Industrial Biotechnology, 15*(2), 52–62.
Tzachor, A., Richards, C. E., & Holt, L. (2021) Future foods for risk resilient diets. *Nature Food, 2*. https://doi.org/10.1038/s43016-021-00269-x.
UN DESA. (2018). *2018 Revision of world urbanization prospects*. Population Division of the UN Department of Economic and Social Affairs.
Vorontsova, A. V., Vorontsova, V. L., & Salimgareev, D. V. (2016). The development of urban areas and spaces with the mixed functional use. *Procedia Engineering, 150*, 1996–2000.
Weisdorf, J. L. (2005). From foraging to farming: Explaining the Neolithic revolution. *Journal of Economic Surveys, 19*(4), 561–586.
Wrigley, C. W., Corke, H., Seetharaman, K., & Faubion, J. (Eds.). (2015). *Encyclopedia of food grains*. Academic.

F

The Future of Reducing Urban Vulnerability with Perspectives of Child Development in Zimbabwe

Lessons from the COVID-19 Pandemic

Gloria Nyaradzo Nyahuma-Mukwashi[1], Tinashe Natasha Mujongonde-Kanonhuwa[2] and Innocent Chirisa[3]
[1]Department for International Development (DFID), Zimbabwe, Harare, Zimbabwe
[2]Department of Rural & Urban Planning, University of Zimbabwe, Harare, Zimbabwe
[3]Department of Demography Settlement and Development, Social & Behavioural Sciences, University of Zimbabwe, Harare, Zimbabwe

Introduction

The Coronavirus Disease of 2019 (COVID-19), whose first cases were recorded in Wuhan City, Hubei Province, China (Phelan et al. 2020; Wilder-Smith and Freedman 2020), has had numerous impacts on people's livelihoods, particularly those of children. Due to people's cross-region travel, the disease has spread to many countries in both the developed and the developing worlds. The African continent has also experienced the rampant spread of the virus, and cases have been recorded in South Africa (being the most affected so far), Zimbabwe, Namibia, and Botswana. This has resulted in countries taking strict measures to curb the spread of the virus, hence the introduction of the national lockdown system and social distancing (Nicola et al. 2020). This development restricts people to staying in their homes and halting work and business, except for critical skills, such as health-care services (Thakur et al. 2020). In efforts to reduce the continued spread of the virus, many governments have enforced issues of social distancing and the banning of unnecessary public gatherings to reduce the continued loss of human lives. These measures have affected children in one way or the other (cf. Rothstein 2015).

Parents and guardians in the informal sector have been the most affected as they have ceased working (ILO 2020a; Kelley et al. 2020), making it difficult for parents to provide for their children. Children have stopped going to school (Ghosh et al. 2020), and this has made children more vulnerable to child neglect, abuse, or abandonment and, in some cases, early child marriages (Ahmad 2020; de Paz et al. 2020). This paper evaluates the impacts of the deadly COVID-19 pandemic on children (good or bad) by also revealing developments made toward bettering the lives of children in the COVID-19 era. Cases have also been drawn from Zimbabwe's urban areas, including Harare and Bulawayo.

Literature Review

The COVID-19 pandemic has had numerous effects on children across the globe. It has been recorded that 1.58 billion children and youths have been affected by the closure of schools during the COVID-19 era; 117 million children may miss out on receiving the life-saving measles vaccines and nutrition services (UNICEF 2020). This indicates that children have been affected in one way or the other by the COVID-19 pandemic at the global and local levels. Lockdown measures have significantly hit on informal workers. In India, 400 million informal workers face deeper poverty and starvation as they have been put out of work (Kelley et al. 2020). The inability of informal workers to fend for their families has a negative effect on children as it becomes difficult to provide the basics like food for their children (Buheji et al. 2020; Peterman et al. 2020; Webb et al. 2020).

Studies have shown that children with parents working in hospitals, or those with parents or guardians undergoing quarantine, spend more time on the Internet or watching television (Auerswald et al. 2006; Thakur et al. 2020). Such children experience loneliness and psychosocial problems, such as low self-esteem (Thakur et al. 2020). As a result, such children may become more anti-social and find it difficult to interact with other children even after the end of the COVID-19 pandemic. Liu et al. (2020) explain that children who have lost parents or caregivers to COVID-19 are more susceptible to psychological problems as they are more likely to suffer from fear and grief due to the loss of a parent or loved one. In Spain, it has been observed that the strict lockdown measures by the Spanish government have resulted in heightened stress levels and anxiety in young children (Semo and Frissa 2020).

The separation of children from caregivers is argued to disturb a child's psychosocial well-being, a very crucial component in a child's development into adulthood (Thakur et al. 2020). In the United States, grandparents have been discouraged from providing childcare, and the sharing of childcare responsibilities with neighbors and friends is now very limited (Alon et al. 2020). The change in children's daily routines also affects their psychosocial development (Greendorfer 1987; Eisenmann and Wickel 2009; Catsambis and Buttaro 2012).

Due to changes in their lifestyle, children undergoing quarantine are argued to be exposed to a plethora of psychological burden, psychosocial stigma, and varied neuropsychiatric manifestations (European Center for Disease Prevention and Control 2020; Ghosh et al. 2020). This comes as a result of the change in a child's lifestyle as they may not see their parents and loved ones. There is therefore a need for hospitals to put in place measures where children may be able to see and communicate with their parents through audio-visual devices (Ghosh et al. 2020) to reduce their stress levels.

Besides being educational centers for children, schools are also a home outside a home as they offer a scope of interaction with fellows and seniors and provision of psychological solace (Ghosh et al. 2020). Not enabling children to attend school would thus have negative psychological effects as they may be forced to change their daily routine and adapt to the new way of living, which would now allow less interaction with peers and teachers.

Evidence has also shown that when children are out of school (e.g., weekends and summer holidays), they become less physically active, endure irregular sleep patterns, and eat less favorable diets, which results in weight gain and loss of cardiorespiratory fitness (Ghosh et al. 2020; Wang et al. 2020b). These negative health effects are argued to be more evident in children who are usually confined to their homes with no outdoor activities and no interaction with same-aged peers during the lockdown (Wang et al. 2020b). This shows that the inability of children to do outdoor activities and to interact with others has negative health and psychosocial effects.

More burden has also fallen on the girl child during the COVID-19 era. This is because society expects the girl child to endure extra responsibilities, such as the carrying out of household chores and taking care of the sick (Samman et al. 2016). This places more burden on girls in case of any sicknesses as they are also expected to care for the sick. The closure of schools is also argued to expose children to early marriages, sexual exploitation, and abuse (Ahmad 2020; de Paz et al. 2020). In this regard, the girl child may suffer heavily at the hand of the COVID-19 pandemic.

However, the lockdown system may present a chance for parents and their children to improve on family ties and bonding. Wang et al. (2020b) argues that parents are important role models for their children, and good parenting skills become critical during the lockdown period. Parents have a role to play in molding children into self-disciplined individuals by engaging in talks with their children as this also allows them to educate the younger generation about the COVID-19 pandemic.

To reduce child vulnerability during the COVID-19 pandemic, many governments across the globe, with the help of non-governmental organizations and the United Nations, have engaged in various activities aimed at supporting families and reducing child vulnerability in different nations. UNICEF has assisted 170,000 children in Cambodia with critical Water Sanitation and Health (WASH) messages and providing emergency water trucking in peri-urban neighborhoods reaching out to 6272 children (UNICEF 2020). In Haiti, with support from UNICEF, the national child support has been ensuring that the 754 residential institutions, hosting 25,000 children, are equipped with the necessary hygiene supplies and the information to mitigate potential COVID-19 outbreaks and preventative measures (UNICEF 2020). In Chile, the government has facilitated the food delivery system to benefit 1,600,000 children. In Argentina, cash transfer programs to benefit minors and children living with disabilities were availed (Gentilini et al. 2020). Children in Belize continue to receive take-home food rations during lockdown (Gentilini et al. 2020). These various program initiatives show efforts by various governments and UNICEF in assisting children during the difficult COVID-19 times.

The effects of the deadly COVID-19 pandemic on children have also had devastating effects in Sub-Saharan Africa. Most countries in Sub-Saharan Africa already have under-resourced health-care systems (International Peace Institute 2009; International Labour Organization 2016). As such, most of the Sub-Saharan countries suffer from malnutrition and child and maternal mortality (Agyepong et al. 2017). The Sub-Saharan health system has limited capacity to absorb the pandemic (Kaseje 2020) and, hence, the many efforts by some of the African nations to reduce the detrimental impacts of the disease that children may possibly suffer while confronted by the deadly pandemic.

In dealing with the issue of child malnutrition in pre-COVID-19 times, many African governments, including Sao Tome and Principe, Mozambique, and South Africa, had engaged in the provision of food and nutritional programs in schools (FAO 2018). This was done with the aim to improve the health of school children (FAO 2018) and reduce the rate of malnutrition in school children. The provision of food supplements in schools was, however, disturbed by the closing down of schools due to the lockdown, resulting in the exposure of many children to hunger (cf. Grantham-McGregor 2005; Lund et al. 2009; Buhl 2010; Drake et al. 2017).

COVID-19 may widen the gap between children in high-income families and those in low-income families due to unavailability of facilities for home-schooling, such as audio-visual systems and good Internet (Ghosh et al. 2020). Many children also lack the required books, a stable residence, computers, smart phones, and access to outdoor leisure activities in both the developed and developing nations (Ghosh et al. 2020). This sets many children at a learning disadvantage, hence widening the learning gap between the rich and the poor (McLoyd 1998; Davie 2000; Comber and Hill 2000; Jeon et al. 2014).

UNICEF has also helped reduce the rate of child vulnerability in Africa. In Tanzania, UNICEF has involved itself in the dissemination of information on how parents can offer their support to children during the COVID-19 era (UNICEF 2020). Cash transfers have been adopted by around 78 countries, including Nigeria, Angola, and Chad (Gentilini et al. 2020). This would help families and children to be provided with their basic needs. There are also other ways to prevent more negative effects on children as a result of the occurrence of the COVID-19 pandemic at the international and regional levels (Ghosh et al. 2020).

In reducing some of the negative effects on children, governments ought to ensure that the right environment exists for online learning and children must be equipped with the basic health knowledge on the disease (Tarimo 1991; Von Lubitz and Wickramasinghe 2006; Yousafzai et al. 2014). To improve the health and physical well-being of children, they need to exercise, create indoor games, and control the rate of watching television (Robinson 1999; Granich et al. 2010; McCurdy et al. 2010; Hesketh et al. 2012). In all this, parental patience becomes very essential in reducing the rate of child vulnerability during the COVID-19 pandemic.

Methodology

The article utilizes the desktop review of literature. Due to technological advancements, many sources can now be found on the Internet for use by researchers (Johnston 2017). To ensure validity and reliability of information, scholarly articles to support arguments were used. Local newspaper articles were used to build cases on the issue of COVID-19 and the children in Zimbabwe's urban areas.

Results

Evidence points out that COVID-19 has had many negative effects on children's well-being and social development at international and regional levels.

Some of the children who heavily relied on schools for nutritional support in the developed and developing worlds now suffer from food insecurity due to closure of schools (FAO 2018; Ghosh et al. 2020). The inability of the beneficiaries to get food from schools poses a threat to the well-being of many students. In reducing child vulnerability, countries, such as Chile, have already started engaging in food delivery programs that targeted children (Gentilini et al. 2020) and hence a positive step toward improving the welfare of children in the host nation.

The strict lockdown measures have increased the stress levels of children (Semo and Frissa 2020), with evidence linking to the loss of cardiovascular fitness and weight gain in children to inactivity born as a result of the inability of children to attend school (Wang et al. 2020a). Psychological challenges in the form of grief and fear have also been observed in children who had lost a parent/caregiver to COVID-19 (Liu et al. 2020). These show the unfriendly consequences that the COVID-19 pandemic has brought in the lives of children. In efforts to reduce the negative effects of the pandemic on children, parents must utilize the lockdown opportunity, by making sure that they bond more with their children by creating more family time (Ghosh et al. 2020), and this gives parents and guardians an opportunity to teach children more about the COVID-19 pandemic.

COVID-19 is believed to attack people with weaker immune systems and the old as compared to the young (Wang et al. 2020b), and this has resulted in discrimination of the old. In the United States, grandparents have been discouraged from providing childcare (Alon et al. 2020). This weakens the family and social ties between grandparents and grandchildren. Children are therefore forced to adopt to a new way of living that contains limited or no interaction with grandparents and neighbors, and this may be stressful for some children as they would be forced to change their daily routines and adjust to the new way of living. The article now directs focus on the issue of COVID-19 and children in Zimbabwe's urban areas of Harare, Bulawayo, and Mutare.

The Case of Harare

In efforts to curb the spread of the COVID-19 pandemic, the Zimbabwean government made arrangements to clear the homeless street children off the streets, not only for Harare but for Bulawayo as well. Fifty-nine homeless children have been placed in the Ruwa Training Center, where they are being provided with essential basic food supplies, clean water, and sanitation services (OCHA 2020b). This reduces child vulnerability for street children as they have been provided with shelter and food during the difficult COVID-19 times.

However, there have been reports that these street children have been running away from these "centers of safety" and going back to the streets on claims that the food they were being provided with was too little (Pindula News 2020; Zimbabwe Voice 2020), hence the increased exposure of these "run-away" street children to the COVID-19 virus.

Harare's informal businesses have largely been affected by the COVID-19 pandemic. Street vending has been a dominant activity in the Harare city center (Mutami and Gambe 2015). The introduction of the lockdown system has restricted informal businesses from operating (ILO 2020b), further making it difficult for urbanites to provide for the needs of their dependent-children and theirs as well.

In assisting vulnerable families, the government has facilitated cash transfers and food distributions from the 30th of March, and 740 people have been provided with food assistance, while 60,000 people have received money transfers (OCHA 2020b). This has significantly reduced the rate of child vulnerability in Harare, as parents and guardians become capable of feeding themselves and their dependent children.

The Harare residents have not been spared from the water woes. Children and adults still queue for hours at boreholes while waiting for their turn to fetch water (Human Rights Watch 2020). This exposes children to stress as they wait long hours to get water for the household's daily use, which further exposes children to the COVID-19 virus as a result of the increased interaction in public places. In reducing the high levels of water shortages in the country, 40 boreholes in Harare have been repaired (OCHA 2020c). This has reduced the level of child vulnerability due to the reduction in the level of the water burden. The number of hours that one has to wait to fetch water has also reduced due to availability of more functional water points within Harare.

The Case of Bulawayo

Bulawayo has confirmed COVID-19 cases and has put in place strict quarantine measures for returnees mainly from Botswana, with the United College of Education and the Bulawayo Polytechnic forming part of the city's major quarantine centers (OCHA 2020c). As of 4 May 2020, the city has hosted 700 Zimbabwe returnees, and these have been placed in mandatory quarantine, and of the people quarantined, 50 were children (All Africa News 2020; Bulawayo 24 News

2020). Sprang and Silman (2013) explain that quarantine and isolation can be very traumatic for children with research showing that those quarantined showed mean post-traumatic stress scores that were four times more than those who had not been quarantined. This is partly contributed to by the lack of recreational facilities that are child-specific in these quarantine centers and the change in daily routines of children in these centers. A combination of these centers would culminate in the increase in stress levels for children.

Children living on the streets in Bulawayo have also been put in alternative care arrangements (OCHA 2020c), with 30 street children between the ages of 13 and 18 being kept at the Jairos Jiri Institution (Bulawayo News 2020). This shows great efforts by the government to provide shelter to the homeless children in the difficult COVID-19 times. Rehabilitation services, such as family-tracing and counseling, have also commenced in these centers, and UNICEF is also assisting these centers with prepositioned recreational kits (OCHA 2020c). These initiatives would assist in reducing the level of child vulnerability in these centers, as children are given the chance to feel comfortable in these facilities.

The Case of Mutare

In the midst of recovering from the destructive effects of Cyclone Idai of March 2019, which left many homeless, the Manicaland Province had to deal with another deadly pandemic that hit the country. Evidence has it that 198 families are still in camps after exposure to the Cyclone Idai disaster and part of the victims are children (OCHA 2020a). Compared to children from affluent families, these children risk lagging behind in terms of education as they may not have access to e-learning.

However, the reintroduction of radio and television lessons, after a two-decade break, will help reduce the learning gaps (Dzenga 2020). Three media companies including the Zimbabwe Broadcasting Corporation, AB Communications, and Zimpapers will provide facilities to record the content to be taught, and radio stations under these companies would then broadcast the lessons (Dzenga 2020). These developments will help in reducing the learning gaps between students in primary school and those in secondary schools.

There is also need for the country to also give attention to other diseases, such as HIV&AIDS, cancer, and malaria. Kaseje (2020) notes that the rainy season in Sub-Saharan Africa came early in the year 2020, which means an anticipated increase in malaria cases, which may coincide with the ongoing COVID-19 pandemic. Zimbabwe has already lost 131 lives to malaria amidst the COVID-19 pandemic (Calvin-Smith 2020), with 3 of the victims being from the Mutare district (Health Times News 2020). The increase in malaria-related deaths could leave more children suffering as they may lose a parent or guardian. This calls for more awareness among health workers, families, and children to understand risks that they may fall prey to during the COVID-19 era.

The Case of Chitungwiza

Many parts of Chitungwiza lack access to clean water. Regardless of the country experiencing lockdown, people still have to go in search of safe water to drink, mainly at community boreholes. Access to clean water is a basic human right (Gleick 1996; ZPP 2020). Thousands of women and school-going-aged children queue at boreholes in Chitungwiza for close to 8 or 9 h and at times spend the whole night in a line waiting for their turn to fetch water in these crowded boreholes/ narrow water wells (Human Rights Watch 2020). Children have also been observed queuing at grocery stores to buy subsidized meal (Muronzi 2020; Nkomo 2020). These scenarios compromise on issues of social distancing and hygiene, which in turn exposes children to danger of contracting the COVID-19 virus.

A research carried by community-based organizations working in urban Chitungwiza revealed that the issue of social distancing and hygiene was only being preached and not being practiced especially in scenarios where people are lacking the much-needed basic commodities (Mackworth-Young et al. 2020). This in turn increases child vulnerability as they risk contracting the virus in these public places. With the help of the United

Nations Children's Education Fund, a total of 20,000 girls and boys in Epworth and Chitungwiza are expected to receive sanitizers, sanitary wear, learning material, buckets, soap, and face towels (OCHA 2020d), and these would assist in reducing the level of child vulnerability in Chitungwiza.

Discussion

Zimbabwean children have been affected in many ways by the existence of COVID-19 cases in the urban landscape. Due to the closure of schools, many children now spend time in their homes (Thakur et al. 2020), while others queue at boreholes in Harare and Chitungwiza to get portable water to drink and at grocery stores to buy subsidized meal (ZPP 2020; Muronzi 2020; Nkomo 2020). This indicates the increased level of (especially from poor families) child labor.

These situations go against the government's mandate to promote social distancing, which makes it difficult for government and health officials to control the spread of the deadly virus. The Zimbabwean government has, however, mandated the police the task to ensure that people wear masks and maintain social distancing in public places (Nkomo 2020). This will in turn reduce exposure of the disease to children in public places. Forty boreholes have also been repaired in Harare to ease the water woes (OCHA 2020c), which is in turn a positive step toward reducing long queues from forming at boreholes.

The closure of informal businesses as a result of the lockdown has made it difficult for parents and guardians to generate income (ILO 2020b), further making it difficult for parents and guardians to provide for themselves and their dependent children. The Government of Zimbabwe has, however, assisted families with cash transfers and food distributions (OCHA 2020b). This has prevented many children from suffering from hunger, as parents and guardians (especially those in the informal sector) have become capable of providing the basics for their dependent children.

The education system has come to a halt, and this has affected the learning progress for most children. While other children from affluent families may have access to online learning, those from poor families may have limited or no access to Internet (Bates 2020), especially those who fell victim to Cyclone Idai and were placed in camps in Mutare. This creates a gap in learning between the rich and the poor. The re-introduction of radio and television lessons will, however, help in reducing the gap between learners (Dzenga 2020). These would be broadcasted on radio, and many children may have access, thus reducing the level of child vulnerability, even for those in camps.

Children living on the streets have been put in Bulawayo's Jairos Jiri (Bulawayo 24 News 2020), while those in Harare have been put in the Ruwa Training Center (OCHA 2020b). The removal of street kids from the streets by the government and securing decent facilities for the children to live in is one step up toward reducing children's exposure to the COVID-19 disease. In these facilities, rehabilitation services in the form of family tracing and counseling practices have commenced (OCHA 2020c). Counseling reduced child vulnerability, and through family tracing, some of the street children may end up being reunited with their own families. However, some the street kids are reported to have been running back to the streets, with reasons that the food they being given in these institutions is too little (Pindula News 2020; Zimbabwe Voice 2020). This is a step back toward ensuring that these homeless children are kept safe and away from easily contracting the virus while on the streets. This calls for strict government measures to keep street children in these "new" homes.

There has also been a concern for government not to only focus on the COVID-19 pandemic at the expense of other diseases, such as HIV&AIDS, tuberculosis, and malaria (AU 2020). The country has lost 131 people to malaria amidst the coronavirus outbreak (Calvin-Smith 2020). The number exceeds that of four COVID-19 deaths recorded in the country as of 1 May 2020 (MoHCC 2020). This calls for the government to balance off and give attention to other possible life-threatening disease

F

outbreaks, such as malaria, HIV&AIDS, and cancer. If people continue to lose lives from other diseases, children risk losing either a parent/guardian, which may in turn increase the number of child-headed families.

Conclusion and Recommendations

It is concluded that, indeed, the COVID-19 pandemic has exposed many children to many harsh and tough conditions (in both the developed and the developing worlds). Many children felt isolated due to school closures, while others felt that the issue of social distancing was emotionally challenging. The closure of schools comes with possibilities for increased number of school dropouts, child marriages, and the lagging behind for the poor and marginalized children who may not have access to the Internet or online learning. The Government of Zimbabwe, in efforts to reduce child vulnerability, has put street children in various institutional facilities where they are provided with basic essentials and counseling services. Cash transfers to the needy have also been facilitated, and this has reduced children's susceptibility to hunger in the event of parents and guardians failing to provide basic essentials for their dependent children. Some of the radio stations have started broadcasting radio lessons targeted at students in primary and secondary schools. These various initiatives all make positive contributions in the lives of children, especially in the wake of the COVID-19 pandemic. This study therefore recommends that, to reduce the level of present and future child vulnerability, there is need for:

- Strict government regulations and penalties, against child offenders.
- Children to be empowered to say "no" to child marriages, child labor, and abuse; a heavy penalty must be imposed on parents and guardians who force children into early marriages.
- Government to work toward revamping its health and education systems in preparation for possible future disasters and allow for the improvement of e-learning facilities for all children in urban and remote areas.

Cross-References

- ▶ From Vulnerability to Urban Resilience to Climate Change

References

African Union (AU). (2020). Impact of the Corona virus (COVID 19) on the African economy. https://www.google.com/url?sa=t&source=web&rct=j&url=https://au.int/sites/default/files/documents/38326-doc-covid-19_impact_on_african_economy.pdf&ved=2ahUKEwjzpdbx8a3pAhUBsXEKHeKzBhlQFjABegQIAxAB&usg=AOvVaw368Oy6ESUTjKFYUJci8r1f. Accessed 12 May 2020.

Agyepong, I. A., Sewankambo, N., Binagwaho, A., Coll-Seck, A. M., Corrah, T., Ezeh, A., Fekadu, A., Kilonzo, N., Lamptey, P., Masiye, F., & Mayosi, B. (2017). The path to longer and healthier lives for all Africans by 2030: The lancet commission on the future of health in the sub-Saharan Africa. *The Lancet, 390*(10114), 2803–2859.

Ahmad, T. (2020, April 15). Social and moral responsibility of the individuals, governments and the protection of human rights in the Coronavirus Pandemic (Covid-19). Governments and the protection of human rights in the Coronavirus Pandemic (COVID-19).

All Africa News. (2020). Zimbabwe: Bulawayo investigates 223 suspected COVID-19 cases but confirms 12. https://allafrica.com/stories/202005040299.html. Accessed 7 May 2020.

Alon, T. M., Doepke, M., Olmstead-Rumsey, J., & Tertilt, M. (2020). The impact of COVID-19 on gender equality (No. w26947) National Bureau of Economic Research.

Auerswald, P. E., Branscomb, L. M., La Porte, T. M., & Michel-Kerjan, E. O. (Eds.). (2006). *Seeds of disaster, roots of response: How private action can reduce public vulnerability.* Cambridge, United Kingdom: Cambridge University Press.

Bates, T. (2020). Emergency online learning and inequity: Developed countries https://www.tonybates.ca/2020/04.20/emergency-online-learning-and-inequity-developed-countries/. Accessed 11 May 2020.

Buheji, M., da Costa Cunha, K., Beka, G., Mavric, B., de Souza, Y. L., da Costa Silva, S. S., Hanafi, M., & Yein, T. C. (2020). The extent of covid-19 pandemic socio-economic impact on global poverty. A global integrative multidisciplinary review. *American Journal of Economics, 10*(4), 213–224.

Buhl, A. (2010). Meeting nutritional needs through school feeding: A snapshot of four African nations. *Global Child Nutrition Foundation*, 1–79. Available at http://citeseerx.ist.psu.edu/viewdoc/citations;jsessionid=777F0183F1833F3625B55A32D9E66D71?doi=10.1.1.656.9697

Bulawayo 24 News. (2020). Respiratory illnesses spike in Bulawayo. https://bulawayo24.com/index-id-news-sc-national-byo-184744.html.

Calvin-Smith, G. (2020). Zimbabwe hit by malaria outbreak amid Covid-19 lockdown. https://ampfrance24.com/en/africa/20200421-malaria-outbreak-zimbabwe. Accessed 24 Apr 2020.

Catsambis, S., & Buttaro, A. (2012). Revisiting "kindergarten as academic boot camp": A nationwide study of ability grouping and psycho-social development. *Social Psychology of Education, 15*(4), 483–515.

Comber, B., & Hill, S. (2000). Socio-economic disadvantage, literacy and social justice: Learning from longitudinal case study research. *The Australian Educational Researcher, 27*(3), 79–97.

Davie, R. (2000). *Combating educational disadvantage: Meeting the needs of vulnerable children*. United Kingdom: Psychology Press.

de Paz, C., Muller, M., Manoz Boudet, A. M., & Gaddis, I. (2020). *Gender dimensions of the COVID-19 pandemic*. Washington, DC: World Bank. © World Bank. https://openknowledge.worldbank.org/handle/10986/33622

Drake, L., Fernandes, M., Aurino, E., Kiamba, J., Giyose, B., Burbano, C., Alderman, H., Mai, L., Mitchell, A., & Gelli, A. (2017). *School feeding programs in middle childhood and adolescence. Chapter 12 in: Child and adolescent health and development* (3rd ed.). The International Bank for Reconstruction and Development / The World Bank: Washington, DC. https://doi.org/10.1596/978-1-4648-0423-6_ch12. 2017 Nov 20. Chapter 12. Available from: https://www.ncbi.nlm.nih.gov/books/NBK525249/.

Dzenga, L. (2020). Zimbabwe: Zim reintroduces radio lessons. The herald. https://allafrica.com/stories/202006040234.html. Accessed 6 Jul 2020.

Eisenmann, J. C., & Wickel, E. E. (2009). The biological basis of physical activity in children: Revisited. *Pediatric Exercise Science, 21*(3), 257–272.

European Center for Disease Prevention and Control. (2020). Coronavirus disease 2019 (COVID-19) in the EU/EEA and the UK–ninth update. Electronic article available at https://www.ecdc.europa.eu/en/publications-data/rapid-risk-assessment-coronavirus-disease-2019-covid-19-pandemic-ninth-update

Food and Agriculture Organisation of the United Nations. (2018). *Regional overview of national school food and nutrition programs in Africa*. FAO. Electronic resource, Available at: https://www.fao.org/policy-support/tools-and-publications/resources-details/en/c/1158936/

Semo, B. W., & Frissa, S. M. (2020). The mental health impact of the COVID-19 pandemic: Implications for Sub-Saharan Africa. *Psychology Research and Behavior Management, 13*, 713–720. https://doi.org/10.2147/PRBM.S264286

Gentilini, U., Almenfi, M., Orton, I., & Dale, P. (2020). *Social protection and jobs responses to COVID-19: A real-time review of country measures. Live document*. Washington, DC: World Bank.

Ghosh, R., Dubey, M. J., Chatterjee, S., & Dubey, S. (2020). Impact of COVID-19 on children: Special focus on psychosocial aspect. *Education, 31*, 34.

Gleick, P. H. (1996). Basic water requirements for human activities: Meeting basic needs. *Water International, 21*(2), 83–92.

Granich, J., Rosenberg, M., Knuiman, M., & Timperio, A. (2010). Understanding children's sedentary behaviour: A qualitative study of the family home environment. *Health Education Research, 25*(2), 199–210.

Grantham-McGregor, S. (2005). Can the provision of breakfast benefit school performance? *Food and Nutrition Bulletin, 26*(Suppl 2), S144–S158.

Greendorfer, S. L. (1987). Psycho-social correlates of organized physical activity. *Journal of Physical Education, Recreation & Dance, 58*(7), 59–64.

Health Times. (2020). Malari kills 131 in Zimbabwe. https://healthtimes.co.zw/2020/04/18/malaria-kills-131-in-zimbabwe/. Accessed 07 May 2020.

Hesketh, K. D., Hinkley, T., & Campbell, K. J. (2012). Children's physical activity and screen time: Qualitative comparison of views of parents of infants and preschool children. *International Journal of Behavioral Nutrition and Physical Activity, 9*(1), 152–161.

Human Rights Watch. (2020). Zimbabwe: Unsafe water raises COVID-19 Risks. Severe water, sanitation crisis undermines pandemic fight. https://www.hrw.org/news/2020/04/15/zimbabwe-unsafe-water-raises-covid-19-risks. Accessed 6 Jun 2020.

International Labour Organization (2016). World employment social outlook 2016: Trends for youth. https://www.ilo.org/wcmsp5/groups/public/@dgreports/@dcomm/@publ/documents/publication/wcms_513739.pdf

International Labour Organization (ILO). (2020a). COVID-19 crisis and the informal economy: Immediate responses and policy challenges. Electronic resource available at: https://www.ilo.org/global/topics/employment-promotion/informal-economy/publications/WCMS_743623/lang--en/index.htm

International Labour Organization (ILO). (2020b). *COVID-19: Guidance for labour statistics data collection. Essential labour force survey content and treatment of special groups (Rev. 1)*. Electronic resource available at https://www.ilo.org/wcmsp5/groups/public/—dgreports/—stat/documents/publication/wcms_741145.pdf

International Peace Institute (IPI). (2009). Underdevelopment, resource scarcity and environmental degradation. Task forces on strengthening multilateral security. Available at https://www.ipinst.org/wp-content/uploads/publications/devresenv_epub.pdf

Jeon, L., Buettner, C. K., & Hur, E. (2014). Family and neighborhood disadvantage, home environment and children's school readiness. *Journal of Family Psychology, 28*(5), 718–731.

Johnston, M. P. (2017). Secondary data analysis: A method of that the time has come. *Qualitative and Quantitative Methods in Libraries, 3*(3), 619–626.

Kaseje, N. (2020). Why sub-Saharan Africa needs a unique response to COVID-19. World Economic Forum. https://www.weforum.org/agenda/2020/03/why-sub-saharan-africa-needs-a-unique-response-to-covid-19/.

Kelley, M., Ferrand, R. A., Muraya, K., Chigudu, S., Molyneux, S., Pai, M., & Barasa, E. (2020). *An appeal for practical social justice in the COVID-19 global response in low-income and middle-income countries*. The Lancet Global Health.

Liu, J. J., Bao, Y., Huang, X., Shi, J., & Lu, L. (2020). Mental health considerations for children quarantined because of COVID-19. *The Lancet Child & Adolescent Health, 4*(5), 347–349.

Lund, F., Noble, M., Barnes, H., & Wright, G. (2009). Is there a rationale for conditional cash transfers for children in South Africa? *Transformation: Critical Perspectives on Southern Africa, 70*(1), 70–91.

Mackworth-Young, C. R. S., Chingono, R., Mavodza, C., McHugh, G., Tembo, M., Dziva Chikwari, C. et al. (2020). 'Here, we cannot practice what is preached': Early qualitative learning from community perspectives on Zimbabwe's response to COVID-19. [Preprint]. *Bulletin of the World Health Organization*. E-pub: 20 April 2020. https://doi.org/10.2471/BLT.20.260224

McCurdy, L. E., Winterbottom, K. E., Mehta, S. S., & Roberts, J. R. (2010). Using nature and outdoor activity to improve children's health. *Current Problems in Pediatric and Adolescent Health Care, 40*(5), 102–117.

McLoyd, V. C. (1998). Socioeconomic disadvantage and child development. *American Psychologist, 53*(2), 185–195.

Ministry of Health and Child Care (MoHCC). (2020). Coronavirus (COVID-19) update: 01 May 2020. Government of Zimbabwe.

Muronzi, C. (2020). "We'll die of hunger first": Despair as Zimbabwe lockdown begins. https://www.aljazeera.com/amp/news/2020/03/die-hunger-despair-zimbabwe-lockdown-begins-200330054919081.html. Accessed 11 May 2020.

Mutami, C., & Gambe, T. R. (2015). Street multi-functionality and city order: The case of street vendors in Harare. *Journal of Economics and Sustainable Development, 6*(14), 124–129.

Nicola, M., Alsafi, Z., Sohrabi, C., Kerwan, A., Al-Jabir, A., Iosifidis, C., Agha, M., & Agha, R. (2020). The socio-economic implications of the coronavirus pandemic (COVID-19): A review. *International Journal of Surgery, 78*, 185–193. https://doi.org/10.1016/j.ijsu.2020.04.018

Nkomo, C. (2020). Zimbabwe: Observing social distancing at Westlea N. Richards a challenge. All Africa News. https://allafrica.com/stories/202004270242.html. Accessed 12 May 2020.

OCHA. (2020a). Humanitarian response plan Zimbabwe 2020. Humanitarian Program Cycle.

OCHA. (2020b). Zimbabwe situational report. Last updated: 8 April 2020. https://reports.unocha.org/en/country/zimbabwe/. Accessed 04 May 2020.

OCHA. (2020c). Zimbabwe situation report. Last updated: 21 Apr 2020. https://reports.unocha.org/en/country/zimbabwe/.

OCHA. (2020d). Zimbabwe situation report. Last updated: 11 June 2020. https://reports.unocha.org/en/country/zimbabwe/.

Peterman, A., Potts, A., O'Donnell, M., Thompson, K., Shah, N., Oertelt-Prigione, S., & van Gelder, N. (2020). Pandemics and violence against women and children. Center for Global Development working paper, 528.

Phelan, A. L., Katz, R., & Gostin, L. O. (2020). The novel coronavirus originating in Wuhan, China: Challenges for global health governance. *JAMA, 323*(8), 709–710.

Pindula News. (2020). The food rations were too little-homeless kids now back on the streets after escaping lockdown safe houses. https://news.pindula.co.zw/2020/04/17/the-food-rations-were-too-little-homeless-kids-now-back-on-the-streets-after-escaping-lockdown-safe-houses/. Accessed 11 May 2020.

Robinson, T. N. (1999). Reducing children's television viewing to prevent obesity: A randomized controlled trial. *JAMA, 282*(16), 1561–1567.

Rothstein, M. A. (2015). From SARS to Ebola: Legal and ethical considerations for modern quarantine. *Indiana Health Law Review, 12*, 227–234.

Samman, E., Presler-Marshall, E., Jones, N., Bhatkal, T., Melamed, C., Stavropoulou, M., & Wallace, J. (2016). *Women's work: Mothers, children and the global childcare crisis*. London: ODI. Obtenido de http://docplayer.net/17690651-Women-s-work-women-s-work-global-childcare-crisis-globalchildcare-crisis.html.

Sprang, G., & Silman, M. (2013). Posttraumatic stress disorder in parents and youth after health-related disasters. *Disaster Medicine and Public Health Preparedness, 7*(1), 105–110.

Tarimo, E. (1991). *Towards a healthy district: Organizing and managing district health systems based on primary health care*. Geneva: World Health Organization.

Thakur, K., Kumar, N., & Sharma, N. (2020). Effect of the pandemic and lockdown on mental health of children. *The Indian Journal of Pediatrics, 87*, 552.

UNICEF. (2020). UNICEF global Covid-19 situation report no. 3. UNICEF. https://www.unicef.org/appeals/files/UNICEF_Global_CoViD19_Situation_Report_No3__1_15_April_2020.pdf

Von Lubitz, D., & Wickramasinghe, N. (2006). Key challenges and policy implications for governments and regulators in a networkcentric healthcare environment. *Electronic Government, An International Journal, 3*(2), 204–224.

Wang, L. S., Wang, Y. R., Ye, D. W., & Liu, Q. Q. (2020a). A review of the 2019 Novel Coronavirus (COVID-19) based on current evidence. *International Journal of Antimicrobial Agents, 55*, 105948.

Wang, G., Zhang, Y., Zhao, J., Zhang, J., & Jiang, F. (2020b). Mitigate the effects of home confinement on children during the COVID-19 outbreak. *The Lancet, 395*(10228), 945–947.

Webb, J. W., Khoury, T. A., & Hitt, M. A. (2020). The influence of formal and informal institutional voids on entrepreneurship. *Entrepreneurship Theory and Practice, 44*(3), 504–526.
Wilder-Smith, A., & Freedman, D. O. (2020). Isolation, quarantine, social distancing and community containment: Pivotal role for old-style public health measures in the novel coronavirus (2019-nCoV) outbreak. *Journal of Travel Medicine, 27*(2), taaa020.
Yousafzai, A. K., Rasheed, M. A., Daelmans, B., Manji, S., Arnold, C., Lingam, R., Muskin, J., & Lucas, J. E. (2014). Capacity building in the health sector to improve care for child nutrition and development. *Annals of the New York Academy of Sciences, 1308*(1), 172–182.
Zimbabwe Peace Project (ZPP). (2020). Human Rights in the midst of the Covid-19 Global Pandemic: Zim in state of neglect. Monthly Monitoring Report.
Zimbabwe Voice. (2020). Street kids escape Harare safe house: The food was little. https://zimbabwevoice.com/2020/04/17/street-kids-escape-Harare-safe-house-the-food-was-little/amp/.

Future of the City-Region Concept and Reality

Implications of Sustainable Infrastructure and Community Linkages

Tafadzwa Mutambisi[1] and Innocent Chirisa[2]
[1]Department of Rural and Urban Planning, University of Zimbabwe, Harare, Zimbabwe
[2]Department of Demography Settlement and Development, Social & Behavioural Sciences, University of Zimbabwe, Harare, Zimbabwe

Definition

Rural hinterland – rural territory

Economic development – activities like farming, mining, and tourism that result in the creation and expansion of household or regional wealth

Introduction

The world has urbanized, but it has not done so exclusively or even mainly in large cities. Almost two billion people, 27% of the world's total population, more so half of the world's urban population, are residents of towns and cities. An additional 3.4 billion people are classified as living in rural areas, or 46% of our planet's inhabitants. The majority of the world's poor, perhaps as many as 70%, live in these towns and small and medium cities and the rural areas more proximate to them, and poverty rates are also higher in small and medium cities than in large urban agglomerations (Berdegué et al. 2014). Therefore, growing urbanization has brought the concept of city-regions as a major factor that has to be considered in development across the globe (Scott 2008). This is because urban development on a massive scale is a major city that expands beyond administrative boundaries to cover small cities, towns, and semi-urban and rural hinterlands, city to city conurbation. The city region concept articulates the relationships between the city and its environments (Davoudi 2008).

Consequences of wide spread and imprecise use of the city-region concept are that the concept can be utilized in describing either a combination or individual territorial units at subnational level sorely excluding small local governments. The core element found in different definitions of the city-region concept is that all city regions have the main city with functional ties to it rural areas or hinterland (Marrazzo 1996). The nature of the linkages between the core city and its hinterland varies from one definition to another, but they generally include a combination of elements such as marketing, economic, travel-to-work, housing market, or retail catchment factors. Identity and the social and cultural domination of the core city are also included as essential elements that make up a city-region (Berdegué et al. 2014). Historically, cities have always depended on a hinterland for the supply of vital goods, while chiefdoms and principalities always needed a regional basis for tribute, defense, and a wide range of other purposes. The logic behind city-region is that planning and organization cannot be limited to the territorial extent of an individual city but must take into account space flows through the city to regional economies. Spaces of flow assist in subordination of spaces of the place, and it characterizes former forms of subnational regionalism (Greenberg 2010).

City-regions are defined by their spatial characteristics, linkages between the city and surrounding regional areas, as well as infrastructure surrounding them. The continuous assessment of the structural integrity of infrastructure systems is critical if there is to be reliable characterization of infrastructure's investment decisions and continued functionality, leading to the sustainability of infrastructure systems (Qi et al. 2019). The concept of sustainable development has been gaining popularity across various disciplines, including infrastructure. To enhance sustainability in the Australian construction industry, the national infrastructure network identified priorities in maintaining economic success and environmental stability.

Infrastructure has played a critical role in the development of many countries. The infrastructure required is both physical, that is, electric cables, cell phone and broadband transmitters, roads, buildings, and water and sewerage pipes, and social-human resource complex. This includes administration, research, ideological functions, education, and so on (Luo and Shen 2008). Sustainability of infrastructure is important in developing city-regions as it determines the efficiency and effectiveness of regional systems. This is very important in the Global South as there exist only two types of cities. The first type is cities coping with informal hyper growth. These cities are characterized by rapid population growth, both through migration and natural increase; an economy heavily dependent on the informal sector; very extensive poverty, with widespread informal housing areas; basic problems of the environment and of public health; and difficult issues of governance (Munyasya and Chileshe 2018). These are typical of the bigger cities in Southern Africa.

The second type is cities coping with dynamic growth. These cities face falling population growth and the prospect of an aging population, rapid economic growth, and emerging environmental problems (Pascariu and Czischke 2015). The territorial approach to governance and planning is untenable given the gravity of the flow of resources and information through the cities. This leads to the traditional city center becoming incapacitated to accommodate expansion of infrastructure and services that are requisite for its nodal functions as the city center becomes physically too small. This leads to expansion of cities beyond their boundaries, and in other cases, urban centers or cities are located in different cities (Pain et al. 2008).

At the same time, organic processes of clustering agglomeration of economic agents enable greater operational flexibility and innovative capacity at a scale that follows economic rather than administrative logic; city administrators and planners must align their functions with these new forms. In these circumstances, the formation of a city-region incorporating all the territory and all the functions into one functional unit is a logical step. Historically, public infrastructure development in many developing countries has been disconnected from management of existing infrastructure assets and has contributed to years of deferred maintenance of existing systems and the contemporary infrastructure crisis (Kitchin and Thrift 2009). The proper management of infrastructure assets over a life cycle affects the integrity and level of service of these systems and thus the infrastructure sustainability. Proper management can include new development and installation as well as maintenance and rehabilitation of existing components. Therefore, this chapter aims to clearly articulate the city-region concept and its impacts on infrastructure as well as community linkages between the city and the surrounding regional areas.

The chapter made use of the qualitative research approach which is based on the descriptive aspect presentation based on the reviews and citations on the social context of the chapter (Creswell 2014). Document analysis, case study research approach, and secondary data were implored in the research. Secondary sources such as books, journals, newspapers, and policy documents were perused and analyzed. This enabled the pap to gain an understanding of the evolution of the city-region concept. The case study approach was vital in assessing a specific area in this chapter concentrating largely on a developed and developing country that has adapted the city-region concept in its development plans to get a clear picture of the concept and its impact.

Theoretical Perspectives on the City-Region Concept and the Infrastructure-Community Nexus

The chapter makes use of the two theories in its quest to highlight the realities of the city-region concept in prompting sustainable development. The theories are growth pole and central place. The growth pole theory sees economic development, or growth, is not uniform over an entire region but instead takes place around a specific pole or around specific key industries which link the development of other industries through direct and indirect effects. These poles may include automotive, aeronautical, agribusiness, electronics, and steel (Bayang 2006). The growth pole theory has direct and indirect effects which imply that the core industry purchases goods and services from its suppliers; this is also termed upstream linked industries. The pole also provides goods and services to its customers (downstream linked industries) (Bayang 2006). Its indirect effects include increase in high demand for specific goods and services by the people who are employed by the main industry; also, other industries linked to the core end up supporting the development and expansion of the area through the establishment of various economic activities such as retail.

In geography, the central place theory is described as a spatial theory that attempts to give reasons as well as to explain distribution patterns, numbers, and sizes of cities and towns around the world. This theory was created by Walter Christaller (1933). He recognized that the economic relationships between cities and their hinterlands or areas further away from the city accumulate in the core cities so as to share ideas and goods. That is, earlier forms of regionalism were based more on a geographical core-periphery model, with a center dominating a surrounding periphery (Bayang 2006). Theoretically, the new regionalism, by contrast, subordinates location for process. By this logic, cities that remain bound into a place-based identity become disconnected from power that is located in the processes of the flows of the circulation of capital (commodities, finance, and people). A shift to a global identity permits planners to identify, orient, and tap into the vast streams of power flow through the cities.

The defining characteristic of the metropolis was considered to be the commuting pattern of the workforce which was assumed to be radial from the periphery to the center(s) where the jobs were located (Uddin et al. 2019). There are four approaches to looking at city-regions: the first one focuses on urban sector analysis and theory; the second emphasizes policy development; the third considers the governmental, planning, and services footprint of a functional region; and the fourth is based on the ecological footprint of an urban system. The revival of the city-region concept is not only a reincarnation of an analytical construct to understand complex spatial relations but also a manifestation of a political move toward a new regionalism (Greenberg 2010). This has been coupled with the rescaling of state intervention to intermediate levels such as the city-region (Scott 1998). Hence, the contemporary relevance and significance of the city-region concept lies in its potential, firstly, to evoke a relational understanding of space and place in policy and practice and, secondly, to encourage researchers to seek new methodologies for capturing the less tangible interconnections that define the virtual contours of what Castells7 called the space of flows (Davoudi 2008).

Globally competitive cities and the city-region concepts are not neutral. They have been gathering in force through their use and distribution used and distributed in the periods of development of capitalism which is understood not merely in its economic aspects but in its social, political, and economic totality. Relational understandings prompt useful questions about the links between employment locations, geography of land and property values, labor and housing markets, transportation systems, and patterns of social segregation, rather than providing simple administrative answers (Hunt and Rogers 2014). First, although labor market area boundaries do not offer a "one-shot" solution for the definition of city-regions, commuting patterns are undeniably important evidence (Coombes 2014). Linkages between the city and its surrounding regions have grown important over the years as the role of cities has grown in importance, and contested neo-liberalization has enhanced this role and has seen the growth of importance of sustainable infrastructure in city-regions.

The future performance of infrastructure systems is mainly concerned with a sustainable maintenance strategy where optimization processes entirely depend on an anticipation of damage nucleation and evolution. Suitable infrastructure is central to the delivery of all the Sustainable Development Goals (SDGs) (Ellingsen and Leknes 2012). It underpins the socioeconomic goals and has impacts on the environmental goals. Sustainable infrastructure underpins the delivery of all the social SDGs. Perhaps most pertinent is infrastructure's role in alleviating poverty (SDG 1), which is crucial to global development, given that 3.5 billion people are still living below the poverty line. Sustainable infrastructure is also important in maintaining community links through a god transport and communication system (Meidani and Ghanem 2015). This is because rural-urban linkages are reciprocal flows of people, goods, services, money, and environmental services. Under certain conditions, aided by geographical proximity, they can lead to interdependence between rural and urban and to the formation of intermediate rural-urban functional areas (territories) that very often cut across administrative boundaries and that encompass a number of rural localities, as well as a few towns and small and medium cities (SDG Knowledge Hub 2020). Ultimately, rural-urban relations result from the combination of the structural transformation that is a key feature of development and the concomitant transformation.

The City-Region, Sustainable Infrastructure, and Community Linkages: Experiences and Cases

Over the years, there have been changing borders of rural and urban regions with a variety of views on the foundational debates on the nature of rural-urban relations since the 1950s (Hussein and Suttie 2016). Policies were based on the proposition that long-term national development required the transfer of surplus of labor and capital from the agricultural sector to urban industries. Authors such as Myrdal (1957) started noticing the relationships between the city and its region by discussing the "spread" and "backlash," "polarization," and "trickle down" effects of industrialization (Bayang 2006). The intellectual and policy tension is on whether urbanization and urban development will pull rural people to conditions of greater opportunities and well-being or, if on the contrary, urban development is parasitic of rural areas and people, leading to countries that are increasingly polarized spatially and socially. For instance, in subnational governance in England, the core fundamentals of the city-region concept have remained unchanged since the formation of the concept by Geddes (1915). Their definitive features reflect the processes that create the development of modern economies and the associated increase in linkages between areas. Classic examples of city-gions are New York or London; these are not just big global cities but critical conducts and nodes for the flows of information, finance, and skills that constitute the global capitalist economic system (Yigitcanlar and Dur 2010).

The city-region concept has gained popularity over the years due to population growth and rapid urbanization in the cities. For instance, in Latin America, there is the distribution of towns and cities of different sizes, identified 986, 394, and 103 functional territories in Mexico, Colombia, and Chile, respectively (Qi et al. 2019). Those that correspond to rural-urban configurations involving one or more towns or small and medium cities and a rural hinterland house 43%, 38%, and 37% of the total population, while deep rural territories that lack an urban nucleus range have 7%, 16%, and 6%. In Brazil, up to 3400 rural and urban municipalities could be part of these rural-urban functional territories, involving 93 million people, 37 million of which live in poverty (Greenberg 2010). The West African Border Markets project of the Center for Economic and Social Research (CEPS/Instead) has identified

11 functional regions, involving cities and rural regions, crossing international as well as national political and administrative boundaries. Indonesia's government has a certain tradition of policies that seek to promote and take advantage of the development of functional rural-urban territories.

A prominent example is the Agropolitan Development Program that seeks to improve the competitiveness of a rural-urban region by improving rural-urban linkages, small and medium agricultural production, and urban-based services in support of the development of the rural hinterland and its economic activities (Schmitt et al. 2015). It is very important to emphasize, however, that the linkages between towns and small and medium cities are not always mutually beneficial. In fact, such relations can be predatory of the rural hinterland, through a number of mechanisms such as exposing rural economic activities to outside competition, financial systems that capture rural savings and lend them mostly for urban consumption or investment, urban bias in local government investments despite the fact that numbers of rural people and rural poverty incidence contribute to enhancing the position of the district or municipality or province in the allocation of central government budget transfers, and inability of some rural dwellers to participate in the more lucrative non-farm activities due to lack of assets or because of social barriers (Stepanova et al. 2020). On average, territories that contained an urban center reduced poverty and increased per capita income faster, which also shows that income or consumption inequality also rises, and it is likely that a proportion of that is due to growing rural-urban income gaps within the rural-urban territory.

In India, small towns with a population of under 50,000 or 100,000, surrounded by a rural hinterland, constitute locally integrated local economies. If so, the number of such places, or territories, would be very large in this country (Rodríguez-Pose 2008). Also, in West Africa, the Africapolis program reports that the increased spread of agglomerations, going from 125 units to 1500 in 70 years, has encouraged the filling of empty spaces, as well as the densification of already urbanized areas. Between 1950 and 2000, the average yearly expansion of the urbanized surface has been 5.1%, as opposed to 4.3% for the population. The average distance separating agglomerations has been divided by three, going from 111 Km to 33 Km for the whole region. There are five methodological approaches that can be used to define city-regions: labor market, housing market, economic activity-based, service district, and administrative. The chapter uses two cases from developed and developing regions to highlight the realities of city-region concepts. These cases are Norway and South Africa.

- **Labor-market** – is measured by journey-to-work.
- **Housing-market** – is measured by those households' search for residential locations (house prices move in tandem).
- **Economic activity-based** – is seen through businesses and business services maybe important in terms of the supply chains and procurement activities of firms.
- **Service district** – is highlighted by the central place theory where customers use the nearest service large number of places frequently used to provide services to smaller numbers of higher-order settlements, more specialized services.
- **Administrative** – administrative regions can be considered a subset of service district. Boundaries are formal and "artificial" – functional areas and strategies are developed within the area.

City-Regions in the European Periphery: The Case of Norway

Nationally, the emergence of city-regions in Norway is related to economic and demographic centralization, which has characterized regional development since the 1970s (Ellingsen and Leknes 2012). Thus, historically, regional policy has been a process of territorial bridging, securing the cohesion of the state by policies overcoming uneven regional development. Cities in Norway have grown significantly in terms of population, economy, and space. Therefore, the city-region concept has been a concept used to help counter the decline in rural areas; the state regional and rural policy-makers have responded with various measures of support, based on Norway's political objective of achieving balanced population development in all parts of the country. The dichotomy of center and periphery has pervaded regional politics and governments, which have focused on development in rural areas rather than promoting a specific industrial policy for urban areas.

The central role of the city is described as being a driver of national and regional economic development. Cities respond to business strategies by developing competitive advantages: centrality in relation to markets, infrastructure, institutional flexibility, and cultural and social variation. The question of how to organize the regional level between the state and the municipalities has been regularly debated in Norway since Parliament since the 1970s. Besides three democratically elected levels of government, the state also has administrative bodies at different regional levels, including the supervision of municipalities by the county governor and other sectors, such as state roads and health in regions that each consist of three to seven counties, creating a patchwork of partly overlapping regionalization (Ellingsen and Leknes 2012). However, the lack of coordination of Norwegian regional policy has resulted in spatial fragmentation of the regional level of government in Norway. The fragmentation of regionalization according to sectoral interests may arguably also reflect territorial politics of divide and rule by the state. The multiplicity of challenges for the city regions of Norway has triggered various responses within the governmental system. One such response is that the largest city regions are assigned an increasingly important role in the national economy (Ellingsen and Leknes 2012).

City-Regions in Latin America: The Case of Fortaleza

Fortaleza is the state capital of Ceará, located in Northeastern Brazil. It belongs to the metropolitan mesoregion of Fortaleza and microregion of Fortaleza, and it is Brazil's 5th largest city and the 12th richest city in the country in GDP. It also has the third richest metropolitan area in the North and Northeast regions. It is an important industrial and commercial center of Brazil, the nation's eighth largest municipal. The city-region concept has been dominant in Latin America due to population growth. The rapid growth of cities in Latin America has led to people and businesses settling in marginalized areas as they have cheaper and more affordable services than the cities. A key feature of urbanization in Latin America is that the rural-urban shift occurred in less than 40 years. This has led to the growth of city-regions in the continent. There has been the development of agglomeration regions such as Maranguape in Fortaleza due to urbanization. Maranguape is a large agglomeration with a population of over 90,000 people. 50,000 of the population resides in the capital. The area lies in the direct sphere of influence of Fortaleza, the metropolis of the Nordeste, and the fifth city in Brazil with a population of over two million. The agglomeration is located 24 km away from Fortaleza. It is a high migratory area as it is in the periphery dominated by a metropolitan center. Maranguape has its own political influence since it is the capital of an administrative district. It offers a number of services to its hinterland or neighboring rural population and acts as a hub for public transport. The importance of city-region was the relief of pressure of city center resources. In Brazil, agglomerations have been given their own administrative powers through decentralization. However, there are still relative constraints in planning for the destruction of centralities, the intense construction of policentralities. The worldwide expansion of metropolitan content is the ongoing challenge of the metropolization of space.

City-Regions in Southern Africa: The Case of Gauteng City-Region

In South Africa, the Gauteng city-region (GCR) is a relatively new concept, although the model has existed and grown for decades in other parts of the world. Integrated development plans have been the primary tool for local government integration of wall-to-wall local government in the 1990s. In the territorial areas of GCR, there are three metropolitan areas and six district municipalities; each district municipality has a number of local municipalities. In developing an integrated approach to the GCR, government has focused on key points of integration starting with transportation and to some extent human settlement at least at the level of a planning framework (Greenberg 2010). These key points include:

(i) Placing emphasis on growing the urban economy as a pragmatic condition for development
(ii) Recognizing that the global spatial location of economic growth increasingly focuses on cities and their immediate surrounding regions
(iii) Observing a new international hierarchy of urban regions competing for economic growth within which the existence of a globally competitive city-region is an important precondition for growth and development
(iv) Motivating that the Gauteng urban region meets the preconditions and characteristics of becoming a globally competitive city-region

The main aim and objective of GCR strategy is to build the Gauteng Province as an integrated region that is highly competitive and as an area where various economic activities that exist in different parts of the province complement each other. This is meant to cement the province as an economic hub of Africa and an internationally recognized global city-region. Urban debates are around the deeper questions of the location of the city-region concept where there exists a high rate of inequalities in the socioeconomic system of the Gauteng Province.

Conclusion and Future Direction

In conclusion, the reality of the city-region concept is that the concept is defined due to its various contexts and regions being applied around the world. The relationship between the region and the city is seen as a growing concern due to the growing populations in the city as well as rapid urbanization. There is need to create sustainable environments where city-regions can co-exist and interact as well as increase its effectiveness and economic efficiency. Creating sustainable environments is complex and relies on long-term and large-scale participation to understand more fully the opportunities and threats to the natural and human-built environment. Firstly, policy-making at the city-region level implies that there is a shift from sectoral to territorial approaches to development. Secondly, there is need for adjustment of the development strategies in many varying contexts so as to come up with greater and better policies that are diverse and innovative. Thirdly, coordination of various institutional sectors is pivotal in development at the city-region scale. Lastly, strategy development at subnational levels encourages the use of bottom-up and participatory approaches. The production of city boundaries fit for governance begins with area definition processes. Area definition should therefore be robust that is based on valid and reliable data about the region known across all areas as well as plausibility that boundaries broadly conform to expectations. Sustainable development is anchored by the triple bottom line of environmental conservation, economic prosperity, and social equity.

Mayors and city planners, local and regional, are now recognizing the social, economic, and environmental opportunities offered through the strengthening of city-region systems. These include:

- Localized production that is urban and peri-urban agriculture for food and income security at household level to reduce market distortions and reduce dependency on imported supplies, new enterprise, and marketing opportunities
- Entry point for awareness raising on health foods and lifestyles
- Resource recovery (urban waste) and climate change adaptation such as designating low-

lying areas and flood plains for agriculture to limit construction and reduce the impact of floods
- Reduced emissions related to food transport and food waste, thus lowering the urban footprint (Berdegué et al. 2014.

Community linkages can be described as reciprocal flows of goods and services, people, money, and environmental services between rural and urban locations. In a narrower sense that we believe to be more fruitful, rural-urban linkages are reciprocal and repetitive flows of goods and services, people, money, and environmental services between specific rural and urban locations that, and to a large extent as a result of those flows, become interdependent and constitute socio-spatial arrangements that we call territories, places with a socially constructed identity (Schejtman and Berdegué 2004).

Access to sustainable infrastructure is the key to the economic as well as social construct of the city-region. Infrastructure such as roads, telecommunications, and housing are vital in city-region development. According to Chirisa (2013), city margins that exist in Africa are extremely difficult to manage and organize, and this problem is escalated by housing manifestations that exist in those areas. Peri-urban areas in many developing regions are characterized by a mixture of planned and unplanned settlements, population growth, insecure land tenure, inadequate health problems, and social and poor service infrastructures. Access to services such as clean and reliable energy is difficult in developing countries, and this has been an ongoing issue in these regions. Developing nations strive to acquire consumer technologies and industrialized economies which have been constrained by various political and economic factors. Traditional methods of transforming stored energy into usable electricity are expensive, and it requires the combustion of fossil fuels. Over the years, research has suggested that climate change is a result of burning fossil fuels and release of toxic greenhouse gases into the Earth's atmosphere.

Grouping together both urban and rural areas sustains links as the regional areas provide carbon sequestration mechanisms which assist in the absorption of harmful gases in the atmosphere; at the same time, cities process clean energy for the regional areas, therefore feeding off each other. Furthermore, infrastructure projects should also be planned to account for transition risks. Infrastructure assets should be designed to be durable and flexible, allowing for easy reconfiguration, deconstruction, and recycling of project components. This also supports the development of circular economies. One of the most important components of the city-region concept in the cities is transport system. They are important as they slow the movement of people, goods, and services between the hinterland and the city. Sustainable transport systems are threatened when transport becomes inefficient, is perceived as unsafe, contributes toward a deteriorating air quality standard, and creates delays and bottlenecks for users inter alia. Urban transport systems in many cities of the developing countries exhibit these shortcomings. Effective transport systems increase a region's competitive advantage in the global market. The links between infrastructure and global development have been recognized for at least the last 25 years. At the same time, we rely upon diverse forms of infrastructure to deliver essential services and support our economies. Human well-being depends upon water and sanitation infrastructure, just as quality education and productivity depend on access to energy. Purposefully planned urban infrastructure including smart public transportation, green and energy-efficient buildings, as well as green spaces is vital to ensure that the world's fast-growing cities are in line with the 2030 Agenda (Weber et al. 2016).

Cross-References

▶ An Overview of the Relationship of the Sustainable Development Goals and Urban and Regional Development
▶ Behavioral Science Informed Governance for Urban and Regional Futures

References

Bayang, A.G.M. (2006). City-region concept body. http://rgdoi.net/10.13140/RG.2.1.2171.8800. 4 August 2020.

Berdegué, J.A., Proctor, F.J., & Cazuffi, C. (2014). *Inclusive rural–urban linkages*. Working paper series 123. Working Group: Development with Territorial Cohesion.

Chirisa, I. (2013). Housing and stewardship in peri-urban settlements in Zimbabwe: A case study of Ruwa and Epworth. http://rgdoi.net/10.13140/RG.2.2.16420.07042. 11 February 2020.

Christaller, W. (1933). *Central places in Southern Germany*, Jena: Fischer.

Coombes, M. (2014). From city-region concept to boundaries for governance: The English case. *Urban Studies, 51*(11), 2426–2443.

Creswell, J.W. (2014). *Research design qualitative, quantitative, and mixed methods approaches* (4th ed.). Thousand Oaks: SAGE Publications. www.sagepublications.com

Davoudi, S. (2008). Conceptions of the city-region: A critical review. *Proceedings of the Institution of Civil Engineers – Urban Design and Planning, 161*(2), 51–60.

Ellingsen, W., & Leknes, E. (2012). The city region as concept, object, and practice. *Norsk Geografisk Tidsskrift – Norwegian Journal of Geography, 66*(4), 227–236.

Geddes, P. (1915). *Cities in evolution: an introduction to the town planning movement and to the study of civics*. London, Williams.

Greenberg, S. (2010). *The political economy of the Gauteng city-region*. Johannesburg: Gauteng City-Region Observatory.

Hunt, D. V. L., & Rogers, C. D. F. (2014). Barriers to sustainable infrastructure in urban regeneration. *Engineering Sustainability*, 15(1), 4–11.

Hussein, K. & Suttie, D. 2016. *Rural-urban linkages and food systems in sub-Saharan Africa: The rural dimension*. Rome: IFAD.

Kitchin, R., & Thrift, N. J. (Eds.). (2009). *International encyclopedia of human geography* (1st ed.). Amsterdam: Elsevier.

Luo, X., & Shen, J. (2008). Why city-region planning does not work well in China: The case of Suzhou–Wuxi–Changzhou. *Cities, 25*(4), 207–217.

Marrazzo, W.J. 1996. The challenge of sustainable infrastructure development. *Environmental Progress, 15*(3): F3–F3.

Meidani, H., & Ghanem, R. (2015). Random Markov decision processes for sustainable infrastructure systems. *Structure and Infrastructure Engineering, 11*(5), 655–667.

Munyasya, B., & Chileshe, N. (2018). Towards sustainable infrastructure development: Drivers, barriers, strategies, and coping mechanisms. *Sustainability, 10*(12), 4341.

Myrdal, G. (1957). *Economic theory and underdeveloped regions*. London: Duckworth.

Pain, K., Harrison, J., Johnson, C., Kenworthy, J., Black, J., & Calder, J. (2008). *City–regions and economic development*. Metropolis Sydney.

Pascariu, S., & Czischke, D. (2015). Promoting urban-rural linkages in small and medium sized cities. Available online: https://urbact.eu/sites/default/files/synthesis_report_urbact_study.pdf

Qi, L., Zeng, F., & Chen, L. (2019). Research on world city evaluation system based on the concept of regional space flow. *IOP Conference Series: Earth and Environmental Science, 267*, 052057.

Rodríguez-Pose, A. (2008). The rise of the "city-region" concept and its development policy implications. *European Planning Studies, 16*(8), 1025–1046.

Schejtman, A., & Berdegué, J. A. (2004). Rural territorial development. Working paper/Rural Territorial Dynamics Program. RIMISP-Latin American Centre for Rural Development; no. 4. Programa Dinámicas Territoriales Rurales. Rimisp, Santiago, Chile.

Schmitt, P., Volgmann, K., Münter, A., & Reardon, M. (2015). Unpacking polycentricity at the city-regional scale: Insights from Dusseldorf and Stockholm. *European Journal of Spatial Development, 59*(2), 1–26.

Scott, A.J. (1998). Global city-regions and the new world system. Available online: http://citeseerx.ist.psu.edu/viewdoc/download?doi=10.1.1.547.689&rep=rep1&type=pdf

Scott, A. J. (2008). *Social economy of the metropolis*. Oxford University Press. http://www.oxfordscholarship.com/view/10.1093/acprof:oso/9780199549306.001.0001/acprof-9780199549306. 4 August 2020.

SDG Knowledge Hub. Capacity-building projects enhance urban, rural resilience. http://sdg.iisd.org/news/capacity-building-projects-enhance-urban-rural-resilience/. 31 July 2020.

Stepanova, N., Gritsenko, D., Gavrilyeva, T., & Belokur, A. (2020). Sustainable development in sparsely populated territories: Case of the Russian Arctic and Far East. *Sustainability, 12*(6), 2367.

Uddin, M. S., Routray, J. K., & Warnitchai, P. (2019). Systems thinking approach for resilient critical infrastructures in urban disaster management and sustainable development. In E. Noroozinejad Farsangi, I. Takewaki, T. Y. Yang, A. Astaneh-Asl, & P. Gardoni (Eds.), *Resilient structures and infrastructure* (pp. 379–415). Singapore: Springer Singapore. http://link.springer.com/10.1007/978-981-13-7446-3_15. 4 August 2020.

Weber, R., Tammi, I., Anderson, T., & Wang, S. (2016). A spatial analysis of city-regions: Urban form & service accessibility, Nordregio working paper 2016:2. Available online: https://www.diva-portal.org/smash/get/diva2:933727/FULLTEXT01.pdf

Yigitcanlar, T., & Dur, F. (2010). Developing a sustainability assessment model: The sustainable infrastructure, land-use, environment and transport model. *Sustainability, 2*(1), 321–340.

Future of Urban Governance and Citizen Participation

George Makunde[1], Valeria Muvavarirwa[2] and Innocent Chirisa[3]
[1]George Makunde Institute, Harare, Zimbabwe
[2]Department of Demography Settlement and Development, Social & Behavioral Sciences, University of Zimbabwe, Harare, Zimbabwe
[3]Department of Demography Settlement and Development, Social & Behavioural Sciences, University of Zimbabwe, Harare, Zimbabwe

Definition

Citizenship – state of being a participating urban dweller
Governance – processes and institutions defining the affairs an area.

Introduction

This chapter examines salient issues that impact governance and citizen participation with the view to streamline some generic approach to urban governance attached to sustainable service delivery. Indeed, the future of urban governance and citizen participation has taken center stage during the turn of a decade or two resulting from a number of global events that have seen so many changes coming in as a result of globalization (Elson 2019; Amavilah et al. 2017). The fact that more people have been able to access education throughout the globe has resulted in those educated being able to know and demand enablers to better life (Kelly 2019; Li et al. 2020). The world has grown into a global village.

In this global village, there is serious interaction and integration among people, institutions, organizations, and countries which has become the new phenomenon. To this end, the interactions and engagements have resulted in open discussion let alone debates on matters to do with better education, more stable jobs, stable economies, more income, reliable social security, better medical and health care, improved housing, good water and sanitation, and above all a better and beautiful environment (Hidayati 2017; Hussain and Arch 2020). These are the pillars that enhance judgment/measurement of whether living standards of a particular country, society, town, or village meet the recommended standards world over. The lack of any one of the above poses problems to governance which may result in irreparable repercussions as people demand what they feel they are worth. Therefore, it is of essence to posit that any governance system derives its authority to govern from the ethos of the politics of the ruling party.

This entails following through a specific and chartered route map, parameters, principles, and economic framework chosen and settled for, which may be classified as capitalism or socialism. These two pillars of development will give the path, theory, and system that define governance and development. In short, the path allows one to reach a goal, and the theory will guide to act while the system provides fundamentals for success (Wang and Fei 2019; Hofmann 2017). These three will also act as determinants of who plays what role, why, when, and how in the whole equation which then brings the component of citizen participation to the fore. It is this part of participation that has seen a lot of countries and institutions including towns and rural areas becoming ungovernable as those governed demand what they view as worth their while. The aim is not solely for the current generation to be happy, but they demand and want guarantee for future generations to have sound growth, good jobs, and enjoyable lives. That guarantee only arises from economic development which fulfils fundamental guarantee to success (Spaiser et al. 2017; Bashir 2018).

Conceptualizing Urban Governance and Citizen Participation

Defining governance in a global village poses a lot of serious questions emanating from the heterogeneity of the sphere that we call the globe. It brings in several issues that hinge on levels and demarcations and classifications of nations, societies, economies, politics, and social networks in the global village. For purposes of this chapter, urban governance shall be defined in the realm of how individuals and institutions, both public and private, manage and plan the common programs of their town. This definition requires one to look at the decision-making process and how such or any decision arrived at is then implemented and managed for the good of the whole.

Theories arising from the definition include rapid urban growth, local economic development, infrastructure and service provision, settlements and spatial planning, urban resilience, gender, social policy, social exclusion, and urban poverty engendering equality versus equity (Singh 2016; Cobbinah and Erdiaw-Kwasie 2018; Narayanaswamy 2020). All the above will then fall into the expansionist theory of urban governance resulting in the need to introduce the component of political economy as a vehicle to enhance the pinnacle of urban governance and citizen participation. Political economy will then enhance the definition by bringing the set of criteria that determine how the nexus of urban dwelling is governed or run on a daily basis (Atkinson and Fulton 2017; Ting 2018).

This idea then brings in the relational political and administrative conduit between the state and the different groups in society as they dwell in an urban scenario. Therefore, it is critical to underline that the success or failure of urban governance and citizen participation hinges on the capacity to integrate and give form to local interests, both public and private, and reinforce local identity. To this end there is no need to overemphasize the relationship between the civil society and the state, between the rulers and the ruled, and finally the government and the governed which at the end shapes the aspect of citizen participation in the governing structures of cities (da Cruz et al. 2019; Foo 2018).

The political economy takes the effect of combining economy, society, and politics as we grapple to understand the concept of urban governance attached, of course, to the participation of the beneficiaries of the outcomes of governance who in this case are the citizens (Parkhurst 2017; Cheema 2020). Franklin Obeng-Odoom in *Governance for Pro-Poor Urban Development: Lessons from Ghana* (2013, p. 39), citing David Harvey, Manuel Castells, and David Drakakis-Smith in their works (Harvey, *Analysis of Economic Interests Shaping Cities* (1973); Castells, "Analysis of State Intervention, Class, Power and Cities"; and Drakakis-Smith, "Analysis of Colonialism and Cities"), made the assertion that:

> Urban economics approach for urban analysis is less appropriate for research in Africa than an alternative political economic approach that emphasizes class, production relationships, capital accumulation, exploitation and conflict, mode of production, the state and collective consumption, inequality and the relationship between the urban crisis.

Therefore, it is imperative for the discourse of urban governance to be looked at from an interdisciplinary perspective to unearth the connectivity of economy, politics, society, and culture in the running of cities in Africa. In that way one will be able to unpack, examine, critique, and evaluate urban governance to the full. This is best defined as political economy inseparable from the governance of any settlement.

Urban Governance and Citizen Participation

The local government system in Zimbabwe derives its operational definition and mandate from the Constitution of Zimbabwe where according to Makunde et al. (2018 , p. 19) urban local government system consists of four types of local authorities, namely, local boards, town councils, municipalities, and cities. The above is in terms of Clause 274 of the Constitution of Zimbabwe Amendment (No. 20), which states that "there are urban local authorities to represent and manage the affairs of people in urban areas throughout Zimbabwe." It goes further to qualify the demarcation as follows: "different classes of local authorities may be established for different urban areas, and two or more different areas may be placed under the management of a single local authority." Clause 264 (Devolution of governmental powers and responsibilities) of the Constitution of Zimbabwe also brings in a critical component of devolution which governs the running and funding of councils in Zimbabwe. It states that:

> . . .whenever appropriate, governmental powers and responsibilities must be devolved to provincial and metropolitan councils and local authorities which are competent to carry out those responsibilities efficiently and effectively'. Further to this the preamble to Chapter 14 of the constitution categorically introduces the democratization of the urban governance system in Zimbabwe by stating that, 'whereas it is desirable to ensure: the democratic participation in government by all citizens and communities of Zimbabwe' and 'the equitable allocation of national resources and the participation of local communities in the determination of development priorities within their areas.

The above citations are meant to illustrate the urban governance scenario in Zimbabwe with the view to introduce insights into how urban governance and citizen participation are practiced in Zimbabwe and giving impetus to the futuristic view we need to propose in terms of this chapter.

The current situation in Zimbabwe in terms of Clause 264 as read with Clause 274 juxtaposed to the enabling act, i.e., the Urban Councils Act [Chapter 29:15], has not been soundly depicting the situation and aspirations of the people, hence several outcries of the lack of appropriate service delivery being experienced across all urban councils (Marumahoko et al. 2020; Makunde et al. 2018).

The situastion in Africa, in general, and Zimbabwe, in particular, shows serious gaps in meeting the constitutional provisions which were very well intended on the onset. The current hashtag this and hashtag that are an indication of discord within the resource distribution systems which have somewhat left the owners of urban governments struggling to enjoy the fruits of their voting franchise they off loaded to the selected few councilors and other leadership (Ronay et al. 2020; Shaw 2018). This has come from the class nature of the state intervention in its or their bid to curb the inadequacies of a market economy. The intention is to give meaning to the collective consumption of urban governance in the form of transport, land, employment opportunities, water, sanitation, waste management, housing, and above all economic development.

Social justice has been thrown out of the fray as no clue is seen in economic interests/policies being generated to shape the cities that urban dwellers want. Demonstrations for services in Cape Town, Johannesburg in South Africa, and downing of tools by doctors and nurses in Zimbabwe have fast become a normal (Bateman 2016; Valiani 2019). Illicit land deals in the cities and urban settlements in Zimbabwe have been the order of the day further negatively impacting on the provision of social amenities and other ancillary needs that fit into the social contract (Chiweshe and Mutondoro 2017; Chiweshe 2017). Those with power and close to power become winners, and they elbow the weak out who unfortunately are the neediest when it comes to the barest minimum needs in life, i.e., water, housing, health services, etc.

This is a sign that certain aspirations of the people who are part of the social contract have not been met. Questions that do arise include the following. How representative are the people thrust in positions of power to further the interests of urban dwellers? How far are the ordinary intended beneficiaries of urban governance policies participating in the policy formulation and decision-making process? How democratic are the mechanisms for recourse in place to ensure remedial action is taken across the globe?

Synthesis

The subject of urban governance and citizen participation has seen a number of emerging issues chief among them being gender mainstreaming, equality and equity and unplanned settlements, urban resilience, climate change, transformation, and change management coming up.

Gender Mainstreaming

Gender mainstreaming encompasses the incorporation of both males and females in decision-making positions and providing access to participation. Females have a history of being ignored and left out in decision-making and have slowly gained entry into decision-making arenas. The absence of women in certain environments means that their voices are not being heard (Africa Development Bank Group 2009). For instance, on the issue of service delivery, females, women and girls, are the most affected in the water sector as they are the ones who fetch water for the family. Without a platform for them to offer their thoughts on the issue, they may tend to demonstrations which are a clear indication of inefficiency in service delivery (Zhu and Peyrache 2017; Moono 2017).

Gender mainstreaming does not just focus on putting items in writing; there is also the issue of following through, making sure that women are being considered in decision-making and are given a platform to participate. When the Zimbabwe Women's Bank was opened, did it really serve its purpose in providing loans to women's businesses? There is need for accountability in dealing with issues of recognizing a sector. Women participation in the economic arena gives a rise in the economic development of a family and the country. Steps have been taken to mainstream women; a women's bank was established; the quota system was written down in the rules and regulation. The question now is has this been effective and been adhered to and what alterations can be made to make the process and the outcome better?

Equality and Equity

Equality and equity are vital in good governance (Walks 2017; Bollens 2018; Carstensen and Ibsen 2019). They propose that people should be treated the same, without exceptions. This relates to the provision of services, where people in high-density areas and those in low-density areas should receive adequate services without discrimination. There is need of fairness in order for there to be good governance (Gorwa 2018; Ernst 2019). Without a sense of fairness, the underprivileged people will find it necessary to protest against the way they are being treated. The unfair treatment leads to the disruption of society and the economy. In the absence of equity and equality, the poorer people find it hard to comply with the rules and try to circumvent them. This leads to the growing informal and illegal sectors such as the transport issue. There are illegal commuter omnibuses that are ferrying people. People are also making illegal electricity and water connections. These are all ways of the people trying to dismiss the regulations imposed on them that they deem unfair. The lack of consistent policies and regulations on how to deal with these issues is of growing concern.

Developing countries are plagued with the concern of inequalities, not just spatial inequalities but inequalities resulting from discrimination of other groups of people. Unequal treatment of people divides the output. The lack of universality prevents participation of the whole and only allows a segment to participate and function (Kingston and Stam 2017; Swapan 2016). Barriers installed due to gender, wealth, and disability exclude other groups of the population from being considered in the development of economies. This perpetuates the growth of poverty and for some, a barrier from escaping poverty traps (Gerber 2004).

There is also the issue of inequality on economic opportunities (Egbulonu and Eleonu 2018; Ward 2017). The opening up of economic opportunities is crucial and should benefit the whole population with entry allowed to anyone willing and not just the wealthy. The barriers to economic opportunities see the rich accumulating more wealth while the poor remain poor. This negatively affects the country's GDP and only benefits the rich. The informal sector is also part of the income-generating machine and should be considered as such. Mechanisms that allow the participation of the informal sector in the issues to do with the national economy need to be inclusive and support the informal industry (Gerber 2004).

The issue of equality is three-pronged, with the first sector being equal life chance where there should be equal access despite differences in factors that cannot be helped such as gender, race, or disability. The other aspect touches on equal concern for needs which encompasses some goods and services that are a basic need and should be accessible to everyone and not just the wealthy. The third prong is growth according to merit, where anyone is eligible for growth in a sector or a promotion due to merit without applying any discriminatory factors. However, this day and age is plagued with corruption and imbalances that do not allow fair treatment and progression of people.

Climate Change

In order to combat climate change that has become a global issue, there is need for a universal definition of the concept. This then calls into question how the whole world is working towards the same goal. The initiative is upon the government to come up with different methods of implementing the regulations that are set up at an international level (Mees et al. 2019; Michaelowa and Michaelowa 2017). The government is responsible for enacting laws that help in the vision to combat climate change (Tanner et al. 2009). Zimbabwe has enacted policies that stop people from cutting down trees and stopping emission of toxic substances into the air and into water sources. It is vital for the government to ensure that the policies are being adhered to and if not, what is being done. The question is now that do wealthy industries find it easier to just pay the fine rather than adhere to the policies, if so, then what can then be done? (Filho 2010).

The government creation of measures to prevent climate change should ensure good governance. Community adaptation to climate change measures hinges upon the way the government establishes the measures (Fraser and Kirbyshire 2017). The initiatives that work as preventive measures need to encourage participation for their success because the people know what they want and have ideas that can be helpful to the government and make sure there is good governance.

Unplanned Settlements

Unplanned settlements are a symptom of poor governance, as it reflects that there is a missing link between the government and the governed (Suhartini and Jones 2019; Wolff et al. 2018). Why do unplanned settlements form? This is usually due to shortage of housing delivery to the people and also the hikes in rent prices that leave people homeless and desperate leading to the creation of illegal settlements. These sites then become hubs for a lot of criminal activity and other illegal activities. These areas are difficult to govern because of difficulty of establishing jurisdiction. If there is an issue of jurisdiction, it becomes difficult to establish who controls what and how in the illegal settlements (Tomar et al. 2017; Weber 2017). The question then becomes who made it possible for people to settle at a location, working out a blame game, whereas the issue should then be why did the people settle here and work out solutions. There is the concern of solving the wrong problem with the wrong solutions as well. This can even exacerbate the living conditions of the urban poor (Nabutola 2007).

It is a basic right to have adequate shelter, which is being infringed upon, and the people take matters into their own hands. This leads to the government taking drastic measures such as demolitions which do not always go so well with the people. This creates an atmosphere where the people do not respect the government because they feel betrayed and harassed, thereby creating a hostile community, which even police men and women are afraid to enter, even if it is for the benefit of the residences of that area. So, the question is who governs those areas that do not want to recognize the police and the government?

Unplanned settlements have no clear pattern as the structure appear organically. This makes the areas un-aesthetical because of sprouting structures and land uses (Magina et al. 2020; Sarkar 2017). This makes the land hard to maneuver, especially when putting infrastructure as upgrading and regularization through servicing the land. However, there is an issue to consider what then can be done to integrate the unplanned settlement into the running city. The responsibility of the upgrading is also another vital issue. Will it be the government or non-benefit organizations? People will demand accountability for actions and progress of the programs (Nabutola 2007).

Urban Resilience

Resilience refers to the ability to adapt to change and remain in functioning condition. There have been numerous talks on encouraging resilient structures (Sharifi 2019; Rahman 2016). This also translates to urban resilience, where the people and their environment are expected to be resilient. The importance of resilience is to ensure the sustainability of cities and their durability. It is vital to create and upgrade cities to be resilient in order to be able to absorb and adjust to any shocks that may hit the urban areas. This can be in the form of natural or economic hits. The people need to feel and actually be heard in deciding on the growth and development of urban areas. However, this is seen as a technical issue that needs experts for the whole process, thereby leaving out the people and blocking any form of participation they may have. This creates areas that are best in theory but hold no regard to the people who did not choose them. The issue is then on what needs to be done in order to incorporate urban resilience in development (Dhakal 2018).

In developing resilient cities that are sustainable and can withstand, absorb, and adapt to natural, economic, and man-made shocks and strains, there is need for good governance. The decision-making in the process of developing resilient countries hinges upon adept, transparent, and confident thinking. The success of urban resilience lies within the success of good governance. The government cannot do this alone and therefore asks other stakeholders for their input (Berkowitz and Kramer 2018; Staddon et al. 2018). The concept of resilience is easily affected by a variety of factors (Meyer and Auricombe 2019). This includes the economic situation of a country.

If the country has a stable and thriving economy, then the development of its features into resilient structures and functions is high and affordable, whereas a struggling economy faces more challenges. The issue is then on deciding when a country is stable enough to initiate such developments and estimating how they will be received. In its success, urban resilience encourages sustainability (Bedi et al. 2014). This becomes difficult for countries that have dwindling economies that are suffering, for example, Zimbabwe, a largely agricultural-based economy. The economy is suffering because of the change in rainfall patterns that has affected the seasons and the harvest. There have also been technological advancements in other parts of the world that are making it difficult to remain rooted and dependent on primary activities. Other countries have moved on to tertiary services and value addition.

Management

Another essential element for there to be good governance is the management. The leadership of the government and various other entities responsible for development and service delivery is crucial for the development of cities, nations, and the whole world. Due to globalization, there have been ever-growing technological advancements (Malik and Kapuria 2020). These have made it necessary for everyone to be part and parcel of the technical world. The management of development issues aligns itself and works with the technology on demand. However, countries like Zimbabwe are always lagging behind when dealing with improvements in technology. This means the country is left behind in development and always has to play catch up with the rest of the world.

The resistance to change of those in power in developing countries also creates friction in governance. The people want to feel confident in who is leading them, and they want to see results; however, the older generations that have power in African countries do not want to pave way for new and younger minds to pick up the mantle and further develop the world. With such resistance, the people are pushing back in the form of rallies and demonstrations that have seen an increase in violence and vandalism that is not good for anyone. Conflict management skills are therefore essential to have, as the way that the situations are handled either appeases the people or further enrages them (Safeena and Velnampy 2017; Fisher et al. 2020).

Good governance is also based on the ability of those in power to accept accountability of the actions that they have taken. It does not just mean those in governmental power but also those in the private sector, nongovernmental organizations, and civil society organizations. Accountability in the form of reports provides a clear picture of what is being done and by whom. This should be the order of the day not to wait until the public cries out that there are injustice and corruption that are taking place. This will be reactive and defeats the purpose of open and honest interactions between the government and those in power with the general population. All these issues are centered on elements of good governance such as transparency, accountability, and participation. The pillars pave the way for all other facets to follow. The application of the elements makes it possible to be a good and working relationship between the governed and the governors.

Governance is multifaceted and does not present one issue at a time but is bombarded by varying aspects. These usually have a ripple effect on other issues directly or indirectly. An issue on climate change-resilient infrastructure cannot be looked upon in isolation but includes other fields and requires a multi-sectorial approach. The wide range of stakeholders involved in sorting out these issues makes good governance a complex aspect that does not just require participation but also accountability (Nachbaur et al. 2020).

Conclusion and Future Direction

The foregoing narratives are a sign of serious gaps that reflect on how governments have somewhat let down the ordinary population in their quest to enjoy the fruits intended by their respective constitutions world over. It is high time that all nations go back to the principles of political economy and apply them for redress to be achieved. The State that the Aristotles and Platos needs to be revisited in order for policy placements and realignment be faithfully applied for the good of urban governance to be felt. For countries like Zimbabwe, the full application of and adherence to the provisions of the constitution is the route to go. Devolution in its full meaning calls those in the echelons of power and authority to apply it to the maximum. With several hashtags becoming a new normal globally with the resultant incessant fights and conflicts happening daily, it is glaringly prudent that nations apply transformational theories and policies to disentangle conflicting players in the urban governance and citizen participation arena. Makunde et al. (2018 , p. 48) defines transformation as "...the movement (changing) from a previous state to a desired state of affairs." The policies and regulations put in place for the inclusion of disadvantaged groups and fines need regular monitoring and evaluation, so as to evaluate the accountability of the management so that the rules and regulations are not just a milestone but are being practiced (Jones 2009).

Cross-References

▶ Behavioral Science Informed Governance for Urban and Regional Futures

References

Africa Development Bank Group. (2009). *Checklist for gender mainstreaming in governance programs*. https://www.afdb.org/en/documents/document/checklist-for-gender-mainstreaming-in-governance-programmes-20012

Amavilah, V., Asongu, S. A., & Andrés, A. R. (2017). Effects of globalization on peace and stability: Implications for governance and the knowledge economy of African countries. *Technological Forecasting and Social Change, 122*, 91–103.

Atkinson, M. M., & Fulton, M. (2017, June). The political economy of good governance. In *ICPP3 conference* (Vol. 32).

Bashir, S. (2018). *Analytical study of community development programs for socio-economic development in Pakistan*. Karachi: University of Karachi.

Bateman, C. (2016). PE hospital turmoil: CEO leaves, nurses snore in patient beds. *SAMJ: South African Medical Journal, 106*(4), 320–320.

Bedi, N., Bishop, M., Hawkins, U., Miller, O., Pedraza, R., Preble, A., & Rico-Rairan, A. (2014). Linking resilience and good governance: A literature review. *Anthós, 6*(1), Article 3. https://doi.org/10.15760/anthos.2014.15.

Berkowitz, M., & Kramer, A. M. (2018). Helping cities drive transformation: The 100 Resilient Cities Initiative. Interviews with Michael Berkowitz, president of 100 Resilient Cities, and Dr. Arnoldo Matus Kramer, Mexico City's Chief Resilience Officer. *Field Actions Science Reports: The Journal of Field Actions, 18*(Special Issue), 52–57.

Bollens, S. A. (2018). Inequality and governance in the metropolis: Place equality regimes and fiscal choices in eleven countries. *Public Administration Review, 78*(6), 931–933.

Carstensen, M. B., & Ibsen, C. L. (2019). Three dimensions of institutional contention: Efficiency, equality and governance in Danish vocational education and training reform. *Socio-Economic Review*, 1–27. https://doi.org/10.1093/ser/mwz012.

Cheema, S. (2020). Governance for Urban Services: Towards political and social inclusion in cities. In: Salim, W and Drenth, M (eds.) Governance for Urban Services (pp. 1–30). Singapore: Springer

Chiweshe, M. K. (2017). Analysis of land-related corruption in Zimbabwe. *Africa Insight, 46*(4), 112–124.

Chiweshe, M. K., & Mutondoro, F. S. (2017). Political corruption and the post 2018 agenda in Zimbabwe. *Journal of Public Administration and Development Alternatives (JPADA), 2*(2), 34–46.

Cobbinah, P. B., & Erdiaw-Kwasie, M. O. (2018). Urbanization in Ghana: Insights and implications for urban governance. In *E-planning and collaboration: Concepts, methodologies, tools, and applications* (pp. 256–278). London: IGI Global.

da Cruz, N. F., Rode, P., & McQuarrie, M. (2019). New urban governance: A review of current themes and future priorities. *Journal of Urban Affairs, 41*(1), 1–19.

Dhakal, S. (2018). The future of urban governance and capacities for resilient cities. *The Asia-Pacific Cities Report*, 2019.

Egbulonu, K. G., & Eleonu, I. S. (2018). Gender inequality and economic growth in Nigeria (1990–2016). *International Journal of Gender and Women's Studies, 6*(1), 159–167.

Elson, A. (2019). The governance of globalization: National and international dimensions. In *The United States in the world economy* (pp. 173–201). Cham: Palgrave Macmillan.

Ernst, A. (2019). How participation influences the perception of fairness, efficiency and effectiveness in environmental governance: An empirical analysis. *Journal of Environmental Management, 238*, 368–381.

Filho, W. L. (2010). Climate change and governance: State of affairs and actions needed. *International Journal of Global Warming, 2*(2), 128.

Fisher, J., Stutzman, H., Vedoveto, M., Delgado, D., Rivero, R., Quertehuari Dariquebe, W., ... Rhee, S. (2020). Collaborative governance and conflict management: Lessons learned and good practices from a case study in the Amazon Basin. *Society & Natural Resources, 33*(4), 538–553.

Foo, K. (2018). Examining the role of NGOs in urban environmental governance. *Cities, 77*, 67–72.

Fraser, A., & Kirbyshire, A. (2017). *Supporting governance for climate resilience: Working with political institutions*. London: Overseas Development Institute.

Gerber, M. (2004). *Promoting equality, including social equity, gender equality and women's empowerment*. 8th Session of the Open Working Group on Sustainable Development Goals, New York, 3–7 Feb 2014.

Gorwa, R. (2018). Towards fairness, accountability, and transparency in platform governance. *AoIR Selected Papers of Internet Research*.

Hidayati, F. (2017). Can decentralization affect public service delivery? A preliminay study of local government's innovation and responsiveness in Indonesia. *JPAS: Journal of Public Administration Studies, 2*(1), 80–86.

Hofmann, B. (2017). Saving liberalism: Governance through Global Hanses. St. Gallen Symposium.

Hussain, S. M., & Arch, B. (2020). Facets of public private debate and discourse of urban service delivery-A case of Juhapura, Ahmedabad.

Jones, H. (2009) *Equity in development: Why it is important and how to achieve it*. https://www.odi.org/sites/odi.org.uk/files/odi-assets/publications-opinion-files/4577.pdf.

Kelly, L. (2019). Barriers and enablers for women's participation in governance in Nigeria. https://opendocs.ids.ac.uk/opendocs/bitstream/handle/20.500.12413/14588/596_Nigerian_Women_Governance.pdf?sequence=1.

Kingston, L. N., & Stam, K. R. (2017). Recovering from statelessness: Resettled Bhutanese-Nepali and Karen refugees reflect on the lack of legal nationality. *Journal of Human Rights, 16*(4), 389–406.

Li, L., Collins, A. M., Cheshmehzangi, A., & Chan, F. K. S. (2020). Identifying enablers and barriers to the implementation of the Green Infrastructure for urban flood management: A comparative analysis of the UK and China. *Urban Forestry & Urban Greening, 126770*.

Magina, F. B., Kyessi, A., & Kombe, W. (2020). The urban land Nexus–challenges and opportunities of regularising informal settlements. *Journal of African Real Estate Research, 5*(1), 32–54.

Makunde, G., Mazorodze, C., Chirisa, I., Matamanda, A., & Pfukwa, C. (2018). Local governance system and the urban service delivery in Zimbabwe: Issues, practices and scope. *International Journal of Technology Management, 3*(1), 13–13.

Malik, S., & Kapuria, C. (2020). The globalization and governance Nexus-evidence from emerging Asia. *IUP Journal of Applied Economics, 19*(1), 7–27.

Marumahoko, S., Afolabi, O. S., Sadie, Y., & Nhede, N. T. (2020). Governance and urban service delivery in Zimbabwe. *Strategic Review for Southern Africa, 42* (1), 41–68.

Mees, H. L., Uittenbroek, C. J., Hegger, D. L., & Driessen, P. P. (2019). From citizen participation to government participation: An exploration of the roles of local governments in community initiatives for climate change adaptation in the Netherlands. *Environmental Policy and Governance, 29*(3), 198–208.

Meyer, N., & Auricombe, C. (2019). Good urban governance and city resilience: An afrocentric approach to sustainable development. *Sustainability, 2019*(11), 5514. https://doi.org/10.3390/su11195514.

Michaelowa, K., & Michaelowa, A. (2017). Transnational climate governance initiatives: Designed for effective

climate change mitigation? *International Interactions, 43*(1), 129–155.
Moono, H. (2017). Exorcising government inefficiency9 through e-systems. *International Growth Centre Blog.*
Nabutola, W. (2007) *Governance issues in informal settlements.* TS 8D – SDI towards Legalizing Informal Development.
Nachbaur, J., Feygina, I., Lipkowitz, E., & Karwat, D. (2020). *Climate change resilience: Governance and reforms. A report.* CPSO Arizona State University.
Narayanaswamy, L. (2020). *Engendering transformative change in international development: by Gillian Fletcher.* Abingdon/New York: Routledge, 2019, 186pp.,£ 115.00 (hbk),£ 39.00 (ebook), ISBN: 9781138575332.
Parkhurst, J. (2017). *The politics of evidence: From evidence-based policy to the good governance of evidence* (p. 182). Chicago: Taylor & Francis.
Rahman, S. (2016). *Encouraging women's contribution to resilient cities* (No. 103270, pp. 1–2). New York: The World Bank.
Ronay, R., Maddux, W. W., & von Hippel, W. (2020). Inequality rules: Resource distribution and the evolution of dominance-and prestige-based leadership. *The Leadership Quarterly, 31*(2), 101246.
Safeena, M. G. H., & Velnampy, T. (2017). Factors influencing integrating conflict management. *Journal of Business Studies, 4*(1), 91–103.
Sarkar, S. (2017). *Conflict ecologies: Gender, genre, and environment in narratives of violent conflict in post-colonial India* (Doctoral dissertation, The George Washington University).
Sharifi, A. (2019). Urban form resilience: A meso-scale analysis. *Cities, 93*, 238–252.
Shaw, E. (2018). 'What dire effects from civil discord flow': Party management and legitimacy breakdown in the Labour Party. In *Labour united and divided from the 1830s to the present.* New Dehli: Manchester University Press.
Singh, A. K. (2016). Engendering budgeting and gender inclusive urban governance in India. *Anveshana, 6*(1), 36–59.
Spaiser, V., Ranganathan, S., Swain, R. B., & Sumpter, D. J. (2017). The sustainable development oxymoron: Quantifying and modelling the incompatibility of sustainable development goals. *International Journal of Sustainable Development and World Ecology, 24*(6), 457–470.
Staddon, C., Ward, S., De Vito, L., Zuniga-Teran, A., Gerlak, A. K., Schoeman, Y., … Booth, G. (2018). Contributions of green infrastructure to enhancing urban resilience. *Environment Systems and Decisions, 38*(3), 330–338.
Suhartini, N., & Jones, P. (2019). Who gains and benefits from the outcomes of formal urban governance for basic urban services. In *Urban governance and informal settlements* (pp. 115–161). Cham: Springer.
Swapan, M. S. H. (2016). Who participates and who doesn't? Adapting community participation model for developing countries. *Cities, 53*, 70–77.
Tanner, T., Mitchell, T., Polack, E., & Guenther, B. (2009). *Urban governance for adaptation: assessing climate change resilience in ten Asian cities. IDS Working Papers, 2009*(315), 01–47. https://onlinelibrary.wiley.com/doi/pdfdirect/10.1111/j.2040-0209.2009.00315_2.x.
Ting, M. M. (2018). *The political economy of governance quality.* Working Paper, Columbia University.
Tomar, S., Kaur, A., Dangi, H. K., Ghawana, T., & Sarma, K. (2017). Fire risk analysis using geospatial approach and mitigation measures for South-West Delhi. *International Journal of Emerging Research in Management and Technology, 8*, 131–137.
Valiani, S. (2019). Public health care spending in South Africa and the impact on nurses: 25 years of democracy? *Agenda, 33*(4), 67–78.
Walks, R. A. (2017). Metropolitanization, urban governance, and place (in) equality in Canadian metropolitan areas. In *Inequality and governance in the metropolis* (pp. 79–106). London: Palgrave Macmillan.
Wang, L., & Fei, Z. (2019, March). The relationship between the core values of socialism in the new era and the finance curriculum. In *2018 8th International Conference on Education and Management (ICEM 2018).* Atlantis Press.
Ward, T. (2017). Inequality and growth: Reviewing the economic and social impacts. *The Australian Economic Review, 50*(1), 32–51.
Weber, B. (2017). Addressing informal settlement growth in Namibia. *Namibian Journal of Environment, 1*, B-26.
Wolff, S. M., Kuch, A., & Chipman, J. (2018). *Urban land governance in Dar es Salaam.* C-40412-TZA-1. IGC Working Paper.
Zhu, M., & Peyrache, A. (2017). The quality and efficiency of public service delivery in the UK and China. *Regional Studies, 51*(2), 285–296.

Future of Urban Land-use Planning in the Quest for Local Economic Development

Tinashe Bobo[1] and Innocent Chirisa[2]
[1]Town Planning Section, Harare City Council, Harare, Zimbabwe
[2]Department of Demography Settlement and Development, Social & Behavioural Sciences, University of Zimbabwe, Harare, Zimbabwe

Synonyms

Land-use – land function; Development – expansion/growth; Planning – organization

Definition

Urbanity – the state of a place or territory being non-rural, and the economy within it being largely non-agricultural

Land-Use Planning: the organization of land so that functions and use are clearly demarcated

Local Economic Development: the structuring of economic development at a local level such that processes and institutions that define such an economy embrace largely the endogenous initiatives

Introduction

The central philosophical perspective of urban land-use planning relates to the modes of intervention in the public domain and the theoretical and epistemological frames of reference among urban planners and urban planning organizations. Urban planners throughout the world are deeply concerned with environmental sustainability and equity and also the profession of urban planning is highly regulatory and technical in practice. The major orientation being on land-use planning that speaks to processes by that land is allocated between competing and sometimes conflicting uses in order to secure the rational and orderly development of land in an environmentally sound manner to ensure the creation of sustainable human settlements. However, there is a new consensus emerging around effective modes of government action in the economic sphere that has significant implications for the reform of the subnational economic development function. There is the realization that it is of essence that functions of local planning and local economic development (LED) be joined and treated inseparably. This chapter is centered on the understanding that urban land-use planning does not exist in isolation and it is necessary to perceive land-use planning as an integral part of the process of national growth and development. Scholars, such as Feser (2014), argue that the urban planning profession should be a robust source of guidance about how best to restructure and manage institutions of LED in the face of an increasingly globalized and competitive economy.

Background and Context

It is undisputable that sustainable development is an essential development paradigm in the twenty-first century. Cities across the globe have adopted sustainable development mechanisms that aim to improve urban development and management and fostering an economically competitive environment so that there is no decrease in welfare and quality of life of urban populace. Professions, such as urban land-use planning, have been central in the understanding and implementation of sustainability principles that have soared variedly across the world, but their key development connotations remain highly environmental and social rather than economic. Mandisvika (2015) argues that cities are also important global hubs of finance, manufacturing, trade, and administration.

Rondinelli (1983) argues that cities offer locations for services that require high population thresholds and large markets to operate efficiently. This is because cities are centers for innovation and diffusion, and they facilitate widespread modernization. Mandisvika (2015) provides statistics that Lagos has 5% of Nigeria's population and has 57% of total value in manufacturing and has 40% of the nation's highly skilled labor. This realization that cities are important both to environmental sustainability as well as to economic sustainability has led to the growing popularity of local economic development (LED). Mandisvika (2015) indicates that LED has the purpose to mobilize the local economic potential by bringing innovation to all its growth dimensions that include infrastructure, local SME's, and their skills, attracting foreign investment, fostering territorial competitiveness, and strengthening local institutions. LED is highly connected to local planning that highlights a major opportunity for the urban land-use planning profession to change the face of urban areas in the future. This chapter tries to discuss the future of urban land-use planning under the impact of LED. Cases will be proffered from places across the world, such as Europe, North America, Asia, and Australia.

Research Methodology

This chapter is based on a wide desktop study that is a secondary type of research. The researcher carried out a desk research through reviewing previous research findings to gain broad understanding on the relationship between urban land-use planning and LED. The research was broadly spread throughout four continents that are Europe, America, Asia, and Australia. The central focus was to understand the future of urban land-use planning under the impact of LED. It was garnered from the research that urban land-use planning does not exist in isolation and it is necessary to perceive land-use planning as an integral part of the process of national growth and development.

Literature Review

Urban land-use planning refers to the process by that land is allocated between competing and sometimes conflicting uses in order to secure the rational and orderly development of land in an environmentally sound manner to ensure the creation of sustainable human settlements. According to Thomas (2001), the process of land-use planning consists of two main functions of development/land-use planning and development control. These two functions are supported by relevant research and mapping that are also major components of the land-use planning process. Land-use planning is an integral part of the process of national growth and development. Among other things, this process seeks to identify, articulate, and satisfy the basic social/human needs of a country's population within the context of available economic/financial resources and technical knowledge. Thomas (2001) observes that people have needs that must be satisfied, and these in the urban areas may include, among others, housing, jobs, education, opportunities for recreation, transport, and basic services like water, electricity, clean air, and health care. These facilities, goods, and services operate at the center of urban land-use planning. According to Namangaya (2014), land-use planning seeks to accommodate these needs within a technical and spatial framework. It is also important to note that urban land-use/spatial planning is one of the crucial tools in attaining economic development of a place (Fainstein 1994). There is much interaction between spatial planning and socioeconomic development that is explained by Namangaya (2014) as a double-edged interaction between urban spatial planning dynamics of economic and social change in an urban region.

The concept of LED is an emerging place-based planning approach to local and regional economic development (Barca et al. 2012). This concept has witnessed widespread popularity and significance across the globe where it is inseparable from the changing world economy and especially the advance of globalization. The central focus of LED is on the wellbeing and development of local residents. The World Bank (2002) maintained that the activity of LED is about local people working together to achieve sustainable economic growth that brings economic benefits and quality of life improvements for all in the community. In the reshaped landscape of development planning, LED has appeared as a novel planning focus in both developed and developing countries, particularly in the context of pervasive trends toward decentralization (Barca et al. 2012). The focal activities in LED are based upon maximizing the comparative advantages of localities and include improving the local business environment, building local skills, cluster development, and encouraging trust and partnerships between the private sector, public sector, public institutions, and civil society (Ruecker and Trah 2007).

LED emerged as a way of responding to industrial cities, such as Manchester in England and to come up with answers to the city's problems. According to Feser (2014), LED did not emerge simply as land-use planning nor did it emerge as a sectorial approach but was concerned with eliminating the impacts of the communist economy approach that led to congestion, underdevelopment, and unbalanced development.

Results

Europe/North America

The evolution of the planning profession in the USA and the UK highlights a desperate need for growth to stem inner city decline in older industrial cities in the 1970s and 1980s. That need shaped urban land-use planning toward regulation and control (Feser 2014). Since then urban land-use planning and LED as government functions have long stood in somewhat ambivalent relation to one another. In the UK, the guidance and control of growth, traditional concerns of the British Statutory Planning System since 1947, had quite suddenly been replaced by an obsession with encouraging growth at almost any cost (Hall 2000, p. 347).

In the USA, the worst excesses of urban renewal programs and the emergence of powerful and undemocratic urban growth coalitions encouraged many planners already instinctive inclination to act as a counterweight rather than enthusiastic partner in economic development. However, the twenty-first century has presented a new consensus worldwide around effective modes of government action in the economic sphere. That's where the concept of LED comes of handy in the development and management of urban landscapes. Feser (2014) argues that this new consensus in the economic sphere is responsible for the widespread reform of the subnational economic development function currently underway in the UK and USA. Feser (2014) further implores the planning profession through its various organizations and practices to develop new models and to train professions who can function effectively under LED. As a result, this chapter tries to explore the future of urban land-use planning under the impact of the rising consensus of LED.

In the same sense, Feser (2014) claims that planning and economic development are intimately related. However the International Economic Development Council (IEDC) that is the largest organization of economic development practitioners in North America, the planning profession is inattentive to land-use policy, its privileging of industrial and commercial development over housing, its lack of focus on the neighborhood scale, and its general tendency to work in a "silo" unconcerned with other urban issues that affect the overall health of cities (Garmise et al. 2008). Realizing the professional divide between local planners and economic developers led to the development of economic development topics in urban planning programs that include things like development finance, employment or manpower planning, economic impact analysis, and regional development theory. These developments took center stage in the 1970s and 1980s in the USA, and the major stride was to teach city planners something about emerging global and local economic dynamics, economic analysis techniques, and potential planning and policy solutions to persistently high unemployment. LED thus promises to become a promising area of professional practice for urban land-use planners.

The development of cities is closely associated with local governments that are argued by Nolon (2011) that with their strong planning visions and clear implementation plans, they may help and guide the development of cities. The economic development of cities can thus be treated as a component of the city's comprehensive land-use plan. This entails creating incentives, such as density bonuses in zoning ordinances or real property tax abatements, and reforming land-use regulations to stimulate needed and feasible redevelopment and revitalization. According to Nolon (2011) land-use planning in the light of economic development should contain the strategies and implementation tools they will use to accomplish their visions, and these include how local land-use regulations, processes, and protocols will be reformed. In the same manner, Nolon (2011) further argues that a good local comprehensive land-use plan should not only guide the physical development of a municipality but also guides economic development, accommodating social, environmental, and regional concerns. This highlights the need to have economic development coordinated with land-use planning strategies. The coordination reflects local, regional, and state-wide concerns with employment, labor, industry, and quality of life. In terms of local planning, this translates into LED.

The coordination of economic development and urban land-use planning in this context is also supported by the realization that urban planners in cities across the world should supplement their economic development knowledge with up-to-date private sector knowledge on local, regional, and national markets, development trends, and available equity and debt financing, including the types of economic development projects that can attract financing. This entails a shift of pure urban land-use planning in cities local authorities to create and implement coordinated, market-realistic jobs and economic development components in their comprehensive plans. Since most of economic data gathering in cities is largely done by the private players, Nolon (2011) argues that it is the ideal time for cities that are thinning out planning staffs that may not have engaged in this type of planning to create a public/private planning partnership that will invest the private sector in the work of cities as they might not have been in the past.

Nolon (2011) equated this to "bringing sophistication in-house" in the sense that this may lead to more precise targeting of city assets and implementation strategies on targeted sectors whose needs align with urban locations in the region. Some of these sectors include medical service providers, food processors, training companies, utility headquarters, green products firms, and services and knowledge-based firms that need the younger, educated workforce seeking urban living or the semiskilled laborers who already live there. In addition to the above, land-use planning in cities should contain an inventory of existing and proposed locations and intensities of various land uses, including public facilities, commercial and industrial facilities, significant natural and environmental resources, and existing housing resources and future housing needs.

Nolon (2011) further highlights an example of Baltimore's comprehensive plan that provides a good example to illustrate what cities can do through the urban land-use planning process. The plan aims to encourage economic growth and to streamline its regulatory system. According to Nolon (2011), the plan states that it "encourages economic growth in port-and defence-related industries as well as six burgeoning employment sectors identified by the Workforce Investment Board by better articulating the development process and ensuring development compatibility in all parts of the city." The urban land-use plan for Baltimore commits to foster economic development by ensuring that local residents are trained and equipped to fill the jobs anticipated in the plan.

The Baltimore plan readjusts its residential land-use "to account for the change in population, aging housing stock, the critical need for moderately priced, quality housing to attract and retain the middle class, the growing market for providing homes for Washington, DC commuters, the expanding market for condominiums, the opportunity to capture an increasing share of the expected 800,000 new residents who will settle in the region by 2020 and enhancing the wonderful mix of architecture, lifestyles and neighborhoods that make Baltimore a premier place to live." The plan, in addition, contains strategies to "capture and encourage biotech job opportunities, create larger tracts of land for commercial or industrial development near transportation centers and connect residents to available employment. Currently the health, medical, financial and construction sectors are large and growing. Education and tourism continue to be strong."

Asian countries have a string focus on the physical development plans, infrastructure, and public programs of assistance. This part of the paper will highlight on the common spatial approaches of governments and economic development issues across Asia. Most governments in Asia have tried to use a variety or wide range of national, regional, and local master or strategic plans to push urban and economic development. These have largely constituted urban land-use planning in Asia. Choe and Roberts (2011) indicate that many of these physical plans, for example, those in Bangladesh and Vietnam, are linked to master plans.

Choe and Roberts (2011) further indicate that many governments in Asia have treated urban land-use planning as a supply-side approach to economic development. This is without considering that cities need to respond more to drivers of demand and not just to drivers of supply to gain competitive advantage. Thus, urban land-use planning across Asia in the sense of master planning has failed due to a plethora of reasons. Some of the reasons are that most of the plans are unrealistic or overly optimistic. They assume that the resources needed to implement the plans will be available; planning is not backed by solid economic and financial analyses, most master plans lack mechanisms for implementing projects for financing or through public-private partnership; and urban master planning has worked in countries like Singapore because the government understands the relationship between planning and the economics of development.

Adding to the shortcomings of master planning, Choe and Roberts (2011) indicate that poor planning and management of land-use and utilities have led to high congestion and underperformance of assets and service delivery that reduces the returns to business and government from their capital investment. Due to such dynamics, more and more business activities have migrated away from the central areas of cities to peripheries as a result. Thus, it is important under the principles of economic development for improvements to be made especially on the functional and spatial productivity of cities or urban areas in Asia for them to attain a more sustainable and prosperous economy. Choe and Roberts (2011) further indicate that most cities in Asia have regional plans or city development plans for the next 20–25 years that determine their land-use planning. These plans contain land-use, social, and physical infrastructure and zoning provisions, but they tend to be rigid and static and not sufficiently responsive to rapid change. Choe and Roberts (2011) highlight that they are often inadequate for sustainable development and this also includes LED.

Countries, such as Singapore, Hong Kong, and China, have been applauded by their use of their well-developed planning systems effectively and enforcing building and land-use regulations to create competitive advantage for business investment. Other countries, such as japan, the Republic of Korea, and Malaysia also have robust planning systems, and city development plans that ensure efficient property markets reduce the risks associated with property rights, clearly define zoning requirements and uses, and reduce the risks from environmental hazards. Choe and Roberts (2011) provide that the strength of the centralized planning systems in these countries is a major factor behind the success of their urban economics. Thus, the future of urban land-use planning follows a stricter planning and development control that makes property markets much more stable and creates a favorable climate for demand-side development in cities throughout the country. However, currently, there are other Asian countries with weak planning that has destabilized property markets and made them less transparent, exposing investors and developers to higher risk.

The future of urban land-use planning in Asia under the tenets of LED has meant that there is need to follow a demand-side approach to economic development. One key aspect is that of favoring the provision of infrastructure. Choe and Roberts (2011) provide that a significant problem for Asian countries is on keeping up with the demand for infrastructure to support economic development. According to an ADB report (Nataraj 2007), Southeast Asian countries will need to spend $412 billion per year between 2007 and 2012 or about 6% of regional GDP, on roads, railways, airports, ports, and electricity. Most of these projects will be targeting urban land-use planning, and for countries, such as India, the infrastructure investment needed during the 5-year period is estimated to be $410 billion.

From the estimates highlighted above, it should be noted that infrastructure investment in more developed Asian countries will mostly be in health, education, and telecommunications, while countries that are less developed give priority of urban land-use planning to investment in roads water supply and basic sanitation. Countries in the second group tend to invest heavily in special economic zones (SEZs), industrial estates, and business parks to encourage multinational corporations to invest in LED. Choe and Roberts (2011) also highlight that in terms of the relationship between urban land-use planning and economic development, most primate cities in Asia and some larger metropolitan cities have to adopt policies to develop new satellite towns as part of a polycentric strategy to decentralize employment and create more specialized economic activities. Thus, more metropolitan plans for Asian cities provide for satellite towns, many of that develop around major industrial or commercial areas. Pudong, on the east side of the Huangpu River

F

in Shanghai, is a major administrative and commercial center. Malaysia has developed Cyberjaya, a new town south of Kuala Lumpur, chiefly for ICT industries. Clark in the Philippines has become a major electronics and logistics center for assembly manufacturing and logistics, taking advantage of the high-quality air and the nearby port facilities at Subic Bay. Bekasi, southeast of Jakarta, is a new town where heavy manufacturing industries are being developed.

Local and regional economic development activities in Australia are organized at the state level rather than nationally. According to Beer and Maude (2002), there are more agencies, more programs, and greater sources of funding operating within the states than empowerment through federal initiatives. There are also state and territory governments that have put some form of local and regional economic development framework in place since the mid-1980s. Beer and Maude (2002) further indicate that in all Australian jurisdictions, the state development activities of the central bureaucracies are complemented by regional development bodies that operate at a regional, that is, sub-state level. There are seven different systems of local and regional economic development in Australia, and these according to Beer and Maude (2002) raise questions of theoretical and practical significance. Several questions can be asked like who has power and responsibility for local and regional development? What types of development assistance are pursued, and how are these programs organized? How are agencies in each state funded? Is there more than one type of economic development agency, and if so, what is the relationship between the various bodies? How have the state frameworks been integrated with federal policy initiatives? Are there better models of local and regional economic development, and what can be learnt from the failings of current systems?

Discussion

The profession and practice of urban land-use planning have traditionally aligned itself toward environmental sustainability and equity. The major thrusts were solely technical and regulatory in practice. However, with the unfolding of events and further development of sustainability concepts, there are huge opportunities where urban land-use planning provides a comparative advantage that is in LED. This chapter maintains that it is important that local planning and LED be joined and treated inseparably. This means that urban land-use planning should in the future be done with an economic development lens where developing a locality would be of high priority. This view was shared by scholars, such as Feser (2014) who argued that urban land-use planning should be robust through guiding the restructuring and management of LED in the face of an increasingly globalized and competitive economy.

Looking at urban land-use planning in continents, such as Europe and America, highlights the traditional purposes of the urban planning profession. The major or central focus of urban land-use planning was to deal with decline and squalor conditions that were widespread in the industrial cities. Scholars, such as Feser (2014), highlight on this background that more focus was on stemming inner city decline, and from then onward urban land-use planning and LED as government functions stood in somewhat ambivalent relation to one another. Such relationship is like that because urban land-use planning was more focused on bringing sanity or aesthetics rather than furthering value or economic development through incorporating economic planning principles. Hall (2000) indicates that due to such professional inclination, many urban land-use planners were not concerned with economic development, and hence they remained cut out of economic growth and development discourse. This, however, has been countered by the recent growth of the 21st consensus in the economic sphere that local areas should drive their economies and make use of already existing local authorities and systems. It is only after some 20 years after the start of the twenty-first century that LED institutions and organizations are sprouting across the world.

The development of stand-alone LED institutions and departments in urban areas is starting to gain traction. However, other scholars like Nolon (2011) argue that local governments through their strong planning visions and clear implementation plans may house LED systems. Nolon (2011) argues explicitly that the economic development of cities that this chapter treats as LED can be treated as a component of cities comprehensive land-use plans. This means that urban areas across the world should start to appreciate a paradigm shift from the traditional urban land-use planning to one that takes into considerations aspects of LED. This has been highlighted above that urban land-use planning that considers LED should create incentives, such as density bonuses in zoning ordinances or real property tax abatements and reforming land-use regulations to stimulate needed and feasible redevelopment and revitalization. It is of paramount significance that local authorities realize that their core business when doing land-use planning is no longer to guide only physical development of their locality but also to guide economic development, accommodating social, environmental, and regional concerns.

Conclusion and Policy Options

This chapter aimed at expounding the future of urban land-use planning under the impact of LED. It was highlighted that urban land-use planning is key in trying to improve the public domain and is a key tool as a theoretical and epistemological frame of reference among urban planners and urban planning organizations. The orientation of urban land-use planning toward environmental sustainability remains valid and valuable in the realization of sustainable development. However, the twenty-first century is calling for urban land-use planning to broaden its horizon toward issues of economic sustainability where the chief target should be local areas. It is understood that exiting systems and local government setups are ideal to implement LED; thus urban land-use planning can work best toward LED. The paper outlined that it is possible to promote economic development through urban land-use planning, and some of the possible strategies include coordinated efforts in creating and implementing market-realistic jobs and economic development components in their comprehensive urban land-use plans.

References

Barca, F., McCann, P. & Rodríguez-Pose, A. (2012). The case for regional development intervention: Place-based versus place neutral approaches. *Journal of Regional Science 52*(1), 134–152. https://doi.org/10.1111/j.1467-9787.2011.00756.x.

Beer, A., & Maude, A. (2002). *Local and regional economic development agencies in Australia. Report prepared for the Local Government Association of South Australia*. Adelaide: School of Geography, Population and Environmental Management, Flinders University.

Choe, K., & Roberts, B. (2011). *Competitive cities in the 21st century: Cluster-based local economic development* (Urban development series). Manila: Asian Development Bank.

Fainstein, S. (1994). *The city builders: Property, politics, and planning in London and New York* (Studies in urban and social change). New Jersey: Blackwell Publishers.

Feser, E. (2014). Planning local economic development in the emerging world order. *TPR, 85*(1). https://doi.org/10.3828/tpr.2014.4.

Garmise, S., Nourick, S., & Thorstensen, E. (2008). *Forty years of urban economic development: A retrospective*. Washington, DC: International Economic Development Council.

Hall, P. (2000). Creative cities and economic development. *Urban studies, 37*(4), 639–649.

Mandisvika, G. (2015). The role and importance of local economic development in urban development: A case of Harare. *Journal of Advocacy, Research and Education, 4*(3), 198–209.

Namangaya, A. H. (2014). Urban spatial planning and local economic development: Comparative assessment of practice in Tanzanian Cities. *International Journal of Business, Humanities and Technology, 4* (6), 20–31.

Nataraj, G. (2007). *Infrastructure challenges in South Asia: The role of public–private partnerships*. Tokyo: Asian Development Bank Institute.

Nolon, J. R. (2011). *Land-use for economic development in tough financial times*. New York: Pace University, Pace Law Faculty Publications.

Rondinelli, D. A. (1983). Implementing decentralization programmes in Asia: A comparative analysis. *Public administration and development, 3*(3), 181–207.

Ruecker, A., & Trah, G. (2007). Local and Regional Economic Development. Towards a common framework for GTZ's LRED Interventions in South Africa. Eschborn, GTZ.

Thomas, D. (2001). The importance of development plans/land-use policy for development control. Prepared for the USAID/OAS Post-Georges Disaster Mitigation Project, Workshop for Building Inspectors.

World Bank. (2002). Local Economic Development – A Primer, Washington DC: World Bank.

G

Garden City

▶ Green Cities in Theory and Practice

Gemstones

▶ The Source Waters of Tanga

Gender

▶ Smart Agriculture and ICT

Gender Blind Urban Planning and Design

▶ Women in Urbanism, Perpetuating the Bias?

Gender Equality

▶ Feminist Planning and Urbanism: Understanding the Past for an Inclusive Future

R. C. Brears (ed.), *The Palgrave Encyclopedia of Urban and Regional Futures*,
https://doi.org/10.1007/978-3-030-87745-3

Gender Inclusive Planning

▶ Understanding Women's Perspective of Quality of Life in Cities

Gender Inequalities in Cities: Inclusive Cities

Nora Libertun de Duren
Inter-American Development Bank, Washington, DC, USA

Synonyms

Cities, urban areas, urban regions; Exclusion, discrimination, segregation; Inclusion, integration

Definition

Discrimination Any distinction, exclusion, or restriction which has the effect or purpose of impairing or nullifying the recognition, enjoyment, or exercise, by any person, of human rights, and fundamental freedoms in the political, economic, social, cultural, civil, or any other field.

Empowerment	The process of gaining access to and developing one's capacities with a view to actively participating in shaping one's own life and that of one's community in economic, socio-cultural, political, and religious terms.
Equal Rights	A truly egalitarian situation in which women and men share equal economic, political, civil, cultural, and social rights.
Gender	The roles, duties, and responsibilities which are culturally or socially ascribed to women, men, girls, and boys.
Gender discrimination	The systematic, unfavorable treatment of individuals on the basis of their gender, which denies rights, opportunities, or resources.
Gender equality	Equal enjoyment of rights and access to opportunities and outcomes, including control of resources, by women, men, girls, and boys.
Gender gap	The differences between women and men in access to and control over resources, especially as reflected in political, intellectual, cultural, or economic attainments and attitudes.

This entry elaborates on women-related challenges in urban space from three major perspectives: housing, mobility, and public space. Women equality refers to a recognition of the different needs and interests of women, which may require a fair redistribution of power, resources, opportunities, and responsibilities. Achieving gender equality in cities is explicitly upheld by the Sustainable Development Goals (SDG) in SDG 5 and in SDG 11 (UN-Habitat 2016). In particular, Goal 11.7 calls for the provision of universal access to safe, inclusive, and accessible, green, and public spaces, for women and children, older persons, and persons with disabilities. Also, the United Nations New Urban Agenda adopted in Quito, Ecuador, in 2016, envisages cities that "achieve gender equality and empower all women and girls by ensuring women's full and effective participation and equal rights in all fields, as well as in leadership at all decision-making levels, by ensuring decent work and equal pay for equal work, or work of equal value, for all women by preventing and eliminating all forms of discrimination, violence and harassment against women and girls in private and public spaces" (13.c).

While these agreements are adopted at the national level, many of the actions therein fall upon local authorities. Nonetheless, women are less likely to be engaged in political life and are underrepresented in elected positions, particularly at the local level. Less than a quarter of parliamentarians are women (World Bank 2017), while less than 10% of mayors are women, and only 25 of the largest 300 world cities have female mayors (City Mayors 2017). Local and national authorities must work closely together to empower women in cities and improve gender equality. Therefore, national, and local governments should make joint efforts to recognize and respond to the different needs, concerns, and interests of all urban residents.

Housing. Women have less access to housing tenure, credit, and well-located housing than men and are disproportionately affected by housing deficits. Only one in four people living in urban areas has access to onsite improved sanitation facilities (UNICEF/WHO 2017). This limitation poses an extra burden on women, who face increased risks of sexual assault when using sanitation facilities located outside of their homes during the night. Also, only three in five people living in urban areas worldwide have access to safe, readily available water at home (UNICEF/WHO 2017), and women and girls are responsible for water collection in 80% of the households without access to water on premises (UN-Women 2018). In addition, inadequate sources of energy, such as wood-burning stoves, also represent a time burden for women, as they spend extra time buying fuel and cooking more often because of limited access to refrigeration (IADB, CAF and UN-Habitat 2020).

Due to differential treatment by law or by custom, women are less likely to own, manage, control, or inherit assets and property (Rakodi 2016). Most land titles are still registered under men's names, and UNICEF estimates that women account only for 25% of the landowners in Latin America (Chant and McIlwaine 2015). Security of tenure correlates with a decline in gender-based violence, either due to a change in men's attitudes (Amaral 2017) or to the fact that women are more inclined to leave abusive relationships (Moser 2017). Security of tenure can also promote women's economic empowerment (OECD 2017) and contribute to the reduction of income inequality (Duflo 2012), as women with housing ownership can access bank loans more easily, which could, in turn, enable them to develop their own businesses and access to financial services (Libertun de Duren 2018). The gender gap is a 7-percentage point in account ownership, an 11-percentage point gap in access to formal savings at financial institutions, and a 3-percentage point gap in access to formal credit (World Bank 2019).

Housing location also affects women development. Mixed land use makes it easier for women to balance paid work with their domestic responsibilities (Taccoli and Satterwhite 2013), as it tends to reduce the distance between housing, workplace, and public services. Conversely, when cities expand without adequate planning, poor households headed by women are the most disadvantaged, as they tend to be in precarious neighborhoods, with limited access to efficient means of public transportation. In Puebla, Mexico, for instance, more than two-thirds of the households in affordable housing units located more than 30 kilometers away from the city are headed by women. Their daily commute takes between 2 and 3 hours, which limits their options for personal and economic development (Libertun de Duren 2017). In addition, these women usually make long journeys at early or late hours, when the frequency of public transportation is very low, increasing their exposure to sexual violence in their daily commute (McIlwaine 2013).

Mobility. Lack of gender-responsive planning can make travelling more complicated, more expensive, and more dangerous for women than for men. Women generally rely more than men on public transportation. In Latin America and the Caribbean, on average, over 50% of public transportation users are women and, in the case of Argentina, women represent over 60% of the public transport users in the city of Buenos Aires (IADB 2017). However, most of the existing public bus routes and sidewalks in the region are not designed with the needs of women in mind (IADB 2015). Typically, women make more multipurpose trips (Stanford University 2018), combining their daily work commute with trips to school, childcare facilities, healthcare centers, and trips for shopping purposes (McGuckin et al. 2005). Data shows that having a young child in the house will increase the number of trips a woman makes, for example, an additional 23% in the case of London, United Kingdom (IADB 2015), and 13% in the case of Buenos Aires, Argentina (Ochoa et al. 2014). Women are also more likely to be victims of gender-based violence on public transportation, which prevents them from enjoying equal access to mobility. For example, in Mexico City, over 65% of women using public transportation have experienced sexual harassment while travelling (IADB 2015).

Public space. Design also needs to promote safe and inclusive spaces that respond to the needs of all citizens. The typical urban locations that cause fear and insecurity to women are dark areas, isolated public parks, empty and poorly lit streets, underground parking lots, and pedestrian underpasses (Libertun de Duren 2020). The lack of adequate sanitation facilities in public spaces also increases women's risk of sexual assault and their exposure to disease. Women tend to experience sexual harassment in public places more frequently than men, with 92% of women in Rabat, Morocco, and 68% of women in Quito, Ecuador; having experienced sexual harassment in public spaces (UN-Women 2017), urban design and interventions that reduce vulnerability to crime and harassment in public spaces become an issue of great relevance for women's equal access to the city.

Conclusions. Urbanization poses specific challenges to women, who have limited access to adequate housing, tend to benefit less from mobility services, and are more exposed to violence in public spaces. Both national and local governments need to address these issues, as these challenges violate human rights, hinder economic opportunities for women, and limit social development for all. At the national level, sound normative frameworks are essential to ensure the basic rights of women, especially regarding housing rights and inheritance laws, land policy and security of tenure, and laws against gender-based violence. At the local level, gender-responsive regulations and interventions are required, particularly in the design and management of transport mobility systems and public spaces. Meaningful participation of women in decision-making processes should be increased in all levels of government while establishing effective, accountable, and transparent governance mechanisms and data information systems to ensure that cities are built to provide opportunities for all.

Cross-References

- ▶ Children, Urban Vulnerability, and Resilience
- ▶ Feminist Planning and Urbanism: Understanding the Past for an Inclusive Future
- ▶ Improving Social Equity and Community Health and Well-Being in Low-Income Suburbs and Regions
- ▶ Social Urbanism
- ▶ Spatial Justice and the Design of Future Cities in the Developing World
- ▶ Sustainable Development Goals
- ▶ Sustainable Development Goals from an Urban Perspective
- ▶ Women in Urbanism, Perpetuating the Bias?

Disclaimer All opinions expressed in this article are the opinions of the author and do not necessarily reflect the views of the Inter-American Development Bank or any of its members

References

Amaral, S. (2017). *Do improved property rights decrease violence against women in India?* Institute for Social and Economic Research, University of Essex.

Chant, S., & McIlwaine, C. (2015). *Cities, slums and gender in the global south: Towards a feminised urban future*. London: Routledge.

City Mayors. (2017). *Largest cities in the world and their mayors (Largest cities with women mayors)*. Accessed 31 July 2018.

Duflo, E. (2012). Women empowerment and economic development. *Journal of Economic Literature, 50*(4), 1051–1079.

IADB, CAF and UN-Habitat. (2020). *Gender inequality in cities*. Prepared for the Urban 20 (U20).

Inter-American Development Bank – IADB. (2015). *The relationship between gender and transport* (p. 72).

Inter-American Development Bank – IADB. (2017, November). *Mujeres y ciclismo urbano: Promoviendo políticas inclusivas de movilidad en América Latina* (p. 75).

Libertun de Duren, N. R. (2017). The social housing burden: comparing households at the periphery and the centre of cities in Brazil, Colombia, and Mexico. *International Journal of Housing Policy, 18*(2), 177–203. https://doi.org/10.1080/19491247.2017.1298366.

Libertun de Duren, N. (2018). *Inclusive cities: Urban productivity through gender equality*. IDB. https://doi.org/10.18235/0001320.

Libertun de Duren, N. R. (2020). Effects of neighborhood upgrading programs on domestic violence in Bolivia. *World Development Perspectives, 19*, 100231. Published online 2020 Jun 24. https://doi.org/10.1016/j.wdp.2020.100231.

McIlwaine, C. (2013). Urbanization and gender-based violence: Exploring the paradoxes in the global south. *Environment and Urbanization Environment and Urbanization, 25*, 65–79.

Moser, C. O. N. (2017). Gender transformation in a new global urban agenda: challenges for Habitat III and beyond. *Environment and Urbanization, 29*(1), 221–236. https://doi.org/10.1177/0956247816662573.

Ochoa, C., Peralta Quiros, T., & Mehndiratta, S. (2014). *Gender, travel and job access: Evidence from Buenos Aires*. World Bank.

Rakodi, C. (2016). Addressing gendered inequalities in access to land and housing. In C. Moser (Ed.), *Gender, asset accumulation and just cities: Pathways to transformation*. ISBN-13: 978-1138193536.

Taccoli, C., & Satterwhite, D. (2013). *Gender and urban change*. Environment and Urbanization.

The Organisation for Economic Co-operation and Development (OECD). (2017). Chapter 11: Women at work: A snapshot of women in the labour force. *The pursuit of gender equality*.

UN-Habitat. (2016). Issue paper 11: Public space – draft. United Nations conference on housing and sustainable urban, Habitat III. Quito.

UN-Women. (2017). *Safe cities and safe public spaces* (p. 24). New York: UN Women Headquarters.
UN-Women. (2018). *Turning promises into action: Gender equality in the 2030 agenda for sustainable development* (p. 124). New York: UN Women Headquarters.
WHO/UNICEF Joint Monitoring Program for Water Supply. (2017). *Progress on drinking water, sanitation and hygiene: 2017 Update and SDG baselines*. Available at https://www.unicef.org/publications/files/Progress_on_Drinking_Water_Sanitation_and_Hygiene_2017.pdf. Accessed 13 June 2018.
World Bank. (2017). *Proportion of seats held by women in national parliaments (%)*. Aaccessed 31 July 2018.
World Bank. (2019). *The little data book on gender 2019*. World Development Indicators. Washington DC: World Bank Group.
McGuckin, N., Zmud, J., & Nakamoto, Y. (2005). Trip-chaining trends in the United States: Understanding travel behavior for policy making. *Transportation Research Record, 1917*(1), 199–204.

Gender Mainstreaming

▶ Feminist Planning and Urbanism: Understanding the Past for an Inclusive Future

Gender Responsive Planning and Design

▶ Women in Urbanism, Perpetuating the Bias?

Gender-Neutral Cities

▶ Women in Urbanism, Perpetuating the Bias?

Geographic

▶ Spatial Demography as the Shaper of Urban and Regional Planning Under the Impact of Rapid Urbanization

Geographic Information System (GIS)

▶ The State of Extreme Events in India

Geography

▶ New Localism: New Regionalism

G

Getting Our Built Environments Ready for an Aging Population

Getting Our Built Environments Ready for an Aging Population

Renard Y. J. Siew
Climate Change & Sustainability, Centre for Governance & Political Studies (CENT-GPS), Kuala Lumpur, Malaysia
Institute for Globally Distributed Open Research and Education (IGDORE), Bali, Indonesia

In 2019, the Population Division of the United Nations Department of Economic and Social Affairs submitted a report as a contribution to the World Assembly on Ageing (UNDESA 2019). The report acknowledged that population aging is unprecedented and that the twenty-first century will witness even more rapid aging that it did in the previous century.

The US Health and Human Services reported that the elderly population numbered 39.6 million in 2009; this represents approximately 12.9% of the US population. It is projected that this will increase to 72.1 million by 2030 in the USA alone. In the Asia Pacific, approximately 12.4% of the population in the region was 60 years or older in 2016, but this is projected to increase to more than a quarter or 1.3 billion by 2050. The United Nations Population Fund (UNFPA) anticipates that the number of elderly population will more than double in Asia, Caribbean, and the Mediterranean. This global phenomenon is pervasive, affecting every man, woman, and child. This phenomenon has profound implications for many facets of human life.

Implications of an Aging Population

In many countries, the total dependency ratio continues to increase. This is largely driven by the fact that the working age population is shrinking and long-standing social conventions such as designated pension age and barriers to women entering workforce. Enabling older people to age in place is a complicated task. It requires comprehensive planning and provision of a wide range of support services in the community as well as the removal of barriers that segregate older people and limit their activities.

Apart from more detailed planning by master planners, other impacts include reduced economic growth due to lesser productivity, as more elderly people are forced into early retirement, and increase in government spending on healthcare and pensions. Those in retirement tend to pay lower taxes because they are usually seen as no longer productive in the workforce. There is also an issue of shortages of skilled labor trained to care for aged patients which is alarming (Milova 2017).

Are Our Built Environments Ready for an Aging Population?

Our society will need to adapt and develop mechanisms and practices in the future to either minimize or mitigate social detachment especially among elderly individuals. To a certain extent, it is argued that there is much less intergenerational support today compared to the past causing recurrent periods of social detachment which leads to an increase in significant health risks. Social innovation is needed at a much faster pace with new concepts such as "age-friendly communities" surfacing suggesting the need for these individuals to remain active agents of their own care.

In a 2018 survey conducted by the American Association of Retired Persons (AARP 2019), 76% of Americans aged 50 and older say they prefer to remain in their current residence, and 77% would like to live in their community as long as possible. While the demand of "aging in place" may be growing, the fundamental question which needs to be asked is whether our built environments are ready to cater for this need. Admittedly, this is an often-overlooked subject even among experienced built environment practitioners.

Over the years, sustainability rating tools (SRTs), for example, Building Research Establishment Environmental Assessment Method (BREEAM), Leadership in Energy & Environmental Design (LEED), Green Star, and the Green Building Index (GBI), have emerged to guide the development of more eco-friendly buildings (Siew et al. 2013; Ding 2012). Despite some of the critique that these tools have received from scholars, the lack of harmonization in terms of the adoption of criteria and standards; the red tape, and additional cost that practitioners will have to incur to implement the criteria, the application of these tools is nothing short of impressive, further propelling the built environment as one of the key sectors to watch.

Recently, however, these tools have also expanded in scope to consider the broader issues within a built environment. The focus has shifted from analyzing the eco-efficiency of just a single building to townships. Naturally, the social dimension becomes more prevalent where there is now a much stronger focus on the quality of life, job creation, accessibility to public transportation, and the attachment that people have for their neighborhood (place-making).

Yet, the aged population has a very different set of needs when compared to a younger demographic. A comparative study of international published reports by governmental agencies and civil society (i.e., World Health Organization, Lifetime Neighbourhoods, American Association of Retired Persons inter alia) revealed similar needs among the aging community (more commonly known as "age-friendly criteria"). Such aspects include the need for an active community engagement program which allows senior citizens to provide feedback, a built environment which promotes multi-generational interaction and cooperation, infrastructure mobility that is age-friendly (e.g., there has been suggestions of carving out areas within a township to be designated just for elderly communities and designed to cater for their needs; a mere 15-second waiting time at a pedestrian crossing may not be feasible for an elderly person who suffers from mobility issues), easy access to healthcare services, housing affordability, as well as built-in safety and security measures.

In a landmark study, Siew (2016) find that there is a huge gap when analyzing mainstream SRTs (de facto framework which guides the development of townships) as many still do not sufficiently emphasize on age-friendly criteria. This jives with anecdotal evidence that most existing built environments and majority of the cities in Asia are not ready to cater for the needs for an aging population.

What Needs to Be Done?

There is still much that needs to be done to address existing gaps, and it is hoped that this paper would be a start to help spur critical discussions in this space. Efforts by the World Health Organization (WHO) to establish a Global Network for Age-Friendly Cities and Communities (GNAFCC) should be recognized as such a platform that helps foster the exchange of experience and mutual learning between global cities and communities (differing in size and location). Such efforts occur within very diverse cultural and socioeconomic context.

It is proposed that mainstream SRTs should consider allocating mandatory credits for age-friendly criteria. This would incentivize important stakeholders, namely, property developers and city planners, to consider and incorporate the needs of the elderly population. Embedding age-friendly criteria in SRTs would also encourage a more meaningful dialogue on the concept of active aging and possibly raise awareness among practitioners on the importance of this issue. Local governments could play a part in this by introducing policies or incentive schemes to encourage property developers to incorporate age-friendly criteria in the design of cities. A critical success factor in the implementation of age-friendly programs is to ensure that there is active engagement from various parties as is the case with Akita City in Japan; Basque Country, Spain; and the City of Brussels, Belgium. Akita City emphasizes three key priorities: the need to involve Akita residents in a leading role; ensure cooperation between private enterprises, administrative organizations, and citizens; and encourage cooperation between all relevant departments in the city government.

Ongoing research needs to be more multi-inter-transdisciplinary. For example, the field of gerontology (the study of aged groups) needs to evolve to also consider how to enhance the built environments for this demographic. The ongoing discussions about integrating the Internet of Things (IoT) into the development of built environment is exciting and one that has immense potential to cater for the needs of the elderly. Already we have seen substantial development of new applications of sensor technology in the home, although this has tended to be tele-health focused. Much more research is needed to study the role of IoT and aging in place that more broadly considers caregiving within the built environment. When the built environment is conducive for the elderly, there is no doubt that half of the battle is already won in combating the challenges of an aging population.

References

American Association of Retired Persons (AARP). (2019). *2018 home and community preferences: A national survey of adults ages 18 plus*. Washington, DC. https://www.aarp.org/research/topics/community/info-2018/2018-home-community-preference.html

Ding, G. K. C. (2012). *Environmental assessment tools, handbook of sustainability management* (pp. 441–471). Singapore: World Scientific Publishing.

Milova, E. (2017). *Four main economic implications of an aging population, and how life extension technologies could solve them*. https://www.lifespan.io/news/4-main-economic-implications-of-an-aging-population-and-how-life-extension-technologies-could-solve-them/

Siew, R. Y. J. (2016). Assessing the readiness of sustainability reporting tools (SRTs) for an age-friendly built

environment. *Journal of Financial Management of Property & Construction, 21*(2), 122–136.
Siew, R. Y. J., Balatbat, M. C. A., & Carmichael, D. G. (2013). A review of building/infrastructure sustainability reporting tools (SRTs). *Smart and Sustainable Built Environment, 2*(2), 106–139.
United Nations Department of Economic & Social Affairs (UNDESA). (2019). *World population ageing 2019*. New York. https://www.un.org/en/development/desa/population/publications/pdf/ageing/WorldPopulationAgeing2019-Highlights.pdf

GGI

► Integrated Urban Green and Grey Infrastructure

Ghost Estates and New Ruins

► Zombie Subdivisions

Global Goals

► Sustainable Development Goals from an Urban Perspective

Global Homelessness: Neoliberalism, Violence, and Precarious Urban Futures

Neoliberalism, Violence, and Precarious Urban Futures

Simon Springer
Centre for Urban and Regional Studies, Dicipline of Geography and Environmental Studies, University of Newcastle, Australia, Callaghan, NSW, Australia

Synonyms

Displaced; Dispossessed; Street-engaged

Definition

Homelessness represents a continuum of accommodation deficiency that extends from a condition of no shelter at all to pronounced housing stress or "emergent" homelessness. The phenomena and lived realities of homelessness and its associated experiences and emotional registers of insecurity are not geographically isolated, but instead play themselves out in communities all across the globe.

The unfolding geographies of violence in cities across the globe are marked by homelessness. Ever more people are losing access to one of life's most basic necessities in the wake of intensifying neoliberal reform. How 'street-engaged' people cope with this alienating experience is paramount to successful responses, which requires examining the lived experiences of the homeless. Studies of homelessness are well represented in the social sciences, yet surprisingly, relatively little attention is paid to the actual experience of being homeless. Future research must begin to offer a more complete picture of the daily lives, activities, and interactions of the homeless as they negotiate the everyday geographies of stigma, exclusion, survival, and care.

Greater insight into how homelessness is experienced is pivotal to any solution, which cannot be achieved without input from those most affected. Thus, the voices of homeless peoples should be placed at the centre of any analysis. Such an approach seems a vital first step in overcoming the invisibleness and silencing that characterizes the current neoliberal moment, a situation wherein governments fail to acknowledge how the social policies and political reforms that arise from competitive market-based strategies actually underscore significant forms of social marginalization. As countries around the world continue to experiment with ongoing economic reform along neoliberal lines, the implications for global society should be clear.

Explanations of homelessness have covered mental illness (Schutt and Goldfinger 2011), drug abuse (Kemp et al. 2006), joblessness and unemployment (Grace et al. 2008), cuts in welfare benefits (Hackworth 2003), gentrification (Smith 1996), the destruction of skid row areas (Howard 2013), family breakdown (Song 2006), property relations (Blomley 2009), and the growing unaffordability of housing (Loftus-Farren 2011). Each of these explanations has played a role in the proliferation of the phenomenon. As appendages of the crisis of capitalism, these understandings of homelessness make it a domain that is ripe for charitable intervention (Jencks 1995), where historically this has come first as tragedy, and then as farce.

Programs and policies addressing homeless people all too often serve to alienate them, while consoling the cognitive dissonance of a global society that has produced this mess, yet fails to take responsibility. Yet instead of charity, Springer (2020a) has argued that mutual aid offers a critical intervention against the helplessness of waiting for state responsibility. While funds that could be allocated to alleviation are channelled into policing, surveillance, and criminalization in further stigmatizing the poor (Amster 2003), mutual aid instead builds a sense of agency and belonging. Mutual aid works to reconfigure the "geography of survival" (Mitchell and Heynen 2009), which describes the spatial relations that "structure not only how people may live, but especially *whether* they may live". Yet the practice of mutual aid has been attacked, notably through the criminalization of food sharing (Heynen 2010), which speaks to the contested notion of 'the right to the city', or more accurately its constriction for homeless peoples (Mitchell 2003).

Lancione (2011) has invigorated the study of homelessness, considering it as a subjective condition that arises from the entanglements of the individual and the city. She argues that official framings of homelessness fail to consider the connections between homeless people and the 'mechanosphere' (i.e., the relation between nature and culture) within the city. This focus on relationality is imperative to the framing of homelessness as it connects it to a broader understanding of geography. Specifically it speaks to a notion of 'integral geographies' (Springer 2016b) that is evident in the work of early anarchist geographers like Élisée Reclus (1894) and Peter Kropotkin (1902/2008). Thinking integrally can potentially move us towards a more empathetic horizon as we come to realize that we are all linked, whereby doing violence to one means doing violence to all. Such an extension of care constitutes a radical politics of homelessness that undermines the exclusion that frames the lives of homeless peoples (Arnold 2004).

Building a greater sense of affinity is vital to reducing the potential of homeless peoples to be rendered as 'bare life' (Agamben 1998), or lives that don't count in the view of authorities (Feldman 2006). Recognizing emotion within the lives of the homeless is crucial to realizing social justice, as it (re)positions homeless people as fully human (Cloke et al. 2008). The hostility of urban environments towards homeless peoples has been largely framed as an outcome of 'the revanchist city' (Wacquant 2009). Less widely acknowledged is a parallel rise of rights-based approaches (Sparks 2010) and the 'urge to care' (Hartnett and Harding 2005), evidenced by a growing number of shelters, hostels, and day centres that offer respite to homeless peoples. While these developments are part of an empathetic geography, there is room for skepticism. Springer (2016a, 2020b) has written about the vile façade that 'rehabilitation centers' represent for Cambodia's homeless, wherein abuse is disguised as kindness. There is also ambiguity within the notion of 'spaces of care', not only for the possibility of hidden violence, but also the potential for neoliberal cooptation (Klodawsky 2009).

Evoking the possibilities of a global framework of understanding homelessness, Smith (1996) speaks of scaling up our concerns. While this is an important geographical vision, Marston et al. (2005) advocate a 'flat ontology,' wherein the immediacy it offers may do more to help as connection is fostered through a horizontalism that networks people as equals. Horizontalism avoids the perils of hierarchical modes of effecting care (Springer 2014), as well as the problems that lurk within these spaces of poverty management (DeVerteuil 2003). Such a grounded ontological premise facilitates deeper reflection on the contested definitions of homelessness, which is crucial if we are to move towards a global frame of understanding (Springer 2000). This reflection is all the more necessary when we move to a global analysis, where cultural understandings of what homelessness entails differ considerably.

Indeed, the normative definition of 'home' should be contested, and recent geographical scholarship confirms that critical thinking around the question of definitions may potentially reveal greater understanding (Brickell 2012). In this vein, Veness (1993) encouraged us to think of the urban poor as often being neither 'homeless' nor 'housed'. What this speaks to is the spectrum of homelessness, from couch surfing to public shelters, and rough sleeping to fully street-engaged. Generic responses to homelessness are consequently inappropriate, and by globalizing our analyses, this does not mean we can offer universal solutions (Speak and Tipple 2006). Rather it allows us to draw attention to contextual specificity and the importance of an articulated view of neoliberalism as it rubs up against existing policy domains and institutional settings.

If we are to improve knowledge that will reduce the scope and potential of the marginalization neoliberalism sows, we need to start recognizing more clearly how those subjected to structural inequalities and a global economic agenda that has abandoned adequate social supports are actually experiencing a from of violence (Springer 2015). Homelessness itself should accordingly be considered as violence. To equate homelessness with violence offers new possibilities for critiquing neoliberalism by directly implicating the state, establishing a basis to explore non-state forms of reciprocity. Specifically, we can look to notions of mutual aid (Kropotkin 1902/2008) as improving care within our communities and on the wider global stage.

By exploring how cooperative forms of sharing are implemented in the everyday lives of homeless people, we may acquire not only a view of how the homeless survive, but also a lens on alternatives to capitalism in general, and neoliberalism in particular. If effective forms of cooperation can be engaged among those who have so little, this example provides hope for global society to be transformed in more collaborative ways. A truly global perspective on homelessness has not yet been adequately realized in the literature. By employing a global frame to the issue of homelessness we would begin to move towards significantly advancing geographical knowledge, with implications for urban studies, sociology, anthropology, development studies, and political science. Sound research carried out in cities experiencing sustained patterns of violence in the form of homelessness is a vital first step in increasing awareness within the academy and beyond.

Cross-References

- ▶ Housing Affordability
- ▶ Spatial Justice and the Design of Future Cities in the Developing World

References

Agamben, G. (1998). *Homo sacer: Sovereign power and bare life*. Stanford: Stanford University Press.

Amster, R. (2003). Patterns of exclusion: Sanitizing space, criminalizing homelessness. *Social Justice, 30*(1), 195–221.

Arnold, K. R. (2004). *Homelessness, citizenship, and identity: The uncanniness of late modernity*. Albany: State University of New York Press.

Blomley, N. (2009). Homelessness, rights, and the delusions of property. *Urban Geography, 30*(6), 577–590.

Brickell, K. (2012). Mapping' and 'doing' critical geographies of home. *Progress in Human Geography, 36*(2), 225–244.

Cloke, P., May, J., & Johnsen, S. (2008). Performativity and affect in the homeless city. *Society and Space, 26*(2), 241.

DeVerteuil, G. (2003). Homeless mobility, institutional settings, and the new poverty management. *Environment and Planning A, 35*, 361–379.

Feldman, L. C. (2006). *Citizens without shelter: Homelessness, democracy, and political exclusion*. Ithaca: Cornell University Press.
Grace, M., Batterham, D., & Cornell, C. (2008). Multiple disruptions: Circumstances and experiences of young people living with homelessness and unemployment. *Just Policy: A Journal of Australian Social Policy, 48*, 23–41.
Hackworth, J. (2003). Public housing and the rescaling of regulation in the USA. *Environment and Planning A, 35*(3), 531–550.
Hartnett, H. P., & Harding, S. (2005). Geography and shelter: Implications for community practice with people experiencing homelessness. *Journal of Progressive Human Services, 16*(2), 25–46.
Heynen, N. (2010). Cooking up non-violent civil-disobedient direct action for the hungry. *Urban Studies, 47*(6), 1225–1240.
Howard, E. (2013). *Homeless: Poverty and place in urban America*. Philadelphia: University of Pennsylvania Press.
Jencks, C. (1995). *The homeless*. Boston: Harvard University Press.
Kemp, P., Neale, J., & Robertson, M. (2006). Homelessness among problem drug users. *Health & Social Care in the Community, 14*, 319–328.
Klodawsky, F. (2009). Home spaces and rights to the city. *Urban Geography, 30*(6), 591–610.
Kropotkin, P. (1902/2008). *Mutual aid: A factor in evolution*. Charleston: Forgotten.
Lancione, M. (2011). *Homeless subjects and the chance of space*. Doctoral dissertation, Durham University.
Loftus-Farren, Z. (2011). Tent cities: An interim solution to homelessness and affordable housing shortages in the United States. *California Law Review, 99*(4), 1037–1081.
Marston, S. A., Jones, J. P., & Woodward, K. (2005). Human geography without scale. *Transactions of the Institute of British Geographers, 30*(4), 416–432.
Mitchell, D. (2003). *The right to the city: Social justice and the fight for public space*. New York: Guilford Press.
Mitchell, D., & Heynen, N. (2009). The geography of survival and the right to the city. *Urban Geography, 30*(6), 611–632.
Reclus, E. (1882–95). In E. G. Ravenstein & A. H. Keane (Eds.), *The earth and its inhabitants: The universal geography, vol I*. London: JS Virtue.
Schutt, R. K., & Goldfinger, S. M. (2011). *Homelessness, housing, and mental illness*. Cambridge, MA: Harvard University Press.
Smith, N. (1996). *The new urban frontier: Gentrification and the revanchist city*. East Sussex: Psychology Press.
Song, J. (2006). Family breakdown and invisible homeless women. *Positions: East Asia Cultures Critique, 14*(1), 37–65.
Sparks, T. (2010). Broke not broken: Rights, privacy, and homelessness in Seattle. *Urban Geography, 31*(6), 842–862.
Speak, S., & Tipple, G. (2006). Perceptions, persecution and pity: The limitations of interventions for homelessness in developing countries. *International Journal of Urban and Regional Research, 30*(1), 172–188.
Springer, S. (2000). Homelessness: A proposal for a global definition and classification. *Habitat International, 24*(4), 475–484.
Springer, S. (2014). Human geography without hierarchy. *Progress in Human Geography, 38*(3), 402–419.
Springer, S. (2015). *Violent neoliberalism: Development, discourse, and dispossession in Cambodia*. New York: Palgrave Macmillan.
Springer, S. (2016a). Homelessness in Cambodia: The terror of gentrification. In K. Brickell & S. Springer (Eds.), *The handbook of contemporary Cambodia*. London: Routledge.
Springer, S. (2016b). *The anarchist roots of geography: Towards spatial emancipation*. Minneapolis: University of Minnesota Press.
Springer, S. (2020a). Caring geographies: The COVID-19 interregnum and a return to mutual aid. *Dialogues in Human Geography, 10*(2), 112–115.
Springer, S. (2020b). The violence of homelessness: Exile and arbitrary detention in Cambodia's war on the poor. *Asia Pacific Viewpoint, 61*(1), 3–18.
Veness, A. R. (1993). Neither homed nor homeless: Contested definitions and the personal worlds of the poor. *Political Geography, 12*, 319–340.
Wacquant, L. (2009). *Punishing the poor: The neoliberal government of social insecurity*. Durham: Duke University Press.

Global Pandemic Fallout

▶ Impact of Universities on Urban and Regional Economies

Global Partnership

▶ Sustainable Development Goals from an Urban Perspective

Global Social Contract

▶ Sustainable Development Goals from an Urban Perspective

G

Global Survey of Food Waste Policies

Curt J. Davis
Independent Scholar, University of Delaware, Newark, DE, USA

Synonyms

Food loss; Food losses and waste; Organic waste

Definition of Food Waste

The term "food waste" is often generically used by the layperson to refer to what scholars and policy experts categorize as *food losses and waste* (FLW). Both food "loss" and "waste" occur across multiple sectors beginning upstream at the farms where food is grown and extending downstream to the homes or restaurants where it is ultimately either consumed or discarded. Definitions of food waste are often ambiguous and inconsistent (Chaboud and Daviron 2017), leaving the specific categorization and qualification of what constitutes as a "loss" opposed to a "waste" open to much interpretation.

In a report that was prepared for the Food and Agriculture Organization (FAO) of the United Nations (UN) – and perhaps the most cited report on "food waste" to date – Gustavsson et al. (2011) define food *losses* as "the decrease in edible food mass throughout the *part of the* supply chain that specifically leads to edible food for human consumption...[taking] place at production, post-harvest and processing stages in the food supply chain" (p. 2; italics original). An important distinction clarified with this definition is that agricultural biomass not intended for direct human consumption, such as crops for biofuels or the vines, plants, or trees on which fruit and vegetables are grown, are not categorized as food loss. Therefore, unharvested edible crops that are left to rot in fields fit into this characterization of "loss." Although this serves as a good starting point, classifications of FLW vary globally in organizations across municipalities, states, provinces, departments, and nations. For example, in the European Union (EU), agricultural produce is not considered "food" until after it has been harvested with the intention of human consumption (Reynolds et al. 2020).

Alternatively, food *waste* – in the narrow sense as a singular component of the FLW acronym – generally includes downstream food that was intended for human consumption but does not ultimately get eaten because of actions – whether voluntary or not – that occur at the retail and/or consumer stages. Examples of this include disposal because of spoilage, storage logistics (i.e., an overabundance of food), taste preferences, or excessive meal portions. In summary, food *loss* happens upstream and is usually attributed to unavoidable logistical inefficiencies during harvest, production, and delivery, whereas food *waste* happens downstream as the result of retailer and/or consumer actions – or inactions – that may or may not be intentional.

The definitions and classifications of food loss and waste become conceptually more challenging when viewed from different linguistical and cultural perspectives, since the very definition of "food" is open to interpretation as seen with the example of the EU. Also, parts of "food" including stalks, seeds, peels, shells, bones and marrow, etc., may be considered edible in some culinary cultures and not in others (Reynolds et al. 2020). In this chapter, the terms food loss and food waste will be employed within the context of each given policy, voluntary action, campaign, etc., as opposed to attempting to transpose each municipality's or nation's "food loss" or "food waste" law to fit one universal definition.

Introduction

The FAO estimates that about a third of all food grown for human consumption is wasted every year around the world (Gustavsson et al. 2011). When considering only food waste, the United Nations Environment Programme (UNEP) estimates that in 2019 approximately 931 million tonnes, including "inedible parts" of downstream waste was generated worldwide. Of this waste, over half came from households (61%), followed by food service (26%) and retail (13%) (Forbes et al. 2021).

Historically, the primary concern over food waste has been the social impacts of food insecurity. It is widely understood that certain regions of the world are relatively food secure, while others experience epochal famines. However, even in higher-income countries, many urban areas experience the tragic reality where mass amounts of food are discarded, while many residents go hungry. Although there have been considerable global achievements in addressing hunger, food insecurity continues to be a mostly solvable global problem in both lesser- and higher-income countries.

Looking beyond these social injustices, there also exists the logistical challenge of the disposal of food waste. Rotting food attracts insects, rodents, and other organisms, and can foster the growth of dangerous bacteria and lead to diseases. Therefore, it requires the same careful planning and infrastructural considerations as managing human waste. Traditional solutions, such as landfills and incinerators, however, may lead to environmental impacts, such as air, soil, and water contamination in local and regional communities.

In addition to direct contamination of the environment that can be calculated immediately, the anaerobic decomposition of organic materials, such as food waste, leads to the release of methane, which has a higher global warming potential than carbon dioxide (United Nations Environment Programme and Climate and Clean Air Coalition 2021). In 2011, the FAO estimated that the total carbon footprint of FLW – including considerations for the land use change required to grow food – was approximately 4.4 gigatons of carbon dioxide equivalence per year, which accounts for approximately 8% of total global anthropogenic carbon emissions (Scialabba 2015). In addition to the climate impacts and environmental degradation, significant amounts of water and energy are required to not only grow, process, and deliver food, but also to dispose of it.

When one considers the issue through the lens of the energy-water-food nexus, the environmental impacts of FLW are even more significant. For example, Cuéllar and Webber (2010) estimated the embedded energy cost in food waste in the USA to be the equivalent of approximately 2% of the annual energy consumption, while Skaf et al. (2021) place estimates at 22 megatonnes of oil equivalence. Even as the world transitions to more sustainable forms of energy production and efficiency, water will continue to be a precious and increasingly scarce resource essential for agriculture. In just the USA alone, the embedded water in food waste is estimated at 11 billion cubic meters annually (Skaf et al. 2021).

Mainstream awareness of the myriad social and environmental impacts related to food waste has grown in the past decade and has been the subject of numerous articles, books, and documentary films (Skaf et al. 2021; Treutwein and Langen 2021). With multiple UN agencies invested in addressing this far-reaching development concern, the 2030 Agenda for Sustainable Development was careful to include target 12.3 of the Sustainable Development Goals (SDGs): "[b]y 2030, halve per capita global food waste at the retail and consumer levels and reduce food losses along production and supply chains, including post-harvest losses" (United Nations, n.d.). Creating goals and targets is an important first step, but they must be followed with strong policy frameworks and action to address the full spectrum of FLW.

Global Food Waste Policies

Just as there is ambiguity and inconsistency with regards to the definitions of food, losses, and waste, the word "policy" is not used consistently across governments, academia, civil society, or the media. Policies can be laws or guidelines that are established and enforced at the federal, state, and local levels. Whereas laws can be enforceable, guidelines are not. In many cases, a federal "policy" is actually more of an obligation to create guidance on an issue, sometimes placing the responsibility for action on states and municipalities. Other times, a policy can refer to a law that is enforceable and punishable by fines, imprisonment, or other legal action.

Rethink Food Waste through Economics and Data (ReFED), a multi-stakeholder American nonprofit organization dedicated to reducing food waste, offers a "Policy Finder" page on its website. At the state level for the USA, it lists three categories of "policy": prevention, recovery, and recycling (ReFED 2021). Focusing primarily on downstream food waste, the subcategories for each policy are as follows:

- Prevention: date labeling
- Recovery: liability protection, and tax incentives
- Recycling: animal feed, organic waste bans, and waste recycling laws

For the federal level, which will be discussed later in the section on North America, it lists three policies related to liability protection, tax incentives, and animal feed (ReFED 2021).

Similarly the EU has a website that offers information on the targets, measures, and acts of its 27 member states toward reducing food waste. The "acts" subcategory not only includes policies and legislative measures, but also information related to *initiatives* to reduce food waste (European Commission n.d.-b). Thus, the classification of what constitutes a food waste "policy" varies depending on the scholar, publisher, organization, reporting agency, and media outlet. This chapter is meant to serve as an overview of global food waste policies in the broadest sense and is not geographically exhaustive. Instead, it will highlight some of the most ambitious and novel food waste awareness and reduction campaigns and policies in select nations representing some of the largest economies and scales of both food production and consumption.

Europe

For years, many highly industrialized and technologically advanced European countries, such as Germany, Italy, France, and the United Kingdom (UK), have been at the global forefront of developing and implementing alternative organic waste treatment technologies, such as anaerobic digestion (Akhiar et al. 2020). The region has also been home to some of the most progressive food waste reduction targets and policies. In line with SDG 12.3, the EU has set a goal of reducing food waste by 50% by 2030 and since 2020 has required that each member state report their annual amount of food waste (Win 2018).

To help coordinate efforts to meet this goal, the EU Platform on Food Losses and Food Waste was formed in 2016. The Platform includes experts from various EU institutions with the goal of shared learning on long-term planning, best practices, and evaluation of FLW trends. The Platform's mandate will run through the end of 2021 after which it will establish a new working group with additional members running from 2022 until 2026 (European Commission n.d.-a). The Platform fits within the EU's Farm to Fork Strategy, which aims to address FLW while ensuring the sustainability of food production, processing, distribution, and consumption. It also contributes to the European Green Deal to confront climate change (European Commission n.d.-c) and the EU's Circular Economy Action Plan.

Almost every EU member state has established a target for food waste reduction (as of October 2021, information for Cyprus, Greece, and Malta was not available and targets for Slovakia and Slovenia were "forthcoming"). Most have some type of plan or program to address food waste through public awareness and education programs, requirements for clear date markings, corporate tax credits and incentives for donated foods, or guidance for retailers and food service industries pertaining to the hierarchy of preferred food redistribution and/or disposal (European Commission n.d.-b).

France has received the most attention in terms of food waste policies as the first nation to implement a food waste ban at the retail level. The 2016 Supermarket Law requires food retailers with a floor plan greater than or equal to 400 m^2 to sign binding contracts with local food banks and other nonprofit organizations that redistribute food to the hungry (Reynolds et al. 2020). Through the provisions outlined in the contract, unsold food that is accepted by the food banks is diverted from landfills. Supermarkets that do not sign contracts face financial penalties. The law also prohibits retailers from purposely tainting or contaminating food in an attempt to prevent people from foraging through dumpsters and it provides legal protection in the event that people get sick from food that was donated in good faith (Bunting Eubanks 2019). Unlike other similar liability protection laws that were created primarily to target hunger, such as in the USA, environmental concerns of food waste were also driving impetuses for the French law (Bunting Eubanks 2019). In addition to implementing a ban on food disposal, the policy also established a hierarchy of strategies to fight food waste.

Italy passed the "Gadda" food waste law in 2016, but unlike the French law, the Italian law offers incentives for food donation rather than penalties. Under the law, retailers who donate unsold food may be eligible for tax deductions from municipalities after they report their donations. In addition, the law also established a multi-sectoral panel of experts to coordinate efforts to curb food wastage (European Commission n.d.-b). Switzerland, Poland, and the Czech Republic have also instituted laws that ban food waste at supermarkets similar to France and Italy (Leket Israel 2020). In addition to national legislation, Italy is also responsible for having a significant global reach in the effort to reduce food waste through the formation of the Milan Urban Food Policy Pact (MUFPP) launched in 2015.

The MUFPP is the first ever international agreement to unite the mayors of urban areas on food policy issues. The Pact offers guidance on a total of 37 actions clustered into six categories – including food waste – related to food policy. As of October 2021, the MUFPP's website lists 217 signatory cities distributed across every region of the world, including 98 in Europe (Milan Urban Food Policy Pact, n.d.).

In addition to EU member states, many non-member states have also taken very pro-active steps toward reducing food waste in the region. The UK has made much progress in raising awareness of food waste and promoting voluntary action by supermarkets. An example of this is the Courtauld Commitment 2030, which is a voluntary agreement of cooperation and collaboration between virtually every actor spanning the food chain in the UK with a shared commitment to reduce food waste, carbon emissions, and water stress (WRAP, n.d.).

Although the UK is no longer a member of the EU, from September 2016 until January 2020 the EU funded the TRiFOCAL London project In the UK's largest urban area. The pilot program facilitated awareness campaigns to reduce food waste, promote food waste recycling, and encourage healthy eating behaviors (WRAP et al. 2020). As an EU-funded program, the final deliverable of the program was to share lessons learned with eleven other cities throughout the member states (Treutwein and Langen 2021).

In addition to the UK, other economically powerful non-EU members are making efforts to reduce FLW. In June 2017, five ministries within the Norwegian government signed an agreement with a dozen food industry organizations to meet SDG 12.3 (Ministry of Climate and Environment 2017). Although the agreement is voluntary, government ministries believe that by extending the agreement to all sectors across the entire food value chain, each actor can make the best decisions on how to reduce waste without imposing disproportionate burdens of responsibility on any one actor. Switzerland has also adopted the SDGs and data on food waste has been collected by the federal government since 2013 when it adopted a measure for its reduction. In 2019, a postulate was adopted that mandated the creation of an action plan on food waste prevention (Federal Office for the Environment 2021).

Asia

With its significantly larger aggregate population and higher rates of income inequality than in Europe, the issue of food insecurity is a much more daunting social issue for many Asian nations. Combined with the environmental impacts related to food production, food waste in Asia is becoming an increasingly relevant issue. However, unlike the EU with its guiding frameworks for member states, Asian nations are left to create their own targets and action plans to confront food waste.

Japan enacted its first national food waste policy in 2001 with the Food Waste Recycling Law, which underwent subsequent revisions in 2007 and 2015 (Reynolds et al. 2020). As the name suggests, the law focuses mainly on reducing food waste through recycling measures with the 2020 burden highest for manufacturers who must recycle 95% of food waste and lowest for restaurants (50%) (Reynolds et al. 2020). In May 2019, the Japanese Diet enacted the Food Loss Act, which calls for greater collaboration between government, business, and consumers (*Japan Enacts Law Calling for "national Movement" to Slash 6 Million Tons of Food Wasted Annually*, 2019). Although the national government does not clearly outline which steps should be taken, it delegates the responsibility of confronting food waste to municipalities (Amoroso 2021) of which five cities have signed on to the MUFPP (Milan Urban Food Policy Pact, n.d.).

China has a new law passed in late April 2021 that targets frivolous over portioning and consumption of food (Liu 2021). Important measures of the 32-clause law are the banning of binge-eating videos, and outlawing restaurants' practices of encouraging customers to order more food than they can eat. Both measures are punishable by fines. In addition, the law allows restaurants to implement surcharges on uneaten food by customers (Liu 2021).

The 2021 law comes after President Xi Jinping launched the "Operation Empty Plate" campaign in 2020. The campaign encourages restaurant diners to order one less plate than normal as a way of avoiding uneaten food. Although the campaign attempts to avoid unnecessary food waste, critics have pointed out that the campaign is overreaching and targets a demographic (lower income bracket) that is already thrifty and averse to waste (Kuo 2020). At the municipal level, four Chinese cities have signed on to the MUFPP, including Beijing and Shanghai (Milan Urban Food Policy Pact, n.d.).

The South Korean government has taken an aggressive stance to curb waste. In 2005, food waste was banned from landfills and by 2013 a compulsory food waste recycling law required households to purchase biodegradable disposal bags (Broom 2019) and pay for food waste disposal by weight. At 95%, South Korea has one of the highest rates of food waste recycling in world with most of it being used to create fertilizer and animal feed (Broom 2019). In addition to the capital city, Seoul, three other cities have signed on to the MUFPP (Milan Urban Food Policy Pact, n.d.).

By some estimates, Indonesia is considered to be the second-largest food waste producer in the world; however, it does not have a national food waste policy (Reynolds et al. 2020). Two Indonesian cities have signed on to the MUFPP (Milan Urban Food Policy Pact, n.d.).

In a comprehensive report on FLW prepared by the World Resources Institute in August 2021, Agarwal et al. (2021) indicate that there is not much in the way of federal policies to address either food loss or waste in India. However, in an effort to encourage food donations to reduce food waste, especially in urban areas, in 2017 the Indian government created a social platform called the Indian Food Sharing Alliance (Agarwal et al. 2021). In some cities, such as Pune – the only city in India to sign on to the MUFPP (Milan Urban Food Policy Pact, n.d.) – communities are using digesters to convert their organic waste to biogas, which is then used to produce community electricity with gas-powered generators (Basu 2021).

Oceania

In 2017, Australia announced their National Food Waste Strategy to reduce food waste by 50% by 2030 with four priority areas: policy support, business improvements, market development, and behavior change (Commonwealth of Australia 2017). The four components of policy support include: measurements for both a baseline calculation of food waste as well as progress toward meeting the goal, identifying the most lucrative areas for investment, the creation of a voluntary reduction program, and enabling legislation that can best support the reduction and recycling/repurposing of food waste (Commonwealth of Australia 2017).

At the state level, New South Wales will begin requiring the separate collection of food and organic waste from high-volume hospitality businesses and supermarkets by 2025, followed by households by 2030 (State of New South Wales through Department of Planning, Industry and Environment 2021). At the municipal level, both Sydney and Melbourne have signed on to the MUFPP (Milan Urban Food Policy Pact, n.d.).

In New Zealand, food waste policy has fallen under the umbrella of the broader Waste Minimisation Act 2008 (Reynolds et al. 2020). There is a Waste Minimisation Fund that supports initiatives to rescue and redirect edible food to needy families, as well as diverting unused fish parts to groups that value them. The fund also supports educational programs to teach people about the impacts of food waste and how to sustainably manage it (Ministry for the Environment 2021). The City of Wellington, which signed on to the MUFPP in 2021 (Milan Urban Food Policy Pact, n.d.), has a Long Term Plan 2018–2028 that includes a commitment to researching sustainable options to reduce food waste (New Zealand Government 2020).

Middle East and Africa

Information on food waste reduction strategies and policies in the Middle East region is sparse (Reynolds et al. 2020), which is not to say that it does not exist. In fact, it is such a problem in parts of the region that by some estimates, the average resident of both the United Arab Emirates (UAE) and the Kingdom of Saudi Arabia (KSA) wastes approximately double the global average of food (Sheldon 2020) with the latter ranking number one in the world in food waste per capita (Reynolds et al. 2020).

In the UAE, the government has not set any food waste policy; however, it has asked hotels and restaurants to take voluntary actions toward its reduction with the use of high-tech software applications (Sheldon 2020). No formal food waste policy has been set by the government of KSA either although an action plan was created in 2014 (Reynolds et al. 2020). As of 2020, the KSA government was considering several proposals for policies including one that would fine diners at restaurants who waste food excessively (Reynolds et al. 2020), similar to the law mandated in China in 2021.

Similar to the "good Samaritan" laws in other countries like the USA, Israel adopted the Food Donation Act in 2018 to protect food donors from liability. Although there are currently no other food waste laws in Israel, they are dedicated to addressing the issue and the National Food Bank of Israel, Leket Israel, has a remarkably comprehensive listing of global food waste policies on their website and offers ample reports and education materials (Leket Israel 2020).

Cities of the region that have signed on to the MUFPP include Dubai, Bethlehem, Kfar Saba, Herzliya, Tel Aviv, and Ramat Gam (Milan Urban Food Policy Pact, n.d.).

Much like the EU, the African Union (AU) provides leadership in the region with regards to goals and targets. In 2013, the AU launched the Agenda 2063 with 20 development goals. Although none of the goals align directly with SDG 12.3, goals 1 (quality of life) and 5 (modern agriculture) connect with SDG 2 (ending hunger and promoting sustainable agriculture for food security and nutrition) (African Union, n.d.). For most of Africa, addressing upstream food loss is a higher priority than addressing downstream food waste. To that end, AU heads of state signed the Malabo Declaration of Accelerated Agricultural Growth and Transformation for Shared Prosperity and Improved Livelihoods in June 2014, which includes a commitment to halve post-harvest crop losses by 2025 over 2014 levels (African Union Development Agency 2016).

Although there is a general lack of FLW laws across most of Africa, some governments have taken steps to facilitate voluntary agreements. An example is the South African Food Loss and Waste Voluntary Agreement, launched in September 2020, in which signatory retailers and food manufacturers must commit to SDG 12.3 (Department of Forestry, Fisheries and the Environment 2020). Despite the slow progress toward food waste policy at the national level, 38 cities in Africa have signed on to the MUFPP (Milan Urban Food Policy Pact, n.d.).

South America

As is the case in Africa, the main impetus for addressing FLW in Latin America is food security. As members of the Community of Latin American and Caribbean States (CELAC), 33 countries in the region have pledged to support the Plan for Food and Nutrition Security and the Eradication of Hunger 2025 (FAO 2014). As part of this plan, the region is committed to reducing FLW by 50% by 2025 (FAO 2017).

Countries with national food waste laws or policies include Argentina, Colombia, Peru (Harvard Law School Food Law and Policy Clinic and The Global FoodBanking Network 2021), and Brazil. Enacted in June 2021, the Brazilian law allows for the donation of safe and edible food to food banks and charities (FAO 2021). Argentina has a similar law that provides protections to donors (Argentina.gob.ar 2019). Although the Chilean government has not enacted any FLW laws, it has created the National Committee for the Prevention of Food Losses and Waste (Harvard Law School Food Law and Policy Clinic and The Global FoodBanking Network 2021). There are 20 MUFPP signatory cities in South America (Milan Urban Food Policy Pact, n.d.).

North America

Food waste is becoming an increasingly relevant issue in North America with numerous awareness campaigns and news stories covering the topic. As members of the Commission for Environmental Cooperation, the governments of Canada, Mexico, and the United States are working together to address FLW (Government of Canada 2020).

Canada, where over 50% of food is wasted annually (Impact Canada 2021), is committed to SDG 12.3. Initiatives to address FLW are organized in the Strategy on Short-lived Climate Pollutants, and the Food Policy for Canada, with the latter sponsoring the national Food Waste Reduction Challenge (Government of Canada 2020). The Challenge offers cash prizes to individuals, organizations, or businesses who develop innovative and replicable solutions to reduce food waste (Impact Canada 2021).

Metro Vancouver has had a ban on organic waste since January 2015, which applies to all residents and businesses. The ban is enforced at the point of disposal through visual inspection and the penalty of a 50% surcharge is applied to waste loads considered to contain excess food waste (Metro Vancouver 2021). In addition to Vancouver, Toronto and Montreal have also signed on to the MUFPP (Milan Urban Food Policy Pact, n.d.).

Mexico has no national FLW policy; however, federal law allows tax deductions on annual income of up to 12% for individuals and business for eligible food donations (Harvard Law School Food Law and Policy Clinic and The Global FoodBanking Network 2021). As one of – if not *the* – largest metropolitan area in North America, Mexico City has signed on to the MUFPP along with Guadalajara and Mérida. Signatory cities in Central America include Guatemala City, San Salvador, Tegucigalpa, and Santa Ana in Costa Rica (Milan Urban Food Policy Pact, n.d.).

The American federal government has committed to address food waste and has established a formal cooperation and coordination agreement between multiple agencies, including the Environmental Protection Agency (EPA), the Food and Drug Administration (FDA), and the Department of Agriculture (USDA) (FDA 2020). Both the EPA and FDA have websites that offer guidance on how to prevent food waste at home and while eating out (EPA 2021; US FDA 2019).

In terms of federal law, the US Congress passed the Bill Emerson Good Samaritan Food Donation Act in October 1996. The law provides federal protection to food donors, such as retailers, restaurants, food service providers, farmers, food banks, and others, from civil and criminal liability if food is donated in "good faith" (USDA, n.d.). Essentially, any food that is deemed to be fit for safe human consumption can be donated without the risk of legal repercussions if the recipient should fall ill because of pathogens, bacteria, etc. Some legal scholars argue that US federal laws were established to only target food insecurity and not the environmental impacts of food waste (Bunting Eubanks 2019). Thus, there is no federal law that prohibits the disposal of food, although such laws do exist at the state and local levels (Schultz 2017).

According to ReFED's policy finder, as of October 2021 five states: California, Connecticut, Massachusetts, Rhode Island, and Vermont, along with six municipalities or counties: Austin, Boulder, Minneapolis and Hennepin County, New York City, San Francisco, and Seattle, have enacted bans on organic waste (ReFED 2021). Most of these bans, such as in Massachusetts (Commonwealth of Massachusetts 2021), target high volume organic waste producers and require that institutions who produce more than a set threshold per week/month divert food waste from landfills to reuse, recycling, composting, or conversion facilities. In addition to a few states making progress with legally binding policies, 13 US cities have signed on to the MUFPP (Milan Urban Food Policy Pact, n.d.).

Summary

The social, environmental, and economic impacts of food waste are clear. Fortunately, many cities, states, nations, and regional alliances are taking action to address the issue. As the world continues to become more urbanized, waste management in general will continue to present challenges to the welfare of cities. Several good examples of successful food donation policies exist, especially in Europe. Elsewhere, food waste bans in places like Seoul offer excellent solutions on how to both manage organic waste in cities, while encouraging home composting and urban farming (Kim 2019). Above all, organizations like the UN and MUFPP offer frameworks for goals and commitments, which can lead to multi-sectoral and even multinational collaborations to achieve success and share lessons learned.

Cross-References

- ▶ Circular Cities
- ▶ Circular Economy and the Water-Food Nexus
- ▶ Circular Economy Cities
- ▶ Circular Water Economy
- ▶ Formulating Sustainable Foodways for the Future: Tradition and Innovation
- ▶ Future Foods for Urban Food Production
- ▶ Green Cities
- ▶ Green Economy Policies to Achieve Water Security
- ▶ Integrating Agriculture, Forestry, and Food Systems into Urban Planning: A Key Step for Future Resilient and Sustainable Cities
- ▶ Planning for Food Security in the New Urban Agenda
- ▶ Sustainable Development Goals
- ▶ Sustainable Development Goals from an Urban Perspective
- ▶ Sustainable Development Goals in Relation to Urban and Regional Development in Japan
- ▶ Urban Food Gardens
- ▶ Urban Policy and the Future of Urban and Regional Planning in Africa
- ▶ Voluntary Programs for Urban and Regional Futures
- ▶ Vulnerability to Food Insecurity Among the Urban Poor in Sri Lanka: Implications for Policy and Practice

References

African Union. (n.d.). *Linking agenda 2063 and the SDGs | African Union*. Retrieved October 31, 2021, from https://au.int/agenda2063/sdgs

African Union Development Agency. (2016). Malabo declaration on accelerated agricultural growth and transformation for shared prosperity and improved livelihoods. African Union Commisssion. https://www.nepad.org/caadp/publication/malabo-declaration-accelerated-agricultural-growth.

Agarwal, M., Agarwal, S., Ahmad, S., Singh, R., & Jayahari, K. M. (2021). Food loss and waste in India: The knowns and the unknowns. *World Resources Institute*. https://doi.org/10.46830/wriwp.20.00106.

Akhiar, A., Ahmad Zamri, M. F. M., Torrijos, M., Shamsuddin, A. H., Battimelli, A., Roslan, E., Mohd

Marzuki, M. H., & Carrere, H. (2020). Anaerobic digestion industries progress throughout the world. *IOP Conference Series: Earth and Environmental Science, 476*, 012074. https://doi.org/10.1088/1755-1315/476/1/012074.

Amoroso, P. (2021, January 30). *How COVID-19 is forcing us to re-examine food waste*. The Japan Times. https://www.japantimes.co.jp/life/2021/01/30/food/covid-19-food-waste-japan/

Argentina.gob.ar. (2019, April 4). *Se reglamentó el Plan Nacional de Reducción de Pérdidas y Desperdicios de Alimentos y la Ley Donal*. Argentina.gob.ar. https://www.argentina.gob.ar/noticias/se-reglamento-el-plan-nacional-de-reduccion-de-perdidas-y-desperdicios-de-alimentos-y-la

Basu, M. (2021, September 3). Waste to watts: India generates green energy from food leftovers. *Reuters*. https://www.reuters.com/article/us-india-climate-change-biogas-idUSKBN2FZ110.

Broom, D. (2019, April 12). *South Korea once recycled 2% of its food waste. Now it recycles 95%*. World Economic Forum. https://www.weforum.org/agenda/2019/04/south-korea-recycling-food-waste/

Bunting Eubanks, L. (2019). From a culture of food waste to a culture of food security: A comparison of food waste law and policy in France and in the United States. *William & Mary Environmental Law and Policy Review, 43*(2), 23.

Chaboud, G., & Daviron, B. (2017). Food losses and waste: Navigating the inconsistencies. *Global Food Security, 12*, 1–7. https://doi.org/10.1016/j.gfs.2016.11.004.

Commonwealth of Australia. (2017). *National food waste strategy: Halving Australia's food waste by 2030* (p. 54). https://www.awe.gov.au/sites/default/files/documents/national-food-waste-strategy.pdf

Commonwealth of Massachusetts. (2021, October). Guide: commercial food material disposal ban. https://www.mass.gov/guides/commercial-food-material-disposal-ban.

Cuéllar, A. D., & Webber, M. E. (2010). Wasted food, wasted energy: The embedded energy in food waste in the United States. *Environmental Science & Technology, 44*(16), 6464–6469. https://doi.org/10.1021/es100310d.

Department of Forestry, Fisheries and the Environment. (2020, September 29). *Consumer Goods Council of South Africa (CGCSA) launches the South African food loss and waste voluntary agreement | Department of Environmental Affairs*. https://www.environment.gov.za/mediarelease/consumergoodscouncil_launches foodloss_wastevoluntaryagreement

EPA. (2021, July 2). *Reducing wasted food at home* [Overviews and factsheets]. https://www.epa.gov/recycle/reducing-wasted-food-home

European Commission. (n.d.-a). *EU platform on food losses and food waste*. EU platform on food losses and food waste. Retrieved October 26, 2021, from https://ec.europa.eu/food/safety/food-waste/eu-actions-against-food-waste/eu-platform-food-losses-and-food-waste_en

European Commission. (n.d.-b). *European food loss and waste prevention hub*. EU food loss and waste prevention hub. Retrieved October 27, 2021, from https://ec.europa.eu/food/safety/food_waste/eu-food-loss-waste-prevention-hub/eu-member-states

European Commission. (n.d.-c). *Farm to Fork strategy*. Retrieved October 26, 2021, from https://ec.europa.eu/food/horizontal-topics/farm-fork-strategy_en

FAO. (2014). The CELAC Plan for Food and Nutrition Security and the Eradication of Hunger 2025. 10.

FAO. (2017, November 13). Ecuador forms a network for the prevention and reduction of food waste | Agronoticias*: Agriculture News from Latin America and the Caribbean | Food and Agriculture Organization of the United Nations*. https://www.fao.org/in-action/agronoticias/detail/en/c/1059434/

FAO. (2021). *FAOLEX database. Law no. 14.016 providing for combating food waste and donating surplus food for human consumption*. https://www.fao.org/faolex/results/details/en/c/LEX-FAOC196025/

FDA. (2020). Formal agreement among EPA, FDA, and USDA relative to cooperation and coordination on food loss and waste. FDA. https://www.fda.gov/food/domestic-interagency-agreements-food/formal-agreement-among-epa-fda-and-usda-relative-cooperation-and-coordination-food-loss-and-waste

Federal Office for the Environment. (2021, February 15). Food waste. https://www.bafu.admin.ch/bafu/en/home/themen/thema-abfall/abfallwegweiser%2D%2Dstichworte-a%2D%2Dz/biogene-abfaelle/abfallarten/lebensmittelabfaelle.html

Forbes, H., Quested, T., & O'Connor, C. (2021). Food waste index report 2021. United Nations Environment Programme. https://www.unep.org/resources/report/unep-food-waste-index-report-2021.

Government of Canada. (2020, December 14). Food loss and waste. https://www.canada.ca/en/environment-climate-change/services/managing-reducing-waste/food-loss-waste.html.

Gustavsson, J., Cederberg, C., & Sonesson, U. (2011). *Global food losses and food waste: Extent, causes and prevention; study conducted for the International Congress Save Food! at Interpack 2011, [16–17 May]*. Düsseldorf: Food and Agriculture Organization of the United Nations.

Harvard Law School Food Law and Policy Clinic, & The Global FoodBanking Network. (2021). The global food donation policy atlas. Mapping the barriers to food donation. http://atlas.foodbanking.org.

Impact Canada. (2021). Food waste reduction challenge: business models. https://impact.canada.ca/en/challenges/food-waste-reduction-challenge.

Japan enacts law calling for "national movement" to slash 6 million tons of food wasted annually. (2019, May 24). The Japan Times. https://www.japantimes.co.jp/news/2019/05/24/national/japan-enacts-law-calling-national-movement-slash-6-million-tons-food-wasted-annually/

Kim, M. S. (2019, April 8). The Country winning the battle on food waste. HuffPost UK. https://www.huffpost.com/entry/food-waste-south-korea-seoul_n_5ca48bf7e4b0ed0d780edc54.

Kuo, L. (2020, August 13). "Operation empty plate": Xi Jinping makes food waste his next target. *The Guardian*. https://www.theguardian.com/world/2020/aug/13/operation-empty-plate-xi-jinping-makes-food-waste-his-next-target

Leket Israel. (2020, November). *Policy tools for reducing food waste and loss, in Israel and around the world – Food waste & rescue in Israel report*. https://foodwastereport2019.leket.org/en/policy-tools-for-reducing-food-waste-and-loss-in-israel-and-around-the-world/

Liu, C. (2021, April 29). *China adopts law against food waste; binge eating, excessive leftovers to face fines*. Global Times https://www.globaltimes.cn/page/202104/1222490.shtml

Metro Vancouver. (2021). *About food scraps recycling*. http://www.metrovancouver.org/services/solid-waste/recycling-programs/food-scraps-recycling/about/Pages/, http://www.metrovancouver.org:80/services/solid-waste/recycling-programs/food-scraps-recycling/about/Pages/default.aspx

Milan Urban Food Policy Pact. (n.d.). Milan urban food policy pact. Retrieved October 28, 2021, from https://www.milanurbanfoodpolicypact.org/

Ministry for the Environment. (2021, August 30). Reducing food waste. Ministry for the Environment. https://environment.govt.nz/what-government-is-doing/areas-of-work/waste/reducing-food-waste/.

Ministry of Climate and Environment. (2017, June 28). *Agreement to reduce food waste* [Nyhet]. Government. No; regjeringen.no. https://www.regjeringen.no/en/historical-archive/solbergs-government/Ministries/kld/news/2017/agreement-to-reduce-food-waste/id2558931/

New Zealand Government. (2020, February 21). Food waste in New Zealand. Wellington City Council. https://wellington.govt.nz/rubbish-recycling-and-waste/reducing-your-waste/reducing-waste-at-home/food-waste-in-nz.

ReFED. (2021, September 13). *U.S. food waste policy finder*. ReFED | Rethink food waste. http://policyfinder.refed.com

Reynolds, C., Soma, T., Spring, C., & Lazell, J. (Eds.). (2020). *Routledge handbook of food waste* (Vol. 1–1 online resource (xxx, 450 p)). Routledge; WorldCat.org. https://public.ebookcentral.proquest.com/choice/publicfullrecord.aspx?p=6012212

Schultz, J. (2017). Fighting food waste. *Legis Brief: National Conference of State Legislatures, 25*(46) https://www.ncsl.org/research/agriculture-and-rural-development/fighting-food-waste.aspx.

Scialabba, N. (2015). Food wastage footprint & climate change. Food and Agriculture Organization of the United Nations. https://www.fao.org/3/bb144e/bb144e.pdf.

Sheldon, M. (2020, March 6). Artificial intelligence to reduce food waste in UAE. NYC Food Policy Center (Hunter College). https://www.nycfoodpolicy.org/food-policy-snapshot-uae-food-waste/.

Skaf, L., Franzese, P. P., Capone, R., & Buonocore, E. (2021). Unfolding hidden environmental impacts of food waste: An assessment for fifteen countries of the world. *Journal of Cleaner Production, 310*, 127523. https://doi.org/10.1016/j.jclepro.2021.127523.

State of New South Wales through Department of Planning, Industry and Environment. (2021). NSW waste and sustainable materials strategy 2041. *Stage, 1*, 2021–2027. 44.

Treutwein, R., & Langen, N. (2021). Setting the agenda for food waste prevention – A perspective on local government policymaking. *Journal of Cleaner Production, 286*, 125337. https://doi.org/10.1016/j.jclepro.2020.125337.

United Nations. (n.d.). *Goal 12 | Department of Economic and Social Affairs*. Goals: 12: Ensure sustainable consumption and production patterns. Retrieved October 6, 2021, from https://sdgs.un.org/goals/goal12

United Nations Environment Programme and Climate and Clean Air Coalition. (2021). *Global methane assessment: Benefits and costs of mitigating methane emissions* (DTI/2352/PA). United Nations Environment Programme. https://www.unep.org/resources/report/global-methane-assessment-benefits-and-costs-mitigating-methane-emissions

US FDA. (2019, November 15). Tips to reduce food waste. FDA. https://www.fda.gov/food/consumers/tips-reduce-food-waste.

USDA. (n.d.). Frequently asked questions about the Bill Emerson Good Samaritan food donation act. United States Department of Agriculture. https://www.usda.gov/sites/default/files/documents/usda-good-samaritan-faqs.pdf

Win, T. L. (2018, April 18). *Europe takes major step to tackle billion-dollar food waste*. Reuters. https://www.reuters.com/article/us-eu-food-waste-law-idUSKBN1HP2SC.

WRAP. (n.d.). The Courtauld commitment 2030 | WRAP. Retrieved October 28, 2021, from https://wrap.org.uk/taking-action/food-drink/initiatives/courtauld-commitment

WRAP, LWARB, & Goundwork London. (2020). TRiFOCAL. transforming city food habits for life. Summary report. http://trifocal.eu.com/wp-content/uploads/2020/01/TRiFOCAL-Summary-Report.pdf.

G

Globalization and Financialization

► Financing: Fiscal Tools to Enhance Regional Sustainable Development

Governance

► Amsterdam's Pathway to Climate Neutrality: Creating an Enabling Environment
► New Localism: New Regionalism

The Governance of Smart Cities

M. Cavada
School of Architecture, Lancaster University, Lancaster, UK

Synonyms

City vision; Liveable cities; Smart cities; Smart (er) cities; Urban Futures;

Definition

How are smart cities governed? This chapter describes the design process for an open and transparent method to govern smart cities. This four-step methodology describes the essential steps into democratic decision-making across the city communities. These steps are Data Assessment, Dissemination, Evaluation, and Smart Design. In each step, a smart tool explores the potential outcomes to support discussion between the participants.

Debating whether a smart solution is appropriate for people is a public issue. Often, solutions that might be considered smart do not yield the desirable results, which should be to the benefit of enhancing liveability in cities (Cavada et al. 2016, 2017). This is a design process of four steps: (i) existing data assessment, (ii) dissemination, (iii) evaluation, and (iv) smart solution design. This process aims to create a design approach in governance for urban areas which can provide clarity and inclusion to the smart vision. For this process, we developed a collaboration between a team of academics, a team from the local governance, and members of the wider academic and policy community, to establish a model of designing the governance of becoming smart for Lancaster, UK. Working together, we explored local issues in Lancaster, concerning the urban environment and related policy. We assessed these policies using the Smart Model Assessment Resilient Tool (SMART) (Cavada et al. 2019) and presented the outcomes to a workshop to provide feedback for designing smart solutions. The intention is to build a design model which can be used for other cities and explore collaboration between academics, local governance, and the wider communities. We anticipate that results will support further collaboration between the city governance and academics aiming to design smartness into urban policy.

Governance and Smart Cities

City governance is multifaceted because governance affects a wide area of service and actors across the city. City governance is a way of managing the city resources, ensuring that the city itself, its organizations, and people can prosper in a safe and secure urban context. Governance should adapt a holistic approach in policy, city infrastructure, and services to reflect the urban context. Governance is needed to organize and manage all systems found in the city, for example, the hard infrastructure (civil and the built environment) and soft infrastructures in cities and the economy are some of them (Cavada et al. 2017). Contemporary cities face challenges often because they developed in a fast pace; often this happened because their fast development exacerbates urban challenges, for example, air pollution, arbitrary construction, and high cost of living. These are some of the issues of contemporary cities. Often, the city governance is concerned with unforeseeable outcomes and is called to solve these through the city governance, or implementing strategies and city policy. City governance aims to overcome these issues by the new city trends, for example, smartness and digitalization, and implement these practices in order to provide solutions and offer benefits to all. Specifically, when it comes to smart cities, the argument is that governance needs to adopt an open and democratic agenda as a way to provide these maximum benefits to people. In the next section, we describe how it is possible to design a process for the governance of smart cities.

A Design Approach into the Governance of Smart Cities

The concept of a smart city is considered a recent advancement, where suitable policy has not been adequately designed to facilitate smartness in a holistic way. Initially, the vision of smartness referred to a city which takes advantage of the digital services provision. Soon, this perception revealed weaknesses in ownership, governance, and implementation of the smart cities' delivery. However, there are also deeper issues revealed in the problem conceptualizing smartness, this is because the private companies involved in the service provision in smartness are focusing primarily on the fiscal outcomes of service provision. Smart cities became part of the consumer economy, and the role of an individual became the receiver of the service. With this in mind, there is a strong criticism when it comes to smartness because of the private services and the high costs of the services which are provided. For example, is everyone able to take part in the digital revolution that smart cities promise to deliver, or is this something that only those who can afford to be part of digitalization able to harness the benefits of?

How to Design Policy for the Governance of Smart Cities

Design for urban policy sets the foundations for understanding the needs and assets of a place before the decision-making on city matters is made by the policy officers. The role of design in urban policy is to demonstrate the appropriate actions needed which can provide benefits to the people and the urban context for prosperous living. It has its emphasis on the communication and relations between all actors in the city, while it aims to create a support system for public participation. Design can support smart cities by setting the appropriate steps, actions, and discussions between all actors in the city. The way to achieve this is to design a four-step approach, which is described in this text, to identify the important parts and provide clarity. This design approach is a new approach to the governance of smart cities, and we summarize it in four steps, which are explained below (Fig. 1).

In the smart cities' agenda, people should be at the heart of the benefits and also be part of the decision-making in the smart city. This model is designed to support the smart concept using a transparent method based on the Smart Model Assessment Resilient Tool (Cavada et al. 2018, 2019) The SMART is developed to provide a deeper understanding of smartness, by assessing these initiatives developed in the smart cities' agenda. The tool is adopted in the design of a model for a process to enhance participation. This is because smartness should change from a service provision agenda towards a participatory

The Governance of Smart Cities, Fig. 1 Design model for the smart cities' governance

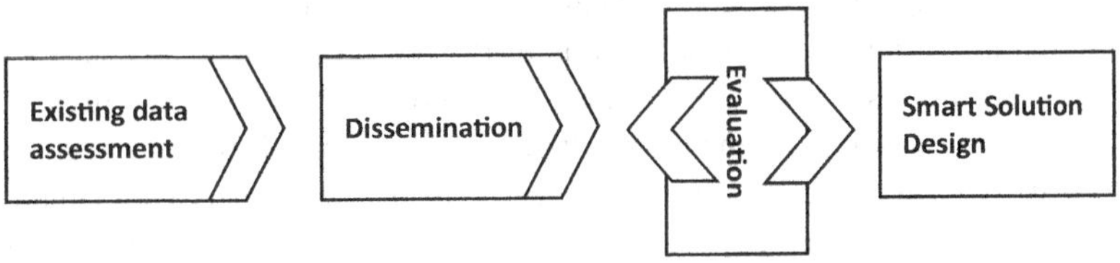

approach, we aim to change the top-down towards the bottom-up approach. The four steps are:

(i) **Existing data assessment**

This is the first step of the design model which starts with an exploration into the current policies in the city. This includes the information that reflects the city operations, for example, the city plan of a city and the collaborations with the city council or the unitarian authority. The city policies can be assessed using the criteria in the SMART. This can provide a critical appraisal across the four lenses (environmental, societal, economy, and governance) of the tool. The process is led by the principal researcher to ensure the clarity of the process along with a research team, where the principal researcher provides guidance and explanation in every step of the assessment. The outcomes of the assessment are collected and presented in a graphical and coherent manner.

(ii) **Dissemination**

In this second step, the results of the assessment are presented to a group of actors in the city; this means initially to the academics and policymakers. The reason being is to bring two different areas of interest in debating research and policy and introduce them in the subject of smartness. Following on, the aim is to bring in wider audiences to the discussion. Dissemination is a clear and transparent process, and the SMART is designed with this in mind, where the criteria, actions, and their KPIs are open and accessible. This is because the dissemination step is designed to be an exchange of critical points in the discussion. In more detail, actors are encouraged to discuss the outcomes of the assessment.

(iii) **Evaluation**

This design model is a continuous process, which might involve rethinking parts of the progression. In this step, we ensure that this is not a linear process of assessment and dissemination, rather a creative process which allows participants to provide feedback and suggestions that can allow for exploration loops in the process. A discussion between policy officers and academics allows consideration of the assessment and also the opportunity to design new practices on smartness, based on the presented outcomes. Therefore, actors through mutual understanding of local influences together with the SMART outcomes could design a smart initiative for the city. This exploration for new actions or initiatives can take place in the evaluation proceeds to reveal which actions might lead to new smart solutions. An evaluation process that can bring some of these issues to the discussion table and re-design parts of the solution or smart initiative.

(iv) **Smart solution design**

A smart solution design or a smart initiative is the development of an idea that, as part of the smart agenda, provides benefits to those who live in cities and the city itself. The SMART is designed to assess whether a number of smart initiatives provide benefits to the city and people. From the discussion between the participants, some key points are decided into which initiatives can support the local context of the city and also how well they satisfy the smart criteria set in the SMART. In the fourth stage of designing the smart process, it is possible to develop a new smart initiative, ensuring that is fulfilling all (or most) of the criteria are met. This could seem a challenging part of the design model. However, the discussion process so far provides understanding both in the local governance and policy and the way of assessing smartness using the SMART. By now, participants are familiar with the usability of the tool and the language used in the smart cities' conceptualization.

The fours-step design model described here is to be implemented in a UK city and document the process and real city outcomes. Lancaster is chosen to implement the design model and also using the SMART. There are two main reasons for this decision: primarily, due to the collaboration between the University of Lancaster, specifically the Beyond Imagination research team, and the Lancaster City Council. This allows communications between both organizations and supports understanding for mutual benefit. Lancaster is considered a medium-size city; therefore, an exploration in policy design for implementing the smart agenda would allow the conversation on local matters an easier process. For example, if we are exploring the smart agenda for bigger cities, Manchester, Birmingham, or even London, it would need a wider exploration of local policy and actors. Additionally, big urban centers have adopted smart city agendas already in their policy and local plans. Inevitably, the SMART assessment would provide criticism on the existing smart agenda and also note how the local agenda. For example, Birmingham and indeed London Smart City Plan are influenced by the digital service provision; this is evident by collaborations with the service providers. Therefore, making suggestions for Lancaster sets a new approach to smartness, as argued by Cavada et al. (2016) based on liveable, low-carbon solutions.

Conclusion: Implementation of the Design Model

This article demonstrates a transparent process for smart cities' governance. It described a design model and explore collaboration between academics, local governance, and the wider community for the subject of smart cities. The four steps of the design model aim to support further collaboration between the city governance and academics aiming to design smartness into urban policy. This is a model which can be used for other cities in their exploration of becoming smart. This process was based on the SMART, a tool developed to assess smartness and used Lancaster City. It described the design for assessing and developing smartness within local actors, specifically a group of policy officers and academics. The model is designed to focus on the usability of the SMART and demonstrate the path for communication between the actor groups. This is important for smart cities, because of the different needs that cities can have, for example, population size or their service provision. Due to communication uncertainties between different professional groups, the design progression allows for evaluation, rethinking the applicability of the actions for the smart city. This allows for considerate decision-making when designing smart initiatives, aimed to bring clarity and provide benefits for the city and those in it. Delivering the smart city is not a panacea for all city problems. However, smart city governance needs to be transparent to allow transparent decision-making on the smart cities' solutions. The four steps described in this document set the basis for communication between actors and allows for the discourse of smart cities. Within the context of service provision, a democratic approach is a novel approach, giving all actors a voice over their local matters. It is also essential for understanding the real needs and wants of people and not those of the service providers. It is anticipated that this is the basis for further research on the governance and participation in smart cities. This is because the design model explained here needs to include all communities, meaning public participation has the opportunity and the tools to be enhanced in the governance process.

Cross-References

- ▶ City Visions: Toward Smart and Sustainable Urban Futures
- ▶ Future of Urban Governance and Citizen Participation
- ▶ Smart City: A Universal Approach in Particular Contexts

Acknowledgment The author gratefully acknowledges the financial support of the Expanding Excellence in England Fund (E3) provided by Research England to Imagination Lancaster.

References

Cavada, M., Rogers, C., & Hunt, D. (2014). *Smart Cities: Contradicting definitions and unclear measures* (Forum, 1–30 November 2014; Sciforum Electronic Conference Series) (Vol. 4, p. f004). https://doi.org/10.3390/wsf-4-f004.

Cavada, M., Hunt, D., & Rogers, C. (2016). Do Smart Cities realise their potential for lower Co_2 emissions?

Institute of Civil Engineers ICE Engineering Sustainability, Theme issue 2016, *169*(6), 243–252. https://doi.org/10.1680/jensu.15.00032.
Cavada, M., Hunt, D., & Rogers, C. F. D. (2017). *Little book of smart cities*. Liveable Cities. Imagination Lancaster. ISBN-13: 978-0704429499.
Cavada, M., Tight, M., & Rogers, C. (2018). Smart Singapore case study: Is Singapore truly smart? Chapter 16. In *Smart City Emergence: Cases from around the world* (1st ed.). Elsevier. ISBN: 9780128161692.
Cavada, M. (2019). *Smart model assessment resilient tool (smart): A tool for assessing truly smart cities*. Ph.D. The University of Birmingham, UK. E-theses.

Governing for Food Security: A Cultural Perspective

Caroline Fabianski[1] and Samantha Suppiah[2]
[1]Independant Researcher, La Seyne sur Mer, France
[2]Possible Futures, Manilla, Philippines

Synonyms

Organization; Food security; Food governance; Policy design

Terms Definition: Key Concepts

- Food system – all activities that permit to achieve food security that is providing access to sufficient, safe, nutritious food for healthy and active life.
- Governance – the act to coordinate a multiplicity of public-private actors towards the achievement of positive outcomes.
- Culture – the diverse norms and values that enable multiple parties to actualize their preferred form of organization.
- Food sovereignty – an approach that put those who produce, distribute, and consume food at the heart of food systems and policies and this in order to not let food systems depend on the demands of markets and corporations.

Introduction

This entry deals with the governance of food systems. It exposes the problematic aspect of current global schemes by providing a picture, an interpretation of the context where the "risk of hunger" breeds and emerges, namely a governance network that functions as a complex system and gives relevance to institutional and organizational design. While design is introduced as the way forward, it requires to properly account for the path-dependent character of the system that makes it resistant to change. To this purpose, Cultural Theory and the Grid-Group Model are introduced. Both deal with the recurring tensions, frictions, and other forms of conflicts that organizing for food security raises. It demonstrates that the design process is challenged by diverse cultures and ways of organizing, accounting for the controversies around the organization of the United Nations Food System Summit (UNFSS) on 23 September 2021 in New York.

The Complex Character of Food Security

Food Security and subsequent "risk of hunger" are not new. Numerous scholars, activists, and practitioners have pointed out the deeper structural and systemic causes of the phenomenon, globalization in particular, to argue that the modern industrial society is radically unsustainable (Koc and Dahlberg, 1999). Many critics shed light on alternative approaches, putting countries like Cuba to the fore because they are not fully integrated into the global food system but seem nonetheless to have developed greater flexibility for self-food reliance. Also, they found that initiatives and agriculture practices supported by local farmers markets and community are ahead of academia, giving prominence to the very local even if it is a nebulous domain – *Does food security rely on local governments or communities? How does it relate to national governments and political coalitions?* These questions pertain to food system governance, which refers in broad terms to the configuration of food systems, and how they are organized. Food systems involve both human and ecological dimensions, and encompass all activities that permit to achieve food security, that is providing access to sufficient, safe, nutritious food for healthy and active life (Van Bers et al., 2019, Hospes and Brons, 2016). Yet, food systems' functioning is increasingly viewed as complex. Indeed food systems are a whole where plural and interconnected components coexist at different levels – global, national, and local, and interact dynamically to trigger feedbacks across multiple scales. In this sense, food systems' structures and processes are often conceived as emergent. This means that the patterns that regulate the organization of food systems cannot be predicted, they become coherent only in retrospect, and they might not continue to repeat (Kurtz and Snowden, 2003). This aspect challenges scholars' and policy makers' capacities to comprehend the dysfunctional character of food systems, which manifests itself through recurring crises such as the world food price crisis of 2007/2008 and 2010 (Hospes and Brons, 2016) and now the COVID-19 pandemic. Subsequently, organizing for food security consists of a wicked problem.

Wicked problems are hard to define due to the inadequacy of the knowledge base available to treat them (Rittel and Webber, 1973, Ritchey, 2013). They often result from policies based on faulty design that downplay the difficulties to make decision in a complex political environment, ignoring the multiplicity of actors and stakeholders involved in the framing of the prob lem as well as its solution (Peters, 2017). From this perspective, rational planning is ill-equipped to address the major social issues of modern life because there is no reliable criteria to assess them. The way forward might consist in working contextually at multiple levels with a range of policy instruments, adopting adaptive management strategies (Head, 2019) that favor learning and continuous evaluation. Recurring food crises are typical examples of wicked problems as government systems fail to prepare, coordinate, and mobilize resources for. They require new type of responses that account for value differences and disagreements about the nature and significance of the problem (Head and Alford, 2015). In this context, food security becomes an issue that transcends scales; as it cannot be solved at the global or national level, only there is a need to consider more local and hyperlocal actions as well. Food security issues and subsequent "risk of hunger," as wicked problems, challenge traditional governance schemes and public intervention. Governance scholars recommend alternative approaches such as designing policies that are more adaptive in scope as they are developed iteratively along an evaluative process that aimed to engage the widest range of stakeholders at different time horizon. Still such approaches do not take into consideration the resiliences that prevent current food systems to change. The objective of this entry is therefore to overcome this shortcoming, applying Cultural Theories and the Grid-Group Model to explore how cultures impact food system governance.

G

The Meaning of Governance

Governance relates to public interventions and more precisely to the reforms of the State that took place in the 1980s which consisted of revising the role of governments (Pollitt and Bouckaert, 2000). The reforms involved the implementation of New Public Management (NPM) principles. According to NPM, the role of governments is to steer rather than to row (Klijn, 2008). Implementation matters were left to separate public agencies and third parties. However the change overlooked the complexities, interdependencies, and dynamics of public services delivery. Command and control mechanisms of bureaucracy have been replaced by more complicated relationships between multiple parties (Klijn & Koppenjan, 2016). Firms, governments, and nonprofits play a role in the delivery of services. Services are jointly produced and no single organization has the power to deliver them alone. Subsequently, there was a need for a renewed analytical framework to understand the procurement of public services. New Public Governance (NPG) (Koppenjan, 2012, Klijn, 2008) based on network theory constitutes such a framework. NPG reckons that public services delivery takes place within networks of public and nonpublic actors. Policies and their implementation are made through a web of relationships between governments, businesses, and civil society actors. From the NPG perspective, the State and the public sector are plural, made of multiple actors and processes. The focus is very much on interorganizational relationships (Osborne, 2006) as governance relies on interdependencies and interactions as means of coordination (Kickert et al., 1997; Kooiman, 1993; Rhodes; 1997). These interactions might be difficult to manage. In contrast with the NPM, network governance and NPG are much more horizontal (Klijn, 2008). Policies and their implementation are co-constructed with a multiplicity of actors and stakeholders through design (Klijn and Koppenjan, 2006). Yet, this design process is characterized by institutional complexity as it faces the set of rules that regulates actors' behavior. There are historic and enduring relationships within networks and design is aimed at changing their institutional characteristics, the system of formal and informal rules that structures interactions.

Back to food systems and subsequent food security issue, this entry assesses how network governance deals with the "risk of hunger" given that it does not only depend on States. Central governments have been hollowed out (Peters and Pierre, 2006) as power has been devolved. Concurrently, the risk, a wicked problem, can be considered as having reached and unprecedented level of complexity and call for adaptive strategies from institutional networks. Again, network governance tends to frame risks in organizational terms, considering two possible responses (Czarniawska, 2009). The first response is planning and relies on prescriptions and actors responsibilities. The second refers to spontaneous organizing as a reaction to a concrete event/crisis that should make who is responsible for what unimportant. In other words, food crises should lead to a joint action within governance networks, triggering institutional change. However, the current COVID-19 pandemic demonstrates that such joint actions did not occur. On the contrary, it seems that network governance remained salient. Governmental systems in general failed to plan for food security and they also struggled to respond concretely to the crisis. This demonstrates the limits of network governance as a theoretical and analytical framework that makes sense of what is happening on the ground, thus the need to complement it with a model that better accounts for culture.

Introducing the Grid-Group Model

Network governance frames the treatment of wicked problems and the risks they involve as organizational issues – how a multiplicity of actors and stakeholders organize and coordinate themselves to achieve food security. This process is subject to insufficient and inadequate knowledge as well as disagreements and conflicts around what constitutes the problem and its solution. This requires attention be paid to the diversity of norms, values, and moral commitments involved, i.e., culture. According to Douglas (1970, 2003), culture is what would enable different parties to actualize their preferred forms of organization. It manifests in specific ways of doing things. Looking at culture permits to understand the governance of food security issues and the risk of hunger, *how they are organized in practice.* This requires to further elaborate on the link between culture and organization as depicted in the Grid-Group Model.

According to Fabianski (2017), the Grid-Group Model presents a typology of organizations coined *Hierarchist, Individualist, Egalitarian, and Fatalist* (also referred as culture). It originates from the study of communities in order to understand, categorize, and *generalize* on how people organize and collaborate in their daily life. Consequently, the four organizations aim at representing the diverse forms of *organizational structures* that human societies present. The Grid-Group Model assumes that the four organizational archetypes coexist in tension in any social setting and display their own perspective or culture that assume different attitudes towards risks. Beyond anthropology and the study of communities, the Grid-Group Model has been applied in diverse contexts such as technology and climate change in order to address risk issues and their governance. In both contexts, the coexistence of multiple cultures (*Hierarchist, Individualist, Egalitarian, and Fatalist*) challenges governance because the cultures adopt a different mindset when it comes to the treatment of risks. Any attempt to deal with the risks at the global level faces resistance. The cultures present different perspectives and considerations of the risks. This leads to different information being held within disagreements on treatment of wicked problems. Therein lies the universal character of the arrangement, i.e., the mechanism that could mitigate the negative impacts of technology, global warming, and any other threats. This is all the more relevant in a connected world because the diversity of cultures translates into organizational terms, posing coordination issues.

Grid-Group is a parsimonious model seeking to represent a universal picture of the society, the whole social order, relying on four different types of social organization and their subsequent cultural rationales (Douglas, 1999). According to Hood (1998), this model captures satisfactorily the essential components of organizational cultures with two dimensions: Grid and Group. Grid-dimension represents the level of autonomy of individuals (individualization) while Group-dimension represents the level encouragement, pressure, or coercion that a group exerts upon individuals to comply (integration). By crossing both dimensions, four types of social organization emerge, strictly distinct from each other, which will respond to different forms of authority. Later the model was enriched and became known as Cultural Theory on the basis of three new assumptions: (1) At the family level as at the national level, all social organization is likely to comprise the four cultures; (2) At the cultural level, each of the four cultures are defined relatively to the others; and (3) The four cultures are in conflict with each other at the society level (Thompson et al., 1990). Consequently, while considering the society and the social context, conflicts are endemic (Werveij et al., 2006). This evolution of the Grid-Group Model permits to consider it as a theory, in the sense that the patterns it describes and explains will recur over time in different collaborative setting.

Implications of the Grid-Group Model

At the societal level, the Grid-Group Model, as four types of social organization, and its subsequent institutional perspective introduce power and legitimacy issues: "*Different decisions-makers worry about different risks – war, pollution, employment, inflation*" (Douglas and Wildavsky, 1983:1). This means practices, actions, and decision-making are not only driven against conflicting interests but also against divergent perceptions of interests. This confirms that controversies are intrinsic to social relationships. In this perspective, risk and uncertainty hold a particular place: risk discourses allow tracking and revealing internal structures and systems. For Douglas (1994), misfortunes and risks are not things but social constructs: "When risk enters as a concept in political debate, it becomes a fearful and menacing thing, like a *flood, an earthquake, or a thrown brick. But it is not a thing; it is a way of thinking, and a hi*ghly artificial contrivance at that" (Douglas, 1994; 46). Douglas' argument does not discuss risks' potential to be real or to harm; it seeks to account for the fact that risks are mobilized within social institutions in the process of holding to account those handling power (Tansey, 2004). Within social institutions, hazards breed political and governmental strategies towards a legitimization of action, referring to how potential dangers generate collective sense-making concerning their treatment, thus it is part of a rhetorical argument, which contributes to the mechanisms of legitimizing action. Concurrently, while institutions disagree with each other, risks turn into a discourse, a persuasive argument against power competition. Here, power could be seen as a synonym of influence, a technological relationship which "*could be* readily exercised if its source is recognized as legitimate by those subject to it" (Johnston, 1981: 469). Risks do not only point out dangers but also refer to disapproved behaviors (Douglas, 1994); social institutions indicate what is morally unacceptable.

G

While the web of power is taken into consideration, risks become a resource to impose order and authority. Considering the divergent attitudes towards risks, it is possible to define the four cultural organizations as follow:

- ***The Isolated:*** fatalist approach to risks, unable to organize themselves to prevent or face risks. The *Isolated* are nonresponsive to risks.
- ***The Individualist:*** the problem of risk is posed as "should we take a risk or not." When question of risks is posed, the *Individualists* calculate and tend to assess the opportunities it represents and therefore their propensity to involve, cooperate, or organize, or they consider the risks as potential negative impacts and therefore their propensity for dissent. The *Individualists* are either risk takers or risk averse.
- ***The Hierarchies*****:** the risks are perceived as an organizational problem. To be dealt with, risks imply appropriate information, resources, and means of coordination. The *Hierarchies* rationalize the treatment of risks and would plan and act for their prevention. In general, *hierarchies* tend to be objective towards risks and to organize to decrease the uncertainty that surround them.
- ***The Egalitarians:*** Risks are the reasons for the existence of *Egalitarians* as a specific group. Risks are seen as external to the Egalitarian group, the threat is such that it legitimizes the *Egalitarians* as a distinct group from the rest of the society. However, in contrast with the hierarchies, who internalize the risks to organize for them, *Egalitarians* tend to use risks to sustain their own boundaries.

Approaching wicked problems from a cultural angle contrasts with the postmodernist argument of the *Risk Society* (Beck, 1992, 1999). Drawing on Giddens, Beck conceptualizes risk and uncertainty as a process, which has reached an unprecedented level of complexity in the "modern world." Risk and uncertainty therefore travel between individual and collective spheres, fostering adaptive strategies from institutions. Conversely, acknowledging Douglas's contribution, this rationale would be misleading because it undermines institutions, as collective entities by giving prominence to initiative of individuals as triggers for structural change. Such evolutionary assumptions deal with social systems as an increasing level of complexity, although the argument does not necessarily contravene the social environment as conceptualized in Grid-Group: on the contrary, institutions persist within complex settings (Ostrander, 2002). However, the resilience of institutions do not suppose a static nature: since they act as a filter for individuals' perceptions, institutions are also sensitive to time and places but they operate a selection of what should be remembered and forgotten – see Douglas's (1987) perspective on history:

> When we look closely at the construction of past time, we find the process has very little to do with the past at all and everything to do with the present. Institutions create shadowed places in which nothing can be seen and no questions asked. They make other areas show finely discriminated detail, which is closely scrutinized and ordered. History emerges in an unintended shape as a result of practices directed to immediate, practical ends. To watch these practices establish selective principles that highlight some kinds of events and obscure others is to inspect the social order operating on individual minds (Douglas, 1987:69)

When power moves from one institutional order to another, some *events* are emphasized and others are forgotten, often intentionally. Such a change defines time lines and the trajectory of history. Such dynamic on risk and temporality applies to the *events* that characterize food security issues, culture influencing the succession of *events* that matters over a sense-making process. This renders the Grid-Group a highly relevant interpretative framework.

After such conceptual clarifications, one should emphasize an important implication of culture for analysis: the Grid-Group model comprises the recognition of unintended effects as the result of collective actions; social organizations and subsequent cultures imply feedback loops that would offer a theoretical background for apparent illogical/irrational patterns (Tansey, 2004) that characterize wicked problems so well. As individuals are social actors, their perceptions are also biased and context-dependent (Olli,

1999). Such findings enable Grid-Group to be used far beyond its primary goal since it retrofits social organizations into analysis via a coherent model, to also provide an interpretative basis for social and organizational change by focusing analysis on tension between the Grid and Group dimensions. However, to make the model operational, there is a need to emphasize that Grid-Group is a relativist model: each culture flourishes in contrast with the others (Douglas, 2003) and hence the model is technically incapable to distinguish one specific social organization regardless of the others. This is part of the interpretative dynamic of the framework. In other words, where conclusions refer to one cultural rationale, they must always been depicted in comparison with the three others. Secondly, the model is drawn to identify social pressures; hence analysis should focus on conditions for change first and then infer conditions for stability. Ultimately, the applicability of the model relies on a fundamental hypothesis: the "world" where analyses apply must be clearly defined and limited in order to be considered as stable; this means it should constitute a terrain where social forms and culture are likely to sustain each other (Douglas, 2005). At this point, there is need to picture the "world" of food security in Grid-Group terms.

So far the entry has presented the characteristic of the complexity of food systems governance. It has framed food security and the subsequent risk of hunger as a wicked problem that poses governance challenges. The entry then defined governance, drawing on network theories and New Public Governance (NPG) that deal with the general transformations of State intervention for public services delivery. It depicts a context where there is no clear chain of command and whose public action depends on the interactions between a multiplicity of actors and stakeholders that transcend scales – global, national, and local, and tends to frame food security as an interorganizational issue that should be solved through policy design. Still, the entry argues that these theories have missed the problematic aspect of governance by overlooking the diversity of cultures and social organizations present in society. To fill this gap, Cultural Theories and the Grid-Group model have been introduced as an analytical framework that makes sense of the recurring conflicts and controversies around food security and interventions to tackle the risk of hunger. Subsequently, the following aims at applying Cultural Theories insight to the current situation, namely the global COVID-19 pandemic. What follows is a reflection on the methodological requirements of the Grid-Group Model.

The COVID-19 Pandemic from the Grid-Group Perspective

To make the Grid-Group Model operational, it is clearly stipulated to first focus on social pressures and change in order to allow the four social organizations – *Hierarchist, Individualist, Egalitarians, and Isolated* – to express themselves and sustain each other, thus a focus on the COVID-19 pandemic as an event. Events (or more specifically "chains of events") offer the opportunity to trace possible divergent stories and context for different sense-making and meaning, demonstrating where institutions connect actions conventionally, or where they are disturbed, raising conflicts and suggesting change. Therefore, the COVID-19 pandemic is seen as a critical *event.* Critical *events* are viewed by practitioners as the conditions for change and opportunities to (re) negotiate norms and values, i.e.: the culture. Faithful to the Grid-Group model, paying specific attention to *Events* permits to question conflicts of interests, need for actions, and potential for governance change. The conditions for governance are unstable, they should be linked to risks, what is at stake but also the implementation of power. As mentioned earlier, cultural institutions express themselves in a rhetorical manner against risks, calling at a diversity of mechanisms to treat them. This process explains how specific governance arrangements emerge, which comprises actions and ways of doing things and give rise to a specific network of actors. This translates in contesting the organizational structures and governance arrangements in place; a process which, to a large extent, refers to power. In other words, such a standpoint assumes that *events* could lead to the emergence of competitive narratives, the *polyphony* that underpins the plurality of governance rationales and practices as depicted by the Grid-Group model. The nature of the account is interpretative (Stake, 1995) in the sense that it provides an understanding of the governance network experience of it.

In this interpretative frame, the idea is to trace *Hierarchist, Individualist, Egalitarians, and Isolated* rationales. To this purpose, the analysis is based on a review of articles and policy briefs produced during the COVID-19 pandemic as well as a one to one conversation with an activist. These materials are treated as narratives (Czarniawska, 2004) that permit to comprehend the network governance of food systems and assess the relative position of the different actors and stakeholders within it. In this respect, *Hierarchist* and *Individualist* are well understood social organizations that often relates to Hierarchy vs. Market. Hierarchy often refers to governmental apparatus, the public sector, and it acts in relation to businesses, the market. Recollecting the reforms of the 1980s and the transformation of governments' intervention, a new form of procurement within network governance has emerged for the delivery of public services: Public-Private Partnership (PPP). PPPs are deemed to overcome the public-private divide to reflect on how the public sector and markets mutually support each other. PPPs are seen as a way to create something new within networks; that is something that could not have been delivered otherwise, hence an innovative character (Klijn & Teisman, 2000; Wettenhall, 2003). They call for collaboration and risk-sharing as well as the development of trust between the different parties involved. However, in practice, PPPs fail to take into consideration the complete picture of the social world where risk debates take place because they downplay two additional social organizations to *Hierarchist* and *Individualist* – the *Egalitarians* and the *Isolated*. And this is precisely what is missing to understand the recurring conflicts around the governance of food systems. The following accounts for the COVID-19 pandemic and the implications in governance terms to highlight how the voices of *Egalitarians* and *Isolated* manifest themselves.

The COVID-19 pandemic has made the structural weaknesses of food systems more apparent. Most vulnerable communities, especially in the Global South, have been the most impacted by the COVID-19 pandemic and food security issues which have resulted in logistic systems collapse and price spikes due to shortages or fear of shortage. According to Blay-Palmer et al. (2021), vendors around the world could not sell their food as markets were closed. There were also impacts from international trade restrictions and countries that cancelled exports of vegetables to preserve domestic supply. In the Pacific Small Island Developing States (PSIDS), COVID-19 measures meant no tourism which shifted the economic focus, especially in food systems as the decline in remittances forced families into debt to be able to eat. COVID-19 has doubled the number of people facing acute food insecurity globally by the end of 2020 (Carey et al., 2020). This also concerns vulnerable and low-income population groups in the Global North such as United States. Cities deserves a specific attention because they entail complex logistical challenges. Rapid urban expansion has reduced cities capacity to produce fresh food, hence cities have become more dependent on distant source of food. There is growing recognition of the need for diversity in the source of food, diversity in the scale at which food is produced and distributed, and diversity in the type of production systems and food enterprises involved. As the majority of people are now living in cities, municipal governments have an important role to play in promoting equitable access to healthy food. This is in this context that the United Nations Food System Summit (UNFSS) was organized on 23 September 2021 in New York.

The UNFSS seeks to set the stage for global food system transformation in order to achieve the United Nations Sustainable Development Goals (UN-SDG) by 2030. The Summit objective is to provide a platform for ambitious new actions, innovative solutions, and plans to transform food systems and leverage these shifts to deliver progress across all of the UN-SDGs. It reckons the centrality of food systems to the entire sustainable development agenda and aims to align stakeholders around a common understanding and narrative of a food system framework as a foundation for concerted action, as well as to catalyze and accelerate bold action for the transformation of food systems by all communities, including countries, cities, companies, civil society, citizens, and food producers. However, the UNFSS was controversial. Hundreds of farmer organizations, indigenous groups, social and economic justice NGO and social movements have rejected and boycotted the UNFSS despite 746 million people suffering from severe food insecurity and an additional 1.25 billion experiencing moderate food insecurity (Mousseau, 2021). Such opposition could be assimilated to cultural conflicts in Grid-Group terms because it refers directly to governance disagreements, an *Egalitarians*' attempt to make their voice heard to organize and reassert their boundaries against the *Hierarchist* and *Individualist* alliance. This attitude is made clear in the arguments they mobilize and a direct reference to power. The *Egalitarians* movement contests the entire setting of the UNFSS. First, they do not consider the actors that lead the summit as legitimate. They do not accept the centering of large corporate businesses that promote monoculture, the use of fertilizers, and a model of fossil fuel-based agriculture at the expense of family farmers, pastoralists, and indigenous communities that are marginalized and forced off their land. Second, the UNFSS is viewed as intentionally exclusive in the sense that it ignored an unprecedented number of petitions, events, public communication, and other advocacy actions that took place around the world. From the *Egalitarians'* perspective, the UNFSS reinforces the dominance of *Hierarchist and Individualist* cultures, ignoring the *Isolated,* here local farmers and indigenous communities that failed to adequately organize in face of the risk, often due to lack of resources or direct instruction from *Hierarchist*'s authorities not to.

The *Egalitarians* describe the *Isolated/Fatalist* as the peasant class, the bottom million of the world who feels the oppression under a capitalist world and whose voice is being ignored, silenced, or suppressed. *Fatalists* are not homogeneous. They are presented as diverse groups that draw on indigenous knowledge and traditional practices but have no guarantee of security of life on the planet. For the *Egalitarians*, these groups are denied. They view the global governance arrangements around food systems as mechanisms to serve Western interests, the ruling class. The allusion to power is clear and is felt by *Isolated/ Fatalist* groups, who are deprived from their capacity to take their fate in their own hands. Arrangements such as the UN-SDGs, the New Urban Agenda (NUA), etc., are not only seen as ineffective but also as a form of hegemonic propaganda that adopts an overly optimistic and positive neoliberal outlook in order to consolidate influence and hence power. *Egalitarians* call for solidarity and argue for alternative modes of governance based on the notion of food sovereignty. Food sovereignty contrasts with the food security approach and consists of putting those who produce, distribute, and consume food at the heart of food systems and policies (Patel, 2009). This is so as to not allow food systems depend on the demands of markets and corporations. Food sovereignty prioritizes local and national economies and empowers peasant and smallholder farmer-driven agriculture based on environmental, social, and economic sustainability. It offers a route to resist and dismantle the current corporate trade and food regime.

Conclusion

This entry has shed a cultural perspective on the governance of food systems. The Grid-Group Model identified how diverse cultures manifest themselves around food systems governance issues. It demonstrates the limits of current arrangements and questions how they could be renewed by taking into consideration the *Egalitarians* and *Fatalists*. In this respect, the design approach is challenged by divergent ways of organizing and power relationships. The way forward might consist in recognizing differences and accepting that there are plural ways. The next step is to reflect on how *Egalitarians* and *Isolated* can be engaged in a healthy manner to enable a constructive treatment of cultural risks. In this respect, the possibility to establish a polycentric governance arrangement which promotes multiple and independent decision-making units as well as mechanisms for conflict resolution should be explored. In this case, the urban and regional level of intervention becomes important, the question being the role and room for maneuver of city officials.

Cross-References

▶ Planning for Food Security in the New Urban Agenda

References

Beck, U. (1999). *World risk society*. Cambridge: Polity Press.

Beck, U. (1992). *Risk society: Towards a new modernity*. London: Sage.

Blay-Palmer, A., Santini, G., Halliday, J., Malec, R., Carey, J., Keller, L., ... van Veenhuizen, R. (2021). City region food systems: Building resilience to COVID-19 and other shocks. *Sustainability, 13*(3), 1325.

Carey, R., Murphy, M., & Alexandra, L. (2020). COVID-19 highlights the need to plan for healthy, equitable and resilient food systems. Cities & health, 1–4.

Czarniawska, B. (2009). *Organizing in the face of risks and threat*. Cheltenham: Edward Elgar Publishing.

Czarniawska, B. (2004). *Narratives in social science research*. Sage.

Douglas. (1987). *How institution thinks*. London: Routledge and Kegan.

Douglas, M., & Wildavsky, A. (1983). *Risk and culture: An essay on the selection of technological and environmental dangers*. London: University of California Press.

Douglas, M. (1994). *Risk and blame: Essay on cultural theory*. London: Routledge.

Douglas, M. (2005). *Grid and group, new development. Workshop on complexity and cultural theories in honour of Michael Thompson*. London: LSE.

Douglas, M. (1999). The evolution of a parsimonious model. *Geo Journal, 47*, 411–415.

Douglas, M. (1970). *Natural symbols*. New York: Pantheon.

Douglas, M. (2003). Being fair to Hierarchists. *University of Pennsylvania Law Review, 151*(4), 1349–1370.

Fabianski, C. J. C. (2017). *Complex partnership for the delivery of urban rail infrastructure project (URIP): How culture matters for the treatment of risk and uncertainty* (Doctoral dissertation, UCL (University College London)).

Head, B. W. (2019). Forty years of wicked problems literature: Forging closer links to policy studies. *Policy and Society, 38*(2), 180–197.

Head, B. W., & Alford, J. (2015). Wicked problems: Implications for public policy and management. *Administration & society, 47*(6), 711–739.

Hood, C. (1998). *The art of the state: Culture, rhetoric and public management*. Oxford: Oxford University Press.

Hospes, O., & Brons, A. (2016). Food system governance: A systematic literature review. *Food Systems Governance*, 13–42.

Johnston, R. (1981). *The dictionary of human geography*. Oxford: Blackwell Reference.

Kickert, W., Klijn, E. H., & Koppenjan, J. (1997). *Managing complex networks: Strategies for the public sector*. London: Sage.

Klijn, E. H. (2008). Governance and governance networks in Europe: An assessment of ten years of research on the theme. *Public Management Review, 10*(4), 505–525.

Klijn, E. H., & Koppenjan, J. (2016). *Governance network in the public sector*. London: Routledge.

Klijn, E. H., & Koppenjan, J. F. (2006). Institutional design: Changing institutional features of networks. *Public Management Review, 8*(1), 141–160.

Klijn, E. H., & Teisman, G. R. (2000). Managing public private partnership: Influencing processes and institutional context of public private partnership. In O. Van Heffen, W. J. M. Kickert, & J. J. A. Thomassen (Eds.), *Governance in modern society: Effect, change and formation of government institutions*. Springer: Dordrecht.

Koc, M., & Dahlberg, K. A. (1999). The restructuring of food systems: Trends, research, and policy issues. *Agriculture and Human Values, 16*(2), 109–116.

Kooiman, J. (1993). *Modern governance: New government-society interactions*. Sage.

Koppenjan, J. (2012). *The new public governance in public service delivery*.

Kurtz, C. F., & Snowden, D. J. (2003). The new dynamics of strategy: Sense-making in a complex and complicated world. *IBM Systems Journal, 42*(3), 462–483.

Mousseau, F. (2021). *People vs. agribusiness corporations: The battle over global food and agriculture governance. Policy brief*. The Oakland Institute.

Olli, E. (1999). Rejection of cultural biases and effects on party preference. In M. Thompson, G. Grendstad, & P. Selle (Eds.), *Science*. London: Routledge.

Osborne, S. P. (2006). The new public governance? *Public Management Review, 8*(3), 377–387.

Ostrander. (2002). *One-and two dimensional models of the distribution of beliefs* in Douglas, M. (2002). Essays in the sociology of perception. Routledge: London.

Patel, R. (2009). Food sovereignty. *The Journal of Peasant Studies, 36*(3), 663–706.

Peters, B. G. (2017). What is so wicked about wicked problems? A conceptual analysis and a research program. *Policy and Society, 36*(3), 385–396.

Peters, B. G., & Pierre, J. (2006). *Governance, government and the state* (pp. 209–222). The state: Theories and issues.

Pollitt, C., & Bouckaert, G. (2000). *Public management reform: A comparative analysis* (Vol. 89). Oxford: Oxford University Press.

Rhodes, R. (1997). *Understanding governance.* Buckingham: Open University Press.

Ritchey, T. (2013). Wicked problems. *Acta morphologica generalis, 2*(1).

Rittel, H. W., & Webber, M. M. (1973). Dilemmas in a general theory of planning. *Policy Sciences, 4*(2), 155–169.

Stake, R. E. (1995). *The art of case study research: Perspectives and practice.* London: Sage.

Tansey, J. (2004). Risk as politics, culture as power. *Journal of Risk Research, 7*(1), 17–32.

Thompson, M., Ellis, R., & Wildavsky, A. (1990). *Cultural theory.* Westview Press, Boulder.

Van Bers, C., Delaney, A., Eakin, H., Cramer, L., Purdon, M., Oberlack, C., ... Vasileiou, I. (2019). Advancing the research agenda on food systems governance and transformation. *Current Opinion in Environmental Sustainability, 39*, 94–102.

Werveij, M., Douglas, M., Engel, C., Hendriks, F., Lohmann, S., Ney, S., Rayner, S., & Thompson, M. (2006). Clumpsy solution for a complex world: The case of climate change. *Public Administration, 84*(4), 817–843.

Wettenhall, R. (2003). The rhetoric and reality of public-private partnership. *Public 105 Organization Review, 3*(1), 77–107.

Government Procurement

▶ Public Procurement for Regional and Local Development

Governments

▶ Internationalization of Cities

Green and Blue Infrastructure

▶ Nature-Based Solutions for River Restoration in Metropolitan Areas

Green and Blue Infrastructure (GBI) in Urban Areas

Luís Valença Pinto[1,2], Carla Sofia Ferreira[1,3,4], António Ferreira[1], Zahra Kalantari[3,4,5] and Paulo Pereira[2]

[1]Research Centre for Natural Resources, Environment and Society (CERNAS), Polytechnic Institute of Coimbra, Coimbra Agrarian Technical School, Coimbra, Portugal

[2]Environmental Management Laboratory, Mykolas Romeris University, Vilnius, Lithuania

[3]Department of Physical Geography and Bolin Centre for Climate Research, Stockholm University, Stockholm, Sweden

[4]Navarino Environmental Observatory, Messinia, Greece

[5]Department of Sustainable Development, Environmental Science and Engineering, KTH Royal Institute of Technology, Stockholm, Sweden

G

Definition

Green and blue infrastructure (GBI) comprises a network of natural and semi-natural areas and includes different environmental features (Jato-Espino et al. 2018). Among other characteristics, GBI contains green (vegetated land) and blue (water) spaces such as parks, gardens, green corridors, wetlands, urban forests, edible gardens, green walls, green roofs, street trees, rivers, lakes, and ponds. However, only areas important for biodiversity and providing goods and services should be considered GBI. For instance, areas covered by intensive land uses, such as tree or crop monocultures, are not GBI (Estreguil et al. 2019). GBI can range from transnational to national, regional, and local, and can encompass rural, peri-urban, and urban environments. It can deliver multiple ecosystem services (ES), such as water and air purification, carbon sequestration, climate regulation, water purification, recreation, and landscape aesthetics. These ES provide critical benefits for human well-being in urban areas and enhance resilience to climate change challenges. GBI is relevant for its contribution to the United Nations' Sustainable Development Goals (SDGs) and the European Union Biodiversity Strategy for 2030.

Acronyms

Green Infrastructure (GI), Blue Infrastructure (BI), Blue-Green Infrastructure (BGI), Urban Green Infrastructure (UGI), Urban Green and Blue Infrastructure (UGBI), United Nations' Sustainable Development Goals (SDGs), Ecosystem Services (ES), Ecosystem Disservices (ESD)

Introduction

The GBI concept has been used more widely in recent years, but it is not clear when it was first used and by whom, with sources conflicting and not easily verifiable (Cullen 2013). However, it can probably be traced back to the early 1990s (Cullen 2013) green infrastructure management is as relevant as management of other human-related infrastructure, such as roads, water, or electricity (MacKay and Reed 1994). According to Mell (2017), GBI is a holistic approach to landscape planning in which MacKay and Reed (1994) synthesized and brought together old ideas, some of them dating back more than 150 years, such as conservation, parkways, greenways, and environmental management. MacKay and Reed (1994) highlighted the contribution of a diverse and representative set of stakeholders for GBI development, including conservationists, recreationists, and businesses, focusing on multifunctionality and multidisciplinarity and not just on nature conservation and connectivity. According to Benedict and McMahon (2002), the GBI concept involves two essential functions: (1) linking green and blue spaces for the benefit of people and (2) preserving and linking natural areas to benefit biodiversity. The use of the term "infrastructure" originated from the need to depart from conventional concepts in order to: (i) change the way people perceive nature, shifting from nature as an amenity, something "pleasant" to have to something necessary or essential to have; (ii) address the sustainability of green areas, shifting the perception from green and blue spaces as self-sustained spaces to considering them as infrastructure that needs to be actively maintained, and restored; and (iii) introduce multifunctionality and connectivity between green and blue spaces, strengthening the notion that they are interconnected and multifunctional (Benedict and McMahon 2002).

GBI is a broad concept in which Cullen (2013) identified four distinct categories, clearly interconnected and sometimes overlapping: (i) GBI for conservation, perhaps the first use of the term, focusing on conservation and land use planning; (ii) GBI for stormwater management, where it is presented as an alternative or an addition to the conventional grey infrastructure for water drainage and sewer pipes, and water treatment plants; (iii) GBI for the urban forest, with arborists and urban foresters increasingly addressing urban forest management under the umbrella of multifunctionality; and (iv) GBI for more sustainable building, which addresses site- and building-specific practices to reduce the environmental impacts of buildings.

GBI for stormwater management is usually associated with the United Nations' Sustainable Development Goals (SDGs). This approach mainly focuses on on-site scale GBI elements, including, e.g., retention basins, bioswales, green roofs, and rain gardens. GBI has been increasingly linked to global sustainability issues, focusing on climate change and urban sustainability (Mell 2017).

GBI and Ecosystem Services: Benefits and Trade-Offs

One of the key aspects of GBI is its multifunctionality and ability to provide multiple ES (Jato-Espino et al. 2018; European Environmental Agency 2019). ES are defined as the goods provided from nature to people (Pereira 2020) that are usually classified as provisioning, regulating, and cultural (Haines-Young and Potschin-Young 2018). GBI can provide primary and secondary benefits (Martín 2020), based on the specific

GBI's overarching goal. For example, dunes' primary benefit is coastal flood protection and biodiversity maintenance, but additional benefits such as rainwater infiltration or aesthetic values are also provided. The scale of GBI plays a vital role in the ES provided. For instance, a small raingarden can provide a set of specific ES such as flood regulation or landscape aesthetical value, while a large urban forest can provide a wider diversity of ES. The diversity and multifunctionality of GBI usually bring a myriad of ES, associated with a set of complex and often nonlinear synergies and trade-offs (Raudsepp-Hearne et al. 2010). The negative effects of ecosystem functions on human well-being, such as diseases, allergens, and poisonous plants and animals, are called ecosystem disservices (EDS) in the scientific literature (von Döhren and Haase 2015). A particular GBI can provide both ES and EDS, e.g., fruit bats provide essential pollination and pest-control ES, but can also be a vector of diseases and parasites (Shapiro and Báldi 2014). In another example, intensive recreational use of urban parks may conflict with aesthetic or spiritual enjoyment. Similarly, increasing provisioning ES (e.g., food or timber) can provoke disservices in regulating (e.g., nutrient cycling or flood protection) and cultural (e.g., recreation) functions.

Quantification and integrated valuation of both services and disservices is fundamental for the full evaluation of ES benefits and trade-offs (Schaubroeck 2017). However, assessing ES synergies and GBI trade-offs can be hindered by several aspects, such as the interaction between distinct GBI in a given area, which can influence how each GBI functions, bringing changes in the ES provided. Additionally, the spatial dimension of GBI, with different GBI types functioning both as individual GBI elements and as subcomponents of other GBI types, raises challenges for full ES assessment. For example, a pond, which is a type of GBI, can be an element of a wider GBI (e.g., urban park), or a stretch of river can be part of the accessible areas associated with a large urban park, providing access to water sports or activities. Furthermore, assessing different types of benefits can require several methodologies, some involving quantitative parameters, others based on qualitative evaluations (Charoenkit and Piyathamrongchai 2019). For instance, biodiversity benefits are sometimes assessed by indicators such as the number of species in a specific area. Simultaneously, the evaluation of emotional well-being might involve population surveys to investigate preferences, motivations, and importance attributed to users' activities (Carmen et al. 2020). Some benefits may also be evaluated by indirect assessments (e.g., through mathematical modeling), such as GBI's impact on climate change mitigation (Alves et al. 2020). However, future climate change may also affect the functioning of GBI and the benefits provided, so assessing GBI may involve a temporal scale. The ability of GBI to supply ES is site and context specific (Ruckelshaus et al. 2016). For planning purposes, cost-effectiveness assessment of GBI is fundamental for comparing different alternatives and to support decision-making. Different approaches and tools have been developed to understand better the cost-benefits of GBI (e.g., Brown et al. 2014; Grêt-Regamey et al. 2015; Burkhard and Maes 2017; Harrison et al. 2018; Ryfield et al. 2019; Havinga et al. 2020; Czúcz et al. 2020).

For GBI planning purposes, comparisons of different GBI options must be performed on assessments based on a mixed social sciences and natural sciences approach (Lyytimäki et al. 2008). A multidisciplinary approach can prioritize objectives based on the agreement between various stakeholders and implement a hierarchy of ecosystem functions (Dorst et al. 2019). This can provide an informed and overarching base for effective land use science and policy.

Typologies of GBI

GBI can be characterized based on a set of different characteristics. The multiple approaches to the concept of GBI and the diverse areas of interest (e.g., urban planning, forestry, resource management, and water management) bring complexity to its definition, with no consensus on a comprehensive classification of GBI (Bartesaghi Koc et al. 2017). Table 1 shows a set of features that

Green and Blue Infrastructure (GBI) in Urban Areas, Table 1 Typological characteristics of green and blue infrastructure (GBI) and their subclasses

Feature	Description	Levels
Type	Basic distinction between types of GBI (Depietri and McPhearson 2017)	(1) Grey-green-blue (2) Green and blue (3) Green (4) Blue
Primary qualifiers	Related to type of GBI in terms of typological analysis (based on Bartesaghi-Koc et al. 2019)	(1) Green open spaces (2) Water bodies (3) Tree canopy (4) Green roofs (5) Vertical greenery systems (6) Grey-green-blue infrastructures
Functional categories	Addresses the main land use functions	(1) Parks, gardens, amenity green spaces, and natural spaces (2) Water bodies (3) Brownfields (4) Agricultural land (5) Forest (6) Link with other solutions (e.g., grey infrastructure) (7) Mixed green and blue elements (containing several functions)
Surface type	Degree of soil sealing (based on Bartesaghi-Koc et al. 2019)	(1) Impervious (2) Mixed (3) Pervious (4) Aquatic
Naturalness	Degree of naturalness of the GBI	(1) Low (mainly urbanized or highly artificial) (2) Medium (green/blue but partially artificial) (3) High (mainly natural setting, no clear human intervention)
Morphology	Morphological category	(1) Artificial elements (2) Semi-natural linear and point-like elements (man-made) (3) Vegetation structure/land cover/water
Vegetation type		(1) No vegetation/aquatic/riparian vegetation (2) Low vegetation (grasses) (3) Medium vegetation (mainly shrubs) (4) High vegetation (mainly trees)
Vegetation distribution		(1) No vegetation/aquatic (2) Aligned (3) Scattered (4) Clustered (5) Compact
Vegetation density		(1) No vegetation/aquatic (2) Low density (3) Medium density (4) High density
Multifunctionality	Degree of diversity in ES provided	(1) Low (provides specific ES/only from one type, e.g., regulation) (2) Medium (provides specific ES from diverse ES types) (3) High (provides several ES)
Scale	Spatial scale of implementation	(1) Local (building, street level) (2) Neighborhood (block) (3) Urban or region (multi-block or higher)
Specialization/ integration level	Related to size and integration with other GBI	(1) Highly specialized (mainly focusing on a single GBI, which can be incorporated into other GBI) (2) Average (can include several GBI elements, and integrated with other GBI elements) (3) Overarching (can include a large set of GBI elements)

Green and Blue Infrastructure (GBI) in Urban Areas, Table 2 Main types of green and blue infrastructure (GBI), their main characteristics (see also Table 1) and ecosystem services (ES) provided. Based on Haase (2015)

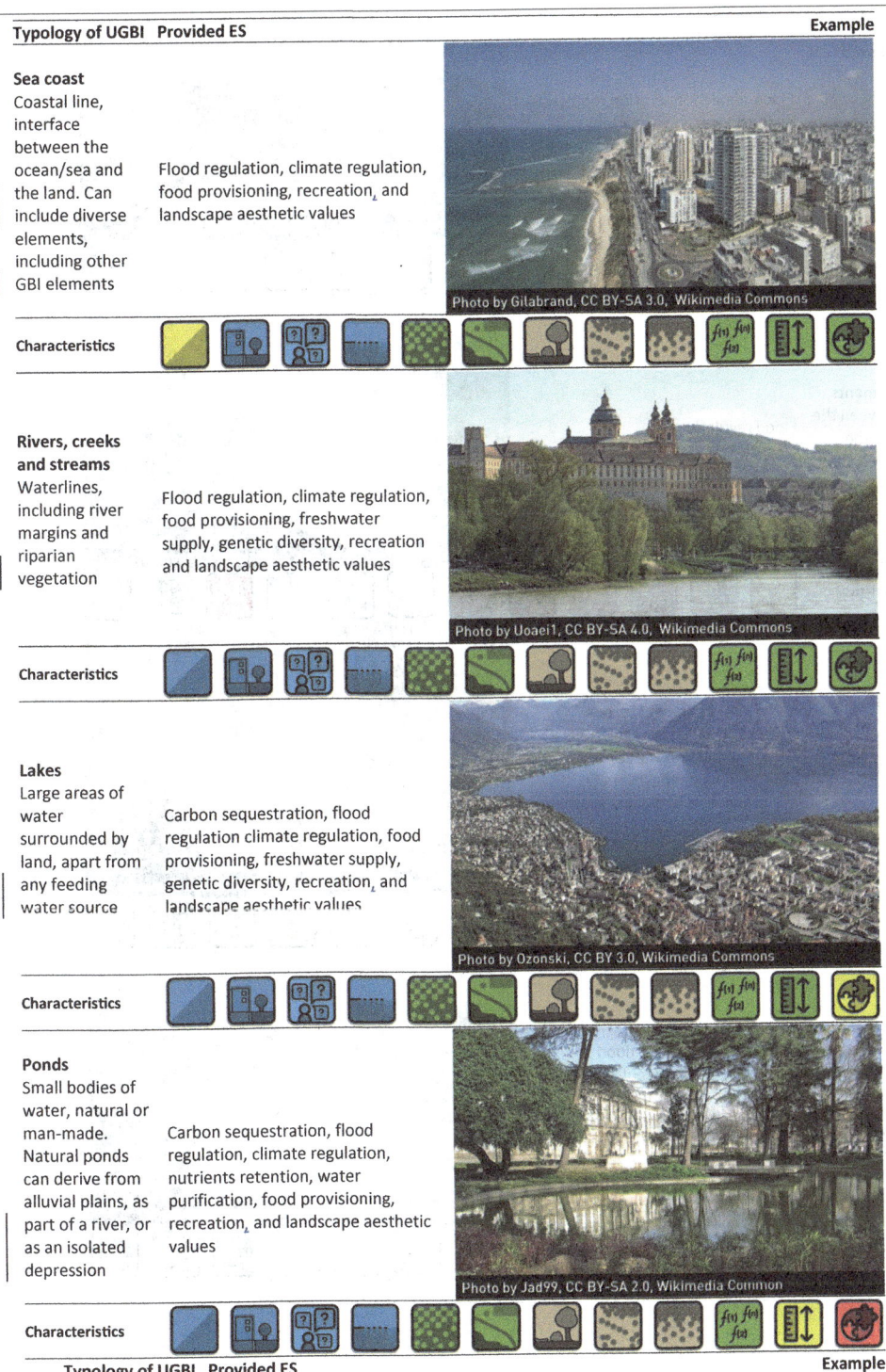

Typology of UGBI	Provided ES	Example
Sea coast Coastal line, interface between the ocean/sea and the land. Can include diverse elements, including other GBI elements	Flood regulation, climate regulation, food provisioning, recreation, and landscape aesthetic values	 Photo by Gilabrand, CC BY-SA 3.0, Wikimedia Commons
Characteristics		
Rivers, creeks and streams Waterlines, including river margins and riparian vegetation	Flood regulation, climate regulation, food provisioning, freshwater supply, genetic diversity, recreation and landscape aesthetic values	 Photo by Uoaei1, CC BY-SA 4.0, Wikimedia Commons
Characteristics		
Lakes Large areas of water surrounded by land, apart from any feeding water source	Carbon sequestration, flood regulation climate regulation, food provisioning, freshwater supply, genetic diversity, recreation, and landscape aesthetic values	 Photo by Ozonski, CC BY 3.0, Wikimedia Commons
Characteristics		
Ponds Small bodies of water, natural or man-made. Natural ponds can derive from alluvial plains, as part of a river, or as an isolated depression	Carbon sequestration, flood regulation, climate regulation, nutrients retention, water purification, food provisioning, recreation, and landscape aesthetic values	 Photo by Jad99, CC BY-SA 2.0, Wikimedia Common
Characteristics		
Typology of UGBI	Provided ES	Example

(continued)

Green and Blue Infrastructure (GBI) in Urban Areas, Table 2 (continued)

Typology of UGBI	Provided ES	Example
Coastal sand dunes Sand dune systems along the coast	Flood regulation, erosion regulation, genetic diversity recreation, and landscape aesthetic values	 Photo by Rúdisicyon, CC BY-SA 4.0, Wikimedia Commons
Characteristics		
River embankments Link between the river and the urban area, built and impervious	Flood regulation	 Photo by Ewan Munro, CC BY-SA 2.0, Wikimedia Commons
Characteristics		
Peat bogs Areas with high groundwater level, organic soils, and natural vegetation	Air quality regulation, carbon sequestration, flood regulation, climate regulation water purification, genetic diversity, recreation, and landscape aesthetic values	 Photo by François Philipp, CC BY-SA 3.0, Wikimedia Commons
Characteristics		
Wetlands Ecosystem flooded by water. Can include, e.g., mangroves, or marshes	Carbon sequestration, flood regulation, climate regulation, water purification, genetic diversity, recreation, and landscape aesthetic values	 Photo by Jupiter3112, CC BY-SA 4.0, Wikimedia Commons
Characteristics		

(continued)

Green and Blue Infrastructure (GBI) in Urban Areas, Table 2 (continued)

Typology of UGBI	Provided ES	Example
Green corridors Open spaces, linear or not, of natural character (e.g., water courses, channels, and woodlands) or that connect different green areas. Designed to ensure the continuity of ecological functions	Carbon sequestration, flood regulation, climate regulation, nutrient retention, food provisioning, recreation, and landscape aesthetic values	 Photo by ThomasKBerl, CC BY-SA 3.0, Wikimedia Commons
Characteristics		
Brownfields Formerly occupied areas (industry, commerce, services, and housing), currently expectant or abandoned	Flood regulation, nutrients retention and recreation	 Photo by Mtaylor848, CC BY-SA 3.0, Wikimedia Commons
Characteristics		
Urban forest Forest areas within urban areas	Air quality regulation, carbon sequestration, flood regulation, climate regulation, erosion regulation, nutrient retention, water purification, noise regulation, food provisioning, timber and biomass provisioning, genetic diversity, recreation, and landscape aesthetics value	 Photo by King Otto, CC BY-SA 3.0, Wikimedia Commons
Characteristics		
Cultivated areas Agriculture areas inside the cities, usually associated to small-scale agriculture, but also including areas of larger dimensions	Carbon storage, flood regulation, climate regulation, water purification, nutrients retention, food provisioning and recreation	 Linda from Chicago, USA, CC BY 2.0, Wikimedia Commons
Characteristics		

(continued)

Green and Blue Infrastructure (GBI) in Urban Areas, Table 2 (continued)

Typology of UGBI	Provided ES	Example
Allotment and community gardens Areas of urban agriculture, created or adapted with proper infrastructures	Carbon storage, flood regulation, climate regulation, water purification, nutrients retention, food provisioning, and recreation	 Photo by Allotmenteer, CC BY-SA 3.0, Wikimedia Commons
Characteristics		
Parks and gardens Green urban areas with semi-natural character, public or private, intended for recreation and the preservation of the natural environment	Air quality regulation, carbon sequestration, climate regulation, flood regulation, nutrient retention, erosion regulation, water purification, recreation, and landscape aesthetics value	 Photo by Cryptic C62, CC BY-SA 3.0, Wikimedia Commons
Characteristics		
Street trees Isolated trees or in group, usually aligned along the streets	Air quality regulation, climate regulation, carbon sequestration, and landscape aesthetic values	 Photo by Glen Bowman , CC BY 2.0, Wikimedia Commons
Characteristics		
Hedgerows Line of trees or shrubs, planted to form a barrier, which may have different purposes (e.g., protection against prevailing winds and separation of property)	Air quality regulation, carbon sequestration, climate regulation, and noise regulation	 Photo by Maurice Pullin, CC BY-SA 2.0, Wikimedia Commons
Characteristics		

(continued)

Green and Blue Infrastructure (GBI) in Urban Areas, Table 2 (continued)

Typology of UGBI	Provided ES	Example
Communitarian compost areas Areas receiving a set of organic solid waste composters, allowing for the treatment of organic waste and producing fertilizer to be used by the inhabitants and local entities	Nutrient's retention and soil fertility	 Photo by Zarateman, CC0, Wikimedia Commons
Characteristics		
Fountains Artificial water fountains located in urban areas	Climate regulation, recreation, and landscape aesthetic values	 Photo by Dietmar Rabich, CC BY-SA 4.0, Wikimedia Commons
Characteristics		
Water channels Artificial watercourses within urban areas, usually connected to the river systems	Climate regulation, recreation, and landscape aesthetic values	 Photo by Nino Barbieri, CC BY-SA 3.0, Wikimedia Commons
Characteristics		
Retention basins Areas designed for the surface retention of runoff generated during precipitation events. It can be a natural and/or artificial drainage infrastructure	Climate regulation, flood mitigation, carbon sequestration, nutrients retention, food provisioning, recreation, and landscape aesthetic value	 Photo by ProfessorX, CC0, Wikimedia Commons
Characteristics		

(continued)

Green and Blue Infrastructure (GBI) in Urban Areas, Table 2 (continued)

Post-mining lakes Mining ponds that can be used for recreational and leisure purposes	Flood regulation, climate regulation, and recreation	 Photo by I. Elkman, CC BY-SA 3.0, Wikimedia Commons
Characteristics		
Greywater recycling Ponds designed for the treatment of grey waters, allowing the subsequent use of water for secondary uses (e.g., irrigation and miscellaneous washes)	Air quality regulation, carbon sequestration, flood regulation, climate regulation, nutrients retention and recreation	 Photo by SuSanA Secretariat, CC BY 2.0, Wikimedia Commons
Characteristics		
Green roofs Artificial system creating green spaces over urban structures such as buildings, bridges, and other structures	Air quality regulation, carbon sequestration, flood regulation, climate regulation, flood regulation, nutrient retention, food provisioning, recreation, and landscape aesthetic values	 Photo by Joalpe, CC BY-SA 4.0, Wikimedia Commons
Characteristics		
Green facades Artificial system establishing vertical green spaces on built facades and other types of construction	Flood regulation, climate regulation, and landscape aesthetic values	 Photo by Nemo, CC BY-SA 4.0, Wikimedia Commons
Characteristics		
Typology of UGBI	**Provided ES**	**Example**

(continued)

Green and Blue Infrastructure (GBI) in Urban Areas, Table 2 (continued)

Permeable pavements Pavements which allow the progressive infiltration of rainwater, commonly used in parking areas or similar	Flood regulation	 Photo by Alexander Eichler, CC BY-SA 3.0, Wikimedia Commons
Characteristics		
Sustainable drainage systems (SUDS) Systems designed to reduce the potential impact of urban (and rural) development projects on rainwater management, through low environmental impact solutions	Flood regulation	 Photo by Luís Pinto
Characteristics	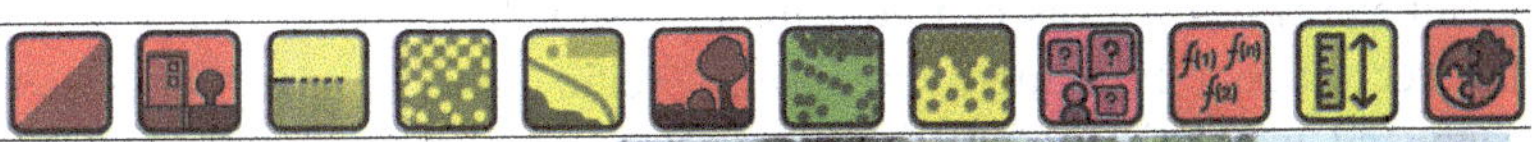	
Rain gardens Spaces designed to collect runoff from impermeable surfaces (e.g., roofs, streets, sidewalks, and car parks), creating the opportunity for the water to infiltrate into the soil	Flood regulation, climate regulation, water purification, and landscape aesthetic values	 Photo by Chris Light, CC BY-SA 4.0, via Wikimedia Commons
Characteristics		
Bioswales Infrastructure consisting of a drainage ditch filled with vegetation, compost, and/or rockfill, designed to remove runoff water pollution	Flood regulation	 Photo by Guzzardo Partnership, CC BY-SA 4.0, Wikimedia Commons
Characteristics	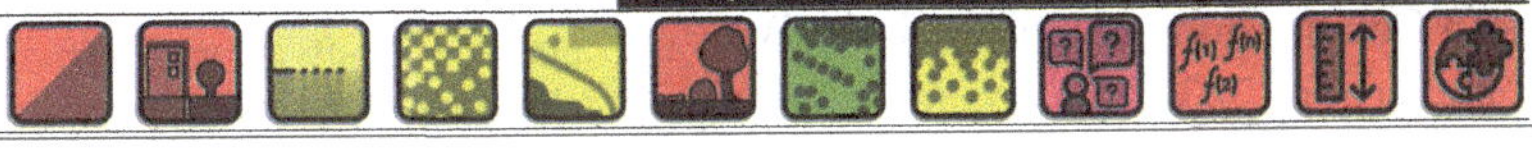	

can be relevant for GBI characterization and evaluation. These include a diversity of areas in terms of functions, scale, coverage type, morphology, degree of naturalness, ES type, vegetation type, distribution, and density. Table 2 lists the most widely used GBI, their relevant characteristics, and the ES they provide.

GBI Relevance to Meet Global Targets

The global population has been growing in recent decades, with the expanding population concentrating in urban areas. In 2018, the urban population represented 55% of the global population, and this value is expected to rise to 68% by 2050 (UN-DESA 2019). Green areas in urban contexts face several challenges, such as degradation (e.g., soil sealing and pollution). However, green areas are increasingly seen as fundamental for society. As such, GBI is essential to support several of the SDGs, particularly: Goal 3 – Good health and well-being, for which GBI areas can contribute cultural and recreational ES, improving users' mental health and well-being; Goal 6 – Clean water and sanitation, for which GBI can provide regulating ES, such as water purification and regulation; Goal 8 – Decent work and economic growth, where GBI design, implementation, and management is able to support opportunities for green jobs; Goal 11 – Sustainable cities and communities, with GBI contributing to universal access to safe, inclusive, and accessible green and public spaces; Goal 13 – Climate action, with GBI contributing to climate regulation, carbon sequestration, and strengthening urban resilience; and Goal 15 – Life on land, for which GBI can contribute regulating and provision ES (United Nations 2015). SDG 11 is also directly linked to the United Nations' New Urban Agenda, launched in 2017, which states that green and public spaces should be accessible to all society (United Nations 2015) and that GBI should "make cities inclusive, safe, resilient, and sustainable" (United Nations 2017). At the European level, the European Commission recognizes, in its Biodiversity Strategy for 2030, the crucial value of investing in nature protection and restoration for both the European Union economy and its inhabitants' mental and physical well-being (European Commission 2020).

Summary

Humanity is facing significant global challenges, especially in urban areas, where degradation is advancing at a relatively fast rate, and pressure on urban natural environments and human well-being is causing major concerns. Green areas are relevant in the urban environment to provide urban dwellers with some ES.

The GBI concept is associated with land use planning at a broad scale, but various specialists use it. GBI can range from entirely natural features to artificial interventions applied at diverse scales, aiming to enhance ES provision. Several goods and services can be provided by the wide variety of GBI typologies (e.g., green roofs and urban forests), due to these spaces' multifunctionality. Specific use of the term "infrastructure" in the concept name is fundamental for acknowledging the importance of green and blue spaces, equating them with other types of infrastructures such as roads, communications, or water systems, considered essential in developed areas.

Acknowledgments This work was produced with the support of PhD grant SFRH/BD/149710/2019 from the Portuguese Foundation for Science and Technology (FCT).

References

Alves, A., Vojinovic, Z., Kapelan, Z., et al. (2020). Exploring tradeoffs among the multiple benefits of green-blue-grey infrastructure for urban flood mitigation. *Science of the Total Environment, 703*, 134980. https://doi.org/10.1016/j.scitotenv.2019.134980.

Bartesaghi Koc, C., Osmond, P., & Peters, A. (2017). Towards a comprehensive green infrastructure typology: A systematic review of approaches, methods and typologies. *Urban Ecosystem, 20*, 15–35. https://doi.org/10.1007/s11252-016-0578-5.

Bartesaghi-Koc, C., Osmond, P., & Peters, A. (2019). Mapping and classifying green infrastructure typologies for climate-related studies based on remote sensing data.

Urban Forestry & Urban Greening, 37, 154–167. https://doi.org/10.1016/j.ufug.2018.11.008.

Benedict, M. A., & McMahon, E. T. (2002). *Green infrastructure: Smart conservation for the 21st century*. Washington, DC: Sprawl Watch Clearinghouse.

Brown, G., Schebella, M. F., & Weber, D. (2014). Using participatory GIS to measure physical activity and urban park benefits. *Landscape and Urban Planning, 121*, 34–44. https://doi.org/10.1016/j.landurbplan.2013.09.006.

Burkhard, B., & Maes, J. (Eds.). (2017). *Mapping ecosystem services*. Sofia: Pensoft Publishers.

Carmen, R., Jacobs, S., Leone, M., et al. (2020). Keep it real: Selecting realistic sets of urban green space indicators. *Environmental Research Letters, 15*, 095001. https://doi.org/10.1088/1748-9326/ab9465.

Charoenkit, S., & Piyathamrongchai, K. (2019). A review of urban green spaces multifunctionality assessment: A way forward for a standardized assessment and comparability. *Ecological Indicators, 107*, 105592. https://doi.org/10.1016/j.ecolind.2019.105592.

Cullen, S. (2013). What is green infrastructure? *Arboricultural Consultancy, 46*, 7.

Czúcz, B., Haines-Young, R., Kiss, M., et al. (2020). Ecosystem service indicators along the cascade: How do assessment and mapping studies position their indicators? *Ecological Indicators, 118*, 106729. https://doi.org/10.1016/j.ecolind.2020.106729.

Depietri, Y., & McPhearson, T. (2017). Integrating the grey, green, and blue in cities: Nature-based solutions for climate change adaptation and risk reduction. In: Kabisch N., Korn H., Stadler J., Bonn A. (eds) Nature-Based Solutions to Climate Change Adaptation in Urban Areas. Theory and Practice of Urban Sustainability Transitions. Springer, Cham. https://doi.org/10.1007/978-3-319-56091-5_6.

Dorst, H., van der Jagt, A., Raven, R., & Runhaar, H. (2019). Urban greening through nature-based solutions – Key characteristics of an emerging concept. *Sustainable Cities and Society, 49*, 101620. https://doi.org/10.1016/j.scs.2019.101620.

Estreguil, C., Dige, G., Kleeshulte, S., et al. (2019). *Strategic green infrastructure and ecosystem restoration – Geospatial methods, data and tools*. Joint Research Center, European Environmental Agency, Luxemburg.

European Commission. (2020). *EU biodiversity strategy for 2030 – Bringing nature back into our lives*. Brussels, Belgium.

European Environmental Agency. (2019). *Tools to support green infrastructure planning and ecosystem restoration*. EEA.

Grêt-Regamey, A., Weibel, B., Kienast, F., et al. (2015). A tiered approach for mapping ecosystem services. *Ecosystem Services, 13*, 16–27. https://doi.org/10.1016/j.ecoser.2014.10.008.

Haase, D. (2015). Reflections about blue ecosytem services in cities. *Sustainability of Water Quality and Ecology, 5*, 77–83. ISSN 2212-6139. https://doi.org/10.1016/j.swaqe.2015.02.003.

Haines-Young, R., & Potschin-Young, M. (2018). Revision of the common international classification for ecosystem services (CICES V5.1): A policy brief. *One Ecosystem, 3*, e27108. https://doi.org/10.3897/oneeco.3.e27108.

Harrison, P. A., Dunford, R., Barton, D. N., et al. (2018). Selecting methods for ecosystem service assessment: A decision tree approach. *Ecosystem Services, 29*, 481–498. https://doi.org/10.1016/j.ecoser.2017.09.016.

Havinga, I., Bogaart, P. W., Hein, L., & Tuia, D. (2020). Defining and spatially modelling cultural ecosystem services using crowdsourced data. *Ecosystem Services, 43*, 101091. https://doi.org/10.1016/j.ecoser.2020.101091.

Jato-Espino, D., Sañudo-Fontaneda, L. A., & Andrés-Valeri, V. C. (2018). Green infrastructure: Cost-effective nature-based solutions for safeguarding the environment and protecting human health and well-being. In C. M. Hussain (Ed.), *Handbook of environmental materials management* (pp. 1–27). Cham: Springer International Publishing.

Lyytimäki, J., Petersen, L. K., Normander, B., & Bezák, P. (2008). Nature as a nuisance? Ecosystem services and disservices to urban lifestyle. *Environmental Sciences, 5*, 161–172. https://doi.org/10.1080/15693430802055524.

MacKay, B., & Reed, N. P. (1994). *Creating a statewide greenways system: Florida greenways commission report to the governor*. Tallahassee: Florida Greenways Commission.

Martín, E. G. (2020). An operationalized classification of nature based solutions for water-related hazards – From theory to practice. *Ecological Economics, 7*, Article number 106460.

Mell, I. C. (2017). Green infrastructure: Reflections on past, present and future praxis. *Landscape Research, 42*, 135–145. https://doi.org/10.1080/01426397.2016.1250875.

Pereira, P. (2020). Ecosystem services in a changing environment. *Science of the Total Environment, 702*, 135008. https://doi.org/10.1016/j.scitotenv.2019.135008.

Raudsepp-Hearne, C., Peterson, G. D., & Bennett, E. M. (2010). Ecosystem service bundles for analyzing tradeoffs in diverse landscapes. *Proceedings of the National Academy of Sciences, 107*, 5242–5247. https://doi.org/10.1073/pnas.0907284107.

Ruckelshaus, M. H., Guannel, G., Arkema, K., et al. (2016). Evaluating the benefits of green infrastructure for coastal areas: Location, location, location. *Coastal Management, 44*, 504–516. https://doi.org/10.1080/08920753.2016.1208882.

Ryfield, F., Cabana, D., Brannigan, J., & Crowe, T. (2019). Conceptualizing 'sense of place' in cultural ecosystem services: A framework for interdisciplinary research. *Ecosystem Services, 36*, 100907. https://doi.org/10.1016/j.ecoser.2019.100907.

Schaubroeck, T. (2017). A need for equal consideration of ecosystem disservices and services when valuing nature; countering arguments against disservices. *Ecosystem Services, 26*, 95–97. https://doi.org/10.1016/j.ecoser.2017.06.009.
Shapiro, J., & Báldi, A. (2014). Accurate accounting: How to balance ecosystem services and disservices. *Ecosystem Services, 7*, 201–202. https://doi.org/10.1016/j.ecoser.2014.01.002.
UN-DESA. (2019). World urbanization prospects: The 2018 revision. https://www.un-ilibrary.org/content/books/9789210043144.
United Nations. (Ed.). (2015). Transforming our world: The 2030 agenda for sustainable development. https://sdgs.un.org/2030agenda.
United Nations. (2017). New urban agenda. https://habitat3.org/the-new-urban-agenda/.
von Döhren, P., & Haase, D. (2015). Ecosystem disservices research: A review of the state of the art with a focus on cities. *Ecological Indicators, 52*, 490–497. https://doi.org/10.1016/j.ecolind.2014.12.027.

Green and Smart Cities in the Developing World

M. Terdiman[1] and T. Angert[2]
[1]The Institute for Environmental Security and Well-being Studies, Jerusalem, Israel
[2]The Institute for Environmental Security and Well-being Studies, Herzliya, Israel

Synonyms

Sustainable city; Smart city; Smart sustainable city; City vision; Urban futures

Definitions

Sustainable city – A city which considers the three dimensions of sustainability – environmental sustainability, social sustainability, and economic sustainability – in its planning along with the creation of a resilient habitat for its present population without harming the ability of future generations to enjoy the same conditions.

Smart city – Has several definitions. Some definitions focus on specific types of Information and Communication Technology (ICT), while others focus on specific challenges, big date for instance, and specific applications such as transportation. It has six points of focus: smart economy, smart mobility, smart environment, smart people, smart living, and smart governance.

Smart sustainable city – Combination of the definitions of a sustainable city and a smart city.

Urban futures – Term that has been used to imagine how cities and urban areas will look like in the long term, how they will function, what infrastructures and government systems will support and coordinate between them and how they are shaped and impacted in the best way by its main stakeholders (civil society, government, businesses and investors, academia and others).

City vision – The formal process of creating an urban vision or a joint and desirable future for a specific city or urban area.

Introduction

Rapid urbanization is one of the unique phenomena of the twentieth and the twenty-first centuries. Nowadays, for the first time in human history, most of the world population resides in cities. According to UN statistics, as of 2018, 55% of the world population resides in cities and it is expected to grow even more to two-thirds by 2050. Furthermore, almost all global population growth until 2050 is predicted to take place in cities which will result in the growth of small and medium-sized cities, an increase in the number of megacities and the creation of more cities in various sizes to lower the population pressure on infrastructure and natural resources in the existing cities.

The main cause for this rapid global urbanization growth has been and most likely will continue to be the economic development which makes cities the hub of economic growth, industrialization, higher education, art, healthcare and innovation, where the well-being and health are the highest. Cities account for more than half of the global gross domestic product. As such, it attracts people from rural areas who are in search of employment and education opportunities. Yet, as a result of this rapid urbanization growth, cities around the world face a variety of interrelated challenges including, but not limited to: mobility, traffic congestion, poverty, inequalities, poor air quality, health risks, housing, socioeconomic inequalities, waste management, urban governance, urban sprawl, environmental degradation, water, food and energy insecurity, poor sewage and sanitation infrastructures, inappropriate land use, urban heat island effects, decreasing well-being and quality of life, etc.

Climate change and COVID-19 have exacerbated these challenges. On the one hand, cities are the main contributors of global carbon emissions and its residents suffer from climate change impacts, such as floods, sea level rise, smog, and natural disasters. On the other hand, cities, as hubs of innovation, may come up with solutions for this issue. Cities are also at the forefront of the COVID-19 crisis since the majority of people live in them. Yet, dealing with COVID-19 has also given an opportunity for the urban governance and residents alike to experience cleaner air, flowing traffic, and more wildlife.

The same challenges, but on a much bigger level, are true to the developing world. The developing world has experienced an unprecedented increase in urbanization levels due to migration from rural areas, high population growth and expansion of cities without any prior planning, which makes it difficult to govern them. Despite the fact that according to the UN (2018), the population living in cities or urban areas was 41% in lower-middle-income countries and 32% in low-income countries, Sub-Saharan Africa has experienced the highest urban growth since the 1990s at an average rate of 3.5% per year and huge megacities, much bigger than the ones in the developed world, have sprung in Asia, Africa, and Latin America. Most of these megacities are capital cities, where the vast majority of paid employment opportunities can be found. Furthermore, it is predicted that over the next three decades most of the global population growth will take place in urban areas in the developing world.

The speed of the urban increase in the developing world presents huge challenges for the state and urban governance. The main challenges include: the rising numbers of the urban poor and risks to the immediate and surrounding environment, to natural resources, to health, to well-being and quality of life, to social cohesion, and to individual rights. Additionally, in Sub-Saharan Africa, the challenges are even more enormous than in Asia and Latin America since many African cities are economically marginalized in the global economy and they grow despite poor economic performance and poor direct foreign investment. This makes it impossible for urban municipalities to provide adequate basic infrastructure and services to its populations.

Due to these challenges, cities all over the world have to develop a comprehensive set of strategies, solutions, and models that address present and future issues taking into account their interrelatedness and the need to prepare for different scenarios that may to be caused by changing climate or environmental conditions, by emerging and new technologies, and/or by another plague like the COVID-19. They also must be based on high-quality statistics and the future patterns and trends of urban change. All the main stakeholders (government, businesses, civil society, academia, etc.) need to be involved in this process. Given the dominance of climate change and sustainability in the discourse throughout the developed world, the models of the sustainable city, the smart city, and the sustainable smart city are increasingly considered as potential answers to the challenges arising from rapid urbanization growth.

These models have also dominated the discourse in the developing world for various reasons including: to increase foreign investments from the developed countries, to achieve personal and local goals, to strengthen their power over cities and megacities that are difficult to manage and control, to secure their power and to find solutions for the challenges created by climate change impacts, etc.

Not much has been written concerning sustainable and smart cities in the developing world.

Hence, the aim of this piece is to describe the development of sustainable cities and smart cities in the developing world. It will first deal with the challenges in conducting research on urbanization and various urban forms and futures in the developing world. Then, it will move on to discuss some relevant definitions of these terms, the challenges and constraints involved in the construction of these kind of cities and to describe few examples from the developing world for three main different kinds of cities: sustainable cities, smart cities, and smart sustainable cities. The summary will look at future trends.

Urban Futures and City Vision

As this piece also deals with future trends and with city visions in the developing world that enabled the creation of sustainable cities, smart cities, and smart sustainable cities, it is important to go some more into detail concerning urban future and city vision.

The abovementioned future predictions and tendencies and its significance are part of preparing for the future. This subdiscipline is called urban futures and city vision and is located under the future thinking and systems research discipline. This discipline does not give certain answers as for how the future will look like, but it provides the ability to analyze, research, raise ideas based on past and present tendencies to prepare for future tendencies and predictions in certain probabilities to which we can prepare ourselves. Throughout history, researchers and philosophers researched different types of settlements and, as a result, thinking about the "ideal city" was developed. As part of this thinking, the idea of a sustainable city, which we use today, was developed in the 1960s and 1970s and penetrated the urban research.

The importance of cities and the fact that most of the global population lives inside them makes it necessary for them to adapt to the challenges faced by its population – like the COVID-19, population density, natural disasters, air pollution, affordable, available, and effective access to water, food, and energy resources – and to develop long-term planning given the fact that these challenges will continue to grow if untreated and without enough preparation.

Urban futures is a term that has been used to imagine how cities and urban areas will look like in the long term, how they will function, what infrastructures and government systems will support and coordinate between them, and how they are shaped and impacted in the best way by its main stakeholders (civil society, government, businesses and investors, academia, and others). Urban future thinking necessitates the abovementioned four main stakeholders in the city to build and develop together a city vision.

The city vision is the formal process of creating an urban vision or a joint and desirable future for a specific city or urban area. It can relate to a single preferable future or to different and alternative urban futures. This topic is part of the science of thinking about the future of cities and includes a variety of future-based methods and tools to assist in the building and development of a city vision. For example, the method of in retrospect, which begins with the definition of the desirable future and afterwards looks at the policies, programs, and tracks which will connect the present with the future.

Thinking and creating a city vision is a challenging process. It necessitates resources, listening, compromises, partnerships, and prioritizing the big goal in order to overcome the conflicting interests and to create a good leadership. Thinking in the city level as we know it today and given the fast urbanization process necessitates an innovative thinking. It also provides a lot of opportunities and shapes the culture, history and the future. It has a lot of impacts like reform and constructive change. It raises questions such as what will be the status of the city in the future, how people from different disciplines will work to develop the vision, how will the economy look like, how will the dynamics between the people look like, and many more questions.

Challenges Facing the Research of Urbanization and Future Development in the Developing World

There are several challenges that face researchers of urbanization and various urban forms and futures in the developing world.

The first and most basic challenge is the lack of a global standard which defines what counts as an urban environment and what are its boundaries. Thus, the definition can vary across countries and even within a single country and can be a function of several criteria: population size, population density, administrative or political boundaries, and definitions and economic function. Hence, new developing sustainable, smart, and sustainable cities can be either a suburb or a village, town, neighborhood, etc. with either fixed or unfixed numbers of people.

Secondly, the lack of reliable and updated demographic data makes it difficult to analyze patterns of urban growth over time. This results in the fact that current urban populations are frequently inaccurate, which, in some cases, makes it difficult to ascertain the real size of the population living in sustainable, smart, and sustainable smart cities.

Third, cities have changed their economic character and spatial form, mixing agricultural and nonagricultural sources of livelihood and, thus, cannot be considered as either urban or rural.

Definitions

Another kind of challenge facing researchers of urbanization in the developing world is that sustainable cities, smart cities, and smart sustainable cities have several definitions. Thus, each sustainable city, smart city, and smart sustainable city serves different goals in different countries and, therefore, has developed its own distinct aspects despite the existence of a common ground between cities in each one of the three groups.

A sustainable city, which is called also ecological city, is a city which considers the three dimensions of sustainability – environmental sustainability, social sustainability and economic sustainability – in its planning along with the creation of a resilient habitat for its present population without harming the ability of future generations to enjoy the same conditions. Other scholars and institutions added two more dimensions to the definition of a sustainable city: political or governance sustainability and physical or built sustainability. Yet, all the definitions for a sustainable city share the three dimensions of sustainability.

According to the UN Sustainable Development Goal 11, the aim of sustainable cities or ecological cities is to "make cities inclusive, safe, resilient and sustainable" by the creation of safe and affordable housing, affordable and sustainable transportation system, inclusive and sustainable urbanization, protection of world's cultural and natural heritage, reduction of the social and economic effects of natural disasters, reduction of the environmental impacts of cities (mostly air quality, waste management, water pollution), and provision of access to safe and inclusive green and public spaces. A broader aim of sustainable cities, according to the Urban 21 Conference in 2000, is "improving the quality of life in a city, including ecological, cultural, political, institutional, social and economic components without leaving a burden on the future generations."

Yet, a smart city has several definitions. Some definitions focus on specific types of Information and Communication Technology (ICT), while others focus on specific challenges, big date for instance, and specific applications such as transportation. Overall, the smart city includes six main points of focus: smart economy, smart mobility, smart environment, smart people, smart living, and smart governance.

According to some definitions, smart cities aim at achieving sustainability goals. One such definition of smart cities is "places where information technology is combined with infrastructure, architecture, everyday objects, and even our own bodies to address social, economic and environmental problems." According to another definition which emphasizes this point even better, a smart city is an "effective integration of physical, digital and human systems in the built environment to deliver a sustainable, prosperous and inclusive future for its citizens."

According to other definitions, smart cities aim at the achievement of other goals, such as addressing public issues, improving the inhabitants' quality of life, making better use of the city's resources, and making the traditional networks and services more efficient. An example for such a definition is: "a 'smart city' is intended as an urban environment which, supported by pervasive ICT systems, is able to offer advanced and innovative services to citizens in order to improve the overall quality of their life." Another definition of this sort is: "a Smart City is a place where the traditional networks and services are made more efficient with the use of digital and telecommunication technologies, for the benefit of its inhabitants and business." Yet, another definition of a smart city is focused on the partnership between the main stakeholders in the city in order to address public issues and says that a smart city is "a city seeking to address public issues via ICT-based solutions on the basis of a multi-stakeholder, municipality-based partnership."

Therefore, smart cities are not necessarily sustainable cities, although some of them are aimed at achieving sustainability while sustainable cities do not necessarily use ICT in order to achieve their goals.

Last but not least, the definition of a smart sustainable city is a combination of the definitions of a sustainable city and a smart city. Thus, the United Nations Economic Commission for Europe and the International Telecommunication Union define a smart sustainable city as "an innovative city that uses ICTs and other means to improve quality of life, efficiency of urban operation and services, and competitiveness, while ensuring that it meets the needs of present and future generations with respect to economic, social, environmental as well as cultural aspects."

Construction of Sustainable and Smart Cities in the Developing World: Difficulties and Challenges

The developing countries face a few difficulties and challenges designing and transitioning towards sustainable and smart cities. These difficulties and challenges result from some interrelated basic factors of each country including: its economic, social, political, security and gender issues, its health system, its technological progress, the environmental challenges it faces, its education system, its history, its interests, its international status and alliances, etc.

The challenges facing the construction of green cities and the fast green growth existing cities in the developing world in general are numerous and interrelated and dealing with them necessitates urgent and substantial comprehensive changes. Some of these challenges are: a lack of stable governance in the national, district and city levels; a lack of ability to set short-term, medium–term, and long-term goals while taking into consideration the needs of the people on the ground; there is not always either a vision or a strategy for the construction of a sustainable city that fits the location; lack of funding and in case there is some funding, a preference to invest in the present in order to deal with the current challenges instead of thinking and planning ahead; regulatory issues in cases of allowing a newly constructed sustainable city to produce its electricity by itself or in cases of introducing new technologies; weak and inadequate infrastructure; poverty; environmental deterioration, unplanned land use; high population density; natural disasters such as frequent floods; air and water pollution; unreliable and unaffordable access to basic natural resources such as water, food, and energy; investment considerations by foreign parties.

Since smart cities are based on ICTs and they have several definitions, there are a host of different challenges facing the transition from a regular city to a smart city in the developing world. One such challenge deriving from the fact that there are several definitions for what is a smart city is the creation of the changing models of the needs, infrastructure, and security services according to the different aims that the smart cities are planned to achieve.

There are other challenges emanating from it. From a planning point of view, since smart cities are built on the infrastructure, systems, and populations of existing cities, these do not always suffice and sometimes are inadequate for the transition to a smart city. Also, not every city and government has the financial capability or the funding to make this transition happen even though they build a plan to do so as the infrastructures are very expensive. During the planning process, the city governance needs to take into account the possibility that there will be a negative impact on the wellbeing of the current city's population during the transition period and after it.

Another challenge related to the planning is the digital gap existing within the city's population. This gap is manifested by different levels of digital skills and literacy and of the ability to use the computer and smart applications. It may create inequity between populations who will be able to use the smart application and those who will not be able to do so.

Moreover, since smart cities use sensors in order to collect, monitor, and analyze information to improve the well-being and the economic situation of the city's and country's population, a lack of internet connectivity, access to internet, power outages, and short bandwidth speed may also be a barrier for the transition to smart cities in some developing countries. Another challenge is the possibility to construct advanced, cheap, and available ICT systems that will be able to have a huge number of users without burdening the system. It corresponds with the UN Sustainable Development Goal 9 which aims at building resilient infrastructure, promoting inclusive and sustainable industrialization and fostering innovation. Furthermore, connecting between different systems owned by different entities in the private and public sectors that, in some cases, compete with each other and creating one network may also be a huge challenge.

Additionally, a smart city includes different ways to decrease carbon emissions while some of the technologies used for that aim emit carbon emissions. Therefore, a smart city will need to find the right balance between the different kinds of technologies and to create infrastructure for clean energies.

Ensuring the human and city's security is also a huge challenge in smart cities. Whereas in the past, the police and army were responsible for the security of countries and cities, nowadays, there are new threats to the human security. One of them is cyber security. Hackers and potential enemies may break into the smart city's system and access sensitive information concerning people, governance systems, etc. This challenge is interrelated to another challenge which is the possibility of harming the people's sense of privacy which may be harmed by the use of cameras to curb crime or by the huge collection of data from smart systems that people use and are exposed too daily and which may necessitate a prior preparation of the city's population for the transition to a smart city.

Last but not least, since the transition to a smart city is a very sophisticated, complicated, and expensive endeavor, in order for it to succeed, it necessitates collaboration between all the stakeholders in the city, public involvement, and participation in the process, securing enough funding resources from local and foreign sources and a stable and visionary government with an ability to plan long term and make decisions accordingly.

General Characteristics of Sustainable and Smart Cities in the Developing World

Despite all these challenges, there are sustainable cities and smart cities throughout the developing world, in Latin America, in Africa, in the Middle East and North Africa, in Southeast Asia, and in South Asia.

Broadly speaking, sustainable cities or ecological cities can be divided into two types: new cities or urban areas that are usually being built by governmental or private companies on a small scale and much larger existing cities or urban areas that transitioned into a sustainable city.

Sustainable cities have various goals. The Sustainable City in Dubai aims to be the first net-zero energy development project in the Emirate of Dubai and to demonstrate Dubai's regional and global pioneering and leading role in dealing with climate change. However, the ecological city of Zenata, which is located in the northeast of Casablanca, aims at finding a solution to alleviate the population density and the unemployment in Casablanca and, thus, it will create employment opportunities, target specifically the middle class, victims of housing shortage or poor housing in the slums while also dealing with climate change and protection of the environment. Curitiba, which is the capital and the largest city in the Brazilian state of Paraná as well the eighth most populous city in Brazil and the largest in southern Brazil, developed in the 1960s a new city vision in order to limit the city sprawl, deal with the fast urbanization process and enable economic growth.

All sustainable cities do their best to integrate social, economic, and environmental sustainability into their development. Yet, they develop different visions according to the aims for which they were constructed.

For example, the Sustainable City in Dubai has been recognized as "the happiest community" in the Gulf States by the Gulf Cooperation Council for the last 3 years. This city includes: a residential area of 500 houses and villas using traditional architecture, which are grouped into five residential clusters, each with their own urban farm; 11 natural biodome, hydroponic, and aquaponic greenhouses systems using passive cooling method with fans and pads; organic farm and individual garden farms for local food production; waste water recycling; solar panels on the roofs of the city's houses and parking areas for the electric cars; biking and jogging trails; a sustainable plaza with shops, restaurants, cafes, offices, and apartments that provide the means for residents to sustain themselves and their families. Its borders, which are lined with more than 2500 trees, act as a buffer zone against pollutants and purify incoming air.

Curitiba is a unique example for an existing city that transitioned to a sustainable few decades before this city vision started to gain dominance. Curitiba is considered one of the world's best examples of urban planning, and in 2007, it was placed third in a list of 15 Green Cities in the world. It is also considered one of the most prosperous cities in Latin America. It is also a member of C40 Cities, which is a global network of megacities that are committed to dealing with climate change and create a future where everyone can thrive. The city has more than 400 square kilometers of public parks and forests with 52 square meters of green space per capita; it is the first city that introduced a planned transportation system, which includes lanes on major streets devoted to a bus rapid transit system which charges one price whatever the distance is and with access for disabled people. Both things together make the air pollution in Curitiba the lowest in Brazil. It also has business incubators designed to help small companies get established and prosper; Crafts Lycée which trains people for professions such as marketing and finance in order to provide jobs and income for the unemployed among the hundreds of thousands of people living in its peripheral towns; a program that exchanges trash and recyclables for bus tokens, food, and cash which makes the city clean.

Another Latin American existing urban space that transitioned to a sustainable city using a very unique and first of its kind model is Curridabat in Costa Rica. Curridabat is the eastern suburb of San Josȇ, the capital of Costa Rica. It developed a unique model called the "sweet city." This model considers everyone living in Curridabat, whether human or not, a city dweller with the bees as the center of urban design. This model overcomes the long-lived antagonism between city and nature that has characterized traditional urban development. The "Sweet City" vision became a public policy in 2016 and the official Curridabat brand. The features of the "Sweet City" are: creation of shade using native trees, excellent waste management, promoting orchards; improving the wellbeing of its residents by balance with the natural environment with the understanding that when the environment is healthy, the residents are also healthier and the children grow up in a natural environment. This model has also served to raise the awareness of its residents for the preservation of the environment. This model has five dimensions linked to well-being: biodiversity, infrastructure, habitat, coexistence, and productivity. It has seven goals: recovery of soil health; improved security in key spaces like parks; enhancement of intermodal transportation and walkability; promotion of conscious eating and mental health; and strengthening of participatory local governance.

Also smart cities can be divided into two types: new cities or urban areas that are usually being built by the government with private-public partnerships on a small scale and much larger existing cities or urban areas that transitioned into a smart city.

Existent cities transitioned to smart cities in order to achieve various goals. For example, the government of the Emirate of Dubai aims at making the city of Dubai into the happiest city on earth and a world-leading city. During the transition process, Dubai developed an integrated futuristic, innovative, and sustainable approach. In 2014, Dubai launched the Smart Dubai Initiative in order to transform itself into a smart city in 2021. The objectives of this initiative included all the six main points of focus of smart cities: smart economy, smart mobility, smart environment, smart people, smart living, and smart governance. This initiative envisions a city where all its resources are optimized for maximum efficiency, where services are integrated seamlessly into daily life and where the people and information are protected. It includes 100 smart initiatives and more than 1000 smart services which results in the transformation of Dubai into a model smart city. All these initiatives are conducted in partnership and involvement of the public and private sectors. One of these initiatives is Dubai 10X Initiative that envisions a Dubai that is ahead of the world by 10 years. Due to its efforts, Dubai has been recognized by the United Nations as a role model for a smart, sustainable, and resilient city.

On the other hand, other cities, such as Medellín in Colombia, wanted to transform itself to a smart city in order to rebuild society with an emphasis on alleviating poverty, isolation, and lack of opportunity that led young people to resort to crime as their best path for success. Therefore, the focus of the transition to a smart city was on the improvement of the everyday life of all citizens. Thus, data gathered on everyday life in Medellín can be used both by the government and by the residents. It is open to everyone. Moreover, Medellín uses other smart applications such as smart mobility system integrating communications, infrastructure and transportation; a prepaid energy management solution; an early warning system for smog and natural disasters; an emergency platform that significantly reduced waiting time for help to arrive; a system for access to education for all; smart technology in the city's administrative services.

Yet, other cities went through this transition to smart cities in order to deal with the challenges created by the rapid urbanization growth. Thus, unlike Dubai, in general, most smart cities in the developing world do not have a comprehensive approach towards the transition into smart cities, but focus on specific issues. For example, Cape Town, the second largest city in South Africa and one of the smartest cities in Africa, launched an action plan to make itself a smart city. This plan is based on digital infrastructure, digital inclusion, E-government, and digital economy. This action plan is focused on harnessing the Internet of Things to foster its development by deploying smart sensors to meet demands for electricity, water management and waste treatment; to optimize agriculture and traffic management and to prevent or deter crimes. The Internet of Things also serves as a means of extending government services and benefits and of providing the population with access to basic healthcare. The city acts transparently with its citizens via open data web portal. The Indian City of Amritsar, which is located in Punjab, and the City of Dehradun, the capital of the state of Uttarakhand, are focused on the development of smart transportation.

Smart cities may face challenges during and after their development in case the planners do not keep the big picture in mind and do not emphasize well-implemented infrastructure and citizen needs. Technology for technology's sake will not create solutions to some of the developing world's cities biggest challenges. A good example for this is Rwanda, which is aiming at creating smart cities. As part of this initiative, the Rwandan government launched a partnership with Nokia for investment in network connectivity and sensor deployment to improve public safety, waste management, utility management, and health care. However, when Kigali, Rwanda's capital, started using bases with free Wi-Fi and cashless payment service, the buses had connectivity issues related to the Korea-built technology's inability to adapt to local conditions. Additionally, there has been criticism around the lack of inclusivity of certain smart cities projects such as Kigali's Smart Neighborhood project, Vision City, which creates a tech-enabled neighborhood with solar power street lamps and free Wi-Fi in the town square since the project ignores the socioeconomic realities of a city where 80 percent of its population lives in slums with monthly earnings below $240.

Another challenge related especially to African smart cities is that they return to an older phenomenon and serve to redefine African countries. In the decades after the end of colonial rule in the 1950s and 1960s, African countries began building new cities from scratch as a way of redefining themselves. These cities, which were then appointed as capitals, were to be centrally located and politically and ethnically neutral. Abuja was constructed in the 1980s to replace Lagos as Nigeria's capital, Dodoma replaced Dar es Salaam as Tanzania's capital in 1974, etc. These new capitals represented a novel kind of civic utopianism. They gave their respective countries a means through which to imagine and pursue better versions of themselves. The drive to build smart cities is this dream carried to its natural conclusion in the IT-dominated twenty-first century. Yet these cities have fallen far short of what was promised. Even on the rare occasion when smart cities are actually completed, they never match up to what was sold at the start – the chance for African countries to showcase that they are on par with the Western and Asian nations that dominate the global tech sector. Nor have African governments been able to fully sell these projects to the investors and public.

Last but not least, smart sustainable cities have being planned and started to be constructed in Nigeria (Eko Atlantic City) and Southeast Asia, where the transition to sustainable smart cities in the joint vision of the Association of Southeast Asian Nations.

Summary

Creating sustainable cities and smart cities has been very dominant in the developing world in order to deal with the challenges of rapid urbanization growth and climate change. In parallel, the newly developed smart sustainable city has started to develop in existing cities, such as Dubai, and in newly built cities in Africa and Southeast Asia. It is probable that this tendency will continue in the future at least in the short and medium terms.

However, the challenges facing cities in the developing world may vary according to its level of development and its ability to govern its population and supply them with their basic needs. Additionally, emerging and future technologies may create new challenges that cannot be grasped today that combined with current ones may create the need to find new solutions and to come up with new models of cities.

Cross-References

▶ Big Data for Smart Cities and Inclusive Growth
▶ City Visions: Toward Smart and Sustainable Urban Futures
▶ Green Cities in Theory and Practice
▶ Making of Smart and Intelligent Cities
▶ Smart(er) Cities
▶ Smart Cities
▶ Smart City
▶ The Sustainable and the Smart City: Distinguishing Two Contemporary Urban Visions

Green Belts

Jonathan Manns
Rockwell, London, UK
UCL, London, UK

Synonyms

Green buffer; Green corridor; Green finger; Green strip; Green wedge

Definition

A green belt is a land use designation which severely restricts the development of land which is typically, but not exclusively, previously undeveloped. Their primary purpose is to prevent the expansion of existing or proposed settlements in a particular location. A green belt may encircle a settlement, as the name suggests, but can also take the form of a buffer, corridor, strip, or wedge.

Introduction

A green belt is a land use policy designation intended to severely restrict the development of typically undeveloped land and thereby to stop the unplanned or unmanaged expansion of existing or proposed settlements by preventing their encroachment onto such land. While the specific reasons for the use of green belts vary from one context to another, their general form, support, and criticism are reflected in almost all circumstances. This briefly summarizes the historical emergence of the concept alongside the widely replicated debate which green belts attract.

Historical Emergence

Consideration of the relationship between a settlement and the area beyond it has existed since humans began to establish non-nomadic lifestyles during the First Agricultural Revolution some 12,000 years ago, prompting questions about ownership, administration, agriculture, and amenity. It consequently receives a degree of historic attention in art and literature, with Thomas More's *Utopia* (1516), for example, making reference to "the country that belongeth to the city" (More, 1516).

The first formal decision to actively restrict the growth of a city by limiting development around the entire periphery was in England in the reign of Queen Elisabeth I (1533–1603). On 6 July 1580, she decreed that a 3-mile-wide *cordon sanitaire* should be established around London which prohibited new building. This was an attempt to limit an outbreak of the plague, but was hard to police and widely ignored.

It was not until proactive attempts at urban planning became more commonplace, particularly from the late eighteenth century onwards, that more considerable and focused attention was given. The catalyst was twofold, being (a) the desire of nation-states, organizations, and individuals to develop entirely new settlements and (b) the rapid expansion of existing urban areas fueled primarily by the Industrial Revolution in Europe.

Planned Green Interventions

The idea of green space encircling a settlement, commonly referred to as a "green girdle" during the nineteenth century, grew in prominence in the 1830s and was initially incorporated into newly planned settlements. In Australia, Colonel William Light's plan for Adelaide in 1837 set aside substantial "Park Grounds." In New Zealand, the New Zealand Company's instructions to its surveyor Captain William Mein Smith, charged with establishing Wellington in 1840, was that "[i]t is indeed desirable that the whole outside of the Town, inland, should be separated from the country by a broad belt of land which you will declare that the Company intends to be public property on condition that no buildings be ever erected upon it" (Cook, 1992). The notion was also incorporated into conceptual new towns. In 1849, James Silk Buckingham put forward proposals for a new "model city" which would include an agricultural town belt (Buckingham, 1849). Buckingham's

G

idea went on to directly inform Ebenezer Howard's notion for new "Garden Cities," which would also be encircled by an agricultural belt (Howard, 1898).

The establishment of new settlements paralleled the expansion of existing ones as the result of rapid urbanization, linked to the Industrial Revolution, which prompted an increasing amount of thought about their condition at a metropolitan scale, on issues related to the living conditions of residents such as traffic congestion, housing standards, air quality, parks, and open space. In 1829, the English landscape architect John Claudius Loudon proposed concentric rings of green to assist London with "the supply of provisions, water and fresh air, and to the removal of filth of every description, the maintenance of general cleanliness, and the despatch of business" (Loudon, 1829). His concentric rings would, as the urban area grew, alternate mile-wide bands of buildings with half-mile-wide zones of accessible country or gardens.

Accessible green space was increasingly introduced into existing urban areas where the opportunity arose, most notably in the form of landscaped thoroughfares. In 1857, Emperor Franz Joseph I of Austria ordered the removal of Vienna's city walls to facilitate connectivity with the city's inner suburbs (*Vorstädte*). They were replaced with the *Ringstraße*, a grand boulevard which both defined the edge of the historic center and served as an accessible area for recreation and amenity. Some two decades later, in North America, simmilar so-called parkways were introduced, with the first being New York's Eastern Parkway, built between 1870 and 1874 and designed by Frederick Law Olmsted and Calvert Vaux. These were popularly rolled-out as part of the City Beautiful movement which saw boulevard streets at the edge of urban centers in the context of beautification of the built environment to enhance moral and civic virtue.

Green Belts and Green Wedges

It was when thinking about improvements to the urban condition of London that the term "green belt" was first coined by the English social reformer Octavia Hill in 1888. "Might it not also be possible," she wrote in her essay "More Air For London," "to secure a green belt to a road newly cut across the country, plant it with tree, and make of it a walking and riding way?". She envisaged connections being made to other protected green space, stating that "besides these two kinds of garden or park, I have always felt that working-men wanted good walks, and whenever I thought about buying land in the suburbs, it always seemed to me that money would go much further if, instead of buying a large square area for a park, one could buy a field path with a good space on each side of it. If this were done, the approach to the commons and parks, the forest, or the heath, might still be these pleasant walking ways, which are so much more refreshing that the train or the omnibus, and bring the sense of country much nearer to the town" (Hill, 1888).

The first specific proposals to be mapped for an existing settlement were for London and came within weeks of each other in 1901, from William Bull and Lord Meath (for London's example, see Manns 2014). There followed a myriad of proposals for designated green belt in various countries throughout the first half of the twentieth century, often emerging alongside proposals for transport improvements. The majority were reliant on the idea of rings of green space, often rippling through a settlement. However, increasing thought was also given to alternative forms that green belts could take, quickly prompting the concept of a "green wedge." First proposed by Bruno Möhring, Rudolf Eberstadt, and Richard Peterson in their third-prize entry for the competition for a Greater Berlin Plan in 1910 (Möhring et al., 1910), these were influenced by the way that larger parks connected into American parkways and were seen as a way of introducing green space into the city itself. The result is that both "belts" and "wedges" received attention as exercises in strategic scale planning that were given impetus by the First and Second World Wars: conflicts which themselves contributed to a greater focus on the health of urban dwellers, not least in terms of their amenity and diet.

Two notable examples of green belts to emerge following the Second World War are those of London and Copenhagen. The former was designated locally from 1955 onwards but had most recently been planned by Patrick Abercrombie. Abercrombie's proposal retained a traditional orbital ring shape, at the edge of the city, supported by green wedges, where the green belt would be "for no further building other than that ancillary to farming" (Abercrombie, 1943) and was "to provide primarily for recreation and fresh food for the Londoner, and to prevent continuous suburban growth" (Abercrombie, 1944). It proved effective at limiting the expansion of the settlement but less effective at supporting public access. Conversely, the latter is notable for its distinctive radial approach, known as the Five Finger Plan (1947), which was developed through the Urban Planning Laboratory with Steen Eiler Rasmussen and Christian Erhardt Bredsdorff. It did not assume any significant population growth and was therefore more focused on the speed of connectivity to the urban center along key economic corridors interspersed with green space. Accordingly, it proved a less effective limit to sprawl but more effectively provided green wedges with greater public accessibility. Both have gone on to influence the design of green belts around the world.

Critique and Defence

The historical emergence of the concept raises questions about the relationship between town and country and the location of new development. These underpin much debate as to the shape and permanence of green belt designations. The iterative way in which the idea emerged, catering to concerns about amenity, connectivity, and agriculture, helps explain the varied forms and purposes which are incorporated beneath the umbrella objective of preventing uncontrolled urban expansion.

While the reason for designating green belt varies from one place to another, the arguments for and against green belts are consistent in most national and local contexts (albeit the strength of debate and case put forward varies in each circumstance).

The concept of a green belt is easily understood and, in most circumstances, popular with the general public. This may in part explain why the debate surrounding the green belt tends to be expressed publicly in binary for and against terms, as opposed to being focused on ways in which the green belt can be either reformed or the same outcomes better achieved. Broadly, the case for revision tends to be made in more evidence-based terms while that against revision tends to be more emotive. In both instances, the case for and against is often expressed in terms which are reactive and defensive as opposed to those which are proactive and postive.

The following five components feature commonly in the cases both *for* and *against* the designation of green belts.

For: They Prevent Urban Sprawl

Urban sprawl occurs when settlements expand in an unplanned or uncoordinated way. Green belts are a clear, simple, and effective prevention against this happening by stopping development from taking place. The result is to encourage development to occur elsewhere, potentially assisting in the regeneration of existing urban areas by increasing the focus of developers on land within a settlement boundary.

For: They Protect the Setting of Historic Towns

As urban areas expand, there is a prospect that they will absorb others which were previously outside their boundary, many of which will have been existing for long periods of time. By preventing development in the area beyond an existing settlement, including a large-scale expansion of those existing settlements within the green belt, they are able to preserve the character and setting of smaller towns and villages.

For: They Provide Amenity and Recreational Space

Green belts ensure that areas of open land are preserved on the edge of, or indeed as wedges within, larger settlements. This means that residents do not need to travel as far to access open space in which to undertake recreational or sporting activities. Engaging with such space has been shown to improve mental and physical well-being. Similarly, food grown within the green belt can be made available locally and closer to a potential marketplace.

For: They Support the Agricultural Industry

The agricultural industry is a key employer in rural areas and performs an important function providing fresh, locally sourced food in addition to contributing to food security. By designating agricultural land as a green belt, the value of this land is maintained at a low level as there can only be limited speculation as to its potential for alternative uses. In preventing the development of the land, it can either continue to be used, or alternatively put to use, for agricultural purposes.

For: They Protect the Environment Without Preventing Development

Green belts contribute to environmental protection and the mitigation of climate change by ensuring that space is available for uses other than those which require development. This might include, for example, the maintenance of forestry or provision of areas of land set aside from farming. This does not come at the expense of development, which simply takes place elsewhere, and does not prevent such development from being either managed or sustainable.

Against: They Increase Land Prices and Reduce Housing Affordability

The key characteristics of any market, no less for land and housing, are that they express a relationship between supply and demand. In situations where there the supply of land for housing is reduced, particularly where demand increases, an affordability crisis ensues which cannot be met except through increased supply. In allowing planning professionals to control the supply of land in any one location, they are artificially influencing the market and, where surplus demand exists, contributing to rising property prices and reduced housing affordability.

Against: They Encourage Less Sustainable Commuting Patterns

While green belts are able to prevent the expansion of settlements, they cannot restrict the desire of people to work within those settlements. It therefore separates the place in which some people live and work, resulting in commuting between the two. Likewise, smaller settlements within the green belt are less able to support the range of services and facilities available in a larger center. For this reason, those who live there will be more reliant on private means of transport, particularly the car, contributing to local traffic congestion.

Against: They Create a "Leapfrogging" of Development

The blanket protection afforded to particular locations by green belts does not remove the market demand for a particular type of development; it merely affects the locations in which that takes place. As such, the designation of green belt does not stop the development from occurring; it merely takes place elsewhere. As not all development can be contained within existing settlements, the result is that development will often occur "beyond" the boundary of the green belt, in an effect termed "leapfrogging." This also extends the functional economy of the larger settlement, to the detriment of "dormitory" settlements elsewhere.

Against: They Prevent the Optimum Use of Land

The process of plan making is predicated on assessing which areas are best suited to development and which are not. In putting a blanket restriction on development, the green belt excludes certain locations from being properly considered as part of this process. Consequently, while there may be sites which are well-suited to meeting the needs of a particular settlement (well-connected to public transport infrastructure, low ecological or visual amenity value, previously developed land, etc.), these may not be able to come forward for development unless removed from the green belt: a decision which could prove unpopular. The additional focus for development within existing urban areas also intensifies the use of those sites, creating pressure for both use and development on existing urban sites and green space, which may be more impactful on local communities or of higher social and amenity value.

Against: They Do Not Benefit the Natural Environment

While it is often suggested that green belt land is of a high landscape or environmental quality, the designation itself is not an acknowledgment of either landscape or environmental quality. These are almost always protected by other policy designations. Instead, green belts simply prevent development from occurring. They do not therefore prevent other activities occurring which may be environmentally damaging, such as infrastructure provision or commercial agricultural practices. Nor do they require enhancements to be made either to secure improvements to the natural environment, to prevent decline or to address challenges such as climate change.

Summary/Conclusion

Green belts have a considerable history and an iterative evolution, underpinned not only by a desire to prevent uncontrolled urban expansion but also to assist with the well-being of those who live in and around the settlements which they serve. While debate exists as to the efficacy of green belts as a policy tool, both criticisms and defenses are made in this context. There is therefore scope for further evolution and future consideration of how green belts might be adapted to better meet the aspirations of those making cases on either side of the debate.

Cross-References

- ▶ Circular Economy Cities
- ▶ Climate Change
- ▶ Peri-urbanization
- ▶ Sustainable Development

References

Buckingham, J. S. (1849). *National evils and practical remedies*. Clifton: Augustus M. Kelley, 1973.

Hill, O. (1888). More air for London. *Nineteenth Century, xxiii*, 181–188.

Howard, E. (1898). *To-morrow! A peaceful path to real reform*. London: Swann Sonnenschein. Second edition as *Garden Cities of To-morrow*. London: Swann Sonnenschein, 1902.

Loudon, J. C. (1829). Hints for breathing places. *Gardeners Magazine*, Vol. 5, pp. 686–690.

Manns, J. (2014). *Green Sprawl: Our current affection for a preservation myth?* London: London Society.

Möhring, B., Eberstadt, R., & Petersen, R. (1910). *Et In Terra Pax*. Berlin.

Ward, J. (1992). Quoted in Cook, W. *Background Report 3: Wellington Town Belt Management Plan Review*. Wellington: Wellington City Council.

Further Reading

Bowie, D. (2017). *The radical and socialist tradition in British planning: From Puritan colonies to garden cities*. London: Routledge.

Elson, M., Walker, S., & Macdonald, R. (1993). *The effectiveness of green belts*. London: HMSO.

Hall, P. (1973). *The containment of urban England*. London: PEP.

Hall, P. (1988). *Cities of tomorrow*. Malden: Blackwell.

Manns, J. (2019). Revived or retired, the green belt must be rethought. *Journal of Urban Regeneration and Renewal, 12*(3), 215–217.

Green Blue Infrastructure

▶ Concepts, Approaches, and Methodologies for Ecological Flood Resilience Assessment: A Review

---

Green Buffer

▶ Green Belts

Green Building

▶ Sustainable Community Masterplan

Green Cities

What Is a Green City?

Ranjan Ritesh and Firoz C. Mohammed
Department of Architecture and Planning,
National Institute of Technology Calicut,
Kozhikode, Kerala, India

Synonyms

Eco-city; Sustainable City; Green Development

There are several definitions of "Green City" based on different literary sources. The definitions are synonymously used as or instead of Green Cities. Table 1 explains some of the keywords and synonyms used by different authors.

- **Eco-city** is an approach for the development of cities. This term was derived from an interdisciplinary study of urban ecology. Various parameters that make an eco-city are environmental technology, bioregionalism, social ecology, economic development at the community level, the green movement, and sustainable development. Also, termed as a sustainable community.
- **Sustainable City** is a city designed for social, economic, and environmental equity while ensuring resilience for the existing population without compromising with the same for the future generation.
- **Green Development** is a concept that aims at managing and improvising the overall quality of water, air and land in urban areas, the relationship with the hinterland and wider systems, as well as the resulting benefits for the environment and residents.

Definition

The meaning of the word "Green" varies to different people. The term is used in a variety of domains like architecture, engineering, business, social science, including urban and regional planning. It has become primarily associated with sustainability and eco-friendliness, and has been adopted by several organizations, both public and private. Based on literature reviews of several sources (Asian Development Bank 2015; Bagaeen 2009; Johnson and Johnson 1979; Kurth et al. 2011; Roseland 1997; UNEP 2011), the green city can be defined as:

> A Green City is a livable and more equitable city with a low ecological footprint, where the citizens get a good quality of life by achieving high environmental quality with responsible social and political decisions. This city is expected to be resilient to natural disasters and epidemics.

Introduction

In a global context, the population is growing at a very exponential rate mostly in developing countries. The majority of people tend to migrate to urban areas. As per current trends (Nations United 2018), the population in urban area is expected to reach 68% by 2050. The adverse impact of urbanization has contributed to intensified global environmental change. This has resulted in various natural disasters in the recent past. The need to balance development with nature has been realized long back. The green city concept was discussed initially in 1979 by Roger T Johnson in his review of the educational research manuscript "Conflict in the Classroom: Controversy and Learning." Since then, the concept has been ever-evolving. From a simple thought of greening the city to now a multidimensional concept that includes various parameters such as greenery by vegetation, green infrastructure, bioregionalism, green economy, liveability, sustainability, disaster resilience, sociopolitical aspects, green technology, etc. New parameters are continuously being added to make green city a holistic concept. There have been multiple definitions of green city and different authors have termed it differently such as sustainable city, eco-city, and green development. The overall concept in every definition has more

Green Cities, Table 1 "Green City": change in dimensions of – What is meant by Green? (Adapted from Pace et al. 2016)

Year	Term introduction	Change in the broad dimension of – What is green?	References
1979	Green City	Multidimensionality – responsible society	Johnson and Johnson (1979)
1997	Eco-city, Sustainable community	Multidimensionality – responsible society	Roseland (1997)
2009	Green City	High environmental performance – human well-being – responsible society	Bagaeen (2009)
2011	Green City	Human well-being – high environmental performance – responsible society	UNEP (2011)
2011	Green City	High environmental performance, human well-being	Kurth et al. (2011)
2015	Green City, Green development	High environmental performance – societal action – responsible policy and multidimensionality	Asian Development Bank (2015)
2020	Green City, Urban nature	High environmental performance – multidimensionality – societal action, urban green, urban agriculture, urban wild area, urban protected area – policy and economy	Artmann (2020)

G

or less been the same and the aim has been to make the city development in balance with nature.

Several international environmental policy goals have been promoted since the United Nations' Rio Earth Summit in 1992, including a reduction of carbon dioxide emissions and green development in cities to reduce environmental impact (Warf 2014). This UN premise has already been adopted by cities throughout the world. It has been explicitly and implicitly mandated that urban settings should reduce their environmental impact since the 1970s, perhaps even earlier than that, if one considers the garden cities of the late nineteenth century (Bauer and Melosi 2012). In the wake of the 1992 Rio Earth Summit, cities increasingly became vocal and engaged in the issue. Further, in the subsequent UN summits, the concerns towards the environment were made louder and heard.

The green brand offers significant ecological advantages over its competitors and is therefore preferred by those who place a high value on ecological responsibility (Grant, 2008). The ecological advantage mentioned here demonstrates the importance of ecology, which emphasizes the interactions among organisms and the environment they live in. As a result, from a corporate branding point of view, green brands represent products that are environment conscious (e.g., an organic product, products that are free of artificial components) or associated with sustainability. Stakeholders in city branding include inhabitants, enterprises, tourists, capitalists, and environmentalists. In contrast, the key stakeholders in promoting and branding a place are tourists, people living there, and businesses (Wang 2019). There has been a discussion where it has been inferred that the green city is a more market-oriented concept and an outcome of branding.

The use of eco-modernization, a market-oriented version of sustainability, has become popular since the 2000s to meet wider economic development goals in urban areas. Efficiencies in energy and resources have become the focus of eco-modernization schemes, which emphasize decoupling environmental degradation from economic growth and commercializing green technologies to resolve it. In recent years, urban sustainability has increasingly been embraced as part of the "Smart City" discourse. This promotes the idea of increasingly dense urbanization and economic expansion if they follow sustainable techniques and designs. This reduces damage to the environment in due course without any makeshifts in contemporary lifestyles or existing sociopolitical structures (Chan 2019).

Green cities are often thought of as engines of sustainable economic growth, sustainable transportation, eco-friendly construction infrastructure, and waste management systems. They can also enable more engaged, socially and environmentally just, and citywide futures that are responsive to citizens. To develop a green city or to transform an existing city into a green city, there have been several pieces of research have been done to develop the parameters to measure greenness, and these conclude to a list of fundamental segments of a green city and implementation strategies to achieve each component.

In this entry, different components of a green city from different authors have been inferred and clubbed based on its applicability. Further, various implementation strategies are devised to achieve green city development.

Fundamental Elements of Green City and Implementation Strategies

To develop a green city or to transform an existing one, it would be important that each strategy is devised for each component of the green city. A gradual transformation of a city into a green city can be achieved through natural resource conservation, regeneration or renewable design, and the restoration of the natural ecosystem. There are six themes in the green movement in discussion and policies around the world, taking into account the characteristics of a city to resolve a problem (Chang 2018). Furthermore, the fundamental elements of green city (Chang 2018; Manea et al. 2014) can be clubbed together to develop implementation strategies to create green city as shown in Fig. 1.

1. **City greening** is a concept that entails the conservation and creation of green spaces and connecting the same to form a green spatial network. Green spaces include natural greenery, such as forests, conservation zones, etc., and built-up green spaces, such as parks, gardens, etc. Further to connecting these spaces by corridors such as walking paths, bicycle paths, greenways, and waterways.
   (a) **Resource optimization**: To prevent the loss of original ecological function due to urbanization and expansion, it involves optimizing available natural resources, green infrastructure, and other green systems. Natural green resources must be protected, and to protect these land areas from being disturbed, land-use restrictions are imposed. To identify and manage the area of ecological production, agricultural land, green corridors, and buffer zones shall be established. Also, to manage buffer zones between rivers and green corridors, preservation of vegetation types and ecological corridors needs to be prioritized.
   (b) **Replenish green resources**: To create artificial green infrastructures in the city, such as nature-oriented parks with leisure and recreational facilities, it is important to develop consolidated green and blue corridors (i.e., green streets and water streams and waterfront). Greenery can also be developed in the existing public buildings and spaces around them.
   (c) **Expand the green covered area**: The greenery percentage in the urban spaces shall be increased at the city level or macro-level, by the development of green corridors (i.e., by planting trees in the street to guide the wind flow in the city). At the micro-level, all urban buildings shall be introduced with vertical gardening and rooftop farming using organic means. This would further improve the green cover and reduce the heat islands in the urban area.
2. **Green water** is a concept that aims at quality improvement and restoration of the natural water cycle. To achieve this, it is important to manage the infiltration of pollutants during surface runoff and increase porous ground coverage, which would reduce the possible flooding, improve groundwater supply, and restore the natural water cycle.

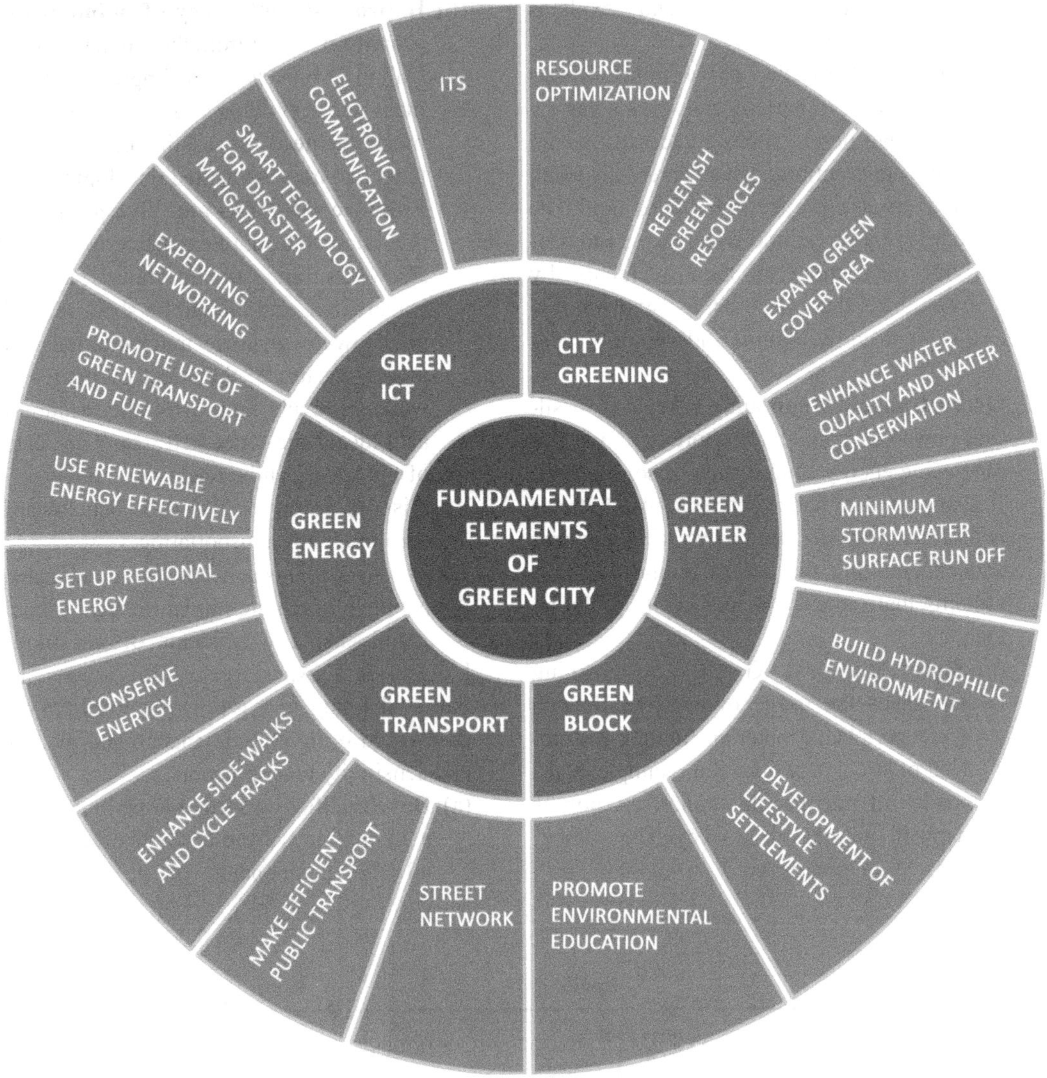

Green Cities, Fig. 1 Fundamental elements of green city: an interpretation by authors based on literature (Chang 2018; Manea et al. 2014)

(a) **Enhance water quality and water conservation**: To improve water efficiency, sanitary sewers must be utilized more and a city water cycle should be established. Water can be utilized effectively within the city by installing a water cycling system that recycles and preserves water and transforms waterways into fisheries and recreational areas. The sewage that is discharged shall be treated in the sewage treatment plant installed at city levels. Also, the natural treatment of artificial wetlands by natural processes achieves twofold purification of water, the vitality of aquatic life, and stimulation of the aquatic ecosystem (Chang 2018).

(b) **Minimum stormwater runoff**: The concept is to increase soil permeability, water storage, rainwater reuse, and sidewalk permeability to recover rainwater and bring it into the water cycle, Figs. 2 and 3 explains the concept graphically.

(c) **Build hydrophilic environment**: Rivers and streams in the city can serve a variety of human needs, such as access to water, flood prevention, wastewater disposal, and disease prevention. Cities could be cooled by rivers and waterways, and provide recreation to their residents by planning the city using hydrophilic/ water-friendly planning ideology (Chang 2018).

3. **Green transportation** aims to replace fossil-fuel-powered public transportation with a renewable energy source such as solar power. It also promotes better connectivity of streets with green practices such as walking, cycling, and encouraging the use of public transport systems that reduces the per capita pollution load.

   (a) **Street network**: Green city planning envisages the usage of eco-friendly and water permeable materials for the construction of street networks (for example, slabs made of permeable concrete, porous asphalt, natural stone, recyclable rubber tiles, tartan boards, etc.)

   (b) **Improve the efficiency of public transportation**: Apart from the emphasizing on walking, bicycle, and improved public mass transit systems, this concept also promotes the need for sustainable transit systems like transit-oriented development.

   (c) **Enhance sidewalks and bike paths**: The provision of a more extensive network of sidewalks and public bike paths is constructed, and the public shall be made more aware of cycle transportation. The increasing use of the street by people by walk or cycle improves the economic activities in the concerned area. Experience of the street can be made more lively, and this can be considered as a step towards zero-emission transportation.

4. **Green blocks**: This concept aims at educating the people by creating miniature ideal models or eco-blocks. The residents of this model block would experience various boons and banes of living in a green city. Furthermore, this would bring attitude change and increase eco-sensitivity among people.

   (a) **Develop lifestyle settlements**: Encourages a 15-minute walkable neighborhood with all the components of a green city.

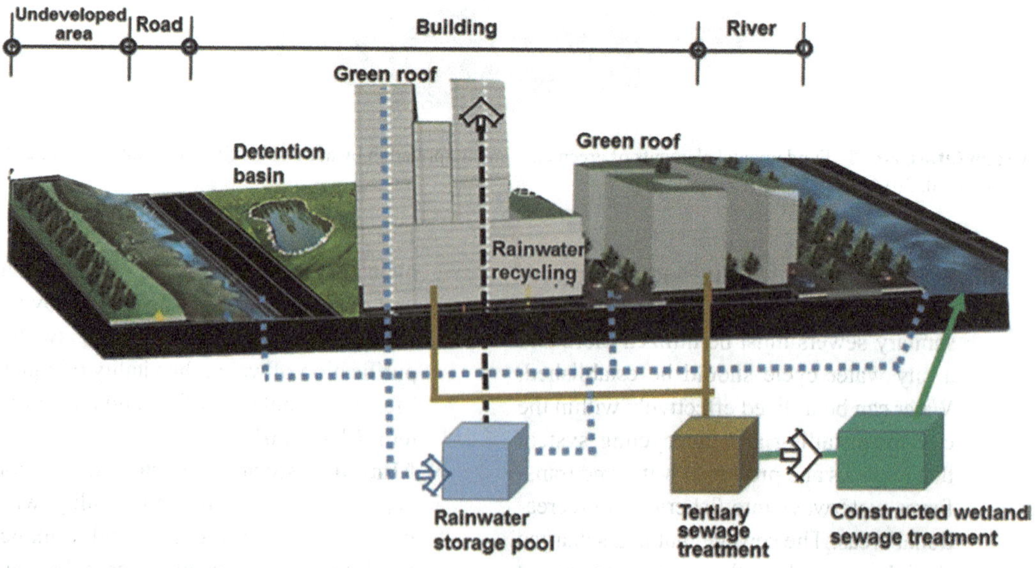

Green Cities, Fig. 2 Concept for stormwater permeability and storage and reuse facilities. (Source: Chang 2018)

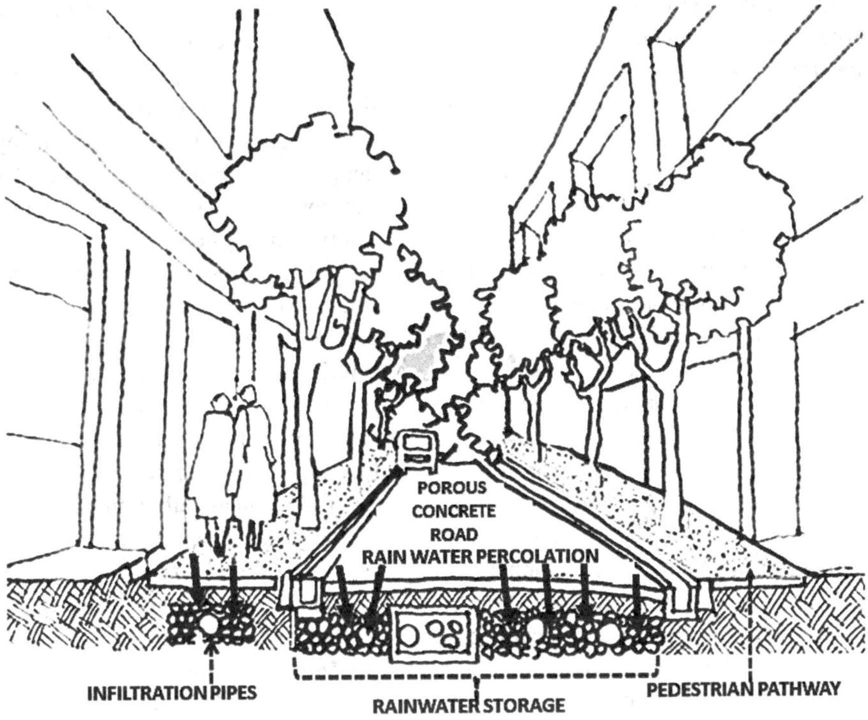

Green Cities, Fig. 3 Development strategies for city green roads. (Adapted from Chang 2018)

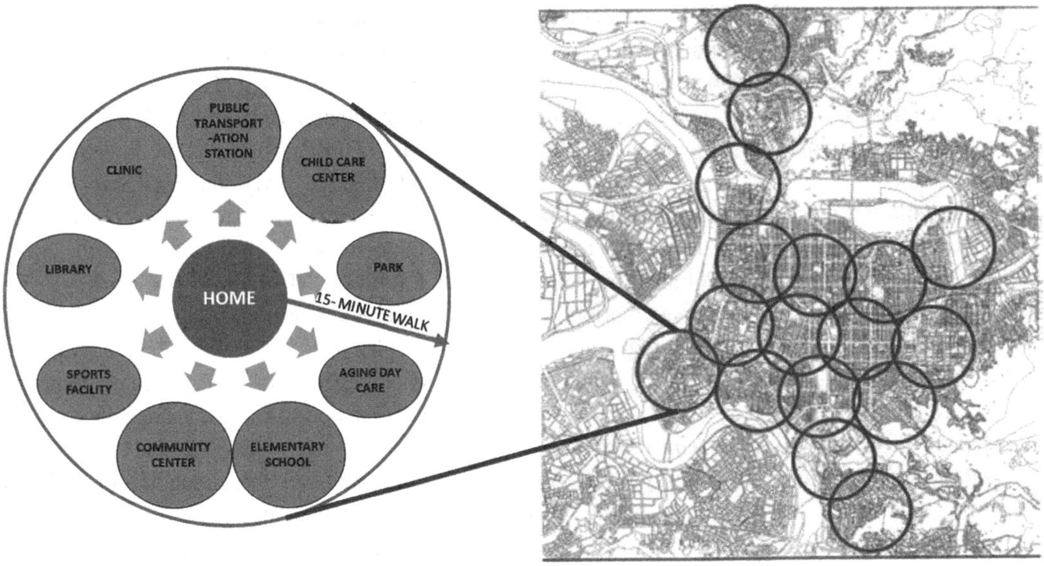

Green Cities, Fig. 4 Concept: a 15-minute walking neighborhood for a community settlement (Chang 2018)

Developing green buildings and ensuring complete recycling of all wastes generated. Figure 4 describes this concept graphically.

(b) **Promote environmentally friendly education**: Education shall include ecological courses as well. This would in turn raise awareness about environmental issues and protect the environment.

5. **Eco-friendly energy**: This aims at ways to reduce greenhouse gas emissions by promoting green buildings, local energy production, and alternate or renewable energy. The best way to save energy is to reduce unnecessary energy consumption in daily life.
   (a) **Conserve energy**: Avoid misuse or wastage of energy by the use of technology (such as using motion sensor light appliances in the corridors). The use of energy-efficient appliances and fixtures can reduce the overall consumption of energy at individual level and then at neighborhood or city level.
   (b) **Set up regional energy**: The creation of an energy demonstration area is intended to motivate and promote the development of regional energy supply systems. Energy sources like hot, cold water, steam, gas, or even solar energy are used to power buildings and equipment.
   (c) **Use renewable energy effectively**: The natural or artificial green spaces, plazas, new and old buildings must be installed with solar power generation systems to suffice their energy requirements. Power generation techniques such as direct burning, transfer of heat, biotransformation, etc., can be used to convert waste into energy.
   (d) **Promote the use of green transportation and fuel**: Encourage the use of new modes of transport such as mass transit to reduce pollution as well as consumption of energy. Newer and cleaner fuels such as hydrogen fuel, solar-powered vehicles, etc., shall be promoted.
6. **Green information and communication technology**: This concept envisages the use of green and smart information and communication technology (ICT) and the internet of things (IoT) based technologies. The benefits of electronic commuting include freedom of choice of location for the individual as well as the company. Video conferencing and real-time communication can reduce traffic and travel time in cities that further contribute to the reduction of air pollution. For prevention of disaster, the application of smart and intelligent technologies, such as remote sensing(RS) and geographic information system(GIS), a global positioning system (GPS), and a management information system (MIS), shall be used to make the process more accurate and efficient. The use of video conferencing and real-time communications reduces the need for physical travel for the people, and this further reduces the traffic on streets (hence, less energy consumption) (Shim et al. 2009).
   (a) **Expediting the construction of wireless and optical cable networks**: Emphasis must be given to make all the places connected to a high-speed internet connection, which would require speeding up of the construction process.
   (b) **Integrate intelligent technologies into disaster relief and prevention**: This can be done by creating and maintaining a database for disaster, a database for real-time data monitoring, disaster mitigation, and the relief repository, the fire rescue operations database, and the disaster-relief energy repository. These repositories can help the experts to use GIS, RS, GPS, and MIS. Furthermore, this facilitates artificial intelligence (AI) and machine learning (ML) technology to analyze and predict the occurrence of the disaster.
   (c) **Facilitate electronic commuting**: This involves communicating digitally without a physical commute. People shall be encouraged to use this mode of communication by performing all office duties reducing the need for physical travel. This contributes to the reduction of pollution due to greenhouse gas emissions. This has proved to be a very successful model during the recent COVID-19 pandemic period.
   (d) **Develop an intelligent transport system**: A satellite-based solution, that helps to reduce traffic congestion, increases safety

and contributes to keeping the environment unharmed. Such transit system may have a central control point (a hub where all transit points are monitored and controlled), as being used in many cities as a smart solution.

Quantification and Measurement of the Greenness of the City

There are multiple factors, theories, and indicators that have been in practice to measure the greenness of a city. The critical dimension to be thought is the challenge imposed due to the continuous evolution of the concept since inception. The applicability of the indicators or theories become obsolete with time, and the need for update arises. For example, postclassical theories such as garden city concept and neighborhood concept had contributed successfully to the making of a green city in the past (where parameters such as safety, security, physical greening, connectivity, walkability, etc., were in practiced then); however, with the advent of time, in the postmodern cities, new emerging concepts such as new urbanism, compact cities, sustainable cities became the trend. These had much wider domains and parameters for the greenness of the city (such as neighborhood development, mixed-use, connectivity, quality of life). Later, during the 1980s, when concerns about climate change were discussed, new parameters concerning pollution and disaster resilience were made important. Of late, during the 1990s, late 2000s, and recent years, the components concerning energy efficiency and resilience of cities have become important variables to measure greenness. Therefore, for green cities, indices and indicators vary from region to region, city to city, authors to authors, and from literature to literature.

Further, based on literature studies, various organizations used different measurement indices and indicators to measure the greenness of the city at a global level (such as green city index, environmental sustainability index, etc.). The approach of these indices is focused towards a specific domain, and its validity is limited to specific regions. Whereas the concept of the green city is advanced to a great extent that covers various other domains and is widely accepted. The organizations and institutions of European countries have been at the forefront of research and development of context-specific measurement indices and indicators. Many European organizations have conducted different case examples and measurement tools for the respective domain. Some of them are listed here: Urban Ecosystem Europe (2006–2007); European Green City Index; European Green Capital Award; and Sustainable Development Goal 11.

Urban Ecosystem Europe 2006–2007 (UEE)

This tool was developed as a joint venture: between the international bank of Dexia and a research consultancy named Ambiente Italia. With the use of UEE, governments can assess the quality of their urban environment using a simplified reporting system. For the first time, UEE was performed in 2006, analyzing the urban environment setting of 26 large European cities from 13 European countries. In 2007, 32 cities from 16 different countries took part in an exercise similar to this. Compared to the first application, the second application had improved in quality and data availability. The UEE was based on a questionnaire using 25 indicators grouped into 6 major themes. These themes are: local action for health and natural common goods; judicious consumption and lifestyle choices; planning, design and better mobility, less traffic; local to global energy and climate change; vibrant, sustainable local economy and social equity, justice, and cohesion; local management towards sustainability and governance (Pace et al. 2016).

European Green City Index 2009 (EGCI)

This was an Economist Intelligence Unit's (EIU) research project, funded by Siemens. As per EIU 2012, EGCI evaluated and compared the environmental impacts of Europe's leading cities. More than 30 cities across 30 countries in Europe were evaluated, using 30 individual indicators per city. These indicators were further divided into 17 quantitative indicators (measuring the current performance of a city) and qualitative indicators (measuring a city's environmental ambitions), under the broad eight groups namely: Carbon dioxide; Energy; Buildings; Transport; Water; Waste and land use; The quality of air and Governance for the environment (Pace et al. 2016).

European Green Capital Award, since 2010 (EGCA)

EGCA was launched in 2008 as a policy tool of the Environmental Directorate General of the Commission. The aim of this policy tool and award was to improve the standards of urban environment. Through this tool, that is based on a theme-based approach for an urban setting, an integrated approach to urban management is encouraged. Every year since 2010, one of the European cities is awarded the title of "European Green Capital." It encourages cities to focus on the environmental aspects of urban planning and to improve their quality of life.

The EGCA assessment criteria are based on three objectives: greenest city; implementation of efficient and innovative measures; and communications with networking. Further, it is divided into 12 equally weighted indicators that are used for the evaluation, namely: climate change mitigation and adaptation; local transport; green urban areas incorporating sustainable land use; nature and biodiversity; ambient air quality; the quality of the acoustic environment; waste production and management; water management; waste water management; eco-innovation and sustainable employment; energy performance; and integrated environmental management.

Each indicator is given a score by the panel of experts based on the data provided by the standardized questionnaire filled in by local authorities. Each indicator is evaluated by an international panel of experts over a period of approximately 5 months through a peer-review process. The evaluators shortlist three or four cities in the first phase; then, in the second phase, they assign scores to each of the indicators. The city that achieves the highest score wins (Meijering et al. 2014; Pace et al. 2016).

Sustainable Development Goal 11, Since 2016 (SDG 11)

During the period of 2016–2030, the UN SDGs will guide global development efforts (Pace et al. 2016; UN Statistical Commission 2016). These goals also include the millennium development goals (MDGs) such as poverty reduction, hunger eradication, and good health. It also consists of environmental targets like industrialization, urbanization, water, and other resources (Choi et al. 2016; Pace et al. 2016). Urban sustainability assessments are expected to involve many factors involving SDG 11 indicators in the coming decades. Therefore, this can be included in the review of green city measurement initiatives.

As a result of an analysis of all the indicators proposed by the four indices mentioned above, the parameters are grouped into 13 categories. These categories were selected based on their relevance and recurrence in two or more indices. These categories are: CO_2; quality of air; energy buildings; transport; water; waste; green areas; land use; acoustic environment; health; safety; education; as well as equity and participation.

Summary/Conclusion

In this chapter, the term "Green City" is used as a metaphor for the conservation of nature along with making it habitable by humans, i.e., about establishing a balance between nature and the development of the built environment. The need of making cities green has been strongly felt due to the fast changes in the global climate. Developing a green city requires collaboration and commitment. Although the idea of the green city is a well-explored concept, there has been a continuous evolution of the topic since its inception. The newer dimensions of the measurement of the greenness of cities are discovered, leaving some of the older parameters incomplete. For example, in the past, green city was about adding urban green spaces to the city. Further with time, other important dimensions were added like pollution control, quality of life, energy efficiency, net-zero city, resilient city, etc. Based on the literature study, a difference in perception among different authors was observed. This contributes as a major setback while compiling the literature. The factors that make a city green and parameters to measure greenness are multidimensional and are still an evolving field. Therefore, the question of what makes a city green is still open for discussion and further research.

To sum up, it can be said that the green city acts as a wonderful measurement mechanism and concept for enhancing sustainability of cities. It establishes a balance between nature and the built environment, thereby promoting a better quality of life for the residents of the society and the world in total.

Cross-References

- ▶ Green and Smart Cities in the Developing World
- ▶ Green Belts
- ▶ Green Cities in Theory and Practice
- ▶ Green Cities: Nature-Based Solutions, Renaturing and Rewilding Cities
- ▶ Stewarding Street Trees for a Global Urban Future

References

Artmann, M. (2020). In J. Breuste (Ed.), *Making green cities*. Springer Nature Switzerland AG. https://doi.org/10.1007/978-3-030-37716-8

Asian Development Bank. (2015). In E. Lewis (Ed.), *Green City development tool kit*. Asian Development Bank. http://www.adb.org

Bagaeen, S. (2009). Green cities: Urban growth and the environment by M. E. Kahn. *Journal of Urban Design, 14*(4), 565–567. https://doi.org/10.1080/13574800903265413

Bauer, J., & Melosi, M. (2012). Cities and the environment. In *A companion to global environmental history* (pp. 360–376). https://doi.org/10.1002/9781118279519.ch20

Chan, C. S. (2019). Which city theme has the strongest local brand equity for Hong Kong: Green, creative or smart city? *Place Branding and Public Diplomacy, 15* (1), 12–27. https://doi.org/10.1057/s41254-018-0106-x

Chang, H.-T. (2018). Green City vision, strategy, and planning. In Z. Shen, L. Huang, K. Peng, & J. Pai (Eds.), *Green City planning and practices in Asian cities: Sustainable development and smart growth in urban environments* (pp. 19–38). Springer International Publishing. https://doi.org/10.1007/978-3-319-70025-0_2

Choi, J., Hwang, M., Kim, G., Seong, J., & Ahn, J. (2016). Supporting the measurement of the United Nations' sustainable development goal 11 through the use of national urban information systems and open geospatial technologies: A case study of South Korea. *Open Geospatial Data, Software and Standards, 1*(1), 1–9. https://doi.org/10.1186/s40965-016-0005-0

Johnson, D. W., & Johnson, R. T. (1979). Conflict in the classroom: Controversy and learning. *Review of Educational Research, 49*(1), 51–69. https://doi.org/10.3102/00346543049001051

Kurth, H, Henze, M, & Burckhardt, E. (2011). Green City Europe – For a better life in European cities. *Green City Europe–for a Better Life in . . .* (pp. 10–11). http://die-gruene-stadt.de/wp-content/uploads/2011/09/ELCA-Tagungsband-in-englischer-Sprache.pdf#page=30

Manea, G., Iuliana, V., Matei, E., & Vijulie, I. (2014). Green cities – Urban planning models of the future. *Cities in the globalizing world and Turkey: A theoretical and empirical perspective, November*, 462–479. https://doi.org/10.13140/2.1.4143.6487

Meijering, J. V., Kern, K., & Tobi, H. (2014). Identifying the methodological characteristics of European green city rankings. *Ecological Indicators, 43*, 132–142. https://doi.org/10.1016/j.ecolind.2014.02.026

Nations United. (2018). *2018 revision of world urbanization prospects*. Department of Economic and Social Affairs; United Nations. https://www.un.org/development/desa/publications/2018-revision-of-world-urbanization-prospects.html

Pace, R., Churkina, G., & Rivera, M. (2016). How green is a " Green City "? A review of existing indicators and approaches. *IASS working paper, December*, 27. https://doi.org/10.2312/iass.2016.035.

Roseland, M. (1997). Dimensions of the eco-city. *Cities, 14*(4), 197–202. https://doi.org/10.1016/s0264-2751(97)00003-6

Shim, Y. H., Kim, K. Y., Cho, J. Y., Park, J. K., & Gyou Lee, B. (2009). Strategic priority of Green ICT policy in Korea: Applying analytic hierarchy process. *World Academy of Science, Engineering and Technology, 58*, 16–20. https://doi.org/10.5281/zenodo.1330973

UN Statistical Commission. (2016). Report of the inter-agency and expert group on Sustainable Development Goal Indicators (E/CN.3/2016/2/Rev.1), Annex IV. In *Report of the inter-agency and Expert Group on Sustainable Development Goal indicators* (Issue E/CN.3/2016/2/Rev.1). https://sustainabledevelopment.un.org/content/documents/11803Official-List-of-Proposed-SDG-Indicators.pdf

UNEP. (2011). Towards a green economy: Pathways to sustainable development and poverty eradication – a synthesis for policy makers. www.unep.org/greeneconomy%0ADisclaimer

Wang, H. J. (2019). Green city branding: Perceptions of multiple stakeholders. *Journal of Product and Brand*

Management, 28(3), 376–390. https://doi.org/10.1108/JPBM-07-2018-1933
Warf, B. (2014). United Nations conference on environment and development. In *Encyclopedia of Geography*. https://doi.org/10.4135/9781412939591.n1173

Green Cities in Theory and Practice

Jon Kellett[1] and Martin Larbi[2]
[1]University of Adelaide, Adelaide, SA, Australia
[2]Kwame Nkrumah University of Science and Technology, Kumasi, Ghana

Synonyms

Biophilic city; Eco-city; Eco-efficient city; Garden city; Resilient city; Sustainable city

Definition

Increasingly, across the globe, cities are referring to themselves as Green. This chapter explores the concept of Green Cities as set out in the academic literature and then examines the claims made by a number of cities, relating these to the theory. It argues that these claims to greenness are often partial in practice and in some instances are driven by the need to project a Green image for economic reasons. In reality, few cities are able to provide hard evidence to support their claims. Nevertheless, a shift towards a greener urban future is vital if cities are to flourish in coming decades. The chapter argues for a stronger evidence base against which cities' claims to green credentials might be rigorously assessed.

Introduction

Cities are widely acknowledged as the cradle of civilization; however, they are also identified as the epicenter of several global challenges. Global efforts in moving towards sustainability are likely to be won or lost in cities (Ahern 2011; Fitzgerald 2010; Wackernagel et al. 2006). In the last century, cities across the globe grew tremendously in population and size, with more than half of the world's population living in cities since 2007 (Seto and Shepherd 2009). This growth trend has serious implications for consumption of natural resources and production of waste. Already, cities consume 75% of the world's natural resources, produce 80% of greenhouse gas (GHG) emissions, and are responsible for more than half of global waste; although they occupy less than 3% of global terrestrial surface (Gago et al. 2013). A corollary of these developments is the rapid depletion of natural resources, inability of the ecosystem to assimilate anthropogenic wastes and emissions, and changes in global and local climate.

Although cities are the main drivers and results of these global phenomena, they also serve as the locus of innovation for technological, social, and economic development (Bettencourt et al. 2007; Shearmur 2012). Cities account for more than 80% of global gross domestic product. In addition, they provide opportunities for commerce, employment, education, healthcare, art and cultural expressions, and recreation, among others. As pointed out by McLaren and Agyeman (2015, p. vii), the idea "*that you can't fix the planet without fixing our cities is obvious, but less obvious is that cities can fix the planet.*" The key question, however, is how cities can minimize their negative environmental and social impacts while promoting a high quality of urban life, and protection of the natural environment.

The concept of urban greenness is highly topical in current debates about city form and growth. Superficially, the Green City may be viewed simply in the light of the extent of its tree canopy and amount of green space. Figure 1 shows the southern suburbs of Adelaide with the city center in the background. Adelaide looks very green because of its low density and abundant street trees but despite these being important factors in city planning and development, they represent just one dimension of greenness. A truly Green City is better viewed as one that treads lightly on the

Green Cities in Theory and Practice, Fig. 1 Looking towards Adelaide CBD from the south showing its extensive tree canopy. (Photo: J Kellett)

earth, an urban community which provides for all of the multiple requirements of urban living such as housing, employment, services, transport, and recreation at little or no cost to the natural environment (Fainstein 2010; Beatley and Newman 2012; Lehmann 2010a). The Green City tag is at the outset, potentially misleading in that it refers to more than simply appearance and land-use attributes.

This chapter seeks to unravel the concept and arrive at a clearer understanding of what makes a city green. First, it explores the literature around the Green City concept, identifying the factors which taken together might define the Green City. It then looks at cities around the world which lay claim to greenness analyzing these claims and relating them to the theory as set out in the literature. There is no attempt to assess or rank these cities but rather to relate the attributes that contribute to their greenness to the established theory. The chapter concludes by discussing the process of transformation and how cities might become green or at least greener.

The Literature on Green Cities

In the early twenty-first century, the Green City concept has gained considerable traction in both academic and policy discourse. It was originally conceived as a metaphor for integrating nature into the built environment (Cohen 2010). Though not new, the Green City concept has evolved over the last two centuries in response to the complex and dynamically changing realities of cities. In response to the challenges posed by progressive industrialization and urbanization, several pioneer urban thinkers such as Ebenezer Howard, Patrick Geddes, Lewis Mumford, and Ian McHarg espoused theories, metaphors, and imageries that aimed to inspire a conscious care for and preservation of nature in the urban landscape. The

Garden City movement, for example, is widely acknowledged as one of the most influential planning ideologies in the twentieth century, with around 700 Garden Cities in 34 countries around the world (Stern et al. 2013). The ideas of these pioneer urbanists are important forerunners of the Green City concept (Fishman 1982; Lehmann 2010a). Their interpretation of what a Green City is largely underlies the conventional view of the latter as one that integrates and promotes intimate contact with nature in the built environment.

In recent decades, enthusiasm for the Green City has been at a high point. However, the concept has become more broadly interpreted beyond its traditional association with the natural environment. Contemporary urbanists reflect on this concept not only in terms of the biotic component of cities, but also the underlying socio-technical systems that influence how cities function. These include factors such as how urban transportation systems are organized, sources of energy, waste management practices, urban form and land-use planning, materiality and architecture of the built environment, and urban governance among others. For example, studies by Beatley (2000), Beatley and Newman (2012), Newman (2010), Newman and Matan (2013), Newman et al. (2017), Low (2005), Kahn (2006), and Lehmann (2010a, b) have been particularly enlightening in demonstrating how the Green City concept could be reconceived to embrace broader visioning towards sustainability beyond the literal application of greenery-based solutions to cities.

Indeed, in recent years, the Green City has become almost indistinguishable from the Sustainable City. The concept of a development paradigm that respects planetary boundaries while promoting the wellbeing of people neatly summarizes the underlying rationale for both ideas, though it can be argued that the Green City carries a stronger environmental focus than the Sustainable City, which includes economic, social, and cultural aspects with differing weights applied in diverse circumstances.

In his seminal book: *Green Urbanism—Learning from European Cities*, Beatley (2000) argues that what makes a city "*green*" is the extent to which the latter is designed to function in ways that make it liveable and sustainable. His study of seven European cities (i.e., Venice, Freiburg, Copenhagen, Vitoria-Gasteiz, Helsinki, Paris, and London) reveals some salient characteristics of a Green City which include: compact and mixed-use development, energy efficiency and renewable energies, passive design, sustainable transport (public transit, electric cars, car sharing, and active modes of transport), circular economy, environmental literacy, and inclusive urban governance (Beatley 2000). Drawing on these European studies, Beatley and Newman (2012) conclude that these attributes provide practical guidelines for cities to think globally and act locally to reduce their impacts on global ecosystems, and promote sustainable, healthy, and vibrant local economies.

Lehmann (2010a) adds to this list of principles by proposing what he identifies as systematic and holistic criteria for assessing Green Cities. His set of principles largely draw on early writings and works of pioneer urbanists such as Baron Haussmann, Ildefons Cerdà, Frederick Olmsted, Ebenezer Howard, Reyner Banham, Lewis Mumford, and Ian McHarg. He lays out 15 core principles of Green Cities most of which are akin to Beatley's criteria. These principles include:

- Designing with climate and context
- Renewable energy for zero CO_2 emissions
- Zero-waste
- Water efficiency
- Landscape, gardens, and urban biodiversity
- Sustainable transport and good public space
- Local and sustainable materials with less embodied energy
- Density and retrofitting of existing districts
- Green buildings and districts, using passive design principles
- Liveability, healthy communities, and mixed-use programs
- Local food and short supply chains
- Cultural heritages, identity, and sense of place
- Urban governance, leadership, and best practice
- Environmental education
- Low-cost sustainability solutions appropriate for cities in developing countries

According to Lehmann (2016), these principles should not be viewed in silos, but regarded as an eclectic mix of principles that require synergy in their theoretical conception and mainstream application.

To anchor these principles in a more concise theoretical framework, Newman (2010) proposes seven "*archetypal cities*" that exemplify, on a conceptual level, some critical features of a Green City, namely, the:

- Renewable energy city
- Carbon neutral city
- Distributed city
- Biophilic city
- Eco-efficient city
- Place-based city
- Sustainable transport city.

It is worth noting that except for the Distributed and Place-based city archetypes which have less resonance with the principles discussed earlier, Newman's taxonomy overlaps significantly with Beatley and Lehmann's reading of a Green City.

In regards to the Distributed city archetype, Newman and Matan (2013) argue for the decentralization of key services such as power, water, and sewage treatment plants to promote the resilience of urban systems and enable cities to reduce their ecological footprint. Here it is possible to discern overlap with another contemporary concept, namely, the Resilient City. Their approach is emphasized in the resilience literature as a means of building redundancy, modularity, and independence into urban systems to ensure that the failure of a part of the system does not result in the failure of the overall system (Tyler et al. 2010). On the place-based city, Newman (2010, p.158) notes that "*the more place-oriented and locally self-sufficient a city's economy is, the more it will reduce its ecological footprint.*" Therefore, the creation of jobs and the supply of basic needs such as food, healthcare, and education should be promoted locally to reduce the need for people to travel. This principle lies at the heart of Howard's quest for a self-contained city where residents would have access to all the basic human needs locally to support a healthy livelihood.

In addition, Low (2005) emphasizes bottom-up engagement and the support of socially shared expectations as fundamental to the building and rebuilding of cities under green principles. The notion of grassroots level participation is not a new proposition. Urban activists such as Jane Jacobs (1961) promoted this idea through intensive and creative discourses to encourage a community-centered approach to urban planning. Strong grassroots participation is seen to encourage collective action and foster voluntary compliance of citizens with new practices that lead to a more sustainable future (Scharpf 2009). Thus, for a Green City to thrive, it must reflect the deeply lived experience and everyday lives of the people.

That said, it is worth noting that human judgments and choices are not always rational. Therefore, regulations are needed to elicit behavior change that helps in achieving desired outcomes. This brings the question of urban governance to the fore. In this respect, Kahn (2006) advances the idea of "*Greener Governance*" as a means of modulating the environmental externalities of people's behavior outcomes. The key implication drawn from this idea is that human behavior needs to be regulated to avoid the so-called *tragedy of the commons*. Expounding on this point, Lehmann (2016) points out that the Green City concept should not only be viewed in light of the visual and tangible attributes of cities, but also how decisions are made and the implications of these decisions for social justice and equity.

In sum, current discourse around the Green City concept evokes a wide range of ideas, images, and interpretations among urbanists, policy makers, and academics. These ideas interrogate how the hard (physical attributes) and soft (underlying governance processes) elements of a city can be reconfigured to deliver desired outcomes. In essence, a Green City should be judged by its accomplishment, not its promise.

While there is no consensus on the definition of a Green City, the existing literature broadly agrees on some salient principles that offer an opportunity for establishing a common ground for discussion and policy-making. Drawing on the analysis

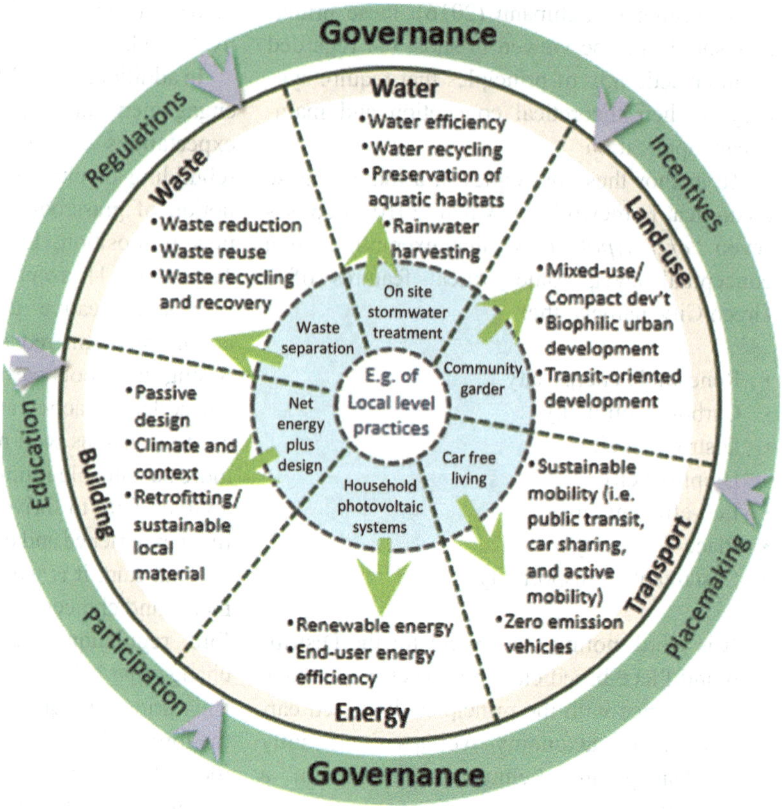

Green Cities in Theory and Practice, Fig. 2 Summary of the criteria of a Green City. (Source: Larbi (2019))

above, these principles can be summarized under six broad criteria, that is, land-use planning, transport, energy, building, waste, and water. As illustrated in Fig. 2 below, the various principles should be applied within a governance framework that provides opportunities for bottom-up engagement, with both carrot (incentives) and stick (regulations) interventions. In addition, it is important to point out that the Green City concept should not be imposed dogmatically, but should be responsive to the specific conditions of its application domain. This is particularly important because cities are complex systems that vary in their social, cultural, political, and physical configurations.

The Reality of Green Cities

Having interrogated the theoretical dialogue concerning Green Cities, it is important to examine urban performance in practice. Just as cities compete to achieve high rankings in international league tables relating to liveability, so greenness is also emerging as a coveted label. Many cities around the world lay claim to green credentials and there is increasing competition to be labeled a Green City.

An Internet search for Green Cities yields multiple results. Various organizations and publications with differing agendas and criteria list their top ten Green Cities. It is immediately noteworthy that the cities included on these lists differ widely in location and scale. For example, London and Reykjavik both appear on multiple lists despite having vastly different characteristics in respect of population size, built area, climatic conditions, transport networks, economy, and energy profile. Some cities such as Amsterdam and San Francisco appear on several rankings lists mainly because they have one significant green characteristic (a high incidence of cycling in the case of

Amsterdam and high level of waste recycling in the case of San Francisco) that marks them out as unusual. Some of the ranking lists are explicit about the criteria they have employed to identify high performing cities and others less so. *Resonance* (Fair 2020), for example, lists the percentages of green space, energy from renewable sources, commuting by public transport as well as walkability (derived using a widely employed assessment tool), the rate of recycling and composting, the incidence of farmers markets, and the concentration of PM10 particles in the air as its indicators of greenness. Only one criterion, water consumption, uses a per capita indicator. *The Green City Times* (undated) list includes a range of criteria relating to renewable energy, public transit, and green space provision and notes that all cities included in its rankings exceed their national Paris Climate Accord Greenhouse Gas (GHG) emissions targets. Some ranking lists are light on the reasons for including certain cities and sometimes these relate more to policy statements or claims made by the cities themselves than actual measurable criteria. Cities that appear on several lists published by different agencies include Amsterdam, Copenhagen, Curitiba, Freiburg, London, and Singapore. Kenworthy (2016) provides a useful discussion of cities that are moving towards eco city development under the UN New Urban Agenda in an address to the ISOCARP conference. This list includes cities of varying scales from North America, Europe, Asia, and Australia. Table 1 describes these cities in respect of a series of 10 characteristics identified by Kenworthy, most of which relate to the land use transport interface, a factor which he argues is critical in making cities more ecologically sustainable.

Cities in different cultures, geographical locations, and climate zones display markedly different characteristics in respect of their topography, built form, density, and land use patterns. Consider, for example, the distinctly different density and physical appearance of Auckland with its extensive low rise suburban spread to the high density concentration of Shanghai. Cities vary enormously in population size, so drawing meaningful comparisons between the 20 million population of Cairo with 800,000 in Copenhagen, may be problematic. It is also important to recognize that there may be benefits attached to recognition as a Green City. These may range from attracting tourism, to businesses and people seeking high quality environments in which to locate. Therefore, there is a risk of greenwashing, making claims to greenness based on unreliable criteria in order to achieve a competitive advantage. For that reason alone, data driven measures and tools are important in assessing such claims. While policy initiatives, infrastructure provision, and identifiable projects are important in moving toward greenness, it is measurable performance in respect of the factors listed in Fig. 2 and Table 1 that may form the best comparative measures for assessing green credentials.

Green Cities in Theory and Practice, Table 1 Green performance factors in selected cities

	Boulder	Freiburg	Perth	Portland	Seoul	Vancouver
Compact		✓		✓		✓
Green environment		✓			✓	✓
De-emphasize freeways	✓	✓		✓	✓	✓
Closed loop environmental technology		✓				
Polycentric		✓				✓
High quality public realm	✓	✓	✓			✓
Legible& permeable			✓			✓
Economic performance		✓				
Inclusive planning	✓		✓	✓		✓
Sustainability based decision making	✓	✓	✓			✓

Source: Derived from Kenworthy (2016)

G

Assessing Greeness

A wide range of tools have been developed to measure sustainability. A review of these is provided by Harsimran and Garg (2019). Several themes emerge, notably that different tools are appropriate to different scales of analysis and that they are invariably partial and often do not integrate multiple factors and interdependencies well. Lowe et al. (2015) also note that many city planning policy documents pay lip service to the usefulness of indicators but either do not include them or contain a weak rationale in respect of the choice of indicators. The impact of development on ecological factors and on social sustainability is also a common weakness of many sustainability assessment approaches. It is noteworthy that many tools such as LEED and BREAM use predictive assessment methods to analyze designs (e.g., the energy performance of a new building). Actual performance evaluation is less common but nevertheless vital in validating these modeling tools. Perhaps the best known evaluation tool is the ecological footprint (Rees and Wackernagel 1996) which uses quantified measures to compare a city's performance in respect of natural resource consumption with global ecological capacity. The technique provides a snapshot of environmental load which the city exerts. It does not take social, equity, or governance factors into account nor does it allow for change in technology, policy, or natural systems. Nevertheless, it provides a useful comparative measure which cities can use to benchmark each other. The ecological footprint measure applied to Vancouver, for example, demonstrates the city occupies a notional area some 200 times larger than its actual physical footprint. There is great variation in the assessments that have been made to date. London, for example, has a footprint 125 times its physical area (Wackernagel et al. 2006). Overall population density and per capita income appear to be two key factors in explaining these differences though many others may be influential.

The ecological footprint uses land as the common measure to allow comparisons to be made between disparate cities. Another useful approach is to relate evidence of greenness to people. This approach overcomes the population size variation problem mentioned above. Just as comparison of carbon emissions is most meaningful when expressed on a per capita basis, consideration of the greenness of any city may be best achieved by assessing its impact on the natural environment using per capita population, per unit of built area, and per annum change measures. Short (2015) makes a case for the development of an index of urban sustainability which would allow like for like comparison in order to assess urban sustainability or greenness on a rational basis. He notes that existing indices are useful but partial, citing the ecological, the carbon and the water footprint, all of which are widely accepted methods, which use per capita measures of sustainability allowing comparison between disparate cities. However, significant problems remain. Firstly, the input data used for these assessments is often not available. National governments, for example, generally collect and analyze GHG emission data on a national or regional/state-based scale. Actual carbon emission data for individual cities is rarely available and calculating it is likely to produce significantly diverse results depending on the methods used (Hamilton et al. 2008). Establishing the water footprint throws up similar issues. Definitional problems relating to boundaries are also important. For example, when calculating transport impacts, the energy and carbon budget of vehicles which cross administrative boundaries requires sophisticated handling. Cities such as Singapore, London, or Dubai all of which serve as major hubs for air travel face significant issues in respect of carbon allocation. Should fuel loaded at these hubs be included in the city's index? In addition to these concerns, as suggested above, there are a range of factors such as green cover, species loss, and resource usage which are not dealt with by the mainstream indices. Some progress is being made in respect of data availability and modeling tools. Australian cities, for example, now have a comparable database of carbon emissions. Similar data for solid waste generation and disposal, passenger miles travelled by varying modes of transport, food miles and water consumption per capita, are all important data, which might allow reliable comparison of the relative sustainability or greenness of cities. Such assessments are as yet partial and patchy.

How Do Theory and Practice Relate?

Contrasting the theoretical discourse about Green Cities with the reality of cities around the world that lay claim to the label, it is immediately apparent that no single city is able to demonstrate the full range of attributes and qualities identified by the main commentators on urban greenness. Few cities, if any, would be able to take the attributes shown in Fig. 2 and provide hard evidence of outstanding performance across all criteria. Table 1 demonstrates this argument in respect of transport related green criteria, with only Freiburg and Vancouver demonstrating positive action against most of the criteria listed. Even the two widely recognized paragons of greenness, Freiburg and Curitiba, appear partial in their approach to these criteria. Freiburg scores highly on its approach to energy and carbon emissions and to an extent, on urban transport. Nevertheless, conflict between government and people particularly over transport, parking, and green space provision has occasionally demonstrated the gap that exists between the prevailing system and the requirements of greenness. Citizen concern about loss of urban greenspace to government sponsored housing development in the city has also become an issue in recent years. Curitiba's urban transportation and approach to waste recycling were exemplary in the 1980s and 1900s though recent economic advances have led to greater levels of private car use and declining public transport ridership. In contrast to Freiburg's grassroots green movement, Curitiba's reforms were a product of an enlightened top down government which ironically was operating under the auspices of a brutal military dictatorship. Singapore, another city which has set out to foster an image of greenness, demonstrates a high degree of central control which has imposed significant restraint on grassroots activism even in the area of environmental protection (Hamnett and Yuen 2019). Furthermore, while the physical appearance of greenness of Singapore is an immediately noticeable characteristic, with extensive green roofs and walls, landscaped highways and extensive tree planting, the city exhibits one of the highest carbon footprints per capita of any world city.

Further to this observation, some Green City initiatives in China have been criticized for engendering social exclusion and economic exploitation (Ghiglione and Larbi 2015). For example, the Dongtan Eco-city which started in 2005, touted a car-free lifestyle, mainstream adoption of renewable energies, zero emissions, and a circular economy. However, the project started poorly. It failed to adhere to some fundamental principles of an Eco-city such as preservation of greenfields and wetlands, as well as to pay regard to social justice and equity (Chang and Sheppard 2013; Sze 2015). The project reflects an eco-dictatorship where little regard is given to citizen participation. This observation is not surprising given the socio-political milieu of China. Indeed, as explained by Ghiglione and Larbi (2015) and Sze (2015), Dongtan may have turned out differently, if political and economic interests had been modulated to interface with the expectations and rights of its residents. In fact, in Dongtan and elsewhere, the "*green concept*" can come across as a selling proposition for attracting high income populations which can lead to the marginalization or neglect of the needs of the broader society (Anguelovski 2016; Walker 2013). So even those cities that lay claim to green credentials are often incapable of providing a comprehensive evidence base in support of their policies.

G

Transitioning Towards Greener Cities

Despite the fact that no city can currently lay claim to be completely green or sustainable, that is not a reason for pessimism, nor does it suggest that the aim is flawed. The relationship between the green city and the sustainable city concepts is worthy of consideration particularly in respect of implementation. The sustainable city may appear to be a broader concept than the Green City in that it contains an overtly human dimension, taking in issues of equality, social justice, and economy as well as the environmental aspect upon which the

Green City focuses. However, in respect of the implementation of green ideas, the experience of Freiburg and Curitiba suggests that elements such as green transport, waste recycling, and green energy are more likely to gain grassroots support if they align with social benefits such as time and cost saving. There is an argument therefore that to achieve the Green City these key human factors need to be incorporated. Indeed, it would be unjust if the application of "green principles" to cities were to leave some people worse off than they were previously. Therefore, an understanding of how social equity considerations interface with the criteria of a Green City to deliver just and inclusive development is critical. Governance, as suggested by Lehmann, appears to be a key concern in moving cities towards a greener future. Strong leadership setting out a long-term vision, as in the case of Curitiba and Singapore, appears important but as suggested above, is not risk free. Competing social and sectoral interests along with income disparities appear to be potential obstacles to achieving green status, which suggest that high level government efforts at poverty alleviation and regulation of economic development, particularly in respect of energy policy are important contextual frameworks which can assist in moving towards the Green City. A clear vision and policy framework forms an important prerequisite. However, two caveats are important here, Firstly, there is no single solution or roadmap to the Green City. Local conditions (climate, city size, politics, wealth) suggest that movement towards greenness is as dependent on unique circumstances as it is on political will. Secondly, flexibiliy, including taking advantage of change both at the macroeconomic and political level and in respect of grassroots inventions, movements, and technical shifts, can be key to allowing change to occur. So, a Green City in Africa, for example, may well display quite different physical, cultural, and governance characteristics from one in Europe or North America. A growing body of literature, which is developing important theoretical explanations that can be tested in the experience of different cities (Anguelovski 2016; Hodson and Marvin 2010; May and Perry 2017), is an exciting aspect of this discussion.

Conclusions

Achieving Green and Sustainable urbanization could go a long way to addressing the environmental and climate crises currently facing the world. There is widespread support amongst academics and policy makers for the concept of the Green City. In most respects, this idea aligns with the Sustainable City concept which has been a focus of debate for some decades. No city can yet claim to be completely Green or Sustainable, though many claim credit in moving towards these ideals. In most cases, the evidence that these cities rely on in claiming green credentials is partial and often unquantified. There is a strong case for insisting on greenness being able to be tested against quantifiable performance measures such as per capita carbon, waste, and pollution emissions. Only then will it be possible to compare performance across cities of different scales and in different locations with any degree of confidence. To achieve this aim, improvements to data collection across diverse sectors are required. The potential benefits include debunking claims of high performance that amount to little more than greenwashing and the ability to set benchmarks and targets to which cities can aspire and thereby demonstrate their green credentials based on reliable evidence. In assessing Green Cities, it is vital to measure actual performance rather than relying on claims and intentions or even on simplistic indicators of volume or counts that do not relate to population or land area. Einstein is credited with pointing out, "*everything that can be counted counts but not everything that counts can be counted.*" The task of achieving a Green City is formidable, the task of demonstrating a city is green is proving just as challenging.

Cross-References

- ▶ Adapting Cities to Climate Change
- ▶ Cities in Nature
- ▶ Green Cities
- ▶ Green Cities: Nature-Based Solutions, Renaturing and Rewilding Cities

- ▶ Integrated Urban Green and Grey Infrastructure
- ▶ Overcoming Barriers in Green Infrastructure Implementation
- ▶ Strategies for Liveable and Sustainable Cities: The Singapore Experience
- ▶ Sustainable Development Goals from an Urban Perspective
- ▶ Toward a Sustainable City

References

Ahern, J. (2011). From fail-safe to safe-to-fail: Sustainability and resilience in the new urban world. *Landscape and Urban Planning, 100*(4), 341–343.

Anguelovski, I. (2016). From toxic sites to parks as (green) LULUs? New challenges of inequity, privilege, gentrification, and exclusion for urban environmental justice. *Journal of Planning Literature, 31*(1), 23–36.

Beatley, T. (2000). *Green urbanism – Learning from European cities*. Washington, DC: Island Press.

Beatley, T., & Newman, P. (2012). *Green urbanism down under: Learning from sustainable communities in Australia*. London: Island Press.

Bettencourt, L. M., Lobo, J., Helbing, D., Kühnert, C., & West, G. B. (2007). Growth, innovation, scaling, and the pace of life in cities. *Proceedings of the National Academy of Sciences, 104*(17), 7301–7306.

Chang, I. C. C., & Sheppard, E. (2013). China's eco-cities as variegated1 urban sustainability: Dongtan eco-city and Chongming eco-island. *Journal of Urban Technology, 20*(1), 57–75.

Cohen, N. (2010). *Green cities an A-to-Z guide*. Thousand Oaks: Sage.

Fainstein, S. (2010). *The just city*. Ithaca: Cornell University Press.

Fair, C. (2020). The World's Greenest cities are our future. https://www.bestcities.org/news/2020/04/22/the-worlds-greenest-cities/. Visited 21/11/2020.

Fishman, R. (1982). *Urban utopias in the twentieth century: Ebenezer Howard, Frank Lloyd Wright, and Le Corbusier*. London: MIT Press.

Fitzgerald, J. (2010). *Emerald cities: Urban sustainability and economic development*. New York: Oxford University Press.

Gago, E. J., Roldan, J., Pacheco-Torres, R., & Ordóñez, J. (2013). The city and urban heat islands: A review of strategies to mitigate adverse effects. *Renewable and Sustainable Energy Reviews, 25*, 749–758.

Ghiglione, S., & Larbi, M. (2015). Eco-cities in China: Ecological urban reality or political nightmare? *Journal of Management and Sustainability, 5*(1), 101–114.

Green City Times. (n.d.). What makes a city a leader in sustainability, and amongst the greenest cities in the world? https://www.greencitytimes.com/10-greenest-cities-in-the-world/. Visited 21/11/2020.

Hamilton, C., Kellett, J., & Yuan, X. (2008). Carbon profiling: An analysis of methods for establishing the local emissions baseline. *Proceedings of the 3rd International Solar Cities* Congress, Adelaide, pp. 331–340.

Hamnett, S., & Yuen, B. K. P. (2019). *Planning Singapore: The experimental city*. London: Routledge, Taylor & Francis Group.

Harsimran, K., & Garg, P. (2019). Urban sustainability assessment tools. *Journal of Cleaner Production, 210*, 146–158.

Hodson, M., & Marvin, S. (2010). Can cities shape socio-technical transitions and how would we know if they were? *Research Policy, 39*(4), 477–485.

Jacobs, J. (1961). *The death and life of great American cities*. New York: Vintage.

Kahn, M. E. (2006). *Green cities: Urban growth and the environment*. Washington, DC: Brookings Institution Press.

Kenworthy, G. (2016). Characteristics of the cities we need, *ISOCARP Conference,* Knowledge for Better Cities, Quito pp 16–46.

Larbi, M. (2019). Green urbanism in contemporary cities: A socio-technical transition analysis, PhD thesis, University of Adelaide.

Lehmann, S. (2010a). *The principles of green urbanism: Transforming the city for sustainability*. London: Earthscan.

Lehmann, S. (2010b). Green urbanism: Formulating a series of holistic principles. *Surveys and Perspectives Integrating Environment and Society & Natural Resources, 3*(2), 2–10.

Lehmann, S. (2016). Transforming the city towards low carbon resilience, YouTube. https://www.youtube.com/watch?v=eQ1blPnibOs. Visited 27/11/2020.

Low, N. (2005). *The green city sustainable homes, sustainable suburbs* (1st ed.). Sydney: UNSW Press.

Low, M., Whitzman, C., Badland, H., Davern, M., Aye, L., Hes, D., Butterworth, I., & Giles-Corti, B. (2015). Planning healthy, liveable and sustainable cities, how can indicators inform policy? *Urban Policy and Research, 33*(2), 131–144.

May, T., & Perry, B. (2017). Knowledge for just urban sustainability. *Local Environment, 22*(1), 23–35.

McLaren, D., & Agyeman, J. (2015). *Sharing cities: A case for truly smart and sustainable cities*. London: MIT Press.

Newman, P. (2010). Green urbanism and its application to Singapore. *Environment and Urbanization ASIA, 1*(2), 149–170.

Newman, P., & Matan, A. (2013). *Green urbanism in Asia: The emerging green tigers*. Singapore: World Scientific.

Newman, P., Beatley, T., & Boyer, H. (2017). Build Biophilic urbanism in the city and its bioregion. In *Resilient cities* (pp. 127–153). Island Press: Springer.

Rees, W., & Wackernagel, M. (1996). Urban ecological footprints: Why cities cannot be sustainable – And why the are key to sustainability. *Environmental Impact Assessment Review, 16*, 223–248.

Scharpf, F. W. (2009). Legitimacy in the multilevel European polity. *European Political Science Review, 1*(2), 173–204.
Seto, K. C., & Shepherd, J. M. (2009). Global urban land-use trends and climate impacts. *Current Opinion in Environmental Sustainability, 1*(1), 89–95.
Shearmur, R. (2012). Are cities the font of innovation? A critical review of the literature on cities and innovation. *Cities, 29*, S9–S18.
Short, J. R. (2015). How green is your city? Towards an index of urban sustainability, *The Conversation,* March 13.
Stern, R. A., Fishman, D., & Tilove, J. (2013). *Paradise planned: The garden suburb and the modern city.* New York: Monacelli Press.
Sze, J. (2015). *Fantasy Islands: Chinese dreams and ecological fears in an age of climate crisis.* Berkeley: University of California Press.
Tyler, S., Reed, S. O., MacClune, K., & Chopde, S. (2010). Planning for urban climate resilience: Framework and examples from the asian cities climate change resilience network (ACCCRN). *Climate Resilience in Concept and Practice, ISET Working Paper 3.* Boulder, Colorado, p. 60.
Wackernagel, M., Kitzes, J., Moran, D., Goldfinger, S., & Thomas, M. (2006). The ecological footprint of cities and regions: Comparing resource availability with resource demand. *Environment and Urbanization, 18* (1), 103–112.
Walker, G. (2013). Inequality, sustainability and capability: Locating justice in social practice. In *Sustainable practices* (pp. 197–212). London: Routledge.

Green Cities: Implementing the Miyawaki Method in Lahore, Pakistan

Ahmad Ahsan
Lahore University of Management Sciences, Lahore, Pakistan

Definition

As cities and towns around the world grow faster, urbanization is heralding a new era of harsh climate change impacts. These impacts can be mitigated by adopting the green city approach, which prioritizes development of environmentally friendly, sustainable green infrastructure. A key element of urban greening is tree plantation and the development of green spaces within urban limits. The Miyawaki method is a multilayered forest regeneration technique requiring little human intervention that uses native vegetation to restore forest ecosystems in various environments within 20–30 years. Named after the Japanese botanist Akira Miyawaki, the approach has been championed around the world from Malaysia to Kenya to Brazil. In recent years, the approach has been extensively explored for implementation in urban areas where dwindling green spaces give birth to a host of urban challenges. In addition to its flexibility, environmental viability, and major role in promoting biodiversity, the Miyawaki method also aims to support disaster risk reduction. Community involvement is identified as a key pillar of the method, thus promoting ownership and sustainability of forests well into the future.

Introduction

For the first time in history, more people now live in urban areas than rural regions. By 2050, more than two thirds of the world's population will live in cities. Cities and urban areas are drivers of activity, contributing around 60% to the global GDP. At the same time, they also contribute about 75% of global carbon emissions (UN Environment 2020). Urbanization has become a reality around the world, including in developing countries which have traditionally been home to agrarian economies. Urbanization has the potential to improve quality of life, socio-economic welfare, education, and health if managed properly. On the other hand, poorly planned urban sprawl has the potential to increase pollution, set back sustainability, create public health challenges, and even promote social exclusion. The uncontrolled growth of cities, particularly in developing countries, makes sustainable urban planning and adequate governance an urgent need.

The transformative nature of urbanization is converging with climate change in an ominous manner. Cities and towns are increasingly bearing the brunt of natural disasters such as urban flooding and smog. Research indicates that the effects of climate change will disproportionately affect urban dwellers in the years to come. Concentrated and intense human activities in cities are magnifying the urban ecological footprint. Increased utilization of poor-quality fuels for transportation, power generation, and industrial purposes, coupled with deforestation, land use changes, and infrastructure development, contributes to increasing environmental pressures. Unchecked population growth necessitates additional housing, urban infrastructure, and a transformation of natural agricultural and farming practices to meet consumer demand. Without effective and sustainable planning, design, and investment, urban dwellers will face a growing number of unprecedented negative impacts, not limited to climate change but also in terms of economic growth, quality of life, urban infrastructure and services, and social instability.

Dwelling in a uniform urban setting from which plantation has been eradicated to make way for infrastructure takes a toll on everyone from schoolchildren to the elderly. The effects of this man-made environment extend beyond the tangible impacts in energy consumption, waste management and sanitation, and pollution control to long-term intangible effects. The sustainable urbanization approach built on a broad development framework can help tackle these critical challenges and promote disaster and climate resilience in tandem. A Special Report on Climate Change and Land Use (IPCC 2019) identified planting forests and protecting existing plantations as the key to limiting global warming.

In addition to promoting green infrastructure, urban greening is emerging as a tried and tested approach to mitigate some of the impact of climate change. A key element of urban greening is tree plantation and the development of green spaces within urban limits. Green spaces not only increase the quality of the urban environment but also enhance local resilience and promote sustainable lifestyles by impacting the health and well-being of the residents. From an ecological standpoint, living greenery is the only "producer" in the ecosystem (Miyawaki 2006a). This establishes a proactive need to regenerate and nurture a rich environment for forests made of native trees that can flourish in the local environment.

Lahore is the second largest city in Pakistan, with a surveyed population of more than 11 million (Pakistan Bureau of Statistics 2018). Urban sprawl is evidenced by the fact that the city's population nearly doubled in merely 20 years (Pakistan Bureau of Statistics 2018). Detailed land cover change analysis of the district is given in the table below.

The aforementioned figures indicate that during the 7-year period from 2010 to 2017, the tree cover in Lahore fell by nearly 75%. In 2019, the Lahore High Court observed in *Farooq v Federation of Pakistan* (2018) that Pakistan had the world's highest deforestation rate. To any citizen witnessing the rapid urban sprawl over the last decade, the disappearance of green spaces – plantations, parks, green belts, and urban gardens – has become synonymous with "development." In addition to unchecked tree cutting for fuel, infrastructure construction, development of new housing "societies," and a shift toward large-scale agricultural production to meet growing demand all contribute to the problem. The Punjab Clean Air Action Plan (PCAAP) identified urbanization and new construction as a major driving force behind deforestation in the province (Environment Protection Department n.d.). Table 1 identifies environmental degradation in terms of reduced agriculture and a steady increase in settlements (built areas). In 2019, around 715 km^2 or 40% of the total area of Lahore was built up, while parks covered only 24 km^2 or 1.2% of the district area. Out of these parks, a maximum of 10% had tree cover (Lahore Development Authority 2019).

Realizing the multiple benefits of urban green spaces such as improving environmental conditions, protecting and improving biodiversity, improving public health, and promoting social engagement and participation, the Government of the Punjab in Pakistan recently launched a tree plantation campaign in Lahore. This campaign, envisioned to generate fully grown urban forests in 20–30 years, is based on the Miyawaki method. Named after the Japanese botanist Akira Miyawaki, the Miyawaki method is a forest

G

Green Cities: Implementing the Miyawaki Method in Lahore, Pakistan, Table 1 Green cities: implementing the Miyawaki method in Lahore, Pakistan (Lahore Development Authority 2019)

Land cover class	1990	2000	2010	2017
Agriculture (percentage of total district area)	1151.7 km^2 (65.4%)	1081 km^2 (61.4%)	972 km^2 (55.2%)	895 km^2 (50.9%)
Water bodies (percentage of total district area)	26.5 km^2 (1.5%)	10.6 km^2 (0.6%)	9.5 km^2 (0.55%)	9.7 km^2 (0.55%)
Tree cover (percentage of total district area)	79 km^2 (4.5%)	51 km^2 (2.9%)	40.6 km^2 (2.3%)	10.6 km^2 (0.6%)
Bare land (percentage of total district area)	356.7 km^2 (20.3%)	283 km^2 (16.1%)	127 km^2 (7.2%)	87 km^2 (4.94%)
Settlements (percentage of total district area)	147 km^2 (8.4%)	335 km^2 (19%)	612 km^2 (34.8%)	759 km^2 (43.1%)

regeneration technique that aims to recreate self-sustaining multilayered indigenous forests on degraded land with little to no human intervention. The goal is restoring forest ecosystems indigenous to the habitat by promoting natural vegetation.

The Miyawaki method has been successfully used to regenerate forests with indigenous species in more than 38 countries. It has been championed as a sustainable, cost-efficient model for environmental protection, disaster prevention, and water source protection. A fundamental requirement of the Miyawaki method is the replication of a native forest. Any species introduced to the area via plantation or by humans is not used. Another core principle of the Miyawaki method is the creation of multilayered communities comprised of tall trees, medium to short trees, and bottom weeds that function together as an entire forest ecosystem. This allows Miyawaki forests to have nearly 30 times the surface area of single-layer lawns and parks, ideally creating a forest 30 times denser than usual. Consequently, the impact in terms of climate change mitigation, environment preservation, and air purification can be 30 times higher as well (Miyawaki 2006a).

Additionally, primary trees (tall trees) in forests establish deep axial roots, thus becoming more resistant to falling which has significant implications in terms of disaster risk reduction. The trees also have successors in waiting underneath them, consequently creating a forest system that can sustain itself semi-permanently. By creating multilayered indigenous forests that cover as much area as possible, the Miyawaki method emerges as a cost-effective, sustainable measure to reduce carbon dioxide and mitigate climate change. A standout feature of the method is that it can be applied in a wide variety of settings, including dense urban areas, mountains, riverfronts, agricultural land, and villages.

In order to restore and reconstruct forests indigenous to the habitat, the Miyawaki method relies on rigorous field investigations of the local vegetation and ecological systems. Close attention must be paid to find the optimum combination of local species, with consideration given to layers, qualities, spacing, and species' associations with each other based on forest surveys. Soil texture must be carefully analyzed to determine water holding capacity and infiltration, root perforation, nutrient retention, and erosion. The Miyawaki method of plantation is designed to restore native vegetation on degraded land by closely planting together seedlings of different native species with well-developed roots in nutrient-rich soil. Upon completion of the field survey and selection of suitable indigenous species, seedlings are planted densely together to resemble the system of a natural forest.

This lays the foundation for promoting those species already thriving in the existing environment which means that extensive soil treatment, fertilizers, and insecticides are not needed.

Moreover, water requirements are in line with the local climate and average rain, eliminating the need to install and maintain expensive water supply systems. These factors make the Miyawaki method a cost-efficient option for large-scale plantations.

In the initial stages following plantation, biomass (such as rice straw), water-retaining materials, and organic fertilizers are used to aid plantation. The forest must be watered once a day – the prevalent water hose and shower method is recommended over installing any irrigation equipment (Citizen Matters 2020). For the first 3 years, it is recommended that monitoring and evaluation be conducted after every 2 months to check the growth of selected species and carry out any required changes. After 3 years, the forests require minimum upkeep and maintenance. No cutting or pruning is required, and dead leaves, flowers, twigs, and wood are allowed to turn into mulch naturally. This creates a rich, fertile top soil layer which promotes plant growth by up to ten times faster than normal and ensures viability and sustainability of the forest in the long run (Citizen Matters 2020). Mulch allows the plants to survive even if they are not watered for up to 40 days; and in the event of a sudden deluge, soil will not be washed away (Miyawaki 2006b). It is also recommended to allow local fauna, including birds and insects, to establish habitats in these forests. The presence of local species and pollinators can help restore natural balance in the ecosystem.

Classical succession assumed that between 150 and 200 years would be required to restore an indigenous forest on barren land. The Miyawaki method proved otherwise by recreating indigenous forests in merely 20–30 years through dense plantation of mixed native vegetation. It was also assumed that a rainforest destroyed intentionally was impossible to restore; however, the Miyawaki method has achieved demonstrable results in restoring tropical rainforests by promoting local vegetation. Akira Miyawaki conducted exhaustive surveys of all types of vegetation growing in cities, industrial areas, natural forests, and rural settlements. His work, in the form of "vegetation maps," proved invaluable in urban and regional planning, land zoning, and land conservation in Japan.

The main participants of forest creation, according to Miyawaki, are citizens themselves. People of all ages participate in tree plantation campaigns, and sponsoring organizations can capitalize on community involvement in urban greening. The World Health Organization (WHO) recommends adopting a dual approach coupling physical improvement with social engagement to promote ownership and utilization of green spaces. Urban green spaces are most sustainable when stakeholders from society, community groups, municipalities, local authorities, and private sectors collaborate. In Japan, for example, the Miyawaki method has been supported by corporations, government, municipal organizations, and civil society, thus becoming a sustainable model of urban greening that encourages community participation.

G

Conclusion

As a fundamental part of green infrastructure, urban greening must be considered a crucial component of the city and regional planning process. Urban greening initiatives must become an essential part of local planning and development frameworks. The fundamental principle behind the Miyawaki method is to densely plant different native trees keeping in view the natural vegetation of the area. The concept of "natural vegetation," i.e. what sort of vegetation a geographical area has the ability to support, is central to the success of regenerating forests. Rather than simply restoring forests in limited locations, this approach allows the creation of native forests based on ecological research to ensure long-term viability and sustainability. Forests that have been regenerated on the basis of potential natural vegetation cost nothing to maintain, are long-lasting, and contribute to disaster risk reduction, environmental conservation, and maintenance of biodiversity.

Cross-References

- ▶ Green Cities
- ▶ Green Cities: Nature-Based Solutions, Renaturing and Rewilding Cities
- ▶ Green Cities in Theory and Practice
- ▶ How Cities Can Be Resilient
- ▶ Innovation to Bring Nature-Based Solutions to Life: Tales of Two Cities

References

Environment Protection Department. (n.d.). The Punjab Clean Air Action Plan. Retrieved October 23, 2020, from https://epd.punjab.gov.pk/system/files/Annex%20D2%20Punjab%20Clean%20Air%20Action%20Plan_0.pdf

Farooq v. Federation of Pakistan, Lahore High Court (HCJ DA 38 2018). Retrieved October 23, 2020, From https://sys.lhc.gov.pk/appjudgments/2019LHC3025.pdf

Intergovernmental Panel on Climate Change. Special Report on Climate Change and Land. (2019, September 1). Retrieved October 23, 2020, from https://www.ipcc.ch/srccl/

Lahore Development Authority. (2019, December 4). *How to increase tree cover of the City from 1% to 15% of the district. Presented at the launch ceremony of the district tree plantation campaign*. Lahore.

Citizen Matters. (2020, January 30). How to make a mini forest with Miyawaki method. Retrieved October 23, 2020, from https://bengaluru.citizenmatters.in/how-to-make-mini-forest-miyawaki-method-34867

Miyawaki, A. (2006a). A Call to Plant Trees. Retrieved October 23, 2020, from https://www.af-info.or.jp/blueplanet/assets/pdf/list/2006essay-miyawaki.pdf

Miyawaki, A. (2006b). *Aiming for the Restoration of a Green Global Environment: Restoration of the Green Environment on the Basis of Field Studies and Research into the Ecology of Vegetation*. Lecture presented at Asahi Glass Foundation's 15th Blue Planet Prize.

Pakistan Bureau of Statistics. (2018). Province Wise Results of Census 2017. Retrieved October 23, 2020, From http://www.pbs.gov.pk/sites/default/files/PAKISTAN%20TEHSIL%20WISE%20FOR%20WEB%20CENSUS_2017.pdf

UN Environment. (2020). Cities and climate change. Retrieved October 20, 2020, from https://www.unenvironment.org/explore-topics/resource-efficiency/what-we-do/cities/cities-and-climate-change

Green Cities: Nature-Based Solutions, Renaturing and Rewilding Cities

Steffen Lehmann
School of Architecture, University of Nevada, Las Vegas, NV, USA

Introduction

There are consequences to the pace of today's urban growth, which is the fastest in human history, including loss of biodiversity, urban heat islands, climate vulnerability, and human psychological changes. Globally, past urban planning decisions, such as the prioritization of the car, have given rise to cities that, but for scattered parks, tend to be divorced from nature.

Urban greening refers to the process of establishing the components of green infrastructure and plants within the built environment. Urban forests, community gardens, plant-festooned buildings, and other renaturing and rewilding efforts can help bolster climate resilience, biodiversity, and even the moods of residents.

Compared to renaturing, rewilding is mostly maintenance-free compared to highly managed parks and gardens. Both strategies, renaturing and rewilding, can help solve three challenges: loss of biodiversity (rewilding has become a powerful strategy to bring back butterflies, insects, birds, and wildlife), urban over-heating (shade and greenery provide coolness and leaves absorb solar radiation using evaporative cooling and photosynthesis), and climate resiliency (treed spaces can serve as carbon sinks).

There are many benefits from urban greenery and a reconnection with nature. Numerous scholars agree that renaturing and rewilding's potential to mitigate climate change and its effects is a core benefit. In general, renaturing and rewilding aim to make cities better and more sustainable for people, plants, and animals. Contact with nature is essential for human existence, urban wellbeing, and good quality of life. A large group of researchers found evidence of the psychological health benefits of nature in urban spaces based on environmental psychology theory on nature contact.

Wilson (1984) and Kaplan and Kaplan (1989) discussed the psychological health benefits of nature in urban spaces based on environmental psychology theory on nature contact. In "The experience of nature," they describe how office workers with a view of nature were happier and healthier at work and how generally exposure to natural environments has proven to lift people's moods and enhance their ability to mentally focus. Ulrich et al. (1991; Ulrich 1993) suggests that looking at nature is beneficial to our brain and introduces stress reduction theory (SRT) to explain emotional and physiological reactions to natural spaces.

Working Definitions

The following part provides short working definitions of the terms nature-based solutions, urban resilience, urban heat island effect, and urban greening.

The term *nature-based solutions* (NBS) refers to the use of nature for addressing environmental, cultural, and societal challenges while increasing biodiversity and balancing urban temperatures of the city cores. According to IUCN (2019), NBS are "actions to protect, sustainably manage and restore natural or modified ecosystems that address societal challenges effectively and adaptively, simultaneously providing human well-being and biodiversity benefits." Nature-based solutions for public and urban spaces play a key role in urban regeneration and improving well-being in urban areas. The solutions can include measures such as green roofs, green walls and screens, and sustainable drainage systems. A *green wall* is a term that encompasses all forms of vegetated wall surfaces, whereas *green screens* consist of steel or plastic mesh on which climber vegetation grows.

Urban greening refers to the process of establishing the components of green infrastructure and plants within the built environment. There is growing appreciation that regreening cities helps to provide viable solutions using and exploiting the properties of natural ecosystems and the services they provide. Municipalities are now looking at how urban areas can adapt their landscapes to better manage the increasing heat stress and to build adaptive capacity. Ecosystem services delivered by city vegetation through healthy street trees, tree-lined avenues, gardens, parks, wetlands, urban forests, green roofs, and living walls are now becoming more appreciated and part of urban master planning. Recent modeling studies by researchers at the University of Surrey found that green roofs could result in a reduction in air pollution of up to 30% to 60%.

Rewilding activities are ecological restoration and conservation efforts aimed at restoring and protecting natural processes and wilderness areas by restoring an area of land to its natural uncultivated and self-regulated state. The term is used especially with reference to the reintroduction of species of wild animals, insects, birds, and flora and fauna that have been driven out or exterminated. Rewilding has significant potential to increase biodiversity, create self-sustainable environments, and mitigate climate change. Passive rewilding aims to reduce human intervention in ecosystems, giving human-cultivated land back to nature and restoring nature.

The *urban resilience* of cities refers to cities' ability to maintain human and ecosystem functions simultaneously over the long term, even during a disaster or crisis, and the capacity to deal with sudden change and continue to develop (Alberti and Marzluff 2004). Similarly, urban resilience, also called *adaptive capacity*, refers to a city's ability to cope with and recover quickly from hardship or crisis. A resilient city is typically one that is prepared and well equipped to contend with and mitigate the multiple effects of climate change, such as heat waves, urban flooding, energy blackouts, and other potential disasters. A resilient city has a robust infrastructural system and can turn even a crisis into a positive development (Mitchell and Harris 2012; Meerow et al. 2016; Lehmann 2018).

The dangerous urban heat island (UHI) effect leads to significantly warmer urban areas compared to surrounding suburban or rural areas, and this temperature difference is usually larger at night than during the day. The UHI effect occurs because the dark surfaces (e.g., black asphalt on roads and concrete on building roofs) absorb and store heat during the day and then release it at night. The main cause of the UHI effect is the modification of land surfaces and material, for instance, concrete roofs that store and trap solar radiation heat during the day. Urban greenery can help reduce this heat gain and the negative impact on human health (Sailor 2014; Lehmann 2015). Therefore, green roofs and facades can best counteract it with plants and vegetation, white or light-colored surfaces (using the Albedo effect to reflect solar radiation), and the use of materials that absorb and store less heat.

The Importance of Applying the Concepts of Regreening and Rewilding of Cities

In recent urban planning and landscape design, rewilding has emerged as an important part of new public parks and gardens. In contrast to highly managed parks and gardens, these rewilding initiatives are leaving allotted spaces mostly uncultivated and self-regulated.

Rewilding activities are ecological restoration and conservation efforts aimed at restoring and protecting natural processes and wilderness areas by restoring an area of land to its natural uncultivated and self-regulated state. The term is used especially with reference to the reintroduction of species of wild animals, insects, birds, and flora and fauna that have been driven out or exterminated.

Rewilding has significant potential to increase biodiversity, create self-sustainable environments, and mitigate climate change. Passive rewilding aims to reduce human intervention in ecosystems, giving human-cultivated land back to nature and restoring nature. Biodiversity blooms in cities where manicured green spaces were once allowed to proliferate as native grasses, shrubs, and wildflowers attract more animals that are diverse, with a diverse flora and fauna, over time. First projects have shown that when self-regulating meadows and areas of biodiversity replace maintenance-heavy monoculture, these urban wilds also become more drought resistant and sequester more CO_2. It is a shift away from a centuries-long tradition of managing and controlling public green spaces and using pesticides.

Rewilding means to let green spaces develop without the interference of humans. It means to let nature take its course over a number of years, so wildlife can flourish – an effective NBS strategy that is not technology dependent and addresses the climate crisis with a minimal number of resources.

A return to the wild for selected underused urban areas can be a powerful way to reintroduce lost biodiversity back into our cities and bring communities into closer connection with nature. Rewilding gardens can create "green lungs" and even improve local economies through nature-based tourism. It is timely to rethink the idea of merely planting trees and to instead support the landscape-scale development of natural forests in and around cities.

Rewilding initiatives can be focused on smaller areas of existing parks. Early urban rewilding initiatives were the Mauerpark in Berlin-Kreuzberg, rewilded areas in Berlin-Tempelhof, and the High Line in New York City, a once-abandoned elevated railway that went wild over decades before being adapted into a blossoming public park. Berlin's former inner-city airport Tempelhof has been turned into a natural oasis and popular public recreation area: Tempelhof Field has been successfully renatured, also offering an efficient carbon sink. The German city of Dessau bought up enough unused properties, brownfield sites, and inner-city land to create a 120-hectare public green zone to return to nature (see Fig. 1). In addition, the grounds of existing housing estates became part of a rewilding project; making them more attractive and improving the lives of the residents.

There are countless other projects: Big Marsh Park, a former steel mill and dumping ground on Chicago's Southeast side that now has hiking

Green Cities: Nature-Based Solutions, Renaturing and Rewilding Cities, Fig. 1 City planners in Dessau rewilded former industrially used areas in the city

trails and an environmental education center to help the population reconnect with nature. In 2021, the Australian city of Brisbane transformed an 18-hole inner-city golf course into a 64-hectare public green space replete with revegetated forests, native bushland, restored water holes, and a lake. Turning a disused golf course back into a swamp and wetland will create a hotspot for local native species and formerly locally extinct flora and fauna.

Nature-Based Solutions

The term nature-based solutions (NBS) refers to the use of nature for addressing environmental, cultural, and societal challenges while increasing biodiversity and balancing urban temperatures of the city cores. The solutions include measures such as green roofs, green walls and screens, and sustainable drainage systems. A green wall is a term that encompasses all forms of vegetated wall surfaces, whereas green screens consist of steel or plastic mesh on which climber vegetation grows.

To introduce the integration of nature-based solutions (NBS) as a strategy in urban planning with the aim to strengthen urban resilience and to slow down the biodiversity decline. Green spaces in cities – big or small – all contribute to the health and wellbeing of residents. However, many cities do not offer residents easy access to green space within the city.

Improving the better distribution of and access to green spaces and extending gardens and parks is likely to deliver a large number of benefits, such as: ecosystem services, better water management for enhanced urban flood control, slowing down the biodiversity loss, contributing to food security, and restoring damaged ecosystems.

Additional green space and NBS help to keep cities cool during heat waves and improve the urban microclimate. When it comes to enhancing urban resilience through applying nature-based solutions (NBS) and regreening strategies, what works in one city may not work in another. However, urban regreening projects generally allow for "repairing" and restoring some of the damage caused to ecosystems while enhancing urban resilience.

Good urban design and planning can make a profound positive contribution to solving the problems related to climate change and societal challenges. The EU Biodiversity Strategy for 2030: *Bringing Back Nature into our Lives* gives a comprehensive list of NBS measures inspired and supported by nature (2019).

Conclusion

Recent research published by Steffen et al. (2018) in their paper "Trajectories of Earth System in the Anthropocene" and the World Cities Report 2020 "The Value of Sustainable Urbanization," published by UN-Habitat (2020), all point in the same direction: our models of urbanization based on GDP growth and increased consumption cannot continue forever and the integration of NBS will play a key role in the future change of these outdated models. The intrinsic value of sustainable urbanization can and should be harnessed for the well-being of all. While we need specializations, most of the complex problems in cities require interdisciplinary teams to resolve them. Clearly, the need exists for longer-term strategies, clarity of policy, and leadership and ambition of government that is followed by action, adopting applied research, and scientific knowledge of NBS as the basis for informed decision-making.

The spatial planning system empowers local authorities to manage land use and set requirements for planning approvals and development decisions. However, there are often wider systemic barriers such as conflicting policy priorities, lack of leadership for policy reform, alignment at local levels, and reduced institutional capacity in local government. This can make it difficult to use planning powers that effectively promote NBS regreening scenarios as part of a whole systems approach. In future, our focus will be on urban ecosystem repair and restoration, and on the application of knowledge that is more scientific: more research and development are still needed in the field of renaturing and sustainable urbanization models (Lehmann 2019).

Urban vegetation represents one of the most considered urban heat mitigation measures for cities and strengthens urban resilience and disaster preparedness; for example, in case of heat waves or urban flooding. Additional tree planting and rewilding initiatives in cities considerably decrease the levels of heat-related illness and the maximum temperatures of UHIs while improving air quality and runoff-water management. There will be significant benefits expected from additional urban greenery and regreening projects that are now on the way worldwide. Vegetation, and, in particular, tree planting and rewilding, offer benefits to the urban climate, health, and well-being by contributing to decreased ambient temperatures, heat-related mortality, and levels of harmful pollutants in the air.

Renaturing and rewilding are effective strategies to improve urban health and wellbeing in cities, improve mental health, and reduce the dangerous Urban Heat Island effect. Nature-based solutions are used to address environmental, cultural, and societal challenges while increasing biodiversity and balancing urban temperatures of cities.

However, it is important to use the right tree species that has the greatest positive impact in cooling and does not require large amounts of water. It is not as simple as the more trees one has in an urban space, the better the air quality will be. Some trees are markedly more effective at filtering pollutants from the air than others. For effective renaturing, it is important to explore which tree species is doing the best job; for example, conifers offer highly effective particulate matter (PM) reduction because they are an evergreen species. It also depends on canopy size, leaf size, and leaf structure.

The next step is to up-scale citywide climate intervention strategies deployed to keep cities cool. Detailed heat maps will be required that help to identify hot spots and forecast urban temperatures for climatic adaptation strategies. More green spaces, green infrastructure, and nature-based solutions (green roofs, bio-filtration swales, and similar "technical solutions") will be required to reduce the damaging effects of urban heat islands. By planting more green spaces and employing smart urban planning techniques, it will also become feasible to reduce the demand for air conditioning.

Research on regreening cities is still in its infancy, and some research gaps have been identified. After examining the literature, it is obvious that there are still knowledge gaps, for example in the areas of practical advice on which NBS might work best, which tree species is most effective in which context, and knowledge gaps on the strategy of rewilding cities that could be examined at the specific intersection of the built environment, unbuilt environment, and nature-based solutions. It is also still unclear in the literature how exactly renaturalization or rewilding are addressed as strategies to strengthen urban resilience; this inter-connectedness will be subject to future research. It has become clear that there are numerous facets of urban greening initiatives involving multiple benefits, sensitivities, and limitations. Nature-based solutions in combination with renaturalization and rewilding strategies can strengthen urban resilience.

References

Alberti, M., & Marzluff, J. (2004). Ecological resilience in urban ecosystems: Linking urban patterns to human and ecological functions. *Urban Ecosystem, 7*(3), 241–265.

European Commission. (2019). *EU biodiversity strategy for 2030: Bringing back nature into our lives.* Brussels: European Commission. A comprehensive list of NBS measures inspired and supported by nature (Compiled by the EU-Commission, with a link to the NBS Atlas). Available online: https://ec.europa.eu/environment/strategv/biodiversitv-strategy-2030_en

International Union for Conservation of Nature (IUCN). (2019). *Global standard for nature-based solutions: A user-friendly framework* (1st ed.). Available online: https://www.iucn.org/theme/nature-based-solutions

Kaplan, R., & Kaplan, S. (1989). *The experience of nature: A psychological perspective*. Cambridge University Press.

Lehmann, S. (2015). Urban microclimates: Mitigating urban heat stress. In S. Lehmann (Ed.), *Low carbon cities: Transforming urban systems* (p. 251). London: Routledge.

Lehmann, S. (2018). Implementing the Urban Nexus approach for improved resource-efficiency of developing cities in Southeast-Asia. *Journal of City, Culture and Society, 13*(6), 46–56. https://doi.org/10.1016/j.ccs.2017.10.003. Elsevier. Accessed 10 Nov 2020.

Lehmann, S. (2019). Reconnecting with nature: Developing urban spaces in the age of climate change, Emerald Open Research, Sustainable Cities Gateway. https://emeraldopenresearch.com/articles/1-2. Accessed 10 Nov 2020.

Meerow, S., Newell, J. P., & Stults, M. (2016). Defining urban resilience. A review. *Landscape Urban Plan, 147*, 38–49.

Mitchell, T., & Harris, K. (2012). *Resilience: A risk management approach. Background note ODI.* London: Overseas Development Institute.

Sailor, D. J. (2014). A holistic view of the effects of urban heat island mitigation. In S. Lehmann (Ed.), *Low carbon cities: Transforming urban systems* (pp. 270–281). London: Routledge.

Steffen, W., Rockström, J., Richardson, K., Lenton, T., Folke, C., Liverman, D., Summerhayes, C., Barnosky, A., Cornell, S., Crucifix, M., Donges, J., Fetzer, I., Lade, S., Scheffer, M., Winkelmann, R., & Schellnhuber, H. (2018). Trajectories of the earth system in the Anthropocene. *Proceedings of the National Academy of Sciences, 115*(33), 8252–8259.

Ulrich, R. S. (1993). Biophilia, biophobia, and natural landscapes. In S. A. Kellert & E. O. Wilson (Eds.), *The biophilia hypothesis*. Washington, DC: Island Press.

Ulrich, R. S., Simons, R. F., Losito, B. D., Fiorito, E., Miles, M. A., & Zelson, M. (1991). Stress recovery during exposure to natural and urban environments. *Journal of Environmental Psychology, 11*, 201–230. https://doi.org/10.1016/S0272-4944(05)80184-7.

UN-Habitat. (2020). *World cities report: The value of sustainable urbanization*, Nairobi, Kenia.

Wilson, E. O. (1984). *The biophilia hypothesis*. New York: Island Press.

G

Green Corridor

► Green Belts

Green Development

► Green Cities

Green Economy

► Green Economy Policies to Achieve Water Security
► Water Security and the Green Economy

Green Economy Policies to Achieve Water Security

Robert C. Brears
Our Future Water, Christchurch, New Zealand

Synonyms

Fiscal tools; Green economy; Nonfiscal tools; Water quality; Water quantity; Water security

Introduction

Economic growth typically refers to an increase in the level of goods and services produced by an economy. It involves the combination of different types of capital to produce goods and services, including produced capital, human capital, natural capital, and social capital. While economic growth has produced many benefits including raising living standards and improving quality of life around the world, it has also resulted in the depletion and degradation of water resources, impacting human health and well-being as well as ecosystem health. In response, many multilateral organizations have called for the development of a green economy that improves human well-being and social equity and reduces environmental degradation. This chapter first defines the concept of the green economy before discussing the role of water in the green economy. Finally, this chapter discusses the range of complementary fiscal and nonfiscal tools available to reduce water security pressures while creating multiple economic, environmental, and social benefits.

Green Economy

The United Nations Environment Programme defines the green economy as "*one that results in improved human well-being and social equity, while significantly reducing environmental risks and ecological scarcities*". As such, it is common for the green economy to be described as low carbon, resource efficient, and socially inclusive. Specifically,

- *Economically*: The green economy secures growth through resource efficiency and sustainable consumption and production patterns.
- *Environmentally*: The green economy preserves the natural capital, invests in natural resources, and mitigates climate change through low-carbon and resource-efficient solutions.
- *Socially*: The green economy improves human well-being, provides decent jobs, reduces inequalities, and tackles poverty (European Commission and UNEP 2019; UNEP 2012).

Green Economy and Water Security

Water affects every aspect of the environment and society and is essential for human well-being. Water is embedded in all aspects of development, including sustaining economic growth in agriculture, industry, and energy generation as well as in health and poverty reduction (Global Water Partnership 2012). As such, the critical component of the green economy is ensuring water security for all users and uses, both human and natural, where water security is the "*capacity of a population to safeguard sustainable access to adequate quantities of acceptable quality water for sustaining livelihoods, human well-being, and socio-economic development, for ensuring protection against water-borne pollution and water-related disasters, and for preserving ecosystems in a climate of peace and political stability.*" Table 1

Green Economy Policies to Achieve Water Security, Table 1 Synergies between the green economy and water security

Characteristics of the green economy	Characteristics of water security
The efficient use of natural resources in economic growth	Ensure enough water for social and economic development
Valuing ecosystems	Ensure adequate water for maintaining ecosystems
Inter-generational economic policies	Sustainable water availability for future generations
Protection of vital assets from climate-related disasters	Balance the intrinsic value of water with its uses for human survival and well-being
Reduce waste of resources	Harness the productive power of water, maintain water quality, and avoid pollution and degradation

Green Economy Policies to Achieve Water Security, Table 2 Impacts of climatic and Nonclimatic trends on water security

Challenge	Impact on water resources
Climate change	• In many countries, climate change is likely to decrease the availability of renewable surface water and groundwater resources significantly, intensifying competition for water resources among users • Climatic extremes will lower water quality, even with conventional treatment processes
Population growth and urbanization	• The world's population is projected to increase by 2 billion people over the next 30 years, from 7.7 billion currently to 9.7 billion in 2050, while 68% of the world's population is expected to live in urban areas by 2050, up from 55% today, increasing water scarcity and ecosystem degradation from excess water withdrawal, eutrophication, and pollution • Peri-urban water competition is likely to increase with further urbanization
Economic growth	• Global demand for water will significantly increase due to manufacturing, industry, and domestic consumption • The global economy will require 55% more water by 2050 compared to 2000 • Household water demand is likely to increase with higher incomes and living standards, including a shift in diet to water-intensive meat and dairy products
Rising demand for energy	• Between now and 2040, the amount of energy used in the water sector will likely double, due to increased desalination, increased large-scale water transfers, and increased demand for wastewater treatment as well as higher levels of treatment • Energy-related water consumption is projected to increase by nearly 60% between 2014 and 2040
Rising for demand food	• By 2050, the world will require at the minimum 60% more food production to maintain current consumption patterns • The volume of global water withdrawn for irrigation will likely increase from 2.6 billion cubic kilometers in 2005–2007 to 2.9 billion cubic kilometers in 2050 • Agricultural intensification will result in excess nitrogen and phosphorus in water, deteriorating water quality and leading to eutrophication and dead zones

provides a summary of the synergies between the green economy and water security (Brears 2016; Global Water Partnership 2012; UN-Water 2013).

Nonetheless, water security is challenged by a variety of trends that impact water quantity and water quality including climate change, population growth and urbanization, economic growth, and rising demand for energy and food as summarized in Table 2 (Arnell et al. 2015; Brears 2016, 2018, 2020; FAO 2015; Harlan et al. 2009; IEA 2016; Kearney 2010; McDonald et al. 2016; PwC 2015; UN-Water 2014; UNESCO 2012; United Nations Department of Economic and Social Affairs 2019a, b; Xie and Ringler 2017). Recognizing these challenges, governments at various levels can facilitate the transition toward a green economy that achieves water security through a range of complementary fiscal and nonfiscal tools that encourage innovation to address environmental problems, boost resource efficiency, and stimulate demand for green technologies (Brears 2017).

Fiscal Tools

Fiscal tools can shift the focus of investments from ones that are resource intensive to those that reduce water security pressures while creating multiple economic, environmental, and social benefits (UNEP 2012; Verma and Gayithri 2018).

Environmental Taxes and Charges

In the past, environmental policy was typically dominated by "command-and-control" regulations that were generally prescriptive and highly targeted, such as banning or limiting substances or requiring certain industries to use specific technologies. Over time, there has been growing interest in environmental taxes and charges due to the many benefits they provide.

Taxes and charges directly address the market failure by "pricing in" environmental costs. The implementation of taxes and charges in environmental policy is based on the existence of environmental externalities that are side effects of processes of production and consumption. Levying a tax or a charge on the cause of environmental damage provides an incentive to the taxpayer to reduce their liability to the tax/charge by reducing the cause of the environmental externalities. An incentive is imparted to the producers/consumers to switch to alternative, less polluting products/processes.

Taxes and charges offer flexibility to consumers and businesses to determine the least-cost way to reduce environmental damage. Most regulatory approaches involve the government specifying how to reduce emissions or waste or who should make the reduction. Similarly, subsidies and incentives for environmentally preferred goods or practices involve the government steering the economy in favor of specific environmental solutions over others. Both approaches involve the government trying to "pick winners," which directs the market in a prescriptive way. This requires significant information about ever-changing conditions and technologies, which enhances the risk of making suboptimal choices. Regulations often result in higher costs than taxes since they force a particular type of technology, even if cheaper alternatives exist. In contrast, taxes and charges allow market forces to determine the least-cost way to reduce environmental damage.

Additional benefits of environmental taxes and charges include the following:

- *Ongoing flexibility to abate*: A target-based or technology-based regulation provides no incentive to abate once the target or technology is met. In contrast, environmental taxes and charges provide a continuous incentive to abate at all levels of pollution, even after significant abatement has been achieved.
- *Improved competitiveness of low-polluting alternatives*: Environmental taxes and charges increase demand for low-emission, low-polluting alternatives. This results in economies of scale that help to make such alternatives more viable, without subsidies.
- *Strong incentives to innovate*: Environmental taxes and charges increase the cost to a polluter of generating pollution, providing incentives for firms to develop innovations and adopt existing ones (ECOTEC 2001; Kreiser et al. 2011; OECD 2011a; PBL Netherlands Environmental Assessment Agency 2012; Sterner and Coria 2002; Verma and Gayithri 2018).

Types of Environmental Taxes and Charges

Environmental taxes and charges come in a variety of types including:

- *Cost-covering charges*: Those who use the environment contribute or cover the cost of monitoring or controlling that use. The level of a cost-covering charge is determined by the service it is intended to deliver or the other purposes which the revenues will support.
- *Incentive taxes*: These are levied for the intention of changing environmentally damaging behavior and without any intention to raise revenues. The success of the tax can be judged by the extent to which initial revenues from it fall, as behavior changes.
- *Revenue-generating taxes*: These may influence behavior but still yield significant revenues over and above those required for related environmental regulation (Brown and Johnstone 2014; ECOTEC 2001; OECD 2011a).

Case: The Netherlands' Environmental Tax Scheme

In the Netherlands, entrepreneurs that invest in environmentally friendly technology

(continued)

may be eligible for the environmental investment deduction (MIA) and arbitrary depreciation of environmental investments (Vamil) schemes. The MIA enables entrepreneurs, across different sectors such as agriculture, industry, and waste processing as well as those who invest in the circular economy and sustainable buildings and so forth, to deduct up to 36% of the investment costs for an environmentally friendly investment on top of their regular investment tax deductions. At the same time, Vamil allows them to decide when to write off 75% of the investment costs. To be eligible for MIA and Vamil, the entrepreneur must be paying income or corporate tax and invest in a business asset that is on the environment list, which includes water storage, water-saving, and water reuse technologies. Companies can also make proposals to include a product or business asset on the next environmental list. To do so, the environmental investment must:

- Have a clear benefit for the environment.
- Be innovative or still have a smaller market share than the conventional product.
- Be more expensive than the environmentally friendly alternative (Netherlands Enterprise Agency 2020).

Case: San Francisco Public Utilities Commission's Variable Rate Structure

San Francisco Public Utilities Commission's (SFPUC) water rate structure consists of two components: a fixed monthly service charge based on meter size and a variable charge which is based on volumetric water usage. The variable rate charge for residential customers is comprised of a two-tier, inclining block rate structure, while non-residential customers are charged a uniform commodity rate (Table 3). A temporary drought surcharge can be imposed on the volumetric portion of all retail water rates when the commission imposes a stage of water delivery reduction per the retail water shortage allocation plan (Stage 1, Stage 2, or Stage 3), with the drought surcharge varying with the stage of drought (Table 4) (SFPUC 2019).

Subsidies

A subsidy is a payment by the government, either directly or through another body, to those that undertake certain activities the government wishes to promote. Subsidies are an effective way for governments to encourage and support the development and consumption of green

Green Economy Policies to Achieve Water Security, Table 3 San Francisco Public Utilities Commission's variable rate charge

Customer type	The volume of water used	Variable rate FY 2020–2021
Single-family residential	First four units per month[a]	$8.68
	All additional units	$10.15
Multiple-family residential	First three units per dwelling unit per month[a]	$8.73
	All additional units	$10.23
Commercial, industrial, and general Uses	For all units[a] of water	$9.81

[a]1 Unit = 1 Ccf of water = 748 gallons

Green Economy Policies to Achieve Water Security, Table 4 San Francisco Public Utilities Commission's drought surcharge

Retail water shortage allocation plan stage	Target usage reduction	Drought surcharge on volumetric water rates
Stage 1	5–10%	Up to 10%
Stage 2	11–20%	Up to 20%
Stage 3	Over 20%	Up to 25%

products. In particular, subsidies are incentive mechanisms which can stimulate enterprise to produce green products as well as encourage consumers to purchase these green products (Ministry for the Environment 2006). However, compared with ordinary products, green products are usually more expensive as they often require higher-level production processes and technology and management experience, and so consumers often have to pay a premium when purchasing green products (Brears 2017; Sachdeva et al. 2015).

The successful development and promotion of green products depend on three conditions. First, strict environmental regulations exist. With strict environmental regulations that state minimum environmental requisites that products need to satisfy, the production of green and high value-added products is promoted through subsidies while resource-intensive, heavily contaminating products decrease. Second, there needs to be intent and technical feasibility of organizations to conduct green product research and development. Governments hope that private sector organizations conduct research and development of new green technology and products. However, these technologies are not yet mature, and the market prospects of the new products are unclear. Therefore, subsidies can reduce the market risk of developing new green technologies and products and encourage long-term green product research and development and promotion. Third, there needs to be public demand for green products. Demand for green products depends on the consumptive level and environmental awareness among members of the public. Subsidies for green products not only stimulate the consumptive demand of consumers but also cultivate awareness of environmental protection and the concept of green consumption. As such, subsidies are often combined with information and communication programs that help customers overcome a lack of information on available green products (Olsthoorn et al. 2017; Zhao and Chen 2019).

In designing subsidy programs, the following issues should be considered:

- Is public support provided only in cases where public goods are expected to be generated, for example, where significant environmental improvements would otherwise not be provided by producers?
- Are public support measures likely to be the most efficient and effective way of reaching an environmental target?
- Has the feasibility and cost of pricing the externality been directly assessed?
- How strong is the substitutability between the subsidized activity and the dirty activities it is supposed to replace?
- Do clear and transparent eligibility criteria exist concerning who is entitled to receive support, and under what circumstance?
- Do existing public environmental expenditure programs have the secondary effect of encouraging additional demand for, or supply of, polluting products over activities over the long term?
- Is public support allocated first to private agents that commit to achieving the most substantial environmental improvement per unit of support? (OECD 2011b)

Overall, subsidies should be innovative and improve environmental performance based on the following principles:

- *Stringent*: How ambitious is the policy target?
- *Predictable*: Does the subsidy reduce investor uncertainty?
- *Flexible*: Does the subsidy allow the public to identify the best green technology to meet the target?
- *Incidence*: Does the subsidy target the environmental objective as carefully as possible?
- *Depth*: Do incentives exist for firms to innovate through a range of potentially ascending objectives? (OECD 2011b)

Case: The Danish Eco-innovation Program
In 2015, Denmark adopted a new law that established the Environmental Technology Development and Demonstration Program (MUDP) for the period 2015–2018. The purpose of MUDP was to support the development and application of new

(continued)

environmental and resource-efficient solutions addressing prioritized environmental challenges, including in the areas of water management such as improving phosphorous recovery from wastewater and the reusing of wastewater for agricultural irrigation. The MUDP was comprised of three pillars:

- *The subsidy scheme*: The subsidy scheme provided funding for the development, testing, and demonstration of new technology to create the foundation for higher environmental standards and/or the possibility to comply with existing regulation using smarter technologies.
- *Innovation partnerships*: The fostering of cooperation and dialogue between private companies, knowledge institutions, and authorities aimed to create better and cheaper environmental solutions.
- *International environmental cooperation*: This initiative aimed to demonstrate Danish companies' environmental solutions to foreign partners and targeted export promotion (Ministry of Environment and Food of Denmark 2020).

Tradable Water Rights

In the context of water resources management, there are two forms of tradable water rights: those that allow a maximum limit of water use and those that consent a maximum limit of water pollution.

Tradable Water Abstraction Rights

Tradable water abstraction rights are used for water quantity management where water rights can be permanent and unlimited (property rights to the water resource) or temporary and limited (transferrable rights to use water). One of the main objectives of water rights trading is to provide an instrument for the reallocation of water rights, so they can be put to more economically beneficial use. In a tradable water abstraction rights program, a regulatory authority sets the water consumption cap – the maximum amount of water abstraction allowed in the hydrological basin – and allocates the corresponding abstraction rights among the users of the basin who can exchange them according to their present and/or future expected water consumption needs. The result is water users who are more efficient and can reduce their consumption will be able to avoid purchasing costly abstraction rights and possibly gain revenue from selling excess rights (Borghesi 2014; Ecologic Institute 2003).

Tradable Water Pollution Rights

Tradable water pollution rights are used for the protection and management of surface water quality. These rights can relate to point, or nonpoint sources and trades can be arranged among different kinds of sources. Under this approach, a responsible authority sets maximum limits of the total allowable emissions of a pollutant, then allocates this total among the sources of the pollutant by issuing permits that authorize industrial plants or other sources to emit a stipulated amount of pollutant over a specified period. The holders of the permits can trade them on a secondary permit market, for example, a polluting point source, which has low pollution abatement costs, can increase its depuration capacity and sell permits to sources which have high clean-up costs. The result is the total cost of reducing pollution is minimized as the depuration effort is carried out by the firm who can meet the objectives with lower costs (Borghesi 2014; Ecologic Institute 2003).

Considerations when Establishing a Tradable Water Rights System

Many considerations should be taken into account when establishing a tradable water rights system, including the following:

- *Efficiency*: Tradable water rights systems should be designed to minimize administrative and transaction costs. Also, trades of water rights should improve the efficient use of water while preventing environmental impacts.
- *Equity*: Measures can be taken in designing water rights systems to promote fairness

among rights holders and other stakeholders and to ensure that broader social and environmental impacts are minimized or prevented, for example, the inclusion of users and other stakeholders in decision-making will generally result not only in better decisions being made but also fairer and more acceptable decisions being made.

- *Transparency*: The transparency of tradable water rights systems can be enhanced by:
  - Setting clear, objective, and verifiable standards for decision-making in connection with the issue, modification, or transfer of water rights
  - Involving water users and other stakeholders in decision-making
  - Establishing clear and effective procedures for the recording and registration of water rights
  - Making sure that information is made available to the public and other water rights holders including ensuring public access to inspect registers of water rights
- *Environment*: Procedures for the sale or trade of water rights should ensure that environmental issues are taken into consideration and that overall water rights trades should have positive environmental impacts (FAO 2006).

Case: Water Markets and Trading in the Murray-Darling Basin, Australia

Water in the Murray-Darling Basin can be bought or sold, either temporarily or permanently. The water is traded on markets within catchments, between catchments, or along river systems, enabling water users to buy and sell water in response to individual needs. Water trade allows water holders to decide when to buy or sell water at a particular time. There are two types of water trade: Permanent trade is the trade of water entitlements ("entitlement trade"), for example, an entitlement holder can sell their water entitlement. In contrast, a temporary trade is the trade of water allocations ("allocation trade"); for example, an entitlement holder can sell their allocation in any season. Water entitlements are rights to an ongoing share of water within the system, and the water market determines the financial value. In contrast, water allocations are the amount of water distributed to water entitlement holders in a given year. Allocations change according to rainfall, inflows into storages, and how much water is already stored. Allocations change throughout the year in response to changes in the system. The Murray-Darling River Basin Authority (MDBA) facilitates fair, consistent, and transparent water trade across the Murray-Darling system, with the authority providing information on water trading and working with Basin State and Territory Governments to ensure their rules comply with the Murray-Darling Basin Plan Trading rules as well as with the irrigator infrastructure operator rules. The Basin State and Territory Governments are responsible for determining water allocations, developing policies and procedures for trade, monitoring water use, developing water resource plans that set the rules for sharing water between users and the environment, and day-to-day trade operations. All the current market and price information is available at the Australian Government's Bureau of Meteorology's water markets dashboard, which provides water-trading information for catchments across Australia. The Bureau of Meteorology also provides monthly reports on water trade statistics for many water entitlement types across the Murray-Darling Basin (Department of Agriculture 2020; Murray Darling Basin Authority 2020).

Payment for Ecosystem Services

Payment for ecosystem services (PES) is a voluntary transaction where a well-defined ecosystem service is bought by an ecosystem service buyer from an ecosystem service provider, with the transaction based on the ecosystem service provider securing ecosystem service provision (conditionality). Conditionality is where payments are dependent on the delivery of ecosystem service benefits. In practice, PES usually involves a series of payments to land or other natural resource managers in return for a guaranteed flow of ecosystem services over-and-above what would otherwise be provided in the absence of payment. Payments are usually based on the implementation of management practices that the contracting parties agree are likely to give rise to these benefits. Most PES schemes target at least one of four ecosystem services: carbon sequestration, watershed services, biodiversity, and/or scenic beauty. PES schemes are most likely to be applied in the following situations:

- Specific land or resource management actions have the potential to increase the supply of a particular ecosystem service (or services).
- There is an apparent demand for these services, and its provision is financially valuable to one or more buyers.
- Land or resource managers can enhance the supply of an ecosystem service or services (Alston et al. 2013; Center for International Forestry Research 2014; Schomers and Matzdorf 2013).

Types of PES Schemes

PES schemes are implemented in a variety of institutional settings and vary in their institutional structures, with their source of funding the primary distinction among PES schemes (Alston et al. 2013). There are three main types of PES schemes:

- *Public payment scheme*: This involves the government paying land or resource managers to enhance ecosystem services on behalf of the public.
- *Private payment schemes*: These are self-organized private deals where beneficiaries of ecosystem services contract directly with service providers.
- *Public-private payment schemes*: These schemes involve both government and private funds paying land or resource managers for the delivery of ecosystem services (Defra 2013).

Payment Mechanism for PES Schemes

There are two main payment mechanisms for PES schemes. Performance-based payments are where payments are made based on the actual provision of the ecosystem service, for example, an improvement in water quality. Input-based payments are where payments are made based on certain land or resource management practices being implemented, for example, the creation of riparian buffers. These payments only eventuate if buyers are willing to accept that specified inputs/activities will result in the provision of the desired ecosystem service (Defra 2013).

Case: Farm-Based Ecosystem Services Pilot in Kansas

General Mills, Kansas Department of Health and Environment (KDHE), and Ecosystem Services Market Consortium (ESMC) have signed a memorandum of understanding to pilot test ESMC's program that rewards farmers for generating environmental assets by improving soil health on their land. Involving farmers in Kansas' Cheney Reservoir watershed, which provides water to more than 400,000 Wichita residents, the pilot project aims to improve water quality as part of the statewide Watershed Restoration and Protection Strategy. The pilot will test ESMC's protocols and procedures to measure and reward the impacts of beneficial agricultural management in an ecosystem services market for agriculture. ESMC's impact-based program will, in turn, pay farmers for increased soil carbon, reduced greenhouse gas emissions, and improved water quality and water use efficiency. Meanwhile, General Mills, KDHE, and ESMC will work with farmers growing row crops in Kansas to improve sustainable agriculture outcomes (ESMC 2020; General Mills 2020).

Green Bonds

A bond is a fixed-income financial instrument for raising capital through the debt capital market. The bond issuer raises a fixed amount of capital from investors over a set period, repaying the capital when the bond matures and paying an agreed amount of interest along the way. Typical bond investors include institutional investors such as pension funds, insurance companies, mutual funds, and sovereign wealth funds. The difference between a green bond and a regular one is that the issuer publicly states that it is raising capital to finance or refinance green projects, assets, or business activities with an environmental benefit. Examples of which include sustainable water and wastewater management initiatives including sustainable infrastructure for clean and/or drinking water, wastewater treatment, sustainable urban drainage systems, and river training and other forms of flooding mitigation. Other examples include initiatives that encourage environmentally sustainable management of living natural resources and land use, including sustainable agriculture and climate-smart farm inputs such as biological crop protection or drip-irrigation (International Capital Market Association 2018; KPMG International 2015; Maltais and Nykvist 2020).

Policy Actions to Increase Green Bond Investment

A range of policy actions can be taken to increase green bond investment; for instance, public sector entities can facilitate private investors to invest in green bonds through capacity building. The public sector can provide educational materials, host workshops, and support market-led initiatives for green bond investor engagement and training. Meanwhile, tax incentives can be used to encourage green bond investment by reducing the cost of capital with a relatively low impact on public finances. For instance, tax incentives can be used to make incomes from bond investments tax-exempt, facilitating a localized market as only investors under the jurisdiction of the country would be eligible for the incentives. Tax incentives can also be used to attract foreign investors into domestic bond markets through preferential withholding tax rates for green bonds (Climate Bonds Initiative 2015).

Case: Nordic Investment Bank's Nordic-Baltic Blue Bond

In 2019, the Nordic Investment Bank (NIB) launched a 5-year SEK 2 million NIB Nordic-Baltic Blue Bond to support the Bank's lending to selected water resources management and protection projects in the Nordic-Baltic region. The region faces challenges to its water environments, particularly from eutrophication, due to high levels of nitrogen and phosphorous discharge. The bond was launched under the NIB Environmental Bond Framework, with the use of proceeds from the transaction being allocated to a separate account for onward disbursement of loans to new wastewater treatment, prevention of water pollution, and water-related climate change adaptation projects in the Nordic-Baltic region. Regarding the NIB Environmental Bond Framework, two main principles govern it:

1. *Financing of new projects*: Proceeds of NIB environmental bonds are only to be used for disbursement of loans to new environmental projects.
2. *Separation of funds*: Bond proceeds are allocated to a separate portfolio, the "Green Fund Pool." Limited mismatches between the point of time of raising funds and disbursements are managed by money market instruments, overall; however, a straight link between proceeds and disbursements of loans for projects is established (Nordic Investment Bank 2014, 2019).

Nonfiscal Tools

A range of nonfiscal tools can be employed to promote green economy-related technologies and services as well as modify the attitudes and behavior of society toward water and related resources to achieve water security.

Regulations

Regulations control behavior and are enforceable through policing institutions with penalties for noncompliance. Regulations encourage or restrict economic activities through the legal system. Standard regulatory tools include the granting of licenses or permits. The purpose of regulations is to encourage certain behaviors over other types of behaviors. Regulations can be used to influence resource efficiency by encouraging production efficiency and reducing the number of by-products. Effective regulations include performance and technology standards to reduce negative externalities when market prices fail to reflect some of the cost of economic activities (Brears 2017). In water resources management, it is common for regulations to be temporary or permanent:

- Temporary regulations restrict certain types of water use during specified times and/or restrict the level of water use to a specified amount, for example, water restrictions during extended dry periods.
- Permanent regulations are typically amendments to building codes or regulations requiring the installation of water-saving devices in all newly constructed or renovated developments (Brears 2016).

Case: The City of Boston's Building Energy Reporting and Disclosure Ordinance

The City of Boston's Building Energy Reporting and Disclosure Ordinance (BERDO) requires Boston's large- and medium-sized buildings to report their annual energy and water use. It also requires buildings to complete a significant energy savings action or energy assessment every 5 years. In 2020, BERDO requires the following buildings to report their annual energy and water usage over the period January 1, 2019, to December 31, 2019:

- Nonresidential buildings that are 35,000 square feet or larger
- Residential buildings that are 35,000 square feet or larger, or have 35 or more units
- Any parcel with multiple buildings that sum to 100,000 square feet or 100 units

To enhance transparency, the City of Boston allows the public to search the publicly disclosed energy and water database. This helps property owners and interested stakeholders to understand how a buildings' performance compares with similar buildings nationally. They also learn, after public disclosure, how the building compares with other Boston buildings (City of Boston 2020).

Ecolabels

Ecolabels are seals of approval given to products that are deemed to have lower environmental impacts than similar products. There are three primary ecolabeling schemes:

- *First-party-labeling schemes*: These are established by individual companies based on their product standards. The standards may be based on criteria related to specific environmental issues. These schemes are also known as "self-declared."
- *Second-party-labeling schemes*: Industry associations establish these ecolabels for their members' products. The members develop certification criteria with verification of compliance done through internal certification procedures within the industry or through external certifying companies.
- *Third-party-labeling schemes*: These are usually established by a public or private initiator, independent from the producers, distributors,

and sellers of labeled products. Products that are certified are then labeled with information to the consumers that the product was produced in an environmentally friendly manner. The label is usually licensed to a producer and may appear on or accompany a product derived from a certified producer (FAO 2001).

Objectives of Ecolabeling Schemes

Ecolabeling has become a useful tool for governments to encourage sound environmental practices as well as for businesses to identify and establish markets for their environmentally preferable products. While ecolabeling schemes may differ across products and countries, they seek to achieve three main objectives:

1. *Protecting the environment*: Environmental conservation and protection is the primary objective with governments and/or non-governmental ecolabeling schemes seeking to influence consumer decisions and encourage the production and consumption of environmentally friendly goods and the provision and use of environmentally friendly services. Some of the specific environmental objectives include the following:
   - Encouraging the efficient use of renewable resources
   - Promoting the efficient use of non-renewable resources
   - Facilitating the reduction, reuse, and recycling of consumer and commercial waste
   - Encouraging the protection of ecosystems and biodiversity
2. *Encouraging environmentally sound innovation and leadership*: Ecolabeling programs offer a market incentive to environmentally innovative and conscious businesses. By offering products and services with ecolabels, these businesses can establish or reinforce a market niche and positive corporate image among consumers. Ecolabeling criteria usually only reward the top environmental performers in a product category, with most schemes gradually and incrementally raising standards to encourage producers and service providers to keep abreast with new and emerging performance improvements.
3. *Building consumer awareness of environmental issues*: Ecolabeling schemes raise consumer awareness of environmental issues and the implications of their choices. To ensure ecolabels are effective, there needs to be educating of consumers on the presence and meaning of ecolabels so that purchasing of eco-labeled products, or comparing ecolabel products with one another when making purchasing decisions, becomes habitual (Global Ecolabelling Network 2004; Iraldo et al. 2020; Song et al. 2019).

Case: Taiwan's Green Mark

The Green Mark, administered by Taiwan's Environmental Protection Administration, is a voluntary ecolabel scheme to encourage companies to manufacture products that have less impact on the environment by reducing waste and promoting recycling. Operating since 1992, the program aims to promote green consumerism by encouraging consumers to select recyclable, low-polluting, resource-saving products. The certification of Green Mark is based on the ISO 14024 eco-friendly principles. It is used as an economic tool to increase products and services that have less impact on the environment. The Green Mark is awarded to products in the top 20–30% of their category, with product categories including dual-flush water-saving toilets, water-saving faucets/devices, water conservation dual-flush toilet retrofit devices, and water fountains. To further promote environmentally friendly products and services, the Type 2 Green Mark certification was developed in 2014 for products not listed in the Green Mark categories. To be eligible

(continued)

for the Type 2 Green Mark, the product or service should comply with specific environmental claims, including that the product or service can save more water than other similar products (Environmental Protection Administration 2020).

Environmental Awards and Recognition

Government agencies can create environmental awards to help raise environmental awareness through businesses and the community and help businesses be publicly recognized for their excellent environmental performance. Environmental awards should be widely promoted in relevant media, for instance, business and industry publications. Environmental recognition awards can also be used to recognize the role of different stakeholders, including individuals as well as entrepreneurs and small businesses who are developing resource efficiency concepts. Award winners then become role models for the wider community. By creating role models, individuals or communities will likely strive to feel connected with the role model and shift their behavior to be in accordance with the norm the role model has created (Brears 2017; OECD 2018).

Case: Indiana State's Governor's Awards for Environmental Excellence

The Indiana Department of Environmental Management Governor's Awards for Environmental Excellence is the state's most prestigious environmental recognition award. They are reserved for the most innovative, sustainable, and exemplary programs or projects that provide significant environmental, economic, and social benefits to the state. Only one award is given per category with any Indiana citizen, business, nonprofit organization, school, university, or government agency eligible to submit a nomination or be nominated for the award. Applications must include a narrative of how the program or project is achieving excellence in its respective category (Table 5) (Indiana Department of Environmental Management 2020).

Green Public Procurement

Public procurement is the acquisition of goods or services by public entities with the primary objective being efficiency and cost-effectiveness. Secondary policy objectives of public procurement are to produce societal benefits such as environmental improvement. Green public procurement is where public authorities procure goods and services with a lower environmental footprint than alternatives that have comparable function and performance. Green public procurement requires a set of environmental criteria, including eco-labels and standards for resource efficiency and emissions intensity (Rainville 2017; Sparrevik et al. 2018). Identifying which product or service should be prioritized can be based on four main factors:

- *Environmental impact*: Selecting those products or services which have a high positive impact on the environment over their life cycle.
- *Budgetary importance*: Green public procurement should focus on areas of significant expenditure within a department.
- *Potential to influence the market*: Green public procurement should focus on areas where there is the most potential to influence the market.
- *Political priorities*: Green public procurement can be used to address a variety of environmental priorities (Table 6) (European Commission 2016; Switch2green 2020).

Green Economy Policies to Achieve Water Security, Table 5 Award categories and narrative

Category	Narrative
Energy efficiency/ renewable resources	The application must describe how it prevents pollution by either reducing energy use or by producing energy using zero- or very-low-emission technologies
Greening the government	The application must describe how individuals, facilities, or agencies within local, county, or state government pursue improvements in the environmental performance within their operations
Land use/conservation	The application must describe how the program or project incorporates innovative and effective methods or practices to preserve or improve land use
Environmental education and outreach	The application must describe how the education or outreach program promotes environmental stewardship or results in enhanced environmental protection
Pollution prevention	This category is for entities that have implemented one or more pollution prevention projects finalized in 2018 or 2019
Recycling/reuse	The application must describe how their recycling or reuse programs use innovative methods to reduce the amount of waste sent for final disposal
Five years of continuous improvement	The application must describe how the business or industry has achieved proven environmental results through a comprehensive environmental management system

Green Economy Policies to Achieve Water Security, Table 6 Examples of environmental priorities addressed by green public procurement

Environmental priority	Example
Deforestation	Green public procurement can address deforestation through the purchase of wood and wood products from legally harvested and sustainably managed forests
Greenhouse gas emissions	Green public procurement can lower greenhouse gas emissions through the purchase of products and services with a lower carbon dioxide footprint throughout their life cycle
Water consumption	Green public procurement can reduce water consumption by choosing more water-efficient fittings
Energy efficiency and resource use	Green public procurement can enhance energy efficiency and resource use through choosing products which are more efficient and implementing environmentally conscious design principles
Air, water, and soil pollution	Green public procurement can reduce pollution by controlling chemicals and limiting the use of hazardous substances
Waste	Green public procurement can reduce waste by specifying processes or packaging which generate less waste or encouraging reuse and recycling of materials
Sustainable agriculture	Green public procurement can encourage sustainable agriculture by purchasing organically produced food

Case: Hong Kong's Green Procurement

Since 2000, the Government of Hong Kong requires its bureaus and departments to consider environmental considerations when procuring goods and services. Specifically, bureaus and departments are encouraged to avoid single-use disposable items and purchase products with improved recyclability, high recycled content, reduced packing, greater durability, and greater energy efficiency or utilize clean technology and/or clean fuels, which, overall, result in reduced water consumption, fewer irritating or toxic

(continued)

Green Economy Policies to Achieve Water Security, Table 7 Green procurement evaluation

	Product A	Product B	Product C
Price $	$$$ (most expensive)	$ (inexpensive)	$ (inexpensive)
Mandatory green specification 1	✓	✓	✓
Desirable green specification 1	X	✓	X
Selected	No	**Yes**	No

substances emitted during installation, or the smaller production of toxic substances, or of less toxic substance. For the purchase of everyday user items, the government has adopted green specifications as mandatory requirements in the tender specifications when the items are available on the market with adequate models and quantities in supply. For new green specifications developed with uncertain market availability, the green specifications will be included in the tender specifications as desirable features. Tenderers are invited to indicate whether their items comply with these green features, and where appropriate submit supporting documents for verification. The tender assessment panel will evaluate tender offers which can meet the mandatory requirements and recommend either the lowest conforming offer or the highest scoring conforming offer for acceptance. Where there are two or more lowest conforming offers which are identical in all respects, the one which could meet the desirable green specifications could be given the preference (Table 7) (Government of Hong Kong Environmental Protection Department 2020). To promote green procurement in general, the Environmental Protection Department has set up a dedicated information portal that, in addition to providing the green specifications adopted by the government, provides examples of best practices, information on overseas and local environmental labeling schemes, and other relevant guidelines and resources.

Education and Awareness

Education of the public is critical in generating an understanding of water scarcity and creating acceptance of the need to implement water conservation programs. In schools, water utilities can educate students on the water cycle and encourage the sustainable use of scarce water resources. At the same time, the public can be informed of the need to conserve water through a variety of instruments. Specifically:

- *School programs*: Water utilities can educate students on water conservation in many ways including school presentations and the distribution of water conservation information and materials that can be used in the school curriculum.
- *Public education programs*: Water conservation messages can be distributed or made available through public service announcements, television commercials, and social media as well as in newspaper articles and advertisements.
- *Public events*: Water utilities can provide customers with information on water conservation and distribute free water-saving devices at public events such as expos, fairs, and conservation workshops.
- *Information in the water bills*: Water bills should be clearly laid out, detailing the volume of usage, rates, and charges. It should also allow customers to compare their current bill with previous bills. Billing inserts with water conservation tips can also be provided (Brears 2016).

G

Case: Singapore World Water Day Programmes and Resources for Tertiary Schools

World Water Day is held annually on March 22, a day that is designated by the United Nations to celebrate the importance of water sustainability. In Singapore, World Water Day is a nationwide celebration by the community, for the community to celebrate the importance of water in March. Schools are encouraged by Singapore's Public Utilities Board (PUB) to organize their own World Water Day event. Schools set a date for their water event in March, decide on a theme for their event (which can be linked to Singapore World Water Day's theme), draft fun activities, and explore how water topics can be incorporated into lesson plans and assembly talks. Schools can share their event on Facebook, Twitter, and Instagram with #SGWorldWaterDay. PUB can help twin schools with other community partners as well as provide water conservation materials for schools that make a request (subject to availability). Schools that are organizing a water-related event can request a variety of materials including:

- *Water conservation A1 panels*: The water conservation A1 panels educate students on various topics, ranging from general water conservation advice to specific subjects such as restoration of biodiversity.
- *Water conservation (standee frames)*: The Standee Frames is an eye-catching set of three standees that support the full range of water conservation A1 panels. The content of the Frames is flexible and can be changed by slotting in different A1 panels set with a variety of water-related topics.
- *Friends of water TV panel*: Share water messages and learn about Singapore's Four National Taps with Friends of Water characters on PUB's 32-inch screen TV.
- *Educational videos on water topics*: The videos educate the students on the Singapore Water Story and simple steps to conserve water resources through animated infographics (Public Utilities Board 2017).

Conclusion

While economic growth has produced many benefits including raising living standards and improving quality of life around the world, it has also resulted in the depletion and degradation of water resources, impacting human health and well-being as well as ecosystem health. In response, many multilateral organizations have called for the development of a green economy that improves human well-being and social equity and reduces environmental degradation. A vital component of the green economy is ensuring water security for all users and uses, both human and natural. Water security is challenged, however, by a variety of trends that impact water quantity and water quality, including climate change, population growth and urbanization, economic growth, and rising demand for energy and food. Recognizing these challenges, governments at various levels can facilitate the transition toward a green economy that achieves water security through a range of complementary fiscal and nonfiscal tools that reduce water security pressures while creating multiple economic, environmental, and social benefits. Regarding fiscal tools, governments at all levels can implement environmental taxes and charges to encourage investments in water-efficient technologies, provide subsidies to water users to encourage the development, testing, and demonstration of new water-efficient technology, develop water-trading markets to maximize the efficient use of water or reduce pollution levels of waterways in the most economically efficient manner possible, develop or collaborate with the private sector on PES schemes that deliver water quality improvements, and issue green bonds for a range of initiatives that protect water quantity and water quality. Governments at all levels can also implement a range of nonfiscal tools to achieve water security including developing regulations to reduce water consumption temporarily during set times of the year or permanently via building code changes, implementing ecolabeling schemes to encourage consumers to purchase water-efficient products, initiating green public procurement policies to lower the environmental footprints of government organizations including reducing water consumption, and creating education and awareness campaigns that encourage the wise use of water.

Cross-References

- ▶ Circular Economy and the Water-Food Nexus
- ▶ Circular Water Economy
- ▶ Water Security and the Green Economy
- ▶ Water Security, Sustainability, and SDG 6

References

Alston, L. J., Andersson, K., & Smith, S. M. (2013). Payment for environmental services: Hypotheses and evidence. *Annual Review of Resource Economics, 5*(1), 139–159. https://doi.org/10.1146/annurev-resource-091912-151830.

Arnell, N. W., Halliday, S. J., Battarbee, R. W., Skeffington, R. A., & Wade, A. J. (2015). The implications of climate change for the water environment in England. *Progress in Physical Geography: Earth and Environment, 39*(1), 93–120. https://doi.org/10.1177/0309133314560369.

Borghesi, S. (2014). Water tradable permits: A review of theoretical and case studies. *Journal of Environmental Planning and Management, 57*(9), 1305–1332. https://doi.org/10.1080/09640568.2013.820175.

Brears, R. C. (2016). *Urban water security*. Chichester/Hoboken: Wiley.

Brears, R. C. (2017). *The green economy and the water-energy-food Nexus*. London: Palgrave Macmillan.

Brears, R. C. (2018). *Blue and Green Cities: The role of blue-green infrastructure in managing urban water resources*. Palgrave Macmillan.

Brears, R. C. (2020). *Developing the circular water economy*. Cham: Palgrave Macmillan.

Brown, Z. S., & Johnstone, N. (2014). Better the devil you throw: Experience and support for pay-as-you-throw waste charges. *Environmental Science & Policy, 38*, 132–142. https://doi.org/10.1016/j.envsci.2013.11.007.

Center for International Forestry Research. (2014). Payments for ecosystem services (PES): A practical guide to assessing the feasibility of PES projects. Retrieved from http://www.cifor.org/publications/pdf_files/Books/BFripp1401.pdf

City of Boston. (2020). Building energy reporting and disclosure ordinance. Retrieved from https://www.boston.gov/departments/environment/building-energy-reporting-and-disclosure-ordinance

Climate Bonds Initiative. (2015). Scaling up green bond markets for sustainable development. Retrieved from https://www.climatebonds.net/resources/publications/scaling-green-bond-markets-sustainable-development

Defra. (2013). Payments for ecosystem services: A best practice guide. Retrieved from https://www.cbd.int/financial/pes/unitedkingdom-bestpractice.pdf

Department of Agriculture, Water, and the Environment. (2020). *Market price information for Murray–Darling Basin water entitlements*. Department of Agriculture, Water, and the Environment.

Ecologic Institute. (2003). The role of tradable permits in water pollution control. Retrieved from https://www.ecologic.eu/sites/files/publication/2015/1872-03_paper_tradablepermits_santiago_final.pdf

ECOTEC. (2001). Study on the economic and environmental implications of the use of environmental taxes and charges in the European Union and its member states. Retrieved from https://ec.europa.eu/environment/enveco/taxation/pdf/ch1t4_overview.pdf

Environmental Protection Administration. (2020). The green mark. Retrieved from https://greenliving.epa.gov.tw/newPublic/Eng/GreenMark/First

ESMC. (2020). General Mills, Kansas Department of Health and Environment and-Ecosystem Services Market Consortium launch farm-based ecosystem services pilot. Retrieved from https://ecosystemservicesmarket.org/2020/01/28/general-mills-kansas-department-of-health-and-environment-and-ecosystem-services-market-consortium-launch-farm-based-ecosystem-services-pilot/

European Commission. (2016). Buying green! A handbook on green public procurement. Retrieved from https://ec.europa.eu/environment/gpp/pdf/Buying-Green-Handbook-3rd-Edition.pdf

European Commission and UNEP. (2019). Switch to an inclusive green economy. Retrieved from https://www.switchtogreen.eu//wordpress/wp-content/uploads/2019/03/5.-Press-Folder.pdf

FAO. (2001). Product certification and ecolabelling for fisheries sustainability. Retrieved from http://www.fao.org/3/y2789e/y2789e00.htm#Contents

FAO. (2006). Modern water rights: Theory and practice. Retrieved from http://www.fao.org/3/a-a0864e.pdf

FAO. (2015). Towards a water and food secure future: Critical perspectives for policy-makers. Retrieved from http://www.fao.org/3/a-i4560e.pdf

G

General Mills. (2020). General Mills launches multi-year regenerative agriculture pilot with wheat growers in Central Kansas. Retrieved from https://www.generalmills.com/en/News/NewsReleases/Library/2020/January/General-Mills-launches-multi-year-regenerative-agriculture-pilot

Global Ecolabelling Network. (2004). Introduction to ecolabelling. Retrieved from https://globalecolabelling.net/assets/Uploads/intro-to-ecolabelling.pdf

Global Water Partnership. (2012). Water in the green economy. Retrieved from https://www.gwp.org/globalassets/global/toolbox/publications/perspective-papers/03-water-in-the-green-economy-2012.pdf

Government of Hong Kong Environmental Protection Department. (2020). Green procurement. Retrieved from https://www.epd.gov.hk/epd/english/how_help/green_procure/green_procure.html

Harlan, S. L., Yabiku, S. T., Larsen, L., & Brazel, A. J. (2009). Household water consumption in an Arid City: Affluence, affordance, and attitudes. *Society & Natural Resources, 22*(8), 691–709. https://doi.org/10.1080/08941920802064679.

IEA. (2016). WEO-2016 special report: Water-energy Nexus. Retrieved from https://webstore.iea.org/weo-2016-special-report-water-energy-nexus

Indiana Department of Environmental Management. (2020). Governor's awards for environmental excellence. Retrieved from https://www.in.gov/idem/partnerships/2457.htm

International Capital Market Association. (2018). Green bond principles. Voluntary process guidelines for issuing green bonds. Retrieved from https://www.icmagroup.org/green-social-and-sustainability-bonds/green-bond-principles-gbp/

Iraldo, F., Griesshammer, R., & Kahlenborn, W. (2020). The future of ecolabels. *The International Journal of Life Cycle Assessment, 25*(5), 833–839. https://doi.org/10.1007/s11367-020-01741-9.

Kearney, J. (2010). Food consumption trends and drivers. *Philosophical Transactions of the Royal Society of London. Series B, Biological Sciences, 365*(1554), 2793–2807. https://doi.org/10.1098/rstb.2010.0149.

KPMG International. (2015). Gearing up for green bonds. Retrieved from https://assets.kpmg/content/dam/kpmg/pdf/2015/03/gearing-up-for-green-bonds-v1.pdf

Kreiser, L., Sirisom, J., Ashiabor, H., & Milne, J. (2011). *Environmental taxation and climate change: Achieving environmental sustainability through fiscal policy.* Cheltenham: Edward Elgar Publishing Limited.

Maltais, A., & Nykvist, B. (2020). Understanding the role of green bonds in advancing sustainability. *Journal of Sustainable Finance & Investment*, 1–20. https://doi.org/10.1080/20430795.2020.1724864.

McDonald, R. I., Weber, K. F., Padowski, J., Boucher, T., & Shemie, D. (2016). Estimating watershed degradation over the last century and its impact on water-treatment costs for the world's large cities. *Proceedings of the National Academy of Sciences, 113*(32), 9117–9122. https://doi.org/10.1073/pnas.1605354113.

Ministry for the Environment. (2006). Market-based approaches to marine environmental regulation: Stage 2: Instrument assessment framework and case study. Retrieved from https://www.mfe.govt.nz/publications/marine/market-based-approaches-marine-environmental-regulation-stage-2-instrument

Ministry of Environment and Food of Denmark. (2020). The Danish eco-innovation program. Retrieved from https://eng.ecoinnovation.dk/the-danish-eco-innovation-program/

Murray Darling Basin Authority. (2020). Water markets and trade. Retrieved from https://www.mdba.gov.au/managing-water/water-markets-and-trade

Netherlands Enterprise Agency. (2020). MIA and Vamil. Retrieved from https://english.rvo.nl/subsidies-programmes/mia-and-vamil

Nordic Investment Bank. (2014). NIB environmental bond framework. Retrieved from https://www.nib.int/filebank/a/1410449130/c14f001e548bdeef346b853a6cd82c2a/3986-NEB_Framework.pdf

Nordic Investment Bank. (2019). NIB issues first Nordic–Baltic blue bond. Retrieved from https://www.nib.int/who_we_are/news_and_media/news_press_releases/3170/nib_issues_first_nordic-baltic_blue_bond

OECD. (2011a). Environmental taxation: A guide for policy makers. Retrieved from https://www.oecd.org/env/tools-evaluation/48164926.pdf

OECD. (2011b). Tools for delivering on green growth. Retrieved from https://www.oecd.org/greengrowth/48012326.pdf

OECD. (2018). *Environmental policy toolkit for SME greening in EU eastern partnership countries*. OECD.

Olsthoorn, M., Schleich, J., Gassmann, X., & Faure, C. (2017). Free riding and rebates for residential energy efficiency upgrades: A multi-country contingent valuation experiment. *Energy Economics, 68*, 33–44. https://doi.org/10.1016/j.eneco.2018.01.007.

PBL Netherlands Environmental Assessment Agency. (2012). *Environmental taxes and Green Growth: Exploring possibilities within energy and climate policy.* PBL Netherlands Environmental Assessment Agency.

Public Utilities Board, Singapore. (2017). Programmes & resources for tertiary schools. Retrieved from https://www.pub.gov.sg/Documents/Tertiary.pdf

PwC. (2015). The world in 2050: Will the shift in global economic power continue? Retrieved from http://www.pwc.com/gx/en/issues/the-economy/assets/world-in-2050-february-2015.pdf

Rainville, A. (2017). Standards in green public procurement – A framework to enhance innovation. *Journal of Cleaner Production, 167*, 1029–1037. https://doi.org/10.1016/j.jclepro.2016.10.088.

Sachdeva, S., Jordan, J., & Mazar, N. (2015). Green consumerism: Moral motivations to a sustainable future. *Current Opinion in Psychology, 6*, 60–65. https://doi.org/10.1016/j.copsyc.2015.03.029.

Schomers, S., & Matzdorf, B. (2013). Payments for ecosystem services: A review and comparison of developing and industrialized countries. *Ecosystem Services, 6*, 16–30. https://doi.org/10.1016/j.ecoser.2013.01.002.

SFPUC. (2019). Rates schedules and fees for water power and sewer service effective with meter readings made on or after July 1, 2019 (FY 2019-20). Retrieved from https://sfwater.org/modules/showdocument.aspx?documentid=7743

Song, L., Lim, Y., Chang, P., Guo, Y., Zhang, M., Wang, X., . . . Cai, H. (2019). Ecolabel's role in informing sustainable consumption: A naturalistic decision making study using eye tracking glasses. *Journal of Cleaner Production, 218*, 685–695. https://doi.org/10.1016/j.jclepro.2019.01.283

Sparrevik, M., Wangen, H. F., Fet, A. M., & De Boer, L. (2018). Green public procurement – A case study of an innovative building project in Norway. *Journal of Cleaner Production, 188*, 879–887. https://doi.org/10.1016/j.jclepro.2018.04.048.

Sterner, T., & Coria, J. (2002). *Policy instruments for environmental and natural resource management*. Florence: Taylor & Francis Group.

Switch2green. (2020). EU green public procurement. Retrieved from https://www.switchtogreen.eu/?p=1527

UNEP. (2012). Green economy: What do we mean by green economy?. Retrieved from https://wedocs.unep.org/bitstream/handle/20.500.11822/8659/-%20Green%20economy_%20what%20do%20we%20mean%20by%20green%20economy_%20-2012Main%20briefing%202012%2D%2DFinal.pdf

UNESCO. (2012). Managing water under uncertainty and risk. Retrieved from http://www.unesco.org/new/fileadmin/MULTIMEDIA/HQ/SC/pdf/WWDR4%20Volume%201-Managing%20Water%20under%20Uncertainty%20and%20Risk.pdf

United Nations Department of Economic and Social Affairs. (2019a). World population prospects 2019: Highlights. Retrieved from https://population.un.org/wpp/Publications/Files/WPP2019_Highlights.pdf

United Nations Department of Economic and Social Affairs. (2019b). World urbanization prospects: The 2018 revision. Retrieved from https://population.un.org/wup/Publications/Files/WUP2018-Report.pdf

UN-Water. (2013). What is water security? Infographic. Retrieved from https://www.unwater.org/publications/water-security-infographic/

UN-Water. (2014). Partnerships for improving water and energy access, efficiency and sustainability. Retrieved from http://www.un.org/waterforlifedecade/water_and_energy_2014/pdf/water_and_energy_2014_final_report.pdf

Verma, R., & Gayithri, K. (2018). Environmental fiscal instruments: A few international experiences. *Margin: The Journal of Applied Economic Research, 12*(3), 333–368. https://doi.org/10.1177/0973801018768974.

Xie, H., & Ringler, C. (2017). Agricultural nutrient loadings to the freshwater environment: the role of climate change and socioeconomic change. *Environmental Research Letters, 12*(10), 104008. https://doi.org/10.1088/1748-9326/aa8148.

Zhao, L., & Chen, Y. (2019). Optimal subsidies for green products: A maximal policy benefit perspective. *Symmetry, 11*(1). https://doi.org/10.3390/sym11010063.

Green Finance

► Financing: Fiscal Tools to Enhance Regional Sustainable Development

Green Finger

► Green Belts

Green Growth

► Water Security and the Green Economy

Green Infrastructure

The Multiple Benefits of Green Infrastructure

G. Osei[1], F. Pascale[1], N. Delle-Odeleye[1] and A. Pooley[2]
[1]Anglia Ruskin University, Chelmsford, UK
[2]Centre for Alternative Technology, Pantperthog, UK

Definition

Green Infrastructure (GI):

GI is a structural approach that utilizes nature, natural process, and mechanisms for infrastructure development and city planning (or land use) while developing the economy and society (Nakamura 2019). GI present multiple benefits and these can be defined as multi-functional structures, producing more than one service to society, economy, and the environment. Benefits can be achieved directly or indirectly through the implementation of GI.

Green Infrastructure: The Multiple Benefits

Green Infrastructure (GI) is a structural approach that utilizes nature, natural process, and mechanisms for development. Despite the name "green," GI often includes natural infrastructure that have blue (water) elements such as ponds, swales, and wetlands, used primarily for sustainable drainage systems (Chow et al. 2014). Due to numerous blue and green features available for GI, multiple benefits can be achieved in areas of increasing grey infrastructure, permitting a more liveable space for wildlife (Plummer et al. 2020). Humans also require this enhancement of liveability due to the discomfort of the earth's changing climate (Blanusa et al. 2019). There are many examples of GI for achieving multiple benefits. Rain gardens and bioswales deliver surface water management and aesthetic pleasure; permeable paving deliver infiltration and friction in icy conditions; green roofs deliver ecological habitat creation and acoustic insulation of a building; tree canopies deliver thermal comfort and place distinctiveness; peatlands deliver wildlife habitat and carbon storage; and more.

The term GI can at times be ambiguous and is likened to the liberal use of the word "sustainability." The lack of clarity is in part due to terms being relatively recent concepts. Their definitions are still evolving but have well accepted characteristics. Both terms are generally linked to achieving environmental, social, and/or economic benefits to differing degrees. GI, however, focuses on the use of natural structures, e.g., trees, meadows, hedges etc., to achieve these benefits. GI is considered a sustainable and resilient form of infrastructure. Conventional infrastructure, sometimes referred to as "grey infrastructure," is commonly used to serve a single purpose. GI, however, aims to not only help manage flood disasters, mitigate urban cooling, and conserve nature, but also deliver on other environmental, social, and economic benefits (Nakamura 2019). Goals for net zero, garden cities, and other sustainable development targets, require multiple benefits of GI; initiating the re-framing of what is meant by infrastructure and city living.

The extensive capability of GI means that their implementation can provide not only direct and indirect gains, but also physical and psychological interventions. The multiple benefits of GI can be summarized into ecosystem services and political/economic services. These services can be further categorized into climate benefits, disaster risk reduction, ecological benefits, social resilience, environmental justice, and economic growth (Fig. 1).

Ecosystem Services

Ecosystem services are a large proportion of the multiple benefits that GI can provide. Within GI implementation, ecosystem services aim to address sustainability and resilience objectives (Calderón-Argelich et al. 2021). These services predominantly focus on environmental and social benefits.

Climate Benefits

Climate action agendas including the United Nations Sustainable Development and the Paris Agreement have led to national and international targets for achieving net zero and carbon neutral by many countries. These targets have caveated for woodland creation and other forms of GI to provide climate resilience on a micro and macro scale. GI provides an environment that is more resilient through reduced vulnerability to natural and human induced hazards and disasters such as floods, heat, fire, and drought (Foster et al. 2011). Examples such as the use of agro-forestry for temperate climates are being implemented as a

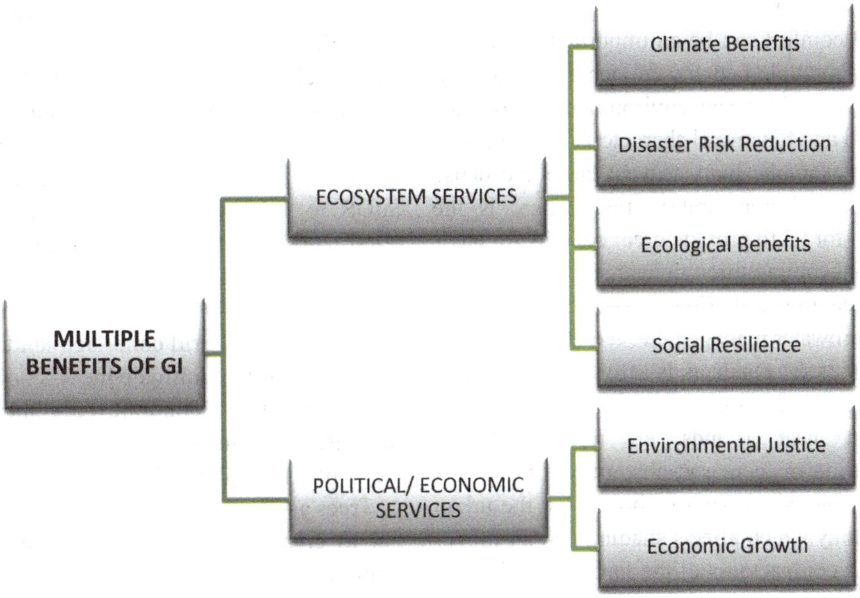

Green Infrastructure, Fig. 1 The multiple benefits of Green Infrastructure

mitigation method due to increased awareness of climate change within the twenty-first century (De Linares 2018).

Air Quality Improvements

Carbon sequestration is highly reflected within the current creations of GI, commonly considered through the delivery of tree lines and woodlands. These trees store different amounts of cardon at different rates based on their age (Chen 2015), with an estimated maximum growth rate for carbon capture of around 30 years. However, tree carbon sequestration often results in an overall low percentage of required carbon capture, within the single figures (Chen 2015; Escobedo et al. 2010); therefore, it is required that GI implementation is considered holistically to provide and calculate optimal carbon sequestration (McPherson et al. 2013; Gill et al. 2007). Sequestration through soil is also possible, as GI implementations such as grasslands can provide higher soil carbon content and microbial biomass whiles reducing the disturbance of ground soil, in contrast to methods such as tradition farming (Bullock et al. 2021), allowing for carbon to remain in the soil as another form of capture. Further, it is relevant to note that peatlands, marshland, and wetlands sequester the largest amounts of carbon but do this over a long period of time.

Additionally, plants are utilized to stabilize, transfer, remove or destroy contaminants in soils and groundwater as a bioremediation process known as the phytoremediation of soil pollutants. This method of soil treatment allows for mitigating and adapting to climate change without compromising the soil fertility (Zgorelec et al. 2020). It should be noted that there are also air quality disservices associated to GI implementation. This disservice is produced through the release of Volatile Organic Compounds (VOCs) and Biogenic Volatile Organic Compounds (BVOCs) from plant (and animal respiration), and organic decomposition (Blanusa et al. 2019). Awareness of potential VOC and BVOC amounts in proposed species of GI implementation results in achieving optimum air quality benefits.

Thermal Comfort

Thermal comfort can be an important strategy in ensuring multiple benefits are achieved from GI, primarily within urban environments. From the structural and functional characteristics of vegetation, for example: hairy leaves; large and dense canopies; and transpiration rates, there is the potential for GI to provide great benefits for environmental cooling through heat and rainfall regulation (Blanusa et al. 2019). When combining GI with stormwater management, the effect of urban heat islands for residents is effectively mitigated (Johnson et al. 2021).

Urban heat islands are created by a combination of increased absorption of sunlight by dark-colored surfaces of buildings, the urban morphology which affects shading and air movement, the compactness of the land-use and travel proximity, and the deficiency in urban green spaces (Aram et al. 2019). GI structures such as hedge rows provide a barrier to heat within outside spaces in urban environments promoting human comfort (Blanusa et al. 2019). The cooling of urban environments through GI implementation are therefore vital to reduce urban heat island effects.

Disservices associated with thermal comfort from GI implementation include excessive shading (Blanusa et al. 2019). Therefore, the benefit of thermal comfort is heightened by considering the location of thermal comfort GI features for suitable height and coverage.

Disaster Risk Reduction

Disaster risk reduction is a valuable factor when considering infrastructure for development (Nakamura 2019, p. 37). The use of GI for disaster risk reduction allows for measurable benefits, such as aquifer recharge as well as pollution reduction, and its use for surface and groundwater flood management has a history in cities with arid climates (Porse 2014). These factors have made GI popular in policy making especially in mitigating flood disaster risk.

Water Management

GI is frequently utilized for flood disaster risk reduction through its use in Sustainable Drainage Systems (SuDS). This term is used in the United Kingdom, with globally similar systems such as the United States' Low Impact Development (USEPA 2006), and Australia's Water Sensitive Urban Design (Brown and Clarke 2007). Modern cities are using GI in SuDS to retain and infiltrate surface runoff (Orsi 2004; Fratini et al. 2012). The issues within cities with surface runoff are increase not by a naturally occurring arid climate, but by the artificially arid concreted land. GI offers the opportunity to mitigate the risks of flooding, while providing economic gain through cost savings, and human health benefits such as aesthetic pleasure (Porse 2014).

Trees as part of the GI offering has the potential to create defenses to stormwater damage through "restoring more natural hydrologic regimes in urban watersheds by intercepting rainfall, delaying runoff, infiltrating, and transpiring captured stormwater" (Kuehler et al. 2017). Furthermore, GI goes beyond its ability to provide reduced disaster risks, as an added benefit to water quality and flood mitigations, GI focused SuDS feature other ecosystem services and socio-cultural benefits (Lawson et al. 2014). As an example, they are also used within school grounds as an added means of education.

Ecological Benefits

Separate to the global benefits of GI implementation, ecological benefits are vital to the overall multi-beneficial character of these types of infrastructure.

Value to Wildlife

GI provides habitat and support for wildlife. Through strategic design the implementation of GI provides wildlife with a habitat devoid of disturbance, fragmentation, and species isolation (Felappi et al. 2020). Disturbance level is minimized through habitat characteristics, providing refuge and limiting human exposure (Tablado and Jenni 2017). Implementing GI networks to connect existing natural spaces have become a helpful measure for adapting areas to the impacts of climate change (Apud et al. 2020). These networks are aimed at enabling wildlife to move within an unrestricted pathway, devoid of place boundary lines. Biodiversity within GI has the potential to support wildlife conservation. The ecological impact of urbanization can be reduced through species conservation, providing environmental gains (Plummer et al. 2020). Large, well-connected landscapes greatly assist biodiversity in GI to support species (Plummer et al. 2020), with these spaces also linked to social preference by such means as colorful planting (Hoyle et al. 2017).

Biodiversity

In identifying the biodiversity richness of a GI site, many have used biodiversity indicators that identify the richness of species including plants, insects, birds, and butterflies (Hoyle et al. 2017; Fuller et al. 2007; Plummer et al. 2020). Through evidence-based decisions, different species are utilized to support various services. However, balancing the levels of biodiversity is important for achieving multiple benefits. Maclvor et al. 2016 notes that limiting biodiversity can improve the survival and performance of inhabiting species within environmentally challenging locations, whiles enhancing biodiversity can provide direct human benefits such as improved aesthetics. Further, a noteworthy disservice is allergens. Evidence on whether the delivery of specific species, such as *Carpinus betulus* hedgerows, cause high allergen content due to flowering profusely will enhance the benefit of diverse vegetation (Blanusa et al. 2019). To deliver multiple ecosystem services appropriate species/cultivar selections are needed with the added benefits of minimizing possible disservices (Blanusa et al. 2019).

Social Resilience

Shifting perceptions of natural GI has allowed for increasing acceptance of "untidied" ecologically aesthetic planting (Hoyle et al. 2017). Where information and education on the adaptability of GI is provided, evidence also shows that individuals opt for more effective multi-beneficial GI options (Derkzen et al. 2017).

Structural Benefit

GI produces direct benefits to society when used as a structural feature. These features include the use of GI as a natural wall aimed at: security, as a barrier; seclusion, for privacy; and screening, for unsightly views (Blanusa et al. 2019). GI walls are effective for security when species such as thorny leaves are used to provide an impenetrable surface. GI walls also benefit from being an aesthetically pleasing way to create privacy in spatial planning both from onlookers and to shield unsightly view. These physical interventions provide a feasible alternative to grey infrastructure with the added benefit of aesthetic value. Additionally, GI structures such as green roofs and hedgerows are utilized for noise reduction, through their ability to function as a sound barrier.

However, invasiveness of species, both vegetative and wildlife, are associated disservices of GI implementation (Blanusa et al. 2019). Income limitations can also hinder social preference to GI, with low socioeconomic areas generally unwilling to pay for GI services due to limited "disposal" income (Derkzen et al. 2017). This income disparity therefore results in limited social benefits from low quality GI within socially significant areas (Amano et al. 2018). Therefore, mitigating these occurrences through robust management processes.

Human Health Benefits

Enhanced human health add to the multiple benefits of GI through physical and mental wellbeing provisions. There is positive wellbeing gained from connecting to nature, with research showing that the psychological benefits are great for those in regions with increased health challenges (Hoyle et al. 2019). Sensory gardens, healing gardens, and therapeutic landscapes are examples of GI features that may produce direct human wellbeing benefits. They encourage a biophilic effect within humans when within the space and can promote multiculturality and social cohesion among the community (Dushkova et al. 2020). The human health benefits of GI also indirectly link to many other social resilience indicators, providing a higher life expectancy and satisfaction, cognitive function, social cohesion, and sense of ownership (Wood et al. 2016; Soga and Gaston 2016).

Community Satisfaction

Moreover, Public spaces such as public parks are beneficial as an "escape" for residents. They are a meaningful way of drawing the communities out of their busy lifestyle (Holland et al. 2007) and into the appreciation of the opposing, natural environment. GI features such as wildflower meadows enhance the offering of these spaces through the aesthetic provision and opportunity to connect with wildlife such as pollinators.

GI can further promote social cohesion between classes with historically famous designs of public spaces, such as the New York Central Park, designed primarily to bridge the class divides felt in urban living. Uses of such concepts as that explored in the 1800s design of Olmsted's Riverside, Chicago, had varying GI elements in shape; size; and variety, proved relevant for community satisfaction as it allowed a diverse urban landscape (Crow et al. 2006) and enhanced the sustainability of the development. Such GI considerations promoted social cohesion, place belonging, and space distinctiveness.

Political/Economic Services

Political and economic benefits are further services that promote the achievement of GI multiple benefits. They go beyond the ecosystems services of GI and demonstrate the infrastructure's capability to produce investible gains. These services are primarily of interest for corporate agendas and political policies. The recognition of the multiple benefits of GI provides the ability for these organizations to relay financial benefits whiles demonstrating corporate social responsibility.

Economic Benefits

The economic benefits of GI are defined by the financial gains that GI can provide. There are measures to monetize the benefits of GI, both direct and indirect benefits, including the financial benefits gained from cost saving. The practical implementation of monetizing GI benefits can be demonstrated globally through examples such as carbon and biodiversity offsetting.

Monetarizing GI Benefits

The multiple benefits of GI may be monetized by various tools such as the TEEB and B£ST from the Netherlands and UK respectively. In general, these tools monetize GI through ecosystem service evaluation. TEEB values in proportion to local adjustments allowing for differences in how monetization is applied, whiles physical performance of GI is favorable within the B£ST tool (Hamann et al. 2020). In the implementation of GI, economic measures of the multiple benefits vary. Hamann et al. (2020) notes that the B£ST tool has been proven to give higher negative impacts than TEEB in some instances. This variation also extends to the differences in the categories used and how monetized values are determined.

GI creation has also allowed for a rise in "green industry" jobs. This is exampled by an increased involvement of ecologist within place making (Maclvor et al. 2016). There are also enhanced educational provisions for those seeking to specialize in the GI related roles. Further to these direct increase to the economy, there is also indirect cost savings resulting in GI implementation. These cost saving include saving to the health system from improve human health and wellbeing, reduced cost for green vs. grey features such as natural barriers instead of concrete walls, and reduced damages to premises through flood mitigation.

A prominent disservice of GI that is considered against economic benefits is the level of labor required for ongoing maintenance (Blanusa et al. 2019). As a result, for multiple benefits of GI to be realized, a healthy maintenance budget is beneficial (Vineyard et al. 2015). This can be achieved where GI has a "management" budget, similar to the "fixing" maintenance budget grey infrastructures require.

Practical Implementation of Monetarizing GI Benefits

Cardon offsetting has been established in many nations, predominantly in the west. This initiative places onus on organizations that produce notable carbon footprints to offset primarily through funding of woodlands. Carbon schemes such as the European Union Emission Trading Scheme (EU ETS), UK ETS, Woodland Carbon Code, Peatland Carbon Code, and CORSIA are in place to regulate the certification and trading of carbon credits on the open market. For example, CORSIA, Carbon Offsetting and Reduction Scheme for International Aviation, is a global offsetting scheme for tackling emissions gained through the process of international aviation. The buying of carbon credits can be undertaken on both a mandatory and voluntary basis.

Within the practice nations such as the UK have formed an actualized the monetarization of GI through the implementation of tools such as the BNG matrix. This ensures the offsetting of biodiversity lose in medium to large projects. The offsetting is gained though onsite habitat creation or offsite funding for new or enhanced habitats to those lost on the original site. Landowners can hence gain funding for the creation and maintenance of habitats that are rich in biodiversity.

Environmental Justice

There is a link between income/race disparities and the distribution of ecosystem services through GI. Commonly GI is low within areas with low socioeconomic groups (Calderón-Argelich et al. 2021). Hence, environmental justice for social and environmental accessibility are necessary in producing wide reaching multiple benefits.

Justice for the Community

Social perceptions and preferences toward green infrastructure are critical in demonstrating to policy makers how this form of infrastructure may be utilized to achieve multiple benefits, with society as part of the decision-making process (Lawson et al. 2014). Engagement with society, in GI implementation, can achieve heightened spatial benefits (Azadi et al. 2011). This engagement requires community input exampled in open ended communication (Golden 2014) to achieve responses for social cohesion of residents (Bélanger et al. 2012; Salone et al. 2017). In GI development, communities often carry a responsible role after the completion of the project in a "hand-over" of responsibilities. GI therefore indirectly promotes the balance of powers sponsoring the voice of vulnerable groups to be considered. The environmental justice of a community is enhanced where urban policy and planning ensure investment in GI that do not reinforce or exacerbate potentially existing environmental injustices against different classes of society (Herreros-Cantis and McPhearson 2021).

Justice for Nature

Enhance implementation of GI supports environmental justice for nature. Increased delivery of GI provides enhanced knowledge and favor for nature and therefore an amplified voice to nature by impacted professionals and citizens. These voices help in widening the political call and increase awareness of the needs, interests, and livelihood of nature (Gray et al. 2020). They change the systems of exploitation and domination of human focused benefits (Meijer 2017). This awareness fosters an appreciation of and respect for nature and their needed representation in economically heavy situations.

The conventional Corporate Social Responsibility process is a driving forces for achieve environmental and social benefits for sustainability in a political/economic setting. Environmental justice is therefore an indirect benefit of GI implementation necessitating the appreciation of the intrinsic value and rights of nature (Gray and Curry 2020) using the observation and "signals" provided by nature (Mathews 2017). The appreciation of GI has enabled the political backing for some large GI structure around the world, such as: legal rights for nature in Ecuador; protection of parts of the Amazon in Colombia; acknowledgement of the Whanganui River as a living being in New Zealand; and the rights to all rivers in Bangladesh (World Economic Forum).

Conclusion

The prospect of achieving numerous benefits through GI implementation has made the concept popular in the twenty-first century, allowing that the same portion of land can be used for multiple benefits at a time (Porse 2014). The benefits can be summarized into ecosystem services and political/economic services. These services are further categorized into climate benefits, disaster risk reduction, ecological benefits, social resilience, environmental justice, and economic growth. Achieving the benefits depend on evidence-based investigation on the location and species required for all the desired effects, opposed to a singular focal objective.

However, there is great reliance on GI to provide all multiple benefits at all times, however it is most vital to have the proper understanding of the case-by-case use of GI. This requires a paradigm shift toward green, away from the grey forms of infrastructure. Sussams et al. (2015) argue that though GI has been charged with mitigating urban heat islands, flood risk management, and ecosystem resilience, these require effective interdisciplinary work through academia, practice, and policy. The multiple benefits of GI therefore initiates the re-framing of what is meant by infrastructure and city living to actualize the goals for net zero, garden cities, and other sustainable development targets.

Delivery of GI is challenged to learn from the mistakes of other sustainable infrastructure initiatives. Passive houses for example have been analyzed to find indicators of shortfalls in their ability to produce their prepositioned advantages (Hasselaar 2008). Therefore, the ability for GI to achieve multiple benefits is not solely on GI itself but requires a multidisciplinary approach to be meet the challenges for effective implementation.

The full scope of a multi-beneficial GI is best met through the understanding of the synergy, trade-off, and interrelation of impacting and impacted systems (Sussams et al. 2015; Schindler et al. 2016; Maclvor et al. 2016). Case-by-case multiple benefits of GI is made effective when a diverse assembly of skilled agents and interested agents partake in the planning and implementation of GI (Schindler et al. 2016). This assembly should include direct specialist in nature such as landscape architects and ecologist to produced well-evidenced GI benefits (Maclvor et al. 2016).

Identifying and addressing disservice, e.g., invasiveness and allergenicity, of GI (Blanusa et al. 2019) is essential for actualizing the associated multiple benefits, yet it is generally accepted that when done well, GI implementation provides a collection of multiple benefits.

Cross-References

- ▶ Beyond knowledge: Learning to cope with climate change in cities
- ▶ Blue-Green Cities: Achieving Urban Flood Resilience, Water Security, and Biodiversity
- ▶ Green and Blue Infrastructure (GBI) in Urban Areas
- ▶ Innovation to Bring Nature-Based Solutions to Life: Tales of Two Cities
- ▶ Multiple Benefits of Green Infrastructure
- ▶ Stewarding Street Trees for a Global Urban Future
- ▶ Sustainable Cities via Urban Ecosystem Restoration
- ▶ Urban Forestry in Sidewalks of Bogota, Colombia

References

Amano, T., Butt, I., & Peh, K. S. H. (2018). The importance of green spaces to public health: A multi--continental analysis. *Ecological Applications, 28*(6), 1473–1480. https://doi.org/10.1002/eap.1748.

Apud, A., Faggian, R., Sposito, V., & Martino, D. (2020). Suitability analysis and planning of green infrastructure in Montevideo, Uruguay. *Sustainability (Switzerland), 12*(22), 1–18. https://doi.org/10.3390/su12229683.

Aram, F., Higueras García, E., Solgi, E., & Mansournia, S. (2019). Urban green space cooling effect in cities. *Heliyon, 5*(4), e01339. https://doi.org/10.1016/j.heliyon.2019.e01339.

Azadi, H., Ho, P., Hafni, E., Zarafshani, K., & Witlox, F. (2011). Multi-stakeholder involvement and urban green space performance. *Journal of Environmental Planning and Management, 54*(6), 785–811.

Bélanger, H., Cameron, S., & De La Mora, C. (2012). Revitalization of public spaces in a working class neighborhood: Appropriation. Identity and the urban imaginary. In H. Casakin & F. Bernardo (2012). *The role of place identity in the perception, understanding, and design of built environments*. Bentham Science.

Blanusa, T., Garratt, M., Cathcart-James, M., Hunt, L., & Cameron, R. W. F. (2019). Urban hedges: A review of plant species and cultivars for ecosystem service delivery in north-west Europe. *Urban Forestry and Urban Greening, 44*. https://doi.org/10.1016/j.ufug.2019.126391.

Brown, R. R., & Clarke, J. M. (2007). *Transition to water sensitive urban design: The story of Melbourne, Australia*. Facility for Advancing Water Biofiltration, Monash University Melbourne.

Bullock, J. M., McCracken, M. E., Bowes, M. J., Chapman, R. E., Graves, A. R., Hinsley, S. A., et al. (2021). Does agri-environmental management enhance biodiversity and multiple ecosystem services?: A farm-scale experiment. *Agriculture, Ecosystems & Environment, 320*, 107582.

Calderón-Argelich, A., Benetti, S., Anguelovski, I., Connolly, J. J. T., Langemeyer, J., & Baró, F. (2021). Tracing and building up environmental justice considerations in the urban ecosystem service literature: A systematic review. *Landscape and Urban Planning, 214*. https://doi.org/10.1016/j.landurbplan.2021.104130.

Chen, W. Y. (2015). The role of urban green infrastructure in offsetting carbon emissions in 35 major Chinese cities: A nationwide estimate. *Cities, 44*, 112–120.

Chow, J., Savić, D., Fortune, D., Kapelan, Z., & Mebrate, N. (2014). Using a systematic, multi-criteria decision support framework to evaluate sustainable drainage designs. *Procedia Engineering, 70*, 343–352. https://doi.org/10.1016/j.proeng.2014.02.039.

Crow, T., Brown, T., & De Young, R. (2006). The riverside and Berwyn experience: Contrasts in landscape structure, perceptions of the urban landscape, and their effects on people. *Landscape and Urban Planning, 75*(3–4), 282–299.

De Linares, P. G. (2018). Comparing urban food systems between temperate regions and tropical regions introducing urban agroforestry in temperate climates through the case of BUDAPEST. *International Journal of Design and Nature and Ecodynamics, 13*(4), 395–406. https://doi.org/10.2495/DNE-V13-N4-395-406.

Derkzen, M. L., van Teeffelen, A. J. A., & Verburg, P. H. (2017). Green infrastructure for urban climate adaptation: How do residents' views on climate impacts and green infrastructure shape adaptation preferences? *Landscape and Urban Planning, 157*, 106–130. https://doi.org/10.1016/j.landurbplan.2016.05.027.

Dushkova, D., & Haase, D. (2020). Not simply green: Nature-based solutions as a concept and practical approach for sustainability studies and planning agendas in cities. *Land, 9*(1). https://doi.org/10.3390/land9010019.

Dushkova, D., & Ignatieva, M. (2020). New trends in urban environmental health research: From geography of diseases to therapeutic landscapes and healing gardens. *Geography, Environment, Sustainability, 13*(1), 159–171. https://doi.org/10.24057/2071-9388-2019-99.

Escobedo, F., Varela, S., Zhao, M., Wagner, J. E., & Zipperer, W. (2010). Analyzing the efficacy of subtropical urban forests in offsetting carbon emissions from cities. *Environmental Science & Policy, 13*(5), 362–372.

Fratini, C.F., Geldof, G.D., Kluck, J., & Mikkelsen, P.S. (2012). Three Points Approach (3PA) for urban flood risk management: A tool to support climate change adaptation through transdisciplinarity and multifunctionality. *Urban Water Journal, 9*(5), 317–331.

Felappi, J. F., Sommer, J. H., Falkenberg, T., Terlau, W., & Kötter, T. (2020). Green infrastructure through the lens of "One health": A systematic review and integrative framework uncovering synergies and trade-offs between mental health and wildlife support in cities. *Science of the Total Environment, 748*, 141589.

Foster, J., Lowe, A., & Winkelman, S. (2011). The value of green infrastructure for urban climate adaptation. *Center for Clean Air Policy, 750*(1), 1–52.

Fuller, R. A., Irvine, K. N., Devine-Wright, P., Warren, P. H., & Gaston, K. J. (2007). Psychological benefits of greenspace increase with biodiversity. *Biology Letters, 3*(4), 390–394. https://doi.org/10.1098/rsbl.2007.0149.

Gill, S. E., Handley, J. F., Ennos, A. R., & Pauleit, S. (2007). Adapting cities for climate change: The role of the green infrastructure. *Built Environment, 33*(1), 115–133.

Golden, S. M. (2014). Occupied by design: Evaluating performative tactics for more sustainable shared city space in private-led regeneration projects. *WIT Transactions on Ecology and the Environment, 191*, 441–452.

Gray, J., & Curry, P. (2020). Ecodemocracy and political representation for non-human nature. In *Conservation* (pp. 155–166). Springer.

Gray, J., Wienhues, A., Kopnina, H., & DeMoss, J. (2020). Ecodemocracy: Operationalizing ecocentrism through political representation for non-humans. *The Ecological Citizen, 3*(2), 166–177.

Hamann, F., Blecken, G., Ashley, R. M., & Viklander, M. (2020). Valuing the multiple benefits of blue-green infrastructure for a Swedish case study: Contrasting the economic assessment tools B£ST and TEEB. *Journal of Sustainable Water in the Built Environment, 6*(4). https://doi.org/10.1061/JSWBAY.0000919.

Hasselaar, E. (2008). Health risk associated with passive houses: An exploration. *Indoor Air, 2008*, 17–22.

Herreros-Cantis, P., & McPhearson, T. (2021). Mapping supply of and demand for ecosystem services to assess environmental justice in New York city. *Ecological Applications, 31*, e02390.

Holland, C., Clark, A., Katz, J., & Peace, S. (2007). *Social interactions in urban public places*. Policy Press.

Hoyle, H., Hitchmough, J., & Jorgensen, A. (2017). All about the 'wow factor'? The relationships between aesthetics, restorative effect and perceived biodiversity

G

in designed urban planting. *Landscape and Urban Planning, 164*, 109–123. https://doi.org/10.1016/j.landurbplan.2017.03.011.

Hoyle, H., Jorgensen, A., & Hitchmough, J. D. (2019). What determines how we see nature? Perceptions of naturalness in designed urban green spaces. *People and Nature, 1*(2), 167–180. https://doi.org/10.1002/pan3.19.

Johnson, D., Exl, J., & Geisendorf, S. (2021). The potential of stormwater management in addressing the urban heat island effect: An economic valuation. *Sustainability, 13*(16), 8685.

Kuehler, E., Hathaway, J., & Tirpak, A. (2017). Quantifying the benefits of urban forest systems as a component of the green infrastructure stormwater treatment network. *Ecohydrology, 10*(3). https://doi.org/10.1002/eco.1813.

Lawson, E., Thorne, C., Ahilan, S., Allen, D., Arthur, S., Everett, G., et al. (2014). Delivering and evaluating the multiple flood risk benefits in blue-green cities: An interdisciplinary approach. *WIT Transactions on Ecology and the Environment, 184*, 113–124. https://doi.org/10.2495/FRIAR140101.

Maclvor, J. S., Cadotte, M. W., Livingstone, S. W., Lundholm, J. T., & Yasui, S.-E. (2016). Phylogenetic ecology and the greening of cities. *Journal of Applied Ecology, 53*(5), 1470–1476. https://doi.org/10.1111/1365-2664.12667.

Mathews, F. (2017). *Ecology and democracy.* Routledge.

McPherson, E. G., Xiao, Q., & Aguaron, E. (2013). A new approach to quantify and map carbon stored, sequestered and emissions avoided by urban forests. *Landscape and Urban Planning, 120*, 70–84.

Meijer, E. R. (2017). *Political animal voices*. Universiteit van Amsterdam [Host].

Nakamura, K. (2019). *Green Infrastructure.* Izumi, T., Shaw, R., Ishiwatari, M., Djalante, R., Komino, T., Sukhwani, V., & Adu Gyamfi, B., eds. (2019). 30 innovations linking Disaster risk reduction with sustainable development goals. Keio University, the University of Tokyo, UNU-IAS, CWS Japan, Japan,: International Institute of Disaster Science (IRIDeS), pp.36

Orsi, J. (2004). *Hazardous metropolis*. [e-book] University of California Press. Available through: google.

Plummer, K. E., Gillings, S., & Siriwardena, G. M. (2020). Evaluating the potential for bird-habitat models to support biodiversity-friendly urban planning. *Journal of Applied Ecology, 57*(10), 1902–1914. https://doi.org/10.1111/1365-2664.13703.

Porse, E. (2014). Risk-based zoning for urbanizing floodplains. *Water Science and Technology, 70*(11), 1755–1763. https://doi.org/10.2166/wst.2014.256.

Salone, C., Bonini Baraldi, S., & Pazzola, G. (2017). Cultural production in peripheral urban spaces: Lessons from Barriera, Turin (Italy). *European Planning Studies, 25*(12), 2117–2137.

Schindler, S., O'Neill, F. H., Biró, M., Damm, C., Gasso, V., Kanka, R., et al. (2016). Multifunctional floodplain management and biodiversity effects: A knowledge synthesis for six European countries. *Biodiversity and Conservation, 25*(7), 1349–1382. https://doi.org/10.1007/s10531-016-1129-3.

Soga, M., & Gaston, K. J. (2016). Extinction of experience: The loss of human – Nature interactions. *Frontiers in Ecology and the Environment, 14*(2), 94–101. https://doi.org/10.1002/fee.1225.

Sussams, L. W., Sheate, W. R., & Eales, R. P. (2015). Green infrastructure as a climate change adaptation policy intervention: Muddying the waters or clearing a path to a more secure future? *Journal of Environmental Management, 147*, 184–193. https://doi.org/10.1016/j.jenvman.2014.09.003.

Tablado, Z., & Jenni, L. (2017). Determinants of uncertainty in wildlife responses to human disturbance. *Biological Reviews, 92*(1), 216–233.

USEPA (US Environmental Protection Agency) (2007). *Reducing stormwater costs through low impact development (LID) strategies and practices*. US Environmental Protection Agency, EPA 841-F-07-006, Washington, DC.

Vineyard, D., Ingwersen, W. W., Hawkins, T. R., Xue, X., Demeke, B., & Shuster, W. (2015). Comparing green and grey infrastructure using life cycle cost and environmental impact: A rain garden case study in Cincinnati, OH. *Journal of the American Water Resources Association, 51*(5), 1342–1360. https://doi.org/10.1111/1752-1688.12320.

Wood, C. J., Pretty, J., & Griffin, M. (2016). A case-control study of the health and well-being benefits of allotment gardening. *Journal of Public Health (Oxford, England), 38*(3), e336–e344. https://doi.org/10.1093/pubmed/fdv146.

Zgorelec, Z., Bilandzija, N., Knez, K., Galic, M., & Zuzul, S. (2020). Cadmium and mercury phytostabilization from soil using miscanthus× giganteus. *Scientific Reports, 10*(1), 1–10.

Green Infrastructure in Metropolis Dimension: Case Study of Llobregat River, Barcelona Metropolitan Area

Marina Matashova
Andorra-LAB, Forward Consulting Group, Barcelona, Spain

Introduction

The modern dynamics of the metropolis characterizes by the transition from a compact model to a dispersed, discontinuous city (Fig. 1a) (Llop 2008). In this situation, the regeneration of large open spaces in the metropolis that correspond, in many cases, to altered natural accidents allows us to speak about a new form of public space on a metropolitan scale, the socio-ecological landscape, as the integrator that allows "projecting the city from the new model of continuity. Thus, it is no longer a question of the traditional compact city ..., but of a new interpretation of the dispersed city in which the new large open space can allow cohesion, make it clear, be the new strategy that defines the shape of the metropolis" (Fig. 1d) (Battle 2011).

As a result of this paradigm shift, the Llobregat River Park project demonstrates an understanding of the river floodplain as a generator of new public space, integrating urban planning activities and acting on the environment's regeneration in the tens of kilometers long (Fig. 2) (MMAMB 2003). Thus, this disturbed territory, identified for a long time as an infrastructural corridor (Fig. 1), in recent decades has undergone a process social and environmental regeneration – to become a vital component of the Barcelona Metropolis Green Structure.

The concept that brings together the new vocations of the metropolitan open space (ecological, social, territorial) and extreme scales has different synonymous denominations such as "Green Structure," "Green Infrastructure," etc.

According to the definition given by the European Commission, "Green Infrastructures" are not equivalent to "Green Areas," the document emphasizes three main criteria: integration in the

G

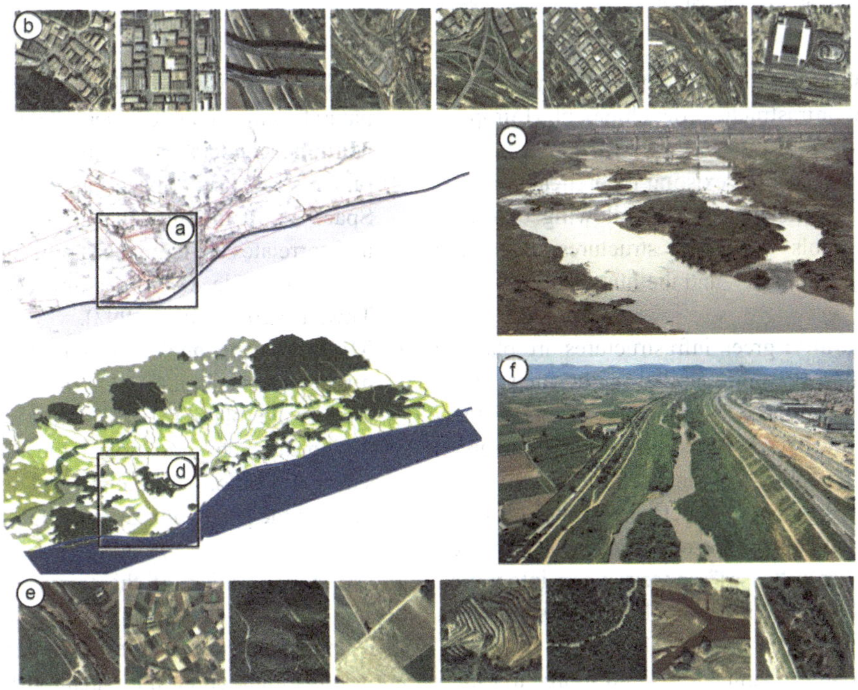

Green Infrastructure in Metropolis Dimension: Case Study of Llobregat River, Barcelona Metropolitan Area, Fig. 1 Low Llobregat, Current State. Own elaboration based on: (**a**) Infrastructure corridor Lower Llobregat in the structure of the Barcelona Metropolis – own elaboration based on A. Font (2004). (**b**) Technogenic and anthropogenic landscapes – source Google Earth. https://earth.google.com. (**c**) Disturbed floodplain landscape – Retrieved from: https://www.amb.cat/. (**d**) Green corridor of Lower Llobregat in the Ecological Matrix of Barcelona – own elaboration based on E. Batlle (2011); (**e**) Natural and agricultural landscapes– source Google Earth. https://earth.google.com. (**f**) Ecological regeneration of a flooded floodplain – Retrieved from: https://www.amb.cat/

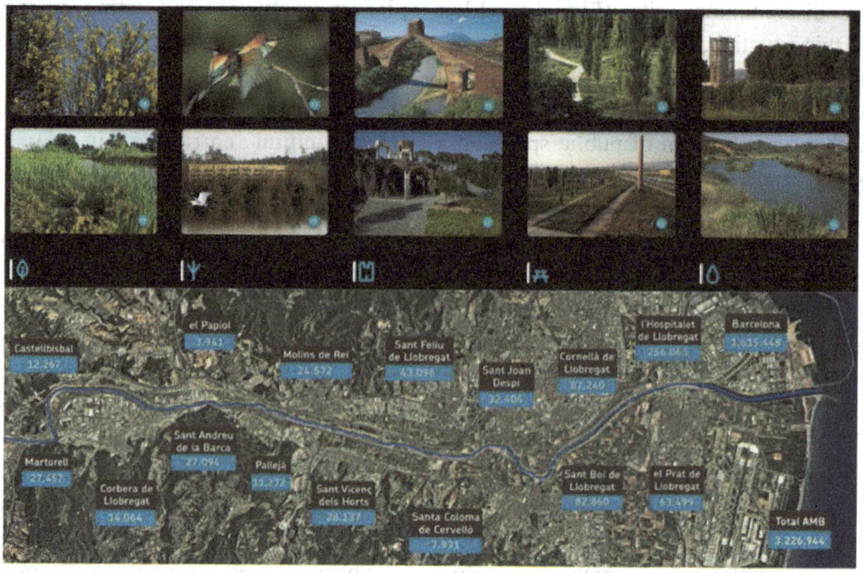

Green Infrastructure in Metropolis Dimension: Case Study of Llobregat River, Barcelona Metropolitan Area, Fig. 2 Llobregat River Park. Retrieved from: https://pemb.cat/. Natural, cultural, historical attractors in the structure of the park. Scheme of walking and cycling routes along the river, indicating the adjacent municipalities and population

urban fabric, environmental function, and its role in social infrastructure development (European Comission 2005).

Summarizing the existing theoretical concepts allows the detection of the following criteria – planning tools – for green structures concerning the territory, the city, and the human scale:

- **Continuity:** green infrastructures are planned intentionally and parallel with other urban infrastructures. The objectives of the continuity of the natural flows restoring and the green spaces systems creating determine the vision from a regional perspective (Forman 1986).
- **Compatibility of vocations** and transition from green "oases" conserving to integrating social and environmental functions in green structures (Orellana 2014).
- **Recycling** marginal areas: a resource to establish green and social infrastructure connectivity (Papers 2008).
- **Soft interventions** and point interventions with a systemic effect: existing structures rethinking with triggering self-regulatory mechanisms of natural components (Clement 2004).
- **Motion scenario** as a way of metropolitan public space use: "Space-Path," "Experienced Space". Soft mobility that imparts coherence and correlates the large metropolis open spaces with the scale of a person (Simkins and Thwaites 2006; Augé 1992).
- **Multiscale approach** – "From satellite to magnifying glass" (Battle 2011): it is necessary to identify the role of the component in the global ecological matrix, to reveal the scenarios of social adaptation at the city level, to take into account the perception from the human scale.

The consideration of the terms "Limit" and "Frontier" in philosophy and urban planning and their application to the case of Lower Llobregat allows us to talk about the transition from a normative approach limiting urban scenarios and natural processes toward the strategic approach in the fluvial areas planning as a frontier – a transitory space between the city and the river and within reach of the interaction synergies of these territories.

However, the morphogenesis of the Lower Llobregat reveals the consolidation process of the following barriers that prevent the accessibility of the Llobregat open space for its public use.

Physical barrier – Embodied in the form of linear infrastructural objects located between the river and adjacent settlements, blocking the continuity of social and natural flows toward the river.

Morphological barrier – Determined by the centripetal logic of the compact municipalities development. As a result, the peripheral riverside areas are excluded from the urban public spaces structure, evolved mainly in the central part.

Mental barrier – Associated with the difficulty of overcoming the inertia of riverside areas perception as a marginal zone and the necessity of new mental patterns grafting redefining its social and environmental value.

The need to resolve these conflicts determines a consideration of the metropolitan open space issue from different points of view: structural, functional, morphological, anthropological.

Structure – Consideration from the flows point of view – strategies that overcome infrastructural barriers, ensuring social accessibility of the riverside territory.

Function – An attempt to overcome the two-dimensional vision of "zoning" and the search for synergy between different zones, options for combining urban development, and public access to the riverside area with measures for the regeneration of nature.

Form/Perception – Morphological and anthropological aspects – forms and types of public spaces in micro, mezzo, and macro scales; strategies for correlating them with the person's scale.

Frontier Sinergies (Structure)

In the current situation characterized by the fragmentation and discontinuity of built-up areas in metropolis, the structural interpretation of the territorial mosaic tiles contributes to promoting soft mobility, giving continuity to the public space of the metropolitan city.

Considering the territory from the point of view of flows, the following flow types can be distinguished: high-speed transit traffic flow in isolated routes; mixed pedestrian and traffic flow in urban fabric; natural – drainage of the territory and biotic flows; and, finally, the type of movement that gives social meaning to the open spaces of the metropolis – soft mobility – pedestrian and bicycle flows in the structure of routes, trails, passages.

According to the English dictionary, "passage" is defined as "a path, channel, or tunnel that allows you to walk along, through or over something" (English Dictionary; Institut per la Ville en movement 2013). This chapter considers the strategies that correspond to these modes applied for soft mobility infrastructure implantation and strategies for overcoming physical barriers.

"Along"

Adapting existing structures for pedestrian use and soft mobility. Implantation of soft mobility infrastructure into territorial structures, which, by definition, have a continuity:

The drainage system of the territory – with the implantation of pedestrian and bicycle routes along the banks of watercourses. The Llobregat river park project thus forms a 40 km long public space – a new version of the central boulevard uniting the municipalities adjacent to the river (Fig. 3a).

Structure of transit transport – integrating a lane for soft mobility in the highway profile. The pedestrianization project for the V-23 highway at the connection to Diagonal Avenue includes soft mobility lanes with landscaping, viewing platforms, and rest areas design. It guarantees pedestrian/cyclist connection between the city center and Llobregat River Park, contributing to its identification as an urban public space (Fig. 3b).

G

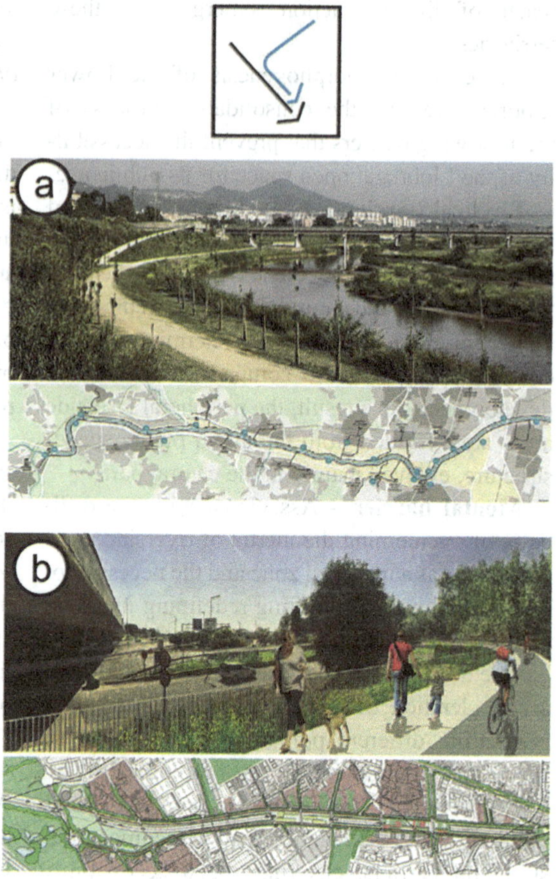

Green Infrastructure in Metropolis Dimension: Case Study of Llobregat River, Barcelona Metropolitan Area, Fig. 3 "Along" logo – own elaboration, M. Matashova (2017). (**a**) Llobregat River Park. Retrieved from: https://pemb.cat/. (**b**) B-23 highway urban intgration and connection with Diagonal street. Retrieved from: https://pemb.cat/

"Through"

A method green axes tracing across the urban settlement through and through, connecting the various urban open spacesin city center, regenerated areas, and metropolitan open spaces out of the consolidated node. Applying this method "Cornellá-Natura" project allowed to restore the connectivity of public spaces, redistribute the recreational intensity, and trigger the mechanism of social adaptation of the urban and natural environment on the city/nature frontier (Fig. 4a).

"Ower"

The method of vertical stratification – involves the formation of bridges at the intersection of the flows and is characterized by the evolution of approaches from the priority of fast and safe movement from A to B point toward the stay on the road scenarios enrichment with speed decrease and soften interventions. In the case of Low Llobregat, this transition is represented by the following space-passages: pedestrian bridges (Fig. 5a), flooded bridges – wade (Fig. 5b), park spaces in flooded areas, where the bridge turns into a resting place and an observation deck, and street furniture and species composition withstand flooding (Fig. 5b).

Integral Vision (Functioning)

There are holistic vision strategies that allow overcoming the limitations of certain types of public spaces that form the modern metropolis:

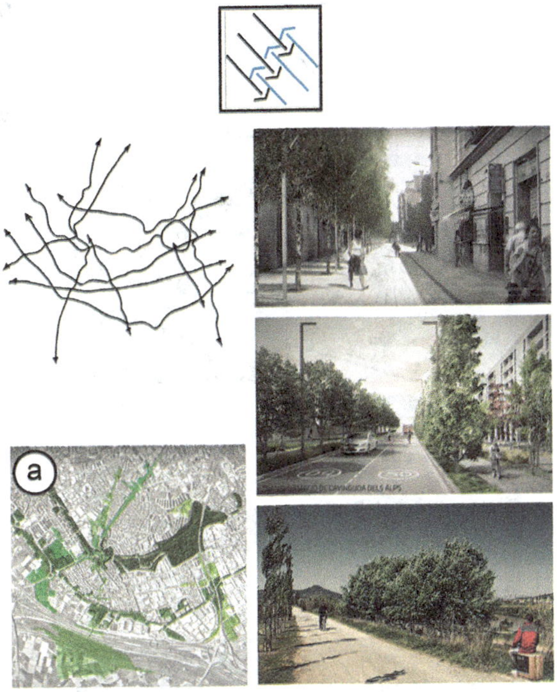

Green Infrastructure in Metropolis Dimension: Case Study of Llobregat River, Barcelona Metropolitan Area, Fig. 4 "Through" logo – own elaboration, M. Matashova (2017). (**a**) "Cornella – Nature" Project Public spaces regeneration and articulation with continuous axes forming. Retrieved from: https://pemb.cat/

- *The discontinuity of the street network outside the compact nuclei between municipalities* (Fig. 6-1),
- *The low intensity of the use of natural and recreational landscapes outside the compact city* (Fig. 6-2),
- *Isolation of public and commercial complexes along transit routes* (Fig. 6-3).

A balanced, integral operation can be achieved through their synergies at an economic, social, and ecological level (Fig. 7).

Method of the Recreational Load Redistribution from the Public Space on the City's Border

Solana Park is the small residential yard on the municipality's border. At the same time, the access from the park to pedestrian bridge connection across the river with the Llobregat River Park makes it possible to compensate for the lack of public spaces in the peripheral zone by implementing sports and recreational scenarios in the structure of metropolitan open space (Fig. 8a).

Method of the Shared Multi-Scale Use of the Highway-Side Shopping and Entertainment Complex

In the case of Cornella municipality, it demonstrates the ability to overcome the isolated and geographic nature of the highway-side commercial nodes. The complex works as an access point to the metropolitan open space, both at the local level – from municipalities and the metropolitan level – through the transit transport network. At the same time, the method allows a new environment to implement some of the complex's recreational outdoor scenarios (Fig. 8b).

Method of Organizing Zones of Ecological Auto-Regeneration with a Public Access Scenario

It involves small-size interventions that trigger self-regulation mechanisms, such as installing

Green Infrastructure in Metropolis Dimension: Case Study of Llobregat River, Barcelona Metropolitan Area, Fig. 5 "Ower" logo – own elaboration, M. Matashova (2017). (**a**) Pedestrian bridge. Retrieved from: https://pemb.cat/. (**b**) Wade. Retrieved from: https://pemb.cat/. (**c**) Observation deck. Retrieved from: https://pemb.cat/

deflectors in a river to form meanders that reduce the flow rate and increase biodiversity. At the same time, the soft mobility infrastructure implantation allows controlling the areas of access depending on the ecologic stability/fragility of the sites (Fig. 8c).

The joint consideration of urban and metropolitan public spaces of the ecological matrix, landscape parks, and urban public spaces allows achieving connectivity and overcoming each type's limitations separately (Fig. 6-4).

Green Infrastructure in Metropolis Dimension: Case Study of Llobregat River, Barcelona Metropolitan Area, Fig. 6 Sectorial Vision – Retrieved from: Google Earth. https://earth.google.com

Public Space on Three Scales (Form and Perception)

Green infrastructure, namely, the park of the Llobregat River (Fig. 6-2), *is a new type of public space in the metropolis, which differs in its appearance from the traditional urban public space* (Fig. 6-1) *and the highway public and shopping complexes* (Fig. 6-3). *Slow mobility shapes it as a space of movement. Thus, the materialization of this type of urbanization is soft interventions that form the infrastructure of access and slow mobility.*

Method of Information Filling of Spaces – Routes

A high density of activity types characterizes the urban public space. The metropolitan open space is characterized by the density of visited environments and a visually perceived kaleidoscope

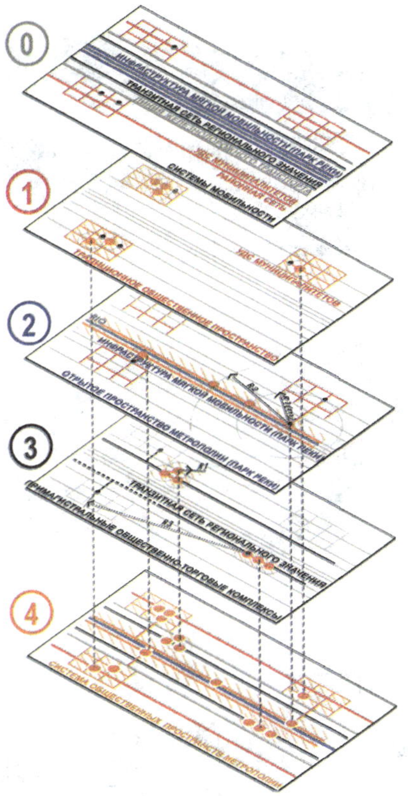

Green Infrastructure in Metropolis Dimension: Case Study of Llobregat River, Barcelona Metropolitan Area, Fig. 7 Integral Vision – own elaboration, M. Matashova (2017)

of landscape components. It implies developing visitors' "reading" skills to perceive recreational, visual, and cultural resources. The Project of Design and Implantation of Signage in the Llobregat River Area aiming to develop an information-filled space. The basis of the Project is cartography: the intervals of signage elements implantation is determined taking into account the routs and speed of a pedestrian, runner, cyclist. The signage informs about access to the park from municipalities, highways, and parking lots to thematic routes, allow access to virtual reality. Thus, a mental landscape map with longitudinal and transverse connections is formed, inviting to "wander" as if through the city's streets but in a larger and more ecological context.

Method of Overlay Scenarios: From Daily Recreation to Vacation Leisure, Park – Transition

The adjoining residential park, connected to the park of the Llobregat River through a pedestrian bridge, allows the implementation of various scenarios and routes: adjoining residential space for low-mobility groups and daily recreation; trails for daily sports and theme routs for a weekend getaway in the park of the Llobregat River (Fig. 8a).

Method of the Implementation of Specific Functional Programs in Open Access Spaces

Access to metropolitan open spaces provides a new spatial resource for sports centers and social infrastructure facilities at the city border. Furthermore, it allows a broader range of new open-air scenarios: regular sports activities (running, trekking, cycling), competitions, festival programs that increase the variety of recreational use and contribute to overcoming the mental barrier in the social adaptation of the metropolitan open spaces (Fig. 8c).

Method of Access Without Approximation

This provides a range of visual resources accessible from viewpoints. Four observation towers are integrated into the structure of Llobregat Park. These vertical dominants provide a transition to a different scale of human perception of a large open space of the metropolis, which contributes to the identification of the space, and at the same time allows one to assess the natural and aesthetic qualities of the landscape at a distance, and thus replace physical access to a number of ecologically sensitive zones with visual, and also create additional visual reference points that structure extended distances (Fig. 8c).

The public space organized in the structure of natural landscapes of the metropolis is not characterized by a clear functional and planning structure typical for the traditional type of urban public

Green Infrastructure in Metropolis Dimension: Case Study of Llobregat River, Barcelona Metropolitan Area, Fig. 8 Green Infrastructure – Retrieved from: https://pemb.cat/

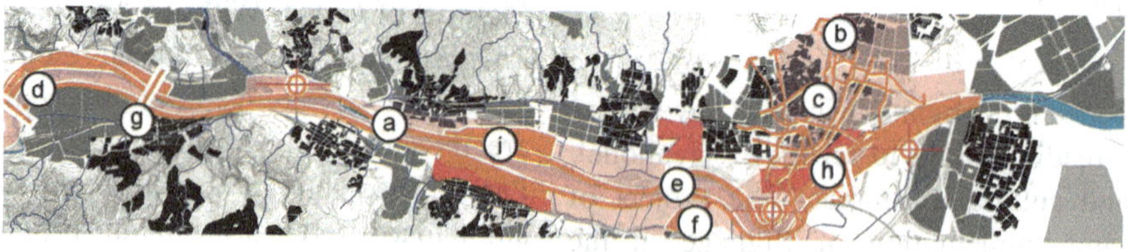

Green Infrastructure in Metropolis Dimension: Case Study of Llobregat River, Barcelona Metropolitan Area, Fig. 9 Own elaboration, M. Matashova (2017)

space. However, interesting synergies arise between them – on the one hand, compact cores form access points to the natural and recreational landscape, but on the other hand get a new varied environment, new content for sports and leisure scenarios (Fig. 9).

The discourse performed shows how, through the strategies, criteria, and methods discussed, Lower Llobregat is becoming a vital component of the green infrastructure of the Barcelona Metropolis.

Cross-References

- ▶ Mainstreaming Blue Green Infrastructure in Cities: Barriers, Blind Spots, and Facilitators
- ▶ Multiple Benefits of Green Infrastructure

References

Augé, M. (1992). *Non-places: Introduction to an anthropology of supermodernity.* Le Seuil, Verso, p. 122.

Battle, E. (2011). *El jardín de la metrópoli. Del paisaje romántico al espacio libre para una ciudad sostenible.* Barcelona: Gustavo Gili.

Clement, G. (2004). *Manifeste du Tiers Paysage.* Paris: Sujet/Objet. 48 p.

English Dictionary. Retrieved from https://www.vocabulary.com/dictionary/passage

Font, A. (Ed.). (2004). *La Explosión de la Ciudad.* Colegio de Arquitectos de. Cataluña. COAC & Forum Universal de la Culturas de Barcelona.

Forman, R. (1986). *Landscape ecology.* Nueva York: Wiley.

Institut pour la ville en mouvement. (2013). Le passage un lien, un lieu, Projet de passages rapport Phase 1, p. 16. Retrieved from: https://www.ville-en-mouvement.com/fr/content/un-lien-un-lieu.

Llop, C. (2008). Paisatges metropolitans: Policentrisme, dilatacions, multiperifèries i microperifèries. Del paisatge clixé al paisatge calidoscopi, Papers no. 47, pp. 8–13.

Matashova, M. (2017). *Del Corredor Infraestructural al Eje De Infraestructura Verde: Análisis de las Intervenciones Proyectuales Realizadas en Bajo Llobregat 2001–2017.* Barcelona: Tesina del Master, UPC.

MMAMB. (2003). Projecte-marc de recuperacio ambiental de l'espai fluvial del Riu Llobregat. http://www3.amb.cat/projectes/projecte_marc_llobregat.pdf.

Orellana, D. (2014). *Redes simples, redes complejas y matrices de tejidos continuos: evolución de la valoración de los espacios abiertos en cuatro casos de estudio: Maryland, Londres, Bolonia y Barcelona.* Barcelona: UPC Tesina Máster Universitario en Urbanismo.

Papers, num 47. (2008). El reto del paisaje en ámbitos metropolitanos. Barcelona.

PEMB. Recuperació social i ambiental de l'espai fluvial del riu Llobregat en el tram metropolità. Retrieved from: https://pemb.cat/

Simkins, I., & Thwaites, K. (2006). *Experiential landscape: An approach to people, place and space.* Nueva York: Routledge. 256 p.

Werquin, A. C., et al. (2005). *Green structure and urban planning. Final report, COST Action C11.* Brussels: European Comission. http://www.greenstructureplanning.eu/COSTC11-book/pdfs/a-Intro.pdf.

Green Infrastructure Planning

▶ Overcoming Barriers in Green Infrastructure Implementation

Green Justice

▶ Public Policies to Increase Urban Green Spaces

Green Spaces

▶ Multiple Benefits of Green Infrastructure

Green Strip

▶ Green Belts

Green Wedge

▶ Green Belts

Greenbelts

▶ Planning for Peri-urban Futures

Greenhouse Gas Emissions (GHG)

▶ The State of Extreme Events in India

Greenspace

▶ Planning for Peri-urban Futures

Greenspace also known as green spaces or parks

▶ Need for Greenspace in an Urban Setting for Child Development

Gross Domestic Product (GDP)

▶ The State of Extreme Events in India

Gross State Domestic Product (GSDP)

▶ The State of Extreme Events in India

Groundwater

▶ Smart Agriculture and ICT

Growth

▶ Furthering the Sustainable Development Agenda by Putting Urban Heritage and Value Extraction at the Center

Growth – Development

▶ Growth, Expansion, and Future of Small Rural Towns
▶ Harare

Growth, De-growth, and Nature-Based Solutions

Shama Perveen
Senior Manager (Water), Ceres, Boston, MA, United States

Synonyms

Blue/green infrastructure; Ecological engineering; Ecosystem services; Ecosystem-based adaptation; Ecosystem-based approach

Definition

Nature-based solutions are defined as actions to protect, sustainably manage, and restore natural or modified ecosystems that address societal challenges effectively and adaptively, simultaneously providing human well-being and biodiversity benefits (IUCN 2020).

Growth, Biophysical Limits and De-growth

We are literally producing nature while we are actively destroying it (Redclift 1987). Many of our discussions on natural resource management are based on society's narrow perspective of *preservation* versus *use*. The Enlightenment thought and ideals that emerged in the seventeenth and eighteenth centuries were based on the paradigms of advancement, progress and rational thought – where nature was used as a force to be dominated and used. In the footsteps of the Enlightenment, the Industrial era further intensified human influence upon the Earth system with humans shaping the biosphere considerably. Between 1800 and 2000, population grew more than sixfold, the global economy about 50-fold, and energy use about 40-fold (McNeill 2001). Land degradation, deforestation, agricultural intensification, and large scale transformation of the hydrological cycle with accelerating number of large dams became the norm.

Intricate, yet often unexamined relationships, exist among our concepts of how we as a society treat nature, approach it from a scientific perspective, and organize our political and social institutions (Bird 1987). In the 1960s, around the time when modern environmental movement was shaping up, a number of critical authors started pointing at runway economic growth as the underlying cause of the ever-increasing environmental problems. Hardin's *Tragedy of Commons* (1968), Boulding's concept of *Spaceship Earth* (1966), Georgescu-Roegen's *Entropy Law and Economic Process* (1971), and Meadows et al.'s *Limits to Growth* (1972, 2004) – all raised concern at nature's ability to sustain the rapidly growing population by continuing to provide the services required to maintain viable human civilizations. In the influential *Tragedy of Commons* discussion, Hardin argues that people are incapable of putting "collective" interests before "private" ones; as a result of which, the resource base is constantly under threat.

The much acclaimed but often criticized *Limits to Growth* model specifically explored the impacts of unbridled use of natural resources and the limits it would impose if exponential growth rates in population and industrial development continued. A common critique of the model was that the estimates were pessimistic and failed to take into account the potential for technological, social, and institutional innovation thus predicting doom and gloom. Mainstream economists also refuted the glum estimates vehemently. Deriving advantage from these discussions, O'Riordan's (1981) *ecocentric* theory proposed that the problem is *not* the balance between the population and resources per se but the *ends* to which resources are put in the pursuit of economic growth. The ecocentric perspective takes issue with the *objectives* of development as well as the *means* – essentially suggesting that not only have we lost our respect for Nature, we have also reached the biophysical limits.

Empirical models subsequently developed to counteract the *Limits to Growth* claims predicted optimistic scenarios but failed to take into account the dynamic and coupled nature of human and environmental systems (Smith and Hargroves 2007). Globally, evidence suggests that natural and social systems behave in nonlinear ways, exhibit marked thresholds in their dynamics, and act as strongly coupled, complex and evolving integrated systems (Liu et al. 2007). The management of critical natural capital was then posited as an important objective of *Sustainable Development* (WCED 1987; Brand 2009). However, given the incompatible views between environment and development; "strong" and "weak" definitions of sustainability were put forth (Neumayer 2003). When natural capital can be substituted by human capital it is called weak sustainability. Strong sustainability, on the other hand, demonstrates that natural capital cannot be viewed as a mere stock of resources but a set of complex systems with biotic and abiotic elements constantly evolving and interacting in ways that determine the ecosystem's capacity to provide a wide array of functions and services (Ekins et al. 2003). The underlying tenets of sustainable development have, however, never really been openly challenged and discussed and the term is often viewed as an oxymoron by some or an attempt to sustain the unsustainable by others.

Rockstrom et al. (2009) proposed a new approach to global sustainability – the concept of *planetary boundaries* as a safe operating space for humanity with respect to functioning of the Earth system. A key aspect of the planetary boundaries concept is that they are inter-related, i.e., transgression in one (e.g., biodiversity, freshwater flows, soil quality) can trigger an abrupt system change when thresholds are crossed (Scheffer et al. 2012). As human actions render ecosystems unable to provide the services, the negative shifts represents loss of resilience. The *threshold* is the point at which a system has lost enough resilience and reorganizes into a new regime characterized by a different set of processes and structures (Groffman et al. 2006).

Five decades since the *Limits to Growth* predictions and the debate on ecosystem limits in a physically finite planet continues. Despite the continued technological optimism, the ecosystems and the services it provides, continues to be the prime contributor to growth. In this context however, the normative concept of "growth" as interpreted and used differently by businesses, governments, economists and environmentalists needs to be examined. While governments and businesses interpret growth in *economic* terms, i.e., financial gains; Environmentalists interpret growth to be *physical* correlating it with resource depletion and environmental damage. Daly (1977) proposes that if we use "growth" to mean quantitative change and "development" as qualitative, then a sustainable steady state economy (SSE) develops where throughput is maintained within ecological limits.

In recent debates on environmental problems and policies thus, the strategy of "de-growth" has appeared as an alternative to the paradigm of economic growth since endless economic growth is not only ecologically unsustainable but also, after a threshold, undesirable. Schneider et al.

(2010) have defined de-growth as, "An equitable downscaling of production and consumption that increases human well-being and enhances ecological conditions." While these authors maintain that de-growth is a *process* whose end goal is a SSE; Kerschner (2010) has explored the relationship between ideas of de-growth, SSE and "strong" sustainability in detail, and concludes that these concepts are complementary. He further argues that countries in the global North must follow a de-growth path to reach a SSE and provide the environmental space needed for the global South to follow a path of decelerating growth. O'Neill (2012) has proposed a framework with biophysical and social indicators in order for countries to determine if de-growth is occurring or how close they are to SSE. More recently, de-growth scenarios are being explored for single countries (D'Alessandro et al. 2020) and bottom-up studies show that high living standards can be maintained with substantially less per capita energy use than currently consumed in affluent countries (Keyßer and Lenzen 2021; Millward-Hopkins et al. 2020).

With natural systems deteriorating beyond a point of no return, an increasing number of perspectives have reflected an anthropocentric view of the management of natural resources, including biodiversity and the environment, with a focus on the benefits that nature may provide for humans. Concepts like Ecosystem based Adaptation (EbA), natural capital, ecosystem services, and nature-based infrastructure have emerged at the center of this narrative. The most recent entry to this discourse is "Nature Based Solutions" (NBS), a concept introduced specifically to promote nature as a means for providing solutions to climate mitigation and adaptation challenges (IUCN 2020). These approaches recognize the complexity of socio-ecological systems and the fact that they are dynamic, and allows for self-reorganization. Tradeoffs, however, can arise depending on the extent and the rate at which ecological sustainability is achieved vis-a-vis other objectives. Alternatively known as synergy, wherein the interaction between one or two actions can lead to an impact greater or less than the sum of individual effects, the tradeoffs can be positive or negative (Mainali et al. 2018). The challenge is to use the gift of our intelligence to find the right relationship to the natural world.

Nature Based Solutions as "No Regret" Options

The term "nature-based solutions" (NBS) was first coined in the late 2000s (MacKinnon et al. 2008) in the context of finding new solutions to mitigate and adapt to climate change effects while simultaneously protecting biodiversity and improving sustainable livelihoods. It marks a perspective shift in the relationship between people and nature, putting explicit emphasis on solutions that use nature to tackle major societal challenges such as climate change (Eggermont et al. 2015). With drivers including the Bonn Challenge, New York Declaration on Forests, Aichi Biodiversity Targets and the CBD post-2020 Biodiversity Framework, more and more countries are pledging to protect and restore their natural ecosystems. NBS represent a powerful way of allowing nations to deliver on international commitments, meet the Sustainable Development Goals with limited finance, and ultimately achieve sustainable and equitable development in a warming world (Martin et al. 2021).

To date, several definitions have been applied to describe NBS (Cohen-Shacham et al. 2016). They are systemic interventions that bring more, and more diverse, nature and natural features and processes and can help address the impacts from climate change, and reverse biodiversity decline and ecosystem service loss (Seddon et al. 2021). One of the most common and widely used definitions of NBS comes from the International Union for Conservation of Nature (IUCN 2016, 2020): "Nature-based Solutions are defined as actions to protect, sustainably manage, and restore natural or modified ecosystems that address societal challenges effectively and adaptively, simultaneously providing human well-being and biodiversity benefits." For the European Commission (2015), NBS are actions inspired by, supported by, or copied from nature and aims to help societies address a variety of environmental, social, and economic challenges in sustainable ways. NBS has evolved from, and are strongly connected to ideas such as natural systems agriculture, ecosystem based adaptation, blue/green infrastructures, natural solution, agroecology, ecosystem-based approaches, ecosystem services, ecological engineering and similar others. While these concepts are all complementary with considerable overlap; NBS can be considered as an umbrella concept with a strong solution-oriented focus and biodiversity lying at its core (Seddon et al. 2019b).

As evidence of the efficacy of NBS as "no-regret" options continues to grow, ecosystems are receiving increasing attention in the international policy fora. The *no-regret* approach is essentially an important part of EbA planning and focuses on maximizing positive and minimizing negative aspects of nature based adaptation strategies (IUCN 2014). The World Bank, UN Development Programme, UN International Strategy for Disaster Relief, and Global Facility for Disaster Reduction and Recovery and others in the development community are all promoting a *no-regrets* approach to climate change and disaster management (Siegel 2002). They are actions taken by households, communities, and institutions and can be justified from economic, social, and environmental perspectives. No-regret actions increases resilience, and the ability of the system to deal with different types of hazards in a timely, efficient, and equitable manner. Specifically targeting poor and at-risk individuals and communities, NBS are flexible no regret options, adept in dealing with and providing a proactive defense against a constantly changing climate and supporting adaptation over the long term. Going beyond the traditional biodiversity conservation and management principles, NBS re-focuses the debate on humans by specifically integrating factors such as poverty alleviation, livelihood, well-being and governance principles.

Conceived as a powerful ally to address the impending global challenges, NBS is suggested to be more sustainable and cost effective compared to the traditional grey engineering solutions and weaves natural features or processes into the built environment to promote adaptation and resilience. Although more research is needed, a rapidly growing evidence base demonstrates that well designed NBS can indeed deliver multiple benefits (Chausson et al. 2020). It is now well documented that watershed protection, and optimizing the performance of traditional infrastructures like dams and levees can not only generate income for local communities but also help meet other critical societal needs like providing green jobs, increased property values, and improvements to public health, including better disease outcomes and reduced injuries and loss of life (FEMA 2021).

Current national policies informing on carbon dioxide emissions have the world on track for 2.3–4.1 °C by 2100 (https://climateactiontracker.org); which may soon require considerably deeper cuts if we are to avoid catastrophic climate impacts. To directly address the interdependent factors that are driving these major global challenges, NBS needs to play a central role. Various studies confirm that NBS could provide around 30% of the cost-effective mitigation that is needed by 2030 to stabilize warming to below 2 °C (Griscom et al. 2017). The Paris Agreement explicitly recognizes nature's role in helping us deal with climate change, calling on all Parties to acknowledge the importance of ensuring the integrity of all ecosystems, including oceans, and the protection of biodiversity. From comparative analyses of the Nationally Determined Contributions (NDCs) submitted to UNFCCC by signatories of the Paris Agreement, it seems that the majority of world's nations (66%) include NBS in some form or other to meet the climate mitigation/adaptation goals (Seddon et al. 2019a).

More recently, NBS have been highlighted in a number of global assessment reports and are a focus of a growing number of major new programs implemented by governments as well as private sector institutions (https://www.nbspolicyplatform.org/). The Leaders Pledge for Nature (http://www.leaderspledgefornature.org/) spearheaded by the UK government, commits signatories to cooperating and holding one another to account in their joint mission to reverse biodiversity loss by 2030. Parallely, hundreds of private sector organizations have made commitments and have formed a coalition called Business for Nature (https://www.

businessfornature.org) to commit to and call on governments to adopt policy to support NBS and help reverse nature loss. Several different international platforms aligned with the Sustainable Development Goals have been formed (e.g., UN Global Compact's CEO Water Mandate; Ceres' Ag water challenge, Act4nature International, 1t.org of World Economic Forum) to help the private sector to reduce the most significant impact of their businesses. This positive framing of nature is strikingly different from past interpretations of nature as an enemy of civilization, something to be tamed and cultivated in order to be useful for human well-being.

Way Forward: Making Vulnerable Ecosystems Resilient

The ubiquitous debate between those who doubt that unlimited economic growth is possible on a finite planet, and those who believe that human ingenuity will eventually overcome all potential limitations to economic growth is ongoing. Recent warnings, however, confirm continued and alarming trends of environmental degradation from human activity, leading to profound changes in essential life-sustaining functions of the Earth, thus calling for a reassessment of the role of growth-oriented economies and the pursuit of affluence (Ripple et al. 2017). The implications of consumption on scarce resources has been confirmed by many through indicators as varied as CO_2 emissions, raw materials input, air pollution, biodiversity loss, nitrogen emissions, and scarce water-energy use. Driven mostly by economic growth, the loss of biodiversity continues unabated as human appropriation of net primary production of biomass increases, and the world input of materials to the economy continues to peak.

Failure to adapt to the effects of climate change is being flagged to be the single greatest threat to health, wealth and well-being around the world (WEF 2019). The ambitious 1.5 °C scenarios (i.e., limiting the temperature increase to 1.5 °C above pre-industrial levels) reported by the Intergovernmental Panel on Climate Change (IPCC 2021) rely on combinations of negative emissions and unprecedented technological change, while assuming continued growth in gross domestic product (GDP). Deemed necessary to support societal well-being, continued GDP growth basically entails increased material and energy consumption and therefore prolonged mitigation challenges. Keyßer and Lenzen (2021) have done a comparative analysis of IPCC's integrated modeling versus the de-growth scenarios and find that even though the political challenges regarding the feasibility of implementing de-growth strategies are substantial, the de- growth scenarios minimize many risks compared to technology-driven pathways. In many scenarios, net negative emissions, achieved either through reforestation and other land-based "natural climate solutions" (Griscom et al. 2017) or negative emission technologies (Rogelj et al. 2019), are required after 2050 to bring the climate back from an overshoot over the climate-change mitigation targets to the specified target level. Recently, the potential of NBS to meet global goals for the reduction in greenhouse gas emissions has been publicized considerably, reflecting the fundamental importance of ecosystems as sources and sinks for CO_2.

Encompassing a broad range of practices, NBS are designed either as a stand-alone measure or in conjunction with grey infrastructure and/or other types of strategies, for example regional or watershed planning, policy making, or economic development. Ecosystem based Adaptation (EbA) is a nature based approach that uses biodiversity and services to help people adapt to the adverse effects of climate change and has significant secondary and long-term benefits. GIZ launched a global project "Mainstreaming EbA – Strengthening Ecosystem-based Adaptation in planning and decision-making processes", to strengthen the ability of decision-makers at international, national and local level to mainstream EbA into policy and planning processes (GIZ et al. 2018). And although there are hundreds of tools and methodologies available to support the integration of EbA into adaptation strategies, information about these tools and instructions on using them are not easily accessible. To fill this gap, a searchable database of over 240 EbA Tools and methods has been developed by IIED, UNEP-WCMC et al. (2019) to help practitioners and policy makers include EbA into climate adaptation planning. In addition to containing tools and methods specifically designed for EbA, the content draws on a variety of disciplines, including climate change adaptation, biodiversity conservation and human development.

What all NBS have in common is that they seek to maximize the ability of nature to provide ecosystem services that help address a human challenge, such as climate change adaptation, food production or disaster risk reduction. While the evidence base is still developing, it is clear that nature-based and hybrid solutions often provide low-risk low-cost solutions protecting against multiple climate related hazards and offer advantages over engineered solutions (Keestra et al. 2018). In contrast, NBS can be more affordable and, if properly implemented, may bring a wide range of ecosystem services especially for the more vulnerable sectors of society. Crucially, ecosystems are more dynamic in that they "leave room for selfreorganization and mutability" and in this way may be more able to withstand and/or adapt to perturbation (i.e., are more resilient) compared to static engineered interventions. There is also growing evidence of the economic benefits of maintaining natural habitats through avoided losses from climate change related disasters. For example, in one study it was found that the presence of intact coastal wetlands in northeast USA protected US $625 million worth of property from direct flood damage during Hurricane Sandy, reducing damages by 20–30% in 50% of affected areas (Narayan et al. 2017). Another recent international analysis revealed that annual expected damages from flooding would double and costs from frequent storms would triple in the absence of reefs globally (Beck et al. 2018).

NBS for sustainable development are not new and there are many examples around the world but mainstreaming, accelerating and scaling up progress in implementing the NBS remains a challenge. This is to some extent the result of lack of knowledge of their benefits and limitations, but more importantly of the challenges of bringing multiple stakeholders together to agree on integrated solutions and developing innovative financing mechanisms to implement such solutions successfully on a wider scale (UNE-DHI et al. 2018). Seddon et al. (2021) have discussed the potentials and pitfalls of NBS framing and urged policymakers, practitioners and researchers to urgently consider the synergies and trade-offs associated with the prioritized ecosystem services under selected NBS. For instance, the Grain-for-Green program in China succeeded in rapidly increasing tree cover to restore degraded agricultural soils, but used mainly fast growing non-native species that have reduced water supply and also resulted in a decrease of 6% in native forest cover as farming was displaced to new areas (Chausson et al. 2020). Complex tradeoffs arising from an investment or policy decision requires more research and assessments to determine the full suite of resulting impacts at multiple geographic scales. Ohashi et al. (2019) found that while afforestation and bio energy carbon capture and storage can cause local biodiversity loss in certain regions, it can achieve net biodiversity benefits at the global level through its contribution to mitigating climate change.

Environmental management is usually a responsive set of techniques rather than a framework for implementing policies proactively. Sustainable development requires a broader view of both economics and ecology than most practitioners in either discipline are prepared to admit, together with a political commitment to ensure that the development is "sustainable" (Redclift 1987). Lately, the growing recognition that GDP is a poor indicator of prosperity and progress has led to a number of major projects globally to investigate alternatives to GDP – those that are more inclusive of environmental and social aspects of progress. These include the European Commission's *Beyond GDP initiative* (http://www.beyond-gdp.eu/), the OECD's project on *Measuring Well Being and Progress* (https://www.oecd.org/statistics/measuring-well-being-and-progress.htm) including the Commission on the Measurement of Economic Performance and Social Progress, World Economic Forum's *Beyond GDP Initiative* (https://www.weforum.org/focus/beyond-gdp) and the World Happiness Report (https://worldhappiness.report/). A number of governments have already moved in this direction and Britain's *National Wellbeing Program* is a leading example of beyond GDP measurement (ONS 2018). It is believed that the shift to new prosperity indicators could bring about transformative impacts directing societies toward sustainability, less consumption-intensive sources of wellbeing, and greater equity.

In efforts to find solutions for a just and a climate resilient transition, the traditional paradigm of growth is being critically (re)evaluated. The impending climate crisis and the unprecedented global crisis brought on by the COVID-19 pandemic adds further to the concern. As scientists draw clear connections between economic growth and environmental degradation, with distinct implications on human health, a systemic change is needed. However, as Harding (1968) has warned – in natural resource policy, no effort aimed at conservation, sustainability or environmental protection, will come without some unintended consequences. While environmental, health and well-being considerations should play a critical role in the choice of a technological option; that is seldom the case. Nature based solutions forms part of an integrated sustainability strategy cutting across sectors and bringing together different forms of transdisciplinary knowledge, with the deciding element being "constraint" rather than "indulgence" to limit the ecological breakdown.

Cross-References

- ► Circular Economy and the Water-Food Nexus
- ► Emerging Concepts Exploring the Role of Nature for Health and Well-Being
- ► From Vulnerability to Urban Resilience to Climate Change
- ► Green and Blue Infrastructure (GBI) in Urban Areas
- ► Green Cities: Nature-Based Solutions, Renaturing and Rewilding Cities
- ► Green Infrastructure
- ► Innovation to Bring Nature-Based Solutions to Life: Tales of Two Cities
- ► Multiple Benefits of Green Infrastructure
- ► Nature-Based Solutions for River Restoration in Metropolitan Areas
- ► Policy and Practices of Nature-based Solutions to Build Resilience in Seoul, Korea
- ► Urban Ecosystem Services and Sustainable Human Well-being

Disclaimer: The views and opinions expressed in the article are those of the author and do not reflect the views of the organization they are affiliated with.

References

Beck, M. W., Losada, I. J., Menéndez, P., Reguero, B. G., Díaz-Simal, P., & Fernández, F. (2018). The global flood protection savings provided by coral reefs. *Nature Communications, 9*, 2186.

Bird, E. A. R. (1987). The social construction of nature: Theoretical approaches to the history of environmental problems. *Environmental Review, 11*, 255–264.

Boulding, K. (1966). The economics of the coming spaceship Earth. In H. Jarrett (Ed.), *Environmental quality in a growing economy* (pp. 3–14). Baltimore: Resources for the Future/Johns Hopkins University Press.

Brand, F. (2009). Critical natural capital revisited: Ecological resilience and sustainable development. *Ecological Economics, 68*(3), 605–612.

Chausson, A., Turner, C. B., Seddon, D., Chabaneix, N., Girardin, C. A. J., Key, I.,. . . Seddon, N. (2020). Mapping the effectiveness of nature-based solutions for climate change adaptation. *Global Change Biology, 26*, 6134–6155.

Cohen-Shacham, E., Walters, G., Janzen, C., & Maginnis, S. (Eds.). (2016). *Nature-based solutions to address global societal challenges*. Gland: IUCN. xiii + 97 pp.

D'Alessandro, S., Cieplinski, A., Distefano, T., & Dittmer, K. (2020). Feasible alternatives to green growth. *Nature Sustainability, 3*, 329–335.

Daly, H. E. (1977). *Steady-state economics: The economics of biophysical equilibrium and moral growth*. San Francisco: W. H. Freeman.

Eggermont, H., Balian, E., Azevedo, J. M., Beumer, V., Brodin, T., & Claudet, J. (2015). Nature-based solutions: New influence for environmental management and research in Europe. *GAIA – Ecological Perspectives for Science and Society, 24*(4), 243–248.

Ekins, P., Simon, S., Deutsch, L., Folke, C., & De Groot, R. (2003). A framework for the practical application of the concepts of critical natural capital and strong sustainability. *Ecological Economics, 44*, 165–185.

European Commission (EC). (2015). Towards an EU research and innovation policy agenda for nature-based solutions and re-naturing cities. *Final report of the Horizon 2020 expert group on 'Nature-based solutions and re-naturing cities'*. Publications Office of the European Union.

FEMA. (2021). *Building community resilience with nature-based solutions. A guide for local communities. June*. Available online: https://www.fema.gov/sites/default/files/documents/fema_riskmap-nature-based-solutions-guide_2021.pdf

Georgescu-Roegen, N. (1971). *The entropy law and the economic process*. Cambridge: Harvard University Press.

GIZ, EURAC, & UNU-EHS. (2018). *Climate risk assessment for ecosystem-based adaptation – A guidebook for planners and practitioners*. Bonn: GIZ.

Griscom, B. W., Adams, J., Ellis, P. W., Houghton, R. A., Lomax, G., Miteva, D. A., ... Fargione, J. (2017). Natural climate solutions. *Proceedings of the National Academy of Sciences, 114*, 11645–11650.

Groffman, P. M., Baron, J. S., Blett, T., Gold, A. J., Goodman, I., Gunderson, L. H., Levinson, B. M., Palmer, M. A., Paerl, H. W., Peterson, G. D., Poff, N. L., Rejeski, D. W., Reynolds, J. F., Turner, M. G., Weathers, K. C., & Wiens, J. (2006). Ecological thresholds: The key to successful environmental management or an important concept with no practical application? *Ecosystems, 9*, 1–13.

Hardin, G. (1968). Tragedy of commons. *Science, 162*(3859), 1243–1248.

IPCC. (2021). *Climate change 2021: The physical science basis. Contribution of Working Group I to the sixth assessment report of the Intergovernmental Panel on Climate Change*. Cambridge University Press (in press).

IUCN. (2014). *Ecosystem based adaptation: Building on no regret adaptation measures*. Technical Paper, 20th session of the Conference of the parties to the UNFCCC and the 10th session of the Conference of the Parties to the Kyoto Protocol, Lima, 1–12 Dec 2014.

IUCN. (2016). *Resolution 6.069: Defining nature-based solutions, WCC-2016-Res-069*. Gland: IUCN.

IUCN. (2020). *Global standard for nature-based solutions. A user-friendly framework for the verification, design and scaling up of NBS* (1st ed.). Gland: IUCN.

Keesstra, S., et al. (2018). The superior effect of nature based solutions in land management for enhancing ecosystem services. *Science of the Total Environment, 610*, 997–1009.

Kerschner, C. (2010). Economic de-growth vs. steady state economy. *Journal of Cleaner Production, 18*, 544–551.

Keyßer, L. T., & Lenzen, M. (2021). 1.5°C degrowth scenarios suggest the need for new mitigation pathways. *Nature Communications, 12*, 2676.

Liu, J., Dietz, T., Carpenter, S. R., Folke, C., Alberti, M., Redman, C. L., ... Provencher, W. (2007). Coupled human and natural systems. *Ambio, 36*, 639–649.

MacKinnon, K., Sobrevila, C., & Hickey, V. (2008). *Biodiversity, climate change and adaptation: Nature-based solutions from the Word Bank Portfolio*. Washington, DC: World Bank. (as cited in Eggermont et al. 2015).

Mainali, B., Luukkanen, J., Silveira, S., & Kaivo-oja, J. (2018). Evaluating synergies and trade-offs among sustainable development goals (SDGs): Explorative analyses of development paths in South Asia and Sub-Saharan Africa. *Sustainability, 10*(3), 1–25.

Martín, E. G., Costa, M. M., Egerer, S., & Schneider, A. (2021). Assessing the long-term effectiveness of nature-based solutions under different climate change scenarios. *Science of the Total Environment, 794*, 148515.

McNeill, J. R. (2001). *Something new under the sun*. New York/London: W.W. Norton, 416 pp. (as cited in Steffen et al. 2007).

Meadows, D. H., Meadows, D. L., Randers, J., & Behrens, W. (1972). *The limits to growth*. London: Pan.

Meadows, D. H, Meadows, D. L, & Randers, J. (2004). A synopsis of limits to growth: The 30 year update. Chelsea Green Publishing Company. pp 338.

Millward-Hopkins, J., Steinberger, J. K., Rao, N. D., & Oswald, Y. (2020). Providing decent living with minimum energy: A global scenario. *Global Environmental Change, 65*, 102168.

Narayan, S., Beck, M. W., Wilson, P., Thomas, C. J., Guerrero, A., Shepard, C. C., ... Trespalacios, D. (2017). The value of coastal wetlands for flood damage reduction in the Northeastern USA. *Scientific Reports, 7*, 9463.

Neumayer, E. (2003). *Weak versus strong sustainability: Exploring the limits of two opposing paradigms*. Northampton: Edward Elgar.

O'Neill, D. W. (2012). Measuring progress in the degrowth transition to a steady state economy. *Ecological Economics, 84*, 221–231.

O'Riordan, T. (1981). *Environmentalism*. London: Pion.

Office of National Statistics (ONS). (2018). Measuring national well-being: Quality of life in the UK. April. Available online: https://www.ons.gov.uk/peoplepopulationandcommunity/wellbeing/articles/measuringnationalwellbeing/qualityoflifeintheuk2018

Ohashi, H., Hasegawa, T., Hirata, A., Fujimori, S., Takahashi, K., Tsuyama, I., ... Matsui, T. (2019). Biodiversity can benefit from climate stabilization despite adverse side effects of land-based mitigation. *Nature Communications, 10*, 5240.

Redclift, M. (1987). Sustainable Development: Exploring the Contradictions. Methuen, London. pp 217. https://doi.org/10.4324/9780203408889

G

Ripple, W. J., et al. (2017). World scientists' warning to humanity: A second notice. *Bioscience, 67*, 1026–1028.

Rockström, J., Steffen, W., Noone, K., Persson, A., Chapin, F. S., Lambin, E., . . . Foley, J. (2009). Planetary boundaries: Exploring the safe operating space for humanity. *Ecology and Society, 14*(2): 32.

Rogelj, J., Huppmann, D., Krey, V., et al. (2019). A new scenario logic for the Paris Agreement long-term temperature goal. *Nature, 573*, 357–363.

Scheffer, M., Carpenter, S. R., Lenton, T. M., Bascompte, J., Brock, W., Dakos, V., . . . Vandermeer, J. (2012). Anticipating critical transitions. *Science, 338*, 344–348.

Schneider, F., Kallis, G., & Martínez-Alier, J. (2010). Crisis or opportunity? Economic degrowth for social equity and ecological sustainability. Introduction to this special issue. *Journal of Cleaner Production, 18*(6), 511–518.

Seddon, N., Sengupta, S., Espinosa, G. M., Hauler, I., Herr, D., & Rizvi, A. R. (2019a). Nature-based solutions in nationally determined contributions: Synthesis and recommendations for enhancing climate ambition and action by 2020. Gland, Switzerland and Oxford, UK: IUCN and University of Oxford

Seddon, N., Turner, B., Berry, P., Chausson, A., & Girardin, C. A. J. (2019b). Grounding nature based climate solutions in sound biodiversity science. *Nature Climate Change, 9*, 84–87.

Seddon, N., Smith, A., Smith, P., Key, I., Chausson, A., Girardin, C., . . . Tuner, B. (2021). Getting the message right on nature based solutions to climate change. *Global Change Biology, 27*, 1518–1546.

Siegel, P. B. (2002). *"No Regrets" approach to decision-making in a changing climate: Toward adaptive social protection and spatially enabled governance.* World Resources Institute Report.

Smith, M., & Hargroves, K. (2007). *Review of limits to growth 3rd update – To mark the 30th anniversary of limits to growth.* Earthscan Publishing.

UN Environment-DHI, UN Environment, & IUCN. (2018). Nature-based solutions for water management: A primer. Available online: https://www.unepdhi.org/wp-content/uploads/sites/2/2020/05/WEB_UNEP-DHI_NBS-PRIMER-2018-2.pdf#:~:text=Naturebased%20solutions%20%28NBS%29%20for%20water%20resources%20management%20involve,water%20infrastructure%20to%20bring%20about%20more%20sustainable%20outcomes.

UNEP-WCMC, IIED, IUCN, & GIZ. (2019). *Ecosystem-based adaptation tools navigator: A searchable database of tools and methods relevant to EbA.* Final version: June 2019.

World Commission on Environment and Development (WCED). (1987). *Report of the world commission on environment and development: Our common future.* Oxford: Oxford University Press.

World Economic Forum (WEF). (2019). *Global risks report 2019* (14th ed.).

Growth, Expansion, and Future of Small Rural Towns

Jeofrey Matai[1], Walter Musakwa[2] and Innocent Chirisa[3]

[1]Department of Architecture and Real Estate, University of Zimbabwe, Harare, Zimbabwe

[2]Future Earth and Ecosystem Services Research Group, Department of Urban and Regional Planning, University of Johannesburg, Johannesburg, South Africa

[3]Department of Demography Settlement and Development, Social & Behavioural Sciences, University of Zimbabwe, Harare, Zimbabwe

Synonyms

Growth – development; Small rural town – growth point; Transformation – change

Definition

Small rural towns – the definition of a town differs from one country to another based on the history of urban development of each country. The definitions can also be based on three criteria: population size, function, and morphology. In Zimbabwe, a town is defined administratively, that is, a settlement designated as urban and demographically as a compact settlement of 2,500 people or more, the majority of whom will be involved in nonfarm employment.

Rural Transformation – rural transformation refers to a comprehensive societal change in which dependence on nonagricultural activities increases while dependence on agricultural activities decreases. Rural transformation is often accompanied by changes in landscape as a result of urbanization of rural places and invasion and conversion of rural places by urban activities.

Spatial Transformation – refers to the changes that take place in space. It is characterized by, among others, the expansion of urban areas, land-use changes, and the reallocation of land resources with consequential effects on rural transformation.

Growth – refers to a gradual development in maturity, age, size, weight, or height. It can also refer to the process or manner in which something grows. In the context of spatial planning and development, growth refers to changes that take place in the size of cities, population, and economic activities, among others.

Introduction

Over the years, the earth has undergone transformations that are attributable to anthropogenic factors and natural phenomena (Viana et al. 2019). Urbanization is one process that is both a product and a result of the transformations that are modifying the earth. Urbanization used to be confined and concentrated to a few large centers, which, due to increased rural-urban migration and natural population increase, grew into large metropolitans. However, owing to several factors, among them the diseconomies of scale, improvements in technology, and government policies, new towns have evolved in rural areas (Bates 1978; Nyandoro and Muzorewa 2017). As is the case with other urban centers, the small towns grow and expand. The population sizes continue to rise, and economic activities increase in size and diversity, further attracting more people and economic activities to towns. The expansion, however, is not without problems. Population growth increases the need for housing and other amenities, agricultural land is converted to urban land, and environmental problems including deforestation and pollution of land, water, and air also increase (Glaeser and Kahn 2005; Viana et al. 2019; Jenberu and Admasu 2020).

In Zimbabwe, cities are growing and the urban population is also increasing. According to the DFID (2017), the urban population growth for Zimbabwe is 0.6% per annum. While the greater proportion of the population in Zimbabwe is living in major cities, there is evidence of increasing and expanding small rural towns (Potts 2018; DFID 2017). This increase in urban population is accompanied by many problems which are not limited to poverty, poor service provision, environmental degradation, and high levels of unemployment (Munzwa and Jonga 2010; Chirisa and Muchini 2011; Chirisa 2013; Muchadenyika 2015). This raises questions on the sustainability of cites. While literature and debates on the sustainability of cities in Zimbabwe are plenty, there is a dearth of literature on the drivers of small rural towns' growth and expansion as well as their sustainability.

To be able to address sustainability issues in small towns, this study examines the factors behind their growth and expansion in the Zimbabwean context to recommend context-specific approaches to deal with the sustainability of small rural towns. Following this introduction section, the chapter provides a conceptual framework for the study and reviews literature on the growth and expansion of small towns. Scholarship and experiences on the growth of small towns are presented in the fourth section of the chapter followed by a discussion on the emerging issues. This is followed by conclusions and recommendations.

Conceptualizing Growth, Expansion, and Sustainability of Small Rural Towns

Urban growth is defined as the increase in the coverage of land that is urbanized. According to Viana et al. (2019), one way through which urban growth takes place is through the urban extension. Bhatta (2010) defines urban growth as a branch of urban geography that focuses on the physical and demographic expansion of cities and towns. Urban growth is a spatial-demographic process that denotes to the enlarged importance of towns and cities as a nucleus of the population in a particular economy and society. Urban growth occurs when the distribution of population changes from being largely hamlet and village-based to being principally town and city-dwelling (ibid.). When urban expansion takes place in a spontaneous or unplanned way, it is called urban sprawl. Urban sprawl, therefore, is a form of urban expansion. The growth and expansion of cities result in the transformation of the land. Durlauf

G

and Blume (2010) use other variables such as population, employment, and density to describe urban expansion. However, these are not always used at once to define urban growth and can differ from country to country.

Viana et al. (2019) categorize urban growth into infills, extensions, and outlying. Outlying growth is characterized by a transformation of land from non-urban to urban that happens away from existing urban areas beyond the urban fringe (Wilson et al. 2003 in Viana et al. 2019). Isolated growth is branded by the urbanization of non-urban areas further away from existing urban areas and is likened to a new house that is surrounded at most by limited urban space (ibid.). Of particular importance to this study is isolation which concerns the development of new urban centers away from the existing urban centers. However, once the new centers develop, they evolve into cities through urban expansion; hence, all forms of urban growth are relevant and cannot be treated in isolation.

Urban expansion involves the physical growth of cities usually horizontal expansion. As such, Mohapatra et al. (2014) link urban expansion to urban sprawl. However, this does not mean to say that urban expansion is associated with unplanned development. Jiao (2015) establishes that the rate of urban expansion is a reflection of its growth rate. This shows that urban growth and urban expansion cannot be separated, giving a thesis that urban expansion is a product of urban growth. The opposite is also true, that is, urban growth is a product of the expansion of cities and towns. However, the growth and expansion of cities are associated with urbanization problems that include, among others, poverty; high rates of unemployment; pollution of air, land, and water; and conversion of agricultural land (Munzwa and Jonga 2010; Chirisa 2013; Mohapatra et al. 2014). These and other problems not stated here raise questions about the future of small rural towns.

Small towns are characterized by, in addition to the small land coverage and population size, the permanence of settlement and excess importance among other things (Vaishar et al. 2015; Steinführer et al. 2016). The latter relates to the economic, administrative, and/or cultural relevance of the town to the inhabitants of the town and the surrounding places. This entails that small towns have the same structural attributes with cities although with varying degrees of intensity and importance. This means to say that services that are required in large cities should also be provided in small towns and the functions of small towns may also be similar to those of large cities as explained by Steinführer et al. (2016). However, Bell, and Jaune (2009) in Vaishar et al. (2016) argue that small rural towns provide a limited number of jobs and services compared to big cities and that they seem to have negligible environmental problems. However, this assertion is dismissed by some scholars who argue that small towns offer significant employment opportunities especially considering that about half of the urban population live in small towns (Wandschneider 2004; Tacoli and Agergaard 2017).

Small towns are also experiencing socioeconomic and environmental problems such as pollution, conversion and succession of agricultural land into urban land, water and sanitation problems, and infrastructural development challenges (Zhang 2002; Mohapatra et al. 2014; Vaishar et al. 2016; Tacoli and Agergaard 2017). These challenges raise the need think, plan, and act sustainably to ensure that the future of small towns provides opportunities for the improvement of people's quality of life.

Evidence confirm that rural towns that are well planned and developed and properly integrated into the space economy have a positive impact to the economy, environment, and well-being of the people living within and outside the towns (Wandschneider 2004). They provide opportunities for the diversification of local economies through the introduction of platforms for practicing non-farm economic activities. Based on this discussion of the relationship among growth, expansion, and sustainability of small towns, the growth and expansion of small cities are inevitable and are providing opportunities for improvement of quality of life for the inhabitants and surrounding communities. However, if the small towns are not well planned and managed and are not integrated into the space economy, they can present challenges to the inhabitants and the surrounding communities. It is therefore important to think about the sustainability of the small towns at a localized scale by understanding the factors behind their growth and expansion and the problems associated with the expansion and exploring the opportunities that the growth and expansion of the small towns present. This is critical for policy formulation and practice. The following section looks into the factors that drive the growth and expansion of small towns as well as the problems posed and the opportunities presented therewith.

The Growth and Expansion of Small Towns

Growth in the context of cities refers to the increase in the land covered by urban activities (Viana et al. 2019), and expansion refers to the extension of urban boundaries (Mohapatra et al. 2014). However, the growth and expansion of towns can also refer to structural changes that include social and economic activities (Durlauf and Blume 2010). The processes of urban growth and expansion are a resultant effect of several driving forces which are mainly anthropogenic although biophysical factors are also responsible (Glaeser and Kahn 2005; Mohapatra et al. 2014; Tacoli and Agergaard 2017; Viana et al. 2019). One of the prime causes of urban growth and expansion is the increase in the urban population. Population growth can be a result of natural increase and rural to urban migration, the most significant one in the recent time being rural to urban migration (Bhatta 2010).

Migration to towns is explained in terms of the push and pull factors. Some of the push factors include climate change which results in reduced agricultural yields for the rural population who then find it better to move to urban centers in search of non-farm activities; shortage of land to cultivate, pushing people to cities in search of livelihood activities; and shortage of amenities. Pull factors are those attractive circumstances in urban areas such as availability of better amenities and employment opportunities. Besides, urban places are perceived as places where money, wealth, and services are concentrated (Bhatta 2010) as opposed to rural places that are characterized by unpredictable environmental conditions such as droughts and floods. As a result, people move to urban areas in search of a better quality of life. However, Bhatta (2010) argues that this growth strains the capacity of cities to provide the services that are required to sustain the growing population. Growing small rural towns across the globe resemble these problems as they grow (see Zhang 2002; Chirisa and Muchini 2011; Manyanhaire et al. 2011; Vaishar et al. 2016).

Expectations about the future demands of development in the absence of spatial planning frameworks also contribute to the growth of towns. In cases where there is a demand for development, development agents make independent decisions that enable them to meet their expectations for the future and anticipated demands. Thus, the government, the private sector and in some cases the local communities take it upon themselves to direct development to meet their demands. The resultant effect is development that is not coordinated, unplanned, and uncontrolled (Harvey and Clark 1995 in Bhatta 2010). This kind of development is biased toward reinforcing and supporting the interests of the elite and powerful people in the towns while negatively affecting the poor and the less powerful groups of society. For example, when independent decisions are made on where to invest, the land is often acquired and converted from rural to urban. The less powerful people living and depending on the land proxy to the town lose the land. This makes them poorer as they would have been stripped off their livelihood source.

Economic growth is also a driver of urban growth and expansion. An increase in the number of employed people and per capita income has an effect of increasing housing demand (Boyce 1963 in Bhatta 2010). This encourages the construction of new houses and infrastructure by developers. The result is often the creation of an uncoordinated and discontinuous urban environment. Economic growth is also linked to the industrialization of the countryside. This attracts people as they seek employment. Besides, these people would require housing resulting in the growth and expansion of towns in the rural places.

Speculation, the expectation for land appreciation, and the land hunger attitude are also drivers of urban growth and expansion (Bhatta 2010). Speculation about future growth and government policies usually results in premature growth of towns and cities and is usually blamed for sprawling development as it encourages withholding of land for development (Clawson 1962 in Bhatta 2010). Similarly, expectations that land proxy to the town may be acquired for urban development cause some landowners to withhold the land in the hope that it will appreciate value. The desire to own land by individuals and institutions also contributes to the growth and expansion of towns. Land grabbing without developing creates parcels of vacant land at the core of towns and often makes infill policies impossible. These factors result in the outward growth of cities and towns, leaving parcels of undeveloped land in between. Physical features also contribute to the expansion of towns. In cases where land is unsuitable to develop, development leapfrogs to places that can be developed resulting in sprawling development.

Government policies also play a role in the growth and expansion of towns. Restrictive policies can result in development moving to other places where control is not restrictive (Barnes et al. 2001 in Bhatta 2010). On the other hand, policies to promote the growth of smaller rural centers can attract development to the small centers through the growth center policy, for example (Jenberu and Admasu 2020). Some scholars (Bhatta 2010; Katsoulakos et al. 2016; Jenberu and Admasu 2020) also argue that the lack of proper planning policies contributes to the expansion of towns. Single-use zoning, for example, results in expansive development of cities. On the other hand, unsuccessful enforcement of land-use plans and policies is cited as one of the reasons for the sprawling development of cities.

The consequences of urban growth and expansion are numerous and interrelated and can be positive or negative although negative impacts tend to override the positive impacts. Urban growth brings with it improved economic production and provides employment opportunities for the unemployed and underemployed (Christiaensen and Kanbur 2017). The growth and expansion of towns also extend services such as sewer and water and services such as education and healthcare facilities to more people. In the context of small towns, growth provides platforms for people to diversify income-generating activities and to link with the wider economy in terms of markets and information (Pedersen 1994; Bhatta 2010; Gibson et al. 2017; Tacoli and Agergaard 2017).

The negative effects are multiple and are often of higher magnitude to offset the benefits of urban growth. Urban growth is believed to sap resources from rural places (see Lipton 1977, 1984). Although Tacoli and Agergaard (2017) dismiss this argument, there is evidence of shrinking rural towns and rural places at the expense of large cities (Mbiba 2017). The continued extraction of resources and the conversion of land from rural to urban as towns expand modify the physical environment (Zhang 2002). The modifications are in most cases unpleasant, for example, increase in the built-up space at the expense of forests and agricultural land. The effects are worse when the towns are poorly planned such that the growth and expansion of towns threaten the environment, health, and quality of life (Bhatta 2010). Loss of farmland often makes people who relied on agriculture poorer.

Global and Regional Experiences in the Growth, Expansion, and Sustainability of Small Towns

The growth and expansion of small towns differ globally with some towns showing positive results while others the opposite. The literature on the role of small towns in contributing to economic growth, in reducing poverty, and in improving the livelihoods of the people within and in surrounding places is expansive (Wandschneider 2004; Vaishar and Zapletalová 2009; Vaishar et al. 2016; Gibson et al. 2017; Nyandoro and Muzorewa 2017; Tacoli and Agergaard 2017). Common to the literature is on the drivers of urban growth and expansion although there are some variations in some, for example, the influence of government policy in the growth of small towns and the impact that these policies have on the growth of towns. Bhatta (2010) argues that government policies can cause leapfrog development where policies on land-use development vary, with some places having strict policies and some less strict policies. Observations by Potts (2018); Manyanhaire et al. (2011); and Nyandoro and Muzorewa (2017) on the influence of policy on the growth of small towns in Zimbabwe differ. In Zimbabwe, the main policy that drives urban growth and expansion is the Growth with Equity (GWE) policy which saw the adoption of the Growth Point policy. Several growth points were established.

Although it is argued that the growth points failed to sustain except for a few (Munzwa and Jonga 2010; Manyanhaire et al. 2011), the policy is responsible for the growth of most rural towns today. Policies such as the Economic Structural Adjustment Program (ESAP) which had resultant effect of retrenchment of many people contributed to the growth of the growth points into the town (using the demographic definition of towns in Zimbabwe). Some of the people who lost their jobs in large cities relocated to the growth points where the cost of living is generally lower (Potts 2018). One of the roles of small rural towns is to provide employment opportunities through the diversification of livelihood activities (Liedholm et al. 1994; Vaishar et al. 2015, 2016; Steinführer et al. 2016; Tacoli and Agergaard 2017). The argument put forward here is that the small towns link the rural people to the global economic space by providing a market to rural products and by providing information on the market and innovations.

These centers are also regarded as the distribution points upon which products from the large cities are sold to the rural people (Wandschneider 2004). Although it can be argued that this exposes the rural population to competition from the global market, this exposure to market and information to the rural population are critical to the growth and expansion of farm and non-farm businesses. The global experiences on the role of small towns are also shared in Zimbabwe. For example, a study by Pedersen (1994) shows that small total towns play a critical role as marketing centers for agricultural products.

Similarly, Liedholm et al. (1994) show that the presence of small rural towns has contributed to the increase in the number of small enterprises and the employment therewith. However, a study by Dalu and Dalu (2019) indicate that while the small towns link rural places to the global market, these small towns are generally shrinking at the expense of the growth of bigger towns. On the other hand, the small towns in Zimbabwe are more of residential towns which are by-passed by-products from the rural hinterlands and, as such, have little or insignificant economic activities (Potts 2018; Munzwa and Jonga 2010; Chirisa et al. 2016; Dalu and Dalu 2019). This shows a variation from general global experience from countries such as India and China.

Small rural towns generally suffer from neglect. There is little commitment financially, technically, and administratively given to small towns across the globe. The literature on small towns shows that studies conducted on small towns indicate that they are often given secondary attention despite being critical human settlements (Liedholm et al. 1994; Pedersen 1994; Zhang 2002; Vaishar and Zapletalová 2009; Steinführer et al. 2016).

Wandschneider (2004) argues that development agencies tend to associate rural development to development in agriculture. This has an effect of limiting the growth of wage employment in rural towns and the development of other sectors that have the potential of driving the growth of local economies. Zimbabwe is not an exclusion to this. Policies that are targeted at rural development are devoid of emphasis in programs that promote the growth and sustainability of rural towns (PlanAfric 2000). As such, the small towns face multiple problems which range from poor planning policies, lack of enforcement of policies, unplanned development, and sprawling development (Zhang 2002; Tacoli and Agergaard 2017). The effect of neglecting small towns ranges from environmental problems, social problems, and economic problems as alluded earlier. When this happens, the sustainability of the small towns is questionable.

Emerging Issues and Synthesis

From the foregoing discussion, several issues can be classified into factors driving urban growth and expansion as well as questions of the sustainability of small towns. The growth of small towns is driven by several economic, social, environmental, and institutional factors. Growth and expansion of small rural towns are therefore a multifaceted process that involves and affects the nature, rate, and direction of growth. Economic policies can have a direct and indirect influence on the growth of small towns. For example, in Zimbabwe, the ESAP contributed to the growth and expansion of small towns in different parts of the country as people relocated from large cities to smaller cities (Potts 2018) and some rural dwellers who were previously dependent on city dwellers moved to small urban centers in search of economic opportunities. Similarly, the desire for urban lifestyle and better amenities also pushes people to emerging urban centers.

Institutional policies that are meant to address spatial disparities also play an influential role in the development, growth, and expansion of small rural towns. Policies that are restrictive to land-use development as is the case with most large cities push people to smaller towns where land-use policies are more relaxed. On the other hand, planning policies and their enforcement are determinants of the rate and nature of urban growth and expansion. It also emerges that small rural towns are critical in linking rural places with the global economy as alluded by Wandschneider (2004); Vaishar et al. (2015); Tacoli and Agergaard (2017); and Jenberu and Admasu (2020). Another emerging issue in the study is that the growth and expansion of small rural towns, although presenting opportunities for the improved quality of life and economy, are associated with problems that range from social, economic, institutional, and environmental. These problems threaten the sustainability of the small towns and the surrounding areas that are dependent or are directly linked to the small towns.

Conclusion and Future Direction

Conclusions made from this study are that the growth and expansion of small rural towns are attributed to several factors that are anthropogenic and biophysical. Factors that drive the growth of small towns are generally similar globally although Zimbabwe has some unique forces that drive the growth of small towns. The growth of small towns is both a product of economic growth and a source of economic growth. The growth of small towns is critical in driving the rural economy by providing employment opportunities and room for diversification of livelihood activities. They also act as a link between the rural areas and large cities and the global space. It is also concluded that the small rural towns are given less attention in planning and management and financial support. Most governments treat rural towns as part of the rural landscape and consider agricultural development as rural development, ignoring the role that small towns play as both market centers and distributors of inputs. This makes the small towns unsustainable. Thus, the future of small rural towns is uncertain if the current approaches to planning and management of small rural towns are not changed to consider the role of small cities and to integrate them into the space economy.

It is recommended that planning approaches such as spatial planning that integrate settlements and the activities of the various players in the growth and management of settlements be adopted to ensure that rural and urban places are considered as a unit. This entails integrating the small towns into the space economy, not as competitors but as facilitators for national economic growth and engines for the growth of rural places. Systems should be put in place to plan for and manage rural and urban places as places that are interconnected by the flow of capital, information, and resources including labor. The recommendation here is to relook into the current institutional arrangements for the planning and management of rural and urban places.

Cross-References

▶ Small Towns in Asia and Urban Sustainability

References

Bates, V. E. (1978). The impact of energy boom-town growth on rural areas. *Social Casework, 59*(2), 73–82. https://doi.org/10.1177/104438947805900202.

Bhatta, B. (2010). *Analysis of Urban Growth and Sprawl from Remote Sensing Data | Basudeb Bhatta (auth.) | download*. Kolkata/Berlin/Heidelberg: Springer. Available online: https://book4you.org/book/892486/fd7989. Accessed 11 Aug 2020.

Chirisa, I. (2013). Solid waste, the "Throw-Away" culture and livelihoods: Problems and prospects in Harare, Zimbabwe. *Journal of Environmental Science and Water Resources, 2*(1), 1–8. https://doi.org/10.5897/JAHR12.036.

Chirisa, I., & Muchini, T. (2011). Youth, unemployment and peri-urbanity in Zimbabwe: A snapshot of lessons from Hatcliffe. *International Journal of Politics and Good Governance, 2*(2), 1–15.

Chirisa, I., Bandauko, E., & Muzenda, A. (2016). Establishment of new satellite towns in Zimbabwe: Old wine in new skins? In *Peri-urban developments and processes in Africa with special reference to Zimbabwe* (pp 105–113). Cham: Springer International Publishing.

Christiaensen, L., & Kanbur, R. (2017). Secondary towns and poverty reduction: Refocusing the urbanization agenda. *Annual Review of Resource Economics, 9*, 405–419. https://doi.org/10.1146/annurev-resource-100516-053453.

Dalu, M. T. B., & Dalu, T. (2019). Determinants of participation in rural non-farm economy in Zvimba District, Zimbabwe. *Journal of Agriculture and Rural Development in the Tropics and Subtropics, 120*(1), 63–70. https://doi.org/10.17170/kobra-20190613555.

DFID. (2017). Available online: https://assets.publishing.service.gov.uk/media/595217e340f0b60a4400003e/ICED_Zimbabwe_Final_Scoping_Report_240217.pdf. Accessed on 3 October 2020.

Durlauf, S. N., & Blume, L. E. (2010). Urban growth. In S. N. Durlauf & L. E. Blume (Eds.), *Economic growth* (pp. 264–269). London: Palgrave Macmillan.

Gibson, J., et al. (2017). For India's rural poor, growing towns matter more than growing cities. *World Development, 98*, 413–429. https://doi.org/10.1016/j.worlddev.2017.05.014. Elsevier Ltd.

Glaeser, E. L., & Kahn, M. E. (2005). Sprawl and urban growth. *SSRN Electronic Journal, 4*(04). https://doi.org/10.2139/ssrn.405962.

Jenberu, A. A., & Admasu, T. G. (2020). Urbanization and land use pattern in Arba Minch town, Ethiopia: Driving forces and challenges. *GeoJournal, 85*(3), 761–778. https://doi.org/10.1007/s10708-019-09998-w. Springer Netherlands.

Jiao, L. (2015). Urban land density function: A new method to characterize urban expansion. *Landscape and Urban Planning, 139*, 26–39. https://doi.org/10.1016/j.landurbplan.2015.02.017. Elsevier B.V.

Katsoulakos, N. M., et al. (2016). Environment and development. In S. P. Inglezakis & V. Inglezakis (Eds.), *Environment and development: Basic principles, human activities, and environmental implications* (pp. 499–569). Elsevier. https://doi.org/10.1016/B978-0-444-62733-9.00008-3.

Liedholm, C., McPherson, M., & Chuta, E. (1994). Small enterprise employment growth in rural Africa. *American Journal of Agricultural Economics, 76*(5), 1177–1182. https://doi.org/10.2307/1243413.

Lipton, M. (1977). *Why poor people stay poor: A study of urban bias in world development*. Temple Smith; Australian National University Press. Available online: https://openresearch-repository.anu.edu.au/handle/1885/114902. Accessed 3 Apr 2019.

Lipton, M. (1984). Urban bias revisited. *The Journal of Development Studies, 20*(3), 139–166. https://doi.org/10.1080/00220388408421910.. Taylor & Francis Group.

Manyanhaire, I. O., Rwafa, R., & Mutangadura, J. (2011). A theoretical overview of the Growth Centre Strategy: Perspectives for reengineering the concept in Zimbabwe. *Sustainable Development in Africa, 13*(4), 1–13.

Mbiba, B. (2017). *DFID Zimbabwe Country Engagement: Final Scoping Report. Report produced for the UK Department for International Development.*

Mohapatra, S. N., Pani, P., & Sharma, M. (2014). Rapid urban expansion and its implications on geomorphology: A remote sensing and GIS based study. *Geography Journal, 2014*, 1–10. https://doi.org/10.1155/2014/361459.

Muchadenyika, D. (2015). Land for housing: A political resource – Reflections from Zimbabwe's urban areas. *Journal of Southern African Studies, 41*(6), 1219–1238. https://doi.org/10.1080/03057070.2015.1087163. Routledge.

Munzwa, K., & Jonga, W. (2010). Urban development in Zimbabwe: A human settlement perspective. *Theoretical and Empirical Researches in Urban Management, 5*(14), 120–146.

Nyandoro, M., & Muzorewa, T. T. (2017). Transition from growth point policy to liberal urban development in

Zimbabwe: The emergence of Ruwa Town, 1980–1991. *The Journal for Transdisciplinary Research in Southern Africa, 13*(1), 1–10. https://doi.org/10.4102/td.v13i1.426.

Pedersen, P. O. (1994). Agricultural marketing and processing in small towns in Zimbabwe-Gutu and Gokwe. In B. Jonathan & P. O. Pedersen (Eds.), *The rural-urban interface in Africa: Expansion and adaptation* (pp. 102–124). Uppsala: The Scandinavian Institute of African Studies. https://doi.org/10.25071/1920-7336.21708.

PlanAfric. (2000). Local strategic planning and sustainable rural livelihoods rural district planning in Zimbabwe: A case study. Available online: https://pubs.iied.org/sites/default/files/pdfs/migrate/7827IIED.pdf. Accessed on 1 October 2020.

Potts, D. (2018). Urban data and definitions in sub-Saharan Africa: Mismatches between the pace of urbanisation and employment and livelihood change. *Urban Studies, 55*(5), 965–986.

Steinführer, A., Vaishar, A., & Zapletalová, J. (2016). The small town in rural areas as an underresearched type of settlement. Editors' introduction to the special issue. *European Countryside, 8*(4), 322–332. https://doi.org/10.1515/euco-2016-0023.

Tacoli, C., & Agergaard, J. (2017). *Urbanisation, rural transformations and food systems: The role of small towns*. https://doi.org/10.1016/j.rser.2013.08.099.

Vaishar, A., & Zapletalová, J. (2009). Small towns as centres of rural micro-regions. *European Countryside, 1*(2), 70–81. https://doi.org/10.2478/v10091/009-0006-4.

Vaishar, A., Šťastná, M., & Stonawská, K. (2015). Small towns – Engines of rural development in the South-Moravian region (Czechia): An analysis of the demographic development. *Acta Universitatis Agriculturae et Silviculturae Mendelianae Brunensis, 63*(4), 1395–1405. https://doi.org/10.11118/actaun201563041395.

Vaishar, A., Zapletalová, J., & Nováková, E. (2016). Between urban and rural: Sustainability of small towns in the Czech Republic. *European Countryside, 8*(4), 351–372. https://doi.org/10.1515/euco-2016-0025.

Viana, C. M., et al. (2019). Land use/land cover change detection and urban sprawl analysis. In P. Hamid Reza & G. Candan (Eds.), *Spatial modeling in GIS and R for earth and environmental sciences* (pp. 621–651). Libson: Elsevier. https://doi.org/10.1016/b978-0-12-815226-3.00029-6.

Wandschneider, T. (2004). Small rural towns and local economic development: Evidence from two poor states in India. *International Conference on Local Development, 44*(0), 0–31.

Zhang, L. (2002). *Ecologizing industrialization in Chinese small towns*. Wageningen University. Available online: https://library.wur.nl/WebQuery/wurpubs/319545. Accessed 11 Aug 2020.

Guerilla Gardening

► Urban Food Gardens

Guidelines

► Public Policies to Increase Urban Green Spaces

Guidelines for Water-Sensitive Informal Settlement Upgrading in the Global South

Tom Sanya[1], Caitlin Anthea Lewis[1] and Itumeleng Mogola[2]
[1]Architecture Planning and Geomatics, University of Cape Town, Cape Town, South Africa
[2]C40 Cities, Benoni, South Africa

Introduction

Freshwater suffuses and is indispensable for the health and vitality of people and ecosystems. The presence of clean freshwater nourishes habitats and signifies ecological health. In *the Garden Cities of Tomorrow*, Ebenezer Howard sought to find a balance between the extremes of the over-crowded polluted industrial city and sparse countryside (Howard 1902). Building on Howard's work, Geddes (1915) argued that nature and decent housing must underpin city design. He maintained that access to nature and decent housing should not just be the privilege of rich industrialists, merchants, mining magnates, and nobles but must be extended to colliers, factory workers, and other laborers. This, he asserted, would improve workers' wellbeing, enhance their productivity and hence guarantee the city's economic vitality. Above all, Geddes argued, a good city can only evolve from a strong civic consciousness and participation.

This progressive proselytizing largely fell to the wayside as the industrial city expanded apace. The link between nature and the city became increasingly forgotten. In this new era, streams were buried in pipes underground, urban green was paved over, and farmland was consumed as cities irrevocably altered natural cycles (Sisolak and Spataro 2011). Cities have become centers of population and production. Urban demands for resources cause exhaustion of renewable and non-renewable resources, and drive the pollution of land, water, and air. In tandem, hordes of the urban poor, particularly in the developing countries, continue to settle in substandard housing in unsanitary informal settlements.

In this critical century, the world is faced with environmental degradation, rapid population growth, accelerated urbanization, and persistent poverty. One billion of the world population today lives in informal settlements mostly in the developing countries of Asia and Africa. In Africa, according to the Organisation for Economic Cooperation and (OECD), the urbanized population is almost 600 million. Of these, more than 230 million stay in informal settlements (i.e., just under 40% of Africa's urban population live in informal settlements). Globally, according to the United Nations, an estimated three billion people will require adequate housing by 2030 (UN 2021 and OECD 2020). The year 2007 was when the majority of the world's population started living in urban areas. By 2030, the majority of people in developing countries will be urbanized (UN -DESA 2021). 70% of the world population will be urbanized by 2060 (World Bank 2020). Most of this urban growth will happen in the developing countries of Asia and Africa. If proactive solutions are not implemented, most of these people could end up in existing or new informal settlements. The extent and growth of Informal settlements manifest socioeconomic and spatial urbanization of abject poverty. Informal settlements present numerous problems, including water scarcity, poor sanitation, ill health, poor infrastructure, and locations on marginal lands and disaster-prone lands. To achieve just and equitable urban development, the problems of informal settlements must be urgently addressed.

Cities make up only about 2% of global land coverage (Cleugh and Grimmond 2011) but generate more than 80% of global GDP (World Bank 2020). Cities are thus largely responsible for consumption-driven source-side and sink-side degradation of hinterland ecosystems. They upset the natural carbon and water balances, drive the demand for mineral and organic raw materials, and spew pollution into land, water, and air. The physical footprint of cities causes paving over of ecosystem and fertile agricultural land. As urban populations and standards of living grow, there is a commensurate increase in demand for water, food, energy, and raw materials. This puts enormous pressure on hinterlands. Therefore, it is precisely in the city that the problem of global environmental degradation is to be arrested and reversed, and catalysts for new regenerative futures ignited. This requires a new urbanization development model centered on social and ecological justice. In all this, upgrading of informal settlements must be priotised because they continue to house significant numbers of population. Using principles of Water Sensitive Design (WSD) a model of socially inclusive and sustainable upgrading of informal settlements is presented in this paper.

Reading Urban and Informal Settlement Problems Through Water

Informal settlements house the most vulnerable urban dwellers. These residents are indispensable for the city's functioning and productivity. The social, economic, and environmental ills of the city converge and are amplified in its informal settlements. Structural violence (Galtung 1969) and slow violence (Nixon 2011) are perniciously internalized by these communities. Low wages and negligence by responsible authorities force the urban poor into substandard housing in marginal lands with poor water and sanitation services in informal settlements. Many informal settlements are located in dangerous areas such as landfills, flood-prone lowlands, and unstable steep slopes.

The World Economic Forum (WEF) places water crises as among the top ten risks both in terms of likelihood and impact (WEF 2019). Water crises occur when there is a significant decline in the available quality and quantity of freshwater, resulting in harmful effects on human health and/or economic activity. This was the case, for example, in the Australian Millennium Drought (Van Dijk et al. 2013), Flint Michigan water pollution (Butler et al. 2016), and the prospect of Day Zero in Cape Town (Taing et al., 2019; Wolski, 2018). Up to 66% of the global population lives under conditions of severe water scarcity for at least one month of the year (Mekonnen and Hoekstra 2016) and 40% of the world's population will be living in seriously water-stressed areas by 2035 (Guppy and Anderson 2017; Addams et al. 2009). There has been a 55% drop globally in available freshwater per capita (Guppy and Anderson, 2017). And yet, global water demand is expected to grow by 50% by 2030. This is according to the United Nations Educational Scientific and Cultural Organization [UNESCO] (UNESCO 2020).

Agriculture is the largest water use sector. But with most people and industries located in the cities, food and agricultural raw materials are largely consumed in cities. As the world urbanizes, water scarcity and stress will increasingly be experienced most acutely in cities, especially in the poorly serviced informal settlements of the Global South. Even as informal settlements struggle with conditions of poor water and sanitation, more affluent residents waste 80% clean potable water on uses for which non-potable water would suffice, for example, toilet flushing and watering lawns (Sisolak and Spataro 2011).

The drivers of water crises include prolonged droughts and growing population and increasing living standards, inefficient water-use patterns, crumbling old centralized infrastructure, and lack of capital to invest in infrastructure as well as institutional failures (Sisolak and Spataro 2011; Rijsberman 2006). Climate change is predicted to negatively impact freshwater supplies due to changes in rainfall patterns and increased evaporation due to global warming (Lüthi et al. 2011; Schewe et al. 2014; Gosling and Arnell 2016).

Globally, 44% of household wastewater is not safely treated – mostly in emerging and developing countries (UN-Water 2021). This pollutes freshwater systems, thus threatening the health of people and the environment and the proper functioning of economies. Inadequate clean water and sanitation for the urban poor in informal settlements converges with other factors into a grand challenge in the water-food-health nexus ((Olsson 2013; Mabhaudhi et al. 2016; Halbe et al. 2015; Walker et al. 2014). Millions of people are faced with increased morbidity and mortality annually due to malnutrition, diarrhea, dysentery, and cholera (Sanya 2020). Through the water cycle, poor sanitation in informal settlements triggers unexpected consequences in the formal sector far away. For example, in South Africa, raw-sewage pollution from poorly serviced informal settlements upstream of the Berg River has resulted in infection of water with malignant bacteria (E.coli and staphylococcus) and contamination of fresh fruit on the downstream river-irrigated commercial farms. This in turn threatens access to international markets putting farm profitability and thousands of jobs at risk (Struyf et al. 2012).

Loss of valuable nutrients (phosphorous and nitrites) from food waste and agricultural systems and their undesirable build up in natural systems via sewage networks causes eutrophication in freshwater bodies (Lüthi et al. 2011). Increased Biochemical Oxygen Demand (BOD) due to untreated organics in water put freshwater biota at risk.

As human water demand expands, environmental water requirements are threatened (Rijsberman 2006; Smakhtin et al. 2004). This results in drying and depletion of groundwater and water bodies (Mekonnen and Hoekstra 2016), endangering of habitats, and destructions of flora and fauna. This is particularly acute in urban areas where buildings, paving, and infrastructure have radically altered the natural water balance (Wong and Eadie 2000). This makes cities more vulnerable to flooding, increases the urban heat island effect and significantly diminishes ecosystem service value.

In 2017, 90% of all disasters were water related (Centre for Research on the Epidemiology of Disasters [CRED] 2017). This includes flooding, landslides, and droughts. The April 2022 deluge in KwaZulu Natal in which hundreds of lives were lost and buildings washed away is a vivid recent example in South Africa. With the majority of the world population already living in cities and urbanization continuing apace, particularly amongst the vulnerable populations of the Global South, disaster preparedness and response must be central to urban design and development. These measures, such as those described in the Sendai Framework (see Pearson and Pelling 2015), should particularly focus on informal settlements, many of which are located on flood-prone lowlands, steep unstable hillsides, and other dangerous areas.

The convergence of rapid urbanization, urban poverty, and increasing natural disasters are begetting existential trepidation that is spurring impatient youth into large riotous movements such as Rhodes Must Fall (Newsinger 2016), Black Lives Matter (Lebron 2017), and Extinction Rebellion (Kinniburgh 2020). The social and environmental problems of today are real and urgent. The moment to act is now. And the site of action must be in cities especially in the informal settlements of the Global South where increasing hordes of frustrated poor people reside. Through the lens of Water Sensitive Design (WSD), this paper outlines guidelines for informal settlement founded in socio-economic, spatial, and environmental justice.

Water: A Framing Medium for Informal Settlement Upgrade (in the Twenty-First Century)

Water is a vital omnipresent substance of all life. The hydrological cycle binds the animate and inanimate, terranean and subterranean, biosphere, and atmosphere (Sanya 2020). Water inextricably interconnects rich and poor, humans and nature. Healthy water systems nourish healthy ecosystems within which communities can enact sustainable thriving livelihoods. Solutions to make our cities just, sustainable, and liveable can be tapped from a deep understanding of the flows of water.

Water-sensitive approaches can undergird transformative measures for urban regeneration and informal settlement upgrading. This requires centering urban design and planning in an ecosystem services framework. The notion of ecosystems was first defined by Tansley (1935). The Millennium Ecosystems Assessment (M.E.A), an initiative of the United Nations Environmental Programme (UNEP), defines ecosystem services as the benefits people obtain from ecosystems in the form of goods and services. Ecosystem services are of four main categories: provisioning, regulating, cultural, and primary services (M.E.A 2005). As the lifeblood of environmental systems, water directly and indirectly underpins all ecosystem services (see Fig. 1).

Water is a distinct *provisioning ecosystem service*. Water supply to society should be of sufficient quantity and of the right quality for different end-uses. The ecosystem service of *water regulation* includes flooding control and purification. Purification covers the range of physical and biological processes for the treatment of water by the environment. *Cultural services*, specifically aesthetics, sense of place, recreational and inspirational value are important objectives for water sensitive urban design (WSUD) – (see Wong and Brown 2009). The above are direct ecosystem services of water. Equally vital, it is evident that water permeates and/or sustains all other ecosystem services, including primary production and the preservation of biodiversity and the intrinsic worth of the environment.

Freshwater ecosystems provide a range of valuable goods and services such as flood protection, fisheries, recreation opportunities, and wildlife sanctuaries (Revenga 2003; Vié et al. 2009). These services are worth trillions of US dollars annually (Carpenter 1997). On the other hand, ecosystem degradation has resulted in the loss of ecosystem services valued at 20 trillion dollars annually (Constanz et al. 1997). It is therefore in society's interests that aquatic ecosystem water requirements are met to preserve these services. This means that in urban and regional planning, ensuring sufficient environmental water

G

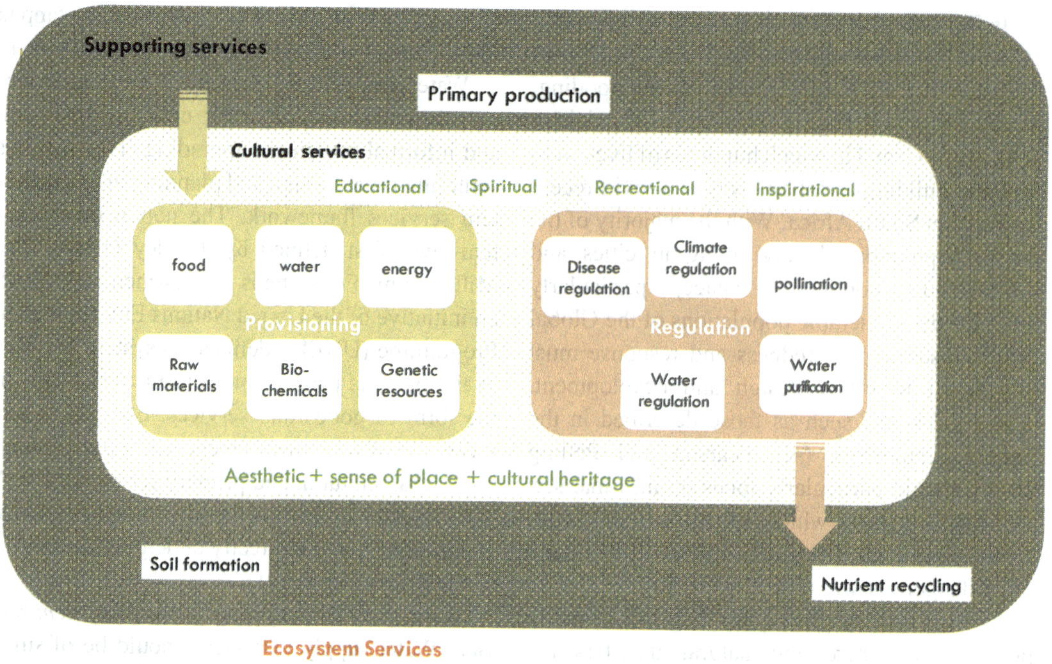

Guidelines for Water-Sensitive Informal Settlement Upgrading in the Global South, Fig. 1 Ecosystem Services: Ecosystems provide multiple values to people. Water is a distinctive ecosystem service. Water is also indispensable for the health of nature and therefore safeguards all other ecosystem services. (The image is based on M.E.A 2015). @Tom Sanya (CC BY 2.0)

availability for hinterlands and watersheds is imperative (see Sanya 2020; Smakhtin et al. 2004).

Urban ecological infrastructure mediates between nature and the urban built environment. Such green infrastructure can be woven into cities through the approach of Water Sensitive Urban Design (WSD). In this approach, existing and new water bodies and greenery are gradually linked into networks of blue-green natural infrastructure (Council 2010; Wong and Eadie 2000). The interlinked linear and zonal ecological infrastructure safeguards and enhances the value of freshwater ecosystems in the city and beyond. When implemented, WSD offers multiple benefits. It increases recreation opportunities such as walking, jogging, cycling, canoeing, and fishing. Blue-green infrastructure in cities is also known to mitigate the urban heat island effect (Brears 2018). Reduced temperatures, opportunities for active recreation, and the pollution-cleansing effects of nature result in overall health benefits for city residents (Steiner 2014). The current COVID-19 pandemic highlights the value of green infrastructure in cities as lungs for the city and as breakout space under lockdown (Lopez et al. 2021; Venter et al. 2021; Yamazaki et al. 2021). In a review of the benefits of WSD, Morgan (2013) identified multiple advantages at variety scales in the city. Amongst others, WSD safeguards water supply, contributes to wastewater treatment, contributes to global warming mitigation, improves human health and wellbeing, enhances economic prosperity, and protects ecosystem health (See Fig. 2). In 2010, the United Nations enshrined the right to safe water and sanitation. WSD can help cities realize this right as well as unlock the multiple of surfacing water.

Beyond the city, WSD protects hinterland water systems and ecologies. Nesting WSD within a broader approach of Integrated Water Resource Management (IWRM) promotes the coordinated development and management of water, land, and related resources for equitable

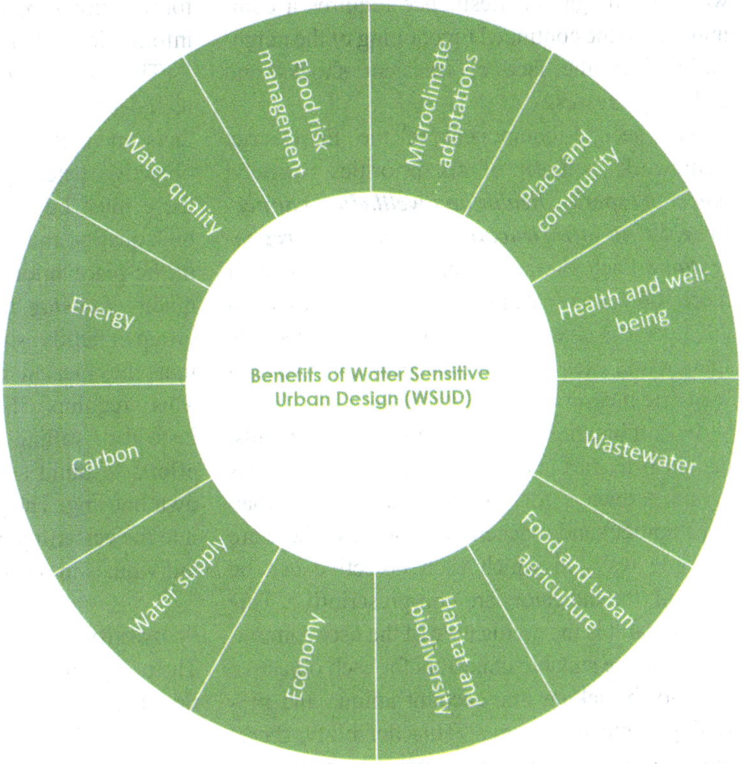

Guidelines for Water-Sensitive Informal Settlement Upgrading in the Global South, Fig. 2 Advantages of Water Sensitive Design (WSD): Designing a city in an ecological green-blue infrastructure grid offers multiple benefits. These benefits are indispensable for transitioning informal settlements into equitable people-centered sustainable EcoDistricts. (The image is based on Morgan et al. 2013). @Tom Sanya (CC BY 2.0)

social and economic development while preserving ecosystem health (Lenton and Muller 2012). Because of prevalent poverty and the unique challenges of the Global South, a livelihoods approach to IWRM is required (Merrey et al. 2005; Braga 2010). WSD and IWRM in the Global South should foreground equitable development, improvement of public health, poverty alleviation, and promotion of racial and gender equality (Khandker et al. 2020; Lande 2015).

EcoDistricts: From Theory to Aspirational Informal Settlement Attributes

There are several approaches to advancing ecological urban neighborhood development. But the EcoDistricts Protocol is amenable to in situ urban regeneration. Inclusive collaboration and development, and long-term reflexive implementation (Ramiller 2019). The EcoDistrict Protocol is aimed at inclusive development and regeneration of people-centered, economically vibrant, and sustainable urban precincts. For these reasons, EcoDistricts Protocol defines a suitable framework for informal settlement upgrading.

Research and development of the EcoDistricts Protocol were initiated in 2012 and engaged a diverse set of expert stakeholders. It involved roundtables, listening sessions, and peer review. The first version of the Protocol was published in 2016. By 2018, the Protocol had been applied to the urban regeneration of over 60 neighborhoods (EcoDistricts 2018).

As defined in the Protocol, the EcoDistricts framework first and foremost requires a commitment to **equity**, **climate protection,** and **resilience**. Equity requires meaningful inclusion and participation, and upliftment of vulnerable groups such as the poor, migrants, old, infirm, disabled frail, women, and children. This is particularly important in Global South Cities where affluent residential enclaves often stand in stark juxtaposition with informal settlements in the same suburbs. Climate protection is aimed at global warming mitigation. Resilience requires a commitment to the continued functioning of the neighborhood in the face of transient shocks and enduring stresses.

The second component of the EcoDistrict framework is a set of six priorities, namely: ***place, prosperity, health and wellbeing, connectivity, living infrastructure, and resource regeneration.*** Each of the six priorities has three or more objective categories to give a total of 20 objective categories (see Fig. 4). To use the framework, each objective category must have at least one measurable indicator.

In combination with the three commitments, the six priorities and the 20 objective categories define pathways for just and sustainable urban development and regeneration. Because there are multiple ways in which each objective can be fulfilled, the indicators are non-prescriptive. This means that it is the prerogative of the user community to define suitable indicators for each objective category by taking into account unique and prevailing contextual factors. This flexibility, therefore, makes the EcoDistricts amenable to informal settlement upgrading in the Global South. Similar to the Nordic Framework for performance-based standards, such non-prescriptive flexibility allows stakeholders with different motives, knowledge, and capacities to engage in productive participatory design and co-creation (Sanya 2016).

In the next section, the priorities and objective categories are described, drawing particular attention to their value for informal settlement upgrading.

Place

The place priority has four objective categories. The ***engagement and inclusion*** objective category is about making civic structures strong and processes inclusive and representative. This priority is also about strengthening sharing networks and programs. In informal settlements, social capital is indispensable to survival and ability to rebuild livelihoods after disasters such as floods and fires. This objective can therefore not be overemphasized for informal settlement upgrading. Networks within the informal settlement and between the settlement and the better-resourced formal city should be enhanced by harnessing information technologies and other mechanisms.

The ***Culture and identity*** objective category includes preservation of heritage places and artifacts, and participation in cultural events. Another objective category is ***public spaces***. Whereby these must be engaging, universally accessible, high quality, and active. A final objective category in the place priority is aimed at achieving high-quality ***housing*** to meet a diversity of income groups, family structures, and needs. Housing must be located close to necessary amenities. This requires that government intervention, instead of stifling initiative, catalyzes individual efforts to build and incrementally upgrade their own housing. The government must therefore tap into the versatility of the private sector and private individuals to meet a diversity of housing needs.

Prosperity

The prosperity priority has three objective categories. The ***access to opportunity*** objective category promotes the reduction of income and race inequality, equitable access to quality education, new career pathways, and access to training. This is very important for informal settlements because they experience high rates of unemployment and underemployment. The ***economic development objective*** category is about retaining jobs in the settlement. In informal settlements, the starting point should be retaining the existing informal jobs, such as evening popup markets, roadside kiosks, and food gardens, etc. But above all, these jobs need to be enhanced through training, microcredit, linking into markets, and establishing synergetic relationships with firms in the formal sector. New jobs should be encouraged by bringing flagship projects through, for example, legislation changes and subsidies. Access to opportunity is very important also for giving people hope, and pathways and ability to contribute to community development. This is necessary to pre-empt crime and social breakdown as the world rapidly urbanizes. The ***innovation*** objective category aims to enhance interaction between entrepreneurs and at growing jobs in non-traditional sectors such as IT, programming, robotics, blockchain, remote working, industry

4.0, green mobility technologies, e-learning, digital art, music, and Non-Fungible Tokens (NFTs), etc. This involves retooling people, offering physical and online spaces for interaction and creation. This allows for melding of ideas, imagination of new products and services, and formation of partnerships.

Health and Well-being

There are four objective categories in the health and well-being priority. The ***active living*** objective category is about increasing well-being through measures such as access to recreational facilities and services, promoting walkability, and cycling. Here the value of WSD blue-green infrastructure is apparent – parks, nature trails, and multifunctional infrastructure (e.g., detention pond which becomes a sports field when dry). Walking under well-shaded paths is much better than working in the hot sun. Vegetation shelters people against harsh winter winds and the hot summer sun hence encouraging people to be outdoors to enjoy fishing, boating, swimming, and hiking – all offerings which can be attained through developing the city along water-sensitive principles. Therefore, instead of locking-in unsustainable spatial structures, WSD can be used to open up these opportunities for active living, particularly considering that inequity in access to urban green is currently significant (Lopez et al. 2021). The ***health*** objective category is about ensuring equitable health outcomes, including life expectancy. This involves ensuring environments are non-polluted and conducive to living. It also involves providing affordable, good quality healthcare within easy reach of informal settlement residents. Furthermore, it requires remediating and regenerating toxic or hazardous disaster-prone environments or moving them away to better lands. This is important because many informal settlements are located on marginal lands like toxic landfills, flood-prone lowlands, and unstable slopes. The objective category of ***safety*** is about creating safe places, particularly for the most vulnerable, such as women and children. The Violence Prevention through Urban Upgrading (VPUU) program in Cape Town is a good example. This program centered on public participation and design for public safety. It not only resulted in safe places but also improved infrastructure and stimulated businesses (Sanya 2016). The next objective category is ***food systems***. This focuses on making healthy and affordable food available to all. WSD is important here because food gardens can be planted as part of the blue-green network and watered with captured rainwater. Such gardens allow the urban poor to supplement their diets with leafy and other vegetables, save money, and even sell some for income. Food production should be encouraged by preserving fertile land, training, availing inputs, and opening up markets for fresh produce. Linking into circular metabolism, recycling of the nutrients in urine, and using compost should also be used to enhance local food gardening under a water-sensitive framework (see Pötz 2016).

Connectivity

The connectivity priority has three objective categories. The ***Street network*** objective category is about supporting all modes of travel and accommodating diverse ages and abilities. The poorest in informal settlements do not own a personal car. They use public transport, walk, and ride bicycles – all environmentally friendly travel modes. The blue-green grid of water sensitive offers opportunities for defining in walking and cycling pathways. Another objective category in connectivity priority is about increasing ***shared mobility***. This is good for the environment and also helps people reduce transport costs. Universal access should be designed into the street networks, so that the aged and infirm, women with prams, and persons with disabilities can move around easily. Multi-modal travel should also be enhanced with walking, cycling, and linking into public transport means. This helps to increase connectivity to other neighborhoods. In the ***digital connectivity*** category, the quality of internet connections is important throughout the district. By providing digital connectivity, spatially and socially marginal communities can find networking spaces in the digital world. This may include availing free or subsided data in local government buildings like administration buildings, clinics, community centers, libraries, and schools. Digital connectivity enhances peer-to-peer learning, cross-boundary socialization, and innovation.

Living Infrastructure

The living infrastructure priority overlaps completely with WSD and blue-green ecological infrastructure grids. The living infrastructure priority has three objective categories. The ***Natural features*** objective category is about protecting ecosystems and their services. The other objective category is protecting and improving ***ecosystem health*** for instance by maximizing the use of water captured onsite, managing water in a natural cycle and ensuring wastewater is fully cleaned onsite before releasing it downstream. Informal settlement upgrading should therefore happen within an overall approach of identifying, increasing, and linking up zonal and linear blue-green infrastructure. A furthermore category is about enhancing the ***connection with nature*** for city residents. Designing a city using a blue-green grid of ecological infrastructure is valuable in this regard. And objective category also calls for integrating natural processes into the built environment, for example, using sustainable urban drainage systems (SUDS).

Resource Regeneration

Resource regeneration priority has three objectives. ***Air*** objective category is about reducing greenhouse gas emissions through energy efficiency and carbon-neutral fuels. It also includes encouraging clean production and use of renewable energies like solar PV, solar thermal by offering subsidies and availing credit, for example. Therefore, new low-cost housing in informal settlements should use passive design principles for achieving indoor environmental comfort, for example, insulation, natural lighting, natural ventilation, passive heating in winter, and solar shading in summer. This objective also involves mitigating the urban heat island effect. As seen above, one advantage of WSD is mitigation of the urban heat island. In the ***Water*** objective category, just as is required in WSD, potable water should be used efficiently, and alternative diverse water sources found for nonportable purposes. This also includes protecting water quality from domestic and industrial wastewater pollution. Here, onsite wastewater treatment such as composting toilets, and neighborhood-level ones are encouraged [see (Tilley 2014) for a full range of examples]. Improving sanitation to make living conditions safe and protect environment is imperative in poorly serviced informal settlements. In this regard, using Bellagio Principles to work with communities in informal settlements to design and implement safe and socially acceptable sanitation is recommended. The ***Land*** category involves remediating contaminated land to be productive. It also involves reducing, reusing, and recycling waste to mitigate pressure on landfills. Overall, in the resource priority, waste recovery should be encouraged with an emphasis on working to approach the natural water, material, and carbon balances.

From Aspirational Objectives to Informal Settlement Upgrading Implementation

The EcoDistricts framework extends Water Sensitive Design, particularly in the social, participatory, and economic dimensions. Arguably, the approach and benefits of WSD as discussed above and as shown in Fig. 2 define a necessary pillar for socially inclusive and ecologically sustainable EcoDistricts shown in Fig. 3. Defining aspirational objectives, as above, is important. The biggest challenge, however, is in implementation. The informal settlement problem is huge and solving it is complex and quintessentially fluid. In implementation, this complexity can engender paralysis, costly mistakes, and even obfuscation. Successful implementation of solutions requires resource mobilization, planning across several spheres and sectors of governance, and collaboration with the private sector, civic organizations, and multiple stakeholder groups. It further involves a commitment to participatory methods so as to tap into the experiential informal knowledge of activists and laypeople and hold governments to account. In addition, implementation requires long-term planning and phasing. To borrow a metaphor from Schön (2017) moving from the ivory tower of aspirational

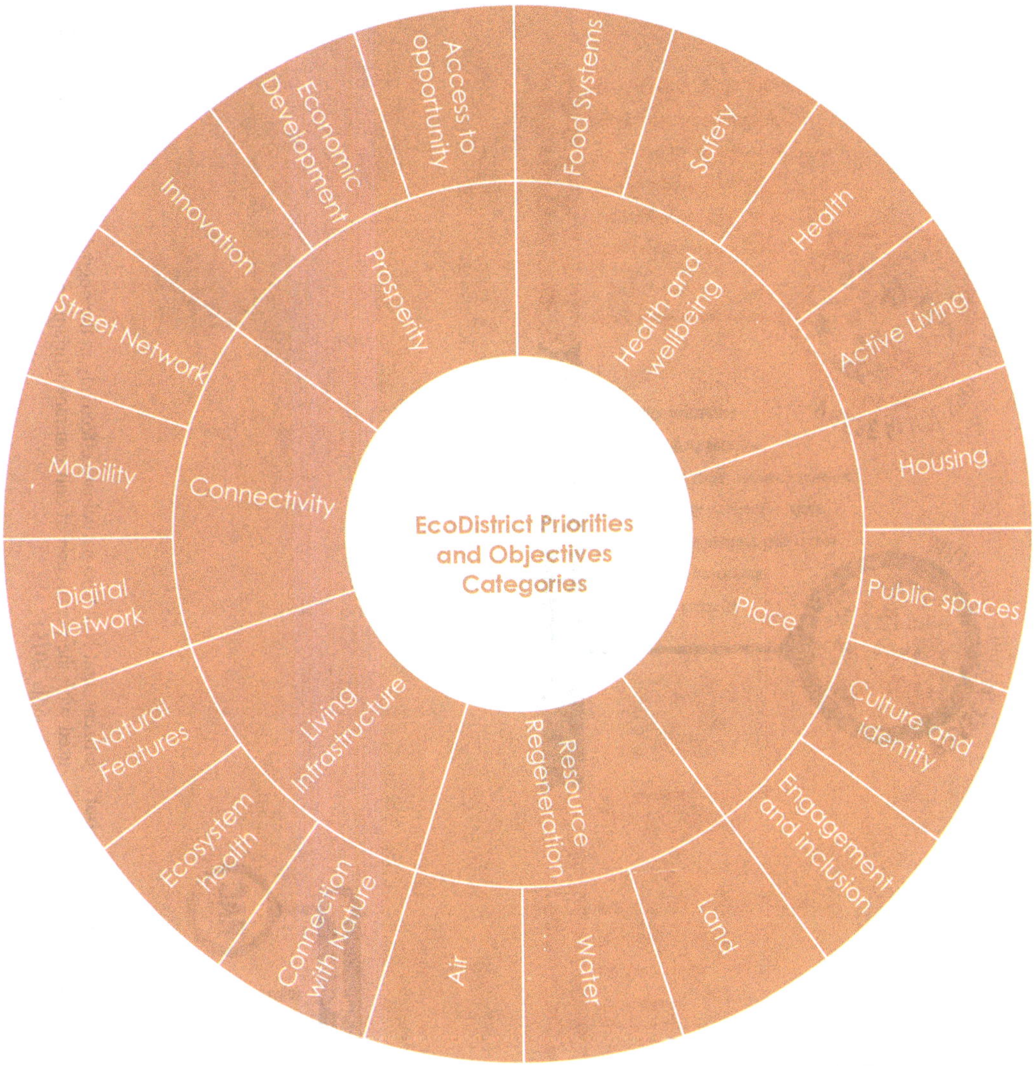

Guidelines for Water-Sensitive Informal Settlement Upgrading in the Global South, Fig. 3 EcoDistricts priorities and objective categories. Many of these 20 objective categories overlap with the benefits of Water Sensitive Design shown as discussed and shown in Fig. 2 above. (The image is adapted from EcoDistricts 2018). @Tom Sanya (CC BY 2.0)

objectives to the "lowly swamp of practice" substantive informal settlement upgrading is fraught, complex, and contingent. Implementation is a dynamic process, with new problems and opportunities continually emerging, melding, and transforming. It, therefore, requires a clear strategy and versatile tactics.

South Africa, perhaps more than any other country, vividly specializes in socioeconomic injustice in the city. Recent apartheid history and prevailing inequity have engendered and perpetuated informal settlements in physically and economically marginal locations of metropoles. The country's post-apartheid constitution is widely described as transformative and post-liberal (Klare 1998) because it is based on rights and justice. Amongst others, the constitution guarantees the rights to decent housing, adequate water and sanitation, and a clean environment. The constitution mandates the government to work for the

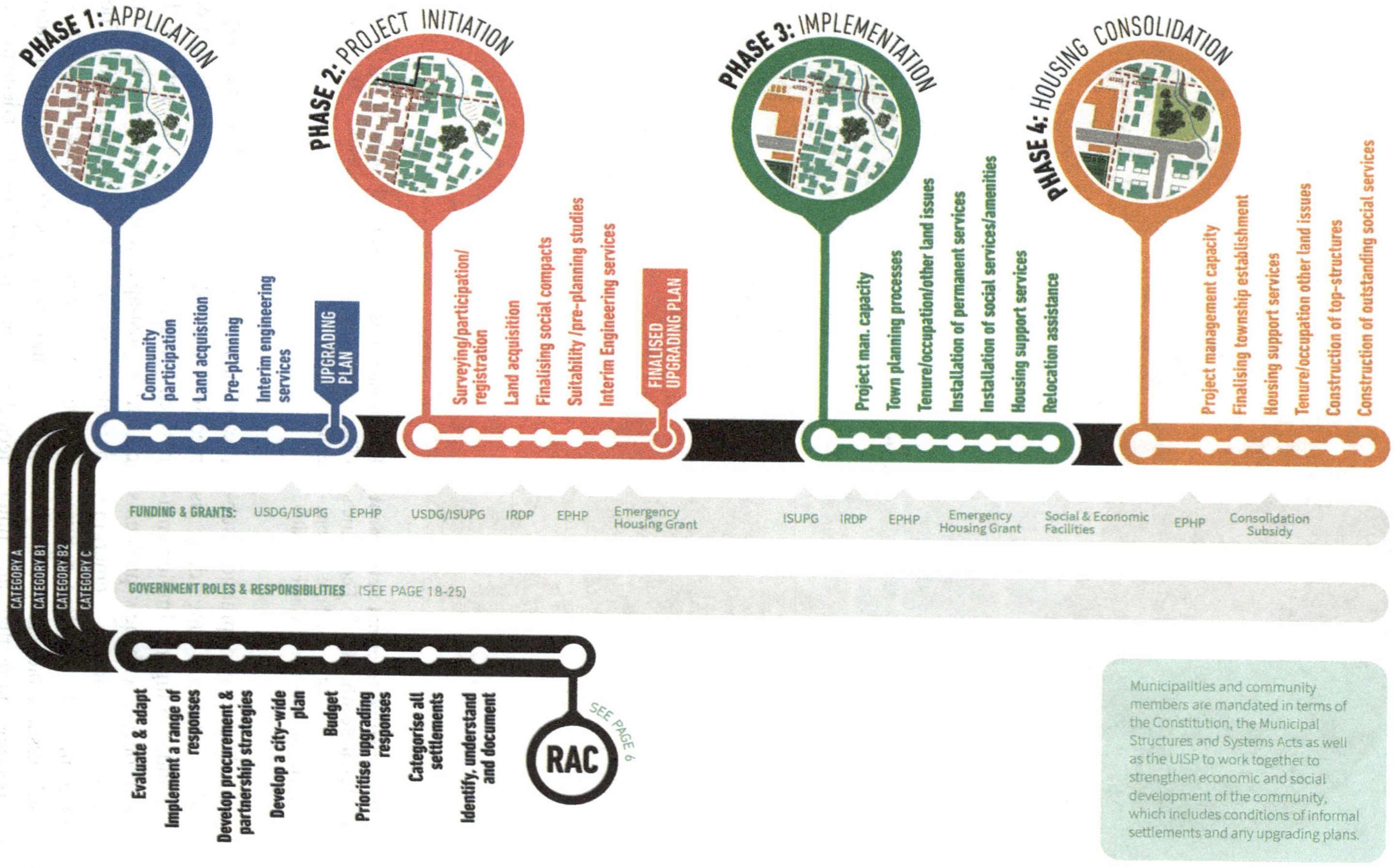

Guidelines for Water-Sensitive Informal Settlement Upgrading in the Global South, Fig. 4 The five-step roadmap for informal settlement upgrading in South Africa. The roadmap highlights legislatively mandated upgrading stages, responsibilities, and resources hence informing and empowering government officials, citizens, and NGOs to participate effectively in the upgrading process. (This image is used with the permission of and gratitude Mirjam van Donk of the Isandla Institute. © Isandla 2019)

progressive realization of socioeconomic rights for all South Africans.

To fulfill this mandate, the government has put in place several acts, policies, programs, and measures. These cascade through the national, provincial, and local spheres of government (see Glazewki and du Toit 2013 for a detailed breakdown). Between 1994 and 2013, just under three million new low-cost housing units were constructed mostly associated with the Reconstruction and Development Programme. But by 2011, the housing backlog was two million and continues to grow at more than 170,000 units per year. New policy leans toward in situ informal settlement upgrading in preference to constructing new low-cost housing. In South Africa, informal settlement upgrading is underpinned by the National Housing Act of 1997 which defines roles and responsibilities of different spheres of government. Set within a transformative constitution, these policies and measures for informal settlement upgrading are progressive but they are mired in the complexity entailed in cooperative governance. Navigating the pathway from the constitutional mandate to substantive informal settlement upgrading is a complex, fluid, and adaptive process involving different spheres of government, various policies, regulations requirements and mandates, and funding instruments. Apart from de facto and legislatively mandated role players, vigilance from communities and their partners is required to ensure legislated role players are held to account. To guide actors in implementation, Isandla (2019) put together an institutional roadmap and guidelines for informal settlement upgrading. The guidelines condense measures, means, funding mechanizations, and formal stages for upgrading in a country that is constitutionally obliged to progressively fulfill socioeconomic rights – including the right to housing. Therefore, the guidelines are an exemplar of how to approach just and sustainable informal settlement upgrading. Isandla condenses the complex implementation constellation into easy-to-navigate guidelines in a roadmap with five steps, namely: rapid assessment review, application, project initiation, implementation, and housing consolidation. These steps are summarized below, respectively (the reader is advised to read the full easy-to-navigate document as it has the appropriate level of detail).

Rapid Assessment Review (RAC)

The first stage is Rapid Assessment Review (RAC). Each municipality is required to undertake a RAC process for all informal settlements in its jurisdiction. Prioritizing the health and safety of the inhabitants, the municipality assigns each informal settlement to one of four upgrading categories:

- Category A: full conventional upgrading can start immediately
- Category B1: can be upgraded but only incrementally
- Category B2: the settlement is hazardous/unsafe and must be relocated, but immediate resettlement is not possible, for example, due to community issues
- Category C: settlement can and must be immediately relocated.

Regardless of classification, the municipality is obliged to provide interim services, such as water and sanitation, until a permanent solution is found.

Only after RAC can upgrading commence, and this is articulated in four phases as described below.

Phase 1: Application

Phase 1 is the Application Phase. The municipality applies for funding to the relevant grant-making authority. Award of funds is preconditioned on RAC, baseline and pre-feasibility study, demographic survey, initial geotechnical survey, census of households including identifying those that must be relocated, work plan, and an estimated budget. Granted funding is largely for the provision of interim services – in line with keeping resident health and safety as a priority. Application usually involves submitting the municipality's upgrading plans, which must reference the relevant sections of the municipality's broader Integrated Development Plan (IDP). Upgrading plans should amongst

G

others include information from RAC but must also justify why certain settlements must be prioritized. This phase also involves desktop scoping of the environmental assessment; the estimated number of households; the estimated number of households to be relocated; identification of illegal immigrants; preliminary work plan for the project implementation; and preliminary budget for the project.

If the application is approved by the relevant authority, the municipality can proceed to Phase 2.

In Phase 2: Project Initiation

Phase 2 involves land acquisition, preparing a full demographic and socio-economic profile, land acquisition, a full geotechnical survey, and the finalization of a business plan. According to Isandla (2019), though estimated to take about a single year, this phase is usually drawn out due to the complexities of implementation. Successful completion of Stage 2 can be a tipping point in the upgrading process.

Phase 3: Implementation

In Phase 3 the municipality must prepare a final upgrading plan and township layout. The plan must align with all other relevant plans. Clear modalities for community participation must be formulated, the work plan revised and extra streams of funding identified. Approval mandate for implementation is vested in the National Department of Human Settlements. Approval may be granted in full or conditional on revisions. After approval, the metro receives funding for the establishment of housing and project management. Permanent infrastructure such as water, sanitation and roads are provided, and social and community facilities are constructed. Housing support centers are established to boost the involvement of households in incremental construction of their dwelling in accordance with their needs, means, and aspirations; This stage may also involve the land acquisition of land and land rehabilitation.

Phase 4: Housing Consolidation

The focus of Phase 4 is the finalization of township establishment, ownership registration, housing construction, and any provision of any outstanding social amenities. A mix of housing options is recommended in terms of investment options, rentals, ownership, sizes, typologies, and so on. In all this, municipalities and communities are mandated to work together.

Conclusion

This is the urban century. The schisms between the affluent city and informal settlements get starker as the world urbanizes, especially in the Global South. The interconnections in the water cycle highlight that the dichotomies that humans frame within society and with nature are untenable. Ignoring informal settlements does not just hurt a significant and growing portion of the world population, but the entire humanity and the planet. Water Sensitive Design provides a lens to approach informal settlement upgrading as a necessity and within a broader Integrated Water Resource Management framework to stitch the formal and informal city and sustain livelihoods and nature. Anchoring Water Sensitive Design in an EcoDistricts framework yields a set of 20 actionable objectives for transforming informal settlements into people-centered, economically vibrant, and sustainable neighborhoods.

There is a huge chasm between inspirational objectives and substantive on-the-ground implementation. Implementation is a messy process that involves many actors at various levels of society with different mandates, means, and motives. Progress is always partial and evolving with ongoing negotiation and contestation of interests. During implementation, new problems and opportunities emerge and ability to work contingently is required. Successful implementation, therefore, requires an explicit commitment to participation and deliberative governance. Such commitment must be founded in social justice and sustainable development, and legislated at the highest levels of government. Regulations, mandates, specific roles, and responsibilities must be institutionalized and resourced appropriately. In all this complex institutional landscape, success is predicated on implementing spheres of government, communities and their partners, and all other stakeholders knowing their specific roles. This chapter has presented a five-step roadmap for informal settlement upgrading drawn from the legislative and policy landscape of South Africa. The roadmap enables informal settlement upgrading actors to bridge the gap between aspiration and substantive change on the ground. The roadmap gives a clear condensed birds-eye view of a complex process and hence enables government officials, active citizens, and activists to deliberatively engage in informal settlement upgrading. Importantly, the informal settlements roadmap highlights legislatively mandated upgrading stages, responsibilities, and resources. This, therefore, empowers communities to participate effectively and hold authorities accountable for the progressive realization of the socio-economic right to sustainable neighborhoods and livelihoods. The five-step roadmap is unique to South Africa, a country with a transformative post-liberal constitution. But the context in other countries is different. However, the key message remains: successful informal settlement upgrading must be grounded in social and environmental justice; must be protected in the apex law; and must be deployed and enacted in legislation, policy, normative guidelines, funding, and oversight. Through the lens of Water Sensitive Design (WSD) and using an EcoDistricts framework, this chapter has highlighted pathways to socially just and environmentally sustainable informal settlement upgrading in the Global South.

Cross-References

- ▶ Blue-Green Cities: Achieving Urban Flood Resilience, Water Security, and Biodiversity
- ▶ Climate Resilience in Informal Settlements: The Role of Natural Infrastructure
- ▶ Disaster Risk in Informal Settlements and Opportunities for Resilience
- ▶ Future of Urban Governance and Citizen Participation
- ▶ Green and Blue Infrastructure (GBI) in Urban Areas
- ▶ Green Cities: Nature-Based Solutions, Renaturing and Rewilding Cities
- ▶ Multiple Benefits of Green Infrastructure
- ▶ Health and the Role of Nature in Enhancing Mental Health
- ▶ Multiple Benefits of Green Infrastructure
- ▶ Nature-Based Solutions for River Restoration in Metropolitan Areas
- ▶ Need for Greenspace in an Urban Setting for Child Development
- ▶ Need for Nature Connectedness In Urban Youth For Environmental Sustainability
- ▶ Rainwater Harvesting for Water Security in Informal Settlements: Techniques, Practices, and Options
- ▶ Urban Food Gardens
- ▶ Urban Heat Islands
- ▶ Water, Sanitation, and Hygiene Question of Future Cities of the Developing World

G

References

Addams, L., Boccaletti, G., Kerlin, M., & Stuchtey, M. (2009). *Charting our water future: Economic frameworks to inform decisionmaking*. New York: McKinsey & Company.

Armitage, N., Fisher-Jeffes, L., Carden, K., Winter, K., Naidoo, V., Spiegel, A., Mauck, B. & Coulson, D. (2014). Water sensitive Urban Design (WSUD) for South Africa: Framework and guidelines. Water Research Commission report no. TT 588/14, ISBN. 978-1-4312-0530-1.

Braga, B. (2010). Integrated urban Water resources management: A challenge into the 21st century. *International Journal of Water Resources Development, 17*(4), 581–599.

Brears, R. C. (2018). Blue-green infrastructure in managing urban Water resources. In *Blue and green cities* (pp. 43–61). London: Springer.

Butler, L. J., Scammell, M. K., & Benson, E. B. (2016). The Flint, Michigan, water crisis: A case study in regulatory failure and environmental injustice. *Environmental Justice, 9*(4), 93–97.

Carden, K., Armitage, N., Fisher-Jeffes, L., Winter, K., Mauck, B., Sanya, T., et al. (2018). *Challenges and opportunities for implementing Water sensitive Design in South Africa*.

Carpenter, S. P. S. (1997). Freshwater ecosystem services. In G. C. Daily (Ed.), *Nature's services: societal dependence on natural ecosystems* (pp. 195–214). Washington D.C.: Island Press.

Centers for Disease Control and Prevention. (2015). Diarrhea: Common illness, global killer. Department of

Health and Human Services. Centre for Research on the Epidemiology of Disasters [CRED] (2017) Disaster Data: A balanced perspective: Floods: Issue No. 48 "Disaster Data: A Balanced Perspective" September 2017.

Cleugh, H., & Grimmond, S. (2011). *Urban climates and global climate change. The future of the world's climate* (2nd ed., pp. 47–76).

Constanz, R., et al. (1997). The value of the world's ecosystem services and natural capital. *Nature, 387*, 253–260.

Council, B. C. (2010). *WaterSmart strategy: Supporting the liveability of Brisbane by managing water sustainably.*

Desa, U. N. (2019). *World population prospects 2019: highlights*. New York: United Nations Department for Economic and Social Affairs.

Ecodistricts (2018). EcoDistricts Protocol (version 1.3): the Standard for Urban and Community Development.

Galtung, J. (1969). Violence, peace, and peace research. *Journal of peace research, 6*(3), 167–191.

Geddes, P. (1915). *Cities in evolution: An introduction to the town planning movement and to the study of civics*. London: Williams.

Glazewski, J. (1999). Environmental Law in South Africa. Butterworth-Heinemann.

Gosling, S. N., & Arnell, N. W. (2016). A global assessment of the impact of climate change on water scarcity. *Climatic Change, 134*(3), 371–385.

Guppy, L., & Anderson, K. (2017). *Global Water crisis: The facts*. Hamilton, UNU-INWEH.

Halbe, J., Pahl-Wostl, C., Lange, A. M., & Velonis, C. (2015). Governance of transitions towards sustainable development–the water–energy–food nexus in Cyprus. *Water International, 40*(5–6), 877–894.

Howard, E. (1902). *Garden cities of tomorrow*. London: S. Sonnenschein &, Ltd. (at Google Books).

Isandla. (2019). *Informal settlement upgrading: An institutional map for NGOs*. Cape Town: Isandla Institute.

Khandker, V., Gandhi, V. P., & Johnson, N. (2020). Gender perspective in Water management: The involvement of women in participatory Water institutions of eastern India. *Water, 12*(1), 196.

Kinniburgh, C. (2020). Can extinction rebellion survive? *Dissent, 67*(1), 125–133.

Klare, K. E. (1998). Legal culture and transformative constitutionalism. *South African Journal on Human Rights, 14*(1), 146–188.

Lande, L. (2015). *Eliminating discrimination and inequalities in access to water and sanitation*. UN-Water.

Lebron, C. J. (2017). *The making of black lives matter: A brief history of an idea*. New York: Oxford University Press.

Lenton, R., & Muller, M. (2012). *Integrated water resources management in practice: Better water management for development*. London: Routledge.

Lopez, B., Kennedy, C., Field, C., & McPhearson, T. (2021). Who benefits from urban green spaces during times of crisis? Perception and use of urban green spaces in new York City during the COVID-19 pandemic. *Urban Forestry & Urban Greening, 65*, 127354.

Lüthi, C., Panesar, A., Schütze, T., Norström, A., McConville, J., Parkinson, J., Saywell, D., & Ingle, R. (2011). *Sustainable sanitation in cities: A framework for action*. Sustainable Sanitation Alliance (SuSanA) & International Forum on Urbanism (IFoU).

Mabhaudhi, T., Chibarabada, T., & Modi, A. (2016). Water-food-nutrition-health nexus: Linking water to improving food, nutrition and health in sub-Saharan Africa. *International Journal of Environmental Research and Public Health, 13*(1), 107.

Mekonnen, M. M., & Hoekstra, A. Y. (2016). Four billion people facing severe water scarcity. *Science Advances, 2*(2), e1500323.

Merrey, D. J., Drechsel, P., de Vries, F. P., & Sally, H. (2005). Integrating "livelihoods" into integrated water resources management: Taking the integration paradigm to its logical next step for developing countries. *Regional Environmental Change, 5*(4), 197–204.

Millennium ecosystem assessment, M. E. A. (2005). *Ecosystems and human well-being* (Vol. 5). Washington, DC: Island press.

Morgan, C., Bevington, C., Levin, D., Robinson, P., Davis, P., Abbott, J., & Simkins, P. (2013). *Water sensitive Urban Design in the UK–ideas for built environment practitioners*. London: Ciria.

Newsinger, J. (2016). Why Rhodes must fall. *Race & Class, 58*(2), 70–78.

Nixon, R. (2011). *Slow violence and the environmentalism of the poor*. Harvard University Press.

OECD/SWAC. (2020). *Africa's urbanisation dynamics 2020: Africapolis, mapping a new urban geography, west African studies*. Paris: OECD Publishing. https://doi.org/10.1787/b6bccb81-en.

O'Farrell, P., Anderson, P., Culwick, C., Currie, P., Kavonic, J., McClure, A., . . . Audouin, M. (2019). Towards resilient African cities: Shared challenges and opportunities towards the retention and maintenance of ecological infrastructure. *Global Sustainability, 2*, E19. https://doi.org/10.1017/sus.2019.16.

Olsson, G. (2013). Water, energy and food interactions – Challenges and opportunities. *Frontiers of Environmental Science & Engineering, 7*(5), 787–793.

Pearson, L., & Pelling, M. (2015). The UN Sendai framework for disaster risk reduction 2015–2030: Negotiation process and prospects for science and practice. *Journal of Extreme Events, 2*(01), 1571001.

Pötz, H., Sjauw En Wa-Windhorst, A., & van Someren, H. (2016). *Urban green-blue grids manual for resilient cities*. In atelierGROENBLAUW.

Ramiller, A. (2019). Establishing the green neighbourhood: Approaches to neighbourhood-scale sustainability certification in Portland. *Oregon. Local Environment, 24*(5), 428–441.

Revenga, C. (2003). *Status and trends of biodiversity of inland water ecosystems*. Secretariat of the Convention on Biological Diversity.

Rijsberman, F. R. (2006). Water scarcity: Fact or fiction? *Agricultural Water Management, 80*(1–3), 5–22. https://doi.org/10.1016/j.agwat.2005.07.001.

Sanya, T. (2016). Participatory design: An intersubjective schema for decision making. *ArchNet-IJAR: International Journal of Architectural Research, 10*(1), 62.

Sanya, T. (2020). Freshwater: Towards a better understanding of a wicked problem. *International Journal of Environmental Science & Sustainable Development, 5*(2), 48–59.

Schewe, J., Heinke, J., Gerten, D., Haddeland, I., Arnell, N. W., Clark, D. B., Dankers, R., Eisner, S., Fekete, B. M., Colón-González, F. J., & Gosling, S. N. (2014). Multimodel assessment of water scarcity under climate change. *Proceedings of the National Academy of Sciences, 111*(9), 3245–3250.

Schön, D. A. (2017). The reflective practitioner: How professionals think in action. Routledge.

Scientific and Cultural Organisation (UNESCO). (2020). The United Nations world water development report 2017.

Sisolak, J., & Spataro, K. (2011). *Toward net zero water: Best management practices for decentralized sourcing and treatment.* Cascadia Green Building Council.

Smakhtin, V., Revenga, C., & Döll, P. (2004). A pilot global assessment of environmental water requirements and scarcity. *Water International, 29*(3), 307–317.

Steiner, F. (2014). Frontiers in urban ecological design and planning research. *Landscape and Urban Planning, 125*, 304–311.

Struyf, E., et al. (2012). *Nitrogen, phosphorus and silicon in riparian ecosystems along the Berg River (South Africa): The effect of increasing human land use* (pp. 597–606). https://doi.org/10.4314/wsa.v38i4.15. Available on website http://www.wrc.org.za.

Taing, L., Chang, C. C., Pan, S., & Armitage, N. P. (2019). Towards a water secure future: Reflections on Cape Town's day zero crisis. *Urban Water Journal, 16*(7), 530–536.

Tansley, A. G. (1935). The use and abuse of Vegetaional concepts and terms. *Ecology, 16*(3), 284–307. http://links.jstor.org/sici?sici=0012-9658%28193507%2916%3A3%3C284%3ATUAAOV%3E2.0.CO%3B2-P.

Tilley, E. (2014). *Compendium of sanitation systems and technologies*. Eawag.

UN-Department of Economic and Social Affairs Statistics Division. (2021). Make cities and human settlements inclusive, safe, resilient and sustainable https://unstats.un.org/sdgs/report/2019/goal-11/

United Nations Educational UN-Water. (2021). Summary progress update 2021: SDG 6 – water and sanitation for all. https://www.unwater.org/publications/summary-progress-update-2021-sdg-6-water-and-sanitation-for-all/

Van Dijk, A. I., Beck, H. E., Crosbie, R. S., de Jeu, R. A., Liu, Y. Y., Podger, G. M., … Viney, N. R. (2013). The millennium drought in Southeast Australia (2001–2009): Natural and human causes and implications for water resources, ecosystems, economy, and society. *Water Resources Research, 49*(2), 1040–1057.

Venter, Z. S., Barton, D. N., Gundersen, V., Figari, H., & Nowell, M. S. (2021). Back to nature: Norwegians sustain increased recreational use of urban green space months after the COVID-19 outbreak. *Landscape and Urban Planning, 214*, 104175.

Vié, J. C., Hilton-Taylor, C., & Stuart, S. N. (Eds.). (2009). *Wildlife in a changing world: An analysis of the 2008 IUCN red list of threatened species*. IUCN.

Walker, R. V., Beck, M. B., Hall, J. W., Dawson, R. J., & Heidrich, O. (2014). The energy-water-food nexus: Strategic analysis of technologies for transforming the urban metabolism. *Journal of environmental management, 141*, 104–115.

Wastewater the Untapped Resource. United Nations Educational, Scientific and Cultural Organization (UNESCO), Paris. Retrieved from http://unesdoc.unesco.org/images/0024/002471/247153e.pdf

Water, U. N. (2021). *Summary progress update 2021: SDG 6 – Water and sanitation for all*. Geneva: Switzerland.

Wolski, P. (2018). How severe is Cape Town's "day zero" drought? *Significance, 15*(2), 24–27.

Wong, T. H., & Brown, R. R. (2009). The water sensitive city: Principles for practice. *Water Science and Technology, 60*(3), 673–682.

Wong, T. H., & Eadie, M. L. (2000). Water sensitive urban design: A paradigm shift in urban design. In 10th world Water congress: Water, the worlds Most important resource (1281). International Water Resources Association.

World Bank. (2020). Urban development. Available online https://www.worldbank.org/en/topic/urbandevelopment/overview#1

World Economic Forum [WEF]. (2019). *The global risks report 2019* (14th ed.). Insight Report.

Yamazaki, T., Iida, A., Hino, K., Murayama, A., Hiroi, U., Terada, T., … Yokohari, M. (2021). Use of urban green spaces in the context of lifestyle changes during the COVID-19 pandemic in Tokyo. *Sustainability, 13*(17), 9817.

The Palgrave Encyclopedia of Urban and Regional Futures

Robert C. Brears
Editor

The Palgrave Encyclopedia of Urban and Regional Futures

Volume 2

H–R

With 418 Figures and 143 Tables

palgrave
macmillan

Editor
Robert C. Brears
Our Future Water
Christchurch, New Zealand

ISBN 978-3-030-87744-6 ISBN 978-3-030-87745-3 (eBook)
https://doi.org/10.1007/978-3-030-87745-3

This Palgrave Macmillan imprint is published by the registered company Springer Nature Switzerland AG.
The registered company address is: Gewerbestrasse 11, 6330 Cham, Switzerland

Preface

The *Palgrave Encyclopedia of Urban and Regional Futures* provides readers (practitioners, academics, researchers, etc.) with expert interdisciplinary knowledge on how urban centers and regions in locations of varying climates, lifestyles, income levels, and stages development are creating synergies and reducing trade-offs in the development of resilient, resource-efficient, environmentally friendly, liveable, socially equitable, integrated, and technology-enabled centers and regions. In particular, the *Palgrave Encyclopedia of Urban and Regional Futures* provides chapters, authored by subject matter experts, on interdisciplinary policies, best practices, lessons learnt, technologies in various stages of development, and case studies of urban centers and regions that aim to decouple economic growth from resource consumption, enhance resilience to climatic extremes, invest in low/zero carbon and smart technologies, lower emissions, reduce economic disparities, improve quality of life, and protect ecosystems and the services they provide for humans and nature.

Christchurch, New Zealand
December 2022

Robert C. Brears
Editor

Acknowledgments

First, I wish to thank Ruth Lefevre and Rachael Ballard for being visionaries who enable Major Reference Works like mine to come to fruition. Second, I wish to thank Anusha Cherian for being an excellent project coordinator. Third, I wish to thank Mum, who has a great interest in the environment and has supported me in this journey.

List of Topics

About the Editor

Robert C. Brears is an international sectoral expert on water for the UN's Green Climate Fund and the World Bank. He is the Editor in Chief of the *Palgrave Handbook of Climate Resilient Societies* and the *Palgrave Encyclopedia of Urban and Regional Futures*. He is the author of 11 books, including the Palgrave Macmillan titles *The Green Economy and the Water-Energy-Food Nexus*, *Blue and Green Cities: The Role of Blue-Green Infrastructure in Managing Urban Water Resources*, *Natural Resource Management and the Circular Economy*, *Developing the Circular Water Economy*, *Developing the Blue Economy*, and *Financing Nature-based Solutions*. He is the founder of Our Future Water, which has knowledge partnerships with World Bank, World Meteorological Organization, and UNEP initiatives.

Contributors

Ayodeji Adeniyi Deception Bay, QLD, Australia

Humera Afaq National University, San Diego, CA, USA

Kristin Agnello Department of Architecture and Planning, Norwegian University of Science and Technology, Trondheim, Norway

Atharv Agrawal University of Toronto, Toronto, ON, Canada

S. Ahilan College of Engineering, Mathematics and Physical Sciences, University of Exeter, Exeter, Devon, UK

Iftekhar Ahmed School of Architecture and Built Environment, University of Newcastle, Callaghan, NSW, Australia

Mubeen Ahmad School of Earth and Environmental Sciences, The University of Queensland, Brisbane, QLD, Australia

Ahmad Ahsan Lahore University of Management Sciences, Lahore, Pakistan

Meredian Alam Sociology and Anthropology Department, Universiti Brunei Darussalam, Gadong, Brunei

Amani Alfarra Land and Water Division, Food and Agriculture Organization of the United Nations, Rome, Italy

Jamal Alibou Department of Civil Engineering, Hydraulic, Environment and Climate, Hassania School of Public Works, Casablanca, Morocco

Sara Alidoust School of Earth and Environmental Sciences, The University of Queensland, Brisbane, QLD, Australia

Angélica Tanus Benatti Alvim Graduate Program in Architecture and Urbanism, Mackenzie Presbyterian University, Sao Paulo, Brazil

P. Ambily Department of Civil Engineering, National Institute of Technology, Calicut, Kerala, India

Grace Andrews Masters Environmental Management, College of Humanities, Arts and Social Sciences, Flinders University, Adelaide, South Australia

Jenna Andrews-Swann School of Liberal Arts, Georgia Gwinnett College, Lawrenceville, GA, USA

T. Angert The Institute for Environmental Security and Well-being Studies, Herzliya, Israel

Shyni Anilkumar National Institute of Technology Calicut, Kozhikode, Kerala, India

Aikaterini Antonopoulou

Hadi Arbabi Department of Civil & Structural Engineering, The University of Sheffield, Sheffield, UK

Md. Arfanuzzaman Food and Agriculture Organization (FAO) of the United Nations, Dhaka, Bangladesh

Felipe Armas Vargas Departamento de Ingeniería de Procesos e Hidráulica, CBI, Universidad Autónoma Metropolitana-Iztapalapa, Ciudad de México, Mexico

S. Arthur Heriot-Watt University, Edinburgh, UK

Hedda Askland University of Newcastle, Callaghan, NSW, Australia

Ditjon Baboci Tirana, Albania

Guy Baeten Urban Studies, Malmö University, Malmö, Sweden

Elham Bahmanteymouri The University of Auckland, Auckland, New Zealand

Nilesh Bakshi School of Architecture, Victoria University of Wellington, Wellington, New Zealand

M. Balasubramanian Centre for Ecological Economics and Natural Resources, Institute for Social and Economic Change, Bangalore, Karnataka, India

Zoran Balukoski School of Geography and Sustainable Communities, The University of Wollongong, Wollongong, NSW, Australia

Jonathan Banfield University of Toronto, Toronto, ON, Canada

Kaya Barry Griffith University, Brisbane, QLD, Australia

Department of Culture and Learning, Aalborg University, Aalborg, Denmark

Matthias Barth Leuphana University, Lüneburg, Germany

Prabal Barua Department of Environmental Sciences, Faculty of Physical and Mathematical Sciences, Jahangirnagar University, Dhaka, Bangladesh

James A. Beckman University of Central Florida, Orlando, FL, USA

Sara Bice Crawford School of Public Policy, The Australian National University, Acton, ACT, Australia

School of Public Policy and Management, Tsinghua University, Beijing, China

S. Birkinshaw University of Newcastle, Newcastle, UK

Stefan Blachfellner The Bertalanffy Center for the Study of Systems Science, Vienna, Austria

Bruno Blanco-Varela Department of Applied Economics, Faculty of Economics, Universidade de Santiago de Compostela, Santiago de Compostela, Galicia, Spain

Tijana Blanusa Royal Horticultural Society, Wisley, UK

University of Reading, Reading, UK

Tinashe Bobo Town Planning Section, Harare City Council, Harare, Zimbabwe

Cherice Bock Portland Seminary of George Fox University, Portland, OR, USA

Antonija Bogadi Department of Urban and Spatial Planning and Research, Technical University Vienna, Vienna, Austria

Simone Borelli Food and Agriculture Organization of the United Nations (FAO), Rome, Italy

Candice Boyd University of Melbourne, Melbourne, VIC, Australia

Christopher T. Boyko Lancaster University, Lancaster, UK

Robert C. Brears Our Future Water, Christchurch, New Zealand

Maria Julieta Brezzo Institutional Relations and Events, Ciudades Globales – CIGLO, Córdoba, Argentina

Katja Brundiers School of Sustainability, Arizona State University, Tempe, AZ, USA

Valerio Alfonso Bruno Università Cattolica del Sacro Cuore, Milan, Italy

Center for European Futures, Naples, Italy

Centre for the Analysis of the Radical Right, Leeds, UK

Felipe Bucci Ancapi Department of Management in the Built Environment, Faculty of Architecture and the Built Environment, Delft University of Technology, Delft, The Netherlands

Felix Bücken Institute of Geography, Osnabrück University, Osnabrück, Germany

Paul Burton Cities Research Institute Griffith University, Gold Coast, QLD, Australia

Alessandro Busà School of Geography, Geology and the Environment, University of Leicester, Leicester, UK

Judy Bush Lecturer in Urban Planning at University of Melbourne, Melbourne, VIC, Australia

Gareth Butler Masters Environmental Management, College of Humanities, Arts and Social Sciences, Flinders University, Adelaide, South Australia

Andrew Butt RMIT University, Melbourne, VIC, Australia

Michael Buxton RMIT University, Melbourne, VIC, Australia

Mohammed Firoz C. Department of Architecture and Planning, National Institute of Technology Calicut, Kozhikode, Kerala, India

Maléne Campbell Department of Urban and Regional Planning, University of the Free State, Bloemfontein, South Africa

Julien Carbonnell Artificial Intelligence on Citizen Engagement, Democracy Studio

M. Cavada School of Architecture, Imagination Lancaster, Lancaster University, Lancaster, UK

Rebecca Cavicchia Department of Urban and Regional Planning, BYREG – Norwegian University of Life Science, Ås, Norway

Lauriane Suyin Chalmin-Pui Royal Horticultural Society, Wisley, UK

The University of Sheffield, Sheffield, UK

Deborah Nabubwaya Chambers Community Health, National University, San Diego, CA, USA

Shenglin E. Chang National Taiwan University, Graduate Institute of Building and Planning, Taipei, Taiwan

Marianna Charitonidou Department of Art Theory and History, Athens School of Fine Arts, Athens, Greece

School of Architecture, National Technocal University of Athens, Athens, Greece

Department of Architecture, ETH Zurich, Zurich, Switzerland

Charles M. Chavunduka Department of Architecture and Real Estate, University of Zimbabwe, Harare, Zimbabwe

Ambika Chawla Urban Climate Innovations, Washington, DC, USA

Fei Chen School of Architecture, University of Liverpool, Liverpool, UK

Andrew Chigudu Department of Demography Settlement and Development, Social & Behavioural Sciences, University of Zimbabwe, Harare, Zimbabwe

Halleluah Chirisa Population Services International Zimbabwe, Harare, Zimbabwe

Innocent Chirisa Department of Demography Settlement and Development, Social & Behavioural Sciences, University of Zimbabwe, Harare, Zimbabwe

Chipo Chitereka Department of Social Work, University of Zimbabwe, Harare, Zimbabwe

N. R. Chithra Department of Civil Engineering, National Institute of Technology, Calicut, Kerala, India

Marcyline Chivenge Department of Demography Settlement and Development, Social & Behavioural Sciences, University of Zimbabwe, Harare, Zimbabwe

Suehyun Cho University of Toronto, Toronto, ON, Canada

Tanya Clark School of Behavioral Sciences, California Southern University, Costa Mesa, CA, USA

M'Lisa Lee Colbert The Nature of Cities, Montreal, QC, Canada

Ramon Fernando Colmenares-Quintero Faculty of Engineering, Universidad Cooperativa de Colombia, Medellín, Colombia

Elif Çolakoğlu Department of Security Sciences, Gendarmerie and Coast Guard Academy, Ankara, Turkey

Michela Conigliaro Food and Agriculture Organization of the United Nations (FAO), Rome, Italy

Sean Connelly University of Otago, Dunedin, New Zealand

A. Contin Politecnico di Milano, Milan, Italy

Rachel Cooper Lancaster University, Lancaster, UK

Samantha Copeland Ethics and Philosophy of Technology, Delft University of Technology, Delft, The Netherlands

João Cortesão Landscape Architecture and Spatial Planning, Wageningen University, Wageningen, The Netherlands

Adriano Cozzolino Center for European Futures, Naples, Italy

Università degli Studi della Campania "Luigi Vanvitelli", Caserta, Italy

Stewart Craine Village Infrastructure Angels, London, UK

Roberta Cucca BYREG – Norwegian University of Life Science, Ås, Norway

Gary Cummisk Central Washington University, Ellensburg, WA, USA

Paul Cureton ImaginationLancaster, Lancaster University, Lancaster, UK

Susan Cyriac Department of Architecture and Planning, National Institute of Technology, Calicut, Kerala, India

Sebastien Darchen School of Earth and Environmental Sciences, The University of Queensland, Brisbane, QLD, Australia

Curt J. Davis University of Delaware, Newark, DE, USA

D. Dawson University of Leeds, Leeds, UK

Evelyne de Leeuw Centre for Health Equity Training, Research and Evaluation (CHETRE), UNSW Australia Research Centre for Primary Health Care & Equity, South Western Sydney Local Health District, Ingham Institute, Sydney, NSW, Australia

Healthy Urban Environments (HUE) Collaboratory, Maridulu Budyari Gumal Sydney Partnership for Health, Education, Research and Enterprise SPHERE, Sydney, NSW, Australia

Ingham Institute for Applied Medical Research, Liverpool, NSW, Australia

Valerio Della Sala Politecnico di Torino (Italy), Interdepartmental Research Centre for Urban Studies (OMERO), Universitat Autonoma de Barcelona (Spain), Turin, Italy

N. Delle-Odeleye Anglia Ruskin University, Chelmsford, UK

Cheryl Desha Cities Research Institute, Griffith University, Brisbane, QLD, Australia

María Mercedes Di Virgilio Instituto de Investigaciones Gino Germani, Universidad de Buenos Aires/ CONICET, Ciudad Autónoma de Buenos Aires, Argentina

Roshini Suparna Diwakar Mahila Housing Trust, New Delhi, India

Timothy J. Dixon School of the Built Environment, University of Reading, Reading, UK

Michelle Duffy University of Newcastle, Callaghan, NSW, Australia

Smart Dumba Department of Demography Settlement and Development, Social & Behavioral Sciences, University Zimbabwe, Harare, Zimbabwe

Nick Dunn Lancaster University, Lancaster, UK

Jenna Dutton Senior Planner – Social Policy, City of Victoria and Research Associate, Center for Civilization, University of Calgary, Calgary, AB, Canada

Vupenyu Dzingirai Department of Community and Social Development, University of Zimbabwe, Harare, Zimbabwe

Charity Edwards Monash University & University of Melbourne, Melbourne, Australia

Huascar Eguino Fiscal Management Division, Inter-American Development Bank (IDB), Washington, DC, USA

Theodore S. Eisenman Department of Landscape Architecture and Regional Planning, University of Massachusetts-Amherst, Amherst, MA, USA

Christina R. Ergler School of Geography, University of Otago, Dunedin, New Zealand

Oscar Escolero Departamento de Dinámica Terrestre y Superficial, Instituto de Geología, Universidad Nacional Autónoma de México, Ciudad de México, Mexico

Javier Esquer Graduate Sustainability Program, Industrial Engineering Department, University of Sonora, Hermosillo, Mexico

G. Everett University of the West of England, Bristol, UK

Caroline Fabianski La Seyne sur Mer, France

Francesco Femia The Center for Climate and Security, an Institute of the Council on Strategic Risks, Washington, DC, USA

Melisha Shavindi Fernando Faculty of Science, Horizon Campus, Malabe, Sri Lanka

Carmen Zuleta Ferrari Food and Agriculture Organization of the United Nations (FAO), Rome, Italy

Carla Sofia Ferreira Research Centre for Natural Resources, Environment and Society (CERNAS), Polytechnic Institute of Coimbra, Coimbra Agrarian Technical School, Coimbra, Portugal

Department of Physical Geography and Bolin Centre for Climate Research, Stockholm University, Stockholm, Sweden

Navarino Environmental Observatory, Messinia, Greece

António Ferreira Research Centre for Natural Resources, Environment and Society (CERNAS), Polytechnic Institute of Coimbra, Coimbra Agrarian Technical School, Coimbra, Portugal

Daniel Fischer School of Sustainability, Arizona State University, Tempe, AZ, USA

Wesley Flannery Urban Planning, School of Natural and Built Environment, David Keir Building, Queen's University Belfast, Belfast, UK

Claudia Fonseca Alfaro Institute for Urban Research, Malmö University, Malmö, Sweden

Mariana Fonseca Braga ImaginationLancaster, Lancaster Institute for the Contemporary Arts (LICA), Lancaster University, Lancaster, Lancashire, UK

Julien Forbat University of Geneva, Institute of Global Health, Geneva, Switzerland

Martin Franz Institute of Geography, Osnabrück University, Osnabrück, Germany

Robert Freestone School of Built Environment, University of New South Wales, Sydney, NSW, Australia

Frances Furio School of Behavioral Sciences, California Southern University, Costa Mesa, CA, USA

Tatiana Gallego Lizon Washington, DC, USA

Emilio Garcia The University of Auckland, Auckland, New Zealand

Birgit Georgi UIA Expert/Strong Cities in a Changing Climate, Egelsbach, Germany

Daniela Getlinger Graduate Program in Architecture and Urbanism, Mackenzie Presbyterian University, Sao Paulo, Brazil

David J. Gilchrist University of Western Australia, Perth, WA, Australia

Brendan Gleeson Monash University & University of Melbourne, Melbourne, Australia

V. Glenis University of Newcastle, Newcastle, UK

Moritz Gold Sustainable Food Processing Laboratory, ETH Zurich, Zurich, Switzerland

Eugenio Gómez Reyes Departamento de Ingeniería de Procesos e Hidráulica, CBI, Universidad Autónoma Metropolitana-Iztapalapa, Ciudad de México, Mexico

Megan Gordon University of Northern British Columbia, Prince George, BC, Canada

Alexa Gower Monash University, Melbourne, VIC, Australia

Sonia Graham School of Geography and Sustainable Communities, The University of Wollongong, Wollongong, NSW, Australia

Institut de Ciència I Tecnologia Ambientals (ICTA), Universitat Autònoma de Barcelona, Barcelona, Spain

Danielle Griego Center for Augmented Computational Design in Architecture, Engineering and Construction, D-BAUG, ETH Zurich, Zurich, Switzerland

Kai Michael Griese Hochschule Osnabrück University of Applied Sciences, Osnabrück, Germany

Carl Grodach Monash University, Melbourne, VIC, Australia

Bern Grush Urban Robotics Foundation, Toronto, Canada

Medhisha Pasan Gunawardena Biodiversity Educational Research Initiative, Colombo, Sri Lanka

Faculty of Science, Horizon Campus, Malabe, Sri Lanka

Hector Manuel Guzman Grijalva Sustainability Graduate Program, University of Sonora, Hermosillo, México

Jochen Hack Technical University of Darmstadt, Section of Ecological Engineering, Institute of Applied Geosciences, Darmstadt, Germany

Perrine Hamel Asian School of the Environment, Nanyang Technological University, Singapore, Singapore

Earth Observatory of Singapore, Nanyang Technological University, Singapore, Singapore

Ben Harris-Roxas School of Population Health, University of New South Wales, Sydney, NSW, Australia

Wolfgang Haupt Leibniz-Insitute for Research on Society and Space, Erkner, Germany

Naomi Hay Australian National University, Canberra, ACT, Australia

Fatime Barbara Hegyi Joint Research Centre – European Commission, Seville, Spain

Hayley Henderson Research Fellow at Crawford School of Public Policy, Australian National University, Canberra, ACT, Australia

Michael Henderson Ramboll Ltd and Oxford Brookes University, London, UK

Cole Hendrigan University of Wollongong and Wollongong City Council, Wollongong, NSW, Australia

Andreas Hernandez Marymount Manhattan College, New York, NY, USA

Victoria Herrmann The Arctic Institute – Center for Circumpolar Security Studies, Washington, DC, USA

Halima Hodzic Food and Agriculture Organization of the United Nations (FAO), Rome, Italy

Karen Horwood The Leeds Planning School, Leeds Beckett University, Leeds, UK

Mette Hotker RMIT University, Melbourne, VIC, Australia

Karin Huber-Heim Circular Economy Forum, Austria, Vienna, Austria

Raisa Binte Huda Department of Geography and Environment, University of Dhaka, Dhaka, Bangladesh

Dan Xuan Thi Huynh School of Economics, Can Tho University, Can Tho, Vietnam

Ligocka Ilona Ministry of Climate and Environment, Warsaw, Poland

Tanya Gottlieb Jacobsen State of Green, Copenhagen, Denmark

Bhanye Johannes Department of Community and Social Development, University of Zimbabwe, Harare, Zimbabwe

Katrina Johnston-Zimmerman THINK.urban, Philadelphia, PA, USA

Kirsty Jones Crawford School of Public Policy, The Australian National University, Acton, ACT, Australia

Alain Jordà Local Development Expert, Manresa, Barcelona, Spain

Gaurav Joshi University of Chinese Academy of Sciences, Beijing, China

Anuja Joy National Institute of Technology Calicut, Kozhikode, Kerala, India

Mahjabin Kabir Adrita Department of Geography and Environment, University of Dhaka, Dhaka, Bangladesh

Zahra Kalantari Department of Physical Geography and Bolin Centre for Climate Research, Stockholm University, Stockholm, Sweden

Navarino Environmental Observatory, Messinia, Greece

Department of Sustainable Development, Environmental Science and Engineering, KTH Royal Institute of Technology, Stockholm, Sweden

Eleni Kalantidou Griffith University, Brisbane, QLD, Australia

Tinashe Natasha Kanonhuhwa Department of Demography Settlement and Development, Social & Behavioral Sciences, University Zimbabwe, Harare, Zimbabwe

L. Kapetas 100 resilient Cities Project, New York, USA

Thomas Karakadzai Department of Demography Settlement and Development, Faculty of Social & Behavioral Sciences, University of Zimbabwe, Harare, Zimbabwe

Abdulrazak Karriem University of the Western Cape, Cape Town, South Africa

Hewa Thanthrige Ashan Randika Karunananda Biodiversity Educational Research Initiative, Colombo, Sri Lanka

Rosemary Kasimba Department of Demography Settlement and Development, University of Zimbabwe, Harare, Zimbabwe

J. O. Kawira County Government of Laikipia, Laikipia, Kenya

Jon Kellett University of Adelaide, Adelaide, SA, Australia

Vlada Kenniff Long Island University, Brookville, NY, USA

Jeffrey Kenworthy Curtin University Sustainability Policy Institute, Curtin University, Perth, WA, Australia

Frankfurt University of Applied Sciences, Frankfurt am Main, Germany

Ganesh Keremane Adelaide, South Australia

Tien Dung Khong School of Economics, Can Tho University, Can Tho, Vietnam

Teng Chye Khoo National University of Singapore, Singapore, Singapore

F. I. Kihara The Nature Conservancy, Nairobi, Kenya

Lorenzo Kihlgren Grandi City Diplomacy Lab, Columbia Global Centers | Paris, Paris, France

C. Kilsby University of Newcastle, Newcastle, UK

Jinhee Kim Centre for Health Equity Training, Research and Evaluation (CHETRE), UNSW Australia Research Centre for Primary Health Care & Equity, South Western Sydney Local Health District, Ingham Institute, Sydney, NSW, Australia

Michael Koh Centre for Liveable Cities, Ministry of National Development, Singapore, Singapore

Victoria Kolankiewicz Faculty of Architecture, Building and Planning, University of Melbourne, Melbourne, VIC, Australia

Weichang Kong The University of Queensland, Brisbane, QLD, Australia

Mrudhula Koshy Norwegian University of Science and Technology, Trondheim, Norway

Maria Kottari School of Transnational Governance, European University Institute, Florence, Italy

Daniel Kozak Universidad de Buenos Aires, Consejo Nacional de Investigaciones CientÃficas y Técnicas (CONICET), Buenos Aires, Argentina

Teresa Kramarz University of Toronto, Toronto, ON, Canada

Tamara Krawchenko University of Victoria, Victoria, BC, Canada

Peleg Kremer Department of Geography and the Environment, Villanova University, Villanova, PA, USA

V. Krivtsov The Royal Botanic Garden, Edinburgh, UK

Arvind Kumar India Water Foundation, New Delhi, India

Gerard Kuperus University of San Francisco, San Francisco, CA, USA

Sigrid Kusch-Brandt Department of Civil, Environmental and Architectural Engineering, University of Padua, Padua, Italy

Faculty of Mathematics, Natural Sciences and Management, University of Applied Sciences Ulm, Ulm, Germany

Ndarova Audrey Kwangwama Department of Architecture and Real Estate, University of Zimbabwe, Harare, Zimbabwe

Oliver Lah Wuppertal Institute for Climate, Environment and Energy, Berlin, Germany

Urban Electric Mobility Initiative (UEMI) a UN-Habitat Action Platform, Berlin, Germany

Khee Poh Lam National University of Singapore, Singapore, Singapore

J. Lamond University of the West of England, Bristol, UK

Martin Larbi Kwame Nkrumah University of Science and Technology, Kumasi, Ghana

Alexander Laszlo The Bertalanffy Center for the Study of Systems Science, Buenos Aires, Argentina

Lucie Laurian School of Planning and Public Affairs, The University of Iowa, Iowa City, IA, USA

Alison Lee Centre for Liveable Cities, Ministry of National Development, Singapore, Singapore

Steffen Lehmann School of Architecture, University of Nevada, Las Vegas, NV, USA

Carlos Leite School of Architecture and Urbanism, Mackenzie Presbyterian University, Sao Paulo, Brazil

Social Urbanism Center, Insper´s Arq.Futuro Cities Lab, Sao Paulo, Brazil

Caitlin Anthea Lewis Architecture Planning and Geomatics, University of Cape Town, Cape Town, South Africa

Nora Libertun de Duren Inter-American Development Bank, Washington, DC, USA

Jade Lindley Law School and Oceans Institute, The University of Western Australia, Crawley, WA, Australia

Yan Liu School of Earth and Environmental Sciences, The University of Queensland, St Lucia, Australia

Adam Loch Centre for Global Food and Resources, School of Economics and Public Policy, Faculty of the Professions, University of Adelaide, Adelaide, SA, Australia

Aynaz Lotfata Department of Geography, Chicago State University, Chicago, IL, USA

Pavel Luksha Global Education Futures, Moscow, Russia

Mengxing Ma Department of Social Work, University of Melbourne, Melbourne, VIC, Australia

Department of Geography, University of Sheffield, Sheffield, UK

Danielle MacCarthy Queen's University Belfast, Belfast, Northern Ireland, UK

Shamiso Hazel Mafuku Department of Architecture and Real Estate, University of Zimbabwe, Harare, Zimbabwe

Kamilia Mahdaoui Hassania School of Public Works, Casablanca, Morocco

Israa H. Mahmoud Laboratorio di Simulazione Urbana Fausto Curti, Department of Architecture and Urban Studies, Politecnico di Milano, Milan, Italy

David Mainenti Palmer iSchool of Library and Information Studies, Long Island University, Brookville, NY, USA

Innocent Maja Faculty of Law, University of Zimbabwe, Harare, Zimbabwe

Soumaya Majdoub Research Group Interface Demography, Department of Sociology, VUB Free University of Brussels, Brussels, Belgium

Brussels Center for Urban Studies (BCUS), Brussels, Belgium

Brussels Interdisciplinary Research Centre for Migration and Minorities (BIRMM), Brussels, Belgium

George Makunde George Makunde Institute, Harare, Zimbabwe

Eleanor Malbon University of New South Wales, Kensington, NSW, Australia

Wendy W. Mandaza-Tsoriyo Department of Rural and Urban Development, Great Zimbabwe University, Harare, Zimbabwe

Manfredo Manfredini School of Architecture and Planning, The University of Auckland, Shanghai University, Auckland, New Zealand

Elton Manjeya Department of Architecture and Real Estate, University of Zimbabwe, Harare, Zimbabwe

Jonathan Manns Rockwell, London, UK

UCL, London, UK

Patrick M. Marchman American Society of Adaptation Professionals/ Climigration Network, Kansas City, MO, USA

Age Mariussen University of Vaasa, Vaasa, Finland

Cecilia Marocchino Food and Agriculture Organization of the United Nations (FAO), Rome, Italy

Andresa Ledo Marques Graduate Program in Architecture and Urbanism, Mackenzie Presbyterian University, Sao Paulo, Brazil

Institute of Urban Design and Planning, Leibniz Universität, Hannover, Germany

Martha Marriner State of Green, Copenhagen, Denmark

Stephen Marshall Bartlett School of Planning, University College London, London, UK

Natalia Martsinovich Department of Chemistry, University of Sheffield, Sheffield, UK

Nesbert Mashingaidze Department of Rural and Urban Development, Great Zimbabwe University, Masvingo, Zimbabwe

Jeofrey Matai Department of Architecture and Real Estate, University of Zimbabwe, Harare, Zimbabwe

Abraham R. Matamanda Department of Urban and Regional Planning, University of the Free State, Bloemfontein, South Africa

Marina Matashova Andorra-LAB, Forward Consulting Group, Barcelona, Spain

Brilliant Mavhima Department of Architecture and Real Estate, University of Zimbabwe, Harare, Zimbabwe

Patience Mazanhi Department of Demography Settlement and Development, Social & Behavioral Sciences, University of Zimbabwe, Harare, Zimbabwe

Chad J. McGuire Department of Public Policy, University of Massachusetts, Dartmouth, MA, USA

Matthew H. McLeskey Department of Sociology, University at Buffalo, State University of New York, Buffalo, NY, USA

Wendy McWilliam School of Landscape Architecture, Faculty of Environment, Society and Design, Lincoln University, Lincoln, New Zealand

Ojilve Ramón Medrano Pérez CONACYT-Centro del Cambio Global y la Sustentabilidad, A.C. (CCGS), Villahermosa, Tabasco, Mexico

Asma Mehan Senior Researcher, CITTA Research Institute, Faculty of Engineering (FEUP), University of Porto, Porto, Portugal

Mahziar Mehan School of Urban Planning, Faculty of Fine Arts, University of Tehran, Tehran, Iran

Prakhar Mehta Digital Transformation: Bits to Energy Lab Nuremberg, School of Business, Economics and Society, Friedrich-Alexander University Erlangen-Nürnberg (FAU), Nuremberg, Germany

Lorena Melgaço Department of Human Geography, Lund University, Lund, Sweden

D. Mendoza Tinoco University of Coahuila, Coahuila, Mexico

Julián Andrés Mera-Paz Faculty of Engineering, Universidad Cooperativa de Colombia, Popayán, Colombia

Magnus Højberg Mernild State of Green, Copenhagen, Denmark

Jessica Ostrow Michel School for Environment and Sustainability, University of Michigan, Ann Arbor, MI, USA

Yoko Mochizuki UNESCO, Paris, France

Itumeleng Mogola C40 Cities, Benoni, South Africa

Mohsen Mohammadzadeh School of Architecture and Planning, Auckland University, Auckland, New Zealand

Abinash Mohanty Council on Energy, Environment and Water (CEEW), New Delhi, India

Mehri Mohebbi Transportation Equity Program, University of Florida (UFTI), Gainesville, FL, USA

Anne Mook University of Georgia, Athens, GA, USA

Eugenio Morello Laboratorio di Simulazione Urbana Fausto Curti, Department of Architecture and Urban Studies, Politecnico di Milano, Milan, Italy

Charlotte Morphet Women and Planning research bursary, Planning, Housing and Human Geography, The Leeds Planning School, Leeds Beckett University, Leeds, UK

Nicky Morrison Western Sydney University, Sydney, NSW, Australia

Sina Mostafavi TU Delft, Delft, The Netherlands

Edmos Mtetwa Department of Social Work, University of Zimbabwe, Harare, Zimbabwe

Tinashe Natasha Mujongonde-Kanonhuwa Department of Rural & Urban Planning, University of Zimbabwe, Harare, Zimbabwe

Manasi R. Mulay Department of Chemistry, University of Sheffield, Sheffield, UK

Grantham Centre for Sustainable Futures, Sheffield, UK

Richard Müller Sustainable Development Institute/Institut udrzatelneho rozvoja, Nitra, Slovakia

Yvonne Munanga Department of Architeture and Real Estate, University of Zimbabwe, Harare, Zimbabwe

Dalia Munenzon College of Architecture, Texas Tech University, Lubbock, TX, USA

Nora Munguia Graduate Sustainability Program, Industrial Engineering Department, University of Sonora, Hermosillo, Mexico

Solomon Muqayi Department of Governance and Public Management, University of Zimbabwe, Harare, Zimbabwe

Cassandra Murphy Department of Psychology, Maynooth University, Maynooth, Ireland

Teagan Murphy University of Maryland, College Park, MD, USA

Brendan Murtagh Urban Planning, School of Natural and Built Environment, David Keir Building, Queen's University Belfast, Belfast, UK

Walter Musakwa Future Earth and Ecosystem Services Research Group, Department of Urban and Regional Planning, University of Johannesburg, Johannesburg, South Africa

Tafadzwa Mutambisi Department of Rural and Urban Planning, University of Zimbabwe, Harare, Zimbabwe

Chipo Mutonhodza Department of Rural and Urban Development, Great Zimbabwe University, Masvingo, Zimbabwe

Valeria Muvavarirwa Department of Demography Settlement and Development, Social & Behavioral Sciences, University of Zimbabwe, Harare, Zimbabwe

Jean Nacishali Nteranya Department of Geology, Faculty of Sciences, Université Officielle de Bukavu (UOB), Bukavu, Democratic Republic of Congo

Anupam Nanda University of Manchester, Manchester, UK

Luzma Fabiola Nava CONACYT-Centro del Cambio Global y la Sustentabilidad, A.C. (CCGS), Villahermosa, Tabasco, Mexico

International Institute for Applied Systems Analysis (IIASA), Laxenburg, Austria

Celeste Nava Jiménez División de Ciencias Económico Administrativas, Campus Guanajuato, Universidad de Guanajuato, Guanajuato, Mexico

Thilini Navaratne Department of Business Economics, Faculty of Management Studies and Commerce, University of Sri Jayewardenepura, Nugegoda, Sri Lanka

Roselin Ncube Women's University in Africa, Harare, Zimbabwe

S. Ncube Heriot-Watt University, Edinburgh, UK

Etienne Nel University of Otago, Dunedin, New Zealand

David Nichols Faculty of Architecture, Building and Planning, University of Melbourne, Melbourne, VIC, Australia

Alejandro Nuñez-Jimenez Sustainability and Technology Group, D-MTEC, ETH Zurich, Zurich, Switzerland

Belfer Center for Science and International Affairs, Harvard University, Cambridge, MA, USA

Gloria Nyaradzo Nyahuma-Mukwashi Department for International Development (DFID), Harare, Zimbabwe

E. O'Donnell University of Nottingham, Nottingham, UK

G. O'Donnell University of Newcastle, Newcastle, UK

Narteh F. Ocansey Water Resources, Freelance, Accra, NA, Ghana

Yukyung Oh King's College London, London, UK

Carolina G. Ojeda Doctorado en Arquitectura y Estudios Urbanos, Pontificia Universidad Católica de Chile, Providencia, Santiago de Chile, Chile

Departamento de Historia, Facultad de Comunicaciones e Historia, Universidad Católica de la Santísima Concepción, Concepción, Chile

Hasan Volkan Oral Faculty of Engineering, Department of Civil Engineering (English), Istanbul Aydın University, Istanbul, Turkey

P. Ortiz International Metropolitan Institute, Madrid, Spain

International Metropolitan Institute, Washington, DC, USA

G. Osei Anglia Ruskin University, Chelmsford, UK

Laura Patricia Otero-Durán Urban Development Institute, Bogotá, Colombia

Maria Pafi Urban Planning, School of Natural and Built Environment, David Keir Building, Queen's University Belfast, Belfast, UK

F. Pascale Anglia Ruskin University, Chelmsford, UK

Maibritt Pedersen Zari School of Architecture, Victoria University of Wellington, Wellington, New Zealand

María Concepción Peñate-Valentín Department of Applied Economics, Faculty of Economics, Universidade de Santiago de Compostela, Santiago de Compostela, Galicia, Spain

Paulo Pereira Environmental Management Laboratory, Mykolas Romeris University, Vilnius, Lithuania

Ben Perks

Shama Perveen Senior Manager (Water), Ceres, Boston, MA, USA

Evi Petersen Institute of Sports, Physical Education and Outdoor Life, University of South-Eastern Norway, Oslo, Norway

Son Phung Department of Civil and Environmental Engineering, Auckland University, Auckland, New Zealand

Francesca Piazzoni

Czarnocki Piotr Ministry of Climate and Environment, Warsaw, Poland

Dorina Pojani The University of Queensland, Brisbane, QLD, Australia

A. Pooley Centre for Alternative Technology, Pantperthog, UK

K. Potter Open University, Milton Keynes, UK

Abdellatif Qamhaieh American University in Dubai, Department of Architecture, Dubai, United Arab Emirates

Md. Anisur Rahman Center for Policy and Economic Research (CPER), Dhaka, Bangladesh

Syed Hafizur Rahman Department of Environmental Sciences, Faculty of Physical and Mathematical Sciences, Jahangirnagar University, Dhaka, Bangladesh

Lakshmi Priya Rajendran The Bartlett School of Architecture, University College London, London, UK

Ritesh Ranjan Department of Architecture & Planning, National Institute of Technology Calicut, Kozhikode, Kerala, India

Andreas Raspotnik High North Center for Business and Governance, Nord University, Bodø, Norway

The Arctic Institute – Center for Circumpolar Security Studies, Washington, DC, USA

Hanna A. Rauf Asian School of the Environment, Nanyang Technological University, Singapore, Singapore

Aaron Redman School of Sustainability, Arizona State University, Tempe, AZ, USA

William E. Rees School of Community and Regional Planning, University of British Columbia, Vancouver, BC, Canada

Christian Reichel University of Applied Sciences for Media, Communication and Management (HMKW), Berlin, Germany

Kimberley Reis Cities Research Institute, Griffith University, Brisbane, QLD, Australia

Catherine E. Richards Center for the Study of Existential Risk (CSER), University of Cambridge, Cambridge, UK

Department of Engineering, University of Cambridge, Cambridge, UK

Lauren Rickards Urban Futures Enabling Capability Platform, RMIT University, Melbourne, Australia

Ritesh Ranjan Department of Architecture and Planning, National Institute of Technology Calicut, Kozhikode, Kerala, India

Alejandra Rivera Vinueza 4CITIES Erasmus Mundus Joint Master Degree (EMJMD) in Urban Studies, Vrije Universitet Brussel (VUB), Brussels, Belgium

Institute for Human Rights and Business (IHRB), Built Environment Global Programme Manager, London, UK

Daniela Rizzi Nature-based Solutions and Biodiversity – Sustainable Resources, Climate and Resilience Team, Freiburg, Germany

Michael Robbins HIPR, New York City, NY, USA

Héctor Rodal Architect and Urban Planner, Barcelona, Spain

Robert Rogerson Institute for Future Cities, University of Strathclyde, Glasgow, UK

Watch Ruparanganda Department of Social Work, University of Zimbabwe, Harare, Zimbabwe

María Carmen Sánchez-Carreira Department of Applied Economics, Faculty of Economics, Universidade de Santiago de Compostela, ICEDE Research Group, CRETUS, Santiago de Compostela, Galicia, Spain

Rami Sabella United Nations Economic and Social Commission for Western Asia, Beirut, Lebanon

Peter Sainsbury School of Medicine, University of Notre Dame, Sydney, NSW, Australia

Samuel Sandoval Solis Department of Land, Air and Water Resources, University of California Davis, Davis, CA, USA

Guido Santini Food and Agriculture Organization of the United Nations (FAO), Rome, Italy

Tom Sanya Architecture Planning and Geomatics, University of Cape Town, Cape Town, South Africa

Hasan Saygın Application, and Research Center for Advanced Studies, Istanbul Aydın University, Istanbul, Turkey

Alice Schmidt Global Health Advisory Service to the European Commission, Mechelen, Belgium

Vienna University of Economics and Business, Vienna, Austria

AS Consulting, Vienna, Austria

Jörg Schröder Institute of Urban Design and Planning, Leibniz Universität, Hannover, Germany

Barbara Schröter Leibniz Centre for Agricultural Landscape Research (ZALF), Working group Governance of Ecosystem Services, Müncheberg, Germany

Lund University Centre for Sustainability Studies (LUCSUS), Lund, Sweden

Kim Philip Schumacher Institute of Geography, Osnabrück University, Osnabrück, Germany

Abel Schumann Organisation for Economic Co-operation and Development, Paris, France

Samad M. E. Sepasgozar School of Built Environment, University of New South Wales, Sydney, NSW, Australia

Alan Shapiro British Columbia Institute of Technology, Vancouver, BC, Canada

Aviram Sharma School of Ecology and Environment Studies, Nalanda University, Rajgir, Bihar, India

Tian Shi Department of Primary Industries and Regions, Adelaide, South Australia

Amna Shoaib Department of City and Regional Planning, Lahore College for Women University (LCWU), Lahore, Pakistan

Yat Shun Kei

Renard Y. J. Siew Climate Change & Sustainability, Centre for Governance & Political Studies (CENT-GPS), Kuala Lumpur, Malaysia

Institute for Globally Distributed Open Research and Education (IGDORE), Bali, Indonesia

David Simon Department of Geography, Royal Holloway, University of London, Egham, UK

Jean Simos Institute of Public Health, Faculty of Medicine, University of Geneva, Geneva, Switzerland

S2D – Health and Sustainable Development, Rennes, France

Neil Sipe School of Earth and Environmental Sciences, The University of Queensland, Brisbane, QLD, Australia

Ben Sonneveld Amsterdam Centre for World Food Studies/Athena Institute, Vrije Universiteit, Amsterdam, The Netherlands

Micol Sonnino The Bertalanffy Center for the Study of Systems Science, Vienna, Austria

Simon Springer Centre for Urban and Regional Studies, Dicipline of Geography and Environmental Studies, University of Newcastle, Australia, Callaghan, NSW, Australia

Janet Stanley Melbourne Sustainable Society Institute, University of Melbourne, Melbourne, Australia

Wendy Steele Centre for Urban Research, RMIT University, Melbourne, VIC, Australia

Justin D. Stewart Department of Ecological Science, Vrije Universiteit Amsterdam, Amsterdam, The Netherlands

Raisa Sultana Department of Geography and Environment, University of Dhaka, Dhaka, Bangladesh

Samantha Suppiah Possible Futures, Manilla, Philippines

Sylvia Szabo Department of Social Welfare and Counselling, University of Seoul, Seoul, South Korea

Gerti Szili College of Humanities, Arts and Social Sciences, Flinders University, Adelaide, South Australia

Bouchra Tafrata Willy Brandt School of Public Policy, University of Erfurt, Erfurt, Germany

Ling Min Tan Department of Civil & Structural Engineering, The University of Sheffield, Sheffield, UK

M. Terdiman The Institute for Environmental Security and Well-being Studies, Jerusalem, Israel

Jacqueline Thomas School of Civil Engineering, The University of Sydney, Sydney, NSW, Australia

M. K. Thomas Rural Focus Limited (RFL), Nanyuki, Kenya

S. Thomas Rural Focus Limited (RFL), Nanyuki, Kenya

C. Thorne University of Nottingham, Nottingham, UK

Karine Tollari Japan Local Government Centre, London, UK

Chiara Tomaselli Consultant – Urban Asset Advisory, Arcadis France, Paris, France

Percy Toriro Municipal Development Partnership for Eastern and Southern Africa, Harare, Zimbabwe

African Centre for Cities, University of Cape Town, Cape Town, South Africa

Isabella Trapani Food and Agriculture Organization of the United Nations (FAO), Rome, Italy

Alejandra Trejo-Nieto Centre for Demographic, Urban and Environmental Studies, El Colegio de Mexico, Mexico City, Mexico

Stella Tsani Department of Economics, University of Ioannina, Ioannina, Greece

Asaf Tzachor Center for the Study of Existential Risk (CSER), University of Cambridge, Cambridge, UK

School of Sustainability, Interdisciplinary Center (IDC) Herzliya, Herzliya, Israel

Zdravka Tzankova Vanderbilt University, Nashville, TN, USA

Kristina Ulm Faculty of Arts, Design and Architecture, University of New South Wales, Sydney, NSW, Australia

Geraldine Usingarawe Department of Architecture and Real Estate, University of Zimbabwe, Harare, Zimbabwe

Luís Valença Pinto Research Centre for Natural Resources, Environment and Society (CERNAS), Polytechnic Institute of Coimbra, Coimbra Agrarian Technical School, Coimbra, Portugal

Environmental Management Laboratory, Mykolas Romeris University, Vilnius, Lithuania

Ellen Van Bueren Department of Management in the Built Environment, Faculty of Architecture and the Built Environment, Delft University of Technology, Delft, The Netherlands

Karel Van den Berghe Department of Management in the Built Environment, Faculty of Architecture and the Built Environment, Delft University of Technology, Delft, The Netherlands

Jeroen van der Heijden School of Government, Victoria University of Wellington, Wellington, New Zealand

School of Regulation and Global Governance, Australian National University, Canberra, ACT, Australia

Wim van Veen Vrije Universiteit, Amsterdam Centre for World Food Studies, Amsterdam, The Netherlands

Lia van Wesenbeeck Vrije Universiteit, Amsterdam Centre for World Food Studies, Amsterdam, The Netherlands

Christopher Vanags Vanderbilt University, Nashville, TN, USA

Kamiya Varshney School of Architecture, Victoria University of Wellington, Wellington, New Zealand

Luis Velazquez Industrial Engineering Department, University of Sonora, Hermosillo, Mexico

Luis Eduardo Velazquez Contreras Sustainability Graduate Program, University of Sonora, Hermosillo, México

T. Vilcan Open University, Milton Keynes, UK

Luiza O. Voinea Urban Planner Certified by The Romanian Register of Urban Planners, Bucharest, Romania

Shreya Wadhawan Council on Energy, Environment and Water (CEEW), New Delhi, India

Sameh N. Wahba The World Bank, Washington, DC, USA

Haiyun Wang School of Earth and Environmental Sciences, The University of Queensland, St Lucia, Australia

Siqin Wang School of Earth and Environmental Sciences, The University of Queensland, St Lucia, Australia

Noelia Wayar National University of Córdoba, Córdoba, Argentina

Oliver Weigel Urban Development Policy Division at the Federal Ministry of the Interior, Building, and Community, Berlin, Germany

Kadmiel H. Wekwete Midlands State University, Gweru, Zimbabwe

Caitlin Werrell The Center for Climate and Security, an Institute of the Council on Strategic Risks, Washington, DC, USA

Andreas Wesener School of Landscape Architecture, Faculty of Environment, Society and Design, Lincoln University, Lincoln, New Zealand

Bettina Wilk ICLEI European Secretariat, Senior Officer for Nature-based Solutions and Biodiversity – Sustainable Resources, Climate and Resilience Team, Freiburg, Germany

Erich Wolff Monash Art, Design and Architecture, Monash University, Melbourne, VIC, Australia

Sam Wong University College Roosevelt, Middelburg, The Netherlands

N. Wright Nottingham Trent University, Nottingham, UK

Junjie Xi

Belinda Young Melbourne Sustainable Society Institute, University of Melbourne, Melbourne, Australia

Asaduz Zaman Centre for Action Research – Barind, Rajshahi, Bangladesh

Asian Development Bank, Dhaka, Bangladesh

Fathima Zehba M. P. Department of Architecture and Planning, Calicut, National Institute of Technology, Calicut, Kerala, India

David Slim Zepeda Quintana Sustainability Graduate Program, University of Sonora, Hermosillo, México

Yuerong Zhang Bartlett School of Planning, University College London, London, UK

Eric Zhao University of Toronto, Toronto, ON, Canada

Metron Ziga University of the Western Cape, Cape Town, South Africa

Monika Zimmermann Urban Sustainability Expert & Former Deputy Secretary General of ICLEI, Freiburg, Germany

Willoughby Zimunya Department of Demography Settlement and Development, University of Zimbabwe, Harare, Zimbabwe

Department of Urban and Regional Planning, University of the Free State, Bloemfontein, South Africa

Michaela Zint School for Environment and Sustainability, University of Michigan, Ann Arbor, MI, USA

Tara Rava Zolnikov School of Behavioral Sciences, California Southern University, Costa Mesa, CA, USA

Department of Community Health, National University, San Diego, CA, USA

[illegible] London, UK

[illegible]

[illegible]

[illegible] Germany

[illegible]

[illegible]

[illegible] Ann Arbor, [illegible] USA

[illegible] Costa Mesa [illegible]

Department of [illegible]

H

Habitat Provisioning

A Regenerative Building Design Approach

Kamiya Varshney, Maibritt Pedersen Zari and Nilesh Bakshi
School of Architecture, Victoria University of Wellington, Wellington, New Zealand

Synonyms

Biodiversity conservation; Refuge; Stepping-stone; Urban biodiversity; Urban ecology

Definition

Ecosystems provide living space for (and are made up of) plant and animal species. These are known as *habitats*. Habitats provide all the resources, biotic and abiotic environmental conditions that an individual plant or animal needs to survive, such as food, water, shelter (protection from predators and weather), and space (to thrive) (TEEB 2010). Different plants and animals need different habitat types that are required to meet all the environmental conditions they need to survive. For an animal, a habitat means everything it needs to find and gather food, accessibility to water, shelter from predators, protection from weather conditions, and a place to successfully reproduce. Similarly, for a plant, a good habitat must provide the right combination of light, air, water, and soil conditions.

In an urban built environment, living organisms primarily find their habitats in blue or green spaces and/or in or on buildings and infrastructures (birds nesting on chimneys, for example). Essentially, they adapt to anthropogenically altered environmental conditions. These habitats are known as *urban habitats*, i.e., shared living spaces for plants, animals, and humans. Each ecosystem provides different habitats for resident and transient populations that are essential to different lifecycle phases of the species they sustain. This ecosystem service is known as *habitat provisioning* (Costanza et al. 1997; TEEB 2010). When aiming to increase urban biodiversity, it is essential to understand habitat requirements, plant-animal and human-nature interactions, and enhance shared urban habitats by strategically designing for regenerative built environments.

Introduction

Core Concepts: Biodiversity, Ecosystems, and Urban Ecosystems

This chapter focuses on how the built environment can facilitate habitat provisioning for urban biodiversity by translating ecological knowledge into an architectural context and address, in part, the global issue of biodiversity loss.

R. C. Brears (ed.), *The Palgrave Encyclopedia of Urban and Regional Futures*,
https://doi.org/10.1007/978-3-030-87745-3

Biodiversity is the complete variety of life at any spatial scale, from microsites and habitat patches to the entire biosphere, including diversity within and between species, and ecosystems (DeLong 1996). It includes three hierarchical levels (CBD 1992; MEA 2005b; Turner 2018):

- *Genetic diversity*: variation in the genetic composition among individuals of same species
- *Species diversity*: a variety of species within a specific area
- *Ecosystem diversity*: a variety of ecosystem types within a specific region, including all the species and their habitats, and physical environment

Urban biodiversity is the diversity of living organisms in and on the edge of an urban area, including genetic diversity (for example, variety of indigenous species in an urban area) and ecosystem diversity (for example, different types of habitats across an urban area) (Müller et al. 2013; Puppim de Oliveira et al. 2014).

Biodiversity loss occurs either through the reduction of diversity per se, or if the potential of biodiversity to provide a particular ecosystem service is diminished (MEA 2005b).

An *ecosystem* is a functional unit comprising of all species (and their habitats) interacting with one another in a physical environment (MEA 2005a). In the urban built environment context, an *urban ecosystem* is composed of the interaction of biological components (the coexistence of plants, animals, humans, and other forms of life) and social components (such as human culture, behavior, and activities) in a physical environment of an urban area (made up of elements such as buildings, roads, infrastructure, and designed landscapes) (Berkowitz et al. 2003). Humans are a dominant component in urban ecosystems, unlike many other ecosystems (Alberti et al. 2003).

The benefits that people derive from nature, such as climate regulation, air purification, habitat provisioning, are known as *ecosystem services* (Costanza et al. 1997; De Groot et al. 2002). MEA (2005a) categorizes ecosystem services into:

- *Provisioning services*: the material benefits humans derive from ecosystems, such as food, water, timber, and fiber.
- *Regulating services*: such as regulating climate, moderation of extremes, regulating disease, and maintaining air, soil, and water quality.
- *Cultural services*: the non-material benefits that are essential for human health and well-being, such as recreation and leisure, aesthetic beauty, and spiritual meaning.
- *Supporting services*: these underpin almost all ecosystem services. For example, habitat provisioning, soil formation, and nutrient cycling.

Urban ecosystems play a significant role in *providing habitats* for biodiversity. For example, migratory birds depend upon finding places of refuge during their journeys. While urban green and blue space is crucial in this regard, building-integrated vegetation could provide further habitat for species, particularly those that are affected by fragmentation and/or land-use changes (Williams et al. 2014).

Habitat Provisioning Ecosystem Service: Translating to the Built Environment Context

Because some ecosystem services (such as pollination and regulation of species diversity) cannot be easily integrated within the physical built context of building or grey infrastructure design, habitat provisioning in the context of built environment design comprises several other ecosystem services: *genetic information; pollination and seed dispersal; fixation of solar energy; biological control; and species maintenance* (Fig. 1) (Pedersen Zari 2018b).

These five ecosystem services are described below:

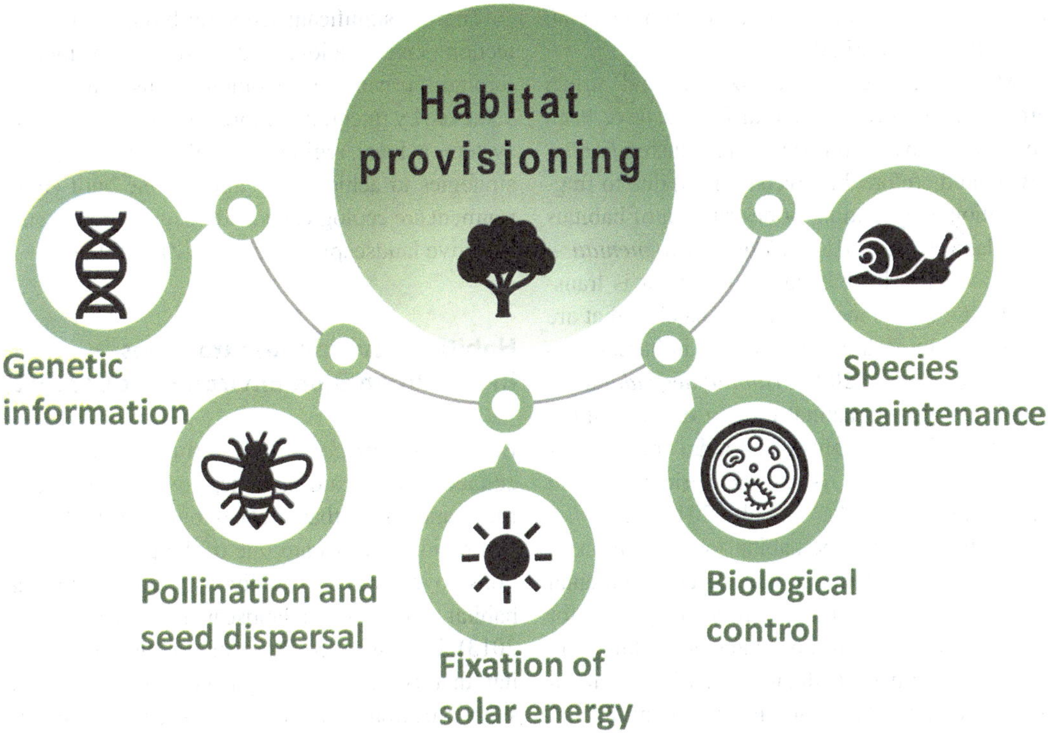

Habitat Provisioning, Fig. 1 Habitat provisioning bundled ecosystem services in built environment context. (Adapted from Pedersen Zari 2018b)

Genetic information facilitates the maintenance of ecological balance and evolutionary processes (De Groot et al. 2010).

Pollination and seed dispersal is the ecosystem service provided by pollinators (such as bees, birds, and bats) for the reproduction of plant populations by transferring the genetic material (Costanza et al. 1997).

Fixation of solar energy is the oxygen-producing benefit of photosynthetic processes occurring within ecosystems that is a fundamental basis for food production on Earth (MEA 2005a).

Biological control is the ecosystem service related to the regulation of pests and vector-borne diseases through trophic relations (De Groot et al. 2010).

Species maintenance refers to the provision of biodiversity, natural selection, and self-organization in ecosystems and plays a vital role in the continuation of all other ecosystem services (Pedersen Zari 2018b, page 122).

Urbanization and Biodiversity: Constraints and Opportunities

Human settlements and biodiversity are interconnected in a feedback loop. Human settlements are a significant driver of biodiversity loss. Land-use change (also resulting in habitat loss), climate change, nitrogen deposition, the introduction of invasive species to ecosystems are some of the key anthropogenic drivers of biodiversity loss (Pedersen Zari 2014). Concurrently, biodiversity loss affects human settlements in terms of climate changes, decreased resilience to changes, and reduced quality or quantity of ecosystem services, adversely impacting ecological health and human well-being (Aerts et al. 2018; Cardinale et al. 2012; Pedersen Zari 2018a).

Moreover, due to urbanization and urban growth, large areas of natural habitat have been converted into grey infrastructure, causing *habitat loss* (and therefore, biodiversity loss) due to fragmentation, degradation, or destruction of habitats (Haddad et al. 2015). *Habitat fragmentation* occurs when a large expanse of habitat is transformed into a number of smaller patches that are isolated from each other, resulting in habitat disintegration (Fahrig 2003). *Habitat degradation* is the decline of the habitat quality due to natural factors (such as droughts and floods) or anthropogenic interventions (such as agriculture and urbanization) (Müller et al. 2013). *Habitat destruction*, an extreme habitat loss condition, is the alteration or elimination of the environmental conditions required by a particular species for survival so that it can no longer be sustained in that particular place (Tilman et al. 1997). Habitat loss is a major threat to the continuation and existence of many species and, therefore, to ecosystems and ecosystem services.

Conversely, increased biodiversity and habitat provisioning have positive impacts on urban ecosystems. Examples include:

- Increased ecological health (e.g., regeneration and provision of ecosystem services, such as purification of air, water, and soil, carbon sequestration nutrient cycling) (Gómez-Baggethun et al. 2013; Varshney et al. 2020.
- Climate change mitigation and adaptation (Biju Kumar and Ravinesh 2017; Swingland 2013)
- Economic benefits (e.g., carbon offset costs and increased workplace productivity) (Liquete et al. 2016).
- Social benefits (e.g., social cohesion and security) (Fuller and Irvine 2010; Sandifer et al. 2015.
- Cultural benefits (e.g., aesthetics, creation of a sense of place) (Clark et al. 2014; Hausmann et al. 2016.
- Human well-being (e.g., psychological and physical health) (Aerts et al. 2018; MEA 2005b)

There is significant scope for biodiversity protection, conservation, and increasing resilience through urban regeneration. This could be achieved by integrating more effectively with ecosystems at local, regional, and global scales. Some strategies to achieve this through the built environment are ecological corridors, green roofs, and sensitive landscaping (Pedersen Zari 2014).

Habitat Provisioning Strategies in the Urban Built Environment Context

Ecological corridors in the form of strategically connected domestic gardens, green walls, green roofs, and other urban green or blue spaces at local and regional scales are one of the potential solutions to conserving biodiversity and providing habitat within urban landscapes (Vergnes et al. 2013). Ecological corridors are the pieces of habitat that connect larger patches of habitat and support ecological connectivity, such as the movement of species for finding basic resources to survive. A regenerative building design approach can aid in considering the interconnectedness of species and ecosystems across local and ecosystem scales and enhance ecological restoration. In a regenerative built environment, buildings are considered to be part of ecosystems with a focus on both human and ecological well-being.

Below are some design strategies which building design professionals may apply to enhance habitat provisioning (Table 1):

- *Providing habitat requirements for keystone species* through hard infrastructure. For example, nesting boxes, ledges, and planters for vegetation. *Keystone species* are those which fundamentally affect aspects of ecosystem structure, functioning, or productivity of habitat on small or large scales (Mills et al. 1993; Power et al. 1996). Loss of keystone species may lead to significant ecosystem changes on broader scales. Mouquet et al. (2013) extended the concept of keystone species to apply to community and ecosystem levels.
- *Adding vegetation to buildings* (walls, roofs, interiors) specifically to support biodiversity

Habitat Provisioning, Table 1 Strategies for habitat provisioning through building design

Bird nesting boxes  Source: simon_redwood, https://www.flickr.com/photos/29654876@N02/6975673734, CC BY-NC-ND 2.0	**Planter boxes**  Source: PermaCultured, https://www.flickr.com/photos/45338605@N08/4967216882, CC BY 2.0
Green wall  Source: Thomas Claveirole, https://www.flickr.com/photos/80318369@N00/277493405, CC BY-SA 2.0	**Multifunctional landscapes**  Source: Deoma12, https://commons.wikimedia.org/wiki/File:Oasis_Terraces_2018.jpg, CC BY-SA 4.0

(continued)

Habitat Provisioning, Table 1 (continued)

Habitat provisioning through green-blue spaces  Source: La Citta Vita, https://www.flickr.com/photos/la-citta-vita/7279717578, CC BY-SA 2.0	**Green roof**  Source: Newtben33, https://commons.wikimedia.org/wiki/File:Green_Roof_Shed.JPG, CC BY-SA 3.0
Increasing ecological space  Source: SteelMaster buildings, https://www.flickr.com/photos/35055618@N07/17298528912, CC BY-SA 2.0	**Connecting people with biodiversity**  Source: Ohalo123, https://commons.wikimedia.org/wiki/File:Ohalo_biophilic_learning_space01.jpg, CC BY-SA 4.0

and reducing habitat fragmentation over a larger scale through the creation of, or integration with, ecological corridors and/or stepping stone habitats (Saura et al. 2014). *Stepping-stones* are the habitat patches in a landscape that provide species with refuge as they travel between other larger patches of habitat (Saura et al. 2014). In this context, building-integrated vegetation could be considered as potential stepping-stones that can connect habitat patches (and reduce fragmentation) through, for example, using green infrastructures such as green roofs, green walls, internal courtyards, and in-ground landscaping.

- *Multifunctional landscape design* to strategically increase the ratio of ecological space to human-only space. This could potentially improve the ecosystem functioning and provisioning of other ecosystem services by evaluating multidisciplinary factors from an ecological perspective, such as analyzing the spatial arrangement, size, shape, connectivity, and ecological functioning of urban landscapes (Lindenmayer et al. 2008; O'Farrell and Anderson 2010).
- *Protection of desirable species* (and their habitat): for example, by providing and improving the habitat requirements of indigenous or endemic species (Breuste 2004). *Indigenous species* are the species that naturally originate and occur in a particular place. Many countries, such as New Zealand and Australia, have developed biodiversity strategies for protecting indigenous and threatened species and their habitats.
- *Control of undesirable species* (and their habitat), especially those that are a potential threat to indigenous biodiversity. Strategies include designing to prevent predators or weeds, and instigating pest management plans (Klotz and Kühn 2010). This is particularly important in places where native species are threatened by relatively newly introduced pests or weeds.
- Restoration of ecosystems through the *reintroduction of habitat* for specific species by recognizing the ecosystem functioning, interrelatedness, and dependencies of a certain place (Chapman and Blockley 2009). An example of this is building seawalls by mimicking natural rocky shores to (re)create coastal habitat (Chapman and Blockley 2009).
- *Amending configuration, shape, and physical conditions* of abiotic elements (hard infrastructure) to support habitat requirements. Design strategies that are compatible between plant, animal, and human needs to facilitate habitat requirements for food, water, and shelter could be considered (Fuller and Irvine 2010).
- *Passive design* may enhance the protection of biodiversity against climate change and increase the resilience of urban environments in terms of energy, water, and food security by using strategies such as wind tunnel effects, noise regulation, shading, and reflective surfaces that impact microclimates for plants, animals, and people (Pitman et al. 2015).
- Using strategies to create or celebrate *human-biodiversity relationships,* for example, through *biophilic design* (a design to include elements that nurture human-nature relationships. (Beatley 2011; Hausmann et al. 2016; Sandifer et al. 2015).
- *Monitoring performance related to habitat provisioning* and restoring the building and its surrounding context over time in response. Biodiversity is dynamic and, therefore, needs monitoring and the use of management plans, particularly in urban contexts where ecosystems are less likely to be entirely self-regulating (Savard et al. 2000).

Habitat Provisioning Dynamics

There is an urgent need for a large multiscale approach to address biodiversity loss. One aspect of this should be seeking to ensure that built environments strategically provide habitat. However, it is important to consider the dynamics of compositional (structural) and process (functional) attributes while designing for habitat provisioning (Lindenmayer et al. 2008). Habitat provisioning through building-integrated vegetation can be improved by understanding the physical and biological environment requirements for different species, their diversity and distribution over space and time, interactions among species, and functional traits (MEA 2005b). Some parameters that influence biodiversity levels in urban built environments that should be considered while designing are:

- *Species diversity and distribution*: Species diversity is the number of species in a community (richness) and the abundance of each species in a community (evenness). According to Savard et al. (2000), a larger amount of species diversity positively affects ecosystem functions. In addition to species diversity, species distribution is important to investigate, to know how the species are distributed in a particular area at a particular time. Species distribution provides more information about how different species interact with each other and with their environment.
- *Plant species selection*: Indigenous species (native to a particular place) are preferable compared to exotic species (non-native) to continue the biodiversity in urban landscapes (Hui and Chan 2011). However, exotic species are either beneficial, neutral, or perhaps detrimental to the indigenous species and are already part of ecosystems in many instances. Some exotic species may contribute to the generation of other ecosystem services. For example, although Pinus radiata (a fast-growing exotic pine species to New Zealand) sequesters carbon faster than New Zealand natives initially (when planted in New Zealand), pine forests are not as beneficial as native ones in terms of biodiversity outcomes. Therefore, the role of different species in the ecosystem must be holistically understood in order to restore ecosystems and also have well-being outcomes for humans and non-humans alike (WCC 2015).
- *Ecosystem functionality*: Ecosystem functions are the biological or system properties, or processes of ecosystems that provide ecosystem services (Costanza et al. 1997). For example, storage, internal cycling, processing, and acquisition of nutrients are the ecosystem functions of nutrient cycling ecosystem service (Pedersen Zari 2018b, p. 149). It is important to understand the ecosystem functioning and the impact of one species on another and how strongly such interactions affect ecosystems (McDonald et al. 2016; MEA 2005b). Some examples of selecting species based on ecosystem functionality are:
  - Nitrogen-fixing plants like legumes versus non-nitrogen-fixing plants (MEA 2005b).
  - Synergistically related species versus conflicting and competing species in terms of predation, parasitism, competition, and facilitation (such as pollination) (Klotz and Kühn 2010).
  - Considering food-web and trophic level interactions.
  - Plant-animal-human interactions (Luck and Smallbone 2010).
- *Connectivity and proximity to surrounding ecologies*: Patch size (small green areas in urban landscape) and connectivity are important factors to support biodiversity in an urban matrix (Farinha-Marques et al. 2011). Species diversity increases with the increase in patch area (Fahrig 2003). Both functional and physical connectivity should be analyzed between patches of vegetation while designing buildings with integrated vegetation. For instance, physical connectivity could be associated with challenging issues like pest dispersal.
- *Area/coverage/density of vegetation*: Vegetation cover, density, tree or shrub height, or volume of vegetation, and degree of shading in vegetation patches are important factors to be investigated for improving the biodiversity levels (Farinha-Marques et al. 2011). For example, the more vegetated green area there is, the more biodiversity there tends to be (Hui and Chan 2011).
- *Substrate type and depth*: Intensive green roofs (substrate depth no less than 15 cm), for example, provide more urban biodiversity compared to extensive green roofs (substrate depth less than 15 cm) (Hui and Chan 2011).

- *The presence of microtopography* (similar to natural cliffs, for example): Varying slopes can enhance species movement, which promotes heterogeneity (Lundholm 2006).
- *Drivers of ecosystem degradation:* The presence of invasive species, over-utilization of resources, environmental pollutions (air, water, soil, noise, light), and contamination are some drivers of ecosystem degradation and can result in poor quality of habitat (McDonald et al. 2016). The factors that lead to environmental degradation should be carefully monitored and considered while designing for urban biodiversity.

Conclusions

Habitat provisioning and biodiversity are intimately related and contribute to ecological and human well-being. Habitat loss affects biodiversity and ecosystems in many ways, such as the decline and extinction of species, decreasing ecosystem resilience, and decreasing supply of ecosystem services. This suggests that it is essential to maintain suitable habitats to ensure the continuation or regeneration of biodiversity, and the provision of ecosystem services, directly or indirectly.

In an urban built environment context, the expansion of built-up areas, meaning less vegetated areas, results in small landscape patches that are largely fragmented from each other. This has a significant impact on ecological connectivity for biodiversity functioning and the provisioning of ecosystem services such as pollination and seed dispersal. Although several measures to address biodiversity loss through the medium of the built environment have been identified, there have been considerable challenges in the acceptance and implementation of those measures in the building industry (Opoku 2019). Therefore, an integrated and regenerative approach is required to execute strategies for providing habitat for urban biodiversity through urban built environment development. Moreover, monitoring and analyzing biodiversity levels at design, construction, maintenance, and deconstruction phases contributes to biodiversity-positive building and regenerative goals.

Cross-References

▶ Ecological Restoration
▶ Green Cities
▶ Nature-Based Solutions

References

Aerts, R., Honnay, O., & Van Nieuwenhuyse, A. (2018). Biodiversity and human health: Mechanisms and evidence of the positive health effects of diversity in nature and green spaces. *British Medical Bulletin, 127*(1), 5–22. https://doi.org/10.1093/bmb/ldy021.

Alberti, M., Marzluff, J. M., Shulenberger, E., Bradley, G., Ryan, C., & Zumbrunnen, C. (2003). Integrating humans into ecology: Opportunities and challenges for studying urban ecosystems. *Bioscience, 53*(12), 1169–1179.

Beatley, T. (2011). Biophilic cities: What are they? In *Biophilic cities* (pp. 45–81). Springer.

Berkowitz, A. R., Nilon, C. H., & Hollweg, K. S. (2003). *Understanding urban ecosystems: A new frontier for science and education.* Springer Science & Business Media.

Biju Kumar, A., & Ravinesh, R. (2017). *Climate change and biodiversity* (Vol. 1). Singapore: Springer. https://doi.org/10.1007/978-981-10-3573-9_5.

Breuste, J. H. (2004). Decision making, planning, and design for the conservation of indigenous vegetation within urban development. *Landscape and Urban Planning, 68*(4), 439–452.

Cardinale, B. J., Duffy, J. E., Gonzalez, A., Hooper, D. U., Perrings, C., Venail, P., Narwani, A., Mace, G. M., Tilman, D., & Wardle, D. A. (2012). Biodiversity loss and its impact on humanity. *Nature, 486*(7401), 59–67.

CBD (Convention on Biological Diversity). (1992). *Sustaining Life on Earth.* https://www.cbd.int/convention/guide/ [accessed on 20 September 2021].

Chapman, M., & Blockley, D. (2009). Engineering novel habitats on urban infrastructure to increase intertidal biodiversity. *Oecologia, 161*(3), 625–635.

Clark, N. E., Lovell, R., Wheeler, B. W., Higgins, S. L., Depledge, M. H., & Norris, K. (2014). Biodiversity, cultural pathways, and human health: A framework. *Trends in Ecology & Evolution, 29*(4), 198–204.

Costanza, R., d'Arge, R., De Groot, R., Farber, S., Grasso, M., Hannon, B., Limburg, K., Naeem, S., O'Neill,

H

R. V., & Paruelo, J. (1997). The value of the world's ecosystem services and natural capital. *Nature, 387*(6630), 253–260.

De Groot, R. S., Wilson, M. A., & Boumans, R. M. (2002). A typology for the classification, description, and valuation of ecosystem functions, goods, and services. *Ecological Economics, 41*(3), 393–408.

De Groot, R. S., Fisher, B., Christie, M., Aronson, J., Braat, L., Haines-Young, R., Gowdy, J., Maltby, E., Neuville, A., & Polasky, S. (2010). Integrating the ecological and economic dimensions in biodiversity and ecosystem service valuation. In *The economics of ecosystems and biodiversity (TEEB): Ecological and economic foundations* (pp. 9–40). Routledge: Earthscan.

DeLong, D. C. (1996). Defining biodiversity. *Wildlife Society Bulletin (1973–2006), 24*(4), 738–749.

Fahrig, L. (2003). Effects of habitat fragmentation on biodiversity. *Annual Review of Ecology, Evolution, and Systematics, 34*(1), 487–515.

Farinha-Marques, P., Lameiras, J., Fernandes, C., Silva, S., & Guilherme, F. (2011). Urban biodiversity: A review of current concepts and contributions to multidisciplinary approaches. *Innovation: The European Journal of Social Science Research, 24*(3), 247–271.

Fuller, R. A., & Irvine, K. N. (2010). Interactions between people and nature in urban environments. In K. J. Gaston (Ed.), *Urban ecology* (pp. 134–171). Cambridge University Press. https://doi.org/10.1017/CBO9780511778483.008.

Gómez-Baggethun, E., Gren, Å., Barton, D. N., Langemeyer, J., McPhearson, T., O'Farrell, P., Andersson, E., Hamstead, Z., & Kremer, P. (2013). Urban ecosystem services. In *Urbanization, biodiversity and ecosystem services: Challenges and opportunities* (pp. 175–251). Dordrecht: Springer.

Haddad, N. M., Brudvig, L. A., Clobert, J., Davies, K. F., Gonzalez, A., Holt, R. D., Lovejoy, T. E., Sexton, J. O., Austin, M. P., & Collins, C. D. (2015). Habitat fragmentation and its lasting impact on Earth's ecosystems. *Science Advances, 1*(2), e1500052.

Hausmann, A., Slotow, R., Burns, J. K., & Di Minin, E. (2016). The ecosystem service of sense of place: Benefits for human well-being and biodiversity conservation. *Environmental Conservation, 43*(2), 117–127.

Hui, S. C. M., & Chan, K. L. (2011). Biodiversity assessment of green roofs for green building design. In *Proceedings of Joint Symposium 2011: Integrated Building Design in the New Era of Sustainability* (pp. 10.1–10.8), 22 November 2011 (Tue). Kowloon, Hong Kong: Kowloon Shangri-la Hotel, Tsim Sha Tsui East.

Klotz, S., & Kühn, I. (2010). Urbanisation and alien invasion. In K. J. Gaston (Ed.), *Urban ecology* (pp. 120–133). Cambridge University Press. https://doi.org/10.1017/CBO9780511778483.007.

Lindenmayer, D., Hobbs, R. J., Montague-Drake, R., Alexandra, J., Bennett, A., Burgman, M., Cale, P., Calhoun, A., Cramer, V., & Cullen, P. (2008). A checklist for ecological management of landscapes for conservation. *Ecology Letters, 11*(1), 78–91.

Liquete, C., Cid, N., Lanzanova, D., Grizzetti, B., & Reynaud, A. (2016). Perspectives on the link between ecosystem services and biodiversity: The assessment of the nursery function. *Ecological Indicators, 63*, 249–257. https://doi.org/10.1016/j.ecolind.2015.11.058.

Luck, G. W., & Smallbone, L. T. (2010). Species diversity and urbanization: Patterns, drivers and implications. In K. J. Gaston (Ed.), *Urban ecology* (pp. 88–119). Cambridge University Press. https://doi.org/10.1017/CBO9780511778483.006.

Lundholm, J. T. (2006). Green roofs and facades: A habitat template approach. *Urban Habitats, 4*(1), 87–101.

McDonald, T., Gann, G., Jonson, J., & Dixon, K. (2016). *International standards for the practice of ecological restoration–including principles and key concepts*. Washington, DC: Society for Ecological Restoration. Soil-Tec, Inc.,© Marcel Huijser, Bethanie Walder.

MEA. (2005a). *Ecosystems and human well-being, general synthesis*. MEA.

MEA. (2005b). *Ecosystems and human well-being: Biodiversity synthesis*. Washington, DC: World Resources Institute.

Mills, L. S., Soulé, M. E., & Doak, D. F. (1993). The keystone-species concept in ecology and conservation. *Bioscience, 43*(4), 219–224.

Mouquet, N., Gravel, D., Massol, F., & Calcagno, V. (2013). Extending the concept of keystone species to communities and ecosystems. *Ecology Letters, 16*(1), 1–8.

Müller, N., Ignatieva, M., Nilon, C. H., Werner, P., & Zipperer, W. C. (2013). Patterns and trends in urban biodiversity and landscape design. In *Urbanization, biodiversity and ecosystem services: Challenges and opportunities* (pp. 123–174). Dordrecht: Springer.

O'Farrell, P. J., & Anderson, P. M. (2010). Sustainable multifunctional landscapes: A review to implementation. *Current Opinion in Environmental Sustainability, 2*(1–2), 59–65.

Opoku, A. (2019). Biodiversity and the built environment: Implications for the Sustainable Development Goals (SDGs). *Resources, Conservation, and Recycling, 141*, 1–7.

Pedersen Zari, M. (2014). Ecosystem services analysis in response to biodiversity loss caused by the built environment. *SAPIENS: Surveys and Perspectives Integrating Environment and Society, 7*(1), 1–14.

Pedersen Zari, M. (2018a). The importance of urban biodiversity–an ecosystem services approach. *Biodiversity International Journal, 2*(4), 357–360.

Pedersen Zari, M. (2018b). *Regenerative urban design and ecosystem biomimicry*. Routledge.

Pitman, S. D., Daniels, C. B., & Ely, M. E. (2015). Green infrastructure as life support: Urban nature and climate change. *Transactions of the Royal Society of South Australia, 139*(1), 97–112.

Power, M. E., Tilman, D., Estes, J. A., Menge, B. A., Bond, W. J., Mills, L. S., Daily, G., Castilla, J. C., Lubchenco, J., & Paine, R. T. (1996). Challenges in the quest for keystones: Identifying keystone species is difficult—But essential to understanding how loss of species will affect ecosystems. *Bioscience, 46*(8), 609–620.
Puppim de Oliveira, J. A., Doll, C. N. H., Moreno-Peñaranda, R., & Balaban, O. (2014). Urban biodiversity and climate change. In B. Freedman (Ed.), *Global environmental change* (pp. 461–468). Springer Netherlands. https://doi.org/10.1007/978-94-007-5784-4_21.
Sandifer, P. A., Sutton-Grier, A. E., & Ward, B. P. (2015). Exploring connections among nature, biodiversity, ecosystem services, and human health and well-being: Opportunities to enhance health and biodiversity conservation. *Ecosystem Services, 12*, 1–15.
Saura, S., Bodin, Ö., & Fortin, M. J. (2014). EDITOR'S CHOICE: Stepping stones are crucial for species' long-distance dispersal and range expansion through habitat networks. *Journal of Applied Ecology, 51*(1), 171–182.
Savard, J.-P. L., Clergeau, P., & Mennechez, G. (2000). Biodiversity concepts and urban ecosystems. *Landscape and Urban Planning, 48*(3–4), 131–142.
Swingland, I. R. (2013). *Capturing carbon and conserving biodiversity: The market approach*. Routledge.
TEEB. (2010). *The economics of ecosystems and biodiversity: Mainstreaming the economics of nature: A synthesis of the approach, conclusions, and recommendations of TEEB*. Ginebra: UNEP.
Tilman, D., Lehman, C. L., & Yin, C. (1997). Habitat destruction, dispersal, and deterministic extinction in competitive communities. *The American Naturalist, 149*(3), 407–435.
Turner, C. (2018). *Climate change and biodiversity*. Scientific e-Resources.
Varshney, K., Pedersen Zari, M., & Bakshi, N. (2020). The intersection of carbon sequestration and habitat provision in built environments: building rating tools comparison. In A. Ghaffarianhoseini, A. Ghaffarianhoseini and N. Nasmith (Eds.), *Imaginable Futures: Design Thinking, and the Scientific Method, 54th International Conference of the Architectural Science Association 2020* (pp. 1145–1154), 26-27 November 2020. Auckland, New Zealand: Auckland University of Technology.
Vergnes, A., Kerbiriou, C., & Clergeau, P. (2013). Ecological corridors also operate in an urban matrix: A test case with garden shrews. *Urban Ecosystem, 16*(3), 511–525.
WCC. (2015). *Our natural capital: Wellington's biodiversity strategy and action plan 2015*. Wellington: Wellington City Council.
Williams, N. S., Lundholm, J., & Scott MacIvor, J. (2014). Do green roofs help urban biodiversity conservation? *Journal of Applied Ecology, 51*(6), 1643–1649.

Harare

Evaluating and Epitomizing Master Plan Performance in an African City in Search of a Positive Future

Percy Toriro[1,2] and Innocent Chirisa[3]
[1]Municipal Development Partnership for Eastern and Southern Africa, Harare, Zimbabwe
[2]African Centre for Cities, University of Cape Town, Cape Town, South Africa
[3]Department of Demography Settlement and Development, Social & Behavioural Sciences, University of Zimbabwe, Harare, Zimbabwe

H

Synonyms

Growth – development; Master plan – comprehensive plan; Infrastructure – utilities/facilities

Definition

Master Plan: a comprehensive plan that covers a whole region or city outlining the goals, strategies for land use, and infrastructure development as envisioned.

Growth – demographic or spatial expansion in/of a specified area or territory.

Urban planning: the technical and political processes that produce decisions that shape a city or town.

Statutory planning: the reference to legislative instrument in shaping the goals and priorities of an area.

Planning law – a set of legal instruments and reference to them in defining the future and management of a given area (town, region, or rural).

Infrastructure – the facilities, utilities that act a conduit for bringing services to given areas (for instance, roads, rails, dams, power lines).

Introduction

Planning has been defined as "the ordering of space" (Kamete 2009). It has also been defined as "all public policies which affect urban and regional development, zoning and land use, or what is often called public production of space" (Yiftachel 1998, p. 2). Planning is concerned with the creation of "orderly environments" that are expected to "enhance living conditions" (Fainstein 2000). It has also been assumed to be "progressive" and "reformist" and is said to mostly pursue a "modernist" agenda (Dear 1986; Hall 1988). Urban planning is a public good that is delivered to achieve the efficient use of land as well as sustainability (Cassidy and Patterson 2008). One supposed virtue of planning is that it is practiced in "public interest" (Faludi 1973). However, "planning as a discipline and as a profession has been increasingly challenged" (Yiftachel 1988, p. 24). The idea of planning being done in public interest has been challenged by several scholars who argue that planners have not exercised their professional discretion in a manner that benefits the public (Watson 2003; Pieterse 2008; Potts 2008; Kamete 2009; Yiftachel 2009). Planning and the production of master and local plans in Zimbabwe are regulated by the law. The law sets both the procedures to be followed as well as the substance of the plans (GoZ 1996). However, there is evidence of planning law having a "poor record in Africa" with accusations of the profession gifting "oppressive regimes, whether colonial or independent, with useful legal mechanism for restricting social and economic opportunities for most people" (Berrisford 2011, p. 215). There are several examples of the use and abuse of laws to advance agendas of the rich as well as improper use of discretion by planning professionals in Africa (Kamete 2009; Toriro 2018).

The laws regulating planning have not been reviewed as regularly as there have been changes to the operating environment. This has been documented as a common problem in many African cities (Berrisford 2011). These laws are generally out of sync with the reality of most African cities such as the realities of poverty, unemployment, food insecurity, and high levels of informality (Battersby 2018; Chen 2012; Crush et al. 2015). Perhaps these laws fail to fit because of the manner in which they were introduced to the different countries. They all emerged from a standard draft that was prepared for use in different colonial cities and were imposed without due consideration of peculiar circumstances (McAuslan 2003).

Several scholars have argued against the efficacy of statutory planning that borrows heavily from the traditional blueprint planning for their failure to deliver plans that are relevant to people's lived experiences (Roy 2005; Chirisa 2008; Kamete 2012). In Africa, creating a vision for cities has not been easy. There is a tendency for a disconnect between visions of officials that tend to be modernist and aspire for "world-class cities" and the rest of the people who are living in poverty (Watson 2014). This disconnect between the intended plans of officials and the generality of the residents has also been described as a "conflict of rationalities" where officials display a detachment from the lived reality of the people (Watson 2003).

Although there is evidence of strict traditional planning in Zimbabwe, this does not appear to have translated to improved services. Zimbabwe has a strong land use planning background that is visible on the orderly layout of most of its urban settlements (Toriro 2007). There is documentation of poor service delivery in many local governments in Zimbabwe, and this has prompted many residents to gang up and find voice in residents associations (Musekiwa and Chatiza 2015). There was a cholera outbreak in Harare during the 2008–2009 period, and 4000 people died as a result of service delivery failure in the water and sanitation sectors (Chatiza et al. 2013). Some of the urban problems are attributed to the country's initial racial policies. Cities were initially planned as temporary homes for black Zimbabweans who were considered rural dwellers expected to return there after their working life in urban areas (Potts 2010). These early policies affected the layout of Harare in significant ways, and these early planning policy challenges were identified as some of the contemporary Harare challenges in the early

1990s (Zinyama et al. 1993). These initial racial policies are reflected in sprawling urban areas that inefficiently utilize infrastructure and land.

Urban planning has been identified as one of the tools that should be used to address the many urbanization challenges confronting Africa today, but values of planning professionals stand in the way. While the United Nations accepts the important role that planning could play, they also observe that planning professionals have not delivered plans that solve common realities of African cities observing rather that as plans are prepared, most cities become exclusive (Tibaijuka 2006). This point of planning making cities exclusive of the poor is graphically made by one scholar who asserts that the planned city "sweeps" out the poor (Watson 2009). Planning in Harare has also been criticized for not embracing inclusive concepts such as the right to the city (Purcell 2013; Rogerson 2016). Maybe this should be addressed by changing the training curriculum of planning professionals at university level so that it reflects the lived reality of the cities in which the planners work (Watson and Odendaal 2013). Perhaps planning in Harare and other African cities must be more sensitive to the many challenges that affect these cities (Parnell et al. 2009). It would appear that the country's planning has not benefitted from the emerging new ways of planning including principles of the New Urban Agenda which further articulates principles embedded in the Sustainable Development Goals (Satterthwaite 2016).

This chapter reviews the performance of master planning in an African city using the case of Harare, Zimbabwe. It used secondary data reposed in Zimbabwe's planning laws, planning guides, and statutory instruments. The existing statutory plans were reviewed. Other published sources on the practice of planning, the theory of planning, and the strengths and weaknesses of planning were also used for purposes of comparison. The secondary data was complimented with interviews of key informants. The key informants were practicing planners and lecturers in planning schools who were purposively selected for their knowledge in the subject matter. They gave insights into the practice of planning in Zimbabwe and their opinions on the performance and continued relevance of the Harare Combination Master Plan. Observations were also made of the relationship between the master plan and developments on the ground.

Background to the Harare Master Plan

This section borrows from the background of the Harare Master Plan to reconstruct a brief history of what became Harare. The settlement was founded as Salisbury by the Cecil John Rhodes-led British South Africa Company (BSAC) as part of the colonization of the modern state of Zimbabwe. The company commissioned the first town plan in 1891 which covered the period 1891–1895. The main purpose of this first plan was the provision of water and sanitation to the settlement. In 1897, the Municipality of Salisbury was established, and racial segregation determined the siting of an "African township" called Mbare a kilometer south of Kopje, the landmark that had motivated the location of Salisbury. The newly established city grew steadily until the period 1937 to 1941 when it rapidly grew by 50% buoyed by the impact of the Second World War. This period saw the expansion of the city into the northern farms such as Avondale and Borrowdale creating new suburbs and promoting the urban sprawl that has continued to characterize Harare. The passing of the 1930 Land Apportionment Act which further institutionalized racial segregation led to the creation of another residential area for blacks known as Highfield with 2500 stands 13 km from the city center. As the city expanded, it faced administrative challenges of unifying the city as the different town boards managed different areas. The 1971 Greater Salisbury Local Authority was established to determine a suitable governance structure and recommended the current setup comprising a central administration and local service centers known as district offices. The setup created a city with a geographical footprint of 550 km^2 (CoH 1991).

The initial planning in the city was driven by sanitary considerations, and there was no formal

H

planning law. The first homegrown planning legislation borrowed from British planning law. It was the Town and Country Planning Act of 1933. This provided for the preparation of Town Planning Schemes, and several of these were prepared. This was followed by a slightly modified Town and Country Planning Act of 1945. An interesting feature of the Town Planning Schemes was that they left out African areas or high-density areas in compliance with racial segregation laws existing at the time. The 1976 Regional Town and Country Planning Act came with significant changes by introducing three types of statutory plans, namely, regional plans, master plans, and local plans. The three tiers defined the scope of plans with regional plans covering a large area such a province(s), master plans covering a whole city, and local plans covering a part of a settlement or a specific subject. The Harare Combination Master Plan was prepared using the 1976 act.

The preparation of the master plan was motivated by a number of factors primarily the "changing requirements and rapid growth especially within the City of Harare" (CoH 1991, p. 9). This was going to be achieved through the production of a coordinated strategic land use development framework for the sustainable development of Harare and its environments over a 15- to 20-year period. The plan set out to address four issues: create policies that addressed important emerging issues caused by rapid urbanization, and consider other national and regional factors; craft relevant development control measures; set out guidance for the preparation of local plans; and develop a strategy for phasing, implementation, and monitoring of development include resource allocation.

In drawing up boundaries for the Harare Combination Master Plan, the combination authority was guided by the need to include a geographical area with a "strong functional relationship" with the City of Harare at the time of plan preparation as well as into the future (CoH 1991, p. 4). This also included the following three objectives: the need to manage and contain the urban sprawl footprint of Harare; the identification of the growing water needs of the city and its environs; and the identification and inclusion of areas with strategic economic connections with the City of Harare. This area was found to cover a geographical area of 4571 km^2 in extent. The 1993 Harare Master Plan was jointly prepared by 11 local authorities, namely, Bromley Ruwa Rural Council, Arcturus Rural Council, Mazowe Rural Council, Harare West Rural Council, Banket-Trelawney Rural Council, Norton-Selous Rural Council, Beatrice-Harare South Rural Council, Harava District Council, Goromonzi-Kubatana District Council, Chitungwiza Town Council, and Harare City Council. Two central government ministries were also included: the Ministry of Local Government, Rural and Urban Development, as well as the Ministry of Natural Resources and Tourism.

The plan intended to "formulate realistic land use proposals related to prevailing socio-economic conditions" (CoH 1991, p. 9). Perhaps the main challenge that the master plan aimed to address was the rapidly growing Harare population. With the advent of political independence, the cessation of war, and the removal of influx control regulations, Harare's population ballooned (CoH 1991; Zinyama et al. 1993; Potts 2010). From a population of 310,360, the population had risen by 24% to 386,040 by 1969. By 1982, the population was at 658,364, a further growth of 70.5% from 1969. In line with the pre-independence rural-urban migration trends, the majority of the population were males in the 20–34-year-old age group (CoH 1991).

The rapidly growing population exerted immense pressure on the city's infrastructure. The master plan study identified water as a major challenge for the city. The study noted that "Harare is ill-located to facilitate future growth economically" as the only way it could guarantee adequate supplies was through recycling of wastewater (CoH 1991, p. 13). It went on to project that even with recycling, the maximum population supporting the existing resource could serve was between 4 and 5 million. The then chief water engineer further indicated that after that, providing water to Harare would become "more expensive than whisky" (CoH 1991, p. 14). In addition to limited water availability, bottlenecks were also identified in its distribution. Water could be provided cheaply within a 15-km radius of the city center. The network could be stretched to a 22.5-km radius beyond which "prohibitive investment" in the network was required.

The study also identified drainage constraints in providing reticulated sewerage to Harare with the southern and western parts of the city easiest to sewer. The eastern areas such as Mabvuku and Tafara could only be expanded after the development of new treatment works on Ruwa River for which a trunk sewer was required. A similar constraint to water was the large cost that would be required to connect areas beyond a 12.5-km radius. To keep infrastructure investments of water and sewerage to affordable levels, there was need to contain urban sprawl. Considering all development potential factors, the identified areas with the highest development potential were:

(a) The Darwendale and Muzururu area to the west
(b) The Waterfalls area
(c) The Gwebi drainage area
(d) Areas adjacent to Chitungwiza

At the time of the master plan preparation, the City of Harare and the country in general had a passenger vehicle deficit caused by the shortage of foreign currency. The country had an estimated motor vehicle demand of 20–25,000 units per annum and had a backlog of 100,000 vehicles. In 1986, the planning area had a vehicle population of 115,000 vehicles. The major public transport provider, the Harare United Omnibus Company operated 640 buses with a monopoly service within a 26-km radius of Harare CBD. The service was overstretched and was described as reaching a crisis situation.

There were also 735 permits for long distance buses plying different inter-city and rural routes out of the Mbare Musika long-distance bus terminus situated a few kilometers south of Harare CBD (CoH 1991). There was a plan for the redevelopment of the sole long-distance bus terminus. The overall road network consisted of a paved portion of 2280 km and a gravel network of 121 km in 1986. Then, there was no significant congestion on most of the roads. There was, however, need to complete the Harare Drive links as well as the airport road links over the railway lines. There was also need for inter-suburban links to avoid traffic passing through the city center. One example was to link High Glen Road from Bulawayo Road roundabout to Old Mazowe Road. The roads infrastructure was maintained using funds raised from vehicle licenses and rates on a 50% from each account basis. Capital funds for road construction are obtained from loans. A proportion of this was obtained from government as a grant to cater for the proportion of national roads in Harare.

Problems of congestion, noise, and air pollution, as well as safety and general inconvenience caused by heavy trucks, were a noted concern. The master plan was expected to plan for heavy vehicle routes which avoided builtup areas. Railway transport also available but with no intracity service. A need was identified to plan for Chitungwiza railway line to link Harare with the dormitory city of Chitungwiza, 25 km southwards. There was a busy container terminal at Rugare Township just 5 km from the city center which handled 80 cargo containers a day and at least 500 a week. There was a wide network of railway sidings serving at least 300 premises in Harare's different industrial sites.

Demand for housing since independence exceeded housing production. This was against a situation of very low national investment in housing. In addition, there were disparities in land availability with 31,561 ha of land having accommodated 39,172 housing units in low-density areas against 5972 ha accommodating 64,850 units in high-density areas from independence in 1980 to the start of studies informing the master plan process in 1987. This translated to a density of 109.46 people per hectare in high density areas against 9.24 people per hectare in low-density areas. Harare averaged 24.66 people per hectare. This compared with New York's density of 300 people per hectare and Cairo's 1000 people per hectare.

Another indicator of housing demand is the housing waiting list. This is a list of prospective

H

homeseekers who register to be allocated a housing plot or a house by the municipality. In 1977, the waiting list for Harare was 16,000. This grew to 29,279 in 1987. Although this shows a phenomenal rise in demand, it may even have understated the real demand as many people do not register on the list due to the many conditions that one must meet to be registered. A study by the master plan team also found 22% of high-density residents living in unapproved outbuildings, another indicator of the high demand for housing in Harare. Although the city was going to avail approximately 55,000 units of stands during the master plan preparation phase, this was still to "fall woefully behind demand" (CoH 1991, p. 23). The under provision for the high- and middle-income brackets of residents was a cause for concern. The fear was that these groups were likely to go for housing provided for lower income groups resulting in a "downward raiding" phenomenon.

Harare had a large land bank of industrial area as well as a relatively well-developed industry. The industries were located in Workington, Southerton, Graniteside, Ardbennie, Beverley, Aspindale Park, as well as the outlying areas of Ruwa, Arcturus, Norton, Mt Hampden, and Chitungwiza. The Harare industries occupied a total of 2137 ha of which almost 66% was developed. The highest industrial development was reported to have taken place during the period 1971–1975 when the isolated Rhodesian government was pursuing self-sufficiency in many areas. During this time approximately 10.7 ha per year were developed for industries compared to 7 ha per year during the period 1981–1983. The master plan technical team estimated an even lower investment during the plan preparation period. The team further estimated that industrial land in Harare would be exhausted by the year 2002 (CoH 1991). At an established estimated employment rate of 134 employees per hectare in industry, there were an estimated 200,000 employees in Harare's industrial areas and potential for an additional 85,000 employees.

Harare's commercial developments comprised of three identified categories: the central shopping area of Harare also known as the CBD; the suburban shopping areas in low-density areas; and the small grocery shops found in the high-density shopping areas. The CBD had the most well-developed and varied commercial services. Most low-density suburban shopping areas were old having been established during the 1950–1960 period and another lot developed in the 1960s when the country experienced a large influx of white migrants. The high-density shopping centers were the least developed typically comprising of grocery shops, bottle stores, and butcheries. Residents could only access low order goods with most goods obtained from the city center.

At the time of the master plan preparation, the majority of the people were employed in the CBD (61.5%), with 22.5% employed in industry, and only 5% employed "elsewhere" (CoH 1991). At the time of the master plan study, there was an estimated potential workforce of 432,390 people. There was a total of 289,140 people in formal employment giving an unemployment rate of approximately 24% in Harare. The rate of job creation growth was 2.6% per annum against a potential workforce growth rate of 5.6% per year.

Most of the employees (64%) lived in the high-density areas, 20% lived in the low-density areas, and the 11% lived in nearby rural areas such as Domboshawa.

Fieldwork done during the plan preparation process indicated that 77.5% of the high-density employees used buses to go to work, as well as 41.5% from low-density areas. About 4.2% of workers cycled to work while another 16.5% walked to work. For 57% of the workers the journey to work was 30 min, while it was 1 h for 34.5% of the workers. Only 7.7% of the workforce had a journey to work of an hour or longer. The situation was, however, observed to be deteriorating with the aging bus fleet.

An important issue that only received scant attention during the master plan preparation was the environment. This is understandable since the studies feeding into the master plan were undertaken before the advent of climate change and global warming as global concerns. Even new local challenges such as pollution, wetland development, and waste management were not yet major challenges. The main weakness of the master plan was a failure to deeply examine and mainstream environmental matters.

Refuse management received some consideration but only from a refuse disposal perspective. The two landfills at Teviotdale and Golden Quarry were receiving 3,000,000m^3 of refuse annually. Both landfills were assessed to have a mean life of only 6 years each. This meant that there had to be plans for new landfills by 1992. To extend the lifespan of the landfills, the master plan encouraged the setting up of incinerators to reduce the volume of waste deposited at the landfills as well as refuse reclamation and recycling. Hazardous waste and air pollution monitoring were also identified as important areas for consideration. More than 40 t of chemical waste and 800,000 liters of cyanide and other waste were being dumped at Teviotdale. Air pollution was still within acceptable levels, and the team indicated that at the time there was no "apparent cause for concern" (CoH 1991).

The master plan team also found a growing need for educational facilities since Harare is dominated by a young population. The same applied to public health facilities such as maternity hospitals and primary care clinics. Colonial development had meant that there were huge disparities in the availability of recreational and other community facilities and the master plan intended to address these. The master plan survey found Harare to have an overprovision of large hotels which were only half full even at large international conferences such as the Non-Aligned Movement conference of 1986. With two international hotels nearing completion, it was concluded that Harare had adequate hotel facilities. Demand was, however, high and underprovided in the low-budget hotel sector where facilities were always full.

The master plan team noted that the death rate in the planning area was falling. However, the city's six cemeteries were already 71% full. There was, therefore, urgent need for identification of new cemetery sites. The team anticipated difficulties in identifying suitable land due to two factors; cemeteries are considered in planning to be "a bad neighbor," and finding an ideal site with good soil conditions suitable for grave depth requirements without polluting adjacent soils was not going to be easy.

For all of the problems that the master plan intended to address that have been discussed above, specific proposals were formulated to address them. The 1987 population was estimated at 1,450,897 and was expected to reach 2,481,509 by the year 2005. To address the rapid population growth, a few proposals were made. The planners realized the need for a national policy that recognized the high population growth rates. It was recommended that population be included in school curricula so as to teach about population control from an early age. Birth control measures were identified as another strategy to control population growth. The master plan included strategies to promote economic growth at surrounding rural service centers to discourage rural to urban migration. It also required policies for employment creation since 35% of the population was under 15 years old.

The plan identified high development potential in Harare South, around Chitungwiza, and in the Gwebi catchment, provided infrastructure was upgraded. The area of highest economic development potential was, however, within a 15-km radius of city center. For the areas with development potential but further away from the CBD, investment in infrastructure such as sewerage works was required. The study also noted a paradox; some land with high urban development potential was also suitable for agriculture. A balance was, therefore, made where the good soils in the northern end of the city would be retained for agriculture and instead expand the city southwards towards the less fertile sandy soils.

To increase water provision, the water treatment works had to be upgraded quickly to provide higher demand for water. To open up new areas around Chitungwiza and Harare South for development, a new sewerage treatment works was proposed on the Mupfure Catchment.

On the road network, there was a proposal to complete a system of ring roads to avoid traffic passing through the center. This included completing the link between airport road and enterprise road over the railway lines. The plan also discouraged the implementation of freeways arguing their implementation in some areas was no longer practical due to new developments along the proposed routes.

The response to the high demand for housing was with an "abysmally low supply"; hence the plan made a raft of proposals including allowing smaller subdivisions in some areas and infill development. These technical proposals would be supported by policies that attracted private and institutional housing developers to bring in higher levels of investments (CoH 1993). The overall goal of the plan was to deliver adequate houses to all income brackets.

Due to the fast-growing population and rising demand for employment, the plan sought to create employment zones in strategic areas. In the city center, the master plan sought to encourage optimum use of land by permitting higher bulk factors. Outside the CBD the plan proposed the establishment of office parks and special industrial areas as well as the revitalization of small suburban shopping centers with the objective of bringing services closer to most people. It also sought to provide for the small-scale sector by encouraging smaller industrial stands and creating conducive conditions for the nurturing and growth of micro businesses and SMEs. Employment could also be encouraged by intensive urban and peri-urban agriculture which was nearer to urban markets.

As already noted, the examination of environmental matters, and the subsequent formulation of proposals was cursory. Proposals were made on conserving and improving the "existing high-quality environment"; planning for refuse incineration to reduce volume of waste; introducing the environmental impact assessment (EIA) as a tool to promote sustainable development; and conserving rivers and water supply sources by imposing buffers along river corridors. On waste management, a transfer station complete with an incinerator was proposed at Golden Quarry Landfill. To avoid urban sprawl, it was proposed to establish a greenbelt around Harare beyond which development would be locked. While some of the identified areas are important, the examination of environmental matters was brief and cursory; hence the proposals were very shallow and lacked detail.

The plan proposed to provide adequate educational and health facilities in line with population growth. Other social and recreational facilities were also going to be provided for including areas for religious and social well-being. There was even an ambitious objective to develop a "coordinated regional open space system" that would be developed into high-quality relaxation and scenic resorts. Also included was a proposal to develop new cemeteries in Donfontein, Mt Hampden, and Gillingham. This was informed by the rapidly filling up of all existing cemeteries.

Progress on Implementation of Proposals: A Discussion

The Harare Master Plan recommended certain corridors for development. These were Harare South, Chitungwiza surroundings, and areas within a 12.5-km radius of Harare city center. It also specifically recommended that areas with good fertile soils be preserved for agriculture so that the city remains food secure. All these proposals were distorted or ignored. Development did occur in the southern direction, but it was all largely unplanned and, in some areas, went against the master plan:

> The southern areas were developed as a corruption of the fast track land reform program (FTLRP). The City of Harare had prepared a local plan known as the Southern Incorporated Areas Local Development Plan No.31. Before this plan could be implemented, the area was occupied by housing activists as part of the FTLRP around the year 2000. Groups of invaders self-allocated the land and planned and allocated it without municipal approval. The planned density in many areas was altered to suit the occupiers' requirements. Proper planning and approvals only happened much later as a politically motivated regularization process. (City Planner 2019)

Chitungwiza's developments occurred about 10 years later around the year 2010. While for some farms the motive for occupying the land was similar to that of the land reform, some of it was done by traditional leaders in the surrounding rural areas and other community leaders. Some occupations were led by corrupt urban activists with political links to the two main political parties, namely, ZANU PF and MDC, as was explained below by an official employed by Manyame Rural District Council which has jurisdiction over most of the land:

> The occupations happened gradually and very secretly for a long time. It was only around the 2013 elections that activists from both parties started openly mobilizing their supporters by promising access to land. They dished out land on the urban fringes where were the demarcations of Chitungwiza Municipality or Manyame RDC were not clear. The occupations extended to rural communal areas and adjacent commercial farms. That is when there was a lot of unsanctioned land parceling. We then reported to the Minister of Local Government who in turn appointed a commission of inquiry to investigate the matter. The committee confirmed our observations but also found that a lot of land within Chitungwiza town had also been quietly allocated in the same manner. (Manyame Official 2020)

Urban development in the proposed corridors was, therefore, largely driven by unauthorized parties leading to a distortion of the plan's intention. In the areas around Chitungwiza, the Government of Zimbabwe ended up protecting the people to whom land had been sold because they had already paid a lot of money to the unauthorized individuals who were described in the investigating committee report as "land barons." The land beneficiaries, however, ended up paying twice for the same land as the local authorities also asked them to pay. While these beneficiaries ended up getting security of tenure after the re-planning and regularization, they lost large amounts of money to the land barons.

The northern and some northwestern parts of Harare had been identified to have fertile soils and had been recommended for continued agriculture utilization. This did not happen as planned. Suburbs such as Mt Pleasant Heights and portions of Good Hope were developed for housing despite the high agriculture productivity. This was partly motivated by short-term profit consideration by some landowners as well as failure to adhere to plan provisions by corrupt planning officials. (Interview with a central government planning agency senior planning official.) Similarly, no greenbelt to demarcate city boundary was put in place; hence applicants well outside the proposed development area continued to be considered for development and were issued with development permits. Harare thus continues to sprawl eating into surrounding rural and commercial farming areas.

Some of the most significant master plan proposals aimed to promote massive housing provision by promoting densification and infill development. Except for infill development, which was probably overdone, densification included vertical development achieved limited success. The intensive development of areas within a 12.5-km zone to optimize the use of existing infrastructure only achieved little success. This would have meant development of new areas as well as redevelopment of identified strategic areas to produce large numbers of housing. A dozen or so blocks of walk-up flats were constructed in the avenue area adjacent to the city center. Even fewer blocks which are beyond six floors were developed in the same area. In that zone, there are still many old detached dwelling houses where there is potential for higher floor area factors. An opportunity to house many more people within walking distance of the city center has thus been unutilized or underutilized.

The policy also provided for the opportunity to subdivide stands to a minimum of 500 m^2 in areas with sewerage reticulation. In the residential areas adjacent to Harare City center such as Eastlea, Belvedere, Milton Park, and Hillside, the minimum subdivision in most plans remains at 1000 m^2. This is despite the master plan opportunity for increased densities. Other suburbs within the prescribed 12.5-km zone such as Highlands, Alexandra Park, Gunhill, Avondale, and Mabelreign have not fully taken advantage of the densification provision. As a result, Harare continues to develop outwards threatening to swallow many surrounding farms and settlements under other jurisdictions.

Perhaps the only policy that has been fully implemented to a point of abuse has been the infill development policy. In many suburbs both high density and low density, undeveloped land has been further subdivided to create new stands. While this has provided housing, the policy has been abused to develop in ecologically sensitive areas such as wetlands and other public open spaces. There are many complaints of developments in wetlands and other open spaces under the guise of implementing the infill policy. (The Harare Wetlands Trust reported that Harare officials were corruptly converting wetlands to housing development.)

Perhaps the weakest link in master plan implementation is the limited or non-implementation of some of the key infrastructure projects that were supposed to drive new development in the master plan area. While development has occurred in some instances beyond the master plan projections, most infrastructure that should have supported that development has not been implemented.

The new areas that the master plan identified as having potential for development did not have infrastructure such as trunk sewers, water storage reservoirs, and even trunk roads. This applies to Harare South, areas around Chitungwiza, and the Gwebi catchment. To date, all these areas are developed with houses that are already occupied: some with poor infrastructure that was not upgraded to accommodate the expansion, others without any basic infrastructure at all. The new areas in the Gwebi catchment in Good Hope and other areas have inadequate water storage and a small sewerage treatment plant at Marlborough. Areas south of the city are worse. Much of Harare South is not connected to both municipal water and sewerage. Areas further south around Chitungwiza are in a similar state of not having both reticulated water and sewerage. They also do not have proper roads and other services. The proposed Mabvuku outfall sewer which was supposed to transport sewerage to the proposed Ruwa River Sewerage Treatment Plant at Lyndhurst remains on the drawing board.

The master plan also identified new dam projects to ensure adequate water supplies in the planning areas. The technical team projected that there would be water shortages by the year 2000. Three dam sites were proposed, Kunzvi and Musami Dams for Harare and Muda Dam for Chitungwiza. None of these projects were implemented yet; the whole planning areas now suffers critical water shortages. Most of the major road projects that were proposed in the master plan have not been done or completed. The airport link with enterprise road, which would decongest the city center, remains incomplete. There are still missing links on the Harare Drive ring road. Although significant portions were constructed in Pomona, Mt Pleasant, and Marlborough, it remains incomplete and cannot perform as intended. Traffic from most areas is still forced to pass through or near the city center continuing to cause bottleneck pollution and congestion. Similarly, the Chitungwiza Railway project remains a pipedream.

As a result of the failure to implement these and other proposed infrastructure projects, Harare now has large sections of the city that do not have water and sewerage. It now also has congestion on many roads particularly on the way to and within the CBD. While there was a backlog in motor vehicle supply at the time of the master plan preparation, there has been an influx of cheap imported used cars that have increased the vehicle population at least fourfold the number during the master planning period. (Interview with an official from the Harare traffic unit.)

The plan had a few proposals to manage Harare's environment. As already indicated, environmental matters did not receive adequate consideration in the master plan. Unfortunately, even the few important clauses that were in the plan were not fully implemented. One of the biggest threats to Harare's environment is the destruction of wetlands despite a proposal to protect water courses and rivers. As a result, there has been a massive invasion of sensitive ecological spaces without due regard in many areas of Harare. Several organizations are now working to protect the remaining important environmental areas out of concern that the plan and the municipality are not doing enough to manage them. (The Harare

Wetlands Trust, Birdlife Zimbabwe, Environment Africa, and other organizations now have programmes to lobby for the protection of wetlands and other open spaces.)

Also not attended to are cemeteries and waste management. Although the Harare Master Plan projected that there was urgent need for new cemetery sites as well as landfills, neither has been done. The landfills were projected to fill up by 1992, yet waste continues to be deposited at Pomona Landfill.

New Imperatives for the New Harare Master Plan

There are several other areas that did not receive attention in the master plan for different reasons. Some of these areas that now constitute current challenges were not major challenges at the time of plan preparation. The informal sector was so small as to be insignificant at the time the studies that inform the master plan were conducted. For example, vending was not a problem for authorities in the city center. The designated markets at the bus termini were sufficient to accommodate all vegetable vendors. Unemployment was within levels that did not push most people to resort to informal employment. While the plan had a few proposals directed at the informal sector, they were not significant because the problem was equally insignificant.

Harare residents were documented as having some of the best services and safety nets in the region (Potts 2010). As a result, most families could afford basic necessities including good food. This has changed significantly. There have been several shocks such as the 1990s Economic Structural Adjustment Programme (ESAP) and other developments that have increased vulnerability and increased unemployment (Potts and Mutambirwa 1998; Tibaijuka 2005). This also increased the proportion of food insecure residents that began to engage in urban agriculture as a survival strategy and grinding own mealie meal to cut costs (Mbiba 1995). In a study conducted by AFSUN in 11 SADC cities in 2009, Harare was among the cities with the highest levels of food insecurity among the urban poor (Tawodzera et al. 2012). Food security is, therefore, a new consideration that was a non-issue in Harare at the preparation of the 1993 master plan. However, any plan prepared will have to address the increasingly dire levels of food insecurity in Harare and other urban areas. The master plan must take a food systems approach to addressing urban food security in Harare.

Traditionally Zimbabwean urban areas including Harare did not tolerate unplanned settlements. The state and the local governments in colonial Rhodesia as well as in independent Zimbabwe were known to ruthlessly deal with unplanned settlements (Patel and Adams 1981; Zinyama et al. 1993). As a result, unlike most African cities, Zimbabwean urban areas did not have notable unplanned areas or slums. This changed with the advent of the FTLRP when landless Zimbabweans in urban areas took advantage of the revolution to claim residential land. This has created many unplanned or unapproved settlements around Harare which present various planning challenges (Toriro 2015). Most of these settlements do not have approved plans for layouts and buildings. They also lack infrastructure such as reticulated water and sewerage. This must be corrected through proper planning and/or regularization. Any new master plan must consider all these settlements if Harare is to return to having properly planned and serviced settlements.

Conclusion and Way Forward

The Harare Master Plan introduced many useful interventions. Housing delivery improved; suburban shopping experiences were revamped, but many infrastructure projects that would have supported the inevitable growth were not implemented. Dams were not built, treatment plants were not expanded significantly, and only a few water reservoirs were built in different areas.

New areas were opened up and developed without supportive frameworks. Most of these settlements lack most basic amenities. The few useful provisions for urban environmental management have not been implemented. This has resulted in the depletion of wetlands and other open spaces. The city has also continued to sprawl outwards in an unsustainable manner. The standards of living have also fallen from those of the first decade of independence. New challenges have emerged that were not in existence when the plan was prepared. There were insignificant levels of informality as most Harare residents were formally employed and regulations were strictly enforced. The structure of the economy has changed significantly from formal and large scale to informal and small scale. There were also events that happened after the master plan was prepared that significantly altered the growth patterns of Harare. These were ESAP and the FTLRP. The 1993 Master Plan was not designed to accommodate these huge impact developments. These are now "stubborn realities" of Harare's urbanity.

It is timely that a new master plan is being considered. The new master plan must certainly consider the "stubborn realities" of Harare which are typical of other Global South cities (Watson 2012). The planning tools must equip officials to manage these. Planning officials removing informal operators from the streets of Harare reported that they were only enforcing existing planning regulations (Rogerson 2016). Harare inherited Salisbury's characteristics of a colonial city, and many such characteristics remain, but these can no longer solve present problems (Rakodi 1995). Perhaps planners must introspect whether it is the Harare that the current residents aspire for.

Cross-References

▶ Local and Regional Development Strategy

References

Battersby, J. (2018). Cities and urban food poverty in Africa. In G. Bhan, S. Srinivas, & V. Watson (Eds.), *The Routledge companion to planning in the global south*. New York: Routledge.

Berrisford, S. (2011). Why it is difficult to change urban planning laws in African countries. *Urban Forum, 22* (3), 209–228.

Cassidy, A., & Patterson, B. (2008). *The planner's guide to the urban food system*. Los Angeles: University of Southern California, School of Policy, Planning, and Center for Sustainable Cities.

Chatiza, K., Makanza, V., Musekiwa, N., Paradza, G., Chakaipa, S., Mukoto, S., & Mushamba, S. (2013). Capacity building for local government and service delivery in Zimbabwe. *Report of the 2013 local government capacity assessment*.

Chen, M.A. (2012). *The informal economy: Definitions theories and policies*. WIEGO working paper no. 1, Cambridge.

Chirisa, I. (2008). Population growth and rapid urbanization in Africa: Implications for sustainability. *Journal of Sustainable Development in Africa, 10*(2), 361–394.

City of Harare. (1991). *Report of study of the Harare Combination Master Plan*. Harare: Harare Combination Master Plan Authority.

City of Harare. (1993). *Written statement of the Harare Combination Master Plan*. Harare: Harare Combination Master Plan Authority.

City Planner. (2019). This is an interview with the City Planner.

Crush, J., Skinner, C., & Chikanda, A. (2015). *Informal migrant entrepreneurship and inclusive growth in South Africa, Zimbabwe and Mozambique*. Cape Town: Southern Africa Migration Programme.

Dear, M. J. (1986). Postmodernism and planning. *Environment and Planning D: Society and Space, 4*(3), 367–384.

Fainstein, S. (2000). New directions in planning theory. *Urban Affairs Review, 35*(4), 451–478.

Faludi, A. (1973). *A reader in planning theory*. New York: Pergamon.

Government of Zimbabwe. (1996). *Regional, town and country planning act [Chapter 29: 12]*. Harare: Government Printers.

Hall, P. (1988). *Cities of tomorrow*. New York: Basil Blackwell.

Kamete, A. Y. (2009). In the service of tyranny: Debating the role of planning in Zimbabwe's urban clean-up operation. *Urban Studies, 46*(4), 897–922.

Kamete, A. Y. (2012). Interrogating planning's power in an African city: Time for reorientation? *Planning Theory, 11*(1), 66–88.

Manyame Official. (2020). This is an interview with the Mhanyame Rural District Council official.

Mbiba, B. (1995). *Urban agriculture in Zimbabwe: Implications for urban management and poverty*. Aldershot: Avebury.

McAuslan, P. (2003). *Bringing the law back: Essays in law and development*. London: Ashgate.

Musekiwa, N., & Chatiza, K. (2015). Rise in resident associational life in response to service delivery decline by urban councils in Zimbabwe. *Commonwealth Journal of Local Governance, 16/17*, 120–136.

Parnell, S., Pieterse, E., & Watson, V. (2009). Planning for cities in the global south: An African research agenda for sustainable human settlements. *Progress in Planning, 72*(4), 195–250.

Patel, D. H., & Adams, R. J. (1981). *Chirambahuyo: A case study in low-income housing*. Gweru: Mambo Press.

Pieterse, E. A. (2008). Urbanisation trends and implications. In *City futures: Confronting the crisis of urban development* (pp. 16–38). London: Zed Books.

Potts, D. (2008). The urban informal sector in sub-Saharan Africa: From bad to good (and back again?). *Development Southern Africa, 25*(2), 151–167.

Potts, D. (2010). *Circular migration in Zimbabwe and contemporary sub-Saharan Africa*. Woodbridge: Boydell and Brewer.

Potts, D., & Mutambirwa, C. (1998). "Basics are now a luxury": Perceptions of structural adjustment's impact on rural and urban areas in Zimbabwe. *Environment and Urbanization, 10*(1), 55–76.

Purcell, M. (2013). The right to the city: The struggle for democracy in the urban public realm. *Policy & Politics, 41*(3), 311–327.

Rakodi, C. (1995). *Harare: Inheriting a settler-colonial city: Change or continuity?* Chichester: Wiley.

Rogerson, C. M. (2016). Responding to informality in urban Africa: Street trading in Harare, Zimbabwe. *Urban Forum, 27*(2), 229–251.

Roy, A. (2005). Urban informality: Toward an epistemology of planning. *Journal of the American Planning Association, 71*(2), 147–158.

Satterthwaite, D. (2016). A new urban agenda? *Environment and Urbanization, 28*(1), 3–12.

Tawodzera, G., Zanamwe, L., & Crush, J. (2012). *The state of food insecurity in Harare, Zimbabwe*. Cape Town: AFSUN.

Tibaijuka, A. K. (2005). *Report of the fact-finding mission to Zimbabwe to assess the scope and impact of operation Murambatsvina*. Nairobi: UN Habitat.

Tibaijuka, A. K. (2006). *The importance of urban planning in urban poverty reduction and sustainable development*. Vancouver: World Planners Congress.

Toriro, P. (2007). Town planning in Zimbabwe: History, challenges and the urban renewal operation Murambatsvina. In K. Kujinga & S. Chingarande (Eds.), *Zimbabwe's development experiences since 1980*. Harare: OSSREA.

Toriro, P. (2015). *Findings of the investigation committee on Caledonia farm*. Harare: Report presented to the Ministry of Local Government, Public Works and National Housing.

Toriro, P. (2018). *Food production processing and retailing through the lens of spatial planning legislation and regulations in Zimbabwe* (PhD Thesis, University of Cape Town).

Watson, V. (2003). Conflicting rationalities: Implications for planning theory and ethics. *Planning Theory and Practice, 4*, 395–408.

Watson, V. (2009). The planned city sweeps the poor away: Urban planning and the 21st century urbanisation. *Progress in Planning, 72*, 151–193.

Watson, V. (2012). Planning and the 'stubborn realities' of the global south-east cities: Some emerging ideas. *Planning Theory, 12*(1), 81–100.

Watson, V. (2014). African urban fantasies: Dreams or nightmares? *Environment and Urbanization, 26*(1), 215–231.

Watson, V., & Odendaal, N. (2013). Changing planning education in Africa: The role of the association of African planning schools. *Journal of Planning Education and Research, 33*(1), 96–107.

Yiftachel, O. (1988). The role of the state in metropolitan planning: The case of Perth, Western Australia, 1930–1970. *Urban Policy and Research, 6* (1), 8–18.

Yiftachel, O. (1998). Planning and social control: Exploring the dark side. *Journal of Planning Literature, 12*(4), 395–406.

Yiftachel, O. (2009). Theoretical notes on gray cities': The coming of urban apartheid? *Planning Theory, 8*(1), 88–100.

Zinyama, L., Tevera, D., & Cumming, S. (1993). *Harare: The growth and problems of the city*. Harare: University of Zimbabwe Publications.

H

Harvesting – Collecting, Harnessing

▶ Rainwater Harvesting for Water Security in Informal Settlements: Techniques, Practices, and Options

Hazard

▶ Role of Disaster Relief Policy in Building Resilient Coastal Regions in the United States

Health

▶ Emerging Concepts Exploring the Role of Nature for Health and Well-Being

Health and the City: How Cities Impact on Health, Happiness, and Well-Being

Alice Schmidt
Global Health Advisory Service to the European Commission, Mechelen, Belgium
Vienna University of Economics and Business, Vienna, Austria
AS Consulting, Vienna, Austria

Synonyms

Urban health and well-being

Definition

Health is a state of complete physical, mental, and social well-being and not merely the absence of disease. The enjoyment of the highest attainable standard of health is a fundamental human right that applies without distinction of race, religion, political belief, and regardless of economic or social condition (WHO 2006). Urban health applies to people who live in urban regions.

The health of individuals and communities is determined by a wealth of factors, including people's individual characteristics and behaviors as well as the socioeconomic and physical environment in which they live. Most of these factors are beyond individuals' direct control.

These "determinants of health" include income and social status; education (low education levels are linked with poor health, more stress, and differences in health-seeking behavior); gender (men and women suffer from different types of diseases at different ages); physical environment, including access to safe water, clean air, safe housing, etc.; working conditions (people in employment are healthier, particularly those who can control their working conditions); social support networks (greater support from families, friends, and communities is linked to better health); culture, i.e., customs and traditions that affect beliefs and behavior; genetics (determining life span and the likelihood of developing certain illnesses); individual behavior and coping skills, including balanced eating, keeping active, tobacco and alcohol consumption, and resilience to stress and challenges; and access to and use of quality health services (WHO 2017).

The social determinants of health overlap with the concept of environmental health. The latter describes the influences of both the natural and the built environment on human health. In cities, the built environment, i.e., the human-made structures such as buildings, parks, roads, etc., that define how humans live, move around, and work, is particularly relevant.

Introduction

The health of its inhabitants is arguably a city's most important asset (WHO 2016a). This is why pandemics, such as COVID-19, present an enormous challenge for urban development. With humanity's growing disregard for nature and fundamental lack of understanding of the value of ecosystem services, such pandemics will become more likely, with potentially grave consequences for urban futures.

Peoples' health and well-being determine their educational progress, productivity, income, and resilience to stress factors. Despite its importance in sustainable development and helping cities thrive, health has been insufficiently integrated into urban policies and planning.

While urbanization has provided unprecedented economic opportunity for large numbers of people, including better access to health services, the rapid, uncontrolled growth of cities is having profoundly negative effects on health, both directly and via effects on environmental degradation and climate change. As urbanization is accelerating, the pivotal role of cities in influencing the health of their inhabitants is greater than ever (WHO 2016a).

A quarter of the world's urban residents – about one billion people – live in informal settlements or slums (WHO 2014). By 2050, 68% of people will live in cities (UN DESA 2018). The number of people living in slum-like conditions is likely to increase significantly given rapid urbanization, particularly in developing countries.

Urban Health: Mitigating Risks and Harnessing Opportunities

Cities provide a wealth of opportunities for their residents. At the same time, they are associated with significant health risks and hazards. Nowhere is health more associated with social determinants than in urban environments. Housing, transport, employment opportunities, social services, education, nutrition, air pollution, violence, and the built environment all play key roles in urban settings, and they interact with each other. It is noticeable that urbanization per se is considered a social determinant of health by the World Health Organization (WHO n.d.-a).

Cities are associated with extremes: occupying 2% of the total land, they account for 70% of GDP, greenhouse gas, and global waste, respectively; and they consume over 60% of global energy (Habitat III 2017). All of these factors affect human health in urban areas more than in rural areas.

As urbanization comes with both opportunities and challenges for human health and well-being, the type and speed of urbanization matters a great deal when it comes to creating an "urban advantage." Given rapid urbanization, particularly in developing countries, this urban advantage is not always realized, particularly where rapid urbanization leads to growth of slums and other types of informal settlements.

Population Density: A Blessing and a Curse

Population density is a key characteristic of urban areas. Population density can be an advantage as investing in services, including in poorer areas, offers high returns on investment because benefits are shared across many people (Ezeh et al. 2016). While population density thus allows urban leaders to create efficiencies in delivering services such as education, healthcare, or transport, it also comes with significant risks.

For example, population density facilitates the rapid spread of infectious diseases such as COVID-19, cholera, Ebola, dengue, or Zika. Moreover, the impact of adverse events, such as contamination of the water supply or natural disasters, as well as air and noise pollution, is amplified in densely populated urban areas.

Health Services: Better Offer, Unequal Access

Provided by a mix of public, private for-profit, and private not-for-profit entities, a country's best healthcare facilities and specialized health services typically concentrate in its biggest cities. Health facilities are thus within easier reach for urban residents, and they tend to offer better quality services where better educated and supported health workers provide them.

However, in many places, urban growth has been so fast that governments and municipalities have been unable to offer even essential services and infrastructure to all residents. Therefore, while tertiary care of excellent quality may be available in a given city in principle, even basic primary healthcare is often limited in poorer areas, for example, in slums where the urban poor concentrate. As a result, the latter often resort to seeking care from unqualified health practitioners, paying for care that is of poor quality and, in some cases, harmful. Combined with insufficient or inadequate regulation and control in health and pharmaceutical markets, this leads to imbalances in healthcare supply and increases hazards such as antimicrobial resistance.

Urbanization Affects Diets and Lifestyles

Urbanization is associated with profound changes in diet and exercise, leading to increased consumption of unhealthy, processed foods and low levels of physical activity. These factors increase the prevalence of obesity and are also associated with increases in type 2 diabetes and cardiovascular disease (Patil 2014). Noncommunicable diseases (NCDs), sometimes called chronic diseases, include strokes, heart attacks, diabetes, respiratory disease, and cancer. Noncommunicable diseases are the leading causes of ill health globally and account for 70% of deaths worldwide (Bennett et al. 2018). While high-income settings have seen a shift from infectious to noncommunicable diseases as a main area for health concerns, most cities in developing countries now face a so-called double burden of disease: they are struggling with high levels of infectious diseases and NCDs at the same time. Some cities are indeed struggling with a "triple burden," where large numbers of injuries are added to the mix. All cities are highly susceptible to pandemics, such as COVID-19.

Air Pollution and Waste: Environmental Influences on Health Abound

Air pollution – principally driven by fossil fuel combustion and exacerbated by climate change – damages the heart, lungs, and every other vital organ. It represents the most direct link between environmental factors and ill health. Even by conservative measures, air pollution kills seven million people annually or close to 20,000 per day (WHO 2018a). A lack of policies to prevent and reduce air pollution contributes to NCDs and thus premature deaths from strokes, heart attacks, cancers, and other issues. Of city dwellers, 90% breathe air that does not comply with the safety standard set by the World Health Organization (WHO 2016b).

Both outdoor and indoor air pollution cause mortality (death) and morbidity (disease), and they interrelate. Urban outdoor air pollution from cars or industrial manufacturing can be an important contributor to indoor air quality, especially in highly ventilated houses or in houses near pollution sources. Similarly, indoor air pollution sources can also be an important cause of urban outdoor air pollution, particularly in cities where many people use biomass fuels or coal for heating and cooking (WHO n.d.-b). Indoor air pollution disproportionately affects women and children (WHO 2018b).

The removal and management of solid waste represents one of the most critical challenges in many cities and especially in slum areas. Uncollected solid waste blocks drains, causes flooding, and can lead to the spread of waterborne diseases. While the proportion of municipal solid waste that gets collected has increased, 20% of urban dwellers are still without collecting services, and the waste that is collected is often not disposed of properly (United Nations 2019). Poorly managed waste and stagnant water provide breeding grounds for the mosquitoes that transmit infectious diseases such as Zika, dengue fever, and malaria.

Urban Health and Inequality: Different Worlds within a City

Growing prosperity in cities leaves behind significant "hidden" urban areas and populations. In other words, rich and poor people live in different epidemiological worlds, even within the same city (Rydin et al. 2012). Living conditions in slums or slum-like surroundings are often poor, with residents having little or no access to urban services – despite these services being readily available to better-off residents in other parts of the city.

Positive and negative influences on health cluster according to geographic and socioeconomic parameters. The failure of authorities to provide essential services (e.g., health, education, and social protection) for the urban poor leads to significant inequalities in health service access and outcomes. Rapid, unplanned urbanization thus also contributes to urban poverty. These factors negatively impact not only on the health of people living in informal settlements but also come with great risks to residents in other areas, for example, by facilitating the spread of infectious diseases.

Migration is a key factor in urban population growth, and it comes with specific health risks. Those who migrate to cities to escape difficult circumstances, such as violent conflict or the effects of climate change, often experience a double jeopardy in cities: preexisting vulnerabilities combine with greater exposure to migration-related stressors. A social and economic gap often emerges between longtime urban residents and migrants.

Many of today's urban poor are not only much worse off than their wealthier fellow citizens, they may even lag behind rural populations. The so-called urban advantage over rural areas is the idea that people are generally better off in cities than in rural areas. However, this urban advantage is not a given for all countries or for all parts of the urban population within a city. Economic growth and associated urban expansion do not necessarily drive improvements in health outcomes. Such an urban advantage needs to be actively created (Rydin et al. 2012).

Mental Health: The Underestimated Value of Greenspace

Besides its complex interaction with physical health, urbanization is a risk factor for mental health. For example, prevalence rates for psychiatric disorders are significantly higher in urban areas compared with rural areas, particularly for mood and anxiety disorders, even when controlling for confounding factors (Peen et al. 2010). The association between urbanicity and risk of schizophrenia is also well established and estimated to be 2.37 times higher in most urban compared to rural environments (Vassos et al. 2012).

Greenspace is one of the factors that greatly distinguishes cities from non-cities and that has an important positive influence on human health. Environmental parameters, such as access to greenspace (as well as blue space) and the way in which they interact with exposure to air pollution and other factors, influence not only physical but also mental health. Access to greenspace during childhood is associated with enhanced mental health (Engermann et al. 2019).

Greenspace can promote mental health by reducing stress and anxiety, supporting psychological restoration, encouraging exercise, improving social coherence, decreasing noise and air pollution affecting cognition and brain development, and improving immune functioning (Engermann et al. 2019). The mental health of people who move from less green to more green areas improves significantly, and this effect persists over time (Alcock et al. 2014). Arguably, greenspace itself provides vital health services (Barton and Rogerson 2017), and it does so largely for free.

Besides reducing heat, clearing the air, improving mental health, and decreasing the risk of flooding, greenspace also helps mitigate noise pollution, which itself is an important risk factor in death and disease. Importantly, access to green and blue space in urban areas can reduce the social isolation of older people, as well as improve community interaction and coherence (EEA 2020). There are economic aspects too: buildings surrounded by greenery require less energy for cooling. Health costs go down where greenspace helps reduce air or noise pollution and thus increases both mental and physical health.

Besides income, health and well-being are major determinants of people's happiness. This this is true for both happiness in terms of perceived quality of life and in terms of more immediate emotions and affects. Access to greenspace too is an important happiness factor. On average, urban residents are happier than people living in rural areas, and this is true for most countries. However, this relationship reverses as countries and cities develop, a phenomenon that can be referred to as the "urban paradox": up to a certain point of economic development, happiness of city dwellers improves. After this point, urban residents may be less happy than their fellow citizens in rural areas, particularly if they live in the largest cities (Helliwell et al. 2020).

Violence is another risk factor in cities that is associated with mental health issues, and it interacts with gender factors. Poor women living in cities face an increased risk of physical, sexual, and psychological violence as well as barriers in accessing health services due to lack of control over family financial resources, child-care responsibilities, and limited decision-making power. Sexual and gender-based violence affects more women in urban than in rural areas due to the configuration of urban spaces, e.g., in slums and insecure environments, and the nature of urban activities, such as increased alcohol or drug use, changing social norms, or the concentration of sex work (McIlwaine 2013).

Urban Health in a Global Context

With 50% and 43% of the population in Asia and Africa, respectively, living in urban areas, these regions are currently less urbanized than the Americas and Europe, where urbanization stands at above 80% and 74%, respectively. Of the increase in the global urban population, 90% will take place in Asia and Africa. It is expected that future increases in the size of the world's urban population will be highly concentrated in a few countries: India, China, and Nigeria alone will account for 35% of urban population growth between 2018 and 2050 (UN DESA 2018).

The number of people living in slums will remain especially high in sub-Saharan Africa where 56% of the urban population already lives in slums and in Southern and Southeast Asia (Ezeh et al. 2016). This is associated with increased health risks. For example, most people living in the cities of sub-Saharan Africa do not have access to networked sewerage systems. Instead, septic tanks, pit latrines, or open defecation are the norm. In densely populated slums, one toilet facility is often shared by many households, sometimes even by hundreds of people (Rydin et al. 2012). Higher infant and neonatal mortality in slums versus rural areas have been reported in Kenya, Ecuador, Brazil, Haiti, and the Philippines. Neonatal, infant, and under-five mortality rates were similar to or higher in urban slums than in rural areas in Bangladesh and Kenya (Ezeh et al. 2016).

Healthy Urban Design: How to Make Cities Future Fit

Cities must invest in environmental determinants of health that are known to yield positive health benefits in a cost-effective manner. These include the following (see WHO 2016a):

- Efficient public transport and cycling networks to lower risks associated with air pollution, reduce traffic deaths and injuries, and promote physical activity (thus lowering the risk of NCDs)
- The prioritization of greenspace (e.g., green belts) to improve air quality, help preserve watersheds, and reduce drinking water contamination, as well as improve mental health and resilience to heat waves and act as a buffer against extreme weather events
- Provision of water, sanitation, hygiene, wastewater, and sewage management to reduce waterborne and infectious illnesses, such as diarrhea and other diseases
- Reuse of resources, recycling, and reduction of solid waste to eliminate the need to burn or bury it, improving air quality and reducing water and soil contamination
- Energy efficiency to retrofit and improve ventilation in buildings, reducing indoor air pollution and carbon emissions

Smart urban design takes health into account at every stage of policy and planning. In some cases, one policy measure can positively affect a number of parameters and sectors, including several determinants of health, at the same time. For example, investing in high-quality public transport reduces air pollution and noise as well as lowers traffic accidents and associated injuries and deaths. Incentivizing people to use public rather than individual motorized transport also reduces sedentary behavior, thus improving health and reducing healthcare costs. Effective transport strategies that start from a bigger picture perspective also take into account the provision of public spaces and green infrastructure. They have the power to influence several health determinants, including by encouraging walking and cycling, improving public safety, and providing safe and pleasant spaces for social interaction. They create win-wins by solving several public health and related policy issues at the same time.

The broader political framework for creating healthy cities and making cities future fit exists. Sustainable Development Goal (SDG) 11 focuses on making cities inclusive, safe, resilient, and sustainable. Several targets of SDG 11 specifically address determinants of health, such as access to basic services and upgrading slums; expanded public transport and improved road safety; reduced deaths and disease from water-related disasters; and improved air quality and waste management.

Another leading global agreement, the New Urban Agenda, was adopted in 2016 to implement the conclusions of the Third UN Conference on Housing and Sustainable Urban Development (Habitat III). It sets out the goals for sustainable development of urban regions within the United Nations' overall SDG framework, emphasizing the important linkages with health.

Summary/Conclusion

The health of urban residents is an indicator of functioning cities. It is both a cause and a consequence of sustainable development. In fact, the good health of its citizens is one of the most effective markers of any city's sustainable development (WHO 2016a). Effective urban planning, infrastructure development, and governance can mitigate risks and harness opportunities, thus promoting the health and well-being of urban populations.

Despite strong evidence of its importance in helping cities thrive, health has not received adequate attention and investment in urban policy making. Sustainable solutions for urban development require consideration of all health determinants and a focus on health equity. These must be addressed through multi-stakeholder action.

Urban Policy and Planning Must Address Health and Equity Head-on

Setting up adequate surveillance systems and both preventive and curative health services in cities is critical to reducing the disease burden and preventing health emergencies, including pandemics such as COVID-19 that easily spread across municipal, regional and national borders. Decisions related to urban planning, finance, and governance can create or exacerbate major health risks, or they can foster healthier environments and lifestyles that reduce or mitigate health risks (WHO 2016a).

If done poorly, urban design can reinforce health inequalities. To address prevailing inequities, it is necessary to go well beyond the provision of health services. To improve health for all urban residents, it is essential to explicitly and systematically consider social protection and focus on the social and environmental determinants of health. Moreover, effective data collection and management systems, including to measure health equity, are crucial. The same is true for engaging communities in the design of health services that maximize accessibility for all urban dwellers.

Multisectoral and Multi-Stakeholder Actions Are Crucial in Addressing the Determinants of Health

Urban health is not a matter that concerns the medical or public health sectors alone. Quite the opposite, human health affects most government sectors and industries, and it links all strata of society. Urban planning that places health upfront offers significant opportunities for improving social determinants of health and health outcomes through measures such as controlling pollution, efficient provision of water and sanitation services, or investments in greenspace. At the same time, multisectoral collaboration is essential for maximizing the potential of healthy urban citizens.

New models of cooperation and cross-sector collaboration are needed to identify synergies and generate actions that result in overall gains to society. Effective urban governance requires intersectoral work characterized by both vertical and horizontal integration. Understanding the barriers to such intergovernmental action is a prerequisite for integrating health into urban development.

Integrating health into urban governance and planning needs a very proactive approach (Irwin and Scali 2010). Urbanization is such an important trend that its health consequences need to be taken into account in all health sector interventions, including outside the urban sphere. By the same token, city representatives must be part of health planning efforts at the national level.

With improved sensitization to urban health issues, the health sector can function as a knowledge leader in such multisectoral dialogue. Cities can become demonstration sites for multisectoral approaches to health at national and international level. Encouraging active participation of multiple stakeholders including communities, in the prevention and control of disease, is essential for effective, sustainable solutions (Schmidt and Jeannée 2018).

H

Addressing Urban Complexity Requires Unconventional Approaches

Cities are complex systems. Urban health outcomes depend on many interactions and feedback loops. In this multi-determinant and multisectoral space, the planning process is challenging. Unintended consequences of policies and programs are to be expected, and they must be tracked. In order to remain in the driving seat, cities must plan ahead for natural growth and migration and invest in a forward-looking manner to ensure that infrastructure and services meet growing future demand. A linear or cyclical planning approach is insufficient in conditions of complexity: urban planning for health should include experimentation and include practitioners and communities in proactive dialogue at eye level.

Acknowledgments The research and preparation of this internal paper for the European Commission by the same author as the present Encyclopedia contribution has informed the development of the latter.

References

Alcock, I. et al. (2014). Longitudinal effects on mental health of moving to greener and less green urban areas. Environmental Science & Technology. 2014 Jan 21;48(2):1247–55. https://doi.org/10.1021/es403688w. Accessed on 29 December 2020 at https://pubmed.ncbi.nlm.nih.gov/24320055/

Barton, J. and Rogerson, M. (2017). The importance of greenspace for mental health. BJPsych Int. 2017 Nov; 14(4): 79–81. Accessed on 16 December 2020 at https://www.ncbi.nlm.nih.gov/pmc/articles/PMC5663018/.

Bennett, J. E., Stevens, G. A., Mathers, C. D., Bonita, R., Rehm, J., Kruk, M. E., et al. (2018). NCD countdown 2030: Worldwide trends in non-communicable disease mortality and progress towards sustainable development goal target 3.4. *The Lancet., 392*(10152), 1072–1088.

Engermann, K. et al. (2019). Residential green space in childhood is associated with lower risk of psychiatric disorders from adolescence into adulthood. February 25, 2019, 19–22. Accessed on 16 December 2019 at https://doi.org/10.1073/pnas.1807504116

European Environment Agency (2020). Healthy environment, healthy lives: how the environment influences health and well-being in Europe. EEA Report 21/2019. ISSN 1977–8449. Accessed on 28 December 2020 at https://www.eea.europa.eu/publications/healthy-environment-healthy-lives?utm_source=POLITICO.EU&utm_campaign=c337ee14d4-EMAIL_CAMPAIGN_2020_09_08_02_30&utm_medium=email&utm_term=0_10959edeb5-c337ee14d4-190657764

Ezeh, A., et al. (2016). The history, geography, and sociology of slums and the health problems of people who live in slums. *Lancet 2017, 389*, 547–558.

Habitat III (2017). Implementing the New Urban Agenda. Powerpoint presentation, 30 January 2017. Accessed on 28 December 2020 at https://www.eukn.eu/fileadmin/Files/Documents/04_Presentation__2_.pptx

Helliwell, John F., Richard Layard, Jeffrey Sachs, and Jan-Emmanuel De Neve, eds. (2020). World Happiness Report 2020. New York: Sustainable Development Solutions Network. Accessed on 14 January 2021 at https://worldhappiness.report/ed/2020/

Irwin, A., & Scali, E. (2010). *Action on the social determinants of health, learning from previous experiences. Social determinants of health discussion paper 1 (debates)*. Geneva: World Health Organization.

McIlwaine, C. (2013). Urbanization and gender-based violence: exploring the paradoxes in the global South. vol 25(1): 65–79. Accessed on 16 December 2020 at https://doi.org/10.1177/0956247813477359.

Patil, R. (2014). Urbanization as a Determinant of Health: A Socioepidemiological Perspective, Social Work in Public Health, 29:4, 335–340. Accessed on 26 July 2018 at https://www.ncbi.nlm.nih.gov/pubmed/24871771

Peen, J et al (2010). The current status of urban-rural differences in psychiatric disorders. Acta Psychiatrica Scandinavica. 2010; 121:84–93. Accessed on 28 July 2018 at https://onlinelibrary.wiley.com/doi/abs/10.1111/j.1600-0447.2009.01438.x

Rydin, Y. et al. (2012). Shaping cities for health: complexity and the planning of urban environments in the 21st century. Lancet 2012; 379: 2079–108. Accessed on 15 November 2020 at https://www.thelancet.com/journals/lancet/article/PIIS0140-6736(12)60435-8/fulltext

Schmidt, A. and Jeannée, E. (2018). Health and the urban poor. Briefing paper, internal working document for EU DEVCO. September 2018, updated July 2020.

UN DESA (2018). World urbanization prospects, the 2018 revision. New York, United Nations Department of Economic and Social Affairs, Population Division, 2018. Accessed on 28 December 2020 at https://esa.un.org/unpd/wup/

United Nations (2019). The Sustainable Development Goals Report 2019. Accessed on 28 December 2020 at https://unstats.un.org/sdgs/report/2019/

Vassos, E. et al. (2012). Meta-Analysis of the Association of Urbanicity With Schizophrenia. Schizophrenia Bulletin vol. 38 no. 6 pp. 1118–1123, 2012. Accessed on 28 July 2018 at https://www.ncbi.nlm.nih.gov/pubmed/23015685

WHO (2006). Constitution of the World Health Organization. Basic Documents, Forty-fifth edition, Supplement, October 2006. Accessed on 29 December

2020 at https://www.who.int/governance/eb/who_constitution_en.pdf
WHO (2014). Urban HEART. Urban Health Assessment and Response Tool. Brochure. Accessed on 26 July 2018 at www.who.int/kobe_centre/measuring/urban-global-report/UrbanHeart_infographics.pdf
WHO (2016a). Health as the Pulse of the New Urban Agenda. UN Conference on Housing and Sustainable Urban Development. Quito. October 2016. Accessed on 15 November 2020 at http://apps.who.int/iris/bitstream/10665/250367/1/9789241511445-eng.pdf
WHO (2016b). WHO Global Urban Ambient Air Pollution Database (update 2016). Accessed on 28 December 2020 at https://www.who.int/phe/health_topics/outdoorair/databases/cities/en/
WHO (2017). Determinants of Health. Q&A, 3 February 2017. Accessed on 15 November 2020 at https://www.who.int/news-room/q-a-detail/determinants-of-health
WHO (2018a). 9 out of 10 people worldwide breathe polluted air, but more countries are taking action. News release on 2 May 2018. Accessed on 15 November 2020 at https://www.who.int/news/item/02-05-2018-9-out-of-10-people-worldwide-breathe-polluted-air-but-more-countries-are-taking-action
WHO (2018b). Indoor Air Pollution, Fact Sheet, Geneva. Accessed on 26 July 2018 at http://www.who.int/news-room/fact-sheets/detail/household-air-pollution-and-health
WHO (n.d.-a). Social Determinants of Health: Urbanization. Accessed on 29 December 2020 at https://www.who.int/social_determinants/themes/urbanization/en/
WHO (n.d.-b). Background information on urban outdoor air pollution. Accessed on 12 January 2021 at https://www.who.int/phe/health_topics/outdoorair/databases/background_information/en/

Health and the Role of Nature in Enhancing Mental Health

Lauriane Suyin Chalmin-Pui[1,2] and Tijana Blanusa[1,3]
[1]Royal Horticultural Society, Wisley, UK
[2]The University of Sheffield, Sheffield, UK
[3]University of Reading, Reading, UK

Synonyms

Mental health; Nature; Physical health; Social health; Urban green infrastructure; Well-being

Definition

Health has been conceptualized in countless ways either as normality, as the absence of disease, as equilibrium, as functionality, as fitness, as resilience, as thriving, as a right, or as a resource. In 1948, the World Health Organization (WHO) adopted a definition of health as a "state of complete physical, mental and social wellbeing and not merely the absence of disease or infirmity." In 1968, the WHO expanded their definition by adding: "to reach a state of complete physical, mental and social wellbeing, an individual or group must be able to identify and to realize aspirations, to satisfy needs, and to change or cope with the environment. Health is, therefore, seen as a resource for everyday life, not the objective of living." This has become one of the most widely used definitions of health as it alludes to three key (and interdependent) domains of health: physical, mental, and social.

While "complete" health seems unrealistic to achieve, this is relative to social, emotional, and psychological circumstances of the life course. For example, people with chronic conditions and disabilities can still be considered healthy if they can function effectively, fulfill their needs, and be resilient to major stresses (Bircher and Kuruvilla 2014; Huber et al. 2011).

Determinants of Health

Creating resilient regional and urban futures is dependent on a population that is in good physical, mental, and social health. There is also recognition that the manner in which we design, build, and inhabit our communities, and how we engage with nature, have deep impacts on our well-being (Frumkin et al. 2011; Roe and McCay 2021).

Physiologically, an individual's health is determined by their physical cells, organs, genetics, and biochemical processes, alongside the availability of water, nutrition, oxygen, and shelter. Yet health is not a purely physical function. Human health and well-being are also influenced by an individual's behavior and lifestyle as well as

H

broader socioeconomic, political, cultural, and technological structures (Lovell 2018). The wider determinants of health are embedded at all scales of built, social, and natural environments (Barton and Grant 2006; Dahlgren and Whitehead 1991). From the global climate to soil quality, from street design to social care, health is determined by a nested variety of factors (Fig. 1).

The places in which we live and work are planned, designed, and built in a way that is increasingly associated with rising incidences of cardiovascular disease, chronic stress and anxiety, and social isolation (Barros et al. 2019; Rautio et al. 2018; Sallis et al. 2020). According to the Global Burden of Diseases and Injury study, mental and addictive disorders affect more than one billion people and are responsible for 19% of all years lived with disability in 2016 (Naghavi et al. 2017; Rehm and Shield 2019). Moreover, the burden of mental diseases may be underestimated because even though mental disorders are not often listed as causes of mortality, they are associated with higher mortality rates and lower life expectancy than the general population. Conscious efforts need to be made to use multi-sectoral expertise to create places and communities that promote good health and prevent ill health.

On the scales of the regions, cities, neighborhoods, and even streets, nature can have beneficial health impacts. Places can foster good health by promoting restoration from stress and attentional fatigue, or they may be intended to heal specific illnesses (Roe and McCay 2021; Townsend et al. 2018; World Health Organization 2016). However, as the global population is increasingly concentrated in urban areas, direct nature experiences are decreasing in quality and quantity for many people (Bratman et al. 2019; Soga and Gaston 2016).

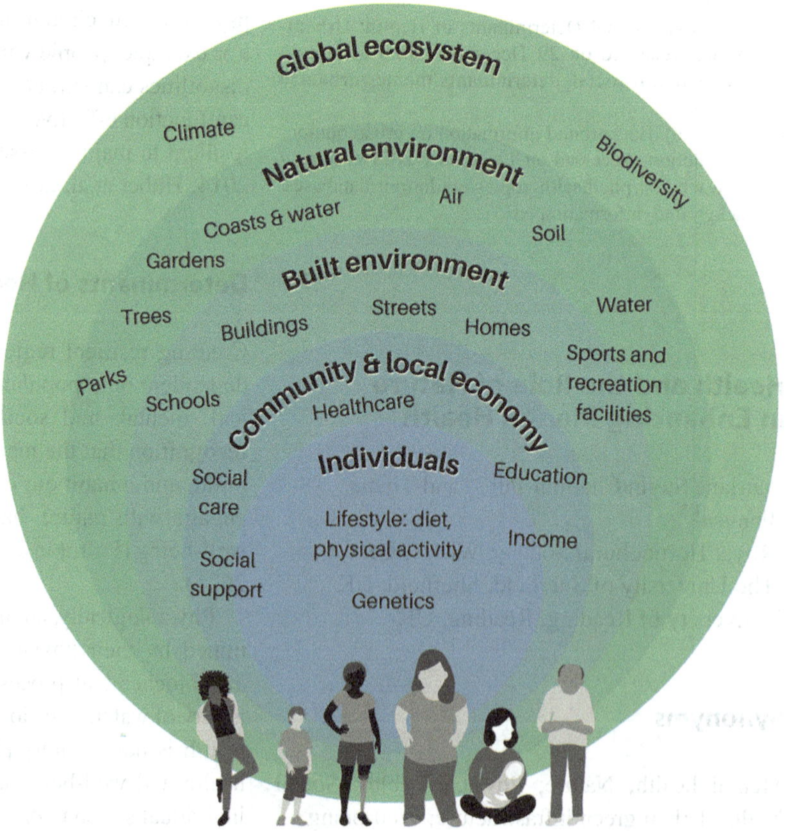

Health and the Role of Nature in Enhancing Mental Health, Fig. 1 The determinants of health. (Adapted from Barton and Grant (2006))

While the physical, mental, and social dimensions of health are not easily separable, this entry focuses on mental health and the role that nature can play to enhance this. Mental health is defined by the WHO (2004) as a state of well-being in which individuals can realize their own potential, can cope with the normal stresses of life, can work productively and fruitfully, and are able to make a contribution to their community.

The Benefits of Nature on Mental Health

There is strong evidence for the benefits of nature on mental health (see Bratman et al. 2019 for a review). Though not medically linked to disease states and mortality, exposure to nature is associated with improved health-related outcomes including physical activity, social contacts, stress physiology, emotional well-being, and cognitive capacity. The strongest consensus is around the association between nature experience and increased psychological well-being, a reduction of risk factors for mental illness, and a lowered burden of mental illness.

Table 1 highlights the range of impacts of nature that have been demonstrated in various contexts. The principal pathways through which nature contributes to health can be conceptualized in three domains: reducing harm, restoring capacities, and building capacities (Markevych et al. 2017). These domains operate individually as well as in complementarity with each other.

Human interaction with nature can be divided into two distinct experiences: those with active contact such as gardening or another activity in a natural setting; and those dealing with more passive interactions such as the view from an office window (Brown and Grant 2005). Mechanisms underlying positive effects include Attention Restoration Theory (Kaplan and Kaplan 1989), Stress Reduction Theory (Ulrich et al. 1991), prospect-refuge theory (Appleton 1975), flow theory (Csikszentmihalyi 1990), and the human brain–gut–microbiome axis (Allen et al. 2017).

The Importance of Multifunctional Green Spaces

Nature in urban, peri-urban, and rural areas exists in many different forms, sizes, on public and private land, and with different functions. Mental health benefits may be accrued from a diversity of accessible greenspace. The most common forms of green spaces are: residential gardens, public parks, allotments, community gardens, botanical gardens, green rooftops, green walls, pocket

Health and the Role of Nature in Enhancing Mental Health, Table 1 Evidenced impacts of nature on mental health

Pathway	Impact	Context	References
Reducing harm	Mental health protection	Nature exposure during childhood acts as a buffer for mental health problems in adulthood	Cherrie et al. (2018), Engemann et al. (2018), and Thygesen et al. (2020)
	Management of mental illness	Symptom reduction of mental health problems (e.g., ADHD and PTSD)	Bettmann et al. (2021), Faber Taylor and Kuo (2011), and Ohly et al. (2016)
Restoring capacities	Emotional well-being	Reduced depression and anxiety, improved mood, and self-esteem	Chalmin-Pui et al. (2021b), Lehberger et al. (2021), and Marselle et al. (2020)
	Cognitive health	Increased mental alertness, memory recall, productivity, reduced cognitive decline	Besser (2021), Moss et al. (2017), Van Hedger et al. (2019), and Yin et al. (2018)
	Improved stress regulation	Improved and healthier indicators of physiological stress such as salivary cortisol, skin conductance, and Heart Rate Variability	Chalmin-Pui et al. (2021c), Kondo et al. (2018), Shuda et al. (2020), and Souter-Brown et al. (2021)
Building capacities	Social well-being	Place attachment, place identity, increased social interaction, increased altruism	Chalmin-Pui et al. (2021a) and Jennings and Bamkole (2019)

parks, rain gardens, street trees, road verges, school gardens, river banks, urban farms, urban forests, and wetlands. All of these are important assets for our global ecosystem as well as for more local built and natural environments.

The composition of those spaces, and how they are managed, can have a significant environmental impact (Cameron et al. 2012). In designing and integrating green spaces with the existing built environment, initial criteria to consider include: accessibility, (perception of) safety, shared spaces and functionality, physical characteristics of the area, microclimate, and planting composition.

The drivers determining planting composition choice should go beyond the traditional considerations of cost, suitability for physical survival on site, and ornamental appeal. There is an increasing understanding of the capacity of plant species (and an overall planting mix) to simultaneously provide multiple other benefits (Blanusa et al. 2019; Cameron and Blanusa 2016). This includes the potential of a green space to support biodiversity, absorb/reduce noise, improve air quality, help with water attenuation and localized cooling. On an individual botanical and physiological level, plant species can vary greatly in the extent of ecosystem service they provide (Cameron and Blanusa 2016). On a garden scale the biggest environmental benefit is likely achieved by increasing overall green cover, maximizing species diversity, and using large-stature and perennial species.

Residential Gardens as an Example of Multifunctional Green Spaces

Gardening is a common leisure activity around the world and thus can have multiple roles and meanings for gardeners. Residential (domestic, private) gardens are an important example of multifunctional green spaces as they are the most readily accessible green spaces for residents, with an estimated 88% of households in Great Britain having access to a private or shared garden (Office for National Statistics 2020). Although small in size individually, the 24 million residential gardens in Great Britain make up a combined area equivalent to approximately 30% of the total urban built-up area (Office for National Statistics 2018). The provision and extent of domestic gardens have also been studied in other national contexts, including in Romania (Badiu et al. 2019), Germany (Wellmann et al. 2020), India (Balooni et al. 2014), Ecuador (Finerman and Sackett 2003), Chile (Reyes-Paecke and Meza 2011), South Africa (King and Shackleton 2020), Belgium (Notteboom 2018), and Spain (Garcia-Garcia et al. 2020).

Research has demonstrated the essential role of private gardens in delivering natural capital that improves human health and well-being. This benefit is independent of the motivation to garden, which could be for leisure, health, professional, creative, or maintenance reasons (Chalmin-Pui et al. 2021b). Gardens with more vegetation are associated with better physiological and psychological stress regulation (Chalmin-Pui et al. 2021c).

Incorporating Gardens into the Public Health Landscape

Green spaces can become assets and resources in a public health system that prescribes nature and nature-based activities to individuals as part of their health and social care. On a national, regional, and city scale, residential gardens could provide a public health benefit by helping people cope with and recover from mental ill-health, and also to keep the general population well (Cameron et al. 2012). "Green social prescriptions" can be tailored to specific needs for the treatment of ill-health or in the context of preventative measures. This could include nature walks, community gardening, or biodiversity conservation activities. The equitable provision of green spaces close to where people live is critical to sustaining a resilient and healthy population.

Cross-References

▶ Health and the City: How Cities Impact on Health, Happiness, and Well-Being

- ▶ Healthy Cities
- ▶ Multiple Benefits of Green Infrastructure
- ▶ Role of Nature for Ageing Populations
- ▶ Urban Health Paradigms
- ▶ Urban Nature

References

Allen, A. P., Dinan, T. G., Clarke, G., & Cryan, J. F. (2017). A psychology of the human brain–gut–microbiome axis. *Social and Personality Psychology Compass, 11*(4), e12309. https://doi.org/10/gft4w2.

Appleton, J. (1975). *The experience of landscape*. New York: Wiley-Blackwell.

Badiu, D. L., Onose, D. A., Niţă, M. R., & Lafortezza, R. (2019). From "red" to green? A look into the evolution of green spaces in a post-socialist city. *Landscape and Urban Planning, 187*, 156–164. https://doi.org/10/ghs8bh.

Balooni, K., Gangopadhyay, K., & Kumar, B. M. (2014). Governance for private green spaces in a growing Indian city. *Landscape and Urban Planning, 123*, 21–29. https://doi.org/10.1016/j.landurbplan.2013.12.004.

Barros, P., Ng Fat, L., Garcia, L. M. T., Slovic, A. D., Thomopoulos, N., de Sá, T. H., Morais, P., & Mindell, J. S. (2019). Social consequences and mental health outcomes of living in high-rise residential buildings and the influence of planning, urban design and architectural decisions: A systematic review. *Cities, 93*, 263–272. https://doi.org/10/gmb5gr.

Barton, H., & Grant, M. (2006). A health map for the local human habitat. *Journal of the Royal Society for the Promotion of Health, 126*(6), 252–253.

Besser, L. (2021). Outdoor green space exposure and brain health measures related to Alzheimer's disease: A rapid review. *BMJ Open, 11*(5), e043456. https://doi.org/10/gkm3qv.

Bettmann, J. E., Prince, K. C., Ganesh, K., Rugo, K. F., Bryan, A. O., Bryan, C. J., Rozek, D. C., & Leifker, F. R. (2021). The effect of time outdoors on veterans receiving treatment for PTSD. *Journal of Clinical Psychology, 77*(9), 2041–2056. https://doi.org/10/gmnqwr.

Bircher, J., & Kuruvilla, S. (2014). Defining health by addressing individual, social, and environmental determinants: New opportunities for health care and public health. *Journal of Public Health Policy, 35*(3), 363–386. https://doi.org/10/f6cx8q.

Blanusa, T., Garratt, M., Cathcart-James, M., Hunt, L., & Cameron, R. W. F. (2019). Urban hedges: A review of plant species and cultivars for ecosystem service delivery in north-west Europe. *Urban Forestry and Urban Greening, 44*, 126391. https://doi.org/10/ghxhp5.

Bratman, G. N., Anderson, C. B., Berman, M. G., Cochran, B., de Vries, S., Flanders, J., Folke, C., Frumkin, H., Gross, J. J., Hartig, T., Kahn, P. H., Kuo, M., Lawler, J. J., Levin, P. S., Lindahl, T., Meyer-Lindenberg, A., Mitchell, R., Ouyang, Z., Roe, J., … Daily, G. C. (2019). Nature and mental health: An ecosystem service perspective. *Science Advances, 5*(7), eaax0903. https://doi.org/10/gf6z8v.

Brown, C., & Grant, M. (2005). Biodiversity and human health: What role in nature for health urban planning. *Built Environment, 31*(4), 326–338. https://doi.org/10/b7qwx2.

Cameron, R. W. F., & Blanusa, T. (2016). Green infrastructure and ecosystem services – Is the devil in the detail? *Annals of Botany, 118*(3), 377–391. https://doi.org/10/f86cbr.

Cameron, R. W. F., Blanusa, T., Taylor, J. E., Salisbury, A., Halstead, A. J., Henricot, B., & Thompson, K. (2012). The domestic garden – Its contribution to urban green infrastructure. *Urban Forestry and Urban Greening, 11*, 129–137. https://doi.org/10/f2qdrw.

Chalmin-Pui, L. S., Griffiths, A., Roe, J., & Cameron, R. (2021a). Gardens with kerb appeal – A framework to understand the relationship between Britain in Bloom gardeners and their front gardens. *Leisure Sciences, 32*, 1–21. https://doi.org/10/gjwd64.

Chalmin-Pui, L. S., Griffiths, A., Roe, J., Heaton, T., & Cameron, R. (2021b). Why garden? – Attitudes and the perceived health benefits of home gardening. *Cities, 112*, 103118. https://doi.org/10/gjvknm.

Chalmin-Pui, L. S., Roe, J., Griffiths, A., Smyth, N., Heaton, T., Clayden, A., & Cameron, R. (2021c). "It made me feel brighter in myself" – The health and well-being impacts of a residential front garden horticultural intervention. *Landscape and Urban Planning, 205*, 103958. https://doi.org/10/ghdg5m.

Cherrie, M. P. C., Shortt, N. K., Mitchell, R. J., Taylor, A. M., Redmond, P., Ward Thompson, C., Starr, J. M., Deary, I. J., & Pearce, J. R. (2018). Green space and cognitive ageing: A retrospective life course analysis in the Lothian Birth Cohort 1936. *Social Science and Medicine, 196*, 56–65. https://doi.org/10/gc2xv3.

Csikszentmihalyi, M. (1990). *Flow: The psychology of optimal experience*. New York: Harper and Row.

Dahlgren, G., & Whitehead, M. (1991). *The main determinants of health model – European strategies for tackling social inequities in health: Levelling up, part 2*. World Health OrganizationRegional Office for Europe

Engemann, K., Pedersen, C. B., Arge, L., Tsirogiannis, C., Mortensen, P. B., & Svenning, J.-C. (2018). Childhood exposure to green space – A novel risk-decreasing mechanism for schizophrenia? *Schizophrenia Research, 199*, 142–148. https://doi.org/10/gfcq2n.

Faber Taylor, A., & Kuo, F. E. (2011). Could exposure to everyday green spaces help treat ADHD? Evidence from children's play settings. *Applied Psychology: Health and Well-Being, 3*(3), 281–303. https://doi.org/10/dp8fkg.

Finerman, R., & Sackett, R. (2003). Using home gardens to decipher health and healing in the Andes. *Medical Anthropology Quarterly, 17*(4), 459–482. https://doi.org/10.2307/3655347.

Frumkin, H., Wendel, A., Abrams, R. F., & Malizia, E. (2011). An introduction to healthy places. In *Making healthy places – Designing and building for health, well-being, and sustainability* (pp. 3–30). Washington, DC: Island Press.

Garcia-Garcia, M. J., Christien, L., García-Escalona, E., & González-García, C. (2020). Sensitivity of green spaces to the process of urban planning. Three case studies of Madrid (Spain). *Cities, 100*, 102655. https://doi.org/10/gj2fg4.

Huber, M., Knottnerus, J. A., Green, L., van der Horst, H., Jadad, A. R., Kromhout, D., Leonard, B., Lorig, K., Loureiro, M. I., van der Meer, J. W. M., Schnabel, P., Smith, R., van Weel, C., & Smid, H. (2011). Health: How should we define it? *BMJ, 343*(7817), 235–237. https://doi.org/10/ccfsst.

Jennings, V., & Bamkole, O. (2019). The relationship between social cohesion and urban Green space: An avenue for health promotion. *International Journal of Environmental Research and Public Health, 16*(3), 452. https://doi.org/10/gg5c7r.

Kaplan, R., & Kaplan, S. (1989). *The experience of nature: A psychological perspective*. Cambridge, UK: Cambridge University Press.

King, A., & Shackleton, C. M. (2020). Maintenance of public and private urban green infrastructure provides significant employment in Eastern Cape towns, South Africa. *Urban Forestry and Urban Greening, 54*, 126740. https://doi.org/10/gj2fhk.

Kondo, M. C., Jacoby, S. F., & South, E. C. (2018). Does spending time outdoors reduce stress? A review of real-time stress response to outdoor environments. *Health and Place, 51*, 136–150. https://doi.org/10/gdpb8c.

Lehberger, M., Kleih, A., & Sparke, K. (2021). Self-reported well-being and the importance of green spaces – A comparison of garden owners and non-garden owners in times of COVID-19. *Landscape and Urban Planning, 212*, 104108. https://doi.org/10/gj2fgn.

Lovell, R. (Ed.). (2018). *Demystifying health* (issue valuing nature paper VNP13). Valuing Nature. https://valuing-nature.net/sites/default/files/documents/Reports/VNN-DemystifyingHealth-Web.pdf

Markevych, I., Schoierer, J., Hartig, T., Chudnovsky, A., Hystad, P., Dzhambov, A. M., de Vries, S., Triguero-Mas, M., Brauer, M., Nieuwenhuijsen, M. J., Lupp, G., Richardson, E. A., Astell-Burt, T., Dimitrova, D., Feng, X., Sadeh, M., Standl, M., Heinrich, J., & Fuertes, E. (2017). Exploring pathways linking greenspace to health: Theoretical and methodological guidance. *Environmental Research, 158*, 301–317. https://doi.org/10/gbvm6t.

Marselle, M. R., Bowler, D. E., Watzema, J., & Eichenberg, D. (2020). Urban street tree biodiversity and antidepressant prescriptions. *Scientific Reports, 10*, 1–11. https://doi.org/10/fqtc.

Moss, M., Earl, V., Moss, L., & Heffernan, T. (2017). Any sense in classroom scents? Aroma of rosemary essential oil significantly improves cognition in young school children. *Advances in Chemical Engineering and Science, 7*(4), 450–463. https://doi.org/10/gjz5p2.

Naghavi, M., Abajobir, A. A., Abbafati, C., Abbas, K. M., Abd-Allah, F., Abera, S. F., Aboyans, V., Adetokunboh, O., Afshin, A., Agrawal, A., Ahmadi, A., Ahmed, M. B., Aichour, A. N., Aichour, M. T. E., Aichour, I., Aiyar, S., Alahdab, F., Al-Aly, Z., Alam, K., . . . Murray, C. J. L. (2017). Global, regional, and national age-sex specific mortality for 264 causes of death, 1980–2016: A systematic analysis for the Global Burden of Disease Study 2016. *Lancet, 390*(10100), 1151–1210. https://doi.org/10/gbxxk6.

Notteboom, B. (2018). Residential landscapes – Garden design, urban planning and social formation in Belgium. *Urban Forestry and Urban Greening, 30*, 220–238. https://doi.org/10/gdg2jb.

Office for National Statistics. (2018). UK natural capital: Ecosystem accounts for urban areas. https://www.ons.gov.uk/economy/environmentalaccounts/bulletins/uknaturalcapital/ecosystemaccountsforurbanareas

Office for National Statistics. (2020). One in eight British households has no garden. https://www.ons.gov.uk/economy/environmentalaccounts/articles/oneineightbritishhouseholdshasnogarden/2020-05-14

Ohly, H., White, M. P., Wheeler, B. W., Bethel, A., Ukoumunne, O. C., Nikolaou, V., & Garside, R. (2016). Attention Restoration Theory: A systematic review of the attention restoration potential of exposure to natural environments. *Journal of Toxicology and Environmental Health. Part B, Critical Reviews, 19*, 1–39. https://doi.org/10/f3ttj3.

Rautio, N., Filatova, S., Lehtiniemi, H., & Miettunen, J. (2018). Living environment and its relationship to depressive mood: A systematic review. *International Journal of Social Psychiatry, 64*(1), 92–103. https://doi.org/10/dqr2.

Rehm, J., & Shield, K. D. (2019). Global burden of disease and the impact of mental and addictive disorders. *Current Psychiatry Reports, 21*(2), 10. https://doi.org/10/dqrz.

Reyes-Paecke, S., & Meza, L. (2011). Jardines residenciales en Santiago de Chile: Extensión, distribución y cobertura vegetal. *Revista Chilena de Historia Natural, 84*(4), 581–592. https://doi.org/10/gf42jm.

Roe, J., & McCay, L. (2021). *Restorative cities: Urban design for mental health and well-being*. London: Bloomsbury Academic.

Sallis, J. F., Cerin, E., Kerr, J., Adams, M. A., Sugiyama, T., Christiansen, L. B., Schipperijn, J., Davey, R., Salvo, D., Frank, L. D., De Bourdeaudhuij, I., & Owen, N. (2020). Built environment, physical activity, and obesity: Findings from the International Physical Activity and Environment Network (IPEN) Adult Study. *Annual Review of Public Health, 41*(1), 119–139. https://doi.org/10/ghf26k.

Shuda, Q., Bougoulias, M. E., & Kass, R. (2020). Effect of nature exposure on perceived and physiologic stress: A systematic review. *Complementary Therapies in Medicine, 53*, 260. https://doi.org/10/gj2fdv.
Soga, M., & Gaston, K. J. (2016). Extinction of experience: The loss of human-nature interactions. *Frontiers in Ecology and the Environment, 14*(2), 85. https://doi.org/10/f8jd9x.
Souter-Brown, G., Hinckson, E., & Duncan, S. (2021). Effects of a sensory garden on workplace wellbeing: A randomised control trial. *Landscape and Urban Planning, 207*, 103997. https://doi.org/10/ghm9q4.
Thygesen, M., Engemann, K., Holst, G. J., Hansen, B., Geels, C., Brandt, J., Pedersen, C. B., & Dalsgaard, S. (2020). The association between residential green space in childhood and development of attention deficit hyperactivity disorder: A population-based cohort study. *Environmental Health Perspectives, 128*(12), 127011. https://doi.org/10/gh76td.
Townsend, M., Henderson-Wilson, C., Ramkissoon, H., & Weerasuriya, R. (2018). Therapeutic landscapes, restorative environments, place attachment, and well-being. In M. van den Bosch & W. Bird (Eds.), *Oxford textbook of nature and public health* (pp. 57–62). Oxford, UK: Oxford University Press.
Ulrich, R. S., Simons, R. F., Losito, B. D., Fioritom, E., Miles, M. A., & Zelson, M. (1991). Stress recovery during exposure to natural and urban environments. *Journal of Environmental Psychology, 11*(3), 201–230. https://doi.org/10/csp47k.
Van Hedger, S. C., Nusbaum, H. C., Clohisy, L., Jaeggi, S. M., Buschkuehl, M., & Berman, M. G. (2019). Of cricket chirps and car horns: The effect of nature sounds on cognitive performance. *Psychonomic Bulletin and Review, 26*(2), 522–530. https://doi.org/10/gj2fd9.
Wellmann, T., Schug, F., Haase, D., Pflugmacher, D., & van der Linden, S. (2020). Green growth? On the relation between population density, land use and vegetation cover fractions in a city using a 30-years Landsat time series. *Landscape and Urban Planning, 202*, 103857. https://doi.org/10/gjkrkv.
World Health Organization. (2004). *Promoting mental health: Concepts, emerging evidence, practice: Summary report.* Department of Mental Health and Substance Abuse in collaboration with the Victorian Health Promotion Foundation (VicHealth) and the University of Melbourne. https://www.who.int/mental_health/evidence/en/promoting_mhh.pdf
World Health Organization. (2016). *Urban green spaces and health – A review of evidence.* København: WHO Regional Office for Europe.
Yin, J., Zhu, S., Macnaughton, P., Allen, J. G., & Spengler, J. D. (2018). Physiological and cognitive performance of exposure to biophilic indoor environment. *Building and Environment, 132*, 255–262. https://doi.org/10/gdcvgg.

Health Environment

▶ Walkable Access and Walking Quality of Built Environment

Health Promotion

▶ Theme Cities Networks

Healthy Cities

H

Evelyne de Leeuw[1,2,3] and Jean Simos[4,5]
[1]Centre for Health Equity Training, Research and Evaluation (CHETRE), University of New South Wales, Sydney, NSW, Australia
[2]South Western Sydney Local Health District Population Health, Liverpool, NSW, Australia
[3]Ingham Institute for Applied Medical Research, Liverpool, NSW, Australia
[4]Institute of Public Health, Faculty of Medicine, University of Geneva, Geneva, Switzerland
[5]S2D – Health and Sustainable Development, Rennes, France

Definition

A Healthy City is one that is continually creating and improving those physical and social environments and expanding those community resources which enable people to mutually support each other in performing all the functions of life and in developing to their maximum potential.

Introduction

Cities and health have gone together since the dawn of civilization (De Leeuw 2017). From the beginnings, the “yin and yang” of this bipole was a dynamic struggle between pathogenesis and salutogenesis (Antonovsky 1987). The modern "Healthy Cities movement" has roots in social systems for health and well-being and their interface with urban planning. It gained traction from the 1960s.

A strong belief emerged in the period of the Enlightenment that certain types of urban planning would be more beneficial to health (salutogenesis) and the prevention of disease (prevent pathogenesis) than others. Maneglier (1990) describes how Voltaire complains about the markets of Paris "established in narrow streets, showing off their filthiness, spreading infection and causing continuing disorders." It took another century before Louis-Napoleon Bonaparte in the middle of the nineteenth century commissioned George Eugene Haussmann (1809–1891) to *aérer, unifier, et embellir* (provide air, unify, and beautify) the great city. The grand boulevards radiating through the city have become emblematic for Parisian charm, but Haussmann clearly had health in mind in designing the infrastructure; in his memoirs he wrote *"The underground galleries are an organ of the great city, functioning like an organ of the human body, without seeing the light of day; clean and fresh water, light and heat circulate like the various fluids whose movement and maintenance serves the life of the body; the secretions are taken away mysteriously and don't disturb the good functioning of the city and without spoiling its beautiful exterior"* (De Moncan and Heurteux 2002).

In Britain, similar reflections led to a surge in urban and public health development. Edwin Chadwick founded, and gained support from, the Health of Towns Association in 1844 (Ashton and Ubido 1991). In 1875, Sir Benjamin Ward-Richardson presented his vision of "Hygeia: A City of Health" to the Social Science Association in Brighton, UK (Cassedy 1962). The urban planning innovations that he identified in his address benefited to no small end from the passing of a series of public health regulations. His utopian city, then, incorporated concepts as clean air, public transport, small community-based hospitals, community homes for the aged and the insane, occupational health and safety, the absence of tobacco and alcohol, and many other advances. His ideas were taken up by others, notably Ebenezer Howard, who proposed and developed the first "garden cities" in Britain in the 1890s – a movement still alive today through New Towns, sustainable communities, or transition town projects (Alexander 2009).

One year after the presentation of "Hygeia: A City of Health," Pasteur and Koch marked the birth of the public health era that has variously been labeled the stage of "germ theory" or "biomedicine" (Davies et al. 2014; Kickbusch 2007). The professed direct causality between a pathogen and an ill health outcome turned out to have much appeal, in fact, to such an extent that larger, more complex, humanistic systems of healthy urban planning devolved to a lower priority. The vision of urban planning suffered a similar fate (Deelstra 1985): from visionary socio-cultural perspectives in the nineteenth century, the urban planning perspective shifted to a structural-physical model in which architectural hardware is all that matters in planning. The three waves of urban planning (from hygienism through functionalism to sustainable planning (Roué Le Gall et al. 2014) map onto the four waves of public health (de Leeuw 2020). The fourth wave in the latter embraces a post-modern networked view of glocal health development.

The Resurgence of Complex Health-Urban Systems

As an echo of nineteenth century socio-cultural ideals, US psychiatrist Leonard Duhl initiated a long-range program development in the National Institute of Mental Health in 1955. His vision was to explore, across disciplines, the impact of the physical environment on human behavior. He used his position and connections to build a network of several hundred multidisciplinarian scholars and practitioners who met regularly as "The Space Cadets" (Martin 2014). Their deliberations enable Duhl to lay the foundations of contemporary Healthy Cities.

In this *"The Urban Condition – People and Policy in the Metropolis"* (Duhl 1963) authors discuss health, housing, ethology, violence, mental health, pathology, planning, and matters that have only re-emerged since the turn of the twenty-first century: climate change, systems thinking, and complexity in the urban environment. The work argues that the city must be viewed as a system in which physical infrastructure (which co-author Deevey (1963) dubs "urbs") and its people and their capabilities (which he calls "civitas") constantly interact dynamically and often unpredictably. Duhl himself, in his introduction to the book, states in no uncertain terms that looking at the parts of the city is possible and often understandable, but in order to make sense of the connection between health and urban dynamics one must see its complexity as an ecological and almost organic whole. Duhl further developed the scholarship at the interface of public health, psychiatry, and urban planning in Berkeley. His work appeared center stage during the 1984 conference in Toronto that celebrated the tenth anniversary of the publication of the Lalonde Report and the consecration of health determinants approach. This landmark Canadian government statement was the first, ever and anywhere, to argue for a shift from health care delivery public policy to integrated policies across *all* determinants of health. The 1984 conference declared the Canadian city's ambition to become a "Healthy City." Duhl presented a keynote. Representatives of the European Office of the World Health Organization (WHO) saw significant similarities between these perspectives and their own efforts to innovate the public health realm (de Leeuw and Harris-Roxas 2016).

In one of the foundation documents of the WHO European Healthy Cities project, Hancock and Duhl (1986) justified the ecological, inclusive, and dynamically complex view of urban health: "Some question the city's ability to initiate and implement health initiatives in the face of a variety of problems that include deterioration of the physical environment, poverty, unemployment, economic stagnation, homelessness, hunger, family violence, and crime and youth alienation. In some respects, cities may be seen as the potential or actual 'victims' of national and international policies – most spectacularly in connection with the threat of nuclear annihilation, more mundanely as a result of social, economic, immigration and other policies. On the other hand, others point to the many real strengths of the city." With foresight, they in fact wrote a prelude to what was later seen as "new urbanism" and endorsed at the dawn of the Urban Century in Habitat-III.

Based on their understanding of the urban fabric, Hancock and Duhl (1986) identify that good cities. . .

- Have a common "gameboard" where everyone comes together to make decisions by a commonly accepted set of rules
- Are multidimensional, yet succeed in relating the various parts to each other
- Are homogeneous and heterogeneous at the same time (the dominant culture accepts new cultures without engulfing them and is enriched by them)
- Have an extensive and redundant network of formal and informal communication linkages, both among its own people and with the outside world
- Can adapt to change, cope with breakdown, repair themselves, and learn from their own experience and that of other cities
- Have a commonly accepted mythology about themselves, in terms of a sense of history, and image of the city as it is today and a vision of what the city should be in the future.

For the European Office of WHO, they suggest a definition of a Healthy City as one

> *. . . that is continually creating and improving those physical and social environments and expanding those community resources which enable people to mutually support each other in performing all the functions of life and in developing to their maximum potential.*

Such Healthy Cities strive to provide:

- A clean, safe, high quality physical environment (including housing quality).
- An ecosystem, which is stable now and sustainable in the long term.
- A strong, mutually supportive and non-exploitative community.

- A high degree of public participation in and control over the decisions affecting one's life, health, and well-being.
- The meeting of basic needs (food, water, shelter, income, safety, work) for all the city's people.
- Access to a wide variety of experiences and resources with the possibility of multiple contacts, interaction, and communication.
- A diverse, vital, and innovative city economy.
- Encouragement of connectedness with the past, with the cultural and biological heritage and with other groups and individuals.
- A city form that is compatible with and enhances the above parameters and behaviors.
- An optimum level of appropriate public health and sick care services accessible to all.
- High health status (both high positive health status and low disease status).

Healthy Cities Then, Now, and in the Future

Healthy Cities, particularly in Europe, have been described often as a "visionary social movement." They drew on the evidence and foresight of visionaries like Duhl, Hancock, Kickbusch, and Tsouros. We surmise that the vision was so strong (and enabled the movement) because it resonated well with deeply held human values that urbanites and their institutions can easily identify with. In most cities, (health) inequity is at your doorstep or around the corner in your gutter. Solidarity is what happens at train stations and in farmer's markets. Sustainability is seen in gardens and parks. Admittedly, urban planning toward "Truman Show"-like gated communities does effectively block out misery and filth (Cunningham 2005) and community action is easily compromised by a "smile and you're happy" subterfuge (Ehrenreich 2010), but that is precisely why urbanites identify with, or at least have emotional connotations with, efforts to strengthen communities, effective and healthy transport, etc.

The Healthy Cities movement has grown since the earliest efforts in the 1980s. By some counts (De Leeuw and Simos 2017) there are over 15,000 localities/jurisdictions around the world that have self-defined as a Healthy City. Of these, as far as we know, the smallest self-declared "Healthy City" is l'Isle-aux-Grues (a community on an island in the St. Lawrence River in Quebec, with around 200 inhabitants). The largest is the conurbation of Shanghai with over 16 million people. Brenner's multi-scalar perspective on urbanity seems to be of particular relevance in understanding these thousands of diverse local health ambitions (e.g., Brenner 2019). By other counts (Rojhani et al. 2019) there are only a few dozen major global cities that have formed a partnership for health. This has happened under the inspiration and funding of the Bloomberg Philanthropies. Elsewhere in this Encyclopedia we are reflecting on global networks of urban jurisdictions that are agglutinating somehow to enhance opportunities for resilience, sustainability, and quality of life (de Leeuw et al. 2020) – all critical determinants of individual and community health.

The Healthy City: A Work in Progress

What *Is* a Healthy City? And Who *Needs* a Healthy City?

In Europe, the European Healthy Cities Network (HCN) consists of "designated" cities (following a process of meeting specific membership requirements) and "accredited" National Networks (supporting larger groups of mostly nation-based local governments). HCN operates in Phases since its inception in 1986. Each Phase lasts ~5 years and pursues specific, commonly agreed, objectives and parameters. Phases have been evaluated with increasingly sophisticated methodological and theoretical toolboxes.

It seems there are many different views, visions, and ambitions for local governments and their communities declaring themselves Healthy Cities as there are cities. This may come as a surprise, particularly for the heavily codified European instance of the global phenomenon.

By the end of the first phase of the WHO/EURO effort, we already discovered more diversity than unity (Draper et al. 1993). For instance, one of the questions we asked was "Why did you join the network?" The responses were interesting:

- "As we are the home of the National School of Public Health, we thought it impossible *not* to join."
- "We needed to create political momentum to invest in sustainable public transport and we thought that Healthy Cities would give us significant clout."
- "Being a Healthy City would enable us to bring back the old grandeur of our city as a spa town and would bring more tourists."
- "We wanted to show off our efforts at urban renewal internationally – Healthy Cities seemed a good forum."
- "We were very happy to be able to join a pan-European city network."

Over the following decades, other motivations emerged, from the most shared (e.g., learning from the others through innovative experiences exchange) to the most unique (underground diplomacy, e.g., between China and Taiwan or Turkey and Cyprus, e.g., Acuto et al. 2017).

Since their "official" launch in 1986 Healthy Cities has evolved and grown naturally and in great diversity. There has been general support for the initiative by WHO in its different regions, but this support has come in different shapes, directions, and times (De Leeuw and Simos 2017). Admittedly, many local healthy city and community initiatives have had very little to do with the vision of a bureaucratic international agency such as WHO; other interests and commitments stimulated local action, for instance, concern about sustainability spurred by things like the Brundtland report and the first Rio conference, and the Québécois network of Villes et Villages en Santé initially developed wholly autonomous from endeavors by the Pan American Health Organization. Indeed, the WHO American Region has consistently nurtured and supported local government action for health (e.g., through resource packs for mayors and other local politicians) but has not engaged in the formalization of networking. The Western Pacific Regional Office of WHO (WPRO) has always had a number of member states where Healthy Cities is high on the agenda, initially with Australia, Japan, and Malaysia (Kuching) a consistent presence. Since the turn of the century this region has formally acknowledged the need to focus on urbanization and health, and guidance documentation and a strategic program to this effect have been developed. WPRO is not proactive in its approach of "designating" cities, but through its networking efforts many cities across the region have signed up as a "Healthy City," for instance, in Cambodia, Mongolia, and Laos, with an abundance of cities in the Republic of Korea. However, there is no formal program that establishes monitored relations between local governments and WPRO. The WHO Eastern Mediterranean Region has over the last 10 years revitalized a fledgling Healthy Cities and Villages program by building on active programs, notably in the Gulf and Iran. Again, no formal designation, support, or recognition has been developed. The WHO African Regional Office has always been active in Healthy Cities and has been supported by a global Francophone network (Europe and Quebec) in developing capacities for Healthy Cities in its French-speaking countries.

Healthy Cities also embrace methodological diversity – they do not exclude a biomedical model of health in favor of a social model. They are not hung up on ownership of their program by the health care sector (e.g., de Leeuw 2015). Goumans (1998) in fact showed for Dutch and British Healthy Cities that ownership can be shared between the education sector, police, parks and recreation, and many more – very much beyond health *care*. Healthy Cities are archetypal for Health in All Policies development.

Healthy Cities is also well implemented in the very reality of local communities and deals with the actual concerns of their populations. There is a truly glocal harmony in Healthy Cities approaches and effectiveness. Participating for many years in national (e.g., France and Greece) or regional

H

(e.g., West Africa) Healthy Cities meetings, we note the similarities between those different countries in their local priorities regarding environmental and social determinants of health issues (e.g., migrants and minorities like Roma integration and health or better quality of green spaces and biodiversity in Europe, healthy markets, or waste management in Africa). These similarities operate independently from the range of wealth and demographics of the municipalities, which presented their case studies in those meetings.

In fact, this last observation leads us to a relevant conclusion: for a municipality to be a Healthy City do not depend on the level of health and environmental status of its community but mainly on its will and involvement to improve them. We call it a "process of continuous improvement." Thus, Healthy Cities is for municipalities the equivalent of what is for companies ISO 9000 regarding their quality process or ISO 14000 for their environmental compatibility.

Cross-References

- ▶ City Visions: Toward Smart and Sustainable Urban Futures
- ▶ Green Cities
- ▶ Health and the City: How Cities Impact on Health, Happiness, and Well-Being
- ▶ The Sustainable and the Smart City: Distinguishing Two Contemporary Urban Visions
- ▶ Theme Cities Networks
- ▶ Urban Health Paradigms
- ▶ Urban Well-Being

References

Acuto, M., Morissette, M., & Tsouros, A. (2017). City diplomacy: Towards more strategic networking? Learning with WHO healthy cities. *Global Policy, 8*(1), 14–22.

Alexander, A. (2009). *Britain's new towns: Garden cities to sustainable communities*. Routledge.

Antonovsky, A. (1987). *Unraveling the mystery of health. How people manage stress and stay well.* San Francisco: Jossey-Bass Publishers.

Ashton, J., & Ubido, J. (1991). The healthy city and the ecological idea. *Social History of Medicine, 4*(1), 173–180.

Brenner, N. (2019). *New urban spaces: Urban theory and the scale question*. Oxford University Press.

Cassedy, J. H. (1962). Hygeia: A mid-Victorian dream of a city of health. *Journal of the History of Medicine and Allied Sciences, 17*(2), 217–228.

Cunningham, D. A. (2005). A Theme Park built for one: The new urbanism vs. Disney design in "The Truman Show". *Critical Survey, 17*(1), 109–130.

Davies, S. C., Winpenny, E., Ball, S., Fowler, T., Rubin, J., & Nolte, E. (2014). For debate: A new wave in public health improvement. *The Lancet, 384*(9957), 1889–1895.

de Leeuw, E. (2015). Intersectoral action, policy and governance in European Healthy Cities. *Public Health Panorama, 1*(2), 175–182.

De Leeuw, E. (2017). Cities and health from the Neolithic to the Anthropocene. In E. De Leeuw & J. Simos (Eds.), *Healthy cities – The theory, policy, and practice of value-based urban planning* (pp. 3–30). New York: Springer.

de Leeuw, E. (2020). Health promotion. Vol. 1, Chapter 2.13. In J. Firth, C. Conlon, & T. Cox (Eds.), *Oxford textbook of medicine* (6th ed., pp. 152–156). Oxford: OUP.

de Leeuw, E., & Harris-Roxas, B. (2016). Crafting health promotion: From Ottawa to beyond Shanghai. *Environ Risque Sante, 15*, 1–4. https://doi.org/10.1684/ers.2016.0921.

De Leeuw, E., & Simos, J. (Eds.). (2017). *Healthy cities – The theory, policy, and practice of value-based urban planning*. New York: Springer.

de Leeuw, E., Simos, J., & Forbat, J. (2020). Urban health and healthy cities today. In D. McQueen (Ed.), *Oxford research encyclopedia of global public health*. Oxford University Press. https://doi.org/10.1093/acrefore/9780190632366.013.253.

De Moncan, P., & Heurteux, C. (2002). *Le Paris d'Haussmann*. Paris: Les éditions du Mécène.

Deelstra, T. (1985). Urban Development & Mental Health: A Planners View. Paper presented at the 'Health in Towns' conference, Council of Europe, Strasbourg, December 3–5, 1985.

Deevey, E. S. (1963). General and urban ecology. Ch. 3. In L. J. Duhl (Ed.), *The urban condition – People and policy in the metropolis* (pp. 20–32). New York/London: Basic Books.

Draper, R., Curtice, L., Hooper, J., & Goumans, M. (1993). Review of the First Five Years: WHO Healthy Cities Project (1987–1992). In *Document EUR/ICP/HSC 644*. Copenhagen: WHO Regional Office for Europe.

Duhl, L. J. (Ed.). (1963). *The urban condition – People and policy in the Metropolis*. New York/London: Basic Books.

Ehrenreich, B. (2010). *Smile or die: How positive thinking fooled America and the world.* Granta books.

Goumans, M. J. (1998). Innovations in a fuzzy domain: Healthy cities and (HEALTH) POLICY DEVELOPMENT IN THE NETHERLANDS AND THE UNITED KINGDOM (Doctoral dissertation, Maastricht University).
Hancock, T., & Duhl, L. (1986). Promoting health in the urban context. WHO healthy cities papers no. 1. FADL Publishers, Copenhagen.
Kickbusch, I. (2007). Health governance: The health society. In D. V. McQueen, I. Kickbusch, & L. Potvin (Eds.), *Health and modernity: The role of theory in health promotion* (pp. 144–161). New York: Springer.
Maneglier, H. (1990). Paris impérial: la vie quotidienne sous le Second Empire. Colin.
Martin, L. J. (2014). Space cadets and rat utopias. The Appendix – Futures of the Past 2(3) August 28, 2014. http://theappendix.net/issues/2014/7/space-cadets-and-rat-utopias. Last accessed 6 Nov 2015.
Rojhani, A., Hung, C., Chew, S., Honeysett, C., Mullin, S., & Karpati, A. (2019). The partnership for healthy cities. Urban health, 293. In S. Galea, C. K. Ettman, & D. Vlahov (Eds.), *Urban health*. Oxford: Oxford Univeristy Press.
Roué Le Gall, A., Le Gall, J., Potelon, J.-L., Cusin, Y. (2014). Agir pour un urbanisme favorable à la santé, concepts & outils. Guide Ecole des Hautes Etudes en Santé Publique (EHESP), Rennes. ISBN: 978-2-9549609-0-6.

Healthy Planning

▶ Planning Healthy and Livable Cities

Healthy Urban Environment

▶ Public Policies to Increase Urban Green Spaces

Healthy Urban Life

▶ Hidden Enemy for Healthy Urban Life

Heritage Cuisines

▶ Formulating Sustainable Foodways for the Future: Tradition and Innovation

Hidden Enemy for Healthy Urban Life

Mehri Mohebbi
Transportation Equity Program Director, University of Florida (UFTI), Gainesville, FL, USA

Synonyms

Active living; Discrimination; Equity; Healthy urban life; Microaggression; Muslim women; Perceived safety; Structural racism; Walkability; Walking behavior

Definition

The world population continues to grow, and by 2050 about 70% of the total population will be residing in urbanized areas. Therefore, improving the quality of urban life is significant as it will impact the majority — more than two-thirds — of the world population's health and well-being through environmental exposures, including increased air pollution and sedentary lifestyles. In addition to the physical environment of cities, social environment plays a pivotal role in urban livability standards. There are instruments that are designed to capture multi dimensional aspects of social environments in relation to the quality of urban life. However, there is a need to update the existing tools to better reflect the convoluted and multi-layered role that social environmental factors play in the lives of urban residents. Specifically, there needs to be a focus on communities who are historically overburdened with health and social inequities (such as racial/ethnic/religious minorities). The role of subtle and obvious forms of discrimination in detaching specific members of urban society is considerable. Becoming discouraged from being active in everyday life and abstaining from bicycling, walking, or taking transit is a threat to urban society, as, for instance, walking is the most available and affordable way to achieve the minimum recommended amounts

H

of physical activity for an individual. This article emphasizes the impact of social factors, such as microaggressions, on urban populations' active living experiences, such as walking.

Introduction

While physical environments and urban planning can create equitable opportunities for healthy living, this is only possible in the absence of social stressors, such as discriminatory behavior. However, it is naïve to solely overemphasize the role of built environment in encouraging healthy urban life — the enormous impacts of social factors must also be underscored. There is a plethora of evidence on how impactful social factors are in discouraging a healthy urban life, even in a built environment which is considered theoretically livable and accessible for all (Ghani et al. 2016; see also Owen et al. 2007). In addition, one's gender, ethnic, and racial identification influence their involvement in social experiences and impact their level of life satisfaction (Lee et al. 2017).

Across the world, cities have been observed as sites of inequality as resources are not equitably distributed, nor could all members of urban society bear the weight of uplifting equity in every aspect of urban life (Tonkiss 2013). Equity is a multifaceted concept that needs to be highlighted both for the public and practitioners; however, the importance of understanding the influence of race, ethnicity, and gender is often left out of plans by policymakers and urban planners (Spain 2014). Such a blind approach has resulted in defining vulnerable parts of urban population as *problems* in contemporary urban settings (Lelandais 2013; see also McKenzie 2015). In response to existing social inequities, there has been a constant call for rights-based approaches to urban development, which centrally address the needs and concerns of historically marginalized groups (such as racial/ethnic minorities, senior citizens, children, and people with special physical and mental abilities) within urban contexts.

To combat this blind spot, there are existing tools that emphasize reframing the concept of healthy urban life. For example, THRIVES, which stands for "Towards Healthy uRbanism: InclusiVe Equitable Sustainable," is a new tool that focuses on awareness about health impacts of new developments, structural barriers, and the urgency of environmental degradation (Pineo 2020). In addition, there exists a considerable number of frameworks and guidelines narrating the hierarchy of social factors impacting one's access to healthy living opportunities and resources (Sallis 2020; see also Mohebbi 2019; Alfonso 2005). A commonly accepted framework is social determinants of health (SDOH), which consists of five domains: (1) economic stability; (2) education access and quality; (3) health care access and quality; (4) neighborhood and built environment; and (5) social and community context (Office of Disease Prevention and Health Promotion 2020). This chapter focuses on the fifth SDOH domain and narrates the story of healthy urban life for a historically marginalized community.

Understanding the human context and types of interactions that an individual encounters on a daily basis is key to reclaiming public spaces for distinct social and cultural groups, while also creating opportunities for multicultural engagement within urban society (Lubitow et al. 2019). Unhealthy social interactions can result in voluntary segregation and isolation for historically marginalized groups. There are different forms of unhealthy and, to a considerable extent, destructive social relations, such as subtle and obvious racist behaviors. To live a healthy urban life, every member of urban society should feel socially comfortable and safe using active living amenities such as urban trails. This chapter introduces a newly developed framework called the *hierarchy of social environment components (HiSEC)* to discuss how subtle forms of racism (e.g., microaggressions) adversely impact the active living routines of historically/systematically marginalized groups (Mohebbi 2019).

Segregation and Healthy Urban Life

While physical inactivity and/or sedentary lifestyle is listed as one of the highest risk factors for mortality (Carbone et al. 2020; see also Lynch and Leitzmann 2017), living a healthy urban life is not easily achievable in many cities across the globe. Although the importance of being physically active is clear, there is a considerable lack of supportive social and built environments, specifically for members of *historically overburdened communities with social and health inequities* (including racial/ethnic minorities, senior citizens, children, and people with special physical and mental abilities). Such inequities are rooted in the existing spatial segregation of urban communities, which decreases the level of access to opportunities and services for individuals coming from a specific race, ethnicity, and/or gender.

Spatial segregation, a common theme linking urban form with access to basic needs (such as physically active spaces and environmental health), hinges on vulnerable populations being residentially separated from advantaged groups. If vulnerable populations were evenly dispersed, they would not suffer from issues such as differential exposures to most environmental hazards or a lack of active living resources. Advances in communication and transportation technology have allowed social groups to move more easily across geographic space, but elite groups have long used residential locations as a way to distance themselves. This pattern has cut across virtually all industrial and post-industrial societies and continues to impact the present-day realities of historically marginalized groups (Musterd 2005). For example, Blacks in the United States and South Africa, as well as ex-untouchable castes in India, experience high levels of segregation. However, even in countries where the divide is less pronounced, some residential differential still exists (van Kempen 2019; see also Sidhwani 2015; Musterd 2005; Dupont 2004).

Several urban movements have attempted to address such pronounced division in urban society. Rights-based innovative solutions, for instance, investigate the correlation between one's social and ethnic identity and their ability to pursue their wills and celebrate their values (Fainstein and Servon 2005). Parallel with rights-based approach, there were numerous other solutions introduced for urban development. For instance, gender-centered movements carefully looked into the influence of an individual's gender identity or sexual orientation on forming their social experiences and defining their social power. Further, urban life is the outcome of social ecology — in other words, it's the result of a series of interconnected but everchanging factors.

Rapid urbanization has caused shifts that influence expectations and perceptions about surrounding built and human environments. Those expectations and perceptions in turn frame urban issues, and often become the benchmark for decisions that either further social isolation or promote social cohesion. In fact, those goals can be simultaneously pursued by separate groups, deepening the feelings of division across the community as a whole. Over time, this either encourages cultural diversity or widens the gap between minoritized groups and the rest of the urban community (Mohebbi 2019).

The challenge is to create socially integrated neighborhoods that increase levels of trust among community members. Lack of trust and experiencing subtle forms of discrimination adversely impact members of minoritized groups. Communicating with minoritized groups to understand their concerns and address their needs is a key component of any effort emphasizing healthy urban life. In most instances, members of those communities are considered a hard-to-reach population not because they're uncooperative, but because researchers, policymakers, or urban planners do not have enough knowledge and experience about how, when, and where members of those communities can be reached.

Reaching Hard-To-Reach Populations

Being defined under the *hard-to-reach population* category indicates a wide range of barriers researchers or policymakers face on reaching members of a community. One of the pivotal barriers a researcher faces is institutional legitimacy (Ellis 2021); a researcher must present their study as legible, impactful, and appropriate to community gatekeepers. In a broader perspective, a researcher's positionality plays a strong role in how they will be seen by the target population.

Although it is still contested, the most widely-accepted solution to access hard-to-reach populations is to collaborate with an insider or be an insider yourself. Several group experiences and/or characteristics can serve insider purposes, such as religious values (Widdicombe 2015; see also Ahmed 2014). However, determining how one could be considered an insider is not as simple as identifying a clear insider dimension.

An insider status can emerge during spontaneous insider moments or can be negotiated during research through constant interaction with research participants (Cui 2014; see also Al-Makhamreh and Lewando-Hundt 2008). An insider's perspective is most valuable when working to gain access to a target population. In practice, however, very few scholars found insider status yield an increased quality of data (Mannay and Creaghan 2016; see also Hassan 2015). Whether the researcher is an insider or an outsider, shared identities such as gender, race, class, sexual orientation, ability, and religion can facilitate access and rapport. However, a lack of shared identities does not preclude successful research. Rather, these differences and considerations must be woven into the data collection process itself.

Insider researchers are still *outsiders* in some ways as the research evolves. There are several ways to overcome barriers that may be caused by an outsider's perspective, such as community immersion. Community immersion involves being with and living among people in the target communities (Mohebbi et al. 2018). While living there, researchers observe interactions and daily social processes. This method allows researchers to build trust as an inside researcher — and facilitates the collection of insider data (Matsuda et al. 2016; see also Katigbak et al. 2015). This approach hints at the importance of creating long-term relationships among communities, planners, and policymakers to:

- Understand communities' concerns
- See the issues from communities' perspective
- Build trust among community members and any outsiders whose daily works intersects the communities' needs (Elefteriadou et al. 2021).

To create a well-connected urban environment that provides a variety of opportunities for social interactions and physical activities, the first step is to understand who is living there, the present state of social cohesion, and which public engagement method/s can ease the process and accelerate the creation of long-lasting positive impacts that uplift health equity.

Conclusion: Healthy Living as a Right, Not a Privilege

Social/community context is one of the five social determinants of health domains. Increasing social support is an optimal goal, specifically in socially segregated contexts where members of disadvantaged groups encounter barriers to enjoying a healthy urban life. Getting the minimum requirement of physical activity is interwoven with the quality of the built environment and social interactions experienced while being physically active outside. In socially segregated areas, walking infrastructures are not equitably present, even though walking is the easiest and most affordable way to connect with nearby opportunities and resources. Further, the health benefits of walking are undeniable. Access to walking infrastructure has been considered a human right, and compact cities with a dense urban fabric and higher level of walkability have been introduced as a way to further social equity (Walljasper 2015; see also Burton 2000). The *hierarchy of social environment components* (HiSEC) (see the Fig. 1), introduced in 2019, emphasizes the impact of societal and group-level factors on an individual's walking behavior (Mohebbi 2019). This behavioral model was developed based on a study addressing Muslim women's active living in southeast Michigan. The model suggests that societal and group-level factors played a more influential role than individual-level factors in Muslim women's decision-making processes surrounding walking behavior. The societal and group-level factors influencing this behavior and, in a broader perspective, the social experiences of these Muslim women, include fear of *otherness* (personal safety impacted by socio-political events) (see the

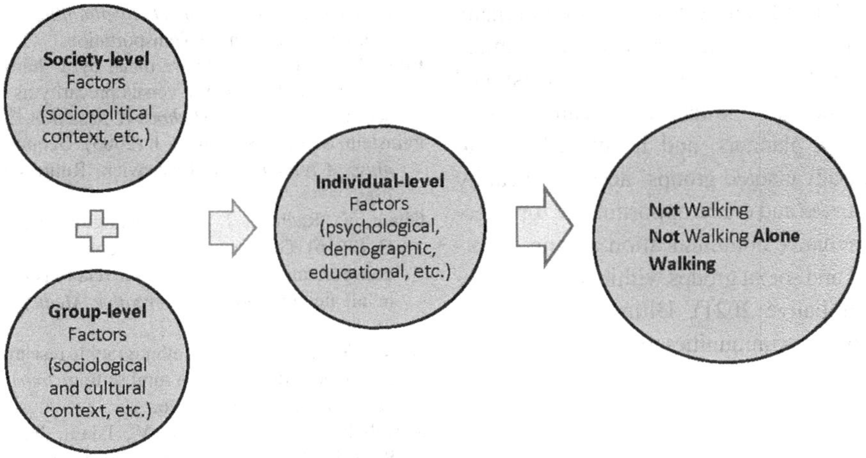

Hidden Enemy for Healthy Urban Life, Fig. 1 The Hierarchy of Social Environment Components (HiSEC) (Mohebbi 2019)

Hidden Enemy for Healthy Urban Life, Table 1 Explaining social environment components of Muslim women's walking behavior

Social environment components	Factors	Key issues
Societal & Group-level Characteristics	Socio-political context	Public misconception about Muslims Being minority even among minorities
	Social acceptance	Public illiteracy on diversity among Muslim community Social discomfort Voluntary segregation (de facto segregation) Negative public attitudes toward Muslim community
	Social safety	Microaggression (e.g., micro-insult and micro-assault) Hate crimes Social mistrust between Muslims and host community

Mohebbi (2019)

Table 1), social acceptance by non-Muslims, and level of social comfort in their neighborhoods (Mohebbi 2018).

In modern cities, connecting to *others* within socially diverse settings is a central issue. *Others* are defined as cultural and/or social groups that celebrate different (even contradictory) values compared to the majority groups. Minoritized groups have suffered from ethnic exclusion, gender stereotypes, and identity crises as a result of being *othered* (Noizet 2020; see also Kimourtzis et al. 2017). While some minoritized groups experience heightened levels of alienation in daily urban life, almost every group experiences it at some level (Coward 2010, 2017). For some, this situation has led to refraining from public spaces where inter-cultural interactions cannot be avoided. Microaggression, when an individual experiences overt forms of racism and discrimination over and over, is one of the most common manifestations of such behavior.

Subtle racial/ethnic/religious discrimination is a threat to public health because it both gradually impacts mental health and also discourages participation in physically active spaces (Devakumar et al. 2020; see also Nadal et al. 2012). Microaggression is the hidden enemy of healthy urban life, and adversely influences urban communities' well-being. The negative impacts increase tremendously for historically marginalized groups (such as Muslim women), whom in many instances are disconnected from or have limited access to healthy living resources. Widespread community-based initiatives can assist policymakers, planners, and health officials to enhance disadvantaged groups' access to healthy living resources and create opportunities for effective and inclusive communication among different cultural and social groups within urban society (PHEAL Initiative 2021). Ultimately, this will lead to stronger communities and healthier living for all.

References

Ahmed, S. (2014). Reflections on conducting research on the 'war on terror': Religious identity, subjectivity and emotions. *International Journal of Social Research Methodology*, 1–14.

Alfonso, M. (2005). To Walk or Not to Walk? The hierarchy of walking needs. *Environment and Behavior (EAB)*, 808–836.

Al-Makhamreh, S. S., & Lewando-Hundt, G. (2008). Researching "at home" as an insider/outsider: Gender and culture in an ethnographic study of social work practice in an Arab society. *Qualitative Social Work*, 9–23.

Burton, E. (2000). The potential of the compact city for promoting social equity. In E. Burton, M. Jenks, & K. Williams (Eds.), *Achieving sustainable urban form* (pp. 19–29). New York: Routledge.

Carbone, S., Ozemek, C., & Lavie, C. L. (2020). Sedentary behaviors, physical inactivity, and cardiovascular health: We better start moving! *Mayo Clinic Proceedings: Innovations, Quality & Outcomes*, 627–629.

Coward, M. (2010). Urbicide: The politics of urban destruction. *Global Discourse*, 186–189.

Coward, M. (2017). In defence of the public city. In R. Atkinson, L. McKenzie, & S. Winlow (Eds.), *Building better societies: Promoting social justice in a world falling apart* (pp. 67–80). Bristol: Policy Press.

Cui, K. (2014). The insider–outsider role of a Chinese researcher doing fieldwork in China: The implications of cultural context. *Qualitative Social Work*, 356–369.

Devakumar, D., Selvarajah, S., Shannon, G., Muraya, K., Lasoye, S., Corona, S., ... Achiume, T. (2020). Racism, the public health crisis we can no longer ignore. *The Lancet*, 112–113.

Dupont, V. (2004). Socio-spatial differentiation and residential segregation in Delhi: A question of scale? *Geoforum*, 157–175.

Elefteriadou, L., Classen, S., Manjunatha, P., Srinivasan, S., Mohebbi, M., Steiner, R., ... Patni, S. (2021). *Transportation mobility assessment and recommendations for smart city planning*. Tallahassee: Florida Department of Transportation.

Ellis, R. (2021). What do we mean by a "hard-to-reach" population? Legitimacy versus precarity as barriers to access. *Sociological Methods & Research*, 1–31.

Fainstein, S. S., & Servon, L. J. (2005). *Gender and planning: A reader*. New Brunswick: Rutgers University Press.

Ghani, F., Rachele, J. N., Washington, S., & Turrell, G. (2016). Gender and age differences in walking for transport and recreation: Are the relationships the same in all neighborhoods? *Preventive Medicine Reports*, 75–80.

Hassan, R. (2015). Issues in ethnographic research: Female researcher's dilemma in a rural setting. *Journal of Gender & Social Issues*, 81–96.

Katigbak, C., Devanter, N. V., Islam, N., & Trinh-Shervin, C. (2015). Partners in health: A conceptual framework for the role of community health workers in facilitating patients' adoption of healthy behaviors. *American Journal of Public Health*, 872–880.

Kimourtzis, P., Kokkinos, G., Papageorgiou, I. V., & Kypriotis, D. (2017). Crisis, otherness and integration. Local policies in South Eastern Europe: The case of Rhodes island. *Advances in Social Sciences Research Journal*, 146–161.

Lee, J., Vojnovic, I., & Grady, S. C. (2017). The 'transportation disadvantaged': Urban form, gender and automobile versus non-automobile travel in the Detroit region. *Urban Studies*, 2470–2498.

Lelandais, G. E. (2013). Citizenship, minorities and the struggle for a right to the city in Istanbul. *Citizenship Studies*, 6–7.

Lubitow, A., Tompkins, K., & Feldman, M. (2019). Sustainable cycling for all? Race and gender-based bicycling inequalities in Portland, Oregon. *City & Community*, 1181–1202.

Lynch, B. M., & Leitzmann, M. F. (2017). An evaluation of the evidence relating to physical inactivity, sedentary behavior, and cancer incidence and mortality. *Cancer Epidemiology*, 221–231.

Mannay, D., & Creaghan, J. (2016). Similarity and familiarity: Reflections on indigenous ethnography with mothers, daughters and school teachers on the margins of contemporary Wales. *Studies in Qualitative Methodology*, 85–103.

Matsuda, Y., Brooks, J. L., & Beeber, L. S. (2016). Guidelines for research recruitment of underserved populations (EERC). *Applied Nursing Research*, 164–170.

McKenzie, L. (2015, January 21). The estate we're in: How working class people became the 'problem'. *The Guardian*.

Mohebbi, M. (2018, December 5). The Discrimination Muslim Women Face: Lessons for City Planning Outreach. *Bloomberg CityLab*.

Mohebbi, M. (2019). *Exploring social/cultural factors that influence the motivation of Muslim women to walk in their neighborhoods* (A case study of detroit metro

area). Cincinnati: University of Cincinnati - Ph.D. Thesis.
Mohebbi, M., Linders, A., & Chifos, C. (2018). Community immersion, trust-building, and recruitment among hard to reach populations: A case study of Muslim women in Detroit metro area. *Qualitative Sociology Review*, 24–44.
Musterd, S. (2005). Social and ethnic segregation in Europe: Levels, causes, and effects. *Journal of Urban Affairs*, 331–348.
Nadal, K. L., Griffin, K. E., Hamit, S., Leon, J., Tobio, M., & Rivera, D. P. (2012). Subtle and overt forms of islamophobia: Microaggressions toward Muslim Americans. *Journal of Muslim Mental Health*, 15–37.
Noizet, H. (2020). Spaces and spatialities in Paris between the ninth and nineteenth centuries: Urban morphology generated by the management of otherness. *Urban History*, 401–420.
Office of Disease Prevention and Health Promotion. (2020, September 25). *Healthy people 2030*. Retrieved from U.S. Department of Health and Human Services: https://health.gov/healthypeople/objectives-and-data/social-determinants-health
Owen, N., Cerin, E., Leslie, E., duToit, L., Coffee, N., Frank, L. D., . . . Sallis, J. F. (2007). Neighborhood walkability and the walking behavior of Australian adults. *American Journal of Preventive Medicine*, 387–395.
PHEAL Initiative. (2021, May 25). *PHEAL (planning for health equity, Advocacy, leadership) principle #2 COMMUNITY-BASED ADVOCACY.* Nationwide Health Equity Initiative, USA. Retrieved from the State of Place: https://www.stateofplace.co/pheal-principle-2
Pineo, H. (2020, October 1). Owards healthy urbanism: Inclusive, equitable and sustainable (THRIVES) – An urban design and planning framework from theory to praxis. Cities & Health, 1–19. Retrieved from https://www.ucl.ac.uk/bartlett/environmental-design/news/2020/jun/new-framework-redefines-healthy-urban-development
Sallis, J. F. (2020). *Neighborhood environment walkability scale (NEWS)*. San Diego, USA.
Sidhwani, P. (2015). Spatial inequalities in big Indian cities. *Economic & Political Weekly*, 55–62.
Spain, D. (2014). Gender and Urban Space. *Annual Review of Sociology*, 581–598.
Tonkiss, F. (2013). *Cities by design: The social life of urban form*. Cambridge: Polity Press.
van Kempen, R. (2019). Social exclusion: The importance of context. In H. T. Andersen & R. van Kempen (Eds.), *Governing European cities*. London: Routledge.
Walljasper, J. (2015, November 18). *Health Benefits of Walking: Make It a Human Right*. Retrieved from HUFFPOST: https://www.huffpost.com/entry/health-benefits-of-walkin_b_8591332
Widdicombe, S. (2015). 'Just like the fact that I'm Syrian like you are Scottish': Ascribing interviewer identities as a resource in cross-cultural interaction. *British Journal of Social Psychology*, 255–272.

Hidden Potential of Wastewater

Martha Marriner and Tanya Gottlieb Jacobsen
State of Green, Copenhagen, Denmark

Introduction

On a global scale, less than half of all wastewater is collected and less than one-fifth is treated (State of Green and Danish Water Forum 2020). Untreated wastewater has resulted in severe environmental degradation and the contamination of many inland and coastal waters around the world. Untreated wastewater is also a public health issue, heightening the risk of outbreaks of diseases such as cholera, dysentery, and polio. Increased urbanization, population growth, and improving economic fortunes mean greater amounts of wastewater are being generated, placing additional stress on the water supply and requiring greater amounts of energy for treatment purposes. Accordingly, the United Nations' Sustainable Development Goal 6 aims to halve the proportion of untreated wastewater discharged into our water bodies by 2030 (United Nations n.d.).

With the water sector accounting for approximately 4% of the world's total electricity consumption (IEA 2018) and wastewater treatment alone responsible for a quarter of this (ibid), there exists considerable potential for energy savings in the water sector, both via energy efficiency measures and energy recovery, which will assist countries in fulfilling climate targets. Furthermore, while the organic content in the wastewater can be utilized as a resource for energy production, the phosphorus present in the water can be used for fertilizer production with several advantages compared to the application of sewage sludge on agricultural land. Finally, the water itself can be treated to such high standards that it can be reused in a number of ways – e.g., for flushing toilets, in laundry machines, or as irrigation for crops.

A deepening appreciation of the potential of wastewater is reflected in the theme of the 2017 edition of the UN's annual world water development report, Wastewater: The Untapped Resource. (WWAP (United Nations World Water Assessment Programme) 2017) The report advocates for a new paradigm in wastewater management, where the focus shifts from viewing wastewater as "waste," and as something to be treated and disposed of, to viewing wastewater as a tool in the transition to a circular economy, where the emphasis is on reusing, recycling, extracting, and recovering as many resources from the water as possible.

One such country exploiting the potential of wastewater is Denmark (State of Green 2020). The Danish water sector has a goal of becoming energy and climate neutral by 2030 (The Confederation of Danish Industry 2020), and the country as a whole has set ambitious targets regarding the reduction of CO_2 emissions on a national level. Furthermore, Denmark has had strict requirements in place regarding the treatment of wastewater since 1987 (The Danish Environmental Protection Agency n.d.). The current focus is on expanding the scope of wastewater treatment plants where, in addition to treating wastewater and thereby protecting the country's ecosystems, they become energy and resource recovery facilities that extract nutrients from the wastewater and produce organic fertilizer and biogas. This entry describes some of the principles guiding the Danish approach and highlights key features of Danish wastewater treatment practices, including ambitions, regulations, and policies. It is intended to give an accurate picture of the state of wastewater in Denmark and provide inspiration for utilities, policy makers, and research institutions on how wastewater can be utilized to solve a number of environmental and climate challenges.

Environmental Crises and the Origins of Denmark's Approach to Wastewater

Increasing water scarcity and population growth highlight the need to treat and appreciate the value of wastewater in order to protect already stressed ecosystems. For Denmark, a string of environmental scandals in the 1970s (Danmarkshistorien 2012) necessitated a new approach to the handling of wastewater and water management as a whole. Like most Western nations, industrial and agricultural production boomed in postwar Denmark. Untreated wastewater was dispersed unchecked into the soil, streams, ocean, and lakes. Not only did this place the future water supply under threat and bathing at beaches and the harbor was frequently forbidden (ibid), the oxygen supply in bodies of water was seriously depleted, leading to the death of fish and fish with high levels of mercury were found (ibid).

As a response to growing concern over environmental degradation, Denmark was the first country (The Danish Environmental Protection Agency 2001) to establish a ministry for the prevention of pollution in 1971, which was subsequently renamed the Ministry of Environmental Protection in 1973. While Denmark's initial strategy in the 1970s was to pump wastewater further out into the ocean, making it less visible and less harmful for populated areas, the first environmental protection law was passed in 1973. Although many of the elements set out in the legislation proved difficult to execute in practice, the framework legislation dictated the establishment of regional plans for wastewater management and the so-called "polluter pays principle."

The 1970s were the decade where the understanding that water is a valued resource and that care should be taken in both the extraction and the use of it, just as emphasis should be placed on cleaning it and releasing it back to nature in the right state, began to take hold in Denmark – both politically and among the wider population.

Wastewater Treatment and the Danish Action Plan for the Aquatic Environment

With less than 20% of all wastewater generated from both industries and households being adequately collected and treated, SDG 6 aims to halve the amount of untreated wastewater discharged into nature by 2030 (United Nations n d). In Europe, 97% of the generated wastewater is collected and subjected to treatment beyond the primary level; however, only 69% is subjected to tertiary treatment, which is considered the most advanced form of treatment (European Environment Agency 2020). Ninety-eight percentage of the wastewater in Denmark is subjected to tertiary treatment, which is primarily carried out by the 110 municipally owned wastewater companies (Zinck 2021).

While wastewater treatment today is highly advanced in Denmark and wastewater treatment and management was indeed becomingly increasingly sophisticated due to the environmental protection legislation enacted in 1973 that spurred investment in wastewater infrastructure, the water pollution crises Denmark experienced in the 1970s continued into the 1980s. As a result, an action plan for Danish waters was introduced in 1987. The first of its kind in Europe, the Action Plan for the Aquatic Environment established stringent criteria for wastewater treatment for key pollutants, organic matter, nitrogen, and phosphorus (Ministry of Environment Denmark and The Technical University of Denmark 2015).

The action plan was implemented in three phases, where the first phase concerned the reduction of nutrient discharges from bigger cities, where biological treatment technologies were developed in Denmark in 1988 (Ministry of Environment Denmark and The Technical University of Denmark 2008) for the removal of nitrogen and phosphorous. The second phase concentrated on building storage tanks to handle stormwater discharges and the final phase aimed to improve wastewater management in rural areas via the establishment of mini treatment plants and the transportation of wastewater to larger, centralized treatment plants.

The action plan was the impetus for a large-scale modernization of wastewater treatment plants in Denmark, leading wastewater tariffs to double between 1985 and 1990 (Nielsen et al. 2019). However, since its inception in 1987 and subject to three revisions, it is estimated that pollution in Danish waters has been reduced by up to 90% (Ministry of Environment Denmark and The Technical University of Denmark 2008). Furthermore, the action plan was adopted in almost its entirety by the EU Urban Wastewater Directive of 1991 (ibid).

In Denmark, aquatic environments have been substantially improved as a result, with cases of salmon breeding in rivers and creeks being recorded – even in very densely populated areas. In addition, the recreational value of the Danish water bodies has also benefitted positively from improved wastewater treatment. In Denmark's biggest cities of Copenhagen and Aarhus, significant investment in modernizing the sewerage system has meant that the old industrial harbor ports, which were finally closed to bathers in 1960 after being periodically closed since the 1930s (The City of Copenhagen and Ministry of Environment Denmark 2008), have been transformed into clean harbor baths to which the cities' residents flock to on sunny days.

In the case of Copenhagen (ibid), the city's municipality built a number of stormwater tanks that can retain excess amounts of rainwater and sewage arising from episodes of heavy rain that would otherwise cause the combined sewerage system to overflow. At present, the capital's citizens can swim in a total of nine different harbor baths. In addition to lower levels of pollution, improved bathing water quality, public health benefits in the form of additional public recreation facilities, and greater biodiversity, Copenhagen's harbor baths have reinvigorated formerly industrial quarters of the city to become attractive, new residential areas.

In terms of financing wastewater treatment, Denmark stands in contrast to the bulk of countries around the world, who view wastewater treatment as a public good to be financed by the state. In Denmark, the polluter pays principle applies to both domestic and industrial users, so that the cost of wastewater collection and treatment are covered by the water tariffs. These costs comprise two-thirds of the overall price consumers pay, which, at an average of EUR 9,6 per cubic meter, means that Denmark has one of the world's highest water prices (Nielsen et al. 2019). The price is based on full cost recovery (ibid), i.e., it covers all of the water utility's costs related to the process of water extraction, distribution, wastewater transportation and treatment, groundwater protection, as well as various taxes. Furthermore, a discharge tax has been implemented, which means that the polluter (in this case the wastewater treatment facility) has to pay a tax on every kilo of discharge of the three key parameters: organic matter, total phosphorous, and nitrogen (State of Green and Danish Water Forum 2020).

Wastewater as a Source of Clean Energy

Treating wastewater has traditionally been an energy-intensive process. The International Energy Agency (IEA) estimates that the water sector accounts for approximately 4% of the world's total electricity consumption and wastewater treatment alone accounts for a quarter of this (IEA 2020). Similarly, water and wastewater treatment processes typically account for 25–40% of a municipal electricity budget (Aarhus Water 2020). Meeting SDG 6, which aims to halve the proportion of untreated wastewater discharged into our water bodies by 2030, could therefore put significant upward pressure on energy demands and municipal energy costs.

However, wastewater actually contains significant amounts of embedded energy. The IEA estimates that if this energy could be harnessed, it could cover more than half of the electricity needs of municipal wastewater utilities (IEA 2020). In Denmark, the energy consumption of wastewater treatment plants has fallen to 1.9% of the country's total energy consumption (State of Green and Danish Water Forum 2020). This has been achieved via a mix of energy efficiency initiatives, process optimization, and energy recovery.

In terms of energy efficiency, online monitoring and energy management systems, replacement of surface aeration by more energy efficient bottom aerators, continuous optimization of pumps and equipment, including the installation of frequency converters, and different operational approaches are responsible for the efficiency gains in Danish wastewater utilities (ibid).

However, the most significant gains to be made in terms of reducing energy consumption will be via energy recovery. By separating the sludge from the wastewater, Danish wastewater utilities can produce biogas that is then used to generate both electricity and heating, which is subsequently sold as green energy to electricity and district heating companies. This is possible as the Danish Water Sector Act allows wastewater companies to produce energy as part of their core services (The Danish water sector law 2009).

Accordingly, a number of the country's largest utilities are managing to recover enough energy to cover almost all of their energy needs. The Marselisborg wastewater treatment plant located in Denmark's second largest city, Aarhus, produced on average 30% more electricity than the amount consumed by the plant itself between 2015 and 2019. At the same time, the treatment plant produced 75% more heat than it consumed, resulting in a total net energy production of 150% (State of Green and Danish Water Forum 2020). In Odense, Denmark's third largest city, the Ejby Mølle wastewater treatment plant produced 188% of the energy the plant consumes (VandCenterSyd 2020). The growing uptake of heat pumps, which at least two utilities in Denmark have instituted (Nielsen et al. 2019), is expected to have a significant impact on the energy consumption of utilities going forward.

Danish wastewater utilities are thus transitioning from being energy guzzlers to energy neutral plants and even net energy producers. The ultimate aim is to achieve an energy neutral water cycle. To achieve this, the energy production from the wastewater treatment plants must be able to cover the energy consumption related to its groundwater extraction, water treatment, water and wastewater transport, as well as wastewater treatment. As the examples mentioned above illustrate, this goal has already been achieved at some of the largest water utilities in Denmark, who now aim to generate enough energy from their wastewater treatment processes to cover the energy consumption of all of their activities across the entire water cycle (State of Green and Danish Water Forum 2020).

Examples of additional measures being considered by selected utilities in Denmark is the potential to recover the heat from the wastewater before it is discharged, which has the additional benefit of reducing the temperature impact on the receiving waters (ibid).

Resource Recovery from Wastewater in the Circular Economy

Considering wastewater as a resource is a relatively new perspective, and Kehrein et al. (2020) argue that the bulk of wastewater treatment plants in Europe still primarily focus on treatment rather than resource recovery (Kehrein et al. 2020). However, by reusing and extracting the resources found in wastewater, it would be possible to lower CO_2 emissions, reduce energy consumption, conserve water resources, and source valuable nutrients.

A shift is nonetheless occurring in both attitudes and practices, where wastewater can be considered an input in the growing shift to a circular economy, where the water is reused, recycled, and upcycled. Traditional wastewater treatment plants in Denmark are morphing into energy and resource recovery facilities, where wastewater is treated to ensure that it does not contaminate the recipients and valuable resources are recovered before it is discharged back into nature. If treated to a high enough standard, the treated water could also be used to irrigate crops or used for household applications such as flushing toilets or in washing machines – the so-called "fit-for-purpose water" approach, where water is treated to a level that will not harm the user, rather than utilizing potable water for purposes other than human consumption. In Denmark, this is currently mostly applied internally in various industries to minimize the company's total water consumption.

Green energy in the form of biogas is used to generate electricity and heating. Scarce nutrients such as phosphate are extracted and then used to produce organic fertilizer, helping meet global demand for fertilizer. Currently, two full-scale plants are working on recovering phosphate from wastewater sludge that can be used as a pure mineral phosphorus fertilizer (struvite). Doing so offers several advantages in comparison to the application of sewage sludge on agricultural land in the form of lower costs, reduced risks of groundwater contamination, a more flexible and easily stored product, etc. (State of Green and Danish Water Forum 2020).

The wastewater treatment plants of the future can be envisaged as production lines, where valuable substances such as phosphorus and ammonium are removed along the way and other products are removed and refined, e.g., organic matter to produce biogas or base chemicals which can be used for high priced products in the pharmaceutical industry, etc. (State of Green 2020). Further research and technological advancements are necessary to make the recovery of some of these elements possible on a commercial scale (ibid).

Vice Chairman of the Danish government's Climate Partnership on Waste, Water and Circular Economy and CEO of the Danish utility Aarhus Water, Lars Schrøder, maps out the current focus for the Danish wastewater sector:

"*Besides removing nutrients from wastewater to a very high standard, treatment plants in Denmark are being converted into resource recovery facilities. Through advanced control strategies it is possible to lower the energy consumption of the facilities while minimising emissions of methane and nitrous oxide, thus reducing the overall carbon footprint of the facilities. The result is not only improved water quality but also a more sustainable future.*" (Nielsen et al. 2019).

For Denmark, a key driver of the shift to resource recovery in the wastewater sector is to contribute to national and international climate and environmental targets. Another key driver relates to the regulatory demands placed on the utilities from the Danish authorities. Water utilities are subject to mandatory benchmarking on operational parameters and face cost-efficiency demands each year from the Danish Competition and Consumer Authority (Danish Competition and Consumer Authority 2019). This is done in order to enhance the competitiveness of the utilities and secure efficient operations, given that the utilities enjoy a natural monopoly, as well as to keep water prices for Danish consumers at sustainable levels (ibid). The Danish Water and Wastewater Association (DANVA) also carries out its own voluntary benchmarking each year (DANVA n.d.)· These practices have helped spur innovation in wastewater treatment optimization and cost-efficient solutions for both the construction and operation of infrastructure. Innovation projects are often based on collaboration across governmental bodies, water utilities, consulting companies, technology suppliers, universities, and research institutions.

Wastewater Infrastructure: A Diversified Approach

Environmental challenges and rapid urban population growth are placing upward pressure on the water and wastewater treatment infrastructure, often necessitating further expansion of sewer networks and the construction of new treatment plants or upgrades to existing ones. Solutions that can adequately dispose of wastewater and safeguard public health at a minimum cost are paramount.

Historically, centralized wastewater treatment plants have tended to meet the needs of urban areas, while decentralized, smaller local treatment plants and low-technology solutions were the most cost-effective option for rural or regional areas. However, this traditional, binary approach is being challenged. Rapid urban growth may require a cluster approach to maintain sewer networks of a manageable size. Furthermore, water scarcity is driving the reuse of wastewater in local areas. Treating wastewater so that it can be used for street cleaning, watering parks, and irrigating crops are all options that can be applied at decentralized treatment plants.

Decentral wastewater treatment therefore now takes place at widely different scales, from clusters within a megacity to scattered individual households in rural areas. Thus, rather than an "either/or" approach, both centralized and decentralized wastewater solutions can ensure effective and cost-efficient treatment through a network of sewer systems designed according to local conditions.

This diversified approach can also be seen in the case of Denmark. Centralized wastewater treatment, where wastewater is pumped from smaller towns and villages to the closest regional center, has become common over the past 10–15 years (State of Green and Danish Water Forum 2020). The main driver for this trend is heightened demand for cost-efficient solutions that reduce operating costs, labor, and maintenance costs, and result in more energy efficient plants. Significantly, the recovery of energy, phosphorus, and other valuable resources from wastewater is currently only feasible at larger, centralized wastewater treatment plants. Doing so is too costly and too complex for smaller plants (ibid).

In tandem with the tendency towards centralization in urban areas, a strengthened focus on wastewater treatment solutions for villages and scattered households in rural areas has been present in the Danish wastewater treatment strategies since 2004 (ibid, p. 18). The need to protect groundwater aquifers or surface waters, which are vital drinking water sources and sensitive to nutrient pollution, has led to the development of a new range of solutions for decentralized treatment. These include both high-tech, prefabricated mini-treatment units for households or small villages, which are typically compact purification systems based on biological processes, where they are mostly confined in tanks and reactors that are covered to prevent spreading of unpleasant odors and simpler, biological sand and gravel filters that are designed for discharge to surface water or for infiltration of treated wastewater into the soil (ibid).

Denmark's diversified approach to wastewater treatment is also applicable when it comes to industrial wastewater. Different sectors may find it more advantageous to treat their wastewater centrally or decentrally, depending on the composition of the wastewater. Discharge of waste from food processing companies is welcomed by wastewater treatment plants in Denmark, (State of Green and Danish Water Forum 2020) as it is carbon rich and enables the plants to increase their biogas production, thereby generating additional energy, as well as improving the biological removal of nitrogen and phosphorous. The more complex composition of wastewater from manufacturing companies, on the other hand, can hamper biological processes and make the sludge unsuitable for use in agricultural fertilizers (ibid). In such instances, treatment or pretreatment at the source may be advantageous or even mandatory. Treating the wastewater from manufacturing at the source allows for the tailored removal of industrial pollutants and requires low investment and low operating costs. It also can make it possible to reuse or recycle the wastewater in the production process and recover and reuse raw materials and chemicals (ibid).

Presently, Denmark's industrial wastewater focus is turning to areas that are poorly regulated. For example, hospital wastewater contains a complex mixture of pharmaceutical residues and other substances that are hazardous to both human health and the environment. Nevertheless, the discharge of this type of wastewater to the public sewer network lacks appropriate regulatory frameworks. Denmark is therefore working on developing new treatment technologies that can be applied to protect the aquatic environment against pharmaceuticals and microplastics (ibid). In one of Denmark's largest hospitals, technology suppliers have collaborated to implement a full-scale wastewater treatment plant that treats the hospital's wastewater. A membrane bio reactor (MBR), ozonation, granulated activated carbon (GAC), and UV treatment remove hazardous pharmaceuticals, estrogenic activity, and pathogens, resulting in an effluent which is discharged directly to a local stream. Doing so has also resulted in significant sewer tax savings (ibid). In Denmark, municipal wastewater treatment plants are also increasingly testing new methods for removal of micropollutants, including pharmaceutical residues. In Brædstrup (ibid, p. 9), ozonation in the main process at the wastewater treatment plant combined with tertiary treatment (also with ozone) was tested and the results are very promising, especially in terms of increased removal of *E. coli* and antibiotic-resistant bacteria as well as antibiotic-resistant genomes.

The Danish Regulatory Framework for Wastewater Treatment

Wastewater in Denmark today is governed through national legislation, water plans, and the EU directives, including the EU Water Framework Directive and the EU Urban Wastewater Treatment Directive.

On a national level, it is the Ministry of Environment that regulates standards for environmental and nature protection, while the Ministry of Climate, Energy and Utilities is responsible for the financial regulation of the Danish water utilities (State of Green 2020).

On a local level, each municipality prepares water supply plans in which it is described how water is supplied to the consumers, how waterworks should operate, and what should happen to private wells, etc.

In 2008, the Water Sector Reform Act was passed. This meant that the task of supplying drinking water and collecting and treating wastewater was moved from the municipalities to water utility companies. These are established as municipal limited companies, where they are fully owned by the municipalities but operate as separate legal entities. They are not-for-profit and operate under a break-even principle based on full cost recovery.

Furthermore, the water and wastewater utilities are subjected to mandatory benchmarking on both environmental and operational parameters and annual cost efficiency requirements from the Danish state. This has helped drive innovation in the sector.

The 70% Reduction in Denmark's CO_2 Emissions by 2030: The Role of Wastewater

Six months after what was dubbed the "green election" (Lykkeberg 2019) took place in Denmark, the Danish government and a broad majority of parties in the Danish parliament passed a legally binding National Climate Act. The Act was passed during the UN Climate Change

Conference, COP 25, on 6 December 2019 (State of Green 2019).

The climate act enshrines in law the goal of reducing Denmark's CO_2 emissions by 70% in 2030 (compared to 1990 levels). Notably, it commits Denmark to assisting other countries in delivering on their own climate targets and engaging on internationally on climate issues. Furthermore, it makes the government responsible for failure to deliver on the 2030 target, with the possibility of ministers being dismissed for lack of progress (Timperley 2019).

A month prior to the climate act being adopted by the parliament, the government appointed 13 climate partnerships (Ministry of Industry, Business and Financial Affairs 2019). The partnerships were composed of leading private sector organizations and captains of industry, where they were asked to formulate solutions from industry's perspective as to how the 2030 goal can be achieved.

The water sector is part of the climate partnership for waste and water, circular economy, and the results of their work was presented in March 2020, where 14 focus areas were outlined. The report outlines the water sector's vision of becoming energy and climate neutral by 2030 and states that it is realistic for the sector to be able to reduce CO_2 emissions by 67% by 2030 (Hermansen, et al. 2020). It is also recommended that efforts are diverted to increasing the export of Danish water technology and international consulting to assist Europe and the global community in reducing CO_2 emissions. Doing so would reduce Europe's emissions by 1.7 million tonnes of CO_2 and the globe's by up to 30 million tonnes, the report estimates (ibid).

In order to achieve an energy and climate neutral sector, the report recommended to focus on three key areas (ibid, p. 34):

1. Increased biogas production at wastewater treatment plants, which can replace energy sourced from fossil fuels
2. Heat production that is based on wastewater and potable water, which can produce energy and replace the use of fossil fuels
3. Afforestation in order to protect the country's groundwater resources and contribute to an increased absorption of CO_2

In addition, Danish utilities have been working consistently with the UN SDGs since their inception in 2015 (Nielsen et al. 2019). A study conducted by Ernst and Young in 2019 showed that 82% of the 199 utilities surveyed were working with the SDGs either by integrating them in the overall strategy for the utility or considering how to integrate them into their strategy (ibid). The Danish Water and Wastewater Association, DANVA has also released an inspiration catalogue detailing how and why the sector can contribute to realizing the SDGs (DANVA 2018).

The Danish water sector has set a common goal of becoming energy and climate neutral by 2030. In 2020, this goal was implemented in the government's national climate plans.

Conclusion

Today, Denmark is home to some of the most cutting-edge wastewater treatment plants in the world. It is pioneering advanced methods of wastewater treatment and energy and resource recovery, viewing itself as an actor with a role to play in meeting not only Denmark's climate and energy targets but also assisting in reducing European and global CO_2 emissions. Denmark itself attributes the success of its wastewater treatment strategies, growing success with resource and energy recovery and improved water quality to unique cooperation between Danish universities, companies, government, and municipalities, where clearly defined roles and responsibilities for each of the different stakeholders have been established (Ministry of Environment Denmark and The Technical University of Denmark 2015). Other practitioners seem to support this argument, stating that wastewater treatment is a complex issue that requires an integrated approach in order to make effective decisions (Bozkurt et al. 2017).

Cross-References

- ▶ Circular Cities
- ▶ Circular Economy Cities
- ▶ Circular Water Economy

References

Aarhus Water. (2020). *Marselisborg WWTP – from wastewater plant to power plant* [online]. State of Green. Available at: https://stateofgreen.com/en/partners/aarhusvand/solutions/marselisborg-wwtp-energy-neutral-water-management/

Bozkurt, H., Gernaey, K. V., & Sin, G. (2017). *Innovative wastewater treatment & resource recovery technologies* (pp. 581–597). IWA Publishing.

Danish Competition and Consumer Authority. (2019). *Economic frameworks.* [online] Available at: https://www.kfst.dk/vandtilsyn/okonomiske-rammer/

Danmarkshistorien. (2012). *Miljøpolitik til lands, til vands og i luften*. [online]. Available at: https://danmarkshistorien.lex.dk/Milj%C3%B8politik_til_lands,_til_vands_og_i_luften

DANVA. (n.d.). *Water in figures*. [online] Available at: https://www.danva.dk/publikationer/vand-i-tal/

DANVA. (2018). *The water sector and the Sustainable Developement Goals*. 1st edition. [pdf] Skanderborg: DANVA. Available at: https://www.danva.dk/media/5698/the-water-industry-and-the-un-sdgs_final_english.pdf

European Environment Agency, (2020). *Urban waste water treatment in Europe. [online]*. Available at: https://www.eea.europa.eu/data-and-maps/indicators/urban-waste-water-treatment/urban-waste-water-treatment-assessment-5

Hermansen, C.A., Schrøder, L. & Petersen, H. G. (2020). *Climate partnership waste and water, circular economy* (1st ed.). Copenhagen: The Confederation of Danish Industry. Available at: https://www.danskindustri.dk/globalassets/dokumenter-analyser-publikationer-mv/pdfer/klimapartnerskaber/afrapportering%2D%2D-klimpartnerskab-for-affald-vand-og-cirkular-okonomi.pdf.

IEA. (2018). *The energy sector should care about wastewater.* [online]. Available at: https://www.iea.org/commentaries/the-energy-sector-should-care-about-wastewater

IEA. (2020). *Introduction to the water-energy nexus.* [online] Available at: https://www.iea.org/articles/introduction-to-the-water-energy-nexus

Kehrein, P., van Loosdrecht, M., Osseweijer, P., Garfí, M., Dewulf, J., & Posada, J. (2020). A critical review of resource recovery from municipal wastewater treatment plants: Market supply potentials, technologies and bottlenecks. *Environmental Science-Water Research & Technology, 6*(4), 877–910. https://doi.org/10.1039/c9ew00905a.

Lykkeberg, A. (2019). Det røde Danmark vinder det grønne valg, nu kommer det svære: At vinde I virkeligheden. *Information*. [online] Available at: https://www.information.dk/indland/leder/2019/06/rune-lykkeberg-roede-danmark-vinder-groenne-valg-kommer-svaere-vinde-virkeligheden

Ministry of Environment Denmark and The Technical University of Denmark. (2008). Modernizing sewers and wastewater systems with new technologies. In *The Danish action plan for promotion of eco-efficient technologies – Danish Lessons* [pdf] (1st ed.). Available at: https://eng.ecoinnovation.dk/media/mst/8051449/Spildevand_baggrundsartikel_print.pdf

Ministry of Environment Denmark and The Technical University of Denmark. (2015). Stricter regulatory goals improve Danish waste water systems. In *The Danish action plan for promotion of eco-efficient technologies – Danish Lessons* [online] (1st ed.). Available at: https://eng.ecoinnovation.dk/media/mst/8051446/Spildevand_appetitvaekker.pdf

Ministry of Industry, Business and Financial Affairs. (2019). *Climate partnerships*. [online] Available at: https://em.dk/ministeriet/arbejdsomraader/erhvervsregulering-og-internationale-forhold/klimapartnerskaber/

Nielsen, A., Sørensen, T. B., Fischer, L., & Larsen, C. (2019). *Water in figures*. 2019 edition. [pdf]. Skanderborg: DANVA. Available at: https://www.danva.dk/publikationer/vand-i-tal/

State of Green. (2019). *During COP25, Denmark passes Climate Act with a 70 per cent reduction target.* [online] Available at: https://stateofgreen.com/en/partners/state-of-green/news/during-cop25-denmark-passes-climate-act-with-a-70-per-cent-reduction-target/

State of Green and Danish Water Forum. (2020). *Unlocking the potential of wastewater: Using wastewater as a resource while protecting people and ecosystems*. 3rd. [pdf] Copenhagen: State of Green. Available at: https://stateofgreen.com/en/publications/unlocking-the-potential-of-wastewater-treatment/.

State of Green. (2020). *Water for smart liveable cities* (1st ed.). [pdf] Copenhagen: State of Green. Available at: https://stateofgreen.com/en/publications/water-for-smart-liveable-cities/.

The City of Copenhagen and Ministry of Environment Denmark. (2008). The Port of Copenhagen – from a heavily polluted industrial port to a clean and thriving aquatic environment. In *The Danish action plan for promotion of eco-efficient technologies – Danish Lessons* [pdf] (1st ed.). Available at: https://eng.ecoinnovation.dk/media/mst/8051440/Havn_baggrundsartikel_print.pdf

The Confederation of Danish Industry. (2020). *Klimapartnerskab for affald, vand og cirkulær økonomi har afleveret deres forslag til regeringen.* [online] Available at: https://www.danskindustri.dk/politik-og-analyser/di-mener/miljoenergi/nyheder-framiljo-og-klima/2020/032/klimapartnerskab-for-affald-vand-og-cirkular-okonomi-har-afleveret-deres-forslag-til-regeringen/

The Danish Environmental Protection Agency. (n.d.). *Action Plan for the Aquatic Environment III 2005–2009*. [online]. Available at: https://eng.mst.dk/trade/agriculture/nitrates-directive/action-plan-for-the-aquatic-environment-iii/

The Danish Environmental Protection Agency. (2001). A historical overview of environmental policy in Denmark. In *Environmental factors and health: The Danish Experience* [online] (1st ed.). Available at: https://www2.mst.dk/udgiv/publications/2001/87-7944-519-5/html/kap04_eng.htm

The Danish Water Sector Law. (2009). *Lov om vandsektorens organisering og økonomiske forhold*. (LOV nr 469 af 12/06/2009). Available at: https://www.retsinformation.dk/eli/lta/2009/469

Timperley, A. (2019). Denmark adopts climate law to cut emissions by 70% by 2030.*Climate Home News*. [online] at: https://www.climatechangenews.com/2019/12/06/denmark-adopts-climate-law-cut-emissions-70-2030/

United Nations. (n.d.). *Goal 6: Ensure access to water and sanitation for all*. [online]. Available at: https://www.un.org/sustainabledevelopment/water-and-sanitation/

VandCenterSyd. (2020). *Der er energi i lortet*. [online] Available at https://www.vandcenter.dk/viden/energi

WWAP (United Nations World Water Assessment Programme). (2017). *The United Nations World Water Development Report 2017. Wastewater: The untapped resource*. [online]. Available at: http://www.unesco.org/new/en/natural-sciences/environment/water/wwap/wwdr/2017-wastewater-the-untapped-resource/

Zinck, A.M. (2021). Email to Iver Høj Nielsen, 15 January.

High Density

► Residential Crowding in Urban Environments

Holistic Management

► *Wadi* Sustainable Agriculture Model, The

Holistic Well-Being

► Systemic Innovation for Thrivable Cities

Home Garden

► Urban Food Gardens

Housing

► Residential Crowding in Urban Environments

Housing Affordability

Ayodeji Adeniyi
Deception Bay, QLD, Australia

Synonyms

Accessible Homes; Affordable Housing; Affordable Shelter; Housing Costs; Housing Stress

Definition

A measure of whether people can afford shelter after housing costs such as rents or mortgages are subtracted from income.

Introduction

Housing affordability is a concept describing whether households can afford shelter. A more comprehensive definition considers if households can afford to rent or own socially acceptable housing while still affording basic living costs. Empirical measures of housing affordability include relative, ratios, budget-based, and residual approaches. The relative measure assesses whether household income is proportional to house prices or rent. The value is given as a mean or median multiple of prices. The benefit of this measure is that a benchmark ratio can be established based on historically sustainable thresholds. For example, Demographia (2019) maintains that liberally regulated markets (minimal regulations and unobstructed competition) have median house prices that do not exceed three times median incomes (median multiple of 3.0). Authorities and analysts use this benchmark to measure the progression of affordability. However, the relative measure does not establish the number of households above or below an affordability threshold (Earl et al. 2017).

Housing Stress

While the relative relationship between house prices/rent and income is useful in some applications, housing affordability can also be conceptualized as the ratio of individuals experiencing housing stress. Housing stress (also known as cost burden) occurs when households spend more on housing-related costs compared to their overall income. There are three types of housing stress: mortgage stress, rental stress, and housing affordability stress (Australian Housing and Urban Research Institute 2018). Mortgage stress is often declared when mortgage repayments and related housing costs exceed 30% of a household's income. Mortgage stress can affect both lower-income homeowners and wealthier buyers. Households may experience mortgage stress due to rampant housing prices, low wages (income-based affordability problems), or an increase in the prioritization of housing aspirations or housing investments (quality-based affordability problems). While surging house prices and low incomes are the most documented factors relating to mortgage stress, households sometimes voluntarily increase their housing-related expenditure. Therefore, mortgage stress does not necessarily signal a decline in financial security.

Similarly, rental stress occurs when household rental payments exceed 30% of a household's income. Both affluent and underprivileged households can experience rental stress since rent can outpace incomes in disadvantaged and advantaged suburbs. Like mortgage stress, some households also voluntarily spend more on rent for aspirational reasons. However, quality-based affordability problems are less prevalent among the renter cohorts.

Moderate-income homeowner households are more likely to consume housing for quality-based reasons. These groups are less likely to be experiencing unaffordability since their consumption exceeds socially acceptable housing norms. Socially acceptable housing satisfies habitation needs and enables convenient access to employment, amenities, services, and educational opportunities. Based on this definition, the bottom 40% of the population is often considered the group deprived of socially acceptable housing needs. Housing affordability stress embodies this principle and is defined as household expenditure of the bottom 40% income earners that exceed 30% of gross household income.

However, while rental market participants are less likely to face quality-based affordability choices, addressing the housing affordability choice problem exclusively by tenure is limited. Different subsets of the population have differing needs. Housing configurations and structures affect the types of people housed: individuals, couples, single-family, multifamily, or shared accommodation. These are not necessarily quality-based priorities.

Budget Standards and Residual Approaches

The most established budget-based measure is the budget standards approach. The approach specifies a basket of market-priced goods consisting of essential housing, food, and other consumer items. This basket of goods then determines the minimum budget that a specific household type requires. Examples of household types include low-income, modest-income, and labor force status (Henman and Jones 2012). The budget standards measure is useful for establishing benchmarks for disposable consumption as well as housing costs.

H

The residual method measures if the household budget remaining after essential food and consumption (disposable housing income) is sufficient for housing-related costs. This evaluation can be determined by comparing the residual prices with the budget standards benchmarks.

Affordable Homeownership and Rental Barriers

In addition to exacerbating housing stress, rising housing costs and relatively stagnant wages have also led to higher entry barriers for prospective renters and homeowners. Entry barriers partly exist because low- and moderate-income earners have fewer housing options to choose. Individuals may also find it hard to access private rentals based on their rental history or lack thereof. Similarly, prospective buyers may have difficulty securing loans based on their credit history. Part-time and casual applicants also face more obstacles accessing home loans than permanent workers. Moreover, it is harder for low- and moderate-income prospective homeowners to raise the initial deposit required for new home loans.

Affordable Housing

Since housing affordability is best expressed as capturing both households' ability to pay for housing and housing suitability, ideally, house prices would be reasonably priced across diverse geographic suburbs, typologies, and tenure types. In contrast, housing stress is more pronounced in some communities than in others. Hence, affordable housing initiatives, spearheaded mainly by governments and nonprofit organizations, deliver lower-priced housing for lower-income and, in some cases, moderate-income households. The exclusive focus on lower- and moderate-income earners is the critical distinction between affordable housing and housing affordability, as recognized by entities such as the parliament of Australia (Thomas 2017) and the Department for Communities and Local Government (2012) of the United Kingdom. However, in the United States, affordable housing is analogous to housing stress (housing expenditure which does not exceed 30% of gross income) (US Department of Housing and Urban Development 2012). If the lower- and moderate-income definition is accepted, then affordable housing is only one small component in the housing affordability discourse since there are other means of achieving affordability.

Affordable housing primarily resolves housing affordability by artificially lowering prices (house prices/rent are subsidized), while other strategies aim to boost income levels or decrease house prices through market forces. Affordable housing includes both means-tested housing (known as social housing) and non-means-tested lower- and moderate-income private housing. Public and community housing are the two forms of social housing. Public housing refers to accommodation provided by government authorities, while community housing refers to housing provided by nonprofit organizations.

Financial Assistance

In addition to supply-side affordable housing schemes (see Section "Affordable Housing"), financial assistance for low- and moderate-income households also improves affordability. This financial assistance enables households to meet their rental, mortgage, or housing deposit obligations. Financial assistance also functions as an effective safety net, allowing households to have enough income remaining for disposable spending. Housing allowances are common in countries such as France, the United Kingdom, and Australia. In the United States, not all eligible for affordable housing vouchers are recipients (Marya 2016). Financial assistance has been found to lower housing stress in contexts where the allowances are widely deployed.

Ownership and Rental Terms

Homeowners and renters can also address affordability by negotiating their ownership or rental terms. Homeowners have mortgage repayment options, although these options are limiting. Borrowers can refinance their home loans, allowing them to change mortgage terms, but they must meet credit requirements, and there are costs involved. Many lenders offer mortgage holidays (also known as payment holidays), which offer a reprieve from repaying all or some of the full amount of home loans for a specified period. However, these loans can only be offered a few times during the entire loan period and are usually only offered to clients who have been customers for a few years.

Renters can control their rental prices by agreeing upon pre-set rents in long-term lease agreements. Yet, long-term contracts are uncommon in some countries, for example, in the United Kingdom and Australia. Moreover, even in countries such as France and Germany, where long-term leases are commonplace, there is a risk that lengthy agreements can preclude renters from the benefits of falling rents (Treanor 2015). However, countries with a prevalence of long-term and open-ended leases (indefinite terms, e.g., month to month) usually have rent controls as well. Rent controls prevent renters from excessive rent increases. Authorities often control rents using initial rent levels and/or rent increases. Initial rent levels can be determined based on predefined agreements, benchmark rates, or the dwelling's present commercial value or capital investment. Ongoing rent increases can also be controlled by stipulating how often rents can be increased and by what margin (OECD 2019).

Housing Affordability and Society

Housing affordability is a multidimensional problem involving financial, social, economic, and environmental pressures. Hence, housing affordability is typically associated with financial security, borrower eligibility, and regulatory controls, such as appropriate zoning and densities, cost-effective construction standards, responsive land release, and desirable property taxes. However, there are differing opinions on what factors are primarily responsible. According to Demographia (2008), unaffordability stems from the failure to loosen regulations, as seen in the New Zealand and Australian markets. The authors mention the slow conversion of Australian urban fringe land into housing (housing supply). Alternatively, Robertson (2007) has described affordability as a product of demand, and high price to income ratios reflects cities with more spending power. Other demand-side arguments include low-interest rates, foreign buyers, and speculation, while another supply-side factor is communities' resistance to density.

Aside from the differences in contributing factors, policymakers, developers, home loan lenders, academics, and housing providers, approach affordable housing with different strategies. Policymakers want to monitor the extent of affordability through time to ensure that problems such as homelessness are alleviated. Developers and housing providers want to determine what share of the low- and moderate-income market they can serve. In contrast, home loan lenders want to assess if buyers can reliably repay their mortgages. Moreover, due to the impact that affordability has on livelihoods, academics are interested in studying how housing affordability is measured, its impact and progression, policy responses, and solutions.

Housing affordability is a tale of both context-specific events and universal features. Rural communities often experience lower-quality dwellings and incomes, undiversified local economies, competition from outside buyers, and difficulty consolidating rural land for housing construction (e.g., in China due to rural land rights and Australia due to planning restrictions). However, rural communities also face the same challenges of regulatory controls, supply shortages, and a lack of affordable housing present in urban environments, although unique to urban communities is the magnitude of housing market speculation, location-driven demand, and foreign buyer interest.

References

Australian Housing and Urban Research Institute. (2018). Mortgage stress, rental stress, housing affordability stress: what's the difference? Understanding different definitions of housing related financial stress. Accessed 15 Jan 2021.

Demographia. (2008). *6th annual demographia international housing affordability survey*. Edited by Wendell Cox and Hugh Pavletich. Belleville: Demographia. Accessed 25 May 2019.

Demographia. (2019). *15th annual demographia international housing affordability survey*. Edited by Wendell Cox and Hugh Pavletich. Belleville: Demographia. Accessed 25 May 2019.

Department for Communities and Local Government. (2012). *National planning policy framework, communities and Local Government*: Crown copyright. Accessed 15 Jan 2021.

Earl, G., Roca, E., Liu, B., Jayawardena, N. I., & Johnson, D. (2017). *volume 1: Review of literature and choice of an affordability measure for inner Brisbane and beyond, housing affordability policy and measures*. Griffith University and The Sustainable Living Infrastructure Consortium. Accessed 12 Jan 2021.

Henman, P., & Jones, A. (2012). *Exploring the use of residual measures of housing affordability in Australia: Methodologies and concepts, AHURI final report*. Melbourne: Australian Housing and Urban Research Institute. Accessed 20 Jan 2021.

Marya, E. (2016). *Rental housing: An international comparison, working paper*. Harvard University: Joint Center for Housing Studies.

OECD. (2019). Rental regulation. OECD Affordable Housing Database: OECD – Social Policy Division – Directorate of Employment, Labour and Social Affairs. Accessed 22 Jan 2021.

Robertson, R. (2007). Housing affordability: Why some cities are so expensive. Accessed 14 Jan 2020.

Thomas, M. (2017). Housing affordability in Australia. Accessed 16 Jan 2019.

Treanor, D. (2015). Housing policies in Europe: M3 HOUSING limited.

U.S. Department of Housing and Urban Development. (2012). Glossary of terms to affordable housing. Accessed 20 Jan 2021.

Housing and Development

Informal Settlements and the SDGs

María Mercedes Di Virgilio
Instituto de Investigaciones Gino Germani, Universidad de Buenos Aires/ CONICET, Ciudad Autónoma de Buenos Aires, Argentina

Synonyms

Adequate housing; Housing upgrading; Informal settlements; Urban agenda; Urban development

Introduction

The Sustainable Development Goals (SDGs), also known as the Global Goals were adopted by the United Nations (UN) to reduce poverty and achieve sustainable development. The SDGs have replaced the Millennium Development Goals (MDGs) from 2016 to 2030, incorporating broader objectives (17 goals) than the MDGs (8 Goals). "Whereas the MDGs have been focused on tackling poverty in poorer countries, the 17 SDGs [apply] to all countries in recognition that transitioning to more sustainable development globally in an interdependent world requires commitments and substantive changes to the status quo by each country, regardless of its current state of economic and social development" (Arfvidsson et al. 2017: 100).

Broadly speaking, the SDGs goals include involving more countries – both developed and developing – expanding sources of funding, increasing the emphasis on human rights and strengthening the involvement of civil society with a more progressive perspective (Kumar et al. 2016).

With the inclusion of Sustainable Development Goal 11 (ODS11) and the associated targets related to urban development, the Agenda for Sustainable Development, and its Sustainable Development Goals (SDGs) have for the first time introduced a goal focusing on urban areas

Housing and Development, Table 1 Sustainable Development Goal 11 and associated targets

11.1 – By 2030, ensure access for all to adequate, safe, and affordable housing and basic services and upgrade slums
11.2 – By 2030, provide access to safe, affordable, accessible, and sustainable transport systems for all, improving road safety, notably by expanding public transport, with special attention to the needs of those in vulnerable situations, women, children, persons with disabilities, and older persons
11.3 – By 2030, enhance inclusive and sustainable urbanization and capacity for participatory, integrated, and sustainable human settlement planning and management in all countries
11.4 – Strengthen efforts to protect and safeguard the world's cultural and natural heritage
11.5 – By 2030, significantly reduce the number of deaths and the number of people affected and substantially decrease the direct economic losses relative to global gross domestic product caused by disasters, including water-related disasters, with a focus on protecting the poor and people in vulnerable situations
11.6 – By 2030, reduce the adverse per capita environmental impact of cities, including by paying special attention to air quality and municipal and other waste management
11.7 – By 2030, provide universal access to safe, inclusive, and accessible, green and public spaces, in particular for women and children, older persons and persons with disabilities
11.a – Support positive economic, social, and environmental links between urban, peri-urban, and rural areas by strengthening national and regional development planning
11.b – By 2020, substantially increase the number of cities and human settlements adopting and implementing integrated policies and plans toward inclusion, resource efficiency, mitigation and adaptation to climate change, resilience to disasters, and develop and implement, in line with the Sendai Framework for Disaster Risk Reduction 2015–2030, holistic disaster risk management at all levels
11.c – Support least developed countries, including through financial and technical assistance, in building sustainable and resilient buildings utilizing local materials

Source: https://www.un.org/sustainabledevelopment/cities/

and housing to the international development agenda (see Table 1).

One of the implications of the adoption of SDG 11 is the recognition of housing as a key factor for the achievement of remaining 16 SDGs (see Table 2). Thereby, the adoption of the 2030 Agenda for Sustainable Development and its 17 SDGs redefined and invigorated the agenda on international housing policy.

Urban Informal Settlements: Definition Proposal

Some of the targets associated with SDG11 are particularly committed to improving informal settlements: SDG11.1, 11.3, 11.5, 116.6, 11.7, and 11.b (see Table 1). They focus on the problem of urban informality, emphasizing the need for spatial organization, accessibility, and design of urban spaces, as well as infrastructures and provision of basic services, together with development policies that promote social cohesion, equality, and inclusion.

> People have lived in parts of the city where housing is precarious for as long as cities have existed. However, it was only with the more recent and widespread phenomenon of rapid urbanization that began in the middle of the 19th century, that informal settlements became a characteristic feature of urban structure. As the number of people arriving and the speed with which they came to the cities, was not matched by an equally rapid provision of adequate housing, informal settlements were quickly formed. One of the main changes that propitiated the formation of irregular settlements was the process of deagrarianization that forced people to migrate from rural areas to cities in a quest for jobs. More recently, from thc latc 1970s onwards, neoliberalism and the withdrawal of the state have had an unprecedented impact on the formation of informal settlements, increasing their number. In the past, rapid urbanization made it possible for migrants to cities to obtain better incomes and improve their living conditions, but this is no longer the case. From the late 1980s over-urbanization has increased poverty and inequality, and informal settlements occupy a large area of developing cities. (Jiménez Huerta 2019: 940)

Informality refers to a relationship of apparent exteriority and/or conflict with the norms and institutions of the State and/or the market. It also refers to "activities that are not regulated by the

Housing and Development, Table 2 Housing's contribution to the SDGs

SDGs	Housing's contribution to the SDGs
1. No poverty	Housing fosters conditions to reduce extreme poverty, ensuring that all people have access to economic resources and basic services
2. Zero hunger	Housing, located in areas with the presence of supply centers, contributes to access to sufficient and nutritious food, in addition to reducing malnutrition conditions in the country
3. Health and well-being	The social determinants of health position housing and neighborhood environments as key to the health of the population. A house built with quality materials, with adequate infrastructure and facilities for heating and ventilation, and with sufficient space, contributes to the reduction of diseases and deaths due to air, water, and soil contamination, as well as to the mental and physical health of its occupants
4. Quality education	Adequately located housing favors proximity to educational, technical, vocational, and higher education and training centers, as well as to educational facilities adapted to the needs of children and adolescents with disabilities and to gender differences
5. Gender equality	Gender A sustainable housing is adapted to the needs of women and girls in their personal development and in the possibility for them to achieve a higher level of well-being
6. Clean water and Sanitation	Housing with appropriate water and sanitation facilities directly supports universal and equitable access to safe drinking water, improved sanitation and hygiene services, improved water quality, and increased use of water resources

(continued)

Housing and Development, Table 2 (continued)

SDGs	Housing's contribution to the SDGs
7. Affordable and clean energy	A house with resource-efficient technology supports universal access to reliable and modern energy services, increases the share of renewable energy, and improves energy efficiency
8. Decent work and economic growth	Sustainable housing promotes decent work in the construction process and boosts new sectors of the economy, protects labor rights, and ensures a safe working environment
9. Industry, innovation and infrastructure	Safe, sustainable, and innovative housing contributes to the modernization and reconversion of industry toward more sustainable processes, as well as to the development of new sectors of the economy
10. Reduced inequalities	Inequality access to sustainable housing, which is a basic social condition that determines people's equality and quality of life, contributes directly to improving the income of the most vulnerable population and promotes economic, social, and political inclusion
11. Sustainable cities and communities	Sustainable housing promotes access to basic services, safe and accessible public transportation systems, and an inclusive and sustainable urbanization process
12. Responsible production and consumption	Housing built with sustainable materials and technologies contributes to the efficient use of natural resources and the rational management of chemicals and waste, which reduces the generation of waste derived from the sector's activity and encourages companies to adopt sustainable practices in their production processes
13. Climate action	Through the efficient use of resources throughout the life cycle of housing, it is

(continued)

Housing and Development, Table 2 (continued)

SDGs	Housing's contribution to the SDGs
	possible to reduce the risks associated with climate change
14. Life under water	Housing located outside risk areas or areas of high ecological value, with adequate water and sanitation infrastructure, directly contributes to the reduction of marine pollution and the protection of marine and coastal ecosystems
15. Life on land	Similarly, to the previous point, well-located housing contributes to the conservation of terrestrial ecosystems and freshwater
16. Peace, justice, and strong institutions	Formulating and implementing inclusive housing policies helps reduce different forms of violence and ensures inclusive, participatory, and representative decision-making that considers the needs of all people
17. Partnerships to achieve the objectives	The formulation and effective implementation of decent housing and habitat policies contributes to the mobilization of resources and the formation of multi-sector and multi-stakeholder alliances to promote improvements in the quality of life of the most vulnerable population and finds novel mechanisms to involve society in the process

Source: Based on Hernández (2019) and Ortíz and Di Virgilio (2020)

State in social environments where similar activities are regulated" (Castells and Portes 1989: 12). In this way, informality emerges in contexts in which the solutions provided by formal and legal land and urban housing markets – including public housing or land service programs with infrastructure – fail to meet the housing needs of a significant number of families residing in cities, either because of their own growth or due to migratory processes. Informal options for access to land and housing usually entail a precarious initial habitat and poor living conditions that may last for at least one generation (i.e., between 20 and 30 years) (Calderón 2006).

In the context of informal housing developments, the mechanisms of access to land and/or housing are neither necessarily nor solely market mechanisms: informal sale and purchase and/or leasing. Social networks also constitute a key mechanism of access to land and housing. Such informal channels to access to land and/or housing in general involve real estate brokers who either do not adhere to the established institutional rules or do not fall under their protection (Feige 1990: 990). In these cases, the prices of land and housing often decrease due to the impossibility of families to prove their ownership of the property (lack of title deeds or equivalent documentation), the lack of services and the progressive development and consolidation of the habitat at the residents' own cost (self-building) (Gilbert and Ward 1985). From this perspective, informality originates when the tenure and/or urban planning situation is not adequate to the rules that regulate the relations of access to and occupation of land and housing. It thus refers to the legal order that regulates social relations and that is expressed spatially in residential location patterns and in the predominant housing situation among different social sectors. Thus, informality is defined by "the way in which the relationship with the land and housing market and the property system is resolved" (Herzer et al. 2008: 176).

The notion of *informality* implies considering not only the way in which low-income sectors resolve the process of access to habitat. It also applies to the relationship with the property system more broadly, either because the good (in our case, land and housing) is produced informally or because, as a good, it might be traded in informal markets. Thus, to speak of informality necessarily entails paying attention to the restrictive role of the formal land and housing market, which remains out of reach for large sectors of the population. As Portes (1999: 27) points out, "the basic difference between the formal and the informal does not lie in the character of the final product or good that is produced, but in the way in which that product is produced or exchanged." This definition of informality does not imply adhering to a dualistic taxonomy on the existence of a formal land and housing market versus an informal one. Rather, it implies that informality is a constituent part of the productive and territorial structure of the city and that it highlights the segmented nature of a single land and housing market that reflects the heterogeneity of the system. It is understood that there are couplings and interrelationships between both (sub) markets: the formal market provides capital and inputs to the informal market, while the goods and services produced by the latter are usually consumed by formal wage earners.

Informal settlements include a wide range of forms; the best known are land occupations, located on the urban periphery. The occupation of undeveloped land is at the basis of urban land production processes. With this concept, we refer to the social process of transformation that land must undergo to be used for urban purposes. This process involves the generation of use value and, therefore, its transformation into a commodity. It also implies carrying out land subdivision activities, opening of streets, laying of networks, etc. As Abramo (2012) points out, in the case of Latin American cities, the occupation of undeveloped land is the dominant form of access to housing for lower income sectors.

Urban Informal Settlements and the ODS11 Agenda

At present, most informal settlements in the world are found in what are now called developing nations.

Faced with the situation of informal settlements, governments have developed multiple strategies. At the beginning in 1970 decay, these initiatives were based on the logic of eradication, promoting the relocation of the inhabitants of settlements to social housing complexes (Jauri 2011). As part of this framework, states promoted the construction and direct distribution of housing through the financing of massive turnkey housing projects (Guevara 2020). Toward the end of the 1970s, public programs shifted toward the provisioning of urbanized lots with sanitation services and minimum housing solutions. Both initiatives, on numerous occasions, involved the eradication of irregular settlements by transferring populations to urban peripheries (Brakarz et al. 2002). Moreover, these programs did not generally achieve the expected results and have been highly criticized by the inhabitants themselves and by experts from various fields. Experts highlighted the elevated social costs that resettlement programs generated for the occupants of informal settlements. These costs included the loss of resources that had been invested in the production and improvement of informal housing, the loss of social networks, and the loss of access to services and sources of income. In addition, critics pointed out that these programs often failed to target or use resources efficiently such that in many cases, the benefits of the programs were captured by middle-class families rather than by poor families. Meanwhile, insufficient resources were allocated in many cases to ensure the effective possibility of scaling up the initiatives (Viratkapan and Perera 2006).

Since the 1980s and very prominently, in the 1990s, governments have adopted a new approach which is still in force today. This approach promotes the establishment of settlements – with the exception of those locations under conditions of environmental risk (Brakarz et al. 2002), through actions of consolidation and progressive improvement of neighborhoods in situ, taking advantage of preexisting constructions where possible and providing the necessary infrastructure and equipment. This approach was supported by international organizations promoting the targeted urbanization of informal areas through the provisioning of infrastructure and basic services (Ochsenius et al. 2016). The pioneering initiatives were aimed at regulating the urbanization process, through the regularization of the domain, guaranteeing access to land ownership. In Latin America, among the initiatives of this generation, the case of Chile stands out as a case where, through different projects and programs, the irregularity of many settlements has been effectively addressed (Clichevsky 2006; Rojas and Fretes Cibils 2010).

Despite some marked successes achieved through such new approaches, the persistence of social problems and the lack of health and education services in informal settlements have limited the results of physical integration. In addition, in the late 1980s, initiatives began to recognize the growing social mobilization around questions of habitat and the importance of involving residents in the design and implementation of neighborhood improvement programs. In this scenario, the municipalities of large urban areas took the initiative by promoting experiences that manage to articulate neighborhood improvement with a broad urban planning process.

> As a result of the experience accumulated with these programs, a consensus has developed around the fact that strategies based on the settlement of populations in the areas already occupied by them is the most socially and economically desirable solution. This leads to the implementation of various modalities of programs, from those restricted to the regularization of irregularly occupied properties, to integrated neighborhood improvement programs in their most comprehensive conception. (Brakarz et al. 2002: 21)

These initiatives go beyond the regularization of lot ownership toward promoting the full incorporation of irregular settlements into the formal city. They also promote investments in the improvement of infrastructure and urban equipment in the neighborhoods. Finally, they develop coordinated programs aimed at mitigating the main social problems of the communities and improving the quality of life in the settlements. Thus, a more comprehensive approach to the improvement of settlements seeks to develop better living conditions among the most disadvantaged populations. Actions like renovating their surroundings through urban planning of settlements, regularization, and architectural and/or urban design are part of these initiatives. In addition, a comprehensive approach entails undertakings such as the construction of basic infrastructures and equipment that not only beautify housing and habitat, but also provide the elements that allow a better integration of the inhabitants to the urban environment, greater social cohesion, and an increase in their quality of life.

The 2030 Agenda for Sustainable Development and the SDG11 provide a new impetus to these guidelines by committing the international community and governments to the improvement of informal settlements.

The Challenge of the Neighborhood Improvement Strategy Within the Framework of the SD's Agenda

It is therefore worth reflecting on the current consensus and the tensions that these experiences still pose, to revisit them considering the challenges of the 2030 Agenda. For this analysis, we will draw on the literature that reviews these experiences in Latin America and Africa, the regions where these initiatives have been most widely developed.

One of the most challenging aspects of neighborhood improvement initiatives is the complexity of scaling up successful interventions. Many programs focus on the neighborhood scale as the basis of diagnosis and solution to problems related to slums. The excessive focus on local territories seems to run up against the dynamics of cities and the close links between the "formal city" and the "informal city" – both of which are part of a single market for land and housing, labor, and consumption. To achieve deeper solutions, then, it seems necessary to broaden the scope and scale of interventions. Broadening the scope has to do with several aspects of the interventions:

(1) The spatial scale of intervention should be extended to a larger territorial scope. It seems necessary to operate based on an integrated vision of habitat problems in the entire territory that presents deficiencies in a city or metropolitan agglomeration. This scale exceeds that of the neighborhood. "The interventions they finance often end up generating islands of good services and formal land tenure in the middle of a large formally urbanized neighborhoods that suffer from deficiencies very similar to those of the settlements targeted by the programs. The search for

poverty targeting, which in many programs translates into operational rules that select settlements where at least 80% of the poor population lives, prevents them from allocating resources to the improvement of formal urbanized areas with deficiencies where a mix of poor families live with lower-middle and low-income families. In search of a greater impact of these interventions, it is necessary to review the scope of the programs to make them more useful in accordance with an integral vision of improving the quality of life in cities" (Rojas and Fretes Cibils 2010: 19).

(2) The scale of financing. As a result of budget limitations, many programs end up benefiting only a few informal settlements, without responding to the more structural problems of the housing sector. In this context, it is worth recalling the experience of PAC-Favela in Brazil. Prior to the implementation of PAC-Favela, social housing projects and improvements by local governments depended solely on municipal resources and, therefore, the scope of such initiatives was limited. The addition of funding from the federal government increased the investment capacity of the municipalities. In addition, the territorial extension of PAC Favela promoted the coordination and articulation of institutions and levels of government (Lonardoni 2016).

(3) The scale and scope of the problems to be addressed. Initiatives have focused on urban aspects, leaving aside key issues in neighborhoods and slums such as the question of economic integration to generate income among the population, legal security, and the prevention of violence. Thus, solutions should be based on an intersectoral and multilevel approach of agencies with responsibility for relevant issues to ensure the exercise of rights and the coordination of budgetary and institutional resources (Rojas and Fretes Cibils 2010).

A second critical aspect is the need to articulate improvement initiatives with other urban policies. Neighborhood improvement programs alone have a limited capacity to respond to housing deficits and housing problems faced by low-income populations. In addition, mass housing production programs have often produced poor quality and poorly located housing, which fuels urban sprawl. They produce a large proportion of low-quality housing, located on the periphery, and often designated for people evicted from precarious neighborhoods, thus exacerbating socio-spatial segregation. It would seem necessary to articulate improvement programs with on-site social housing production initiatives, which would allow combining the benefits of improvement with the need for a progressive reduction in the quantitative housing deficit. The articulation of neighborhood upgrading initiatives with rental and housing policies is essential to prevent the formation of new settlements. This change in the political agenda, characterized by broader policies, also implies articulating neighborhood improvement policies with resources to improve the quality of existing housing or to improve the urban environment with the provision of adequate services. In this sense, another Achilles heel of improvement programs seems to be the availability of land for the incorporation of land for commercial and productive use, considering diverse housing typologies and functions. Comprehensiveness is also associated with the systematic incorporation of the ecological dimension and climate change in the design of interventions (Becerril 2019). Finally, more holistic policies involve strengthening the technical capacities of local governments to develop and implement quality design projects, in multi-sector and multi-scale partnerships, with a wider range of financial mechanisms and improvements in governance (Magalhães 2016b).

Third, it is worth highlighting the need for federal (national) frameworks to support and guide local initiatives to improve neighborhoods and slums. "Colombia, in this respect, was innovative, and its experience indicates the direction that housing policies should take to tackle the housing problem in all its dimensions and significantly improve urban life. Bogota and Medellin led their own slum upgrading strategies and designed their own integrated, large-scale programs based on an integrated urban intervention approach. Their starting point was acknowledging the important role played by informal markets in supplying housing for the poor and electing slum upgrading as a central issue on the policy agenda and a target for public expenditure. They institutionalized a broader, long-term commitment to upgrade and structure informal settlement areas, while at the same time promoting social and community development. As part of city policy in Medellin, each project was used as an opportunity to enhance connectivity, introduce new urban open space networks, and create new nodes with public facilities, enhancing the periphery overall urban dynamic. These actions reflect a broader city vision. The successful experiences of Bogota and Medellin offer inspiring examples for the [Latin America] region about how to use national established legal frameworks and planning instruments as a platform for large-scale, integrated low-income housing policy, slum upgrading, and land policy operations. Unfortunately, those experiences were not scaled up" (Magalhães 2016a: 113).

Fourth, it seems that multi-stakeholder coordination is a key element to ensure the viability and sustainability of the initiatives. Experiences show that the private efforts of families to improve the neighborhood and their homes are clearly not enough. Facilitating support from different levels of government appears to be fundamental to making initiatives sustainable. In this sense, sustained and long-term social agreements between residents and authorities are a sine qua non condition for the viability and durability of actions. For example, the sponging and relocation processes associated with upgrading can be regulated successfully if neighborhood leaders are actively involved and consensus is built around them. These leaders tend to be legitimate interlocutors with government agencies and officials, especially those at the local level. It is a recurrent strategy to incorporate these territorial representatives in regularization processes to "pave" the way for public intervention in the neighborhoods (Di Virgilio et al. 2012). Likewise, different types of neighborhoods upgrading programs can be implemented to respond to the needs of the populations living in informal settlements, for example, housing upgrading for rental or purchase purposes, based on existing consensuses (Danso-Wiredu and Midheme 2017). In general, such articulations must be sustained in the program, i.e., in the intervention, and programs should guarantee the capacity to transversally unite institutional and extra-institutional efforts aimed at building citizenship. That is to say, the active commitment of the policy to the accessibility of rights involved in the recognition and construction of citizens (Arias and Sierra 2019). In this regard, Brakarz poits out: "the conclusion of works in a settlement should not mean the end of governmental concern for that community. The objectives of urban and social integration are only achieved in the medium term with the continuity of social actions and with an adequate operation and maintenance of urban infrastructures and equipment, particularly of the drinking water, drainage, sewage and garbage collection systems" (Brakarz et al. 2002: 87).

Fifth, the programs must confront operational challenges head-on. Improvement operations require adequate organizational schemes for execution. The different types of works, systems, and services implemented demand extra coordination and cooperation efforts in the execution of the programs. "The work of coordinating the execution of a program that involves investments in multiple sectors and faces delicate problems of relationship with the communities (when executed in neighborhoods that are inhabited) presents a considerable technical challenge. It requires technical implementation teams that include various specialties; an intense inter-institutional coordination effort; good implementation control and monitoring mechanisms; and adequate information systems" (Rojas 2009: 169).

Finally, linked to the previous point, the temporality of the execution of the works is not the same temporality that organizes the life and transformation of the neighborhoods. Settlements and slums are not static urbanizations. On the contrary, if anything characterizes these urbanizations, it is their dynamism and permanent change. This implies that different temporalities and dynamics can be recognized in each of the neighborhoods. Sometimes they are associated with neighborhood-specific processes – for example, when population growth results in the occupation of adjacent land. Others are related to broader processes of the region or the city – for example, the rise of migratory movements or the forced eradication of population from other areas of the city. The different stages of land occupation and population dynamics (population increase due to the influx of migratory flows, vegetative growth of the original population, etc.) force the permanent redefinition of improvement initiatives. This applies both to the preparation of censuses and lists of beneficiaries, and to the general planning of the future neighborhood (location of public spaces and collective facilities, layout of the main arteries, etc.). The progressive occupation of vacant land also reduces the margins for adequate urban management (Di Virgilio et al. 2012). Likewise, changes in the administration at any of the levels of government involved also tend to damage or delay programs, due to changes in priorities, authorities, technical teams, etc.

Cross-References

- ▶ An Overview of the Relationship of the Sustainable Development Goals and Urban and Regional Development
- ▶ Circular Cities
- ▶ City Financing and Social Urbanism in Latin America: The Importance of Good Fiscal Management
- ▶ Disaster Risk in Informal Settlements and Opportunities for Resilience
- ▶ Gender Inequalities in Cities: Inclusive Cities
- ▶ Habitat Provisioning
- ▶ Housing Affordability
- ▶ Housing and Development
- ▶ Local and Regional Development Strategy
- ▶ Social Urbanism
- ▶ Spatial Justice and the Design of Future Cities in the Developing World
- ▶ Sustainable Development Goals
- ▶ Sustainable Development Goals from an Urban Perspective
- ▶ Urbanization, Planning Law, and the Future of Developing World Cities

References

Abramo, P. (2012). La ciudad com-fusa: Mercado y producción de la estructura urbana en las grandes metrópolis latinoamericanas. *Eure, 114*(38), 35–69.

Arfvidsson, H., Simon, D., Oloko, M., & Moodley, N. (2017). Engaging with and measuring informality in the proposed Urban Sustainable Development Goal. *African Geographical Review, 36*(1), 100–114.

Arias, A., & Sierra, N. (2019). La accesibilidad en los tiempos actuales. Apuntes para pensar el vínculo entre los sujetos y las instituciones. *Margen, 92*. http://www.margen.org/suscri/margen92/arias-92.pdf

Becerril, H. (2019). The long-term effects of housing policy instrumentation: Rio de Janeiro's case from an actor–network theory perspective. *Housing Studies, 34*(2), 360–379.

Brakarz, J., Greene, M., & Rojas, E. (2002). *Ciudades para todos: La experiencia reciente en programas de mejoramiento de barrios*. Washington, DC: Banco Interamericano de Desarrollo.

Calderón, J. (2006). *Mercados de tierras urbanas, propiedad y pobreza*. Lima: SINCO Editores.

Castells, M., & Portes, A. (1989). World underneath: The origins, dynamics and effects of the informal economy. In M. Castells, A. Portes, & L. Benton (Eds.), *The informal economy: Studies in advanced and less developed countries*. Baltimore/London: The Johns Hopkins University Press.

Clichevsky, N. (2006). *Regularizando la informalidad del suelo en América Latina y el Caribe: Una evaluación sobre la base de 13 países y 71 programas*. Santiago: CEPAL – Naciones Unidas.

Danso-Wiredu, E. Y., & Midheme, E. (2017). Slum upgrading in developing countries: Lessons from Ghana and Kenya. *Ghana Journal of Geography, 9*(1), 88–108.

Di Virgilio, M. M., Arqueros, M. S., & Guevara, T. (2012). Conflictos urbanos en los procesos de regularización de villas y asentamientos informales en la región metropolitana de Buenos Aires (1983–2011). *Urban NS04*, pp. 43–60.

Feige, E. (1990). Defining and estimating underground and informal economies: The new institutional economics approach. *World Development, 18*(7), 989–1002.

Gilbert, A., & Ward, P. M. (1985). *Housing, the state, and the poor: Policy and practice in three Latin American cities*. Cambridge, UK: Cambridge University Press.

Guevara, T. (2020). Movimientos populares, Nueva Agenda Urbana, derecho a la ciudad e integración socio-urbana. In M. M. Di Virgilio & M. Perelman (Eds.), *Desigualdades urbanas en tiempos de crisis*. Santa Fe: UNL.

Hernández, F. (2019). *Vivienda, clave para cumplir ODS: ONU*. Centro Urbano. https://centrourbano.com/vivienda/vivienda-contribuye-los-objetivos-desarrollo-sostenible/

Herzer, H., Di Virgilio, M. M., & Rodríguez, M. C. (2008). ¿Informalidad o informalidades? Hábitat popular e informalidades urbanas en áreas urbanas consolidadas. *Pampa: Revista Interuniversitaria de Estudios Territoriales, 4*, 85–112.
Jauri, N. (2011). Las villas de la ciudad de Buenos Aires: Una historia de promesas incumplida. *Questión. Revista Especializada en Periodismo y Comunicación, 1*(29). https://perio.unlp.edu.ar/ojs/index.php/question/article/view/565
Jiménez Huerta, E. R. (2019). Informal settlements. In A. M. Orum, D. R. Judd, B. R. Roberts, M. G. Cabeza, & C. P. Pow (Eds.), *The Wiley-Blackwell encyclopedia of urban and regional studies* (pp. 940–943). Hoboken: Wiley-Blackwell.
Kumar, S., et al. (2016). Millennium development goals (MDGs) to sustainable development goals (SDGs): Addressing unfinished agenda and strengthening sustainable development and partnership. *Indian Journal of Community Medicine, 4*, 1–4.
Lonardoni, F. (2016). From mass public housing to a twin-track approach. In F. Magalhães (Ed.), *Slum upgrading and housing in Latin America* (pp. 31–60). New York: BID.
Magalhães, F. (2016a). Shifting gears. In F. Magalhães (Ed.), *Slum upgrading and housing in Latin America*. New York: BID.
Magalhães, F. (2016b). *Slum upgrading and housing in Latin America*. New York: BID.
Ochsenius, F., Carman, M., Lekerman, V., & Wertheimer, M. (2016). Políticas hacia villas y casas tomadas de la ciudad de Buenos Aires: Tensiones entre la inclusión y la exclusión. *Revista INVI, 31*(88), 193–215.
Ortíz, C., & Di Virgilio, M. M. (2020). *Laboratorios de Vivienda (LAVs) Asentamientos precarios y vivienda social: Impactos del covid-19 y respuestas [working paper]*. UHPH/Cities Alliace. https://discovery.ucl.ac.uk/id/eprint/10106201/1/lav_impactos_de_la_crisis_del_covid-19_en_asentamientos_workingpaper_vf.pdf
Portes, A. (1999). La economía informal y sus paradojas. In J. Carpio, E. Klein, & I. Novacovsky (Eds.), *Informalidad y exclusión social*. Buenos Aires: SIEMPRO/OIT/FCE.
Rojas, E. (2009). *Construir ciudades: Mejoramiento de barrios y calidad de vida urbana*. Washington, DC: BID.
Rojas, E., & Fretes Cibils, V. (2010). Construir ciudadanía para una mejor calidad de vida. In E. Rojas (Ed.), *Construir ciudades: Mejoramiento de barrios y calidad de vida urbana*. Washington, DC: BID.
United Nations. (n.d.). *Goal 11: Make cities inclusive, safe, resilient, and sustainable*. Sustainable Development Goals. https://www.un.org/sustainabledevelopment/cities/
Viratkapan, V., & Perera, R. (2006). Slum relocation projects in Bangkok: What has contributed to their success or failure? *Habitat International, 30*(1), 157–174.

Housing Costs

► Housing Affordability

Housing Stress

► Housing Affordability

Housing Upgrading

► Housing and Development

How Cities Can Be Resilient

Emilio Garcia
The University of Auckland, Auckland, New Zealand

Introduction: Rankings and Resilience

Discrepancies and vagueness about the definition of resilience, along with incoherencies in its measurements, are all factors that make the understanding of resilience in the built environment more complicated (Garcia 2020). Resilience is not a word frequently used in everyday conversation, a factor that makes it even more enigmatic. For planners, urban designers, or architects the challenge for the application of resilience to the built environment is that the concept has not been created with cities in mind. Moreover, there are many understandings of resilience coming from engineering, psychology, and ecology (Garcia and Vale 2017). Regardless of the fuzziness about resilience, institutions and firms have made use of the concept of resilience in a more direct way, as they pick up a definition that is suitable for their purposes and build rankings accordingly (Savills Resilient Cities 2021 2021; The EIU 2019: 34). In this way, rankings work as instruments to apply resilience in built environments, and their results become an example of how cities can be resilient.

In principle, there is no need for ranking resilient cities if they are all assumed as inherently resilient (Garcia 2017). Cities around the world have historically gone through social problems, economic crises, and natural hazards. Aleppo, in Syria, is around 8000 years old and still in the same place. Tokyo has been through multiple fires, earthquakes, and tsunamis, and its metropolitan area with more than 37 million residents keeps on growing in the same place (Garcia et al. 2014). Cities in Nepal, Indonesia, India, Bangladesh, and Philippines have been affected year after year by seasonal flooding, but their citizens have managed to find ways to keep their cities working while absorbing the impact of social and natural hazards (Gautam 2017). All cities have resilience because they have people and people have the capacity to learn and to adapt. They extend these qualities to their built environments by changing them in order to persist. Without people, built environments are ruins whose changes only show the pass of time through the deterioration of materials (Garcia et al. 2021).

Resilience starts in communities and individuals, who self-organize to solve problems. Changes in built environments happen as a result of that problem-solving demand to overcome adverse situations, to satisfy new lifestyles, different interests, and needs. In this way cities can adapt and learn from previous experiences. People make cities to be complex and adaptive, which provides an inherent resilience to all of them (Garcia 2017). From this point of view, building more urban infrastructure does not necessarily equate to more resilience for a city. Assuming that all cities, as complex adaptive systems, have an inherent resilience capacity, why would cities need to care about resilience?

History can prove that cities have resilience but that they are not resilient to everything, all the time, in all circumstances. The resilience of built environments is not infinite. Resilience has limits. This is explained in the theory of resilience through the adaptive cycle, which explains that changes happen in cycles, involving different phases of development and including release, which is equivalent to major change (Gunderson and Holling 2002). Many civilizations have disappeared. Empires, countries, cities, and villages have historically collapsed with some of them reorganizing after collapse and others disappearing (Garcia et al. 2021). Even though all cities have resilience they are exposed to different problems and situations that make them more or less vulnerable. Pacific islands, like Vanuatu and Kiribati, are at risk of disappearing due to climate change–induced sea-level rise. There is no affordable infrastructure or magic engineering trick that can save them. They will have to migrate to other countries (Dawson 2017). For these reasons the resilience of vulnerable communities to social or natural hazards is something that deserves attention.

The institutional perception of resilience tends to oversell it by confusing being resilient with having resilience and completely forgetting that all cities have resilience capacity. This situation can be very useful in the competition between cities, a situation that can be reflected in the creation and diffusion of rankings (White and Kitchin 2021). City rankings are the type of information that can go viral on the Internet and make people hit the bottom "like" on social media posts. This can have the effect of increasing the reputation of cities positively ranked or stigmatizing cities poorly ranked (Erdi 2019).

Rankings of resilient cities are not as popular and abundant as other city rankings related to the smartest and most sustainable cities in the world (McManus 2012). Nonetheless, the published resilient rankings have had an impact on the media. One of the most cited rankings was the Grosvenor ranking of resilience cities, which has been previously analyzed by Garcia and Vale (2017) and is no longer available on the website. The Safe City Index, which contains a resilience section, reached more than 100 million people through social media and produced 400 news stories covered by global media (Economist Media Group 2021).

Rankings get the attention of people and have the power to influence policies (Carrera Portugal 2019). Regardless of the rationale behind the rankings or their accuracy, they still matter for people, firms, and institutions. Using rankings to compare cities is an old practice that has taken off since the 2000s (Acuto et al. 2021a: 364). A resilient ranking can be useful for firms that are interested in knowing where to invest and particularly in forecasting how safe it is to investment in a city (Verisk Maplecroft 2021). Therefore, rankings can inform researchers about how resilience is perceived and used by firms and, in this way, identify expectations about resilient cities.

The "ranked resilience" of cities can be used to understand the institutional perception of resilience and its implications. This is important because rankings are linked to the ability of cities to compete with each other (Giffinger et al. 2010). One of the implications of city competitiveness is the influence that institutions and the private sector have in setting goals for the urban design and the planning of cities (Acuto et al. 2021a). The information collected in this chapter helps to explore what cities would need to be recognized as resilient. How could rankings be used to understand what cities would have to do to be resilient?

The following sections highlight some relevant points about resilience related to its theoretical understanding, especially the definitions and characteristics of resilience that can help to explain how the term will be used. These definitions and characteristics of the concept will be introduced in the methodology and used in the analysis of resilience rankings.

Resilience and Sustainability

The first step toward resilient cities should focus on a development that is more sustainable. The OECD recognizes the link between resilience and sustainability by defining resilience cities as cities that can "absorb, recover and prepare for future shocks [...] promote sustainable development, well-being and inclusive growth" (OECD 2021). The problem in this definition is that resilient cities are portrayed as perfect cities, something totally unattainable. No city can be prepared for whatever comes in the future, neither can it promise to deliver such a level of confidence so as to face and succeed against uncertain shocks. To promote sustainable development is, in the definition usually given, something equally vague and very general as a goal. The reality about sustainability for cities is much simpler. Actually, there is no rocket science about it. Countries and cities involved in the Paris Agreement need to reduce their emissions by 70% by 2030 (UNCC 2021). Perhaps if cities, particularly in developed countries, followed a path toward a more sustainable environment, they wouldn't need to build any extra resilience (Chicca et al. 2018).

The popularity gained by resilience is arguably linked to the failure of stopping climate change by sustainable means (Garcia 2020). The wealthier part of the world is responsible for more significant carbon dioxide emissions that contribute to climate change issues and that impact on the least wealthy countries (Garcia et al. 2021). The excessive emissions of a few countries and cities impact in the present and near future a significant majority (Chancel and Piketty 2015). While one wealthy part of the world is planning and investing in infrastructure to "build resilience" to potential threats, the rest is obliged to be resilient with scarce resources. This situation creates an undeniable link between resilience and sustainability. Trying to be "resilient" without caring about sustainability is like throwing gasoline on a fire and expecting the fire truck to arrive soon enough.

These subtle differences are relevant because they tend to be misunderstood at institutional and private levels. Organizations involved in ranking have the power to define which city is resilient and which one is not. This becomes a guideline with which to define how cities can be resilient and recognized as such. The analysis of rankings of resilient cities proposed in this chapter can help to highlight some of these misunderstandings (Acuto et al. 2021b).

Ecological and Engineering Resilience

The understanding of resilience has two possibilities: the ecological and the engineering approach (Gunderson 2000). The basic difference is in how they understand and frame the relationship between changes, disturbances, and the stability of a system, for example, a city (Holling 1973). In the engineering approach disturbances threaten stability, and therefore it is important to get back to normal as quickly as possible (Pimm 1984). In urban terms, the faster the recovery, the more resilient a city is. Time is key, and changes are minimized so the city persists. In the ecological approach, the focus is not on the disturbance but on the system itself, in the urban field, cities, towns, or neighborhoods. The goal is to create the conditions for the persistence of a system. It is about acknowledging change as part of persisting and moving forward (Garcia and Vale 2017). The theoretical differences matter a lot. An engineering analysis of the earthquake that hit Christchurch, New Zealand, in 2011 could focus in assessing the time it takes for the government to rebuild the city or the time it takes for its economy to get back on track. However, an ecological approach applied to cities would focus on what the city of Christchurch did to keep on working while the CBD was closed. This could imply investigating the role of the green and gray infrastructure of the city to buffer the impact of the earthquake (Garcia 2017), or analyzing the shift of commercial activities from the CBD to the periphery, or highlighting the emergence of new spaces like the new market. Depending on the resilience approach used, the subject of analysis changes and produces discrete results that are not necessarily complementary.

Resilience Has No Ethical Goals

Resilience has no ethical goals; it is just a theory for understanding change in complex systems (Garcia and Vale 2017). It can be desirable or undesirable. Crime, poverty, and inequality are very resilient; they keep on adapting and changing and can be very persistent in neighborhoods and cities. That resilience is undesirable. The resilience to sea-level rise or to economic crises is the kind of resilience that is desirable. On the contrary, sustainability has goals based on ethical principles and decisions. Polluting less benefits everybody. Therefore, a resilient city does not imply a good or better city unless what is being resilient to what else is specified. However, a sustainable city is always a positive quality.

Specific Resilience: The Resilience of What, to What, for Whom

Regardless of the approach in resilience, engineering, or ecology, there are two variables in common: a system that is threatened (countries, cities, towns, neighborhoods) and threats (disturbances, hazards, shocks, or crises). The system can vary in size, number of elements, and scales in the same way that there can be multiple threats, but the comparison between systems and threats remains the same, and the resilience of something to something else. This way of framing resilience follows a simple formula: "the resilience of what to what?" (Carpenter et al. 2001). Meerow and Newell (2019) suggested that the resilience for whom, what, when, where, and why should be added to conceiving and implementing projects concerning urban resilience. The analysis of rankings done in this chapter has followed the structured approach of looking for the resilience of what, to what and for whom.

Resilience and Time: Past, Present, and Future

There are always before- and aftershock scenarios in a resilience assessment (Garcia 2017). It is in the history of a system where the identity is developed (McGrath and Lei 2021). The resilience of a system is shown after going through, at the least, one crisis, but not before (Roggema et al. 2021). One can say New York has shown resilience to Hurricane Sandy, but it would be hard to understand a statement like, "New York is resilient to future hurricanes," since these events have not occurred yet. Therefore the assessment of the resilience of a system, whether engineering or ecological, is always a measurement of its past, never its future.

The relationship between resilience and time is useful to give external validation to resilience rankings.

Resilience and Hazards in the Built Environment

One of the expectations about resilience is helping cities to deal with natural disasters. Terms like "bouncing back" (Aldunce et al. 2014), "building back better" (Fan 2013), "recovery" (Cimellaro 2016) are usually used to describe what resilience can do for cities after a natural hazard strikes. Natural hazards are one of the worst nightmares for the built environment, especially big cities where people and resources are clustered, exposed, and more or less vulnerable depending on social and environmental factors (Cutter 2020). Natural hazards can be geological, like earthquakes; hydrological, like river flooding; meteorological, like strong winds; and biological, like viruses and bacteria (Milanović 2020). Flooding and earthquakes are particularly significant for cities. Flood risk "threatens more people than any other natural catastrophe [...] river-flooding poses a threat to over 379 million residents. That is more than the 283 million inhabitants potentially affected by earthquakes and the 157 million people at risk from strong winds" (Sundermann et al. 2013: 11).

However, while the world population is becoming increasingly urbanized and natural disasters happen more frequently, global deaths resulting from natural disasters have diminished drastically since 1930s (see Fig. 1). This could give the impression that communities, on average, are getting better at avoiding deaths from natural disasters. Are they becoming more resilient?

Social hazards are also important for cities, but they are less associated with resilience (Rómice et al. 2020). Humans are the main cause of social hazards (Emdad Haque and Etkin 2007). They involved a wide range of problems from personal issues to relationships between people, hierarchy, institutions, power, and money (Hutter et al.

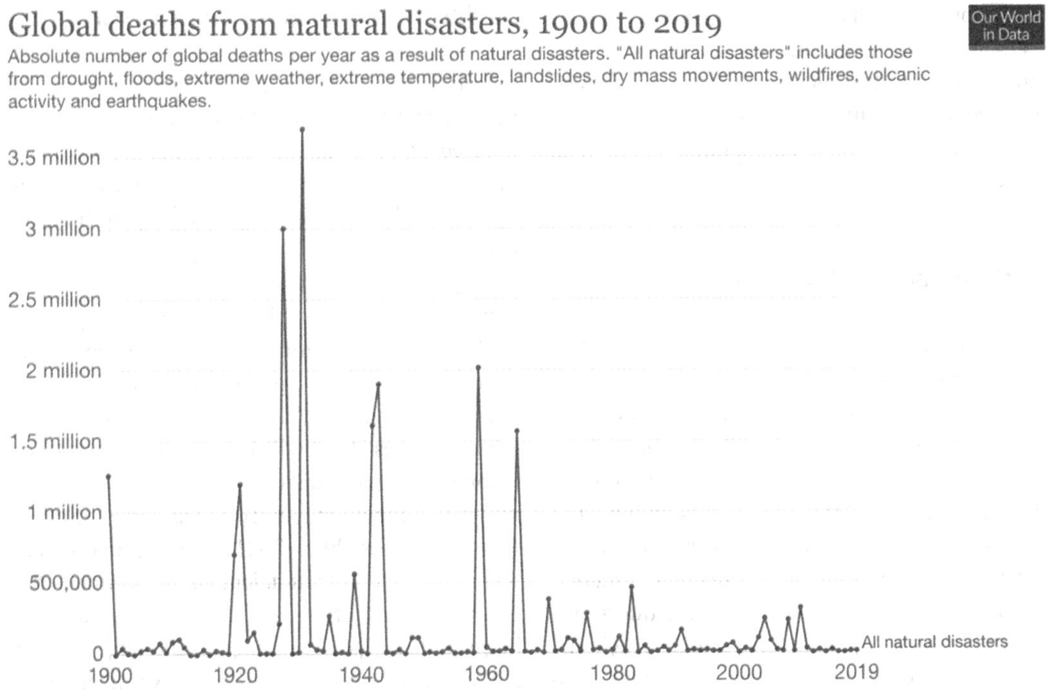

How Cities Can Be Resilient, Fig. 1 Global deaths from natural disasters from 1900 to 2019 (Ritchie and Roser 2014)

H

2013). The built environment as a product of communities is deeply related to social hazards (Lizarralde et al. 2015). Built environments can cause displacements and discrimination within communities (Ye and Aldrich 2021) as well as affect the environment through pollution and emissions (Garcia et al. 2021). Changes in the built environments can also affect the identity of a group, their sense of belonging, and sense of place (Muminović et al. 2014). One of the reasons for social hazards being given less attention than natural hazards could be the problem of linking social hazards and deaths. It is easier to prove that earthquakes killed people and much more difficult to prove that a financial crisis did the same. However, millions can be affected by an economic crisis. Natural hazards tend to be measured by their economic losses, which creates a link between social and natural factors (Meyer et al. 2013). Economic crises can be potentially bigger than the economic losses caused by a natural disaster (Cambridge Centre for Risk Studies 2015). From this economic point of view, investing in resilience could be a good and sensitive approach considering that the return on investment is significant (Harris et al. 2020). This is an important element that should also be assessed when comparing rankings to see what matters most in resilience, rankings, people, or money.

Methodology and Methods

This study is a comparative analysis between two resilient rankings and their similarities with other international rankings across different topics. The subjects of analysis are the Savills Resilient Cities Index (SRCI) and a ranking produced with information included in the Safe Cities Index. The selection of these rankings is linked to relevance and popularity. Following the resilience theory, the analysis is organized around three questions: the resilience of what?, to what?, and for whom? These three questions are used to analyze both resilient rankings separately. Every question is explored in one step. The objective is looking for matches between the top ten resilient cities and other rankings, across different topics. After analyzing both rankings separately using the resilience of what, to what, for whom, the results are compared.

The method uses matrices to count the number of matching cities between two or more rankings. Every matrix is created to count the number of cities found in other rankings (if any) that are also included in the resilient rankings. The number of matching cities between one resilient ranking and another ranking is counted in separate columns. The subtotal is calculated as a ratio of matching cities divided by the total number of cities ranked, which are always ten, although these are not always the same ten cities. The result is, therefore, a number between 0 and 1. In every matrix, multiple categories and rankings are analyzed. All subtotals are averaged to produce final results per category. Table 1 shows an example of one matrix. In the analysis section only subtotals and totals per category are shown to better fit into the space available.

Savills Resilient Cities Index (SRCI)

Savills is a property agent originally founded in the UK that has grown globally with offices distributed in 70 countries around the world. They provide a wide range of services from asset management, advice for buying or selling, to finance, and even guidance into the planning process. Management of data related to the real estate markets is the bread and butter of the company.

The research department of the firm, the Savills World Research team, produced the Savills Resilient Cities Index 2021. Savills defines resilient cities as cities that "will be able to withstand or embrace the many disruptive forces facing global real estate today and in 10 years' time" (Savills Resilient Cities 2021 2021). They propose that resilient cities "attract talent and encourage the innovation that drives city and personal wealth" because these factors impact on the decision that real estate investors make (Savills Resilient Cities 2021 2021). Therefore, resilient cities are cities attractive to real estate investors. The top ten most resilient cities in the Savills Resilient Cities Index

How Cities Can Be Resilient, Table 1 Sample of a matrix used in the analysis of resilient rankings

	Housing markets		Economy		Technology	
	CBRE's most expensive housing markets	Housing markets average per square foot	Top ten GDPs	Global Financial Centres Index	IMD-SUTD Smart City Index Report	IESE Smart Cities Ranking
	Hong Kong	Hong Kong	Tokyo	New York	Singapore	London
	Munich	Singapore	New York	London	Helsinki	New York
	Singapore	Paris	Los Angeles	Shanghai	Zurich	Paris
	Shanghai	Munich	Seoul	Tokyo	Auckland	Tokyo
	Shenzhen	London	London	Hong Kong	Oslo	Reykjavík
	Beijing	Shenzhen	Paris	Singapore	Copenhagen	Copenhagen
	Vancouver	Shanghai	Osaka	Beijing	Geneva	Berlin
	Los Angeles	Beijing	Chicago	San Francisco	Taipei	Amsterdam
	Paris	Berlin	Moscow	Shenzhen	Amsterdam	Singapore
	New York	Milan	Shanghai	Zurich	New York	Hong Kong
New York	1	0	1	1	1	1
Los Angeles	1	0	1	0	0	0
London	0	1	1	1	0	1
Tokyo	0	0	1	1	0	1
San Francisco	0	0	0	1	0	0
Paris	1	1	1	0	0	1
Seoul	0	0	1	0	0	0
Boston	0	0	0	0	0	0
Washington DC	0	0	0	0	0	0
Dallas	0	0	0	0	0	0
Subtotal	3	2	6	4	1	4
Subtotal	0.30	0.20	0.60	0.40	0.10	0.40
Total	0.25		0.50		0.25	

are New York, Los Angeles, London, Tokyo, San Francisco, Paris, Seoul, Boston, Washington DC, and Dallas (see Table 2). The index ranked 500 cities on "their economic strength, demographics, the knowledge economy and technology, environmental resilience, as well as the depth of their real estate markets" (Savills 2021: 9). The economic strength measures how wealthy cities are in terms of their population size, economic activity, and spending, with GDP growth and GDP per capita as key variables. Curiously cities with the highest forecast GDP growth are expected to be in Vietnam, India, and China, but none were ranked in the top 20 resilient cities. In the knowledge economy area, they highlight that technology drives wealth and high-paid jobs; therefore cities that host big tech companies, with high-ranked universities that are competitive in the production of patents, tend to score higher in the ranking. The environment, social, and governance dimension is interested in the use of renewables, food and water security, natural resources, and electric vehicles. However, they declare that the data is scarce and the ranking is done at country scale, with most of the Scandinavian countries leading the ranking for use of renewables, although again Scandinavian cities are scarcely represented in the top 20 final ranking. The real estate metrics refer to the size of the real estate market. The resilience is apparently measured by comparing total investments, change in

How Cities Can Be Resilient, Table 2 SRCI's top ten resilient cities, extracted from Savills Resilient Cities 2021 (2021)

Rank	Top ten resilient cities	Real estate	Economic strength	Knowledge economy and technology	Environment social governance
1	New York	Los Angeles	New York	Seoul	Norway
2	Los Angeles	New York	Tokyo	London	Sweden
3	London	Paris	Los Angeles	New York	Denmark
4	Tokyo	San Francisco	London	Boston	Finland
5	San Francisco	London	San Jose	Paris	Iceland
6	Paris	Seoul	Shenzhen	Daejeon	Switzerland
7	Seoul	Tokyo	San Francisco	Berlin	Netherlands
8	Boston	Dallas	Shanghai	Washington DC	New Zealand
9	Washington DC	Boston	Beijing	San Francisco	Canada
10	Dallas	Atlanta	Washington DC	Los Angeles	Germany

relation to the last year, and cross-border share of investment. This means that cities whose real estate markets keep on growing (or that have lost less than other markets), while reducing the spread of benefits to other cities, are seen as more resilient.

The Resilience of What: Housing Markets, Economy, or Technology?

This section compares the SRCI ranking against other rankings that can be included in the same categories used by Savills, namely, real estate, economic strength, technology, and environment. In the "Housing Markets" category, rankings chosen are used to show similarities with the most expensive housing markets. The "Economy" category used top ten GDPs and most importantly global financial centres rankings to show similarities with cities that have robust economies. The "Technology" category includes two smart city rankings, since technology is one of its pillars (Savills 2021), noting that these rankings are very different. Table 3 shows the results of the matrices comparing the top ten cities in the SRCI ranking with other rankings.

The top ten GDPs is the most similar ranking to the SRCI with six matches (0.6) (see Table 3). Figure 2 shows that the category "Economy" contains a bigger percentage of the top ten cities as included in the SRCI. This result can be used to infer that the GDP of cities is the variable that has more weight and the one that seems to be deciding the resilience of "what" is being analyzed. It is important to highlight that as the GDP is linked to the housing market and the development of technology, these could be considered as subsystem within the category "Economy."

The Resilience to What: Housing Markets, Economy, and Environmental Impacts

This section compares threats in the categories previously used. The environmental category was added in this section since it is an impact and a threat. The category "Housing Markets" includes two rankings that are related to threats to real estate markets, namely, real estate bubbles and housing affordability. The section "Economies and finances at risk" uses city rankings of financial risk and GDP at risk to see if the top ten cities in the SRCI are also the riskiest. The environmental impact is compared through the carbon footprint and CO_2 emissions.

The matrix in Table 4 shows that the most important threats are economic and financial. This is consistent with the subject of analysis,

How Cities Can Be Resilient, Table 3 The resilience of what: housing markets, economy, or technology?

Housing Markets		Economy		Technology	
CBRE's Most expensive housing markets[a]	Housing markets[b] average per square foot	Top-ten GDPs[c]	Global Financial Centres Index[d]	IMD-SUTD Smart City Index Report[e]	IESE Smart Cities Ranking[f]
Hong Kong	Hong Kong	Tokyo	New York	Singapore	London
Munich	Singapore	New York	London	Helsinki	New York
Singapore	Paris	Los Angeles	Shanghai	Zurich	Paris
Shanghai	Munich	Seoul	Tokyo	Auckland	Tokyo
Shenzhen	London	London	Hong Kong	Oslo	Reykjavík
Beijing	Shenzhen	Paris	Singapore	Copenhagen	Copenhagen
Vancouver	Shanghai	Osaka	Beijing	Geneva	Berlin
Los Angeles	Beijing	Chicago	San Francisco	Taipei	Amsterdam
Paris	Berlin	Moscow	Shenzhen	Amsterdam	Singapore
New York	Milan	Shanghai	Zurich	New York	Hong Kong
0.30	0.20	0.60	0.40	0.10	0.40
0.25		0.50		0.25	

Cells in gray are matching cities with SRCI
[a] CBRE Research (2021)
[b] CBRE Group & Tatler Asia (2021)
[c] World Population Review (2021a)
[d] Morris et al. (2020)
[e] IMD & SUTD (2020)
[f] IESE Business School (2020)

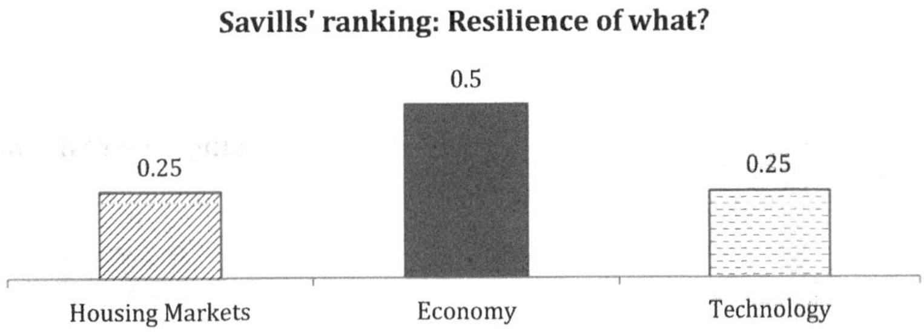

How Cities Can Be Resilient, Fig. 2 The resilience of what? (SRCI)

the resilience of the economy. Therefore it makes sense that the most important threats are risks to the economy. When cities have large GDPs they also have more at stake and therefore more to lose, factors that increase their risk. Figure 3 shows that environmental impact is also an equally important threat. Almost half of the top ten cities (0.4) within the SRCI match cities ranked within the top ten emitters. This is not a coincidence. The link between GDP and emissions has been documented before. The more you have the more you have to lose and the more you can impact on the planet. The need for resilience increases with the GDP of the city and its impact on the planet. A statistical analysis of the relationship between threats in housing markets, economies, and environmental impact deserves to be the subject of further research. Figure 3 suggests that

How Cities Can Be Resilient, Table 4 The resilience to what?: housing market, economies, finances, and environmental impact

Housing Markets		Economies and finances		Environmental impact	
UBS' Global Real Estate Bubble Index[a]	Demographia Housing Affordability 2021[b]	Lloyds' City risk Index[c] (Financial and Economic)	Cambridge Global Risk Index[d] (by GDP at risk)	Carbon footprint by Urban Cluster[e]	CO_2 Emissions[f]
Munich	Hong Kong	New York	Tokyo	Seoul	Tokyo
Frankfurt	Vancouver	Sao Paulo	New York	Guangzhou	Incheon
Toronto	Sydney	Moscow	Manila	New York	Kaohsiung
Hong Kong	Auckland	Tokyo	Istanbul	Hong Kong	New York
Paris	Toronto	Istanbul	Taipei	Los Angeles	Seoul
Amsterdam	Melbourne	Mexico City	Osaka	Shanghai	Hong Kong
Zurich	San Jose, CA	Madrid	Los Angeles	Singapore	London
Vancouver	San Francisco	Los Angeles	Baghdad	Chicago	Taiyuan
London	Honolulu	Buenos Aires	London	Tokyo	Rotterdam
Tokyo	Los Angeles	London	Shanghai	Riyadh	Houston
0.30	0.20	0.40	0.40	0.40	0.40
0.25		0.40		0.40	

Cells in gray are matching cities with SRCI
[a]Holzhey and Skoczek (2020)
[b]Cox (2021)
[c]Cambridge Centre for Risk Studies (2019)
[d]Cambridge Centre for Risk Studies (2018)
[e]Moran et al. (2018)
[f]Nangini et al. (2019)

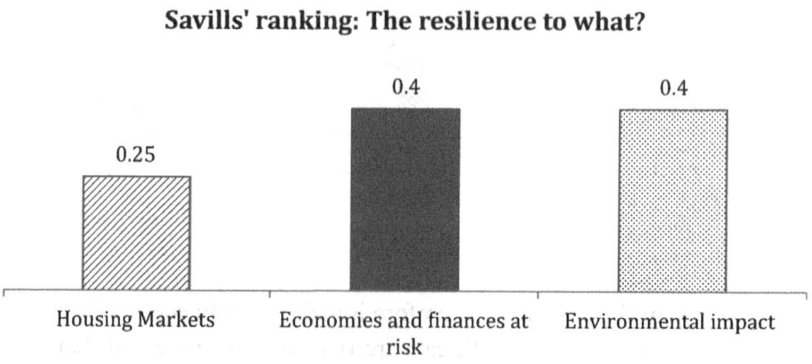

How Cities Can Be Resilient, Fig. 3 The resilience to what? (SRCI)

the resilience of the economy of a city (what) should include more than the analysis of threats to the economy of a city (to what) and acknowledge the equally important impact of environmental issues.

The Resilience for Whom: Population or Wealthy Population?

This section tries to unfold whose resilience is at play in the top ten cities of the SRCI. The

How Cities Can Be Resilient, Table 5 The resilience for whom?: population and wealthy population

Resilience for population		Resilience for wealthy population	
Cities by population size[a]	%GDP at Risk[b]	Top ten GDPs[c]	Cities by wealthy population[d]
Tokyo	Manila	Tokyo	New York
Delhi	Rosario	New York	Tokyo
Shanghai	Taipei	Los Angeles	Hong Kong
Sao Paulo	Xiamen	San Francisco	Los Angeles
Mexico City	Kabul	Paris	London
Dhaka	Port au Prince	Osaka	Paris
Cairo	Kathmandu	Chicago	Chicago
Beijing	Santo	London	San Francisco
Mumbai	Ningbo	Dallas	Washington DC
Osaka	Hangzhou	Shanghai	Dallas
0.10	0.00	0.70	0.80
0.05		0.75	

Cells in gray are matching cities with SRCI
[a]World Population Review (2021b)
[b]Cambridge Centre for Risk Studies (2015)
[c]World Population Review (2021a)
[d]Imberg and Shaban (2020)

approach is very general, but it tries to show in a polarized way whose population benefits from the resilience of economies (what) to financial and economic threats (to what). The category "Resilience for population" uses two top ten rankings: one related to population and the other related to percentage of GDP at risk (see Table 5). The "Cities population" is used to check if the most resilient cities according to SRCI contain some of the biggest agglomerations of population. The percentage of GDP at risk describes how important the GDP that one city has is at risk when compared with its total GDP. Cities in developing countries could have smaller GDPs than cities in developing countries. However, cities in developing countries can have bigger percentage of their total GDP at risk. Therefore their GDPs can be smaller, but in comparison, they have much more to lose. Moreover the assets at risk in cities in developing countries tend to be cheaper than cities in developed countries. However when a family loses their house, it means everything for them regardless of the price of the house. In the category "Resilience for wealthy population," two rankings are used: top ten cities' GDPs per capita and cities by wealthy population. In this way both extremes are represented, the population that has more to lose against the wealthiest population.

Table 5 shows that the most similar ranking to the SRCI are the rankings related to wealth and population. None of the top ten cities in the SRCI matches any of the top ten cities that have more to lose in terms of percentage of GDP at risk (see % GDP at risk in Table 5). Figure 4 shows that the differences between the two categories are roughly eight to one. The number of matching cities between the top ten cities in the SRCI and the wealthy rankings is the highest (seven and eight cities out of ten) for all the rankings compared so far. However, this result is coherent with the identity of the SRCI ranking that is about cities that attract and concentrate personal wealth. After all resilience is a tool for analyzing change and not necessarily a referent for moral and ethics goals. This point has been discussed and highlighted as a concrete difference between resilience and sustainability, with the latter having concrete goals that imply ethical decisions.

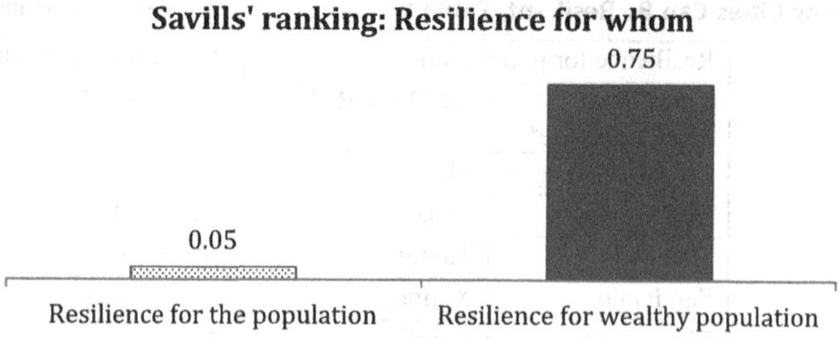

How Cities Can Be Resilient, Fig. 4 Resilience for whom? (SRCI)

How Cities Can Be Resilient, Table 6 Resilient ranking derived from EIU's report

Rank	Cities	Damage	Assets	Preparation	Average
1	Singapore	97.5	100	99.2	98.90
2	Tokyo	94.6	98.1	98.9	97.20
3	Chicago	94.1	95.6	98.4	96.03
4	Osaka	96.4	92.4	98.4	95.73
5	Amsterdam	96.7	90.2	99.2	95.37
6	San Francisco	94.1	92.6	98.4	95.03
7	Toronto	93.6	92.4	98.9	94.97
8	Dallas	93.7	92	98.4	94.70
9	Melbourne	92.9	92.4	97.9	94.40
10	Sydney	92.9	92.4	97.9	94.40

Case Study 2: Resilience Cities from the Economist Intelligence Unit (EIU)

The Safe Cities Index is a report produced by the Economist Intelligence Unit, which is sponsored by NEC Corporation (Nippon Electric Company). Since 2017 the index has ranked 60 cities across 57 indicators. In the 2019 version of the Safe Cities Index the top ten safest cities were Tokyo, Singapore, Osaka, Amsterdam, Sydney, Toronto, Washington DC, Copenhagen, and Seoul. In the 2019 version it also contains a section dedicated to resilience indicators. The report proposes, "Resilience is about avoiding, mitigating or responding to potential shocks"(The EIU 2019: 34), and in the case of cities, resilience is inseparable from safety (p43). Shocks include natural and cultural hazards, like natural disasters, cyber attacks, and terrorism. The authors of the report recognized the challenge of deciding on what the threat is that they need to prepare for. They emphasize that the assets have to be flexible but more details about which types of assets are included and what flexible means in the context is not clear. The main idea is to understand resilience as a synonym of being ready and prepared to deal with shocks. The report does not contain a ranking of resilient cities but presents scores across different categories. There are three categories: damage and threat multipliers, which refers to risk; relevant assets, which relates to infrastructure and money; and preparation, which alludes to planning. By averaging scores across the three dimensions for the 60 cities analyzed, it is possible to rank them and select the top 20 cities. Table 6 shows a ranking of cities by averaging the scores across the different categories analyzed in the report.

The analysis suggests that in order to be resilient cities needs to have a good infrastructure to deal with earthquakes and floods and to have planning policies to coordinate responses. An interesting point is the report highlights the relevance of transparency for planning but the absence of sustainability measurements. The report suggests that resilience does not necessarily have to be costly. However, they show that

How Cities Can Be Resilient, Table 7 The resilience to what? Population and economic losses

Population potentially affected			Economic losses	
Metro Area by population potentially affected by storm[a]	Population affected by all risks[a]	Top ten cities by population affected (earthquake)[a]	Top ten cities by Expected loss (Flooding)[b]	Top ten cities by expected loss (Earthquake)[b]
Pearl-River Delta	Tokyo	Tokyo	Sao Paulo	Lima
Tokyo	Manila	Jakarta	London	Tianjin
Manila	Pearl-River Delta	Manila	Osaka	Tehran
Osaka	Osaka	Los Angeles	Paris	Istanbul
Taipei	Jakarta	Osaka	Delhi	Los Angeles
Shantou	Nagoya	Teheran	Los Angeles	Taipei
Nagoya	Kolkata	Nagoya	Chicago	Beijing
Mumbai	Shanghai	Lima	New York	Chengdu
Chennai	Los Angeles	Taipei	San Francisco	San Francisco
Tainan-Kaohsiung	Tehran	Istanbul	Singapore	Jakarta
0.2	0.2	0.2	0.4	0.1
0.2			0.25	

[a]Sundermann et al. (2013)
[b]Cambridge Centre for Risk Studies (2015)

wealthy countries tend to score higher than poor countries in terms of safety and transparency, while poor countries are more exposed to hazards and corruption. Unfortunately they forget to compare the environmental impact of wealthy countries. Perhaps the report assumes that cities do not need to be sustainable to be resilient. The following section analyzes and compares the top ten cities in the assembled EIU ranking against a range of rankings ordered in categories following the structure of the resilience of what, to what, for whom.

The Resilience of What: People or Money?

The potential impact of natural hazards is usually assessed in rankings through the potential people affected and/or potential economic losses. The present section tries to identify whether the EIU ranking is closer to cities that rank in the top ten of populations potentially affected by natural hazards or with rankings whose cities are in the top ten for potential economic losses. In the category "Population potentially affected" three ranking were used (see Table 7). Two of them rank cities whose populations can be affected by specific threats, namely storms and earthquake. The remaining ranking in the category ranks cities by population exposed to multiple natural hazards. In the category "Economic losses" two ranking are included: one that ranks cities by expected loss due to flooding and the other which ranks cities by expected loss due to earthquakes. The main goal is to see how exposed to hazards are the top ten cities in the EIU ranking and whether the ranking favors the resilience of the population affected or expected losses.

Table 7 shows that the matching cities between the EIU ranking and the rankings in the category of "Population affected by hazards" are very consistent with two matching cities (out of ten) per ranking, giving an average score of 0.2. However, differences between rankings in the category "Economic losses" are more evident. If all rankings are compared, the ranking of "Cities by expected losses" due to flooding is the most similar to the EIU's resilience ranking with four cities matching (out of ten). In the other four rankings

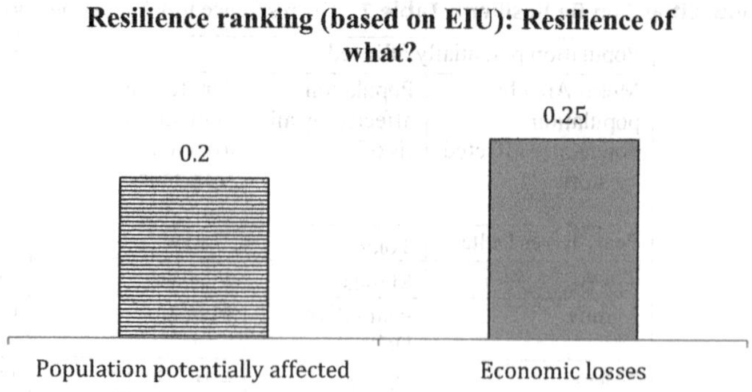

How Cities Can Be Resilient, Fig. 5 Resilience of what? (EIU)

the representation of the top ten cities included in the EIU ranking is very poor. One of the biases with the information included in rankings of population affected is that it doesn't cover small cities from small countries that can cease to exist. The other point to consider is related to economic losses. Economic losses in cities from developing and developed countries should be considered separately since they can have different impacts on the economy.

Figure 5 shows that values of average matching cities are low but very similar between categories. These can have different interpretations. Categories chosen might not be relevant to testing or unfolding the intentions behind the EIU's ranking. Alternatively, by averaging scores from each ranking, differences are hidden. The low percentage of matching cities could also be related to the lack of exposure of cities in the top ten EIU ranking to single or multiple natural hazards. This could be the case if a ranking is particularly focused on social and political hazards, like wars, terrorism, and cyber attacks. The question arising is can cities be ranked as the most resilient to shocks without being the cities most threatened by natural hazards?

The Resilience to What: General or Specific Threats?

This section digs deeper into the relationship between threats and resilient cities as ranked in the EIU ranking. The threats considered in the rankings selected are natural hazards since they are susceptible to increase due to climate change, and the risks these pose to cities have been documented. Following the resilience framework proposed by Walker and Salt (2012) two categories are compared, the resilience to general and specific threats. The resilience to general threats uses rankings that focus on people or assets potentially exposed to multiple natural hazards. This means that cities considered in the top ten of these rankings can be the most threatened by a number of natural hazards. The Verisk Maplecroft Report (2021) predominantly ranked cities from India and Indonesia in the top ten. Nonetheless, the report assessed multiple hazards across many countries but tried to warn firms interested in investing in India and South East Asia that natural hazards and climate change could create serious risks. The resilience to specific threats includes rankings of port cities only and assesses the exposed assets to one specific threat, flooding, since this is the variable that affects more people and assets worldwide and that also occurs more frequently. By having the two categories, links with specific and general resilience can be checked in the EIU ranking.

Table 8 shows that the general and specific threats categories are poorly represented in the EIU's resilience ranking with an average of 1 match per ranking within and across categories. The majority of the most threatened cities in the world are in developing countries. Figure 6 shows that the top ten cities in the EIU ranking are slightly more included in rankings related to general than specific threats (0.13 against 0.1). Without considering the Verisk Maplecroft

How Cities Can Be Resilient, Table 8 Resilience to what? General and specific threats

Resilience to general threats			Resilience to specific threats	
People potentially affected [a]	Cities vulnerable to at least 3 natural hazards [b]	Risk from environmental hazards [c]	Top ten port cities by asset exposure [d]	Flood average annual losses [e]
Tokyo	Tokyo	Jakarta	Miami	Guangzhou
Manila	Mexico City	Delhi	New York	Miami
Pearl-River Delta	Osaka	Chennai	New Orleans	New York
Osaka	Dhaka	Surabaya	Osaka	New Orleans
Jakarta	Karachi	Chandigarh	Guangzhou	Mumbai
Nagoya	Kolkata	Agra	Shanghai	Nagoya
Kolkata	Istanbul	Meerut	Mumbai	Tampa
Shanghai	Manila	Bandung	Kolkata	Boston
Los Angeles	Los Angeles	Aligarh	Alexandria	Shenzhen
Tehran	Tianjin	Kanpur	Ho Chi Minh City	Osaka
0.2	0.2	0	0.1	0.1
0.13			0.1	

[a]Sundermann et al. (2013)
[b]Gu et al. (2015)
[c]Choudhury (2021)
[d]Nicholls et al. (2008)
[e]Hallegatte et al. (2013)

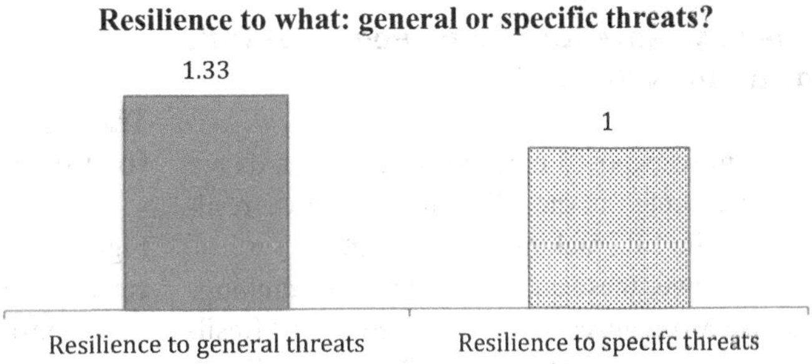

How Cities Can Be Resilient, Fig. 6 Resilience to what? (EIU)

Report (2021), whose top ten is dominated by Indian and Indonesian cities, the average of matching cities within the category general threats could be double that of specific threats. Even in this case the similarities between the top ten cities in the EIU ranking and general threats are still low. As mentioned in the previous section, a low average match between resilient cities and threats could be biased by the types of hazard taken into consideration (social or natural). Perhaps natural hazards were not even considered by the EIU. In any case, the mismatch between ranking of the most threatened cities in the world and cities ranked as the most resilient in the world is significant. In the present comparison, the cities ranked as the most resilient in the EIU ranking do not include the majority of the cities under multiple or specific threats. Why should cities be pushed to be resilient if the most resilient cities in the world are not being harassed by any important threats? How can cities be ranked as the most resilient in the

How Cities Can Be Resilient, Table 9 Resilience for whom? Population at risk and wealthy population

Resilience for the population at risk		Resilience for the wealthy population	
Metro Area by population potentially affected by storms[a]	People potentially affected by all risks[a]	Top-ten GDPs[b]	Cities by wealthy population[c]
Pearl-RiverDelta	Tokyo	Tokyo	New York
Tokyo	Manila	New York	Tokyo
Manila	Pearl-RiverDelta	Los Angeles	Hong Kong
Osaka	Osaka	San Francisco	Los Angeles
Taipei	Jakarta	Paris	London
Shantou	Nagoya	Osaka	Paris
Nagoya	Kolkata	Chicago	Chicago
Mumbai	Shanghai	London	San Francisco
Chennai	Los Angeles	Dallas	Washington DC
Tainan-Kaohsiung	Tehran	Shanghai	Dallas
0.2	0.2	0.5	0.4
0.20		0.45	

[a]Sundermann et al. (2013)
[b]World Population Review (2021a)
[c]Imberg and Shaban (2020)

world if they are not the most troubled in the world?

The Resilience for Whom: People at Risk or Wealthy People?

If the most exposed cities to natural hazards are not represented in the EIU ranking, whose resilience is the EIU ranking considering?

To answer these questions a group of rankings has been organized into two categories: "Resilience for the population at risk" and "Resilience for the wealthy population" (see Table 9). In the first category, the goal is to see the extent to which cities ranked in the top ten of the largest population at risk are also in the top ten rankings of the EIU. The other category uses top ten cities ranked by GDP per capita and wealthy population.

Table 9 shows that the majority of the cities with the largest populations at risk (to single or multiple natural hazards), with the exception of Osaka, Tokyo, and Los Angeles, are mostly in developing countries. The wealthiest cities are all in developed countries. Figure 7 shows that cities in the category "Resilience for the wealthy population" match, on average, almost half of the top ten resilient cities in the EIU ranking, and it is double the average matching cities in rankings related to people at risk.

The Resilience of What, to What, for Whom in SRCI and EIU Rankings

Figure 8 shows the highest scores in the three categories analyzed (resilience of what, to what, for whom) for the SRCI. The average across categories is around 0.5. The highest matching averages between rankings are found in the resilience for whom, represented by the wealthy population. Looking at the highest scores in each category is possible to reassemble the resilience questions of what, to what, and for whom. The top ten cities in the SRCI match closer with rankings that highlight the resilience of the economy of cities (of what) to financial risks and environmental impacts (to what) for wealthy populations (for whom).

Figure 9 shows the highest average scores in each category for the analysis of the top ten cities in the EIU ranking. The resilience for whom,

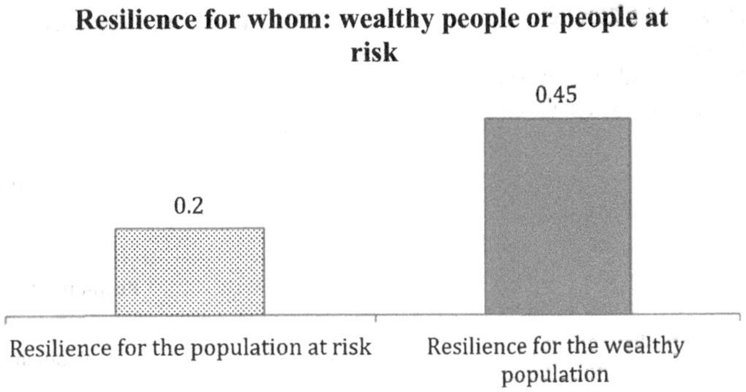

How Cities Can Be Resilient, Fig. 7 Resilience for whom? (EIU)

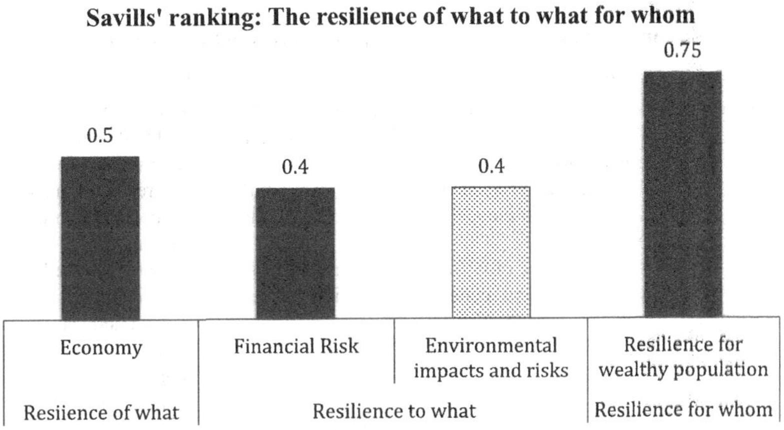

How Cities Can Be Resilient, Fig. 8 The resilience of what, to what, for whom (SRCI)

illustrated by the resilience for the wealthy, is the highest average score (0.45) across all the categories. The economic losses due to natural hazards and the exposure to general hazards represent the resilience of what, to what. The top ten resilient cities in the EIU ranking could be framed as the resilience of economic losses (of what) to general threats (to what) for wealthy populations (for whom).

Table 10 presents a comparison between the SRCI and the EIU rankings. It shows that both rankings are associated to the economy of the cities ranked and their economic challenges.

Discussion: Resilience and Wealth

Figure 10 presents a summary of the highest scores in the categories related to the resilience of what, to what, and for whom for both rankings, SRCI and EIU. Figure 10 shows that magnitudes increase toward the left side of the chart, the resilience of whom, which is characterized in both cases by the resilience for the wealthy population. Wealth is the most sensitive category. If cities want to be in the top ten of most resilient cities, they must have an important population of wealthy people and a robust economy. Wealth is the precondition for the resilience of cities. Without wealth, the chance to be ranked in the top ten of resilient cities is reduced because the resilience for whom is more significant than the resilience of what to what. The comparison between rankings permits the highlighting of the importance of the resilience for whom, something usually obliterated by the analysis of threats or the system harassed by them.

The popularity of resilience cities will keep on growing because of the competition between cities. The study of rankings is helpful in unveiling key messages contained in the perception of resilience by firms and institutions. This analysis permits an understanding of the expectations of institutions and firms about resilient cities.

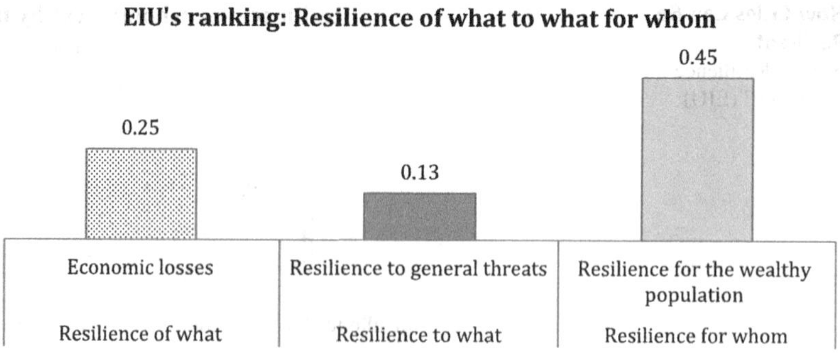

How Cities Can Be Resilient, Fig. 9 Resilience of what, to what, for whom (EIU)

How Cities Can Be Resilient, Table 10 Comparison between SRCI and EIU rankings

Rankings	Savills	EIU
Differences	Resilient cities are cities attractive to investors. Resilience to social hazards. Resilience of real estate markets to threats. Cities ranked as resilient tend to be more similar to GDP city rankings than cities with a real estate bubble. The "resilience of what, to what, for whom" is very specific and related to economies, financial risks, and wealth.	Resilience is associated with safety, being ready to deal with social and natural hazards. Social and natural hazards are considered. The resilience analyzed is, in theory, the general resilience with multiple factors measured in each ranking. The cities ranked as resilient match poorly (10%) with the cities ranked as highly threatened by general and specific natural hazards. Rankings related to wealth and economic losses are more similar to the resilience ranking than rankings related to general and specific threats
Similarities	The definition of resilience is linked to an engineering approach. The "resilience of what" is predominantly related to the economy and economic losses. The economic impact is more important than losing people. The "resilience for whom" tends to match rankings related to wealth. Sustainability is not measured or assessed.	

Figure 10 clearly shows that the goal of resilient cities is to protect the economy, finance, and the wealth of the population that owns this.

The analysis showed that the richest cities are also ranked in the top ten cities in terms of emissions and carbon footprint. Rich cities will need to decide whether to keep on competing to be the most resilient or to reduce their impacts on the planet. If they choose to keep on growing economically to protect the wealth of a few, risks and the need for more resilience will also keep on growing. Framed in this way, the resilience represented by these rankings is a measurement of a maladaptive cycle. Cities compete to grow their economies to be more prepared and more resilient. In doing so they increase their impact on the environment and the exposure to threats, which in turn demands bigger investments and more economic growth. Getting richer and wealthier increases the chance for cities to be ranked as resilient. Therefore, if cities want to be ranked as resilient, they must find a way to get richer as soon as possible. If they are already rich, they must use resilience to protect their wealth and economy. After all, this is what makes them ranked in the top ten of resilient rankings.

Conclusion: How Cities Can Be Resilient

The analysis shows that comparison between rankings can be useful to highlight contradictions. The relationship between resilience and sustainability fits into this category. The first step toward

How Cities Can Be Resilient,
Fig. 10 Resilience and wealth

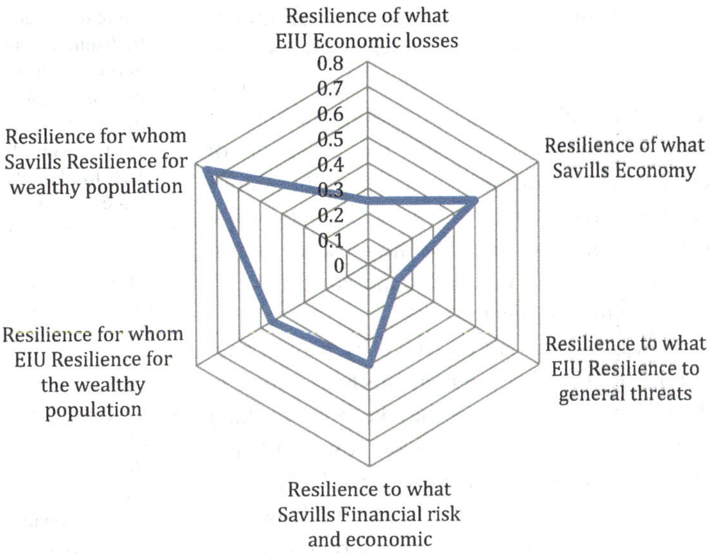

resilient cities should be avoiding the processes and drivers of change that can lead to a faster collapse (Garcia and Vale 2017). The focus of planning policies and urban design strategies should be about tackling the root of problems and avoiding creating bigger ones (Anique and Golousova 2021). The OECD proposed that resilience and sustainability should be linked. However, the only link between resilience and sustainability observed in the analysis presented in this chapter shows that the most resilient cities can also be ranked among the bigger emitters. One interpretation is to assume that the most resilient cities do not have to be the most sustainable. Another way of reading this result is by implying that some of the most resilient cities need to be extremely resilient because they keep on increasing their own problems through their emissions. This understanding can be used to make a division between resilience in developing and developed countries. In developed countries there is more budget to be resilient. Therefore, if a city contributes to exacerbating the problems they are trying to battle, the need for more investments in resilience needs to follow. More problems mean bigger investments in infrastructure, which contributes to increasing the GDP of a country. As shown in the analysis, the top ten cities by GDP per capita are very similar to the most resilient cities. However, cities in developing countries will need to enhance their resilience to overcome threats exacerbated by the impact of wealthy cities on the environment plus their own vulnerabilities. Therefore, from this point of view, cities have more chance to become resilient if they are not sustainable but wealthy.

The comparative analysis between rankings is also useful for challenging and unveiling the subject of analysis of the resilience ranked. When analyzing the resilience of what to what, a theme that emerges is that the resilience to social hazards, especially economic problems, is more important for cities than natural hazards. Even though natural hazards can kill more people, this is less relevant for the resilience rankings than the money that a city can lose. For these rankings, protecting money is more relevant than protecting citizens. The analysis of resilient rankings permits revealing a bold message to institutions and governments of cities: only wealthy cities with rich people can be resilient.

Another way of looking at the result of the resilience rankings is by analyzing the discrimination behind them. The analysis of the resilient rankings does not show much hope for cities in poor countries. The assumption derived from the ranking is that to recover from shocks, cities need two things: money and a robust infrastructure, which means more money. Poor cities do not have these resources, and they will not have them in the near future. They are doomed and relegated from the rankings, which might impact on their prestige and their capacity to compete against other cities. Very few cities ranked as resilient are also considered within the top ten most hazardous cities. With the exception of Osaka and Tokyo, the top ten most hazardous cities are predominantly in developing countries. How can cities be ranked as the most resilient if they are not even the most exposed to threats? What are they resilient to?

Poor cities that are pressured by natural hazards have none of the requirements to be ranked in the resilient rankings, but they keep on working and surviving. These cities have low GDPs per capita, deprived infrastructure, volatile finances, and economies with hard limitations. However, they find ways to keep on buffering multiple threats with scarce resources. Is this not the definition of resilience?

Perhaps rankings of resilience measure cities that would like to be more resilient while the most resilient cities of the world just keep on working without the need of being ranked.

References

Acuto, M., Pejic, D., & Briggs, J. (2021a). Whose city benchmarks? the role of the critical urbanist in comparative urban measuring. *International Journal of Urban and Regional Research, 45*(2), 389–392.

Acuto, M., Pejic, D., & Briggs, J. (2021b). Taking city rankings seriously: Engaging with benchmarking practices in global urbanism. *International Journal of Urban and Regional Research, 45*(2), 363–377.

Aldunce, P., Beilin, R., Handmer, J., & Howden, M. (2014). Framing disaster resilience: The implications of the diverse conceptualisations of "bouncing back". *Disaster Prevention and Management, 23*(3), 252–270.

Anique, J., & Golousova, E. S. (2021). Evaluation of challenges faced by urban planners. A study of ecological construction and sustainability in built environment. *Весенние дни науки.—Екатеринбург, 2021*, 692–699.

Cambridge Centre for Risk Studies. (2015). *World cities risk 2015–2025.* Retrieved October 11, 2021 from https://www.jbs.cam.ac.uk/wp-content/uploads/2020/08/crs-lloydsworldcities-pt1-overviewresults.pdf

Cambridge Centre for Risk Studies. (2018). *Lloyd's city risk index executive summary.* Retrieved October 11, 2021 from https://cityriskindex.lloyds.com/wp-content/uploads/2018/06/Lloyds_CRI2018_executive%20summary.pdf

Cambridge Centre for Risk Studies. (2019). *Lloyd's city risk index global analysis of finance, economic and trade risks*. Retrieved October 11, 2021from https://cityriskindex.lloyds.com/wp-content/uploads/2019/12/Lloyds_city-risk-index_finance-economic-and-trade-risks_v3.pdf

Carpenter, S., Walker, B., Anderies, J. M., & Abel, N. (2001). From metaphor to measurement: Resilience of what to what? *Ecosystems, 4*(8), 765–781.

Carrera Portugal, A. (2019). The role of city rankings in local public policy design: Urban competitiveness and economic press. *Global Media and China, 4*(2), 162–178.

CBRE Group, & Tatler Asia. (2021). *Most expensive residential property markets worldwide in 2020.* Retrieved October 7, 2021 from https://www.statista.com/statistics/1040698/most-expensive-property-markets-worldwide/

CBRE Research. (2021). *Global living 2020.* Retrieved October 7, 2021 from https://www.cbreresidential.com/uk/sites/uk-residential/files/CBRE-Global%20Living_2020_Final.pdf

Chancel, L., & Piketty, T. (2015). *Carbon and inequality: From Kyoto to Paris Trends in the global inequality of carbon emissions (1998–2013) & prospects for an equitable adaptation fund.* World Inequality Lab. Retrieved October 23, 2021 from https://wid.world/document/chancel-l-piketty-t-carbon-and-inequality-from-kyoto-to-paris-wid-world-working-paper-2015-7/

Chicca, F., Vale, B., & Vale, R. (Eds.). (2018). *Everyday lifestyles and sustainability: The environmental impact of doing the same things differently.* London: Routledge.

Choudhury, S. (2021). *Asia accounts for 99 out of 100 top cities facing the biggest environmental risks.* Retrieved from October 7, 2021 https://www.cnbc.com/2021/05/14/asian-cities-face-the-biggest-environmental-risks-study.html

Cimellaro, G. P. (2016). *Urban resilience for emergency response and recovery: Fundamental concepts and applications.* Cham: Springer International Publishing AG.

Cox, W. (2021). *2021 Demographia international housing affordability.* Retrieved October 11, 2021 from http://www.demographia.com/dhi.pdf

Cutter, S. L. (2020). Community resilience, natural hazards, and climate change: Is the present a prologue to the future? *Norwegian Journal of Geography, 74*(3), 200–208.

Dawson, A. (2017). *Extreme cities: The peril and promise of urban life in the age of climate change.* London/New York: Verso.

Economist Media Group. (2021). NEC: The safe cities index 2019. Retrieved 16 December, 2021 from https://thoughtthatcounts.economist.com/showcase/nec-safe-cities-index

Emdad Haque, C., & Etkin, D. (2007). People and community as constituent parts of hazards: The significance of societal dimensions in hazards analysis: People, community and resilience: Societal dimensions of environmental hazards. *Natural Hazards (Dordrecht), 41*(2), 271–282.

Erdi, P. (2019). *Ranking: The unwritten rules of the social game we all play*. New York: Oxford University Press.

Fan, L. (2013). *Disaster as opportunity? Building back better in Aceh, Myanmar and Haiti*. Humanitarian Policy Group, Overseas Development Institute.

Garcia, E. (2017). Between grey and green: Ecological resilience in urban landscapes. *Landscape Review, 17*(2).

Garcia, E. J. (2020). Ecological resilience and the built environment. In R. C. Brears (Ed.), *The Palgrave handbook of climate resilient societies* (pp. 1–19). Cham: Springer International Publishing.

Garcia, E. J., & Vale, B. (2017). *Unravelling sustainability and resilience in the built environment*. London/New York: Routledge.

Garcia, E. J., Muminović, M., Vale, B., & Radović, D. (2014). The role of green spaces for the resilience of a city. In *New urban configurations* (pp. 865–871). Amsterdam: IOS Press.

Garcia, E., Vale, B., & Vale, R. (2021). *Collapsing gracefully: Making a built environment that is fit for the future*. Cham: Springer Nature Switzerland.

Gautam, D. (2017). Assessment of social vulnerability to natural hazards in Nepal. *Natural Hazards and Earth System Sciences, 17*(12), 2313–2320.

Giffinger, R., Haindlmaier, G., & Kramar, H. (2010). The role of rankings in growing city competition. *Urban Research & Practice, 3*(3), 299–312.

Gu, D., Gerland, P., Pelletier, F., & Cohen, B. (2015). *Risks of exposure and vulnerability to natural disasters at the city level: A global overview*. Retrieved October 10, 2021 from https://population.un.org/wup/Publications/Files/WUP2014-TechnicalPaper-NaturalDisaster.pdf

Gunderson, L. (2000). Ecological resilience and application. *Annual Review of Ecology and Systematics, 31*, 425.

Gunderson, L. H., & Holling, C. S. (2002). *Panarchy: Understanding transformations in human and natural systems*. Washington, DC: Island Press.

Hallegatte, S., Green, C., Nicholls, R. J., & Corfee-Morlot, J. (2013). Future flood losses in major coastal cities. *Nature Climate Change, 3*(9), 802–806.

Harris, J. A., Denyer, D., Harwood, S., Braithwaite, G., Jude, S., & Jeffrey, P. (2020). Time to invest in global resilience. *Nature (London), 583*(7814), 30.

Holling, C. S. (1973). Resilience and stability of ecological systems. *Annual Review of Ecology and Systematics, 4*, 1–23.

Holzhey, M., & Skoczek, M. (2020). *USB global real estate bubble index*. Retrieved October 11, 2021 from https://www.ubs.com/content/dam/static/noindex/wealth-management/ubs-global-real-estate-bubble-index-en-1509323.pdf

Hutter, G., Kuhlicke, C., Glade, T., & Felgentreff, C. (2013). Social resilience in hazard research and planning. *Natural Hazards (Dordrecht), 67*(1).

IESE Business School. (2020). *IESE cities in motion index 2020*. Retrieved October 10, 2021 from https://media.iese.edu/research/pdfs/ST-0542-E.pdf

Imberg, M., & Shaban, N. (2020). *A decade of wealth. A review of the past 10 years in wealth and a look forward to the decade to come*. Retrieved October 10, 2021 from https://www.wealthx.com/wp-content/uploads/2020/05/Wealth-X_A-Decade-of-Wealth_May-2020.pdf

IMD, & SUTD. (2020). *Smart cities index 2020*. Retrieved October 7, 2021 from https://www.imd.org/globalassets/wcc/docs/smart_city/smartcityindex_2020.pdf

Lizarralde, G., Valladares, A., Olivera, A., Bornstein, L., Gould, K., & Barenstein, J. D. (2015). A systems approach to resilience in the built environment: The case of Cuba. *Disasters, 39*, s76–s95.

McGrath, B., & Lei, D. (2021). The embodied multisystemic resilience of architecture and built form. In *Multisystemic resilience* (pp. 603–624). Oxford: Oxford University Press.

McManus, P. (2012). Measuring Urban Sustainability: the potential and pitfalls of city rankings. *Australian Geographer 43*(4), 411–424. https://doi.org/10.1080/00049182.2012.731301

Meerow, S., & Newell, P. (2019). Urban resilience for whom, what, when, where, and why? *Urban Geography, 40*(3), 309–329.

Meyer, V., Becker, N., Markantonis, V., Schwarze, R., Bergh, J. C. J. M., Bouwer, L. M., et al. (2013). Review article: Assessing the costs of natural hazards – state of the art and knowledge gaps. *Natural Hazards and Earth System Sciences, 13*(5), 1351.

Milanović Pešić, A. (2020). Natural hazards: Interpretations, types, and risk assessment. In W. Leal Filho, A. M. Azul, L. Brandli, P. G. Özuyar, & T. Wall (Eds.), *Climate action. Encyclopedia of the UN sustainable development goals*. Cham: Springer.

Moran, D., Kanemoto, K., Jiborn, M., Wood, R., Többen, J., & Seto, K. C. (2018). *Global gridded model of carbon footprints (GGMCF)*. Retrieved October 11, 2021 from http://citycarbonfootprints.info/

Morris, H., Mainelli, M., & Wardle, M. (2020). *The global financial centres index 28*. Long Finance: Financial Centre Futures. Retrieved October 11, 2021 from https://www.longfinance.net/media/documents/GFCI_28_Full_Report_2020.09.25_v1.1.pdf

Muminović, M., Garcia, E. J., Vale, B., & Radović, D. (2014). Resilient assemblages: The complex identity of Nezu in Tokyo. In *New urban configurations* (pp. 807–813). Amsterdam: IOS Press.

Nangini, C., Peregon, A., Ciais, P., Weddige, U., Vogel, F., Wang, J., et al. (2019). A global dataset of CO2

H

emissions and ancillary data related to emissions for 343 cities. *Scientific Data, 6*(1), 180280.
Nicholls, R., Hanson, S., Herweijer, C., Hallegatte, S., Corfee-Morlot, J., Chateau, J., & Muir-Wood, R. (2008). *Ranking port cities with high exposure and vulnerability to climate extremes: Exposure estimates.* Retrieved October 11, 2021 from https://www.oecd-ilibrary.org/docserver/011766488208.pdf?expires=1634512286&id=id&accname=guest&checksum=E9D09DD2F7B1C3B401BCB8F191F3949E
OECD. (2021). *Resilient cities.* Retrieved October 23, 2021 from https://www.oecd.org/cfe/regionaldevelopment/resilient-cities.htm
Pimm, S. (1984). The complexity and stability of ecosystems. *Nature, 307*(5949), 321–326.
Ritchie, H., & Roser, M. (2014). *Natural disasters.* Published online at OurWorldInData.org. Retrieved October 24, 2021 from: https://ourworldindata.org/natural-disasters
Roggema, R., Tillie, N., Keeffe, G., & Yan, W. (2021). Nature-based deployment strategies for multiple paces of change: The case of Oimachi, Japan. *Urban Planning, 6*(2), 143–161.
Rómice, O., Porta, S., & Feliciotti, A. (2020). *Masterplanning for change: Designing the resilient city.* London: Riba Publishing.
Savills. (2021). *Impacts the future of global real estate. evolve: How changing cities are shaping how we work and live our lives.* Retrieved October 7, 2021 from https://www.savills.com/impacts/Impacts3_pdfs/SavillsImpacts2021.pdf
Savills Resilient Cities 2021. (2021). Retrieved July 24, 2021 from https://www.savills.com/impacts/market-trends/resilience-ranked.html
Sundermann, L., Schelske, O., & Hausmann, P. (2013). *Mind the risk a global ranking of cities under threat from natural disasters.* Retrieved October 11, 2021 from https://www.swissre.com/dam/jcr:1609aced-968f-4faf-beeb-96e6a2969d79/Swiss_Re_Mind_the_risk.pdf
The EIU (The Economist Intelligence Unit). (2019). *Safe cities index 2019. Urban security and resilience in an interconnected world.* Retrieved October 11, 2021 from https://safecities.economist.com/wp-content/uploads/2019/08/Aug-5-ENG-NEC-Safe-Cities-2019-270x210-19-screen.pdf
UNCC (United Nations Climate Change). (2021). *The Paris agreement.* Retrieved October 23, 2021 from https://unfccc.int/process-and-meetings/the-paris-agreement/the-paris-agreement
Verisk Maplecroft. (2021). *Environmental risk outlook 2021.* Retrieved October 23, 2021 from https://www.maplecroft.com/insights/analysis/environmental-risk-outlook-2021/#report_form_container
Walker, B. H. & Salt D. (2012). Resilience practice building capacity to absorb disturbance and maintain function. Washington, DC: Island Press.
White, J. M., & Kitchin, R. (2021). For or against 'The business of benchmarking'? *International Journal of Urban and Regional Research, 45*(2), 385–388.
World Population Review. (2021a). *Richest city in the world.* Retrieved from October 7, 2021 https://worldpopulationreview.com/world-city-rankings/richest-city-in-the-world
World Population Review. (2021b). *World city populations 2021.* Retrieved from October 7, 2021 https://worldpopulationreview.com/world-cities?utm_medium=website&utm_source=archdaily.com
Ye, M., & Aldrich, D. P. (2021). How natural hazards impact the social environment for vulnerable groups: an empirical investigation in Japan. *Natural Hazards, 105*, 67–81.

How Cities Cooperate to Address Transnational Challenges

Lorenzo Kihlgren Grandi
City Diplomacy Lab, Columbia Global Centers | Paris, Paris, France

Definitions

City diplomacy – On the international scene, a growing number of cities worldwide are emerging as actors capable of designing and conducting their own strategies. City diplomacy represents the practice through which municipal governments enter into a wide range of international partnerships with their peers and other foreign actors to pursue local interests and universal values.

Transnational challenges – From climate change to migratory flows, from growing inequalities to pandemics: the primary challenges humanity faces today feature a dual nature, i.e., transnational in scope and overwhelmingly urban in impact. The first aspect indicates that the broadest possible multilateral collaboration is crucial to address these challenges effectively. The second aspect, inherently linked to the global trend of urbanization, implies that such multilateral cooperation necessitates the active involvement of municipalities.

Introduction: From Undergone Urban Internationalization to City Diplomacy

The COVID-19 pandemic represents the latest dramatic example of how our era is characterized by an unprecedented pervasiveness across human society of increasingly serious challenges. Challenges that find in urban communities their most stark expression.

This trend is undoubtedly linked to the fact that in the twenty-first century, the urban population has surpassed the rural population for the first time in human history (UNDESA 2019). In addition to this quantitative aspect, the root causes of these challenges are closely linked to the globally widespread contradictions of urban development. The driving role of urban centers, especially the larger ones, in their national economy (OECD 2019) matches with the concentration of the most evident and marked social inequalities and greenhouse gas emissions; the opportunities in terms of employment and rights that cities offer to migrants and asylum seekers coexist with the most prominent challenges in terms of integration and cohesion; while several municipal governments have proven to be effective in fostering a climate of innovation, urban residents experience the most deleterious effects of the impact of digital technologies on sectors such as the real estate market and the labor market (Habitat III 2015; UN-Habitat. 2016; UNHCR 2020). Finally, while it is in cities that health facilities are concentrated, it is within their boundaries that 90% of COVID-19 infections have been recorded, due essentially to the promiscuity and close contacts typical of urban space, and in particular of the lower-income neighborhoods (OECD 2020; United Nations 2020).

Awareness of the close link between urbanity and major transnational challenges is leading an increasing number of municipal governments to develop strategies to move from passively undergoing these dynamics to equipping themselves with tools to manage them. Paralleling the "glocal" nature of these challenges – i.e., simultaneously local in impact and global in diffusion – municipal governments have developed a wide range of ways to join forces internationally to define and implement strategies to address these issues locally.

This practice is known as city diplomacy. Such a term refers precisely to the ability of cities to forge international partnerships to strengthen their capacity to pursue local goals and promote universal values on a regional and planetary scale (Kihlgren Grandi 2020a).

Origin and Evolution of City Diplomacy

The ability of cities, *de jure* or *de facto*, to draw up diplomatic strategies autonomously from the foreign policy of their own nations represents one of the primary evolutions of the international relations system in force. If the complex and stratified system of the European Middle Age had allowed various cities to establish advantageous and lasting partnerships, the latter had been made impossible by the reform of the international system implemented by the Peace of Westphalia in 1648. In this context, international relations were attributed exclusively to countries. This quickly led to the exclusion of cities from international relations, marking the end of what can be called the forerunner of modern city networks, namely the Hanseatic League (Take 2017).

The return of cities to international relations found suitable conditions only 275 years later. In 1913 the "International Congress of the Art of Building Cities and Organising Community Life" was held in Ghent, then the host of the International Exhibition. In this congress took place the foundation of modern city diplomacy.

Welcoming his colleagues from around the world, the mayor of the Flemish city Emile Braun called on them to collaborate internationally to "deliberate on the major problems that arise from the universal nature of the conditions of present-day life, which are more or less the same everywhere" (UCLG 2013). As a result, the congress led to the creation of the International Union of Cities (renamed in 1928 as International Union of Local Authorities-IULA), the first global city network, whose aim was precisely to encourage technical cooperation between city administrations to respond to shared challenges.

This vocation for problem-solving finds its potential in two characteristics shared by all municipalities in the world. Despite differences in size and political and socio-economic conditions, cities worldwide are both the political institution closest to citizens and the main provider of public services. These two aspects have proven to be essential, respectively, in the strategic definition and concrete application of the diplomatic action of cities.

The institutional and physical proximity between local leaders and the population determines that the political success of the former – and their reelection, in democratic and decentralized systems – is directly linked to the ability to respond to the needs and hopes of the latter, both deeply impacted by the aforementioned major international challenges. Since the main instrument for dealing with these expectations lies in the capacity to deliver public services – social care, traffic management, public transportation, public lighting, waste management, water supply, cultural heritage management, etc. – it is clear that urban policies, much more than national ones, are, in fact, open to direct evaluation by citizens.

The result is an additional stimulus for municipal administrations to equip themselves with all possible tools to strengthen their capacity to respond to citizens' calls. Hence the interest in entering into international partnerships with foreign cities to benefit from the knowledge and best practices sharing and collaborate to develop new methodologies and services.

It is worth noticing that, besides focusing on all traditional municipal actions, an international collaboration for digital technologies has recently emerged, often referred to as smart city diplomacy. Namely, this methodology has enabled the spread of platforms to enable data communication and facilitate bottom-up decision-making processes. Examples of this are the decision-making platforms created by Barcelona (*Decidim*) and Paris (*Lutèce*), both of which have now been adopted by a number of other cities in line with this collaborative spirit of city diplomacy, as well as the knowledge exchange practices of the African smart cities project ASToN and the Nordic Smart Cities Network (Kihlgren Grandi 2020a).

Geographic Distribution

In over 100 years of existence, the spread of such practice has gone hand in hand with the process of administrative decentralization. In fact, the latter appears to be a fundamental prerequisite for the international action of cities to be considered an expression of the local community, and as such genuine diplomacy – and not a mere execution of the foreign policy drawn up by the central government.

The numbers of city diplomacy, particularly in terms of participation in international networks, thus unsurprisingly illustrate an uneven global distribution, with a concentration in regions whose political and institutional conjuncture has favored administrative decentralization. A study on city networks by the UCL City Leadership Laboratory published in 2017 showed that most city networks (107 out of 200) had their headquarters in Europe, and 43.5% of these networks operated on a European regional scale. In addition, the cities with the highest number of network memberships were European (Barcelona in first place with 23 regional and global network memberships) (Acuto et al. 2017).

However, it is plausible to expect that three dynamics will contribute to a rebalancing of this dimension:

1. The decentralization process is underway in the regions least represented by city diplomacy, i.e., Africa and Asia.
2. The spread of knowledge and capacity-building processes resulting from intercontinental partnerships in a North-South or triangular North-South-South key.
3. The support for the international action of local governments by regional intergovernmental organizations, such as the African Union, Mercosur, ASEAN, and the UN Regional Commissions.

However, it should be pointed out that the figures for city diplomacy do not consider municipalities' effective capacity to define and put into practice international strategies independently of their respective central governments. This aspect refers to the relationship between foreign policy and city diplomacy, which takes on extremely diverse expressions worldwide depending on the level of democracy and decentralization.

At the level of international and domestic law, it is evident that the Westphalian principle of central government control over the definition and implementation of foreign policy objectives remains in place. In the eyes of international law, cities are largely conceived of as subsets of the central power. Hence, as the latter bears responsibility for the former's potentially illicit conduct, it is clear why every nation in the world ensures that its cities comply with international obligations and the guidelines of national foreign policy.

While all national systems postulate such hierarchical inferiority of city diplomacy to national diplomacy, decentralized and democratic ones allow cities – and often other local authorities – to independently formulate their own priorities for international action and join an international association of local authorities. The latter right is formalized in the Council of Europe's European Charter of Local Self-Government (European Charter of Local Self-Government 1985).

This capacity does not detract from the fact that central governments in democratic and decentralized countries have mechanisms for preventing (through prior approval requirements) or blocking (through direct intervention) municipal actions deemed harmful to state interests or obligations. However, this has not averted the formation of municipal coalitions on a national scale that express open criticism of certain aspects of national foreign policy. For example, this is the case with the sanctuary cities movement, an expression of a welcoming approach to international migration, the opposition to the arms race proper to "nuclear-free zones" (Leffel 2018), or the desire to abide by certain international conventions neglected by the central government, as happened with the declaration "We are still in" signed by over 290 American cities following the Trump government's decision to abandon the Paris Agreement (We Are Still In 2021).

In the case of authoritarian and centralized countries, it is evident that such a discrepancy cannot occur. On the contrary, the boundaries between city diplomacy and national foreign policy become very difficult to draw in such contexts. Where political autonomy in defining priorities for international action and the instruments to pursue them is lacking, it is appropriate to speak of foreign policy by proxy, a dynamic observed in the Chinese case (Klaus and Curtis 2020). Nevertheless, cities and local authorities in authoritarian and centralized countries can enjoy a certain level of autonomy in the technical component of city diplomacy, for example, by participating actively in exchanges of best practices and the introduction of joint projects in the framework of twinning agreements or city networks. One example of this is the active participation of Chinese cities in various environmental networks such as C40 and ICLEI, through which several Chinese mayors were, for example, present at COP 26 in October and November 2021 in Glasgow, in the absence of president Xi Jinping (Muglia 2021).

The Instruments of City Diplomacy

At both the global and regional levels, city diplomacy is currently carried out in the context of bilateral and multilateral cooperation, through both of which cities demonstrate their ability to link local development to cross-national dynamics.

Bilateral Cooperation

From the post-World War II period until around the 2010s, the spread of city diplomacy has mainly been characterized by its bilateral dynamics, expressed through cooperation partnerships often referred to as twinning or sister cities agreements.

Through the conclusion of such agreements, the twinned cities commit themselves to developing a path of collaboration and solidarity in the long term, exchanging municipal management practices, promoting each other, and responding jointly to structural and unforeseen challenges of urban development.

Several considerations drive the identification of one's partner city abroad. The phenomenon developed in the immediate aftermath of the

Second World War between European cities with a clear reconciliation and solidarity objective – the first twinnings linked German cities with English and French ones (Handley 2006; Defrance 2008) – and was often established based on previous historical and cultural relationships (Kihlgren Grandi 2020a, pp. 11–12, 126). By the mid-1950s, twinnings had already spread globally, including a myriad of solidarity-drive agreements between cities in the North and cities in the South. The latest wave of agreements debuted in the 1990s and featured a clear goal of joint economic development, explaining the proliferation of ties with cities in emerging economies, particularly in East Asia.

Whatever the origin of these agreements, their potential to generate tangible impacts on the population is largely linked to the ability of the respective administrations to identify shared development priorities to pursue together. This is the case of the partnership between Rotterdam and Mexico City concerning the management of downpours, which affect both cities. The collaboration between the two cities has focused on the "Water Squares" technology developed by the Dutch city and refined by collaborating on its adoption in the Mexican capital (Ilgen et al. 2019).

Similarly, bilateral collaborations between cities linked by migration processes have become widespread. This phenomenon, referred to as co-development (or, less frequently, as migration for development), involves development cooperation projects designed and carried out in collaboration with migrant communities (Alvarez Tinajero and Sinatti 2012). As the project thusly implemented benefits from the knowledge and skills of the migrant community, the latter enjoys further integration and empowerment in the host community.

A famous example is the "microjardin" project between the twin cities of Milan and Dakar, carried out in three steps between 2006 and 2016 and which allowed to strengthen the food security and safety of more than 4000 families in the Senegalese capital. The activities widely involved and empowered the Senegalese community in Milan, namely its cultural associations, who became partners with the municipality on subsequent projects (Comune di Milano 2015).

Finally, it is worth mentioning the support provided by the European Union to bilateral partnerships between cities in its member countries and those in other continents. The most famous example is represented by the International Urban Cooperation (IUC) program, active since 2017 and through which 165 cities were involved in urban development and the fight against climate change as of early 2021. In early 2021, the new phase of the project, renamed International Urban and Regional Cooperation Program (IURC), was launched. In such a step, the bilateral dimension will be combined with multilateral collaboration in the form of thematic clusters and a global network.

Multilateral Cooperation

Although less known to the general public than twinning arrangements, cross-border multilateral action has become the main driver of city diplomacy.

Over the past few years, there has been a broad and decisive trend in favor of activities carried out in "municipal promiscuity" (Handley 2006) – i.e., with two or more foreign partners – to the point that some cities have decided to abandon bilateral projects altogether and concentrate on them. A notable example is the aforementioned city of Ghent, in Flemish Belgium. This trend is based precisely on the realization that the main urban challenges are shared across borders, and as such, can be effectively addressed through three types of activities, i.e., city networks, multilateral projects, and advocacy campaigns.

It is undoubtedly city networks that have led the expansion of city diplomacy over the past few decades. They consist of a group of municipal administrations belonging to different countries, whose collaborative activity tends to have an open time horizon and benefits from the coordination of a permanent administrative structure, most often housed in one of the member municipalities.

It is possible to categorize city networks according to two criteria:

- Geographical criterion, depending on whether they are open to membership from cities

around the world or exclusively from a specific region
- Thematic criterion, depending on whether they promote collaboration across all fields of municipal action or focus on a particular topic

Among the best-known examples are the global and multi-purpose United Cities and Local Government (UCLG), created in 2004 by merging the aforementioned IULA with its major competitor, the United Towns Organization (UTO), the global environmental networks C40 and ICLEI, the regional and multi-purpose EUROCITIES network, and the regional and thematic ASEAN Smart Cities Network.

The high number of networks, which rose from 50 in 1985 to more than 200 today (Acuto et al. 2017), is due precisely to the realization that today's major urban challenges cross national boundaries. However, while the relative ease of creation and management and the non-binding nature of these multilateral initiatives have encouraged their spread, the limited coordination among them has led to much overlap and duplication. This is particularly evident in environmental policies, to which more than 50 networks of cities are dedicated (Teles 2016; Acuto et al. 2017).

In addition, most city networks do not sanction their members' little or no participation in projects and campaigns, as long as they pay their membership fees. As a result, several networks tend to simultaneously feature a nucleus of active and motivated cities and a relatively passive group of members (Kern and Bulkeley 2009).

Alongside the networks, multilateral projects of limited duration have multiplied in recent decades. The spread of such international projects is mainly due to the numerous calls for proposals issued by global and regional organizations, development assistance agencies and banks and, to a lesser extent, the city networks themselves.

The fact that these projects are concentrated in Europe is due to a large number of calls for proposals from the European Union, divided into those also open to cities (Europe for Citizens, Horizon Europe, Interreg Europe, Erasmus+, among others) and those conceived exclusively for them, such as URBACT. The EU identifies these calls as a tool for strengthening European cohesion between local communities and as an instrument for concrete advances in citizens' quality of life through the empowerment of municipal administrations.

The European Union is also a promoter of multilateral projects between its cities and those located on other continents. This is the case, for example, of the Mediterranean City-to-City Migration Project (2015–2018 and 2018–2021), aimed at promoting improved migration governance at the urban level in the Mediterranean region (European Union 2020), and the "International City Partnerships" project (2021–2023) aimed at promoting collaboration for green and inclusive recovery with cities in Canada, South Africa, Singapore, South Korea, Taiwan, and Hong Kong (European Union 2021). Moreover, the European methodology has also inspired multilateral projects in other continents. An example is ASToN (2020–2022), a project financed by the French Development Agency (AFD), through which 11 African cities are collaborating to develop digital technologies capable of promoting sustainability and inclusion on an urban scale (ASToN Secretariat 2020).

Advocacy Campaigns

The spectrum of multilateral activities is completed by advocacy, which has become increasingly widespread in recent years, especially among cities in democratic and decentralized regimes. It consists of the practice of defining and broadly disseminating a shared political vision among cities on issues of common interest and aims to influence both international public opinion and national and international decision-makers. The spread of this practice can be traced back to the desire of cities to influence the governance framework of major urban transnational challenges, which generally lies in the hands of nations. According to a slogan repeatedly employed by proponents of this approach, cities should therefore be granted a "chair at the global table" (Metropolitan District of Quito et al. 2016), so to represent their interests in official negotiations on matters with urban impact.

These campaigns have contributed to the emergence of the global notoriety of several mayors of large cities, authors of a progressive and multilateralist positioning about the protection of social rights and the fight against climate change. Mayors who have achieved the most significant notoriety through strong positioning on transnational issues include the mayors of Paris Anne Hidalgo, Los Angeles Eric Garcetti, London Sadiq Khan, Bogota Claudia López Hernández, and Freetown Yvonne Aki-Sawyerr in relation to the fight against climate change, the mayors of Barcelona Ada Colau and Montreal Valérie Plante regarding the fight against inequality, the mayor of Amsterdam Femke Halsema and the mayor of Buenos Aires Horacio Rodríguez Larreta regarding sustainable economic development, and the mayor of Milan Giuseppe Sala regarding the need to ensure a post-pandemic revival that is economically, socially, and environmentally sustainable.

Networks are among the main promoters of these campaigns, notably in the form of declarations. Examples include the *Cities for Adequate Housing* (2018) (UCLG 2018) and *Right to the City for Women* (UCLG 2019) declarations. Moreover, when declarations are promoted by networks with a limited number of members due to stringent membership criteria, they are frequently open to non-member cities. For example, this is the case with C40's Clean Bus Declaration (2015), which affirms the willingness of cities across the world to support the deployment of green public transportation (C40 2015).

While the content of such advocacy campaigns obviously varies considerably, they generally share the intent to highlight two main characteristics of the municipal approach: multilateralism and pragmatism. These two dimensions are well summarized in the famous quote by former New York City mayor Michael Bloomberg: "while nations talk, cities act" (Florida 2012). This claim received considerable impetus during the COVID crisis. In the early months of the pandemic, when the main bodies devoted to international collaboration, starting from the UN and the European Union, were struggling to foster a joint solution to the crisis, more than 60 online trackers and monitors for exchanging best practices were put in place by a wide array of organizations directly or indirectly involved in city diplomacy (UCLG et al. 2020). Parallelly, city networks deployed several advocacy campaigns – as did C40 (on a global scale) and Eurocities (on a European scale) – to call on national leaders to act in line with the values of international solidarity and collaboration, often referring to the initially largely ignored calls to this effect by the leadership of intergovernmental organizations (Eurocities 2020; Watts 2020).

Furthermore, city networks have demonstrated their willingness to coordinate with each other and with nongovernmental and nonprofit organizations to achieve broader advocacy outcomes. An example of this is the "Race to Zero" campaign, launched at the 2019 UN Climate Action Summit and containing a pledge to reach net-zero by 2050 at the latest. The more than 1000 cities and local governments signing of the campaign, a result of the joint efforts of C40, ICLEI, and the Global Covenant of Mayor, was presented by these networks as one of the key urban contributions to the shift to a decarbonized economy at COP26 in Glasgow (C40 2021).

Conclusions: The Challenges Ahead

Despite being an increasingly mediatized, thriving practice and an object of growing focus by research institutions, city diplomacy, with its little more than 100 years of age, represents a relatively recent dynamic on the international scene. Not surprisingly, in the evolution of its boundaries, the ambition of cities and their networks collides with a number of structural challenges.

The first of these challenges appear to be related to city diplomacy's communication and awareness-raising. If in the first decades, city diplomacy was justified essentially as a moral imperative responding to goals of friendship and solidarity between peoples – to which, as we have seen, we owe the first significant expansion of twinning agreements – this is hardly politically sufficient today. As the French case demonstrates, one of the main obstacles to the spread of this practice is the mayors' difficulty in presenting the concrete impact of this practice on residents and local actors. When the impact is not clearly measured and communicated, city diplomacy lends itself to easy criticism from the local opposition, generally centered around the accusation of waste of public money and "political tourism" of mayors and officials on missions abroad (Kihlgren Grandi 2020b).

However, two dynamics are helping to make the reporting of this practice more effective. First, in the vast majority of today's calls for projects issued by government organizations, development banks, national agencies, and the third sector, impact assessment is generally a prerequisite. Secondly, a growing number of city networks have deployed mechanisms to map and assess their impact. This practice is led by the environmental networks, which in return have started enjoying considerable visibility among the general public. A noteworthy example is the decision of the Global Covenant of Mayors for Climate and Energy, which has over 11,000 cities in 140 countries as members (November 2021), to launch in January 2019 the Common Reporting Framework in order to support its cities in quantifying their local climate change impacts in a harmonized and transparent manner (Global Covenant of Mayors 2019). Moreover, estimates from Global Covenant show that the mentioned "Race to Zero" campaign has the potential to reduce global emissions by at least 1.4 gigatons annually by 2030 (C40 2021).

However, it seems clear that further efforts will be needed to extend this approach to most city networks, which still provide little evidence of the tangible impact of their actions – unless they are supported by external funding that formally requires it. Certainly, this objective is complicated by the dynamics of duplication and overlap between networks of cities, particularly in light of the fact that different levels of local government acting on the same territory may have different affiliations, making it particularly complex to quantify the contribution of each to identical goals.

However, it is clear that the main obstacle to the international ambitions of cities lies in the aforementioned architecture of international relations. It remains the case that only states are responsible for defining and approving, through negotiation, treaties aimed at addressing the great challenges of humanity, which are, as already mentioned, transnational in nature and largely urban in impact.

Although a "seat at the global table" is requested by different types of non-state actors, the demand by cities benefits from advantages related to the very nature of municipal governments. In fact, unlike other actors, the governance of municipalities is based in most states around the world on a popular mandate conferred through elections. Moreover, they represent, as mentioned above, the primary providers of public services. These elements appear not to have eluded international organizations, whose official acts concerning cities and local governments have seen a transition from identification as civil society organizations to recognition as true actors (Acuto et al. 2021).

In spite of this recognition, and in spite of increasing visibility of city diplomacy in media and scholarly literature, cities are still far from obtaining the much sought-after inclusion within multilateral decision-making processes.

If, therefore, such a "chair" has yet to be formally offered to cities, two opportunities for official recognition of the role of cities in international relations have emerged. These are Urban 20 (U20), the stakeholder engagement group of the G20 launched in 2017 to gather the major cities of the world's largest economies, and the Forum of Mayors, created in 2019 by the UN Economic Commission for Europe (U20 2021; Forum of Mayors 2020). In both cases, the voice of cities expressed in these contexts is officially presented to the relevant intergovernmental bodies (G20 and UNECE Committee on Urban Development, Housing, and Land Management), potentially influencing the respective decision-making processes.

While still not allowing cities to negotiate and vote on agreements that are binding on the international community, the inclusion of cities in these processes testifies to the willingness of the respective international institutions to recognize the contribution of cities in terms of both vision and the concrete application of international strategies approved by nations.

The mere existence of such instances is, in turn, contributing to the creation of an alliance between cities and international intergovernmental organizations, fostering a wide-ranging reflection on a bolder integration of cities in the international decision-making processes of the future.

Cross-References

- ▶ Age-Friendly Future Cities
- ▶ An Overview of the Relationship of the Sustainable Development Goals and Urban and Regional Development
- ▶ Big Data for Smart Cities and Inclusive Growth
- ▶ Collaborative Climate Action
- ▶ Healthy Cities
- ▶ Internationalization of Cities
- ▶ Networking Collaborative Communities for Climate-Resilient Cities
- ▶ Planning for Food Security in the New Urban Agenda
- ▶ Theme Cities Networks
- ▶ Urban Health Paradigms
- ▶ Voluntary Programs for Urban and Regional Futures

References

Acuto, M., Decramer, H., Morissette, M., Doughty, J., & Ying, Y. (2017). *City networks: New frontiers for city leaders* (UCL City leadership lab report). London: University College London.

Acuto, M., Kosovac, A., Pejic, D., & Jones, T. L. (2021). The city as actor in UN frameworks: Formalizing "urban agency" in the international system? *Territory, Politics, Governance*, 1–18. https://doi.org/10.1080/21622671.2020.1860810.

Alvarez Tinajero, S. P., & Sinatti, G. (2012). *Migration for development: A bottom-up approach*. EC-UN Joint Migration and Development Initiative. http://www.migration4development.org/sites/m4d.emakina-eu.net/files/jmdi_august_2011_handbook_migration_for_development_0.pdf

ASToN Secretariat. (2020). ASToN baseline study. https://aston-network.org/wp-content/uploads/2021/05/Aston-BaselineStudy-VA-210430-JB.pdf

C40. (2015). *C40 Clean Bus Declaration*. C40. http://c40-production-images.s3.amazonaws.com/other_uploads/images/884_C40_CITIES_CLEAN_BUS_DECLARATION_OF_INTENT_FINAL_DEC1.original_EC2.original.original.pdf?1479915583

C40. (2021). From LA to Bogotá to London, global mayors unite to deliver critical city momentum to world leaders tasked with keeping 1.5 degree hopes alive at Glasgow's COP26 (Press release). https://www.c40.org/news/cop-26-cities-race-to-zero/

Comune di Milano. (2015). Progetti di cooperazione internazionale. http://mediagallery.comune.milano.it/cdm/objects/changeme:20771/datastreams/dataStream3924617956805694/content

Defrance, C. (2008). Les jumelages franco-allemands. Aspect d'une coopération transnationale. *Vingtième Siècle. Revue d'Histoire, 99*(3), 189–201. https://doi.org/10.3917/ving.099.0189.

Eurocities. (2020, June 9). *European cities respond to the coronavirus crisis: Frankfurt – Solidarity with twin city Milan*. Live updates COVID-19. https://covidnews.eurocities.eu/2020/06/09/frankfurt-solidarity-with-twin-city-milan/

European Charter of Local Self-Government, ETS No.122. (1985). https://rm.coe.int/CoERMPublicCommonSearchServices/DisplayDCTMContent?documentId=090000168007a088

European Union. (2020). *Mediterranean City-to-City Migration (MC2CM) – Phase II: Action fiche*. European Union. https://ec.europa.eu/trustfundforafrica/sites/euetfa/files/action_fiche_revised.docx.pdf

European Union. (2021). International Urban and Regional Cooperation (IURC): North America. https://www.iurc.eu/wp-content/uploads/2021/04/iurc-north-america-brochure.pdf

Florida, R. (2012, June 13). What if mayors ruled the world? *Bloomberg CityLab*. https://www.bloomberg.com/news/articles/2012-06-13/what-if-mayors-ruled-the-world

Forum of Mayors. (2020). *Geneva Declaration of Mayors*. UNECE. https://unece.org/sites/default/files/2021-03/Mayors%20declaration%20booklet%20-%20ver.4.pdf

Global Covenant of Mayors. (2019). *Climate emergency: Unlocking the urban opportunity together [2019 impact report]*. Brussels: Global Covenant of Mayors.

Habitat III. (2015). Issue paper on inclusive cities (Habitat III issue papers). http://habitat3.org/wp-content/uploads/Habitat-III-Issue-Paper-1_Inclusive-Cities-2.0.pdf

Handley, S. (2006). *Take your partners – The local authority handbook on international partnerships* (No. 10, LGIB international reports). London: Local Government International Bureau.

Ilgen, S., Sengers, F., & Wardekker, J. A. (2019). City-to-city learning for urban resilience: The case of water squares in Rotterdam and Mexico City. *Water, 11*(5), 983. https://doi.org/10.3390/w11050983.

Kern, K., & Bulkeley, H. (2009). Cities, Europeanization and multi-level governance: Governing climate change through transnational municipal networks. *Journal of Common Market Studies, 47*(2), 309–332. https://doi.org/10.1111/j.1468-5965.2009.00806.x.

Kihlgren Grandi, L. (2020a). *City diplomacy*. Basingstoke: Palgrave Macmillan. https://doi.org/10.1007/978-3-030-60717-3.

Kihlgren Grandi, L. (2020b). *Le nouveau rôle international des villes (et pourquoi il faut l'encourager)*.

Terra Nova. http://tnova.fr/notes/le-nouveau-role-international-des-villes-et-pourquoi-il-faut-l-encourager
Klaus, I., & Curtis, S. (2020, July 6). Ties that bind: China's BRI and City diplomacy in a shifting world order [text]. *ISPI*. https://www.ispionline.it/it/pubblicazione/ties-bind-chinas-bri-and-city-diplomacy-shifting-world-order-26852
Leffel, B. (2018). Animus of the underling: Theorizing city diplomacy in a world society. *The Hague Journal of Diplomacy, 13*(4), 502–522. https://doi.org/10.1163/1871191X-13040025.
Metropolitan District of Quito, Government of Mexico City, European-Latin American Cooperation Alliance among Cities (AL-LAs), Global Taskforce of Local And Regional Governments, & United Cities and Local Governments (Eds.). (2016). A seat at the global table: Local governments as decision-makers in world affairs. https://cdn.theatlantic.com/assets/media/files/a_seat_at_the_global_table.pdf
Muglia, A. (2021, November 14). La diplomazia (agile) delle città. *Corriere Della Sera*, pp. 16–17.
OECD. (2019). *OECD regional outlook 2019: Leveraging megatrends for cities and rural areas*. Paris: OECD. https://doi.org/10.1787/9789264312838-en.
OECD. (2020). *COVID-19 cities policy responses*. OECD policy responses to coronavirus (COVID-19). https://read.oecd-ilibrary.org/view/?ref=126_126769-yen45847kf&title=Coronavirus-COVID-19-Cities-Policy-Responses
Take, I. (2017). The Hanseatic League as an early example of cross-border governance? *Journal of European Integration History, 23*(1), 71–96. https://doi.org/10.5771/0947-9511-2017-1-71.
Teles, F. (2016). *Local governance and inter-municipal cooperation*. New York: Palgrave Macmillan.
U20. (2021). Urban 20 calls on G20 to empower cities to ensure a green and just recovery. An official communiqué from Urban 20 (U20). Urban 20. https://www.urban20.org/wp-content/uploads/2021/06/U20-2021-Communique-Final.pdf
UCLG. (2013). 100 Years: Testimonies. https://www.uclg.org/sites/default/files/libro%20centenario-web%20(1).pdf
UCLG. (2018). *Cities for adequate housing*. Municipalist Declaration of Local Governments for the Right to Housing and the Right to the City. https://citiesforhousing.org/
UCLG. (2019). The right to the city for women. https://www.uclg-cisdp.org/sites/default/files/EN%20-%20The%20Right%20to%20the%20City%20for%20Women.pdf
UCLG, Metropolis, & LSE Cities. (2020). *COVID-19 monitors of relevance to urban and regional governance*. Analytics note no. 1; Emergency Governance Initiative for Cities and Regions. https://www.lse.ac.uk/cities/Assets/Documents/EGI-Publications/Analytics-Note-01.pdf
UNDESA. (2019). *World urbanization prospects: The 2018 revision*. New York: United Nations.
UN-Habitat. (2016). *Urbanization and development: Emerging futures*. Nairobi: UN-Habitat.
UNHCR. (2020). *Forced displacement in 2019: Global trends*. UNHCR. https://www.unhcr.org/5ee200e37.pdf
United Nations. (2020). COVID-19 in an urban world. https://www.un.org/sites/un2.un.org/files/sg_policy_brief_covid_urban_world_july_2020.pdf
Watts, M. (2020, April 16). *Cities unite to tackle COVID-19 as president Trump attacks the World Health Organisation*. Medium. https://c40cities.medium.com/cities-unite-to-tackle-covid-19-as-president-trump-attacks-the-world-health-organisation-and-bcaa7e65e086
We Are Still In. (2021, November 30). *We Are Still In*. https://www.wearestillin.com

Human Rights

▶ Social Urbanism: Transforming the Built and Social Environment

Human Security

▶ Water Security and Its Role in Achieving SDG 6

Hybrid (Green-Grey) Infrastructure

▶ Integrated Urban Green and Grey Infrastructure

Hydrographic

▶ Water Policy in the State of Tabasco

Hygiene - Cleanliness

▶ Water, Sanitation, and Hygiene Question of Future Cities of the Developing World

H

I

ICT

▶ Smart Agriculture and ICT

Ideology–Thought

▶ Neoliberalism and Future Urban Planning

IGGI

▶ Integrated Urban Green and Grey Infrastructure

Illegal International Movement

▶ Transnational Crimes: Global Impact and Responses

IMD – India Meteorological Department

▶ Adapting to a Changing Climate Through Nature-Based Solutions

R. C. Brears (ed.), *The Palgrave Encyclopedia of Urban and Regional Futures*,
https://doi.org/10.1007/978-3-030-87745-3

Impact of Universities on Urban and Regional Economies

Nicky Morrison
Western Sydney University, Sydney, NSW, Australia

Synonyms

Global pandemic fallout; Local economic development; Localized price effects; Universities expansion; Urban and regional economies

Definition

This entry examines the role that universities play in urban and regional economies and the pressures they face to expand in order to remain internationally competitive. It then sheds light on how universities' expansion plans can have adverse localized price effects. It presents several lessons for urban and regional planning authorities to consider. First, the potential of universities to generate localized economic growth should not be underestimated. If surrounding housing supply remains inelastic, universities' expansion plans will create further upward pressure on housing prices. Second, the strategy of a university to concentrate its resources in particular urban areas, while attractive to certain businesses, hold potential downsides, including putting pressure on existing social and urban infrastructure. It is therefore important that planning authorities work closely together to ensure that both housing stock and associated infrastructure expand to mitigate any adverse impacts from a university's growth strategies. The entry concludes with a word of caution. The fallout of the global pandemic has left many universities facing major organizational restructuring, casting doubt on their ability, and need, to expand their physical footprints.

Introduction

The role of universities as engines or drivers of local economic development has now become well-established. Numerous studies have centered on individual university's anchor role in local economies, drawing extensively on examples from the United States and Europe (Huffman and Quigley 2002; Glasson 2003; Haktan 2014). At the same time, heightened international competition has put pressure on universities to expand (Morrison 2013). To sustain research excellence and compete for national and international recognition, universities, world-wide, have focused on extensive capital works programs, expanding their campuses, teaching and research facilities (Paradeise and Thoenig 2013). Urban and regional planning authorities have also faced pressures from national governments to accommodate universities' expansion plans, with their growth strategies considered nationally and internationally significant, and beyond local concerns (Morrison and Szumilo 2019). As Kitson (2010, p. 4) suggests, "places need embedded economic actors, and universities are one of the most important."

As such, the position of universities within urban and regional economies has radically shifted, from a more idealistic position focused on the creation of knowledge to an increasingly instrumentalist position in which they are now explicitly part of the system of governance (Abel and Deitz 2011). Governance, for Stoker (2002, p. 3) revolves around "the capacity to get things done through collective action in the realm of public affairs, in conditions where it is not possible to rest on recourse to the state." In this context, universities are now increasingly required to act on behalf of the state, even though in principle they are autonomous entities.

Little attention, however, has been given to the dynamic between universities' expansion plans and adverse, localized price effects. Universities are spatially fixed. By their nature, their expansion emulates from their founding location, unlike "footloose" businesses that can close original premises and relocate to different jurisdictions for financial reasons. Thus, universities' increasingly global ambitions exercise an ever-stronger localized impact (Morrison and Szumilo 2019).

The purpose of this entry is two-fold. First, it draws together the premise from previous research: that universities play a crucial role in urban and regional economies. Secondly, it sheds light on this question of universities' localized price effects. Here, the entry draws on and highlights key lessons from earlier research carried out Morrison and Szumilo (2019). These key lessons will be of relevance to planning authorities, as well as to universities themselves. Although focused within the United Kingdom, this research is undeniably applicable for the story of universities across the world. Finally, the entry concludes with a note of caution, and a further instruction. Following the global pandemic, universities across the globe are having to scrutinize and reassess their expansion strategies. Planning authorities, universities, and industry will now need to collaborate more than ever to work out a combined strategy going forward.

Critiquing the Role of Universities in Local Economic Development

Measuring the Expanded Value of Universities

Conventional approaches to valuing the economic activity generated by universities have tended to focus on direct employment or expenditure effects, coupled with calculations as to the multiplier effect to capture associated additional indirect and induced outcomes (Bleaney 1992).

Universities are often among the top half dozen employers in their respective local economies. Considered in conjunction with these multipliers, universities invariably comprise an important economic power in local income and employment improvements. However, commentators documenting the role of universities have also demonstrated how their influence goes well beyond these standard effects, as a direct result of the ways in which they help build knowledge and skills, that is, the human capital, of a local region (Kumar et al. 2014; Lawton-Smith and Charles 2016; Beneworth and Fitjar 2019). For Fischer and Varga (2003), higher levels of human capital generate both greater levels of economic activity, and more rapid economic growth. These spillover effects then further boost the region's demand for high skilled, talented workers. University graduates themselves tend to be more productive through higher working hours and earn higher wages; they have higher employment rates and higher lifetime incomes compared to lower-skilled professionals (Kumar et al. 2014).

Universities also deepen the skills and knowledge – again, the human capital – of the residents of the local region in which it is situated, and then in turn the local economic activities in which such local graduates are employed (Ackers and Gill 2005). The volume and orientation of graduates at higher Master and Doctoral levels can then further enhance this process, not just in relation to local economic development but also in terms of wider benefits to society. As Addie (2017) contends, universities therefore create a "generation of capabilities." Although high rates of migration among graduates may undercut these effects within a specific local economy (Doutriaux 2003), local benefits to the supply of human capital are still an apparent result of the overall knowledge production function of (such local) universities (Cattaneo et al. 2017).

Universities' Role in Fostering Innovation

The knowledge dispersal and research functions of universities are critical to innovation. As globalization continues to intensify competition across all industries and sectors, innovation has emerged as a critical factor in supporting business growth. It means there is also now an increased recognition of the role that universities play in promoting innovation, particularly through co-location and collaborations with industries where they are based (Laursen et al. 2011; Kumar et al. 2014), and of the traditional dependency of local economies on the skills of their local university graduates and researchers. Private sector companies often locate themselves close to reputable tertiary education centers to take advantage of research collaboration opportunities and spillover effects and gain access to a skilled labor force (Guerrero and Urbano 2014).

The main advantage of industry and university collaboration is in the sharing of new knowledge and ability to source funds for equipment, research assistants, start-up research projects, and the ability to test practical applications of theories. These collaborations are mutually reinforcing, allowing the university and industry to retain their respective competitive positions, raise revenue or profit margins, and increase their share in international markets (Beneworth and Fitjar 2019).

The Growth of Entrepreneurial Universities

One particular outcome of the processes described above is that governments and policy makers, again around the world, have been encouraging what have once been seen as "traditional" universities, focusing on conventional educational programs, to become entrepreneurial; with an emphasis now on stimulating innovation, co-creating research with end users, notably private sector businesses, and launching new graduate entrepreneurs of the future (McAdam and McAdam 2008; Goldstein 2010; Smith and Bagchi-Sen 2012). For Clark (1998), five elements are required if universities are to achieve such transformations: (i) a strengthened "steering core" within the internal organization of the university; (ii) an "expanded developmental periphery" in how the university interacts with its environment; (iii) a "diversified funding base," less dependent on government funds and drawing on other sources such as from industry; (iv) a "stimulated academic heartland" disseminating high quality outputs; and (v) an "integrated entrepreneurial and innovative culture" permeating every aspect of the university.

Most universities undertake entrepreneurial activities to improve urban, regional, as well as national economic performance, and, to financially uplift the university. The success of such activities will depend on the significance given to both internal elements, such as the university's status and policy, and external elements such as public policy and regional conditions (Smith and Bagchi-Sen 2012). For Shattock (2010), the development of an embedded university, in effect, pushes local economic development, which in turn serves as a platform for development of the university, creating a reinforcing motivational force.

The Expansion of Global Research Centers

In the local and global "market-place," a university wanting to become or remain competitive must match the offer of its competitors. Competitive attractors then include the availability of facilities and size of the research community. In this international "arms-race," even existing best universities will soon be surpassed by others and be under pressure to expand (Wildavsky 2012). Financial futures will depend on attracting the best talent from around the world, whether researchers or students (Morrison and Szumilo 2019), and success in this regard will impact a university's position in international league tables, involving a series of metrics from research output to students' academic performance (THE 2021a). These tables in turn form the basis of many individuals' decisions about which institution to attend (Boliver 2013).

There are several ways in which universities can gain or maintain an international competitive advantage. A key strategy is through investing in state-of-the-art research facilities. The creation of a global research center is attractive not just because good facilities are themselves desirable, and to both researchers and students. Such centers can also create a positive spillover effect whereby businesses are attracted to a locality where research fits their needs. Individuals are then also thereby encouraged to the university, enthused by the prospect of nearby jobs (Morrison and Szumilo 2019).

Further, a consequence of the associated knowledge spillovers is that academics, in conjunction with those associated industries, can create and develop new ideas faster. Interdisciplinary research is much easier in an environment where experts from different disciplines are easy to contact (Fischer and Varga 2003). Productivity increases because of the ability to share not only facilities but also ideas and experience: clear agglomeration benefits exist from having a large and diverse community of scholars. Again, the resultant knowledge production and new technologies attract new firms and talented workers to the region and help existing firms expand and innovate.

Universities As Catalysts for High-Tech Clusters

There are now many documented cases where a university has played a lead and active role in the initiation and evolution of high-tech clusters and which in turn have played a major part in stimulating local economic growth (Lazzeretti and Tavoletti 2007; McAdam and McAdam 2008; Smith and Bagchi-Sen 2012). A key finding here is that such universities have developed strong and embedded linkages with the local system of governance and have become highly responsive to industry needs. Matching with the needs of employers and an associated prioritizing of research endeavors has become increasingly important part of the new "entrepreneurial" university. This dynamic has then also been extended to their perceived, and actual, role in the local community, with universities providing civic culture, social equity, and sustainability programs, among other actions in direct response to local needs and concerns (Glasson 2003; Breznitz and Feldman 2012).

There are number of ways that universities develop industry linkages. These include specific technological initiatives (for example, advanced manufacturing practices, technology transfer programs, techno-parks, and spin-off companies), student and graduate placements, community projects (for example, community lifelong courses, continuing education programs, and the support of the cultural vitality of the community), and the promotion of green initiatives. The number and type of initiatives vary by university. Their role in this regard has been well documented in the USA and Europe (Lawton-Smith and Glasson 2017; Haktan 2014); and the success stories of Stanford University and Silicon Valley, the Massachusetts Institute of Technology (MIT) and Route 12, and the Research Triangle Area in North Carolina are well studied (Huffman and Quigley 2002; Goldstein 2005; Hsu et al. 2007).

Factors Affecting Knowledge Spillovers

The ability for a university to transmit its knowledge base will profoundly affect its ability to then impact its local environment and determine its role in a regional innovation system. Several factors will contribute to this "spillover" ability. These include the university's founding mission, institutional context, and prior experience with commercial activity. A university's founding mission and the orientation of its entrepreneurship endeavors will also vary depending on the size of the university, its catchment area, and the local and regional context (Goldstein 2010).

Here, the direction or orientation a university takes is often also aligned with political agenda (at central and local levels). Studies which have distinguished between entrepreneurial universities and more traditional research universities have identified the former as having a more important and active role in urban and regional economies. Drawing on examples from the USA and Europe, entrepreneur universities are also more successful in socioeconomic development, thereby making possible additional, often considerable, benefits to wider society (Goldstein 2010; Lawton-Smith and Charles 2016; Addie 2017).

A Research Gap: Understanding Localized Price Effects

The new entrepreneurial university now invests heavily in facilities that will attract talented newcomers into its "home" locality, support the establishment of new global research centers, and enable collaborative activities with local industry; all with a goal of local area stimulus. But while the business rationale for such university expansion exists, there is limited research on any unintended adverse price effects from this expansion, particularly within any local (generally urban) area with restricted geographical and physical access to immediate other complementary or alternate economies for its residents to draw on. A particular example is where employment demand pressures far outstrip housing supply in constrained locations, adding pressure onto existing housing stock and creating localized price effects (Morrison 2013). Understanding these dynamics warrants active investigation within any process or policy to establish new university structures aimed at the overall enhancement of such local economies.

Universities' Global Research Ambitions and Their Localized Price Effects

To begin to fill in this research gap on the localized price effects of university expansion, Morrison and Szumilo (2019) adopted a novel spatiotemporal model to explore the impact of universities' investment in global research facilities on local house prices. Publicly available statistics of housing transactions in five selected locations in the United Kingdom (UK) were used to build up a picture of impact over time.

The five locations were all cities dependent on university-related businesses: Oxford, Cambridge, Durham, Exeter, and York. A key finding was that in all five locations, there was indeed a strong correlation between house price growth and the establishment or expansion of those universities' global research centers. This correlation was most noticeable in the immediate area, with impact diminishing with distance. In Cambridge, for example, the results show that the restricted size of the housing stock and the growth of the University of Cambridge's two global research centers had created a spatial price effect of price increases that has not been observed in comparable cities in the UK.

Further research is necessary to fully explain why this relationship between the spatially weighted number of research centers and house prices was found to be especially strong in Cambridge. Yet the speed with which the University of Cambridge's research facilities have grown and the lagged response by local planning authorities in supplying housing land are likely to be key contributory factors. The University of Cambridge has effectively been able to use the political clout from its significant role in the local economy and as a major landowner to push regional and city planners toward growth strategies that meet their needs to expand their research facilities (Morrison 2010a, 2013). Yet at the same time, Cambridge's limited housing availability and concomitant high housing costs has remained (Morrison 2010a, b; Boddy and Hickman 2016).

Key Lessons

There are several lessons that planning authorities in other locations can take away from this piece of research. First, the potential of universities to generate localized economic growth should not be underestimated. Viewed from one angle, all five UK universities analyzed are success stories. Their investment in state-of-the-art research facilities has undoubtedly attracted newcomers. The University of Cambridge, which has concentrated its resources on expanding its biomedical and sciences facilities, has created attractive opportunities for related businesses that clearly benefit from co-locating and collaborating with top academics in their field. Most notably, the multinational pharmaceutical company, AstraZeneca, relocated its corporate headquarters and global research and development facilities to Cambridge in 2013 to co-locate alongside the university's new biomedical campus, bringing with it, thousands of new jobs. New workers have continued to move into Cambridge from around the world despite relatively high living costs.

Lessons of caution, however, need to be drawn. Oxford and Cambridge, for instance, are cities with a particularly high proportion of jobs in education within the local economy. In 2018, 28.9% of jobs in Oxford were in education. In Cambridge the number was 23.6%. Both figures are well above the UK national average of 8.9% (Morrison and Szumilo 2019). At the same time, both cities are witnessing significant housing affordability problems. With demand far outstripping new housing supply, low-income earners and the unemployed are the ones being priced out of each city. The University of Cambridge's recent expansion plans have been especially rapid, contributing to localized (increased) house price effects. It is therefore important that universities work alongside planning authorities to ensure that growth plans are sustainable, including ensuring the establishment (and co-creation) of concurrent solutions to address housing affordability problems in their immediate vicinities. The University of Cambridge's new urban quarter (named Eddington) in Northwest Cambridge provides an exemplary case in point. Heralded as the university's largest capital project to date, on completion the scheme will comprise a total of 5000 new housing units, with 1500 for market housing, 1500 dedicated to university key workers, and 2000 to postgraduate students.

That said, the University of Cambridge has also committed to expand its global research centers, to further strengthen the positive agglomeration benefits for each, along with the university's international competitiveness (Boddy and Hickman 2016). This includes the expansion of one site to accommodate one of the largest global science and medical research campuses. On completion, the Cambridge biomedical campus is predicted to create as many as 8000 new jobs by 2026. It will occupy almost the same footprint as the University already does within the city (Cambridge University Hospitals NHS Foundation Trust 2021). Another site has become a premier location for physical sciences and technology. Expansion of this site, in West Cambridge, includes several buildings accommodating the Departments of Physics, Chemical Engineering and Biotechnology, and part of Electrical Engineering Department (University of Cambridge 2021). Associated economic development within these two areas will further encourage incoming labor attracted by the growth of employment prospects and wages (Cambridge Ahead 2017). Yet if housing supply remains inelastic, this will not only create a further upward pressure on house prices but will also have a spatial effect of pushing lower income earners and the unemployed further away from Cambridge over the long term.

Two further lessons can be drawn from this study, in reference to the specific experiences of Cambridge and Oxford. In Oxford, which boasts an even higher percentage of jobs in education than in Cambridge, there was a weaker correlation between the university's global research centers and localized house price effects than found in Cambridge. This difference is likely to be due to the University of Oxford dispersing its university jobs more evenly around the whole city, as compared to the University of Cambridge, which has created specialized research clusters in two specific localities (Morrison and Szumilo 2019).

This finding ties in to the second important lesson. The strategy of the University of Cambridge to concentrate its resources in particular areas, while proven attractive to certain businesses such as AstraZeneca, also holds potential downsides that planning authorities, as well as universities, must recognize. The cost of agglomeration is that there can also be a tipping point beyond which further growth will yield declining economic benefits by putting pressure on existing social and urban infrastructure; an area may suffer from overcrowding or congestion, for example. It reiterates the importance of universities to work closely with planning authorities to ensure both housing stock and associated infrastructure expand to meet the need for sustainable and inclusive growth for the whole area. Importantly, this will need to include a wide range of housing types that caters for both existing residents and newcomers. In doing so, adverse impacts from a university's expansion plans can be mitigated, ensuring that everyone in the locality can benefit from a university's role in local economic development.

Conclusions

Universities have become far more than centers of education. By virtue of their status as landowners and local employers, universities have an economic impact that now often matches their contribution to research and learning (Morrison 2013). Universities, however, do not have total control over their own destinies. The growing pressure for accountability in public funding of academic research has made the responsiveness of universities to industry demands a critical remit. At the same time, politicians and policy makers are keen to better understand the importance of universities, particularly in the development of urban and regional economies. Entrepreneurial activities will therefore continue to be part of this instrumentalist positioning, in which a university's mission is converging with primarily private sector interests (Abel and Deitz 2011). Attention will remain on ways to strengthen university–industry collaborations, with universities praised for contributing to the up skilling of the labor market, the host locality's innovation network, and to the generation of mutual assets for innovation (Shattock 2010).

Yet while planning authorities covet the presence of universities on the basis of their ability to bring not just stability but growth in the local economy, a word of caution is now required. The attractiveness of any university is partly dependent on the quality of life in the surrounding area. While a high quality of life is a key factor with respect to international mobility destination choices, high housing costs increasingly limit the location choices of newcomers, restricting their ability to take advantage of these benefits (Morrison and Szumilo 2019).

The adverse consequences of the global pandemic on the university sector, world-wide, have been immediate, and are anticipated to endure for many years. The fall-out has left universities, particularly those in more developed countries and dependent on international student revenues, urgently exploring how they can continue to fund their growth strategies. Reduced reliance on such revenues and a restricted ability to recruit talented workers have called on universities to interrogate their role within not just local but national economies – and on where their strengths and weaknesses will lie in the future in responding to an ever-increasing cut-throat competitive global environment. Strategic policy choices must be made to varying degrees by each university to mitigate predicted financial losses (Marshman and Larkins 2020).

The establishment of university global research centers, with commensurate capital works programs and activities are likely to be called into question. Many universities are facing major organizational restructuring and have been compromised in terms of their research missions and growth strategies (Robinson 2021; DePietro 2021; THE 2021b). Moreover, the pandemic has laid bare the future of smart working and the growth of online learning. Work and education are likely to become more flexible in terms of locations, hours, and shared responsibilities, facilitated not just through physical but increasingly digital infrastructure. This trend in turn raises doubt on the need for universities to continue to expand their physical footprints.

Further research, and particularly case study research, is needed to chart the role of universities in urban and regional economies, and in the intricate university–industry linkages, post the global pandemic. Measuring the different tangible inputs, outputs, outcomes, and, in particular, the localized price effects, will need to comprise a central component of this research agenda, and continue to demonstrate the complexity of these relationships. Moreover, this agenda will also require a greater collaboration among policy makers, industry, and universities to respond to the significant economic uncertainties and growth challenges ahead.

Cross-References

- ▶ City Visions: Toward Smart and Sustainable Urban Futures
- ▶ Financing: Fiscal Tools to Enhance Regional Sustainable Development
- ▶ Future of Urban Governance and Citizen Participation
- ▶ Future of Urban Land-use Planning in the Quest for Local Economic Development
- ▶ Internationalization of Cities
- ▶ Local and Regional Development Strategy
- ▶ Regulation of urban and regional futures
- ▶ Smart Cities
- ▶ The Urban Planning-Real Estate Development Nexus
- ▶ Urban and Regional Leadership

References

Abel, J., & Deitz, R. (2011). The role of colleges and universities in building local human capital, Federal Reserve Bank of New York. *Current Issues in Economics and Finance, 17*(6), 1–7.

Ackers, L., & Gill, B. (2005). Attracting and retaining 'early career' researchers in English higher education institutions 1. *Innovations, 18*(3), 277–299.

Addie, J. P. D. (2017). From the urban university to universities in urban society. *Regional Studies, 51*(7), 1089–1099.

Beneworth, P., & Fitjar, R. (2019). Contextualising the role of universities to regional development: Introduction to the special issue. *Regional Studies Regional Science, 6*(1), 331–338.

Bleaney, M. (1992). What does a university add to its local economy? *Applied Economics, 24*, 305–311.

Boddy, M., & Hickman, H. (2016). The 'Cambridge Phenomenon' and the challenge of planning reform. *Town Planning Review, 87*(1), 31–52.

Boliver, V. (2013). How fair is access to more prestigious UK universities? *The British Journal of Sociology, 64*(2), 344–364.

Breznitz, S., & Feldman, M. (2012). The larger role of the university in economic development: Introduction to the special issue. *The Journal of Technology Transfer, 37*(2), 135–138.

Cambridge Ahead. (2017). Press Releases. http://www.cambridgeahead.co.uk/. Accessed 06 Apr 2017. 23rd January 2017 report: http://www.cambridgeahead.co.uk/wp-content/uploads/2017/01/PRESS-REL-new-Cambridge-Growth-data-FINAL-230117.docx

Cambridge University Hospitals NHS Foundation Trust. (2021). http://www.cuh.org.uk/corporate-information/about-us/our-profile/our-partners/cambridge-biomedical-campus

Cattaneo, M., Malighetti, P., Meoli, M., & Paleari, S. (2017). University spatial competition for students: The Italian case. *Regional Studies, 51*(5), 750–764.

Clark, B. (1998). *Creating entrepreneurial universities: Organisational pathways of transformation*. Guildford: IAU Pergamon/Elsevier.

DePietro A. (2021). *Here's a look at the impact of coronavirus (COVID-19) on colleges and universities in the US* forbes.com/sites/andrewdepietro/2020/04/30/impact-coronavirus-covid-19-colleges-universities/?sh=1a17baf961a6

Doutriaux, J. (2003). University-industry linkages and the development of knowledge clusters in Canada. *Local Economy, 18*(1), 63–79.

Fischer, M. M., & Varga, A. (2003). Spatial knowledge spillovers and university research: Evidence from Austria. *The Annals of Regional Science, 37*(2), 303–322.

Glasson, J. (2003). The widening local and regional development impacts of the modern universities – A tale of two cities (and North-South perspectives). *Local Economy, 8*(1), 21–37.

Goldstein, H. (2005). The role of knowledge infrastructure in regional economic development: The case of the research triangle. *Canadian Journal of Regional Science, 28*(2), 199–223.

Goldstein, H. (2010). The 'entrepreneurial turn' and regional economic development mission of universities. *The Annals of Regional Science, 44*(1), 83.

Guerrero, M., & Urbano, D. (2014). Academics' start-up intentions and knowledge filters: An individual perspective of the knowledge spillover theory of entrepreneurship. *Small Business Economics, 43*(1), 57–74.

Haktan, S. (2014). The role of universities in local economic development: A case of TRA2 region in Turkey. *Research Journal of Business Management, 1*(4), 448–459.

Hsu, D., Roberts, E., & Eesley, C. (2007). Entrepreneurs from technology-based universities: Evidence from MIT. *Research Policy, 36*(5), 768–788.

Huffman, D., & Quigley, J. (2002). The role of the university in attracting high tech entrepreneurship: A Silicon Valley tale. *The Annals of Regional Science, 36*, 4-3-419.

Kitson, m. (2010). *The connected university: Driving recovery and growth in the UK economy.* London: NESTA.

Kumar, V., Mumari, A., & Saad, M. (2014). *The role of universities in local economic development: A literature review.* Conference: 4th international COSINUS conference, Bordj-Bou-Arreridj, Algeria: Innovation systems and the new role of universities, at University of Bordj-Bou-Arreridj, Algeria.

Laursen, K., Reichstein, T., & Salter, A. (2011). Exploring the effect of geographical proximity and university quality on university–industry collaboration in the United Kingdom. *Regional Studies, 45*(4), 507–523.

Lawton-Smith, H., & Charles, D. (2016). Universities and local economic development: An appraisal of the issues and practices *Local Economy,* 5.

Lawton-Smith, H., & Glasson, J. (2017). UK university models of technology transfer in a global economy. *University Technology Transfer* (pp. 199–221). London: Routledge.

Lazzeretti, I., & Tavoletti, E. (2007). Higher education excellence and local economic development: The case of the entrepreneurial university of Twente. *European Planning Studies, 13*(3), 475–493.

Marshman, I., & Larkins, F. (2020). *Modelling individual Australian universities resilience in managing overseas student revenue losses from the COVID-19 pandemic.* https://melbourne-cshe.unimelb.edu.au/lh-martin-institute/fellow-voices/modelling-individual-australian-universities-resilience-in-managing-overseas-student-revenue-losses-from-the-covid-19-pandemic

McAdam, M., & McAdam, R. (2008). High tech start-ups in University Science Park incubators: The relationship between the start-up's lifecycle progression and use of the incubator's resources. *Technovation, 28*(5), 277 290.

Morrison, N. (2010a). Securing key worker housing through the planning system. In S. Monk & C. Whitehead (Eds.), *Making housing more affordable: The role of intermediate tenures.* New Jersey: Wiley-Blackwell, Jersey City.

Morrison, N. (2010b). A green belt under pressure: The case of Cambridge, England. *Planning Practice and Research, 25*(2), 157–181.

Morrison, N. (2013). Reinterpreting the key worker problem within a university town: The case of Cambridge, England. *Town Planning Review, 34*(6), 721–742.

Morrison, N., & Szumilo, N. (2019). Universities' global research ambitions and their localised effects. *Land Use Policy, 85*, 290–301.

Paradeise, C., & Thoenig, J. C. (2013). Academic institutions in search of quality: Local orders and global standards. *Organization Studies, 34*(2), 189–218.

Robinson, I. (2021). *Protecting UK universities from COVID-19.* https://www.timeshighereducation.com/hub/hsbc/p/protecting-uk-universities-covid-19

Shattock, M. (2010). The entrepreneurial university: An idea for its time. *London Review of Education, 8*(3), 263–271.

Smith, H. L., & Bagchi-Sen, S. (2012). The research university, entrepreneurship and regional development: Research propositions and current evidence. *Entrepreneurship & Regional Development, 24*(5–6), 383–404.

Stoker, G. (2002). Governance as theory: Five propositions. *International Social Science Journal, 50*(155), 17–28.

Times Higher Education (THE). (2021a). *World University Rankings.* https://www.timeshighereducation.com/world-university-rankings

THE. (2021b). *The impact of coronavirus on higher education.* https://www.timeshighereducation.com/hub/keystone-academic-solutions/p/impact-coronavirus-higher-education

University of Cambridge. (2021). West Cambridge. http://www.westcambridge.co.uk/project/location-and-site-context

Wildavsky, B. (2012). *The great brain race: How global universities are reshaping the world.* Princeton University Press.

I

Improved Water Sources

▶ Meeting SDG6: Ensuring Safe Drinking Water for All in Rural India

Improving Social Equity and Community Health and Well-Being in Low-Income Suburbs and Regions

Frances Furio[1], Tara Rava Zolnikov[1,2] and Tanya Clark[1]
[1]School of Behavioral Sciences, California Southern University, Costa Mesa, CA, USA
[2]Department of Community Health, National University, San Diego, CA, USA

Introduction

Health, an essential component of well-being, is also an essential component of overcoming the harmful or negative effects of poverty and social disadvantage (Braveman and Gruskin 2003). Research has shown that socioeconomic variables directly influence the health of individuals and larger populations (Sena et al. 2007). Poverty and deprivation are interlinked with poor health outcomes (Eades 2000), shorter life expectancies (Marmot and Wilkinson 2005), serious illnesses, and premature death (Wilkinson and Marmot 2003). Improving health and sequential well-being of individuals requires the consideration of social equity and the various barriers to health faced by those living in low-income suburbs and regions. Having a positive impact on community health also requires the consideration of advocacy and improvement efforts, including the ways in which public health knowledge can move into political action (Marmot and Wilkinson 2005).

Social Equity and Public Policy

Equity has been described as an ethical principle that is closely related to principles of human rights (Braveman and Gruskin 2003). As an ethical principle, social equity has emerged as an ethical component of public administration (Frederickson 2005). When social equity emerged as a component of public administration in the 1960s, experiences related to discrimination in society were widespread (Frederickson 2005). A theory of social equity was developed in 1968 as a component of public administration, and in 1997, a standing panel on social equity was established by the congressionally chartered National Academy of Public Administration to create a forum for deliberations on topics related to equality, justice, and fairness in public administration (Frederickson 2015).

Since this time, public administration efforts have continued to develop. Three pillars in public administration have been widely discussed in literature: efficiency, economy, and equality (Frederickson 2015). Social equity has been described as an important component of the third pillar (Frederickson 2015). Social equity values are related to the fairness of organizations, the management of organizations, and the delivery of public services (Frederickson 2015). Obligations of social equity include efforts related to administering laws fairly, proactively furthering equality, and engaging in moral leadership (Frederickson 2005).

When discussing social equity, context of inequity is important (Sena et al. 2007). This context includes where inequity takes place, what is causing inequity, and other determinants related to inequity (Sena et al. 2007). Social equity, as a focus and a movement, is interested in considering who organizations are well-managed for, who they are efficient for, and who they are economical for (Frederickson 2015). There is also consideration of who public services are either more or less delivered for (Frederickson 2015). Social equity has been recognized as multidimensional, in that it is comprehensive and includes various intersecting factors and variables simultaneously (Pascual et al. 2014). Among the dimensions are the political context and surrounding social conditions, which could include education, gender, or power dynamics (Pascual et al. 2014), as well as equality in relation to race, ethnicity, language, sexual orientation, certain mental and physical conditions, and other economic circumstances (Frederickson 2005). These existing markers could influence an individual's advocacy, influence, and recognition (Pascual et al. 2014).

In regard to public policy and public administration, social equity is a core topic of growing importance (Frederickson 2015). Interestingly, while social equity has grown in importance within public administration, inequality has only continued to increase (Frederickson 2005). Policies that privilege an unequal distribution of wealth and the continued production of this unequal distribution have directly contributed to inequality among various regions and groups and repercussions for populations with less privilege (Sena et al. 2007). In this way, the playing field has shifted toward privileged groups and away from the underprivileged groups (Frederickson 2005).

Community Health and Well-Being

In the nineteenth century, crowded cities and poor sanitation practices and access to contaminated water contributed to disease outbreaks, like pneumonia, tuberculosis, typhoid, and cholera (Bentley 2014). Gradual public health movements helped advance knowledge on these issues, eventually resulting in public health policies and legislation (Bentley 2014). Moreover, access to healthcare has significantly improved over time. Healthcare has been described as a major determinant of health, and while healthcare has been a conversation in government, there is still a significant proportion of society that continues to lack access to comprehensive healthcare (Hug 2015). In fact, access to healthcare is one of the many areas where inequality within communities is most evident (Frederickson 2005).

To date, there are many significant gaps that have yet to be sufficiently addressed (Hug 2015). One of the most persistent problems has been barriers related to uninsurance (Hug 2015). Problems also include the distribution of policies and services, quality of healthcare services, fairness of procedures, financing of healthcare, allocation of healthcare resources, services related to disease management and preventative care, and outcomes for different population groups (Hug 2015). These are some of the significant disparities related to healthcare, and these disparities are not decreasing (Hug 2015). These types of health disparities are examples of social injustice (Wilkinson and Marmot 2003). It is for these reasons that healthcare social equity efforts should include consideration of distribution, fairness, and quality (Hug 2015).

Fast forward to the twenty-first century, inequality continues to be evident in virtually all aspects of society: jobs, housing, and transportation (Frederickson 2005). This situation moves beyond these aspects and contributes to adverse health effects as well, as poor social and economic circumstances impact an individual's health throughout their lifetime (Wilkinson and Marmot 2003). Long-term, social and psychological circumstances can do significant damage to health (Marmot and Wilkinson 2005). Research has shown that social factors play a significant role in contributions to health inequity, including communicable and non-communicable diseases (Marmot and Wilkinson 2005). Psychosocial demands and other vulnerability factors contribute to and influence psychobiological stress responses, including neuroendocrine, autonomic, and immune responses (Marmot and Wilkinson 2005).

All around the world, people with fewer opportunities and privilege face more illnesses and shorter life expectancies (Wilkinson and Marmot 2003). This is likely due to a variety of reasons, including poor access to healthcare, decreased access to quality food, and low health literacy rates (Hug 2015). As social inequity has continued to increase, researchers have seen a growing need for informed health policies that take into account the examination of the relationship between health implications and social and economic policies (Wilkinson and Marmot 2003). Equity in health between different social groups should be a priority (Dahlgren and Whitehead 1991). A view of public health should move beyond individual people, diseases, and symptoms to include all determinants of health (Bentley 2014).

Ultimately, the focus on social determinants related to health is an essential component to understanding and effectively addressing these disparities (Marmot and Wilkinson 2005). These determinants of health include age, sex, education, employment, social hereditary factors, social influences, community influences, living conditions, working conditions, socioeconomic conditions, cultural conditions, environmental conditions, and individual lifestyle factors (Bentley 2014). There are also many challenges related to environmental determinants of health and equity, including shortages in clean water, climate change, and oil dependency (Bentley 2014). Other issues related to health equity that must be considered include adequate nutrition and access to adequate housing (Hug 2015). Even with this information, there is limited consideration of social determinants of health in public policy and public administration efforts (Hug 2015).

Poverty and the Health Gap

Poverty has been described as a persistent moral issue (Frederickson 2005). Social equity, especially in relation to poverty, has been described as a moral stance on the issue (Frederickson 2005). The gap is continuing to grow between those that are poor and the rest of society (Frederickson 2005). Residents of neighborhoods in low-income areas are directly involved with inequality, especially in relation to opportunity (Frederickson 2005). Poverty is often linked to various forms of disadvantage: poorer education, insecure employment, fewer family assets, living in poor housing, and other difficult circumstances (Wilkinson and Marmot 2003). People subject to low-income, or even moderate-income, suburbs and regions are less likely to have their voices heard by government officials (Frederickson 2005).

Health is an essential component of an individual's overall well-being (Braveman and Gruskin 2003). Even further, health is an essential component of overcoming the other harmful or negative effects of poverty and social disadvantage (Braveman and Gruskin 2003). Those at risk of experiencing inequality related to community health include, but are not limited to, migrant populations, elderly populations, younger populations, unemployed populations, and uneducated populations (Dahlgren and Whitehead 1991). People experiencing poverty, which already socially disadvantages individuals, face further disadvantages related to health outcomes (Braveman and Gruskin 2003).

Aspects related to poverty and deprivation are interlinked with poor health outcomes (Eades 2000). Research has shown that those living in more deprived areas have shorter life expectancies (Marmot and Wilkinson 2005). People who are considered lower on the "social ladder" are more likely to experience serious illness and premature death (Wilkinson and Marmot 2003). Poverty is so linked to health disparities that researchers described the issue as the *poverty-health complex* over 50 years ago (Kosa 1969). There is a significant division among different socioeconomic groups in regard to health and well-being (Dahlgren and Whitehead 1991). Studies have shown that people in low socioeconomic conditions have worse access to healthcare than people with greater incomes in higher socioeconomic conditions (Hug 2015). It is this magnitude of inequity that demonstrates the importance of achieving equity related to community health (Dahlgren and Whitehead 1991).

Advocacy and Improvement Efforts

Past efforts to incorporate social equity as part of a larger political agenda have included using neutral definitions, raising awareness, and involving fact-based information in policy-making processes (Dahlgren and Whitehead 1991). The history of equity in health is rooted in theories of justice and fairness (Bentley 2014). Health equity brings up the importance of enforcing civil rights (Hug 2015). Actively pursuing social equity requires policy-makers to consider the interests of the less advantaged and the less privileged members of their communities (Frederickson 2005). It requires policy-makers to actively look at the experiences of those with less opportunity (Frederickson 2005). Issues related to race, gender, and other inequalities should be actively engaged (Frederickson 2005). Achieving good health will involve addressing education disparities, financial insecurity, unemployment, and poor housing situations (Wilkinson and Marmot 2003). It has been stated that "public administration should be all about seeing after the interests of minorities and the poor" (Frederickson 2005, p. 38).

To work toward equity in health, everyone must collectively work toward eradicating systemic disparities that impact health (Braveman and Gruskin 2003). This includes working toward eradicating the barriers related to social determinants of health (Braveman and Gruskin 2003). Ignoring the powerful determinants of health inextricably linked to social inequity would also be ignoring the most significant social justice issues facing societies today (Wilkinson and Marmot 2003). The differences in health equity between advantaged and disadvantaged groups need to be assessed and evaluated to understand how best to create change that exists between these gaps (Braveman and Gruskin 2003). Comparing these differences allows practitioners to assess whether policies are bringing us closer or further from social equity and justice in community health and well-being (Braveman and Gruskin 2003). In order to do so, the disparities between those with underlying social advantage and social disadvantage must be addressed (Braveman and Gruskin 2003).

Social justice is also linked to environmental justice (Bentley 2014). From an ecological perspective, equity and justice are not only environmental concerns but also social concerns (Bentley 2014). This perspective asserts that there is no division between nature and culture; examining the connection between the two aids in the exploration of the various interactions and interactions that occur (Bentley 2014). In urban areas, we see the impact of socio-ecological factors, especially in relation to the reproduction of various social inequities and environmental inequities (Bentley 2014). From this perspective, it could be helpful to consider public health in a way that acknowledges people as part of a larger ecosystem (Bentley 2014).

On a local, national, and global level, principles of action include improving the conditions of daily life for those in low-income areas but also tackling the structural inequities present in our communities (Marmot 2008). This would include addressing the inequitable distribution of money, resources, and resulting power (Marmot 2008). It would also include measuring and evaluating the problem and action, increasing the awareness, training, and knowledge base regarding social equity and social determinants of health (Marmot 2008).

Efforts related to social equity start in the day-to-day tasks of public administrators (Frederickson 2005). Outcomes related to national policies can be focused on local families, neighborhoods, and cities (Frederickson 2005); that said, efforts should be made locally, it is argued, because the consequences seen from social inequity are often seen locally (Frederickson 2005). At the local level, public administrators are in a position where they can either influence new policies being developed or implement the policies that are already established in a way that addresses the effects of poverty and promotes opportunity (Frederickson 2005).

It has been recommended that those interested in social equity, and committed to the efforts that pursue it, should focus on specific causes and then enlist in the organizations that are likely to influence just and fair policies (Frederickson 2005). There are several traits that may aid in the effectiveness of policy-makers and policy administrators. People working within public policy and public administration are said to need wit, courage, and knowledge to make progress related to social equity in communities (Frederickson 2005). Money, organization, persistence, and determination could also aid in these efforts (Frederickson 2005).

Conclusion

It is difficult to ignore the vast and ever-growing disparities related to community health, especially considering the many inequities faced by those living in low-income suburbs and regions. Research has continued to stress the importance of focusing on social determinants related to health as factors that are inextricably linked to social equity and community health efforts. Improving the health and well-being of individuals, especially those living in low-income suburbs and regions, will require social and political action and advocacy efforts that directly consider and actively address the many social justice issues still relevant in communities. Locally, public officials and public administrators can focus on action by influencing new policies or implementing existing policies that address the effects of poverty on health and promoting opportunity among those living in low-income suburbs and regions.

Cross-References

- ► Need for Greenspace in an Urban Setting for Child Development
- ► Need for Nature Connectedness in Urban Youth for Environmental Sustainability

References

Bentley, M. (2014). An ecological public health approach to understanding the relationships between sustainable urban environments, public health and social equity. *Health Promotion International, 29*(3), 528–37.

Braveman, P., & Gruskin, S. (2003). Theory and methods: Defining equity in health. *Journal of Epidemiology and Community Health, 57*(4), 254–258.

Dahlgren, G., & Whitehead, M. (1991). *Policies and strategies to promote social equity in health.* Stockholm: Institute for Future Studies, 14.

Eades, S. J. (2000). Reconciliation, social equity and indigenous health. *Aboriginal and Islander Health Worker Journal, 172*(10), 468–469.

Frederickson, H. G. (2005). The state of social equity in American public administration. *National Civic Review, 94*(4), 31–38.

Frederickson, H. G. (2015). *Social equity and public administration: Origins, developments, and applications*. Routledge: Taylor and Francis Group, Abingdon: Oxfordshire.

Hug, R. W. (2015). Social equity, health, and health care. In *Justice for all: Promoting social equity in public administration, 1, 1–44, Routledge: Taylor and Francis Group, Abingdon: Oxfordshire.*

Kosa, J. E. (1969). *Poverty and health: A sociological analysis*. Harvard University Press, Cambridge: MA.

Marmot, M. (2008). *Closing the gap in a generation: Health equity through action on the social determinants of health: Commission on social determinants of health final report.* Geneva: World Health Organization Commission on Social Determinants of Health, 372(9650), 1661–1669.

Marmot, M., & Wilkinson, R. (2005). Social determinants of health. *The Lancet, 365*(9464), 1099–104.

Pascual, U., Phelps, J., Garmendia, E., Brown, K., Corbera, E., Martin, A., Gomez-Baggethun, E., & Muradian, M. (2014). Social equity matters in payments for ecosystems services. *BioScience, 64*(11), 1027–1036.

Sena, R. R., Seixas, C. T., & Silva, K. L. (2007). Practices in community health toward equity contributions of Brazilian nursing. *Advances in Nursing Science, 30*, 343–352.

Wilkinson, R. G., & Marmot, M. (2003). *Social determinants of health: The solid facts*. World Health Organization, Geneva.

Inclusive Cities

► Women in Urbanism, Perpetuating the Bias?

Increasing Young People's Environmental Awareness

Javier Esquer[1], Nora Munguia[1] and Luis Velazquez[2]

[1]Graduate Sustainability Program, Industrial Engineering Department, University of Sonora, Hermosillo, Mexico

[2]Industrial Engineering Department, University of Sonora, Hermosillo, Mexico

Introduction

With the release of the Agenda 2030 (United Nations 2015), and the 17 Sustainable Development Goals (SDG), the concern for environmental problems, like climate change crisis and water pollution, has aroused. Moreover, even the current COVID-19 pandemic has been linked to Climate Change, and thus, to sustainability, highlighting both biodiversity's vulnerabilities and humanity's vulnerabilities (Perkins et al. 2020). Youth involvement in sustainable development topics has been crucial. The way environmental initiatives are put in practice provides a way to demonstrate that sustainability is not a merely theoretical issue and that barriers in doing so must be addressed to deter them properly (Płonka and Dacko 2019). Additionally, with the increasing use of information and communication technologies (ICTs), education and environmental awareness among young people has been strengthened (Estevao Goulart 2017).

A Way of Visualizing Young's People Thoughts Environment

Since children need to interact with others through different stages such as childhood, adolescence, and youth, well-being is subject to various factors such as the education received and the economic difficulties that the person has – family, pressure from friends, and social networks, among other aspects (Birch et al. 2020). According to education experts, propelling young people toward a positive education will provide them with increased resilience, hope, and gratitude (Trask-Kerr et al. 2019). Furthermore, it leads to forming a school culture that seeks to improve safety and well-being and give importance to relationships, happiness, and caring for the environment.

This situation leads to forming a school culture that seeks to increase safety and well-being and give importance to relationships, happiness, and caring for the environment. Young people have become increasingly disillusioned with politics and political parties over several decades. They are now much more likely to be interested in issues than join political parties and are interested in a wide range of subjects – from housing to crime to the environment (Kiraz and Firat 2016). That is why it is necessary to take advantage of that environmental concern to awaken a way of life.

Young people will be the generations to face the problems derived from being indifferent to what happens around us in terms of the environment; this concern must include today's young people as well as the not-so-young. In the conditions that we leave the planet, it will be relevant for its immediate future, according to partner organizations for the Convention on Biological Diversity in 2018 (Chawla 2020). Such a treaty deals with the need to connect people with their environmental surroundings and thus undertake actions for its conservation. Although this report focuses on childhood, it nevertheless applies to all ages. In the case of young people, they have the challenge of moving toward more sustainable futures that allow them to create communities with greater empathy and attitude toward the environment (Shafiei and Maleksaeidi 2020).

Informally today, the so-called "influencers" could be a support in which young people are reflected and tend to imitate their performance and thus help to awaken or increase environmental awareness. One of the most notorious cases is Greta Thunberg, an environmental activist of Swedish origin who was invited to COP 24 in 2018 to reduce global warming (Sutter and Davidson 2018). Greta, although has received also a lot of criticism from a sector of society (Zraick 2019), is a young woman who has awakened the interest of young people. She calls to hold protest marches and participate with organizations such as Fridays for Future Scotland to promote the fight against climate change.

It is known that today other influencer organizations promote actions to awaken that environmental awareness, mainly in young people. On their website, Ecovidrio (2019) lists a series of young people dedicated to promoting and executing actions aimed at raising environmental awareness in different aspects such as reforestation, climate change, sustainable fashion, and recycling, in other topics related to the environment. Most influencers use social networks as their communication tool, so this type of communication has gained strength. As a result, the messages to promote environmental awareness about the various ecological problems are disseminated faster and very short (Mallick and Bajpai 2018).

The Role of Education in Increasing Sustainable Environmental Awareness

According to Partanen-Hertell et al. (1999), environmental awareness (EA) is defined as a combination of elements that interact to achieve environmental awareness. These elements are knowledge, motivation, and skills that provide strategies that distinguish each of them, and together they form all the aspects that determine environmental awareness. In the case of knowledge, it could be clearly defined as knowledge of facts, truths, or principles. People's knowledge about their environment is essential for developing their EA. The motivation to improve the environment is based on values and attitudes (Bamberg and Möser 2007).

When environmentally aware individuals encounter an external physical stimulus, they may realize the potential for some action. If their worldview and values support pro-environmental efforts, they are motivated to make environmentally friendly choices. Based on an individual's knowledge and skills and current opportunities to act, this motivation may manifest in pro-environmental actions (Harju-autti and Kokkinen 2014).

Environmental awareness is one of the components included in ecological literacy, consisting of five aspects: knowledge, awareness, behavior, involvement, and attitude (Jannah et al. 2013). On the other hand, the indicators of ecological awareness defined by the OECD are described by:

(a) Awareness of environmental problems represents how informed students are about current ecological issues.
(b) Perception of environmental problems, which represents how concerned students are about ecological problems.
(c) Environmental optimism: The students' belief that their actions or human actions can contribute to maintaining and improving the environment.

Environmental awareness is an integral part of the movement's success. By teaching our friends and family that the physical environment is fragile and indispensable, we can begin to solve the problems that threaten it (Partanen-Hertell et al. 1999).

Environmental education plays a fundamental role in influencing the change of attitudes and ecological behavior, awakening positive initiatives toward the preservation and protection of the environment and the search for solutions to environmental problems. From primary education, it is possible to trigger that involvement of the little ones and revive in them that responsibility toward environmental care. With the decree of the decade of education for sustainable development (ESD) by UNESCO in 2005, skills, abilities, and knowledge were developed to change society toward sustainable lifestyles for citizens. As a result, there is expected participation toward awakening conscience toward protecting the environment (Buckler and Creech 2014).

According to the United States Environmental Protection Agency (EPA 2001), environmental education seeks to integrate people and the environment they develop, solving problems and applying action measures to improve the environment. In addition, to understand in a more conscious and informed way and thus be more responsible in the decisions and actions that are taken. The components of awareness, sensitivity, and knowledge, and understanding that environmental education must face include the attitudes, participation, and skills required to achieve it.

Environmental Education was officially instituted during the 1970s; several important international conferences addressed Environmental Education needed to be taught at an early age and continue for all the following courses of the school trajectory (Tsekos et al. 2012). Teachers or instructors play a significant role in environmental education since they influence their students. For example, Tuncer et al. (2009) argued that teachers create environmentally literate students. In addition to ecological knowledge, they must have positive attitudes toward the environment and be concerned about environmental problems.

In many cases, the foundations of primary education are taught directly or indirectly on environmental issues. Therefore, schools play a crucial role in culture and education; therefore, it is necessary to take advantage of their high potential to be used efficiently (Loughland et al. 2002). At the same time, the children learn and understand together because of the care for the environment or nature. Moreover, understanding that environmental problems have adverse consequences on the personal health of human beings and, secondly, affect the economy, society, and culture. Therefore, what is sought is to create public concern about environmental hazards. These problems are, in many cases, directly related to human activities, and they are affected by environmental changes (Arani et al. 2016).

In the 1990s, the concern that most concerned educators were the misuse of resources, which is why they brought discussions on environmental education to the fore, later becoming the forerunner of education for sustainable development (Aikens et al. 2016). UNESCO, for its part, made a global effort to promote education for sustainable development (ESD) and try to improve the curricula during the so-called decade. The decade of ESD is followed by the Global ESD Action Program (GAP), which began in 2015 and ended in 2019, contributing substantially to the 2030 plan to scale up ESD action in all academic areas and disciplines. More recently, the UN implemented 17 sustainability goals in 2015. Specifically, Goal four focused on ensuring access to quality education for all children. Some academics affirm that education for sustainable development is the most important of the 17 SDGs.

Higher education institutions (HEIs) are responsible for transmitting generation by generation concern for the environment and feeding, through the curricula, sustainable attitudes and practices for their lives (Chinedu et al. 2018). Universities also play an essential role in developing "sustainable solutions," which has enormous implications for university research (Cortese 2003). On the other hand, education for sustainable development promotes academic learning. It applies knowledge to encourage activities and conduct research for the benefit of the general population that helps solve long-term problems.

Among the activities carried out are those that work to implement sustainability in academic study plans and those focused on university sustainability. During the last decade, universities witnessed enormous progress in sustainability through curricula and campus actions and have led outcomes like conferences, articles, or documents focused on making a sustainable campus, presented as success stories in different parts of the world (Vaughter et al. 2013).

Integrating education for sustainable development within the school curriculum is increasingly relevant (Haensly et al. 1985). That is why both curricular and extracurricular activities seek to integrate an education focused on environmental issues as activities that have to do with concern for the environment. Curricular activities, for their part, have helped to incorporate and highlight environmental topics learned in class into their daily life. In contrast, extracurricular activities, regularly known as supplementary activities, have been shown to enormously help create this environmental concern (Amran et al. 2019).

The Role of Social Media in Environmental Awareness

Nowadays, it is not only common to talk about Information and Communication Technologies (ICT) as well as social networks, or social media, but, even more, a large part of the population is involved with them. ICT and social media are profoundly interconnected and can be used to trigger social changes, including issues like economic development, political progress, cultural change, and social revolution (Life Learners 2018).

First, *information and communication technologies* (ICT) can be defined as "a diverse set of technological tools and resources used to transmit, store, create, share or exchange information. These technological tools and resources include computers, the Internet (websites, blogs and emails), live broadcasting technologies (radio, television and webcasting), recorded broadcasting technologies (podcasting, audio and video players, and storage devices) and telephony (fixed or mobile, satellite, visio/video-conferencing, etc.)" (UNESCO 2009).

Second, the term *social media* involves a wide spectrum of many things where people can either create or share information with other people or interact with them (Marshall 2018). Social media has multiple definitions that, in some way, are closely related. For instance, according to Brunskill (2015), they may refer to "Internet technologies that allow people to connect, communicate and interact in real time to share and exchange information (text, photographs, images, video or audio files)." Likewise, Carr and Hayes (2015) describe social media as "Internet-based channels that allow users to opportunistically interact and selectively self-present, either in real-time or asynchronously, with both broad and narrow audiences who derive value from user-generated content and the perception of interaction with others."

Globally, there are more than 3.8 billion users (Dollarhide 2021) on countless social media platforms where, every day at every time, people of all ages and cultures interact and share information of many kinds. The most popular of these platforms are listed in Table 1.

Like many things in our society, information technologies might be a double-edged sword with their pros and cons. On the one hand, we can mention that some disadvantages are ethical issues derived from their misuse (Rauf 2021). For instance, the psychological attachment that people have to social media (Brunskill 2015) is considered "an essential part of adolescent life," producing undesired health behaviors (Buda et al. 2021). This includes addiction to playing computer games, "being online," pathological gambling, as well as a negative impact on school education and adolescent sleep length (Wojdan et al. 2021). All of these aspects may contribute also to social isolation coupled with environmental destruction (Threadgold 2012). Social media can also trigger unnecessary consumption through many sponsored posts and ads popping up along the different platforms and apps that promote a variety of goods and services (Sertdurak 2020).

On the other hand, despite the difficulties that ICTs and social media may pose, for this manuscript, we want to highlight their advantages, particularly for topics related to sustainable development and environmental awareness. In this sense, some of the sustainability reports of the main social media, or environmental publications about them were browsed, leading to the following key points (Table 2):

Social media can be a low-cost, effective tool to encourage the thinking and decision-making of new technology users to be inclined toward certain behaviors (Zhang et al. 2021); including the ability to influence how users understand sociopolitical messages on social media affecting how the source of sociopolitical news is considered within society (Wilkins et al. 2021). Some of the key players making an important role in sending those messages are the so-called social media influencers (SMIs), who "represent a new type of independent third-party endorser who shapes audience attitudes through blogs, tweets, and the use of other social media" (Freberg et al. 2011). In addition, these influencers sometimes consider themselves as a moral reference when supporting a particular case, either favoring some rights or being against what they think is inappropriate (Marshall 2018).

Increasing Young People's Environmental Awareness, Table 1 Most popular social media platforms (data are billions of users)

1. Facebook (2.85)	7. TikTok (0.732)	13. Kuaishou (0.481)
2. YouTube (2.29)	8. QQ (0.606)	14. Pinterest (0.478)
3. WhatsApp (2.00)	9. Douyin (0.600)	15. Reddit (0.430)
4. Instagram (1.38)	10. Telegram (0.550)	16. Twitter (0.397)
5. Facebook Messenger (1.30)	11. Sina Weibo (0.530)	17. Quora (0.300)
6. Weixin/ WeChat (1.24)	12. Snapchat (0.514)	

Source: Own elaboration based on Statista (2021)

Increasing Young People's Environmental Awareness, Table 2 Sustainability and/or environmental statements related to social media platforms

Social media (Source)	Main sustainability and/or environmental statements related to their users
(a) Facebook (2020)	Facebook is a company that owns Facebook app, Oculus, Portal, Instagram, Workplace, Messenger, Novi, and WhatsApp. In their Sustainability Report 2020 they have the following statements: Beyond doing our part to reduce our environmental footprint, our approach is to accelerate access to authoritative information and encourage positive action on climate through our core products and services, while working with others to scale solutions that help create a healthier planet for all. Internal and external stakeholder perspectives are important to our sustainability journey. We regularly have formal and informal meetings and conversations with different stakeholders, including the people who use our programs and technologies, colleagues, communities, suppliers, industry peers, nongovernmental organizations (NGOs), policymakers, and investors. These conversations help inform our sustainability programs and advance our progress. Using Facebook's platform, products, and services to promote and enable a more equitable, safe, and healthy society. Empowering action on key sustainability issues and mitigating negative impacts through core products and services.
(b) Google (2020)	Google's core products and platforms are Android, Chrome, Gmail, Google Drive, Google Maps, Google Play, Search, and YouTube. We're driving positive environmental impact throughout our value chain in five key ways: designing efficient data centers, advancing carbon-free energy, creating sustainable workplaces, building better devices and services, and empowering users with technology.
(c) Twitter (2020)	We believe that public conversation is better when as many people as possible can participate. With that philosophy in mind, we work hard to promote healthy conversations on our service. Creating our environmental sustainability strategy has brought

(continued)

Increasing Young People's Environmental Awareness, Table 2 (continued)

Social media (Source)	Main sustainability and/or environmental statements related to their users
	together passionate Tweeps* across the company to determine how we can most responsibly operate our business and use our platform to make a positive environmental impact around the world. [*Tweep is a person who uses the Twitter online message service to send and receive tweets (Merriam-Webster 2021)]
(d) Weixin/ WeChat (Chen 2020)	Research results show that Chinese WeChat users' environmental information-sharing behavior is motivated by both egoistic factors (self-presentation, information seeking, and socializing) and altruistic factors (awareness of consequences and ascription of responsibility). From the perspective of egotistic motivations, this study has identified three types of gratification with positive influence: The desire for entertainment, self-presentation, and having social interactions with others. From the perspective of altruistic motivations, Chinese WeChat users' awareness of consequences and ascription of responsibility significantly influence the antecedents of the environmental information-sharing behavior. First, the results show that awareness of consequences affects people's attitudes, perceived subjective norms, and personal norms regarding sharing behavior. Chinese WeChat users' attitudes, subjective norms, and personal norms toward sharing environmental information are vital factors driving their behavioral intention.
(e) TikTok	We work to maintain an environment where everyone feels safe and welcome to create videos, find community, and be entertained. We believe that feeling safe is essential to feeling comfortable expressing yourself authentically, which is why we strive to uphold our Community Guidelines by removing accounts and content that violate them. Our goal is for TikTok to remain a place for inspiration, creativity, and joy (Tiktok 2021).

(continued)

I

Increasing Young People's Environmental Awareness, Table 2 (continued)

Social media (Source)	Main sustainability and/or environmental statements related to their users
	In just a few scrolls, users can learn about climate change, sustainability practices, and conscious shopping. Designers, climate change experts, and vintage stores have used the platform to share tips, tricks, and facts about saving our planet. While some may use TikTok as an entertaining, humorous escape from the chaos of the world, others are using the app as an educational source. Conscious consumerism is a common thread among these videos – the idea that consumers need to understand the true impact of their purchases. This activism is centered around social and environmental change. TikTok is predominately popular among Gen-Z – the generation that will undoubtedly carry the weight of the climate crisis on their shoulders (Gurry 2020).
(f) Pinterest (Swasti 2019)	In fact, many Pinners are inspired to get out of their comfort zone and try a more eco-conscious lifestyle for the first time, with searches for "sustainable living for beginners" up by 265% since last year. Top Sustainability Searches on Pinterest: Sustainable living, Sustainable architecture, Sustainable fashion, Sustainable packaging, Sustainable home, Sustainability in childcare, Self-sustainable living, Sustainable development, Sustainable energy, Sustainable products.
(g) Snapchat (Snap Inc., 2021)	At Snap, our mission is to contribute to human progress by empowering people to express themselves, live in the moment, learn about the world, and have fun together. In doing so we lead with our values – kind, smart and creative – but it always begins with kindness. We design kindness into the way we run our company, how we build our products, and how we treat our people, society, and the planet. That's how we think about Environmental, Social, and Governance (ESG), and we set our ESG priorities to match. At Snap we recognize that we hold a unique position and the opportunity to engage with our community of Snapchatters to catalyze change for the future. We know that reducing our impacts is important, but leveraging our platform to raise awareness and catalyze change on environmental issues can have an even bigger impact. To increase environmental awareness and engage with our Snapchat community of 280 million daily active users we have partnered with the UN Environment Programme and other organizations to create in-app activations that educate our community and provide them with resources to become a part of the solution.

Source: Own elaboration based on the several sources cited

Wielki (2020) proposed five categories for influencers regarding their motivation to take action: idols, experts, lifestylers, artists, and activists. The last one includes those people concerned with topics like environmental protection. Although activists usually conduct actions "offline," that is, in the field, nowadays, social media have enabled mobilization and coordination of such efforts by connecting with an enormous amount of the population (Cammaerts 2015; Sertdurak 2020). That is why advocates for environmental protection causes use social media more frequently as a leading means to increase communities' knowledge and commitment to supporting their initiatives by addressing technical issues as well as society's structure and culture (Lallana 2014).

In the end, sustainability and social media are two megatrends where the latter are crucial in raising sustainability awareness (Lee et al. 2021), particularly for young people (Vanko and Zaušková 2019). Innovative ways to use social media and ICTs can be helpful to increase environmental consciousness and a sustainable lifestyle by addressing local knowledge that is valuable for informed environmental planning decisions (Krätzig and Warren-Kretzschmar 2014). In this sense, Souter (2012) describes different levels or orders of ICTs impact on sustainability:

- First-order (or direct) effects are related to manufacturing ICTs devices, including computers and mobile phones. These usually tend to be more negative than positive due to the carbon-related emissions of the production processes.
- Second-order (or indirect) effects are related to how those ICTs are used. In contrast, these usually tend to be ultimately more favorable due to the dematerialization, by using more virtual goods, and increasing energy and human efficiency as a result of technological improvements.
- Third-order (or societal) effects are the combined results of many people using ICTs over the medium-to-long term in ways that alter how economies and societies work.

Conclusion

As stated by Płonka and Dacko (2019), the future for the younger generations must be designed with high-level talent, from elementary school toward higher education, seeding a systems approach toward sustainability solutions for the extremely complex environment, society, and economic problems through a variety of scientific disciplines.

Cross-References

- ▶ Environmental Education and Non-Governmental Organizations
- ▶ Sustainability Competencies in Higher Education

References

Aikens, K., McKenzie, M., & Vaughter, P. (2016). Environmental and sustainability education policy research: A systematic review of methodological and thematic trends. *Environmental Education Research, 22*(3), 333–359. https://doi.org/10.1080/13504622.2015.1135418.

Amran, A., Perkasa, M., Satriawan, M., Jasin, I., & Irwansyah, M. (2019). Assessing students 21st century attitude and environmental awareness: Promoting education for sustainable development through science education. *Journal of Physics: Conference Series, 1157*(2), 022025. https://doi.org/10.1088/1742-6596/1157/2/022025.

Arani, M. H., Bagheri, S., & Ghaneian, M. T. (2016). The role of environmental education in increasing the awareness of primary school students and reducing environmental risks. *Journal of Environmental Health and Sustainable Development, 1*(1), 9–17.

Bamberg, S., & Möser, G. (2007). Twenty years after Hines, Hungerford, and Tomera: A new meta-analysis of psycho-social determinants of pro-environmental behaviour. *Journal of Environmental Psychology, 27*(1), 14–25. https://doi.org/10.1016/J.JENVP.2006.12.002.

Birch, J., Rishbeth, C., & Payne, S. R. (2020). Nature doesn't judge you – How urban nature supports young people's mental health and wellbeing in a diverse UK city. *Health & Place, 62*, 102296. https://doi.org/10.1016/J.HEALTHPLACE.2020.102296.

Brunskill, D. (2015). The dangers of social media for the psyche. In A. Bennet (Ed.), *Social media : Global perspectives, applications and benefits and dangers*. Nova Science Publishers, Inc..

Buckler, C., & Creech, H. (2014). *Shaping the future we want : UN decade of education for sustainable development (2005–2014): Final report*. United Nations Educational, Scientific and Cultural Organization.

Buda, G., Lukoševičiūtė, J., Šalčiūnaitė, L., & Šmigelskas, K. (2021). Possible effects of social media use on adolescent health behaviors and perceptions. *Psychological Reports, 124*(3), 1031–1048. https://doi.org/10.1177/0033294120922481.

Cammaerts, B. (2015). Social media and activism. In R. Mansell & P. Hwa (Eds.), *The international encyclopedia of digital communication and society* (pp. 1027–1034). Wiley-Blackwell. https://doi.org/10.1002/9781118767771.wbiedcs083.

Carr, C. T., & Hayes, R. A. (2015). Social media: Defining, developing, and divining. *Atlantic Journal of Communication, 23*(1), 46–65. https://doi.org/10.1080/15456870.2015.972282.

Chawla, L. (2020). Childhood nature connection and constructive hope: A review of research on connecting with nature and coping with environmental loss. *People and Nature, 2*(3), 619–642. https://doi.org/10.1002/PAN3.10128/SUPPINFO.

Chen, Y. (2020). An investigation of the influencing factors of chinese wechat users' environmental information-sharing behavior based on an integrated model of UGT, NAM, and TPB. *Sustainability, 12*(7), 2710. https://doi.org/10.3390/su12072710.

Chinedu, C. C., Wan-Mohamed, W. A., Ogbonnia, & Abdurrahman, A. (2018). A systematic review on education for sustainable development: Enhancing Tve Teacher Training Programme. *Journal of Technical Education and Training, 10*(1), 109. https://doi.org/10.30880/jtet.2018.10.01.009.

I

Cortese, A. D. (2003). The critical role of higher education in creating a sustainable future. *Planning for Higher Education, 31*(3), 15–22.

Dollarhide, M. (2021, August 31). *Social media definition: Sharing ideas & thoughts*. Investopedia – Small Business; Dotdash publishing. https://www.investopedia.com/terms/s/social-media.asp

Ecovidrio. (2019). *Reciclaje de envases de vidrio en España | Ecovidrio*. https://www.ecovidrio.es/

EPA. (2001). *An organizational guide to pollution prevention*. U.S. Environmental Protection Agency.

Estevao Goulart, E. (2017). Cultural and educational aspects of using social media: A study with undergraduate students. *Estudios Sobre Las Culturas Contemporáneas, 23*(3), 27–40.

Facebook. (2020). *Sustainability report 2020*. Facebook.

Freberg, K., Graham, K., McGaughey, K., & Freberg, L. A. (2011). Who are the social media influencers? A study of public perceptions of personality. *Public Relations Review, 37*(1), 90–92. https://doi.org/10.1016/j.pubrev.2010.11.001.

Google. (2020). *Google environmental report 2020*. Google.

Gurry, L. (2020, October 10). *TikTok: An unexpected leader In sustainability*. https://curiosityshots.com/tiktok-and-sustainability/.

Haensly, P. A., Lupkowski, A. E., & Edlind, E. P. (1985). The role of extracurricular activities in education. *The High School Journal, 69*(2), 110–119.

Harju-autti, P., & Kokkinen, E. (2014). A novel environmental awareness index measured cross-nationally for fifty seven countries. *Universal Journal of Environmental Research and Technology, 4*(4), 178–198.

Jannah, M., Halim, L., Meerah, T. S. M., & Fairuz, M. (2013). Impact of environmental education kit on students' environmental literacy. *Asian Social Science, 9*(12), 1–12. https://doi.org/10.5539/ass.v9n12p1.

Kiraz, A., & Firat, A. (2016). Analyzing the environmental awareness of students according to their educational stage. *Researchers World: Journal of Arts, Science and Commerce, VII*(2), 15–25. https://doi.org/10.18843/rwjasc/v7i2/02.

Krätzig, S., & Warren-Kretzschmar, B. (2014). Using interactive web tools in environmental planning to improve communication about sustainable development. *Sustainability, 6*(1), 236–250. https://doi.org/10.3390/su6010236.

Lallana, E. C. (2014). *Social media for development*. United Nations Asian and Pacific Training Centre for Information and Communication Technology for Development.

Lee, J. H., Wood, J., & Kim, J. (2021). Tracing the trends in sustainability and social media research using topic modeling. *Sustainability, 13*(3), 1–23. https://doi.org/10.3390/su13031269.

Life Learners. (2018, June 7). *Relationship between Social Media and ICT – Life Learners Limited*. https://lifelearners.ng/relationship-between-social-media-and-ict/

Loughland, T., Reid, A., & Petocz, P. (2002). Young people's conceptions of environment: A phenomenographic analysis. *Environmental Education Research, 8*(2), 187–197. https://doi.org/10.1080/13504620220128248.

Mallick, R., & Bajpai, S. P. (2018). Environmental awareness and the role of social media. In S. Narula, S. Rai, & A. Sharma (Eds.), *Impact of social media on environmental awareness* (pp. 140–149). IGI Global. https://doi.org/10.4018/978-1-5225-5291-8.ch007.

Marshall, C. (2018). *Writing for social media*. BCS Learning & Development Ltd.

Merriam-Webster. (2021). *Tweep | Definition*. https://www.merriam-webster.com/dictionary/tweep

Partanen-Hertell, M., Harju-Autti, P., Kreft-Burman, K., & Pemberton, D. (1999). *Raising environmental awareness in the Baltic Sea area*.

Perkins, K. M., Munguia, N., Ellenbecker, M., Moure-Eraso, R., & Velazquez, L. (2020). COVID-19 pandemic lessons to facilitate future engagement in the global climate crisis. *Journal of Cleaner Production, 290*, 125178. https://doi.org/10.1016/j.jclepro.2020.125178.

Płonka, A., & Dacko, M. (2019). Secondary school youth and the idea of sustainable development – Opinions and attitudes. *Scientific Papers of Silesian University of Technology – Organization and Management Series*, 139, 489–502.

Rauf, A. A. (2021). *New moralities for new media ? Assessing the role of social media in acts of terror and providing points of deliberation for business ethics*. 229–251. https://doi.org/10.1007/s10551-020-04635-w.

Sertdurak, N. E. (2020, November 18). *The position of social media in sustainability: An obstacle or a step in reaching the goals?* . The Sustainable Development Watch. https://sdwatch.eu/2020/11/the-position-of-social-media-in-sustainability-an-obstacle-or-a-step-in-reaching-the-goals/

Shafiei, A., & Maleksaeidi, H. (2020). Pro-environmental behavior of university students: Application of protection motivation theory. *Global Ecology and Conservation, 22*, e00908. https://doi.org/10.1016/J.GECCO.2020.E00908.

Snap Inc. (2021). *2021 CitizenSnap report*. Snap Inc.

Souter, D. (2012). *ICTs, the Internet, and sustainability: A discussion paper* (Issue May).

Statista. (2021, July 1). *Most popular social networks worldwide*. Social Media & User-Generated Content; Statista. https://www.statista.com/statistics/272014/global-social-networks-ranked-by-number-of-users/

Sutter, J., & Davidson, L. (2018, December 17). *Teen tells climate negotiators they aren't mature enough*. CNN. https://edition.cnn.com/2018/12/16/world/greta-thunberg-cop24/index.html

Swasti, S. (2019, February 7). *How Pinterest inspires people to live a more sustainable lifestyle*. https://newsroom.pinterest.com/en/node/5601.

Threadgold, S. (2012). "I reckon my life will be easy, but my kids will be buggered": Ambivalence in young people's positive perceptions of individual futures and their visions of environmental collapse. *Journal of Youth Studies, 15*(1), 17–32. https://doi.org/10.1080/13676261.2011.618490.

Tiktok. (2021, July 1). *Tiktok Transparency Report*. https://www.tiktok.com/safety/resources/transparency-report-2020-2?lang=en

Trask-Kerr, K., Chin, T. C., & Vella-Brodrick, D. (2019). Positive education and the new prosperity: Exploring young people's conceptions of prosperity and success. *Australian Journal of Education, 63*(2), 190–208. https://doi.org/10.1177/0004944119860600.

Tsekos, C. A., Christoforidou, E. I., & Tsekos, E. A. (2012). Planning an environmental education project for kindergarten under the theme of "The Forest.". *Review of European Studies, 4*(2), 111–117. https://doi.org/10.5539/res.v4n2p111.

Tuncer, G., Tekkaya, C., Sungur, S., Cakiroglu, J., Ertepinar, H., & Kaplowitz, M. (2009). Assessing pre-service teachers' environmental literacy in Turkey as a mean to develop teacher education programs. *International Journal of Educational Development, 29*(4), 426–436. https://doi.org/10.1016/J.IJEDUDEV.2008.10.003.

Twitter. (2020). *Global impact report*. Twitter.

UNESCO. (2009). Guide to measuring information and communication technologies (ICT) in education. In *UNESCO Institute for Statistics (UIS) – technical paper 2* (issue 2). United Nations Educational, Scientific and Cultural Organization – UNESCO Institute for Statistics. https://doi.org/10.15220/978-92-9189-078-1-en

United Nations. (2015). Transforming our world: The 2030 agenda for sustainable development. In *United Nations*. United Nations. https://doi.org/10.1007/s13398-014-0173-7.2.

Vanko, M., & Zaušková, A. (2019). Raising public awareness of eco-innovations through social media. *Proceedings of the European Conference on Innovation and Entrepreneurship, ECIE, 2*(2010), 1076–1085. https://doi.org/10.34190/ECIE.19.167.

Vaughter, P., Wright, T., McKenzie, M., & Lidstone, L. (2013). Greening the ivory tower: A review of educational research on sustainability in post-secondary education. *Sustainability, 5*(5), 2252–2271. https://doi.org/10.3390/SU5052252.

Wielki, J. (2020). Analysis of the role of digital influencers and their impact on the functioning of the contemporary on-line promotional system and its sustainable development. *Sustainability, 12*(17). https://doi.org/10.3390/su12177138.

Wilkins, D. J., Livingstone, A. G., & Levine, M. (2021). One of us or one of them? How "peripheral" adverts on social media affect the social categorization of socio-political message givers. *Psychology of Popular Media, 10*(3), 372–381. https://doi.org/10.1037/ppm0000322.

Wojdan, W., Wdowiak, K., Witas, A., Drogoń, J., & Brakowiecki, W. (2021). The impact of social media on the lifestyle of young people. *Polish Journal of Public Health, 130*(1), 8–13. https://doi.org/10.2478/pjph-2020-0003.

Zhang, W., Chintagunta, P. K., & Kalwani, M. U. (2021). Social media, influencers, and adoption of an eco-friendly product: Field experiment evidence from rural China. *Journal of Marketing, 85*(3), 10–27. https://doi.org/10.1177/0022242920985784.

Zraick, K. (2019, September 24). *Greta Thunberg, after pointed U.N. speech, faces attacks from the right*. The New York Times. https://www.nytimes.com/2019/09/24/climate/greta-thunberg-un.html

INDCs – Intended Nationally Determined Contributions

▶ Adapting to a Changing Climate Through Nature-Based Solutions

Independent Mobility

▶ Pre-schoolers and Sustainable Urban Transport

India Meteorological Department (IMD)

▶ The State of Extreme Events in India

Indian Summer Monsoons (ISM)

▶ The State of Extreme Events in India

Indigenous Cosmologies

▶ New Orbital Urbanization

Industrial Symbiosis

Towards More Resource-Efficient Industrial Systems

Sigrid Kusch-Brandt
Department of Civil, Environmental and Architectural Engineering, University of Padua, Padua, Italy
Faculty of Mathematics, Natural Sciences and Management, University of Applied Sciences Ulm, Ulm, Germany

Synonyms

Eco-industrial network; Industrial synergy; Networked eco-industrial system

Definition

Industrial symbiosis describes synergistic business relationships between traditionally separated industrial entities, resulting from networked endeavors to collaborative manage material or energetic resources, infrastructures, capacities, or know-how. Nonindustrial organizations can be included as well. By establishing mutually beneficial interlinkages between participants, such a network achieves more efficient use of materials, energy, or other resources, and thus accomplishes both higher business profit and reduced adverse impacts on the environment. Interfirm valorization of by-products and wastes is one common feature of industrial symbiosis.

Introduction

In natural ecosystems, a symbiosis captures a close interaction of living species. As an analogy, industrial symbiosis refers to complexly interlinked companies or other organizations in an industrialized environment. The widely cited definition of Chertow (2000) states: "The part of industrial ecology known as industrial symbiosis engages traditionally separate industries in a collective approach to competitive advantage involving physical exchange of materials, energy, water and by-products." Lombardi and Laybourn (2012) have specified: "[Industrial symbiosis] engages diverse organizations in a network to foster eco-innovation and long-term cultural change. Creating and sharing knowledge through the network yields mutually profitable transactions for novel sourcing of required inputs, value-added destinations for non-product outputs, and improved business and technical processes."

Since the beginning of the twenty-first century, when the OECD (Organization for Economic Cooperation and Development) recognized industrial symbiosis as an innovative and powerful element to advance green growth, the creation of networked eco-industrial systems has evolved into a prioritized strategy on policy agendas to achieve more sustainable management of natural resources (Kusch-Brandt 2020). As an example, the European Commission, embedded in its flagship Roadmap to a Resource Efficient Europe, has recommended the implementation of industrial symbiosis to EU member states.

Main Characteristics of Industrial Symbiosis

The most essential element of industrial symbiosis is the establishment of cooperative interactions between business actors which create win-win situations for all participants. Companies or other industrial entities are important actors of a symbiotically functioning network in an industrial environment, but the networked system can also include other actors such as municipalities.

A prevalent type of interlinkage between members of an industrial symbiosis is the value-added cooperative management of material and energy, which involves interfirm exchange of physical resources. This typically includes cases where materials regarded as waste by one entity become valuable resources for networked entities (Chertow and Ehrenfeld 2012). Other elements with potential to unlock synergies are the sharing of infrastructures, capacities, or expertise. Pooling of resources, networked usage of technical equipment or facilities, joint treatment of emissions, joint mitigation of risks, and sharing of information and skills between business actors are some examples (Lombardi and Laybourn 2012; Martin et al. 2015; van Berkel 2010). The symbiotic interactions can be classified into two main types (Kosmol 2019): (1) exchange-based business relationships (e.g., reuse of wastes and by-products) and (2) sharing-based business relationships (e.g., shared usage of utilities, infrastructures, services, logistics, knowledge). Physical flows of material are often in focus when analyzing networked eco-industrial systems, while it is more recent that sharing of information and knowledge has been recognized as an element to promote industrial symbiosis; in particular, ICT tools (information and communication technologies) can ease networked interactions and thus support business success (Kosmol 2019; Maqbool et al. 2018).

Spatial proximity, a requirement of a symbiosis between living species in nature, facilitates industrial symbiosis, especially by reducing transport costs, but it is neither necessary nor sufficient for industrial symbiosis (Lombardi and Laybourn 2012). Network members can be located at different geographical sites and, depending on the type of interaction and the economic implications, might even be in far distance from each other (Vladimirova et al. 2018). Spatial proximity is a crucial decision criterion regarding a shared usage of equipment or physical infrastructure, or an exchange of resources with low technical or economic transportability. As an example, supply of steam can only be implemented over short distances. For by-products or wastes with low market value per unit of mass (e.g., food industry residues or construction and demolition waste), transport distances have a significant impact on the economic viability of valorization schemes, but for materials with high market value, such as rare earth elements, a short transport distance is less relevant compared to better business opportunities at a more remote location (Domenech et al. 2018).

Industrial symbiosis creates dynamic networks with quantitatively and qualitatively strongly connected participants who cooperatively unlock synergies in the system to increase value creation (Boons et al. 2014). This goes beyond traditional approaches of considering single economic entities as potential contract partners for specific transactions (e.g., delivery of waste for recycling). While the single entity remains an autonomous business actor, it also responds to the needs of network members to enable cooperative solutions; thus, a coevolution of participants occurs. Process steps to modify a by-product might be required (Domenech et al. 2018), or changes in manufacturing schemes or management procedures.

Benefits of Symbiotic Industrial Relationships

Industrial symbiosis makes a strong contribution to sustainable development (Laybourne and Lombardi 2012; UNEP and UNECE 2016) and more circular economies (Dong et al. 2021); environmental benefits, positive economic impacts, and socially beneficial impacts are achieved (Domenech et al. 2018). Business actors primarily benefit from economic advantages related to lower costs for resources or facilities and from new business opportunities. At the same time, shared usage of facilities, networked management of resources, and valorization of by-products and wastes means less burden to the environment; it saves primary resources and reduces emissions (Chertow and Ehrenfeld 2012). This translates into lower societal costs to manage environmental pollution and wastes. Increasing economic prosperity becomes decoupled from environmental degradation. Furthermore, network-oriented business practices enhance the innovation capacity and resilience of industrial systems, and thus facilitate adaptations to changing environments (Kusch-Brandt 2020).

Challenges and Research Perspectives

Several hundreds of industrial symbiosis cases are documented in literature (Kusch-Brandt 2020); the best-known example is the Danish

Kalundborg industrial network which has been evolving since the 1960s (Chertow 2000), but case studies from different regions are available (Aissani et al. 2019; Chertow and Ehrenfeld 2012). Nevertheless, the share of business actors who are actively participating in industrial symbiosis schemes is low; in Europe, only 0.1% out of 26 million companies have been reported to engage in such schemes (Maqbool et al. 2018). Enabling factors, especially factors to support long-term success of networked eco-industrial systems, are not yet fully understood (Maqbool et al. 2018; Lombardi et al. 2012). Each network is unique and dynamic (Boons et al. 2014; Morales and Diemer 2019), and must respond to constantly changing environments (legislation, market dynamics, business strategies of single participants, organizations leaving or joining, etc.). For robust networks, Chertow and Ehrenfeld (2012) observed characteristics of complex adaptive systems.

Further research is required to comprehensively capture the full complexity and dynamics of eco-industrial networks and to better frame the success elements. Some relevant elements can already be outlined. Robust symbiotic partnerships have developed in a bottom-up process, where an initial cooperation of few actors has gradually expanded to create a broader network (Chertow 2007; Domenech et al. 2018). Spontaneous interactions encourage long-lasting relationships, while overplanning might negatively affect network dynamics (Vladimirova et al. 2018). ICT tools can promote symbiotic interactions (Kosmol 2019; Domenech et al. 2018). However, most proposed ICT solutions failed to remain operational for longer periods (Maqbool et al. 2018); incompatibility of information flows is one possible explanation.

Trust is an essential component of successful eco-industrial networks. It is a prerequisite to establish a close "mental distance" of participants (Lombardi and Laybourn 2012). Trust can unfold directly between business actors or via an independent body which offers evaluations, coordination, and network support (Vladimirova et al. 2018). Some form of institutionalization of the industrial symbiosis along with an independent evaluation is essential for sustainable success (Chertow 2007; Chertow and Ehrenfeld 2012; Domenech et al. 2018; Lombardi et al. 2012). A newly created platform or an affiliation to a public or private competence center can fulfill these functions (Kusch-Brandt 2020). Essential roles of such a body are coordination, knowledge building, proactive identification of innovation potential, independent and detailed analysis of economic viability of possible interactions, and the value-added integration of interests of participants.

Conclusion

Industrial symbiosis makes a strong contribution to achieving more resource-efficient and more sustainable industrial systems. Its implementation reflects activities of business entities which engage in collaborative management of material or energetic resources, infrastructures, or know-how. Such complex schemes of networked interactions between autonomous economic entities differ significantly from traditional business practices, where single entities rather try to reduce potential influence by other market actors. The benefits of industrial symbiosis are well known, and many successful cases are documented in literature. Nevertheless, the share of companies involved in such schemes is still low. Enabling factors have not yet been fully understood, but some key elements can be highlighted. Networked eco-industrial systems must respond to constantly changing environments. Bottom-up processes where initial interactions have dynamically expanded into a wider network are among the most robust initiatives documented in literature. Trust plays an essential role and can be supported through the creation or involvement of an independent body.

Cross-References

▶ Circular Economy
▶ Green Economy
▶ Resource Efficiency

References

Aissani, L., Lacassagne, A., Bahers, J. B., & Le Feon, S. (2019). Life cycle assessment of industrial symbiosis: A critical review of relevant reference scenarios. *Journal of Industrial Ecology, 23*(4), 972–985.

Boons, F., Spekkink, W., & Jiao, W. (2014). A process perspective on industrial symbiosis – Theory, methodology, and application. *Journal of Industrial Ecology, 18*(3), 341–355.

Chertow, M. R. (2000). Industrial symbiosis: Literature and taxonomy. *Annual Review of Energy and Environment, 25*, 313–337.

Chertow, M. R. (2007). "Uncovering" industrial symbiosis. *Journal of Industrial Ecology, 11*(1), 11–30.

Chertow, M., & Ehrenfeld, J. (2012). Organizing self-organizing systems – Towards a theory of industrial symbiosis. *Journal of Industrial Ecology, 16*(1), 13–27.

Domenech, T., Doranova, A., Roman, L., Smith, M., & Artola, I. (2018). *Cooperation fostering industrial symbiosis: Market potential, good practice and policy actions*. Brussels: European Commission.

Dong, L., Liu, Z., & Bian, Y. (2021). Match circular economy and urban sustainability: Re-investigating circular economy under Sustainable Development Goals (SDGs). *Circular Economy and Sustainability*. https://doi.org/10.1007/s43615-021-00032-1.

Kosmol, L. (2019). Sharing is caring – Information and knowledge in industrial symbiosis. In *2019 IEEE 21st conference on business informatics (CBI)* (pp. 21–30). Moscow, Russia: IEEE. https://doi.org/10.1109/CBI.2019.00010.

Kusch-Brandt, S. (2020). Industrial symbiosis: Unlocking synergies to achieve business advantages and resource efficiency. In F. W. Leal, A. M. Azul, L. Brandli, S. A. Lange, & T. Wall (Eds.), *Industry, innovation and infrastructure. Encyclopedia of the UN Sustainable Development Goals*. Cham: Springer. https://doi.org/10.1007/978-3-319-71059-4_110-1.

Laybourne, P., & Lombardi, D. R. (2012). Industrial symbiosis in European policy. *Journal of Industrial Ecology, 16*(1), 11–12.

Lombardi, D. R., & Laybourn, P. (2012). Redefining industrial symbiosis – Crossing academic-practitioner boundaries. *Journal of Industrial Ecology, 16*(1), 28–37.

Lombardi, D. R., Lyons, D., Shi, H., & Agarwal, A. (2012). Industrial symbiosis – Testing the boundaries and advancing knowledge. *Journal of Industrial Ecology, 16*(1), 2–7.

Maqbool, A. S., Mendez Alva, F., & Van Eetvelde, G. (2018). An assessment of European information technology tools to support industrial symbiosis. *Sustainability, 11*, 131.

Martin, M. N., Svensson, N., & Eklund, M. (2015). Who gets the benefits? An approach for assessing the environmental performance of industrial symbiosis. *Journal of Cleaner Production, 98*, 263–271.

Morales, M. E., & Diemer, A. (2019). Industrial symbiosis dynamics, a strategy to accomplish complex analysis: The Dunkirk case study. *Sustainability, 11*, 1971.

UNEP, UNECE. (2016). *GEO-6 Assessment for the pan-European region*. Nairobi: United Nations Environment Programme.

van Berkel, R. (2010). Quantifying sustainability benefits of industrial symbioses. *Journal of Industrial Ecology, 14*(3), 371–373.

Vladimirova, D., Miller, K., & Evans, S. (2018). *Lessons learnt and best practices for enhancing industrial symbiosis in the process industry*. SCALER Project Report. http://www.scalerproject.eu

Industrial Synergy

▶ Industrial Symbiosis

Industry Self-Regulation

▶ Voluntary Programs for Urban and Regional Futures

Informal Settlements

▶ Housing and Development

Infrastructure – Utilities/ Facilities

▶ Harare

Inheritance

▶ Furthering the Sustainable Development Agenda by Putting Urban Heritage and Value Extraction at the Center

Innovation

► Sustainable Development Goals and Urban Policy Innovation

Innovation Eco-systems

► Philanthropy in Sustainable Urban Development: A Systems Perspective

Innovation to Bring Nature-Based Solutions to Life: Tales of Two Cities

Birgit Georgi
UIA Expert/Strong Cities in a Changing Climate, Egelsbach, Germany

Synonyms

Green infrastructure; Renaturing cities; Urban green; Urban greenery; Urban greenspace

Definitions

Nature-based solutions (NBS) refer to solutions that are inspired and supported by nature, which usually provide simultaneously environmental, social, and economic benefits and help to build resilience. They bring more nature and natural features and processes into cities and regions (EC n.d-b., adapted).

Green and blue infrastructure are natural and seminatural areas on land and water with other environmental features designed and managed to deliver a wide range of ecosystem services such as climate mitigation and adaptation, water purification, air quality, or space for recreation to improve environmental conditions and therefore citizens' health and quality of life (EC n.d-a., adapted).

Over the last years, implementing nature-based solutions has become a popular concept in many cities around the globe. They deliver many benefits for a better urban climate, health, and well-being or simply nature in cities. Hence, they are increasingly appreciated for their ability to raise the attractiveness and livability of cities. However, while they are generally appreciated, the concrete planning and broader implementation of nature-based solutions is still hampered by a couple of challenges that cannot be solved by business-as-usual approaches. Tackling the challenges, Greater Manchester in the UK and Breda in the Netherlands have developed different innovative ideas. They are currently implementing and testing these in their projects *IGNITION – Innovative financing and delivery of natural climate solutions* in Greater Manchester and *GreenQuays – Urban River Regeneration through Nature Inclusive Quays*, which are funded by the European Union's *Urban Innovative Actions* program. The municipalities' general approach was to get out of business-as-usual planning and delivery procedures and think differently in order to boost nature in their cities.

Urban Density Versus Urban Green

The center of European cities is usually built up densely. From an environmental perspective, such urban density is a desired feature of cities as it enables them to be very efficient on energy use for heating or cooling of houses, mobility, and minimizes the use of space per resident. On the downside, the high coverage by buildings, stone, and asphalt heats them up further in hot summers, prevents natural drainage of rain and stormwater, and allows little space for urban green areas. A higher degree of greening and urban density at the same time appears contradicting.

To bring more green infrastructure into its very dense medieval city center, the city of Breda has decided to reverse its inner-city development of the 1950s and 1960s when the river section

Innovation to Bring Nature-Based Solutions to Life: Tales of Two Cities, Image 1 Visualization of nature inclusive quay. (Copyright: GreenQuays Municipality of Breda)

flowing through the center had been covered in order to establish a two-lane road and a parking garage. Since the 1990s, the mindset has changed again and nature in the city has been increasingly valued. Step by step, the city restores its harbor and old channels. The water will however run between steep quay walls, which would not leave much space for green, hence the initial design included only a few pieces of greenery. Residents did not agree with these initial plans and wanted more urban greenery. The municipality has taken up that request from citizens and started to experiment with different elements and forms of nature-based solutions to make the area greener. As part of the broader river Mark restoration project, *GreenQuays* is developing and testing different quay wall designs – different type of bricks, mortar, joints, bonds, and wall pattern, which shall provide stability and, at the same time, enable nature to invade the construction. Herbaceous plant species are set into the joints and niches; and their thriving and spreading is monitored to find the most appropriate plants and best supporting construction patterns and material. In the nursery, trees have been grown in curved pipes; and the bended trees have then been planted into containers integrated in the vertical quay walls. These solutions have been implemented and monitored in a small pilot. In a next step, these will be upscaled in a real-life pilot of 175 m length that will be established in 2022–2023 and inspire in the wider renewal of the river Mark (Image 1).

A Patchwork of Public and Private Areas

Urban areas are a patchwork of public and private plots. Just fostering green infrastructure on public land may not suffice to capitalize desired effects such as reducing heat stress in summer, drain or store stormwater, or provide equally accessible green space to all resident groups as the public plots may not be situated at the places most needed and not properly connected. In addition, private engagement could supplement public financing of urban green, which is of particular importance, when public budgets for the development and maintenance of green areas is tight. Including private plots however requires working with a multitude of stakeholders.

In the wider urban development, Breda also faces the problem of having a patchwork of public and private plots and the need for more green in the inner city. Hence beyond the river Mark renewal and other public projects, Breda has mobilized private houseowners to green their facades and courtyards as an important supplement to public greening. Through this municipal program, houseowners receive advice on small private gardens and greening elements. After the first ones have started, the idea is picking up and becoming more and more visible in the streets, making the city greener overall and linking to the river section.

The *IGNITION* project of Greater Manchester follows a different approach to overcome the patchwork of public and private areas to boost green infrastructure. The lead partner, the Greater Manchester Combined Authority saw that increasing public green space through publicly funded grants alone would not be sufficient to achieve their ambitious targets for an uplift of urban green infrastructure by 2038 to make the city climate resilient. Private action and new financing schemes and business models are needed. The focus of the project is, therefore, on (private) developers and building investors. It explores how this audience can be convinced to use more nature-based solutions instead of engineered "grey" solutions for climate resilience.

Private Investors Hesitate to Apply Nature-Based Solutions

Private landowners, investors, and developers show, however, often a low interest in applying nature-based solutions and prefer known conventional engineering solutions instead. Reasons are various.

Research as part of *IGNITION* among potential investors' awareness, knowledge, and confidence revealed gaps on knowledge about costs and benefits of different nature-based solutions and a lack of trust in their effectiveness as the solutions are new to them. Their and their clients' main interest is typically on getting high returns for their investment. With little or no knowledge on the functioning and the real benefits of nature-based solutions they fear the risk in applying these new solutions. In a first step, IGNITION developed a publicly accessible *evidence base for nature-based solutions* to show the different options and the range of benefits and costs. The second step is the implementation of a Living Lab at Salford University, which combines different nature-based solutions, measures their performance on diverse benefits, and thus delivers real data and tangible experience in the immediate vicinity. The first stage of the Lab has been inaugurated in autumn 2020 (Image 2) and the implementation and operation of the whole Lab will be finalized in 2021.

Based on this information and data, IGNITION started to develop innovative funding streams and business models for private partners and/or public-private partnerships, including also health or highway authorities, for co-investment as well as water charge–based solutions. Once developed, the knowledge shall be managed at a newly established Climate Adaptation Services Company (CASCo) and the different funding streams used as pipelines to roll out the business models and serve large-scale investment in green infrastructure far beyond *IGNITION*.

Conclusions: Lessons Learnt

A main success factor for developing and implementing innovative nature-based solutions in both cities has been a favorable attitude for innovation in the city, the region, their responsible authorities, and from politicians. There is a

Innovation to Bring Nature-Based Solutions to Life: Tales of Two Cities, Image 2 Small green wall in the raingarden of the Living Lab Salford. (Copyright: IGNITION Living Lab, University of Salford)

willingness to step out of usual administrative procedures and experiment with new and innovative pathways. This includes the acceptance to fail in parts; rather seeing this as an opportunity to learn and develop even better solutions instead of real failure.

Co-creation with citizens and other stakeholders that could have a stake in the developed solutions is another important element for success. Both cities have organized co-creation workshops for various stakeholder groups – citizens groups, students, service providers and maintenance staff, developers, and investors – to co-design the solutions.

Finally, innovative projects need an innovation-minded partnership. If team members are curious for other partners' knowledge, input, and to exchange ideas, further innovative solutions may come up. In Breda, for example, researchers had worked on bricks, mortar, and construction pattern for stable quay walls before, but have been challenged now to integrate nature. In return, the design team has learned about the different qualities of brick-and-mortar combinations, and the partners caring for nature bring forward all the ecological requirements. They find solutions in joint discussions, while each partner broadens its knowledge.

Cross-References

▶ Green Cities: Nature-Based Solutions, Renaturing and Rewilding Cities
▶ Overcoming Barriers in Green Infrastructure Implementation

References

EC. (n.d-a). *Green infrastructure – Environment – European Commission*. https://ec.europa.eu/environment/nature/ecosystems/index_en.htm. Accessed 17 Feb 2021a.

EC. (n.d-b). *Nature-based solutions*. Text. European Commission – European Commission. https://ec.europa.eu/info/research-and-innovation/research-area/environment/nature-based-solutions_en. Accessed 17 Feb 2021b.

Further Readings

GreenQuays. https://www.greenquays.nl/

IGNITION Greater Manchester. http://www.ignitiongm.com/

Living Lab Salford. https://www.youtube.com/watch?v=4z-G-VU-Z1g

UIA Green Quays. https://uia-initiative.eu/en/uia-cities/breda-call4

UIA IGNITION. https://www.uia-initiative.eu/en/uia-cities/greater-manchester

Integrated Farming Systems

▶ *Wadi* Sustainable Agriculture Model, The

Integrated Green Grey Infrastructure

▶ Integrated Urban Green and Grey Infrastructure

Integrated Green-Grey Infrastructure

▶ Integrated Urban Green and Grey Infrastructure

Integrated Greening of Grey Infrastructure

▶ Integrated Urban Green and Grey Infrastructure

Integrated Planning

▶ Land Use Planning Systems in OECD Countries

Integrated Urban Green and Grey Infrastructure

Andreas Wesener and Wendy McWilliam
School of Landscape Architecture, Faculty of Environment, Society and Design, Lincoln University, Lincoln, New Zealand

Synonyms

Combined green and grey infrastructure; Coupled green and grey infrastructure systems; GGI; Hybrid (green-grey) infrastructure; IGGI; Integrated green grey infrastructure; Integrated green-grey infrastructure; Integrated greening of grey infrastructure

Definition

Integrated urban green and grey infrastructure combines “green” (natural; living) and “grey” (human-made; anthropogenic) infrastructure constituents to provide facilities in support of key (ecosystem) services and functions to urban communities. Integration may occur in the form of green infrastructure adding improved or additional services or functions to grey infrastructure, or grey infrastructure increasing the functionality of green infrastructure. It may also occur along a green-grey continuum based on the proportional composition of green or grey constituents within integrated green and grey infrastructure systems. Integration occurs at three spatial scales (*scale of integration*): first, at the **intracomponent scale** when one infrastructure component includes both green and grey constituents (e.g., a biofiltration facility); second, at the **intercomponent scale** when two components, one grey and one green, work together to provide particular services or functions (e.g., a levee protecting a wetland from coastal floods); and finally, at the **system scale** when multiple green, grey, and/or green-grey components establish a system (e.g., rain gardens, planters, trees, street furniture, and car parks working together to provide street calming services). *Type of integration* refers to how green-grey infrastructure constituents are connected with each other. **Physical integration** means that infrastructure components and/or systems are physically connected (e.g., a water conveyance system). **Functional integration** means that green, grey, and/or green-grey infrastructure components and/or systems together provide services or function(s) without being physically connected (e.g., street trees, planters, and raingardens along a street providing bird habitat without being physically connected).

Introduction

The Oxford English Dictionary defines *infrastructure* as "the basic physical and organizational structures (e.g., buildings, roads, power supplies) needed for the operation of a society or enterprise" (Stevenson and Waite 2011). While the term originated from the French in the nineteenth century, in support of the need for railroad construction, the number and complexity of infrastructure systems considered essential for the operation of urban communities continues to change through time. For example, urban infrastructural components including road systems, fortifications, wells or water pipes predate the industrial city and have been indispensable for the functionality, safety, and survival of urban populations for many centuries (Schott 2015). However, additional and more complex systems such as gas, water, electricity, or telecommunication networks have developed and been viewed as essential. Such systems were largely engineered, using "grey" human-made materials.

Beginning in the 1990s, additional services and functions provided by nature were identified by scholars within cities. For example, natural systems are uniquely suited to provide carbon sequestration to slow the rate of climate change, mitigate harsh microclimates, cleanse the air and water, produce food, and provide habitat in support of biodiversity and food production. In recognition of the importance of natural systems in cities in support of these services and functions, the term *green infrastructure* was developed and its essential role in cities promoted since the mid-1990s (Seiwert and Rößler 2020). There are many definitions of green infrastructure (e.g., Boyle et al. 2014; Wright 2011), and it is notable that a duality of green infrastructure concepts as either being "visually green" and/or "sustainably green" has occurred. This has led to a distinction between (visually) "ecological/natural" types of green infrastructure (e.g., parks, trees, forests, grasslands, etc.) and "engineered" types including cycle paths, storm water channels, energy efficient buildings, etc. Combinations of visual and sustainable types of infrastructure such as green walls and roofs have been proposed (Mell 2013). However, the above examples illustrate that definitions of "green" infrastructure that are fundamentally distinct from "grey" infrastructure are not always useful. Most recently, the benefits of both grey and green, and in particular, their combined benefits, are being recognized. In response, new notions of *green-grey infrastructure* types and compositions have emerged.

Integrated Urban Green and Grey Infrastructure Concepts and Definitions

The origins of green-grey urban infrastructure integration concepts go back to Davies et al. (2006) who proposed a *green-grey continuum* (Fig. 1) that includes both "green" and "grey"

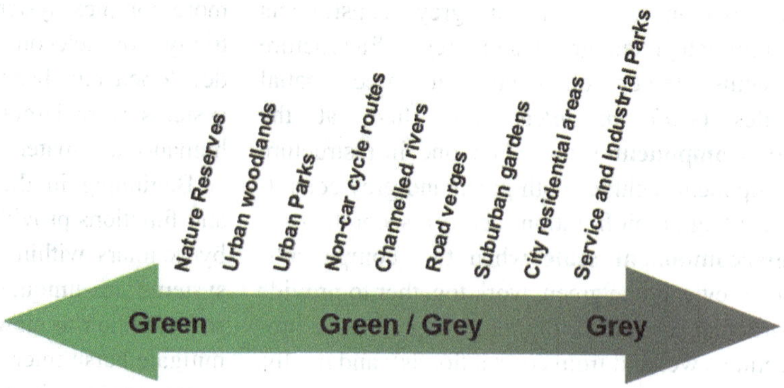

Integrated Urban Green and Grey Infrastructure, Fig. 1 The grey-green continuum (Davies et al. 2006, p. 3). (Permission to use the figure in this publication has been granted)

infrastructure elements. Their concept depicts predominantly "green" (natural; living) infrastructure systems (e.g., nature reserves and urban woodlands) at one end and predominantly "grey" (man-made; anthropogenic) ones (e.g., industrial parks and inner city residential areas) at the other end. The "green/grey" spectrum in the middle depicts hybrid or semi-natural infrastructure elements or systems such as cycle routes, road verges, or suburban gardens.

In the years following Davies et al.'s (2006) publication, notions of green-grey infrastructure integration, interaction, and development have been considered by several authors who have promoted their benefits for human health and wellbeing (Svendsen et al. 2012), their contribution to reducing risk from environmental disasters (Denjean et al. 2017), and other ecosystem services (Tiwary and Kumar 2014), such as contribution to reducing sewer overflows (Alves et al. 2016).

The European Union-funded international research project *Green Surge* (2013–2017) advanced the concept of integrated green-grey infrastructure further, defining it as "the interaction and links between urban green infrastructure and other urban structures" (Hansen et al. 2017, p. 23). Their definition led to a recognition of the physical and functional synergies between green and grey infrastructure that has the potential to provide a wider array of services and functions than green or grey infrastructure alone. However, their concept is limited by its focus on complementing or replacing (existing) grey infrastructure with green infrastructure rather than integrating the advantages of both. Likewise, Naylor et al. (2017) define integrated green and grey infrastructure, or "IGGI," as the "greening of hard infrastructure that cannot be replaced with softer green (or blue-green) solutions [. . .] IGGI therefore sits between entirely 'green' and entirely 'grey' options along a continuum of engineering approaches [. . .] and in simple terms is 'greening the grey'" (p. 3). Similarly, Firth et al. (2020, p. 1763) only seem to recognize the biodiversity-related services of IGGI, arguing it is "a new conservation strategy that involves biodiversity enhancement of hard infrastructure that cannot be replaced with green solutions [. . .] to improve multifunctionality; in particular the ecological value of hard infrastructure." The focus of integration is on improving (new or existing) grey infrastructure through its replacement with green, for example, by incorporating a green roof or wall into a building.

Depietri and McPhearson's (2017) notion of "hybrid" green-grey infrastructure provides a reverse conceptual approach. They define hybrid infrastructure as a system "where a technological or built infrastructure complement the service delivered by a green or blue infrastructure" (p. 102). The authors place their concept within a social-ecological-technological systems framework (SET) that combines engineered components with those provided by nature. Examples are artificial wetlands that include engineered (grey) components such a levee to provide flood protection, or water filtration or conveyance systems such as bioswales.

However, the most detailed conceptual discussion on integrated urban green-grey infrastructure has been provided by Aleksandrova et al. (2019) in the context of residential street retrofits. The authors assess the ratio of grey and green components and/or functions of street-based infrastructure within a green-grey infrastructure continuum, similar to Davies et al. (2006), in order to define green-grey infrastructure more specifically:

> *Green-grey infrastructures are components and systems that have areas or volumes of 25 percent or more of either nature and human-made materials, or, where it cannot be determined from the study, comparable proportions of green or grey components (e.g. rain gardens). Potentially, networks could shift from one type to the other as their infrastructure components are retrofitted through time (e.g. a storm-water management system along a street could change from grey to green-grey infrastructure when pipes (grey infrastructure components) are replaced with bio-retention facilities (green-grey components).* (Aleksandrova et al. 2019, p. 7)

The authors distinguish between scale and type of integration. With regard to the scale of integration, there are, on the one hand, individual components that are made up of green and grey elements (e.g., an individual biofiltration facility). On the other hand, there are green, grey, or green-grey components that may work together at different spatial scales, from individual streets, to neighborhoods to cities and regions, as an integrated green-grey infrastructure system. For example, "biofiltration facility (green-grey infrastructure), planter, planting area and trees (green infrastructure), and street furniture and car parks (grey infrastructure) working together to provide street calming services" (p. 8; Fig. 2). With regard to the type of integration, the authors distinguish between physical types of integration, in which green-grey infrastructure components or networks are physically connected, and functional types of integration where green-grey infrastructure

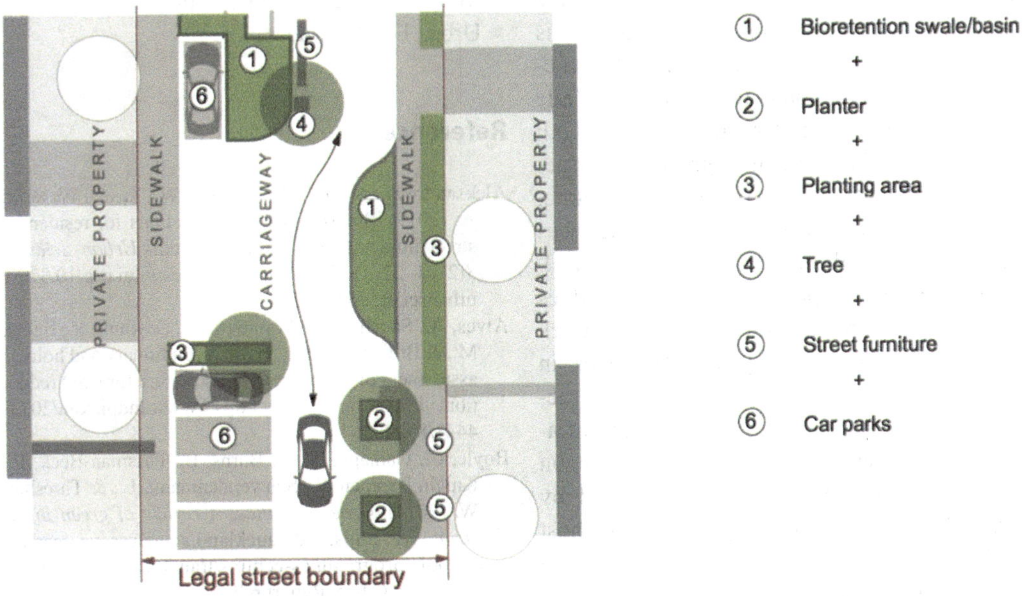

Integrated Urban Green and Grey Infrastructure, Fig. 2 Integrated green-grey infrastructure in a residential street (system scale; functional integration) (Aleksandrova et al. 2019, p. 9). (Permission to use the figure in this publication has been granted)

components or systems work together by contributing to particular services, however, without necessarily being physically linked with each other (Fig. 2).

Summary/Conclusion

Notions of integrated urban green and grey infrastructure have emerged in the literature since the early 2000s in response to the experienced limitations of a binary categorization of infrastructure types into "grey" and "green." In response, urban infrastructure types have been conceptualized as a continuum including "green" (natural; living), "grey" (man-made; anthropogenic), and hybrid or semi-natural infrastructure elements or systems (Davies et al. 2006). Literature-based concepts of integrated urban green and grey infrastructure share the insight that green and grey elements or functions may work together and complement each other, rather than compete. However, a consensus regarding its definition among scholars is still to develop, with three slightly different perspectives currently at play. The first perspective follows a "greening the grey" approach where green infrastructure is used to support new or existing grey infrastructure by adding additional functionality. While this perspective integrates functions of both grey and green infrastructure, green infrastructure acts like an "add-on" to grey infrastructure, rather than an equal partner in support of services and functions (Firth et al. 2020; Naylor et al. 2017). The second perspective acknowledges that different infrastructure types can work together in an integrated system. However, in contrast to the first perspective, grey infrastructure is seen to improve (existing) green infrastructure services (Depietri and McPhearson 2017). The third perspective, put forward by Aleksandrova et al. (2019), presents the most detailed and systematic approach to integrated urban green and grey infrastructure. Instead of focusing on which infrastructure type complements or adds to another, the authors develop a framework of grey, green, and green-grey infrastructure integration based on the proportional composition of green or grey constituents or elements at both component and system levels (*scale of integration*). In addition, the authors define the *type of integration* as either "physical," where infrastructure components and/or systems are physically connected, or "functional," where different infrastructure components and/or systems work together without being physically linked with each other.

The above definition of integrated urban green and grey infrastructure is a general one. The literature review shows that green-grey infrastructure concepts relate to a variety of spatio-functional situations, for example, wetlands, stormwater management systems, streetscapes, or building design. Thus, more specific contextual definitions for particular integrated urban green and grey infrastructure may evolve over time.

Cross-References

- ▶ Green and Blue Infrastructure (GBI) in Urban Areas
- ▶ Multiple Benefits of Green Infrastructure
- ▶ Urban Food Gardens

References

Aleksandrova, K. I., McWilliam, W. J., & Wesener, A. (2019). Status and future directions for residential street infrastructure retrofit research. *Urban Science, 3*(2), 1–22. (article No. 49).. https://doi.org/10.3390/urbansci3020049.

Alves, A., Sanchez, A., Vojinovic, Z., Seyoum, S., Babel, M., & Brdjanovic, D. (2016). Evolutionary and holistic assessment of green-grey infrastructure for CSO reduction. *Water, 8*(9), 402. https://www.mdpi.com/2073-4441/8/9/402.

Boyle, C., Gamage, G. B., Burns, B., Fassman-Beck, E., Knight-Lenihan, S., Schwendenmann, L., & Thresher, W. (2014). *Greening cities: A review of green infrastructure*. https://cdn.auckland.ac.nz/assets/creative/schools-programmes-centres/transforming%20cities/Greening_Cities_Report.pdf

Davies, C., MacFarlane, R., McGloin, C., & Roe, M. (2006). *Green infrastructure planning guide*. http://www.greeninfrastructurenw.co.uk/resources/North_East_Green_Infrastructure_Planning_Guide.pdf

Denjean, B., Altamirano, M. A., Graveline, N., Giordano, R., van der Keur, P., Moncoulon, D., Weinberg, J., Máñez Costa, M., Kozinc, Z., Mulligan, M., Pengal, P., Matthews, J., van Cauwenbergh, N., López Gunn, E., & Bresch, D. N. (2017). Natural assurance scheme: A level playing field framework for green-grey infrastructure development. *Environmental Research, 159*, 24–38. https://doi.org/10.1016/j.envres.2017.07.006.
Depietri, Y., & McPhearson, T. (2017). Integrating the grey, green, and blue in cities: Nature-based solutions for climate change adaptation and risk reduction. In N. Kabisch, H. Korn, J. Stadler, & A. Bonn (Eds.), *Nature-based solutions to climate change adaptation in urban areas*. Heidelberg: Springer.
Firth, L. B., Airoldi, L., Bulleri, F., Challinor, S., Chee, S.-Y., Evans, A. J., Hanley, M. E., Knights, A. M., O'Shaughnessy, K., Thompson, R. C., & Hawkins, S. J. (2020). Greening of grey infrastructure should not be used as a Trojan horse to facilitate coastal development. *Journal of Applied Ecology, 57*(9), 1762–1768. https://doi.org/10.1111/1365-2664.13683.
Hansen, R., Rall, E., Chapman, E., Rolf, W., & Pauleit, S. (2017). *Urban green infrastructure planning: A guide for practitioners*. http://greensurge.eu/products/planning-governance/UGI_Planning_Guide_Sep_2017_web.pdf
Mell, I. C. (2013). Can you tell a green field from a cold steel rail? Examining the "green" of green infrastructure development. *Local Environment, 18*(2), 152–166. https://doi.org/10.1080/13549839.2012.719019.
Naylor, L. A., Kippen, H., Coombes, M. A., Horton, B., MacArthur, M., & Jackson, N. (2017). *Greening the grey: A framework for Integrated Green Grey Infrastructure (IGGI)* [Technical report]. http://eprints.gla.ac.uk/150672/
Schott, D. (2015). Stadt und Infrastruktur: Einleitung. *Informationen zur modernen Stadtgeschichte, 45*(1), 5–16.
Seiwert, A., & Rößler, S. (2020). Understanding the term green infrastructure: Origins, rationales, semantic content and purposes as well as its relevance for application in spatial planning. *Land Use Policy, 97*, 104785. https://doi.org/10.1016/j.landusepol.2020.104785.
Stevenson, A., & Waite, M. (2011). Infrastructure. In *Concise Oxford English dictionary*.
Svendsen, E., Northridge, M. E., & Metcalf, S. S. (2012). Integrating grey and green infrastructure to improve the health and well-being of urban populations. *Cities and Environment, 5*(1), 3.
Tiwary, A., & Kumar, P. (2014). Impact evaluation of green–grey infrastructure interaction on built-space integrity: An emerging perspective to urban ecosystem service. *Science of the Total Environment, 487*, 350–360. https://doi.org/10.1016/j.scitotenv.2014.03.032.
Wright, H. (2011). Understanding green infrastructure: The development of a contested concept in England. *Local Environment, 16*(10), 1003–1019. https://doi.org/10.1080/13549839.2011.631993.

Integrating Agriculture, Forestry, and Food Systems into Urban Planning: A Key Step for Future Resilient and Sustainable Cities

Simone Borelli, Michela Conigliaro, Isabella Trapani, Cecilia Marocchino, Guido Santini, Halima Hodzic and Carmen Zuleta Ferrari
Food and Agriculture Organization of the United Nations (FAO), Rome, Italy

Background

Urban population first exceeded the rural one in 2007, and it is projected that cities will host more than 68% of the global population by 2050 (UNDESA 2019). This rapid urban growth, combined with growing demand for food and basic services, poses major challenges to local governments aiming to ensure decent living conditions for all citizens.

Although cities consume 70% of the global food supply, the level of food insecurity and malnutrition in urban and peri-urban communities remains high (FAO 2020). At the same time, the high prevalence of unhealthy diets and the lack of physical activity in urban areas have not only led to rise of overweight and obesity but also to an increase in noncommunicable diseases such as diabetes, cardiovascular diseases, and cancer (Bodicoat et al. 2014; Global Panel 2017; UN-HABITAT and WHO 2020). This seriously undermines the health of urban communities and is especially burdensome in low and middle-income countries and across marginalized communities (Global Panel 2017).

I

Furthermore, population growth leads to urban sprawl at the expense of natural and seminatural green spaces (including agricultural land, forests, and parks), affecting the important services they provide to cities such as food production, air and water quality, biodiversity conservation, and wood and woodfuel supply, as well as opportunities for recreation, physical activity, and psychological restoration (Artmann et al. 2019; Bren d'Amour et al. 2017; UCCRN 2018; van Vliet 2019). This loss also undermines the capacity of cities to adapt to the harmful effects of climate change, as natural ecosystems play a key role in preventing and mitigating the impact of floods and landslides, and in moderating the impact of heatwaves and the urban heat island effect (Depietri et al. 2012; FAO 2019a; Zimmermann et al. 2016).

To address these challenges, national and subnational governments need to rethink the way in which they plan urban and peri-urban areas, identifying more inclusive, resilient, and green pathways to environmental, social, and economic well-being. The transformation of food systems and the expansion of green spaces in cities and peri-urban areas offer an important opportunity to increase the overall health and resilience of urban communities and to create livable urban environments (WHO 2017). For this to happen, urban planners need to go beyond their professional boundaries and promote a holistic and multidisciplinary approach (Cabannes and Marocchino 2018).

Barriers and Challenges to the Integration of Green Spaces and Food Systems in Urban Planning

Various experiences suggest that agriculture, forestry, and food systems can be mainstreamed into urban planning by connecting sectors and actors, and linking different spatial scales (Douglas et al. 2017; Alexandra and Norman 2020; Cabannes and Marocchino 2018). Local and subnational governments have initiated significant changes, recognizing the importance of prioritizing green spaces and food systems in their urban development agenda. There is also growing interest in the direct connections between quality public green spaces and the livability, safety, sustainability, and health of the city (Barton 2009; WHO 2016, 2017; UN-HABITAT and WHO 2020).

While a number of cities have already adopted novel approaches to include food systems and green spaces in their urban agenda, their long-term sustainability and integration in urban and territorial planning still presents many challenges (Bush 2020; Cabannes and Marocchino 2018; Haaland and van Den Bosch 2015). Despite their crucial role, local governments are often not sufficiently recognized as key players, and thus, not adequately supported by national governments with policy and regulatory instruments (FAO 2019b). In addition, they often have limited financial and technical capacities to face these challenges, as well as inadequate understanding of urban food systems (Tefft et al. 2020). This is sometimes exacerbated by corruption and conflicts of interest.

To achieve sustainable urban transformation, local and regional governments need to work across departments, institutional and administrative boundaries, and in collaboration with other actors such as civil society, private sector, and academic institutions (UN-HABITAT 2018). However, many urban administrations work in silos, with health, environment, infrastructure, and food usually being separate portfolios. These jurisdictions do not necessarily work seamlessly, and they run the risk of compartmentalization. Cities may also have limited capacity to promote meaningful integration of urban planning and design at different scales, from neighborhood to territorial and regional planning levels (Cabannes and Marocchino 2018).

Also, urban and regional planners and leaders often lack specialized knowledge on how to integrate multifunctional green spaces such as agriculture and forests as well as food systems into the planning processes. To be meaningful, their planning and implementation need to be integrated across many jurisdictions and benchmarked against other cities (ARUP 2021).

From the spatial standpoint, scarcity of available land exacerbates the competition among different uses such as agriculture, green spaces, housing, infrastructure, and industry. As urban and peri-urban areas expand, the increased demand for land and the changes in land use put pressure on land tenure arrangements, which are often customary or informal (Dadashpoor and Ahani 2019). Since land uses like housing, industry, and other infrastructure have a higher economic value than agriculture and green spaces, local authorities must duly consider land tenure arrangements, ensuring that the needs of citizens are met and that trade-offs among different land uses and priorities are managed (Lyu et al. 2022). Moreover, a territorial and landscape approach to natural resource management can provide a spatial framework to protect and manage the natural and built environment of cities and territories as well as fostering functional relationships between rural and urban areas and their effect on the provision of ecosystem goods and services such as renewable energy, air, and water. Forests and trees, for example, can act as natural barriers to protect city infrastructure and agricultural land from natural and environmental hazards and risks from climate shocks (UN-Habitat 2015).

Reinforcing urban-rural linkages and promoting a territorial approach is crucial for integrating agriculture, forestry, and food systems into urban planning. Strengthening synergies between urban and rural areas can address the challenge of supplying food, water, energy, and wood to urban consumers by more effectively linking urban centers with their "catchment areas." Thereby, neighboring cities can foster the complementarity of actions among themselves, creating rural income opportunities and using agro-ecological zoning and landscape approaches to ensure sustainable development (FAO 2017, 2018). Thus, the often limited urban-rural synergies and the lack of regional integrated planning on food systems and ecosystem services call for improvement of multilevel governance with a meaningful engagement of local authorities. These mechanisms play a fundamental role in promoting policy coherence and fostering the integration of green spaces and food systems within urban planning (Cabannes and Marocchino 2018).

Urban planners and designers also need to gain a better understanding of the connections between different spatial scales and their multiple relations, especially regarding the integration of the food-water-energy nexus into urban planning. Currently, subject matter specialists often design infrastructural interventions independently, without considering their mutual interactions (Rahman 2015; Zhou et al. 2014). However, energy, water, food, infrastructure, and urban green spaces cannot be considered in isolation as they have many interactions with the surrounding built and natural environment (Antrop 2004). Thus, neighborhoods need to be designed in a holistic manner, integrating energy, water, food access, housing, and urban greening, and promoting a circular economy. Above all, an optimal solution must include social equity in the planning and design process (UN-HABITAT 2018).

An additional challenge, mainly in low-income countries, is the existence of the informal food system. Street vendors, informal markets, and informal urban and peri-urban agriculture play a significant role in responding to citizens' need and in reaching the poor in informal settlements and underserved areas. They increase accessibility and affordability of food, contributing critically to food and nutrition security as well as to the local economy. However, if not properly regulated, informality in agriculture and retail sectors may pose serious food safety hazards, due to untreated wastewater use, poor hygiene in food handling, and inadequate food storage and conservation. While local administrations often deal with informality in a restrictive way, promoting the use of public spaces for these activities, where they already exist, can present an opportunity to transform a perceived problem into a solution (Cabannes and Marocchino 2018).

From Policy to Practice: Experiences From Cities

In order to gain practical insights into the integration of agriculture, forestry, and food systems into urban planning, semistructured interviews were conducted with key informants in a diverse range of cities: Baltimore (USA), Quito (Ecuador), Chengdu (China), Ljubljana

I

(Slovenia), and Kisumu (Kenya). The interviews focused on cities' experience of integrating urban and peri-urban agriculture, forestry, and food systems in urban and territorial planning. The reasons behind their interventions ranged from building resilience to external threats such as climate change and pandemics and improving food and nutrition security to creating livable and economically vibrant cities for urban residents. For some cities, the call for action resulted from recognizing the need to protect the natural environment and achieve food security and improved nutrition for all through expanded green spaces and the promotion of sustainable food systems, including urban and peri-urban agriculture, better food environments, as well as the prevention and management of food waste.

Kisumu

In Kisumu County, the main drivers for integrating food systems, forestry, and agriculture into urban planning were rapid urbanization, population growth, and food insecurity in urban areas. The effects of climate change exacerbate these challenges, placing measures for climate adaptation high on the urban agenda. Another important driver for action was the lack of green spaces and green recreation zones within the county. To this aim, the local government is integrating agriculture and forestry into the urban fabric through the newly proposed Local Physical and Land Use Development Plan. In particular, recognizing the low urban forest cover, Kisumu is working on planting new trees in the city's five main parks. Moreover, the county aims to create livable and sustainable urban spaces through additional pedestrian walkways and benches for citizens, better integrating the Victoria Lakefront into the city, beautifying the city by planting grass and flowers, and incorporating agriculture and trees into urban renewal plans. Synergies between forestry and farming are created by planting fruit trees (mango, guavas, and avocados) within the urban area. Additionally, the county promotes small-scale farming and commercial urban gardening for economic empowerment of vulnerable groups, especially for women and unemployed youth in informal settlements. As a starting point for integrating food systems in urban planning, Kisumu is in the process of developing a holistic food systems strategy recognizing food systems as an important driver to address food security nutrition and poverty challenges in urban areas. Existing food system challenges are being addressed in relation to the integration of urban and peri-urban agriculture, market spaces, food outlets, and informal street food vendors within the urban planning process.

Chengdu

The City of Chengdu is working on the integration of agriculture, forestry, and food systems into urban planning to increase livability and urban resilience. This is happening both in response to the national priority that China has been giving to food security and improved nutrition, and to the recognition that promoting agriculture and green spaces is a way to foster urban-rural integration. The city follows a people-centered strategy, moving from an "industry-city-people" approach to a "people-city-industry" and from the idea of a "park in a city" to "city in a park." Chengdu's spatial planning fully respects and protects the city's geography and natural landscape, striving for a balance between urban, ecological, and agricultural space. The city is divided into five functional zones with diversified development objectives based on their respective resources and capabilities. As China's first "Park City,", Chengdu aims to increase the amount of green spaces as well as the livability and esthetic of the city, with all the related environmental, social, and economic benefits. As such, the city plans to create a greenway system around and across the whole city, including green infrastructure leveraging air currents to create a more comfortable urban climate. Through greening and ecological restoration, Chengdu also intends to establish an urban carbon sequestration mechanism. In addition, food supply and rural development play an important role in urban planning, and the city adopts mechanisms to avoid the transformation of farmland to other land use. Furthermore, Chengdu has developed innovative spatial patterns through the creation of 86 urban-rural integration units, breaking administrative boundaries, and establishing five rural revitalization corridors in the city. Agriculture is also being integrated into other industries such as commerce, art, tourism, and sports, and modern agricultural industry development is accelerated through a systems approach across the value chain.

Baltimore

While the integration process in Baltimore already started in 2010 with the hiring of a Food Policy Director, the turning point were the 2015 protests that raised the question of the city's role in emergency situations. This was reiterated during the COVID-19 crisis, in which the city could significantly contribute to increasing the resilience of the local food system and overall food security. Baltimore's urban development plans are largely guided by the Sustainable Development Goals, while applying a particular equity lens. Urban food systems are integrated into the Master Plan with urban agriculture and green spaces being incorporated through amending the zoning and building codes. This includes repurposing underutilized industrial and brownfield sites for food-related business and activities, both as a form of work creation and to enhance the sustainability of the local food system. The city also aims to increase its tree canopy from the current 20% to 40% and to create a comprehensive Urban Forest Management Plan. Also, the city has introduced a number of policies to support procurement of local produce as well as equitable access to quality housing, green spaces, and other physical and non-physical infrastructure and services crucial for the overall resilience of the city. Finally, through its Master Plan, Baltimore aims to ensure that all residents are within 1.5 miles of grocery stores offering healthy and fresh produce as well as of other health and social services. In particular, Baltimore's Grocery Store Initiative intends to attract healthy market options to the inner city in order to combat the challenge of food deserts.

Ljubljana

Since 2005, Ljubljana has been working on the integration of food production and forestry into urban development, primarily focusing on greening and bluing the city, to foster sustainability and tourism. The push to become greener came from the mayor, largely in preparation for the European Green City award. However, sustainability and the image of a "green city" have soon become part of Ljubljana's identity, which has led to sustained action and public participation by its citizens over the years. Indeed, two-thirds of the total metropolitan area of Ljubljana are rural, and there are around 800 small active farms within the city boundaries. Ljubljana aims to increase self-sufficiency by ensuring the flow of quality agricultural and forestry goods from the surrounding areas and strengthening urban-rural-linkages. In its Urban Master Plan, 83% of the future city development is directed toward repurposing existing brownfields and underutilized sites, thus addressing the challenge of competing interests and needs for urban land. The city's Environmental Protection Programme guards and enhances the natural environment and biodiversity by raising environmental awareness, increasing green spaces, and through urban beekeeping. Through these interventions, Ljubljana aims to increase job opportunities along the food chain, increase the production and consumption of local produce, enhance nutrition and environmental education, ensure more resilient food supply and value chains, and better match consumers' needs to decrease surplus production and food waste.

Quito

Quito's Development Plan 2012–2022 calls for an equitable, sustainable, and participatory city and envisages a "green Quito" to improve environmental quality and mitigate the impacts of climate change. The city integrates food into urban development through a strong urban agriculture program, consisting of training and education, support for community farms, and a center for research and development of urban agriculture. Sustainable food and nutrition are at the core of Quito's Vision 2040, which aims for greater food system resilience and increased and equal access to nutritious food. It includes a comprehensive urban green network integrated into city planning, with green corridors connecting creeks, squares and parks, and vacant spaces incorporated into the public space. Urban agriculture should form an integral part of the urban landscape. The plan also incorporates improved management of food and organic waste, resilient and sustainable food supply chains, an inclusive food economy, and stronger urban-rural interlinkages. To achieve this proposal, the city uses territorial strategies to facilitate metropolitan food production and distribution, specifically focusing on the promotion of organic production and markets. Moreover, Quito fosters synergies between resilience and climate change as well as food and nutrition security by developing a territorial plan that includes indicators related to food systems and green spaces.

Lessons Learned

While the starting point for integrating green spaces and food systems in urban planning might differ from city to city, based on their distinct socioeconomic and environmental conditions, there is a general recognition that integrating agriculture, forestry, and food systems provides an important contribution to the health and well-being of urban dwellers as well as to overall urban resilience.

Sectoral integration is a key aspect of the manner in which the five interviewed city administrations operate. In fact, integration is a key issue in urban planning, as the effective planning of a city requires policies and laws aimed at harmonizing the range of interests by developing and strengthening a common vision and collaborative actions.

To address existing challenges of food insecurity and climate change, the municipality of Kisumu aims to integrate agriculture, food systems, and green spaces into urban planning by ensuring that the departments of lands, physical planning, housing and development, utility, roads, environment, water, engineering, and private sector/economic affairs work closely together. Chengdu also involves all municipal departments in the implementation of the city's overarching masterplan. Ljubljana holds weekly interdepartmental meetings to facilitate sectoral integration and collaboration, and to foster strong horizontal governance between the environmental, housing, transport, health, and education sectors. Within the municipality of Baltimore, inter-agency collaboration is used as a crucial instrument for mainstreaming food into the agenda of all city agencies and leveraging the city's internal food system expertise. It also ensures that food systems and zoning are strongly connected through agriculture, green spaces, and forestry within Baltimore's 12 municipal plans. Likewise, Quito's Secretariat of Economic Development facilitates cross-sectoral exchange along strategic axes to collaborate on integrated solution. The current plans for developing an integrated master plan will be the first inter-secretariat experience in the city.

The engagement of and collaboration between a wide range of actors and different sectors in a municipality allow for increased knowledge sharing and breaking down silos, leading to more effective and inclusive solutions in addressing food access issues and policy barriers.

Baltimore, for example, has created a Food Policy Action Coalition that includes over 60 members representing residents, businesses, universities, nonprofits, and other relevant actors. The coalition works on a wide range of food-related issues, such as access to nutritious food and obesity, and is critical in developing inclusive food policies and managing food retail. To ensure that policies and solutions are inclusive, a group of resident food equity advisors, representing marginalized and minority groups, works closely with food policy officials in the municipality. Public participation in Kisumu is enshrined into its constitutional framework, with a designated directorate ensuring and encouraging public participation in the planning process. The county adopts a multidisciplinary approach, involving relevant technical advisors as well as the private sector, academia, and civil society organizations in every step of the planning process. With a specific focus on integrating food systems into urban development, Kisumu has established a multiactor county-level food governance mechanism called Food Liaison Advisory Group (FLAG), including thematic subgroups. The FLAG advises local authorities on food-related issues and on the development of a holistic food systems strategy currently under development. It includes representatives from different county's departments, producer associations, vendor organizations, civil society, and academic institutions. Ljubljana works on fostering urban-rural linkages by organizing and providing space for regular interaction between civil society and decision makers. These meetings have resulted in important knowledge exchange, for example, on access to local food and space for recreation during the COVID-19 pandemic, and they have facilitated engagement of a full range of actors in planning processes. In 2018, Quito launched the multistakeholder governance platform "Pacto Agroalimentario de Quito," consisting of representatives of civil society, the private sector, academia, international cooperation, and local, provincial, and national government.

Integrating different actors in urban planning is a challenging process, and good governance and institutional support play an essential role in this.

Kisumu's efforts to integrate agriculture, forestry, and food systems into urban planning benefited from the ongoing process of institution alizing frameworks and embedding policies into the law. Moreover, political will and the development of partnerships have been crucial. Ljubljana also received strong institutional support from the city's top leadership, which proved essential for its progress. Furthermore, great public participation and support underpinned the push to create a green city identity and the further development in this direction. Acknowledging the state of the city's food system and its challenges, Baltimore brought in specialized expertise and moved the food policy division to the planning department to strengthen the integration of food into urban planning.

Concluding Remarks

Urban policy coherence at city, territorial, and national levels is fundamental in order to build an enabling environment for local governments to take action and integrate food systems and green infrastructure into the planning process. To support an integrated approach, sectoral and spatial policies need to be linked. Moreover, urban governments need to contribute to the design of national policies related to food security, nutrition, and environmental challenges and ensure linkages with urban and territorial planning. This entails inter alia strengthening the technical, regulatory, institutional, and financial capacities of local administrations. In this respect, it is crucial to raise awareness of the role of local governments and provide a clear understanding of why food systems and green spaces should be an integral part of any urban planning agenda as tools to address food security, nutrition, and environmental challenges. As the interviews with the five diverse cities showed, good governance, institutional support, and political will are of great importance for the success of sustainable urban transformation.

Multilevel governance and multistakeholder collaboration should be at the center of urban planning processes, leveraging inclusive and integrated decision-making by bringing together civil society, academia, the private sector, and government representatives to work on integrated planning approaches. Indeed, the successful integration of food systems or urban and periurban agriculture and forestry into urban planning is usually connected to the existence of collaborative and cross-sectoral governance mechanisms such as food policy councils, food labs (Cabannes and Marocchino 2018). The experience of the interviewed cities revealed that multistakeholder governance platforms such as the Food Liaison Advisory Group in Kisumu, the "Pacto Alimentario de Quito," the Food Policy Action Coalition in Baltimore, and their specific subworking groups could become unique spaces for innovation, inclusive development, and enhanced accountability.

Moreover, the city interviews demonstrated the great importance of sectoral integration and interdepartmental collaboration. Urban planning and design require a good understanding of the connections between different spatial scales and their multiple relations. Taking into account interlinkages between energy, water, food, infrastructure, and urban green spaces as well as the circular economy should be central to integrated urban planning and design. Thus, city-level development plans should be complemented by land use plans, zoning regulations, and integrated infrastructure design at different scales, from neighborhood to district level. Furthermore, reinforcing urban-rural linkages and promoting the territorial approach are crucial for integrating urban and peri-urban agriculture, forestry, and food systems in urban and territorial planning and strengthening synergies between urban and rural areas.

Thus, integrating food systems and green infrastructure into urban planning does not depend so much on the entry point. What is at stake, that is to say a systemic plan that will be sustainable over time, is related to the capacity of the planning process to gradually address the different challenges and connect the different dots in a coherent, comprehensive, and systemic manner (Cabannes and Marocchino 2018).

In conclusion, there is an increasing need to leverage place-based, participatory, and integrated solutions for mainstreaming food systems and green spaces in urban and territorial planning and design that cut across different scales and administrative boundaries. The role of urban planners will be crucial to connect different actors, sectors, and spatial scales. They need to adopt systemic thinking and work in strong collaboration with other actors and across various systems to work toward ensuring the health and well-being of urban dwellers. City networks, UN-Agencies, and academic institutions can also play an important role in producing evidence, raising awareness, and developing capacity of policy makers, city planners, and local stakeholders to support local planning processes. By signing the Milan Urban Food Policy Pact (2015), over 200 mayors have acknowledged the strategic role of cities in developing sustainable food systems and promoting healthy diets. In 2018, FAO and RUAF launched the City Region Food Systems Programme to support local government in assessing climate risk in local food systems and plan measures to build resilience (FAO 2019a). In 2018, FAO and the Arbor Day Foundation launched the Tree Cities of the World Programme, a recognition scheme that acknowledges cities striving to make their cities greener, healthier, and happier places. Today, more than 120 cities have already joined the program. Furthermore, through the Framework for the Urban Food Agenda (FAO 2019b), FAO supports local governments in mainstreaming food systems into local planning and actions. Launched in 2020, The FAO's Green Cities Initiative, launched in 2020 (FAO 2020), supports local governments to leverage multifunctional green spaces for strengthened urban-rural linkages, access to healthy food from sustainable agrifood systems, and climate change mitigation and adaptation.

As more and more evidence of the role of urban agriculture, urban forestry, and urban food systems in building the resilience of cities in the face of external shocks is gathered and shared through different channels, it is expected that an increasing number of cities will adopt integrated planning approaches, ensuring access to both nutritious food and quality green spaces for all, leaving no one behind.

Cross-References

▶ Future Foods for Urban Food Production
▶ Urban Food Gardens

Acknowledgments We would like to thank the following persons for their availability to be interviewed and provide their insights on the integration of forestry, agriculture, and food systems in the planning process of their respective cities: Holly Freishtat (Baltimore); Maruška Markovčič (Ljubljana); Shulang Fei, Bowen Sheng, and Nan Wang (Chengdu); Geofrey Ochieng, Evans Gichana, and Girlchrist Okuom (Kisumu); and Alexandra Rodriguez (Quito).

References

Alexandra, J., & Norman, B. (2020). The city as forest – Integrating living infrastructure, climate conditioning and urban forestry in Canberra, Australia. *Sustainable Earth, 3*(1), 1–11.

Antrop, M. (2004). Landscape change and the urbanization process in Europe. *Landscape and Urban Planning, 67*(1–4), 9–26.

Artmann, M., Inostroza, L., & Fan, P. (2019). Urban sprawl, compact urban development and green cities. How much do we know, how much do we agree? *Ecological Indicators, 96*, 3–9.

Barton, H. (2009). Land use planning and health and well-being. *Land Use Policy, 26*(Supplement 1), S115–S123.

Bodicoat, D. H., O'donovan, G., Dalton, A. M., Gray, L. J., Yates, T., Edwardson, C., Hill, S., Webb, D. R., Khunti, K., Davies, M. J., & Jones, A. P. (2014). The association between neighbourhood greenspace and type 2 diabetes in a large cross-sectional study. *British Medical Journal Open, 4*, e006076.

Bren d'Amour, C., Reitsma, F., Baiocchi, G., Barthel, S., Güneralp, B., Erb, K. H., Haberl, H., Creutzig, F., & Seto, K. C. (2017). Future urban land expansion and implications for global croplands. *Proceedings of the National Academy of Science, 114*(34), 8939–8944.

Bush, J. (2020). The role of local government greening policies in the transition towards nature-based cities. *Environmental Innovation and Societal Transitions, 35*, 35–44.

Cabannes, Y., & Marocchino, C. (Eds.). (2018). *Integrating food into urban planning*. London/Rome: UCL Press/FAO.

Dadashpoor, H., & Ahani, S. (2019). Land tenure-related conflicts in peri-urban areas: A review. *Land Use Policy, 85*, 218–229.

Depietri, Y., Renaud, F. G., & Kallis, G. (2012). Heat waves and floods in urban areas: a policy-oriented review of ecosystem services. *Sustainability Science, 7*(1), 95–107.

Douglas, O., Lennon, M., & Scott, M. (2017). Green space benefits for health and well-being: A life-course approach for urban planning, design and management. *Cities, 66*, 53–62.

FAO. (2017). *The state of food and agriculture: Leveraging food systems for inclusive rural transformation*. Rome.

FAO. (2018). *The state of food and agriculture 2018. Migration agriculture and rural development*.

FAO. (2019a). *City region food systems programme: Reinforcing rural-urban linkages for climate resilient food systems*. Rome.

FAO. (2019b). *FAO framework for the Urban Food Agenda. Leveraging sub-national and local government action to ensure sustainable food systems and improved nutrition*. Rome.

FAO. (2020). *Green cities action programme. Building back better*. Rome.

Global Panel. (2017). *Urban diet and nutrition: trends, challenges and opportunities for policy action* (Policy brief no. 09). London: Global Panel for Agriculture and Food Systems for Nutrition.

Haaland, C., & van Den Bosch, C. K. (2015). Challenges and strategies for urban green-space planning in cities undergoing densification: A review. *Urban Forestry and Urban Greening, 14*(4), 760–771.

Lyu, Y., Wang, M., Zou, Y., & Wu, C. (2022). Mapping trade-offs among urban fringe land use functions to accurately support spatial planning. *Science of the Total Environment, 802*, 149915.

Rahman, M. A. U. (2015). Coordination of urban planning organizations as a process of achieving effective and socially just planning: A case of Dhaka city, Bangladesh. *International Journal of Sustainable Built Environment, 4*(2), 330–340.

Tefft, J., Jonasova, M., Zhang, F., & Zhang, Y. (2020). *Urban food systems governance – Current context and future opportunities*. Rome/Washington: FAO and The World Bank.

UCCRN. (2018). The future we don't want: How climate change could impact the world's greatest cities [online]. https://www.c40.org/other/the-future-we-don-t-want-homepage. Accessed 23 Dec 2021.

UN-Habitat. (2015). *International guidelines on urban and territorial planning*. Nairobi.

UN-HABITAT. (2018). *Leading changes. Delivering the New Urban Agenda through urban and territorial planning*. Nairobi.

UN-HABITAT, & WHO. (2020). *Integrating health in urban and territorial planning: A sourcebook*. Geneva: UN-HABITAT and World Health Organization.

United Nations Department of Economic and Social Affairs, Population Division. (2019). *World urbanization prospects: The 2018 revision (ST/ESA/SER.A/420)*. New York: United Nations.

van Vliet, J. (2019). Direct and indirect loss of natural area from urban expansion. *Nature Sustainability, 2*(8), 755–763.

WHO. (2016). *Urban green spaces and health – A review of evidence*. Geneva: World Health Organization.

WHO. (2017). *Urban green space interventions and health. A review of impacts and effectiveness*. Geneva: World Health Organization.

Zhou, D., Zhao, S., Liu, S., Zhang, L., & Zhu, C. (2014). Surface urban heat island in China's 32 major cities: Spatial patterns and drivers. *Remote sensing of environment, 152*, 51–61.

Zimmermann, E., Bracalenti, L., Rubén Piacentini, R., & Inostroza, L. (2016). Urban flood risk reduction by increasing green areas for adaptation to climate change. *Procedia Engineering, 161*, 2241–2246.

Websites

ARUP. (2021). Integrated planning for complex cities. https://www.arup.com/perspectives/integrated-planning-for-complex-cities. Accessed 23 Dec 2021.

I

Integrating Sustainability into Construction Project Management

Renard Y. J. Siew
Climate Change & Sustainability, Centre for Governance & Political Studies (CENT-GPS), Kuala Lumpur, Malaysia
Institute for Globally Distributed Open Research and Education (IGDORE), Bali, Indonesia

Definition

1. Sustainable Development Goals (SDGs) – encompasses a total of 17 Goals and 169 targets agreed by 193 member countries
2. Project Management Body of Knowledge (PMBOK) – a body of knowledge encompassing key project management areas
3. Sustainability Reporting Tools (SRTs) – a set of sustainability reporting frameworks, standards, ratings, and indices
4. Green Project Management (GPM) – a movement that advocates for the embedment of principles and value-based methods in project management
5. Project Sustainability Maturity Levels (PSML) – a framework to measure sustainability maturity in projects

Background

Project management can be broadly defined as the application of processes, methods, skills, knowledge, and experience to achieve certain set objectives within agreed parameters. Projects are unique in that they are not part of a routine operation but rather a specific set of operations designed to accomplish the objectives. The *Project Management Body of Knowledge* (PMBOK) often discussed in the literature typically encompasses 10 core areas, namely, integration; scope; time management; cost; quality; human resource; communication; risk; procurement; and stakeholder management (PMI 2020). The benefits of applying project management tools and practices are well-documented. Practitioners have cited that the use of project management has led to increased productivity and efficiency as well as visible reduction in cost and workload. This is backed with substantial evidence in the literature attempting to measure both qualitative and quantitative benefits of project management (see Lappe and Spang 2014; Gomes and Romão 2016).

In September 2015, an ambitious set of Sustainable Development Goals (SDGs) was introduced to the world to set a new direction for both developed and developing countries. The SDGs cut across diverse areas such as the need to end poverty (Goal 1); the need for access to clean water and sanitation (Goal 6); focus on responsible consumption and production (Goal 12); promote climate action (Goal 13); and develop sustainable cities and communities (Goal 11) just to name a few. One hundred ninety-three member countries collectively agreed on the 17 SDGs which consist of 169 targets. To ensure that this becomes a success, it is argued here that the SDGs or thinking around sustainability should be integrated into every decision-making tool including project management practices and existing tools. With the increasing demand by project stakeholders for more transparency on sustainability performance (i.e., the need for adaptable and more resilient buildings and infrastructure; incorporating water security into built environments), there is now a more urgent need to integrate sustainability into existing project management especially by construction practitioners.

Yet, to date, there is very little discussion on how sustainability can be embedded into existing practices. This paper aims to explore this subject further and outline recommendations to enhance current practice.

Sustainable Project Management

Few scholars have attempted to integrate sustainability thinking into project management. A large majority of them are focused on the development of sustainability criteria that could potentially be relevant to project success (see Mavi and Standing 2018; Silvius 2017; Siew 2015a). To a certain extent, such criteria is also reflected in existing sustainability reporting tools (SRTs) such as Leadership in Energy and Environmental Design (LEED), Building Research Establishment Environmental Assessment Method (BREEAM), and Green Star or the Green Building Index (GBI) although these tools have been critiqued for predominantly focusing on the environmental dimension with little consideration for other social aspects of sustainability.

There is also the emergence of the concept of Green Project Management (GPM 2019), a movement which advocates for the embedment of principles and value-based methods in project management. GPM offers training and relevant certification to interested individuals and organizations.

Project Sustainability Maturity Levels (PSML)

In recent years, the concept of project maturity levels has also emerged in the project management field. Siew et al. (2016) proposed a method for measuring project sustainability maturity levels. Being able to measure levels of sustainability attainment in projects will help project owners to better identify their strengths and weaknesses in their current practices. It might be worth outlining the summary of the framework here (see Table 1) as this may be useful application for construction practitioners.

The PSML model proposed was developed based on relevant PMBOK areas and consist of six main criteria, namely, integration, scope, cost, human resources, communication, and procurement. These criteria are further subdivided into 15 sustainability subcriteria. The proposal is to

I

Integrating Sustainability into Construction Project Management, Table 1 PSML Model (Siew et al. 2016)

Criteria	Subcriteria
Integration	• Project development – strategic planning to create a consistent document that can be used to guide the execution of sustainability practices (i.e., the focus areas of the SDGs)
	• Project execution – the execution of sustainability practices or the SDGs
	• Change control – the management of factors that create changes in the sustainability practices of a project
	• Information systems – the use of digitalization or technology to ensure that sustainability information/data (i.e., water, waste, energy consumption etc.) are captured in a database
Scope	• Business requirements – the incorporation of sustainability practices/ SDGs in business related requirements
	• Technical requirements – the incorporation of sustainability practices in the technical requirements of the project (i.e., requirements of green building/infrastructure projects)
	• Deliverables – the development of a detailed sustainability project scope; work breakdown structure
	• Scope change – the incorporation of additions, changes, and deletions to the scope of sustainability
Cost	• Estimation of the costs of resources needed to deliver sustainability practices (i.e., installation of more energy or water efficient products; waste management processes; improving living conditions of workers; enhancing health and safety practices)
Human resources	• Resource planning – the planning of human resources required to deliver sustainability practices
	• Recruitment process – the hiring process of sustainability professionals
Communication	• Planning – how sustainability information is conveyed to stakeholders (see Siew 2015a)
	• Sustainability reporting – the collection and reporting on sustainability progress of a project
Procurement	• Planning – the consideration of sustainability issues in procurement and the entire supply chain (i.e., is there a set of sustainability criteria that is used in pre-screening vendors)
	• Contracts management – the development of contract clauses to ensure that sustainability targets are delivered

measure each subscriterion using four levels of maturity level 1, ad hoc; level 2, defined; level 3, managed; and level 4, integrated (depending on the level of attainment). While fuzzy sets mapped to linguistic terms was proposed to measure the maturity level of each of these subcriteria (Siew et al. 2016), it should be noted that this is not the only way of measuring maturity levels (for non-technical personnel, a simple use of the Likert survey would suffice). The term sustainability practices is used interchangeably to refer to the focus areas of the SDGs.

The Project Sustainability Maturity Level (PSML) model is as follows:

The perspective offered is a very simplistic view of how a single project can be managed with sustainability in mind. However, the reality is that practitioners are often confronted with multiple projects at once. The management of multiple projects is often referred to as portfolio management. In another contribution, Siew (2015b) proposed two methods to account for sustainability in construction project portfolio management, namely, through (i) screening development of a set of sustainability related project criteria and (ii) optimal portfolio selection encouraging practitioners to use an efficient frontier to facilitate the selection of an optimal portfolio of projects from a sustainability perspective. This arguably may have been too advanced and technical now for use by practitioners; but it certainly is a way forward to ensure optimum planning and efficient use of resources to ensure sustainability outcomes are met by construction practitioners.

Recommendations

A few recommendations are proposed to further enhance the integration of sustainability thinking into project management practices. Firstly, given the proliferation of sustainability frameworks (i.e., United Nations 10 Principles, SDGs, life cycle analysis, green rating tools; Green Project Management), standards (i.e., IS0 14,001; OHSAS 18001), and reporting guidelines (i.e., Global Reporting Initiative (GRI); integrated reporting <IR>) which seems to be diverging there must instead be a coordinated effort globally to converge, harmonize, and integrate them with existing project management tools. Ideally, apart from ensuring relevance to construction industry practitioners, integration efforts should focus on the delivering the aspirations of the SDGs as they have been collectively agreed upon by 193 member countries.

Secondly, a professional network needs to be established to create a platform for the sharing of best sustainability practices within project management. The platform should serve to build up capability of practitioners to be well-versed with sustainability matters. Although GPM seems to be well-positioned to take on this role, it has not become mainstream. This is especially needed in the construction industry as it is well-known to be fragmented by nature and incorporating sustainability thinking is an opportunity to "connect the dots" and facilitate integration across multiple knowledge areas.

Thirdly, to encourage adoption of sustainable project management practices, an incentive mechanism should be set up for construction practitioners. This could be in the form of credit points available as is the case with green rating tools. There needs to be increased access to project green funding for buildings and infrastructure. This is an important aspect to consider as one of the common barriers to the adoption of green practices is the cost of entry (Siew 2015c). Many practitioners claim that investing in green and renewable technology/practices are often more expensive, and this either presents a financial burden to them or make them less competitive compared to their peers.

References

Gomes, J., & Romão, M. (2016). Improving project success: A case study using benefits and project management. *Procedia Computer Science, 100*, 489–497.

Green Project Management (GPM). (2019). *Our story*. https://www.greenprojectmanagement.org/about/our-story. Accessed 18 Aug 2020.

Lappe, M., & Spang, K. (2014). Investments in project management are profitable: A case study- based analysis of the relationship between the costs and benefits of

project management. *International Journal of Project Management, 32*(4), 603–612.
Mavi, R. K., & Standing, C. (2018). Critical success factors of sustainable project management in construction: A fuzzy DEMATEL-ANP approach. *Journal of Cleaner Production, 194*, 751–765.
Project Management Institute (PMI). (2020). *What is project management?* https://www.pmi.org/about/learn-about-pmi/what-is-project-management. Accessed 27 Aug 2020.
Siew, R. Y. J. (2015a). Health and safety communication strategy in a Malaysian construction company: A case study. *International Journal of Construction Management, 15*(4), 310–320.
Siew, R. Y. J. (2015b). Integrating sustainability into construction project portfolio management. *KSCE Journal of Civil Engineering, 20*(1), 101–108.
Siew, R. Y. J. (2015c). Briefing: Integrated reporting- challenges in the construction industry. *Proceedings of the ICE-Engineering Sustainability, 168*(1), 3–6.
Siew, R. Y. J., Balatbat, M. C. A., & Carmichael, D. G. (2016). Measuring project sustainability maturity level-a fuzzy based approach. *International Journal of Sustainable Development, 19*(1), 76–100.
Silvius, G. (2017). Sustainability as a new school of thought in project management. *Journal of Cleaner Production, 166*, 1479–1493.

Integrative Design Process

▶ Sustainable Community Masterplan

Intermediate Towns

▶ Small Towns in Asia and Urban Sustainability

Internally Displaced Persons

▶ Multi-stakeholder Partnerships to Support Climate Migrants in Fragile Cities

International Law

▶ Transnational Crimes: Global Impact and Responses

International Partnerships

▶ Multi-stakeholder Partnerships to Support Climate Migrants in Fragile Cities

International Relations

▶ Internationalization of Cities

Internationalization of Cities

From Spectators to Participants

Noelia Wayar[1] and Maria Julieta Brezzo[2]
[1]Extension Secretary, National University of Córdoba, Córdoba, Argentina
[2]Institutional Relations and Events, Ciudades Globales – CIGLO, Córdoba, Argentina

Synonyms

Decentralized cooperation and political strategy; Governments; International relations; Local development; Political science and development; State

Definitions

Internationalization of cities: a multidisciplinary and dynamic process, including the networking and interdisciplinary work of local and regional governments whose objective is to promote the development of subnational unity.

International action: when a subject takes a role beyond the local.

Decentralized cooperation: a variant of traditional cooperation involving subnational governments, especially cities and a new approach in cooperation relations that seeks to establish direct links with local representative bodies and stimulate their own capacities to project and lead development initiatives with the direct participation of interested population groups, taking into account their interests and their points of view on development

Local development: a process of transformation of the local economy and society, aimed at overcoming the existing difficulties and challenges, which seeks to improve the living conditions of its population. It should be considered case by case, as the needs and demands of each city and territory are different; the capabilities of the inhabitants, companies, and the local community change; and, in addition, each community has its own views about the priorities that development policies should incorporate.

Political strategy: the implementation of actions designed, measured, and adapted to the context from the government to achieve a certain objective

Introduction

International relations can no longer be exclusively understood as the external linkage of central governments. In a world of complex realities, numerous relevant and influential actors have emerged and are beginning to develop capacities and carry out international activities, such as NGOs, companies, and subnational governments. The latter, and in particular cities, have considerably increased their role on the international scene, developing networking, decentralized cooperation, and positioning of citizens and the territory. Those who govern these local units are recognized as counterparts in international forums for debate on issues of the international agenda.

With the processes of decentralization of governments, globalization, and growing urbanization, local governments have been taking responsibilities that go beyond their traditional role. Currently, the administration of a municipality is not limited to the provision of services but is in charge of promoting local development and, to that end, the interactions of a city beyond its local limits is essential.

The importance of local economic development, its multidimensionality, and the increasingly wide acceptance of its geographic-territorial scale have stimulated local governments to take responsibility for a large number of issues. According to Ponce Adame (2018), this necessarily motivates the revaluation of the municipal government space and the different public and social actors.

The new responsibilities and the issues on the global agenda that impact the local territory led local governments to seek alternatives and innovative instruments for international cooperation and linkage. In this sense, Díaz Castro (2018) proposes the irreversible internationalization of local governments. Irreversible in a globalized, urban, and interconnected world, local internationalization is a necessary practice for any substate entity that wishes to provide solutions to the problems of its territory. This is why we will analyze the relationship between cities, local governments, and their international actions.

Main Text

Every city is different. At government level, practices and procedures have been developed to address the needs of the citizens and the territory, therefore each one has its own reality. This is why the challenge of developing internationalization processes and strategies has not been addressed in the same way by all cities; some governments still consider their international relations as a set of isolated actions, and others are leading strategic processes of international ties that are part of the local development plan.

The prevailing system of government in each country is decisive in the degree of internationalization (those states with federal systems of government are more likely to allow international relations to their subnational units).

In relation to this, it is very likely that if the central government does not promote internationalization practices at the local level, cities will have difficulties in developing a strategy. Collaboration between the two levels of government will depend mainly on two factors: the political affinity between them and the existence of a legal framework that regulates the international actions of cities.

Generally speaking, central governments have one of the following positions: they allow international action, they object to it, or they are indifferent. Legal stability and a regulatory framework are essential to encourage the long-term development of these actions.

How Is a City Internationalized?

Internationalizing a city does not mean internationalizing a government, but rather a territory and its citizens. For this reason, the chosen strategy should not be a unilateral "top-down" strategy, but it should be agreed with other levels of government and institutions of the city in order to promote governance as the managing strategy.

Cities seek spaces for plural debate guided by the government's strategic plan, giving political influence to international action. This will force local governments to develop mechanisms for raising awareness and participation, and to listen to all actors in the city to reach a consensus on the best way to "go out into the world." The development of multi-stakeholder dynamics is essential to transform the international actions of a government into sustainable and legitimized public policies.

To go from strategy into practice, governments must professionalize themselves and have a team that not only has training in the subject but also, fundamentally, has the ability to work across all areas of the local government. The tools for the internationalization of cities are multiple and the selection of a toolkit that best suit the objectives will depend on the guiding strategy.

These strategies for the internationalization of cities are based on the concept of decentralized cooperation. According to Eduardo Gana (1996), the traditional concept of international cooperation is that: *"solutions that are good for one country can be good for the majority."* This perception has favored for decades the formulation of all-encompassing solutions, projects of wide applicability, and the imposing of the same guidelines to all countries.

Descentralized cooperation, as a different concept in relation to the traditional cooperation, is "*a new approach in cooperation relations that seeks to establish direct links with local representative bodies and stimulate their own capacities to project and lead development initiatives with the direct participation of interested population groups, taking into account their interests and their points of view on development*" (González Parada 1998).

UN-Habitat has expanded the use of the term "city-to-city cooperation," abbreviated as C2C. As this concept is endorsed at the international level, it has encouraged the development of actions from local governments and currently acts as a sort of "legal framework" that enables their action.

Among the most common tools for the internationalization of cities, according to Eugene D. Zapata Garesché (2007), we can name:

1. Horizontal collaboration and exchange of experiences: a relationship that favors dialogue and collaboration rather than substitution, subordination, or competition between two or more local governments. Its advantage is that it mobilizes and transforms the government much more profoundly than mere financial transactions.
2. Twinning and bilateral projects: Through them, friendly relations have been established between two cities or regions, establishing institutional ties between governments, promoting closeness between their communities, cultural exchanges, and support for development projects.
3. Reciprocity: Mutual interest and benefits. Reciprocity is the mutual interest to cooperate and the possibility of deriving benefits from cooperation on both sides.

The emergence of unconventional actors within international relations increased in recent decades. One of the main reasons that led cities to internationalize were several economic crises that affected different countries during the 1980s and 1990s. This led local governments to start designing new action plans for their economic recovery having those previous tools as a starting point.

Currently, cities seek to promote international cooperation, to improve their economic relations, and thus be able to offer their citizens infrastructure, services, and technology that meet their needs. By taking action on the local level, international cooperation is intervening in the agendas of local and regional governments.

Networking is one of the main pillars on which the internationalization of cities is built.

Manuel Castells developed the concept of network society: "…is *one whose social structure (human organizational agreements) is made up of networks (interconnected nodes) powered by information and communication technologies.*" The characteristics of this network society are flexibility, adaptability, and survival capacity, all fundamental to unleash the potential of society and to achieve a technological paradigm that has given way to the Information Age.

In addition, the network society is a global society. This means that all the processes take place in global networks of a dominant social structure. All basic activities that make up human life in every corner of the planet are organized in global networks. "*The social structure is global, but most of the human experience is local in both a territorial and cultural sense*" (Borja et al. 1998).

Political actors began to expand their networks, generating new nodes that are centered in the cities. This concept by Castells applied to cities, and what we have been proposing, relates in an interesting way with the concept of *smart city*, where cities are intertwined with technology and the search for optimization of processes, even influenced by urban planning and architecture, and solutions to needs that have more and more points in common and are shared by other cities in the world.

When we talk about the internationalization of cities, we are talking about a node, as defined by Castells, where we can transversely observe: territory, citizenship, technology, economy, culture, architecture and urbanism, law, politics, relationships and links, social development, and welfare. On this line, Mariana Borrell developed the concept of urban competitiveness within the frame of the international projection of cities in the context of globalization. Generally, a competitive city has to be attractive and allow for international investments and events, has modern infrastructure, and offers technology and services in various areas up to international standards. A competitive city is also innovative in its architectural developments and urban appearance, has good communication networks with its physical and virtual environment, implements local policies of exchange with its counterparts on other countries, and has an active and diversified international presence, among other determinants.

So, what are cities looking for when going international?

First, we must consider that the ultimate goal of a government that is internationalizing is to positively impact citizens. This is done through new opportunities for the city, better conditions for the territory, the construction of an improved city model that allows comprehensive economic, social, and cultural development benefiting everyone. Otherwise, international actions would be meaningless.

The second point to consider is local development. Local development policies define their actions with a territorial approach. The local development strategy should be considered case by case, as the needs and demands of each city and territory are different, the capabilities of the inhabitants, companies, and the local community change, and, in addition, each community has its own views about the priorities that development policies should incorporate. The development of a city or territory requires that public and private actors execute their investment programs in a coordinated manner (Vazquez-Baquero 2009).

The third motivation is the positioning of the city. A city internationally recognized, either for its natural characteristics, the services it offers, its own assets, or its successful administration, is an attractive city for tourists, investors, or to be a model to other cities. The political leader has a fundamental role in this matter. Internationalizing a city in an intelligent way implies legitimizing an administration with all local actors and making its political figure visible.

In addition to this "proactive" attitude, there is also a reaction to the global demand: cities are being called to contribute.

The currently pandemic of the Covid-19 has generated a rebirth of nationalisms and a strong protagonism of the National States; however, the cooperation actions have been strengthened and the mayors, using these tools, have been recognized as fundamental actors to manage the crisis. The mayors of certain cities have taken on an even greater role, confronting national governments with the discourse that: no one knows the population as we do, decisions will be local. Something very similar to what was being discussed in the debates on global issues. In this crisis management, the international action, the strength of the mayor's role, and the international support serve those who have to make decisions to take the lead, lead and manage a health crisis without precedents but also to position themselves as undisputed leaders in the near future.

International Agendas Are Calling for Cities to Intervene

The agendas for development that we started to see since 2015, and the main change that is apparent, is the multiplicity of actors that are considered relevant for the fulfillment of those agendas.

When we speak of agendas, we refer to the 2030 Agenda, the Paris Agreement, the Sendai Framework Agreement, the New Urban Agenda, and the Financing for Development Agenda.

Specifically, the 2030 Agenda and the sustainable development goals are the result of multiple processes of participation by organizations of different types and representatives of various spheres from countries on all five continents. The result was a multi-stakeholder agenda.

In turn, the inclusion of SDG 11 regarding sustainable cities is the product of pressure from organized cities that made themselves heard. This is, perhaps, the best example of the importance of cities on the agenda. On the other hand, the Habitat III meeting in Quito and the New Urban Agenda are another example that, in this era, cities became protagonists.

This situation is not only desirable but it is necessary: the proximity that a Major has with constituents brings detailed knowledge of the needs of the population and the impact on the territory of the decisions of the Agenda made in the large international forums, whether positive or negative. As an example, we will give a brief account of the Syria Program in Argentina and its relationship with the local government of the city of Cordoba.

The Syria Program in Argentina: Actions of the Municipality of Córdoba

During 2017, the conflicts in Syria escalated. Many people decided to emigrate in search of better living conditions.

The large Syrian-Lebanese community in the city of Córdoba, Argentina, invited for some to choose this Argentine city to settle.

They did so within the framework of the "Special Humanitarian Visa Program for Foreigners affected by the conflict in the Syrian Arab Republic," called the Syrian Program, whose objective is, through the processing of visas for humanitarian reasons, to facilitate the entry into the country of the people fleeing the conflict in Syria, allowing them to build a new life in Argentina or to settle temporarily until the end of the conflict.

The Syria Program is the product of a positive alliance between government and civil society, and receives financial assistance and technical support from the international community.

The program is under the responsibility of the central government, and local governments celebrated this decision, but the migrants arrived in the city in a precarious situation: very little resources, a cumbersome emotional burden, and no knowledge of the local language.

The Municipality of Córdoba with Major Ramón Mestre, and the Syrian Consulate in Córdoba met to define how they will face this international problem of local impact.

They decided that it was urgent to collaborate for the inclusion of migrants, and the most important obstacle was the language, so they formed an alliance with the Faculty of Languages of the

National University of Córdoba so that all the people arriving in Córdoba as refugees receive free Spanish language and culture classes. This approach would give them the practical tools to enable intercultural communication with the host population.

This is a clear example of an issue on the global agenda (refugees), of measures taken by national governments, in this case the argentine central government, that have a direct impact on the territory, Córdoba, that had to be addressed by the Major of the city.

How Do You Implement the New Urban Agenda in the Territories?

The New Urban Agenda presents a paradigm shift based on the science of cities; establishes standards and principles for planning, construction, development, administration, and improvement of urban areas, with five main pillars of application: national urban policies, urban legislation and regulations, urban planning and design, local economy and municipal finances, and local implementation. It is a resource for this common ideal to be realized from all levels of government, from central to local, civil society organizations, private sector, groups of interested parties and people who consider that the urban spaces of the world they are their "home." The New Urban Agenda incorporates a new recognition of the correlation between good urbanization and development. It highlights the links between good urbanization and job creation, opportunities to generate income and improved quality of life, which should be included in all urban renewal policies and strategies. This further highlights the connection between the New Urban Agenda and the 2030 Agenda for Sustainable Development, in particular Goal number 11, which deals with sustainable cities and communities.

We have seen that local and regional governments have more and more responsibilities and assume new commitments. However, it is difficult for cities to have all the necessary resources to face the new challenges of the Development Agendas.

According to UCLG (United Cities and Local Governments), within the framework of *the Global Report on Local Democracy and Decentralization (GOLD IV)* and recommendations of the Global Taskforce of Local and Regional Governments prepared during the Habitat III process and included in the Bogotá Commitment and in the Action Agenda, local governments should follow these recommendations:

1. Improve the strategic planning capacities of local and regional governments.
2. Promote the dynamic and autonomous participation of civil society to co-create cities and territories.
3. Take advantage of integrated urban and regional planning to shape the future of cities and territories.
4. Guarantee access to quality infrastructure and quality basic services to everyone.
5. Develop local economic opportunities for the creation of quality employment and promote social cohesion.
6. Place the "Right to the City" at the center of urban and territorial governance.
7. Lead the transition to resilient and low-carbon cities and regions.
8. Promote local heritage, creativity, and diversity through people-centered cultural policies.

Political views will always be present: internationalization of cities is used as a positioning tool for a political figure. Participation in international forums and advocacy on global agendas are desirable scenarios for any politician, and this is not a bad thing, but it is necessary to promote discussions and consensus, so that the strategy does not only serve a partisan purpose but also allows for development of the city that can be maintained over time even through changes in administration and different approaches to internationalization.

Internationalization of cities is the result of a consensus among the most diverse institutions in a territory; to internationalize a city is not to internationalize a government but a territory. The ultimate goal of any process of this type must be focused on the citizen. Local development and best opportunities for everyone must be the focus of these strategies and, therefore, as a practical matter, must be within the city's development plan. The teams in charge of carrying out the challenging task of internationalizing the city must be trained to promote a new multi-sector and multi-level governance system that promotes citizen participation.

Conclusion

Although cities predate nation states, international dynamics before the twenty-first century did not consider them valid actors in the global system. This has changed in recent decades, showing processes of internationalization of cities, creating multi-stakeholder spaces where they are recognized as necessary voices to be able to fully implement different global agendas that require a local counterpart.

Some factors that have made this trend possible can be summarized as: the growing process of urbanization, the decentralization of central governments, the globalization process, and the new development agendas.

The coronavirus has also deepened on the traditional cooperation actions, promoting in cities to generate concrete actions of change preparing themselves for the post-pandemic: redefining city schemes, promoting healthier practices, generating more committed environmental policies, working in alternative initiatives of urban mobility, and in all this, cooperation is fundamental not only for the new challenges that cities will face but to manage the treatment of pre-existing problems in cities, which were deepened as a consequence of Covid-19.

In relation to the Agendas, it is important to note that there is a double cause-and-effect relationship between their contents and local governments: local governments have worked hard to have an impact on these agendas because they understand that the issues discussed there directly affect the cities. But the agendas, in turn, recognize and request nontraditional actors to participate.

Faced with this challenge, the internationalization of cities should be considered as a multidisciplinary dynamic, including networking and interdisciplinary work of local and regional governments. The concepts of multi-level and multisector governance should be key while designing an international strategy for a city.

Cities must assume the enormous responsibility of persevering on the fight for a spot at the decision-making table; they must take charge of the important task of raising awareness among governments and teams about the importance and need in today's world of cities that are global, integrated, and influential: build the global from the local.

References

Barquero, A. V. (2007). Desarrollo endógeno. Teorías y políticas de desarrollo territorial. Investigaciones Regionales. *Journal of Regional Research,* (11), 183–210.

Borja, J., Castells, M., Belil, M., & Benner, C. (1998). Local y global: la gestión de las ciudades en la era de la información (Vol. 5). Madrid: Taurus.

Borrell, M. (2012). La proyección internacional de las ciudades en la globalización: una revisión del concepto de competitividad urbana.

del Huerto Romero, M. (2004). La cooperación descentralizada: nuevos desafíos para la gestión urbana. *Urbano, 7*(9), 76–85. Recuperado a partir de http://revistas.ubiobio.cl/index.php/RU/article/view/545

Gana, Eduardo (1996). Las relaciones económicas entre América Latina y la Unión Europea: el papel de los servicios exteriores, CEPAL, Santiago de Chile.

Garesché, E. D. Z. (2007). Guidelines for the international relations of local governments and decentralised cooperation between the European Union and Latin America. 1. Practical manual for the internationalisation of cities. Observatorio. ONU Habitat - Nueva Agenda Urbana. http://uploads.habitat3.org/hb3/NUA-Spanish.pdf

González Parada, J. R. (1998). Cooperación descentralizada. Un nuevo modelo de relaciones Norte-Sur, IUDC/UCM, Madrid.

Las Competencias, D. P. Y. M., & La Esfera, E. N. (2017). Los gobiernos locales en la internacionalización del territorio. Coordinación General: Pablo costamagna, Mariana foglia, 3. (This one is related to Diaz Castro 2018)

Ponce Adame, Esther y otros (2018). Teoría y práctica de la cooperación internacional para el desarrollo. Ciudad de México: Honorable Cámara de Diputados, 2018

The Interplay of Intersectionality and Vulnerability Towards Equitable Resilience

Learning from Climate Adaptation Practices

Aynaz Lotfata[1] and Dalia Munenzon[2]
[1]Geography Department, Chicago State University, Chicago, IL, USA
[2]College of Architecture, Texas Tech University, Lubbock, TX, USA

Introduction

Recently, planning and implementing a community-driven, equitable climate preparedness process got worldwide attention (e.g., Murray and Poland 2020). Community-driven planning empowers and enables those facing the most significant climate risks (Kirkby et al. 2018). Equity should be the center of any adaptive approach, and it should not reduce to the equal spread of vulnerability among groups (Reckien et al. 2017). This study suggests considering the root causes of inequity in neglected intersectional identities to understand who will be most sensitive to climate change impacts. We live in the context of emerging identities that are often not involved in planning and design processes. Intersectionality of ethnicity, race, religion, gender, sexual orientation, nationality, (dis)ability, political affiliation, relationship status, profession, and socio-economic status define urban dynamics (Evans et al. 2018). However, the lack of risk adaptation infrastructure in terms of diversity within urban planning practices inhibits the long-term adaptation (Wei et al. 2015). Urban practices need to develop a multiple-level approach to model equities at the intersection of multiple social identities. A multiple-level system carries the sense of commonalities (Opach et al. 2020) where social, cultural, economic, political, ecological capitals of a geographical location are in harmony with individuals' and groups' needs without loading strains on natural resources and causing social inequities and exclusions. In this context, the primary concern is to expand pre-hazard and post-hazard equity in adaptation to minimize vulnerabilities. In the global context, urban practices considered equity 52% in planning and 59% in implementation, and in equity planning, limited marginalized groups are included (GAMI 2021). In this regard, to what extent are historically marginalized social groups included in planning (procedural equity) or documented as intended participants/beneficiaries *(distributive justice)* of adaptation responses left out unanswered (Araos et al. 2021; Meerow et al. 2019). Similarly, though adaptation associated with urban resilience is debated frequently, there is still no explicit understanding of the resilience in cities and who benefits the most from the adaptation. The United Nations Human Settlements Programme (UN-Habitat) highlights that "global agendas having resilience as a key concept will ensure that the call for sustainable and resilient cities leaves no one behind" (UN-Habitat 2018).

The evolving identities create unforeseen discrimination and inequities, especially in areas with poor adaptation strategies to risks. The integrative adaptation research framework can enable incremental-transformative adaptive governance (Wilson et al. 2020; Few et al. 2017; Pelling 2011). In this rein, building self-organized communities (Yang and Jiang 2020), focusing on target groups, encouraging local participation, and co-production development are four pathways toward adaptive responses to risks and minimizing vulnerabilities (ECDCP 2017). These pathways raise significant questions about adaptive responses: "who is responding?", "what is the type of adaptation responses?", "are adaptive responses reducing climate change risks?", "is local and indigenous knowledge considered in adaptive response planning?", "What is the cost of adaptation"?, "is adaptation cross-cutting themes?", "who is financing adaptive response?", and "what is depth, scope, speed of adaptive response?". Thereby, vulnerability should not be limited to a set of predefined conditions of population that derive from the historical and prevailing cultural, social, environmental, political, and economic conditions. Vulnerability is an uncertain social phenomenon (Kuran et al. 2020). It should be rediscovered in the interplay of equity, adaptation, and resilience because most studies homogenized identities of vulnerable groups without considering that urban context is diverse and dynamic, leading to unequal exposure to disasters (Vickery 2018; Blennow et al. 2019). In risk management studies (Chisty et al. 2021; Kuran et al. 2020; Lovell et al. 2019; Schuller et al. 2015), intersectionality is used as an approach and tool to consider diverse values and preferences in resilience planning and to foster community well-being (Ryder and Boone 2019; Prohaska 2020; Collins and Bilge 2020). The intersectionality outlook can contribute to studying societal processes in a fine-grained way, though it can yet homogenize members to a specific group.

In this study, the equity with the engagement of intersectionality perspective was qualitatively studied and debated in Climate Adaptive Practices to uncover the differences in vulnerability, adaptive responses, and factors that prevent community resilience. While the notion of resilience underlines the long-term goals, the predetermined qualification of vulnerability contradicts a long-term mission of resilience thinking while individuals' interests, values, and identities change. Hence, this study incorporates intersectional equity and vulnerability into resilience planning. In this regard, first, we argued Resilience Thinking and the interplay of Equity, Adaptation, and Community Resilience engaged with intersectionality. Second, we designed the framework to read the Urban Adaptation Projects regarding Equity, Intersectionality, and Community Resilience. Finally, we developed the adaptive framework regarding intersectional perspectives and recommendations that scholars and practitioners may implement in practice. We seek to add value to current debates rather than propose an alternative framework.

Resilience Thinking

Resilience is widely used across various disciplines, including science, sociology, economics, and engineering (Larimian et al. 2020; Saja et al. 2019). It was derived from the Latin word "resilio," meaning "to jump back" (Klein et al. 2003), and it emerged in the early 1970s in the ecological systems (Walker et al. 2004). It evolved from its initial definition (the ability to bounce back to equilibrium following external stressors) into a more comprehensive theory where adaptation and transformability are vital components (e.g., Bindu and Subha 2021; Olazabal 2017). Adaptability addresses the capacity of the system to learn and adjust in response to disturbance into the stable domain (e.g., Broeke et al. 2017). Transformability is the capacity to evolve into the fundamental new system/condition when the current situation does not afford the changing demands and needs (Olazabal 2017).

Resilience is defined by the dynamic interactions of slow (e.g., identities, preferences, and climate) and fast-changing (e.g., markets and weather variation) variables (Zanotti et al. 2020). Change is naturally occurring and ignoring or resisting change reduces resilience and increases the vulnerability of a system (e.g., community) to risks and disasters. Feedback loops between social dynamics (e.g., intersectional identities) and stakes define the pathway of change. Risks and social dynamics are indeed in a cyclic relationship. In other words, the occurrence of risks enables and exacerbates socioeconomic risks and vice versa. Social dimensions (e.g., social equity and justice) get attention when the questions of where, who, when, where, and why were debated in areas at natural hazard risks.

The origin of resilience thinking bridges natural and social sciences by using the same terminology referring to social systems and ecological systems. The transference of the resilience concept from the ecological sciences to social systems (Davidson 2010) is a barrier to understanding differences in ecological and social systems' essence (Adger 2000). Accordingly, a system-based approach on social dimensions of resilience was debated mainly in material terms, for example, access to resources, physical infrastructure, and urban network (Dwiartama and Rosin 2014), and ignored changing habits, beliefs, and identities of the communities. Further studies on reflection of resilience thinking on social systems with taking intersectional perspective in the center of attention are needed to move social dimensions of resilience beyond communities' material and physical structure toward equitable resilience. There is not yet a clear definition and consensus for resilience. Resilience as a bouncing back system in the face of risks (e.g., Hoegl and Hartmann 2021), adoption of resilience definition as a transformative evolutionary process (e.g., Komninos et al. 2019), and/or fuzzy understanding of resilience (e.g., Schaefer et al. 2020) create an ambiguous situation to find an explicit solution for inequities and vulnerabilities. Overall, resilience depends on identifying variables that define the system's dynamic, determining the risks that may destabilize these variables, and illustrating variables' thresholds (Kwak et al. 2021) while interscale cooperative mobilizing of stakeholders can maintain the system from moving toward undesirable situations (Rahmani et al. 2021).

Intersectional Equity and Community Resilience

The way to engage intersectional equity with the concept of resilience is ambiguous. The overarching goal of intersectionality is to study the power distribution and understand the mechanism of domination and oppression within intersected identities causing inequities and making individuals vulnerable (Chisty et al. 2021; Anguelovski et al. 2020; Crenshaw 1989). Given that resilience is a transformative evolutionary process, intersectional equity should be built on the positive and negative feedbacks from changing social dynamics in different stages of risk (pre-risk, during risk, and post risk). Likewise, it is critical to keep in mind that the context of risk is the bedrock of unforeseen risks which may cause intersectional vulnerabilities if the adaptive response is not made in timely depth and speed.

The identification of equity and public well-being predominates the following components: access to material needs, health, security, good social relations, and freedom of choice (e.g., Sung and Phillips 2018). This predetermined identification creates a barrier to reaching social equity, collective well-being, mutual trust, and a sense of belonging (e.g., Jiang and Zhen 2021) in the context of risks. A community should not function as a closed system (Zhao et al. 2017) where intersectional equity is undermined. In recent years, the intersectional perspective has indeed expanded outside the field of race, such as sustainable development (Ryder and Boone 2019), environmental hazards (Fletcher 2018), and disaster vulnerability (Vickery 2018).

The intersectional perspective centers attention on building community equity beyond one dimension leading to building homogeneous societies. It does encourage the combination of several factors, such as race, sex, ethnicity, race, disability, and immigration status (e.g., Buckingham-Hatfield 2000). Intersectional equity is based on differences in individual lives, collective and social practices. In the context of intersectionality, communities with different individuals' and groups' identities in different sizes build inclusive relationships using co-production as a tool (Fig. 1). Chambers et al. (2021) state the six pathways of co-production which enable intersectional equity. The pathways are: 1. They are finding solutions for equity within socio-ecological dynamic interactions (Frank et al. 2017). 2. Enabling marginalized groups to self-identify their own needs, priorities, and portrayal can help to mitigate misrepresentation (Fitzgibbons and Mitchell 2019). 3. Planning practices can include disempowered groups and individuals in the planning process using a

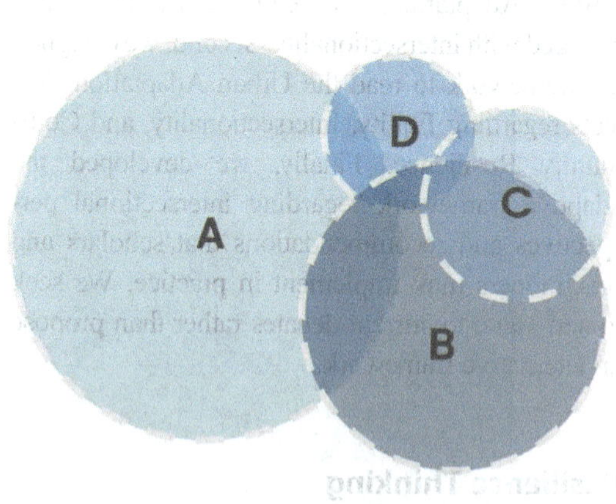

The Interplay of Intersectionality and Vulnerability Towards Equitable Resilience, Fig. 1 Using co-production approach among communities with different identities and sizes (Community A,B,C,D's Social, cultural, political, and economic capitals of each community and their geographical locations)

participatory approach to breaking pre-framed power (Eidt et al. 2020). 4. Reframing power within the incremental-transformative governance framework. 5. navigating differences and 6. reframing agency.

The intersectional perspectives illustrate how socially constructed identities and differences are re(produced) in response to disasters and hazards (Hopkins 2019). In addition, how different social groups are impacted by dangers and their impacts on, effects of disasters and threats (Thomas et al. 2019). Traditionally, identifying vulnerability was a one-sided process, the impact of anthropogenic or natural hazards and/or combination of these two on communities. This definition is primarily used in risk assessment and management studies (e.g., Jena et al. 2020) as unmanageable, undesirable, and uncertain conditions. At the same time, on another side, vulnerability is socially constructed, and its degree depends on cultural, geographical, and environmental variables imposed on specific individuals (Bene et al. 2016). The type and extent of vulnerability depends on who is exposed to the risks (e.g., Alizadeh et al. 2018). The intersectional perspective calls attention to more narrowed observations on community members than the general terms "people with disability are vulnerable" and "women are vulnerable." The sensitivity and lack of adaptation capacity should be included as a measure of vulnerability (Cardona et al. 2012).

Understanding resilience and equity under the dynamic nature of intersectionality centers strain on climate adaptive actions. It highlights that the traditional predetermined approach in understanding vulnerability and limits social dimensions of resilience to physical infrastructure and urban flows (Raicu et al. 2019; Da Silveira et al. 2018) can mislead, cause delays in adaptation, and extend recovery time. In addition, environmental justice highlights the importance of social dynamics and equitable processes in achieving climate change adaptation as equity debated in three categories 1. "distributive justice" – distribution of burdens and allocation of benefits across society (i.e., who suffers from climate risks, who benefits from adaptation responses?); 2. "procedural justice" – processes of representation and participation in decision-making (i.e., how are adaptation actions prioritized?); 3. "recognition justice" – who is recognized as a legitimate actor and how are their needs and interests acknowledged and included (i.e., whose concerns matter?) (Singh et al. 2021).

The public policies tend to statically group people while dealing with hazards. The intersectional perspective opens avenues to challenge the concepts of resilience and vulnerability in urban planning and public policies, which ignores differentials and the way these differentials interact in the real world. In addition, traditional vulnerability definitions should evolve so that individuals and communities' vulnerability, robustness, and resilience are in change both in terms of impacts and outcomes of hazards (Alvarez and Evans 2021). The degree of vulnerability and resilience depends on the individual capacity (Kuran et al. 2020). The capacity generally identifies with health, access to resources, income, and good social relations. However, access to all these depends on the spatial and socioeconomic distribution of resources and dynamics of socially constructed context. The various social factors, norms, values, cultures, and social relations shape the responsibilities and roles critical to consider in risk assessment and flourishing individuals' well-being in terms of environmental and social equity.

Thereby, intersectionality with urban practices requires understanding individual perceptions about their living environment and uncovering marginalized people and how they are marginalized to develop sustainable mitigation strategies. Likewise, we should keep in mind that while people are vulnerable to some hazards, they might be resilient for others. In addition, becoming deep into individual needs should not cause us to neglect the social dimension. In this regard, a co-production approach initiated as a qualitative approach implemented across scales (micro and macro) to develop an action-based approach to building relational well-being and social equity. Social equity should support both individualistic and social actions and provide a valuable framework to understand people's motivation and behavior in short and longer terms (Hunter 2005).

We propose an assessment of community well-being in the context of resilience and social equity through the lens of an intersectional perspective to uncover qualitative differences in vulnerability. In this context, resilience and its opposite side, vulnerability, are both socially constructed. Resilient and vulnerable groups/communities define through their pre-event social, economic, and cultural factors, which can cause a lack of access to resources and exclusion if resources do not manage appropriately in the broader context. For example, income, livelihood, education, health, and residential location define vulnerability. The level of access to all these can cause inequity, and it also can change over time and add more complexity.

The most vulnerable people – people with disabilities, women, children, older persons, minority and indigenous groups, LGBT, persons with chronic health conditions, and others who are contextually marginalized live with existing inequalities – require the highest levels of resilience.

Intersectionality theories examine the diversity and interconnected nature of such systems of oppression and exclusion. Therefore, the question of intersectionality is inherent to the equitable delivery of urban services as anticipated in the SDGs (Stuart and Woodroffe 2016).

This entry uncovers intersectional approaches within climate adaptation practices to vulnerability reduction and resilience-building. The intersectional approach studies the historical, social, cultural, and political contexts to explain that social groups are neither homogeneous nor static. Indeed, using an intersectional approach contributes to understanding vulnerability and resilience in an implementable way. Understanding changing social dynamics within a community is critical to reducing temporal lag or timely response and recovery from hazards. There is no single approach or defined set of methods for seeking intersectional understandings of vulnerability and resilience relating to climate change and natural hazards. More research on intersectional approaches to vulnerability reduction and resilience building is required, mainly qualitative and contextual analysis, to fully understand how inequalities intersect and affect people in different contexts (Fig. 2).

The Interplay of Intersectionality and Vulnerability Towards Equitable Resilience, Fig. 2 Explaining broader ways in conceptualizing literature study

Vulnerability and Resilience in Practice: Studying Urban Adaptation Projects

Conceptual Model

Adaptation should reduce vulnerability and/or increase adaptive capacity, especially of the most vulnerable and those most at risk to climate change

We selected three guideline documents between the years of 2017 and 2019. They are 1. Community-Driven Climate Adaptation Planning developed by the National Association of Climate Resilience Planners (NACRP), 2. Urban Sustainability Directors Network (USDN), and 3. the NAACP Environmental and Climate Justice Program (ECJP) to study how these projects define vulnerability, incorporate intersectionality across the evaluation process, and integrate it in the strategies development process. To analyze the applicability of such strategies for vulnerable communities, we reviewed the City of Providence's Climate Justice Plan, Creating an Equitable, Low-Carbon, and Climate-Resilient Future, and the framework supported "An Urban Sustainability Directors Network Innovation Fund Project."

To assess the context of intersectional resilience, we propose to focus on Community Resilience as the scale that can accommodate both individual identities, social dynamics, and geography. Although operating at this scale, we acknowledge that each community operates and is impacted by policies and practices prescribed at a higher governmental level. The focus on community resilience provides access to the approach of Co-production as defined in the literary review above.

The literary review outlines several theoretical approaches to intersectionality, resilience, environmental justice, and community well-being. The scholarly study shows that the current academic discourse around climate adaptation is focused on infrastructure and city-wide action and is insufficient for equitable and effective long-term transformation. The following review of guideline documents and climate adaptation reports explicitly created for community-based organizations shows that the convergence of all these issues is already happening in practice.

Vulnerability is analyzed through the spatial condition of exposure to risk, the social dynamic in play, and individual identities and variables. All can be viewed as constructed by past social and power dynamics through the framework of Environmental Justice. When these elements are in play, they exacerbate and reinforce each other, and the introduction of external risk amplifies the existing challenges. Therefore, many of the projects reviewed here identify well-being as an objective to address these layers in depth. The co-production pathways identified by Chambers et al. (2021) are used as the framework to analyze and deliver well-being in these marginalized communities. And a mechanism as trust-building utilized to respond to the need for participation, power and agency building, and most importantly, the negotiation of differences. Figure 3 illustrates the conceptual framework to study the selected Urban Practices of this study.

"*Equitable climate preparedness planning strives to fairly distribute the benefits and burdens of climate change and climate actions through a community-driven planning process that empowers those most affected to shape the decisions that will impact their lives.... Traditional planning practices are not properly designed to understand and address how racial and social inequities intensify climate risk. As a result, climate adaptation plans may inadvertently perpetuate or exacerbate existing racial and social inequities*" (USDN 2017a, b). In this study, the community-driven urban practices are studied by questioning three critical points:

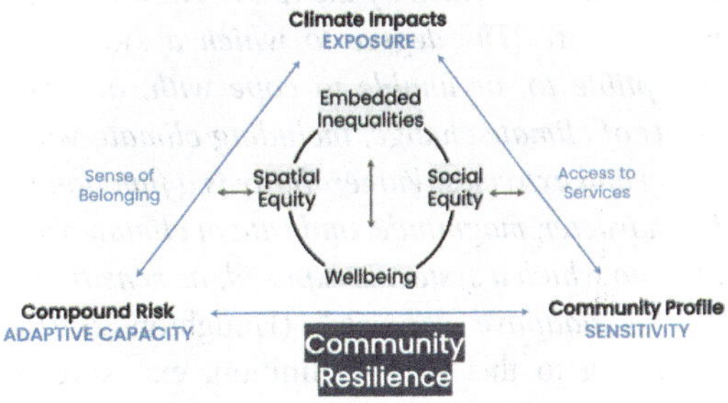

The Interplay of Intersectionality and Vulnerability Towards Equitable Resilience, Fig. 3 A model to study the current urban practices

- The premise of the projects studied.
- Point of inception – Which issues are framed as consensus at the starting point of these documents?
- Why do they pursue a community-driven process?

Overlapping and Exacerbating Impacts and Vulnerabilities

The reviewed projects and guideline documents offer strategies to achieve equitable climate adaptation for communities that identify as vulnerable. Historic inequities in representation, capital, and services distribution, and spatial planning (planning decisions and investment in infrastructure that created the current built environment) are the core of the risk created by climate impacts. The Guide to Equitable, Community-Driven Climate Resilience Planning (USDN 2017a, b) defines vulnerability as the combination of three categories:

- Climate hazard (the exposure to the event)
- Risk factors (physical and demographic conditions)
- Contributing causes (systemic socio and economic conditions)

These categories resonate with the vulnerability definition provided by the IPCC Third Assessment Report "*The degree to which a system is susceptible to, or unable to cope with, adverse effects of climate change, including climate variability and extremes. Vulnerability is a function of the character, magnitude, and rate of climate variation to which a system is exposed, its sensitivity, and its adaptive capacity*" (Houghton 2001). According to this IPCC definition, exposure is the degree of hazard, sensitivity, the impacted resources, and the scale of risk to the built environment; adaptive capacity, the system's ability to adjust and cope. This paper's reviewed documents expand exposure, sensitivity, and adaptive capacity to include well-being and social cohesion variables. Thus, they identify components that can provide the framework for transformative actions.

Contributing causes of inequity can be outlined as cascading variables that exacerbate individual and community sensitivity and directly affect adaptive capacity. By framing these aspects of livelihood and well-being as crucial adaptation components, these projects emphasize the role of urban planning, policy, and governance in creating equitable resilience. The Guide to Equitable, Community-Driven Climate Resilience Planning (USDN 2017a, b) describes how climate risks combined with economic insecurity can foster multiple challenges to individual livelihood (Fig. 4). For example, daily weather conditions can impact one's access to employment when combined with the state of transit and transportation infrastructure. Similarly, individuals who are required to work long hours outdoors are more subjected to the health impacts of cold/hot weather.

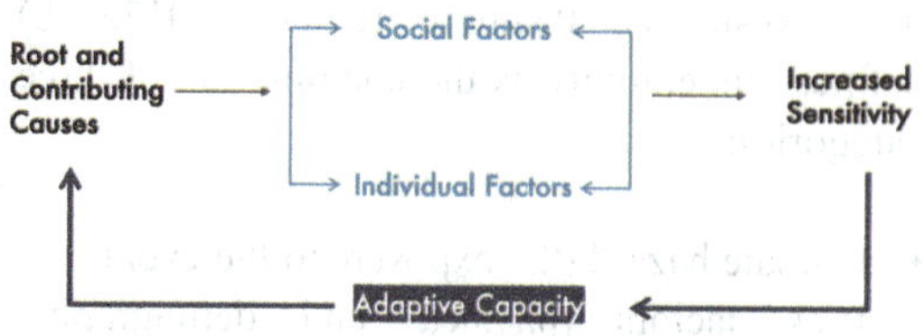

The Interplay of Intersectionality and Vulnerability Towards Equitable Resilience, Fig. 4 Overlapping vulnerabilities between social inequity and adaptive capacity

Moreover, neighborhoods burdened with subgrade utility systems will be more susceptible to power outages and risk residents' exposure to cold or heat. In low-income and underserved communities, any unplanned monetary expenses on health services or access to employment can result in their long-term adaptive capacity. The factors leading to disproportionate vulnerability to climate change are outlined below (Table 1) with the summary of indicators and variables as identified in the studies projects. Notably, this document combines social factors with climate hazards. While traditionally, exposure and climate only account for extreme heat, flooding, wildfires, etc., this document adds expected rising utility and food costs due to the changing environment to the list of hazards, thereby reinforcing the connection between fundamental well-being needs as part of the adaptation conversation.

Although each of the projects listed above has a slightly different approach to discussing, analyzing, and studying community vulnerabilities, they all identify systemic and root causes and social risk factors that, if addressed, can be the levers of resilience and well-being. The Guide to Equitable, Community-Driven Climate Resilience Planning singles out racial and ethnic inequality as a root cause to health, education, and natural hazard disparities. The document emphasizes the disproportionate burden on communities of color created by "*structural and institutional racism in our economic, government, and social systems*" (USDN 2017a, b). For example, racial segregation historically resulted in neighborhood disinvestment and political disenfranchisement. Planning policy in the service of racial segregation and inequity has perpetuated neighborhoods exposed to more significant climate impacts and are more vulnerable (Hoffman et al. 2020).

Social factors that result from these root causes are both spatial and social. They range from access to green space and affordable services to language barriers. Individual and biological factors are the demographic and health data of each resident driving personal sensitivity. While age, gender, and other factors are given, health is deeply connected to the root causes listed.

The Interplay of Intersectionality and Vulnerability Towards Equitable Resilience, Table 1 Survey of identified vulnerability variables

Name of document	Climate impact (Exposure)	Contributing and root causes (Adaptive Capacity)	Risk factors -social (Sensitivity)	Risk factors -individual (Sensitivity)
The Guide to Equitable, Community-Driven Climate Resilience Planning	Heat, coastal flooding, wildfire and air, rising cost of living (utility and food)	Concentrated neighborhood disinvestment, Racial segregation, poverty, income inequality, lack of living wage jobs, gaps in educational opportunities and attainment	Residential location, ability to afford basic necessities and resources, access to affordable and quality housing, access to reliable and affordable transportation, access to affordable health care, access to green spaces, green infrastructure, and tree cover, linguistic isolation, social cohesion	Age, chronic and acute illnesses, mental and physical disabilities, overall health status
Community-Driven Climate Resilience Planning: A Framework, 2017, by National Association of Climate Resilience Planners	Drought, pollution	Poverty, racial disparities, educational divestment, rate of incarceration	Access to affordable, healthy food; affordable energy	Preventable diseases
Our Communities Our Power, Advancing Resistance and Resilience in Climate Change Adaptation, Action toolkit, 2019 by the NAACP (ECJP)	Agricultural yields, sea level rise, and extreme weather	Systems, policies, programs/services, protocols, and governance/decision making	Income/wealth, employment, literacy, education, housing stock, insurance status, etc.	Age, gender, race, preexisting health condition
The City of Providence's Climate Justice Plan, creating an equitable, low-carbon, and climate resilient future, Fall 2019, By the City of Providence	Air pollution, sea level rise, flooding	Structural racism, economic inequality, extractive economy	Income, representation, transportation, housing, community safety, policing practices, services	Chronic illness, gender, race

According to Hoffman's study (Hoffman et al. 2020) – at the same time as practices of rezoning, urban renewal, and redlining pushed communities of color to live in less desirable areas in low-lying lands and next to polluting industries, municipalities, and local governments reduced investments fostering spatial deterioration and individual burdens. These actions lead communities residing in vulnerable geographic locations to develop biological factors, such as chronic illnesses, mental trauma, and physical disabilities. Illuminating sensitivities through individual and social factors and root causes provides a multilayered framework to climate adaptation that can overlap and resolve the three paths of climate resilience preparedness, recovery, and adaptation. How to translate these factors to actions and strategies will be discussed in the following section.

As many root causes become institutionalized and geographically grounded, projects such as the Community-Driven Climate Resilience Planning: A Framework (USDN 2017a, b) emphasize community co-production to identify local social factors with specific adaptive strategies. Their work aims to address the social aspects of organic and healthy food access, energy systems, and political representation, with a unique focus on building capacity for governance. Two of the case studies documented from California – PODER and Rooted in Resilience – demonstrated a community and youth-led vulnerability analysis. The team interviewed residents about their livelihood challenges, priorities, and climate-related concerns. The goal of such an approach, according to the report, was to identify community and culturally specific needs and develop a collaborative narrative for the future. Community surveys and in-depth conversations were used to identify both the root causes and risk factors.

The analysis of personal and community narratives helps incorporate intersectionality in the planning toward equitable adaptation. The NAACP Environmental and Climate Justice Program (ECJP) provides a guide for juxtaposing several risk factors to determine "*intersectional relationships in systems, communities and individual lives*" (NAACP 2019). Their action toolkit, Our Communities Our Power, Advancing Resistance and Resilience in Climate Change Adaptation, categorizes most individual risk factors as static that require consideration when proposing strategies and social risk factors as vulnerabilities to address. For example, for a group representing several intersecting risk factors, the document shows that during Hurricane Katrina, the most impacted group was "low-income, African American women" (Rhodes et al. 2020). Studying the root causes of high injury and mortality rates should focus on the smallest possible geographic area and specific cultural norms, circumstances, and local issues. While this method is grounded in fieldwork and community outreach, similar has been done in academia with the combination of "inter-categorical intersectional analyses" and neighborhood-level data (Alvarez and Evans 2021). The detailed and comprehensive analysis and categorization of personal narrative can set the ground to multi-scalar adaptation strategies focusing on emergency preparedness, services and education, and urban climate adaptation infrastructure design. The guideline documents studied here show the overlap between Environmental Justice and analysis of intersectional issues per individuals and communities and the need to propose multilayered solutions to improve well-being and promote climate adaptation.

Well-Being as an Applicable Goal in Implementing Equitable Climate Adaptation

To in-depth investigate how intersectional and inter-categorical approaches of analyzing vulnerabilities to climate hazards and how are translated into actions to achieve adaptive capacity, we studied The City of Providence's Climate Justice Plan, Creating an Equitable, Low-Carbon, and Climate-Resilient Future (City of Providence Office of Sustainability 2019), and the Innovation Funded Project by the USDN, with Facilitating Power and the National Association of Climate Resilience Planners (USDN 2017a, b). The fund's goal was to strengthen communities' ability to influence municipal decision-making and form equity-focused committees with community leaders. These committees were formed to foster community-driven governance models through co-production and trust-building between communities and City leadership and staff. Successful integration of equity into municipal climate action plans was the desired result of the effort. In this research, we examined the method applied by the Providence plan and the USDN process to address intersectional issues in communities through co-production and trust-building and the crucial role well-being plays in their goal setting and solution development.

The concept of community is framed by a geographic area and shared culture (local, ethnic, or racial) as defined by the guideline documents studied above. However, community vulnerabilities can result from societal processes, prevailing discrimination, and individual identities and conditions. Therefore, creating an inclusive committee and community-driven process should reflect the overlap of both definitions. The Providence Climate Justice Plan identifies vulnerability as a predefined reality for the City's frontline communities. According to the plan's glossary, Frontline Communities are ethnic and racial minorities impacted by "*the crises of ecology, economy, and democracy,*" focusing on the foreign-born, incarcerated, and LGBTQ, hence merging individual and social factors with root causes. In line with the framing of vulnerable communities in the Environmental Justice discourse, according to Jeff Turrentine, frontline communities "*have historically shouldered a disproportionate share of our collective environmental burden, while simultaneously not always benefiting from policies to reduce pollution and recent dire climate warnings have exacerbated the issue*" (Turrentine 2019).

The Providence Racial and Environmental Justice Committee (e.g., REJC), established in 2016 under the Office of sustainability, developed several projects focused on integrating equity into environmental and climate adaptation projects. The REJC's mission is to identify the intersectional risk factors created by the root causes of personal and institutional racism, identify ecological priorities and concerns, and propose long-term strategies for collaboration between the city and the communities. A core objective is to develop a "Critical Analysis of Race and Intersectionality," draw the line between lived experiences, inequities, and root causes, and develop solutions that address structural racism. To include both geography-based communities and population with other social and individual factors, the REJC comprises ten members from local low-income and BIPOC communities. More specifically, "*four of the seats are neighborhood-based, five are topical, and one is held for a representative of one of the tribes that originally inhabited the territory that Providence occupies*" (USDN 2017a, b). To foster accountability and deep democracy, each committee member has a group of at least ten community members with whom they meet before each committee meeting to report and discuss progress. This iterative process and structure can ensure greater inclusion and participation of various identities in a population. As a result, it enables a comprehensive understanding of intersectional challenges and the development of holistic solutions. The REJC receives a stipend for their role in the committee to respect and value their time and effort. In addition, the committee is supported by city agencies representatives as listeners and the Office of Sustainability staff with stewarding the focus on racial and environmental equity.

The vision REJC developed prioritizes social and individual well-being and includes housing, access and services, transit, safety, and equitable policing practices. The REJC report emphasizes the vision of well-being with the term – "*El Buen Vivir: living well without living better at the expense of others. The fundamental human right to clean, healthy, and adequate air, water, land, food, education, transportation, safety, and housing. Just relationships with each other and with the natural world, of which we are a part*" (REJC 2016). REJC widens the definition of well-being beyond the economic and individualistic perspective (Hunter 2005) toward the collective good. This collective conception includes the public commons as services, infrastructure, environment, and relationships between individuals.

Co-production and trust-building are the means to promote this vision, with the unpacking of lived experience, incorporating community rights and self-determination into city practices, expanding capacity, and avenues for decision making. The co-production approach assists the progress of collaboration between individual residents, communities, and the municipal staff. Bovaird and Loeffler argue it is a pragmatic exchange between the population and the public sector that can create greater social and environmental value (Bovaird and Loeffler 2012). The USDN Innovation Fund Project Grant used a tool developed by Rosa González of Facilitating Power, in collaboration with Movement Strategy Center called The Spectrum of Community Engagement to Ownership (Gonzalez and Facilitating Power 2019). The Spectrum examines opportunities to introduce community participation and impact municipal decision-making practices. This engagement and collaborative planning process start from the status quo of marginalization that must be addressed by giving the community agency a level of autonomy to have tangible influence. The Spectrum ranks participation from zero to five, respectably from ignoring, marginalization to community ownership.

Most climate adaptation projects cited in the USDN Community Ownership document are located between 2 – consult and 3 – involving; the public sector often has the tool to collect input from the community but not enough willingness to ensure the implementation of priorities. This tool can be used for the design of a process and the analysis of specific conditions. Analysis of collaboration between the community and local government can include members participating in policy implementation and city practices to recognize policy challenges and capacity limitations. Shared indicator analysis can boost implementation capacity. Building community networks and having a clear purpose can help create the conditions needed for collaborative governance (commitment, resourcing, capacity, and trust-building). The community must be able to participate in clear and transparent decision-making processes. It is critical to foster a multidirectional learning environment where community members provide crucial input to city staff on policies and procedures. City staff includes vital information to community members on impacts and community assets. A collaborative process and the ability to act generate levels of trust in the development of an equitable decision-making protocol. Equitable decisions cultivate accountability between community and government.

Root causes of climate vulnerability such as racial discrimination and systemic disinvestment foster distrust and suspicion between the residents and the local governments. In line with the vision for well-being offered by the REJC, the NAACP toolkit document views Social Cohesion as a core lever toward adaptation. According to the NAACP Toolkit, Social Cohesion is the relationships and links the residents with each other and their built environment; it is the cooperation and trust-building against exclusion and toward well-being (NAACP 2019). Ideally, a cohesive society will have the collective capacity to cope and respond to extreme events and hazards. However, beyond the structure of a specific geographic community, the question in climate adaptation planning goes back to the relationship with the local governance systems. The scale of transformation required for adaptation is grander than what can be accomplished at the local level. It necessitates a relationship with the public sector and access to policy-making mechanisms and public funding.

Hence, the dynamics between the public sector and communities are a significant variable in effective adaptation. The vulnerabilities identified by factors or root causes have been created and perpetuated partially by the same institutions with which communities and individuals are now expected to collaborate. Distrust is one of the most significant barriers in pursuing inclusive and equitable processes, framed by the USDN document From Community Engagement to Ownership (USDN 2017a, b). Additional challenges are embedded in bureaucratic practices that make it difficult to build alliances. Trust-building is a communication practice used to bridge these gaps by creating personal and direct relationships and creating pathways to ensure strategies and policies are implemented. Conversations between public sector officials and community members must unpack root causes, discuss hard truths (USDN 2017a, b), and be willing to make structural changes and increase pathways for communication and accountability. A prerequisite is thus identifying who is vulnerable and who can adapt and why people are vulnerable and why they hold differential, adaptive capacities (Thomas et al. 2019).

Community and individual capacity to participate in collaborative design and planning processes are an additional hurdle toward effective adaptation. Frontline communities, burdened by injustice and inequities, are limited in their ability to participate and address the complex issues they are facing. To build equitable participation and trust, the process will need to propose a "*community resourcing strategy*" (USDN 2017a, b). Institutional capacity is required to create pathways for communications and engagement with community leaders. And the institutional capacity provides the resources to move forward solutions and demonstrate implementation and goodwill. Defaulting on moving the decisions forward risks trust and "reinforces public disillusionment" (Hernández and Franco 2021). Thus, a successful process provides a response to the three Environmental Justice components (Singh et al. 2021) and addresses the integration of community members in governmental procedures and practices, their ability to access and prescribe the distribution of funds, and the recognition of these community members and their respective community's needs and priorities.

Proposed Strategies

The co-production approach establishes the framework for transformation at the governance level by building relationships, providing opportunities for dialogue, and access to training and funding. Other strategies address well-being and spatial issues like health, economy, housing, energy, transportation, and preparedness. Table 2 categorizes these actions based on root causes and risk factors to overlay with the vulnerability analysis methodology. Activities and strategies that address root causes (and not a specific climate hazard) with policy or infrastructure have better chances of increasing the effectiveness of the proposed plan for the long term.

Discussion: From Solutions to Transformative Actions

In the last several years, climate adaptation expanded from FEMA's five mission areas: prevention, protection, mitigation, response, and recovery (FEMA 2020). While infrastructure and emergency management offer solutions to reduce exposure, policy and governance actions aimed at long-term systemic change can address the root causes of frontline communities' marginalization and, consequently, foster community resilience.

Equitable community resilience is built through multi-scalar actions at the local community levels and in the long term on the city level. Emergency management and preparedness, large-scale infrastructure and spatial design, community-based organizational actions and capacity building, and the reversal of disinvestment by health-focused policies should all be carried out concurrently. Transformative actions address root causes and social and individual risk factors by offering to address the built environment and the mechanisms

The Interplay of Intersectionality and Vulnerability Towards Equitable Resilience, Table 2 Adaptive strategies and intersecting vulnerabilities

Issue	Scale and responsibility	Type of action	Contributing and root causes (Adaptive Capacity)	Risk Factor (Sensitivity)	Climate impact (Exposure)	Time frame of impact
Governance and capacity building	City + community	Co-production	Representation and recognition			Near term +long
Preparedness	City	Infrastructure		Health	Heat, storm, flood	Near term
Green spaces	Community	Infrastructure	Neighborhood disinvestment, Racial segregation, poverty		Heat, storm, flood	Long term
Transportation	City	Infrastructure		Health, access, income		Long term
Housing	City	Policy + infrastructure		Displacement, income		
Energy	City	Infrastructure		Health, age	Heat, storm, flood	
Health	City	Policy			Air pollution	
Economy	Community	Capacity		Low income, education		

that created it. All these necessitate political will to transform the governance mechanism, as demonstrated by the spectrum, funding to implement large-scale infrastructure changes, and culturally and locally sensitive actions based on community priorities and needs.

In the framework of this study, we only reviewed a limited number of guideline documents and plans – these case studies aimed to address equitable resilience and propose strategies for frontline communities. However, as the analysis by the fund showed, within the limits of political will and existing planning frameworks, they were only able to achieve limited progress towards inclusion. It leaves the need to analyze other examples of more long-term and successful climate adaptation plans that could execute transformational actions over time and perhaps test their adaptive capacity in the face of exposure. Future studies should focus on adaptive capacity in communities' overtime. Large-scale resilience infrastructure plans face implementation challenges due to the complexity of funding, policy, regulations, and governance. Therefore, the political will and local bottom-up activism should be studied through the lens of co-production.

Conclusion

This study aimed to understand how intersectional approaches are incorporated in climate adaptation planning – vulnerability assessment and solution development. Results show intersectionality works at the level of the community and public sector. The analysis of vulnerabilities through root causes and personal and social factors helps uncover historical inequities and latent layers of factors that can reduce the effectiveness and adaptive capacity. Based on the projects we examined, we concluded that intersectionality in vulnerability assessment is leveraging co-production to incorporate intersectional identities into climate planning and the power of well-being as a vision and a tool. Our research connects ideas from the literature about better equity, environmental justice, and intersectionality and demonstrates their applicability in practice.

References

Adger, W. N. (2000). Social and ecological resilience: Are they related? *Progress in Human Geography, 24*(3), 347–364.

Alizadeh, M., Alizadeh, E., Asadollahpour, K. S., Shahabi, H., Beiranvand Pour, A., Panahi, M., Bin Ahmad, B., & Saro, L. (2018). Social vulnerability assessment using Artificial Neural Network (ANN) model for earthquake hazard in Tabriz City, Iran. *Sustainability, 10*(10), 3376.

Alvarez, C. H., & Evans, C. R. (2021). Intersectional environmental justice and population health inequalities: A novel approach. *Social Science & Medicine, 269*, 113559, ISSN 0277-9536.

Anguelovski, I., et al. (2020). Expanding the boundaries of justice in urban greening scholarship: Toward an emancipatory, antisubordination, intersectional, and relational approach. *Annals of the American Association of Geographers, 1–27*, Routledge. https://doi.org/10.1080/24694452.2020.1740579.

Araos, M., Jagannathan, K., Shukla, R., Ajibade, I., Coughlan de Perez, E., Davis, K., Ford, J.D., Galappaththi, E. K., Grady, C., Hudson, A. J., Joe, E. T., Kirchhoff, C. J., Lesnikowski, A., Gabriela Nagle Alverio, Miriam Nielsen, Ben Orlove, Brian Pentz, Diana Reckien, A. R. Siders, Ulibarri, N., van Aalst, M., Zulfawu Abu, T., Agrawal, T., Berrang-Ford, L., Bezner Kerr, R., Coggins, S., Garschagen, M., Harden, A., Mach, K. J., Marshall Nunbogu, A., Spandan, P., Templeman, S., & Turek-Hankins, L. L. (2021). Equity in human adaptation-related responses: A systematic global review. *One Earth, 4*(10), 1454–1467, ISSN 2590-3322, https://doi.org/10.1016/j.oneear.2021.09.001.

Bene, C., Al-Hassan, R. M., Amarasinghe, O., Fong, P., Ocran, J., Onumah, E., Ratuniata, R., Tuyen, T. V., McGregor, J. A., & Mills, D. J. (2016). Is resilience socially constructed? Empirical evidence from Fiji, Ghana, Sri Lanka, and Vietnam. *Global Environmental Change, 38*, 153–170.

Bindu, C. A., & Subha, V. (2021). *IOP Conference Series: Materials Science and Engineering, 1114 012045.*

Blennow, K., Persson, E., & Persson, J. (2019). Are values related to culture, identity, community cohesion and sense of place the values most vulnerable to climate change? *PLoS One, 14*(1), e0210426.

Bovaird, T., & Loeffler, E. (2012). From engagement to co-production: The contribution of users and communities to outcomes and public value. *VOLUNTAS: International Journal of Voluntary and Nonprofit Organizations, 23*(4), 1119–1138.

Broeke, T., van Voorn, G. A., Ligtenberg, G., & Molenaar, J. (2017). Resilience through adaptation. *PLoS One, 12*(2), e0171833.

Buckingham-Hatfield, S. (2000). *Gender and environment*. London: Routledge.

Cardona, O. D., van Aalst, M. K., Birkmann, J., Fordham, M., McGregor, G., Perez, R., Pulwarty, R. S., Schipper,

E. L. F., & Sinh, B. T. (2012). Determinants of risk: Exposure and vulnerability. In C. B. Field, V. Barros, T. F. Stocker, D. Qin, D. J. Dokken, K. L. Ebi, M. D. Mastrandrea, K. J. Mach, G.-K. Plattner, S. K. Allen, M. Tignor, & P. M. Midgley (Eds.), *Managing the risks of extreme events and disasters to advance climate change adaptation. A Special Report of Working Groups I and II of the Intergovernmental Panel on Climate Change (IPCC)* (pp. 65–108). Cambridge/New York: Cambridge University Press.

Chambers, J. M., Wyborn, C., Ryan, M. E. et al. (2021). Six modes of co-production for sustainability. *Nature Sustainability*.

Chisty, M. A., Dola, S. E. A., Khan, N. A., & Rahman, M. M. (2021). Intersectionality, vulnerability and resilience: Why it is important to review the diversifications within groups at risk to achieve a resilient community. *Continuity & Resilience Review, 3*(2), 119–131.

City of Providence Office of Sustainability. (2019). *The City of Providence's climate Justice plan*. Creating an equitable, low-carbon, and climate resilient future. Accessed 5 Nov 2019 from http://www.providenceri.gov/wp-content/uploads/2019/10/Climate-Justice-Plan-Report-FINAL-English-1.pdf

Collins, P. H., & Bilge, S. (2020). *Intersectionality*. Wiley.

Crenshaw, K. (1989). *Demarginalizing the intersection of race and sex: A black feminist critique of antidiscrimination doctrine, feminist theory, and antiracist politics, 1989* (p. 139). University of Chicago Legal Forum.

Da Silveira, M. E. M., et al. (2018). Environmental Justice and climate change adaptation in the context of risk society. In F. Alves, F. W. Leal, & U. Azeiteiro (Eds.), *Theory and practice of climate adaptation. Climate change management*. Cham: Springer.

Davidson, D. J. (2010). The applicability of the concept of resilience to social systems: Some sources of optimism and nagging doubts. *Society & Natural Resources, 23*(12), 1135–1149.

Dwiartama, A., & Rosin, C. (2014). Exploring agency beyond humans: The compatibility of actor-network theory (ANT) and resilience thinking. *Ecology and Society, 19*(3), 28.

Eidt, C. M., Pant, L. P., & Hickey, G. M. (2020). Platform, participation, and power: How dominant and minority stakeholders shape agricultural innovation. *Sustainability, 12*(2), 461.

Evans, L. R., Williams, D. R., Onnela, J.-P., & Subramanian, S. V. (2018). A multilevel approach to modeling health inequalities at the intersection of multiple social identities. *Social Science & Medicine, 203*, 64–73, ISSN 0277-9536.

Federal Emergency Management Agency (FEMA). (2020). *National preparedness goal. FEMA.gov*. Access link: https://www.fema.gov/emergency-managers/nationalpreparedness/goal#:~:text=Capabilities%20to%20Reach%20the%20Goal,tied%20to%20a%20capability%20target. Accessed 25 Oct 2021.

Few, R., Morchain, D., Spear, D., et al. (2017). Transformation, adaptation and development: Relating concepts to practice. *Palgrave Communications, 3*, 17092.

Fitzgibbons, J., & Mitchell, C.L. (2019). Just urban futures? Exploring equity in "100 Resilient Cities". *World Dev, 122*:648–659. https://doi.org/10.1016/j.worlddev.2019.06.021.

Fletcher, A. J. (2018). *More than women and men: A framework for gender and intersectionality research on environmental crisis and conflict*.

Frank, B., Delano, D., & Caniglia, S. (2017). Urban Systems: A socio-ecological system perspective. *Sociology International Journal, 1*(1), 00001.

Global Adaptation Mapping Initiative (GAMI)-Accessed-Link: https://globaladaptation.github.io/

Gonzalez, R., & Facilitating Power. (2019). The spectrum of community engagement to ownership. *Movement Strategy Center*. Access link: https://movementstrategy.org/resources/the-spectrum-of-community-engagement-to-ownership/. Accessed 24 Oct 2021.

Guide To Equitable, Community Driven Climate Preparedness Planning(ECDCP). (2017). Link https://www.usdn.org/uploads/cms/documents/usdn_guide_to_equitable_community-driven_climate_preparedness-_high_res.pdf

Hernández, F., & Franco, Á. (2021). Urban spaces of fear and disillusionment. In A. Scribano, L. M. Camarena, & A. L. Cervio (Eds.), *Cities, capitalism and the politics of sensibilities*. Cham: Palgrave Macmillan.

Hoegl, M., & Hartmann, S. (2021). Bouncing back, if not beyond: Challenges for research on resilience. *Asian Business & Management, 20*, 456–464.

Hoffman, J. S., Shandas, V., & Pendleton, N. (2020). The effects of historical housing policies on resident exposure to intra-urban heat: A study of 108 US urban areas. *Climate, 8*, 12.

Hopkins, P. (2019). Social geography I: Intersectionality. *Progress in Human Geography, 43*(5), 937–947.

Houghton, J. T. (2001). *Climate change 2001: The scientific basis: Contribution of working group I to the third assessment report of the intergovernmental panel on climate change*. Cambridge: Cambridge University Press.

Hunter, L. M. (2005). Migration and environmental hazards. *Population and Environment, 26*, 273–302.

Jena, R., Pradhan, B., Beydoun, G., et al. (2020). Seismic hazard and risk assessment: A review of state-of-the-art traditional and GIS models. *Arabian Journal of Geosciences, 13*, 50.

Jiang, Y., & Zhen, F. (2021). The role of community service satisfaction in the influence of community social capital on the sense of community belonging: A case study of Nanjing, China. *Journal of Housing and the Built Environment*.

Kirkby, P., Williams, C., & Huq, S. (2018). Community-based adaptation (CBA): Adding conceptual clarity to the approach, and establishing its principles and challenges. *Climate and Development, 10*(7), 577–589.

Klein, R. T. J., Nicholls, R. J., & Thomalla,F. (2003). Resilience to natural hazards: How useful is this concept?. *Global Environmental Change Part B: Environmental Hazards, 5*(1), 35–45, https://doi.org/10.1016/j.hazards.2004.02.001.

I

Komninos, N., Kakderi, C., Panori, A., & Tsarchopoulos, P. (2019). Smart city planning from an evolutionary perspective. *Journal of Urban Technology, 26*(2), 3–20. https://doi.org/10.1080/10630732.2018.1485368.

Kuran, C. H. A., Morsut, C., Kruke, B. I., Krüger, M., Segnestam, L., Orru, K., Nævestad, T. O., Airola, M., Keränen, J., Gabel, F., Hansson, S., & Torpan, S. (2020). Vulnerability and vulnerable groups from an intersectionality perspective. *International Journal of Disaster Risk Reduction, 50*, 101826, ISSN 2212-4209.

Kwak, Y., Deal, B., & Mosey, G. (2021). Landscape design toward urban resilience: Bridging science and physical design coupling sociohydrological modeling and design process. *Sustainability, 13*(9), 4666.

Larimian, T., Sadeghi, A., Palaiologou, G., & Schmidt, R., III. (2020). Neighborhood social resilience (NSR): Definition, conceptualisation, and measurement scale development. *Sustainability, 12*(16), 6363.

Lovell, E., Twigg, J., & Lung'ahi, G. (2019). *Building resilience for all: Intersectional approaches for reducing vulnerability to natural hazards in Nepal and Kenya*.

Meerow, S., Pajouhesh, P., & Miller, T. R. (2019). Social equity in urban resilience planning. *Local Environment, 24*(9), 793–808.

Murray, S., & Poland, B. (2020). Neighborhood climate resilience: Lessons from the lighthouse project. *Canadian Journal of Public Health, 111*, 890–896.

National Association for the Advancement of Colored People(NAACP). (2019). *Our communities our power, advancing resistance and resilience in climate change adaptation, Action toolkit*, 2019 By the NAACP Environmental and Climate Justice Program (ECJP)- Access link: https://naaee.org/sites/default/files/equity_in_resilience_building_climate_adaptation_indicators_final.pdf

Olazabal, M. (2017). Resilience, sustainability and transformability of cities as complex adaptive systems. In S. Deppisch (Ed.), *Urban regions now & tomorrow. Studien zur Resilienzforschung*. Wiesbaden: Springer.

Opach, T., Scherzer, S., Lujala, P., & Ketil Rød, J. (2020). Seeking commonalities of community resilience to natural hazards: A cluster analysis approach. *Norsk Geografisk Tidsskrift - Norwegian Journal of Geography, 74*(3), 181–199.

Pelling, M. (2011). *Adaptation to climate change: From resilience to transformation*. Routledge.

Prohaska, A. (2020). Still struggling: Intersectionality, vulnerability, and long-term recovery after the Tuscaloosa, Alabama USA Tornado. *Critical Policy Studies*, 1–22.

Racial and Environmental Justice Committee (REJC). (2016). *Equity in Sustainability.* City of providence. Access link: https://providenceri.gov/wp-content/uploads/2017/02/Equity-and-Sustainability-SummaryReport-2-20-reduced.pdf. Accessed 24 Oct 2021.

Rahmani, M., Lotfata, A., Khoshnevis, S., & Javanmardi, K. (2021). medRxiv 2021.09.13.21263435; https://doi.org/10.1101/2021.09.13.21263435.

Raicu, S., Rosca, E., & Costescu, D. (2019). Resilience of urban technical networks. *Entropy, 21*(9), 886.

Reckien, D., Creutzig, F., Fernandez, B., Lwasa, S., Tovar-Restrepo, M., Mcevoy, D., & Satterthwaite, D. (2017). Climate change, equity and the sustainable development goals: An urban perspective. *Environment and Urbanization, 29*(1), 159–182.

Rhodes, J., Chan, C., Paxson, C., Rouse, C. E., Waters, M., & Fussell, E. (2020). The impact of hurricane Katrina on the mental and physical health of low-income parents in New Orleans. *The American Journal of Orthopsychiatry, 80*(2), 237–247.

Ryder, S., & Boone, K. (2019). Intersectionality and sustainable development. In W. Leal Filho, et al. (Eds.), *Gender equality, encyclopedia of the UN sustainable development goals*.

Saja, A. M. A., Goonetilleke, A., Teo, M., & Ziyath, A. M. (2019). A critical review of social resilience assessment frameworks in disaster management. *International Journal of Disaster Risk Reduction, 35*, 101096, ISSN 2212-4209.

Schaefer, M., Thinh, N. X., & Greiving, S. (2020). How can climate resilience be measured and visualized? Assessing a vague concept using GIS-based fuzzy logic. *Sustainability., 12*(2), 635. https://doi.org/10.3390/su12020635.

Schuller, M., Manyen, P. A., & Konsa, F. N. (2015). Intersectionality, structural violence, and vulnerability before and after Haiti's earthquake. *Feminist Studies, 41*(1), 184–210.

Singh, C., Lyer, S., New, M. G., Few, R., Kuchimanchi, B., Segnon, A. C., & Morchain, D. (2021). Interrogating 'effectiveness' in climate change adaptation: 11 guiding principles for adaptation research and practice. *Climate and Development*, 1–15.

Stuart, E., & Woodroffe, J. (2016). Leaving no-one behind: can the sustainable development goals succeed where the millennium development goals lacked?. *Gend Dev, 24*(1), 69–81.

Sung, H., & Phillips, R. G. (2018). Indicators and community well-being: Exploring a relational framework. *International Journal of Web Based Communities, 1*, 63–79.

Thomas, K., Hardy, R. D., Lazrus, H., Mendez, M., Orlove, B., Rivera-Collazo, I., Roberts, J. T., Rockman, M., Warner, B. P., & Winthrop, R. (2019). Explaining differential vulnerability to climate change: A social science review. *WIREs Climate Change, 10*(2), e565.

Turrentine, J. (2019). *A roadmap for frontline communities*. NRDC. Access link at: https://www.nrdc.org/stories/roadmap-frontline-communities. Accessed 24 Oct 2021.

UN-Habitat. (2018). *Resilience*. https://unhabitat.org/urban-themes/resilience/

Urban Sustainability Directors Network (USDN). (2017a). *From community engagement to ownership*. USDN. Access link: https://www.usdn.org/uploads/cms/documents/community_engagement_to_ownership_-_tools_and_case_studies_final.pdf. Accessed 24 Oct 2021.

Urban Sustainability Directors Network (USDN). (2017b). Access Link: https://www.usdn.org/uploads/cms/documents/usdn_guide_to_equitable_community-driven_climate_preparedness-_high_res.pdf
Vickery, J. (2018). Using an intersectional approach to advance understanding of homeless persons' vulnerability to disaster. *Environmental Sociology, 4*(1), 136–147.
Walker, B., Hollin, C. S., Carpenter, S. R., & Kinzig, A. (2004). Resilience, adaptability and transformability in social-ecological systems. *Ecology and Society, 9*.
Wei, Y. M., Wang, K., Wang, Z. H., et al. (2015). Vulnerability of infrastructure to natural hazards and climate change in China. *Natural Hazards, 75*, 107–110.
Wilson, R. S., Herziger, A., Hamilton, M., et al. (2020). From incremental to transformative adaptation in individual responses to climate-exacerbated hazards. *Nature Climate Change, 10*, 200–208.
Yang, M., & Jiang, P. (2020). Socialized and self-organized collaborative designer community-resilience modeling and assessment. *Research in Engineering Design, 31*, 3–24.
Zanotti, L., Ma, Z., Johnson, J. L., Johnson, D. R., Yu, D. J., Burnham, M., & Carothers, C. (2020). Sustainability, resilience, adaptation, and transformation: Tensions and plural approaches. *Ecology and Society, 25*(3), 4.
Zhao, L., Li, H., Sun, Y., Huang, R., Hu, Q., Wang, J., & Gao, F. (2017). Planning emergency shelters for urban disaster resilience: An integrated location-allocation modeling approach. *Sustainability, 9*(11), 2098.

Intersectional Feminism

▶ Feminist Planning and Urbanism: Understanding the Past for an Inclusive Future

Intersectionality – Complex

▶ At the Intersection and Looking Ahead

Interstitial Spaces

▶ Zombie Subdivisions

Involvement

▶ Closing the Loop on Local Food Access Through Disaster Management

IPCC – The Intergovernmental Panel on Climate Change

▶ Adapting to a Changing Climate Through Nature-Based Solutions

IUCN – International Union for Conservation of Nature

▶ Adapting to a Changing Climate Through Nature-Based Solutions

J

Just Energy Transition

► European Green Deal and Development Perspectives for the Mediterranean Region

Justice – Fairness, Equity

► Spatial Justice and the Design of Future Cities in the Developing World

R. C. Brears (ed.), *The Palgrave Encyclopedia of Urban and Regional Futures*,
https://doi.org/10.1007/978-3-030-87745-3

L

Lack of Political Will

▶ Water Policy in the State of Tabasco

Land and Buildings Development

▶ The Urban Planning-Real Estate Development Nexus

Land Use – Land Organization

▶ Zooming Regions into Perspective

Land Use Land Cover (LULC)

▶ The State of Extreme Events in India

Land Use Planning Systems in OECD Countries

Trends, Practices, Innovations, and Reforms

Tamara Krawchenko[1] and Abel Schumann[2]
[1]University of Victoria, Victoria, BC, Canada
[2]Organisation for Economic Co-operation and Development, Paris, France

Synonym

Integrated planning; Land use regulation; Planning; Spatial planning; Zoning regulation

Definition

Land use planning involves determining and regulating land uses on the basis of desired social, economic, environmental, and cultural outcomes. In most countries, local authorities adopt strategic land use plans alongside detailed plans that contain zoning regulation and ordinance and upper orders of government develop strategic plans and policy guidelines with spatial implications to coordinate the territorial development of an entire region or of the whole nation.

R. C. Brears (ed.), *The Palgrave Encyclopedia of Urban and Regional Futures*,
https://doi.org/10.1007/978-3-030-87745-3

Introduction

The 38 member countries of the Organization for Economic Cooperation and Development (OECD) are a collection of the world's most advanced economies. As such, this subset of countries lends themselves to some degree of comparability, despite their different structures encompassing federal, quasi-federal, and unitary states. Across OECD countries, it is mostly subnational, and in particular, local governments that are responsible for land use policies, while upper-level governments set the overarching framework that governs the system. The strong involvement of local governments can be explained on normative as well as on practical ground. From a normative perspective, land is fundamental to daily life in communities and local issues should be decided upon locally based on the principle of subsidiarity. From a practical perspective, most land use decisions are highly context-dependent and most higher-level governments lack the necessary local knowledge to make informed decisions. However, many land use decisions by local governments have important impacts that can be felt far beyond their jurisdictions and many upper-level governments aim at playing a coordinating role to ensure coherent and integrated planning across territories – with varying degrees of success.

This entry provides an overview of land use planning systems and reforms in OECD countries. First, it gives a brief summary of land use changes in 32 OECD countries, including how land consumption has kept pace with population growth or decline. Second, it outlines land use planning systems in the OECD across the national, regional, metropolitan, and local scales. Third, it provides an overview of innovative practices and recent reforms. This entry draws on research and findings of the two-part volume: *The Governance of Land Use in OECD Countries* (Part 1) and *Land Use Planning Systems in the OECD* (Part 2), including a survey of 32 OECD member countries (OECD, 2017a, 2017b). The 32 surveyed OECD countries are: Australia, Austria, Belgium, Canada, Chile, Czech Republic, Denmark, Estonia, Finland, France, Germany, Greece, Hungary, Ireland, Israel, Italy, Japan, Korea, Mexico, the Netherlands, New Zealand, Norway, Poland, Portugal, Slovak Republic, Slovenia, Spain, Sweden, Switzerland, Turkey, the United Kingdom, and the United States.

Land Use Trends in the OECD

Built-up land constitutes only a small share of all land in OECD countries. In all OECD countries except Belgium and the Netherlands, it is below 10% of the total land mass and often below 1%. Even in regions that the OECD defines as urban, the total share of developed land is usually below 20%. While the total amount of developed land is relatively small, its share has increased since 2000 in all countries – but at vastly different growth rates.

Among OECD countries, Norway saw the highest rate in growth of built-up land between 2000 and 2014 with an increase of 30%, followed by Mexico (29.8%) and Finland (27.4%). At the opposite end of the scale, the Slovak Republic (9.2%), the UK (8.7%), and Japan (7.5%) have had the lowest growth in developed land over the same time period. While all countries recorded growing areas of built-up land, in approximately a third of OECD countries, the growth rate of developed land was lower than the population growth rate. In those countries, the area of developed land per capita declined, implying that on a per capita basis, land use has become more sparing.

However, built-up land is a highly imperfect measure of the environmental and cultural impact of land use. The small share of built-up land in most countries means that even significant growth in built-up areas does not necessarily have major impacts. If built-up areas were to increase from 2% to 3% of the total land mass of a country, the overall environmental impact could be minor. However, if the growth occurs in a scattered and sprawling fashion that disrupts natural habitats and destroys landscapes, its impact is much larger than becomes apparent from the numbers. Thus, the impact of land use depends on various factors that are still hard to quantify and measure.

Land Use Planning Systems in the OECD

Multiple sets of legal instruments determine how land can be used. Land is governed by legislation that determines the rights associated with it, such as property rights and expropriation rights, and also the obligations associated with its use. Regulatory policies, such as environmental regulations and building code regulations provide further detail on rights and obligations in specific contexts. Fiscal policies create varying financial obligations depending on the type of land use and influence land use by altering the incentives associated with it. While any of these instruments can have differential effects in different local contexts, none of the instruments is usually used with the primary objective to steer land use in a specific context.

Spatial and land use plans are the key policy instrument to influence the use of land in specific locations. Generally, they exist at multiple scales – national, regional, and most importantly, local. They either provide the permitted land uses for a specific area directly or they set out how land uses should be decided (often while also providing indicative guidance on intended uses). At a first glance, land use planning systems across the OECD look quite similar (OECD, 2017a). Upper-level governments generally provide the framework laws that set out the planning system and enact environmental legislation, while local governments make decisions about detailed land uses. In practice, the governance of land use can vary greatly, even within countries, let alone across them. Depending on planning practices and legal interpretations, the role of spatial and land use plans as well as the role of planners themselves varies widely. In some places, there are informal partnerships between the many actors involved in the governance of land use, while in others, there is a distinct hierarchy between levels of planning, and the institutions involved operate on the basis of statutorily defined roles.

The OECD land use governance survey has identified 229 unique types of spatial and land use plans in the 32 surveyed countries. This number does not count plans separately that exist several times in a country. The responsibility for the 229 plans is approximately evenly split between the national, regional, and the local governments on average (across all countries). In several countries (e.g., Austria, the Netherlands), the same type of plan can be enacted by several levels of government. Most plans are enacted by the level of government that corresponds to their geographical scope. However, almost a third of all plans are enacted by a higher level of government for the jurisdiction of a lower level of government.

This description of the planning system does not capture the land use planning practices of Indigenous nations and communities. Indigenous peoples are defined by the United Nations as those who inhabited a country prior to colonialization, and who self-identify as such due to descent from these peoples, and belonging to social, cultural, or political institutions that govern them. Across 14 OECD countries, there are more than 38 million Indigenous peoples. Land rights are crucial to the maintenance of the collective identity of Indigenous groups and in many countries, First Peoples have their own land management regimes. For example, Canada's First Nations Land Management Regime. The following descriptions of planning systems do not include Indigenous land use planning.

National Plans

More than three-quarters of all national plans (plans covering the entire national territory) contain policy guidelines and more than 60% contain strategic plans. Few national plans are binding for private landowners, but a majority of them (68%) is legally binding for the planning policies of subnational governments and other public authorities (Whether a plan is in practice binding for subordinate plans is a different question. Often, the language of national and other high-level plans is vague and therefore leaves subordinate planning authorities with a wide range of discretion.). Boundary plans are rare in national plans and only 24% include them. When they are included, they usually target specific high priority areas, such as national plans. Countrywide boundary plans at the national level such as Israel's National Master Plan No. 35 are very rare across the OECD. It is noteworthy that there are ten

OECD countries in which the national government neither prepares any general spatial or land use plan (except for thematically narrow sectoral plans such as a national transport plan) nor any general guidelines on land use: Austria, Belgium, Canada, France, Italy, New Zealand, Spain, Sweden, the United Kingdom, and the United States. Most national plans are approved through a regulatory decision, but seven plans (in six countries: Greece, Hungary, Mexico, Portugal, Slovenia, and Turkey) are approved by a vote of the national parliament. All plans that are approved by a vote of parliament are legally binding.

Reflecting the trend of integrated planning, most plans cover a broad range of policy fields. Of the 29 national plans that refer to a list of 6 thematic areas (transport, environment, housing, industry, commerce, and agriculture), 76% cover three or more of those policy fields and only 24% cover only one or two of them. Most common are transport and the environment, followed by housing, industry, and commerce. Even agriculture – the least common thematic area – is still discussed in more than half of the national level spatial plans.

Regional Plans

Sixty-six regional and 5 subregional plans have been identified across the OECD. Subregional plans are distinguished from regional plans by the fact that they cover only a part of an administrative region, while covering a larger territory than local or metropolitan plans. Given their small number and given that their characteristics do not otherwise differ systematically from regional plans, they included under regional plans. Just as for national plans, a majority of regional plans contain general policy guidelines (70%) and elements of strategic plans (75%). Compared to national plans, they are somewhat more oriented to strategic planning and focus less on providing policy guidelines. The share of plans that contain elements of boundary plans is 32%, which is only marginally higher than in the case of national plans. Even when boundary plans are included in regional plans, they are rarely central elements. Legally, regional plans tend to have a weaker status than national plans: 34% of regional plans are legally binding for subordinate plans and allow few or no exceptions compared to 46% for national plans. Likewise, the share of plans that is neither legally binding nor incentivized is 28% at the regional level compared to 18% at the national level. Approximately three-quarters of all regional plans are approved by regional governments, whereas the remaining quarter is mostly approved by national governments.

Metropolitan and Inter-municipal Plans

Dedicated metropolitan and inter-municipal plans are rare in OECD countries. Only 11 types of such plans were identified. Some plans, such as France's Territorial Coherence Scheme (SCoT) or the Metropolitan Area Plan in Korea, are prepared for every metropolitan area of a country. However, others are prepared only for a single metropolitan area (e.g., the Finger Plan for Copenhagen, the Auckland Plan, the Budapest Priority Region Plan, and the London Plan).

Local Plans

Local plans are the most common plan type and give the highest level of detail. Almost all local plans contain boundary plans and only around a third of them also provide general policy guidelines or strategic planning. The vast majority (87%) of all local plans are legally binding for land owners and 52% of them allow no or only rare exemptions. The relatively large share of strictly binding plans is a consequence of the regulatory role that local plans play in the planning process. Compared to national and regional plans, local plans are more frequently approved through the vote of an elected body (e.g., the municipal assembly).

Innovative Practices and Recent Reforms

Planning systems show a strong institutional persistence. As of 2016, the median time evolved since the last major reform of the planning systems in the 32 surveyed countries was 37 years. Across OECD countries, major reforms to planning systems are rare, especially considering that many of the reforms that occurred in the 1990s and 2000s happened in eastern European countries as part of their transition from socialism to democratic market economies. Aside from those reforms, most land use governance systems in the OECD were established prior to 1970. The persistence of formal planning systems contrasts to the fluidity with which planning practices can change. Thus, despite the persistence of overarching planning systems, there is a degree of flexibility in current land use governance regimes. The same frameworks can accommodate considerable differences in planning practice if this is supported by the involved actors.

While major reforms to land use governance systems are rare, smaller reforms that concern specific elements of planning systems occur frequently. For example, Chile has introduced regional land use plans (Orellana Ossandón et al., 2020). France has adopted regional plans that integrate previously separate sectoral plans: the Planning, Sustainable Development and Territorial Equality Regional Plan (Schéma Régional d'Aménagement, de Développement Durable et d'Egalité du Territoire, SRADDET) places regions as lead actors in the field of planning and sustainable development and adds a requirement for a region to develop a specific plan on the prevention and management of waste (Ministère de la cohésion des territoires et des relations avec les collectivités territoriales, 2020). Greece has simplified its complex planning system by replacing and abolishing existing plans (Vitopoulou & Yiannakou, 2018). In Ireland, a major reform at the regional level replaced eight Regional Authorities with three Regional Assemblies. Consequently, the old Regional Planning Guidelines have been replaced with Regional Spatial and Economic Strategies that are supposed to be more comprehensive spatial documents (Williams & Nedović-Budić, 2020). In Israel, different national master plans have been merged into a single plan (Feitelson, 2018). In Italy, ten provinces have been given the status of metropolitan cities (2015) and have adopted metropolitan plans (Tomàs, 2020). In Poland, the Metropolitan Association Act of 2015 has introduced a new planning instrument, the Framework Study for Metropolitan Areas (Smętkowski et al., 2020). Thus, there have been changes to aspects of the planning system in many OECD countries, including the structures of local and regional governance.

Many of these reforms aim at making planning more integrated and cross-sectoral, while increasing the responsiveness of the planning process. These steps have the potential to overcome some of the major challenges facing the planning system, such as sectoral divisions and cumbersome processes that are too slow to react to change. However, they demand different institutional mechanisms and ways of working.

While large-scale changes to planning systems are uncommon in OECD countries, the Netherlands stands out as recently having adopted ambitious reforms. In 2016, the Netherlands adopted an Environment and Planning Act (Omgevingswet) which has two primary objectives: (1) to ensure a safe, healthy, and high-quality physical environment and (2) to ensure the effective management, use, and development of the physical environment to fulfill social functions (Rijksoverheid, 2021). The Environment and Planning Act merges 26 separate acts into 1; merges 120 orders in Council into 4; and simplifies over 100 ministerial regulations in order to create greater coherency. Environmental plans will replace the structural visions at each level of government. Existing zoning plans will be transferred to the environmental plan and local governments will have a period of 10 years to transform them. The Environment and Planning Act is expected to enter into force on 1 January 2022.

A major rationale underlying these reforms is the need for a more flexible and innovative land use planning system (Jansen, 2020). Flexible approaches to land use management can be structured in a number of different ways. For example, they might entail the establishment of specific zones in a community which are more open to experimentation and temporary uses. With greater planning flexibility, there are fewer rules about how land is used and each project is judged on the basis of its own merit, typically framed by overarching guidelines and objectives about community needs and aspirations. Under such systems, more effort needs to be put in upfront in order to collaboratively define projects and reach consensus between investors, developers, governments, residents, and other actors (OECD, 2017c). This can breed experimentation and innovation, and respond in a more timely way to emerging trends and needs. An important caveat is that more flexibility need not be embraced everywhere. For example, historical districts and environmentally sensitive areas often need more stringent rules than transitional spaces such as brownfield sites.

Flexible planning systems, such as in the Netherlands, differentiate places according to their need for protection and differentiate planning procedures accordingly. Increasing the flexibility in the planning system inevitably implies that planners exert less direct control over land use. Without complementary measures, this would increase the risk of uncontrolled development, potentially leading to undesired outcomes such as more sprawl, inefficient transport systems, and incompatible land uses in close proximity. As greater flexibility imposes fewer restrictions on land use, it is essential that private actors can be compelled to pursue desirable patterns of development out of their own interest. Planners need to engage proactively with land owners to persuade them to develop land in desirable ways. Yet, such efforts will only be successful if land owners face the right incentives. Many public policies provide incentives for land use such as tax policies. At the moment, these incentives are rarely used to influence land use actively. Increasing the flexibility of planning systems and still achieving desired spatial outcomes requires that public policies are more effectively used to set the right incentives. In particular, fiscal policies, which are currently considered to be outside the domain of spatial and land use planning, must be used more effectively.

In order to implement more flexible planning systems, a high degree of capacity is needed at the local level, since a broader range of considerations has to be taken into account in the decision-making process. Furthermore, decision makers have to be accountable and need to be trusted by the public in order to ensure that land use decisions are accepted even by those who would prefer different outcomes and planners need to engage more proactively with landowners and other stakeholders. Flexible planning systems also need effective monitoring and evaluation to ensure that key objectives are achieved and if necessary, adjust policies.

Conclusion

How land is used is connected to some of the most important policy issues of our time: sustainable development, reducing territorial inequalities, economic development, and the rights of future generations to name but a few. Land use is also highly persistent, which implies that any decisions taken today will have consequences for decades. Given this, there is an urgency to ensure land is managed efficiently, equitably, and responsively. Spatial and land use planning at all levels of government faces considerable pressures in the future, as some areas face population growth, others decline, and shifting policy priorities such as climate change pose new challenges.

Spatial and land use policies encounter the same kinds of silos, sectoral divisions, and intra-government coordination issues as other policy areas, thus detracting from their effectiveness. However, they also face distinct scalar challenges – with many issues such as economic development, transportation, climate change mitigation and adaptation, and landscape protection extending across multiple administrative boundaries and scales. Efforts to integrate planning across policy sectors can make planning more effective. However, they also create new challenges for planners and other policy makers, as they demand more coordination across policy sectors to be effective.

As a consequence, spatial planning is increasingly complex and interdisciplinary, requiring approaches that cross sectoral boundaries. As the purview of what planners think about and seek to respond to has expanded, so too has the scale at which planning operates. Many issues, such as landscape protection, housing, transportation and water, are best tackled at scales beyond that of local municipal boundaries. However, governance at the metropolitan level is often weak with few statutory instruments and a strong reliance on voluntary collaboration, thus adding a further dimension of coordination needs. Concurrently, there is a continuing need for vertical cooperation, with local, regional and national governments, thus further increasing the coordination needs. The exceptional range of public actors involved in the governance of land use bears witness to transversal nature of land. It affects a multitude of issues from the very local to the global scale.

Cross-References

▶ Toward a Sustainable City

References

Feitelson, E. (2018). Shifting sands of planning in Israel. *Land Use Policy, 79*, 695–706. https://doi.org/10.1016/J.LANDUSEPOL.2018.09.017

Jansen, F. J. (2020). De Omgevingswet: wetgeving als symbool of communicatieve wetgeving? *Recht Der Werkelijkheid, 40*(2), 44–64. https://doi.org/10.5553/RDW/138064242019040002004

Ministère de la cohésion des territoires et des relations avec les collectivités territoriales. (2020, December 30). Schémas régionaux d'aménagement et de développement. Retrieved July 16, 2021, from Ministère de la cohésion des territoires et des relations avec les collectivités territoriales website: https://www.cohesion-territoires.gouv.fr/schemas-regionaux-damenagement-et-de-developpement

OECD. (2017a). *Land-use planning systems in the OECD: Country fact sheets*. OECD, Paris, France.

OECD. (2017b). The governance of land use in OECD countries: Policy analysis and recommendations. http://www.oecd-ilibrary.org/urban-rural-and-regional-development/the-governance-of-land-use-in-oecd-countries_9789264268609-en

OECD. (2017c). The governance of land use in the Netherlands: The case of Amsterdam. In *OECD*. https://doi.org/10.1787/9789264274648-en.

Orellana Ossandón, A., Arenas Vásquez, F., & Moreno Alba, D. (2020). Ordenamiento territorial en Chile: nuevo escenario para la gobernanza regional [Land planning in Chile: New scenario for regional governance]. *Revista de Geografía Norte Grande, 77*, 33–49. Retrieved from http://siedu.ine.cl/index.html

Rijksoverheid. (2021). Omgevingswet. Retrieved 16 July 2021 from https://www.rijksoverheid.nl/onderwerpen/omgevingswet

Smętkowski, M., Celińska-Janowicz, D., & Romańczyk, K. (2020). Metropolitan areas in Poland as a challenge for urban agenda at different territorial levels. *Urban Book Series*, 139–163. https://doi.org/10.1007/978-3-030-29073-3_7

Tomàs, M. (2020). Metropolitan revolution or metropolitan evolution? The (dis)continuities in metropolitan institutional reforms. *Metropolitan Regions, Planning and Governance*, 25–39. https://doi.org/10.1007/978-3-030-25632-6_2.

Vitopoulou, A., & Yiannakou, A. (2018). Public land policy and urban planning in Greece: Diachronic continuities and abrupt reversals in a context of crisis. *27*(3), 259–275. https://doi.org/10.1177/0969776418811894.

Williams, B., & Nedović-Budić, Z. (2020). Transitions of spatial planning in Ireland: Moving from a localised to a strategic national and regional approach. *Planning Practice and Research*, 1–21. https://doi.org/10.1080/02697459.2020.1829843

Land Use Regulation

▶ Land Use Planning Systems in OECD Countries

Land–Real Estate

▶ Neoliberalism and Future Urban Planning

Landscape Fragmentation

▶ The Challenges for Wildland-Urban Interfaces (WUI) in Metropolitan Areas: Reducing Fire Risk, Providing Employment Opportunities, and Preserving Natural Habitat

Landscaping

▶ Urban Forestry in Sidewalks of Bogota, Colombia

Land-Use – Land Function

► Future of Urban Land-use Planning in the Quest for Local Economic Development

Latin America Peripheries

► New Forms of Shared Governance and Local Action Plan in Socially Vulnerable Settlements

Law

► Regulation of Urban and Regional Futures

Law – Legal Instruments

► At the Intersection and Looking Ahead

Lead Exposure in US Cities

Matthew H. McLeskey
Department of Sociology, University at Buffalo, State University of New York, Buffalo, NY, USA

Synonyms

Lead poisoning; Lead risks; Plumbism; Threat of lead exposure

Definition

In cities across America, parents worry about how to keep their children safe from lead exposure. Initially used in industrial and building products, lead poses numerous dangers to childhood health and well-being. Pathways to lead exposure for children include contact with paint, dust, water, and soil found in older homes, especially those built before 1950 when lead paint was most common. Federal regulations banned lead in residential paints in 1978, though its traces linger throughout urban centers (National Center for Healthy Housing 2019; Markowitz and Rosner 2014). Worldwide, children remain at risk of lead exposure from multiple sources as well, such as from contaminated soil from battery recycling and mining operations (World Health Organization 2019).

Exposure to lead in dilapidated housing poses significant health risks to young children under age six because their small, developing bodies cannot absorb it. Consequently, it negatively impacts children's organ systems and neurological development. This results in a host of issues which affects children's well-being and life outcomes: lower IQ, attention deficits, behavioral issues, and slowed physical development. Experts use a lead level based on the US population of children ages 1–5 years who are in the top 2.5% of children when tested for blood lead levels, with a testing threshold of 5 micrograms per deciliter of lead in blood (Center for Disease Control and Prevention 2021). In 2014, the Center for Disease Control and Prevention found that 4.2% of children test, or half a million, had elevated blood lead levels of 5 (μg/dL) or greater (Frostenson 2017).

Due to its association with the causes and consequences of deindustrialization, medical and practitioner communities have known for decades how the lead-based paint used in the nation's homes and other buildings contain the power to harm children who sucked on lead-painted windowsills, toys, cribs, and woodwork (Needleman and Landrigan 1981; Warren 2000). Yet, lead-based paint still covers the walls of houses, apartment buildings, and workplaces across the United States and remains in place almost four decades later, especially in disinvested neighborhoods where mostly low-income African-American and Hispanic children reside (Center for Disease Control and Prevention 2020).

Current Controversies

The Flint, Michigan, lead water crisis was covered throughout international news media in 2015 after it was discovered that Flint residents were being literally poisoned by the water running through their pipes and out of their faucets. The crisis began the year before when Flint city government switched its water source from the Detroit water system to the Flint River as a cost-saving solution, but proper measures to ensure the older piping would not leach lead into the water were not taken. The local water authority was completing a pipeline from Lake Huron but had not completed it and opted to temporarily use the Flint River until completion of the pipeline. The result of this decision would prove disastrous for the Flint community (Clark 2018). Over a half-decade later, the impact of this incident remains long-lasting for the affected residents and serves as a warning sign about the health risks caused by derelict infrastructure.

After the decision to switch the water source, residents showed concern about the worrisome color and odor of their tap water. A majority-African-American city with high poverty levels, Flint residents with pre-existing medical conditions began suffering symptoms such as skin rashes and hair loss and began to approach local and state authorities at meetings with containers of dark-colored, obviously unsafe drinking water to bring more attention to the issue (Vella 2021). A study conducted by Virginia Polytechnic Institute and State University on the water quality in Flint found 40% of Flint homes had elevated blood lead levels (Edwards et al. 2015). Of the thousands of adults and children exposed to lead in Flint, almost 9000 children were under age six (Strauss 2019). Scholars have pointed out how the Flint case represents failure at various levels of government to protect their citizens and enforce the regulations which already existed, as well as the need for a more robust regulatory framework that ensures protection of vulnerable populations (Doyon-Martin 2020).

Lead exposure from paint found in dilapidated homes in neighborhoods across America's cities proves a similarly troubling case. Living in dilapidated rental housing in a disinvested neighborhood means navigating chipping paint from repeatedly moving the windows up and down in warmer months or flaking paint from the ceiling which falls, often unnoticed, into the nearby soil outside where children play. In the United States, various levels of government spend up to $15 billion each year to handle new lead poisoning cases, including remediating lead paint in 38 million housing units and cleaning countless tons of soil contaminated by exterior paint flaking off and dissolving into the ground below (Zaleski 2020).

The case of Freddie Gray, Jr. is one of the most infamous news stories involving the consequences of lead poisoning of lead paint. In 2015, at age 25, Gray died from police brutality in the back of a police vehicle following an arrest for alleged possession of an illegal knife in Baltimore; he was comatose and died from spinal cord injuries 7 days after his arrest. His death inspired protests and riots throughout the city (Rector 2017). A life characterized by classroom struggle and run-ins with law enforcement were in part caused by the house he grew up in – Gray and his siblings were all diagnosed with lead poisoning as children and received an undisclosed settlement against their property owner in a 2008 lawsuit (McCoy 2015).

Other Issues and Social Effects

Childhood lead poisoning undermines generations of learning potential and exacerbates racial disparities in income, education, and health. Threat of lead exposure can prove detrimental to the material well-being for urban residents, even when they manage to avoid lead exposure.

Therefore, lead exposure also exacerbates pre-existing racial and economic inequalities, disrupts the lives of low-income, often minority residents, and even puts small-scale landlords in a bind (Muller et al. 2018). Tenants navigate to find – and keep – safe, healthy housing amidst the threat of lead poisoning in the post-industrial city as landlords struggle to maintain and renovate their rental units.

While research on the physiological consequences of lead poisoning abounds in health science fields, social scientists recently analyzed the spatial concentration of lead poisoning to more deeply understand the causes and consequences of urban inequalities caused by a legacy of lead (Liévanos et al. 2021; Sampson and Winter 2016; Moody et al. 2016). For example, statistical analyses of neighborhood-level lead exposure in Chicago and Detroit, respectively, found a strong relationship between spatial concentration of ethnic minorities and prevalence of lead poisoning via elevated blood lead levels in children along with increased proximity to industrial facilities and hazardous waste sites. This form of environmental injustice results from structural barriers such as socioeconomic inequality, racial segregation, housing quality, and toxic exposure. Therefore, childhood lead poisoning can be seen as intertwined with other economically and racially concentrated social problems such unemployment and segregation. As a result, generations of children face a lifetime of prolonged disadvantage compared to more well-off peers.

Scholars have noted how social scientists have not fully uncovered the extent to which threat of lead exposure strains disinvested urban communities (Muller et al. 2018). For families concerned with keeping their children safe, threat of lead exposure compounds preexisting inequalities such as employment scarcity, emotional stress, and rent burden. Even worse, it also creates new ones, such as ill and/or disabled children, healthcare hardship, and even housing relocation due to conflicts with their landlord. For landlords, keeping older housing lead safe proves daunting because lead often resides in paint and minor rehabilitation can transfer lead paint from walls to the floor via dust. These unsafe landscapes of urban America result in housing situations burdensome for both tenants and landlords. Thus, lead exposure exacerbates a number of urban inequalities.

Understanding tenants' and landlords' experiences with lead poisoning provides deeper insight into the broader texture of urban poverty. Lead poisoning does not just burden tenants – it overwhelms landlords. Potentially lead poisoned housing occurs not necessarily due to landlord neglect; a landlord may need to evict tenants upon learning about their unit's lack of habitability. Lead poisoning also persists as a housing problem contributing to urban inequalities due to variations in the strength of lead abatement policies across cities and the complexities of local, county, state, and federal governance, including the ability to enforce compliance with lead-based paint regulation, detection, and remediation. Understanding the role of the threat of lead poisoning in the reproduction of urban poverty requires further understanding into how Desmond and Bell (2015: 23) conceptualize urban-based environmental inequalities such as lead poisoning: "a housing condition, neighborhood effect, and legal problem worthy of serious inquiry."

Conclusion

According to a recent UNICEF (2020) study, up to one in three children worldwide may suffer from lead poisoning due to the lingering traces of lead gas, paint, and piping and, in a more global context, the informal recycling of lead acid batteries. Exposure to environmental compounds such as lead can have effects for multiple generations, exacerbating hardship across a lifetime and widening social disparities. In the context of the United States, lead abatement regulations vary widely across cities even with similar demographics and economic landscapes. Abatement implementation and housing code compliance remain less enforced in marginalized communities, particularly those ignored by gentrification, expanding how landlords impact stratification processes in urban centers (Rosen 2020). Stable housing and a healthy environment remain one of the basic necessities of life and can also be generally the most expensive, especially for low-income families. Housing quality and stability and the policies which allow or prevent access to these basic tenets of individual and family well-being remain a rich yet untapped area of social science inquiry.

Cross-References

▶ Housing Affordability

References

Center for Disease Control and Prevention. (2020). *Health effects of lead exposure*. Center for Disease Control and Prevention. https://www.cdc.gov/nceh/lead/prevention/health-effects.htm

Center for Disease Control and Prevention. (2021). *Blood lead levels in children*. Center for Disease Control and Prevention. https://www.cdc.gov/nceh/lead/docs/lead-levels-in-children-fact-sheet-508.pdf

Clark, A. (2018). *The poisoned city: Flint's water and the American urban tragedy*. New York: Metropolitan Books.

Desmond, M., & Bell, M. (2015). Housing, poverty, and the law. *Annual Review of Law and Social Science, 11*, 15–35. https://doi.org/10.1146/annurev-lawsocsci-120814-121623

Doyon-Martin, J. (2020). The Flint water crisis: A case study of state-sponsored environmental (in)justice. In A. Brisman & N. South (Eds.), *Routledge international handbook of green criminology* (pp. 317–332). New York: Routledge.

Edwards, E., Roy, S. & Rhoads, W. (2015). Lead testing results for water tested by residents. Flint has a very serious lead in the water problem. Flint Water Study. http://flintwaterstudy.org/information-for-flint-residents/resultsfor-citizen-testing-for-lead-300-kits/

Frostenson, S. (2017, April 27). 1.2 million children in the U.S. have lead poisoning. We're only treating half of them. *Vox*. https://www.vox.com/science-and-health/2017/4/27/15424050/us-underreports-lead-poisoning-cases-map-community

Liévanos, R. S., Evans, C. R., & Light, R. (2021). An intercategorical ecology of lead exposure: Complex environmental health vulnerabilities in the Flint water crisis. *International Journal of Environmental Research and Public Health, 18*(5), 2217–2240. https://doi.org/10.3390/ijerph18052217

Markowitz, G., & Rosner, D. (2014). *Lead wars: The politics of science and the fate of America's children*. Berkeley: University of California Press.

McCoy, T. (2015, April 29). Freddie Gray's life a study on the effects of lead paint on poor blacks. *The Washington Post*. https://www.washingtonpost.com/local/freddie-grays-life-a-study-in-the-sad-effects-of-lead-paint-on-poor-blacks/2015/04/29/0be898e6-eea8-11e4-8abc-d6aa3bad79dd_story.html

Moody, H., Darden, J. T., & Pigozzi, B. W. (2016). The racial gap in childhood blood levels related to socioeconomic position of residence in metropolitan Detroit. *Sociology of Race and Ethnicity, 2*(2), 200–218. https://doi.org/10.1177/2332649215608873

Muller, C., Sampson, R. J., & Winter, A. S. (2018). Environmental inequality: The social causes and consequences of lead exposure. *Annual Review of Sociology, 44*, 263–282. https://doi.org/10.1146/annurev-soc-073117-041222

National Center for Healthy Housing. (2019). *Lead*. National Center for Healthy Housing. https://nchh.org/information-and-evidence/learn-about-healthy-housing/lead/

Needleman, H., & Landrigan, P. J. (1981). The health effects of low-level lead exposure. *Annual Review of Public Health, 2*, 277–298. https://doi.org/10.1146/annurev.pu.02.050181.001425

Rector, K. (2017, September 13). Freddie Gray case: DOJ won't charge Baltimore police officers. *The Baltimore Sun*. https://www.baltimoresun.com/news/crime/bs-md-ci-doj-decline-charges-20170912-story.html

Rosen, E. (2020). *The voucher promise: "Section 8" and the fate of an American neighborhood*. Princeton: Princeton University Press.

Sampson, R. J., & Winter, A. S. (2016). The racial ecology of lead poisoning: Toxic inequality in Chicago neighborhoods, 1995–2013. *Du Bois Review: Social Science Research on Race, 13*(2), 261–283. https://doi.org/10.1017/S1742058X16000151

Strauss, V. (2019, July 3). How the Flint water crisis set back thousands of students. *The Washington Post*. https://www.washingtonpost.com/education/2019/07/03/how-flint-water-crisis-set-back-thousands-students/

UNICEF. (2020, July 29). A third of the world's children poisoned by lead, new groundbreaking analysis says. https://www.unicef.org/press-releases/third-worlds-children-poisoned-lead-new-groundbreaking-analysis-says

Vella, E. (2021, April 22). What happened to. . .Flint Michigan water crisis. *Global News*. https://globalnews.ca/news/7754252/what-happened-to-flint-michigan-water-crisis/

Warren, C. (2000). *Brush with death: A social history of lead poisoning*. Baltimore: John Hopkins University Press.

World Health Organization. (2019). *Lead poisoning and health*. World Health Organization. https://www.who.int/news-room/fact-sheets/detail/lead-poisoning-and-health

Zaleski, A. (2020, January 2). The unequal burden of urban lead: Decades after federal regulations banned the use

L

of the deadly metal in paint, gasoline, and plumbing, the effects of lead continue to be felt across America's cities. *Bloomberg CityLab*. https://www.bloomberg.com/news/articles/2020-01-02/undoing-the-legacy-of-lead-poisoning-in-america

Additional Reading

Hanchette, C. L. (2008). The political ecology of lead poisoning in eastern North Carolina. *Health & Place, 14*(2), 209–216. https://doi.org/10.1016/j.healthplace.2007.06.003

Morello-Frosch, R., & Lopez, R. (2006). The riskscape and the color line: Examining the role of segregation in environmental health disparities. *Environmental Research, 102*(2), 181–196. https://doi.org/10.1016/j.envres.2006.05.007

Pauli, B. J. (2019). *Flint fights back: Environmental justice and democracy in the Flint water crisis*. Cambridge, MA: MIT Press.

Lead Poisoning

► Lead Exposure in US Cities

Lead Risks

► Lead Exposure in US Cities

Learning Ecosystems

► Systemic Innovation for Thrivable Cities

Learning Outcomes

► Sustainability Competencies in Higher Education

Lifetime Neighborhood

► Age-Friendly Future Cities

Lifewide Learning

► Systemic Innovation for Thrivable Cities

Linear Economy

► Circular Economy Cities

► Resource Effectiveness in and Across Urban Systems

Link/Relationship

► The Urban Planning-Real Estate Development Nexus

Livability

► Understanding Women's Perspective of Quality of Life in Cities

Livable Communities

► Multiple Benefits of Green Infrastructure

Liveable Cities

► The Governance of Smart Cities

Local Action Plan

► New Forms of Shared Governance and Local Action Plan in Socially Vulnerable Settlements

Local and Regional Development Strategy

Alain Jordà
Local Development Expert, Manresa, Barcelona, Spain

Definition

Managing a territory, either a region or a city, is a very complex issue. As this Encyclopedia of Urban and Regional Futures shows, there is a very wide range of fields (mobility, housing, waste, economy, sustainability, security, poverty, etc.) to manage, and, moreover, they are to be managed for a community of people that are free to drive their lives as they want at every moment.

Two big problems territorial development face are, on one side, the need for a continuous development action along 10–20 years and, on the other side, the need to get all local stakeholders involved and committed with the development of their land.

To overcome these problems, the solution is to get a territorial strategy, i.e., a long-term vision and a plan to make it come true. This strategy must be unique, point to an excellence goal for the territory, and have the consensus of all stakeholders in order to be able to generate synergies between them.

Introduction

Managing a territory, either a region or a city, is a very complex issue. In fact, it's much more complex than managing a private company. And that's so complex because you not only need to manage simultaneously a very wide range of fields (mobility, housing, waste, economy, sustainability, security, poverty, etc.), but you need to do it for a community of people that are free to drive their lives as they want at every moment.

The challenges a territorial leader must deal with is not only to keep the city/region working on normally (this is management of the everyday life) but to transform it by building an outstanding place that offers a high quality of life to all its local citizens (and this, **managing the future, is what we call development**).

What it's to be understood as "development" is a true complete transformation of the city/region. It's a change in every sphere of the territory, and, therefore, it's a change that requires working along 10, 15, or 20 years.

To say that territorial development is about transformation means that it's not only about economic development, nor only about urban change, nor only about social inclusion, nor only about sustainability. It's about an integral transformation of the city or the region involving all these four fields. And this is why we need a strategy for the whole territory, a strategy that will set guidelines for every field of local development.

Two Big Problems of Development

When understanding the word "development" in this way, every territory will deal with two big problems (Jordà 2019):

First of all, you will need to guarantee that those development efforts will keep on for one or two decades, and that means that it should take place along several political changes.

Secondly, most changes will not be possible if you don't get direct involvement of citizens and local stakeholders. For instance, you cannot reduce water consumption, improve litter recycling, or increase commuting by public means if citizens don't contribute through their positive attitude. Moreover, public administrations financial capacities are quite limited, and they will need private investments from local stakeholders in order to boost local development.

So, two big questions need to be answered:

1. How to guarantee that consecutive local governments will keep on furthering local development efforts in the same direction?
2. How to guarantee that local stakeholders and citizenship will support those development efforts through their investments and commitment?

The Ideal Scenario You Need to Reach

The answer to those two questions is to get a local strategy. However, it's important to agree on what is a proper local strategy and which are the requirements it must fulfill.

The first quality of a strategy to be successful is **to be unique**. This means that a strategy is a goal to reach and it's not only a desideratum. Therefore, all those territories that define their strategy as "we want to be an inclusive, competitive, vibrant and sustainable place" don't have a true strategy. This kind of sentences may be good desiderata, but as far as they are obviously valid for any territory and they have a complete lack of focus, they will never be able to play the role of a territorial strategy.

Moreover, when a strategy consists of "our city aims at being an inclusive, competitive, vibrant and sustainable city," chances are that any development plan based on such statements will produce four disconnected sectorial plans: one for social inclusion, one for a competitive economy, one for a vibrant city, and the last one for a carbon-free city. But getting four disconnected sectorial plans has very few to see with an integrated city strategy plan since these four plans will not only be disconnected from one another, but they may easily try to pull the city in four different directions instead of generating synergies between them.

The aim of a strategy is to be an engine to transform the territory. And to do so, **the strategy needs to point to a goal of excellence for the territory**. This excellence goal is then to be built jointly for, maybe, 10–20 years time by all local stakeholders. Through this transformation process, the city will become outstanding. Some examples of those goals may be either "we produce paramount cheese for our country" or "we provide the best adventure experience to our visitors" or "we are the national center of excellence on metal-mechanics."

This kind of goals is able to excite local stakeholders and move them to invest in the project of jointly building the goal. Moreover, any of these examples promotes excellence and will be able to bring either talent and business or tourists, or all together to the site.

There's another key requirement for a territorial strategy to succeed. It needs **to generate full consensus** around it. Let put it on a different way: the best territorial strategy is the one that all local stakeholders are keen on pushing and investing in.

So, a strategy is not only worth to show the long-term path to follow by the territory, but it's also what keeps both, local stakeholders and citizenship, working together and joining efforts and resources, and what generates **synergies** between them in order to create the new future their strategy outlines.

Finally, a strategy will guarantee as well the continuation of the transformation efforts along different local governments through the join will of local stakeholders to reach the common goal the strategy determines.

How to Get This Ideal Situation in Your City/Region?

To get an effective territorial strategy will require using the right methodology in the process of defining it. Let's see which are the main rules to follow.

The way to do it properly is by all local stakeholders defining the strategy jointly through a methodology including a gradual process of proposals, debate, and consensus.

This is quite different from the old way of defining city strategy plans when a team of consultants was in charge of it, and they did it by studying local socioeconomic data, getting some interviews with "key" local people then building a proposal to fit all those requirements, and, finally, submitting this proposal to citizen assemblies to discuss that proposal.

The main concern of using this methodology is that local stakeholders don't assume the result as being "their own" proposal. As a result, municipality or regional government stays alone at developing the plan. No stakeholders really contributed to it simply because they didn't feel that it was "their" plan but just the government plan.

What we need instead is a plan that the whole city feels it is "their own" plan. For that, they must feel they have really contributed, and then they will stay in control on the deployment of the plan.

To get such a result, we will split the process of defining the development plan in two steps. The first step is defining the long-term vision city aspires to. This will be contained in a ten-page document, and it must be defined, jointly, by local stakeholders. The second step will be the regular definition of the plan for each territorial field (sustainability, economy, social, infrastructures, etc.). The difference is that now, each sector will define their plan starting from the long-term vision for the territory which is common to all territorial sectors. This means that, for instance, the economy development plan will be designed with a focus on getting the long-term vision.

Let's make a short list of guidelines for the definition of the long-term vision:

1. **A collaborative leadership**. The role of the local government is to announce the beginning of a process to define a local development plan and to call everybody to contribute. After that, the government will provide the means for this process to develop smoothly. Its contribution to the development plan will be chanelled as that of any other stakeholder. The final development strategy is not to be fixed by government but, instead, by a debate between all local stakeholders. Another responsibility included in the "collaborative leadership" concept is to manage to get an agreement with all local political forces in order to keep the development plan out of the political debate.
2. **Starting with a white sheet of paper**. It's of a paramount importance to start this process, from scratch, with all local stakeholders at the table. No preliminary proposal should be presented by local government. The "white sheet of paper" is to be filled with all stakeholders through an open debate and reaching consensus. Moreover, this is the only way to produce an innovative strategy.
3. As far as the strategy is not to be defined by any consultant, a city needs to involve territorial intelligence – a selected group of representative people – and work with them through a procedure aimed at generating **territorial innovation**. Territorial innovation is key to provide innovative ideas for the future of the territory.
4. Using **an "ad hoc" participative process**. The proposal issued from point 3 is to be shared and enriched through a number of participative meetings with different local sectors. This part of the process will spread the basic proposal, validate it, and get more ideas and projects to enrich it.

What Else After That? What Are the Next Steps?

At this point the territory needs to set up some kind of **mixed development agency** to manage all the development process from that moment on. "Mixed" means that all local stakeholders must be represented in the management body of this agency.

The first job of this development agency will be to manage the second part of the process which is defining sectorial plans.

After that, the agency will be in charge of coordinating any project contributing to the territory strategy. Since there will be projects from any local stakeholder, the agency will be in charge of coordinating all of them and ensuring all those projects fit into the local strategy and avoiding both, duplicities and shortcomings.

Summary/Conclusion

As managing a territory is a very complex issue, a strategy is required to transform it if we want to build an outstanding place offering a high quality of life to all its local citizens.

This strategy, to be effective, needs to be unique, to point to an excellence goal and to generate full consensus among local stakeholders. Under these conditions, the strategy will be able to excite local stakeholder in order for them to commit and contribute to "their" territorial project.

The way to get such an effective strategy is by using a collaborative leadership, starting from a "white sheet of paper," generating territorial innovation and sharing the project with citizenship through a participative process.

L

References

Jordà, A. (2019). Desarrollo Local y Territorial, Guía para Políticos y Técnicos. Editorial Letrame. ISBN 978-84-17897-32-1

Local Content

► European Green Deal and Development Perspectives for the Mediterranean Region

Local Development

► Internationalization of Cities

Local Economic Development

► Impact of Universities on Urban and Regional Economies

Local Food Traditions

► Formulating Sustainable Foodways for the Future: Tradition and Innovation

Local Governance

► Participatory Governance for Adaptable Communities
► Philanthropy in Sustainable Urban Development: A Systems Perspective

Local Knowledge

► Beyond Knowledge: Learning to Cope with Climate Change in Cities

Localized Price Effects

► Impact of Universities on Urban and Regional Economies

Localizing SDGs

► An Overview of the Relationship of the Sustainable Development Goals and Urban and Regional Development

Localizing Sustainable Development Goals (SDGs) Through Co-creation of Nature-Based Solutions (NBS)

Towards an Assessment Framework for Local Governments

Israa H. Mahmoud[1], Eugenio Morello[1], Daniela Rizzi[2] and Bettina Wilk[2]
[1]Laboratorio di Simulazione Urbana Fausto Curti, Department of Architecture and Urban Studies, Politecnico di Milano, Milan, Italy
[2]ICLEI European Secretariat, Senior Officer for Nature-based Solutions and Biodiversity – Sustainable Resources, Climate and Resilience Team, Freiburg, Germany

Working Definitions

Nature-based Solutions

According to the European Commission's definition (See also https://ec.europa.eu/info/research-and-innovation/research-area/environment/nature-based-solutions_en) (2015), nature-based solutions (NBS) are solutions that are "*inspired and supported by nature, which are cost-effective, simultaneously provide environmental, social and economic benefits and help build resilience. Such solutions bring more, and more diverse, nature and natural features and processes into cities, landscapes, and seascapes, through locally adapted, resource-efficient and systemic interventions. Nature-based solutions must therefore benefit biodiversity and support the delivery of a range of ecosystem services.*"

Sustainable Development Goals

The 17 global Sustainable Development Goals were introduced in 2015 by the United Nations General Assembly as part of a new global development agenda to be achieved by the year 2030. They comprise 169 targets addressing the developmental challenges facing the world including economic growth, urbanization, poverty, inequality, climate change, environmental degradation, peace and justice, see https://sdgs.un.org/goals.

Introduction: NBS, Co-creation and the SDGs, a Possible Triangulation

NBS have the aim to address many urban challenges such as climate change, biodiversity loss, urban heat island, and deforestation; those challenges put pressures on human health and well-being to natural capital depletion, and the security of food, water as well as energy. The EU's research & innovation (R&I) policy has been investing in NBS research and implementation, considering their superior potential to optimize the synergies between nature, society, and the economy (Faivre et al., 2017). NBS address specific demands or challenges, and at the same time seek to maximize other environmental, social, and economic co-benefits for health, the economy, society, and the environment (European Commission, 2015, 6). Considering the need for evidence on the realistic effectiveness of NBS, research, innovation, and demonstration Horizon 2020 projects in cities have been contributing to set up urban living laboratories (ULLs) for innovation, experimentation, and testing of good practices, methods, and tools with the view to exploit a range of ecological, social, and economic co-benefits for all (Chausson et al., 2020; Seddon et al., 2020).

The Horizon 2020 EU's research framework program accounts for over 240 million euros of investment in research and innovation in the field of NBS related projects (Davies et al., 2021, 54). Around 20 projects with specific focus on NBS were launched from 2014 to 2020, among which there are "Research and Innovation Actions" (RIAs) and "Innovation Actions" (IAs). Being RIAs projects that include basic and applied research, technology development and integration, testing and validation on a small-scale prototype in a laboratory or simulated environment, whereas IAs may include prototyping, testing, demonstrating, piloting, large-scale product validation, and market replication. At least 12 of the spearheaded NBS projects are dedicated to activities and research in cities (urban and peri-urban areas), among which eight are "Innovation Actions" and four are "Research and Innovation Actions" (Somarakis et al., 2019). In brief, those projects endeavor to explore how NBS work in different urban contexts with differing political, social, cultural, institutional, environmental, and economic situations (Dushkova & Haase, 2020).

In 2015, the United Nations General Assembly formally adopted the universal, integrated, and transformative 2030 Agenda for Sustainable Development, along with a set of 17 Sustainable Development Goals and 169 associated targets. The relationship between NBS and the SDGs (SDG reference is generally omitted since we refer to the official website here, https://sdgs.un.org/goals) has since then been highlighted in a number of publications, highlighting that NBS can be directly relevant to SDGs (Lo, 2016; Dudley et al., 2017; Vasseur et al., 2017) or even deliver on all 17 SDGs (Osieyo, 2020). Publications also advocate for NBS as cost-effective and no-regret solutions to address the complex task of meeting SDGs at a local scale in the long term (Acharya et al., 2020). Nonetheless, the recognition that NBS can contribute to SDGs on the local scale is present in only few examples from literature and practice recently (Beceiro et al., 2022; Schmidt et al., 2022).

The "SDG wedding cake," see Fig. 1, conceptualized by the Stockholm Resilience Institute (Folke et al., 2016), presents a holistic view of

Localizing Sustainable Development Goals (SDGs) Through Co-creation of Nature-Based Solutions (NBS), Fig. 1 The "SDG Wedding-Cake" by Stockholm Resilience Centre (https://www.stockholmresilience.org/research/research-news/2016-06-14-the-sdgs-wedding-cake.html) (left side) and an image (right side) by the University of Oxford under its "Nature-Based Solutions Initiative (NbSI)" on the fundamental importance of NBS to the hazards and impacts of climate change, using the SRC's framework to highlight the interconnection of NBS with all SDGs. (Sources: Folke et al., 2016; IUCN, 2020)

the SDGs, in which the prosperity and well-being of societies are depicted as dependent on the health of the planet. The cake's concept moves away from a sectorial approach in which social, economic, and ecological development are seen as separate parts, instead putting forward an interconnection between NBS and all SDGs as if economies and societies are embedded parts of the biosphere. The model highlights the intertwined nature of social-ecological systems and suggests the biosphere as the basis for sustainable development. The authors introduce the notion of "biosphere stewardship," raising the challenge of stewardship in tune with the biosphere as critical for sustainable development. The cake's concept advocates that focusing primarily on human well-being and social resilience while remaining disconnected from the biosphere and its stewardship is not a recipe for long-term sustainability. By framing biosphere stewardship as a process of engaging people to collaborate with shared visions across different levels and scales. Among few, Folke et al. (2016) and Kabisch et al. (2019) also provide an initial framework for a triangulation between NBS, co-creation, and SDGs.

Literature Review: NBS, Co-creation, and SDGs

NBS are increasingly recognized as a feasible means to address urban sustainability challenges and climate change actions (e.g., SDG 11 and 13), see also IUCN French Committee (2019). On a practical level there is an urge to engage a variety of stakeholders in an inclusive and efficient collaborative co-creation framework for NBS planning and implementation (Mahmoud & Morello, 2020). Particularly, a recent pan-European research and innovation projects' stream (*Demonstrating innovative nature-based solutions in cities*) is focusing on the application of NBS to address social inclusiveness (SDG 10) and shared governance challenges within urban regeneration processes (https://cordis.europa.eu/programme/id/H2020_SCC-02-2016-2017). This requires a paradigm shift from a theoretical to a practical framework of collaborative decision-making processes. Further, there is a missing link between practical collaborative processes and the SDG goals. Since the focus of these projects is primarily on pilot demonstration of NBS and experimentation with different planning approaches, ULLs are the preferred approach and arena to plan, design, and implement NBS while placing citizens at the center of decision-making mechanisms and processes (Bulkeley et al., 2018; Zingraff-Hamed et al., 2020).

SDGs and How Localized NBS Contribute to Their Attainment

The SDGs stand out as a holistic framework to provide structure and direction to nations, regions, and cities toward a sustainable future. Since the adoption of the UN SDGs in 2015, progress has been made but there is still a lot to be done for cities and gaps to be filled (De Maio et al., 2020). While there are many existing sustainability frameworks, the SDGs provide the most comprehensive and integrated approach for tracking sustainable development targets. In fact, SDGs present a major milestone for the local-to-global conversation by making room for local demands to be voiced at the global stage.

Although SDGs are set for national governments, they are quite relevant for local governments, since the establishment of partnerships with multiple stakeholders has been recognized as a crucial component of strategies linked to the 2030 Agenda, and indeed, "most SDGs will not be achievable without local level support" (Schuthof et al., 2019, 3). However, limitations to localizing SDGs in regional and urban development must also be acknowledged; these include political power, limited public finances, low institutional capacities to work across departments, absence of intergovernmental and multi-level cooperation and multi-stakeholder participation (Trejo-nieto, 2021). These limitations make the co-creation approach to implement SDGs very bumpy within ULLs and local governments.

In literature, NBS multifunctionality and ability to deliver several environmental, economic, and societal co-benefits such as social cohesion, awareness on biodiversity, GI ecosystem service provisioning, and human health and well-being (European Commission, 2020) which make them suited for indirectly or directly addressing all SDGs (Gómez Martín et al., 2020). Specifically, NBS are directly relevant to SDG 1 (no poverty), SDG 2 (food security), SDG 3 (health and well-being), SDG 6 (clean water and sanitation), SDG 7 (affordable and clean energy), SDG 11 (sustainable cities and communities), SDG 12 (responsible consumption and production), SDG 13 (climate action), SDG 14 (conservation and sustainable use of oceans, seas, and marine resources), and SDG 15 (protection, restoration, and promotion of sustainable use of terrestrial ecosystems), see Vasseur et al., 2017; Wendling et al., 2018; Cohen-Shacham et al., 2019. In sum, given the multifunctional character of NBS, these will indirectly contribute to all 17 SDGs, see also (Mahmoud, et al., 2022).

Co-creation, ULLs, and SDGs

Co-creation is understood as *"the systematic engagement of all relevant stakeholders from the start to the end of a project (and beyond) towards ensuring a smooth urban transition"* (Beck, 2018). What differentiates co-creation from more traditional forms of stakeholder engagement is the intensity of citizen involvement and the influence of societal actors in and on processes (Frantzeskaki & Rok, 2018; Menny et al., 2018). As an emerging form of collective urban governance and experimentation, ULLs are set up to address sustainability challenges and opportunities created by rapid urbanization. ULLs involve citizens in co-creation processes to increase social acceptance, foster NBS place-based ownership, support and plant the seed for co-implementation and co-maintenance of NBS (Malmberg et al., 2017; Breukers & Duneworks, 2017). In fact, active engagement of all relevant stakeholders from the very beginning of the planning process and throughout is likely to produce mutually valued outcomes (i.e., vision narratives, new understandings of problems and opportunities, etc.), and can thus build ground for trust, shared responsibility, and ownership of the NBS infrastructure (Voorberg et al., 2015; Pauleit et al., 2019).

ULLs have different goals and are initiated by various actors, forming different types of partnerships (Voytenko et al., 2016). ULLs are therefore where SDGs can be pursued locally and closest to citizens, who are critical partners to implement sustainability on the ground. In fact, *"ULLs are advanced as an explicit form of place-based interventions delivering sustainability goals for cities"* (Bulkeley et al., 2016; Menny et al., 2018). As comprehensive as SDGs are, they demand sectoral integration and require the transformation of managerial systems and systematic urban planning policies. And it is precisely in ULLs that diverse actors (public bodies, civil society, private actors) are brought together to experiment co-creation processes while addressing urban place-based challenges (Connop et al., 2015; von Wirth et al., 2019). ULLs can also be an important arena to give insight into experimentation with integrated approaches across sectors and departments.

Moreover, ULLs operate as sites where co-creation methods can be directly tested with end users and through which learning loops can take place in real time (Mahmoud et al., 2021). Allowing for joint dynamics of stakeholder engagement processes at the local level, ULLs contribute with co-production of knowledge to foster transition towards more collaborative governance structures in cities, encouraging more balanced power distribution, local leadership, and ownership via community involvement (Lund, 2018; Xie & Bulkeley, 2020).

Diversity of NBS types and scales brings in different timelines within an ULL co-creation pathway (Mahmoud & Morello, 2020). For instance, within the same ULL there could be a variety of "NBS in place" that requires different execution timelines and skills for co-design and co-implementation. In fact, NBS timelines widely differ: the spatial-temporal implementation tends to be the most challenging dimension because it varies according to the type of NBS (e.g., building-scale interventions, public space interventions, water body systems, transport linear infrastructure, natural areas, and ecological habitat interventions), see (Morello et al., 2019). These differences drive the need to be addressed with various implementation techniques and processes to overcome long-term maintenance timelines and development responsibilities.

In this sense, co-creation pathways and methodologies present an alternative trajectory to get NBS implemented in ULLs in a way that is inclusively shared and collectively governed; especially regarding the localization of SDGs in urban regeneration processes, some factors are to be considered. **Firstly**, NBS are living systems that continuously evolve and require caretaking and maintenance; hence, the importance of a shared co-management and collaborative governance of the interventions. **Secondly**, integrating NBS within existing urban regeneration dynamics requires the involvement of a multiplicity of stakeholders, as well as the activation of a solid and complex shared governance model, see (Mahmoud & Morello, 2021). **Lastly**, consolidated groups of interests and associated stratification of practices, as well as different memories of local communities, call for an inclusive approach to decision-making, which can bring a diversity of perspectives into the design process and thus takes a longer time to develop.

To sum up, on the one hand, the abovementioned complexity is related to the suitability of NBS, and the specific impact generated by a given NBS in place to be measured against the SDGs. On the other hand, the node to these co-creation pathways applied for NBS delivery remains the impact measurement against SDGs. In the following sections, this entry explores the co-creation methodology and co-creation pathways applied within the CLEVER Cities project to foster NBS implementation.

Identifying the Knowledge Gaps Around Co-creation of NBS in Relation to the SDGs

Why integrate SDGs in NBS co-creation? A possible answer is: Aligning with the universally recognized targets and speaking the common language of the 2030 Agenda can help break silos in the agenda of local governments and boost partnerships, leading to concrete societal transformative change (Kirsop-Taylor et al., 2021). Also, the SDGs provide a common framework of targets and thus, impacts to be achieved by co-creation of NBS.

Several knowledge gaps identified by the authors make it difficult to harness the full potential of co-created NBS in relation to the SDGs. That is due to the novelty of the co-creation topic within projects for NBS implementation (only 3 European-funded projects started in 2017, 6 in 2018, and 2 in 2019), see also (Carlotta et al., 2020: 19). However, recently a remarkable effort has been carried out to connect different SDGs to NBS objectives by Somarakis et al. (2019, 29,30).

The first hindrance arises from translating SDG targets to NBS implementation in practice and related local actions, since the integration of local needs, priorities, local capacities, and expectations are critical factors to determine SDG attainment (Tosun & Leininger, 2017). (Keeys & Huemann, 2017; Kotsila et al., 2020; Xie & Bulkeley, 2020). **The second knowledge gap** is a methodological one. There is an absence of standardized co-creation frameworks in investigated methods and practical guidances in all NBS projects (no one size fits all) (Kruger et al., 2018; Forde, 2020); most H2020 projects with a focus on co-created NBS pilot projects establish their own co-creation pathways and possible indicators for monitoring the impacts of their own co-creation processes, see (Mahmoud & Morello, 2018). Multiple differing co-creation frameworks are used, hence, there is a lack of a "universal" technical language in the co-creation discourse, especially as highlighted in the latest NBS sister projects recurrent EC NBS Task Force meetings. (Starting March 2020, a specific Task Force (VI) was formed between sister projects, these Tasks Forces are promoted by the EC "Network Nature" in order to facilitate the NBS projects collaboration, they meet every two months. The focus of this TF(VI) in specific is to uptake a common approach on co-creation processes that involve multiplicity of stakeholders and involve citizens in the whole process of NBS co-governance and implementation.) This methodological gap was identified by assessing numerous online workshops, publications, and meetings in addition to active participation in different TFs.

Further, for local governments to design relevant approaches and programmes and consequently evaluate methods to effectively deliver on the SDGs, guidance is needed on how to localize targets on a city scale and thus, derive required actions. Local authorities and municipalities are lacking an analysis framework that associates local co-creation processes, stakeholder validation, and impacts of NBS in relation to the SDGs (Beceiro et al., 2020). Currently, a set of consolidated performance and impact indicators are missing, so no streamlined analysis of success (or failure) against the SDGs can be performed. In most cases, partners of ongoing Horizon2020 NBS projects are the ones to assess their co-creation processes in terms of effectivity considering their local contexts and urban regeneration planning framework, which makes it difficult to establish a comparable framework.

This relates to the third knowledge gap: the lack of a concrete set of KPIs that analyze co-creation processes impacts and associated success or failure, added values as a work-in-progress and not just after finalization. That is off-course logic since the NBS H2020 projects only embedded co-creation pathways as a novel innovation policy during the latest 2 or 3 years; hence, it needs to be considered that there is no consolidated knowledge at this point but an ongoing effort to cover the knowledge gap. Scholars often look at the impact of NBS in terms of monitoring the performance of environmental, societal, and economic challenges, which is in correlation to the SDGs (Colléony & Shwartz, 2019; Connop et al., 2020; Lam et al., 2020). However, only a few recent references look at the SDGs while also reviewing the methodological co-creation frameworks (Kabisch et al., 2019; Dumitru et al., 2020).

Another shortcoming is the standardization of co-creation processes with a proliferation of methods and guidance (Kruger et al., 2018; Forde, 2020). Not all NBS are "co-creatable" in the same way, nor are the same efforts required during an NBS implementation process in terms of stakeholder engagement and co-management with the groups of interest.

Lastly, **there is a research and evidence gap** on the added value of co-creation processes for NBS design, co-implementation, and co-monitoring, and thus, of the co-creation impacts on NBS delivery and even more, of the sustainability of NBS, as well as of how co-creation processes ultimately contribute to achieving SDGs.

A Methodological Approach: Toward Structuring a Co-creation Assessment Framework for Local Governments Related to SDGs

A quick search query on Scopus (NBS AND SDGs) revealed a number of 105 papers over the last 10 years focusing on NBS and SDGs. (The query was limited to titles, abstracts and keywords of the publications then corrected manually: 32 out of 105 in the year 2020 and 69 records in 2021, 13 records in 2022 till the submission of manuscript.) Predominant research foci are environmental, societal, and economic impacts of NBS with an intrinsic link to the SDGs (Colléony & Shwartz, 2019; Connop et al., 2020; Lam et al., 2020). However, research and respective papers focusing on SDGs in relation to co-creation is still scarce. Co-creation in relation to SDGs was only addressed in 42 documents, (Co-creation term is also used in business and educational ambits; hence an elaboration was needed to review queries accuracy.) out of which only three focused on NBS. In other databases, such as Springer, co-creation features prominently in public policies and social sciences (80 documents). However, no correlation with either NBS or SDGs could be found in any of the publications when limited to the domain of environmental and social policies (Elsevier, 2015) as of date.

In this research, the co-creation phases developed in the CLEVER Cities project are the base for creating links with the SDG targets. In this project, a list of potential co-creation KPIs are developed, which will be used in this entry as a hinge point for creating an assessment framework for local governments to evaluate the "added value" of inclusive shared governance models in NBS implementation and urban regeneration planning.

The present tentative model for an assessment framework considers five macro categories (see Linking co-creation phases to the SDGs) as related to the phases of the co-creation pathway within CLEVER Cities as an anchor to identify the correlation between co-creation and SDGs. In this model, SDGs and their associated targets relevant to co-creation processes were identified through the perspective of desired impacts to be generated by "NBS in place," such as reducing inequality, increase well-being, increase social cohesion and inclusivity, etc.

CLEVER Cities Co-creation Framework in Five Phases: Methodological Testing and Implementation

In this work, the authors analyze where (in which phases) and how the SDGs come into play along the NBS co-creation pathway. For the purpose of pragmatism, this entry refers only to the CLEVER Cities co-creation framework, which represents a modular and complete pathway that covers all the potential phases of collaboration and stakeholder engagement around NBS (Mahmoud & Morello, 2018, 2020; Morello et al., 2018). For instance, co-creation could be implemented on a variety of NBS delivery phases, such as co-design, co-implementation, co-monitoring, and co-development. Since, literature argues that co-creation has emerged in connection to the participatory approach in public policies and human-centered design, where it is about empowering people in decision-making processes and working practices (Voorberg et al., 2015; Jansen & Pieters, 2017; Frantzeskaki, 2019). Hence, it was considered that co-creation tools for NBS implementation in practice encounter a variety of user types and possible platforms to bring together citizens, stakeholders, researchers, academia, and policymakers to address collective urban challenges.

In the CLEVER Cities' pathway, co-creation goes beyond running simple collaborative workshops. In the latest article, Basnou et al. (2020, 1) refer to the need for a well-planned co-design process and engagement strategy that supports inclusive participation and social learning through enabling dialogue, co-production of knowledge, and equity in urban and territorial planning processes. In fact, in the CLEVER Cities framework, the complete sequenced co-creation pathway (See CLEVER Cities

Co-creation Guidance website, https://clevercitiesguidance.wordpress.com/) represents a flexible structure to be put in practice in the ULL, i.e., a place-based medium for fostering inclusive urban regeneration processes. Hence, the process of implementing NBS according to a collaborative co-design process that involves stakeholders from the start until the end requires the development of guidelines drawing special attention on achieving more social inclusiveness, and consequently, enhance shared governance and collaborative decision-making mechanisms.

The CLEVER Cities phases of co-creation are five, see Fig. 2, **area (a) Co-creation phases** for an exhaustive illustration of the co-creation phases and coverage of stakeholder mapping and engagement of citizens, co-designing, co-implementation, co-monitoring, and co-development of NBS.

Why ULLs Are the Field of NBS Implementation in CLEVER Cities

ULLs involve citizens in co-creation processes as a means to increase social acceptance, foster support, and plant the seed for co-implementation and co-maintenance of the NBS (Malmberg et al., 2017; Breukers & Duneworks, 2017). Active engagement from the very beginning is likely to produce mutually valued outcomes and can thus build ground for trust, responsibility, and ownership of the NBS infrastructure (Malmberg et al., 2017; Hansen et al., 2019). Following (IAP2, 2014; Emerson & Nabatchi, 2015), stakeholder engagement can range from information, consultation, involvement, and collaboration into actual empowerment. It differs with regard to the extent of power, willingness and influences stakeholders have on decision-making processes and on the development of the final solution. In CLEVER Cities, this engagement process starts gradually in the so-called urban alliance phase, which is the initial phase of identifying city-wide targets to be achieved throughout the partnership activation. In this sense, it is very relevant within the CLEVER Cities co-creation pathways to consolidate mature and inclusive Urban Innovation Partnerships (UIP), designed to host community leaders, local associations, local SMEs, etc.

L

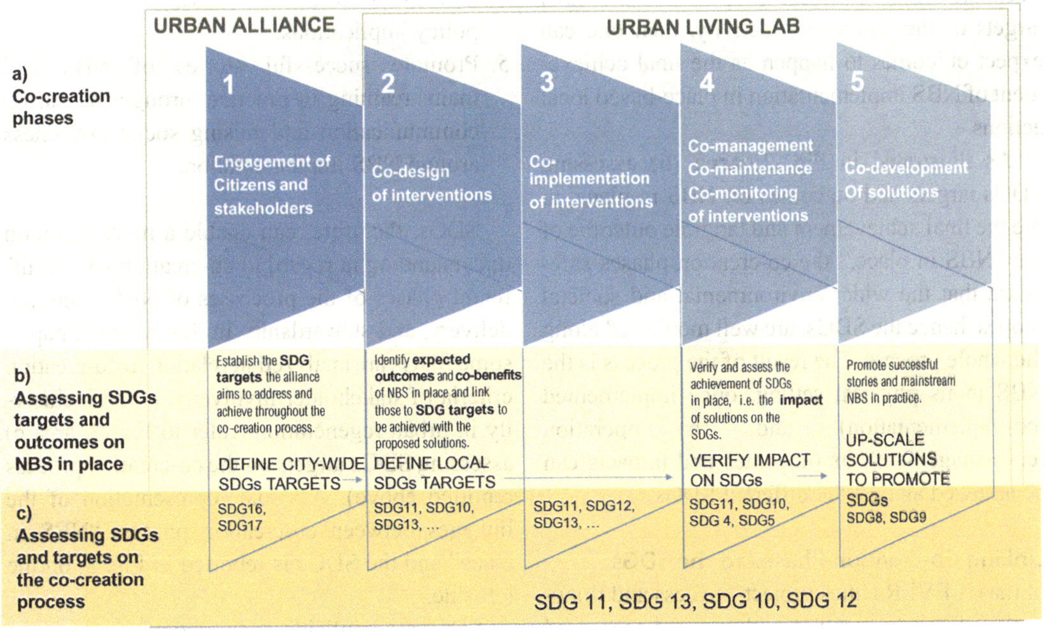

Localizing Sustainable Development Goals (SDGs) Through Co-creation of Nature-Based Solutions (NBS), Fig. 2 Co-creation phases and SDGs correlations. (Source: The Authors)

Results: A Possible Assessment Framework for the NBS Co-creation Process and to Assess Tangible Outcomes of "NBS in Place" Against Achieving the SDGs

The CLEVER Cities' co-creation pathway contribution to the SDGs can be investigated from the perspective of process itself or from the standpoint of generated impacts. In terms of process the efficacy of the process itself in embedding the SDGs along NBS implementation can be assessed, as well as the of the level of inclusiveness process toward a real shared governance, where no one is left behind. In terms of impact, co-created "NBS in place" can be evaluated regarding the tangible outcomes and impacts of the process (Many efforts are done by CLEVER Cities sisters projects on planning assessment, for instance see also https://connectingnature.eu/innovations/impact-assessment) such as co-benefits generated by specific actions. These two dimensions, the process to implement NBS on one side, and the tangible outcomes and impacts of "NBS in place" on the side, are strictly related. In fact, if the process explicitly emphasizes the SDGs as value proposition and final targets of the co-creation activity, then one can expect outcomes to happen as the final achievement of NBS implementation in place-based local actions.

As illustrated in Fig. 2, **area (b)** assessing SDGs targets and outcomes on NBS in place, to see the final achievement and tangible outcome of the "NBS in place," the co-creation phases safeguard that the wider environmental and societal scopes, hence the SDGs, are well monitored along the whole process. The result of the process is the NBS in its physical setting: once implemented (co-implementation) and in operation (co-management), its outcomes and impacts can be assessed against the different SDGs.

Linking Co-creation Phases to the SDGs

In the CLEVER Cities project, success and drawbacks of the co-creation pathway are measured against five major macro indicators areas (This categorization is a work in progress by the first two authors, it was requested by the EC in review meetings to showcase of the process co-creation success and added value. The categories are mainly based on the development of the co-creation pathway implementation in FR Cities and the outcomes of the project till September 2021, three workshops were carried out to confirm which indicators are most common to all three FR cities.):

1. Stakeholder engagement in the urban alliance following the ladder of engagement (inform, consult, involve, collaborate, empower) as in Arnstein (1969).
2. Timeline and duration of engagement: In which stage of co-creation are citizens/community members/local groups and other stakeholders involved? Is there a continuity of engagement during the project lifetime?
3. Co-creation pathway governance: Flexibility and resilience of the process against experimentation and adaptation to shocks and hazards, as well as transparency of the operational process.
4. Measurable outcomes and added values of co-creation in decision-making processes and policy implications.
5. Promote successful stories of NBS and mainstreaming in practice through measuring communication and raising social awareness around NBS implementation.

SDGs, therefore, can enable a more common understanding in regard to co-creation within different phases of the processes of NBS planning, delivery, and stewardship. In this research paper, some SDGs are analyzed in relation to co-creation criteria of stakeholder involvement and inclusivity in urban regeneration (refer to Fig. 2, area c) assessing SDGs targets on the co-creation process reported above). A visual representation of the linkages between co-creation phases, "NBS in place" and the SDGs is reported in Fig. 3, on the left side.

SDGs are implicitly addressed within the overall approach to participatory decision-making and

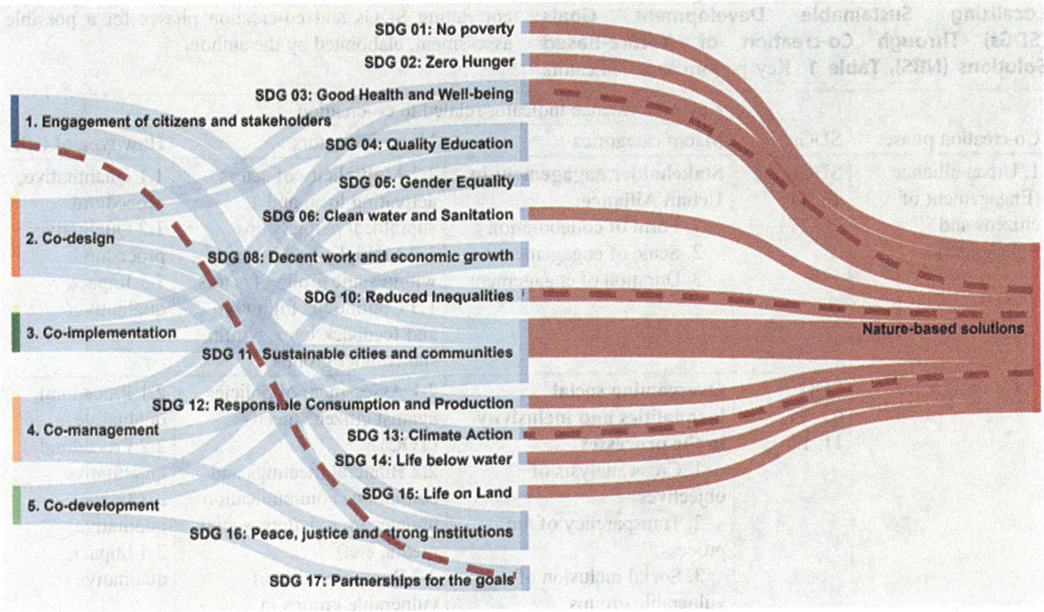

Localizing Sustainable Development Goals (SDGs) Through Co-creation of Nature-Based Solutions (NBS), Fig. 3 Alluvial diagram connecting the SDGs to the co-creation phases (on the left) and the "NBS in place" (on the right). In red dotted line, example of connections between Co-creation phases and SDGs on the left. Example of connection between SDG 03, 10, 13, and NBS on the right. (Source: The Authors)

shared governance. The overarching nature of SDG 16 and SDG 17, for instance, in relation to the partnerships and decision-making processes is the most correlated ones to the co-creation phase of fostering urban alliance (e.g., creating alliances and partnerships around NBS awareness such as nature forums). Furthermore, within the ULLs, the SDG 10, SDG 11, and SDG 13 relate to overcoming social inequalities and inclusivity in NBS processes inserted within a wider climate action framework. Considering responsible consumption and production, SDG 12 is addressed in co-implementation and co-management phases, whereby local assets (financial resources, skills, material resources) can be exploited. SDG 4 and SDG 5 are addressed within the co-management and the co-maintenance phases of NBS, because they relate to building knowledge and creating awareness around the relevance of nature in cities (hence improving quality education), and also offer job opportunities as well as reduce gender inequalities (e.g., citizen caretaking of NBS in public space is an example of knowledge and skill building around nature). In regard to upstreaming successful solutions to strengthen sustainable communities (SDG11) and boosting the local economy through NBS (SDG8), the final phase of co-development addresses this, see Fig. 3. (In Fig. 3, Not all SDGs seem to have a direct link to assess the impact of NBS in place. In this case it is NA.)

Discussion: Proposed Co-creation Key Performance Indicators

Based on the topics illustrated above, two tentative lists of main KPIs are reported below. In Table 1, each phase of co-creation in relationship to the SDGs is crossed, resulting from the previous Fig. 3, toward dividing the KPIs in two main categories: (1) **Macro** categories, which are directly related to SDGs and (2) **Micro** indicators, which mainly relate to the impact generated by the NBS throughout the co-creation phase. Specifically, Table 1 addresses the fulfilment of the

Localizing Sustainable Development Goals (SDGs) Through Co-creation of Nature-Based Solutions (NBS), Table 1 Key performance indicators correlating SDGs and co-creation phases for a possible assessment, elaborated by the authors

Co-creation phases	SDGs	Key performance indicator related to co-creation Macro categories	Micro indicators	How/type of KPI
1. Urban alliance (Engagement of citizens and stakeholders)	SDG 10, 11, 16, 17	**Stakeholder engagement in Urban Alliance:** 1. Form of collaboration 2. Scale of engagement 3. Duration of engagement	1.1 Multiplicity of actors activating local and supralocal partnerships 1.2 Multiplicity of roles within same group of actors 1.3 Continuous follow-up and feedback loops within stakeholder groups	1.1 Quantitative, procedural 1.2 Qualitative, procedural 1.3 Impact, qualitative
2. Co-design	SDG 03, 04, 11, 12	**Overcoming social inequalities and inclusivity in the processes** 1. Cross analysis of objectives 2. Transparency of the process 3. Social inclusion of vulnerable groups 4. Flexibility and adaptability	2.1 Assessment of policies against citizen's needs (TOC) 2.2 Bilateral meetings and continuous communication updates (newsletters, social media, etc.) 2.3 Representation of vulnerable groups in decision-making 2.4 Experimentation through NBS implementation to change plans	2.1 Procedural, qualitative 2.2 Procedural, quantitative 2.3 Impact, qualitative 2.4 Impact, qualitative
3. Co-implementation	SDG 04, 08, 11	**Co-creation pathway governance:** 1. Development of a specific plan related to place-based challenges 2. Exploitation of expertise through shared governance	3.1 Tailor placed-based NBS implementation to their desired co-benefits 3.2 Encouraging citizens to get hands on their NBS ownership and be the main decision-makers	3.1. Impact, qualitative 3.2. Procedural, qualitative
4. Co-management	SDG 03, 04, 11	**Measurable outcomes** and verification of added impact 1. Building Trust with citizens and build legacy 2. Encourage new partnerships and ownership	4.1 Added values of the co-creation process, what is the residual left in place after the co-design ends 4.2 Spill overs measured in new partners interested to take place and catalyze the management process.	4.1 Impact, qualitative 4.2 Impact, quantitative
5. Co-development	SDG 08, 11	**Promote Successful stories of NBS and Mainstreaming in practice** 1. Assessment of potential upscaling and Mainstreaming 2. Replication success factors 3. Communicate success and failures equally	5.1 Financial revenues generated by NBS upscaling and mainstreaming 5.2 Increase of frequency of uses of the 'NBS in place' and social recognition by citizens. 5.3 Increase communication and social awareness around NBS	5.1 Impact, quantitative 5.2 Impact, qualitative 5.3 Impact, quantitative

SDGs within the co-creation process, while Table 2 evaluates the outcomes and impacts of the "NBS in place" against the SDGs. The KPIs are then sub-divided by two different typologies: procedural and impact (This subdivision is decided by the authors with respect to the

Localizing Sustainable Development Goals (SDGs) Through Co-creation of Nature-Based Solutions (NBS), Table 2 Suggested indicators for evaluating the "NBS in place" – Some examples of indicators on SDGs 03, 10, 13, and 15

SDG	Targets	Indicators	Examples of NBS typologies
SDG 3	Contribution to mental health	Physically and visually accessible NBS in open spaces [leaf area index, nr. f trees, m2 of tree canopy]	Tree-lined streets Green walls Fruit trees
	Contribution to physiological health	Physically accessible NBS in open spaces [leaf area index, nr. of trees, m^2 of tree canopy]	Tree-lined streets Urban forests
	Contribution to a healthy diet	NBS providing food grown locally [nr. of trees, Kg. of grown food, CO_2 emissions of grown food]	Community gardens Fruit trees Espalier fruit trees
SDG 10	Empower and promote the social, economic, and political inclusion of all, irrespective of age, sex, disability, race, ethnicity, origin, religion, or economic or other status	Lowering number of people living below 50% of median income, by age, sex, or disabilities	Community gardens Green roofs
SDG 13	Reduce urban rainwater runoff (heavy precipitation hazard)	Volume of water absorption through vegetation roots and permeable soil [m3]	Trees Urban forests Green roofs Green parking lots
	Mitigate outdoor climate (extreme temperatures hazard)	Measured temperature reduction [°C]	Trees Urban forests Green walls
SDG15	Enhance biodiversity	Species, biodiversity [nr. of species reported]	Urban forests Ecological corridors Beehives Butterfly oasis

Many NBS projects are working on the providing NBS catalogues that satisfy the NBS categorization that correspond to their standards and needs (technological, social, productive, etc.), see http://www.labsimurb.polimi.it/nbs-catalogue/ and https://urbinat.eu/nbs-catalogue/

evaluation carried out with the cities and the consortium seemed to have possibility to measure procedure as part of the co-creation impacts.) measured either quantitatively or qualitatively, as explained after Table 1.

It is to be noted that the proposed subdivision is mainly assessing the generated outcomes from the CLEVER Cities' co-creation pathway process and the associated measurable co-benefits throughout the process itself. The main idea was to divide the assessment framework into stages that look at different types of indicators, those that are of an operational nature, such as the ones related to impact, and the ones associated to procedures. Finally, it was also relevant to examine the indicators at different co-creation phases and according to data collection types (*qualitative or quantitative*). However, a final and more consolidated assessment framework is still to be defined with the rest of the CLEVER Cities consortium. Currently, the assessment framework is being co-developed in a living document with involved cities and is subject to changes over time. A prospective scoreboard for each city to be able to assess its co-creation process in terms of inclusivity and success or failure is the final aim. (Possible Limitations: The SDG dimension was not identified during the initial development of CLEVER Cities project, the gain of relating all this to SDGs is mainly the added value of speaking the common language across different cities and to the EASME with collaboration of cities local governments.)

Table 2 provides an example for linking SDGs, targets, indicators with different NBS typologies. This relationship suggests how different NBS typologies differently respond to environmental (biosphere), social, and economic challenges. In the impact assessment of a certain type of NBS, it is not always possible to establish linkages to all SDGs. After all, some SDGs are only relevant in specific contexts: for instance, a NBS can respond to SDG1, SDG2 (and to some extent also to SDG5) in lower income countries but these SDGs might not be addressed in richer ones. Moreover, not all types of NBS can be easily co-created (such as highly technological solutions or large-scale infrastructure). Future work will better investigate the relationship between the performance of a solution, the impact on the SDGs, as well as consider the context from a cost-benefit perspective.

Conclusions: Toward a Co-creation Assessment Framework for Local Governments Connecting NBS and SDGs

This entry has discussed the triangulation among NBS, co-creation, and the SDGs, from the implementation of the framework in the CLEVER Cities project. It shows that establishing stronger links between NBS and the SDGs in the entire co-creation process is essential because the SDG language has become universal by gaining increased popularity and application in different domains since its promotion in late 2015. Hence, speaking the SDG language benefits from the opportunity to align urban planning processes with the engagement of the civic society. This makes it easier to establish partnerships and collaborating on the same targets (win-win condition i.e., SDG 17). Moreover, it is easier to assess and quantify the progress and success of processes and outcomes through indicators converging to the 2030 Agenda because it has its established targets to be locally measured in many municipalities' strategic plans.

Considering the CLEVER Cities' co-creation methodological pathway, a challenge is recognized in addressing and referring to the SDGs and targets. Therefore, this entry generates a basic understanding of when and how inclusive NBS-led urban regeneration implicitly intersects with the SDGs. Since anchoring the NBS co-creation framework in the 2030 Agenda is essential, a set of macro categories and micro–Key Performance Indicators are indicated as a new framework of assessment putting together co-creation of NBS and the SDGs.

Such a conceptual framework should enable players to assess the contribution of both the implementation process of NBS on one side and the tangible outcomes and impacts of the NBS to SDGs, on the other side. In order to reach this purpose, these KPIs are a starting point for developing a practical assessment tool for local governments. These shall help local governments to (1) reflect the SDGs in co-creation and assess the performance of ULLs and inclusive shared governance toward the 2030 Agenda; and (2) measure the impact of tangible outcomes of the implemented NBS, to be evaluated on the basis of the expected SDGs as established in the co-design phase.

This entry also uncovers the knowledge gap of missing links between indicators on SDGs and NBS. The idea is to develop an easy-to-use assessment instrument for co-creation processes and impact indicators that can measure co-creation effectiveness for NBS delivery from the planning side. The entry establishes indicators for reflecting on SDGs while affording a view of the impacts of NBS within ULLs/co-creation on the practice side.

Another important aspect is showcasing the plurality of indicators that each and every project could adopt to assess their own co-creation processes to identify the major different impact originating from the co-benefits of "NBS in place." Nowadays many projects are reviewing initial objectives and revaluating desired impacts after the COVID-19 Pandemic situation, which raises the relevance of evaluating co-creation processes to assess the dynamic changes and adjustments in terms of citizen engagement, such as online co-design activities moving to online platforms. This is also linked to the dynamic evolution of co-benefits, which increases with higher NBS effectiveness and related collaborative processes.

Lastly, the importance of the NBS capacity for addressing SDGs is highlighted, which is highly dependent on NBS multifunctionality and on the local contexts of co-creation processes. Nonetheless, engaging stakeholders in the first stages of NBS design and implementation is key. The effectivity of co-creation processes is also highly influenced by recognizing the relevance to perform multiple stakeholder engagement within different local and supralocal partnerships.

Recommendations for Future Research

1. Exploration of the opportunity of the co-creation task forces to set up a common language.
2. Interpretation of the content on existing NBS platforms to make the connection with the SDGs.
3. Proposal of a new framework of investigation across projects rather than isolated evaluation of results (evidence-based approach).
4. Development of a co-creation KPIs' framework that is transversal amongst projects and eventually represents the shared governance fundaments already developed for CLEVER Cities.

Cross-References

- ▶ An Overview of the Relationship of the Sustainable Development Goals and Urban and Regional Development
- ▶ Cities in Nature
- ▶ Green Cities: Nature-Based Solutions, Renaturing and Rewilding Cities
- ▶ Sustainable Development Goals

Acknowledgment Authors want to thank CLEVER Cities Project team to support different reflections on the actual co-creation processes.

Funding This document has been prepared in the framework of the European project CLEVER Cities. This project has received funding from the European Union's Horizon 2020 innovation action program under grant agreement no. 776604. The sole responsibility for the content of this publication lies with the authors.

References

Acharya, P., Gupta, A. K., Dhyani, S., & Karki, M. (2020). New pathways for NbS to realise and achieve SDGs and post 2015 targets: Transformative approaches in resilience building. In S. Dhyani et al. (Eds.), *Nature-based solutions for resilient ecosystems and societies* (pp. 435–455). Springer.

Arnstein, S. R. (1969). A ladder of citizen participation. *Journal of the American Institute of Planners, 35*, 216–224.

Basnou, C., Pino, J., Davies, C., et al. (2020). Co-design processes to address nature-based solutions and ecosystem services demands: The long and winding road towards inclusive urban planning. *Frontiers in Sustainable Cities, 2*, 1–4. https://doi.org/10.3389/frsc.2020.572556.

Beceiro, P., Brito, R. S., & Galvão, A. (2020). The contribution of NBS to urban resilience in stormwater management and control: A framework with stakeholder validation. *Sustainability, 12*, 2537. https://doi.org/10.3390/su12062537.

Beceiro, P., Brito, R. S., & Galvão, A. (2022). Assessment of the contribution of Nature-Based Solutions (NBS) to urban resilience: Application to the case study of Porto. *Ecological Engineering, 175*, 106489. https://doi.org/10.1016/j.ecoleng.2021.106489.

Beck, S. (2018). Urban transitions: Public anthropol a borderless. *World, 2018*, 314–350. https://doi.org/10.2307/j.ctt9qdb3w.17.

Breukers, S., & Duneworks, Y. J. (2017). Step-by-step guide for co-production and co- creation of Nature-based Solutions. Retrieved from https://www.nature4cities.eu/_files/ugd/55d29d_7880df6bd3ca41b4aaca7000534724a8.pdf

Bulkeley, H., Coenen, L., Frantzeskaki, N., et al. (2016). Urban living labs: Governing urban sustainability transitions. *Current Opinion in Environment Sustainability, 22*, 13–17. https://doi.org/10.1016/j.cosust.2017.02.003.

Bulkeley, H., Marvin, S., Palgan, Y. V., et al. (2018). Urban living laboratories: Conducting the experimental city? *European Urban and Regional Studies, 26*(4), 317–335. https://doi.org/10.1177/0969776418787222.

Carlotta, F., Martina, P., Martina, B., & Sjoerdje, V. H. (2020). *Handbook of sustainable urban development strategies, EUR 29990*. Luxembourg: Publications Office of the European Union.

Chausson, A., Turner, B., Seddon, D., et al. (2020). Mapping the effectiveness of nature-based solutions for climate change adaptation. *Global Change Biology, 26*(1), 6134–6155. https://doi.org/10.1111/gcb.15310.

Cohen-Shacham, E., Andrade, A., Dalton, J., et al. (2019). Core principles for successfully implementing and upscaling nature-based solutions. *Environmental Science & Policy, 98*, 20–29. https://doi.org/10.1016/j.envsci.2019.04.014.

Colléony, A., & Shwartz, A. (2019). Beyond assuming co-benefits in nature-based solutions: A human-centered approach to optimize social and ecological outcomes for advancing sustainable urban planning. *Sustainability, 11*, 4924. https://doi.org/10.3390/su11184924.

Connop, S., Nash, C., Elliot, J., Haase, D., & Dushkova, D. (2020). Nature-based solution evaluation indicators: Environmental Indicators Review. Retrieved from

L

https://connectingnature.eu/sites/default/files/images/inline/CN_Env_Indicators_Review_0.pdf

Connop, S., Vandergert, P., Eisenberg, B., et al. (2015). Renaturing cities using a regionally-focused biodiversity-led multifunctional benefits approach to urban green infrastructure. *Environmental Science & Policy, 62*, 99–111. https://doi.org/10.1016/j.envsci.2016.01.013.

Davies, C., Chen, W. Y., Sanesi, G., & Lafortezza, R. (2021). The European Union roadmap for implementing nature-based solutions: A review. *Environmental Science & Policy, 121*, 49–67. https://doi.org/10.1016/j.envsci.2021.03.018.

De Maio, S., Kuhn, S., Esteve, J. F., & Prokop, G. (2020). Indicators for European cities to assess and monitor the UN Sustainable Development Goals (SDGs). Retrieved from https://www.eionet.europa.eu/etcs/etc-uls/products/etc-uls-report-2020-08-indicators-for-european-cities-to-assess-and-monitor-the-un-sustainable-development-goals-sdgs

Dudley, N., Ali, M., & Kettunen, K. M. (2017). Protected areas and the sustainable development goals. *Parks, 23*, 9–12. https://doi.org/10.2305/IUCN.CH.2017.PARKS-23-1.en.

Dumitru, A., Frantzeskaki, N., & Collier, M. (2020). Identifying principles for the design of robust impact evaluation frameworks for nature-based solutions in cities. *Environmental Science & Policy, 112*, 107–116. https://doi.org/10.1016/j.envsci.2020.05.024.

Dushkova, D., & Haase, D. (2020). Not simply green: Nature-based solutions as a concept and practical approach for sustainability studies and planning agendas in cities. *Land, 9*, 19. https://doi.org/10.3390/land9010019.

Elsevier. (2015). *Sustainability science in a global landscape*.

Emerson, K., & Nabatchi, T. (2015). *Collaborative governance regimes, public man*. Washington, DC: Georgetown University Press.

European Commission. (2015). *Towards an EU research and innovation policy agenda for nature-based solutions & re-naturing cities*.

European Commission (2020), Directorate-General for Research and Innovation, Bulkeley, H., Naumann, S., Vojinovic, Z., et al., Nature-based solutions : state of the art in EU-funded projects, Freitas, T.(editor), Vandewoestijne, S.(editor), Wild, T.(editor), https://data.europa.eu/doi/10.2777/236007

Faivre, N., Fritz, M., Freitas, T., et al. (2017). Nature-based solutions in the EU: Innovating with nature to address social, economic and environmental challenges. *Environmental Research, 159*, 509–518. https://doi.org/10.1016/j.envres.2017.08.032.

Folke, C., Biggs, R., Norström, A. V., et al. (2016). Social-ecological resilience and biosphere-based sustainability science. *Ecology and Society, 21*(3). https://doi.org/10.5751/ES-08748-210341.

Forde, C. (2020). Co-production and co-creation: Engaging citizens in public services. *VOLUNTAS: International Journal of Voluntary and Nonprofit Organizations, 31*, 454–455. https://doi.org/10.1007/s11266-019-00182-9.

Frantzeskaki, N. (2019). Seven lessons for planning nature-based solutions in cities. *Environmental Science & Policy, 93*, 101–111. https://doi.org/10.1016/j.envsci.2018.12.033.

Frantzeskaki, N., & Rok, A. (2018). Co-producing urban sustainability transitions knowledge with community, policy and science. *Environmental Innovation and Societal Transitions, 29*, 47–51. https://doi.org/10.1016/j.eist.2018.08.001.

Gómez Martín, E., Giordano, R., Pagano, A., et al. (2020). Using a system thinking approach to assess the contribution of nature based solutions to sustainable development goals. *Science of the Total Environment, 738*, 139693. https://doi.org/10.1016/j.scitotenv.2020.139693.

Hansen, R., Olafsson, A. S., van der Jagt, A. P. N., et al. (2019). Planning multifunctional green infrastructure for compact cities: What is the state of practice? *Ecological Indicators, 96*, 99–110. https://doi.org/10.1016/j.ecolind.2017.09.042.

IAP2. (2014). IAP2' s public participation spectrum. *Int Assoc Public Particip. 2014*.

IUCN. (2020). *IUCN global standard for nature-based solutions: A user-friendly framework for the verification, design and scaling up of NbS* (1st ed.). Cham: Gland.

IUCN French Committee. (2019). *Nature-based solutions for climate change adaptation & disaster risk reduction* (Vol. 23). Paris: IUCN French Committee.

Jansen, S., & Pieters, M. (2017). *The 7 principles of complete co-creation*. Amsterdam: Bis Publishers.

Kabisch, S., Finnveden, G., Kratochvil, P., et al. (2019). New urban transitions towards sustainability: Addressing SDG challenges (research and implementation tasks and topics from the perspective of the scientific advisory board (SAB) of the joint programming initiative (JPI) Urban Europe). *Sustainability, 11*, 2242. https://doi.org/10.3390/su11082242.

Keeys, L. A., & Huemann, M. (2017). Project benefits co-creation: Shaping sustainable development benefits. *International Journal of Project Management, 35*, 1196–1212. https://doi.org/10.1016/j.ijproman.2017.02.008.

Kirsop-Taylor, N., Russel, D., & Jensen, A. (2021). Urban governance and policy mixes for nature-based solutions and integrated water policy. *Journal of Environmental Policy and Planning, 0*, 1–15. https://doi.org/10.1080/1523908X.2021.1956309.

Kotsila, P., Anguelovski, I., Baró, F., et al. (2020). Nature-based solutions as discursive tools and contested practices in urban nature's neoliberalisation processes. *Environment and planning E: Nature and space, 4*(2), 252–274. https://doi.org/10.1177/2514848620901437.

Kruger, C., Caiado, R. G. G., França, S. L. B., & Quelhas, O. L. G. (2018). A holistic model integrating value co-creation methodologies towards the sustainable development. *Journal of Cleaner Production, 191*, 400–416. https://doi.org/10.1016/j.jclepro.2018.04.180.

Lam, D. P. M., Martín-López, B., Wiek, A., et al. (2020). Scaling the impact of sustainability initiatives: A typology of amplification processes. *Urban Transformations, 2*(1), 1–24. https://doi.org/10.1186/s42854-020-00007-9.

Lo, V. (2016). Synthesis report on experiences with ecosystem-based approaches to climate change adaptation and disaster risk reduction. Technical Series No.85. Secretariat of the Convention on Biological Diversity, Montreal, 106 pages. Retrieved from https://www.cbd.int/doc/publications/cbd-ts-85-en.pdf

Lund, D. H. (2018). Co-creation in urban governance: From inclusion to innovation. *Scandinavian Journal of Public Administration, 22*, 3–17.

Mahmoud, I. H., & Morello, E. (2018). Co-Creation Pathway as a catalyst for implementing Nature-based Solutions in Urban Regeneration Strategies Learning from CLEVER Cities framework and Milano as test-bed. Urbanistica Informazioni., 278(Special issue), 204–210. https://re.public.polimi.it/retrieve/handle/11311/1079106/348151/2018_Mahmoud-Morello_XI INU_sessione n3.pdf

Mahmoud, I. H., & Morello, E. (2020). Are Nature-based solutions the answer to urban sustainability dilemma? The case of CLEVER Cities CALs within the Milanese urban context. Atti Della XXII Conferenza Nazionale SIU. L'Urbanistica Italiana Di Fronte All'Agenda 2030. Portare Territori e Comunità Sulla Strada Della Sostenibilità e Della Resilienza, 1322–1327. http://media.planum.bedita.net/78/be/Atti_XXII_Conferenza_Nazionale_SIU_Matera-Bari_WORKSHOP_3.1_-Planum_Publisher_2020.pdf

Mahmoud, I. H., & Morello, E. (2021). Co-creation Pathway for Urban Nature-Based Solutions: Testing a Shared-Governance Approach in Three Cities and Nine Action Labs. In A. Bisello et al. (Ed.), Smart and Sustainable Planning for Cities and Regions (pp. 259–276). Springer International Publishing. https://doi.org/10.1007/978-3-030-57764-3

Mahmoud, I. H., Morello, E., Ludlow, D., & Salvia, G. (2021). Co-creation pathways to inform shared governance of urban living labs in practice: Lessons from three European projects. *Frontiers in Sustainable Cities, 3*, 1–17. https://doi.org/10.3389/frsc.2021.690458.

Mahmoud, I. H., Morello, E., Lemes de Oliveira, F., & Geneletti, D. (2022). Nature-based Solutions for Sustainable Urban Planning (1st ed.; I. H. Mahmoud, E. Morello, F. Lemes de Oliveira, & D. Geneletti, eds.). https://doi.org/10.1007/978-3-030-89525-9

Malmberg, K., Vaittinen, I., Evans, P., Schuurman, D., Ståhlbröst, A., & Vervoort, K. (2017). Living Lab Methodology Handbook. Zenodo. https://doi.org/10.5281/zenodo.1146321

Menny, M., Voytenko Palgan, Y., & McCormick, K. (2018). Urban living labs and the role of users in co-creation. *Gaia, 27*(68–77), 10.14512/gaia.27.S1.14.

Morello, E., Mahmoud, I., & Colaninno, N. (2019). *Catalogue of nature-based solutions for urban regeneration*. http://www.labsimurb.polimi.it/nbs-catalogue/

Morello, E., Mahmoud, I., Gulyurtlu, S., et al. (2018). CLEVER Cities Guidance on co-creating nature-based solutions: PART I – Defining the co-creation framework and stakeholder engagement. *Deliverable, 1*(1), 5.

Osieyo, M. A. (2020). Building a new relationship between people and nature. Retrieved from https://wwfeu.awsassets.panda.org/downloads/nature_in_all_goals_2020.pdf

Pauleit, S., Andersson, E., Anton, B., et al. (2019). Urban green infrastructure – Connecting people and nature for sustainable cities. *Urban Forestry & Urban Greening, 40*, 1–3. https://doi.org/10.1016/j.ufug.2019.04.007.

Schmidt, S., Guerrero, P., & Albert, C. (2022). Advancing sustainable development goals with localised nature-based solutions: Opportunity spaces in the Lahn river landscape, Germany. *Journal of Environmental Management, 309*, 114696. https://doi.org/10.1016/j.jenvman.2022.114696.

Schuthof, R., Kuhn, S., Morrow, R., & Kotler, A. (2019). *15 pathways to localise the sustainable development goals*. ICLEI.

Seddon, N., Chausson, A., Berry, P., et al. (2020). Understanding the value and limits of nature-based solutions to climate change and other global challenges. *Philosophical Transactions of the Royal Society B, 375*, 20190120. https://doi.org/10.1098/rstb.2019.0120.

Somarakis, G., Stagakis, S., & Chrysoulakis, N. (2019). *ThinkNature nature-based solutions handbook*. ThinkNature project funded by the EU Horizon 2020 research and innovation programme under grant agreement.

Tosun, J., & Leininger, J. (2017). Governing the interlinkages between the sustainable development goals: Approaches to attain policy integration. *Global Challenges, 1*, 1700036. https://doi.org/10.1002/gch2.201700036.

Trejo-Nieto A. (2021) An Overview of the Relationship of the Sustainable Development Goals and Urban and Regional Development. In: Brears R. (eds) The Palgrave Encyclopedia of Urban and Regional Futures. Palgrave Macmillan, Cham. https://doi.org/10.1007/978-3-030-51812-7_57-1

Vasseur, L., Horning, D., Thornbush, M., et al. (2017). Complex problems and unchallenged solutions: Bringing ecosystem governance to the forefront of the UN sustainable development goals. *Ambio, 46*, 731–742. https://doi.org/10.1007/s13280-017-0918-6.

von Wirth, T., Fuenfschilling, L., Frantzeskaki, N., & Coenen, L. (2019). Impacts of urban living labs on sustainability transitions: Mechanisms and strategies for systemic change through experimentation. *European Planning Studies, 27*, 229–257. https://doi.org/10.1080/09654313.2018.1504895.

Voorberg, W. H., Bekkers, V. J. J. M., & Tummers, L. G. (2015). A systematic review of co-creation and co-production: Embarking on the social innovation journey. *Public Management Review, 17*, 1333–1357. https://doi.org/10.1080/14719037.2014.930505.

Voytenko, Y., McCormick, K., Evans, J., & Schliwa, G. (2016). Urban living labs for sustainability and low carbon cities in Europe: Towards a research agenda.

Journal of Cleaner Production, 123, 45–54. https://doi.org/10.1016/j.jclepro.2015.08.053.
Wendling, L. A., Huovila, A., zu Castell-Rüdenhausen, M., et al. (2018). Benchmarking nature-based solution and smart city assessment schemes against the sustainable development goal indicator framework. *Frontiers in Environmental Science, 6*, 69. https://doi.org/10.3389/fenvs.2018.00069.
Xie, L., & Bulkeley, H. (2020). Nature-based solutions for urban biodiversity governance. *Environmental Science & Policy, 110*, 77–87. https://doi.org/10.1016/j.envsci.2020.04.002.
Zingraff-Hamed, A., Hüesker, F., Albert, C., et al. (2020). Governance models for nature-based solutions: Seventeen cases from Germany. *Ambio, 50*(8), 1610–1627. https://doi.org/10.1007/s13280-020-01412-x.

Low Earth Orbit

► New Orbital Urbanization

Low-Carbon Transport

Policies to Encourage Cycling in Sprawling Cities

Weichang Kong and Dorina Pojani
The University of Queensland, Brisbane, QLD, Australia

Introduction

Urban sprawl has been extremely detrimental for cities (Brueckner 2000). It has led to an excessive reliance on cars for travel (Rubiera Morollón 2015), with knock-on effects on congestion, health, energy consumption, and pollution (Habibi and Asadi 2011). With a climate breakdown looming, policy makers everywhere are now seeking and implementing strategies to reverse sprawl and automobility. To this end, many places have prioritized cycling alongside walking, ride-hailing, and public transport (Ogilvie et al. 2011). Cycling is expected to provide environmental, economic, and social benefits to both cyclists and society (Handy et al. 2014). However, the rates of utilitarian cycling (as opposed to recreational cycling) remain abysmal in most places (Dixon et al. 2018).

Common natural environment barriers to cycling include frosty, scorching, and/or muggy weather, precipitation (rain and snow), and a hilly topography (An et al. 2019; Bean et al. 2021; Lee and Pojani 2019). In sprawling cities, large distances between destinations present a built environment barrier (An et al. 2019). In addition, insufficient and/or low-quality designated infrastructure undermines cycling – especially among novice cyclists and more safety-conscious persons such as women and the elderly. This is understandable given that cyclists are among the most vulnerable road users, and severe injuries and/or fatalities can occur in a collision between a moving bicycle and a car (Deffner et al. 2012; Jacobsen and Rutter 2012). In many places, car drivers lack experience in interacting with cyclists (Deffner et al. 2012) and, at least in some contexts, are known to behave aggressively towards cyclists (Johnson et al. 2014).

Some sociocultural factors are at play too. For example, in very status-conscious settings, the wealthy shun eco modes in favor of luxury cars. Here, cycling is seen as an activity for children or for the poor (Ashmore et al. 2018; Daley and Rissel 2011; Li et al. 2019). In some settings, the "mamil" image ("middle-aged men in lycra," meaning men dressed in body-hugging spandex clothing who ride expensive racing bicycles at high speeds) is problematic. Mamil are seen as arrogant or irresponsible, and as an impediment to both drivers and pedestrians. As such, they do not help promote the status of cycling in society.

Problems can also lie with the institutions in charge of land-use and transport planning. Where these are uncoordinated, overly bureaucratic, or simply unfriendly to cycling, adequate cycling guidelines and infrastructure are typically missing (ECMT 2004). Earmarked funding is rarely provided, and cycling is constantly at risk of being shortchanged in favor of other modes, which are perceived as more deserving (Pojani et al. 2018).

Only a handful of larger Northern European and East Asian cities have managed to achieve high proportions of cycling for transport. Also, in a few smaller university towns or working-class enclaves, cycling is normalized as part of the everyday culture (Aldred and Jungnickel 2014). The cycling policies and programs which have been learned from these places are reviewed below. The review is structured in accordance with a conceptual framework formulated by Methorst et al. (2010) and Harms et al. (2015), which considers three types of strategies: "hardware" (concrete infrastructure and vehicles), "software" (policies such as education and communication), and "orgware" (institutional settings).

Strategies to Encourage Cycling

Hardware

Build Segregated Bicycle Path Networks and Bicycle-Friendly Intersections

CROW (2017) – the most influential design manual for bicycle traffic – outlines five design principles for successful cycling infrastructure: *cohesion*, *directness*, *safety*, *comfort*, and *attractiveness*. Segregated cycling paths, which separate bicycles from vehicular traffic through physical barriers, are considered as the best way to ensure the safety of cyclists along busy traffic roads (Cohen 2013). Where possible, cycling-only streets (or cycling highways) can be created. Cycling networks must be fully integrated and interconnected within cities (Pucher and Buehler 2007). As the crucial (and most accident-prone) links of the cycling network, intersections must also be carefully designed to reduce the risks of collisions and achieve seamlessness (Monsere et al. 2020). So far, the Netherlands has the most extensive urban network of segregated cycling paths in the world (37,000 kilometers), relative to the population (Dutch Cycling Embassy 2021).

Design Weather-Proof Cycling Infrastructure

Weather is an important determinant for bicycle ridership; on the other hand, cycling can make a significant contribution to reducing global warming and other deteriorating weather conditions (An et al. 2019). Planning cycling infrastructure must take weather into consideration (An et al. 2019). In regions where summers are very hot, weather-proof cycling infrastructure involves covered paths, tree-shaded paths, and even artificially cooled paths. In cold winter conditions, heated paths are an option.

While the concept of weather-proofing cycling infrastructure is relatively new, several countries have already implemented it as a part of their cycling programs. For example, the Netherlands has built a heated cycling path that connects two neighboring cities (Wageningen and Arnhem) to keep snow and ice away from the street surface (Boffey 2018). Singapore has been creating covered and shaded cycling and pedestrian paths (Urban Redevelopment Authority and Land Transport Authority 2018), and Qatar has proposed a 35 km cooled cycling path in Doha, powered by solar energy (Naparstek 2006).

Launch Bikesharing Schemes

Bikesharing schemes – publicly or privately owned and operated – have become popular worldwide (Bremmer 2018; Ma et al. 2020; Pojani et al. 2020). They offer users the convenience of cycling without the costs, space requirements, and maintenance responsibilities associated with owning a bicycle (Shaheen et al. 2010). Shared bicycles can be used for full door-to-door trips or, in countries where bicycle ownership is high, they can be used to complete the first/last mile of public transit trips. During the COVID-19 pandemic, bikesharing systems have been more resilient than public transportation (Teixeira and Lopes 2020). The most recent schemes are dockless, and the membership cards are integrated with public transport passes.

Encourage the Uptake of Electric Bicycles

E-bikes and power-assisted bicycles have grown in popularity as the technology has become more affordable to a mass of urbanites (de HaaS et al. 2021). Compared to conventional push bicycles, electric bicycles can travel faster and require less physical effort. These features make them more suitable for longer distances (e.g., in sprawling cities) and hilly and hot settings. Also, they allow less fit persons to ride with ease.

European countries have adopted a number of cost-effective programs to encourage the uptake of e-bikes and power-assisted bicycles, including free hires and purchase subsidies. These have been offered by municipalities, companies, or large nonprofit institutions such as universities. To name a few examples, in Eindhoven (the Netherlands), a program called "try a pedelec" allowed customers to hire an e-bike at no cost for a set period before purchasing. In Totnes (the UK), a neighborhood-scale e-bike hire trial provided 10 e-bikes to be circulated among community members. And in Vorärlberg (Austria), e-bike purchases were subsidized between 2009 and 2011 as part of the "Landrad" project (Cairns et al. 2017).

Provide Secure and Weather-Proof Bicycle Parking

The provision of secure, comfortable, and covered bicycle parking is essential to cyclists (Pucher and Buehler 2008b), and it is often associated with more bicycle commuting (Buehler 2012). In addition to serving cyclists, the provision of bicycle parking facilities minimizes community complaints prompted by randomly scattered bicycles (Pucher and Buehler 2008b). Well-designed, simple-to-use, and easy-to-find bicycle parking facilities should be provided not only in apartment buildings and dormitories but also near transit centers, workplaces, and shopping malls. The physical form of bicycle parking can vary from simple racks in the open air to guarded storage boxes in enclosed garages, depending on the location and users. But in all cases, sufficient surveillance – by staff or cameras – is required to reduce bicycle theft (Van der Spek and Scheltema 2015). Reasonable parking fees can offset the cost of security.

Apply Traffic Calming in Residential Neighborhoods

On residential streets, where traffic levels are low and/or fully segregated bicycle paths are unfeasible, traffic calming can be adopted instead (Pucher and Buehler 2008a). Traffic calming involves road design and traffic regulation measures such as speed limits (in some cases as low as "footpace"), traffic circles, speed bumps, chicanes, pavement markings, narrow roadways, and cycling-friendly signage (Furth 2012). Evaluation studies have also shown that traffic calming can reduce cycling collisions and injuries (Pucher and Buehler 2008a). In the Netherlands, the "woonerf" (or "living street") concept is a successful example of integrating multiple traffic calming measures so that residential streets can be safely shared by motorists, cyclists, and pedestrians (YOURS 2012).

Integrate Cycling with Other Transport Modes

Integrating cycling with other modes, in particular public transport, is crucial to achieving seamless, multimodal trips (Veryard and Perkins 2018). To this end, cities should prioritize cycling routes leading to transit stations (Jonkeren et al. 2021). Stations should be equipped with bicycle parking facilities, a range of bicycle services (such as repair or wash shops), and changing rooms for cyclists (Kager and Harms 2017). Bicycles should be accommodated on buses, trams, and metros – for example, in designated spaces (e.g., rail carriages or bus front racks) or time periods (ECMT 2004). It is also useful to stimulate the purchase of folding bicycles which take up less space on transit – through programs similar to those delineated above to encourage the uptake of e-bikes and power-assisted bicycles. For commuters who do not own bicycles, an option is the provision of rental or shared bicycles at transit stations (Kager and Harms 2017).

Software

Improve the Image of Cycling

Normalizing cycling as a mainstream (rather than niche) activity is a key challenge. Cities must work to improve the image of both cycling and cyclists. This requires overcoming a variety of pejorative stereotypes. Social marketing campaigns could make a significant contribution by breaking psychological barriers (Daley and Rissel 2011). They should aim at promoting the positive aspects of cycling, which connect to people's core values (e.g., fitness, wellbeing, sustainability) and inspire positive visions of low-energy, zero carbon cities. The costs and benefits of different travel options should be publicized so that residents can reevaluate their travel behaviors based on transparent information. It is also useful to talk about climate change so that urban dwellers come to regard it as an immediate risk that requires action in the form of travel modification (APS 2017). A range of different activities, displays, and events – ranging from billboards, posters, and flyers to concerts, exhibitions, and workshops – should be designed to target specific socioeconomic and age groups (Deffner et al. 2012). Among youth, cycling should be framed as a trendy activity.

Provide Training Programs for Novice Cyclists

Cyclists need to possess a certain level of skill in order to cycle more efficiently, confidently, and safely on urban roads. Training courses should be offered for people who want to cycle but are afraid of accidents. Courses can provide information around traffic rules and regulations for all modes but also include group rides which can help participants overcome the initial fear of venturing on the roads on a bicycle. Importantly, courses are more effective when they target specific neighborhoods or demographic groups. For example, a course in Belgium aims to encourage daily cycling for health among the elderly; a course in the Netherlands focuses on teaching migrant women how to cycle as a way to integrate in Dutch society; and in many countries "safe routes to school" programs target children (ECMT 2004).

Offer Road Safety Education Courses for Drivers

In contemporary sprawling cities, drivers need to learn to cooperate with bicycles as well as an increasing number of other micromobility modes, such as electric scooters and monowheels (Yang et al. 2020). To create a safe traffic environment for all, drivers' education in sharing the road with others, beginning at the licensing stage, is crucial (Deffner et al. 2012). Seemingly minor details, such as techniques to safely change lanes, turn, or open car doors when exiting a parked car, can make a big difference in terms of cycling safety (Johnson et al. 2013).

Orgware

Define the Roles of Different Levels of Governments

The roles and responsibilities for cycling as transport should be clearly defined at every level of government. Cycling should be treated the same as other vital infrastructure, such as water, sanitation, and telecommunication. A long-term national vision and commitment to cycling is key to success. It is important to establish a well-integrated cycling policy framework at a national level, which can guide and coordinate the implementation of cycling projects at the regional and/or local level (ECMT 2004).

Adopt Pro-cycling Legislation and Regulation

National governments should firmly ground cycling in the transport policy framework and set aside funds earmarked specifically for cycling (ECMT 2004). These funds need to be made transparent (Pojani et al. 2018). The United Nations Environment Programme recommends that, from now on, 20% of transport funding be spent on cycling (UNEP 2016). Meanwhile, bicycle-related facilities – lanes, parking lots, intersections, and signage – should be standardized in the national regulations to ensure equity between cyclists and other road users (ECMT 2004). Pro-cycling legislation should address safety concerns without overburdening cyclists. For example, laws that require cyclists to wear helmets – adopted in Finland, Sweden, and Australia (LeBlanc et al. 2002) – are counterproductive and lead to reductions in the number of cyclists (Clarke 2012). On the other hand, Dutch laws which presume that the driver is at fault and bears the burden of proof in case of a car–bicycle collision empower cyclists and lead to much more careful driving (Cycling in the Netherlands n.d.).

Encourage Broad Public Participation

Public involvement in the decision-making process regarding cycling can have a significant impact on outcomes (Gil et al. 2011). Planning processes should not be dominated by either cycling-averse drivers or highly experienced cyclists. Policy makers and transport planners should make an effort to include (existing and potential) cyclists of all ages, genders, and abilities and be open to listening to their concerns, experiences, and suggestions. Planners should use their judgment in handling NOMS (Not On My Street), an attitude which refers to residents and/or business owners who are pro-cycling in theory but in practice oppose the installment of bicycle paths in their vicinity for fear of losing on-street parking (Butterworth and Pojani 2018).

Conclusion

While most cities – especially sprawling ones – are far behind in making cycling a centerpiece of their transport systems, a variety of strategies are available which can be adopted. No single strategy can work in isolation. For example, building a cycling path that only connects two destinations or launching a bikesharing scheme in a place that lacks all cycling infrastructure will not help. While infrastructure has a crucial role to play, it needs to be complemented by education and awareness-raising campaigns, as well as pro-cycling legislation and budgeting. In sum, a combination of "hardware," "software," and "orgware" will be needed to promote cycling in sprawling contemporary cities.

The *hardware* includes building segregated bicycle networks and bicycle-friendly intersections, designing weather-proof cycling infrastructure, launching bikesharing schemes, encouraging the uptake of electric bicycles, providing secure and weather-proof bicycle parking, applying traffic-calming in residential neighborhoods, and integrating cycling with other transport modes. The *software* comprises improving the image of cycling, providing training programs for novice cyclists, and offering road safety education courses for drivers. Finally, the *orgware* involves defining the roles of different levels of governments, adopting pro-cycling legislation and regulation, and encouraging broad public participation. The ultimate goal is to make cycling "irresistible" while making cars less convenient (Pucher and Buehler 2008b).

References

Aldred, R., & Jungnickel, K. (2014). Why culture matters for transport policy: The case of cycling in the UK. *Journal of Transport Geography, 34*, 78–87. https://doi.org/10.1016/j.jtrangeo.2013.11.004.

An, R., Zahnow, R., Pojani, D., & Corcoran, J. (2019). Weather and cycling in New York: The case of Citibike. *Journal of Transport Geography, 77*, 97–112. https://doi.org/10.1016/j.jtrangeo.2019.04.016.

Ashmore, D. P., Pojani, D., Thoreau, R., Christie, N., & Tyler, N. A. (2018). The symbolism of "eco cars" across national cultures: Potential implications for policy formulation and transfer. *Transportation Research Part D: Transport and Environment, 63*, 560–575. https://doi.org/10.1016/j.trd.2018.06.024.

Australian Psychological Society (APS). (2017). Climate change empowerment handbook. https://psychology.org.au/getmedia/88ee1716-2604-44ce-b87a-ca0408dfaa12/climate-change-empowerment-handbook.pdf

Bean, R., Pojani, D., & Corcoran, J. (2021). How does weather affect bikeshare use? A comparative analysis of forty cities across climate zones. *Journal of Transport Geography, 95*, 103155. https://doi.org/10.1016/j.jtrangeo.2021.103155.

Boffey, D. (2018, April 10). Europe's longest heated cycle path to connect Dutch cities. The Guardian. https://www.theguardian.com/world/2018/apr/10/europes-longest-heated-cycle-path-to-connect-dutch-cities

Bremmer, D. (2018). Public transport bicycle achieves new record: 4 million journeys in a year. https://www.ad.nl/binnenland/ov-fiets-haalt-nieuw-record-4-miljoen-ritten-in-een-jaar~a42932bb/?referrer=https%3A%2F%2Fen.wikipedia.org%2F

Brueckner, J. K. (2000). Urban Sprawl: Diagnosis and remedies. *International Regional Science Review, 23*(2), 160–171. https://doi.org/10.1177/016001700761012710.

Buehler, R. (2012). Determinants of bicycle commuting in the Washington, DC region: The role of bicycle parking, cyclist showers, and free car parking at work. *Transportation Research Part D: Transport and Environment, 17*(7), 525–531. https://doi.org/10.1016/j.trd.2012.06.003.

Butterworth, E., & Pojani, D. (2018). Why isn't Australia a cycling mecca? *European Transport, 69*(69).

Cairns, S., Behrendt, F., Raffo, D., Beaumont, C., & Kiefer, C. (2017). Electrically-assisted bikes: Potential impacts on travel behaviour. *Transportation Research Part A: Policy and Practice, 103*, 327–342. https://doi.org/10.1016/j.tra.2017.03.007.

Clarke, C. F. (2012). Evaluation of New Zealand's bicycle helmet law. *New Zealand Medical Journal, 125*(1349).

Cohen, E. (2013). Segregated bike lanes are safest for cyclists. *Canadian Medical Association Journal, 185*(10), E443–E444. https://doi.org/10.1503/cmaj.109-4468.

CROW. (2017). *Design manual for bicycle traffic*. The Netherlands: CROW (The National Information and Technology Platform for Transport, Infrastructure and Public Space).

Cycling in the Netherlands. (n.d.). In Wikipedia. https://en.wikipedia.org/wiki/Cycling_in_the_Netherlands

Daley, M., & Rissel, C. (2011). Perspectives and images of cycling as a barrier or facilitator of cycling. *Transport Policy, 18*(1), 211–216. https://doi.org/10.1016/j.tranpol.2010.08.004.

de Haas, M., Kroesen, M., Chorus, C., Hoogendoorn-Lanser, S., & Hoogendoorn, S. (2021). E-bike user groups and substitution effects: Evidence from longitudinal travel data in the Netherlands. *Transportation, 1-26*. https://doi.org/10.1007/s11116-021-10195-3.

Deffner, J., Hefter, T., Rudolph, C., & Ziel, T., (Eds). (2012). Handbook on cycling inclusive planning and promotion. Capacity development material for the multiplier training within the mobile2020 project. https://ec.europa.eu/transport/sites/default/files/cycling-guidance/mobile_2020_more_biking_in_small_and_medium_sized_towns_of_central_and_eastern_europe_by_2020.pdf

Dixon, S., Irshad, H., Pankratz, D. M., & Bornstein, J. (2018). The Deloitte City mobility index. *Deloitte Insights*. https://www2.deloitte.com/content/dam/Deloitte/cn/Documents/consumer-business/deloitte-cn-consumer-city-mobility-index-en-180613.pdf

Dutch Cycling Embassy. (2021). Best practices Dutch cycling. https://www.dutchcycling.nl/downloads/DCE%20Best%20Practices%20Dutch%20Cycling.pdf

European Conference of Ministers of Transport (ECMT). (2004). Implementing sustainable urban travel policies: Moving ahead. https://doi.org/10.1787/9789282103296-en.

Furth, P. G. (2012). Bicycling infrastructure for mass cycling: A transatlantic comparison. In J. Pucher & R. Buehler (Eds.), *City cycling* (pp. 105–140). Cambridge, MA: MIT Press. https://doi.org/10.7551/mitpress/9434.003.0009.

Gil, A., Calado, H., & Bentz, J. (2011). Public participation in municipal transport planning processes – The case of the sustainable mobility plan of Ponta Delgada, Azores, Portugal. *Journal of Transport Geography, 19*(6), 1309–1319. https://doi.org/10.1016/j.jtrangeo.2011.06.010.

Habibi, S., & Asadi, N. (2011). Causes, results and methods of controlling urban sprawl. *Procedia Engineering, 21*, 133–141. https://doi.org/10.1016/j.proeng.2011.11.1996.

Handy, S., van Wee, B., & Kroesen, M. (2014). Promoting cycling for transport: Research needs and challenges. *Transport Reviews, 34*(1), 4–24. https://doi.org/10.1080/01441647.2013.860204.

Harms, L., Bertolini, L., & Brömmelstroet, M. T. (2015). Performance of municipal cycling policies in medium-sized cities in the Netherlands since 2000. *Transport Reviews, 36*(1), 134–162. https://doi.org/10.1080/01441647.2015.1059380.

Jacobsen, P. L., & Rutter, H. (2012). Cycling safety. In J. Pucher & R. Buehler (Eds.), *City cycling* (pp. 141–156). Cambridge, MA: MIT Press. https://doi.org/10.7551/mitpress/9434.003.0010.

Johnson, M., Newstead, S., Oxley, J., & Charlton, J. (2013). Cyclists and open vehicle doors: Crash characteristics and risk factors. *Safety Science, 59*, 135–140. https://doi.org/10.1016/j.ssci.2013.04.010.

Johnson, M., Oxley, J., Newstead, S., & Charlton, J. (2014). Safety in numbers? Investigating Australian driver behaviour, knowledge and attitudes towards cyclists. *Accident Analysis & Prevention, 70*, 148–154. https://doi.org/10.1016/j.aap.2014.02.010.

Jonkeren, O., Kager, R., Harms, L., & te Brömmelstroet, M. (2021). The bicycle-train travellers in the Netherlands: Personal profiles and travel choices. *Transportation, 48*(1), 455–476. https://doi.org/10.1007/s11116-019-10061-3.

Kager, R., & Harms, L. (2017). Synergies from improved cycling-transit integration: Towards an integrated urban mobility system. *International Transport Forum Discussion Papers*. https://doi.org/10.1787/ce404b2e-en.

LeBlanc, J. C., Beattie, T. L., & Culligan, C. (2002). Effect of legislation on the use of bicycle helmets. *CMAJ, 166*(5), 592–595.

Lee, Q. Y., & Pojani, D. (2019). Making cycling irresistible in tropical climates? Views from Singapore. *Policy Design and Practice, 2*(4), 359–369. https://doi.org/10.1080/25741292.2019.1665857.

Li, X., Chen, H., Shi, Y., & Shi, F. (2019). Transportation equity in China: Does commuting time matter? *Sustainability, 11*(21), 5884. https://doi.org/10.3390/su11215884.

Ma, X., Yuan, Y., Van Oort, N., & Hoogendoorn, S. (2020). Bike-sharing systems' impact on modal shift: A case study in Delft, the Netherlands. *Journal of Cleaner Production, 259*, 120846. https://doi.org/10.1016/j.jclepro.2020.120846.

Methorst, R., Monterde, I., Bort, H., Risser, R., Sauter, D., Tight, M., & Walker, J. (Eds.). (2010). *COST 358 pedestrians' quality needs, PQN final report*. Cheltenham: Walk21.

Monsere, C. M., McNeil, N. W., & Sanders, R. L. (2020). User-rated comfort and preference of separated bike lane intersection designs. *Transportation Research Record: Journal of the Transportation Research Board, 2674*(9), 216–229. https://doi.org/10.1177/0361198120927694.

Naparstek, A. (2006). Mist-cooled bike paths being built in Qatar. StreetsBlog. https://nyc.streetsblog.org/2006/08/10/mist-cooled-bike-paths-being-built-in-qatar/

Ogilvie, D., Bull, F., Powell, J., Cooper, A. R., Brand, C., & Mutrie, N. (2011). An applied ecological framework for evaluating infrastructure to promote walking and cycling: The iConnect study. *American Journal of Public Health, 101*(3), 473–481. https://doi.org/10.2105/ajph.2010.198002.

Pojani, D., Kimpton, A., Corcoran, J., & Sipe, N. (2018). Cycling and walking are short-changed when it comes to transport funding in Australia. The Conversation. https://theconversation.com/cycling-and-walking-are-short-changed-when-it-comes-to-transport-funding-in-australia-92574

Pojani, D., Chen, J., Mateo-Babiano, I., Bean, R., & Corcoran, J. (2020). Docked and dockless public bike-sharing schemes: Research, practice and discourse. In C. Curtis (Ed.), *Handbook of sustainable transport* (pp. 129–138). https://doi.org/10.4337/9781789900477.00025.

Pucher, J., & Buehler, R. (2007). At the frontiers of cycling: Policy innovations in the Netherlands, Denmark, and Germany. Monograph. *World Transport Policy & Practice, 13*(3), 8–56.

Pucher, J., & Buehler, R. (2008a). Cycling for everyone. *Transportation Research Record: Journal of the Transportation Research Board, 2074*(1), 58–65. https://doi.org/10.3141/2074-08.

Pucher, J., & Buehler, R. (2008b). Making cycling irresistible: Lessons from the Netherlands, Denmark and Germany. *Transport Reviews, 28*(4), 495–528. https://doi.org/10.1080/01441640701806612.

Rubiera Morollón, F., González Marroquin, V. M., & Pérez Rivero, J. L. (2015). Urban sprawl in Spain: Differences among cities and causes. *European Planning Studies, 24*(1), 207–226. https://doi.org/10.1080/09654313.2015.1080230.

Shaheen, S. A., Guzman, S., & Zhang, H. (2010). Bikesharing in Europe, the Americas, and Asia. *Transportation Research Record: Journal of the Transportation Research Board, 2143*(1), 159–167. https://doi.org/10.3141/2143-20.

Teixeira, J. F., & Lopes, M. (2020). The link between bike sharing and subway use during the COVID-19 pandemic: The case-study of New York's Citi bike. *Transportation Research Interdisciplinary Perspectives, 6*, 100166. https://doi.org/10.1016/j.trip.2020.100166.

United Nations Environment Programme (UNEP). (2016). Global outlook on walking and cycling. https://www.unep.org/resources/report/share-road-global-outlook-walking-and-cycling-october-2016

Urban Redevelopment Authority and Land Transport Authority. (2018). Walking and cycling: design guide. https://www.ura.gov.sg/Corporate/Guidelines/Active-Mobility/-/media/

Van der Spek, S. C., & Scheltema, N. (2015). The importance of bicycle parking management. *Research in Transportation Business & Management, 15*, 39–49. https://doi.org/10.1016/j.rtbm.2015.03.001.

Veryard, D., & Perkins, S. (2018). Integrating urban public transport systems and cycling. *ITF Roundtable Reports*. https://doi.org/10.1787/bd177112-en.

Yang, H., Ma, Q., Wang, Z., Cai, Q., Xie, K., & Yang, D. (2020). Safety of micro-mobility: Analysis of E-scooter crashes by mining news reports. *Accident Analysis & Prevention, 143*, 105608. https://doi.org/10.1016/j.aap.2020.105608.

Youth For Road Safety (YOURS). (2012). The Dutch 'Woonerf' – An example of safe road spaces. http://www.youthforroadsafety.org/news-blog/news-blog-item/t/the_dutch_woonerf_an_example_of_safe_road_spaces

M

Mainstreaming Blue Green Infrastructure in Cities: Barriers, Blind Spots, and Facilitators

Hayley Henderson[1], Judy Bush[2] and Daniel Kozak[3]
[1]Research Fellow at Crawford School of Public Policy, Australian National University, Canberra, ACT, Australia
[2]Lecturer in Urban Planning at University of Melbourne, Melbourne, VIC, Australia
[3]Professor at Universidad de Buenos Aires, Researcher at Consejo Nacional de Investigaciones Científicas y Técnicas (CONICET), Buenos Aires, Argentina

Introduction

As the world becomes increasingly urbanized, the integration of Blue Green Infrastructure (BGI) into cities and towns is gaining focus and importance. In essence, BGI can help address complex urban problems, including a range of urban environmental issues, as well as contributing to residents' health and well-being and providing biodiversity habitat. Despite its growing theoretical currency, take-up of BGI in practice has faced difficulties associated with shifting away from business as usual. This entry offers a broad overview of the barriers, blind spots, and facilitators for mainstreaming BGI in cities.

The entry draws on case study research conducted in cities from both the Global South and Global North, as well as from a review of literature on BGI and related initiatives, like Nature-based Solutions and Ecosystem Adaptation, in cities. It commences by providing a description of the theoretical and applied genealogy of BGI and some of its recognized characteristics. Then, the entry focuses on the unique qualities required to mainstream BGI in urban water and green space management as an innately dispersed and decentralized approach. Here, and in line with an ethic of decolonization, we seek to draw special attention to the traditional, nonlinear and plural approaches to managing water and ecosystems in cities occurring around the world. We underscore the importance of framing this concept in history and the different trajectories of human–nature relations, as well as the importance of encouraging context-specific approaches and knowledge exchange on equal terms between regions and cultures. Then, we provide a brief overview of some frameworks utilized to evaluate capacity for and delivery of NbS, as well as a detailed examination of cross-cutting barriers, blind spots, and facilitators for mainstreaming BGI.

R. C. Brears (ed.), *The Palgrave Encyclopedia of Urban and Regional Futures*,
https://doi.org/10.1007/978-3-030-87745-3

Background: Conceptual and Applied Genealogy of Blue Green Infrastructure

The idea behind the concept of BGI points to the recognition of the innate capacities of green space and water, and the ecosystems in which they are immersed, to produce environmental benefits and to enhance the quality of life of people living in cities. In contrast to the historical and conventional management of stormwater, with an emphasis on gray infrastructure (We refer to *gray infrastructure* as conventional rainwater infrastructure, generally underground, centralized, and impermeable.), BGI responds both to a demand to improve environmental quality in cities and to the limitations of traditional solutions, by taking advantage of the geomorphic features of natural systems (Kozak et al. 2020).

The term "Blue Green Infrastructure" first appeared in the English-speaking world in the 2000s (Lamond and Everett 2019: 1), around the same time that the concept of *Trame Verte et Bleue* (TVB) was coined in France, stressing the protection, consolidation, and production of biodiversity corridors along the watercourses and green spaces that cross cities and metropolitan regions (Vimal et al. 2012). Put simply, BGI is a network of vegetation, watercourses, and bodies of water that provide ecosystem services. These services include carbon sequestration; the moderation of the Urban Heat Island effect, and the regulation of temperature in general; improvements in air quality through the use of the phytoremediation capacity of urban vegetation; noise reduction; restitution or establishment of biodiversity corridors; and greater control in the management of both quantity and quality of stormwater runoff; among many other benefits for human populations including increased access to green and blue recreational spaces and improved health outcomes.

Some of the components of BGI are parks, nature reserves, green corridors, rivers, streams, lagoons, wetlands, bio-retention reservoirs, and floodable parks. They also include simple traditional public works, such as tree-lined boulevards and gardens, to more sophisticated, but still low-tech interventions, such as vegetated depressions designed to capture and filter rainwater (Tayouga et al. 2016: 2–3), green swales, and other bio-infiltration devices (Zellner et al. 2016: 116–117). They can be of public or restricted access, and they can be located on public or private lands. One of the most widespread definitions of BGI describes it as a:

> Strategically planned network of natural and semi-natural areas with other environmental features designed and managed to deliver a wide range of ecosystem services. (JNCC 2019: 5) (In earlier documents by the European Commission, the same definition is used for the concept of Green Infrastructure: http://ec.europa.eu/environment/nature/ecosystems/index_en.htm)

More recently, the concept of "Nature-based Solutions" (NbS) has been proposed. NbS is an umbrella term for ecosystem-based approaches to addressing a range of societal challenges. The definition for NbS offered by the European Commission highlights the necessity for cost-effectiveness in NbS implementation:

> Solutions that are inspired and supported by nature, which are *cost-effective*, simultaneously provide environmental, social and *economic* benefits and help build resilience. Such solutions bring more, and more diverse, nature and natural features and processes into cities, landscapes and seascapes, through locally adapted, resource-efficient and systemic interventions (emphasis added). (European Commission 2020: 5)

BGI is often understood as one of the central approaches to implementing NbS, and the two terms are often used interchangeably. In fact, a range of terms in addition to BGI and NbS are used interchangeably by practitioners and researchers, including "ecological infrastructure," "soft engineering," "living architecture," and "integrated urban water management" (Moosavi et al. 2021). Among this diversity of terminology, we utilize BGI in this entry due to its familiarity and adoption broadly among both researchers and practitioners including in non-European contexts. Furthermore, it is strategically important to talk about "infrastructure" when positioning BGI mainstreaming practices and debate in the context of traditional infrastructure.

The inclusion of the economic dimension at the core of the definition can be problematic – in a similar way to which the idea of "economic sustainability" can be problematic when it is merely understood as "economic viability" (Kozak and Romanello 2012: 14–15) – but it has the advantage of raising the issue of the quantification and monetization of NbS as well as the fundamental requirement of cost-effectiveness when promoting its implementation as public policy. Sometimes BGI can be costly to implement and often is more cost effective over the long-term. This is one of the areas where progress is most needed, however, because the multiple co-benefits of NbS (and BGI) and long timeframes for delivering co-benefits, like increasing biodiversity, require relatively complex evaluations compared to traditional gray infrastructure, which seeks to deliver singular or fewer benefits. This can be one of the factors that causes resistance to mainstreaming NbS (and BGI), as discussed in Sect. 4 of this entry about the barriers and blindspots facing BGI mainstreaming.

The study of BGI, NbS, and ecosystem-based approaches is currently one of the most fertile fields of inquiry in relation to the search to maximize synergies between ecosystem health and human well-being (OECD 2020). There is also a steady increase in political commitment in the international arena and across financing opportunities, which both reflect growing urgency associated with climate change and increasing knowledge about addressing complex urban problems (FEBA 2017). Nevertheless, despite the growing interest in these approaches at a global level – supported by a solid academic consensus based on its multiple benefits – recent studies coincide in pointing out that their implementation has been limited so far (OECD 2020; Kapos et al. 2019; Browder et al. 2019). Predominantly, BGI and NbS still do not represent the priority option for national governments, local authorities, or the private sector in the allocation of resources and the planning of public policies and budgets, which are all still dominated by the logic of conventional gray infrastructure (OECD 2020).

Conceptualizing "Mainstreaming" for BGI

In order to analyze the barriers, blind spots, and facilitators for mainstreaming BGI it is first necessary to frame what "mainstreaming" means in this entry. BGI will be recognizable as a mainstream service when it is discussed and implemented on equal terms with conventional gray infrastructure. This requires reversing the current default; it involves the idea that BGI can offer effective solutions to many complex urban problems, like flooding in high density areas or within vulnerable communities, and improved solutions to others, like water quality. We argue that while both gray infrastructure and BGI will continue to be important components of the total green space and urban water management system, mainstreaming BGI denotes that it will occupy a central role as a structurally significant solution, with gray infrastructure as a complement in some scenarios (e.g., to remove excess flood waters during extreme events).

There are different approaches to achieve this mainstreaming scenario. Some authors on mainstreaming stress the importance of internalizing environmental goals into other sectoral policy areas to create synergy effects and cost-savings (e.g., Cowling et al. 2008; Runhaar et al. 2018). Furthermore, theoretical frameworks have been developed and applied to test pathways to mainstreaming different environmental objectives, from urban ecology to BGI such as integrated urban stormwater management. For example, a transitions approach was used to study the trajectory of sustainable urban stormwater management in Melbourne, Australia from niche to regime practices (e.g., Rogers et al. 2015). A number of these studies demonstrate that there is an ongoing important role for gray infrastructure, though BGI can become a substantial element of future infrastructure systems that deliver enhanced ecosystem services and many co-benefits, for example, in relation to health outcomes for vulnerable groups like children (Kabisch et al. 2017).

While evidence offers hope for mainstreaming possibilities, there are challenges involved. Also, there is a significant difference between what it meant to scale-up gray infrastructure in the past as a centralized and structural solution for urban water management relative to defining a path for mainstreaming BGI in the future. We identify three key factors to consider. First, it is necessary to account for its multifunctional purpose and many co-benefits. Second, framing increased use of BGI requires a shift to dispersed and decentralized management. Lastly, given the diversity of contexts, it is important to adopt diversity in types of interventions and in this regard, there are lessons to be learnt from different approaches between regions around the world.

In terms of multifunctionality and co-benefits, BGI presents new challenges for developing business cases that are competitive in contexts with traditional metrics because rather than isolating a single problem and designing a uni-functional solution (e.g., flooding resolved with larger drains), BGI often seeks to address multiple problems – from delivering air and water quality improvements to flood management and open space provision, which can occur at different times and over different scales. In this regard, it is important that regulatory frameworks and project decision-making are geared to accommodate this paradigm change.

Second, BGI as a structural solution is comprised of multiple, dispersed interventions that are ultimately connected in a network for multiplier effects. Part of this systems approach requires new, sustainable interventions, but also removing harmful elements of the system, like unnecessary impermeabe surfaces, and avoiding developments that worsen outcomes, like unnecessary land clearing or soil works in sensitive catchments. In this shift away from centralized water management to planning for total system and water cycle benefits, the dispersed interventions also require a decentralized management approach (Bettini et al. 2015). On the one hand, this requires a framework for planning and evaluating mainstreaming efforts that move beyond short-term fixes to consider long-term benefits of synergies within ecosystems, including building resilience and biodiversity outcomes. On the other hand, mainstreaming BGI requires governance arrangements that engage diverse stakeholders. This presents a suite of unique barriers and facilitators, which are discussed in the next sections.

Thirdly, given that a BGI approach requires locally specific responses to environmental and urban conditions, there are many more alternative solutions that can be considered, which adds a layer of complexity to design, planning, monitoring and evaluation of projects. For example, while rainwater gardens can be a common feature of BGI, their size, design, vegetation, and materials can differ greatly. For this, it is important to upscale the learning about alternative experiences around the world and to ensure BGI designs respond to local context.

In particular, mainstreaming BGI at a global scale requires decentering Global North to broaden knowledge about plural and nonlinear approaches adopted in the Global South. At present, high-quality evidence and academic literature about experiences and advances in BGI exists on Global North cities, frequently those like Copenhagen, Singapore, or Washington, D.C., which appear highest in sustainability rankings like the Siemans Green City Index and Arcadis's Sustainability Index (e.g., Brears 2018). However, in a search of nearly 50,000 publications on urban ecology, Shackleton (2021) found that only 31% included the Global South. While the Global North focus in current studies produces new knowledge vital to advancing sustainability objectives in cities, most of the world's population lives in the Global South where qualities about cities, their inhabitants and modes of governance differ from the Global North. These factors, which are often tied to colonial legacies, need to occupy an elevated position in research to mainstream BGI at a global scale. Currently, not only are there fewer studies published on the Global South, but often the findings do not seem to penetrate debate in the Global North (Watson 2009). Overcoming legacies of colonialism that have included the transfer of urban planning and governance modes from the Global North to the Global South, requires that the particular characteristics of green space and urban water management in Global South cities take center stage in developing new conceptual insights.

Specifically, when conceptualizing the mainstreaming of BGI there are a range of factors that, while present in some Global North cities, are particularly prevalent and problematic for cities of the Global South. First, urbanization has historically been directed toward coasts as a result of centuries of colonialism (and resource extraction), which exposes urban populations in the Global South to sea level rise and greater risks associated with flooding (Myers 2021; Shackleton 2021). Secondly, another legacy is segregation in many cities, where the poor have been relegated to live in parts of cities that have fewer environmental services – such as parks – and that are at higher risk of environmental change and quality problems, including industrial pollution or landslides (Dobbs et al. 2021). In this regard, mainstreaming BGI requires a focus on risk reduction in Global South settings. Thirdly, invasive species were introduced and continue to afflict ecosystems in the Global South.

Lastly, there are enduring impacts on the urban policymaking and governance settings that require additional attention. In particular, informality is an important dynamic of urbanization in the Global South, and it exists across policy domains, from housing to employment. Diversity in approaches to managing urban water is another central feature of cities, from top-down interventions funded by multilateral organizations to smaller, bottom-up approaches that harness local social networks and activism, particularly in informal settlements where networked infrastructure is often absent. While also true in other parts of the world, traditional customary appoaches to governance may be more common. Overall, in many Global South contexts there is likely to be greater reliance on community-based approaches, which are orientated to survival where poverty is entrenched; here the importance of these efforts and their interconnections is central to mainstreaming BGI. Finally, colonialism shaped urban governance arrangements with norms and laws that too often are taken for granted and tend to project Western urban forms and the exclusion of community participation, especially active resistance. Some of these factors are contemplated below in the exploration of barriers, blind spots, and facilitators to mainstreaming BGI, though we suggest that there is significant scope to review relevant existing findings and to study further the plural interventions that highlight BGI potential in the Global South.

Barriers and Blind Spots to Mainstreaming Blue Green Infrastructure

Despite growing evidence about the efficacy of BGI to deliver sustainable urban water management solutions, there are barriers and blind spots to implementation. There are biophysical challenges, such as the type of soil or slope of the territory (Sarabi et al. 2020), though overwhelming research suggests that the main difficulties to mainstreaming BGI are political and institutional (Bettini et al. 2015; Brown 2005; Croeser et al. 2021; Sarabi et al. 2020; Uittenbroek 2016). Ramping up the delivery of BGI necessitates politico-institutional dynamics that can comprehend and be adaptive to evolving, complex human–nature systems. Existing barriers are "rooted in social and organizational cultures, practices, and processes, and hence are difficult to overcome" (Dhakal and Chevalier 2017: 172). The following Table 1 summarizes the main barriers and blind spots uncovered in our literature review and case study work, which are grouped across the following categories: technical, financial, regulatory, prioritization and leadership, governance, and coordination.

Limitations in Technical Capacities

The first barrier relates to technical capacity within institutions responsible for planning, designing, implementing, maintaining, and monitoring BGI. Among existing professionals, there is less experience and knowledge about newer BGI technologies relative to more traditional approaches (Bettini et al. 2015; Qiao et al. 2018)

Mainstreaming Blue Green Infrastructure in Cities: Barriers, Blind Spots, and Facilitators, Table 1 Summary of barriers for mainstreaming BGI in cities

Technical	Shortage of professionals in general (e.g., hydraulic engineers) Limited awareness and expertise related to BGI (newer technology). Difficulties in knowledge transfer between places due to diverse local conditions Lack of interdisciplinary and "boundary spanning" skills necessary to design and evaluate the multifunctionality of BGI Challenges in measuring long-term performance and co-benefits, for example, biodiversity outcomes Dominance of conventional engineering perspectives and preference for tried approaches Longer timeframes for planning BGI given multifunctionality and multi-actor involvement from the design stage through to maintenance and monitoring
Financial	Difficulties in developing competitive business cases (monetizing multiple, system-wide, and long-term benefits, including difficult-to-value benefits) Institutional funding for stormwater and urban greening tends to sit with municipalities and is limited relative to higher levels of government Major impediments are current contracting models: public works and maintenance are often centralized and combined, relying on singular interventions as opposed to BGI with distinct elements and more localized maintenance requirements Current funding and financing models tend to focus on single or a few objectives. There are barriers to expanding these approaches to incorporate environmental goals together with other funding priorities (i.e., shelter, food, health, basic sanitation infrastructure)
Regulatory	Norms and laws can prohibit BGI or create unreasonable liability for governments and private landowners or developers Lack of global design or engineering standards for BGI Conflicting mandates or confusing provisions, e.g., approvals, processes, or unclear authority Lack of regulatory interaction between policy areas BGI absent from some aligned policy areas, such as health or housing Vertical integration problems, where conflicts or inconsistencies may exist between municipal, provincial, and national regulations Strong regulatory measures may exist but may not be implemented to their full potential
Leadership and prioritization	Sustainability outcomes can be overshadowed by or considered contrary to resolving other urgent concerns (e.g., building housing or roads) Hesitancy to support BGI as a novel or perceived "luxury" approach when there are traditional, short-term fixes to pressing problems, like flooding Disconnect between short-term actions in short-term political cycles and long-term sustainability planning and partnerships for BGI Financial constraints open opportunities for lobbying and private interests, diminishing public sector support for green issues Power of lobbying generally on leadership and interests conflict with sustainability objectives
Governance and coordination	Horizontal integration barriers within governments, including fragmentation or overlap of relevant roles, goals, timeframes, and responsibilities between divisions. Sectoral language differences can compound these functional challenges Vertical coordination difficulties exist between levels of government responsible for the management of some bioregions and ecosystems Incongruent biophysical and jurisdictional boundaries Multi-jurisdictional management faces challenges in sourcing funding (e.g., for specific basin committees or resource authorities) and in defining leadership and responsibilities, from design through to maintenance In dual and three-tiered systems of government there are power imbalances and difficulties in working toward policy coherency, especially when they are administered by different political parties or coalitions Challenges to engaging nongovernment stakeholders from early design stages through to delivery and maintenance. To date there is also a lack of recognition of customary approaches and essential direct actions of citizens and social movements already underway that constitute BGI, especially in the GS Knowledge about the benefits and co-benefits of BGI is not generally not widespread in the community. In fact, there can be perceived negative outcomes from the interaction between humans and nature (e.g. associated with floods, water-borne diseases, and contamination), especially in GS

and in many cases, especially in the Global South, there is a shortage of professionals in general. One of the challenges associated with designing and delivering BGI is that it requires nuanced attention to local conditions and knowledge and experiences from one place may not be applicable to others (Qiao et al. 2018). Given the multi-functional character of BGI, design and delivery work require interdisciplinary skills and "trans-boundary actors skilled in speaking the language of different groups" (Sarabi et al. 2020) that include participatory approaches to engage stake-holders and system thinking about ecosystem-wide approaches. Furthermore, given relatively new BGI approaches and that they differ greatly between contexts, there are generally more uncertainties about their performance within a wider system (catchment) (Drosou et al. 2019) compared to more common and accepted gray infrastructure. In this regard, and compared to traditional approaches, like piped stormwater drainage systems, there are difficulties in sourcing information about, measuring, and adequately recognizing the multiple co-benefits over the long-term, from environmental (e.g., biodiversity, climate change adaptation) and social (e.g., health outcomes in vulnerable groups) to economic measures (e.g., green jobs, new maintenance businesses).

In addition to the aforementioned technical issues, studies have also identified "technocratic path dependencies" (Brown and Farrelly 2009) associated with the dominant, gray infrastructure approach which prioritizes engineering knowledge. Engineering professionals are central to gray and BGI design, for example in terms of flood protection and water quality. However, where a gray infrastructure approach seeks to pigeonhole water issues, BGI requires greater complexity in systems thinking and multi-actor governance, thereby necessitating collaboration and power-sharing among disciplines like environmental officers, community engagement professionals, urban planners, and hydraulic engineers. Given these complexities, often longer timeframes are needed in technical planning work, which falls short of expectations given gray infrastructure planning precedents. There are also transaction costs (Mekala and Hatton MacDonald 2018) associated with this work, for example in searching for, organizing, processing, and analyzing data.

Often in the Global South, while information may exist it can be poorly integrated for decision-making, for example on climate trends (e.g., see case study from Argentina by Janches et al. 2014). This is especially true in Global South contexts where pressures associated with high rates of urbanization, informality and vulnerability place limitations on governance capacity and limit opportunities for experimentation over tried and tested gray infrastructure. In this regard, risk aversion and a preference for continuing to apply tried approaches permeates the private sector in both the GS and GN as well where uncertainties can raise concerns about potential financial losses (Davies and Lafortezza 2019; Kabisch et al. 2016).

Financial Impediments

While studies have shown that BGI can offer more cost effective or equally affordable solutions for urban water management compared to traditional gray infrastructure, BGI faces greater funding constraints (Dhakal and Chevalier 2017). In part, the limitations on funding BGI are related to its incipient stage of development. Business cases are harder to develop for this new approach, especially when BGI seeks to produce multiple benefits over the long term, which are not traditionally measured in gray infrastructure proposals. Presently, direct provisioning services tend to be measured while other ecosystem services and co-benefits (e.g., cultural, recreational, health) are not considered, there is a lack of indicators for monitoring, particularly "intangible" benefits, or there is a lack of data on associated costs and outcomes, which increases the risk profile of BGI for decision-makers both in the public and private sectors (Raymond et al. 2017). In this regard, project decision-making is set on a path of tested solutions, so that innovative proposals find it difficult to permeate, let alone change decision-making criteria.

Furthermore, in many places institutional funding for stormwater management is at the local level, which is limited and, while some large water projects are funded by provincial and national level governments, often these funds are dedicated to gray infrastructure projects without significant consideration of sustainability objectives. In a similar vein, general revenue is often dedicated to specific needs like housing and food policy, especially in the Global South, and again, these proposals often don't contemplate complementary sustainability outcomes that could be achieved by incorporating BGI. In fact, in some Global South contexts, studies have found that limited funding "is the primary challenge in adopting BGI concepts" (Drosou et al. 2019). This challenge flows through to contracting models and processes, which tend to combine public works and maintenance and which tend to be preformulated for centralized and single-purpose traditional infrastructure approaches.

The underlying impediment is that traditional gray infrastructure investments in cities are usually designed with a single objective and that financing and funding are designed for this. For example, in the case of the culverted stormwater system, the objective is to remove excess rainwater as quickly and as far as possible from the city. Consequently, the evaluation by means of cost-benefit analysis is usually relatively simple. The cost is mainly that of its construction and maintenance, and the benefit is to what extent this one-dimensional objective is achieved. On the contrary, in the case of BGI, due to the inherent characteristic of producing multiple socio-environmental benefits, the evaluation is necessarily more complex. BGI responses to urban drainage management, for example, seek to replicate the natural mechanisms of absorption, retention, and expansion, with the aim of solving stormwater drainage closer to the site of origin. In other words, they seek to recover the natural interaction of the watercourse bed with its natural geoformations, typically these include floodplains and their associated forms (Kozak et al. 2021). Unlike traditional hydraulic-engineering, which historically focused on the volume of water to be displaced, the objective of BGI is not only to address the problem of the quantity of water, but also to prioritize its quality, along with its potential to generate quality public space, urban amenities, and ecosystem services. For this reason, it is essential to carry out sophisticated analyses, where the indirect benefits or co-benefits produced are assessed, as well as quantified and monetized when possible.

Investment in BGI in the management of the rainwater system, following the above example, not only generates benefits in terms of reducing flood risk, but also produces a multiplicity of socio-environmental co-benefits that must be considered for its comparison with conventional gray infrastructure to be accurate. In other words, it is essential that the indicators used to assess NbS also include those that seek to measure its co-benefits, for example, in terms of healthcare savings as a result of a decrease in respiratory diseases due to a decrease in air pollution because of the phytoremedial action of the vegetation involved in BGI projects. Figure 1 schematically shows the monetized costs and benefits of a NbS project in Valencia, Spain, which involves a new green corridor that connects existing green areas and offers a new path for nonmotorized mobility under the shade of green pergolas (GrowGreen 2019). The economic benefits identified include: gray water treatment costs and irrigation costs with drinking water avoided, improvements in air quality, carbon sequestration, among others.

Figure 2 below shows the results of another project, including the quantification of some of the same co-benefits identified in Fig. 1. The name of the project is *Corredores Verdes* (Green Corridors) and it was implemented with 36 distinct interventions by the Government of Medellín in Colombia between 2016 and 2019, within the framework of Horizon 2020's Urban GreenUP (2019). The response of the project to two climate change scenarios was evaluated: the first RCP 2.6 (increase in global mean temperature at 2100 between 0.9 °C and 2.3 °C); and the second RCP 8.5 (increase in global mean temperature at 2100 between 3.2 °C and 5.4 °C). The second one is the business-as-usual scenario. In Scenario 1, it was estimated that *Corredores Verdes* would contribute to 49.5 fewer days above the temperature threshold set as risky. Consequently, it would avoid 513 deaths per year and 33,919 for the period 2020–2030 if this action were projected to 12% of Medellín. In economic terms, the impact of this single action would be equivalent to USD 153 million for avoided healthcare costs and USD 10,166 million for the 2020–2030 period. In Scenario 2, the reduction in temperature

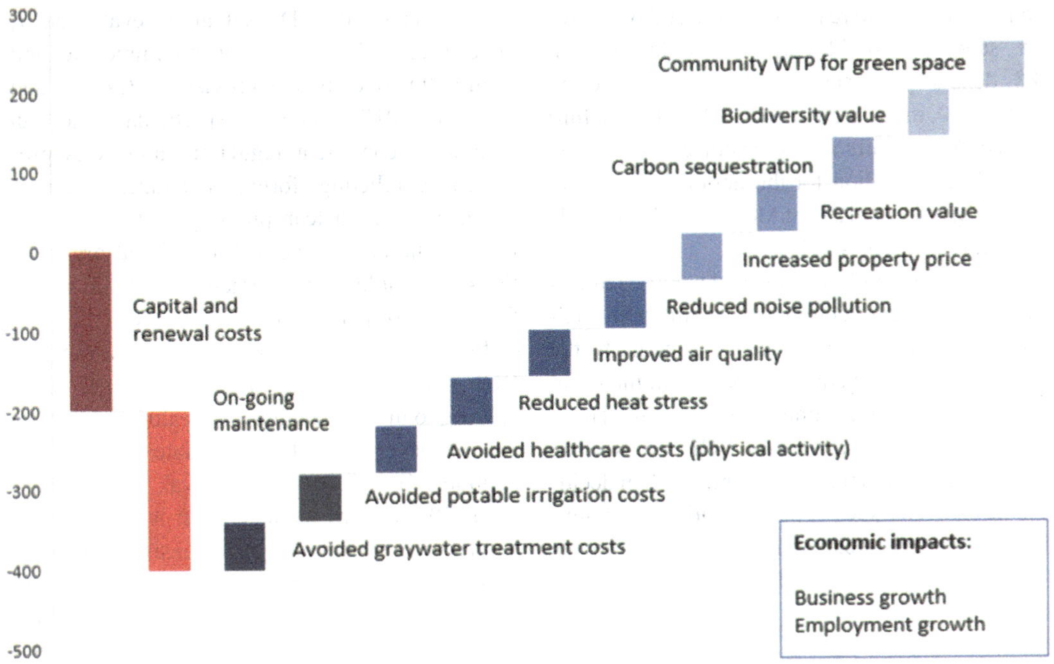

Mainstreaming Blue Green Infrastructure in Cities: Barriers, Blind Spots, and Facilitators, Fig. 1 Illustration of costs and multiple benefits of NbS project in Valencia, Spain. (Source: GrowGreen 2019, quoted in Whiteoak 2020: 208)

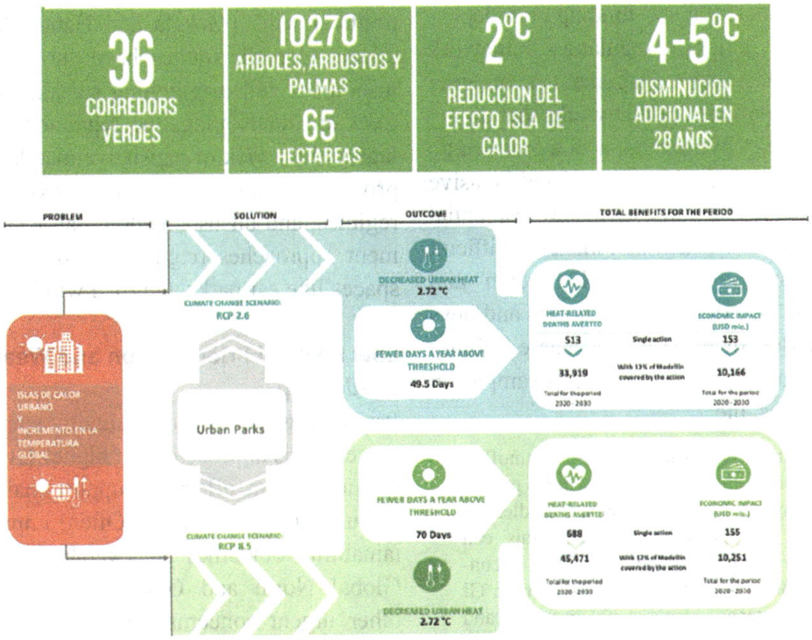

Mainstreaming Blue Green Infrastructure in Cities: Barriers, Blind Spots, and Facilitators, Fig. 2 Quantification of co-benefits derived from the program "Corredores Verdes" in Medellín. (Sources: "Estrategia de renaturalización para Medellín," Alcaldía de Medellín 2019a: 29; "Plan de Acción Climática Medellīn 2020–2050," Alcaldía de Medellín 2019b: 145; https://www.plataformaarquitectura.cl/cl/921605/medellin-crea-30-corredores-verdes-para-mitigar-el-calentamiento-urbano)

would contribute to reaching 70 fewer days above the same threshold, which would prevent 688 deaths per year (45,471 for the period 2020–2030, in the projection to 12% of Medellín) and USD 155 million for health costs avoided (USD 10,251 million for the period 2020–2030, in the projection to 12% of Medellín) (Alcaldía de Medellín 2019b: 144–145).

Carrying out multidimensional analyses of this type undoubtedly represents a significant challenge. It is much more complex to quantify the indirect benefits generated by NbS than the usual ones from traditional gray infrastructure. However, in order to mainstream BGI and NbS so they are considered the default option when deciding the characteristics of public works, it is essential to remove financial and funding impediments for ease of public and share management approaches.

Inadequate Regulatory Framework (Not BGI Friendly)

In chorus with the abovementioned barriers associated with technocratic and funding path dependencies, in many cities the regulatory framework that guides urban water management favors gray infrastructure solutions or other works that do not deliver improved environmental outcomes (e.g., the promotion and use of introduced and invasive species for ornamental purposes in new residential developments) to an extent that makes it difficult to promote mainstreaming BGI (Brown and Farrelly 2009). In extreme cases, norms and laws can prohibit BGI or place unreasonable liability on the developer or landowner. For example, in the United States, the:

> Common Enemy Rule regards stormwater runoff as a common enemy and allows a landowner to protect his/her land by any means necessary, regardless of the possible consequences to others...this can encourage hydrologic disruptions, which is contrary to the concept of GI. As discussed earlier, GI regards stormwater as a resource, not an enemy, and utilizes it on-site by reestablishing the disrupted hydrology. (Dhakal and Chevalier 2017)

In other circumstances, it is uncommon to find design or engineering standards for BGI either as standalone frameworks or within existing policy (Croeser et al. 2021; Dhakal and Chevalier 2017; Qiao et al. 2018). In fact, it is common to find "conflicting or confusing provisions" (Dhakal and Chevalier 2017). Some of the particular issues can include "inconsistent regulatory approvals processes, conflicting formal mandates amongst organisations, unclear property rights, and the lack of authority/power of operational organisations to implement alternatives" (Brown and Farrelly 2009). Evidence also suggests that the different regulatory frameworks relevant to BGI – from stormwater management and landscaping to urban planning and health – lack interaction, have conflicting mandates or do not include BGI solutions (Qiao et al. 2018; Sarabi et al. 2020). Vertical integration is also a challenge, where conflicts or inconsistencies may exist between municipal, provincial, and national levels (Sarabi et al. 2020). In other places, for example in many Australian cities, strong regulatory measures exist to underpin BGI; however "they are not used to their full potential... The failings may be occurring as a result of the discretionary nature, enforcement mechanisms and political will" (Mekala and Hatton MacDonald 2018). Overall, studies have shown that while discourses and overarching sustainability objectives are developing, the regulatory framework lags and the lack of legislative mandates tends to privilege path dependency, existing policy regimes, and unsustainable urban water management approaches (e.g., maximizing impervious spaces like car parks and road widths).

Low Political Prioritization and Weak Leadership

Prioritizing BGI means assigning special importance to this approach from higher levels of government in an effort to support mainstreaming throughout organizations. Often, improving sustainability outcomes are overshadowed in both Global North and Global South contexts by other urgent concerns, for example relating to housing needs or road building, which are seen as singular objectives without exploring sustainability synergies. As Brown succinctly stated, "stormwater is a low political priority" (Brown 2005), with the exception of crisis moments when the typically invisible management of this resource is overwhelmed by floods that harm people and property In a detailed case study conducted in Indonesia, researchers also found that public servants were faced with prioritization challenges whereby BGI was conceived as a luxury approach when pressure was high to address basic needs (Drosou et al. 2019). In their study, respondents stated that BGI was viewed as "buying healthy organic rice" when "people are starving and are penniless," which also led to the statement "Do not mention green infrastructure, when we cannot run the grey infrastructure" (Drosou et al. 2019). While the theory is not to offer a luxury approach, the multifunctionality and long-term results of BGI are harder to convey and appreciate alongside quick fix or traditional solutions to specific problems in cities, like park landscaping or stormwater drainage. This kind of disconnect between short-term actions and long-term sustainability goals is common in many contexts and one of the major blind spots to be addressed (Kabisch et al. 2016; Sarabi et al. 2020).

Across the Global North and South, researchers have consistently found that "political support is lacking" and that leadership on advancing a BGI agenda can be obstructive or inadequate (Croeser et al. 2021). That political leaders and policy decision-makers can be hesitant to support BGI is because of some of the previously mentioned factors, including that it is a relatively new approach, there is some uncertainty about measuring the costs and benefits, and because often the benefits are reaped long term (Sarabi et al. 2020). Short-term fixes are more palatable in contexts with high turnover in administrations and their leadership. Often, long-term visions with measurable indictors are lacking (Brown and Farrelly 2009). In a study of Polish cities, Kronenberg found that financial constraints create asymmetries in decision-making that tend to favor private development interests and creates a lack of public sector support for green issues (Kronenberg 2015). Other researchers have also highlighted the power of lobbying on leadership where interests conflict with sustainability objectives (Runhaar et al. 2018) and run against a sense of urgency in adopting BGI, despite their potential to address a range of urban challenges from improved health outcomes and food security to vulnerability risk reduction and biodiversity conservation.

Governance and Coordination Challenges

There are multiple and overlapping governance challenges relating to the coordination of stakeholders necessarily involved in BGI design and delivery. Firstly, it is common to uncover problems to horizontal integration of policy areas within governments. Horizontal integration refers to the material and communicational linkages that are made between government divisions. For operational reasons, governments departments are separated into different areas of service provision, for example from roads and rail to water supply and stormwater management. Some of the blind spots for mainstreaming BGI relate to subtle differences in the ways responsibility are defined and shared among departments, as well as areas of overlap between these sectoral "silos" (Brown 2005), which may be disguised behind sectoral language differences (Kabisch et al. 2016). In some country contexts, like Australia and Argentina, local governments are creations of provincial government and tend to operate as state or provincial implementation authorities. In these instances, vertical coordination framed around sector priorities can be stronger than horizontal, cross-departmental integration in the same level of government. The fragmentation of responsibilities within governments presents operational difficulties for mainstreaming BGI (Dhakal and Chevalier 2017) because BGI doesn't fit within a single sector but rather requires the input of multiple sectors to maximize its multifunctional purpose. For example, some of the challenges include the way BGI can be linked to the specific objectives and policy timeframes of sectors, which "operate on the basis of distinct visions, goals, legal structures and ways of thinking" (Sarabi et al. 2020).

Secondly, there are common challenges associated with vertically coordinating different levels of government as well as between neighboring jurisdictions of the same level. One of the innate challenges to managing urban water courses is that their biophysical and ecological systems often do not correspond with the jurisdictional boundaries of local or even single provincial governments (in many cases there are also international relations required). Therefore, mainstreaming BGI requires interjurisdictional governance arrangements that can coordinate the actions of different institutions. In these contexts, one of the first challenges is to define the leading or responsible authority, or a system of shared responsibility with good communication and collaboration processes. Unclear leadership is a common problem, which leads to unclear definition of responsibilities (Qiao et al. 2018). This problem is common in managing large urban river basins, where initiatives can become delayed for long periods of time while public actors negotiate their stakes and responsibilities, everything from plan design to post-implementation maintenance. For example, in Argentina a number of basin committees have been established to lead planning efforts; however they often don't have the mandate or funding to implement plans, leaving questions around which public authority is responsible for sector plans and tasks. This problem becomes acute when transitioned away from a predominantly gray infrastructure system to advance BGI as the number of stakeholders grows. In fact, some meta-studies of literature have found that "the most commonly identified impediment was the lack of a coordinated institutional framework" in sustainable urban water management (Brown and Farrelly 2009).

Furthermore, coordination between municipal governments and between different provincial or regional governments is not often promoted, for example coordinated approaches to planning are not budgeted for. Vertical coordination between levels of government can be challenged by power imbalances and differing priorities, for example in statutory reach or funding scope. For example, a local government may seek to implement an innovative BGI project, while higher authority with regulatory or funding power may have conflicting policy frameworks (Croeser et al. 2021). Challenges in vertical coordination are manifested in fragmented natural resource management and dispersal of resources between different sectors and levels of government (Drosou et al. 2019).

Thirdly, there are challenges to engaging non-government stakeholders in the BGI design and delivery process. As Dhakal and Chevalier (2017) so clearly concluded in their analysis of barriers for green infrastructure (GI) in the USA, current governance arrangements were:

> designed for and is adept at governing centralized gray infrastructure, therefore it inherently supports gray, not GI. As opposed to GI, which is a decentralized approach requiring the involvement of many stakeholders, the existing governance is centralized and exclusively technocratic (p.175).

The implementation gap associated with BGI can largely be linked to this governance gap that requires collaboration between new and different actors so as to design, implement, and maintain multifunctional solutions (Frantzeskaki et al. 2020). Research has shown that both private sector stakeholders and civil society actively involved in sustainability planning translates to increasing awareness, build trust and behavior change favorable to sustainability outcomes, including willingness to pay (Fünfgeld 2010; Qiao et al. 2018). This is important in supporting the paradigm shift away from gray infrastructure or to utilizing gray infrastructure as a complement in some scenarios (e.g., to remove excess flood waters during extreme events) toward mainstreaming BGI, which requires tangible contact and a relationship of care with blue and green spaces in the city, on private and public land. For example, private actors "can help to manage stormwater on-site and reduce stormwater runoff, and also to solve the challenges of land and funding shortages" (Qiao et al. 2018). Currently, research from different contexts from across the Global North and South like Argentina, Australia, Poland, and Indonesia show that the pressure exerted on public managers is to maintain unsustainable approaches, for example to cutting grass short to present manicured lawns or to keep streams culverted and out of sight, many times in an effort to reduce costs or perceived risks to private interests (Brown 2005; Drosou et al. 2019; Kronenberg 2015).

Not only are there challenges to engaging nongovernment stakeholders to effect a mainstreaming process of BGI policy-making and implementation, there are also challenges in recognizing the customary and other existing nongovernment advances in BGI within formal policy arrangements. The direct actions of citizens and organized movements are responsible for bringing about significant change, especially in some Global South contexts (Miraftab 2009) where governance arrangements can be weaker at including diverse voices and dissent. Here, bottom-up approaches are not only a viable alternative but can prove to be essential in addressing urban water problems like flooding or to managing green spaces and trees (Dobbs et al. 2021; Drosou et al. 2019).

Finally, there is a particular challenge when engaging communities and developers on BGI in terms of some gaps in knowledge about potential environmental services as well as a notion of environmental disservices. In this regard, mainstreaming BGI also confronts a barrier to acceptance because cultural and experiential factors lead some people to associate negative outcomes from the interaction between humans and nature. This is true across the world and is often associated with an increasing distance between humans and nature, so that small incursions of nature can be seen as problems (like grass emerging through cement cracks). However, this sentiment is especially common in the Global South, where higher exposure to environmental disservices like floods, seawater penetration of potable water reserves, water-borne diseases, insect plagues, and contamination is more frequent, especially for vulnerable populations. It is more common in Global South contexts for communities to seek the removal of environmental resources, like trees (Kronenberg 2015) or to drain streams so as to minimize maintenance needs and reduce risks. Studies have also shown that it is also true in Global North contexts that attitudes toward environmental resources including BGI can be affected by "cognitive factors" like fear or lack of awareness, for example about maintenance (Dhakal and Chevalier 2017). In both the Global South and North, perceptions about environmental resources like BGI are affected by values, traditions, and prior experiences in each context. In some circumstances, negative perceptions can create high levels of anxiety or fear, for example regarding possible flood events, the presence of animals or rubbish (Shackleton 2021).

Many of the aforementioned issues associated with technical, financial, regulatory, leadership, and governance constraints combine to entrench path dependencies within delivery organizations. "Institutional inertia" has been identified as responsible for the slow uptake of BGI and other NbS (Brown and Farrelly 2009). Delivery organizations are locked into historical patterns that favor unsustainable approaches to managing urban water and green spaces. While strong barriers exist to mainstreaming BGI within institutions, there are also signs of promise and factors that facilitate changes in patterns of decision-making. The following section provides an overview of the factors that facilitate the scaling up of BGI.

Facilitators

With increasing experience in the implementation of BGI, from small-scale pilots and experiments to more comprehensive and larger scale approaches, the key elements that address the barriers to BGI implementation and facilitate mainstreaming are increasingly being analyzed and identified in both research and practice. Facilitation factors mediate between internal (within organizations) and external (stakeholders and wider context) factors (Ibrahim et al. 2020). The key facilitators can be grouped into four broad categories: policy dimensions; knowledge sharing and collaboration; intermediaries and boundary organizations; and leadership and champions.

Policy Dimensions

Policy dimensions that facilitate mainstreaming include the presence of high level objectives for BGI in global and national frameworks. Examples of international treaties and commitments that support BGI implementation include the Sendai Framework for Disaster Risk Reduction (2015), Agenda 2030 and the Sustainable Development Goals (SDGs) (2015), the Paris Agreement of the UN Framework Convention on Climate Change (2015), and the New Urban Agenda (NUA) (2016). These international frameworks include both implicit and explicit goals and objectives for inclusion of BGI in cities. For example, Sustainable Development Goal 11 Sustainable Cities and Communities includes the target "By 2030, provide universal access to safe, inclusive and accessible, green and public spaces, in particular for women and children, older persons and persons with disabilities." Directives from both international, national and subnational levels can actively influence or even mandate BGI implementation at city or local government scales.

Knowledge Sharing and Collaboration

Cities are increasingly aware of the need to enhance resilience to climate impacts, including resilience of urban infrastructure, buildings, social structures, and institutions (Frantzeskaki et al. 2019). Frantzeskaki et al. (2020) highlighted the effectiveness of "targeted and tailored" capacity building programs for sharing knowledge and skills for implementation of nature-based solutions, as well as creating large-scale collaborative learning and research platforms. Knowledge sharing and collaboration can effectively address technical and skills deficiencies and barriers, particularly as pilot and experimental approaches to BGI implementation mature and the results of both implementation successes and failures can be shared. Both research and practitioner monitoring and reporting can contribute to these knowledge sharing roles.

Intermediaries and Boundary Organizations

Intermediaries and boundary-spanners or organizations can play key facilitation roles by linking across policy domains and between different levels of government to support both vertical and horizontal policy integration (Frantzeskaki et al. 2020; Herslund et al. 2018; Runhaar et al. 2018). Intermediary organizations, including nongovernment global networks and associations can also support knowledge sharing and skills development (Frantzeskaki et al. 2020). International nongovernment organizations acting as BGI intermediaries include ICLEI (e.g., Sustainable Urban Resilient Water for Africa or SUReWater4Africa), United Cities and Local Governments (UCLG), C40 network and the 100 Resilient Cities (100RC) network (funded by the Rockefeller Foundation). Likewise, there are several international research collaborations (e.g., CONEXUS, Naturvation) that are bringing together knowledge, experience, data, and case studies from a range of contexts spanning Global North and Global South. In research on the development of a metropolitan-scale urban forest strategy in Melbourne Australia, the role of intermediaries was highlighted (Frantzeskaki and Bush 2021). Intermediaries, both individuals as well as key organizations, played key roles in "linking between sectors, across different levels of government and between disciplines and policy domains" (Frantzeskaki and Bush 2021). Indeed, an "ecology of intermediaries" was identified, with a range of actors adopting a range of roles (that shifted across the strategy development process) to support metropolitan-scale planning and implementation (Frantzeskaki and Bush 2021).

Leadership, Champions

Strong leaders and influential champions can support and facilitate uptake of BGI approaches, and implementation of demonstration projects can effectively communicate and inspire wider interest and action (Bush 2020; Herslund et al. 2018). City planners can act as "policy entrepreneurs," creating "enabling space for innovation to scale up" (Frantzeskaki et al. 2020: 8). In this regard, policy entrepreneurs can facilitate horizontal mainstreaming between departments within an organization or level of government. Further, policy entrepreneurs may also be influential in utilizing key events, such as droughts, flooding, or other climate-related impacts, as "windows of opportunity" to prompt action (Runhaar et al. 2018). For example, Herslund et al. (2018) highlighted that with water shortages highlighted on a local community's agenda, linking BGI to water supply could motivate increased BGI implementation. It should be noted however that while such events may temporarily increase action or while individuals may generate increased interest in pursuing BGI because of their particular interest or education, "mainstreaming" requires continued and concerted efforts that need to be underpinned by institutionalized responses (Runhaar et al. 2018).

The four categories of facilitation often interact to create synergistic and strengthened responses (Ibrahim et al. 2020). Interactions and synergies include, for example, between knowledge and policy factors, with knowledge sharing on effective implementation underpinning development of evidence-based policy making (Frantzeskaki et al. 2020). Likewise, the intermediaries identified in Frantzeskaki and Bush's (2021) case study also contributed to knowledge sharing, collaboration, and to leadership and championing to build political support for BGI implementation. As implementation of BGI increases in a range of contexts, and addressing a range of urban sustainability challenges, there will be increased opportunities for learning from and replicating these approaches, including both the technical, as well as collaborative governance dimensions. This learning and replication is underpinned by effective knowledge sharing, communication and collaboration mechanisms from local to global scales.

Finally, it is important to note the scaling potential generated by pilots across these facilitating factors. Pilot projects can support relationship development that bridges organizational fragmentation as well as developing links to non-government actors, helping to raise awareness about BGI but also facilitating the co-construction of knowledge necessary for design and delivery. While partnerships can be important and the role of intermediaries can play a role, government is the key actor in driving pilot opportunities for novel BGI and through whole-of-project monitoring and evaluation, the public sector can provide risk-reducing information about the performance of BGI. There are multiple opportunities for BGI from small-scale initiatives in stormwater management to large-scale developments such as stadiums or social housing where BGI can be applied and showcased.

Conclusions

This entry provides an overview of the barriers, blind spots, and facilitators to mainstreaming Blue Green Infrastructure in cities. While significant opportunities exist to deliver multifaceted solutions to complex urban problems through BGI, the barriers to uptake are entrenched and revolve around politico-institutional trajectories and path dependencies that reproduce business-as-usual infrastructure solutions. This entry highlights five categories of barriers and blind spots relating to technical capacities, financial impediments, regulatory frameworks, leadership, and governance. Mounting pressure to build more resilient communities as well as incremental change occurring to deliver BGI with recognized co-benefits are opening up spaces of opportunity for BGI mainstreaming, for example where boundary-spanning policymakers build new linkages between fragmented areas of government or leaders mobilize regional and international networks to build political support for sustainability outcomes. However, change to date has been slow and there is great need to continue building the case for mainstreaming BGI. Strengthening the facilitating factors identified in this entry offers a pathway to support mainstreaming. Advancing a BGI agenda globally necessitates building and sharing knowledge between regions, in particular by centering the advances and learning provided from the experience of Global South cities where

M

most of the world's population live and where most BGI experiences go under-recognized.

Cross-References

- ▶ Blue-Green Cities: Achieving Urban Flood Resilience, Water Security, and Biodiversity
- ▶ Green Cities: Nature-Based Solutions, Renaturing and Rewilding Cities
- ▶ Integrated Urban Green and Grey Infrastructure
- ▶ Urban Greening and Green Gentrification

Acknowledgments We would like to acknowledge Prof. Sara Bice who originally planned to co-author this entry with us, though circumstances outside of her control meant that she was unable to contribute as an author. We greatly appreciate her contribution as a peer reviewer of our draft and support of this work.

Part of the research carried out in Argentina to inform this chapter received funding from the European Union's Horizon 2020 research and innovation programme under grant agreement no. 867564 for the CONEXUS project.

References

Alcaldía de Medellín. (2019a). *Estrategia de renaturalización para Medellín*. Informe Técnico. Available in: https://cdn.locomotive.works/sites/5ab410c8a2f42204838f797e/content_entry5ab410faa2f42204838f7990/5ad0b06574c4837def5d27e9/files/Climate_Action_Plan_Medellin.pdf?1618309681.

Alcaldía de Medellín. (2019b). *Plan de Acción Climática Medellín 2020–2050. Acción por el clima*. Informe Técnico. Available in: https://www.medellin.gov.co/irj/go/km/docs/etc/PortalMedellin/archivos/medio_ambiente/Publicaciones/Documentos/2021/PAC_Medellin_Libro_Digital.pdf.

Bettini, Y., Brown, R. R., de Haan, F. J., & Farrelly, M. (2015). Understanding institutional capacity for urban water transitions. *Technological Forecasting and Social Change, 94*, 65–79. https://doi.org/10.1016/j.techfore.2014.06.002.

Brears, R. (2018). *Blue and green cities: The role of blue-green infrastructure in managing Urban water resources*. London: Palgrave Macmillan.

Browder, G., Ozment, S., Rehberger Bescos, I., et al. (2019). *Integrating green and gray: Creating next generation infrastructure*. Washington, DC: World Bank and WRI.

Brown, R. R. (2005). Impediments to integrated urban stormwater management: The need for institutional reform. *Environmental Management, 36*(3), 455–468. https://doi.org/10.1007/s00267-004-0217-4.

Brown, R. R., & Farrelly, M. A. (2009). Delivering sustainable urban water management: A review of the hurdles we face. *Water Science and Technology, 59*(5), 839–846. https://doi.org/10.2166/wst.2009.028.

Bush, J. (2020). The role of local government greening policies in the transition towards nature-based cities. *Environmental Innovation and Societal Transitions, 35*, 35–44. https://doi.org/10.1016/j.eist.2020.01.015.

Cowling, R. M., Egoh, B., Knight, A. T., O'Farrell, P. J., Reyers, B., Rouget, M., . . . Wilhelm-Rechman, A. (2008). An operational model for mainstreaming ecosystem services for implementation. *Proceedings of the National Academy of Sciences, 105*(28), 9483–9488. https://doi.org/10.1073/pnas.0706559105%.

Croeser, T., Garrard, G. E., Thomas, F. M., Tran, T. D., Mell, I., Clement, S., . . . Bekessy, S. (2021). Diagnosing delivery capabilities on a large international nature-based solutions project. *npj Urban Sustainability, 1*(1), 32. https://doi.org/10.1038/s42949-021-00036-8.

Dhakal, K. P., & Chevalier, L. R. (2017). Managing urban stormwater for urban sustainability: Barriers and policy solutions for green infrastructure application. *Journal of Environmental Management, 203*, 171–181. https://doi.org/10.1016/j.jenvman.2017.07.065.

Dobbs, C., Vasquez, A., Olave, P., & Olave, M. (2021). Cultural Urban ecosystem services. In C. M. Shackleton, S. S. Cilliers, E. Davoren, & M. J. D. Toit (Eds.), *Urban ecology in the global south* (pp. 245–264). New York: Springer International Publishing.

Drosou, N., Soetanto, R., Hermawan, F., Chmutina, K., Bosher, L., & Hatmoko, J. U. D. (2019). Key factors influencing wider adoption of blue–green infrastructure in developing cities. *MDPI Water, 11*(6), 1234–1254.

European Commision. (2020). *Biodiversity and Nature-based Solutions. Analysis of EU-funded projects.* Report by Naumann, S. & Davis, M., Directorate-General for Research and Innovation. Avalaible in: https://op.europa.eu/en/publication-detail/-/publication/d7e8f4d4-c577-11ea-b3a4-01aa75ed71a1.

FEBA (Friends of Ecosystem-based Adaptation). (2017). *Hacer que la soluciones basadas en la naturaleza sea eficaz: un marco para definir criterios de cualificación y estándares de calidad*, GIZ, Bonn, Alemania, IIED, Londres, Reino Unido, y UICN, Gland, Suiza.

Frantzeskaki, N., & Bush, J. (2021). Governance of nature-based solutions through intermediaries for urban transitions – A case study from Melbourne, Australia. *Urban Forestry & Urban Greening, 64*, 127262. https://doi.org/10.1016/j.ufug.2021.127262.

Frantzeskaki, N., McPhearson, T., Collier, M. J., Kendal, D., Bulkeley, H., Dumitru, A., . . . Pintér, L. (2019). Nature-based solutions for Urban climate change adaptation: Linking science, policy, and practice communities for evidence-based decision-making. *BioScience, 69*(6), 455–466. https://doi.org/10.1093/biosci/biz042%.

Frantzeskaki, N., Vandergert, P., Connop, S., Schipper, K., Zwierzchowska, I., Collier, M., & Lodder, M. (2020). Examining the policy needs for implementing nature-based solutions in cities: Findings from city-wide transdisciplinary experiences in Glasgow (UK), Genk (Belgium) and Poznań (Poland). *Land Use Policy, 96*, 104688. https://doi.org/10.1016/j.landusepol.2020.104688.

Fünfgeld, H. (2010). Institutional challenges to climate risk management in cities. *Current Opinion in Environmental Sustainability, 2*(3), 156–160. https://doi.org/10.1016/j.cosust.2010.07.001.

GrowGreen. (2019). A partnership for greener cities to increase liveability, sustainability and business opportunities. Available in: http://growgreenproject.eu/

Herslund, L., Backhaus, A., Fryd, O., Jørgensen, G., Jensen, M. B., Limbumba, T. M., … Yeshitela, K. (2018). Conditions and opportunities for green infrastructure – Aiming for green, water-resilient cities in Addis Ababa and Dar es Salaam. *Landscape and Urban Planning, 180*, 319–327. https://doi.org/10.1016/j.landurbplan.2016.10.008.

Ibrahim, A., Bartsch, K., & Sharifi, E. (2020). Green infrastructure needs green governance: Lessons from Australia's largest integrated stormwater management project, the River Torrens Linear Park. *Journal of Cleaner Production, 261*, 121202. https://doi.org/10.1016/j.jclepro.2020.121202.

Janches, F., Henderson, H., & MacColman, L. (2014). Working Report: Urban Risk and Climate Change Adaptation in the Reconquista River Basin of Argentina. Lincoln Institute of Land Policy. Accessed 15.08.22 https://www.lincolninst.edu/es/publications/working-papers/urban-risk-climate-change-adaptation-reconquista-river-basin-argentina.

JNCC (Joint Nature Conservation Committee). (2019). *Roadmap towards a blue green infrastructure manual. Bridging the knowledge gap in the field of Blue Green Infrastructures*. Available in: https://data.jncc.gov.uk/data/354f40aa-1481-4b7f-a1eb-82c806893409/BGI-Manual-Report.pdf

Kabisch, N., Frantzeskaki, N., Pauleit, S., Naumann, S., Davis, M., Artmann, M., … Bonn, A. (2016). Nature-based solutions to climate change mitigation and adaptation in urban areas: Perspectives on indicators, knowledge gaps, barriers, and opportunities for action. *Ecology and Society, 21*(2). https://doi.org/10.5751/ES-08373-210239.

Kabisch, N., van den Bosch, M., & Lafortezza, R. (2017). The health benefits of nature-based solutions to urbanization challenges for children and the elderly – A systematic review. *Environmental Research, 159*, 362–373. https://doi.org/10.1016/j.envres.2017.08.004.

Kapos, V., et al. (2019). *The role of the natural environment in adaptation*. Rotterdam/Washington, DC: Background Paper for the Global Commission on Adaptation, Global Commission on Adaptation.

Kozak, D., & Romanello, L. (2012). *Sustentabilidad II: Criterios y normativas para la promoción de sustentabilidad urbana en la CABA*. Buenos Aires: Ediciones CPAU.

Kozak, D., Henderson, H., de Castro Mazarro, A., Rotbart, D., & Aradas, R. (2020). Blue-Green Infrastructure (BGI) in dense Urban watersheds. The case of the Medrano Stream Basin (MSB) in Buenos Aires. *Sustainability., 12*(6). https://doi.org/10.3390/su12062163.

Kozak, D., Henderson, A., Rotbart, D., & Aradas, R. (2021). Beneficios y desafíos en la implementación de Infraestructura Azul y Verde: una propuesta para la RMBA. In D. Zunino Singh, V. Gruschetsky, & M. Piglia (Eds.), *Pensar las infraestructuras en Latinoamérica* (pp. 223–244). Buenos Aires: Editorial Teseo.

Kronenberg, J. (2015). Why not to green a city? Institutional barriers to preserving urban ecosystem services. *Ecosystem Services, 12*, 218–227. https://doi.org/10.1016/j.ecoser.2014.07.002.

Lamond, J., & Everett, G. (2019). Sustainable blue-green infrastructure: A social practice approach to understanding community preferences and stewardship. *Landscape and Urban Planning, 191*. https://doi.org/10.1016/j.landurbplan.2019.103639.

Mekala, G. D., & Hatton MacDonald, D. (2018). Lost in transactions: Analysing the institutional arrangements underpinning Urban green infrastructure. *Ecological Economics, 147*, 399–409. https://doi.org/10.1016/j.ecolecon.2018.01.028.

Miraftab, F. (2009). Insurgent planning: Situating radical planning in the global south. *Planning Theory, 8*(1), 32–50. https://doi.org/10.1177/1473095208099297.

Moosavi, S., Browne, G. R., & Bush, J. (2021). Perceptions of nature-based solutions for Urban water challenges: Insights from Australian researchers and practitioners. *Urban Forestry & Urban Greening, 57*, 126937. https://doi.org/10.1016/j.ufug.2020.126937.

Myers, G. (2021). Urbanisation in the global south. In C. M. Shackleton, S. S. Cilliers, E. Davoren, & M. J. D. Toit (Eds.), *Urban ecology in the global south* (pp. 27–50). New York: Springer International Publishing.

OECD. (2020). Nature-based solutions for adapting to water-related climate risks, OECD Environment Policy Paper No. 21.

Qiao, X.-J., Kristoffersson, A., & Randrup, T. B. (2018). Challenges to implementing urban sustainable stormwater management from a governance perspective: A literature review. *Journal of Cleaner Production, 196*, 943–952. https://doi.org/10.1016/j.jclepro.2018.06.049.

Raymond, C. M., Frantzeskaki, N., Kabisch, N., Berry, P., Breil, M., Nita, M. R., … Calfapietra, C. (2017). A framework for assessing and implementing the co-benefits of nature-based solutions in urban areas. *Environmental Science & Policy, 77*, 15–24. https://doi.org/10.1016/j.envsci.2017.07.008.

Rogers, B. C., Brown, R. R., de Haan, F. J., & Deletic, A. (2015). Analysis of institutional work on innovation trajectories in water infrastructure systems of Melbourne, Australia. *Environmental Innovation and Societal Transitions, 15*, 42–64. https://doi.org/10.1016/j.eist.2013.12.001.

Runhaar, H., Wilk, B., Persson, Å., Uittenbroek, C., & Wamsler, C. (2018). Mainstreaming climate adaptation: Taking stock about "what works" from empirical research worldwide. *Regional Environmental Change, 18*(4), 1201–1210. https://doi.org/10.1007/s10113-017-1259-5.
Sarabi, S., Han, Q., Romme, A. G. L., de Vries, B., Valkenburg, R., & den Ouden, E. (2020). Uptake and implementation of nature-based solutions: An analysis of barriers using interpretive structural modeling. *Journal of Environmental Management, 270*, 110749. https://doi.org/10.1016/j.jenvman.2020.110749.
Shackleton, C. M. (2021). Ecosystem provisioning Services in Global South Cities. In C. M. Shackleton, S. S. Cilliers, E. Davoren, & M. J. D. Toit (Eds.), *Urban ecology in the global south* (pp. 203–226). New York: Springer International Publishing.
Tayouga, S. J., Sara, A., & Gagné, S. A. (2016). The socio-ecological factors that influence the adoption of green infrastructure. *Sustainability., 8*, 3–17.
Uittenbroek, C. J. (2016). From policy document to implementation: Organizational routines as possible barriers to mainstreaming climate adaptation. *Journal of Environmental Policy & Planning, 18*(2), 161–176. https://doi.org/10.1080/1523908X.2015.1065717.
Urban GreenUP. (2019). *Renaturing Urban plans to mitigate the effects of climate change*. Available in: wwww.urbangreenup.eu/
Vimal, R., Mathevet, R., & Michel, L. (2012). Entre expertises et jeux d'acteurs: La trame verte et bleue du Grenelle de l'environnement. *Natures Sciences Sociétés, 20*, 415–424. https://doi.org/10.1051/nss/2012043.
Watson, V. (2009). Seeing from the south: Refocusing Urban planning on the Globe's central Urban issues. *Urban Studies, 46*(11), 2259–2275. https://doi.org/10.1177/0042098009342598.
Whiteoak, K. (2020). Market challenges and opportunities for NBS. In: Wild, Bulkeley, Naumann, Vojinovic, Calfapietra y Whiteoak (Eds.) Nature-based solutions. State of the art in EU-funded projects. Bruselas: European Commission. Available in: https://op.europa.eu/en/publication-detail/-/publication/8bb07125-4518-11eb-b59f-01aa75ed71a1.
Zellner, M., Massey, D., Minor, E., & Gonzalez-Meler, M. (2016). Exploring the effects of green infrastructure placement on neighborhood-level flooding via spatially explicit simulations. *Computers, Environment and Urban Systems, 59*, 116–128. https://doi.org/10.1016/j.compenvurbsys.2016.04.008.

Maintainability

▶ Furthering the Sustainable Development Agenda by Putting Urban Heritage and Value Extraction at the Center

Making of Smart and Intelligent Cities

Challenges and Options for Africa

Wendy W. Mandaza-Tsoriyo[1], Geraldine Usingarawe[2], Abraham R. Matamanda[3] and Innocent Chirisa[4]
[1]Department of Rural and Urban Development, Great Zimbabwe University, Harare, Zimbabwe
[2]Department of Architecture and Real Estate, University of Zimbabwe, Harare, Zimbabwe
[3]Department of Geography, University of the Free State, Bloemfontein, South Africa
[4]Department of Demography Settlement and Development, Social & Behavioural Sciences, University of Zimbabwe, Harare, Zimbabwe

Definition

Sustainability: capacity to live longer under enduring pressures.

Africa: the continent of the majourity of the global black population

Intelligent city: city where digital technologies are the major facilitator for business and social activities.

Introduction

Smart and intelligent cities produce the solutions for large populations that live in areas largely supported by technology. The overall aim of implementing smart and intelligent cities is to mitigate the problems associated with urbanization. This is especially important in the African continent, where urbanization rates are high and the population tends to be concentrated in capital cities only or in a few urban centers (Chirisa et al. 2014). The United Nations (2014) has noted a worldwide trend during which cities have become central to the transformation of the world economy. Urbanization has dramatically changed energy consumption patterns, water and sanitation needs, food security, and the mobility of human beings (UN-HABITAT 2009; Freire et al. 2014). In response to challenges, emerging schools of thought envision good and sustainable cities that make the most of ICT to boost the flexibility of leaders to make decisions about the allocation of resources and improving the delivery of services. The Intelligent Building Institution described an intelligent building as one which integrates various systems in a quest to effectively manage the resources in a coordinated mode to maximize technical performance; investment and operating cost savings; and flexibility (Deakin and Al Waer 2011; Deakin 2012; Ghaffarianhoseini et al. 2016, p. 338; Ghaffarianhoseini et al. 2018).

There are many definitions of smart cities, and the International Telecommunication Union (2014, p. 1) defines a smart city as "an innovative town that uses ICTs [Information and Communication Technologies] and alternatives to boost quality of life, the efficiency of urban operations and services and competitiveness, while making certain that it meets the needs of present and future generations with respect to economic, social and environmental aspects." The "smart city" concept is underpinned by the conception that rational and optimal decisions, as well as intelligent behaviors can be attained in city environments through the utilization of knowledge and communication technologies (ICTs) (Komninos 2008, 2009; Komninos et al. 2013; Tella 2018).

It is interesting to note that the terms smart cities and intelligent cities are used interchangeably. According to Gardner (2000), it ought to be remembered that "intelligent" may be a broader and also a holistic term than "smart," both technically and linguistically. The smart city and intelligent city concept are not an exercise without merit (Glasmeier et al. 2003; Kitchin 2014); it can address the challenges faced by the African urban areas identified in this chapter. In as much as smart cities are supported by technology, the first consideration should be human capital, which should become an intrinsic component of the equation to achieving smart cities (Hollands 2008, p. 315), since people are a critical factor in any successful community, enterprise, or venture. The critical thing about information technology is not the capacity which it has to create smart cities. Rather, what is important is that there is a need for communication, which ought to be part of a social, economic, and cultural development. This communication and manipulation of the environment are engineered and operated by humans, who have to be intelligent and receptive to the technology, otherwise all the initiatives will be in vain (Wang et al. 2016).

This chapter makes use of reports from different organizations and events such as Smart Africa, Deloitte, United Nations (UN), and Transform Africa Summit (May 2017) as well as articles by authors who are practicing. It also makes reference to case studies of countries outside Africa. When analyzing the challenges to and opportunities for sustainable urban development in the cities featured in the case studies, one might ask how sustainable development in these cities is or could be. The "how" refers to measurable criteria used when assessing their performance. In an effort to provide standardized measures of different levels of urbanization over time, researchers make use of the WDR agglomeration index methodology, which incorporates GIS data and analyses involving travel time rasters, density rasters, and other biophysical and infrastructure variables gathered at national level (Li et al. 2016; Siba and Sow 2017).

After identifying the challenges that African countries face because of urbanization, this chapter provides possible solutions. First, we examine global trends relating to smart cities, that is, how other countries outside Africa are managing their population growth of which African countries are trying to draw lessons from. The chapter will also include the methodology that was used in studies to gather information. Finally, there is a presentation of case studies within African countries, namely, Kenya, South Africa, Zimbabwe, Sudan, and Rwanda.

Smart and Intelligent Cities: A Review

Cities are complex entities that require effective and efficient management (Geyer and Rihani 2010). Urbanization is the one major component that makes the cities complicated. Africa and Asia are projected to urbanize faster than other continents (UN-HABITAT 2014b). Ironically, such urbanization is associated with poverty, lack of economic or industrial growth, and above all poor governance (Chirisa et al. 2014). All sorts of problems manifest in African cities, and they take different forms and dimensions, a situation which makes these cities complex (Myers 2011). It is thus critical to consider urbanization in the face of poverty and poor governance. Adopting policies that have been used in the Global North may not be the best solution for Africa, warns Watson (2009), who highlights that Africa is unique and therefore needs her own recipe to development. In as much as the African city is complex, just like any city across the world, the smart and intelligent city concept may be useful in untangling some of the challenges the continent faces. However, it remains unclear as to which manual works best for African cities (African Union 2013).

The move towards intelligent buildings began in the early 1980s in the USA, and for the past three decades, intelligent buildings (IBs) were only conceptual frameworks representing future buildings (Castells 2010). However, intelligent buildings are now influencing the policies for design and development of future buildings (Girardet 2015). The rationale is that urban areas must be influenced by intelligent buildings so as to promote smart growth, green development, and healthy environments (Choon et al. 2011; Pearson et al. 2014). Ghaffarianhoseini et al. (2016) have pointed out that the intelligent buildings are a focal point that embraces green, sustainable, smart, and other related attributes. Smartness and technology awareness are one of the key performance indicators of intelligent buildings. The goals of intelligent cities include the reduction of operational costs through efficient energy management and "user-oriented" improvement in safety, health, and well-being (Silva et al. 2012; Cempel and Mikulik 2013).

Due to overcrowding, cities can present major challenges for national and native governments in scaling up their public infrastructures and services (i.e., education, transportation, water and energy supply) and meeting strict environmental standards to provide wealthy and sound living conditions, and in relation to efficient and effective health care; security and protection systems; accessible social, inventive, and cultural networks; and a large variety of products and services. Densely populated cities also present new challenges for organizations in the design and management of their operations. Therefore, intelligent buildings were designed to embrace the fundamental roles of addressing the requirements of the users in terms of functional and sensory needs; use of smart technology to allow for security and monitoring to aid facilities management; and ensuring sustainability with regards to energy, water, and waste through incorporation of smart design.

Intelligent and smart cities leverage ICT capacity by integrating urban and technological infrastructure (e.g., streets, utilities, and telecom networks) and digitalizing traditional services (such as online land registry, energy smart metering, and electronic healthcare records). This will enable urban authorities to integrate online services that are not connected to a network and reach citizens at lower costs and better quality (Layne and Lee 2001; Ho 2002; Moon 2002; West 2004; Carter and Bélanger 2005; Nam and Pardo 2014). The development of smart cities creates new opportunities for organizations to rethink their operations models. Smart cities seek to address issues of urbanization, economic development, and the technological needs of its inhabitants and visitors. They provide an area to live, learn, and work in; a space to develop innovative businesses and entrepreneurial activity within the digital arena; as well as healthcare provision – a crucial issue in Africa. The terms intelligent and smart are complementary; it is all about using the ever-growing potential to optimize the performance and impact of buildings and cities. During the Industrial Revolution, cities were the engines of economic development, innovation, and interaction (Jacobs 1969, 1984; Perlman et al. 1998). Following the problems of the industrial revolution, there was a move towards the sustainable development of cities. The aim of sustainable development is to meet people's needs in such a way that the environment does not suffer (WECD 1987; McGranahan and Satterthwaite 2003).

We then explore how countries outside of Africa have made efforts to roll out intelligent cities projects as providers of smart services to ensure the safety of their habitats and visitors. In as much as smart cities are supported by technology, there is a need to understand that such cities are not places where technology is produced, as Silicon Valley, for example. Smart cities act as receptacles of technology or places where technology is deployed to facilitate the functioning of the city (Glasmeier et al. 2003). In the face of complexity of the city and its myriad of challenges, technology is applied systematically to address these problems and ultimately make the city a safe haven for the urbanites.

There are many smart cities around the world from which lessons can be drawn. The following are some of the smart cities where different forms of technology have been used to ensure the well-being of the citizens:

- Masdar town in the United Arab Emirates has developed in a desert and makes use of renewable energy and clean technology (Kingsley 2013).
- In South Korea, the City of Songdo has, since 2003, been able to embrace technology in efforts to improve the lives of the citizens. In attempts to develop smart cities, initiatives implemented include free Wi-Fi for the city and the use of sensors to monitor temperature, energy use, and traffic flow. In efforts to control traffic in Songdo, digital road signs help update citizens (EYETRO Digital 2018).
- In Europe, Barcelona and Amsterdam have also made great strides in formulating and implementing smart city initiatives. These initiatives include the retrofitting of the cities' buildings (Hollands 2015; Kirby 2013).

If one is to consider the example of India, its government established the Smart Cities Mission focused on "'sustainable and inclusive development' resulting in the creation of a replicable model which will act like a lighthouse to other wishful cities" (Government of India 2015). The mission has presented a list of smart solutions to be implemented: e-governance and national services, waste management, water management, energy management, urban mobility, as well as telemedicine and tele-education, incubation/trade facilitation centers, and skill development centers. Further, the mission focuses on four strategic parts of "area-based development," namely, improvement of the city (retrofitting), city renewal (redevelopment), town extension (greenfield development), and pan-city initiative within which those solutions will be implemented, encompassing larger parts of a city. With an endowment of Indian Rupees 3205 crores (The Economic Times 2016), the Smart Cities Mission is "meant to outline examples which will be replicated both in and outside the smart city, catalysing the creation of comparable smart cities in numerous regions and parts of the country" (Government of India 2015).

In the example of Singapore, in addition to the big glass skyscrapers, a system called the SingPass, which allows Singaporean citizens to access government services with a single, unified login, has been launched. The SingPass has become a digital identification system to allow access to more than just the services it was intended for initially but now works with banks to mechanically fill out forms. It is expected that in the future, SingPass is expected to automatically bill ERP road users, and it is already rolling out as a way to securely store medical information so that doctors and nurses will simply access it. SingPass is placing Singapore on a firm footing for the future, and for any nation attempting to develop itself into a smart society, a national ID service such as SingPass is essential. The Housing & Development Board (HDB) of Singapore is also creating the Punggol Eco-Town in the north-east, a part of Singapore Island. The area was modelled in 3D before any construction was put in place. For that reason, architects were ready to

M

predict the wind flow within the area and create adaptations on the buildings so that zones clear of the shore would still get good airflow, therefore minimizing the use of air conditioners.

Realities in African Cities: Cases and Experiences

African countries are presently in the early stages of the urbanization process. Africa was the least urbanized region in the world in 2015, with only 40% of sub-Saharan Africa's population staying in cities, and is now the second-fastest urbanizing region in the world (after Asia). Population experts predict that by 2020, Africa will be on top. Given this rapid growth, now is the time for African policymakers to incorporate smart cities into their urbanization strategies. Since some African countries currently lack telecommunication cable installations, African cities will have to install the newest available ICT technology, eliminating the costs associated with removing or upgrading present ICT infrastructure. During the Transform Africa Summit in May 2017, there were initiatives to incorporate information and communication technologies to drive socioeconomic development in Africa. The summit brought together 300 African mayors in addition to the usual heads of state and ministers. At the summit, the Rwandan government unveiled the Smart Cities Blueprint, a framework aimed to extend the adoption of ICT-driven initiatives in cities across Africa. The blueprint was intended to inspire African countries to think progressively in the face of urban expansion. In addition, the blueprint also encouraged African countries to seek nontraditional sources of financing such as public-private partnerships and smart bonds to fund their projects. Following is a presentation of case studies of African countries such as Kenya, Ethiopia, Rwanda, Zimbabwe, Sudan, and South Africa and how those countries have made efforts to launch smart cities.

Kenya

The population of Nairobi, the capital of Kenya, grew from 8,000 people in 1901 to at least 4.5 million in 2011, and it is estimated that by 2025 the city will accommodate 5.8 million people (Omwenga 2011). Owing to its economic dominance within the country, Nairobi emerges as the hub for industry, education, and culture in Kenya and therefore continues to attract more people (Chirisa et al. 2018). The city is also home to the offices of a number of international organizations such as the International Labour Organization, the World Health Organization, the World Bank, and the International Monetary Fund. This strengthens the significance of Nairobi as a diplomatic, commercial, and cultural city for Africa at large. The large numbers of people that the city has to accommodate population as a result of its status were not envisaged when the city was established. In an attempt to keep pace with the demands imposed on the city of Nairobi, there have been initiatives by the central government and local authority to adopt smart initiatives. Initiatives towards attaining the smart city in Nairobi include great strides in mobile money and payment systems and building infrastructure aimed at supporting the digital economy. In recognition of these efforts, Nairobi has been named as the most intelligent city in Africa, for 2 consecutive years, 2015 and 2016 (Versa 2017).

Other smart initiatives in the country include the development of Konza City, which will be situated approximately 60 km from the center of Nairobi. This city will replicate the Silicon Valley and will be home to Kenya's blooming tech companies. It ought to be able to collect data from smart devices and sensors embedded within the urban environment, such as roadways, buildings, and other assets. Roadway sensors will be monitoring pedestrian and automobile traffic and adjust traffic light timings accordingly to optimize traffic flows. The city is going to be connected to Nairobi through a high-speed train. People operating and living in Konza will have access to a variety of real-time data-driven material such as traffic maps, emergency warnings, and quality information on energy and water consumption. In as muh as such plans are underway, there are a multiplicity of challenges which stand in the way of and may stifle the success of the implementation of the smart city initiatives in Kenya. Firstly, Nairobi has a large network of infrastructure, which is, however, dispersed and disconnected (Wanyonyi 2017). Secondly, there is a lack of knowledge among citizens relating to the technology, although the use of mobile phones facilitates the adoption of the smart city concept. Funding is also another challenge that complicates the success of implementing the smart city concept in Nairobi. Mwaniki et al. (2016) highlight that the success of smart cities initiatives in Nairobi have been impeded by challenges such as a weak manufacturing sector, the dominance of the informal economy.

Zimbabwe

Harare, the capital city of Zimbabwe, is home to much of the country's urban population. According to the 2012 national census, approximately 16% of the country's population resides in Harare (ZimStat 2013). As a result, Harare experiences serious urban crises Potts (2006a, b), which are manifested in the near collapse of the public service delivery, particularly electricity, pharmaceuticals, transport, and housing (Munzwa and Jonga 2010; Musemwa 2014). In Harare, housing challenges are evident in the proliferation of slums with restricted or no access to roads, electricity, water, and sanitation (Muchadenyika 2015). Through the *Harare Strategic Plan 2012–2025*, Harare seeks to become a world-class city through the adoption of smart city initiatives. The strategic plan spells out that it will achieve this objective through human capital reforms, which focus on human factor issues. Information technology is also placed at the fore of the plan, with the rationale being that ICT is fast becoming a management tool that has the ability to redress most problems haunting Harare. Technology will also be used to influence and transform urban form and urbanization processes (City of Harare 2012). The smart initiatives have been used in the transportation sector, where solar power has been used to power street lights and traffic lights across Harare. The same applies in some areas where increased use is made of solar to pump water. The introduction of the e-billing system by the City of Harare has been a great stride towards smart city status, and this has resulted in over 30,000 residents registering to the system (EyeTro Digital 2018). On the other hand, the City of Harare has also introduced the geographic information system department, which will be responsible for managing geospatial data for Harare. The electronic traffic management system that has been introduced by the council also facilitates the technological monitoring and management of transport system in the city. However, enormous challenges have impeded the success of the smart city initiatives in Harare. Mismanagement, corruption, and disputes within urban councils have led to the failure of cities to provide urban services such as water and sanitation, housing, public transport, and refuse collection (Muchadenyika 2014). Brain drain is another challenge that has affected the success of the smart city initiatives in Harare, since a large number of competent personnel have gone abroad in search of greener pastures. On the other hand, the existence of a massive informal sector in the city has also limited the success of the smart city initiatives as it has become difficult, for example, to monitor the operations of the informal transport operators.

Sudan

Khartoum, the capital city of Sudan, is a primate city. Due to its economic dominance and vast educational opportunities, the city attracts many people from rural areas and other small towns and cities. International migrants and refugees also contribute to the population influx in Khartoum. If this trend continues, the population of Khartoum is expected to reach 10 million in the near future (Petterson 2017). In an effort to cope with the city's growing population, many national strategic plans have been established over the past six decades. The plans mostly emphasized the importance of extending services to rural areas located on the fringes of the city, of accompanying investment to extend the efficiency of agricultural production, and of microfinance support for urban families and of support to key industries (Seif El Islam 2012). The plan additionally aimed to control the environmental pollution generated by population increase by stipulating that no new project can be executed unless an environmental impact assessment is carried out. The challenges associated with the implementation of the smart city concept in Khartoum are multifaceted. Among these challenges are the city's fragile economy, ineffective government institutions, environmental issues, conflict, and economic and political interests, which adversely affect the implementation of the smart city concept (Pantuliano et al. 2011; Seif El Islam 2012; UN-HABITAT 2014a). This has led to uncontrolled conurbation and misuse of land.

Rwanda

Rwanda is one among the pioneers of smart city engineering in the continent of Africa. Modernizing Kigali, the capital city of Rwanda, was part of a wider effort by the Rwanda government to extend and simplify access to public services. The government of Rwanda established the Irembo platform providing e-government services that allow its citizens to complete public processes online, including registering for driving exams and requesting birth certificates. The country involved the private sector in its goal towards creating smart cities. Recently, the Internet of Things (IoT) project was unveiled. It seeks to link up and connect citizens and organizations in Kigali, as part of the smart city initiatives (Hoy 2017), and contribute to efforts relating to environmental monitoring, smart farming initiatives, as well as a "smart" bus equipped with satellite Internet meant to facilitate connectivity for remote communities (Hoy 2017). Kigali's smart neighborhood project, Vision City, was also developed to create a neighborhood in which technologies would enable the provision of services such as street lamps powered by solar and free Wi-Fi in the town (Fremy 2017). The government of Rwanda has also partnered with Nokia and SRG around the deployment of smart city technology. The partnership focuses on the investment in network connectivity and the deployment of sensors to improve the well-being of the city's citizens.

Rwanda's smart city initiative has not been without its challenges. For instance, in 2016, the city started rolling out buses with free Wi-Fi. However, the buses have had connectivity problems owing to the Korea-built technology's inability to adapt to native conditions of the town (Muvunyi 2017; Hoy 2017). Critics also argued that the projects launched did not take into account the socioeconomic realities of the city, where 80% of its population lived in slums with lower monthly earnings. Rwandan planners responded to the critics stating that affordable housing was going to be developed in the later phases of the project. Moreover, there was also the possibility that such initiatives would be funded by international organizations such as a United Arab Emirates firm that is set to invest $50 million to facilitate the implementation of the Smart Kigali initiative (Mwai 2017).

South Africa

Cape Town is South Africa's second largest city in terms of population. Cape Town is the 10th most populous city in Africa, with an estimated population of 3.8 million people (UN-HABITAT 2014b). The city has been experiencing rapid urbanization mainly owing to population increase (City of Cape Town 2014). Such rapid urbanization calls for the provision of different kinds of infrastructure, including transportation systems, water infrastructure, housing, and energy (Chirisa et al. 2018). Cape Town is among the best places to do business on the continent because the South African government continues to implement thoughtful planning and technology intended to draw in businesses and improve the lives of its citizens. The city contributes approximately 11% to the gross domestic product in South Africa.

A number of smart initiatives have been introduced under the Cape Town Smart City strategy to enable it to keep pace with the rapid rate of urbanization and associated urban challenges. The city has introduced several initiatives to address different issues pertaining to the development of the city. Firstly, in an effort to minimize urban crime and violence in the city, CCTV for surveillance purposes have been used in Cape Town, with at least 560 cameras installed and located in different areas around the city. Public Wi-Fi, which was rolled out in 2016, has been a great stride in ensuring that citizens remain connected and have the ability to connect to the internet. In an effort to promote e-governance, which is also an intrinsic component of smart cities, an open data portal was launched in Cape Town in 2015. This portal made it possible for the citizens in Cape Town to pay their utility bills, apply for municipal services, licenses and permits, report crimes, or offences and request municipal services via the Internet. Lastly, the smart grid, which is currently being explored, is proving to be effective in monitoring the usage of water and electricity. Data on the consumption of the services that are being provided is captured electronically, allowing the government to plan a better way to invest new resources in the city, and all the data of the town registers from its citizens is made available to everyone through the open data portal website. Taxify, which allows people to request for a taxi through a smartphone, was established as a competitor of Uber in Cape Town. It provided the same platform of connecting drivers with passengers. As seen in the case studies presented above, the smart city strategy in Cape Town is not without its challenges either. Corruption is rife in Cape Town, and this tends to be a major drawback to the sustainability of the city's smart initiatives. Dordley (2018) reports that Cape Town was ranked the third most corrupt city in South Africa for the year 2017. Such levels of corruption include embezzlement of public funds, which compromises service delivery.

Conclusion and Future Directions

The foregoing case studies revealed that in Africa, efforts are being made to introduce smart city projects as key performance indicators of intelligent. In a 2015 issue paper, UN-Habitat urged town planners to avoid viewing smart cities as the final product but rather a way of minimizing transport costs, reducing the costs of service delivery, and maximizing land use, to ensure that the city reduces congestion, creates spaces for recreational uses, enhances service delivery, and improves the quality of life. ICTs will also continue to evolve. Town planners and policymakers should keep the big picture in mind when promoting smart cities, concentrating on well-developing infrastructure, and meeting citizen needs.

Several problems faced by some of the countries were identified during the research, principally that technology alone will not solve some of Africa's cities biggest challenges, which include high-cost, low-quality, and inaccessible services. The success of smart towns also depends on the government implementing the right policies (Moore 2014; Neirotti et al. 2014; Petrolo et al. 2014; Kummitha and Crutzen 2017). For example, Singapore benefited from close cooperation between the government and the business world. Moreover, the major problem of the urbanization process will concentrated in one region, in this case, the capital cities, as seen in the examples of Zimbabwe and Ethiopia. The research has revealed decentralization serves to promote social change through urban governance. However, it has failed in Zimbabwe, owing to corruption. By way of example, to acquire a driver's license, one must either be related to someone who operates in the relevant offices or pay a bribe to those conducting the tests. If not, one cannot pass the tests. The fight between political parties to take charge of the urban constituency and the management of urban affairs continues at the expense of urban service delivery. Currently, service delivery is being used as a source and resource for political agency. It can also be seen that the control of intelligent or smart cities and buildings is not only restricted and should rather be shaped by interactions among different-sized enterprises, nongovernmental organizations (NGOs), and local governments, and the entire process needs the more effective involvement of citizens in decision-making (Bolívar 2015).

This chapter sought to explore challenges that are being faced, particularly in Africa. Since intelligent cities are the smart providers of services, they do so by rolling out smart city projects, which are to be used to measure the performance of intelligent cities. It emerged from the study that if the smart cities are properly launched and well managed, they will mitigate environmental and socioeconomic problems and improve the inhabitant's quality of life. It can be concluded that if we academics wish to see smart town development and rhetoric address bigger issues and pursue more socially relevant uses of new technology

M

applications, then it is our responsibility to conduct the collaborative research that helps we the practitioners understand what the technology can and cannot do; what the application use conditions are (proximity; density); and how far off applications are from widespread marketability (Kitchin 2015). Options for future direction are that if people are to be persuaded to return to, or remain in, the rural areas, there must be a redress in the imbalance between the rural and urban areas.

Cross-References

- ▶ City Visions: Toward Smart and Sustainable Urban Futures
- ▶ Smart City: A Universal Approach in Particular Contexts
- ▶ The Sustainable and the Smart City: Distinguishing Two Contemporary Urban Visions

References

African Union. (2013). Agenda 2063 vision and priorities. Available online: http://agenda2063.au.int/en//vision

Bolívar, M. P. R. (2015). Smart cities: Big cities, complex governance? In *Transforming city governments for successful smart cities* (pp. 1–7). Cham: Springer.

Carter, L., & Bélanger, F. (2005). The utilization of E-government services: Citizen trust, innovation and acceptance factors. *Information Systems Journal, 15* (1), 5–25.

Castells, M. (2010). The rise of the network society. *The Information age: Economy, society, and culture* (2nd ed., Vol. 1). Oxford, UK: Wiley-Blackwell.

Cempel, W. A., & Mikulik, J. (2013). Intelligent building reengineering: adjusting life and work environment to occupant's optimal routine processes. *Intelligent Buildings International, 5*(1), 51–64.

Chirisa, I., Matamanda, A., & Bandauko, E. (2014). Ruralised urban areas vis-à-vis urbanised rural areas in Zimbabwe: implications for spatial planning. *Regional Development Dialogue, 35*, 65–80.

Chirisa, I., Mukarwi, L., & Matamanda, A. R. (2018). Sustainability in Africa: The service delivery issues of Zimbabwe. In: S. J. Garren (Ed.), *The Palgrave handbook of sustainability* (pp. 699–715). Cham: Palgrave Macmillan.

Choon, S. W., Siwar, C., Pereira, J. J., Jemain, A. A., Hashim, H. S., & Hadi, A. S. (2011). A sustainable city index for Malaysia. *International journal of sustainable development and world ecology, 18*(1), 28–35.

City of Cape Town (2014). Smart Living Handbook. Available online: http://resource.capetown.gov.za/documentcentre/Documents/Procedures,%20guidelines%20and%20regulations/CCT_Smart_Living_Handbook.pdf

City of Harare. (2012). *City of Harare strategic plan 2012–2025*. Harare: City of Harare.

Deakin, M. (2012). Intelligent cities as smart providers: CoPs as organizations for developing integrated models of eGovernment services. *Innovation: The European Journal of Social Science Research, 25*(2), 115–135.

Deakin, M., & Al Waer, H. (2011). From intelligent to smart cities. *Intelligent Buildings International, 3*(3), 140–152.

Dordley, L. (2018) MEC threatens to shut down taxi ranks amid violence. *Highbury Media*. https://www.capetownetc.com/news/mecthreatens-shut-taxi-ranks-amid-violence/

EYETRO Digital (2018). Will Harare achieve being a smart city by 2025? Available online: https://www.eyetrodigital.com/2018/01/19/harare-smart-city/

Freire, M.E, Lall, S & Leipziger, D, 2014. *Africa's urbanisation: Challenges and opportunities* (Working paper no. 7). Washington, DC: World Bank.

Fremy, L. (2017). Smart Kigali trying to give a brighter future to Rwanda. Available online: https://atelier.bnpparibas/en/smart-city/article/smart-kigali-give-brighter-future-rwanda

Gardner, H. E. (2000). *Intelligence reframed: Multiple intelligences for the 21st century.* New York: Hachette UK.

Geyer, R., & Rihani, S. (2010). *Complexity and public policy: A new approach to 21st century politics, policy and society.* London: Routledge.

Ghaffarianhoseini, A., Berardi, U., AlWaer, H., Chang, S., Halawa, E., Ghaffarianhoseini, A., & Clements-Croome, D. (2016). What is an intelligent building? Analysis of recent interpretations from an international perspective. *Architectural Science Review, 59*(5), 338–357.

Ghaffarianhoseini, A., AlWaer, H., Ghaffarianhoseini, A., Clements-Croome, D., Berardi, U., Raahemifar, K., & Tookey, J. (2018). Intelligent or smart cities and buildings: a critical exposition and a way forward. *Intelligent Buildings International, 10*(2), 122–129.

Girardet, H. (2015). *Creating regenerative cities.* New York: Routledge.

Glasmeier, A., Wood, L., & Kleit, A. (2003). *Broadband Internet service in rural and urban Pennsylvania: A commonwealth or digital divide?* Pennsylvania: EMS Environment Institute Pennsylvania State University, Centre for Rural Pennsylvania.

Government of India (2015) *Smart cities – Transformnation: Mission statements & guidelines*. New Delhi: Government of India, Ministry of Urban Development.

Ho, K. C. (2002). Globalization and Southeast Asian urban futures. *Asian Journal of Social Science, 30*(1), 1–7.
Hollands, R. G. (2008). Will the real smart city please stand up? Intelligent, progressive or entrepreneurial? *City, 12* (3), 303–320.
Hollands, R. G. (2015). Critical interventions into the corporate smart city. *Cambridge Journal of Regions, Economy and Society, 8*, 61–77.
Hoy, K. 2017. Smart Kigali: An IoT project to transform Rwanda. Available online: https://www.idgconnect.com/abstract/27057/smart-kigali-an-iot-project-transform-rwanda
Jacobs, J. (1969) The economy of cities. New York: Vintage Books.
Jacobs, J. (1984) Cities and the wealth of nations. New York: Random House.
Kingsley, P. (2013). Masdar: The shifting goalposts of Abu Dhabi's ambitious eco-city. *Wired Magazine*.
Kirby, T. (2013, April 18). City design: Transforming tomorrow. *The Guardian*.
Kitchin, R. (2014). The real-time city? Big data and smart urbanism. *GeoJournal, 79*(1), 1–14.
Kitchin, R. (2015). Making sense of smart cities: Addressing present shortcomings. *Cambridge Journal of Regions, Economy and Society, 8*, 131–136.
Komninos, N. (2008). *Intelligent cities and globalisation of innovation networks*. New York: Routledge.
Komninos, N. (2009). Intelligent cities: Towards interactive and global innovation environments. *International Journal of Innovation and Regional Development, 1* (4), 337–355.
Komninos, N., Pallot, M., & Schaffers, H. (2013). Special issue on smart cities and the future internet in Europe. *Journal of the Knowledge Economy, 4*(2), 119–134.
Kummitha, R. K. R., & Crutzen, N. (2017). How do we understand smart cities? An evolutionary perspective. *Cities, 67*, 43–52.
Layne, K., & Lee, J. (2001). Developing fully functional E-government: A four stage model. *Government Information Quarterly, 18*(2), 122–136.
Li, F., Nucciarelli, A., Roden, S., & Graham, G. (2016). How smart cities transform operations models: A new research agenda for operations management in the digital economy. *Production Planning and Control, 27*(6), 514–528.
McGranahan, G., & Satterthwaite, D. (2003). Urban centers: an assessment of sustainability. *Annual review of environment and resources, 28*(1), 243–274.
Moon, M. J. (2002). The evolution of e-government among municipalities: Rhetoric or reality? *Public Administration Review, 62*(4), 424–433.
Moore, T. (2014). From intelligent to smart cities. *Australian Planner, 51*(3), 290–291.
Muchadenyika, D. (2014). Contestation, confusion and change: Urban governance and service delivery in Zimbabwe (2000–2012). Unpublished MAdmin thesis. Cape Town: University of the Western Cape.
Muchadenyika, D. (2015). Slum upgrading and inclusive municipal governance in Harare, Zimbabwe: New perspectives of the urban poor. *Habitat International, 48*, 1–10.
Munzwa, K. M., & Jonga, W. (2010). Urban development in Zimbabwe: A human settlement perspective. *Theoretical and Empirical Researches in Urban Management, 4*(14), 120–146.
Musemwa, M. (2014). *Water, history and politics in Zimbabwe: Bulawayo's struggles with the environment* (pp. 1894–2008). Trenton: Africa World Press.
Muvunyi, L. (2017). Rura to push stalled smart Kigali project. Available online: http://www.theeastafrican.co.ke/rwanda/News/Rura-to-push-stalled-Smart-Kigali-project%2D%2D-/1433218-4091126-78j75cz/index.html
Mwai, M. (2017). UAE firm to invest $50m in Kigali 'smart city' initiative. *The New Times*. Available online: http://www.newtimes.co.rw/section/read/212435
Mwaniki, D., Kinyanjui, M., & Opiyo, R. (2016). Towards smart economic development in Nairobi: Evaluating smart city economy impacts and opportunities and challenges for smart growth. In T. M. Vinod Kumar (Ed.), *Smart economy in smart cities*. Singapore: Springer.
Myers, G. (2011). *African cities: Alternative visions of urban theory and practice*. London: Ze Books.
Nam, T., & Pardo, T. A. (2014). The changing face of a city government: A case study of Philly311. *Government Information Quarterly, 31*, S1–S9.
Neirotti, P., De Marco, A., Cagliano, A. C., Mangano, G., & Scorrano, F. (2014). Current trends in Smart City initiatives: Some stylised facts. *Cities, 38*, 25–36.
Omwenga, M. (2011) Integrated transport system for liveable city environment: A case study of Nairobi Kenya, 47th ISOCARP Congress 2011. Available online: http://www.isocarp.net/data/case_studies/2022.pdf
Pantuliano, S., Buchanan-Smith, M., Metcalfe, V., Pavanello, S., & Martin, E. (2011). *City limits: Urbanisation and vulnerability in Sudan. Khartoum case study*. London: HPG.
Pearson, R. G., Stanton, J. C., Shoemaker, K. T., Aiello-Lammens, M. E., Ersts, P. J., Horning, N., ... & Akçakaya, H. R. (2014). Life history and spatial traits predict extinction risk due to climate change. *Nature Climate Change, 4*(3), 217–221.
Perlman, J., Hopkins, E., & Jonsson, A. (1998). Urban solutions at the poverty/environment intersection. *The Mega-Cities Project, Publication MCP, 18*.
Petrolo, R., Loscri, V., & Mitton, N. (2014). Towards a smart city based on cloud of things. *Proceedings of the WiMobCity – International ACM MobiHoc workshop on wireless and mobile technologies for smart cities*.
Petterson, D. (2017). Turning Africa's megacities into smart Cities Available online: http://www.infrastructurene.ws/2017/08/29/turning-africas-megacities-into-smart-cities/
Potts, D. (2006a). 'All my hopes and dreams are shattered': Urbanization and migrancy in an imploding economy—The case of Zimbabwe. *Geoforum, 37*, 536–551.

M

Potts, D. (2006b). 'Restoring order'? Operation Murambatsvina and the urban crisis in Zimbabwe. *Journal of Southern African Studies, 32*(2), 273–291.

Seif El Islam, M. (2012). How planning can be wizard to solve third world cities problems? How to use the magic of planning to solve greater khartoum problems? 48th ISOCARP Congress 2012.

Siba, E., & Sow, E. (2017 November, 1) Smart city initiatives in Africa. Available online: https://www.brookings.edu/blog/africa-in-focus/2017/11/01/smart-city-initiatives-in-africa/

Silva, T. H., de Melo, P. O. S. V., Almeida, J. M., & Loureiro, A. A. (2012). Social media as a source of sensing to study city dynamics and urban social behavior: Approaches, models, and opportunities. In M. Atzmueller, A. Chin, A. Helic, & A. Hotho (Eds.), *Ubiquitous Social Media Analysis* (pp. 63–87). Heidelberg: Springer, Berlin

Tella, O. (2018). Agenda 2063 and its implications for Africa's soft power. *Journal of Black Studies, 49*(7), 714–730.

The Economic Times. (2016, Feb 29).Budget 2016: Over Rs 7,290 crore allocated for AMRUT, 'smart cities' in budget, *The Economic Times*.

UN-HABITAT. (2009). *Planning sustainable cities: Global report on human settlements 2009*. London: Earthscan.

UN-HABITAT. (2014a). *The state of African cities: Re-imagining sustainable urban transitions*. Nairobi: UN-HABITAT.

UN-HABITAT. (2014b). *Sudan's report: For United Nation's 3rd conference and sustainable urban development, (Habitat III), 2016*. Nairobi: UN-HABITAT.

United Nations. (2014). *World urbanization prospects: The 2014 revision-highlights*. UN, Department of Economic and Social Affairs Population Division, (ST/ESA/SER.A/352).

Versa, O. (2017). Smart cities in Africa: Nairobi and Cape Town. Available online: https://www.howwemadeitinafrica.com/smart-cities-africa-nairobi-cape-town/58209/

Wang, X. Y., Peng, Z., Yu, X. Q., Zhao, J., & Wang, L. (2016). Intelligent architecture design and research based on smart city. *Mechanics and architectural design: Proceedings of 2016 international conference*, Suzhou, Jiangsu, China, 156–161.

Wanyonyi, P. (2017). Smart cities: The case of Nairobi. Available online: http://www.nairobibusinessmonthly.com/smart-cities-the-case-for-nairobi/

Watson, V. (2009). Seeing from the South: Refocusing urban planning on the Globe's central urban issues. *Urban Studies, 46*(11), 2259–2275.

West, D. M. (2004). E-government and the transformation of service delivery and citizen attitudes. *Public Administration Review, 64*(1), 15–27.

World Commission on Environment and Development. (1987). *Our common future, World commission on environment and development* (The Brundtland commission). Oxford: Oxford University Press.

ZimStat (Zimbabwe National Statistics Agency). (2013). *Census 2012: Provincial report Harare*. Harare: ZimStat.

Managed Retreat

► Climate-Induced Relocation

Management/Guidance

► Metropolitan Discipline: Management and Planning

Managing Africa's Urban Flooding Challenges from the Bottom Up: A View from Ghana

Narteh F. Ocansey
Water Resources, Freelance, Accra, NA, Ghana

Rainfall and Its Benefits

Rainfall as part of the hydrological cycle is a natural phenomenon that brings freshwater to sustain life. Rains recharge groundwater; fill up lakes, streams, and rivers; and aid in plant growth. Runoffs resulting from rainfall bring rich nutrients downstream and facilitate migration of aquatic species. Coastal wetlands are also recharged by runoff and provide habitat for some ocean species to spawn! Land modifications have been part of human evolution to support man's need for shelter, food, energy, transportation, and the likes. When these interactions are not properly managed, the benefits of rains as runoff are impaired, and with it comes devastating floods.

Floods

Flood is an overflow of water that submerges land that is usually dry as a result of rainfall. The way land is developed can have dramatic impact on the volume and speed of runoff, due to the loss of retention and interception characteristics of the catchment. Anthropogenic factors such as overgrazing, land compaction, deforestation, building on slopes, and an increase in the area of impermeable surface, i.e., roads and concrete, can all dramatically increase flood risk downstream. Loss of wetlands through legal or illegal reclamation may also contribute to worsening the flood situation.

Reasons for Flooding

Accra's flooding is driven by two main factors, climate change (CC) and increased urbanization (impervious surface), among other hydrological factors. The Intergovernmental Panel on Climate Change (IPCC) predicts increased short intense rainfall and frequency of flooding due to climate change. With Accra experiencing increasing frequency of minor season flooding in October, this prediction may already be happening. Hydrological factors that affect runoff within the catchment area are the size, topography, shape, aspect, geology, and soil. High intense rainfall on impervious surfaces causes faster catchment response time and increases peak runoff as in a hydrograph, leading to floods.

Urbanization and Flooding

Africa's rapid urbanization comes with the challenges of urban sprawl – a situation that worsens the response mechanisms to flooding. Without basic drainage infrastructure in the suburbs, runoff water finds its own levels wreaking havoc on neighborhoods at the slightest rainfall. According to the National Spatial Development Framework (2015), urbanization is increasing at a rate of 7.3% p.a. But within the Odaw catchment in Accra, urbanization peaks at 3.2% p.a., while runoff volume is increasing at 2.4% p.a., according to a recent model result. Confirms the inability the already inefficient drainage system to be easily overwhelmed, resulting in frequent flash floods.

Ordinarily this would require a corresponding yearly expansion of the drainage infrastructure, something that is simply unsustainable. To alleviate the problem, the approach has been to desilt or line drains and streams. Though such maintenance schemes are necessary, its effectives has been met with scorn by the public over the years. More so drainage masterplans that consider population growth rate for the designs underestimate future runoff rates with serious implications for the coping capacity of the system.

The much talked about building on floodplains and wetlands is a phenomenon of urban sprawl, uncontrolled housing development, and slums. Occupants of such lands are the poor and vulnerable who bear the most impact when their homes are flooded. They are driven to this extent as a result of the huge housing deficit and a lack of equitable access to land. Settlers, mostly migrants from the rural areas, use such habitats to hunt for economic opportunities provided by the city.

Short-Term Solutions

In the short term, firstly, what is needed is the political will to demolish structures sited in floodplains and make way for the river. Compensation could be paid to structures with permits but are sited wrongly in drainage channels. Secondly the ministries, sector organizations, and local authorities must implement the numerous policy documents gathering dust on the shelves that calls for sustainable management of water. The state must provide these institutions with the necessary resources including capacity building for them to deliver on their mandate.

Thirdly there is an urgent need to develop a planning policy guidance for hillside development because of the direct impact of slope on catchment response time and peak runoff volume on downstream settlements. Accra's expansion has now reached the Akwapim Hills, and its continued expansion poses a high risk to the city which is downstream.

Long-Term Solutions

In the long term, a more holistic and risk-based approach is required that gives priority to water in all its forms and underlying factors, starting off by addressing issues in the land tenure system that ensures equitable access to land and title, without which land use planning is severely constrained. Investment in housing by government is also key to reap the benefits of planned cities that enhance livability for its people. Adoption of more sustainable approaches to the management of urban stormwater that mimic natural drainage conditions like the best management practice (BMP) and green infrastructure (GI) as best practice from the USA and other parts of the world. We can take it a step further by requesting drainage impact and flood risk assessment for all properties as part of the local authority's building permitting process provided by civil engineering firms.

Conclusion

Urbanization as with population growth is inevitable, but this should not exacerbate the impact of flooding on cities and its inhabitants. To tackle this challenge, we need to properly regulate the land sector, plan and control land use, make room for water, provide mass housing, and adopt best practices such as GI. Within the GI policy, we should regulate runoff such that pre-development runoff does not exceed post-development runoff from the site. Ultimately to succeed will require strong leadership with a vision and a plan which is data driven and adequately funded. In the endless cycle of nature, the rains will come, and so will the flood waters rise; therefore, preparedness remains the key to cities' resilience.

Cross-References

- ▶ Blue-Green Cities: Achieving Urban Flood Resilience, Water Security, and Biodiversity
- ▶ Mainstreaming Blue Green Infrastructure in Cities: Barriers, Blind Spots, and Facilitators
- ▶ Multiple Benefits of Green Infrastructure

References

Government of Ghana. (2015). *Ghana National Spatial Development Framework 2015–2035*. Vol. II. Retrieved from http://www.luspa.gov.gh/files/NSDF%20Final%20Report%20Vol%20II%20Final%20Edition_TAC%201.pdf

Managing the Risk of Wildfire Where Urban Meets the Natural Environment

Janet Stanley and Belinda Young
Melbourne Sustainable Society Institute, University of Melbourne, Melbourne, Australia

Synonyms

Wildfire; Bushfire; Forest fire

Introduction

The world is facing a convergence of multiple, serious challenges that appear to be beyond human ability to modify or solve. High on the list is climate change and its association with extreme events. This brief paper discusses the interface between one of these extreme events, wildfire and its interface with other challenges, associated with cities and regions, that have become more pronounced over the past few decades.

Although wildfires have always been present in Australia, they have recently grown in size, intensity, and number. Official records are kept on these fires, but they greatly underestimate the number that occur, by a factor of around five. Recent years have also shown a new pattern where many fires are now occurring, often concurrently. In Australia, over the 5 months, from September 1, 2019 to February 29, 2020, 960,041 fires were observed by satellites (Global Forest Watch https://www.globalforestwatch.org/topics/fires/). These fires resulted in 18 million hectares being burned, 6000 buildings destroyed, a billion animals killed, and greenhouse gas emissions almost doubling Australia's annual emissions. Similar disasters are happening in the USA and Europe and have occurred in countries without a history of severe fires, such as Scandinavian countries (in 2018) and Alaska (in 2019) (Stanley et al. 2020).

The Increasing Number of Wildfires and Their Location

Two issues are important to understand about why these wildfire disasters are becoming more common. Firstly, the relationship between wildfire severity and climate change. Rising temperatures, changes in rainfall patterns, drought, and low humidity and soil moisture levels have resulted in the drying of vegetation, thus increasing its flammability. The creation of such weather conditions has been predicted for many years by scientists (see, for example, Hughes 2015). The evidence to support these predictions have been strongly growing over the past decade, such that it is now irrefutable (IPCC 2021).

Secondly, there is a need to understand why these fires are igniting. The great majority of fires (at least 85%) start as a result of human activity. This can be due to purposeful or malicious lighting; reckless or accidental lighting, such as burning off rubbish on a fire danger day, sparks from machinery, and powerlines lighting a fire; or prescribed burns, to reduce forest fuel, turning into a wildfire. Nonhuman ignitions may result from naturally occurring events, such as volcanic eruptions, spontaneous combustion in natural ground litter, and lightning, the latter accounting for 0.55% of fires recorded in Western Australia over the period 2004–2012 (Plucinski 2014). The severity of present-day fires now being experienced has seen increased occurrences of fires creating other fires. This may be due to traveling embers and very hot, large fires changing the localized weather by creating strong wind events and increasing the risk of lightening, often now not associated with rainfall (Romps et al. 2014).

There are three location types where fires frequently start, each with differing characteristics. Fires occur in a built environment, with ignitions attributed to deliberate lighting (e.g., revenge and insurance claim) or accidental lighting (e.g., wiring faults). Fires start on the interface between the urban and natural environment, mostly due to human ignition. In the USA research has shown that human-caused ignitions account for 97% of the residential homes threatened by wildfire (Cattau et al. 2020). The third area is the occurrence of fires in forested areas, where the US study found that about 59% of wildfires in this location were linked to human-caused ignitions. This research also points out other important findings on human-lit fires. When compared with forested lands, human-lit wildfires double the length of the fire season, double the area burned in the forested areas during the shoulder fire seasons, and spread faster than lightning-lit fires.

The Cause of Wildfires

The proportion of wildfires that are purposefully lit is not widely recognized, being possibly 40–60% of all fires. Juveniles aged 14–20 years are considered the largest group of people associated with purposefully lit fires. Other groups of concern include younger children and unemployed males over 30 years of age (Stanley et al. 2020), the latter group more likely to be involved in fire lighting on high fire danger days.

Those involved in fire lighting (maliciously or recklessly) are usually males with troubled and/or socioeconomically disadvantaged backgrounds. They are likely to have a history of absent fathers, child abuse, or neglect, and engage in a number of antisocial activities with other youth, such as vandalism, drug use, and absenteeism from school (Dolan and Stanley 2010). Youth who commit arson are predominantly males, who may engage in substance abuse and other criminal behaviors, including theft, criminal damage, and vandalism. These young people may be motivated to light a fire due to boredom, displaced anger, peer pressure, excitement/thrill seeking, an interest in fire which may be a family pattern of behavior, or may have an intellectual disability (Willis 2004) (Stanley et al. 2020).

Returning to the broader issues of cities and regions, there are a number of trends that, it is argued, are linked to wildfire. The first of these is population growth, with about 83 million people being added to the world's population every year (United Nations Department of Economic and Social Affairs 2017). Thus, if problems that lead to fire-lighting are not being resolved, the number of fires being lit will also be increasing. As well as population grows, there is a trend for rural to urban population movement. An outcome of this is city growth and accompanying urban sprawl in many countries, such as in Southern Europe, Northern America, and Australia.

Population pressure and the need for affordable housing has commonly led to poor urban planning that allows urban sprawl into urban/rural interface and peri-urban areas (see ▶ Improving Social Equity and Community Health and Well-Being in Low-Income Suburbs and Regions). The pace of this land use change has left many governments playing catch-up with infrastructure, services, and jobs, needed by people who move into these newly opened up areas of housing. There is also a movement of people into interface areas, often older people or established families, who are looking for a lifestyle change in a healthier environment. The population growth in the interface areas has resulted in 43% of all residential development in the USA occurring in peri-urban areas since the 1990s, exacerbating wildfire risk (Radeloff et al. 2005; Radeloff et al. 2018; Syphard et al. 2007). This pattern is predicted to double in size by 2030 (Theobald and Romme 2007). Australian interface areas can also be very large, said to extend about 160 kilometres around Melbourne's perimeter (Buxton et al. 2011).

Urban sprawl associated with affordable accommodation attracts young families with lower incomes, who also may experience poor socioeconomic well-being. Research on this issue in Melbourne found that there were poorer health outcomes, lower levels of trust, earlier school leaving, and less post-school education in high growth areas, when compared to other parts of the city (Brain et al. 2019). The outer, and fastest growing suburban areas of Melbourne have the lowest number of jobs per 1000 residents and the poorest levels of public transport than found in other parts of Melbourne. Thus, it would not be unexpected to find that many of the urban/rural interface areas in Melbourne have high levels of youth unemployment, underemployment, and disengagement from work and education, such that the numbers rise to one-third of the youth population in some locations. The collision of these two problems, high fire risk and poor urban planning reducing youth well-being, is increasing the risk of fire ignitions, a trend that is clearly evident in many countries.

Youth experiencing disadvantage, disillusionment, and boredom create a ripe breeding ground for lighting fires (Tomison 2010). The pattern of lighting commonly fits the behavioral patterns of youth. Fires are lit close to their residential location in an isolated bush setting, often a comfortable bicycle ride away and commonly in unsupervised times such as Saturday nights, and after school, when parents may be at work.

Urban and Land Use Planning

Related issues are found in regional areas where poor transport, in particular, is leading to diminished opportunities for youth and high unemployment levels for many. This presents a problem with risk of fires on the edges of rural towns, but also in forested areas. Dumped cars and rubbish are often a preferred means of fire-lighting, commonly occurring within 24 h of dumping, and frequently leading to a wildfire (Sibthorge and Lowrey 2018; US Fire Administration 2014). Malicious fires tend to occur along roads that provide access points into areas for dumping, campsites, and illicit activities. Isolation from opportunities, and community interactions and values, can also extend to the older unemployed male seeking attention and excitement through fire-lighting. Unfortunately, research and documented cases in Australia and the USA recognize that this opportunity may arise through joining the fire brigade and lighting fires, to become a "hero" by reporting and extinguishing the fire.

The "lifestyle" seekers that move into natural environmental areas present a slightly different problem, where the fire risks relate more to accidental or reckless-type fires. These forms of ignitions are often associated with small farms, where burning rubbish, and machinery use on a high fire danger day, can cause sparks. The interface areas with small farms can have a fairly high residential population in these areas. Buxton et al. (2011) reported a population of about 300,000 people in an interface area of 12,000 square kilometers. Prescribed burns to reduce the vegetation fuel loads are commonly undertaken in peri-urban and forested areas. Such burns may be associated with smoke pollution risks and not uncommonly result in the development of a wildfire.

Wildfire Prevention

It is critical that robust policies are urgently set in place to reduce carbon emissions and the growing climate change conditions that facilitate catastrophic wildfires. Other actions that target the reduction of fire ignitions in the first place are significantly overlooked internationally. There is a widespread heavy reliance on environmental modification, a form of mitigation or planning designed to reduce fuel loads should a fire occur. Prescribed burning is based on the premise where once a fire occurs, lower amounts of flammable material will reduce the severity of a wildfire. Apart from findings suggesting that this assumption is not always correct in all conditions and areas, it overlooks actions that could be taken to prevent the ignition in the first place.

Insights and actions in relation to wildfire prevention could be achieved if an integrated land use and planning approach is undertaken (Alexander 2020; March et al. 2020a; Stanley 2020; Stanley et al. 2017). The starting point is engaging with communities and maintaining their ongoing involvement and decision-making. Part of this process is through the use of social media to facilitate a dialogue between locals and the range of government authorities, businesses, non-government organizations, and others (Young 2021) (see ► Collective Emotions and Resilient Regional Communities). There is a need to understand the interrelationships across a wide range of environmental, social, and economic issues and how changes in one policy area will impact many other areas (Stanley et al. 2017).

Once all elements are on the table, then strategic policy needs to be elaborated and aligned across the multiple levels of impact, at city, regional, local government, and community levels. Decisions are made according to the desired outcomes, presumably prioritizing safety, health, and well-being. Trade-offs need to be carefully elucidated and considered, local communities having a very important role in this decision-making, as many decisions around wildfire are grounded in value perspectives (Stanley et al. 2020). Vital to this process are monitoring and evaluation of outcomes, and research and development to increase understanding and improve responses.

A coordinated response to the prevention of wildfire that realizes improvements in land use strategies addresses the multiple impacts associated with the consequences of fire. These include government costs for fire suppression and rehabilitation, personal costs in recovery, growth in inequality, health problems due to smoke inhalation, and loss of the natural environment, with many follow-on consequences around the loss of ecosystem services.

Some examples of improved land use planning are now given as an illustration of the approaches that could be taken to reduce the number of wildfire ignitions. The approaches range from high level and large to the local and specific.

Crime prevention and good mental health outcomes necessitate the importance of providing a "sufficient" level of services and infrastructure to enable all people to reach the capability to be what they wish (Sen 1992). By way of example, research has shown that the failure to offer youth accessibility to education, jobs, and particularly community and social contacts increases their social isolation and diminishes their well-being and opportunities in life (Stanley et al. 2019). The association between disadvantage and crime is recognized in the environmental criminology and psychology literature (see, for example,

Baumeister et al. 2005). The passage to this outcome is commonly through combinations of lowered self-esteem, anger, frustration, emotional denial (cutting off feelings), and cognitive impairment (not thinking well).

A range of other options exist that include siting critical infrastructure such as hospitals away from high-risk areas, building designs that are able to combat most fire events, and safe associated evacuation places (March et al. 2020b). Situational crime prevention addresses specific planning that structures specific sites so that the configuration diminishes opportunities for fire-lighting (Cozens and Christensen 2011). This harks back to Jacob's (1961) "eyes on the street" ideas, so that people can observe behavior through presence (a people-friendly place, seats, and shades), lighting put in places, and reducing areas of unoccupied and neglected buildings and rubbish. It will also be the provision of places for youth, such as sports facilities, places where they may like to meet and use, that are also accessible by active transport.

Conclusion

Society is fast approaching a time when the political failure to address both climate change and growing inequality will lead to faster growing catastrophic disasters, along with a reduced opportunity to correct these issues as time passes (IPDD 2021). There are many prevention actions that could, and should, be done now, to at least reduce the daunting outcome for society that is fast approaching. An important option is to review and adjust the urban planning associations with wildfire ignitions.

Cross-References

- ▶ Collective Emotions and Resilient Regional Communities
- ▶ Improving Social Equity and Community Health and Well-Being in Low-Income Suburbs and Regions

References

Alexander, J. (2020). Burning bush and disaster justice in Victoria, Australia: Can regional planning prevent bushfires becoming disasters? In A. Lukasiewicz & C. Baldwin (Eds.), *Natural hazards and disaster justice: Challenges for Australia and its neighbours* (pp. 73–88). Singapore: Palgrave, Macmillan.

Baumeister, R., DeWall, C., Ciarocco, N., & Twenge, J. (2005). Social exclusion impairs self-regulation. *Journal of Personality and Social Psychology, 88*(4), 589–604.

Brain, P., Stanley, J. K., & Stanley, J. R. (2019). *Making the most of our opportunities: First report to the Municipal Association of Victoria*. Melbourne: NIEIR and Stanley and Co.

Buxton, M., Haynes, R., Mercer, D., & Butt, A. (2011). Vulnerability to bushfire risk at Melbourne's urban fringe: The failure of regulatory land use planning. *Geographical Research, 49*(1), 1–12.

Cattau, M., Wessman, C., Mahood, A., & Balch, J. (2020). Anthropogenic and lightning-started fires are becoming larger and more frequent over a longer season length in the U.S.A. *Global Ecology and Biogeography, 29*(1), 20. Jan.

Cozens, P., & Christensen, W. (2011). Environmental criminology and the potential for reducing opportunities for bushfire arson. *Crime Prevention and Community Safety, 13*(2), 119–133.

Dolan, M., & Stanley, J. (2010). Risk factors for juvenile firesetting, *Advancing Bushfire Arson Prevention in Australia, report from collaborating for change: Symposium Advancing Bushfire Arson Prevention in Australia*, held in Melbourne, 25–26 March, 2010, Monash University and Australian Institute of Criminology, pp. 31–32.

Hughes, L. (2015). *The burning issue: Climate change and the Australian bushfire threat*. Sydney: Climate Council.

IPCC (Intergovernmental Panel on Climate Change). (2021). Climate change 2021: The physical science basis: Summary for policymakers, 7 August.

Jacobs, J. (1961). *The death and life of great American cities*. New York: Random House.

March, A., Nogueira de Moraes, L., & Stanley, J. (2020a). Dimensions of risk justice and resilience: Mapping urban planning's role between individual versus collective rights. In A. Lukasiewicz & C. Baldwin (Eds.), *Natural hazards and disaster justice: Challenges for Australia and its neighbours* (pp. 93–111). Singapore: Palgrave, Macmillan.

March, A., Nogueira de Moraes, L., van Delden, H., Stanley, J., Riddell, G. A., Dovers, S., Beilin, R., & Maler, H. (2020b). Urban planning and natural Hazard risk reduction: Critical frameworks for best practice, November Bushfire & Natural Hazards CRC.

Plucinski, M. (2014). The timing of vegetation fire occurrence in a human landscape. *Fire Safety Journal, 67*, 42–52.

Radeloff, V. C., Hammer, R. B., & Stewart, S. I. (2005). Rural and suburban sprawl in the US Midwest from 1940 to 2000 and its relation to forest fragmentation. *Conservation Biology, 19*, 793–805.

Radeloff, V. C., Helmers, D. P., Kramer, H. A., Mockrin, M. H., Alexandre, P. M., Bar-Massada, A., Butsic, V., Hawbaker, T. J., Martinuzzi, S., Syphard, A. D., & Stewart, S. I. (2018). Rapid growth of the US wildland-urban interface raises wildfire risk. *Proceedings of the National Academy of Sciences, 115*(13), 3314–3319. https://doi.org/10.1073/pnas.1718850115.

Romps, D., Seeley, J., Vollaro, D., & Molinari, J. (2014). Projected increase in lightning strikes in the United States due to global warming. *Science, 346*, 851–854.

Sen, A. (1992). Capability and wellbeing. In M. Nussbaum & A. Sen (Eds.), *The quality of life* (pp. 30–53). Oxford: Clarendon Press.

Sibthorge, C., & Lowrey, T. (2018, November 4). Residents urged to create bushfire survival plans as firefighters work to bring Canberra blaze and spot fires under control. *ABC News*, https://www.abc.net.au/news/2018-11-04/crews-still-trying-to-get-canberra-fire-under-control/10463648

Stanley, J. (2020). How a failure in social justice is leading to higher risks of bushfire events. In A. Lukasiewicz & C. Baldwin (Eds.), *Natural hazards and disaster justice: Challenges for Australia and its neighbours* (pp. 205–217). Singapore: Palgrave, Macmillan.

Stanley, J. K., Stanley, J. R., & Hansen, R. (2017). *How great cities happen: Integrating people, land use and transport*. Edward Elgar.

Stanley, J., Stanley, J., Balbontin, C., & Hensher, D. (2019). Social exclusion: The role of mobility and bridging social capital in regional Australia. *Transportation Research Part A: Policy and Practice, 125*, 223–233.

Stanley, J. R., March, A., Ogloff, J., & Thompson, J. (2020). *Feeling the heat: International perspectives on the prevention of wildfire*. Delaware: Vernon Press.

Syphard, A. D., Volker, C. R., Jon, E. K., Todd, J. H., Murray, K. C., Susan, I. S., & Roger, B. H. (2007). Human influence on California fire regimes. *Ecological Applications, 17*(5), 1388.

Theobald, D., & Romme, W. (2007). Expansion of the US wildland–urban interface. *Landscape and Urban Planning, 83*(4), 340–354.

Tomison, A. (2010). Bushfire arson: Setting the scene. In: J. Stanley, & T. Kestin (Eds.) *Advancing bushfire arson prevention in Australia, report from collaborating for change: Advancing bushfire arson prevention in Australia symposium*, Melbourne, 25–26 March

United Nations Department of Economic and Social Affairs. (2017). World population prospects 2017. Department of Economic and Social Affairs of the United Nations Secretariat, Population Division.

Mandate

► Regulation of Urban and Regional Futures

Master Plan – Comprehensive Plan

► Harare

Master Planned Estates and the Promises of Suburbia

Paul Burton
Cities Research Institute Griffith University, Gold Coast, QLD, Australia

Synonyms

Community; Sprawl; Suburban; Suburbanization; Suburbia; Sustainability

Definition

Suburban development or peripheral urbanization has resulted in a built form that has dominated the major cities of North America, Australia, and the UK for almost one hundred years. While early forms involved little more than detached housing, more recently suburban development has claimed to involve the creation of master-planned communities. These promise to reconcile a desire for residential and social detachment with community development and ecological sustainability. This chapter explores critically the achievement in practice of this reconciliation.

Introduction

Suburban forms of residential development have been popular in many countries for hundreds of years and have become a dominant built form in the USA, Canada, Australia, and the UK. While "peripheral urbanisation" in many parts of the Global South involves residents building their own houses and creating their own neighborhoods, in these late capitalist countries of the Global North suburban built environments are produced more often by large developers. They are often supported and encouraged in this by government planners responsible for metropolitan or subregional growth management. In trying to reconcile the need for new houses and jobs with the timely provision of infrastructure, while simultaneously trying to protect and preserve ecosystems and highly valued natural environments, peripheral urbanization presents many challenges. A common feature of this approach in the Global North is to approach them as exercises in the creation of "master planned communities." These promise to achieve the reconciliation described above, but at a more localized scale and to form part of a coherent and sustainable metropolitan settlement pattern. This entry provides a brief review of the historical trajectory of suburban development, focusing on the emergence of master planned communities and looking critically at whether or not they deliver on their promise of a suburban good life.

A Brief History of Suburbs

As Walks (2013) observes, while we make increasing reference to suburbia, suburbanization, and to suburbs themselves, there has been no concomitant increase in the clarity of definition and conceptualization of these terms. Of course, this lack of precision is not peculiar to the suburban. Saunders' (1981) attempt to use social theory to answer "the urban question" reflected something of revival of interest among urban theorists toward the end of the second half of the twentieth century, but this has not been maintained. There has, however, been a steady stream of scholarly interest in similarly suburban places, including "technoburbs" (Fishman 1987), "boomburbs" (Lang and LeFurgy 2007), "metroburbia" (Knox 2008), "post-suburban" landscapes (Lucy and Phillips 1997), or even as "exopolis" within the "post-metropolis" (Soja 2000).

Within this growing body of scholarship on suburbs, processes of suburbanization, and forms of suburban life, some texts stand out as exemplary accounts of the origins of suburban places and ways of life. Robert Fishman's *Bourgeois Utopias* provides an excellent account of the emergence of the first suburbs around London from the middle of the eighteenth century onward as an emerging middle class, whose wealth came from manufacturing in and trade between cities, sought refuge from a city life that was becoming increasingly unpleasant, unhealthy, and dangerous. While suburbia was, in Fishman's words, "the collective creation of the Anglo-American middle class: the bourgeois utopia'" (1987: p. *x*), it did not appear overnight, inspired by the singular vision of a great architect. Instead, it evolved on the basis of trial and error to reflect the developing aspirations, values, and priorities of this emergent class. Nor was it the only way in which they chose to live, in Paris, for example, the *bourgeoisie* decided to stay in the center of the city, transform its built form, and drive the urban proletariat to the fringes into what are now the infamous Parisienne *banlieues*. It is interesting to speculate on the spread of this French solution around the globe if they had been a more significant imperial and colonial power: Might North American and Australian cities have embraced the wide boulevards and grand apartments introduced in Paris by Haussmann and, been less enamored of the low density, detached suburbs that now dominate their metropolitan landscapes?

Another foundational book on suburban development is Robert Fogelsong's *Bourgeois Nightmares*, which charts meticulously the imposition from the late nineteenth century of restrictive covenants and other impositions on the ways that suburban subdivisions could be developed. It is noteworthy that while some of these were underpinned by racist values (when they were used to exclude "non-Caucasians" from owning or even investing in these schemes), they were introduced primarily for economic reasons. In short, these restrictive covenants gave owners and investors a degree of certainty that the value of their investment would not be undermined by bad neighbors, inappropriate design, or poor construction quality. Restrictions and limitations that are now the bedrock of most planning schemes designed and implemented by governments and their agencies were introduced by private developers using the force of common law. The epitome of this approach was the Palos Verdes Estate on the Southern Californian coast, developed according to plans produced by Olmsted Brothers of Massachusetts, in which the owners of lots had to submit their plans for assessment and approval by the Palos Verdes Art Jury. Approval required a high degree of conformity with height limits, site coverage, building materials, and the prevailing style of Californian architecture, and the subsequent use of the building and maintenance of the property was tightly regulated by the Home Association (Fogelsong 2005: 16).

Modern suburbs have, therefore, always been highly planned by public or private means in order to achieve conformity, order, and a degree of exclusivity, not just for their own sake but because they serve to maintain and enhance property values. An important question when we consider contemporary master planned communities is whether the standards imposed by these different planning regimes have been and are still as good as they could be.

The Rise of the Master Planned Community and Its Promise

As Cheshire (2012) notes, while contemporary developers of master planned estates or communities often attract the opprobrium of critical scholars for,

> . . .lacking ethics, aggressively pursuing profits at the expense of powerless individuals, engaging in environmental vandalism by building on pristine farmland or open space, and responsible for a development aesthetic that ranges from dull, standardised, and 'ticky-tack' to ostentatious 'McMansions' that are excessive consumers of energy.

some of their earliest proponents, such as Ebenezer Howard, are more often lauded as visionaries and social reformers. Howard's seminal work, *To-morrow: a Peaceful Path to Real Reform*, published in 1898, builds on his concern with urban overcrowding and its consequences and refers to the conflicting attractions of "the town magnet" and "the country magnet," each pulling in different directions and offering their own mix of opportunities and problems. His resolution was to imagine a place that offered the best of both worlds, where people would be attracted by "the town-country magnet." His idea for a garden city embodied many of the key principles of what we would now call sustainable development, where cooperative governance structures reinvest rents into local infrastructure including concert and lecture halls, theaters, libraries, museums, and hospitals. Beautiful gardens complement parkland and waste is used by small-scale agricultural producers based on the fringe of the city. Howard's plans were based also on a detailed economic case for managing land values and rents and ensuring that residents earned wages that enabled them to live comfortably and to sustain local businesses.

Many of Howard's ideals and principles remain popular and form the mainstay of contemporary manifestations of the desire to blend the best of urban and rural life in master planned communities on the fringes of our cities (Gwyther 2005). But do they deliver on their promises? A growing body of research on suburban environments and lifestyles should provide some answers to this question.

The Lived Experience

Critics of those who apply very broad concepts like "neoliberalism" to explain virtually every aspect of contemporary life (e.g., Peck et al. 2018) typically argue for more empirically based explorations of "actually existing neoliberalism" that reflect better the spatial and temporal variations in how neoliberalism plays out on the ground. This approach can also be applied productively to the concept of "suburbia," so that we develop a more sophisticated understanding and appreciation of the actual existence of different suburban landscapes and ways of life. A number of studies of contemporary Australian suburban landscapes and lifestyles have been carried out in recent years that give some insights into "actually existing suburbia," including some of the level of satisfaction of suburban residents with their environment and some that compare master planned estates with other forms of suburban development. Cheshire et al. (2003) examined a number of large, privately developed master planned estates in five Australian states and compared them with surrounding suburbs. Among their findings of note is that while a desire for a safer environment and to live "among a better class of people" are important reasons given by many for moving to a master planned estate, once there they do not form community bonds that are significantly stronger than in other areas. Sometimes there is an initial but temporary sense of community, but this can be explained by the fact that early residents sometimes see themselves as pioneers and are often encouraged to attend community events hosted by the developer.

Walters and Rosenblatt (2008: 410) concluded their study of sense of community in master planned estates by suggesting,

> the property developer is able to successfully invoke the idea of this idealised version of community in an environment where it is unlikely to ever come to fruition...its realisation is not critical to residents' wellbeing...[because]...of the ability of the late modern residents of suburbia to make the 'cognitive leap' from the reality of their everyday social lives...

In other words, the allure and promise of suburban living continues for many prospective suburbanites and among developers, even if the reality does not match the promise. The developers of new master planned communities continue typically to sell the chance to own property (even if many residents now live in houses rented from investors) in place offering an active, healthy lifestyle in a thriving community, with local jobs and education opportunities. This is echoed by Cheshire (2012: 198) who suggests,

> ...developers invest more than just a significant sum of money into master planning since they also invest the symbolic value of their brand and reputation as well.

Shifting our attention from the experiences of suburban residents to the physical landscapes and buildings themselves, an enduring attraction of suburban life has been the ownership of a detached house surrounded by land sufficient for children to play safely, for entertaining friends, and perhaps to grow food. While master planned suburban communities also provide communal parks, playgrounds, sports fields, and other greenspace, the private and typically fenced backyard or garden has been of paramount social and cultural importance. Although the postwar suburbs of the mid-twentieth century often provided gardens sufficient to grow enough food to keep a typical family in fresh fruit and vegetables (if they were so inclined), this space and indeed usage has dwindled to the point that few suburban families now have the inclination, the time, or the space to achieve this degree of self-sufficiency.

Lot sizes in contemporary master planned communities vary, depending on the location and target market. In Australia, in the larger communities that emphasize affordability as well as proximity to major employment centers, these lots typically range from 250–650 m^2 with the most common mid-sized lots of 400 m^2 currently (2021, Queensland) selling for around \$200,000. House and land packages in this range cost around \$450,000 for a four-bedroom, single-story home with an integrated garage, which will typically occupy at least half of the lot with a mandatory 6m setback from the road frontage. As a consequence, one of those traditional features of suburban living – the backyard or back garden – is often very small. Hall (2010) has described the "life and death of the Australian backyard" and noted that while an emotional attachment to the idea of the traditional "quarter acre block" (approximately 1000 m^2) continues, it is now rarely seen outside upmarket and more expensive developments or in ones situated far from urban centers, where land is still relatively cheap.

Could They Be Better?

One of the most highly valued features of suburban developments built in the postwar years was the lot size. As described above, the quarter acre block was commonplace and served as a benchmark for many years, providing room for children to play, families to entertain their friends and neighbors, cultivation of fruit and vegetables and perhaps the keeping of poultry, and even room to extend the house. The enlargement of the original house through rear and side extensions allowed a degree of scaling up from a relatively affordable beginning, whereas many contemporary suburban developments begin with a high level of site coverage (or plot ratio) on smaller lots and leave little room for expansion or modification. This tends to limit the capacity of homes to meet changing household demands as children arrive or leave, older relatives want to move in, or homeworking expectations change.

High levels of car dependency continue to be a significant feature of most contemporary suburban developments, including master planned communities. Without good public transport facilities and services, and without significant employment centers in the locality, there is often a need for most if not all adult members of the household to have a car of their own. This in turn exposes households that have stretched themselves financially to buy a suburban property to fuel cost increases, as Dodson and Sipe (2008) demonstrated with their VAMPIRE Index. And, as we may not yet have reached "peak car size (It was recently reported that, for the first time, just over half of all new vehicles sold in Australia were some forms of SUV, including soft-roaders, off roaders, and "faux wheel drives" it becomes increasingly difficult for all of these vehicles to be parked on the lot, whether in a garage or on the driveway. On street parking can then be a significant problem in some of these communities, limiting the ability of children to play in the street (even if they were inclined to do so), impeding waste collection and emergency service vehicles and doing little to create a pleasant environment rather than an informal parking lot.

There is growing evidence that new suburban developments are beginning to include a greater range of housing types that cater for a more diverse range of household types, such as townhouses (row houses), duplexes, and small, low-rise apartment blocks. But detached houses still dominate, and it is possible that as more of these are owned by investors, they will be rented by groups of unrelated sharers. While this can indicate a welcome increase in household and resident diversity, there is still evidence in some countries of stigma attaching to homes occupied by renters rather than owners (e.g., Arthurson 2004).

There is also some evidence of master planned communities being designed to have a minimal impact on the landscape, eschewing extensive earthworks and vegetation clearance, insisting through local covenants and bylaws that each dwelling touches the ground lightly, and requiring a range of energy minimization features such as solar alignment, generous eves, on-site water capture and energy generation, and the use of recycled and sustainable building materials (See Currumbin Ecovillage at: https://theecovillage.com.au/about/). There is certainly scope for more on-site water and waste treatment, energy production and distribution, and digital connectivity, even in more conventional master planned suburban estates.

In conclusion, one can ask whether contemporary master planned communities will stand the test of time and become, like many of their postwar equivalents, highly desirable and more expensive places in the future (Burton 2015)? Related to this is the question of whether they represent the very best in planning and urban design, demonstrating unequivocally the state of the art of master planning? Do they really help build a sense of community spirit, encourage healthy and active lifestyles, and provide flexible spaces for changing household needs, including the opportunity to work well from home? Time will tell and it usually takes decades before the success or otherwise of substantial master planned developments is apparent, but at present there is no clear and unequivocally positive answer to these questions.

Conclusions

Suburban landscapes and lifestyles have become increasingly popular in many countries, especially but not exclusively in those of the Global North. But the definition and conceptualization of these places and lifestyles, as well as the people who choose to live in them, have expanded as well to the point that there is a danger that the terms lose their descriptive power and analytical purchase. If suburban areas now contain rich and poor people, are no longer (typically and thankfully) ethnically segregated enclaves, contain medium-rise apartment blocks as well as detached houses, and can be significant places of employment as well as dormitory settlements, then we might ask: Which places are not suburban? Typically, the answer is that only the places known as downtowns, or CBDs, or city centers are excluded. Even small towns in rural areas might have entire neighborhoods bearing the hallmarks of suburbia: detached dwellings, punctuated only occasionally by a convenience store or fuel station.

Many scholars have risen to the challenge of defining suburbia more precisely (eg Jackson 1985; Hayden 2003; and Davison 2013) and a major contribution to the broad field of suburban studies has been made in recent years by Roger Keil of York University in Canada, including as Series Editor for the University of Toronto Press series on Global Suburbanisms. However, an enduring challenge faced by academic suburban studies has been its engagement with a distinctly political project of advocating for suburban development in the face of what is often perceived to be an unhealthy preoccupation with urban questions and an unwarranted emphasis by governments at all levels on investment in urban infrastructure. This "project" (which is by no means a concerted and coordinated movement) has emerged in response a belief that a somewhat unholy alliance of architects, planners, urban studies academics, and other members of a metropolitan elite class not only hold jaundiced and prejudicial views about suburbia but have succeed in influencing governments to skew their investment decisions in favor of urban facilities such as mass transit systems, at the expense of suburban infrastructure. Furthermore, planning policies and metropolitan growth management strategies that seek to limit peripheral expansion and encourage higher density development within cities are held to be economically inefficient and contrary to the wishes of the majority who are happy to live in ever-expanding suburban environments.

In order to know whether proposed new master planned communities on the fringes of our cities deliver on their promises, respecting the legacy of Howard and other early town planners, we must continue to apply the highest scholarly standards to our research and be wary of being used, perhaps unwittingly, as uncritical promoters of contemporary bourgeois utopias.

Cross-References

- ▶ Changing Paradigms in Urban Planning 2000–2020
- ▶ Cities in Nature
- ▶ Green Belts
- ▶ Peri-urbanization
- ▶ The Challenges for Wildland-Urban Interfaces (WUI) in Metropolitan Areas: Reducing Fire Risk, Providing Employment Opportunities, and Preserving Natural Habitat

References

Arthurson, K. (2004). From stigma to demolition: Australian debates about housing and social exclusion. *Journal of Housing and the Built Environment, 19*(3), 255–270.

Brenner, N., & Theodore, N. (2002). Cities and the geographies of "Actually existing neoliberalism". *Antipode, 34*(3), 349–379.

Burton, P. (2015). The Australian good life: The fraying of a suburban template. *Built Environment, 41*(4), 504–518.

Cheshire, L. (2012). *Master plan developers, international encyclopaedia of housing and home* (pp. 195–199). Elsevier.

Cheshire, L., Wickes, R., & White, G. (2003). New suburbs in the making? Locating master planned estates in a comparative analysis of suburbs in South-East Queensland. *Urban Policy and Research, 31*(3), 281–299.

Davison, G. (2013). The suburban idea and its enemies. *Journal of Urban History, 39*(5), 829–847.

Dodson, J., & Sipe, N. (2008). Shocking the suburbs: Urban location, homeownership and oil vulnerability in the Australian city. *Housing Studies, 23*(3), 377–401.

Fishman, R. (1987). *Bourgeois utopias: The rise and fall of suburbia*. New York: Basic Books.

Fogelsong, R. (2005). *Bourgeois nightmares: Suburbia, 1870–1930*. New Haven: Yale University Press.

Gwyther, G. (2005). Paradise planned. *Urban Policy and Research, 12*(1), 57–72.
Hall, T. (2010). *The life and death of the Australian backyard*. Collingwood: CSIRO Publishing.
Hayden, D. (2003). *Building suburbia: Green fields and urban growth, 1820–2000*. New York: Pantheon Books.
Jackson, K. (1985). *Crabgrass frontier: The suburbanization of the United States*. New York: Oxford University Press.
Keil, R. (2018). *Suburban planet*. Cambridge: Polity Press.
Knox, P. (2008). *Metroburbia, USA*. New Brunswick: Rutgers University Press.
Lang, R., & LeFurgy, J. (2007). *Boomburbs: The rise of America's accidental cities*. Washington, DC: The Brookings Institute.
Lucy, W., & Phillips, D. (1997). The post-suburban era comes to Richmond: City decline, suburban transition, and exurban growth. *Landscape and Urban Planning, 36*(4), 259–275.
Peck, J., Brenner, N., & Theodore, N. (2018). Actually existing neoliberalism. In D. Cahill, M. Cooper, M. Konings, & D. Primrose (Eds.), *The Sage handbook of neoliberalism* (pp. 3–15). London: Sage.
Saunders, P. (1981). *Social theory and the urban question*. London: Hutchinson and Co..
Soja, E. (2000). *Postmetropolis: Critical studies of cities and regions*. Oxford: Blackwells.
Walks, A. (2013). Suburbanism as a way of life, slight return. *Urban Studies, 50*(8), 1471–1488.
Walters, P., & Rosenblatt, T. (2008). Cooperation of co-presence? The comforting idea of community in a master planned estate. *Urban Policy and Research, 26* (4), 397–413.

The Mediatized City

The Structural Transformation of Emplaced Relationality and the Resilience of Depoliticized Urban Commons

Manfredo Manfredini
School of Architecture and Planning, The University of Auckland, Shanghai University, Auckland, New Zealand

Synonyms

- Network City (Mediatized City)
- Planetary City (General Urbanisation)
- Recombinant City (Reassembling Urbanity)
- Uneven Spatial Development (Unequal Development)
- Extractive Urbanism (Exploitative Urbanism)
- Transcalar Urbanization (Discontinuous Urbanization)

Definition

Mediatization is considered a general descriptor for the complex process by which media constitute the *molding forces* (Hepp 2012) of our mutually constitutive social, cultural, and spatial dimensions. In our advanced digital age, its pervasion among the concrete and agential elements of our daily life has produced a reality–virtuality continuum in which hegemonic forces have exacerbated the unevenness of the abstractive and relational dynamics (Rosa 2013). *City* is understood as a meta-descriptor for a type of settlement undergoing deep transformations which progressively accomplishes the general *planetary urbanization* (Brenner 2014). In our urban age, the fading of conventional geographical determinations that produces an urban–rural continuum facilitates the domination projects of leading powers overcoming locally established social pacts to implement a progressively extractive, transcalar, and uneven more-than-spatial spatial development. Accordingly, this entry proposes the *mediatized city* as a topos for the elaboration of the discourse on equality and spatial justice that redresses the reading of the *space of flow* of the *network society* (Castells 2010) by focusing on the threats and opportunities of the crisis of urban politicizing territoriality.

Introduction

The increasingly rapid transformation of our technological framework and the continuous growth of social inequalities and power imbalances are progressively fragmenting the social and spatial fabrics of the city, profoundly reforming the constitution and spatialization processes of urban communities. The relations between actual and potential are reframed by a *reality–virtuality continuum* that pervades all dimensions of our lives, deterritorializing appropriations, associations, and stabilization of emplaced concrete objects, networks and agents. Severe restructuring of the traditional geographical and temporal determinations make relationality progressively unstable and *distracted*. Criticism has associated this restructuring to multidimensional social and spatial issues including *depleted transindividuation* (Stiegler 2012), *revanchist depoliticization* (Low and Smith 2006), *embourgeoisement* (Harvey 2005b), *inequitable public entrepreneurialism* (Madanipour 2019), *strategic decommoning* (Stavrides 2015), *biopolitical securitization* (Schuilenburg 2018), *disciplinary mediatization* (Han 2017), and *de-individuating hegemonic timescape* (Hassan 2020). Studies on spatial justice have linked it with adverse dynamics of *brutal enclosing* (Sennett 2017) *antagonistic politics* (Mouffe 2008), *ontogenetic despatialization* (Harvey 2005b, 2008), *control by privatization* (Kohn 2004) negation of the *Right to the city and centrality* (Harvey 2008; Lefebvre 1968; Marcuse 2014; Mitchell 2003; Purcell 2002).

Through an analysis of key drivers of the structural transformation of the systems that regulate the opposition between urban commons and enclosures, this entry illuminates the emerging and complex modes of territorial production by restructuring. It addresses the changing production of critical central nodes of urban systems that are progressively subject to acceleration of cycles of deterritorializations and reterritorializations moderated through articulated strategies of abstraction by transnational hegemonic financial actors supported by sweeping economic liberalization policies. It explores how the emerging forms of collective territorialization of communities, no longer bound by geographic boundaries enable novel constitutive processes across space and time, and use new technologies to form mutable and mobile assemblages of infrastructure, activation, and agents. Firstly, the discussion considers the effects of the acceleration of all processes on relational synchronization and the capacity of hegemonic actors to implement abstractive fictions; secondly, it establishes the relations between this fictional production and rampant globalization by elaborating on how double abstraction processes successfully sustain financially extractive enclosures; thirdly, it elaborates upon the combined impact of arrhythmic fictionality and abstracted redistributions of the sensible on the formation of counterforces of sociospatial transindividuation that expands the potential of the mediatized planetary city to reactivate processes of *reconfigurative othering* (Rancière 2010) and *counter-desubjectification* (Hardt and Negri 2017, p. 28).

Acceleration and Arrhythmia of Abstractive Fictions

In the unrelenting acceleration of our lifeworld, the exponential increase of the speed of change of each aspect of urban life steadily augments our daily productive, distributive, consuming, and associative capacities both in their individual and combined realms. However, the benefits of these improvements are associated with a critical growth of the complexity of their underlying systems. The enormous accretion of instrumental, cultural, relational, and affective capability progressively exceeds the limits of our capacity driving a progressive technological, cognitive, social, and psychological desynchronization. The accelerated rhythms of our urban ecosystems produce an arrhythmia that pushes embedded practices and organizations towards *fatal disorder* (Lefebvre 2004). Incoordination, maladjustment, and dissociation expand triggering unprecedented disruptions of social norms, networks, and territorialities across the various components of the socius. The intensification of this disorder has been directly correlated to the growth of socioeconomic inequalities and linked to the growing pace and strength of financial, democratic, ecological, and psychosocial crises occurring in the contemporary city (Hardt and Negri 2017; Rosa et al. 2017, pp. 60–64). The crises compromise the resilience of the urban body, exposing its components to *objective abstraction*: the substantive removal of individualities from stabilized conditions that allow understandings, interpretations, and actions articulated through transindividuative relations, knowledge, and practices (Connolly 2000; Rosa 2013). The abstracted body has values removed, dispossessed, reappropriated, and reformed from "forms and practices of human cooperation and sociality that are external to them" (Mezzadra and Neilson 2019, p. 138).

The uneven distribution of the abstractive agency allows hegemonic actors to profit from it and reverse crises into opportunities to expand their power. These opportunities are generated by implementing complex resource-intensive machines that harness the annihilation of stabilized values, conventions, and relations and restructure situated ecosystems at a foundational level, concentrating affluence and diffusing and deepening deprivation, and increasing social segregation and spatial injustice. The machines can control abstraction by establishing enclosed systems with *dynamically stable conditions* (Rosa et al. 2017) where ad hoc governance resynchronizes, spatially and temporally, the internal systemic motility. The enclosures promptly adapt, transition, and transmutate values, conventions, and relations through metastabilization dynamics that offset the crises with other perturbating processes of reassembling by fragmenting, overcoding by decoding, and prorupting by disrupting. The introduced foundational contradictions expand the unevenness of the broader systems, fostering the conflicting and incongruous dynamic instability of their socio-spatial development. The local disjunction of the enclosures – compensated by their integration within translocal networks of analog elements – produces permanent *rescaling* and *scale bending* (MacKinnon 2010) with a multiscalarity (Peck 2017c) that exceeds the geographical domain of stabilized collectivities, disempowering them from controlling their urban processes and practices.

With the development of the apparatuses of dynamic stabilization, *semifixed thick territorial insertions* of various sizes and forms pervade urban areas, introducing continuous remodulations of their spaces and contributing to the increasing translocalization of people, goods, and technologies. The uneven and chaotic diffused *horizontal metropolis* (Viganò et al. 2017), with hyper-connective infrastructures of the *space of flow* era (Castells 2005), exacerbates its incoherent and unstable "landscape of equalization and differentiation" (Smith 1982, p. 142). Increasingly transient mediations between unstable hegemonic apparatuses, which have instituted "common or connective processes, in conversation across multiple sites" (Peck 2015, p. 168), dissipate embedded social, environmental, and cultural systems, imposing transcalar transpositions on their constitutive elements (Jones et al. 2016) with irreconcilable rhythms and flows.

From a cultural perspective, these disrupted landscapes have complex patterns of consensus-production functional to strategic geopolitical financial remapping of hegemonic powers. Free-market logics articulate contingent narratives composed of dynamically stable fictions which promptly and conjuncturally adapt to self-legitimize their disruptive effects. Cunning "construction of set[s] of relations between sense and sense, between things that are said to be perceptible and the sense that can be made of those things" (Rancière 2013, p. 1) institute the fictions. Locally and temporally situated associations severely disrupt the civicness of their environment, eradicating any space for political dissensus and subjectification. Severe policing identifies and dismisses any agent that may challenge or obstruct the alienating processes they have put in place (Rancière 1999). Arrhythmia reigns and desynchronization *rules by expulsion* (Sassen 2014). The institution of ultrascalar, metamorphic, and mobile hegemonic spatial regimes resynchronizes in crescendo networks of severely policed fictions of growing reassociation, restabilization, and de-translocalization. Fictions impose *dressage-like rhythms* (Lefebvre 2004) that, by equating losing the pace to dismissal, sustain the growth of dissensus-free metastable regimes where the modulations of the adaptivity imperative moderate growing intrinsic contradictions and protect the systems from the very same incongruities they exacerbate.

Financialization and Double Abstraction of Extractive Enclosures

In our market-dominated urban era, an essential imperative of capital, *reproduction through compound-rate growth* (Harvey 2014, pp. 222–245), drives a general process of desynchronization-by-acceleration. This process is intensified by the limits posed by the material world to capital expansion, which can neither exceed the basic financialization of everything – a process in its rapid making – nor exponentially increase the value of the concrete asset. To comply with its imperative, capital shifts its operational area to an immaterial domain: the abstract component of its ecosystem. With the creation of nonproductive abstract machines to exploit the *perpetual mobility of capital* (Marx 1973, p. 130), operations with fictitious capital, like financial derivatives, opens limitless trade development and value creation. Bracketing the constitutive core part of capital's reproduction process, *money–money*, allows the hindrances of the material components of the whole process *commodity–money–money–commodity* (de Angelis 2017, pp. 174–181) to be lifted.

The recent progress of these machines for the *abstract financialization of everything* has upheld the exorbitant growth in value realization and degraded the mainly urban material asset. The differential between the increasing valuation of the purely financial and the concrete basis has gradually marginalized but not removed the cardinal role of the latter (Sassen 2017). As demonstrated by triggering the global financial crisis of 2008, the effects of the sudden collapse of property value triggered enormous losses in the global economy, starting from the transnational market for derivatives based either on direct real estate acquisition or indirect third-party financing.

The progressive absorption into the financial sphere of the material asset and thus its marginalization generate two main issues. On the one hand, it increases the power unbalance between the two main groups of stakeholders of that asset, those in control of the global financial machines and their *customers*, severely aggravating the indebtedness and dependency of the latter. De-collectivization and de-privatization reduce the freedom and autonomy of large swaths of society to produce their own individual and common space (Harvey 2005a, 2009, pp. 69–116, 2019). On the other hand, it enhances the penetration into urban processes of the abstractive financial logic that prompts the neglect of sociospatial dynamics and material conditions that have no direct influence on monetary value, deflecting the focus of urban governance from welfare problems (Harvey 2010; Sassen 2005). The combination of these issues has severe effects, the intensity of which increases in parallel with the penetration in any given domain of their abstract machines: the more these machines extend their control over the city's economy, the more the care for equality and autonomy of its communities loses centrality in its development. In other words, the deeper the "annihilation of space by time" (Harvey 1990; Marx 1973, p. 499), the more the *reorganization of time by space* of the abstractive networked enterprises (Castells 2010, pp. 407–408) obliterates the urban commons, disintegrating both the internal and reciprocal relations between all physical, social, and cultural dimensions of the urban space.

Spatially, the alliance between apparatuses of capital reproduction and neoliberal regimes of all sorts characterizes the framework for deploying the *extractive urbanism models* (Sassen 2005). This alliance prevents or moderates crises generated by the neglect of the concrete base of the financial machines. Negotiable regulatory frameworks, ad hoc means, and medium- to long-term public–private partnership programs increasingly erode the broader institutional frameworks determined by social pacts. Disestablishing reforms concerted through collective struggles during the twentieth century enables this system to minimize or offload risks and maximize the gains deriving from crises of the "marginal" concrete material and convertible bases that are inseparably coupled with the core financial ones. Increasingly *entrepreneurial* urban governance paradigms push local administrations to compete aggressively to attract global finance capital. Sassen's (2005) comparative analyses of global cities, Harvey's (2010) studies on the 2008 global credit crunch, and Peck's (Peck 2017a, b; Peck and Whiteside 2016) work on the municipal bankruptcy of Detroit and Atlantic City eminently describe the progressive effectiveness and breadth of these models.

The territorial emplacements of extractive models inscribe in space the indiscriminate system of relations of global capitalism that steers it (Brenner 2014; Sassen 2018a; Smith 2008). Fragments of the ecology of flow that reach all dimensions of social production pervade our cities with elements of very different scales. These fragments are semi-randomly located along transportation corridors throughout central cities, suburbs, exurbs, towns, and rural areas reifying their exogeneity to any reality on the *planetary* scale (Brenner 2014). Embodiments in multiscalar forms negate any commons and simulate them within enclosures with abundant idiosyncratic and multivalent combinations of near-public, quasi-public, and pseudo-public spaces. Ambivalent openings between highly asymmetrical territories make boundaries concurrently hard and impenetrable, soft and welcoming. At a larger scale, these elements include (a) centers of conspicuous consumption, such as the mini-cities of Unibail-Rodamco-Westfield; (b) extensive comprehensive urban precincts for the elites, such as the Hudson Yards in New York of Related Companies, Oxford Properties, and Morgan Stanley; and (c) outright private cities, such as the South Korean Songdo of Gale International and POSCO. At a smaller scale, they comprehend emplacements with various functions and formats, often deceptively merged in the city's body, such as semi-gated office parks, residential communities, and gray or green infrastructural networks and nodes.

The extractive enclosures disrupt stabilized territorial organization through restructuring processes that privatize what is public and redefine what is private (in some cases semi-publicizing it). By annihilating space, they produce urban landscapes with a proliferation of commercial, residential, and recreative rhizomatic enclosure systems that determine formally dissociated patterns of functionally mismatching locales with idiosyncratic multiscalar punctuations. They scale up and trans-scale what Rowe and Koetter (1978) described as *colliding fields of internal coherence* that, together with the resultant *interstitial debris,* have made up the city throughout history. They subsume the centralities of public life by conflating and polarizing the networks of urban amenities within their controlled domains. They redistribute the social infrastructure, creating spatial discontinuities, displacements, and inaccessibilities that are particularly acute in the fragmented network of the detritic space where socioeconomic deprivation is at its highest (Low and Smith 2006; Soja 2010). They institute *bordered geographies of centrality* with a general territorial disruption-by-recombination that compromises the intrinsically integrated, i.e., relational, collaborative, and multidimensional, nature of the urban production and amplifies the longstanding crisis of the urban commons. This compound crisis exacerbates the entrenched social justice problems across all well-being areas, such as the socialization, culture, and health of progressively disenfranchised collectivities (de Angelis 2017). The abstract machines operate by introducing a new level of spatial abstractions. Enclosures with fabricated spatial constructs not only prevent people's grasp of the financially speculative reality underpinning them by disguising their perception and spatial practices, but also introduce a *double abstraction.*

Abstracted parts of the city are swiftly de-relocalized, remobilized, and recombined to constitute agile workable *ultrascalar units* for translocal people, organizations, and processes that relinquish their emplaced sociospatial relationality. The compensation for the spatialities that these enclosures contribute to obliterating is designed to bracket the perception of their relational system within their reconciliatory fictions, offering short-circuited sets of desirable elements and relations. These sets present imperfect and ill-arranged associations that appear redeeming in contrast to the complementary fiction that exposes, as deviated and wrong, what they disrupt outside them. The precession of the double abstracted over the negated operates through drastic reductions of idealized public institutions and paradigms. The salvific rule of the market forces that once relied on basic abstraction emplacing analog mirrored representational imaginaries has now moved to the derivative abstraction with which hyperreal simulacra envelop the whole edifice of representation with the "liquidation of all referentials" (Baudrillard 1994, p. 2). In other words, the ability of the capital to establish workable spatial relations in the total absence of embedded historical dynamics and collective concertations relies on its capacity to transition between the simulations of simple-real reproduction to simulations of abstract-real simulacra of derivatives. These simulacra, paraphrasing Debord (1983), are systems that restructure society in aggregations making dissociation the base of relationality. They isolate and bring individuals together in "pseudo-[meta]communities" (p. 96) ruled by (dis)order, (dis)unity, and (dis)placement "through new mechanisms of control and constant conflict" (Hardt and Negri 2004, p. xiii) moderated by meticulously synchronized transduction.

In the malls' sector, typical instances of the former kind are static themed ensembles, such as denatured Italianated concoctions of piazzas, palazzi, porticoes, fountains, parterres, and cafes; while responsive metamorphic associations of interactive, hyper-connected open spaces, buildings, and interfaces are characteristic of the latter. The double abstraction reverses the simulated practices of collective production of the city with *mutable immobile* configurations that operate a secondary restabilization of the destabilized centralities to continuously remap, splinter, dissociate, dislocate, and segregate their domains. Ever-expanding networks of *adaptive immobiles* impose themselves as normative models that concurrently exacerbate and counter the dynamic desynchronization of deterritorialized and translocalized urbanites. Abstracted citizens are pacified by alienation and left in disjoined perspectival positions amid an overpowering "plurality of new antagonism[s]" (Laclau 1990).

Discursive fictions no longer replace decoded systems of sociospatial relationality of each place with permanent overcoding. Rather they steadily annihilate and rewrite narratives, themes, characters, and settings through "a strategy of the real, of the neoreal and the hyperreal that everywhere is the double of a strategy of deterrence" (Baudrillard 1994, p. 7). Segregation also negates biases, inequalities, and conflicts, and, consistently, reintroduces them through another negation: the reformulation and resynchronization by the spectacle of the diversity of abstracted retention. While the primary negation operates through dissimulation of missing elements, the secondary materializes hyperreal redistributions of sensible parties that transform the lack of inherited and protensive relationality into eventful choreographed fictions delivering experiential (in)authenticity. Metastable places in permanent becoming live on eventfulness. Static generic cities of consumption are replaced by transformative cities of locally and instantly rebalanced relations between center and periphery, public and private, commons and enclosures that abstractly reconnect places, individuals, routes, and histories (Laclau 1996; Laclau and Mouffe 1985). The affirmation-by-hegemony of these fictions relies on complex mechanisms for the production of the civic derivative where the system makes "itself material in an organ which assumes itself to be the representative of a unitary people" (Laclau and Mouffe 1985, p. 187). Yet, paradoxically, these fictions expose a peculiar character of our *post-civil society* (Dehaene and de Cauter 2008): the capacity of fabricated appearances to frame the disrupted reality of the city as a failed collective oeuvre, induce compliant and complacent depoliticized behaviors that turn the citizen into an accomplice, and sanction their dictated neglect and abandonment of both the public space and the public sphere (Low 2016, p. 296).

Conclusion: Mediatization and the Formation of Counterforces on Transindividuation

Mediatization is the core engine of the accelerating pervasion of urban fiction. The development of digital technology has profoundly contributed to the abstraction of spatial configurations, functions, and meanings of the entire city. New media, precisely the digital, radically modify people's conceptions, practices, and actions by leading a structural shift from the modern *mode of production* to a postmodern *mode of prosumption* (Ritzer and Jurgenson 2010). Digital enhancements of prosumption – the intimate combination of production and consumption – boost the disruptive desynchronization of the general acceleration by redistributing and restructuring all tangible and intangible means, practices, and rhythms of both production and reproduction.

The historical distinction of processes of production and consumption has dwindled and introduced an unprecedented condition: the subversion of the canonic *production-distribution-[realization-]consumption* order with a novel *consumption-[realization-]distribution-production* process that does not supplant the former but instead integrates and overlaps with it, seamlessly, in circulation loops with alternated sequences (Ritzer 2019). Information technology has generalized this reaffirmative subversion by comprehensively suffusing, in each act of everyday life, multiple arrhythmic digitally augmented, more-than-spatial, and multichronic dimensions of spatial engagement.

Acts of consumption pertaining to the recreative sphere constitute practices of prosumption that involve execution of labor profoundly involved in determinant processes of production of goods and services for power and financial accumulation. The activity on social media epitomizes this dual process. On sites and applications such as Facebook and Instagram, where X-realities and mixed temporalities have comprehensive desynchronizing effects, the digital labor of the consumer is concurrently *relational-reproductive*, and *alienating-productive*. A good example is the interplatform content creation through photographs and reviews of food and beverage service experiences, where consumption generates invaluable commercial data which is collected by location and expense tracking and recording systems and processed by AI systems for marketing use (Fuchs 2014; Manfredini 2022b).

Prosumption in the media environment is acutely affected by the fatal arrhythmia of abstract financialization. A profound restructuring led by dominant actors has complemented the subsequent *spatial fixes* that reorganize the city (Harvey 2001) with *digital spatial fixes* (Greene and Joseph 2015) where these actors have succeeded in tightening their hegemony by establishing a strict oligopolistic form of *platform capitalism* (Srnicek 2017). The digital spatial fixes have increasingly involved substantive parts of both the online and offline practices of the individuals' everyday lives, rapidly proceeding towards their control. Through this process – identified in the information and marketing trade as *identity resolution* – the personal domains are dissected, reordered, and reformed into coherent subjects using algorithmic computational systems (Brusseau 2019, 2020).

Conforming to the logic of expulsion that generates the metastable enclosures, these identities are created through abstract determinations of pseudo-integrities that fit the construction of the widening landscape of fiction. Surveillance capitalism reconstructs identities of sublated relationality through modulated desynchronizations and resynchronizations processes. Artificial intelligence politicizes technology by repoliticizing the individual through technological processing of their subjectification. Interplatform engines control the abstractive reidentification parametrically, policing, curating, and supporting how the individual conceives the past, present, and future with *technologies of the self* (Brusseau 2019).

Desynchronized and abstracted individualities are resynchronized and reconstituted in omnipresent digital identities with "integral" retention of the *relational past* by services such as Facebook Timeline. The individual is made into a fragile dynamically stabilized *dividual* (Foucault 1982, p. 5) that is made exploitable not only for its orchestrated co-prosumption, but rather, more importantly, for the collaborative production of the fictions. The *dual abstraction* operates, firstly, by creating fictive identities in the form of pseudo-digital twins that cover reality with a coherent narration and compensate for its dissipation; secondly, through the recombination of such identities that *derivatively reaffirms* the negation of their multiple becoming by homogenizing and adapting them to ever-evolving constructed

M

profiles. Momentary reconciliations of the abstracted individual resurrect – paraphrasing Baudrillard – where the *embedded subject and collectivities* have disappeared.

This problem affecting the individual applies to cities too. The city cannot be substituted by a system of abstracted derivatives that suppresses its relational nature and "cuts us off from those channels that precisely enable us not to distinguish the frame from the person moving about within it" (Latour 2012).

A real-time twin city produced with advanced technologies is not even a simulation. In it, "everything ... remains invisible, everything and, over and above everything, the city taken as a whole" (Latour 2012, p. 91). The abstraction of a city inhibits the engagement with what Latour (2012) defines as its *plasma*, and only produces *membra disjecta* that are not related to their natural, social, and discursive contexts and detract from their political dimension (p. 93). The "smart city" idea epitomizes how collectivities are induced to ignore the process of spatial production, potentially offsetting the benefits of technology by leaving the built environment increasingly controlled by multinational interests, algorithms, and unconstrained avarice (Cuthbert 2020). The criticality of these systems is supreme. The risk involved in their misuse is enormous, as demonstrated at the time of this writing by the move of Google to temporarily disable its Maps live traffic data in Ukraine for the safety of local communities after consulting local state authorities.

Financially, the mediatized cities are decisive for processes functional to hegemonic organizations since they facilitate their externalization and wide-ranging divestment of the fixed production capital. In the city of dynamically resynchronizing enclosures, the "democratic" redistribution (i.e., personally acquired and managed) of means of production (Hardt and Negri 2017) enacted by dominant actors of integrated conventional and platform capitalism is selectively directed to the strict public of accomplices. This redistribution, however, is both residual and strategic. It is residual, since it does not involve the core means that operate on their derivatives but only the basic asset. It is strategic, since it mobilizes and coopts latent forces to expand the unfinished financialization. The after-hours farmers or night markets in parking lots of shopping centers are prosaic and yet relevant examples of this phenomenon. These markets, while seemingly subverting the corporate retail logics, sustain the value of the commercial enclosures strengthening their reputation, centrality, and referentiality.

However, mediatization sustains the emancipatory capacity granted with the redistribution of the means of production. Digitalization enables further reappropriation of these (residual) means and exposes the enclosures' contradictions and expands the prosumers' autonomous connective and creative capacity. Mobile internet liberates new relational spaces for commoning by widening spaces that suspend the imposed law and order, resurging forces of associative production that reactivate processes of *reconfigurative othering* (Rancière 2010) and *counter-desubjectification* (Hardt and Negri 2017, p. 28). Emerging ultra-rapid feedback loops are moderated by real-time subversive machines with the amplitude, efficacy, and agency granted by the digital systems. These loops recreate conditions of "agonistic pluralism" (Mouffe 2016) that transform consenting abstraction into disagreement and conflictual differentiation. This heterogeneity fosters the production of concrete counter-elements that disrupt pseudo-identities, associations, and centralities. Multiplicity generates relational antisystems that decode the (dis) connected externally policed relations of the thick territorial insertions, and, after taking possession and commoning their mechanisms of (de)synchronization, makes available rich mixes of creative instrumentalities for social production. Insurgent discourses extract and reassociate narratives and liberate powers of imagination and desire that establish acculturating associations.

Counter-transterritorializing and counter-deterritorializing processes affirmatively negate the double abstractive paradigms, overturning them into frameworks for immediate and self-determining immanence. The transcendence of the immediate locality enables the subverted apparatuses to sustain eminent relational and anti-extractive social assemblages and commoning practices (Manfredini 2022a, b). Global interconnectedness, permanence, and continuity of instruments, relational systems, and agendas enable cross-pollination, experimentation, and sharing of collectively created innovation. By constituting the practical conditions for radical reconstitution of revolutionary advanced trans-"mutable mobiles" (Latour 2005), reappropriated urban centralities foster the alliance between counterhegemonic and nonhegemonic forces to form productive conflictual networks of equivalence "built up collectively by co-equal groups who have chosen to work in concert" that can "be substantially centralized without a leading class" (Purcell 2012, p. 519). The subversion of the thick territorial insertions instates open transnational networks creating a "space with new economic and political potentialities ... for the formation of new types of presences, including transnational identities and communities" (Sassen 2018b, p. 27). Necessary instrumentalities are liberated by exploiting the internal contradictions of the systems and made available to counterforces for the formation of independent assemblages with reconstructive and recreative seeding machines for sociospatial reproduction. Independent associations for sharing, reciprocal altruism, and collaboration in unique and conjunctural sociality of being together, separately and diachronically demonstrate their transformational potential in overturning the alienating space (Harvey 2018) into surplus space (Amin 2008). This surplus space, surreptitiously produced at the core of ultrascalar urban simulacra through basic commoning practices, forms new assemblies that convert the pseudo-ordinary ordinary into differential extraordinary. Emergent fleeting digitally networked commons radically reconfigure the hegemonically reterritorialized domains, making egalitarian logic emerge from "interruptions, fractures, irregular and local" (Rancière 1999, p. 137), and presenting a "political being-together" that overturns the regimes that negate the exercise of the Right to the City. Paraphrasing Marx, only when relationality has become world relationality and has as its basis large-scale production, only when all nations are drawn into the competitive struggle, is the permanence of the acquired productive forces assured.

Cross-References

- ▶ Augmented Reality: Robotics, Urbanism, and the Digital Turn
- ▶ Big Data for Smart Cities and Inclusive Growth
- ▶ City Visions: Toward Smart and Sustainable Urban Futures
- ▶ Conceptualizing the Urban Commons
- ▶ Digital Twin and Cities
- ▶ Digitalization, Urbanization, and Urban-Rural Divide
- ▶ How Cities Can Be Resilient
- ▶ New Cities
- ▶ Public Space
- ▶ Social Urbanism: Transforming the Built and Social Environment
- ▶ Urban Futures: Pathways to Tomorrow
- ▶ Urban Resilience

References

Amin, A. (2008). Collective culture and urban public space. *City, 12*(1), 5–24.

Baudrillard, J. (1994). *Simulacra and simulation*. University of Michigan.

Brenner, N. (Ed.). (2014). *Implosions/explosions: Towards a study of planetary urbanization*. Jovis Verlag.

Brusseau, J. (2019). Ethics of identity in the time of big data. *First Monday*. https://doi.org/10.5210/fm.v24i5.9624.

Brusseau, J. (2020). Deleuze's postscript on the societies of control: Updated for big data and predictive analytics. *Theoria, 67*(3), 1–25.

Castells, M. (2005). The network society: From knowledge to policy. In M. Castells & G. Cardoso (Eds.), *The network society: From knowledge to policy* (pp. 3–21). Center for Transatlantic Relations.

Castells, M. (2010). *The rise of the network society*. Blackwell.

Connolly, W. (2000). Speed, concentric cultures, and cosmopolitanism. *Political Theory, 28*(5), 596–618.

Cuthbert, A. (2020). Eliot's insight – The future of urban design. *Journal of Urban Design, 25*(1), 11–14.

de Angelis, M. (2017). *Omnia sunt communia: On the commons and the transformation to postcapitalism*. Zed Books.
Debord, G. (1983). *Society of the spectacle*. Black and Red.
Dehaene, M., & de Cauter, L. (2008). Notes. In M. Dehaene & L. de Cauter (Eds.), *Heterotopia and the city: Public space in a postcivil society* (pp. 22–29). Routledge.
Foucault, M. (1982). The subject and power. *Critical Inquiry, 8*(4), 777–795.
Fuchs, C. (2014). Digital prosumption labour on social media in the context of the capitalist regime of time. *Time & Society, 23*(1), 97–123. https://doi.org/10.1177/0961463X13502117.
Greene, D., & Joseph, D. (2015). The digital spatial fix. *Triple-C, 13*(2), 223–247.
Han, B.-C. (2017). *Psychopolitics: Neoliberalism and the new technologies of power*. Verso.
Hardt, M., & Negri, A. (2004). *Multitude: War and democracy in the age of empire*. Penguin Press.
Hardt, M., & Negri, A. (2017). *Assembly*. Oxford University Press.
Harvey, D. (1990). Between space and time: Reflections on the geographical imagination. *Annals of the Association of American Geographers, 3*, 418–434.
Harvey, D. (2001). Globalization and the spatial fix. *Geophische Revue, 2*(3), 23–31.
Harvey, D. (2005a). *A brief history of neoliberalism*. Oxford University.
Harvey, D. (2005b). The political economy of public space. In *The politics of public space* (pp. 17–34). Routledge.
Harvey, D. (2008). The right to the City. *New Left Review, 53*, 23–40.
Harvey, D. (2009). *Spaces of global capitalism: A theory of uneven geographical development*. Verso.
Harvey, D. (2010). *The enigma of capital and the crises of capitalism*. Profile Books.
Harvey, D. (2014). *Seventeen contradictions and the end of capitalism*. Profile Books.
Harvey, D. (2018). Universal alienation. *Journal for Cultural Research, 22*(2), 137–150. https://doi.org/10.1080/14797585.2018.1461350.
Harvey, D. (2019). The enigma of capital and the crisis of capitalism. *Estado & Comunes, Revista de Politicas y Problemas Públicos*. https://doi.org/10.37228/estado_comunes.v1.n1.2013.9.
Hassan, R. (2020). *The condition of digitality: A postmodern Marxism for the practice of digital life*. University of Westminster Press.
Hepp, A. (2012). *Cultures of mediatization*. Polity Press.
Jones, J. P., Leitner, H., Marston, S. A., & Sheppard, E. (2016). Neil Smith's scale. *Antipode, 49*(1), 138–152.
Kohn, M. (2004). *Brave new neighborhoods: The privatization of public space*. Routledge.
Laclau, E. (1990). *New reflections on the revolution of our time*. Verso.
Laclau, E. (1996). *Emancipation(s)*. Verso.
Laclau, E., & Mouffe, C. (1985). *Hegemony and socialist strategy: Towards a radical democratic politics*. Verso.
Latour, B. (2005). *Reassembling the social: An introduction to actor-network-theory*. Oxford University Press.
Latour, B. (2012). Introduction: Paris, invisible city: The plasma. *City, Culture and Society, 3*(2), 91–93.
Lefebvre, H. (1968). *Le Droit à la ville [The right to the city]*. Anthropos.
Lefebvre, H. (2004). *Rhythmanalysis: Space, time and everyday life*. Continuum.
Low, S. (2016). Public space and diversity: Distributive, procedural and interactional justice for parks. In G. Young & D. Stevenson (Eds.), *The Ashgate research companion to planning and culture* (pp. 295–310). Routledge.
Low, S., & Smith, N. (Eds.). (2006). *The politics of public space*. Routledge.
MacKinnon, D. (2010). Reconstructing scale: Towards a new scalar politics. *Progress in Human Geography, 35*(1), 21–36.
Madanipour, A. (2019). Rethinking public space: Between rhetoric and reality. *Urban Design International, 24*, 38–46.
Manfredini, M. (2022a). Affirmatively reading deterritorialisation in urban space: An Aotearoa/New Zealand perspective, in Territories. In A. M. Brighenti & M. Kärrholm (Eds.), *Territories, environments, governance: Explorations in territoriology*. Routledge.
Manfredini, M. (2022b). Envisioning urban commons as civic assemblages in the digitally augmented city. A critical urbanism exploration of counterhegemonic individuation in the age of networked translocalism, multiassociative transduction and recombinant transculturalism. In A. Taufen & Y. Yang (Eds.), *The Routledge handbook of sustainable cities and landscapes in the Pacific Rim*. Routledge.
Marcuse, P. (2014). *City analysis of urban trends, culture, theory, policy, action reading the right to the City*. https://doi.org/10.1080/13604813.2014.878110.
Marx, K. (1973). *Grundrisse*. Penguin Books & New Left Review.
Mezzadra, S., & Neilson, B. (2019). *The politics of operations*. Duke University Press.
Mitchell, D. (2003). *The right to the city: Social justice and the fight for public space*. Guilford.
Mouffe, C. (2008). Public spaces and democratic politics. In *Highrise–common ground. Art and the Amsterdam Zuidas area* (pp. 135–156). Valiz.
Mouffe, C. (2016). Democratic politics and conflict: An agonistic approach. *Política Común*. https://doi.org/10.3998/pc.12322227.0009.011.
Peck, J. (2015). Cities beyond compare? *Regional Studies, 49*(1), 160–182.
Peck, J. (2017a). Transatlantic city, Part 1: Conjunctural urbanism. *Urban Studies, 54*(1), 4–30.
Peck, J. (2017b). Transatlantic city, Part 2: Late entrepreneurialism. *Urban Studies, 54*(2), 327–363.

Peck, J. (2017c). Uneven regional development. In D. Richardson, N. Castree, M. F. Goodchild, A. Kobayashi, W. Liu, R. A. Marston, & K. Falconer Al-Hindi (Eds.), *International encyclopedia of geography: People, the earth, environment, and technology* (pp. 7271–7291). Wiley-Blackwell.

Peck, J., & Whiteside, H. (2016). Financializing Detroit. *Economic Geography, 92*(3), 235–268. https://doi.org/10.1080/00130095.2015.1116369.

Purcell, M. (2002). Excavating Lefebvre: The right to the city and its urban politics of the inhabitant. *GeoJournal, 58*(2–3), 99–108.

Purcell, M. (2012). Gramsci is not dead: For a both/and approach to radical geography. *ACME: An International Journal for Critical Geographies, 11*(3), 512–524.

Rancière, J. (1999). *Disagreement*. University of Minnesota Press.

Rancière, J. (2010). *Dissensus: On politics and aesthetics*. Continuum.

Rancière, J. (2013). In what time do we live? *European Graduate School, 4*, 1–9. https://quod.lib.umich.edu/p/pc/12322227.0004.001?view=text;rgn=main.

Ritzer, G. (2019). Prosumption: Contemporary capitalism and the 'new' prosumer. In F. F. Wherry & I. Woodward (Eds.), *The Oxford handbook of consumption* (pp. 75–93). Oxford University Press.

Ritzer, G., & Jurgenson, N. (2010). Production, consumption, prosumption. *Journal of Consumer Culture, 10*(1), 13–36.

Rosa, H. (2013). *Social acceleration: A new theory of modernity*. Columbia University Press.

Rosa, H., Dörre, K., & Lessenich, S. (2017). Appropriation, activation and acceleration: The escalatory logics of capitalist modernity and the crises of dynamic stabilization. *Theory, Culture and Society, 34*(1), 53–73.

Rowe, C., & Koetter, F. (1978). *Collage city*. MIT Press.

Sassen, S. (2005). The global city: Introducing a concept. *The Brown Journal of World Affairs, 11*(2), 27–43.

Sassen, S. (2014). *Expulsions: Brutality and complexity in the global economy*. Harvard University Press.

Sassen, S. (2017). Predatory formations dressed in wall street suits and algorithmic math. *Science, Technology & Society, 22*(1), 6–20.

Sassen, S. (2018a). Embedded borderings: Making new geographies of centrality. *Territory, Politics, Governance, 6*(1), 5–15.

Sassen, S. (2018b). The global city: Strategic site, new frontier. In L. Ferro, M. Smagacz-Poziemska, M. V. Gómez, S. Kurtenbach, P. Pereira, & J. J. Villalón (Eds.), *Moving cities: Contested views on urban life* (pp. 11–28). Springer.

Schuilenburg, M. (2018). *The securitization of society: Crime, risk and social order*. New York University Press.

Sennett, R. (2017). The open city. In T. Haas & H. Westlund (Eds.), *In the post-urban world: Emergent transformation of cities and regions in the innovative global economy* (pp. 97–105). London.

Smith, N. (1982). Gentrification and uneven development. *Economic Geography, 58*, 39–155.

Smith, N. (2008). *Uneven development: Nature, capital, and the production of space*. University of Georgia Press.

Soja, E. (2010). *Seeking spatial justice*. University of Minnesota Press.

Srnicek, N. (2017). *Platform capitalism*. Polity Press.

Stavrides, S. (2015). Common space as threshold space: Urban commoning in struggles to re-appropriate public space. *The Foot, 9*(1), 9–19.

Stiegler, B. (2012). Relational ecology and the digital pharmakon. *Culture Machine, 13*, 1–19. https://culturemachine.net/wp-content/uploads/2019/01/464-1026-1-PB.pdf.

Viganò, P., Cavalieri, C., Barcelloni Corte, M., Arnsperger, C., & Lanza, E. (2017). Rethinking urban form: Switzerland as a "horizontal metropolis". *Urban Planning, 2*(1), 88–99.

Meeting SDG6: Ensuring Safe Drinking Water for All in Rural India

Aviram Sharma
School of Ecology and Environment Studies, Nalanda University, Rajgir, India

M

Synonyms

Clean water; Drinking water standards; Improved water sources; Rural areas; Safe drinking water; SDG 6; Water for all; Water quality

Definition

United Nations Sustainable Development Goal (SDG) 6 aims to ensure availability and sustainable management of water and sanitation for all by 2030. Despite tremendous efforts and progress made in the water sector, globally, two billion people lack access to safe drinking water. Most of the population without proper access to safe drinking water is primarily from the Global South. India is one of the major countries, which is lagging in meeting the SDG 6 targets. Millions of Indians do not have yet access to safe drinking water. Compared to urban India, the population without access to safe drinking water is higher in rural India. Against this backdrop, the entry reviews the interventions and progress made in ensuring safe drinking water for all in rural India. In addition, the entry charts out the challenges for meeting SDG 6 in rural India.

Introduction

The rate of urbanization in India has increased over the last three decades. Yet, according to the 2011 census, around 68% of the Indian population is rural. In other words, more than 833 million Indians resides in rural parts of the country. The rural economy and society have witnessed major changes over the last few decades. The rural sector's contribution to the overall GDP has gone down, and rural areas are facing multiple vulnerabilities and threats (Bardhan 1999; Pandey 2020). Against this backdrop, the government has planned multiple interventions to address socioeconomic challenges faced by the rural populace. Improving the rural drinking water infrastructure and governance is one such major public policy intervention. The production, distribution, and consumption of drinking water have significantly changed over the years in rural areas (Biswas and Mandal 2010; Srikanth 2009). Yet, the drinking water supply in rural areas is understudied in the academic literature. Socioeconomic, cultural, and environmental aspects of rural water supply and drinking water economies, in general, is inadequately explored in the existing literature. The majority of the studies primarily deals with the monitoring of water pollution in high contamination zones (such as areas affected by excessive fluoride, arsenic). More importantly, the debate on drinking water quality (especially in rural areas) is still confined among small academic and policy groups. The drinking water quality is not comprehensively and adequately analyzed at the regional levels. Due to domestic priorities and international commitment to meet SDG 6 (Roy and Pramanick 2019; Rayasam et al. 2020; Sharma 2021), the focus has returned on quality aspects of drinking water and ensuring safe water for all (GoI 2021a). Currently, the government aims to improve service levels through piped schemes and move away from the earlier policy focus on only expanding access through handpumps and other means (Hutchings et al. 2017).

Water quality issues were treated as local concerns during the colonial era and decades after the independence. Hardly there is any literature which talks about the water quality issues in the rural parts of the country during the 1950s and 1960s. By the 1970s, water quality contaminations started to be reported from different parts of rural India. In the last five decades, water quality concerns have severely aggravated in different regions. Earlier, biological contaminants were the major water quality threats. After the 1970s, chemical pollutants became a significant threat, besides the biological contaminates. During this period, India started setting regulatory institutions (such as Central Pollution Control Board – CPCB, State Pollution Control Boards – SPCB) and designing policies for monitoring, controlling, and preventing water pollution and managing groundwater and surface water sources (such as Water (Prevention and Control of Pollution) Act 1974, the Environment Protection Act of 1986, Coastal Zone Regulation of 1991, and many others) (Sharma 2017). During the later decades, the production of water for drinking purposes has shifted from open wells to hand pumps and eventually to bore wells and piped water supply in rural India. However, the transition is incomplete, and there is a wide disparity in water production and distribution for drinking purposes at the household and regional levels (GoI 2021a). Around 80% of India's rural drinking water needs are met by groundwater (Srikanth 2009), and the groundwater aquifers in different regions of India face multiple threats. For instance, 66 million people are at risk due to high fluoride, and 10 million people are at risk due to high arsenic; similarly, large populations are at risk due to high iron and excess nitrate; moreover, coastal areas face excessive salinity (Susheela 1999; Srikanth 2009; Biswas and Mandal 2010). Along with the poor quality of groundwater, overextraction is widely reported (Biswas and Mandal 2010) from different parts of India.

Policies and Interventions in Rural Drinking Water Sector: "Safe Water" as a Mirage

In the water policy discourse, increasing access to "improved" drinking water sources remained a significant concern in India. Several programs and interventions were devised over the last seven decades to achieve that (Srikanth 2009). In 1954, the Government of India introduced the first program named the National Water Supply and Sanitation Programme to improve access to safe water. In the second phase, the Government of India introduced Accelerated Rural Water Supply Programme (ARWSP) in 1972–1973 to increase the coverage of drinking water supply in the rural areas of India (Shrivastava 2013), which continued for a decade. In 1986, the Government of India converted it into mission mode and introduced the National Drinking Water Mission (NDWM), which was renamed as Rajiv Gandhi National Drinking Water Mission in 1991 (Shrivastava 2013). During this period, the government attempted to increase access to safe drinking water by installing hand pumps and bore wells for rural water supply. The first National Water Policy was formulated in 1987. The second National Water Policy of the Government of India was released in 2002 and the third in 2012. All these policies emphasized on providing "safe" drinking water.

Before the dedicated interventions planned by the state institutions started for improving water access in rural areas, predominantly, the drinking water was dealt at individual and community levels. For most of this period, the policy discourse did not debate or defined water quality in scientific terms. Access to handpumps and bore wells were considered as a proxy for safe drinking water. In 1999, the Department of Drinking Water Supply was created for overlooking drinking water supply in rural areas of the country under the Ministry of Rural Development. During this phase, the rural drinking water supply schemes witnessed major changes. The National Rural Drinking Water Programme emphasized increasing the coverage at the household level rather than the habitation level. It advocated for increasing the role of local elected bodies in water supply schemes (Bandyopadhyay 2016).

In rural areas, drinking water is still predominantly supplied by the Public Health Engineering Department (PHED) in most of the states. Apart from PHED, other organizations entrusted to supply drinking water includes locally named Statutory Agencies and *Panchayati Raj* (a system of local self-government) Institutions. Even after all this investment, it has been reported that around 90% of rural households depend on untreated surface or groundwater to meet their drinking water requirements (Srikanth 2009). According to some estimates, nearly 11,330 million US $ was invested in India for improving water supply in rural areas of India through government schemes until 2007 (Bassi and Kabir 2016). In recent years, India reasserted its commitment to ensuring safe drinking water to all (Roy and Pramanick 2019; Rayasam et al. 2020). To speed up and accelerate the process to meet SDG 6, the Government of India launched *Jal Jeevan* (Water is Life) Mission, which aims to provide functional household tap connection to every household by 2024 (GoI 2021a). Till now, around 40% of the rural population has access to piped water (GoI 2021a). The government wish to reach 100% by 2024 (GoI 2021a). Yet, what is meant by "safe" drinking water is not explicitly explained in the policy discourse and made mandatory by law.

The Drinking Water Quality Discourse: What Is Safe Water?

Even though water quality has been acknowledged as a significant concern by scientists and policymakers (Srikanth 2009; Bandyopadhyay 2016), yet it has not become a major public health concern in public forums. There are no mandatory drinking water standards applicable at the national level (Sharma 2017). Different organizations (such as the Department of Rural Water Supply, Ministry of Urban Development, Indian Council of Medical Research – ICMR, CPCB, Bureau of Indian Standards – BIS) promote different water quality standards for drinking water (Lodhia 2006). Except for the BIS standards for packaged drinking water and natural mineral water, none of them is mandatory (Sharma 2017). Courts in India recognize that access to safe drinking water is a fundamental right under Article 21 of the constitution and the onus to provide safe drinking water is on the state. Even the recent policy intervention (*Jal Jeevan* Mission) has not made it mandatory. In the absence of mandatory standards, water quality concerns are primarily managed individually and in a piecemeal manner in urban India (Bhaduri et al. 2015). Urban India is increasingly dependent on water purification technologies to negate the drinking water quality concerns. Water purification technologies are employed at a much lower rate in the rural parts of the country (Biswas and Mandal 2010).

In the scientific arena, systematic water quality monitoring at the institutional level started after the creation of the CPCB in 1974. Initially, microbial contaminations were the major threat and diseases such as diarrhea, dysentery, and cholera were rampant in rural India. After the 1980s, chemical contaminants became the most significant threat, along with biological contaminants. The early cases of arsenic poisoning were reported in the 1980s. Unlike arsenic, fluorosis was identified as long back as 1937, but state agencies attempted to tackle it not before 1986–1987. Shankar et al. (2011) reported that, according to government figures, out of 593 districts in India, high fluoride is reported from 203 districts, iron from 206 districts, salinity from 137 districts, nitrate from 109 districts, and arsenic from 35 districts in 2006. The situation has further aggravated since then.

During the last two decades, radioactive contamination of groundwater has been reported from few places in India, specifically Punjab. Apart from that, pollution from fertilizers, agrochemicals, and industrial effluents are other significant contaminants. In addition, several emerging contaminants (endocrine disrupters and traces of pharmaceuticals) are reported, which will pose a more significant threat to human health and the economy in the coming decades. Limited resources are available at the individual and institutional level in the rural areas to address these mega challenges. The ill impacts of many of these contaminants emerge slowly over time and have chronic impacts.

In terms of scientific studies, more toxicological and epidemiological studies are required to assess the ground level situation, especially in the context of emerging pollutants (traces of different drugs in water) in rural areas. No guidelines exist for many emerging pollutants, even in many developed countries (Schriks et al. 2010). In such a situation, how the policy actors in India will address these concerns will decide the health of millions of rural populace. In addition, the combined effects of different pollutants on human health are often not well documented, which requires further studies.

The "safety" and "quality" aspects of drinking water must be explicitly defined, and quality standards should be made mandatory. Moreover, rather than treating the standards and guidelines set by different regulatory institutions as static entities, the standards need to be periodically revised based on available scientific knowledge, technological know-how, and public health concerns.

Conclusion

The policy discourse on water quality in India is quite perplexing. The scientific community and government institutions are aware of the threats and risks posed by contaminated water sources (Srikanth 2009; Biswas and Mandal 2010; Sharma 2017). Yet, the public policy interventions to address these challenges have remained ad hoc and incomprehensive. The water quality standards prescribed by regulatory institutions are treated merely as guidelines, and state agencies supplying water in rural areas are not compelled to follow them. The judiciary emphasizes on providing safe drinking water. However, even the judiciary does not underscore the need to define the safe limits (both minimum and maximum permissible limits) for different water contaminants and adhering mandatory standards (Sharma 2017).

The water quality monitoring programs, which were started in the 1970s, are still not able to cover the depth and breadth of the country. Under the *Jal Jeevan* Mission, the federal government has restarted emphasizing water quality monitoring and surveillance. The government expects the drinking water suppliers (PHED and Rural Water Supply Departments) to monitor the quality and the local communities and local self-rule institutions to do the surveillance. The government has devised steps to display the water quality data using digital platforms. In addition, regulatory mechanisms and supporting institutions for water testing have been strengthened. Yet, the number of testing labs opened by the government of India (around 2000 laboratories) (GoI 2021b) are inadequate for covering the entire country, given the size and the population pressure. For instance, Bihar, with 40,172 villages and a rural population of around 9,74,87,502, has only 122 testing laboratories (JJM-WQMIS 2021). Further, the majority of the laboratories are not certified by the National Accreditation Board for Testing and Calibration Laboratories of India (JJM-WQMIS 2021). Inadequate water quality testing infrastructure and low institutional capacity are even reported from urban India (Sharma 2018; Rayasam et al. 2020); in such situations, the challenges for small town and rural India is even more significant.

Independent studies would be required to evaluate how effective and transformative are the various government-community initiatives and public-private partnership schemes planned under the *Jal Jeevan* Mission. However, given the digital divide, weak capacity of the state, and vast numbers of the marginalized and disenfranchised population, the challenges are monumental. The fault lines and stark inequalities between urban and rural and between well-off and marginalized sections of Indian society in terms of access to water were blatantly revealed during the current pandemic. For instance, the World Health Organization and Ministry of Health and Family Welfare advised frequent handwashing during the current pandemic. Such advice is practically impossible to follow for the majority of the marginalized sections living in the urban slums, small towns, and rural India due to lack of access to an adequate amount of water (International Commission of Jurist 2020).

Robust infrastructure to monitor, regulate, and manage the vast water resources at the regional and local level is still missing and drastically varies from region to region. More third-party studies are required to assess the ground-level situation. In addition, more studies from the social science and public health perspectives are required to develop a comprehensive understanding.

The quality and quantity are tied with each other. The quantity aspect (55 liters per capita per day) is clearly explained in *Jal Jeevan* Mission; however, the quality debate is still loosely defined (GoI 2021b). Besides, water stress is also emerging as an imminent threat (Biswas and Mandal 2010), and thus, protecting and managing the overexploited water sources is essential. Pure technological fixes are not going to address this complex problem. For example, certain water purification technologies (such as reverse osmosis – RO) help solve water quality concerns, but in return, they may further exacerbate the water stress at the local level (Bhaduri et al. 2015). The wastage of water is still quite massive in RO, posing a threat to groundwater aquifers in water-stressed areas. It is crucial to move away from short-term techno-managerial fixes. The state needs to invest in developing long term, effective, and participatory governance models and robust supporting institutions to meet the targets of SDG 6 and sustain it in the longer run.

Cross-References

- ▶ Green Economy Policies to Achieve Water Security
- ▶ Small Towns in Asia and Urban Sustainability
- ▶ WASH (Water, Sanitation, and Hygiene): Infrastructure as a Measure of Sustainable Development
- ▶ Water Security and Its Role in Achieving SDG 6
- ▶ Water Security, Sustainability, and SDG 6

References

Bandyopadhyay, S. (2016). Sustainable access to treated drinking water in rural India. In *Rural water systems for multiple uses and livelihood security* (pp. 203–227). Elsevier.

Bardhan, P. (1999). *The political economy of development in India*. Oxford University Press.

Bassi, N., & Kabir, Y. (2016). Sustainability versus local management: Comparative performance of rural water supply schemes. In *Rural water systems for multiple uses and livelihood security* (pp. 87–115). Elsevier.

Bhaduri, S., Sharma, A., & Talat, N. (2015). Growth of water purification technologies in the era of 'regulatory vacuum' in India. *Current Science, 108* (8), 1421–1423.

Biswas, P. K., & Mandal, K. (2010). Drinking water in rural India: A study of deficiency, quality and some social implications. *Water Policy, 12*(6), 885–897.

GoI. (2021a). *Jal Jeevan mission*. New Delhi: Department of Drinking Water & Sanitation, Ministry of Jal Shakti, Government of India.

GoI. (2021b). *Jal Jeevan mission note*. New Delhi: Department of Drinking Water & Sanitation, Ministry of Jal Shakti, Government of India.

Hutchings, P., Franceys, R., Mekala, S., Smits, S., & James, A. J. (2017). Revisiting the history, concepts and typologies of community management for rural drinking water supply in India. *International Journal of Water Resources Development, 33*(1), 152–169.

International Commission of Jurist. (2020). COVID-19 pandemic in India: The right to water. A briefing paper. Geneva: International Commission of Jurists.

JJM-WQMIS. (2021). Jal Jeevan mission water quality management information system: Dashboard. New Delhi: Department of Drinking Water & Sanitation, Ministry of Jal Shakti, Government of India. Accessed from https://neer.icmr.org.in/website/main.php. Accessed on 20 July 2021.

Lodhia, S. (2006). *Quality of drinking water in India: Highly neglected at policy level* (Working paper no. 11). Ahmedabad: CFDA.

Pandey, P. (2020). RRI's commitment to care and vulnerability of agrarian systems: The 'problem' of rice straw burning in India. *Science, Technology and Society, 25* (2), 240–255.

Rayasam, S. D., Rao, B., & Ray, I. (2020). The reality of water quality monitoring for SDG 6: A report from a small town in India. *Journal of Water, Sanitation and Hygiene for Development, 10*(3), 589–595.

Roy, A., & Pramanick, K. (2019). Analyzing progress of sustainable development goal 6 in India: Past, present, and future. *Journal of Environmental Management, 232*, 1049–1065.

Schriks, M., Heringa, M. B., van der Kooi, M. M., de Voogt, P., & van Wezel, A. P. (2010). Toxicological relevance of emerging contaminants for drinking water quality. *Water Research, 44*(2), 461–476.

Shankar, P. V., Kulkarni, H., & Krishnan, S. (2011). India's groundwater challenge and the way forward. *Economic and Political Weekly, 46* (2), 37–45.

Sharma, A. (2017). Drinking water quality in Indian water policies, laws, and courtrooms: Understanding the intersections of science and law in developing countries. *Bulletin of Science, Technology & Society, 37*(1), 45–56.

Sharma, A. (2018). Science-based mandatory standards and the implementation gap: The case of bottled water regulations in India. *Current Science, 114*(1), 29–33.

Sharma, A. (2021). Aligning regional priorities with global and national goals: Constraints on the achievement of Sustainable Development Goal 6 in rural Bihar, India. *Area Development and Policy, 6*(3), 363–373.

Shrivastava, B. K. (2013). Mitigation of naturally occurring fluoride in drinking water sources in rural areas in India: An overview. *Journal of Water Sanitation and Hygiene for Development, 3*(3), 467–478.

Srikanth, R. (2009). Challenges of sustainable water quality management in rural India. *Current Science, 97*, 317–325.

Susheela, A. K. (1999). Fluorosis management programme in India. *Current Science, 77*(10), 1250–1256.

Megacity Region

► Participatory Planning: A Useful Tool for the Development of Sustainable Mega-City Regions

Megaconstellation

► New Orbital Urbanization

Mental Health

► Health and the Role of Nature in Enhancing Mental Health

Metabolic City

► Circular Cities

Meteorological conditions' distortion and surface water resources in Sri Lanka

► Climate Change and Surface Water Resources in Sri Lanka

Metropolitan Discipline: Management and Planning

A New Paradigm for a New Dimension

P. Ortiz[1,2] and A. Contin[3]
[1]International Metropolitan Institute, Madrid, Spain
[2]International Metropolitan Institute, Washington, DC, USA
[3]Politecnico di Milano, Milan, Italy

Synonyms

Dimension/Size; Discipline/Practice; Management/Guidance; Paradigm/Method; Planning/Coordination

The Metropolitan Fact

Definitions

Urban growth addresses a new age phenomenon, which started in the second half of the twentieth century: rapid growth of urban population and especially the explosion of the largest urban agglomerates, metropolis, and megapolis. The definition of metropolis or megapolis is not yet well defined nor accepted. Scholars have focused on two approaches: quantitative and morphological. The quantitative perspective (Eurostat, 2016, 2020; UN-Habitat, 2020) defines megapolises as systems that have reached 10 million-plus inhabitants. Currently, there are about 25 megapolises, depending on how they are measured. The threshold definition of metropolises, however, has been discussed in more depth. The most commonly used is the 1 million inhabitants' threshold: a simple figure, which has historically been linked to the fact that for 5000 years only, three cities had reached that figure (Rome, Beijing, and London). Today there are 500 metropolises, and they house almost 25% of world population. Nevertheless, the demographic increase and the new dimension of inhabited areas have introduced theoretical problems that have not yet been overcome. The management of these new urban realities is also problematic due to the fact that the unitary management of the city of the past is no longer viable. An operational definition of the complexity achieved should take into account a very varied combination of geographical and historical elements, which makes it impossible to propose a purely quantitative definition (Fig. 1).

In terms of the morphological approach, and for the sake of a standard procedure that can be applied to research, scholars tend to conceptualize metropolises as large conurbations: urban continuums that have reached a significant size. The problem with this definition is that the real metropolis extends far beyond the border of that continuum. UN -Habitat has recently published a map of this conflict (Fig. 2).

Nevertheless, the increase in the cities' resident population has often not led to a similar increase in typically urban activities and functions. The city, therefore, has turned out to be only larger with elementary demographic concentrations, but not a placc where the urbanization of the population also corresponds to the greater complexity of social, economic, and cultural relations typical of what over time has been defined as a city. In recent times, the term has been taken up again to describe a metropolitan region, understood as a portion of determined territory, where a city of considerable size performs superordinate and coordinating fusions concerning networks of medium and small cities in the same geographical area. Without denying the interest of other definitions as partial expressions of the metropolitan phenomenon (size, morphology, etc.), a structural definition is much more convincing.

Some important multilateral studies (Angel et al., 2012) define a metropolis as a local government area comprising the urban area as a whole

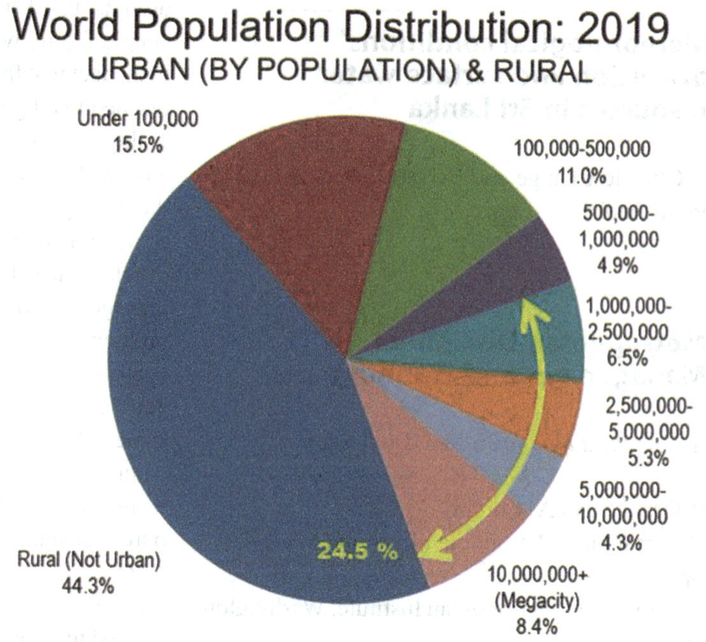

Metropolitan Discipline: Management and Planning, Fig. 1 World population distribution: 2019. Urban (by population) and rural. (Source: United Nations 2018)

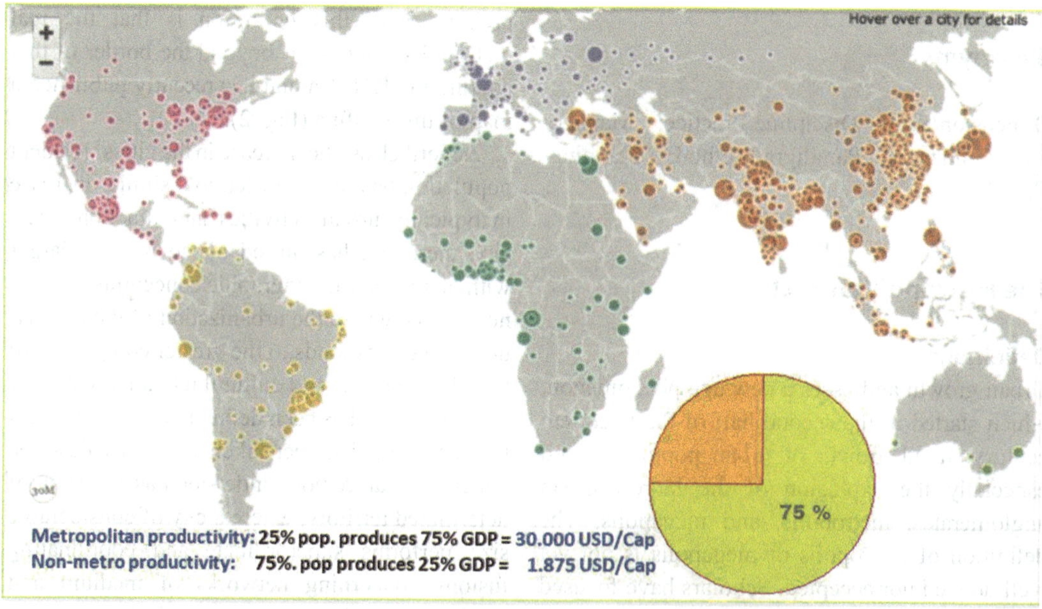

Metropolitan Discipline: Management and Planning, Fig. 2 Urban Agglomeration and metropolitan area. (Source: UN Habitat 2020. Global State of Metropolis 2020 – Population Data Booklet)

and its primary commuter areas to share a daily significant movement of people and goods. Usually the two areas have different densities and also different commuting facilities.

The largest daily flows of commuters into a city were recorded in the northern Italian greater city of Grande Milano ("Greater Milano") and in the French capital, Paris, where in 2011 the local workforce was supplemented, on average, by more than a million commuters. There were more than half a million commuters arriving in the Portuguese, Spanish, and Belgian capitals of Greater Lisbon, Madrid (2008) and Bruxelles/Brussels (2008). The flux of people in the metropolitan area must also be related to the day and nighttime life of the metropolis. Eurostat defines a metropolitan region, by including, additionally to demographic statistics, data such as crimes recorded by police, by economic account; intellectual property rights; business demography; transport statistics; and labor market statistics.

The definition of the "metropolitan fact" must start from a different perspective. The contemporary metropolis enables new lifestyles and behaviors that are in some measure universal and homogeneous, supported by real-time virtual communications. Nevertheless, the issue of the metropolis cannot be reduced to a matter of quantitative data. The scale shift has also introduced a qualitative shift in human interrelations with and within space, as a result from the emergence of the out-of-scale spaces. The idea of measure has always been closely related to the human scale which was comparable with the urban fabric and density parameters of the historical urban settlements. Today, however, this kind of relationship, typical of the local scale or scale of proximity, is embedded in a larger context of mass mobility of people and goods. This implies a greater order of magnitude in the interactions between individuals and groups in space and time. Geographer David Harvey (1989), described this condition as "time-space compression." It is imperative to understand the complex relation between the physical context and the metropolitan patterns of settlement and reconsider the relationships between nature and the built space in the entire scale spectrum as the accessible and inhabitable landscape has to be thought of as a metropolitan public realm. In order to define the metropolitan approach, it is essential to recognize the paradigm shift from the urban to the metropolitan scale, thus considering the contemporary metropolis a "net-city."

According to D. G. Shane (2005) the net-city is a multicentered network system of different sizes, functioning as a whole through a network of physical and virtual infrastructures. In this polycentric system, however, not only the nodes and edges of the network are relevant. According to authors such as Terry McGee (2009), Edward Soja (2000), Neil Brenner (2014), and many more, it is a hybrid territory where the urban and the rural define a seamless heterogeneous landscape.

During the past century, growth models related to economic efficiency have spread the so-called "international" style that often gave a homogeneous look and layout to the newly rising cities. Nature has been reduced to a thin surface of leisure space. The standardized work routine, which started with the industrial revolution and was consolidated with the invention of the conveyer belt, also influenced the idea of standardized quality of life. As metropolises grew and the population became more and more diverse, it became necessary to embrace the complexity in quality of life and thus the shaping of the metropolises through the acknowledgement of the "glocal," the co-existence of global and local conditions. The shift from the paradigm of a polycentric model to the network of the net-city created the need to deal with space in between the networks, which must be reconceptualized with new meaning and structure and new images in the metropolitan era.

These differences bring about significantly different phenomena, behavior, discipline analysis, knowledge, and management criteria. Extreme urbanization suggests the ponderous question of how to deal with the metropolitan complexity to attain the goal of sustainable growth, achieving the well-being of the population in the postcolonial, Anthropocene era. Metropolitan complexity refers to the issues of the contemporary metropolitan system: social and economic inequality; the fragility of environmental ecosystems in relation to the global climate change; the emergence of the political idea of the metropolitan dweller as a global citizen; the preservation of cultural heritage; the postcolonial identity; and the governance and policy issues. These questions cannot be addressed with a single, static, and traditional disciplinary approach, but, rather, they require a comprehensive and multidisciplinary vision in order to be understood.

The Metropolitan Discipline

The Need for Metropolitan Discipline

Size matters. The metropolitan dimension, which is 10 times larger than the urban one, produces a radically different phenomenology compared to the urban one. Those differences concern the four metropolitan components: governance, economy, social, and physical. The specificities of the metropolitan dimension are not yet set out, and, as they differ, it is necessary to approach them in a specific perspective, addressing and responding to those differences, which are in no way urban.

The metropolitan phenomenon is quite recent. The explosion of car ownership in the postwar economic boom of the 1950s and 1960s provided large urban masses a level of mobility never experienced before. Although mass public transport based on the commuter train was promoted in metropolises like London and Paris since the middle nineteenth century, creating satellite towns along these commuter rails, the diffusion of the private car was the event that allowed towns a level of freedom never experienced before. In the USA, this sprawl tendency started in the 1930s, but it has now expanded globally and the rate of car ownership is growing exponentially in the developing economies, making this phenomenon global (Fig. 3).

It takes time to develop a discipline specific to the metropolitan phenomenon; 70 years, from the 1950s onwards, might not have been enough time to do so. However, the lack of understanding of these differentials and the reluctance of the academic world to accept the rise of a new discipline might have played a certain role in this delay. This lack of discipline is producing disrupting consequences. Metropolises are being run as large cities. Metropolitan plans are drafted as regulatory urban plans. The metropolitan disjointed governance system, far from the unitary governance of municipalities, makes those plans inapplicable and useless, if not harmful. The transport problems of the metropolises approached with urban-sized transport models, such as the BRT or the bicycle, are obviously inadequate for the 50 km metropolitan distances. The car is king in absence of adequate metropolitan scale policies.

In terms of the environmental issue, the specificity of the current city is a multipolar way of local growth/transformation involving increasingly large areas based on the effectiveness (performance) of infrastructure networks. If the natural environment is present in the city, while not being the city itself, it becomes peripheral. The metropolis is a discontinuous system; therefore the environment becomes quintessential to the nature of the metropolis. The environment is interstitial to the urban fabric and cannot be separated from it. It is the structure of the

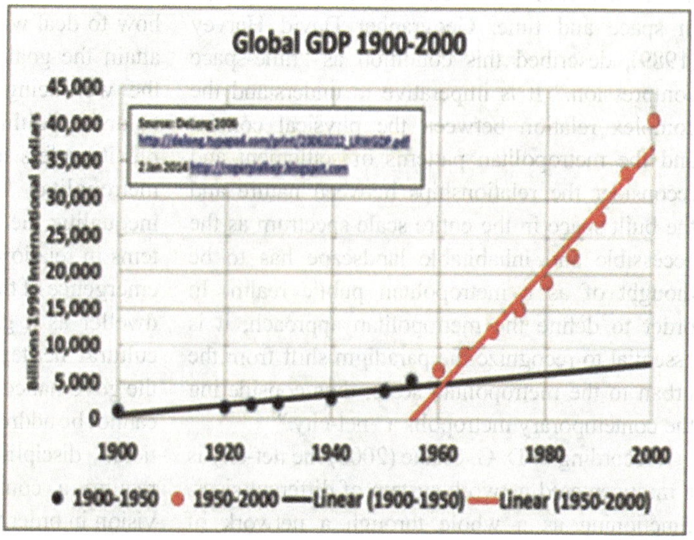

Metropolitan Discipline: Management and Planning, Fig. 3 Global GDP. (Source: http://rogerpielkejr.blogspot.com/2014/01/global-poverty-rates-and-economic-growth.html)

metropolis. This is not fully understood, and the consequence is a considerable destruction of the environmental assets through an unrestricted and uncontrolled expansion of the urban fabric into a metastatic invasive continuum. The recognition of the net-city "space in between" allows to discover new settlement patterns that are beyond the dichotomy of urban and rural patterns. It has introduced the totally new possibilities of shifting between different scales and time that require new spatial practices, social behaviors, and organizational structures. This change, additionally, fostered engagements of new spatial agencies such as private, collective, and public organizations, universities, and families in the interactions, among global and local forces, challenging fixed administrative boundaries at different scales and requiring innovative forms of institutional organization and planning.

The green-gray infrastructure is the metropolitan form, the architectonics (the structure) of the physical reality of the metropolitan scale. The metropolitan city with the green-gray metropolitan infrastructure and its networks of medium and small cities constitute the urban vertebrae of an eco-region. Regarding the physical metropolitan dimension, the goal of the metropolitan discipline is to structure the metropolitan form, identifying the metropolitan dynamics that have generated the vulnerability of territories and recognizing their shortcomings. It proposes a project based on a metabolic vision (maintenance, improvement, or transformation) of the life cycle of the city, which determines the metropolitan biography (Contin, 2016) over time (Fig. 4).

The absence of a discipline is producing a lack of epistemological awareness regarding new theoretical disciplinary questions and harmful management that will be the heritage of an unaware generation to the next, just as the urban errors of the overcrowded and uncontrolled industrial expansion of early nineteenth century. In European cities, this is a heritage still felt in the most conflictive areas of modern cities, 200 years later. The errors in metropolitan management of present metropolises are likely to be endured for centuries to come. This can only be curtailed by the discipline that would appropriately address this management.

The Metropolitan Discipline Versus Urban or Regional Disciplines

The question on the relationship between citizens (the society), politics, economic models, and the spatial dimension has been analyzed by Spangenberg and Bonniot (1998) and discussed during the Lisbon Conference (2000) that dealt with the European Space of Knowledge within the seventh Framework Programme of Research. Within the metropolitan discipline, the metropolitan dimensions as fields of actions of metropolitan projects are divided in four categories following the prism of sustainability: physical, social, economy, and governance.

The Physical Dimension

The physical dimension is the most radically different component. Working with this component requires a design effort to close the gaps between the ideal city (principles) and the real city (metropolitan issues) with metropolitan projects. While urban fabrics require a continuum to provide enjoyment of the street scene and experience, the efficiency of infrastructures, and transportation for proximity to basic social facilities and commercial services, the metropolitan fabric, quite the opposite, needs to be discontinuous in order to allow the environmental interstitial network and, consequently, the rise of the urban nuclei that must remain distinctively separated and connected through mass transport (train commuters).

The metropolitan discipline, according to the Metro Matrix paradigm, defines five physical components linked to the main functions of a city. The green and gray infrastructure are a continuous system, whereas the three components of housing, industrial facilities, and social activities are discontinuous elements that still function as a network in the city (Fig. 5).

Within the metropolitan discipline, these five components cannot remain separate but need to be integrated both spatially and conceptually. The main issues to consider in the green and gray infrastructure are:

- Ground design concerning environmental knowledge and risk, water management

Metropolitan Discipline: Management and Planning, Fig. 4 Aerial view of Torreon. (Source: A. Contin)

- Mobility and permeability utilizing and sharing data
- Public realm as a continuously evolving cultural landscape with new a process of inclusion and new functions that generate the green economy

The main issues to consider among the discontinuous components are:

- Social housing: affordability; policy and technology related to sustainable construction methods; smart devices used in interior environment
- Industrial activities: sustainable economy, 4.0 industry, research and education
- Social activities: welfare and cultural heritage

As we have already said, the green and the gray infrastructure must be conceived as a new integrated infrastructure: the green-gray infrastructure that constitutes the strength of the metropolitan region and city. This concept of space relevant to the new situation of metropolitan life needs to be immediately perceived and understood as a principle of mapping and possible mental map even on a metropolitan scale. The tonality of the social behavior key becomes metropolitan and no longer rural or urban.

From the Großstadt to the Metropolitan Models The metropolitan structure is "a set of urban units," as opposed to a single unit. Edward Leman in 2001 set up the diagrams of four

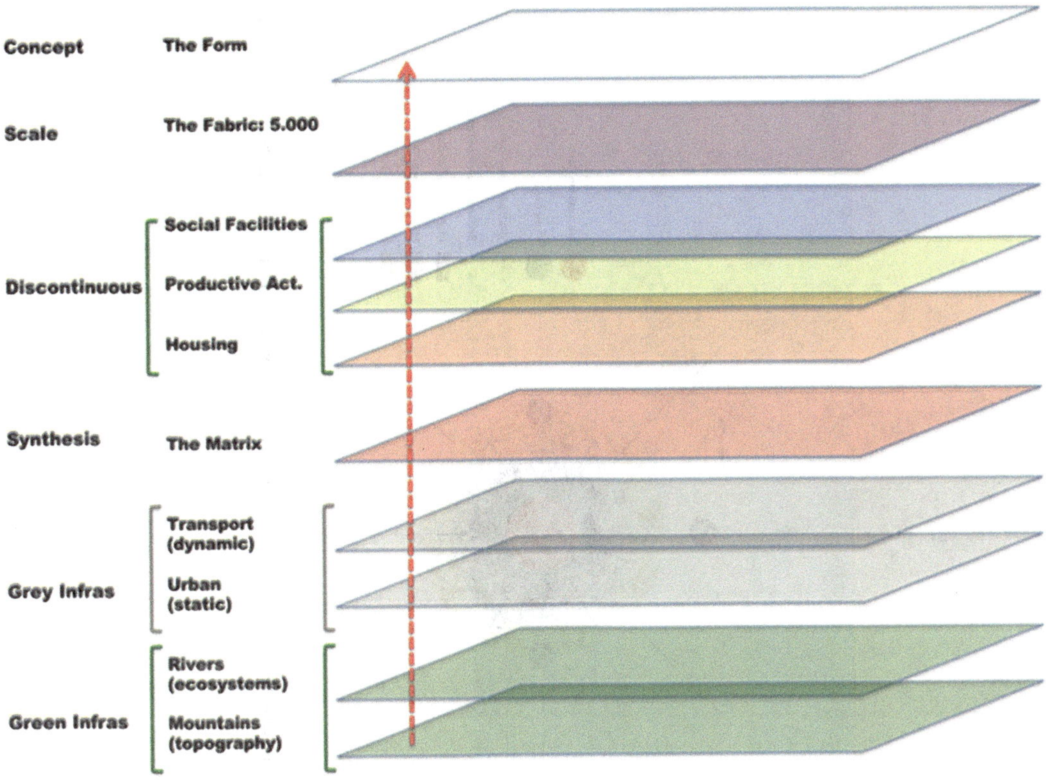

Metropolitan Discipline: Management and Planning, Fig. 5 The metro toolkit. The five physical dimension components. (Source: P. Ortiz)

metropolitan typologies: the first monocentric, the last polycentric. By describing the monocentric structure, he was in fact describing what a metropolis is not and should not be. He was describing a city overgrown into a metropolis, not understanding that a metropolis actually has a different DNA compared to a city (Fig. 6).

The problem arose long ago. Each discipline should base itself on history, not because history itself would legitimize or justify anything in the future but because every cultural question in relation to a specific theme can be identified in the history of the discipline. History is the ideal place to start re-evaluating the original answers, by once more asking the questions in these present times.

Based on the theory of practice presented in the Renaissance texts and on the principles of military art, the necessary doubling of the city begins to take shape in the 1500s. That growth took place in two ways: the construction of a new city parallel to the first and the multiplication of settlements, according to a preferred direction. The Großstadt developed in the last century, consuming its territory. The paradigmatic models of the Anglo-Saxon multiplication were born with New York and London. The growth of New York, in particular, developed within a model that provided strips of fast lines (the avenue) intersected by slow streets that introduced "villages," completely erasing agriculture from the city. Jane Jacobs (1961)'s argument, who in 1961 published *The Death and Life of Great American Cities*, and the subsequent Lynch (1996) studies tried to remedy – through new analytical tools – this far-ranging cancellation of geography and landscape from American cities.

The alternatives to this model of growth were the centripetal forms of Vienna first, Paris (with its plaza-hinge points) and Barcelona (with the diagonal inlaying in the territory) later. Vienna, in

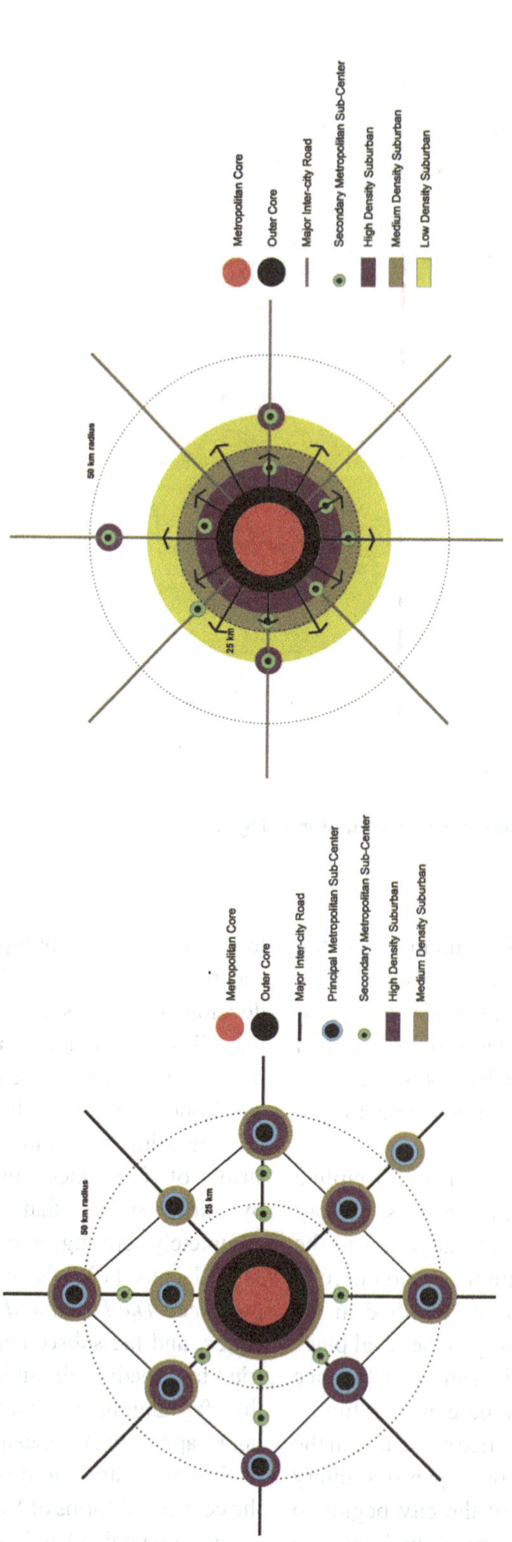

Metropolitan Discipline: Management and Planning, Fig. 6 Monocentric urban structure and polycentric metropolitan structure. (Source: Edward Leman, chreod ltd., 2001)

particular, represents a point of no return: in 1910, Otto Wagner was invited by the Columbia University to give a talk at the International Congress of Urban Art in New York, where he decided to present the study "Die Großstadt" (The Metropolis) whereby he imagined Vienna extended by unlimited growth. In 1892, the Vienna Municipality launched a competition for the elaboration of the new Vienna plan, which was won by Otto Wagner, thanks to a project whose theme was *Artis sola domina necessitas*. The project was based on the expansion of the city starting from the center. According to Wagner, the planning and creation of a dense network of public transport and large and straight streets is fundamental to such a big city: it allows prolonging the radial lines and the construction of 20 of what he called *Stellen* or district centers, equipped with the infrastructure and services needed in a modern metropolis. The size of each *Stelle* corresponds to that of an old village: 2 km × 2 km. Thus the size of a new metropolitan digit was established.

In 1933, Walter Christaller (Christaller, 1933) developed the central place theory presented in the work "The central location in Southern Germany," in which he theorized a distribution of the jurisdictional hierarchy structure in relation to the geographical structure of the territory and in relation to three variables: the courts, the markets, and the mobility network (time). What matters in Christaller's model is permanence or the territorial "image" of the model. This outlines a new relationship between the elements – needed for the metropolitan dimension – and the land necessarily deformed because of their impact. This will be the problem of the future metropolis. This approach, which is included in the field of localization theories, was inspired by the concept of urban hierarchy and identified the rules by which it is possible to interpret urban systems, explaining size, frequency, and distance of urban centers at each hierarchical level. (Unlike Christaller, in the Metro Matrix model, Ortiz Castano (2014) provides the relationship between two infrastructure networks, train and highway, industry, housing, and services).

In his American period, Hilberseimer's urban theory evolved through a complex review of previous studies on the Big City and through a verification of the dimension of the city and the architectural dimension (Hilberseimer, 2014). The progress of his research is developed primarily around the theme of the relationship between city and territory, introducing then the issue of the regional city dimension and, consequently, the multiple scale theme. Hilberseimer changed his point of view between 1920 and 1944. In 1920, the discussion centered on the compact city and was expressed in the vertical city paradigmatic figure.

In 1944, the decentralization and regional dimension of the city, expressed through the linear city paradigmatic figure, were the main subjects of study (Hilberseimer, 2012). Within the new regional pattern, (Hilberseimer, 1949) the decentralization produced by the exodus from the industrial city was in line with the desire to return to nature. In the plan for Chicago, in 1948, working on a proposal for urban reorganization, based on the first hypothesis for the rail network, Hilberseimer developed a paradigmatic ideal form of a metropolis merged into the landscape. The widespread urban expansion in the territory was conceived as a linear structure of horizontal links: motorways and railways. Hilberseimer criticized the introduction of the car and its crucial role in the development of urban sprawl. The car accelerated the process of decentralization that was not yet based on adequate principles of planning. In Chicago, therefore, he started from the reorganization of road and rail traffic and developed solutions that defined a new relationship between road and block, reconstructing the relationship between residence, work, and leisure. His effort was aimed at understanding the new relationship that links the anthropometric scale with the territorial scale. The impossibility of defining a relationship between the two scales determined the model of settlement units: independent units of settlement, limited in size, containing all the elements necessary for an individual town, defined in its function. The prototypes of settlement units are related to a low density; the relationship between the different areas of the city are all detailed in numerous diagrams and plans.

In his plan, the backbone of the unit is the main traffic artery separating the different city functions. The main themes developed by Hilberseimer were zoning (seeking the best shape for the relationship between work, housing, and leisure); the role of the two infrastructures structuring the plan; the hierarchy of roads to solve the traffic issue; the proposal of the introduction of nature in the city; and the different types of residential buildings. The definition of the theoretical system of the urban sprawl was related to the idea of a flexible plan that was despite the fact that Hilberseimer discarded the idea of public spaces, whose solution was left to the natural elements. The Broadacre City of Frank Lloyd Wright, 1958, seemed to be the alternative.

The 1950s, with Jacobs, Lynch (1960, 1996) in the 1960s, 1980s, and up to the 2000s with Venturi et al. (1972), Frampton (1983), and the geographer Gottmann (1961), saw the beginning of the theoretical critique of the huge growth city model, deemed unable to produce urbanity. In 1961, in his text *"Megalopolis: the urbanized Northeastern Seaboard of the United States,"* Gottmann reveals the emergence of a new urban form that grows linearly with a multicentric structure and multiple concentration of functions for the 700 kilometers between Boston and Washington via the largest cities: NY, Philadelphia, Baltimore, with a density, then, of 40 million inhabitants.

The empirical approach of Appleyard (Appleyard et al., 1964) and Lynch's "*The view from the road*," (1964), and Venturi's "*Learning from Las Vegas*," (1972), although with different approaches, introduced a pragmatic perspective that was excluded from a theoretical discussion in the European academy. Frampton, 1999, in *Megaform as Urban Landscape*, asserted that the breaking point in the contemporary urban narrative begins when, in the twenty-first century, the industrial city expands, breaking the city that lay within the walls. These two facts, understood as an anomaly or the urban paradigm puzzle, marked the need for a radical change of the traditional "set of methodological rules, interpretive hypothesis and explanatory models and define practices." It was a real revolutionary breakdown, until the completion of a new shared metropolitan discipline.

In 1986, Rem Koolhaas (2006) presented the "Ville Nouvelle" for Melun-Sènart. This project proposed three main features:

- City as expression of bigness
- Control on the morphology from a non-morphological point of view, working on the problem of designing accidental relationships within a complex system
- Work of assembly functions that take a linear form as a basic element, recognizing its structuring value

The Contemporary Metropolitan Models: The Modernity of the Urban Form as a Critique to the Megalopolis Traditionally urban history acknowledges three main stages of the urban form that Cedric Price (Hardingham, 2016) in the Taskforce Project of 1982 envisioned as breakfast egg dishes: the hard-boiled, the fried, and the scrambled city. Drawing from the ironic reading of Price, a group of planners at ISOCARP (International Society of City and Regional Planners) in 2001 named them, respectively, "archi-città," "cine-città," and "tele-città." The late stage of the urban form is described as the "tele-città" to stress, according to Grahame Shane (2005), the role of television as a cohesive medium of a dispersed physical reality. Indeed, the ancient Greek prefix "tele-" (τηλε- meaning "far") accentuates the distance and physical separation of urban elements. The city comes in the form of a scrambled egg where it is no longer possible to recognize neither yolk nor egg white, but it is rather an amorphous agglomeration of diverse pieces and fragments spread across vast tracts of the territory forming an urban archipelago (L. Hilberseimer 1944, The new city. Principles of planning, Chicago: Paul Theobald). That triad obviously refers to the three regulatory models proposed by Lynch in 1981. In "*The good city form*," (Lynch, 1996) the latest model, the eco-city, can be seen as the fulfillment of the effort to de-structure the Metropolis, which begun in the 1960s, which was motivated by a profound criticism of the primacy of the structure of movement as the foundation of the metropolitan dimension, as well as a profound criticism of the cancellation of relationship with nature.

The metropolitan scale cannot simply be a matter of flow system, but it is also linked to the adapted space distribution. The space must be described in terms of categories describable at every scale, namely, controllable and significant in relation to the scale of belonging (the real form of every scale). Every space derives its meaning from its structure, location, and size, which also shapes its identity. To produce, therefore, a sense of identity even at the metropolitan scale, it is essential to design a recognizable landscape of high-quality housing, with a robust civic image even at the metropolitan scale.

David Grahame Shane in 2005 in his text Recombinant Urbanism, and in 2011 with the book *Urban Design Since 1945: A Global Perspective*, emphasizes the fact that the contemporary model of net-city is the new paradigm that defines the transition from the metropolitan dimension to the mega city and the meta cities. Thereafter, the condition of net-city leads to the creation of a new discipline, landscape urbanism, which will identify the environmental issue as crucial in the re-conceptualization of those "spaces in between" or "spaces of the differences" Shane had called the "body space" of the new urban dimension.

Today, due to the growth of the landscape urbanism discipline, landscape is interpreted as an urban ecosystem and designed according to the two concepts of resilience and robustness. It is a landscape that is always seen within a continuous metabolic process. Starting from its exploitation, at the time of its birth, it leads to a reorganization aimed at storing and then subsequently to a release, so that to a death or destruction corresponds the creation of some of its parts, leading today to its renewal by means of the use of new systems of energies. The environmental issue, too, as mentioned, is addressed in particular by the American School of McHarg (1969) "*Design with Nature*" and today by Waldheim 2006, *The Landscape Urbanism Reader*, and Terry McGee (2009), "*The spatiality of Urbanization: The Policy Challenges of Mega -Urban and Desakota Regions of Southeast Asia*."

The metropolitan territory is a hybrid territory. According to McGee, there are two stages of development: in a first phase, when the urban subjects do not yet have the economic strength to implement a concrete development, an archipelago system, regulating a water and energy system loop, can survive on-site in a more local dimension. This is, however, to prepare the territory to be coupled at a later time to the typical size of the metropolitan model, which is characterized by the form of a net, a grid. This form is more characteristic of the metropolitan dimension off-site and envisages the possibility of the coexistence of models at different scales, linked to different economies. The idea of a hybrid metropolitan landscape and a related hybrid architectural entity is relevant because the body dimension is already fundamental in order to experience the metropolitan models. The body, though small, is essential to understanding the real impact of the gigantic metropolitan projects and the meanings of certain regional organizations on the living spaces. To take this into account, the metropolitan urbanity is shaped by places in which it is possible to find the permanence of the scale value, and different times, too.

For this reason too, the metropolitan discipline, together with landscape urbanism, takes on a fundamental role: to study the size of the net-city and especially the hybrid landscape that lies between the networks and is to this day rarely conceptualized as an acceptable living space that can be considered the true realm of the contemporary metropolitan public good (Fig. 7).

The Metro-Matrix Naturally evolved metropolises are such because they are in key positions that provided them with a location advantage. These key positions are most generally the point of interface between two ecosystems: land and sea, river crossing, and mountain ridge passage. Controlling that point makes them the metropolises they have become. Those ecosystems have, most of the time, a linear border, coast, river, and ridge, that provides them with a main directionality and a cross directionality of the gradients of intensity to that linear feature: this forms a reticular system, as opposed to a hexagonal one. Metropolises are never in a featureless

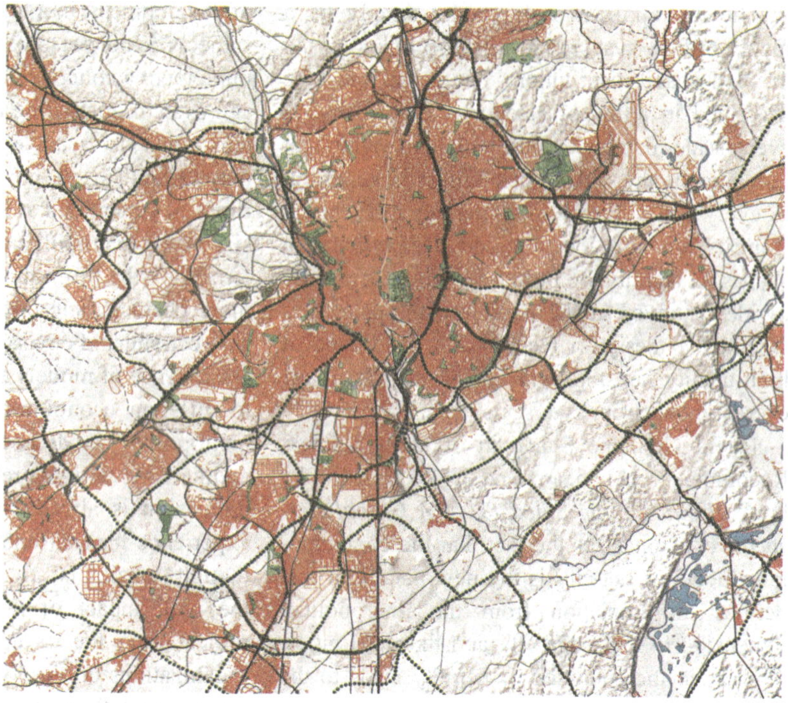

Metropolitan Discipline: Management and Planning, Fig. 7 Metro-matrix: Madrid 1996. (Source P. Ortiz)

plain. That is why they are not hexagonal either: they are reticular.

The matrix is a way to explain the birth of a metropolis in relation to the rule of its physical shape without resorting to an economic model. It is a mental paradigm that comes from a deep ambition to synthesize, demonstrating a strong will towards a geometric choice. By changing the existing shape of the city, the city articulation changes among its different units, and the risk leads to a fogging of the existing relationships. Blocks in the metropolitan fabric no longer exist, substituted by fields of practicability. The form of the city has opened to the line: no longer, therefore, streets and blocks, no longer the interpenetration of a microcirculation in the center, but rather the articulation between different orders of magnitude. The question of the metropolitan project will concern the definition of the unit of intervention and how micro intervention should be articulated at the higher macroscale: what persists, and what changes? Thus, the search for the definition of a metropolitan potential.

It is a tool for action. The metro-matrix presents a diachronic, dimorphic, and diastatic structure which is stable and complete in time, form, and scale. It provides the clues for what is missing or what is needed. As in an isostatic truss structure, each knot and each bar is necessary. If any are missing the structure fails. In the metro-matrix, if a bar or a knot is missing, the structure underperforms, and these should be provided. The metro-matrix allows for redundancies. Inherited wrong decisions must be considered as possible unbuilt yet wrong investments but politically committed. The metro-matrix includes these deviations, although it focuses on minimizing redundancies and allowing for unwanted alterations to the core of the structure.

The metro-matrix immediately allows prioritizing the main strategic projects necessary to the metropolis. It is an integrated tool for action even though political and administrative action is mostly disjointed and incrementalistic. Comprehensive strategic and structural plans (see difference in attached document) would be welcome but are not imperative. As a matter of fact, metro-matrix proposals have been implemented around the world as independent projects without the backing of an integrated plan.

The metro-matrix methodological tool puts the metropolitan geographical structure under the accordingly located set of urban units. The metro-matrix developed for Madrid in the 1996 Metropolitan Plan owes its approach to the 1929 New York regional plan that sets a double reticulum in New Jersey and Long Island-Connecticut and to the Milton Keynes Plan that expands the nineteenth-century urban grid to an organic 1-mile-square reticulum. If metropolises have two perpendicular directionalities imposed by the geographical structure of their setting, that natural reticulum has a different mesh width depending on the geological conditions. The metropolis should be planned according to its mesh width and underlying DNA. London Green Grid 2012, Ruhr's regional plan, and the Gran Paris approach with the Rocade are all developed, thanks to this reticular theory.

Governance Dimension

One of the problems present-day metropolises suffer from is that the management and planning is produced applying urban solutions to an object that is not urban and requires a different set of knowledge.

Phenomenology of Metropolitan Contexts: Relationships Between Scales The metropolitan discipline analyzes the scale relationships between the different levels of the territory that make up the metropolization phenomenon. There are four basic relational levels of net-city determincd by:

- Interaction between the intercontinental metropolis
- Interaction between the metropolitan scale and the different regions
- Interaction between the urban centers and the suburbs
- Collaborative interactions between neighborhoods within the urban scale

Starting from these situations, the metro-discipline determines some integrated projects, which deal with interface spaces and related functions able to allow the possible scale relationships. These happen in all metropolitan contexts that are:

- Near-spontaneous sprawls next to regional roads. Hybrid Ground (Rural-Urban)
- The ground as amorphous support of the infrastructural network (gray and web) hyper-planned. Strategy for a new coexistence between old and new portions of the city (intimate time) and the planning of hyper-infrastructure networks
- The natural ground. Eco Armature, porous space, connected and permeable
- The object that absorbs the ground. Metropolitan epicenters as public realm: hybrids or distributed

A specific jurisdiction able to define the scale problem setting is defined at each level. For each environment, the operation necessary to build the metropolitan shape or to reshape must be outlined.

The Peer-to-Peer Dialogue Among Institutions Municipalities have a single government that rules the whole metropolis. In constitutional terms, this is called a unitary system although several approaches to the implementation of a unitary system either as centralized or decentralized are present. On the other hand, in the unitary system power is held at the center, with different levels of delegation. That is not the case of metropolises.

Metropolises cover many municipalities, each with their own sovereign capacity to make decisions. The relevant investments required by the metropolises, however, cannot be performed with the meagre municipal budgets, often only 1% of the public investments. They are performed by the central state, with a central budget. The national ministries become the main actors of metropolitan management. Additionally, there are various agencies for continuous structural systems, such as environment and transport, removed from discontinuous systems, housing, productive activities, and social facilities, when the legislative branch of government realizes these systems cannot be managed following a disjointed incrementalistic approach. This disaggregated reality does not have a constitutional structure, despite the need for it.

The two systems are radically different. The unitary system has a pyramidal structure of power

M

where sovereignty is at the apex, with various delegations below. The circular analogy is also valid, with power at the center and peripheral delegations. Aristotle defined this power structure as *Potestas*. The participatory dialogue between center and periphery is defined as bottom-up top-down popular participation.

The metropolitan disaggregated structure in need of a system is a peer-to-peer dialogue, as each of the institutions involved has its own competences and independent responsibilities. The dialogue among all the actors needs to be constant and reiterated. The metropolitan region diagram of the metro matrix is a matrixial structure where each member represented in the columns and rows must establish communication with all others in the intersections of rows and columns. The dialogue should be carried out through negotiations, intelligence and compromise, and not by imposition. Aristotle called this approach *Authoritas* (Fig. 8).

Metropolises require a specific form of government to be able to address problems that cannot be solved by aggregation of disjointed municipal policies. This can only be one of three types. Constitutional discipline provides three alternatives: confederal, federal, and unitary.

Most national states (and international multilaterals) promote the confederation of municipalities for metropolitan management. Confederations in history have not lasted long. However, their limited efficiency, regarding coordination and dialogue, might be in the intention of national governments afraid of the importance that metropolises have in the national economy, and the political power they might acquire with a federal devolved or unitary decentralized system.

Economy Dimension

Metropolises are enormously powerful in economic terms. Urban economics or urban development deals with dimensions similar to national economics in need of policies as complex and extensive as national macroeconomics (Fig. 9).

The absolute gross domestic product (GDP) of metropolises is similar to the national ones and often superior to the GDP of many entire nations. In a comparative analysis made for UN Habitat in 2018, out of 195 nations in the world, Tokyo proves to be the 15th "nation" and New York the 16th, ranking higher than Mexico and 180 other nations. Madrid is in a range similar to Peru or Colombia, and Mumbai ranks higher than Pakistan, etc. Out of 100 most productive territorial units in the world, 46 are metropolises. Out of the next 100, the first half are metropolises. It is not possible to compare metropolises to cities. They are similar to nations, and as such, they have to be economically managed as nations (Fig. 10).

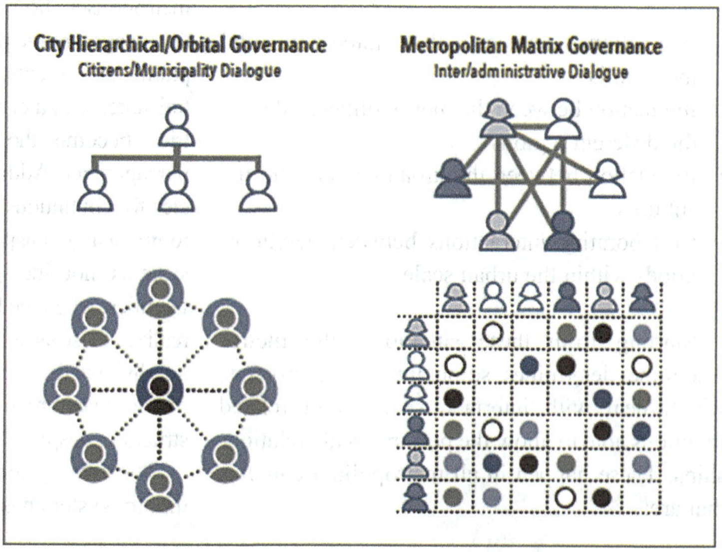

Metropolitan Discipline: Management and Planning, Fig. 8 Metropolitan governance matrix. (Source: Interamerican Development Bank – Steering the metropolis, 2017)

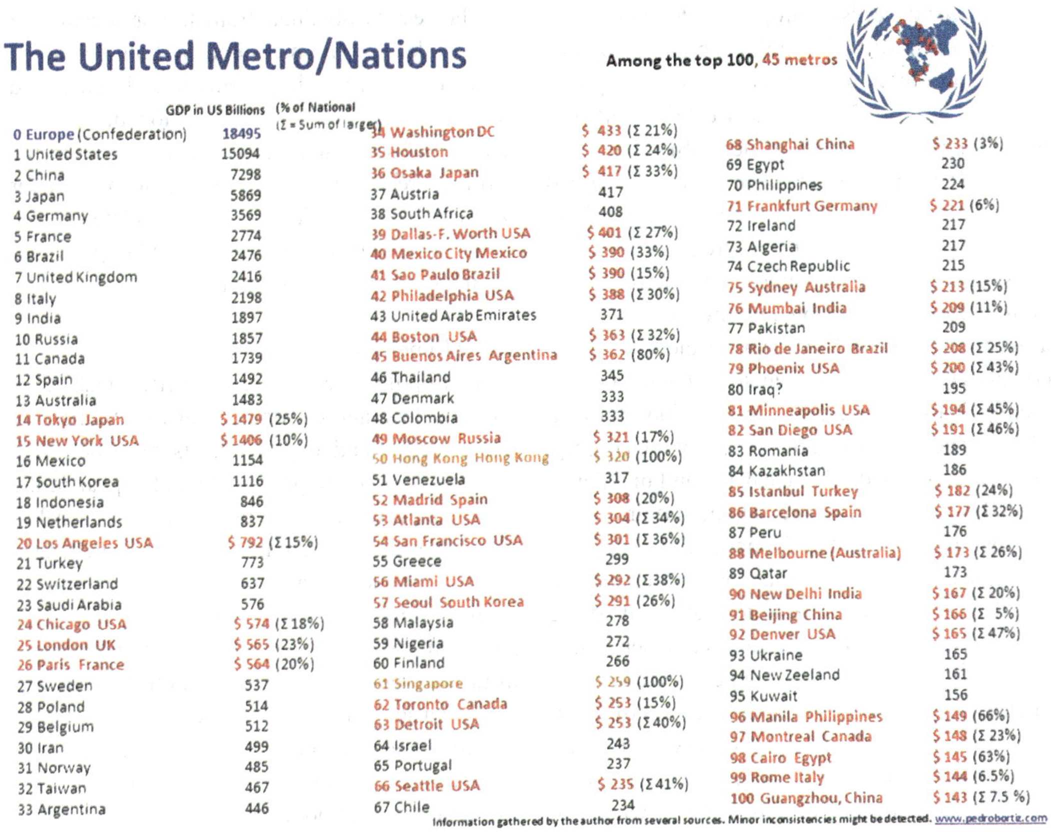
The United Metro/Nations
Among the top 100, 45 metros
GDP in US Billions
(% of National
(Σ = Sum of larger)
0 Europe (Confederation) 18495
1 United States 15094
2 China 7298
3 Japan 5869
4 Germany 3569
5 France 2774
6 Brazil 2476
7 United Kingdom 2416
8 Italy 2198
9 India 1897
10 Russia 1857
11 Canada 1739
12 Spain 1492
13 Australia 1483
14 Tokyo Japan $ 1479 (25%)
15 New York USA $ 1406 (10%)
16 Mexico 1154
17 South Korea 1116
18 Indonesia 846
19 Netherlands 837
20 Los Angeles USA $ 792 (Σ 15%)
21 Turkey 773
22 Switzerland 637
23 Saudi Arabia 576
24 Chicago USA $ 574 (Σ 18%)
25 London UK $ 565 (23%)
26 Paris France $ 564 (20%)
27 Sweden 537
28 Poland 514
29 Belgium 512
30 Iran 499
31 Norway 485
32 Taiwan 467
33 Argentina 446
34 Washington DC $ 433 (Σ 21%)
35 Houston $ 420 (Σ 24%)
36 Osaka Japan $ 417 (Σ 33%)
37 Austria 417
38 South Africa 408
39 Dallas-F. Worth USA $ 401 (Σ 27%)
40 Mexico City Mexico $ 390 (33%)
41 Sao Paulo Brazil $ 390 (15%)
42 Philadelphia USA $ 388 (Σ 30%)
43 United Arab Emirates 371
44 Boston USA $ 363 (Σ 32%)
45 Buenos Aires Argentina $ 362 (80%)
46 Thailand 345
47 Denmark 333
48 Colombia 333
49 Moscow Russia $ 321 (17%)
50 Hong Kong Hong Kong $ 320 (100%)
51 Venezuela 317
52 Madrid Spain $ 308 (20%)
53 Atlanta USA $ 304 (Σ 34%)
54 San Francisco USA $ 301 (Σ 36%)
55 Greece 299
56 Miami USA $ 292 (Σ 38%)
57 Seoul South Korea $ 291 (26%)
58 Malaysia 278
59 Nigeria 272
60 Finland 266
61 Singapore $ 259 (100%)
62 Toronto Canada $ 253 (15%)
63 Detroit USA $ 253 (Σ 40%)
64 Israel 243
65 Portugal 237
66 Seattle USA $ 235 (Σ 41%)
67 Chile 234
68 Shanghai China $ 233 (3%)
69 Egypt 230
70 Philippines 224
71 Frankfurt Germany $ 221 (6%)
72 Ireland 217
73 Algeria 217
74 Czech Republic 215
75 Sydney Australia $ 213 (15%)
76 Mumbai India $ 209 (11%)
77 Pakistan 209
78 Rio de Janeiro Brazil $ 208 (Σ 25%)
79 Phoenix USA $ 200 (Σ 43%)
80 Iraq? 195
81 Minneapolis USA $ 194 (Σ 45%)
82 San Diego USA $ 191 (Σ 46%)
83 Romania 189
84 Kazakhstan 186
85 Istanbul Turkey $ 182 (24%)
86 Barcelona Spain $ 177 (Σ 32%)
87 Peru 176
88 Melbourne (Australia) $ 173 (Σ 26%)
89 Qatar 173
90 New Delhi India $ 167 (Σ 20%)
91 Beijing China $ 166 (Σ 5%)
92 Denver USA $ 165 (Σ 47%)
93 Ukraine 165
94 New Zeeland 161
95 Kuwait 156
96 Manila Philippines $ 149 (66%)
97 Montreal Canada $ 148 (Σ 23%)
98 Cairo Egypt $ 145 (63%)
99 Rome Italy $ 144 (6.5%)
100 Guangzhou, China $ 143 (Σ 7.5 %)
Information gathered by the author from several sources. Minor inconsistencies might be detected. www.pedrobortiz.com

Metropolitan Discipline: Management and Planning, Fig. 9 The united metro nation list. (Source P. Ortiz)

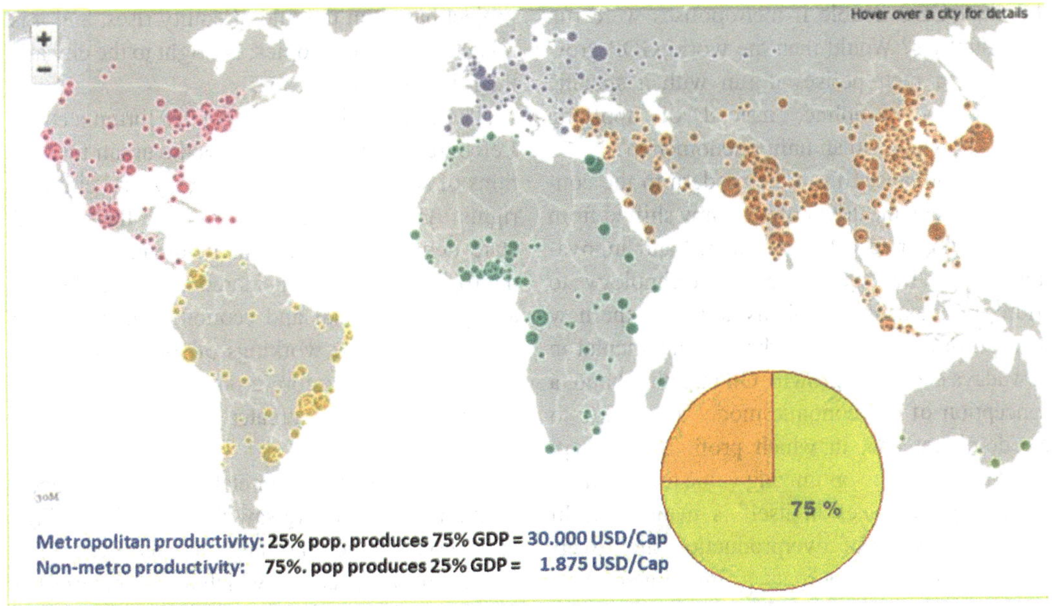
Hover over a city for details
75 %
Metropolitan productivity: 25% pop. produces 75% GDP = 30.000 USD/Cap
Non-metro productivity: 75%. pop produces 25% GDP = 1.875 USD/Cap

Metropolitan Discipline: Management and Planning, Fig. 10 The benefit of the metropolitan productivity. (Source: P. Ortiz)

The world needs metropolises to work. The 500 metropolises with 1 million-plus inhabitants that account for almost 25% of the world's population account for 75% of the world's GDP. The world GDP is close to 70 trillion USD. The GDP per capita for metropolitan inhabitants is 30,000 USD. The GDP per capita for non-metropolitan inhabitants is 1.875 USD. Metropolitan inhabitants are 16 times more productive than non-metropolitan inhabitants are. It could be argued that metropolises deprive the rest of the country of resources for their sole benefit and that metropolises are in a dominant position, drawing off the benefit produced by the rest of the country, going even beyond. That cleavage and unbalance could be considered inefficient and unfair.

Nevertheless, the fact is that national country governments are responsible for this unbalance, and they are proving unable to manage it. If the economy of metropolises were run independently, they would probably achieve an even greater productivity and a more substantial amount of transfers to the rest of the country for their development and service provision. The living proof of the efficiency of a metropolis, when run as a country economy, is Singapore. The GDP per capita of Singapore is 65,000 USD, more than twice of the metropolitan average. Would the metropolitan GDP double if metropolises were run like countries? Would then the world GDP grow by 75% in metropolises if run with a specific metropolitan discipline, instead of applying urban knowledge to an unfit phenomenon?

As we advanced from the modern to the contemporary era, the driver of economy shifted from land and labor to entrepreneurship and innovation. The new combination of technology to address the existing problems and meet the new demand of an evolving market is fundamental in nowadays economic growth. On the other hand, a conception of an economic model alternative to the dominant one, in which profit is still kept as a value, but as an optimizing value that includes a plurality of values in itself is mandatory. In the current state of overproduction and overconsumption, it is important to investigate alternative models, such as circular or green economy, within the metropolitan discipline.

The results obtained from the application of new models can be measured by both economic indicators (increase in product added value) and environmental indicators (reduction of the production of solid and liquid pollutants, or in the form of greenhouse gases). In a growing number of areas of application, indicators related to the social sphere are also considered and in particular the increase in job opportunities.

Social Dimension

In metropolitan contexts, citizens, inhabitants, city dwellers, and commuters move into the new public realm and experience its novelties as a stimulus to actively participate and appear in the new spaces of relationships acquiring new roles. The ways of life that the metropolis generates have a psychological and social dimension or even basically, an anthropological side. This social dimension has become an essential element in urban theory today. The relationship between the social space-time project, marked in the working life, and the individual space-time project programmed on citizens' agendas must match. Communities that do not wish to or cannot afford to be included in this metropolitan space-time project are dissolved either by dilution of their identities or by marginalization. The need for a project based on the rule of equity rises, and it is related to two main topics: the right to the city and the right to the landscape.

The right to the city, originally formulated by Lefebvre (1974), re-establishes the urban foundations of seeking justice, democracy, and citizens' rights. After centuries during which the national state defined citizenship and human rights, the metropolitan city is considered as a special space and place of social and economic advantage, a focal point for the workings of social power and hierarchy, and therefore a powerful battleground for struggles seeking greater democracy, equality, and justice.

As Harvey (2005) later elaborated in his liberal formulations, Lefebvre saw the normal workings of everyday metropolitan life as generating unequal power relations, which in turn manifested themselves in inequitable and unjust distributions of social resources across the space of the metropolis. The demand for greater access to social power and valued resources by those most disadvantaged by inequitable and unjust geographies defined the struggle to reclaim the manifold rights to the city. The aim, at least from a liberal and egalitarian point of view, is to gain greater control over the forces shaping urban space, in other words, to reclaim democracy from those who have been using it to maintain their advantaged positions. A conceptual leap in the context of metropolitan studies suggests the inclusion of the rights to landscape and lifestyle within this framework. Social management of metropolises requires a different approach, both endogenous and exogenous and both in terms of human resources and social resources.

The metropolis, as a whole, with its population mobility cannot be socially approached in a disjointed way as a municipality. A social municipal policy has the effect of attracting or expelling population in need from or to the other municipalities around the metropolis. However, this transfer phenomenon does not mean that social techniques must change dramatically, at least, not on the service provision for the most in need among the human resource stock of the metropolis. It means that a metropolitan government is needed to extend social policies to the whole metropolis and avoid pockets of wealth or deprivation if not applied consistently.

The metropolitan social discipline differs when it is integrated with other metropolitan components that have their own discipline, for instance, economic or physical components. The economy of the metropolis requires a different management from the approach adopted for cities. Cities most commonly do not have a management of their economy as such. Nations do as well as metropolises.

Human Resources Social management of metropolises, when related to their economic management, features two specific elements unknown to cities. Metropolises must have an economic strategy with a 40-year time horizon. It needs to take into account the role and rank of the metropolis into a globalized network of metropolises, what that metropolis should specialize on, which products, and which markets it needs to target. The wealth of the metropolis and thus its capacity to address its inhabitants' needs depend on that. Metropolitan culture and education have to foresee all needs 20 years ahead and set targets to meet them: subsidizing education with public funds to target social benefits. In this way, capitalizing on human resource stock of the metropolis responds to social and economic needs.

Social Resources Capitalizing on human resources is not enough. Sharing the same value system promotes a metropolitan sociocultural context. Cities within the metropolises do not have differences in social resources. Cities within the national context do, but these differences are mostly reflected in the value of their elected officials.

This social resource element, within the social component, can also be defined as collective intelligence. Collective intelligence (Malone and Bernstein 2015) is the capacity of a group, or a metropolis, to make the right decision within a reasonable span of time. It is not related to the accumulation of knowledge of the human resource stock. It is related to the system of values that define the interactions among its inhabitants. It is not defined by knowledge, but by ethics, not morale. Cities do not promote that. Metropolises have to. This element will define the future success of the metropolis in a globalized world and not the number of its roads or the number of buses or housing deficit figures. Those are only a consequence of the collective intelligence of the metropolis. Governance must represent the ethical axis of the metropolis development, where the social and the individual projects coincide.

Metropolitan Discipline Integrated Approach

All these aspects, components, branches, sectors, elements, and factors need to be included in an integrated approach for practical purposes, i.e., they need to work together to make the metropolis function (Fig. 11).

Biology provides us with a helpful analogy when trying to understand a complex matter such as a metropolis. The Chilean biologists Humberto R. Maturana and Francisco J. Varela

Metropolitan Discipline: Management and Planning, Fig. 11 The genome diagram. (Source P. Ortiz)

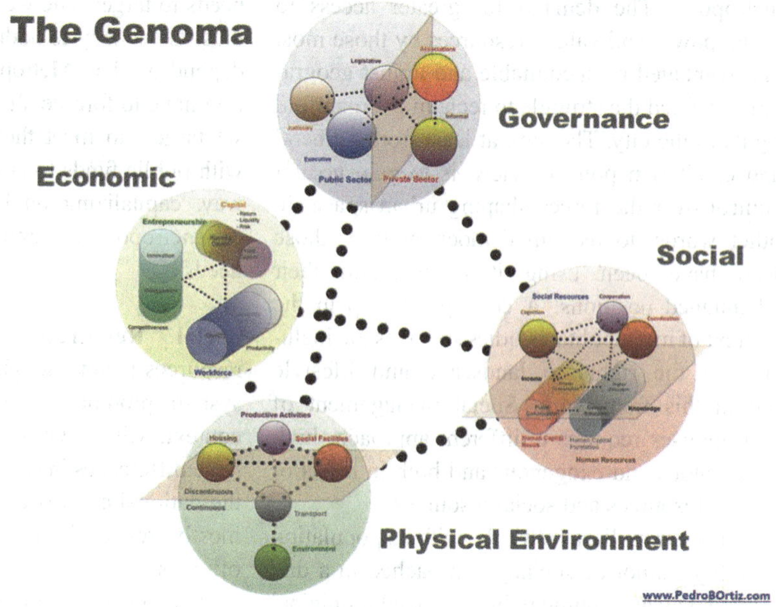

(2001) explain genome and phenome as the two fundamental structures in understanding the complexity of genetic studies. A genome is a complete set of genetic material of a living being. In a broad sense, the genome defines the principles, the very essential rules, and the ideas that constitute an entity. In our case, the basic conditions, rules, and principles that make up a contemporary metropolis is a genome.

On the other hand, in the field of biology, a phenome is a set of phenotypes expressed by organisms. Each phenotype has a basis in its genotype, but it may be influenced by the environment. It is possible to apply the genome's principles and rules to all metropolises. However, as much as the principles and rules are essential, the way that they react in an environment with specific conditions (phenome) also has a significant importance in the way we perceive complexity in reality. It is important to focus simultaneously on both genome and phenome to investigate something as complex as a metropolis.

The metropolitan genoma (Ortiz 2021) is the integration of the different disciplines into a single DNA model for the metropolis. The metropolitan managers need to understand them all at the same time in order to be able to forecast the interaction between components. They must detach the implications of any proposed policy and evaluate the possible need to complement or address any negative collateral effect and if the result of all this is going to produce positive or negative effects on the external world.

Space. Metropolises are on an intermediate scale between cities and countries. They must be integrated transversally within the different scales. Metropolises unlike cities behave in many ways as countries do. They have their scale of intervention, and this is not to be mistaken with the urban one. The disciplines of the lower scales are well defined: architecture generally operates at a 1/50 scale. Also urban design works on a 1/500 scale. Urbanism was developed in the eighteenth century in Latin-European countries, and town planning started in the nineteenth century in England and the USA. Their procedural scale is mostly around the 1/5.000. The global scale is on the other extreme of the physical component, and it involves disciplines as macroeconomics and geopolitics. The scale of operation is 1/5 million. Nations have long tried an integrated approach for their allocation of investments and infrastructures. Their scale of operations was on a 1/500.000 scale, depending on the size of the country. The metropolitan scale ranges between the urban one, 1/5.000, and the national one,

1/500.000; the metropolis is in the range of 1/50.000 (Fig. 12).

The study of metropolitan complexity is always related to scale interactions. Metropolitan complexity produces specific morphologies that determine the metropolitan context. It is fundamental to define these morphologies because the metropolitan approach will lead to integrated projects dealing with the interface-spaces and related functions to allow any possible scale relations.

The metropolitan context often does not belong to the traditional urban morphology and therefore results in being abandoned, unserved, precarious, and therefore, undesired. Studying and implementing the context within the metropolitan practice is mandatory, and addressing such context requires an integrated approach that deals with all dimension of the metropolis (Fig. 13).

Time. Metropolitan actions must be diachronically integrated in time, not just synchronically. They need to exert a consecutive effect connecting synergies in time. Long-term planning (30-year horizon) also needs to be compatible with the programing objectives of a 4-year political mandate. The long term has to be compatible as well with the medium term, and the impact investments produce should result in the highest multiplier effect. Variable geometry is required in metropolitan planning, especially when political tendencies rule consecutively, and each has the legitimate democratic right to introduce its own peculiar approach, backed by the citizens' will or vote, even beyond the mandate term. Those periodical changes require a sliding horizon approach to ensure mandate adaptation, without losing sight of long-term planning (Fig. 14).

Dialogue is essential, and it is a constituent part of the governance system and of the collective intelligence of a metropolis. Governance and design must interact. Disintegration is the price of nonintegration. Integrating the scales needs integrated transport modes to be included in the

Metropolitan Scale

Spatial Scales and Disciplinary approach

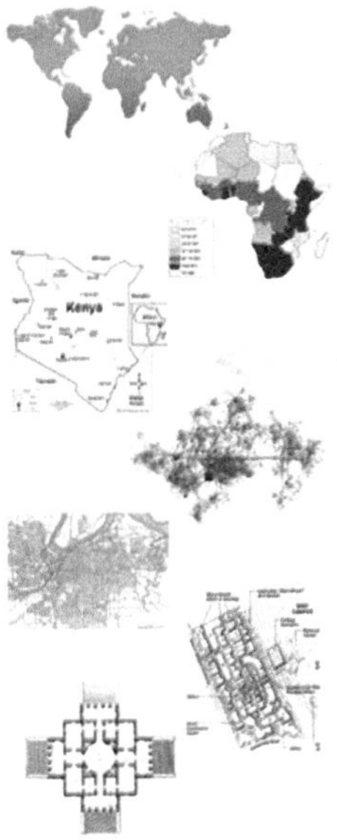

Scale	Discipline	Knowledge areas
1: 50.000.000	Geopolitics	UN. NATO. US. EU. IMF, OPEP,
1: 5.000.000	'Regional' (Continental) Politics	UN. OAS. WB. BID. ADB. AFDB.
1: 500.000	National Development	Politics, Economics, Sociology, Environment, Geography
1: 50.000	Metropolitan Planning	Economics, Sociology, Infrastructure, Environment, Utilities
1: 5.000	Urban Planning	Housing, Industrial, Services, Commerce, Transport, Environment
1: 500	Urban design	Space, Volume, Semiotics, Engineering,
1: 50	Architecture	Light, Space, Texture, Materials, Structure, Installations, Budget

© www.PedroBOrtiz.com

Metropolitan Discipline: Management and Planning, Fig. 12 The scale of planning. (Source: P. Ortiz)

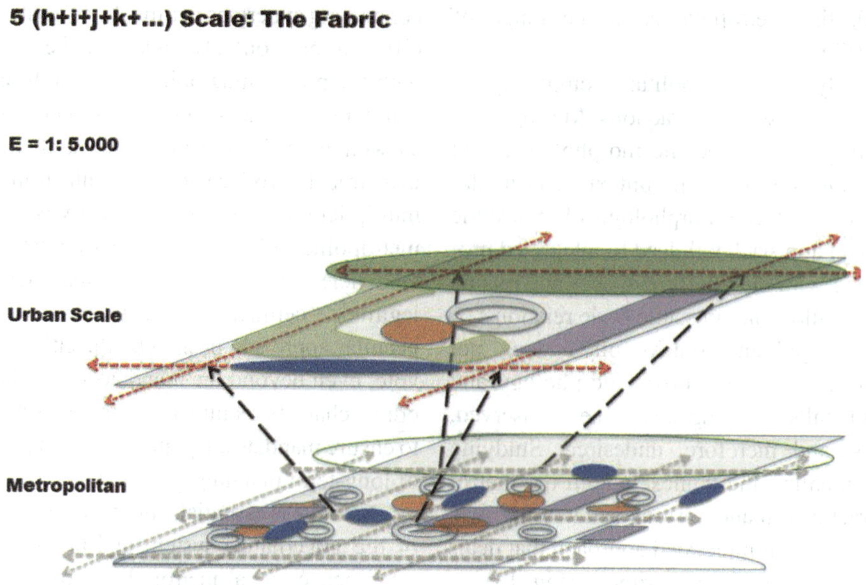

Metropolitan Discipline: Management and Planning, Fig. 13 Urban and metropolitan scale connection. (Source: P. Ortiz)

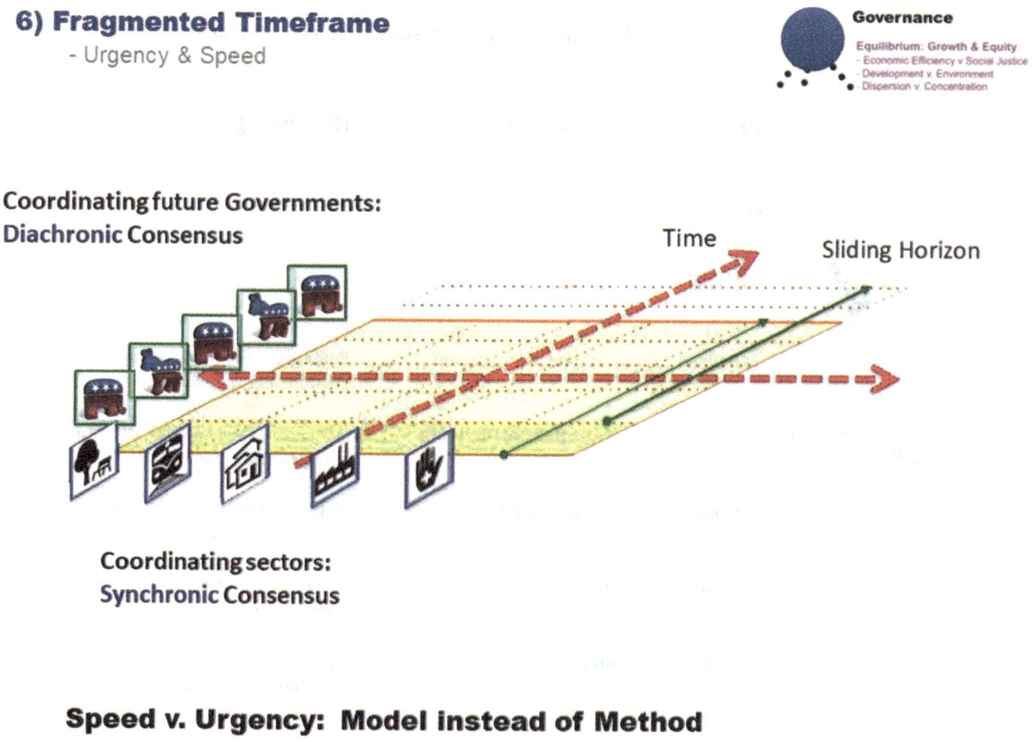

Metropolitan Discipline: Management and Planning, Fig. 14 Diachronic and synchronic consensus. (Source: P. Ortiz)

system in order to support the transport chain. The integrated alternative includes the local scale mode (last-mile bicycle), the metropolitan scale mode (commuter train stations and bicycle parking), and the urban scale mode (BRT) to reach the city center and the final destination in 40 min. The disintegrated alternative, instead, involves a significant loss of time due to the long distances to be covered to reach public transport and long waiting times.

Interconnection of Projects and Policies

As mentioned earlier, policies and projects are closely interlinked. Commuter trains must comprise urban centralities to serve and be served. If both are not combined in time and space, the result will be twofold:

- The commuter trains will serve empty destinations. Ridership will be very low and cost of service maintenance very high and unaffordable if considered in an opportunity cost analysis. The service will be discontinued or, if political failure needs to be concealed, maintained at the minimum frequency to reduce losses.
- The urban centralities with no available public transport will either attract car ridership resulting in the congestion of the insufficient access roads, or will not attract projects/investments from the private sector, and the original infrastructure investment effort will be wasted.

This kind of disjointed incrementalism approach produces the actual anarchic result of metropolises. Long-term negative externalities of this disintegrated approach are not being addressed. The price will be extremely high in the space of a century.

The Metropolitan Approach to the Impact of Foreseeable Complexity in World's Governance

If metropolises become efficient, equitable, sustainable, and politically stable, this is going to have consequences for the world. The requirement is to provide for self-governance attributions.

Metropolitan Governance

Metropolises require nation-like competences to run nation-like economies suffering from nation-like social and sustainability problems. The national model is threefold: nations can only be confederations, federations, or unitary systems. Confederations do not work in the long term. In confederations, sovereignty resides with its members that can stay or leave at any time (Brexit). The confederative approach to metropolitan government sits the mayors to a table and comprises three stages of development: collective intelligence, New York new school (Fig. 15)

- Cognition, when they share their policies
- Coordination, when their knowledge alters and adapts their local policies
- Cooperation, when they decide to do something together, also by creating an agency for its management. Never applicable for political decisions

Unitary systems can be centralized or decentralized:

- Centralized, when central ministries make all decisions. The lack of local knowledge and cognizance as well as intersectoral discoordination at local level makes them inefficient. The subsidiarity principle is not applied.
- Decentralized or delegated, when the central government creates local delegations. These delegations play the role of coordinating at local level the actions of the central ministries.

The centralized system does not work for metropolises as the decisions of central ministries often lack coordination and efficiency. The decentralized system does work when the national president appoints a metropolitan governor to coordinate the national budget (the only one able to address metropolitan-sized issues) in their metropolitan implementation. This role should have authority over the ministers in order to be able to coordinate. A vice president could undertake this kind of task.

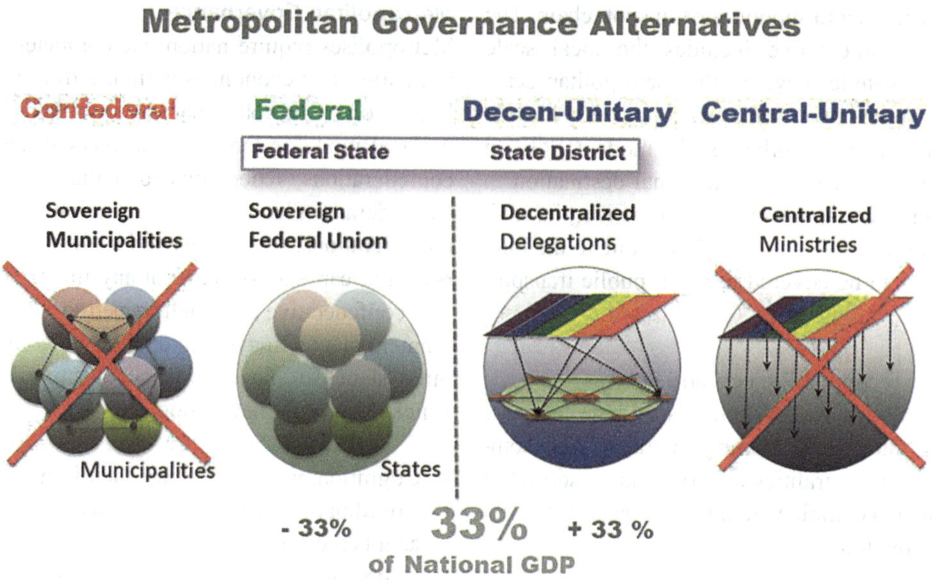

Metropolitan Discipline: Management and Planning, Fig. 15 Confederations, federations, or unitary system comparison. (Source: P. Ortiz)

Federations are systems where the sovereignty is at the center but where competences and political decisions are shared with the state governments. This system is called devolution, as opposed to unitary decentralization. There are two kinds of statehood (Fig. 16):

- Territorial statehood: Most federations favor this approach. States represent geographical or historical units whose decision autonomy has been preserved at some extent within the federation.

  In this context, the role of the metropolis within the state raises the same issues as in the relation between the metropolis within the nation. The economic impact of the metropolis for the state is almost overwhelming. A delegation from the state governor or the president is thus pertinent.
- Metropolitan statehood: Some new federation constitutions, like the German one in 1949 or the Spanish quasi-federative system in 1978 recognized, this new phenomenon and provided a single statehood for some metropolises. That is the case of Berlin, Hamburg, and Bremen in Germany and Madrid in Spain.

It is interesting to point out that the main difference between the unitary decentralized and the federative metropolitan systems is who appoints or elects the governor. In the unitary system, it is the national president; in the federative it is the state electorate. Both are equally democratic; however, the difference is that governors appointed by the president will represent the national interests of the democratically elected president. When the governors are elected by the people, they will defend the interests of the state. Both are legitimate interests; however, the preference will have to be defined in terms of the local sociocultural context. As a matter of fact, the solutions discussed in each country are closely related to their specific socioeconomic context.

Metropolitan Finance

A metropolis can be managed in two ways. The first way involves making decisions within the competences entrusted to the administration by the law and imposing them; the second way considers the budget you can spend to promote development and negotiate with other institutions where both can invest. The metropolitan project critically rethinks aspects capable of producing

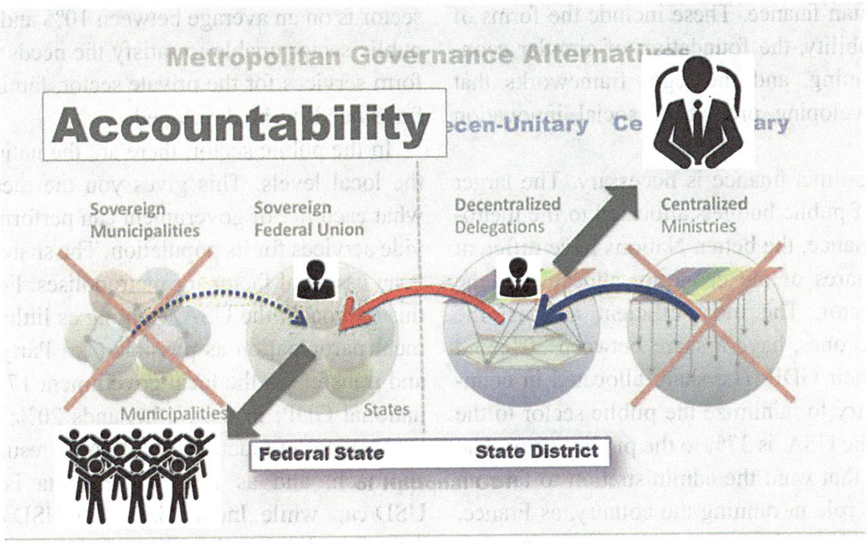

Metropolitan Discipline: Management and Planning, Fig. 16 Metropolitan governance alternatives. (Source: P. Ortiz)

only one adjustment tool on the sustainability issue. Instead, a good metropolitan financial instrument should be able to solve the conflicts between state, regional governments, and global private international firms, no longer on a local scale but on a metropolitan scale.

In fact, usually, the investigation of metropolitan regions from the "economic" point of view starts from the analysis of the "genetic principles" of the city combined with the companies' territorial location factor. The metropolitan city is referred to as a "diffused city" where the fragmentation of urban functions moves in parallel with other fragmentations, from the social one to the labor market one. And, in the end, a fragmentation of the actors of the urban setting occurs, with territories overlapping the boundaries and the areas of decision-making competence. The result is a sprawling metropolis that blends in with different urban territorial systems to create a new structure, a settlement continuum, "spread," yet always interconnected.

Therefore, it is necessary to focus on the rationalization of the new models of territorial organization and to plan from a local and global sustainability perspective, citizens' prosperity, and leveraging the study of the best systems of governance. The metropolitan project develops a critical discourse on the academic literature that considers the main "genetic principles" of the new cities as urban agglomeration and accessibility, where urban agglomeration concerns urbanization economies, advantages for firms and households settled in a city (i.e., presence of infrastructures, public and private services, amenities, etc.). The factors influencing the companies' location are geographical-infrastructural factors, economic-financial factors, and others (i.e., regulatory, environmental, labor market factors but also "softer" factors such as personal considerations and "mental maps").

Among these factors, in particular, the aim of the metropolitan discipline is the analysis of the reasons underlying the definition of the new border (real and virtual) that marks the difference between center and periphery, that is to say, the accessibility of services. This question is relevant since it introduces the issue of periferization caused by the lack of service accessibility. Starting from the consideration that the new world order and the geopolitical competition concern the dispute around the demarcation of a border at various possible scales, from the global hegemony to the control of urban spaces, the metropolitan discipline needs to explore some topics related to projects and policies referred to metropolitan finance. These include the forms of urban mobility, the foundation of circular economy planning, and the legal frameworks that allow developing project of social innovation (Fig. 17).

Metropolitan finance is necessary. The larger the part of public budgets allocated to the metropolitan finance, the better. Nations have different budget shares of the economy allocated to the public sector. The most efficient nations, the developed ones, have a share between 30% and 50% of their GDP. The share allocated in countries that try to minimize the public sector to the most, as the USA, is 37% to the public sector. The countries that want the administration to take an important role in running the country, as France, allocate 56%. In these circumstances, we need to understand that it will be very difficult to foster development in countries like India, or Guatemala, where the share of spending for the public sector is on an average between 10% and 13%. A public sector unable to satisfy the needs and perform services for the private sector, families, and firms needs to be developed.

In the public sector, there are the national and the local levels. This gives you the measure of what each tier of government can perform to provide services for its population. The share amount is an essential factor for metropolises. Following this approach, the USA, wanting as little government participation as possible (Tea Party), raises and transfers to the local government 17% of the national GDP; in the Netherlands 20%; in India 0.13%; and in Guatemala 1%. As a result of this approach, and as USA GDP/capita is 60,000 USD/cap while India's is 1,700 USD/cap, the American local authorities will have more than 10,000 USD/cap to spend for each of its citizens, while the Indian one will only have 2.2 USD/cap. The difference in GDPs is 35 times higher, but the

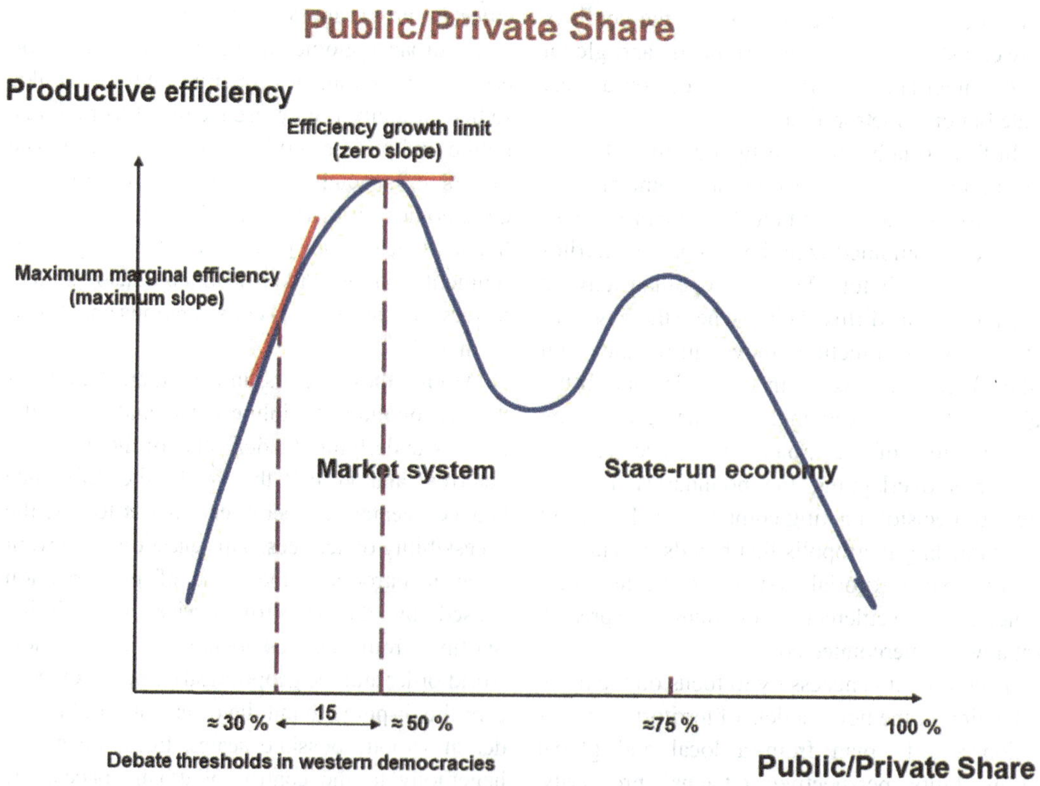

Metropolitan Discipline: Management and Planning, Fig. 17 Productive efficiency and public/private shared benefit. (Source: P. Ortiz)

public budget is 4,600 times higher. And this is a country that does not believe in the importance of the public sector (Fig. 18).

How can metropolises raise the budget necessary not only to fulfil their basic duties but also to promote efficiency and foster all the metropolis aspects of potential development without forgetting the principle of equity? Experts on local finance suggest that if public budget range should be between 30% and 50% of the GDP, the share of local budgets should be in a range between 20% and 40% of the public budget, with an average of 30%. That share would amount to 13% of the national GDP. Knowing the national GDP/cap is possible to determine the accumulated local budget. You just need to multiply the 0.13 GDP/cap per population figure. This approach is close to the world's standard policies. However, it is not the same everywhere. If metropolises are essential for the development of nations, budget is essential to the development of metropolises.

Global Governance

The balance of the world is going to be changed. An integrated metropolitan government will make metropolises much more efficient and economically powerful. Singapore is a good example. The weight of countries in the global balance of power will also depend on the economic presence of their metropolises in this global scenario.

Nevertheless, the world as a whole is going to be transformed, not just the single countries. Actually, there are 500 metropolises with 1 million plus inhabitants. They produce 75% of world GDP. They are 16 times more productive than the rest of the world population. This is enough to have their power recognized, both nationally and internationally. The metropolises that will be able to increase their efficiency will be the ones that will become the driving force of world economic power and cultural change. Their economic and political international presence will shape world policies. Should we reconsider the share of international presence between nations and metropolises?

Metropolitan and National Dialogue

The power of autonomous metropolitan regions creates an issue that needs to be addressed. What should be their responsibility towards the rest of the nation they belong to? Metropolises would

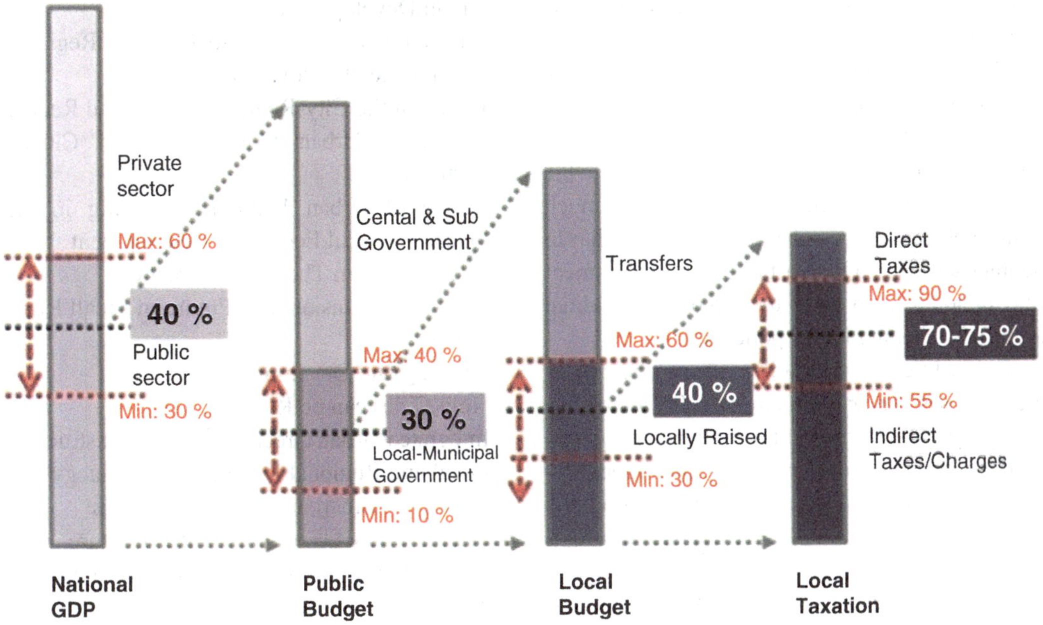

Metropolitan Discipline: Management and Planning, Fig. 18 Municipality budget. (Source: P. Ortiz)

become centers of cultural production and power for these nations. The nations would be the providers, as any other nation could be. Nations would have to endure the metropolitan footprint. This creates a serious unbalance both in social and environmental terms that is unacceptable. And as we learn from most independentist movements and parties, this unbalance grows because the wealthier they represent will get even more wealthy if they do not have to bear the burden of those in need. Generally, the wealthy do not want to support the ones that are not.

How can we address this issue? There are two ways: increasing productivity and increasing social transfers. It is a two-pronged policy. The way to increase productivity in non-metropolitan nations is by investing capital and technology, together with education. This investment will not be as efficient as those made within the metropolis, but it promotes social equity as a byproduct to justify this choice. This has to be performed by a higher tier of the government; it needs to be at a national level.

Culture will be transformed. The non-metropolitan areas, due to a metropolitan project able to define new patterns to bind urban to rural, will be able to increase the value of their products, and this will result in increased benefits and wealth.

The way to increase social transfers is by increasing taxes on the higher metropolitan productivity and benefits, in order to build up a social policy and to increase the provision services to the deprived population that deserves them as much as the metropolitan population do. Taxation has to be imposed from a higher tier of the government (the national one) able to avoid metropolitan lobbies weakening its tax policies.

The political responsibility cannot be left to metropolises. Their long-term benefits are different from the markets' short-term benefits. Metropolitan discipline proves collective intelligence by fostering a path, towards a federal metropolis or a unitary decentralization, that will make way for a long term efficient, equitable, sustainable, and well-balanced governance for both nations and metropolises (Fig. 19).

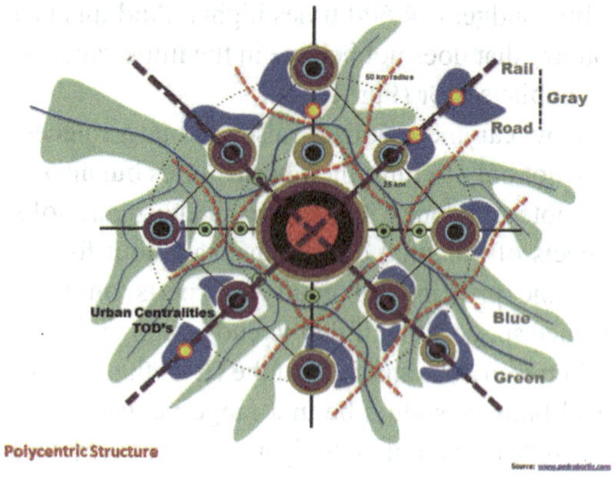

Metropolitan Discipline: Management and Planning, Fig. 19 The metropolitan digit. Green-gray infrastructure and settlements integration. (Source: P. Ortiz)

Cross-References

- ▶ Changing Paradigms in Urban Planning 2000–2020
- ▶ City Visions: Toward Smart and Sustainable Urban Futures
- ▶ Connecting Urban and Regional Innovation Ecosystems to Enhance Competitiveness
- ▶ Education for Inclusive and Transformative Urban Development
- ▶ Financing: Fiscal Tools to Enhance Regional Sustainable Development
- ▶ Future of the City-Region Concept and Reality
- ▶ Future of Urban Governance and Citizen Participation
- ▶ Future of Urban Land-use Planning in the Quest for Local Economic Development
- ▶ Green Cities in Theory and Practice
- ▶ Growth, Expansion, and Future of Small Rural Towns
- ▶ Housing and Development
- ▶ How Cities can be Resilient
- ▶ Integrated Urban Green and Grey Infrastructure
- ▶ Local and Regional Development Strategy
- ▶ Making of Smart and Intelligent Cities
- ▶ Multiple Benefits of Green Infrastructure
- ▶ Neoliberalism and Future Urban Planning
- ▶ New Cities
- ▶ Planning for Peri-urban Futures

- ▶ Spatial Demography as the Shaper of Urban and Regional Planning under the Impact of Rapid Urbanization
- ▶ Sustainable Urban Mobility
- ▶ Transportation and Mobility
- ▶ Urban and Regional Leadership
- ▶ Urban Densification and Its Social Sustainability
- ▶ Urbanization, Planning Law, and the Future of Developing World Cities

References

Appleyard, D., Lynch, K., & Myer, J. (1964). *The view from the road*. Cambridge MA: MIT Press.

Brenner, N. (2014). *Implosions/explosions: Towards a study of planetary urbanization*. Berlin: Jovis.

Christaller, W. (1933). *Die zentralen Orte in Suddeutschland*. Jena: Gustav Fischer. (Translated (in part), by Baskin C.W., (1966) as Central Places in Southern Germany, Prentice Hall.

Contin, A. (2016). *The narrative structure of the agro-urban metropolitan territory. The metropolis as Hypertext for the history of the XXI century: A network of middle cities as an operational topography*. Berlin: Springer.

Eurostat. (2016). Urban Europe: Statistics on cities, towns and suburbs – Working in cities. Retrieved from https://ec.europa.eu/eurostat/statistics-explained/pdfscache/50937.pdf

Eurostat. (2020). Metropolitan Region Dataset. Retrieved from https://ec.europa.eu/eurostat/web/metropolitan-regions/data/database

Frampton, K. (1983). Towards a critical regionalism: Six points for an architecture of resistance. *The Anti-Aesthetic Essays on Postmodern Culture, 1*, 16–31. Seattle: Bay Press. https://monoskop.org/images/archive/0/07/20150505143904!Foster_Hal_ed_The_Anti-Aesthetic_Essays_on_Postmodern_Culture.pdf.

Gottmann, J. (1961). *Megalopolis: The urbanized northeastern seaboard of the United States*. New York: The Twentieth Century Fund.

Hardingham, S. (2016). *Cedric Price, Works 1953–2003, A forward-minded retrospective*. Montreal: Architectural Association (AA) and the Canadian Centre for Architecture (CCA).

Harvey, D. (1989). *The condition of postmodernity*. Oxford: Blackwell.

Harvey, D. (2005). *A brief history of neoliberalism*. Oxford: University Press.

Hilberseimer, L. (1944). *The new city. Principles of planning*. Paul Theobald: Chicago.

Hilberseimer, L. (1949). *The new regional pattern: Industries and gardens, workshops and farms*. Chicago: P. Theobald.

Hilberseimer, L. (1993). *Das neue Frankfurt 1926–1931* (G. Grassi, Ed.). Torino: Dedalo Edizioni. (Original work published 1929).

Hilberseimer, L. (2012). *Metropolis architecture*. New York: GSAPP Columbia University Graduate School of Architecture, Planning and Preservation. (Original work published 1920).

Hilberseimer, L. (2012). New City. Principles of Planning, Madrid: HardPress Publishing

Hilberseimer, L., (2014). Metropolisarchitecture, Edited by Richard Anderson. Translated by Richard Anderson and Julie Dawson. Afterword by Pier Vittorio Aureli, New York: Columbia University Press

Jacobs, J. (1961). *The death and life of great American cities, (2016)*. New York: Random House.

Koolhaas, R. (2006). in El Croquis OMA/(Issues 53+79) (English and Spanish Edition) (Spanish) Hardcover – August 1, 2006.

Lefebvre, H. (1974/1991). *The production of space* (D. Nicholson-Smith, Trans.). Oxford: Basil Blackwell.

Lynch, K. (1960). *The image of the city*. Cambridge, MA: MIT Press.

Lynch, K. (1996). *A theory of good city form. (R. Melai.). Progettare la città. La qualità della forma urbana*. Milano: Etaslibri. (Original work published 1981).

Malone, T. W., & Bernstein, M. S. (2015). *Handbook of collective intelligence*. Cambridge, MA: MIT Press.

Maturana, H. R., & Varela, F. J. (2001). *Autopoiesi e cognizione. La realizzazione del vivente*. Venezia: Editore Marsilio (collana Saggi. Critica).

McGee, T. (2009). The spatiality of urbanization: The policy challenges of mega-urban and Desakota Regions of Southeast Asia. In *Proceeding of UNU-IAS 161*, Vancouver

McHarg, J. (1969). *Design with nature*. London: Wiley.

Ortiz Castano, P. B. (2014). *The art of shaping the Metropolis*. New York: Mc Graw Hill.

Ortiz, P., (2021). The metropolitan genome. In: Contin, A., Giordano, P. and Nacke, M. (eds.).Training for education, learning and leadership towards a new metropolitan discipline. Inaugural book. Buenos Aires: CIPPEC.

Shane, G. D. (2005). *Recombinant urbanism: Conceptual modeling in architecture, urban design, and city theory*. Chichester: Wiley.

Shlomo Angel, Jason Parent, Daniel L. Civco, and Alejandro M. Blei (2012). Atlas of urban expansion, Lincoln Institute of Land Policy, Cambridge Massachusetts.

Soja, E. W. (2000). *Postmetropolis: Critical studies of cities and regions*. Los Angeles: Blackwell Publishing. Ed. spagnola (a cura di) Hendel Veronica, Cifuentes Monica, Postmetropolis. Estudios criticos sobre las ciudades y las regiones, Traficantes de sueños, Madrid, 2008.

Spangenberg, J., & Bonniot, O. (1998). *Sustainable indicators – A compass on the road towards sustainability* (Wuppertal paper No. 81). Wuppertal: Wuppertal Institute.

Un-Habitat. (2020). Global State of Metropolis - Population Data Booklet. Un-Habitat, HS/013/20E. Retrieved

from https://unhabitat.org/global-state-of-metropolis-2020-%E2%80%93-population-data-booklet
United Nations. (2018). World Urbanization Prospects. Population Dynamics. Retrieved from: https://population.un.org/wup/
Venturi, R., Scott Brown, D., & Izenour, S. (1972). *Learning from Las Vegas*. Cambridge, MA: MIT Press.
Waldheim, C. (2006). *The landscape urbanism reader*. New York: Princeton Architectural Press.

Micro Utility Devices

► Personal Delivery Robots: How Will Cities Manage Multiple, Automated, Logistics Fleets in Pedestrian Spaces?

Micro, Small and Medium Enterprises (MSMEs)

► The State of Extreme Events in India

Microaggression

► Hidden Enemy for Healthy Urban Life

Microgrids

► Faith Communities as Hubs for Climate Resilience

Migration – Movement

► Transnational Migrants on the Margin

Millennium Development Goals

► Sustainable Development Goals from an Urban Perspective

Mitigation

► From Vulnerability to Urban Resilience to Climate Change

Mixed Land Use

► Urban Structure and Its Impact on Mobility Patterns: Reducing Automobile Dependence Through Polycentrism

Mobility

► Amsterdam's Pathway to Climate Neutrality: Creating an Enabling Environment
► Pre-schoolers and Sustainable Urban Transport

Mobility – Movement

► Transportation and Mobility

Moving Towards Sustainable, Liveable, and Care-Full Urban Environments: Pre-schoolers' Rights and Visions for Planning Just, Socially, and Ecologically Integrated Cities

Christina R Ergler
School of Geography, University of Otago, Dunedin, New Zealand

Synonyms

Care ethics; Care-full cities; Child-friendly cities; Participatory; Sustainability; Urban; Young children

Definition

Children under the age of five are often denied participation in sustainable planning initiatives. If they get the opportunity to share their vision for liveable environments, they plan care-full cities that are just and socially and ecologically integrated. In their visions for these cities full of care, the young, pre-literate children demonstrate a deep and holistic knowledge about how cities should function to be safe and socially and physically connected by creating destinations, amenities, and services available for all ages and abilities. Moreover, pre-schoolers create not only cities based on their own experiences, but cities that ensure liveability, flourishing, and well-being for humans and non-humans. This entry outlines why it is important to involve young, pre-literate children in planning initiatives, what such participation could look like and what we can learn from young children when it comes to creating inclusive, sustainable, and liveable cities.

Introduction

Pre-school-aged children are the current and future users of urban environments. But they are traditionally not consulted on voicing their experiences and visions for sustainable, socially equitable, and liveable urban environments. One reason for this omission is the commonly – but falsely – held view that pre-schoolers lack the competency to reflect on large-scale environments. Thus, they are silenced in discussions on creating environmentally friendly, liveable, and socially just urban environments; they are not viewed as an important stakeholder group and thus denied the opportunity to participate in the planning for sustainable urban environments that speak to their vision of liveability.

Why Is It Important to Value Pre-schoolers' Rights to Participate in Sustainable City Planning Initiatives?

First, the social and physical environments children are growing up in impact on their social, physical, and mental health and well-being in the short and long term. For example, emissions from cars or manufacturing industries expose them to air, noise, and water pollution that can be harmful to young bodies. Similarly, the way neighborhoods are designed can provide for nutritious and affordable local foods and encourage safe and active transport routes that reduce harmful emissions for people and the planet. Neighborhoods can inspire young children's imagination and invite safe play and exploration; thus, how the environment is designed can increase social and nature interactions with positive effects on bodily, social, and mental health. But young children often do not have a voice in the decision-making process on how these environments are shaped or should be shaped. They are often denied consultation, as adults view them as hard to work with and incompetent in voicing their experience beyond their immediate surroundings, such as places traditionally associated with young children like their home, playgrounds, and kindergartens. In recent years, planners and policy-makers started to honor the United Nations Convention on the Rights of the Child (UNCRC) and began to move away from a deficit discourse about young children's capabilities, fostering their consideration as an important stakeholder group by inviting adults to rethink their approaches to eliciting their ideas and not to measure their contribution through adult-centric ways.

Second, children of all ages and abilities have – according to UNCRC – the right to be consulted on and to shape the issues affecting their life (UNICEF 1995). A majority of all UN member states have signed the Convention since it was first ratified in 1989, and, by doing so, UNCRC began to advance children's standing in society by advocating for their social, cultural, educational, health, economic, political, and civil rights and outlining government and societal responsibilities and obligations. UNCRC, in other words, paved the way for a rethinking about children and their capabilities. Participating in urban issues is thus their right as citizens and a legal duty clearly sketched out by this Convention. Nonetheless, this shift in seeing children as competent social actors in their own right has not been translated into the integration of young pre-literate children's ideas in city planning per se. Even the further endorsement by the United Nations General Comment No. 7 (UN Committee 2006) to consult with and foster the participation of especially young, pre-literate children has not resulted in the integration of their voices more widely. They remain framed as incompetent, unresponsive actors, and thus they are denied their citizen rights.

Third, ensuring their citizen rights through participation empowers young children and is a more democratic process. Viewing them as valuable contributors to city planning has positive effects for the participating child and the liveability of communities on a local, national, and global scale. Incorporating pre-literate children's ideas for change can transform existing urban environments and their shortfalls as it moves away from the adult-centric visions for sustainable urban environments. If outcomes are attuned to young children's everyday experiences, values, and needs, spaces like parks, playgrounds, and transport infrastructure can be improved and reimagined. Moreover, when pre-literate children's input for change is taken seriously, they not only experience meaningful participation and learn to value democratic processes, but also their self-esteem is enhanced, and they usually gain valuable skills (e.g., voicing and structuring their thoughts in a capability-appropriate way, making sense of complex urban processes).

Fourth, to ensure justice and equity in societies it is pertinent to incorporate children of all ages and abilities as active citizens. However, this implies that different generations work together as a team and value the contribution of anyone. In this way, participation is viewed as a process of negotiation and co-learning and thus moves away from existing adult-child relationships that are tainted by unequal power structures. Bringing the experiences, values, and practices of different generations, abilities, cultures, and ethnicities together can create more inclusive and equitable working relationships. In turn, these relationships foster sustainable, liveable urban environments, as diverse voices and needs are heard in the transformation process.

What Concepts Can Adults Draw on for Planning Just, Socially, and Ecologically Integrated Cities with Children?

Two helpful conceptual frameworks exist that try to ensure children grow up, are valued, and can be part of shaping the future of their cities in a meaningful way: the child-friendly city initiative and the concept of care-full cities.

A child-friendly city is a "city, town, community or any system of local governance committed to fulfilling child rights [...] It is a city or community where the voices, needs, priorities and rights of children are an integral part of public policies, programmes and decisions" (UNICEF 2018, p. 1010). Thus, the child-friendly city concept is based on a right-based approach and informed by ideas of social justice and aims to foster children and young people's long-term well-being but also wants to ensure their rights as citizens by providing adequate access to:

- Health and basic services
- Safe and secure environments
- Natural environments and green spaces
- Peer and intergenerational gathering places
- Getting around in their communities
- Participation in the design of and planning for their communities
- Being able to learn about and care for their physical and human environments

This concept has evolved from a resolution passed at the second United Nations Conference on Human Settlement (Habitat II) and subscribes to realizing children's rights as articulated in UNCRC, but the concept has also the underlying premise that cities that work for children are a liveable space for all generations. Nonetheless, children and young people are often accused of just focusing on their own needs rather than the needs of all city dwellers. To avoid the ambiguity of the term "child-friendly," researchers and planners began to turn to the concept of care-full cities, which recenters the focus on care and care relationships in urban environments.

The care-full city concept has developed from the feminist ethics of care literature and focuses on care as transformative practice to move beyond self-interest, to recognize the lived and experienced realities of – and to facilitate reciprocal care relationships between – the social and material environment. A very influential publication by Fisher and Tronto (1990, p. 40) defines these caring relationships as "a species activity that includes everything that we do to maintain, continue, and repair our world so that we can live in it as well as possible. That world includes our bodies, ourselves, and our environment, all of which we seek to interweave in a complex, life sustaining web." This definition outlines a reciprocal care relationship between different species and the material world that is practiced as taking care of, caring for/care giving, care receiving, and caring with. As a consequence, seemingly mundane actions such as volunteering, gardening, or playing along a sidewalk are conceptualized as part of the care negotiating activities, and things like mini-libraries, colorful flowers, or empty bottles facilitate, foster, mediate, and co-constitute care relationships such as being healed and becoming attuned to the need for care. The definition points to an ontological embedding in more than human-human ontologies but also has a strong alignment to ideas of justice that questions how everyday caring practices in cities mirror values of fairness, redistribution, recognition, equity, and democracy. Thus, a just and caring city prioritizes the notion of "care" in the development, functioning, and relationship the city has with the people who live there. Children in this context are then conceptualized both as care agents who practice diverse forms of caring about, caring for, and acting in caring ways towards the social, material, and natural world and as recipients of care by the non-human world and through diverse social rights and responsibilities.

How Can Pre-schoolers Experiences in, Thoughts About, and Visions for Care-Full Cities Be Elicited?

To clearly value and incorporate young children's voices in sustainable urban planning, a culture shift is needed that responds to the UN's obligation to ensure children's genuine participation in all issues affecting their lives. This means that adults need to acknowledge that pre-literate children have the right to seek, receive, and communicate their ideas and views through any media they feel comfortable with. Often participatory, child-centered and so-called "child-friendly" methods are the go-to tool kit for eliciting young children's voices. These methods include but are not limited to photovoice, story-telling activities and puppet plays, drawings, crafting models, and mapping activities or go-along walks. All methods have in common that they are tailored to the interests, needs, and capabilities of the young participants and are often based on a rights-based approach. This approach moves away from adultist perspectives and current deficit discussions and is based on the following principles:

- Pre-schoolers are bearers of human rights.
- Pre-schoolers are current and not just future citizens.
- Pre-schoolers are competent meaning makers on issues affecting their lives.
- Pre-schoolers have the right for age- and capabilities-appropriate methods to be used that are robust, rigorous, and sensitive to cultural and social contexts.
- Pre-schoolers are worth listening to without pre-establishing answers or outcomes, and thus adults should not force their knowledge and categorizations on pre-schoolers and thereby pre-empt outcomes.

What Can Adults Learn from Pre-schoolers About Planning and Creating Sustainable, Liveable Urban Environments?

If pre-schoolers get the possibility to reveal their visions for a sustainable, liveable city, studies showed that they think holistically about the needs of urban dwellers at all ages and also care for the needs of non-human inhabitants. They create their version of equitable and integrated sustainable environments.

M

Studies have showed that they expect to find in their cities at least all basic amenities, such as health services, government (e.g., council, police, fire brigade), and educational services, but also facilities that can enrich the mind (churches, libraries) and body (pools, sport facilities) as well as those that allow for spending time in nature and with friends and families (e.g., parks, walkways). They incorporate services and amenities of a city that ensure inhabitants' social and physical well-being. This also includes well-designed traffic infrastructure such as safe crossings for pedestrians, traffic lights, and street lighting. Greening and coloring their cities are also important for children. They include trees, parks, flowers, and many different animals and insects in their cities. For them, cities need to ensure the liveability of many different species. By ensuring biodiversity in cities, they intuitively create diverse haptic elements, soundscapes, and colorscapes but also smellscapes that they and other human beings can enjoy. However, to ensure these non-humans can live an appropriate life, the young children advocate for protection, conservation, and the cleaning up of urban environments. In other words, these studies show that preschoolers connect themselves with their city and its people, infrastructures, and all other non-living things in a bodily and cognitive manner that mirrors young children's care agency as complex, messy, and relationally performed within the emotional, ecological, and material context they live and act in.

Young, pre-literate children create cities that are a complex, life-sustaining web fostering the flourishing and well-being of all life. In other words, the young children shared how they envision how caring relationships can be played out in their cities. They build cities that aim to enhance vibrancy, socialization, and intergenerational and interspecies inclusion. By integrating these ideas of a sustainable urbanism, young children also demonstrate a far more holistic understanding of the complex socio-nature relationships than they have often been credited for. For them, cities need to be just, socially and ecologically integrated and inclusive, sustainable, and liveable urban environments on the micro- and macroscale. In doing so, they combine the social and ecological dimensions of sustainability which traditionally are looked at separately in the sustainability framework. Pre-literate children express a deep interconnectedness of humans and non-humans in their cities and thus highlight the importance of this relationship for well-being, flourishing, and liveability in cities.

Conclusion

Young children see matters and notice things that adults tend to overlook, but their expertise is seldom acknowledged, let alone used, in urban planning processes. As young children demonstrate a layered understanding of places and people, they are able to envision alternatives for their immediate environments. Listening to and valuing young children's environmental experiences and perspectives are essential to support their well-being but also for creating sustainable, resilient environments. How they are treated by the city influences their life chances and their well-being as young children, which in turn affects their well-being later in life.

Children's firm inclusion in sustainable city discussions will recenter debates around maintaining, continuing, and repairing an unequal world and opens up a new understanding of the interrelatedness of the human and non-human city spheres and their meaning for well-being, mutuality, and liveability. Children of any age or ability have the right to access urban resources and a right to meaningful participation full of care in the creation, transformation, and development of urban environments – environments that care for the well-being of human and non-human inhabitants and speak to their needs, be it through formal planning and development projects or their everyday practices. The aim to enable the well-being of humans and non-humans needs to be viewed both as a process and outcome of care. However, to ensure children can exercise their rights in and to cities, we need to respect, value, and ensure their rights; researchers, planners, and urban policy-makers need to develop ways of integrating their views, experiences, and suggestions to actually create sustainable, liveable urban environments that care for all living and non-living things in cities.

Cross-References

- ▶ Age-friendly Future Cities
- ▶ Children, Urban Vulnerability, and Resilience
- ▶ Green Cities
- ▶ Healthy Cities
- ▶ Need for Greenspace in an Urban Setting for Child Development
- ▶ Participatory Governance for Adaptable Communities
- ▶ Pre-Schoolers and Sustainable Urban Transport
- ▶ Public Space
- ▶ Toward a Sustainable City
- ▶ Walkable Access and Walking Quality of Built Environment

References

UN Committee on the Rights of the Child (CRC). (2006, September 20). *General comment No. 7 (2005): Implementing child rights in early childhood*, CRC/C/GC/7/Rev.1. Available at: https://www.refworld.org/docid/460bc5a62.html. Accessed 7 Oct 2021.

UNICEF. (1995). *United Nations Convention on the Rights of the Child.*

UNICEF. (2018). *Child friendly cities and community handbook*. Geneva/New York: United Nations Children's Fund.

Further Reading

Ataol, Ö., Krishnamurthy, S., & van Wesemael, P. (2019). Children's participation in urban planning and design: A systematic review. *Children, Youth and Environments, 29*(27), 47.

Bernard van Leer Foundation. (2017). *Building better cities with young children and families: How to engage our youngest citizens and families in city building – A global scan for best practices*. The Hague: Bernard van Leer Foundation.

Bessell, S. (2017). Rights-based research with children: Principles and practice. In R. Evans, L. Holt, & T. Skelton (Eds.), *Methodological approaches*. Singapore: Springer Singapore.

Blades, M., Blaut, J. M., Darvizeh, Z., Elguea, S., Sowden, S., Soni, D., Spencer, C., Stea, D., Surajpaul, R., & Uttal, D. (1998). A cross-cultural study of young children's mapping abilities. *Transactions of the Institute of British Geographers, 23*, 269–277.

Clark, A. (2010). *Transforming children's spaces: Children's and adults' participation in designing learning environments*. Hoboken: Taylor & Francis.

Clark, A. (2011). Multimodal map making with young children: Exploring ethnographic and participatory methods. *Qualitative Research, 11*(3), 311–330.

Derr, V., & Tarantini, E. (2016). "Because we are all people": Outcomes and reflections from young people's participation in the planning and design of child-friendly public spaces. *Local Environment*, 1–22.

Ergler, C. (2016). Children's participation. In M. A. Peters (Ed.), *Encyclopedia of educational philosophy and theory* (pp. 1–6). Singapore: Springer Singapore.

Ergler, C., Smith, K., Kotsanas, C., & Hutchinson, C. (2015). What makes a good city in pre-schoolers' eyes? Findings from participatory planning projects in Australia and New Zealand. *Journal of Urban Design, 20*(4), 461–478. https://doi.org/10.1080/13574809.2015.1045842.

Ergler, C., Freeman, C., & Guiney, T. (2020). Walking with pre-schoolers to explore their local wellbeing affordances. *Geographical Research*.

Ergler, C., Freeman, C., & Guiney, T. (in press). Pre-schoolers' vision for liveable cities: Creating 'care-full' urban environments. *Tijdschrift voor economische en sociale geografie*.

Eriksson, C., & Sand, M. (2018). Belonging in transience: Vocal mapping for a commuting preschool practice. *Emotion, Space and Society, 29*, 1–8.

Fisher, B. & Tronto, J. (1990). Toward a feminist theory of caring. In: Abel, E. K. & Nelson, M. K. (eds.) Circles of care: Work and identity in women's lives. Albany: State University of New York.

Freeman, C., & Cook, A. (2019). *Children and planning*. London: Lund Humphries.

Freeman, C., Ergler, C., & Guiney, T. (2017). Planning with preschoolers: City mapping as a planning tool. *Planning Practice & Research, 32*(3), 297–318. https://doi.org/10.1080/02697459.2017.1374790.

Power, E. R., & Williams, M. J. (2020). Cities of care: A platform for urban geographical care research. *Geography Compass, 14*, e12474.

Templeton, T. N. (2018). 'That street is taking us to home': Young children's photographs of public spaces. *Children's Geographies*, 1–15.

Multidimensional Urban Development

▶ Social Urbanism: Transforming the Built and Social Environment

M

Multi-functionality

► Multiple Benefits of Green Infrastructure

Multi-level Governance

► Collaborative Climate Action

Multiple Benefits of Green Infrastructure

Amna Shoaib
Department of City and Regional Planning, Lahore College for Women University (LCWU), Lahore, Pakistan

Synonyms

Green spaces; Livable communities; Multi-functionality; Parks; Spatial distribution; Sustainable development; Urban planning

Definition

The rapid increase in the urban population has made it difficult for city authorities and policymakers to ensure sustainable development and quality of life in cities. Urban green infrastructures are multi-functionally planned and managed green spaces that deliver an array of benefits to the social, health, environmental, and economic sectors. It offers facilities in different dimension for climate change adaptation, sustainable development, and urban resilience. This chapter briefly highlights the diverse benefits of green infrastructure in an urban setting.

Introduction

Urbanization is growing at an unprecedented rate with 68 percent of the world's population predicted to live in cities by the year 2050 (United Nations 2018). The massive increase in urban population has a varying extent on the overall development of cities. These impacts are distributed unevenly in the world and influencing the urban infrastructure and well-being of citizens in diverse parameters (Peterson 2017). Governments and policymakers are trying to provide a healthy environment in cities to enhance the quality of life for residents.

Many countries like Sweden, Singapore, France, and Norway successfully revolutionized their urban development by incorporating features of green cities in planning (see section "Green Cities"). However, developing countries are not giving required attention to the soft infrastructure measures which exacerbates the threats of climate change, air pollution, mental health, etc. There is a dire need to include green infrastructure in developmental planning to improve sustainability and resilience in cities. There are various benefits of green infrastructure that can create beneficial and long-term impacts.

Green Infrastructure

Green infrastructures (GIs) are planned spatial regions with interconnectivity to environmental elements, natural spaces, open areas, and landscape features (European Commission 2013; Lovell and Taylor 2013; Naumann et al. 2011). These natural structures are more capable to fulfill social, economic, and environmental objectives than gray infrastructure in cost-efficient and resilient ways (Landscape Institute 2009). Green infrastructure initially emerged in the 1990s in the United States as a result of increased concern for unmanageable urban sprawl (Benedict and McMahon 2006), but afterward, it developed as an important policy measure to maintain sustainable development and smart growth (Hansen and Pauleit 2014; SEP 2012).

Green infrastructure is connected to spatial planning with strong linkages to landscape architecture and ecology. The various theoretical considerations of GI's principles have developed the two main concepts of connectivity and multi-functionality (Pauleit et al. 2011; Roe and Mell 2013). Connectivity includes a range of environmental features with natural and semi-natural areas (SEP 2012), whereas multi-functionality refers to the diverse results in combining ecological, economic, and social parameters of green infrastructures (Ahern 2011; Hansen and Pauleit 2014). Multi-functionality is the key to GI which helps in developing sustainable land uses (Roe and Mell 2013). Urban planners and policymakers now consider green infrastructures as a tangible solution for sustainable planning and climate change mitigation (Lovell and Taylor 2013).

GI can be developed in various places including man-made elements, like green roofs, aqueducts on highways, marshlands, and mangroves, as well as on any natural and semi-natural regions in rural, urban, and marine land. Green infrastructures can provide a range of functions on both land and water surfaces with multiple benefits. Although biodiversity stayed as the core mission of green infrastructures for a long time, but it has numerous other advantages as well, besides serving as a perseveration tool (European Commission 2013).

Benefits of Green Infrastructure

Green infrastructure has the most quantifiable benefits in the environmental sector. It plays a vital role in ensuring a healthy and safe environment with enhanced biodiversity or habitat options, climate interventions, and better air quality. Social benefits are the range of non-profitable parameters people acquire from GI, including improved housing facilities, enhanced aesthetic value or sense of place, and improved public services. Accessible green spaces advance the health and well-being sector by enhancing physical and mental well-being,

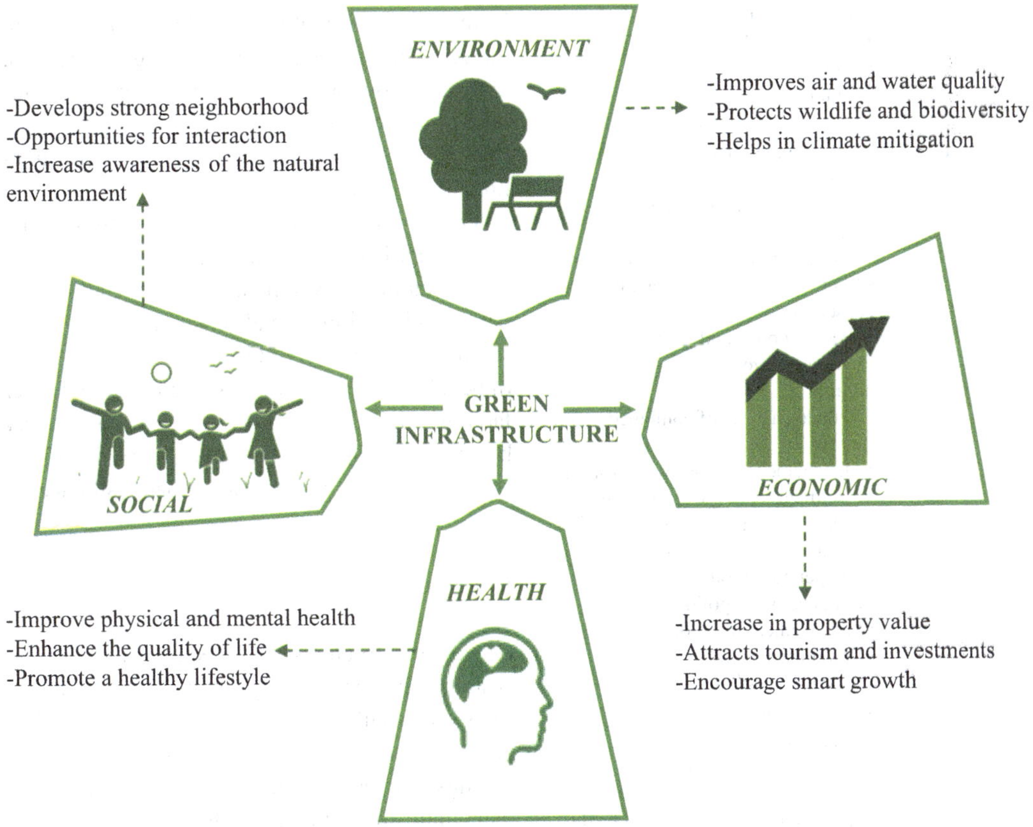

Multiple Benefits of Green Infrastructure, Fig. 1 Framework of GI's benefits and multi-functionality

M

Multiple Benefits of Green Infrastructure, Table 1 Sectoral benefits of green infrastructure

Types of benefits	Description
Economic	• Increase in property value and tax revenue • Creation of green jobs • Enhanced marketability • Better retail sales • Community benefits with local economic development • Upgradation of vital urban centers • Promoting tourism opportunities • Reduction in infrastructure, development, and healthcare cost • Minimize energy usage for cooling
Social	• Improvement in housing quality • Better access to public services • Improvised quality of life and public participation • Reduce crimes • Stronger community cohesion and development • More spaces for social gatherings • More recreational opportunities and nature interaction • Sense of ownership • Efficient land uses • Cultural representation of community • Encourage sustainable development
Environment	• Mitigate the effects of urban heat island • Upgrade biodiversity and improve ecosystem • Reduction in downstream erosion • Improvement in water quality and conservation • Prevent floods and give protection for storm surge and other natural hazards • Reduce chances of combined sewer runoffs • Management of stormwater runoff • Minimizes ecological footprint • Helps in climate change adaptation and mitigation measures • Temperature moderation • Reduced noise pollution • Improvement in air quality • Carbon sequestration • Improved aesthetics with ample landscape and townscape options
Health and well-being	• Increases life expectancy • Reduced health inequality • Improvement in physical activity • Promotion of psychological health • Enhance productivity • Stress reduction • Boost cognitive and mental well-being

Sources: Burley (2018), CNT (2010), EEA (2011), Heckert and Rosan (2018), Kim and Song (2019), Law et al. (2017), Lovell and Taylor (2013), Mathey et al. (2015), Parker (2017), Sarkara et al. (2018), Semeraro et al. (2018), SEP (2012), Suppakittpaisarn et al. (2017), The Scottish Government (2012), Tiwary et al. (2016), Wan et al. (2018), Xiao (2018)

eliminating health inequalities, and increasing life expectancy.

Green infrastructure helps in improving quality of life and economic health which in turn attracts businesses and economic opportunities in the area. This economic growth can lead to the establishment of profitability in local markets, an increase in property prices near green spaces, increased tourism, and business opportunities with crime reduction. Figure 1 shows the illustrative representation of the benefits and multi-functionality of green spaces.

Green infrastructures are now getting more importance because of the diverse benefits in multiple sectors. They emphasize health and well-being and environmental, economic, and social segments of society. Table 1 discusses the range of benefits in detail. GI has the potential to deliver facilities in various ways including community benefits, climate change adaptation, livability, sustainable development, and urban resilience.

Conclusion

Green infrastructures are effective in enhancing sustainability and quality of life in urban areas. They are usually found in the form of open fields, parks, green corridors, and other nature-based settings. Green infrastructures are famous for their multi-functionality in diverse fields including social, health, economic, and environmental sectors with manifold advantages. GI has the capacity to become a highly efficient policy instrument with the ability to provide benefits at diverse scales with sustainable and viable solutions.

Cross-References

▶ Green Cities

References

Ahern, J. F. (2011). From fail-safe to safe-to-fail: Sustainability and resilience in the new urban world. *Landscape and Urban Planning, 100*, 341–343.

Benedict, M., & McMahon, E. (2006). *Green infrastructure: Linking landscapes and communities*. Washington, DC: Island Press.

Burley, B. A. (2018). Green infrastructure and violence: Do new street trees mitigate violent crime? *Health & Place, 54*, 43–49.

Center for Neighborhood Technology (CNT). (2010). *The value of green infrastructure: A guide to recognizing its economic, environmental and social benefits*. Chicago: Center for Neighborhood Technology (CNI).

European Commission. (2013). *Communication from the commission to the European Parliament, the council, the European economic and social committee and the Committee of the Regions. Green infrastructure (GI) – Enhancing Europe's natural capital*. Brussels: European Commission.

European Environment Agency (EEA). (2011). *Green infrastructure and territorial cohesion: The concept of green infrastructure and its integration into policies using monitoring systems*. Copenhagen: European Environment Agency (EEA).

Hansen, R., & Pauleit, S. (2014). From multi-functionality to multiple ecosystem services? A conceptual framework for multi-functionality in green infrastructure planning for urban areas. *Ambio, 43*, 516–529.

Heckert, M., & Rosan, C. D. (2018). Creating GIS-based planning tools to promote equity through green infrastructure. *Frontier Built Environment, 4*, 1–5.

Kim, D., & Song, K. S. (2019). The multifunctional benefits of green infrastructure in community development: An analytical review based on 447 cases. *Sustainability, MDPI, 11*, 3917.

Landscape Institute. (2009). *Green infrastructure: connected and multifunctional landscapes. Landscape Institute position statement*. London: Landscape Institute.

Law, E. P., Diemont, S., & Toland, T. R. (2017). A sustainability comparison of green infrastructure interventions using energy evaluation. *Journal of Cleaner Production, 145*, 374–385.

Lovell, S. T., & Taylor, J. R. (2013). Supplying urban ecosystem services through multifunctional green infrastructure in the United States. *Landscape Ecology, 28*, 1447–1463.

Mathey, J., Rößler, S., Banse, J., Lehmann, I., & Bräuer, A. (2015). Brownfields as an element of green infrastructure for implementing ecosystem services into urban areas. *Journal of Urban Planning and Development, 141*, 1–13.

Naumann, S., Davis, M., Kaphengst, T., Pieterse, M., & Rayment, M. (2011). *Design, implementation and cost elements of green infrastructure projects*. Brussels: European Commission.

Parker, J. A. (2017). *Survey of park user perception in the context of green space and city liveability: Lake. Claremont, Western Australia*. Master's thesis. Murdoch University, Perth, Australia.

Pauleit, S., Liu, L., Ahern, J., & Kazmierczak, A. (2011). Multifunctional green infrastructure planning to promote ecological services in the city. In *Urban ecology, patterns, processes, and application* (pp. 272–285). Oxford: Oxford University Press.

Peterson, A. L. (2017). *Exporting strategies for urban livability: Examining Copenhagen, Denmark as a model city for quality of life generated through urban design*. Senior Theses, Trinity College, Hartford. Trinity College Digital Repository.

Roe, M., & Mell, I. (2013). Negotiating value and priorities: Evaluating the demands of green infrastructure development. *Journal of Environmental Planning and Management, 56*, 650–673.

Sarkara, S., Butcher, J. B., Johnson, T. E., & Clark, C. M. (2018). Simulated sensitivity of urban green infrastructure practices to climate change. *Earth Interactions, 22*, 1–37.

Science for Environment Policy (SEP). (2012). *The multifunctionality of green infrastructure*. Brussels: European Commission.

Semeraro, T., Pomes, A., Del Giudice, C., Negro, D., & Aretano, R. (2018). Planning ground based utility scale solar energy as green infrastructure to enhance ecosystem services. *Energy Policy, 117*, 218–227.

Suppakittpaisarn, P., Jiang, X., & Sullivan, W. C. (2017). Green infrastructure, green stormwater infrastructure, and human health: A review. *Current Landscape Ecology Reports, 2*, 96–110.

The Scottish Government. (2012). *Making the most of communities' natural assets: Green infrastructure*. Scotland: The Scottish Government.

Tiwary, A., Williams, I. D., Heidrich, O., Namdeo, A., Bandaru, V., & Calfapietra, C. (2016). Development of multi-functional streetscape green infrastructure using a performance index approach. *Environment Pollution, 208*, 209–220.

United Nations, Department of Economic and Social Affairs, Population Division. (2018). *World urbanization prospects: The 2018 revision, highlights*. New York: United Nations.

Wan, C., Shen, G. Q., & Choi, S. (2018). The moderating effect of subjective norm in predicting intention to use urban green spaces: A study of Hong Kong. *Sustainable Cities and Society, 37*, 288–297.

Xiao, X. D. (2018). The influence of the spatial characteristics of urban green space on the urban heat island effect in Suzhou Industrial Park. *Sustainable Cities and Society, 40*, 428–439.

Multi-stakeholder Partnerships to Support Climate Migrants in Fragile Cities

Ambika Chawla
Urban Climate Innovations, Washington, DC, USA

Synonyms

Cities in low-and-middle income countries; Climate displaced persons; Internally displaced persons; International partnerships; Urban centers; Weak governance

Definition

Globally, increasing numbers of climate migrants are arriving to "fragile cities" in low- and middle-income countries. Multi-stakeholder partnerships with other actors can facilitate much needed support for climate migrants living and working in these fragile urban centers. This chapter introduces the Durable Solutions Initiative (DSI) and the Protection Return Monitoring Network (PRMN) in Somalia as case studies of multi-stakeholder partnerships to support climate migrants living in fragile urban contexts.

Introduction

Every year, millions of people around the world are forced to abandon their lands, livelihoods, and communities due to the effects of climate change. And, the rate of climate-induced migration is increasing – with most of this migration taking place in the form of rural-urban migration within countries.

According to the recent World Bank report "Groundswell Preparing for Internal Climate Migration," the number of internal climate migrants could reach more than 143 million by 2050, with the majority of these climate migration "hotspots" occurring in the regions of Sub-Saharan Africa, Latin America, and South Asia. Most will be forced from their homes by extreme weather, while others will move due to slow-onset climate-related events, such as desertification and sea level rise (https://www.worldbank.org/en/news/infographic/2018/03/19/groundswell%2D%2D-preparing-for-internal-climate-migration).

Kayly Ober, Senior Advocate for the Climate Displacement program at Refugee International, points out that climate induced migration will likely pose a number of unique challenges for the international community.

"If you are displaced due to climate-related factors, you don't have the same types of protection provisions as a traditional refugee. Also, the scale and scope of the displacement is novel in the face of repeated and increasingly intense climate-related factors. Finally, we need to remember that climate change is not just a driver of migration, but even a factor in protracted displacement where migrants do not have the ability to return home," says Ober.

Pablo Escribano, the Latin American specialist on Migration and Climate Change for the International Organization for Migration (IOM), believes that in addition to regional hotspots for climate migration, we will also see an increasing number of "urban hotspots" for climate induced migration.

According to Escribano, sea level rise will have strong implications for cities like Jakarta, Bangkok, and Dhaka, while in Latin America, the cities of Rio de Janeiro, Lima, La Paz, and Mexico City will experience migration pressure from sea-level rise, melting glaciers and other climate-change effects. Rapidly urbanizing cities in Africa, such as Lagos, Luanda, and Kinshasa are also considered to be urban hotspots for climate migration.

Climate Migrants Arrive in Fragile Cities

The majority of these climate migrants are moving to rapidly expanding cities in low- and middle-income countries. These are under-resourced cities with weak governance, limited in their capacities to provide social services and infrastructure to their residents.

Urban development expert Robert Muggah has dubbed these areas as "fragile cities." As the co-director of the Igarape Institute in Brazil, Muggah developed eleven indicators that determine urban fragility, such as crime, inequality, lack of access to services, climate change threats, among others.

City fragility is not a steady state but occurs due to an aggregation of risks and stresses that can result in extreme vulnerability for city residents. Fragility intensifies when city governments are not able to provide their residents with basic services, infrastructure, or security. Cities in Somaliland, the Congo, Syria, Sudan, and Libya are considered to be exceedingly fragile, while cities in Europe and Australia have lower degrees of fragility.

Challenges Climate Migrants May Face When They Arrive in Fragile Cities

Climate migrants are likely to face many challenges upon arrival to fragile urban settings, such as having limited access to services, adequate housing, infrastructure, and security. In addition, migrants tend to engage in the informal economy of a city, which is characterized by unsafe working conditions, low wages, and long working hours.

"Major metropolitan cities have not done a great job of integrating newly arrived migrants – so they will find themselves in precarious situations, in precarious settlements which exposes them to different types of vulnerabilities," says Ober.

For example, fragile cities struggle to provide adequate housing for their residents. In the mega-city of Nairobi, approximately 60% of the city's population reside in informal slum settlements, often with no running water and poor sanitation (Wamukoya 2020).

In Latin America, it is estimated that 25% of the population lives in informal settlements. In Rio de Janeiro, for example, 1.5 million people live in informal settlements, also known as "favelas," where they struggle to access basic services and sanitation (World Bank 2017; Castano 2020). Rampant crime, violence and even homicidal violence is common in Rio, where an average of five people are killed in a day, often due to clashes with the police (BBC News 2019).

Migrants in fragile urban contexts often do not have access to basic services, such as healthcare and education. Healthcare is often too costly for new migrants. In addition, new arrivals to cities rarely possess the kind of official documentation in their new places of work, which makes it difficult for them to access municipal healthcare clinics.

According to Canh Toan Vuh, a program director for the nongovernmental organization ISET – the Institute for Social and Environmental Transition in Vietnam – one of the major challenges for migrants in Vietnam, particularly temporary migrants who are in need of social services in large cities such as Hanoi and Ho Chi Minh City, is the residential registration system.

"You need to be registered as a permanent resident in order to have access to basic social services. If you are not registered, then your kids cannot go to a public school, which is much cheaper than a private school," he says.

Multi-stakeholder Partnerships to Support Climate Migrants in Fragile Cities

These challenges, which will intensify as increasing numbers of persons move to fragile urban settings due to a variety of climate-related factors, demonstrate a clear need for multi-stakeholder partnerships of all kinds. The power of multi-stakeholder collaboration comes from the different approaches of various actors (government, grassroots, civil society, and business) and the resources they bring to collectively bring to work towards solving pressing challenges – in this case, the challenge of increasing waves of climate migrants arriving to under-resourced urban centers.

By forging partnerships with international humanitarian organizations, NGO's, research institutions, UN agencies, private companies, other city governments, and their respective national governments, fragile city governments can strengthen their capacities to effectively deal with their most pressing urbanization and climate migration challenges.

City governments can build partnerships with international humanitarian organizations with the goal to deliver humanitarian assistance to climate migrants. Research finds that the delivery of humanitarian assistance is more effective when strategies for multi-stakeholder partnerships are developed early, prior to a crisis or emergency which requires an immediate response (Grunewald et al. 2011).

Sister city partnerships between cities in the developed and developing world can facilitate much needed funding for migrant inclusion programs in fragile cities. Sister Cities International is an example of a nonprofit organization which can strengthen the capacities of fragile city governments through the building of institutional partnerships between cities, knowledge sharing, and via grant programs which can provide resources for migrants living in cities.

Fragile city governments can also build bridges with global research programs with the aim to fill gaps in knowledge on the dynamics of climate migration to cities, urban trends, and city governance, particularly given the fact that due to their financial constraints, local governments are often limited in their capacities to collect data (IISD 2019). Because data/information on climate migration to cities is often limited in accuracy, greater knowledge can help city officials design policies which can better support climate migrants living and working in fragile urban contexts.

According to Pablo Ferrandez of the Internal Displacement Monitoring Center (IDMC), a think tank based in Geneva, multi-stakeholder partnerships play a crucial role in the gathering of data and information about IDP's for his organization.

"When it comes to research, we work directly with some key partners. For example in Ethiopia and Somalia, we work with the IOM (International Organization for Migration), the government of Ethiopia, as well as some local universities," says Ferrandez.

"We start with the people affected – internally displaced persons and host communities- and from there, we build up the agenda, collaborating with national governments, UN agencies, NGOs, academia, and research centers."

Partnerships with the private sector can bring much-needed investments in the form of humanitarian assistance for climate migrants. Private companies are increasingly playing a more active role in humanitarian responses in urban contexts. Over the years, multinational corporations such as Ericsson, TNT, UPS, IKEA, and Google have provided different forms of humanitarian assistance for persons living in urban areas (Chawla 2017).

For this next section, we will look at the Durable Solutions Initiative (DSI) and the Protection Return and Monitoring Network (PRMN) in Somalia as examples of multi-stakeholder initiatives to support climate migrants in fragile cities in Somalia.

Case Study 1: The Durable Solutions Initiative (DSI) in Somalia

Currently, there are approximately 2,648,000 million displaced persons in Somalia, with the largest concentration, around half a million, living in the capital city of Mogadishu (IDMC 2020; Yarnell 2019). These internally displaced persons arrive in fragile cities in Somalia due to conflict, violence, food insecurity, and climate-related factors, such as drought and flooding.

For the last decade, Somalia has experienced protracted drought, and more recently, heavy flooding. Last year, heavy rains in Somalia from October to December led to severe flooding, resulting in the displacement of 370,000 people.

According to the Humanitarian Response Plan 2020 Somalia, over the years, erratic weather patterns ranging from severe drought to torrential rains have left hundreds of thousands of people in Somalia homeless (United Nations Office for the Coordination of Humanitarian Affairs 2020). "Before the floods, 300,000 people had already been displaced by drought and conflict in 2019, adding to the 2.6 million internally displaced people living in 2,000 sites across Somalia," the Plan states.

Upon arrival to the fragile city of Mogadishu, or to the smaller cities of Kismayo and Baidoa, these climate migrants often find themselves living in challenging situations in one of the many IDP camps that border the city center.

In 2016, the Durable Solutions Initiative (DSI) was created as a joint effort led by the Federal Government of Somalia and the United Nations, with the aim to strengthen the capacities of government at all levels to provide long-term solutions for IDPs, including climate migrants, with the hope that 1 day, they will no longer have special protection needs and they will become self-reliant.

Teresa del Ministro, the Durable Solutions coordinator for Somalia, explains that the "DSI has been a collective effort linked to global trends of the limitations of humanitarian approaches to deal effectively with IDPs. With that trend increasing worldwide, it appeared that multi-stakeholder partnerships are needed at all levels."

The DSI is a multi-stakeholder partnership which has triggered a collective effort from various donors, such as the EU, DANIDA (Danish International Development Agency), DFID (Department for International Development), the Peace Building Fund, the World Bank, as well as UN agencies, sections of the government in Somalia, and humanitarian organizations.

Isabelle Peter, the Durable Solutions Coordination Officer in Somalia, shares that the DSI is considered innovative due to its "bottom-up" approach. "A participatory, locally owned approach is one of the programming principles for the DSI, which were developed with the government and the DSI community to guide interventions for durable solutions in Somalia," she says.

One example of a community-led initiative which was implemented in direct support to the Durable Solutions Initiative (DSI) is the Midnimo I project ("unity" in Somali), which received initial funding from the Peacebuilding Fund and was led by UNDP, with the IOM and UN-Habitat as partners.

Midnimo is a peacebuilding initiative which took place in Southern and Central Somalia where displaced communities, including climate migrants, women and youth groups, engaged in "visioning" exercises with their host communities and city/national government officials with the aim to articulate their short-term needs, as well as ideas to move towards self-reliance.

Representatives of the displaced communities would come up with priorities for infrastructure investments or other types of investments for community projects. If a project didn't have funding for these priorities, the government would then convene other actors and ask for their support.

According to an evaluation report by the IOM, the Midnimo project created short-term employment opportunities for IDPs, led to the construction of community infrastructure projects, the establishment of a land commission, and contributed to improved relations between host and displaced communities (https://www.un.org/peacebuilding/sites/www.un.org.peacebuilding/files/documents/somalia_2018_project_evaluation_mid-term.pdf).

Teresa del Ministro and Isabelle Peter agree that the long-term success of the DSI will depend upon overcoming a number of challenges, such as ensuring that there are sufficient resources for community-led initiatives, overcoming the obstacles of coordination, and strengthening the capacities of fragile city governments.

"The need for strengthening the capacities of cities is huge. Stronger capacities are needed in human resources in city planning. There is a need to have financial resources available. Developing the skills and knowledge of people who are equipped to deal with challenges in cities is needed," says Peter.

Case Study 2: The Protection Return and Monitoring Network (PRMN) in Somalia

The Protection and Return Monitoring Network (PRMN) is a multi-stakeholder partnership in Somalia which is led by the United Nations High Commissioner for Refugees (UNHCR), who partners with the Norwegian Refugee Council (NRC), and civil society groups throughout the country. Through collaborative efforts, these actors collect data, information, and trends about internal displacements across the country, sharing this information on a monthly basis via an online dashboard.

PRMN gathers different types of data about IDP's in Somalia, such as the drivers for internal displacements – like conflict, drought, and flooding – as well as the numbers of persons displaced due to these drivers. PRMN presents its data in the form of an online dashboard (https://unhcr.github.io/dataviz-somalia-prmn/index.html) and "flash reports" which provide an overview of the latest trends in internal displacements in Somalia. The data is then used by city/national government officials, UN agencies, the media, and international humanitarian organizations, with the hope that it can help them better support IDPs.

PRMN also gathers "protection incidences" related to displacement which encompass the range of experiences which violate a person's rights, such as rape or torture. Finally, PRMN staff collaborate with displaced populations to identify their priority needs.

During the 2017 drought, for example, PRMN provided vital data and information about drought-related internal displacements, as well as the priority needs for these IDP's. PRMN staff conducted extensive interviews with IDP's upon arrival, with the goal to highlight their immediate needs (i.e., food, healthcare, and shelter). As a result, the international community was able to provide a much more effective humanitarian response in Somalia during the drought.

Joseph Jackson, the UNHCR coordinator for PRMN, shares the ways in which the initiative is a collaborative effort between NRC and UNHCR:

"We do joint monitoring—if there is an incidence of displacement, we work together to verify the incident. Next, every quarter we come together, review the data, and come up with a joint analysis of trends, which is then shared with the humanitarian community. Finally, we produce flash reports which report incidences that have the potential to undermine humanitarian efforts."

Mustafa Ghulam, a Food Security Specialist for the Norwegian Refugee Council, explains that local civil society organizations in the country are engaged in the PRMN initiative through coordination meetings where leaders of displaced communities share their concerns with NRC staff.

"We have networks with the leaders of local, displaced communities. NRC provides technical support, training them on gathering data on their communities," he says. "We have local partners who are collecting data on a daily basis. We organize coordination meeting where we select one topic in every meeting and we talk about it."

In terms of some of the challenges the PRMN experiences, for one, Somalia is a country characterized by conflict and insecurity. "We have not been able to access all of the areas because of insecurity. There are dangerous areas controlled by militant groups," says Jackson.

Next, there have been challenges with the issuing of joint press releases. As a result, each organization now issues its own press release.

Ghulam emphasizes the important role of technology for overcoming any coordination challenges between the partners active in PRMN. "Technology has helped us a lot. These days we are connected—we can organize meetings with zoom and skype. You have to have a coordinated response. While it is better to go and meet in person, if you cannot meet, we conduct meetings through the use of technology."

Challenges and Recommendations

Research on multi-stakeholder partnerships as an innovative form of governance is relatively nascent and there is a need to better understand the challenges that may arise through these cooperative arrangements, particularly at the local level, which some researchers have called "place-based multi-stakeholder partnerships" (Orozco et al. 2018).

To date, most of the research examining the challenges associated with multi-stakeholder partnerships have focused on national and transnational partnerships with less of an understanding of the challenges associated with place-based multi-stakeholder collaboration, particularly in the context of the Global South. Place-based multi-stakeholder partnerships are those which are focused on addressing local-level issues and which draw upon local assets and knowledge to design solutions to local challenges (Bellafontaine and Wisner 2011).

One notable study (2018) on place-based multi-stakeholder collaboration, which was led by a team of Mexican researchers, analyzed 38 multi-stakeholder socio-environmental projects throughout rural and urban areas in Mexico, with the aim to gain a more comprehensive understanding of some of the key challenges associated with place-based multi-stakeholder partnerships. According to the study, some of the key challenges included the following: (a) the divergent visions and interests across stakeholders; (b) inadequate project planning and management; (c) lack of funding; (d) weak governance, insecurity, and violence at the national level; (e) and inadequate communication and among the participants engaged in the multi-stakeholder partnership.

The research team, in consultation with the participants engaged in the partnerships, recommended a series of strategies for fostering improved place-based multi-stakeholder collaboration. For one, participants proposed the need for strengthening project management, such as improving strategies for evaluating projects while increasing accountability, ensuring that planning and evaluation takes place in a participatory manner, improving access to funding opportunities, and improving the management of human and financial resources.

Second, the need for a common vision was mentioned, with emphasis on engaging underrepresented communities, for developing formal agreements among the participants in the partnership, building shared knowledge among participants, developing conflict prevention mechanisms, and for strengthening strategic partnerships, primarily with academia and government.

Third, capacity-building and tools for improving communication were mentioned, with participants recommending the building of a common language, the use of diverse communication media, the improving of transparency and empathy in communication, and the organization of forums and other types of meeting venues.

Fourth, participants proposed strategies in the area of forms of organization and community institutions, such as the equitable distribution of power in communities, enhanced decision-making mechanisms including those for conflict-resolution, and respect for a diverse range of viewpoints among community members. Finally, they mentioned strategies for overcoming challenges related to governance and security and proposed recommendations for the prevention of crime and violence, as well as ideas to reduce the bureaucracy which commonly characterizes government institutions.

Conclusion

With the coming years, climate induced migration to "urban hot spots" will only intensify. As it does, multi-stakeholder partnerships can help fragile city governments deliver a more effective humanitarian response in times of crisis, while also empowering IDPs to move towards greater self-reliance by offering training, services, and other resources that can help them fully integrate into mainstream society.

The Durable Solutions Initiative (DSI) and the Protection and Return Monitoring Network (PRMN) in Somalia are two examples of multi-stakeholder partnerships which have been effective in providing much needed support to climate migrants living in fragile cities in Somalia. The DSI, in particular, gives internally displaced persons a voice in decision-making processes that shape their future and offers a model for other cities around the world that are, or soon will be, in similar circumstances.

In conclusion, fragile city governments around the world cannot address the manifold challenges associated with internal climate migration alone. While initial research has taken place on the challenges associated with place-based multi-stakeholder partnerships, and recommendations for strengthening these partnerships, much more research is needed on the challenges associated with multi-stakeholder partnerships at the city level, particularly in the context of the Global South. Hopefully, greater knowledge in this area can help decision-makers in fragile cities to design more effective policies and programs to support climate migrants living and working in their respective cities.

Cross-References

▶ Climate-Induced Relocation
▶ Collaborative Climate Action
▶ From Vulnerability to Urban Resilience to Climate Change
▶ Future of Urban Governance and Citizen Participation
▶ Participatory Governance for Adaptable Communities

References

BBC News. (2019). Rio violence: Police violence reach record high in 2019. https://www.bbc.com/news/world-latin-america-51220364

Bellafontaine, T., & Wisner, R. (2011). *The evaluation of place-based approaches. Questions for further research* (p. 33). Toronto: Policy Horizons Canada.

Castano, AEH. (2020, Apr 6). In the slums of Rio, communities have to choose between hunger and coronavirus. *ABC News*.

Chawla, A. (2017). Climate-induced migration and instability: The role of city governments. In *One earth future foundation (OEF) research*. Denver, Colorado.

Data from the Internal Displacement Monitoring Center (IDMC). https://www.internal-displacement.org/countries/somalia. Retrieved on October 10, 2020.

Grunewald, F., Boyer, B., Kauffman, D., & Patinet, J. (2011). Humanitarian aid in urban settings: Current practice, future challenges. Groupe URD report, December 2011. Available at https://www.scribd.com/document/179808610/Humanitarian-aid-in-urbansettings-current-practice-future-challenges

IISD. (2019). Experts examine multi-stakeholder partnerships' challenges, potential. https://sdg.iisd.org/news/experts-examine-multi-stakeholder-partnerships-challenges-potential/

Orozco, B. A., et al. (2018). Challenges and strategies in place-based multi-stakeholder collaboration for sustainability: Learning from experiences in the global south. *Sustainability, 10*, 3217.

United Nations Office for the Coordination of Humanitarian Affairs (2020). Humanitarian response plan Somalia 2020. Humanitarian Programme Cycle 2020. OCHA (United Nations Office for the Coordination of Humanitarian Affairs).

Wamukoya, M. (2020). The Nairobi urban health and demographic surveillance of slum dwellers, 2002–2019: Value, processes, and challenges. *Global Epidemology, 2*, 100024.

World Bank. (2017). The cities of the future in Latin America: Fewer cars, fewer youth. https://www.worldbank.org/en/news/feature/2017/10/05/ciudades-del-futuro-en-america-latina

Yarnell, M. (2019). Durable solutions in Somalia: Moving from policies to practice for IDP's in Somalia. *Refugees International.*

Municipal

▶ Closing the Loop on Local Food Access Through Disaster Management

Muslim Women

▶ Hidden Enemy for Healthy Urban Life

N

NAPCC – National Action Plan on Climate Change

▶ Adapting to a Changing Climate Through Nature-Based Solutions

National Disaster Management Authority of India (NDMA)

▶ The State of Extreme Events in India

Natural Foodways

▶ Formulating Sustainable Foodways for the Future: Tradition and Innovation

Nature

▶ Health and the Role of Nature in Enhancing Mental Health

Nature Connectedness

▶ Emerging Concepts Exploring the Role of Nature for Health and Well-Being

Nature Relatedness

▶ Emerging Concepts Exploring the Role of Nature for Health and Well-Being

Nature-Based Interventions, Planning with Nature, Green/Soft Engineering

▶ Nature-Based Solutions for River Restoration in Metropolitan Areas

Nature-Based Solutions

▶ Urban Forestry in Sidewalks of Bogota, Colombia

R. C. Brears (ed.), *The Palgrave Encyclopedia of Urban and Regional Futures*,
https://doi.org/10.1007/978-3-030-87745-3

Nature-Based Solutions (NbS)

▶ Policy and Practices of Nature-Based Solutions to Build Resilience in Seoul, Korea

Nature-Based Solutions for River Restoration in Metropolitan Areas

The Example of Costa Rica

Jochen Hack[1] and Barbara Schröter[2,3]
[1]Technical University of Darmstadt, Section of Ecological Engineering, Institute of Applied Geosciences, Darmstadt, Germany
[2]Leibniz Centre for Agricultural Landscape Research (ZALF), Working group Governance of Ecosystem Services, Müncheberg, Germany
[3]Lund University Centre for Sustainability Studies (LUCSUS), Lund, Sweden

Synonyms

Ecological engineering; Ecological restoration; Ecosystem-based adaptation; Ecosystem-based disaster risk reduction or building with nature; Ecosystem-based management; Green and blue infrastructure; Nature-based interventions, Planning with nature, green/soft engineering

Definition

Nature-based solutions (NBS) are "actions which are inspired by, supported by or copied from nature" (Commission 2015), or more specifically, actions that (i) alleviate a well-defined societal challenge, (ii) utilize ecosystem processes, and (iii) are embedded within viable governance models (Albert et al. 2019). The term is used as an umbrella term for existing concepts such as ecological restoration, ecological engineering, green and blue infrastructure, ecosystem-based management, ecosystem-based adaptation, ecosystem-based disaster risk reduction, or building with nature.

River restoration refers to a variety of ecological, physical, spatial, and management measures aimed at restoring the natural state and functioning of the river system in support of biodiversity, recreation, flood management, and landscape development (ECRR 2021).

NBS for river restoration all include in-stream and off-stream measures that initiate or accelerate the recovery of degraded, damaged, or destroyed river ecosystems. Examples for in-stream measures are wetland restoration, reconnection of seasonal streams, revitalization of floodplains, or the reestablishment of alluvial forest. Off-stream measures NBS for river restoration contribute to a more natural water balance and flow regime as well as control water quality and erosion, such as bioretention areas, bioswales, infiltration trenches, raingardens, green roofs, or other elements of green stormwater infrastructures and water-sensitive urban design.

Introduction

In addition to the urbanization process itself, the current climate and environmental crises confront cities with a variety of challenges. In former times, rivers were the prerequisite for larger settlements as they supplied drinking water, enabled agricultural production, and provided the waterway for trading. With the foundation and development of cities, rivers have been further transformed by humans, e.g., to increase their function as drainage, waste receiver, or to produce energy (Brown et al. 2009). These transformations have led to undesirable ecological effects over time, such as increased erosion, decrease and pollution of groundwater resources, increase in flood probability, and decline in fisheries and biodiversity as well as loss of esthetics and recreational functions (Sabater et al. 2018). In consequence, rivers are one of the most affected ecosystems within cities which makes it difficult for them to adapt to the changes in the river and flood regime due to the climate and ecological crises (Pletterbauer et al. 2018). Pollution, eutrophication, salinization, missing wastewater treatment, and the scarcity of clean water constitute major challenges in cities (Haase 2015), with huge impacts on health, well-being, and high economic costs to sustain people's quality of life (Vörösmarty et al. 2010). Particularly in megacities of the Global South, there is an unsustainable way of using water resources (Niemczynowicz 2009) although rivers often represent the only remaining ecosystems providing habitat and ecosystem services within cities (Hack et al. 2020).

NBS for river restoration in response to these challenges can be distinguished into measures implemented within the river corridor (in-stream) and measures implemented in the river basin (off-stream, basin-wide). Especially in strongly urbanized river basins, the reestablishment of a more natural water balance in form of off-stream measures is often necessary to achieve flow regimes (reduced runoff volumes and peak flows, augmented base flows) that are similar to a pre-development state (Walsh et al. 2005). This implies measures to reduce surface runoff through a reduction of surface sealing and to increase infiltration and evapotranspiration by providing storage in soils and water bodies as vegetated space (Collentine and Futter 2016). Common concepts that support the reestablishment of a more natural water balance and flow regimes are water-sensitive urban design, urban green infrastructures, sustainable urban drainage systems, or low-impact development. Besides these measures aiming at a quantitative hydrological change, there is also a need to improve runoff from urban areas qualitatively through in-stream measures. Not all measures that improve quantitative aspects of the urban water balance also have a qualitative impact. Measures need to be specifically designed to have a positive water quality impact. Different filtration and vegetated systems as NBS are used to improve runoff quality before infiltrating into the ground or discharging it to a water body. In addition to NBS within the drainage area of a river, measures within the river corridor (in-stream) are essential for river restoration in urban areas. These measures are related to the reestablishment of longitudinal and lateral connectivity – e.g., through the removal of weirs, the establishment of fish passes, or the reconnection of floodplains – as well as to a general improvement of the morphology and ecological communities of rivers and their corridors.

NBS for river restoration can be applied in a spectrum of ecosystem conditions, from artificial to managed to largely intact ecosystems. Eggermont et al. (2015) distinguish three NBS types differing in the level of engineering or management applied to biodiversity and ecosystems, the number of services to be delivered and stakeholder groups targeted, and the likely level of maximization of the delivery of targeted services. These types range from no or minimal intervention in ecosystems, via implementation of management approaches that develop sustainable and multifunctional ecosystems and landscapes up to creating new ecosystems. They are beneficial over purely technical solutions as they minimize negative side effects and instead create co-benefits for people and nature (Albert et al. 2019), as they can serve different purposes (multifunctionality) and impact on various challenges such as climate regulation, biodiversity conservation, enhanced food security, and improved livelihoods (Watkins et al. 2019).

Moreover, NBS are considered important for combating climate change, safeguarding biodiversity, advancing ecosystem restoration, and enabling a green recovery in the aftermath of the SARS-COV-2 pandemic (Albert et al. 2021).

Nature-Based Solutions for River Restoration in Metropolitan Areas in Latin America

There are several challenges and opportunities for the implementation of NBS for river restoration in metropolitan areas, in particular in Latin America, a region with a high share of urban population, a long history of urban growth (Pérez Rubi and Hack 2021), and more biodiversity hotspots than in other regions (Dobbs et al. 2019). Challenges are, among others:

1. Cities are constantly growing without proper planning. Settlers arrive and construct their

housing often informally, often at unsuitable sites without appropriate infrastructure. Thus, in already urbanized areas, space availability for NBS measures is often poor (Pérez Rubi and Hack 2021) as housing, industrial, and transport development is prioritized while public green spaces are largely neglected (Fluhrer et al. 2021), unequally distributed, or of poor quality depending on the residents' socioeconomic status (Dobbs et al. 2019).

2. Planning and governance are characterized by ineffective governmental institutions, lack of transparency, poorly defined tenure regimes, and lack or ineffectiveness of planning tools (Dobbs et al. 2019). Planning departments do not exist or are understaffed, and information or records regarding the existing public infrastructure are often missing (Pérez Rubi and Hack 2021). This implies that NBS design and implementation have to be retrofitted to an existing often chaotic and not well-planned infrastructure.
3. There is a strong difference between formal and informal institutions. While formal rules and laws for sustainable water management exist – in Costa Rica, e.g., the protection zone of rivers establishes a strip of 15 min in urban areas without tourism projects, real estate, or productive activities – in reality, settlements do not respect this rule. Monitoring and enforcement are time-consuming and protracted (Pérez Rubi and Hack 2021).
4. As the better developed infrastructure of larger cities in Latin America is available mainly in wealthier districts, justice issues of NBS implementation are present from the beginning. This aggravates findings for recent studies, in general, that NBS in urban areas tend to foster social exclusion, neoliberalist growth, and unequal power relations regarding class, race, ethnicity, gender, or sexuality (Toxopeus et al. 2020; Cousins 2021; Sekulova et al. 2021). As the management of green areas usually depends on municipal revenues and homeowner access to resources, municipalities and neighborhoods with lower income generally have few, poorly vegetated urban green spaces (Dobbs et al. 2019).
5. Although the financing situation of NBS is difficult already in the Global North, as the example from Germany shows (Droste et al. 2017), it is even more limited in Latin American cities. Municipalities are on a tight budget to invest and often have other investment priorities (Chapa et al. 2020; Neumann and Hack 2020).
6. Responsibilities and resources for environmental and infrastructure issues are split between different branches of administration lacking integration. While environmental management resides typically in the Ministry of Environment, decisions regarding infrastructure are made in the Ministry of Planning and Finance. Therefore, NBS are not always considered in early stage-planning and procurement processes for infrastructure (Watkins et al. 2019). This is reflected on different administrative levels. In addition, the environment departments usually have less financial resources at their disposal than the infrastructure-engineering departments.
7. Integration of social, environmental, and economic policies hardly exists (Varis et al. 2006), and NBS need to be better mainstreamed into policy, legislation, and regulations (Watkins et al. 2019). The full spectrum of stakeholders for decision-making is not included, and synergies between distinct municipal departments are not established. By revising and reformulating their current regulatory plans, municipalities should capitalize more comprehensively on the multiple benefits provided by NBS (Pérez Rubi and Hack 2021).
8. The population is rarely involved in problem solving and decision-making; establishment and monitoring of public acceptance is neglected in governance processes (Wantzen et al. 2019).

However, there are also strong opportunities for successful NBS implementation in Latin America:

1. NBS are already considered or incorporated into national policy, legislative, regulatory, or organizational arrangements (Watkins et al. 2019). The general guidelines for the

incorporation of resilience measures in public infrastructure in Costa Rica (MINAE 2020), e.g., define NBS as important principle.

2. The recent Escazú Agreement among the countries of Latin America and the Caribbean to establish regional transparency and environmental standards regarding information, participation, and access to justice such as access to environmental information, participation rights, rights of citizens affected by resource exploitation to sue, and protection mechanisms for environmental activists (ECLAC 2018).
3. Latin America in general has a strong and organized civil society that is engaged in community–based environmental management (Sattler et al. 2016). These organizations could act as a powerful actor for NBS implementation at the local level.
4. Municipalities have a lot of scope to adapt and use existing policies. For instance, they can use local funds from stormwater fees and loan programs combined with other institutions' funding to stimulate the adoption of NBS. Further, they can implement incentive mechanisms in local policies, such as grant programs, rebates, and tax abatements for the development of NBS, as well as regulations that enhance a constant knowledge transfer between citizens and urban planners (Neumann and Hack 2020).
5. While in comparison to Europe or the USA in Latin America rivers are commonly more polluted and suffer from a bad water quality due to solid waste disposal in the river corridor, poor wastewater treatment, and stormwater management, they are less morphologically changed than rivers in Europe. For NBS implementation, this is an advantage, as structural changes in the river such as flood plain disconnections, artificial embankments and river beds, dams, or dikes are more difficult to treat than water quality challenges (Beißler and Hack 2019).
6. North-South and South-South city cooperation agreements and networks to exchange experiences and learn from each other are becoming increasingly important.

Examples for Implementation of Nature-Based Solutions for River Restoration in Costa Rica

Costa Rica serves as a good example to illustrate the implementation of NBS for river restoration in metropolitan areas. It is a country noted for its natural wealth and conservation efforts through protected wildlife areas, which are highly visited by foreign and local tourism. In contrast, in almost all urban areas, and especially in the Greater Metropolitan Area of the capital of San José (GAM), Costa Rica still owes a debt to its inhabitants for the conservation of nature and the benefits it provides. The GAM is the largest urban agglomeration in Costa Rica, composed of 31 municipalities located in the provinces of San José, Heredia, Cartago, and Alajuela. It has a population of approximately 2.7 million people – more than half of Costa Rica's total population – and covers an area of 2044 km^2, a little more than 4% of the country's territory, making it the most urbanized region with the highest population density (Potthast and Geppert 2019). At the same time, the GAM is the center of the country's most important economic activities with the main service, industry, and infrastructure sectors, as well as the headquarters of state institutions.

Three exemplary instruments for promoting NBS for river restoration in Metropolitan Areas of Costa Rica will be presented: The Hydrological Code, the initiative of Interurban Biological Corridors, and real-world lab prototypes.

An important process for the promotion of NBS in urban areas in Costa Rica has been initiated by the Federated College of Engineers and Architects of Costa Rica of Engineers and Architects of Costa Rica (*Colegio Federado de Ingenieros y de Arquitectos de Costa Rica*; CFIA). The CFIA created a Joint Commission (*Comisión Paritaria para la Creación del Código Hidrológico*) to develop a **Hydrological Code** for the country (CITEC 2021). The principal objective of the Hydrological Code is to cope with the flood-generating impact of rapid urbanization. The code intends to establish a common national framework for all municipalities to deal with stormwater issues more comprehensively. It

N

recognizes the insufficiency of traditional stormwater systems to cope with ever increasing volumes of urban runoff and the deteriorating impact that these systems have on river ecosystems. Furthermore, the code explicitly includes green infrastructures as NBS that use vegetation and soil to maintain or recover hydrological processes such as interception, ponding, infiltration, percolation, evaporation, and transpiration – processes which are eliminated or diminished by urban development. After approval of the code, the commission intends to link it as a complement to the technical standards for design and construction of drinking water supply, sewerage, and sanitation, to provide training, and to extend the scope of the code to road infrastructures.

In the GAM, urban sprawl and loss and degradation of urban green spaces jeopardize connectivity and affect the formation of the network of biological corridors. The contamination of urban water bodies by sewage, solid waste dumps, invasion of protected areas, and logging of riparian forests are other substantial challenges faced. However, in recent years, many public, private, and civil society initiatives and efforts have been developed that seek to change this situation. One of these initiatives is the **Interurban Biological Corridors** (IBC). According to Decree 40.043 of MINAE (MINAE 2016), IBCs are "urban territorial extensions that provide connectivity between landscapes, ecosystems, and modified and natural habitats that interconnect micro-watersheds and green spaces or wild protected areas. These spaces contribute to the maintenance of biodiversity, enabling migration, dispersion of flora and fauna species, and include cultural, socioeconomic and political dimensions." As part of the IBCs, strategies are designed and implemented by local committees, participatory platforms with representation from civil society, private enterprise, state institutions, and municipalities.

Today, IBC initiatives exist along different urban rivers and across several municipalities. They were initiated as a response to the growing concern about highly polluted and degraded microwatersheds. This situation culminated in 2007 with a complaint of unconstitutionality and the ruling of the Constitutional Court to public institutions to seek an effective, articulated, and multisectoral solution to eliminate the sources of contamination in the Grande de Tárcoles river basin. In 2019, there were six official IBCs in the GAM, in addition to several initiatives in the process of creation and officialization that in the future will be included in the map of the National Biological Corridors Program (Potthast and Geppert 2019).

Currently, a 3-year pilot plan in the Torres and María Aguilar IBC microwatersheds is being implemented to systematize experiences and replicate these actions in other urban microwatersheds and apply this strategy in at least two sub-basins or microbasins, with a total impact on at least six water bodies by 2030 (MINAE 2020). As a first result of the application of this instrument in the sub-basin of the María Aguilar River was the establishment of an intermunicipal agency that supports the harmonization of regulatory plans and stormwater norms among the participating municipalities and is provided with an own technical team. The activities of the IBC aim to improve the environmental quality of the space, facilitate recreation and enjoyment of the natural elements, as well as promote ecosystem regeneration and the efficiency of the city's metabolic flow through targeted interventions within the so-called Life Network (Red de Vida). These interventions include the conditioning, construction, and improvement of: ecological paths and trails, river parks, viewpoints and observation points for flora and fauna, watershed stabilization works (such as gabions and others), runoff retardation ponds and water farms, urban orchards and nurseries, and pedestrian and wildlife bridges.

The third exemplary instrument is the implementation of NBS as **prototypes in a real-world lab** in the Metropolitan Area of San José, by a 5-year transdisciplinary research project financed by the German Federal Ministry of Education and Research (SEE-URBAN-WATER 2021). The real-world lab provides physical space and a socioeconomic context representative for testing of nature-based solutions resulting. It serves for joint knowledge generation and synthesis as well as a basis for knowledge transfer and upscaling of the tested nature-based solutions. The project started in 2018 by selecting the Quebrada Seca-Río Burío watershed, a highly urbanized watershed of the Metropolitan Area with severe urban flooding and river contamination problems (Neumann and Hack 2020). The watershed, which is part of the Tárcoles River Basin that drains most of the Metropolitan Area, is of special political importance as the Constitutional Court of Costa Rica ordered municipalities and other relevant institutions to strictly consider urban flood mitigation and river restoration in future urban development.

In order to select a typical residential neighborhood as a real-world lab for detailed study of NBS implementation potential and prototype implementation, the five municipalities sharing the watershed were asked to present proposals for potential NBS for river restoration sites including specifications about the type of NBS in a participatory process. Based on the quality of the proposal and the guaranteed support of the municipality to support the implementation of prototypes, a real-world lab in the District of Llorente of the Municipality of Flores was established (Chapa et al. 2020).

The real-world lab represents an already urbanized residential area of 25 ha with about 2500 inhabitants of mostly low-income households with less than 175 USD per month. Therefore, a retrofitting with NBS as adaptation of existing infrastructure and alternative uses of space is pursued. With a participatory co-design approach, it was aimed for the development of context-adapted NBS prototypes and the establishment of a shared vision for transformation. The initial field work-based assessment of the potential for urban NBS consisted of four steps: (1) a detailed site analysis and co-design process including consultations of local community leaders, street markets, residents' workshops, and interviews leading to an identification of suitable types of NBS, general placement locations, and dimensions as well as vegetative design aspects; (2) the establishment of design criteria and placement strategies to achieve a high degree of multifunctionality; (3) the development of spatial typologies based on street network and open space characteristics allowing upscaling through replication of NBS in other areas of the watershed with similar spatial typology, i.e., implementation potential for NBS such as available space and land use patterns; and based on the steps before, (4) a spatial suitability assessment of NBS elements to reveal the specific NBS implementation potential for the real-world lab (Fluhrer et al. 2021).

The results of this assessment allowed a detailed high-resolution hydrological modeling of the flood-mitigating effect (Towsif Khan et al. 2020) and microclimate-regulating impact of NBS in this particular area (Wiegels et al. 2021). Furthermore, it enabled the identification of suitable sites and guided the implementation of four exemplary NBS prototypes to address the prevailing socioecological challenges of river contamination by untreated greywater discharge, urban flooding, and a general lack of green spaces. In the beginning of 2020, three prototypes for household greywater treatment of different degrees of decentralization (household front in responsibility of the house owner, street block, and neighborhood level in responsibility of the municipality) as well as one prototype for stormwater storage were constructed (see Fig. 1). The performance of these prototypes is still being monitored and evaluated; however, the planning and construction process already revealed several challenges regarding the implementation of NBS in densely urbanized areas (Chapa et al. 2020). For instance, the involvement of multiple stakeholders and the assignment of new responsibilities to achieve a context-adapted and multifunctional design of NBS are necessary but have proven difficult since demands and objectives vary among stakeholders. In addition, NBS represent new solutions that require flexibility and openness from municipal authorities. The operation and maintenance of the NBS prototypes result in new tasks related to the control of clogging and sedimentation of waste and stormwater entries as well as vegetation care. However, the realization of NBS prototypes initiated an important learning process about implementation and maintenance

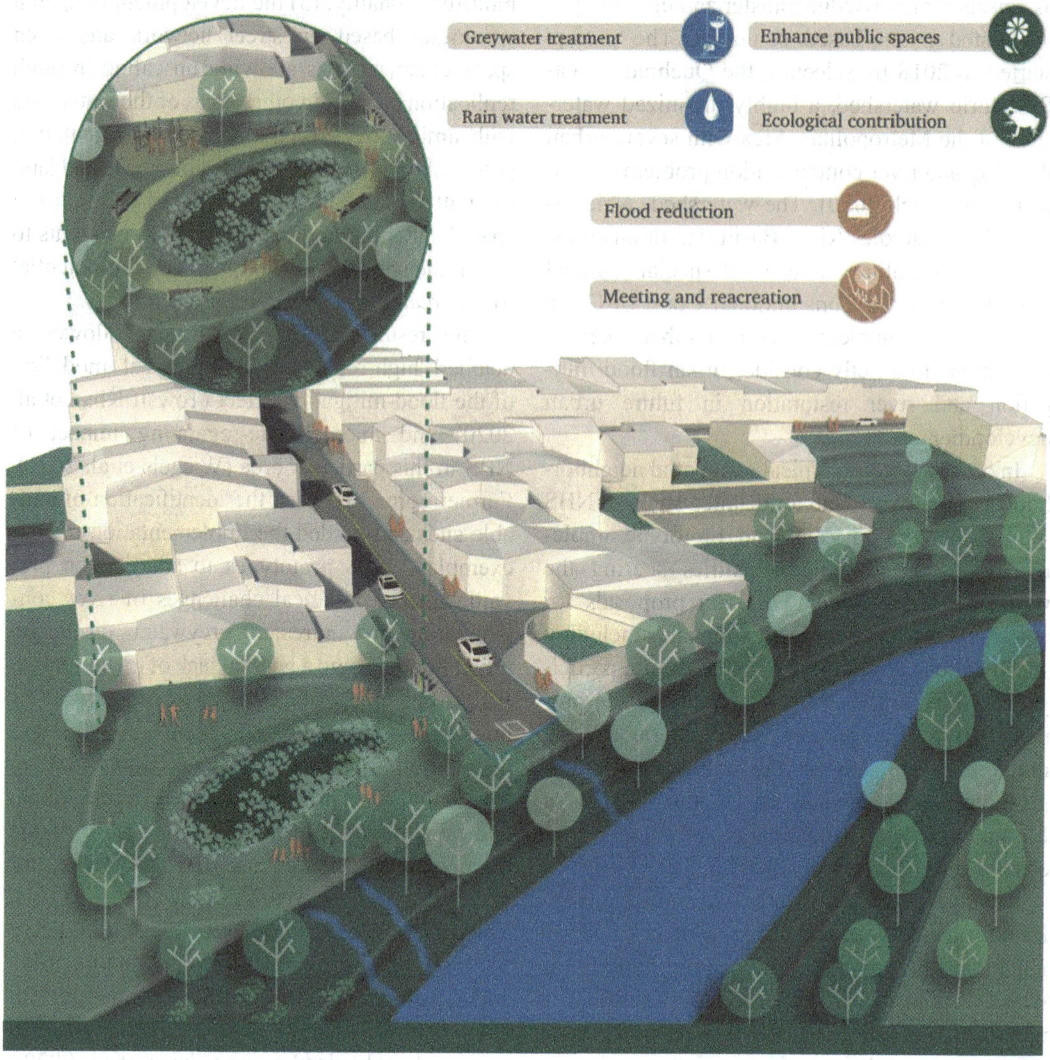

Nature-Based Solutions for River Restoration in Metropolitan Areas, Fig. 1 Illustration of a NBS prototype for waste and storm water treatment of residential areas realized in a real-world lab. (Source: SEE-URBAN-WATER)

challenges for a larger socioecological transformation (from neighborhood to watershed level).

The assessment of NBS potential, the implementation and operation of prototypes, provided also important insights for the evaluation of the upscaling potential in the form of replicating measures at similar sites within the watershed. The NBS potential identified in the real-world lab was used to develop and model upscaling scenarios based on similarities of spatial typologies in other parts of the watershed (Chen et al. 2021). The modeling of upscaling scenarios supports the identification of most effective NBS types at the watershed level and the contribution of retrofitted NBS in already developed areas to river restoration. In addition to retrofitted NBS, potential sites for multifunctional NBS as flood retention measures along the river corridor at the watershed scale have been identified. Since space is very limited in already urbanized areas, also unbuilt areas will be needed for NBS to restore the rivers of the Metropolitan Area.

Summary/Conclusion

Recently, the design and implementation of NBS has gained attention for making cities more sustainable and resilient. Latin America is a region with continuous high growth rates for urban areas which struggles with many water-related issues due to unstructured planning of infrastructure. The fact that rivers are more polluted and have a low water quality but are not extensively structurally changed presents an advantage for successful river restoration measures. Also, several instruments already exist or are being developed such as rules similar to Costa Rica's Hydrological Code, protection areas such as interurban biological corridors, or the testing of prototypes in real-world labs. Stressing the co-benefit potential of NBS and engaging the public in the design, their implementation may help to overcome some structural obstacles in the region such as social exclusion, economic inequalities, environmental injustice, and unequal power relations. In this sense, Latin America, and Costa Rica in particular, bears the potential to become a pioneer of NBS for urban river restoration in the Global South.

Cross-References

- ▶ Artificial Urban Wetlands
- ▶ City Visions: Toward Smart and Sustainable Urban Futures
- ▶ Climate Resilience in Informal Settlements: The Role of Natural Infrastructure
- ▶ Green and Blue Infrastructure (GBI) in Urban Areas
- ▶ Green and Smart Cities in the Developing World
- ▶ Green Cities
- ▶ Green Cities: Implementing the Miyawaki Method in Lahore, Pakistan
- ▶ Green Cities: Nature-Based Solutions, Renaturing and Rewilding Cities
- ▶ Green Cities in Theory and Practice
- ▶ Green Infrastructure in Metropolis Dimension: Case Study of Llobregat River, Barcelona Metropolitan Area
- ▶ Healthy Cities
- ▶ Innovation to Bring Nature-Based Solutions to Life: Tales of Two Cities
- ▶ Integrated Urban Green and Grey Infrastructure
- ▶ Managing Africa's Urban Flooding Challenges from the Bottom up: A View from Ghana
- ▶ Multiple Benefits of Green Infrastructure
- ▶ Policy and Practices of Nature-Based Solutions to Build Resilience in Seoul, Korea
- ▶ Resilient Urban Climates
- ▶ Small Water Retention Measures in Haluzice Gorge

Acknowledgments We thank Gonzalo Pradilla Villamizar for a friendly review of an earlier version of this entry. We further acknowledge the funding that we received from the German Federal Ministry of Research and Education (Grant ID: 01UU1704 and 01UU1601B).

References

Albert, C., Hack, J., Schmidt, S., & Schröter, B. (2021). Planning and governing nature-based solutions in river landscapes: Concepts, cases, and insights. *Ambio*. https://doi.org/10.1007/s13280-021-01569-z

Albert, C., Schröter, B., Haase, D., Brillinger, M., Henze, J., Herrmann, S., . . . Matzdorf, B. (2019). Addressing societal challenges through nature-based solutions: How can landscape planning and governance research contribute? *Landscape and Urban Planning, 182*, 12–21. https://doi.org/10.1016/j.landurbplan.2018.10.003

Beißler, M. R., & Hack, J. (2019). A combined field and remote-sensing based methodology to assess the ecosystem service potential of urban rivers in developing countries. *Remote Sensing, 11*(14), 1697.

Brown, R. R., Keath, N., & Wong, T. H. F. (2009). Urban water management in cities: Historical, current and future regimes. *Water Science and Technology, 59*(5), 847–855. https://doi.org/10.2166/wst.2009.029

Chapa, F., Pérez, M., & Hack, J. (2020). Experimenting transition to sustainable urban drainage systems – Identifying constraints and unintended processes in a tropical highly urbanized watershed. *Water, 12*(12), 3554.

Chen, V., Bonilla Brenes, J. R., Chapa, F., & Hack, J. (2021). Development and modelling of realistic retrofitted nature-based solution scenarios to reduce flood occurrence at the catchment scale. *Ambio*. https://doi.org/10.1007/s13280-020-01493-8

CITEC. (2021). *Comisión Paritaria para la creación de un Código Hidrológico [WWW Document]. Webpage Colegio de Ingenieros Técnologos*. Retrieved from https://www.citec.or.cr/2021/03/12/comision-permanente-reglamentacion-sobre-el-ejercicio-profesional/. Accessed 11 June 21.

N

Collentine, D., & Futter, M. N. (2016). Realising the potential of natural water retention measures in catchment flood management: Trade-offs and matching interests. *Journal of Flood Risk Management*. https://doi.org/10.1111/jfr3.12269

Commission, E. (2015). *Towards an EU research and innovation policy agenda for nature-based solutions & re-Naturing cities (Final report of the horizon 2020 expert group on nature-based solutions and re-Naturing cities)*. Retrieved from Brussels, Belgium: Retrieved from – https://publications.europa.eu/en/publication-detail/-/publication/fb117980-d5aa-46df-8edc-af367cddc202

Cousins, J. J. (2021). Justice in nature-based solutions: Research and pathways. *Ecological Economics, 180*, 106874. https://doi.org/10.1016/j.ecolecon.2020.106874

Dobbs, C., Escobedo, F. J., Clerici, N., de la Barrera, F., Eleuterio, A. A., MacGregor-Fors, I., . . . Hernández, H. J. (2019). Urban ecosystem Services in Latin America: Mismatch between global concepts and regional realities? *Urban Ecosystem, 22*(1), 173–187. https://doi.org/10.1007/s11252-018-0805-3

Droste, N., Schröter-Schlaack, C., Hansjürgens, B., & Zimmermann, H. (2017). Implementing nature-based solutions in urban areas: Financing and governance aspects. In N. Kabisch, H. Korn, J. Stadler, & A. Bonn (Eds.), *Nature-based solutions to climate change adaptation in urban areas: Linkages between science, policy and practice* (pp. 307–321). Springer International Publishing.

ECLAC. (2018). Regional agreement on access to information, public participation and justice in environmental matters in Latin America and the Caribbean. Economic Commission for Latin America and the Carribean. Retrieved from https://repositorio.cepal.org/bitstream/handle/11362/43583/1/S1800428_en.pdf

ECRR. (2021). *What is river restoration?* Retrieved from https://www.ecrr.org/River-Restoration/What-is-river-restoration

Eggermont, H., Balian, E., Azevedo, J. M. N., Beumer, V., Brodin, T., Claudet, J., . . . Le Roux, X. (2015). Nature-based solutions: New influence for environmental management and research in Europe. *GAIA – Ecological Perspectives for Science and Society, 24*(4), 243–248. https://doi.org/10.14512/gaia.24.4.9

Fluhrer, T., Chapa, F., & Hack, J. (2021). A methodology for assessing the implementation potential for retrofitted and multifunctional urban Green infrastructure in public areas of the global south. *Sustainability, 13*(1), 384.

Haase, D. (2015). Reflections about blue ecosystem services in cities. *Sustainability of Water Quality and Ecology, 5*, 77–83. https://doi.org/10.1016/j.swaqe.2015.02.003

Hack, J., Molewijk, D., & Beißler, M. R. (2020). A conceptual approach to modeling the geospatial impact of typical urban threats on the habitat quality of river corridors. *Remote Sensing, 12*(8), 1345.

MINAE (2016). Decreto No 40.043 – Regulación del Programa Nacional de Corredores Biológicos. Ministerio de Ambiente y Energía - República de Costa Rica. San José, Costa Rica.

MINAE (2020). Decreto No 42465-MOPT-MINAE-MIVAH – Lineamientos generales para la incorporación de las medidas de resiliencia en infraestructura pública. Ministerio de Ambiente y Energía - República de Costa Rica. San José, Costa Rica.

Neumann, V. A., & Hack, J. (2020). A methodology of policy assessment at the municipal level: Costa Rica's readiness for the implementation of nature-based-solutions for urban Stormwater management. *Sustainability, 12*(1), 230.

Niemczynowicz, J. (2009). Megacities from a water perspective. *Water International, 21*(4), 198–205. https://doi.org/10.1080/02508069608686515

Pérez Rubi, M., & Hack, J. (2021). Co-design of experimental nature-based solutions for decentralized dry-weather runoff treatment retrofitted in a densely urbanized area in Central America. *Ambio*. https://doi.org/10.1007/s13280-020-01457-y

Pletterbauer, F., Melcher, A., & Graf, W. (2018). Climate change impacts in riverine ecosystems. In S. Schmutz & J. Sendzimir (Eds.), *Riverine ecosystem management: Science for governing towards a sustainable future* (pp. 203–223). Springer International Publishing.

Potthast, M., & Geppert, S. (2019). Corredores Biológicos Interurbanos: Fusionando el capital construido y el capital natural de la ciudad. *Ambientico – Revista Trimestral sobre la Actualidad Ambiental, 272*, 5–12.

Sabater, S., Bregoli, F., Acuña, V., Barceló, D., Elosegi, A., Ginebreda, A., . . . Ferreira, V. (2018). Effects of human-driven water stress on river ecosystems: A meta-analysis. *Scientific Reports, 8*(1), 11462. https://doi.org/10.1038/s41598-018-29807-7

Sattler, C., Schröter, B., Meyer, A., Giersch, G., Meyer, C., & Matzdorf, B. (2016). Multilevel governance in community-based environmental management: A case study comparison from Latin America. *Ecology and Society, 21*(4). https://doi.org/10.5751/ES-08475-210424

SEE-URBAN-WATER. (2021). SEE-URBAN-WATER [WWW Document]. Webpage of SEE-URBAN-WATER.

Sekulova, F., Anguelovski, I., Kiss, B., Kotsila, P., Baró, F., Palgan, Y. V., & Connolly, J. (2021). The governance of nature-based solutions in the city at the intersection of justice and equity. *Cities, 112*, 103136. https://doi.org/10.1016/j.cities.2021.103136

Towsif Khan, S., Chapa, F., & Hack, J. (2020). Highly resolved rainfall-runoff simulation of retrofitted Green Stormwater infrastructure at the Micro-watershed scale. *Land, 9*(9), 339.

Toxopeus, H., Kotsila, P., Conde, M., Katona, A., van der Jagt, A. P. N., & Polzin, F. (2020). How 'just' is hybrid governance of urban nature-based solutions? *Cities, 105*, 102839. https://doi.org/10.1016/j.cities.2020.102839
Varis, O., Biswas, A. K., Tortajada, C., & Lundqvist, J. (2006). Megacities and water management. *International Journal of Water Resources Development, 22*(2), 377–394. https://doi.org/10.1080/07900620600684550
Vörösmarty, C. J., McIntyre, P. B., Gessner, M. O., Dudgeon, D., Prusevich, A., Green, P., ... Davies, P. M. (2010). Global threats to human water security and river biodiversity. *Nature, 467*(7315), 555–561. https://doi.org/10.1038/nature09440
Walsh, C. J., Roy, A. H., Feminella, J. W., Cottingham, P. D., Groffman, P. M., & II, R. P. M. (2005). The urban stream syndrome: Current knowledge and the search for a cure. *Journal of the North American Benthological Society, 24*(3), 706–723. https://doi.org/10.1899/04-028.1
Wantzen, K. M., Alves, C. B. M., Badiane, S. D., Bala, R., Blettler, M., Callisto, M., ... Zingraff-Hamed, A. (2019). Urban stream and wetland restoration in the global south – A DPSIR analysis. *Sustainability, 11*(18), 4975.
Watkins, G. G., Silva Zuniga, M. C., Rycerz, A., Dawkins, K., Firth, J., Kapos, V., ... Amin, A.-L. (2019). Nature-based solutions: Scaling private sector uptake for climate resilient infrastructure in Latin America and the Caribbean. Retrieved from https://publications.iadb.org/publications/english/document/Naturebased_Solutions_Scaling_Private_Sector_Uptake_for_Climate_Resilient_Infrastructure_in_Latin_America_and_the_Caribbean.pdf.
Wiegels, R., Chapa, F., & Hack, J. (2021). High resolution modeling of the impact of urbanization and green infrastructure on the water and energy balance. Urban Climate 39100961. https://doi.org/10.1016/j.uclim.2021.100961

NbS – Nature-Based Solutions

► Adapting to a Changing Climate Through Nature-Based Solutions

NDMA – National Disaster Management Authority of India

► Adapting to a Changing Climate Through Nature-Based Solutions

Need for Greenspace in an Urban Setting for Child Development

Tara Rava Zolnikov[1,2], Frances Furio[2] and Tanya Clark[2]
[1]National University, Department of Community Health, San Diego, CA, USA
[2]California Southern University, School of Behavioral Sciences, Costa Mesa, CA, USA

Synonyms

Greenspace also known as green spaces or parks

Definition

Greenspace – an undeveloped area or landscape, typically in an urban setting.

Worldwide, the amount of people living in urban areas is approximately 55% (United Nations [UN] 2018). This number is projected to increase to include nearly 70% of the total world population by 2050 (UN 2018). This large growth will result in expansion of urbanized areas, which focuses on an industrial transformation as the demography shifts from rural to urban settings (Huang et al. 2018). These modifications can have population effects; for example, affected health and well-being have been linked to living in urban settings (Nutsford et al. 2013; Mitchell 2013; Carter and Horwitz 2014). Research suggests that access to greenspace can help mitigate adverse health effects that can result from living in urban areas (Mitchell 2013; Carter and Horwitz 2014).

Greenspace is defined as vegetation in urban environments and can include, but is not limited to, parks, open spaces, gardens, trees, and more (Kabisch and Haase 2013). In addition to the visual pleasure of a nature-based setting, this environment can also minimize urban noise pollution, improve air quality by absorbing and shielding particulate matter, and prevent heat stress with tree shade (Margaritis and Kang 2016; Nowak

N

et al. 2006; Hartig et al. 2014; Lee et al. 2016). These types of areas are beneficial because they offer considerable health benefits to people who have access to them; these benefits include providing areas for physical activity, social interaction, community attachment, and generally improved well-being (Hillsdon et al. 2006; Gardsjord et al. 2014; Seeland et al. 2009; Arnberger and Eder 2012; Tzoulas et al. 2007; Barbosa et al. 2007). That said, not all greenspace is created equal; research has confirmed the need for ecological diversity (e.g., various fauna and flora) to help foster health benefits, like psychological restoration (Wood et al. 2018) (Fig. 1).

Greenspace has an influence on child development, as this type of environment provides activity to promote both physical and cognitive health (Sallis et al. 2012; Cervero and Kockelman 1997; McCormick 2017). Children surrounded by more greenspace are less susceptible to psychological effects of everyday stressors, and in fact, it has been found to decrease cortisol concentrations in children (Wells and Evans 2003; Roe et al. 2013). Greenspace has the ability to promote attention restoration, memory, and competence, improve social interaction, boost self-esteem, increase self-discipline and self-regulation, moderate stress, and improve behavior (e.g., including decreasing symptoms of ADHD), decreased psychiatric disorders, and is associated with higher standardized test scores (McCormick 2017; Engemann et al. 2019) (Fig. 2). Other research has confirmed an overall high quality of life regarding health outcomes with greenspace use, confirming a positive impact for child development (Dadvand et al. 2015).

This type of information confirms a myriad of ways to improve child development in urban settings, which can often be disadvantaged because of an increase in externalized behavior (e.g., aggression, cheating, stealing, and disobeying rules), exposure to at-risk settings (e.g., areas

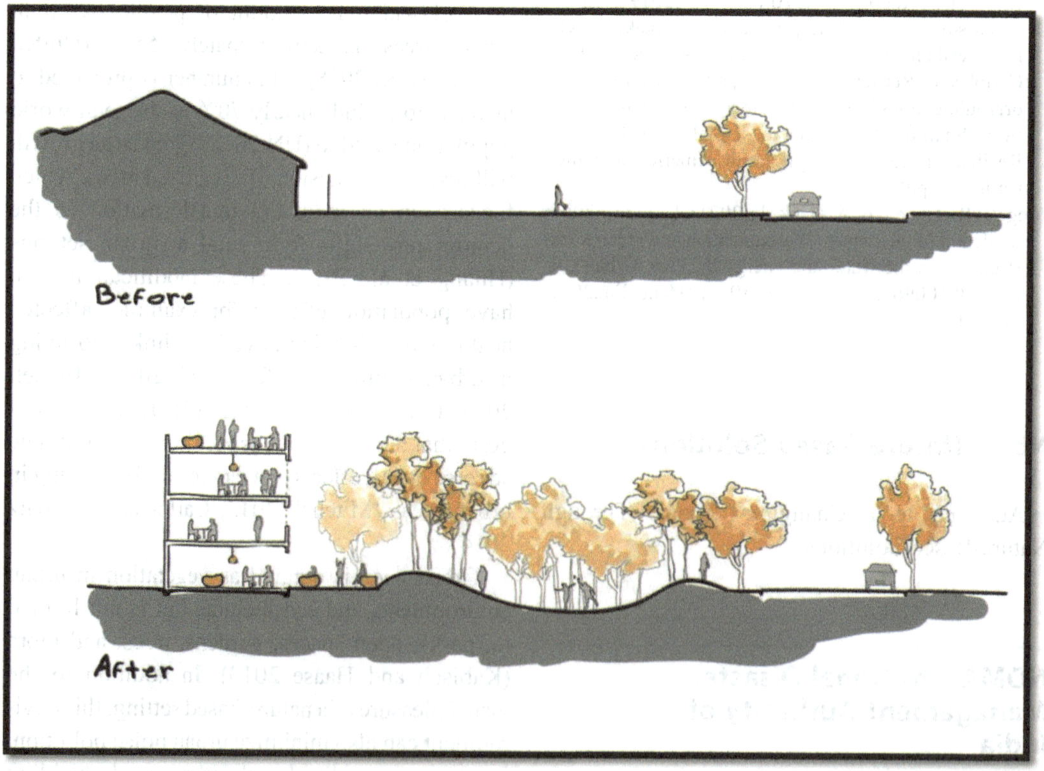

Need for Greenspace in an Urban Setting for Child Development, Fig. 1 Greenspace versus improved greenspace. (Source: Bolleter and Ramalho 2019)

Greenspace and Child Development Outcomes
✓ Improved attention
✓ Physical activity and exercise
✓ Emotional and behavioral development
✓ Self-confidence, self-esteem, and self-regulation
✓ Social skills and interaction

Need for Greenspace in an Urban Setting for Child Development, Fig. 2 The benefits of greenspace on child development outcomes

with high rates of crime, violence, delinquency, substance use, abuse, and poverty), and other possible areas of influence, like poor access to resources and other social and economic forces (Hope and Bierman 1998; Fotso et al. 2012).

Health and well-being of populations in urban settings has become a major focus of public health efforts. Arguably, this is more important in children, whose childhood experiences affect subsequent health status across the life span (Hertzman 1999). Thus, interventions improving adverse developmental outcomes in early life can have long-lasting effects on cognitive and socio-emotional development in children. For example, chronic stress in childhood can cause later impairment in learning, behavior, and physical and mental health (Shonkoff et al. 2012); it is known that access to greenspace can mitigate stressors and could contribute to improved outcomes experienced in later life as well as during childhood (Wells and Evans 2003). This example highlights the importance of greenspace as an intervention to improve childhood outcomes and development in urban settings. As such, using improved biodiverse greenspace should be recognized as a healthy tool to improve childhood outcomes despite possible negative effects of urbanization.

Cross-References

▶ Need for Nature Connectedness in Urban Youth for Environmental Sustainability

References

Amberger, A., & Eder, R. (2012). The influence of green space on community attachment of urban and suburban residents. *Urban Forestry & Urban Greening, 11*(1), 41–49.

Bolleter, J., & Ramalho, C. E. (2019). *Greenspace-oriented development: Reconciling urban density and nature in suburban cities*. Cham: Springer Nature.

Barbosa, O., Tratalos, J. A., Armsworth, P. R., Davies, R. G., Fuller, R. A., Johnson, P., & Gaston, K. J. (2007). Who benefits from access to green space? A case study from Sheffield, UK. *Landscape and Urban Planning, 83*(2–3), 187–195.

Carter, M., & Horwitz, P. (2014). Beyond proximity: The importance of green space useability to self-reported health. *EcoHealth, 11*, 322–332. https://doi.org/10.1007/s10393-014-0952-9.

Cervero, R., & Kockelman, K. (1997). Travel demand and the 3Ds: Density, diversity, and design. *Transportation Research. Part D, Transport and Environment, 2*(3), 199–219.

Dadvand, P., Nieuwenhuijsen, M. J., Esnaola, M., Forns, J., Basagaña, X., Alvarez-Pedrerol, M., et al. (2015). Green spaces and cognitive development in primary schoolchildren. *Proceedings of the National Academy of Sciences, 112*(26), 7937–7942.

Engemann, K., Pedersen, C., Arge, L., Tsirogiannis, C., Mortensen, P., & Svenning, J. C. (2019). Residential green space in childhood is associated with lower risk of psychiatric disorders from adolescence into adulthood. *Proceedings of the National Academy of Sciences, 116*(11), 5188–5193. https://doi.org/10.1073/pnas.1807504116.

Fotso, J. C., Madise, N., Baschieri, A., Cleland, J., Zulu, E., Mutua, M. K., & Essendi, H. (2012). Child growth in urban deprived settings: Does household poverty status matter? At which stage of child development? *Health & Place, 18*(2), 375–384.

Gardsjord, H. S., Tveit, M. S., & Nordh, H. (2014). Promoting youth's physical activity through park design: Linking theory and practice in a public health perspective. *Landscape Research, 39*(1), 70–81.

Hartig, T., Mitchell, R., Vries, D. S., & Frumkin, H. (2014). Nature and health. *Annual Review of Public Health, 35*, 207–228. https://doi.org/10.1146/annurev-publhealth-032013-182443.

Hertzman, C. (1999). The biological embedding of early experience and its effects on health in adulthood. *Annals of the New York Academy of Sciences, 896*(1), 85–95.

Hillsdon, M., Panter, J., Foster, C., & Jones, A. (2006). The relationship between access and quality of urban green space with population physical activity. *Public Health, 120*(12), 1127–1132.

Hope, T. L., & Bierman, K. L. (1998). Patterns of home and school behavior problems in rural and urban settings. *Journal of School Psychology, 36*(1), 45–58.

Huang, N. C., Kung, S. F., & Hu, S. C. (2018). The relationship between urbanization, the built environment, and physical activity among older adults in Taiwan. *International Journal of Environmental Research and Public Health, 15*(5), 836. https://doi.org/10.3390/ijerph15050836.

Kabisch, N., & Haase, D. (2013). Green spaces of European cities revisited for 1990–2006. *Landscape and Urban Planning, 110*, 113–122.

Lee, H., Mayer, H., & Chen, L. (2016). Contribution of trees and grasslands to the mitigation of human heat stress in a residential district of Freiburg, Southwest Germany. *Landscape and Urban Planning, 148*, 37–50. https://doi.org/10.1016/j.landurbplan.2015.12.004.

Margaritis, E., & Kang, J. (2016). Relationship between green space-related morphology and noise pollution. *Ecological Indicators, 72*, 921–933. https://doi.org/10.1016/j.ecolind.2016.09.032.

McCormick, R. (2017). Does access to green space impact the mental well-being of children: A systematic review. *Journal of Pediatric Nursing, 37*, 3–7.

Mitchell, R. (2013). Is physical activity in natural environments better for mental health than physical activity in other environments? *Social Science & Medicine, 91*, 130–134. https://doi.org/10.1016/j.socscimed.2012.04.012.

Nowak, D. J., Crane, D. E., & Stevens, J. C. (2006). Air pollution removal by urban trees and shrubs in the United States. *Urban Forestry & Urban Greening, 4*, 115–123. https://doi.org/10.1016/j.ufug.2006.01.007.

Nutsford, D., Pearson, A. L., & Kingham, S. (2013). An ecological study investigating the association between access to urban green space and mental health. *Public Health, 127*(11), 1005–1011.

Roe, J. J., Thompson, C. W., Aspinall, P. A., Brewer, M. J., Duff, E. I., Miller, D., Mitchell, R., & Clow, A. (2013). Green space and stress: Evidence from cortisol measures in deprived urban communities. *International Journal of Environmental Research and Public Health, 10*, 4086–4103.

Sallis, J. F., Floyd, M. F., Rodríguez, D. A., & Saelens, B. E. (2012). Role of built environments in physical activity, obesity, and cardiovascular disease. *Circulation, 125*(5), 729–737.

Seeland, K., Dübendorfer, S., & Hansmann, R. (2009). Making friends in Zurich's urban forests and parks: The role of public green space for social inclusion of youths from different cultures. *Forest Policy and Economics, 11*(1), 10–17.

Shonkoff, J. P., Garner, A. S., Siegel, B. S., Dobbins, M. I., Earls, M. F., McGuinn, L., et al. (2012). The lifelong effects of early childhood adversity and toxic stress. *Pediatrics, 129*(1), e232–e246.

Tzoulas, K., Korpela, K., Venn, S., Yli-Pelkonen, V., Kaźmierczak, A., Niemela, J., & James, P. (2007). Promoting ecosystem and human health in urban areas using Green Infrastructure: A literature review. *Landscape and Urban Planning, 81*(3), 167–178.

United Nations [UN]. (2018). *68% of the world population projected to live in urban areas by 2050, says UN.* https://www.un.org/development/desa/en/news/population/2018-revision-of-world-urbanization-prospects.html

Wells, N. M., & Evans, G. W. (2003). Nearby nature a buffer of life stress among rural children. *Environment and Behavior, 35*(3), 311–330.

Wood, E., Harsant, A., Dallimer, M., Cronin de Chavez, A., McEachan, R. R. C., & Hassall, C. (2018). Not all green space is created equal: Biodiversity predicts psychological restorative benefits from urban green space. *Frontiers in Psychology, 9*, 2320. https://doi.org/10.3389/fpsyg.2018.02320.

Need for Nature Connectedness in Urban Youth for Environmental Sustainability

Tanya Clark[1], Tara Rava Zolnikov[1,2] and Frances Furio[1]
[1]School of Behavioral Sciences, California Southern University, Costa Mesa, CA, USA
[2]Department of Community Health, National University, San Diego, CA, USA

Introduction

Environmental sustainability is one of the most significant global challenges in the twenty-first century (World Health Organization [WHO] 2015; Barrera-Hernández et al. 2020). Educating individuals on how to engage in pro-environmental behavior is crucial for solving environmental degradation issues (Biggar and Ardoin 2017; Ernst and Theimer 2011). Experts believe that environmental education (EE) for youth is a key factor in mitigating this dire scenario (Barrera-Hernández et al. 2020). Research shows that connecting children and adults with the natural environment is a powerful environmental sustainability intervention (Martin et al. 2020). Additionally, studies show that nature connectedness is positively correlated to well-being, further reinforcing the power of this intervention to influence ecological change (Barrable and Booth 2020).

Environmental education is a widespread mechanism for introducing youth to the complexities of environmental sustainability (Liefländer et al. 2013; Kopnina 2012). From 1994 to 2013, the number of published studies on measurable student outcomes in EE grew steadily and continues to increase (Ardoin et al. 2017). EE research is a relatively new field and in general environmental attitudes, knowledge, and behavior are traditional outcomes (Ardoin et al. 2018). Research shows that the scope of EE activities is broad, ranging from traditional group workshops to outdoor geo games (Hoang and Kato 2016; Schneider and Schaal 2018).

Research suggests that the optimal way to establish a connection to nature for the purpose of inducing pro-environmental behavior may not be by means of cognitive knowledge but rather via emotion, appreciation, and sustained contact (Lumber et al. 2017). A large body of EE research has concentrated on environmental knowledge or attitudinal outcomes, while less studies have focused on the promotion of nature connectedness (Liefländer et al. 2013). However, the Nature in Self Scale has been used to show that EE participation can culminate in a strong short-term rise in nature connectedness (Liefländer et al. 2013).

As such, enhancing one's internal connection to nature should be considered a primary focus for EE (Frantz and Mayer 2014) and in particular a goal for children's EE (Otto and Pensini 2017). Nature connectedness in EE merits special consideration among urban youth who are deprived of experiential opportunities within the natural environment (Martin et al. 2020).

Environmental Education Overview

For decades, EE has been recognized worldwide as a valid method for developing environmental sustainability (Maurer and Bogner 2019). Interacting with nature can improve all aspects of well-being, which has been comprehensively reviewed by understanding associations between exposure to green space (e.g., any aspect of vegetation) (Houlden et al. 2018). Moreover, some research suggests that nature connectedness is a greater predictor of pro-environmental behavior than is exposure to nature (Martin et al. 2020).

History

Environmental sustainability initiatives gained momentum in the 1970s (Wright 2004). In October of 1977, the United Nations Educational, Scientific and Cultural (UNESCO) Intergovernmental Conference took place in Tbilisi, Georgia (Gillett 1977). The high-level international conference included interdisciplinary professionals who acknowledged the need for environmental education (Gillett 1977). At that time, participants addressed potential curricula and research objectives (Gillett 1977). The outcome was the Tbilisi Declaration, a call to action for the remediation of environmental issues (Hungerford and Peyton 1980). The group called for educational institutions to play a role in environmental sustainability efforts via environmental education opportunities specifically those that would aid in the understanding of the relationship between humans and the environment (Wright 2004). In 1994, the North American Association for Environmental Education (NAEE) further addressed EE by suggesting best practices via environmental education guidelines and offered additional tools and language for the field (McCrea 2006).

Environmental Sustainability, Well-Being, and Health Outcomes

Research broadly supports the supposition that human contact with nature is positively connected to the inclination to behave in environmentally sustainable ways (Martin et al. 2020; Beery and Wolf-Watz 2014). Moreover, research shows that psychological connectedness with nature may actually increase such behavior in the presence of other EE methods (Martin et al. 2020). Many contend that nature and the human mind are intimately linked and that this connection holds primary importance to both human and planetary health (Krčmářová 2009). Subsequently it has been suggested that the peril of being severed from the natural world is the development of action and outlook injurious to both human and environmental health (Mayer and Frantz 2004; Pritchard et al. 2020).

The nature connectedness construct encompasses a sense of emotional affinity for the natural world and being a fundamental part of nature (Mayer and Frantz 2004). Several studies have found a positive link between nature connection and well-being in adults and children (Capaldi et al. 2014; Howell et al. 2011; Nisbet and Zelenski 2013). Finally, data shows that nature connectedness is a reliable predictor of well-being (Howell et al. 2013, 2011; Mayer and Frantz 2004; Nisbet and Zelenski 2013; Capaldi et al. 2014). Additionally, research indicates that connectedness to nature influences sustainable behaviors, which, in turn, result in happiness (Barrera-Hernández et al. 2020).

The multifaceted nature of environmental issues makes environmental education challenging as well as multipronged (Ardoin et al. 2017). Research suggests that nature connectedness can promote interest in factors related to environmental sustainability which points to the importance of using this intervention with nature-impoverished urban youth. Researchers consider childhood as the window for development of values and beliefs (Eccles and Wigfield 2002), and data indicates that adult nature connection and pro-environmental behavior may be rooted in early developmental years (Wells and Lekies 2006). In fact, access to green space can mitigate stressors and improve child development and could contribute to improved outcomes experienced in later life as well as during childhood (Wells and Evans 2003).

Nature Connectedness

The scientific literature offers different definitions for nature connectedness or connectedness to nature. Mayer and Frantz (2004) define it as a personality trait that facilitates an emotional connection to the natural world. Zylstra and colleagues (2014) explain connectedness to nature as a stable cognitive, affective, and experiential state characterized by awareness of the interconnections between self and nature. Nisbet et al. (2009) use the term nature relatedness to describe a relatively stable characteristic across the lifetime manifested by appreciation and understanding of the interconnection between human beings and nature (Barrera-Hernández et al. 2020).

In turn these related constructs have variously been referred to as nature connectedness (Mayer and Frantz 2004), inclusion of nature in self (Schultz 2002), emotional affinity toward nature (Müller et al. 2009), and nature relatedness (Nisbet et al. 2009). Regardless of the terminology, the fundamental theory is similar across the constructs: a subjectively perceived connection to the nonhuman elements of the natural environment (Capaldi et al. 2014).

Researchers have measured these constructs using the Connectedness to Nature Scale (Mayer and Frantz 2004), the Inclusion of Nature in Self

Scale (Schultz 2001), and the Nature Relatedness Scale (Nisbett et al. 2009; Maurer and Bogner 2019). Research shows that these measures correlate strongly with each other, as well as show similar correlations with measures of well-being and environmental beliefs and behaviors (Capaldi et al. 2014). Acknowledged limitations to date of the most widely used nature connection scales include limitations in the age range of validated measures, a lack of self-report measures for children younger than 8 years of age, and the possibility that these measures impose an artificial ceiling effect that prevents the measurement of changes in highly connected individuals (Barrable and Booth 2020).

Examples of Interventions

Researchers theorize that human disconnection from the natural environment can accelerate the decimation of the planet (Barrable and Booth 2020), and research supports this supposition with evidence that shows the natural environment can influence pro-environmental attitudes and behaviors (Hartig et al. 2001). For instance, attending an EE workshop can increase elementary school age students' knowledge about ecological practices (Hoang and Kato 2016). Research has shown that connectedness to nature can increase the pro-environmental outcomes of a nature contact intervention (Lumber et al. 2017). Compelling evidence indicates that nature contact by itself may not be sufficient to influence environmental sustainability behavior but that psychological connection to the environment in addition to contact has a greater influence on pro-environmental behavior (Capaldi et al. 2014; Pritchard et al. 2020).

Research with children demonstrates a significant relationship between nature connectedness and sustainable behavior, which consequently impacted levels of happiness (Barrera-Hernández et al. 2020). Being in a natural environment at least one time per week is associated with an increase in pro-environmental behavior such as buying environment-friendly products and recycling (Hartig et al. 2001). Other research has confirmed an overall high quality of life regarding health outcomes with green space use, confirming a positive impact for child development (Dadvand et al. 2015). Nature connectedness as an EE intervention is especially germane and valuable with urban youth deprived of direct nature contact at a time when youth are spending less time in nature (Ernst and Theimer 2011) and younger children are spending very little time outdoors (Barrable and Booth 2020).

Areas that can offer green space and sequential nature connectedness include schoolyards, playgrounds, community gardens, lawn-covered common areas, or public seating areas in public plazas or malls and even vacant lots. These areas all can provide environmental education and help foster change through future generations.

Conclusion

Research shows that nature connectedness influences pro-environmental behaviors (Whitburn et al. 2019) and that it is correlated to overall improved well-being (Mayer and Frantz 2004). At a time when half of the world's residents live in nature-poor urban settings, nature connectedness should be considered an important EE tool for environmental sustainability (Barrable and Booth 2020). Connectedness to nature represents one's subjective sense of mutuality with the natural environment and has been made operative with a number of scales including the most methodically analyzed instrument, the Connectedness to Nature Scale (Mayer and Frantz 2004; Olivos et al. 2011). In particular, connectedness to nature has been shown to be a superior driver of ecological behavior, versus the introduction of environmental knowledge, in elementary school age youth (Otto and Pensini 2017). There has been increased effort in the last few decades to connect children to nature via EE, and nature connectedness is a significant means of doing so for the substantial segment of the population with severely limited access to the natural world (Ernst and Theimer 2011).

Furthermore, data suggests that EE influences outcomes beyond its scope to include a heightened sense of well-being (Capaldi et al. 2014; Pritchard et al. 2020). Both nature contact and connectedness to nature are linked with various health and sustainability-related outcomes although more research is called for as evidence has been fragmented (Barrera-Hernández et al. 2020; Martin et al. 2020). This is particularly true in children, which is a population that can offer considerable change for the future by understanding how the environment plays such an integral role in quality of life (Zolnikov et al. 2020 in press). Ultimately, this generation has the ability to carry forward learned behaviors and promote environment-focused action.

Because research shows a link between the health of the planet and the health of the individuals who reside on it (Krčmářová 2009; Barrable and Booth 2020; Capaldi et al. 2014), it is difficult not to appreciate an environment-promoting intervention that comes with a potential side effect of enhanced individual well-being. Increasing environmental sustainability, while at the same time enhancing one's own well-being, is not a chimera but likely a seamless and sustaining cycle, an ecological win-win situation.

References

Ardoin, N. M., DiGiano, M. L., O'Connor, K., & Podkul, T. E. (2017). The development of trust in residential environmental education programs. *Environmental Education Research, 23*(9), 1335–1355.

Ardoin, N. M., Bowers, A. W., Roth, N. W., & Holthuis, N. (2018). Environmental education and K-12 student outcomes: A review and analysis of research. *The Journal of Environmental Education, 49*(1), 1–17.

Barrable, A., & Booth, D. (2020). Increasing nature connection in children: A mini review of interventions. *Frontiers in Psychology, 11*, 492.

Barrera-Hernández, L. F., Sotelo-Castillo, M. A., Echeverría-Castro, S. B., & Tapia-Fonllem, C. O. (2020). Connectedness to nature: Its impact on sustainable behaviors and happiness in children. *Frontiers in Psychology, 11*, 276.

Beery, T. H., & Wolf-Watz, D. (2014). Nature to place: Rethinking the environmental connectedness perspective. *Journal of Environmental Psychology, 40*, 198–205.

Biggar, M., & Ardoin, N. M. (2017). More than good intentions: The role of conditions in personal transportation behaviour. *Local Environment, 22*(2), 141–155.

Capaldi, C. A., Dopko, R. L., & Zelenski, J. M. (2014). The relationship between nature connectedness and happiness: A meta-analysis. *Frontiers in Psychology, 5*, 976. https://doi.org/10.3389/fpsyg.2014.00976.

Dadvand, P., Nieuwenhuijsen, M. J., Esnaola, M., Forns, J., Basagaña, X., Alvarez-Pedrerol, M., … & Jerrett, M. (2015). Green spaces and cognitive development in primary schoolchildren. *Proceedings of the National Academy of Sciences, 112*(26), 7937–7942.

Eccles, J. S., & Wigfield, A. (2002). Motivational beliefs, values, and goals. *Annual Review of Psychology, 53*(1), 109–132.

Ernst, J., & Theimer, S. (2011). Evaluating the effects of environmental education programming on connectedness to nature. *Environmental Education Research, 17* (5), 577–598.

Frantz, C. M., & Mayer, F. S. (2014). The importance of connection to nature in assessing environmental education programs. *Studies in Educational Evaluation, 41*, 85–89.

Gillett, M. (1977). The Tbilisi declaration: Conference on environmental education. *McGill Journal of Education.*

Hartig, T., Kaiser, F. G., & Bowler, P. A. (2001). Psychological restoration in nature as a positive motivation for ecological behavior. *Environment and Behavior, 33*(4), 590–607.

Hoang, T. T. P., & Kato, T. (2016). Measuring the effect of environmental education for sustainable development at elementary schools: A case study in Da Nang city, Vietnam. *Sustainable Environment Research, 26*(6), 274–286.

Houlden, V., Weich, S., Porto de Albuquerque, J., Jarvis, S., & Rees, K. (2018). The relationship between greenspace and the mental wellbeing of adults: A systematic review. *PloS One, 13*(9), e0203000.

Howell, A. J., Dopko, R. L., Passmore, H. A., & Buro, K. (2011). Nature connectedness: Associations with well-being and mindfulness. *Personality and Individual Differences, 51*(2), 166–171.

Howell, A. J., Passmore, H. A., & Buro, K. (2013). Meaning in nature: Meaning in life as a mediator of the relationship between nature connectedness and well-being. *Journal of Happiness Studies, 14*(6), 1681–1696.

Hungerford, H. R. & Peyton, R. B. (1980). A paradigm for citizen responsibility: Environmental action. *Current Issues.*

Kopnina, H. (2012). Education for sustainable development (ESD): The turn away from 'environment'in environmental education? *Environmental Education Research, 18*(5), 699–717.

Krčmářová, J. (2009). EO Wilson's concept of biophilia and the environmental movement in the USA. *Klaudyán: Internet Journal of Historical Geography and Environmental History, 6*(1/2), 4–17.

Liefländer, A. K., Fröhlich, G., Bogner, F. X., & Schultz, P. W. (2013). Promoting connectedness with nature through environmental education. *Environmental Education Research, 19*(3), 370–384.

Lumber, R., Richardson, M., & Sheffield, D. (2017). Beyond knowing nature: Contact, emotion, compassion, meaning, and beauty are pathways to nature connection. *PLoS One, 12*(5), e0177186.

Martin, L., White, M. P., Hunt, A., Richardson, M., Pahl, S., & Burt, J. (2020). Nature contact, nature connectedness and associations with health, wellbeing and

pro-environmental behaviours. *Journal of Environmental Psychology, 68*, 101389.
Maurer, M., & Bogner, F. X. (2019). How freshmen perceive Environmental Education (EE) and Education for Sustainable Development (ESD). *PLoS One, 14*(1), e0208910.
Mayer, F. S., & Frantz, C. M. (2004). The connectedness to nature scale: A measure of individuals' feeling in community with nature. *Journal of Environmental Psychology, 24*(4), 503–515.
McCrea, E. J. (2006). Leading the way to environmental literacy and quality: National guidelines for environmental education. *Environmental Education and Training Partnership.*
Müller, M. M., Kals, E., & Pansa, R. (2009). Adolescents' emotional affinity toward nature: A cross-societal study. *Journal of Developmental Processes, 4*(1), 59–69.
Nisbet, E. K., & Zelenski, J. M. (2013). The NR-6: A new brief measure of nature relatedness. *Frontiers in Psychology, 4*, 813.
Nisbet, E. K., Zelenski, J. M., & Murphy, S. A. (2009). The nature relatedness scale: Linking individuals' connection with nature to environmental concern and behavior. *Environment and Behavior, 41*(5), 715–740.
Olivos, P., Aragonés, J. I., & Amérigo, M. (2011). The connectedness to nature scale and its relationship with environmental beliefs and identity. *International Journal of Hispanic Psychology, 4*(1), 5–19.
Otto, S., & Pensini, P. (2017). Nature-based environmental education of children: Environmental knowledge and connectedness to nature, together, are related to ecological behaviour. *Global Environmental Change, 47*, 88–94.
Pritchard, A., Richardson, M., Sheffield, D., & McEwan, K. (2020). The relationship between nature connectedness and eudaimonic well-being: A meta-analysis. *Journal of Happiness Studies, 21*(3), 1145–1167.
Schneider, J., & Schaal, S. (2018). Location-based smartphone games in the context of environmental education and education for sustainable development: Fostering connectedness to nature with Geogames. *Environmental Education Research, 24*(11), 1597–1610.
Schultz, P. W. (2001). The structure of environmental concern: Concern for self, other people, and the biosphere. *Journal of Environmental Psychology, 21*(4), 327–339.
Schultz, P. W. (2002). Inclusion with nature: The psychology of human-nature relations. In *Psychology of sustainable development* (pp. 61–78). Boston: Springer.
Wells, N. M., & Evans, G. W. (2003). Nearby nature a buffer of life stress among rural children. *Environment and Behavior, 35*(3), 311–330.
Wells, N. M., & Lekies, K. S. (2006). Nature and the life course: Pathways from childhood nature experiences to adult environmentalism. *Children Youth and Environments, 16*(1), 1–24.
Whitburn, J., Linklater, W. L., & Milfont, T. L. (2019). Exposure to urban nature and tree planting are related to pro-environmental behavior via connection to nature, the use of nature for psychological restoration, and environmental attitudes. *Environment and Behavior, 51*(7), 787–810.
World Health Organization. (2015). Health in 2015: From MDGs, millennium development goals to SDGs, sustainable development goals.
Wright, T. (2004). Definitions and frameworks for environmental sustainability in higher education. *Higher Education Policy.*
Zylstra, M. J., Knight, A. T., Esler, K. J., & Le Grange, L. L. (2014). Connectedness as a core conservation concern: An interdisciplinary review of theory and a call for practice. *Springer Science Reviews, 2*(1), 119–143.

Neighborhood

▸ The Social and Solidarity Economy

Neighbourhood

▸ Closing the Loop on Local Food Access Through Disaster Management

N

Neither Rural Nor Urban: A Critical Review of the Fringe Dynamics of Settlements

Susan Cyriac[1], Mohammed Firoz C[1] and Lakshmi Priya Rajendran[2]
[1]Department of Architecture and Planning, National Institute of Technology, Calicut, Kerala, India
[2]The Bartlett School of Architecture, University College London, London, UK

Introduction

Fringe regions form the confluence of the transition zones of urban and rural areas with a mix of both characteristics. Such regions can be easily detectable by their form, with various shades of neither rural nor urban features found in the immediate outskirts of cities. Different literature refers to these areas as the peri-urban, suburban, rurban, rural-urban continuum, composite settlements, gragara, and desakota settlements, depending on the region's size, scale, and settings. The fringe regions span various scales ranging from a city's peri-urban areas to a matrix of the intercontinental areas. The complex forms of such developments are often polycentric. The lack of official recognition and definition of such regions is a significant cause of concern in the fast urbanizing world.

The majority of the countries differentiate rural and urban based on a strict dichotomous definition, which is often the farther ends of a rural-urban spectrum. The countries develop their definition based on their geographical and policy-specific attributes. Hence, they vary widely from country to country. The lack of official recognition of the fringe regions thereby causes problems with the country-specific urban definitions. The fringe areas get classified as either urban or rural even though they are neither urban nor rural. The following section examines the existing criteria for classifying urban areas.

The Existing Classification of Urban and Rural Areas

Globally, settlements are classified as urban and rural for ease in policy purposes. Such a classification is often used for administrative advantages, governance, funding, and resource allocation. Each country formulates its urban definition based on the parameters most significant to its context. Hence, it is not possible for a universal classification of urban areas. The urban definitions of most countries were formulated at least half a century earlier, while there existed stereotypical differences between the urban and rural areas. According to Hugo et al. (2001), conventionally, the significant divergences among urban and rural were in the economy, occupational structure, education levels and provisions, politics, information, accessibility to services, demography, ethnicity, and migration levels. The comparison of rural and urban, hence varies considerably in dimensions.

The urban definitions used for administrative purposes typically employ criteria that best define the country's urban characteristics. The definition and the criteria used varies widely from country to country. While 29% of the countries use a single criterion for defining urban, 24% use a combination of four criteria (United Nations 2018). The most common criteria for determining urban is the population, followed by the status designated to the administrative area. The other prominent criteria include employment in the primary sector, infrastructure and amenities, population density, and urban characteristics. A majority of the countries follow the dichotomous rural and urban classification. The areas that are not urban by definition are considered rural by the majority of the countries. However, there are traces of intermediate classifications other than the dichotomous classifications. For example, Spain has classified urban, intermediate, and rural areas, while New Zealand has three urban classifications – main, secondary, and minor urban areas apart from rural areas. Israel has three different types of urban and rural classes.

The definitions used for policy purposes typically differ from the dichotomous administrative classification. These are more specific and could vary based on the targeted sector of the policies. For instance, in the USA, the Department of agriculture categorizes rural areas on five criteria: demographic data, socioeconomic aspects, jobs, income, and county classifications (John and Flood 2019). The US Department of Health and Human Services identifies nonmetro counties, areas within metros with a density of fewer than 35 persons per sq. mile, etc., as rural (Federal Office of Rural Health Policy 2021). The criteria selection thus differs based on the thrust area of the policy. The policy definitions need not necessarily follow any administrative boundaries.

It can be observed that the rural and urban definitions differ considerably across the globe. It makes cross-country comparisons difficult as the definitions vary significantly in terms of criteria and thresholds involved. For instance, countries such as Ecuador, Kenya, and Liberia uses a sole criterion of population threshold of 2000 to define areas as urban, while Japan uses four criteria of population, dwelling density, occupation pattern, and built-up area to determine urban areas. The population threshold for classifying urban settlements in Japan is 50,000 (United Nations 2018). Certain countries such as Indonesia use even sophisticated definitions based on infrastructure and amenities for urban areas. Hence, the following section attempts to understand the variations in the urban definitions in select countries.

Country-Specific Urban Definitions

The country-specific urban definitions vary widely for different countries. Some use the administrative criterion alone whereas others use a variety of criteria for the urban and rural classifications. Some countries have just two categories – urban and rural, while other countries have various subdivisions. A comparison of the definition and classification of a few countries is provided to illustrate the existing differences.

Nicaragua follows a dichotomous rural and urban classification based on administrative area, population size, and infrastructure and amenities. The areas with a sparse population and concentration of fewer than 1000 inhabitants without meeting the urban conditions are classified as rural. Urban areas have an administrative or departmental and regional center. The minimum population is 1000, and the locality should have urban characteristics such as paved roads, electricity, and industrial and commercial centers (United Nations 2018).

India being a large and highly populous country has more extensive geography with emerging forms of urbanizations (van Duijne and Nijman 2019) classified based on population, density, occupation, and administrative criteria. In India, there are four broad categories of settlements with three urban classifications. The first one is the statutory towns, which are defined based on administrative status (a municipality, corporation, cantonment board, or a notified town area committee). The second category of settlements are called as census towns, which are urban areas satisfying the three criteria of a population of 5000 persons, a density of 400 persons per square kilometer, and at least 75% of the main male working population in the nonagricultural activities (Sircar 2017). The third category of classification is the Outgrowths, which are urban villages or parts contiguous to statutory towns that possess urban facilities. The fourth category of settlements are all the nonurban areas, which are classified as rural.

Countries such as Japan and the USA have high population thresholds for classifying areas as urban. The USA differentiates urban areas as "urbanized areas" and "urban clusters." It is based on the population size, density, dwelling type, and land use. The population threshold of urbanized areas is a minimum of 50,000. All the other areas are categorized as rural areas. On the other hand, Greece has three classifications for which the chief criteria are the administrative status of the area and the population size. The urban areas have 10,000 or more inhabitants, while areas between 2000 and 9999 are semi-urban areas (United Nations 2018). The less populated areas are categorized as rural areas.

Indonesia follows a scoring system of three criteria to classify settlements as urban or rural. For a settlement to be classified as urban, the criteria of population density, percentage of agricultural households, and number/availability of urban facilities within a threshold distance are scored. The settlements with a cumulative score of more than ten are categorized as urban and those scoring less than ten are categorized as rural settlements.

Thus, there exists differences in thresholds and criteria of classification across various countries. The population size, population density, administrative status, occupation, and infrastructure and amenities are the most occurring criteria for classification. However, the thresholds and the requirements vary widely amid countries, thereby making it difficult for comparisons. Of late, attempts are on the anvil for a universal urban definition with the help of satellite imagery and census information from the countries, such as the degree of urbanization (Dijkstra et al. 2020). Such a definition can help to compare and benchmark the urbanization levels of various countries. There is a need to define settlements to be further classified to assist planning, administration, and governance.

A significant drawback of the country-specific dichotomous definitions is the unrecognized fringe areas. It results from faulty classification, which often forms the basis for resource allocation, infrastructure provision, policies, planning, and management strategies. Thus, there is a need to recognize fringe regions and classify them based on their degree of urbanization. Fringe areas vary based on the number of urban cores associated with the area. The following section explores the monocentric and polycentric development patterns and the fringe areas associated with these.

Monocentric and Polycentric Development Patterns

The development pattern of the urban system is determined by the functions performed, the scale, connectivity, and location. Monocentric city forms were most prominent till the 1970s. These had a significant Central Business District (CBD) forming the core. The concentric zone theory by E.W.Burgess was based on the monocentric cities depicting the core-periphery relationships. Monocentric urban settlements' boundary regions and hinterlands have blurred rural and urban characteristics, eventually merging to surrounding rural settlements. These blurred boundaries are synonymously called the ex-urban (Hussain 2017) and Peri-urban settlements (Iaquinta and Drescher 2000). The other forms of peripheral urban development, which lead to blurring urban and rural characteristics, include edge cities, exurbia, and extended metropolitan regions (Hugo et al. 2003). Examples of such monocentric cities include New York and Tokyo. However, with the diversification of economic activities, improved transportation, and technological advances, the monocentric patterns gave way to polycentric urban systems (Fig. 1).

The polycentric urban systems vary widely in scale from edge cities to world cities. They are predominantly of four types based on the scale (You 2017). The first one, Edge cities, is an intra-urban pattern developed due to the dispersion of population and service. The second category is the metropolitan areas with multiple centers, satellite cities, and the rural hinterland that depends on the core. The polycentric urban region, which is a matrix of different cities, forms the third category. The cities are spatially and administratively independent entities and are devoid of a primate city. Megalopolis, network cities, multinucleated metropolitan regions, city-regions, and world cities are the other terminologies used to indicate such regions. The fringe regions of the monocentric and polycentric development patterns have various terminologies. The following section attempts to familiarize and differentiate the nomenclature associated with the fringe dynamics.

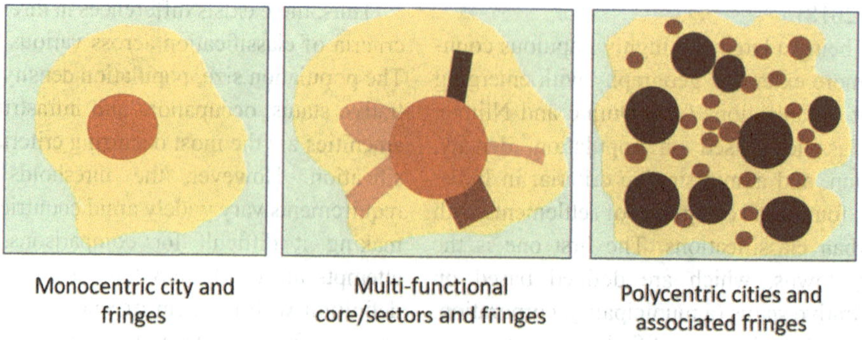

Neither Rural Nor Urban: A Critical Review of the Fringe Dynamics of Settlements, Fig. 1 Schematic representation of monocentric and polycentric cities

An Overview of Definitions of Various Forms of Fringe Development

The fringe region occurs in various scales depending on the geographical settings, availability of transportation facilities, demographic characteristics, and urban growth. Most of the terminology used to denote fringe regions are synonymous and are devoid of a universally applicable definition. The main terminologies for fringe regions include peri-urban, suburban, rurban, rural-urban continuum, composite settlements, gragara, and desakota settlements. The following sections attempt to define these fringe regions and bring out the similarities and differences of these terms. Table 1 finally provides an overview of the various terms associated with RUC settlements.

Peri-Urban

Peri-urban is a term with no concrete definition. It could refer to a specific geography or a process. Geographically, it relates to the transition areas in the periphery of cities with mixed urban and rural characteristics. It is a process by which the rural areas near cities acquire urban attributes with changes in socioeconomic activities. The peri-urban areas have a higher dependence on agriculture compared to the urban areas. There is a possibility that the peri-urban areas gradually get converted to urban areas when the city grows spatially. The Peri-Urban Land Use Relationships (PLUREL) project defines peri-urban as "discontinuous built development containing settlements of each less than 20,000 population, with an average density of at least 40 persons per hectare (averaged over 1 km cells)" (Piorr et al. 2011). However, the definition might not be applicable on a global level. The peri-urban definition needs to consider the geographical settings of the region under consideration. Figure 2 illustrates the building footprint and terminologies associated with peri-urban areas.

Suburb

The terminology suburban often refers to areas that are beyond the core areas of the city. The

Neither Rural Nor Urban: A Critical Review of the Fringe Dynamics of Settlements, Table 1 Features of various RUC terminologies

Term	Urban connection	Demographic/land use component	Geographic component	Temporal component
Rural-urban continuum	Matrix of urban and rural areas	Mixed land uses	Small cities and large villages coexist with no clear boundary	Urban sprawl expansion, good connectivity
Peri-urban	Urban periphery	Mixed land uses	Interface of rural and urban areas	Expansion of urban areas
Suburb	Outskirts of urban cores	Predominantly residential	Within the periphery of cities or locations with good connectivity to the urban core	Affinity to transportation network
Rurban	A mix of large rural and small urban areas	Less than 20,000 and higher densities	Composed of towns, small towns, and large villages	Emerges as urban in developing countries
Composite settlements	Predominantly rural	Mixed land use offering services to surrounding rural areas	Collection of villages with a prominent village as a service center	
Gragara	A mix of urban and rural	Large population, high density, mixed land uses	Stretch of varied economic activities with agricultural land uses. Adjacent towns within 3 to 22 miles	Affinity to the transportation network, change in occupation pattern
Desakota	Matrix of urban and rural areas	Large population, high density, mixed land uses	Mixed land uses with agricultural land parcels, predominantly wet agriculture	Intensification of urbanization in developing countries

N

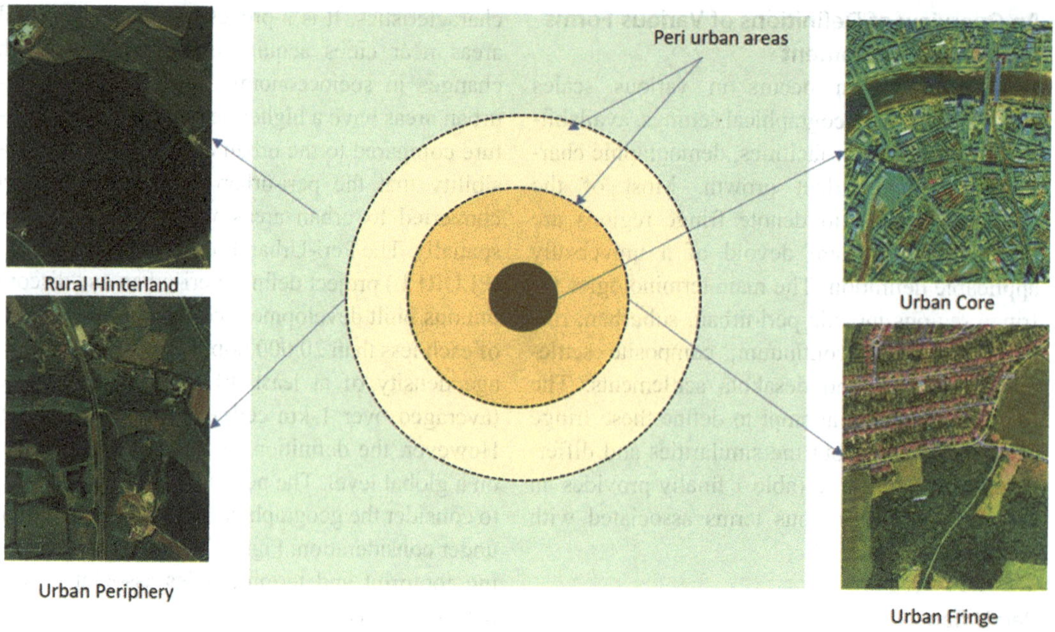

Neither Rural Nor Urban: A Critical Review of the Fringe Dynamics of Settlements, Fig. 2 Illustration of peri-urban areas of monocentric cities. (Source: compiled from Google Maps, 2021)

suburbs are also known as "Edge Cities," denoting their geographical proximity to the urban cores. They are predominantly residential, though in specific contexts have mixed land use. Geographically, the suburban areas can be located outside the city's core or can be a separate area easily accessible from the city. The proximity to an urban center, the advances in transportation facilities, and relaxed zoning regulations outside the urban centers have contributed to the suburban landscape. Exurbia is a form of a suburb, which houses the high-income group. Figure 3 represents the schematic representation of a suburb and examples of suburban regions.

Desakota

Terry McGee coined the term "Desakota" in 1991 to denote the mix of urban and rural spatial structures occurring in Asian countries. Initially, McGee coined the term "Kotadesa" while conducting researches in Indonesia and other Asian countries. The typical characteristics of Desakota regions include: a population with smallholder cultivation of rice, an increase in nonagricultural activities, good connectivity resulting in increased mobility of the people, a mix of land uses such as cottage industries, suburban development interspersed with agriculture, development of technology, increase in female labor force participation, and formation of "grey zones" were informal activities group.

Based on the drivers of the economy, there are three distinct variations of desakota (Fig. 4). Japan and South Korea have the first type of desakota pattern characterized by declining rural settlements and agricultural population accompanied by an increasing income level. The second type of desakota pattern (found in Java, parts of China, Kolkata region of India) also shifts from agricultural to nonagricultural activities. The settlements have urban characteristics in terms of connectivity and infrastructure provisions. The third type of desakota has low economic growth and high density, similar to the urban settlements, and is found in Bangladesh, Java, Kerala, etc.

The different types of desakota are determined by the pace of transformation of the economy and spatial manifestations (McGee 1991). A distinct feature of the desakota pattern is the presence of agricultural land use. Large population size and

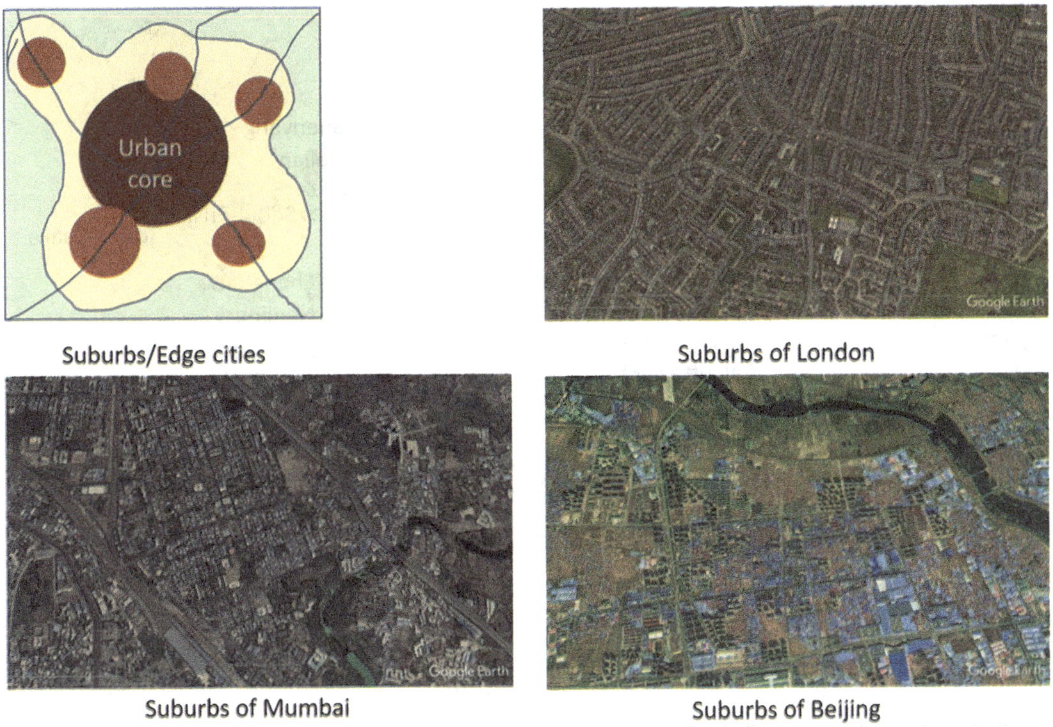

Neither Rural Nor Urban: A Critical Review of the Fringe Dynamics of Settlements, Fig. 3 Schematic representation of suburbs and examples. (Source: compiled from Google Maps, 2021)

density are also features of desakota development. The urbanization process of desakota settlements is different from the classic urban models. In the desakota pattern, it is not easy to distinguish the urban and rural boundaries. The urban expansion of these areas is not necessarily due to suburbanization; instead, it is due to the location of economic activities. These regions have a mix of industrial, residential development amid the agricultural land uses.

The Desakota system is an interlinked urban and rural livelihood in terms of communication, transport, and economic activities. Schematic representation of the Desakota settlements is indicated in Fig. 5.

Rural-Urban Continuum (RUC)

The rural-urban continuum refers to the merging of rural and urban areas, thereby eradicating the assumption of a clear-cut distinction between both entities. The RUC settlements range from a city and its suburbs to large-scale corridors extending beyond international boundaries. Megacities, metropolises, ribbon developments, and corridor developments are a few examples of such recognized regions with settlements that possess rural and urban characteristics. Demographic particulars depend on the scale and settings of the RUC.

A study on global cityscapes (Isabel et al. 2016) identified different forms of urban-rural settlement distribution and concisely classified them based on similarity in morphology, travel time, infrastructure, the extent of the area, and the number of major cities in the formation. Based on the morphology of these settlements, the polycentric fringes can be categorized broadly into three patterns (Fig. 7).

The first category of the RUC settlements can be broadly called Polycentric Cities (Fig. 6a), formed by multiple monocentric cities. Functional Urban regions (Hall 2009), Urban Field (Friedmann and Miller 1965), and Larger Urban Zone (Dijkstra 2009) are the other terminologies used to describe such morphology. The terms

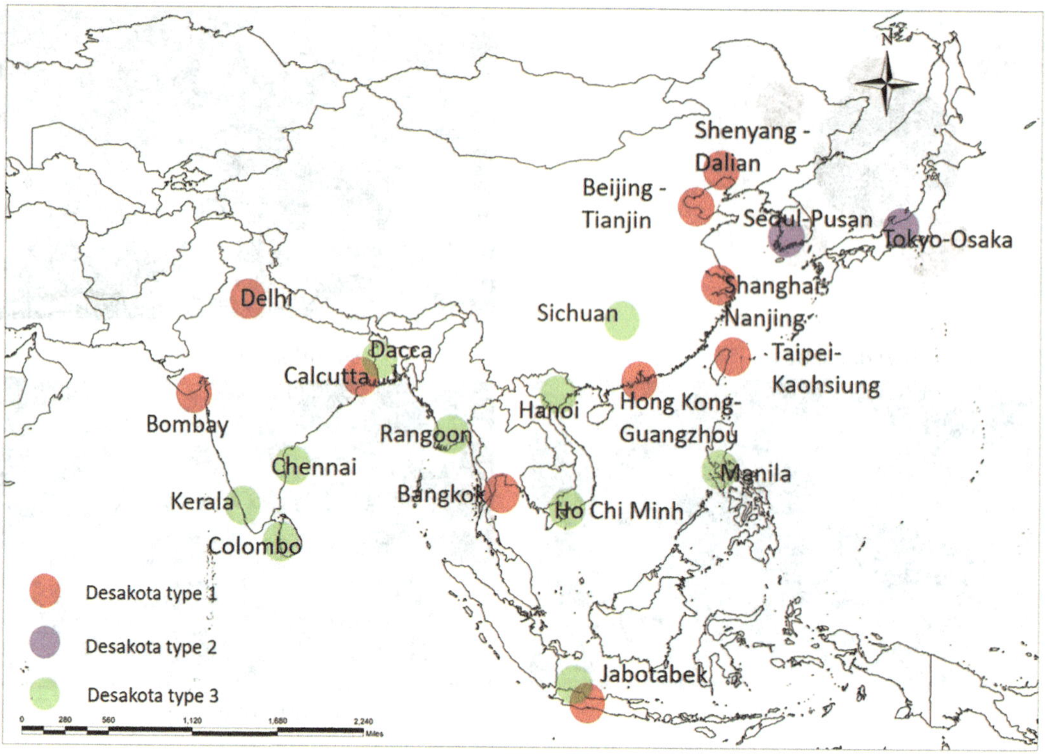

Neither Rural Nor Urban: A Critical Review of the Fringe Dynamics of Settlements, Fig. 4 Three types of desakota pattern in Asia. (source: adapted from McGee 1991)

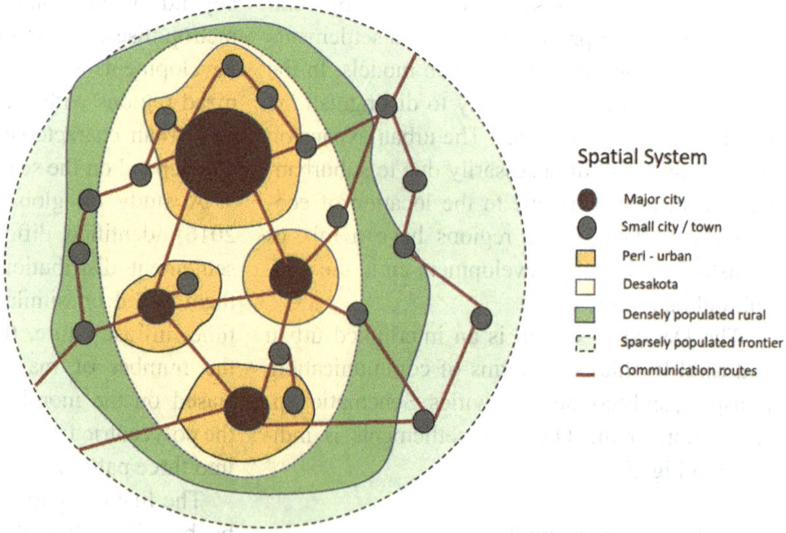

Neither Rural Nor Urban: A Critical Review of the Fringe Dynamics of Settlements, Fig. 5 Illustration of Desakota settlements. (Source: adapted from McGee 1991)

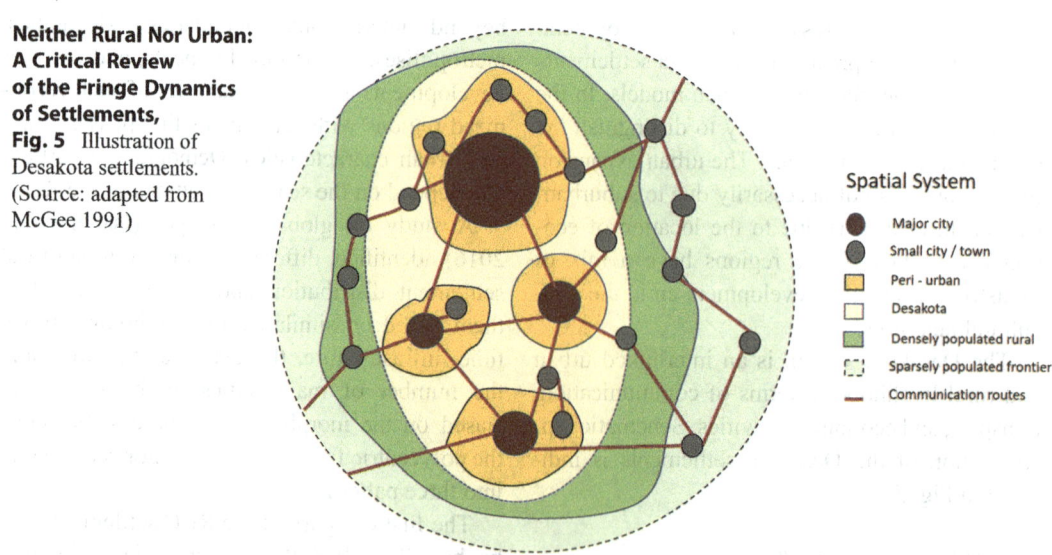

Megacity Region, Megalopolis, Mega-(Urban) Region, and Mega-Metropolitan Region also represent a combination of Rural-Urban tracts spreading over a vast geographic region. The Jean Gottmann Bos-Wash region is a Megalopolis (Bos-Wash: The heavily populated Megalopolis extending from Boston to Washington). The Pearl River Delta in China, Tokyo-Kobe in Japan, and the Nile Valley in Africa are examples of the megalopolis. “Megapolitan” is also a similar cityscape with distances extending from 50 to 200 miles and two or more prominent cities (Georg et al. 2016). “Metropolitan Agglomeration, Metropolitan Region, and Metroplex” are large-scale RUC settlements having two or more major cities and connectivity in the form of a high-speed road or rail.

The second category of RUC settlements is the Ribbon development settlements (Fig. 6b). “Ribbon Development” and “urban corridors” form a linear pattern of RUC stretches with a significant difference in scale. Ribbon development is a linear stretch of RUC settlements within a 1-h travel time (up to 100 kilometers) in proximity to a single city. Simultaneously, urban corridors are linear RUC settlements that stretch more than several hundred kilometers, with high-speed road and rail connectivity (Georg et al. 2016). Urban corridors are the most complex of cityscapes and comprise a network of settlements of varying complexities. They often stretch beyond international boundaries. Bangkok and Hangzhou are examples of ribbon development, while Bos-Wash, Blue Banana (Blue Banana: The European Megalopolis is a corridor spread over Western and Central Europe), and BESETO are examples of urban corridors (Georg et al. 2016) (BESETO: Extended urban area extending from Beijing via Seoul to Tokyo). In India, such urban corridors though of more minor scales, are emerging along the Mumbai-Pune Highway, Bangalore-Chennai Highway, and many other locations connecting major urban centers.

The third category of Morphology of RUC settlements is the Desakota Settlements (McGee 2013) (Fig. 6c). Such settlements are regional forms with a more equitable mix of urban and rural settlements, occasionally with a higher inclination to the rural characteristics as discussed in the previous section.

Other Related Terminologies

Composite Settlements: Composite settlements are predominantly rural. A predominant village that offers services and facilities to a group of surrounding villages function as a composite settlement. The village will have a mix of rural and urban activities. The main village thus acts as a central place. Firoz (2015), in his studies, has used the term composite settlements to describe the rural-urban continuum settlements of Kerala. The settlements with mixed urban-rural characteristics, which cannot be differentiated, are termed composite settlements.

Umland: Umland is a term coined by the French geographer Ander Allix in 1914 to denote areas linked to an urban center socially, culturally, and economically. It is typically used for urban centers in the inland region, which grows in all directions. Urban field, the sphere of influence, urban catchment area, hinterlands, etc., are standard terms that refer to Umland.

Gragara: Gragara combines the words “gramam,” meaning rural, and “nagaram”

N

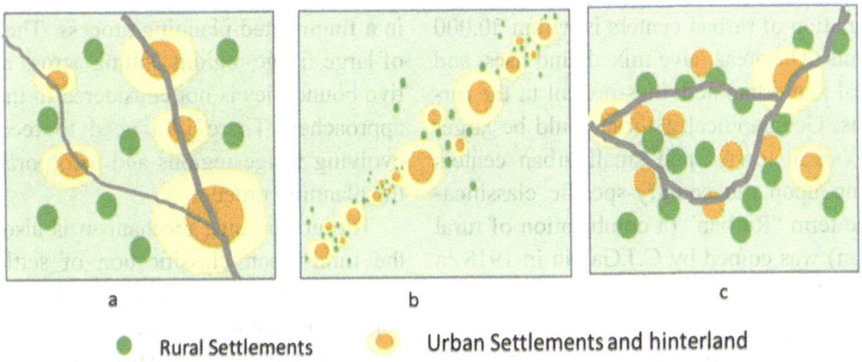

Neither Rural Nor Urban: A Critical Review of the Fringe Dynamics of Settlements, Fig. 6 Broad morphology of polycentric rural-urban continuum settlements implying urban. The terminology describes the RUC settlements found in the southern state of Kerala in India, which has unique rural-urban continuum settlements that occur along the coastal stretch. The landscape and economy of the settlements of Kerala have both rural and urban components that set it apart from the rest of the Indian states. The villages and towns have comparable population sizes and densities. It is not easy to distinguish between an urban or rural area in Kerala.

Ruralopolis: The RUC with a high concentration of rural settlements than urban is also called "Ruralopolis" settlements (Qadeer 2000). Ruralopolis regions are primarily formed due to urbanization by implosion and are found primarily in South Asian countries like India, Pakistan, Bangladesh, and China (Qadeer 2004). In South Asian countries, rural settlements have a high population and density with an advanced agro-based economy. The small landholding patterns create increased pressure on land with a growing demand for services and infrastructure development (Mohammed Firoz 2015).

Rurban: Rurban term was first used in 1915 to indicate a mix of rural and urban characters. During the initial periods, the term denoted a residential area with farming activities. However, in developing countries such as India, the term refers to emerging small towns under the governance of rural local bodies (Pranav and Kumar 2016). The process of rurbanization is the increase in population and urban-related activities in the rural spaces close to the large urban areas. These areas tend to have increasing population size and density. Often the population of rurban centers is within 20,000 inhabitants. The areas have mixed land uses, and both rural and urban activities prevail in the rurban areas. Geographically, these could be larger villages, small towns, and small urban centers depending upon the country-specific classification. The term "Rurban" (a combination of rural and urban) was coined by C.J.Galpin in 1918 in his book titled "Rural Life."

The fringe dynamics in various scales and forms occur in various geographies. However, most countries continue to follow a strict dichotomous rural and urban classification. It often results in classifying the in-transition settlements into either urban or rural, which is inappropriate. Myriad forms of RUC settlements of varying scales occur all across the world. Hence, it is time to acknowledge such regions and the rural and urban spectrum that unfolds in these regions. However, most countries continue to classify settlements as urban or rural (Hugo et al. 2003). It results in the inappropriate classification of settlements with mixed urban and rural characteristics as either category depending on the country-specific definition (Mohammed Firoz et al. 2014; Mohammed Firoz 2015). The following section explores the problems associated with following a dichotomous classification in RUC settlements.

Problems Associated with Dichotomous Classifications

The planning process, especially in the Asian countries, is based on the dichotomous rural and urban classification though they have RUC development patterns. Often, the urban plans prepared are limited to the urban boundaries. Instead of a holistic regional planning approach, the focus of physical planning is limited to urban areas. The fringe areas with similar urban populations and densities often get classified as rural and become regions of chaotic development.

While mega-urban regions in Asia have physical planning taking care of the infrastructure for the well-defined urban administrative boundaries, the economic and social process in the fringe regions is often neglected. Countries like India follow decentralized governance, which results in a fragmented planning process. The formation of large fringe regions cutting across administrative boundaries is not considered in the planning approaches. There is a need to recognize the evolving fringe regions and incorporate them in the planning process.

The governance mechanism is also based on the rural-urban classification of settlements. It results in a fragmentation of services as the concerned authorities in charge of resources (transportation, water, sewerage, etc.) have the power only in their jurisdiction. The extension of resources and infrastructure to the fringe area is often neglected, resulting in chaotic development.

The urban classified local bodies have higher staff allocations, while the rural local bodies with similar demographics will have fewer human resources. The urban local bodies can avail higher taxes owing to their classification. However, it becomes a burden for the rural local bodies to raise taxes equivalent to the urban local bodies, as they have to perform almost all functions of an urban local body.

The lenient byelaws of the fringe regions, which are often classified as rural areas, promote uncontrolled growth and haphazard developments. The lack of zoning regulations results in rampant land conversion, often detrimental to the environment. The lack of proper planning results in poor infrastructure and amenities, further reducing the environmental quality. Hence, the fringe regions should adopt a spectrum of urban ranges based on urbanization instead of a dichotomous classification. It will help in the resource allocation and planning process. There exist various approaches that move away from the typical dichotomous classification followed by various countries and organizations. The following section discusses a few methods for classifying settlements apart from the usual dichotomous rural and urban classification.

Classification Approaches Contrary to the Dichotomous Classification

The dichotomous classification often ignores the rural-urban fringe areas. Hence, several organizations have started certain attempts primarily for policy purposes to acknowledge such ignored fringe areas into the settlement classification process. Some of the most prominent ones are discussed here.

Classifications in Australia

Australia developed the index of accessibility to classify settlements to indicate accessibility in nonmetropolitan areas. The index is known as the "Accessibility/Remoteness Index of Australia" (ARIA) (GISCA 2001). Service centers are identified and categorized into four levels based on their population sizes according to the classification. ARIA scores for the settlements are calculated based on the road distance between each populated locality in Australia and their nearest service centers. The methodology helped categorize settlements into five ranging from highly accessible to very remote settlements, which helped frame appropriate policies.

Classifications in the USA

For census purposes, the areas are delineated into geographical regions. Urban areas have a minimum population of 50,000 within urbanized areas, and urban clusters have inhabitants between 2500 and 50,000. Detailed criteria include population density, high degree of impervious land cover, contiguity of the urban fabric, etc. The territories that fall outside the urban areas are categorized as rural.

The United States Department of Agriculture (Economic Research Service (ERS)) recognizes rural as a combination of open countryside, rural towns with a population less than 2500, and urban areas with 2500–49,999 inhabitants that are not part of metropolitan areas. For assisting the Federal programs in the USA, the rural areas are further classified based on Rural-Urban Continuum codes, urban-influence codes, natural amenities scale, and the ERS typology codes that categorize rural regions by their economic and policy types.

The rural-urban continuum codes subdivide the metro regions into three subcategories based on the population size and the nonmetro regions into six categories based on the degree of urbanization and proximity to the metro area. The urban-influence codes differentiate counties by population size of the metro area. As per the urban-influence codes, the nonmetro categories are subdivided into ten categories depending on the proximity to urban areas. The Natural amenities scale measures the physical characteristics and is constructed combining the measures of climate, water area, and topography. The country typology codes capture the economic and social aspects of the regions.

Regional Classifications by Organization of Economic Cooperation and Development (OECD)

The regions of OECD member countries are classified broadly into three regions – predominantly urban, intermediate, and predominantly rural. The criteria include a minimum density of 150 people per square kilometers (excluding Japan and Korea, which has a density threshold of 500 persons per square kilometer), the percentage of the population living in the rural local units, and the size of urban centers (OECD 2011).

Regional Classifications by European Statistical Office (EUROSTAT)

The EUROSTAT broadly categorizes the entire region into Nomenclature of Territorial Units for Statistics (NUTS), a hierarchical division of the territory for statistical and analytical purposes (Fig. 7). NUTS 1 is a broad classification of the major socioeconomic regions, while NUTS 2 subdivides areas as basic regions for policy application. A more detailed small region for specific diagnoses is classified as the NUTS 3 region.

At the NUTS 3 level, there is urban-rural, metropolitan, and coastal typology. The predominantly urban, the intermediate, and the predominantly rural regions are the different classes of urban-rural typology. At the local level, the classifications are based on the degree of urbanization and functional urban areas. The functional urban areas comprise cities and their commuting zones.

Classification by the Degree of Urbanization

The need for a uniform definition of cities, towns, semi-dense areas, and rural areas was felt by the United Nations. In 2020, the methodology known as the new degree of urbanization was formulated (Dijkstra et al. 2020). The methodology uses remote sensing data and census data. Accordingly, settlements are classified as urban centers, core and semi-dense areas, and rural areas. An urban center should comprise contiguous grid cells of 1 sq. km with 1500 population density and a population of 50,000. The contiguous cells with a population of 5000 and 300 persons per sq. km are an urban cluster, and the rest of the cells are rural grid cells. The advantage is the recognition of rural-urban continuum regions and the universal applicability of the method.

The fringe developments and rural-urban continuum regions differ in size, scale, and composition based on the associated urban core. The urban cores could be monocentric and polycentric based on the number of urban centers and related functions. The following section provides an insight into the fringe dynamics associated with the urban centers.

Conclusion

The increasing urbanization in the world has resulted in the emergence of various types of fringe regions. The initial monocentric structure gave way to highly complex polycentric systems

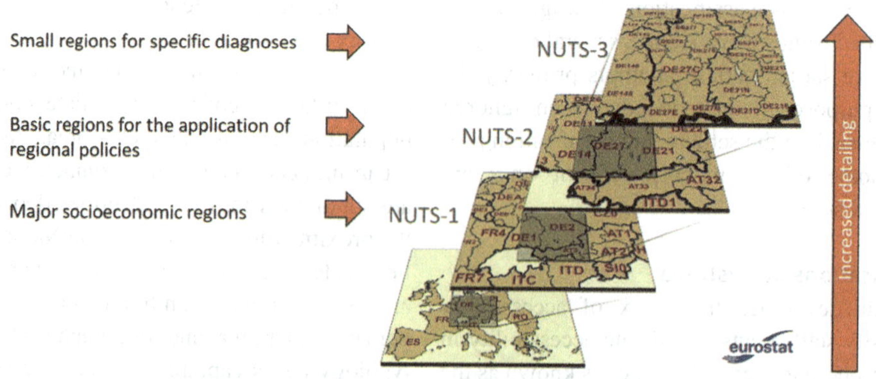

Neither Rural Nor Urban: A Critical Review of the Fringe Dynamics of Settlements, Fig. 7 NUTS classification. (Source: adapted from (Eurostat 2021))

of megaregions. The associated fringe regions thus acquired a mix of both rural and urban characteristics. However, countries differentiate their rural and urban areas based on the most appropriate criteria by which the urban is defined in the context of each country. They often rely on a dichotomous classification for ease in differentiating settlements. Often the country-specific categorization, while differentiating urban and rural, fails to acknowledge the formation of rural-urban continuum regions. Most often, the RUC settlements get erroneously classified as either urban or rural. The misclassification is associated with several problems affecting resource allocation, governance, planning, management, and the environment.

The fringe regions are known by various terminologies such as peri-urban, suburban, megalopolis, urban corridors, desakota, etc. The multiple scales and composition of fringe developments and classification systems applied to differentiate the areas based on accessibility, density, and contiguity are mentioned in the study. Often, the fringe regions get neglected and remain primarily unrecognized due to the lack of clear definitions and classifications. The study attempts to put forward the various associated terminologies and identify their differences to reduce ambiguities while using the terminologies. It is necessary to acknowledge the fringe developments according to the context, composition, and scale and classify them based on their degree of urbanization for better management, resource allocation, and planning of the region.

Cross-References

- ▶ Future of the City-Region Concept and Reality
- ▶ Growth, Expansion, and Future of Small Rural Towns
- ▶ Peri-urban Regions
- ▶ Peri-urbanization
- ▶ Planning for Peri-urban Futures
- ▶ Small Towns in Asia and Urban Sustainability
- ▶ Spatial Demography as the Shaper of Urban and Regional Planning Under the Impact of Rapid Urbanization
- ▶ Urban Densification and Its Social Sustainability
- ▶ Urban Structure and Its Impact on Mobility Patterns: Reducing Automobile Dependence Through Polycentrism
- ▶ Urbanization, Planning Law, and the Future of Developing World Cities

References

Dijkstra, L. 2009. Metropolitan regions in the EU. *Regional Focus*. http://ec.europa.eu/regional_policy/sources/docgener/focus/2009_01_metropolitan.pdf

Dijkstra, L., Florczyk, A. J., Freire, S., Kemper, T., Melchiorri, M., Pesaresi, M., & Schiavina, M. (2020). Applying the degree of urbanisation to the globe: A new harmonised definition reveals a different picture of global urbanisation. *Journal of Urban Economics, November*, 103312. https://doi.org/10.1016/j.jue.2020.103312.

Eurostat. 2021. NUTS – nomenclature of territorial units for statistics – Eurostat. European Commission. 2021. https://ec.europa.eu/eurostat/web/nuts/background

Federal Office of Rural Health Policy, Heath Resources and Services Administration. (2021). Defining Rural Population | Official Web Site of the U.S. Health Resources & Services Administration. 2021. https://www.hrsa.gov/rural-health/about-us/definition/index.html

Firoz, M. (2015). *Reclassification of the Typlogy and pattern of composite settlements : A case of Kerala, India*. Kharagpur: Indian Institute of Technology.

Friedmann, J., & Miller, J. (1965). The urban field. *Journal of the American Institute of Planners, 31*(4), 312–320. https://doi.org/10.1080/01944366508978185.

Georg, I., Blaschke, T., & Taubenböck, H. (2016). New spatial dimensions of global cityscapes: From reviewing existing concepts to a conceptual spatial approach. *Journal of Geographical Sciences, 26*(3), 355–380. https://doi.org/10.1007/s11442-016-1273-4.

GISCA, Department of Health and Aged Care and the National Key Centre for Social Applications of Geographical Information Systems. (2001). Measuring remoteness : Accessibility/remoteness index of Australia (ARIA). Occasional Papers: New Series No. 14. https://www1.health.gov.au/internet/main/publishing.nsf/Content/health-historicpubs-hfsocc-ocpanew14a.htm

Hall, P. (2009). Looking backward, looking forward: The City region of the mid-21st century. *Regional Studies, 43*(6), 803–817. https://doi.org/10.1080/00343400903039673.

Hugo, G., Champion, A., & Lattes, A. (2001). *New conceptualisation of settlement for demography:*

N

Beyond the rural/urban dichotomy. Paris: IUSSP. no. June.

Hugo, G., Champion, A., & Lattes, A. (2003). Toward a new conceptualization of settlements for demography. *Population and Development Review, 29*(2), 277–297. https://www.jstor.org/stable/3115228.

Hussain. (2017) Rural urban continuum and the necessity of integrated planning. Geographyandyou.Com. 2017. https://www.geographyandyou.com/rural-urban-continuum-necessity-integrated-planning/

Iaquinta, D. L., & Drescher, A. W. (2000). *Defining Peri-urban: Understanding Rural-Urban Linkages and Their Connection to Institutional Contexts*. IRSA.

Isabel, G., Thomas, B., & Hannes, T. (2016). New spatial dimensions of global cityscapes : From reviewing existing concepts to a conceptual spatial approach. *Journal of Geographical Sciences, 26*(3), 355–380. https://doi.org/10.1007/s11442-016-1273-4.

John, P. L. C., & Flood, J. (2019). *What is rural?* USDA, National Agricultural Library. 2019. https://www.nal.usda.gov/ric/what-is-rural.

McGee, T. G. (1991). The emergence of 'Desakota' regions in Asia: Expanding a hypothesis. In N. Ginsberg (Ed.), *The extended Metropolis: Settlement transition in Asia* (pp. 3–26). Honolulu: University of Hawaii Press.

McGee, T. G. (2013). The emergence of 'Desakota' regions in Asia: Expanding a hypothesis. In N. Brenner (Ed.), *Implosions/Explosions*. https://doi.org/10.1515/9783868598933-010.

Mohammed Firoz, C. (2015). *Reclassification of the typology and pattern of composite settlement systems: A case of Kerala, India*. Kharagpur: Indian Institute of Technology.

Mohammed Firoz, C., Banerji, H., & Sen, J. (2014). A methodology to define the typology of rural urban continuum settlements in Kerala. *Journal of Regional Development and Planning, 3*(1), 49–60. https://econpapers.repec.org/RePEc:ris:jrdpin:0025.

OECD, Directorate for Public Governance and Territorial Development. (2011). OECD Regional Typology. http://www.oecd.org/gov/regional-policy/42392595.pdf

Pauchet, M., & Oliveau, S. (2008). Kerala: A Desakota? European Population Conference. 2008. https://halshs.archives-ouvertes.fr/halshs-01140931/document

Piorr, A., Ravetz, J., & Tosics, I. (2011). *Peri-Urbanisation in Europe: Towards a European policy to sustain urban-rural futures*. Frederiksberg: Forest & Landscape, University of Copenhagen. 144.

Pranav, N., & Kumar, K. (2016). Rurban Centres : The new dimension of urbanism. *Procedia Technology, 24*, 1699–1705. https://doi.org/10.1016/j.protcy.2016.05.198.

Qadeer, M. A. (2000). Ruralopolises : The spatial organisation and residential land economy of high-density rural regions in South Asia. *Urban Studies, 37*(9), 1583–1603. https://doi.org/10.1080/00420980020080271.

Qadeer, M. A. (2004). Urbanization by implosion. *Habitat International, 28*(1), 1–12. https://doi.org/10.1016/S0197-3975(02)00069-3.

Sircar, S. (2017). 'Census towns' in India and what it means to be 'urban': Competing epistemologies and potential new approaches. *Singapore Journal of Tropical Geography, 38*(2), 229–244. https://doi.org/10.1111/sjtg.12193.

United Nations, International Labour Organization. 2018. Inventory of Current Definitions of Rural/Urban. 2018. http://ilo.org/stat/Areasofwork/Compilation/WCMS_389373/lang%2D%2Den/index.htm

van Duijne, R. J., & Nijman, J. (2019). India's emergent urban formations. *Annals of the American Association of Geographers*. https://doi.org/10.1080/24694452.2019.1587285.

You, Y. (2017). The classification of Urban Systems: A review from monocentric to polycentric. *Advances in Economics, Business and Management Research, 42*. https://doi.org/10.2991/isbcd-17.2017.1.

Neoliberalism and Future Urban Planning

An Inquiry into Africa's Readiness

Brilliant Mavhima[1], Charles M. Chavunduka[1] and Innocent Chirisa[1,2]
[1]Department of Architecture and Real Estate, University of Zimbabwe, Harare, Zimbabwe
[2]Department of Demography Settlement and Development, Social & Behavioural Sciences, University of Zimbabwe, Harare, Zimbabwe

Synonyms

Ideology–thought; Neoliberalism–capitalism; Land–real estate

Definitions

Neoliberalism is an ideology that for economies to grow, government interference should be removed and allow for free markets.

Ideology is a firm belief in a process or direction of thinking.

Keynesianism is an ideology that posits that governments should be central for sustainable economic developments.

Modernism is an ideology that there is a single truth and the quest of research is to find it.

Postmodernism is an ideology that truth is objective and there exist many truths.

Urban planning is a process of designing, organizing, and management of urban settlements.

Land market entails exchange of land rights by buyers and sellers.

Introduction

What will become of urban planning in the future given the existing neoliberal wave? The passing of Keynesian economics and the introduction of neoliberalism in the1960s meant that the government's central role in the development of economies and distribution of resources (the invisible hand) was reduced. A *laisses faire* economy was created. In urban planning, the evolution of planning paradigms has continued starting with a move from rational planning (modernist planning) to other urban planning paradigms (postmodernism) like disjointed incrementalism, mixed scanning, and feminism, among others. While the planning approaches have been increasing, the role of the central government remains prominent. This is discordant of the existing global economic agenda of neoliberalism.

Urban planning operates in a dynamic environment. Neoliberalism is a stage in the process of the urban planning environmental dynamism. Existing literature indicates how urban planning has remained a tool of the state in distribution of resources (Fincher and Iveson 2008; Friendly and Stiphany 2019; Yang and Cai 2020). The process makes the state a central agent in the process of urban planning; this resonates with Keynesian economics that emphasizes on the significance of the state (Tily 2016). Global economic trajectory has reached the stage of neoliberalism that reduces the role of the state in the process of urban planning and management of land markets (Ostry et al. 2016). This raises issues that include, if the environment dictates the need for less government intervention and increased efficiency, what is to happen to urban planning.

Urban planning, as a profession, derives its relevance from the state; the central notion behind neoliberalism is to reduce the power of the state. Neoliberalism undercuts the sole purpose of urban planning as a tool of correcting market failures. Baeten (2012) states urban planning is a prerequisite of neoliberalism as decision on land, and investments can only be functional if regulated by planning institutions. The process of neoliberalism presents planners with a wide array of challenges. This chapter centers on seeking to answer the following questions:

- What are the key tenets of the concept of neoliberalism?
- Beyond neoliberalism, what might come and how can urban planning prepare for it?
- What is urban planning and what are the existing advancements in urban planning towards meeting the demands of the existing neoliberal environment?
- How have other countries managed their urban planning systems towards responsiveness to neoliberalism?

Neoliberalism and Urban Planning: A Review

Neoliberalism has been described as one of the most successful agenda across the world (Springer 2016; Whyte and Wiegratz 2016; Ruano 2017; Slobodian 2020). Its success in permeating or aspects of human lives has brought about various questions as to what it entails. In a paper by Venugopal (2015) the following questions were raised:

> Does neoliberalism imply a contraction of the state vis---a---vis the market, or just a different kind of state that promotes and works at the behest of markets? Is neoliberalism a depoliticized and technocratic fetishization of the market or is a deeply political agenda of class rule and neo---colonial domination? Is it a Leviathan that bludgeons its way around the world or is it a far more subtle, mutating, localized, contingent force that works by transforming individual subjectivities? Is

N

neoliberalism an absolute final state of being, or is it a relative category, describing a direction of travel?

The further development and overeating of the term neoliberalism have created a high degree of ambiguity as to what entails neoliberalism. This has risen various controversies as to what the concept entails. Existing literature points to the position that the term neoliberalism keeps changing definitions from paper to paper. Turner (2008, p. 2) points that what neoliberalism stands for is both confusing and confused. This creates a problem for urban planning as the focus moves away from fairness and redistribution towards profit maximizations. The question remains, what is neoliberalism? Brenner et al. (2010, p. 330) define neoliberalism as a historically specific, unevenly developed, hybrid, patterned tendency of market disciplinary regulatory restructuring. The purpose of neoliberalism is to shift from government to private and market-oriented strategies (Slobodian 2020). This process moves away from Keynesian economics that focuses on welfare state economics towards economic liberalization. This process views the government as an intrusion in contractual agreements between individuals (Seixas 2017).

The drive behind neoliberalism is to provide administrative efficiency, entrepreunualism, as well as giving economic freedom. Current neoliberalism is characterized by economic globalization, international capital mobility fewer business restrictions, extended property rights, and devolution of property rights, among other characteristics. While there are conceptualizations of what neoliberalism entail, existing literature indicates that the idea of neoliberalism is context defined and no country has had the full features of neoliberalism. Each country implements what is appropriate for the area.

In urban planning, Sager (2013) indicates that the attitude of urban planners is to embrace neither market nor politics. Urban planners have always looked at the planning profession as having professionally good solutions (Kafka 2018). In this process, spatial planners argue that market should transfer some responsibilities to the government. This is a political process. While this seems clear, Sager (2013) indicates that the consequences of neoliberalism on urban planning is still unknown. The only clear position is that the neoliberal wave led to the cutting of public budgets leading to many plans being shelved (Lovering 2007). The concept of neoliberalism however is very significant in planning as it helps in describing political trends as well as transformations that occur in the planning working environment. The term neoliberalism is a term that is critical in understanding the urban conditions in line with social democracy and the regulations that exist. Planning is viewed as a distortion of market mechanisms as it's a threat to private sector.

Planning activities often reflect the prevailing values and processes. The neoliberal waves that have rocked the world have brought with its various stresses on planning. Kotz (2002), in the context of England, highlight that the governments that have existed in England have been moving towards moving away from public interests and discretionary urban planning and using market forces in making planning decisions. Houghton et al. (2012) in the context of continued change in planning argued that planning is a political football that is constantly reshaped and reoriented every time a need arises. In England, the government tried to relax the regulation of planning through introduction of short-term letting and permitted development rights. These allowed people to convert office space into residential without the application or planning process (Kotz 2002). This approach was to remove unnecessary planning bureaucracies and allow market forces to flow and dominate. This weakened the position of the planner in terms of development planning options as well as the dominance of public interest drive in planning (Sosnowki 2019). This is so as planners had not permitted that kind of development since the tourist destination nature of London would slowly lead to people converting their residential areas to hotel accommodation.

In a broader context, Mackay (2020) indicates that the drive to smart cities growth has furthered the concept of neoliberalism. The argument is that the principles driving the smart cities is to develop high-density mixed-use developments and reduce sprawling. This drive has led to policy driven gentrification, where planning old houses are destroyed and replaced with new high rental and high-density developments (ibid.). This resonates with the principles of neoliberalism that indicates profit maximization. This can be seen with the Halifax Centre Plan that has replaced old residential buildings with new commercial buildings. The central notion in the plan has been to focus on brown field redevelopments with the intention of maintaining urban financial health and competitiveness.

Planners have adopted and entrepreneurial approach amid the context of pervasive neoliberalism. Fox-Rogers and Murphy (2016) indicate that planners have moved away from intervention and development control-oriented planning and moved towards private development interests. The need to portray a pro-development approach in planning is for planners to demonstrate their relevance (Fox-Rogers and Murphy 2016). This approach has permeated the planning schools of thought with Bengs (2005) indicating that the purpose of communicative planning is to allow for the facilitation of neoliberalism and real estate markets. Planning decisions tend to mirror political decisions (Sager 2018).

Regardless of existing literature indicating how planning has been facilitating neoliberalism, Wright (2013) brings a different perspective. The argument is that neoliberals view urban planning as a distortion to land markets and an increase to costs as it creates unnecessary bureaucracies. This makes planning an enemy of neoliberal ideals. What planning dies is to make various considerations before approval of a plan as well as allocation of a development permit. This is unnecessary in the context of neoliberalism as it emphasizes on building once the resources are available. Having looked at how planning and neoliberalism interact, it is clear that neoliberalism is not universally applicable, as such the planners' approach operates in that context. In Africa, the planner's approach towards market-driven ideals indicates a different trajectory to that of developed nations. The discussion above indicates that planners have taken an entrepreneurial approach in their duties which is different to that of planners in the African continent.

The idea of neoliberalism in planning in Africa came around the 1990s with the introduction of the structural adjustment programs (Bruff 2016). The period saw the reduction of state interventions with the state taking the role of the gatekeeper of the neoliberals, towards involving those excluded by neoliberalism. The emphasis of development planning was towards cities and securing of investments hence the introduction of the concept of the "world of cities." With this happening, the question is, what happened to planning in Africa?

The coming of the idea of a "world of cities," coupled with other factors, has seen the rise in urbanization in Africa that has created a series of planning challenges like informality, pressure on basic services among others. Watson and Agbola (2013) indicate that planning remains a state tool that has been used to manage the developments that are occurring. The planners in Africa present an image of technical that are assumedly apolitical (Cobbinah and Darkwah 2017). This is in contrast with a deeper and critical look at planning in Africa that presents deep involved in politics as it deals with political interests. This makes planning one profession that has been responding to neoliberalism, since politicians have been the ones driving the neoliberalist ideology.

The availability of planners as a political tool has made the planning process less of a process of adapting to neoliberalism approach and more of a tool for state repression (Grooms and Frimpong Boamah 2018). Planning tools in Africa are used to whip people in line with the political ideology prevailing (Watson and Agbola 2013). An example is the *Operation Murambatsvina* (restore order) that occurred in Zimbabwe in the year 2005. This was an eviction process of all informal settlement dwellers following the British developed spatial planning act, the Regional Town and Country Planning Act [Chapter 29:12]. This case indicates that while other planners around the globe follow neoliberalism and reorienting to maximizing profits, planning in Africa remains a political tool that is used when it is necessary for the state.

The existence of planning as a political tool means that planning has not been developing as per the development needs of the African population. The bureaucracies in the development of spatial plans and other planning frameworks remain central in planning in African countries. This directly contradicts the ideas behind the concepts of neoliberalism that emphasizes the significance of efficiency (Ruano 2017). As a result of the fixed nature of planning and planning instruments in Africa, the continent has been marred with illegal developments with informality being evident in all sectors of African life (Watson and Agbola 2013). The private property sector has been booming in Africa; however, most of the developments have been done informally. This is so as the tool that exist for planning has also been unresponsive and remained outdated (Mbiba 2019). Thus, Watson and Agbola (2013) indicate that most developments in Africa have been occurring in a completely non-planned and non-transparent manner.

Having tools and a planning environment that is not evolving, spatial planning in Africa operates in a very dynamic neoliberal environment. The developments in Africa are inspired by developed nations. The cities in Africa takes the shape of their colonial masters with Zimbabwean and Zambian cities having a British taste, while Mozambique has an ambience of Portuguese cities (Mbiba 2019). Following this, the informality that has been created by the neoliberal wave that is in Africa is "wished away by national governments" (Watson and Agbola 2013, p. 3). When planning is engaged to deal with the issues, it is often slow and bureaucratic in a way that things change before the plan is implemented (Mbiba 2019).

The central idea in neoliberalism remains unimplemented in the planning process of urban settlements in Africa. This is so as the master plans in Africa are almost drawn by central governments (Watson and Agbola 2013, p. 4). This mean that the role of the state remains central in the development of African cities. The state continues to be the sole planning organization instead of the adjudicator as advocated for by the neoliberalist principles. Thus, regardless of the market demands, the government implements what it thinks is proper (Grooms and Frimpong Boamah 2018).

The coming of neoliberalism has reduced the role of the state and therefore to some extent incapacitated and weakened the spatial planning toolkit to deal with problems in urban Africa (de Satgé and Watson 2018). This has also changed the focus of urban planning from creating decent places for people into developing places for elite consumption and production (Grooms and Frimpong Boamah 2018). This process has led to failure of urban planning to respond to the growth in Africa. However, the private developers have taken over development in Africa, trying to maximize profit in development. An example is Zimbabwe that engaged private land developers giving them 40% of all alienated land (land ear marked for development) (Paradza and Chirisa 2017). This mean that development is carried out through private developers and financiers, thereby speeding up the land development process and cutting down some of the land development bureaucracies.

Emerging Issues and Synthesis

The existing literature has highlighted what planning and neoliberalism are. How they are supposed to interact and how they interact. Neoliberalism has changed planning approach of planners in developed countries, while the planning in developing countries remains state centered. From the scholarly debates raised above, the following issues emerge:

Neoliberalism Is Here

One key issue that raises is that neoliberalism is here. The concept of having a central focus on efficiency and profit orientation has become central in global development. The existence of neoliberalism in different countries and the features that exist in different countries are contrary. No country has all the characteristics of neoliberalism. However, different features of neoliberal states exist in every country (Wright 2013). The most common characteristic in existence is the slow removal of the state in urban planning activities and the perception of the state as the bureaucrat that slows down the development projects.

Regardless of the strand of neoliberalism that focus on state involvement, in African countries and developing world at large, state involvement is still emphasized very much. However, the key feature of neoliberalism in African and developing nations is the sudden focus on private developments. This approach has seen the state having to be avoided in some processes of land development, thereby improving efficiencies in the process. Furthermore, this has led to commodification of land which was once viewed as a free product under socialistic approaches in state-led developments (Steel et al. 2019).

Planning Has Been Responding Differently Depending with Context

The coming of neoliberalism and the acknowledgement of its existence has impacted the urban planning processes. Spatial planners in developed nations have changed their approaches from being just state tools to becoming entrepreneurs (Fox-Rogers and Murphy 2016). Planning approaches and the evolution of planning approaches have shifted from achieving social good towards maximization of profits. Planning standards and master plans in developed countries have been revised to allow for permission of development without applications being necessary (Sager 2016). This has increased the efficiencies in the process of private developments.

In developing nations, planning remains a state tool. Planning continues to implement the master plans and planning regulations that were developed by the colonial masters. This approach means that planning in most developing nation is becoming a hindrance to development particularly that which is driven by the millennial forces of neoliberalism (Choplin 2017). In this context, planners operate in a vacuum, pretending to be unaware of the neoliberal wave, resisting development all under pretense of serving the social good.

Cities' Planning in Africa Remains under the State

The African urban planning processes are still largely state driven. The government, through various departments administer land the spatial planning processes and governs the laws behind development control (Watson and Agbola 2013). This makes planning a state process that is used to deal with market failure through resource redistribution. This goes against the principles of neoliberalism which requires a free market approach with the state have little to no involvement. Social good in this context is not a consideration (ibid.).

The continued existence of planning under the state means that the processes continue to be old and filled with red tape (Watson and Agbola). State-owned planning institutions focus more on the needs and demands of the political muscle in the country (ibid.). This mean that planning tools and regulations are engaged when the politicians feel the need to (see the case of Operation Restore Order in Zimbabwe 2005). In this context, neoliberalism becomes a distance force that only influences urban planning practices when allowed to.

Bureaucracy and Slow Planning Process

Having planning operating under the state in African countries means that the urban planning processes are slow and filled with red tape (Watson and Agbola 2013). This is against the principles of neoliberalism and market orientation that the globe has taken. Neoliberalism advocates for efficient planning process which mean less time in planning permit acquisition. This has not been the case with planning processes in Africa taking years to process. A case is that of master plans that expire before they get approved and finalized (Choplin 2017). The slow processes in the context of neoliberalism have brought about informality as a new evil within the context of urban planning.

Informality Is a Sign of Urban Planning Resistance to Neoliberalism

In a neoliberal world, demands, tastes, and needs evolve quicker and according to global development trajectories. Concepts evolve and people adopt them quickly (Carmody and Owusu 2016). Informality and the existing of illegal development can be pointed to as changes in tastes that urban planning resisted to follow. Urban planning rules and regulations have been stagnant in Africa (ibid.) while human needs have evolved. This mean that human needs have been rendered illegal with the failure of urban planning to change to the neoliberal wave. The sudden need for financial security through investment in urban land has prompted the need for home ownership, hence a sudden rise in informal developments and informal settlements.

What Then Is the Purpose of Planning?

The question that then rises from the central issues that came is to the purpose of urban planning in the context of neoliberalism. Urban planning in Africa has slowed down the process of neoliberalism. The process and practice of urban planning seem to take a state-centered orientation to development planning. This in turn forces development to favor political ideology, thus making the planning processes restrictive than permissive to development. Acquiring permits for developments take longer than necessary; hence most developments get permits when they are long overdue. This leads to developments that are less and less attractive compared to other nations. The idea of competition comes in from neoliberalism that has been coupled with globalization to create a global competition for cities into the development of what are called global cities.

Conclusion and Future Direction

The purpose of the chapter was to identify the issues surrounding the concept of neoliberalism and urban planning towards mapping the future of urban planning in the context of the neoliberal wave. The chapter indicated that neoliberalism is an ideology that has taken over the political space and is driven by profit maximization, efficiency, and free market approach. This goes against the central tenets of the concept of urban planning that emphasizes the significance of the state in correcting marketing failures and pushing forward the idea of public interest. Urban planners in developed nations have enhanced the concept of neoliberalism through relaxation of development permission process, thereby minimizing the influence of state. On the other hand, developing nations still have the state being central in all the planning processes. This makes it very slow and bureaucratic, thereby reducing the competitiveness of the products of the planning process. The question then becomes what needs to be done for planning in developing nations to improve in the context of neoliberalism.

Urban Planning Liberalization

The first stage is the liberalization of the process of urban planning. Urban planning needs to be given a high degree of autonomy from the political nitty-gritties. The process of planning is technical and yields better results if not influenced by personal interests. While planning cannot be termed apolitical, it is significant to keep political activities at bay and planning to operate according to the forces that determine the planning environment to develop competitive urban developments.

Indigenous Planning Statutes

Urban planning in most African nations is determined by the rules and regulations set by colonial masters. The rules were designed in a way that would create hospitable environments for colonialists and not the indigenous people. The rules and standards have become too high (Tibaijuka 2005) for Africans. In a neoliberal environment where Africans need properties to act as collateral in their investments, people have turned to private developers for land. The highness of the standards is because of the levels of development of the receiver of the regulations and the ones who drafted them. The colonial masters came from the nations that are developed whose approach towards neoliberalism is different to that of the recipient which is developing.

Follow What we Need to Follow

Urban planning in Africa needs to follow what it needs to follow in the context of neoliberalism. Neoliberalism has brought with its various concepts that are being sold by superior nations. For African nations to compete in this context, there is need to become unique and avoid overflowing with the tide. Previous experience of flowing with the tide has seen the African nation being clients to every negative externality of the developed nations.

Cross-References

▶ Global Homelessness: Neoliberalism, Violence, and Precarious Urban Futures

References

Baeten, G. (2012). Neoliberal planning: Does it really exist?. In: T. Taşan-Kok, & G. Baeten (Eds.), *Contradictions of neoliberal planning: Cities, policies, and politics* (Vol. 102). Dordrecht: Springer Science & Business Media. Springer.

Bengs, C. (2005). Planning theory for the naive? *European Journal of Spatial Development*, July 2005. Retrieved from http://www.nordregio.se/en/European-Journal-of-Spatial-Development/Debate/

Brenner, N., Peck, J., & Theodore, N. (2010). After neoliberalization?. *Globalizations, 7*(3), 327–345.

Bruff, I. (2016). Neoliberalism and authoritarianism. In *Handbook of neoliberalism* (pp. 135–145). London: Routledge.

Carmody, P., & Owusu, F. (2016). Neoliberalism, urbanization and change in Africa: The political economy of heterotopias. *Journal of African Development, 18*(1), 61–73.

Choplin, A. (2017). African urban subalternity: Hegemonic planning, subaltern practices and neoliberal citizenship (Nouakchott-Mauritania). *TERRITORIO, 81*, 28–30.

Cobbinah, P. B., & Darkwah, R. M. (2017). Urban planning and politics in Ghana. *GeoJournal, 82*(6), 1229–1245.

de Satgé, R., & Watson, V. (2018). African cities: Planning ambitions and planning realities. In *Urban planning in the global south* (pp. 35–61). Cham: Palgrave Macmillan.

Fincher, R., & Iveson, K. (2008). *Planning and diversity in the city: Redistribution, recognition and encounter*. Basingstoke: Macmillan International Higher Education.

Fox-Rogers, L., & Murphy, E. (2016). Self-perceptions of the role of the planner. *Environment and Planning. B, Planning & Design, 43*(1), 74–92. https://doi.org/10.1177/0265813515603860.

Friendly, A., & Stiphany, K. (2019). Paradigm or paradox? The 'cumbersome impasse' of the participatory turn in Brazilian urban planning. *Urban Studies, 56*(2), 271–287.

Grooms, W., & Frimpong Boamah, E. (2018). Toward a political urban planning: Learning from growth machine and advocacy planning to "plannitize" urban politics. *Planning Theory, 17*(2), 213–233.

Houghton, S., Yamada, E., & Yamada, E. (2012). *Developing criticality in practice through foreign language education*. Oxford: Peter Lang.

Kafka, K. (2018). Urban planners in Poland. Practicing the urban planning profession in Poland and other European countries. *Technical Transactions, 11*(5), 39–51.

Kotz, D. M. (2002). Globalization and neoliberalism. *Rethinking Marxism, 14*(2), 64–79.

Lovering, J. (2007). The relationship between urban regeneration and neoliberalism: Two presumptious theories and a research agenda. *International Planning Studies, 12*(4), 343–366.

Mackay, F. (2020). Dilemmas of an academic feminist as manager in the neoliberal academy: Negotiating institutional authority, oppositional knowledge and change. *Political Studies Review*, 14789299209 58306.

Mbiba, B. (2019). Planning scholarship and the fetish about planning in southern Africa: The case of Zimbabwe's operation Murambatsvina. *International Planning Studies, 24*(2), 97–109.

Ostry, J. D., Loungani, P., & Furceri, D. (2016). Neoliberalism: oversold. *Finance & Development, 53*(2), 38–41.

Paradza, P., & Chirisa, I. (2017). Housing cooperative associations in Zimbabwe: A case study of current housing consortium in Budiriro, Harare. *Journal of Public Policy in Africa, 4*(2), 159–175.

Ruano, R. (2017). De-linking from neoliberalism. *The American Papers*, p. 103.

Sager, T. (2013). *Reviving critical planning theory: Dealing with pressure, neo-liberalism, and responsibility in communicative planning*. New York: Routledge.

Sager, T. (2016). Activist planning: A response to the woes of neo-liberalism? *European Planning Studies, 24*(7), 1262–1280.

Sager, T. (2018). Planning by intentional communities: An understudied form of activist planning. *Planning Theory, 17*(4), 449–471.

Seixas, B. V. (2017). Welfarism and extra-welfarism: A critical overview. *Cadernos de Saúde Pública, 33*, e00014317.

Slobodian, Q. (2020). *Globalists: The end of empire and the birth of neoliberalism*. Cambridge, MA: Harvard University Press.

Sosnowski, P. (2019). Urban planner as the profession of public trust. *Central European Review of Economics & Finance, 29*(1), 23–33.

Springer, S. (2016). Neoliberalism. In *The Routledge research companion to critical geopolitics* (pp. 169–186). London: Routledge.

Steel, G., Van Noorloos, F., & Otsuki, K. (2019). Urban land grabs in Africa? *Built Environment, 44*(4), 389–396.

Tibaijuka, A. K. (2005). *Report of the fact-finding mission to Zimbabwe to assess the scope and impact of Operation Murambatsvina*. Nairobi: UN-HABITAT.

Tily, G. (2016). *Keynes's general theory, the rate of interest and Keynesian' economics*. Amsterdam: Springer.

Turner, R. (2008). *Neo-liberal ideology*. Edinburgh: Edinburgh University Press.

Venugopal, R. (2015). Neoliberalism as concept. *Economy and Society, 44*(2), 165–187.

Watson, V., & Agbola, B. (2013). *Counterpoints: Who will plan African cities*. London: Africa.
Whyte, D., & Wiegratz, J. (Eds.). (2016). *Neoliberalism and the moral economy of fraud*. London: Routledge.
Wright, I. (2013). 'Are we all neoliberals now? Urban Planning in a neoliberal era' unpublished manuscript prepared for the 49th ISOCARP Congress. Available at: http://www.isocarp.net/Data/case_studies/2412.pdf
Yang, Q., & Cai, Y. (2020). Housing property redistribution and elite capture in the redevelopment of urban villages: A case study in Wuhan, China. *Journal of Cleaner Production, 262*, 121192.

Neoliberalism–Capitalism

▶ Neoliberalism and Future Urban Planning

Networked Eco-industrial System

▶ Industrial Symbiosis

Networking Collaborative Communities for Climate-Resilient Cities

M'Lisa Lee Colbert
The Nature of Cities, Montreal, QC, Canada

Submission

Twenty-First Century Urban Expansion

Urban expansion is expected to reach unprecedented levels in the coming decades. UN Habitat has predicted that by 2050, two thirds of the world's total population will be living in a city (Habitat 2016; Elmqvist, et al. 2018). This is approximately an increase of 2.4 billion people at a rate of growth that is equivalent to building a city with the population of London every 7 weeks. To facilitate this growth, humanity will urbanize an area of 1.2 million km^2 which is larger than the country of Colombia (McDonald et al. 2018a). Moreover, cities are responsible for over 70% of global greenhouse gas emissions and, at the same time, 90% of urban areas are situated on coastlines, making the majority of the world's population increasingly vulnerable to climate change (Elmqvist 2019).

Given this reality, cities will play an increasingly important role in global sustainability efforts. Most notably, an increasingly central role where it comes to climate resilience. For example, we see this shift manifest itself in the many international efforts cities around the world participate in, such as international urban networks like C40, Innovate4Cities, International Council for Local Environmental Initiatives (ICLEI), The Nature of Cities (TNOC), European Association for Local Democracy (ALDA), Global Covenant of Mayors, The Urban Knowledge-Action Network (UKAN) and various other urban networked initiatives. This networked approach is befitting because cities as infrastructure at their core are networked flows of capital, goods, people, and data that are becoming ever denser and more interconnected globally as cities grow and continue to internationalize (Batty 2017; Comunian 2011). Some even liken the "world city network as the 'skeleton' upon which contemporary globalization has been built (Taylor and Derudder 2015)."

It is increasingly clear to many of the world's mayors and municipal officials that cities will be the site where the battle for climate resilience will be won or lost (Guterres 2019), and that this battle is an incredibly complex one with various social, political, economic, technological, and biophysical processes and variables that exist across many spatial and temporal scales (Keeler et al. 2019a, b; McPhearson et al. 2016a; Bai et al. 2016). Within any city, all of these processes interact with each other creating challenges for us in recognizing and understanding interactions, synergies, and potential trade-offs and co-benefits between each component so that we can make changes and develop solutions (Bai et al. 2020). This being the likeness of our cities, so to must it be the character of the solutions we orchestrate to work within them (Bai et al. 2016, 2020). This is the incredible power of networks.

The Concept of Urban Networks

In the context of urban sustainability challenges, the concept of networks can take on multiple meanings, and encompass a range of social and institutional structures. Urban networks can be both local and global in scale, and can broadly be defined as a collection of things: people, organizations, materials, infrastructures, intangibles, etc., that come together as a system of things on the basis of shared relationships and interactions that may be intangible or tangible and exist within, around, or concerning the city. The idea of the networked city refers to the multiple infrastructure networks that support the functioning of modern cities (Monstadt and Schramm 2017), and the role of cities as global nodes in the flow of capital and information (Sassen 2004, 2013), and their role in connecting other networks and forming urban corridors (Pflieger and Rozenblat 2010; Peng and Bai 2019; Shove 2016; Taylor et al. 2015; Yeung 2020). They can facilitate cross-city learning, support inter-connection and cross-fertilization across diverse urban contexts to promote innovative urban practices (Bai et al. 2019; Acuto 2018; Elmqvist, et al. 2018; Nagendra et al. 2018) Bai et al. 2017; Elmqvist et al. 2017; Pflieger and Rozenblat 2010; Peng and Bai 2019; Keeler et al. 2019a, b). Urban networks are important because they facilitate exchange among citizens, diverse disciplines, and urban practitioners that can leverage local action into global impacts. Within these complex, multidimensional urban systems, actors, processes, and governance structures are intrinsically and intricately connected (Bai 2019). Including, where it comes to these systems integrating and interacting with the natural world.

The Urban-Nature Divide

These increasingly global social, environmental, cultural, and economic links in local places are coupled with rapid urbanization affecting mega cities to mid-sized cities, that are further perplexing human-nature relationships in urban areas, leaving in its midst international challenges for integrating and sustainably developing nature in cities and urban green infrastructure (Hillel; Colbert et al. 2020; McDonald et al. 2019). The city is inseparable from the natural world, yet the current relationship is growing more and more worrisome as the sustainability gap grows ever wider between city life and nature.

A recently conducted global nature assessment called Nature in the Urban Century reminds us that natural habitats play an important role in climate mitigation by, among other things, sequestering and storing carbon in their biomass. The assessment quantifies how much carbon dioxide would be released as a result of natural habitat lost due to urban growth between now and 2030. It found that if occurring as forecasted in a business-as-usual scenario, urban growth would destroy natural habitats that are responsible for storing an estimated 1.19 billion metric tons of carbon, or 4.35 billion metric tons of carbon dioxide. This is roughly equivalent to the annual carbon dioxide emissions from 931 million cars (McDonald et al. 2018a, b). This is an immense challenge on a global scale that necessitates the participation and action of many diverse actors from various geographies around the global. Thus, strategies will need to include many smaller but concentrated actions in local places and across scales, geographies, governments, people, and natural habitats to solve. If networked, the solutions we come up with have the best chance of achieving resilience in many cities because of the access they will have to cross-fertilizing further solutions, and resource sharing for maximum impact.

Furthermore, in many cases, the best global nature-based solutions are those that are enthusiastically supported locally. So, getting people and communities onboard to support efforts that reconnect humans and nature and adaption and mitigation challenges related to urbanization, is critically important. This could be to retrofit old areas, or plan new ones in cities. For example, in many shrinking cities, especially those affected by the loss of coal-based heavy industries, there are empty brownfield sites and vacant lots which are often discarded as wasted spaces, but instead should be valued for their potential to enhance urban biodiversity while satisfying resident's needs for nature and greenspace (Hillel; Colbert et al. 2020; Rega-Brodsky et al. 2018). Re-uses of such land in the UK, including London's Queen Elizabeth Park, home to the 2012 Olympic Games, and the Carbon Landscape wetlands of Greater Manchester are examples of this (Douglas 2019).

The urban network framework has the organizing potential to better are approach to the various challenges we face, including those listed here; closing the gap on the urban-nature divide and helping humans live sustainably and achieve resilience in spite of twenty-first century urban expansion. Urban life and the scale at which it operates is huge in many respects, but it is at the same time a collection of many small things going on all at once. Underneath a cities skin are many people, living in smaller barrios, cartiers, neighborhoods, and communities among and intrinsically a part of urban complexity every day. As climatic changes continue to occur, creating strong communities in cities will be a critical part of urban resilience. Networks function wonderfully for gathering cities, organizing goods, collecting data and information, etc., yet this approach can also be an important formation for organizing communities and citizens inside cities to bring about change for resilient cities. When communities and people network, they harness the power of something big to achieve changes in smaller local places that then in turn create global impact for resilience.

Networks are Tools for Communities Too

Networks can help give small community initiatives a large impact. The movement and energy given to each local community effort within a network of things helps to energize small local efforts. They can enhance relations between like members, and help them find their commonality to strengthen their community (Lombardi et al. 2020). The networked approach can move them from a more passive position where they may be acting on their own, and perhaps facing more hurdles because of it, to working with many other groups in various other places to accomplish the same thing. Examples of this are spread far and wide including, Climate Action Network, Global Shapers, Fridays for Future, Extinction Rebellion, Green Peace, Library for Food Sovereignty, Agriculture Biodiversity Community, and The Carbon Underground, to name a few. This networked approach affords them all kinds of tools, such as lobbying potential, visibility, resource sharing, knowledge-exchange, etc., that they would not have had without the network. The network concept is not new. It exists as a feature in many areas of our life. Take for example the sprawling instantaneous social networks we all participate in daily. Networks in this way are utilized by Fortune 500 tech companies like Facebook, Instagram, TikTok, and others to network individuals all around the world. In this world, this action is sometimes referred to as 'converting a passive asset' or harnessing 'collective potential' (Manville 2016). Networking urban community efforts essentially function the same way, and thus have the same potential to be wildly successful.

In many ways local community initiatives already consist of all the things they need to embark on this approach. To recall the definition of urban networks, it is at its core a collection linked on the basis of shared relationships and/or interactions. A community on the other hand can broadly be defined as a group or unit of people and or including other species that are linked together by something in common (Merriam Webster 2021). Sometimes this is a location, interest, need, objective, etc., or all of these things at the same time. Urban communities in the broadest sense are the backbone of all human life in a city. We form communities, whether it's for public purposes, like technology communities, science, or creative/artistic communities, or for personal purposes like a group of friends, writing groups, amateur sports leagues, etc., because we are social creatures. We like to be in and among each other and we need this to be happy and healthy (Aristotle 1984; Plato 1968). Cities grow and thrive on the basis of that reality first and foremost (Nature Human Behavior 2018; Jacobs 1961). Without people and this social aspect for all our many needs and wants, we wouldn't have the city (Jacobs 1961). In fact, in many cities, the need for community continues to be the single most challenging obstacle to following the safety regulations that demand isolation and distance from each other as we continue to deal with the COVID-19 crisis. It's a very unnatural thing for us. Having to spend time alone without our families, and communities. It even made many other aspects of human life, such as mental health issues in our lives skyrocket because it is difficult to cope alone (WHO 2021; Chiappini 2020).

Yet, we are also faced with increased challenges when it comes to creating community and networking urban communities for climate resilience in cities because of the nature of the modern capitalist system.

Capitalism Versus Collaboration in the City

What we are facing now adays is a great social, ecological, and economic paradox between the ecological changes we need to adapt to, and the capitalist systems upon which all modern cities and nations depend (Klein 2014). On a large scale, this impacts all networks and systems in cities, but on a smaller scale it also impacts communities and individual people, as capitalism has had a profound impact on how we act and behave together. It has affected how we collaborate and how we build community. In the Wealth of Nations, Smith deciphers that capitalism is based on the idea that humans are inherently selfish, and that all our human behavior can be predicted to, at some point, boil down to this fact. So, by designing a system that works with that reality, we can help the human species flourish (Smith 2002). In many ways, this actually works for humanity. For example, capitalism created economies of scale that allow more people than ever to benefit from goods and services that were once reserved for hierarchies of people such as aristocrats and landowning classes. This gave many people greater space in society and access to increased agency and independence to live according to their interests, and not have to remain confined to the classes and castes they were born into via earlier generations (DuBoff 1990; Heilbroner 1993). Capitalism fueled innovations, progress, and industrialization that afford many the modern conveniences we enjoy today. Yet, we also know that capitalism causes great in equity (Du Bois 1923; Dubois 1935; Dubois 1939; Douglas 2019) drives unsustainable consumption and production (Marx 1977; DuBoff 1990; Heilbroner 1993), and distracts efforts to develop climate solutions because of its preoccupation with profit accumulation (Klein 2014). Yet, this is difficult to change, because capitalism operates according to an entrenched set of powers.

Power is and has been substantiated by many things in society. It can be sourced from beauty, status, money, intelligence, knowledge, etc. but at its core power is essentially the human capacity and ability of our faculties, or something that attaches or relates to them that we can then use and carry around with us throughout life to exert our will. This aids us in survival, distinguishing differences, and creating value among our peers (Heilbroner 1995; Foucault 1972). Like being social creatures, this we need to. Power under capitalism is profit. Under this system power is achieved by accumulating endless growth. Thus, more and more profit. Profit ensures economic survival, and political and social status as the valued attribute, and thus we go to great lengths to obtain it. For example, in business, the smaller the producer you are, the more you aspire to grow The bigger the producer you are the more you have to grow to protect what you have. The most important rule is that profit growth is infinite. So, the system inherently creeps into every facet of our lives to create space for profit. Somethings that generations before us would never have imagined could become products or consumables, such as intangible things like, identities, sound, or communities, or tangible things like water, land, soil, rain, sunshine, etc., are all things we have found a way to commercialize and capitalize on under capitalism. There is seemingly no limit, even in the face of vast environmental degradation.

This is part of what makes the Anthropocene such a historical moment in time because, at a planetary level, humans have actually managed to change, shift, and alter the course of planetary balance by the actions and decisions we make as we live due to our systems (Lenton 2015; Lenton et al. 2008; Rockström et al. 2009a, b; Scheffer 2009; Folke et al. 2016). Many of these actions are driven by the preoccupations of the capitalist system as it is currently. Thus, human action is both responsible for the climate crisis in having perpetuated this instability with our unsustainable systems, ever growing unsustainable consumption and production, urban sprawl, pollution, etc., yet as demonstrated through the work on networks, our climate solutions very much depend on the human capacity, ability, and willingness and to act together to change.

One persistent challenge at a social level is that capitalism has changed how we function and relate to one another in our human communities. If solutions are dependent upon our action, decision making capacities and behavior, then understanding and acknowledging capitalisms impact on these aspects of our lives is crucial. Hannah Arendt in The Human Condition argues that public space –the only spaces in cities and societies that people can come to be together – have long been filled with producers and not people. She reminds us that the power that holds the public market together and in existence is not the potentiality that springs up between people when they come together in action and speech, but a combined "power of exchange" which each of the participants acquired in isolation (Arendt 1998). This is essentially dehumanizing (Marx 1977; Heilbroner 1993). This dehumanization has ultimately caused a deep human frustration (Marx 1977). The more we live human lives where the focus is to consume, produce, and build profit the more frustrated we become, the more individualized, hyper-specialized, and the less oriented toward community and the other we become.

Our roles in society as the consumer and the producer can in many ways help us develop solutions for climate change, such as becoming an ethical, conscious or sustainable consumer or producer. We can collectively lower consumption and production, waste, pollution, etc., if we all do less of these things as conscious consumers, and consider how our products are made or our farms are farming as ethical producers. Yet, capitalism is still designed on the premise that we are selfish. It continues to shorten spaces for collaboration and drives us into divisive siloes due to how we must act, behave and decide, to function under it. Given the state of things, we will very much need human cooperation to solve the climate crisis. Thus, we need to start rewarding and embracing this ability and capacity to collaborate and build community among us that supports ecological sustainability and regenerate human nature connections.

Conclusion: Building Networks and Re-humanizing the City for Climate Resilience

Bringing the human back into the city may just be a big part of the social solution to our resilience problem. Coming together, not a 'togetherness' that Jacobs denounces as the sense that if anything is shared, all must be shared in a city (Jacobs 1992), but accordingly with her, an active relationship among societies members that extends from a nurtured sentiment that we are in this together, and we will need to work together (The Cooperative Human 2018). Under this framework, the goal toward greater and greater profit is a hinderance and at the very worst end of the spectrum, completely impossible. To be in effective communities together where we can count on togetherness as part of our climate mitigation and adaptation solutions –whether these be crisis administration, mutual aid, or resource sharing realities like we witness with water in the city of Cape town, these community efforts and city initiatives cannot be organized upon the goal of profit accumulation and remain sustainable and effective. Keeping the human species healthy and together in societies where we need and desire to be is a much more logical organizing principle. This human need and desire to form communities is one of our biggest strengths as a species. And this togetherness has produced incredible things over the ages. In art and culture, community is at the core of whole movements and works of art. Together we have pushed the boundaries of aesthetic, design, and imagination to invent and create wonderful things: bending, curving, sky-scraping infrastructure of all shapes and sizes; lush environments rich in color and texture; master plans and ideas that have formed regions, etc. Our species has done phenomenal things, and we are at a moment in time where we are being called upon to do what may be our greatest feat yet. Collaboration and effectively networked communities, organized by principles other than profit accumulation, and driven by respect and inclusion for earths ecosystems will be critical to achieving human resilience in cities.

References

Acuto, M. (2018). Global science for city policy. *Science, 6372*, 165–166.

Arendt, H. (1998). *The human condition* (p. 349). Chicago: The University of Chicago Press.

Aristotle. (1984). *The politics* (p. 284). London: The University of Chicago Press.

Bai, X., et al. (2016). Defining and advancing a systems approach for sustainable cities. *Current Opinion in Environmental Sustainability, 23*, 69–78.

Bai, X., McPhearson, T., Cleugh, H., Nagendra, H., Tong, X., Zhu, T., & Zhu, Y. G. (2017). Linking urbanization and the environment: Conceptual and empirical advances. *Annual Review of Environment and Resources, 42*, 215–240.

Batty, M. (2017). *The new science of cities*. Cambridge, MA: MIT Press.

Comunian, R. (2011). Rethinking the creative city: The role of complexity, networks and interactions in the urban creative economy. *Urban Studies, 48*(6), 1157–1179. https://doi.org/10.1177/0042098010370626.

Douglas, A. J. (2019). *W.E.B. Du Bois and the critique of the competitive society*. Athens: The University of Georgia.

Du Bois, W. E. B. (1923). The superior race (An essay). *The Smart Set: A Magazine of Cleverness, 70*(4), 55–60.

Du Bois, W. E. B. (1935). *Black reconstruction in America, 1860–1880*. New York: Henry Holt and Company.

Du Bois, W. E. B. (1939). The Negro scientist. *The American Scholar, 8*(3), 309–320.

Elmqvist, T., Andersson, E., Gaffney, O., & McPhearson, T. (2017). Sustainability and resilience differ. *Nature, 546*, 352.

Elmqvist, T., Bai, X., Frantzeskaki, N., Griffith, C., Maddox, D., McPhearson, T., ... Watkins, M. (Eds.). (2018a). *Urban planet: Knowledge towards sustainable cities*. Cambridge: Cambridge University Press.

Elmqvist, T., et al. (Eds.). (2018b). *Urban planet: Knowledge towards sustainable cities*. London: Cambridge University Press.

Elmqvist, T., Andersson, E., Frantzeskaki, N., et al. (2019). Sustainability and resilience for transformation in the urban century. *Nat Sustain, 2*, 267–273.

Folke, C., Biggs, R., Norström, A. V., Reyers, B., & Rockström, J. (2016). Social-ecological resilience and biosphere-based sustainability science. *Ecology and Society, 21*, 41.

Foucault, M. (1972). *The archaeology of knowledge: A discourse on language* (p. 245). New York: Pantheon Books.

Guterres, A. (2019). *C40 world mayors summit*. UNFCCC: Copenhagen. Public Talk. https://unfccc.int/news/guterres-cities-are-where-the-climate-battle-will-largely-be-won-or-lost.

Heilbroner, R. (1993). *21st century capitalism* (p. 175). New York: W.W Norton and Company.

Heilbroner, R. (1995). *The worldly philosophers: The lives, times and ideas of the great economic thinkers* (p. 365). New York: Simon and Schuster.

Hess, C., & Ostrom, E. (Eds.). (2011). *Understanding knowledge as a commons: From theory to practice*. Boston: MIT Press.

Jacobs, J. (1992). *The death and life of great American cities* (p. 458). New York: Vintage Books.

Keeler, B. L., Hamel, P., McPhearson, T., et al. (2019a). Social-ecological and technological factors moderate the value of urban nature. *Nat Sustain, 2*, 29–38.

Keeler, B. L., Donahue, M., McPhearson, T., Hamann, M., Donahue, M., Prado, K. M., Arkema, K., Bratman, G., Brauman, K., Finlay, J., Guerry, A., Hobbie, S., Johnson, J., MacDonald, G., McDonald, R., Neverisky, N., & Wood, S. (2019b). Social-ecological and technological factors moderate the value of urban nature. *Nature Sustainability, 2*, 29–38.

Kirby, P. (2013). Transforming capitalism: The triple crisis. *Irish Journal of Sociology, 21*(2), 62–75.

Klein, N. (2014). *This changes everything: Capitalism vs. the climate* (p. 566). New York: Simon and Shuster.

Lenton, T. M. (2015). *Earth system science. A very short introduction* (p. 153). Oxford: Oxford University Press.

Lenton, T. M., et al. (2008). Tipping elements in Earth's climate system. *Proceedings of the National Academy of Sciences of the United States of America, 105*, 1786–1793.

Lombardi, M., Lopolito, A., Andriano, A. M., Prosperi, M., Stasi, A., & Iannuzzi, E. (2020). Network impact of social innovation initiatives in marginalised rural communities. *Social Networks, 63*, 11–20.

Manville, B. (2016). Six principles for a networked community strategy. Forbes. Web. Accessed 1 Sept 2021.

N

Marx, K. (1977). *Capital Volume One* (p. 1141). New York: Vintage Books.
McDonald, R., et al. (2018a). Nature in the urban century assessment. https://doi.org/10.13140/RG.2.2.11429.14563.
McDonald, R. I., Colbert, M.'. L., Hamann, M., Simkin, R., & Walsh, B. (2018b). *Nature in the urban century.* Washington, DC: The Nature Conservancy.
McPhearson, T., Parnell, S., Simon, D., Gaffney, O., Elmqvist, T., Bai, X., Roberts, D., & Revi, A. (2016a). Scientists must have a say in the future of cities. *Nature, 538*, 165–166.
McPhearson, T., Pickett, S. T. A., Grimm, N. B., Niemelä, J., Alberti, M., Elmqvist, T., Weber, C., Haase, D., Breuste, J., & Qureshi, S. (2016b). Advancing urban ecology toward a science of cities. *Bioscience, 66*(3), 198–212.
Monstadt, J., & Schramm, S. (2017). Toward the networked city? Translating technological ideals and planning models in water and sanitation Systems in Dar es Salaam. *International Journal of Urban and Regional Research, 41*(1), 104–125.
Nagendra, H., Bai, X., Brondizio, E. S., & Lwasa, S. (2018). The urban south and the predicament of global sustainability. *Nature Sustainability, 1*(7), 341.
Peng, Y., & Bai, X. (2019). Scaling urban sustainability experiments: Contextualization as an innovation. *Journal of Cleaner Production*. 227:21.
Pflieger, G., & Rozenblat, C. (2010). Introduction. Urban networks and network theory: The city as the connector of multiple networks. *Urban Studies, 47*(13), 2723–2735.
Plato. (1968). *The republic of Plato* (p. 487). New York: The Perseus Books Group.
Rega-Brodsky, C. C., Nilon, C. H., & Warren, P. S. (2018). Balancing urban biodiversity needs and resident preferences for vacant lot management. *Sustainability, 10*, 1679.
Rockström, et al. (2009a). *Nature, 461*, 472–475.
Rockström, et al. (2009b). *Ecology and Society, 14*(2), 32.
Sassen, S. (2004). The global city: Introducing a concept. *Brown J World Affair, 11*, 27.
Sassen, S. (2013). *The global city: New York, London, Tokyo*. Princeton University Press.
Scheffer, M. (2009). *Critical transitions in nature and society.* Princeton University Press.
Shove, E. (2016). Infrastructures and practices: Networks beyond the city. In O. Coutard & J. Rutherford (Eds.), *Beyond the networked city: Infrastructure reconfigurations and urban change in the north and south* (pp. 242–258). London: Routledge.
Smith, A. (2002). *The wealth of nations*. Oxford, UK: Bibliomania.com Ltd.. Web. Accessed Sept 1 2021.
Taylor, P. J., & Derudder, B. (2015). *World city network: A global urban analysis*. Routledge.
The cooperative human. Nat Hum Behav. 2018 Jul;2(7): 427–428. https://doi.org/10.1038/s41562-018-0389-1.
Yeung, Y. M. (2020). *Globalization and networked societies: Urban-regional change in Pacific Asia*. University of Hawaii Press.

New Cities

Victoria Kolankiewicz[1], David Nichols[1] and Robert Freestone[2]
[1]Faculty of Architecture, Building and Planning, University of Melbourne, Melbourne, VIC, Australia
[2]School of Built Environment, University of New South Wales, Sydney, NSW, Australia

Synonyms

Company town; New town; Planned city; Satellite city; Satellite town

Definition

New cities are purposefully designed tracts or centers of urban development which are located beyond the periphery of an existing metropolitan area. They are typically planned and constructed to accommodate population growth with a view to address metropolitan primacy. The creation of new cities can reflect national or state-level planning policies. In some instances, they are also private sector-led real estate projects. New cities can also be purpose-built to accommodate a workforce in a settlement proximate to resources or industry. The opportunity to plan anew can provide opportunities to redress known urban ills, in addition to applying novel approaches to social and physical planning.

Overview

New cities or towns are a longstanding aspiration of modern urban planning globally in representing the best opportunity to realize social, economic, environmental, and design ideals unencumbered by the inherited constraints of legacy cities. The nature of the ideal community has evolved since the late nineteenth century but there are recurring preoccupations with normative notions of self-containment, amenity, sociability, density, mobility, and shape (Reiner 1963). Implementation has been more discontinuous and dependent on either regional, state, or national government policies to redistribute population for more "balanced" settlement, entrepreneurial moves to seek profit through new spatial opportunities for production and consumption, and public-private partnerships to share infrastructural and running costs. Just like early utopian thinking which they channel, the principles upon which new cities are designed and settled are responsive to the cultural, institutional, and urban conditions of the period in history which spawn them.

Broad themes underpinning the genesis of new towns can thus be identified internationally. European new towns of the 1920s and 1930s captured socialist ideals of reform and nation-building. British new towns of the 1940s sought to provide an increased quality of life amidst rapid population growth. In China in the 1950s, new towns were urban satellites linked to industrialization policies. In the 1960s in the United States, new settlements sprung forth primarily as speculation-driven real estate projects. In Australia, new town programs were developed by both federal and state governments in the 1970s to address the twin fears of metropolitan congestion and suburban sprawl. In the USSR in the 1980s, new town building continued apace to colonize underpopulated hinterlands in pursuit of a national settlement strategy. Clearly many 'new towns' are actually now quite 'old' and as they have aged they have had to address new economic, demographic and environmental issues not unlike those of the larger settlements they once challenged (Sies et al. 2019).

A recent global enumeration of twentieth and twenty-first planned century new towns (defined as urban places of at least 30,000 population) identifies a total of around 700 (Peiser and Forsyth 2021). Their categorization denotes the multiplicity of primary rationales: capital city, industrial, metropolitan deconstruction, military/research, reconstruction, resettlement, resort, and satellite. Many proposals have ultimately failed to thrive, with planners having failed to anticipate the complexities and limitations of forced or incentivized population dispersal. Nevertheless, globally there is a notable correlation between cycles of construction and the pace of urbanization, with the current frontier in fast-growing developing countries (also known as the Global South).

Early Extra-Urban Settlements

There were numerous historical precedents for an orderly decentralization of population centers located outside of existing cities. Initial forays came at the tail-end of the second Industrial Revolution when the effects of industrialization were being keenly felt by urban dwellers: density, pollution, and competing and conflicting land-uses threatened the quality of life of working-class citizens. This coincided with the emergence of the town planning profession. Drawing on the work of Graham Taylor, Douglass (1925) in the mid-1920s distinguished between these suburban spaces of production as industrial satellites versus more conventional dormitory communities.

Private sector interest arose in the creation of pleasant, healthful, and hygienic environments with the understanding that workers would benefit physically and ideologically from the beneficence of these spaces compared to overcrowded slums as the norm for working-class industrial families. Welfare capitalists' investments in environmental aesthetics paid dividends for a while. However even pioneering showpiece developments beyond the metropolitan fringe from the 1880s like the company town of Port Sunlight, established by the Lever Brothers outside Liverpool, and Pullman, south of Chicago, for a railway coach building business faced their fair share of industrial turmoil and worker disgruntlement over the cost of living (Rees 2012). Nevertheless, near the turn of the twentieth century, as-built environments, they captured the financial feasibility of comprehensive new community building (Creese 1966).

The Garden City

The garden city ideal was not the singular new city progenitor but nevertheless proved enormously influential (Ward 2016). Ebenezer Howard's garden city manifesto (1898) was an Anglocentric pathway to "a better and brighter civilization" through planned new communities embodying the best of urban and rural living with none of the disadvantages and in time developing into polycentric "social cities." Within commuting distance of London, both Letchworth Garden City from the Edwardian era and Welwyn Garden City post World War I are remarkable products of a private movement to demonstrate a new form of urban living in compact, walkable communities with ample provision of public open space woven through tracts of human-scale housing.

Both Letchworth and Welwyn served as formative influences for new cities, towns, and – much to the displeasure of the ardent garden city advocates – suburbs built in the interwar years in many countries. Robert A.M. Stern's et al. (2013) monumental *Paradise Planned* records some 900 of them, many at a city-scale. Many of these developments were seminal in their own right. The template for Radburn, New Jersey, defined "the garden city for the motor age" with an innovative plan by Clarence Stein and Henry Wright articulating independent circulation for pedestrians and vehicles with generous communal green spaces (Miller 2002, p. 18). While both Radburn and the garden city movement have long been eclipsed as cutting edge urban design, their inspiration is still traceable into innovative community plans today, many under the banner of New Urbanism (Talen 2005). Moreover, bodies such as the British Town and Country Planning Association descended from Howard's lobby group remain a vigorous advocate of neo-garden city planning at both the community and site planning levels, as does the Prince of Wales in both conceptual and practical ways such as Poundbury and Nansledan. The British Government has latterly endorsed the garden settlement concept as a sustainable and affordable urban development paradigm (Smith and Pratt 2017).

The Metropolitan Region

A fertile application of new town thinking is in the penumbral shadow of metropolitan areas. Here the intent was less to forge a radically new approach to urban development but to rationalize the inevitability of outward extension along tidier, more ordered, and liveable lines. By the mid-1920s, the satellite town was cast by British town and country planning advocates like Charles Purdom (1925) as a strategic compromise between a metropolitan dependence and securing some civic and corporate identity without succumbing to a suburban subtopia. Raymond Unwin was an influential theorist and policy shaper who transposed Ebenezer Howard's thinking into a kind of planet-and-moon urban morphology articulated by green belts in perpetuity (Unwin 1912).

These ideas took hold in Europe through the work of Ernst May and others in guiding the peripheral extension of cities. The most memorable expression of this thinking was Patrick Abercrombie's Greater London Plan of 1944 (Van Roosmalen 1997). One of the most famous applications was in Stockholm with new rail-based communities such as Vällingby dubbed ABC Towns, an acronym for *arbete*, *bo*, and *centrum*, the Swedish words for work, dwell, and center (Hall and Tewdwr-Jones 2019). These communities helped point the way toward the more recent formulation of transit-oriented development.

Of the satellite town projects successfully undertaken, anticipated scales of population and development were often not achieved or the development area was wholly absorbed into existing metropolitan areas. However, in some instances the original spatial vision if not strategic planning intent remains recognizable. Notable examples are the new towns designed and constructed around Paris since the mid-1960s, such as Cergy-Pontoise and Evry. These settlements have absorbed both population and employment growth to mitigate the problems of the French capital as a large monocentric urban region (Desponds and Auclair 2017). They have also avoided the social disadvantage and tensions evident in peripheral Parisian suburbs.

Purposeful New Cities

While politically centralized space economies such as Russia have conflated economic development and population redistribution through new city construction at national and regional scales, elsewhere the catalyst for non-metropolitan initiatives has come from opportunities presented by agricultural, mining, industrial, and defense-related development. These settlements have long delivered opportunities for planners to apply best-practice or emerging knowledge in a relatively blank slate setting. Lyon architect Tony Garnier helped to legitimize this activity with his 1904 ideal scheme for a *Cité Industrielle*.

In the United States, until stymied by the Great Depression then World War II, some corporations invested significantly in the design of company towns in isolated locations. Some of the same leading practitioners responsible for progressive metropolitan projects were engaged to avoid the disorder and grimness of spontaneous settlements (Crawford 1995). Two "model cities" from 1915 were Morgan Park built by US Steel and now a residential neighborhood of Duluth, Minnesota, and Tyrone, now a ghost town in New Mexico, designed by Bertram Goodhue. While the physical standard of living was enhanced in such settlements, their economic and social sustainability was usually compromised by narrow economic bases and paternalistic corporate governance.

In Canada, planned resource extraction and processing towns are found in regions where the interests of the mining sector and government coincide (Bone 1998). These differ greatly with respect to specialization, form, and overall sustainability. Kitimat, built in the early 1950s to house workers in an aluminum smelter plant in British Columbia, is a prime example of a community well-planned to counter its isolation. Clarence Stein's design sought to overcome the limitations of a "company town" in crafting a human-scale environment oriented around open space corridors, unimpeded pedestrian movement, neighborhood units, and a central core supporting commerce and civic life. Kitimat effectively realized the ambitions and qualities of Radburn "to a larger scale and in a frontier context" (Cross 2016, p. 11).

Regional Australia has its share of smaller planned communities linked to resource development opportunities (Freestone 2010). American expatriate Walter Burley Griffin who had won an international design competition in 1912 for the federal capital city of Canberra was also commissioned to lay out new and expanded towns as service centers for a major irrigation project in southern New South Wales. The coal-mining town of Yallourn in Victoria in the 1920s was modeled on Letchworth Garden City. The secretive town of Woomera captured advanced site planning ideas when opened in the 1940s to service a rocket testing range in outback South Australia. A network of corporate settlements, often with innovative designs not all well received in the harsh arid environment, accompanied the opening up of the massive iron ore deposits of Western Australia from the 1970s (Newton 1985). The prospect of a significant counter-urbanization trend was regardless dampened by mining companies reverting to temporary housing to service the preferred fly in-fly out employment model.

Remote new cities share one distinguishing factor: a life-cycle that inherently limits the scope for their continued use and re-use (Lucas 1971). Such towns also experience transferals of ownership and governance and suffer from over-specialization which constrains opportunities for a normalized demographic profile.

Speculative New Cities

Mainstream urban developers have been more enamored with new city opportunities within commuting distance of major metropolitan areas, either in response to official planning strategies for accommodating future population growth or by making the best of their own opportunities. The speculative end of the private enterprise spectrum has long been attracted by the potential of new cities to offer immense riches from previously undeveloped rural land. In Australia, there were early bold moves to leverage large "new city" areas adjacent to the apparent security of government enterprises in hitherto remote areas and resulting mostly in unrecoverable losses by small investors (Freestone and Nichols 2010).

In the absence of major direct national government investment in new cities in the Global North, the market has filled the gap. In the USA, more substantial enterprises like the 1960s Lake Havasu City in Arizona with its designer schooled in theme parks and expositions were similarly based on government infrastructure – it was located on the banks of a new dam – and, famously, incorporated the iconic London Bridge into its design as a "conversation starter" and tourist attraction (Wildfang 2005). The most substantial and acclaimed schemes dating from this era are large master-planned stand-alone communities such as Reston, Columbia, and Irvine, all impressive and popular as commercial ventures but lacking a strong foundation in public policies (Rapkin 1967). As environmental values were more consciously incorporated in such developments, pathways to more sustainable smart growth became evident (Forsyth 2005).

The Future Pace of New Cities

Today, new cities are still proposed and proceeding worldwide, envisioned to serve a multitude of purposes. The visions are updated for current and emerging new challenges notably climate change adaptation. Songdo International Business District, a privately-owned "smart city" completed in 2015 is a case in point; its urban form and functions align with sustainability principles, and intended to demonstrate the capacity of new cities to address contemporary crises and economic opportunities with responsive and conscious design. When seeking to pre-empt market demand, new cities serve as investment vehicles to further private sector financial interests, but often to the exclusion of populations already residing in regions earmarked to become new cities (Moser 2020).

Hundreds of new cities have been established globally in recent decades, generally comprehensively master-planned, and predominantly in the Global South. Countries such as India and China have committed to national settlement policies to address urbanization, promote economic development, and stimulate employment growth within a pro-market environment. New towns become practical instruments for managing the implications of a continuing and inexorable shift from rural to urban living. A growing concern in India is the rise of urban agglomerations outside any local authority administration (Sharma 2021).

In China, new cities planned at high densities are important components of growing articulated city clusters in metropolitan and regional settings (Bonino et al. 2019). When the pace of development has momentarily overtaken demand, "ghost cities" have resulted, revealing that one may plan and build an ideal city, but it takes careful strategizing to understand how to populate these places (Shepard 2015). Physical planning – of street layouts, public open space, zoning, and infrastructure – comprises the simpler portion of the equation. More complex are the drivers of human movement that underpin voluntary detachment from existing places to pursue a new urban frontier.

New city projects have also emerged across Africa in private sector-led ventures, with urban forms emphasizing qualities of containment and self-sufficiency (van Noorloos and Kloosterboer 2018). These projects anticipate significant population increase in addition to rural-to-urban migration, whilst simultaneously addressing the incapacity of government to deliver spaces for habitations. They may additionally invoke the use of a magnetic land-use or purpose to attract specific populations and investors, such as the knowledge-oriented Konza Technology City in Kenya or the extractive industrial city of Takoradi in Ghana.

Conclusion

New cities present a fascinating counterpoint to existing settlements. They are an opportunity to realize utopian ideals for living into practice, perhaps even discovering the limitations of best-practice concepts. They illustrate that magnetism – that is, practice or purpose – is a crucial force in enticing populations to abandon established settlements. This is underlined by the many stillborn schemes such as Stonehouse, the Scottish new town development to follow Milton Keynes, discontinued on the ascension of the Thatcher government to power in 1979 (Smith 1978) and Monarto, in South Australia, scuttled by crisis-inducing population projections found to be over-optimistic (Walker et al. 2015). Nonetheless, even projects that never really left the drawing-board can offer learnings and lessons for planning in practice.

There is no one master narrative beyond the planner's dream of the *tabula rasa*. They are, increasingly, vehicles for speculation and investment vulnerable to both gains and losses. They warrant analysis from a contemporary standpoint as cities grow and become denser. Should existing metropolises, with all their attendant weaknesses arising from organic growth, continue to receive investment and intervention – or is it better to begin anew? The timeless existential rationale for new towns remains; they are "the material structure of potentialities and aspiration, and the frontier between two worlds: the present and the future" (Wakeman 2016, p. 307).

Cross-References

► Growth, Expansion, and Future of Small Rural Towns

References

Bone, R. M. (1998). Resource towns in the Mackenzie Basin. *Cahiers de Géographie du Québec, 42*(116), 249–259. https://doi.org/10.7202/022739ar.

Bonino, M., Governa, F., Repellino, M. P., & Sampieri, A. (Eds.). (2019). *The city after Chinese new towns: Spaces and imaginaries from contemporary urban China*. Birkhäuser: Basel.

Crawford, M. (1995). *Building the Workingman's paradise: The Design of American Company Towns.* New York: Verso.

Creese, W. L. (1966). *The search for environment: The garden city, before and after.* New Haven: Yale University Press.

Cross, B. (2016). Modern living "hewn out of the unknown wilderness": Aluminum, city planning, and Alcan's British Columbian industrial town of Kitimat in the 1950s. *Urban History Review/Revue d'histoire Urbaine, 45*(1), 7–17.

Desponds, D., & Auclair, E. (2017). The new towns around Paris 40 years later: New dynamic centralities or suburbs facing risk of marginalisation? *Urban Studies, 54*(4), 862–877.

Douglass, H. P. (1925). *The suburban trend.* New York: The Century.

Forsyth, A. (2005). *Reforming suburbia: The planned communities of Irvine, Columbia and The Woodlands.* Berkeley: University of California Press.

Freestone, R. (2010). *Urban nation: Australia's planning heritage.* Melbourne: CSIRO Publishing.

Freestone, R., & Nichols, D. (2010). Town planning and private enterprise in early twentieth century Australia. *History Australia, 7*(1), 05.1–05.24. https://doi.org/10.2104/ha100005.

Hall, P., & Tewdwr-Jones, M. (2019). *Urban and regional planning* (6th ed.). London: Routledge.

Howard, E. (1898). *Tomorrow: A peaceful path to real reform.* London: Swan Sonnenschein.

Lucas, R. (1971). *Minetown, milltown, railtown: Life in Canadian communities of single enterprise.* Toronto: University of Toronto Press.

Miller, M. (2002). Garden cities and suburbs: At home and abroad. *Journal of Planning History, 1*, 6–28.

Moser, S. (2020). New cities: Engineering social exclusions. *One Earth, 2*(2), 125–172. https://doi.org/10.1016/j.oneear.2020.01.012.

Newton, P. W. (1985). Planning new towns for harsh arid environments: An evaluation of Shay Gap and Newman mining towns, Australia. *Ekistics, 311*, 180–188.

Peiser, R., & Forsyth, A. (Eds.). (2021). *New towns for the twenty-first century: A guide to planned communities worldwide.* Philadelphia: University of Pennsylvania Press.

Purdom, C. B. (1925). *The building of satellite towns.* London: J.M. Dent and Sons.

Rapkin, C. (1967). New towns for America: From picture to process. *Journal of Finance, 22*, 208–219.

Rees, A. (2012). Nineteenth-century planned industrial communities and the role of aesthetics in spatial practices: The visual ideologies of Pullman and Port Sunlight. *Journal of Cultural Geography, 29*(2), 185–214. https://doi.org/10.1080/08873631.2012.680816.

Reiner, T. A. (1963). *The place of the ideal community in urban planning.* Philadelphia: University of Pennsylvania Press.

Sharma, N. S. (2021). A tale of new cities: The main challenge before govt's aim to develop eight new cities. *Economic Times,* 20 September 2021. https://economictimes.indiatimes.com

Shepard, W. (2015). *Ghost cities of China: The story of cities without people in the world's most populated country.* London: Zed Books.

Sies, M.C., Gournay, I. & Freestone. R., eds. (2019). *Iconic Planned Communities and the Challenge of Change.* Philadelphia: University of Pennsylvania Press.

Smith, R. (1978). Stonehouse – An obituary for a new town. *Local Government Studies, 4*(2), 57–64. https://doi.org/10.1080/03003937808432733.

Smith, L., & Pratt, A. (2017). *Briefing paper on garden cities, towns and villages.* London: House of Commons.

Stern, R. A. M., Fishman, D., & Tilove, J. (2013). *Paradise planned: The garden suburb and the modern city*. New York: Monacelli Press.

Talen, E. (2005). *New urbanism and American planning: The conflict of cultures*. New York: Routledge.

Unwin, R. (1912). *Nothing gained by overcrowding*. London: Town and Country Planning Association. [2013 edition published by Routledge].

van Noorloos, F., & Kloosterboer, M. (2018). Africa's new cities: The contested future of urbanisation. *Urban Studies, 55*(6), 1223–1241. https://doi.org/10.1177/0042098017700574.

Van Roosmalen, P. (1997). London 1944: Greater London plan. In K. Bosma & H. Hellinga (Eds.), *Mastering the city: North-European town planning 1900–2000* (Vol. 2, pp. 258–265). The Hague: NAI.

Wakeman, R. (2016). *Practicing utopia: An intellectual history of the new town movement*. Chicago: University of Chicago Press.

Walker, P., Grant, J., & Nichols, D. (2015). Monarto's contested landscape. *Landscape Review, 16*(1), 20–35. https://doi.org/10.34900/lr.v16i1.864.

Ward, S. V. (2016). *The peaceful path: Building garden cities and new towns*. Hatfield: University of Hertfordshire Press.

Wildfang, F. B. (2005). *Lake Havasu City*. Mount Pleasant: Arcadia Publishing.

New Forms of Shared Governance and Local Action Plan in Socially Vulnerable Settlements

Carlos Leite[1], Angélica Tanus Benatti Alvim[2], Jörg Schröder[3], Andresa Ledo Marques[2,3] and Daniela Getlinger[2]

[1]Graduate Program in Architecture and Urbanism, Mackenzie Presbyterian University and Insper Arq.Futuro Cities Lab, Sao Paulo, Brazil

[2]Graduate Program in Architecture and Urbanism, Mackenzie Presbyterian University, Sao Paulo, Brazil

[3]Institute of Urban Design and Planning, Leibniz Universität, Hannover, Germany

Synonyms

Latin America peripheries; Local action plan; Slums upgrading; Social urbanism; Urban governance.

Definition

Transition peripheries: The entry presents two innovative dimensions in socially vulnerable settlements upgrading programs: first, the shared governance process which is developed with a strong presence of organized civil society through the articulation of a nongovernmental organization (NGO) network and the local community; second, a local action plan – Neighborhood Plan – developed with the community. The entry presents the case study of Jardim Lapena, a socially vulnerable settlement in the periphery of the city of Sao Paulo, Brazil, in terms of advances, limitations, and lessons learned, as well as discussions for potential replicability in other cities in the Global South.

Peripheries in Transition

The increase in socio-spatial inequalities, in many cities around the world, is becoming a major topic in scientific debate. Urban planning can be seen in a promising position to contribute to the evolving research in this context, not only through novel empiric evidence of the comprehensive access to urban life and urban space in concrete situations, but also by means of innovative methodologies in a "project-and development-oriented" approach of "activist researchers" (Scaffidi et al., 2019). In particular, a territorial notion of targeted urban planning projects in scale of neighborhoods not only extends to multiscalar implications in urban networks but also approaches challenges of climate change and social fragmentation with a "cosmopolitan mission" (Schröder et al., 2021), leading to a reformulation of concepts, models, and tools for urban transition to sustainability: (1) ecological transition to face climate change, "which refers to the process by which humans incorporate nature into society" (Bennett & Belshaw, 1976) and urban planning and design adaption as a social and spatial process; and (2) a sustainable and inclusive transformation that includes social, cultural, spatial, ecological, and economic aspects and their interrelation. Effectively, sectorial transition processes (in energy, industry, or agri-food) as well as mitigation and adaptation to hazards and changed nature in climate change are increasingly seen as deeply interconnected with social inclusiveness and collaboration to become effective as well as ethical.

In the comprehensive vision of urbanism, fragility, as a spatial manifestation of division and impact of nature, is to be tackled not only in an ecological and social dimension, but overall, also as a risk for cities as a living space. The right to the city manifests itself not only to the important provision of housing, basic services, and safety, but also the crucial possibility to participate and collaborate that facilitates a right to the future. Still, the share of urban population living in informal settlements and slums – marginalized and fragile urban areas – counts to one billion[1] and is constituting a major challenge for sustainability goals, as referred on United Nations Sustainable Development Goal 11 (UNSDG 11), to make cities and human settlements inclusive, safe, resilient, and sustainable (UN, 2015).

The focus on critical urban situations, as brought forward with this entry, involves a trans-scalar perspective on processes of spatial fragility and social inequality that not only affects specific areas, but also are a danger for the future of entire urban systems, urban life, and urban economy. Effectively, the concept of peripheries as a manifestation of marginalization can be set not only in a transversal perspective across different contexts (metropolitan, intermediate, and rural), in order to discover possibilities to transfer concepts and approaches, but also in a dynamic sense of changing interaction and linkages between different innovative spatial and social phenomena that offer relevant potentials for local sustainable development.

The marginal situations, as particularly promising fields to enhance own resilience and contribute to systemic transition, can be understood with Sennett's concept of liminal space that is at the borders of standard cultural and political perception: Porosity and borders "create liminal space; that is, space at the limits of control, limits which permit the appearance of things, acts, and persons unforeseen, yet focused and sited." In opposition to the attributes of closed systems – fixed shape, equilibrium, and integration – he characterizes liminal space even as crucial for an open system as "non-linear, and within that frame range from path-dependency to the patterns of chance" (Sennett, 2018), in interaction between spatial creation and social behavior that together form urban agency. Liminal space thus can become a place and a disseminator for innovation toward resilience – in many fields of community and knowledge building for urban action, e. g., for public space, water facilities, new models of social housing, use of renewable resources, sustainable mobility, new living-working models, circular economy, food cycles, etc.

[1] http://www.un.org/sustainabledevelopment/cities/

The Context

The patterns of rapid urbanization that occurred during the twentieth century have left over one billion people living in marginalized and fragile urban areas around the world, mostly within informal settlements and slums of developing nations. Now, in the early decades of the twenty-first century, the scenario is yet more somber, under the threats of potential adverse impacts of climate change, and more recently, the devastating impacts of the Covid-19 pandemic.

Similar to other countries in Latin-America, Brazil has presented intense urbanization process, mainly from the late twentieth century onward, when peripheries started to be densely occupied. Peripheries in large cities were occupied by houses built by their own householders, in illegal allotments and slums, mostly due to intense internal migration from North-eastern Brazil towards the South-eastern region of the country. The city of Sao Paulo – the biggest in South America with 12.3 million inhabitants – recorded 11.4% of the total population living in slums (IBGE, 2010).

The precariousness of urban space in Brazilian cities is allied to the struggles of social movements that have called, since the 1960s, for urban reform and the introduction of instruments aimed at reducing socio-spatial inequalities. It was only in the beginning of the twentieth century that the instruments for participatory processes were implemented, which originated in the Federal Constitution of 1988, known as the Citizen Constitution. Its regulation, however, only occurred in 2001, with the establishment of the City Statute Federal Law which enhanced the cities Master Plans together with the implementation of a set of social inclusion instruments and the Neighborhood Plans.

In Sao Paulo, the urban regulatory framework instituted in 2014, the Strategic Master Plan (PDE), is a fundamental part of promoting a more inclusive and sustainable city. In this context, stands out the Neighborhood Plan, a social instrument designed for a local scale that enables, mainly, the technical and political confrontation of conflicts that arise in the place, enabling aspects of social inclusion. However, the city had only one experience developed until the emergence of Jardim Lapena Neighborhood Plan in 2018, which is discussed in this entry (CEPESP/FGV & FTS, 2018).

Concerning slum-upgrading programs, the pioneering case in Latin America, the Favela-Bairro program, implemented in the city of Rio de Janeiro in the 1990s, did not have specific and integral local action plans or Neighborhood Plans developed with the participation of the local community, altogether with the technical support of the municipality (Leite et al., 2020).

Thus, the problem of qualifying the poor peripheries unfortunately remains far from being resolved or having consistent and long-lasting programs.

In essence, the problems are the following: (i) a strong number of slums and a housing deficit of more than 500,000 units in the city of Sao Paulo, the majority on the peripheries; (ii) a lack of local plans developed, despite the existence of the legal framework (Neighborhood Plan); and (iii) a lack of innovative local shared governance models (Fig. 1).

This entry presents the innovative case of Jardim Lapena territory, on the eastern periphery of the city, based on the two following significant criteria:

- The existence of an innovative governance model, based on the presence in the territory, for 12 years, of a robust NGO, the Fundacao Tide Setubal (FTS), which provided governance shared with the local community, an unprecedented factor in the city
- The development of a Neighborhood Plan in 2018, based on PDE urban instruments, another unprecedented factor in a slum in this context.

The Case Study Analytical Framework

The case study of Jardim Lapena was analyzed from three main axes, having the methodological support of the study of "Slum upgrading: Lessons learned from Brazil" (IADB, 2012), seeking to highlight three important points: evaluating the effective participatory dimension and the aspect of multilevel governance; assessing the advances and limitations of the Neighborhood Plan; and highlighting urban qualities and positive ecological effects under technical feasibility. The analytical framework developed is shown in Table 1.

Constituting an urban agency proposal, these three key dimensions for case study evaluation can be confirmed and extended further, as follows.

For the axis of governance and institutional aspects (1), urban agency needs to be built on the transition process built in interaction between bottom-up initiatives and associative networks – as strong points of true urban regeneration projects – with directing frameworks, policies, and public intervention that facilitate to extend projects into processes of change (Schröder et.al., 2021). "Collaborative networks, advanced models of city-making, and social innovation initiatives are expressed in a rediscovered adaptive capacity of urban communities and, as a sort of 'local resilience', involved in bridging the dramatic gap of public welfare by offering a view of sharing and circularity..." (Lino, 2018). Thus, public stakeholders, civic organizations, and economic actors need to develop coherent activities as common engagement and shared "spatial strategy" (Healey, 2007).

For the axis of social and participatory aspects (2), toward action-oriented projects, urban agency is shown to rely not only on narratives as analytical

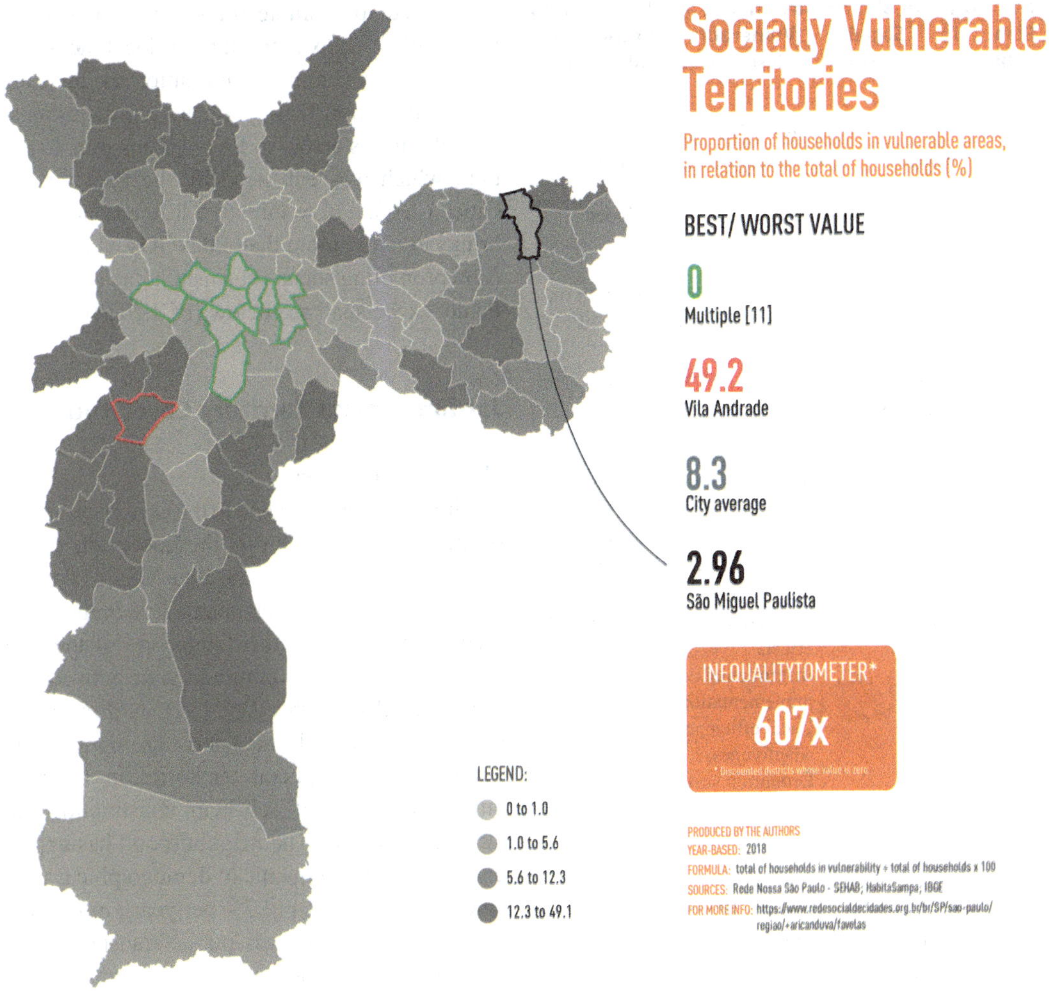

New Forms of Shared Governance and Local Action Plan in Socially Vulnerable Settlements, Fig. 1 Map of Inequality showing the proportion of slums in the total number of households, by districts of Sao Paulo city. Jardim Lapena territory is located inside the peripheral Sao Miguel Paulista District (Data from RNSP 2019)

tool, in order to grasp cultural resources and lay the ground to enhance social capital, but also on narratives as a projective tool (Salmon, 2007) implying a critique of scripts and indicators that lead to an "over-determination both of the city's forms and its social functions" (Matthey, 2014). Thus, narrative can be seen as a basis for action-oriented projects that are necessary as guidance to implement public policies effectively. Furthermore, trans-sectoral cooperation - within concrete and comprehensive urban projects - is crucial for interventions in fragile territorial contexts, bringing together multiple public policies, for example, social housing, public spaces, water sanitation, risk management, poverty, education, health, mobility, etc., and linking them to diverse civic initiatives involving multiple sectors (intersectorial social action).

For the axis of urban and physical aspects (3), a place-based approach has been proven to offer analytical access to specific forms of spatial and social capital, and foremost to their potentialities in a specific context, leading to place-based strategies that "consider economic, social, political, and institutional diversity in order to maximize both the local and the aggregate potential" (Barca et al., 2012) as well as giving feedback for the further development of the – necessary and facilitating – policy frameworks, since in a place-based perspective.

New Forms of Shared Governance and Local Action Plan in Socially Vulnerable Settlements, Table 1 Three axes of analysis for the Jardim Lapena case study

Axes	Elements
(1) governance and institutional aspects	Consensus, support, and mobilization Intersectorality developed Efficiency of multilevel governance Influence of local planning on the further development of planning frameworks
(2) social and participatory aspects	The neighborhood plan Local social capital strengthened Effective participatory dimension in planning and implementation Intersectoral social action Social work in support of interventions
(3) urban and physical aspects	Urban qualities and ecological effects Instrumental coherence in implementation Local efficiency and adaptiveness Economic sustainability

Source: prepared by the authors

The three key dimensions – in combination – are further referred to the concept of social urbanism (Leite et al., 2020), articulating a specific framework of urban policies and instruments in Sao Paulo.

The strategy of organizing the territory in a more inclusive way seeks to align local regulations and innovative urban instruments to concrete transformations in the territory, even if incrementally, which can be called here as a continuous and incremental process of "social urbanism." Social urbanism here is understood as a reformist urbanism, assuming the existence of a strong, fully standardized, and democratic State, which defines the rules for urban development with social participation and control, where there is a call for other private agents and civil society organizations act and assist in the construction of the city as a territory for collective, democratic, and inclusive use, with emphasis on policies.

The differential that is one of the pillars of the investigation of this entry is local governance established to promote transformation in urban vulnerable contexts through a local action of active participation by community residents and technical support that culminated in the articulation of various actors in favor of this transformation, which was enhanced by the Pact for Just Cities (PJC), a new initiative in the country that brings together more than 20 NGOs whose focus is on promoting social urbanism in the city of Sao Paulo, as we shall see below (PCJ 2020).

Jardim Lapena Territory Case Study

The Urban Context

Jardim Lapenais is located in the eastern periphery of the city of Sao Paulo in the floodplain area of the Tiete River, within the Sao Miguel Paulista region. Total 12,000 inhabitants live in the settlement also known as the "underside of the train line" to describe the confined area between the railway tracks and the Tiete river. Jardim Lapena presents the typical characteristics found in the landscape of peripheral territories in Brazilian cities, the result of self-help construction and land subdivisions. The neighborhood has experienced an intense dynamic of demographic growth since the early 2000s (2.7% per year), mostly due to its proximity to mass public transport, good road accessibility, proximity to the center of Sao Miguel's district, and its wide range of public facilities. This growth rate is much higher than those observed in the city in the same period, within the context of peripheries of Brazilian cities, that is, irregular occupation of distant and environmentally hazardous areas (flood-prone area in this case), by the poor population through self-help construction as the only form of family housing, due to the low production of affordable housing by local governments (Fig. 2) Getlinger (2021).

Governance and Institutional Aspects: Analysis Axis 1

The Jardim Lapena slum has some important differential elements in its form of social organization. Being a relatively small territory limited by physical barriers, it has been unable to expand

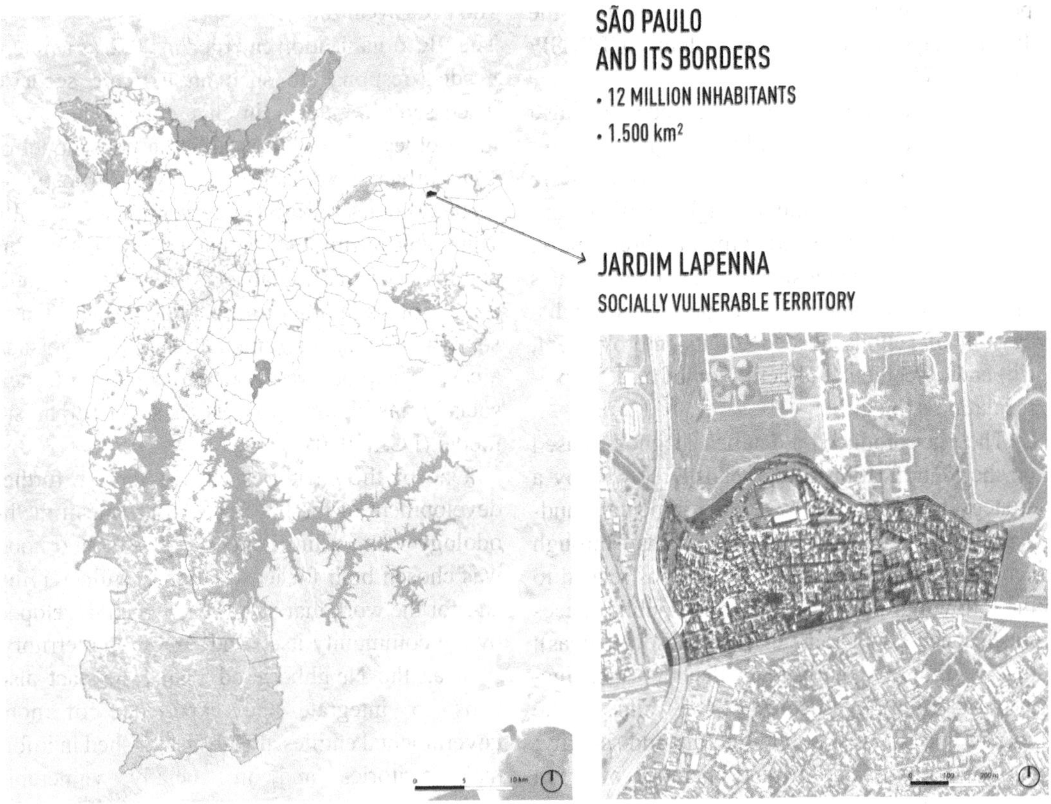

New Forms of Shared Governance and Local Action Plan in Socially Vulnerable Settlements, Fig. 2 Jardim Lapena and its urban context: territory of social vulnerability contained by built and natural barriers

N

physically, thus facilitating the emergence of the neighborhood community association, the first form of social organization in the area.

The arrival of a strong nonprofit institution, Fundacao Tide Setubal (FTS)[2] – with focus on child education – associated with the historical struggle for better living conditions in the area, strengthened the organization of the local community. One of the most important initiatives of this partnership is Galpao ZL, a place for training and actions that foster innovation, transformation, and entrepreneurship, which has been active for over 12 years (CEPESP/FGV & FTS, 2018).

This organization has managed to gradually achieve various collective community facilities over the past years: two Child Education Centers, a public school, a Basic Health Unit, a municipal reading center, and a Child and Adolescent Center.

The Neighborhood Urban Plan

In 2018, the local community got organized in order to tackle housing deficit, social, environmental problems through the Lapena Residents' Forum, a social mobilization group with a significant record of achievements. With FTS financial support, the forum hired a technical consultant from the Center for Policy and Economy of the Public Sector of the Getulio Vargas Foundation – important academic institution on public

[2]With over a decade of work, Tide Setubal Foundation (FTS) is an NGO that works to encourage initiatives that promote social justice and the sustainable development of urban peripheries and contribute to addressing socio-spatial inequalities in large cities, in articulation with various agents of civil society, research institutions, the State, and the market. https://fundacaotidesetubal.org.br/tide-setubal-foundation/

policies – to assist in the development of the Jardim Lapena Neighborhood Plan (CEPESP/FGV & FTS, 2018).

The construction of the Plan aimed at tracing a common future as well as urban design actions for a two square kilometer area. Priorities were defined by the community through a participatory process which revolved around three major moments: (i) diagnosis of the main issues affecting the neighborhood, (ii) proposals and discussion of actions needed for its improvement, (iii) and agreement and definition of basic implementation strategies.

The plan produced a detailed diagnosis based on the identification of demands, followed by a strategy to promote urban, environmental, landscape, and housing improvements through actions, investments, and interventions related to the following: mobility and safety; public spaces (open spaces and green, leisure, and social areas); macro drainage; garbage collection; cleaning, afforestation and gardening of sidewalks, public spaces, and squares; public lighting and security; and accessibility, supply, and operation of urban and social facilities.

It was organized into the four following challenges: to strengthen an active and effective community organization process; to promote environmental qualification of the neighborhood; to improve the walkability and connectivity of the routes, prioritizing road safety and the adequacy of open spaces focusing on early childhood: infants, toddlers, and caregivers; and finally, to protect and qualify existing facilities with actions related to infrastructure (basic sanitation) and use public facilities (turning it attractive and accessible) in order to promote decent living conditions for the residents (CEPESP/FGV & FTS, 2018).

It is important to highlight that the development of the Neighborhood Plan relied on a high level of community engagement, with several different views and perspectives. Another point is that while the medium- and long-term projects were elaborated and planned, there was an active mobilization of the population for short-term actions aimed at improving the neighborhood, such as cleaning and maintenance of public spaces.

The PJC Organization

The PJC organization emerged in 2020 as a much-needed response to such an extreme scenario discussed previously in this entry. The pact's main objective was to establish a new model of local urban governance and civic participation methodologies for urban interventions in socially vulnerable territories. For that, twelve NGOs, plus community and affordable housing organizations, joined forces in order to expand the reach of civil society organizations and their capillarity in such areas, in a pioneering strategy to manage their social transformation assets, a social urbanism model (PCJ, 2020).

One of the territories chosen for the further development of the Pact's social urbanism methodology was Jardim Lapena. The neighborhood was chosen both for its situation of vulnerability and for the work that had already been developed by the community itself and FTS in the territory, such as the Neighborhood Plan. The pact also aims to integrate the experience of non-governmental entities already established in informal territories and in socially vulnerable territories with actions of the formal public sector in those areas (PCJ, 2020).

The shared governance model establishes a thorough democratic participation environment that promotes wide community participation and places community demands in the center of the process, from the development of proposals through governance of public facilities. The expected outcome is the establishment and consolidation of good governance practices based on civil society in order to protect successful models and guarantee the continuation of such projects during local government transition periods. Therefore, the pact's role is to look beyond inequalities in multiple spheres and assist entities that already deal with local dynamics, thus superseding sector-oriented state-planning hierarchy. While integrated into official actions, this model gives voice to the residents in order to safeguard initiatives with transformation potential. It is widely understood that urban qualification projects cannot, and should not, occur abruptly, but through an incremental approach, collectively constructed and with metrics that start from the horizontal relations in these territories. So, in this model, a mix of concurrent bottom-up and top-down initiatives feedback each other through a progressive cycle. As in other contemporary cases, the local community collaborative workshop series has become an excellent instrument in the process of building trust as well as new forms of governance.

The objective of the pact was to combine social mobilization, technical support, and political articulation with the aim of improving the quality of life of the local community. For that, the PJC, the organization that has been in action since April of 2020, utilizes the following methodology: (i) the development of a participation methodology to evaluate the territories, (ii) the execution of a participation diagnosis, and (iii) the formulation of guidelines for the elaboration of projects to integrate sector policies (shared governance) (PCJ, 2020).

Discontinuity has been a constant and strong problem in the implementation of public policies in Brazilian cities. Due to this fragility of the public power, the PJC is based on the interpretation that the transformation of the territories cannot depend exclusively on municipal administrations; there needs to be a movement from civil society itself as a key agent and guardian of long-term projects. It is important to emphasize that the collaborative governance model does not invalidate the role of the State, but seeks to build some common patterns of action and a space for sharing and engagement of different actors in solving complex problems of the cities. In this sense, it is relevant to verify that after 1 year, the organization has grown with almost 30 NGOs currently active in the Jardim Lapena area. Moreover, other cities have increasingly demonstrated interest in replicating the model in their vulnerable territories.

Social and Participatory Aspects: Analysis Axis 2

Within the framework of the PJC, Jardim Lapena became one of the three territories chosen for priority action of the Sao Paulo City Hall. The choice of territories was made based on the technical diagnosis and participatory analysis of the areas. The technical diagnosis involved the construction of indicators for monitoring, context analysis, access to public services, quality of these services, and frequency of use. The participation analysis started with the identification and formulation of problems, challenges, practices, and opportunities of each territory by means of participation methodologies such as interviews, questionnaires, and conversation rounds.

The program is structured through the five following working groups organized under specific themes (PCJ, 2020):

- Social Participation: formulation of methodology and implementation of the participatory process, consultations with the population, systematization of proposals, and feedbacks
- Urban Design and Physical Intervention: diagnosis and formulation of urban interventions, project development, and implementation monitoring
- Improvement and Integration of Public Services: diagnosis and guideline development for improvement and integration of public services
- Shared Governance: development of shared governance models for planning and governance in the territory within the several NGOs from Pact for the Just Cities, FTS, and the neighborhood community association
- Indicators and Monitoring of Impact: development of models for evaluating the impacts of interventions

Urban and Physical Aspects: Analysis Axis 3

Social Urbanism Local Plan

The physical component of the Neighborhood Plan emerged in 2020 in the context of the PJC organization, to establish a pedestrian-oriented network between public spaces and facilities, that included specific urban design for the early childhood.

The project consisted of the implementation of a qualified pedestrian-oriented network

N

interconnecting existing social, educational, sports, health, cultural, and leisure facilities, as well as reclaimed residual existing open spaces in order to turn them into new public spaces for leisure, especially for children.

The two following complementary phases were developed:

- Phase 1: walkability, public spaces, and connectivity network (Fig. 3)
- Phase 2: open spaces and the Mutirao square (Figs. 3, 4 and 5)

A landscaped route interconnects main facilities that focus on early childhood. Two public squares located near schools and day care center were redesigned to meet the needs of children.

Traffic-calming strategies are also included in this project as an opportunity to meet the community's wishes to guarantee everyone the right to walk and enjoy sidewalks safely. Sidewalks have been expanded and redesigned, which allowed the creation of three new small squares, adding quality and connected public spaces to the existing Praca do Mutirao. All public and non-governmental facilities are currently connected through an integrated and fully refurbished route, with shaded places for resting, playing, and meeting along the way.

Finally, the elements analyzed in the three research axes in an integrated manner provided relevant actions:

- Monitoring the municipal budget-planning process, ensuring the necessary resources to make the actions agreed in the Neighborhood Plan feasible
- Training for community managers, including conflict mediation
- Monitoring the proposals of the Neighborhood Plan in a continuous and transparent way with the community

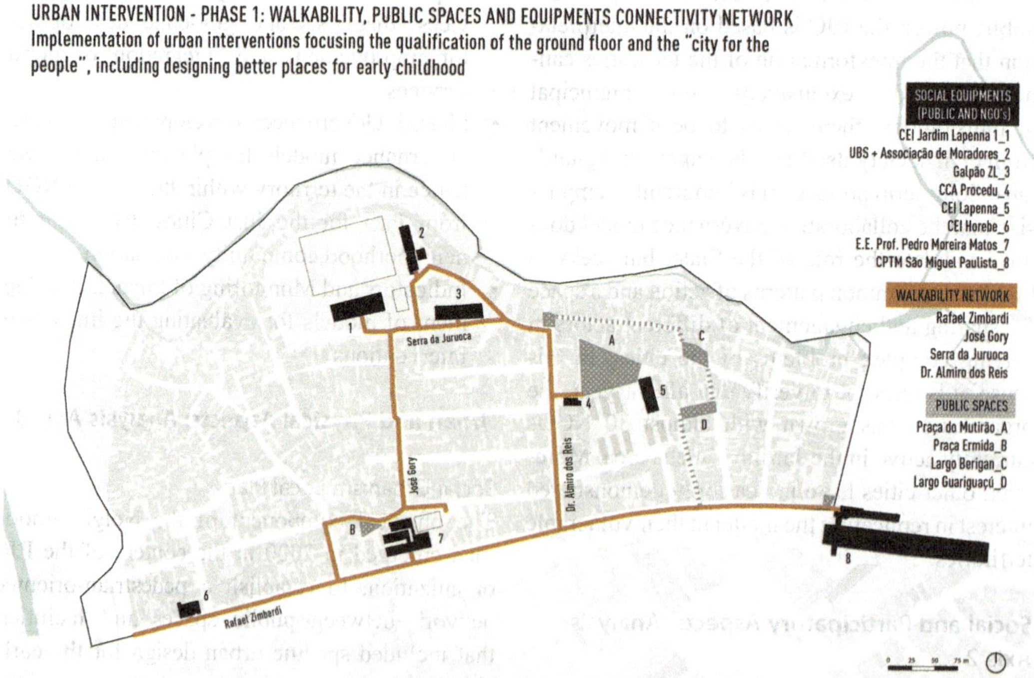

New Forms of Shared Governance and Local Action Plan in Socially Vulnerable Settlements, Fig. 3 Urban interventions, phase 1 and 2: pedestrian-oriented routes connecting public spaces and social facilities, and proposal for new traffic calming strategies and Mutirao square (PCJ, 2020)

New Forms of Shared Governance and Local Action Plan in Socially Vulnerable Settlements, Fig. 4 Urban interventions, phase 1 and 2: pedestrian-oriented routes connecting public spaces and social facilities, and proposal for new traffic calming strategies and Mutirao square (PCJ, 2020)

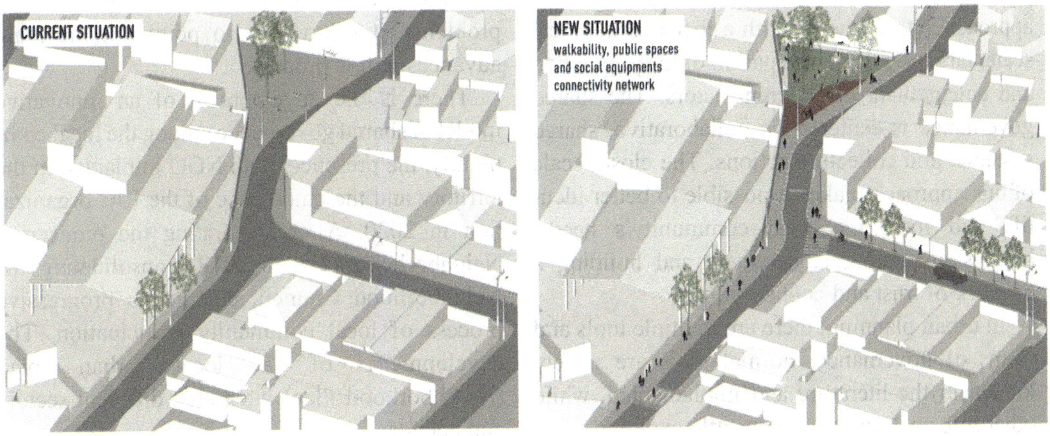

New Forms of Shared Governance and Local Action Plan in Socially Vulnerable Settlements, Fig. 5 Urban interventions, phase 1 and 2: pedestrian-oriented routes connecting public spaces and social facilities, and proposal for new traffic calming strategies and Mutirao square (PCJ, 2020)

Advances and Limits on New Forms of Shared Governance and Local Action Plan

There is a worldwide consensus that one of the biggest challenges of the contemporary urban world are territories of high social vulnerability where millions of poor people are left anonymously on the margins of society, especially in the cities of the Global South. Furthermore, it has been widely argued that coexistence in cities cannot be effective if socio-spatial disparities prevail.

In the context of decades of increasing informal settlements, several slum-upgrading approaches have been implemented in varying conditions in the Global South to improve the living conditions of these communities (Yeboah et al., 2021). Within the Brazilian context, there are examples of well-known slum-upgrading policies and practices, such as the Favela Bairro Program in Rio de Janeiro (Lara, 2013).

But, despite the success of these important cases in providing improvements, they did not have specific and integral local action plans or Neighborhood Plans developed within the local community participation altogether with the technical support of the municipalities. Besides, the programs suffer from discontinuity and interruptions, due to the Brazilian reality of discontinuity of public policies and lack of investments in peripheral areas. One of the major difficulties in ensuring continuity and improvements in slum-upgrading processes commonly pointed in literature is related to governance aspects.

Overcoming these challenges in socially vulnerable settlements implies in setting up innovative forms of shared governance and local action plan which are understood as decentralized approaches that emerge with a focus on the local scale, targeting the participation of the population and engagement of different actors. The shared governance is centered on collaborative, shared systems, and collective actions. The closer scale of this approach makes it possible to better identify and meet the local community's needs, besides generating engagement and building a network of trust and cooperation.

In urban planning, there are multiple tools and forms of governance. Formal tools are widely known in the literature and implemented, while other informal models, often little explored, continue to evolve. Articulating these multiple forms of governance can generate innovation and bring positive transformations to urban planning (Carmona, 2017).

In the Brazilian context, the case of Jardim Lapena shows a robust advance in innovation in relation to shared urban governance, mainly. This experience reflects how local aspirations for the improvement of living conditions can be addressed through a well-organized community and a robust presence of organized NGOs with two essential objectives: (i) development of a shared governance model and (ii) development of a local urban plan that integrates and territorializes policies and programs.

The Jardim Lapena community took the pioneering initiative to develop a Neighborhood Plan through a completely bottom-up movement. Therefore, the importance of the Jardim Lapena Neighborhood Plan is twofold: the construction of collective knowledge; and the development of clear and effective methodologies for the application of the instruments of the neighborhood plan. These two measures allowed the effective participation of the local community in the establishment of local urban policies and in the implementation of the urban plan.

The analysis of the ongoing process in Jardim Lapena points to the sustainable evolution of an innovative process of incremental social urbanism. The analysis on the three axes addressed in this work can be summarized in some key points of advances and limits faced so far.

Considering this case study as a reference and with potential lessons for other cities with similar problems, it is possible to point out the main advances in this process.

There is the development of an innovative model of shared governance among the local community, the presence of an NGO implanted in the territory and the emergence of the PJC organization in 2020. After elaborating the pioneering Neighborhood Plan in 2018, consolidating the Neighborhood Council, emerged a progressive process of local community participation. The development of a local urban plan (Neighborhood Plan) integrated different sectors and agendas.

The implementation of urban interventions since 2018 has continuously promoted the qualification of public spaces with a focus on pedestrian paths, safety, and better spaces for early childhood.

Incremental transformations are underway. It is known that slum-upgrading actions cannot occur abruptly, but rather through an incremental approach, built collectively and with metrics based on horizontal relationships in these territories. So, in this model, combined simultaneous bottom-up and top-down initiatives feed each other along a progressive cycle, and this process has taken place.

Despite these advances, there are still two complex structural challenges, both involving important forms of financing. There is a large housing deficit, including affordable rent: It is estimated that more than 1000 residential units are needed to be built for families currently living in unsanitary conditions, adjacent to a polluted stream and subject to flooding. And there is a demand for the implementation of drainage infrastructure to solve serious flooding problems; so far, microdrainage systems have been adopted in the urban intervention, phase 1 of the project.

The intention in the next phases of the project is to build affordable housing together with the basic infrastructure, articulating and integrating housing complexes with the fabric of the city, overcoming the erroneous tradition in Brazilian and Latin American cities of building them isolated from the city, without the desirable integration.

With the support of the FTS, the Neighborhood Council is currently articulating with various sectors of the municipal and state government to find structural and lasting solutions related to drainage infrastructure and, on the other hand, seeking to hire consultants specialized in popular housing systems to develop a plan composed of housing construction techniques combined with innovative financing models.

Final Considerations

In the cities of the Global South where the issue of inequality is already challenging, questions such as social inclusion, lack of access to land with basic infrastructure (water and energy supply networks, sanitation, public and social facilities, adequate public transport, etc.), and decent housing for all are greatly worsened by the current pandemic. In the context of decades of increasing informal settlements, several slum-upgrading approaches have been implemented in varying conditions in the Global South to improve the living conditions of these communities.

However, one of the major difficulties in ensuring continuity and improvements in slum-upgrading processes is related to governance aspects and to the necessity of better articulation between micro and macro policies in urban planning. Perhaps the key point of a successful and sustainable transformation is related to innovative forms of shared governance focused on the local scale, closer of the local community's needs, and targeting on the engagement of different actors in a continuous incremental process.

In this scenario, decentralized approaches emerge with a focus on the local scale, targeting the participation of the population and engagement of different actors, such as the so-called local governance that is centered on collaborative, shared systems and collective actions. The closer scale of this approach makes it possible to better identify and meet the local community's needs, besides generating engagement and building a network of trust and cooperation.

Some cases in developing countries demonstrate the importance of local governance for effective transformation of vulnerable territories, such as the case of Quezon City (Philippines), which, based on a bottom-up approach, relied on the active participation of NGOs and civil society groups to mobilize resources for slum upgrading, including funds from the World Bank and the Asian Development Bank, and the case of Kibera, in the city of Nairobi, Kenya, which used a hybrid approach from the partnership between the Kenyan Government, UN-Habitat, community groups, and a local NGO to transform the territory (Meredith et al., 2021).

The remarkable experience and accomplishments of the Jardim Lapena case study demonstrate the importance of strong local organized communities to effectively tackle problems in socially vulnerable territories. Not only there have been real spatial gains and improvement of public quality of life, but also it allowed for the community to quickly adapt to the new demands imposed by the pandemic. The experience of Jardim Lapena is based on the active participation of organized civil society and the fundamental role played by a place-based robust NGO for more than 12 years. This articulation between

N

the third sector, civil society, and municipality has culminated in a participatory process that transforms the local community itself into an agent of change for its territory, which in an organized manner leads, in conjunction with other agents, a process of struggle and demands for a more inclusive and sustainable territory.

This innovative movement gains strength and feedbacks on the PJC's initiative. These initiatives are essential for developing countries that have a significant part of their population living in precarious conditions. These spaces can become transition peripheries, as recommended in this entry. This transition takes place through the joint action of different agents, based on participatory methodologies, and also on local development, as shown in the case study.

The immense challenge of continuing slum-upgrading programs and actions in the cities of the Global South can receive new possibilities through the innovative model of shared governance, combined with a Neighborhood Plan and urban transformation that are consistent, participatory, dynamic, and constantly updated as in the case of Jardim Lapena territory.

Acknowledgments This work was supported by the CAPES PRINT to Mackenzie Presbyterian University under Grant 745884P.

References

(PCJ) Pacto pelas Cidades Justas. (2020). *Pacto pelas Cidades Justas*. https://cidadesjustas.org.br.

(RNSP) Rede Nossa Sao Paulo. (2019). *Map of Inequality 2019*. https://www.nossasaopaulo.org.br/wp-content/uploads/2019/11/Mapada_Desigualdade_2019_apresentacao.pdf.

Barca, F., McCann, P., & Rodriguez-Pose, A. (2012). The case for regional development intervention: Place-based versus place-neutral approaches. *Journal of Regional Science, 52*(1), 134–152.

Bennett, J., & Belshaw, C. (1976). *The ecological transition*. Oxford: Pergamon.

Carmona, M. (2017). The formal and informal tools of design governance. *Journal of Urban Design, 22*(1), 1–36. https://www.tandfonline.com/doi/full/10.1080/13574809.2016.1234338.

CEPESP/FGV and FTS (Fundacao Tide Setubal). (2018). *Plano de Bairro Jardim Lapena: A Rota Para Um Territorio de Direitos*. Sao Paulo: Fundacao Tide Setubal. https://fundacaotidesetubal.org.br/publicacoes/plano-de-bairro-jardim-lapenna/#boletim-modal.

Getlinger, D. (2021). *Plano de ação local comoelemento de integração e territorializaçãode políticaspúblicas: o caso do Jardim Lapena*. Sao Paulo: Tese de doutorado. UniversidadePresbiterianaMackenzie.

Healey, P. (2007). *Urban complexity and spatial strategies*. Oxford: Routlegde.

IADB (Inter-American Development Bank). (2012). *Slum upgrading: Lessons learned from Brazil. Edited by Fernanda Magalhaes and Francesco di Villarosa*. Washington: Inter-American Development Bank. https://publications.iadb.org/en/slum-upgrading-lessons-learned-brazil.

IBGE (Instituto Brasileiro de Geografia e Estatística). (2010). *CensoDemografico 2010*. Resultados. https://censo2010.ibge.gov.br/resultados.html.

Lara, F. L. (2013). Favela upgrade in Brazil: A reverse of participatory processes. *Journal of Urban Design, 18*(4), 553–564.

Leite, C., Acosta, C., Militelli, F., Jajamovich, G., Wilderom, M., Bonduki, N., Somekh, N., & Herling, T. (2020). *Social urbanism in Latin America: Cases and instruments of planning, land policy and financing the City transformation with social inclusion*. Springer. https://doi.org/10.1007/978-3-030-16012-8.

Lino, B. (2018). In-transition processes. In J. Schröder, M. Carta, M. Ferretti, & B. Lino (Eds.), *Dynamics of periphery. Atlas of emerging creative and resilient habitats* (pp. 96–99). Jovis.

Matthey, L. (2014). *Building up stories: Sur l'actionurbanistique à l'heure de la société du spectacle intégré*. Geneva: A-Type.

Meredith, T. et al. (2021). *Partnerships for Successes in Slum Upgrading: Local Governance and Social Change in Kibera, Nairobi* (pp. 237–255). https://link.springer.com/10.1007/978-3-030-52504-0_15.

Salmon C. (2007) Storytelling: La machine á fabriquer des histoires et á formater les esprits. Paris, La Decouverte

Scaffidi F., Lopez, F. M., Mottee L., & Sharkey M. (2019) The role of activist researchers in urban and regional planning. In: *AESOP 2019 conference proceedings* (pp. 311–323).

Schröder, J., Carta, M., Scaffidi, F., & Contato, A. (Eds.). (2021). *Cosmopolitan habitat. A research agenda for urban resilience*. Berlin: Jovis.

Sennett, R. (2018). *Building and dwelling: Ethics for the City*. New York: Farrar, Straus and Giroux.

UN. (2015). *Transforming our world: The 2030 agenda for sustainable development resolution adopted by the general assembly on September 25, 2015, A/RES/70/1*. United Nations General Assembly.

Yeboah, V., Asibey, M. O., & Abdulai, A.-S. J. (2021). Slum upgrading approaches from a social diversity perspective in the global south: Lessons from the Brazil, Kenya and Thailand cases. *Cities, 113*, 103164. https://linkinghub.elsevier.com/retrieve/pii/S0264275121000627.

The New Leipzig Charter: From Strategy to Implementation

Oliver Weigel[1] and Monika Zimmermann[2]
[1]Urban Development Policy Division at the Federal Ministry of the Interior, Building, and Community, Berlin, Germany
[2]Urban Sustainability Expert & Former Deputy Secretary General of ICLEI, Freiburg, Germany

Definition

The New Leipzig Charter is a key policy document on integrated urban development in Europe and beyond, prepared by the German EU Presidency in 2020 and adopted in November 2020 by all ministers responsible for urban affairs in EU member States.

Summary

Following the first Leipzig Charter from 2007, the New Leipzig Charter (2020) presents key principles for an integrated, sustainable and equitable urban development with *just, green, and productive cities*. The New Leipzig Charter also defines conditions for urban development following these principles, which include urban policy for the common good, integrated approach, participation and co-creation, multi-level governance, and a place-based approach.

The New Leipzig Charter is led by the subsidiarity principle, which points out that decisions should be taken by the lowest reasonable level of government and thus allocates relevant responsibilities to the local level.

Communities as a Driver and a Subject of Global Change

On November 30, 2020 under the German Presidency of the Council of the European Union, a policy document was resolved in order to strengthen the strategic, participatory and integrated urban development policies in Europe. Prior to this resolution by the ministers of the EU member states responsible for urban development policy, there had been an intense two-year working phase, with a discussion between governments and countless stakeholders on the proposed principles and recommendations for action.

The insight that global and local levels are now more closely interwoven than ever before was one driver along the path to a new Leipzig Charter. It is becoming increasingly clear that challenges such as climate change, scarcity of resources, the simultaneity of growth and contraction, migration movements, pandemics, and digitization have a direct and visible influence on our cities and towns. These challenges also harbor risk of new types of social, economic, and ecological conflicts and disparities in our societies. And at the same time, they offer new opportunities that need to be seized actively.

Urban areas will play a decisive role in the "century of the cities" when it comes to designing the future; they will be key drivers of the transformation in favor of sustainability, because they are at the same time drivers and subjects of global changes. This shows what responsibility cities and towns bear as regards the development of our societies the world over. It also highlights what opportunities arise from urbanization if it is the local level that emphatically defines it and gives it a sustainable thrust.

Needless to say, this insight is not new. In Europe, at the latest after 1990 both the economic and the socio-political framework conditions of spatial development changed completely. At the European, national, and regional levels, this transformation process created winners and losers. This tended to be measured in terms of core economic indicators and almost always resulted in massive migratory movements, the likes of which Europe had not seen since the end of World War II. At the latest after reunification, these trends of course ere especially intense in Germany.

This trend in Europe was intensified by globalization which, by dint of the technological boost

N

delivered by increasing networking and digitization of the world of work, impacted an ever-greater number of areas of everyday life. During this period, the central role of cities in actively meeting the challenges and design of the future of our societies became ever more apparent and there was no longer any way to ignore the fact that cities were unable to master these challenges alone. Urban development therefore increasingly came to be viewed as a task for all levels of government.

For this reason, Germany together with its European partners used the opportunity given by its 2007 Presidency of the Council of the European Union and slightly less than 3 years of the EU's eastwards expansion to draw up a pan-European document on urban development, and by resolving it at a meeting of ministers in Leipzig to strongly signal the importance of modern and responsible urban development. The "Leipzig Charter on the Sustainable European City" as resolved on May 24, 2007, defines the principles for a strategic, integrated and participatory urban development policy. In it, the member states that resolved it commit to realizing an integrated urban development policy in a multilevel government approach. The objective is to promote focused urban development programs in all member states, to describe the strengths and weaknesses of the cities and neighborhoods based on an analysis of the current situation, and to define consistent development objectives for the whole urban area. In the process, through an integrated approach the idea is to coordinate and spatially focus the use of funds by public and private sector players. This objective – and it was very strongly shaped by the favorable experiences Germany had made with promoting urban planning and the principle of subsidiarity – set high standards for future urban development policy in Europe.

Urban Trends and the Advancement the Leipzig Charter of 2007

Political, technological, social, and economic developments in Europe since 2007 have been extraordinarily dynamic. The finance and banking crisis led to distortions in the international financial markets, in the course of which the differences in income and wealth in part became massive, and that includes the trends between EU member states.

Today, the need to spare resources, the climate, and the environment is so obvious that it defines political debate in many countries. Realizing social and ethnic integration – and they need primarily to be addressed at the communal level – generates enormous tasks for cities and communities. At the same time, society in many European countries and indeed in many parts of the world continues to grow older, and this means our societies face even greater and new challenges – as the pandemic has just demonstrated with great acuity. And the statement that the very first iPhone came on the (German) market only 6 months after the Leipzig Charter was resolved presumably suffices to highlight just how technological challenges and potentials have evolved in urban development over the last 14 years.

Some things have also changed in the charter's "political environment". While the Leipzig Charter 2007 was still one of the very few international urban development documents, today a raft of international agreements influences urban development policy – world-wide, in Europe, nationally, and specifically "on site" in communities. Of note in this regard are the Sustainable Development Goals (SDGs) of the UN Agenda 2030 resolved in 2015 or the Paris Climate Protection Agreement (2016). The UN's New Urban Agenda resolved in 2016 in Quito (Habitat III) directly references urban development policy just as does the (European) Pact of Amsterdam with its Urban Agenda for the EU; it was likewise resolved in 2016.

In other words, there were countless reasons both why the Leipzig Charter needed to be advanced in substantive and political terms and for making its resolution a core aspect of the 2020 German Presidency of the Council of the European Union in the field of urban development.

The first work on the new document commenced in 2017, and from 2018 onwards a very broad participation process was developed and put into action. A total of 12 national and European meetings elaborated and discussed themes from June 2018 onwards and supported the actual work on the texts. Between 50 and 75 urban development experts from all state levels, including from the European Commission, science and research, associations, NGOs and clubs, took part in these national and European meetings. The results of the meetings were consistently inserted into the document after each meeting and were communicated and politically evaluated on an ongoing basis in the framework of both the committees of the German National Urban Development policy (the Board of Trustees, the National Urban Development Policy Working Party) and those entailing collaboration between the member states in Europe (meetings of the Urban Development Group, Director Generals and Ministers).

The New Leipzig Charter Describes the Transformative Power of Cities for the Common Good: Just, Green, and Productive Cities Are the Goal

The Leipzig Charter has grown, albeit less in size and above all in terms of the depth of the statements it makes. Initially it refers to the five basic principles of good governance: These are the principle of focusing on the common good, of participation and of co-creation, the integrated approach to urban development, the principle of multilevel governance and the place based approach.

These basic principles are applied to the three dimensions of the city: the just, the green, and the productive city. These are central fields of sustainable action and need to be thought of as one. Digitization is a (dominant) cross-cutting topic in this context with immense impacts on the other dimensions, and a purpose in itself.

The just city: It guarantees equal opportunities and justice for all, irrespective of gender, socioeconomic status, age, and origin. The just city does not exclude anyone. It offers every single person the opportunity to integrate into society.

The green city: It contributes to the battle against global warming and to high environmental quality as regards the air, water, and earth, and to a sustainable use of space. This also includes access to green and leisure zones, climate-neutral energy supplies, the use of renewable resources, implementation of energy efficiency measures, and climate-constant and CO2-neutral buildings – alongside many other characteristics of a green city.

The productive city: It rests on a broad-based business community that generates jobs and a sound financial bedrock for sustainable urban development. As attractive, innovative, and competitive business hubs, cities require skilled labor, social, technological, and logistics infrastructures, as well as affordable free space. An innovation-friendly environment that creates opportunities for local and regional production is also key.

The above-mentioned tasks need to be mastered in all the city quarters. For that reason, the New Leipzig Charter has abandoned focusing solely on disadvantaged urban quarters and now focuses with its strategical approaches on the quarter, the community, and its functionally interwoven city/surrounding regions. In line with their respective function, all these levels require a place-based approach. In many respects, they are interlinked and interdependent. Strategies that do not consider all the spatial dimensions run the risk of not being sufficiently focused (Fig. 1).

The New Leipzig Charter defines the preconditions that need to be fulfilled in order to pursue a successful urban policy. Cities require sufficient financial and personal resources as well as an appropriate legal position in the sense of subsidiarity.

Five key principles of good urban governance should define urban development policy in Europe:

Focus on the common good: Communities should act in the interest of the general population and therefore make available services and infrastructure geared to the common good. These should be inclusive, affordable, safe, and available to all. The services and infrastructure geared to the common good include healthcare facilities, social services,

N

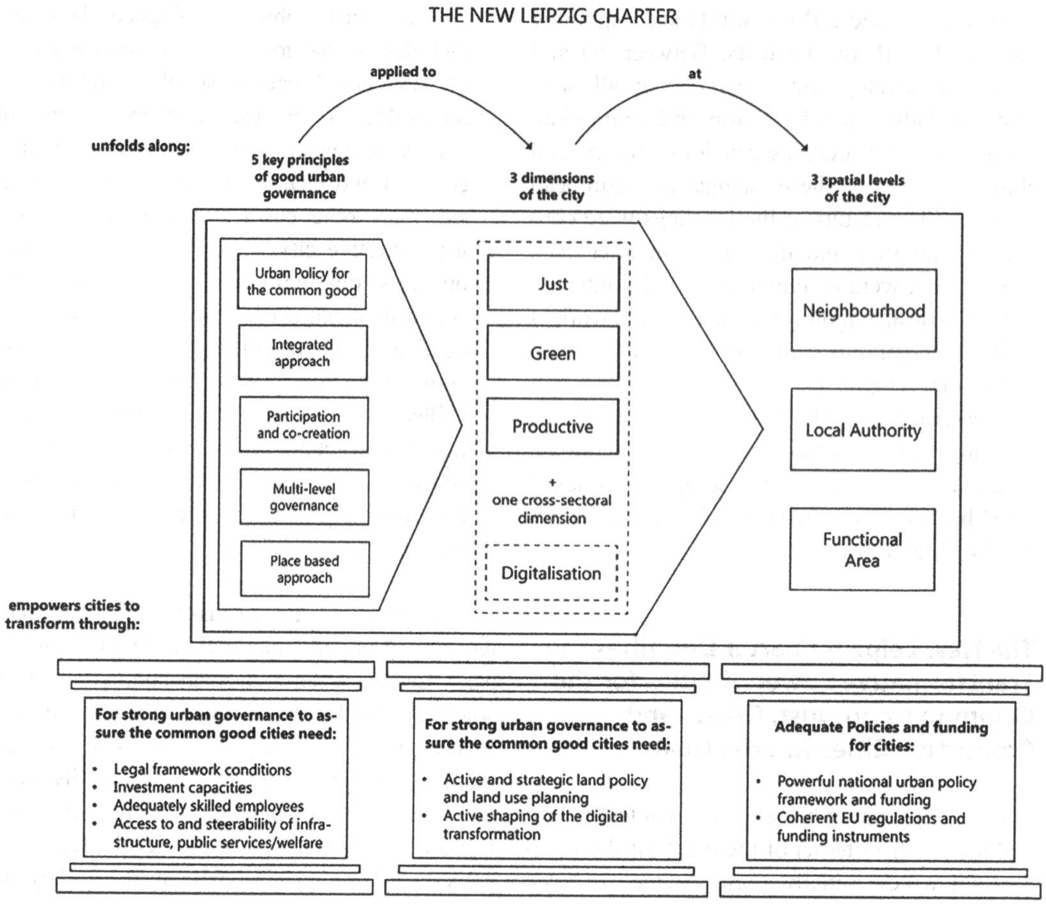

The New Leipzig Charter: From Strategy to Implementation, Fig. 1 Schematic representation of the New Leipzig Charter

education, cultural offerings, housing, water and energy utilities, waste management, public local transport, and digital information and communications systems. Also important are high-grade public spaces as well as green and blue infrastructures, likewise the preservation and revitalization of our built heritage *("Baukultur")*. Good urban development policy is capable of harmonizing public, business, and private interests.

An integrated approach: All fields of urban development policy should be coordinated in spatial, sectoral, and temporal terms. This integrated approach is based on the simultaneous and justly balanced consideration of all concerns and interests of relevance to urban development. For this reason, it should bundle the different, in part contradictory interests and align them. Cities should compile integrated and sustainable urban development concepts.

Participation and coproduction: The integrated approach calls for the involvement of all economic players, the broad general public, and other stakeholders to ensure their concerns and specialist knowledge is onboarded. A public involvement in urban development processes should include all urban actors. This also serves to strengthen local democracy. Citizens should be given a voice wherever urban development processes impact their daily lives.

Multi-level cooperation: All levels of the political administration – local authorities, regional bodies, metropolitan councils, and the national, European, and global levels – are in some way or other responsible for the future of our cities. The foundation for this is provided, among other things, by the principle of subsidiarity, according to which the competence for regulation should always be "as low as possible and as high as necessary."

Place based approach: It enables urban transformation from within and reduces local socio-economic differences.

The principles formulated in the New Leipzig Charter and designed to guide sustainable urban development are closely bound up with the practices of urban development policy and serve to promote the latter. For example, an active and strategic policy on land and soil as well as land use planning are especially important, as in many cities there is only a limited amount of surface area available, and this frequently triggers conflicts of interest. Communities need sustainable, transparent, and just strategies for land use and soil policy. These include land ownership and the management of land use by the local authorities. Important in this context are:

Polycentric settlement structures with appropriate densities and compactness in urban and rural areas: Among other things, the objective must be to reduce the routes from homes to work to leisure time, training/education facilities, stores, and services. This minimizes traffic volumes inside and between cities and reduces the need for mobility, which in turn dampens sprawl; as a result, areas dedicated for transportation can then be scaled back.

Fostering cross-administrative and cross-border cooperation: In this way, city/country relationships are factored into the equation to pre-empt and prevent sprawl.

Reducing spatial footprints: The focus will be on renewal and the comprehensive revitalization of urban districts in order to limit areas covered in asphalt/concrete.

Offsetting spatial use: It means that areas asphalted/concreted over or otherwise transformed must be offset by other areas where the city's biodiversity flourishes, fostering climate-neutral, robust, and eco-friendly urban development and improving air quality.

Designing and managing safe public spaces: Places that are freely accessible to all citizens and offer them a healthy environment in which to live.

Mixed urban spaces: Places that are characterized by mixed usages. As a result, new forms of production and enterprises are supported and a green, creative, service-oriented economy initiated.

These "basic pillars" lay the foundations for the New Leipzig Charter. They must be implemented through national urban development policies. In order to strengthen cities' ability to act, states must consider urban development policy a task to be discharged by government as a whole (Fig. 2).

From Document to Lived Practice . . .

Neither the "old" nor the New Leipzig Charter was designed to be a program of clear actions. Its objective was and is to improve the framework conditions for successful, sustainable, integrated, and participatory urban development. For this reason, one decisive step in improving the framework conditions for successful urban development in the respective EU member states was when the ministers committed themselves to

The New Leipzig Charter: From Strategy to Implementation, Fig. 2 Adoption of the New Leipzig Charter on November 30th, 2020, State Secretary Anne Katrin Bohle

implementing urban development policies of their own in line with the Charter's principles.

The Charter's success is expressed, in other words, initially and above all in the national urban development policies that the member states have introduced. Moreover, another objective is to prompt the European Commission to gear their instruments to the aspects of a strategic, integrated, and participatory urban development policy using a multi-level approach.

. . .in Germany

With the introduction of Germany's *National Urban Development Policy* as a joint initiative by the Federal government, the states, and local authorities, in 2007 the country took a far-reaching step toward a strategic, integrated, and participatory urban development policy.

One objective of this national urban development policy is to consistently advance the instruments of urban development. These instruments include systematic support for urban planning by the Federal and state governments, support for experimental formats for innovative action by communities, inculcating "good practices," and a conscious exchange of knowledge between the worlds of science, research, administration, and politics. As a result, urban development policy at all levels can swiftly respond to challenges and current opportunities can be exploited. In this respect, the annual federal congresses of the National Urban Development Policy, the University Open Days, and Round Tables are all now firm components of urban development practice in Germany – and beyond (Fig. 3).

With its diverse instruments, the National Urban Development Policy provides an innovation space in which the New Leipzig Charter can be realized in Germany. It champions the complexity of the city, its sub-spaces, and the way it is embedded in a regional context. Both focus strongly on an alignment to the common good, transparency, close links to citizens and civil society, and local participation. And.

. . . in the EU

In the other member states that have traditionally possessed strong urban development policies (such as France or the Netherlands), the self-commitment triggered by the Charter has led to said policies being taken forward. The process that prompted the "old Leipzig Charter" persists to this day. For example, in 2015 Poland introduced its National Urban Development Policy (see on this: BMVBS 2012 and BBSR 2017). Romania is currently working on a concept that is destined to translate the principles of the New Leipzig Charter into a national multi-level approach.

In terms of the cooperation between the member states, with the Pact of Amsterdam resolved during its 2016 Presidency of the EU Council, the Netherlands created a key milestone along the route to a European culture of urban development (see on this: EUKN 2017). And the European Commission itself has, by adding the "Urban Dimension" to the Structural Funds, likewise made an important contribution to realizing the objectives of the Leipzig Charter.

. . . and Outside Europe

One insight of recent years is that it no longer suffices to construe urban development in national or European terms. Rather, the unabated continuation of urbanization world-wide and the increasing social and ecological challenges, above all in developing, emerging, and transformation countries, creates immense political pressure to act. Both political control and planned management often lag behind rapid urbanization in these regions. The administrative actors involved often are not sufficiently networked and lack the integrated approaches necessary to combat the complex set of issues involved. In order to handle the structural challenges worldwide, the UN's New Urban Agenda defines a series of actual possible global solutions: developing and realizing urban policy measures at all levels, strengthening urban management

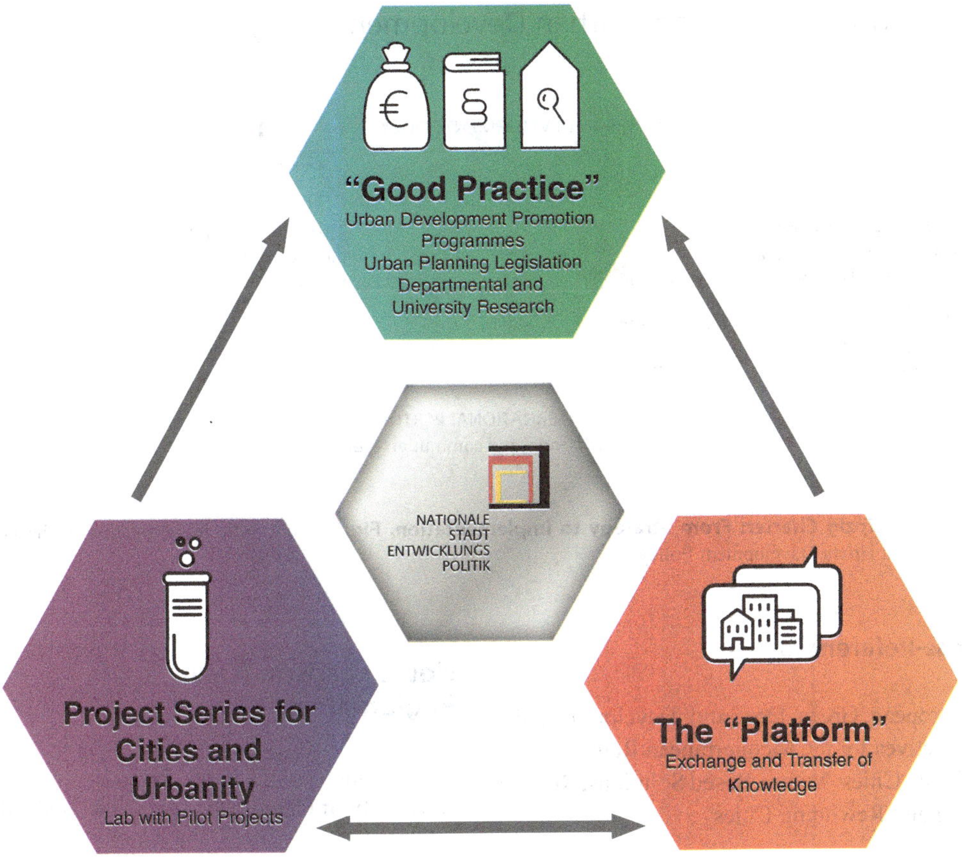

The New Leipzig Charter: From Strategy to Implementation, Fig. 3 The three pillars of the National Urban Development Policy

structures and processes, promoting long-term and integrated urban and spatial planning, and supporting an effective, innovative, and sustainable financial framework.

It is here that German practices can and should provide an international example. In the form of integrated urban development, the German system relies on an instrument that can support national governments and local authorities in pursuing the principles of the New Urban Agenda and the SDGs in the local, spatially adapted setting. And the international exchange of procedures and methods as well as joint learning together with partners from other cultural and institutional contexts effectively renews urban development practices inside Germany and ensures they are advanced on an ongoing basis (Fig. 4).

It is always decisive in this regard to ensure that the local citizens participate in developing the ideas and reaching decisions. In the spatial context, integrated urban development focuses on optimally linking and coordinating different usages and interests, and thus supports SDG 11 with its focus on sustainable cities and communities.

Urban development is a permanent task. That has always been the case, and the considerably faster cycles of economic, social, and ecological developments today also make this apparent to the broader general public. This is an opportunity to draw greater attention to this important political field and in this way arrive at enduring sustainable solutions for our cities.

The New Leipzig Charter: From Strategy to Implementation, Fig. 4 Schematic representation of the German International Urban Development Policy

Cross-References

- ▶ European Green Deal and Development Perspectives for the Mediterranean Region
- ▶ Green Cities: Nature-Based Solutions, Renaturing and Rewilding Cities

References

English text of the Charter: https://ec.europa.eu/regional_policy/sources/docgener/brochure/new_leipzig_charter/new_leipzig_charter_en.pdf

Background Information

https://www.nationale-stadtentwicklungspolitik.de/NSPWeb/EN/Initiative/Leipzig-Charter/leipzig-charter_node.html

https://ec.europa.eu/regional_policy/en/information/publications/brochures/2020/new-leipzig-charter-the-transformative-power-of-cities-for-the-common-good

https://www.europarl.europa.eu/RegData/etudes/ATAG/2020/652020/EPRS_ATA(2020)652020_EN.pdf

In German: https://www.nationale-stadtentwicklungspolitik.de/NSPWeb/DE/Initiative/Leipzig-Charta/Neue-Leipzig-Charta-2020/neue-leipzig-charta-2020.html

European Knowledge Network (Ed.) (2017). *One year pact of Amsterdam – EUKN Report*. The Hague.

New Localism: New Regionalism

Sean Connelly and Etienne Nel
University of Otago, Dunedin, New Zealand

Synonyms

Agency; Geography; Governance; Place; Scale; Space; Territory

Definitions

New localism	Localism initially emerged in the nineteenth century as a local state response to local level development challenges and needs. In the late twentieth century it re-emerged, in a new political-economic context in which local places, as a result of the "hollowing out of the state" from the 1980s and neoliberalism, were obliged to engage in locally driven economic and social responses to

	prevailing crises (after Goetz and Clarke 1993). This in turn encouraged recognition of the role of unique locality or place-based economic action.
New regionalism	New regionalism emerged as a response to the demise of Keynesian economics to understand the way that processes of regionalism and regional development occurred in the context of a devolved state. As a result, greater attention is placed on the endogenous characteristics of regions, their resources, capacities, and social capital, in shaping regional outcomes. Based on a new regionalism analysis, regional futures are driven less by top-down government interventions and result more from the negotiated intersection of diverse institutions, networks, and partnerships within regions and beyond.

Introduction

Space and place are, in many ways, the foundation of the discipline of Geography. Despite the often elusive definitional nature of these terms, their basic premises have helped guide and shape how and at what levels geographers have engaged with the world through their applied and theoretical research (Johnston 1991; Massey 1994; Hubbard et al. 2002). Space and place variously exist in absolute terms, as points of reference, or as elements which are socially constructed and which have agency (Agnew 2011). Invariably, the nature of that engagement has led to questions pertinent to another of geography's core tenants, that of scale, and it is with references to the scalar dimension of research into place and space that conceptual framing has evolved around the applied and theoretical notions of localities and localism, and regions and regionalism. This chapter seeks to engage with these latter concepts.

Both localities and regions exist as definable geographic and political entities, but they have also been accorded agency and relativity. They exist in a conceptual sense with a high degree of fluidity regarding their meaning and associations. Interpretations have varied from seeing them as locations with fixed coordinates to that of being constructs which shape and, in turn, are shaped by social relations, society, and power dynamics (Massey 1984; Cox and Mair 1991), to newer interpretations associated with the politics of power and neoliberalism in the governmentality sense, as espoused by "new localism" (Jones 2019) and "new regionalism" (MacLeod 2001). According to Clarke (2013, p. 493), localities are defined by the "the scale or range of experiences, defined by (the) day-to-day activities of people in the ordinary business of their lives." Similarly, regions and regionalism provide a response to the universalizing claims of globalization based on a particular mobilization of cultural, economic, and political sub-national divisions based on (re-) territorialization, identity and belonging (Tomaney 2017).

While we can analyze localities and regions and the processes occurring within them at a theoretical level, they also have a high degree of association with government and governance structures. Over the course of human history, political processes have oscillated between questions of local autonomy and central control. At an applied policy level, as Jones (2019) argues, there has been an on-going policy debate and pendulum shifts in the UK, in particular, over the last 200 years from support for regions and regionalism to support for localities and localism. This reality both complicates and enlivens the interest in these topics as the negotiation between place and power creates and refines understanding of these terms and their associations, as the following sections will illustrate.

Localities, Localism, and New Localism

While the terms "local" or "location" have been used in common English for centuries, from the

1980s, "localities" became a defined sub-field of interest for geographers interested in "understanding the process of socio-economic restructuring and the role of place and spatial variation within it" (Gregory 2009, p. 425). This built on an earlier interest in the local and local action (i.e., what was termed "localism"), a practice which sought to address urban socio-economic needs, which had emerged in the nineteenth century. Localism was a response to deprivation and the need for places to be sustainable with "municipal local states" emerging with a focus on issues of local water, sanitation, and taxation (Hogan and Lockie 2013; Beel et al. 2017; Jones 2019). This local policy response persisted until the rise of the welfare state. The demise of the latter from the 1970s helped catalyze a renewed interest in localities and the potential of localism.

Foci of interest in the concept of localities from the 1980s have included the study of the nature, causes and consequences of the spatial differentiation which localities reflect, and how they have changed over time. Nuanced understandings of localities emerged, with Cooke (1989, p. 3) detailing that at one level localities are a "descriptive term for the place where people live out their lives," but they are also "the sum of local energy and agency . . . they are not passive or residual but, in varying degrees centres of collective consciousness." Massey's (1984, 1994) work on the social and economic processes which interact in localities was critical in helping to develop an understanding of the key links between place, people, and power within localities. Over time, debates became increasingly refined, recognizing, as Jones (2019) clearly articulates, the importance of issues such as absolute and relative space, processes of "locality making," and associations with uneven development. From a theoretical perspective, divergences in understanding emerged regarding whether localities and the "localism" they were associated with should be seen as a political expression of the spatial division of labor and political culture, or as a natural way of life and expression of place, or as a place of struggle to defend local autonomy (Clarke 2013).

Cox and Mair's (1991) work on the notion of "localities as agents," helped shape a growing recognition of what was then regarded as "*new* localism" in which local places, as a result of the hollowing out of the state from the 1980s and neoliberalism, were obliged to engage in locally driven economic and social responses to prevailing crises (Goetz and Clarke 1993). This, in turn, encouraged recognition of the role of unique locality, or place-based economic action, led by local leaders which has variously referred to as "urban entrepreneurialism" (Harvey 1989) and Local Economic Development (Blakely 1989). These actions can be variously understood as either a response to capitalism's hegemonic and destabilizing tendencies or as a triumph of market-based individualism. Whatever the outcome, processes of globalization, neoliberalism, a shift from a social welfare to a market focus in many countries, new public management, and what Clarke (2009) terms as the "new politics of scale" helped accord new attention to "localities" and the economic and social processes either occurring within them or devolved to them, i.e., "new localism." In parallel, growing interest in the role of cities and city regions in a globalized economy, has drawn attention to local processes and the place of the "local" in the "global" (Beel et al. 2017).

This understanding of localities and new localism was contested from the 1990s with authors such as Marvin and Guy (1997) pointing out that new localism was grounded on a partial understanding of the local environment, which reified the city and created "myths" about the potential benefits of locally driven action. In a similar vein, Brenner and Theodore (2002) have pointed to the limitations and vulnerabilities of local action in an age of deregulation and hence have challenged "new localism" as being elitist and ambiguous. In reviewing the recent applied history of "new localism," Haughton and Allmendinger (2013) commented that since the 1990s, particularly in Britain, it has been used as a justification by politicians to convert local development successes into "recipes" for local economic restructuring and devolution, despite the limitations of such approach, not least the failure to consider contextual factors which impact development outcomes. In order to be more effective, there have been calls for advocates of localism to recognize and engage with participatory governance (Evans et al. 2013a) and to avoid the post-political associations which the concept can engender.

In recent years, interest in localities and localism has resurfaced, in part in response to state thinking in the EU and the UK as detailed below, but in part also by continued academic interest in localities as exemplified by the call made by Jones and Woods (2013) to "return to locality" and for a "new localities" research approach with an emphasis on processes of "locality making." Added impetus and interest in localities, from a policy and political perspective, came from the Barca Report (2009) for the EU which argued the case for "place based development" leading to a greater interest in place sensitive interventions and support from EU development agencies.

Within UK politics, recent attachment to the concepts of devolution and localism, albeit with a very defined political agenda, have given increasing prominence to localities and the processes occurring in them. The passage of the Local Government Act of 2011 endorsed a local growth agenda, building on the 2010 Local Growth White Paper (Clarke and Cochrane 2013; Jones 2019). The entrenchment of this agenda has led to the recognition of what, in the literature is now referred to as "*New* New Localism" (Jones 2019). This has led to a situation in which, according to Collins (2016), both "optimism and anxiety" now shape urban policy in an era of neoliberalism and austerity. This has meant that "localities" and "localism" do not exist just as theoretical constructs employed by geographers, but, in a very real way, the concepts have been co-opted by politicians to justify and enact policies of neoliberal management and devolution. Moving on from an initial belief that localism was about local action to address local needs (Hogan and Lockie 2013), it now, in a political and applied sense, according to Evans et al. (2013b, p. 405), refers to "the devolution of power and/or functions, and/or resources away from central control towards front-line managers, local democratic structures, local institutions and local communities," with a strong emphasis on partnership formation between the local state and the private sector (Jones 2019). The critique from Australia is that, in practice, while localism can lead to bottom-up decision-making, local empowerment, and resourcing, localism, pursued with the goal of self-sufficiency, can simply lead to "regulatory dumping, responsibility shifting, (and) under resourcing" (Hogan et al. 2014).

As a net result, while localities exist in geographic space as places of local action, identity, and change, from a political perspective localism, or more specifically its new hybrid "new new localism" has evolved from independent local action to a situation of devolution to "empower" local leaders with associate economic and political overtones. As Jones and Woods (2013) argue, academic interest in localities and localism waned prematurely in the 1990s, and these concepts remain valid, albeit contested in both applied and theoretical geography, justifying continued interest in the concepts.

Regions, Regionalism, and New Regionalism

In much the same way as the conceptual understanding of localities and localism has evolved over time, so too have understandings of the next higher scalar construct, that of regions and regionalism evolved. As noted below, at one level, regions can be seen as static entities often with resource and political determinants. That said they can also be seen as spaces with agency and fluidity which leads to an engagement with the concept of regionalism (Tomaney 2002). Interest in regions has, in no small measure, been encouraged by their continued recognition and national or supranational support for them, not least from the EU (Barca 2009).

Debates about regionalism are dependent on definitional constructs of "the region" and how the purpose and process of regional development occurs in the context of uneven development at a sub-national scale (Hodge and Robinson 2013). The scope and the placement of boundaries that differentiate one region from another or other scales are tied to identity and belonging, and these constructs provide meaning to material and symbolic spaces (Hudson 2007; Tomaney 2002). As such, regions and regional thinking are not inherent, but rather are imagined and created, based on jurisdictional boundaries, internal profiles, or notions of difference. As Peck (2002) notes, the way regions are defined, and by whom, are the outcome of multi-scalar processes that are shaped by preexisting economic activities, normative values about "leading or lagging" and assemblages of identities, cultures, experiences, and aspirations. Thus the region can be defined as a heterogeneous contested space consisting of cross-cutting institutional agencies, partnerships, and interests (Allen and Cochrane 2007).

The purpose of the region and the role it plays in processes of economic development has also shaped conceptual and applied debates that have evolved over time. In the colonial Empire building period, the purpose of regions were shaped primarily by environmental conditions that served as the basis for investment or extraction decisions (Paasi et al. 2018). The region served as a functional and sectorally defined space for the purpose of resource extraction from distant places for home markets. These cores and peripheries were the basis of Innis's staples theory of development in the Canadian context (Watkins 1963). During this period, regional thinking was dominated by identification of and support for key resources and industries as the primary driver for development decisions. Investments in people and infrastructure were made primarily to facilitate resource and industry needs (e.g., company towns, transportation infrastructure, etc.) that reflected Fordist development patterns (Moulaert et al. 1988). As such, regions were seen as passive spaces, subject to lock-in and path dependence. Regions were the site where development happened, primarily driven by state intervention, to support the development and exploitation of resources and industries (Scott and Storper 1992). This notion of the "region as container," persisted throughout the twentieth century, as states intervened to support regional economies by either deliberately placing functions in underperforming regions or through smoke-stack chasing attempts at attracting investment and industry to regions to support developmental agendas (Markey et al. 2005).

The demise of Keynesian economics and the reduced role of the state required a new way of thinking about the processes of regulation of economic activity at the regional scale (MacLeod and Jones 1999), as shifts from productivist, to post-productivist and to multifunctional regional economies (Wilson 2001, 2009; Argent 2002, 2011; Woods 2007) unsettled and fragmented the role and purpose of the region. In the midst of economic restructuring, there was increased recognition that regionalist interventions were needed to address the negative impacts of the roll out and roll back of neoliberalism (Peck and Tickell 2002), while others noted how factors endogenous to regions were playing a more important role in shaping regional competitiveness (Kitson et al. 2004). Recognition of these processes and related challenges led to calls for the analysis of regional development policies and programs to be more socially embedded in local spaces of production and consumption (Rossi 2004). As such, "*new regionalism*" emerged as a response.

New regionalism can be characterized by a shift in focus away from resources and industries, with greater emphasis placed on the way that the internal characteristics and meanings of regions shape and are shaped by activities beyond the region. From a new regionalism perspective, the region is no longer a passive recipient, but rather an active site where issues of local economic development, community development, quality of life, social capital can all play a role in promoting, maintaining or altering regional outcomes (Keating 1998; Jones and MacLeod 1999). The reduced role of the state has led to greater devolution, resurgence in territorial identity, and greater attention to place-based regional development opportunities (Paasi et al. 2018). However, the initial efforts to account for new regionalism were critiqued for not sufficiently accounting for the relational components of power, ideas, concepts, and resources across regions and scales (Paasi 2002) and the failure to place regional change within the broader political context characterized by state retrenchment (Lovering 1999).

For example, Lovering (1999) identified the problem of policy capture, based on critical debates about how the "region" was used by powerful actors at a range of scales to rearticulate and re-conceptualize policy interventions to suit their needs. In their view, new regionalist approaches failed to account for the way that external forces were still the primary actors in shaping regional development outcomes. However, for Markusen (1999), the problem with new regional approaches was their limited policy engagement and the lack of a developmental agenda for regional policy that suggested that regional policy development was too focused on economic efficiency (and a broader neoliberal agenda) and less on the more normative goals of equity, democracy, social and spatial justice, and human rights.

Rossi (2004) addressed some of these critiques in their review of the Mezzogiorno region of Italy, by highlighting how paying greater attention to power and spatial justice in regional analysis can illustrate how regional economies in depressed areas are potentially much more competitive and dynamic at the local level. According to Rossi (2004), the task of regional development policy is to activate and mobilize local capacities and resources through collaboration between the local state and regional organizations that explicitly acknowledge the social and spatial inequality that persists in order to maximize endogenous development potential. A critical new regionalism was called for to question how loose and temporary agglomerations of regional activity often present a veneer of institutional support, what MacLeod (2001) refers to as a mask of social capital or institutional thickness that serves to explain buy-in to regional futures.

In response to these critiques, more recent approaches to new regionalism place greater attention on how different perspectives, conflicts, and tensions come together to shape regions and regional thinking. As such, regions are recognized as being fluid, rather than static and shaped by dynamic relations across places and scales (Markey 2011; Zirul et al. 2015), and how regions are networked across multiple spatial, scalar, and territorial geographies. Rather than privileging one reading of the region, regions and their development opportunities are shaped by the intersection of the functional components of past iterations of the region with reference to path dependence and comparative advantage and with context specific endogenous factors of local institutions, local capacity, and local resources (Markey et al. 2006). It is the intersection of place-based and extra-local factors that provide opportunities for thinking about regional policy development that is contextually specific, yet informed by broader assemblages of social, economic, environmental, and political relationships across scales (Vodden et al. 2019).

As such, these approaches to new regionalism highlight the fuzziness of the way regions are defined, how boundaries are delineated and how their purpose is articulated based on a "contingent 'coming togetherness' or assemblage of proximate and distant social, economic and political relationships" (Jonas 2012, p. 263). In so doing, more recent regional analysis highlights how changing power relations, institutions, and diversified markets reduce the emphasis on a single resource as a defining characteristic of a region and open space for more diverse actors to play a greater role in shaping regional futures. These new assemblages of regional activities, values, norms, and functions (Lewis et al. 2013) lead to a more open, less bounded, less fixed, and less path dependent understandings of regions and regionalism, one that is more open to diversified ways of defining the region and its purpose.

Conclusion

As the above discussion suggests interest in and understanding localities and regions, and the parallel processes of localism and regionalism have both evolved and waxed and waned over time. These have been areas of on-going debate, not least because of the political overlay with which they are associated and linked notions of identity, agency, and power. The concepts of new localism and new regionalism provide insight into broader debates in geography, by drawing our attention to the processes by which place(s) are bounded and/or unbounded. There remains active debate about the degree to which they valorize place and local and regional actors in the face of powerful global processes. However, they provide useful insights into the way politics are embedded in scalar analysis and action in understanding the way local and regional futures play out, for and by whom. As Jones and Woods (2013 intimate, these concepts remain of value and are deserving of further attention.

Cross-References

► Public-Private Partnerships: The Danish Way of Turning Climate Change Measures into Policies and Long-Term Commitments

References

Agnew, J. (2011). Space and place. In J. Agnew & D. Livingstone (Eds.), *Sage handbook of geographical knowledge* (pp. 316–330). Los Angles: SAGE.

Allen, J., & Cochrane, A. (2007). Beyond the territorial fix: Regional assemblages, politics and power. *Regional Studies, 41*(9), 1161–1175.

Argent, N. (2002). From pillar to post? In search of the post-productivist countryside in Australia. *Australian Geographer, 33*(1), 97–114.

Argent, N. (2011). Trouble in paradise? Governing Australia's multifunctional rural landscapes. *Australian Geographer, 42*(2), 183–205.

Barca, F. (2009). *An agenda for a reformed cohesion policy: A place based approach to meeting European Union challenges and expectations*. Brussels: DG Regio.

Beel, D., Jones, M., & Jones, I. R. (2017). *City-region building and Geohistorical matters. In reanimating regions*. London: Routledge.

Blakely, E. (1989). *Planning local economic development: Theory and practice*. Newbury Park: SAGE.

Brenner, N., & Theodore, N. (2002). Preface: From the "new localism" to the spaces of neoliberalism. *Antipode, 34*(3), 341–347.

Clarke, N. (2009). In what sense 'spaces of neoliberalism'? The new localism, the new politics of scale, and town twinning. *Political Geography, 28*(8), 496–507.

Clarke, N. (2013). Locality and localism: A view from British human geography. *Policy Studies, 34*(5–6), 492–507.

Clarke, N., & Cochrane, A. (2013). Geographies and politics of localism: The localism of the United Kingdom's coalition government. *Political Geography, 34*, 10–23.

Collins, T. (2016). Urban civic pride and the new localism. *Transactions of the Institute of British Geographers, 41*(2), 175–186.

Cooke, P. (1989). *Localities: The changing face of urban Britain*. London: Unwin.

Cox, K. R., & Mair, A. (1991). From localised social structures to localities as agents. *Environment and Planning A, 23*(2), 197–213.

Evans, M., Stoker, G., & Marsh, D. (2013a). In conclusion: Localism in the present and the future. *Policy Studies, 34*(5–6), 612–617.

Evans, M., Marsh, D., & Stoker, G. (2013b). Understanding localism. *Policy studies, 34*(4), 401–407.

Goetz, E. G., & Clarke, S. E. (Eds.). (1993). *The new localism: Comparative urban politics in a global era*. SAGE.

Gregory, D. (2009). *The dictionary of human geography*. Chichester: Wiley Blackwell.

Harvey, D. (1989). From managerialism to entrepreneurialism: The transformation in urban governance in late capitalism. *Geografiska Annaler: Series B, Human Geography, 71*(1), 3–17.

Haughton, G., & Allmendinger, P. (2013). *Spatial planning and the new localism*. London: Taylor and Francis.

Hodge, G., & Robinson, I. M. (2013). *Planning Canadian regions*. Vancouver: UBC Press.

Hogan, A., & Lockie, S. (2013). The coupling of rural communities with their economic base: Agriculture, localism and the discourse of self-sufficiency. *Policy Studies, 34*(4), 441–454.

Hogan, A., Cleary, J., & Lockie, S. (2014). *Localism and the policy goal of securing the socio-economic viability of rural and regional Australia*. London: Routledge.

Hubbard, P., Bartley, B., Fuller, D., & Kitchin, R. (2002). *Thinking geographically: Space, theory and contemporary human geography*. London: Continuum.

Hudson, R. (2007). Regions and regional uneven development forever? Some reflective comments upon theory and practice. *Regional Studies, 41*(9), 1149–1160.

Johnston, R. J. (1991). *Geography and geographers*. London: Edward Arnold.

Jonas, A. (2012). Region and place: Regionalism in question. *Progress in Human Geography, 36*(2), 263–272.

Jones, M. (2019). *Cities and regions in crisis*. Cheltenham: Edgar Elgar.

Jones, M., & MacLeod, G. (1999). Towards a regional renaissance? Reconfiguring and rescaling England's economic governance. *Transactions of the Institute of British Geographers, 24*, 295–313.

Jones, M., & Woods, M. (2013). New localities. *Regional Studies, 47*(1), 29–42.

Keating, M. (1998). *The new regionalism in Western Europe: Territorial restructuring and political change*. Cheltenham: Edward Elgar.

Kitson, M., Martin, R., & Tyler, P. (2004). Regional competitiveness: An elusive yet key concept. *Regional Studies, 38*(9), 991–999.

Lewis, N., Le Heron, R., Campbell, H., Henry, M., Le Heron, E., Pawson, E., Perkins, H., Roche, M., & Rosin, C. (2013). Assembling biological economies: Region-shaping initiatives in making and retaining value. *New Zealand Geographer, 69*(3), 180–196.

Lovering, J. (1999). Theory led by policy: The inadequacies of the 'new regionalism' (illustrated from the case

of Wales). *International Journal of Urban and Regional Research, 23*(2), 379–395.
MacLeod, G. (2001). New regionalism reconsidered: Globalization and the remaking of political economic space. *International Journal of Urban and Regional Research, 25*(4), 804–829.
MacLeod, G., & Jones, M. (1999). Reregulating a regional rustbelt: institutional fixes, entrepreneurial discourse, and the 'politics of representation'. *Environment and Planning D: Society and Space, 17*(5), 575–605.
Markey, S. (2011). *A primer on new regionalism. Canadian regional development: A critical review of theory, practice, and potentials* (pp. 1–8). Retrieved from http://cdnregdev.ruralresilience.ca/wp-content/uploads/2013/03/primernewregionalism-markey.pdf
Markey, S., Pierce, J., Vodden, K., & Roseland, M. (2005). *Second growth: Community economic development in rural British Columbia*. Vancouver: UBC Press.
Markey, S., Halseth, G., & Manson, D. (2006). The struggle to compete: From comparative to competitive advantage in Northern British Columbia. *International Planning Studies, 11*(1), 19–39.
Markusen, A. (1999). Fuzzy concepts, scanty evidence, policy distance: The case for rigour and policy relevance in critical regional studies. *Regional Studies, 33*(9), 869–884.
Marvin, S., & Guy, S. (1997). Creating myths rather than sustainability: The transition fallacies of the new localism. *Local Environment, 2*(3), 311–318.
Massey, D. (1984). *Spatial divisions of labour: Social structures and the geography of production*. London: Macmillan International Higher Education.
Massey, D. (1994). *Space, place and gender*. Cambridge: Polity Press.
Moulaert, F., Swyngedouw, E., & Wilson, P. (1988). Spatial responses to Fordist and post-fordist accumulation and regulation. *Papers in Regional Science, 64*, 11–23.
Paasi, A. (2002). Place and region: Regional worlds and words. *Progress in Human Geography, 26*(6), 802–811.
Paasi, A., Harrison, J., & Jones, M. (2018). *Handbook on the geographies of regions and territories*. Cheltenham: Edward Elgar Publishing.
Peck, J. (2002). Political economies of scale: Fast policy, interscalar relations, and neoliberal workfare. *Economic Geography, 78*(3), 331–360.
Peck, J., & Tickell, A. (2002). Neoliberalizing space. *Antipode, 34*, 380–404.
Rossi, U. (2004). New regionalism contested: Some remarks in light of the case of the Mezzogiorno of Italy. *International Journal of Urban and Regional Research, 28*, 466–476.
Scott, A. J., & Storper, M. (1992). Industrialization and regional development. In *Pathways to industrialization and regional development* (pp. 15–28). London: Routledge.
Tomaney, J. (2002). The evolution of regionalism in England. *Regional studies, 36*(7), 721–731.
Tomaney, J. (2017). Regionalism. In D. Richardson, N. Castree, M. F. Goodchild, A. Kobayashi, W. Liu, & R. A. Marston (Eds.), *International encyclopedia of geography: People, the earth, environment and technology*. https://doi.org/10.1002/9781118786352.wbieg0112.
Vodden, K., Douglas, D. J., Markey, S., Minnes, S., & Reimer, B. (2019). *The theory, practice and potential of regional development: The case of Canada*. London: Routledge.
Watkins, M. H. (1963). A staple theory of economic growth. *Canadian Journal of Economics and Political Science/Revue Canadienne de Economiques et Science Politique, 29*(2), 141–158.
Wilson, G. (2001). From productivism to post-productivism ... and back again? Exploring the (un)changed natural and mental landscapes of European agriculture. *Transactions of the Institute of British Geographers, 26*, 77–102.
Wilson, G. A. (2009). The spatiality of multifunctional agriculture: A human geography perspective. *Geoforum, 40*(2), 269–280.
Woods, M. (2007). Engaging the global countryside: Globalization, hybridity and the reconstitution of rural place. *Progress in Human Geography, 31*(4), 485–507.
Zirul, C., Halseth, G., Markey, S., & Ryser, L. (2015). Struggling with new regionalism: Government trumps governance in northern British Columbia, Canada. *Journal of Rural and Community Development, 10*(2), 136–165.

New Orbital Urbanization

Charity Edwards and Brendan Gleeson
Monash University & University of Melbourne, Melbourne, Australia

N

Synonyms

Extended urbanization; Indigenous cosmologies; Low Earth orbit; Megaconstellation; Planetary urbanization; Satellite infrastructure

Definition

A new site of urbanization – low Earth orbit – increasingly structures our lives through the extension of urban informatics and sensing infrastructure circling the world. Such processes are conspicuously absent from many urban and regional debates however, which remain focused on human settlement patterns across the land surface of the Earth. Consideration of an expanding array of devices, equipment, and monitoring systems should extend beyond simply seeing their role in planetary-scale militarization, surveillance overreach, and/or global computational capacity. The technological capture of wider landscapes prefigures their eventual urban transformation, which must be a cause for democratic oversight. The proliferation of visible craft and signal pollution increases space debris risks, impacts scientific observations, causes real harm to the cosmos, and ultimately constrains what futures are possible for urban and regional communities around the Earth.

Introduction

Debates on the role of technology in urban and regional futures have till now focussed on human settlement patterns across the land surface of the Earth. However, beyond and, indeed, above this, an expanding array of sensory devices, satellite equipment, and monitoring systems structure everyday experience around the globe. That is, urbanization extends outwards around the world and now upwards into the atmosphere. Visions of entrepreneurial innovation in outer space often, however, obscure the implications of ongoing technological changes *up there* on lives experienced on Earth today. These trajectories operate in low Earth orbit but relate to terrestrial conditions – and cities in particular – in new and rapidly unfolding ways. Thus, while scholars now recognize that urban processes extend beyond the traditional container of the city as forms of "extended urbanisation" (Brenner, 2013), the role of low Earth orbit and its sensing networks in supporting and restructuring the city is less well understood. Much scholarship considers the objects of low Earth orbit in terms of planetary-scale militarization, surveillance overreach, and/or global computational capacity, but relatively little interrogates them as an extension of urban processes into the Earth's atmosphere.

Significantly, the technological capture of wider landscapes prefigures their urban transformation. Impacts of extended urbanization therefore become particularly urgent where technologies engulf planetary scale operations, escaping urban and regional oversight. A clear example is SpaceX's new venture, *Starlink*, which is deploying many thousands of small but bright satellites to support high-density urban internet bandwidth and increased spread of low-cost global communications. At the same time, proliferating visible craft and signal pollution will increase space debris risks, impair scientific observation, and cause significant harm to dark sky environments (Walker et al., 2020). This entry will illustrate the increasing urbanization of orbital space: a project rooted in Cold War politics but with surprising new trajectories in the twenty-first century. In particular, the discussion will highlight where an emerging apparatus of urban informatics meets a largely disregarded sensing infrastructure circling the world, with attendant implications for urban and regional futures.

Extending Urbanization

As argued by geographer Matthew Gandy (2014) and others, cities are just *one* type of urbanization. While it is true that a city may appear as a singular object, neatly delineated on a map of the Earth, the material and technological limits of urban areas are far less clear-cut. Providing a counterpoint to the narrow focus that sociologists Hillary Angelo and David Wachsmuth (2015) term "methodological cityism," a diverse range of scholarship now recognizes urban processes extend well beyond the traditionally conceived container of the city, and into regional and rural ways of life as they expand through networks of capital and consumption around the planet. This is reflected in the diverse work of urbanists Neil Brenner and Christian Schmid (2011), planners such as Ananya Roy (2016), geographers like Martin Arboleda (2014) and Stephen Graham (2016), and even media and technology scholars including Jennifer Gabrys (2016) and Shannon Mattern (2017). The revelation of this extended urbanization in our "global urban age" has limits however, tending to position remote environments such as the ocean, Amazonian forests, and outer space as fully outside "the city" and therefore (somehow) not party to urbanizing processes.

An Increasing Array

As urbanization extends outwards around the world, so too does it expand upwards – and beyond the Earth. Outer space imaginaries and visions of entrepreneurial innovation often however obscure the implications of ongoing technological changes *up there* for lives *down here* on Earth. In turn, much discussion of technologies in urban and regional futures tends toward a limited range of scenarios: "smart cities," public surveillance concerns, and the role of drone vehicles on the ground and in the air. Above all this though is an expanding array of sensory devices, satellite equipment, and monitoring systems that structure everyday experience. These technologies operate in orbit around our planet but direct earthbound phenomena, especially cities, in new and unfolding ways. Despite now acknowledging that urbanization does extend into spaces that are decidedly *not* cities, increasingly urbanizing remote environments such as low Earth orbit – and its associated sensing networks that both support and reconfigure the city – are less well understood. To correct this gap, Stephen Graham (2016: 27) in particular draws attention to ways that orbiting satellites completely organize contemporary life despite "their apparent removal from the worlds of earthly politics."

Remote Capture

The technological capture of wider landscapes on Earth prefigures their urban transformation. This concern becomes particularly urgent where technologies escape both terrestrial and politically bound oversight through planetary-scale operations. The capabilities of remote sensing and worldwide observation grew dramatically following the launch of *Sputnik I* in 1957. Indeed, Hannah Arendt used the figure of this satellite to project an emerging planetary society in the prologue to her classic text, *The Human Condition* (1998 [1958]). Anthony Giddens (1999) later claimed this first artificial orbiting object marked the beginnings of contemporary globalization (rather than simply a Cold War nexus of technology and nation-state interests). Surveillance from on high continued to expand in new and surprising ways via the proliferation of multinational satellite TV corporations and cheap broadcasting in the 1980s. Media scholar Lisa Parks (2005) argues however that the wild array of today's orbiting technologies stem largely from policy changes enacted by US President Clinton's administration during the 1990s: making military-grade high-resolution satellite imagery available for commercial sale by private companies. A resulting lucrative combination of increased satellite coverage, planetary broadcasting reach, and networked computational capacity allowed for the intensification of activities as diverse as policing civil society across continents (Rothe & Shim, 2018), navigating archaeological excavation at a distance (Parcak, 2019), searching for weapons of mass destruction from a position of aerial safety (Perkins & Dodge, 2009), and burgeoning land consolidation and real estate speculation (Desai, 2019; Juergens & Meyer-Heß, 2021). As Jennifer Gabrys (2016) reminds us, this planet is a *changeable* thing. That is, the Earth does not just exist to be translated via devices: rather, it is always in the process of becoming *through* and *with* technologies that encircle it.

Beyond the Critical Gaze

The success of endless real-time observation, ordering, and re-ordering of vast territories from a distance relies on co-opting a planetary envelope known as the "low Earth orbit (LEO) region." Extending up to 2000 km above the Earth, this altitude enables circling objects to sweep the globe in just over 2 hours and complete 11 full orbits a day (Inter-Agency Space Debris Coordination Committee, 2007). A close geocentric orbit provides cost-efficient stability and energy expenditure for most artificial objects launched into space. However, as noted by aerospace physicists Nicholas Crisp et al. (2020), new proposals by companies such as Space X now also seek to exploit a closer and hitherto avoided zone of "very low Earth orbit (VLEO)." VLEO is located from 100 km to 450 km above the Earth's surface and provides additional space and mobility in a now crowded orbital "marketplace." This extreme form of diminishing returns strategy recalls similar efforts by mining companies intent on extracting polymetallic nodules from financially marginal deep seabed territories (Miller et al., 2018).

An urbanizing LEO obviously manifests via material objects like satellites, spacecraft, and space stations. As at August 2021, the United Nations Office for Outer Space Affairs' *Index of Objects Launched into Outer Space* lists 7859 such items currently orbiting Earth, and the years of 2020–2021 represent more than a 20% increase on previous launches. Similarly, SpaceX's new "megaconstellation" (Witze, 2018: 25), *Starlink*, will station over 10,000 small satellites in the VLEO-LEO range to create high-density urban internet bandwidth and a universalized spread of low-cost global communications. Invisible processes of wireless networking and intensive cloud computing that together enable the Internet of Things (IoT) (Li et al., 2021) are thus another significant feature of new forms of orbital urbanization. By way of illustration, the World Economic Forum (2020) positions the expansion of IoT as fundamental to wider societal recovery in the wake of COVID-19. In doing so, they have already partnered with corporations alongside more than 200,000 cities and governments to embed integrated technologies in city planning, remote working, vaccine delivery, and governance frameworks. Uneven and unexpected convergences result where the instruments of urban data become entangled with a disregarded sensing infrastructure circling the world.

Into (Not So Much) Darkness

Space X sought to launch 60 satellites every fortnight during 2020, and while many astronomers remain concerned about their damaging impacts (McDowell, 2020), discussion of such orbiting artifacts and their signal pollution has been conspicuously absent from contemporary urban debates. Once LEO missions are complete, providers are supposed to "de-orbit" objects (including defunct satellites, abandoned payload carriers, solid motor effluents, derelict rockets, and some intentionally released waste). However, a majority of orbital debris remains in LEO, and is only increasing in size and scope as urbanizing processes continue to extend into the atmosphere. Debris over 10 cm in diameter are tracked by NASA through the US Space Surveillance Network and, by late 2020, numbered more than 23,000 objects, or the equivalent of 8 metric tons of material orbiting the Earth at approximately 8 km per second. While some nations have guidelines on managing space debris in orbit, no international agreement exists to police these recommendations. It is clear that increasing orbital urbanization brings substantial risks with it. The potential for uncontrolled space debris collisions and ongoing contamination of critical scientific observations is exacerbated, along with significant harm enacted upon dark sky environments and nocturnal species already under pressure from earthbound urban processes and industrial-scale light pollution.

The increasing urbanization of spaces beyond the Earth's surface also reinscribes everyday lives with long patterns of harm by extending settler-colonial violence and extractive structures into the plural worlds of many Indigenous and First Nations communities. Indigenous cosmologies typically reject the delineation of Earth from outer space and dominant Western assumptions of extra-planetary environments as either empty or non-sentient: that is, they resist reordering orbital space through the fiction of *terra nullius* logics. As it is, recent research co-produced by Yolŋu and non-Indigenous authors on Bawaka Country (2020) in northeastern Arnhemland, Australia, demonstrates the ongoing role of ethical obligations between Sky Country and ancestral (yet still agential) beings who reside there. Through this text, the Bawaka Collective explain how actions that modify Sky Country – such as the expansion of contemporary sensing networks – disrupt the (ongoing) dwelling places of their kin and damage (already existing) knowledges of the dark sky. Significantly, Yolŋu recognizes such moves fundamentally erase relations with other people and beings throughout the cosmos and thus constrain the types of futures imaginable for all.

Conclusion

A new site of urbanization – low Earth orbit – structures our lives on the planetary surface. Here, an emerging apparatus of urban informatics meets a largely disregarded sensing infrastructure circling the world. Contemporary visions of entrepreneurial extraplanetary innovation obscure the implications of ongoing changes *up there,* while the critical regard of such processes is conspicuously absent from many debates on urban and regional futures. An expanding array of devices, equipment, and monitoring systems increasingly direct everyday experience around the globe. The operation of LEO objects exceeds much current scholarship, which focuses on their role in planetary-scale militarization, surveillance overreach, and/or global computational capacity. Critically, such artifacts escape the traditionally conceived container of the city, spreading into regional and rural ways of life in mostly invisible ways as yet another form of "extended urbanisation." The technological capture of wider landscapes prefigures their eventual urban transformation, which must be a cause for political and policy oversight, as well as urban enquiry. It is urgent that the mounting impacts of low Earth orbit intrusions are understood and managed before their urban trajectories are fixed. Proliferating visible craft and signal pollution increase space debris risks, deleteriously impact scientific observations, and cause significant harm to dark sky environments. They also serve to reinscribe settler-colonial violence and extractive practices into the cosmos, and significantly constrain what futures are possible for urban and regional communities across the Earth.

Cross-References

- ▶ Augmented Reality: Robotics, Urbanism, and the Digital Turn
- ▶ City Visions: Toward Smart and Sustainable Urban Futures
- ▶ Regulation of Urban and Regional Futures

References

Angelo, H., & Wachsmuth, D. (2015). Urbanizing urban political ecology: A critique of methodological Cityism. *International Journal of Urban and Regional Research, 39*(1), 16–27. https://doi.org/10.1111/1468-2427.12105.

Arboleda, M. (2014). Implosions/explosions: Towards a study of planetary urbanization. *Area,* 46(3), 339–340. https://doi.org/10.1111/Area.12092.

Arendt, H. (1998 [1958]). *The human condition* (2nd Ed.). University Of Chicago Press.

Bawaka Country Including, Mitchell, A., Wright, S., Suchet-Pearson, S., Lloyd, K., Burarrwanga, L., & Maymuru, R. (2020). Dukarr Lakarama: Listening to Guwak, talking Back to space colonization. *Political Geography, 81*. https://doi.org/10.1016/J.Polgeo.2020.102218.

Brenner, N. (2013). Theses on urbanization. *Public Culture, 25*(1), 85–114. https://doi.org/10.1215/08992363-1890477.

Brenner, N., & Schmid, C. (2011). Planetary urbanisation. In M. Gandy (Ed.), *Urban constellations*. Berlin: Jovis Verlag.

Crisp, N. H., Roberts, P. C. E., Livadiotti, S., Oiko, V. T. A., Edmondson, S., Haigh, S. J., & Schwalber, A. (2020). The benefits of very low earth orbit for earth observation missions. *Progress in Aerospace Sciences, 117*, 100619. https://doi.org/10.1016/J.Paerosci.2020.100619.

Desai, R. S. (2019). Afterlives of orbital infrastructures: From Earth's high orbits to its high seas. *New Geographies: Extraterrestrial, 11*, 39–45.

Gabrys, J. (2016). *Program earth : Environmental sensing technology and the making of a computational planet*. University Of Minnesota Press.

Gandy, M. (2014). Implosions/explosions : Towards a study of planetary urbanization. In N. Brenner (Ed.), *Where does the city end?* (pp. 86–89). Berlin: Jovis.

Giddens, A. (1999). Runaway World. *Bbc Reith Lectures* [Lecture Recording]. Bbc.

Graham, S. (2016). *Vertical: The city from satellites to bunkers*. London: Verso.

Inter-Agency Space Debris Coordination Committee. (2007). *Space Debris Mitigation Guidelines*. (Iadc-02-01). Vienna: Un Office For Outer Space Affairs Retrieved From http://Www.Unoosa.Org/Documents/Pdf/Spacelaw/Sd/Iadc-2002-01-Iadc-Space_Debris-Guidelines-Revision1.Pdf

Juergens, C., & Meyer-Heß, M. F. (2021). Identification of construction areas from Vhr-satellite images for macroeconomic forecasts. *Remote Sensing, 13*(13), 2618. Retrieved From https://www.Mdpi.Com/2072-4292/13/13/2618.

Li, C., Zhang, Y., Xie, R., Hao, X., & Huang, T. (2021). Integrating edge computing into low earth orbit satellite networks: Architecture and prototype. *Ieee Access, 9*, 39126–39137. https://doi.org/10.1109/Access.2021.3064397.

Mattern, S. (2017). A city is not a computer. *Places Journal*. https://Doi.Org/2010.22269/170207

Mcdowell, J. C. (2020). The low earth orbit satellite population and impacts of the Spacex Starlink constellation. *The Astrophysical Journal, 892*(2), 1–10. https://doi.org/10.3847/2041-8213/Ab8016.

Miller, K. A., Thompson, K. F., Johnston, P., & Santillo, D. (2018). An overview of seabed mining including the current state of development, environmental impacts, and knowledge gaps. *Frontiers in Marine Science, 4*, 418. Retrieved From https://www.Frontiersin.Org/Article/10.3389/Fmars.2017.00418.

Parcak, S. (2019). *Archaeology from space: How the future shapes our past*. Henry Holt And.

Parks, L. (2005). *Cultures in orbit: Satellites and the televisual*. Durham, Usa: Duke University Press.

Perkins, C., & Dodge, M. (2009). Satellite imagery and the spectacle of secret spaces. *Geoforum, 40*(4), 546–560. https://doi.org/10.1016/J.Geoforum.2009.04.012.

Rothe, D., & Shim, D. (2018). Sensing the ground: On the global politics of satellite-based activism. *Review of International Studies, 44*(3), 414–437. https://doi.org/10.1017/S0260210517000602.

Roy, A. (2016). What is urban about critical urban theory? *Urban Geography, 37*(6), 810–823. https://doi.org/10.1080/02723638.2015.1105485.

Walker, C., Hall, J., Allen, L., Green, R., Seitzer, P., Tyson, A., Joachim, P. (2020). *Impact of satellite constellations on optical astronomy and recommendations toward mitigations*. Retrieved From Tucson, Usa: https://Noirlab.Edu/Public/Products/Techdocs/Techdoc003/

Witze, A. (2018). The quest to conquer Earth's space junk problem. *Nature, 561*(7721), 24–26. https://doi.org/10.1038/D41586-018-06170-1.

World Economic Forum. (2020). *Shaping the future of the internet of things and urban transformation* [Press Release]. Retrieved From https://Www.Weforum.Org/Platforms/Shaping-The-Future-Of-The-Internet-Of-Things-And-Urban-Transformation

New Town

► New Cities

New Towns

► Small Towns in Asia and Urban Sustainability

Non-communicable Diseases (Non-infectious Diseases)

► Epidemiological Shifts in Urban Bangladesh

Nonfiscal Tools

► Green Economy Policies to Achieve Water Security

Non-state Market-Driven (NSDM) Governance Systems

► Voluntary Programs for Urban and Regional Futures

Nudging

► Behavioral Science Informed Governance for Urban and Regional Futures

O

OECD – Organization for Economic Cooperation and Development

▶ Adapting to a Changing Climate Through Nature-Based Solutions

Organic Engineering Systems

▶ Future Foods for Urban Food Production

Organic Waste

▶ Global Survcy of Food Waste Policies

Organization

▶ Governing for Food Security: A Cultural Perspective

Organized Crime

▶ Transnational Crimes: Global Impact and Responses

Overcoming Barriers in Green Infrastructure Implementation

Building Governance's Adaptive Capacities as a Valuable Approach

Antonija Bogadi
Department of Urban and Spatial Planning and Research, Technical University Vienna, Vienna, Austria

Abbreviations

ES Ecosystem services
UGI Urban green infrastructure

Synonyms

Adaptive governance; Ecosystem-based adaptation; Green infrastructure planning

Definition

Urban green infrastructure (UGI) — It is a strategic and spatial approach to landscape and urban planning. It is a network of patches and corridors of land that are planned and managed for biodiversity conservation,

R. C. Brears (ed.), *The Palgrave Encyclopedia of Urban and Regional Futures*,
https://doi.org/10.1007/978-3-030-87745-3

water resources, land protection, recreation, cultural use, urban development control, and climate change mitigation and adaptation. A green infrastructure network can consist of networks of greenways, open spaces, greenbelts, urban greening, cultural landscapes, urban open spaces, ecological networks, agricultural land, and another unbuilt land that could support vegetation.

Ecosystem-based adaptation — It emerges from the theory and practice of climate change adaptation and ecosystem services. Millennium Ecosystem Assessment (MA 2005) and the Economics of Ecosystem Services and Biodiversity (TEEB 2010) grouped **ecosystem services (ES)** in four major categories: provisioning (water supply, material supply), regulating (regulation of biophysical/biotic/physiochemical environment, flow regulation), habitat, and cultural services (symbolic, intellectual, and experimental).

Introduction

As we move to a highly urbanized future characterized by social and climate uncertainties, there is a need to prepare cities to take on new drivers of change by offering various solutions to the problems they will be facing. Direct climate change impacts can be ameliorated to a significant extent by the implementation of green infrastructure, through the ecosystem services it provides.

To address real-world problems in implementing such ecosystem-based solutions, we need to be better at recognizing, and utilizing, the dynamic relationships and transformations that complex processes of governing the recourses for effective UGI undergo.

For green infrastructure to be effective, it should be implemented at multiple scales, gradually, and flexibly, allowing adjustments as climate change and other changes materialize.

The same characteristics that qualify GI as an effective spatial adaptation tool within urban regions, i.e., GI's multifunctional and multi-scalar properties, create difficulties for mainstreaming GI into adaptation planning. These characteristics create problems in organizing intervention areas, jurisdictional and stakeholders' coordination, and trade-offs in economic benefits and land use quality.

Within those circumstances, it is necessary to strategically plan GI for short-term and long-term benefits, while remaining adaptive for changes over time. This kind of responsibility goes beyond the conventional methods of planning.

This chapter relies on a premise that enhancing the adaptive capacity of governance is one of the important answers to the above-stated challenges. Managing diversity and redundancy; creating optimal connections between actors; enabling constant experimentation, learning, and participation processes; and identifying variables and managing feedbacks are processes through which those capacities can be built.

Urban Green Infrastructure Implementation: Circumstances and Needs

When discussing circumstances around urban green infrastructure implementation, the topics of the scarcity of spatial recourses, jurisdictional incoordination, lack of cooperation between stakeholders behind generating ecosystem services, and trade-offs are crucial to be taken into account.

Scarcity of Spatial Recourses and Competing Urban Priorities

Densification processes in towns and cities potentially endanger urban green spaces. This process is a major obstacle to developing urban areas for alternative land uses, like urban green infrastructure elements. That makes green space planning and management challenges, especially in areas under densification, since important ecosystem services are supposed to be delivered by limited green spaces. As pressure has grown within expanding and densifying urban areas, urban green space has to compete with high land prices and increasing densities of development and urban infrastructure, which often results in squeezing nature out of our cities, followed by unwanted social and health impacts.

One of the greatest difficulties for municipalities is to introduce a priority to implement UGI into an already resource-stretched environment, especially over or along with popular social policies like housing provisions (Barthel et al. 2005). Environmental conservation most often has a lower priority than other areas of city development and is not yet effectively integrated across complementary departments and initiatives (Wilkinson et al. 2013).

Jurisdictional Incoordination and Scale Mismatch

A further difficulty in UGI planning is that it must simultaneously function on various scales, in synergy and harmony. Andersson et al. suggest three spatial scales of ecological processes: local scale green areas, city-scale green networks, and a regional scale green infrastructure (Fig. 1).

Because of this GI's multi-scalar and multifunctional properties, its planning has to be carried through initiatives at different scales, from individual parcels to community, regional, and state levels. Furthermore, green infrastructure planning has to incorporate ecological, social, and economic/abiotic, biotic, and cultural functions of green spaces, which need to be physically and functionally connected at different scales and from different perspectives.

Green infrastructure includes the usage of all kinds of urban space, e.g., natural and seminatural areas, water bodies, and public and private green spaces, such as parks and gardens. As a consequence, green infrastructure planning also seeks the integration and coordination of urban green with other urban infrastructures in terms of physical and functional relations, e.g., built-up structure, transport infrastructure, and water management system.

The most common problem in effective UGI governance is a mismatch between scales of ecosystem processes on the one hand and scales of management on the other.

The main challenge in UGI governance appears either because of fragmented governance, where several jurisdictions exist within the city with overlapping and often uncoordinated responsibilities, or because ecosystem functioning does not align with administrative boundaries. Furthermore, governmental decisions across scales have long-term consequences extending beyond the period for which elected officials are responsible.

For instance, municipal managers and urban planners have a wider landscape perspective and are mostly focused on protected land and urban parks, while local actors in so-called developed land don't mind ecological processes beyond their focus area. Those scale mismatches are resulting in understudied city-scale networks and neglected resources for providing ecosystem services in urban areas (Colding and Barthel 2013).

Lack of Cooperation Between Stakeholders Behind Generating Ecosystem Services

Responsibility for ecosystem services provision and UGI is usually divided between governmental bodies operating at the citywide, regional, and state scales, nonprofit organizations, research institutions and universities, private businesses, informal community groups, and individual volunteers operating in ecological regions, across cities, and in specific neighborhoods.

Furthermore, green infrastructure planning is based on knowledge from different disciplines, such as landscape ecology, urban and regional planning, and landscape architecture and developed in partnership with various local authorities, stakeholders, and final users, all with different understandings, motivations, and needs for UGI, so effective UGI demands exceptionally communicative and inclusive planning and management environment.

High population density and limited space are creating the need for multifunctional (green) spaces, so various interests and objectives are

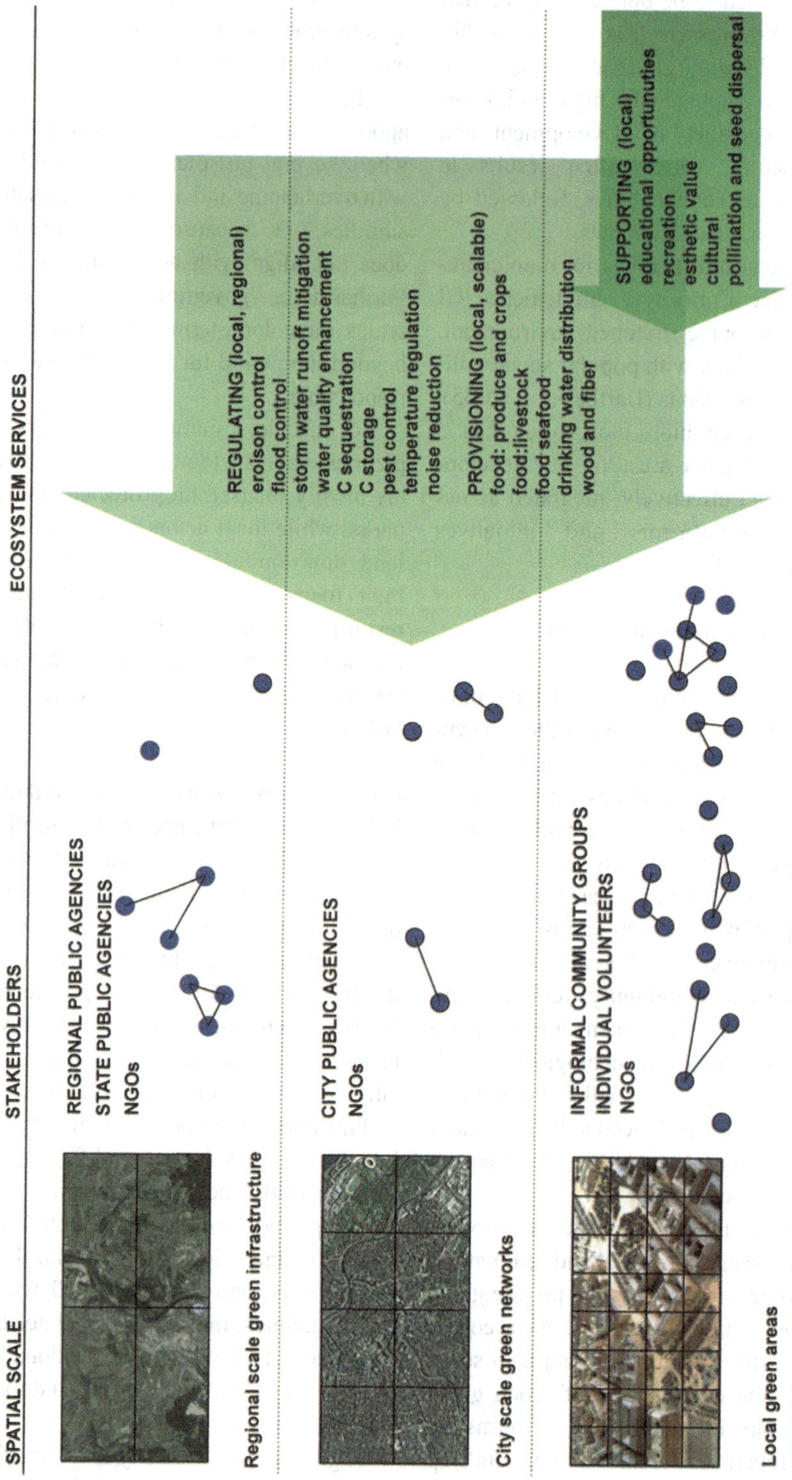

Overcoming Barriers in Green Infrastructure Implementation, Fig. 1 Multifunctional and multi-scalar properties of green infrastructure linked to the urban spaces with the capacity for ecosystem services provision.

Different stakeholders may have conflicting priorities even on the same ecosystem services. Also, cities do not necessarily have the political commitment or fiscal and institutional capacity to govern effective UGI. For all cities, the challenge is how to coordinate all the abovementioned stakeholders and their actions for effective ecosystem services provision.

For example, regional and city-level organizations have particular interests that are long-termed and larger-scaled and are neglecting the potential of small scale patches and corridors in the built-up areas. Cultural ecosystem services, such as recreation, aesthetics, and cultural heritage, are often prioritized in planning, design, and management of urban green spaces (Colding and Barthel 2013).

On the other hand, local civic organizations and agencies are using the available space for their current and mostly short-termed needs and don't have a broader concept of knowledge about the whole spectrum of ecosystem services. They are mostly interested in provisioning types of ecosystem services, e.g., urban agriculture and community gardening and home gardens.

Regarding the lack of understanding and knowledge about ES, stakeholders are most often lowering the effectiveness of those services. The reason for this is that ES depends on carefully planned functional connectivity based on a knowledge of spatial interdependencies, e.g., pollination, pest control, moderation of extreme events, carbon sequestration, and storage, which are functioning and interacting simultaneously on multiple spatial scales and that is rarely understood by stakeholders concerned in urban environmental issues.

Trade-Offs in the Context of Spatial Planning and Provision of Ecosystem Services

Despite the desire to produce "win-win solutions" in spatial planning, they seem to be rare in environments where decision-makers work within trade-offs and hard choices, finding ways to organize land use to fulfill diverse requirements of society, such as the needs of local residents, the viability of economy, and maintaining environmental quality.

In the context of spatial planning and ES, trade-offs can be seen as "land-use or management choices that increase the delivery of one (or more) ecosystem service(s) at the expense of the delivery of other ecosystem services" (derived from TEEB 2010).

This occurs when the co-use is impossible, e.g., housing development vs. nature conservation, when two or more desired ES either can't be delivered at the same time, e.g., agriculture vs. flood control, or when the ES benefits are unequally distributed over different stakeholders, e.g. maintaining traditional landscapes vs. tourism (Quintas-Soriano et al. 2016).

Knowledge about ES trade-offs can't be generalized, because ES trade-offs are influenced by social, economic, institutional, and ecological factors, which are always context-specific.

Place-based studies that focus on the local specificities of trade-off mechanisms, involving local knowledge, are often the most efficient and reliable way to study these ES trade-offs, but such studies are still rare.

A better understanding of the underlying causes and mechanisms for trade-offs can help to predict where and when trade-offs might take place; encourage dialogue, learning, and trust between stakeholders; and help to obtain more equitable outcomes.

Designing Adaptive Governance Processes for Establishing Effective UGI

Rhodes (1997) describes the fragmentation of the capacity of the traditional authorities to influence the urban system as the shift from government to governance. This fragmentation increases the need for governments to cooperate with a wide array of other actors and factors influencing outcomes. Governance includes navigating relationships and dynamics between the authorities on various levels, the civil society, and the private sector, all with sometimes conflicting and overlapping intentions.

"Adaptive governance," with its ability to adapt and even thrive in changing social and ecological circumstances, is seen as one of the most promising models of governance.

A well-functioning adaptive governance system is in tune with the dynamics of the ecosystems under management and with external perturbations, uncertainty, and surprise, like changes in climate, disease outbreaks, hurricanes, global market demands, subsidies, and governmental policies. Well-functioning adaptive governance systems can use barriers and disturbances as opportunities to transform into more desired states (Fig. 2).

For adaptive governance to work, self-organization of stakeholders and collective action have to be understood as vitally important, and this is influenced by the ability of the involved parties to agree on resource-related problems and resource status (Ostrom 2005).

Holling's idea of the "adaptive renewal cycle" provides a valuable context for the analysis and design of adaptive governance processes, an area of research under rapid development with implications for resilient and sustainable development.

The adaptive renewal cycle is a heuristic model, generated from observations of ecosystem dynamics, of four phases of development driven by discontinuous events and processes. There are periods of exploitation phase, of growing stasis and rigidity, periods of readjustments and collapse, and phase of renewal.

The five practical ways forward for building adaptive capacities in governance's processes to be able to go through the adaptive cycle, derived from a review of the literature and presented below, include maintaining diversity and redundancy, directing connectivity, controlling slow variables and feedbacks, encouraging learning and experimentation, and broadening participation.

Guideline 1: Maintaining the Right Amount of Diversity and Redundancy Among Actors

A variety of organizational forms, e.g., government departments, nongovernmental organizations, and community organizations, with overlapping domains of authority can provide redundancy and diversity. Various organizational forms provide diversity because organizations with different sizes, cultures, funding mechanisms, and internal structures provide more ways to respond to change and disturbance.

Also, greater actors' diversity of knowledge and experience leads to a greater range of possible responses and of finding creative solutions to changes or disturbances that threaten ecosystem services. In this regard, there is a growing need to deliberately increase the diversity of knowledge by combining various types of knowledge, such as local ecological knowledge and scientific knowledge.

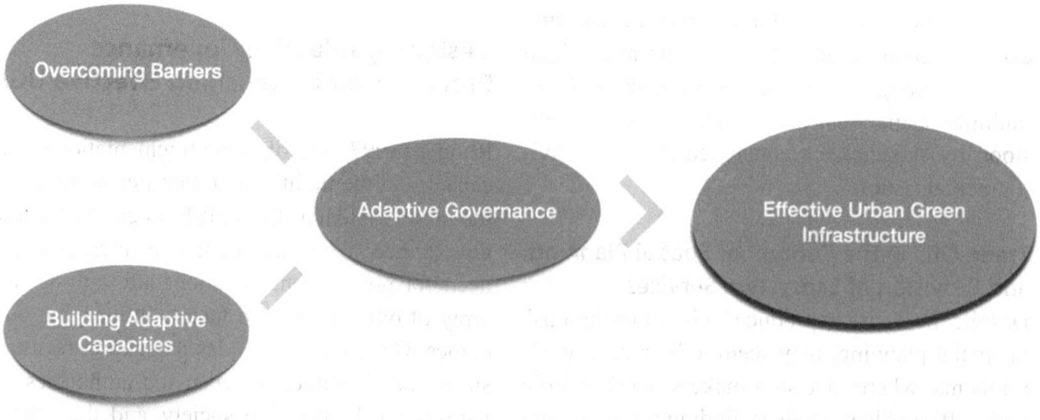

Overcoming Barriers in Green Infrastructure Implementation, Fig. 2 Overcoming barriers and building adaptive capacities of governance for effective UGI

Historical (Walker 2007) and social movement research (Ernstson et al. 2008) suggest that urban green areas that engage a high range of interest and user groups have more chances to be efficient, thanks to a more efficient exchange of knowledge and skills and increasing potential for adequate action.

On the other hand, very high levels of diversity and/or redundancy can cause negative impacts, like stagnation. For example, a high diversity of interests, preferences, and understandings of climate change impacts among nations has been identified as a bottleneck in climate negotiations.

High redundancy in management organizations also increases administrative costs, coordination, and other types of transaction costs, the possibility for contradictory regulations, which may hinder UGI governance.

Guideline 2: Optimizing Connectivity Between Social Actors

Increasing connectivity between different social groups gives individuals access to new information, various "outside" perspectives and ideas, and development of trust and reciprocity that can increase governance's adaptive capacity.

More specifically, improving stakeholders' cooperation can be generated through increasing the level of communication between stakeholders by connecting the governance through spatial scales, by increasing the level of knowledge of ES and scale-dependent responsibilities among all actor groups, and by increasing the diversity of stakeholders on a particular site.

When developing instruments for optimizing connectivity, it is important to keep in mind that highly connected networks may also limit social learning, reduce the options to find optimal solutions, and reduce the capacity for experimentation. For example, when homogenization of norms occurs, explorative ability drops, leading to a dead-end situation in which actors believe they are doing the right actions, while they are ineffectively managing ecosystems (Bodin and Norberg 2005).

Guideline 3: Identifying Key Slow Variables and Managing Feedbacks in All Scales

Identifying key slow variables and managing feedbacks to avoid UGI performance thresholds or facilitate transformations is crucial for increasing adaptive capacities of UGI governance. Designing governance processes that can act on knowledge about key slow variables in urban ecosystems is therefore critical.

Appropriate action to manage feedbacks may not occur for a variety of reasons, even in the environments where knowledge and monitoring information exist. For instance, although key slow variables and feedbacks of climate change are known, actors' diverse and competing interests can restrict the coordinated response to avoid potential climate-driven regime shifts.

Another obstacle to properly react to changes is that governance institutions are often structured to operate on shorter timeframes than the timeframes over which important changes in controlling slow variables might occur, so these variables are treated as constant and ignored.

Guideline 4: Stimulating Experimentation and Learning Processes

To enable adaptation to change in governance processes, we need to constantly update existing knowledge and maintain ecosystem services through the phases of disturbance and change. This can be achieved by learning and experimentation, which means creating new knowledge and allowing the re-evaluation of existing values and new understandings.

Power asymmetries in social systems have an important impact on how learning is harnessed. For instance, relatively more powerful actors can dominate poorly implemented learning processes and assert the influence of their knowledge, thereby covering up other voices within communities. Likewise, powerful organizations on a national level can restrain the potential contribution of learning and innovation at the local scale.

In adaptive governance, a key focus tends to include knowledge sharing and re-evaluating across scales. This cross-scale focus of learning and experimentation is pursued because developing adaptive governance prioritizes the creation of new social norms and cooperations toward nested organizational structures and bridging organizations to match the scale of decision-making and responsibilities to the scale of ecological processes.

Guideline 5: Enabling Fair and Continuous Participation

Participation, seen as the active engagement of relevant stakeholders in the governance process, can play a positive role in supporting transparency, knowledge sharing, trust-building, the legitimacy of decisions, and learning. Participation can range from just informing stakeholders to complete change of power division, occurring in some or all stages of a decision-making process, from identifying problems and goals to implementing policies, monitoring results, or evaluating outcomes. Participatory processes generally aim to involve actors who can contribute in some way to the management of urban green infrastructure, by providing either knowledge or services, management practices, funding, or political support.

While participation can enhance the adaptive capacities of the governance, its quality depends on participants, the participation process, and the social and institutional environment. These factors are interdependent and are also context-dependent, and, if not well understood, participation can undermine or compromise adaptive capacities.

For example, participation can allow some stakeholders to use their influence at the expense of others, by increasing their power or influence within the participation process. Likewise, participation can cause high connectivity among different actors in the system, who then start to use the same information gained through those new links and develop "single-mindedness."

Also, there is a tendency for the scientists to do the science first, or governmental agencies to develop the agenda first, and then present it to the different groups to choose from already established frameworks.

Participation can also deepen already bad stakeholder relationships. For example, dealing with the contested resource in an environment where the legal authority is unclear may fully degrade relationships between stakeholders. Another negative impact of participation is the danger that too many participatory schemes lead to degradation of the governance process as the community experiences "consultation fatigue" (de Vente et al. 2016).

Conclusion/Summary

Ecosystem-based adaptations to the various effects of climate change in cities have attracted a great deal of interest in recent years.

This chapter discusses the circumstances around urban green infrastructure planning and implementation and synthesizes main governance barriers in implementing ecosystem-based adaptation into urban areas. In sum, the fundamental problem lies in the fact that urban areas with the potential to provide ecosystem services are often overseen or insufficiently used, due to conflicting jurisdictional and stakeholder coordination, land-use priorities, and trade-offs.

Based on the overview of the main challenges for UGI implementation, the key insight is that the capacity of urban ecosystems to produce ecosystem services depends on collaborative urban planning and design that can integrate formal and informal institutions, knowledgeable stakeholders at various urban scales, social networks, and diverse interest groups in urban ecosystem management. Such collaborative urban planning and design should also have the capacity to cope with the change, like climate change and rapid urbanization.

Those opportunities can be rephrased into the term "adaptive capacities building," and in that format used for the strategical design of "adaptive governance," the structure that has capacities to deal with existing and future challenges.

As "adaptive governance" pursues UGI plans, it has to enhance its adaptive capacity across multiple sectors to address social, environmental, and economic vulnerabilities, via maintaining the right degrees of diversity, redundancy, and connectivity among actors, by being able to identify key slow variables and managing feedbacks at multiple scales, stimulating experimentation and learning processes, and enabling continuous participation.

Cross-References

- ▶ Mainstreaming Blue Green Infrastructure in Cities: Barriers, Blind Spots, and Facilitators
- ▶ Multiple Benefits of Green Infrastructure

References

Barthel, S., Colding, J., Elmqvist, T., & Folke, C. (2005). History and local management of biodiversity-rich, urban, cultural landscape. *Ecology and Society, 10*, 10.

Bodin, Ö., & Norberg, J. (2005). Information network topologies for enhanced local adaptive management. *Environmental Management, 35*(2), 175–193.

Colding, J., & Barthel, S. (2013). The potential of 'urban green commons' in the resilience building of cities. *Ecological Economics, 86*, 156–166.

de Vente, J., Reed, M., Stringer, L., Valente, S., & Newig, J. (2016). How does the context and design of participatory decision making processes affect their outcomes? Evidence from sustainable land management in global drylands. *Ecology and Society, 21*(2), 24.

Ernstson, H., Sörlin, S., & Elmqvist, T. (2008). Social movements and ecosystem services – The role of social network structure in protecting and managing urban green areas in Stockholm. *Ecology and Society, 13*(2), 39.

MA (Millennium Ecosystem Assessment). (2005). *Ecosystems and human wellbeing: Synthesis*. Washington, DC: Island Press.

Ostrom, E. (2005). *Understanding institutional diversity*. Princeton: Princeton University Press.

Quintas-Soriano, C., Castro, A. J., Castro, H., & García-Llorente, M. (2016). Impacts of land use change on ecosystem services and implications for human well-being in Spanish drylands. *Land Use Policy, 54*, 534–548.

Rhodes, R. A. (1997). *Understanding governance: Policy networks, governance, reflexivity and accountability*. Open University Press.

TEEB. (2010). *The economics of ecosystems and biodiversity: Mainstreaming the economics of nature: A synthesis of the approach, conclusions and recommendations of TEEB*. TEEB.

Walker, C. (2007). Redistributive land reform: For what and for whom? In L. Ntsebeza & R. Hall (Eds.), *The land question in South Africa: The challenge of transformation and redistribution* (pp. 132–151). Cape Town: HSRC Press.

Wilkinson, C., Sendstad, M., Parnell, S., & Schewenius, M. (2013). Urban governance of biodiversity and ecosystem services. In *Urbanization, biodiversity and ecosystem services: Challenges and opportunities* (pp. 539–587). Springer. Jones et al. 2012.

An Overview of the Relationship of the Sustainable Development Goals and Urban and Regional Development

Alejandra Trejo-Nieto
Centre for Demographic, Urban and Environmental Studies, El Colegio de Mexico, Mexico City, Mexico

Synonyms

Development Goals; Localizing SDGs; Sustainability; Territorial development

Definitions

Urban and Regional Development

Development is a process by which a societal problem is solved by implementing a systematic process of change. Regional and urban development is a broad and multidimensional concept that refers to the complex space-time dynamics of regions and cities changing their social, economic and environmental welfare positions.

Sustainable Development Goals

The 17 global Sustainable Development Goals were introduced in 2015 by the United Nations General Assembly as part of a new global development agenda to be achieved by the year 2030. They comprise 169 targets addressing the developmental challenges facing the world including economic growth, urbanization, poverty, inequality, climate change, environmental degradation, peace and justice.

O

Introduction

Faced with longstanding social, economic and environmental challenges including urbanization, poverty, inequality, economic stagnation, unemployment, energy transitions and climate change, the international community has applied itself to improving development globally. The Sustainable Development Goals (SDGs) were published in 2015 by the United Nations (UN) as part of a renewed global development agenda. The 17 goals, with their 169 specific targets, are very comprehensive in scope, covering all the policy domains that are critical to sustainable growth and development (UN 2015). Following the eight Millennium Development Goals (MDGs), which aimed to eradicate extreme poverty and hunger, promote gender equality and reduce child mortality from 2000 to 2015, the scope and breadth of the Agenda 2030 for Sustainable Development and its SDGs represent the greatest challenge to developed and developing countries to date (OECD 2020). The SDGs are widely used as the basis of social, economic, and environmental policies and action around the world, and most countries have implemented programs and activities aiming to achieve their development compromises (Trejo Nieto 2017).

Despite the SDGs' significant potential for activating macro-level development processes, there are multiple specific subnational processes both for taking in and fully understanding the Agenda 2030 and for its implementation. Cities and regions have a crucial role in achieving the SDGs, as they have core policy responsibilities that are central to sustainable development. The OECD (2020) estimates that at least 100 of the SDGs' 169 targets cannot be reached without the proper engagement and coordination of local and regional governments. The global development agenda, in turn, is relevant to subnational development approaches which have been applying sustainability principles to improve social, economic, and environmental standards in cities and metropolitan regions.

Because regions and cities play a crucial role in achieving sustainable development, the adoption of a territorial perspective of development, or "localization," is seen as a necessary condition for accomplishing the Agenda 2030. Localization implies, among other things, public policies based on a multilevel governance approach, and requires the participation of the appropriate subnational actors. The relevance of subnational territories in global development was intensely debated during the preparation of the Agenda 2030. The demands of subnational authorities and other stakeholders resulted in the incorporation of SDG 11 on "Sustainable Cities and Communities." SDG 11 addresses making urban settlements inclusive, safe, resilient and sustainable (Ojeda Medina 2019). The formulation of SDG 11 uncovers the many existing and crosscutting opportunities for achieving development goals via cities and urbanization: urbanization is capable of stimulating many aspects of global development and plays a critical role in facilitating balanced territorial development, while cities are well-recognized as centers of growth, innovation and investment, and play a leading role in driving industrialization and economic growth in both developed and developing countries. Cities are also places that can connect, territorially speaking, most of the SDGs because they include an urban component (UN-Habitat 2018). Regions, too, have a distinctive role in the implementation and achievement of the SDGs, particularly as intermediaries between the national and the local level (OECD 2020). Regional governments can help to alleviate the burden of national government, especially in following up and reviewing development progress (Messias 2017).

This chapter presents a framework for understanding the opportunities and synergies as well as the potential challenges and contradictions inherent in the interaction between the SDGs and urban and regional development. It offers insights into how the visions, goals and targets of the global development agenda promote and support the development of cities and regions, while cities and regions in turn support the progression of global development. It discusses the often complex, contradictory and manifold associations between the urban-related SDGs targets and other development targets, making an important contribution to informing national, regional and urban development policies. Mexico City is used as a short case study to exemplify a local approach for sustainable development.

Agenda 2030 and the Sustainable Development Goals

At the Millennium Summit in 2000, 184 UN Member States made a commitment to the reduction of poverty and the improvement of human development in the world. The Millennium Declaration of the General Assembly issued at this summit enunciated the eight MDGs as part of a common framework for action and global cooperation for development, to be accomplished by 2015 (UN 2015). Several evaluations of the MDGs' implementation, delimitation, structure and scope were made prior to the 2015 deadline. Even where important progress had been made, the costs and benefits were unevenly distributed among countries, and some problems became more relevant as the deadline for reaching the MDGs approached. Climate change, armed conflict, security, economic growth, migration and inequality were identified as the most significant pending development issues not included in the MDGs. This evaluation of the MDGs led to the planning and establishment of the Post-2015 Agenda or Agenda 2030 with its 17 SDGs, building on lessons learned from the MDGs and aiming to close the gaps left by the Millennium Agenda (Trejo Nieto 2017). The year 2015 was the turning point in the evolution of the socio-economic and political priorities that shape international development. Four major UN international conferences were held that year: the Third World Conference on Disaster Risk Reduction, which adopted the Sendai Framework for Disaster Risk Reduction 2015–2030; the Third International Conference on Financing for Development, which adopted the Addis Ababa Action Agenda; the 2015 United Nations Conference on Climate Change (COP 21), which adopted the Paris Climate Agreement on limiting global warming; and the UN's SDG Summit launching the Agenda 2030 for Sustainable Development. These conferences defined the structure and components of a global development framework that aimed not only to eradicate poverty but also to halt and reverse the adverse effects of climate change and promote sustainable, comprehensive and inclusive development (Kanuri et al. 2016).

At its Summit in New York on September 25–27 2015 the UN assembly approved the document "Transforming our World: The Agenda 2030 for Sustainable Development" which formalizes the post-2015 development framework. The member states ratified the 2030 Agenda for Sustainable Development with its 17 SDGs to be fulfilled in 15 years. All the goals and their 169 - targets were the result of long negotiation and apply to all countries, while recognizing their different priorities and different levels of development (UN 2015). Figure 1, below, presents the SDGs. The Agenda 2030 involves broader objectives, strategies and policies than those of the Millennium Agenda, aiming to create inclusive economic and social development and environmental sustainability (CEPAL 2018). The SDGs seek to generate more sustainable and resilient societies by focusing on the most pressing global challenges: poverty, inequality, climate change, economic growth and employment, urbanization, environmental degradation, and peace and justice. Although the SDGs are not legally binding, they are a normative framework via which countries at different stages of development may achieve national and local sustainability (Krellenberg et al. 2019). The Agenda particularly emphasizes the measurement of performance and results against a set of indicators to assess achievement of the SDGs and gather lessons and recommendations (Ojeda Medina 2019).

The Agenda 2030 has been described as inclusive (i.e., leaving no one behind), universal (applying to all countries regardless of income level) and comprehensive (it addresses economic, social and environmental aspects of development). In addition to expanding the scope and vision of progress and well-being, the Agenda outlines a new development architecture defined by multilevel and multi-stakeholder governance (PNUD 2017). It recognizes that many issues are no longer exclusive to nation-states and supranational institutions, and that the 17 SDGs and their

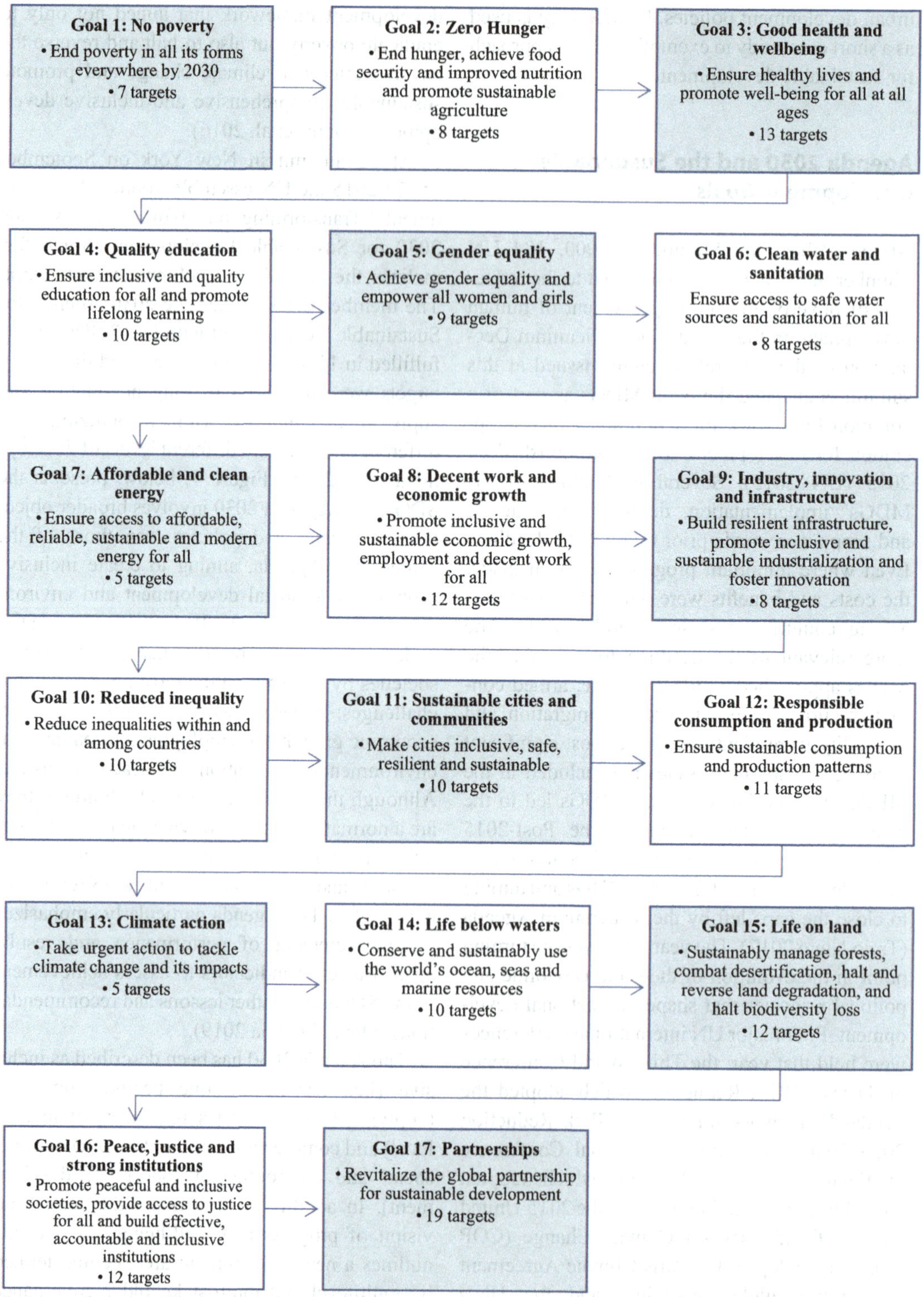

An Overview of the Relationship of the Sustainable Development Goals and Urban and Regional Development, Fig. 1 The Sustainable Development Goals. (Source: UN 2015)

169 goals must be achieved globally, nationally and subnationally, and so compliance with them requires the participation of governments and multiple subnational public and private actors (GTFLRG et al. 2016).

Localizing or territorializing the Agenda 2030 is an approach that makes explicit the significant links between the SDGs and regional and urban development and incorporates multilevel governance approaches to development. Localizing the SDGs takes subnational contexts into account from establishing the objectives and targets to determining the means of implementation, and uses indicators to measure and monitor progress. Localization also means that local and regional governments can support the achievement of the SDGs with action from below while the SDGs provide a framework for local development policy (UCLG 2018).

On the other hand, global recognition of the fundamental role of cities and regions in the achievement of development goals is manifest in a specific SDG on making cities and human settlements inclusive, safe, resilient and sustainable (SDG 11: Sustainable cities and communities), contributing to the identification of intersections between the Agenda 2030 and subnational development. In SDG 11 the development community recognize that sustainable urban development and management are crucial to the quality of all people's lives.

A Territorial Approach to the Sustainable Development Goals

The fundamental role of regional and local governments, local communities and the private and non-governmental sectors in fulfilling their obligations in relation to the new international development framework is now widely recognized (Reddy 2016; Oosterhof 2018). Because the accomplishment of the SDGs require concrete actions at different territorial levels then local, regional, and national actors' participation determines the successful adoption and implementation of the SDGs (PNUD 2017). Focusing efforts toward sustainable development on cities and regions is not only a practical requirement but also a strategic choice, as meeting sustainability goals in cities presents several opportunities and challenges (Vaidya and Chatterji 2020).

Even though urban areas occupy only a small proportion of the global landmass, they have a disproportionate impact on development that can be exploited for substantive gain in the fight against poverty, inequality and climate change (Kanuri et al. 2016). Cities are key to achieving the global agenda for sustainable development (CBD et al. 2017), and the SDGs offer a basic model for defining the scope and context of development, setting city-specific targets, contributing to urban planning and improving the clarity of local development activities (Finnveden and Gunnarsson-Östling 2017; UN-Habitat 2018). Regions, too, have a distinct role in the implementation and accomplishment of the Agenda 2030 as intermediaries and interlocutors between the national and local levels, especially in the alignment of policy priorities across levels of government and the enhancement of multilevel governance. Regions can channel resources and funding in the form of public investment for implementation of the SDGs (OECD 2020). Regions and cities are not only beneficiaries but also a significant component of the development effort, and their commitment is critical to defining and achieving results because they are in a position to transform the Agenda 2030 into concrete and efficient implementation. They can approach the goals and targets in an ad hoc way, integrating them into their own contexts and adapting them to their specific realities and needs (GTFLRG et al. 2016).

By actively encouraging the implementation of a territorial approach to the SDGs during the preparation of the Agenda 2030, different local sectors achieved explicit recognition of their roles as both recipients and key actors in development. The culmination of their work is reflected in SDG 11: "Make cities and human settlements inclusive, safe, resilient and sustainable," which explicitly relates to the urban and regional agenda (Ojeda Medina 2019). Yet, almost all the SDGs imply local action and consequences. Subnational appropriation and execution of the agenda implies carrying out a process that has been called localization. Localization is the approach to articulate the global, national, and local agendas for development.

Generally, localization is the process of restricting something to a particular place (Patole 2018). Localizing the SDGs involves authorities and other interested parties in defining, implementing and monitoring local strategies with the intention that they will contribute to achieving the Agenda's goals and targets (UCLG 2018). At the same time, it takes into consideration the characteristics of subnational contexts for achieving the Agenda 2030 (Kanuri et al. 2016). From this perspective, localization represents implementation from below in cities and human settlements (Kanuri et al. 2016; Reddy 2016; Ojeda Medina 2019). Localization places cities and regions, their governments and other subnational actors at the center of addressing the Agenda 2030 (Messias et al. 2019). While local and regional governments' responsibilities vary from country to country, they are often in charge of key areas of public policy and are in a strong position to identify and respond to gaps and need for the successful implementation of the SDGs (Oosterhof 2018).

Localizing the SDGs means the adoption of policies and solutions to the local context (Reddy 2016), involves the implementation of supranational policies in projects at the appropriate subnational level to guarantee the provision of goods, services, infrastructure, and capacities to the population (Patole 2018) and requires political will to interact with communities and find local solutions to global challenges and objectives (UCLG 2018). On the other hand, localization has become a framework for subnational planning, for local and regional governments' allocation of resources, and for local implementation of development policy (Ojeda Medina 2019). Thus the normative framing of the SDGs and its translation to specific territorial contexts through localization supports cities and regions' efforts toward sustainability (Krellenberg et al. 2019).

The starting point for localizing the SDGs is an inclusive informative process aimed at fostering local actors' awareness of what the SDGs mean. Subsequent steps include the establishment of a local agenda for the development goals, an implementation plan, and the design of indicators and a system for recording, monitoring and evaluating progress (Kanuri et al. 2016; Ojeda Medina 2019). Ultimately, localization should contribute to strengthening local capacity to formulate and implement public policy that adequately responds to societal needs in a particular territory. According to Reddy (2016), this line of action represents a significant change in terms of recognizing the benefits of a local approach to development and promoting sustainable development from below. PNUD (2017) proposes a comprehensive and not necessarily linear process comprising three major components: implementation, facilitation of institutional arrangements, and capacity-building, as illustrated in Fig. 2.

The universal, inclusive, and comprehensive nature of the Agenda 2030 requires collaboration by multiple levels of government in all sectors to accelerate the implementation of the SDGs. Oosterhof (2018) sees the process of localizing the SDGs as ideally a top-down and bottom-up process that improves vertical and horizontal policy coherence and governance, and as an integral part of a multilevel development governance system. Multilevel governance is understood as the decision-making system that defines and implements public policies produced by a collaborative relationship, whether vertical (between different levels of government including national, federal, regional and local) or horizontal (across different stakeholders). OECD (2020) underlines that the SDG framework favors multilevel governance by triggering vertical coordination across local, regional and national governments. Moreover, it fosters horizontal coordination across different sectors in cities, regions and countries as well as stakeholders' engagement to promote a holistic development approach.

Multilevel governance of the Agenda 2030 explicitly calls for national governments and international institutions to collaborate with local and regional governments in the implementation of the SDGs (Fig. 3). Multilevel governance is a powerful strategy for successful urban and regional development as the impact of traditional

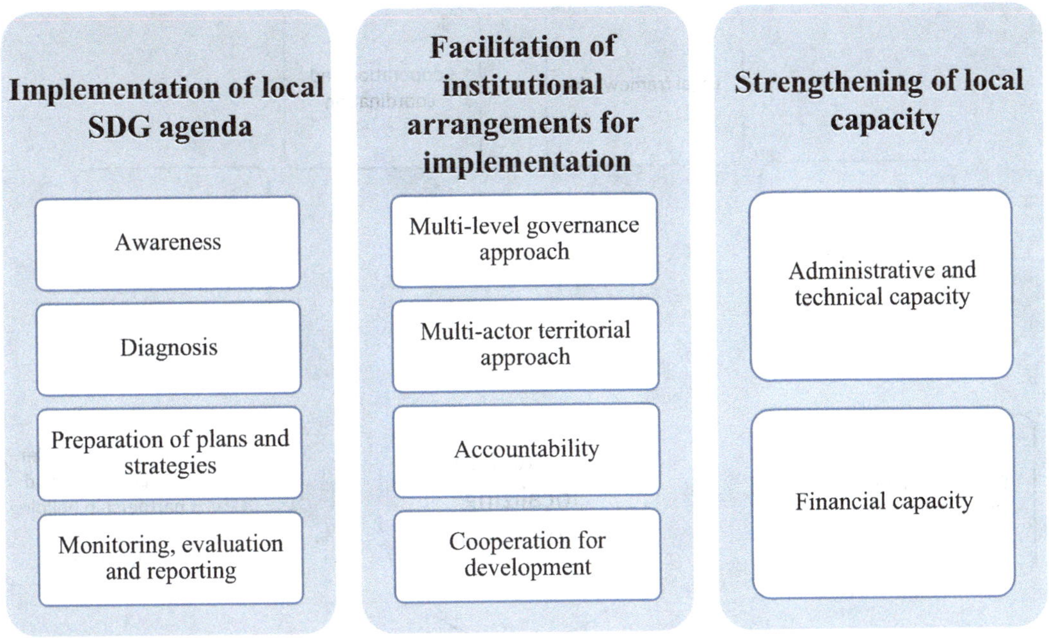

An Overview of the Relationship of the Sustainable Development Goals and Urban and Regional Development, Fig. 2 Main components of SDG localization. (Source: Based on PNUD 2017)

An Overview of the Relationship of the Sustainable Development Goals and Urban and Regional Development, Fig. 3 Multilevel governance for the implementation of Agenda 2030

top-down territorial development approaches has been questioned due to their inability to promote a more balanced territorial development. Persistent spatial disparities have threatened national growth prospects in both developed and developing economies, increased political instability and eroding the potential for sustainable development. Although top-down development initiatives are widely implemented, this multilevel approach sees the enormous potential of the bottom-up approach to achieve sustainable development objectives. Subnational spheres of government are often in a better position to facilitate the mobilization of stakeholders in local development, especially civil organizations, the private sector, and local communities in conjunction with national and international governments and organizations (Niestroy 2014).

However, the limits to localization must be acknowledged. Regional and urban governments face a number of constraints to promoting sustainable development. These include weak political and fiscal power, imperfect decentralization, limited public finances and financing, low institutional capacity, the absence of intergovernmental and multilevel cooperation and coordination, and inability to promote multi-stakeholder participation and the generation of partnerships between

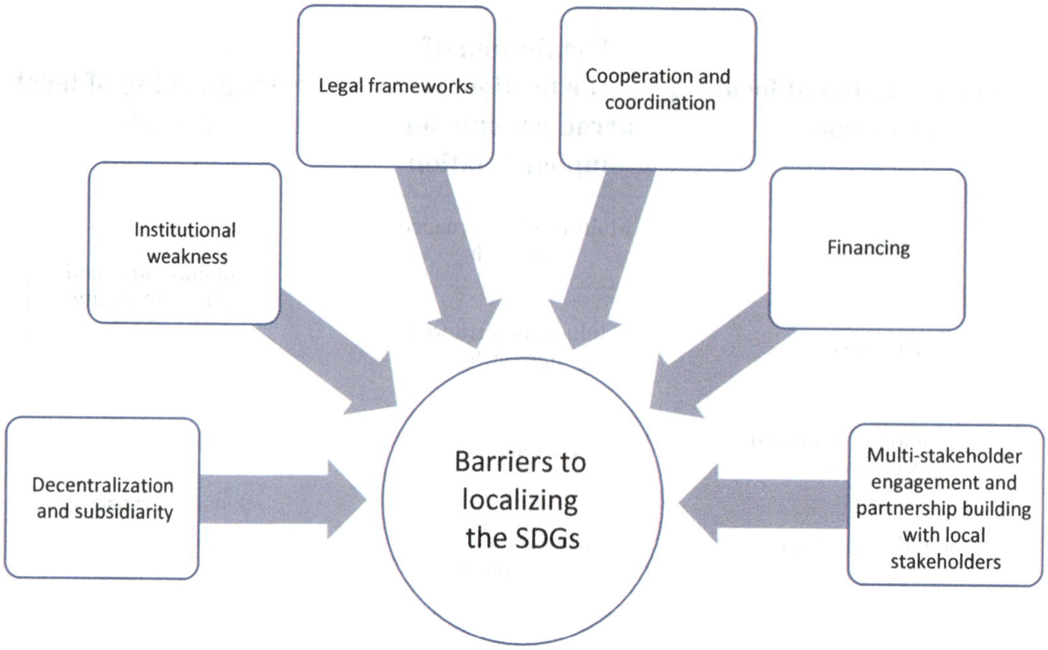

An Overview of the Relationship of the Sustainable Development Goals and Urban and Regional Development, Fig. 4 Barriers to localizing the SDGs

local actors (Fig. 4). Such constraints vary widely across territories within and across nations, especially in developing countries, imposing severe barriers to development globally (Kanuri et al. 2016; Reddy 2016).

Interlinkages Between SDG 11 and Other SDGs

The SDGs came into effect in an increasingly urban world. Since 2007 more than half the world's population has been living in cities, rising in 2018 to around 55%, and an estimated 70% by 2050 (Trejo Nieto 2020). Moreover urban areas produce 85% of global gross domestic product (Vaidya and Chatterji 2020). As the global population becomes increasingly urban, a country's cities largely determine its successes and failures. With their growing populations and economic activity, cities are recognized as growth hubs that can unlock development potential and drive sustainability. If well planned and developed, cities can promote economically, socially, and environmentally sustainable societies. However, the way in which urbanization is rapidly unfolding is deeply problematic and accompanied by various problems including housing needs, waste management, water unavailability, heavy traffic, pollution, environmental change, poverty, slums and irregular settlements, inequality, disasters, unemployment, segregation, and the deterioration of public safety. For instance, cities contribute 70% of both global energy consumption and global carbon emissions (UN-Habitat 2018). These problems can lead to the decline and stagnation of urban functions and services, affecting the development of cities, regions, and nations (WEF 2018).

As explained before, a number of initiatives have emerged to underline cities and regions' contributions to sustainable development, including the territorial approach, or localization of the SDGs. More importantly, with the introduction of SDG 11, urban development has become a prominent goal of Agenda 2030. SDG 11 involves ten targets (see Fig. 5, below), which guide the urban vision of sustainable and inclusive development

11.1 By 2030, ensure access for all to adequate, safe and affordable housing and basic services and upgrade slums
11.2 By 2030, provide access to safe, affordable, accessible and sustainable transport systems for all, improving road safety, notably by expanding public transport, with special attention to the needs of those in vulnerable situations, women, children, persons with disabilities and older persons
11.3 By 2030, enhance inclusive and sustainable urbanization and capacity for participatory, integrated and sustainable human settlement planning and management in all countries
11.4 Strengthen efforts to protect and safeguard the world's cultural and natural heritage
11.5 By 2030, significantly reduce the number of deaths and the number of people affected and substantially decrease the direct economic losses relative to global gross domestic product caused by disasters, including water-related disasters, with a focus on protecting the poor and people in vulnerable situations
11.6 By 2030, reduce the adverse per capita environmental impact of cities, including by paying special attention to air quality and municipal and other waste management
11.7 By 2030, provide universal access to safe, inclusive and accessible, green and public spaces, in particular for women and children, older persons and persons with disabilities
11.A Support positive economic, social and environmental links between urban, peri-urban and rural areas by strengthening national and regional development planning
11.B By 2020, substantially increase the number of cities and human settlements adopting and implementing integrated policies and plans towards inclusion, resource efficiency, mitigation and adaptation to climate change, resilience to disasters, and develop and implement, in line with the Sendai Framework for Disaster Risk Reduction 2015-2030, holistic disaster risk management at all levels
11.C Support least developed countries, including through financial and technical assistance, in building sustainable and resilient buildings utilizing local materials

An Overview of the Relationship of the Sustainable Development Goals and Urban and Regional Development, Fig. 5 SDG 11 and its targets. (Source: UN 2015)

by introducing transformative change in cities addressing their urgent problems, from providing affordable housing to universal access to public transport and public spaces, reducing their environmental impact and the impacts of disasters, inequality and segregation, protecting their cultural heritage, and building resilient societies. Based on this focus, the following attributes of a city are crucial in order to achieve SDG 11: inclusivity, safety, resilience, sustainability and participatory, integrated and sustainable planning. Housing policies, strategies, and legislation occupy a central place in the articulation of most urban solutions, and in ensuring cities' sustainable development (UN-Habitat 2018).

SDG 11's targets present an integrated view of sustainable urban development, with a strong orientation toward urban planning and policies (Krellenberg et al. 2019). Target 11.A in particular denotes the importance of strengthening regional development planning to support positive economic, social, and environmental links between urban, peri-urban, and rural areas and assumes that cities and regions can address a range of sustainable development domains through regulatory, fiscal, planning and policy instruments (Vaidya and Chatterji 2020).

According to Vaidya and Chatterji (ibid.) the SDGs are based on a systems approach with an emphasis on cross-linkages among several interrelated sustainability and developmental issues. Some of these connections that can help progress on the development goals but also trade-offs that can prevent or even worsen the performance among other SDGs (Finnveden and Gunnarsson-Östling 2017). For instance, there are synergies between promoting economic growth (SDG 8), ending poverty (SDG 1) and promoting equality (SDG 10). On the other hand, higher levels of CO_2 emissions – resulting from increased economic activity (SDG 8) – work against combating climate change (SDG 13) and the sustainable use of natural resources (SDG 15). Key policy questions arise especially regarding trade-offs, such as how to support the transition to more sustainable and responsible consumption and production (SDG 12) while contributing to economic growth and job creation (SDG 8) and reducing negative environmental impact (SDGs 13, 14 and 15). Whereas identifying synergies among the SDGs can help to promote policy integration in areas that are traditionally sectoral, pinpointing trade-offs is essential to trigger policy coordination mechanisms (OECD 2020).

SDG 11 has robust connections with several other SDGs. Therefore, it is not enough to focus on sustainable urban development via SDG 11 alone, as other goals also include significant urban concerns (Finnveden and Gunnarsson-Östling 2017). Issues related to affordable housing and inclusive planning in SDG 11 are closely linked to SDG 1 on eradicating poverty. Specific targets in other goals clearly fall within the interests of SDG 11, such as equal access to resources by all people (SDG 1), reducing the number of road accident fatalities (SDG 3), access to safe water and sanitation for all (SDG 6), reducing waste generation (SDG 12) and strengthening capacity for resilience and adaptation to climate hazards (SDG 13) (Vaidya and Chatterji 2020). UN-Habitat (2018) finds that SDG 11 is directly related to at least 11 other SDGs, and that 30% of indicators within the Agenda 2030 can be measured at the urban level (Fig. 6).

While Finnveden and Gunnarsson-Östling (2017) claim that all SDGs are relevant to urban areas, according to the Cities Alliance (2015) not all of the linkages are clearly defined and most can only be identified through interpretation. Such interpretation is problematic because can be

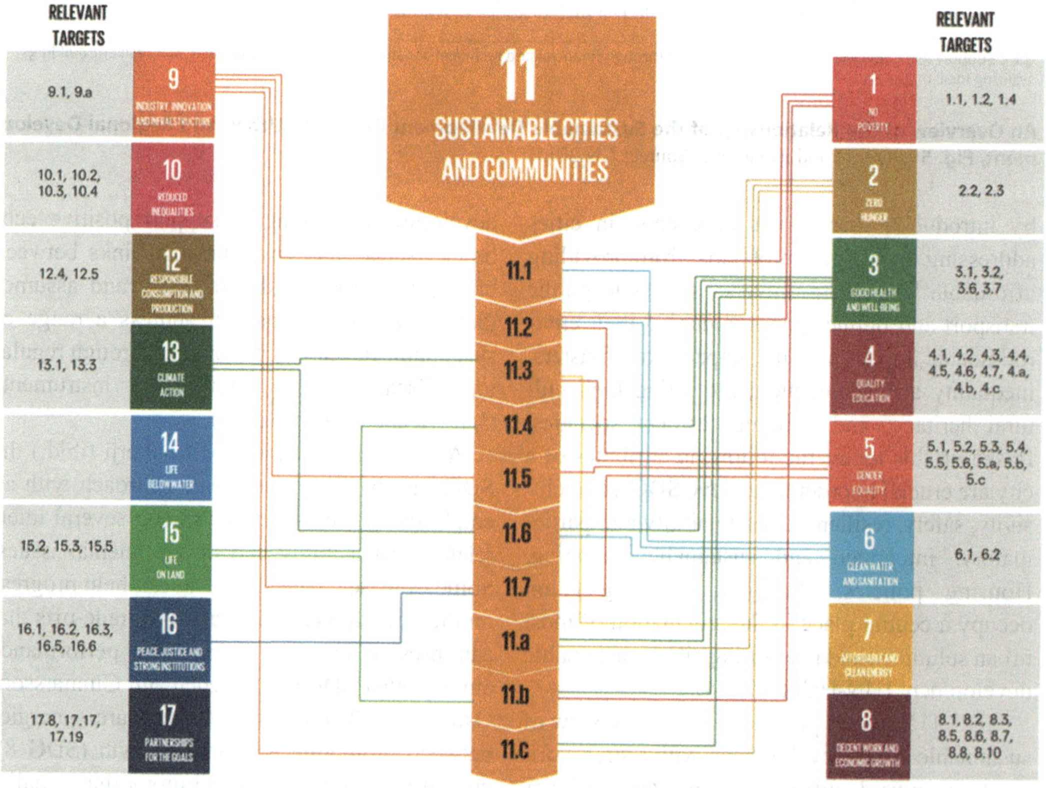

An Overview of the Relationship of the Sustainable Development Goals and Urban and Regional Development, Fig. 6 SDG 11 and the Agenda 2030 framework. (Source: Based on UN-Habitat 2018)

highly context-specific. In a general way, the targets linked to climate change, financing, sustainable production and consumption, inequality, infrastructure and basic services, gender-based violence, food security and nutrition, and migration are clearly linked to SDG 11 targets. Table 1 indicates the most direct interlinkages between SDG 11 and all other SDGs.

According to the OECD (2020), in many policy domains cities and regions are the more appropriate scale for unpacking the complexity of synergies and trade-offs among SDGs and managing and tailoring concrete solutions to specific places. Vaidya and Chatterji (2020) argue that because trade-offs between conflicting goals and priorities in Agenda 2030 are negotiated at various levels of governance, the priorities assigned to economic, environmental, and social concerns vary according to local context. For this reason different cities and regions need to establish different methods and approaches such as network analysis, participatory policymaking and matrix approaches to analyze the synergies and trade-offs among the SDGs (OECD 2020).

In addition to strong ties with certain SGDs, SDG 11 is complemented by other important urban agendas. Global urban policy frameworks such as COP 24, the Paris Agreement, the New Urban Agenda-Habitat III and numerous national urban policies are closely related to and aligned with SDG 11's focus on urban centers to achieve long term developmental objectives and create direct and tangible sustainability and inclusiveness (Vaidya and Chatterji 2020). SDG 11 is particularly connected to the New Urban Agenda, a specialized framework to drive urban sustainability. Habitat III articulates governance frameworks and financing to focus on what needs to be done to ensure that cities and human settlements are vehicles of inclusive and sustainable development (UN-Habitat 2018). With SDG 11 and the global urban and environmental agendas the world faces the urgent need to halt uncontrolled urban sprawl, reverse the growth of urban slum populations, institute safe and efficient urban transport systems, improve urban environments through creating safe public spaces, manage air pollution and solid waste, as well as promote sustainable buildings, ecosystem corridors and consumption and production patterns.

Agenda 2030 for Regional and Urban Development: The Case of Mexico City

Mexico City is one of the biggest and most complex capital cities in the world. It is also the most important demographic, political, economic, financial and cultural urban nucleus in Mexico's urban system, and one of the largest urban centers in Latin America and the world. It is often depicted as a case of urban gigantism with chaotic urban development, like many metropolises in the Global South (Gilbert et al. 2016; Trejo Nieto 2020). Its metropolitan population is calculated to exceed 21 million, and its challenges result from many derivatives of demographic concentration, urban growth and expansion such as pollution, traffic, waste management, scarce water supply, lack of affordable and adequate housing, transport and mobility, poverty, segregation, etc. In addition, Mexico City has long faced a myriad of natural and climate challenges including earthquakes, floods, droughts, heat waves and landslides, due to its geographical location.

Policy agendas and programs recognize the environmental, social and economic challenges that the city and its metropolitan region face. For instance, since the 1990s when the United Nations labelled it as the most polluted city, Mexico City has implemented strong local climate action plans and has participated actively in international forums, reporting on its commitments and policies such as the Climate Action Local Strategy 2014–2020 and the Climate Action Program 2014–2020, which together define the broad guidelines of the city's climate strategy and corresponding action plans (CBD et al. 2017). The city government has scheduled programs and actions to advance sustainable development, addressing the Agenda 2030 in its planning by including the SDGs in its development plans. This puts the city in the vanguard of localizing the SDGs in Mexico.

In 2017 the Government of Mexico City, in conjunction with academia and international and

An Overview of the Relationship of the Sustainable Development Goals and Urban and Regional Development, Table 1 Explicit links between SDG 11 the other SDGs

SDG	Relationship to SDG 11
Goal 1: No poverty	With humanity becoming increasingly urban, poverty is following and is often represented by the rise in the number of slum dwellers in cities across developing countries who lack access to basic services and adequate housing. Security of land tenure is key for the provision of services in urban areas, and also offers a foundation for access to a basic means of production
Goal 2: Zero hunger	Food security is linked to several SDG 11 targets which cover issues of nutrition, agriculture and food production, rural-urban linkages, food waste, productivity, the impact of pollution associated with cities, and consumption patterns. Sustainable urbanization that considers the need for agricultural land is a requirement for attaining SDG 2
Goal 3: Good health and wellbeing	Good health and cities are strongly linked, as health is often affected by location. Rapid and unplanned urbanization leads to more road traffic accidents and environmental and health hazards that affect the health of city dwellers. Integrated urban planning, access to basic services, decent and affordable housing reduce the incidence of non-communicable diseases and limit environmental hazards such as air pollution and dangerous traffic, contributing to better health
Goal 4: Quality education	Access to equitable, good-quality education for the urban poor and those facing vulnerability (e.g., women) contributes to making cities inclusive and sustainable. Education may help slum dwellers acquire the skills they need to find decent work, which in turn contributes to improving their living conditions
Goal 5: Gender equality	Gender equality and empowerment are linked to SDG 11 through access to and safety in public spaces, access to and use of basic infrastructure, and participation in local governance and decision-making. Promoting the inclusion and empowerment of women helps to create inclusive and sustainable cities
Goal 6: Clean water and sanitation	Clean water and sanitation are connected to SDG 11 through Target 11.6, which calls for the reduction of per capita environmental impact of cities by reducing air pollution and managing city waste better. Effective urban planning is crucial to ensuring access to safe drinking water, sanitation and hygiene, and to improving the quality and sustainability of water resources. The achievement of SDG 6 will promote better housing through the upgrading of slums and reduce the number of people affected by water pollution
Goal 7: Affordable and clean energy	Access to clean and efficient energy systems is critical to the development of safe, resilient, inclusive, and sustainable human settlements, allowing them to grow and perform efficiently. SDG 11 creates conditions for the achievement of SDG 7 through access to more sustainable transport, housing, urban planning, reduced pollution, and climate change mitigation. In contrast, unsustainable patterns of urban consumption may contribute to environmental degradation in various forms and increase the direct consumption of energy
Goal 8: Decent work and economic growth	Cities are positive and potent forces for addressing sustainable economic growth and prosperity as they drive innovation, consumption and investment. Inclusive and sustainable cities are key to achieving SDG 8 through innovation, entrepreneurship, job creation, and greater productivity. In turn, inclusive and sustainable economic growth will promote inclusive and resilient cities with better housing and urban planning, and access to basic services
Goal 9: Industry, innovation and infrastructure	Investment in infrastructure and the application of innovative technology are critical to achieving urban development. Industrialization and innovation are key to making cities safe and sustainable
Goal 10: Reduced inequality	Inequality is highly prevalent in cities, but cities are also best positioned to address this through better opportunities for employment, addressing affordable housing challenges, providing better spaces for inclusion, accessible transport, etc. Poor urban planning, design, and governance can exacerbate the exclusion and marginalization of vulnerable people

(continued)

An Overview of the Relationship of the Sustainable Development Goals and Urban and Regional Development, Table 1 (continued)

SDG	Relationship to SDG 11
Goal 12: Responsible consumption and production	SDG 11 contributes to achieving SDG 12 through the efficient management of natural resources and the treatment and safe disposal of toxic waste and pollutants. Cities that use their resources efficiently in an innovative manner increase their productivity and reduce their environmental impact, offering residents greater consumption choices and sustainable lifestyles. Integrated city planning that reduces sprawl can improve consumption patterns. Standards for buildings, energy and transport can help reduce embedded energy as well as a city's material footprint. In turn, sustainable consumption and other patterns promote inclusive, resilient, and sustainable cities by reducing latent stressors
Goal 13: Climate action	Cities both contribute to climate change and are particularly vulnerable to it and to the impact of natural disasters. SDG 13 is key for achieving the sustainability elements of SDG 11. In turn, SDG 11 offers many opportunities to develop climate change mitigation and adaptation strategies, especially through environmentally sustainable and resilient urban development and by ensuring responsible urban development plans and policies
Goal 14: Life below water	SDG 11 has a direct positive impact on the achievement of Goal 14 through the proper management of waste generated by cities, which can pollute the sea. Achieving SDG 14 in turn reinforces sustainable urban planning and resilient settlements, as much urban development occurs along the coast to take advantage of the economic advantages and opportunities presented by proximity to the sea
Goal 15: Life on land	SDG 11 contributes to achieving SDG 15 by promoting sustainable urbanization, better urban planning, development of green infrastructure, safe management and treatment of waste, and protection of the world's natural heritage. In turn, SDG 15 contributes to developing sustainable cities and human settlements through advocating for nature-based solutions and disaster risk reduction
Goal 16: Peace, justice and strong institutions	SDG 11 will only be achieved if there is peace and effective governance and financial and institutional resources for their implementation. As humanity becomes increasingly urban, the kind of urban societies we build will greatly shape progress toward the SDG's 2030 deadline. Urban crime, violence, and insecurity must be addressed to achieve both SDG 11 and SDG 16. Corruption and illicit financial flows increasingly occur in cities and in many urban development efforts. Peaceful, inclusive, and sustainable cities rely on strong institutions
Goal 17: Partnerships	Goal 11 will only succeed if there are strong partnerships for sustainable urban development, involving a wide network of actors including international organizations, UN Member States, international and regional associations of cities, NGOs, the private sector, specialized funding bodies, goodwill ambassadors and civil societies, and National Commissions

Source: Based on UN-Habitat (2018)

civil society organizations, initiated various actions to promote compliance with the SDGs and demonstrate the alignment of local objectives, actions and achievements with those of the Agenda 2030. The city government instituted a Council for the Follow-up of the Agenda 2030 for Sustainable Development in Mexico City consisting of different city government agencies and chaired by the Head of Government. Operational guidelines and four technical committees for the thematic monitoring of all SDGs, with the exception of SDG 14 – Life below water – which is out of alignment, were created. The Council designed a mechanism that contrasts the objectives of the institutional programs available to the local public administration with the 169 SDGs targets and relates the city government's actions to the objectives and targets of the Agenda 2030. The General Development Plan for Mexico City 2013–2018 was aligned to a total of 126 SDG targets. The government also introduced the Government Management Monitoring and

Evaluation System to identify and follow up progress toward achieving the SDGs. In October 2017 the Social Development Evaluation Council included plans for social programs aligned with the Agenda 2030 (Gobierno de la Ciudad de Mexico 2018).

Based on the model proposed by the federal government in its Guide to Incorporating the Agenda 2030 in the Preparation of State and Municipal Development Plans, the city government designed technical committees for the economy, society, the environment, and alliances to monitor the SDGs (Fig. 7). This framework systematizes the alignment of local development with the Agenda 2030 and specific programs for each SDG are usually incorporated.

The core principle of the Government of Mexico City's development policies for 2019–2024 is sustainability. The most recent government has explicitly engaged with the Agenda 2030, incorporating sustainable development into the new Development Program of Mexico City 2019–2024 which is based on the global development agenda. Deep-rooted inequality, violence, and resilience are the greatest policy challenges, and the program is organized into six main areas: equal rights; sustainable city; more and better mobility; Mexico City-cultural capital; zero aggression and better security; and innovation and transparency. Each of these has specific objectives in response to the global development agenda (Government of Mexico City 2019). Mexico City's resilience strategy, published as part of the 100 Resilient Cities Initiative, also addresses the city's development challenges with a five-pillar structure incorporating elements of planning

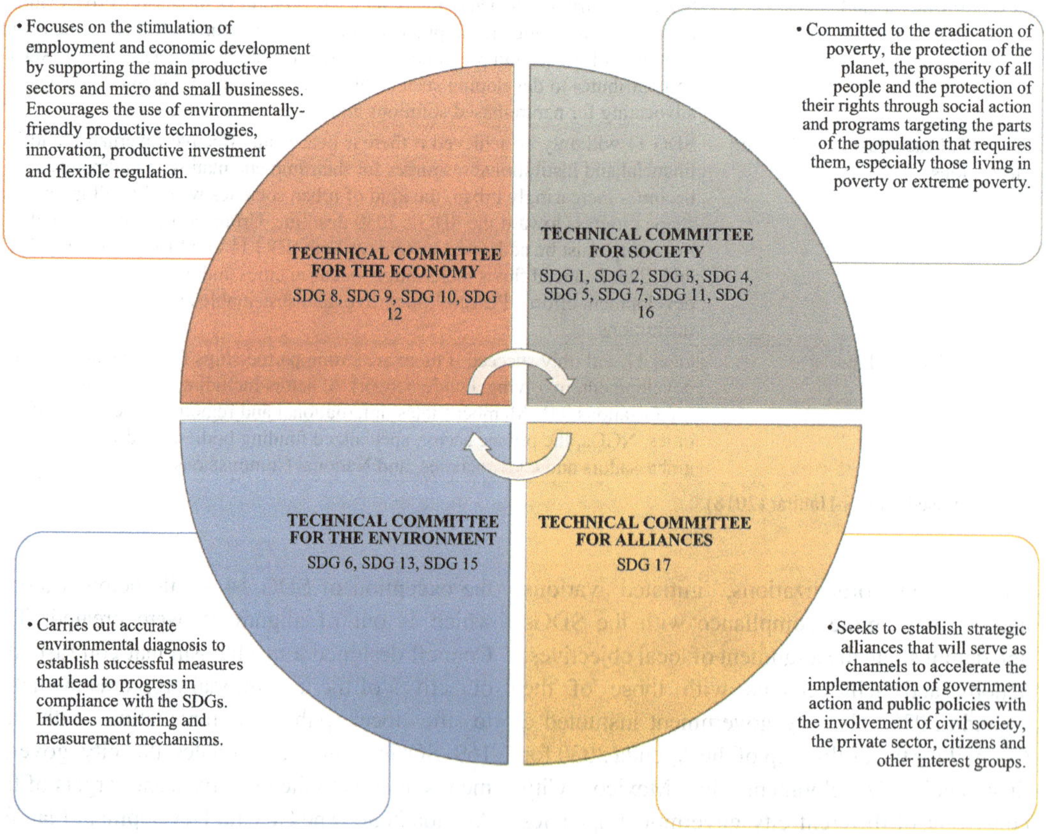
• Focuses on the stimulation of employment and economic development by supporting the main productive sectors and micro and small businesses. Encourages the use of environmentally-friendly productive technologies, innovation, productive investment and flexible regulation.
• Committed to the eradication of poverty, the protection of the planet, the prosperity of all people and the protection of their rights through social action and programs targeting the parts of the population that requires them, especially those living in poverty or extreme poverty.
TECHNICAL COMMITTEE FOR THE ECONOMY
SDG 8, SDG 9, SDG 10, SDG 12
TECHNICAL COMMITTEE FOR SOCIETY
SDG 1, SDG 2, SDG 3, SDG 4, SDG 5, SDG 7, SDG 11, SDG 16
TECHNICAL COMMITTEE FOR THE ENVIRONMENT
SDG 6, SDG 13, SDG 15
TECHNICAL COMMITTEE FOR ALLIANCES
SDG 17
• Carries out accurate environmental diagnosis to establish successful measures that lead to progress in compliance with the SDGs. Includes monitoring and measurement mechanisms.
• Seeks to establish strategic alliances that will serve as channels to accelerate the implementation of government action and public policies with the involvement of civil society, the private sector, citizens and other interest groups.

An Overview of the Relationship of the Sustainable Development Goals and Urban and Regional Development, Fig. 7 Local framework for aligning with the Agenda 2030 in Mexico City. (Source: Based on Gobierno de la Ciudad de Mexico 2018)

for urban and regional resilience, sustainability and inclusiveness: (1) Foster regional coordination; (2) Promote water resilience as a new paradigm to manage water in the Mexico basin; (3) Plan for urban and regional resilience; (4) Improve mobility through an integrated, safe and sustainable system; (5) Develop innovation and adaptive capacity (CBD et al. 2017).

Regular reviews must be conducted to ensure that goals and actions are evaluated and updated. Programs for and results of the localization of the Agenda 2030 in Mexico City are registered progressively, with sustainability at the core of the city's planning practice. Nevertheless, there have been pressing issues to deal with as development action is implemented. Like many other cities in the world that are actively implementing a sustainable development agenda, Mexico City has had to consider alternative financing for its development plans and programs, and was the first city in Latin America to issue green bonds. The introduction of green bonds as a source of finance was fostered by international organizations, national government policies and strategies, and then adopted by city governments to mobilize capital for climate-change-related investment that promotes sustainable development. One of the main drivers behind the issuance of city green bonds has been the gradual transition of public policy and legislation toward sustainability. For a city to issue green bonds, it must be creditworthy to raise capital in the financial market. A local Mexican development bank, Nacional Financiera, issued its first green bond in November 2015, paving the way for Mexico City's first green bond worth 1000 million Mexican pesos (around 46 million euros) in 2016. This contributed to the sustainability agenda by meeting the city's urgent need for sustainable infrastructure and buildings, renewable energy, energy and water efficiency, wastewater management, pollution prevention and control, the conservation of biodiversity, and adaptation to climate change (CBD et al. 2017).

Notably, most of the SDGs have become benchmarks for sustainable city action, and the local SDG strategy has helped to consolidate a sustainable administration model of which social inclusion, environmental action, economic growth, and the strengthening of alliances are fundamental pillars. However local progress on the SDGs requires models of governance that attract and coordinate the efforts of all the local stakeholders, including citizens. In the case of this extended city, a metropolitan approach to sustainability is an urgent need. Despite significant effort in Mexico City and other Mexican cities and regions, continuing the sustainability model successfully will require further endeavor and partnerships between government, multilateral development banks, and the private sector.

Summary

In 2015 a new global agenda to achieve sustainable development by 2030 was unanimously adopted by the 193 UN Member States. The Sustainable Development Agenda, or Agenda 2030, consists of 17 Sustainable Development Goals (SDGs) with 169 targets which follow and expand on the Millennium Development Goals, a previous development agenda to end poverty agreed in 2000. The 17 SDGs address the interconnected economic, social, and environmental elements of sustainable development and convey the world's complex development challenges. Universality, inclusiveness and integrality are the guiding principles of the Agenda, which applies to all countries independently of their level of development.

The Agenda recognizes that most people live and work in cities and that urbanization is continuing across the world, with 70% of the global population expected to be urban dwellers by 2050. The inclusion of SDG 11 as a development goal addressing cities and human settlements highlights the importance of urbanization in sustainable development and commits national governments to support increased sustainability, inclusivity, and resilience in their subnational territories. However, cities and regions matter for much more than SDG 11 because they are at the front line of sustainable development as they can translate the Agenda's goals into plans for local progress.

The Agenda creates two-way interlinkages between urban/regional development and the

SDGs: the SDGs provide a significant guide to locally-tailored and sustainable regional and city development planning; the Agenda 2030 can facilitate sectorally integrated urban planning as the SDGs are interlinked by nature; the SDGs can support more integrated multilevel planning (central, regional, and local) and enhance horizontal cooperation among stakeholders; and they are an important means of information, communication, and accountability for local governments. To take advantage of this link, national governments are localizing the SDGs in a territorial approach with four main objectives: inform and communicate the global development approach to subnational stakeholders; implement development plans and policies from below; measure and monitor cities and regions' progress toward realizing the SDGs; and analyze how cities and regions use the SDGs to design policy and communicate and coordinate with upper levels of government. Effective localization of the SDGs can support the identification of best practice and lessons learnt from international experiences. Other benefits of using the SDGs to promote sustainable urban and regional development include the systematic implementation of planning measures, effective use of scarce resources, reduction of trade-offs between development objectives, and reinforcement of local capacity to formulate, implement, and evaluate development plans and policies.

On the other hand, cities and regions are basic territorial units that make the advancement of the global development agenda possible. In many cases cities and regions have significant competency in policy areas underlying the SDGs such as economic productivity and employment creation, water, housing, transport, infrastructure, land use, and climate change. Therefore local and regional governments can have a direct impact on the implementation of the majority of the 169 targets. Cities and regions can reinforce the interlinkages and integrated approaches across sectors and goals and facilitate access to local data and knowledge. However, it is important to identify the limits and barriers to local actors' and especially governments' implementation of the development agendas. Structural challenges to localization include uneven decentralization and limited access to finance, the need to strengthen cooperation between governments and local actors (civil society, industry, and academia) and an appropriate governance approach.

The relevance of cities and regions in improving sustainability imposes a necessary multilevel approach to governance where the global, the national, and the local reinforce one another. The need for collaboration and coordination across many sectors poses unique opportunities and challenges for international agencies and national and subnational governments. Multilevel governance is important to reinforce the interlinkages between urban and regional development and the SDGs. While national governments are necessary for the formulation of local plans aligned with national SDG implementation strategies, they cannot achieve the goals of the Agenda alone. Moreover, local and regional governments often participate in the design of national SDG implementation strategies and lead consultations with communities. They can be present on national councils and committees for the Agenda 2030 which makes possible the systematic involvement of local authorities in SDG implementation strategies.

Different proposals on integrating subnational spaces into the theory and practice of the Agenda 2030 and the SDGs have been implemented around the world. In Mexico, the second most important Latin American country, Mexico City is pioneering the localization of SDGs and the integration of a local sustainable development agenda. The problems it faces are gigantic in both nature and magnitude as it is one of the largest metropolitan areas in the world. Financing solutions and policies for sustainable development are a key issue for such huge urban centers. Mexico City has introduced innovative planning and financing tools such as green bonds. While considerable efforts have been made to address sustainability challenges and important progress in urban development has been made, monumental challenges to realizing the Agenda 2030's far-reaching vision still remain.

References

CBD, et al. (2017). Connecting cities and communities with the SDGs. CBD, ECLAC, FAO, ITU, UNDP, UNECA, UNECE, UNESCO, UN Environment, UNEP-FI, UNFCCC, UN-Habitat, UNIDO, UNU-EGOV and WMO. Geneva.

CEPAL. (2018). *La Agenda 2030 y los Objetivos de Desarrollo Sostenible: una oportunidad para América Latina y el Caribe. Objetivos, metas e indicadores mundiales*. Comisión Económica para América Latina. https://www.cepal.org/es/publicaciones/40155-la-agenda-2030-objetivos-desarrollo-sostenible-oportunidad-america-latina-caribe

Cities Alliance. (2015). Sustainable Development Goals and Habitat III: Opportunities for a successful New Urban Agenda. Cities Alliance Discussion Paper, No. 3.

Finnveden, G., & Gunnarsson-Östling, U. (2017). Sustainable development goals for cities. In Bylund, J. (ed), *Connecting the dots by obstacles? Friction and traction ahead for the SRIA urban transitions pathways*. Brussels: JPI Urban Europe Symposium. http://urn.kb.se/resolve?urn=urn:nbn:se:kth:diva-206339

Gilbert, L., Khosla, P., & De Jong, F. (2016). Precarización y crecimiento urbano en la zona metropolitana de México. *Espacialidades,* 6(2), 5–32. http://espacialidades.cua.uam.mx/ojs/index.php/espacialidades/article/view/133

Gobierno de la Ciudad de Mexico. (2018). *Informe de la agenda 2030 para el desarrollo sostenible en la Ciudad de México, 2030 CDMX*. Mexico City: Gobierno de la Ciudad de Mexico, Coordinación General de Modernización Administrativa. http://www.monitoreo.cdmx.gob.mx/statics/g_apoyo/Libro_2030CDMX.pdf

Government of Mexico City. (2019). *Innovation and rights; A program to advance sustainable development in Mexico City*. Government of Mexico City. https://www.cgaai.cdmx.gob.mx/storage/app/uploads/public/5df/7d6/d97/5df7d6d973e49264453690.pdf

GTFLRG, UN-HABITAT, & UNDP. (2016). *Roadmap for localising the SDGs: Implementation and monitoring at subnational level*. Global Taskforce of Local and Regional Governments, UN Habitat, and United Nations Development Programme. https://sustainabledevelopment.un.org/content/documents/commitments/818_11195_commitment_ROADMAP%20LOCALIZING%20SDGS.pdf

Kanuri, C., Revi, A., Espey, J., & KuhlKanuri, H. (2016). *Getting started with the SDGs in Cities: A guide for stakeholders*. Sustainable Development Solutions Network.

Krellenberg, K., Bergsträßer, H., Bykova, D., Kress, N., & Tyndall, K. (2019). Urban sustainability strategies guided by the SDGs. A tale of four cities. *Sustainability, 11*(1116), 1–20.

Messias, R. (2017). *SDGs at the subnational level: Regional governments in the voluntary national reviews*. Brussels: Network of Regional Governments for Sustainable Development and United Regions Organization. https://www.regions4.org/wp-content/uploads/2019/06/R4_SDGsatSubnationalLevel2017-1.pdf

Messias, R., Grigorovski Vollmer, J., & Sindico, F. (2019). *Localizing the SDGs: Regional governments paving the way: 2018 Report*. Brussels: Regions4 for Sustainable Development and University of Strathclyde Centre for Environmental Law and Governance. https://www.regions4.org/wp-content/uploads/2019/06/Localizing-the-SDGs.pdf

Niestroy, I. (2014). *Sustainable Development Goals at the subnational level: Roles and good practices for subnational governments* (SDplanNet Briefing Note). Sharing Tools in Planning for Sustainable Development. https://www.iisd.org/sites/default/files/publications/sdplannet_sub_national_roles.pdf

OECD. (2020). *A territorial approach to the sustainable development goals: Synthesis report* (OECD urban policy reviews). Paris: OECD Publishing. https://doi.org/10.1787/e86fa715-en.

Ojeda Medina, T. (2019). El Rol Estratégico De los Gobiernos Locales y Regionales en la Implementación de la Agenda 2030: Experiencias Desde la Cooperación Sur-Sur y Triangular. *OASIS, 31*, 9–29. https://doi.org/10.18601/16577558.n31.03.

Oosterhof, P. D. (2018). *Localizing the SDGs to accelerate the implementation of the 2030 Agenda for Sustainable Development: The current state of Sustainable Development Goal Localization in Asia and the Pacific*. Asian Development Bank, Governance Briefs, No. 3, 1–14. https://www.adb.org/publications/sdgs-implementation-2030-agenda-sustainable-development

Patole, M. (2018). Localization of SDGs through disaggregation of KPIs. *Economies, 6*(1), 15.

PNUD. (2017). *Articulación de redes territoriales para el Desarrollo Humano Sostenible: Resumen 2015–2016*. Brussels: Programa de las Naciones Unidas para el Desarrollo.

Reddy, P. S. (2016). Localising the SDGs: The role of local government in context. *African Journal of Public Affairs, 9*(2), 1–15.

Trejo Nieto, A. (2017). Crecimiento Económico e Industrialización en la Agenda 2030: Perspectivas para México. *Problemas del desarrollo, 48*(188), 83–112. http://www.scielo.org.mx/scielo.php?script=sci_arttext&pid=S0301-70362017000100083&lng=es&tlng=es

Trejo Nieto, A. (2020). *Metropolitan economic development. The political economy of urbanisation in Mexico*. London: Routledge. https://doi.org/10.4324/9780429456053.

UCLG. (2018). *Towards the localization of the SDGs. local and regional government's report to the 2018 HLPF,*

O

No. 2. Barcelona: United Cities and Local Governments. https://www.gold.uclg.org/sites/default/files/Towards_the_Localization_of_the_SDGs.pdf

UN. (2015). *Draft outcome document of the United Nations Summit for the Adoption of the Post-2015 Development Agenda: Draft resolution submitted by the President of the General Assembly*. New York: United Nations. https://digitallibrary.un.org/record/800852

UN-HABITAT. (2018). *Tracking progress towards inclusive, safe, resilient and sustainable cities and human settlements: SDG 11 synthesis report – High level political forum 2018*. New York: UN. https://doi.org/10.18356/36ff830e-en.

Vaidya, H., & Chatterji, T. (2020). SDG 11 sustainable cities and communities. In I. B. Franco et al. (Eds.), *Actioning the global goals for local impact* (pp. 173–185). Springer Singapore. https://doi.org/10.1007/978-981-32-9927-6_12.

WEF. (2018). *The global risks report*. Geneva: World Economic Forum. http://www3.weforum.org/docs/WEF_GRR18_Report.pdf

P

Paradigm/Method

▶ Metropolitan Discipline: Management and Planning

Paris Agreement

▶ Policy and Practices of Nature-Based Solutions to Build Resilience in Seoul, Korea

Parks

▶ Multiple Benefits of Green Infrastructure

Parks and Green Space

▶ Policy and Practices of Nature-Based Solutions to Build Resilience in Seoul, Korea

Participate

▶ Closing the Loop on Local Food Access Through Disaster Management

Participative

▶ Participatory Planning: A Useful Tool for the Development of Sustainable Mega-City Regions

Participatory

▶ Moving Towards Sustainable, Liveable, and Care-Full Urban Environments: Pre-schoolers' Rights and Visions for Planning Just, Socially, and Ecologically Integrated Cities

Participatory Governance for Adaptable Communities

Lessons from India's Informal Settlements

Roshini Suparna Diwakar
Mahila Housing Trust, New Delhi, India

Synonyms

Community leadership; Democratic processes; Local governance; Urban governance; Urban poor

R. C. Brears (ed.), *The Palgrave Encyclopedia of Urban and Regional Futures*,
https://doi.org/10.1007/978-3-030-87745-3

Definitions

Participatory governance	The process wherein citizens engage in policy-making through deliberative practices. It enables the citizens to engage in the entire decision-making process – proposal, design, implementation, monitoring and evaluation, and redesign – and empowers them to collaborate with the state more substantively than in conventional democratic governance.
Adaptable communities	Spatially defined communities that are responsive to the changes that are caused by a shift in policy, environment, or development practices.
Informal settlements	Term that is broadly used to define those settlements in urban spaces that are not planned or formalized by city or state structures. Informal settlements include slums and unauthorized settlements or what is commonly known as squatter settlements on public and private land and resettlement colonies or settlements which are relocated by the government for development projects.

Introduction

The decade of the 2020s began with a global pandemic forcing a re-evaluation of existing social policies and approaches toward community development. Countries world over have had to take stock of not only the ability of their respective healthcare system to respond to such a crisis but also the very structures that govern public life. It has become evident that local communities are at the center of social life and that the strengthening of these ecosystems is critical in order to effectively respond to such crises. In India, local government institutions became *the* access point to basic resources for marginalized communities. These state structures bridged the gap between isolated settlements and those beyond immediate geographies. They were challenged unequivocally, and stark inadequacies were unraveled. The role of nongovernmental organizations (NGOs) and community-based organizations (CBOs) grew during this phase, as they collaborated with state machinery and communities to facilitate access to resources and protect the fundamental right to food, healthcare, shelter, and livelihood. In spaces where community as well as state leadership and networks were strong, the response to the locally emerging needs was more efficient.

With the impact of climate change becoming increasingly evident, the need to respond to environmental and health crises on an immediate as well as systemic way will only grow. In order for communities to be able to adapt to these evolving realities, thrive, and flourish, local leadership must be developed. Solutions that are contextual and sustainable can emerge when communities participate in a dialogue with policy-makers and are included in decision-making processes. This is particularly true for marginalized communities whose realities are often misunderstood, invisibilized, and deprioritized by those with decision-making powers and continue to grapple with fundamental issues of access to basic amenities (Chattopadhyay 2015).

The shift toward creating more participatory institutions has already occurred on paper in India through the 73rd and 74th Constitutional Amendments, though the process of inclusion has had varying degrees of success in reality (Patel et al. 2016).

Framework for Participation

India, in the early 1990s, witnessed two major, intertwined shifts in policy; the first was the liberalization and privatization of the economy and, the second, in keeping with global trends, the emergence of local democratic institutions in order to increase citizen participation and promote "good governance" practices (Chattopadhyay 2015; Fischer 2012; Dupont 2007; Daly and Silver 2008).

The 73rd and 74th Constitutional Amendments mandated the institutionalization of a third tier of government – the Panchayati Raj institutions in rural areas, the Nagar Panchayats in transitional areas, and the municipal councils/corporations in urban areas, expanding the existing center and state government structures. Most critically, the 74th Amendment directed the state to create Ward Committees (WCs) as the lowest, directly elected body that would represent the local needs of the community and streamline service delivery. It provided for the reservation of seats on the committees for representatives of marginalized castes and tribes as well as women. The devolution of power has been strengthened by the creation of financial structures that enable these institutions to fulfill their functions (Sachdeva 2011).

More than 25 years later, the adoption of the Amendment has been varied. While some states have had successes in creating effective and functional institutions, others' attempts have been piecemeal and inadequate. A major concern that emerged was the further marginalization of those who lack social capital by the local elite (Patel et al. 2016; Dupont 2007). This has been a common experience of participatory governance processes not only in India but across the globe.

In the Indian context, the accelerated increase in the urban population, especially as a result of migration from rural areas, has led to significant inequities across socioeconomic indicators (Shaw 2012). The expansion and creation of cities was coupled with the growth of informal settlements that accommodated the urban poor (ibid.). In addition to the income and job insecurity that the urban poor face in the informal economy, they are confronted by the threat of eviction, lack access to entitlements and dignity, and struggle to break out of the poverty cycle due to sociopolitical marginalization. Their access to their democratic representatives is often restricted to the 5-year election cycle wherein they are viewed as a "vote bank" to tap into (Auerbach 2019).

Thus, participatory governance structures that truly include everyone can help these communities break out of the poverty cycle, meet their immediate demands, and become empowered citizens within the polity. There has been significant literature in the "democracy and development" academic space that reiterates the claims made by the proponents of participatory governance (Fischer 2012). For the communities in India's urban informal settlements, participatory governance is a means to access resources and meet their needs, as well as an end in itself – the emergence of an empowered citizen.

Community Action Groups (CAGs): A Model for Local Leadership

Origin of the CAG Model

The Gujarat Mahila Housing SEWA Trust (MHT) is a nongovernmental organization that emerged out of the Self-Employed Women's Association (SEWA) in the mid-1990s. The goal of the organization was to address the habitat issues of urban poor women. One of its first programs was the Slum Networking Project (SNP) introduced by the Ahmedabad Municipal Corporation (AMC) in the state of Gujarat. The project aimed at the upgradation of services in the urban poor settlements of the city (Jhabvala and Brahmbhatt 2020). Central to the project was community participation and contribution. Residents of these settlements would contribute financially toward its development and collectively decide which common services they wanted to prioritize. This required substantial dialogue and coordination among the residents. To facilitate this process, MHT enabled the creation of Community Action Groups (CAGs) that constituted 8–12 women residents chosen from among and by the women members of every 250 households. These CAG members worked with the AMC by sharing information, bringing community members on board, and articulating the priorities of the community. They mobilized the community to participate in the process and co-create changes within their settlements. This flagship program, launched just 2 years after the adoption of the 74th Amendment, is an oft-cited example of participatory governance and community engagement.

Process

When the organization begins a relationship with a new community, the first goal is to identify immediate needs and build trust with its members.

Trust can be built when common threads between the community and the external stakeholder are highlighted. This process takes months, with frequent interaction critical for relationship-building. Over the course of this period, women residents participate in meetings organized by MHT and share the concerns of the community. As trust gets built, women community members identify 8–12 leaders from among them to form the CAG. Once the CAG members are selected, they are trained in "individual and basic services," "formation of CAG," "map sharing and participation," "urban local bodies (ULBs) structure," "health and hygiene," "maintenance of services," "introduction to materials for construction of basic services," "solid waste management" (SWM), "basic accounting," "pre-monsoon planning," and "climate change." This training is the foundation on which they begin engagement with multiple stakeholders, specifically various ULBs, to advocate for their rights.

The mobilization of the community and their acknowledgement of the importance of participating in this process is critical. CAGs meet monthly or when issues are raised by community, and solutions are sought and next steps determined during the meeting. Coordination among the members ensures transparency within the group and makes the follow-up process easy.

As the CAGs develop a collective identity that is recognized within the community and by other actors of the system, their ability to negotiate and make demands increases. Networks get built over time, and the leadership capabilities of the CAG members are realized, enabling them to strategically navigate their position and fulfill their needs within the system.

Over time, CAG members are empowered to mentor a new group of leaders within their community and expand their geographical reach to neighboring settlements that lack such local leadership structures. Engagement with other informal settlements and solutioning for local problems provides the CAG leaders with the opportunity to identify patterns across contexts. These patterns of issues are organized as broader demands and policy recommendations for the planning and development of the city. Two leaders are chosen from every CAG by their members to represent them at a city-level platform. This organically creates a bottom-up, people-centric approach toward planning and engages marginalized communities in a dialogue with those in power. They themselves recognize a shift in thinking beyond the slum to broader trends that impact the lives of the urban poor in informal settlements.

The range of issues being addressed widens over time as well. While the initial demands are pressing in nature and what would be considered "low-hanging fruit," the successes from having these demands met lead to an expansion in the role of the CAG and a focus on deep-rooted problems.

Unlike some other local leadership models, the CAG does not focus on issues on an individual level or only those that affect the members of the group; the cornerstone of the model is to participate and create transparency and accountability with local government. What makes the model unique is also that it is issue agnostic and responds to the communities' needs as they evolve and emerge.

Underlying this process is the slow creation and accumulation of social capital by the CAGs.

Building Social Capital

The term "social capital" became popular in the late 1980s to the early 1990s, and its origin, characteristics, and manifestations have been debated widely in academia (Poder 2011). Various camps have emerged to deliberate whether it rests with an individual or a collective, its positive and negative effects, and if it can truly be termed "capital" when compared to other forms such as physical and human capital (ibid.).

The CAG model leans on Putnam's approach toward social capital, identifying it as a resource that is held collectively. It is created over time, incrementally, through repeated interactions, the development of relationships and networks, and the building of trust (ibid.).

There are two factors that commence the process of social capital formation among the CAG members: (1) the belief that participation will lead to change and (2) that collective action can be more powerful than individual action. The latter is particularly true for these marginalized communities who lack access to resources and have limited political salience at an individual level. However, when collectives are formed, the "strength in numbers" increases their bargaining power and enables them to begin the process of building networks.

For the newly formed CAGs, the first step to building social capital is developing an identity and creating networks within the community and among group members. "Bonding" social capital enables the CAGs to identify and prioritize the needs of the community, coalescing various sub-identities under common threads (Daly and Silver 2008). This process creates a collective identity of the CAG that is recognized and valued within the community. The CAGs then form linkages with other stakeholders such as local government administrators, elected representatives, and private service delivery agents. This "bridging" social capital develops networks with external stakeholders and creates intergroup cooperation (ibid.).

Both "bonding" and "bridging" social capital are built by the CAGs not through the training that they receive but by acting on the needs expressed by the community. It is developed incrementally, over time, tacitly as other actions, interactions, and partnerships are undertaken. This built social capital, in turn, acts as a resource to expand the network, address complex issues, and delve deeper into the advocacy for participatory governance structures.

In the context of the Community Action Groups, social capital is first manifested when services are delivered efficiently and in a time-bound manner. As interactions with the government increases, a mutual respect is developed between the CAGs and the state, and collaborative partnerships are created. This accelerates the expansion of the network, linking the CAGs to other external actors. In time, the relationship becomes mutually productive, with the CAGs bridging the gap between the local government and the community.

The course of social capital creation is initiated by an external stakeholder, in this case MHT, whose role as the catalyst evolves as social capital accumulates.

Role of the Catalyst

The role of the catalyst is vital for the creation of social capital and, consequently, enabling participatory governance. While local leadership and social capital does exist in these informal settlements, they are usually geared toward addressing cultural and religious issues, focusing on the salience of those identities, rather than common developmental problems. The catalyst, an external stakeholder, can thus create the conditions required for the generation of collective social capital.

Training individual leaders is often an easier process than building collective leadership. However, the institutionalization of local leadership is possible only when structures that are not dependent on individuals are constructed. For participatory governance to be effective, participation must be inclusive. The creation of new hierarchies within the communities wherein individual leaders consolidate social capital will only further entrench them in existing inequities.

For the CAG model to be adopted, therefore, the catalyst must focus on the development of collective leadership. This is a long process that is not determined by specific projects undertaken but rather through an organic empowerment of the group. The catalyst must be equipped to invest in the community and their needs for long durations and continue to facilitate the evolution of the CAGs as they move toward self-actualization.

A good indicator for the built capacities of the group is when it approaches the catalyst with issues emerging from the community itself, often beyond the purview of the catalyst, or when they are able to identify and resolve problems by partnering with other stakeholders without the catalyst linking the two actors. At this stage, the position of the catalyst changes to equate that of any other external stakeholder engaged in the system.

The withdrawal of the catalyst or change in its position must occur organically, without disrupting the rest of the network. The social capital of the CAG will enable the structure to remain intact as the catalyst realigns its position.

Enablers for the Model

The creation of a CAG is enabled when there is an urgent need for local leadership within the community. This requires the acknowledgement of issues that exist within the community and the belief that engaging with the process will create change. While the latter can be created through dialogue with the catalyst, the former must exist for participation. The community must believe that participation will pay off in order for them to engage (Fischer 2012).

Another initial enabler is the focus on "low hanging fruit" or issues that are urgent and will be concretely realized. Concerns raised at the start are often common across identities within the community, creating trust in both the catalyst and the CAG when these are addressed.

When the collective is built, leaders within the community invest in ensuring its sustainability. Their focus is on institutionalizing the model by not focusing on individual leadership and developing collective social capital, as earlier stated.

Most significantly, as issues get addressed and the network grows, the CAG develops a sense of purpose. This becomes a crucial driver for their continued engagement.

Barriers to the Model's Success

The agenda behind the catalyst's entry into the community is an important determinant for the model's success. When a top-down approach is taken by the catalyst without consulting the community on their priorities, the community has little incentive to engage deeply in the process. Mobilizing marginalized communities to take action requires significant investment of time and resources by the catalyst, something it is often unable to do due to project durations and donor commitments.

As a model that empowers the marginalized within marginalized communities (women), contextual barriers emerge. The mobility of women and active participation in "public life" is controlled by the men in the family or other women who are positioned higher in the hierarchy. Breaking out of these patriarchal molds and pushing the boundaries of engagement is a slow, nonlinear process.

The overshadowing of collective identity by individual leaders can hamper the growth of the CAG and prevent the institutionalization of the structure. While the individual leader might be successful in meeting the immediate demands of the community, dependence on the individual is not sustainable, and leadership will collapse once they are gone.

Informal settlements in urban India are congested, densely populated, and heterogeneous in sociocultural identities. Thus, there are several conflicting interests that emerge within these communities along with their own models of leadership. Once Community Action Groups are created, they are often threatened by other local leadership, especially those connected to political parties. This leads to internal conflicts and can stifle the smooth functioning of the CAG.

Conclusion

The Community Action Group (CAG) model effectively illustrates how enabling marginalized communities and facilitating the creation of social capital among these groups to collaborate with and monitor the government can produce structures for participatory and responsive governance. Over the past two decades, CAGs have engaged in the implementation and monitoring of local policies that apply to their contexts. The model is operational in nine cities across India, leading to the creation of 883 CAGs.

However, in order to truly create participatory governance, there is a need for them to be included in the design process. Participation of some CAG leaders in Ward Committees in Ahmedabad provides hopes for others to similarly engage. This, nevertheless, remains a vision for the future rather than a current reality.

The Indian Constitution lays out principles for local participation in urban spaces in the 74th Amendment. However, the realization of the philosophy behind the Amendment remains varied. If the CAG model were to be legalized by states and incorporated into their existing structures, the foundation for participatory governance would be laid. This proposal, however, is likely to meet with resistance from existing hierarchies that would be forced to realign along more equitable lines and redistribute resources justly.

The success of the CAG model lies in the manner of network creation and expansion. Even as they hold the government accountable and emphasize transparency, the CAGs are not confrontational. The duality of bonding and bridging social capital enables them to articulate the common needs of their diverse communities and act as the link between the community and local government.

This entire process requires time and investment not just from the community but an external catalyst. The catalyst will facilitate the creation of an ecosystem that will respond to crises as they emerge – allowing for communities to adapt and strengthen their democratic powers.

References

Auerbach, A. M. (2019). *Demanding development: The politics of public goods provision in India's urban slums* (Cambridge Studies in Comparative Politics). Cambridge: Cambridge University Press.

Chattopadhyay, S. (2015). Contesting inclusiveness: Policies, politics and processes of participatory urban governance in Indian cities. *Progress in Development Studies, 15*(1), 22–36. https://doi.org/10.1177/1464993414546969.

Daly, M., & Silver, H. (2008). Social exclusion and social capital: A comparison and critique. *Theory and Society, 37*(6), 537–566. https://doi.org/10.1007/s11186-008-9062-4.

Dupont, V. (2007). Conflicting stakes and governance in the peripheries of large Indian metropolises – An introduction. *Cities, 24*(2), 89–94. https://doi.org/10.1016/j.cities.2006.11.002.

Fischer, F. (2012). Participatory Governance: From theory to practice. Oxford Handbooks Online, 1–18. https://doi.org/10.1093/oxfordhb/9780199560530.013.0032.

Jhabvala, R., & Brahmbhatt, B. (2020). *The city-makers*. Gurugram: Hachette India.

Patel, S., Sliuzas, R., & Georgiadou, Y. (2016). Participatory local governance in Asian cities. *Environment and Urbanization ASIA, 7*(1), 1–21. https://doi.org/10.1177/0975425315619044.

Poder, T. G. (2011). What is really social capital? A critical review. *The American Sociologist, 42*(4), 341–367. https://doi.org/10.1007/s12108-011-9136-z.

Sachdeva, P. (2011). *Local government in India* (1st ed.). Delhi: Pearson India.

Shaw, A. (2012). *Indian cities*. New Delhi: Oxford University Press.

Participatory Irrigation Management: Barind Model – A New Sustainable Initiative

Asaduz Zaman
Center for Action Research-Barind, Rajshahi, Bangladesh
Asian Development Bank, Dhaka, Bangladesh

Introduction

Participatory Irrigation Management

J. Raymond Peter (Executive Director, International Network on Participatory Irrigation Management, Washington, D.C.) states the term participatory irrigation management refers to the participation of users – the farmers – in the management of the irrigation system. Further he refers the handbook on PIM that defines participatory irrigation management as the involvement of the irrigation users in all aspects of irrigation management, and at all levels. All aspects include planning, operation and maintenance, financing, decision, rules, and the monitoring and evaluation of the irrigation system. All levels include the primary, secondary, and tertiary levels. In the governance paradigm, PIM can be considered as a partnership between governments, agencies, and users (Raymond Peter 2004).

Mark Svendson and others stated participation in irrigation management by water users can take a wide variety of forms. Farmers can be involved in various system management functions, planning, design, operations, maintenance, rehabilitation, resource mobilization, and conflict resolution. Moreover, they can be involved in these functions at various system levels: from the field channel to entire system. Almost all irrigation

P

systems have some involvement by water users in system management. When people speak of introducing "Participatory Irrigation Management" (PIM), they are thus usually referring to a change in the level, mode, or intensity of such participation that would increase farmer responsibility in management process. PIM is designed to shift the financial burden for irrigation services from the agency to the users (Svendsen et al. 1997). This idea is being supported by Mr. Adrian Laycock, as he stated, privatization started to become politically fashionable in the late 1980. In this context, privatization means off-loading government ownership or responsibility for operation into the private sector, either to the farmers themselves or to an intermediate private subcontractor. But is it just a means of off-loading responsibility from a government line management system that can't cope, or is it really to benefit the farmers? (Laycock 2011).

Sandra Ricart and others show how stakeholder engagement in irrigation systems shapes hydrosocial territories: (1) by reducing tension between stakeholders, (2) by redirecting regional planning and strategy, (3) by highlighting water crises, (4) by decentralizing water responsibilities, and (5) by integrating values and beliefs from different stakeholders. Stakeholder engagement is one of the main characteristics of the shift from governmental to nongovernmental ownership, management, and administration of water resources and services. Stakeholder engagement implies a combination of *collaboration* – which involves cooperation to achieve goals of efficiency, equity, and sustainability in water resources and comprehension – which is made up of forces, systems, and mechanisms consisting of the ability to put oneself in the place of the other, sharing social identity, and promoting work together to achieve particular outcomes at different scales.

Analysis of PIM

A major weakness that continues to plague the productivity of large irrigation schemes is the lack of efficient and sustainable Maintenance Operation and Management (MOM). As a consequence, the infrastructure of these schemes is degraded and needs rehabilitation and modernization. Other reasons include inadequate government financing, lack of beneficiary empowerment and engagement in MOM, and limited capacity of public agency resulting in weak service delivery. Specific issues are the: (i) inadequacy of budget to support system MOM; (ii) lack of distinction between annual, periodic, or emergency maintenance of a system; and (iii) poor cost recovery from the water management groups.

Water management is very limited with minimal attention to how the scarce water resources are allocated, and most schemes only meet a small portion of their target production. Many farmers turn to the use of groundwater to support limited and irregular water supplies. The mix of irrigation from surface and groundwater makes evaluation difficult and masks deficiencies of the surface water supplies.

During the past 20 years, substantial efforts were made to improve irrigation MOM through introduction of new model of participatory irrigation management (PIM). PIM proved generally successful on small and medium schemes but it has yielded limited results on large schemes. The variable performance of PIM in improving irrigation MOM is internationally documented and private sector participation through public private partnership (PPP) is seen as an interesting alternative approach. It has demonstrated promising results in few developing countries such as Brazil, Morocco, and Ethiopia but is still to be developed in Asia. The irrigation systems in Asia are characterized by densely populated farmers doing subsistence agriculture. The farm size among Asian farmers are small as opposed to large commercial farms in the west; farmers are generally poor and usually augment incomes through non-agriculture-related activities. Inequities among farmers in terms of farm size, access to credit and markets, caste structures make irrigation in Asia very complex (Raymond Peter 2004). However, in Bangladesh, this issue of cast structure is not evident especially in irrigation sector.

Water User Association (WUA)

There are Water User Association (WUA) and cooperative groups in government documents but practically is not in effective form especially in most of the irrigation schemes in South East Asia. If there is any WUA in any irrigation scheme, that is only in paper. If one asks the farmers, the president or secretary of the WUA are for how many years, the reply would be from the beginning of formation of WUA. The problem is the stakeholders as farmers are of very heterogeneous – there are rich and poor farmers, educated and illiterate farmers, own land owners and share croppers or lessee farmers. Practically, it is quite a critical issue. Moreover, this Water Users Association is advocated by most of the studies but possibly one aspect is overlooked or bypassed, the remunerations to office bearers of the WUA. How one can expect they will render their services properly? It is believed those office bearers of WUA or so-called cooperatives manage certain benefits monetarily or other ways (own land irrigated by the cost of other water users). However, this unholy practice is being followed in most of the WUA and cooperatives of irrigation schemes.

Barind Multipurpose Development Authority (BMDA)

The Barind Multipurpose Development Authority (BMDA) is an autonomous authority reporting to the Ministry of Agriculture. BMDA has taken an innovative approach to management of rural infrastructure, particularly the distribution irrigation water and management of irrigation systems. The part of greater Rajshahi, Dinajpur, Ranqpur, and Boqra District of Bangladesh, and the Indian territorial Maldah District of West Bengal is geographically identified as the Barind Tract. The hard red soil of these areas distinguishes it from of the other parts of the country. A typical dry climate with comparatively high temperatures prevails in the Barind area from November to May. The total cultivable area is 600,000 ha of which 84% is single cropped, 13% is double cropped, and the rest is triple cropped. The Barind Multipurpose Development Authority (BMDA) was established under the Ministry of Agriculture in 1992 through the Secretarial power of the concerned Ministry. However, Parliament passed the Act in October 2018 and now covers 16 districts; 124 upazilas, 1094 unions, and 20,153 mouzas, i.e., whole of northwest part of Bangladesh as shown in Fig. 1.

BMDA Management

BMDA is managed by a board chaired by an appointee of the Prime Minister. The Executive Director (Chief Executive Officer) of the Authority is a member and also secretary of the Board. The Board is constituted as follows (Zaman 2013b):

I. Chairman – Appointed by Prime Minister
II. Member – Deputy Inspector General of Police, Division
III. Member – Deputy Commissioners of the districts
IV. Members – Three members representatives from the farmers
V. Member Secretary – Executive Director, BMDA.
VI. Advisor – Members of the Parliament of BMDA jurisdiction

Irrigation Development

There are around 16,072 irrigation equipment (tube well – 15,553 and surface water lift pump – 519) under BMDA management and all are governed by BMDA Zone (field) Offices. Most important aspect not a single irrigation equipment is in idle condition for technical, financial, or political reasons. BMDA pays employee salary, wages, all allowances, retirement benefits, and support costs for its around 1036 regular staff members, 375 muster roll staff, and 16,072 pump operators on hourly basis (no work no remuneration), i.e., total around 17,483 engaged manpower, O&M costs of all irrigation equipment, and transports, irrigation equipment's electricity bills. BMDA bears all the above costs without seeking unreliable funding sourced from the government. The detail of irrigation equipment and its use is shown in Table 1.

However, the year-wise use of number of irrigation equipment and corresponding irrigated area are shown below in the graph, Fig. 2.

P

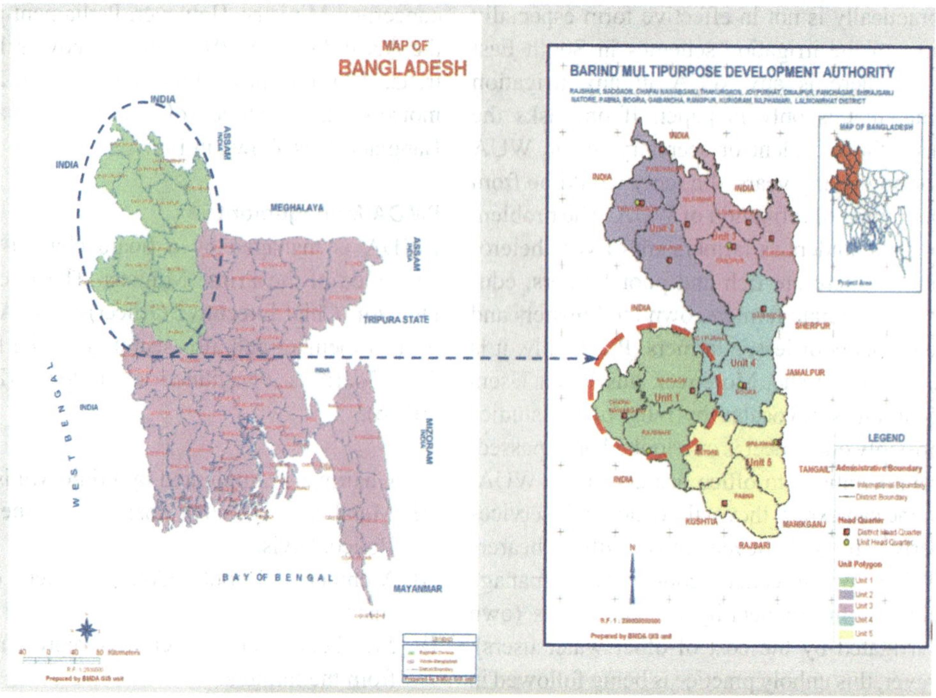

Participatory Irrigation Management: Barind Model – A New Sustainable Initiative, Fig. 1 Map of Barind Authority

Participatory Irrigation Management: Barind Model – A New Sustainable Initiative, Table 1 Status of present irrigation equipment and its use

Year	Used nos. of irrigation equipment		Total nos. of irrigation equipment	Irrigated area (Ha.)	Realized irrigation charges (Million BDT)	Realized irrigation charges (Million USD)
	Tube well	Surface water lift				
2018–2019	15,553	519	16,072	540,240	1561.80	19.52

Prepaid Metering System and Smart Cards

Despite its success, BMDA does not follow the principles of the Guidelines for Participatory Water Management (GPWM) nor does it follow the participatory processes that global lessons claim are necessary for sustainability. A significant feature of the BMDA operation is the prepayment, by farmers, for irrigation water.

In response to the universal concern for irrigation sustainability, that O&M will fail through a lack of user raised funding, BMDA has introduced two innovative concepts for collection of the irrigation charge (IC) – a prepaid metering system. The prepaid metering system provides every farmer with a user card. The user card provided by BDMA has photo ID, name, and a user number. The card is loaded with credit by paying cash at any BMDA office or to an accredited dealer. The card is then inserted in a slot at the pump station and water is pumped automatically with the charge levied against the credit on the card. The meter, usage, and pump delivery is checked every day or every 2 days by a BMDA official in a similar manner as for the coupon system. Card uploading stations (vending stations) are connected electronically to

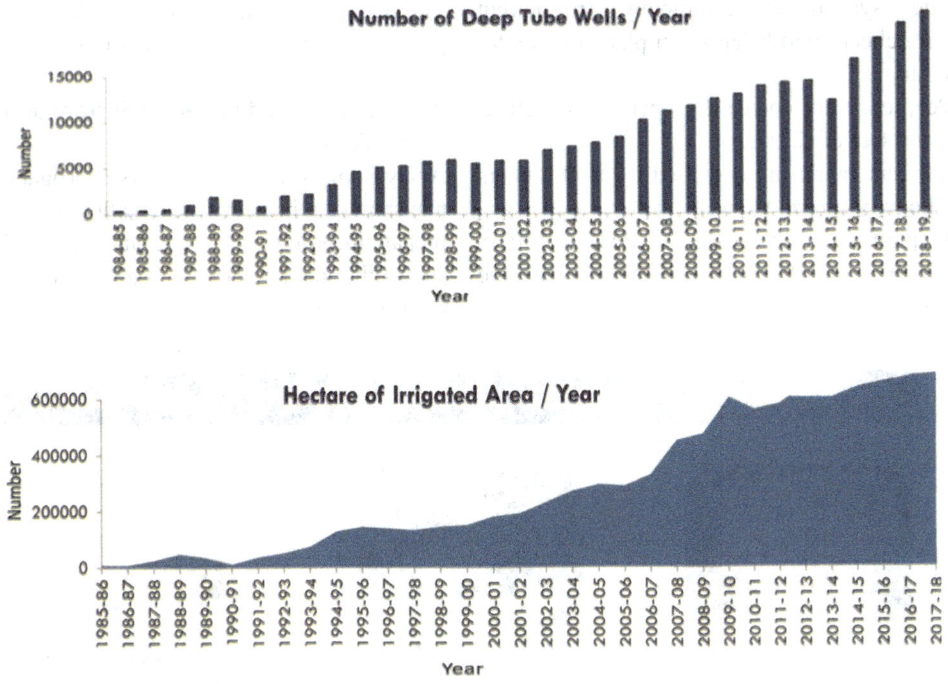

Participatory Irrigation Management: Barind Model – A New Sustainable Initiative, Fig. 2 Year-wise deep tube well and irrigated area (hector)

a central BMDA operator and all data can be monitored remotely. The process works using technology similar to that used for loading credit to cell phones. The initial concern for the introduction of a high technology electronic system was dismissed on viewing the system in operation. Farmers were using the card with confidence and the procedure was working. It fails when there is no electricity, but then so do the electrically powered pumps (Fig. 3).

Mode of Operation of Prepaid Meter and Card

- Each farmer is issued with a prepaid *User* card which when introduced into a prepaid *Meter* enables the pump to start and water delivered until such time as the card is removed or its credit expires. The amount debited to the card is proportional to the pumping duration and therefore volume.
- Farmers recharge their cards using a handheld *Mobile Vending Unit* (MVU) kept by a *Dealer* who collects farmer payments. The Dealer recharges his MVU credit whenever required from a *Vending Station* (VS) at the local BMDA office after depositing the recharge amount into the BMDA bank account.
- The prepaid meter system which currently extends over 16 districts is managed by BMDA. Repairs to the system, for example for the 16,072 m, are done under contract with a private company (currently Sanakosh Associates Ltd). The system was supplied and installed by Wasion Group, China (Zaman 2013a).
- All the pumps under BMDA are electric pumps. There are issues of reliable electricity but the farmers manage an informal backup system pumping from ponds and khals. Farmers understand the limitations of electricity and some schemes are not fully planted due to the limitations of the number of hours of electricity.

Benefits of Prepaid Meter and Smart Card

- All water provided is paid for in advance.
- There is no opportunity to bypass the meter.

- The system is completely transparent with checks and balances in place to counter fraud.
- People cannot coerce the operator to deliver water free of charge.
- Farmers cannot be exploited by land owners who may control the well.
- Prepaid water charges go to BMDA coffers which support the sustainability of the BMDA.

The findings of benefits after installation of prepaid meter could be seen in the following Table 2.

Impact of Prepaid Metering System and Smart Cards

Analysis of different parameters of irrigation system shows a miraculous picture after installation of prepaid metering system and smart cards. Table 3 shows the impact before and after installation of prepaid meters.

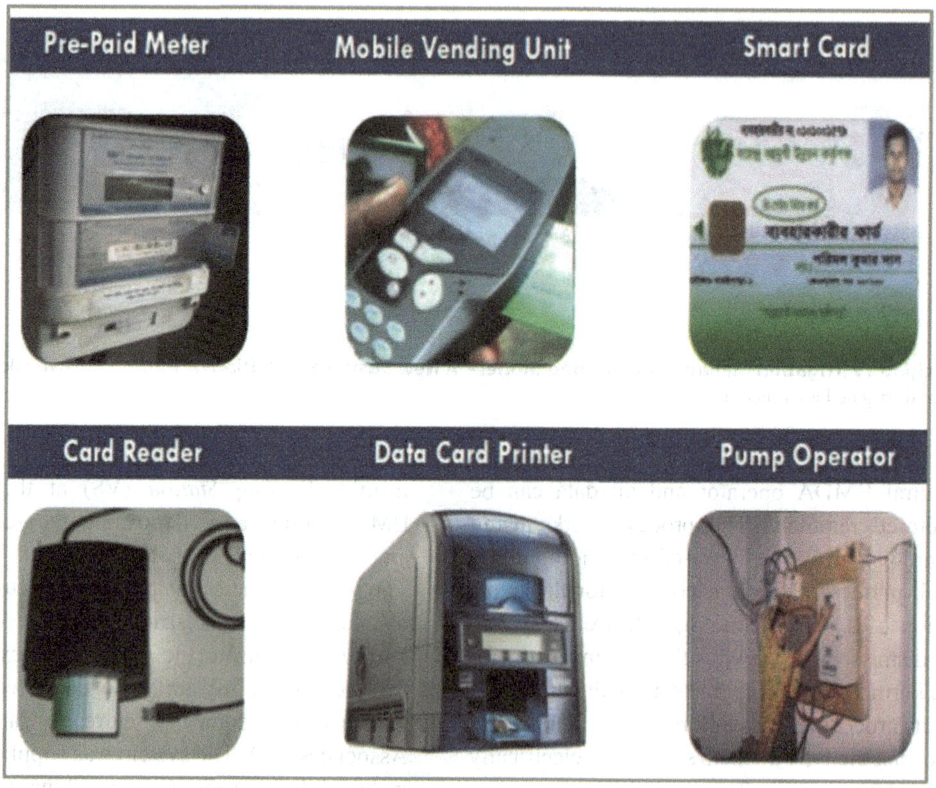

Participatory Irrigation Management: Barind Model – A New Sustainable Initiative, Fig. 3 Components of prepaid meter

Participatory Irrigation Management: Barind Model – A New Sustainable Initiative, Table 2 Findings of benefits after installation of prepaid metering system

Component	Without Prepaid meter	With Prepaid meter	Comparison
Irrigation cost/ha (BDT)	11,040	5440	51% improvement
Irrigation cost/ha (USD)	138	68	51% improvement
Average water use/ha (inch)	82	59	28% improvement
Water required/kg of Boro rice production (l)	3400	2250	34% improvement
Average yield/ha (kg)	6084	6602	34% improvement
Average earning/ha of Boro rice (BDT)	31,200	44,480	43% improvement
Average earning/ha of Boro rice (USD)	390	556	43% improvement

Assessment of Irrigation Charges

The irrigation charges per irrigation equipment under BMDA management are calculated by a committee of engineers. The committee calculates and recommend for consideration of the BMDA board. Once BMDA board approves the rate of irrigation charges, become effective for implementation in the field. The rate of irrigation charges to be such that all with the intention to keep Barind Irrigation as a sustainable model and different stakeholders feel satisfied. However, the following parameters are being considered for the assessment of irrigation charges.

$$\text{Irrigation charge} = EC + POR + RMC + MCB + DC + SO + MVC + VAT$$

where,

EC = electricity cost per hour KWh
POR = pump operator remuneration
RMC = repair and maintenance cost
MCB = maintenance cost of buried pressure pipe system
DC = depreciation cost of machine and equipment
SO = scheme operational staff cost
MVC = mobile vendor's commission
VAT = value-added tax

The present rate of irrigation charge as been assessed and approved for irrigation equipment are as follows. The per hour irrigation charges are as follows for ground water lifting and surface water lifting and those are variable on the basis of discharge from the pumps are shown in Tables 4 and 5. Per hour irrigation charge is dependable on main aspect, i.e., rate of electricity tariff.

Financial Viability

Financial viability of irrigation project is the key factor for achieving the sustainable irrigation management system. The chart in Fig. 4 shows the year-wise earnings versus expenditures

Participatory Irrigation Management: Barind Model – A New Sustainable Initiative, Table 3 Impact before and after installation of prepaid metering system

Parameter	Before prepaid meter	After prepaid meter	Comparison
Irrigated area/well in ha	30	39	30% increase
Number of water users/well	70	89	27% increase
Irrigation charges/well in BDT	254,960	286,560	12% increase
Irrigation charge/well in USD	3187	3582	12% increase
Annual operating hours/well (avg.)	2884	3132	9% decrease
Annual electricity bill/well in BDT	129,040	117,360	9% decrease
Annual electricity bill in USD	1613	1467	9% decrease

Participatory Irrigation Management: Barind Model – A New Sustainable Initiative, Table 4 Present rate of irrigation charges per hour for tube wells pumping

Varied discharge capacities for tube wells				
Unit	14 Lt./Sec.	15–21 Lt./Sec.	22–28 Lt./Sec.	29–56 Lt./Sec.
Per hour rate in BDT	Tk. 85/	Tk. 100/	Tk. 110/	TK. 125/
Per hour rate in US$	$1.06	$1.25	$1.37	$1,56
Price per 1000 m^3 in BDT	Tk. 1687	Tk. 1323	Tk.1091	Tk. 620
Price per 1000 m^3 in US$	$21.08	$16.54	$13.64	$7.75

Note: These rates are effective from 1 February 2018 and the USD to BDT exchange rate was set at USD 1 = BDT 80 at the time

Participatory Irrigation Management: Barind Model – A New Sustainable Initiative, Table 5 Present rate of irrigation charges per hour for surface water lift pumping

Unit	Single lift pumping 28–56 Lt./Sec.	Double lift pumping 28–56 Lt./Sec.
Per hour rate in BDT	Tk. 125	Tk. 160
Per hour rate in US$	$1.57	$ 2.00
Price per 1000 m^3 in BDT	Tk. 1240.00	Tk. 776.70
Price per 1000 m^3 in US$	$15.50	$9.71

Note: These rates are effective from 1 February 2018 and the USD to BDT exchange rate was set at USD 1 = BDT 80 at the time

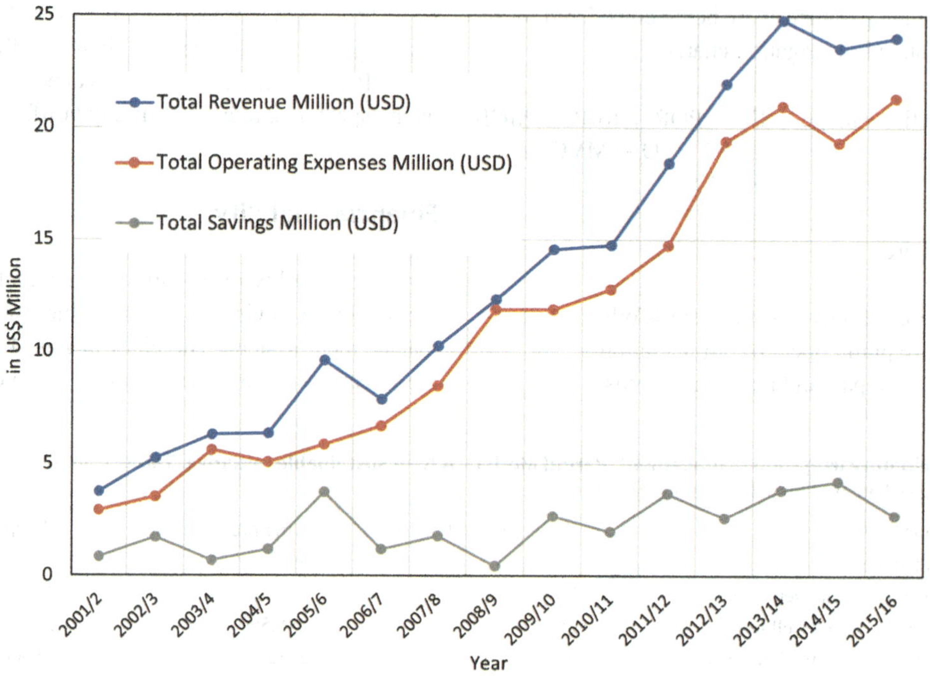

Participatory Irrigation Management: Barind Model – A New Sustainable Initiative, Fig. 4 Year-wise revenue earning versus expenditures

of the authority. Regarding financial viability Dr. Tushaar Shah who wrote in one of his publication after visiting and interviewing the farmers in 2014, *"The revenue s earned are used for O&M and for expanding the system. Studies suggest that this system is financially self-sustaining. . . .The institutional arrangement is incentive -compatible. The Barind's socio-economic impacts are deep and wide. In a region where farmers found it hard to grow one crop, today they grow up to three crops annually. Some 1.5 million small farmers have benefited and many more will benefit when the remaining 450,000 ha are brought into the tube well programme."* (Shah 2014) Similarly, Tonkin + Taylor did a field study of Barind in 2016 under Swiss Red Cross sponsorship and wrote "BMDA in Barind has a highly capable and well organized management system with 100% revenue return on water pricing" (2016). These study reports show that Barind Irrigation Management system is a self-sustained model.

Participatory Irrigation Management: Barind Model – A New Sustainable Initiative, Fig. 5

Appendix-A

IWMI International Water Management Institute

HOME ABOUT IWMI REGIONAL PROFILES RESEARCH & IMPACT TOPICS PUBLICATIONS TOOLS & RESOURCES MEDIA CENTER

Media Center

Search Website vacancies feedback sitemap contacts sign-up

SHARE

India Water Week

8-12 April, 2013 - Vigyan Bhawan, New Delhi

Boomtime in Barind: a model for India's irrigators to follow?

IWMI's Anna Deinhard talks to Dr Asaduz Zaman about the innovative approach his organization took to sustainable irrigation management in Bangladesh, which he recently presented at Asia Water Week in Manila, Philippines.

Testing of a tube well to determine needed head and capacity of the pump.
Photo: Asaduz Zaman

In the Barind region of Bangladesh rainfall is highly seasonal and poverty is endemic. Farmers with few resources were only able to grow one crop per year. In the late 1980s the Barind Multipurpose Development Authority was set up to explore ways of improving livelihoods through sustainable farm irrigation. New rice varieties that could grow in a relatively dry environment were introduced. To reduce water loss through evaporation an extensive network of tube wells and a pipe system were sunk into the ground as a replacement for surface irrigation channels.

Farmers still had to pay for access to water but cash transactions were replaced with a pre-paid swipe-card system for pump usage. This helped avert corruption and led to more optimal water usage, allowing more farmers access to groundwater and an expansion of irrigable land by over a fifth. Aside from poverty elimination, the project's major driver, other positive changes were observed. The electricity network, for instance, that was set up to run the pumps led to improved household electrification and the operation of a rice mill.

IWMI: The *Barind model* of irrigation management has proved such a success that the ADB is now planning to replicate it in other regions. What was Barind like before the project, and what changes did you achieve?
AZ: The farmers used to be very dependent on rainfall to grow a (variety of) monsoon rice crop. Consequently farm laborers (the poorest residents in the region, and therefore a key target for the project) could only work for around 4 months each year and then had to move elsewhere to sustain their lives. In addition rainfall in Barind used to be the lowest in the country and desertification started to become a problem.

Now after the implementation the farm labourers seem actually happier than land owners. They have work through the year and are even able to bargain. The demand for female labour has risen a lot too for certain agricultural practices.
Desertification is unheard of today. Irrigation wells are now performing a dual purpose for drinking water supply through pipe distribution system with overhead tank.

What were the challenges farmers had to deal with?
Well, first of all farmers did not have experience in how to grow seeds for Boro (the new dry season rice variety), which we introduced as an alternative to monsoon rice. It is a totally different procedure, so famers had to be trained. Knowledge of crop diversification and high value crops were also a totally new concept to the farmers.

In your presentation you highlighted the "Pre-paid-meter system. What is special about it and how does it work?
From the beginning of the project, we had the intention to avoid cash money and introduce a prepaid system for the irrigation system. Initially we started a coupon system, printing coupons from a government press. But even though, we ended up having problems with corruption and found the coupons being replicated on a local press. The police had to intervene. It was then that I started thinking about the swipe cards.

How was the system perceived by the farmers and other users?
At first some of the MPs and also some of the farmers seemed a bit reluctant. However soon after introduction of the new system those farmers came back and said, "This is good. We are saving money, water and time."

Do you think the model could be replicated in, for instance, India?
Back in 2004-05 when we introduced the pre-paid meter it still seemed very high tech, but I would say it is very easy to introduce nowadays. I am confident the model could be replicated in India for irrigation purposes.

In your opinion which are critical factors for the success of a replication?
I think the Irrigation project authority should introduce the provision of incentives and penalties for good and bad performances respectively. Then there are matters of infrastructure, all wells should be electrified. The area should be brought under a buried plastic pipe water distribution system. But these factors are not impossible to achieve.
Last but not the least I think the chief executive of a successful irrigation project should have a clear idea about people, land and water.

About the interviewee:
Dr. Asaduz Zaman from Bangladesh has 40 years of work experience in groundwater and surface water irrigation management. He worked with the Barind Multipurpose Development Authority (BMDA), where he was responsible for the implementation of an irrigation project known as "Barind model".
Presently he is working with the Asian Development Bank (ADB) on a replication of this model.

P

Conclusions with Salient Features

Financial Cost Recovery

– The prepaid smart card system ensures 100% collection from farmers and by being totally transparent eliminates rent seeking and corruption. It is also fair with farmers paying according to volume of water they use.
– The use of electric power for pumping greatly reduces pumping costs due to the government's subsidy. In most areas covered by the BMDA, load shedding is not a major problem.
– Farmers themselves provide an informal backup of water supplies by pumping from ponds and khals.
– The BMDA prepaid smart card system requires electric powered pumps and power supply for the meters.
– With a buried pipe distribution system, the pump runs for 10–20 s after a farmer's user card is withdrawn (the amount being charged to the card). Another card needs to be inserted within this time before the pump shuts down. Once a pump stops, it cannot be restarted for 3–5 min to avoid damage to the motor.
– BMDA undertakes a number of cost recovery/income generation activities which help support the financing of the authority as well as provide direct and indirect benefit to farmers.

Water Use Efficiency

– While the prepaid card system can work with open channels, operational losses are significant as the canal prism empties/fills when water is rotated to different outlets. A buried pipe system remains full. Pipeline diameters used vary from 6 to 10 inches (160–250 mm).

– Farmer irrigation service is good with a direct link between volume supplied and cost. It is claimed that water use efficiency has greatly improved as farmers minimize their costs.

Scheme Size

– The full pumped flow (typ. 28–56 l/s) is used by one farmer who may split the flow so that field crop damage is avoided.

Cropping

– With the prepaid user card and buried pipe system, farmers irrigate a variety of crops, including tomato, and boro – paddy.

Backup Power

– Despite frequent power cuts, the Barind does not provide any backup power supply. Farmers individually or collectively provide some backup supply from ponds and khals (drains).

Concluding Remark

In conclusion, the author was asked by IWMI representative in an interview during India Water Week on 8–12 April, 2013, and published in the IWMI's under the title "Boomtime in Barind: a model for India's irrigators to follow?" (2013) is being enclosed as Appendix A. Based on the field experience in Barind for more than 20 years, author strongly believes that the Chief Executive of successful irrigation project should have a clear idea about water, soil, and people.

Appendix A

See Figure 5

References

International Water Management Institute (IWMI), India Water Week. (2013). *Boomtime in Barind: A model for India's irrigators to follow?* An Interview with Dr. Asaduz Zaman 8–12 April.

Laycock, A. (2011). *Irrigation systems: Design, planning and construction*. CABI.

Raymond Peter, J. (2004). *Executive Director, International Network on Participatory Irrigation Management* (p. 6). Washington, DC: Participatory Irrigation Management (PIM), Newsletter.

Shah, T. (2014). *Groundwater governance and irrigated agriculture* (TEC Background Paper no. 19). Global Water Partnership.

Svendsen, M., Trava, J., & Johnson, S. H., III. (1997). *Participatory irrigation management: Benefits and second generation problems*. Cali: World Bank & IIMI.

Tonkin+Taylor, prepared for Swiss Res Cross. (2016, August). The Integrated Water Resource Management in the BarindTract, Bangladesh.

Zaman, A. (2013a). *Replication of Barind model to Muhuri Project May*. Paper presented at IWMI Delhi.
Zaman, A. (2013b). *Barind: A paradigm of sustainable irrigation management for Bangladesh and beyond*. Paper presented in Asia Water Week.

Participatory Planning: A Useful Tool for the Development of Sustainable Mega-City Regions

Valerio Della Sala
Politecnico di Torino (Italy), Interdepartmental Research Centre for Urban Studies (OMERO), Universitat Autonoma de Barcelona (Spain), Turin, Italy

Synonyms

MCR: Megacity region; Participatory: Participative, Shared, Collaborative, Community; Planning: Zoning, Programming, Design; Social: Community; Stakeholders: Partner

Definition

Megacity region (MCR): A megacity region (MCR) involves a cluster of highly networked urban settlements grouped into one or more cities. One of the prerogatives of mega city regions is the distance between one location and another.

Sustainable model: A sustainable development model encompasses a holistic vision that ensures that the principles of socio-economic development are respected in respect of the natural resources in the project area.

Participatory planning: Participatory planning is that active development process that includes citizens, associations, and local authorities within the planning lines.

Introduction

Megacity regions (MCR) created a big urban transformation, which allows a new critical perspective concerning public services and management measures. The evolution of the MCRs in Europe implies a critical reflection on the social sustainability of the model. Furthermore, in relation to the UN indications within the 2030 agenda and the Rio de Janeiro agreement in 1992, citizen participation can be a new tool for supporting local decisions. The composition of a megaregion made up of a constellation of urban or metropolitan centers suggests a new approach in the design of the transport network on the territory. A regional development model that can integrate participation in planning processes will be able to respond in a timely manner to sustainability agreements.

The Megacity Region: A Brief Introduction

The phenomenon of the polycentric region of megacities (MCR) is emerging in almost all the most urbanized parts of the world. The long decentralization process of traditional cities has evolved in the extension of metropolitan areas. The new form of the polycentric region involves and incorporates five cities or more. In some cases, such as Japan, the constellations reach up to 15 cities. Cities in MCR are separated in institutional processes but integrated through a network of infrastructures in the new economic dimension of the entire region. Places in MCR can either function as single entities or as global and therefore more complex identities.

The consideration of a polycentric region inevitably involves a rethinking of railway lines, motorways, and the flow of people within the area. This complex urban form is a phenomenon that has become of greater interest in the twenty-first century because of its important implications for the sustainable development of the region.

Otherwise, recent UN estimates show that by 2050, 80% of the world's population will live within urbanized areas. European Commission research, Polynet, analyzing the functioning of

8 polycentric regions in Europe, hypothesizes that the increasing share of population and decentralization of production processes will lead to a decrease in trade within the central city, increasing the movement of workers within the regional area.

Polycentricity mainly refers to the tendency of central cities to suppress lower order cities, placing them within the space of flows, advanced by Castells.

Indeed, researches conducted by Kloosterman in 2011, and Taylor in 2003, suggest that Europe's polycentric urban regions may conflict with the sustainability goals set out in the Rio de Janeiro agreement. In this contention, the concept of polycentricity essentially depends on the scale of intervention considered by public administrations.

As advanced by Sassen in 1991, global cities are defined in terms of exchanges of information with the outside world. Therefore, polycentric cities should be considered in terms of exchanges of information with the inside.

The real challenge for planners in the next decade will be to adopt sustainable development strategies that involve participatory planning by citizens.

Participatory planning seems to be one of the useful elements that can reduce the polarization processes of megacity regions.

Polycentric Megacity Region and Sustainability

Sustainable development is promoted by governments and supragovernmental institutions as an objective of spatial planning. Unfortunately, within the day-to-day processes of governments, sustainability is at likelihoods with national and global development plans.

As Hall states, polycentricity has implications for environmental sustainability (Hall 2006).

Polycentricity fuels regional commuting and travel that is not always achievable through regional public transport. Currently, functional polycentricity linked to the evolution of metropolitan clusters is justified through the promotion of new jobs and new infrastructure interventions.

Polycentricity, observed as a simple ecological planning process that reduces commuting distances, can have the opposite effect for the whole region, increasing commuter flows and thus stimulating the use of the private car rather than public transport. A fragmented urban development model reduces the efficiency of the information economy advanced by Castells (1989).

Moreover, the changing role of metropolitan centers may suggest a change in the global city competitions proposed by Sassen (1991). Also, the concentration of global functions within MCRs will inevitably lead to external pressures that will influence the decision-making process of territorial governance.

For example, in Central Belgium, the Regional Plan of Flanders develops a polycentric urban plan that is sponsored as a "concentrated devolution" capable of strengthening the urban structure that fits into European urban networks. The growth of urbanized territories within megaregions should not be identified as a sustainable planning practice. The creation or development of new networks should only be sponsored if they are complementary to each other, otherwise sustainable development will not be promoted. In this way, the development of the territory becomes a practice of sustainability that can be replicated in other territories with a similar production history. The complementarity of centers can certainly encourage sustainable development.

On the other hand, the polarization of urban centers, observed in Radstad, England, Belgium, and Germany, will only contribute to an increase of fragmented urbanism and commuting flows across the territory. Failing the sustainability practices identified above. In addition, cities embedded in global networks will want to benefit from the same economic structures as other cities in order to achieve a stable economy.

Thus, the concentration of centers becomes an essential component of regional economic sustainability, as well as being less harmful to the environment.

Participation Planning: A New Future

Over the last century, with the repositioning of socioeconomic relations, the megacity region

(MCR) has become a model for the research and evolution of geographical space.

The fragmentation of areas, the division of processes, and the differentiation of institutional levels have led to an increase in the flow of people within mega-areas. The increase in the exchange of goods and the flow of people inevitably entails a major investment in transport infrastructure.

The composition of a megaregion made up of a constellation of urban or metropolitan centers suggests a new approach to the design of the transport network on the territory.

Agreeing with Hall, megacity regions can be defined as a set of functional and neighboring urbanized centers (Hall 2006). The proximity of these centers is essential for them to be functional megacity regions.

The creation and development of a complex structure such as megacity regions require a totally different governance approach compared to the one observed in the second industrial revolution. In urban studies, the concept of governance spent refers to the collective action of the private sector, civil society, and local stakeholders.

In the last 10 years, participatory planning practices have demonstrated to be an excellent tool for defining priorities within the territory. Involving citizens and local stakeholders in decision-making processes ensures acceptance of public works and private investments in the territory.

Participatory planning as a new form of governance needs strong community collaboration and acceptance before it can be implemented in the territory.

According to Hajer, collective action is increasingly taking place in an institutional vacuum, of which governance is merely a logical outcome (Hajer 2009).

Collective action should be considered as the fundamental tool for decision-making processes within the territory. In particular, the participation of citizens in choosing priorities is becoming increasingly important within urban centers.

The intensification of contrasts between cities and regions in relation to economic investment in the territory has contributed to the evolution of territorial competition between urban areas.

The public policies of urban centers in relation to housing, spatial organization, and infrastructure investment have become increasingly complex processes which, given the fragmentation of the public sector, now inevitably require private participation for their implementation (Innes et al. 2011).

Conclusion

Observing the phenomenon of megacity regions, we can affirm that polycentricity has no direct relevance for sustainable development. On the other hand, the adoption of participatory planning practices can help reduce forms of inequality on the territory, contributing to equitable and sustainable development.

Collective action enables a strengthening of group values and identity in one's own community. This strengthening will be crucial for the future of our cities and regions. Public participation in decision-making processes has now become a central element in respect of the principles of inclusion and integration of individuals within democratization processes.

Participatory planning forms have become a reality that contrasts the different policies adopted by institutions in traditional models of territorial development. Public administrations should position themselves as independent, listening to the needs of all stakeholders included in the decision-making process.

Participation appears to be a fundamental element for the expansion and implementation of democratic processes within one's own territory (Kooiman 1993).

According to Healey (2008), the communicative/collaborative planning model helps dialogue with stakeholders and increases citizen consent. Although public participation cannot solve the institutional processes related to democratic planning, participatory planning is a fundamental element for territorial development. Citizen participation in decision-making processes can provide a useful tool for urban policies, setting new priorities within neighborhoods and cities. Citizen participation in neighborhoods is one of the crucial elements for the future of our cities.

The change of scale, from global to local, will inevitably be the challenge for cities for the next 20 years. Therefore, agreeing with Hall in relation to the unsustainability of the megacity region model, participation could be a key element in shifting the focus to the holistic aspects of sustainability.

Finally, through participatory planning practices, a spatial justice that is able to reduce socio-economic diversity can be achieved.

Cross-References

- ▶ Digital City Modeling and Emerging Directions in Public Participation in Planning
- ▶ Future of Urban Land-use Planning in the Quest for Local Economic Development
- ▶ Metropolitan Discipline: Management and Planning
- ▶ Participatory Governance for Adaptable Communities

References

Castells, M. (1989). The Informational City: Information Technology, Economic Restructuring and the Urban-Regional Process. Oxford:Blackwell.

Hajer, M. (2009). Authoritative Governance: Policy Making in the Age of Mediation. Oxford: Oxford University Press.

Hall, P. (2006). *The polycentric metropolis learning from mega-city regions in Europe*. London: Earthscan.

Healey, P (Ed.). (2008). Interface: Civic engagement, spatial planning and democracy as a way of life. *Planning Theory & Practice, 9*, 379–414.

Innes, J., Booher, D. & Di Vittorio, S. (2011). Strategies for megaregion governance. *Journal of the American Planning Association, 77*(1), pp. 55–67.

Kooiman, J. (1993). Modern Governance: New Government–Society Interactions. London: Sage.

Sassen, S. (1991). The Global City. New York, London, Tokyo. Princeton, New Jersey: Princeton University Press.

Partner

▶ Participatory Planning: A Useful Tool for the Development of Sustainable Mega-City Regions

Peace–Conflict Security

▶ Water Security and Its Role in Achieving SDG 6

Pedestrian Density Flow

▶ Walkable Access and Walking Quality of Built Environment

Peer Learning

▶ Beyond Knowledge: Learning to Cope with Climate Change in Cities

Perceived Safety

▶ Hidden Enemy for Healthy Urban Life

Perception and Reality for Sustainable Irrigation System with Micro-irrigation

Asaduz Zaman
Centre for Action Research – Barind, Rajshahi, Bangladesh
Asian Development Bank, Dhaka, Bangladesh

Introduction

Water being a necessity for crop production is one of the most important natural resources for sustaining human life on earth. Now, in order to feed the growing population and to further increase farm incomes and livelihood of farmers, the overall agricultural production needs to grow exponentially. One of the key ways to boost overall agricultural production is to implement better soil-water management techniques that would provide better access to irrigation water, without actually increasing the stress on available natural source of water.

Role of Irrigation Water

It has been found that irrigation water alone claims an increase in production by 25–40% depending upon soil moisture condition. With improved farming practices, seeds, fertilizer, pest control, and processing, the output rates could be increased by several times. However, the contributions of the latter are positive when water is ensured. Water for agriculture, thus, is an essence for optimum and sustained productivity of land and therefore, for the farmer a security for economic sustenance (My Agricultural Information Bank 2019).

Crop Credit

Farmers would need the important inputs for achieving optimum yield of crops; these are (i) irrigation, (ii) seeds, (iii) fertilizer, and (iv) pesticides. Farmers need irrigation water and s/he is in need to buy water from private shallow and or deep tube well owners, or from public sector deep tube well. In any case irrigation water is not freely available, farmers have to borrow money from formal or informal money lenders. Informal money is readily available but cost of borrowed fund is exorbitant. On the other hand, formal loan would be comparatively cheaper but required paper document is not available with its updated status. Sufferings are much higher in this type of formal loan from the state-owned financial institutions. Both has merits and demerits – farmers agree whatever the rate of interest for borrowing the money even at a higher rates. Same as for high quality seeds, fertilizer, and pesticides – these commodities are available in the markets but either farmer is to buy at cash payment or at loan with much higher rate of interest – farmers are compelled to agree whatever the rate of interest of fund as s/he needs these without delay (Siddiqui 1975).

Under the above scenario, the four inputs (irrigation, seeds, fertilizer, and pesticides) are applied inputs. These have direct impact on crop yield. We can consider another input as catalytic and or facilitating this one is crop credit. It does not have direct impact on plants/crops but has very important role to make available those applied inputs. This could be self-explanatory from the following flow diagram shown in Fig. 1.

Micro-irrigation may be defined as the application of water at low volume and frequent interval under low pressure to plant root zone. Besides the land, water also an important factor in the progress of agriculture. Since water is the limiting factor today, we must utilize it properly and derive optimum benefit from the used amount as much possible. The expansion of area under irrigation is essential for obtaining increased agriculture production required to feed growing population.

The expansion could be done only by additional development conservation and efficient management of the available water resources, i.e., use of micro-irrigation means application of

P

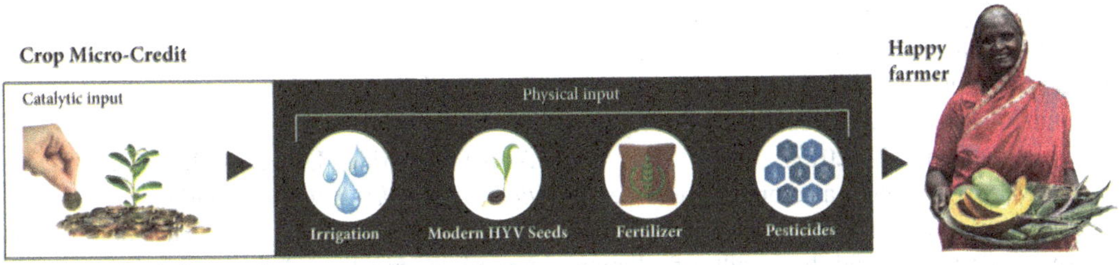

Perception and Reality for Sustainable Irrigation System with Micro-irrigation, Fig. 1 Crop microcredit flow chart

optimum water according to plant requirement. This could be achieved by introducing advanced and sophisticated methods of irrigation, viz., drip irrigation, sprinkler, etc.

Why Micro-irrigation?

Globally, it is well established that micro-irrigation technologies increase crop yield, save water, improve crop quality, enhance the fertilizer/chemical application efficiency, conserve energy, reduce labor cost, improve pest management, increase feasibility of irrigating difficult terrains, improve suitability of problem soils, improve tolerance to salinity, etc. In micro-irrigation, supply of optimum quantity of water in the form of tiny streams, fine spray, or continuous drops mitigates water loss due to evaporation and on account of seepage and percolation. This further reduces waterlogging and improves soil health. Consequently, there is an increase in productivity and the quality of produce, thereby leading to a rise in the overall farm incomes. Micro-irrigation technology is promoted primarily for the following reasons: (1) as a means to save water in irrigated agricultural land; (2) as an initiative to increase farm incomes and reduce poverty; and (3) to enhance the food and nutritional security of rural households. Also substantial dependence on rainfall makes cultivation a high risk and less productive activity, so assured irrigation and in-situ moisture conservation encourages farmers to invest more in farming technology and inputs that lead to productivity enhancement and increased farm income. Further, the rate of return vis-à-vis farm productivity from investment in drip irrigation is observed to be relatively higher than that of sprinkler irrigation and can be as high as 150%. Understandably, the minimum payback period has been found to be less than 1 year and the maximum to be 2–3 years in both drip and sprinkler method (NITI 2017).

Nowadays, drought is a common phenomenon due to impact of global climatic change. The cause of drought could be result of climatic change cycle and also man-made disastrous situation like overexploitation of both surface and groundwater resources without considering environmental impact. Introduction of micro-irrigation may help to some extent to mitigate the environmental adverse impact through judicious use of water resources. However, Professor Milos Holy Ph.D. SC, Director of the Institute for Irrigation and Drainage, Technical University, Prague, in the FAO paper No. 8 provided a range of classification of climatic region on the basis of annual precipitation, shown below.

Classification of climatic region on the basis of annual precipitation:

Desert climate	annual rainfall less than 120 mm
Arid climate	120–250 mm
Semiarid climate	259–500 mm
Moderately humid climate	300–1000 mm
Humid climate	1000–2000 mm
Very humid climate	more than 2000 mm

At first sight, the mean annual precipitations in arid regions and the moisture requirement of cultivated plants clearly indicate a high degree of water use by agriculture. But it is high not only in arid regions. It is also substantial in regions with high precipitations if these occur outside the vegetation period or if they are unevenly distributed (Milos Holy 1971).

Very recently FAO, UNICEF, IFAD, WFP, and WHO published a document that shows highest magnitude of agricultural damage and loss in agriculture by climate-related disasters shown in Fig. 2. With judicious use of both surface and groundwater resources along with the role of micro-irrigation comes into picture to mitigate the adverse effect of drought.

Perception

As per Cambridge University Dictionary *Perception, a thought, belief, or opinion, often held by many people and based on appearances.* Regarding micro-irrigation, there is an existing perception among most of the end-users (farmers), field engineers, and the policy makers are in confusing

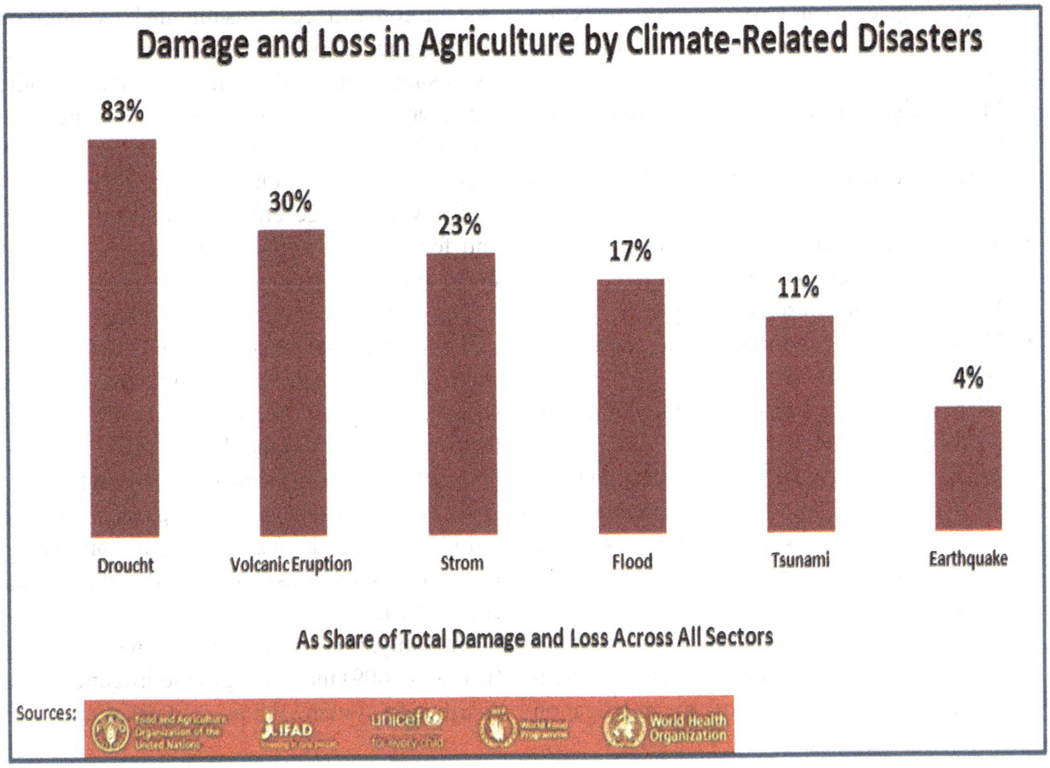

Perception and Reality for Sustainable Irrigation System with Micro-irrigation, Fig. 2 Damage and loss in agriculture by climate-related disasters

status. The reasons may be they do not have the orientation or experience about the advanced modern irrigation system. Correctly, it is well said by John Maynard Keynes, the famous British Economist (5 June 1883–21 April 1946), "The problem lies not so much in developing new ideas, but in escaping from old ones."

The existing perceptions in the field are as follows:

- Micro-irrigation system is so sophisticated.
- Micro-irrigation system is of very high-tech.
- Micro-irrigation system is very complicated.
- Micro-irrigation cannot be operated by common farmer.
- Its maintenance is beyond capability of rural people.
- Micro-irrigation is not suitable for heavy soil, like clay and clay loam.
- Micro-irrigation is not feasible/suitable for paddy cultivation.
- Micro-irrigation is feasible/suitable only for orchard, vegetables, potato, and wheat.
- Micro-Irrigation system is of very high cost involvement.

Reality

Reality*, the state of things as they are, rather than as they are imagined to be* (source: Cambridge University Dictionary). In reality, micro-irrigation has many positive aspects to attain sustainable irrigation management system. Salient features are noted below for better understanding of micro-irrigation.

- Micro-irrigation can help to achieve 50–80% water use efficiency.
- Micro-irrigation combining with prepaid metering system and smart card can achieve 70–90% water use efficiency.

- Micro-irrigation can cover much bigger command area in comparison to other irrigation systems.
- Micro-irrigation technologies increase crop yield.
- Micro-irrigation saves water and improves crop quality.
- Micro-irrigation enhances the fertilizer/chemical application efficiency.
- Micro-irrigation conserves energy.
- Micro-irrigation reduces labor cost and improve productivity.
- Micro-irrigation increases feasibility of irrigating difficult terrains.
- Micro-irrigation supply of optimum quantity of water in the form of tiny streams, fine spray, or continuous drops mitigates water loss due to evaporation and on account of seepage and percolation.

These above points are proven facts from field observations. With usage of micro-irrigation system, conveyance loss is minimal. Evaporation, runoff, and deep percolation are also reduced by using micro-irrigation methods. Another water-saving advantage is that water source with limited flow rates such as small water wells can be used. Micro-irrigation provides significantly higher water usage efficiency due to proximity and focused application. Water usage efficiency under various irrigation systems in India is shown in Fig. 3 in a recent publication (2016).

It is evident that micro-irrigation can show how stakeholder engagement in irrigation systems shapes hydrosocial territories: (1) by reducing tension between stakeholders, (2) by redirecting regional planning and strategy, (3) by highlighting water crises, (4) by decentralizing water responsibilities, and (5) by integrating values and beliefs from different stakeholders (Sandra Ricart et al. 2018).

In reality, it is observed from a research study (Kashem 2009) that underground irrigation water distribution system with prepaid meter and smart

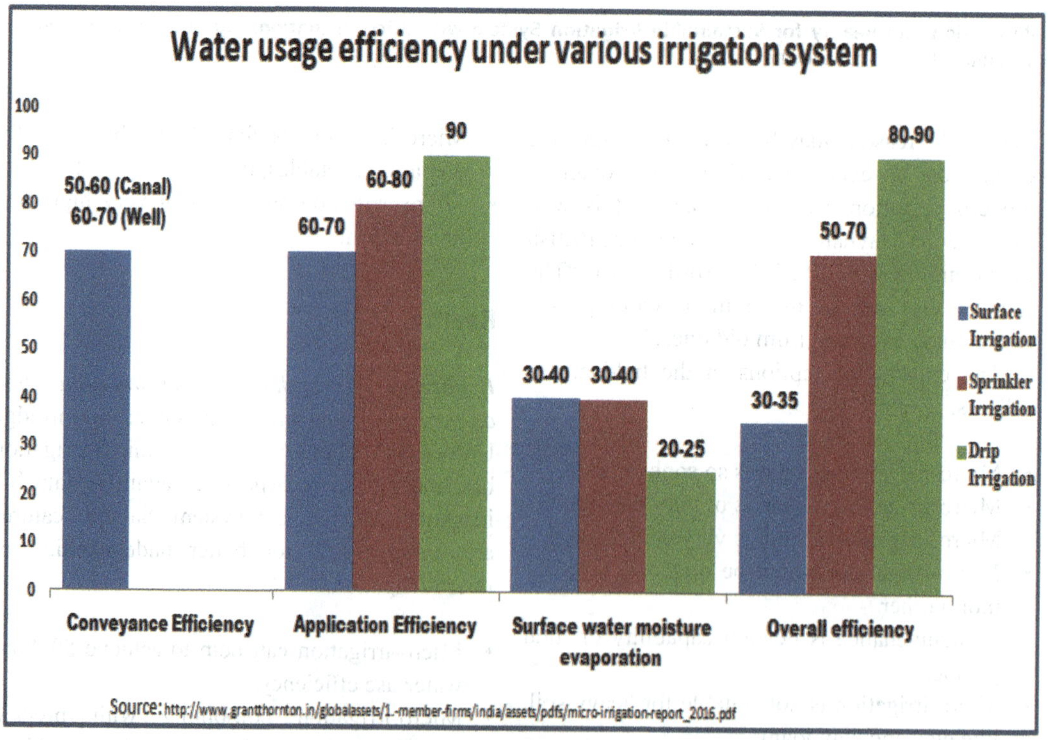

Perception and Reality for Sustainable Irrigation System with Micro-irrigation, Fig. 3 Water usage efficiency under various irrigation systems in India

card saves around 40% of irrigation water in comparison to flood irrigation from deep well. It is assessed that micro-irrigation with reliable water source and prepaid metering system including smart card could save around 50–90% of irrigation water. And also being considered is the best for achieving sustainable irrigation system and its management. In conclusion, once the public sector conceives the idea of sustainable irrigation system with modernized micro-irrigation facilities and collaborate with private sector, it is going to work effectively in the irrigation sector through judicious use of water resources (Kashem 2009).

Global Scenario

Arjen Y. Hoekstra and Ashok K. Chapagain – Globalization of Water, Sharing the Planet's Freshwater Resources, 1988, expressed the virtual-water content of various products in terms of liter of water per kilogram (kg) of product (Hoekstra and Chapagain 1988) is shown in Table 1.

Let us think about Indian subcontinent (India, Bangladesh, Nepal, Pakistan, and Sri Lanka) where rice is the main crop to attain food self-sufficiency. In fact, rice is the main staple cereal crops. Table 1 shows that around 3400 l of water is required for production of 1 kg of rice. If we can increase the water use efficiency to around 60–70%, then a huge volume of water could be saved and much bigger command area could be covered with same amount of water resources.

Perception and Reality for Sustainable Irrigation System with Micro-irrigation, Table 1 Global average virtual-water content of selected products (per unit of product)

Product	Virtual-water content (liters)
1 kg rice	3400
1 kg wheat	1300
1 kg maize	900
1 potato (100 g)	25
1 tomato (70 g)	13
1 orange (100 g)	50
1 apple (100 g)	70
1 cup of tea (250 ml)	35
1 cup of coffee (125 ml)	140
1 kg cotton	11,000

Micro-irrigation in Bangladesh

Since long, there are micro-irrigation system with sprinklers in tea gardens and army conducted livestock grazing field. Very recently under Barind Authority, there more than 35 sites of drip irrigation at low discharge well for low water consumptive crops like vegetables (cabbage, tomato, chili) and orchards. It is expected to boost up the micro-irrigation system under Barind Authority especially in the zone of water stressed both in respect of surface and groundwater resources. An installation of drip irrigation is shown in Fig. 4 in one of the dug well in Barind.

Barind Authority

The Barind Multipurpose Development Authority (BMDA) is an autonomous authority reporting to the Ministry of Agriculture. BMDA has taken an innovative approach to management of rural infrastructure, particularly the distribution irrigation water and management of irrigation systems.

In response to the universal concern for irrigation sustainability, that O&M will fail through a lack of user raised funding, BMDA has introduced an innovative concepts for collection of the irrigation service charge (ISC). The most important component to be implemented was the introduction of the prepaid meter with smart cards for tube well irrigation. This really proved to be the miracle solution towards a more stable and efficient sustainable irrigation management. Other new ways were also implemented. All wells were electric ones that eased the operation and maintenance. Underground plastic pipe water distribution system was managed for all areas that housed wells.

The ruling principle was to provide the best possible service to farmers with reasonable and affordable service charges (Zaman 2013).

The Prepaid Metering System provides every farmer with a user card. The card is loaded with credit by paying cash at any BMDA office or to an accredited dealer. The card is then inserted in a

Perception and Reality for Sustainable Irrigation System with Micro-irrigation, Fig. 4 Drip irrigation in Barind

slot at the pump station and water is pumped automatically with the charge levied against the credit on the card. Card uploading stations (vending stations) are connected electronically to a central BMDA operator and all data can be monitored remotely. The process works using technology similar to that used for loading credit to cell phones. The benefits of the prepaid system, either coupon or electronic card are (Zaman 2014):

(i) All water provided is paid for in advance.
(ii) There is no opportunity to bypass the meter.
(iii) The system is completely transparent with checks and balances in place to counter fraud.
(iv) People cannot coerce the operator to deliver water free of charge.
(v) Farmers cannot be exploited by land owners who may control the well. Prepaid water charges go to BMDA coffers which support the sustainability of the BMDA.

Irrigation systems based on the following principles:

(i) Water is available when, and in the quantities, required all year round.
(ii) All repairs and maintenance costs are borne by BMDA.
(iii) Technical support is provided free of charge to the farmers.
(iv) All electricity used by pumps or other irrigation equipment is paid for by BMDA.
(v) Pump operators are remunerated by BMDA on an hourly basis.
(vi) The prepaid coupons and electronic card reloading are sold by dealers who are paid on a commission basis, and dealers retain a 5% commission for selling the coupons. Commission on card reloading was reported to be 2.5%.
(vii) BMDA pays employee salary and support costs for its 1036 staff members without seeking unreliable funding sourced from the government.

Sustainability

Sustainability *is the process of maintaining change in a balanced fashion, in which the exploitation of resources, the direction of investments, the orientation of technological development and institutional change are all in harmony and enhance both current and future potential to meet human needs and aspirations.* Once surface irrigation from canal and other water bodies were the most popular and dependable irrigation system. Due to scarcity of water at the intake of canals and gradual increase of pumping units on the other hand depleting of the water resources due to siltation of the canal and water bodies, the initial sustainable condition does not exist anymore. Same way groundwater abstraction has been initiated during mid-1960s; it was quite fine till around 2000, with huge increments of tube wells, but it is now having huge crisis. Presently, spacing of wells and numbers are being strongly monitored to control the groundwater abstraction. Now proposal being considered at the policy level to increase the water use efficiency to cover the existing command area and also to extend more area under controlled irrigation to achieve food self-sufficiency to feed the growing millions. Under this consideration, policy makers and relevant engineers are thinking to switch over for micro-irrigation to avoid the natural havoc/disastrous situation of the country. That is why micro-irrigation (both sprinkler and drip irrigation) is being considered seriously but its technology is not that popular like surface canal or underground pipe for irrigation water distribution system. Moreover, there is certain land terrain where canal or tube well irrigation is not that much suitable and or feasible but micro-irrigation could be best possible solution. These need to be demonstrated in the field, and let other engineers and policy makers have a look physically.

Conclusion and Recommendation

It is assessed that micro-irrigation with reliable water source and prepaid metering system including smart card could save around 50–90% of irrigation water. And also being considered is the best for achieving sustainable irrigation system and its management. In conclusion, once the public sector conceives the idea of sustainable irrigation system with modernized micro-irrigation facilities and collaborate with private sector, it is going to work effectively in the irrigation sector through judicious use of water resources.

All study says micro-irrigation saves water of sources may be underground and surface. Less water will be required to pump to cover the projected command area. This is correct, on the other hand, less water for irrigation will cause less seepage and percolation means less recharge of the groundwater resources. Dr. Rashid mentioned in his Ph.D. thesis “Conclusions to the following specific water management aspect/issues are drawn from the research study: 1. In Boro Season (February to May) 55% of the applied water needed for ET (Evapotranspiration) and the rest 45% is lost from field as seepage and percolation. Likewise, in Transplanted Aman (T.Aman) season (Kharif-II, July to October) 45% of the applied water in the rice field was needed for ET and the rest 55% is lost from the field as seepage and percolation. Therefore, combining the results of two seasons together it can be concluded that irrespective of seasons 50% of the irrigation water is needed for ET and the rest 50% is lost through S and P annually from double cropped rice field.” (Rashid 2005)

It means there will be obviously less recharge to the groundwater when less amount of irrigation water sprayed over bigger command area through micro-irrigation system. Evapotranspiration remains same. Question will there be a sustainable condition continue in those command area in the concerned areas. This dilemma/issues need to be addressed through proper in-depth study.

References

Hoekstra, A. Y., & Chapagain, A. K. (1988). *Globalization of water*. Sharing the Planet’s Freshwater Resources.

P

Irrigation Association of India, and Federation of Indian Chambers of Commerce & Industry. (2016). *Accelerating growth of Indian agriculture: Micro irrigation an efficient solution* (Strategy Paper – Future Prospects of Micro Irrigation in India).
Kashem, A. (2009). *The economics of prepaid irrigation: Barind Multipurpose Development Authority (BMDA) perspective in Bangladesh*. Ph.D. thesis.
Milos Holy. (1971). *Water and the environment* (Irrigation and Drainage Paper No. 8). FAO.
My Agricultural Information Bank. (2019). Online http://www.agriinfo.in/default.aspx?page=topic&superid=8&topicid=2234
NITI. (2017). AAYOG-guidelines -micro irrigation through public private partnership "From the Source to the Roots". Draft Concept note, October, 2017.
Rashid, A. (2005). *Groundwater management for rice irrigation in Barind Area of Bangladesh*. Ph.D. thesis.
Sandra Ricart, C., et al. (2018). How to improve water governance in multifunctional irrigation systems? Balancing stakeholder engagement in hydrosocial territories. *International Journal of Water Resources Development*. https://doi.org/10.1080/07900627.2018.1447911.
Siddiqui, M. F. A. (1975). *An infra-structure for development in Bangladesh-Irrigation*. Paper presented in Integrated Rural Development in November 29–December 3 1975, Dacca.
Zaman, A. (2013). *Barind: A paradigm of sustainable irrigation management for Bangladesh and beyond*. Paper presented at ADB's water week on 12 March.
Zaman, A. (2014, June 23). New management initiatives for sustainable irrigation management.

Peri-urban – City-Edge

► Strategies for Taming the City

Peri-urban Regions

Benefits and Threats

Michael Buxton
RMIT University, Melbourne, VIC, Australia

Synonyms

Benefits; Rural development; Threats; Urban development

Definition

Billions of people enjoy the values and use the resources of peripheral urban areas. The considerable benefits to humanity and nature from these regions constitute a global resource but are being incrementally reduced by development. No global institution and few intergovernmental processes have responded adequately to the global threat to peri-urban resources. No international peri-urban policies exist. The future of peri-urban areas is left to states and cities with the result that peri-urban policy is fragmented or nonexistent.

Peri-urban regions hold high strategic, spatial, social, economic, and environmental significance. They fulfill personal and social needs, allow urban dwellers to relate to countryside, and provide preventative mental and physical health and other social services. They contain a range of natural resources, produce much of the world's food, provide a range of ecosystem services, often contain valued landscapes and significant reserves of habitat and biological diversity, and are important water catchment areas.

Their traditional rural characteristics are intrinsic to their value but are threatened by forces that are unprecedented in scale and type. The principal threats remain urbanization, small lot subdivision, and the uncontrolled proliferation of commercial and other urban related land uses in rural landscapes. These and inadequate governance systems interact with emerging new factors, principally climate change, and threaten to overwhelm the resilience of peri-urban areas globally through catastrophic nonlinear change.

Introduction

Peri-urban regions are those areas on the urban periphery into which cities expand or which cities influence (Houston 2005; Burnley and Murphy 1995). Such regions are commonly taken to extend up to 150 km from a city edge into landscapes which are predominately rural or semirural and often include townships of varying sizes. The extent of the influence of a metropolitan area on its hinterland generally is defined by commuting time to the nearby metropolis or by spatial characteristics such as the density of rural settlement and types of land uses.

The spatial characteristics of peri-urban areas vary considerably between regions and countries. These variations can complicate the designation of land as peri-urban. The designation of urban growth boundaries to expanding cities is a common way of defining the inner boundary. But, often, both the inner and outer boundaries of peri-urban areas can be indistinct. The edges of cities may be characterized by a range of urban uses which merge with rural residential developments and traditional rural activities. Development may occur in a linear fashion along transport routes or be more haphazard. Often development "leapfrogs" rural land placing pockets of urban activity in rural land and making the identification of a clear urban edge difficult. The identification of outer boundaries can be even more difficult. Amenity farming, rural residential development, intensive agriculture, and a diverse range of other uses such as forestry and rural industries eventually merge in indistinct ways with broadscale agriculture.

Such varying spatial characteristics of peri-urban land are intricately related to types of land uses. This entry examines these relationships to identify both the benefits of and threats to peri-urban areas. The entry also examines the governance arrangements which can protect peri-urban areas and which themselves often constitute threats to the maintenance of traditional rural land uses.

Protected or Multifunctional Regions

Commentators have taken two contrasting positions on which peri-urban land uses are regarded as beneficial or as threats. These positions can broadly be described as "preservationist,", expressed as "monofunctional" land uses, and "heterogeneous" expressed as "multifunctional" land uses. A preservationist approach will seek to permit only nonurban land uses and encourage the retention of traditional rural uses. It will regard three types of land uses as the principal threats: metropolitan and regional city expansion, small lot rural subdivision, and urban-related land uses. A heterogeneous approach, in contrast, will seek to extend the type of permitted land uses, and include many urban-related uses, and will regard the three land use types as opportunities.

This debate is often focused on the notion of an urban green belt. A green belt is a defined non-urban or rural belt of land around the periphery of an urban area intended to permanently protect rural areas from urbanization. Green belts are structural components of the peri-urban zone situated closest to metropolitan settlements, while more distant land forms outer peri-urban areas. Many variations to the green belt concept exist. Some surround cities, others parts of cities often by delineating wedges of land between urban land. Parkways, or nonurban corridors linking areas of public land, are common. Large urban parks often form open areas which supplement the exclusion of urban uses from privately owned land.

Traditionally, the fundamental element in the green belt idea is that the outward spread of urban settlement must be halted. Rydin and Myerson (1989: 476), for example, argue that "green belts are about defining the boundary between urban and rural areas. Indeed, the operation and implementation of green belt policy is boundary definition." Hall (1973: 52) argued that the "notion of urban containment, and the notion of rural preservation, come together in the concept of the green belt." Some advocates, such as Mumford (1961), have argued that a green belt needs no further justification than its own existence serving to halt the outward spread of cities. Mumford (1961: 595) argued that outward urban growth must be contained if the benefits and values of a functioning city and countryside are to be maintained and that both are threatened by the "extension of vast masses of sub-urban. . .housing. . .into open country."

Competing Ideologies

A number of underlying perspectives affect how spatial relationships are viewed and influence the importance placed on varying drivers of peri-urban change. An urban perspective emphasizes the powerful impacts a city exerts on its peri-urban areas and concentrates on the needs of the city, regarding nearby nonurban areas as the means to satisfy urban needs by providing land and resources. The proximity of peri-urban land to urban markets creates pressures for change. The types of urban influence include the effects on peri-urban population size or type and distribution, tourism and accommodation developments, land price, agriculture, habitat and biological diversity, and landscapes. A constantly recurring impact is that peri-urban land becomes regarded as an urban land bank awaiting transformation into suburbia (Angelo 2016). A rural perspective, in contrast, regards the peri-urban zone as invaded countryside threatened by urban expansion which can offer opportunities by introducing new income and skills (Bunce 1994).

Another contrasting conceptual framework concerns the process of how change occurs. Many researchers draw from Conzen's model to propose a typology of concentric rings where the urban influence on peri-urban areas lessens in an orderly and predictable manner as distance from the urban fringe increases (Burnley and Murphy 1995; Hart 1991). This picture presents a continuum of progressive, measured change across landscape as a gradation of peri-urban land uses in an orderly spatial pattern of gradually lower populations and lot numbers along with fewer urban-related uses. In contrast, a second typology emphasizes that change may not be uniform but a disorderly pattern of land uses in a heterogeneous, multifunctional land use mosaic. Urban-related uses appear out of sequence with agricultural and other traditional rural uses in no apparent pattern. A multitude of different lot sizes in a differentiated subdivision pattern underlies the existence of commercial, retail, and industrial uses, residential, rural-residential, and farming lots, and multiple other types of land uses (Daniels 1990; Nelson 1999; Audirac 1999; Allen 2003).

This entry now turns to examining the proposed benefits of peri-urban areas before outlining the perceived threats.

The Value of Amenity

Amenity attracts people to peri-urban areas from nearby metropolitan centers both as visitors enjoying natural and cultural features and as residents. Natural amenity factors such as forests, coastal environments, high-quality landscapes, and water resources are common peri-urban features. Cultural factors such as heritage architecture often reinforce natural attractions (Argent et al. 2010). In turn, the location of peri-urban areas close to metropolitan centers enables ready access by residents to urban facilities particularly for work, entertainment, and contact with family and friends. Amenity is the value given by residents and visitors to attributes which are esthetically attractive, and to access to services (Argent et al. 2010). Early commentators identified two types of amenity, "amenity resources" such as climate, coastal areas, and high-quality landscapes (Perloff and Wingo 1964), and cultural resources (Moss 1987). Amenity can be defined objectively in terms of physical conditions such as natural features, heritage buildings, and infrastructure, and subjectively as personal perceptions, tolerance, diversity, and safety.

More recently, much attention has been paid to the relationships between spatial factors, amenity, and human well-being. The notion of well-being also can be defined in terms of both objective indicators such as life span, levels of person interaction and participation, income, and health and sustenance, and subjective ones such as fulfillment and happiness (Stanley et al. 2013). There is clear evidence of the positive relationships between amenity, well-being, and personal and community health outcomes (Rogerson et al. 2019). People living in rural landscapes experience increased contact with nature and report higher well-being compared to urban dwellers, even controlling for other variables such as income (White et al. 2013). Peri-urban residents and urban dwellers alike gain both physical health benefits and improved psychological well-being from rural landscapes through such activities as outdoor recreation, inspiration, and wildlife viewing (Gibbons et al. 2014). Factors such as quality landscapes, plant species, and wildlife improve mood and self-esteem (Clark et al. 2014), social cohesion (Rogerson et al. 2019), promote calm (Aspinall et al. 2015), and so reduce stress (Ulrich 1981).

Peri-urban Food Production

Cities usually are located in areas rich in natural resources, often on or near fertile soils, in benign climates, and with access to abundant water resources suitable to a wide variety of agricultural production. Often few substitutes exist for the type of food production most suitable to such conditions. Peri-urban agriculture is advantaged further by being located close to urban markets, and using sophisticated transport and communication networks to convey food rapidly and efficiently to distribution points for regional, national, and international use, at relatively low cost. The proximity of metropolitan centers provides other advantages, such as access to labor and food-processing plants, and the potential of using recycled water for intensive agriculture.

Renewed interest in the importance of peri-urban agriculture occurred after the global food price crises of 2007–2008 and led to an "explosion of interest in food studies" (Moragues-Faus and Morgan 2015: 275). This interest has been expressed in the sustainability of existing food production models and urban and peri-urban agriculture, focused particularly on food sources and diversity, planning for public health and threats to food systems. Peri-urban areas are the traditional food sheds of cities (Zasada et al. 2017) and make major contributions to world food supply. They contribute high percentages of widely practiced crop production globally, such as rice, maize, and wheat, particularly in developing countries (Bryce 2016). Their locational advantages often lead to forms of intensive agriculture with many times the financial return per hectare than broadscale farming. Peri-urban land is 1.77 times more productive than the global average and contains over 60% of irrigated croplands (Bren d'Amour et al. 2017). Intensive agriculture continues to be practiced often immediately adjacent to or amid urban settlements in both developing and developed countries. Such productive areas often become part of valued landscapes valued by residents for their varied appearance and contributions to food supply.

Peri-urban agricultural production also often dominates certain food sources. For example, these areas provide 86% of the US fruit and vegetables and 63% of its dairy products (AFT 2014). Canadian peri-urban food production provides a successful case study of the ways that agricultural production is integrated with an urban greenbelt. The Vancouver Agricultural Land Reserve is a 4.72 million ha area which includes 61,000 ha of metropolitan agricultural land. This area produces 27% of the value of British Columbia's agriculture on 1.5% of provincial land. The Ontario government has taken a similar approach, preserving a green belt of 730,000 ha where innovative agricultural uses flourish. Australian peri-urban areas are particularly important contributors to food supply. Melbourne's area provides 41% of the city's food needs providing 47% of the state's vegetables, 67% of its eggs, and almost all its chicken, lettuce, herbs, asparagus, and berry fruit (Sheridan et al. 2015). The Sydney region provides almost 100% of New South Wales' Chinese cabbages and sprouts, 80% of fresh mushrooms and 91% of spring onions and shallots, 55% of meat, 40% of eggs, and 38% of dairy (Sinclair 2015).

Biodiversity and Natural Resources

Peri-urban areas make important contributions to the reserves of the earth's life support systems of plants, animals, other organisms and genetic strains within species and ecosystems particularly at this time of continued global assault on biological diversity. Globally, the contribution of peri-urban areas to biodiversity protection varies, but the collective impacts and benefits are profound. The protection of habitat and other natural resources provides a variety of additional benefits such as the provision of ecosystem services, recreational spaces, and their contributions to human health. These benefits vary between countries. Many European countries have extensively modified natural landscapes particularly around cities though even here large urban parkland, modified peri-urban forest and wetland areas, pockets of original or modified habitat, or culturally applied flora such as hedgerows can provide considerable benefits to biodiversity. Peri-urban agriculture has led to widespread modification of habitat around many Asian cities though the picture is more mixed around many cities in Africa, South America, North America, and Australia. The location of cities in resource-rich regions means that rich reserves of biodiversity often remain in heterogeneous habitats but that such reserves contain high proportions of threatened species at particular risk. For example, over 50% of threatened species occur in and around the fringes of Australian cities (Ives et al. 2016).

An ecosystem approach (UNCBD 2010) builds on the notion of interconnected benefits from the preservation of nature by providing a framework for the integrated management of land, water, and living resources. It can seek to provide ecosystem services for human use in the form of those that are supporting (such as soil), products (such as food), regulating (such as pollination), or cultural (such as health benefits and spiritual renewal). This approach is aimed primarily at ecosystem conservation through the maintenance of abiotic, biological, ecosystem, and biochemical resources. Globally, large numbers of urban regions are reevaluating urban and peri-urban biodiversity because of their inherent contribution to the reserve of global biodiversity and for their human benefits. Typically, peri-urban biodiversity protection is often incorporated into integrated urban and regional planning through both the maintenance and restoration of habitat.

Many urban and regional metropolitan centers rely wholly or partially on peri-urban sources for water supply. Water in peri-urban regions is stored and consumed for domestic, agricultural, and industrial purposes and is also used for recreational and the maintenance of natural aquatic environments and ecosystems. It is also a vital resource for nonurban water use, such as irrigated agriculture. Urban water and wastewater reuse is increasingly advantaging peri-urban agriculture by separating peri-urban rural water use from dependence on large riverine irrigation systems which are subject to climate change and overallocation.

This entry now turns to threats to the future of peri-urban areas. Opinions are divided on whether particular factors are threats or opportunities. Attitudes toward each issue considered here are influenced by more fundamental values which in turn affect the way evidence is evaluated.

The Threat of Urban Development

All outward urban expansion extends into peri-urban areas, and this growth is the greatest threat to traditional peri-urban landscapes and land uses. Urban influences are commonly regarded as demographic, economic, and environmental. Some researchers focus on the suburbanization process arising from the phenomenon of cities continually absorbing their fringe areas and creating new urban frontiers (Nelson and Sanchez 1999; Golledge 1960). Others concentrate on a broader peri-urban belt extending well beyond the metropolitan fringe, and on differences between peri-urban and urban areas (Houston 2005). Ford (1999: 307) argues that "in-flows from the metropolitan area are not solely the result of extended suburbanization" and points to a complex interacting range of causes of peri-urban growth, including growth from rural and metropolitan areas and from within peri-urban regions.

The physical areas of urban settlement in both developed and developing countries are expanding much faster than populations are increasing, consuming disproportionate amounts of land. Between 1990 and 2015, urban expansion in relation to urban population growth in developed countries increased 1.5 times and in developing countries by 3.5 times. The extent of urban areas in East Asia expanded annually by 7.2%, Southeast Asia 5.7%, and sub-Saharan Africa 5.1%. By 2030, cities are expected to cover three times as much land as they did in 2000, with much of the expansion occurring in relatively undisturbed key biodiversity-threatened areas and on productive agricultural land. This process of global urban expansion into peri-urban areas is leading to new settlement forms growing outward from urban peripheries often disconnected from the main urban settlements (UN Habitat 2020).

Many cities have established urban growth boundaries which often form a fixed edge to settlement. These are most apparent in Europe and North America but also exist into some Asian and other cities. One quarter of US metropolitan areas use an urban growth boundary often coupled with a green belt (Dawkins and Nelson 2002) while 15 green belts cover 14% of England. However, the world will further urbanize over the next decade, from 56.2% or 4.3 billion people in 2020 to 62% or 5.4 billion by 2036 and 68% by 2050. By mid century, 96% of urban growth will occur in the less developed regions of East Asia, South Asia, and Africa with 35% of world urban population increase from 2018 to 2050 occurring in India, China, and Nigeria (UN Habitat 2020). Over two-thirds of the populations of both China and India will eventually be housed in continuous urban regions (Friedmann 2010).

But the fastest growing settlements are the small and medium cities with less than 1 million inhabitants containing 59% of global urban population in 2020 (UN Habitat 2020; UN DESA 2019). Over 50% of Europe's population lives in city-regions of over 100,000 people (Briquel and Collicard 2005). High growth rates are being recorded on the fringes of many German and Mediterranean cities in a trend to detached single family houses in peri-urban hinterlands. Urban sprawl is now characteristic of many dispersed settlements as new edge metropolitan suburbs have emerged leading to spatial differentiation and the socioeconomic segregation of populations in a way that is characteristic of the US urban pattern, although less so (Hoffmann-Martinot 2004). For example, peri-urban rings occur around 230 large urban centers in France. The population of peri-urban rings of France's 12 major cities grew by an average 52% from 1968 to 2011, almost double the rate for their urban areas. One quarter of the French population or over 15 million people now live on a peri-urban land area of 38% of the nation (Cusin et al. 2016).

In the United States, peri-urban land contains about one-third of the US population (Audirac 1999) and land area (Nelson 1999). Thirty per cent of all new housing in the United States is built there. Between 1990 and 2010, 12.7 million more houses were built in the wildland-urban interface (WUI), or peri-urban regions, housing 25 million more people (Pierre-Lewis and White 2018). Even though the WUI comprises less than one-tenth of the land area of the United States, 43% of all new houses were built there in the 20 years to 2010 on an area of 189,000 km^2, an area larger than Washington State. Between 1982 and 2012, more than 42 million acres, including at least 24 million acres of agricultural land, were developed on peri-urban land in the United States (Daniels and Payne-Riley 2017). Nelson (1999: 137) states that much of the United States may eventually no longer "be distinguishable as either urban or rural, being instead characterized mostly as low density, exurban development." The same trends are apparent in parts of Western Europe.

Threats from Small Lot Subdivision and Urban-Related Uses

The introduction of small lot rural subdivision and urban-related uses also threatens traditional rural attributes. Smaller rural-residential lots remove much agricultural land from production. Copland covers 12% of the Earth's ice-free surface, and 3.2% of this, or 46 million ha, is expected to be urbanized by 2030, with 84% of losses expected in Asia and Africa (Bren d'Amour et al. 2017). Controversy continues over the impacts of urban expansion on food production. In the United States, for example, some commentators argue that the US urban area of 3% of the total US land area exerted little impact on food production (Gottlieb 2015; Vesterby et al. 1994). However, high-value land comprises 17% of the US land area, and most land lost is in peri-urban areas, making the effects on prime agricultural land disproportionately severe. Between 1992 and 2012, 31 million acres was taken from production, much of this loss irreplaceable. The costs of such loss are considerable. A study of the costs of urban development scenarios to food production in the

Melbourne peri-urban area, for example, showed a 20-year reduction in the net present value of the cumulative decrease in Gross Regional Product under a moderate urban sprawl scenario of $1.33 billion (in 2014–2015 dollars) (Deloitte Access Economics 2016).

Another dispute between advocates of protection or development is over whether peri-urban land should be regarded as monofunctional (reserved for homogeneous land uses) or heterogeneous (allowing multifunctional uses). Sometimes, industrial, commercial, tourism, retail, business, and other urban-related uses are permitted well into the peri-urban zone. These uses are incompatible with the preservation principle and are a de facto breaching of an urban growth boundary. Many commentators oppose a preservationist ideology and argue for policy flexibility allowing for multiple land uses including substantial additional housing and other development (Sturzaker and Mell 2017). But many rural land uses are far from homogeneous. Traditional landscapes, while excluding urban uses, are often managed for a diverse range of uses with environmental, social, and economic benefits. Much agricultural land protects remnant natural values such as vegetation and wetlands. Even intensive farming which has altered such features often retains high landscape value. Rural production through such uses as aquaculture and forestry is often significant, and other innovative nonurban rural activities are encouraged (Buxton and Butt 2020).

Both consumption and production landscapes can substantially alter traditional peri-urban landscapes. The progressive subdivision of productive land into smaller "rural-living" lots for amenity dwellers "consumes" the landscape for amenity purposes at the expense of agricultural and other rural uses. The closer settlement associated with commodified landscapes also often leads to loss of environmental features such as habitat and can be part of a process of "creative destruction" which creates a new economy as it annihilates the rural landscapes previously formed (Tonts and Greive 2002). Many commentators have shown how maintaining rural heterogeneity can protect peri-urban landscapes by linking environmental preservation with innovative farming and other rural practices (Adell 1999). Similarly, broadscale agricultural production can lead to devastating environmental impacts, necessitating the removal of habitat and even land forms, and leading to the monopolization of water. The extensive use of greenhouse technology, in areas such as the south of Spain, and other changing patterns of intensified agricultural practice threaten biodiversity, landscapes, stream flow, and aquifers. Such technology is emerging as a global threat to traditionally integrated peri-urban landscapes.

Nonlinear Environmental Threats

Cities and their peri-urban areas coexist in a reciprocal relationship. A future which values rural hinterlands will differ radically from one where urban settlements gradually expand and often merge. The prosperity and livability of cities in this century are likely to lie in their connections with their regions. As humans decrease the resilience of biophysical systems, progressively smaller disturbances can move the system beyond its threshold into a different regime with a different structure and function (Walker and Salt 2006). This change may be gradual and linear, or it may be nonlinear. Peri-urban resilience therefore can be assessed by an evaluation of system elements and complex interactions to indicate the capacity to withstand or adapt to radical systems shocks. This capacity is determined by the strength of the pressures from nearby urban areas, the ability of the functioning peri-urban system to respond to change, and the impact of the ruling governance system.

The possibility of rapid, nonlinear change once critical thresholds are exceeded requires anticipatory planning to prevent or adapt to such change. Peri-urban areas are among the most susceptible on earth to likely fundamental change. Anthropogenic climate change reinforces the notion of uncertainty as global and regional climates change. Climate models predict higher temperatures, sea level rise, greater climate variability, and more extreme events along with a wide range of environmental, social, and economic impacts. Peri-urban areas will be among the areas most affected by bushfires, flooding, coastal erosion, and extreme events such as storms. Yet large-scale settlement of land continues globally on the fringes of cities in areas which climate change will make more dangerous, exponentially increasing risk. Often the most vulnerable citizens are placed in these increasingly risky locations. Extreme events will require retreat and resettlement of large numbers of people in many countries from increasingly high-risk and ultimately uninhabitable peri-urban areas.

Uncertainty from climate change and related factors such as energy supply and food security increases the importance of peri-urban regions, and the need to protect their assets and values and to plan their future. The societies able to adapt most successfully to change will be those which previous generations have advantaged by retaining the greatest possible decision-making potential. Uncertainty about the future requires the exercise of caution and the retention of future options over future land uses. Energy and food supply chains are likely to be subjected to increased costs and disrupted. Pressure on global food resources will increase, advantaging cities with regional peri-urban food supplies. Fresh produce may already travel thousands of kilometers to urban consumers. The amount of food shipped between countries increased fourfold in the last four decades of the twentieth century, and the trend is increasing (Weekes 2006). The fact that "once developed, land stays in urban uses' (Vesterby et al. 1994) leads to 'an "option value" argument that tends to support the preservationist point of view" (Gottlieb 2015: 5).

Governance Threats

Many peri-urban areas are detrimentally affected by fragmented governance and inadequate policy responses. Many governance models influence the relationships between land use and interacting spatial, social, natural resource, economic, and environmental factors. Regulatory policy is common, as in the UK greenbelts, while regional policy and planning provisions have protected many peri-urban areas in Germany and elsewhere in Europe. Legislation and zones governing land use are used to establish urban growth boundaries and protect greenbelts. Regulatory controls affecting peri-urban land have proved particularly controversial in green belts because they are "a major planning intervention in rural land economics" which limit or remove development rights and affect rural land uses (Taylor 2019: 463).

In Portland, Oregon, several municipalities have cooperated to establish a growth boundary and peri-urban zones to achieve orderly, sequential land release over a long time period. Some authorities and agencies have used other techniques, such as the purchase of property rights. US government agencies and trusts have preserved over 5 million acres at a cost of US$5 billion to purchase development rights over peri-urban land to protect farmland in ~50 metropolitan counties (AFT 2014). By 2015, 28 states, almost 100 counties, and over 500 land trusts were involved in such programs (Daniels and Payne-Riley 2017). Alternatively, some municipalities and governments have relied on land purchase to protect designated parts of peri-urban areas. Boulder, Colorado, for example, established a land purchase fund in 1967, financed by a sales tax, and has purchased over 18,000 ha at a cost of US$200 million. Large areas of green belt land have been purchased progressively in Britain. For example, the London County Council purchased 55 square miles (142 km^2) of land for the London Green Belt. Australian state governments and municipalities also have a long record of purchasing or reserving peri-urban land for parkland.

Legislation and planning controls, such as zones, which mandate allowable land uses provide more certain direction over allowable peri-urban uses and developments than discretionary policies intended to guide decisions. However, to be consistent with the preservation ethic, the content of legislation and planning provisions should allow only nonurban uses and seek to control quasi-industrial rural uses, such as extensive shed-based agriculture. Many planning provisions limit the effects of regulation by allowing some urban-related uses. Controls over subdivision also vary significantly, with much small rural lot subdivision allowed, so limiting the rural nature of the peri-urban zones.

Conclusion

Perhaps the greatest threat to peri-urban planning is a neoliberal ideology which has come to dominate land use planning in many countries. The reduction in the role of government has led to eliminating or reducing controls over land use and development coupled with the lack of long-term anticipatory planning. This process had led globally to the loss of public benefits of peri-urban areas. Fragmented governmental and sectoral arrangements often accompany neoliberal governance, characterized by vertically disintegrated systems of national, regional, state, and local government coupled with horizontally fragmented sectoral planning. Cross-sectoral planning integrates spatial planning regimes with the management of land, water, and natural resources and requires the reciprocal consideration of sectoral impacts. Applied on a regional scale, it provides an effective model for the maintenance of peri-urban values.

Land tenure is a key independent factor in a complex network of interacting variables and reciprocal relations. The regional significance of peri-urban regions collectively constitutes a global resource which benefits humanity and nature. Yet their future is determined by incremental local and regional decisions with little or no regard to the cumulative impact of decisions. Only a global body, such as the United Nations, can recognize the global benefits and threats and adopt measures which can form the context for international action to protect these vital areas.

References

Adell, G. (1999). *Theories and models of the peri-urban interface: A changing conceptual landscape, strategic environmental planning and management for the peri-urban interface research project*. London: Development Planning Unit, University College London.

AFT (American Farmland Trust). (2014). *Fact sheet: Why save farmland?* Washington, DC: Farmland Information Center, American Farmland Trust.

Allen, A. (2003). Environmental planning and management of the peri-urban interface: perspectives on an emerging field. *Environment & Urbanization, 15*(1), 135–147.

Angelo, H. (2016). From the city lens toward urbanisation as a way of seeing: Country/city binaries on an urbanising planet. *Urban Studies, 54*(1), 158–178. https://doi.org/10.1177/0042098016629312.

Argent, N., Tonts, M., Jones, R., & Holmes, J. (2010). Amenity-led migration in rural Australia: A new driver of local demographic and environmental change? In G. Luck, R. Black, & D. Race (Eds.), *Demographic change in Australia's rural landscapes* (pp. 23–44). Dordrecht: Springer.

Aspinall, P., Mavros, P., Coyne, R., & Rove, J. (2015). The urban brain: Analysing outdoor physical activity with mobile EEG. *British Journal of Sports Medicine, 49*, 272–276. https://doi.org/10.1136/bjsports-2012-091877.

Audirac, I. (1999). Unsettled views about the fringe: rural-urban or urban-rural frontiers?. In O. Furuseth and M. Lapping (Eds.), *Contested Countryside: The Rural Urban Fringe in North America* [pp. 7–22]. Aldershot: Ashgate.

Bren d'Amour, D., Femke, C., Reitsma, F., ... Seto, K. (2017). Future urban land expansion and implications for global croplands. *Proceedings of the National Academy of Sciences of the United States of America, 114*(34), 8939–8944. https://doi.org/10.1073/pnas.1606036114.

Briquel, V., & Collicard, J. (2005). Diversity in the rural hinterlands of European cities. In K. Hoggart (Ed.), *The city's hinterland: Dynamism and divergence in Europe's peri-urban territories* (pp. 22–44). Aldershot: Ashgate.

Bryce, E. (2016) Growing mega-cities will displace vast tracts of farmland by 2030, study says. *The Guardian*, 28 December.

Bunce, M. (1994). *The countryside ideal: Anglo-American images of landscape*. London: Routledge.

Burnley, I., & Murphy, P. (1995). Exurban development in Australia and the United States: Through a glass darkly. *Journal of Planning Education and Research, 14*, 245–254. https://doi.org/10.1177/0739456X9501400402.

Buxton, M., & Butt, A. (2020). *The future of the fringe. The crisis in peri-urban planning*. Melbourne: CSIRO Press.

Clark, N., Lovell, R., Wheeler, B., Higgins, S., Depledge, M., & Norris, K. (2014). Biodiversity, cultural pathways and human health: A framework. *Trends in Ecology and Evolution, 29*, 198–204. https://doi.org/10.1016/j.tree.2014.01.009.

Cusin, F., Lefebvre, H., & Sigaud, T. (2016). The peri-urban question: A study of growth and diversity in peripheral areas in France. *Revue Française de Sociologie, 57*, 641–679. https://doi.org/10.3917/rfs.574.0641.

Daniels, T. (1990). Policies to preserve prime farmland in the USA: A comment. *Journal of Rural Studies, 6*(3),

331–336. https://doi.org/10.1016/0743-0167(90)90087-O.
Daniels, T., & Payne-Riley, L. (2017). Preserving large farming landscapes: The case of Lancaster County, Pennsylvania. *Journal of Agriculture, Food Systems, and Community Development, 7*(3), 67–81. https://doi.org/10.5304/jafscd.2017.073.004.
Dawkins, C., & Nelson, A. (2002). Urban containment policies and housing prices: An international comparison with implications for future research. *Land Use Policy, 20*(5), 771–794.
Deloitte Access Economics. (2016). *The economic contribution of Melbourne's foodbowl: A report for the foodprint Melbourne project*. Melbourne: University of Melbourne.
Ford, T. (1999). Understanding population growth in the peri-urban region. *International Journal of Population Geography, 5*, 297–311. https://doi.org/10.1002/(SICI)1099-1220(199907/08)5:4<297::AID-IJPG152>3.0.CO;2-O.
Friedmann, J. (2010). Place and place-making in cities: A global perspective. *Planning Theory and Practice, 2*, 149–165. https://doi.org/10.1080/14649351003759573.
Gibbons, S., Mourato, S., & Resende, G. (2014). The amenity value of English nature: A hedonic price approach. *Environmental and Resource Economics, 57*(2), 175–196. https://doi.org/10.1007/s10640-013-9664-9.
Golledge, R. (1960). Sydney's Metropolitan Fringe: A Study in Urban-Rural Relations. *Australian Geographer, 7*(6), 243–255.
Gottlieb, P. (2015). Is America running out of farmland? *Choices, 30*(3), 1–7.
Hall, P. (1973). *The containment of urban England*. Oxford: George Allen & Unwin.
Hart, J. (1991). The perimetropolitan bow wave. *The Geographical Review, 81*, 35–51.
Hoffmann-Martinot, V. (2004) *Towards an Americanization of French metropolitan areas*? Paper presented at the Department of Political Science, Universidad Autonoma de Madrid and International Relations and at the International Metropolitan Observatory Meeting Pole Universitaire de Bordeaux, 9–10 January.
Houston, P. (2005). Revaluing the fringe: Some findings on the value of agricultural production in Australia's peri-urban regions. *Geographical Research, 43*(2), 209–223. https://doi.org/10.1111/j.1745-5871.2005.00314.x.
Ives, C., Lentini, P., Threlfall, C., Ikin, K., Shanahan, D., … Kendal, D. (2016). Cities are hotspots for threatened species. *Global Ecology and Biogeography, 25*(1), 117–126. https://doi.org/10.1111/geb.12404.
Moragues-Faus, A., & Morgan, K. (2015). Reframing the foodscape: The emergent world of urban food policy. *Environment and Planning A., 47*(7), 1558–1573. https://doi.org/10.1177/0308518X15595754.
Moss, L. (1987). *Santa Fe, New Mexico, post-industrial amenity-based economy: Myth or model?* Edmonton/Santa Fe: Alberta Ministry of Economic and Trade & International Cultural Resources Institute.
Mumford, L. (1961). *The city in history: Its origins, its transformations and its prospects*. London: Secker and Warburg.
Nelson, A. (1999). The exurban battleground. In O. Furuseth and M. Lapping (Eds.), *Contested Countryside: The Rural Urban Fringe in North America* [pp. 137–150]. Aldershot: Ashgate.
Nelson, A., & Sanchez, T. (1999). Debunking the exurban myth: A comparison of suburban households. *Housing Policy Debate, 10*(3), 689–709.
Perloff, H., & Wingo, L. (1964). Natural resource endowment and regional growth. In J. Friedmann & W. Alonso (Eds.), *Regional development and planning: A reader* (pp. 215–239). Cambridge, MA: MIT Press.
Pierre-Lewis, K., & White, J. (2018). Americans are moving closer to nature, and to fire danger. *New York Times*, 15 November.
Rogerson, M., Gladwell, V., Pretty, J., & Barton, J. (2019). Landscape and well-being. In M. Scott, N. Gallent, & M. Ghartzios (Eds.), *Routledge companion to rural planning* (pp. 495–507). London: Routledge.
Rydin, Y., & Myerson, G. (1989). Explaining and interpreting ideological effects: A rhetorical approach to green belts. *Environment and Planning D, Society & Space, 7*, 463–479. https://doi.org/10.1068/d070463.
Sheridan, J., Larsen, K., & Carey, R. (2015). *Melbourne's foodbowl: Now and at seven million*. Melbourne: Victorian Eco-Innovation Lab, University of Melbourne.
Sinclair, I. (2015, 9–11 December). Growing food in a residential landscape. In *State of Australian cities conference*. Gold Coast: Queensland.
Stanley, J., Birrell, R., Brain, P., Carey, M., Duffy, M., … Wright, W. (2013). *What would a climate-adapted settlement look like in 2030? A case study of Inverloch and Sandy Point*. Gold Coast: National Climate Change Adaptation research Facility.
Sturzaker, J., & Mell, I. (2017). *Green belts: Past, present, future?* New York: Routledge.
Taylor, L. (2019). The future of green belts. In M. Scott, N. Gallent, & M. Ghartzios (Eds.), *Routledge companion to rural planning* (pp. 458–468). London: Routledge.
Tonts, M. and Greive, S. (2002). Commodification and creative destruction in the Australian rural landscape: the case of Bridgetown, Western Australia. *Australian Geographical Studies 40*(1), 58–70. https://doi.org/10.1111/1467-8470.00161.
Ulrich, R. (1981). Natural versus urban scenes: Some psychophysiological effects. *Environment and Behavior, 13*, 523–556. https://doi.org/10.1177/0013916581135001.
UN Habitat. (2020). *World cities report. The value of sustainable urbanization*. Nairobi: United Nations Settlements Programme.
UNCBD (UN Convention on Biological Diversity). (2010). *Year in review 2010*. Montreal: UN Environmental Programme.
United Nations, Department of Economic and Social Affairs, Population Division. (2019). *World urbanization prospects: The 2018 revision* (ST/ESA/SER.A/420). New York: United Nations.

P

Vesterby, M. Heimlich, R. Krupa, K. (1994). *Urbanization of Rural Land in the United States*. Agricultural Economic Report No. 673. Washington, DC: US Department of Agriculture.

Walker, B., & Salt, D. (2006). *Resilience thinking: Sustaining ecosystems and people in a changing world*. Washington, DC: Island Press.

Weekes, P. (2006). This meal has travelled all over the world. *The Sunday Age*, 7 March.

White, M., Alcock, I., Wheeler, B., & Depledge, M. (2013). Would you be happier living in a greener urban area? A fixed-effects analysis of panel data. *Psychological Science, 24*, 920–928. https://doi.org/10.1177/0956797612464659.

Zasada, I., Schmutz, U., Wascher, D., Kneafsey, M., … Piorra, A. (2017). Food beyond the city: Analysing foodsheds and self-sufficiency for different food system scenarios in European metropolitan regions. *City, Culture and Society, 16*, 25–35. https://doi.org/10.1016/j.ccs.2017.06.002.

Peri-urban Spaces

► The Challenges for Wildland-Urban Interfaces (WUI) in Metropolitan Areas: Reducing Fire Risk, Providing Employment Opportunities, and Preserving Natural Habitat

Peri-urbanization

David Simon
Department of Geography, Royal Holloway, University of London, Egham, UK

Synonyms

Urban edge; Urban–rural fringe; Urban–rural interface

Definition

Peri-urbanization is a process of dynamic change in land-use and livelihoods affecting the perimeter of growing or stable urban areas. Unlike the static and simplistic urban–rural dichotomy, a peri-urban interface (PUI) constitutes a zone of transition between rural and urban status. The breadth, composition, and physical location of a PUI around any given urban area usually vary in different directions at a particular moment and temporally, as the leading edge of the interface may move outwards at a different rate from that of the inner edge, where areas become effectively urban or suburban. Originally formulated in the context of rapidly growing African and Asian cities, the concept of peri-urbanization has recently been applied globally, as researchers also discern (re)new(ed) processes of change and complexity in many parts of the global North, for which traditional conceptions of urban fringe or edge are now inadequate. Some researchers claim that suburbanization is a preferable term for these processes, but this essay demonstrates the important differences between them, both conceptually and empirically. Meanwhile, recent research into complex and sustainable ancient urban agglomerations on different continents within which peri-urban agro-hydraulic systems were integral may hold important lessons for current sustainable urban transitions.

Introduction

The term peri-urbanization gained currency in the 1990s to describe the complex processes of change playing out on the fringes of urban areas in many parts of the world. The inadequacy of the overly simplistic rural–urban dichotomy to distinguish urban from nonurban areas was particularly evident in rapidly growing towns and cities of Africa, Asia and some small island states over the preceding decades. This situation still applies today.

While some urban expansion occurs in an orderly and planned manner, much takes place in piecemeal fashion, with individual developments often ignoring or bypassing official mechanisms and processes. This commonly creates conflicts between incompatible land uses; generates resource degradation and environmental pollution through unregulated discharges of waste; makes retrospective infrastructure and service provision both difficult and more costly; and provides profound governance and planning challenges.

Unscrupulous developers and landowners stand to make large profits but many residents and new migrants may also benefit from more affordable and perhaps appropriate housing and facilities than would be permitted officially.

The result is often an "untidy" zone of transition, of in betweenness, that is neither any longer physically or functionally rural nor fully urban – hence the term "peri-urban." Initial attempts to capture this conceptually strove to determine empirical parameters by measuring the width of the peri-urban zone in different cities. This proved immensely difficult and showed that there was no uniformity even around a single city. This diversity reflected a combination of circumstances, such as physiographic conditions (topography, (un-)suitability for construction); accessibility; land tenure arrangements and the willingness or otherwise of particular landholding communities and chiefs, private individuals or public authorities to sell, permit or tolerate settlement by newcomers.

Such attempts also proved futile because – measurement problems aside – they were by definition static and completely missed the essence of peri-urbanization, which is highly dynamic change. Hence a peri-urban interface (PUI) is most appropriately understood as a zone or interface in transition from rural to urban appearance, land-use mix, functions, and status. As a city grows, previously rural land and communities become peri-urban on the outer perimeter or fringe of the zone, while the inner edge may move outwards too, as some areas become fully urban. Since these processes differ markedly in rate and extent, even around a particular city over the same time period, the width of the PUI may grow or shrink simultaneously in different parts of the zone. Even slow-growing or essentially stable cities may have active PUIs.

A key stimulus to this perspective was provided by the British Government's Department for International Development (DFID), which financed a decade-long PUI research program (1995–2006), with a suite of projects centered on Hubli-Dharwad in India and Kumasi in Ghana. The objective was to understand the complexity of changing livelihoods of the poor in a comparative perspective and hence draw generalizable lessons to inform more effective peri-urban sustainable natural resource utilization and planning in the global South (Mbiba and Huchzermeyer 2002; Simon et al. 2004; McGregor et al. 2006; Mattingly 2009). Even in the very different and rapidly urbanizing Chinese context, where dense settlement in largely anthropogenic agrarian landscapes has a long history, the PUI approach aids understanding of new urban forms and also processes of socio-ecological adaptation (Lin 2006; Friedmann 2011; Abramson 2016; Li et al. 2018).

Extending the Concept Historically

Given this particular origin, peri-urbanization is thought of predominantly as a distinctive present-day phenomenon but examination of recent archaeological evidence derived from remote sensing has challenged this assumption. Indeed, it appears to have occurred in many different ancient urban agglomerations worldwide under diverse social-ecological conditions, forming the underpinning of sustainable urban agro-ecosystems, from which current efforts to promote urban sustainability transitions could learn much (Simon 2008; Simon and Adam-Bradford 2016; Smith 2010; Barthel and Isendahl 2013; Sinclair et al. 2016).

Current Controversies

Until the mid-2000s, various terms had been used to denote the urban-rural interface elsewhere in the world. Some had local origins while others reflected traditional academic conceptions dating from post-World-War-II urban containment, land-use planning, and green belt policies that these zones were stable or static in many European countries, for instance (Simon 2008). However, (neo-)liberalization of planning controls, coupled with profound demographic change and immigration has blurred many of these lines, both literally and figuratively, and given greenbelt development new impetus. Many of these urban fringes also now display more complex and dynamic land-use mixes, even in conditions of only modest urban and population growth. Hence, there has been a growing tendency since the mid-2000s to frame analysis of European, American, and Japanese cities in terms of PUIs (Qviström 2013; Simon and Adam-Bradford 2016: 63–4; Sorensen 2016). Indeed, this represents a still all-too-rare example of analytical globalization and cross-fertilization from the global South to the global North.

Another debate is attempting to displace peri-urbanization and the PUI concept with suburbs (e.g., Mabin et al. 2013; Keil 2018). Its proponents argue that suburbanization more accurately describes the middle-class residential expansion on the fringes of many cities in the global South, often driven by large-scale and even transnational property development firms offering the growing indigenous middle classes and returning diasporic community members westernized visions of low-density, high amenity living. Sometimes, it is led by individual households buying or leasing plots and developing them piecemeal on the urban fringe (e.g., Mercer 2017, 2020).

However, such perspectives fail to acknowledge the diverse forms and processes that may co-exist and reveal even contradictory or countervailing patterns in the same PUI. Second, longstanding "indigenous" residents of small villages being encroached and their land progressively converted from agriculture to concrete culture might see things differently from middle-class outsiders, not least in terms of whether they individually benefit or lose through land transactions and associated processes of accumulation and dispossession. Over recent years, I have observed peri-urbanization in Ununio, a small coastal fishing and agricultural village in a different part of the-then outer fringe of metropolitan Dar es Salaam from where Mercer (2017) undertook her research, and it has been very different from what might reasonably be called suburbanization. Yet just a short distance away, planned new developments, complete with infrastructure, could appropriately be described as outlying suburbs.

Suburbanization has very particular connotations, not least of overwhelmingly residential land-use, with historically specific origins and translations from metropolitan to colonial and imperial territories (Bartels et al. 2020). In the global South, suburbs were reimagined by the indigenous elites and new middle classes to symbolize their status and modernity and to affirm their place in a globalizing world. Recent research in the USA also demonstrates the complexity, diversity, and fluidity of boundaries and concepts, providing a warning against the imagining and application of universalizing terminology (Garner 2017; Lichter and Ziliak 2017).

Systematic research around Kumasi, Ghana, in the early 2000s also demonstrated the continuum of conditions along a transect through the peri-urban interface. At the outer edge, agricultural villages were beginning a transition in livelihoods profiles and land uses, while at the inner extreme villages comprised fully urbanized areas which outsiders could not distinguish from the surrounding suburbs but which the residents still firmly identified as constituting their villages with clear boundaries (Simon et al. 2004, 2006). Village councils under chiefs and elders held sway, albeit with reduced remits since their land allocation function had been lost through land sales and sometimes formal registration of land title. Longitudinal peri-urban research in Accra, the Ghanaian capital, reveals similar processes (Gough and Yankson 2006; Owusu 2008). Such processes have continued subsequently, with complexity, dynamism and diversity of prevailing conditions of all kinds still very much in evidence (Bartels et al. 2018, 2020; Gough et al. 2019). Such multilocal comparative research reveals the diversity of built forms, processes and overlapping meanings coexisting in dynamic peri-urban interfaces. This may change further over time but in many contexts, peri-urban and suburban are currently not mutually exclusive categories.

Conclusions

Peri-urbanization refers to changing mixes of land-use, livelihoods, and lifestyles during processes of urbanization, and is most appropriately conceived of as a dynamic continuum between rural and urban, over both space (both directionally and with increasing distance from the city) and time. Suburbanization may be one component, or an outcome, of peri-urbanization where the dominant land-use is residential. However, other parts of a PUI might emerge as predominantly commercial, light or heavy industrial, service-oriented, or remain mixed.

Although developed in the dynamic urban contexts of the global South, the term has acquired global relevance in understanding the complex dynamics now again found on the margins of often older urban areas in Europe, North America, and Japan. It is also now proving helpful in understanding ancient urban forms and processes around the world, from which lessons may be learned for current efforts to promote urban sustainability. Nevertheless, its diverse applications defy elaborate theorization.

Cross-References

- ▶ Food Security
- ▶ Green Cities
- ▶ Green Infrastructure
- ▶ Nature-Based Solutions
- ▶ Urban Agriculture

References

Abramson, D. B. (2016). Periurbanization and the politics of development-as-city building in China. *Cities, 53*(2), 156–162. https://doi.org/10.1016/j.cities.2015.11.002.

Bartels, L. E., Bruns, A., & Alba, R. (2018). The production of uneven access to land and water in peri-urban spaces: De facto privatisation in greater Accra. *Local Environment, 23*(12), 1172–1189. https://doi.org/10.1080/13549839.2018.1533932.

Bartels, L. E., Bruns, A., & Simon, D. (2020). Towards situated analyses of uneven peri urbanization: An (urban) political ecology perspective. *Antipode, 52*(5), 1237–1258. https://doi.org/10.1111/anti.12632.

Barthel, S., & Isendahl, C. (2013). Urban gardens, agriculture, and water management: Sources of resilience for long-term food security in cities. *Ecological Economics, 86*, 224–234. https://doi.org/10.1016/j.ecolecon.2012.06.018.

Friedmann, J. (2011). Becoming urban: Periurban dynamics in Vietnam and China. *Pacific Affairs, 84*(3), 425–434. https://www.jstor.org/stable/23056183.

Garner, B. (2017). "Perfectly positioned": The blurring of urban, suburban, and rural boundaries in a southern community. *The Annals of the American Academy of Political and Social Science, 672*, 46–63. https://doi.org/10.1177/0002716217710490.

Gough, K. V., & Yankson, P. W. K. (2006). Conflict and cooperation in environmental management in peri-urban Accra, Ghana. In D. McGregor, D. Simon, & D. Thompson (Eds.), *The peri-urban interface: Approaches to sustainable natural and human resource use* (pp. 196–210). London: Earthscan.

Gough, K. V., Yankson, P. W. K., Wilby, R., Amankwaa, E. F., Abarike, M. A., Codjoe, S. N. A., Griffiths, P. L., Kasei, R., Kayaga, S., & Nabilse, C. K. (2019). Vulnerability to extreme weather events in cities: Implications for infrastructure and livelihoods. *Journal of the British Academy, 7*(S2), 155–181. https://doi.org/10.5871/jba/007s2.155.

Keil, R. (2018). *Suburban planet: Making the world urban from the outside in*. London: Polity Press.

Li, C., Wang, M., & Song, Y. (2018). Vulnerability and livelihood restoration of landless households after land acquisition: Evidence from peri-urban China. *Habitat International, 79*, 109–115. https://doi.org/10.1016/j.habitatint.2018.08.003.

Lichter, D. T., & Ziliak, J. P. (2017). The rural–urban interface: New patterns of spatial interdependence and inequality in America. *The Annals of the American Academy of Political and Social Science, 672*, 6–25. https://doi.org/10.1177/0002716217714180.

Lin, G. C. S. (2006). Peri-urbanism in globalizing China: A study of new urbanism in Dongguan. *Eurasian Geography and Economics, 47*(1), 28–53. https://doi.org/10.2747/1538-7216.47.1.28.

Mabin, A., Butcher, S., & Bloch, R. (2013). Peripheries, suburbanisms and change in sub-Saharan African cities. *Social Dynamics, 39*(2), 167–190. https://doi.org/10.1080/02533952.2013.796124.

Mattingly, M. (2009). Making land work for the losers; policy responses to the urbanisation of rural livelihoods. *International Development Planning Review, 31*(1), 37–53. https://doi.org/10.3828/idpr.31.1.3.

Mbiba, B., & Huchzermeyer, M. (2002). Contentious development: Peri-urban studies in sub-Saharan Africa. *Progress in Development Studies, 2*(2), 113–131. https://doi.org/10.1191/1464993402ps032ra.

McGregor, D., Simon, D., & Thompson, D. (Eds.). (2006). *The peri-urban interface: Approaches to sustainable natural and human resource use*. London: Earthscan. xxiv + 336 p.

Mercer, C. (2017). Landscapes of extended ruralisation: Postcolonial suburbs in Dar es Salaam, Tanzania. *Transactions of the Institute of British Geographers, 42*(1), 72–83. https://doi.org/10.1111/tran.12150.

Mercer, C. (2020). Boundary work: Becoming middle class in suburban Dar es Salaam. *International Journal of Urban and Regional Research, 44*(3), 521–536. https://doi.org/10.1111/1468-2427.12733.

Owusu, G. (2008). Indigenes' and migrants' access to land in peri-urban areas of Accra, Ghana. *International Development Planning Review, 30*(2), 177–198. https://doi.org/10.3828/idpr.30.2.5.

Qviström, M. (2013). Searching for an open future: Planning history as a means of peri-urban landscape analysis. *Journal of Environmental Planning and Management, 56*(10), 1549–1569. https://doi.org/10.1080/09640568.2012.734251.

Simon, D. (2008). Urban environments: Issues on the peri-urban fringe. *Annual Review of Environment and Resources, 33*, 167–185. https://doi.org/10.1146/annurev.environ.33.021407.093240.

Simon, D., & Adam-Bradford, A. (2016). Hybrid planning in the peri-urban interface: Applying archaeology and contemporary dynamics for more sustainable, resilient cities. In B. Maheshwari, V. P. Singh, & B. Thoradeniya (Eds.), *Balanced urban development: Options and strategies for liveable cities* (pp. 57–83). Dordrecht/Frankfurt am Main/London: Springer. https://doi.org/10.1007/978-3-319-28112-4_5.

Simon, D., McGregor, D., & Nsiah-Gyabaah, K. (2004). The changing urban–rural interface of African cities: Definitional issues and an application to Kumasi, Ghana. *Environment and Urbanization, 16*(2), 235–247. https://doi.org/10.1177/095624780401600214.

Simon, D., McGregor, D., & Thompson, D. (2006). Contemporary perspectives on the peri-urban zones of cities in developing countries. In D. McGregor, D. Simon, & D. Thompson (Eds.), *The peri-urban interface: Approaches to sustainable natural and human resource use* (pp. 3–17). London: Earthscan.

Sinclair, P., Isendahl, C., & Stump, D. (2016). Beyond rhetoric: Towards a framework for an applied historical ecology of urban planning. In *Oxford handbooks online* (p. 20). https://doi.org/10.1093/oxfordhb/9780199672691.013.34.

Smith, M. E. (2010). Sprawl, squatters and sustainable cities: Can archaeological data shed light on modern urban issues? *Cambridge Archaeological Journal, 20*(2), 229–253. https://doi.org/10.1017/S0959774310000259.

Sorensen, A. (2016). Periurbanization as the institutionalization of place: The case of Japan. *Cities, 53*(2), 134–140. https://doi.org/10.1016/j.cities.2016.03.009.

Personal Delivery Robots are Sometimes Called: Personal Delivery Devices

► Personal Delivery Robots: How Will Cities Manage Multiple, Automated, Logistics Fleets in Pedestrian Spaces?

Personal Delivery Robots: How Will Cities Manage Multiple, Automated, Logistics Fleets in Pedestrian Spaces?

Bern Grush
Urban Robotics Foundation, Toronto, Canada

Synonyms

Personal delivery robots are sometimes called: personal delivery devices; Sidewalk robots; Delivery robots; Delivery bots; Sidewalk drones; Micro utility devices

Introduction

The company currently leading the last-mile logistics market with small, personal robotic deliveries claims to have made 1.5 million deliveries as of May 2021, with one-third of this number achieved in the first 5 months of 2021. This same company claims to "make more than 80,000 road crossings every day" (Edwards 2021).

This latter number should give pause to any city transportation planner. A claimed achievement of 500,000 deliveries over 5 months, implies an average of 3310 trips per day, 80,000 daily crossings imply 24 crossings per trip. Assuming 12 crossings each way on a round trip, this means the average trip would be a bit under a mile – certainly a reasonable distance for such service. What is troubling is the potential for pedestrian, cycling, and automobile interactions at those crossings should these 80,000 daily crossings swell to several billion per day, worldwide, in a few years. See Fig. 1.

As of 2021, the worldwide volume of such deliveries made by all providers will likely be at least double the numbers claimed by the single, lead provider. When express delivery companies combine this technology with micro-warehousing and local lock boxes – all technologies in various

Personal Delivery Robots: How Will Cities Manage Multiple, Automated, Logistics Fleets in Pedestrian Spaces?, Fig. 1 A robot (pink, center) crossing a busy intersection in Toronto in 2021. (Image courtesy tinymile.ai)

stages of development and deployment and with very few remaining *technical* barriers – the daily number of trips and their related street crossings could potentially rival the number of pedestrians and bikes at these crossings in locations of greatest demand. This would be especially true in cities with less-active and more auto-oriented populations, where the relative, local numbers of such robots could easily exceed pedestrian volumes.

As will be developed in this entry, robotic services in public spaces, especially last-mile delivery, promise enormous benefits for our cities and its citizens if deployed in ways that are clean, safe, and comfortable (see Table 1). They could also pose a challenge to urban livability if left unmanaged as we did street parking for most of the twentieth century, and which most cities still manage so poorly.

The promise made for last mile delivery (and similarly for other services) by sidewalk robots is that they would replace larger vehicles such as congestion-causing stepvans from express-delivery companies or would avoid "mov[ing] a 2-pound burrito in a 2-ton car" (serverobotics.com). Most would agree that small, quiet, slow, electric delivery devices would be more desirable than internal combustion step vans, but a shift from goods movement on the roadway to goods movement on the pedestrian footway portends new implications that are yet fully understood.

This entry will attempt to address that.

Personal Delivery Robots: How Will Cities Manage Multiple, Automated, Logistics Fleets in Pedestrian Spaces?, Table 1 An example set of planning and design principles. (Transport for London 2020)

Safe	The public realm should be safe to use at all times of day and for people to feel safe to spend time in
Inclusive	All walking environments should adhere to the principles of inclusive design by ensuring that they are accessible to, and useable by, as many people as reasonably possible without the need for special adaptation or specialized design
Comfortable	Designated walking areas should allow unhindered movement for pedestrians by providing sufficient space
Direct	Facilities should be positioned to provide convenient links between major walking trip attractors
Legible	Features should be consistent and easy to understand for all pedestrians to know intuitively how to navigate within a space
Connected	Walking networks should have a high density of route options to suit pedestrians needs
Attractive	Walking environments should be inviting for pedestrians to pass through or spend time in

Robots Operating in Public Spaces

There are several aspects to consider when planning or regulating commercial and service robots operating in public places. These range from technology, social and urban impacts, through traffic and safety impacts, as well as standards and monetization issues. This entry section provides a cursory overview.

Table 1 lists the design principles developed by Transport for London for pedestrian networks comprising publicly accessible footways and other public spaces where people are permitted to walk (Transport for London 2020). The principles listed in this table were developed for the full spectrum of pedestrian circumstances, without consideration of the introduction of robotic devices sharing these spaces. Nonetheless such planning principles, where they exist, would directly impact future regulations regarding delivery robots operating in these spaces.

Technology Capability

The fundamental design purpose of a robot is to perform a defined task effectively. To that end, robotics is an engineering discipline – at its core, mechatronics, but inclusive of many others including industrial, human factors, and artificial intelligence. Matters of user safety as well as time and cost efficiency are of first order in applications such as factory, agriculture, warehousing, and mining. In those well-understood environments, proximate humans are usually working in collaboration with, or are trained to be wary and respectful of, robotic operations. In this entry, however, we are focused on robots to be deployed in public spaces at a considerable, non-line-of-sight distance from responsible human oversight while surrounded by noninvolved humans (sometimes referred to as "incidentally co-present persons" (InCoPs).

The proximity of noninvolved humans, such as nearby or passing pedestrians, adds a considerable level of complexity to the core system issues of human safety, operating time, and cost efficiency. Consider also that some pedestrians may be vulnerable due to age, ability, health, or distraction, while others might engage in mischief or vandalism.

Add to this challenge the fact that robots operating in public spaces are inclined to be on sidewalks, pathways, laneways, parking lots or bike lanes almost all of which have discontinuities, gradients, curbs, and narrow passages, as well as fixed and transient obstacles. These spaces assume ambulatory, flexible, meandering, erratic human users at speeds between 0 and 8 kph.

Hence, the physical design of a robot to plough snow, pick litter, or transport a meal will be sized and shaped to its task and its intelligence design configured to navigate the pathways it is expected to encounter.

To address the intelligence challenge, robotics designers approach this with similar technologies and solutions as are being applied to the problem of the autonomous vehicle. Hence, these machines variously use multiple cameras, LIDAR, HD-maps, AI software, and teleoperators. A key design goal for any of these robotics designers is to determine the sensor and software configuration sufficient to maximize the ratio of robots to teleoperator. At this writing, the leading practitioners can operate two or three robots per human teleoperator, depending on pathway conditions.

Using the SAE scale (SAE International 2021), current robots operating in public space, delivering groceries or food, for example, are operating at Level 2 (one teleoperator fully attentive to one robot) or Level 3 (one teleoperator attending to a small number of robots that need only occasional interventions). The design goal is to achieve Level 4 (multiple teleoperators share-managing large fleets that are 5, 10, 20, times more numerous than the human operator collective). In such highly automated fleets, each robot would seldom demand attention; hence a very high level of fleet automation is required to match such a high level of robot operation.

Level 3 capability is already achievable in some, limited public spaces, and these machines are beginning to alleviate last-mile delivery issues in those selected environments. To become a sustainable industry – i.e., to become pervasive – such environments must be made more numerous and be managed to maintain a profitable ratio of robots per operator.

A key motivating factor for this segment of the robotics industry is that the current costs of last mile delivery are very high. The COVID pandemic exaggerated the impact of e-commerce, food, and grocery delivery and has promoted both supply and demand for final-mile delivery (Grush 2021a). In 2020–2021, supply has meant the expansion of gig delivery operators as well as accelerated innovation and investment by companies such as Amazon and FedEx, and numerous startups such as Starship and Serve Robotics (a recent Uber spin-off).

The fact that Level 3 sidewalk robotics is within reach, means profitable delivery operations are also within reach from the machine-sidewalk-teleoperator perspective. But that is still insufficient for a workable (governable) robotic delivery system to operate at scale in an urban, public environment. Remember, also, deliveries are only one of the many potential tasks such machines will be able to perform in these same public spaces.

With this logic, the greatest design challenge we face is the safe, societally acceptable operation of multi-operator, mixed-purposed, variably scheduled fleets of robots in a partially managed or often poorly managed, urban spatial environment.

After considering some social, urban, and traffic matters in the remainder of this section, the second section of this entry will outline a proposed solution to this challenge "A System to Manage Sidewalk Robots".

Social Impacts

Sidewalk robots (Fig. 2) arrive for us at a complex juncture. After well over a century of the influence of the automobile in determining how humans are regulated and channeled to walk in constrained public spaces that are spatially and speed dominated by lethal machines, we are just beginning to claw back space for cycling and vulnerable road users. The needs of the latter have been largely, discounted until recently. Municipalities have begun to take seriously competing demands for scarce sidewalk real estate, but aging infrastructure poses challenges and complexity when designing to accommodate these smaller vehicles. We are also just beginning to add micro-mobility forms at scale, to design a (very) few "complete streets," and to promote metrics such as "walkability" or concepts such as the "15-minute city." At the same time that urban areas and populations grow, the average size and personal ownership ratios of motorized vehicles is at least sustained and, in aggregate, continues to grow, even if peaking among some populations. Urban space demanded by automobiles is not abating.

Add to that the social distancing demands of COVID-19 and our cities have become effectively denser with mobile humans, devices, and machines. One example is the increased incursion of bicycles and micromobility devices sharing

Personal Delivery Robots: How Will Cities Manage Multiple, Automated, Logistics Fleets in Pedestrian Spaces?, Fig. 2 These are four of many tens of delivery robots in small-scale commercial or trial use today. They range 68–91 cm (length), 53–71 cm (width), 55–147 cm (height, without flag), 33–136 kg (gross weight). and 5–24 km/h (max speed) (Dimensions). Larger and faster delivery robots are planned for roadway use. Ambulatory (legged) robots are being developed to handle stairs. Innovation is only just beginning. (Illustration commissioned by the author.)

sidewalks and multiuse trails in many cities. I have experienced this personally numerous times in both Toronto and Montreal.

We must ask ourselves: "Have sidewalk robots arrived at the best possible time, or at the worst possible time?" The answer, of course, is that it is up to us. If we are to welcome these robots into our pedestrian and active-transportation spaces, then how will relative accessibility for all existing parties including these new devices be delineated and ensured (Clamann and Bryson 2021) (Grush 2021b)?

Jeremy Hsu describes both the upsides and downsides of sidewalk robots. On the upside, he quotes from a study: "By moving the last leg of deliveries from the road to the sidewalk, cities could reduce congestion and eliminate the parking problem entirely." And on the downside, he asserts: "When the pavement gets more crowded, even robots rolling along at walking speeds will face challenges which will get worse in US cities with narrow sidewalks." He concludes: "Given these obstacles, sidewalk delivery robots are not necessarily destined to win the future" (Hsu 2019).

While it is unnecessary to take a fixed position on any of these matters, there will clearly be cities that will see small, well-behaved robots as a way to promote the fortunes of local retail devastated by box-stores, ecommerce and the pandemic, to encourage small, quiet, slow, electric deliveries as a way to reduce automotive travel for short-haul shopping (Grush 2021c), or to cope with rising congestion from e-commerce and food deliveries.

Kristen Thomasen outlines three views of public space that might guide a regulator of robots on sidewalks: Communal Public Square, Regulated and Orderly Public Square, or State-Owned Property (Thomasen 2020). Depending on how these views influence relevant regulations, robots would be governed locally in more or less restricted ways.

It should also not be surprising if pedestrian, accessibility, cycling, and/or labor advocates demand limits on the use of such systems. The next decade will see much social discourse about the deployment of these technologies. It will become a goal of urban planners and municipal managers to govern the deployment of such technology somewhere between the extreme approaches of outright banning or complete laissez-faire. Paraphrasing Alanna Coombes:

> *To thrive we need community, business and political agreement on who has what rights at the kerb, footway and crossing. In turn, these rights need to be turned into clearly defined priorities that meet the needs of citizens, including those traditionally excluded, and businesses. Public space as we are contemplating for robotic traffic must be inclusive and protective of community and artistic expression and livability. These vital public spaces – like the city centers in which they exist – need to adapt to the needs of current and future generations, addressing their economic, social, community needs and their wellbeing.* (Coombes and Grush 2022)

Urban Planning

Closely related to these social issues is urban planning – an instrument that can be used to further social progress, especially its civil aspects. In the same way that active transportation and micro-mobility modalities, as well as livability and climate have become central concerns for planners over the past decade or more, the sustainability issues of urban freight and especially e-commerce and food delivery have grown in volume and urgency.

For any municipality wishing to deploy robots for a public or commercial purpose, it will be necessary to confirm that the sidewalks, pathways, and crosswalks to be used can accommodate those robots without violating applicable accessibility guidelines. For example, a sidewalk would need to be wide enough to permit a pedestrian in an accessible mobility device or aided by a service dog to pass a delivery device at most locations on the pathway. As well, if there were places such that a robot must stand aside for a wheelchair to pass, then sufficient waiting space is needed for such robots, and that remaining narrow passages need to be very short to minimize robot-pedestrian standoffs. Further complicating accessibility challenges is the proliferation of pet friendly establishments. Will robots be able to distinguish between a pet and a service animal? For example, if an algorithm depends on the ability of a human to act in a specified manner when a robot is approaching, how will that algorithm need to consider pedestrians with sight loss that may not be able to see an approaching robot and act accordingly – or at least not be alarmed or confused? Micromobility operators have begun to seriously consider adopting acoustic vehicle alerting systems similar to those required of electric/hybrid cars but little standardization exists with regard to such alerting systems.

Many other matters such as maximum in-line or cross gradients and surface conditions are important. Many of these are already incorporated in draft standards under development (Grush 2021b). One critical goal is to ensure that all urban pathways and trails intended for robot use be sufficiently specified, organized, and maintained to meet applicable accessibility guidelines, *even while supporting the intended robot traffic volumes*. This latter point implies the planner has some way to know or control the dynamic volume of robotic traffic, which is one of the intentions of the above referenced standard.

Ideally, the innovators of these devices and the deployers of the expected fleets would engage in meaningful dialogue with other road, bicycle, and sidewalk users, but too little such dialogue has thus far taken place. That may be understandable from the innovators' perspective partly because they are still determining what is feasible to innovate, but it may also be partly due to the sense that it is often easier to apologize later rather than ask now. That would be especially true if a new industry were able to accrue sufficient demand to act as a defensive buffer, later.

Regardless of one's urban-moral stance in this, the advocacy opportunity for pedestrians, vulnerable road users, cyclists, micro-mobility, and all other active transportation users is to press for attention now. The opportunity is to collaborate with those logistics companies that will deploy robotics to lobby for coherent urban spaces and approaches for active and small-device mobility corridors. The basis of this collaboration would be municipal monetization of robotic commerce to fund, manage, and maintain such pathways and corridors sufficient to both accessibility requirements and commercial needs. This would imply universal design at an appropriate scale.

In the zero-sum game currently being played out between motor vehicle and active modes, the sidewalk robot has properties of both. Several of the US State Senate bills that authorize a class of sidewalk robots called personal delivery devices define these robots as pedestrians in terms of the rules they must follow and the rights of way that motor vehicles must grant them in turn. At the same time, they are a special "pedestrian" class that must always grant rights-of-way to human pedestrians (Grush 2020) (Kingson 2021). This arrangement may not be suitable in the future case of sidewalk robots deployed for fire, police, or Emergency medical services (e.g., ambulance) work.

In any city where robotic goods delivery scales dramatically, the urban mobility space that is now split between motor vehicles and active modes may need to be repartitioned. How we currently segregate space among pedestrians, cyclists, street parking, transit lanes, and moving motor vehicles has grown increasingly fragmented, and in some places bordering on the irrational. The unintended consequence of shifting last-mile goods delivery from curbside stepvans onto sidewalk robots may be to force the re-rationalization of urban mobility space.

The most critical matter planners must face is the conundrum between the possible, the probable, and the preferable. Should planners find ways to accommodate sidewalk robots, or should they take advantage of the potential of sidewalk robots, and other forms of vehicle automation, to build the city they prefer?

P

Traffic Management

Even if mobility spaces that mix robotic service modes and active-mobility modes were optimally designed, we cannot avoid consideration of traffic management within those spaces. As well, we do not yet have a full and common traffic code that guides all these mobility forms to cohabit and cooperate on these pedestrian pathways and crosswalks. Standards for many of these issues are being developed. One will be mentioned in section "A System to Manage Sidewalk Robots" of this entry.

One of the most important matters will be sidewalk congestion wherein a large number of robots among pedestrians may become unworkable and frustrate pedestrian passage.

Urban planners can take a lesson from the last half-century in recalling the unfortunate but often necessary behavior of goods delivery vehicles in their parking behaviors as they frequently blocked traffic through lanes and more recently bicycle lanes for lack of loading and unloading infrastructure proximate to the delivery point.

As an industry, express delivery has often had little choice but to rely on infractions, citations, and traffic court as a "part of doing business" in larger cities. In this unwitting traffic management agreement, motor traffic, cyclists, and shipment receivers bear these costs. Wherever goods movement will be relocated to the sidewalk, any time that these vehicles cannot be accommodated within the existing space, the delivery operator will necessarily resort to some scheme to complete its delivery.

Urban transportation managers have been able to survive the problem of courier parking in our cities for several decades by relying on double parking, citations, and traffic court. Any other solution was deemed to incur more trouble, more complexity, and even higher costs.

If sidewalk robots are deployed at scale, such a violation-citation-payment approach would become even more unworkable.

The key point here is that an adherence to universal design principles is insufficient to manage a volume of robots that would overwhelm pedestrian traffic, unless there were independent lanes for robots – itself an unaffordable solution except in very few cases. Physical urban planning alone cannot eliminate this issue; such traffic requires dynamic, digitalized control.

Roadway traffic management incorporates universal patterns: traffic signs, signals, circles, speeds, parking areas, protocols, enforcement, citations, etc. Details may differ in each jurisdiction, and some at each intersection. Traffic management rules, or "robot orchestration" will differ slightly at each location, as well, and all of the signs and signals will be communicated via digital maps in real time. This will provide the opportunity needed for both congestion control and monetization (Grush and Coombes 2022).

A System to Manage Sidewalk Robots

Section "Robots Operating in Public Spaces" of this entry briefly highlights some of the technical, social, urban, and traffic management issues relevant to the large-scale introduction of mobile, human-scale robots for a variety of service purposes – especially final-mile delivery services – on public sidewalks, bike lanes, footpaths, and crosswalks.

To manage multiple automated fleets of mobile robots in pedestrian spaces, we will break the problem into four loosely connected layers: Regulatory, Orchestration, Fleet, and Machine.

The top, regulatory layer is provided by local, regional, and/or state and national regulations. These would be designed for local purposes and demands, and could be expressed in digital form using standard data definitions and procedural elements of ISO/DTS 4448 (Grush 2021b).

In this section of the entry, we are concerned with the second layer, designed for the digital orchestration of regional robot traffic. We assume that the two lower, digital layers for fleet and machine management preexist the orchestration layer. All three of these layers are in the "connected vehicle" domain.

Problem Definition

A high-level problem statement for orchestrating multiple, concurrent fleets of robots would be:

Orchestrate the flow of an arbitrary number of robots from an arbitrary number of fleets of robots that comprise an arbitrary number of machines assigned independent tasks, with independent schedule constraints, within a mapped "operational design domain" including dynamically changing traffic volume constraints.

(a) Each fleet of robots has an independent operator with an independent fleet-operating system (fleet layer)
(b) Robots in each fleet navigate safely and collaboratively among any proximate robot, human, or obstacle (machine layer)
(c) Data at the orchestration layer can only be communicated to or from the fleet layer

(d) The orchestration layer cannot communicate with any robot
(e) Latency between layers is effectively zero from a traffic management perspective. This means that any instruction or constraint from the orchestration layer can reach a robot by way of its respective fleet layer within 2 or 3 s.

Digital Management Layers

Figure 3 illustrates these four key system layers. As stated, we are focused only on the orchestration layer. This layer is the robot traffic control system that a region or municipality may employ to govern usage, manage congestion, and monetize public infrastructure for commercial usage in what are usually pedestrianized operating design domains. The same system structure could be used to manage loading and unloading goods and passengers from robotic vehicles at the curb.

The fleet layer in Fig. 3 is populated by fleet managers, each of which comprises software representing the business of a single entity, say an express delivery company or a de-icing fleet. This system is agnostic in regard to how elements of the fleet layer assign work to its robots, how many fleet managers are active, or how many active robots a fleet manager has deployed.

The machine layer (Fig. 3, bottom) is only populated by robots, and may represent multiple machine models, perform multiple task types, be made by multiple manufacturers, and utilize multiple software platforms. Coordinating each subfleet within whatever constraints are passed down from the orchestration layer is the problem of the respective fleet operator and is assumed to be handled by software that is arranged to manage that communication. The orchestration layer never communicates with the machine layer. This simplifies system management and helps to keep private the business of the fleet operator. This entry is not concerned with this layer, or the one above it.

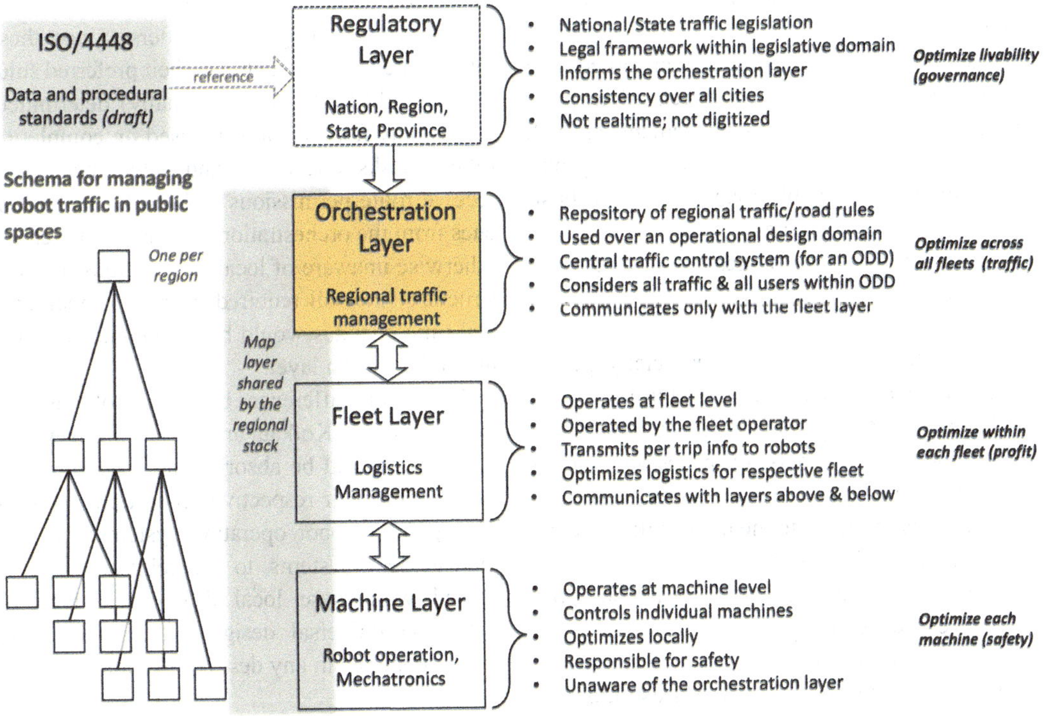

Personal Delivery Robots: How Will Cities Manage Multiple, Automated, Logistics Fleets in Pedestrian Spaces?, Fig. 3 In this schema for managing robot traffic in public spaces, there are four loosely connected layers. The Machine, Fleet, and Orchestration layers would be fully digitalized and operate in the "connected vehicle" domain

The Orchestration Layer

The orchestration layer (Fig. 3) is a fleet-independent, ground traffic control system that addresses several local (regional) matters especially gross positioning related to traffic distribution. This is distinct from micro-navigation matters critical at the machine layer or logistics and task optimization matters at the fleet layer. The central essence of the orchestration layer is an intelligent, dynamic, constraint-aware routing engine. It also manages real-time information distribution regarding local (block-face) rules some of which may change dynamically. These rules may also include user fees.

The only potential override by the orchestration layer would be the assignment of a route for traffic distribution reasons that might not be the route that the fleet layer for the respective operator would have derived. However, because the orchestration layer is enabled to use pricing to manage congestion, a fleet operator could have multiple route choices, including its own time-optimized, distance-optimized, or charge-optimized choice(s).

The orchestration layer is also concerned with parameterizing robot behaviors related to positional and shy-distancing management, as well as deferential, social behaviors regarding proximate humans, vulnerable road users, pets, businesses, and other machines.

The purpose of the orchestration layer is to maximize:

- Accessibility for all users, especially pedestrians, including vulnerable road users
- Traffic flow (congestion, rights-of way)
- Acceptance of robots in pedestrian spaces (human comfort, robot social behaviors)
- Efficiency in regard to the use of infrastructure within the relevant ODD
- Fleet operator awareness of local conditions for the intended trip

and to minimize:

- Pedestrian confusion, alarm, or frustration
- Spatial conflicts and or congestion
- Unexpected navigation or access barriers for the fleet operator

This layer is generally unconcerned with robot task-related information. Exceptions to this would be in regard to carriers of hazardous goods, robots executing emergency-related tasks, and certain public works tasks such as snowplowing or litter-picking. A traffic authority would likely wish to manage certain types of task features for such cases.

The data exchanged between the orchestration and fleet layers comprises a few dozen elements such as:

- Trip data: Origin, destination, time, actual max speed, etc.
- Robot data: Size, weight, max speed set, equipment capabilities, registration number, etc.
- Trip contract: Several elements indicating agreement to several spatial and pathway travel rules, human-robot communications, and expected conditions regarding surface conditions and weather resiliency, and more.

To summarize, the orchestration layer is concerned about social, urban, livability, infrastructural, and congestion matters. Since these matters are locally structured, their preferred solutions should be locally (regionally) determined. Hence, any fleet operator focused on completing robotic tasks effectively and safely need only receive route permissions and machine behavior cues from the orchestration layer and may remain otherwise unaware of local expectations or infrastructure since all required instruction including real-time changes would be communicated from the orchestration layer.

Hence, the differences between a robot getting from A-to-B in Kolkata and getting from A-to-B in Berlin, would be absorbed in, and communicated from, their respective orchestration layers. This permits robot operating systems and fleet management systems to be relatively unconcerned with these local differences, focusing rather on universal design matters that allow them to operate in any desired location.

Orchestration Process Overview

The communication process (Fig. 4) between a regional orchestration layer and any fleet layer

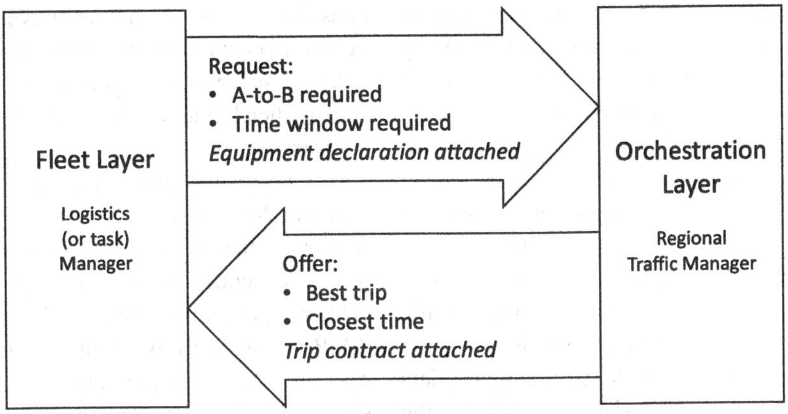

Personal Delivery Robots: How Will Cities Manage Multiple, Automated, Logistics Fleets in Pedestrian Spaces?, Fig. 4 A standardized set of messages are required to negotiate a trip contract between a fleet operator and the regional authority managing the ODD. This simplified figure shows only the fundamental concept of a trip request for a declared device (equipment), and an offer of a trip contract (Grush 2021b)

operating in that region would consist of several messages concluding in a "trip contract."

The contents of a trip contract include several dozen elements (depending on the number of path segments in the trip. In addition to an agreed pathway and time window, these include:

- Information such as "narrowest passage" or "steepest gradient" on each path segment
- Instructions such as "use the right (or left) side" of a crosswalk for each crosswalk between path segments
- Constraints such as maximum speeds or weight per segment and crosswalk
- Equipment provisions such as lights, flags, sounds, etc.
- Financial data such as the user fee per segment on this contract

Draft definitions for the data intended for use in such trip contracts are included in a draft technical standard (ISO/DTS 4448) (Draft technical standard (DTS) 2022). All ISO/DTS 4448 data elements are provided with a default to guide orchestration level set-up. Any element could be set to suit the needs of a local authority.

Many elements are local map-associated parameters (such as max speed, max weight, travel left, travel clockwise) that would be determined, stored, and exchanged within mapping regimes suitable to the local governing authority, and converted to the standard trip contract format.

Many elements are variable over geography, such as maximum cross gradient, or maximum pathway roughness. Other elements may change dynamically, even in real time, depending on local capabilities. These might be user fees affected by congestion or surface friction related to temperature or precipitation.

Note that the draft standard indicted here, while providing metrics, formats and defaults, does not require that every element be updated in real time. Clearly, there will be differences in the deployed capability of regional orchestration level systems depending on local requirements. Any shortcomings will necessarily be absorbed at the lower system levels for fleet and machine (Fig. 3).

Ancillary System Components for a Robot Orchestration Process

It would be possible to simply set up an orchestration system, apply the ISO/DTS 4448 defaults, and expend only a modicum of effort on real time management. In fact, this will be the likely approach until the number of operators, robots, and trips begins to grow, and a degree of congestion management is required. The assumption is that if public-space robotic technology becomes as pervasive as its visionaries promise, then fleet coordination and congestion management will clearly become an issue, if accessibility, livability, and pedestrian advocates don't lobby for other constraints first. This implies that the success of this pending industry within our cities and our active transportation and vulnerable users will rely on a highly competent and socially acceptable fleet orchestration capability.

In order to learn how to adjust appropriate system parameters to consistently optimize the orchestration layer (Fig. 3), data will be needed about pathway conditions and congestion factors. To maintain good order, data will be needed about infractions such a speed or rogue robots (without trip contracts). The management of some of the social behaviors such as auditability and visibility of prescribed robot sounds and lights will also require observation data.

Some of this data may come from the robots themselves, some may come from proximate robots (although that would seem a very difficult approach), and some may come from data capture through sensors on fixed IoT networks.

An enforcement capability will be required, but would be defined locally. Enforcement and its digitalization will depend on the depth of capability embedded in the orchestration layer.

Conclusion

The problem of managing deployed robots is very different from the pure mechatronics problem of designing a robot to perform a particular, well-described task. The easiest cases are factory robots performing a repetitive assembly task in a fixed, bounded space ("caged robots"). The next harder case is an automated mobile robot (AMR) to pick-move-place loads from one spot to another in a factory, warehouse, or mine setting or to plough or spray a given agricultural area.

As we move to the problem of roboticized work in public spaces, we introduce at once the need for robotics management platforms to consider untrained, non-attentive, and noninvolved persons; persons of varying abilities; and complex spaces comprising urban sidewalks with less predictable and highly variable surfaces and barriers.

As we deploy in this new environment with multiple fleets from multiple vendors performing multiple categories of tasks, we engage one of the most difficult, nonmilitary applications for AMRs. This is the context for the vision of using AMRs at scale from a plurality of independent operators for multiple, mundane tasks within shared urban spaces that include pedestrians of every physical, sensory, and cognitive ability.

The value of the services that are envisioned to be provided by these robots is extraordinarily high from livability and commercial perspectives. The opportunity provided by competent fleet orchestration is to manage these services in a way that maximizes their value and minimizes their threat. That much is obvious.

What is less obvious is how to go about doing that. This entry has presented such a proposal but because the related standard is itself still in draft form, this proposal must also be read as a draft.

Cross-References

- ▶ Age-Friendly Future Cities
- ▶ City Visions: Toward Smart and Sustainable Urban Futures
- ▶ Disruptive Mobility: Sharing Electric Autonomous Vehicles (SEAVs) Reshape Our Future Cities
- ▶ Getting Our Built Environments Ready for an Aging Population
- ▶ Green and Blue Infrastructure (GBI) in Urban Areas
- ▶ Green Cities
- ▶ Improving Social Equity and Community Health and Well-Being in Low-Income Suburbs and Regions
- ▶ New Cities
- ▶ Public Space
- ▶ Regulation of Urban and Regional Futures
- ▶ Residential Crowding in Urban Environments

- ▶ Shrinking Towns and Cities
- ▶ Smart Cities
- ▶ Smart City
- ▶ Smart Densification
- ▶ Smart(er) Cities
- ▶ Sustainable Urban Mobility
- ▶ Systemic Innovation for Thrivable Cities
- ▶ The Governance of Smart Cities
- ▶ Transport Resilience in Urban Regions
- ▶ Transportation and Mobility
- ▶ Urban Futures
- ▶ Urban Resilience: Moving from Idealism to Systems Thinking
- ▶ Urban Structure and Its Impact on Mobility Patterns: Reducing Automobile Dependence Through Polycentrism

References

Clamann, M., Bryson, M. (2021). Sharing spaces with robots: The basics of personal delivery devices. An info-brief from the Pedestrian and Bicycle Information Center. https://www.pedbikeinfo.org/cms/downloads/PBIC_InfoBrief_SharingSpaceswithRobots.pdf

Coombes, A., & Grush, B. (2022). Digitization, automation, operation, and monetization: The changing management of sidewalk and Kerb 2000–2025. In J. R. Vacca (Ed.), *Smart Cities policies and financing: Approaches and solutions*. Elsevier.

Dimensions. https://www.dimensions.com/collection/autonomous-delivery-vehicles

Draft technical standard (DTS) ISO 4448 "Ground-based automated mobility" is planned for release in multiple parts in 2022 or after.

Edwards, D. (2021). Demand for delivery robots has 'quadrupled' in past year, says Starship Technologies. https://roboticsandautomationnews.com/2021/05/19/demand-for-delivery-robots-has-quadrupled-in-past-year-says-starship-technologies/43328/

Grush, B., & Coombes, A. (2022). Digitization, automation, operation, and monetization: Standardizing the management of sidewalk and Kerb 2025–2050. In J. R. Vacca (Ed.), *Smart Cities policies and financing: Approachesss and solutions*. Elsevier.

Grush, B. (2020). Are we doing justice to sidewalk robots? https://www.linkedin.com/pulse/we-doing-justice-sidewalk-robots-bern-grush/

Grush, B. (2021a). The Last Block: Towards an international standard to regulate and manage sidewalk robots https://citm.ca/wp-content/uploads/2021/02/Harmonize-Mobility_The-Last-Block_21.02.01.pdf

Grush, B. (2021b). Making Room for Robots: A draft ISO technical standard for ground-based automated mobility: Loading and unloading at the kerbside and footway. http://endofdriving.org/wp-content/uploads/2021/07/Making-Room-for-Robots-final.pdf

Grush, B. (2021c). Footway Robots and Business Improvement Areas. http://endofdriving.org/wp-content/uploads/2021/05/Footway-Robots-and-Business-Improvement-Areas-Grush-final-2021-04-30.pdf

Hsu, J. (2019). Out of the Way, Human! Delivery Robots Want a Share of Your Sidewalk. https://www.scientificamerican.com/article/out-of-the-way-human-delivery-robots-want-a-share-of-your-sidewalk/

Kingson, J. (2021). Sidewalk robots get legal rights as "pedestrians". https://www.axios.com/sidewalk-robots-legal-rights-pedestrians-821614dd-c7ed-4356-ac95-ac4a9e3c7b45.html

SAE International. (2021). Taxonomy and definitions for terms related to driving automation systems for on-road motor vehicles. J3016_202104. https://www.sae.org/standards/content/j3016_202104/

Transport for London. (2020). The planning for walking toolkit: Tools to support the development of public realm design briefs in London. Part B: Planning & Design Principles

Thomasen, K. (2020). Robots, regulation, and the changing nature of public space. https://commons.allard.ubc.ca/fac_pubs/633/

Philanthropy

▶ Philanthropy in Sustainable Urban Development: A Systems Perspective

Philanthropy in Sustainable Urban Development: A Systems Perspective

Stefan Blachfellner and Micol Sonnino
The Bertalanffy Center for the Study of Systems Science, Vienna, Austria

Synonyms

Civic society; Community foundations; Foundations; Innovation eco-systems; Local governance; Philanthropy; Sustainable development; Systemic innovation; Systems thinking; Urban development

Definition

The growing literature on the evolving nature of philanthropy has increasingly explored the influence of foundations in urban spaces. Philanthropic foundations have specifically entered cities as bargaining actors in both local and global levels of policymaking, influencing the corresponding urban and regional governance. This brings about the necessity of an interdisciplinary knowledge stemming from the ethics of philanthropy, urban development, and complex systems studies.

While defining philanthropic foundations proves to be a challenging task due to frail legal frameworks, some common characteristics such as the state-society context or the relationship between private interests and the public good help define these organizations. Secondly, critical assessment of foundations in their role of aiding capital's dominance and eventually supporting socioeconomic status quo is pivotal to understanding their influence in democratic settings.

The increasing role and influence of philanthropy in international development agendas, as well as the impact of foundations on local policymaking networks, has been associated with issues of legitimacy, transparency, and accountability which can be analyzed both historically and in contemporary perspectives.

Systems theory approaches can be applied to grasp how foundations can balance private interests and public good intentions to enable systemic innovation for change. Foundations can offer a relevant cross-sectoral function for the mitigation of societal and environmental issues.

Introduction

Challenging questions arising from the study of philanthropy relate to what defines foundations, and what differentiates them from other forms of charitable organizations. Moreover, philanthropy has been at the center of a loaded discussion around questions of capital accumulation, democratization of capital, and corresponding influence on the public interest exercised by private assets.

As philanthropic foundations grow worldwide in both number and size (Reich 2018; Toepler 2018; Donors and Foundations Networks in Europe AISBL & European Foundation Centre 2021), their roles and purposes change alongside. As a consequence, philanthropy as a field of study has matured into an interdisciplinary subject. Adopting an interdisciplinary outlook, specifically, allows a comprehensive scrutiny of the feasibility of philanthropic action.

Recent literature has particularly underlined the role of foundations in urban settings from a perspective of policy making and governance. While co-governance theories welcome the growing involvement of donors in policy making practices, a critical analysis of the role undertaken by philanthropy in these networks should assess their legitimacy of action (Pill 2019; Thomson 2021).

This entry positions the perspectives of ethics and politics of philanthropy in the field of systemic urban sustainable development. The conceptual framework stems from system's perspectives: the goal is to provide readers with an all-encompassing understanding of the modern role of foundations.

Terminology and Conceptual Frames

While research on philanthropy has a long tradition, the field has gradually broadened alongside a global foundation boom which initially sparked in the 1990s and further increased in the early 2000s (Reich 2018; Toepler 2018). An accelerating wealth accumulation and the intergenerational transfer of this wealth were some of the reasons behind this enormous growth (Toepler 2018).

Governments have sought to mirror this growth by reforming the legal and fiscal frameworks of foundations (Toepler 2018) with different results in Western democracies (Jung et al. 2018), China (Chan and Lai 2018), or Russia (Jakobson et al. 2018).

Many studies consequently attempted to rationalize the variegated results by creating a conceptual framework of foundations and analyzed the specific differences in the legal definition as well as the institutional and regulatory structures of foundations (Jung et al. 2018 cite the report by the Rockefeller Philanthropy Advisors & Marshall Institute 2017; and the work of the European Foundation Centre 2017). This was done particularly in the international comparative scholarship (Anheier 2018; Jung et al. 2018; Toepler 2018).

What distinguishes philanthropic foundations from other charitable institutions? This question has been at the core of research endeavors in the field (Anheier 2010, 2018; Anheier and Daly 2007; Fleishman 2009), resulting in some defining characteristics:

Capital: One common trait of foundations is to possess a permanent asset. Anheier (2001) is commonly quoted for defining foundations as: owning an asset; being non-membership based, private, and relatively permanent; showing an identifiable organizational structure with a nonprofit mission; serving a public purpose (Anheier 2001). Prewitt (2006) further specifies that the permanent endowment is "not committed to a particular institution or activity, provides a grantmaking capacity reaching across multiple purposes and into the indefinite future" (p. 355).

Frail legal definition: A second commonly quoted remark of foundations is that they are in practice often characterized by feeble legal frameworks and unrestricted organizational forms (Anheier and Daly 2007). Jung et al. (2018) remark how no legal distinction is to be found between foundations and other charitable institutions in the United Kingdom. In other countries on the other hand, the United States among all, foundations exist as consequences of federal income taxation.

Charitable purpose and tax-related benefits: All foundations, whether private or not, exist to serve the public interest. Charitable activity in the United States is under two tax incentives: tax exemption for non-profit organizations, including tax on investment income for public charities; and tax deductibility for donations to public charities and private foundations (Reich 2018). Public charities receive assets from the government or the public, while private foundations receive income from donors or their endowment. Both public and private philanthropy are excluded from property taxation (Reich 2018).

These incentives make for a common spark of controversy in the field. Reich (2018) explains: private funds can support the public good through taxation or charitable donations. For the latter, people who wish to give away money for charity in the United States are incentivized through tax advantages. This money makes for foregone tax revenue for citizens. This charitable contribution deduction was created in the US congress in 1917. Many other countries follow this subsidy structure, with the exception of Sweden.

To navigate other aspects of distinction among foundations themselves, scholars have thus referred to further characteristics. For a more comprehensive mapping of notable themes used to differentiate foundations, see Jung et al. 2018.

Asset management for charitable actions: Part of the capital accumulated by foundations is legally bound to be spent in charitable actions through a redistribution payout. The percentage varies across countries and according to the purpose of each foundation. It usually sums up to a yearly 2–6% of the foundation's total assets, varying from country to country (DAFNE and EFC 2021) In the United States, this is 5% of the 12-month average fair market value of the foundation's endowments (Afik and Katz 2019; Reich 2018).

One second major distinction within asset management is whether foundations are of the "grantmaking" or "operating" type. Grantmaking foundations create grants to enable third parties to achieve the foundation's mission. Operating foundations directly utilize their assets to operate their own charitable programs (Toepler 2018). While the United States or Sweden are predominately grant-making, Spain and Italy make for examples of mostly operating-type foundations (Toepler 2018).

Asset management both for revenue activities for the foundation's sustainment and for charitable activities are often unreleased data (Mc-Gregor-Lowndes and Williamson 2018; Toepler 2018).

Asset ownership and sustainment: Foundations can originate their assets from inheritance or collect them from different sources (Toepler 2018). Whether the donor is still alive influences the corresponding asset management (Jung et al. 2018) in terms of mission – how much the original purpose is subject to change, and redistribution – how much of the original capital is intended to be preserved or spent.

Regardless of the origins of these assets, foundations own a capital that can either be spent in a one-off grant or sustained through revenues to be distributed over time. In US-American style philanthropy, grantmaking foundations have shown to be more independent on external donations than operating foundations in Europe (Rey-Garcia 2018; Toepler 2018). Regulations introduced in 2004 in China instead formally state that only government-backed foundations should have a fundraising status for their operations (Chan and Lai 2018; Toepler 2018).

Relation to state, civil society, and business: Different studies have investigated the relationship between philanthropy and the state and the related implications (Bies and Kennedy 2019; Jung et al. 2018, among others). Many underline how liberal welfare regimes such as in the United States result in wider responsibilities for the nonprofits and philanthropy: private resources help in filling the gaps left by governments in public service supply (Jung et al. 2018; Toepler 2018). Elsewhere, foundations act on the side rather than as substitution of the state, acting as revenue sources for the government, for example, in Russia or China (Chan and Lai 2018; Jakobson et al. 2018), or as adjunct to government funding in Europe or Australia (Jung et al. 2018; Toepler 2018).

Purpose, structure, and behavior: Foundations can have different interests and missions, from relief to advocacy, or even research; can have different board compositions; and belong to different sectors (Jung et al. 2018). All in all, Anheier (2018) proposes a "foundation triangle" to cluster foundations in combinations of purposes, approaches, and role sets. As main purpose categories Anheier proposes relief, protection, and/or change. These can be pursued through grantmaking and/or operating approaches, to adopt roles of innovation, complementarity, substitution, or build-out (2018).

Understanding the Implications of Foundations

Foundations can be broadly defined as organizations owning a permanent asset bound to charity action and subject to tax subsidies. This comes with implications. While philanthropic action, intended as the activity undertaken to donate assets, has existed for millennia, critical analysis on the legitimacy of this action focuses on the restrictions that should be posed to avoid issues of power. A certain tension exists between preserving the corpus of a foundation in contrast to distributing the assets for social causes.

Reich (2018) and Toepler (2018) enquire whether the current legal and normative framework of low accountability and transparency can be deemed legitimate and justifiable. To the goal of *understanding* foundations, the following characteristics sum up critical nodes within philanthropy:

Foundations as a tool for exerting power: Anheier and Daly (2007) describe foundations as an inherently political tool, as they make for "private agendas in public arenas outside direct majoritarian public control" (p. 3). Foundations exert a plutocratic power (Reich 2018) when they direct private assets toward a public purpose without scrutiny exercised through democratic voting.

Non restricted: To whom and how foundations release funds are mostly unrestricted processes. Reich (2018) estimates that subsidies for charitable contributions replaced at least $50 billion of US federal tax revenues in 2016. As tax expenditures, these funds would have been subject to a degree of scrutiny: instead, charitable activities of foundations can be directed based on the donors wishes under virtually no transparency obligations.

Foundations' assets are by design planned to remain permanent and "endure across generations" (Reich 2018): there is also no formal duty for the foundation to change the original mission. Reich (2018) also suggests that intergenerational transfers ought to be incentivized to the purpose of precautionary strategies for future generations, and that failing to do so locks up charitable assets in otherwise non-pressing missions.

Accountability: Philanthropic foundations, while exerting power, are not formally subject to accountability and transparency (Reich 2018). Foundations lack accountability on the goal-setting processes, as no accountability system toward donors or grantees is required by law.

The foregone tax revenue is not accounted for, with many arguing that donors' preferences usually favor elitist goals. One additional critic is also that foundations exhibit and reinforce conservative biases (Thomson 2021).

Legitimacy: Toepler (2018) summarizes legitimacy questions as arising from a tension between the autonomy of donors' intent and the duty to serve the public interest as incentivized by tax privileges. The loosely defined obligation put on foundations is that they should use assets for the public good. Legitimacy claims question whether foundations are a proper tool for this scope, and whether the current framework is suitable for a proper accumulation and then redistribution of private wealth (Heydemann and Toepler 2006). Legitimacy of foundations overall also relates to questions on its relation to the government and role in civil society.

Critics also question whether donors should receive incentives to return part of the wealth accumulated, or if this wealth should be taxed in the first place. With *philanthrocapitalism* (Bishop 2008) in addition, critics focus of the growing tendency of big philanthropies of using market logics to advance the common good (Montero 2018); while with philanthro-policy making, critical analysis is put on the increasing tendency of foundations in reinforcing neoliberal tendencies and thereby worsening the condition of marginalized groups (Rogers 2011; Thomson 2021).

Overall, philanthropic foundations can be understood in their characteristics – what Toepler (2018) calls the question of definition; and in their implication – what can be called the question on their purpose, power, and role in a democratic society. Applying a systemic approach to these questions is useful to understand how philanthropy can be refined to enable sustainable development and systems change.

Applying Systems Approaches in Philanthropy to Foster Sustainable Development

When adopting a systems perspective, philanthropy can take a particular role in the sustainable development of cities, which became the focal space for innovation policies. (See chapter ► "An Overview of the Relationship of the Sustainable Development Goals and Urban and Regional Development")

Systems thinking and practice require foundations to appreciate and incorporate "Intersectionality as a Reality" (Grady 2020) to tackle complex problems that require the utmost effort and collaboration on numerous levels.

In a *Nested Systems view* (See chapter ► "Systemic Innovation for Thrivable Cities"), foundations are themselves at the intersection of the social systems and political systems, as well as political systems and economic systems. Their systems properties allow to dismantle silos across the sectors (Grady 2020) and aid synergies, for example, by building collaborations between governmental agencies, or between NGOs and businesses, expanding systems boundaries, as well as influencing any power dynamic that affects communities. As such foundations become a crucial actor in systemic strategies for change, independent of their critical role as *funders*.

In an environment of rapid global change, hardly any organization will be able to handle all the innovation and adaptation on its own. Systemic challenges such as climate change, the depletion of natural resources, and the achievement of the Sustainable Development Goals are multidimensional challenges.

Foundations and philanthropy can therefore shift their purpose, structure, and role from sole funders of particular causes or projects to initiators, co-creators, and facilitators of innovation ecosystems, advancing strong leverages in systems change.

Innovation ecosystems are constantly evolving, from actors, activities, and means of production as well as institutions and their existing and newly formed relationships, which are important for the innovation performance of an actor or a group of actors (Granstrand and Holgersson 2020). In this sense, innovation ecosystems as evolutionary systems are comparable to natural ecosystems and not only as a metaphor.

Ecosystem as a concept originated in the science of ecology (Tansley 1935). It can be defined by recycling flows of nutrients along pathways composed of living subsystems (Shaw and Allen 2018). They interact in a wide variety of ways, from symbiosis to collaboration to competition. Odum incorporated 1953 Ludwig von Bertalanffy's General Systems Theory into the epistemological framing of ecosystems (Marin 1997). A variant of the theory of complex systems developed in ecological approaches, mainly adopted by sociology of organizations and political science. These approaches essentially apply ecosystem ideas and related key concepts of evolutionary change to social reality (Baum 1996; Hannan and Freeman 1977).

In summary all ecological approaches emphasize (1) the dynamic character of interdependencies and interaction between social actors, (2) the multiplexity of relations between the components of these systems, and (3) the existence of multiple and relatively autonomous layers and levels in such systems, and emergent relations between these levels, as well as (4) co-evolution and co-evolutionary interaction structures, which are not just restricted to cooperation and competition, but involve a spectrum of relations (Schneider and Bauer 2007). Stemming from systems science, this ecological perspective has important implications for the study of systems change, adaptation, and innovation and be a useful basis for growth in the field of philanthropic action.

Innovation and its ecosystems are about creating value. Unlike biological ecosystems, an innovation ecosystem comprises an intertwined web of multi-layered relationships through which relevant knowledge and resources flow in the context of sustainable co-creation of value. The term has gained popularity in academia, policy, and business. Development and innovation are enabled in between open systems through related actors who together create an enabling environment that continuously mobilizes assets, is resilient to external shocks, and even has the ability to regenerate itself. All actors can leverage ecosystem relationships for greater value creation by utilizing the synergies and network effects that arise from the complementarity of players. In terms of systems thinking, the whole is greater than the sum of its parts.

Achieving and sustaining development outcomes depends on the ability of a wide range of interconnected actors – governments, civil society, the private sector, universities, individual entrepreneurs, and others – to work together effectively. The effectiveness of each part of the innovation ecosystem is influenced by other parts of the system. Key stakeholders must be brought together from the beginning to jointly develop strategies and translate ideas into activities. More resilient ecosystems are possible when diversity is recognized as a strength. This applies not only to the selection of actors, but also to processes. Exploratory transdisciplinary research and development is needed to identify patterns, to enable more applied research, and to scale innovations up. Efforts must not be focused on one area, but rather on funding and optimizing the whole.

A neutral and independent role for curation is crucial in rallying these various stakeholders to an ambitious common goal. Foundations are well situated for this demanded role, if they are emphasizing their intersectoral capabilities in purpose, structure, and behavior, or support organizational structures which fulfill this role of intersectoral curation.

In light of critical assessments such as a study by the Austrian association of charitable foundations together with the Federal Ministry Republic of Austria for Climate Action, Environment, Energy, Mobility, Innovation and Technology (2021), which concluded that only 3% of all charitable foundations worldwide are involved in climate mitigation actions and environmental sustainability, as well as further findings by the Worldwide Initiatives for Grantmaker Support (WINGS), the Charities Aid Foundation (CAF), and the OECD Development Centre, which stated that at the current rate, it would take more than three centuries for philanthropy to close the SDG funding gap, foundations are consulted to emphasize strengthening philanthropy's ecosystem of support (Bellegy et al. 2019; Knight 2018).

By adopting system-oriented strategies, foundations are shaping the political debate, acting as mediators, influencing the financial market, and building civil societies to affect interdependent and cross-sectoral mitigation efforts.

Shaping the Political Debate Through Advocacy

Goss and Berry (2018) laid out the rationale of how foundation can influence public policy through advocacy. As interest groups, foundations can pursue lobbying as well as other indirect political pressures to regulators. Both studies by Suárez et al. (2018) and Goss and Berry (2018) conclude that foundations are somewhat limited in the directly opposing legislation through lobbying but own a diverse set of tools to pressure politicians and regulator to promote social change and voice diversity – including public information campaigns, research, and coalitions.

Luers (2013) brings the example of climate advocacy through knowledge sharing to pressure for political support. Suárez (2012) underlines that smaller, community oriented, socially focused foundations favor advocacy over other strategies for social change. In the East-Asian context, Bies and Kennedy (2019) contend that a tension exists between the need for a social stronger sector, and the desire of the state's to limit advocacy.

Acting as Mediators: Networks, Structures, and Field Coordination

Philanthropic foundations can adopt different roles as mediators. Williamson and Leat (2021) group these in four themes: group making, capacity and relationship development, temporal bridging, and intermediation: "foundations mediate, intercede, intervene, arrange, broker, buffer, and filter" (p. 2). To intermediate corresponds to form partnerships that can initiate innovation and change (Williamson and Leat 2021).

Influencing Market Dynamics

Foundations manage over 1 trillion USD in assets globally, with an average spend rate of 10% and expenditures of 150 billion USD per year, which indicates the actual amount spent for charitable purposes in relation to the foundation's assets (Gehringer 2020).

While several reports still emphasize either the role of philanthropy as venture capital to create early markets (McKinsey & Company 2021) or funders for R&D and innovation in the public interest (European Commission 2005), foundations need to shift their overall asset management strategies toward divestment investments.

DivestInvest philanthropy, which is concerned with fossil fuels divestment and climate solutions reinvestment (McGregor-Lowndes and Williamson 2018), advocates for this change. "Nearly 200 foundations worldwide have committed to divest, many thousands have not, and the largest American foundations with even explicit missions to combat climate change have opted not to divest" (Williams 2019, 2020). Spending funds on tackling the climate crisis while investing in industries causing climate change is a systemic failure. Asset anxiety, portfolio constraints, and the decision structures in small asset teams and foundation boards (Williams 2019, 2020) are still hindering the needed directional switch for impactful systems change leveraged through the influence foundations have on the financial market, despite the growing evidence that fossil-free or sustainable oriented investments can also bring high yields.

DivestInvest (2021) is a global network of individual investors and organizations supporting the agreement made by governments in Paris at COP21 while protecting their own investment returns. Their pledge requires not just divesting from fossil fuel companies, but also committing at least 5% of assets to climate solutions. Even cities have joined the pledge with, for example, the C40 network of mayors of nearly 100 world-leading cities, including Auckland, Berlin, Bristol, Cape Town, Copenhagen, London, Los Angeles, Milan, New Orleans, New York, Pittsburgh, Rio de Janeiro, Seattle, Vancouver, and many more.

Foundations aid this process, as C40 is supported by the Sainsbury Family Charitable Trust, providing a practical guide for institutional investors to DivestInvest (Harrison 2018), addressing Universities, faith groups, foundations, charities with endowments, family offices, corporate, organizational, and local authority pension funds, as well as city funds.

Building Strong Civil Societies

By supporting grassroot organizing, foundations can aid in building a diversified civil society made of multiple voices. When foundations build social capital, invest in communities, and provide institutional spaces, this contribution enhances the capacity of civil society (Gerzon 1995). Pill (2019) further expands on the role of philanthropy by providing the different conceptualizations of civil society – the work of Kaldor (2003) is also particularly insightful.

In practice, the four categories complement each other (Delanoë et al. 2021). Wu (2021) has underlined in her analysis of 539 annual reports of US community foundations that some strategies were adopted more often than others: in particular, community foundations often engage in capacity building, partnering, and policy engagement. Graddy and Morgan (2006) add that the choice of strategies will depend on the age of the organization, stability of community, as well as professionalization of the field and presence of national competitors.

Community Philanthropy: System's Perspectives on Urban Foundation's Legitimacy

As discussed, foundations are far from being a monolithic group (Martin 2004). This makes it hard to properly analyze their growing prominence in urban policy making (Ravazzi 2016).

Enormous funding is dedicated to city projects: foundations are main contributors in shaping the direction and priorities of global urban development (Fuentenebro and Acuto 2021). Particularly, philanthropic foundations are prominent actors in local policy making and governance processes, thanks to their ability to catalyze resources and creating networks (Ravazzi 2016; Williamson and Leat 2021)

Urban foundations are also prominent in urban economic development (Giloth 2019). This is particularly visible in the development of metropolitan cities, historically coupled with philanthropic involvement (Fuentenebro and Acuto 2021; Henthorn 2018), both in Euro-Northamerican contexts and otherwise. By rationalizing funding into systematic, goal-oriented projects, foundations have been involved in providing cities with modern urban services and thus realized their visions of what the identity of these cities should be (Henthorn 2018).

Pill (2019) and Giloth (2019) underline that the action of philanthropic foundations in cities reflects interpretations of state-society power relationships – in which "civil society" complements the actions of the state in providing services. The inclusion in policy making of philanthropic foundations might be a reflection of a democratic vision of decision making, or "lack of faith in centralized decision making" (Martin 2004, p.395) – the shift of governance to private, capital-intensive actors comes in any case with questions of legitimacy posed earlier in this entry.

One practical pathway for improving the democracy of philanthropy is visible in community foundations. In Europe alone, there are "around 900 community foundations in 23 countries" (ECFI 2021). They are defined as independent, autonomously acting, non-profit foundations with a large number of donors, with the loose goal of serving the common good in a defined area. Intangible resources such as volunteers are included in the assets, so that community foundations combine "long-term, financial sustainability with personal commitment" (Verband für gemeinnütziges Stiften 2021).

Community foundations share these characteristics (ECFI 2020):

- Their mission is broadly defined but specific to a geographically defined area (city, district, province).
- Contributions mostly originate within the community.

- They are mostly grantmaking.
- They have multi sectoral local boards.
- They own permanent assets.

First, the local focus of community foundations implies that their work is under ongoing local scrutiny, which thereby improves the transparency process of the foundation (Mesik and Owen 2008). Local expectations expand accountability horizontally, requiring foundations to be accountable to the community, to donors, and to grantees. Giloth (2019) argues that the local focus of urban community foundations fills the gap potentially left by the urban fragmented civic leadership. Particularly, community foundations take on the role of "'system actors' that have a wide view of stakeholders, geography, assets, and convening power, and as 'social entrepreneurs' that can translate ideas, plans, and strategies into long-term action" (p.160).

Some examples of community foundations include the work of Charles Stewart Mott Foundation. Through the Civil Society, Education and Environment programs, the foundation supports non-profits to build resilient civil societies and community empowerment through education, health, or research as few examples. The community foundation concept is also expanding in all continents except Antarctica, summing up to more than 42 countries (Mott Foundation 2004).

All in all, community foundations can be deemed fit to solve some of the legitimacy questions as they combine civic commitment to proper funding (Giloth 2019). Moreover, by owning permanent endowments, community foundations can ensure financial sustainability. Unlike other associations, foundations are required by law to permanently dedicate the endowments to charitable action (Mesik and Owen 2008; Verband für gemeinnütziges Stiften 2021). This can become a competitive advantage in comparison to other socially oriented associations: this is if the permanent assets are devoted to meeting ever changing intergenerational needs rather than the goals set by one original donor.

There are also critical nodes within community foundations (Mesik and Owen 2008):

- Communities might not always own skilled human resources.
- There might be low participation from the local community in participating in the foundation's efforts.
- The legal environment of a country might hamper the capacities of the foundation to accumulate and distribute financial resources.

Nonetheless, community foundations can be a suitable institution for the application of the four systemic-oriented strategies presented earlier. These strategies would enable a role for foundations that goes beyond grantmaking to strengthen the legitimacy premises questioned earlier in this entry.

Conclusion

The field of philanthropic studies is maturing, with a wealth of approaches questioning the discourse around foundations. This entry has exposed some of the legitimacy questions behind foundations: lack of transparency and regulations; hierarchy and permanency of donors' intentions; and the potential conflict between heterogeneous needs and some elitist funding preferences (Reich 2018). As philanthropy joins the policy making table in urban settings, new issues in governance legitimacy arise.

Nickel and Eikenberry (2009) and Reich (2018), among others, have warned about the undemocratic premises of foundations that act for social change, when these institutions adopt elitist goals and marketized strategies. Far from representing the marginalized, some aspects of the legal framework of foundations aid a depoliticized discourse of policy making and hamper democracy.

System approaches can shine a light on how to increase the beneficial potential of urban foundations. Foundations can be relevant beyond their funding activities and promote a diversification of voices. Shaping the political debate, acting as mediators, influencing the stock market, and building civil societies are some of the main strategies foundations can adopt to affect interdependent and cross-sectoral efforts. Insights from an ecosystems perspective help to overcome silos and achieve synergies across actors involved. A system's perspective can thereby also enhance the benefits that the legal peculiarities of foundations can bring, which include independence from political and market forces, and risk inclination aiming for social transformation.

Finally, adopting systemic strategies aids with a common purpose of dampening the dominance of capital: as visible in philanthrocapitalism, elitist interests can otherwise prevail policy making through wealth and connections (Rogers 2011; Thomson 2021).

Understanding the role of foundations in local and international urban agendas becomes crucial to the goal of decentralizing authority, creating valuable networks, and local participation in democracy.

Cross-References

- ▶ An Overview of the Relationship of the Sustainable Development Goals and Urban and Regional Development
- ▶ Building Community Resilience
- ▶ Building Resilient Communities Over Time
- ▶ City Financing and Social Urbanism in Latin America: The Importance of Good Fiscal Management
- ▶ Collaborative Climate Action
- ▶ Connecting Urban and Regional Innovation Ecosystems to Enhance Competitiveness
- ▶ Education for Inclusive and Transformative Urban Development
- ▶ Financing: Fiscal Tools to Enhance Regional Sustainable Development
- ▶ Future of Urban Governance and Citizen Participation
- ▶ Multi-stakeholder Partnerships to Support Climate Migrants in Fragile Cities
- ▶ Networking Collaborative Communities for Climate-Resilient Cities
- ▶ Participatory Governance for Adaptable Communities
- ▶ Sustainable Development Goals
- ▶ Systemic Innovation for Thrivable Cities
- ▶ Unpacking Cities as Complex Adaptive Systems
- ▶ Urban and Regional Leadership

References

Publications

Afik, Z., & Katz, H. (2019). Reconsidering the philanthropic foundation minimum payout policy under a "new normal". *Journal of Policy Modeling, 41*(2), 219–233. https://doi.org/10.1016/j.jpolmod.2018.09.004.

Anheier, H. (2001). *Foundations in Europe: A comparative perspective*. London School of Economics and Political Science, Centre for Civil Society.

Anheier, H. K. (Ed.). (2010). *American foundations: Roles and contributions*. Brookings Institution Press.

Anheier, H. K. (2018). Philanthropic foundations in cross-National Perspective: A comparative approach. *American Behavioral Scientist, 62*(12), 1591–1602. https://doi.org/10.1177/0002764218773453.

Anheier, H., & Daly, S. (2007). *The politics of foundations*. London: Routledge.

Baum, J. (1996). Organizational ecology. In S. Clegg, C. Hardy, & W. R. Nord (Eds.), *Handbook of organization studies* (pp. 77–114). London: Sage.

Bies, A., & Kennedy, S. (2019). The state and the state of the art on philanthropy in China. *Voluntas: International Journal of Voluntary and Nonprofit Organizations, 30*(4), 619–633. https://doi.org/10.1007/s11266-019-00142-3.

Bishop, M. (2008). *Philanthrocapitalism: How the rich can save the world and why we should let them*. A & C Black.

Chan, K. M., & Lai, W. (2018). Foundations in China: From statist to corporatist. *American Behavioral Scientist, 62*(13), 1803–1821. https://doi.org/10.1177/0002764218773444.

Fleishman, J. L. (2009). *The foundation: A great American secret; how private wealth is changing the world*. PublicAffairs.

Fuentenebro, P., & Acuto, M. (2021). The gifted city: Setting a research agenda for philanthropy and urban governance. *Urban Studies*. https://doi.org/10.1177/00420980211024158.

Gehringer, T. (2020). Corporate foundations as partnership brokers in supporting the United Nations' sustainable development goals (SDGs). *Sustainability, 12*(18), 7820. https://doi.org/10.3390/su12187820.

Gerzon, M. (1995). Reinventing philanthropy. Foundations and the renewal of civil society. *National Civic Review, 84*, 188–195. https://doi.org/10.1002/ncr.4100840304.

Giloth, R. (2019). Philanthropy and economic development: New roles and strategies. *Economic Development Quarterly, 33*(3), 159–169. https://doi.org/10.1177/0891242419839464.

Goss, K. A., & Berry, J. M. (2018). Foundations as interest groups. *Interest Groups Advocacy, 7*, 201–205. https://doi.org/10.1057/s41309-018-0044-2.

Graddy, E. A., & Morgan, D. L. (2006). Community foundations, organizational strategy, and public policy. *Nonprofit and Voluntary Sector Quarterly, 35*(4), 605–630. https://doi.org/10.1177/0899764006289769.

Grady, H. (2020). How philanthropy must address the climate emergency. *Stanford Social Innovation Review*. https://doi.org/10.48558/Q3AD-F668.

Granstrand, O., & Holgersson, M. (2020). Innovation ecosystems: A conceptual review and a new definition. *Technovation, 90–91*, 102098. https://doi.org/10.1016/j.technovation.2019.102098.

Hannan, M. T., & Freeman, J. (1977). The population ecology of organizations. *American Journal of Sociology, 82*(5), 929–964. http://www.jstor.org/stable/2777807

Henthorn, T. C. (2018). Building a moral metropolis: Philanthropy and city building in Houston. *Texas. Journal of Urban History, 44*(3), 402–420. https://doi.org/10.1177/0096144214566951.

Heydemann, S., & Toepler, S. (2006). Foundations and the challenge of legitimacy in comparative perspective. In K. Prewitt, M. Dogan, S. Heydemann, & S. Toepler (Eds.), *The legitimacy of philanthropic foundations: United States and European perspectives* (pp. 3–26). New York: Russell Sage Foundation.

Jakobson, L. I., Toepler, S., & Mersianova, I. V. (2018). Foundations in Russia: Evolving approaches to philanthropy. *American Behavioral Scientist, 62*(13), 1844–1868. https://doi.org/10.1177/0002764218778089.

Jung, T., Harrow, J., & Leat, D. (2018). Mapping philanthropic foundations' characteristics: Towards an international integrative framework of Foundation types. *Nonprofit and Voluntary Sector Quarterly, 47*(5), 893–917. https://doi.org/10.1177/0899764018772135.

Kaldor, M. (2003). The idea of global civil society. *International Affairs, 79*(3), 583–593. https://doi.org/10.1111/1468-2346.00324.

Luers, A. (2013). Rethinking US climate advocacy. *Climatic Change, 120*, 13–19. https://doi.org/10.1007/s10584-013-0797-1.

Marin, V. (1997). General system theory and the ecosystem concept. *Bulletin of the Ecological Society of America, 78*(1), 102–104.

Martin, D. G. (2004). Nonprofit foundations and grassroots organizing: Reshaping urban governance. *The Professional Geographer, 56*(3), 394–405. https://doi.org/10.1111/j.0033-0124.2004.05603008.x.

McGregor-Lowndes, M., & Williamson, A. (2018). Foundations in Australia: Dimensions for international comparison. *American Behavioral Scientist, 62*(13), 1759–1776. https://doi.org/10.1177/0002764218773495.

Mesik, J., & Owen, D. (2008) *Community foundations how to series: Getting started with a Community Foundation* (Social Development Notes No. 112). Washington, DC: World Bank. https://openknowledge.worldbank.org/handle/10986/11156

Montero, S. (2018). Leveraging Bogotá: Sustainable development, global philanthropy and the rise of urban solutionism. *Urban Studies, 57*(11), 2263–2281. https://doi.org/10.1177/0042098018798555.

Nickel, P. M., & Eikenberry, A. M. (2009). A critique of the discourse of marketized philanthropy. *American Behavioral Scientist, 52*(7), 974–989. https://doi.org/10.1177/0002764208327670.

Pill, M. C. (2019). Embedding in the city? Locating civil society in the philanthropy of place. *Community Development Journal, 54*(2), 179–196. https://doi.org/10.1093/cdj/bsx020.

Prewitt, K. (2006). *Legitimacy of philanthropic foundations: United States and European perspectives: United States and European perspectives*. Russell Sage Foundation. http://gbv.eblib.com/patron/FullRecord.aspx?p=4416956

Ravazzi, S. (2016). Philanthropic foundations and local policy making in the austerity era: Does urban governance matter? *Lex Localis – Journal of Local Self-Government, 14*(4), 917–935. https://doi.org/10.4335/14.4.917-935(2016).

Reich, R. (2018). *Just giving*. Princeton University Press. https://doi.org/10.1515/9780691184395.

Rey-Garcia, M. (2018). Foundations in Spain: An international comparison of a dynamic nonprofit subsector. *American Behavioral Scientist, 62*(13), 1869–1888. https://doi.org/10.1177/0002764218773452.

Rockefeller Philanthropy Advisors, Marshall Institute. (2017). *The theory of the foundation: European Initiative 2016*. London. Retrieved from (quoted by Jung et al. 2018).

Rogers, R. (2011). Why philanthro-policymaking matters. *Society, 48*(5), 376–381. https://doi.org/10.1007/s12115-011-9456-1.

Schneider, V. & Bauer, J. M (2007). Governance: Prospects of complexity theory in revisiting systems theory. *Annual conference of the Midwest Political Science Association*, Chicago, 14 April 2007.

Shaw, D. R., & Allen, T. (2018). Studying innovation ecosystems using ecology theory. *Technological Forecasting and Social Change, 136*, 88–102. https://doi.org/10.1016/j.techfore.2016.11.030.

Suárez, D. F. (2012). Grant making as advocacy: The emergence of social justice philanthropy. *Nonprofit Management and Leadership, 22*(3), 259–280. https://doi.org/10.1002/nml.20054.

Suárez, D. F., Husted, K., & Casas, A. (2018). Community foundations as advocates: Social change discourse in the philanthropic sector. *Interest Groups Advocacy, 7*, 206–232. https://doi.org/10.1057/s41309-018-0039-z.

Tansley, A. G. (1935). The use and abuse of vegetational concepts and terms. *Source: Ecology, 16*(3), 284–307.

Thomson, D. E. (2021). Philanthropic funding for community and economic development: Exploring potential for influencing policy and governance. *Urban Affairs*

P

Review, 57(6), 1483–1523. https://doi.org/10.1177/1078087420926698.
Toepler, S. (2018). Toward a comparative understanding of foundations. *American Behavioral Scientist, 62*(13), 1956–1971. https://doi.org/10.1177/0002764218773504.
Williamson, A. K., & Leat, D. (2021). Playing piggy(bank) in the middle: Philanthropic foundations' roles as intermediaries. *Australian Journal of Public Administration, 80*(4), 965–976. https://doi.org/10.1111/1467-8500.12461.
Wu, V. C. S. (2021). Community leadership as multidimensional capacities: A conceptual framework and preliminary findings for community foundations. *Nonprofit Management and Leadership, 32*(1), 29–53. https://doi.org/10.1002/nml.21467.

Weblinks

Austrian association of charitable foundations, Federal Ministry Republic of Austria for Climate Action, Environment, Energy, Mobility, Innovation and Technology. (2021). Stiftungen als Akteure für Umwelt- und Klimaschutz. https://umwelt.gemeinnuetzig-stiften.at/stiftungen-umwelt/circular-economy/ Accessed 31 Jan 2022.
Bellegy, B.; Mapstone, M.; Pavone, L. (2019). The role of philanthropy for the SDGs is not what you expect. https://oecd-development-matters.org/2019/02/05/the-role-of-philanthropy-for-the-sdgs-is-not-what-you-expect/ Accessed 31 Jan 2022.
Delanoë, E., Gautier, A.; Pache, A. (2021). What can philanthropy do for the climate? Strategic pathways for climate giving. https://www.alliancemagazine.org/analysis/what-can-philanthropy-do-for-the-climate-strategic-pathways-for-climate-giving/ Accessed 31 Jan 2022.
DivestInvest. (2021). We are accelerating the clean energy transition. https://www.divestinvest.org/ Accessed 31 Jan 2022.
Donors and Foundations Networks in Europe AISBL (DAFNE) and European Foundation Centre AISBL (EFC). (2021). Comparative Highlights of Foundation Laws: The Operating Environment for Foundations in Europe 2021. https://philea.issuelab.org/resource/comparative-highlights-of-foundation-laws-the-operating-environment-for-foundations-in-europe-2021.html Accessed 1 Dec 2021.
European Commission. (2005). Giving more for research in Europe: The role of foundations and the non-profit sector in boosting R&D investment. https://ec.europa.eu/invest-in-research/pdf/download_en/rec_5_7800_giving_4_051018_bat.pdf Accessed 31 Jan 2022.
European Community Foundation Initiative. (2020). *Connecting community foundations with the SDGs*. Association of German Foundations. https://www.communityfoundations.eu/fileadmin/ecfi/knowledge-centre/Knowledge_Database/ECFI-guide-Connecting_Community_Foundaitons_with_the_SDGs_-2020.pdf. Accessed 31 Jan 2022.
European Community Foundation Initiative. (2021). *Community foundations in Europe*. ECFI. https://www.communityfoundations.eu/community-foundations-in-europe.html?L=0. Accessed 31 Jan 2022.
European Foundation Centre. (2017). The EFC launches the Institutional Philanthropy Spectrum. http://www.efc.be/news/the-efc-launches-the-institutional-philanthropy-spectrum/. Accessed 31 Jan 2022.
Harrison, T. (2018). How to divest invest: A guide for institutional investors. https://www.c40knowledgehub.org/s/article/How-to-Divest-Invest-A-guide-for-institutional-investors?language=en_US. Accessed 31 Jan 2022.
Knight, B. (2018). What makes a strong ecosystem of support to philanthropy? https://wings.issuelab.org/resource/what-makes-a-strong-ecosystem-of-support-to-philanthropy.html Accessed 31 Jan 2022.
McKinsey & Company. (2021). It's time for philanthropy to step up the fight against climate change. https://www.mckinsey.com/business-functions/sustainability/our-insights/its-time-for-philanthropy-to-step-up-the-fight-against-climate-change Accessed 31 Jan 2022.
Mott Foundation. (2004). Community foundations expanding globally. https://www.mott.org/news/articles/community-foundations-expanding-globally/ Accessed 31 Jan 2022.
Verband für gemeinnütziges Stiften. (2021). *Verband für gemeinnütziges Stiften | Gemeinsamstiften*. https://www.gemeinnuetzig-stiften.at/gemeinsam-stiften Accessed 31 Jan 2022.
Williams, T. (2019). Major climate funders are still invested in fossil fuels. Why is that? https://www.insidephilanthropy.com/home/2019/12/19/major-climate-funders-are-still-invested-in-fossil-fuels-why-is-that. Accessed 31 Jan 2022.
Williams, T. (2020). As top foundations resist divesting from fossil fuels, what might change their minds? https://www.insidephilanthropy.com/home/2020/1/13/as-top-foundations-resist-divesting-from-fossil-fuels-what-might-change-their-minds. Accessed 31 Jan 2022.

Physical Activity

▶ Emerging Concepts Exploring the Role of Nature for Health and Well-Being

Physical Health

▶ Health and the Role of Nature in Enhancing Mental Health

PIB – Press Information Bureau

▶ Adapting to a Changing Climate Through Nature-Based Solutions

Place

▶ New Localism: New Regionalism

Places of Worship

▶ Faith Communities as Hubs for Climate Resilience

Planetary Urbanization

▶ New Orbital Urbanization

Planned City

▶ New Cities

Planning

▶ Feminist Planning and Urbanism: Understanding the Past for an Inclusive Future
▶ Land Use Planning Systems in OECD Countries

Planning – Organization

▶ Future of Urban Land-use Planning in the Quest for Local Economic Development
▶ Planning for Food Security in the New Urban Agenda
▶ Urban Policy and the Future of Urban and Regional Planning in Africa

Planning for Food Security in the New Urban Agenda

Marcylene Chivenge[1], Tafadzwa Mutambisi[2], Chipo Mutonhodza[3], Innocent Maja[4], Roselin Ncube[5], Percy Toriro[6] and Innocent Chirisa[7]
[1]Department of Demography Settlement and Development, University of Zimbawbe, Harare, Zimbabwe
[2]University of Zimbabwe, Harare, Zimbabwe
[3]Department of Rural and Urban Development, Great Zimbabwe University, Masvingo, Zimbabwe
[4]Faculty of Law, University of Zimbabwe, Harare, Zimbabwe
[5]Women's University in Africa, Harare, Zimbabwe
[6]Municipal Development Partnership for Eastern and Southern Africa, Harare, Zimbabwe
[7]Department of Demography Settlement and Development, Social & Behavioural Sciences, University of Zimbabwe, Harare, Zimbabwe

Synonyms

Planning – organization; Food security – food governance; Policy – direction

Definition

Food security – the state of adequacy of food, physically, socially, or economically to households and community.

New Urban Agenda – an action-oriented document that mobilizes Member States and other key stakeholders to drive sustainable urban development at the local level.

Urban Planning: technical and political process that defines the use and management of land use, design, and future activities in space.

Policy: decision-makers' outcomes that shape processes of governance.

Management: the aspect of ordering processes and structures toward functionality and achieving desired goals.

Introduction

Increasing rates of urbanization bring pressure on global food systems. As the world's cities expand, there is increased food insecurity, thereby becoming home to an increasing number of malnourished people (Cordell et al. 2009; Battersby and Watson 2019). Food security is a complex and multidimensional problem that needs cross-sectional and broad responses. For people to realize their right to food, they should be able to enjoy other human rights such as the right to adequate housing, the right to property, and the right to freedom from discrimination. Therefore, food security is a human rights issue, and this has led to the adoption of the New Urban Agenda which was adopted in October 2016. The set of non-binding principles and commitments will guide the efforts around urban development through the 20 years to 2036. A new role for towns and cities across the world has been given from the implementation of the New Urban Agenda (NUA) (Dubbeling et al. 2016). The NUA recognizes urban food security as an important component of sustainable urban development. The city region concept in the NUA raises crucial concepts on food systems in relation to urban and territorial planning.

According to the New Urban Agenda, adopted in Quito in October 2016, which will guide implementation of the Sustainable Development Goals (SDG), food security and nutrition are declared as important fundamentals of urban and territorial sustainability (United Nations 2016). It is also recognized that food systems are closely related to other local and regional government sectors, including organic waste management, public health, transport, markets, and enterprise creation in the food system, consumption and food insecurity/malnutrition, land use planning, and climate change adaptation strategies, among others (Maye 2019; Kuylenstierna et al. 2019). This also highlights that the realization of the right to food is the responsibility of the state. There is a need for the state to provide an environment where people can freely produce and procure adequate and sufficient food for themselves and their families. Therefore, the NUA recognizes that there is a need for a food security and law to be part of governments' agenda in making sure that food security is achieved.

This chapter seeks to unpack matters on the role of NUA in reversing the alarming trend in food and nutrition insecurity, the main causes of malnutrition and constraints for people to improve their nutrition status and how can it respond to such matters. Desktop research was used for this study through document and literature review. Detailed and comprehensive search for applicable and related studies on urban food security and the New Urban Agenda was used. Online publications including books, peer-reviewed journal articles, and policy briefs were reviewed. The chapter first presents the introduction which explores the aim of the chapter. This is followed by a background of food insecurity and the development of the NUA, global and regional literature on food security in urban areas and issues raised in the NUA. The chapter then presents the methodology followed by analysis and discussion of results. The chapter concludes by offering recommendations to urban food security and addressing policy gaps.

The Food Agricultural Organization (FAO) has welcomed the New Urban Agenda, agreed upon by countries at the Habitat III conference in Quito, Ecuador, in 2016. It aims to address the diverse challenges posed by urbanization which are crucial in achieving sustainable development and eradicating hunger (Seidenbusch et al. 2018). The New Urban Agenda places food security and nutrition at the center of urban sustainable development (Watson 2016; Battersby and Watson 2019). The agenda represents a fundamental step towards linking urban and rural communities in the planning and development of food systems grounded in a territorial approach that provides food security and improved nutrition for all. New Urban Agenda has been described as a vision for a better and greener urban future, where everyone has access to the benefits of urbanization. Achieving sustainable urban development requires the implementation of sustainable food systems (Ballamingie et al. 2020). For this reason, FAO, together with other partners, made substantive efforts in the negotiation process of the agenda to ensure that a clear reference to food security and nutrition was incorporated in the outcome declaration of the Habitat Conference until 2036. Food security relates to the availability and adequate access at all times to sufficient, safe, nutritious food to maintain a healthy and active life (Watson 2009). Food security analysts look at the combination of the following four main elements: food availability, access, stability, and utilization. Urbanization and climate change have affected availability of food.

Food Security in the New Urban Agenda: A Review

Food system governance and planning have for a long time been absent in urban planning and city policy-making. Nevertheless, city, metropolitan, and national governments have started prominently and actively to take part in local, national, and international dialogue on the future of urban food and nutrition security (Van Veenhuizen 2006; Watson 2016; Earle 2016). International declarations by support organizations and cities call for an increased focus on the role that cities and city regions play in enhancing food security, local economic development, and resilient and sustainable development of urban areas (Monaco et al. 2017). In the New Urban Agenda, adopted in Quito in October 2016, food security and nutrition are mentioned as key elements of urban and territorial sustainability. It is also recognized that food systems are closely related to other local and regional government sectors, including organic waste management, public health, transport, markets, and enterprise creation in the food system, consumption and food insecurity/malnutrition, land use planning, and climate change adaptation strategies, among others (MacNairn 2014).

Urban and territorial planning can be defined as a decision-making process aimed at realizing economic, social, cultural, and environmental goals through the development of spatial visions, strategies, and plans (Graute 2018). It also involves the application of sets of policy principles, tools, institutional and participatory mechanisms, and regulatory procedures. The New Urban Agenda is an action-oriented document that provides the global principles, policies, and standards required to achieve sustainable urban development, to transform the way cities are constructed, managed, and operated (Caprotti et al. 2017; Garschagen et al. 2018). It guides the efforts around urbanization for a wide range of actors including nation states, city and regional leaders, funders of international development, the private sector, the United Nations programs, and civil society for the next 20 years from 2016.

The key components of the New Urban Agenda that will provide strategic direction for the successful transformation of our cities are urban policies, urban governance, and urban planning and design. Developing and implementing urban policies that promote cooperation among local-national government and build multi-stakeholder partnerships enable them to achieve sustainable integrated urban development (Evans et al. 2016). The outcomes in terms of quality of urban settlement depend on the set of rules and regulations that are framed and effected (Lucan et al. 2015). Strengthening urban governance and legislation will provide directive to the urban development and the necessary stimulus to municipal finance (Peter and Yang 2019; Cruz 2017). Authorities should strengthen urban and territorial planning to best utilize the spatial dimension of the urban form and deliver the urban advantage (Minaker et al. 2011).

The NUA-related guideline document on urban-rural linkages makes the connection between food security and the city-region concept: Settlements and their surrounding rural and peri-urban areas need to acknowledge their territorial interdependence for guaranteeing a sustainable food supply for all (Garrett and Ruel 2018).

P

In this context, the relevance of city-region food systems and regional approaches needs to be recognized. Regional food systems are able to create food self-sufficiency, are more environmentally friendly, and foster regional employment (United Nations 2016). The promotion of linkages between towns and their hinterlands is seen as necessary to achieve urban food security. UN-Habitat (2017, p. 42) identifies, as an entry point for sustainable urban food systems, the flows of goods and services that promote sustainable local and regional food supply chains; especially regarding the linkages between urban agglomerations and the relevance of peri-urban areas and the hinterland in producing and supplying food for urban areas (FAO 2017).

There is a growing awareness of the extent of food insecurity and poor nutrition in urban areas, particularly in lower-income countries, and how it is interrelated with a wide range of other development issues (Casper 2007). The Food and Agricultural Organization's (FAO) survey of 146 countries in 2014–2015 found that 50% of urban populations in the least developed countries were food-insecure, compared with 43% in rural areas. This reached 70–95% of the population in urban informal settlements around the world (Garschagen et al. 2018; Koop and van Leeuwen 2017). Food expenditure is a significant component of the most low-income household budgets, usually in competition with other basic needs and priorities, thus contributing directly to urban poverty (Minaker et al. 2011). The food system is a major source of urban employment and livelihoods, especially in the areas.

There is concern over national governments ability to produce national statements on food policy and connecting them to national urban policies. For regional and local governments, there is concern if they are prioritizing food security in plans, for instance, by provision for urban agriculture and local markets. Plans and policies aim to build new urban-rural linkages and give due consideration to protecting the best quality agricultural land from urban development (Pothukuchi and Kaufman 1999). Professional planners and their associations are rethinking the knowledge and skills required of members and how to support the acquisition of knowledge on relatively new topics like food security, carbon reduction, and climate change mitigation and adaptation. Therefore, in spatial or territorial planning, there is a need for community participation in the decision-making process. It is important for communities who have difficulties in accessing food to voice out their needs to relevant authorities. This is important as food security cannot be met when the targeted population needs are not met. Participation should be incorporated throughout a program work plan, including into any baseline assessment, program design and targeting, implementation, and monitoring and evaluation. Any law or policy reform process should be participatory and inclusive.

Food security and the law are important in fulfilling the Sustainable Development Goals (SDG2) of 2015. The right to adequate food is a basic human right that was recognized by the Universal Declaration of Human Rights (1948) (Hussein and Suttie 2016). The policy recognized that food security is a component that is important in achieving an adequate standard of living. When the International Covenant on Economic, Social, and Cultural Rights entered into force in 1976, the declaration of human rights became legally binding. Ending of hunger was a policy action taken up by many countries through decisive national action and international and regional cooperation. Countries need to establish food control and safety policies in order to fulfil their obligations towards the health of their citizens (Pswarayi et al. 2014). Legal frameworks over food security are important as food security is critical to human development (Mudimu 2004). The ability of communities to feed themselves partly depends on national and international food markets, the demand for land, and the accessibility of natural resources. The poor as a vulnerable group are mostly the ones that face food security issues in various regions. Law over food security assists the marginal groups to be able to present their interests to policy-makers (Shava et al. 2009). Upholding the right to food requires that people understand their rights and can access claim mechanisms and receive timely and fair outcomes. However, in Africa political interference is a major problem for policy implementation. This has threatened food security in many African countries.

Practical and Reality Aspects of Food Security in the New Urban Agenda

The New Urban Agenda (2016) views the city-region idea as important for achieving inclusive and sustainable cities and balanced urban and regional territorial development and, second, in certain urban food security and food policy positions which argue that the "city-region food systems" approach is key to addressing the issue of urban food insecurity. Over the years, the city-region construct has been challenged from an analytical as well as a normative perspective in the fields of planning and regional economy, as well as in the field of urban food security. In addition, both these lines of critical thinking have drawn attention to the important influence of context and "place" and have questioned the validity of applying single-model analytical constructs and solutions in all parts of the world.

The New Urban Agenda (NUA), accepted at the Habitat III Conference in 2016, is the focal point of the global urban agenda until 2036 (UNHABITAT 2017). Habitat III included many municipal and civil society representatives, and, as a result, the NUA places more emphasis on subnational actors than the SDGs (Parnell 2016). It also strikes an optimistic tone about the potential for sustainable cities to optimize the benefits of new technologies and models of inclusive governance for the conservation of natural resources, the preservation of eco-systems, and the promotion of equitable growth (Yam et al. 2016; Crush 2017). Although there was some resistance to including food security in the NUA, sustained lobbying by various nongovernmental agencies saw it named in 12 of the 175 articles of the document. In most cases, food or food and nutrition security are simply included in the lists of desirable public goods, services, and outcomes. One section stands out for advocating a broader focus on urban food security (Crush et al. 2017).

The integration of food security and the nutritional needs of urban residents, particularly the urban poor, in urban and territorial planning, in order to end hunger and malnutrition is promoted (Webb and Rogers 2003; Watson 2009). There is coordination of sustainable food security and agriculture policies across urban, peri-urban, and rural areas to facilitate the production, storage, transport, and marketing of food to consumers in adequate and affordable ways in order to reduce food losses and prevent and reuse food waste. There is further promotion of the coordination of food policies with energy, water, health, transport, and waste policies, maintain the genetic diversity of seeds, reduce the use of hazardous chemicals, and implement other policies in urban areas to maximize efficiencies and minimize waste (WHO 2016). While a focus on promoting the integration and needs of urban residents is promising, the policy solution primarily ties food security to production and the reduction of food waste rather than to the full spectrum of actions that would promote food-secure cities. Urban and territorial planning can contribute to ensuring the right to adequate and nutritious food.

Food has always influenced the development and structure of cities: roads connect the fields where food is produced to the various markets where it is sold for consumption (Sonnino 2009). Planning systems that developed in the 1940s in countries that had experienced food shortages have sought to protect agricultural land from urban development. In contrast, in land-rich countries, the planning systems have been more permissive of such land conversion. Land is important to food security as the world population depends on it for survival. Food security is highlighted in SDG 2 which is to end hunger, achieve food security and improved nutrition, and promote sustainable agriculture.

The NUA points to the need to use urban spatial frameworks and urban planning and design to strengthen food system planning and urban resilience. Improving food security and nutrition is one of the many benefits that it attaches to well-distributed networks of open, multipurpose, safe, inclusive, accessible, green, and quality public spaces (Raja et al. 2017). This notes that urbanization is having profound impacts and necessitates re-examination of the ways in which cities are provisioned with food and water. It calls for an integrated approach to urban food policies, connecting the economic, social, and environmental aspects locally, while also linking the local to the regional and national and international scales. In particular it notes the scope for urban and peri-urban agriculture to protect and integrate biodiversity into city region landscapes and food systems, thereby contributing to synergies across food and nutrition security, ecosystem services, and human well-being.

National policy should set the framework for the protection of land for agriculture. Many cities have developed over time in areas of high-value agricultural land production and good soil quality. Often such land will be in peri-urban locations and under pressure from urban expansion. It should not just be left to the market to decide whether this land should be consumed by urban expansion, especially low-density development, as there are strategic issues at stake (Koop and van Leeuwen 2017). Where it is feasible to enforce compact city policies and densification of low-density areas, these measures should be applied to contain the loss of farmland, especially as other benefits in increasing urban sustainability also accrue. In rapidly urbanizing countries where the capacity to plan and enforce regulation may make such approaches infeasible, peri-urban agriculture, even if only on a temporary basis, is likely to be a viable use among new urban residents.

Insofar as this is a labor-intensive sector, agriculture can provide employment to groups who find it hard to access conventional jobs – e.g., teenagers, the elderly, and in some cases women. Therefore, urban and territorial planning practices should regard urban agriculture as a potential resource for creating urban prosperity and employment in situations where significant value added can be achieved and alternative urban uses are comparatively unproductive. This will often involve schemes for stimulating urban agricultural productivity. This can entail not only making space for such activity but also recognizing and legalizing urban agriculture and cooperation to overcome some of the potential hazards of urban agriculture such as poor composting or soil exhaustion (Dubbeling et al. 2017). Urban agriculture can also assist in the management of organic waste which can be used as a form of nutrient. By nurturing and sustaining local food networks, planners can boost local incomes and reduce the "food miles" travelled between producer and consumer, thereby contributing to reducing carbon emissions.

FAO has welcomed the New Urban Agenda and recognizes that progress in addressing the diverse challenges posed by urbanization will be key in achieving sustainable development and eradicating hunger. The New Urban Agenda places food security and nutrition at the center of urban sustainable development. The agreement represents a fundamental step towards linking urban and rural communities in the planning and development of food systems grounded in a territorial approach that provides food security and improved nutrition for all (Cohen and Habron 2018; Chan et al. 2020). Joan Clos, Executive Director of UN Habitat described the New Urban Agenda as "a vision for a better and greener urban future, where everyone has access to the benefits of urbanization." Achieving sustainable urban development will require the implementation of sustainable food systems (Riccaboni and Cavicchi 2019; Valentini et al. 2019). For this reason, FAO, together with other partners, made substantive efforts in the negotiation process of the agenda to ensure that a clear reference to food security and nutrition was incorporated in the outcome declaration of the Habitat Conference, an event which usually only takes place every 20 years.

Urbanization represents one of the most rapid and profound shifts in human history. By 2050, most of the world's population – two thirds – will live in towns or cities, but these urban centers will still depend strongly on rural communities for the provision of food (Battersby and Watson 2019). As such, food security strategies need to foster new synergies, which incorporate the necessary links between urban, peri-urban, and rural areas, Benítez stressed. Sustainable and inclusive urban economies must better engage and interact with rural economies, especially since small-scale farmers produce most of the food consumed in developing countries, including in their cities. It is necessary to go beyond the traditional dichotomy between urban and rural areas. City dwellers cannot be considered as mere consumers, and rural communities must not be seen exclusively as producers.

Urban and rural planning must be integrated in ways that foster sustainable food production including a reduction of food waste in cities – currently one third of food produced globally is lost or wasted. Habitat III has reaffirmed the increasing political momentum that recognizes the interlinkages between food security and nutrition and urban agendas (Battersby 2017). This was also underscored during the Second Mayors' Summit on the Milan Urban Food Policy Pact held in Rome on 14 October 2016, in which concrete commitments were announced by the leaders of several cities. This is also in line with recent development in which many cities across the globe are engaged in food-related initiatives, developing food planning strategies such as food charters, food policies, projects on school catering, urban gardening, food waste management, strengthening urban-rural linkages, as well as developing urban and peri-urban agriculture and forestry and green infrastructure (Halliday 2019; Sonnino 2016).

Emerging Issues

The New Urban Agenda recognizes that the eradication of hunger and poverty is a global challenge and is an indispensable requirement for sustainable development. It also offers inclusive and sustainable development. For instance, women's vulnerability to food insecurity comes which is caused by poor access to and control over land and productive resources due to discrimination in both laws and social customs is eradicated, thus promoting human development. According to FAO, the agenda aims at helping countries eradicate hunger – which still affects nearly 800 million people worldwide and malnutrition through an integrated and multi-sectoral approach targeting both urban and rural areas. The interrelationship between urbanization, the nature of urban poverty, and the relationship between governance, poverty, and the spatial characteristics of cities and towns through a focus on urban food systems and the dynamics of urban food poverty is crucial in understanding relevance of policies in addressing food insecurity (Cordell et al. 2009; Dempsey et al. 2011). Incorporating urban food security and food systems into the NUA through promoting agricultural production is the main solution to food security as well as favoring local (city-region) food supplies over international food supply chains. Addressing urban food security and food systems in the NUA is needed to acknowledge regional difference across the world.

The concept of urban and territorial planning is related to urban food security particularly (although not only) through the concept of the city-region and urban-rural links. The nature of approaches to city-region planning in these documents show little cognizance of the long history of theoretical and policy debate on city-regions – in urban economic development theory as well as in planning theory (Crush and Riley 2017). Adopting older approaches to city-region thinking allowed a greater alignment with current productionist and localist policies on urban food security but avoided newer understandings of urban economies in a globalized world, newer regional planning arguments, and the importance of contextual difference.

The NUA presents shifts in thinking on urban food security policy as well as on the concept of the city-region. The New Urban Agenda was adopted with a vision for global urban development, recognizing that while rapid urbanization poses many challenges, it also presents an opportunity for innovation (Barnett and Parnell 2017). It calls on national and local governments to ensure that cities fulfil the social function of providing equal access to basic goods and services, among them notably food security and nutrition. It addresses the urban poor specifically, stating for the promotion of the integration of food security and the nutritional needs of urban residents, particularly the urban poor, in urban and territorial planning, in order to end hunger and malnutrition (Crush and Riley 2018). The agenda is not prescriptive in how to realize its ideals (neither should it be), so the fulfilment of the objectives depends largely on the commitment of individual Member States. While it is a critical step forward, this does raise some matters about how policies ultimately will be implemented and whether they will truly address the needs of the urban poor.

Conclusion and Recommendations

This chapter sought to unpack the role of the NUA in reversing the alarming trend in food and nutrition insecurity, the main causes of malnutrition, and constraints for people to improve their nutrition status and more pressing, how can it respond to such matters. More than half of the world's population lives in cities, and the number of urban citizens is expected to grow exponentially over the coming decades. Cities have become engines of development and spaces for political participation and social interaction. However, urbanization also poses several challenges to human security. The magnitude and pace of migration have led to the inability of cities to provide adequate public utilities, such as security, education, sanitation, and water and food provision. In the context of growing urbanization, urban poverty, and climate change impacts, the importance of urban food security and urban food systems is increasingly recognized by local and national governments, as well as international actors. There is also a growing understanding that urban development and food systems cannot be decoupled from rural development given the multiple impacts that urban areas have on their surroundings. Therefore, land reform that ensures tenure security and access to land for those who depend on land for their livelihood, particularly the rural poor, women, and other marginalized groups, is critical when implementing food security policies.

The New Urban Agenda acknowledges the need for integrated urban and territorial development and recognizes the centrality of food security and nutrition in planning for sustainable cities. It promotes the creation and maintenance of well-connected and well-distributed networks of open, multipurpose, safe, inclusive, accessible, green, and quality public spaces. It also works towards improving the resilience of cities to disasters and climate change, including floods, drought risks, and heat waves. In addition, it is critical for improving food security and nutrition, physical and mental health, and household and ambient air quality, to reducing noise and promoting attractive and loveable cities, human settlements, and urban landscapes and to prioritizing the conservation of endemic species. It is concluded that the interrelationship between urban growth and the nature of urban poverty and their relationship with governance and poverty call for understanding the spatial characteristics of cities and towns. It is recommended that focusing on urban food systems and the dynamics of urban food poverty is crucial in understanding relevance of policies in addressing food insecurity. There is need for a holistic approach in dealing with food insecurity. Urban areas should not be put in isolation, hence the importance of urban and territorial planning. There is also need to promote accountability in legislators and policymakers. The human rights framework brings out accountability of states and governments (who are the duty bearers) towards the rights holders. This needs governments to be answerable to its human rights obligations.

References

Ballamingie, P., Blay-Palmer, A., Knezevic, I., Lacerda, A., Nimmo, E., Stahlbrand, L., & Ayalon, R. (2020). Integrating a food systems lens into discussions of urban resilience. *Journal of Agriculture, Food Systems, and Community Development, 9*(3), 1–17.

Battersby, J. (2017). Food system transformation in the absence of food system planning: The case of supermarket and shopping mall retail expansion in Cape Town, South Africa. *Built Environment, 43*(3), 417–430.

Battersby, J., & Watson, V. (2019). The planned 'city-region' in the New Urban Agenda: An appropriate framing for urban food security? *Town Planning Review, 90*(5), 497–518.

Barnett, C., & Parnell, S. (2017). Spatial rationalities and the possibilities for planning in the New Urban Agenda for Sustainable Development. In: Bhan, G., Srinivas, S., & Watson, V. (Eds.), *The Routledge companion to planning in the Global South*. London: Routledge.

Caprotti, F., Cowley, R., Datta, A., Broto, V. C., Gao, E., Georgeson, L., ... & Joss, S. (2017). The New Urban Agenda: key opportunities and challenges for policy and practice. *Urban research & practice, 10*(3), 367–378.

Casper, J. K. (2007). *Agriculture: The food we grow and animals we raise*. New York: Chelsea House.

Chan, E. Y. Y., Ho, J. Y., Wong, C. S., & Shaw, R. (2020). Health-EDRM in international policy agenda III: 2030 Sustainable Development Goals and New Urban Agenda (Habitat III). In *Public health and disasters* (pp. 93–114). Singapore: Springer.

Cohen, M., & Habron, G. (2018). How does the new urban agenda align with comprehensive planning in US cities? A case study of Asheville, North Carolina. *Sustainability, 10*(12), 4590.

Cordell, D., Drangert, J. O., & White, S. (2009). The story of phosphorus: Global food security and food for thought. *Global Environmental Change, 19*(2), 292–305.

Crush, J. O. N. A. T. H. A. N., & Riley, L. I. A. M. (2017). *Urban food security, rural bias and the global development agenda* (Discussion Paper no. 11, pp. 1–11). Waterloo: Hungry Cities Partnership.

Crush, J. (2017). *Informal migrant entrepreneurship and inclusive growth in South Africa, Zimbabwe and Mozambique (No. 68)*. Cape Town: Southern African Migration Programme.

Crush, J., Frayne, B., & McCordic, C. (2017). Urban agriculture and urban food insecurity in Maseru, Lesotho. *Journal of Food Security, 5*(2), 33–42.

Cruz, J. (2017). *Fiscal contracts and local public services: Bridging tax justice and inclusive cities for the New Urban Agenda* (Research briefing note, PSI local and regional government sector). Geneva: PSI. http://www.world-psi.org/sites/default/files/documents/research/psi_research_brief-_fiscal_contracts_and_local_public_services.pdf

Crush, J., & Riley, L. (2018). Rural bias and urban food security. In: Battersby, J., & Watson, V. (eds.), *Urban food systems governance and poverty in African cities* (p. 290). London: Taylor & Francis.

Dempsey, N., Bramley, G., Power, S., & Brown, C. (2011). The social dimension of sustainable development: Defining urban social sustainability. *Sustainable Development, 19*(5), 289–300.

Dubbeling, M., Bucatariu, C., Santini, G., Vogt, C., & Eisenbeiß, K. (2016). *City region food systems and food waste management. Linking urban and rural areas for sustainable and resilient development*. Eschborn: Deutsche Gesellschaft für Internationale Zusammenarbeit (GIZ) GmbH.

Dubbeling, M., Santini, G., Renting, H., Taguchi, M., Lançon, L., Zuluaga, J., ... & Andino, V. (2017). Assessing and planning Sustainable City region food systems: Insights from two Latin American cities. *Sustainability, 9*(8), 1455.

Earle, L. (2016). Urban crises and the new urban agenda. *Environment and Urbanisation, 28*(1), 77–86.

Evans, B. M., Rosenfeld, O., Elesei, P., Golubchikov, O., Badynia, A., Saliez, F., ... & Küsters, C. (2016). Habitat III regional report on housing and urban development for the UNECE region: Towards a city-focused, people-centred and integrated approach to the New Urban Agenda. Available online: https://unece.org/DAM/hlm/documents/Publications/HabitatIII-Regional-Report-Europe-Region.pdf. Accessed on 22 October 2020.

FAO. (2017). Strong rural-urban linkages are essential for poverty reduction. Available online: http://www.fao.org/3/a-i7904e.pdf. Accessed on 23 October 2020.

Garrett, J., & Ruel, M. (2018). Background paper (Annex 3: Linking Nutrition with Sustainable Cities and Communities-SDG 11: Make cities and human settlements inclusive, safe, resilient and sustainable). Available online: https://cgspace.cgiar.org/bitstream/handle/10568/99266/Annex%203_Garrett_2018.pdf?sequence=1. Accessed on 21 October 2020.

Garschagen, M., Porter, L., Satterthwaite, D., Fraser, A., Horne, R., Nolan, M., ... & Schreiber, F. (2018). The New Urban Agenda: From vision to policy and action/will the New Urban Agenda have any positive influence on governments and international agencies?/informality in the New Urban Agenda: From the aspirational policies of integration to a politics of constructive engagement/growing up or growing despair? Prospects for multi-sector progresson city sustainability under the NUA/approaching risk and hazards in the New Urban Agenda: A commentary/follow-up and review of the New Urban Agenda. *Planning Theory & Practice, 19* (1), 117–137.

Graute, U. (2018). International guidelines on urban and territorial planning handbook. Available online: https://unhabitat.org/international-guidelines-on-urban-and-territorial-planning. Accessed on 21 October 2020.

Halliday, J. (2019). Cities' strategies for sustainable food and the levers they mobilize. In *Designing urban food policies* (pp. 53–74). Cham: Springer.

Hussein, K., & Suttie, D. (2016). *Rural-urban linkages and food systems in sub-Saharan Africa: The rural dimension*. Rome: IFAD.

Koop, S. H., & van Leeuwen, C. J. (2017). The challenges of water, waste and climate change in cities. *Environment, Development and Sustainability, 19*(2), 385–418.

Kuylenstierna, J., Barraza, H. J., Benton, T., Larsen, A. F., Kurppa, S., Lipper, L., & Virgin, I. (2019). Food security and sustainable food systems. Available online: https://www.mistra.org/wp-content/uploads/2019/04/

P

mistra_bp_-food_security_2019.pdf. Accessed on 21 October 2020.

Lucan, S. C., Maroko, A. R., Patel, A. N., Gjonbalaj, I., Abrams, C., Rettig, S., Elbel, B. & Schechter, C.B. 2015. Changein urban food enviroment: Storefront sources of food/drink increasing over time, and not limited to 'food stores' and restaurants. https://jandonline.org/article/S2212-2672(18)30802-5/fulltext.

MacNairn, I. (2014). *Zimbabwe food security brief.* Harare: International Development Famine Early Warning Systems Network (FEWS NET).

Maye, D. (2019). 'Smart food city': Conceptual relations between smart city planning, urban food systems and innovation theory. *City, Culture and Society, 16*, 18–24.

Minaker, L. M., Fisher, P., Raine, K. D., & Frankd, L. D. (2011). Measuring the food environment: From theory to planning practice. *Journal of Agriculture, Food Systems, and Community Development, 2*(1), 65–82.

Monaco, F., Zasada, I., Wascher, D., Glavan, M., Pintar, M., Schmutz, U., & Sali, G. (2017). Food production and consumption: City regions between localism, agricultural land displacement, and economic competitiveness. *Sustainability, 9*(1), 96.

Mudimu, G. (2004). Zimbabwe food security issues paper for forum for food security in Southern Africa. http://www.odi.org.uk/food-security-forum.

Peter, L. L., & Yang, Y. (2019). *Urban planning historical review of master plans and the way towards a sustainable city.* Dar es Salaam: Frontiers of Architectural Research.

Pothukuchi, K., & Kaufman, J. L. (1999). Placing the food system on the urban agenda: The role of municipal institutions in food systems planning. *Agriculture and Human Values, 16*(2), 213–224.

Pswarayi, F., Mutukumira, A. M., Chipurura, B., Gabi, B., & Jukes, D. J. (2014). Food control in Zimbabwe: A situational analysis. *Food Control, 46*, 143–151. Elsevier Ltd.

Parnell, S. (2016). Defining a global urban development agenda. *World Development, 78*, 529–540.

Raja, S., Morgan, K., & Hall, E. (2017). Planning for equitable urban and regional food systems. *Built Environment, 43*(3), 309–314.

Riccaboni, A., & Cavicchi, A. (2019). Innovation for sustainable food systems: Drivers and challenges. In *Achieving the sustainable development goals through sustainable food systems* (pp. 131–140). Cham: Springer.

Seidenbusch, E.; Villamarin, F.; McMahon, D.; Husain, S.; Islam, S. (2018). *The Urban Agenda: Meeting the food and nutrition security needs of the urban poor. SNV nutrition paper.* The Hague: SNV Netherlands Development Organisation.

Shava, S., O'Donoghue, R., Krasny, M. E. & Zazu, C. (2009). Traditional food crops as a source of community resilience in Zimbabwe, 4(1). https://www.researchgate.net/publication/228618389_Traditional_food_crops_as_a_source_of_community_resilience_in_Zimbabwe?enrichId=rgreq-0cb78f4077d6758e0b9f7f1fd11edccc-XXX&enrichSource=Y292ZXJQYWdlOzIyODYxODM4OTtBUzoxMDI2NjklNzAwMTkzMzdAMTQwMTQ4OTc0MTU4Ng%3D%3D&el=1_x_3&_esc=publicationCoverPdf

Sonnino, R. (2009). Feeding the city: Towards a new research and planning agenda. *International Planning Studies, 14*(4), 425–435.

Sonnino, R. (2016). The new geography of food security: Exploring the potential of urban food strategies. *The Geographical Journal, 182*(2), 190–200.

The New Urban Agenda. (2016). The new urban agenda. Available online: https://habitat3.org/the-new-urban-agenda/. Accessed on 23 October 2020.

United Nations. (2016, 17–20 October). New Urban Agenda. Quito declaration on sustainable cities and human settlements for all. In Proceedings of the united nations conference on housing and sustainable urban development, Quito, Ecuador.

UN-Habitat. (2017). Trends in urban resilience 2017. Available online: https://unhabitat.org/trends-in-urban-resilience-2017. Accessed on 20 October 2020.

Valentini, R., Sievenpiper, J. L., Antonelli, M., & Dembska, K. (Eds.). (2019). *Achieving the sustainable development goals through sustainable food systems.* Cham: Springer International Publishing.

Van Veenhuizen, R. (2006). *Cities farming for the future: Urban agriculture for green and productive cities.* Silang: International Institute of Rural Reconstruction and ETC Urban Agriculture.

Watson, V. (2009). 'The planned city sweeps the poor away…': Urban planning and 21st century urbanisation. *Progress in Planning, 72*(3), 151–193.

Watson, V. (2016). Locating planning in the new urban agenda of the urban sustainable development goal. *Planning Theory, 15*(4), 435–448.

Webb, P., & Rogers, B. L. (2003). *Addressing the "in" in food insecurity.* Food and Nutrition Technical Assistance Project Academy for Educational Development. Available online: http://www.regionalstudies.org/wp-content/uploads/2018/12/Vanessa-Watson-1.pdf. Accessed on 24 October 2020.

World Health Organization. (2016). Health as the pulse of the new urban agenda: United Nations conference on housing and sustainable urban development, Quito, October 2016.

Yam, K., Tan, T., Doyle, R., Clos, J., Gutiérrez, F., Chan, M., … & Mena, M. (2016). Our Planet-October 2016: Urban solutions: Making cities strong, smart, sustainable. Available online: https://wedocs.unep.org/rest/bitstreams/18379/retrieve. Accessed on 23 October 2020.

Planning for Health

▶ Planning Healthy and Livable Cities

Planning for Peri-urban Futures

Food, Landscape, and Agricultural Systems

Andrew Butt
RMIT University, Melbourne, VIC, Australia

Synonyms

Greenbelts; Greenspace; Urban hinterland

Definition

Planning in peri-urban regions is concerned with the management of that region just beyond the urban boundary, particularly around larger cities. The concern with food, agriculture, and landscape has typically emerged through an understanding of the inherent impacts of urban expansion on agriculture as a cultural and economic system, on food production and availability in the city's hinterland, and the values applied by urban populations to surrounding nonurban landscapes, whether as recreational spaces, for biodiversity, or as natural resources systems such as urban water supply.

Introduction

The processes of mass urbanization since the nineteenth century, and those continuing today, have resulted in various policy concerns for the impacts of urban growth on surrounding rural areas. These peri-urban areas are locations of change and influence led by the social and economic processes of urbanization, even when they do not necessarily exhibit the physical features of urbanization.

Contemporary policy and regulatory concern for these locations, including through land use planning systems, emerged in various jurisdictions from the early twentieth century. Key features of the such approaches remain today and are typically focused on the twin processes of containing and managing urban expansion into peri-urban areas and managing (or preventing) the transition of land use from rural activities to urban generated land uses.

Containing Urbanization

Approaches to contemporary peri-urban planning include the seemingly *anti-urban* objectives of the late nineteenth century, with influential examples such as Ebenezer Howard's (1898) vision for the agrarian urbanism of "garden cities" and Patrick Abercrombie's (1926) concerns for sprawl, "ribbon" development and dormitory satellite were framed in response to seemingly ceaseless urban growth and its impacts on rural culture and production. Similar themes are evident in Cole and Crowe (1937), through a survey of Europe and North America that outlines an emerging interest in planning for rural land resources in the face of urbanization, and similar reactions to contemporary urbanization becoming operationalized through examples such as "green belts" in many metropolitan plans throughout the twentieth century.

These concerns came to be operationalized through well-known examples including London's Green Belt (1938–1955), Copenhagen's "Finger Plan" (1947), Ottawa's Green Belt (1955), and other examples, each with different regulatory and tenure approaches and policy goals. These and more recent examples have typically developed to coincide with periods of significant urbanization, both as population growth and as "sprawl" led by growth in car ownership and suburban housing models. These processes include political imperatives to address housing affordability and evident preferences for peri-urban lifestyles in many city regions. They are trends that continue to emerge in many cities today, including increasingly in parts of South and East Asia and in Africa where examples of peri-urbanization as a model of enabling urban settlement and affordable housing create conflicts with existing rural land use (Tian and Guo 2019; Abass et al. 2018).

Food and Agriculture

Food production has long been a key role of peri-urban areas. Traditionally, these areas (coupled with production within the city) were vital feature of supplying urban populations. Since the 1970s, and specifically after the 1996 *Rome Declaration* (FAO 1996), broadening notions of food security have emerged with increasingly concern for sustainability and resilience of food systems that have influenced research agenda and land use policy in many cities.

Critical concerns in planning for food and peri-urban futures relate to the impacts of agricultural land loss and the changing nature of agriculture in these areas. Not only is land "lost" to urban uses, but the changing nature of land markets and the increasing introduction of potentially competing land uses often lead to changing agricultural systems and declining investment. Planning systems have typically responded by restriction nonagricultural land uses with muted success and a limited range of policy instruments (Lapping 2006).

Some approaches to policy protecting agricultural systems within peri-urban areas are framed within a sociocultural protection agenda. This includes the role of broader agricultural support system (for example, within the EU) to maintain farming in peri-urban regions, as well as newer market-orientated models such as community supported agriculture (CSA) and direct-sales models.

Managing Landscapes

Peri-urban paces have typically performed a range of roles beyond farming, many connected to urban activities and services. Policy approaches have included the maintenance of urban water supply, the roles of peri-urban areas for tourism and recreation, and the use of peri-urban areas for noxious land uses. They are also areas where biodiversity is often threatened by urbanization processes, and those areas where productive, natural landscapes (for example, urban water catchments) are vital features of peri-urban landscapes and the policy approaches to protecting them.

The management of peri-urban landscapes within planning systems often also includes the management of particular forms of nonurban settlement from urbanization. The maintenance of a nonurban character has not always precluded forms of urban generated housing and activities, and this continues to be a point of tension in many city regions where development generated by lifestyle opportunities or housing affordability create function change and incremental land use impacts.

Planning Futures for Peri-urban Regions

Planning approaches to peri-urban land management demonstrate both continuity and adaptation. There is an increasing interest in addressing the consequences of urbanization, including urban sprawl, as well as recognizing the risks and vulnerabilities presented by long and stretched food systems, and the opportunity costs of effective management of peri-urban regions. Emerging issues for planning the futures of these regions relate to the tensions inherent in the increasingly multifunctional roles provided by and anticipated for these regions: food production, tourism, biodiversity, and others. Adapting to climate change has presented increased urgency in recognizing the potential of peri-urban place within city regions, and the risks of ongoing physical and functional urbanization. Policy and regulatory frameworks increasingly recognize the need for these areas to serve a range of purposes and to be considered not simply as future urban spaces but as critical to the sustainability of city regions.

Cross-References

- ► Climate-Resilient Technologies and Innovations for Sustainable Agriculture, Improved Landscape, and Food Security
- ► Integrating Agriculture, Forestry, and Food Systems into Urban Planning: A Key Step for Future Resilient and Sustainable Cities
- ► Peri-Urban Regions
- ► Peri-Urbanization

References

Abass, K., Adanu, S. K., & Agyemang, S. (2018). Peri-urbanisation and loss of arable land in Kumasi Metropolis in three decades: Evidence from remote sensing image analysis. *Land Use Policy, 72*, 470–479.

Abercrombie, P. (1926). *The preservation of rural England: The control of development by means of rural planning*. London: Liverpool University Press/Hodder and Stoughton.

Cole, W., & Crowe, H. (1937). *Recent trends in rural planning*. New York: Prentice-Hall.

FAO. (1996). *Declaration on world food security*. Rome: World Food Summit, UN Food and Agriculture Organisation.

Lapping, M. (2006). Rural policy and planning. In P. Cloke, T. Marsden, & P. Mooney (Eds.), *Handbook of rural studies* (pp. 104–122). London: SAGE.

Tian, L., & Guo, Y. (2019). *Peri-urban China: Land use, growth, and integrated urban–rural development*. Routledge.

Planning Healthy and Livable Cities

Sara Alidoust
School of Earth and Environmental Sciences, The University of Queensland, Brisbane, QLD, Australia

Synonyms

Healthy Planning, Planning for Health

Definition

A healthy city promotes the physical, mental, and social health of residents, through the provision of healthy housing and urban infrastructure, social inclusion and equity, climate change adaptation/mitigation, and environmental management, creating resilient communities and integrated land-use planning.

Introduction

Human health is affected by a complex interplay of factors outside the individual sphere. Today, the definition of health has shifted from a mere medical perspective (absence of disease and infirmity) to a balance between the three health dimensions: physical, mental, and social (WHO 1948). With more than half of the world's population living in cities, urban planning is a critical enabler of active healthy lifestyle and can promote health among individuals and communities.

This chapter provides a holistic view of the role of urban planning in promoting health (Fig. 1). It discusses how planning decisions play a central role in the social, mental, and physical health of residents through the provision of healthy housing and urban infrastructure and promoting inclusive and equitable societies. Climate and environmental management can also create or exacerbate health risks for populations. In addition, urban planning can foster healthier environments for people through creating resilient communities. These, however, cannot be achieved without integrated land-use planning and embedding health in all urban planning and policy decisions, as discussed in the next sections.

Housing and Urban Infrastructure

The health impacts of housing can be analyzed at three levels: housing quality (e.g., housing material (Loughnan et al. 2015), design (Wågø et al. 2016), and occupant density (Caffaro et al. 2019)); housing policy (e.g., homeownership (Arcury et al. 2015) and housing stability (Tran and Van Vu 2018)), and housing market (e.g., housing affordability (Marquez et al. 2019)).

Among the potentially health-promoting urban infrastructure are public transport (Giles-Corti et al. 2016) and active mobility that facilitate active modes of travel, such as walking and cycling, and thus engender physical activity and social engagement (Alidoust et al. 2018) and are associated with less risk of obesity and lower rates of chronic diseases (Müller-Riemenschneider et al. 2013). Accessible, affordable, and safe community-based health facilities (Pearce et al. 2006); public amenities (e.g., libraries, community spaces, and local sporting clubs) (Alidoust and Bosman 2019; Alidoust et al. 2019, 2020);

P

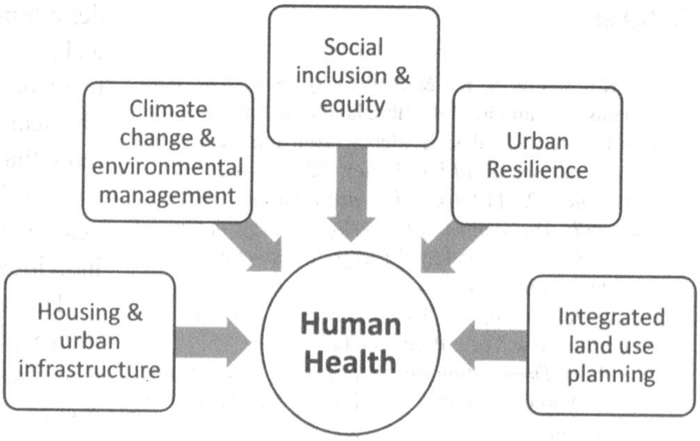

Planning Healthy and Livable Cities, Fig. 1 A holistic view of different mechanisms through which urban planning impacts human health

and green spaces (Sugiyama et al. 2010) also benefit health by encouraging physical activity and social interaction and promoting mental health.

Climate Change and Environmental Management

Climate change is a significant health threat to the world's population. Key health impacts of climate change relate to extreme weather events; threats to food, water, and air quality; the spread of infectious diseases; and mental illnesses (Haines and Patz 2004; McMichael et al. 2003; Séguin 2008). The health impacts of climate change are adverse among vulnerable populations, including older people, socially disadvantaged, Indigenous population, and people with disabilities (Séguin 2008). Living in high-risk environments, such as coastal zones and flood plains with weak public health infrastructure, also places populations of developing countries in a more vulnerable position in facing the health risks of climate change (Haines and Patz 2004).

Cities need to address the changing climate in two ways: first, designing urban environments that contribute less to the causes of climate change (e.g., renewable energy solutions); second, developing climate-resilient settlements that adapt to the increasing impacts of climate change (e.g., preparing cities for extreme weather events) (Costello et al. 2009). Well-designed climate change mitigation and adaptation strategies not only reduce health risks but also improve health outcomes for communities through the achievement of health co-benefits (Ebi et al. 2012) (e.g., supporting active transport not only reduces greenhouse gas emissions but also encourages physical and social activities (Frank et al. 2010)).

Social Inclusion and Equity

An inclusive and equitable environment is a key determinant of the health of individuals. In addition to individual socioeconomic position (e.g., gender, race/ethnicity, age), characteristics of the places in which people live impact their access to equitable opportunities (e.g., education, employment, health care, and public transport) and thus affect health inequities (Diez Roux and Mair 2010; Northridge and Freeman 2011; Turrell et al. 2007). Both urban physical and social environments need to embrace and address the diverse needs of urban populations. A fairer distribution of health-promoting opportunities can be achieved through enhanced political power and participatory planning processes that involve vulnerable groups in decision-making and implementation processes (Northridge and Freeman 2011).

Urban Resilience

Urban resilience is crucial to protect the health of residents. It is defined as "the ability of a system, community, or society exposed to hazards to resist, absorb, accommodate, and recover from the effects of a hazard in a timely and efficient manner, including through the preservation and restoration of its essential basic structures and functions" (UNISDR 2011). Building urban resilience requires a strategic approach, focused on long-term management of urban resources and potential risks they may face (D'Amico and Currà 2018). The risks could be chronic (e.g., from climate changes and poverty) or shocking (e.g., natural disasters) (Capolongo et al. 2018).

Given the rapid population growth and climate change, food security is a key determinant of urban resilience. Food security is not just about the availability of food. It is also about the more complex issues of food adequacy, secure access, and stable food supply (Armar-Klemesu 2000). Urban agriculture is an important component of the urban food system which contributes to food security, particularly for vulnerable populations (Eigenbrod and Gruda 2015).

Integrated Land-Use Planning

Incorporating health in land-use planning and development decisions requires a holistic approach and systems thinking (Pineo et al. 2020). An integrated planning approach helps to avoid the problems resulting from governments operating within traditional sectoral silos, such as fragmented plans, policies, and implementation programs (Holden 2012). Integrated land-use planning takes health measures through a range of strategies, with a focus on promoting the availability and accessibility of health-supportive opportunities. For example, the built environment can be shaped to strategically support the provision of healthy food (e.g., fresh food and farmers market) in local communities and inhibit fast-food outlets in close proximity of schools (Kent 2012). Integrated land-use and transport planning, mainly through the provision of dense, mixed-use, and transit-oriented developments, can also further enhance active travel opportunities and promote health (Giles-Corti et al. 2016).

Creating integrated policies that promote the health of residents requires effective partnerships and collaboration between stakeholders, including the public and community sectors (Holden 2012; Innes and Booher 2000). Additionally, there is a need for further engagement between public health, land-use planning, housing, transport, and economic development decision-makers in order to develop proactive, healthy urban plans (Carmichael et al. 2013). Health measures need to be embedded in planning and policy processes, institutions, and networks, over time to result in trusted relationships and collaborations among diverse sectors in cities (Pineo et al. 2020).

Cross-References

▶ Building Community Resilience
▶ Healthy Cities
▶ Housing Affordability
▶ Urban Climate Resilience

References

Alidoust, S., & Bosman, C. (2019). Planning for healthy ageing: How the use of third places contributes to the social health of older populations. In J. Dolley & C. Bosman (Eds.), *Rethinking third places*. Cheltenham: Edward Elgar Publishing.

Alidoust, S., Bosman, C., & Holden, G. (2018). Talking while walking: An investigation of perceived neighbourhood walkability and its implications for the social life of older people. *Journal of Housing and the Built Environment, 33*(1), 133–150.

Alidoust, S., Bosman, C., & Holden, G. (2019). Planning for healthy ageing: How the use of third places contributes to the social health of older populations. *Ageing and Society, 39*(7), 1459–1484.

Alidoust, S., Bosman, C., & Holden, G. (2020). The "Hothouse Community" and its implications for older age residents. *Journal of Urban Policy and Research, 38*(2), 132–149.

Arcury, T. A., Trejo, G., Suerken, C. K., Grzywacz, J. G., Ip, E. H., & Quandt, S. A. (2015). Housing and neighborhood characteristics and Latino farmworker family well-being. *Journal of Immigrant and Minority Health, 17*(5), 1458–1467.

Armar-Klemesu, M. (2000). Urban agriculture and food security, nutrition and health. In *Growing cities, growing food. Urban Agriculture on the Policy Agenda* (pp. 99–118).

P

Caffaro, F., Galati, D., Loureda, M. V. Z., & Roccato, M. (2019). Housing-related subjective well-being in Turin (Italy) and Havana (Cuba): Dimensions and prediction. *Applied Research in Quality of Life, 14*(1), 273–285.

Capolongo, S., Rebecchi, A., Dettori, M., Appolloni, L., Azara, A., Buffoli, M., ... D'Amico, A. (2018). Healthy design and urban planning strategies, actions, and policy to achieve salutogenic cities. *International Journal of Environmental Research and Public Health, 15*(12), 2698.

Carmichael, L., Barton, H., Gray, S., & Lease, H. (2013). Health-integrated planning at the local level in England: Impediments and opportunities. *Land Use Policy, 31*, 259–266.

Costello, A., Abbas, M., Allen, A., Ball, S., Bell, S., Bellamy, R., ... Kett, M. (2009). Managing the health effects of climate change: Lancet and University College London Institute for Global Health Commission. *The Lancet, 373*(9676), 1693–1733.

D'Amico, A., & Currà, E. (2018). Urban resilience in the historical centres of Italian cities and towns. Strategies of preventative planning. *Journal of Technology for Architecture Environment and Behavior, 15*, 257–268.

Diez Roux, A. V., & Mair, C. (2010). Neighborhoods and health. *Annals of the New York Academy of Sciences, 1186*, 125–145.

Ebi, K., Berry, P., Campbell-Lendrum, D., Corvalan, C., & Guillemot, J. (2012). *Protecting health from climate change: Vulnerability and adaptation assessment*. Geneva: WHO.

Eigenbrod, C., & Gruda, N. (2015). Urban vegetable for food security in cities. A review. *Journal of Agronomy for Sustainable Development, 35*(2), 483–498.

Frank, L. D., Greenwald, M. J., Winkelman, S., Chapman, J., & Kavage, S. (2010). Carbonless footprints: Promoting health and climate stabilization through active transportation. *Preventive Medicine, 50*, S99–S105.

Giles-Corti, B., Vernez-Moudon, A., Reis, R., Turrell, G., Dannenberg, A. L., Badland, H., ... Stevenson, M. (2016). City planning and population health: A global challenge. *The Lancet, 388*(10062), 2912–2924.

Haines, A., & Patz, J. A. (2004). Health effects of climate change. *Journal of American Medical Association, 291*(1), 99–103.

Holden, M. (2012). Is integrated planning any more than the sum of its parts? Considerations for planning sustainable cities. *Journal of Planning Education Research, 32*(3), 305–318.

Innes, J. E., & Booher, D. E. (2000). Indicators for sustainable communities: A strategy building on complexity theory and distributed intelligence. *Planning Theory and Practice, 1*(2), 173–186.

Kent, J. (2012). Healthy planning. In S. Thompson & P. Maginn (Eds.), *Planning Australia: An overview of urban and regional planning* (pp. 381–408). Melbourne: Cambridge University Press.

Loughnan, M., Carroll, M., & Tapper, N. J. (2015). The relationship between housing and heat wave resilience in older people. *International Journal of Biometeorology, 59*(9), 1291–1298.

Marquez, E., Francis, C. D., & Gerstenberger, S. (2019). Where I live: A qualitative analysis of renters living in poor housing. *Health & Place, 58*, 102143.

McMichael, A. J., Campbell-Lendrum, D. H., Corvalán, C. F., Ebi, K. L., Githeko, A., Scheraga, J. D., & Woodward, A. (2003). *Climate change and human health: Risks and responses*. Geneva: WHO.

Müller-Riemenschneider, F., Pereira, G., Villanueva, K., Christian, H., Knuiman, M., Giles-Corti, B., & Bull, F. C. (2013). Neighborhood walkability and cardiometabolic risk factors in Australian adults: An observational study. *BMC Public Health, 13*(1), 1–9.

Northridge, M. E., & Freeman, L. (2011). Urban planning and health equity. *Journal of Urban Health, 88*(3), 582–597.

Pearce, J., Witten, K., & Bartie, P. (2006). Neighbourhoods and health: A GIS approach to measuring community resource accessibility. *Journal of Epidemiology Community Health, 60*(5), 389–395.

Pineo, H., Zimmermann, N., & Davies, M. (2020). Integrating health into the complex urban planning policy and decision-making context: A systems thinking analysis. *Palgrave Communications, 6*(1), 1–14.

Séguin, J. (2008). *Human health in a changing climate: A Canadian assessment of vulnerabilities and adaptive capacity*. Ottawa: Health Canada.

Sugiyama, T., Francis, J., Middleton, N. J., Owen, N., & Giles-Corti, B. (2010). Associations between recreational walking and attractiveness, size, and proximity of neighborhood open spaces. *American Journal of Public Health, 100*(9), 1752–1757.

The United Nations Office for Disaster Risk Reduction (UNISDR). (2011). *Global assessment report on disaster risk reduction: Revealing risk, redefining development*. Geneva: United Nations International Strategy for Disaster Reduction.

Tran, T. Q., & Van Vu, H. (2018). A microeconometric analysis of housing and life satisfaction among the Vietnamese elderly. *Quality & Quantity, 52*(2), 849–867.

Turrell, G., Kavanagh, A., Draper, G., & Subramanian, S. (2007). Do places affect the probability of death in Australia? A multilevel study of area-level disadvantage, individual-level socioeconomic position and all-cause mortality, 1998–2000. *Journal of Epidemiology and Community Health, 61*(1), 13–19.

Wågø, S., Hauge, B., & Støa, E. (2016). Between indoor and outdoor: Norwegian perceptions of well-being in energy-efficient housing. *Journal of Architectural and Planning Research, 15*, 326–346.

World Health Organization (WHO). (1948). *Preamble to the constitution of the WHO as adopted by the international health conference, New York, 19–22 June 1946*. Geneva: WHO.

Planning Law – Legislative Instruments Guiding Planning

▶ Urbanization, Planning Law, and the Future of Developing World Cities

Planning/Coordination

▶ Metropolitan Discipline: Management and Planning

Plant-Based Protein

▶ Future Foods for Urban Food Production

Plumbism

▶ Lead Exposure in US Cities

Policies for a Just Transition

Tamara Krawchenko[1] and Megan Gordon[2]
[1]University of Victoria, Victoria, BC, Canada
[2]University of Northern British Columbia, Prince George, BC, Canada

Synonyms

Equitable Transitions; Sustainable Economies

Definition

The protection of workers' rights, livelihoods, and communities that are negatively impacted by the transition to sustainable production and a post-carbon economy. The "just transitions" framework entails a political imperative, a policy goal, and a set of practices to equitably manage this transition.

Introduction

The concept of a "just transition" is at once an agenda for sociotechnical transformation; a call for social, economic, and environmental justice; and a political imperative to manage the decarbonization of economies. At its core, it speaks to the importance of recognizing the rights and needs of workers and communities impacted by polluting and carbon-intensive industries and their transitions.

The concept of a just transition emerged in the American labor movement in the 1970s and subsequently gained prominence through the work of unions and social activists in other countries and globally. For an overview of the genealogy of just transitions and the importance of national and international unions to its proliferation, see the work of Morena, Krause, and Stevis et al. (2020). While the term was initially used to draw attention to the right to an environmentally safe (toxic-free) workplace, it has increasingly come to focus on how to secure the rights of workers and communities in the transition away from carbon-intensive industries.

Solidifying its global prominence, the 2015 Paris Agreement of the United Nations Framework Convention on Climate Change included the proviso that signatory nations would take into account "the imperatives of a just transition of the workforce and the creation of decent work and quality jobs in accordance with nationally-defined development priorities" (UNFCC 2015). From its labor movement origins, academic and think tank work on just transitions has proliferated in the past decade (Pai et al. 2020). Despite this, the term remains contested terrain. This article briefly outlines the three main interpretations of *justice* in "just transitions" followed by key issues and knowledge gaps important for regional and urban futures.

P

What Constitutes a *Just Transition*?

There Are Three Main Interpretations: Jobs-Focused, Environment-Focused, or Society-Focused

Justice or equity is conceived in distributional, procedural, and restorative terms (McCauley and Heffron 2018; Newell and Mulvaney 2013). When it comes to sustainability transitions, *distributional* justice is concerned with how different groups benefit or experience impacts of changes; *recognitional* justice identifies interest groups and rights holders who may be implicated; and *procedural* justice is concerned with governance – that is, who is included in decision-making and how (Bennett et al. 2019).

The literature reveals three main ideas and approaches around just transitions, each in turn stressing a "jobs-focused," "environment-focused," or "society-focused" view. These interpretations are grounded in the central question: "justice for whom?". Each approach has implications for *where* and *to whom* governments have focused their policy supports and investments.

Jobs-Focused Just Transitions

The "jobs-focused" interpretation of just transition advocates primarily for workers and communities that have been impacted by environmental and climate policies. This approach originates with the labor movement – i.e., social-democratic unions in regions that rely on carbon-intensive industries and resource extraction (Evans and Phelan 2016). From this lens, the imperative for a just transition focusses on the uneven distribution of costs and benefits between resource-dependent workers and communities and the industries and governments who profit from production. Advocates from this lens emphasize the importance of involving unions, workers, and communities in discussions around industrial transition and change in order to increase *procedural* justice.

The "jobs-focused" interpretation of just transitions is the most common among the three types and has been a focus of government policies and programs to address the immediate concerns of impacted workers and communities. It is aligned with a "differentiated responsibility" approach wherein states and capital have a responsibility towards workers impacted by environmental regulations (Stevis and Felli 2014). These reactive supports typically include temporary income support (e.g., bridge to retirement or reemployment), retraining or re-education initiatives, and support for communities in the form of project or diversification funding (Mertins-Kirkwood 2018). These policies have been adopted in Canada, Australia, and the United States in response to transitions in the coal sector.

Of the three interpretations, the "jobs-focused" interpretation is the narrowest. Advocates for a job-focused just transition could be seen as problematic from a distributional, procedural, and recognitional justice perspective as it fails to simultaneously account for other societal challenges. Many other groups (e.g., marginalized populations in impacted communities) may not be perceived as having experienced "direct" impacts (e.g., creating recognitional justice issues), thus excluding them from decision-making forums (i.e., creating procedural justice challenges), and limiting their access to the benefits of transition supports (creating distributional justice issues). Given the primary emphasis on jobs and employment in this interpretation, less consideration is placed on securing "green jobs" for displaced workers (e.g., conversions of coal plants to natural gas plants). Some may question whether this constitutes a truly just outcome, as those who take an "environment-focused" interpretation of just transition insist that a transition can only truly be "just" if it helps progress toward a net zero-carbon economy and facilitates the end of fossil fuel dependence.

Environment-Focused Just Transitions

The "environment-focused" interpretation of just transition evaluates transitions based on the primary objective of enabling the shift to a zero-carbon economy. This approach examines just transition from a socio-technical standpoint, examining the production and consumption patterns of a sector (Meadowcroft 2009; Newell and Mulvaney 2013). From this lens, a just outcome is one wherein communities and workers who rely on carbon-intensive industries shift to green jobs and low-carbon solutions. Advocates recognize that environmental policies and climate actions can harm jobs and the economy if not well managed and point to opportunities in renewable energy, energy efficiency, adaptation projects, and other "green" solutions. Advocates focus on ecological justice, noting the dangers of carbon lock-in (Goddard and Farrelly 2018; Stevis and Felli 2014). This interpretation mirrors a "shared solutions" approach which involves dialogue, mutual understanding, and shared solutions between trade unions and international organizations in the context of climate negotiations (Stevis and Felli 2014). When considering distributional justice, it acknowledges the role that fossil fuel extraction and development have directly benefited a few and yet directly contributes to anthropogenic climate change which has a nearly universal negative impact. Environment-focused just transitions seek to build allyship with workers in fossil fuel-dependent regions by finding green alternatives, thus helping to reduce resistance to climate action (Evans and Phelan 2016). This is the core concern of those who subscribe to the "jobs vs. environment" narrative.

A critical look at an environment-focused interpretation of just transition reveals a number of possible shortcomings. In many fossil fuel-dependent regions, particularly in coal regions, community and worker culture and identity are deeply connected to the industry; thus they may be inherently resistant to the idea of low carbon transition due to hegemonic discourses and political beliefs (Evans and Phelan 2016). Furthermore, advocates for an environment-focused just transition may not consider the practicality of green solutions or the reality of green jobs, given their location, the nature of the work and its compensation, and the transferability of skills.

Society-Focused Just Transitions

"Society-focused" interpretations of just transitions take the broadest view, with the widest application of solutions. This lens sees just transition measures as a means to uplift and support workers, communities, and society at large. Advocates for a society-focused interpretation of just transition include a broad range of interests and advocate for system-wide transformation (Bennett et al. 2019). This approach engages more heavily with environmental and energy justice and advocates for universal equity and justice by addressing inequities at the national and subnational scales (e.g., marginalized communities facing disproportionate harms of resource development) and globally (e.g., energy poverty in developing countries). This interpretation is often used by social justice organizations and in the international governance arena (e.g., ILO, UNEP). This lens stresses the importance of the United Nations Sustainable Development Goals to achieving a just transition (Delina and Sovacool 2018; McCauley et al. 2019; Pai et al. 2020).

While the society-focused interpretation of just transitions has the potential to have the widest and greatest impact, there are few concrete examples and many unanswered questions about how it could be achieved. While the philosophical ideas underpinning radical system upheaval are important to engage with, it leaves few practical and concrete solutions to implement. A macro-level view of low-carbon transition and subsequent just transition measures may leave out important political and cultural nuances at the regional, subregional, or community level that may hinder progress. Furthermore, the goal of a "society-focused" interpretation of just transition requires an examination of other problems in society (e.g., gender and racial inequality, poverty reduction, etc.), and thus, it is much more challenging to measure progress and arrive at tangible just outcomes.

From Concept to Implementation: Knowledge Gaps and Key Considerations

The concept of just transition has evolved from its labor movement roots. There are a growing number of policies and initiatives focused on achieving just outcomes for workers, communities, the environment, and society. However, there remain a number of key considerations and research gaps in understanding how to create pathways to pragmatic and implementable solutions for governments and stakeholders.

What Is the Right Jurisdictional Scale to Manage Just Transitions? Top-Down Versus Bottom-up Approaches

More attention must be given to the scale at which sustainability transitions are occurring and the required level of government that must intervene to achieve a just transition. Many attempts to examine transitions away from fossil fuels have been initiated at a level of government that does not hold jurisdiction over solutions. For example, the findings from a state-initiated panel examining pathways to achieve targets of 50% renewable energy in Queensland, Australia, were contingent on federal policy support, thus limiting their effectiveness (Goddard and Farrelly 2018).

While the literature suggests that a just process must involve top-down and bottom-up collaboration and support to achieve optimal results, more research on successful examples of such coordination is needed (Evans and Phelan 2016; Weller 2019). Much of the literature on just transitions has focused on the global scale (Healy and Barry 2017). Yet transitions and related policy actions are regionally differentiated and regionally focused (Mertins-Kirkwood and Hussey 2020). Scale must be considered when examining options for low-carbon solutions in energy transitions (Goddard and Farrelly 2018).

What About Community Priorities and Concerns of Rural Places?

The concerns and priorities of workers often overlap with community concerns. Given that the majority of case studies on the topic of just transitions or sustainability transitions examine coal communities where workers are typically more deeply connected to place, it is often logical that workers and communities have common interest. However, a hyper-fixation on measures for workers means that community needs risk being overshadowed. The solutions for workers are not necessarily the same solutions for communities. For example, in Canada, the Alberta government has supported the coal phaseout by reimbursing the moving expenses of coal workers leaving to find new jobs – this same policy exacerbates community population loss and economic decline (Mertins-Kirkwood and Deshpande 2019). In order to achieve a just transition for communities, many advocate that new jobs should provide opportunities for the displaced workforce in the community wherever possible (Government of Canada 2019).

The added challenges faced by rural and small-town places (e.g., fewer services and amenities, less diversified economies, greater reliance on personal transportation, rural poverty, etc.) mean that solutions for communities must take a place-based approach to addressing specific challenges, determining unique assets, and considering priorities (Weller 2019). Tensions exist between rural and urban places where resource-dependent economies contribute royalties to support the economic well-being of the region. The economic stability of the region also influences receptiveness to climate action. Rural areas experiencing rapid development have also been found to be more in favor of conservation than those living in weak economies (Mayer 2018).

Communities (rural or otherwise) are complex ecosystems composed of unique subgroups and stakeholders who are also impacted by transitions on a social and personal level. In addition to workers in the impacted sector, families, businesses, service providers, contractors, local governments, nonprofit organizations, and marginalized groups have unique challenges as a result of transitions. The "ripple effects" can be wide reaching and extend beyond challenges related to economic diversification and employment. For example, historic examples of industrial transition have demonstrated that communities have faced population out-migration; increases in poor health and mental health conditions; increase in social issues such as poverty, addiction, domestic violence, and housing market slumps; increased cost of living; decline in available service providers; and other severe consequences that threaten their vitality (Cooling et al. 2015; Government of Canada 2019). Through careful consideration for procedural, distributional, and recognition elements of justice, just transitions could be achieved for communities by taking a place-based approach, while simultaneously reducing resistance to climate action and forwarding environmental objectives and progress (Newell and Mulvaney 2013). For example, full collaboration and engagement between communities and decision-makers, support for community led-initiatives, and socioeconomic analysis can help foster greater community support and buy-in to transitions (McCauley and Heffron 2018). Just outcomes must consider communities as unique units of study within greater socio-technical energy transition and sustainability transformations.

Not All Are Treated and Affected Equally: How to Include and Address the Needs of Marginalized Groups?

A significant portion of the literature on just transitions notes the importance of examining distributional, procedural, and recognitional justice in order to ensure measures do not perpetuate existing in equalities in society (Goddard and Farrelly 2018; Healy and Barry 2017). Many scholars advocate for there to be a more explicit consideration for social justice when managing just transformations as "societal transformations at any scale are shaped by, and will shape, the distribution of wealth, opportunities, and privileges afforded to different social groups" (Bennett et al. 2019, p. 3). White males dominate jobs in the fossil fuel sector (Pollin and Callaci 2019), therefore narrow attempts to find solutions for a displaced fossil fuel workforce may exclude underrepresented groups (Rosemberg 2010). Some literature suggests that solutions for displaced workers should also seek to provide opportunities for other groups facing employment barriers and strive toward gender and racial parity. This is particularly important in industries where there is uneven representation of groups in society which creates uneven distribution of benefits from transition solutions. For example, a study in the United States found that non-whites were underrepresented in the solar energy sector (Mayer 2018). This indicates there must be a concerted effort to address structural barriers and attract individuals to these sectors and professions, such as through affirmative action (Pollin and Callaci 2019). An equity analysis of the coal transition in Canada notes that public investments should be aimed at the community level to extend the benefits of transition to marginalized groups and potentially create opportunities for women, Indigenous peoples, racialized, and immigrant workers in these communities (Mertins-Kirkwood and Deshpande 2019).

An environmental justice analysis looking at the distribution of the harms of industrial development suggests that efforts to achieve a just transition could be an opportunity to address injustices experienced by marginalized communities. For example, the in the United States, energy production with higher pollution is typically found in socially deprived area with high levels of poverty and minority populations (McCauley et al. 2019). Just transition is an opportunity to address inequities caused by climate change, pollution, or environment degradation, given patterns of environmental racism (Newell and Mulvaney 2013).

Who Is Responsible?

To determine whether outcomes are just, it is important to determine who is responsible for enacting a just transition. There is a clear belief that governments have an ethical responsibility to support workers when the cause of transition is environmental regulations, policies, or government decisions (Pai et al. 2020), particularly when looking at just transitions from a jobs-focused lens. In the United States, residents blamed environmental policies as directly contributing to the decline of coal, a sentiment echoed by conservative politicians, although market forces also had an influence (Mayer 2018). Responsibility for managing and implementing just transitions becomes less clear when transition is caused by climate change or market forces. In an assessment of community responses to government intervention regarding coal transition in Australia, there was no difference in level of support for policies to support coal miners if transition was caused by market or by environmental regulations (Mayer 2018). It becomes even less clear when climate change is the reason behind transition. More research is needed on this topic as many industries are highly vulnerable to the impacts of climate change (Rosemberg 2010).

Nongovernmental actors have a role to play. For example, unions have been instrumental in ensuring fair distribution of workforce transitions (Healy and Barry 2017). While tensions between environmentalists and unions have existed in the past (largely as a result of the "environment vs. jobs" narrative); however, labor groups are now taking on new roles as advocates for environmental policies and see just transition as a vehicle for more equitable outcomes for workers (Stevis and Felli 2014). Labor organizations have pointed to the role that industry must play in helping with priority hiring of displaced workers (Canadian Labour Congress 2000).

Critically, the right composition of interests should be represented with adequate resources to fulfil responsibilities (Goddard and Farrelly 2018). Ultimately, it is important to have legitimate processes to achieve a just transition and to ensure proper procedural justice and ensure workers and unions are included in decision-making processes around transition measures and are involved in proactive industrial planning (Goddard and Farrelly 2018).

Can We Have Green Jobs and Good Work?

Economists predict that there will be a net growth in jobs as a result of decarbonization, including in transport, construction, buildings, renewables, agriculture, and technological development (International Trade Union Confederation 2017). A significant portion of workers will reach retirement age in the coming decades, and an estimated 85% of the fossil fuels jobs in the United States will be phased out through attrition (Pollin and Callaci 2019). This will help displaced workers and new workers entering the workforce find jobs. However, green jobs created may not be located in the same regions as jobs lost. This can create challenges, as regions need replacement economic activities and workers are often deeply connected to their communities, as witnessed in the Canadian coal context (Government of Canada 2019). Opportunities in replacement sectors in the industry may not offer the same kind of long-term employment associated with fossil fuel sector jobs. For example, wind and solar energy systems can create temporary construction jobs, but this does not lead to the kind of permanent employment in the surrounding communities compared to coal-fired electricity generation (Government of Canada 2019).

The element of good, decent work is also often left out of discussions around just transitions. New jobs and opportunities for displaced workforce and newly created green jobs must consider the interests and priorities of workers that are high-quality, attractive to people who lose employment in traditional industries, and maintain prevailing wage standards and labor agreements (Cha 2017). For example, veterans in the United States found returning from the Cold War were enrolled in the Defence Reinvestment and Conversion Initiative where they were placed in low wage and did not use skills from defense career (Pollin and Callaci 2019) Furthermore, the loss of coal jobs in Alberta means the loss of many unionized jobs in a province with already low rates of unionization (Mertins-Kirkwood and Hussey 2020). Workers may not be interested in new opportunities available for them. Labor organizations have also noted that new jobs should consider and use workers' existing skills wherever possible (Alberta Federation of Labour 2017; Canadian Labour Congress 2000). Attention should also be paid to opportunities that address skill shortages in other industries (Rosemberg 2010). For example, over half of construction workers will retire over the next decade (Mertins-Kirkwood and Deshpande 2019).

How Is Power Redistributed?

The political economy of fossil fuel-dependent regions influences the initiation, implementation, and success of efforts to achieve a just transition. Large fossil fuel corporations often have great influence over government decision-making by shaping energy policies and influencing energy transition options (Healy and Barry 2017). In Australia, industry lobby groups were successful in halting government funding for sustainability and renewable energy investments, given their reliance on affordable and reliable coal-fired electricity (Goddard and Farrelly 2018). In Canada, the Alberta government was not legally required to compensate companies as part of the phaseout of coal, yet they negotiated "Off Coal" Agreements totaling $1.36 billion with the three impacted power utilities as to not dissuade investment in the region (Mertins-Kirkwood and Hussey 2020). This was also due to the Alberta government's dependence on these private companies as the electricity market was deregulated in 1995 (Mertins-Kirkwood and Hussey 2020). These negotiations reflect the political economy of the provinces energy market which relies on a small number of private companies which provide an essential service, giving companies leverage in these negotiations (Mertins-Kirkwood and Hussey 2020). The lasting effects of a neoliberal policy era have also meant that governments are less willing to oversee a just transition as energy production and consumption are often shared with or delegated to the private sector (Newell and Mulvaney 2013).

The power of the fossil fuel industry can influence the level of ambition of climate policies, as well as workers' perceptions of transition. The hegemony around fossil fuel extraction and its importance to the economy are often renewed and defended in communities and among workers (Evans and Phelan 2016). Some companies have drawn on the "jobs vs. environment" narrative as a way to exploit the employment-related anxiety of workers who fear for their livelihoods (Goddard and Farrelly 2018; Pai et al. 2020). Furthermore, a study found that corporate media was proven to perpetuate this narrative, privileging economic considerations when discussing coverage of pipeline controversies (Hackett and Adams 2018). In order for energy transitions to be "just," carbon lock-in must be avoided as it is associated with several forms of injustice (Goddard and Farrelly 2018; Healy and Barry 2017). Divestment campaigns are noted in the just transition literature as grassroots means to challenge, redistribute, and rebalance power (Healy and Barry 2017).

Is There a Robust Social Infrastructure?

A common theme in the literature is the need for more robust social infrastructure, greater state intervention, and departure from neoliberalism and austerity policy regimes. Just transition should lead to greater state intervention as neoliberal economics have failed to deliver just outcomes (McCauley and Heffron 2018). Social protections could help reduce resistance to transition and "cushion" its impacts by ensuring guaranteed access to income supports and training (Healy and Barry 2017; Rosemberg 2010). For example, funding for rural, small town, and single-industry-reliant communities for diversification could support additional jobs and opportunities and create a need that displaced workers could fill. A full employment economy – one where there is a an abundance of decent jobs available for all people seeking work – would make it easier for workers to find and choose appropriate replacement jobs (Pollin and Callaci 2019). Furthermore, access to social services, including mental health and health supports, could help address some of the social impacts of transition (e.g., stress, mental illness, domestic violence, addictions) (Cooling et al. 2015; Government of Canada 2019).

How Does Community and Worker Identity Matter?

Worker identity and political beliefs can have a significant influence over receptiveness to climate action and transition (Mayer 2018). Conservative political views in industrial regions have been shown to contribute resistance to decarbonization (Healy and Barry 2017). Fossil fuel development has often been the source of economic stability and wealth regions for hundreds of years, particularly in coal producing regions (Evans and Phelan 2016). There is a strong sense of belonging to the places coal sector employees live and work, and these identities often form over generations (Pai et al. 2020). It is also essential to avoid grouping workers in the resource-based, industrial, or fossil fuel sectors into a homogenous group, as there are distinct identities, characteristics, and values of workers and communities in different sectors. In the oil and gas sector, there are higher rates of labor mobility, the work is more seasonal and transitory, and workforce demographics are somewhat different when compared to coal. These factors all shape beliefs about transition and willingness to think about change and transition, hence influencing the dialogue around "just transitions." It is important to acknowledge this identity and strategically manage as "effective climate action isn't likely to happen without a strong base among labour, and workers' rights aren't likely to flourish without people challenging the power of extractivist capital" (Hackett and Adams 2018, p. 30).

Low-carbon transition and sustainability transitions can be seen as threatening this identity, as well as livelihoods. There are potentially lessons to be honed from regions who have successfully redefined their economic identity, such as Ruhr Valley in Germany where inhabitants were deeply connected to the steel and coal manufacturing in the region but successfully shifted to a technology and service economy (Alberta Federation of Labour 2017). More research is needed to understand whether ideology and political identity have over individual's receptiveness to just transition measures (e.g., hesitation to accept "government handouts"; political alignment incompatible with big government/welfare state). Workers identity, political beliefs, and values must be understood and not perceived as being dismissed in order to counter trends such as the rise of right-wing populism (Hackett and Adams 2018). Just transition measures must consider the context of communities and people who depend on the sector.

Policy Learning Across Sectors?

Gaps exist in the just transition literature on sustainability transitions as they are almost exclusively focused on fossil fuels. There are lessons to be learned in other natural resource sectors where significant transformations have taken place (e.g., fisheries, agriculture, forestry), as well as other intensive emissions, trade exposed industries, or extraction and processing activities (e.g., automotive sector, mining). Most of the case studies pertaining to just transition have been in coal regions as there has been a global push to move away from coal creating momentum rendering coal uneconomic in many regions (Evans and Phelan 2016). However, there are examples that can be drawn from other cases of industrial transition, such as the forest sector in British Columbia which frequently experiences change due to boom-and-bust cycles. The province has responded in the past with initiatives for displaced workers, such as Forest Renewal BC which was an initiative that sought to find new jobs for displaced forest sector workers in silviculture and environmental restoration (Cooling et al. 2015). Furthermore, a case study of fisheries in British Columbia provides important lessons for understanding sound governance in transition management and integrating procedural justice considerations (Bennett et al. 2019).

When and for How Long?

A critical part of achieving just outcomes is understanding the timing and duration required to allow workers and communities to absorb information, accept the new circumstances, and prepare for the future. A framework for transitioning fossil fuel-dependent workers and regions in the United States suggests that a two-decade timeframe to achieve a 40% reduction in the oil and gas sectors and 60% reduction in coal is required to achieve climate targets (Pollin and Callaci 2019). However, this is based on a steady rate of contraction and does not account for sudden economic shocks which have historically resulted in added challenges and stress for workers and communities (Pollin and Callaci 2019). Long-term planning is essential to ensuring governments and states enacting transition are avoiding unnecessary costs from inaction (Pai et al. 2020). In energy transitions, replacement technology should also be readily available to better respond to sudden shifts and transitions to facilitate continued, reliable energy access (Delina and Sovacool 2018).

Both proactive and reactive policy supports are needed to help ensure just outcomes: proactive measures helping to maximize long-term benefits and reactive aimed at minimizing the harms of transition (Mertins-Kirkwood 2018). A transition managed proactively would involve more investments in the community and sustainable growth to create jobs and opportunities for both the displaced workforce and other groups in society (Mertins-Kirkwood 2018) Temporary supports that respond to immediate challenges are still needed in combination with a proactive approach.

More research is needed on proactive transition management as there are few examples internationally. One of the critical factors that hinder governments ability to manage transition proactively are the nature of electoral cycles, as priorities and policies often change with change of governments (Hackett and Adams 2018). There is a need to secure multi-partisan support for just transition and low-carbon approaches so that initiatives designed to help communities prepare over the long term survive a change of government (Goddard and Farrelly 2018).

Comprehensive, Multi-scalar, and Integrated Approaches Are Needed

In order to achieve a just transition, the question of "justice for whom" must be central to avoid leaning too heavily on jobs-, environment-, or society-focused solutions. A "just outcome" is different for different groups and stakeholders. More effort must be put into ensuring policies, and programs aimed at achieving a just transition have the greatest impact and benefit the widest cross section of people and address unique challenges. There is also a need to actively and purposefully draw from and incorporate other kinds of justice (e.g., social, environmental, climate, and ecological) to develop a comprehensive just transition plan. Gaps in the knowledge base reveal important considerations, such as the scale at which transitions take place, and the degree of intervention must take place (e.g., active, passive, transformative, minimalist) (Goddard and Farrelly 2018). A vision for a just transition is one where the move to a zero-carbon economy leads to "a future where all jobs are green and decent, emissions art at net zero, poverty is eradicated, and communities thriving and resilient" (International Trade Union Confederation 2017).

Conclusion

Achieving a just transition requires concerted action and effort across disciplines, industries, sectors, and jurisdictions. It requires a combination of proactive and reactive elements. While some actions are more "minimalist" in nature and fit within existing structures, a more transformative approach may be required. The nature of the policy and programs is not the only important factor – the level of ambition, timing, and depth of resources allocated toward these efforts have the ability to facilitate or hinder success.

Cross-References

▸ Amsterdam's Pathway to Climate Neutrality: Creating an Enabling Environment

References

Alberta Federation of Labour. (2017). Getting it right: A just transition strategy for alberta's coal worker. Accessed 5 Feb 2021 from https://digital.library.yorku.ca/yul-1121960/getting-it-right-just-transition-strategy-albertas-coal-workers

Bennett, N. J., Blythe, J., Cisneros-Montemayor, A. M., Singh, G. G., & Sumaila, U. R. (2019). Just transformations to sustainability. *Sustainability, 11*(14), 3881. https://doi.org/10.3390/su11143881.

Canadian Labour Congress. (2000). *Just transition for workers during environmental change*. Ottawa: Canadian Labour Congress.

Cha, J. M. (2017). A just transition: Why transitioning workers into a new clean energy economy should be at the center of climate change policies. *Fordham Environmental Law Review, 29*(2), 196–220. https://doi.org/10.2307/26413303.

Cooling, K., Lee, M., Daub, S., & Singer, J. (2015). *Just transition – Creating a green social contract for BC's resource workers*. Retrieved from www.policyalternatives.ca

Delina, L. L., & Sovacool, B. K. (2018). Of temporality and plurality: An epistemic and governance agenda for accelerating just transitions for energy access and sustainable development. *Current Opinion in Environmental Sustainability, 34*, 1–6. https://doi.org/10.1016/j.cosust.2018.05.016.

Evans, G., & Phelan, L. (2016). Transition to a post-carbon society: Linking environmental justice and just transition discourses. *Energy Policy, 99*, 329–339. https://doi.org/10.1016/j.enpol.2016.05.003.

Goddard, G., & Farrelly, M. A. (2018). Just transition management: Balancing just outcomes with just processes in Australian renewable energy transitions. *Appl Energy, 225*, 110. https://doi.org/10.1016/j.apenergy.2018.05.025.

Government of Canada. (2019). *Task force: Just transition for Canadian coal power workers and communities*. Gatineau: Government of Canada.

P

Hackett, R., & Adams, P. (2018). *Jobs vs the environment? Mainstream and alternative media coverage of pipeline controversies | Corporate Mapping Project*. Retrieved from https://www.corporatemapping.ca/jobs-vs-environment/

Healy, N., & Barry, J. (2017). Politicizing energy justice and energy system transitions: Fossil fuel divestment and a "just transition". *Energy Policy, 108*, 451–459. https://doi.org/10.1016/j.enpol.2017.06.014.

International Trade Union Confederation. (2017). *Just transition – Where are we now and what's next? A guide to national policies and international climate governance*. Retrieved from https://www.ituc-csi.org/just-transition-where-are-we-now

Mayer, A. (2018). A just transition for coal miners? Community identity and support from local policy actors. *Environmental Innovation and Societal Transitions, 28*, 1–13. https://doi.org/10.1016/j.eist.2018.03.006.

McCauley, D., & Heffron, R. (2018). Just transition: Integrating climate, energy and environmental justice. *Energy Policy, 119*, 1–7. https://doi.org/10.1016/j.enpol.2018.04.014.

McCauley, D., Ramasar, V., Heffron, R. J., Sovacool, B. K., Mebratu, D., & Mundaca, L. (2019). Energy justice in the transition to low carbon energy systems: Exploring key themes in interdisciplinary research. *Applied Energy, 233–234*, 916–921. https://doi.org/10.1016/j.apenergy.2018.10.005.

Meadowcroft, J. (2009). What about the politics? Sustainable development, transition management, and long term energy transitions. *Policy Sciences, 42*(4), 323–340. https://doi.org/10.1007/s11077-009-9097-z.

Mertins-Kirkwood, H. (2018). *Making decarbonization work for workers policies for a just transition to a zero-carbon economy in Canada*. Retrieved from https://www.policyalternatives.ca/publications/reports/making-decarbonization-work-workers

Mertins-Kirkwood, H., & Deshpande, Z. (2019). Who is included in a just transition? Considering social equity in Canada's shift to a zero-carbon economy. Retrieved from https://www.policyalternatives.ca/publications/reports/who-is-included-just-transition

Mertins-Kirkwood, H., & Hussey, I. (2020). A top-down transition: A critical account of Canada's government led phase-out of the coal sector. In E. Morena, D. Krause, & D. Stevis (Eds.), *Just transitions: Social justice in the shift towards a low-carbon world* (pp. 172–197). https://doi.org/10.2307/j.ctvs09qrx.14.

Newell, P., & Mulvaney, D. (2013). The political economy of the "just transition". *Geographical Journal, 179*(2), 132–140. https://doi.org/10.1111/geoj.12008.

Pai, S., Harrison, K., & Zerriffi, H. (2020). *A systematic review of the key elements of a just transition for fossil fuel workers*. Ottawa: Smart Prosperity Institute, WP, 20-04.

Pollin, R., & Callaci, B. (2019). The economics of just transition: A framework for supporting fossil fuel–dependent workers and communities in the United States. *Labor Studies Journal, 44*(2), 93–138. https://doi.org/10.1177/0160449X18787051.

Rosemberg, A. (2010). Building a just transition the linkages between climate change and employment. *International Journal of Labour Research, 2*(2), 125–161.

Stevis, D., & Felli, R. (2014). Global labour unions and just transition to a green economy. *International Environmental Agreements: Politics, Law and Economics, 15*(1), 29–43. https://doi.org/10.1007/s10784-014-9266-1.

Stevis, D., Morena, E., & Krause, D. (2020). Introduction: The genealogy and contemporary politics of just transitions. In E. Morena, D. Krause, & D. Stevis (Eds.), *Just transitions: Social justice in the shift towards a low-carbon economy* (pp. 1–31). London: Pluto Books.

UNFCC. (2015). Adoption of the Paris agreement. Accessed 5 Feb 2021 from https://unfccc.int/sites/default/files/english_paris_agreement.pdf

Weller, S. A. (2019). Just transition? Strategic framing and the challenges facing coal dependent communities. *Environment and Planning C: Politics and Space, 37*(2), 298–316. https://doi.org/10.1177/2399654418784304.

Policy – Direction

► Planning for Food Security in the New Urban Agenda
► Strategies for Taming the City
► Urban Policy and the Future of Urban and Regional Planning in Africa

Policy and Practices of Nature-Based Solutions to Build Resilience in Seoul, Korea

Yukyung Oh
King's College London, London, UK

Synonyms

Carbon neutrality; Nature-based solutions (NbS); Paris Agreement; Parks and Green Space; Urban resilience

Definition

This entry outlines how existing and new policies and practices regarding nature-based solutions have been developed in the Republic of Korea, to enhance its resilience to coupled environmental, social, and economic challenges. It suggests how the Republic of Korea has prepared for achieving the goals of the Paris Agreement, from the Framework Act on Low Carbon, Green Growth, to the Carbon Neutrality and Green Growth Act, along with the recent Action Plan 2021 to implement the 2050 Carbon Neutrality Strategy. It also considers the upcoming strategy of the Nature-Based Solution for Reducing and Adapting to Greenhouse Gas Emissions. Prior to the release of the strategy, Korea already developed NbS-related policies and practices at a city level (e.g., the Cheonggyecheon River Project, preservation of Royal Tombs as urban cemeteries, and Tancheon Ecological Landscape Conservation Area and its blue-green grid extension), while taking into account more citizen science approaches in urban greening projects. The chapter provides more sound evidence of South Korean NbS, not just focusing on one or two functions of green spaces, but on their multifunctionality.

Introduction

Nature-based solutions (NbS) become crucial in achieving the goals of the Paris Agreement, limiting global warming to well below 2 °C, preferably to 1.5 °C, above preindustrial levels. As countries should communicate in the plans for climate action, that is, nationally determined contributions (NDCs), to achieve the goals, it has become more crucial to set long-term strategies, while building resilience to climate change impacts. In this sense, NbS have been highlighted as a means to build urban resilience to climate change impacts, as it can encompass environmental, social, and economic challenges (Bush and Doyon 2019). NbS aim to "protect, sustainably manage, and restore natural or modified ecosystems, that address societal challenges (e.g., climate change, food security or natural disasters) effectively and adaptively, simultaneously providing human well-being and biodiversity benefits" (Cohen-Shacham 2016, p. 5). Even though the concept is not a new term but an umbrella concept, expanding from similar concepts such as green-blue infrastructure and ecosystem-based adaptation, more applications and criteria for NbS-related practices should be clarified, so that practitioners can adopt the concept to achieve the Paris Agreement goals. For instance, the International Union for Conservation of Nature (IUCN) released the first Global Standard for Nature-based Solutions, which is a self-assessment for users to adopt and apply NbS, along with the following eight criteria: (1) NbS effectively address societal challenges; (2) design of NbS is informed by scale; (3) NbS result in a net gain to biodiversity and ecosystem integrity; (4) NbS are economically viable; (5) NbS are based on inclusive, transparent, and empowering governance processes; (6) NbS equitably balance trade-offs between achievement of their primary goal (s) and the continued provision of multiple benefits; (7) NbS are managed adaptively, based on evidence; (8) NbS are sustainable and mainstreamed within an appropriate jurisdictional context (IUCN 2020). Such criteria can be useful for countries to prepare an NbS-related regulations and strategies.

Green-blue urban grids, which can be called green-blue infrastructure, with good quality, are crucial for improving urban resilience in a city. Urban resilience refers to "the capacity of a city and its urban systems (social, economic, natural, human, technical, physical) to absorb the first damage, to reduce the impacts (changes, tensions, destruction or uncertainty) from a disturbance (shock, natural disaster, changing weather, disasters, crises or disruptive events), to adapt to change and to systems that limit current or future adaptive capacity" (Ribeiro and Pena Jardim Gonçalves 2019, p. 4). In general, urban resilience is closely related to "the quality of buildings, the effectiveness of land use planning, the quality and coverage of key infrastructure and services, the effectiveness of early warning systems and public response measures," along with income levels of households and administrative capacities for supporting the socially vulnerable (IPCC 2014, p. 548). To handle diverse environmental, social, economic challenges in a sustainable manner (European Commission 2015), the concept of

NbS can be introduced to make Seoul become a resilient city to those challenges. As the South Korean Government has realized the necessity for preparing strategies for low-carbonization of cities and land and achieving carbon neutrality by 2050, by maximizing the function of ecosystems via the expansion and restoration of carbon sinks (for sequestering, absorbing, and storing carbon), the NbS approach is expected to serve a more crucial role in implementing carbon neutrality policies and strategies.

This chapter attempts to demonstrate how NbS concepts have been implemented in Seoul, the capital of the Republic of Korea, with a dynamic history of destruction during the Korean War, and much urban development since then. As of 2020, according to the Seoul Population Density (per district) Statistics of the Seoul Metropolitan Government, the entire population was 9,911,088, the entire area recorded 605.23 km^2, and the population density was 16,376 (persons/km^2). Parks and green spaces in Seoul were managed within the state-led urban planning policy framework from 1962 to 1993, as the Seoul Mayors were appointed by the central government during the period. Yet, since 1993, under the mayors elected by the people, parks and green space policies have been developed based on citizen science.

As Seoul is preparing for providing a long-term Strategy of the Nature-Based Solution for Reducing and Adapting to Greenhouse Gas Emissions with more details, this entry is significant in providing a more coherent point of view of parks and green space policies, as well as NbS-related cases. This entry illustrates the development of the national Climate Change Adaptation Plans, as well as policies, strategies, the legal foundation for carbon neutrality, and NbS-related plans at a national level (section "Roadmap Towards a Carbon Neutral Country"). Section "Seoul's Green Space Policy and Status for Adopting NbS Approaches" handles parks and green area-related policies and practices related to NbS at a city level. Section "Case Studies of Nature-Based Solution " presents two NbS-related case studies in Seoul: Seolleung and Jeongneung Royal Tombs, and Tancheon Ecological Landscape Conservation Area.

Roadmap Toward a Carbon Neutral Country

After its entry into the United National Framework Convention on Climate Change in 1993, the Republic of Korea has established governmental department-led systems and strategies for responding to climate change: four Comprehensive Plans for Combating Climate Change between 1999 and 2012, the establishment of the Countermeasure Committee for Climate Change in 2001, and the operation of a working organization for responding to climate change under the Prime Minister's Office (Planning Office for Climate Change of Prime Minister's Office 2008). The country enacted the Framework Act on Low Carbon, Green Growth in 2010, in a situation where a new engine for growth was needed while facing such challenges as global economic crisis, climate change, and energy issues. As stated in Article 1, the Act aims to achieve economic development by preparing a base for low carbon and green growth, and for new growth drivers via green technology and industries. Ultimately, the goal is to improve the quality of life of the people by making the balance between the economy and environment, and by realizing a low-carbon society.

Under Article 40 of the Act, the government should develop or implement a basic plan for mitigating climate change impact for a planning period of 20 years, while handling such matters as tendencies and forecasts of climate changes in Korea and other countries, changes in GHGs in the atmosphere, status, and outlook of the emission and absorption of GHGs, setting medium- and long-term targets for reducing GHG emissions, cooperation at national and international levels for coping with climate change, etc. The Act was followed by South Korea's first national climate change adaptation measure, the "Comprehensive Climate Change Adaptation Plan" in 2008. Based on Article 48 (4), as well as Article 38 (1) of the Enforcement Decree of the Framework Act on Low Carbon, Green Growth, the Ministry of Environment announced the first National Climate Change Adaptation Plan – the first legal plan, in October 2010 – and shall develop adaptation measures for each five-year period and deal with the abovementioned matters, while consulting with the heads of relevant central administrative agencies.

National Climate Change Adaptation Plans have been developed in phases based on the abovementioned Act, further leading to the contribution to carbon neutrality by 2050. Table 1 indicates its planning period, vision, objectives, structure, tasks, a list of science-based risks, and examination and evaluation for each climate change adaptation plan. The plans have shown more detailed tasks and risks, while taking more science-based and participatory approaches, starting from a broader scale of adaptation plan (i.e., the Comprehensive Climate Change Adaptation Plan) to the recent Plan. The most remarkable point is that the Plans have not solely led by one specific department, but diverse government ministries in charge of corresponding tasks and risks. In addition to local governments' higher role in implementing tasks at a local scale, public participation has been more encouraged to establish countermeasures for adaptation.

As an effort to move beyond existing adaption and mitigation approaches, South Korea has striven to provide a more advanced legal foundation, as well as policies and strategies for achieving carbon neutrality by 2050. For instance, the Carbon Neutrality and Green Growth Act was passed by the country's National Assembly in September 2021. Seoul has prepared for providing an advanced legal foundation, in order to achieve the 2050 goal and cut GHG emissions by 40% from its 2018 levels, from its previous target of 26.3%, by 2030. Even though the Act is replaced by the Framework Act on Low Carbon, Green Growth from 25 March 2022, relevant master plans and adaptation plans are effective until they are completed under the Framework Act.

Prior to this lawmaking approach, in 2020, Korea announced the 2050 Carbon Neutrality Promotion Strategy, and the Long-term Low Greenhouse Gas Emission Development Strategies (LEDS), in order to accelerate to achieve net-zero GHG emissions by 2050 in December 2020. The 2050 Carbon Neutrality Promotion Strategy, the country's net zero roadmap, has the vision of simultaneously achieving carbon neutrality, economic development, and better quality of life, via a "proactive response" rather than "adaptive reduction" with four policy directions (Government of Korea 2020b):

1. Low carbonization of the economic structure (Corresponding tasks: Accelerated energy transition, Innovation of high-carbon industrial structure, Transition to future mobility, and Low carbonization of cities and land)
2. Creation of innovative low-carbon industry ecosystems (Corresponding tasks: Fostering a new promising industry, Building the foundation of the innovation ecosystem, and Activation of a circular economy)
3. Fair transition to a carbon-neutral society (Corresponding tasks: Protection of vulnerable industries and classes, Realization of community-oriented carbon neutrality, and Raising public awareness of a carbon-neutral society)
4. Reinforced systemic foundation for carbon neutrality (Corresponding tasks: Finances, Green finance, R&D, and International cooperation via the reinforcement of carbon price signal and establishment of foundation for more investment in carbon-neutral sectors)

LEDS is the 2050 Carbon Neutral Strategy of the Republic of Korea. It is a long-term vision for responding to climate change, since "all Parties should strive to formulate and communicate long-term low greenhouse gas emission development strategies, mindful of Article 2 taking into account their common but differentiated responsibilities and respective capabilities, in the light of different national circumstances" (Article 4.19 of the Paris Agreement) to the UNFCCC by 2020. The strategy has five main factors for a sustainable and green society, while specifying sectoral strategies in energy, industry, transportation, building, waste, farming, and carbon sinks: "(1) Expanding the use of clean power and hydrogen across all sectors, (2) Improving energy efficiency to a significant level, (3) Commercial deployment of carbon removal and other future technologies,

Policy and Practices of Nature-Based Solutions to Build Resilience in Seoul, Korea, Table 1 National Climate Change Adaptation Plans from the Comprehensive to the Third Plan

	Comprehensive Climate Change Adaptation Plan	The First National Climate Change Adaptation Plan	The Second National Climate Change Adaptation Plan	The Third National Climate Change Adaptation Plan
Planning Period	2009–2030	2011–2015	2016–2020	2021–2025
Vision	Establishment of a safe society and support for green growth through adaptation to climate change	Establishment of a safe society and support for green growth through adaptation to climate change	Establishment of a safe and happy society for the people through adaptation to climate change	Realization of a climate-safe nation for the people
Objectives	Short term (– 2012) : Reinforcement of adaptive capacities to climate change in a comprehensive and systematic manner Long-term (– 2030) : Lower climate change risks and realization of opportunities	–	Lower risks and realization of opportunities from climate change	- Higher climate resilience in all social sectors in preparation for a 2 °C increase in global temperature - Promotion of science-based adaptation by establishing climate monitoring and forecasting infrastructure - Realization of adaptation mainstreaming in which all adaptation actors participate
Structure	- Establishment of climate change risk assessment system - Implementation of climate change adaptation programs in six sectors (i.e., ecosystem, water management, health, disaster, adaptation industry, and energy SOC) - Securing domestic and international cooperation, and institutional foundation	- Promotion of corresponding tasks for each sector (i.e., seven sectors including health, disaster, agriculture, forest, marine/ fisheries, water management, and ecosystem) - 13 central ministries involved (later 14 ministries) including the Ministry of Environment, Ministry of Health and Welfare, Ministry of Land, Infrastructure, and Transport, and Ministry of Agriculture, Food and Rural Affairs, as well as 70 experts in relevant fields - Adaptation-based measures 1. Climate change monitoring and prediction	- Establishment of the plan led by the government (20 ministries involved) and experts - Four Policies 1. Scientific risk management 2. Construction of a safe society 3. Securing industrial competitiveness 4. Sustainable management of natural resources - Implementation Basis 1. Preparation for implementation basis in Korea and other countries	- Three policy-based tasks (232 policies), core strategies (improvement of climate resilience, protection of the vulnerable, activation of citizen participation, and responses to the new climate system), and public-oriented tasks in eight sectors - All implementation actors (e.g., experts, industry, local communities, civil society, and youth groups) in climate change adaptation, participated in governance forums, a working council on climate change adaptation, online public hearings, and discussions on preparing public-oriented

(continued)

Policy and Practices of Nature-Based Solutions to Build Resilience in Seoul, Korea, Table 1 (continued)

	Comprehensive Climate Change Adaptation Plan	The First National Climate Change Adaptation Plan	The Second National Climate Change Adaptation Plan	The Third National Climate Change Adaptation Plan
		2. Adaptation industry/energy 3. Education, public relations, and international cooperation		countermeasures for establishing measures
Selection of tasks	–	–	- Priority selection of ministries' projects - Combination of short- and long-term measures, without high-level administrative planning	- Discovering customized implementation tasks per risk - Specific action plans for 5 years in line with the Basic Plan for Climate Change Response
Construction of a list of science-based risks	–	–	Since 2019, the Korean government constructed a list of national climate risks, based on expert surveys (87 risks in eight sectors)	Establishment of science-based (research paper DB, and causal loop diagram) national risks (84 risks in six sectors)
Examination and Evaluation	- Each ministry establishes and implements an implementation plan every year, evaluating the results and submitting the plan for the next year to the government department in charge - The department reports the aggregated results to the Climate Change Countermeasures Committee - The department constructs a database of annual implementation plans and results of each department, to build a comprehensive and systematic national inventory of adaptation policies and practices, and then prepares a national climate change adaptation assessment report	- Organization of the "Working Council of Government Departments on Climate Change Adaptation" to make major decisions on adaptation measures and collect expert opinions - Operation of working committees for adaptation of relevant ministries and local governments (consultation and evaluation of the performance of adaptation measures) - Supplementation of the basis for implementing adaptation measures (e.g., legal revisions and system improvement) - Operation of the Korea Adaptation Center for Climate Change	- Examination and evaluation of the operation of expert forums -Performance-oriented examination and evaluation of detailed implementation tasks * Based on Article 38 (1) of the Enforcement Decree of the Framework Act on Low Carbon, Green Growth, performance for 2016 was first examined in 2017, followed by consistent monitoring on tasks in 2018 and 2019	- Examination and evaluation of the operation of the citizen evaluation group - Implementation of public-oriented policies, and examination and evaluation of public perceptions

Source: Government of Korea (2008, 2010, 2015, 2020a), restructured by the author. N.B. Unless otherwise specified, all translations from the Korean original are the author's own

(4) Scaling up the circular economy to improve industrial sustainability, (5) Enhancing carbon sinks" (The Government of the Republic of Korea 2020, p. 9).

The Ministry of Environment announced the so-called Action Plan 2021 so as to implement the 2050 Carbon Neutrality Strategy in March 2021. In September 2021, the Ministry also held the Nature-Based Solution Forum to achieve the 2050 Carbon Neutrality in September 2021. The implementation actors (government, national agencies, academia, civic organizations, etc.) participated in the forum, to discuss the expansion of sustainable carbon sinks through NbS and preparation of comprehensive management measures, solutions to social problems. The Ministry has considered the solution as an approach to effectively and flexibly solve social issues such as climate change by protecting and sustainably managing ecosystems. Such recognition can be found in the Action Plan, as it indicated the strategy of the Nature-Based Solution for Reducing and Adapting to Greenhouse Gas Emissions, which categorized items and details for nature/ecology-based strategies. The items encompass: (1) GHG reduction based on Nature and Ecology, (2) expansion and management of ecosystem protection areas, (3) ecological restoration of the entire land, (4) promotion of ecosystem services in urban areas, (5) climate change adaptation based on Nature and Ecology, and (6) (Other) infrastructure (see Table 2). Yet the final version of the long-term strategy with more details is planned to be released in December 2021, after

Policy and Practices of Nature-Based Solutions to Build Resilience in Seoul, Korea, Table 2 Categorized items and details (draft) for nature/ecology-based strategies for GHG reduction and adaptation

Category	Details (Draft)
GHG reduction based on Nature and Ecology	More carbon sinks via forest utilization (Korea Forest Service)
	More carbon sinks in the national environment, such as cities, rivers, and wetlands
	Carbon absorption using marine ecological resources such as wetlands (Ministry of Oceans and Fisheries)
Expansion and management of ecosystem protection areas	Expansion of terrestrial protected areas
	Expansion of coastal protected areas (Ministry of Oceans and Fisheries)
Ecological restoration of the entire land	Restoration of river ecosystems
	Restoration of marine and wetland ecosystems (Ministry of Oceans and Fisheries)
	Restoration of terrestrial ecosystems
	Restoration of species
Promotion of ecosystem services in urban areas	Improvement of water circulation
	Creation of urban micro-ecosystem
Climate change adaptation based on Nature and Ecology	Creation of a water environment in response to the climate crisis
	Ecosystem management responding to the climate crisis
	Coastal disaster prevention and management (Ministry of Oceans and Fisheries)
	Response to agricultural and fishery disasters (Ministry for Food, Agriculture, Forestry and Fisheries, and Ministry of Oceans and Fisheries)
(Other) Infrastructure	Research and Development
	Monitoring and database construction
	Reflection of national and local government plans
	Discovery of industries using NbS
	Reinforced international cooperation

In the case of the restoration of terrestrial ecosystems, expansion of terrestrial protected areas, and the creation of urban micro-ecosystems, there will be more details in cooperation with relevant ministries such as the Ministry of Land, Infrastructure and Transport, the Ministry of Oceans and Fisheries, and the Korea Forest Service
Source: Ministry of Environment (2021), Translated by the Author

collecting opinions from all walks of life including the Forum. In sum, the South Korean government has provided the legal framework and action plans to implement carbon neutrality strategies, along with the foundation for developing NbS strategies to address social challenges in a more integrated manner.

Seoul's Green Space Policy and Status for Adopting NbS Approaches

Even though South Korea does not have specific laws and regulations relevant to green infrastructure – the preexisting term of NbS – there are regulations that have led to the creation, conservation, and expansion of green spaces, which is the crucial and basic element of green infrastructure. The most relevant of these are: the Act on Urban Parks, Green Areas, etc.; the Act on Special Measures for Designation and Management of Development Restriction Zones; the Landscape Act; the Natural Environment Conservation Act; the National Land Planning and Utilization Act; and the Natural Park Act.

Considering different socioeconomic and environmental contexts, the definition of green space should be clarified. In particular, South Korea employs terms of parks, green spaces, or green areas in environmental and land planning regulations and policies. According to the Act on Urban Parks, Green Areas, Etc., "the term of green areas refers to a greenbelt provided for in subparagraph 6 (b) of Article 2 of the National Land Planning and Utilization Act, which is designated according to an urban or county (*gun*) management plan provided for in Article 30 of the same Act to improve the urban landscape by preserving or improving the natural environment and preventing environmental pollution and natural disasters in urban areas." Green areas or spaces can be divided into buffer, landscaping, and connected green areas. Each type has its own objective. Buffer green areas serve to prevent air pollution, noise, odor, and other various disasters such as accidents or natural disasters. Landscaping green areas offer a better urban landscape by restoring or improving damaged natural environments, or preserving natural environments in urban areas. The so-called connected green area serves to organically connect parks, rivers, and mountains in a city, and to provide city dwellers with leisure and relaxation via walking spaces (Ministry of Land Infrastructure and Transport 2021). These are the main types of green space known/planned in South Korea.

When it comes to green space-related policies and practices in Seoul, the city has so far focused more on the expansion of parks, which is still the current priority (i.e., creation of smaller-scale of parks), rather than the qualitative development of green spaces. In the late 1980s, the city started to secure the health of ecosystems, for instance, by designating ecological landscape conservation areas. More recently, the Seoul Metropolitan Government has shown a tendency of focusing on degraded urban green spaces, fair distribution of green and park areas, and more citizen participation for leading urban greening projects. On the riverine island of Nanjido, for example, a recovery project sought to create an ecological park out of a landfill, which had long occupied the island. The city also launched urban greening campaigns to encourage and support citizens in planting flowers or trees within living zones, as well as rooftop gardens.

Institutionally, 1993 saw an important change. From 1962 to 1993, parks and green spaces in Seoul were managed within the state-led urban planning policy framework, as the Seoul Mayors were appointed by the central government during this period. In 1993, however, when the mayor became a democratically elected position, parks and green space policies have been developed based on citizen participation. As part of an effort to introduce the concept of connected green and blue infrastructure, the Seoul Metropolitan Government released the first Master Plan for parks and green spaces, "2030 Seoul Parks and Green Space Master Plan" in 2015.

This plan constituted a milestone in green space policy for Seoul. It suggests a vision of parks and green spaces for 2050 Seoul while being consistent with the Seoul 2030 Master Plan (see Fig. 1), a plan for a long-term urban development based on the Act on Planning and

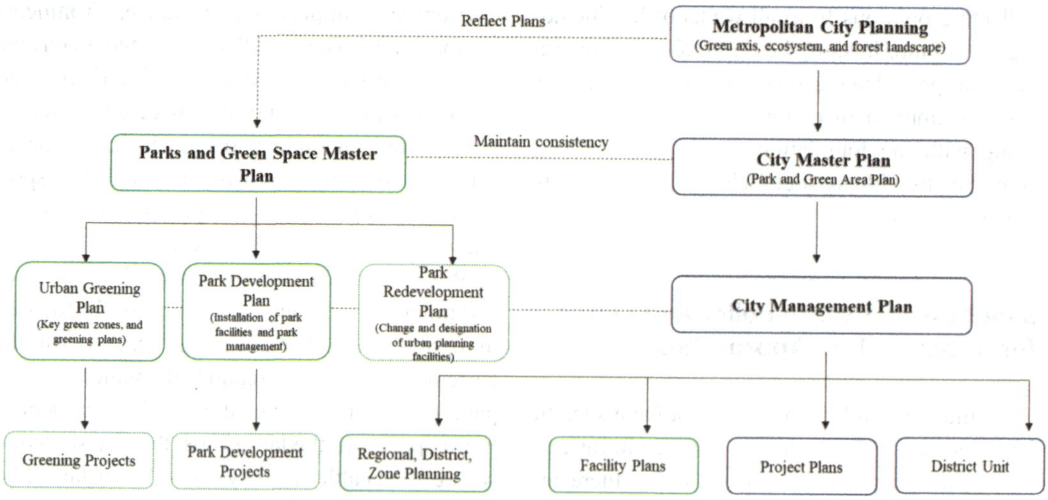

Policy and Practices of Nature-Based Solutions to Build Resilience in Seoul, Korea, Fig. 1 Status of 2030 Seoul Parks and Green Space Master Plan. (Source: translated by the Author, and adapted from the 2030 Seoul Parks and Green Space Master Plan (Seoul Metropolitan Government (2015)), The reuse of this Report is authorized under a Creative Commons Attribution-NonCommercial-NoDerivs 3.0 Unported (CC BY-NC-ND 3.0) license, https://creativecommons.org/licenses/by-nc-nd/3.0/deed.en)

Use of National Territory. Under the Act on Urban Parks, Green Areas, etc., this plan was developed based on citizen participation (e.g., citizens' public hearing, advice from the city park committee, and deliberation of a city planning committee). It suggests a policy direction for developing urban environments, which includes providing citizens with spaces for relaxation and recreation, accessible parks with higher satisfaction levels. It also includes a framework for parks and green spaces encompassing a management system, setting criteria or altered measures for urban parks in line with changing urban management plans, and, importantly, a program to identify all sorts of green areas in order to create a central database. Finally, the 2030 Plan proposed measures to restore damaged parks and green spaces through the analysis of disconnected green spaces, as well as connecting green spaces or creating green networks, and restoring damaged ecosystems (Seoul Metropolitan Government 2015). It is important to note that the City Master Plan continues to determine the alteration and designation of park areas and urban planning facilities: if the contents of the Parks and Green Space Master Plan are different from those of the City Master Plan, the latter has priority (Seoul Metropolitan Government 2015).

Seoul has achieved a middling rank/score on indicators such as quality of public life, psychological comfort and satisfaction, and physical fitness which are provided by public green spaces. As seen in Fig. 2, Seoul shows a fairly reasonable proportion of public green space (27.8%); Istanbul (2.2%), Taipei (3.4%), and Bogota (4.9%) have less than 5% of public green spaces, whereas Oslo has the largest public green space with 68%. Some cities such as Helsinki (40%), Hong Kong (40%), Stockholm (40%), Shenzhen (40.9%), Vienna (45.5%), Sydney (46%), and Singapore (47%) record higher public green spaces than others. Public green space can serve as a tool for comparing green space status and characteristics among cities, as it contributes to the better quality of public life, in terms of psychological comfort and satisfaction, and physical fitness (Morar et al. 2014; Nasution and Zahrah 2012).

In detail, regarding Seoul, according to the Statistics on the Status of Green Spaces in Seoul in 2019, the city had 7862 green spaces with 15,696,410 m^2. These spaces are highly diverse in composition. They encompass several categories: so-called reserved green spaces (i.e., protected or conserved space on which development is

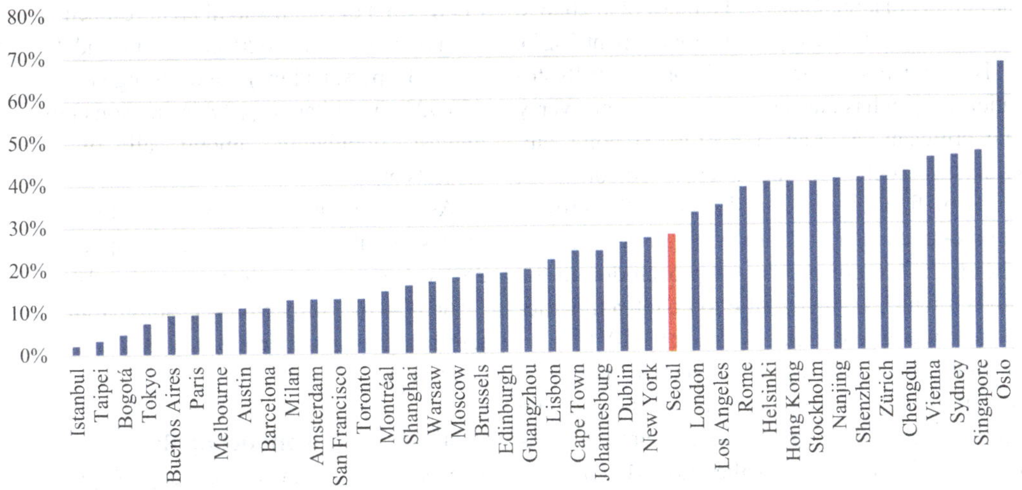

Policy and Practices of Nature-Based Solutions to Build Resilience in Seoul, Korea, Fig. 2 Public Green Space Proportion of World Cities (parks and gardens). (Source: Author's work, based on data from World Cities Culture Forum, © Copyright World Cities Culture Forum 2021, http://www.worldcitiescultureforum.com/data/of-public-green-space-parks-and-gardens)

limited); standard green spaces (including green spaces on plazas and streets); green spaces in road median strips (dividers); green wall establishments; forest areas; riverside landscaping; rest areas; areas around subway vents; areas around apartment buildings and schools; establishment of water-friendly spaces; and a rest category, "others." Of these, "standard" green spaces accounted for the largest category, with 2560 spaces amounting to a total 5,156,593 m^2, whereas the category of areas around subway vents was the smallest, at 67 spaces amounting to a total 16,219 m^2.

Case Studies of Nature-Based Solution

Compared to green spaces, nature-based solutions have only recently entered South Korea's policy toolbox. There are two well-known NbS-related projects in South Korea, both focused on rivers: the Four Major Rivers Restoration Project (Oct. 2009–Dec. 2012 over a total river length of 1266 km) and the Cheonggyecheon River Project (project period: July 2003–Dec. 2005, total length: 5.84 km). They are interesting to review briefly because they are so well-known.

The two projects differ greatly in size and scope. The Cheonggyecheon River is in reality a short stream running through northern Seoul, while the Four Major Rivers are South Korea's biggest rivers. The Cheonggyecheon River project, which involved restoring a stream that had been covered up with concrete and as a result had run dry, has nevertheless attracted by far the most international media attention. It was primarily a socioeconomic and cultural project of urban renewal, rather than an environmental policy. Nonetheless, the Cheonggyecheon's restoration has had several environmental benefits in terms of more biodiversity, better air quality (NO: 51.0 to 46.0PPB, and PM: 71.0 μg/m^3 to 60.0 μg/m^3) (K. T. Kim and Song 2015), and mitigation of the urban heat island effect (around 0.4 °C decrease on average) (K. T. Kim and Song 2015; Y. H. Kim et al. 2008). Also, fauna and flora in the stream increased from 342 in 2006 to 552 in 2017; the number of plants (e.g., *Chionanthus retusa, Salix koreensis Andersson,* and *Miscanthus sacchariflorus*) (160 in 2006 to 375 in 2017), the number of fish (e.g., *Zacco platypus, Pungtungia herzi,* and *Cyprinus carpio*) (21 in 2006 to 25 in 2017), the number of birds (e.g., *Anas platyrhynchos, Ardea cinerea,* and *Egretta alba modesta*) (19 in 2006 to 43 in 2017), the number of benthic macroinvertebrates (27 in 2006 to 40 in 2017), the number of terrestrial insects (113 in 2006 to 69 in 2017), and the number of amphibians and reptiles (2 in 2006 to 0 in 2017) (Seoul Metropolitan Government 2021).

By comparison, the Four Major Rivers Restoration Project has faced much greater controversy and criticism. As Lah et al. (2015) point out, criticisms center on the unclear outcomes in terms of flood prevention, lowering the growth of blue-green algae, and improving water quality. The strong top-down policy approach during the project, which involved little to no citizen participation, was also criticized (Lah et al. 2015).

Nonetheless, the Cheonggyecheon and Four Major Rivers restoration projects can be interpreted as NbS that intervene in natural environments through physical alteration. By contrast, this entry focuses on NbS with less physical alternations. It therefore considers two different and less well-known case studies: Seolleung and Jeongneung Royal Tombs, and Tancheon Ecological Landscape Conservation Area. The target sites are located south of the Han River that divides Seoul in two, specifically in the districts (*gu*) of Gangnam and Songpa. These districts account for 6.53% (39.5km^2) and 5.6% (33.87km^2) of Seoul's total area, respectively. As for population, in 2020 Gangnam had 544,055 (13,773 persons/km^2) and Songpa recorded 673,926 inhabitants (19,896 persons/km^2) the highest population among all of Seoul's districts at.

As shown in Fig. 3, Gangnam-*gu* had the highest number and area of green spaces among 25 local government districts in Seoul (411, and 1,566,004 m^2, respectively). Songpa-*gu* has the fourth-highest number and area (332, and 991,527 m^2, respectively).

Seolleung and Jeongneung Royal Tombs

The Seolleung and Jeongneung Royal Tombs (Seonjeongneung, in the Korean abbreviation) are a complex of three tombs from the Joseon Dynasty (1392–1910). They are examples, albeit ceremonial ones, of urban cemeteries: urban green spaces with both cultural and natural features. South Korea is not unique in considering cemeteries as green spaces. Cemeteries are regarded as green infrastructure with cultural history in Scandinavian countries, and as green spaces with

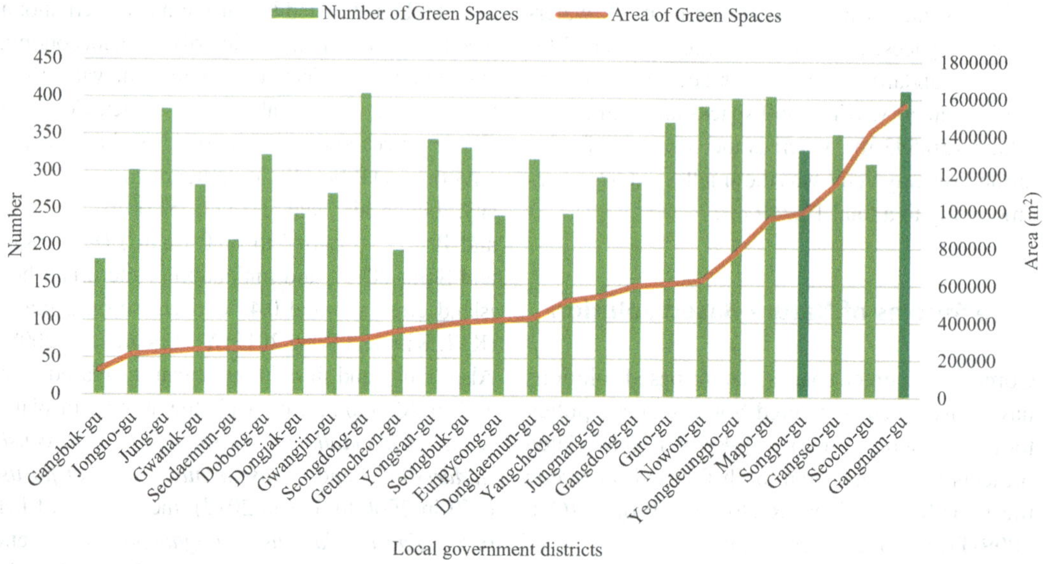

Policy and Practices of Nature-Based Solutions to Build Resilience in Seoul, Korea, Fig. 3 Number and area of green spaces by local government district in Seoul in 2019. (Source: Author's work, based on data from "the Statistics on the Status of Green Spaces in Seoul in 2019 (in Korean)" by Landscape Planning Division, Green Seoul Bureau of Seoul Metropolitan Government, The reuse of this data is authorized under a Creative Commons Attribution- Attribution 2.0 Korea (CC BY 2.0 KR) license, Creative Commons — Attribution 2.0 Korea — CC BY 2.0 KR)

recreational functions in Denmark (Nordh and Evensen 2018). As a multifunctional landscape, urban cemeteries have social and recreational values (Al-Akl et al. 2018), and have a complex structure with such elements as burial space, vegetation, and spatial element (Anna and Ewa 2020).

There are several types of urban cemeteries in Seoul, but royal graveyards are unique as they have historical, cultural, and environmental features. There are 40 Royal Tombs of the Joseon Dynasty (the Dynasty lasted 518 years from 1392 to 1910, producing 27 kings) in Seoul, built by the twenty-seven kings of the Joseon dynasty to honor the royal ancestors and their achievements, and based on the principles of geomancy. They are spread over 18 locations in Seoul and its suburbs. The tombs were inscribed on the UNESCO World Heritage List in 2009 (Cultural Heritage Administration 2021).

Among the burial complexes, Seonjeongneung harbors the tombs of King Seongjong, the ninth Joseon ruler (1469–1494), his Queen Jeonghyeon, and their son King Jungjong, the eleventh ruler (1506–1544). In the twentieth century, Seonjeongneung underwent much change and suffered extensive damage. These include reclamation before and after the Korean War, and even conversion to fields and farms. Although public access and cultivation were prohibited in the area of the tomb, during the Japanese colonial period (1910–1945), the Japanese Governor-General of Korea allowed the land reclamation of the low-lying areas inside the tomb area, leading to a conversion of the forest into cultivated land. To be specific, the area was reduced by approximately 85% over approximately 45 years (1,711,917.36 m^2 in 1960, 198,813 m^2 in 1973, and 240,588.43 m^2 in 2008) due to intensive development of neighboring areas in the district, as well as an urban development plan (Ahn 2016).

The 1980s saw a change to sustainable management, with the planting of trees such as *Alnus japonica Steudel* or *Pinus Densiflora*. This urban space now serves as a crucial urban green infrastructure in terms of cultural, recreational, historical, and environmental benefits. In terms of tree species and green area, the area has 85.7% of historical landscape forest, and 8.9% of sacred space; 94% of the total area is covered by green space, with *Quercus* spp. (more than 50%), *Pinus Densiflora* (20%), *Zelkova serrata* (11%), and other tree species (Ahn 2016). It also provides visitors – mainly workers from the neighboring business districts – with a place to rest, mediate, and walk in the highly compact city. It also doubles as a history education site. Seonjeongneung also has a significant mitigating effect on the urban heat island. According to Kim et al. (2016), which targeted the Seolleung and Jeongneung area between 2003 and 2005, the maximum temperature difference between urban and green areas was about 2.9 °C in summer, and the minimum of 1.7 °C in winter; it underpins the cooling effect of this royal tomb.

Furthermore, this type of green space is now protected to prevent further damage or preserve the sacred area, under Article 13 of Cultural Heritage Protection Act: "an administrative agency in charge of the authorization, permission, etc. of construction works shall examine whether such construction works are likely to affect the preservation of designated cultural heritage before granting authorization, permission, etc. for the construction works. In such cases, the administrative agency concerned shall consult the relevant experts, as prescribed by Presidential Decree" (Article 13 (2)): "the scope of the preservation area shall be within 500 meters from an outer boundary, in consideration of the cultural, artistic, academic, and scenic value of the relevant designated cultural heritage, its surrounding environment, and other necessary matters for the protection of cultural heritage: Provided, That where construction works implemented in an area 500 meters away from an outer boundary of designated cultural heritage are deemed to affect the cultural heritage due to its characteristics, locational conditions, etc., the scope thereof may be set in excess of 500 meters" (Article 13 (3)).

Tancheon Ecological Landscape Conservation Area

Tancheon is a river in the Greater Seoul area and tributary of the Han River. The length of the stream is 35.6 km, from the city of Yongin in Gyeonggi Province (surrounding Seoul) to districts of Songpa and Gangnam districts. In particular, its lower reach is known as the habitats of migratory birds during the winter season, aquatic species, and wild fauna (e.g., little erget (*Egretta garzetta*) and gray heron (*Ardea cinerea*) (Choi et al. 2017)). During the period through to the turn of the century, prior to its designation as a Landscape Conservation Area, the water quality of the stream deteriorated due to reckless development in the surrounding areas, which produced major inflows of domestic sewage and oil and sand from construction sites. However, since the 2000s, the Seoul Metropolitan Government has implemented projects to improve river environments in order to restore the physical properties and ecological functionality of damaged rivers. The Tancheon area is one of the most successful cases of the project, and continuous monitoring is in progress.

In detail, the Seoul Metropolitan Government designated several Ecological Landscape Conservation Areas in which there are high ecological importance and special value for conservation due to the abundant biodiversity, so as to protect those areas from any contamination and damage (pursuant to the Natural Environment Conservation Act, and the Natural Environment Conservation Ordinance). Among 17 designated areas (in total 4,807,327m^2), Tancheon (a tributary of the Han River) Ecological Landscape Conservation Area, which was designated in 2002, has served as a place to protect the ecosystem and nature.

According to Lee et al. (2010), in the target areas of 6.7 km (1,404,636 m^2) from Tanchoen Daegok Bridge to Tancheon Second Bridge, there were 308 florae (e.g., *Miscanthus sacchariflorus, Phragmites australis, Pennisetum alopecuroides,* and *Echinochloa crus-galli.*; *Salix koreensis Andersson, Salix gracilistyla Miq,* and *Robinia pseudoacacia L.*); 5 families and 7 species for mammals; 25 families and 33 species for birds; 3 families and 4 species for amphibians; 3 families and 4 species for reptiles; 58 families and 147 - species for insects; 5 families and 168 species for fish; 22 families and 28 species for benthic macroinvertebrates.

Beside ecological restoration and conservation, this area has been used as a space for recreational or other physical activities for residents. However, as this area has been designated as an Ecological Landscape Conservation Area, this section was disconnected to other causeways for the last five decades. In 2021, this area is now connected to other causeways, completing a unique "circular walking path" along rivers, namely the "Songpa Dulle-gil" of a total length of 21 km. The path was created without damaging the ecological landscape conservation area, along with four ecology observatories to monitor or watch (migratory) birds and spaces for relaxation. Rather than the top-down approach, the path was created based on citizen participation, after presentations for residents, collection of expert opinions in relevant fields to environmental science, birds, and landscape; during and after the project, residents voluntarily participated in monitoring the project to create a nature-friendly walking path. The connection project is continuously expanding to other areas to activate local economy as well.

Conclusion

This entry attempted to suggest how the Republic of Korea has prepared for achieving the goals of the Paris Agreement: the Framework Act on Low Carbon, Green Growth, development of Comprehensive/ National Climate Change Adaptation Plans, 2050 Carbon Neutrality Promotion Strategy, the Carbon Neutrality and Green Growth Act to replace the previous Act, and the Action Plan 2021 to implement the 2050 Carbon Neutrality Strategy, as well as upcoming strategy of the Nature-Based Solution for Reducing and Adapting to Greenhouse Gas Emissions. As Seoul has more than 24% of the country's total population, this entry targeted parks and green area-related policies and practices related to NbS in Seoul, while providing NbS-related case studies to contribute to building more sound evidence of NbS cases, not just focusing on more or two functions of green spaces, but on their multifunctionality.

Focusing on the period prior to the release of the strategy of the Nature-Based Solution, this entry attempted to present how Korea has already developed NbS-related policies and practices (e.g., the Cheonggyecheon River Project, preservation of Royal Tombs as urban cemeteries, and Tancheon Ecological Landscape Conservation Area and the extended connection of blue-green infrastructure). Seoul still takes a strong state-led approach in promoting greening projects, but citizen science approaches are more found in urban greening projects, which enable residents or citizens to become an owner in the city.

Even though Korea has developed a great diversity of NbS-related policies and strategies, there is a lack of integrated and coordinated policies, along with fragmented and arbitrary measures, to achieve the Paris Agreement goals of limiting global warming to well below 2 °C, preferably to 1.5 °C, above preindustrial levels. In this sense, the upcoming NbS strategy could provide an opportunity to integrate parks and green space policies and strategies.

References

Ahn, H. (2016). *Location and ecological interpretation of Forest of Royal Tombs of the Joseon dynasty (in Korean)* (Master). Seoul: Korea Univeristy.

Al-Akl, N. M., Karaan, E. N., Al-Zein, M. S., & Assaad, S. (2018). The landscape of urban cemeteries in Beirut: Perceptions and preferences. *Urban Forestry & Urban Greening, 33*, 66–74. https://doi.org/10.1016/j.ufug.2018.04.011.

Anna, D., & Ewa, K.-B. (2020). How to enhance the environmental values of contemporary cemeteries in an urban context. *Sustainability, 12*(6), 2374. Retrieved from https://www.mdpi.com/2071-1050/12/6/2374.

Bush, J., & Doyon, A. (2019). Building urban resilience with nature-based solutions: How can urban planning contribute? *Cities, 95*, 102483. https://doi.org/10.1016/j.cities.2019.102483.

Choi, J.-K., Choi, M.-K., & Lee, G.-Y. (2017). The process of river landscape for 10years in tan-chun ecological landscape reserve. *Journal of the Korean Society of Environmental Restoration Technology, 20*(6), 107–115.

Cohen-Shacham, E., Walters, G., Janzen, C., & Maginnis, S. (2016). Nature-based Solutions to address global societal challenges. Gland, Switzerland: IUCN

Cultural Heritage Administration. (2021). *Royal Tombs of the Joseon Dynasty, World Heritage*. Retrieved from http://english.cha.go.kr/cop/bbs/selectBoardArticle.do?ctgryLrcls=CTGRY209&nttId=58002&bbsId=BBSMSTR_1205&uniq=0&mn=EN_03_01

European Commission. (2015). *Towards an EU research and innovation policy agenda for nature-based solutions & re-Naturing cities: Final report of the horizon 2020 expert group on nature-based solutions and re-Naturing cities*. Luxembourg: Publications Office of the European Union.

Government of Korea. (2008). *Comprehensive Climate Change Adaptation Plan (in Korean)*.

Government of Korea. (2010). *The 1st National Climate Change Adaptation Plan(2011 - 2015) (in Korean)*.

Government of Korea. (2015). *The 2nd National Climate Change Adaptation Plan (2016 -2020) (in Korean)*.

Government of Korea. (2020a). *The 3rd National Climate Change Adaptation Plan(2021-2025) (in Korean)*.

Government of Korea. (2020b). *2050 Carbon Neutral Promotion Strategy of the Republic of Korea (in Korean)*.

IPCC. (2014). Urban Areas. In *Climate change 2014: Impacts, adaptation, and vulnerability*. Cambridge/New York.

IUCN. (2020). *Global standard for nature-based solutions. A user-friendly framework for the verification, design and scaling up of NbS*. Gland: IUCN.

Kim, K. T., & Song, J. (2015). The effect of the Cheonggyecheon restoration project on the mitigation of urban Heat Island (in Korean). [The effect of the Cheonggyecheon restoration project on the mitigation of urban Heat Island]. *Journal of Korea Planning Association, 50*(4), 139–154. https://doi.org/10.17208/jkpa.2015.06.50.4.139.

Kim, Y. H., Ryoo, S. B., Baik, J. J., Park, I. S., Koo, H. J., & Nam, J. C. (2008). Does the restoration of an inner-city stream in Seoul affect local thermal environment? *Theoretical and Applied Climatology, 92*(3), 239–248. https://doi.org/10.1007/s00704-007-0319-z.

Kim, G.-H., Lee, Y.-G., Lee, D.-G., & Kim, B.-J. (2016). Analyzing the cooling effect of urban green areas by using the multiple observation network in the Seonjeongneung region of Seoul, Korea (in Korean). [Analyzing the cooling effect of urban green areas by using the multiple observation network in the Seonjeongneung region of Seoul, Korea]. *Journal of Environmental Science International, 25*(11), 1475–1484. Retrieved from http://www.dbpia.co.kr/journal/articleDetail?nodeId=NODE09105350.

Lah, T. J., Park, Y., & Cho, Y. J. (2015). The four major rivers restoration project of South Korea: An assessment of its process, program, and political dimensions. *The Journal of Environment & Development, 24*(4), 375–394. https://doi.org/10.1177/1070496515598611.

Lee, S.-D., Shim, J.-H., Ryu, S.-M., Bae, D.-Y., Park, S.-Y., Lee, J.-C., ..., Park, H.-M. (2010). *Assessment of physical habitat and biodiversity in Tancheon ecological landscape conservation area using GIS*. Retrieved from.

Ministry of Environment. (2021). *The 2021 carbon neutrality implementation plan of the Ministry of Environment*.

P

Ministry of Land Infrastructure and Transport. (2021). *Green Area*. Retrieved from https://www.molit.go.kr/USR/policyData/m_34681/dtl?id=85
Morar, T., Radoslav, R., Spiridon, L. C., & Păcurar, L. (2014). Assessing pedestrian accessibility to green space using GIS. *Transylvanian Review of Administrative Sciences, 10*(42), 116–139.
Nasution, A. D., & Zahrah, W. (2012). Public open space's contribution to quality of life: Does privatisation matters. *Asian Journal of Environment-Behaviour Studies, 3*(9), 59–74.
Nordh, H., & Evensen, K. H. (2018). Qualities and functions ascribed to urban cemeteries across the capital cities of Scandinavia. *Urban Forestry & Urban Greening, 33*, 80–91. https://doi.org/10.1016/j.ufug.2018.01.026.
Planning Office for Climate Change of Prime Minister's Office. (2008). *Comprehensive plan for combating climate change.*
Ribeiro, P. J. G., & Pena Jardim Gonçalves, L. A. (2019). Urban resilience: A conceptual framework. *Sustainable Cities and Society, 50*, 101625. https://doi.org/10.1016/j.scs.2019.101625.
Seoul Metropolitan Government. (2015). 2030 Seoul parks and green space master plan.
Seoul Metropolitan Government. (2021). Ecological Condition, Introduction of Cheonggyecheon River. Retrieved from https://www.sisul.or.kr/open_content/cheonggye/intro/animals.jsp
The Government of the Republic of Korea. (2020). *2050 carbon neutral strategy of the Republic of Korea: Towards a sustainable and green society.*

Policy Design

► Governing for Food Security: A Cultural Perspective

Policy Learning

► Beyond Knowledge: Learning to Cope with Climate Change in Cities

Political Science and Development

► Internationalization of Cities

Polycentrism

► Urban Structure and Its Impact on Mobility Patterns: Reducing Automobile Dependence Through Polycentrism

Population Studies

► Spatial Demography as the Shaper of Urban and Regional Planning Under the Impact of Rapid Urbanization

Poverty

► Smart Agriculture and ICT

Poverty Reduction

► Sustainable Development Goals from an Urban Perspective

The Practice of Resilience Building in Urban and Regional Communities

Naomi Hay[1] and Eleni Kalantidou[2]
[1]Australian National University, Canberra, ACT, Australia
[2]Griffith University, Brisbane, QLD, Australia

Definition

Resilience building goes beyond the traditional notion of bouncing back, despite being perceived as a mechanism of recovery. It responds to vulnerability caused by structural injustices, economies of growth and political agendas, by creating conditions of adaptation related to present and forthcoming risks. Instead of placing the burden of rebuilding on the individual, resilience building focuses on a systemic preparation to adapt, which entails a careful evaluation of local, regional, and global conditions. These are examined by taking into consideration the environmental, geopolitical, and socioeconomic parameters that shape them.

Introduction

This chapter outlines an approach to resilience building in the face of increasing global vulnerabilities by recognizing as essential the integration of a community-driven resilience strategy. Exponential population growth, urban sprawl, displacement, pandemics, and environmental degradation combined with natural and anthropocentric disasters, including climate change, generate devastating environmental, geopolitical, and socioeconomic implications. Some challenges can be endured (repeated flooding, cyclones, storm surges, fire risk, or prolonged years of drought), some can be mitigated for or alleviated (building levee banks, strengthening sea walls, flood mitigation, constructing dams, and changing building codes and practices), while other challenges may be solved through a process of adaptation (changing farming practice and crop rotation, recycling water, and harvesting energy alternatively). Beyond these, there are evolving problems brought by the escalating effects of climate change including rising sea levels, rising temperature, and heat islanding in urban zones, making cities potentially uninhabitable in the future. Such issues make evident the need to extend the efforts of mitigating and adapting in order to include preventative measures such as managed resettlement and urban relocation. Adopting extensive interventions requires risk mapping, infrastructural planning, a careful evaluation of resources, and most importantly, community resilience rooted in interdependence.

Resilience and the Urban/Regional/Rural Fabric

Urbanization has become the dominant mode of settlement. The global population currently stands at 7.8 billion, with population growth heavily concentrated in urban zones and more than half of the world's residents living in cities today. This figure is forecasted to increase to 70 per cent by 2050 (United Nations 2021). Contemporary forms of urbanization – at all scales – are in conflict with the environment. To elaborate, the urban is not conceived as a town or city of a specific size or form, but as a phenomenon that encompasses all, forming and structuring spaces; it is complex and difficult to define and delimit. Urban zones can be the commercial center for primary industries (agriculture, mining, forestry), heavily industrialized with secondary industries (power generation, manufacturing, technology), or reliant upon service sectors (tourism, hospitality, education). An urban zone may also be defined in the context of its population density or as a center for administration and governance. Larger and wealthier cities utilize vast amounts of energy and imported resources while they export waste, which results in a flow-on causal effect of environmental degradation in distant regions. Even the smallest, remotest village does not remain untouched, as the urban fabric stretches globally, searching for resources necessary to feed, clothe, transport, accommodate, and provide labor for it to function. Henri Lefebvre (2003) describes the urban fabric as encompassing rural life. In this context, the urban is not limited to megacities, capital cities, or large regional centers, but refers to an urban mode of existence that has additionally invaded agrarian life, villages, and towns of all shapes and sizes.

To interpret the urban, regional, and rural as transitional, a conceptualization of space as dynamic and subject to deterritorialization and reterritorialization under a state of constant negotiation is necessary. Gilles Deleuze and Félix Guattari's (1987) understandings of smooth space (nomadic space) and striated space (sedentary space) demonstrate how space exists only in combination of both, constantly being transformed from one to another (Deleuze and Guattari 1987). As such, urbanization undergoes continuous reconfiguration driven by a process of spatial and cultural fragmentation – examples include

P

growth leading to simultaneous impoverishment of working-class communities, gentrification shifting disenfranchised populations to city fringes, and development further marginalizing suburbs without access to services. Under the weight of uncontrolled expansion, some cities reach critical mass, with failing infrastructure, services, and job markets being unable to provide basic living standards, having as an outcome the migration of urban denizens to regional centers. Conversely, other regional and rural areas' populations are declining as a consequence of dying industries and struggling local businesses, with young people moving to the cities in search of employment, education, and access to better facilities.

As populations shift and reconfigure, communities become increasingly transient and disconnected. Disparities – both socioeconomic and temporal – between the rural/regional and the city are growing. Localities are divided by time, with rural/regional communities often lagging in services, infrastructure, educational opportunities, and employment opportunities. In addition to the ramifications of population movement and infrastructural dissimilarities between localities, a globalized economy further complicates the possibility for environmental, economic, and social resilience. Among the ramifications of globalization is the growing impact of transnational corporations on state governance, and the latter's diminishing connection, contribution, and responsibility toward local communities and their future generations (Bauman 1998).

Resilience at Risk

As human settlements continue to expand, so do levels of risk. Fast disasters, such as earthquakes, floods, cyclones, bush fires, and industrial accidents, are immediate, situated and can be responded to rapidly. On the other hand, slow disasters such as the effects of climate change, pollution, global pandemics, and socioeconomic decline pose a different level of complexity, large and expensive to tackle. These are usually micromanaged by treating their symptoms while disregarding their causes and long-term impact. In this respect, risk needs to be locally examined, with the lived experience of smaller communities being at the epicenter of evaluating exposure to harm. Having said this, communities should not be viewed in isolation, given their existence in an increasingly globalized world. According to Ulrich Beck, ongoing threats against natural and social ecologies pose risks that necessitate institutional preparedness (2006), which so far has not been adopted by traditional disaster management programs. The latter emphasize short-term recovery, in order for the community to preserve their known way of living. To add to the complexity of state or institutional responses, the current level of global geopolitical risks requires a coordinated approach to address challenges faced by global communities, such as the COVID-19 pandemic. The traditional strategy to undertake such issues involves separating them into recognizable problems that may be "solved" by individual governments and experts at various levels, thereby making the task somewhat manageable. However, as the recent experience of a pandemic has shown, the relationality and complexity of global problems demand a systemic modus operandi that goes beyond compartmentalizing them and temporarily contesting them.

From the existing responses to disaster management, the United Nations Sendai Framework for Disaster Risk Reduction 2015–2030 identifies four main priority framework areas: understanding the risk; strengthening disaster risk governance to manage disaster risk; investing in disaster risk reduction for resilience; and enhancing disaster preparedness for effective response and to build back better in recovery, rehabilitation, and reconstruction (UNDRR 2015). This framework responds to the need for holistic disaster risk management at all levels to reduce the number of people affected by disasters and reduce global economic loss. It additionally promotes the implementation of integrated policies towards inclusion, resource efficiency, mitigation adaptation to climate change, and resilience to disasters (United Nations 2021). On an institutional level, this approach aims to strengthen engagement and levels of involvement from all levels of community, including women, children, seniors, Indigenous peoples, migrants, coupled with commitment from private business, not-for-profit, and educational sectors to support the interface between science and policymaking (UNDRR 2015). Despite the principles and guidelines of the framework being promising, governments have traditionally adopted reactive instead of preventative strategies when it comes to disaster management. Repetitive patterns of rebuilding in areas impacted by floods and bushfires in Western settings as well as expanding coastal cities demonstrate the reluctance of adopting policies related to reconfiguring settlements in urban, regional, and rural contexts at-risk. The involvement of people to be, or already affected by disasters is critical considering their input based on lived experience. Traditional knowledge, local socio-economic circumstances, and community idiosyncrasies have to be part of every discussion about identifying and mitigating risk. In this regard, grassroots initiatives around the globe have been actively seeking more citizen participation in decision and policymaking.

The Practice of Resilience Building

Resilience is often described as the ability to bounce back or recover in times of adversity or stress. The term is dealt with in different contexts: environmental resilience in the scientific realm; economic resilience in the business realm; social resilience in the realms of government policy; and resilient infrastructure, buildings, and disaster recovery in the context of urban planning. The term ecological resilience emerged into prominence in the 1970s to describe a system's capacity to maintain itself or recover in the event of stress or disruption to the system (The Rockefeller Foundation and Arup 2015). Crawford Stanley Holling (1973) introduced integrated theories of ecology and systems thinking while linking these concepts to adaptive management. A decade later, Aaron Wildavsky applied resilience thinking to risk management, public policy, and budgeting to develop a framework of resilience principles (Pelling 2003). Since then, the term resilience has been applied widely across ecology, political science, psychology, sociology, engineering, economics, business administration, and more recently urban planning, disaster planning, international development, and complex adaptive systems. However, definitions, theories, and methods of understanding resilience vary significantly across disciplines (Martin-Breen and Anderies 2011; McAslan 2010; Ungar 2018). Some definitions focus upon coping, survival, recovery, and maintenance. Others add adaptation and transformation into the resilience equation (Resilience Alliance 2021; UNHabitat 2021). The Stockholm Resilience Centre (2021) takes this a step further by seeing resilience building as an opportunity for encouraging renewal and future innovation. Other approaches frame resilience in the context of system-based approaches, focusing upon individual subsystems within the larger system of the city. Still, relationality and interdependency are minimally considered between systems at varying scales and their governing structures (The Rockefeller Foundation and Arup 2015). Following this, issues to be examined involve the connection of the urban to the regional and the rural in terms of transport infrastructure, access to food, facilities, and resources, the institutional jurisdictions that govern this interaction, and the limitations deriving from the lack of citizen participation in decision-making. Given the vast number of people living in urban centers, there has been far less focus on community resilience at the level of the regional and the rural/local. Another limitation of the existing approaches is in making attempts to measure resilience through the creation of vulnerability indexes, followed by the proposal of resilience frameworks and/or toolkits for administrators to apply directly to their communities. Even though these are useful tools to raise awareness within communities, create dialogue, and propose guidelines for action, it is critical that community resilience does not get diminished into ticking boxes in the name of expediency.

Community resilience has been defined as hard and soft. Hard resilience refers to the strength of structures and institutions when placed under pressure, and soft resilience denotes the ability of a community to recover without fundamental changes to its structure and function (Moench 2009, p. 256). To ensure the steadfastness of hard resources, which include public buildings, housing, facilities, infrastructure, and services, it is crucial for these to be combined with the scaffolding of soft resources such as employment, social economies, education, and information sharing. Tactical integration of both in a proactive plan is needed to strengthen long-term resilience. In terms of the implementation of merging hard and soft resources, there has been an emphasis on the former when it comes to designing and planning built environments. Nonetheless, examples connected to recurring drought, flooding, or conflicts have shown that it can take years or decades for communities to recover, making visible chronic issues that directly hurt soft resources involving local economies and employability. To enhance community resilience, strategic planning needs to go beyond the design of effective evacuation centers, safe construction practices, early warning systems, and so forth, and encompass a wider range of initiatives to reduce vulnerabilities. Such measures include strengthening the political agency of neglected urban, regional, and rural areas, financial scaffolding for developing nations, and giving communities the means to respond to vulnerabilities identified from within. Assessment strategies for hard resilience of communities at large require a detailed mapping of infrastructural weak points and exit plans in case of disaster, while for soft resilience, they demand a multi-stakeholder evaluation of socioeconomic and political factors interfering with communities' ability to cope following disaster. As a case in point, UN-Habitat promotes resilience building from a multiple hazard's perspective, with a focus upon both natural and human-made crises, and an objective to ensure capacity for rapid recovery from crisis events (UNHabitat 2021). To achieve this, effective collaboration between community, government, non-government organizations, and private sectors has to be established, along with significant investments in long-term social resilience via programs with continuing funding. Considering the political instability both developed and developing countries are experiencing, in addition to a growing precariousness driven by climate change, it is difficult for plans like this to be implemented and sustained for an extended period of time.

Besides institutional directives and infrastructural changes to pursue resilience building, Judith Rodin (2014) argues that resilience building can also be learned as a concept and developed as a practice. Rodin identifies five essential characteristics that an entity with resilience should have: awareness, diversity, integration, self-regulation, and adaptation. To obtain these, three phases of resilience building are necessitated, characterized by readiness, responsiveness, and revitalization. Furthermore, to respond to disruption depends not only on readiness but upon social cohesion with friends, neighbors, colleagues, and strangers supporting each other and providing help in conditions of emergency. Social cohesion offers an element of resilience that influences responsiveness under stress, deriving from strong community ties and social relationships. This encourages a united front when confronted by a crisis (Rodin 2014). Tony Fry (2015, p. 60) also argues that resilience is not about maintaining business as usual, nor a product of policy or institutions of civil society; instead, it is the "product of the collective actions of people who are bound together by their culture and a common will to survive by transcending the conditions that threaten." This thinking recognizes resilience as an underpinning of communal self-sufficiency that diminishes the dependency on state interventions. Such reasoning applies to reducing vulnerabilities caused by both fast and slow disasters, supporting communities to face risk at multiple levels. Each community has unique strengths and the capacity to identify weaknesses based on its locality, political, and socioeconomic makeup. A preemptive planning strategy individualized and designed within each community is essential in risk reduction, mitigation, and regeneration, where decisions can be made in the best interest of the community as a whole. In light of this, resilience is a process, not an end goal, reaching beyond the level of recovery and towards activating awareness, preparedness, and adaptation so as to fully engage with risk at all levels. Resilience building is about learning from the past, facing and overcoming challenges of the present, and prefiguring for and adapting to challenges of the future.

Conclusion

As risk both immediate and unfolding, becomes more complex, systemic planning is required so as to move beyond the limitations of current frameworks. As highlighted in this chapter, resilience is more than recovery. Moving beyond economic rehabilitation and simulated stability, resilience is a process that involves a thorough examination of what the next steps should be to ensure survival and well-being in urban and regional communities. It necessitates a reconsideration of adaptation by prioritizing community building on the ground, and a detailed mapping of the available resources that can be maintained, relinquished, or redesigned. To accomplish this, a careful examination of contemporary infrastructures and everyday practices is needed, in order to identify these that should be preserved under the pressure of present and upcoming crises; the unsafe, which should be left behind; and the ones that should be reconfigured to accommodate new uses.

Moreover, resilience should be grounded in acknowledging the precarious state of the local within the global and identifying cultural practices and lived experience that can provide lessons for adaptability. In doing so, there must be a joint assessment by institutions and communities of fast and slow disasters and their impact on the interconnected urban, regional, and rural. In contrast to the contemporary illusion of stability, building resilience does not have a telos. Instead of focusing on an end goal, it lies in a process of constantly re-evaluating infrastructure, habitation, migration, resettlement, and communal interdependence. The latter, as an essential component of resilience, can only derive from relationships forming through participation in decision-making and self-organizing operational units with corresponding responsibilities. This necessitates opening lines of communication, authentically engaging with, and actively involving the community in reconceptualizing a vision of what an inherently resilient future could look like.

Cross-References

▶ Community in a Changing Climate: Shaping Urban and Regional Futures

References

Bauman, Z. (1998). *Globalization: The human consequences*. New York/Cambridge: Columbia University Press/Polity Press.

Beck, U. (2006). Living in the world risk society. *Economy and Society, 35*(3), 329–345.

Deleuze, G., & Guattari, F. (1987). *A thousand Plateaus: Capitalism and Schizophrenia* (B. Massumi, Trans.). Minneapolis: University Of Minnesota Press.

Fry, T. (2015). *City futures in the age of a changing climate*. Oxon/New York: Routledge.

Holling, C. S. (1973). Resilience and stability of ecological systems. *Annual Review of Ecology and Systematics, 4*, 1–23.

Lefebvre, H. (2003).*The urban revolution* (R. Bononno, Trans.). Minneapolis: University of Minnesota Press.

Martin-Breen, P., & Anderies, J. M. (2011). *Resilience: A literature review*. https://opendocs.ids.ac.uk/opendocs/handle/20.500.12413/3692

McAslan, A. (2010). *The concept of resilience: Understanding its origins, meaning and utility*. https://www.flinders.edu.au/content/dam/documents/research/torrens-resilience-institute/resilience-origins-and-utility.pdf

Moench, M. (2009). Adapting to climate change and the risks associated with other natural hazards: Methods for moving from concepts to action. In E. L. F. Schipper & I. Burton (Eds.), *The Earthscan reader on adaptation to climate change* (pp. 249–282). London/Sterling: Earthscan.

Pelling, M. (2003). *The vulnerability of cities: Natural disasters and social resilience*. London/Sterling: Earthscan Publications.

Resilience Alliance. (2021). *Resilience*. https://www.resalliance.org/resilience

Rodin, J. (2014). *The resilience dividend: Being strong in a world where things go wrong*. Philadelphia: Public Affairs Perseus Books Group.

Stockholm Resilience Centre. (2021). What is resilience? https://www.stockholmresilience.org/research/research-news/2015-02-19-what-is-resilience.html.

The Rockefeller Foundation, and Arup. (2015, December). *City resilience framework*. http://www.100resilientcities.org/resources

P

UNDRR: United Nations Office for Disaster Risk Reduction. (2015). *Sendai framework for disaster risk reduction 2015–2030.* https://www.undrr.org/publication/sendai-framework-disaster-risk-reduction-2015-2030
Ungar, M. (2018). Systemic resilience: Principles and processes for a science of change in contexts of adversity. *Ecology and Society, 23*(34). https://doi.org/10.5751/ES-10385-230434.
UNHabitat. (2021). *Resilience and risk reduction.* https://unhabitat.org/topic/resilience-and-risk-reduction
United Nations. (2021). *Sustainable development goals: Goal 11: Make cities inclusive, safe, resilient and sustainable.* https://www.un.org/sustainabledevelopment/cities/

Precedent

► Closing the Loop on Local Food Access Through Disaster Management

Pre-schoolers and Sustainable Urban Transport

Christina R. Ergler
School of Geography, University of Otago, Dunedin, New Zealand

Synonyms

Active transport; Busing; City; Environmental literacy; Independent mobility; Mobility; Transport modes; Walking; Young people

Definition

Children under the age of 5 are often considered dependent travelers. Thus, their transport needs, experiences, and expectations for a sustainable transport system are less well understood. The few studies that directly work with young children to elicit their mobility experiences show that transport systems are not attuned to the needs of young travelers and their families, but being on the move and using diverse transport modes plays a major role for young children's identity formation and environmental literacy. This entry outlines why it is important that not only the transport system but cities in general need to be more attuned to the needs of preschoolers and their families to foster geographical imagination and sustainability.

More than a billion children are growing up in urban environments. Depending on age and capabilities, they either reach destinations independently or accompany parents and caregivers to go shopping, visits to the doctor or hairdresser, to see friends and family members, or to attend play and educational facilities. Destinations in urban environments are diverse, and their availability and quality vary between cities and countries, but how these destinations are reached depends on how far they are from the home of the child, the design of the urban environment, what transport opportunities are available, and the particular transport norms and practices in families and the city. However, which transport mode children and their families use has a major impact on the environment and children's health and well-being. For example, in a car-dominant transport system, air quality is often low, CO2 emissions and traffic-related injuries are high, and children are less likely to engage in recommended activity levels on a daily basis. Moreover, even when young children walk or wheel in such an environment, they are, due to their height, exposed to fumes from cars, buses, and other motorized transport. Their faces when they are in a stroller are at the same level as the exhaust pipes.

While primary school children's use of different transport modes, their travel experiences, and their preferences for and understanding of a sustainable urban transport system have been the focus of many studies, younger pre-literate children are hardly being considered as transport users; they are viewed as dependent travelers. Thus, their use, experiences, and visions for getting around (more sustainably) in the urban environment is less well researched.

Studies that have looked at young children and transport more widely can be divided into three broad categories:

1. Mobility access and urban design
2. Mobility experiences leading to identity formation and connections with the environment
3. Knowledge about transport systems

Studies in the first category show that mobility access is a major barrier. Transport hubs or road crossings are not attuned to the needs of small walkers or parents with prams. Routes are described as unsafe, and the young children themselves even indicate a lack of accessible destinations within walking distance. To get around more easily in their views, they suggest building wide, safe sidewalks and introducing traffic calming measures and bike lanes in their local environment. Studies in the second category reveal how travel experiences by car or bike, on foot, or on the train shape young children's sense of belonging and contribute positively or negatively to their identity formation. Other studies in this category revealed which social, natural, and built environmental factors contribute to young children being and feeling well in their neighborhood, while others unpacked the embodied and sensory experiences of pre-schoolers being on the move and their playful encounters with their human and non-human surroundings. These studies clearly show that being on the move for young children is more than just getting to a destination; they value the journey as much as reaching a destination. Studies in the third category investigate young children's understanding and knowledge of the environmental impact of various modes of transport.

All studies in these three categories show that young children's experiences of being on the move leave a remarkable impact on their identity formation, contribute to their environmental literacy, and build the basis for forming relationships and connections with their environment. Their mobility experiences shape and are shaped by their and their families' norms and attitudes towards different transport modes. Moreover, longitudinal studies show that early childhood experiences shape transport norms and practices at later life stages. Adults can learn much from young children about how their use of the existing transport system and getting around in the urban environment shapes young children's visions and embodiment of existing, often motorized, transport systems. These findings also show, however, that urban environments that aim to be more liveable and be explored by walking, cycling, and other public transport systems need to work for caregivers and parents, too. Making the environment inviting and easy for parents and caregivers more generally allows new ways for children to be exposed to and experiment with alternative transport modes and thus for the installation of an early sustainable transport imagination.

The Bernard van Leer Foundation has developed three premises to ensure cities work for young children and foster geographical imagination and sustainability, with the underlying premise being that the transport system also needs to work for young children by offering different sustainable modal choices. They advocate:

1. Designing a city for caregiving
2. Creating a 15-minute neighborhood, as proximity matters
3. Thinking "babies [and young children]" as a universal design principal

In a nutshell, this means putting the needs of young children and their families first. When the city – and thus the transport system connecting different destinations – works for young children and their caregivers, it will work for all ages and abilities. The Foundation proposes not only creating safe and comfortable cities and transport systems for adults and older children, but making it easy for anyone to use and get around the city by, for example, including benches on sidewalks to be able to stop at when breaks are needed for young walkers, or providing easy-to-board, convenient, and electric buses. Young children are extremely vulnerable to traffic and thus are often dependent on adults to take them places, but young children also enjoy exploring and playing en route, so sidewalks, pathways, and the transport systems (and cities) need to be safe, clean, and interesting enough for them, so that parents do not use the car as the go-to transport mode. To achieve a more sustainable city in which the transport system plays an important part, the Foundation suggests ensuring that services families frequently use, like education, health, shopping, and leisure facilities, are located within a walkable 15-minute range. The aim is that walking and cycling may become the dominant, enjoyable, and accessible transport forms, as active transport modes reduce emissions and improve air quality, while also increasing young children's environmental literacy and having the possibility to enrich their future sustainable transport imaginations.

Cross-References

- ▶ Healthy Cities
- ▶ Low-carbon Transport
- ▶ Need for Greenspace in an Urban Setting for Child Development
- ▶ New Cities
- ▶ Policies for a Just Transition
- ▶ Sustainable Urban Mobility
- ▶ Understanding Women's Perspective of Quality of Life in Cities
- ▶ Urban Futures
- ▶ Urban Well-Being
- ▶ Walkable Access
- ▶ Youth and Public Transport

Further Reading

Bernard van Leer Foundation. (2017). *Building better cities with young children and families: How to engage our youngest citizens and families in city building – a global scan for best practices*. The Hague: Bernard van Leer Foundation.

Borg, F., Winberg, T. M., & Vinterek, M. (2019). Preschool children's knowledge about the environmental impact of various modes of transport. *Early Child Development and Care, 189*(3), 376–391.

Clement, S., & Waitt, G. (2018). Pram mobilities: Affordances and atmospheres that assemble childhood and motherhood on-the-move. *Children's Geographies, 16*(3), 252–265.

Ergler, C. R., Freeman, C., & Guiney, T. (2020). Preschoolers' transport imaginaries: Moving towards sustainable futures? *Journal of Transport Geography, 84*, 102690. https://doi.org/10.1016/j.jtrangeo.2020.102690

Ergler, C., Smith, K., Kotsanas, C., & Hutchinson, C. (2015). What makes a Good City in pre-schoolers' eyes? Findings from participatory planning projects in Australia and New Zealand. *Journal of Urban Design, 20*(4), 461–478. https://doi.org/10.1080/13574809.2015.1045842.

Ergler, C., Freeman, C., & Guiney, T. (2020). Walking with pre-schoolers to explore their local wellbeing affordances. *Geographical Research*.

Gustafson, K., & van der Burgt, D. (2015). 'Being on the move': Time-spatial organisation and mobility in a mobile preschool. *Journal of Transport Geography, 46*, 201–209.

Karsten, L. (2014). From Yuppies to Yupps: Family Gentrifiers consuming spaces and re-inventing cities. *Tijdschrift voor Economische en Sociale Geografie, 105*(2), 175–188. https://doi.org/10.1111/tesg.12055.

Smith, K., & Kotsanas, C. (2014). Honouring young children's voices to enhance inclusive communities. *Journal of Urbanism: International Research on Placemaking and Urban Sustainability, 7*(2), 187–211. https://doi.org/10.1080/17549175.2013.820211.

Press Information Bureau (PIB)

▶ The State of Extreme Events in India

Private Regulation

▶ Voluntary Programs for Urban and Regional Futures

Private Standards

▶ Voluntary Programs for Urban and Regional Futures

Programming

▶ Participatory Planning: A Useful Tool for the Development of Sustainable Mega-City Regions

Progression

▶ Furthering the Sustainable Development Agenda by Putting Urban Heritage and Value Extraction at the Center

Progression Nexus

► The Urban Planning-Real Estate Development Nexus

Proptech: Issues for the Future

Anupam Nanda
University of Manchester, Manchester, UK

Introduction

Technology is now omnipresent in our lives. Since the arrival of the Internet, all aspects of economic and financial activities have been shaped by solutions powered by digital technology. Affordable access to the Internet has brought widespread use cases and solutions to many problems. The real estate sector has also evolved gradually with each new wave of digital technology. With improvements in digital technology through the introduction of artificial intelligence (AI), machine learning, and block chain technology, the ability to improve transaction processes and understand a property's value and impact better is creating significant investment opportunities across all subsectors of real estate. All these technological interventions in real estate fall under so-called "proptech" solutions.

As a sector, real estate is somewhat slow to adjust to new technologies due to the complex interplay of various actors, regulated practices, and processes. However, the change is inevitable and it is spreading with certainty as new technologies have the ability to offer effective solutions to several typical challenges in real estate. For example, technology can assist in reducing expenses, increasing operational efficiencies, eliminating planning and production delays, identifying needs and development areas, evaluating performance, creating benchmarks, and understanding the risks and uncertainties of property investment better. While traditional real estate service providers and many start-ups are beginning to offer products and services using proptech platforms, several areas need much more attention to be able to fully exploit the technological opportunities. It is very crucial to recognize that real estate, as a sector, should embrace technologies to solve or at least better address societal challenges such as climate change, inequalities, affordability, etc. that are linked to the sector.

In this chapter, the author briefly visited areas of proptech applications and highlight the issues needing attention in the future.

Areas of Proptech Applications

For a detailed discussion of the following application areas, see Nanda (2019).

New and Enhanced Uses of Physical Spaces

With technology, new and enhanced uses of physical space are emerging, along with new ideas of revenue generation. Several new uses are based on the shared economy concepts. Proptech applications are enabling several tech-driven new and shared uses. Unused or less used spaces can be made commercially viable with an appropriate revenue model. This is also partly driven by changing tastes and preferences around work and urban living.

Some of these shared use concepts are growing popular due to demand-supply mismatches and the inability to afford a property due to overpriced markets especially in urban areas with a significant concentration of employment opportunities. There are now technology platforms that can enable sharing of information about available spaces and facilitate peer-to-peer matching much more efficiently. Technology can help match suitable co-occupants and generate revenue from unused spaces. Co-living can release unused stock of spaces and potentially help alleviate supply constraints, especially in cities with high property prices and rents.

With technology, common issues of sharing economy such as trust, security, and privacy can be effectively addressed to improve customer satisfaction. Airbnb is an example. Security, physical depreciation, and revenue uncertainty can be addressed by technology solutions with suitable background checks and real-time information flow. However, the Airbnb model faces concerns of overcrowding and shifting stock from long-term tenancies to more lucrative short-term rentals. Several cities have started devising policies to limit short-term rentals. Another concern is around the possibility that the local residents may get displaced and priced out, especially in high-demand Airbnb locations. Tenant turnover is typically high. Airbnb can also push up rents, which can exacerbate affordability in places where supply is heavily constrained. However, a combination of appropriate policy measures and technological innovation can address some of these concerns.

Similarly, a flexible workplace is another area with significant scope for technology applications with more workers are working without a fixed office allocation, and younger generations are seeking more flexible working arrangements. This trend has sparked significant investment interest in recent years. Airbnb, WeWork, LiquidSpace, and others have all successfully implemented the flexible working principle and developed commercially viable solutions. Moreover, the Covid-19 pandemic has made working from home and hybrid working much more commonplace and it is a trend, which may see a long-run shift. Technology is key to enabling flexible working.

Facilitating Property Search and Transaction Processes

A unique attribute of real estate is information asymmetry (The real estate market can be characterized by several distinctive attributes such as (a) territoriality, (b) heterogeneity, (c) longevity, (d) information asymmetry, (e) transaction costs, (f) externality, and (g) socio-political influence (see Nanda, 2019 for a detailed discussion)). In a typical real estate transaction, the information set differs in quality and quantity across the parties involved. An implication of such asymmetry in information is uncertainty, which leads to speculative behavior by the stakeholders involved in the property transaction. Digital technology has the potential to address information asymmetry, right from the property information as part of the starting point in the property search process to the progression through various stages toward successful completion. In many ways, not only the buyers and sellers but also the intermediaries (such as brokers/agents, surveyors, solicitors, mortgage advisors, etc.) can benefit from a more efficient information flow across various stakeholders. A key example of the application of digital technologies is property listing portals for buying, selling, and renting. Listing portals such as Rightmove and Zoopla are prime examples.

Internet-based services are now commonplace around the world. In recent times, the embedding of third-party information and tools (e.g., integrated mapping service from Google Earth) has added many additional layers of contextual information to facilitate property search and decision-making. Moreover, online tours of the property, as well as virtual and augmented reality applications are now enhancing the search process and customer experience in a significant way. Augmented reality will be able to convey search results to any device connected to 3G, 4G, and 5G networks. Current and future technological platforms can facilitate frictionless interactions between sellers and buyers, as well as automated updating of information, forecasting, and simulation techniques. This has also allowed platforms to offer innovative search processes. Purplebricks, for example, facilitates direct seller-to-buyer transactions at a predetermined price. In near future, a majority of home searches will likely be conducted online, with minimal face-to-face human interactions. Such changes not only cut the monetary costs of property transactions but also reduce the nonmonetary costs (e.g., time and stress) associated with a typical property transaction.

Apart from producing search results, the property listing portals also collect a wealth of information, which can be used for creating an in-depth understanding of the property industry, market trends, demand and supply changes, property values, and shifts in consumer attitudes. The wealth of information can be further analyzed to facilitate the development of new products and services through applications of machine learning and associated data science tools. It can enable more informed policy responses by the regulators and industry bodies. It can also help with impact analysis and evaluation of the effectiveness of the policies.

In terms of property valuation services, more detailed and granular information about a property's performance and other locational attributes can help improve the quality of benchmarks. There may lead to fewer risks and uncertainties in terms of a property's valuation and as a result, the gap between buyers' offer and sellers' asking prices may reduce, resulting in potentially less intense or lengthy bargaining and faster completions. This may significantly aid in a more efficient market clearing and potentially reduce speculative activities by consumers and investors.

Another area of significant potential is the application of blockchain technology in contracting. The blockchain framework can be used to record and manage transactions. By being decentralized and based on "trustless," distributed ledger technology, it can effectively remove old centralized sources, which are costly in terms of monetary and nonmonetary costs, as well as eliminate their potential for abuse and mismanagement. Some parts of transaction processing could benefit from blockchain applications. Direct buyer-seller contracts, land acquisitions, registration and titling, as well as environmental and title searches are examples of potential application areas for integrated smart contracting. However, it is important to note that such integration platforms should be based on sustainable energy sources, which can bring and integrate all relevant datasets and tools within a secure, trusted, and energy-efficient manner.

Innovations in Property Finance Market

The current mortgage finance mechanisms are complex to navigate and are often long drawn. The mortgage approval procedure is extremely process driven and typically creates many dependencies. However, there are several components of mortgage market operations such as formal application, evaluation, affordability checks, underwriting, and approval, and upon mortgage disbursement, regular servicing can become more automated and error free. Technology start-ups and traditional lenders can collaborate to develop a more efficient mortgage market process within a collaborative ecosystem. This could reduce three typical risks in the mortgage market, i.e., default and foreclosure risks for lenders, prepayment risk for lenders, and equity and interest rate risks for borrowers. The advantages are significant, with greater risk-reward matching to reduce investment uncertainties, controlling speculative activities, effective capital recycling, and reduced market volatility. A common matching and engagement platform for fund managers and investors can significantly reduce investment delays and release capital more efficiently. Nonbanking lending platforms can be established. For example, LendInvest provides short-term financing to various stakeholders (e.g., intermediaries, landlords, and developers) and help with streamlined mortgage application and loan servicing processes while operating through a nonbanking organizational framework.

Issues for the Future

While the areas of applications mentioned above are going through further development and will experience more innovations in the coming years, several aspects need careful consideration for proptech to have desirable impacts.

- **Social responsibility** – tackling societal challenges such as climate change, inequality, affordability, rapid urbanization, aging population, and mass migration must be the key focus of all innovations. This is due to our world now facing several critical and existential challenges. Businesses must look beyond short-run revenue goals and focus on long-term solutions through innovations. Technology (especially digital technology) can provide such solutions. Real estate is no different and, as a sector, real estate has much to contribute to shaping a sustainable future. The property market has a very significant carbon footprint and therefore, any technological interventions to reduce carbon emissions from the processes of production and consumption (property development, management, and operation) can go a long way in contributing substantially toward the net-zero goals. Another very important area is inequality and affordability. While housing is a basic need, a sizeable section of our society around the world cannot meet that basic need. Across many major markets around the world, affordability is a major concern, and many young aspiring homeowners are priced out of the housing market. Technological solutions such as the production of affordable homes using new materials, robotics, 3D printing, etc. as well as information technology to understand and analyze affordability needs can provide viable solutions. Proptech solutions should also focus on these issues in the developing world. With rapid urbanization across much of the developing world and major economies (such as the BRICS countries), technology-based property market solutions can address the key challenges.

- **Ethical standards** – proptech solutions tend to collect and analyze a huge amount of proprietary data about the people, place, and property. This is crucial for developing and providing solutions. However, this also comes with the issues of data protection, legal use, and other ethical concerns, which must be addressed. The ethical concerns are not just about the data but also about the processes. Transparency in data handling and governance structures that are accountable and robust is also crucial. As the proliferation of proptech solutions is intensifying, these concerns must be central to devising viable business models. It would also require development of standards that can incorporate the issues and concerns from all relevant perspectives.
- **Industry acceptance and collaborations** – proptech solutions bring two quite distinct industry sectors together – real estate and technology. While the real estate sector adjusts slowly to changes, the technology sector is fast in adopting and adapting to new trends and changes. However, it is important to note that a viable proptech solution should be able to bring both sectors together in a balanced manager with a focus on the property market challenges as that is where the opportunities lie.
- **Education and skill building** – proptech solutions demand an understanding of both the real estate and technology sectors. Skill sets are different for these two sectors but those are important to bring together within a complementary framework. Future real estate professionals and leaders will require new skill sets to navigate the world of real estate, new technologies, and big data. With customers becoming more familiar with technology platforms, fundamental data science and technology competencies will be required. Investing in building those skills would be crucial for scaling up the solutions and encouraging further innovations. While it may not be possible to become experts in both sectors by a real estate professional, a level of competency to be able to appreciate the use cases and their implications would be an important attribute for success. All of this would involve development of new instructional methods, new materials, and training programs.

Concluding Remarks

While the last decade has seen a great deal of interest in using technology to provide better and innovative property market solutions, much needs to be developed and improved further to be able to fully exploit the opportunities that technology can offer to address the perennial challenges in the property market. Moreover, significant societal challenges that are related to the property sector exist. Technology should be embraced to address those challenges.

Selected Reading List

Acemoglu, D., & Pascual, R. (2018). The race between man and machine: implications of technology for growth, factor shares, and employment. *The American Economic Review, 108*(6), 1488–1542.

Bernstein, E. S., & Turban, S. (2018). The impact of the 'open' workspace on human collaboration. *Philosophical Transactions of the Royal Society B: Biological Sciences, 373*(1753), 20170239.
Fuster, A., Plosser, M., Schnabl, P., & Vickery, J. (2019). The role of technology in mortgage lending. *The Review of Financial Studies, 32*(5), 1854–1899.
Nanda, A. (2019). *Residential real estate: urban & regional economic analysis*. Routledge, Abingdon, Oxon: New York, NY.
Saiz, A. (2020). Bricks, mortar, and proptech: The economics of IT in brokerage, space utilization and commercial real estate finance. *Journal of Property Investment & Finance, 38*(4), 327–347.

Prosperity

▶ Sustainable Development Goals from an Urban Perspective

Provincial

▶ Closing the Loop on Local Food Access Through Disaster Management

Provincial Towns

▶ Small Towns in Asia and Urban Sustainability

Provision

▶ Closing the Loop on Local Food Access Through Disaster Management

Public

▶ Closing the Loop on Local Food Access Through Disaster Management

Public Acquisition

▶ Public Procurement for Regional and Local Development

Public Administration

▶ Regulation of Urban and Regional Futures

Public Awareness of and Participation in Municipal Solid Waste Management in Urban Areas of the Mekong River Delta, Vietnam

Tien Dung Khong[1], Adam Loch[2] and Dan Xuan Thi Huynh[1]
[1]School of Economics, Can Tho University, Can Tho, Vietnam
[2]Centre for Global Food and Resources, School of Economics and Public Policy, Faculty of the Professions, University of Adelaide, Adelaide, SA, Australia

P

Introduction

Vietnam is rapidly industrializing driving increased manufacturing, residential areas, commodity demand, and material and energy consumption (Ministry of Natural Resources and Environment of Vietnam 2020). These increases have facilitated the expansion and growth of manufacturing, business, and service industries; all of which have contributed positively to the country's socio-economic development. However, this rapid growth generates a large amount of garbage, particularly solid waste such as air pollution, sludge from wastewater, and/or discarded materials from industrial, commercial, mining, agricultural or community activities.

According to the Vietnamese government, the country collected over 33,167 tons of solid waste last year alone. Of this regular solid waste, the total amount collected and treated as per national technical standards and regulations was approximately 27,067 tons (accounting for 81%). Thus, there was about 6,100 tons of solid waste that was collected but not treated. This untreated waste, together with the significant amount of uncollected solid waste, has typically gone on to pollute the environment, and the volume of untreated waste is forecast to increase from 24% to 30% over the next five years (Government of Vietnam 2020). This is due to the fact that the planning and classification of municipal solid waste (MSW) generated by current households has become increasingly problematic because of limited landfill resources and urban environmental challenges associated with landfills. For example, almost all solid waste collected in the Mekong River Delta (MRD) is concentrated, transshipped, and transported to landfill that does not meet integrated dumping- ground specifications. Further, the collection, transportation, treatment, and disposal of solid waste inevitably results in many environmental concerns including loss of urban beauty, traffic congestion, stench, leachate, and greenhouse gas emissions (Ministry of Natural Resources and Environment of Vietnam 2018, 2020). Environmental pollution treatment and risk management associated with solid waste, therefore, poses an urgent environmental protection requirement for Vietnam and, potentially, many other developing economies.

As an alternative approach, one feasible solution may be to classify MSW at the source; that is, separate waste into different classes and then handle them differently in each case. If implemented at household level via a proper management system and technology the management, collection, class-typing and reuse of solid waste can result in significant economic rewards, environmental protection, and resource conservation for the country (Ko et al. 2020). However, the high expense of MSW classification at the household level also presents other obstacles (Government of Vietnam 2020) which may deter effective implementation. To improve household recognition and adoption of MSW management services it is critical to understand what motivates people to willingly pay for such programs and engage with the classification of MSW at the household level. This research contributes to an existing body of knowledge on willingness-to-pay (WTP) research for improved waste services – where developing country contexts remain understudied – by utilizing a contingent valuation methodology (CVM) and choice experiments (CE) to determine WTP levels and drivers that encourage people to implement change that can address the effects of increased waste on environmental quality.

The Setting

Overview of the Domestic Solid Waste Sorting Program

Based on the 3R (Reduce – Reuse – Recycle) management model, previous studies suggest that a solid waste separation program at the household level is necessary to increase the efficiency of waste management activities. Waste reduction activities in Vietnam and some developing countries are inextricably linked to waste management activities and policies. Waste management is the primary responsibility of local governments (Schübeler et al. 1996). Solid waste management is a complex process that requires organizational capacity and collaboration between numerous private and public sector stakeholders. While solid waste management activities positively impact public health and environmental protection, this practice is still unsatisfactory in most developing country cities (Ferreira and Marques 2015; Jin et al. 2006).

To overcome this issue, numerous waste-reduction policies have been implemented including institutional and control policies (Slack et al. 2009), solid waste regulations, and penalties for violations. However, there are many examples where institutional policies related to MSW have not resulted in compliance or improved environmental quality (Stafford 2002). Instead, individual-level incentives may make market-based policies more effective than institutional-level policies (Driesen 2006) shifting the focus for policy-makers. Negative incentives may include paying sales taxes or adhering to a "*pay as you throw*" policy, as well as paying a fee based on the volume of solid waste discharged. Positive incentives may include sponsorship opportunities or tax breaks for individuals or organizations that reduce their waste (Gellynck et al. 2011). Some specific policies such as deposit return systems combine both positive and negative incentives to try and affect change (Wagner and Arnold 2008). Numerous communities have also implemented voluntary policies such as voluntary participation in recycling (Palatnik et al. 2014). However, while there are numerous policy tools available for solid waste management, their effectiveness varies by community. For instance, the "*pay as you throw*" policy may be ineffective in some developing countries due to a lack of household solid waste collection programs (Longe et al. 2009). As a result, policies and activities aimed at reducing household solid waste generation should take into account the unique characteristics of each municipality or country.

In response to these limitations, Zhuang et al. (2008) have stated that segregating MSW at the source may effectively reduce waste. In response to these findings Germany, the United States of America, and Japan have adopted MSW as part of their waste management systems. On the back of this change, municipal solid waste in Japan was reduced by 69% (Ministry of Environment of Japan 2014) suggesting the usefulness of such approaches. Other evidence indicates that solid waste separation at the source has aided in reducing solid waste discharged into the environment by reusing and recycling a large amount of solid waste.

Although household MSW classification has not been conducted on a large scale or per regulations in Vietnam, some households have classified and sold scraps from their solid waste, which are favorable practices for recycling activities to reduce MSW in Vietnam. On the other hand, each household currently pays about US$0.66 to US$0.87 per month for waste collection services (Ministry of Natural Resources and Environment 2020). With this payment level, the government is expected to enact a policy of compensating for losses with a sizable sum of money. Simultaneously, households are accountable for this compensation under the principle of "*polluters must pay*". So studying the factors affecting participation in the garbage classification program is critical to developing practical policy recommendations. In this chapter, we seek to examine the drivers of effective MSW segregation for one Vietnamese city in an effort to inform a policy of change with good prospects for implementation and continuance over the longer-term.

The State of Solid Waste Management in the Mekong River Delta

According to reports from local Departments of Natural Resources and Environment, MSW collected accounts for 88.3% and 49% of waste generated in urban and rural areas, respectively. However, only 71% of the MSW collected is treated. As a result, a substantial amount of generated solid waste remains uncollected and untreated. The amount of generated MSW that is not collected, together with the amount of MSW that is not collected and is not treated and the amount of MSW that is not properly treated, are the primary sources of environmental pollution. According to the Ministry of Natural Resources and Environment (2020), the impact of solid waste can be seen in the loss of urban space, the pollution of land, water, air, and greenhouse gas emissions (Fig. 1).

Review of the Literature

Appropriate solid waste management benefits the environment and public health (Ministry of Natural Resources and Environment of Vietnam 2020). Individuals have varying preferences for solid waste management programs designed to increase their solid waste management capacity and reduce their solid waste. Preferences are challenging to value because they lack a market price (i.e. the compose non-market values). Generally, non-market value is estimated using non-market valuation methods. The fundamental tenet of these methods is the consumer's willingness to pay or willingness to accept (Bateman et al. 2002) as a

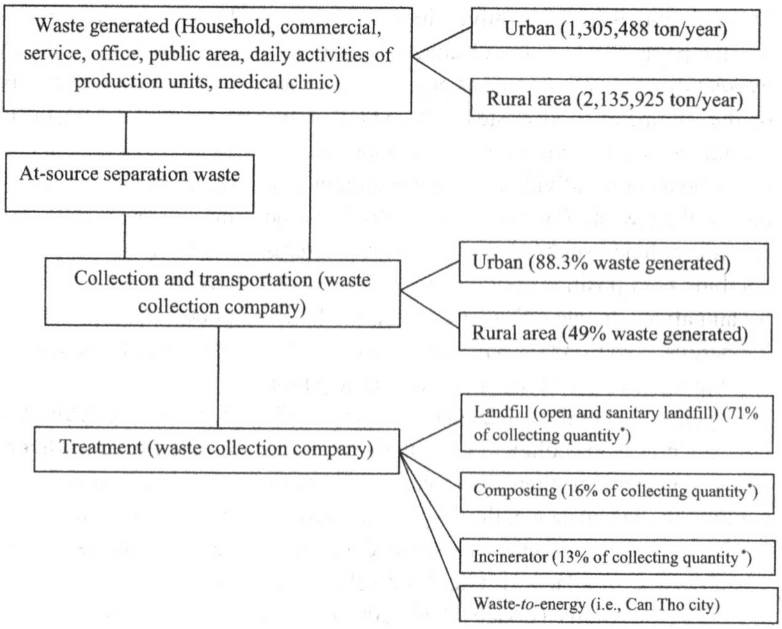

Public Awareness of and Participation in Municipal Solid Waste Management in Urban Areas of the Mekong River Delta, Vietnam, Fig. 1 The situation of solid waste management in the Mekong River Delta. (Note: *national data. Source: Compiled from reports of local Departments of Natural Resources and Environment of provinces in the Mekong River Delta)

basis for economic valuation. Non-market pricing methods include direct and indirect (or preference expression) approaches. Finding an equivalent concept for environmental goods such as improved waste management, on the other hand, is more complex (Carson 1998; Barkmann et al. 2008). Anex (1995), and Struk and Pojezdná (2019) used travel-cost-methods (TCM) to explore the costs associated with waste collection and movement, while Arimah (1996) used the hedonic price method (HPM) to estimate people's willingness to pay for improved service quality in solid waste management. However, these calculated values are unique and applicable only to the programs they were designed for. Thus, in order to fully value a non-market good outside of those contexts, the preference expression method must be used.

Preference statement methods, such as CVM and CM (Adamowicz et al. 1998), are direct valuation techniques based on individual assumptions. Both can reveal a person's actual preference for environmental goods through their behavior in a hypothetical market (Hanley et al. 1998a, b). To achieve this individuals are surveyed regarding their willingness to pay for environmental goods or accept compensation for the loss of those goods. The primary advantage of the preference statement approach is that it can be used to price any good or service with relatively little data (Diafas 2016). Additionally, the preference statement method has been deemed appropriate for estimating an environmental goods' indirect use and non-use values (Freeman 2003), which provides us with a firm rationale for employing the preference statement method in solid waste management program valuation studies.

A review of previous studies including Damigos et al. (2016), Maimoun et al. (2016), Ferreira and Marques (2015), Gillespie and Bennett (2013), Yusuf et al. (2007), Aadland and Caplan (2006), and Othman (2007) shows that randomization of the assignment of subjects to treatment groups, choice alternatives, and/or payment vehicle values is frequently used to examine an individual's preference for solid waste management. Moreover, we see that CVM approaches are quite commonly applied, under which non-market commodity pricing obscures the characteristics of the goods in question. By contrast, applications of the CM method in the study of waste management is regarded in the research literature such as Massarutto et al. (2019), Tarfasa and Brouwer (2018), Fukuda et al. (2018), Yuan and Yabe (2015), Czajkowski et al. (2014), Adeoti and Obidi (2010), Othman (2007), Sakata (2007),

Jin et al. (2006) as a novel contribution, where its application may serve as a baseline for comparison with other CM-related applied research. The CM method's fundamental theory is Lancaster's (1966) theory of multi-property consumption, which enables CM to ascertain the properties of non-market goods that are of interest to consumers. In support of comparing the two approaches, Adamowicz et al. (1998) established that CM is superior to CVM because it accurately describes the trade-offs between attributes. CM also avoids the bias and problems associated with surveys that use dichotomous questions that can be answered with a simple "yes" or "no" (Hanley et al. 2002). Thus, by incorporating the CM method into the valuation study of the solid waste management program, information about the program's attributes that are of interest to the public can be ascertained. Furthermore, the information serves as a necessary foundation for informing and proposing policies to enhance the quality of solid waste management services, thereby reducing solid waste and protecting the environment (Pearce et al. 2006).

Methodology

Survey Instrument Design

The design process for both the CVM and CM survey instruments began with an exhaustive review of available literature and initial discussions with waste management experts and resident groups. Information gathered was then used to determine the necessary framing data in each survey case and the attribute and payment card details. The survey questionnaire was then pre-tested with 20 randomly selected households to ascertain the questionnaire's format and content validity.

Respondents were first introduced to the proposed MSW program during the CVM survey. They were then informed that the program's objective was to reduce landfill solid waste by promoting waste avoidance and reduction methods (e.g., recycling, reusing). They were also informed as to how the MSW program would operate in practice via source separation. Then, respondents were offered a *cheap-talk* script as an *ex-ante* bias correction to remind them to consider financial limits and describe the payment amount as if it was real. The respondents were then asked if they would be willing to pay for the proposed program using open-ended questions based on five possible monthly payment bid values: 20,000 Vietnamese Dong (VND), VND50,000, VND80,000, VND110,000, and VND130,000 (which were then equivalent to US $0.87/month, US$2.18/month, US$3.49/month, US$4.81/month, and US$5.68/month, respectively (i.e., US$1 was equal to VND22,890 on June 30th, 2021). These bid values were based on additional discussions with local waste managers and public officials from urban joint-stock companies who serve as the local authority responsible for waste management in each province in Vietnam. The final bid values reflected the expected range of low and high-cost changes. Finally, respondents were asked to provide socio-demographic information about themselves and their perceptions of the benefits of the proposed MSW program.

When developing the CM questionnaire, the attributes of non-market goods and their levels of characteristics were identified (Pearce et al. 2006). The number of attributes considered was determined by the study's number of observations (Bateman et al. 2002). When numerous observations are collected, a study can reveal larger numbers of the characteristics of an environmental good. When the number of observations is limited, studies should focus on approximately four, five, or six characteristics. In this case, the study was focused on ascertaining the willingness of 380 respondents to pay for an MSW management service improvement program in MRD. As such, we limited our focus to four attributes. The MSW management service improvement program's four attributes were then divided into appropriate levels based on expert advice which included the percentage of MSW recycled (0%, 5%, 10%, and 15%); the rate at which CO_2 emissions are reduced (0%, 5%, 10%, and 15%); the number of MSW types categorized: not classified, two groups (recycled and the rest), three separation class groups (recycled, organic, and the rest);

P

and the monthly fee that would have to be paid by households for improved MSW management services (US$0.87/month, US$2.18/month, US $3.49/month, US$4.81/month, and US$5.68/ month). These attributes were in part based on the Ministry of Natural Resources and Environment's (2016, 2017, 2018) report on the state of solid waste management in Vietnam and research findings from Massarutto et al. (2019), Tarfasa and Brouwer (2018), Fukuda et al. (2018), Yuan and Yabe (2015), Czajkowski et al. (2014), and Jin et al. (2006). Therefore, the attributes estimated in this study were expected to be highly accurate and representative of the population.

Finally, 25 packages of choice-groups were created from the four attributes and their respective levels using an orthogonal combination. These packages were included in each of the five versions of the questionnaire. Each questionnaire also contained five distinct packages: the first questionnaire (version 1) contained the first five options; the second questionnaire (version 2) contained the sixth to tenth options; the third questionnaire (version 3) contained the 11th to 15th options; the fourth questionnaire (version 4) contained the 16th to 20th options. The fifth questionnaire (version 5) contained options 21 to 25. Each option package corresponds to option A in each questionnaire, including options A, B, and C. Option B was a substitute. In all cases, C is the option to retain the status quo. In other words, options A and B are the presumptive MSW management service quality improvement programs, as described by four distinct properties. Option C corresponded to the current MSW management service quality level, stating that there is no need to improve the quality of MSW management services. Table 1 provides an example questionnaire choice-set.

Collecting Data

Primary data for this study were gathered through direct interviews with 380 respondents from three major cities in the MRD including Can Tho (urban grade 1), Long Xuyen (An Giang province – urban grade 2), and Ca Mau (Ca Mau province – urban grade 3). According to a report published by the Ministry of Natural Resources and Environment (2016), urbanization is a significant factor contributing to the sharp increase in MSW. Thus the criteria for stratified sampling should reflect the region and population characteristics. In other urban areas of developing countries, stratified samples are also frequently used (see, for example, Chaudhry et al. 2007). Here we followed a stratified sampling approach so that the findings of this study might be applied to the remainder of the MRD region.

Public Awareness of and Participation in Municipal Solid Waste Management in Urban Areas of the Mekong River Delta, Vietnam, Table 1 Option design for the survey

Properties	Option A	Option B	Option C
The rate of MSW being recycled	Reduce 10% 	Reduce 5% 	**Do not** prefer either A or B
The rate of CO_2 emissions are reduced	Reduce 15% 	Reduce 15% 	
The number of types of MSW is classified	Recycling and remaining 	Non-classified 	
Fees for MSW management services	US$0.87/month	US$2.18/month	
Please tick only 1 of 3 options	□	□	□

Data analysis

Contingent Valuation Approach

The CVM method is based on the random utility theory of Luce (1959) and McFadden (1974) using the indirect utility function of households from the consumption of MSW management services and has the following form:

$$V(p, qi, M, \varepsilon) \quad (1)$$

where p is the price vector, q is the number of goods sold, M is income, and ε is the random error. If the quality of the MSW management service is improved, the utility of the household will be:

$$V(q_1, M, \varepsilon) \geq V(q_0, M, \varepsilon) \quad (2)$$

The probability of a household choosing an improved MSW management service is:

$$Pr[Yes] = Pr[V(q_1, \ M - t_k, \ \varepsilon_1) \geq V(q_0, \ M, \ \varepsilon_0)] \quad (3)$$

Assuming that the utility function is linear:

$$v(q_i, M) + \varepsilon_i \quad (4)$$

It is possible to write a formula to calculate the probability of the choice *Yes* is:

$$Pr[Yes] = Pr[v(q_1, \ M - t_k,) - v(q_0, \ M) + \varepsilon_1 - \varepsilon_0 \geq 0] \quad (5)$$

The household will choose *Yes*, when the sum of utility changes,

$$\Delta U = v(q_1, M - t_k,) - v(q_0, M) \quad (6)$$

and the difference of error, $\eta = \varepsilon_I - \varepsilon_O$, is larger than 0. Probability can be written as:

$$Pr[Yes] = Pr[\eta \geq -\Delta U] \quad (7)$$

Based on probability theory, Eq. (8) is then used:

$$Pr[Yes] = Pr[\eta \geq -\Delta U] = 1 - F\eta(-\Delta U) \quad (8)$$

In which, $F\eta$ is the probability density function (PDF) of η. To satisfy the condition of symmetric distribution here is:

$$(x) = 1 - F(x) \quad (9)$$

It is assumed to have an asymmetric distribution. The probability can be written as

$$Pr[Yes] = F\eta(\Delta U) \quad (10)$$

The cumulative density function (CDF) gives the probability of the observations having an eigenvalue of ΔU. The PDF shows the probability of an observation with an eigenvalue of ΔU. These two functions are similar in distribution. The relationship between these functions is F_η (ΔU) =, where $f(x)$ is the probability density function. Thus, the probability of observing the eigenvalues of ΔU is also the area under the PDF curve.

The model is estimated to have a maximum likelihood. If I_k represents the answer to the kth observation,

$$\text{with} \quad I_k = 1 : Pr[Yes] = Pr[I_k = 1] = Pr[\eta_k \leq \Delta U_k] = F_\eta(\Delta U_k) \quad (11)$$

$$\text{with} \quad I_k = 0 : Pr[No] = Pr[I_k = 0] = 1 - Pr[\eta_k \leq \Delta U_k] = 1 - F_\eta(\Delta U_k) \quad (12)$$

The likelihood function is written as:

$$L = \prod_{k=1}^{N} Pr\ [I_k = 1] Pr[I_k = 0] = \prod_{k=1}^{N} [F_\eta(\Delta U_k)]^{I_k} [1 - F_\eta(\Delta U_k)]^{1-I_k} \quad (13)$$

where N is the number of observations. If the log rational function is taken, this would get the rational function as:

$$\log\ L = \sum_{k=1}^{N} I_k \ln F_\eta(\Delta U_k) + (1 - I_k) \ln (1 - F_\eta(\Delta U_k)) \quad (14)$$

Based on probability theory, we then determine the CDF and estimate it using maximum likelihood estimation to find the values of the corresponding coefficients. A parametric method is used to estimate the mean and median of willingness to pay using the bid coefficient and variable coefficients related to attitude and other socioeconomic characteristics of households. The Logit model was used in this study to estimate the coefficients of these variables. The Logit model is a commonly used approach for estimating the cumulative density function when the random error has a normal distribution. The Logit model is presented in the following manner.

$$P_i = F(x_i'\beta) = \frac{e^{x_i'\beta}}{1 + e^{x_i'\beta}}, \quad (15)$$

where

$$x_i'\beta = \beta_0 + \beta_1 x_1 + \beta_2 x_2 + \beta_3 d_1 + \beta_4 x_3 + \beta_5 d_2 + \beta_6 d_3 + \beta_7 d_4 + \beta_8 d_5 \quad (16)$$

The probability of agreeing to pay the increased MSW management service fee is the dependent variable (Y). This variable has two values: $Y = 1$ if the respondent is willing to pay, and $Y = 0$ if the respondent is not willing to pay. The bid (X_1), the monthly fee per household for increased MSW management services, is one of the independent variables. These rates are based on government and local expert estimates for unsubsidized MSW services. The respondent's age is the variable age (X_2) (in years). The dummy variable male (D_1) has two values: $D_1 = 1$ if the respondent is male, and $D_1 = 0$ if the respondent is female. Edu (X_3) is the number of years the respondent has spent in school (in years). Higher-educated respondents are more likely to understand the benefits of increased MSW management services and the environmental harm caused by solid waste. As a result, they tend to pay higher for any program changes. Income D_2 (dummy variable) holds two values: $D_2 = 1$ if the respondent's monthly income is US $385.44 or more (this is the level of a deduction for taxpayers based on Law No. 26/2012/QH13 issued by the National Assembly of the Socialist Republic of Vietnam on November 22, 2012), and $D_2 = 0$ when less than US$385.44. The demand theory for environmental goods, in this case, assumes that as income rises so does the demand for environmental quality (Lewis and Tietenberg 2019). The non-classified variable (D_3 – dummy variable) has two values: $D_3 = 1$ if the respondent does not recycle MSW by classifying it, and $D_3 = 0$, if otherwise. Finally, urban settings were divided into types, which differed in terms of population size and density, non-agricultural labor rates, and architectural or infrastructure amenities (where type 1 > type 2 > type 3). The dummy variable Urbantype2 (D_4) has two values: $D_4 = 1$ if the respondent lives in an urban city type 2 and $D_4 = 0$, if otherwise. Finally, the dummy variable Urbantype3 (D_5) has two values: $D_5 = 1$ if the respondent is in an urban city type 3 and $D_5 = 0$, if not.

Choice Model approach

The CM approach is founded on Lancaster's (1966) multi-criteria utility theory combined with Thurstone's (1927) random utility theory. According to the random utility theory, an individual consumer's utility is composed of observable and unobservable components. Their assessment of product attributes determines the observable (quantifiable) component of an individual's utility. The unobservable portion is unpredictable and is determined by the individual's preferences. When an individual i consumes product j, that individual utility function is as follows:

$$U_{ij} = V_{ij} + e_{ij} = V(Z_{ij}, S_i) + e(Z_{ij}, S_i) \quad (17)$$

where V denotes the observable component. V_{ij} is a vector representing the degree of the attributes Z of the product j and the economic, social, and behavioral characteristics (S) of the respondent i, where e denotes the unobservable component. When presented with a choice set of numerous products with varying attributes, consumers will select the product that provides them with the most significant utility (max U). The probability that an individual i prefers product j to any other product m equals $U_j > U_m$. To be more precise, the probability of selecting product j of individual i (P_{ij}) will be as follows:

$$P(i) = P(U_{ij} > U_{im}) = P(V_{ij} + e_{ij} > V_{im} + e_{im});\ \forall m \in C \quad (18)$$

Assuming that the random component e_{ij} follows a homogeneous, independent and identical distribution (IID), and follows a Gumbell or Weibull distribution, the probability that a specific choice h is chosen is estimated by the multinomial logit (MNL) model as follows:

$$P(i) = \frac{EXP(V_{ij})}{\sum_{j \in C} EXP(V_{ij})} \quad (19)$$

The linear equation of utility for the choice of the jth product is written as:

$$V_{ij} = ASC + \beta_1 Z_1 + \beta_2 Z_2 + \beta_3 Z_3 + \cdots + \beta_k Z_k \quad (20)$$

Where k denotes the order of the product attributes, the coefficient β can be negative or positive, unique for each attribute, and is "valued" according to each individual's subjective preferences. The coefficient β will vary between subgroups of individuals within a population but will remain constant among individuals within each subgroup. Although there are numerous methods for eliminating mismatches between choices and improving model fit, this research employed the MNL model to minimize error and produce the most accurate results.

Based on formula (20), the utility function of choice A, B, and C in each set of questions can be presented as follows:

Option A $V_1 = ASC + \beta_1 fee + \beta_2 waste + \beta_3 co2 + \beta_4 sep2 + \beta_5 sep3$

Option B $V_2 = ASC + \beta_1 fee + \beta_2 waste + \beta_3 co2 + \beta_4 sep2 + \beta_5 sep3$

Option C $V_3 = \beta_1 fee + \beta_2 waste + \beta_3 co2 + \beta_4 sep2 + \beta_5 sep3$

where V_j denotes the utility function for choice j, and ASC denotes a constant for each choice in the model. The marginal willingness to pay (MWTP) for improved properties via the proposed MSW program is estimated by the marginal rate of substitution between the non-monetary attribute parameter $\beta_{\text{non - monetary attribute}}$, and the monetary attribute factor $\beta_{\text{monetary attribute}}$ follows:

$$MWTP = -\frac{\beta_{\text{non-monetary attribute}}}{\beta_{\text{monetary attribute}}}$$

The coefficients $\beta_{non\ -\ monetary\ attribute}$ and $\beta_{monetary\ attribute}$ are estimated from the MNL model. The variables included in the MNL model are the attributes of the MSW management service quality improvement program. These variables include the variable *fee*, variable *waste*, *co2*, and two separation variables *sep2* and *sep3*. The variable *fee* represents the charging attribute for increased MSW management service stated as the monthly payment the household will incur for the MSW service (US$/month). The variable *waste* (%) represents the percentage of recycled solid waste. The variable *co2* represents an attribute of reduced CO_2 emissions (%) from changes to waste management. In contrast, the variables *sep2* and *sep3* represent numeric attributes of the possible types that MSW may need to be separated into (recycled, organic, residual). *Sep2* is a dummy variable that takes two values: $sep2 = 1$, if the number of MSW types is classified into two types, recycled and remaining, and $sep2 = 0$ if the number of MSW types is not classified into any type (unclassified). *Sep3* is another dummy variable with two values: $sep3 = 1$ if the number of MSW types is classified into three categories (recycled, organic, and residual), or $sep3 = 0$ if the MSW is not classified into those three categories.

Results and Discussion

Descriptive Statistics

Based on the survey data, most respondents are women (70%), while men comprised only 30% of the study sample. In addition, the survey data demonstrates that the survey sample is quite diverse in age, occupation, and income. The

P

respondents' ages ranged from 18 to 88 years and averaged around 50 years. Respondents work in various occupations including private business owners, government employees, home traders, housewives, and retirees. Entrepreneurs and housewives account for 33% and 26% of respondents, respectively. Furthermore, respondents include employees of private companies (3%), state employees (6%), and retired individuals (10%). The majority of respondents earn less than US$386 per month. The results show that the greater the proportion of respondents with higher incomes, the lower the proportion. The majority of respondents have completed education levels 1, 2, and 3 (coded for primary school, junior high school, and high school, respectively), and only a small percentage are unable to attend school.

In the survey, respondents were asked to estimate the amount of their individual household's recycled waste. There were 255 out of 380 households in the survey currently conducting recycling activities. However, that recycling amounts to approximately 124 g per day per household. By comparison, MSW generation amounts to approximately 1842 g per day per household, implying that the average recycling volume of each household is less than 10%. This figure is even higher when only recycling households are included in the data (185 g). It is notable that Can Tho City has the lowest recycling rate as compared to the less urbanized areas (Long Xuyen and Ca Mau) in MDR (Tables 2 and 3).

When considering the reasons for recycling or not recycling, the lowest recycling rate of households in higher urbanization levels (Can Tho) compared to the other less urbanized cities can be explained. Respondents reported that recycling activities are time-consuming, require space, necessitate the use of containers, and are harmful to one's health. Significantly though, one of the primary reported reasons to recycle is to generate additional revenue. Policymakers can use this revenue-earning opportunity to increase motivation when implementing the MSW development program on a national scale. To facilitate the adoption of this program, additional information, such as those mentioned earlier, should be provided to households that have not yet been recycling. A social marketing campaign may be

Public Awareness of and Participation in Municipal Solid Waste Management in Urban Areas of the Mekong River Delta, Vietnam, Table 2 The total quantity of recycled waste by surveyed households (grams/day/household)

	Mean	Std. Dev.	Min	Max	Diff. CT-LX	CT-CM	LX-CM
Can Tho – CT ($n = 146$)	44.26	71.37	0	428.57	−9.2851***	−5.9887***	1.5490^{ns}
Long Xuyen – LX ($n = 120$)	193.99	178.30	0	666.67			
Ca Mau – CM ($n = 114$)	154.82	208.04	0	766.67			
Total ($n = 380$)	124.71	170.64	0	766.67			

Source: Survey data, 2019
Note: *, **, and *** are statistically significant at 10%, 5%, and 1%, respectively, and *ns* are not statistically significant

Public Awareness of and Participation in Municipal Solid Waste Management in Urban Areas of the Mekong River Delta, Vietnam, Table 3 Quantity of recycled waste (involving scrap selling only)

	Mean	Std. Dev.	Max	Diff. CT-LX	CT-CM	LX-CM
Can Tho – CT ($n = 84$)	76.93	79.72	428.57	−8.8714***	−5.8926***	0.8383^{ns}
Long Xuyen – LX ($n = 92$)	250.86	164.74	666.67			
Ca Mau – CM ($n = 78$)	226.28	217.10	766.67			
Total ($n = 255$)	185.06	11.23	766.67			

Source: Survey data, 2019
Note: *, **, and *** are statistically significant at 10%, 5%, and 1%, respectively, and *ns* are not statistically significant

necessary to encourage the households to recycle and to counter the belief that it is a waste of time and only provide little additional income (Table 4).

Analysis of CVM and CM Data for an Improved Management Program

CVM Approach Result

The findings shown in Fig. 2 suggest that approximately 66% of the households are willing to participate in MSW management programs. The proportion of respondents' willingness to pay tended to decrease as the bid rate increased. For example, most respondents agree to participate in the lower bid (95% for the bid US$0.87). The participation rates, however, decrease to 78%, 65%, and 40% at bid levels of $2.18, US$3.49, and US$5.68/month, respectively. These findings are consistent with economic demand theory, which states that increased prices will have a negative effect on the demand for goods or

Public Awareness of and Participation in Municipal Solid Waste Management in Urban Areas of the Mekong River Delta, Vietnam, Table 4 Stated reasons to recycle and not to recycle

Reason		Most important	Second important	Third important
Experienced recycling households ($n = 254$)	It takes not too much time	47 (18.50%)	28 (11.02%)	64 (25.30%)
	Doesn't take up much space	48 (18.90%)	73 (28.74%)	53 (20.95%)
	Provides additional income	72 (28.35%)	65 (25.59%)	59 (23.32%)
	Improves environment quality	47 (18.50)	70 (27.56%)	48 (18.97%)
	Does not affect health	40 (15.75%)	15 (5.91%)	29 (11.46%)
Non-experienced recycling households ($n = 126$)	Takes time to collect	26 (20.63%)	20 (15.87%)	26 (20.63%)
	Requires storage space	14 (11.11%)	25 (19.84%)	36 (28.57%)
	Requires containers	8 (6.35%)	23 (18.25%)	9 (7.14%)
	Money from this activity is small	59 (46.83%)	31 (24.60%)	21 (16.67%)
	Don't think scrap can be sold	5 (3.97%)	15 (11.90%)	13 (10.32%)
	Negative effects to health	3 (2.38%)	3 (2.38%)	15 (11.90%)

Source: Survey data, 2019

P

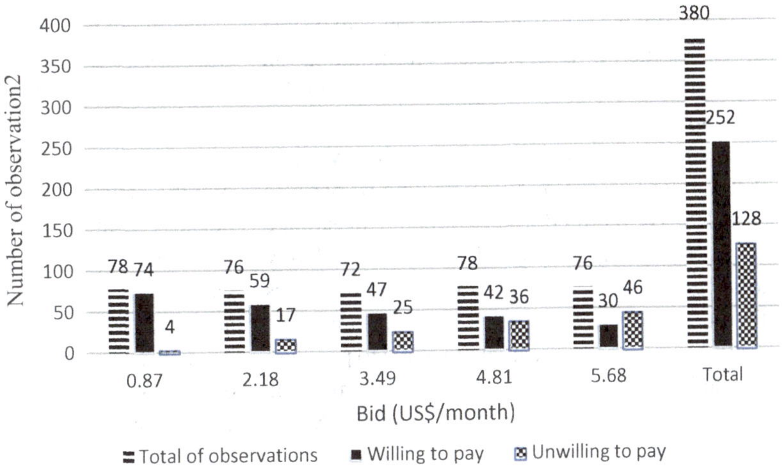

Public Awareness of and Participation in Municipal Solid Waste Management in Urban Areas of the Mekong River Delta, Vietnam, Fig. 2 The proportion of willingness to pay for improved MSW management services. (Source: Survey data, 2019)

services. However, the most critical finding here is the high level of household willingness to participate in the program.

Table 5 below summarizes the Logit model results analyzing the influence on the decision to participate in the program to improve the quality of MSW management services. The respondents were classified into three groups based on their level of urbanization (urban-type group). Can Tho is home to the most urbanized households, while Long Xuyen is home to the least urbanized. The primary objective of this stratification was to ascertain the effects of urbanization on WTP levels and the determinants of these levels. Interestingly, some differences in determinants across three areas with varying degrees of urbanization were observed. The distinction was that variables such as education and income were statistically significant at Can Tho only for those with a higher level of education and income.

Additionally, while the classified variable was not statistically significant in the three separate models, a 10% significance level was observed in the combined mode. It highlights the distinctions among those experienced with waste-separating activities and others. When considering the bid variable, as previously demonstrated by Rahji and Oloruntoba (2009), Pek and Othman (2010), and Altaf and Deshazo (1996), the higher payments required from the program will therefore decrease the proportion of participants in the program, with a statistically significant at the 1% level result in each of the separated and combined models. As a result, the appropriate bid level elicited from the following discussion is expected to contribute to policymakers formulating an appropriate payment level when establishing this program.

Respondents who did not separate MSW stated that they did so due to a lack of time or because sorted MSW required a large amount of storage space. This group of respondents was more likely to suggest their willingness to pay higher prices for the services than others. Respondents in Type 2 and 3 cities were more likely to agree to pay than respondents in Type 1 cities. This may be explained by the fact that existing MSW management systems in Type 1 cities are more thorough than those in other urban settings. As a result, respondents in Type 2 and 3 cities have higher expectations that the implementation of increased MSW management services will positively contribute to environmental improvement.

The WTP mean value derived from both CVM and CM estimates is presented in Table 6. As shown, the mean value varies between US$3.87 and US$6.61. It is critical to emphasize that both approaches share these values. As a result, increased confidence was gained in the payment value provided by this research, which can contribute to the program's implementation. Interestingly, the area with the lowest level of urbanization has a higher value than the area with the highest level of urbanization. This may be explained by the fact that Can Tho City piloted a waste separation program, and thus residents have some experience with the process. However, this program was not successful for several reasons including poor quality of final services and unfamiliarity among many households. As a result, local governments should consider these factors to ensure success when implementing a new program.

CM Approach Result

While there are minor differences among the three groups the determinant outcomes are consistent across the three models. For instance, households care about the properties of recycled solid waste and the rate at which CO_2 is reduced. However, the attribute that reduces CO_2 emissions is more appealing to people because the marginal willingness to pay for the property (US$9.62/month) is more than double the marginal willingness to pay for the property (US$5.09/month), thereby increasing the rate of re-payment (US$5.09/month). The coefficient for the variable fee is negative, indicating a negative correlation between household payment decisions and the variable fee. It is notable that in the CVM approach, as the bid level increases, the probability of the improvement program receiving WTP decreases. As a result, this is another point at which the two approaches are consistent. Thus,

Public Awareness of and Participation in Municipal Solid Waste Management in Urban Areas of the Mekong River Delta, Vietnam, Table 5 The determinants of willingness to pay for improvement of the MSW management program

	Can Tho		Long Xuyen		Ca Mau		Combined group	
Variables	Coef.	*z* value	Coef.	*z* value	Coef.	*z* value	Coef.	*z* value
Constant	2.316155*	1.87	5.237349***	3.54	2.6637*	1.87	2.5517***	3.26
Bid	−0.307716***	−5.20	−0.335705***	−4.48	−0.0234***	−3.02	−0.0292***	−7.60
Age	−0.0248744^{ns}	−1.52	−0.0394147*	−2.05	0.0010^{ns}	0.05	−0.0235**	−2.24
Male	−0.4082008^{ns}	−0.89	0.5342377^{ns}	0.97	0.1463^{ns}	0.27	0.1117^{ns}	0.39
Edu	0.0892957*	1.48	−0.0337889^{ns}	−0.50	0.0309^{ns}	0.44	0.0368^{ns}	1.00
Income	0.000115*	1.93	0.0001458^{ns}	1.41	0.0000531^{ns}	0.69	0.0001**	2.43
Non-classified	0.5061549^{ns}	1.17	0.4776936^{ns}	0.84	0.6977685^{ns}	1.11	0.5625*	1.94
Urbantype2							1.1760***	3.57
Urbantype3							1.6972***	4.89
Log-likelihood	−76.235		−57.1785		−49.793		−186.685	
LR Chi^2 (8)	48.17		38.41		15.06		112.21	
Prob > Chi^2	0.0000		0.0000		0.0198		0.0000	
Pseudo R^2	0.2401		0.2514		0.1313		0.2311	
Observation	146		120		114		380	

Source: Survey data, 2019

Note: *, **, and *** are statistically significant at 10%, 5%, and 1%, respectively, and *ns* are not statistically significant

P

Public Awareness of and Participation in Municipal Solid Waste Management in Urban Areas of the Mekong River Delta, Vietnam, Table 6 Willingness to pay mean value estimated from CVM and CM

		MWTP	Lower bound	Upper bound	ASL
Can Tho	CVM	3.87	3.29	4.51	0.0000
Long Xuyên	CVM	4.79	4.18	5.71	0.0000
Ca Mau	CVM	6.54	5.34	10.76	0.0006
Combined group	CVM	4.87	4.48	5.40	0.0000
	CM	4.61	4.11	5.11	

Note: ASL is the significance level for hypothesis tests: H_0: WTP $<= 0$, H_1: WTP > 0

the more the MSW management program is improved toward gradually increasing the rate of recycled MSW and decreasing its CO_2 content, the more likely people will participate. However, the fee for the improvement program should be reasonable, as this is one of the factors that will reduce citizen participation.

Individuals are interested in the attributes examined in this study, as demonstrated in Table 7. Ca Mau, again has the highest mean WTP across all attributes. This result is consistent with the findings of the CVM. This result may reflect the perceptions and characteristics of households engaged in MSW separation. As discussed previously, households with lower education and income levels may not anticipate many environmental benefits (or attributes) from the MSW program, which is a behavioral change initiative. Moreover, the past pilot program in more urbanized areas was shown to be ineffective. As a result, residents of this area have less reason to participate in this program. What is more interesting is that the current fees are not enough to cover solid waste collection, transportation, and treatment. The total cost of all of these stages is approximately US$20.53 per year and person (World Bank 2018). As a result, the government's budget has been strained. According to the Vietnam Ministry of Finance (2015), state budget expenditures on MSW management have nearly doubled from US$272 million in 2010 to US $498 million in 2015. The research also revealed that people's concerns about waste recycling are largely in line with Prime Ministerial Directive 33 (2020) suggesting that positive opportunities for participation and change at the household level are possible, if implemented correctly.

Concluding and Policy Implications

According to the research results reported here, one factor that influences people's willingness to pay is the rate of recycled MSW. People are willing to pay an extra fee if more MSW is recycled, reducing the amount of MSW that must be treated and discharged into the environment. Apart from increasing the amount of recycled MSW, CO_2 emissions are a factor that households considered and discussed during the survey process. The study results also indicate that households are willing to pay for and accept the program if it successfully reduces CO_2 emissions. Thus, when implementing the program, it is critical to emphasize these two benefits to increase households' willingness to pay. Apart from 72.2% who agreed, 27.8% continue to oppose the MSW program's classification. As a result, it is necessary to overcome the barriers to increase support for MSW classification at the source and improve participation in the program to improve the quality of household waste management services. On the other hand, the time factor is one reason that contributes to the source's disapproval of household MSW classification. Approximately 39.6% of households do not support the MSW classification. To them, it is time-consuming and has a negative impact on their work and personal lives. As a result, local governments may need to run a social marketing campaign to guide and encourage participation in the MSW program. Moreover, in daily practice, households frequently do not classify MSW but instead combine them all. Consequently, collaboration among relevant government agencies is vital to further implement solid waste separation practices such as the use of tanks

Public Awareness of and Participation in Municipal Solid Waste Management in Urban Areas of the Mekong River Delta, Vietnam, Table 7 Determinant of households' willingness-to-pay for source-separation of solid waste

	Can Tho		Long Xuyen		Ca Mau		Combined group	
Variables	Coeff.	*z* value	Coeff.	*z* value	Coeff.	*z* value	Coeff.	*z* value
Fee	−0.1470***	−6.80	−0.1019***	−6.08	−0.0608***	−6.43	−0.0886***	−11.15
Waste	11.2842***	3.44	9.9846***	3.03	10.6546***	4.29	10.2312***	6.81
CO_2	43.6008***	5.78	8.2191**	2.38	18.5381***	5.64	19.3538***	7.81
Sep2	−0.1180^{ns}	−0.45	0.4534*	1.85	1.0184***	4.00	0.4887***	3.79
Sep3	−0.9830**	−2.34	0.6499**	2.20	1.4489***	4.55	0.4887***	2.80
Log-likelihood	−314.7123		−308.6301		−336.2377		−1029.4816	
Observations	146		120		114		380	
Chi^2	163.45		167.31		64.99		348.07	
Prob > Chi^2	0.0000		0.0000		0.0000		0.0000	

Source: Survey data, 2019
Note: *, **, and *** are statistically significant at 10%, 5%, and 1%, respectively, and *ns* are not statistically significant

P

with clear and distinguishable colors and symbols. There are also advertising and recommendation activities to assist people in understanding and implementing MSW classification through images and charts.

References

Aadland, D., & Caplan, A. J. (2006). Curbside recycling: Waste resource or waste of resources? *Journal of Policy Analysis and Management: The Journal of the Association for Public Policy Analysis and Management, 25*(4), 855–874.

Adamowicz, W., Boxall, P., Williams, M., & Louviere, J. (1998). Stated preference approaches for measuring passive use values: Choice experiments and contingent valuation. *American Journal of Agricultural Economics, 80*(1), 64–75.

Adeoti, A., & Obidi, B. (2010). Poverty and preference for improved solid waste management attributes in Delta-State, Nigeria. *Journal of Rural Economics and Development, 19*(1623-2016-134902), 15–33.

Altaf, M. A., & Deshazo, J. R. (1996). Household demand for improved solid waste management: A case study of Gujranwala, Pakistan. *World Development, 24*(5), 857–868.

Anex, R. P. (1995). A travel-cost method of evaluating household hazardous waste disposal services. *Journal of Environmental Management, 45*(2), 189–198.

Arimah, B. C. (1996). Willingness to pay for improved environmental sanitation in a Nigerian City. *Journal of Environmental Management, 48*(2), 127–138.

Barkmann, J., Glenk, K., Keil, A., Leemhuis, C., Dietrich, N., Gerold, G., & Marggraf, R. (2008). Confronting unfamiliarity with ecosystem functions: The case for an ecosystem service approach to environmental valuation with stated preference methods. *Ecological Economics, 65*(1), 48–62.

Bateman, I. J., Carson, R. T., Day, B., Hanemann, M., Hanley, N., Hett, T., & Sugden, R. (2002). *Economic valuation with stated preference techniques: A manual.*

Carson, R. T. (1998). Valuation of tropical rainforests: Philosophical and practical issues in the use of contingent valuation. *Ecological Economics, 24*(1), 15–29.

Chaudhry, P., Singh, B., & Tewari, V. P. (2007). Non-market economic valuation in developing countries: Role of participant observation method in CVM analysis. *Journal of Forest Economics, 13*(4), 259–275.

Czajkowski, M., Kądziela, T., & Hanley, N. (2014). We want to sort! Assessing households' preferences for sorting waste. *Resource and Energy Economics, 36*(1), 290–306.

Damigos, D., Kaliampakos, D., & Menegaki, M. (2016). How much are people willing to pay for efficient waste management schemes? A benefit transfer application. *Waste Management & Research, 34*(4), 345–355.

Diafas, I. (2016). *Estimating the Economic Value of forest ecosystem services using stated preference methods: The case of Kakamega Forest, Kenya*. Doctoral dissertation, Niedersächsische Staats-und Universitätsbibliothek Göttingen.

Driesen, D. (2006). Economic instruments for sustainable development. In: B. J. Richardson, S. Wood (Eds.), *Environmental law forsustainability.* Portland: Hart Publishing

Ferreira, S., & Marques, R. C. (2015). Contingent valuation method applied to waste management. *Resources, Conservation and Recycling, 99*, 111–117.

Freeman, M. (2003). *The measurement of environmental and resource values: Theory and methods.* Washington, DC: Resources for the Future.

Fukuda, K., Isdwiyani, R., Kawata, K., & Yoshida, Y. (2018). Measuring the impact of modern waste collection and processing service attributes on residents' acceptance of waste separation policy using a randomised conjoint field experiment in Yogyakarta Province, Indonesia. *Waste Management & Research, 36*(9), 841–848.

Gellynck, X., Jacobsen, R., & Verhelst, P. (2011). Identifying the key factors in increasing recycling and reducing residual household waste: A case study of the Flemish region of Belgium. *Journal of Environmental Management, 92*(10), 2683–2690.

Gillespie, R., & Bennett, J. (2013). Willingness to pay for kerbside recycling in Brisbane, Australia. *Journal of Environmental Planning and Management, 56*(3), 362–377.

Government of Vietnam. (2020). Directive No. 33/CT-TTg of the Prime Minister: On strengthening the management, reuse, recycling, treatment and reduction of plastic waste. Hanoi.

Hanley, N., MacMillan, D., Wright, R. E., Bullock, C., Simpson, I., Parsisson, D., & Crabtree, B. (1998a). Contingent valuation versus choice experiments: Estimating the benefits of environmentally sensitive areas in Scotland. *Journal of Agricultural Economics, 49*(1), 1–15.

Hanley, N., Wright, R. E., & Adamowicz, V. (1998b). Using choice experiments to value the environment. *Environmental and Resource Economics, 11*(3), 413–428.

Hanley, N., Wright, R. E., & Koop, G. (2002). Modelling recreation demand using choice experiments: Climbing in Scotland. *Environmental and Resource Economics, 22*(3), 449–466.

Huang, B., Wang, W., Bates, M., & Zhuang, X. (2008). Three-dimensional super-resolution imaging by stochastic optical reconstruction microscopy. *Science, 319*(5864), 810–813.

Jin, J., Wang, Z., & Ran, S. (2006). Comparison of contingent valuation and choice experiment in solid waste management programs in Macao. *Ecological Economics, 57*(3), 430–441.

Ko, S., Kim, W., Shin, S.-C., & Shin, J. (2020). The economic value of sustainable recycling and waste management policies: The case of a waste management crisis in South Korea. *Waste Management, 104*, 220–227.

Lancaster, K. J. (1966). A new approach to consumer theory. *Journal of Political Economy, 74*(2), 132–157.

Lewis, L., & Tietenberg, T. (2019). *Environmental economics and policy*. Routledge.

Longe, E. O., Longe, O. O., & Ukpebor, E. F. (2009). People's perception on household solid waste management in Ojo Local Government Area in Nigeria. *Iranian Journal of Environmental Health Science and Engineering, 6*, 209–216.

Luce, R. D. (1959). *Individual choice behavior a theoretical analysis*. Wiley.

Maimoun, M. A., Reinhart, D. R., & Madani, K. (2016). An environmental-economic assessment of residential curbside collection programs in Central Florida. *Waste Management, 54*, 27–38.

Massarutto, A., Marangon, F., Troiano, S., & Favot, M. (2019). Moral duty, warm glow or self-interest? A choice experiment study on motivations for domestic garbage sorting in Italy. *Journal of Cleaner Production, 208*, 916–923.

McFadden, D. (1974). Conditional logit analysis of qualitative choice behavior. In P. Zarembka (Ed.), *Frontiers in econometrics*. New York: Academic.

Ministry of Environment of Japan. (2014). *History and current state of waste management in Japan*. Tokyo: Ministry of Environment of Japan.

Ministry of Natural Resources and Environment of Vietnam. (2016). *Report on the current state of the national environment for the period 2011–2015*. Hanoi: Publishing House of Natural Resources – Environment and Maps of Vietnam.

Ministry of Natural Resources and Environment of Vietnam. (2017). *Report on the current state of the national environment 2016*. Hanoi: Publishing House of Natural Resources – Environment and Map of Vietnam.

Ministry of Natural Resources and Environment of Vietnam. (2018). *Report on the current state of the national environment 2017*. Hanoi: Publishing House of Natural Resources – Environment and Maps of Vietnam.

Ministry of Natural Resources and Environment of Vietnam. (2020). *Report on the state of the national environment in 2019*. Hanoi: Dan Tri Publishing House.

Othman, J. (2007). Economic valuation of household preference for solid waste management in Malaysia: A choice modeling approach. *International Journal of Management Studies (IJMS), 14*(1), 189–212.

Palatnik, R., Body, S., Ayalon, O., & Shechter, M. (2014). *Greening household behaviour and waste* (OECD Environment Working Paper 76).

Pearce, D., Atkinson, G., & Mourato, S. (2006). *Cost-benefit analysis and the environment: recent developments*. Organisation for Economic Co-Operation and Development.

Pek, C. K., & Othman, J. (2010). *Household demand for solid waste disposal options in Malaysia*.

Rahji, M. A. Y., & Oloruntoba, E. O. (2009). Determinants of households' willingness to pay for private solid waste management services in Ibadan, Nigeria. *Waste Management & Research, 27*(10), 961–965.

Sakata, Y. (2007). A choice experiment of the residential preference of waste management services – The example of Kagoshima city, Japan. *Waste Management, 27*(5), 639–644.

Schübeler, P., Christen, J., & Wehrle, K. (1996). *Conceptual framework for municipal solid waste management in low-income countries* (Vol. 9). St. Gallen: SKAT (Swiss Center for Development Cooperation).

Slack, R. J., Gronow, J. R., & Voulvoulis, N. (2009). The management of household hazardous waste in the United Kingdom. *Journal of Environmental Management, 90*(1), 36–42.

Stafford, R. (2002). *The role of environmental stress and physical and biological interactions on the ecology of high shore littorinids in a temperate and a tropical region*. Doctoral dissertation, University of Sunderland.

Struk, M., & Pojezdná, M. (2019). *Non-market value of waste separation from municipal perspective*. In Proceedings of 17th International Waste Management and Landfill Symposium (Sardinia 2019).

Tarfasa, S., & Brouwer, R. (2018). Public preferences for improved urban waste management: A choice experiment. *Environment and Development Economics, 23*(2), 184.

Thurstone, L. L. (1927). A law of comparative judgment. *Psychological Review, 34*(4), 273.

Vietnam Ministry of Finance. (2015). Financial resources for environmental protection for the period 2011–2015, orientation for the period 2016–2020. The 4th National Environment conference, 2015.

Wagner, T., & Arnold, P. (2008). A new model for solid waste management: an analysis of the Nova Scotia MSW strategy. *Journal of Cleaner Production, 16*(4), 410–421.

World Bank. (2018). *Report on solid and industrial hazardous waste management assessment options and action areas to implement the national strategy*. Hong Duc Publishing House. https://documents1.worldbank.org/curated/en/352371563196189492/pdf/Solid-and-industrialhazardous-waste-management-assessment-options-and-actions-areas.pdf

Yuan, Y., & Yabe, M. (2015). Residents' preferences for household kitchen waste source separation services in Beijing: A choice experiment approach. *International Journal of Environmental Research and Public Health, 12*(1), 176–190.

Yusuf, S. A., Salimonu, K. K., & Ojo, O. T. (2007). Determinants of willingness to pay for improved household solid waste management in Oyo State, Nigeria. *Research Journal of Applied Sciences, 2*(3), 233–239.

P

Public Buying

▶ Public Procurement for Regional and Local Development

Public Governance

▶ Regulation of Urban and Regional Futures

Public Policies

▶ European Green Deal and Development Perspectives for the Mediterranean Region

Public Policies to Increase Urban Green Spaces

Case studies from Belgium, Netherlands, and Germany

Luiza O. Voinea
Urban Planner Certified by The Romanian Register of Urban Planners, Bucharest, Romania

Synonyms

Approach, Green justice, Guidelines, Healthy urban environment, Strategy

Introduction

The European Union has multiple initiatives (papers, statements, strategies, programs, and tools) aimed not only to mobilize cities, urban agglomerations, and metropolitan areas in achieving the objectives on environmental issues, biodiversity, and climate change, but also to promote good practices among Member States (e.g., Covenant of Mayors for Climate & Energy, Green City Accord, European Green Capital Network, European Green Leaf, and European Green City Tool). The 7th General Union Environment Action Programme to 2020 entitled "Living well, within the limits of our planet" confirms that urban agglomerations are facing an increased level of air, water, and soil pollution, stating that it is expected that more European cities will implement sustainable policies in urban planning and development (Endl and Berger 2014). This movement is strongly encouraged by the European funds available in this purpose. One of the main objectives of this European program was not only to promote a sustainable urban environment, within the limits of the natural environment, but also to evaluate the performances achieved by various urban agglomerations on environmental issues, considering their specific economic, social, and territorial context. The interest in ensuring a healthier urban environment is reinforced by the fact that according to the statistics, in 2018, over 70% of the European Union population lived in urban areas, following an upward trend.

Another European program aligned with these objectives is "The environment action programme to 2030" based on the "European Green Deal." It focuses on measures aimed to reduce air pollution, to ensure health protection for the European community, to support biodiversity, and to promote climate-resilient settlements. In addition, the European Green Deal through the European Urban Initiative offers assistance to the urban areas interested in aligning their development strategies toward a more sustainable approach.

These European objectives exert an increased pressure on regional and local authorities that have to find the tools and resources necessary to improve the urban environment, in the context of a less predictable dynamic of the urban settlements.

One potential way to improve air quality in urban areas is by increasing the green surfaces that can filter pollution from the air, but this becomes very difficult considering that the public domain on which public authorities can develop this type of projects has been significantly reduced due to the development pressures.

There is a significant change in the balance between public and private domain, many urban sites being subject to privatization, followed by the development of new public facilities using public-private partnerships or only private resources, as a condition for the building permits.

This entry focuses on the alternative solutions proposed by the local authorities from three different countries, in order to expand the green areas within the city, overcoming the public budget limitation and free public land availability.

The case studies were selected based on four criteria: the urban-sprawl phenomenon and urban expansion to the surrounding territories, the green space per capita compared to the European average, the level of air pollution and how it evolved during the last years of urban expansion, and the innovative public policies regarding green areas.

Following the European Union objective to become "climate resilient" until 2050 and to promote a healthier and sustainable urban environment, the central and local authorities have to identify ways to increase the green areas within the cities and urban agglomerations, to ensure a coherent urban development and to improve the access to public facilities, evenly distributed in the territory. The three case studies represent examples of alternative urban policies on green areas, with a high level of transferability to other Member States with similar urban challenges.

Case Study from Belgium

Belgium is among the European countries with the highest weighted urban proliferation and dispersion according to a report that made a comparison analysis between 2006 and 2009 (European Environment Agency 2016). Despite this territorial pressure, the level of air pollution in urban areas has decreased in recent years, as presented by the European Environment Agency. In 2018, the levels of BaP, PM 2.5, and PM 10 were declared 0.0; the only values that still affect the urban environment are those of NO2 (with 2.3% of the urban population affected) and O3 (with 51.2% of the urban population affected) (European Environment Agency 2020a).

Ghent is a member of the European Green Capital Network (an initiative of the European Commission) along with other cities that have won the European Green Capital Award or have qualified among the finalist cities of this competition. Ghent faced many administrative and urban challenges, especially starting from 1977 when the Belgian Government decided to annex the surrounding urban settlements to the main city, resulting a high increase of the population (from 140.000 to 249.000) and a larger urban area that required an integrated approach (Boussauw 2014).

The total population has a very dynamic character as in 2000 there were about 224.180 inhabitants, while in 2015 there were already 253.270 inhabitants, resulting in an increase of 12% in just 5 years.

By 2030, the population may reach 269.000 inhabitants (Department of Urbanism and Spatial Planning 2018). This demographic pressure on the urban territory hardens the local authority mission to support a sustainable and healthy urban environment.

According to the estimations, 10,800 new homes will be built by 2030, residential areas that need access to all public facilities, including green spaces within walkable distances (Department of Urbanism and Spatial Planning 2018).

Reducing Hardened Surfaces from Public or Private Properties in Ghent

Policy Background

Inside the city of Ghent, the green areas cover 38.8% of the total administrative territory (Arcgis 2017), resulting in 15.55 sqm of green areas per capita (2019) (Maes et al. 2019), slightly below the European average value. The hardened surfaces (constructions, pavements, and infrastructure) cover 46% of the administrative territory with a higher percentage in the central of around 80%.

One of the main causes leading to the urban heat island phenomenon is the large hardened surface, significantly reducing green areas with a high impact on the rainwater absorption capacity and also on the air quality. In order to reduce this phenomenon, the local authorities proposed ways to control the new hardened surfaces through the urban policies defined in the Ghent Climate Adaptation Plan 2016–2019 (City of Ghent 2016).

Starting from 2004, after signing the European Covenant of Mayors, the local authorities started to develop strategies aimed to create a less vulnerable city to climate change and a healthier environment for its inhabitants. Ghent was among the first cities of Flanders that aligned to the European Covenant of Mayor objectives and was selected as a pilot project through the European initiative entitled "Cities adapt."

Proposal and Implementation

In order to properly address the environmental issues and integrate the European objectives, the local authorities developed Ghent Climate Adaptation Plan. The second action plan for 2016–2019 defined the principles that have to be followed in urban design and planning activity. The first principle is related to the prevention of more sealed soil by controlling the maximum hardened surface for each urban development. The second and third principles are focusing on greening the city by using all the available surfaces and by developing the green-blue network (City of Ghent 2016).

One of the measures proposed by this action plan was to upgrade the existing public spaces by reducing the hardened surfaces to the minimum required, not to impact its functionality. In addition, the local administration is trying to identify new areas that can become greener in order to compensate new urban developments (infrastructure and buildings).

The canal banks were redesigned with new green areas aimed to support the green blue infrastructure of the city. New restrictions have been added also for private investments, as the mandatory site plan that has to follow the principle of reducing the hardened surfaces and the solutions for rainwater management. The real estate developers have the opportunity to compensate the impact of their new investment by designing green facades and terraces (also considered as a small-scale green network) and improving the urban environment in their surrounding area. These requirements are not only designed exclusively for large residential developments, but also for individual homes whose courtyards and facade gardens are in most cases covered with impervious pavements for car parking or garden pavilions. The new measures were integrated in the Local Urban Regulation as a mandatory approach for all public or private entities. The new investments should focus more on reducing the hardened surfaces and replacing them with green areas or permeable paving solutions, that may improve the overall urban environment.

The local authority has also introduced financial facilities for the owners or developers interested in redesigning the facade gardens or green terraces based on these principles. For implementing green terraces, the authorities grant 31 Euro/sqm for private entities and 25 Euro/sqm for public entities with a maximum value of 25.000 Euro/property (City of Ghent 2016).

The green terraces and facades bring multiple benefits both to the urban environment and the owners that integrate them in their investments: additional thermal and sound insulation with lower costs for the cooling and heating system, improved urban microclimate and landscape, air filtration, and less pressure on the urban rainwater drainage network as the green terraces have an additional retention capacity that may be exploited.

The third action plan developed for 2020–2025 reconfirmed the principles defined in the previous action plan and proposed a target of 15% impervious surfaces reduction for each redesigned public area and new measures meant to control the new urban developments in terms of the green areas that will be provided and their support for the climate challenges. For example, the surface built for rainwater retention must cover at least 7% of the total permeable area in all investments. The local authorities also decided upon implementing green facades and terraces on public buildings, in order to promote them among private investors (City of Ghent 2020).

The objectives stated in the last action plan are more specific; the local authorities want to offer financial support for 365 facade gardens every year, 1500 sqm of green facades on both public and private buildings, and 1500 new trees on street alignments, as well as for developing new public green areas (City of Ghent 2020).

There are many initiatives coordinated by local authorities to encourage these actions among urban actors. For example, The Great Ghent Breakout Programme provides participants with the necessary tools to remove the additional impervious surfaces from their gardens, while considering alternative landscape plans offered as models by the local authority. There are also workshops and information campaigns organized for the same purpose and pilot projects developed in key places with very good visibility and impact for the community members.

Policy Evaluation

The evolution of the total green areas and impervious surfaces and pavements is measured periodically through a monitoring system coordinated by Roads, Bridges and Waterways Service.

Between 2014 and 2016, 754 private facade gardens were redesigned in order to reduce the hardened surfaces, and starting from 2002, 54.640 sqm of green terraces was implemented.

The local authorities continue to promote these actions in order to increase the green areas within the city, to improve the urban microclimate and design a healthier, more sustainable, and climate-resilient city; acknowledging that the solutions to the new urban challenges are not limited to the public domain or buildings, all the actors should join their forces for the same common objective, a better urban environment.

Case Study from Netherlands

In Europe, there are two main clusters defined considering the urban-sprawl phenomenon and territorial dispersion. The first one is localized in the North-Eastern part of France, Belgium, the Netherlands, and the West of Germany, while the second one is located in the United Kingdom.

The Netherlands ranks first in the European Union considering the weighted urban proliferation indicator, according to a report from 2009 (European Environment Agency 2016).

The public authorities are facing a high demographic and territorial pressure, and their urban policies aim to support a sustainable development and maintain a balance between the built-up area and the green surfaces. The percentage of built-up area, related to the national territory, places the Netherlands in the second position on the European level, as mentioned in the statistics from 2009, with 12% (European Environment Agency 2016).

Although the urban sprawl is increasing, this phenomenon did not have any impact on the air quality from urban areas. The pollution level is low for all monitored indicators, as follows: NO2 with 0.7% of the urban population affected and O3 with 7.2% of the urban population affected, while the other pollutants are 0.0, considering the statements from 2018 (European Environment Agency 2020b).

Rotterdam is one of the cities where the demographic pressure was a constant challenge for the authorities. The city was included in the urban agglomeration of Rotterdam and in the conurbation of Randstad that currently has 7 mil. Inhabitants. The urban agglomeration population is estimated at 1.012.007 inhabitants with a regular increase of 0.2% per year since 2015 and a forecast to reach 1.063.885 inhabitants by 2035 (World Population Review 2021). Furthermore, Rotterdam is part of the C40 Cities Climate Leadership Group, a transnational network coordinated by the Mayor of Toronto (Mees and Driessen 2011).

P

Multifunctional Green Roofs in Rotterdam

Policy Background

In the context of all economic, social, and territorial pressures, the local authorities had to find solutions to coordinate this urban growth in a sustainable way, assuring that the new development will have a proper access to all urban facilities, including green areas.

In Rotterdam, the public green areas cover 16.4% of the administrative territory (Arcgis 2017), resulting in 23.85 sqm per capita (2019) above the average of 18 sqm per capita registered at European level (Maes et al. 2019), and the surface covered with water represents 34.9% (Syahid et al. 2017).

Rotterdam, especially the central area is considered very compact, with large hardened surfaces and minimal availability for new public green facilities. The city was developed based more on the principle of densification in the districts that have access to urban facilities and less on territorial expansion, as in other urban agglomerations.

Proposal and Implementation

Rotterdam was one of the first cities in the Netherlands that had initiatives toward a better urban environment and a climate-resilient settlement. The policy on green terraces was launched in 2008 through the program "Up on the roof Rotterdam," being considered an important approach due to the multiple functions that the terraces can integrate: vegetable and fruit gardens, recreational areas, rainwater tanks, support for biodiversity (system of urban steppingstones), air filtering, and other potential roles. There are also significant advantages for large private developments, as green terraces have a positive impact on the buildings' energy consumption for the heating and cooling system; the owners may reuse the collected rainwater and may design attractive social or commercial areas on top of their buildings, using a surface that was given less importance in the past.

The policy is supported by the total surface of terraces in Rotterdam that exceeds 18 sq. km, offering multiple possibilities of design according to the classification made by local authorities: green terraces (for air purification and improved urban landscape), blue terraces (for rainwater retention, with less pressure on the public networks), yellow terraces (for renewable energy production), red terraces (for social functions), orange terraces (for projects related to mobility and transport), purple terraces (for private residential areas), and gray terraces (for technical functions, air conditioners, and chimneys).

Residents are also encouraged to implement mixed terraces, by combining more urban functions (Municipality of Rotterdam 2021a). In order to fully understand the potential of this measure, the local authorities made a territorial analysis on the existing buildings in Rotterdam in the scope of integrating different roles on their terraces (Atelier GroenBlaw 2021).

This measure is stimulated by financial facilities (cofinancing scheme, coupons, subsidies, and reduced taxes), as converting the terraces is still optional for private owners or developers. On the long term, the authorities aim to transform this measure into a mandatory condition for all buildings in Rotterdam (Gemeente Rotterdam 2019a). The subsidies for greening the terraces were defined at 15 Euro/sqm, but starting from 2020 the authorities wanted to change the financial scheme.

The strategies and planning documents developed for Rotterdam integrate this approach of promoting multiple functions on building terraces, in order to be aligned to the same vision, following three major public objectives: climate resilience, energy security, and increased attractiveness for the business environment: Rotterdam City Vision 2030, Action Plan Rotterdam goes green, Rotterdam's Climate Proof Programme (RCP), and other documents developed in the same period of time.

The new Land Development Act has introduced the possibility to not only demand, from the private developers, new green spaces with public access integrated in their investments, designed and implemented with private resources, but also to define clear areas for rainwater retention on 10% of their total land (Mees and Driessen 2011). The new act allowed the distribution of urban costs from the new developments, between public and private entities. The local authorities also plan to integrate the obligation to design the new building terraces according to the city strategy.

Policy Evaluation

In 2019, there were around 400.000 sqm of green terraces in Rotterdam (Municipality of Rotterdam 2021b), and the objective on short term established through the Programmaplan Multifunctionele Daken 2019–2022 was to implement between 60.000 and 80.000 sqm of green terraces until 2022, although the local authority believes that this target can be reached annually.

Every year, an external agency, contracted by the public authority, makes a comparative analysis of the building terraces that were aligned with new city vision, based on satellite images, aerial photographs, and GIS maps (Gemeente Rotterdam 2019b).

Case Study from Germany

Germany is in the sixth position in Europe based on the weighted urban proliferation coefficient. In order to succeed in controlling this accelerated phenomenon, the central authority have defined, within the National Sustainability Strategy for Germany – 2002, one objective related to the maximum urban growth rate allowed. The rate was established at 30 ha/day of new built-up areas for the entire country, a difficult challenge considering that the growth rate in 1990 was around 130 ha/day. Currently, the urban growth rate is 63 ha/day, more than double compared to the authority objective (European Environment Agency 2016).

Although Germany is facing a high urban growth rate, the air quality in urban areas has been improved, except for the O3 pollution that still affects 74.6% of the total urban population (European Environment Agency 2020c).

The urban agglomeration of Berlin has a current population of 3.566.791 inhabitants, with a constant increase of 0.36% per year between 2007 and 2018, followed by a decrease in the growth rate to 0.13% per year. A population of 3.75 mil. Inhabitants by 2030 is expected, if the same upward trend is maintained (World Population Review 2021b). The largest city of Germany is part of Berlin – Brandenburg metropolitan region with a population of 5.8 mil. Inhabitants.

Between 2000 and 2011, there was an important increase in residential areas, from 18.023 ha to 21.044 ha, while the green areas were not developed with the same intensity, growing from 9087 ha in 2000 to 9677 in 2011.

The demographic evolution will lead to the building of 137.000 new homes (Ecosystem service implementation and governance challenges). The local authorities need to coordinate the urban growth without affecting the green infrastructure.

Allotment Gardens in Berlin

Policy Background

In Berlin, the green areas cover 39.7% of the total administrative territory (Arcgis 2017), approximately 23,73 sqm of green space per capita (2019) (Maes et al. 2019).

Even though the average green area per capita exceeds the European Union average of 18 sqm per capita, the public green facilities are not evenly distributed in the territory. There are large differences between the districts from Berlin, with green areas starting from 6 sqm per capita to 35 sqm per capita in the periphery (Green justice).

The allotment gardens have a history of more than 150 years in Germany, with their first aim to provide food security for low-income families. Their role was enriched by adding new urban functions to the existing one, such as recreational and leisure areas, becoming a representative place for the communities. In addition, the allotment gardens support the local biodiversity and a healthier urban environment. In 2019, there were 877 allotment gardens in Berlin, covering an area of 2903 ha (3% of the administrative territory) (Senate Department for the Environment, Transport and Climate Protection).

Proposal and Implementation

The allotment gardens are protected by not only two laws (Federal Act on Small Gardens – 1983, Federal Building Code – 1960), but also by the general urban plans. According to the current regulations, they can only have a recreational, noncommercial, and food production purpose. Their minimum size is 400 sqm with a maximum area for the permanent or temporary buildings of 24 sqm (Drilling et al. 2016). The allotment gardens bring multiple benefits to the surrounding communities: They give the inhabitants the possibility to produce their own food, to spend their free time close to the nature, and to have the opportunity to become members in a specialized association that can initiate them in gardening (Drescher 2001).

The allotment gardens are threatened by the real estate pressure, as they are considered favorable sites for new developments, especially due to

P

their locations. In present, just 6% of the allotment gardens are protected through the urban plans with clear urban functions that cannot be changed by developers (Zimbler 2001).

The gardens can be implemented on public lands, sites owned by public institutions, churches, private companies, or private owners (Gröning 1996).

The free lands, on which the owners do not have plans to invest on short term, have the opportunity to become semipublic areas or allotment gardens, by signing a contract between the land owner and the user, similar to the conditions on public lands (Federal Ministry for the Environment, Nature Conservation, Building and Nuclear Safety 2018).

This represents an opportunity for the local authorities to expand the green areas and the allotment gardens on free private lands with benefits for all actors involved: The landowners get a rent for allowing this facility on their property, users can have a garden close to their home, and the local administration can redirect the public funds that would have been used for land acquisition, to other priorities.

Due to a high demand for allotment gardens, the local authorities are currently looking for ways in which they can redistribute the areas so that more users can benefit from it.

Policy Evaluation

The allotment gardens are monitored and organized by the German Leisure Garden Federation, an organization financed by the annual contribution paid by each garden user. The central authority aims to support the maintenance and development of new allotment gardens, as they are aware of their social, ecological, and educational role. The practices inside the gardens take care of the environment and promote crops diversity. On long term, the administration wants to define a set of guidelines for all the allotment garden users, including workshops and educational programs (Federal Ministry for the Environment, Nature Conservation, Building and Nuclear Safety 2018).

The area covered with allotment gardens was significantly reduced due to the social and economic pressure; for this reason, the associations are trying to include them in the urban plans, so that there will be restrictions for building on these lands.

Conclusion

The case studies revealed several public policies through which the urban green areas may be extended, without limiting to the financial possibility of the authority to buy new lands for these public facilities. The practices detailed in this research paper are based on the fact that the development of a sustainable and climate-resilient city is not the exclusive task of public authorities; it is the responsibility of all urban actors to participate in creating a healthier urban environment.

The solutions based on public-private partnerships or subsidized support for private investments can represent valuable alternatives to the projects dependent entirely on public resources. The public budget is limitless, and so is the availability of land for new public projects, especially due to social and economic pressure. There are many cases in which the local authorities have to decide upon the approval of a new residential, commercial area or the maintenance of public green spaces in the same configuration, with no possible response for the increase demand in new residential spaces.

One potential solution may be the use of free private lands with benefits for all actors involved, so that the unused sites can serve the community as parks, recreational areas, allotment gardens, or other similar functions. Another solution can be represented by the green terraces that are using an area that has not been taken into account until now. The policies on dehardening the surfaces can also be easily stimulated by the local authorities with minimal public resources required, compared to the acquisition of new properties inside the city.

All these measures depend on a good collaboration between the actors involved, a good coordination from the public entity, and periodic evaluations and adjustments of the public policy according to the feedback received from the community and the results.

Cross-References

- ▶ City Visions: Toward Smart and Sustainable Urban Futures
- ▶ Climate Resilience in Informal Settlements: The Role of Natural Infrastructure
- ▶ Urban Food Gardens
- ▶ Urban Greening and Green Gentrification

References

Arcgis. (2017). *Share of green urban areas*. Available: https://www.arcgis.com/apps/MapSeries/index.html?appid=42bf8cc04ebd49908534efde04c4eec8%20&embed=true

Atelier GroenBlaw. (2021). *Green roofs Rotterdam, The Netherlands*. Available: https://www.urbangreenbluegrids.com/projects/rotterdam-the-netherlands/

Boussauw, K. (2014). *City profile: Ghent*. Belgium.

City of Ghent. (2016). *Working towards a climate-robust city. Ghent climate adaptation plan 2016–2019*. City of Ghent.

City of Ghent. (2020). *2020–2025 Climate Plan*. City of Ghent. Department of Urban Development Environmental and Climate Service.

Department of Urbanism and Spatial Planning. (2018). *Room for all ghenteneers. 2030 structural vision in a nutshell*. Department of Urbanism and Spatial Planning.

Drescher, A.W. (2001). *The German allotment gardens – A model for poverty alleviation and food security in Southern African cities*? Available: https://www.cityfarmer.org/germanAllot.html

Drilling, M., Ponizy, L., & Gledych, R. (2016). The idea of allotment gardens and the role of spatial and urban planning. In: Bell S, Fox Kämper R, Keshavarz N, et al. (eds) Urban Allotment Gardens in Europe. London and New York: Routledge, pp. 35–61.

Endl, A., & Berger, G. (2014). *The 7th environment action programme: Reflections on sustainable development and environmental policy integration*. European Sustainable Development Network.

European Environment Agency. (2016). *Urban sprawl in Europe*. Joint EEA-FOEN report. Publications Office of the European Union.

European Environment Agency. (2020a). *Belgium – Air pollution country fact sheet*. Available: https://www.eea.europa.eu/themes/air/country-fact-sheets/2020-country-fact-sheets/belgium-air-pollution-country

European Environment Agency. (2020b). *Netherlands – Air pollution country fact sheet*. Available: https://www.eea.europa.eu/themes/air/country-fact-sheets/2020-country-fact-sheets/netherlands

European Environment Agency. (2020c). *Germany – Air pollution country fact sheet*. Available: https://www.eea.europa.eu/themes/air/country-fact-sheets/2020-country-fact-sheets/germany

Federal Ministry for the Environment, Nature Conservation, Building and Nuclear Safety. (2018). *White paper: Green spaces in the city*. Federal Ministry for the Environment, Nature Conservation, Building and Nuclear Safety (BMUB).

Gemeente Rotterdam. (2019a). *Naar een Rotterdams Daklandschap*. Programma voor Multifunctionele Daken. Gemeente Rotterdam.

Gemeente Rotterdam. (2019b). *Rotterdam gaat voor groen*. Gemeente Rotterdam.

Gröning, G. (1996). *Politics of community gardening in Germany*. Available: http://userpage.fu-berlin.de/~garten/Texte/Groening.html

Maes, J., Zulian, G., Günther, S., Thijssen, M., & Raynal, J. (2019). *Enhancing resilience of urban ecosystems through green infrastructure (EnRoute)*. Publications Office of the European Union.

Mees, H., & Driessen, P. (2011). Adaptation to climate change in urban areas: Climate-greening London, Rotterdam, and Toronto. *Climate Law*. Available: https://www.researchgate.net/publication/228760305_Adaptation_to_climate_change_in_urban_areas_Climate-greening_London_Rotterdam_and_Toronto/citations

Municipality of Rotterdam. (2021a). *Multifunctional roofs*. Available: https://www.rotterdam.nl/wonen-leven/multifunctionele-daken/

Municipality of Rotterdam. (2021b). *Green roofs*. Available: https://www.rotterdam.nl/wonen-leven/groene-daken/

Senate Department for the Environment, Transport and Climate Protection. Allotment Gardens. Data and Facts. Available: https://www.berlin.de/senuvk/umwelt/stadtgruen/kleingaerten/en/daten_fakten/index.shtml.

Syahid, N. C., Lissandhi, N. A., Novianti, K., & Reksa, A. F. A. (2017). *Sustainable cities in the Netherlands: Urban Green Spaces Management in Rotterdam*. Available: https://www.researchgate.net/publication/337058027_Sustainable_Cities_in_the_Netherlands_Urban_Green_Spaces_Management_in_Rotterdam

World Population Review. (2021). *Rotterdam Population 2021*. Available: https://worldpopulationreview.com/world-cities/rotterdam-population

Zimbler, L. R. (2001). *Community Gardens on the Urban Land Use Planning Agenda*. Experiences from the United States, Germany, and the Netherlands. Available: http://www-sre.wu-wien.ac.at/neurus/Zimbler.pdf

Public Policy

▶ Role of Disaster Relief Policy in Building Resilient Coastal Regions in the United States

P

Public Private Partnership (PPP)

► The State of Extreme Events in India

Public Procurement for Regional and Local Development

María del Carmen Sánchez-Carreira[1] and María Concepción Peñate-Valentín[2]
[1]Department of Applied Economics, Faculty of Economics, Universidade de Santiago de Compostela, ICEDE Research Group, CRETUS, Santiago de Compostela, Galicia, Spain
[2]Department of Applied Economics, Faculty of Economics, Universidade de Santiago de Compostela, Santiago de Compostela, Galicia, Spain

Synonyms

Government procurement; Public acquisition; Public buying; Public purchasing; Public tendering

Definition

Public procurement (PP) is the acquisition of products, services, and works by the public sector from the private sector (Arrowsmith 2003). This concept embraces the purchase both by any level of government (national, regional, or local), and by any of its entities (in particular, state-owned enterprises) (OECD 2019).

Introduction

The public sector is a major economic player, which performs different roles. Among all of them, literature commonly focuses on its diverse functions as provider. However, the demand from the public sector cannot be neglected. In fact, global procurement accounts for around 9.5 trillion dollars, according to the World Bank Global Public Procurement Database. The size of total PP ranges from 13% to 20% of Gross Domestic Product (GDP), depending on the country. Thus, PP means 11.8% of GDP as average in the OECD countries in 2017 (OECD 2019). Moreover, the importance of PP varies by area. Hence, health is the area that concentrates the highest percentage of the total PP in the OECD in 2017. Other relevant areas are infrastructure, education, social protection, and defense (OECD 2019). The government is an important customer in many activities and countries, and many times it is also the main buyer and user, as it happens in innovative fields (Overmeer and Prakke 1978; von Hippel 1986; Gregersen 1988; Dalpé et al. 1992; Walker and Brammer 2009; Lember et al. 2014; Edquist 2015; Horner and Alford 2019).

Potential of Public Procurement as Policy Tool

The relevance of the public sector as buyer is not merely restricted to its direct purchase power, but the indirect effects stand out. Thus, besides acting as a procurer and user of existing innovations in the market and the innovations resulting from its early demand, the public sector may create new needs and markets and contribute to the development of new and/or strategic activities through the demonstration effect, the creation of jobs, and the increase in wealth. In this sense, PP can act as policy tool since it addresses different policy goals (OECD 2019). Strategic public procurement refers to its use to reach other policy objectives. In this way, economic, environmental, and social goals underline. Therefore, strategic public procurement aims at promoting regional and local development, circular economy, innovation, or social inclusion (OECD 2019; Sánchez-Carreira et al. 2019; Sánchez-Carreira 2020).

There are different types of PP, based on the used classification (Sánchez-Carreira 2020), which can be the final goal, the result, the users, the description of the needs, or criteria used in the tenders.

Although the role of the public sector as consumer has been used along the history, mainly to procure technology in some areas, it gains relevance since the change of the twenty-first century, due to its widespread use as horizontal policy tool (Edler 2010; Lember et al. 2014). Thus, demand-side initiatives and policies have been implemented at international, national, regional, and local levels. The recent support of the EU to public procurement should stand out through different initiatives. The development of the Regional Smart Specialization Strategies provides an opportunity for using public procurement, in particular, in the EU given that they are a requirement for the access to regional funds in the programming period 2014–2020. At the local level, initiatives such as smart, sustainable, or circular cities use public procurement, as well as the provision of healthy and local food at schools or hospitals or the waste management. In addition, the Sustainable Development Goals (SDGs) within the framework of the 2030 Agenda for Sustainable Development launched by the United Nations in 2015 include the promotion of sustainable procurement practices in accordance with national policies and priorities as one of the targets of the SDG 12 Ensure sustainable consumption and production patterns.

This contribution focuses on the relationship between PP and territorial development, in particular, in the regional and local fields. It provides some ideas and insights for the debate about the trade-off between efficiency and regional development.

Regional and Local Development in the Focus

Public procurement as policy tool can promote growth and regional development. It should be underlined that the majority of the PP is undertaken by regional or local governments. Thus, more than 60% of global PP is executed at the subcentral levels. In the case of the OECD countries, subcentral levels mean 63,3% of total PP in 2017 (OECD 2019). Therefore, it may be an opportunity to foster regional development and the participation of local enterprises, mainly small and medium-sized enterprises (SMEs) (Sánchez-Carreira et al. 2019). Another reason for the implementation of public procurement at subnational levels is the fact that regions, cities, and other localities are relevant in the delivery of public services (Uyarra et al. 2017; Sánchez-Carreira et al. 2019). A critical issue concerns the availability of resources and the capabilities for the design and implementation of PP initiatives, due to the complexity of the process (Uyarra et al. 2017). Nonetheless, this cannot be understood as a very restrictive barrier, because all the levels of government need to build competences as well as training to apply PP. In this way, it may be interesting to start with small initiatives at that local or regional level to experiment with the policy tool, taking advantage of the knowledge of local actors and interactions (Sánchez-Carreira et al. 2019).

Along the history, different countries have used public procurement to develop national sectors and local enterprises, outstanding the experiences of the USA, Japan, or Sweden in computer, telecommunications, defense, or aerospace industries (Overmeer and Prakke 1978; Weiss and Thurbon 2006; Lember et al. 2014; Horner and Alford 2019). Nevertheless, the purpose of regional goals or national purposes is controversial, due to its conflicting relationship with economic efficiency and even with the legal framework. The institutional framework plays a key role to drive PP. In fact, it offers opportunities as policy tool, although it presents limitations. Among the difficulties, it is worth noting the existence of different institutional levels with competences in this area, which involves the need of coordination. This complexity is especially present in the European Union (EU), based in a multilevel organization. Moreover, regulation can become an obstacle for the utilization of PP as an instrument for regional development, as is the case in the EU, where competition and nondiscrimination by nationality of enterprises are the guiding principles.

The main contested issue about the use of PP as an instrument of regional development concerns the decision between choosing the best offer, regardless of the location of the enterprises or contributing to industrial or technological development at the nearest area (local, regional, or even national level). This discussion deals with prioritizing the appropriateness and quality of the project in accordance with the requirements of the tender, or the promotion of local companies (Rolfstam 2016). It leads to the lively debate about efficiency and territorial cohesion. Several authors (such as Rolfstam 2016) consider appropriate to choose based on the project. This fact may lead to most of the tenders being awarded to enterprises outside the region or the country. Nevertheless, even in this situation, local enterprises can benefit from a collaboration with other enterprises. Thus, they gain access to external resources and may build competences and learn from this experience (Saastamoinen et al. 2018). Moreover, this collaboration can make easier the access to the local knowledge and capacities (Sánchez-Carreira et al. 2019).

The public sector, as procurer, can foster regional development, participation of SMEs, or industrial and innovation development. However, this driving role of the public sector from the demand side depends mainly on the local characteristics of the supply side (productive structure, capabilities, knowledge base, innovation, or competitive advantages) (Rothwell 1984; Uyarra et al. 2017; Sánchez-Carreira et al. 2019). In addition, the role of the public sector as demand driver is critical in less developed regions and countries (Rothwell 1984; Sánchez-Carreira et al. 2019). A key aspect is that the awarded enterprises meet the requirements of the public demand. The supply-base of a region may lack the needed resources, capabilities, and skills to fulfill the tender. In these cases, local enterprises would not participate in PP. It should be noted that there is a low level of participation of SMEs and that local enterprises are often in disadvantaged conditions for a global competition or even with specialized enterprises. This is particularly relevant in the field of innovation (Sánchez-Carreira et al. 2019). In this sense, policy measures that reduce the barriers (external and internal) faced by enterprises (and mainly SMEs) to participate in PP processes are advised. In this line, some of the implemented measures in different countries are dividing the bids into lots, fostering consortia, setting minimum PP budgets for SMEs or for the same region (Edler and Georghiou 2007; Loader 2013; Uyarra et al. 2014; Flynn and Davis 2016; Tammi et al. 2017). Otherwise, the regional disparities would increase, and other policies will be focused on this issue, allocating more public resources. It should be highlighted that the participation in PP enhances the competences of the enterprises, increases their recognition and image. As a result, the enterprises may acquire competitive advantages.

Therefore, a relevant issue for achieving better results from public procurement concerning regional development is the coordination among supply and demand policies (Edler and Georghiou 2007; Guerzoni and Raiteri 2015; Uyarra 2016; Sánchez-Carreira et al. 2019). Thus, supply policies may contribute to develop the productive and innovative capacities of local enterprises, upgrading their competences. This effort is more relevant in peripheral areas to become PP in a real opportunity for development. In this sense, the concrete design and implementation of the tool is crucial to achieve other policy goals, such as regional or local development. The anticipation of the needs from the public sector can help to build and develop the needed competences by the enterprises of the area.

Summary/Conclusions

Public procurement focuses on the role of the public sector as relevant consumer. In addition, it may achieve other strategic policy goals, such as regional and local development, which is precisely the issue tackled here. PP poses opportunities but also challenges for regional development, which are critical for less developed areas. The institutional framework and regulation, the non-conflicting combination of goals and criteria for selection, the features of the supply-base, and the coordination between demand and supply sides are the main insights for the lively debate that leads to the tradeoff between efficiency and cohesion. New research is needed to study this issue in-depth, due to the scarce literature, besides some interesting case-studies, mainly regarding sustainability, and to propose the best practices for leveraging the opportunities of PP for territorial development, achieving efficient result at the same time.

Cross-References

- ▶ Innovation to Bring Nature-Based Solutions to Life: Tales of Two Cities
- ▶ Local and Regional Development Strategy
- ▶ New Localism: New Regionalism
- ▶ Regulation of Urban and Regional Futures
- ▶ Sustainable Development Goals
- ▶ Sustainable Development Goals from an Urban Perspective
- ▶ The Sustainable and the Smart City: Distinguishing Two Contemporary Urban Visions
- ▶ Urban and Regional Leadership

References

Arrowsmith, S. (2003). *Government procurement in the WTO*. La Haya: Kluwer Law International.

Dalpé, R., Debresson, C., & Xiaping, H. (1992). The public sector as first user of innovations. *Reseach Policy, 21*(3), 251–263.

Edler, J. (2010). Demand oriented innovation policy. In R. Smits, S. Kuhlmann, & P. Shapira (Eds.), *The theory and practice of innovation policy. An international research handbook* (pp. 177–208). Cheltenham: Edward Elgar.

Edler, J., & Georghiou, L. (2007). Public procurement and innovation – resurrecting the demand side. *Research Policy, 36*(7), 949–963.

Edquist, C. (2015). *Innovation-related public procurement as a demand-oriented innovation policy instrument* (Papers in Innovation Studies. Paper n° 2015/28). Lund: CIRCLE Lund University.

Flynn, A., & Davis, P. (2016). Firms' experience of SME-friendly policy and their participation and success in public procurement. *Journal of Small Business and Enterprise Development, 23*(3), 616–635.

Gregersen, B. (1988). Public sector participation in innovation systems. In C. Freeman & B. A. Lundvall (Eds.), *Small countries facing the technological revolution* (pp. 262–278). London: Pinter Publishers.

Guerzoni, M., & Raiteri, E. (2015). Demand-side vs. supply-side technology policies: Hidden treatment and new empirical evidence on the policy mix. *Research Policy, 44*(3), 726–747.

Horner, R., & Alford, M. (2019). The roles of the state in global value chains. In S. Ponte, G. Gereffi, & G. Raj-Recihert (Eds.), *Handbook in global value chains* (pp. 555–569). Northampton: Edward Elgar Publishing.

Lember, V., Kattel, R., & Kalvet, T. (2014). Public procurement and innovation: theory and practice. In V. Lember & T. Kalvet (Eds.), *Public procurement, innovation and policy* (pp. 13–34). London: Springer.

Loader, K. (2013). Is public procurement a successful small business support policy? A review of the evidence. *Environment and Planning C: Government and Policy, 31*, 39–55.

OECD. (2019). *Government at a Glance 2019*. Paris: OECD.

Overmeer, W., & Prakke, F. (1978). *Government procurement policies and industrial innovation: Report prepared for the six countries programme on government policies towards technological innovation in industry*. Delft: Strategic Surveys TNO.

Rolfstam, M. (2016, April 13–15). Concerning support for SME's as suppliers of public health tech innovation: Some reflections and case evidence. In *20th international research society on public management conference*, Hong Kong.

Rothwell, R. (1984). Creating a regional innovation-oriented infrastructure: The role of public procurement. *Annals of Public and Cooperative Economics, 55*(2), 159–172.

Saastamoinen, J., Reijonen, H., & Tammi, T. (2018). Should SMEs pursue public procurement to improve innovative performance? *Technovation, 69*, 2–14.

Sánchez-Carreira, M. C. (2020). New opportunities for government procurement. In A. Farazmand (Ed.), *Global encyclopedia of public administration, public policy, and governance*. Springer. https://doi.org/10.1007/978-3-319-31816-5_1522-1.

Sánchez-Carreira, M. C., Peñate-Valentín, M. C., & Varela-Vázquez, P. (2019). Public procurement of innovation and regional development in peripheral areas. *Innovation: The European Journal of Social Science Research, 32*(1), 119–147.

Tammi, T., Reijonen, H., & Saastamoinen, J. (2017). Are entrepreneurial and market orientations of small and medium-sized enterprises associated with targeting different tiers of public procurement? *Environment and Planning C: Politics and Space, 35*(3), 457–475.

Uyarra, E. (2016). The impact of public procurement of innovation. In J. Edler, P. Cunningham, A. Gök, & P. Shapira (Eds.), *Handbook of innovation policy impact* (pp. 355–380). Cheltenham: Edward Elgar.

Uyarra, E., Edler, J., Garcia-Estevez, J., Georghiou, L., & Yeow, J. (2014). Barriers to innovation through public

procurement: a supplier perspective. *Technovation, 34* (10), 631–645.
Uyarra, E., Flanagan, K., Magro, E., & Zabala-Iturriagagoitia, J. M. (2017). Anchoring the innovation impacts of public procurement to place: the role of conversations. *Environment and Planning C: Politics and Space, 35*(5), 828–848.
von Hippel, E. (1986). Lead users: A source of novel product concepts. *Management Science, 32*, 791–805.
Walker, H., & Brammer, S. (2009). Sustainable procurement in the United Kingdom public sector. *Supply Chain Management, 14*(2), 128–137.
Weiss, L., & Thurbon, E. (2006). The business of buying American: Public procurement as trade strategy in the USA. *Review of International Political Economy, 13* (5), 701–724.

Public Purchasing

► Public Procurement for Regional and Local Development

Public Realm

► Public Space
► Stewarding Street Trees for a Global Urban Future

Public Space

Fei Chen, Francesca Piazzoni, Junjie Xi, Yat Shun Kei and Aikaterini Antonopoulou
School of Architecture, University of Liverpool, Liverpool, UK

Synonyms

Public realm; Public sphere; Urban open space

Definition

Public space refers to an urban space that is freely accessible to all people. Accessibility here refers not only to the ability to physically enter public space, but also the ability to participate in, and take decisions about, the activities taken place in it (Carmona et al. 2003; Shaftoe 2008). Conventional categorizations of public space include urban parks, civic plazas, streets, markets, playgrounds, community open spaces, waterfronts, greenways, and indoor public venues.

Public space is a physical entity associated with the political concept of *public sphere* (Low and Smith 2006). It can also be referred to as *public realm* when the physical aspect and the political aspect coincide. Understandings of public space have recently moved away from the dichotomy of public and private, with public space being recognized as bearing various degrees of publicness. Dimensions of publicness are further discussed below.

Introduction

Public space forms a large proportion of urban space in our cities today and is vital for the city's future. Ideally, public space provides venues for social gathering, communication, social activities, expression of political views, which are beneficial for people's quality of life. Public space also contributes to the city's ecological system, aesthetics, and economy. Ensuring good quality public space is the key for the development of more sustainable and resilient cities.

This chapter briefly reviews the origin and development of the concept of public space. It focuses on debates of contemporary public space which include the commodification of public space and its consequences on social justice, the temporary uses of public space that helps fulfill the changing demands of societies, and the digital mediation of public space, which also bear implications for equity. We conclude by reflecting on how urban design and planning can respond to the challenges posed by these debates, suggesting that research should extend to cover non-Western countries and regions.

Defining Public Space

Canonical texts associate the origins of public spaces with democracy and participation, using the Greek agora and the Roman forum as examples (Zucker 1959). Scholars critiqued these "origins" narrative by noticing how those allegedly "ideal" Western European public spaces in fact precluded access to women and enslaved people (Arendt 1958; Low and Smith 2006; Sennett 1970).

Early literature suggested that dichotomies between public and private spaces were embedded in bourgeois cities where the "private" was neatly defined and protected by property laws (Benjamin 1979, cited from Parker 2015, 17). Walter Benjamin (1979) argued that in cities where market dynamics were not the main driver of urban development, boundaries between public and private spaces were less defined. Researchers have started to investigate the concept of public space in those contexts, such as in China. The "public" referred to apparatuses of states, the market, and anything in-between, for example, civil society (Warner 2002). Thus, the "public" can have multiple meanings across the public-private spectrum.

This observation has prompted critics to explore and assess the so-called *publicness of public space*. For some, the publicness was defined by access, agency (who controls what happens in space?), and interests (who benefits from the activities that take place?) (Benn and Gaus 1983; Akkar 2005); others stressed the question of ownership of the space, access, and social interaction or use (Kohn 2004; Marcuse 2004). Based on their study of several public spaces in London, Carmona and Wunderlich (2012) suggested that access is the main, if not the only aspect that truly determines the publicness of public space. Varna and Tiesdell (2010) proposed a *Star* model of publicness, assessing public space in terms of control (presence of formal rules), civility (maintenance and facilities), animation (social interaction), physical configuration, and ownership.

While important, these models tend to evaluate public spaces almost exclusively based on their physical qualities and from the perspective of designers, planners, and managers. This approach has been criticized by those who see public space as inseparable from social structures and human agency (Low and Smith 2006). Public space is understood as a product of social practice and social relations (e.g., Lefebvre and Nicholson-Smith 1991; Mitchell 2003; Madanipour 2003; Kohn 2004). The relationship between public space and urban life is dynamic, reciprocal, and ever changing (Loukaitou-Sideris and Banerjee 1998), which makes analyses of public space context-dependent and complex. Within the lively debates over the future of public space, there are recurring concerns over questions of privatization, surveillance, and equity. The next section addresses some of these concerns, focusing on the commodification and social justice, temporary use, as well as mediation and representation of public space.

Commodification and Social Justice

Since the 1990s, postindustrial cities experienced a shift from a manufacturing to a service economy, the growing commercialization of science and technology, and the emergence of professional classes (Bell 1976). As a result, public spaces, which traditionally resulted from economic and urban growth, increasingly carry the burden of generating income through tourism and place "branding" (Gospodini 2006; Ashworth and Voggd 1990). This is evident in efforts in heritage preservation, the creation of "creative clusters," and large-scale urban regenerations which are typically mobilized to attract capitals.

In Britain, examples include the Southwark in London, the docks in Liverpool, and MediaCity in Salford. In Europe, there is the Hafen City in Hamburg or the Bjørvika area in Oslo (Burda and Nyka 2017). Such projects are frequently realized through public-private partnership, with the resultant public spaces being commercialized and privatized. Often, these developments advance the economic interests of the elites, which may conflict with the wider public good. The spatial and political ramifications of privatization have received considerable attention

P

(e.g., Minton 2006). We trace some of these debates below, highlighting recent efforts for countering exclusion.

Public spaces are supposed to be opened to everyone, allowing strangers to encounter one another and democracy to take form (Arendt 1958; Goffman 1963). But opportunities to access and use public spaces are not equally distributed among the residents of a city. Henri Lefebvre's (1968) concept of the *Right to the City* famously captured the political consequences of spatial exclusion. Highlighting how capitalist relations make urban centers increasingly hostile to the poor, Lefebvre argued that all those who inhabit a city should instead have the right to use and produce its spaces. Scholars of the 1970s and 1980s further explored links between public space and social exclusion. For example, Richard Sennett (1970) argued that modern transformations of public spaces prevent people from interacting with each other, therefore impeding societies to "mature." William Whyte (1988) drew attention on how the design of urban forms contributes to banishing traditionally marginalized groups who are designated as "undesirables."

Feminist critics of the 1990s expanded discourses on the *Right to the City.* They demonstrated how people face exclusion not only based on social class, but also because of their gender, race, health, and sexuality (Fraser 1990; Young 1990). Building on this consideration, Leonie Sandercock (2003) suggested that a more just urban condition should entitle all residents to the *Right to Difference*, or the ability to participate in the physical and social production of space. Scholars of the same period focused on detailing how built environments amplify inequities by rendering specific uses – and users – unwelcome (Banerjee 2001; Loukaitou-Sideris and Banerjee 1998). Regulations sanction behaviors such as loitering or selling merchandise, de facto preventing some people from using public space (Austin 1994; Blomley 2007). Privatizations and commodification keep out those who cannot afford to pay to access a space (Boyer 1993; Sorkin 1992). The symbolic construction of landscapes normalizes ideas of who are the "appropriate" users, making those who look different appear as if they do not belong (Hayden 1995; Zukin 1991).

If public spaces can exacerbate inequities, they can also provide people with opportunities to counter oppression. Since the 2000s, a robust scholarship has analyzed *Everyday Urbanisms*, or the spatial appropriations by which the users of a city satisfy needs (Chase et al. 1999) and make a city their own (Low 2000; Hou 2010). While not all the interactions that occur in public space necessarily ease social frictions (Amin and Thrift 2002), a consensus has emerged that ensuring freedom of access and usage to as many people as possible can help increase mutual respect among strangers (Anderson 2011; Kohn 2004).

These considerations have prompted new reflections on urban design as a transformative force for social change. While fully aware that design cannot solve all problems, critics have suggested that spatial transformations can help reverse oppressive relations of power (Tonkiss 2013; Loukaitou-Sideris 2020). *Just Urban Design* has recently emerged as a framework that urges architects and planners to center the voices of excluded people, facilitating their insurgent practices of self-empowerment (Goh et al. 2022; Low and Iveson 2016).

Temporary Use of Public Space for Changing Social Needs

Public spaces are produced, shaped, and given meaning by various human activities, temporary appropriations, and subjectivities. Some of the uses of public space are not planned, but unexpected, spontaneous, and momentary to meet different users' needs. Researchers refer to this quality as the "looseness of space," or the adaptability of public spaces to accommodate a diverse urban life and to facilitate vitality of cities (Franck and Stevens 2006). Scholars suggested that the temporary dimensions of public space are often neglected in the design and management of space, but are nonetheless vital for the functioning of public space (Carmona et al. 2003; Thwaites et al. 2007).

Public space offers possibilities for assembling temporary installations of portable, mobile, and demountable structures. These structures can serve a variety of purposes including hosting cultural and entertainment events, commercial exchange, markets, political activities, and other types of gathering. The COVID19 pandemic has demonstrated the enormous potential of such transportable, reversable structures set up in public spaces for testing and curing COVID patients (Fang et al. 2020). For example, in Wuhan, China, the public space by the Wuhan International Conference and Exhibition Centre was rapidly constructed to accommodate medical beds and facilities. Temporary drive-in stations and tents were set up in many cities for COVID tests.

Not only state-owned, but also private-owned spaces can be temporarily occupied to satisfy needs. As scholars noted (e.g., Bishop and Williams 2012; Madanipour 2018), temporary installations in privately owned spaces can regenerate areas and create opportunities in periods of economic stagnation (Berwyn 2013; Steele 2013). Vacant sites, according to Madanipour (2018, 13), were produced by "the cyclical nature of capitalism and its recurring crises of overproduction," but can be temporarily used by creative or cultural industries while waiting for planned regeneration. These uses can add positive value to the space and the community. For example, in the UK, a charity organization called *Meanwhile Foundation* helps people deliver temporary uses legally and in a cost-effective manner.

Beyond these positive aspects, scholars have also pointed out that temporary use of spaces can become a branding exercise for corporate organizations and pose financial risk to smaller businesses with less capital security (Martin et al. 2019). Moreover, as pop-up parklets, food markets, and art installations have become popular tools to regenerate "neglected" areas, most of these interventions gear towards the tastes of white-collar professionals, further excluding disadvantaged urbanites (e.g., Bostic et al. 2016; Munoz 2019).

Technological Mediations of Public Space

In the context of digitization and global media, digital infrastructure interacts with many of the traditional elements of public space, producing new meanings of it. The technological mediations of public space are underpinned by theorizations of the image of the city in the age of electronic communication (Sorkin 1992; Zukin 1995), by conceptualizations of the city as a medium (Kittler and Griffin 1996; Mumford 1961), by future urban imaginaries based on telecommunication networks (Mitchell 1995; Graham and Marvin 1997; Wigley 2001), by cinematic constructions of urban space (Koeck 2013; Penz and Lu 2011), and by urban experiences shaped by technological devices embedded in urban infrastructure (Shepard 2011; Townsend 2013). Models such as the "networked city," the "media city," and the "smart city" are celebrated for their endless potential, openness, and performativity. Meanwhile, the citizens of these cities oscillate between the promise of infinite freedom and the concern of continuous surveillance.

As the city is digitally recorded and reproduced via the Internet, our understanding of public space is shaped not only by the material presence of buildings and landscapes, but also and increasingly by their digital representations. This has significant impact on how politics play out in public space and on how collective identities are formed. Paul Virilio (1994, 62) argues that video recording and information technologies have marked the end of a logic of public representation and the beginning of a "paradoxical logic," where the real-time image acts upon the object represented and, in this way, subverts the very concept of reality. With social communication taking place upon these recorded perspectives, the public image replaces the public space (Virilio 1994, 64) and the screen absorbs the public realm, becoming the new "city-square" (Virilio 1991, 25–7).

The ever-expanding number of images circulating across different media and personal devices may give us the illusion that the space of the city is brought directly before our eyes. However, it leaves us with an unclear sense of the reality of things as well as with limited capacity for immediate action. Especially in spaces of conflict, it is important to look closely at, and to engage with the places and people that are obscured, distorted, overwritten, or left behind the representations of public space (Boyer 1996, 138). This crisis, therefore, calls local bodies, policy makers, and designers to situate themselves between the complex layers of public space and their representation so that they create more inclusive spaces.

Summary/Conclusion

This chapter has presented a nonexclusive list of debates on contemporary public space focusing particularly on the social impact of these various phenomena: the commercialization and privatization of public space, its temporary use, and its mediation and virtualization. All three phenomena point to similar questions of justice and equity. They call for efforts to allow marginalized people to access and participate in social activities, benefit from them, and to be represented in physical as well as virtual spaces.

Design and planning cannot solve all social problems. But socially conscious urban design can provide a better chance for public spaces to overcome hurdles against these challenges. There is no one-size-fit-all design solution. As we argued in this chapter, the production of public space is highly context-dependent, being imbricated with multiple social, physical, and political relations. The meaning of public space is fulfilled through the entire lifecycle of planning, development, management, representation, exchange, and use (Carmona 2014; Madanipour 2018). Attention should be paid to particular cities and particular public spaces within them.

We also recognize that studies on public spaces have been mostly focusing on cities in Western Europe and North America. In the last decade, public spaces in East Asia, the Global South, and Eastern Europe have attracted more scholarly attention, but work on these regions is still scarce. Although similar phenomena of spatial change such as privatization and gentrification can be found in those areas, the difference in attitude towards state intervention (via design and planning) poses new challenges and opportunities (Hee et al. 2008; Sadowy and Lisiecki 2019). The public spaces in those areas, as reflections of different attitudes towards capitalism and neo-liberalism, call for alternative design solutions and approaches.

Cross-References

▶ Feminist Planning and Urbanism: Understanding the Past for an Inclusive Future
▶ Gender Inequalities in Cities: Inclusive Cities
▶ Spatial Justice and the Design of Future Cities in the Developing World
▶ Urban Forestry in Sidewalks of Bogota, Colombia

References

Akkar, M. (2005). The changing 'publicness' of contemporary public spaces: a case study of the Grey's monument area, Newcastle upon Tyne. *Urban Design International, 10*(2), 95–113.

Amin, A., & Thrift, N. (2002). *Cities: Reimagining the urban*. Cambridge: Polity.

Anderson, E. (2011). *The cosmopolitan canopy: Race and civility in everyday life*. New York: Norton.

Arendt, H. (1998 [1958]). *The human condition*. Chicago: University of Chicago Press.

Ashworth, G. J., & Voogd, H. (1990). *Selling the city: Marketing approaches in public sector urban planning*. London: Belhaven Press.

Austin, R. (1994). "An honest living": Street vendors, municipal regulation and the black public sphere. *The Yale Law Journal, 103*, 2120–2131.

Banerjee, T. (2001). The future of public space: Beyond invented streets and reinvented places. *Journal of the American Planning Association, 67*(1), 9–24.

Bell, D. (1976). *The coming of post-industrial society*. London: Penguin.

Benjamin, W. (1979). *One-way street and other writings*. Frankfurt: Suhrkamp Verlag.

Benn, S., & Gaus, G. (1983). *Public and private in social life*. London: Croom Helm.

Berwyn, E. (2013). Mind the gap: creating opportunities from empty space. *Journal of Urban Regeneration & Renewal, 6*(2), 148–153.

Bishop, P., & Williams, L. (2012). *The temporary city*. London: Routledge.

Blomley, N. (2007). How to turn a beggar into a bus stop: Law, traffic and the function of the place. *Urban Studies, 44*(9), 1697–1712.

Bostic, R., Kim, A. M., & Valenzuela, A. (2016). Contesting the streets, vending and public space in global cities. *Cityscape: A journal of Policy Development and Research, 18*, 3–10.

Boyer, C. (1993). The City of illusion: New York's public places. In P. L. Knox (Ed.), *The restless urban landscape* (pp. 111–126). Englewood Cliffs: Prentice Hall.

Boyer, C. M. (1996). *CyberCities: Visual perception in the age of electronic communication*. New York: Princeton Architectural Press.

Burda, I., & Nyka, L. (2017). Providing public space continuities in post-industrial areas through

remodelling land/water connections. *Materials Science and Engineering, 245*, 1–6.
Carmona, M. (2014). The place-shaping continuum: A theory of urban design process. *Journal of Urban Design, 19*(1), 2–36.
Carmona, M., & Wunderlich, F. (2012). *Capital spaces: The multiple complex public spaces of a global city* (1st ed.). London: Routledge. https://doi.org/10.4324/9780203118856.
Carmona, M., Heath, T., Oc, T., & Tiesdell, S. (2003). *Public places urban spaces: The dimensions of urban design*. London: Architectural Press.
Chase, J., Crawford, M., & Kaliski, J. (1999). *Everyday urbanism*. New York: Monacelli Press.
Fang, D., Pan, S., Li, Z., Yuan, T., Jiang, B., Gan, D., Sheng, B., Han, J., Wang, T., & Liu, Z. (2020). Large-scale public venues as medical emergency sites in disasters: Lessons from COVID-19 and the use of Fangcang shelter hospitals in Wuhan, China. *BMJ Global Health, 5*(6), e002815. https://doi.org/10.1136/bmjgh-2020-002815.
Franck, K., & Stevens, Q. (2006). *Loose space: Possibility and diversity in urban life*. London: Routledge.
Fraser, N. (1990). Rethinking the public sphere: A contribution to the critique of actually existing democracy. *Social Text, 25*(26), 56–80.
Goffman, E. (1963). *Behaviour in public places*. New York: Free Press.
Goh, K., Loukaitou-Sideris, A., & Mukhija, V. (2022). *Just urban design: the struggle for a public city*. MIT Press.
Gospodini, A. (2006). Portraying, classifying and understanding the emerging landscapes in the post-industrial city. *Cities, 23*(5), 311–330.
Graham, S., & Marvin, S. (1997). *Telecommunications and the city: Electronic spaces, urban places*. London: Routledge.
Hayden, D. (1995). *The power of place: Urban landscapes as public history*. Cambridge, MA: MIT Press.
Hee, L., Schrieffer, Su, N., & Li, Z. (2008). From post-industrial landscape to creative precincts: Emergent spaces in Chinese cities. *International Development Planning Review, 30*(1), 249–266.
Hou, J. (2010). *Insurgent public space*. New York: Routledge.
Kittler, F. A., & Griffin, M. (Trans.). (1996). The city is a medium. *New Literary History, 27*(4), 717–729.
Koeck, R. (2013). *Cine-scapes: Cinematic spaces in architecture and cities*. London: Routledge.
Kohn, M. (2004). *Brave new neighbourhoods: The privatisation of public space*. London: Routledge.
Lefebvre, H. (1968 [1996]). The right to the city. In E. Kofman & E. Lebas (Eds.), *Writings on cities* (pp. 147–159). Cambridge: Wiley-Blackwell.
Lefebvre, H., & Nicholson-Smith, D. (1991). *The production of space* (Vol. 142). Oxford: Blackwell.
Loukaitou-Sideris, A. (2020). Responsibilities and challenges of urban design in the 21st century. *Journal of Urban Design, 25*(1), 22–24.
Loukaitou-Sideris, A., & Banerjee, T. (1998). *Urban design downtown: Poetics and politics of form*. Berkeley: University of California Press.
Low, S. (2000). *On the plaza: The politics of public space and culture*. Austin: University of Texas Press.
Low, S., & Iveson, K. (2016). Propositions for more just urban public spaces. *City, 20*(1), 10–31.
Low, S., & Smith, N. (Eds.). (2006). *The politics of public space*. London: Routledge.
Madanipour, A. (2003). *Public and private spaces of the city*. London: Routledge.
Madanipour, A. (2018). Temporary use of space: Urban processes between flexibility, opportunity and precarity. *Urban Studies, 55*(5), 1093–1110.
Marcuse, P. (2004). The threat of terrorism and the right to the city. *Fordham Urb. LJ, 32*, 767.
Martin, M., Deas, I., & Hincks, S. (2019). The role of temporary use in urban regeneration: Ordinary and extraordinary approaches in Bristol and Liverpool. *Planning Practice & Research, 34*(5), 537–557.
Minton, A. (2006). *The privatisation of public space*. London: RICS.
Mitchell, W. J. (1995). *City of bits: Space, place, and the infobahn*. Cambirdge: MIT Press.
Mitchell, D. (2003). *The right to the city: Social justice and the fight for public space*. New York: Guilford press.
Mumford, L. (1961). *The city in history: Its origins, its transformations, and its prospects*. New York: Harcourt, Brace & World.
Munoz, L. (2019). Cultural gentrification: Gourmet and Latinx immigrant food trucks vendors in Los Angeles. *Journal of Urban Cultural Studies, 6*(1), 95–111.
Parker, S. (2015). *Urban theory and the urban experience: Encountering the city*. London: Routledge.
Penz, F., & Lu, A. (Eds.). (2011). *Urban cinematics understanding urban phenomena through the moving image*. Bristol: Intellect Books.
Sadowy, K., & Lisiecki, A. (2019). Post-industrial, post-socialist or new productive city? Case study of the spatial and functional change of the chosen Warsaw industrial sites after 1989. *City, Territory and Architecture, 6*(4). https://doi.org/10.1186/s40410-019-0103-2.
Sandercock, L. (2003). *Cosmopolis II: Mongrel cities in the 21st century*. London: Continuum.
Sennett, R. (1970). *Uses of disorder*. New York: Knopf.
Shaftoe, H. (2008). *Convivial public spaces: Creating effective public places*. London: Earthscan.
Shepard, M. (2011). *Sentient city: Ubiquitous computing, architecture, and the future of urban space*. Cambridge: MIT Press.
Sorkin, M. (1992). *Variations on a Theme Park: The new American city and the end of public space*. New York: Hill and Wang.
Steele, J. (2013). How 'meanwhile' came to the high street. *Journal of Urban Regeneration and Renewal, 6*(2), 172–175.
Thwaites, K., Porta, S., Romice, O., & Greaves, M. (Eds.). (2007). *Urban sustainability through environmental design: Approaches to time-people-place responsive urban spaces*. New York: Taylor & Francis.
Tonkiss, F. (2013). *Cities by design: The social life of urban form*. Cambridge: Polity Press.

Townsend, A. M. (2013). *Smart cities: Big data, civic hackers, and the quest for a new utopia*. New York: W. W. Norton & Company.
Varna, G., & Tiesdell, S. (2010). Assessing the publicness of public space: The star model of publicness. *Journal of Urban Design, 15*(4), 575–598. https://doi.org/10.1080/13574809.2010.502350.
Virilio, P. (1991). translated by Moshenberg, D. *The lost dimension*. New York: Semiotext(e).
Virilio, P. (1994 [1989]). translated by Rose, J. *The vision machine*. New York: Semiotext(e).
Warner, M. (2002). Publics and Counterpublics. *Public Culture, 14*(1), 49–90. https://www.muse.jhu.edu/article/26277
Whyte, W. H. (1988). *City: Rediscovering the center*. New York: Doubleday.
Wigley, M. (2001). Network fever. *Grey Room, 1*(4), 82–122.
Young, I. M. (1990). *Justice and the politics of difference*. Princeton: Princeton University Press.
Zucker, P. (1959). *Town and square: From the agora to the village green*. New York: Columbia University.
Zukin, S. (1991). *Landscapes of power: From Detroit to Disney world*. Berkeley: University of California Press.
Zukin, S. (1995). *The cultures of cities*. Cambridge, MA: Blackwell.

Public Sphere

► Public Space

Public Tendering

► Public Procurement for Regional and Local Development

Public-Private Partnerships: The Danish Way of Turning Climate Change Measures into Policies and Long-Term Commitments

Magnus Højberg Mernild
State of Green, Copenhagen, Denmark

Transitioning to sustainable economies and achieving net zero cannot be done by governments alone. Forging just and financially viable paths requires conscientious and concerted action, where citizens, corporates, and governments work in tandem. A challenging task that calls for effective ways of integrating climate change measures into policies, strategies, and binding commitments.

Following Denmark's general election in 2019, the Danish Government established 13 industry-specific Climate Partnerships with the private sector (The Prime Minister's Office 2019). The Climate Partnerships, which later expanded to 14, are an instrumental lever in realizing Denmark's 2030 climate target as set forth in the country's Climate Act (The Danish Ministry of Climate, Energy & Utilities 2019). Passed in 2020 by 167 of the 179 members of the Danish parliament (State of Green 2021, p. 5), the Act obligates the sitting government to work to reduce Denmark's greenhouse gas emissions by 70% by 2030 compared to 1990 levels and toward net zero by 2050 at the latest (Danish Energy Agency 2022a).

The 14 Climate Partnerships span all sectors of the Danish economy – from energy, finance, and maritime to construction, aviation, and agriculture. Each partnership is chaired by industry leaders such as Ørsted, Maersk, Danfoss, Novozymes, PensionDenmark and consists of a range of representatives from a specific industry, where they are tasked with formulating recommendations as to how their sector can contribute to reducing emissions and reaching the 2030 goal. The partnerships have a dual objective: (1) to make the world more sustainable, and (2) to ensure green growth in Denmark (State of Green 2021, p. 8). Essentially, Denmark has delegated responsibility for making its economy more sustainable to the industries themselves.

Collectively, the Climate Partnerships have produced more than 400 recommendations (The Danish Government 2021), many of which the Danish Government has already implemented through a number of agreements contained within the government's climate action plan. Further, the Danish Minister of Finance has asked the Climate Partnerships to select the most urgent decisions and initiatives that will assist in stimulating the economy in terms of growth and employment, as well as contribute to the green transition.

The most well-known recommendation hitherto is the proposal to build the world's first artificial energy island. Instigated by the climate partnership on energy, a broad majority of the Danish parliament in June 2020 legislated the creation of an energy hub the size of 18 football pitches in the North Sea as part of the Danish Climate Act (Danish Energy Agency 2022b). The island, which will have an installed capacity of up to 10 GW, will be able to power approximately 10 million European households with green energy and eventually fuel heavy transport in a sustainable way through Power-to-X (ibid.). Located 80 km from the Jutlandic peninsula, the North Sea energy island is Denmark's largest infrastructure project ever. The island will be owned by a public-private partnership, where the Danish State will own at least 50.1% (The Danish Energy Agency 2021).

Other recommendations include the partnership for food and agriculture which has put forth solutions with the potential to reduce the sector's climate impact by 62% in 2030 (State of Green 2021, p. 42). Similarly, the construction industry's proposals would see the sector reduce its emissions by approximately 5.8 t of CO_2e yearly (State of Green 2021, p. 23). Other partnerships have also identified how the upscaling of existing technologies within energy, water, life science, and production facilities can bring sub-sectors close to climate neutrality in 2030. Although the approach is forward-looking, the recipe is based on past experiences with effective public-private partnerships.

A Partnership Approach with 50 Years of Experience

When the 1973 oil crisis struck, Denmark was one of the OECD countries most dependent on oil for its energy supply. With more than 90% of its national energy supplies based on imported oil (Danish Energy Agency (2012), the surging prices hit a key nerve in the society and marked a turning point in Denmark's energy policies. Virtually overnight, the crisis spurred a push to diversify the national energy mix, and the establishment of a new regulatory regime was considered a precondition for a successful reorientation of the energy sector. A key factor in the success of Denmark's decision to heighten its energy security has been its approach to policymaking, where long-term agreements with a broad consensus across the political spectrum and the support of industry are devised on energy and environmental issues. This means that policies remain unchanged even when the government changes, and the ensuing political stability has helped secure continuous investment and solutions to many of the nation's sustainable development challenges.

Another key factor is Denmark's strong tradition for entering into public-private partnerships, which has allowed shifting Danish governments to enact regulations and programs with the support of business and industry, thereby ensuring their successful implementation and adherence. The high levels of societal trust that exist in Denmark, and an open, dialogue-based approach to solving challenges such as climate change have reinforced the partnership approach. While the public sector provides ambitious long-term goals and stable framework conditions, the private sector supplies innovation and solutions to achieve the visions. As such, the Danish Model seeks to reflect the constant need for growth and innovation while, at the same time, cushion the inevitable setbacks that societies face on unfamiliar paths.

Decoupling Economic Growth from Energy Consumption

Since 1980, Denmark has managed to decouple its economic growth from its overall energy consumption. In four decades, the national GDP has more than doubled while energy consumption has only increased by 6% (State of Green 2022). Over the same period, water consumption decreased by 40% (Ibid.). The numbers prove that it is possible to create growth without using more energy.

The Danish concept of public-private partnerships has been a catalyst in securing the country's green economy and reconciling economic growth with ambitious green policies. Contributing approximately 12% of the total Danish exports, the share of Denmark's green energy technology exports has grown by more than 40% since 2010 (Danish Energy Agency, Confederation of Danish Industry, Danish Energy, Wind Denmark, Danish District Heating Association 2021). When it comes to research and development in green technologies, the rewards of a whole-of-society approach also stand out. No other OECD country displays a similar development of green technology as measured by patent applications (Danish Ministry of Business. Industry and Financial Affairs 2021). The legacy of long-term political commitments and industry assuming responsibility for solving societal challenges is reflected in the maturity of the country's sustainable financing sector. In 2019, Denmark's pension industry committed to invest more than USD 55 billion in green initiatives toward 2030 (The Danish Government's Climate Partnerships 2021a). Only 2 years down the road, this target appears to be safely within reach. Nationally, Denmark also allocated 59% of the EU COVID-19 recovery funds to green initiatives (The European Commission 2022), against the required 37%.

National Reductions, Global Inspiration

While the Danish learnings and its 14 Climate Partnerships provide solid arguments to enter into national as well as global partnerships, the contribution to achieving net zero globally is minimal. Denmark only accounts for 0.1% of global emission (Danish Ministry of Climate, Energy and Utilities 2020). Therefore, Denmark aims to provide inspiration and experiences from its own route to lowering emissions to spur other countries to embark on a similar path and deepen their commitments, thus heightening its overall impact.

The conclusion to Denmark's learning curve over the past five decades provides another valuable lesson. Namely, investing in renewable energy, water, energy efficiency, and resource optimization makes good economic sense. Recent projections show that every time one gigawatt of offshore wind is set up in Denmark, 14,600 jobs are secured in Danish companies (Danish Shipping, Danish Energy and Wind Denmark 2020). Today, 76,000 Danes hold green jobs out of a national workforce of some 2.8 million people (Danish Ministry of Business 2021b). Looking ahead, realizing the Danish 2030 climate target of reducing its greenhouse gas emissions by 70% by 2030 may create up to 290,000 green jobs (Danish Energy 2020).

In sum, Denmark's transition shows that green business is good business. As such, the Danish transition shows that public-private partnerships can accelerate green growth as an effective way of turning climate change measures into policies and long-term commitments.

Cross-References

▶ Community Engagement and Climate Change: The Value of Social Networks and Community-Based Organizations
▶ Multi-stakeholder Partnerships to Support Climate Migrants in Fragile Cities

References

Danish Energy. (2020). *Beskæftigelseseffekter af investeringerne i den grønne omstilling* [analysis]. https://www.danskenergi.dk/sites/danskenergi.dk/files/media/dokumenter/2020-11/Arbejdskraftanalyse_Beskaeftigelseseffekter-af-investeringerne-i-den-groenne-omstilling.pdf

Danish Energy Agency. (2012). *Energy policy in Denmark*, p. 6, [publication]. https://www.ft.dk/samling/20121/almdel/KEB/bilag/90/1199717.pdf

Danish Energy Agency, Confederation of Danish Industry, Danish Energy, Wind Denmark, Danish District Heating Association (2021), *Eksport af energiteknologi og -service 2020* [analysis]. https://presse.ens.dk/pressreleases/eksporten-af-energiteknologi-faldt-i-2020-3094178

Danish Ministry of Business. Industry and Financial Affairs (2021). *Redegørelse om virksomheders grønne omstilling*, [publication]. file:///C:/Users/MagnusHpercentC3percentB8jbjerg/Downloads/redegoerelse-om-virksomheders-groenne-omstillingpercent20(1).pdf

Danish Shipping, Danish Energy and Wind Denmark. (2020). *Socioeconomic impacts of offshore wind*, p. [study]. https://www.danishshipping.dk/en/press/news/new-report-offshore-wind-secures-thousands-of-jobs/

State of Green. (2021). *Climate Partnerships for a greener future*. [publication]. https://climatepartnerships2030.com/

State of Green. (2022), *The Danish vision,* [online]. https://stateofgreen.com/en/the-danish-green-vision/

The Danish Energy Agency. (2021). *Tender-preparing partial agreement regarding the long-term framework of a call for tenders and ownership of the energy island in the North Sea* [binding agreement]. https://ens.dk/sites/ens.dk/files/Energioer/tender-preparing_partial_agreement_of_1_september_2021.pdf

The Danish Energy Agency. (2022a). *Danish Climate Policies, [online].* https://ens.dk/en/our-responsibilities/energy-climate-politics/danish-climate-policies

The Danish Energy Agency. (2022b). *Denmark's Energy Islands*, [online]. https://ens.dk/en/our-responsibilities/wind-power/energy-islands/denmarks-energy-islands

The Danish Government. (2021). Denmark's recovery and resilience plan, [publication]. https://fm.dk/media/18771/denmarks-recovery-and-resilience-plan-accelerating-the-green-transition_web.pdf

The Danish Government's Climate Partnerships. (2021). Sector road map, *Klimapartnerskab for finanssektoren*, [publication]. https://em.dk/media/14287/sektorkoereplan-for-klimapartnerskab-for-finanssektoren.pdf

The Danish Ministry of Business, Industry and Financial Affairs, (2021a). *sektorkøreplan, Klimapartnerskab for finanssektoren*, [publication]. https://em.dk/media/14287/sektorkoereplan-for-klimapartnerskab-for-finanssektoren.pdf

The Danish Ministry of Business, Industry and Financial Affairs. (2021b). *Redegørelse om virksomheders grønne omstilling,* [publication]. https://em.dk/media/14278/redegoerelse-om-virksomheders-groenne-omstilling.pdf

The Danish Ministry of Climate, Energy and Utilities (2020). *Climate Programme 2020 - Denmark's Mid-century, Long-term Low Greenhouse Gas Emission Development Strategy,* [publication]. https://unfccc.int/sites/default/files/resource/ClimateProgramme2020-Denmarks-LTS-under-thepercent20ParisAgreement_December2020_.pdf

The Danish Ministry of Climate, Energy and Utilities. (2019). *Denmark's Climate Act.* [publication]. Available at: https://kefm.dk/Media/1/D/aftale-om-klimalov-af-6-december-2019percent20FINAL-a-webtilgpercentC3percentA6ngelig.pdf

The European Commission. (2022). *Denmark's recovery and resilience plan*, [publication]. https://ec.europa.eu/info/business-economy-euro/recovery-coronavirus/recovery-and-resilience-facility/denmarks-recovery-and-resilience-plan_en

The Prime Minister's Office. (2019). *The Danish Government's Climate Partnerships* [press release]. https://www.stm.dk/presse/pressemeddelelser/regeringens-klimapartnerskaber/

P

Q

Quality of Life

▶ Theme Cities Networks

R. C. Brears (ed.), *The Palgrave Encyclopedia of Urban and Regional Futures*,
https://doi.org/10.1007/978-3-030-87745-3

R

Rainwater – Precipitation

▶ Rainwater Harvesting for Water Security in Informal Settlements: Techniques, Practices, and Options

Rainwater Harvesting for Water Security in Informal Settlements: Techniques, Practices, and Options

Thomas Karakadzai[1], Abraham R. Matamanda[2] and Innocent Chirisa[3]
[1]Department of Demography Settlement and Development, Faculty of Social & Behavioral Sciences, University of Zimbabwe, Harare, Zimbabwe
[2]Department of Geography, University of the Free State, Bloemfontein, South Africa
[3]Department of Demography Settlement and Development, Social & Behavioural Sciences, University of Zimbabwe, Harare, Zimbabwe

Synonyms

Harvesting – collecting, harnessing; Rainwater – precipitation; Security – provisioning; Slum – informal settlement

Definition

Informal settlement – a slum, being a settlement without adequate facilities

Rain water harvesting – methods and techniques of collecting and storing water from the atmosphere

Water security – ensuring water resources are available for human and related consumption always

Introduction

The term "informal settlement" generally refers to urban settlements that develop outside the legal systems intended to record land ownership and tenure and enforce compliance with regulations relating to planning and land use, built structures, and public health and safety (Revi et al. 2014). Parkinson, Tayler, and Mark (2007, p. 173) have noted that informal settlements fall outside formal laws and regulations on land ownership, land use, and buildings. The informal settlements are illegally formed and therefore not officially part of the city resulting in an exclusion from urban services. Characteristics of informal settlements vary within and across countries, but most settlements have inadequate basic services such as water, sanitation, electricity, waste management, drainage, and roads (World Bank Group 2015). Utilities

R. C. Brears (ed.), *The Palgrave Encyclopedia of Urban and Regional Futures*,
https://doi.org/10.1007/978-3-030-87745-3

underprovide these services, partly because they do not have a clear obligation to serve informal settlements and in some cities do not have the authority to do so. However, many subsequently gain recognition from the local authorities, often as a result of political patronage. Where authorization to deliver services does exist, utilities tend not to prioritize extending services because they are technically, legally, and commercially more challenging to serve relative to formal urban communities (World Bank 2017). Challenges are greater for extending services to settlements with insecure land tenure, those in peri-urban areas that may be more remote or outside of formal utility service districts, and those on land that is technically challenging to reach with traditional infrastructure.

In expanding informal neighborhoods of cities in sub-Saharan Africa (SSA), sustainable management of storm- and wastewater drainage is fundamental to improving living conditions (Mulligan et al. 2020). This has ignited the debate on how to optimally combine green or natural infrastructure, traditional "gray" infrastructure, and "blue" infrastructure which mimics natural solutions using artificial materials. Planners have advocated for small-scale, niche experiments with these approaches in informal settings, in order to learn how to navigate the intrinsic constraints of space, contested land tenure, participation, and local maintenance. Parkinson et al. (2007) note the need to consider the development of city-wide drainage system in order to maximize the effectiveness and efficiency of investments in drainage. Water security is the availability of water, in adequate quantity and quality, to sustain a well-balanced (or sustainable) development. These conditions seem to be more and more challenged considering the growing populations, the global urbanization, and the related environmental consequences.

Curbing water scarcity problems in informal settlements is a top priority for economic and social development. This chapter aims at assessing how informal settlements have embraced rainwater harvesting (RWH) for water security across the globe. In spite of rainwater harvesting being widely promoted as a panacea for the growing drinking water crisis in different settlements, informal settlements have rarely implemented the system. The existing water security strategies commonly pivot around supply-side initiatives to mitigate scarcity, forecasted population growth, or anticipated climate change (McDonald et al. 2014). The rationale of this chapter draws upon the debate regarding the need for more practically relevant tools that can be used for mainstreaming water harvesting techniques in informal settlements. The chapter is organized as follows: The first part presents the introduction, theoretical perspectives, and literature review on rainwater harvesting experiences around the globe. The presentation and analysis of the study findings follow. The last part of the chapter presents a summary of the arguments raised in the chapter.

Theoretical Perspectives

The ecological modernization theory (EMT) provides a useful framework for studying adoption of new ideas and technologies (White 2011). The theory contends that private sector innovation, governance, and consumption reforms mutually influence each other, exerting a positive influence on the environment. The household purchase of a rainwater harvesting system appears idiomatic. The direct effect of pro-environmental technological innovation is hinged on the process by which an innovation is communicated through certain channels over time among the members of a social system. Nevertheless, calls to "rationalize" lifestyle or to reduce consumption are generally received reluctantly by the community, and with the ongoing drought, community use of water is increasingly rationed. This calls for diffusion of innovative ideas by adopting actor-centered criterion theory which asserts that adoption is facilitated by criteria centered on the actor. In the context of urban planning, Do-It-Yourself Urbanism is a fluid concept that refers to small-scale interventions "in which urban residents take it upon themselves to do what cities will not, or cannot, to address urban issues" (Finn 2014, p. 381). This encompasses adoption of new technologies but also new ways of knowing or ideas. However, with new technology, there is a possibility for new values to arise within the society which are incompatible with the old or for the society to come in conflict with the different expectations of the new approach/methodology.

Globally, people are seeking solutions to water insecurity, and the harvesting of rainwater offers potential in many contexts (Button 2017). As populations continue to grow, and as climate variability and change affect water availability and the frequency and severity of extreme events, achieving and maintaining water security is a fundamental development challenge. Water, food, and energy are inextricably linked security concerns and form a critical nexus for understanding and addressing development challenges. The security of water is even more difficult to improve in conflict areas, which suffer from degraded water infrastructure, limited human and financial capacities, and weak institutions (USAID 2018). However, the effect on water security differs regionally depending on a number of factors including geographic location and features, conditions of water availability and utilization, demographic changes, existing management and allocation systems, legal frameworks for water management, existing governance structures and institutions, and the resilience of ecosystems (UN-DESA 2014).

Multidisciplinary approaches and cross-sectoral policies are needed to address water issues underlying human security, for instance, rainwater harvesting. Rainwater harvesting is among the specific adaptation measures that the water sector needs to undertake to cope with future climate change (Kahinda et al. 2010). The art of rainwater harvesting has been practiced since the first human settlements, being particularly developed in the dry regions of the planet given the need to expand the sources of water for human consumption in these areas (Barron 2009). Rainwater harvesting has been practiced in various informal settlements across different countries around the globe.

In Mexico, land-use policy – both urban and environmental – has ignored, or at least failed to include explicitly in its plans and regulations, a strategy for managing informal settlements and has also not defined land reserves for relocating poor groups in the future (Aguilar and Santos 2011). DIY Urbanism intervention was adopted and then formally institutionalized in Tucson as DIY Urbanism actions transition from the informal to the formal realm. The initiative inspired countries and institutions such as the United States and World Bank for their ground-breaking green infrastructure initiatives and approaches to dealing with urban water scarcity (World Bank 2017). The case study helps contribute to better understanding the shape of urban rainwater policies and practices (Soler 2018) and the potential utilization of informal DIY Urbanism rainwater harvesting as a form of practical authority for achieving sustainable urban water management and water democracy.

Germany has seen strong uptake of RWH technologies as reported by Partzsch (2009) with 80,000 installations per annum and a total industry value of 340 million Euros. In the United Kingdom, rainwater harvesting (RWH) systems have traditionally been installed at domestic residences for the single objective of providing a non-potable water supply for use in toilets, for laundry facilities, and for garden irrigation (British Standards Institution 2013).

In Asia and Africa, rainwater harvesting has been widely used for non-potable applications in countries such as Malaysia, India, Kenya, and South Africa. Interest in rainwater harvesting (RWH) has rapidly grown in Kenya over the last 30 years. RWH projects have been carried out in an effort to provide long-term solutions to water resource problems. RWH projects were meant to provide water for domestic needs and agriculture and commercial purposes which resulted in strengthening individual and community fabric by reducing livelihood insecurities (Black et al. 2012). In sub-Saharan Africa, the potential of RWH is yet to be fully recognized as there is insufficient awareness and support of RWH as a water supply solution and a lack of local capacity for implementation. Multitudes of indigenous and recently developed rainwater harvesting techniques are used in different parts of SSA. Some of these indigenous techniques have been introduced and are being widely applied in the drylands of Western Asia (Oweis et al. 2004).

The techniques and modes of application, however, differ regionally. The best experiences in one country have the potential to be adapted in another country which has similar problems of water scarcity. To convince and attract more development partners and avoid skepticism about the significance of the technology in SSA, an overview of the contemporary research findings on the best experiences with rainwater harvesting is essential. Oweis and Hachum (2006) outlined the most important water harvesting practices and their field performance for Western Asia and North Africa, but no such review exists for SSA.

Villages in North India, where water is scarce due to low rainfall, have collected rainwater in large ponds or wells for use by the community for centuries, particularly for watering crops (Button 2017). The city of Mumbai, India, has struggled to provide water to 60% of the population who are living in poverty in informal settlements, with low – and decreasing – water access and low resilience to water shortage (Srivastava and Echanove 2014). The responsibility for securing water resources in Mumbai's middle-class households has been shifted on to the residents themselves, via rainwater harvesting technologies installed for individual households. Several governments of Southern countries have taken the initiative to scale up community-based RWH approaches, and networks have been established between Southern and Northern civil society organizations, governments, private sector, and research institutes to support and promote upscaling of RWH (European Environment Agency 2009).

Most scholars conducting empirical evaluations on rainwater harvesting have opted to cover their studies on issues such as sizing and performance of rainwater harvesting techniques (Campisano and Modica 2012; Imteaz et al. 2011, 2012), socioeconomic constraints and opportunities of developing a rainwater harvesting approach (Kahinda and Taigbenu 2011), and their integration in runoff models in urban areas (Kim et al. 2012) or in decision support systems at the watershed scale (Kahinda et al. 2008; Kahinda and Taigbenu 2011). These studies demonstrated the practice of rainwater harvesting is used in both humid and well-developed regions, in response to the increased green building practices supporting the smart use of water (Jones and Hunt 2010).

Campisano et al. (2017) conducted a study on urban rainwater systems in Africa, Asia, Australia, America, and Europe. A major finding is that the degree of rainwater harvesting system implementation and technology selection are strongly influenced by economic constraints and local regulations. Despite many countries setting many design protocols, recommendations are still often organized only with the objective of conserving water without considering other potential benefits associated with the multiple-purpose nature of rainwater harvesting. The literature has indicated that inappropriate design and material selection results in lower microbial quality of harvested rainwater (Fewtrell and Kay 2007). Sanches Fernandes et al. (2015) conducted a study on rainwater harvesting systems for low demanding applications in Mirandela in Portugal. The study revealed how the design of a rainwater harvesting methodology in regions of temperate climate can deal with water deficit events in this type of rainwater harvesting system and sporadic climate. The study concludes that the dimensioning of rainwater harvesting systems for non-potable uses ought to consider the demand fraction defined as the ratio between the annual demand and the annually collected rainwater as this would improve economic advantages (Dile et al. 2013).

Elder and Gerlak (2019) conducted a study on interrogating rainwater harvesting as Do-It-Yourself (DIY) Urbanism in the United States (Tucson, Arizona). The study interrogates rainwater harvesting as a DIY Urbanism intervention adopted and then formally institutionalized in Tucson as DIY Urbanism actions transition from the informal to the formal realm. An examination of rainwater harvesting as a proposed sustainable urban water management practice for a water-stressed city with aging infrastructure and a growing population particularly vulnerable to a changing climate was done. The study finds out that rainwater harvesting contributes to the vision of an urban utopia for some local activists and NGO actors, as well as for those working within the municipal bureaucracy, despite rainwater harvesting being viewed as an attack on social order and private resource rights (Elder and Gerlak 2019).

Kimani et al. (2015) have evaluated rainwater harvesting technologies and the factors contributing to adoption of the technologies in the ASAL areas with Makueni County within Kenya's Eastern Region. Various rainwater harvesting technologies are used within Makueni County including micro-catchment rainwater harvesting technologies (i.e., earth dams, sand/subsurface dams, shallow wells, rock catchment structures, systems) and rooftop rainwater harvesting and micro-catchment rainwater harvesting technologies for agricultural farming. Lee et al. (2016) discuss the potential of rainwater harvesting in Malaysia under the dynamic climate. Rainwater harvesting emerged as one of the measures to enhance the resilience of human society toward water shortage problem in Malaysia; however, it is yet to be mainstreamed into the national water and climate change policy as an adaptation strategy to climate change, especially in urban areas where water resources are fast depleting due to rapid increase in population and water consumption (Lee et al. 2016). The study suggested the need for inter-ministerial and multi-stakeholders' cooperations as way to promote the development of rainwater harvesting in Malaysia to be authoritative and organized as an alternative water resource.

The most common examples pertain to the use of harvesting systems in low demanding scenarios around a household or business that do not require potable water, such as irrigation, vehicle washing, or toilet flushing. There are however a very limited number of studies focused on mainstreaming rainwater harvesting approaches in informal settlements, the reason why a common-sense approach has not yet been defined for the dimensioning of the system reservoirs.

Experiences in Rainwater Harvesting for Water Security in Informal Settlements

Rainwater harvesting is the collection, conveyance, and storage of rainwater for an intended use (Malesu et al. 2007). The method of rainwater harvesting (RWH) encompasses the accumulation and deposition of rainwater for domestic, commercial, and biomass production use instead of allowing it to run off. Gois et al. (2014) note that rainwater harvesting systems are able to simultaneously address the water scarcity problem and reduce the dependence on conventional water supplies such as piped water, taps, and boreholes. Rainwater harvesting has now much wider perspectives, being studied in a variety of environmental contexts, such as sustainable economic growth and rainfed agriculture, the harmonious functioning of ecosystems, flood control in urban areas, and quality control of surface urban runoff during rainfall (Dile et al. 2013). The way rainwater is harvested can be classified into various categories as shown in Table 1.

Rainwater Harvesting Typologies and Institutions Behind Them

There are two types of rainwater harvesting which include roof water harvesting and ground catchment harvesting. Roof water harvesting is the practice of capturing the rainfall from roofs, diverting it through gutters and drains, and storing the water in tanks of various sizes for later use. The method of harvesting water from roofs is highly determined by the nature of the roofing materials used and type of settlement. Unlike in informal settlements where the structures or buildings are made of shacks, roof harvesting in planned settlements provides a secure, safe, and convenient source of water to satisfy domestic needs (Black et al. 2012). However, in the design of roof water harvesting technology, there is a need to take into consideration factors such as the collection surface (catchment), guttering, and storage (Biazin et al. 2011).

Ground catchment harvesting is the capture and storage of rainwater and water runoff either

Rainwater Harvesting for Water Security in Informal Settlements: Techniques, Practices, and Options, Table 1 Rainwater harvesting categories

Category	Description	Examples
Macro-catchment technologies	This is a system that involves the collection of runoff from large areas which are at an appreciable distance from where it is being used (Biazin et al. 2011). These technologies handle large runoff flows diverted from surfaces such as roads, hillsides, and pastures	Hillside sheet/rill runoff utilization, rock catchments, sand and earth dams
Micro-catchment technologies	Micro-catchment rainwater harvesting systems are designed to collect runoff from a relatively small catchment area, mostly 10–500 m^2, within the farm boundary (Biazin et al. 2011). The systems collect runoff close to the growing crop and replenish the soil moisture (Kimani et al. 2015). The technique is mainly used for growing crops such as maize, sorghum, groundnuts, and millet	Zai pits, strip catchment tillage, contour bunds, semicircular bunds, and meskat-type system (Kimani et al. 2015) Zai pits are a grid of planting pits which is dug across plots that could be less permeable or rock-hard; organic matter is sometimes added to the bottom of the pits (Mupangwa et al. 2006). Zai pits are practiced in countries of West Africa (Burkina Faso, Mali, Niger) and East Africa Contouring (stone/soil bunds) Contours are made of stones or sometimes earthen bank of 0.50–0.75 cm height which is often piled on a foundation along the contour in a cultivated hillslope (WOCAT 2007) Contouring is practiced in East Africa (Kenya, Ethiopia, Tanzania), West Africa (Burkina Faso), and South Africa
		Terracing (Fanya Juu, semicircular and hillside terraces) Bunds in association with a ditch, along the contour or on a gentle lateral gradient, are constructed in different forms. The Fanya Juu terraces are different from many other terrace types in that the embankment is put in the upslope position East Africa (Kenya, Ethiopia, Tanzania) (WOCAT 2007)

Source: Biazin et al. (2011)

in a pond structure or in the soil profile itself (Malesu et al. 2007). Ground catchment harvesting consists of various in situ catchment techniques, which include harvesting ponds, pans, and dams. However, the structures vary in makeup and size, ranging from small manually dug farm ponds to large community earth and sand dams (Malesu et al. 2007). Contour ridges are also used to capture and store rainwater falling between each bund or from a larger upslope catchment as well. This technique is most often intended for agriculture or pasture land regeneration but can also be used for forestation projects.

Institutions cover the wide range of organizations – as well as policies and legal instruments that guide, govern, and possibly constrain everyday water management decisions (USAID 2018). Beyond the performance of water entities, it is often useful to analyze the entire water sector and how it is organized and operates. Issues of accountability and transparency in service provision are also a critical topic for assessing management capacities. The community-based organization groups have proved to be key institutions when dealing with rainwater harvesting. Rainwater harvesting is being taken up by citizens around the world through both informal bottom-up practices in cities like Tijuana, Mexico; Barcelona, Spain; Melbourne; and Mumbai, India. The evolving community of actors have pushed for rainwater harvesting on various levels contributing to the institutionalization of different ideas about its implementation (Soler 2018). For instance, the rainwater harvesting movement in Tucson started at a grassroots level with a few local activists installing rainwater harvesting features in their respective communities.

The Do-It-Yourself (DIY) Urbanism intervention was adopted and then formally institutionalized in Tucson as DIY Urbanism actions transition from the informal to the formal realm. The initiative inspired countries and institutions such as the United States and World Bank for their groundbreaking green infrastructure initiatives and approaches to dealing with urban water scarcity (World Bank 2017). In Mumbai, India, the inhabitants are being made responsible for securing their own water supplies via rainwater harvesting technology, which is increasingly installed in new housing (Button 2017). This shifts responsibility for water provision from city authorities to private households. As a result, community-based organizations influenced policy transformation through the institutionalization of rainwater harvesting in city and county ordinances, policies, and general plans (Word Bank 2017).

Nongovernmental organizations provide opportunities for community members in maneuvering rainwater harvesting process. Shipek (2018) notes that nongovernmental organizations trigger, teach, and develop co-op programs which demonstrate how successful rainwater harvesting projects can be done. For example, in Kibera informal settlement in Nairobi, Kenya, rainwater harvesting projects are mostly built by the support of NGOs. Nongovernmental organizations address multisectoral and multidisciplinary aspects in rainwater harvesting projects, which is an essential ingredient for project success.

Local authorities play a pivotal role in the implementation of rainwater harvesting projects. The local authorities provide a list of policies and guidelines for installing a rainwater collection and utilization system, planning and design, eco-efficiency in water infrastructure for public buildings, and urban stormwater management.

Population growth, climate change, and increasing stress on water resources evidence the urgency of adopting more sustainable urban water management practices such as rainwater harvesting. Cities such as Tucson, Arizona, started rainwater harvesting as a way to improve the sustainability of water provision in arid urban areas. It was revealed that the rainwater harvesting practice was successful through a bottom-up approach (Elder and Gerlak 2019).

Many settlement households face additional barriers to connecting to utility-supplied water. The utilities have different purviews and legal restrictions on providing services to formal areas, and there is a greater variation in how utilities serve informal settlements across countries. This has resulted in most informal settlements across Africa (Kenya, South Africa, and Zimbabwe) and Asia (India) facing challenges of water accessibility and resorting to purchasing water from mobile water suppliers.

Rainwater harvesting in informal settlements especially roof water harvesting is still at its infant stage which might have proven a solution to water scarcity. The lack of complementary infrastructure such as roof tops and gutters for harvesting rainwater tends to exacerbate negative health impacts of missing WASH services. The shacks are made of either plastics or wood material which are also prone to destruction during rainy season and could not sufficiently harvest roof water. Poor water extraction methods from the storage facilities as a result of improper designs lead to safety and health issues. The type of structures which is found in most settlements renders it ineffective for roof water harvesting in informal settlements. In Hopley Farm, in peri-urban Harare in Zimbabwe, water challenges are acute with residents averaging RTGS 10 per 20 liters bucket of water.

However, most informal settlements practice micro-catchment rainwater harvesting techniques which include pitting, contouring, terracing, and micro-basins despite the techniques having different names in different regions and minor differences in design and use. The planting of seedlings in shallow pits (5–15 cm deep and 10–30 cm wide) at intervals of 40–70 cm is being practiced in informal settlements of Zimbabwe and South Africa. The basin and mulch technique are an innovative water conservation technique used in Zimbabwe and South Africa to reduce total runoff to zero and soil evaporation considerably, thus improving crop yields.

In practice, the implementation of rainwater harvesting in informal settlements has remained a challenge due to lack of regulations governing rainwater harvesting practices. There are currently no polices and regulations at the national or local levels governing rainwater harvesting for non-potable use in informal settlements, and the practices vary widely from one location to another. The practice of deploying participatory processes in implementing rainwater harvesting practices remains unexplored in most informal settlements.

Conclusion and Future Direction

The chapter set out to provide a road map for mainstreaming rainwater harvesting systems in informal settlements. The discussion has noted that rainwater harvesting in most informal settlements across the globe, including those in Zimbabwe, still lags far behind in terms of implementation of rainwater harvesting systems. The chapter has established that context-specific approaches, enactment of government regulation and policies, and collaboration ties are key tools through which informal settlements could mainstream rainwater harvesting techniques as a way of dealing with water scarcity. Rainwater harvesting has a potential to make a significant contribution toward livelihoods in informal settlements by proper harvesting, storing, and management of the little water that falls. Various rainwater harvesting technologies (RWHTs) might be used within informal settlements including micro-catchment rainwater harvesting technologies (i.e., earth dams, sand/subsurface dams, shallow wells, rock catchment structures, systems) and rooftop rainwater harvesting and micro-catchment rainwater harvesting technologies for agricultural farming.

There is a need to establish and strengthen local-level institutions as a way of improving rainwater harvesting approaches. This entails the setting up of community-level institutional structures for facilitating planning and implementation of rainwater harvesting approaches in informal settlements. Elder and Gerlak (2019) note that citizen action through Do-It-Yourself (DIY) Urbanism strategy pushes local governments to adopt and institutionalize rainwater harvesting sustainable initiatives. The community-based groups contribute toward building the much-needed social cohesion for groups to better engage in decision-making regarding accessing improved rainwater harvesting technologies. The economic feasibility of rainwater harvesting technology is dependent on various factors despite of it being built and installed in various countries (Badiru and Omitaomu 2007). The reliability of rainwater harvesting technology is largely dependent on and affected by the socioeconomic and infrastructure conditions, the area of the roof, the rainfall intensity, and the hydrological setting. The success of rainfall harvesting depends upon the frequency and amount of rainfall; therefore, it is not a dependable water source in times of dry weather or prolonged drought.

Support through policing by the government enables facilitation of rainwater harvesting practices both at institutional and sociopolitical levels. Campisano et al. (2017) augment that the development of a wider-scale rainwater harvesting which is sustainable requires improved support at institutional and sociopolitical levels as a way to incentive tools and social acceptance. Implementation of rainwater harvesting approaches needs to be context specific guided by the local government policing and regulations. There should not be a one-size-fits-all approach when dealing with rainwater harvesting as different contexts require different approaches of rainwater harvesting. Melville-Shreeve et al. (2016) support that modelling tools and methodologies have been developed over the last 20 years to facilitate the evaluation (and design) of rainwater harvesting systems in different countries. In practice, operationalizing rainwater harvesting necessitates some limitation of spatial extent but should at least reflect on the implications of these designations, cross-scalar interactions, and how fostering resilience at one spatial scale affects those at others (Meerow and Newell 2019). There is a need to understand the political context, decision-making processes, and powerbrokers that define the rainwater harvesting agenda and to carefully consider underlying motives.

The perception of rainwater harvesting system as a purely technocratic approach has created a barrier between stakeholder nodes in terms of collaborating on similar projects and implementing cross-departmental interventions (Ziervogel et al. 2016). Kimani et al. (2015) identify the major technical constraints toward the adoption and success of rainwater harvesting systems such as inadequate guidelines on the construction of rainwater harvesting systems especially in the informal settlements. There is inadequate technological transfer to the beneficiaries (in cases of donor-funded projects) as well as lack of training programs on rainwater harvesting for stakeholders (beneficiaries' artisans) which often result in poor technical selection and usage of local materials in construction of rainwater harvesting systems. For example, residents of Makueni County who harvest rainwater at household level indicated that they had never been trained on rainwater harvesting, health and sanitation and food and nutrition storage properties. Improper training on rainfall data simulation leads to improper sizing of rainwater harvesting storage facilities. Technical aspects of the projects including gutter selection and methods of fixation, fixing taps, tank construction valves, and operation and maintenance guidelines are not fully understood nor issued to the community on the commissioning of the project.

References

Aguilar, A. G., & Santos, C. (2011). Informal settlements' needs and environmental conservation in Mexico City: An unsolved challenge for land-use policy. *Land Use Policy, 28*, 649–662.

Badiru, A. B., & Omitaomu, O. A. (2007). *Computational economic analysis for engineering and industry* (pp. 61–71). Florida, USA: CRC Press.

Barron, J. (2009). Chapter 1: Introduction: Rainwater harvesting as a way to support ecosystem services and human well-being. In J. Barron (Ed.), *Rainwater harvesting: A lifeline for human well-being* (pp. 1–3). York/Stockholm: Stockholm Environment Institute/Stockholm Resilience Centre.

Biazin, B., Sterk, G., Temesgen, M., Abdulkedir, A., & Stroosnijder, L. (2011). Rainwater harvesting and management in rainfed agricultural systems in sub-Saharan Africa – A review. *Physics and Chemistry of the Earth*. https://doi.org/10.1016/j.pce.2011.08.015.

Black, J., Malesu, M., Oduor, A., Cherogony, K., & Nyabenge, M. (2012). Rainwater harvesting inventory of Kenya an overview of techniques, sustainability factors, and stakeholders. World Agroforestry Centre (ICRAF). PO Box 30677 – 00100 Nairobi, Kenya.

British Standards Institution. (2013). *BS 8515:2009 +A1:2013 – Rainwater Harvesting Systems – Code of Practice*. London: BSI.

Button, C. (2017). Domesticating water supplies through rainwater harvesting in Mumbai. *Gender and Development, 25*(2), 269–282. https://doi.org/10.1080/13552074.2017.1339949.

Campisano, A., & Modica, C. (2012). Optimal sizing of storage tanks for domestic rainwater harvesting in Sicily. *Resources, Conservation and Recycling, 63*, 9–16.

Campisano, A., Butler, D., Ward, S., Burns, M. J., Friedler, F., DeBusk, K., Fisher-Jeffes, L. N., Ghisi, E., Rahman, A., Furumai, H., & Han, M. (2017). Urban rainwater harvesting systems: Research, implementation and future perspectives. *Water Research, 115*, 2017.

Dile, Y. T., Karlberg, L., Temesgen, M., & Rockström, J. (2013). The role of water harvesting to achieve sustainable agricultural intensification and resilience against water related shocks in sub-Saharan Africa. *Agriculture, Ecosystems and Environment, 181*, 69–79.

EEA (European Environment Agency). (2009). Water Resources across Europe–Confronting Water Scarcity and Drought. Report No. 2/2009. Available online: https://www.eea.europa.eu/publications/water-resources-across-europe/file. Accessed on 18 June 2018.

Elder, A. D., & Gerlak, A. K. (2019). Interrogating rainwater harvesting as Do-It-Yourself (DIY) Urbanism. *Geoforum, 104*(2019), 46–54.

Fewtrell, L., & Kay, D. (2007). Microbial quality of rainwater supplies in developed countries: a review. *Urban Water J, 4*(4), 253–60.

Finn, D. (2014). DIY urbanism: Implications for cities. *Journal of Urbanism: International Research on Placemaking and Urban Sustainability, 7*(4), 381–398.

Gois, E. H. B., Rios, C. A. S., & Constanzi, R. N. (2014). Evaluation of water conservation and reuse: A case study of a shopping mall in southern Brazil. *Journal of Cleaner Production, 96*, 263–271.

Imteaz, M. A., Ahsan, A., Naser, J., & Rahman, A. (2011). Reliability analysis of rainwater tanks in Melbourne using daily water balance model. *Resour Conserv Recy, 56*(1), 80–86. https://doi.org/10.1016/j.resconrec.2011.09.008.

Jones, M. P., & Hunt, W. F. (2010). Performance of rainwater harvesting systems in the southeastern United States. *Resour Conserv Recycl, 54*, 623–629.

Kahinda, J. M., & Taigbenu, A. E. (2011). Rainwater harvesting in South Africa: Challenges and opportunities. *Physics and Chemistry of the Earth, 36*, 968–976.

Kahinda, J. M., Lillie, E. S. B., Taigbenu, A. E., Taute, M., & Boroto, R. J. (2008). Developing suitability maps for rainwater harvesting in South Africa. *Journal of Physics and Chemistry of the Earth, 33*, 788–799.

Kimani, M. W., Gitau, A. N., & Ndunge, D. (2015). Rainwater harvesting technologies in Makueni county, Kenya. *Research Inventy: International Journal of Engineering and Science, 5*(2), 39–49.

Kim, H., Han, M., & Lee, J. Y. (2012). The application of an analytical probabilistic model for estimating the rainfall–runoff reductions achieved using a rainwater harvesting system. *Science of the Total Environment, 424*, 213–218.

Lee, K. E., Mokhtar, M., Mohd Hanafiah, M., Abdul Halim, A., & Badusah, J. (2016). Rainwater harvesting as an alternative water resource in Malaysia: Potential, policies and development. *J Clean Prod, 126*, 218–222.

Malesu, M., Oduor, R., & Odhiambo, O. (2007). *Green water management handbook: Rainwater harvesting for agricultural production and ecological sustainability*. Technical Manual No. 8 Nairobi, Kenya: World Agroforestry Centre (ICRAF), Netherlands Ministry of Foreign Affairs. 219p.

McDonald, R. I., Weber, K., Padowski, J., Flörke, M., Schneider, C., Green, P. A., et al. (2014). Water on an urban planet: Urbanization and the reach of urban water infrastructure. *Global Environmental Change, 27*(1), 96–105. https://doi.org/10.1016/j.gloenvcha.2014.04.022

Meerow, S., & Newell, J. P. (2019). Urban resilience for whom, what, when, where, and why? *Urban Geography, 40*(3), 309329. https://doi.org/10.1080/02723638.2016.1206395

Mulligan, J., Bukachi, V., Campbell-Clause, J., Jewell, R., Kirimi, F., & Odbert, C. (2020). Hybrid infrastructures, hybrid governance: new evidence from Nairobi (Kenya) on green-blue-grey infrastructure in informal settlements. Anthropocene. https://doi.org/10.1016/j.ancene.2019.100227

Mupangwa, W., Love, D., & Steve Twomlow, S. (2006). Soil–water conservation and rainwater harvesting strategies in the semi-arid Mzingwane Catchment, Limpopo Basin, Zimbabwe. *Physics and Chemistry of the Earth, 31*, 893–900.

Oweis, T., & Hachum, M. (2006). *Water harvesting for improved rainfed agriculture in the dry environments integrated water and land management program*. Aleppo, Syria: International Center for Agricultural Research in the Dry Areas (ICARDA).

Oweis, T., Hachum, M., & Bruggeman, A. (2004). *Indigenous water harvesting systems in West Asia and North Africa*. Aleppo, Syria: International Centre for Agricultural Research in the Dry Areas ICARDA.

Parkinson, J., Tayler, K., & Mark, O. (2007). Planning and design of urban drainage systems in informal settlements in developing countries. *Urban Water Journal, 4*(3), 137–149.

Partzsch, L. (2009). Smart regulation for water innovation – The case of decentralized rainwater technology. *Journal of Cleaner Production, 17*, 985–991.

Peter Melville-Shreeve, P., Ward, S., & Butler, D. (2016). Rainwater harvesting typologies for UK Houses: A multi criteria analysis of system configurations. *Water, 8*, 129.

Revi, A., Satterthwaite, D., Aragoń-Durand, F., Corfee-Morlot, J., Kiunsi, R. B. R., Pelling, M., Roberts, D., Solecki, W., Pahwa Gajjar S., & Sverdlik, A. (2014). Chapter 8: Urban areas in field. In: C. B., V. R. Barros, D. J. Dokken, K. J. Mach, M. D. Mastrandrea, T. E. Bilir, M. Chatterjee, K. L. Ebi, Y. O. Estrada, R. C. Genova, B. Girma, E. S. Kissel, A. N. Levy, S. MacCracken, P. R. Mastrandrea and L. L. White (eds.), *Climate change 2014: Impacts, adaptation, and vulnerability. Part A: global and sectoral aspects. Contribution of working group II to the fifth assessment report of the intergovernmental panel on climate change* (pp. 535–612). Cambridge and New York: Cambridge University Press.

Sanches Fernandes, L. F., Terêncio, D. P., & Pacheco, F. A. (2015). Rainwater harvesting systems for low demanding applications. *The Science of the Total Environment, 529*, 91–100. https://doi.org/10.1016/j.scitotenv.2015.05.061. PMid:26005753.

Shipek, C. (2018). Personal interview. 14 November 2018. Tucson.

Soler, N. G. (2018). Rain and the city: Pathways to mainstreaming rainwater harvesting in Berlin. *Geoforum, 89*, 96–106.

Srivastava, R., & Echanove, M. (2014). 'Slum' is a loaded term. They are homegrown neighbourhoods. The Guardian.

UN-DESA. (2014). World Urbanization Prospects, the 2014 revision. Percentage urban and urban agglomerations by size class. United Nations. Department of Economic and Social Affairs. Population Division. https://esa.un.org/unpd/wup/Maps/CityDistribution/CityPopulation/2014/City/Urban/high.png. Accessed 11 Jan 2015.

USAID. (2018). Thinking and working politically through applied political economy analysis. Center of Excellence on Democracy, Human Rights, and Governance.

White, I. (2011). Rainwater harvesting: theorising and modelling issues that influence household adoption. *Water Science and Technology, 62*(2), 370–377.

WOCAT (World Overview of Conservation Approaches and Technologies). (2007). *Where the land is greener – Case studies and analysis of soil and water conservation initiatives worldwide* (Eds: H. Liniger & W. Critchley.). FAO, Rome.

World Bank. (2017). Transforming water-scarce cities into water-secure cities through collaboration. The World Bank. http://www.worldbank.org/en/news/feature/2017/05/15/water-scarce-cities-initiative

Ziervogel, G., Cowen, A., & Ziniades, J. (2016). Moving from adaptive to transformative capacity: Building foundations for inclusive, thriving, and regenerative urban settlements. *Sustainability (Switzerland), 8*(9).

Further Reading

Abers, R. N., & Keck, M. E. (2013). *Practical authority: Agency and institutional change in Brazilian water politics*. Oxford: Oxford University Press.

da Costa Pacheco, P. R. (2017). A view of the legislative scenario for rainwater harvesting in Brazil. *Journal of Cleaner Production, 141*, 290–294.

Domènech, L., & Saurí, D. (2011). A comparative appraisal of the use of rainwater harvesting in single and multifamily buildings of the Metropolitan Area of Barcelona (Spain): Social experience, drinking water savings and economic costs. *Journal of Cleaner Production, 19*, 598–608.

Douglas, G. C. C. (2014). Do-It-Yourself urban design: The social practice of informal "Improvement" through unauthorized alteration. *City and Community, 13*(1), 5–25.

Environment Agency. (2008). Harvesting rainwater for domestic uses: An information guide. Available online: http://www.highland.gov.uk/NR/rdonlyres/F512E7E7-6C8D-4036-ACFF-1DA761006DF8/0/RainwaterHarvestingforDomesticUse.pdf. Accessed 10 July 2015.

Gato-Trinidad, S., & Gan, K. (2014). Rainwater tank rebate scheme in Greater Melbourne, Australia. *Aqua, 63*(8), 601.

Ghimire, S. R., & Johnston, J. M. (2015). Traditional knowledge of rainwater harvesting compared to five modern case studies. In K. Karvazy & V. Webster (Eds.), *World Environmental and Water Resources Congress 2015* (pp. 182–193). Reston: American Society of Civil Engineers.

Harvey, D. (2008). The right to the city. *New Left Review, 53*, 23–40.

Helmreich, B., & Horn, H. (2009). Opportunities in rainwater harvesting. *Desalination, 248*, 118–124.

König, K. W., Gnadlinger, J., Han, M., Hartung, H., Hauber-Davidson, G., Lo, A., & Qiang, Z. (2009). Chapter 6: Rainwater harvesting for water security in rural and urban areas. In J. Barron (Ed.), *Rainwater harvesting: A lifeline for human well-being* (pp. 44–55). York/Stockholm: Stockholm Environment Institute/Stockholm Resilience Centre.

Lefebvre, H. (2000). *Writings on cities*. Malden: Blackwell Publishers.

Mays, L., Antoniou, G., & Angelakis, A. (2013). History of water cisterns: Legacies and lessons. *Water (Switzerland), 5*, 1916–1940.

Municpal Corporation of Greater Mumbai (MCGM). (2008). *Brihanmumbai Mahanagarpalika: Rain Water Harvesting*. Mumbai: MCGM.

Rahman, A., Keane, J., & Imteaz, M. A. (2012). Rainwater harvesting in Greater Sydney: Water savings, reliability and economic benefits. *Resources, Conservation and Recycling, 61*, 6–21.

Rahul, S., & Echanove, M. (2014). "Slum" is a loaded term. They are homegrown neighbourhoods. *The Guardian*, Friday 28 November.

Rockström, J., Karlberg, L., Wani, S. P., Barron, J., Hatibu, N., Oweis, T., Bruggeman, A., Farahani, J., & Qiang, Z. (2010). Managing water in rainfed agriculture: The need for a paradigm shift. *Agricultural Water Management, 97*, 543–550.

Sturm, M., Zimmermann, M., Schütz, K., Urban, W., & Hartung, H. (2009). Rainwater harvesting as an alternative water resource in rural sites in central northern Namibia. *Physics and Chemistry of the Earth, 34*, 776–785.

Tomaz, P. (2005). *Aproveitamento de água da chuva para áreas urbanas e fins não potáveis*. São Paulo: Navegar. (in Portuguese).

Rapidity

▶ Transport Resilience in Urban Regions

Readiness

▶ Transport Resilience in Urban Regions

Recovery

▶ Transport Resilience in Urban Regions

Recreation

▶ Toward a Sustainable City

Recreational Opportunities

▶ Toward a Sustainable City

Recreational Space

▶ Toward a Sustainable City

Redundancy

▶ Transport Resilience in Urban Regions

Refuge

► Habitat Provisioning

Regenerative City

► Circular Cities

Regenerative Design

► Sustainable Community Masterplan

Region – Precinct

► Zooming Regions into Perspective

Regional

► Closing the Loop on Local Food Access Through Disaster Management

Regional Development

► Financing: Fiscal Tools to Enhance Regional Sustainable Development
► Senegalese Ecovillage Network

Regional Eco-systems of Innovation

► Connecting Urban and Regional Innovation Ecosystems to Enhance Competitiveness

Regions

► Sustainable Development Goals and Urban Policy Innovation

Regulation of Urban and Regional Futures

Jeroen van der Heijden
School of Government, Victoria University of Wellington, Wellington, New Zealand
School of Regulation and Global Governance, Australian National University, Canberra, ACT, Australia

Synonyms

Public governance, Public administration, Mandate, Directive, Law, Statute

Definition

Whether we like it or not, regulation will be essential to achieve sustainable, inclusive, and healthy urban and regional futures. This chapter explores the interdisciplinary regulatory literature to understand what sorts of regulation are available to achieve these goals (from government mandates to self-regulation by firms), and what outcomes we may expect from different regulatory interventions. The chapter challenges populist antiregulation calls for less regulation and more reliance on the self-correcting capacity of the free market, but it also explains why many existing regulatory systems are not fit-for-purpose to accelerate the transition toward the urban and regional futures we desire. In short, the chapter argues that we do not need less or more regulation. Rather, we need different regulation.

There is no disagreement that the transition toward sustainable, inclusive, and healthy urban and regional futures is a typical collective action problem (Bulkeley and Betsill 2003; Castán Broto

2017; Johnson 2018). Collective action problems are those social dilemmas where individuals are better off to cooperate but are not doing so because they have diverging or even conflicting interests (Olson 1965). Regulation is – contrary to claims made by populist anti-regulation rhetoric – one of the most promising means to overcome collective action problems, including at the urban and regional levels (Jagers et al. 2020). Of particular promise is that it is well-trailed, has a proven track record of effectiveness, and provides for considerable versatility (Van der Heijden 2014).

Regulation for urban and regional futures can take many forms. For example, governments can mandate the sorts of behavior and actions that individuals and organizations need to follow to achieve desired futures, and thus bypass the process of cooperation altogether. They can forbid behavior and action that clashes with the desired futures (or reward behavior and action that contributes to it), but without mandating the sorts of behavior and actions that need to be followed. This effectively gives individuals and organizations some freedom to behave and act as they see fit. Governments can also regulate the process of cooperation between individuals and organizations to ensure that collective action toward urban and regional futures is taken.

That having been said in favor of relying on government-led regulation to govern urban and regional transformations, it comes with challenges as well. In what follows, I will first touch on the pros and cons as they relate to the development, implementation, and performance of government-led regulation for urban and regional futures. Following, I will provide some examples of regulation for urban and regional futures that move well beyond the traditional government-led, command-and-control model that we often have in mind when we think about regulation.

Advantages of Government-Led Regulation for Urban and Regional Futures

Oppressive tyrants, benevolent monarchs, and democratically elected governments all have used regulation to govern the development, use, and maintenance of cities and regions. A well-known, classic example is the Codex Hammurabi from around 1800 BC (one of the oldest preserved sets of written laws). In the codex, King Hammurabi mandates, among others, that "If a builder builds a house for some one, and does not construct it properly, and the house which he built falls and kills its owner, then that builder shall be put to death" (King 2004, 21). Such regulation of what we would now call building health and safety, but also government-led regulation for city planning and even regional development, have been traced in ancient Chinese, Greek, Roman, and Muslim law.

Hammurabi's code is an example of what is known as *lex talionis*, or law of retribution – often thought of as "an eye for an eye" (Duff 2008). While government-led regulation has evolved considerably since, three characteristics of this ancient approach to regulation have remained. First is the prescriptive nature of much government-led regulation that stipulates what needs to be done ("command"). Second is that government regulators can enforce regulation with the full power of law ("control"). Third is that often government-led regulation sets a penalty for noncompliance and, effectively, aims at compliance through fear of the consequences of noncompliance ("deterrence"). While government-led "command-and-control" regulation and its accompanying deterrence-orientation makes many people cringe these days, it is a model that has served human civilizations very well for millennia (Fukuyama 2012, 2015).

One of the advantages of government-led regulation is that it allows for treating those regulated equally and consistently under the law across time and space. A ban on the use of drinking water to water the garden or wash cars during a dry spell, for example, makes no exemptions because of people's status or financial position, and it can come in force every time a city or even a region is hit by a dry spell. Another advantage is that enforcement will be executed without a profit motive, meaning that government regulators have (in theory) no perverse incentives to find violations or delay regulatory approval processes.

For example, a municipal building inspectorate will likely charge a set fee for a specific class of building inspections, irrespective whether it concerns a complex retrofit of a heritage building or an off-the-shelf newly built building. Finally, at least in modern democracies, government regulators are held accountable for their actions and the regulations they introduce. This opens opportunities for those who disagree with the regulator or its regulations to hold the regulator to account. Millennia of positive experiences with regulation explain why there is so much of it in the governance of cities and regions (Taylor 2013).

For example, the vast amount of direct regulation that today governs regions and cities in Europe and Northern America has its origins largely in the late nineteenth Century when the detrimental impacts of rapid industrialization (and later globalization) became clear: water and air pollution in cities and beyond, unstoppable sprawl of industrial areas and infrastructure into greenfields, and, at that time, a rapid growth of slum areas to house the working poor (Garvin 2014; Le Corbusier 1929). In that period, government's role in society was that of a "nightwatchman" and in that role it fell back on addressing the new societal and environmental challenges with the tool it knew so well: direct regulatory interventions (Hodge et al. 2014). This allowed government bureaucrats to "rationally" quantify and divide the new challenges into "manageable" parts, and mandate detailed and enforceable conduct (including product and process criteria) often described in a highly technical language (Torfing 1998).

Disadvantages of Government-Led Regulation for Urban and Regional Futures

That all having been said in favor of government-led regulation for the development of cities and regions, it comes with disadvantages as well. First is that government regulation takes much time to develop and implement, and even more time for its results to become visible (Van der Heijden 2014). Considering the urgency of many societal risks, this is a problem faced by governments around the globe. Yet, it is particularly challenging for governments in rapidly developing economies. Here cities and regions are growing fast, sometimes at a speed of 5–10% new buildings and infrastructure annually (World Bank 2017). Experience suggests that governments in rapidly developing economies can take decades to develop, implement, and enforce sufficient regulatory requirements, resulting in a lag between urban and regional development and regulatory response. This could lock cities and regions into, among others, resource inefficiencies, high levels of greenhouse gas emissions, and weak (urban) resilience (Rosenzweig et al. 2018).

Second, cities in developed economies develop too slowly for (new) regulatory interventions for future development to be meaningful. Here governments face a different lag: at an annual rate of urban and regional development of about 2%, it may take 40–70 years for new regulation to transform all buildings and infrastructure (World Bank 2016). More problematically, when new government-led regulation is introduced, existing buildings, infrastructure, and land use in cities and regions are often exempted from compliance. This process, known as "grandfathering," is typically required because governments cannot infringe on (i.e., change or alter) the existing property rights held by the owners, users, and other beneficiaries of these buildings, infrastructures, and lands (van Straalen et al. 2018). By means of illustration, as a result of grandfathering existing buildings, 80% of existing Australian houses are exempted from compliance with building regulations introduced since the 1980s (Yates and Bergin 2009).

Third, as society has become more complex, and government regulators followed suit, we are now witnessing situations of overly complex and sometimes conflicting regulation that hampers rather than supports the transition to safe, just, and sustainable urban and regional futures (Van der Heijden 2014). Critics argue that government regulators will always be behind on the communities, firms, and industries they seek to regulate, and that regulation often aims at restoring harm done rather than preventing it in the first place (Fairman and Yapp 2005). They point at the risk of legalism when the proliferation of government-led regulation leads to over-regulation, which may strangle competition and entrepreneurship in the market (Bardach and Kagan 1982). Equally problematic is that public goals often cannot be expressed in technical regulatory standards, which subsequently makes the enforcement of regulations difficult and expensive (Baldwin et al. 2012).

Fourth, while I have been using the term "government-led regulation" so far, much regulation is (partially) developed or implemented by private sector individuals and organizations. This holds particularly for the regulation of cities and regions (Van der Heijden 2017). Examples include private sector inspector that monitor (and sometimes enforce) building codes, food standards, and health and safety regulation; and, private standard setting organizations such as ISO (International Organization for Standardization) or the German TUV (*Technischer Überwachungsverein*; Technical Inspection Association). Such private sector regulatory intermediaries are typically not publicly accountable (as are government regulators), and they often are profit oriented (or at least operate in an environment where financial losses are not acceptable). Such essential differences may undermine the exact advantages of government-led regulation that were explored in the preceding section (Drahos 2017).

Regulation for Urban and Regional Futures: Good or Bad?

At this point you may wonder: To what extent is regulation promising (or not) for governing the transition toward sustainable, inclusive, and healthy urban and regional futures? Will the advantages outweigh the disadvantages, or not? The answer to those questions is not easy and, in part, they remain to be seen. Still, when thinking about these questions, it is important to acknowledge that in the twenty-first century regulation means something quite different from what it has meant over the last century – and perhaps over the last millennia (Levi-Faur 2011) when our cities and regions have seen remarkable levels of development and expansion (Sassen 2012). To understand what regulation means today and what it will mean for urban and regional futures, I briefly explore the *regulation and governance literature* that has emerged over the last few decades; and area of scholarship that bridged public administration, political science, and legal studies (for those interested in a broader exploration of this literature, see among others: Baldwin et al. 2012; Drahos 2017; Levi-Faur 2011).

Within this literature it is often argued that toward the end of the twentieth Century, regulation has stepped out to the shadow of public policy (Baldwin et al. 2012). Regulation is no longer seen as the final phase of the policy process – the operationalization of policy goals through regulations and their monitoring and enforcement. Rather, regulation has become a dominant government function, with some scholars arguing that in a wide range of countries the welfare state has morphed into a *regulatory* state, "a vast array of new independent institutions and new regulatory powers that seek to constrain and guide the behaviour of actors in the economy" (Humperson 2012, 267). Within this regulatory state, the delivery of public and welfare services is increasingly outsourced, privatized, or contracted out, but strongly regulated by government (Majone 2016). In short, the traditional image of government-led command-and-control regulation does (often) no longer reflect regulatory reality.

At the same time, government regulators and private sector regulatory intermediaries have embraced new forms of regulation that seek to overcome the challenges of the traditional model of command-and-control regulation (Baldwin et al. 2012). For example, prescriptive regulation is increasingly replaced with performance-based regulation. This latter form prescribes a performance to be achieved (say, carbon neutral housing) but allows individuals and organizations freedom to find a way to comply (say, housing that produces more energy than it consumes, housing fully developed from recycled materials, or purchasing carbon credits to off-set the carbon emissions) (May 2011). Likewise, self-regulation by firms, co-regulation between public regulators and private parties, and surrogate regulation by citizens has in many areas replaced government-led regulation (Holley et al. 2012).

Likewise, government regulators and private sector regulatory intermediaries have moved away from considering deterrence as the most suitable (and sometimes only) way to achieve compliance with regulation (Van Rooij and Sokol 2021). Positive incentives such as rewards for desired behavior (e.g., subsidized solar panels, or reduced inspection regimes for restaurants that have a long track record of food-safety compliance) have become common practice, as have information campaigns about the necessity of a regulatory intervention and individuals and organizations can comply with it (e.g., the various Covid-19 campaigns ran by many countries in 2020 and 2021 that explained the need of social distancing) (Cawthorn et al. 2021). Also, insights from the behavioral sciences are increasingly used to develop regulation that is aware of (and seeks to overcome) our inbuilt human biases such as hyperbolic discounting (our inclination to choose immediate rewards over rewards that come later in the future even if the immediate rewards are smaller – a bias that often hampers people to take climate action) (Van der Heijden 2020).

What Future for the Regulation of Urban and Regional Futures?

Arguably the most important insight provided by the regulation and governance literature is that regulation has transitioned from being an important *tool* for government to operationalize public policy and overcome collective action problems, to being an important *process* for doing so (Lodge and Wegrich 2012; Windholz 2018). Combined with the various insights discussed in the preceding sections, it then makes sense to conceptualize the sort of regulation we will likely see for the governance of urban and regional futures as "an *intentional*, and often *structured* and *sustained*, *process* implemented by an individual or collective ('regulator') to *direct* the behaviour of other individuals or collectives ('targets') to achieve a *predefined* aim through a variety of *interventions* ('instruments' and 'strategies') that typically include standard setting, monitoring, enforcement, and retribution or rewards" (Van der Heijden forthcoming, original emphasis).

When applying this type of regulation of urban and regional futures, the regulator can be a public party (governments), a private party (firms, NGOs, non-profits), or a combination of public and private parties. Likewise, the targets of regulation can be public or private parties, or combinations of these. In addition, regulation of the development of cities and regions often not only seeks to steer the behavior of targets, but also seeks to meet the needs of beneficiaries of regulation (minority groups, society at large, more-than-human entities, etc.). Finally, regulatory interventions range from those that make desirable behavior attractive (positive incentives such as encouragement and benefits), and the undesirable behavior unattractive (negative incentives such as dissuasion and punishment) or even impossible (restrictions such as barriers and bans). Put differently, the future of the regulation of urban and regional futures will likely be characterized by regulatory innovation.

Indeed, if we look around the globe, we see a mushrooming of innovative regulatory interventions for urban and regional futures. To name a few (all from the field of building energy efficiency to indicate how even for this single collective action problem many regulatory solutions are available):

- *Tokyo's Cap and Trade Program*, introduced in 2007 by the Tokyo Metropolitan Government. The program sets mandatory targets for greenhouse gas emissions at building level (a cap). Those who produce lesser emission than their allowance can trade their surplus with those who emit more than their allowance. The program incentivizes, among others, owners of existing buildings to retrofit their property in order to reduce their emissions (see, for example, Nishida et al. 2016).
- *GreenMark*, introduced in 2005 by the Government of Singapore. This is a building rating system to evaluate the environmental impact (including energy efficiency) of existing and

new buildings. Depending on a buildings' performance, it will be given a higher or lower rating. All buildings in Singapore have to meet a minimum rating, but the Government of Singapore incentivizes property owners, users, and builders to achieve higher ratings, often through increased levels of energy efficiency (see, for example, Dell'Anna and Bottero 2021).

- *Energy Performance of Buildings Directive (EPBD)* introduced in 2010 (and amended in 2018) by the European Commission. The Directive sets minimum energy performance requirements for new buildings, but not for existing ones. It does, however, require that in European countries an *energy performance certificate* must be issued when a building is sold or rented. It is expected that (prospective) buyers or tenants of existing buildings will use the energy performance certificates to guide their decision between available property – in other words, the certificates are introduced to directly inform and influence the "demand side" of the property market and to indirectly influence its supply side (see, for example, Hogeling and Derjanecz 2018).
- *Property Assessed Clean Energy (PACE)* introduced originally in 2005 in Santa Cruz Country, but now applied throughout the United States. It seeks to help property owners to find finance for energy retrofits (and now a broader set of retrofits). The model is operationalized differently across jurisdictions, but typically, a public organization borrows money from a finance provider and lends it to a property owner. This allows the public organization to set the requirements the property owner must meet to get access to the funds. The debt is typically tied to the property and recouped through (a newly introduced) property tax (see, for example, Kirkpatrick and Bennear 2014).

To conclude, particularly since the 1980s, regulation has become an exceptionally versatile approach to overcome collective action problems. Surely, it will not be a panacea for all collective action problems that stand in the way of achieving sustainable, inclusive, and healthy urban and regional futures. Yet, considering the available evidence, its advantages will likely outweigh its disadvantages.

Cross-References

- ▶ City Visions: Toward Smart and Sustainable Urban Futures
- ▶ Housing Affordability
- ▶ Internationalization of Cities
- ▶ Neoliberalism and Future Urban Planning
- ▶ Policies for a Just Transition
- ▶ Public Policies to Increase Urban Green Spaces
- ▶ Unpacking Cities as Complex Adaptive Systems
- ▶ Urban and Regional Leadership

References

Baldwin, R., Cave, M., & Lodge, M. (2012). *Understanding regulation: Theory, strategy and practice* (2nd ed.). Oxford: Oxford University Press.

Bardach, E., & Kagan, R. A. (1982). *Going by the book: The problem of regulatory unreasonableness*. Philadelphia: Temple University Press.

Bulkeley, H., & Betsill, M. (2003). *Cities and climate change: Urban sustainability and global environmental governance*. London: Routledge.

Castán Broto, V. (2017). Urban governance and the politics of climate change. *World Development, 93*(1), 1–15.

Cawthorn, D.-M., Kennaugh, A., & Ferreira, S. (2021). Beyond command and control: A rapid review of meaningful community-engaged responses to covid-19. *Ambio, 50*(4), 912–821.

Corbusier, L. (1929). *The city of to-morrow and its planning*. London: J. Rodker.

Dell'Anna, F., & Bottero, M. (2021). Green premium in buildings: Evidence from the real estate market of Singapore. *Journal of Cleaner Production, 286*(1), 1–14.

Drahos, P. (Ed.). (2017). *Regulatory theory*. Canberra: ANU Press.

Duff, A. (2008). Theories of punishment. In P. Cane & J. Conaghan (Eds.), *The new oxford companion to law* (pp. 970–971). Oxford: Oxford university Press.

Fairman, R., & Yapp, C. (2005). Enforced self-regulation, prescription, and conceptions of compliance within small businesses: The impact of enforcement. *Law and Policy, 27*(4), 491–519.

Fukuyama, F. (2012). *The origins of political order: From prehuman times to the french revolution*. New York: Farrar, Straus, and Giroux.

R

Fukuyama, F. (2015). *Political order and political decay: From the industrial revolution to the globalization of democracy*. New York: Farrar, Straus, and Giroux.
Garvin, A. (2014). *The american city: What works and what doesn't*. New York: Mc GrawHill.
Hodge, G. A., Maynard, A., & Bowman, D. (2014). Nanotechnology: Rhetoric, risk and regulation. *Science and Public Policy, 41*(1), 1–14.
Hogeling, J., & Derjanecz, A. (2018). The 2nd recast of the energy performance of buildings directive (epbd). *REHA Journal, 2018*(2), 70–72.
Holley, C., Gunningham, N., & Shearing, C. (2012). *The new environmental governance*. London: Routledge.
Humperson, E. (2012). Auditing regulatory reform. In D. Oliver, T. Prosser, & R. Rawlings (Eds.), *The regulatory state: Constitutional implications* (pp. 267–282). Oxford: Oxford University Press.
Jagers, S., Harring, N., Lofgren, A., Sjostedt, M., Alpizar, F., Brulde, B., . . . Steffen, W. (2020). On the preconditions for large-scale collective action. Ambio, 49(7), 1282–1296.
Johnson, C. (2018). *The power of cities in global climate politics: Saviours, supplicants or agents of change?* London: Palgrave Macmillan.
King, L. W. (2004). *The code of hammurabi*. Whitfish: Kessinger Publishing.
Kirkpatrick, A. J., & Bennear, L. (2014). Promoting clean energy investment: An empirical analysis of property assessed clean energy. *Journal of Environmental Economics and Management, 68*(2), 357–375.
Levi-Faur, D. (2011). *Handbook on the politics of regulation*. Cheltenham: Edward Elgar.
Lodge, M., & Wegrich, K. (2012). *Managing regulation: Regulatory analysis, politics and policy*. New York: Palgrave Macmillan.
Majone, G. (2016). The evolution of the regulatory state: From the law and politics of antitrust to the politics of precaution. In A. Burgess, A. Alemanno, & J. O. Zinn (Eds.), *Routledge handbook of risk studies* (pp. 2016–2228). London: Routledge.
May, P. (2011). Performance-based regulation. In D. Levi-Faur (Ed.), *Handbook on the politics of regulation* (pp. 373–384). Cheltenham: Edward Elgar.
Nishida, Y., Hua, Y., & Okamoto, N. (2016). Alternative building emission-reduction measure: Outcomes from the Tokyo cap-and-trade program. *Building Research & Information, 44*(5–6), 644–659.
Olson, M. (1965). *The logic of collective action: Public goods and the theory of groups*. Cambridge, MA: Harvard University Press.
Rosenzweig, C., Solecki, W., Romero-Lankao, P., Mehrotra, S., Dhakal, S., & Ali Ibrahim, S. (2018). *Climate change and cities: Second assessment report of the urban climate change research network*. Cambridge: Cambridge University Press.
Sassen, S. (2012). *Cities in a world economy* (4th ed.). Los Angeles: SAGE.
Taylor, P. (2013). *Extraordinary cities: Millennia of moral syndromes, world-systems and city/state relations*. Cheltenham: Edward Elgar.
Torfing, J. (1998). *Politics, regulation and the modern welfare state*. London: Macmillan Press.
Van der Heijden, J. (2014). *Governance for urban sustainability and resilience: Responding to climate change and the relevance of the built environment*. Cheltenham: Edward Elgar.
Van der Heijden, J. (2017). Brighter and darker sides of intermediation: Target-oriented and self-interested intermediaries in the regulatory governance of buildings. *Annals of the American Academy of Political and Social Science, 670*(1), 207–224.
Van der Heijden, J. (2020). Urban climate governance informed by behavioural insights: A commentary and research agenda. *Urban Studies, 57*(9), 1994–2007. https://doi.org/10.1177/0042098019864002.
Van der Heijden, J. (forthcoming). The politics of regulation: Mapping a half-century of debates (1970–2020). In F. Sager, A. Ladner, & A. Bastianen (Eds.), *Handbook on the politics of public administration*. Cheltenham: Edward Elgar.
Van Rooij, B., & Sokol, D. (Eds.). (2021). *The Cambridge handbook of compliance*. Cambridge: Cambridge University Press.
van Straalen, F., Hartmann, T., & Sheenan, J. (Eds.). (2018). *Property rights and climate change: Land use under changing environmental conditions*. London: Routledge.
Windholz, E. (2018). *Governing through regulation: Public policy, regulation and the law*. Abingdon: Routledge.
World Bank. (2016). *Regenerating urban land : A practitioner's guide to leveraging private investment*. Wasnghinton, DC: World Bank.
World Bank. (2017). *Africa's cities: Opening doors to the world*. Washington, DC: World Bank.
Yates, A., & Bergin, A. (2009). *Hardening Australia: Climate change and national disaster resilience. Special report 29*. Canberra: Australian Strategic Policy Institute.

Renaturing Cities

► Innovation to Bring Nature-Based Solutions to Life: Tales of Two Cities

Renewable Energy

► Circular Water Economy

Residential Crowding in Urban Environments

Interventions Through Planning and Design

Raisa Binte Huda and Raisa Sultana
Department of Geography and Environment, University of Dhaka, Dhaka, Bangladesh

Synonyms

Cross-cultural adaptation; Crowding; Environmental psychology; High density; Housing

Definitions

Residential crowding is a cognitive phenomenon rather than objective reality, it folows then that it can be manipulated and ameliorated by the built environment (Sherrod and Cohen 1979). **Communal housing** refers to high-density dwellings shared by multiple non-kin residents, for example, the *Goshiwons* of Seoul (Yoo et al. 2019) and *Mess* of Dhaka (Real and Jameel 2021) where typical residents are single and young-adult students or workers who usually have migrated to the city. While **Goshiwons** are hostel-like and allot a separate room to each resident, even though extremely small, **Mess** is an informal arrangement where an apartment is rented and shared by multiple tenants, often allocating a single room to 2–3 residents.

Introduction

The United Nations Universal Declaration of Human Rights, Article 25, states that suitable housing is a universal right (United Nations 1948, art. 25). An essential facet of "suitable housing" is adequate space. In fact, in many countries around the world, a lack of sufficient space is considered "housing deprivation" (Filandri and Olagnero 2014).

Due to rapid population growth and drastic urbanization, there's an increasing pressure on housing resources in cities. Despite all flaws, high-density living is inevitable in rapidly densifying city centers. High-density by itself has not been conclusively linked to ill-health, but crowding has been, and even though density is a prerequisite of crowding, it is not deterministic: Not all high-density settings lead to experiences of crowding. While the former is spatial-inadequacy, the latter *is a negative experience of the socio-psychological inconveniences as a result of increasing proximity to others caused by spatial-limitation.* This is why, crowding is classified as a socio-environmental stressor as opposed to just an environmental stressor (Halpern 1995; Cohen et al. 1986).

Crowding has deleterious impacts in any situation, but what makes "residential" crowding particularly detrimental is that it occurs in a primary setting (Stokols 1976), where maximum control is expected, where the most intimate rituals are performed, and where the social ties at stake are neither transitory nor anonymous. Residential crowding and consequent social-withdrawal fares even worse for migrants since they already bear a broken social-network. Moreover, for living-arrangements like *Goshiwons* and *Mess* that are becoming increasingly common in urban areas, there are no familial ties among the residents to buffer against crowding (Levy and Herzog 1974; Silverstein et al. 2006).

As the number and type of such residences will only continue to increase in future cities, it is imperative that planning and design principles are informed by a clear understanding of what aspects of high-density cause residential crowding (i.e., antecedents to crowding), what are its impacts on residents' physical and mental health, and how these can be mediated by modifying the built environment.

The present chapter is a descriptive one. It aims to systematically record available knowledge across diverse disciplines and thus produce new knowledge by virtue of connections made. More precisely, it attempts to answer the following:

R

1. What are the socio-psychological inconveniences (i.e., antecedents) that lead to an experience of crowding?
2. What are the consequences of residential crowding?
3. How can planning and design buffer the antecedents and consequences of residential crowding?

Antecedents to Residential Crowding

Any situation of spatial limitation involves an increased proximity to others, which in turn can lead to several inconveniences: goal-blockage, infringement of personal space, frequent and unwanted social interactions, etc. All these arouse in the individual a sense of inadequacy of space, that is, it makes the high-density situation more salient, leading to an experience of "crowding."

Loss of Control and Overload of Social Stimuli

The over-load model hypothesizes that "crowding" is experienced when increased proximity to others leads to a loss of control over the frequency and intimacy of social interactions, resulting in an overload of unwanted interactions (Baum and Koman 1976; Zlutnick and Altman 1972; Esser 1972; Altman 1975; McCarthy and Saegert 1979; Saegert 1978).

Baum and his colleagues compared two dorms with similar density (space allocated per person), but which differed in design: In the corridor design, 32–34 students shared an undifferentiated hallway with a common bathroom and lounge, while in the suite design, 4–6 students had bedrooms clustered around a shared area which housed a bathroom and lounge, and that opened into a hallway (see Baum and Valins 1977 for floor plans). Residents of the corridor design reported more uninvited interactions in the hallways, were more hostile and withdrawn, and reported more experiences of crowding. Those with bedrooms closer to the bathroom or lounge reported more intense feelings of crowding. The authors hypothesized that this was due to overstimulation and loss of control over social interactions.

Since functionally central facilities were in a common area, the arrangement of space itself forced unwanted interactions on the residents in the corridor design. Furthermore, the public nature of the large undifferentiated hallway right outside bedrooms was not conducive for group-development. Shared commonality of the hallway increased residents' daily exposure to and possibility of interaction with strangers and those they knew by face at most (Cohen 1978). The increased unpredictability in the shared corridor design caused residents to withdraw instead, shifting any group-formation exclusively to outside the dorm where interactions could be better controlled (Baum et al. 1978).

The suite-design, on the other hand, clustered residents into smaller groups and provided them with a semi-private area that could easily be regulated ("defensible space," Newman 1972), making casual interactions better controlled and therefore less "scary." Since only 4–6 students would use the shared lounge, it was also easier for residents to grasp each other's comfort and intimacy thresholds, leading to a more positive residential experience.

These studies strongly suggest that it is the *loss of control and resultant overload of unwanted and unpredictable interactions* that are antecedents to crowding rather than density per se (Baum and Paulus 1987; Baum and Davis 1980; Baum and Valins 1977).

Behavioral Constraints and Infringement of Personal Space

The behavioral constraint model postulates that crowding is experienced by an individual when he realizes that his behavioral choices have been restricted due to the proximity of other people. This experience of crowding is exacerbated when the infringement occurs in what the individual deems his "personal space" (Proshansky et al. 1970; Stokols 1972).

This model extends from and combines Brehm's theory of psychological reactance (Brehm 1966) and Sommer's concept of "personal space" (Sommer 1969). It points out that a person's reactance against crowding will be especially aggressive if his behavioral restriction is due to an invasion of his privacy by others. High-density that would otherwise be alright can lead to an experience of crowding if privacy is required for some behavioral activity. This explains why intrusion is particularly frustrating, persistent, and difficult to resolve in residences than in workplaces or malls (Stokols 1972).

The role of infringement of privacy as an antecedent to crowding has been explored extensively in prison-crowding studies. For example, Cox et al. (1984) found that enclosed sleeping-accommodations alone significantly alleviated negative impacts of crowding, while Wells et al. (1965) found that the simple placement of cubicles reduced negative impacts of crowding in open prison-dormitories.

Furthermore, Karlin et al. (1978) found that dorm-residents living as groups of three in rooms designed for two were more stressed, frustrated, and had lower GPAs. The authors hypothesized that this was due to "goal blockage" (Sundstrom 1978): the "extra" third person increased the proximity of residents, threatening behavioral freedom and thus goal-attainment, such as studying in peace and without distractions. When loss of behavioral choice is associated with others (i.e., intrusion) rather than spatial inadequacy, reactance is more aggressive and feelings of crowding, more intense (Baron and Rodin 1978).

A common goal or clear hierarchy such as in military would act as buffers as seen in Smith and Haythorn (1972). However, the urban residents of *Goshiwons*, *Mess*, and similar typology of housing have tasks which are essentially individualistic. Thus, goal-blockage is common, leading to experiences of crowding.

Residential Crowding, and Stress: Personal and Social Consequences

When there's stress, humans cope. If coping is unsuccessful, prolonged exposure to stressors can affect health. Unfortunately, even when coping is successful, if the coping-mechanism is unhealthy, it can lead to problems of its own.

Social Withdrawal and Breakdown of Social Ties

A common mechanism of coping with crowding is through "social withdrawal": avoiding shared spaces where interactions take place, avoiding eye contact, setting up anti-social cues such as plugging in earphones, "sneaking" in and out of the residences, etc. (Baum and Davis 1980; Baum and Valins 1979; Evans and Lepore 1993; Lepore et al. 1990, 1991; Evans et al. 1989).

Even though effective at coping with crowding (Evans et al. 2000), social withdrawal comes with a steep price: It changes a person's social-information processing abilities, whereby he/she learns to tune out interpersonal cues. As a result, residents not only retain less information about their housemates but will also be unable to realize any cue for help. This breaks down socially supportive relationships among the residential occupants (Evans and Lepore 1993; Evans et al. 1989; Lepore et al. 1991).

Typical residents of *Goshiwons* and *Mess*, who are new in the city, are especially vulnerable when inter-residential bonds sour since they have a weaker support-system in the city to begin with. Additionally, the students and young adult workers who usually live in such housing-type have less time to form and nurture relationships outside their residences due to busy schedules and long commutes.

Ironically, a perceived negative social climate in the residence (caused by the coping mechanism itself) sensitizes a person to cross-situational crowding (Stokols 1978) as well as residential crowding (Baum et al. 1975) creating a vicious cycle of withdrawal, alienation, and crowding stress.

Overgeneralization of Social-Withdrawal and Social Alienation

Overgeneralization occurs when occupants of crowded residences employ social withdrawal toward interactions even in situations where crowding and/or high density is non-existent. For example, Baum and Valins (1977) reported that residents of the crowded, corridor design displayed greater avoidance behavior towards interactions even outside the dorms. They were also happier to be ignored in the hallways (Reichner 1974).

Furthermore, when Baum and Valins (1979) asked students from both dorm-designs to wait alone and with an unrevealed confederate in a waiting-room, their behavior was similar when waiting alone. However, students from the corridor design showed dramatically more avoidance behavior when waiting with the confederate: seats were taken up at greater distances, facial regard was less, and significantly more discomfort was reported.

Residents from crowded homes were also less likely to accept help when in need and equally less likely to offer help to someone in need, that is, they were predisposed to avoiding communication even when it was counter-productive and exhibited lesser altruistic tendencies (Bickman et al. 1973; Evans and Lepore 1993).

Such retreat from fellow humans does not bode well. There is widespread consensus that social support protects individuals from risk (Holahan and Moos 1981; Cohen and Wills 1985). Émile Durkheim (1997) had come to the same conclusion in *Suicide* that a person who does not communicate with others and who has a deficiency of social contact is vulnerable to mental ill-health and suicide.

Learned Helplessness

Despite employing social withdrawal to cope, it is impossible that residents experiencing crowding can control every interaction. When they fail to control interactions frequently, "learned helplessness" results (Seligman 1975). Learned helplessness occurs when a person "learns" that his actions (withdrawal in this case) do not affect the stressor. This overgeneralizes to post-crowding situations and can lead to depression, passivity in the face of all stressors, motivational defects, and reluctance to act even when it is possible to control events (Rodin 1976; Sherrod 1974). Baum et al. (1978) reported that residents of long-corridors who experienced crowding were less likely to question about an ambiguous experiment to follow and showed greater pessimism and in general, greater helplessness.

Interventions Through Planning and Design

At the onset of the chapter, we have mentioned that while high residential density is a physical reality, the experience of residential crowding is a "perceived relationship between the self and the environment" (Sherrod and Cohen 1979), subject to manipulation by design.

It is important to clarify before we progress that successful design and planning interventions need not completely dispose of the antecedents to crowding, but simply provide residents with a choice of avoiding them, even if it's only a "perceived" choice. In other words, successful interventions need only establish a "perception" of control (Sherrod and Cohen 1979; Schopler and Walton 1974). This is because the "possibility" of control alone can buffer against crowding stress: Sherrod (1974) exposed subjects to high and low densities in a study. A few under high-densities were then informed that they could, if they wanted, go to a less-crowded room to work (i.e., the subjects were provided with a behavioral choice of escape and therefore a perceived control over crowding). Although nobody actually left the room, those that had "perceived control" performed significantly better on post-experiment measures of frustration. Glass and Singer (1972) had concluded similarly, although with different stressors.

Perceived control is an illusion with real consequences. As such, crowding interventions can target reinstating perceived control in situations where establishing "true" control over interactions is improbable. Last but not the least, interventions must not only target ameliorating the antecedents to crowding, but also its negative consequences.

Our understanding of crowding is not yet at a stage where we can conclusively suggest design and planning programs. However, what we do have are well-informed hypotheses that can guide design and planning toward certain avenues rather than others (Fig. 1).

Architectural Mediations

Some architectural factors can intervene directly, such as the size and number of windows in a

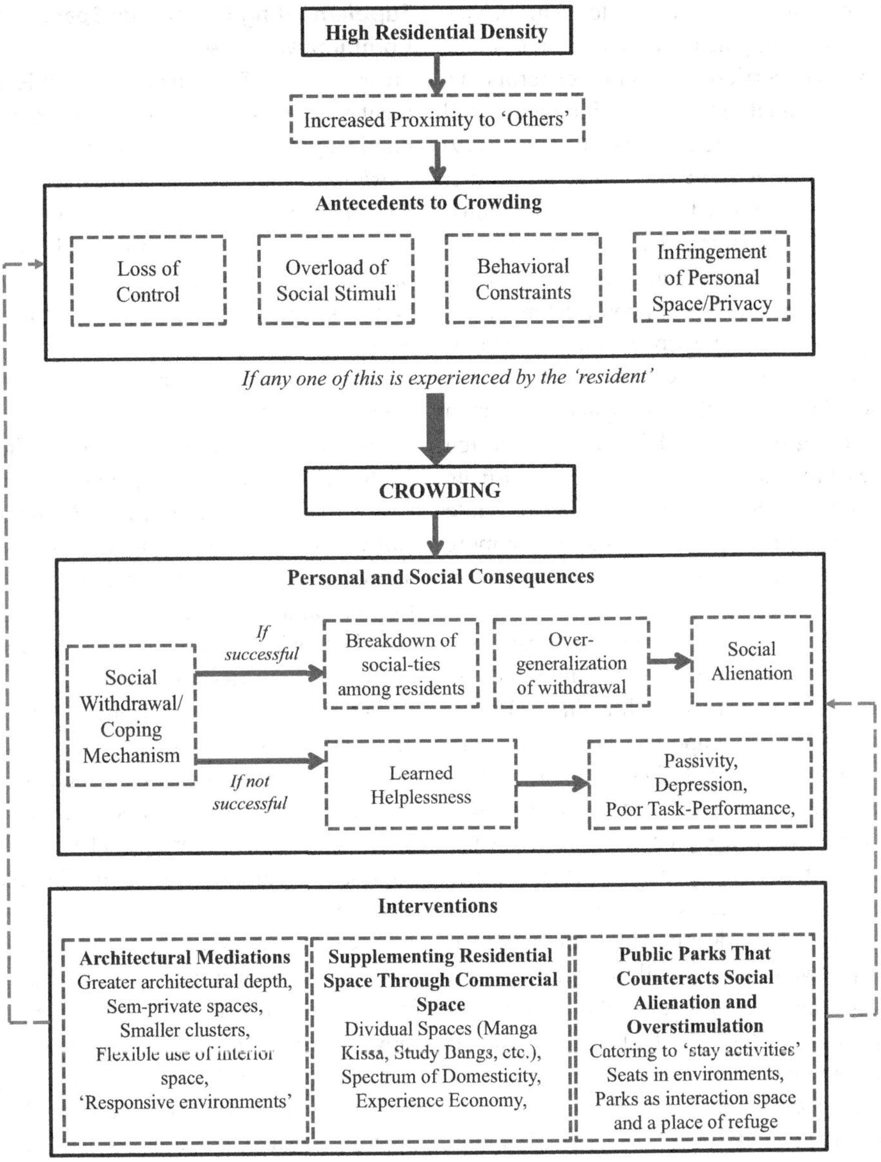

Residential Crowding in Urban Environments, Fig. 1 Density, crowding, and design

residence. Schiffenbauer et al. (1977) found that natural light made rooms seem less crowded. Furthermore, it was found that residents of rooms with more perceived light, also got along better with roommates. Thus, an architectural factor not only has the potential to affect crowding but also its consequences.

Other architectural factors intervene more indirectly. For example, increased architectural depth can provide alternatives to social withdrawal as a way of coping. Evans et al. (1996) found that residents of crowded housing with greater depth – "number of spaces one must pass through to get from one point in a structure to another" – were less likely to withdraw from housemates, because it enabled residents to alter the degree of physical separation between themselves and others, more.

The importance of small-scale, semi-private spaces was highlighted by Baum and Valins (1979) in the corridor vs. suite experiments: The smaller spatial extent is conducive to small group-formations (Sherrod and Cohen 1979) where interactions and intimacy are better established and regulated, decreasing anonymity, and fostering friendships. This can be achieved by avoiding long, undifferentiated corridors and by decreasing the number of people catered to by a semi-private or recreational space, such as keeping a separate lounge for each floor in a high-rise. The hierarchy of space (Newman 1972; Zimring et al. 1982) in the suite design further ensures that intrusion into more private spaces (like bedrooms) are not required, while also not forcing residents to interact in the more uncontrollable "for all" space among strangers (such as in hallways).

When space is limited within a residence for a separate semi-private area, or when multiple residents share a single room such as in a *"Mess,"* flexibility of usage is important instead of "hard architecture" (Sommer 1974). Altman (1975) advocated for "responsive environments" that can be altered for privacy or togetherness. Japanese architecture has pioneered an unusually "fluid" usage of space. (Michelson 1970; Nishi and Hozumi 1996): Room sizes are modified using paper-walls, which allow light to pass through and thus avoid a claustrophobic situation. The large space of the main hall can be modified for different usage; some walls can be moved, and rooms can be temporally joined.

Desor (1972) and Baum et al. (1974) found evidence that partitions reduced feelings of crowding and intrusion of personal space. Not surprisingly, the type of screen did not quite matter (half-way or full, translucent, or opaque) since humans respond to markers of territoriality (Altman 1975): The screen itself acts as a signal to others that they are about to invade personal space and thus makes them more reflective about doing so.

Another modification, although not economically feasible, would be to design into a residence a nook where residents can "escape to" when overwhelmed (Sherrod and Cohen 1979).

Supplementing Residential Space Through Commercial Space

"Escape rooms" are neither possible to implement without structural overhaul, nor is it economically lucrative for many housing types like *Goshiwons* and *Mess* which cater to those who wish to keep rent at a minimum. But the need for a private spot to escape to is indisputable, and often "bathrooms" satisfy this need in many homes when designated "escape rooms" are not available.

Regarding this, an interesting urban-phenomenon to explore are "dividual spaces" (Almazan and Tsukamoto 2006; Choe et al. 2016) that materialized separately in Tokyo and Seoul, such as *manga kissa, PC bangs, study bangs,* etc. These are commercialized spaces in the public realm that offer traditionally domestic facilities and the privacy, intimacy, and comfort associated with one's home at an extremely cheap rate by the hour (Rybczynski 1986).

The facilities are the central attraction rather than any luxury or décor, as in hotels. In fact, spatial-layouts are standardized and décor is kept simple. These ensure what is called an "environment-function fit"(Sherrod and Cohen 1979): a domestic, banal environment for domestic, banal activities.

The *manga kissa* of Tokyo, for example, offers small cubicles for drinking coffee, eating, browsing the internet, reading magazines, watching movies, etc. *Study-bangs* or *gongubangs* in Seoul offer space for studying, in groups or by oneself. What separates these *gongubangs* from libraries is that in the former, a space, albeit small, is allocated to everyone, inhibiting intrusion of personal space, in contrast to a library's "openness" and lack of clear demarcation of spatial-boundaries. Not only does this make the space easier to regulate but the smaller-scale also reproduces the comfortable "domestic-compactness." Essentially, such space creates an experience where it is as if one's own personal study-table simply happens to be outside the home (and needs to be paid for by the hour).

These spaces can then be said to extend the domestic "residential" sphere into the public realm, blurring the distinction between the two.

Interestingly, such spaces do not use sociofugal design (Osmond 1957) and create isolation. Instead, privacy is enforced by the participants themselves. Since it is a space people go to for privacy, any interaction between strangers is unexpected and ignoring one another is legitimized, curbing any mental taxation and guilt that results when avoiding a housemate who naturally expects some form of acknowledgment.

There is actually a different kind of socialization taking place in these dividual spaces, one that leans more towards "passive contact" rather than the active, community-building, involved communication occurring in the West's "third places" (Oldenburg 1989). While the presence of others can be "felt," heard, and even seen in these spaces, the "contact" never intrudes upon the user's personal time and space. In other words, it provides privacy without isolating. (See "liminiality" and "communitas" in Turner 1967; "mechanical solidarity" in Durkheim 2014 and Jacobs 1961 and Gehl 2010 for passive contact.)

The extension of domesticity into the public realm perhaps emerged in high-density cities like Tokyo and Seoul to compensate for the lack of residential space, and their potential for cross-cultural replication is undeniable, given the advent of the "experience economy" (Pine and Gilmore 1998). Existence of such space would ultimately create a behavioral choice: to escape into a semi-private commercial space for retreat and as demonstrated by Sherrod's experiment, it is less important that the individual actually escapes than that escape is simply made possible.

Public Parks That Counteracts Social Alienation and Overstimulation

Magaziner (1988) provided ecological evidence that the highest mental institution admission came from a combination of "high household – high external density" and "low household – low external density." This pattern suggests that residential density can be offset by the opportunity to escape to a relatively low external density, and that the adverse effects of living-alone can be offset by a busy neighborhood.

This highlights the simultaneous roles public parks can play as a place of refuge from overstimulation and a place of social contact, buffering the individual against isolation and alienation.

Extensive literature exists on designing "good" public parks. Due to a limitation of space, we want to focus just one aspect: seats. Seats are essential for "stay/passive activities" like sitting, reading, people-watching, conversing, etc., which are the most preferred and most common activities in parks (Woolley 2003; SimõesAelbrecht 2016; Gehl 2010, Chen et al. 2016). In fact, when Gehl and Svarre (2013) doubled the number of seats in a public space, the number of people seated, doubled as well.

When people are involved in stay activities, they linger. Otherwise, they would simply pass through. Not only does this increase the chances of socialization for those who are staying (Mehta 2007), but it also indirectly affects others by enabling "passive contacts" and by allowing them to be among other people (Abu Ghazzeh 1999; Gehl 2010; Whyte 1980; Madanipour 2003; Cattell et al. 2008; Matsuoka and Kaplan 2008; SimõesAelbrecht 2016). Opportunities for active contact and the passive contacts that do occur are especially important to the migrant population who know fewer people and have few opportunities to meet new people.

Thus, to cater to stay activities across a "spectrum of socialization" from those who seek "public solitude" to those who want to "hang-out," a variety of seats need to be available. For the former, Gehl (2010) suggests long benches and step-like ledges as these enable people to stay close, but still discourage intrusion by making everyone face the same way. Another suggestion is to create enclosure-styled seats around planters: This too enables people to stay close while reducing unwanted-interaction since everyone faces a different direction (Cooper and Francis 1998). For more social stay-activities, Gehl (2010) suggests clustering benches into a "talkscape" while Crankshaw (2009) emphasizes that seats be arranged in a way that maximizes eye-contact. Cooper et al. (1998) recommends concave seating that bends inwards for groups of twos and threes.

Perhaps the best way of ensuring variety is by designing seats into the environment: Along with formal seats, it is equally important to provide lots of flat surfaces (such as boulders, fountains, steps, etc.) that can be turned into make-shift seats.

Unfortunately, the importance of spaces that encourage interaction and community-formation far exceeds the capacity of this chapter. In the limited space, we wanted to highlight that when intervening with residential crowding, it is important to target curbing causes of crowding experiences and, perhaps more importantly, the personal and social consequences of it.

Conclusion

Increasing population and urbanization have put a mounting pressure on urban housing resources. Faced with huge urban masses, there's been a rapid proliferation of high-density housing shared among non-kins. The consequent spatial-inadequacy results in a closer-proximity among housemates, which can become a source of frustration, causing experience of crowding. This leads to social-withdrawal, breakdown of socially supportive networks, alienation, and passivity. Since crowding is essentially a cognitive phenomenon, it can be mediated through the built environment. Architectural mediations, dividual spaces that supplement limited residential space, and public parks that perform dual functions as a place of refuge and social contact hold potential as means of alleviating residential crowding in urban centers. While high-density is often an unavoidable reality, the negative experience of crowding is not. The acceptance of this potential is the start of recognizing the behavioral influence of design and planning and by doing so, of redefining them in relation to urban-residential crowding and stress.

Cross-References

► Housing Affordability
► Spatial Demography as the Shaper of Urban and Regional Planning Under the Impact of Rapid Urbanization
► Toward a Sustainable City

References

Abu-Ghazzeh, T. M. (1999). Housing layout, social interaction, and the place of contact in Abu-Nuseir, Jordan. *Journal of Environmental Psychology, 19*, 41–73.

Almazan, J., & Tsukamoto, Y. (2006). Tokyo public space networks at the intersection of the commercial and the domestic realms study on dividual space. *Journal of Asian Architecture and Building Engineering, 5*(2), 301–308.

Altman, I. (1975). *The environment and social behaviour: Privacy, personal space, territory and crowding*. Monterey, CA: Brooks/Cole.

Baron, R. M., & Rodin, J. (1978). Personal control as a mediator of crowding. In A. Baum, J. Singer, & S. Valins (Eds.), *Advances in environmental psychology (vol 1)*. Hillsdale, NJ: Erlbaum.

Baum, A., & Paulus, P. B. (1987). Crowding. In D. Stokols & I. Altman (Eds.), *Handbook of environmental psychology* (Vol. 1).

Baum, A., & Valins, S. (1979). Architectural mediation of residential density and control: Crowding and the regulation of social contact. *Advances in Experimental Social Psychology, 12*, 131–175. https://doi.org/10.1016/S0065-2601(08)60261-0.

Baum, A., & Davis, G. (1980). Reducing the stress of high density living: An architectural intervention. *Journal of Personality and Social Psychology, 38*, 471–481.

Baum, A., & Koman, S. (1976). Differential response to anticipated crowding: Psychological effects of social and spatial density. *Journal of Personality and Social Psychology, 34*, 526–536.

Baum, A., & Valins, S. (1977). *Architecture and social behavior: Psychological studies of social density*. Hillsdale, NJ: Erlbaum.

Baum, A., Aiello, J., & Calesnick, L. (1978). Crowding and personal control: Social density and the development of learned helplessness. *Journal of Personality and Social Psychology, 36*(9), 1000–1011.

Baum, A., Harpin, R., & Valins, S. (1975). The role of group phenomena in the experience of crowding. *Environment and Behavior, 7*, 185–197.

Baum, A., Riess, M., & O'Hara, J. (1974). Architectural variants of reaction to spatial invasion. *Environment and Behavior, 6*, 91–100.

Bickman, L., Teger, A., Gabriele, T., McLaughlin, C., Berger, M., & Sunaday, E. (1973). Dormitory density and helping behavior. *Environment and Behavior, 5*(4), 465–490. https://doi.org/10.1177/001391657300500406.

Brehm, J. W. (1966). *A theory of psychological reactance*. Academic Press.

Cattell, V., Dines, N., Gesler, W., & Curtis, S. (2008). Mingling, observing, and lingering: Everyday public spaces and their implications for well-being and social relations. *Health and Place, 14*, 544–561.

Chen, Y., Liu, T., & Liu, W. (2016). Increasing the use of large-scale public open spaces: A case study of the North Central Axis Square in Shenzhen, China. *Habitat International, 53*, 66–77.

Choe, S., Almazan, J., & Bennett, K. (2016). The extended home: Dividual space and liminal domesticity in Tokyo and Seoul. *Urban Design International, 21*, 298–316. https://doi.org/10.1057/udi.2016.10.

Cohen, S., Evans, G. W., Stokols, D., & Krantz, D. S. (1986). *Behavior, health, and environmental stress*. Boston, MA: Springer. https://doi.org/10.1007/978-1-4757-9380-2.

Cohen, S. (1978). Environmental load and the allocation of attention. In A. Baum & S. Valins (Eds.), *Advances in environmental research*. Hillsdale, NJ: Erlbaum.

Cohen, S., & Wills, T. A. (1985). Stress, social support, and the buffering hypothesis. *Psychological Bulletin, 98*(2), 310–357.

Cooper Marcus, C., & Francis, C. (1998). *People places: Design guidelines for urban open space* (2nd ed.). New York: Wiley, 384 pp.

Cox, V. C., Paulus, P. B., & McCain, G. (1984). Prison crowding research: The relevance for prison housing standards and a general approach regarding crowding phenomena. *American Psychologist, 39*, 1148–1160.

Crankshaw, N. (2009). *Creating vibrant public spaces: Streetscape design in commercial and historic districts*. Washington: Island Press.

Desor, J. (1972). Toward a psychological theory of crowding. *Journal of Personality and Social Psychology, 21*.

Durkheim, E. (1997). *Suicide: A study in sociology* (G. Simpson and J. A. Spaulding, trans.). Free press. (Original work published in 1951)

Durkheim, E. (2014). The division of labor in society (W.D. Halls, trans.). Simon and Schuster. (Original work published in 1893).

Esser, A. (1972). A biosocial perspective on crowding. In J. Wohlwill & D. Carson (Eds.), *Environment and the social sciences: Perspectives and applications*. American Psychological Association.

Evans, G. W., Palsane, M. N., Lepore, S. J., & Martin, J. (1989). Residential density and psychological health: The mediating effects of social support. *Journal of Personality and Social Psychology, 57*, 994–999.

Evans, G. W., & Lepore, S. J. (1993). Household crowding and socialsupport: A quasi-experimental analysis. *Journal of Personality and Social Psychology, 65*, 308–331.

Evans, G. W., Lepore, S. J., & Schroeder, A. (1996). The role of interior design elements in human responses to crowding. *Journal of Personality and Social Psychology, 70*(1), 41.

Evans, G. W., Rhee, E., Forbes, C., Allen, K. M., & Lepore, S. J. (2000). The meaning and efficacy of social withdrawal as a strategy for coping with chronic residential crowding. *Journal of Environmental Psychology, 20*(4), 335–342. https://doi.org/10.1006/jevp.1999.0174.

Filandri, M., & Olagnero, M. (2014). Housing inequality and social class in Europe. *Housing Studies, 29*(7), 977–993.

Fitch, J. M. (1973). *American building: The historical forces that shaped it*. New York: Schocken Books.

Gehl, J., & Svarre, B. (2013). *How to study public life*. Washington.DC: Island Press.

Gehl, J. (2010). *Cities for people*. Washington: Island Press.

Glass, D. C., & Singer, J. E. (1972). *Urban stress: Experiments on noise and social stressors*. New York: Academic Press.

Halpern, D. (1995). *Mental health and the built environment: More than bricks and mortar?* (1st ed., p. 70). London, Taylor and Francis.

Holahan, C. J., & Moos, R. H. (1981). Social support and psychological distress: A longitudinal analysis. *Journal of Abnormal Psychology, 90*(4), 365–370. https://doi.org/10.1037//0021-843x.90.4.365.

Jacobs, J. (1961). *The death and life of great American cities*. New York: Vintage Books.

Karlin, R. A., Epstein, Y. M., & Aiello, J. R. (1978). A setting specific analysis of crowding. In A. Baum & Y. Epstein (Eds.), *Human responses to crowding*. Hillsdale, N.J.: Lawrence Erlbaum Associates.

Lepore, S. J., Evans, G. W., & Schneider, M. L. (1991). Dynamic role of social support in the link between chronic stress and psychological distress. *Journal of Personality and Social Psychology, 61*, 899–909.

Lepore, S. J., Merritt, K., Kawasaki, N., & Mancuso, R. (1990). *Social withdrawal in crowded residences. Paper presented at the meeting of the Western Psychological Association*. Los Angeles, CA.

Levy, L., & Herzog, A. N. (1974). Effects of population density and crowding on health and social adaptation in the Netherlands. *Journal of Health and Social Behavior, 15*, 228–240.

Madanipour, A. (2003). *Public and private spaces of the City*. London: Routledge.

Matsuoka, R. H., & Kaplan, R. (2008). People needs in the urban landscape: Analysis of landscape and urban planning contributions. *Landscape and Urban Planning, 84*, 7–19.

Magaziner, J. (1988). Living density and psychopathology: A re-examination of the negative model. *Psychological Medicine, 18*(2), 419–431. https://doi.org/10.1017/S0033291700007960.

McCarthy, D. P., & Saegert, S. (1979). Residential density, social overload, and social withdrawal. In J. R. Aiello & A. Baum (Eds.), *Residential crowding and design*. Boston, MA: Springer. https://doi.org/10.1007/978-1-4613-2967-1_5.

Mehta, V. (2007). Lively streets determining environmental characteristics to support social behavior. *Journal of Planning Education and Research, 27*, 165–187.

Michelson, W. (1970). *Man and his urban environment: A sociological approach*. Reading, Mass: Addison-Wesley.

Newman, O. (1972). *Defensible space: Crime prevention through urban design*. New York: Macmillan.

Nishi, K., & Hozumi, K. 1996 [1985] What is Japanese architecture? A survey of traditional Japanese architecture. Tokyo: Kodansha International.

Oldenburg, R. (1989). *The great good place*. New York: Marlowe and Company.

R

Osmond, H. (1957). Function as the basis of psychiatric Ward Design. *Mental Hospitals, 8*, 23–30.
Pine, B. J., & Gilmore, J. H. (1998). Welcome to the experience economy. *Harvard Business Review, 76*(4), 97–105.
Proshansky, H., Lttelson, W., & Rivlin, L. (1970). Freedom of choice and behavior in a physical setting. In H. Proshansky, W. Ittelson, & L. Rivlin (Eds.), *Environmental psychology: Man and his physical setting*. New York: Holt, Rinehart, and Winston.
Real, H.R.K., & Jameel, R. (2021, May 27). "What It's like to be a student tenant in Dhaka". *The Daily Star*, Retrieved from https://www.thedailystar.net/shout/news/what-its-be-student-tenant-dhaka-2099569
Reichner, R. (1974) On being ignored: The effects of residential group size on social interaction. Unpublished Masters' thesis, State University of New York at Stony Brook.
Rodin, J. (1976). Density, perceived choice, and response to controllable and uncontrollable outcomes. *Journal of Experimental Social Psychology, 12*, 564–578.
Rybczynski, W. (1986). *Home: A short history of an idea*. New York: Penguin Books.
Saegert, S. (1978). High density environments: Personal and social consequences. In A. Baum & Y. Epstein (Eds.), *Human responses to crowding* (p. 1978). Hillsdale, NJ: Erlbaum.
Schiffenbauer, A., Brown, J., Perry, P., Shulack, L., & Zanzola, A. (1977). The relationship between density and crowding: Some architectural modifiers. *Environment and Behavior, 9*, 3–14.
Schopler, J., & Walton, M. (1974). The effects of structure expected enjoyment, and participants' internality-externality upon feelings of being crowded. Unpublished manuscript, University of North Carolina at Chapel Hill, 1974.
Seligman, M. E. P. (1975). *Helplessness: On depression, development, and death*. San Francisco: Freeman.
Sherrod, D. R., & Cohen, S. (1979). Density, personal control, and design. In J. R. Aiello & A. Baum (Eds.), *Residential crowding and design*. Boston, MA: Springer. https://doi.org/10.1007/978-1-4613-2967-1_15.
Sherrod, D. R. (1974). Crowding, perceived control, and behavioral aftereffects. *Journal of Applied Social Psychology, 4*, 171–186.
Silverstein, M., Cong, Z., & Li, S. (2006). Intergenerational transfers and living arrangements of older people in rural China: Consequences for psychological well-being. *The Journals of Gerontology Series B: Psychological Sciences and Social Sciences, 61*(5), S256–S266.
SimõesAelbrecht, P. (2016). 'Fourth places': The contemporary public settings for informal social interaction among strangers. *Journal of Urban Design, 21*(1), 124–152. https://doi.org/10.1080/13574809.2015.1106920.
Smith, S., & Haythorn, W. W. (1972). Effects of compatibility, crowding, group size, and leadership seniority on stress, anxiety, hostility, and annoyance in isolated groups. *Journal of Personality and Social Psychology, 22*(1), 67–79. https://doi.org/10.1037/h0032392.
Sommer, R. (1969). *Personal space: The behavioral basis of design*. Englewood Cliffs, NJ: Prentice-Hall.
Sommer, R. (1974). *Tight spaces: Hard architecture and how to humanize it*. Englewood Cliffs, NJ: Prentice-Hall.
Stokols, D. (1972). A social-psychological model of human crowding phenomena. *The Journal of the American Institute of Planners, 6*, 72–83.
Stokols, D. (1976). The experience of crowding in primary and secondary environments. *Environment and Behavior, 8*, 49–86.
Stokols, D. (1978). A typology of crowding experiences. In A. Baum & Y. Epstein (Eds.), *Human response to crowding*. Hillsdale, NJ: Erlbaum.
Sundstrom, E. (1978). Crowding as a sequential process: Review of research on the effects of population density on humans. In A. Baum & Y. Epstein (Eds.), *Human response to crowding*. Hillsdale, NJ: Erlbaum.
The United Nations. (1948). Universal declaration of human rights.
Turner, V. (1967). *Forest of symbols*. Ithaca, NY/London: Cornell University Press.
Valins, S., & Baum, A. (1973). Residential group size, social interaction, and crowding. *Environment and Behavior, 5*, 421–439.
Wells, B. W. P. (1965). The psycho-social influence of building environment: Sociometric findings in large and small office spaces. *Building Science, 1*(2), 153–165.
Whyte, W. H. (1980). *The social life of small urban spaces*. New York: Project for Public Places, 125 pp.
Woolley, H. (2003). *Urban Open Spaces*. London: Spon Press. https://doi.org/10.4324/9780203402146.
Yoo, H.-Y., Yang, J.-W., & Kim, J.-S. (2019). A study on the direction of improvement by analyzing the characteristics of Goshiwons for urban regeneration in deteriorated residential blocks. *Journal of Asian Architecture and Building Engineering, 18*(5), 392–403. https://doi.org/10.1080/13467581.2019.1661254.
Zimring, C., Weitzer, W., & Knight, R. C. (1982). Opportunity for control and the designed environment: The case of an institution for the developmentally disabled. In A. Baum & J. Singer (Eds.), *Advances in environmental psychology* (Vol. 4). Hillsdale, NJ: Erlbaum.
Zlutnick, S., & Altman, I. (1972). Crowding and human behavior. In J. F. Wohlwill & D. H. Carson (Eds.), *Environment and the social sciences: Perspectives and applications*. Washington, DC: American Psychological Association.

Resilience

► From Vulnerability to Urban Resilience to Climate Change
► Role of Disaster Relief Policy in Building Resilient Coastal Regions in the United States

Resilience – Absorption

► At the Intersection and Looking Ahead

Resilience – Absorptive Capacity

► Children, Urban Vulnerability, and Resilience

Resilient Cities

► Urban Climate Resilience
► Urban Resilience

Resilient City

► Green Cities in Theory and Practice

Resilient Rural Electrification for the Twenty-First Century

Stewart Craine
Village Infrastructure Angels, London, UK

Definitions

HDI	Human Development Index
kW	Kilowatt
kWh	Kilowatt-hour
SDGs	Sustainable Development Goals

Introduction

Herein is presented a model to link access to energy for the world's poorest communities to a measurement tool of poverty alleviation, the Human Development Index, so as to give a clear and rational approach to resource allocation and investments, both public and private, that helps to eliminate energy poverty, to take a large step towards eradicating energy poverty overall, and to help the worlds' poor form more resilient communities that are ready to face the challenges of the twenty-first century. In summary, the analysis shows that $1500 of energy access technologies per household can contribute 35% of what is needed to lift the poor out of poverty overall (the balance coming from non-energy interventions), but that small appliances like lighting, radios, and TVs contribute relatively little, and the bulk of development impact comes from cleaner cooking and community-scale productive uses, which is not the primary focus of energy access investments at this time. This requires around 220 W of power and 0.8 kWh/day of electricity access per day for each household if an improved biomass cookstove is used and considerably more if electric cooking is used. Annual incomes are also expected to increase by $500/year.

Energy and Poverty Reduction

A world free of poverty can be, and should be, an achievement of humanity in the twenty-first century. This target is enshrined in the Sustainable Development Goals that aim to achieve this by 2030. As with many goals set to eradicate poverty in its many forms, the SDGs will probably not be achieved by 2030, similar to the worlds' challenge to decarbonize the economy to combat climate change by 2030 or soon thereafter. However, 2050 is a goal for both that should be achievable with the resources available to humanity, and millions of people are devoting their careers to these goals.

SDG7 is the goal focused on clean energy, including ensuring that all people have access to electricity and clean cooking technologies. At this time, approximately 750 million people lack access to electricity, and around 2 billion people are using polluting, inefficient biomass cookstoves and/or open fires. No country has lifted its people out of poverty with a high level of rural electrification. In modern development language, the level of development a nation has achieved is represented by the Human Development Index, where the least developed countries have HDI values of 0.35–0.55, developing countries have a value of 0.55–0.7, emerging economies have values of 0.7–0.8, and developed economies have values higher than 0.8. Thus, access to clean energy should aim to help countries (or smaller subnational regions) achieve an HDI of over 0.7 if they are to achieve the overall objective of the SDGs, which is to eliminate poverty.

Figure 1 shows that there is a strong correlation between HDI and both the access to electricity and the use of energy. The relationship to access to electricity is fairly linear in nature, with few countries achieving and HDI of 0.7 with reaching 80–90% access to electricity. The relationship to energy consumption is quite different and highly non-linear, showing that a small amount of additional energy use leads to a rapid increase in HDI, whereas large increases in energy use by high-HDI countries leads to very little improvement in HDI.

A helpful characteristic of HDI is that it can also be scalable, applied not only to a nation but also to a single house, or a single village, or a single district/province/area. This can help give encouraging results at the local scale that would

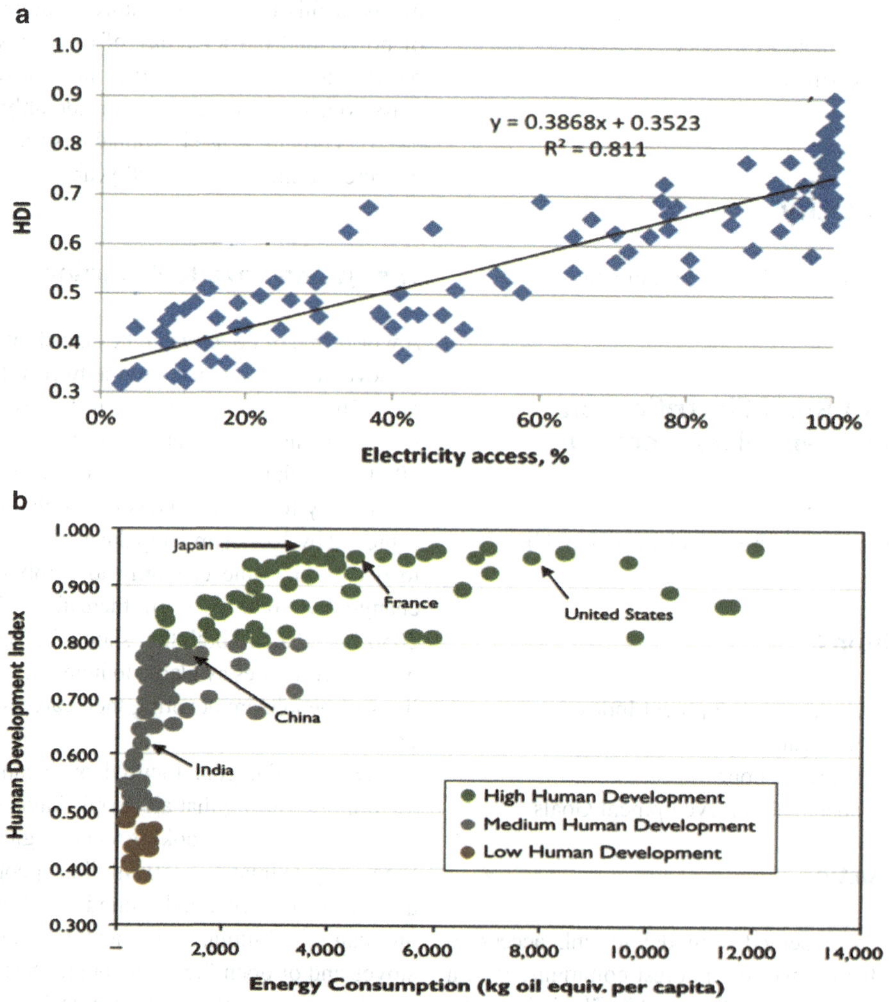

Resilient Rural Electrification for the Twenty-First Century, Fig. 1 Relationship between energy and the Human Development Index. (**a**) Access to electricity vs HDI (Battacharyya 2012). (**b**) Energy consumption vs HDI (Lobner 2017)

be lost in the noise of national-level statistics – sometimes one must look at just a few trees before finding a way to help a whole forest. Thus, it may help to see how these national statistics can translate into a guide for helping an individual household.

The country that has used the least energy to achieve an HDI of over 0.7, Samoa, consumes 319 kg of oil per capita, which is around 3800 kWh of raw energy and, after conversion losses typical of cooking, power generation, and motive engine power of 60–70%, useful energy of around 1350 kWh per person per year. It may be also noted that in developed countries, residential household use of total energy consumption is around 20–30% of national energy use (with the balance being for commercial, industrial, and government sectors), which gives around 400 kWh/person/year of useful energy. This might suggest at least 2000 kWh/year or 5.5 kWh/day of cooking, electricity, and fuel energy "needs" to be consumed for a five-person household to reach an HDI of 0.7 and escape from poverty.

By comparison, a household recently connected to the grid typically consumes 0.5–2 kWh/month, mostly for lighting and small electronic appliances (radio, fan, TV). At the other end of the scale, households in countries such as the UK, the USA, and Australia consume around 50–75 kWh/day of electricity, gas, and motor fuel, of which 30–40% is for motor fuel, 30–40% is for heating/cooling/cooking, and 20–30% is for other electrical appliances (around 10–15 kWh/day). Using similar proportions, it may be suggested that 1–2 kWh/day (20–40% of 5.5 kWh/day) of electricity "needs" to be consumed by the poor to power their way out of poverty. Using more energy-efficient appliances than the average rich household could reduce this.

The targeted improvement in energy access for the global poor has to be achieved among a turbulent global situation, both in the physical world and the economic world. Climate change is making food harder to grow, water harder to find, and coastal areas harder to live in. While the global poor faced many challenges in the past, there will likely be more pressure than ever on them in the coming decades, so helping them to be resilient to these shocks is more important than ever before. The cost of inaction will likely be needless deaths on a scale not yet imagined – the current pandemic is already starting to hint at how global inequalities to essential services can impact the global poor more than the rich and that the rich will tend to secure their own future before taking care of others. With this natural human trait in mind, it is therefore important to not only help the global poor to become resilient as soon as possible but also require as little resources from the rich by which to achieve this, as there is no guarantee such resources will be made available.

As Fig. 2 shows, the increased use of energy to improve living standards sadly, but unsurprisingly, also mimics the increase in unsustainable ecological footprints. Thus, it is not necessarily desirable that the highest level of human development be strived for, if it comes at the cost of a rapidly increasing unsustainable ecological footprint. However, Fig. 2 also shows that this footprint is unlikely to be exceeded at HDI = 0.7, so development impact should be prioritized higher. The poorest households on the planet did not create most of the ecological crisis, so if they need a diesel generator as the least-cost technical option to deliver a given service like milling or as a backup power source to solar when the sun doesn't shine, clean energy ideology should not burden them, or funders, with a solution that costs twice as much, as long as the "dirty" cheap option does not cause excessive harm to the user or the environment.

It is also clear that even "highly developed" economies have further work to do to truly develop as responsible global citizens – to reduce their ecological footprint while maintaining (or improving) their already high standards of living, which means reducing the ecological footprint of their energy consumption (i.e., increased use of cleaner fuels) and/or using less energy to enjoy the same services (i.e., increase use of energy-efficient technologies), as well as other non-energy-related ecological footprint reductions (e.g., eat less meat). Most modern science also indicates that this has to be done quite soon, before 2050, if we are to avoid existential threats to our survival as a species. However, those actions are not the focus of this analysis –

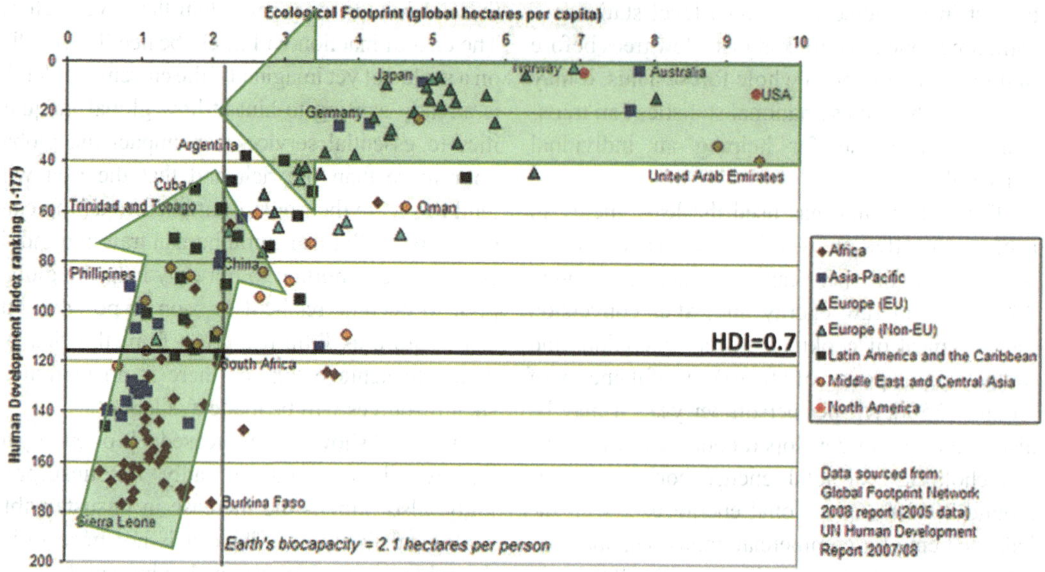

Resilient Rural Electrification for the Twenty-First Century, Fig. 2 Ecological footprint vs HDI (Wikipedia 2008)

lifting the poorest communities out of poverty within the same timeframe is the key focus.

Ultimately, the end goal is that all countries have a medium-to-high HDI and 100% access to sustainable and efficient forms of energy, as shown by the arrows in Fig. 2. The richest people in the world have so far burned their way to wealth with delayed but real ecological impacts verging on catastrophic, but perhaps the poorest 1–2 billion people can attain similar living standards by 2050 while following a more sustainable energy pathway and, in doing so, even educate some of the rich as to what is now possible with modern technologies. Leapfrogging development steps is already apparent in many developing countries, such as mobile phones being a cheaper faster option than landline telephones, and local clean energy solutions may be a similar cheaper and faster alternative to large, centralized fossil-fuel-burning power stations that lifted richer countries out of poverty in the last century.

The Current Situation

Energy poverty has many forms that appear clearly in the daily routine of the global poorest 1–2 billion people. These are all well known – kerosene lamps and candles provide unsafe and polluting lighting instead of electric lamps; wood and charcoal are used for cooking, creating indoor air pollution that makes respiratory disease a bigger global killer than malaria and AIDs combined and causes a significant share of global deforestation; unenergized water supply results in people having to collect their water manually from wells, taps, springs, and rivers, without having any guarantee against water contamination; agriculture without energy means a lack of pumps for irrigation (or if there are pumps, many of them use non-sustainable diesel fuel); and a reliance on diesel-fueled mills or manual labor (mostly by women and girls) to process crops by hand. Energy poverty is not only limited to the home – a lack of electricity in clinics, schools, and other community infrastructure also results in substandard services being available.

The lack of access to modern technology thus places many hours of unnecessary labor onto the poor, which has an opportunity cost in terms of both alternative income-generating activities that cannot be pursued and more time spent in school, robbing students (often girls) of a full education, and causes significant health impacts that contributes to decreased life expectancy. These three elements – income, education, and life expectancy – are what make up the Human Development Index, and thus an improvement in energy services that can improve these three elements will also improve living conditions and thus will reduce poverty. There are other elements of a modern, dignified life that are not covered by the Human Development Index and would also reduce poverty, as covered better by Bhutan's Gross National Happiness Index, but for simplicity and alignment to recognized poverty indicators, the link between HDI and energy poverty will be the main framework used here.

Building More Resilient Communities

Resilience is a wide-reaching concept. Communities, rural and urban, can experience many different types of shocks, which can also disrupt any energy system that serves these communities. Natural hazards include hurricanes, storms, earthquakes, drought, fires, landslides, and, to a lesser degree, severe cold (most of the poorest 1–2 billion live near the equator). Other shocks can include health pandemics (Ebola, coronavirus, cholera), political unrest, terrorist attacks, and economic shocks (high fuel prices, logistical supply issues, hyper-inflation). It is not possible to build an affordable electricity and energy supply system that is completely impervious to such risks, but lessons from recent years can lead to some helpful guidelines:

- **Use local fuels** – the longer the supply chain to secure the fuel required to make electricity, the bigger the risk, particularly if the fuel has to be imported from other countries. The most abundant local fuels are generally renewable – solar energy, hydropower, wind, biomass, and sometimes biogas from animal dung or other sources.
- **Use clean fuels** – most fossil fuels like coal, gas, diesel, and kerosene are not clean fuels, and there are less and less funding and investment for financing projects using these fuels. Some may still be of important use to ensure 100% reliability when combined with clean energy sources as the main source of power (i.e., a hybrid power station). CO_2 emissions are not the only challenge – smoke from cookstoves leads to extremely high indoor air pollution, which also needs to be improved.
- **Generate power locally** – after all major natural disasters have run their course, powerlines are often broken and sometimes even power generation stations. Many utilities globally are increasingly looking at installing local power stations (often solar+storage) instead of rebuilding aged, damaged powerlines.
- **Include system redundancy** – a core principle of energy system planning is having N + 1 redundancy in the system, so that if one source of power becomes unavailable (planned or unexpectedly), another power source can ramp up to meet the demand. This helps keep systems greater than 99.9% reliable (less than 10 h of system outages per year) or better. A system with only one power source may struggle to be even 95% reliable, leading to 18 days of outages per year or more.

Maximizing all of these characteristics, while keeping the service affordable, is not an easy challenge and sometimes may create a conflict – a system that is 100% renewable would be ideal, but achieving reliability and system redundancy might result in unreasonable costs of a very large battery that could be more cost-effectively solved by the inclusion of a small amount of non-renewable energy into the energy system, like a diesel backup generator. This non-renewable technology could also be selected carefully so that it could be converted to being renewable at a later date (e.g., using biodiesel or bioethanol or biogas or even straight vegetable/algae oil) as those technologies mature and become cost-effective and practically deployable at a micro-scale in such villages (or are part of a larger more centralized network at the provincial or national level).

R

From Poverty to Resiliency: Setting Goals

Using HDI to guide village improvements leads to a focus on the three key parts of the HDI formula –

improved income, improved life expectancy, and increased years of education. In the dry and boring language of international development such as the Logical Framework Analysis or Theory of Change, this may mean the desired Impact of poverty-free resilient communities requires these three Outcomes to be attained, which in turn require certain Inputs to the communities, delivered as project Tasks/Activities, that result in Outcome-aligned Outputs. The Inputs/Tasks/Activities and Outputs will be discussed later as we work our way backwards from our desired Impact to the actions that hopefully result in that Impact being achieved – for now, we focus on the three HDI-driven Outcomes.

The HDI formula shown in Fig. 3. The least developed countries have an HDI around 0.4, meaning that each of the three main elements of the formula also has a value of 0.4. This means a life expectancy of $0.4 \times (85 - 20) + 20 = 46$ years (the lowest national averages are currently 55–60 years (HDI of 0.54–0.62)), but sub-national population segments could have much lower average life expectancy than the national average such as rural populations or those suffering from chronic but curable illnesses.

Similarly, an HDI of 0.4 for the Educational Index means only having and expecting 6–7 years of schooling. Countries with the least schooling now average just 1.5–3 years of schooling (HDI of 0.05–0.1) showing that this parameter needs greater improvement than life expectancy.

$$\text{Life Expectancy Index (LEI)} = \frac{(LE-20)}{(85-20)}$$

$$\text{Educational Index (EI)} = \frac{(MYSI+EYSI)}{2}$$

$$\text{Mean years of schooling Index} = \frac{MYS}{15}$$

$$\text{Expected Years of schooling Index} = \frac{EYS}{18}$$

$$\text{Income Index (II)} = \frac{[\ln(GNIpc)-\ln(100)]}{[\ln(75000)-\ln(100)]}$$

$$HDI = \sqrt[3]{(LEI * EI * II)}$$

Resilient Rural Electrification for the Twenty-First Century, Fig. 3 Human Development Index formula, where LE = life expectancy, MYS = mean years of actual schooling, EYS = expected years of schooling, GNIpc = gross national income per capita, in US$

An HDI of 0.4 for the Income Index of around \$1400/person/year or \$7000/year for a five-person household. These values include the Purchasing Power Parity (PPP) factor, which averages around 3 for low-income countries, so the absolute (nominal) income is around \$467/person/year and \$2333 per household. When many men and women in extreme poverty can only bring home \$1–2/day of income, only \$500–1000/year is likely to be earned for the household resulting in a nominal earning of \$100–200/person or PPP earning of \$300–600, which converts to an HDI of 0.17–0.27. Even putting children to work to earn \$3–5/day in total will only earn \$1000–2000/year nominally for the household, or \$600–1200/person PPP gives an HDI of just 0.27–0.38. The minimum national average value in the United Nations statistics is \$660/person/year PPP for Burundi (HDI of 0.28), implying at least \$3300/year/household PPP, but this will hide the extremes within the country, because even the richest households in Burundi are included in this national average.

Non-cash income is likely also included, so if a household can grow enough food to feed itself (5×2000 kcal per day = 10,000 kcal/day/house = 2.5 kg/day/house of food assuming 4 calories per gram), then even a subsistence household with no cash income is "earning" \$1/day of food grown to survive, assuming it is worth \$0.40/kg at local market prices, and thus \$365/house/year or \$73/capita nominal or \$220/capita PPP (HDI of 0.12), 65% lower than the lowest national average from Burundi. Hence, the challenge for income is as almost as great as the challenge for improving the years of education.

This breakdown of the three components of HDI is illustrated for each country in Fig. 4, clearly showing that for the poorest countries, life expectancy is less of a contributor to a low HDI compared to a lack of years in school or low income, which are similar, though education is marginally the worst performing component. This difference may help better allocate resources.

Therefore, clear targets for each of the three HDI components can be set if an HDI of 0.7 is to be attained. For life expectancy, the increase needed is the least of the three components, and

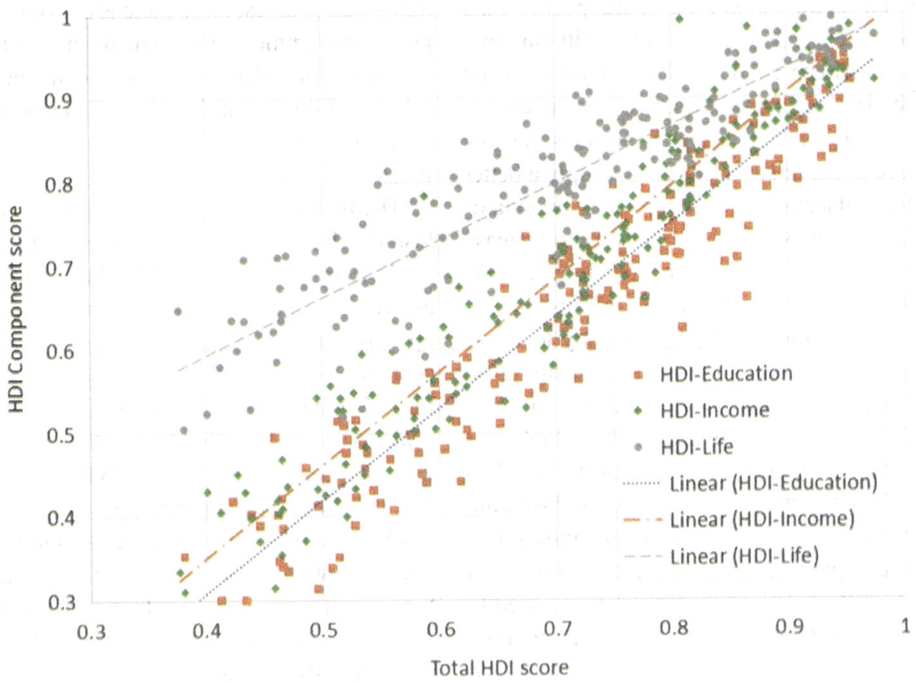

Resilient Rural Electrification for the Twenty-First Century, Fig. 4 Component breakdown of HDI values per country

the target is 65.5 years life expectancy. No country has lifted its people out of poverty with an average life expectancy of less than 64 years. According to the United Nations data, 38 countries currently are below this 65.5-year threshold, considerably fewer than the 75 below HDI of 0.7, but subnational regions within countries that pass the 0.7 HDI threshold may have local life expectancy lower than the 65.5 and 0.7 thresholds, such as in deeply rural areas and poorly serviced villages. This highlights the need for good subnational HDI and component statistics, which are not as easily available. Similarly, an HDI score of 0.7 gives targets of 11.5 years of mean/expected schooling and $10,200/person PPP of income ($3,400 nominal) or $50,000 PPP per five-person household ($17,000 nominal).

Given that improvements in schooling and income are likely to naturally also increase life expectancy, the data suggests that the HDI for life expectancy will likely be above that of income and education, so even attaining an HDI of 0.65 for these two components could be sufficient, giving lower targets of 10.5 years of schooling and $1500/person PPP ($500 nominal) or $37,000/household PPP ($12,300 nominal) of income, assuming life expectancy reached an HDI of 0.81 which is a target of 72.8 years (that 84 countries have not yet reached).

Around 75 countries currently have an HDI below 0.7, and are therefore the primary targets of poverty alleviation, but it is important to also map where subnational regions also fail to attain this score and need active assistance. These countries and regions with an HDI of less than 0.7 will very likely be identical to those that suffer energy poverty – have failed to achieve at least 90% access to electricity and have a very high proportion of households that use open fires or poor-quality, inefficient cookstoves, relying on wood and charcoal for most of their energy need.

The link between rural electrification and increased income has always been high on the agenda, even if not always proven to occur. The link between energy and health and thus life expectancy has also been fairly well explored, particularly by the improved cookstove industry who highlight that respiratory diseases from indoor air pollution kill far more people than AIDS or malaria, yet receives little health funding. The link between energy and education has been peripheral at best and non-existent in many energy access programs. While some data exists to show increased hours of study that might imply a better quality of education, there is probably no data or program strategies that deliberately try to increase the years of schooling as a result of energy access.

However, energy plays a role in this issue – children are taken out of school mostly to work, helping with household tasks and/or helping on the farm, processing crops by hand (beating rice, shelling corn, grinding flour), fetching fuelwood and water, or earning a little money in urban slums. This is especially true for girls. Installing productive uses of energy in communities like mills, water pumps, and even electric cookstoves can reduce the hours of manual labor required to survive and add value to crops and/or grow more crops, increasing income, reducing the workload for children, and potentially keeping them in school for as little as $1–10/month. This could then possibly become one of the cheapest and easiest ways to increase the HDI of a village.

The First Technology Steps Towards Poverty Reduction and Resilience

With three clear goals set for improving income, education, and health, specific energy technologies can be assessed for their potential to help achieve these goals. Improved access to energy alone will not achieve the goals, if there is no school or teachers available in a deeply rural village or urban slum, increasing years of education become impossible, as might increasing life expectancy if there is no health clinic or clean water source, or increasing incomes if accessing markets is extremely difficult (no road nearby or for isolated small islands). However, building this non-energy infrastructure without improving access to energy, which can be found in many locations, is also just as likely to result in failure to build resilience and reduce poverty. A coordinated multi-sectoral integrated approach is required, something the private sector is poorly motivated to invest in, so there is a strong need for public government involvement in putting integrated district plans in place and financing the infrastructure that has long-term benefits ill-suited to the private sectors' comparatively short-term focus.

The first use of electricity for a household has historically usually been lighting, though in the modern age, charging a mobile phone may be seen as even more important. In the twentieth century, lighting was mostly designed based on incandescent bulbs – in Nepal, the typical design load was four 25 W bulbs or 100 W of morning and evening peak power per household, operating for 6–8 h per day, thus consuming 600–800 Wh/day. Added to this are often small appliances often run by batteries like radios and torches that consumed 3–5 W each, followed by larger 10–50 W appliances like fans, music stereos, and TVs. This may total 1kWh/day of energy and 150–200 W of peak load per household if older-style inefficient appliances are used. However, incandescent bulbs were replaced first by fluorescent lamps that took 25% less power and energy to deliver the same amount of light and then more recently by white LEDs that need 90% less power (10 W) than bulbs and just 60 Wh/day of energy, easily provided by a solar panel of just 15 W. A 5–10 W mobile phone charger used for 2 h per day would use another 10–20 Wh/day.

Similar to the improvement in lighting technologies, cathode ray analogue TVs were first replaced by digital flat panel TVs with fluorescent backlights and a liquid crystal display and then by plasma and LED technology. Efficiency improved, but so did screen size, and a large modern TV often still consumes the same 30–50 W that a small CRT TV does and sometimes considerably more (100–200 W). That said, an entry-level smaller modern TV of 12–20 inches can consume as little as 5–15 W, yielding a 50–90% saving, while 30–90-inch TVs consume 30–150 W as shown in Fig. 5.

These first appliances are sometimes referred to as entry-level technologies on the "energy ladder," which has higher power consumption technologies as one climbs up the energy ladder, as shown in Fig. 6. In some cases, like

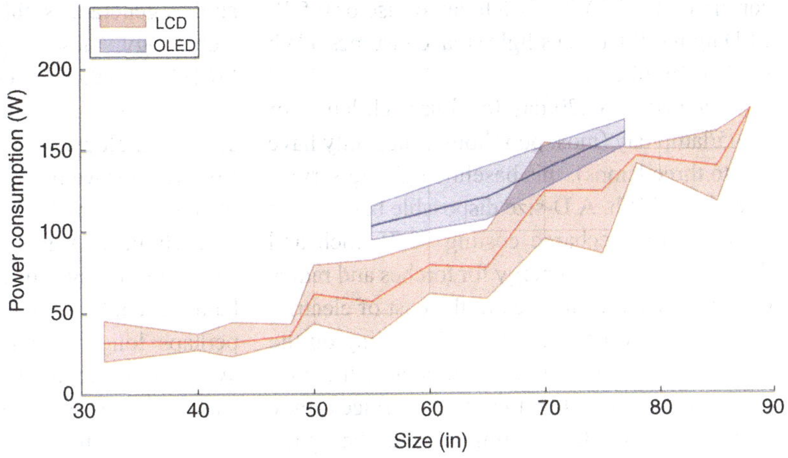

Resilient Rural Electrification for the Twenty-First Century, Fig. 5 Power consumption of modern TVs (Di Giovanni 2021)

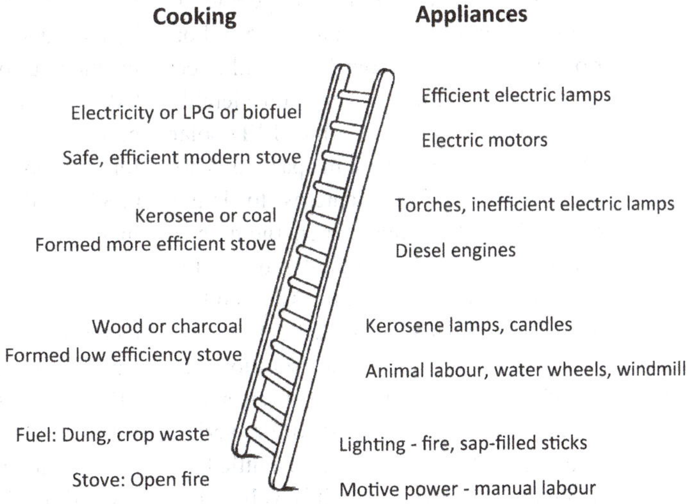

Resilient Rural Electrification for the Twenty-First Century, Fig. 6 The basic energy ladder

communication, several historical steps can be leapfrogged so the poor can immediately start using the best technology available. Similar more advanced energy ladders can be formed which include other modern applied forms of energy, such as for washing clothes (from manual labor to heat pump DC inverter efficient washing machines), transport (from walking to electric cars and bikes), refrigeration (from cellars and zeer pots to modern refrigerators and freezers), ironing (from hot rocks to electric irons), and more. With each step up the energy ladder, manual labor decreases, value can be added to village products, and people can be more productive and efficient, the basis of the industrial revolution and wealth creation.

The baseline cost of these basic first energy services like lighting and TVs is also important to establish as is the technical energy efficiency. Except for the poorest households that use fire for lighting to avoid a cash expenditure on lighting, most households without electricity use kerosene lamps, candles, and battery-based torches for lighting, typically costing \$1–2/week or \$50–100/year. Compared to a household income of \$500–2000/year, this can be 5–10% of the household budget. Charging a mobile phone and buying batteries for torches and radios can increase this even further. A non-pressurized kerosene lamp or a single candle gives only 30–50 lumens of light, about the same as a 2–3 W incandescent bulb (15 lumens/watt) that consumes 12–18 Wh per 6 h day of use or 0.5 W LED light (100 lumens/light) that consumes 3 Wh per 6 h day of use.

At a cost of $0.20/day for four such lamps or $0.05/lamp/day (most poor households only have one to three lamps), this baseline lighting service costs $2–4/kWh. A D-size disposable battery or a mobile phone recharge costing $0.25 each and delivering 25 Wh of energy for torches and radios costs $10/kWh. Compared to the cost of electricity which is $0.10–$0.50/kWh depending on the country and technology used, it is clear the poor pay a huge cost for the poor services they received, often 10–100 times more than they should be paying.

Collectively, around 750 million people or 150 million households spend around $15 billion per year on these services, about 15% of all international aid spent on the poor. This is considerable buying power, and a range of white LED solar lanterns have been developed in the past 15 years from single 0.3 W $5–10 lanterns to larger 1.5–3 W $25–50 lanterns that can charge mobile phones and larger 5–20 W multiple-lamp systems with two to five lamps and sometimes a radio that costs $50–200. Compared to $1–2/week of current expenditure, the modern solar solution can pay itself off in just 5–100 weeks, from a few months to 2 years. With extra costs of delivering these technologies, this might become a little longer, from 3 months to 3 years, but it is well within the desirable zone for the private sector to get involved and mobilize more capital than the public sector can by itself. This mobilization has accelerated greatly since 2005 as both mobile phones and solar white LED lamps became more and more affordable to off-grid villages in developing countries, and now these two technologies alone are a multi-billion dollar industry for the poorest 1–2 billion people.

The Impact of the First Technologies on Poverty Reduction

The quantified impact on poverty alleviation of these first small appliances can now be evaluated, via an estimated increase in HDI. While impact studies generate useful data on the impact of each technology, these impacts are generally never brought together under a single poverty alleviation framework that the HDI model offers. For each technology, an attempt can be made to quantify the improvement in income, education (years of schooling), and health (life expectancy).

Lighting is a good example. A modern solar LED lighting system using good-quality lithium batteries could be expected to last 5–10 years and perhaps longer if replacement parts are easily available. Over this time, assuming a household was spending $50–100 (nominal) on kerosene lamps or candles, phone-charging services provided by others, and batteries for radios and torches, $250–1000 of expenditure can be saved, by a product costing $50–200, giving a net boost in income of $200–800 over 5–10 years or $40–80/year nominal or $120–240 PPP. At a baseline household income of $500/year nominal for the poorest of the poor, this results in a 0.002–0.005 increase in HDI using the Income Index equation, and for a higher baseline of income of $1000/year nominal, the increase in Income HDI is 0.001–0.002 (and is less of an increase in total HDI). Given our income target was to increase nominal (real) incomes to $12,300–17,000/year/household from $500 to $1000, $40–80 is only 0.24–0.71% of the desired income increase, and increased income is only one third of the work needed to increase total HDI. Similarly, the desired increase of HDI from around 0.4 to 0.7 is a difference of 0.3, and a change of 0.001–0.005 in Income HDI is only a 0.1–0.4% contribution to the total desired change in total HDI.

Eliminating kerosene lamps and candles with small 5–20 W solar home system is thus a clear contributor towards reducing poverty, but does not address 99.6% of total poverty alleviation needs. Meeting the 7th Sustainable Development Goal of access to energy for everyone should result in a considerably higher contribution towards poverty alleviation than just 0.4%. The Millennium Village Project by the renowned economist Jeffrey Sachs and others attempted a fully integrated package of poverty-alleviation technologies and interventions which yielded some modest successes and estimated that energy infrastructure represented 10–30% of the total package financing. Similarly, other studies have suggested that infrastructure for the poor amounts to 50% of the financing gap that is required to complete all 17 Sustainable Development Goals.

Therefore, it is suggested that no less than 20–40% of the total HDI uplift target should be seen from the elimination of energy poverty. For the poorest countries that have an HDI of 0.3 and need an uplift of 0.4, this is an HDI increase of 0.08–0.16. The increase in Income HDI of up to 0.005 for kerosene lamp elimination with small solar LED lighting systems, which translates to up to 0.0017 increase in total HDI, amounts to 1–2% of the total HDI increase one might expect from energy poverty elimination. Even the inclusion of radio, torch, TV, and entertainment typically found in small solar home systems only increases this to a maximum energy HDI share of 3%. In other words, 97% of energy access impacts are still yet to be delivered to the poor after the kerosene lamps, candles, fire-based lights, and disposable batteries have been eliminated. It was also earlier estimated that 1–2 kWh/day (or less, with the most efficient appliances) would be needed, whereas 5–20 W solar systems generate only 25–100 Wh, 1–10% of the total expected energy needed. The lower end of this energy consumption range is consistent with the share of energy HDI calculated.

In short, solar lighting alone should not be considered "access to energy" – this is a betrayal to the poor if the claim is being made that the Sustainable Development Goals have been attained simply because a few watts of LED lights have been installed in a house, but there is still no power in the school, at the clinic, for water pumps, and no mill to process crops. At least ten times as much access to energy (50–200 W or 250–1000 Wh) is required for energy poverty elimination to contribute its fair share towards total poverty elimination. The energy access community has a multi-tiered definition of "access to energy," with Tier 0 being the existing situation, Tier 1 being 3–50 W of solar and 12–200 Wh of energy for basic needs, while Tier 2 is for 50–200 W and 200–1000 Wh of energy. Thus, Tier 1 should not be considered "access to energy," and Tier 3 (200–800 W and 1000–3400 Wh) may be more than is required. Tier 2 only focuses on household appliances, so "Tier 2+" is needed for energy access to be attained.

Climbing the Ladder of Energy Access for True Resilience

That "lighting only" is not "energy access" was recognized decades ago by the microhydro minigrid sector in Nepal. Early projects in the latter half of the twentieth century were mostly lighting-only projects, but in the last 20–30 years, it has been recognized that these had limited development impact without also including other non-lighting "productive" uses of energy that reduce manual labor and increase income. The reason this happened far earlier for microhydro-based access to energy projects is that in the twentieth century, solar energy was too expensive to do anything more than power some lights and charge some phones. However, in the past 5–10 years, solar (without battery storage) has now become as cheap as hydropower, around $0.05/kWh, but is still more expensive than hydro at around $0.15–0.30/kWh once battery storage is included such that solar energy can be used outside daytime hours when it is produced.

This is still cheaper than the current default technology for the workhorse of productive energy for the poor – the diesel engine. If kerosene is the workhorse fuel of the poor for lighting, then diesel is the next fuel to aim to displace, mostly used in diesel generators for electricity generation and in diesel engines that provide motive power for water pumps, mills, and a range of other mechanically driven equipment. A diesel engine or generator generally consumes 1 liter of fuel and delivers 3–4 kWh of energy, so if the cost of that fuel is $1/liter, the fuel cost alone of diesel energy is $0.25–0.33/kWh. Capital and operating costs should also be added. Diesel engines do not last long and incur considerable operation and maintenance costs, so once these are factored in, the cost of diesel energy is around $0.30–0.40/kWh.

This is for on-site use of diesel energy and excludes distribution to distant loads via a minigrid, which can increase costs to $0.50/kWh, as is commonly found in many small island economies that rely on diesel generators for most or all of their electricity.

In Nepal, almost every microhydro had a mill as the first big step up the energy ladder, particularly for increasing the daytime use of energy to complement the morning and evening peak loads. Daytime uses of solar power will also be helpful in solar minigrids or stand-alone systems, as it will use solar power as it is being made, vastly reducing or eliminating the need for battery storage and thus making solar energy cheaper for appliances used in the daytime. A mill is a community-focused energy asset, which will be assessed in more detail after household-level individually owned appliances are assessed.

Other smaller steps up the household energy ladder would include those appliances found in medium-sized (20–80 W, 100–400 Wh/day) solar home systems and many grid/minigrid connections, mostly fans, TVs, and entertainment systems. It is less clear that such technologies make any increase in income, education, or health, though they undoubtedly add to customer happiness and an improved feeling of living conditions, which is why customers gladly pay for them. There may be some instances of a negative effect on HDI, as hours spent watching TV reduces the time spent on more productive activities.

Therefore, without strong evidence to the contrary, it is assumed that fans, TVs, and entertainment systems do almost nothing to decrease an HDI-defined level of poverty. Perhaps a fourth factor is needed to be added to the HDI framework to capture improvements in happiness or a contribution to the soul – spending time on religion, art, music, and social gatherings small or large similarly has little measurable impact on income, years of education, and life expectancy, but are important parts of a fulfilling life. However, this is beyond the scope of this analysis.

One exception to the low impact on poverty that small appliances have could be a computer, particularly with Internet connection, as a major step up the ladder from a mobile phone. A computer is capable of allowing its owner to connect to global markets and generate considerable income for a very modest amount of energy (around 100–200 Wh/day) – just ask the well-educated and well-paid energy consultants in the energy access industry, including the author! Efforts like "One Laptop per Child" have attempted to demonstrate this high impact, but the time required to see the end effects exceed the average length of a development project, and thus are likely never truly measured or captured. "One Laptop per Household" might have a more immediate effect if adults can use the tool to boost incomes.

Higher up the household energy ladder are appliances like refrigerators and freezers, clothes washing machines, dishwashers, irons, water pumps, air conditioners, heaters, power tools, and cooking devices (stovetops, kettles, fry pans, toasters, microwaves, etc.). Gas cookers and heaters can also be efficient options, though with power emission cleanliness than renewable electricity. Such technologies require 200–500 Wh/day to operate and will thus deliver the bulk of HDI impact. Clothes washing machines can save 1–3 h of manual labor per wash compared to washing clothes by hand or 2–6 h per week assuming two loads of washing per household. Such time savings can be spent doing more farming, fishing, or making handicrafts or other income-generating activities. The earning capacity of the poor is around $0.5–1/h, so as an example, a washing machine doing 200 cycles per year could create time savings worth $100–250 per year.

Similarly, water pumps and gas or electric cookstoves that replace wood-burning cookstoves can replace around 1 h/day each for fetching water and fuelwood or, in urban areas, can replace cash expenditure on cooking fuel and purchased water which typically totals $0.25–0.50/day (and thus is also worth $100–200/year). Irrigation pumps can increase farm output up to 50–100%, and thus too also income, and can allow a switch from staple crops worth $0.30/kg to higher-value horticultural crops worth $0.50/kg or more, with increased production per hectare also adding to increased income. Refrigeration and freezers can help dairy farmers and fishermen similarly, increasing output by 50–100% and reducing fruit and vegetable spoilage of by 10–30%. Carpentry tools and sewing machines can also reduce manual labor and open up new income-generating possibilities.

These "high-power" (100–1000 W) household appliances may not have a 3-month payback period like LED lights, but collectively could increase household incomes by $500–5000/year nominal ($1500–15,000 PPP), which is 4–40% of the desired income boost of $12,300–17,000/year nominal that increases the Income HDI to 0.7. The upper half of this range is consistent with the aim of energy access technologies helping to deliver 20–40% of the total HDI increase that is required to eliminate poverty.

However, the other two components – education and health – must also be increased. A strong driver of taking children out of school before they complete the desired 10–11 years of schooling, particularly girls, is to help their parents with daily chores and to earn income. If access to modern technologies can save 2–4 h per day of manual labor, parents will have less need for help from children with daily chores. If income has increased 100% or more, there will be less pressure for children to work. The mean years of schooling in countries with HDI of less than 0.7 averages 4–6 years and needs to rise to 10.5, while expected years of schooling averages 9–11 years and needs to be 10.5 years, so the clear aim is to ensure actual schooling years meet this expectation, adding 4–6 years. If energy access technologies help with 20–40% of this target by achieving 1–3 years more schooling due to reduced manual labor and increased income, the target boost in Education-HDI will be achieved. Programs like SELCO's Light for Education program placed a solar battery charging station at schools, so that only by bring a battery to school two to three times per week could a household get the benefit of electric light and phone charging (Abra 2011). This model proved very successful, with over 600 schools funded by corporations in India and tens of thousands of students no longer using kerosene lamps for studying.

An increase in Life Expectancy-HDI is also desirable. The current average for low-HDI countries is 55–60 years for the poorest countries and 60–70 years in less poor countries. Therefore, some countries need no help to hit the target of 65.5 years, while some require 5–10 years of increased life expectancy. If energy access technologies can help deliver 20–40% of this increase, a 1–4-year increase needs to be attained. It is likely that with increased income, 50–100% more food, better access to clean pumped water, reduced indoor air pollution and a properly powered local health clinic, such an increase in lifetime expectancy is highly likely.

Lastly, it is also important to realize that energy access technologies are not only about increasing access to individually owned appliances but also community-/village-scale technologies. A mill (or mills) to process the major crops in a village does not need to be individually owned – each kW of milling power can process around 30 kg of grain per hour, which can likely meet the subsistence needs of 20 subsistence-level households that eat 1.5 kg of uncooked staple crops each day. This may reduce to serving five to ten households if they have excess crops to sell in the market to generate income. For a mill operating 5 h per day, 25–100 households can be served for each kW of milling power installed. Hence, each household needs around 10–40 W of milling power and 10–40 Wh of energy, which is not much, about the same as for LED lighting and small household appliances.

A mill run from clean energy (hydro, solar, etc.) can displace diesel mills and/or 0.5–1 h per day of manual labor shelling corn cobs, hulling rice with sticks, or grinding flour with rocks. Expenditure on such services from a diesel or electric mill averages $1–2/month (including travel costs to and from the mill) so $12–24/year. However, time savings can have considerably higher value of $100–200/year/household, similar to the cash savings of eliminating kerosene lamps, candle, and disposable batteries. Mills also add value to crops, helping to boost incomes – for example, a local coconut oil press could help villagers make $3/kg coconut oil instead of $0.30/kg dried coconut meat (copra) or could make cassava flour instead selling raw cassava roots. If vacuum packing is added, the shelf life of cassava flour can increase from 1 day to 1 year, boosting food security particularly in the aftermath of natural disasters when crops are often destroyed, improving resilience as well as incomes.

Other examples of community energy access technologies include water pumps that could be shared between several farmers for irrigation to increase the amount of food available and washing machines that could be shared in a laundromat form rather than individually owned. Sharing high power appliances can make them more affordable if the capital cost is shared between several households and can lead to higher utilization of the appliance, decreasing the cost of the service. Most of the benefits will target increasing incomes, but other shared energy access technologies like better electricity access and appliances at the school and at the health clinic will have a far more direct impact on the Education-HDI and Life Expectancy-HDI components. As these are non-financial benefits that investors find hard to value, public funding will likely be more appropriate than private sector investment.

Starting the Journey Towards the Future's Resilient Poverty-Free Village

It may be helpful to describe how an HDI-focused energy access program would be implemented, and the outcomes the village would be expected to experience, complete with an energy budget, a financial construction (capex) budget, and an investment analysis via a simple payback period calculation. This may also help predict the public and private capital mix that may be required.

The first step is to size a typical scale of project. A primary school typically has around 300–500 students in Africa, and a typical household may have 3 children, 2 of which are of school age, so a school may serve 150–250 households. Thus, an "average village" can be considered of 200 households (1000 people), though a single health clinic can often be serving 10 times as many households and people (but may be easily overloaded when a pandemic strikes).

Table 1 shows the suite of solar-powered all-electric energy access technologies that could yield the desired boost in overall HDI, starting from a baseline of 55 years life expectancy, 6 years of mean and expected education, and income per capita of $500/year nominal ($1500 PPP) which equates to $2500/year/household nominal ($7500 nominal) for a five-person household. This is an overall baseline HDI of 0.43 and is a typical value for a least developed country (or a poorer-than-average region of a medium developed country).

Some technologies are individually owned by households, while others are community-scale items that have multiple users per installation. Some appliances are used primarily at nighttime or early morning and thus must operate primarily from a battery, while others operate in the daytime and so can use a more minimal amount of battery storage. The seven community technologies average 20 W per household and total 140 W, similar to other small appliances used on an individual household, but cooking power needed is far higher than any of these, and thus both electric cooking and improved biomass cookstoves have been assessed.

The best-performing improved biomass stoves, in terms of fuel efficiency and emissions reduction, are fan-forced stoves, usually using a thermoelectric module that converts some heat into electricity to run the fan. However, no improved biomass cookstove (without a chimney or exhaust fan) has yet delivered indoor air pollution levels that pass the World Health Organization (WHO) permissible limits. A biomass stove also does not reduce fuelwood consumption completely as an electric stove would achieve. Thus, it is expected that only 50% of the health and fuelwood cost (or time) savings would be realized by improved biomass cookstoves, but they are far cheaper than an electric cookstove, so are likely to be more cost-effective. That said, fuelwood and charcoal consumption by 2 billion people consumes at least 500 million kg of wood per year, which is equivalent to around 50% of deforestation, so maximizing fuelwood reductions is a very important goal.

An LPG or bioethanol stove has not been explicitly included as an option, but most certainly is a potential solution. These fuels still

Resilient Rural Electrification for the Twenty-First Century, Table 1 List of energy access technologies to eliminate energy poverty

Item	Description	Usage	Time of use	Power, W
LED lighting	4 × 1 W 100 lumen lamps	Household	Evening	4
Phone charger	10 W 2A for 2 phones	Household	Evening	10
Radio	5 W for small radio	Household	Continuous	5
TV	18 inch efficient TV	Household	Evening	10
Entertainment	Music System, VCR/DVD	Household	Evening	50
Electric cooking	Two-pot 500 W cooker	Household	Daytime	500
Washing machine	10 kg capacity, used 3 times/day	Community	Daytime	200
Mill for crops	Multifunctional, 30 kg/h	Community	Daytime	750
Water pumping system	100 W, 10 k L/day, incl. well, pipes, tank	Community	Daytime	100
School power	3 kWp solar system	Community	Daytime	3000
Health clinic power	3 kWp solar system	Community	Continuous	3000
Shops, commercial	1 kWp solar system	Community	Continuous	1000
Government	1 kWp solar system	Community	Daytime	1000
OR				
Improved wood cooking	Thermo-electric stove, 50% fuel saving	Household	Daytime	2

create CO_2 emissions, though less than a traditional non-improved biomass stove, so are hence similar to an improved cookstove, but can pass the WHO indoor air pollution limits and thus have higher health benefits similar to an electric stove. The cost of an LPG or bioethanol stove is higher initially than an improved cookstove but yields less savings, as the ongoing fuel cost is only 0–25% cheaper than buying wood or charcoal rather than 30–50% cheaper as would be for an improved biomass cookstove – perhaps $0.30/day instead of $0.40/day. The lifetime cost of an LPG or bioethanol stove over the same 10-year life of a solar electric stove is likely around $100/year or $1000, plus around $100 initially for a 2-burner stove and gas bottle deposit, so is likely to cost as much or more than a 500 W solar electric stove and far more than a $50–100 improved biomass cookstove. Hence, it is likely to be a higher cost option and does not have higher benefits, so is likely a less-than-optimum solution.

However, the lower capital cost and high operating cost of a LPG or bioethanol stove compared to a solar electric stove may make it much easier to finance for end customers, as less long-term credit would be required, which makes it more investable for private sector financing. It is also a highly desirable form of cooking by end customers, considerably more so than an improved biomass cookstove, and faster to cook with, yielding important time savings. Solar electric cookstoves will also not function well on rainy days unless a big (and expensive) battery is included and is perhaps not well suited to some forms of cooking like grilling and high-heat frying. Cooking is a complex issue, and the reality is that multiple cookstove solutions will likely be required (called “stove stacking”) to ensure lowest cost but reliable cooking services are available.

Table 2 shows the power and energy demand for current efficient technologies available in the marketplace, as well as the estimated expenditure savings or increased income (cash or by valuing time savings). Small appliances owned by the household (user) collectively have around a $100/year/household benefit, requiring 40–80 W of power and almost 200 Wh of energy. Cooking with electricity can save as much again, but requires far more power, 500 W or 1kWh/day, assuming 2 kg/day of cooked food at 400 Wh/kg (though energy-efficient insulated pots can halve this to 200 Wh/kg). Improved cookstoves reduce fuelwood consumption by 50%, but still give some indoor air pollution, so have half the benefits, but cost less.

Resilient Rural Electrification for the Twenty-First Century, Table 2 Power and energy demand and financial benefits

Item	Power, W	Hours of use	Energy, Wh/day	Battery, Wh	Users per appliance	Power per user	Energy per user	$/year/user increase (nominal)
LED lighting	4	6	24	24	1	4	24	$60
Phone charger	10	4	40	40	1	10	40	$25
Radio	5	6	30	18	1	5	30	$15
TV	10	4	40	40	1	10	40	$5
Entertainment	50	1	50	50	1	50	50	$5
Electric cooking	500	2	1000	500	1	500	1000	$100
Washing machine	200	3	600	200	5	40	120	$50
Mill for crops	750	4	3000	750	20	38	150	$25
Water pumping System	100	4	400	100	3	33	133	$225
School power	3000	6	18,000	3000	200	15	90	$5
Health clinic power	3000	6	18,000	10,500	1000	3	18	$5
Shops, commercial	1000	6	6000	3500	100	10	60	$30
Government	1000	6	6000	1000	1000	1	6	$1
	TOTALS PER HOUSEHOLD (W and Wh/day)					**719**	**1761**	**$551**
Improved wood cooking	2	3	6	2	1	2	6	$50
	TOTALS PER HOUSEHOLD (W and Wh/day)					**221**	**767**	**$501**

Overall, this suite of energy access technologies requires 220–720 W of power, depending on whether cooking is included or not, and 0.8–1.8kWh/day of energy. The resultant benefit is estimated at up to $500/household/year nominal ($1500 PPP), with irrigation likely to contribute almost half of the financial benefit, assuming the target user is a rural farmer. A similar benefit for fishermen may be realized from local cost-effective refrigeration and ice-making, but a similar benefit for the urban poor is hard to identify. For every 20–50 households, a shared basic "productive use centres" of shared "heavy duty" items would be made available including portable water pumps, a laundromat, and mills for crops, similar to the famous "kirana" corner stores all across India. Over time other technologies might be added in some Centres, such as carpentry power tools, brick-making, drying, packaging and bakery tools, an Internet-enabled business center, transport vehicle hire, and possibly banking and postal facilities.

These community-scale shared productive use technologies average around 20 W per household, comparable to a TV, and collectively require 140 W, considerably less power than electric cooking and not much more than the 100 W for individually owned household non-cooking appliances, but with three times more financial benefit ($340/year) as well as additional significant non-financial health and education benefits that household appliances do not create (other than cleaner cookstoves). Table 3 summarizes the estimated poverty-alleviation impact of each technology within the three-component HDI framework, with each technology's contribution above the baseline quantified. As noted earlier, basic household appliances including lighting and TVs only

Resilient Rural Electrification for the Twenty-First Century, Table 3 HDI impact of energy access technologies

Item	$/year/user increase (nominal)	Years schooling increase per user	Lifespan increase per user	Income-HDI	Education-HDI	Health-HDI	Total HDI
LED lighting	$60	0.0	0.0	0.004	0.000	0.000	0.001
Phone charger	$25	0.0	0.0	0.002	0.000	0.000	0.001
Radio	$15	0.0	0.0	0.001	0.000	0.000	0.000
TV	$5	0.0	0.0	0.000	0.000	0.000	0.000
Entertainment	$5	0.0	0.0	0.000	0.000	0.000	0.000
Electric cooking	$100	0.5	1.0	0.006	0.031	0.015	0.018
Washing machine	$50	0.5	0.0	0.003	0.031	0.000	0.013
Mill for crops	$25	0.5	0.0	0.002	0.031	0.000	0.012
Water pumping system	$225	0.5	1.0	0.013	0.031	0.015	0.021
School power	$5	1.0	0.0	0.000	0.061	0.000	0.023
Health clinic power	$5	0.0	2.0	0.000	0.000	0.031	0.008
Shops, commercial	$30	0.0	0.0	0.002	0.000	0.000	0.001
Government	$1	0.0	0.0	0.000	0.000	0.000	0.000
	$551	**3.0**	**4.0**	**0.032**	**0.183**	**0.062**	**0.098**
Improved wood cooking	$50	0.3	0.5	0.003	0.015	0.008	0.009
	$501	**2.8**	**3.5**	**0.030**	**0.168**	**0.054**	**0.089**

deliver 10–15% of the total HDI impact. To eliminate poverty and raise the HDI from the baseline of 0.432 to 0.7, a total HDI rise of 0.268 is required, and this model shows that this suite of energy services can likely increase total HDI by almost 0.1, which is more around 35% of the total HDI rise target, within the 20–40% share expected from energy access interventions. It is also clear that the Education-HDI makes up most of this improvement, followed by health, and the doubling of income from $500/year/household to $1000 contributes little.

Table 4 estimates the installed capital cost of each of the technologies and notes the percentage HDI contribution towards the overall target rise in HDI.

Other non-energy interventions would provide the other 65% of the poverty alleviation contribution, such as toilets, water filtration, health and sanitation education, training and availability of teachers, health workers, agricultural farmer support, microfinance, insurance, disaster relief, and other key services. A lot of the remaining progress would also have to come from the Income-HDI component, and it is likely that this will still require additional energy inputs.

These tables also allow a comparative evaluation of each technology on its overall cost-effectiveness from a total HDI impact perspective and a narrower financial-only analysis of the payback period of each technology, comparing the capital cost to the customer benefits. The result suggests that public investments in school and health clinic power have the most impact on poverty alleviation, 10 times more so than productive uses and 50 times more so than household appliances, assuming the expected benefits are indeed realized (1 year of increased school attendance

R

Resilient Rural Electrification for the Twenty-First Century, Table 4 Capital cost and cost-effectiveness of energy access technologies

Item	% of 20% target	% of 40% target	% of total target	Capex	Capex per user	Capex/ HDI	User payback period, years
LED lighting	2.3%	1.2%	0.5%	$50	$50	$10,640	0.8
Phone charger	1.0%	0.5%	0.2%	$25	$25	$12,659	1.0
Radio	0.6%	0.3%	0.1%	$20	$20	$16,837	1.3
TV	0.2%	0.1%	0.0%	$200	$200	$503,849	40.0
Entertainment	0.2%	0.1%	0.0%	$150	$150	$377,886	30.0
Electric cooking	32.9%	16.6%	6.7%	$750	$750	$11,138	7.5
Washing machine	23.3%	11.7%	4.8%	$600	$120	$2516	2.4
Mill for crops	22.3%	11.2%	4.6%	$1750	$88	$1915	3.5
Water pumping system	37.5%	18.9%	7.7%	$900	$300	$3903	1.3
School power	41.8%	21.0%	8.6%	$5000	$25	$292	5.0
Health clinic power	14.9%	7.5%	3.1%	$7000	$7	$229	1.4
Shops, commercial	1.2%	0.6%	0.2%	$2500	$25	$10,562	0.8
Government	0.0%	0.0%	0.0%	$2000	$2	$25,167	2.0
	178%	**90%**	**36%**		**$1762**	**$4829**	**3.2**
Improved wood cooking	16.5%	8.3%	3.4%	$100	$100	$2952	2.0
	162%	**81%**	**33%**		**$1112**	**$3355**	**2.2**

from a properly functioning school and 2 years of increased life expectancy from a properly functioning clinic).

The next most impactful set of technologies is the community-scale productive uses, which are 5–10 times more impactful than household appliances. Shops and government office energy solutions may be of less benefit, as shops tend to benefit the shop owner at the expense of other community members rather than bring new money into the village, and governments may take as much money from villagers as consumption and income taxes as they bring to the village in the form of subsidies, free capacity-building programs, and free/subsidized services (e.g., schools).

The lower the payback period, the more attractive it might be to the private sector to finance, though it must be noted that "customer benefit" is not the same as "willingness to pay" – a TV may have little financial benefit but would be very popular, while an improved biomass cookstove can often have a payback period of 1 year or less but still struggles to be an attractive purchase to users. Technologies that have a simple payback period of more than 1 year will probably struggle to sell to end customers for cash and will need financing over 1 or more years. In the commercial deployment of such energy access technologies by the private sector, the actual financing period, and thus investment period, will need to be longer than these simple payback periods, to generate the investors' target return on capital invested, cover operating costs, and cover customer default of credit offered. This can easily double the payback periods, so if an investor is not willing to lend to the poor for more than 3 years (quite normal in microfinance), a user payback of more than 1.5–2 years is unlikely to be commercially acceptable for that private sector capital, and public sector capital may be partly or fully needed.

Thus, Table 4 suggests that small household appliances, water pumps, and shops may be suited to private investment, while cooking, washing machines, mills, schools, health clinic, and government office power will need public grants/subsidies or low-interest loan finance involved from national governments or international aid programs. As these technologies mature over time, just as happened for LED lighting systems, costs will decrease and payback periods will reduce, allowing more private sector capital financing, reducing the burden on limited public capital. However, if the overall program has a simple payback period of 2–3 years, there is a clear indication that 4–6-year investments in the world's poorest villages could bring them 50% out of poverty, doubling baseline incomes, increasing school attendance by 3 years to an average finishing age of 15–16 years, and improving life expectancy by up to 4 years, and total HDI improves from 0.43 to 0.55.

Given that on-grid large power station projects selling power to a dubiously creditworthy power utility are typically financed over a 15–30-year investment horizon, it is not unreasonable to suggest that, were the same applied to dubiously creditworthy off-grid poor households, energy poverty could be eliminated. The current infrastructure investment market is more than $3000 billion/year, while an investment of $1500/household for 20 million households per year would total, over the next 8 years to 2030, just $30 billion/year, just 1% of the global total. If every infrastructure investor "risked" 1% of their portfolio on the poorest 1–2 billion people, the Sustainable Development Goal 7 of energy access for all could be achieved.

Completing the Journey Towards a Resilient Poverty-Free Village

To complete the journey out of poverty, nominal incomes will have to increase from $1000/household (nominal) to $17,000/household, a far bigger increase than achieved in the energy access model but is a realistic target as it is typical of the average salary of two workers (the households' mother and father) in countries such as Indonesia, Ecuador, and Egypt that have an HDI of 0.7 and have access to electricity of 90–100%. Years of schooling will also have to increase 1–2 more years and life expectancy another 6–7 years. As noted earlier, the income target could be reduced to $12,300/household if the life expectancy component exceeds its target. However, an increase in real income of $10,000–15,000/year/household is a huge challenge, not least limited by the assets owned by a typical low-income household.

Most are small-scale rural poor farmers who own on average a 0.5 hectare plot of land that currently yields 1000–2000 kg/ha but with irrigation may yield 2000–4000 kg/ha. The price of rice, wheat, and other grains in such markets is generally around $0.30/kg, and the first 500 kg is needed to feed the family, meaning for low-fertility land (or poor farming practices) there is no excess at all to sell (known as subsistence agriculture), and at best there is 1500 kg of excess crop to sell from an average smallholder farm. This brings $500–1000/year of increased income, far below what is needed to achieve the Income-HDI target. Only a farmer that can produce 30–50 tonnes of excess crop, and thus owns around 10 hectares, can farm his way out of poverty.

One strategy is to abandon farming low-value staple crops and focus on horticultural fruits and vegetables or oil-bearing plants like palm and coconut. Fruits and vegetables can yield 25–100 tonnes per hectare. If these similarly sell for $0.30/kg, income can reach $4000–15,000 from a 0.5 hectare plot, which is on target.

Another example: A well-tended coconut plantation would fit 100 trees per hectare and yield around 100 nuts per tree every year, generating $500 per 0.5 hectare plot from unprocessed whole nuts at $0.10/nut and similar income from processing dried copra meat. However, if 1 L of coconut oil can be extracted from every 15 nuts using a local oil expelling mill, over 300 L of oil can be made from the 0.5 hectare plot, which if high quality can be sold at $3/liter can yield $900/year, and the leftover coconut meal can be used for animal feed, adding perhaps $600 more income, thus tripling income to $1500/year. However, this is still well short of the target, so intercropping between trees with high-value crops like vanilla, kava, or good-quality cocoa, or even grazing of beef, can help boost incomes towards the target.

Another alternative to farming higher-value crops is that in sunny rural areas with plenty of space, non-farming options can be considered. 40 years ago, the city of Shenzhen in China which now has 10 million people and tens of thousands of factories was just a small town surrounded by rice paddies. Those farmers mostly now lease (or sold) their land to factory owners and property developers, and few farms are left. The children of those farmers now drive expensive cars and barely have to think about working.

Such industrial development is not the panacea for every rural village, but farming the sun is one interesting option. A 0.5 hectare plot of land can fit 500 kW of solar panels that would generate around 800,000 kWh/year of energy that can be sold to the city at around $0.03–0.05/kWh (the same as existing coal or gas power plants sell energy for). This can yield $24,000–40,000, far more than the target, but a rural farmer could never afford the $250,000–500,000 investment to build the solar power station.

A creative approach would be for a professional solar power station developer to initially own the plant and finance construction but transfer 2% to the farmer each year over 25 years until the end of the project life, at which time the farmer would own 50% of the project and be able to invest in the next project. Over this time, the farmers' income would grow to $12,000–20,000/year, escaping the poverty cycle. In fact, in just the first year, the farmer would likely earn as much from his 2% ownership of the project as he would earn from farming all year.

In infrastructure investing language, this could be called a Build-Own-Operate-Slow-Transfer (BOOST) model, and the BOOST investor would receive a 2.7% lower IRR on investment by slowly giving away half of the solar asset to the landowners. If debt investors in the project lowered their interest rate by just 1.3% (e.g., from 10% to 8.7%) because they wanted to help finance poverty reduction, the developer would earn exactly the same IRR as owning 100% of the project but paying the higher rate of interest.

If lenders want a faster end to poverty, 17 years is possible with a 3% annual transfer rate which needs a 2.1% reduction in interest rates, or 12 years with 4% transfer rate and 2.8% interest rate reduction, or 10 years with a 5% transfer rate and a 3.5% interest rate reduction. The pace of poverty reduction can therefore be accelerated to be achievable within a single generation so children of farmers that live in poverty today do not have to continue to remain in the poverty cycle, the interest rate reduction required in the BOOST model is affordable to the rich funders of such projects, particularly those that specialize in "concessional finance" such as multilateral banks (World Bank, African Development Bank, etc.).

Another way to look at the potential of farming the sun is that just 2–4% of the farmers' land covered by 10 kW of solar panels would earn the same $500–1000/year as farming the entire 0.5 hectare plot of land. In some markets, the amount of fertilizer, water, and power subsidies given to poor farmers actually exceeds the value of the crops they grow, giving a negative economic return on the public funds invested. Making farmers self-sufficient by farming the sun would reduce the need for such subsidies. Even Australian farmers have commented in the media that farming the sun is more profitable and probably more reliable than farming the earth, particularly as soils have been depleted over many decades and both rain and irrigation water become more scarce. In India, farmers are already starting to sell excess power back to the grid from 2–10 kW solar water pumps.

Objections to farming the sun may be raised via the food-vs-fuel debate, but if such projects are located on the least fertile land, which is where most of the global poor can be found, there should be less resistance to the concept. Of course, without a grid connection to these rural areas to allow power to be sold from rural villages back to the city, the potential is lost, so this may be one of the best ways to justify the extension of the grid to deeply rural areas – not to sell power to the poor, but to buy it from them and sell it to richer consumers in the city. The power from one average smallholders' 0.5 hectare 500 kW solar farm would likely meet the energy needs of 100–200 households in the city.

Education is also key to income generation. The older children of rural farmers can, with a good education and access to modern technologies and information, secure good non-farming (but not necessarily non-rural) jobs in the next 10 years that continue to help assist with lifting their communities out of poverty. Earning $50/day from such work is not unreasonable and can generate $12,500 of real income, which is the target. Knowing how to operate a computer and write emails and having a working knowledge of English, French, and/or Spanish will help. Many children can learn language basics informally from television programs, but more structured free programs in their own language run on computers would probably offer a higher-quality learning process. It is never too early to start, and one step towards spending development money in rural villages is to make contact and then secure their assistance to validate satellite imagery locations of buildings in latitude/longitude around the village, so that accurate development plans can be formed. With increased "working from home" trends showing what is possible, and the higher health risks of living in a densely populated city, creating rural services and knowledge sector jobs with good communications infrastructure could be a key strategy to reducing rural poverty.

The Cost of Traditional Power Distribution

The capital costs so far have assumed on-site power generation using solar and battery energy storage. Should microhydro be available thanks to the correct water and topographical resources, this can be cheaper than solar+storage, as no battery investment is required, but on-site power generation would not be possible, so the cost of transporting power via poles and wires from the powerhouse to the point of use must also be included. With a solar minigrid, this cost of traditional distribution of power must be added to the on-site solar+storage costs, because the same solar panels and batteries are located at the powerhouse, but the power must still be transferred to customers. This is a non-negligible cost that increases the capital cost and payback periods considerably, particularly for very low levels of power consumption.

A typical distribution system has low voltage lines at the same voltage as is used in the house and thus no need for transformers to change voltages up or down. For an AC power system of 100–400 V, the maximum distance such systems can reach is generally 500–1000 m before voltage drops along the wire become excessive (more than 10%). Therefore, in minigrids or grid extension projects that extend further than 500–1000 m, a medium voltage line using 3–33 kV transformers will be used. Such lines are have taller, stronger poles and are more expensive. A low-voltage line is typically $5000–10,000 per km ($5–10/meter), while MV lines are typically $15,000–45,000 per km ($15–45/meter). Most rural households are spaced 25–50 m apart, so a low-voltage line adds $125–500 per household to the capital cost, while the addition of MV lines can increase this to $200–800/household. If households have only been spending $5–10/month on energy services, the payback period on these lines adds 1–5 years to the simple payback periods listed in Table 4. Even if such projects succeed in doubling rural incomes as shown in Table 2, it is still fairly unlikely that investors will agree to invest based on 50% non-existent projected income.

Such distribution networks often last for 20 years once built, unless hurricanes/typhoons/earthquakes knock them down, which is becoming increasingly common. Traditionally, each powerline has to be sized for the maximum power demand that each household will draw, because there is no battery storage at or near the house to provide power for surges in demand, only one centralized battery at the center of the village.

Therefore, if an investor has only a short tolerance for the return of their capital (3 years or less as is normal for microfinance or most impact investors), building solar minigrids is simply commercially infeasible – no solar (or even micro-hydro) minigrid is likely to have a simple payback period of 1.5–2 years (which grows to 3 years after the cost of financing and revenue collection). Investors with a 5–10-year investment horizon may find a minority of minigrids will fit their criteria, but 10–15-year investment periods are more likely required (5–7-year simple payback periods). Such investment periods are also typical for large grid-connected renewable energy power generation projects, so the capital is there from infrastructure investors, but such investors are used to $50–500 million projects, not micro projects in rural village. They also expect a single customer, the power utility, to take all the power and pay all the revenue, but in an off-grid project, there can be dozens, hundreds, or even thousands of end customers to manage, which is not something a classical infrastructure investor is keen to be involved with (though is a natural fit for microfinance organizations and investors).

Therefore "micro infrastructure" will need a special class of investors, who are comfortable with end customer retail lending to the poorest 1–2 billion people on the planet (off-taker risk) but have a long-term investment horizon that most investors lack. As noted earlier, it needs just 1% of the existing infrastructure industry to specialize in micro infrastructure to achieve poverty-alleviation goals.

It is also fairly obvious that the lower the demand for energy services that a new customer (residential or business or government) can afford, the less justification there can be for extending expensive powerlines to the customer. Table 4 suggests that to bring a full suite of energy services to a village will cost $1100–1800 per household before the cost of powerlines are included, delivering 220–720 W of peak power per household and delivering 0.75–1.75 kWh/day of energy. Over the 20-year lifespan of the solar panels and powerlines, and allowing 40% extra cost for future battery replacements at 10–15 years, each customer will use 5500–12,500 kWh of energy (depending if solar electric cooking is included), giving a basic cost of $0.19/kWh with electric cooking and $0.28/kWh without. This cost will increase 50–100% once the cost of operations, investors' return on capital, default, and repairs are included.

This is the basic on-site power generation cost of energy before powerlines are added, which increase capital costs $100–250 small, dense low voltage networks and $200–500 per household for larger and less dense medium voltage networks. If electric cooking is included, powerlines will add $0.02–0.04/kWh, and if electric cooking is excluded, the powerlines will add $0.02–0.09/kWh, depending on the density of buildings and size of the network. Again, at least 100% needs to be added to these basic costs for commercially viable costs over 20 years and possibly 200%. Therefore, powerlines can add $0.05–0.30/kWh to the basic $0.19–0.28/kWh cost of power generation.

The lower end of this range ($0.24/kWh) can be attractive, but the higher end of the range ($0.58/kWh) is more expensive than the current diesel minigrids ($0.50/kWh) or on-site diesel power generation ($0.35–0.40/kWh as noted earlier). It is also a fairly typical cost of energy from solar minigrids being developed today, which then cry out that they need subsidization to deliver affordable services. Perhaps fewer poles and wires and more on-site power generation is the answer. This is not just an issue for access to energy in developing countries – in Western Australia, poles and wires that served rural farms and small towns are being removed after bushfires and are being replaced with more reliable and far cheaper on-site 10–1000 kW solar/diesel hybrid power systems.

For a small household using only basic appliances, a 20–50 W solar home system costing $200–400 can cost the same as installing 50–100 m of low-voltage powerline to reach the household. Even a 500–600 W solar system with lights, TV, cooker, and other appliances will only cost $1200–1500, probably similar to 100–150 m of voltage powerline construction. There is thus no rational reason to build powerlines for more than 100 m per household connected for low-revenue customers likely to pay $5–15/month. Higher demand productive loads like mills, schools, clinics, and shops could justify longer lines, but still not likely more than 1–2 km per 1–3 kW load. Similar logic applies to the extension of existing grids. If more than $1500 is being spent per household on a rural electrification project, it has probably been poorly designed and costed or has a high proportion of non-residential loads.

One innovation can help modify these powerline viability guidelines. If energy storage is decentralized to the customer, instead of centralized, the power delivered to customers in a minigrid no longer needs to be the peak demand, but just the total daily energy demand spread over 24 h. This can reduce the peak demand by 50–80%, meaning that thinner conductors can deliver backup power to on-site batteries. This is where the direction of power systems is headed even in rich, developed countries, where electrifying cars can help build the load for existing utilities as on-site 3–10 kW solar panel installations on a Western household meet the majority of smaller appliance loads and maybe even most cooking loads, but only on sunny days, and the main grid still needs to trickle in backup power. Thus, "skinny grids" with distributed energy storage may be a significant change to powerline network design in the twenty-first century. During a disaster when powerlines are knocked down by strong winds or an earthquake, on-site energy storage (and perhaps some on-site solar power generation) can help provide local power to households, business, and government while the traditional powerlines are rebuilt. Underground powerlines are more resilient and can offset weather risks, but are expensive.

At the smallest village level, several companies like Okra Solar and PowerBlox are already building smart 200–2000 W solar controllers that can allow interconnection of solar home systems via a simple powerline to the neighbor (AC or DC voltage, the latter being safer but cannot send power as far). Early results suggest that this allows 10–30% of power generated to be circulated locally and used, rather than being wasted and unused as would happen in an isolated solar home system. If the extra cost of these lines and smart controllers is less than oversizing a solar home system by 10–30% ($100–400 per household depending on the energy demand of each house, particularly electric vs non-electric cooking), then these "mesh grids" make rational sense. If households are spaced more than 50–100 m apart, this solution is unlikely to be optimum, even at a low cost of $5/m.

Spatially Planning Solutions for Resilient Villages

If all buildings and loads are mapped accurately on a map, these guidelines and the distance between buildings can then help planners rapidly assess the most optimum mix of on-site solar+-storage systems, minigrids, and grid extensions to build. For each cluster of N households, it is recommended that the maximum line length should be limited to 100 m × N, so if a household is 140 m from the grid or minigrid, it should be serviced with a solar home system, and if a cluster of 40 households is more than 4 km from the nearest other minigrid or main grid, an isolated minigrid should be built (assuming that the spacing of households within the village is less than 100 m). As demand grows over the years, some systems may be viably interconnected that could not be justified initially, and cost reductions in certain technologies may also change the optimum system design.

Rather than forecasting what the demand might be in 10–20 years as was done last century, building a power system with upgradeable components, responsive appliances, distributed power generation, and energy storage is a more cost-effective way to design in the twenty-first century, even if more complicated, and ultimately also cheaper. Power system design is now not only restricted to highly expensive dedicated software and high-voltage power system engineers, but can be designed using open-source free GIS mapping software by anyone who understands the technologies available and the most cost-effective strategies to employ, particularly for extra-low system voltages (below 48–60 V DC). There is slowly a larger body of software and expertise becoming available and a growing group of helpful experts.

Summary/Conclusion

The twenty-first century will see, and is seeing, a decentralization of traditional electricity networks and at the same time a decarbonizing of power generation to reduce CO_2 emissions that are causing potentially catastrophic climate change. At the same time, those who have not been served by the twentieth-century power infrastructure should not be left behind, and access to energy for all is as essential as improvements in existing power networks. The Sustainable Development Goals to eliminate poverty for the poorest 1–2 billion people cannot be completed without eliminating energy poverty (SDG7), so a model that directly links energy access to poverty alleviation helps quantify each technology used for energy access, and it contribute towards the elimination of poverty. Despite its limitations, a three-component measure of poverty reduction that adds up to an overall Human Development Index (HDI) has been used to measure improvements in income, education, and health.

The benefits of access to lighting, TVs, fans, radios, and other small appliances, the primary result of current expenditures in off-grid energy, are found to be contributing just 2% of the total poverty alleviation impact and require up to 80 W of power per household. However, improving cookstoves can contribute 10–20% of development impact, but the bulk of impact is delivered by "productive" uses of energy, particularly community-scale technologies such as mills, shared water pumps, washing machines for laundromats, and power for schools, health clinics, shops, and government offices. These can provide 80–90% of poverty alleviation impacts but also require more power and energy, up to 140 W per household if cooking is improved with more efficient, lower-emission thermoelectric biomass cookstoves, and require an additional 500 W if solar electric cooking is used, giving respective totals of 220 W and 720 W of power and 0.8 kWh and 1.8 kWh per day of energy consumption.

From a baseline of $500/household/year of real income, education of 6 years schooling, and an average life expectancy of 55 years for those most in need, who are mostly rural smallholder farmers, this suite of energy access technologies that costs $1100–1750 to implement should result in an a doubling of income to $1000/year/household (assuming the same crops are grown), 3 years of additional schooling due to reduced manual workloads and a properly working school, and 3–4 years longer lifespan due to more food to eat, better incomes, reduced indoor air pollution, and properly working health clinic.

The full suite of technologies has a simple payback of 2–3 years which may require 4–6 years of financing to cover commercial operational costs and returns to private sector investors. This is probably too long for most investors, who prefer a maximum exposure of 3 years, so the model also helps isolate which technologies are best suited for rapid paybacks and which others will need public capital. Adding the cost of powerlines can increase these costs by 10–50% depending on the demand for energy and the local population density, so minigrids and grid extension will be affordable in some instances, but otherwise isolated solar+storage home systems will be the least cost option. In no instance is the cost expected to exceed $1500 per household, so for 150 million off-grid households (750 million people) that lack access to electricity today, the total cost is not expected to exceed $225 billion, considerably lower than the estimates of $400–600 billion made by the International Energy Agency which are regularly promoted by the United Nations, Sustainable Energy for All, and other industry leaders. Over 10–15 years, this would average $15–22 billion/year, which seems feasible given baseline expenditure of the worlds' off-grid households is already $15 billion/year spent on very poor-quality services.

It is thus hoped that this HDI-based model of delivering energy access demonstrates how efficient appliances and rational delivery of clean energy solutions can deliver such essential services for considerably less cost than experts currently expect and quantifiably demonstrates that eliminating energy poverty delivers around 35% of the required increase in the Human Development Index that would lift the least developed regions and countries out of poverty. The remaining poverty alleviation effort would come from non-energy interventions and, to attain the required rise in income, may require a substantial shift in historical practices, such as changing from staple crops to higher-value crops, or farming the sun instead of the land by installing solar power in the low-productivity farms and selling the energy to the city, or by relying on the education of the next generation to provide a higher non-farming income.

References

Abra, P. (2011). Title. Retrieved October 9, 2021, from https://economictimes.indiatimes.com/initiative-watch/solar-entrepreneur-harish-handes-solar-electric-light-company-taps-rural-schools-homes/articleshow/11129802.cms

Battacharyya, S. (2012). Energy access programmes and sustainable development: A critical review and analysis. *Energy for Sustainable Development*. Retrieved October 2, 2021, from https://www.semanticscholar.org/paper/Energy-access-programmes-and-sustainable-A-critical-Bhattacharyya/3a82c99c65eecbe56d2f79e5bdcfd34c4fd84a9a

Di Giovanni, N. (2021). *OLED and LED TV power consumption and electricity cost*. Retrieved October 9, 2021, from https://www.rtings.com/tv/learn/led-oled-power-consumption-and-electricity-cost

Lobner, P. (2017, November 26). *Human activities are contributing to global carbon dioxide levels, but possibly not in the way you think they are*. USA: Lycean Group. Retrieved October 2, 2021, from https://lynceans.org/tag/human-development-index

Wikipedia (2008, November 25). *Human development vs ecological footprint*. Retrieved October 2, 2021, from https://en.wikipedia.org/wiki/File:Human_Development_vs_Ecological_Footprint.jpg

Resilient Urban Climates

▶ Urban Climate Resilience

Resource

▶ Closing the Loop on Local Food Access Through Disaster Management

Resource Effectiveness in and Across Urban Systems

Hadi Arbabi and Ling Min Tan
Department of Civil & Structural Engineering, The University of Sheffield, Sheffield, UK

Synonyms

Carbon emissions; Circular economy; Environment; Linear economy; Resource efficiency; Urban systems

Definition

Resource effectiveness metrics quantify city-wide material and energy efficiency. In a thermodynamic formulation of material and energetic flows within and across cities, effectiveness of resource utilization and conversion reflect the efficiency of the city as a consumer and producer, respectively. These dimensionless metrics are based on the ratios of successfully utilized (effectiveness of utilization) or exported (effectiveness of conversion) biophysical resources to the total resources extracted or imported into the city, all measured in units of exergy.

Introduction

Cities are centers of economic growth but also responsible for ever higher resource consumption and greenhouse gases emissions. Rapid urbanization due to increasing human population and resource-intensive economic activities have drawn concerns for the future of urban sustainability (Seto et al. 2017; Krausmann et al. 2017). Often described as thermodynamically open systems, cities rely on intake of resources and are heavily dependent on flows of resources and energy from their external environment to avoid stagnation (Carmona et al. 2021). This resource reliance raises a key question: how *effectively* do cities consume the resources available to them?

Main Approaches to Measurement

In the 1960s, Wolman undertook a thought experiment to estimate the material needs of a typical American city by assembling per capita resource consumption and waste generation figures using available national statistics (Wolman 1965). Drawing on ecological metaphors, he used the term *urban metabolism* in describing the resource input and waste output sustaining cities. Since Wolman, under the umbrella of "urban metabolism," a variety of methods have been developed. These facilitate the measurement and/or estimation of the quantity of materials and energy imported, exported, stocked, and consumed in cities. Many case studies have been undertaken over the last few decades quantifying resource flows in various cities. Table 1 summarizes the most prominent of these approaches within the academic literature.

Material Flow Analysis (MFA)

Material flow analysis as a method relies on a spatiotemporally defined system boundary across which an assessment of the flows and stocks of resources can be analyzed using a mass/energy conservation approach (Brunner and Rechberger 2004). The method is often used to track resource streams across these spatial and temporal boundaries providing a measure of the demand for resources and pace of development. Such a quantification of the inbound and outbound urban resource streams is meant to contribute towards an understanding of how the urban environmental and economic functions interact with the city's surroundings (Bancheva 2014).

Applications of the MFA can take two overall forms based on the treatment of the data used. In top-down approaches, resource flows are estimated using economy-wide, and often national, inflow and outflow statistics collected annually over a given period. For subnational system boundaries, the national statistic is often treated using appropriate population or economic scaling factors. Depending on the availability of the aggregate statistics, resource streams in MFA studies can be subcategorized based on specific economic activities for which they have been recorded. The resource intensity of these economic sectors relative to their economic output can then inform their resource productivity (Fischer-Kowalski et al. 2011).

The bottom-up methods, on the other hand, approach data collection through survey samples. Surveys allow for constructing inventories of products and tracing the stocks and streams of resources embedded in their life-cycle from extraction to their eventual disposal (Brunner and Rechberger 2004). These inventories often contain quantities of resources normalized against suitable indicators, for example, population, area, gross domestic product, etc., coded as *material intensities*. These average characteristics estimated from the survey samples can then be used to extrapolate for material embedded in flows and stocks across other systems of various sizes but similar compositions. They can also be used in more dynamic formulations of the MFA that use demand-driven models to examine past and future

Resource Effectiveness in and Across Urban Systems, Table 1 Broad summary of methods frequently used in quantifying resource flows in cities

Method	What it includes	What it measures	What it is lacking
Material flow analysis	Total material/energy stocked in cities and crossing urban boundaries	Material input, consumption, production, and waste emissions in kilograms of material	Linear system description, mismatched measurement units, lacks quantification of a differential in quality of resource streams
Input-output analysis	Direct and indirect interactions and interdependencies between different sectors	Impact of shocks, disruptions, and ripple effects throughout the system	Rely on monetary supply and use tables requiring large quantity of survey data to be collected frequently
Ecological network analysis	Input-output analysis with an emphasis on network structure of sectoral interactions and interdependencies	Nature of resource interactions between sectors including control and dependence relationships	
Emergy and exergy approaches	Extends the other methods providing unified units of measurement for embodied energy content or its thermodynamic quality	Unified energy content of resources in Joules	Difficult to estimate or unified conversion factors for waste and socioeconomic resource streams

material use and its effects through time (Augiseau and Barles 2017; Müller 2006; Müller et al. 2014).

MFA is, however, limited by its linear and simplified nature. System conceptualization, particularly in top-down MFA, follows black-box definitions that simply estimate throughflow using the net differences in total inputs and outputs. On the other hand, bottom-up approaches can suffer from the unavailability of the intensive resources that are needed for data collection and the time it often takes. More importantly, the implementations of MFA often ignore differences in quality of the resource streams. Such quality differences are crucial when considering material transformation processes that suffer thermodynamic degradation (Barles 2010; Kennedy et al. 2007).

Input-Output Analysis (I/O)

Input-output analysis techniques date back to Leontief's formulation of an input-output model which is used to analyze industrial interdependencies in an economy (Leontief and Strout 1963; Leontief 1986). The method requires development of input-output tables, which are comprehensive adjacency matrices that contain the flows of intermediate goods and services between industries, that is, industry by industry, and their final sale and purchase within an economy (OECD 1999). Constructing IO tables requires meticulous record keeping. The method is thus most reliably used in studying national and global economic systems where data is more readily available. At lower spatial scales, however, multiregional input-output models can be assembled and have been shown to be useful tools in analyzing trade links across interconnected systems (Bruckner et al. 2012).

Most applications of I/O in urban metabolism rely on monetary flows as a proxy for the physical resources exchanged between industries within the system (Bailey et al. 2004a, b). However, extensions of I/O methodology have been implemented to broaden the applications. Environmental-extended input-output is one such extension that enables evaluation of the associated environmental impacts of industrial exchanges. These include system-wide effects of the extraction of natural resources or the carbon emissions associated with industrial interactions (Kitzes 2013). The main difficulty in using I/O rests with its strict requirements for data and its format. Limitations of data that are recorded and are available at city-level often pose constraints on the applicability of the method.

Ecological Network Analysis (ENA)

Ecological approaches to urban metabolism expand on the implied analogies between urban processes and those of ecosystems. This allows for using methods originally developed to study ecosystems and food webs to model complex interactions among processes in and across cities. In doing so, parallels are made between components and their interactions in urban systems, for example, industrial sectors, and those in food webs and ecological networks, that is, various species. In an ecological paradigm, the overall behavior of the system is dictated by the complex interactions of its internal components (Bai 2016; Huang 1998; Newman 1999).

From a theoretical perspective, the ecological network analyses build on the same core concepts as the I/O analysis developed by Leontief (Hannon 1973). However, unlike I/O, the end goals are not so much in studying the ripple effects through the system but rather in the nature of the relationships between different system components as a function of their direct and in-direct interactions. ENA, additionally, allows examining the dynamics that influence the formation of these resource flows between different components in an urban resource network. Due to the background of the methods, these are often articulated as a function of hierarchical relationships between components mirroring those seen in natural ecological pyramids with apex predators towards which the majority of overall trophic resources flow (Bodini et al. 2012; Li et al. 2018, 2011, 2012).

A number of perspectives can be attained using ENA methods. Functional analysis allows the quantification of the total system throughflow much like I/O (Fath and Borrett 2006; Zhang et al. 2010, 2014). Utility analysis allows allocation of metabolic relationships to any component pairs based on their reciprocal flows (Fath and Patten 1998). These are used to determine whether different industrial sectors, or different cities when studying flows of material between cities rather than within them, exhibit competitive, exploitative, or mutually beneficial resource interactions (Tan et al. 2018). Finally, a control allocation analysis allows quantification of the degree to which different sectors exert control over the resource-input others or how different sectors depend on the resource-output of others (Chen and Chen 2015; Schramski et al. 2007).

ENA has been widely used and is considered an effective assessment toolkit for examining urban and regional resource flows (Chen and Chen 2015; Fan et al. 2017; Li et al. 2012; Yang et al. 2014). Implementing ENA, however, suffers from the same difficulties as I/O. At city-level large amounts of data are required in a similar format as is required by I/O. Additionally, studies that do not directly use monetary I/O tables as a proxy face additional difficulties in collecting granular data. The difficulties lie in sourcing data that is both measured in consistent and comparable units and meets the required format in an I/O table for flows of different resources.

Emergy and Exergy Approaches

Emergy and exergy approaches have been developed as means by which to address the problem of comparability of units used in measuring resource streams of different qualities. Emergy as a method was developed within the ecological tradition. It seeks to unify resource measurement by estimating the total *embodied energy* embedded in a resource stream in terms of the solar energy equivalent needed for its creation (Odum 1988, 1996). In principle, this would provide for directly comparable resource streams in both quantity and quality using a single objective unit of measurement. In reality, however, the method can become severely limiting as a function of prior agreement on and ease by which solar energy conversion factors can be defined and estimated for flows of complex resources outside a strictly ecosystem context (Zhang et al. 2015).

Exergy, on the other hand, has its roots in the thermodynamic principles of irreversibility and work availability. In such contexts, it is defined as "the maximum theoretical useful work obtained if a system is brought into thermodynamic equilibrium with the environment by means of processes in which the system interacts only with this environment." (Sciubba 2001; Sciubba and Wall 2007) As such it retains not only the energetic content of resource streams, but also its thermodynamic quality. While estimation of exergy can face similar difficulties as emergy with regard to conversion factors, *chemical equivalent* conversion factors can be used to provide estimations for different resources based on their required primary energy input (Szargut 1989). In addition to industrial resources, exergy-based approaches have also been adapted for quantification of nonenergetic resources, for example, labor and direct capital flows (Sciubba 1999, 2005). In this way, exergy has been more successful than emergy in providing a unified framework. As a unit of measurement for both quantity and quality of resources, exergy can be integrated and used within the other previously mentioned approaches to urban metabolism (Ayres et al. 1998; Gong and Wall 2001; Lozano and Valero 1993; Finnveden and Östlund 1997).

As with emergy, exergetic approaches can still face difficulty when compiling data for complex systems that encapsulate a large variety of physical and energetic resources. These include the recurring concerns about the appropriateness of the conversion factors used when converting flows into an exergetic framework. More specifically, both emergy and exergetic accounting are still lacking a unified approach to the quantification of waste products (Zhang 2013).

Mathematical Description

Answering the question of how effectively cities do consume the resources available to them requires a mixed use of the reviewed approaches. Particularly helpful is the ability of the exergy-based formulation to keep track of quantity and quality of resource streams. *Exergy destruction*, as is often used in the context of describing urban processes, expresses the usefully dissipated part of a resource stream. This is in contrast with the wasted portions of flow streams that due to thermodynamic irreversibilities are not used nor can be directly recovered (Nicolis and Prigogine 1977). Resource effectiveness of urban systems can then be evaluated through comparisons of the destroyed and wasted resources to the total inflows of resources and energy into the system.

A Network Model of Urban Systems

System abstractions used in MFA and I/O to represent cities can be thought of as a directed network of N nodes and E edges. In such networks, nodes can be representative of industrial sectors within a city with edges taking the place of monetary, physical, or energetic flows between them. They can, more broadly, be stand-ins for any such similar roles depending on the context of the domain of study, for example, regional resource flows or international monetary interactions.

For each node i, F_{ij} represents the resource flow passed on from it to node j with Δ_i representing the resource in/outflows that cross the boundary of the overall system, for example, the city's boundary. X_i^U, constitutes part of F_{ij} that is successfully utilized at i, for example, exergy destroyed. Meanwhile, X_i^W denotes the portion lost to thermodynamic irreversibilities. Disutility factors, λ and ϕ, account for process efficiencies that dictate how successful a process is in using available resources and maintain the conservation of energy across the model. More specifically, λ controls the amount of exergy successfully *destroyed,* and ϕ reflects the portion that is irrecoverably lost to waste for each resource stream F_{ij}. In the majority of model formulations, these processes and their efficiencies are characteristics of the node inside which they take place (Arbabi et al. 2020; Tan et al. 2019). Figure 1 shows a schematic representation of such network arrangements.

Expanding this formulation across all nodes and edges would give the system-wise overall resources balance as

$$\sum_i^N \sum_{j,j\neq i}^N F_{ij} + \sum_i^N X_i^U + \sum_i^N X_i^W - \sum_i^N \sum_{j,j\neq i}^N F_{ji} + \sum_i^N \Delta_i = 0.$$

Overall Resource Effectiveness and Balance

Resource effectiveness in each process and across the system as a whole depends not only on the efficiency by which the transformations are performed, that is, combined effects of λ and ϕ, but also the intended purpose of a sector. Urban systems and their processes can exhibit different behaviors and qualities when regarded as consumers of resources or their producers/transformers. For cities, the two aspects as consumers and conversion engines are captured by effectiveness of resource utilization, $\epsilon_U := \frac{\sum_i^N X_i^U}{\sum_i^N \Delta_i^+}$, and

R

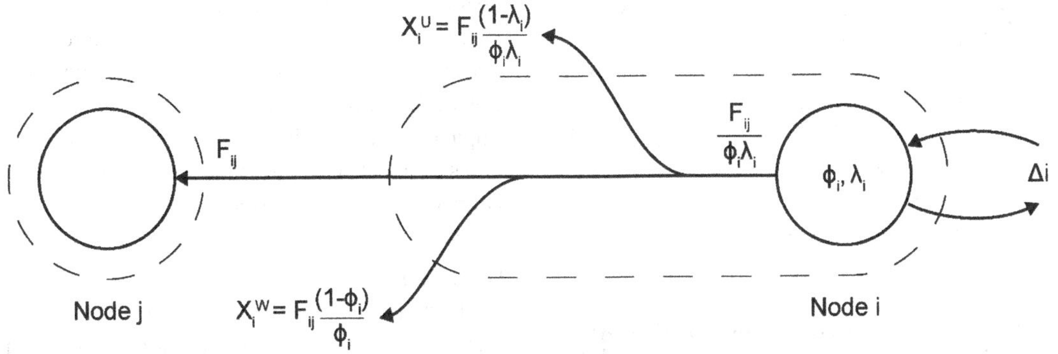

Resource Effectiveness in and Across Urban Systems, Fig. 1 Schematic showing a node pair and the resource flow between them broken down in exergetic terms to its utilized, wasted, and exported components. (Adapted from Arbabi et al. 2020)

effectiveness of resource conversion, $\epsilon_C := \frac{\sum_i^N \Delta_i^-}{\sum_i^N \Delta_i^+}$, where $\sum \Delta^+$ denotes the incoming resources imported into the city and $\sum \Delta^-$ represents those that have been exported outside the city for use in other cities or countries.

Both metrics are dimensionless indicators of performance that measure either the successful exergy destruction or the total exergy of useful product export, inclusive of the capital funds generated in a socially extended framework, per total urban resource requirement. Close examination of the energy conservation equation reveals that the ability of cities to be efficiently self-sufficient in their consumption, that is, values of ϵ_U closer to unity, and their ability to be efficient producers, that is, ϵ_C closer to unity, are at odds. This trade-off between the two aspects of cities is demonstrated in Fig. 2.

The overall magnitude of resource effectiveness of cities can then be captured as $R := \sqrt{\epsilon_U^2 + \epsilon_C^2}$ measuring both producer and consumer capabilities. As such, its value provides a system-wide performance metric for using and transforming resources available. The tension between the consumer/producer behavior of the overall system can be captured as angle $\theta := arctan\left(\frac{\epsilon_U}{\epsilon_C}\right)$, where the system is more dominantly a producer with $\theta < 45^\circ$ and is exhibiting more dominant consumer tendencies with $\theta > 45^\circ$.

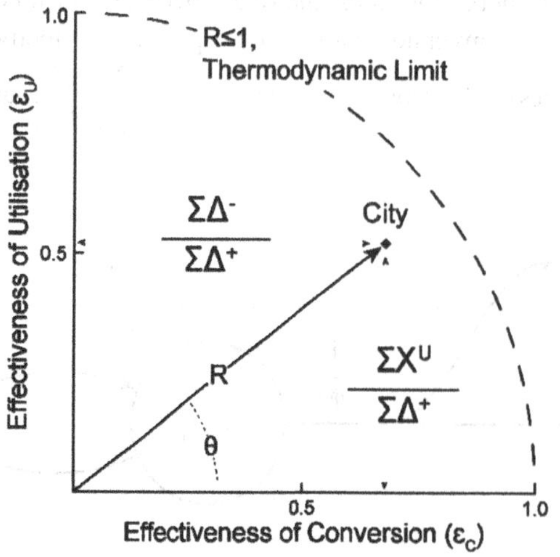

Resource Effectiveness in and Across Urban Systems, Fig. 2 Effectiveness diagram showing balance between resource utilization and conversion, the overall resource effectiveness, R, the overall effectiveness balance, θ, and the city's thermodynamic limit for conservation of energy

Application of Resource Effectiveness in an Urban System

The main use of resource effectiveness metrics is to provide a clear understanding of the role of various economic sectors in cities and explore how this affects their needs and prospects for future growth. Such an understanding of how effective cities are in using their resources facilitates a decentralization of urban resource policy and a focus on sector-specific economic strategies and urban planning informed by the unique urban characteristics of each city (Tan et al. 2021).

Additionally, open system network models that underlie effectiveness assessment can be expanded to include nested representation of sectors in cities and their interactions across cities. Multiscale approaches, as shown in Fig. 3, would enable a thorough investigation of the cross-sector relationships and interdependencies between cities to identify the key channels of resource intake into the system and the external risks the system is exposed to. These range from disruptions due to climate change and sea level rise to changes to the infrastructure, for example, transport, facilitating resource flows. For instance, identifying the possible hazards causing disruptions to resource connections of the urban network can suggest suitable precautionary actions to secure the resource linkages in the supply chains and sustain proper functions of the urban system. Insights of this nature impact regulatory decisions on how sectors and cities connect with one another and the resource connectivity in and across cities.

Data Requirements

Understanding cities and measuring how effective they are at resource consumption are data intensive. Modeling cities as open systems within an exergetic framework that allows estimation of resource effectiveness metrics requires a minimum of the following data types to be available beforehand or capabilities in estimating such information from other available datasets.

Minimum data input requirements are:

- Records of cross-boundary physical resource imports and exports in terms of their mass to estimate overall system boundary flow
- Records of virgin resource extraction through local production activities in terms of their mass or energetic content
- Monetary input-output tables and supply-and-use tables detailing the intensity of interactions between economic sectors
- Employment and labor data for industrial sectors in terms of number of employees, total hours worked, and wages
- Greenhouse gases emission intensity factors for industrial product output and domestic energy use

A UK Example

The example here outlines the resource effectiveness of the 38 functional urban areas building up the urban system in Great Britain. Figure 4 shows both the estimated values of ϵ_U and ϵ_C and the trajectories of R and θ between 2000 and 2010. The widespread tendency for cities to exhibit consumer-like behaviour is clear particularly on panels B and C.

Finally, while effectiveness metrics are informative for management of individual cities, they also provide a means for the assessment of wider urban networks as a whole. Examination of clustering and similarity patterns in the resource-use behaviors across the urban system enables identification of common characteristics that can be addressed in system-wide resource allocation planning. For the system of cities in Great Britain as an example, the individual temporal trajectories in Fig. 4 underlie five fairly distinct consumer/producer characteristics, as shown in Fig. 5.

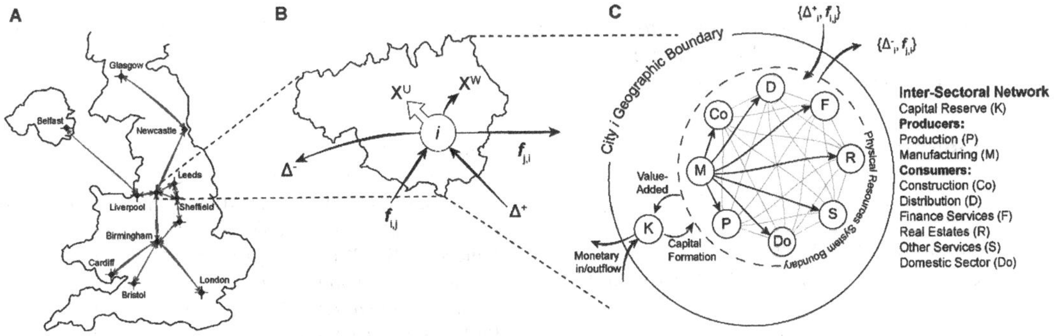

Resource Effectiveness in and Across Urban Systems, Fig. 3 Schematic of an inter-urban flow network (**a**), aggregated flows over a city (**b**), and detailed inter-sectoral physical and financial flows within a city, with those of manufacturing highlighted (**c**) as a nested multiscale resource model

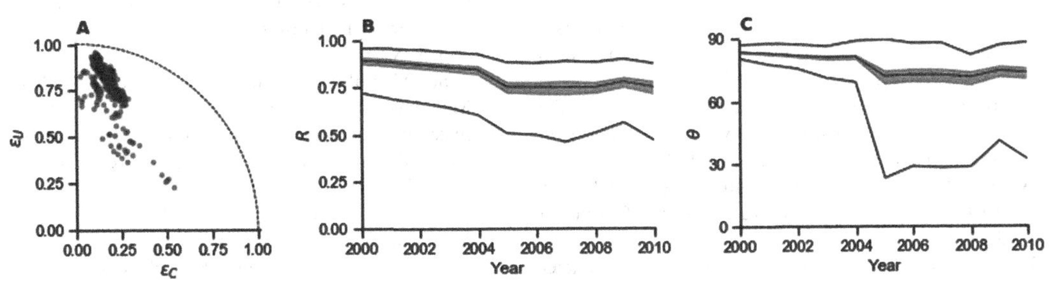

Resource Effectiveness in and Across Urban Systems, Fig. 4 Annual estimates of the effectiveness of resource utilization and conversion for the period 2000–2010 (**a**), annual trend showing mean and its 95% CI (shaded area), minimum, and maximum of the overall resource effectiveness (**b**), and resource balance (**c**). (Adapted from Tan et al. 2021)

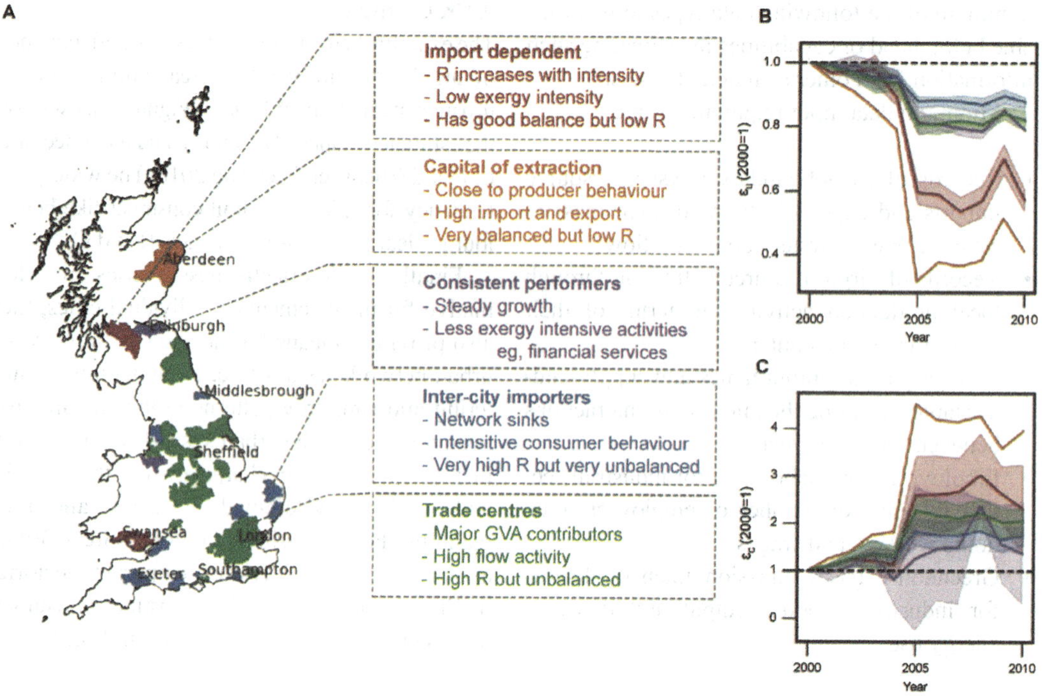

Resource Effectiveness in and Across Urban Systems, Fig. 5 Map of urban clusters by resource effectiveness behavior (**a**) and indexed variations of mean ϵ_U (**b**), and ϵ_C (**c**) for each cluster with their standard deviations. (Adapted from Tan et al. 2021)

Cross-References

- ▶ Circular Cities
- ▶ Circular Economy Cities
- ▶ Sustainable Development Goals from an Urban Perspective

References

Arbabi, H., et al. (2020). On the use of random graphs in analysing resource utilization in urban systems. *Royal Society Open Science, 7*, 200087.

Augiseau, V., & Barles, S. (2017). Studying construction materials flows and stock: A review. *Resources, Conservation and Recycling, 123*, 153–164.

Ayres, R. U., Ayres, L. W., & Martinás, K. (1998). Exergy, waste accounting, and life-cycle analysis. *Energy, 23*, 355–363.

Bai, X. (2016). Eight energy and material flow characteristics of urban ecosystems. *Ambio, 45*, 819–830.

Bailey, R., Allen, J. K., & Bras, B. (2004a). Applying ecological input-output flow analysis to material flows in industrial systems: Part I: Tracing flows. *Journal of Industrial Ecology, 8*, 45–68.

Bailey, R., Bras, B., & Allen, J. K. (2004b). Applying ecological input-output flow analysis to material flows in industrial systems: Part II: Flow metrics. *Journal of Industrial Ecology, 8*, 69–91.

Bancheva, S. (2014). *Integrating the concept of urban metabolism into planning of sustainable cities: Analysis of the Eco^2 cities initiative*. DPU working paper 36.

Barles, S. (2010). Society, energy and materials: The contribution of urban metabolism studies to sustainable urban development issues. *Journal of Environmental Planning and Management, 53*, 439–455.

Bodini, A., Bondavalli, C., & Allesina, S. (2012). Cities as ecosystems: Growth, development and implications for sustainability. *Ecological Modelling, 245*, 185–198.

Bruckner, M., Giljum, S., Lutz, C., & Wiebe, K. S. (2012). Materials embodied in international trade – Global material extraction and consumption between 1995 and 2005. *Global Environmental Change, 22*, 568–576.

Brunner, P. H., & Rechberger, H. (2004). *Practical handbook of material flow analysis*. Boca Raton: CRC/Lewis.

Carmona, L. G., Whiting, K., Wiedenhofer, D., Krausmann, F., & Sousa, T. (2021). Resource use and economic development: An exergy perspective on energy and material flows and stocks from 1900 to 2010. *Resources, Conservation and Recycling, 165*, 105226.

Chen, S., & Chen, B. (2015). Urban energy consumption: Different insights from energy flow analysis, input–output analysis and ecological network analysis. *Applied Energy, 138*, 99–107.

Fan, Y., Qiao, Q., & Chen, W. (2017). Unified network analysis on the organization of an industrial metabolic system. *Resources, Conservation and Recycling, 125*, 9–16.

Fath, B., & Borrett, S. (2006). A MATLAB® function for network environ analysis. *Environmental Modelling & Software, 21*, 375–405.

Fath, B. D., & Patten, B. C. (1998). Network synergism: Emergence of positive relations in ecological systems. *Ecological Modelling, 107*, 127–143.

Finnveden, G., & Östlund, P. (1997). Exergies of natural resources in life-cycle assessment and other applications. *Energy, 22*, 923–931.

Fischer-Kowalski, M., et al. (2011). Methodology and indicators of economy-wide material flow accounting. *Journal of Industrial Ecology, 15*, 855–876.

Gong, M., & Wall, G. (2001). On exergy and sustainable development – Part 2: Indicators and methods. *Exergy, An International Journal, 1*, 217–233.

Hannon, B. (1973). The structure of ecosystems. *Journal of Theoretical Biology, 41*, 535–546.

Huang, S.-L. (1998). Urban ecosystems, energetic hierarchies, and ecological economics of Taipei metropolis. *Journal of Environmental Management, 52*, 39–51.

Kennedy, C., Cuddihy, J., & Engel-Yan, J. (2007). The changing metabolism of cities. *Journal of Industrial Ecology, 11*, 43–59.

Kitzes, J. (2013). An introduction to environmentally-extended input-output analysis. *Resources, 2*, 489–503.

Krausmann, F., et al. (2017). Global socioeconomic material stocks rise 23-fold over the 20th century and require half of annual resource use. *Proceedings of the National Academy of Sciences of the United States of America, 114*, 1880–1885.

Leontief, W. (1986). *Input-output economics*. Oxford: Oxford University Press.

Leontief, W., & Strout, A. (1963). Multiregional input-output analysis. In *Structural interdependence and economic development* (pp. 119–150). London: Palgrave Macmillan

Li, S., Zhang, Y., Yang, Z., Liu, H., & Zhang, J. (2012). Ecological relationship analysis of the urban metabolic system of Beijing, China. *Environmental Pollution, 170*, 169–176.

Li, J., Huang, G., & Liu, L. (2018). Ecological network analysis for urban metabolism and carbon emissions based on input-output tables: A case study of Guangdong province. *Ecological Modelling, 383*, 118–126.

Liu, G. Y., Yang, Z. F., Chen, B., & Zhang, Y. (2011). Ecological network determination of sectoral linkages, utility relations and structural characteristics on urban ecological economic system. *Ecological Modelling, 222*, 2825–2834.

Liu, G. Y., Yang, Z. F., Su, M. R., & Chen, B. (2012). The structure, evolution and sustainability of urban socioeconomic system. *Ecological Informatics, 10*, 2–9.

Lozano, M. A., & Valero, A. (1993). Theory of the exergetic cost. *Energy, 18*, 939–960.

Müller, D. B. (2006). Stock dynamics for forecasting material flows – Case study for housing in the Netherlands. *Ecological Economics, 59*(1), 142–156.

Müller, E., Hilty, L. M., Widmer, R., Schluep, M., & Faulstich, M. (2014). Modeling metal stocks and flows: A review of dynamic material flow analysis methods. *Environmental Science & Technology, 48*, 2102–2113.

Newman, P. (1999). Sustainability and cities: Extending the metabolism model. *Landscape and Urban Planning, 44*, 219–226.

Nicolis, G., & Prigogine, I. (1977). *Self-organization in nonequilibrium systems: From dissipative structures to order through fluctuations*. New York: Wiley-Blackwell.

Odum, H. T. (1988). Self-organization, transformity, and information. *Science, 242*, 1132–1139.

Odum, H. (1996). *Environmental accounting: Emergy and environmental decision making*. New York: Wiley.

OECD. (1999). *The OECD input-output database*. https://www.oecd.org/industry/ind/2673344.pdf

Schramski, J. R., et al. (2007). Indirect effects and distributed control in ecosystems: Distributed control in the environ networks of a seven-compartment model of nitrogen flow in the Neuse River Estuary, USA – Time series analysis. *Ecological Modelling, 206*, 18–30.

Sciubba, E. (1999). Exergy as a direct measure of environmental impact. *Advanced Energy Systems Division (Publication) AES, 39*, 573–581.

Sciubba, E. (2001). Beyond thermoeconomics? The concept of extended exergy accounting and its application to the analysis and design of thermal systems. *Exergy, An International Journal, 1*, 68–84.

Sciubba, E. (2005). From engineering economics to extended exergy accounting: A possible path from monetary to resource-based costing. *Journal of Industrial Ecology, 8*, 19–40.

Sciubba, E., & Wall, G. (2007). A brief commented history of exergy from the beginnings to 2004. *International Journal of Thermodynamics, 10*, 1–26.

Seto, K. C., Golden, J. S., Alberti, M., & Turner, B. L. (2017). Sustainability in an urbanizing planet. *Proceedings of the National Academy of Sciences, 114*, 8935–8938.

Szargut, J. (1989). Chemical exergies of the elements. *Applied Energy, 32*, 269–286.

Tan, L. M., Arbabi, H., Li, Q., Sheng, Y., Densley Tingley, D., Mayfield, M., & Coca, D. (2018). Ecological network analysis on intra-city metabolism of functional urban areas in England and Wales. *Resources, Conservation and Recycling, 138*, 172–182. ISSN 0921-3449, https://doi.org/10.1016/j.resconrec.2018.06.010.

Tan, L. M., Arbabi, H., Brockway, P. E., Densley Tingley, D., & Mayfield, M. (2019). An ecological-thermodynamic approach to urban metabolism: Measuring resource utilization with open system network effectiveness analysis. *Applied Energy, 254*, 113618.

R

Tan, L. M., Arbabi, H., Densley Tingley, D., Brockway, P. E., & Mayfield, M. (2021). Mapping resource effectiveness across urban systems. *npj Urban Sustainability, 1*, 1–14.

Wolman, A. (1965). The metabolism of cities. *Scientific American, 213*, 178–190.

Yang, D., Kao, W. T. M., Zhang, G., & Zhang, N. (2014). Evaluating spatiotemporal differences and sustainability of Xiamen urban metabolism using emergy synthesis. *Ecological Modelling, 272*, 40–48.

Zhang, Y. (2013). Urban metabolism: A review of research methodologies. *Environmental Pollution, 178*, 463–473.

Zhang, Y., Yang, Z., Fath, B. D., & Li, S. (2010). Ecological network analysis of an urban energy metabolic system: Model development, and a case study of four Chinese cities. *Ecological Modelling, 221*, 1865–1879.

Zhang, Y., Zheng, H., & Fath, B. D. (2014). Analysis of the energy metabolism of urban socioeconomic sectors and the associated carbon footprints: Model development and a case study for Beijing. *Energy Policy, 73*, 540–551.

Zhang, Y., Yang, Z., & Yu, X. (2015). Urban metabolism: A review of current knowledge and directions for future study. *Environmental Science & Technology, 49*, 11247–11263.

Resource Efficiency

► Circular Economy Cities

► Resource Effectiveness in and Across Urban Systems

Resource Recovery from Human Excreta in Urban and Regional Settlements

Jacqueline Thomas[1] and Moritz Gold[2]

[1]School of Civil Engineering, The University of Sydney, Sydney, NSW, Australia

[2]Sustainable Food Processing Laboratory, ETH Zurich, Zurich, Switzerland

Definitions

Biosolids	The solid component of treated or partially treated wastewater or fecal sludge
Blackwater	Mixture of human excreta and flush water along with anal cleansing water (if water is used for cleansing) and/or dry cleansing material (e.g., toilet paper), generally from a toilet
Effluent	The liquid component of treated or partially treated human excreta
Greywater	Water that contains human hygiene waste (e.g., shower water and dish washing water) but not human excreta and is considered waste
Fecal sludge	Human excreta that is collected and contained in an on-site sanitation system
Fecal sludge management	The chain of processes that occur when managing fecal sludge from on-site systems, including deposition, containment, emptying, transport, and treatment
Human excreta	Feces and urine excreted by humans
Open defecation	When in the absence of a toilet humans defecate in the open environment
Pit latrines	Type of on-site sanitation system where human excreta is deposited in a hole the ground
Septic tanks	Type of on-site sanitation system that consists of constructed tank system that is designed to treat human excreta on-site
Sewer system	A centralized connection of pipes that collects and transports wastewater to a treatment plant
Sewage	Wastewater that is collected by a sewer system
Sanitation	All the systems that capture, contain, treat, and dispose of human excreta
Toilet front-end	The part of toilet where people deposit their excreta (pedestal toilet, swat toilet, or hole)
Toilet back-end	The part of toilet where human excreta is contained, transferred, and/or treated

Wastewater All the components of water that are released as waste from a household, predominantly consisting of greywater and blackwater. Captures both sewer and onsite sanitation systems

Introduction

Sanitation Challenges

It is estimated that 4.2 billion people do not have access to a safely managed sanitation systems (WHO and UNICEF 2019). Sanitation systems that are not safely managed leads to exposure to human excreta, which contributes to an estimated 829,000 deaths per annum attributed to diarrheal diseases and helminth infections (Pruss-Ustun et al. 2019) and pollution of aquatic ecosystems (Wear et al. 2021).

Despite the health and environment threats from unsafely managed human excreta, there has been limited development in the sanitation technology needed to contain, treat, transport, dispose, or reuse excreta from human settlements. Looking back through history, there are only four major sanitation technology jumps made by human civilization:

1. Approximately 1000 BC in ancient Greece, communal latrines were first used;
2. By 100 BC, the Romans had developed centralized basic sewer networks that transported human excreta from communal baths and private houses (Wald 2016);
3. In England in 1775, flush water seal latrines were first patented which placed an S-shaped bend between the toilet the sewer network to block smell and flies (Goodyer 2012).
4. In the late nineteenth century, wastewater treatment plants were designed to help reduced fecal-oral disease transmission (such as Cholera) by partially treating the combined blackwater and greywater (sewage) transported by sewers before it was discharged to water bodies.

The latrines and sewer networks used today are fundamentally based on the original eighteenth and nineteenth century designs. Further, the most commonly used wastewater treatment plant technologies are limited and designed to simply remove large solids and floating fats and oils; termed primary wastewater treatment. Oceans and bodies of surface water, including rivers and lakes, are then used to dispose of the treated sewage with the intention to dilute it.

The issue of piped sewer systems is only a fraction of the sanitation problem. Globally 2.7 billion people do not have access to a centralized sanitation system and in-stead use on-site sanitation systems such as septic tanks and pit latrines (Strande et al. 2014). On-site systems are predominantly used by rural populations in higher income countries and by the majority of people in lower income countries. The front-end toilet is most commonly a pedestal, squat plates or a hole in the ground and depending on water availability cistern flush, pour-flush or dry system is common (Fig. 1). The most common back-end infrastructure for an on-site system is either a pit (partially lined or unlined), soak away pit or a septic tank. The human excreta with/without water, cleansing material and solid wastc (e.g., rubbish) collected in on-site sanitation systems is called fecal sludge. The treatment chain for faucal sludge is referred to as fecal sludge management. Fecal sludge differs from sewage as it has a much higher portion of total solids, which is a by-product of using less water in the on-site systems and water leaching from the pit or pervious tank into the surrounding area. Recently published global models calculate that 48% of wastewater, from both sewer and on-site systems, are disposed of directly to the environment without any treatment (Jones et al. 2021).

For an on-site sanitation system to be safely managed the faucal sludge must be contained or treated on-site or when the system is full it needs to be removed, transported, and treated off-site. Given that generally lower income urban and more rural communities rely on on-site sanitation systems, there are significant financial and capacity barriers to constructing safe back-ends that effectively contain or treat fecal sludge on-site. Further, the options to safely extract the faucal sludge, transport it, and treat it off-site are often severely inadequate due to prohibitive cost, lack

Resource Recovery from Human Excreta in Urban and Regional Settlements, Fig. 1 Three common toilet front-end types: (**a**) Cistern flush toilet from Samoa, (**b**) Ceramic squat plate pour flush latrine with metal bucket from India, and (**c**) A hole outlined with bricks in the earthen cover of a pit latrine with a broken plastic bucket as a cover. (Copyright – original images taken by Jacqueline Thomas)

of a trained workforce, and very limited treatment facilities. To reduce the fillings rates and extend the life of back-end systems there are significant incentives to build systems that allow the liquid fractions of fecal sludge to rapidly leach into the surrounding environment. For many households who have an on-site system, the only realistic solutions when the system does become full are to either:

1. Unsafely empty the back-end by hand and bury the extracted fecal sludge in ditches or pits close to the original toilet or dump the sludge into drains or surface water bodies that are within close proximity
2. Abandon the old system and build a new one, which leaves the accumulated fecal sludge to leach further, or
3. Start sharing another toilet or practicing open defecation if the back-end cannot be emptied or a new one cannot be built due to cost or space restrictions.

There is an estimated 673 million people globally who practice open defecation (WHO and UNICEF 2019) and while progress has been made, more effort is needed to eliminate this practice in human settlements. Open defecation areas, uncontained fecal sludge and unsafely disposed fecal sludge can leach into surrounding soil contaminating groundwater and the environment which has significant human and environmental health consequences.

Human Health Impacts

Human excreta carries very high-loads of micro-organisms and if an individual is carrying an infection then disease causing micro-organisms (pathogen) are excreted in exceptionally high quantities. These microbial pathogens from many individuals are concentrated in sewage, fecal sludge and open defecation areas. Common fecal-oral diseases which result from pathogen transfer from human excreta include:

1. Viral infections due to norovirus and Hepatitis A
2. Bacterial infections due to *Shigella*, *Salmonella typhi* (Typhoid Fever), and *Vibrio cholerae* (Cholera)
3. Protozoal infections due to *Giardia* and *Cryptosporidium*

4. Helminth infections due to hookworm and roundworm

Humans are exposed to excreta via four main pathways:

1. Either via direct hand contact with human feces via unhygienic sanitation practices.
2. Via ingestion as they drink or bath in water contaminated with human excreta.
3. Grow their crops using human excreta or planting crops in sanitation leach zones.
4. From flies that land on human excreta and then transfer it to food or other surfaces.

Even in high income countries, infections with fecal-oral diseases are common. For example in the United States of America (USA), the Centre for Disease Control and Prevention estimates that 1 in 44 people get sick from a waterborne disease every year and in the year 2014 the top causes of infection were norovirus (1.3 million infections) and Giardiasis (415,000 infections) (Collier et al. 2021).

Environmental Impacts

There is a growing understanding of the environmental impacts of unsafely managed sanitation for aquatic ecosystems. Human excreta has high levels of nutrients, such as nitrogen (N), phosphorous (P) and potassium (K) and can also contain heavy metals and micro-pollutants such as pharmaceuticals like synthetic estrogen (used in the oral contraceptive pill). Oceans and surface water bodies such as rivers, streams and lakes are frequently are polluted with large quantities of human excreta due to direct sewer discharging, illegal dumping or leaching from unsafe sanitation systems. In low income countries the lack of regulation enforcement, limited financial resources, and the high number of unsafe sanitation systems results in very high levels of excreta entering surface water bodies (Fig. 2). But even in developed countries there are significant challenges, for example, in the United States of America is it estimated that annually 4.5 million m^3 of untreated wastewater from sewer systems is discharged directly into rivers due to overloaded or by-passed wastewater treatment plants (Wear et al. 2021).

Coral reefs are one type of ecosystem that is highly sensitive to changes in nutrients and the presence of pollutants. Research on coral reefs has revealed the negative impacts of human excreta on coral growth and reproduction rates, along with increasing rates coral diseases (Wear and Thurber 2015). Similar impacts are known for other fresh-water aquatic ecosystems.

Benefits of Resource Recovery

Given that human excreta is high in valuable nutrients and energy, there exists a valuable opportunity to re-capture those components and beneficially reuse them. There is some evidence that creating a value chain, which transforms excreta from a waste product to a product of value, can contribute to sanitation improvements (Diener et al. 2014).

Production of Agricultural Inputs

Human excreta and respective wastewaters has varying concentrations of three important nutrients; nitrogen (N) and phosphorous (P) and potassium (K). The nutrient composition of human excreta is closely linked to diet, with higher levels of nitrogen seen in people who consume more protein from animal sources. In agriculture crops grow best when there is an optimal nutrient ratio of NPK. Due to intensive cropping practices and using sub-optimal agricultural land, many soils around the world are deficient in NPK and have reduced organic carbon concentrations. To ensure agricultural productivity, farmers routinely apply an NPK fertilizer to plants. However, the sources of NPK and in particular P are globally limited, with some estimations that the world will run out of bio-available P within 50–100 years (Cordell et al. 2009). In low income countries, use of commercial NPK fertilizers is already too expensive for many small holder farmers. The reuse of properly treated human excreta, where the NPK and carbon elements are maintained has been demonstrated to both improve soil health and crop productivity. The methods for using human

Resource Recovery from Human Excreta in Urban and Regional Settlements, Fig. 2 Excreta and solid waste polluted rivers and streams are a common site in many countries: (**a**) a river in India which is not flowing with green algae growing in the stagnant water pools indicating high nutrient levels, (**b**) a main river flowing through Kathmandu, Nepal with high excreta loads and (**c**) a stream flowing through an informal community in Dar es Salaam, Tanzania carrying high excreta loads to the coastline and coral reefs. (Copyright – original images taken by Jacqueline Thomas)

excreta in agriculture has been practiced in some cultures for centuries but is rejected as unsafe in many other cultures. An example of a culture that has long used untreated human excreta in agriculture is Vietnam in South East Asia, however these agricultural practices do have measurable negative health impacts with elevated levels of protozoal soil transmitted helminth infections (Pham-Duc et al. 2013). In addition to direct application of treated human excreta another method to produce agricultural inputs is the conversion of human excreta into protein-rich feeds for farmed animals. The primary means of conversion is feeding mixed human excreta and organic waste to Black Solider Fly (BSF) larvae which naturally feed on decaying organic matter (see case studies section). As the world's population continues to climb and there is ever increasing pressure on food availability and natural resources, it is of paramount importance the reuse of human excreta is adopted to create a circular economy for nutrients.

Production of Energy Inputs

There is now clear global acceptance of climate change and motivation to tackle the causes. The uncontrolled decomposition of unsafely managed human excreta can contribute to climate change by releasing potent greenhouse gases such as methane and the disruption of aquatic ecosystems. In order to achieve the ambitious carbon dioxide reduction targets, alternative bio-based energy sources can be produced from human excreta. Human excreta has a high calorific value of 17 MJ/kg dry solids and can be used as an energy source via two established conversion pathways; (1) drying and combustion/pyrolysis and (2) biological conversation to biogas (Diener et al. 2014). When sewage or fecal sludge is dried to produced dried biosolids it can be used as a combustible solid fuel in industrial kilns (e.g., cement industry) (Gold et al. 2017). In many low-income countries, wood and charcoal are main sources of cooking fuel used by households and this drives rapid deforestation. Dried fecal sludge has been successfully converted to biochar briquettes that can replace charcoal and wood used for cooking (Gold et al. 2018). The second pathway involves harvesting energy from human excreta by using microbiological digestion in the absence of oxygen (anaerobic digestions) to form combustible biogas for cooking. Given human excreta's high energy profiles and the necessity to adopt alternative cleaner energy sources, there exists an opportunity to upscale human waste into energy programs. The benefits of reusing human excreta from urban and regional settlements are clear; however, there remains various challenges to overcome in order to optimize the appropriate resource recovery technologies and ensure safe reuse.

Resource Recovery Processes

There are a number of different treatment technologies for human excreta. The proceeding section will describe six common treatment technologies where resources can be directly recovered to produce agricultural and energy inputs.

Conventional Wastewater Treatment Plants

In most high income countries sewer networks transport sewage for urban or regional populations of settlements. In most cases, these sewers are also connected to a conventional wastewater treatment plant. Treatment plants have four main levels of treatment:

1. Pre-treatment – removes gross solids by screening or settling (e.g., screen and grit chamber),
2. Primary treatment – removes gross solids plus readily settleable solids by primary sedimentation (e.g., Imhoff tank),
3. Secondary treatment – removes most solids and biological oxygen demand by biological and/or chemically assisted treatment (e.g., activated sludge treatment),
4. Tertiary (advanced) treatment – removes nutrients and microorganisms by biological processes and/or chemical precipitation along with disinfection using ultraviolet light and/or chlorination. It can also include activated carbon adsorption of organics such as pharmaceuticals.

Each treatment level increases efficiencies of removal and higher treatment levels require more complex systems with larger energy and operational demands (Fig. 3). In many countries, the minimum treatment level required by regulations is pre-treatment followed by primary treatment. Wastewater treatment plants produce large quantities of treated effluent which are generally disposed of into surrounding water bodies, like lakes and oceans. There are a number of different configurations and designs for secondary and tertiary treatment systems depending on the quality and quantity of sewage and the effluent discharge requirements. Some purpose designed advanced tertiary wastewater treatment can treat wastewater

a) Sewage pre-treatment - Nepal
b) Primary treatment tanks - South Africa
c) Secondary treatment aeration - Samoa

Resource Recovery from Human Excreta in Urban and Regional Settlements, Fig. 3 Conventional centralized wastewater treatment plants: (**a**) the sewage pre-treatment stage of a wastewater treatment plant in Nepal, (**b**) primary treatment clarifying tanks in South Africa, (**c**) the secondary treatment aeration stage of a wastewater treatment plant in Samoa. (Copyright – original images taken by Jacqueline Thomas) to sufficient safety to be reused as recycled water. Urban areas with recycled water dual system, generally use the recycled water for non-potable purposes such as flushing toilets or watering gardens. If the highest levels of wastewater treatment are applied then it is possible to produce treated effluent that is safe enough to drink. In water-vulnerable countries, such as Singapore, wastewater is recycled and directly used for drinking water. Each level of wastewater treatment produces varying qualities and quantities of sludge. Sludge needs to be disposed of or further treated and re-used. Examples of treatment include solar drying and anaerobic digestion. Commonly treated sludge results in biosolids that can be used for non-food agricultural applications, such as in forestry plantations. However, the rates of sludge reuse depend entirely on local policies, regulations and demand for the product. It is important to note that wastewater treatment plants are designed to only treat sewage, which has a low portion of total solids as the human excreta is heavily diluted by large quantities of greywater. A conventional wastewater treatment plant cannot be used to treat the high total solid waste streams of fecal sludge from on-site sanitation systems (pit latrines and septic tanks).

Anaerobic Biological Digestion and Biogas Capture

Anaerobic biological digestion utilizes natural microbial processes to digest wastewater in the absence of oxygen. The process of anaerobic digestion is a commonly used in many treatment systems including anaerobic wastewater treatment ponds and septic tanks. If the anaerobic digestion takes place in a sealed digester or contained space this allows for the capture of the biogas that is produced by the process. Anaerobic digesters are commonly simple gravity fed domes or tanks and can be designed to service a single household or be part of large centralized wastewater treatment plans (Fig. 4). Biogas is made up of 50–70% methane gas and is valuable energy source as it is highly combustible and can be used for cooking, bottled or combusted to produce heat or electricity (Bond and Templeton 2011). During biological digestion a large fraction of both the nutrients and carbon in the raw material is consumed and converted by the microbial populations. The benefits of this process are that it produces effluent and sludge that has much lower nutrient profiles allowing safer release into the environment. However, for the purposes of reuse the final effluent and digested sludge have

a) Anaerobic biodigester construction - Tanzania
b) Urine diversion - Tanzania
c) Biochar production - Tanzania

Resource Recovery from Human Excreta in Urban and Regional Settlements, Fig. 4 Decentralized systems for human excreta treatment and resource recovery in Tanzania: (**a**) a 10 m^3 anaerobic biodigester (biogas dome) under construction, (**b**) the back-end of a household urine diverting toilet with urine collecting in the bucket and feces composting in the concrete vaults, (**c**) a simple metal kiln to produce biochar from dried fecal sludge combined with corn husk. (Copyright – original images taken by Jacqueline Thomas)

much lower nutrient and energy profiles but is still classed as valuable. From a safety perspective, the biological digestion does not significantly reduce microbiological pathogen loads. Hence, there are also restrictions and limitations on how the final effluent and sludge can be used in agriculture. Generally, treated sludge and effluent meet the guideline values to be reused in agricultural practices where it is applied to non-edible plants (e.g., trees) or where there will no contact with the edible parts of crops (e.g., sub-surface irrigation of banana palms). Further treatment of sludge (e.g., by compositing) or effluent (e.g., by wastewater treatment) can enable use for food production and higher levels of water reuse.

Urine Diversion and Composting

When looking at human excreta, urine and feces are generally combined in the sanitation system, but the two have very different properties that warrant handling them separately. Firstly, urine as it is filtered by the kidneys is generally free from pathogens, as long as the person excreting does not have any type of endocrine infection. Urine contains the highest portions of available nutrients as N and P (Simha and Ganesapillai 2017). If handled correctly, urine can be diluted and used with minimal treatment directly as a liquid fertilizer in agriculture. Secondly, feces always contain high loads of potentially infectious microbes and always requires treatment prior to reuse. When urine and feces are combined, the high nitrogen content of urine results in higher levels of ammonia species which can inhibit the natural digestion processes for feces. For these reasons, sanitation systems have emerged that separate the urine from the feces, commonly referred to as urine-diverting toilets (Fig. 4). In such toilets, urine is generally deposited in a hole in the toilet front end and captured separately from the feces which are deposited in a separate hole for collection. The systems are normally dry or low-water systems and additional organic matter (e.g., rice husk or saw dust) are added to feces to prevent smell and/or aid the composting process. The urine is recommended to be stored in a sealed container for 1 month prior to dilution and direct reuse on the roots of crops that are not going to be directly eaten (e.g., banana palms). Direct reuse of the feces requires the chamber to be sealed for 12 months to allow for breakdown of organic matter and inactivation of pathogens. However, there are potential risks with viable protozoan helminth eggs persisting in the sludge and hence, the types of reuse are generally restricted to areas without human contact. Urine diverting toilets have been constructed all over the world. However, in cultures that are not accustomed to reusing human excreta when urine or feces containment vessels become full and the community does not want to empty or use the products. This frequently leads to the latrines being abandoned.

Sludge Settling, Dewatering, and Drying

Settling, dewatering, and drying are common treatment processes for wastewater from both sewer and onsite sanitation systems. The processes generally involves separating liquid and solid fractions to facilitate further treatment along separate treatment chains. Optimal separation is dependent on good knowledge of the wastewater characteristics and appropriate and often costly infrastructure. For settling, the treatment process involves settling and compacting of solids from sewage or fecal sludge. This is normally achieved in large settling tanks or channels. Centrifugation and screw-presses are common technologies for dewatering. Settling and dewatering efficiency can be increased by the use of conditioners and coagulants (Gold et al. 2016). Drying processes use either solar radiation or additional heating processes to evaporate the liquid leaving the solids. Common systems to achieve drying for sludge are solar sludge drying beds where concentrated sludge from sewage or fecal sludge is pumped or deposited into shallow drying beds and left to dry over a number of weeks or months (Mrimi et al. 2020). The final microbiological safety of the dried sludge is dependent on the final percentage moisture and the time that it has been left to dry. The lower the percentage moisture and the longer the time period of solar or heat exposure the lower the concentrations of pathogenic viruses and bacteria. However, the eggs from helminths worms are normally not disinfected by solar drying but can be heat inactivated by heated drying processes above 50 °C degrees (Naidoo et al. 2020). The concentrations of helminth worm eggs need to be strongly considered before reuse. One benefit of this process is that the nutrient levels remain high and the dried sludge has high calorific and carbon profiles for use in agriculture as biosolids or as combustion feedstock to produce energy.

Biochar Production

The production of biochar is a similar process to making charcoal which requires the combusted (carbonization) of a carbon source in the absence of oxygen. Dried fecal sludge can be carbonized to produce biochar either via hydrothermal processes or via slow pyrolysis (Fig. 4). After combustion, the biochar is generally a mixture of small and large burnt fragments. Often the biochar is then mechanically ground to produce fine biochar. The ground biochar can be combined with a binding agent and then extruded to make biochar briquettes that can replace charcoal made from trees and be used for cooking. However, research has shown that carbonization results in a lower calorific value of the final fuel and hence directly using dried fecal sludge might be a preferable process for energy recovery (Andriessen et al. 2019). Alternatively, the ground biochar can be directly applied as a soil conditioner in agriculture. Biochar is known to enhance the soil health by increasing carbon sequestration and moisture retention which can lead to increased crop productivity (Novak et al. 2016). Due to the high temperatures, carbonization generally inactivates all pathogens, but it does produce greenhouse gas emissions. Given that deforestation is such a rampant problem in low-income countries, dried and carbonization fecal sludge presents an alternative cooking fuel source that needs to be further explored.

Insect Bioconversion

The larvae of many Dipteran flies naturally feed on decaying organic matter such as human excreta. The Black Soldier Fly (BSF) (*Hermetia illucens L.*) that can be encountered (especially in pit latrines and on food waste) in tropical and sub-tropical areas between the latitudes of 40°S and 45°N is currently the most promising species for waste conversion. The fly lays its eggs in the vicinity of decaying organic matter. Following hatching, the hatchlings start feeding on the waste and convert it into larval body mass. At ambient temperatures of around 28 °C and nutrient rich waste such as human excreta, BSF fly larvae grow from 0.027 mg to above 200 mg within 2 weeks. Waste nutrients contents, moisture levels, and temperatures are important metrics of efficient conversion. BSF larvae waste conversion has become a popular bioconversion technology due the high market value of larval products as animal feed (e.g., pets, poultry, fish). When commercially or intensively farmed, the larvae are collected before pupation and either fed directly or dried and/or defatted to increase the protein content. At the time of writing, one large-scale commercial BSF facilities in Nairobi (see case study below), Kenya, is successfully producing larval-based animal feeds with human feces and municipal organic solid waste (ratio approximately 40% feces: 60% municipal organic solid waste). Even BSF larvae bioconversion decreases the concentrations of certain pathogens in human feces, several pathogens including helminth eggs are transferred from waste to larvae, requiring heat treatment (e.g., boiling, sun drying) before use (Lalander et al. 2013). The BSF larvae growth performance on low-nutrient wastes such as fecal and wastewater sludge can be increased by mixing with municipal organic wastes or agri-food byproducts (Gold et al. 2020). Depending on heavy metal concentrations in the sludge, larvae grown on wastewater sludge are often unsafe for feed production due to bioaccumulation of heavy metals, but non-food applications (e.g., lipids for biofuel, chitin, and chitosan) should be considered.

Case Studies

Resource Recovery from Fecal Sludge in Kathmandu, Nepal

This fecal sludge treatment plant was constructed in Mahalaxmi Municipality of Kathmandu in 2016. The treatment facility was designed to service the on-site sanitation systems (pit-latrines and septic tanks) of the 2365 households and

10 post 2015 earthquake relief camps in the municipality (BORDA 2017). The plant was funded by donor funds and constructed on 300 m^2 of land provided by a local orphanage in Lubhu, Lalitpur. The treatment plant is run by the municipality and can treat 6 m^3 of fecal sludge per week and which is delivered to the treatment plant via vacuum tankers. The first stage of treatment is where the solid and liquid fractions of the fecal sludge are allowed to separate by settling in the feeding tank (Fig. 5). The separated solid fraction is then treated in an anaerobic biogas digester then the sludge is dried on a planted sludge drying bed. The more liquid fraction of the fecal sludge from the feeding tank is treated in an anaerobic baffled reactor followed by a planted gravel filter and finally the treated liquid component is used for irrigation of crops. The treatment plant produces treated wastewater for irrigation, biogas for cooking, and dried fecal sludge biosolids for agricultural re-use. This decentralized treatment system is a successful example of resource recovery from fecal sludge for an urban settlement and is well supported by key stakeholders.

Resource Recovery Using Black Solider Flies (BSF) in Nairobi, Kenya

In Nairobi, Kenya, through their "fresh-life" brand, the waste management franchise Sanergy provides more than 2500 toilets to 100,000 people daily (Sanergy 2021). Since 2015, Sanergy converts human feces collected from these toilets with agri-food byproducts and municipal organic solid waste into animal feed and compost using BSF larvae. A facility commissioned in 2020 consists of waste pre-processing (e.g., mixing of the different waste streams), fly rearing, a nursery, waste treatment, and product harvesting and refinement and can process up to 72,000 tons per year and fly rearing provides a steady supply of BSF eggs (Dortmans et al. 2021). In the nursery, for 5 days, larval hatchlings are nurtured on a high-quality diet to young larvae with a weight of around 1 mg (Fig. 6). On long beds, the young larvae then convert the waste mixture into a compost-like residue, thereby increasing their weight above 200 mg. Following harvesting and separation, the larvae are washed, dried and sold under the "KuzaPro" brand to feed millers and animal farmers. Some part of the residue is briquetted with saw dust and combusted in an industrial kiln to provide heat for maintaining optimal temperatures (around 28 °C) for larval-based waste bioconversion. The remaining residue is converted into the "Evergrow" compost and sold to farmers as fertilizer. This centralized facility employs over 100 staff and is successful example for re-establishing the connection between sanitation and agricultural production by creating valuable high-quality inputs.

a) Sludge loading into below ground anaerobic biodigesters - Nepal

b) Sludge dewatering and drying - Nepal

c) Crops grown with treated biosolids - Nepal

Resource Recovery from Human Excreta in Urban and Regional Settlements, Fig. 5 Fecal sludge treatment system with anaerobic digestion in Kathmandu, Nepal: (**a**) the sludge loading tank (far left back corner) transfers fecal sludge to two underground anaerobic digesters, (**b**) the digested sludge is pumped into drying beds to dry (**c**) dried fecal sludge (biosolids) is mixed with soil to grow crops. (Copyright – original images taken by Jacqueline Thomas)

a) "Fresh life" Toilet - Kenya
b) Preprocessing of human faeces - Kenya
c) Female Black Soldier Fly (BSF)

Resource Recovery from Human Excreta in Urban and Regional Settlements, Fig. 6 In Nairobi, Kenya, Sanergy operates toilets with a franchise model. (**a**) Feces are collected separately at the toilet front-end. (**b**) Before insect-based bioconversion, the human feces are mixed with other organic wastes. (**c**) The waste is treated using larvae of the black soldier fly (BSF). (Copyright – original images taken by Moritz Gold, image on the right is taken from commons.wikimedia.org)

Current State of Resource Recovery from Human Excreta

There are currently no accurate global estimates for the quantities of wastewater (either sewage or onsite fecal sludge) that are reused as inputs for agriculture and/or energy production. Data from high income countries give an indication of the quantities of biosolids that are reused from conventional wastewater treatment plants. For example, in the USA in 2019 there were an estimated 4.75 million dry metric tons of biosolids produced of which 2.44 million tons (51%) was used in land applications with 1.44 million tons of this portion used on agricultural land (EPA 2019). However, his data does not capture reuse from wastewater treatment plants supplying settlements of less than 10,000 people or from private facilities. In low-income countries, national data sets are generally absent and reuse initiatives tend to be more decentralized and operate on an individual household basis. For example in South Africa a review of anaerobic digestion technologies found only a fraction were used to treat human excreta (Mutungwazi et al. 2018), hence providing reuse for only a tiny fraction of the population. Many initiatives for human excreta reuse are driven by the research and development and the not-for-profit sector. Many are not yet commercially viable or operating at national scales. For resource recovery from human excreta to become mainstream, there needs to be significant investment in monitoring and expanding the current scale of resource recovery activities. Still if this type of resource recovery is to have an assured future in sustainable urban and regional settlements, then a number of challenges need to be overcome.

Challenges in Resource Recovery

In order to increase rates of resource recovery from human excreta four main challenges need to be addressed: (1) societies lack of acceptance, (2) absence of an enabling policy environment, (3) current limited economic viability, and (4) scarcity of a trained sanitation workforce. If the primary challenge of addressing societies lack of acceptance can be overcome with education and awareness, then this will lead to demand which will in turn enhance political will to create supportive polices and the market for the human excreta products will grow. If there is a market for reuse products, then the economics of investing in resource recovery will look more favorable. Further, as the industry grows then there will be incentives to train the required workforces. Action toward increasing societies acceptance of reusing excreta has been partly initiated by the global mobilization around the Sustainable Develop Goals (SDG). In particular SDG 6 – Clean Water and Sanitation, which outlines targets to safely manage human excreta by significantly increasing levels of wastewater treatment and reuse. Combating climate change is another motivating factor and the wider population is becoming increasingly aware of the need to change individual behaviors to protect future life on this planet. This is fortunate, as overcoming common human repulsion to reusing human excreta will be necessary to enable more mainstream reuse technologies to come in to operation. Communication regarding the safety of specific reuse technologies and the products will be necessary to gain public support. Further research is required on all types of human excreta reuse products to measure the exact benefits and risks associated with both their production and applications. Quality research must also inform the creation of appropriate policies and regulations for the reuse of human excreta. In many countries at this point in time, there are inadequate or nonexistent national policies and regulations for the human excreta reuse industry. The absence of enabling policies also discourages private and public sector investment in the sector. The absence of enabling policies is one of the largest challenges that needs to be overcome and collective efforts lead by appropriate multi-lateral organizations can lead the way in the creation of standards, policies, and guidelines for the world to follow. The Bill and Melinda Gates Foundation have been leading innovation in this sector as they have heavily invested in the Reinvent the Toilet Challenge and global standards since 2011. Creating an industry for the reuse of human excreta is not without risk, but all stakeholders need to have the courage to invest and find solutions to mitigate the problems that will be encountered.

Conclusion

Population growth is producing ever increasing quantities of human excreta and placing unsustainable demands on our worlds finite resources. To protect both human health and the health of our environment, human excreta cannot be viewed as a waste product any longer. As fertilizer nutrients become increasingly scarce and the demand for sustainable energy sources grows, there will be greater incentives for urban and regional settlements to invest in resource recovery from human excreta for their residence. Bold political and scientific leadership is needed to steer the course and ensure that the right technologies are invested in that are both safe and sustainable.

References

Andriessen, N., Ward, B. J., & Strande, L. (2019). To char or not to char? Review of technologies to produce solid fuels for resource recovery from faecal sludge. *Journal of Water Sanitation and Hygiene for Development, 9*(2), 210–224. https://doi.org/10.2166/washdev.2019.184.

Bond, T., & Templeton, M. R. (2011). History and future of domestic biogas plants in the developing world. *Energy for Sustainable Development, 15*(4), 347–354. https://doi.org/10.1016/j.esd.2011.09.003.

BORDA. (2017). Faecal sludge treatment plant in Kathmandu Valley Nepal. Retrieved from https://www.borda-sa.org/faecal-sludge-treatment-plant-in-kathmandu-valley-nepal/

Collier, S., Deng, L., Adam, E., Benedict, K., Beshearse, E., Blackstock, A., … Beach, M. (2021). Estimate of burden and direct healthcare cost of infectious waterborne disease in the United States. *Emerging Infectious Diseases, 27*(1), 140–149. https://doi.org/10.3201/eid2701.190676.

Cordell, D., Drangert, J.-O., & White, S. (2009). The story of phosphorus: Global food security and food for thought. *Global Environmental Change, 19*(2), 292–305. https://doi.org/10.1016/j.gloenvcha.2008.10.009.

Diener, S., Semiyaga, S., Niwagaba, C. B., Muspratt, A. M., Gning, J. B., Mbéguéré, M., … Strande, L. (2014). A value proposition: Resource recovery from faecal sludge – Can it be the driver for improved sanitation? *Resources Conservation and Recycling, 88*, 32–38. https://doi.org/10.1016/j.resconrec.2014.04.005.

Dortmans, B., Egger, J., Diener, S., & Zurbrügg, C. (2021). *Black soldier fly biowaste processing – A step-by-step guide*. Dubendorf: Eawag: Swiss Federal Institute of Aquatic Science and Technology.

EPA. (2019). *Biosolids annual report*. Washington, DC. Retrieved from https://www.epa.gov/biosolids

Gold, M., Dayer, P., Faye, M., Clair, G., Seck, A., Niang, S., … Strande, L. (2016). Locally produced natural conditioners for dewatering of faecal sludge. *Environmental Technology, 37*(21), 2802–2814. https://doi.org/10.1080/09593330.2016.1165293.

Gold, M., Ddiba, D. I. W., Seck, A., Sekigongo, P., Diene, A., Diaw, S., … Strande, L. (2017). Faecal sludge as a solid industrial fuel: A pilot-scale study. *Journal of Water Sanitation and Hygiene for Development, 7*(2), 243–251. https://doi.org/10.2166/washdev.2017.089.

Gold, M., Cunningham, M., Bleuler, M., Arnheiter, R., Schonborn, A., Niwagaba, C., & Strande, L. (2018). Operating parameters for three resource recovery options from slow-pyrolysis of faecal sludge. *Journal of Water Sanitation and Hygiene for Development, 8*(4), 707–717. https://doi.org/10.2166/washdev.2018.009.

Gold, M., Cassar, C. M., Zurbrugg, C., Kreuzer, M., Boulos, S., Diener, S., & Mathys, A. (2020). Biowaste treatment with black soldier fly larvae: Increasing performance through the formulation of biowastes based on protein and carbohydrates. *Waste Management, 102*, 319–329. https://doi.org/10.1016/j.wasman.2019.10.036.

Goodyer, J. (2012). Thinking outside the thunderbox. *Engineering & Technology, 7*, 52–55.

Jones, E. R., van Vliet, M. T. H., Qadir, M., & Bierkens, M. F. P. (2021). Country-level and gridded estimates of wastewater production, collection, treatment and reuse. *Earth System Science Data, 13*(2), 237–254. https://doi.org/10.5194/essd-13-237-2021.

Lalander, C., Diener, S., Magri, M. E., Zurbrugg, C., Lindstrom, A., & Vinneras, B. (2013). Faecal sludge management with the larvae of the black soldier fly (Hermetia illucens) – From a hygiene aspect. *Science of the Total Environment, 458*, 312–318. https://doi.org/10.1016/j.scitotenv.2013.04.033.

Mrimi, E. C., Matwewe, F. J., Kellner, C. C., & Thomas, J. M. (2020). Safe resource recovery from faecal sludge: Evidence from an innovative treatment system in rural Tanzania. *Environmental Science: Water Research & Technology, 6*, 1737–1748. https://doi.org/10.1039/C9EW01097A.

Mutungwazi, A., Mukumba, P., & Makaka, G. (2018). Biogas digester types installed in South Africa: A review. *Renewable and Sustainable Energy Reviews, 81*, 172–180. https://doi.org/10.1016/j.rser.2017.07.051.

Naidoo, D., Archer, C. E., Septien, S., Appleton, C. C., & Buckley, C. A. (2020). Inactivation of Ascarisfor thermal treatment and drying applications in faecal sludge. *Journal of Water Sanitation and Hygiene for Development, 10*(2), 209–218. https://doi.org/10.2166/washdev.2020.119.

Novak, J. M., Ippolito, J. A., Lentz, R. D., Spokas, K. A., Bolster, C. H., Sistani, K., … Johnson, M. G. (2016). Soil health, crop productivity, microbial transport, and mine spoil response to biochars. *Bioenergy Research*, 1–11. https://doi.org/10.1007/s12155-016-9720-8.

Pham-Duc, P., Nguyen-Viet, H., Hattendorf, J., Zinsstag, J., Phung-Dac, C., Zurbrügg, C., & Odermatt, P. (2013). Ascaris lumbricoides and Trichuris trichiura infections associated with wastewater and human excreta use in agriculture in Vietnam. *Parasitology International, 62*(2), 172–180. https://doi.org/10.1016/j.parint.2012.12.007.

Pruss-Ustun, A., Wolf, J., Bartram, J., Clasen, T., Cumming, O., Freeman, M. C., … Johnston, R. (2019). Burden of disease from inadequate water, sanitation and hygiene for selected adverse health outcomes: An updated analysis with a focus on low- and middle-income countries. *International Journal of Hygiene and Environmental Health, 222*(5), 765–777. https://doi.org/10.1016/j.ijheh.2019.05.004.

Sanergy. (2021). Sanergy. Retrieved from http://www.sanergy.com/

Simha, P., & Ganesapillai, M. (2017). Ecological sanitation and nutrient recovery from human urine: How far have we come? A review. *Sustainable Environment Research, 27*(3), 107–116. https://doi.org/10.1016/j.serj.2016.12.001.

Strande, L., Ronteltap, M., & Brdjanovic, D. (2014). *Faecal sludge management: Systems approach for implementation and operation*. London: IWA Publishing.

Wald, C. (2016). The secret history of ancient toilets. *Nature, 533*(7604), 456–458. https://doi.org/10.1038/533456a.

Wear, S. L., & Thurber, R. V. (2015). Sewage pollution: Mitigation is key for coral reef stewardship. In A. G. Power & R. S. Ostfeld (Eds.), *Year in ecology and conservation biology* (Vol. 1355, pp. 15–30). Oxford: Blackwell Science Publ.

Wear, S. L., Acuna, V., McDonald, R., & Font, C. (2021). Sewage pollution, declining ecosystem health, and cross-sector collaboration. *Biological Conservation, 255*, 9. https://doi.org/10.1016/j.biocon.2021.109010.

WHO & UNICEF. (2019). *Progress on household drinking water, sanitation and hygiene 2000–2017: Special focus on inequalities*. Retrieved from: https://www.unicef.org/reports/progress-on-drinking-water-sanitation-and-hygiene-2019

Resource-Efficient City

► Circular Cities

Resource-Efficient Logistics

► Sustainability Transition and Climate Change Adaption of Logistics

Response

► Collective Emotions and Resilient Regional Communities

Responsibility to Prepare and Prevent

► Responsibility to Prepare and Prevent (R2P2): Applying Unprecedented Foresight to Addressing Unprecedented Climate Risks

Responsibility to Prepare and Prevent (R2P2): Applying Unprecedented Foresight to Addressing Unprecedented Climate Risks

Francesco Femia and Caitlin Werrell
The Center for Climate and Security, an Institute of the Council on Strategic Risks, Washington, DC, USA

Synonyms

Climate change; Climate security; Conflict; Responsibility to prepare and prevent; Security; Water security

Introduction

The web of intersecting security risks that challenges human civilization today is more complex and challenging than any other stage of history. Some of these security risks, such as tensions among powerful political entities and disputes over geographical boundaries, have been with human civilization for millennia. Other risks, such as nuclear weapons and cyber threats, are relatively recent. In some cases, as with rapid climate change, the risks to human society are unprecedented. However, the tools human society possesses today also give it an unprecedented foresight. This is a primary feature that differentiates the twenty-first century from past periods of disruption – the ability to harness scientific and technological tools to better predict and prepare for a range of plausible future scenarios, including climate change.

In this context, nation-states and intergovernmental security institutions have a responsibility to use their enhanced predictive capacities to manage and minimize climate change and other foreseeable risks. This combination of “unprecedented risk” and “unprecedented foresight” creates the case for a “Responsibility to Prepare and Prevent (R2P2)”– a responsibility to build a resilient world order against a more dangerous yet more reliably foreseeable future.

Unprecedented Risks

The relatively stable climatic period geologists called the Holocene (beginning at approximately 11,701 BP) is making way for a new epoch: the Anthropocene (Waters 2016). The Anthropocene is characterized by human-induced changes in the climate that are happening at an extremely rapid rate in terms of geologic and civilizational time and are unprecedented in history (Solomon 2007). These changes – including the melting of the glaciers and polar icecaps, extreme rainfall variability, and sea-level rise – are all changes that disrupt the foundations of the sociopolitical institutions that form the basis of civilization. Simply put, these changes affect the basic resources that support human livelihoods, nations, and the global order those nations participate in (Werrell and Femia 2016). The implications of a rapidly changing climate, coupled with other demographic, economic, and technological shifts, contribute to an era of unprecedented risk. However, some of those same dynamics – particularly rapid technological change – have also contributed to unprecedented foresight.

Unprecedented Foresight

Despite the unprecedented risk of climate change, there is a small silver lining that provides the foundation for the R2P2 framework. Namely, climate change, especially when compared to other drivers of international security risks, can be modeled with a relatively high degree of certainty. Consider, for instance, that the first accurate climate change model is from 1967, half a century ago, and for the most part, the climate is changing as the model predicted. A political scientist in 1967 would have had a much more difficult time predicting the current international security landscape (Siegel 2017). Other climate models have also shown prescient prediction capabilities (Cowtan et al. 2015). Strikingly, where inaccuracies have occurred, they have often been characterized by an underestimation of the rate and severity of change, showing a milder picture than what eventually emerged (Allison et al. 2009; Stark 2017). Subsequent technological and scientific refinements have led to more complex models and ultimately a strong record of accurate predictions of the rate and scale of global climatic changes under emissions scenarios that ultimately materialized. While significant uncertainties in predicting local-scale climatic changes and ecological interactions remain, existing projections from climate models and Earth observations paint a fairly clear picture of what the future holds for the global climate, which provides a strong basis for governments and societies to plan accordingly.

A Responsibility to Prepare and Prevent Framework (R2P2)

The combination of unprecedented risks and an unprecedented ability to forecast such risks creates a clear responsibility for governments and intergovernmental institutions to prepare for unavoidable changes and prevent the potentially unmanageable ones. The transnational and cross-sectoral nature of climate change risks demand a comprehensive approach that is adaptable to unique local and regional circumstances, but this approach should be clearly articulated and systematized into goals and principles that nations and intergovernmental institutions can adopt, measure, and promote, in order to avoid the paralysis that such complex risks can create. In this context, an R2P2 agenda should adhere to the overarching goal of "climate-proofing" security institutions at all levels of governance (local, national, regional, and international) in order to increase the capacity of states to absorb and reduce climatic stresses. This climate-proofing should consist of six core principles: routinizing, integrating, institutionalizing, and elevating attention to climate and security issues, as well as developing rapid response mechanisms and developing contingencies for unintended consequences.

Routinization: Climate change is happening now and affects nearly all aspects of society, yet that reality is not reflected in the routine activities of most governance bodies responsible for security. Doing so would help break climate change out of its traditional cage within the environment and development ministries and broaden the aperture of security institutions to include this complex risk.

Institutionalization: How climate change impacts security is not deeply understood within and across governments. In this context, the issue requires institutional centers to conduct climate security analysis and inform decision-makers. Creating institutional centers (or leveraging existing institutions in civil society) to collect and interpret information, using the best analytical tools available and then regularly delivering recommendations for action to decision-makers would go a long way in increasing preparedness for such eventualities and strengthen efforts for conflict prevention.

Elevation: In some cases, warnings related to the security risks of climate change are delivered to governments by analysts, but not at a high enough level. This is often based on a particular issue not being prioritized within a government or intergovernmental institution, or the issue not being presented in a fashion that appropriately contextualizes the risks as they pertain to other geostrategic priorities. In this context, elevating such issues within governing bodies is critical for ensuring preparedness.

Integration: In order to ensure that climate and security issues are not treated as a special interest concern, security institutions should integrate climate change trends into their analyses of other critical security priorities. This is the "just add climate" approach, justified by the nature of the threat and the simple fact that changes in the climate, acting as a threat multiplier, will affect the entire geostrategic landscape. For example, the questions of how climate change intersects with health security, conflict, international terrorism, nuclear proliferation, and maritime security are all critically important but may be missed if such analysis sits solely in the kind of specialized centers described above.

Rapid response: Though the approaches above are designed to facilitate preventive solutions, there will undoubtedly be future cases of climate-exacerbated dynamics that demand immediate attention from the security community. Developing scaled warning systems that identify long, medium, and short-term risks and that include clear "triggers" for emergency action on climate and security would help ensure that foreseeable events are acted upon with commensurate levels of urgency. This is particularly important for anticipating low probability/high impact risks and creating a governance capacity to prepare for "unknown, unknowns" or "black swans" (Femia et al 2011).

Contingencies for unintended consequences: Despite best efforts, unintended consequences of solutions to these risks may inevitably arise. Governments should seek to identify these potential eventualities and develop contingencies for addressing them. For example, unilaterally deployed geoengineering solutions to climate change, particularly in the absence of international norms to regulate their use, could result in new and unpredictable disruptions to climate, water, food, and energy systems. These are foreseeable possibilities that security institutions can identify and attempt to prevent sooner rather than later. Facilitating or institutionalizing cross-sectoral coordination to hedge against these unintended consequences would be a good start.

Conclusion

The window of opportunity to strengthen security governance in a world of rapid and unprecedented climate change is narrowing. Stalled or delayed actions may result in diminishing returns and, in the worst-case scenarios, difficult and perhaps inhumane choices in the face of continued strains on natural resources and political will. This scenario is foreseeable. Technological developments have given us climate models, and predictive tools, that enhance human society's ability to anticipate and mitigate risks. These tools must be better utilized and integrated into international, regional, national, and local security institutions in order to manage this new world. That foresight renders the realization of a Responsibility to Prepare and Prevent (R2P2) agenda both practically and morally essential.

Cross-References

▶ New Localism: New Regionalism

References

Allison, I., et al. (2009). *The Copenhagen diagnosis, 2009: Updating the world on the latest climate science.* The University of New South Wales Climate Change Research Centre (CCRC), Sydney, Australia, p. 60, http://www.copenhagendiagnosis.com

Cowtan, K., et al. (2015). Robust comparison of climate models with observations using blended land air and ocean sea surface temperatures. *Geophysical Research Letters, 42*, 6526–6534. https://doi.org/10.1002/2015GL064888.

Femia, F., Parthemore, C., & Werrell, C. (2011). The inadequate US response to a major security threat: Climate change. *The Bulletin of the Atomic Scientists*, July 20, 2011, available at: http://thebulletin.org/inadequate-us-response-major-security-threat-climate-change

Siegel, E. (2017). The first climate model turns 50 and predicted global warming almost perfectly. *Forbes*, March 15, 2017, https://www.forbes.com/sites/startswithabang/2017/03/15/the-first-climate-model-turns-50-and-predicted-global-warming-almost-perfectly/#5b3a8afa6614

Solomon, S., et al. (2007). Summary for Policymakers. In: *Climate change 2007: The physical science basis,' Contribution of working group I to the fourth*

R

assessment report of the Intergovernmental Panel on Climate Change (IPCC), Cambridge, United Kingdom and New York, NY, USA, Cambridge University Press. NASA, Climate Change: Vital Signs of the Planet: Evidence, http://climate.nasa.gov/evidence/

Stark, A. (2017). *Climate models underestimate global warming by exaggerating cloud "Brightening, Lawrence Livermore National Laboratory"*, April 7, 2017. https://www.llnl.gov/news/climate-models-underestimate-global-warming-exaggerating-cloud-brightening

Waters, C. N., et al. (2016). The anthropocene is functionally and stratigraphically distinct from the Holocene. *Science, 351*(6269)., http://science.sciencemag.org/content/351/6269/aad2622.

Werrell, C., & Femia, F. (2016). Climate change, the erosion of state sovereignty, and world order. *The Brown Journal of World Affairs, 22*(2), April 1, 2016. https://www.brown.edu/initiatives/journal-world-affairs/222-spring%E2%80%93summer-2016/climate-change-erosion-state-sovereignty-and-world-order

Responsible and Equitable

▶ Sustainable Development and Responsible Tourism: The Grijalva-Usumacinta Lower River Basin

Responsiveness

▶ Transport Resilience in Urban Regions

Restored Urban Wetlands

▶ Artificial Urban Wetlands

Right to Water

▶ Water Security and Its Role in Achieving SDG 6

Robotics

▶ Augmented Reality: Robotics, Urbanism, and the Digital Turn

Robustness

▶ Transport Resilience in Urban Regions

Role of Disaster Relief Policy in Building Resilient Coastal Regions in the United States

Chad J. McGuire
Department of Public Policy, University of Massachusetts, Dartmouth, MA, USA

Synonyms

Climate change; Coastal management; Disaster relief; Hazard; Public policy; Resilience

Definition

The United States has a long history of providing financial assistance to those it deems have suffered through no fault of their own. In the parlance of federal disaster assistance, this has generally meant the harm is caused by "natural" phenomenon outside human agency. When the harm is seen as "human-caused," for example, economic recessions, there has been a consistent reluctance to provide financial relief to those impacted. This chapter explores this dynamic through the lens of coastal climate change adaptation. Questions explored include the following. Should the impacts of climate change (e.g., increased storm frequency and intensity) be seen as entirely "natural" from a disaster relief perspective. And if so, what impact does this have on a sense of individual and group responsibility toward climate change? Alternatively, if seen as "human-caused," does that increase individual responsibility for the impacts (e.g., home damage due to coastal storm surge) and thus preclude taxpayer subsidized relief for those impacted? The ultimate question to be asked is what role does defining climate-related coastal damage as "natural" versus "human-caused" for purposes of federal disaster relief policy have on developing resilient coastal regions? The answers to these questions are relevant in informing policy as it attempts to address how to approach proactive resiliency planning along our built coastlines.

Introduction

The United States, like many nations with coastlines, has the majority of its population and its major cities located in coastal regions. Currently, well over 50% of the US population lives in a coastal region, an area representing less than 20% of the total US land mass. Coinciding with population, the largest cities in the United States are also located in coastal regions which is similar to global trends where almost half of the world population lives within 100 miles of, and 18 of the largest 25 cities globally are located on, a coastline (Griggs 2017). There are historical and logistical reasons for the correlation between human population centers, economic activity, and coastal areas (see McGuire 2020).

US coastal areas are dynamic and subject to storm-related risks. Like many coastal nations, the United States has developed a number of policies to deal with coastal risks. One such policy is federal disaster assistance. The concept is deceptively simple. When a natural event occurs, such as a coastal storm surge that causes damage, the federal government – in conjunction with state and local governments – provides both financial and logistical support. The United States has been providing disaster assistance since its inception as a nation. Drought, hurricanes, tornadoes, fires, and other hazards have regularly been the impetus for direct financial payments from the federal government. The key in triggering the funding is the finding of the event to be "natural": not caused by, or related to, human activity. Historically, when a hazardous event occurs, for example, a coastal hurricane where many homes and other human developments suffer damage, the event is deemed to be natural and resources are readily made available (Dauber 2013; CRS 2020a).

One issue that arises in the context of this national disaster relief program in the United States is the impact and influence of this relief on local decision-making. For example, in the context of coastal development, does the existence of federal disaster relief, and similar national programs, influence local perceptions of risk? Are homeowners more or less likely to internalize the existing and emerging risks of coastal living in an era of climate change and associated sea level rise? And importantly, if climate change and the associated sea level rise continue to increase the risks of coastal living, what is the effect of subsidizing the costs of redeveloping these areas after storms on future decision-making?

From a policy standpoint, these questions center around fault attribution. As a matter of operational principle, when a hazardous event is deemed natural, the US government does not attribute the cause to humans and readily provides substantial financial relief. However, when fault is attributed to human actions, no matter how unrelated to the acts of those claiming harm, the US government is recalcitrant to provide relief. A prime example is the global economic crisis that began around 2008. By now, a number of key actions, mainly by private financial institutions, have been shown to be accountable for triggering the crisis (Williams 2010). And the United States, like many other countries, found this fact to be important in setting policy that was hesitant to provide generous and unconditional relief to citizens suffering as a result of the crisis.

The causes and effects of climate change are relevant to have an honest assessment of fault attribution, at least in terms of relationships between hazardous events and human activities. The objective evidence is clear that human activities are primarily responsible for observed climate change (IPCC 2014). In addition, that evidence is also clear that climate change is positively influencing the frequency and intensity of storm activity, particularly observed in most coastal regions of the United States (USGCRP 2018). Finally, observed sea level rise, which threatens many coastal areas with inundation from coastal storm events, is directly correlated with observed climate change. Collectively, this suggests that human responsibility for coastal storms, and their resulting damage, exists and is increasing.

If coastal storms and their resulting damage can be attributed to humans through climate change, then should this fact change the calculus by the US government when it deems such storm events as "natural" or otherwise outside of human responsibility? And if it changes that calculus, what does this mean for coastal communities? There is little doubt a pullback in current federal disaster assistance would fundamentally change coastal asset valuations in many areas of the United States.

Alternatively, what are the effects of continued federal support for coastal hazardous events in the face of mounting evidence showing those events are directly linked to human activities? What problems does this present for a government wishing to gain public acceptance and support for initiatives related to climate resiliency? Does the willingness to support coastal community development, and postdisaster redevelopment, in high-risk areas prohibit meaningful public internalization of the risks associated with climate change? These questions are explored in greater detail below in an attempt to address how to best approach proactive resiliency planning along our built coasts.

Examining the Concept of Fault in Federal Disaster Relief

In the United States, there is a long history of providing public assistance and relief to those who have suffered loss *through no fault of their own*. The question of fault is at the heart of public assistance in the United States, including what activities qualify for assistance, the amount of assistance available, and whether the assistance is contingent on meeting special criteria, for example, self-help.

Unemployment compensation, run through federal and state governments, is an example of public benefits provided through a specific event: the loss of employment. But in order to qualify for unemployment benefits in the United States, the loss must generally be shown to be through no fault of the individual claiming benefits; they cannot have voluntarily left the job or have been validly fired for their own conduct. The benefits are not available simply for finding oneself unemployed, but rather only when someone finds themselves unemployed for reasons beyond their own actions or inactions. When someone finds themselves unemployed through forces outside of their personal control in the United States, the government will step in and provide financial assistance.

But aid for unemployment is not unlimited. Rather, it is limited in the amount of aid and the length of time aid is offered. Often the amount of aid provided is a lower percentage of actual wages, and the duration is often limited to a period less than 1 year. For most states in the United States, the person must prove they are seeking active employment as a condition of receiving unemployment benefits. And if a job is offered, even if suboptimal to the prior position and/or prior pay, the person must accept the job or risk losing the unemployment benefit (O'Leary et al. 2020).

The limits placed on unemployment assistance are best contextualized as conditions to aid, both in qualifying for that aid, and even after initial qualification, maintaining that qualification for a limited period of time. For example, to initially qualify, an application cannot be personally responsible for losing their employment. If this hurdle is overcome, then there generally is a myriad of conditions and limitations placed on maintaining qualification. And, ultimately, benefits are exhausted after a discrete period of time – generally less than 1 year – whether or not the person has been reemployed. These kinds of limitations exist in other areas of government benefits including disability claims, welfare, housing, and many other areas of publicly provided assistance.

The concept of personal actions limiting qualification for benefits, or fault, can be conceptually thought of as a continuum when looking at the principles of providing public assistance. First, as noted, there can be no fault found by the person initially claiming the right to the benefit. A person finds themselves in an economic environment where their industry is experiencing downsizing and the employer must cut jobs due to the overarching economic conditions presented. A loss of one's job under these circumstances is reasonably seen as beyond the fault of the person seeking unemployment assistance. But if that person fails to actively seek new employment, then the continuum of fault can find them responsible for their continued unemployment. Public assistance was available to the person for the job loss when fault did not attach to the person for the loss, but fault reattaches and cuts off the availability of public assistance when the person fails to actively seek new employment opportunities. The assistance is conditional not only on the no-fault loss of employment, but also on actively seeking reemployment.

The concept of a continuum of fault, where fault is assessed on actions over time, is similar to a legal principle in the United States of a duty to mitigate the harm one experiences (Slovenko 1998). If one is wronged through the fault of another, and that harm causes acute damages that can be remedied with action, then failure to act by the aggrieved can be a basis of limiting the liability of the person who created the harm. As a simple example, if Person A accidently harms Person B in a way that results in a cut, the failure of Person B to seek normal medical assistance to heal the cut does not make Person A responsible for all injuries that may arise from both the cut and the failure to properly care for the cut. Person A can only be responsible for the reasonable harm caused by the cut, not for subsequent harm caused by a failure to seek normal care for the cut. This is similar to the job seeker who lost their job through no fault of their own but has failed to take reasonable steps to look for alternative employment. As described above, public unemployment assistance in the United States suggests it is available for the initial loss of wages caused by the job loss, but not for any loss of wages caused by the job seeker failing to reasonably look for alternate employment opportunities.

In most areas of publicly funded assistance in the United States, the assistance is intended to be limited in duration: a stop-gap measure to prevent extensive harm caused by an acute event. It is not meant to provide long-term, uninterrupted benefits. There are a few exceptions, including federal retirement (Social Security) and medical (Medicare) benefits, both of which are pension-like systems of public benefits requiring individuals to pay into a system during their working years in order to qualify for long-term benefits when they get to a determined age. Similar benefit plans exist in other federal and state pension systems in the United States. But in all instances, these benefits are not given as a matter of right. Individuals must qualify by paying into these systems for a minimum number of years: They are qualified entitlement systems (CRS 2020c, d).

The principle of fault, as described above, can be used to examine federal disaster assistance policy in the United States. Using the continuum principle, one can look at federal disaster assistance from multiple perspectives to examine whether the concept of fault applies, and if it does, when it is triggered and to what extent does it impact coastal management policies. At the center of any such analysis is some examination of the extent to which coastal storms and sea level rise resulting from a changing environment are being attributed to human activities in US coastal policy. This transition in thinking is critical as it establishes the basis for critiquing a "no-fault" policy principle when it comes to providing financial assistance for harm caused by the coastal effects of climate change.

Natural Hazards and Natural Disasters: Similarities and Distinctions

Disaster relief policy in the United States may best be understood by distinguishing between two terms: *natural hazard* and a *natural disaster*. Natural hazards, generally speaking, are the kinds of events in nature that happen regularly around the world. Natural disasters, however, are a political designation meant to connect an event to the need for government intervention that goes beyond existing structures and institutions available to respond to the hazardous event.

There are natural and recurring forces (for example, mountain snowmelt in the spring) that regularly induce flooding in flood-prone areas. For centuries, humans have learned to live and adapt to these conditions. If the flooding event is "normal" (within expected averages) and causes little to no harm because the local municipality has managed its development and activities around the expected flooding, then the event, while hazardous in relation to nonflooding conditions, is managed in a way that does not cause unexpected and widespread harm. There may be losses incurred, but they are generally supported by existing government and private market structures, such as public emergency response apparatuses and private hazard insurance to cover the specific losses.

These types of events are generally considered natural hazards. They occur on a somewhat expected basis and to an expected extent. For example, the State of Florida in the United States can expect to be impacted by at least one tropical storm or hurricane emanating from the Atlantic Ocean every year. Florida is located in an area that supports the conditions and pathways for hurricanes. Coastal municipalities understand this general risk and prepare with a mix of emergency response planning, land use restrictions, and homebuilding requirements meant to mitigate the expected risk. Individual homeowners who finance the purchase of their property are often required to pay for hazard insurance to mitigate against expected risks. Private insurance companies, in turn, develop policies with acceptable premiums that only pay for losses deemed to occur within these expected hazard conditions. Events deemed natural hazards can certainly cause loss, but that loss is one that is expected in both scope and scale.

Natural disasters are events that are similar to natural hazards, but the scope and scale of the event is beyond accepted expectations. Generally, this is measured in the intensity of the activity and the amount of damage it causes. Using the examples of hurricanes in Florida, the State may be prepared for, and expect, an average of a few hurricanes per year that are of low to moderate intensity, somewhere between a Category 1 and 2 on the hurricane intensity scale with 1 being the lowest intensity and 5 being the highest. If Florida experiences a Category 4 or 5 hurricane, and depending on its characteristics (size, speed, etc.), the event itself can be destructive in a way that is beyond expectations. And as such, it can often lead to damages that far exceed the built capacity of the state and municipal governments, as well as the private insurance industry, based on those expectations. In such instances, Florida may look to the federal government for additional assistance in responding to the damage and resulting human need. It does so by making a formal request that the federal government declare a natural disaster. If declared by the federal government, a plethora of additional aid (monetary, logistics, emergency management, redevelopment, etc.) is provided above and beyond the existing state and private resources (CRS 2020a).

These examples highlight the major differences between a natural hazard and natural disaster designation. Almost every natural disaster arises from events that would be normally deemed natural hazards. But natural disasters are probably best represented as natural hazard events exceeding the expected scope and duration of similar events in that area. In this way, every time a natural disaster is declared, it suggests an event has occurred that exceeds the normal expectations or risk from a hazardous event in that area. If Florida requests multiple natural disaster declarations every year based on hurricanes that exceed expectations, then that would suggest the hurricanes impacting Florida differ from the past, and certainly differ from the state's expectations about the frequency and intensity of such storms. That in itself is informative, especially if Florida's expectations are based on a long history of previous storms. The new information suggests Florida is experiencing an average increase in storm activity and intensity in recent years relative to the history that helped to formulate its current expectations. In addition, if the frequency and intensity of hurricanes continues for the foreseeable future, then it may be reasonable to suggest Florida needs to reassess its expectations about what is the "norm" for hurricane risk in the state on an annual basis.

If Florida were to update its expectations about hurricane storm risk, then it would need to internalize that new risk into its state and local planning. All the important aspects of coastal management would need to be updated. Emergency management, building codes, land use planning like inundation and exclusion zones, and setback requirements would all need to be updated to reflect the additional risk. In addition, such updating would impact private and public hazard insurance. If the risk of loss increases, then the cost to insure against that loss would also likely increase assuming variables such as real estate valuations in risky areas remain stable. And maybe most importantly, internalizing the additional risk would begin to normalize that new level of risk, potentially removing such events from the opportunity to be declared natural disasters because they now exist in the realm of expected background natural hazards. Losing lucrative federal disaster assistance by incorporating increased hazards into background expectations may create a disincentive for coastal states to update their risk assessment.

There are important considerations when thinking about how the existing policy paradigm in the United States, particularly federal disaster relief, might create disincentives to coastal states when assessing and updating coastal risks. This is an area worthy of significant research potential. But there is a fundamental question to be considered in contemplating the existing paradigm of federal disaster relief, and that is the effect of that existing policy on meaningfully incorporating climate change into coastal management planning.

The Impacts of Climate Change on Federal Disaster Relief

Currently, the US federal government is providing more disaster assistance than it ever has in its history. According to the Congressional Research Service (CRS 2020b), since records began on federal disaster relief spending in 1964, the federal government has spent a total of just over US $435 billion in 2020 adjusted dollars. Of that US $435 billion, US $117 billion (≈26% of total) was appropriated (planned) to cover disaster relief requests. The remaining US $318 billion (≈74% of total) was supplemental appropriations that had to be passed to cover actual expenses incurred after natural disaster designations. Importantly, in the most recent decade (2010–2020), annual appropriations totaled just over US $83 billion, or approximately 71% of the US $117 billion dollars appropriated since 1964, showing the federal government appropriating more money toward disaster relief as a result of increasing and costly declared disasters. In the same last 10 years, the federal government spent over US $123 billion in supplemental appropriations, approximately 150% more than the planned appropriations.

Over the last decade, the combined annual appropriations (US $83 billion) and supplemental appropriations (US $117 billion) come to total spending of US $200 billion, which is almost as much as was spent between 1964 and 2009 controlling for inflation. And this spending correlates to higher numbers of major disaster declarations (averaging approximately 60 per year) and the highest number of catastrophic disaster declarations (37 total) over the past 15 years. In sum, the US government is declaring more natural disasters and spending more money to help remediate those disasters than it has in its history.

Spending more on federal disaster relief correlates closely with the most recent science on climate change. The United States Global Change Research Program (USGCRP 2018) has outlined the current state of coastal waters in the United States relative to the observed and predicted impacts of climate change. The assessment indicates that, overall, coastal environments in the United States are already at risk. Increases in frequency, depth, and extent of tidal flooding due to sea level rise have been observed across all coastal regions of the nation. In addition, higher storm surges and increased frequency of heavy precipitation events exacerbate the risk. The risks to private property owners and coastal communities are identified to be significant as chronic high-tide flooding leads to higher mitigation costs while simultaneously lowering property values. Effectively, the financial costs of coping with the impacts of persistent flooding will strain coastal municipal budgets. Lower property values lead to less tax revenue, which amplifies the strain on most coastal budgets dealing with the fallout from coastal property and infrastructure damage.

The national assessment is consistent with global observations and assessments conducted by the Intergovernmental Panel on Climate Change (IPCC 2014) and, importantly, correlates to the increased declaration of, and spending on, natural disasters by the United States as summarized earlier. And all assessments clearly identify human activities as the main driver of observed climate change. This presents an array of important questions for coastal management policy in the United States. Primary among them is to what degree does the current framework of federal disaster relief policy influence the development of resilient coastal regions?

Issues in Attaching Human Responsibility to Natural Disasters

Current disaster relief policy in the United States identifies most natural hazard events that cause unexpected damage as natural (nonhuman) activities. Thus, the "no fault" presumption dominates the policy approach to disaster relief: substantial and unconditional compensation regardless of individual circumstances. For coastal areas, as summarized and described by McGuire (2020), this creates a set of federal policies that act as a de facto, zero-premium insurance net. In essence, coastal regions can maximize development opportunities that do not account for the emerging and growing dangers of climate change. If an event occurs, and causes substantial damage beyond those covered under existing public and private policy paradigms, then federal disaster assistance provides the resources to not only compensate for the substantial damage and redevelopment, but also to catalyze new development.

Bringing forward the continuum of fault conceptual framework described earlier, federal disaster relief policy does not currently attach human responsibility to events it categorizes as natural disasters. But as noted earlier, natural disasters are, effectively, natural hazard events that cause more damage than anticipated. So, high-intensity storms – the very storms that are observed and predicted under climate change observations – end up having the exact characteristics of events that qualify for disaster relief. In this manner of reasoning, it can be argued that US disaster relief policy applies mainly, if not fully, to events that can be linked directly to human-induced climate change.

By not attaching human responsibility to natural disaster declarations, current US policy supports a vision of coastal areas that does not internalize the full risks of climate change. This then reinforces preexisting coastal state emergency response, development, and planning paradigms. And there is little compulsion for coastal states and municipalities to unilaterally change this orientation. Recalling the example of the State of Florida described earlier, its current set of expectations about coastal development and management is premised on the existing policy environment. It relies on federal government intervention for damages that go beyond its current expectations. It has little incentive to unilaterally alter its expectations and internalize the increased coastal risks due to climate change because, under the current paradigm, the federal government is responsible for those additional risks. Voluntarily internalizing those risks would arguably make Florida primarily responsible, including financially responsible, for those added risks.

One cascading effect of the current policy environment is the inability to meaningfully internalize the additional risks of climate change. While, generally speaking, the United States supports the existence, impacts, and risks of climate change, as well as humans as the major cause of that change (see USGCRP 2018), it has yet to formally internalize that reality into many of its existing national policies. This is certainly true of the current national disaster relief policy. And it is reasonable to assume that coastal states and municipalities in the United States will continue to develop and manage their coastlines in a manner that externalizes the risks of climate change so long as the federal government fully subsidizes that risk through its current policy measures.

If the United States were to change its current national disaster relief policy by attaching human responsibility to all manner of coastal storms, effectively forcing the internalization of climate change risk at US coastlines, then there would likely be significant reverberations for coastal economies. A good deal of research has been done on the dynamics of coastal economies in the United States relative to existing national support and subsidy mechanisms including federal disaster assistance. Some of this work deals with market supports, such as housing prices, that are dependent on existing subsidies. Other work focuses on the influence of national disaster assistance and other financial mechanisms on coastal economic development postdisaster: so-called disaster economics.

Ouzad and Kahn (2019) have done a substantial review of the literature on the US housing market's equilibrium pricing of natural disaster risk. What the literature shows is a strong price support for the financing of coastal development. They find the demand side of the real estate market is directly supported by existing subsidies such as national disaster relief and publicly subsidized flood insurance. But they also find that, through mortgage origination in particular, government guarantees of conforming coastal home mortgages result in an increased rate of mortgage originations immediately after a disaster. In essence, the disaster stimulates coastal development. This conforms to research conducted by Elliott and Clement (2017), which showed that disaster relief funding by the federal government not only stimulated the redevelopment of damaged and destroyed homes in hazardous coastal areas, but also stimulated new development in these same areas at the same time redevelopment was occurring. The mix of direct government relief funding, subsidized development loans, and government-backed guarantees on private mortgages provided the mechanisms for housing price support and new development in these coastal areas.

Removal of these existing subsidies could have devastating effects on the economies of coastal regions. Recent research shows the current level of demand for coastal real estate, including current market prices, is supported by government programs that externalize the risk of loss through a mix of national subsidies including, importantly, national disaster relief (Bakkensen and Barrage 2017; Ortega and Taspinar 2018; Zhang and Leonard 2019). And for many coastal regions, real estate demand is the foundation of their economies. Receipts from real estate taxes form the basis of public spending on education, emergency services, and associated civic activities.

A Failed Attempt to Internalize Risk

There is one recent example where the federal government has attempted to remove existing subsidies for coastal living, and in the process allow the increasing risks of climate change to be internalized into the private market: flood insurance. As summarized by Knowles and Kunreuther (2014), the United States currently implements a national public flood insurance program. The program is run by the federal government because private insurance companies were unwilling to provide flood insurance to homes that were at heightened risk of flooding. The private companies determined risk-based premiums would be too high for current and prospective homeowners, creating substantial downside pressure on the coastal real estate market. As a compromise, and because it was already providing "free" flood insurance through its national disaster relief program, the federal government developed a highly subsidized public flood insurance program. Those seeking financing, or in certain cases federal disaster assistance, and living in high-risk coastal areas were required to purchase and maintain public flood insurance. The premiums are set by the federal government and are well below private market rates.

After Superstorm Sandy in 2012, a highly destructive coastal storm along the East Coast of the United States, the national flood insurance program faced a substantial deficit because the claims far exceeded premium payments held in reserve (CRS 2013). The US federal government had to raise the national public flood insurance program's borrowing capacity from US $1.5 billion to over US $30 billion to cover the immediate claims made on the national flood insurance program. And this amount was completely separate from the tens of billions of US dollars allocated to disaster relief by the federal government in supplemental appropriations.

Faced with skyrocketing costs and the prospects of a future where events like Superstorm Sandy occurred more regularly, the US Congress passed bipartisan legislation known as the Biggert-Waters Flood Insurance Reform Act in 2012. The law removed many of the subsidies inherent in the NFIP, including a phasing out of premium subsidies for second homes, businesses, and properties suffering repeat losses and the recalculation of flood zones based on updated information including observed sea level rise. These reforms were not holistic in addressing the externalization of risk through heavily subsidized public flood insurance. For example, they did not include market-based insurance premiums for primary residences. The subsidy for second homes and other properties persisted, lowering incrementally toward a market rate over a 5-year period. And there were numerous exceptions to these reforms.

The Biggert-Waters reforms to the NFIP received immediate public backlash. The attempts to begin to internalize the risks of coastal living through higher flood insurance premiums were highly unpopular. Many coastal homes were presented for sale while the demand for these homes waned, particularly in the immediate aftermath of Superstorm Sandy's unprecedented devastation in coastal New Jersey and New York. Over 50 years of highly subsidized and rarely enforced coastal flood insurance policies institutionalized the expectations of the affected population, such that special interests converged to push back the implementation of Biggert-Waters. In 2014, an amendment to Biggert-Waters – the aptly named Menendez-Grimm Homeowner Insurance Affordability Act – was passed and effectively rolled back all of the risk-internalization amendments of Biggert-Waters.

Conclusion: A Path Forward

The example of Biggert-Waters provides some evidence of the difficulty in establishing new US coastal policy that attempts to internalize the emerging risks to coastal areas in a time of climate change. The United States carries a long history of incentivizing and supporting coastal development. In addition to subsidized flood insurance and government-backed mortgage guarantees, the federal government has consistently provided unlimited disaster relief assistance under a no-fault premise when coastal hazards cause unanticipated harm. These subsidies, and more, have collectively aided in making coastal regions a safe bet for investing and living. And the by-products of this history are evident today. More humans in the United States live in coastal regions where the majority of economic activity occurs.

Climate change is altering this historical bargain. Coastal regions have always been dynamic, subject to hazards such as coastal hurricanes and storm surges that can bring flooding and damage to low-lying areas. But the frequency of those events and the extent of damage they bring are increasing. And it is clear that increase is directly related to human activity. This places the United States, and likely other nations in similar circumstances, into a difficult position. The reason is existing policy, particularly national disaster relief policy, has always treated coastal storm events as "natural," and thus outside of human responsibility. And this is important because, as has been discussed, the United States generally places significant restrictions on providing federal assistance for harm it deems is derived through human behavior.

In order for US policy to move forward meaningfully in helping to build resilient coastlines, it has to internalize the risks associated with climate change. In the parlance of national disaster relief policy, it must move from an assumption that coastal storm activity is not the fault of humans to the reality that the coastal storms of today and tomorrow are directly related to human activities. In this way, the United States can begin to meaningfully remove the implicit and explicit subsidies it has created that make coastal living seem less risky than it actually is in many areas of the country. But it is no easy task. As the Biggert-Watters example shows, even incremental attempted reforms to public flood insurance that marginally increase the costs of coastal property ownership are met with stiff public and special interest resistance.

Even though the path is difficult, there are clear characteristics in existing policy that can help inform a way forward for the United States in building resilient coastlines. Clearly the expectations of federal assistance without conditions have created a scenario where risk can be *almost fully* externalized by coastal regions. If development opportunities are questionable at the local level, but the risks of loss can be mostly mitigated through federal safety nets like disaster assistance, then the decision-making scales will likely balance in favor of development. All of the benefits of development accrue to the local coastal municipality, while most of the costs can be deferred to the future storm event and, then, mostly mitigated through federal assistance.

The federal government can change this calculus. Referring to the continuum of fault concept discussed earlier, it may choose to still provide disaster assistance when a hazard event occurs. But it can also choose to limit the amount of support it provides. Rather than providing unlimited funds, it can place caps or conditions on how funding occurs. Bringing forward the example of one's duty to mitigate loss, the government can review losses after an event and then exclude future aid to the most at-risk areas. This makes future development of risky areas less likely by the coastal community if they know federal aid will not be available. It also places the onus on future decisions, not decisions that have already occurred.

There are many other examples of how the US government can force the increasing risks of climate change on coastal communities. Fundamentally, however it chooses to do so, it must remove the no-fault, nonhuman-related assumptions about coastal hazards it currently imbues into its national disaster relief policy. By doing so, it will remove a major hurdle for coastal regions in the United States in developing more resilient policies.

Cross-References

▶ From Vulnerability to Urban Resilience to Climate Change
▶ How Cities Can Be Resilient

References

Bakkensen, L. A., & Barrage, L. (2017). *Flood risk belief heterogeneity and coastal home price dynamics: Going under water?* (Bureau of Economic Research Working Paper Series: 23854). https://www.nber.org/papers/w23854

Congressional Research Service (CRS). (2013). *The National Flood Insurance Program: Status and remaining issues for Congress*. https://fas.org/sgp/crs/misc/R42850.pdf

Congressional Research Service (CRS). (2020a). *Congressional primer on responding to and recovering from major disasters and emergencies*. https://fas.org/sgp/crs/homesec/R41981.pdf

Congressional Research Service (CRS). (2020b). *The disaster relief fund: Overview and issues*. https://fas.org/sgp/crs/homesec/R45484.pdf

Congressional Research Service (CRS). (2020c). *Social security primer*. https://fas.org/sgp/crs/misc/R42035.pdf

Congressional Research Service (CRS). (2020d). *Medicare primer*. https://fas.org/sgp/crs/misc/R40425.pdf

Dauber, M. L. (2013). *The sympathetic state: Disaster relief and the origins of the American welfare state*. The University of Chicago Press.

Elliott, J., & Clement, M. T. (2017). Natural hazards and local development: The successive nature of landscape transformation in the United States. *Social Forces, 96* (2), 851–876. https://doi.org/10.1093/sf/sox054.

Griggs, G. (2017). *Coasts in crisis: A global challenge*. University of California Press.

Intergovernmental Panel on Climate Change (IPCC). (2014). *AR5 synthesis report: Climate change 2014*. https://www.ipcc.ch/report/ar5/syr/

Knowles, S. G., & Kunreuther, H. C. (2014). Troubled waters: The National Flood Insurance Program in historical perspective. *The Journal of Policy History, 26* (3), 327–353. https://doi.org/10.1017/S0898030614000153.

McGuire, C. J. (2020). Issues in developing and implementing an active hazard framework for coastal climate resiliency planning. In R. Brears (Ed.), *The Palgrave handbook of climate resilient societies* (pp. 1–21). Springer Nature. https://doi.org/10.1007/978-3-030-32811-5_108-1.

O'Leary, C. J., Barnow, B. S., & Lenaerts, K. (2020). Lessons from the American federal-state unemployment insurance system for a European unemployment benefits system. *International Social Security Review, 73*(1), 3–34. https://doi.org/10.1111/issr.12226.

Ortega, F., & Taspinar, S. (2018). Rising sea levels and sinking property values: Hurricane Sandy and New York's housing market. *Journal of Urban Economics, 106*, 81–100. https://doi.org/10.1016/j.jue.2018.06.005.

Ouzad, A., & Kahn, M. E. (2019). *Mortgage finance in the face of rising climate risk*. (National Bureau of Economic Research Working Paper Series: 26322). https://www.nber.org/papers/w26322

Slovenko, R. (1998). Duty to minimize damages. *The Journal of Psychiatry & Law, 26*(4), 579–594. https://doi.org/10.1177/009318539802600415.
United States Global Change Research Program (USGCRP). (2018). *Impacts, risks, and adaptation in the United States: Fourth national climate assessment, volume II*. https://nca2018.globalchange.gov/
Williams, M. T. (2010). *Uncontrolled risk: The lessons of Lehman Brothers and how systemic risk can still bring down the world financial system*. McGraw-Hill.
Zhang, L., & Leonard, T. (2019). Flood hazards impact on neighborhood house prices. *The Journal of Real Estate Finance and Economics, 58*(4), 656–674. https://doi.org/10.1007/s11146-018-9664-1.

Role of Nature for Ageing Populations

Danielle MacCarthy
Queen's University Belfast, Belfast, Northern Ireland, UK

Responding to population ageing, which the UN identifies as one of the four global "megatrends" shaping the twenty-first century (Messerli et al. 2019), may require fiscal, societal, and spatial overhaul. The Organisation for Economic Co-operation and Development states that the population share of adults aged 65 years old and over is expected to rise to around 25% in 2050 across its member states. It takes place simultaneously with three other major megatrends: population growth, international migration, and urbanization. This positions cities and urban areas as a central and focal point for multiple pressures. Currently, there are efforts to address twentieth-century urbanization patterns, which in large part were driven by economic and growth factors and paid little attention to the human scale or the conservation and protection of the natural and ecological networks in cities. In terms of human health, this impacted human habitats and may in part contribute to rising levels of non-communicable diseases (NCDs), alienation, and loneliness, affecting many ageing adults in the twenty-first century. There is growing recognition that the leading causes of disease and death, including heart disease, cancer, cerebrovascular disease, chronic lower respiratory disease, and injury, can be exacerbated by elements within the built environment which contribute to sedentary lifestyles and unhealthy environments. Examining the role of nature for older adults in this context means that while it cannot be a panacea for society's ills, there is a strong and growing evidence base to suggest its positive role for the promotion of health and wellbeing (in addition to ecologically enriching cities and mitigating against climate change as environmentally cities have been sites for degradation and loss of natural environment). Most commonly, urban greenness and urban nature is understood in terms of its benefits for ageing populations, and these benefits have been reported globally and exist across a range of areas including mental health and therapeutic benefits, physical activity, facilitating social cohesion, political activism, and recent work in the field of reminiscence. Across the life cycle, people of all ages experience an innate desire to connect and engage with nature (Ulrich 1983; Wilson 1984; Baxter and Pelletier 2019). For older adults, beyond preference and engagement alone, the widespread health benefits of nature are becoming better understood.

Health Benefits

One of the main functions that nature can provide for older adults is how it aids health and wellbeing. In some instances, differential impacts for older adults have been found and may exceed the health benefits of nature for other groups in cities. A well-established study by Maas et al. (2006) found that, in particular, the elderly, along with youth, and post-secondary educated residents in large cities seem to benefit more from the presence of green areas in their living environments and reported better health. These findings hold true for self-reported health indicating the important role of nature for subjective wellbeing. However, other studies have demonstrated the significant role greenness plays in its relationship to longevity. An early study in the field by Lewis and Booth (1994) found that people living in built-up areas with access to gardens and other green spaces had a lower prevalence of psychiatric morbidity as compared to people in built-up areas with no such access. A couple of decades later, Crouse et al. (2017) conducted a large-scale Canadian cohort study examining 1,265,000 individuals which showed significantly decreased risk of mortality in the range of 8–12% from all causes of death for participants with increased greenness around their residence. The theoretical basis for establishing the cause-effect relationship of nature to health is still not understood fully. However, we know there are several pathways which nature operates on. Markevych et al. (2017) identify three pathways; these are mitigation, restoration, and instoration and can be a useful framework to examine the role of nature (Fig. 1).

Reducing Harm

A recent study from Hong Kong (HK) (Sun et al. 2020) found that the effect of interaction between greenness and air pollution on respiratory mortality for HK elders meant that older adults residing in greener areas are less susceptible to acute air pollution. This demonstrates mitigation as a function of nature and the role it plays for urban dwelling older adults. Another impact of urbanization, in which older adults are particularly vulnerable, is urban heat island (UHI) effects. Multiple studies conclude that green spaces and water bodies can reduce urban heat islands. Gomez-Martinez (2021) show that larger green spaces were significantly cooler and that size can explain almost 30% of temperature variability. Green spaces with higher vegetation index values were also significantly cooler, and the relationship between greenness and temperature strengthened over time. Both air pollution and UHI are global

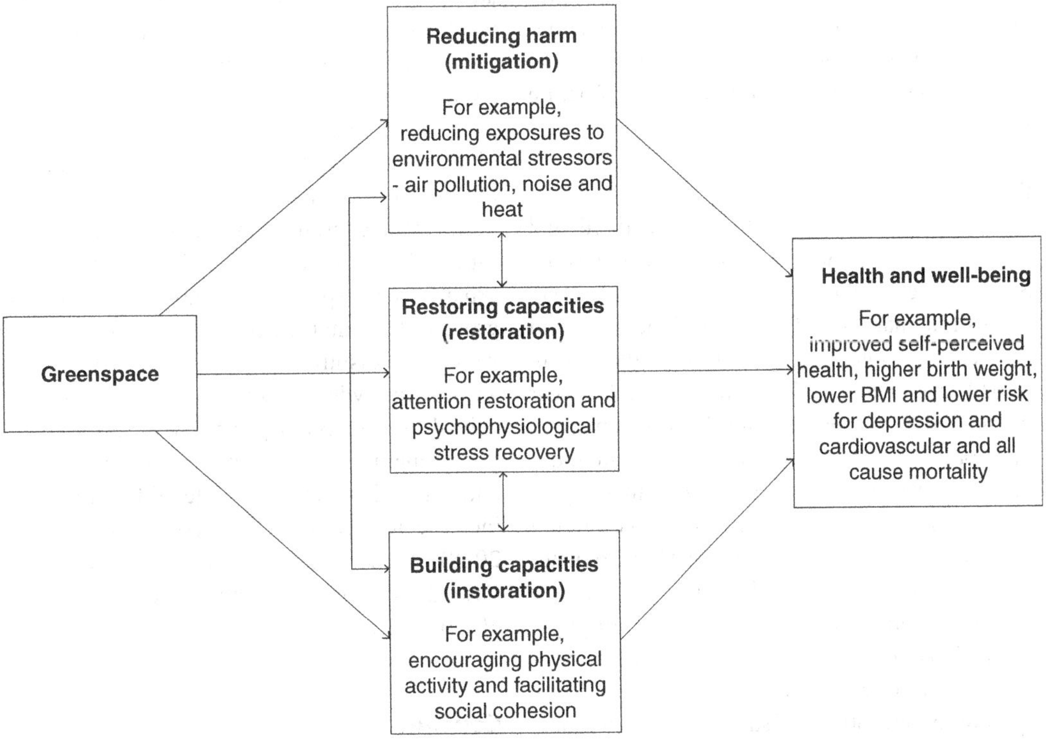

Role of Nature for Ageing Populations, Fig. 1 Three domains of pathways linking green space to positive health outcomes. The arrows represent hypothetical patterns of influence, with specific pathways in each domain potentially influencing one or more specific pathways in the other domains (Markevych et al. 2017)

problems that are likely to grow because of population and urbanization expansion and consequently have detrimental impacts on ageing populations.

Restoring Capacities

A second pathway that Markevych et al. (2017) identify is restoration. Building on early seminal work by Kaplan and Kaplan (1989), establishing the role that nature plays for restoration and stress recovery, there are a number of ways that exposure to, and affiliation with, nature have shown to support mental health for older adults. These include nature's ability to reduce stress, create positive affective states, and improve cognitive functioning. Pun et al. (2018) found a direct association of greenness with perceived stress among older adults. de Keizer et al. (2019) report that higher residential surrounding greenness was associated with slower cognitive decline including tests which assessed short-term memory and verbal fluency over a 10-year follow-up longitudinal study. While it is noted that there are inconsistent findings in the field, much of the research points to positive effects of restoration of nature for older adults.

Building Capacities

Instoration is the third pathway identified and involves physical activity and social cohesion. Surrounding greenness has been shown to play an instrumental role for physical activity and physical functioning and exercise. Dalton et al. (2016) examined 15,672 people over a 7.5-year period and found that neighborhood green space may protect physical activity decline in older adults. de Keijzer et al. (2019) state that higher residential surrounding greenness and living closer to natural environments contribute to better physical functioning at older ages. Nature may play an important role in the promotion of physical activity for older adults; however, this relationship can often be mediated by other factors relating to perceptions of safety in the neighborhood or social capital (Hong et al. 2018). Older adults are more likely to experience shrinking social networks in their community due to deteriorating physical and cognitive ability that can lead to functional limitation and mobility decline. Green space may play an important role for improving older adults' social capital and related health outcomes (Frank et al. 2010).

Aside from health benefits for *all* older adults, research looking at the important role nature plays for the subset of the population of older adults who suffer from dementia is also gaining traction. Recent work in Australia by Astell-Burt et al. (2020) has found that increased tree canopy may help to lower the risk of Alzheimer's, and other studies highlight the role of nature in reducing the risk of dementia.

Disservices

While the above research has identified the benefits that nature confers for older adults, not all studies support these findings. Some research has not found positive effects of green space for older adults, and some research highlights benefits which hold indirect relationships for older adults and can depend on co-occurring active lifestyles (Astell-Burt et al. 2013). In some cases, urban green spaces can be considered a disservice for older adults, dependent on conditions of use. Palliwoda and Priess (2021) examined disservices by age groups and found older persons especially felt more disturbed by other users of urban green spaces compared to other age groups. Criminal activity and safety aspects were more disturbing, in addition to barbecuing and overcrowded parks which were often perceived as negative aspects among respondents. Poorly maintained urban nature such as tree roots and vegetation can be considered hazardous and increase perceived likelihood of falls (Hong et al. 2018). Disservices may also be experienced through tree pollen allergy risks (Aerts et al. 2021).

Conclusion

This chapter briefly outlines how nature can be understood to benefit ageing populations. The CV-19 pandemic has exposed inequities around access to green space, where vulnerable populations such as older people and those in low socioeconomic areas are particularly affected (Robinson et al. 2021). As global populations face shifting demographic structures and ageing populations, there is substantial evidence demonstrating that urban nature and landscape architecture is a powerful tool to improve the human condition and health. One key theme that emerges throughout the research is the importance of where nature is situated, that is, surrounding nature. For both the very young and old (18–60-year age group), the neighborhood vicinity becomes the most relevant criterion for observed health benefits (Chaudhury et al. 2016). While the field of urban design has been recognized to have a primary role for healthy ageing, the integration of nature within this urban environment has yet to be fully recognized as equally important in supporting general health and wellbeing. Further research needs to address the unique and multiple pathways that nature offers in this health-nature relationship and in particular for this age group.

Cross-References

- ▶ Age-Friendly Future cities
- ▶ Health and the role of nature in enhancing mental health

References

Aerts, R., Bruffaerts, N., Somers, B., Demoury, C., Plusquin, M., Nawrot, T. S., & Hendrickx, M. (2021). Tree pollen allergy risks and changes across scenarios in urban green spaces in Brussels, Belgium. *Landscape and Urban Planning, 207*, 104001.

Astell-Burt, T., Feng, X., & Kolt, G. S. (2013). Mental health benefits of neighbourhood green space are stronger among physically active adults in middle-to-older age: Evidence from 260,061 Australians. *Preventive Medicine, 57*(5), 601–606.

Astell-Burt, T., Navakatikyan, M. A., & Feng, X. (2020). Urban green space, tree canopy and 11-year risk of dementia in a cohort of 109,688 Australians. *Environment International, 145*, 106102.

Baxter, D. E., & Pelletier, L. G. (2019). Is nature relatedness a basic. Human psychological need? A critical examination of the extant literature. *Canadian Psychology/Psychologie Canadienne, 60*(1), 21.

Chaudhury, H., Campo, M., Michael, Y., & Mahmood, A. (2016). Neighbourhood environment and physical activity in older adults. *Social Science & Medicine, 149*, 104–113.

Crouse, D. L., Pinault, L., Balram, A., Hystad, P., Peters, P. A., Chen, H., ... Villeneuve, P. J. (2017). Urban greenness and mortality in Canada's largest cities: A national cohort study. *The Lancet Planetary Health, 1*(7), e289–e297.

Dalton, A. M., Wareham, N., Griffin, S., & Jones, A. P. (2016). Neighbourhood greenspace is associated with a slower decline in physical activity in older adults: A prospective cohort study. SSM-population health, 2, 683–691.

de Keijzer, C., Tonne, C., Basagaña, X., Valentín, A., Singh-Manoux, A., Alonso, J., ... Dadvand, P. (2018). Residential surrounding greenness and cognitive decline: A 10-year follow-up of the Whitehall II cohort. *Environmental Health Perspectives, 126*(7), 077003.

de Keijzer, C., Tonne, C., Sabia, S., Basagaña, X., Valentín, A., Singh-Manoux, A., & Dadvand, P. (2019). Green and blue spaces and physical functioning in older adults: Longitudinal analyses of the Whitehall II study. *Environment international, 122*, 346–356.

Frank, L. D., Sallis, J. F., Saelens, B. E., Leary, L., Cain, K., Conway, T. L., & Hess, P. M. (2010). The development of a walkability index: Application to the Neighborhood quality of life study. *British Journal of Sports Medicine, 44*(13), 924–933.

Gomez-Martinez, F., de Beurs, K. M., Koch, J., & Widener, J. (2021). Multi-temporal land surface temperature and vegetation greenness in urban green spaces of Puebla, Mexico. *Land, 10*(2), 155.

Hong, A., Sallis, J.F., King, A.C., Conway, T.L., Saelens, B., Cain, K.L., Fox, E.H. and Frank, L.D., (2018). Linking green space to neighborhood social capital in older adults: The role of perceived safety. *Social Science & Medicine, 207*, pp.38-45.

Kaplan, R., & Kaplan, S. (1989). The experience of nature: A psychological perspective. Cambridge university press.

Lewis, G., & Booth, M. (1994). Are cities bad for your mental health? *Psychological Medicine, 24*(4), 913–915.

Maas, J., Verheij, R. A., Groenewegen, P. P., De Vries, S., & Spreeuwenberg, P. (2006). Green space, urbanity, and health: How strong is the relation? *Journal of Epidemiology & Community Health, 60*(7), 587–592.

Markevych, I., Schoierer, J., Hartig, T., Chudnovsky, A., Hystad, P., Dzhambov, A. M., ... Fuertes, E. (2017). Exploring pathways linking green space to health: Theoretical and methodological guidance. *Environmental Research, 158*, 301–317.

Messerli, P., Murniningtyas, E., Eloundou-Enyegue, P., Foli, E. G., Furman, E., Glassman, A., & van Ypersele, J. P. (2019). Global sustainable development report 2019: The future is now–science for achieving

R

sustainable development. United Nations publication issued by the Department of Economic and Social Affairs

Palliwoda, J., & Priess, J. (2021). What do people value in urban green? Linking characteristics of urban green spaces to users' perceptions of nature benefits, disturbances, and disservices. *Ecology and Society, 26*(1): 28

Pun, V. C., Manjourides, J., & Suh, H. H. (2018). Association of neighborhood greenness with self-perceived stress, depression and anxiety symptoms in older US adults. *Environmental Health, 17*(1), 1–11.

Robinson, J. M., Brindley, P., Cameron, R., MacCarthy, D., & Jorgensen, A. (2021). Nature's role in supporting health during the COVID-19 pandemic: A geospatial and socioecological study. *International Journal of Environmental Research and Public Health, 18*(5), 2227.

Sun, S., Sarkar, C., Kumari, S., James, P., Cao, W., Lee, R. S. Y., ... Webster, C. (2020). Air pollution associated respiratory mortality risk alleviated by residential greenness in the Chinese elderly health service cohort. *Environmental Research, 183*, 109139.

Ulrich, R. S. (1983). Aesthetic and affective response to natural environment. In *Behavior and the natural environment* (pp. 85–125). Boston: Springer.

Wilson, E. O. (1984). *Biophilia*. Harvard University Press.

Role of Urban Agriculture Policy in Promoting Food Security in Bulawayo, Zimbabwe

Metron Ziga and Abdulrazak Karriem
University of the Western Cape, Cape Town, South Africa

Synonyms

Urban food production

Definition

Urban agriculture is "an industry located within (intra-urban) or on the fringe (peri-urban) of a town, a city or a metropolis, which grows or raises, processes and distributes a diversity of food and non-food products, (re-) using largely human and material resources, products and services found in and around that urban area, and in turn supplying human and material resources, products and services largely to that urban area" (Mouget 2000, p. 11). The differences between urban and rural agriculture are not determined by location but by how urban agriculture is embedded in the urban system (de Zeeuw 2004). Urban agriculture policy, therefore, refers to the rules and regulations used to govern the practice of urban agriculture.

Introduction

Food security in many African countries has often been perceived as a rural rather than urban phenomenon. Crush and Frayne (2010, p. 7) describe urban food insecurity as an "invisible crisis," which is ignored by planners and policy makers for more visible urban problems like unemployment. In Zimbabwe, 67.79% of the population was living in rural areas in 2019 (World Bank 2021), and as a result, urban food security issues have been missing on the national policy agenda. The central government has been rolling out food security policies with a strong rural bias such as the national nutrition strategy, the food and nutrition security policy, the Zimbabwe Agenda for Sustainable Socio-Economic Transformation (ZimAsset), Command Agriculture Program, and the Transitional Stabilization Program in the past 5 years. The recent Command Agriculture aims to improve smallholder farmer productivity through the provision of seeds, fertilizers, chemicals, and tillage services (Chisoko and Zharare 2017; Mazwi et al. 2019). These programs exclude low-income urban communities, who in some cases experience higher levels of food insecurity as compared to rural areas (Tacoli et al. 2013).

Urban agriculture is a possible livelihood diversification strategy, which can potentially alleviate urban food insecurity for low-income communities. The practice of urban agriculture varies between low-income areas in different African cities and the significance of the practice differs from city to city (Crush et al. 2011). Urban agriculture can play an important role in increasing household incomes through the sale of surplus produce (Kortright and Wakefield 2011;

Moyo 2013; Jongwe 2014). The income derived through the sale of surplus produce can be used to purchase more nutritious food to promote dietary diversity. Unfortunately despite the positive prospects it proffers, urban agriculture has largely not been tolerated in most African cities as it is often viewed as a rural activity, which detracts from the modern city image (Simatele and Binns 2008). Food insecurity in some African cities can be attributed to the lack of or poor food system governance (Moragues-Faus et al. 2017; Smit 2016). Policies geared at improving the food and nutrition security of low-income communities cannot be efficient unless they promote both income generation and food production measures. There is a paucity of data on urban agricultural policies and their efficacy in the Southern African context.

This chapter seeks to assess the role of the Bulawayo Urban Agriculture Policy in promoting food security in the city. Urban agriculture has always been practiced in Bulawayo from the pre-independence era, but there was no existing policy regulating the practice until the early 2000s. The chapter demonstrates how an enabling policy environment can play a pivotal role in improving the livelihoods and food security needs of low-income communities. Following this introduction is a brief background on the urban food insecurity in Zimbabwe. This is followed by an analysis of the Bulawayo Urban Agriculture policy and the challenges of its implementation in one low-income ward of the city.

Urban Food Insecurity in Zimbabwe

Food security is achieved "when all people, at all times, have physical, social and economic access to sufficient, safe and nutritious food which meets their dietary needs and food preferences for an active and healthy life" (FAO 1996). The pillars of food security are accessibility, utilization, safety, and stability. In urban areas, food security is more of an access than availability issue. Supermarket shelves may be overflowing with food but for the urban poor access can be a challenge if the food is expensive. In low-income countries like Zimbabwe, about two-thirds of urban residents rely on low and irregular incomes which makes food access to be a challenge (Tacoli et al. 2013). Additionally, the urban poor spend more than half of their incomes on food (Frayne et al. 2009; Crush et al. 2011; Oxfam 2014; Anand et al. 2019) when they also have to pay for housing, transport, and other necessities. In Zimbabwe, the urban food crisis is exacerbated by a currency crisis, limited employment opportunities, mismanagement of funds, and recurrent droughts (Muronzi 2019).

The increase in urban poverty and food insecurity in the country dates as far back as the early 1990s when the government introduced the International Monetary Fund (IMF) and World Bank-led Economic Structural Adjustment Programme which was accompanied by massive retrenchments and loss of livelihoods. The already dire economic situation was further worsened by poor macroeconomic policies, which have been accompanied by a massive de-industrialization of the economy, hyperinflation (which led to the abandonment of the Zimbabwean dollar in 2009), and political upheavals (Kutiwa et al. 2011; Tawodzera et al. 2016). The former breadbasket of Southern Africa is now a net food importer. Zimbabwe's economic crisis and resultant food price inflation have forced many urban dwellers to turn to urban agriculture as a food security coping strategy (Kutiwa et al. 2011; Moyo 2013; Pedzisai et al. 2014). Urban agriculture has been intensifying as households seek to adapt to the unstable economic environment, which is threatening the sustainability of urban livelihoods. Smart et al. (2015) contend that in cases of extreme economic hardship and crisis, urban agriculture plays an important role in promoting household adaptation and coping. This is linked to the resilience theory, which stipulates that the lack of an economic and employment mainstay is a catalyst for alternate livelihood strategies (Dawley et al. in Smart et al. 2015).

The rapid growth of urban agriculture in Zimbabwean cities has been witnessed through anecdotal evidence and empirical research. Urban households in Zimbabwe's major cities have been reported to be growing different crops and rearing chickens, ducks, and pigeons for both subsistence and commercial purposes (Dhewa 2015). This growth in urban agriculture has been promoted by both national and local government's recognition of the instrumental role it plays in promoting the livelihoods of urban people. This recognition by different levels of government and NGOs has been pivotal in the promulgation of Municipal policies regulating urban agriculture such as the Bulawayo Urban Agriculture Policy (Moyo 2013). These policies have been necessitated by the recognition that it is the core responsibility of governments "to protect, or at least insulate, the populace, and in particular the poor" from food insecurity, which is a national issue (Haysom 2007, p. 124). Urban agricultural supportive municipal policies have contributed to the unprecedented growth of the livelihood strategy in urban areas of Zimbabwe.

The Bulawayo Urban Agriculture Policy

Zimbabwe's economy is largely agro-based in nature. For many years, urban agriculture in Zimbabwe has been practiced without any formal recognition in urban planning and development (Hadebe and Mpofu 2013). Agriculture has always been viewed as a rural activity, hence there is no clear national policy regulating urban agriculture in the country (Kutiwa et al. 2011). Local governments govern all activities within their jurisdiction including agricultural production, processing, and marketing. A number of statutory instruments such as the Regional Town and Country Planning Act of 1976 and the Urban Councils Act of 1995 guide local governments in regulating urban agriculture (Moyo 2013, p. 130). Significant strides in urban agriculture policy were made after the Nyanga Declaration on Urban agriculture in 2000, where the Urban Councils Association committed to supporting all forms of urban agriculture in the country (Moyo 2013; Toriro 2019). Following the Nyanga Declaration, Ministers of local Government from Tanzania, Kenya, Malawi, Swaziland, and Zimbabwe met in Harare in 2003. The Harare Declaration resolved that urban agriculture was an important livelihood strategy which needed to be promoted by creating an enabling environment (Chaminuka and Makaye 2015). The Harare Declaration influenced the drafting of the National Environmental Draft Policy which provides guidelines to local governments on how they can integrate and coordinate urban agriculture in the cities (Kutiwa et al. 2011). These declarations and policy instruments were fundamental in the development of the Bulawayo Urban Agriculture Policy.

The Bulawayo City Council (BCC) implemented its first Urban Agriculture Policy Guidelines in 2000. Participatory methodologies were not used in the design of the policy guidelines, hence it was mandatory to revise and come up with a solid policy document which was approved on the 7th of February 2008 (BCC 2008). The policy was prepared by a multi-stakeholder forum comprising of Agriculture and Extension services personnel, academicians, NGOs, and technical officers from the BCC. The Forum was established in October 2005 under the auspices of the Cities Farming for the Future (CFF) Program funded by the Resource Center on Urban Agriculture and Food Security Foundation (RUAF). The policy was developed in order to "...legalise, regulate and facilitate access to land and water for urban agriculture; to alleviate poverty, promote economic development and sustainable use of the environment thereby ensuring food security and surplus produce for income as well as guaranteed good nutrition for its citizens in the light of the HIV/AIDS scourge" (BCC 2008, p. 20). The policy also aims to improve the food and nutrition security of low-income earners and vulnerable groups, which include the widowed, the sick, caregivers, single parents, the unemployed, and the elderly (Hadebe and Mpofu 2013). The policy further recognizes urban agriculture as an important land use and economic activity.

Urban agricultural projects in Bulawayo are practiced under different circumstances and arrangements (BCC 2008). The first arrangement occurs in designated or zoned areas. The designated areas are plots of mostly over two hectares designated through the City's Master plans and are located in largely peri-urban and low-density residential areas. Urban agriculture is also practiced in special consent areas, which are usually low-density residential areas. Permits are required for practicing agriculture in these special consent areas with health and environmental conditions. Special consent is required for the rearing of more than 20 chickens (a smaller number is freely permitted in all residential areas), which also is determined by property size.

The city owns two productive commercial farms and sets aside land for allotment gardens in high-density areas. There are currently 12 allotment gardens which mainly target the underprivileged who are selected by Social Workers in the Department of Housing and Community Services. The city also has projects such as Gum Plantation with 1100 plots of 5000 m^2 which utilizes wastewater and the Khami School leavers and Cooperatives which trains school leavers and cooperatives on urban agriculture. The allocation of plots at gum plantation is done per ward by councilors. The final and most dominant arrangement of urban agriculture in the city is unauthorized agriculture. This is the most prevalent yet seasonal type of urban agriculture often practiced by low-income earners. It is carried out on land that was planned, surveyed, and even serviced or awaiting development in some cases. It can also be along roadsides, under electricity power lines, and along streams.

The Bulawayo Urban Agricultural Policy addresses the challenges of urban agriculture in the city including water, land, finance, and the lack of an institutional framework. The City Council pledges to avail land for urban agriculture through techniques which include reserving the land it owns, negotiating with private landowners, and allowing residents to farm in formerly undesignated areas. Residents are allowed to cultivate creeping crops such as groundnuts and beans along roads and encourage proliferous orchards along stream banks. The city council, working together with its partners such as the Agricultural Extension and Services Department and NGOs are to provide training to urban farmers to promote sustainable farming practices.

Drescher et al. (2000) argue that the critical focus of any urban agriculture policy should be food security and nutrition, health, and the urban environment and urban planning. The Bulawayo Urban Agriculture Policy focuses on these three critical areas. The urban agriculture policy was mainly framed to mobilize the city's residents to appreciate urban agriculture as a livelihood strategy, which promotes food security and income generation (BCC 2008). The Health and environment aspects of the policy are guided by the Public Health Act and the City's by laws. Urban farmers are to seek expert advice when using recycled water for irrigation and avoid cultivation on stream banks, verges, and swamps. The city's master plan and all council's urban spatial plans integrate the practice of urban agriculture; urban planners are encouraged to adhere to the plans.

Case Study

One low income ward was purposively sampled to assess the efficacy of the Bulawayo Urban Agriculture Policy. Ward 28 also known as Cowdray Park is a high density suburb. High density suburbs in Bulawayo are highly compact and are characterized by small houses in plot sizes which vary from 190 m^2 to 490 m^2 (Magwaro-Ndiweni 2011). According to the Zimbabwe poverty Atlas of 2015, ward 28 had the second highest poverty prevalence rate in the city of about 42% (ZimStats 2015). Urban agriculture is a prevalent livelihood strategy in Cowdray Park as the city of Bulawayo is grappling with de-industrialization and the resultant high unemployment levels.

The Bulawayo Urban Agriculture Policy makes provisions for land access, training, and technical advice for urban farmers. However, there is a gap between the policy on paper and its application on the ground. The urban farmers in Cowdray Park were not aware of the existence of the city's urban agriculture policy or the legality of urban agriculture. This even extended to the Resident Association's chairperson and representatives who are supposed to be the connecting link between residents and the local councilor. When the Residents Association representative was asked about the city's regulations particularly on

Bulawayo's thriving poultry sector, she responded as follows:

> It's not allowed, from my understanding it is not really allowed...it is not allowed but due to the current economic situation people are now doing this poultry production to improve their household well-being, it is a source of employment for many and the city council cannot just ask them to cease their operations. (Residents Association Representative)

The key informant from Zimbabwe Democracy Trust (ZDT), a local NGO working with urban farmers, expressed concern that some of the urban farmers did not have knowledge on the city's authority guidelines regulating the practice of urban agriculture:

> Some people are not even aware that such guidelines exist. There are no measures taken to ensure that they really follow the guidelines, except maybe discouraging stream bank cultivation through writing the warning on water bill receipts...The city does not have the capacity to monitor more than 170,000 households in the city. Unless the neighbors register complaints with the counsellor or city council nothing is done to monitor compliance with the guidelines. (ZDT Field Officer)

Urban farmers did not receive any technical or financial support from the City Council or NGOs. The farmers relied on their own finance for inputs or they asked their friends or relatives to lend them some money. The lack of financial support for urban agriculture in the City militates against its potential of significantly contributing to sustainable urban development. The local government is not capable of financing urban agriculture as a result of the country's economic crisis. Unlike instances in Cuba where the state played a proactive role in urban agriculture and South Africa where NGOs like Abalimi Bezekhaya and Soil for Life support urban gardeners, urban agriculture is an individual household matter in Cowdray Park. In Johannesburg, South Africa, urban farmers are trained by extension workers, NGOs and by other farmers (Malan 2015). The Urban Agriculture Policy of the City of Cape Town as an example legitimizes all public support for UA in Cape Town, such as the provision of free public land, fencing and infrastructure, and inputs; although some argue that it is good on paper, but not in practice (Olivier 2015). The Bulawayo Urban Agriculture Policy makes provisions for the training of urban farmers, but this has not been put into practice in Cowdray Park.

The study participants might not have been aware of the Bulawayo Urban Agriculture Policy, but they did acknowledge that the city council is now more tolerant of the practice of urban agriculture. Prior to the enacting of the policy, crops in undesignated areas would be slashed down by City council rangers. The national position on urban agriculture is that farmers should not be harassed, but a lack of clear regulations or laws leaves producers in uncertainty. In 2005 for example, crops were destroyed in some areas as a result of the central government's Operation *Murambatsvina* (remove filth) even though the Bulawayo City Council had Urban Agriculture Policy guidelines dating back to the year 2000. Ward councilors are also taking advantage of urban farmer's ignorance of policy to gain political mileage by appearing as if their leadership advocacy is allowing the city authorities to turn a blind eye to agricultural activities. The state of food security of low-income communities in the city can be improved if there is adequate support for urban agriculture from various stakeholders, but the city's policies present entry point opportunities.

Conclusion

The practice of urban agriculture in Bulawayo is acting as a livelihood strategy which promotes food security and income generation. The Bulawayo Urban Agriculture Policy has facilitated the growth and expansion of the sector in the city. There are, however, challenges in the practical implementation of the policy due to lack of awareness, economic crisis, and recurrent droughts. There is a need for the central government to regulate urban agriculture, and play an active role as it does in rural agriculture. NGOs, local government, and the Central government in Zimbabwe should learn from the success stories of countries such as Cuba to fully support urban agriculture. This will also be possible if there is a conducive economic environment. The BCC needs to engage with communities and conduct awareness campaigns so that urban farmers can become familiar with the provisions of the policy. If the Bulawayo Urban Agriculture policy is fully implemented, it can help in ameliorating the food security challenges faced by low-income earners in the city.

References

Anand, S., et al. (2019). Urban food insecurity and its determinants: A baseline study of Bengaluru. *Environment and Urbanization, 31*(2), 421–442.

Bulawayo City Council. (2008). *Revised urban agriculture policy*. Bulawayo: Bulawayo City Council. Retrieved from https://foodsystemsplanning.ap.buffalo.edu/gsfp-policy/urban-agriculture-policy-bulawayo-zimbabwe/.

Chaminuka, N., & Makaye, P. (2015). The resilience of urban agriculture in the face of adversity from city authorities: The case of Mkoba. *Global Journal of Human-Social Science, 15*(3), 15–22.

Chisoko, G., & Zharare, H. (2017). Demystifying command agriculture. Retrieved from https://www.herald.co.zw/demystifying-command-agriculture/

Crush, J., & Frayne, B. (2010). *The invisible crisis: Urban food insecurity in Southern Africa* (AFSUN series, 1). Cape Town: AFSUN.

Crush, J., Horvoka, A., & Tevera, D. (2011). Food insecurity in Southern African cities: The place of urban agriculture. *Progress in Development Studies, 11*(4), 285–305.

De Zeeuw, H. I. (2004). *The development of urban agriculture, some lessons learnt*. Key note paper for the International conference "urban agriculture, agro-tourism and City region development" Beijing 10–14 October 2004.

Dhewa, C. (2015). Rapid Growth of Urban Farming in Harare, Zimbabwe. [Online] Available at http://www.cityfarmer.info/2015/11/03/rapid-growth-of-urbanfarming-in-harare-zimbabwe/ [Accessed on 28 February 2020]

Drescher, A. W., Nugent, R., & de Zeeuw, H. (2000). Urban and Peri-urban agriculture on the policy agenda, FAO/ETC joint Electronic Conference August 21 -September 30, 2000. Retrieved from http://www.fao.org/3/X6091E/X6091E.htm

FAO. (1996). *Rome declaration on world food security and world food summit plan of action*. World food summit 13–17 November 1996. Rome: FAO.

Frayne, B., Battersby-Lennard, J., Fincham, R. & Haysom, G. (2009). Urban Food Security in South Africa: Case Study of Cape Town, Msunduzi and Johannesburg. Development Planning Division Working Paper Series No 15, Midrand: DBSA.

Hadebe, B. L., & Mpofu, J. (2013). Empowering women through improved food security in urban centers: A gender survey in Bulawayo urban agriculture. *African Educational Research Journal, 1*(1), 18–32.

Haysom, G. (2007). Urban agriculture and food security. In J. V. Braun (Ed.), *Towards a 709 healthy and sustainable world food situation 2020–2050*. Zurich: International Research 710 Food Policy Institute.

Jongwe, A. (2014). Synergies between urban agriculture and urban household food security in Gweru City. *Journal of Development and Agricultural Economics, 6*(2), 59–66.

Kortright, R., & Wakefield, S. (2011). Edible backyards: A qualitative study of household food growing and its contributions to food security. *Agriculture and Human Values, 28*, 38–53.

Kutiwa, S., Boon, E., & Devuyst, D. (2011). Urban agriculture in low-income households of Harare: An adaptive response to economic crisis. *Journal of Human Ecology, 32*(2), 85–96.

Magwaro- Ndiweni, L. (2011). Contestation in the use of residential space: House typologies and residential land in Bulawayo, Zimbabwe. *African Review of Economics and Finance, 3*(1), 40–56.

Malan, N. (2015). Urban farmers and urban agriculture in Johannesburg: Responding to the food resilience strategy. *Agrekon, 54*(2), 51–75.

Mazwi, F., et al. (2019). Political economy of command agriculture in Zimbabwe: A state-led contract farming model. *Agrarian South: Journal of Political Economy, 8*(1–2), 232–257. https://doi.org/10.1177/2277976019856742.

Moragues-Faus, A., Sonnino, R., & Marsden, T. (2017). Exploring European food system vulnerabilities: Towards integrated food security governance. *Environmental Science & Policy, 75*, 184–215.

Mouget, L. J. A. (2000). *Urban agriculture: Definition, presence, potentials and policy challenges* (Cities feeding people series, report 31). International Development Research Centre (IDRC): Ottawa

Moyo, P. (2013). Urban agriculture and poverty mitigation in Zimbabwe: Prospects and obstacles in Bulawayo townships. *Journal of Human Ecology, 42*(2), 125–134.

Muronzi, C. (2019). Zimbabwe's food crisis: 'Food security is national security'. Retrieved from https://www.aljazeera.com/economy/2019/11/28/zimbabwes-food-crisis-food-security-is-national-security

Olivier, D.W. (2015). The physical and social benefits of urban agriculture projects run by non-governmental organisations in Cape Town. Doctoral dissertation, Stellenbosch: Stellenbosch University.

Oxfam. (2014). *Hidden hunger in South Africa. The faces of hunger and malnutrition in a food secure nation*. Oxford: Oxfam International.

Pedzisai, E., Kowe, P., Matarira, C. H., Katanha, A., & Rutsvara, R. (2014). Enhancing food security and economic welfare through urban agriculture in Zimbabwe. *Journal of Food Security, 2*(3), 79–86.

R

Simatele, D. M., & Binns, T. (2008). Motivation and marginalization in African urban agriculture: The case of Lusaka, Zambia. *Urban Forum, 19*(1), 1–21.
Smart, J., Nel, E., & Binns, T. (2015). Economic crisis and food security in Africa: Exploring the significance of urban agriculture in Zambia's Copperbelt province. *Geoforum, 65*, 37–45.
Smit, W. (2016). Urban governance and urban food systems in Africa: Examining the linkages. *Cities, 58*, 80–86.
Tacoli, C., Thanh, H. X., Owusu, M., Kigen, L., & Padgham, J. (2013). *The role of local government in urban food security.* London: IIED.
Tawodzera, G., Riley, L., & Crush, J. (2016). Following the crisis: Poverty and food security in Harare, Zimbabwe. *Journal of Food and Nutrition Disorders, 5*(5), 1–10.
Toriro, P. (2019). Urban food production in Harare, Zimbabwe. In J. Battersby & V. Watson (Eds.), *Urban food systems governance and poverty in African cities.* London, New York: Routledge.
World Bank. (2021). Rural population (% of total population). Retrieved from https://data.worldbank.org/-indicator/SP.RUR.TOTL.ZS
ZimStats. (2015). Zimbabwe Poverty Atlas. Small Area Poverty Estimation: Statistics for Poverty Eradication. Harare: Zimbabwe National Statistics Agency.

Rural

► Challenges of Delivering Regional and Remote Human Services and Supports
► Collective Emotions and Resilient Regional Communities

Rural Areas

► Meeting SDG6: Ensuring Safe Drinking Water for All in Rural India

Rural Development

► Peri-urban Regions

Rural Transformation – Rural Change

► Spatial Planning Under the Impact of Urbanization and Rural Transformation in Zimbabwe: A Review of Theories, Philosophies, and Practices

Rural-Urban Continuum Settlements: Selected Case Studies

Susan Cyriac[1], Mohammed Firoz C[1] and Lakshmi Priya Rajendran[2]
[1]Department of Architecture and Planning, National Institute of Technology, Calicut, Kerala, India
[2]The Bartlett School of Architecture, University College London, London, UK

Definition

Rural-Urban Continuum (RUC) – The term was coined by Robert Redfield in 1947 while studying the industrial developments in Mexico. RUC emphasizes that a sharp breaking point of rural or urban is diminishing with the increasing urbanization trends. The rural-urban continuum can be defined as the various degrees of urban ranging from the urban core toward its hinterlands and further to the remote areas. It is the combination of a series of urban centers together with their fringe regions. The scale of the RUC could differ based on the number and characteristics of the urban core associated. The prominent forms of RUC settlements include the peri-urban areas, ribbon developments/urban corridors, desakota, and Ruralopolis. An extended synonym of the RUC is the Ruralopolis, which is defined as "an extended rural region of urban-level population density" (Qadeer 2000). Apart from the high population density, the other attributes of Ruralopolis include corridors of homestead settlements, agrarian economy, small land holding, etc.

Introduction

The typology of rural-urban continuum (RUC) settlements is broadly determined by the number of urban cores and hinterlands. Based on the number of urban cores, the two types are monocentric and polycentric. An urban center with a single prominent primate city forms the monocentric urban system. Multiple urban centers amalgamated either based on homogeneity or functionality form polycentric settlement groups. The rural-urban continuum settlements are associated with both monocentric and polycentric settlements. They occur in various sizes and patterns depending on the associated urban structure. The RUC associated with monocentric cities often occurs in the fringe areas where the rural and urban areas meet. The predominant formations of RUC along a monocentric urban system are the peri-urban development and suburbs.

The polycentric urban system varies widely in scale and comprises of complex urban networks. Such polycentric large-scale RUC patterns sometimes even cross the international boundaries. These could be continuous or discontinuous landmasses. The poly-centric region has a few prominent cities and many smaller cities and mostly lacks a primate city. It is synonymous with city region, megalopolis, polynucleated metropolitan regions, etc. The major urban corridors such as the BosWash and the metropolitan region of Ruhr in Germany are examples of polycentric urban regions. They are mainly classified into four distinct types: edge cities at the intra-urban scale, metropolitan area at the intraregional scale, polycentric urban regions at the regional scale, and world cities at the global scale (Fig. 1) (You 2017). Physical connectivity is not a significant criterion that determines the network of world cities. Cities such as New York, Frankfurt, and London comprise a network despite their proximity. These regions do not follow any particular morphology; instead, they are primarily a functional network group. Apart from these four forms, the two other polycentric RUC systems with more influence on rural characteristics are the Ruralopolis and desakota pattern settlements.

Desakota is a form of RUC which has settlements with a mix of urban activities spread across the rural settlements, thereby making the settlements acquire a blend of rural and urban characteristics. In such settlements, a clear demarcation of the urban or rural boundary is difficult. Ruralopolis, on the other hand, are regions that are predominantly rural with higher densities comparable to the metropolitan cities. These developments are mainly occurring in the Asian and African continents. The higher densities with an agrarian economy make these settlements more rural.

The chapter presents four different case studies showcasing the various types of RUC. These case studies were chosen based on their prominence in literature. The first study discusses polycentric urban regions with three cases representing three different continents:

- The continuous BosWash corridor of the USA, which is the largest megalopolis in the USA, as identified by Jean Gottmann
- The fragmented Blue Banana corridor of Europe, which is the most significant corridor in terms of area (444,883 sq. km) (Georg and Taubenböck 2015)
- The BESETO corridor from China to Japan via South Korea, which is one of Asia's most prominent and emerging corridors, spanning three countries

R

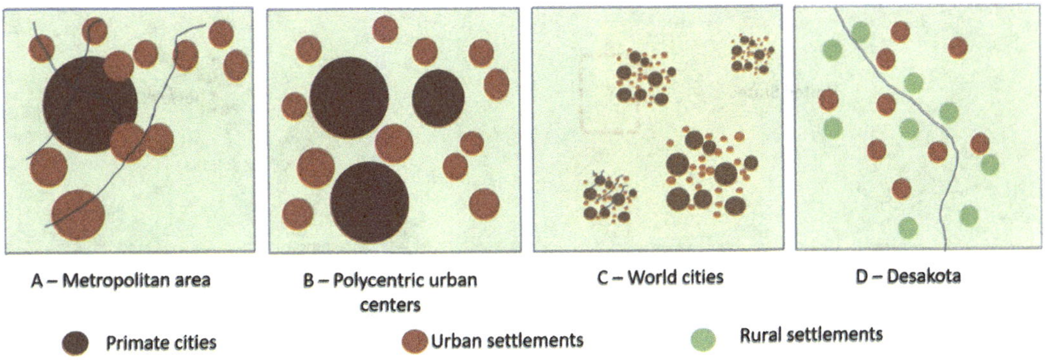

Rural-Urban Continuum Settlements: Selected Case Studies, Fig. 1 Schematic representation of polycentric urban systems and associated RUC

The second case study regions are the desakota rural-urban continuum settlement systems. Here, the three types of the desakota regions, type I (Tokaido in Japan), type II (Pearl River Delta region of China), and type III (Java, Indonesia), are discussed. The third case region is the Rural Urban Continuum settlements of Kerala, India, discussed as an extended type III desakota region. This region is taken as a case study because of its evident uniqueness of truly blurred urban and rural boundary extending up to a 650 KM stretch. Also, much literature has not acknowledged this case region. The Ruralopolis as a concept is explored as the fourth case study region on the developing Asian countries, mainly the settlements of Bangladesh and Pakistan.

The Polycentric Urban Corridors

The Northeast Megalopolis (BosWash Corridor)

The BosWash corridor is an example of a contiguous polycentric urban center at the regional level. The French geographer, Jean Gottmann, in 1961, first identified the connection of the cities via a matrix of the suburban areas. It is a contiguous stretch of major urban centers and is the most urbanized megalopolis in the USA. The corridor connects Boston to Washington DC along the Atlantic Coast and includes New York, Newark, Philadelphia, and Baltimore. The other popular nomenclature for BosWash consists of the Northeast corridor, Boston-Washington corridor, Eastern seaboard, and the Atlantic Seaboard (Fig. 2).

The historical factors that favored the growth of the region include the proximity to Europe, climatic conditions, good connectivity, navigable rivers, and the proximity to the coastal areas. The territory gained prominence with the capital established in Washington D.C., in 1800. It enabled a spillover of the infrastructure and facilities to the nearby regions, and the region is a major center of administration and financial activities. BosWash is an economic powerhouse of the world. It houses 47.6 million population and has an economic output of USD3650 billion, making it the most significant economic contributor, and the GDP output is more than that of Brazil or the UK (Florida 2019).

The 1950s saw densification of the urban cores in the BosWash region. During that period, the megalopolis housed one-fourth of the population of the country (32 million). By 2000, the population of BosWash corridor increased to 49 million, with a density as high as 930 persons per square mile. However, suburban counties started getting densified during this period (Rennie Short 2009).

The transport connectivity also contributed to the formation of the region. The emerging cities in the region were interconnected by a railroad. Later on, these were complemented by other transport modes such as air travel and road networks. The

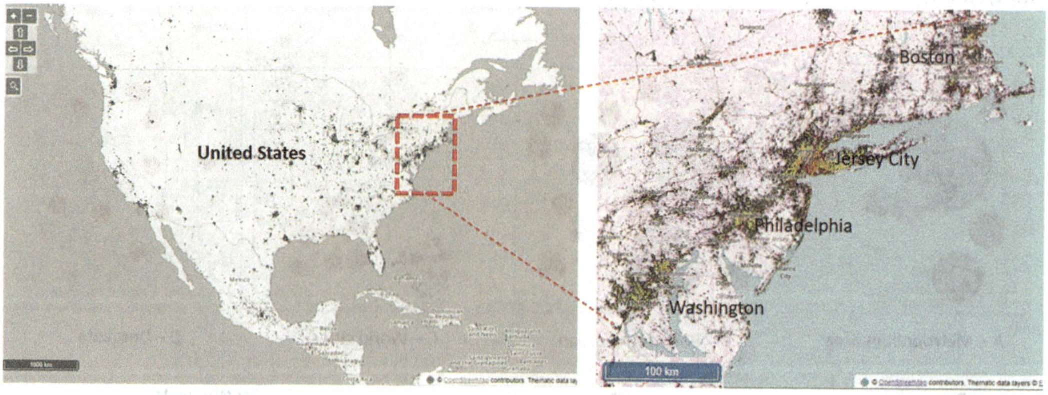

Rural-Urban Continuum Settlements: Selected Case Studies, Fig. 2 BosWash corridor. (Source: modified from (Oswald and Ames 2009, GHSL 2021))

establishment of the highway system in 1956 enabled vehicle-oriented growth and resulted in the creation of massive suburbs in the hinterlands of the cities, thus contributing to the development of the BosWash corridor.

The significant economic driver in the region used to be the industrial sector. However, it got replaced by the service sector of late. The major service sectors in the megaregion are information technology, financial activities, professional and business services, education, and health services. The region has a high concentration of wealthy population, which increased property prices and cost of living. It resulted in the displacement of the people toward the suburbs resulting in sprawls. This also resulted in social inequality in the region (Allen 2017).

Apart from the historic, transportation, and economic drivers, the region thrived by government initiatives. Subsidized housing and the provision of highways in the region promoted the development of the area. Joint development at the megaregion scale could provide further impetus to the growth of the region. However, consideration of the environment is a crucial aspect that needs to be taken care of in this megaregion.

The Blue Banana Corridor

The Blue Banana corridor is an example of a fragmented polycentric urban corridor. The idea of Blue Banana was conceptualized in 1989 by Roger Brunet and his team of geographers based on the economic strength of Europe. The corridor stretches from Northwest England to Northern Italy. The main cities located along the corridor are London, Manchester, Rotterdam, Hague, Amsterdam, Brussels, Genoa, Switzerland, and Milan (Fig. 3).

According to geographers, the Blue Banana dated back to Roman times, and it was the corridor along which the industrial revolution spread. It is a densely populated area, a polycentric city region

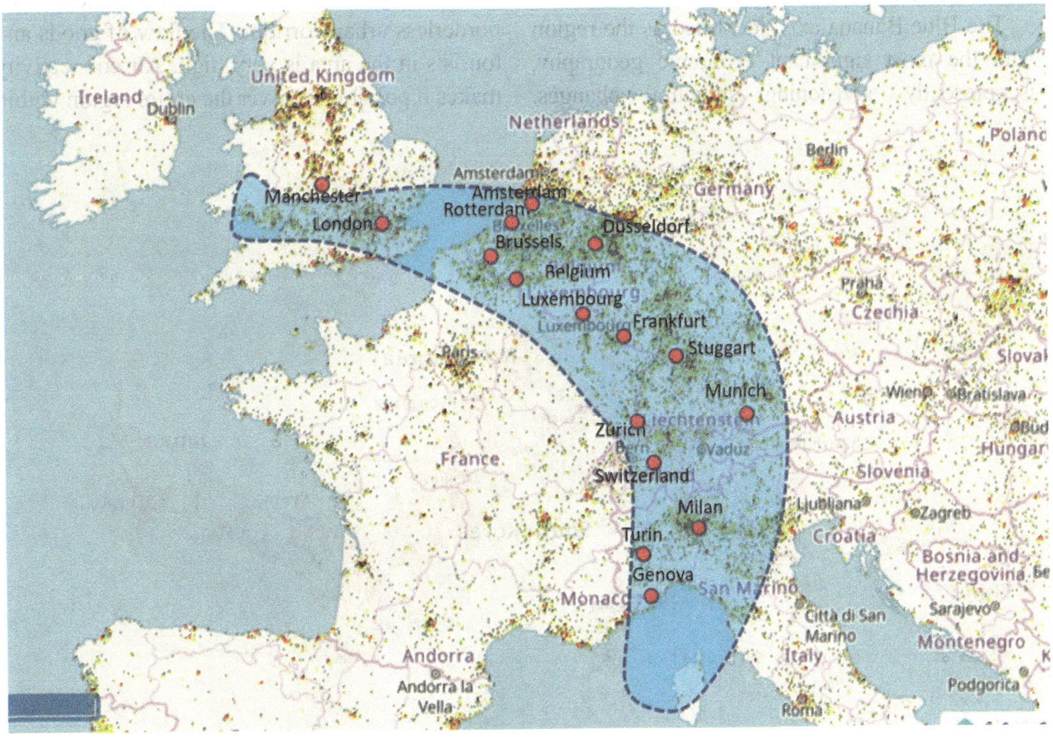

Rural-Urban Continuum Settlements: Selected Case Studies, Fig. 3 The Blue Banana (Source: modified from (GHSL 2021))

with large- and medium-sized cities that housed almost 40% of the European Union in 1996 (Hospers 2003). The corridor has a well-developed physical infrastructure and dense transportation networks. The region acts as the cultural and educational hub of Europe.

The Blue Banana corridor forms the backbone of Western Europe. The urban corridor comprises industries and service activities, and it measures between 1500 and 1700 kilometers. The region has good connectivity through airport hubs and seaports. It has the headquarters of the European Union and houses the International Court of Justice located in the region. The corridor has a favorable temperate climate as well. It is rich in mineral resources and has fertile soil. In the nineteenth century, the region was the center of world industry and finance. Unlike the success of the BosWash corridor, the Blue Banana concept had spatial competition from the surrounding areas with other models such as the Mediterranean arc (Golden Banana), the Alpine Furrow, and the Bunch of Grapes or the European Grape (Jacobs 2014; Faludi 2015).

The Blue Banana axis, identified as the region with the most significant economic geography, is presently envisioning structural changes. The service sector is increasing in the region, which had a dominant industrial base. Hospers (2003) indicates that despite the changes, the corridor will continue to emerge as the main economic backbone of the European Union for a while.

BESETO (Beijing to Tokyo Via Pyongyang and Seoul)

BESETO is the extended urban area from Beijing via Seoul to Tokyo (BESETO), including 112 cities stretching over a 15,000 km strip of high-density land. The region connects the most developed areas of the respective countries (Keum 2000). It joins four corridors of China (Bohai Rim corridor), North Korea (Shinuiju-Kaesong corridor), South Korea (Seoul-Pusan corridor), and Japan (Fukuoka-Tokyo). The total population of BESETO exceeds 98 million. BESETO provides an opportunity to study further about metropolitan regions beyond national borders (Fig. 4).

Researchers use the term Ecumenopolis (coined by C.A. Doxiadis) to indicate the borderless urban corridor. The flow of goods and tourists in the area is very high. Air connectivity makes it possible to cover the entire region within

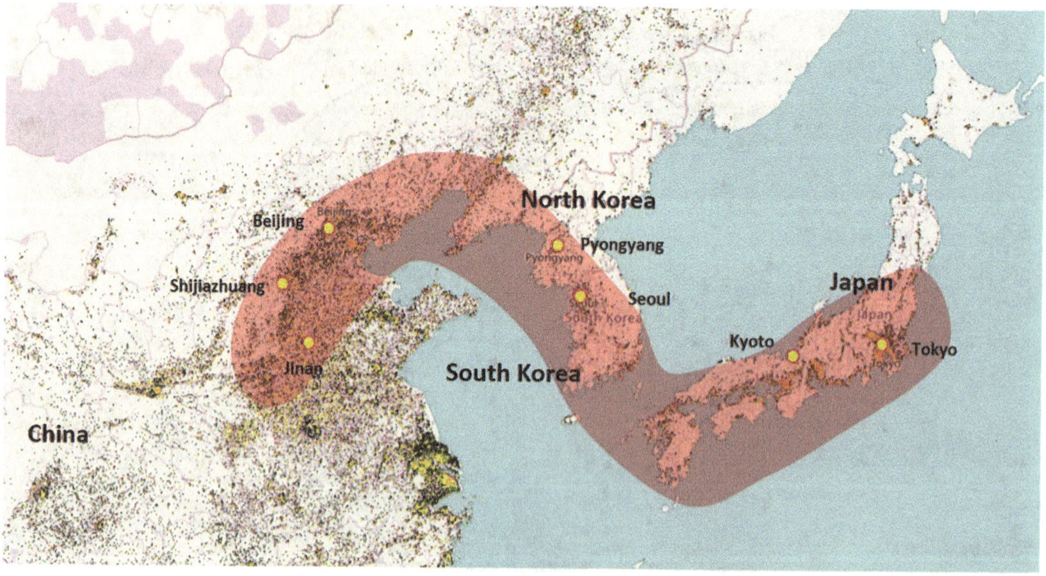

Rural-Urban Continuum Settlements: Selected Case Studies, Fig. 4 BESETO corridor. (Source: adapted from ("The Transnationalization of Urban Systems: The BESETO Ecumenopolis" 1996; modified from GHSL 2021)

2 hours. The major urban centers of these regions have good connectivity comprising of extensive rail and road networks.

Professor Choe Sang proposed the BESETO in 1991 that enabled inter-city cooperation by connecting the capital cities of three major countries of Northeast Asia (Keum 2000). The mayors of the three cities initiated the approaches for collaboration in the BESETO region. The sectors of cooperation identified included urban management, economy, technology, and environment. The association also envisioned the institutionalization of the cooperative scheme.

However, there are few hurdles for the effective establishment of the BESETO. The urban centers involved have different political and economic structures, and coordination among the three countries is complex. The language and cultural differences might also hinder the progress. The way forward for the BESETO corridor is an interdependence amidst the counterparts. The strengthening of inter-modal transport will also help to strengthen the corridor.

Desakota Regions

The term desakota was coined in 1990 by Terry McGee while studying the urbanization of Indonesia. The term is a combination of Indonesian "desa" implying village and "kota" implying city (Bahasa language of Indonesia). The Asian urbanization witnesses the spread of desakota phenomenon which are areas with a mix of urban and rural characteristics (Guldin 1996). According to McGee "the desakota developments have six characteristics. These are:

- A large population of small holder cultivators
- An increase in non-agricultural activities
- Extreme fluidity and mobility of population
- A mixture of land uses with agricultural parcels co existing with cottage industries, suburban development
- Increased work force participation of the female labour force; and
- Grey zones where informal and illegal activities regroup." (WordSense 2021)

Based on the differences in characteristics, there are three types of desakota regions (McGee 2013). The following examples indicate the different typologies of desakota.

Desakota Type I

The first type of desakota has declining rural settlements, land uses, and agricultural population owing to a shift to the urban centers. The pattern has an increase in income. The agricultural land use is primarily due to the protection policies established by the government. Though these regions have rural landscapes, the primary activity is nonagricultural. Japan and South Korea have this desakota pattern. One such type I desakota is the Tokaido region. Tokaido was developed based on the concept of the megalopolis proposed by Jean Gottmann. The Tokyo-Osaka corridor connecting the urban cores is an example of a polycentric mega-conurbation. Historically, Tokaido joined the two capitals of Japan dating back to the feudal era. With the industrial revolution, Tokaido became a highly industrialized area housing the major heavy industries. It had good connectivity by transport systems of a well-connected network of roads, airports, ports, and rails. By 1965, the megaregion housed 50 million people (50% of the total population of Japan) with an average density of 1400 persons per square kilometer (Sorensen 2019). The region thrived on being the prime center of economics and politics of Japan.

The corridor is 600 kilometers in length and includes the urban centers of Tokyo, Kobe, Nagoya, and Osaka. Areas outside the urban cores have high density due to the rapid industrialization in the region. The farmland conversions, the following policy changes that protected the agricultural sector, and the rapid industrialization resulted in a desakota development in the megaregion (Hebbert 1992). The desakota region has a mix of housing, industry, farmland, and urban services. Even the primate urban cores of Tokyo have a sizeable amount of farmland (Hebbert 1992). These characteristic features led to the categorization of this region as desakota (Fig. 5).

However, all these developments came with a series of environmental crises in Japan. During the

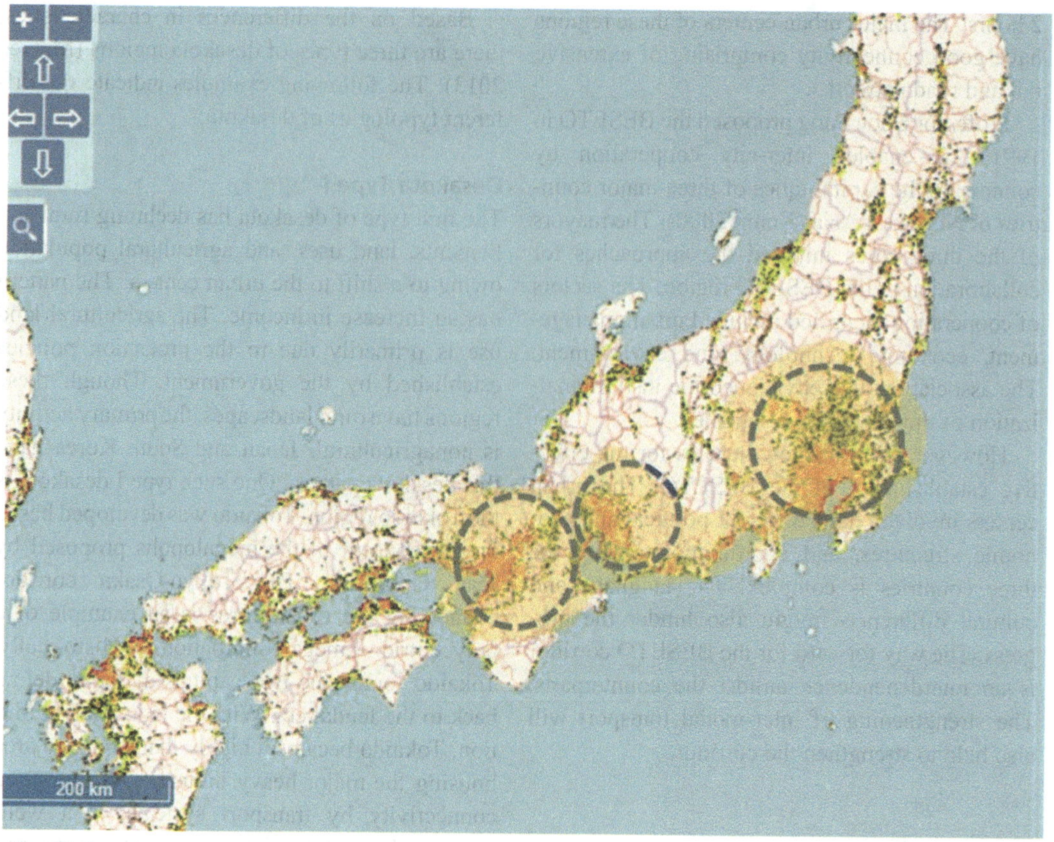

Rural-Urban Continuum Settlements: Selected Case Studies, Fig. 5 The Tokyo-Osaka corridor. (Source: modified from (GHSL 2021))

1960s, Japan was one of the most polluted countries globally (Murphey and Murphey 1984). It created massive protests, and community development movements to protect the environment began to flourish. During this period, the planning process of Japan was concentrated only on the urban centers, while the peripheral areas were left to development without severe restrictions. For instance, the developer's contribution to the public was overlooked in the suburban areas (Sorensen 2019).

Though it was once a thriving conurbation, Tokaido is at present an example of a shrinking megaregion. The demographic decline of the region already impacts the economy. It provides a valuable suggestion that urban form and infrastructure capacity development need to be flexible. It also suggests that urban structures should be flexible and developable after the peak growth period of a conurbation/megaregion is passed.

Desakota Type II

The second desakota pattern is characterized by its shift from agricultural to nonagricultural activities in the urban core areas and the adjacent locations. These settlements have urban characteristics in terms of connectivity, infrastructure provision, and income levels with rapid economic growth. The Kolkata (Calcutta) region in India, Jabotabek in Java, parts of China, etc. are examples of this type of development.

Pearl River Delta in China is an example of a type II desakota region. It is a low-lying delta formed by the Pearl River estuary, where the river meets the South China Sea. Historically the maritime silk road began here. The population of the region was below 10 million during 1978 and comprised of several medium-sized cities embedded in a network of water bodies dominated by mostly villages. The region's estimated population for 2019 is 58 million, indicating the rapid population growth in the area (EROS 2021) resulting out of China's post-1978 economic reforms leading to the inflow of foreign direct investment. Presently the region is considered one of the largest urban areas in size and population, and it comprises 11 municipalities, including Hong Kong, Guangzhou, Shenzhen, Jiangmen, and Macau. It has eight harbors. Presently, the megaregion is an industrial hub that manufactures items ranging from communication and computer equipment to toys and ceramics, thereby resulting in a change in the occupation profile of the residents. The cities of Guangzhou and Shenzhen act as service sectors, and Hong Kong functions as the global city (Bie et al. 2015).

Despite being an industrial hub, large-scale agricultural activities still exist in the region to cater the needs of the urban markets surrounding the Pearl River Delta region. It results in the coexistence of agriculture with manufacturing and service sectors, thereby reducing the rural-urban dichotomy in the region. The transformation is most prominent along the transportation corridors. The continuous stretches of integrated activities result in the expansion of towns that engulf the villages on the way (Fig. 6).

The urbanization process of the Pearl River Delta by Zhou in 1995 indicated four distinct but interlinked processes. These are the urbanization of boundaries between rural and urban areas, small towns, villages, and economic and technological development districts (Eliyu Guldin 1996). The rural communities outside the major cities are urbanized to cater the residential and service needs of the nearby growing towns.

Desakota Type III

High-density regions with low economic growth characterize the third type of desakota. These locations are near the smaller urban centers with slow economic growth and have a high population growth rate and low productivity in all sectors of the economy. According to McGee, Bangladesh, Sichuan Basin in the interior of China, the Jogjakarta region of Java, etc. are examples of type III desakota. Though Kerala in South India is categorized as type III desakota, it has certain distinct peculiarities. Hence, this region shall be discussed separately.

Jakarta metropolitan area is the largest urban area of Indonesia. It includes the cities of Jakarta, Bogor, Depok, Tangerang, and Bekasi. The population of the region in 1980 was 11.4 million,

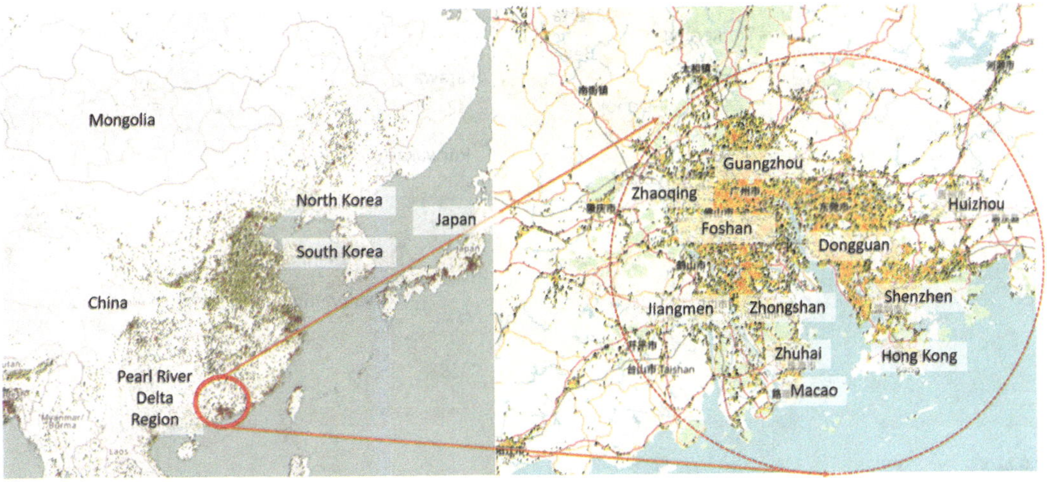

Rural-Urban Continuum Settlements: Selected Case Studies, Fig. 6 Pearl River Delta region (Source: modified from (GHSL 2021))

which increased to 34 million by 2018 (Rustiadi et al. 2021). The region is the administrative and economic center of Indonesia. The Jakarta region includes the urban area of Jakarta, the adjoining cities, and the hinterlands that exhibit the desakota type of development. The region accounts for more than 25% of the GDP of Indonesia (Fig. 7).

The urbanization pattern was different in the Jakarta region as compared to the other Western cities. The extended metropolitan region had a rural population engaged in agricultural activities. However, the change in economic activities led by industrialization and the improved road networks started drawing labor to other activities. The urbanization led to the blurring of the distinction between the urban and rural areas. It also led to the development of a cheaper transportation network. Industries were also located in the region, increasing economic activities in the region. Jakarta had region-centric urbanization (Mcgee and Greenberg 1992). Typically the urban expansion was observed from the city core to the extended metropolitan region. The corridors that emerged from the urban center had the desakota pattern of urbanization where nonagricultural activities and agricultural activities coexist.

The urban expansion of Jakarta follows the transportation corridors. The core urban areas grow and connect to the suburban areas, smaller urban areas, and other metropolitan areas along the major highways. However, the desakota pattern of urbanization, which results in blurred rural and urban zones, has affected the peri-urban regions. The desakota regions, according to McGee, are crucial zones that are the resource base for the core urbanized areas. Considering the potential of the Jakarta megaregion, the government is promoting infrastructure development and industries while ignoring the transitional agricultural activities. It has led to several environmental challenges to the region.

Rural-Urban Continuum Settlements of Kerala

Kerala in India has a unique settlement pattern which is a classic example of the rural-urban continuum, especially along the western coastal stretch. Although the Western coastal plains of India have a

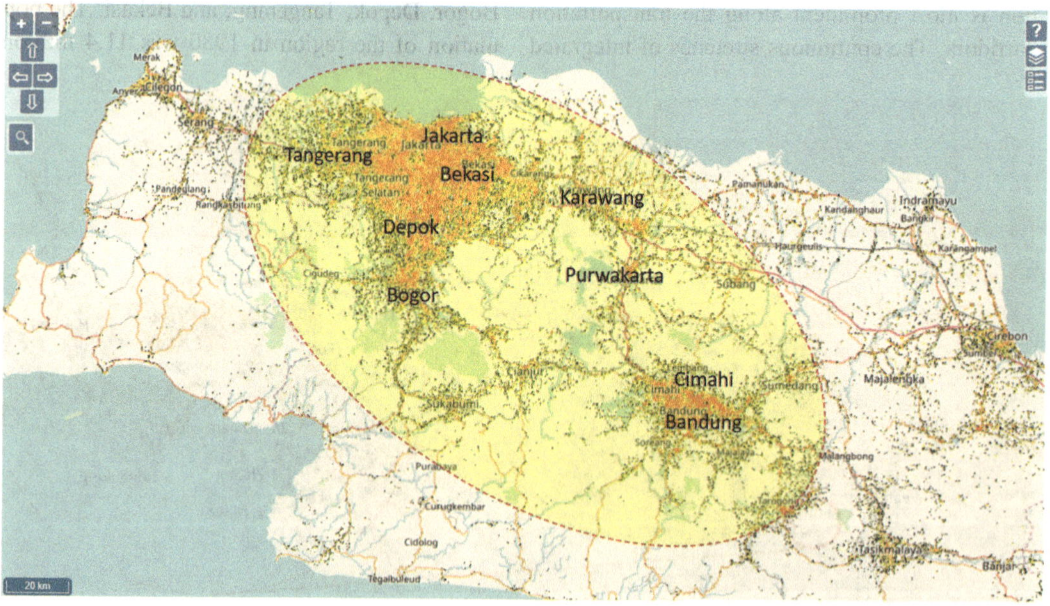

Rural-Urban Continuum Settlements: Selected Case Studies, Fig. 7 Jakarta region (Source: modified from GHSL, 2021)

distinct RUC settlement pattern extending from the state of Goa to Kanyakumari in Tamil Nadu, a uniform and concentrated distribution exists in coastal Kerala, mostly (Firoz C 2015; Firoz C et al. 2014).

The width of Kerala state from west to east varies from 11 to 124 kilometers, with a length of 560 kilometers along the coast in the north-south direction (ENVIS Centre 2014). As per the 2011 census of India, the area of the state is 38,863 square kilometers, which constitutes 1.2% of the geographical area of India, with a population of 33.4 million people. Kerala is the third densely populated state of India, with an average density of 859 persons per square kilometer. Figure 8a indicates the location of Kerala in India and the desakota regions of Kerala. Similar RUC patterns occur along major highways connecting adjacent cities. The settlement system is interconnected though dispersed and forms a linear dense agglomerated stretch (Firoz C et al. 2014). Widely acclaimed as a desakota (type III) settlement model, economic activities encompassing various sectors occur throughout the settlements in most areas. For instance, rural-urban knots performing commercial activities are present throughout most transportation nodes in Kerala (Ansari 1970). The dispersed settlements are devoid of a distinct core and lack any specific nodality (Sreekumar 1990). The influence of topography and physical features has contributed significantly to the settlement pattern of Kerala (Ansari 1970).

The availability of urban facilities in rural areas due to equitable resource allocation and the presence of all sectors of the economy throughout the settlements results in urban characteristics to the rural areas. Simultaneously, intermittent stretches of agricultural land parcels and dispersed homestead settlement patterns along the transportation corridors give rural characteristics to urban areas. Hence, the resulting urban system in Kerala is unique compared to other parts of India. The equitable distribution of higher-order facilities resulted in minimal migration from rural areas to urban areas within the state, unlike other Indian

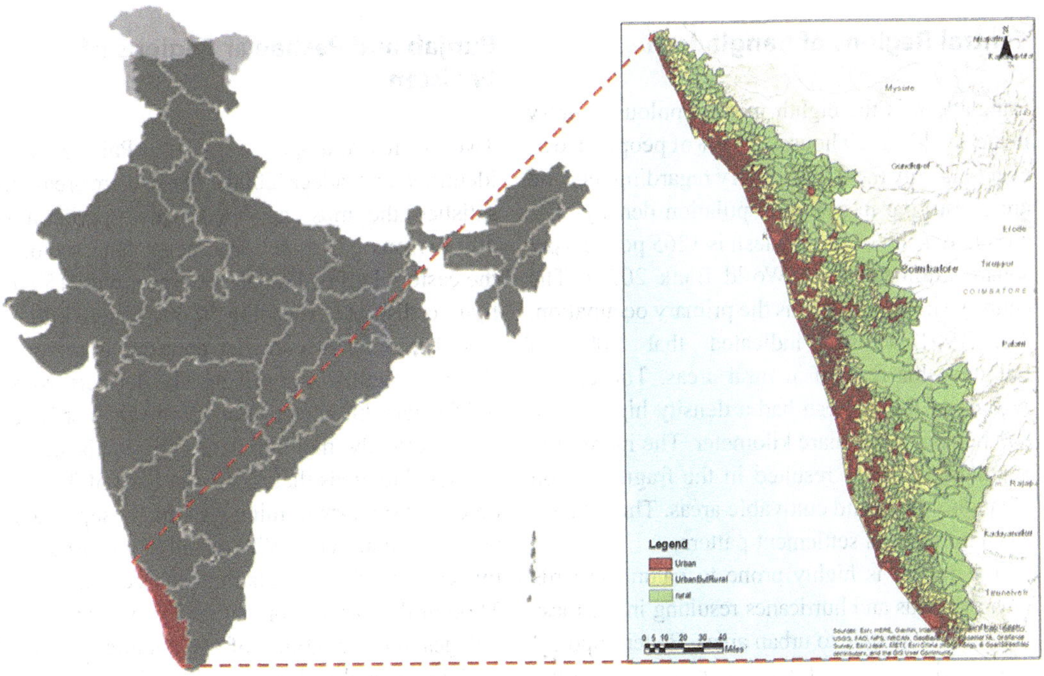

Rural-Urban Continuum Settlements: Selected Case Studies, Fig. 8 (**a**) Location of Kerala in India, (**b**) census classification of settlements of Kerala

states (Firoz C et al. 2014). Apart from the peripheral outlook, both rural and urban settlements have almost similar densities and population structures. Qadeer indicates that Kerala has a mix of desakota and ruralopolitan characteristics (Qadeer 2000).

Ruralopolis Settlements

Ruralopolis is defined as the form of urbanization that occurs by the in-place growth of population occurring in the third world countries which leads to the formation of ruralopolis settlements by Qadeer (2000). The increased density results in ribbon development, especially along the transportation corridors. These high-density regions are seldom recognized as urban, as the primary occupation of the residents is agriculture. Ruralopolis is a region with uniform characteristics and may have settlement systems that are unique for the region. For the case study, we are exploring the ruralopolis found in Bangladesh, Pakistan, and India.

Central Regions of Bangladesh

Bangladesh is the eighth most populous country in the world which houses 2.11% of people. However, it is only the 92nd country regarding the land area, resulting in a high population density. The current density of Bangladesh is 1265 persons per square kilometer (The World Bank 2020). The country has agriculture as the primary occupation. The 1991 census indicated that 80% of Bangladeshis resided in rural areas. The central regions of Bangladesh had a density higher than 800 persons per square kilometer. The increasing population density resulted in the fragmentation of landholdings and cultivable areas. The villages had a homestead settlement pattern.

The region is highly prone to natural hazards such as floods and hurricanes resulting in the mass migration of people to urban areas. Better opportunities and facilities at the urban centers further triggered the migration to urban centers. The rural system had a further setback due to the high population density; the increasing labor force could not be absorbed in the traditional agricultural system, and villages had significantly low non-agricultural activities. The transportation network connecting to the towns was inadequate, and hence daily commutation was at a low level, increasing the migration.

The planned development was limited to individual urban centers and master plans catered to these areas. The improved basic amenities, infrastructure facilities, and services that were an impetus of the migration resulted in the inadequacy of services, shortage of housing for the low-income groups, law and order problems, and deteriorating environmental conditions in Bangladesh (Roberts and Kanaley 2006).

The rural population with densities as high as the significant urbanized regions of the world require more facilities and services that cater to the people. The base for the high-density hinterland needs expansion (Fig. 9). The increasing density pattern of the region will inevitably result in structural changes to the settlement pattern (Qadeer 2000).

Punjab and Peshawar Regions of Pakistan

Two distinct ruralopolis regions in Pakistan were identified by Qadeer (2000). These were areas that satisfied the most accepted density criteria of 400 persons per square kilometer. One region is the eastern half of Punjab province, with 15 contiguous districts exceeding 50,000 square kilometers. The other is a smaller contiguous region in Peshawar valley comprising six districts within 9500 square kilometers (Karrar and Qadeer 2013).

Historically the population in Pakistan was restricted to areas that receive at least 500 mm of precipitation. Dry farming was the most prevalent agricultural activity. With the advent of irrigation by canals, the population started spreading. Though the earlier population lived in nucleated villages, the availability of water sources resulted in the dispersal of homesteads and hamlets. The villages in Punjab province have a star-shaped pattern, and the node is where several roads

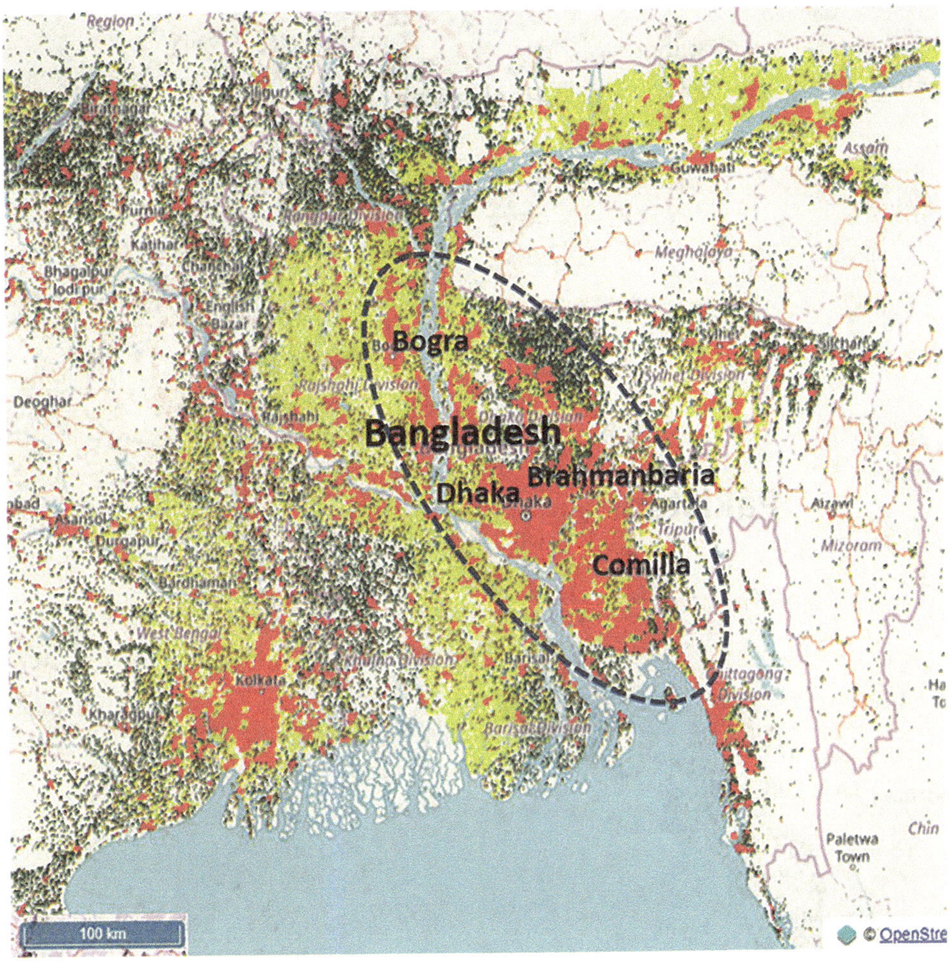

Rural-Urban Continuum Settlements: Selected Case Studies, Fig. 9 Ruralopolis regions of Bangladesh in 1991 identified by Qadeer 2000. (Source: modified from GHSL 2021)

converge. Houses are located along the roads and radiate from the node giving the rural and urban areas a stellar shape ("Functions and Pattern of Rural Settlements," n.d.).

Pakistan had a fivefold increase in its population from 1951 to 2012, turning the cities into metropolitan regions and the villages into ruralopolitan regions (Kugelman et al. 2014). The country experiences two forms of urbanization: the natural growth of cities and towns in an organized format and the population by an implosion characterized by high-density rural settlements. The ruralopolis is showing increasing tendencies of urbanization as the village is turning to urban activities. Further, the rural sprawl fuses with the exurban developments. Qadeer (Kugelman et al. 2014) estimates that if the ruralopolis is considered, the percent of the urban population in Pakistan can exceed 60%.

The high urbanization requires to be dealt with caution in Pakistan. It should be steered away from the agricultural lands into settlements designated for habitation. The necessary urban infrastructure, facilities, and services should be provided for both the ruralopolis regions and the urban areas. Concerning the urbanization policy in Pakistan, Qadeer identifies few institutional prerequisites that include ensuring adequate services and quality of life, responsive local governance, appropriate planning, and building rules (Fig. 10).

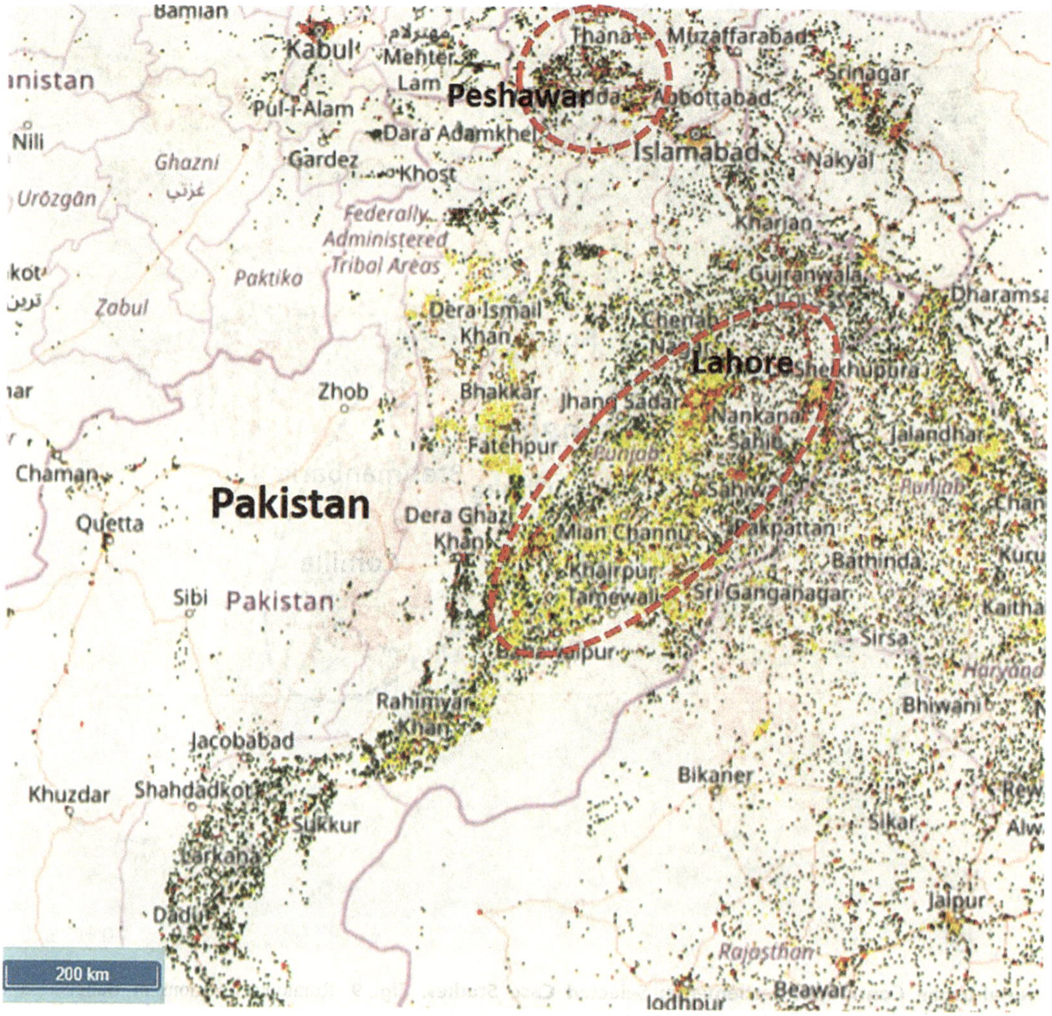

Rural-Urban Continuum Settlements: Selected Case Studies, Fig. 10 Ruralopolis regions of Pakistan, 1998 as identified by Qadeer 2000. (Source: modified from GHSL 2021)

Conclusion

The rural-urban continuum and the polycentric urban centers with mix of urban and rural characteristics are evolving in various forms globally. The interface of urban and rural areas occurs in various scale and different proportions. These emerging geographies that even cross international boundaries need to be recognized. The future research areas will be acknowledging and planning for such large polycentric, urban rural mixed regions.

However, each region has its challenges. Most of the regions deal with an expanded area that does not often follow the administrative boundaries. The integration of the institutional mechanism for megaregions and the provision of facilities, infrastructure, access to health care, housing, etc. are challenges for the evolving megaregions. While the urbanized polycentric megaregions deal with environmental challenges, the regions with high-density rural populations suffer from a lack of facilities, institutional mechanisms, and rapid changes in land use apart from unsustainable ecological practices. The actions that safeguard the environment should go hand in hand with increasing the region's economy.

Cross-References

- ▶ Connecting Urban and Regional Innovation Ecosystems to Enhance Competitiveness
- ▶ Future of the City-Region Concept and Reality
- ▶ Growth, Expansion, and Future of Small Rural Towns
- ▶ Peri-Urban Regions
- ▶ Peri-Urbanization
- ▶ Planning for Peri-Urban Futures
- ▶ Spatial Demography as the Shaper of Urban and Regional Planning Under the Impact of Rapid Urbanization

References

Allen, D. J. (2017). *Lost in the transit desert: Race, transit access, and suburban form*. https://doi.org/10.4324/9781315667027.

Ansari, J. H. (1970). A study of settlement patterns in Kerala. *Ekistics, 30*(180), 427–435. https://www.jstor.org/stable/43616449.

Bie, J., de Jong, M., & Derudder, B. (2015). Greater pearl river delta: Historical evolution towards a global city-region. *Journal of Urban Technology, 22*(2), 103–123. https://doi.org/10.1080/10630732.2014.971575.

Eliyu Guldin, G. (1996). Desakotas and beyond: Urbanization in southern China. *Ethnology, 35*(4), 265–283.

ENVIS Centre, Kerala. (2014). State of environment and related issues. *Kerala State Council for Science, Technology and Environment*. http://www.kerenvis.nic.in/Database/Industry_829.aspx.

EROS, Earth Resources Observation and Science Center. (2021). Pearl River Delta, China. *Earth Shots, USGS*. https://eros.usgs.gov/image-gallery/earthshot/pearl river-delta-china. Accessed 1 Oct.

Faludi, A. (2015). The 'blue Banana' revisited. *European Journal of Spatial Development, 1*(56).

Florida, R. (2019). The real powerhouses that drive the world's economy. *Bloomberg CityLab*. https://www.bloomberg.com/news/articles/2019-02-28/mapping-the-mega-regions-powering-the-world-s-economy.

Functions and Pattern of Rural Settlements. (n.d.). http://gcwk.ac.in/econtent_portal/ec/admin/contents/87_18GC507_2020121706185771.pdf

Georg, I., & Taubenböck, H. (2015). *Identifying urban corridors: Unified concept and global analysis*. CEUR Workshop Proceedings. http://ceur-ws.org/Vol-1598/paper11.pdf

Georg, I., Blaschke, T., & Taubenbock, H. (2018). Are we in Boswash yet ? A multi-source geodata approach to spatially delimit urban corridors. *International Journal of Geo-Information, 1–19*. https://doi.org/10.3390/ijgi7010015.

GHSL, Global Human Settlement Layer. (2021). *Global visualisation*. https://ghsl.jrc.ec.europa.eu/visualisation.php

Guldin, G. E. (1996). Desakotas and beyond: Urbanization in southern China. *Ethnology, 35*(4), 265–283.

Hebbert, M. (1992). Seb-Biki amidst Desakota: Urban Sprawl and Urban Planning in Japan. *Planning for Cities and Regions in Japan*.

Hospers, G. J. (2003). Beyond the blue banana? Structural change in Europe's geo-economy. *Economic Geography*, 76–85. https://doi.org/10.1007/BF03031774.

Jacobs, F. (2014). *The blue banana – The true heart of Europe – Big think*. https://bigthink.com/strange-maps/the-true-heart-of-europe-nil-the-blue-banana/

Karrar, H. H., & Qadeer, M. A.. (2013). *Urbanization of everybody and social sustainability*. https://www.wilsoncenter.org/sites/default/files/ASIA_140502_Pakistan'sRunaway Urbanizationrpt_0530.pdf#page=30

Keum, H. (2000). Globalization and inter-city cooperation in Northeast Asia. *East Asia*. https://doi.org/10.1007/s12140-000-0029-y.

Kugelman, M., Haider, M., Haque, N. U., Hussain, N., Tahir, A., Iqbal, A., Kugelman, M., Nishtar, J. C. S., Chishtie, F., Qadeer, M. A., & Siddiqui, T. (2014). *Pakistan's runaway urbanization: What can be done?* Edited by Michael Kugelman.

Map of China, Beijing, Hong Kong. (2021). https://www.geographicguide.com/asia/maps/china.htm. Accessed 21 Oct.

McGee, T. G. (2013). The emergence of 'Desakota' regions in Asia: Expanding a hypothesis. In N. Brenner (Ed.), *Implosions/Explosions*. https://doi.org/10.1515/9783868598933-010.

Mcgee, T. G., & Greenberg, C. (1992). The emergence of extended metropolitan regions in ASEAN: Towards the year 2000. *ASEAN Economic Bulletin, 9*(1), 22–44. https://doi.org/10.1355/AE9-1B.

Mohammed Firoz, C. (2015). *Reclassification of the typology and pattern of composite settlement systems: A case of Kerala, India*. Kharagpur: Indian Institute of Technology.

Mohammed Firoz, C., Banerji, H., & Sen, J. (2014). A methodology to define the typology of rural urban continuum settlements in Kerala. *Journal of Regional Development and Planning, 3*(1), 49–60. https://econpapers.repec.org/RePEc:ris:jrdpin:0025.

Murphey, R., & Murphey, E. (1984). The Japanese experience with pollution and controls. *Environmental Review, 8*(3), 284–294. https://doi.org/10.2307/3984327.

Oswald, M., & Ames, D. (2009). Evaluating the current state of the BOSWASH transportation corridor and indicators of resiliency. https://ntrl.ntis.gov/NTRL/dashboard/searchResults/titleDetail/PB2014104129.xhtml

Pauchet, M., & Oliveau, S. (2008). Kerala: A Desakota? *European Population Conference*. https://halshs.archives-ouvertes.fr/halshs-01140931/document

R

Qadeer, M. A. (2000). Ruralopolises: The spatial organisation and residential land economy of high-density rural regions in South Asia. *Urban Studies, 37*(9), 1583–1603. https://doi.org/10.1080/00420980020080271.

Rennie Short, J. (2009). The liquid city of megalopolis. *The New Blackwell Companion to the City*, 77–90. https://doi.org/10.1002/9781444395105.ch3.

Roberts, B., & Kanaley, T. (2006). Urbanization and sustainability in Asia, case studies of good practice. https://www.adb.org/sites/default/files/publication/27965/urbanization-sustainability.pdf

Rustiadi, E., Pravitasari, A. E., Setiawan, Y., Mulya, S. P., Pribadi, D. O., & Tsutsumida, N. (2021). Impact of continuous Jakarta megacity urban expansion on the formation of the Jakarta-Bandung conurbation over the rice farm regions. *Cities, 111*. Elsevier Ltd. https://doi.org/10.1016/j.cities.2020.103000.

Sorensen, A. (2019). Tokaido megalopolis: Lessons from a shrinking mega-conurbation. *International Planning Studies, 24*(1), 23–39. https://doi.org/10.1080/13563475.2018.1514294.

Sreekumar, T. T. (1990). Neither rural nor urban: Spatial formation and development process. *Economic and Political Weekly, 25*(35/36), 1981–1990. https://www.jstor.org/stable/4396713.

The Transnationalization of Urban Systems: The BESETO Ecumenopolis. (1996). *Emerging World Cities in Pacific Asia*. http://www.nzdl.org/cgi-bin/library?e=d-00000-00%2D%2D-off-0envl%2D%2D00-0%2D%2D%2D%2D0-10-0%2D%2D-0%2D%2D-0direct-10%2D%2D-4%2D%2D%2D%2D%2D%2D-0-1l%2D%2D11-en-50%2D%2D-20-about%2D%2D-00-0-1-00-0%2D%2D4%2D%2D%2D%2D0-0-11-10-0utfZz-8-00&cl=CL1.2&d=HASH0119908b862414c69769a396.7.3.4&x=1

The World Bank, Open Data. (2020). *Population density – Bangladesh*. https://data.worldbank.org/indicator/EN.POP.DNST?end=2020&locations=BD&start=1961&view=chart

WordSense, Online Dictionary. (2021). Desakota. https://www.wordsense.eu/desakota/. Accessed 21 Oct.

You, Y. (2017). The classification of urban systems: A review from monocentric to polycentric. *Advances in Economics, Business and Management Research, 42*. https://doi.org/10.2991/isbcd-17.2017.1.

9783030877446VOL02

The Palgrave Encyclopedia of Urban and Regional Futures

Robert C. Brears
Editor

The Palgrave Encyclopedia of Urban and Regional Futures

Volume 3

S–Z

With 418 Figures and 143 Tables

palgrave
macmillan

Editor
Robert C. Brears
Our Future Water
Christchurch, New Zealand

ISBN 978-3-030-87744-6 ISBN 978-3-030-87745-3 (eBook)
https://doi.org/10.1007/978-3-030-87745-3

This Palgrave Macmillan imprint is published by the registered company Springer Nature Switzerland AG.
The registered company address is: Gewerbestrasse 11, 6330 Cham, Switzerland

Preface

The *Palgrave Encyclopedia of Urban and Regional Futures* provides readers (practitioners, academics, researchers, etc.) with expert interdisciplinary knowledge on how urban centers and regions in locations of varying climates, lifestyles, income levels, and stages development are creating synergies and reducing trade-offs in the development of resilient, resource-efficient, environmentally friendly, liveable, socially equitable, integrated, and technology-enabled centers and regions. In particular, the *Palgrave Encyclopedia of Urban and Regional Futures* provides chapters, authored by subject matter experts, on interdisciplinary policies, best practices, lessons learnt, technologies in various stages of development, and case studies of urban centers and regions that aim to decouple economic growth from resource consumption, enhance resilience to climatic extremes, invest in low/zero carbon and smart technologies, lower emissions, reduce economic disparities, improve quality of life, and protect ecosystems and the services they provide for humans and nature.

Christchurch, New Zealand
December 2022

Robert C. Brears
Editor

Acknowledgments

First, I wish to thank Ruth Lefevre and Rachael Ballard for being visionaries who enable Major Reference Works like mine to come to fruition. Second, I wish to thank Anusha Cherian for being an excellent project coordinator. Third, I wish to thank Mum, who has a great interest in the environment and has supported me in this journey.

List of Topics

About the Editor

Robert C. Brears is an international sectoral expert on water for the UN's Green Climate Fund and the World Bank. He is the Editor in Chief of the *Palgrave Handbook of Climate Resilient Societies* and the *Palgrave Encyclopedia of Urban and Regional Futures*. He is the author of 11 books, including the Palgrave Macmillan titles *The Green Economy and the Water-Energy-Food Nexus*, *Blue and Green Cities: The Role of Blue-Green Infrastructure in Managing Urban Water Resources*, *Natural Resource Management and the Circular Economy*, *Developing the Circular Water Economy*, *Developing the Blue Economy*, and *Financing Nature-based Solutions*. He is the founder of Our Future Water, which has knowledge partnerships with World Bank, World Meteorological Organization, and UNEP initiatives.

Contributors

Ayodeji Adeniyi Deception Bay, QLD, Australia

Humera Afaq National University, San Diego, CA, USA

Kristin Agnello Department of Architecture and Planning, Norwegian University of Science and Technology, Trondheim, Norway

Atharv Agrawal University of Toronto, Toronto, ON, Canada

S. Ahilan College of Engineering, Mathematics and Physical Sciences, University of Exeter, Exeter, Devon, UK

Iftekhar Ahmed School of Architecture and Built Environment, University of Newcastle, Callaghan, NSW, Australia

Mubeen Ahmad School of Earth and Environmental Sciences, The University of Queensland, Brisbane, QLD, Australia

Ahmad Ahsan Lahore University of Management Sciences, Lahore, Pakistan

Meredian Alam Sociology and Anthropology Department, Universiti Brunei Darussalam, Gadong, Brunei

Amani Alfarra Land and Water Division, Food and Agriculture Organization of the United Nations, Rome, Italy

Jamal Alibou Department of Civil Engineering, Hydraulic, Environment and Climate, Hassania School of Public Works, Casablanca, Morocco

Sara Alidoust School of Earth and Environmental Sciences, The University of Queensland, Brisbane, QLD, Australia

Angélica Tanus Benatti Alvim Graduate Program in Architecture and Urbanism, Mackenzie Presbyterian University, Sao Paulo, Brazil

P. Ambily Department of Civil Engineering, National Institute of Technology, Calicut, Kerala, India

Grace Andrews Masters Environmental Management, College of Humanities, Arts and Social Sciences, Flinders University, Adelaide, South Australia

Jenna Andrews-Swann School of Liberal Arts, Georgia Gwinnett College, Lawrenceville, GA, USA

T. Angert The Institute for Environmental Security and Well-being Studies, Herzliya, Israel

Shyni Anilkumar National Institute of Technology Calicut, Kozhikode, Kerala, India

Aikaterini Antonopoulou

Hadi Arbabi Department of Civil & Structural Engineering, The University of Sheffield, Sheffield, UK

Md. Arfanuzzaman Food and Agriculture Organization (FAO) of the United Nations, Dhaka, Bangladesh

Felipe Armas Vargas Departamento de Ingeniería de Procesos e Hidráulica, CBI, Universidad Autónoma Metropolitana-Iztapalapa, Ciudad de México, Mexico

S. Arthur Heriot-Watt University, Edinburgh, UK

Hedda Askland University of Newcastle, Callaghan, NSW, Australia

Ditjon Baboci Tirana, Albania

Guy Baeten Urban Studies, Malmö University, Malmö, Sweden

Elham Bahmanteymouri The University of Auckland, Auckland, New Zealand

Nilesh Bakshi School of Architecture, Victoria University of Wellington, Wellington, New Zealand

M. Balasubramanian Centre for Ecological Economics and Natural Resources, Institute for Social and Economic Change, Bangalore, Karnataka, India

Zoran Balukoski School of Geography and Sustainable Communities, The University of Wollongong, Wollongong, NSW, Australia

Jonathan Banfield University of Toronto, Toronto, ON, Canada

Kaya Barry Griffith University, Brisbane, QLD, Australia

Department of Culture and Learning, Aalborg University, Aalborg, Denmark

Matthias Barth Leuphana University, Lüneburg, Germany

Prabal Barua Department of Environmental Sciences, Faculty of Physical and Mathematical Sciences, Jahangirnagar University, Dhaka, Bangladesh

James A. Beckman University of Central Florida, Orlando, FL, USA

Sara Bice Crawford School of Public Policy, The Australian National University, Acton, ACT, Australia

School of Public Policy and Management, Tsinghua University, Beijing, China

S. Birkinshaw University of Newcastle, Newcastle, UK

Stefan Blachfellner The Bertalanffy Center for the Study of Systems Science, Vienna, Austria

Bruno Blanco-Varela Department of Applied Economics, Faculty of Economics, Universidade de Santiago de Compostela, Santiago de Compostela, Galicia, Spain

Tijana Blanusa Royal Horticultural Society, Wisley, UK

University of Reading, Reading, UK

Tinashe Bobo Town Planning Section, Harare City Council, Harare, Zimbabwe

Cherice Bock Portland Seminary of George Fox University, Portland, OR, USA

Antonija Bogadi Department of Urban and Spatial Planning and Research, Technical University Vienna, Vienna, Austria

Simone Borelli Food and Agriculture Organization of the United Nations (FAO), Rome, Italy

Candice Boyd University of Melbourne, Melbourne, VIC, Australia

Christopher T. Boyko Lancaster University, Lancaster, UK

Robert C. Brears Our Future Water, Christchurch, New Zealand

Maria Julieta Brezzo Institutional Relations and Events, Ciudades Globales – CIGLO, Córdoba, Argentina

Katja Brundiers School of Sustainability, Arizona State University, Tempe, AZ, USA

Valerio Alfonso Bruno Università Cattolica del Sacro Cuore, Milan, Italy

Center for European Futures, Naples, Italy

Centre for the Analysis of the Radical Right, Leeds, UK

Felipe Bucci Ancapi Department of Management in the Built Environment, Faculty of Architecture and the Built Environment, Delft University of Technology, Delft, The Netherlands

Felix Bücken Institute of Geography, Osnabrück University, Osnabrück, Germany

Paul Burton Cities Research Institute Griffith University, Gold Coast, QLD, Australia

Alessandro Busà School of Geography, Geology and the Environment, University of Leicester, Leicester, UK

Judy Bush Lecturer in Urban Planning at University of Melbourne, Melbourne, VIC, Australia

Gareth Butler Masters Environmental Management, College of Humanities, Arts and Social Sciences, Flinders University, Adelaide, South Australia

Andrew Butt RMIT University, Melbourne, VIC, Australia

Michael Buxton RMIT University, Melbourne, VIC, Australia

Mohammed Firoz C. Department of Architecture and Planning, National Institute of Technology Calicut, Kozhikode, Kerala, India

Maléne Campbell Department of Urban and Regional Planning, University of the Free State, Bloemfontein, South Africa

Julien Carbonnell Artificial Intelligence on Citizen Engagement, Democracy Studio

M. Cavada School of Architecture, Imagination Lancaster, Lancaster University, Lancaster, UK

Rebecca Cavicchia Department of Urban and Regional Planning, BYREG – Norwegian University of Life Science, Ås, Norway

Lauriane Suyin Chalmin-Pui Royal Horticultural Society, Wisley, UK

The University of Sheffield, Sheffield, UK

Deborah Nabubwaya Chambers Community Health, National University, San Diego, CA, USA

Shenglin E. Chang National Taiwan University, Graduate Institute of Building and Planning, Taipei, Taiwan

Marianna Charitonidou Department of Art Theory and History, Athens School of Fine Arts, Athens, Greece

School of Architecture, National Technocal University of Athens, Athens, Greece

Department of Architecture, ETH Zurich, Zurich, Switzerland

Charles M. Chavunduka Department of Architecture and Real Estate, University of Zimbabwe, Harare, Zimbabwe

Ambika Chawla Urban Climate Innovations, Washington, DC, USA

Fei Chen School of Architecture, University of Liverpool, Liverpool, UK

Andrew Chigudu Department of Demography Settlement and Development, Social & Behavioural Sciences, University of Zimbabwe, Harare, Zimbabwe

Halleluah Chirisa Population Services International Zimbabwe, Harare, Zimbabwe

Innocent Chirisa Department of Demography Settlement and Development, Social & Behavioural Sciences, University of Zimbabwe, Harare, Zimbabwe

Chipo Chitereka Department of Social Work, University of Zimbabwe, Harare, Zimbabwe

N. R. Chithra Department of Civil Engineering, National Institute of Technology, Calicut, Kerala, India

Marcyline Chivenge Department of Demography Settlement and Development, Social & Behavioural Sciences, University of Zimbabwe, Harare, Zimbabwe

Suehyun Cho University of Toronto, Toronto, ON, Canada

Tanya Clark School of Behavioral Sciences, California Southern University, Costa Mesa, CA, USA

M'Lisa Lee Colbert The Nature of Cities, Montreal, QC, Canada

Ramon Fernando Colmenares-Quintero Faculty of Engineering, Universidad Cooperativa de Colombia, Medellín, Colombia

Elif Çolakoğlu Department of Security Sciences, Gendarmerie and Coast Guard Academy, Ankara, Turkey

Michela Conigliaro Food and Agriculture Organization of the United Nations (FAO), Rome, Italy

Sean Connelly University of Otago, Dunedin, New Zealand

A. Contin Politecnico di Milano, Milan, Italy

Rachel Cooper Lancaster University, Lancaster, UK

Samantha Copeland Ethics and Philosophy of Technology, Delft University of Technology, Delft, The Netherlands

João Cortesão Landscape Architecture and Spatial Planning, Wageningen University, Wageningen, The Netherlands

Adriano Cozzolino Center for European Futures, Naples, Italy

Università degli Studi della Campania "Luigi Vanvitelli", Caserta, Italy

Stewart Craine Village Infrastructure Angels, London, UK

Roberta Cucca BYREG – Norwegian University of Life Science, Ås, Norway

Gary Cummisk Central Washington University, Ellensburg, WA, USA

Paul Cureton ImaginationLancaster, Lancaster University, Lancaster, UK

Susan Cyriac Department of Architecture and Planning, National Institute of Technology, Calicut, Kerala, India

Sebastien Darchen School of Earth and Environmental Sciences, The University of Queensland, Brisbane, QLD, Australia

Curt J. Davis University of Delaware, Newark, DE, USA

D. Dawson University of Leeds, Leeds, UK

Evelyne de Leeuw Centre for Health Equity Training, Research and Evaluation (CHETRE), UNSW Australia Research Centre for Primary Health Care & Equity, South Western Sydney Local Health District, Ingham Institute, Sydney, NSW, Australia

Healthy Urban Environments (HUE) Collaboratory, Maridulu Budyari Gumal Sydney Partnership for Health, Education, Research and Enterprise SPHERE, Sydney, NSW, Australia

Ingham Institute for Applied Medical Research, Liverpool, NSW, Australia

Valerio Della Sala Politecnico di Torino (Italy), Interdepartmental Research Centre for Urban Studies (OMERO), Universitat Autonoma de Barcelona (Spain), Turin, Italy

N. Delle-Odeleye Anglia Ruskin University, Chelmsford, UK

Cheryl Desha Cities Research Institute, Griffith University, Brisbane, QLD, Australia

María Mercedes Di Virgilio Instituto de Investigaciones Gino Germani, Universidad de Buenos Aires/ CONICET, Ciudad Autónoma de Buenos Aires, Argentina

Roshini Suparna Diwakar Mahila Housing Trust, New Delhi, India

Timothy J. Dixon School of the Built Environment, University of Reading, Reading, UK

Michelle Duffy University of Newcastle, Callaghan, NSW, Australia

Smart Dumba Department of Demography Settlement and Development, Social & Behavioral Sciences, University Zimbabwe, Harare, Zimbabwe

Nick Dunn Lancaster University, Lancaster, UK

Jenna Dutton Senior Planner – Social Policy, City of Victoria and Research Associate, Center for Civilization, University of Calgary, Calgary, AB, Canada

Vupenyu Dzingirai Department of Community and Social Development, University of Zimbabwe, Harare, Zimbabwe

Charity Edwards Monash University & University of Melbourne, Melbourne, Australia

Huascar Eguino Fiscal Management Division, Inter-American Development Bank (IDB), Washington, DC, USA

Theodore S. Eisenman Department of Landscape Architecture and Regional Planning, University of Massachusetts-Amherst, Amherst, MA, USA

Christina R. Ergler School of Geography, University of Otago, Dunedin, New Zealand

Oscar Escolero Departamento de Dinámica Terrestre y Superficial, Instituto de Geología, Universidad Nacional Autónoma de México, Ciudad de México, Mexico

Javier Esquer Graduate Sustainability Program, Industrial Engineering Department, University of Sonora, Hermosillo, Mexico

G. Everett University of the West of England, Bristol, UK

Caroline Fabianski La Seyne sur Mer, France

Francesco Femia The Center for Climate and Security, an Institute of the Council on Strategic Risks, Washington, DC, USA

Melisha Shavindi Fernando Faculty of Science, Horizon Campus, Malabe, Sri Lanka

Carmen Zuleta Ferrari Food and Agriculture Organization of the United Nations (FAO), Rome, Italy

Carla Sofia Ferreira Research Centre for Natural Resources, Environment and Society (CERNAS), Polytechnic Institute of Coimbra, Coimbra Agrarian Technical School, Coimbra, Portugal

Department of Physical Geography and Bolin Centre for Climate Research, Stockholm University, Stockholm, Sweden

Navarino Environmental Observatory, Messinia, Greece

António Ferreira Research Centre for Natural Resources, Environment and Society (CERNAS), Polytechnic Institute of Coimbra, Coimbra Agrarian Technical School, Coimbra, Portugal

Daniel Fischer School of Sustainability, Arizona State University, Tempe, AZ, USA

Wesley Flannery Urban Planning, School of Natural and Built Environment, David Keir Building, Queen's University Belfast, Belfast, UK

Claudia Fonseca Alfaro Institute for Urban Research, Malmö University, Malmö, Sweden

Mariana Fonseca Braga ImaginationLancaster, Lancaster Institute for the Contemporary Arts (LICA), Lancaster University, Lancaster, Lancashire, UK

Julien Forbat University of Geneva, Institute of Global Health, Geneva, Switzcrland

Martin Franz Institute of Geography, Osnabrück University, Osnabrück, Germany

Robert Freestone School of Built Environment, University of New South Wales, Sydney, NSW, Australia

Frances Furio School of Behavioral Sciences, California Southern University, Costa Mesa, CA, USA

Tatiana Gallego Lizon Washington, DC, USA

Emilio Garcia The University of Auckland, Auckland, New Zealand

Birgit Georgi UIA Expert/Strong Cities in a Changing Climate, Egelsbach, Germany

Daniela Getlinger Graduate Program in Architecture and Urbanism, Mackenzie Presbyterian University, Sao Paulo, Brazil

David J. Gilchrist University of Western Australia, Perth, WA, Australia

Brendan Gleeson Monash University & University of Melbourne, Melbourne, Australia

V. Glenis University of Newcastle, Newcastle, UK

Moritz Gold Sustainable Food Processing Laboratory, ETH Zurich, Zurich, Switzerland

Eugenio Gómez Reyes Departamento de Ingeniería de Procesos e Hidráulica, CBI, Universidad Autónoma Metropolitana-Iztapalapa, Ciudad de México, Mexico

Megan Gordon University of Northern British Columbia, Prince George, BC, Canada

Alexa Gower Monash University, Melbourne, VIC, Australia

Sonia Graham School of Geography and Sustainable Communities, The University of Wollongong, Wollongong, NSW, Australia

Institut de Ciència I Tecnologia Ambientals (ICTA), Universitat Autònoma de Barcelona, Barcelona, Spain

Danielle Griego Center for Augmented Computational Design in Architecture, Engineering and Construction, D-BAUG, ETH Zurich, Zurich, Switzerland

Kai Michael Griese Hochschule Osnabrück University of Applied Sciences, Osnabrück, Germany

Carl Grodach Monash University, Melbourne, VIC, Australia

Bern Grush Urban Robotics Foundation, Toronto, Canada

Medhisha Pasan Gunawardena Biodiversity Educational Research Initiative, Colombo, Sri Lanka

Faculty of Science, Horizon Campus, Malabe, Sri Lanka

Hector Manuel Guzman Grijalva Sustainability Graduate Program, University of Sonora, Hermosillo, México

Jochen Hack Technical University of Darmstadt, Section of Ecological Engineering, Institute of Applied Geosciences, Darmstadt, Germany

Perrine Hamel Asian School of the Environment, Nanyang Technological University, Singapore, Singapore

Earth Observatory of Singapore, Nanyang Technological University, Singapore, Singapore

Ben Harris-Roxas School of Population Health, University of New South Wales, Sydney, NSW, Australia

Wolfgang Haupt Leibniz-Insitute for Research on Society and Space, Erkner, Germany

Naomi Hay Australian National University, Canberra, ACT, Australia

Fatime Barbara Hegyi Joint Research Centre – European Commission, Seville, Spain

Hayley Henderson Research Fellow at Crawford School of Public Policy, Australian National University, Canberra, ACT, Australia

Michael Henderson Ramboll Ltd and Oxford Brookes University, London, UK

Cole Hendrigan University of Wollongong and Wollongong City Council, Wollongong, NSW, Australia

Andreas Hernandez Marymount Manhattan College, New York, NY, USA

Victoria Herrmann The Arctic Institute – Center for Circumpolar Security Studies, Washington, DC, USA

Halima Hodzic Food and Agriculture Organization of the United Nations (FAO), Rome, Italy

Karen Horwood The Leeds Planning School, Leeds Beckett University, Leeds, UK

Mette Hotker RMIT University, Melbourne, VIC, Australia

Karin Huber-Heim Circular Economy Forum, Austria, Vienna, Austria

Raisa Binte Huda Department of Geography and Environment, University of Dhaka, Dhaka, Bangladesh

Dan Xuan Thi Huynh School of Economics, Can Tho University, Can Tho, Vietnam

Ligocka Ilona Ministry of Climate and Environment, Warsaw, Poland

Tanya Gottlieb Jacobsen State of Green, Copenhagen, Denmark

Bhanye Johannes Department of Community and Social Development, University of Zimbabwe, Harare, Zimbabwe

Katrina Johnston-Zimmerman THINK.urban, Philadelphia, PA, USA

Kirsty Jones Crawford School of Public Policy, The Australian National University, Acton, ACT, Australia

Alain Jordà Local Development Expert, Manresa, Barcelona, Spain

Gaurav Joshi University of Chinese Academy of Sciences, Beijing, China

Anuja Joy National Institute of Technology Calicut, Kozhikode, Kerala, India

Mahjabin Kabir Adrita Department of Geography and Environment, University of Dhaka, Dhaka, Bangladesh

Zahra Kalantari Department of Physical Geography and Bolin Centre for Climate Research, Stockholm University, Stockholm, Sweden

Navarino Environmental Observatory, Messinia, Greece

Department of Sustainable Development, Environmental Science and Engineering, KTH Royal Institute of Technology, Stockholm, Sweden

Eleni Kalantidou Griffith University, Brisbane, QLD, Australia

Tinashe Natasha Kanonhuhwa Department of Demography Settlement and Development, Social & Behavioral Sciences, University Zimbabwe, Harare, Zimbabwe

L. Kapetas 100 resilient Cities Project, New York, USA

Thomas Karakadzai Department of Demography Settlement and Development, Faculty of Social & Behavioral Sciences, University of Zimbabwe, Harare, Zimbabwe

Abdulrazak Karriem University of the Western Cape, Cape Town, South Africa

Hewa Thanthrige Ashan Randika Karunananda Biodiversity Educational Research Initiative, Colombo, Sri Lanka

Rosemary Kasimba Department of Demography Settlement and Development, University of Zimbabwe, Harare, Zimbabwe

J. O. Kawira County Government of Laikipia, Laikipia, Kenya

Jon Kellett University of Adelaide, Adelaide, SA, Australia

Vlada Kenniff Long Island University, Brookville, NY, USA

Jeffrey Kenworthy Curtin University Sustainability Policy Institute, Curtin University, Perth, WA, Australia

Frankfurt University of Applied Sciences, Frankfurt am Main, Germany

Ganesh Keremane Adelaide, South Australia

Tien Dung Khong School of Economics, Can Tho University, Can Tho, Vietnam

Teng Chye Khoo National University of Singapore, Singapore, Singapore

F. I. Kihara The Nature Conservancy, Nairobi, Kenya

Lorenzo Kihlgren Grandi City Diplomacy Lab, Columbia Global Centers | Paris, Paris, France

C. Kilsby University of Newcastle, Newcastle, UK

Jinhee Kim Centre for Health Equity Training, Research and Evaluation (CHETRE), UNSW Australia Research Centre for Primary Health Care & Equity, South Western Sydney Local Health District, Ingham Institute, Sydney, NSW, Australia

Michael Koh Centre for Liveable Cities, Ministry of National Development, Singapore, Singapore

Victoria Kolankiewicz Faculty of Architecture, Building and Planning, University of Melbourne, Melbourne, VIC, Australia

Weichang Kong The University of Queensland, Brisbane, QLD, Australia

Mrudhula Koshy Norwegian University of Science and Technology, Trondheim, Norway

Maria Kottari School of Transnational Governance, European University Institute, Florence, Italy

Daniel Kozak Universidad de Buenos Aires, Consejo Nacional de Investigaciones CientÃficas y Técnicas (CONICET), Buenos Aires, Argentina

Teresa Kramarz University of Toronto, Toronto, ON, Canada

Tamara Krawchenko University of Victoria, Victoria, BC, Canada

Peleg Kremer Department of Geography and the Environment, Villanova University, Villanova, PA, USA

V. Krivtsov The Royal Botanic Garden, Edinburgh, UK

Arvind Kumar India Water Foundation, New Delhi, India

Gerard Kuperus University of San Francisco, San Francisco, CA, USA

Sigrid Kusch-Brandt Department of Civil, Environmental and Architectural Engineering, University of Padua, Padua, Italy

Faculty of Mathematics, Natural Sciences and Management, University of Applied Sciences Ulm, Ulm, Germany

Ndarova Audrey Kwangwama Department of Architecture and Real Estate, University of Zimbabwe, Harare, Zimbabwe

Oliver Lah Wuppertal Institute for Climate, Environment and Energy, Berlin, Germany

Urban Electric Mobility Initiative (UEMI) a UN-Habitat Action Platform, Berlin, Germany

Khee Poh Lam National University of Singapore, Singapore, Singapore

J. Lamond University of the West of England, Bristol, UK

Martin Larbi Kwame Nkrumah University of Science and Technology, Kumasi, Ghana

Alexander Laszlo The Bertalanffy Center for the Study of Systems Science, Buenos Aires, Argentina

Lucie Laurian School of Planning and Public Affairs, The University of Iowa, Iowa City, IA, USA

Alison Lee Centre for Liveable Cities, Ministry of National Development, Singapore, Singapore

Steffen Lehmann School of Architecture, University of Nevada, Las Vegas, NV, USA

Carlos Leite School of Architecture and Urbanism, Mackenzie Presbyterian University, Sao Paulo, Brazil

Social Urbanism Center, Insper´s Arq.Futuro Cities Lab, Sao Paulo, Brazil

Caitlin Anthea Lewis Architecture Planning and Geomatics, University of Cape Town, Cape Town, South Africa

Nora Libertun de Duren Inter-American Development Bank, Washington, DC, USA

Jade Lindley Law School and Oceans Institute, The University of Western Australia, Crawley, WA, Australia

Yan Liu School of Earth and Environmental Sciences, The University of Queensland, St Lucia, Australia

Adam Loch Centre for Global Food and Resources, School of Economics and Public Policy, Faculty of the Professions, University of Adelaide, Adelaide, SA, Australia

Aynaz Lotfata Department of Geography, Chicago State University, Chicago, IL, USA

Pavel Luksha Global Education Futures, Moscow, Russia

Mengxing Ma Department of Social Work, University of Melbourne, Melbourne, VIC, Australia

Department of Geography, University of Sheffield, Sheffield, UK

Danielle MacCarthy Queen's University Belfast, Belfast, Northern Ireland, UK

Shamiso Hazel Mafuku Department of Architecture and Real Estate, University of Zimbabwe, Harare, Zimbabwe

Kamilia Mahdaoui Hassania School of Public Works, Casablanca, Morocco

Israa H. Mahmoud Laboratorio di Simulazione Urbana Fausto Curti, Department of Architecture and Urban Studies, Politecnico di Milano, Milan, Italy

David Mainenti Palmer iSchool of Library and Information Studies, Long Island University, Brookville, NY, USA

Innocent Maja Faculty of Law, University of Zimbabwe, Harare, Zimbabwe

Soumaya Majdoub Research Group Interface Demography, Department of Sociology, VUB Free University of Brussels, Brussels, Belgium

Brussels Center for Urban Studies (BCUS), Brussels, Belgium

Brussels Interdisciplinary Research Centre for Migration and Minorities (BIRMM), Brussels, Belgium

George Makunde George Makunde Institute, Harare, Zimbabwe

Eleanor Malbon University of New South Wales, Kensington, NSW, Australia

Wendy W. Mandaza-Tsoriyo Department of Rural and Urban Development, Great Zimbabwe University, Harare, Zimbabwe

Manfredo Manfredini School of Architecture and Planning, The University of Auckland, Shanghai University, Auckland, New Zealand

Elton Manjeya Department of Architecture and Real Estate, University of Zimbabwe, Harare, Zimbabwe

Jonathan Manns Rockwell, London, UK

UCL, London, UK

Patrick M. Marchman American Society of Adaptation Professionals/ Climigration Network, Kansas City, MO, USA

Age Mariussen University of Vaasa, Vaasa, Finland

Cecilia Marocchino Food and Agriculture Organization of the United Nations (FAO), Rome, Italy

Andresa Ledo Marques Graduate Program in Architecture and Urbanism, Mackenzie Presbyterian University, Sao Paulo, Brazil

Institute of Urban Design and Planning, Leibniz Universität, Hannover, Germany

Martha Marriner State of Green, Copenhagen, Denmark

Stephen Marshall Bartlett School of Planning, University College London, London, UK

Natalia Martsinovich Department of Chemistry, University of Sheffield, Sheffield, UK

Nesbert Mashingaidze Department of Rural and Urban Development, Great Zimbabwe University, Masvingo, Zimbabwe

Jeofrey Matai Department of Architecture and Real Estate, University of Zimbabwe, Harare, Zimbabwe

Abraham R. Matamanda Department of Urban and Regional Planning, University of the Free State, Bloemfontein, South Africa

Marina Matashova Andorra-LAB, Forward Consulting Group, Barcelona, Spain

Brilliant Mavhima Department of Architecture and Real Estate, University of Zimbabwe, Harare, Zimbabwe

Patience Mazanhi Department of Demography Settlement and Development, Social & Behavioral Sciences, University of Zimbabwe, Harare, Zimbabwe

Chad J. McGuire Department of Public Policy, University of Massachusetts, Dartmouth, MA, USA

Matthew H. McLeskey Department of Sociology, University at Buffalo, State University of New York, Buffalo, NY, USA

Wendy McWilliam School of Landscape Architecture, Faculty of Environment, Society and Design, Lincoln University, Lincoln, New Zealand

Ojilve Ramón Medrano Pérez CONACYT-Centro del Cambio Global y la Sustentabilidad, A.C. (CCGS), Villahermosa, Tabasco, Mexico

Asma Mehan Senior Researcher, CITTA Research Institute, Faculty of Engineering (FEUP), University of Porto, Porto, Portugal

Mahziar Mehan School of Urban Planning, Faculty of Fine Arts, University of Tehran, Tehran, Iran

Prakhar Mehta Digital Transformation: Bits to Energy Lab Nuremberg, School of Business, Economics and Society, Friedrich-Alexander University Erlangen-Nürnberg (FAU), Nuremberg, Germany

Lorena Melgaço Department of Human Geography, Lund University, Lund, Sweden

D. Mendoza Tinoco University of Coahuila, Coahuila, Mexico

Julián Andrés Mera-Paz Faculty of Engineering, Universidad Cooperativa de Colombia, Popayán, Colombia

Magnus Højberg Mernild State of Green, Copenhagen, Denmark

Jessica Ostrow Michel School for Environment and Sustainability, University of Michigan, Ann Arbor, MI, USA

Yoko Mochizuki UNESCO, Paris, France

Itumeleng Mogola C40 Cities, Benoni, South Africa

Mohsen Mohammadzadeh School of Architecture and Planning, Auckland University, Auckland, New Zealand

Abinash Mohanty Council on Energy, Environment and Water (CEEW), New Delhi, India

Mehri Mohebbi Transportation Equity Program, University of Florida (UFTI), Gainesville, FL, USA

Anne Mook University of Georgia, Athens, GA, USA

Eugenio Morello Laboratorio di Simulazione Urbana Fausto Curti, Department of Architecture and Urban Studies, Politecnico di Milano, Milan, Italy

Charlotte Morphet Women and Planning research bursary, Planning, Housing and Human Geography, The Leeds Planning School, Leeds Beckett University, Leeds, UK

Nicky Morrison Western Sydney University, Sydney, NSW, Australia

Sina Mostafavi TU Delft, Delft, The Netherlands

Edmos Mtetwa Department of Social Work, University of Zimbabwe, Harare, Zimbabwe

Tinashe Natasha Mujongonde-Kanonhuwa Department of Rural & Urban Planning, University of Zimbabwe, Harare, Zimbabwe

Manasi R. Mulay Department of Chemistry, University of Sheffield, Sheffield, UK

Grantham Centre for Sustainable Futures, Sheffield, UK

Richard Müller Sustainable Development Institute/Institut udrzatelneho rozvoja, Nitra, Slovakia

Yvonne Munanga Department of Architeture and Real Estate, University of Zimbabwe, Harare, Zimbabwe

Dalia Munenzon College of Architecture, Texas Tech University, Lubbock, TX, USA

Nora Munguia Graduate Sustainability Program, Industrial Engineering Department, University of Sonora, Hermosillo, Mexico

Solomon Muqayi Department of Governance and Public Management, University of Zimbabwe, Harare, Zimbabwe

Cassandra Murphy Department of Psychology, Maynooth University, Maynooth, Ireland

Teagan Murphy University of Maryland, College Park, MD, USA

Brendan Murtagh Urban Planning, School of Natural and Built Environment, David Keir Building, Queen's University Belfast, Belfast, UK

Walter Musakwa Future Earth and Ecosystem Services Research Group, Department of Urban and Regional Planning, University of Johannesburg, Johannesburg, South Africa

Tafadzwa Mutambisi Department of Rural and Urban Planning, University of Zimbabwe, Harare, Zimbabwe

Chipo Mutonhodza Department of Rural and Urban Development, Great Zimbabwe University, Masvingo, Zimbabwe

Valeria Muvavarirwa Department of Demography Settlement and Development, Social & Behavioral Sciences, University of Zimbabwe, Harare, Zimbabwe

Jean Nacishali Nteranya Department of Geology, Faculty of Sciences, Université Officielle de Bukavu (UOB), Bukavu, Democratic Republic of Congo

Anupam Nanda University of Manchester, Manchester, UK

Luzma Fabiola Nava CONACYT-Centro del Cambio Global y la Sustentabilidad, A.C. (CCGS), Villahermosa, Tabasco, Mexico

International Institute for Applied Systems Analysis (IIASA), Laxenburg, Austria

Celeste Nava Jiménez División de Ciencias Económico Administrativas, Campus Guanajuato, Universidad de Guanajuato, Guanajuato, Mexico

Thilini Navaratne Department of Business Economics, Faculty of Management Studies and Commerce, University of Sri Jayewardenepura, Nugegoda, Sri Lanka

Roselin Ncube Women's University in Africa, Harare, Zimbabwe

S. Ncube Heriot-Watt University, Edinburgh, UK

Etienne Nel University of Otago, Dunedin, New Zealand

David Nichols Faculty of Architecture, Building and Planning, University of Melbourne, Melbourne, VIC, Australia

Alejandro Nuñez-Jimenez Sustainability and Technology Group, D-MTEC, ETH Zurich, Zurich, Switzerland

Belfer Center for Science and International Affairs, Harvard University, Cambridge, MA, USA

Gloria Nyaradzo Nyahuma-Mukwashi Department for International Development (DFID), Harare, Zimbabwe

E. O'Donnell University of Nottingham, Nottingham, UK

G. O'Donnell University of Newcastle, Newcastle, UK

Narteh F. Ocansey Water Resources, Freelance, Accra, NA, Ghana

Yukyung Oh King's College London, London, UK

Carolina G. Ojeda Doctorado en Arquitectura y Estudios Urbanos, Pontificia Universidad Católica de Chile, Providencia, Santiago de Chile, Chile

Departamento de Historia, Facultad de Comunicaciones e Historia, Universidad Católica de la Santísima Concepción, Concepción, Chile

Hasan Volkan Oral Faculty of Engineering, Department of Civil Engineering (English), Istanbul Aydın University, Istanbul, Turkey

P. Ortiz International Metropolitan Institute, Madrid, Spain

International Metropolitan Institute, Washington, DC, USA

G. Osei Anglia Ruskin University, Chelmsford, UK

Laura Patricia Otero-Durán Urban Development Institute, Bogotá, Colombia

Maria Pafi Urban Planning, School of Natural and Built Environment, David Keir Building, Queen's University Belfast, Belfast, UK

F. Pascale Anglia Ruskin University, Chelmsford, UK

Maibritt Pedersen Zari School of Architecture, Victoria University of Wellington, Wellington, New Zealand

María Concepción Peñate-Valentín Department of Applied Economics, Faculty of Economics, Universidade de Santiago de Compostela, Santiago de Compostela, Galicia, Spain

Paulo Pereira Environmental Management Laboratory, Mykolas Romeris University, Vilnius, Lithuania

Ben Perks

Shama Perveen Senior Manager (Water), Ceres, Boston, MA, USA

Evi Petersen Institute of Sports, Physical Education and Outdoor Life, University of South-Eastern Norway, Oslo, Norway

Son Phung Department of Civil and Environmental Engineering, Auckland University, Auckland, New Zealand

Francesca Piazzoni

Czarnocki Piotr Ministry of Climate and Environment, Warsaw, Poland

Dorina Pojani The University of Queensland, Brisbane, QLD, Australia

A. Pooley Centre for Alternative Technology, Pantperthog, UK

K. Potter Open University, Milton Keynes, UK

Abdellatif Qamhaieh American University in Dubai, Department of Architecture, Dubai, United Arab Emirates

Md. Anisur Rahman Center for Policy and Economic Research (CPER), Dhaka, Bangladesh

Syed Hafizur Rahman Department of Environmental Sciences, Faculty of Physical and Mathematical Sciences, Jahangirnagar University, Dhaka, Bangladesh

Lakshmi Priya Rajendran The Bartlett School of Architecture, University College London, London, UK

Ritesh Ranjan Department of Architecture & Planning, National Institute of Technology Calicut, Kozhikode, Kerala, India

Andreas Raspotnik High North Center for Business and Governance, Nord University, Bodø, Norway

The Arctic Institute – Center for Circumpolar Security Studies, Washington, DC, USA

Hanna A. Rauf Asian School of the Environment, Nanyang Technological University, Singapore, Singapore

Aaron Redman School of Sustainability, Arizona State University, Tempe, AZ, USA

William E. Rees School of Community and Regional Planning, University of British Columbia, Vancouver, BC, Canada

Christian Reichel University of Applied Sciences for Media, Communication and Management (HMKW), Berlin, Germany

Kimberley Reis Cities Research Institute, Griffith University, Brisbane, QLD, Australia

Catherine E. Richards Center for the Study of Existential Risk (CSER), University of Cambridge, Cambridge, UK

Department of Engineering, University of Cambridge, Cambridge, UK

Lauren Rickards Urban Futures Enabling Capability Platform, RMIT University, Melbourne, Australia

Ritesh Ranjan Department of Architecture and Planning, National Institute of Technology Calicut, Kozhikode, Kerala, India

Alejandra Rivera Vinueza 4CITIES Erasmus Mundus Joint Master Degree (EMJMD) in Urban Studies, Vrije Universitet Brussel (VUB), Brussels, Belgium

Institute for Human Rights and Business (IHRB), Built Environment Global Programme Manager, London, UK

Daniela Rizzi Nature-based Solutions and Biodiversity – Sustainable Resources, Climate and Resilience Team, Freiburg, Germany

Michael Robbins HIPR, New York City, NY, USA

Héctor Rodal Architect and Urban Planner, Barcelona, Spain

Robert Rogerson Institute for Future Cities, University of Strathclyde, Glasgow, UK

Watch Ruparanganda Department of Social Work, University of Zimbabwe, Harare, Zimbabwe

María Carmen Sánchez-Carreira Department of Applied Economics, Faculty of Economics, Universidade de Santiago de Compostela, ICEDE Research Group, CRETUS, Santiago de Compostela, Galicia, Spain

Rami Sabella United Nations Economic and Social Commission for Western Asia, Beirut, Lebanon

Peter Sainsbury School of Medicine, University of Notre Dame, Sydney, NSW, Australia

Samuel Sandoval Solis Department of Land, Air and Water Resources, University of California Davis, Davis, CA, USA

Guido Santini Food and Agriculture Organization of the United Nations (FAO), Rome, Italy

Tom Sanya Architecture Planning and Geomatics, University of Cape Town, Cape Town, South Africa

Hasan Saygın Application, and Research Center for Advanced Studies, Istanbul Aydın University, Istanbul, Turkey

Alice Schmidt Global Health Advisory Service to the European Commission, Mechelen, Belgium

Vienna University of Economics and Business, Vienna, Austria

AS Consulting, Vienna, Austria

Jörg Schröder Institute of Urban Design and Planning, Leibniz Universität, Hannover, Germany

Barbara Schröter Leibniz Centre for Agricultural Landscape Research (ZALF), Working group Governance of Ecosystem Services, Müncheberg, Germany

Lund University Centre for Sustainability Studies (LUCSUS), Lund, Sweden

Kim Philip Schumacher Institute of Geography, Osnabrück University, Osnabrück, Germany

Abel Schumann Organisation for Economic Co-operation and Development, Paris, France

Samad M. E. Sepasgozar School of Built Environment, University of New South Wales, Sydney, NSW, Australia

Alan Shapiro British Columbia Institute of Technology, Vancouver, BC, Canada

Aviram Sharma School of Ecology and Environment Studies, Nalanda University, Rajgir, Bihar, India

Tian Shi Department of Primary Industries and Regions, Adelaide, South Australia

Amna Shoaib Department of City and Regional Planning, Lahore College for Women University (LCWU), Lahore, Pakistan

Yat Shun Kei

Renard Y. J. Siew Climate Change & Sustainability, Centre for Governance & Political Studies (CENT-GPS), Kuala Lumpur, Malaysia

Institute for Globally Distributed Open Research and Education (IGDORE), Bali, Indonesia

David Simon Department of Geography, Royal Holloway, University of London, Egham, UK

Jean Simos Institute of Public Health, Faculty of Medicine, University of Geneva, Geneva, Switzerland

S2D – Health and Sustainable Development, Rennes, France

Neil Sipe School of Earth and Environmental Sciences, The University of Queensland, Brisbane, QLD, Australia

Ben Sonneveld Amsterdam Centre for World Food Studies/Athena Institute, Vrije Universiteit, Amsterdam, The Netherlands

Micol Sonnino The Bertalanffy Center for the Study of Systems Science, Vienna, Austria

Simon Springer Centre for Urban and Regional Studies, Dicipline of Geography and Environmental Studies, University of Newcastle, Australia, Callaghan, NSW, Australia

Janet Stanley Melbourne Sustainable Society Institute, University of Melbourne, Melbourne, Australia

Wendy Steele Centre for Urban Research, RMIT University, Melbourne, VIC, Australia

Justin D. Stewart Department of Ecological Science, Vrije Universiteit Amsterdam, Amsterdam, The Netherlands

Raisa Sultana Department of Geography and Environment, University of Dhaka, Dhaka, Bangladesh

Samantha Suppiah Possible Futures, Manilla, Philippines

Sylvia Szabo Department of Social Welfare and Counselling, University of Seoul, Seoul, South Korea

Gerti Szili College of Humanities, Arts and Social Sciences, Flinders University, Adelaide, South Australia

Bouchra Tafrata Willy Brandt School of Public Policy, University of Erfurt, Erfurt, Germany

Ling Min Tan Department of Civil & Structural Engineering, The University of Sheffield, Sheffield, UK

M. Terdiman The Institute for Environmental Security and Well-being Studies, Jerusalem, Israel

Jacqueline Thomas School of Civil Engineering, The University of Sydney, Sydney, NSW, Australia

M. K. Thomas Rural Focus Limited (RFL), Nanyuki, Kenya

S. Thomas Rural Focus Limited (RFL), Nanyuki, Kenya

C. Thorne University of Nottingham, Nottingham, UK

Karine Tollari Japan Local Government Centre, London, UK

Chiara Tomaselli Consultant – Urban Asset Advisory, Arcadis France, Paris, France

Percy Toriro Municipal Development Partnership for Eastern and Southern Africa, Harare, Zimbabwe

African Centre for Cities, University of Cape Town, Cape Town, South Africa

Isabella Trapani Food and Agriculture Organization of the United Nations (FAO), Rome, Italy

Alejandra Trejo-Nieto Centre for Demographic, Urban and Environmental Studies, El Colegio de Mexico, Mexico City, Mexico

Stella Tsani Department of Economics, University of Ioannina, Ioannina, Greece

Asaf Tzachor Center for the Study of Existential Risk (CSER), University of Cambridge, Cambridge, UK

School of Sustainability, Interdisciplinary Center (IDC) Herzliya, Herzliya, Israel

Zdravka Tzankova Vanderbilt University, Nashville, TN, USA

Kristina Ulm Faculty of Arts, Design and Architecture, University of New South Wales, Sydney, NSW, Australia

Geraldine Usingarawe Department of Architecture and Real Estate, University of Zimbabwe, Harare, Zimbabwe

Luís Valença Pinto Research Centre for Natural Resources, Environment and Society (CERNAS), Polytechnic Institute of Coimbra, Coimbra Agrarian Technical School, Coimbra, Portugal

Environmental Management Laboratory, Mykolas Romeris University, Vilnius, Lithuania

Ellen Van Bueren Department of Management in the Built Environment, Faculty of Architecture and the Built Environment, Delft University of Technology, Delft, The Netherlands

Karel Van den Berghe Department of Management in the Built Environment, Faculty of Architecture and the Built Environment, Delft University of Technology, Delft, The Netherlands

Jeroen van der Heijden School of Government, Victoria University of Wellington, Wellington, New Zealand

School of Regulation and Global Governance, Australian National University, Canberra, ACT, Australia

Wim van Veen Vrije Universiteit, Amsterdam Centre for World Food Studies, Amsterdam, The Netherlands

Lia van Wesenbeeck Vrije Universiteit, Amsterdam Centre for World Food Studies, Amsterdam, The Netherlands

Christopher Vanags Vanderbilt University, Nashville, TN, USA

Kamiya Varshney School of Architecture, Victoria University of Wellington, Wellington, New Zealand

Luis Velazquez Industrial Engineering Department, University of Sonora, Hermosillo, Mexico

Luis Eduardo Velazquez Contreras Sustainability Graduate Program, University of Sonora, Hermosillo, México

T. Vilcan Open University, Milton Keynes, UK

Luiza O. Voinea Urban Planner Certified by The Romanian Register of Urban Planners, Bucharest, Romania

Shreya Wadhawan Council on Energy, Environment and Water (CEEW), New Delhi, India

Sameh N. Wahba The World Bank, Washington, DC, USA

Haiyun Wang School of Earth and Environmental Sciences, The University of Queensland, St Lucia, Australia

Siqin Wang School of Earth and Environmental Sciences, The University of Queensland, St Lucia, Australia

Noelia Wayar National University of Córdoba, Córdoba, Argentina

Oliver Weigel Urban Development Policy Division at the Federal Ministry of the Interior, Building, and Community, Berlin, Germany

Kadmiel H. Wekwete Midlands State University, Gweru, Zimbabwe

Caitlin Werrell The Center for Climate and Security, an Institute of the Council on Strategic Risks, Washington, DC, USA

Andreas Wesener School of Landscape Architecture, Faculty of Environment, Society and Design, Lincoln University, Lincoln, New Zealand

Bettina Wilk ICLEI European Secretariat, Senior Officer for Nature-based Solutions and Biodiversity – Sustainable Resources, Climate and Resilience Team, Freiburg, Germany

Erich Wolff Monash Art, Design and Architecture, Monash University, Melbourne, VIC, Australia

Sam Wong University College Roosevelt, Middelburg, The Netherlands

N. Wright Nottingham Trent University, Nottingham, UK

Junjie Xi

Belinda Young Melbourne Sustainable Society Institute, University of Melbourne, Melbourne, Australia

Asaduz Zaman Centre for Action Research – Barind, Rajshahi, Bangladesh

Asian Development Bank, Dhaka, Bangladesh

Fathima Zehba M. P. Department of Architecture and Planning, Calicut, National Institute of Technology, Calicut, Kerala, India

David Slim Zepeda Quintana Sustainability Graduate Program, University of Sonora, Hermosillo, México

Yuerong Zhang Bartlett School of Planning, University College London, London, UK

Eric Zhao University of Toronto, Toronto, ON, Canada

Metron Ziga University of the Western Cape, Cape Town, South Africa

Monika Zimmermann Urban Sustainability Expert & Former Deputy Secretary General of ICLEI, Freiburg, Germany

Willoughby Zimunya Department of Demography Settlement and Development, University of Zimbabwe, Harare, Zimbabwe

Department of Urban and Regional Planning, University of the Free State, Bloemfontein, South Africa

Michaela Zint School for Environment and Sustainability, University of Michigan, Ann Arbor, MI, USA

Tara Rava Zolnikov School of Behavioral Sciences, California Southern University, Costa Mesa, CA, USA

Department of Community Health, National University, San Diego, CA, USA

S

Sacred Sites

► Faith Communities as Hubs for Climate Resilience

Safe Drinking Water

► Meeting SDG6: Ensuring Safe Drinking Water for All in Rural India

Sanctuary Cities and Its Impact on Quality of Life of Its Citizenry

James A. Beckman
University of Central Florida, Orlando, FL, USA

Introduction

In a broader and larger sense, the concept of a regional government (city, county, or state) actively ignoring federal laws goes back to the very beginning the history of the United States. For example, early in U.S. history, Congress passed legislation collectively known as the Alien and Sedition Acts in 1798. The Alien Act enabled the President of the United States to deport any alien deemed by the President too dangerous to the United States. The Sedition Act criminalized the speaking or writing of "false, scandalous, and malicious" statements against the Congress or the President. These acts were very polarizing in the United States and viewed as political acts meant to silence or deport critics of the governing Federalist party of President John Adams and weaken support for the opposition party (the Democratic-Republican party) led by Vice President Thomas Jefferson. In response, some areas of the country refused to enforce these laws. The state legislatures of Virginia and Kentucky went further in passing resolutions (collectively known as the Virginia and Kentucky Resolutions) which, among other things, purported to allow state legislatures to ignore the Alien and Sedition Acts and declare them to be null and void.

The idea of a city or state being able to nullify and/or not enforce federal laws emerged at many junctures throughout the history of the United States. For instance, in the so-called nullification crisis of 1832, South Carolina promulgated an Ordinance of Nullification, which acted to void a new federal tariff law and preclude its enforcement in South Carolina. In the Twenty-First Century, after the terrorist attacks on the United States by Al-Qaeda on September 11, 2001, Congress promulgated a sweeping set of laws (collectively entitled the "Patriot Act" that (1) restricted some civil liberties and (2) gave law enforcement

R. C. Brears (ed.), *The Palgrave Encyclopedia of Urban and Regional Futures*,
https://doi.org/10.1007/978-3-030-87745-3

increased powers (Ewing et al. 2015). In response, hundreds of cities announced that they would not honor these new police powers and restrictions on the civil liberties of its citizenry. In the 2 years after the Patriot Act was enacted in October 2001, over 300 local governments (primarily cities) announced their opposition and general non-compliance with the Patriot Act (Maxwell 2004). Several state legislative bodies passed resolutions in opposition of the Patriot Act as well (Beckman 2007). These are earlier examples of a type of "sanctuary city" in the United States. However, in the last 20 years, "sanctuary cities" in the United States have come to specifically connote and reference those jurisdictions (typically cities and counties) that refuse to cooperate with federal authorities in the deportation of undocumented immigrants living in those jurisdictions (Kopan 2018).

As of 2021, there is no definitive singular definition of what constitutes a "sanctuary city" (Ingraham 2017). Broadly speaking, and consistent with previous acts of non-compliance of regional governments to the dictates of federal laws throughout United States history (as briefly illustrated above), a sanctuary city could be any regional government or municipality that does not prosecute individuals under certain federal laws, whether those laws are those like the Alien and Sedition Acts of 1798, the federal tariff act of 1833, the Patriot Act of 2001, or modern immigration laws, and provides "sanctuary" from prosecution. However, most properly in the modern parlance and usage, a "sanctuary city" refers to a city (or a county or even state government) that "limits its cooperation with federal immigration enforcement agents in order to protect low-priority immigrants from deportation..." (America's Voice 2019; see also, Shultz 2020). United States Immigration and Customs Enforcement (ICE) refers to sanctuary cities wherein "local authorities refuse to hand over illegal immigrants to federal agents to deportation" (Ingraham 2017). In his report entitled "The Effects of Sanctuary Policies on Crime and the Economy," Tom K. Wong expands the definition slightly by explaining that "[s]anctuary counties – as defined by this report – are counties that do not assist federal immigration enforcement officials by holding people in custody beyond their release date" (Wong 2017).

A sanctuary city is not an area where undocumented, illegal immigrants may somehow be recognized as a lawful alien/resident under United States law. Rather, a sanctuary city connotes a municipality wherein the municipal authorities will not cooperate with ICE when an undocumented illegal immigrant runs afoul of local municipal laws and ordinances (usually minor infractions like traffic violations) and will not turn that individual over from city/county officials to the federal government for deportation hearings. According to a well-publicized report on sanctuary cities, ICE has classified 608 different counties/cities as being sanctuary jurisdictions (Wong 2017). The propriety of sanctuary cities increasingly become fodder for the political and culture wars in the United States with ascendency of Donald Trump as a political candidate for President of the United States in 2015. During the subsequent years of the Trump administration (2016–2020), Trump repeatedly and derogatorily referenced sanctuary cities as being "incubators of crime," "dangerous hotbeds of criminal activities," and locations that will result "in so many needless deaths" (for presumably following a practice that makes the jurisdiction a sanctuary city) (Ingraham 2017).

The practice that leads some cities to be labelled "sanctuary cities" generally is as follows: An undocumented illegal immigrant will be pulled over for a minor infraction such as driving without a license, driving with a broken taillight, running a red light, speeding or some other relatively minor misdemeanor offenses. In a non-sanctuary city area, after the local authorities process the individual for whatever minor misdemeanor offense they have committed, the undocumented individual's information is put into a federal database that is shared with ICE, ICE puts a "hold/detainer" on the individual, and then the person is kept in custody until an ICE agent can arrive, transfer this individual into federal custody, and initiate deportation procedures (Demby 2017). It is not outside the realm of possibility that the undocumented immigrant will be held for local authorities for days (again, for example, for a minor traffic infraction) and then held for much longer periods of time by ICE while ICE deportation proceedings are commenced. By contrast, in a so-called sanctuary city," the city police department will release the undocumented individual after he or she has been "cleared of charges, posted bail, or completed jail time" for the underlying offense for which he/she was initially arrested (America's Voice 2019). No coordination or cooperation with ICE will occur and deportation proceedings will obviously not be commenced.

Basic Reasons Why Sanctuary Cities are Safer

First, the workload of city police departments is decreased when they only need to focus on enforcing their applicable municipal laws (and not be involved in a process that converts the local law enforcement into federal immigration agents for purposes of identifying and processing undocumented immigrants). In sanctuary cities, police therefore have more time to focus on more serious crimes and investigate and prosecute them accordingly. This better protects the general public.

Second, "cities, communities, and law enforcement want undocumented immigrants to trust the police...[i]n order for the police to be most effective at their jobs" (America's Voice 2019). Undocumented immigrants are more likely to talk to the police, provide witness statements, provide leads in cases, and provide other cooperation, but only if they are not petrified that such cooperation with the police will lead to their eventual ICE detention and deportation. Further, undocumented immigrants who are victims of crimes are also reluctant to come forward for the same reason. Thus, according to the Houston Chief of Police, the number of Hispanic-Americans reporting that they are potential victims of a rape decreased 42.8% in 2018 and reporting on other serious crimes dropped 13% (America's Voice 2019). Put plainly, undocumented women will not report rape, sexual assault and/or domestic violence for fear of then being deported for being an "illegal." This phenomenon has created an issue of potential human rights abuses in those non-sanctuary cities. This may explain why the leading police union and advocacy group in the United States, the Fraternal Order of Police, have expressed support for the existence and operation of sanctuary cities. (America's Voice 2019). Other major law enforcement groups, like The Major Cities Chiefs Association (which represents the 68 largest law enforcement agencies in the United States), and the International Association of Chiefs of Police, have strongly supported the operation of sanctuary cities and that the failure to do so would have major detrimental effects to the local communities and waste state and local law enforcement in the enforcement of civil immigration laws (Ingraham 2017). Indeed, the Major Cities Chiefs Association stated that mixing the job expectations of local police with the enforcement of immigration efforts "would result in increased crime against immigrants and in the broadest community, create a class of silent victims and eliminate the potential for assistance from immigrants in solving crimes or preventing future terroristic acts" (Wong 2017; see also, Ferrell et al. 2006).

Evidence of Increased Quality of Life in Sanctuary Cities

One study showed that *non*-sanctuary cities had a 15% increase in crime rate as compared to sanctuary cities (Wong 2017; Ingraham 2017). Phrased another way, according to a study conducted in 2017, the average large urban area experienced 654 fewer crimes per 100,000 residents when that metropolitan area is a sanctuary city (Ingraham 2017). While speculative, one could reasonably argue that freeing up municipal police departments to police and enforce local laws (and not serve as quasi-immigration agents) frees up local police resources and allows for those police departments to spend their time thwarting and deterring crime by spending more time in neighborhoods, thereby decreasing criminal activity. In his report on sanctuary cities, Tom Wong also found that even in smaller counties and rural neighborhoods, "crime rates were also lower for sanctuary areas" (Ingraham 2017).

Further, the benefits of living in a sanctuary city extend beyond lower crime rates. According to Wong, "economies are stronger in sanctuary counties – from higher median household income, less poverty, and less reliance on public assistance to higher labor force participation, higher employment-to-population ratios, and lower unemployment" (Wong 2017). More specifically, Wong in his research and subsequent report, came to the following specific findings to the benefits of living in a sanctuary city/county beyond decreased crime rates, including as follows:

Medial household annual income is, on average, $4,353 higher in sanctuary counties compared to nonsanctuary counties.

The poverty rate is 2.3% lower, on average, in sanctuary counties compared to nonsanctuary counties.

Unemployment is, on average, 1.1% lower in sanctuary counties compared to nonsanctuary counties.

[T]here is significantly less reliance on public assistance in sanctuary counties compared to nonsanctuary counties.

While the results hold true across sanctuary jurisdictions, the sanctuary counties with the smallest populations see the most pronounced effects (Wong 2017).

A Brief Comment on United States Federal Law

Trump issued an Executive Order in January 2017, trying to clamp down on sanctuary cities. However, this order was blocked by a federal judge and then a bevy of other federal courts, and the law appears to support the conclusion that Trump's attempt to impose a restriction on sanctuary cities by executive order was outside the scope of his powers. Trump also threatened to deprive sanctuary cities of federal funds that might be used by the cities for social services, transportation infrastructure, and other endeavors. However, these efforts by the Trump administration have been ineffective and considered to be outside the powers of the constitutional powers of his office.

As a matter of United States constitutional law, the United States Congress may certainly utilize its constitutionally enumerated "funding power" to place restrictions upon states, counties and cities who are to be recipients of the federal funding. Phrased another way, the United States Congress may place certain policy restrictions on State governments (or other political subdivisions of the State) that must be adhered to before the State may receive the federal funding (Voisard 2018). The United States Supreme Court has ruled this practice to be constitutional in the seminal case *South Dakota v. Dole* (1987). However, this is a constitutional power that belongs to the United States Congress and *not* the President (Dietry 2018; see also, Voisard 2018). Furthermore, this power of the United States Congress (and again, not the President) is subject to several other constitutional limitations, such as the requirement that the conditions imposed by the Congress for States to follow to receive federal funds must be unambiguous in exactly what is being required of the State governments by the Congress (Dietry 2018). Finally, Congress's imposed conditions on the "receipt of federal funds [should be done] in a way reasonably calculated to address. . .a purpose for which the funds are expended" (Dietry 2018). Phrased another way, for example, the federal government may not constitutionally deprive educational funding or funds for road and highway infrastructure as a way to coerce a state to cooperate with the federal government on immigration reporting and/or enforcement.

Conclusion

As the above data illustrates, sanctuary jurisdictions are generally safer and economically stronger. Sanctuary jurisdictions allow law enforcement to be for more effective and efficient in enforcing its local laws without having to be "deputized" into policing federal immigration policies as well. The extra time saved by not becoming entangled in immigration enforcement efforts saves valuable time and resources to pursue and enforce laws and policies important to that local jurisdiction. Additional local benefits in sanctuary cities accrue in additional areas as diverse as education and support of children. To the extent sanctuary cities/counties are not complicit with ICE in breaking up family units, those otherwise intact family units will generally contribute to a stronger local economy and put less stress on schools and child-care institutions when the immigrant family is left intact. As Professor Tom Wong has put it, "if you deport the breadwinner [of an immigrant family], that leaves families more economically vulnerable. That means that these economically vulnerable families are more reliant on public assistance" (Ingraham 2017).

References

America's Voice. (2019). Immigration 101: What is a sanctuary city? https://americasvoice.org/blog/what-is-a-sanctuary-city/. Accessed June 2021.

Beckman, J. A. (2007). *Comparative legal approaches to homeland security and anti-terrorism*. Aldershot, Hampshire: Ashgate Publishing Limited.

Demby, G. (2017). Why sanctuary cities are safer. *NPR,* January 29, 2017. https://www.npr.org/sections/codeswitch/2017/01/29/512002076/why-sanctuary-cities-are-safer. Last accessed June 2021.

Dietry, B. (2018). Are sanctuaries safe? A legal analysis of the campaign to defund sanctuary cities. *4 U. Cent. Fla Dep't Legal Stud. L.J. [1], 111-120.*

Ewing, W., Martinzez, D., & Rumbaut, R. (2015). The criminalization of immigration in the United States. *American Immigration Council*. https://www.americanimmigrationcouncil.org/research/criminalization-immigration-united-states. Accessed June 2021.

Ferrell, C. E., & Major Cities Chiefs Association. et al. (2006) M.C.C. Immigration Committee Recommendations for Enforcement of Immigration Laws by Local Policy Agencies, *Major Cities Chiefs Association*, https://majorcitieschiefs.com/pdf/news/MCC_Position_Statement.pdf.

Ingraham, C. (2017). Trump says sanctuary cities are hotbeds of crime; Data says the opposite. *Washington Post*, January 27, 2017.

Kopan, T. (2018). What are sanctuary cities, and can they be defunded? *CNN Politics,* March 26, 2018. https://www.cnn.com/2017/01/25/politics/sanctuary-cities-explained/index.html. Accessed June 2021.

Maxwell, B. (2004). *Homeland security: A documentary history*. Washington, DC: CQ Press.

Shultz, D. (2020). Crime did not surge when California became a 'sanctuary state.' *Science/AASS*. https://www.sciencemag.org/news/2020/02/crime-did-not-surge-when-california-became-sanctuary-state. Last accessed June 2021.

Voisard, A. (2018). Cities, states resist – And assist – Immigration crackdown in new ways. *Pew Institute*. https://www.pewtrusts.org/en/research-and-analysis/blogs/stateline/2018/08/03/cities-states-resist-and-assist-immigration-crackdown-in-new-ways. Last accessed June 2021.

Wong, T. (2017). The effects of sanctuary policies on crime and the economy. *Center for American Progress*. https://www.americanimmigrationcouncil.org/research/criminalization-immigration-united-states. Last accessed June 2021.

Sanitation – Facilities for Cleanliness

▶ Water, Sanitation, and Hygiene Question of Future Cities of the Developing World

SAPCC – State Action Plan on Climate Change

▶ Adapting to a Changing Climate Through Nature-Based Solutions

Satellite City

▶ New Cities

Satellite Infrastructure

▶ New Orbital Urbanization

Satellite Town

▶ New Cities

Scale

► New Localism: New Regionalism

SDG 6

► Meeting SDG6: Ensuring Safe Drinking Water for All in Rural India

Secure

► Closing the Loop on Local Food Access Through Disaster Management

Security

► Responsibility to Prepare and Prevent (R2P2): Applying Unprecedented Foresight to Addressing Unprecedented Climate Risks

Security – Provisioning

► Rainwater Harvesting for Water Security in Informal Settlements: Techniques, Practices, and Options

Self-Regulation

► Voluntary Programs for Urban and Regional Futures

Self-Sustainable City

► Circular Cities

Senegalese Ecovillage Network

Andreas Hernandez
Marymount Manhattan College, New York, NY, USA

Synonyms

Appropriate Technology; Community Development; Ecovillages; Environmental Movement; Regional Development

Definition

The Senegalese Ecovillage Network brings together hundreds of villages in a heterogeneous movement that seeks community-led development by taking the best of West African village life and combining this with green technologies and recuperation of soils and forests. This network has reframed the notion of ecovillages coming from the Global North, and in the process also reframed ideas of West African rural development.

Introduction

Leaders and activists of the Senegalese Ecovillage Network report that the ecovillage model provides a framework to engage the interrelations of culture, economy, technology, and environment, to promote materially and culturally better ways of living over the long term. The ecovillage framework they say is not prescriptive, but orients innovative approaches to protracted problems. They assert that this holistic framework is highly resonant with West African traditional worldviews and provides an effective tool for development that respects traditional village culture while opening to the world and introducing technology (Illieva and Hernandez 2018). The Global Ecovillage Network (GEN) defines an ecovillage as "... an intentional, traditional or urban community that is consciously designed through locally owned participatory processes in all four dimensions of sustainability (social, culture, ecology and economy) to regenerate social and natural environments" (GEN 2021, p. 1). Ecovillages in Senegal have developed projects as diverse as: solar power grids, extensive permaculture gardens, biogas, and solar cookers not reliant on scarce wood fuel, reforestation, reintroduction of dry crops such as millet, and water pumps and tanks that extend growing seasons. They have also developed value-added enterprises and worked to construct new markets (Illieva and Hernandez 2018; Joubert and Dregger 2015). According to the United Nations Development Programme (UNDP) (2021), in at least one ecovillage, years of outmigration have reversed as young people return to new opportunities in villages.

Network leaders claim that Northern ecovillages are often focused on creating community and ecologically viable worldviews and spiritual systems. African villages, they argue, already possess these social and cultural resources, and seek to bring in "clean modern technologies to uplift living conditions" while recuperating the environments upon which villages depend (Illieva and Hernandez 2018, p. 25). Ecovillages in the Global North are generally smaller initiatives "created around shared values and projects that relate to the journey towards a more sustainable society." Ecovillages in the Global South tend to "comprise communities in which local leaders understand the threat that economic globalization poses to the health of their communities and are seeking to wrest back some measure of control over their cultural, ecological and economic resources" (Dawson 2011, p. 5).

History of the Network

Villages in Senegal face dire environmental conditions, which are intertwined with difficult historic social conditions. In the north of the country, the Sahara is arriving where forests existed 60 years ago. Deforestation by colonial powers, villages, and companies have left impoverished landscapes. Organizations such as USAID and the Chinese Government have advocated for and subsidized chemical and water-intensive rice production to sell nationally and for export, poisoning rivers, and mining soils. This constellation of factors has impoverished villages and contributed to hunger, outmigration, and social breakdown.

The Senegalese Ecovillage Network began in the traditional fishing village of Yoff in coalition with the Ithaca Ecovillage and the third international EcoCity Conference, which was held there in 1995. Yoff was organizing to defend their livelihood and culture from land grabbing by public and private entities as Dakar expanded to encircle and subsume traditional villages. Through the internal successes of what became EcoYoff, the ecovillage framework began to spread organically to villages in ecologically diverse regions of Senegal. This framework set the foundations for innovative responses, outside of both traditional village modalities as well as mainline development pathways. Out of these initial experiences emerged the Senegal Ecovillage Network (GENSEN).

Government officials, including a President, took note in the early 2000s and launched the Ministry of Ecovillages, which later became the National Agency for Ecovillages (ANEV) with the project of transitioning half of the country's 28,000 villages into ecovillages. ANEV seeks to involve and support the villages with development assistance that villages request. This includes interventions such as implementing solar power, providing seeds, infrastructure for irrigation, and technical support.

The formal power structures of villages vary between elected mayors and hereditary chiefs. In both cases, villages have taken on the ecovillage framework usually with the leadership, or at least with the strong support of these formal village positions. Thus, government resources are leveraged directly toward ecovillage development at the village level, as villages make this a political focus. Village leadership is also then able to formally interact with federal organs, particularly with ANEV.

With these forms of institutionalization, funds from the UNDP and other international donors became available, creating a split in the Movement. GENSEN frayed and the movement split into two heterogeneous wings. One part of the Movement asserts that the community-led dimensions of ecovillage development are essential, and direct government intervention weakens community agency, creating a situation that looks like other government-led development efforts. The other part of the Movement insists that Government and international aid provides access to crucial and expensive technologies (such as solar power) and infrastructures (such as irrigation), and that villages remain agents in this relationship, participating in decisions of what interventions or resources will be provided. Locally led NGO's have formed to coordinate ecovillage activities. For example, the Network for Ecovillage Emergence and Development in the Sahel (REDES) works with five ecovillages in the North of Senegal and brokers relationships with international partners and donors. GEN created GEN Africa, which has become the overarching and unifying organization to which most ecovillages may relate. In 2014, GEN and the Government of Senegal held a Global Ecovillage Summit in Dakar. Although the goal of 14,000 villages remains distant, hundreds of villages are adopting aspects of the African ecovillage model, often in coalition with ANEV or NGOs, creating one of the most successful grassroots development efforts on the continent. The model is spreading to neighboring countries such as Mali and Democratic Republic of Congo (Illieva and Hernandez 2018; Joubert and Dregger 2015).

Case Studies

The village of Mbackombel is considered one of the more elaborated villages within the ANEV network and has transformed village life through creative use of a solar powered microgrid – particularly around issues of food insecurity. Among numerous benefits, this grid powers pumps to store water, expanding the growing season. This stored water is the basis for new climate-friendly permaculture gardens, reforestation projects, and fish ponds. Young girls have traditionally been charged with the task of fetching water, often hours away. The stored water system has freed up the time of many young girls who are now able to attend school. The villagers of Mbackombel have also concentrated on bioconstruction projects, using locally available materials for permanent structures, and on creating rainwater storage ponds. Women in the village developed solar cookers and highly efficient clay rocket stoves – decreasing dependency on fuels. The solar microgrid has also transformed education and global connections for villagers. The school has a computer room with rechargeable laptops, and adults are able to connect to the wider world through internet (Illieva and Hernandez 2018; Joubert and Dregger 2015).

The village of Guédé Chantier has focused on community power and participation, mobilizing the cultural value of *jokkere endam* (primacy of community solidarity) as the basis of innovative projects. Guédé Chantier is also the home of the multi-village REDES NGO. Women's groups in this village at the edge of the Sahara Desert have led the ongoing transition to organic agriculture focused on climate appropriate crops. These women seek to better use and protect water in the cultivation of diverse healthy food destined first for consumption by village families (instead of cash crops for national markets or for export). Women's groups are at the same time developing value-added processing and preservation for their fruits and vegetables, which they sell in local markets. The community constructed the Center of Genetic Resources for seed saving and sharing – seeds are distributed freely to farmers who also receive training in regenerative farming. US and Senegalese students intern to support these projects. The youth-led Association of Ecoguardians engages wider communities in education initiatives, sponsors large-scale cleanups, and provides training programs for youth populations. The Ecoguardians use community theater as a key modality for social transformation addressing questions of childhood malnutrition, STD's, chemical poisons in the fields, and plastic litter (Illieva and Hernandez 2018; Joubert and Dregger 2015).

Summary

Villages in diverse geographies of Senegal continue to use the ecovillage model as a holistic tool for bringing together the best of West African village life with green technologies. Ousmane Pame, Director of GEN Africa sums up his ecovillage vision:

> *The kind of development we are looking for is a development based on our traditions, our cultures and personalities. And also we would like to open up to the modern world at the same time.* ("Senegalese Ecovillage Movement" 2019; 3:11)

References

Dawson, J. (2011). *From Islands to networks: An exploration of the history – And a glimpse into the future – Of the ecovillage movement*. Totnes: Schumacher Press.

Global Ecovillage Network. (2021). *What is an ecovillage?* Available online: https://ecovillage.org/projects/what-is-an-ecovillage/

Ilieva, R., & Hernandez, A. (2018). Scaling-up sustainable development initiatives: A comparative case study of agri-food system innovations in Brazil, New York, and Senegal. *Sustainability, 10*(11), 4057. https://doi.org/10.3390/su10114057.

Joubert, K., & Dregger, L. (2015). *Ecovillage: 1001 ways to heal the planet*. Axminster: Triarchy Press.

Senegalese Ecovillage Movement. (2019). Available online: https://www.youtube.com/watch?v=tdcQIcj112E

United Nations Development Programme (UNDP). (2021). *L'Ecovillage de Mbackombel sort de l'ombre*. Available online: http://www.sn.undp.org/content/senegal/fr/home/ourwork/environmentandenergy/successstories/-l_ecovillage-de-mbackombel-sort-de-lombre.html

Shantytowns

► Disaster Risk in Informal Settlements and Opportunities for Resilience

Shaper

► Spatial Demography as the Shaper of Urban and Regional Planning Under the Impact of Rapid Urbanization

Share

► Closing the Loop on Local Food Access Through Disaster Management

Shared

► Participatory Planning: A Useful Tool for the Development of Sustainable Mega-City Regions

Shared Urban Codes

► Urban Commons as a Bridge Between the Spatial and the Social

Shared Urban Conventions

► Urban Commons as a Bridge Between the Spatial and the Social

Shared Urban Practices

► Urban Commons as a Bridge Between the Spatial and the Social

Shared Urban Resources

► Urban Commons as a Bridge Between the Spatial and the Social

Shift in Atmospheric Conditions and Surface Water Resources in Sri Lanka

► Climate Change and Surface Water Resources in Sri Lanka

Shrinkage

► Shrinking Towns and Cities

Shrinking Towns and Cities

Etienne Nel
University of Otago, Dunedin, New Zealand

Synonyms

Decline; Shrinkage; Urban

Definition

> an urban area...that has experienced population loss, economic downturn, employment decline and social problems as symptoms of a structural crisis. (Martinez-Fernandez et al. 2012a, p. 214)

Introduction

Urban shrinkage is gaining recognition as one of the most significant processes impacting towns and cities in many parts of the world – primarily in the Global North, and to lesser degree in parts of the South (Martinez-Fernandez et al. 2016; Haase et al. 2017). The 10,000 year long history of human settlement has generally been associated with steady and sustained urban growth, albeit there have been occasional instances or phases of decline associated with a range of incidents such as warfare, disease, and resource depletion. What is distinctive about the current phase of urban history is that in large parts of the world, the coalescence of economic change, globalization, deindustrialization, falling birth rates, and migration have led to a situation in which shrinking towns and cities are not isolated phenomenon, but rather occur extensively as a result of structural change. In the case of Eastern Europe in particular, the postcommunist era has been directly associated with deindustrialization, population movement, and urban shrinkage (Silverman 2020). It has been estimated that 10% of urban centers globally and 40% of the those in the developed world have experienced some degree of population shrinkage in recent decades (Wolff and Wiechmann 2018). The extensive and drawn-out nature of urban decline has taken place in places as diverse as large cities in the US Rust Belt such as Detroit, small town regions such as rural Japan, and large tracts of Eastern Europe. Hollander and Nemeth (2011) note that 25% of cities globally with more than 100,000 people have experienced some degree of shrinkage. In the case of China Long and Gao (2019) note that 180 of the country's 653 cities shrunk between 2000 and 2010.

Despite the extensive nature of small town and city shrinkage – referred to collectively in this chapter as urban shrinkage – as Haase et al. (2014) argue, it is one of the most undertheorized and neglected urban phenomena. This, despite the fact that, as Mamadouh and van Wageningen (2016, p. 154) point out in the case of Europe, "population shrinkage is set to become the new normal. Not only in rural areas but certainly also in a large number of European towns and cities." This chapter examines what is urban shrinkage, how extensive it is, what responses are emerging, and what are the current issues being raised by academics and by policymakers.

Defining Urban Shrinkage

There is no uniform definition about what urban shrinkage means. This is because its causes, nature, impact, and longevity vary across time and space. In many cases attempts at definitions are complicated by differing rates of decline and the fact that in some instances decline is temporary and may even be followed by regrowth. That said, the sheer scale of recent change which has been observed suggests that, despite definitional challenges, we are witnessing a significant urban transformation that justifies greater attention and attempts to better understand what is happening. In broad terms most authors refer to urban shrinkage (or shrinking cities or towns) as situations where urban places experience both sustained population loss and economic decline for several years. While some authors associate shrinkage only with places above a certain size, other authors do not use preexisting size as an indicator (Rocak 2020). One attempt at developing a standardized definition was made by the Shrinking Cities International Research Network, which defined a shrinking city as a place with more than 10,000 people which had experienced population loss for more than two years and which also faced structural economic challenges (in Schackmar 2020).

While many places appear locked into a trajectory of long-term demographic and economic decline, such as former mining and industrial centers, it has also been noted that in some places decline has ceased and a new norm has been reached, while in the case of a few others there are instances of regrowth – such as Leipzig and Liverpool (Haase et al. 2021). The latter tend to be restricted to larger centers which are close to prosperous cities, which are attractive to in-migrants and commuters, centers which attract new rounds of investment (Wolff et al. 2017; Haase et al. 2018). In smaller, more isolated mono-economies reversal seldom occurs. This has led Stryjakiewivz and Jaroszewska (2016) to postulate that shrinkage can be "short-term," "long-term," or "episodic," while in a similar vein Wolff and Weichmann (2018) have identified places which are "continuously shrinking," those which are experiencing "episodic" shrinkages and those which experienced temporary shrinkage but which have stabilized at a new norm. This reality complicates both how we understand what urban shrinkage means and also whether it needs to be responded to or whether "market forces" will determine a new outcome.

How Widespread Is It?

Eastern Europe has the most extensive incidence of urban shrinkage in the world, with Wolff and Weichmann (2018) noting that shrinkage is the reality for over 30% of towns and cities. That is said, as Martnez-Fernandez et al. (2012a) note urban shrinkage is a global phenomenon and evidence of its existence can be found on every continent. In the USA it has been noted that between 1980 and 2010 41% of rural communities – places with population between 5000–10,000 – experienced a period of contraction (Wuthnow 2018). In Australia the comparable figure is 48% of all local government areas with more than 10,000 people (Martinez-Fernandez et al. 2012a), while in China with its many large cities, as noted above, over 180 cities out of 653 have recorded population loss (Long and Gao 2019).

Critique

While yardsticks of population size and economic activity are relatively easy to measure, less measurable but equally serious features associated with shrinkage are processes of sociocultural transformation, well-being challenges, the effects of out-migration, particularly of the most educated, and of aging, all of which impact on the vitality and future prospects for such places (Rocak 2020). Silverman (2020, p. 6) extends this analysis pointing out that shrinking places are not homogenous but within them social distinctions are growing.

Other concerns include the lack of single definitions of what constitutes urban shrinkages which in part reflects the varied nature of such processes around the world. However of greater concern is the general absence of defined, logical responses to urban shrinkage, as is examined in the next section, with Berglund (2020) additionally arguing that there is a need to consider governance and political economic thinking in both the understanding of and the responses to shrinkage.

Responses to Urban Shrinkage

Sustained urban shrinkage has emerged in various parts of the world as a real challenge to urban planners and governments. This is because the conventional expectation has been that one needs to continually plan for urban expansion and growth, not decline. Planners are now challenged, in the case of shrinking cities not to plan for growth, but rather to reduce the housing stock, services, and social facilities (Stryjakiewivz and Jaroszewska 2016). This has, in recent year, spawned thinking and practice around concepts such as "smart decline," "planning for less," "beyond growth," and "right-sizing," whereby the concept of unabated growth is replaced by active policies to reuse abandoned urban facilities for new uses, or to turn abandoned spaces into green space and urban farms. Parallel interventions seek to reduce housing stock and to find new economic activities in places where, often, jobs have been lost and the remaining populations are aging (Hollander and Nemeth 2011; Hummel 2015; Hollander 2018; Leick and Lang 2018). In the case of "beyond growth" authors such as Leick and Lang (2018) argue there is a need to look inward at local leadership and social innovation where traditional market-led growth is unlikely. There are also the political issues associated with the cost of continually providing services to declining areas, and that of managing expectations.

While some countries, such as Poland and Germany, have tried to implement state-led responses to these challenges, and the EU have actively investigated what responses could be initiated (Rink et al. 2012), in other parts of the world the earlier "withdrawal of the state" means that declining communities often have to look inward to their own resources and capacities to negotiate new futures for their community, based on principles of self-reliance, a process which generates very mixed outcomes (de Noronha Vaz et al. 2013). In Hospers' (2014) view policymakers in Europe have responded either through trivializing the phenomenon or alternatively, trying to counter it through interventions, accepting it and facilitating downsizing, or utilizing it, in the sense of using it as a testing ground for new urban norms. As a way forward Sousa and Pinho (2015) argue that planners need to be more flexible in their approaches to planning to accommodate what for many places is the new norm. In the case of the OECD (in Martinez-Fernandez et al. 2012b), looking at sustainable planning, creating employment opportunities for the retired, integrating migrants, and assuring public financial support are seen as critical steps to initiate.

Conclusion

As the literature examined reveals, there are certain difficulties associated with defining what urban shrinkage is given the variation in experience, not least because of the reality that some places have experienced regrowth. The phenomenon of shrinking towns and cities, nonetheless, is a distinctive reality in many parts of the world. In all likelihood as birth rates fall around the world and economic activity selectively privileges or abandons particular localities, the reality and effects or urban shrinkage will become more pronounced. This presents both a challenge and an opportunity for policymakers to rethink conventional notions of urban planning and growth and also to embrace interventions which recognize political economic realities and which respond to local needs and realities, drawing on and working with local capacity (Berglund 2020).

Cross-References

▶ Green Cities
▶ Smart City

References

Berglund, L. (2020). Critiques of the shrinking cities literature from an urban political economy framework. *Journal of Planning Literature, 35*(4), 423–439.

de Noronha Vaz, T., van Leeuwen, E., & Nijkamp, P. (2013). *Towns in a rural world*. Farnham: Ashgate.

Haase, A., Rink, D., Grossmann, K., Bernt, M., & Mykhnenko, V. (2014). Conceptualizing urban shrinkage. *Environment and Planning A, 46*(7), 1519–1534.

Haase, A., Wolff, M., Špačková, P., & Radzimski, A. (2017). Reurbanisation in postsocialist Europe-a comparative view of Eastern Germany, Poland, and the Czech Republic. *Comparative Population Studies-Zeitschrift für Bevölkerungswissenschaft, 42*, 353–389.

Haase, A., Wolff, M., & Rink, D. (2018). From shrinkage to regrowth: The nexus between urban dynamics, land use change and ecosystem service provision. In *Urban transformations* (pp. 197–219). Cham: Springer.
Haase, A., Bontje, M., Couch, C., Marcinczak, S., Rink, D., Rumpel, P., & Wolff, M. (2021). Factors driving the regrowth of European cities and the role of local and contextual impacts: A contrasting analysis of regrowing and shrinking cities. *Cities, 108*, 102942.
Hollander, J. (2018). *A research agenda for shrinking cities*. Cheltenham: Edward Elgar Publishing.
Hollander, J. B., & Németh, J. (2011). The bounds of smart decline: A foundational theory for planning shrinking cities. *Housing Policy Debate, 21*(3), 349–367.
Hospers, G. J. (2014). Policy responses to urban shrinkage: From growth thinking to civic engagement. *European Planning Studies, 22*(7), 1507–1523.
Hummel, D. (2015). Right-sizing cities: A look at five cities. *Public Budgeting & Finance, 35*(2), 1–18.
Leick, B., & Lang, T. (2018). Re-thinking non-core regions: Planning strategies and practices beyond growth. *European Planning Studies, 26*(2), 213–238.
Long, Y., & Gao, S. (2019). Shrinking cities in China: The overall profile and paradox in planning. In *Shrinking cities in China* (pp. 3–21). Singapore: Springer.
Mamadouh, V., & Wageningen, A. (2016). *Urban Europe. Fifty tales of the city.* Amsterdam: Amsterdam University Press.
Martinez-Fernandez, C., Audirac, I., Fol, S., & Cunningham-Sabot, E. (2012a). Shrinking cities: Urban challenges of globalization. *International Journal of Urban and Regional Research, 36*(2), 213–225.
Martinez-Fernandez, C., Kubo, N., Noya, A., & Weyman, T. (2012b). *Demographic change and local development: Shrinkage, regeneration and social dynamics.* Paris: OECD Publishing.
Martinez-Fernandez, C., Weyman, T., Fol, S., Audirac, I., Cunningham-Sabot, E., Wiechmann, T., & Yahagi, H. (2016). Shrinking cities in Australia, Japan, Europe and the USA: From a global process to local policy responses. *Progress in planning, 105*, 1–48.
Rink, D., Rumpel, P., Slach, O., Cortese, C., Violante, A., Bini, P. C., & Krzystofik, R. (2012). *Governance of shrinkage: Lessons learnt from analysis for urban planning and policy.* Leipzig: Helmholtz Centre for Environmental Research.
Rocak, M. (2020). *The experience of shrinkage. Exploring social capital in the context of urban shrinkage* (Doctoral Dissertation, Nijmegen).
Schackmar, J. (2020, September). Smart cities–A new revitalisation approach for shrinking cities? In *SHAPING URBAN CHANGE–Livable City regions for the 21st century. Proceedings of REAL CORP 2020, 25th International Conference on Urban Development, Regional Planning and Information Society* (pp. 739–750). CORP–Competence Center of Urban and Regional Planning.
Silverman, R. M. (2020). Rethinking shrinking cities: Peripheral dual cities have arrived. *Journal of Urban Affairs, 42*(3), 294–311.
Sousa, S., & Pinho, P. (2015). Planning for shrinkage: Paradox or paradigm. *European Planning Studies, 23*(1), 12–32.
Stryjakiewicz, T., & Jaroszewska, E. (2016). The process of shrinkage as a challenge to urban governance. *Quaestiones Geographicae, 35*(2), 27–37.
Wolff, M., & Wiechmann, T. (2018). Urban growth and decline: Europe's shrinking cities in a comparative perspective: 1990–2010. *European Urban and Regional Studies, 25*(2), 122–139.
Wolff, M., Haase, A., Haase, D., & Kabisch, N. (2017). The impact of urban regrowth on the built environment. *Urban Studies, 54*(12), 2683–2700.
Wuthnow, R. (2018). *The left behind: Decline and rage in rural America.* Princetown: Princetown University Press.

Sidewalk Drones

► Personal Delivery Robots: How Will Cities Manage Multiple, Automated, Logistics Fleets in Pedestrian Spaces?

Sidewalk Robots

► Personal Delivery Robots: How Will Cities Manage Multiple, Automated, Logistics Fleets in Pedestrian Spaces?

Slum – Informal Settlement

► Rainwater Harvesting for Water Security in Informal Settlements: Techniques, Practices, and Options

Slums

► Disaster Risk in Informal Settlements and Opportunities for Resilience

Slums Upgrading

► New Forms of Shared Governance and Local Action Plan in Socially Vulnerable Settlements

Small Rural Town – Growth Point

► Growth, Expansion, and Future of Small Rural Towns

Small Towns in Asia and Urban Sustainability

Aviram Sharma
School of Ecology and Environment Studies, Nalanda University, Rajgir, Bihar, India

Synonyms

Intermediate Towns; New Towns; Provincial Towns; Small Urban Centers; Sustainability Challenges

Definition

Around 60% of the global urban population resides in urban spaces having less than 500,000 population. Yet, the scholarly literature predominately focuses on megacities and large urban centers. The urban studies literature does not adequately engage with small towns' socioeconomic and environmental dynamics. To fully comprehend the broad contours of twenty-first-century urbanization, it is crucial to understand the emerging socio-environmental dynamics and urban sustainability challenges of small towns. In this background, this entry engages with small towns in Asia, one of the most rapidly urbanizing regions in the world, and charts out the urban sustainability challenges emanating from such spaces.

Introduction

More people live in urban areas at the global level in today's time than rural (Caprotti and Yu 2017; UN-Habitat 2020). It is expected that the global urban population will keep registering growth in the coming decades. The urban Asian population bypassed the 50% mark in 2020, whereas it is expected that the urban African population will reach the 50% mark by 2035 (UN-Habitat 2020). Interestingly, by 2018, Asia had 20 out of the 33 megacities (with a population of 10 million or more) (UN 2019). Asia, with a 2.3 billion urban population (out of 4.3 billion global urban population), accounts for 54% of the global urban population (UN-Habitat 2020). Caprotti and Yu (2017) argue that the geography of urbanization has shifted from the Euro-American context toward Asia in recent years.

At the global level, most of the urban population resides in urban spaces having less than 1 million population; megacities (more than 10 million population) and large cities (between 5–10 million population) only account for 13% (529 million) and 8% (325 million) of the total urban population (UN 2019). Almost half of the urban population resides in urban spaces having less than 500,000 inhabitants (UN 2019). Similar to the global trend, more urban Asians live in small and medium-sized towns than megacities. Roughly around 60% of the urban Asian population resides in urban areas with less than 1 million people (UN-Habitat 2020). Yet, the scholarly literature primarily focuses on megacities, large cities, and large-scale regional urbanization (on meta- and hypercities). Nonmetropolitan urban areas and especially small and medium-sized towns and urban spaces are relatively understudied in the literature (Bell and Jayne 2009; Sharma 2015; Guin 2019). Scholars have attempted to revive the study of the different facets of small towns (Tan 1986; Ofori-Amoah 2007; Bell and Jayne 2009; Shaw 2013; Denis and Zérah 2017) in both the Global North and the Global South. Such attempts emanating from the global south in recent years were termed as the "second urban turn" in the academic debates, and it contributed in "provincializing the global urban studies" (Guin 2019; Mukhopadhyay et al. 2020). Asian urban transition is happening at an unprecedented rate and in the "era of the environment," when global warming, climate change, and resource depletion have come out as major threats.

The urbanization pattern and rate of urbanization vary across different parts of Asia. There are countries having 100% urban population, such as Singapore and Kuwait. At the same time, few others have more than 90% urban population, such as Japan (92%) and Israel (92%). Besides, several others have less than 30% urban populace, such as Afghanistan (25%), Nepal (20%), and Sri Lanka (18%) (Table 1).

The number of small towns in Asia varies from region to region. Small towns are of varied kinds, and they perform different roles, such as growth centers, a bridge between rural and urban areas, administrative units, strategic sites, company towns, private towns, religious centers, and many other functions (Tan 1986; Dahiya 2012; Shaw 2013; UN 2019; Guin 2019; Mukhopadhyay et al. 2020). Before delving further, it is crucial to understand how small towns are defined and categorized. The definition of small towns and rural/urban areas varies in terms of space (different countries) and time, and there is no universally accepted definition (Guin 2019; Tong et al. 2020). The definitions given by different institutions and nations account for factors such as population size, population density, patterns of economic activities, level of infrastructure, or combination of these variables or are based on even additional criteria (such as administrative decisions).

Shaw (2013) categorized small towns and small cities in India based solely on the population criteria. She categorized small towns as urban spaces with less than 100,000 population and small cities as urban spaces with populations between 100,000 and 500,000. Guin (2019) argues that the Census of India does not explicitly define small towns but defines cities as urban spaces with more than 100,000 population. Thus, urban spaces having a population of less than 100,000 can be categorized as small towns. Tong et al. (2020) argue that the concept of small towns in China does not have a clear population limit, and there are multiple interpretations of small towns. Scholars have argued that the population-based definition of "urban" spaces does not often do justice with the emerging patterns of urbanization (Guin 2019). Different alternative indexes exist to define small towns (Tan 1986; Guin 2019; Tong et al. 2020). Given the broad scope of this review, we will consider urban spaces with less than 500,000 population as small towns in this study. Understanding the socio-environmental dynamics of small towns is paramount to interpret the emerging urban patterns and the related sustainability challenges.

Small Towns in Asia and Urban Sustainability, Table 1 Countries in Asia based on the level of urbanization

Regions	Most urbanized countries (above 80%)	Predominantly urban countries (between 51% and 80%)	Predominantly rural countries (between 30% and 50%)	Least urbanized countries (below 30%)
East Asia	Singapore, China, Hong Kong SAR, China, Macao SAR, Japan, and Republic of Korea	China, Indonesia, Thailand, Malaysia, North Korea, and Brunei Darussalam	Philippines, Vietnam, Timor-Leste, Laos, and Myanmar	Cambodia
West Asia	Kuwait, Israel, Saudi Arabia, Qatar, Oman, Jordan, Bahrain, and United Arab Emirates	Iran, Turkey, Iraq, Cyprus, Syria, Lebanon, Georgia, Armenia, and Azerbaijan	Yemen	Nil
South Asia	Nil	Nil	India, Bangladesh, Pakistan, Bhutan, and Maldives	Afghanistan, Nepal, and Sri Lanka
Central Asia	Nil	Mongolia, Kazakhstan, and Turkmenistan	Kyrgyzstan, Uzbekistan	Tajikistan

Source: World Bank (2019)

S

Typology of Small Towns in Asia

The economy, demographic trends, infrastructural developments, and governance mechanisms in the thousands of small towns in Asia drastically vary from one region to another. There are marked differences among the small towns of the most urbanized countries in Asia and the predominantly rural and least urbanized countries (refer to Table 1). West Asia and East Asia are predominantly more urban than South Asia and Central Asia. Extreme poverty and inequality, a higher percentage of slum population, and dependence on the informal economy are generally high in many Asian cities. Such trends are more emphatic in the small and medium level towns of Asia's least urbanized and predominantly rural economies (Dahiya 2012).

Some small towns are emerging as nodes of economic growth and are well integrated in the global economy (due to globalization) through different networks. The local economy of many other towns is facing loss and stagnation, a general decline in population, and even losing importance at the regional and national level (Guin 2019; Tong et al. 2020). In many small towns in Asia, high population density, lack of robust financial base, infrastructural deficit, and rapid unplanned and unregulated growth of built areas have created challenging situations (Dahiya 2012). Small towns in the global south are uniquely placed in the sustainability debate; they face both the burden of development (pollution-related challenges) and underdevelopment (lack of access to safe water and sanitation) (Véron 2010). The relationship between state, market, and community is complex in many small towns and is in flux. In this background, it is crucial to understand the different varieties of small towns in Asia.

Typologies of small towns are drawn by using insights from the configurations of local economy, political and social dynamics, demographic trends, and administrative and environmental rationalities employed to govern small towns. These categorizations are not watertight and exclusionary, and they do overlap with each other. None of the small towns support only one dominant function; instead, they fulfill multiple socioeconomic and political roles.

Administrative Centers: Over the years, many of the small towns have primarily emerged as the sites of local administration/government, such as county-level administrative units or sub-prefecture level towns (Xinhua in Hubei province of China), district towns (Sheikhpura in Bihar province in India), subdistrict headquarters (Karnaphuli in Chittagong District of Chittagong Division of Bangladesh), or block headquarters (Hasilpur Tehsil in Bahawalpur district of Punjab province in Pakistan). The urban growth and infrastructural development in small towns hosting administrative units primarily happen through state-led institutions in the early years. Many small towns eventually emerge as sites of urban development, with the growth of administrative units, educational institutions, health infrastructure, and related institutions and activities. Over the period, the local economy grows, and livelihood opportunities diversify in these places – eventually giving rise to a higher and complex level of urbanization.

Economic Centers: These categories of towns represent a significant portion of the overall number of small towns. Historically, these towns have supported manufacturing (textile, automobile), trading, and diverse industrial activities (Mukhopadhyay et al. 2020). Some of them developed as market towns (agrarian markets – such as Bihar Sharif in India), commercial centers, transportation hubs, and export centers (Hubli in Karnataka, India). These small towns have often shown strong entrepreneurial and innovative capabilities (such as Tiruchengode from India). These towns play a crucial role in the regional economy, and some of them emerge as large urban centers (Tirupur and Coimbatore in India). Market towns in India are often called as *Bazaar*; in China, they are known as *Jizhen*. In many countries, small towns are resource-based towns (dependent on exploitation and processing of agricultural, mineral, forest, and other natural resources) and act as local growth engines (Tong et al. 2020). These market towns and resource-based towns are mostly well integrated into the regional and national economy. Besides, they also serve their immediate rural hinterlands. Many of these small towns are increasingly getting integrated into the global trade networks. Whereas in some regions, they are also facing decline and shrinkage.

Historical/Cultural/Tourist Towns: Many of the culturally, historically, and religiously sacred spaces in Asia developed as tourist towns over the years (Bodh Gaya, Mathura, Rishikesh, and Rameswaram in India, Ayutthaya in Thailand, and Huế in Vietnam). Such towns' material and social base revolves primarily around the famous institutions (monuments/landmarks/sites) present in the vicinity. The local economy of such towns is often well entrenched in the broad religious and tourist circuits and displays its unique national and international characteristics. With the increase in the intensity of touristic activities, many small tourists (Darjeeling in India) and historic towns (Bukhara in Uzbekistan), especially located in ecologically fragile regions, are facing different kinds of vulnerabilities (such as resource depletion, biodiversity loss, and livelihood-related threats).

New Towns: Over the last several decades, varieties of new towns were planned and developed in many parts of Asia (Islamabad in Pakistan, Chandigarh, Faridabad, Kalyani in India, Alavi, Ramin, Ramshahr, Tis, Binalood, and many more from Iran) for different strategic purposes (Ziari 2006). Some of them turned up into large cities over the years, whereas many did not flourish as expected (Ziari 2006). In the last few decades, new private towns (Lavasa and Aamby Valley City from India, King Abdullah Economic City from Saudi Arabia) have become the new addition in this category. These new privately planned towns are often categorized or proclaimed as eco-city (Masdar City in UAE, Gwanggyo in South Korea, and Dompak in Indonesia) and smart cities (Gujarat International Finance Tec-City (GIFT) city) by many. Several smart cities (for example, Dholera and Rajarhat in India) are initially planned as small and medium-sized settlements. These smart cities are supposed to scale up and address the "challenges of urbanization." Besides strategic new towns (both public and privately planned), many new towns are also getting added up by rapidly urbanizing rural areas, witnessing a rapid increase in population and economic activities (Mukhopadhyay et al. 2020).

Urban Sustainability in Small Towns of Asia

Urban sustainability has emerged as a major concern in megacities and large urban centers and is studied in the literature. However, concerns related to urban sustainability and environmental dynamics of small towns are understudied, which are argued to be at the frontline of transition (Mukhopadhyay et al. 2020). The new forms of economic and social organizations emerging in small towns are significantly altering the traditional patterns of resource use and are putting stress on the local environmental resources.

Urban sustainability can be studied from different perspectives. One group emphasizes on focusing on environmental considerations while defining urban sustainability whereas others argue for taking a holistic approach and the need to balance environmental, economic, and social concerns (Maclaren 1996; Verma and Raghubanshi 2018). In this entry, urban sustainability is analyzed by weighing intergenerational equity, intragenerational equity, protection of natural resource base, moving away from nonrenewable resources to renewables, energy efficiency, economic vitality, individual and community well-being, and fulfillment of basic human needs (Maclaren 1996; Verma and Raghubanshi 2018).

The ecological footprint and carbon footprint are low in most small towns, as the urban populace primarily relies on nearby places to meet their material needs. Consumption and use of material-intensive capital goods are still low in most small towns of Asia. However, it has started picking up in some of the advanced and emerging economies. In addition, unsustainable and exhaustive resource use is emerging as a concern in some of the small towns (Wujiang and Huludao in China, Tiruppur and Manesar in India, and Gazipur in Bangladesh) in the rapidly industrializing regions. Several small towns are well integrated into the global economic and trade networks and emerging as manufacturing and export hubs. The products and services produced from such places are circulating at the national and global level; however, the environmental harm produced while manufacturing diverse material-intensive commodities is mainly faced by local dwellers. Environmental rationalities are not well integrated into the urban policy responses at the grassroots level due to limited governance capabilities, lack of regulations, and low environmental commitments of local institutions. Apart from resource depletion, growing waste generation and lack of proper waste management are becoming a big challenge in many small towns. Vulnerable populations face disproportionally more environmental harm in these transition towns. Thus, intergenerational equity and intragenerational equity are difficult to achieve.

Lack of economic vitality and governance issues are faced in many small towns in Asia. National policies often focus more on megacities, which contributes significantly more to the national GDP (Gross Domestic Product). Infrastructure deficit (housing and others), financial limitations, and governance challenges often lead to environmental challenges related to underdevelopment. For instance, there exists remarkable disparity among different areas (core and peri-urban areas) in Darkhan (having a population of around 74,454) in Mongolia in terms of access to water and sanitation (Sigel et al. 2012). Individual and community well-being severely gets curtailed in such circumstances.

In a nutshell, globalization-led changes, industrialization, deindustrialization, political and economic reconfigurations, social inequalities, and environmental factors (resource overuse and climate-induced factors) significantly influence the socio-environmental sustainability of small towns. The growth and shrinkage of small towns are dependent on a combination of these factors, and so is the question of overall urban sustainability. How different actors plan new interventions, coordinate their efforts at multiple levels, and implement policies and plans at the local level will determine how Asian societies will address these complex challenges.

Conclusion

Small towns have started attracting the attention of the academic community in several countries in Asia in recent years. Yet, most studies still focus on analyzing the socioeconomic and demographic dimensions, and environmental dimensions remain understudied. Infrastructure deficit, high population density, stark inequalities, weak financial base, and dysfunctional governance are often reported from many small towns in Asia as a critical challenge for achieving urban sustainability. However, the situation varies from region to region, and there are numerous success stories (eco-towns of Japan and South Korea, such as Iida and Yōkaichi in Japan) too. The highly diverse Asian countries (Japan to Bangladesh and Bahrain to the Philippines) present great scope to understand the small towns from the sustainability lens. Rather than merely analyzing the small towns from the rural-urban lens (with demographic and economic focus), it is high time small towns are analyzed using multiple lenses to develop a comprehensive understanding of the fundamental changes happening at the local level.

Cross-References

- ▶ City Visions: Toward Smart and Sustainable Urban Futures
- ▶ Growth, Expansion, and Future of Small Rural Towns
- ▶ Peri-urbanization
- ▶ Shrinking Towns and Cities
- ▶ Sustainable Development Goals from an Urban Perspective

References

Bell, D., & Jayne, M. (2009). Small cities? Towards a research agenda. *International Journal of Urban and Regional Research, 33*(3), 683–699.

Caprotti, F., & Yu, L. (Eds.). (2017). *Sustainable cities in Asia*. Routledge.

Dahiya, B. (2012). Cities in Asia, 2012: Demographics, economics, poverty, environment and governance. *Cities, 29*, S44–S61.

Denis, E., & Zérah, M.-H. (Eds.). (2017). *Subaltern Urbanisation in India: An Introduction to the Dynamics of Ordinary Towns*. New Delhi: Springer.
Guin, D. (2019). Contemporary perspectives of small towns in India: A review. *Habitat International, 86*, 19–27.
Habitat, U. N. (2020). World Cities Report 2020: The Value of Sustainable Urbanization.
Maclaren, V. W. (1996). Urban sustainability reporting. *Journal of the American Planning Association, 62*(2), 184–202.
Mukhopadhyay, P., Zérah, M. H., & Denis, E. (2020). Subaltern Urbanization: Indian Insights for Urban Theory. *International Journal of Urban and Regional Research, 44*(4), 582–598.
Ofori-Amoah, B. (2007). *Beyond the metropolis: Urban geography as if small cities mattered*. University Press of America.
Sharma, A. (2015). Sustainable and socially inclusive development of urban water provisioning: A case of Patna. *Environment and Urbanization ASIA, 6*(1), 28–40.
Shaw, A. (2013). Emerging Perspectives on Small Cities and Towns. In R. N. Sharma & R. S. Sandhu (Eds.), *Small Cities and Towns in Global Era: Emerging Changes and Perspectives* (pp. 36–53). New Delhi: Rawat Publications.
Sigel, K., Altantuul, K., & Basandorj, D. (2012). Household needs and demand for improved water supply and sanitation in peri-urban ger areas: the case of Darkhan, Mongolia. *Environmental Earth Sciences, 65*(5), 1561–1566.
Tan, K. C. (1986). Small towns in Chinese urbanization. *Geographical Review, 76*(3), 265–275.
Tong, Y., Liu, W., Li, C., Zhang, J., & Ma, Z. (2020). Small towns shrinkage in the Jilin Province: A comparison between China and developed countries. *PLoS One, 15* (4), e0231159.
United Nations. (2019). *World Urbanization Prospects 2018. Highlights, Department of Economic and Social Affairs*. Population Division.
Verma, P., & Raghubanshi, A. S. (2018). Urban sustainability indicators: Challenges and opportunities. *Ecological Indicators, 93*, 282–291.
Véron, R. (2010). Small cities, neoliberal governance and sustainable development in the global south: A conceptual framework and research agenda. *Sustainability, 2* (9), 2833–2848.
World Bank (2019). World Bank Open Data. From World Bank.
Ziari, K. (2006). The planning and functioning of new towns in Iran. *Cities, 23*(6), 412–422.

Small Urban Centers

▶ Small Towns in Asia and Urban Sustainability

Small Water Retention Measures in Haluzice Gorge

Richard Müller
Sustainable Development Institute/Institut udrzatelneho rozvoja, Nitra, Slovakia

Introduction

Haluzice Gorge is situated 260 m above sea level within administrative boundaries of Haluzice Village in the Western Slovakia about 25 km south from city of Trenčín. It belongs to orographic system of Outer Carpathians and orographic subsystem of White Carpathians. Haluzice Gorge was declared a Nature Monument in fourth degree of protection in 1963. It has a total area of 35,000 m^2. The Gorge was formed by deep erosion activity of the Haluzice Stream. In the upper part on dolomite-limestone flysch bedrock, erosion gradually excavated a 30 m deep and 100 m wide gorge, mostly without vegetation that gradually expanded. Due to weather conditions, fine sandy material from the steep sidewalls eroded and formed embankment cones on the foot of gorge walls. Heavier surges of water transported material to lower parts where the debris created silt that clogged fertile land in the village of Štvrtok and retroactive erosion threatened Haluzice part of the village. On the slopes were formed ridges and ditches that created new gorges. The largest was established in the area Pod kostolom. It has destroyed part of the old cemetery and watchtower of fortified Romanesque church from the thirteenth century (Figs. 1 and 2).

Title of the Project

Stabilization of the Haluzice Stream and erosion protection of the Haluzice Gorge.

Type of the Project

The Haluzice Gorge belongs to Haluzice Village Cadastre, Nové Mesto nad Váhom District and

S

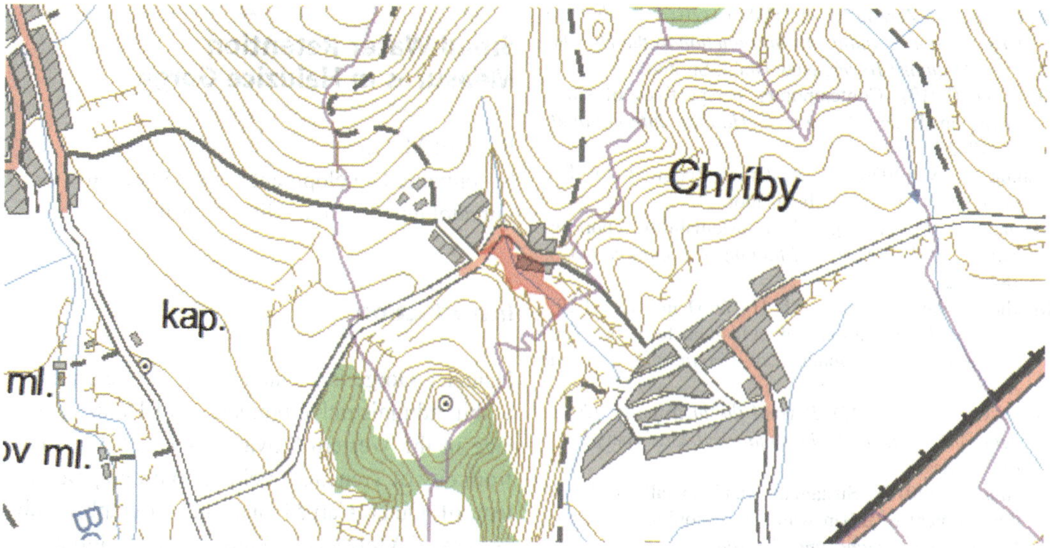

Small Water Retention Measures in Haluzice Gorge, Fig. 1 Location of the Nature Monument Haluzice Gorge. (Source: https://old.uzemia.enviroportal.sk/main/detail/cislo/43)

Small Water Retention Measures in Haluzice Gorge, Fig. 2 Location of Nature Monument Haluzice Gorge. (Source: Google Earth)

Trenčiansky Region in Western Slovakia. Its total area is 3.5 hectares. Stabilization of the Haluzice Stream and erosion protection of the Haluzice Gorge began in 1926. There were applied comprehensive technical measures comprising stone-wall weirs and stone steps. Nontechnical measures included horizontal barriers made of wood sill to slow water flow, tree and shrubs planting on terraces reinforced by baskets made of willow tree to reduce soil erosion. Forest trail Haluzice Gorge was opened in 2011 with several information panels. In addition, Lesy SR incorporated Haluzice Gorge in its network of significant forest areas. These are monuments connected with the history of forestry in Slovakia in terms of natural, construction, technical, and artistic values.

Specific Objective of the Project (Program, Investment)

Technical and nontechnical measures constructed in the Haluzice Gorge are multipurpose, ranging from flood protection, water retention during dry periods, soil erosion reduction, forest ecosystem protection, protection of cultural heritage, and local road to education and recreation. Weirs and stone steps retain silt, caused by erosion in the upper part of the steam, as well as store high water levels during storms and slow runoff and safely drain water.

Legal Basis of the Undertaken Measures

Haluzice Gorge was declared as Nature Monument by the decision of the Educational and Cultural Commission of District Peoples Committee in Trenčín No 40 of 31 October 1963. The Regional Authority in Trenčín, by its generally binding decree No 1/2003 of 27 June 2003 effective from 1 August 2003, established its fourth degree of protection. No protection zone was declared. Haluzice Gorge does not belong to any European network of nature-protected areas and is not a private protected area. Currently, State Nature Conservancy White Carpathians Protected Landscape Area manages the Nature Reserve Haluzice Gorge. The number of basic map sheet in the scale 1: 50,000 of the Slovak Republic maps is 35–14; the standard name is Haluzická súteska. Subject of protection is about 50 m deep epigenetic valley – gorge formed by deep erosion of the Haluzický Stream that originates in the area of little resistant flysch rocks outside the Nature Monument. On the other hand, Haluzice Gorge is geologically a small island of resistant Mesozoic limestone. All activities have to be in line with limits of fourth degree of protection according to Act 543/2002 on Nature and Landscape Protection.

Connections with Other Programs

The area is a part of a spatial plan of large territorial unit of Trenčín Region that was developed in 1998. Three state institutions administer different uses of the Haluzice Gorge. Slovak Water Management Enterprise manages Halizický Stream, Lesy SR, Trenčín Branch technical works (weirs and steps) as well as forestland, and, finally, State Nature Conservancy manages the Nature Reserve. The area belongs to the Váh River basin and its respective river subbasin management plan, prepared in 2009 by Water Research Institute.

Organizational and Financial Basis

Construction of the technical and nontechnical measures started in 1926–1927. General restoration works were undertaken in 1960, 1965, 1979, and 1980. In 2011, LESOSTAV NITRA restored weirs in the right side arm of the gorge. Slovak Water Management Enterprise administers the Haluzický Stream that flows through the Gorge. Technical works on the Haluzický Stream, stone weirs, and stone steps belong to the state property managed by State Enterprise Lesy SR, Trenčín Branch. Similarly, Lesy SR, Trenčín Branch, manages and administers forestland in the Gorge, surrounding the Haluzický Stream.

Technical Description

Prof. Leo Skatula (1889–1974) prepared and realized a project of stabilization of the Haluzický Stream between 1926 and 1927. Prof. Skatula was a forester, aware of the importance of soil protection and water management function of forests. He is the author of the project, which stabilized slopes of the Haluzice Gorge and its side arms and prevented flooding and transport of debris downstream. The complex stabilization system comprised 8 stonewall weirs and 17 stone steps in the lower part of the Gorge. The slopes were stabilized by afforestation through terraces reinforced by willow fences. Shrubs, trees, and creeper plants were planted on the floor, sides, and bare spots. During the following years, trees, shrubs, and creepers gradually filled the whole surface of the Gorge. The last 4 m high weir was constructed in the end in order to stop erosion and protect Haluzice Village, as well as a connecting local road with Štvrtok Village. Total length of modified Haluzický Stream including the Gorge is 1.5 km. Currently, there are six functional weirs in the

S

Gorge on the Haluzický Stream, four are in the right side arm. The purpose of the weirs is to retain silt, caused by erosion in the upper part of the steam as well as retention of high water levels during storms. In addition, 17 stone steps slow runoff and safely drain water.

Measures Impact Assessment (Tangible and Intangible Benefits)

Small water retention measures serve different purposes. Their main purpose is protection against erosion and floods. Other benefits include water storage during dry periods for fauna and flora, recreation, education, and protection of cultural heritage. In addition to forest educational trail, the area is a popular destination due to the ruins of the fortified Romanesque church built in 1240, located near the Gorge. In the past, one of the towers and parts of fortification walls have been destroyed by erosion of main slopes of the Gorge.

Attachments and References (Figs. 3, 4, 5, 6, 7, 8 and 9)

Small Water Retention Measures in Haluzice Gorge, Fig. 3 Ruins of Haluzice church and lower part of the Haluzice Gorge under the church. (Photo credits: (Richard Müller). (Forests of the Slovak Republic. Significant Forest Area Haluzice Gorge [internet]. Banská Bystrica: Forests of the Slovak Republic, state enterprise; 2020. Available from https://www.lesy.sk/lesy/pre-verejnost/kam-do-prirody/vyznamne-lesnicke-miesta/zoznam/haluzicka-tiesnava.html Slovak Environment Agency. State Register of Specially Protected Nature Areas in the Slovak Republic [internet]. Banská Bystrica: Slovak Environment Agency; 2016. Available from https://old.uzemia.enviroportal.sk/main/detail/cislo/43)

Small Water Retention Measures in Haluzice Gorge, Fig. 4 Lower part of the Haluzice Gorge with 4 m high stone weir. (Photo credits: Richard Müller)

Small Water Retention Measures in Haluzice Gorge, Fig. 5 Woody debris retained by the stone weir in lower part of the Gorge and information panel of the significant forest trail. (Photo credits: Richard Müller)

Small Water Retention Measures in Haluzice Gorge, Fig. 6 Middle part of the Gorge with stone weir and woody sill to reduce water flow. (Photo credits: Richard Müller)

Small Water Retention Measures in Haluzice Gorge, Fig. 7 Middle part of the Gorge with woody sill to reduce water flow and lush vegetation. (Photo credits: Richard Müller)

Small Water Retention Measures in Haluzice Gorge, Fig. 8 Upper part of the gorge with two stone weirs. (Photo credits: Richard Müller)

Small Water Retention Measures in Haluzice Gorge, Fig. 9 Side valley of the gorge with two stone weirs. (Photo credits: (Richard Müller). Scans of documents of the Haluzice Gorge Nature Monument http://uzemia.enviroportal.sk/main/detail/cislo/43. Significant forest area Haluzice Gorge: http://www.lesy.sk/showdoc.do?docid=894)

Smart Agriculture and ICT

Sam Wong
University College Roosevelt, Middelburg, The Netherlands

Synonyms

Climate smart agriculture; Developing countries; Evidence-based research; Gender; Groundwater; ICT; Poverty; Solar-powered irrigation systems; Solar pumps.

Definitions

- Climate-smart agriculture adopts an integrated approach to managing landscapes that tackles intertwined issues, such as food security, rising population, and climate change (World Bank website)
- Information and communication technologies (ICTs) include all forms of communication innovations, such as the Internet, mobile phones, software, and social networking, that help users store, access, retrieve, and transmit information digitally (Food and Agriculture Organization website)

– Solar-powered irrigation system, also known as photovoltaic pump system, is defined as a device to convert solar energy into electrical energy by solar panels so as to move water from one place to another (Wong, 2019)

Introduction

Solar-powered irrigation system (hereafter SPIS) is an exemplar of smart-climate agriculture. It has been praised for its "transformational" potential in tackling the trilemma of existing global problems: how to increase food production to feed rising population, without compromising farmers' resilience to changing climate, and without causing negative impact on the environment (Lefore et al., 2021)? International development agencies, such as the World Bank, the United Nations Development Program (UNDP), and the Food and Agriculture Organization (FAO), have all been promoting SPIS in developing countries. In Bangladesh and India alone, 1.5 million units of SPIS are expected to be installed by 2025 (Bastakoti et al., 2019).

In the light of the rising popularity of SPIS, this paper is intended to evaluate the effectiveness of SPIS in three particular aspects: (1) changing farming productivity and practices; (2) the well-being of poor farmers; and (3) the environmental impact. This paper will focus on developing countries only since 800 million populations living in poor countries still rely on farming, and SPIS could play a significant role in livelihood improvement (World Bank, 2013). To evaluate the success and limitations of SPIS, this paper carries out an intensive literature review and provides evidence from a wide range of sources, including academic journal papers and international development reports.

Impact on Food Production and Farming Practices

Literature tends to agree that SPIS is successful in increasing crop yields, raising food quality, and diversifying the choice of farm produce (Bolwig et al., 2020; Schmitter et al., 2018). Gutpa (2019), for example, compares and contrasts two groups of farmers in Rajasthan, India – one group adopts SPIS, whereas the other does not – between 2011 and 2015. His research suggests that the SPIS user group marks a 20% increase in crop productivity than their counterpart.

The increase of productivity could be explained by the availability of constant and reliable supplies of water (surface and underground water), thanks to solar pumps. This helps mitigate the uncertainty of rainfall patterns, and makes all-season cultivation possible, especially in dry seasons. Climate-smart devices, such as humidity monitoring and pest control, provide favorable conditions for crops. Additionally, before the introduction of solar energy, an inadequate coverage and reliable electricity supply and expensive diesels have long restricted the agricultural growth in developing countries.

The solar pump interventions have also led to a higher level of crop diversification. Irrigation enables extra cropping seasons per year, which allows farmers to cultivate higher-valued and more nutritious crops, such as fruits and vegetables (Zande et al., 2020). In the case study of Rajasthan, Gupta (2019) finds out that the production of fruits and vegetables is increased by 33% after the adoption of SPIS.

Despite the positive outcomes, not everyone is optimistic about the development of SPIS. Closas and Rap (2017) are skeptical about the reliability of the research findings related to productivity. They point out that most studies are conducted in a highly controlled, project-based environment. They question whether some farmers may sell their solar panels straight away, rather than investing in SPIS in the long run. Affordability is another hot-debated issue. The initial costs of SPIS remain high when compared to diesel technology (Bastakoti et al., 2019). The World Bank and other development agencies have long called for a reduction of diesel subsidies in order to make solar more competitive. Nevertheless, little action has been taken and there is a lack of political will. Without proper financial arrangements, there are concerns that SPIS adopters may end up in debt. Inadequate maintenance and repair support will also undermine the success of SPIS.

Impact on Poor People's Well-Being

The rising productivity, explained in the previous section, means that SPIS adopters can sell more products to the market, and consequently generate more incomes to farmers. Research by Hossaine and Karim in Bangladesh (2020) and Gupta in India (2019) both support this claim. When compared to non-adopters, Hossaine and Karim discover that SPIS adopters make a significant increase of incomes by 27%. Similarly, Gupta suggests that farmers with SPIS show a 23% difference in annual profits when compared to the non-SPIS counterparts. The additional incomes enable poor households to buy supplementary food items, such as corns, rice, and sorghum, as well as higher-cost, but nutritious items, such as fish and cooking oil. The diversification of the family's diet helps improve the nutrition intakes of poor households (Alaofe et al., 2019).

Apart from incomes, the literature also demonstrates that SPIS makes a significant improvement in people's livelihoods in other aspects. The use of solar energy has made lives safer as diesel is inflammable and smoky. The increased availability of food also enhances food security and nutritional intake, especially to children. According to the research by Bouzidi and Campana (2020), the rising incomes of farmers in Algeria have potential in slowing down the process of depopulation of rural areas.

That said, the capital-intensive nature of SPIS has raised concerns. In their separate studies, Gutpa (2019) and Hossain and Karim (2020) both discover that the SPIS adopters tend to be richer, more educated, and more likely to be self-employed. They are more able to afford the running costs of solar pumps. As a result, Jain and Shahidi (2018) are worried that SPIS could widen the existing gap of inequality between the rich and the poor within communities.

Another social change, brought by SPIS, is gender improvement. Water collection is often considered a task to women and girls. In Sub-Saharan Africa, for example, they spend 33 minutes on average to complete one round trip of water collection in rural areas (UNICEF website). Thanks to solar pumps, obtaining water from local sources is possible, and women are free to perform other income-generating activities and girls can go to schools to receive education. Women farmers also prefer solar to diesel-based technology because the former is less demanding on women's physical strength (Schmitter et al., 2018).

The empowering nature of SPIS has encouraged more governments and organizations to target solar pumps on women and to make SPIS more inclusive to women (Lefore et al., 2019). For instance, female farmers in Nepal are offered financial support to take up solar pumps. They will also get a further 10% discount if the land right is transferred to women. However, some academics are not certain if SPIS is effective in challenging the structural gender inequalities. For example, Wong (2019) raises concerns about whether women could enjoy the equal shares of the rising farming productivity as a result of SPIS intervention. Gugerty and Kremer (2008) also suggest that not all women in developing countries are better off since social interventions tend to attract women who are richer, better educated, and wealthier.

Impact on the Environment

The literature points to the fact that SPIS is effective in reducing pollution and cutting CO2 emissions. Before solar pumps were introduced, farmers in developing countries relied on diesel-powered pumps. Diesel pumps contaminate water with chemicals which affect surrounding vegetation (Melikyan, 2018). Comparing the life-cycle assessment of SPIS and diesel pumps, FAO (2018) finds out that SPIS helps the reduction of greenhouse gas emissions by 98%. All these environmental advantages provide a strong justification for the replacement of nine million diesel pumps in India alone (Jain & Shahidi, 2018).

That said, SPIS is not problem-free. The possibility of over-extraction of water, especially underground water, caused by solar pumps, has raised serious concerns (Bastakoti et al., 2019; Shah et al., 2018). Once the solar pump systems are set up, the marginal costs of extracting an extra unit of water is very small. As a result, farmers are incentivized to expand their farmland or to sell extra water to other farmers in order to compensate the high initial costs of SPIS. The potential damage of the depletion of water level has called for a tighter regulation on groundwater extraction, such as limiting the size of pump and raising farmers' awareness, before a wide implementation of SPIS in developing countries. Yet, scholars, such as Lefore et al. (2021), highlight the challenges of monitoring and regulation.

Another environmental concern of SPIS is the waste management of solar panels. The production of PV panels involves silicon and some toxins. Having a better plan in handling solar waste is crucial to the success of SPIS (FAO, 2018).

Conclusions

This paper has conducted an intensive literature review and has demonstrated, with evidence, that SPIS could be effective in raising agricultural productivity, increasing farmers' incomes, achieving gender equalities, and reducing CO2 emissions in developing countries. However, this paper has also raised serious concerns about the potential environmental problems of SPIS in water over-extraction and solar panel wastes. Other issues, such as affordability, maintenance, and the potential widening gaps between the rich and the poor in communities, deserve a closer scrutiny before SPIS is implemented in developing countries in a large scale.

This paper has also highlighted the methodological challenges in understanding the impact of SPIS because most studies are conducted in a very controlled, project-based environment. More qualitative and in-depth research is needed to examine how SPIS changes the daily interactions of community members from the political, sociological, and environmental perspectives.

Cross-References

▶ Renewable Energy

References

Alaofe, H., Burney, J., Naylor, R., & Taren, D. (2019). The impact of a solar market garden programme on dietary diversity, women's nutritional status and micronutrient levels in Kalalé district of northern Benin. *Public Health Nutrition, 22*(14), 2670–2681.

Bastakoti, R., Raut, M., & Thapa, B. R. (2019). *Adoption of solar-powered irrigation pumps. Experiences from the Eastern Gangetic Plains. Water Knowledge Note prepared for the South Asia Water Initiative technical assistance project Managing Groundwater for Drought Resilience in South Asia.* World Bank: Washington D.C.

Bolwig, S., Baidoo, I., Danso, E. O., Rosati, F., Ninson, D., Hornum, S. T., & Sarpong, D. B. (2020). *Designing a sustainable business model for automated solar-PV drip irrigation for smallholders in Ghana* (UNEP DTU Working Paper Series 2020, No.1). Denmark: Copenhagen.

Bouzidi, B., & Campana, P. E. (2020). Optimization of photovoltaic water pumping systems for date palm irrigation in the Saharan regions of Algeria: Increasing economic viability within multiple-crop irrigation. *Energy, Ecology and Environment, 51.*

Closas, A., & Rap, E. (2017). Solar-based groundwater pumping for irrigation: Sustainability, policies, and limitations. *Energy Policy, 104*(3), 33–37.

FAO. (2016). *Aquastat.* Rome: FAO.

Food and Agriculture Organization (FAO) website. Information and Communication Technologies. available at: http://aims.fao.org/information-and-communication-technologies-ict.

Food and Agriculture Organization of the United Nations. (2016). *Aquastat.* Rome: FAO.

Food and Agriculture Organization of the United Nations. (2018). *The benefits and risks of solar-powered irrigation – A global overview.* Rome: FAO.

Gugerty, M., & Kremer, M. (2008). Outside funding and the dynamics of participation in community association. *American Journal of Political Science, 52*(3), 585–602.

Gupta, E. (2019). The impact of solar water pumps on energy-water-food nexus: Evidence from Rajasthan, India. *Energy Policy, 129*, 598–609.

Hossain, M. and Karim, A. (2020) *Does Renewable Energy Increase Farmers' Well-being? Evidence from Solar Irrigation Interventions in Bangladesh.* ADBI Working Paper 1096. Tokyo: Asian Development Bank Institute.

Jain, A. and Shahidi, T. (2018) *Adopting solar for irrigation. Farmers' perspectives from Uttar Pradesh.* Council on Energy, Environment and Water.

Lefore, N., Giordano, M., Ringler, C., & Barron, J. (2019). Sustainable and equitable growth in farmer-led irrigation in sub-Saharan Africa: What will it take? *Water Alternatives, 12*(1), 156–168.

Lefore, N., Closas, A., & Schmitter, P. (2021). Solar for all: A framework to deliver inclusive and environmentally sustainable solar irrigation for smallholder agriculture. *Energy Policy, 154*, 112313.

S

Melikyan, L. (2018) *Outcome evaluation in the practice of environment and energy outcome evaluation, UNDP Sudan under CPD 2013-2017*. Final report. Available at: file://rsr-ws3-08/Staff/s.wong/Downloads/Sudan%20EECC%20Outcome%202%20evaluation%20final%20report%20Febr%202_CPD%202013-2017%20(1).pdf
Schmitter, P., Kibret, K. S., Lefore, N., & Barron, J. (2018). Suitability mapping framework for solar photovoltaic pumps for smallholder farmers in sub-Saharan Africa. *Applied Geography, 94*, 41–57.
Shah, T., Rajan, A., Rai, G., Verma, S., & Durga, N. (2018). Solar pumps and South Asia's energy groundwater nexus: Exploring implications and reimagining its future. *Environmental Research Letters, 13*(115003), 1–12.
UNICEF website. Collecting water is often a colossal waste of time for women and girls. Available at www.unicef.org/press-releases/unicef-collecting-water-often-colossal-waste-time-women-and-girls.
Wong, S. (2019). Chapter 21: Decentralised, off-grid solar pump irrigation systems in developing countries – Are they pro-poor, pro-environment and pro-women? In P. Castro, A. Azul, W. Leal, & F. Azeiteiro (Eds.), *Climate change-resilient agriculture and agroforestry: Ecosystem services and sustainability* (pp. 367–382). Berlin: Springer.
World Bank. (2013). *Implementing agriculture for development. Agriculture action plan 2013-2015*. Washington D.C.: World Bank.
World Bank website. Climate-smart agriculture. Available at: www.worldbank.org/en/topic/climate-smart-agriculture.
Zande, G., Amrose, S. & Winter, A. (2020) *Evaluating the potential for low energy emitters to facilitate solar-powered drip irrigation in Sub-Saharan Africa*. Paper presented at the Design 2020 International Design (Virtual) Conference.

Smart Cities

► Augmented Reality: Robotics, Urbanism, and the Digital Turn
► The Governance of Smart Cities
► Understanding Urban Engineering

Smart City

► Green and Smart Cities in the Developing World
► The Sustainable and the Smart City: Distinguishing Two Contemporary Urban Visions

Smart City: A Universal Approach in Particular Contexts

Elham Bahmanteymouri
The University of Auckland, Auckland, New Zealand

Introduction

This entry investigates and analyzes the meaning and operation of the "smart city" as a popular and growing phenomenon in the world. To explore this prevailing approach, the entry first explains the method of analysis as based on two philosophical concepts of universalism and particularism. It then examines how the smart city approach has been universally defined and implemented and how it has responded in different particular contexts and cases.

Universalism and Particularism

This entry applies the philosophical dialogue between universalism and particularism to explain how the approach of the smart city as a prevailing universal approach has been diversely operating in different particular contexts. The dichotomy or dialogue between universalism and particularism goes back to the classical ancient philosophy of God and later to the philosophy of science. However, it also has been utilized as an approach to research methodology by providing a more comprehensive understanding of the different dimensions of a phenomenon, as well as the similarities and differences between the context-dependent characteristics of the phenomenon (Torfing 1999).

In the case of the smart city, it is important to understand that although the smart city generally has been accepted as a universal approach with specific principles and models, it has been implemented and/or challenged by urban planning institutions and governments in different contexts, diversely based on their particular characteristics. This methodological approach helps the understand that even though the smart city approach is becoming the dominant view of urban management and planning, it is not possible to standardize its principles as a homogenized prescription for all cities around the world. On the contrary, it should be utilized as a flexible context-dependent strategy that includes diverse features and technologies to solve problems and improve the wellbeing of citizens.

Smart City: An Evolving Universal Approach

There is no standard and generally accepted definition of "smart city." As Picon (2018, p. 270) argues, "the smart city belongs partly to the imagination." The utilization of Information and Communication Technologies (ICTs), blockchain, Artificial Intelligence (AI), as well as advanced technological tools and algorithms to analyze Big Data in urban planning and urban design is a growing universal phenomenon. The term "smart city" has been widely used to describe strategies, policies, and methods of applying these advanced technologies in different dimensions of urban planning, urban management, and urban design. The main objective behind the concept of the smart city is sustaining the economy of a city at its high efficiency and consequently providing quality life for all citizens. Based on a review of the literature (Harrison et al. 2010; Kitchin 2015; Marsal-Llacuna and Segal 2016), this entry suggests that a successful smart city can be identified through specific characteristics including:

- Policy making and land use planning for sustainable and smart urban economic development and urban land development
- Facilitating collaborative planning using digital platforms
- Saving costs through automation
- Increasing the quality of life by tracking and maintaining the quality of the built environment and managing disasters
- Forecasting and reducing risks through tracking the data about activities, pollution, and externalities
- Easing connectivity and flows of information
- Providing flexibility at work and in living spaces
- Facilitating travel through smart transportation
- Enabling smart buildings

New technologies are changing economic, political, and social relations and consequently the pattern of the built environment and cities (Bahmanteymouri 2021). Technologies are taking control of the future of our society in many aspects – from government and defense systems to finance and banking, as well as the built environment and urban areas. Many of the new digital platforms and technologies are considered not only disruptive to the official institutional system or established norms (e.g., Airbnb and Uber), but they also have been operating as rivals or even threats to governments and central banks (e.g., cryptocurrencies) (Shrier 2020). Digital technologies are taking over and running cities sometimes beyond the control of governments and official planning institutions.

Bremmer (2021) maintains that "Technology companies are not just exercising a form of sovereignty over how citizens behave on digital platforms; they are also shaping behaviours and interactions." He suggests that political scientists should make more advanced approaches to understanding the impacts of Big Tech on societies and political events, as well as economic decisions and policies.

However, the collaboration between technological companies and governments and the use of technological advancements in serving states/governments is a growing phenomenon, and regulators and state institutions are gaining more experience in dealing with what was previously considered disruptive. The question is how governments can use the opportunities and take advantage of technologies to provide services for cities and citizens and manage and deal with problems instead of resorting to rivalry with technological companies.

A Universal Theory: A Correlation Between Market Failures and Technological Advancement

There is a strong correlation between the technological advancement and the economic logic of capitalist market failures. From the Marxian (Marx and Engels 2008 [1848]) and Schumpeterian points of view (2009 [1947]), capitalism needs creative destruction and incessant technological advancements to overcome its inherent limitations, contradictions, and imbalances. Creative destruction refers to destroying previous relations, jobs, and technologies and replacing them with new areas of technology. The digital economy is a mode of the technically facilitated economy that emerged mainly to overcome the global economic crisis of 2007–2008 (Bahmanteymouri 2021). New technological innovations and digital platforms are growing as solutions to the capitalist (neoliberal) economic crisis.

Following Schumpeterian philosophy, Mazzucato (2021) suggests the new approach of *mission economy* to overcome the limitations and failures of capitalism by using technological advancements and setting new macro-political economic objectives such as the Green New Deal and climate change strategies. The concept of technological advancements is not neutral; rather, it is closely attached to the operation of the global system of capitalism and its economic crisis. Since early 2020, the COVID-19 pandemic, as both symptom and cause of the economic crisis, has accelerated the use of digital technologies and platforms, specifically transnational e-commerce such as Amazon (Lee 2020) and communications platforms such as Zoom (Kranjec 2021).

A Classification of Technologies That Makes a City Smart

Based on this explanation, the questions are: What areas and what types of advanced technologies and digital platforms are or should be considered as the best tools to achieve the objectives of a smart city? Can governments control and use technologies in the specific ways they need to make a city productive in terms of the concept of the smart city?

Since technologies such as digital platforms and AI have been used for different purposes that have overlapped with each other, including but not limited to the sharing economy, participatory planning, and traffic management and transportation, a neat classification of the different dimensions of the smart city is difficult. Despite this challenge, this entry identifies the following areas that digital technologies have been operating in to make the concept of the smart city meaningful. This categorization helps in the analysis of how planning and governments, as well as private businesses and citizens, have been engaged in the operation of smart cities:

1. Digital platforms:
   - (i) The sharing economy digital platforms
   - (ii) Communication platforms
   - (iii) Participatory platforms
   - (iv) Social media platforms (participatory impacts and social media platforms including Facebook, Twitter, and TikTok)
2. Artificial Intelligence (AI), algorithms, and Internet of Things (IoT)
3. Blockchain

Smart City: Implementations and Characteristics in Particular Contexts

Whether digital platforms operate with or are independent from government organizations, they are becoming new regulators and influential actors in the space of the built environment. It is important to note that the majority of digital platforms and technologies have been developed by private and non-government companies rather than by governments. Even if governments control and/or direct and utilize the power of digital platforms and new technological companies, the power of digital technologies is in their growth as institutions that are able to shift social, economic, and political relations from the physical world to digital space. An investigation of the function of the smart city in particular contexts assists in understanding the importance of the political, social, and economic power of digital and advanced technologies and how different actors including technology firms, governments, and citizens are involved in the smart city.

This section explores how planning and other government institutions have applied, implemented, and/or challenged different digital platforms and companies in particular contexts.

Digital Platforms

Digital platforms operate based on specific criteria. The first and most important criterion is critical mass, that is, an adequate number of users, to the point that a platform operates as a self-sustaining system and consequently achieves the benefits of *network effects*. Network effects means that the high number of users of a platform increases the value and utility of the platform (Shapiro et al. 1998), providing benefit for both the platform providers and users. From the classic economics point of view, economies of scale also explains that more users (higher demand) means more benefits and a lower cost of controlling the market for platform providers. Since critical mass is an important criterion for the operation of digital platforms, transnational platforms have shown themselves to be more successful and able to survive economic/political fluctuations and difficulties. Obviously, access to the internet and advanced digital devices such as smartphones are a requirement for building critical mass. In particular, social networking platforms such as Facebook, Twitter, and Instagram, as well as service provider platforms such as Airbnb and Uber, are successful examples of the effect of critical mass and size at the transnational level. However, these digital platforms have been implemented differently in particular contexts.

The growing power of digital platforms and their influence on cities, known as *platform urbanism* (Bauriedl and Strüver 2020; Sadowski 2020), have changed traditional relationships and interactions between actors and the operation of urban institutions. While digital platforms have shaped a new structure for social, economic, and political interactions between citizens, businesses, and government organizations, these platforms have become a new and important infrastructure for cities.

The Sharing Economy: Digital Platforms

Airbnb and Uber are the most popular examples of platforms within the sharing economy. Such platforms have been considered as disruptive by governments and the official planning institutions, especially in the early and growing phases of these platforms. It is now accepted that Airbnb has been disruptive to the long-term rental housing market, creating a shortage and lifting prices. In addition, Airbnb has created negative externalities for cities (Gurran and Phibbs 2017; Bahmanteymouri 2021; Bahmanteymouri and Haghighi 2020).

Planning organizations and governments have responded with different approaches to short-term rental accommodations, such as Airbnb. In Singapore, for example, Airbnb is subject to strict limitations and taxation. In other countries, there is minimal intervention, while in some cases Airbnb is regarded as complementary to tourism, such as in Denver, London, and Queenstown in New Zealand (Bahmanteymouri 2021). Sharing economy platforms such as Airbnb initially utilized populist slogans such as "sharing against ownership" and "making income from idle assets" and are supported as a progressive alternative to neoliberal capitalism by some leftist approaches (Rifkin 2014). More recently, with the expansion of the Airbnb platform and the challenges that creates for cities, Airbnb has been criticized by the majority of researchers as a monopoly that provides benefits for homeowners while facilitating sharing consumption (Oskam 2019).

The term "disruptive mobility" has been widely used for many transportation digital platforms such as those promoting car sharing, such as Cityhop, ride-sourcing, such as Uber, and ridesharing, such as UberPOOL, as well as Shared Autonomous Vehicles (SAV). The term "disruptive mobility" shows that these digital platforms are regarded as a new phenomenon in cities that should be used and/or responded to with new policies and through a different attitude from the traditional approach to transportation. In particular, it has been pointed out that this group of platforms provide a new mode of transport that is neither private nor public. Many research projects have encouraged governments to use them as a complement to public transportation and an opportunity to respond to the demand for public transportation (Katzev 2003). Mohammadzadeh (2021 p. 5) argues that using SAVs has a strong correlation with the economic situation and the incomes of individuals and households; lower-income groups who are the most frequent public transport users have a strong future intention to use SAVs, while car owners would mostly prefer to privately own an AV. Therefore, sharing digital platforms need more investigation and potentially can be used by the planning system as an important index of smart cities, depending on the local regulations and situations.

Communication Platforms

The COVID-19 pandemic has resulted in digital platform disruption expanding to education, work, shopping, as well as healthcare sectors. A number of digital communication platforms have been available for quite some time, for example, Skype (since 2003), Ovoo (2017), Zoom (since 2012), and WhatsApp (since 2009). However, pre-pandemic, they remained underused due to entrenched attitudes and habits. During the lockdowns and restrictions of the pandemic, these platforms have been helpful and even compulsory for people who must continue to work and live. How these platforms could change the nature of social and economic relations and land use planning needs more time and future investigations (Bahmanteymouri 2021). One point that is currently clear is that the more we use these platforms, the more problems and challenges emerge that need improvement for any future use (Beighton 2021; Vandenberg and Magnuson 2021). It is too early to understand how they may change the particular context of cities.

Participatory Platforms

Facebook group pages are familiar to the majority of internet users around the world. Many communities have utilized Facebook's facilities and capacities to create pages exclusively for residents of a neighborhood or community. In fact, although Facebook, as a global commerce platform, benefits from network effects and critical mass, it also provides a specific design for communities to create locally based pages owned by residents for different purposes including sharing information about a specific district and finding solutions to their problems.

Although one of the criteria for a financially productive digital platform is achieving transnational critical size to take advantage of network effects and economies of scale, some digital platforms that lack this criterion still operate fruitfully at a limited local size. As Filipe (2021) explains, some of the local non-corporate platforms that are built for participatory purposes have successfully been created, implemented, and maintained at geographically restricted scales, such as neighborhood, community, and/or city districts. Two worthy examples of participatory digital platforms are Gebiedonline and Decidim.

Gebiedonline is a not-for-profit digital platform that exemplifies cooperative digital platforms and has been used in Amsterdam for many civil society activities since 2016. The platform is supported functionally and financially by various levels of government while the Amsterdam Municipality is also a member of the platform. However, decision-making, maintenance, and dissemination of the platform are controlled by the members. The platform began in 2012 "when local resident and IT specialist Michel Vogler was asked by a neighbourhood community in IJburg to help them find a way to share information online. The online platform he built (halloijburg.nl) over the years became the Amsterdam best practice for online local community support" (Neven 2017). Without any centralized organization, this successful practice created a model for other communities in Amsterdam. In January 2016, six community representatives founded the new cooperative platform Gebiedonline (area online) to support 20 local communities (in the cities of Amersfoort, Amsterdam, and Gouda) (Neven 2017).

Gebiedonline can be considered as a noble example of bottom-up urban management. Filipe (2021, p. 5) explains that Gebiedonline presents "platform cooperativism" that allows users "to create non-monetary value" from the platform. Gebiedonline offers a communication website for citizens, government, and other parties to share information, work together, address their communities' problems, collectively discuss and find potential solutions and improve their local areas, and even facilitate participatory policy making.

Decidim is another example of a non-corporate participatory digital platform that was created following the 2008 economic downturn and social unrest in Barcelona. Decidim is a FLOSS-based (free/libre open-source software) platform that was initially developed by Barcelona's municipality. It is currently used as a free open-source platform on which to practice participatory democracy within cities and organizations (Decidim n.d.). The platform provides a virtual participatory space for planning and policy making from the early stage of raising a problem, to the final stage of implementation at the local and community scale. Beyond social media, it provides a space for professional debate with evidence, data, and analysis by experts and citizens – leaving the government at only the initial stage of this process. Decidim is participatory, not just from the user-end but also in its technical dimensions; it is designed based on "decentralised crowd intelligence to improve technopolitical processes" (Filipe 2021 p. 7). Decidim's GitHub (GitHub is an international platform that provides opportunity for collaborative coding and software development) has also been developed collaboratively – not by every citizen but instead by a few members who are familiar with coding. Nevertheless, coding and technological development are openly available for all if they are interested in this dimension of the platform.

Some scholars, including Filipe (2021), believe that non-corporate digital platforms such as platform cooperativism and FLOSS may create an alternative to platform/digital capitalism. These types of platforms also have the capacity to go beyond clicktivism and/or slacktivism, which is the main function of social media platforms such as Facebook and Twitter. These platforms develop a more transparent, meaningful, and fruitful digital space based on collective debates, cooperation, and problem-solving. More importantly, in-person meetings, surveys, in-depth understanding and analysis of the problems, and consultations with experts are complementary and necessary for the effective operation of non-corporate digital platforms.

Social Media Platforms (Participatory Impacts and Social Media Platforms: Facebook, Twitter, TikTok, and Public Opinion)

Although social media platforms are the source of many problems, from technical to political issues, including distortions of reality and fake and bot accounts, posts, and tweets (automated accounts/tweets), they are popular and many people are using them as the source of information rather than for personal communication. In many democracies or semi-democracies today, politicians' ability to gain followers on Facebook and Twitter unlocks the money and political support needed to win office (Bremmer 2021). Even worse, these platforms have taken the place of in-depth analysis of cities' problems and urban issues, policy making, and political decisions. Through slacktivism and populist ideologies create political movements and change behaviors. There is a large amount of literature, news, and movies about the manipulative and unethical operations of social media. Facebook and Twitter are familiar examples of the abuse of data and algorithms to influence users' behaviors during political events. As van Raalte et al. (2021) argue:

> The accumulation of social media followers, likes, and retweets not only helps support the populist's claim that they represent "the voice" of "the people." They also help consolidate the appeal of a particular political "brand," whether in the form of a political party brand or an individual political persona. Populist communication is a leadership style that surmounts the division between the Left and the Right to occupy the entire political spectrum. With the use of emoji, memes, slogans, and personal messages, populist leaders seek to imitate the social media behavior of their followers in order to create an illusory symmetry of communication with them.

Analyzing data from Twitter for a period of around 1 year, Ferrara (2020) discovered how the strategies of the US political parties structured voters' behaviors in the 2020 US election. Bartley et al. (2021, p. 65) analyzed how digital media platforms are reshaping our habits; for example, they realized that "Twitter's timeline curation algorithms skew the popularity and novelty of content people see and increase the inequality of their exposure to friends' tweets. In fact, algorithmic curation of content systematically distorts the information people see." Chen et al. (2021, p. 15) investigated the role of Twitter in the 2020 US election and concluded that tweets from "accounts most likely to be bots [outnumbered] tweets from accounts that [were] most likely human for both the Republican and Democratic parties."

Artificial Intelligence (AI), Algorithms, and Internet of Things (IoT)

AI, algorithms, and IoT are used in different dimensions of urban planning and design, from land use planning and zoning, big data analytical platforms, transportation, to even (re)shaping the behaviors of citizens. "Internet of things (IoT) is a promising solution to connect and access every device through internet" (Smys et al. 2020, p. 190). IoT helps devices to connect to other devices automatically without any human intervention. It is gaining increasing attention from planning based on its ability to provide services in urban areas, specifically the provision of sensors that record pollution and traffic congestion problems.

One of the most popular aspects of the smart city is the presence of urban artificial intelligence (AI) capable of thinking and acting in an unsupervised manner. AI can be used in smart cities for various applications – from urban management and support for decision and policy making to the provision of public services for citizens and urban development. AI can facilitate efficiency gains, democratic engagement in decision-making, and economic and environmental sustainability. It may also be run in the form of smart sharing mobility platforms and autonomous vehicles, self-driving cars, robots, and city brains that are taking the management of urban services as well as urban governance out of the hands of humans.

Cugurullo (2020) analyzed the implementation of AI in a case from Abu Dhabi and concludes that "technological innovation in the field of AI is progressing rapidly, and the social sciences and humanities should not lag behind." An interdisciplinary perspective is vital to understanding the ideological, political, and economic impacts and dimensions of AI on cities and human relations.

The application of AI and algorithms in urban areas for the purpose of urban planning and management has many challenges:

- It usually involves collecting, managing, and dealing with big data. There are many issues involved in working with big data from the technical aspect, such as the need for high-performance computers, and the ethical aspect, including the security of private data collected from interactions as well as humans' behaviors.
- Another ethical dimension is the black-box effect (Lampathaki et al. 2021) which is created by a self-learning AI algorithm. The black-box effect can produce and reproduce bias and lead to unfair decisions.
- The creative destruction effects of digital capitalism are more obvious through the operation of AI in the job market and some jobs will be displaced by AI (Pellegrin et al. 2021).
- Urban planning and design as well as regulatory frameworks should adapt to AI (Gavaghan et al. 2021, p. 32).

Blockchain

Blockchain is known as decentralized and Distributed Ledger Technology.

> It keeps records of transactions across many computers.... Every transaction creates a block (an equation) that is linked to other blocks in a tree-like structure. It is not possible to chop down the middle of the tree unless you cut the entire sequence of blocks. This system of blocks guarantees the security of the chain. (Shrier 2020, pp. 15–21)

Bitcoin arose out of this system of blockchain out of the 2008 financial crisis as "a direct response to the threat of central banking and quantitative easing ... [because] existing banking systems are not sophisticated enough to solve new problems (Shrier 2020, p. 2).

Blockchain has become attractive to the financial sector through cryptocurrencies such as Bitcoin and Ethereum. It also has been implemented for governance because it has the capacity to create a decentralized democratic network that is transparent for all members and allows them to create and share information.

Recently, blockchain has become popular in urban planning. Govela (2018) explains that blockchain can be used for specific aspects of cities, including the following:

- Blockchain supports governance and democratic participation in decision-making as well as government accountability. Blockchain presents a transparent and decentralized system that makes it an ideal technology for a citizen-centered system of governance. It increases the trust of citizens by providing the ability to check the interactions, financial situation, and transactions of the government. Using the blockchain system also increases the accountability of governments because their operations are traceable and observable by citizens. In addition, it provides governments with more accurate and detailed information and data about business opportunities, the employment rate, and other economic data that are crucial for planning and decision making. Fiorentino and Bartolucci (2021) explain how blockchain can provide a platform for a sharing economy within governance. Muth et al. (2019) explain that the BBBlockchain system is a set of smart contracts, deployed by the Ethereum blockchain that facilitates all the different dimensions of urban development in Germany – from citizen participation to land use and urban land value, as well as transactions.
- Blockchain provides a reliable record of property and land use. All transactions and the details of the ownership status of land and property can be recorded through blockchain systems. These systems thus provide information for taxation and land use planning institutions, as well as for the provision of infrastructure and public services. Dubai is one of a number of cities that have applied a blockchain strategy to track property and landownership records.
- Blockchain supports infrastructure and urban services. In particular, blockchain provides a smart system for energy production and consumption. It automatically adjusts the energy system and electricity distribution in the most efficient way to respond to citizen demand. Applying the circular economy approach, blockchain also has been used for waste management in France and China (Taylor et al. 2020).
- Blockchain provides an ecosystem of values. All the mechanisms of planning and the provision of urban infrastructure are possible through a system of different types of value in urban areas. Blockchain provides a reliable and comprehensive ledger for all interactions and values that are produced in cities. "In Berlin, there are several commercial platforms involved in building blockchain applications, programming Smart Contracts and connecting them to real world data, applying DLT to pre-existing databases, modular toolboxes, creating blockchain internet networks that connect parallel networks, citizen ownership of data, facilitating payments with cryptocurrencies, and producing market forecasts" (Govela 2018, p. 14).

The UN is also involved in Afghanistan's blockchain land registry tool that records earth information, value, and ownership of lands and properties (Konashevych 2021).

Challenges and Future Opportunities

The universal theory of the new technological innovations – IoT, AI, and digital platforms – reveals that features of the smart city are growing as solutions to the capitalist (neoliberal) economic crisis. However, they will offer only temporary and ephemeral spaces that emerge and take the place of previous technology, as is the case every time there is an economic downturn (Bahmanteymouri and Haghighi 2020).

Digital platforms and new technologies are developing in new areas of our cities, connecting people and institutions while changing relationships through virtual spaces and algorithms. Governments, policy makers, and planning institutions (as well as the education sector) are struggling to adapt to these technologies and use them for different purposes in particular contexts.

The European Commission has presented a new regulatory package in Brussels that fines internet platforms for illegal content and controls high-risk AI applications. The EU is also calling for technology-focused industrial policies – including billions of euros of government funding – to encourage new approaches to pooling data and computing resources. The goal is to develop local or global alternatives to big platforms that are grounded in local values and which consider climate change and protection of the environment, promote workers' rights, combat child and forced labor, and expand resilient and sustainable supply chains (The EU 2021).

> A bipartisan push for regulation in the United States rewards "patriotic" companies that deploy their resources in support of national goals. The government hopes that a new generation of technology-enabled services for education, health care, and other components of the social contract will boost its legitimacy in the eyes of middle-class voters. (Bremmer 2021)

China's government also has regulated digital spaces and the companies "against the growing strength of the platform companies and their charismatic leaders. Another rationale for the move, however, is that the new enforcement will make Chinese digital platform companies more competitive and stimulate innovation" (Wheeler 2021).

There are challenges to fully achieving the wellbeing objectives of the smart city. The first and the most important challenge is abuse of power. What will be a matter of concern in the future is that technological power provides political power and control over resources and wealth. Competition and collaboration between governments and tech companies and platforms in ruling over people's behaviors, desires, public and private assets, and wealth generate challenging ethical dimensions of the concept of the smart city that should be seriously considered (Bahmanteymouri and Mohammadzadeh 2021).

Applying digital technologies in different aspects of urban planning and design requires a range of new skills for planners, new sets of development policies, regulations, and protocols focused on social relations protection, as well as work protocols. Major investment in education is required to provide equal access to technological learning opportunities for all. Finally, we need to think about needs in different contexts and smart city indexes that support citizens' wellbeing. What role can technology and the smart city play within the health system to prevent the next pandemic and provide wellbeing for all? Is the use of technologies to control and surveil all aspects of life ever ethical? These are the questions that must be answered in the future.

References

Bahmanteymouri, E. (2021). Evaluating the impacts of the digital economy on land use planning. In Y. Yang & A. Taufen (Eds.), *The Routledge handbook of sustainable cities and landscapes*. Routledge.

Bahmanteymouri, E., & Haghighi, F. (2020). Airbnb as an ephemeral space: Towards an analysis of a digital heterotopia. In S. Ferdinand, I. Souch, & D. Wesselman (Eds.), *Heterotopia and globalisation in the twenty-first century*. Routledge. Retrieved from https://www.taylorfrancis.com/books/9780429290732

Bahmanteymouri, E., & Mohammadzadeh, M. (2021). The emerging autonomous Smart City and its impacts on planning and power relations in late capitalism. In M. Gunder, K. Grange, & T. Winkler (Eds.), *Handbook on planning and power* (1st ed.). Edward Elgar Publishing.

Bartley, N., Abeliuk, A., Ferrara, E., & Lerman, K. (2021). Auditing algorithmic bias on twitter. In *13th ACM web science conference 2021* (pp. 65–73). https://doi.org/10.1145/3447535.3462491

Bauriedl, S., & Strüver, A. (2020). Platform urbanism: Technocapitalist production of private and public spaces. *Urban Planning, 5*(4), 267–276.

Beighton, C. (2021). *Zoom fatigue: Online communication platforms and technologies of the self.* The UC Research Repository. https://repository.canterbury.ac.uk/item/8x3y5/zoom-fatigue-online-communication-platforms-and-technologies-of-the-self

Bremmer, I. (2021). The technopolar moment: How digital powers will reshape the global order. *Foreign Affairs*. Retrieved 26 Oct 2021 from https://www.foreignaffairs.com/articles/world/2021-10-19/ian-bremmer-big-tech-

global-order?utm_source=Sailthru&utm_medium=email&utm_campaign=FA%20This%20Week_102221_The%20New%20Cold%20War&utm_term=FA%20This%20Week%20-%20112017
Chen, E., Deb, A., & Ferrara, E. (2021). # Election2020: the first public Twitter dataset on the 2020 US Presidential election. *Journal of Computational Social Science*, 1–18. Volume 4, issue 3. https://doi.org/10.1007/s42001-021-00117-9
Cugurullo, F. (2020). Urban artificial intelligence: From automation to autonomy in the smart city. *Frontiers in Sustainable Cities, 2*, 38.
Decidim. (n.d.). General description and introduction to how Decidim works. Retrieved 18 Nov 2021, from https://docs.decidim.org/en/features/general-description/
Ferrara, E. (2020). # Covid-19 on Twitter: Bots, conspiracies, and social media activism. arXiv preprint arXiv: 2004.09531.
Filipe, R. (2021). The unexpected persistence of non-corporate platforms: The role of local and network embeddedness. *Digital Geography and Society, 2*. https://doi.org/10.1016/j.diggeo.2021.100020.
Fiorentino, S., & Bartolucci, S. (2021). Blockchain-based smart contracts as new governance tools for the sharing economy. *Cities, 117*, 103325.
Gavaghan, C., Knott, A., & Maclaurin, J. (2021). *The impact of artificial intelligence on jobs and work in New Zealand*. University of Otago. Retrieved 21 Nov 2021 from https://www.otago.ac.nz/caipp/otago828396.pdf
Govela, A. (2018). *Blockchain, a tool for metropolitan governance?* Metropolis Observatory, World Association of the Major Metropolises. Issue 5.
Gurran, N., & Phibbs, P. (2017). When tourists move in: How should urban planners respond to Airbnb? *Journal of the American Planning Association, 83*(1), 80–92.
Harrison, C., Eckman, B., Hamilton, R., Hartswick, P., Kalagnanam, J., Paraszczak, J., & Williams, P. (2010). Foundations for smarter cities. *IBM Journal of Research and Development, 54*(4), 1–16.
Katzev, R. (2003). Car sharing: A new approach to urban transportation problems. *Analyses of Social Issues and Public Policy, 3*(1), 65–86.
Kitchin, R. (2015). Making sense of smart cities: Addressing present shortcomings. *Cambridge Journal of Regions, Economy and Society, 8*, 131–136.
Konashevych, O. (2021). 'GoLand Registry' case study: Blockchain/DLT adoption in land administration in Afghanistan. In *DG. O2021: The 22nd annual international conference on digital government research* (pp. 489–494).
Kranjec, J. (2021). Zoom retains pandemic gains and doubles half-year revenue. Retrieved 13 Nov 2021 from http://www.koreaittimes.com/news/articleView.html?idxno=108929
Lampathaki, F., Agostinho, C., Glikman, Y., & Sesana, M. (2021). Moving from 'black box'to 'glass box'Artificial intelligence in manufacturing with XMANAI. In *2021 IEEE international conference on engineering, technology and innovation (ICE/ITMC)* (pp. 1–6). IEEE.
Lee, D. (2020). Amazon's advertising business booms in pandemic. https://www.ft.com/content/095d73d5-a7a6-4acc-9dcc-9ee3e3d1fff4
Marsal-Llacuna, M., & Segal, M. (2016). The intelligenter method (I) for making "smarter" city projects and plans. *Cities, 55*, 127–138.
Marx, K., & Engels, F. (2008 [1848]). *The communist manifesto*. Edited by David Harvey and Samule Moore. London: Pluto Press.
Mazzucato, M. (2021). *Mission economy: A moonshot guide to changing capitalism*. Penguin UK.
Mohammadzadeh, M. (2021). Sharing or owning autonomous vehicles? Comprehending the role of ideology in the adoption of autonomous vehicles in the society of automobility. *Transportation Research Interdisciplinary Perspectives, 9*, 100294–100304.
Muth, R., Eisenhut, K., Rabe, J., & Tschorsch, F. (2019, September). BBBlockchain: Blockchain-based participation in urban development. In *2019 15th international conference on eScience (eScience)* (pp. 321–330). IEEE.
Neven, R. (2017). Gebiedonline. Retrieved 16 Nov 2021 from https://amsterdamsmartcity.com/updates/project/gebiedonline-q401tgsl
Oskam, J. A. (2019). *The future of airbnb and the 'sharing economy'*. Channel View Publications.
Pellegrin, J., Colnot, L., & Delponte, L. (2021). *Research for REGI committee – Artificial intelligence and urban development*. Brussels: European Parliament, Policy Department for Structural and Cohesion Policies.
Picon, A. (2018). Urban infrastructure, imagination and politics: From the networked metropolis to the smart city. *International Journal of Urban and Regional Research, 42*, 263–275. https://doi.org/10.1111/1468-2427.12527.
Rifkin, J. (2014). *The zero marginal cost society: The internet of things, the collaborative commons, and the eclipse of capitalism*. St. Martin's Press.
Sadowski, J. (2020). Cyberspace and cityscapes: On the emergence of platform urbanism. *Urban Geography, 41*(3), 448–452. https://doi.org/10.1080/02723638.2020.1721055.
Schumpeter, J. (2009 [1947]). *Can capitalism survive?* London: HarperCollins.
Shapiro, C., Carl, S., & Varian, H. (1998). *Information rules: A strategic guide to the network economy*. Harvard Business Press.
Shrier, D. (2020). *Basic Blockchain: What it is and how it will transform the way we work and live*. Robinson.
Smys, S., Basar, A., & Wang, H. (2020). Hybrid intrusion detection system for Internet of Things (IoT). *Journal of ISMAC, 2*(4), 190–199.
Taylor, P., Steenmans, K., & Steenmans, I. (2020). Blockchain technology for sustainable waste management. *Frontiers in Political Science, 2*, 15.

The European Union (The EU). (2021). EU-US trade and technology council inaugural joint statement report 29 September 2021. Retrieved 21 Nov 2021 from https://ec.europa.eu/commission/presscorner/detail/en/STATEMENT_21_4951

Torfing, J. (1999). *New theories of discourse*. Malden: Blackwell.

Van Raalte, A., Maeseele, P., & Phelan, S. (2021). Twitter as a right-wing populist's playground: The algorithmic populism of Dutch political party 'Forum voor Democratie' and leader Thierry Baudet during their political rise. *Discourse, Context & Media, 44*, 100549.

Vandenberg, S., & Magnuson, M. (2021). A comparison of student and faculty attitudes on the use of zoom, a video conferencing platform: A mixed-methods study. *Nurse Education in Practice, 54*, 103138.

Wheeler, T. (2021). China's new regulation of platforms: a message for American policymakers (September 14, 2021). Retrieved 21 Nov 2021 from https://www.brookings.edu/blog/techtank/2021/09/14/chinas-new-regulation-of-platforms-a-message-for-american-policymakers/

Smart Densification

Mubeen Ahmad, Sebastien Darchen, Sara Alidoust and Neil Sipe
School of Earth and Environmental Sciences, The University of Queensland, Brisbane, QLD, Australia

Synonyms

Smart Density

Definition

Smart densification is a metric that holistically integrates the quantitative (percentage/unit-to-area-ratios) and qualitative (context, perception, user experience) dimensions. The definition extends beyond the numerical relationship between physical space and the occupants who use that space. It incorporates the qualitative aspects of context, user perception, and experience to determine the appropriate configuration and scope of density for planning practice.

Introduction

In policy and planning practice, density is utilized to regulate land use. The current debate in the literature highlights the continual pursuit and rise in advocacy for achieving higher densities supported through policy. The literature also highlights a widening gap between the actual calculable levels of density, and the perception and attitudes surrounding it. At one end is the discrepancy and overemphasis of utilizing density only through a single measurable ratio lens which Schropfer and Menz (2019) argue results in unintended consequences for the built environment. At the opposite end is the view that density calculations should be more comprehensive (Bolleter and Ramalho 2020; McFarlane 2016) and include qualitative and contextual elements – to maximize the density utility. Thus, a new definition of density is needed that overcomes these issues.

Existing Density Definitions

Density is defined as the number of units allocated in each area. The standard density formula consists of the numerator (the number of units) and the denominator (land area). An issue with most density definitions is that they use averages that are not indicative of wider characteristics of a development (Aurand 2010; Jenks and Dempsey 2005; Burton 2000). Moreover, depending on the kind of density required, no one accepted measure is used. When calculating density, variables comprising physicality, occupancy, and the coverage of spaces play a key role.

Figure 1 illustrates the variables in existing density definitions. They are organized to show the prevailing interrelationships that influence how densities are calculated. These overlapping aspects subsequently impact development proposals and enable decision-makers, from architects to developers, planners, and policymakers, to calibrate and apply densities at different scales.

The applicability of these density measures in planning practice brings about further nuances on how density is used. From the perspective of policymakers, the density measures commonly used in local and state planning codes are a

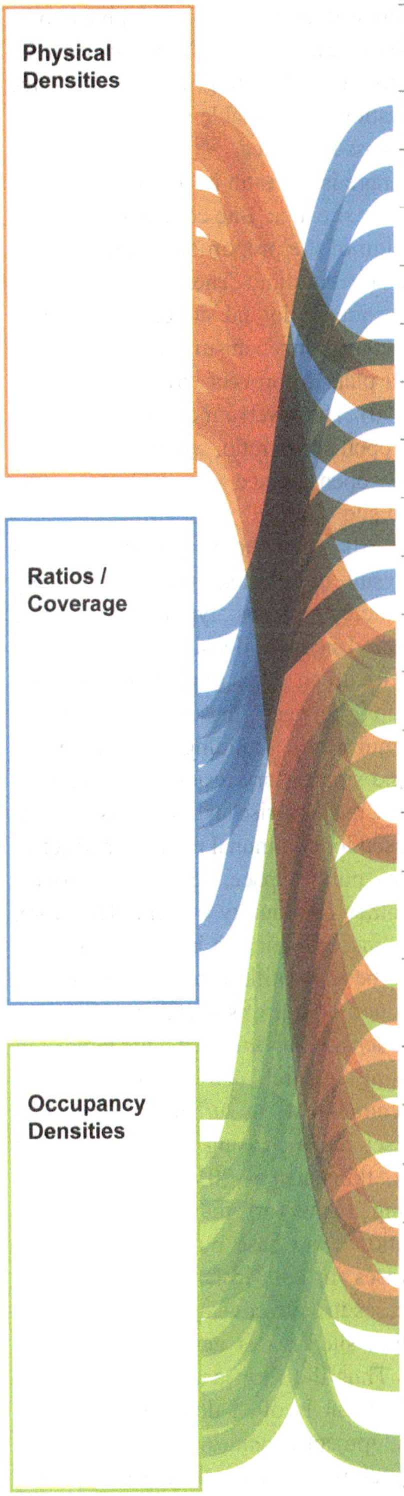

Density Units & Typology *(in order of micro to macro scale)*	Calculations, Measurement Units, Scope
Setbacks: Side Distance between two buildings on separate parcels	Provides a measurement to ascertain building bulk in imperial/metric units.
Setbacks: Front Kerbs	Denotes the "spatial experience" of the setback. Measured in Imperial / Metric Units the distance from the frontal facade of the building to the kerb line.
Setbacks: Front Parcels	Denotes the "spatial experience" of the setback. Measured in Imperial / Metric Units the distance from the frontal facade of the building to the property line.
Heights: Building on Parcel / Site Plot	Provides a measure of the building "intensity", usually perscribed in zoning regulations. Measured in imperial / metric units for individual parcels / site plot area.
Coverage: Impervious Surface Block	(Ground Floor Building Footprint Area) + (Built-up Area) / (Block Land Area)
Coverage: Impervious Surface Parcel	(Ground Floor Building Footprint Area) + (Built-up Area) / (Parcel / Site Plot Area)
Coverage: Building Block	(Ground Floor Building Footprint Area) / (Block Area)
Coverage: Building Site Coverage (or coverage ratio)	(Ground Floor Building Footprint Area) / (Parcel / Site Plot Area)
Floor-Area Ratio	(Floor Area of all floors of the building) / (Parcel / Site Plot Area)
Scale: Net Residential Density at City or Metropolitan Level	(Dwelling Unit / Residential Population) / (Allocated Residentail Land at City / metropolitan scale)
Scale: Metropolitan Density	(Dwelling Unit / Residential Population of Metropolitan Scale) / (Total Area)
Scale: City Density	(Dwelling Unit / Residential Population of City Developed Area) / (Total City Developed Area)
Gross Neighbourhood Density	(Dwelling Unit / Residential Population of Neighbourhood) / (Total Neighbourhood Area)
Net Neighbourhood Density	(Dwelling Unit / Residential Population of Neighbourhood) / (Total Neighbourhood Area excl. areas with city-wide uses)
Net Street Density	(Dwelling Unit / Residential Population of Neighbourhood) / (Total Residential Unit Area based on typology and inclusive of public streets)
Net Neighbourhood Residential Density (Building Type)	(Dwelling Unit / Residential Population) / (Total Area that consists of residential unit based on typology)
Net Neighbourhood Residential Dwelling (Population Density)	(Dwelling Unit / Residential Population) / (Total Land Area that consists of residential unit)
Block Density (Partial)	(Dwelling Unit / Residential Population) / (Definitive Sub-Area of a Block)
Block Density (Full)	(Dwelling Unit / Residential Population) / (Full Extent of a Block Area)
Parcel / Site Plot Density	(Dwelling Unit / Residential Population of Total Floor Area) / (Total Parcel / Site Plot Area)
Occupancy Density	(Individual Habitable Dwelling's Total Number of Occupants) / (Building Total Floor Area)
Habitable Rooms per Hectare/Unit	(Number of Rooms in a dwelling used for living purposes) / (Building Total Land Area)

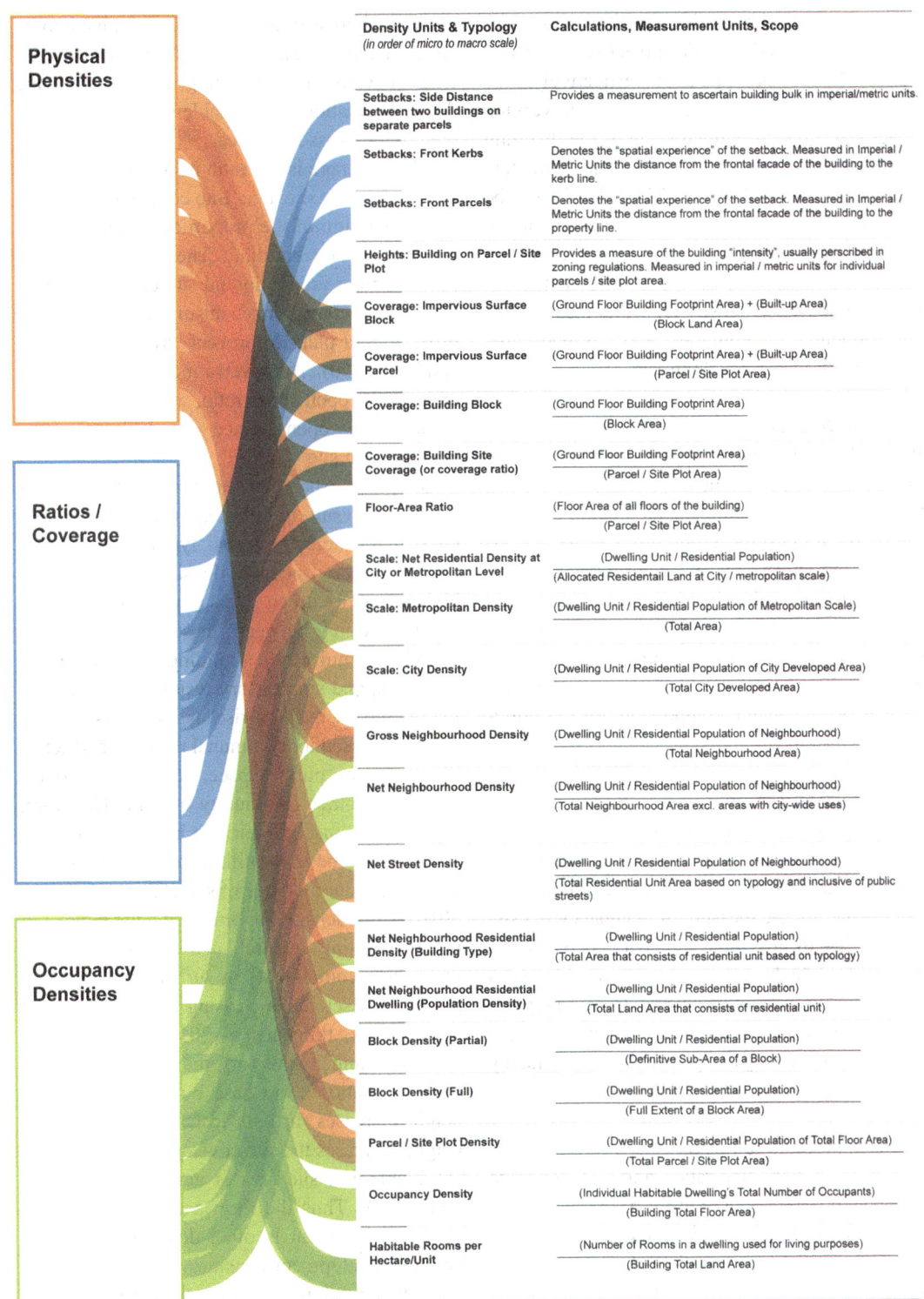

Smart Densification, Fig. 1 Current definitions of density. (Source: Adapted from multiple sources (Atkinson et al. 2014; Forsyth 2003; Friedman 2007; Schropfer and Menz 2019)

S

combination of plot-specific densities and wider neighborhood residential densities (Scoffham and Vale 1996; Vale and Vale 2010). Effectively, they aim to prescribe dwelling densities at varying scales starting from a microscale of the dwelling and increase to the neighborhood and ultimately to the macro-scale of the city. Thus, density can be appropriated for the kind of development envisioned at each scale – that is, gross/net density of plots for specific residential development at the plot, neighborhood density for wider masterplans and residential communities, and city-wide area densities for newer developments.

Physical Densities

Physical densities indicate measurements made in quantitative terms to denote the concentration and the ratio of built structures in an area. In calculating physical densities, a building can be considered an individual unit and determined to be of any form, in any location, and of any program such as a residential unit, office space, and other mixed-use typology. Decision-makers employ physical densities to establish spatial needs for various programmatic components for the buildings and volumetric massing.

Occupancy Densities

Occupancy densities are the ratio of the floor area of an individual building unit to the number of occupants in that unit. The occupancy rate is considered the inverse measure of occupancy density and is used to determine the space available for occupants. A higher occupancy rate would mean a larger liveable area for occupants. In planning policies, the regulation of minimum occupancy rate is often used in building design to enhance the quality of habitable spaces (Atkinson et al. 2014).

Site and Floor-Area Ratios

Variances in floor-area-ratios and site coverage results in differences in built form. In preliminary design documentation and development outlays, floor-area-ratio is commonly referred to as the standard indicator for land-use regulation, zoning, and development. Floor-area-ratio limits are often prescribed in planning codes to regulate the extent of construction in a project. Measured as the ratio of the total gross floor area of a development to its site area, the entire area within the perimeter of the exterior walls of the building, including service ducts, circulation, and useable spaces, is considered of the gross floor area.

Site coverage is the ratio of the building footprint area to its site area. Site coverage is considered a measure of the proportion of the site area covered by the building. The major similarity between floor-area-ratio and site coverage is that it is often a control mechanism used primarily in urban master plans to prevent "over-build" (Vale and Vale 2010). The inverse measure of site coverage is the open space ratio, which specifies the amount of open space available on the development site.

Similarly, urban developments of similar densities can result in different urban forms. Ng (2010) illustrates various typologies ranging from the same residential density but in different high to low rise urban forms (Ng 2010). There are varying forms between these two scenarios, including multistory buildings, medium-rise buildings, and single-story residential dwellings. In this context, the high-rise layout results in space that can be used for communal facilities and activities, such as public space, recreational zones, parks, and other community centers. Therefore, efficient land-use planning is imperative if such spaces are to be fully utilized because inefficiencies can produce a range of societal issues (Tallon 2010). High density has attracted wide interest in urban planning research and is considered an important component in urban development plans to offset the increasing pressure on land availability (Atkinson et al. 2014).

As building heights vary, they may increase the floor-area-ratio at the base but decrease as the height increases. In both instances, the building height decreases the site coverage. From a practical standpoint, site area is generally limited in inner cities. Thus the form and structure in the urban environment is often determined by predefined development densities (Friedman 2007). Parallel to the above density measures is the inclusion of front setback amounts and roadside kerb provisions relative to plot dimensions and adjacent alleyway space in-between structures and buildings. Effectively, this impacts the floor and plot area ratios, which influence how much of the design and development parameters permit aggregate floor areas for developers. In such instances, developers prefer to utilize site plot densities, building-site-coverage, and floor-area-ratios (Forsyth 2003) to collaborate and coordinate with architects using the same terminology to produce a design based on the required dwelling density.

Practical Applications of Density

To determine which density measure to use, calculations should include quantitative and qualitative factors such as the characteristics of the development, the stage at which the proposed development is in the planning process, and aspects of the surrounding areas where the development is located.

Boyko and Cooper (2009) have raised issues regarding net density as the default measurement. Rudlin and Falk (1999) argue that net residential density fails to consider variances in land capacities – specifically within mixed land-use – and that such calculations do not provide any useful guidance for evaluating qualitative aspects regarding the viability of public services such as public and active transport.

The discourse on how density is perceived (Ellis 2004; Jenson 1966; Pun 1994) reflects a paradox that exists in the wider debate, which impacts the kind of development that is pursued. As Fig. 2 shows, spatial dimensions and sense of urban spaces influence the attitudes toward densities. As discussed by Berke and Kaiser (2006), density is perceived differently across urban landscapes, and public attitudes are indicative of the inclination toward lower densities as opposed to the negative connotations associated with higher densities (Berke and Kaiser 2006; Burgess and Jenks 2000; Feinstein 2006; Gleeson and Low 2000; Williams 2000).

The various density definitions provide an array of possible uses, and planning policies are typically based on the understanding that density affects everyone in the same manner (Churchman 1999). However, there is little support for the notion that

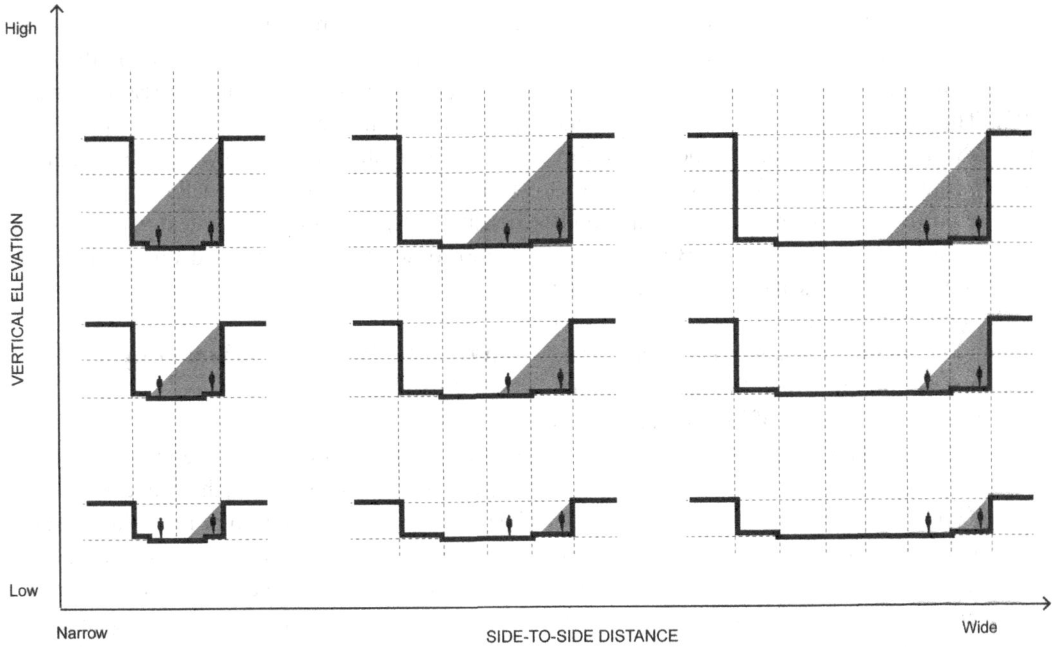

Smart Densification, Fig. 2 Spatial dimension and sense of space. (Source: Author, adapted from multiple sources Ellis 2004; Pun 1994; Vale and Vale 2010)

the public would prefer higher densities (Hall 1999). Key impediments and implications have emerged due to the contested nature of density, particularly the need to integrate the qualitative and quantitative dimensions so that the qualities of the physical environment, the public's perceptions and needs are considered (Churchman 1999). With the plethora of quantitatively based definitions, our understanding of density needs further clarification as it relates to qualitative measures. These can be grouped into three areas: context, perceptions, and user experience.

Context

The primary aspect that influences density is context. Calculating density depends on both subjective and objective definitions. However, the psychological, social, economic, and technological dimensions of density may be at odds with one another (Churchman 1999). For instance, an area may not have the same meaning to everyone. For example the prevailing context and attitudes among different population groups will influence the size of the household. Each context will determine which dimensions of density are relevant. Complexity increases when different scales and locations are compared – because, with varying degrees of contexts, the density would be different even if it measured the same aspect (Alexander 1993). The notion is that higher densities equal higher sustainability or even higher liveability (Doberti and Giordano 2007; Rao 2007) – although this depends upon the prevailing values both in social and cultural contexts (Bramley et al. 2010; Raman 2010). This understanding is crucial since it perceives densities through the lens of context. Therefore, a meaningful effect in one place may be of great significance in another location, primarily because of its varying degrees of social and cultural context.

Perception

The second aspect influencing density is perception. Spatial and social constraints are perceptual dimensions that need to be considered when considering density. How people perceive density informs what emotional responses they emit and how they behave (Dave 2011; Glass and Singer 1972; Grammenos 2011; Rodin et al. 1978; Cohen and Sherrod 1978; Sherrod 1974). Both spatially and socially, people have predispositions that influence their interaction in that situation. In calculating density, the social dimension is comprised of a person's interaction with other peers.

In contrast, the spatial dimension is comprised of how a space places limitations on the ability to move around and have contact with others. In addition, how people perceive the environment is influenced by a host of other factors. Forsyth (2003) argues that perception can be linked to density through the building form, typology, and aesthetic design and appeal.

User Experience

The third aspect influencing density relates to user experience. User experience is intricately linked with our perceptions (Raman 2010). The user experience of density through the lens of a social and psychological perspective is usually "negative" (Cohen and Gutman 2007) because higher densities can lead to "overcrowding" (Rao 2007) and lack of privacy (Bolleter and Ramalho 2020; Hernandez et al. 2018; Haarhoff et al. 2016; Mees 2009; Su 2011). While privacy and overcrowding can be addressed through design, the distinction between privacy/crowding and density is such that one does not increase the other. Gordon and Ikeda (2011) illustrate that an increase in density does not mean overcrowding because perceived density is not the same as crowding in social gatherings. Therefore, an individual's experience of density is dependent on the context and their perception of what privacy or crowdedness is to them.

Smart Densification: Advancing the Definition

Following on from the above definitions of density and their use in planning, the concept needs to be revised. It further needs to be re-evaluated such that it incorporates the complexities of everyday living into the quantitative side of density calculation along with consideration for the qualitative dimensions. To this end, advancing the definition of density is proposed in this chapter, coined as "Smart Density," offering ways to elucidate the overall concept.

The analysis, as discussed above, highlights that density is not only a significant spatial concept but also encompasses more than a simple quantitative calculation. However, density also has qualitative aspects involving context, perception, and user experience. Figure 3 shows how the current understanding of density comprises the quantitative dimensions. Smart density builds on the existing definition by adding a much larger ring with new dimensions. These new dimensions make density more viable in practice by including perception and public needs, along with the qualitative dimensions and versatility. Thus, smart density integrates both quantitative and qualitative dimensions, allowing decision-makers to consider all three dimensions (context, perception, and user experience) holistically and dynamically (Dave 2010).

Application of Smart Density in Planning Policy and Practice

This section brings together the dimensions of density in a checklist (see Fig. 4) to assist policymakers and practitioners in calculating smart density. Applying the calculation based on the dimensions outlined above allows for more appropriate density levels.

First, policies need to include both quantitative and qualitative elements of density. There is a need for a thorough understanding of community and other relevant stakeholder expectations as well as the needs of the built and natural environment (Chan 1998; Cheng 2010; Churchman 1999; Cohen and Gutman 2007; Dave 2011). Smart densification advances the existing definition, enabling planners to formulate policies that reflect the practicalities of built environment design. For example, effective density policies should integrate density with other quality of life indicators. Equipped with a combination of quantitative and qualitative dimensions, smart density can provide building forms and articulation that reflect the context and knowledge about peoples' preferences and perceptions.

Second, density should have greater responsiveness to contextual elements. Figure 1 illustrates the shared aspirations of density across all scales of development. This is important to consider because defining density on a common platform that can be used across different levels of government is helpful in avoiding confusion in terminology and measurement. Smart density

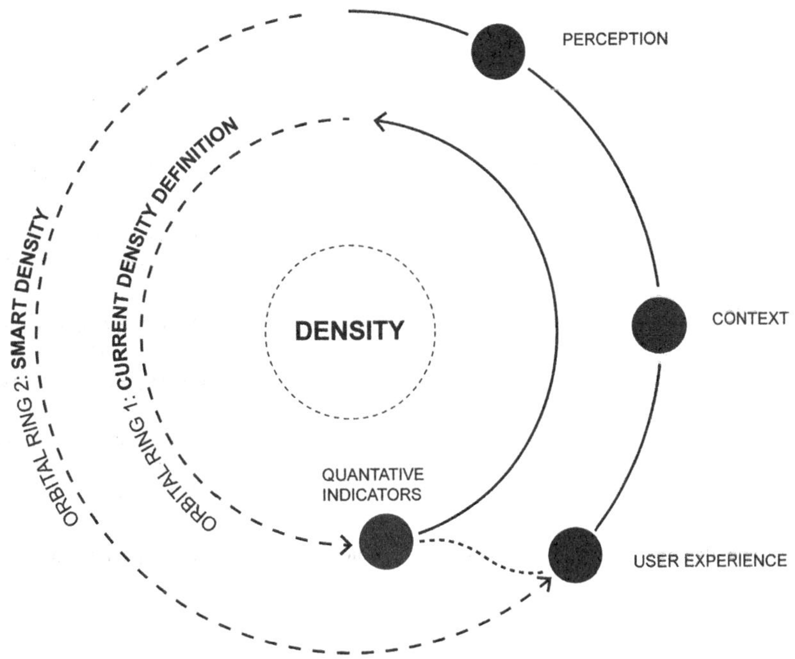

Smart Densification, Fig. 3 Advancing density definition: current understanding and smart density. (Source: Author)

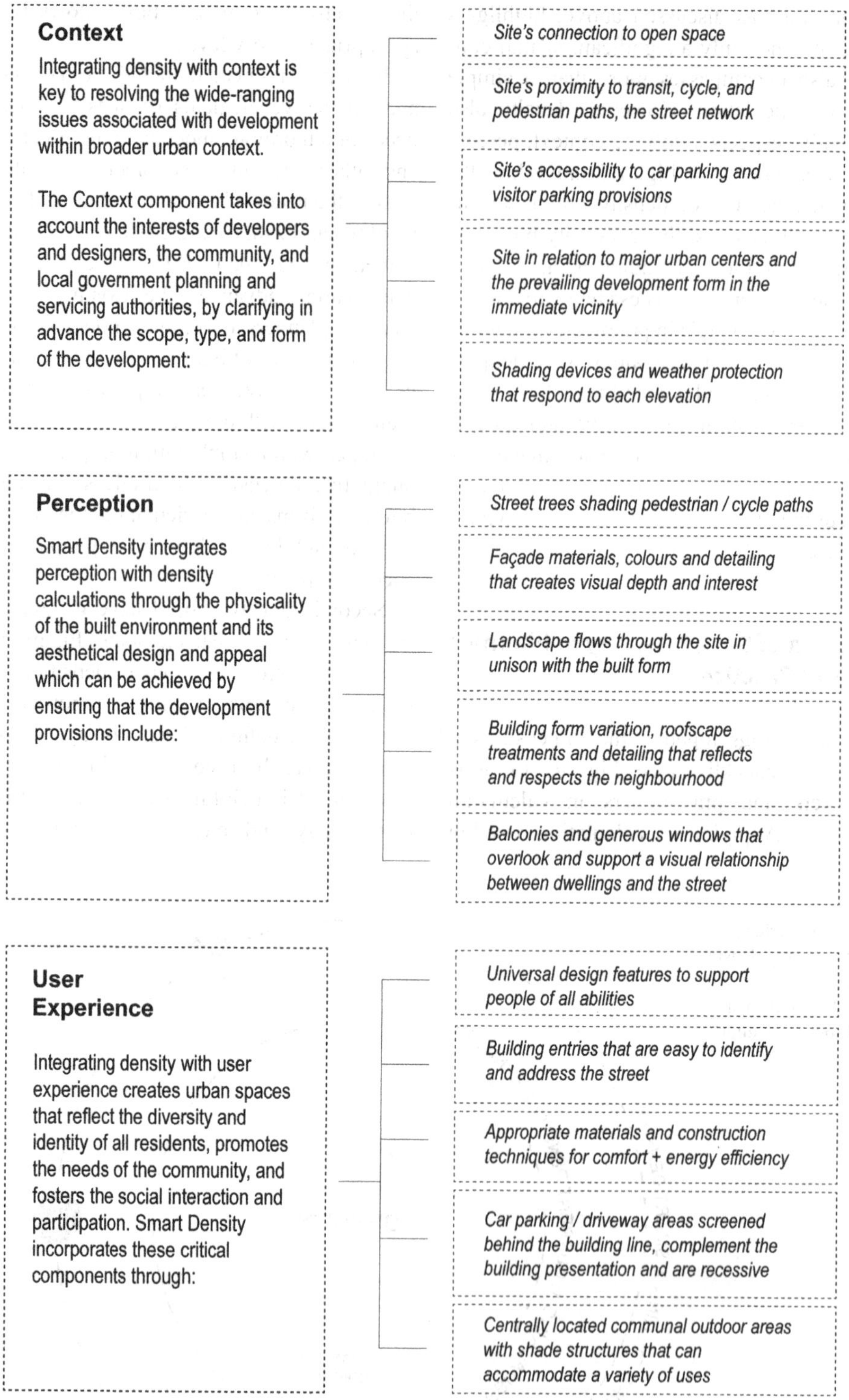

Context
Integrating density with context is key to resolving the wide-ranging issues associated with development within broader urban context.
The Context component takes into account the interests of developers and designers, the community, and local government planning and servicing authorities, by clarifying in advance the scope, type, and form of the development:
Site's connection to open space
Site's proximity to transit, cycle, and pedestrian paths, the street network
Site's accessibility to car parking and visitor parking provisions
Site in relation to major urban centers and the prevailing development form in the immediate vicinity
Shading devices and weather protection that respond to each elevation
Perception
Smart Density integrates perception with density calculations through the physicality of the built environment and its aesthetical design and appeal which can be achieved by ensuring that the development provisions include:
Street trees shading pedestrian / cycle paths
Façade materials, colours and detailing that creates visual depth and interest
Landscape flows through the site in unison with the built form
Building form variation, roofscape treatments and detailing that reflects and respects the neighbourhood
Balconies and generous windows that overlook and support a visual relationship between dwellings and the street
User Experience
Integrating density with user experience creates urban spaces that reflect the diversity and identity of all residents, promotes the needs of the community, and fosters the social interaction and participation. Smart Density incorporates these critical components through:
Universal design features to support people of all abilities
Building entries that are easy to identify and address the street
Appropriate materials and construction techniques for comfort + energy efficiency
Car parking / driveway areas screened behind the building line, complement the building presentation and are recessive
Centrally located communal outdoor areas with shade structures that can accommodate a variety of uses

Smart Densification, Fig. 4 Criteria for application of smart density for planning practice contributes to a neighborhood's unique identity because it can be embedded through building components that conform to a coherent architectural form that enhances the street. Additionally, policies should not be so deterministic that it prevents incorporation of newer solutions for location-specific density-based issues (Haarhoff et al. 2016).

Third, there is a need for greater versatility in how density is conceptualized. Currently, densities are typically the number of dwelling units per area (Aldred 2010). However, density should consider additional land uses, typologies, and forms. Figure 1 shows that density consists of various typologies related to the built and natural form and has a dynamic relationship with people's attitudes, perceptions and experiences. Together these impact the design of the built form. Thus, smart density enhances the quality of development and increases functionality by incorporating connections to open space, communal outdoor areas, and public/active transport. Through combinations of these elements, smart density provides a benchmark for policymakers to use in the development of density policies and codes.

Smart density is a valuable instrument because it can be applied at a range of scales to help generate the design of sustainable urban environments. Determining appropriate densities can be achieved by using the parameters in the checklist below. These criteria incorporate the qualitative considerations and cover the variety of aspects that play an important role in achieving appropriate densities.

Conclusion

The aims of this chapter are twofold – to illustrate the many and complex definitions of density, and to propose smart density as a reconceptualized definition. In addressing these two aims, this chapter consisted of a review of the definitions of density and how they can be improved to integrate both quantitative and qualitative dimensions, utilizing contextual elements. By advancing the definition, smart densification enables the development of efficient and appropriate policies that are conscious of contextual attitudes. This updated definition will help to regulate land use by providing more guidance to developers and ensuring better planning outcomes.

Cross-References

▶ Urban Densification and Its Social Sustainability

References

Aldred, T. (2010). *Arrest development: Are we building houses in the right places?* London: Centre for Cities.

Alexander, E. (1993). Density measures: A review and analysis. *Journal of Architectural and Planning Research, 10*(3), 181–202.

Atkinson, G., Dietz, S., Neumayer, E., & Agarwala, M. (Eds.). (2014). *Handbook of sustainable development* (2nd ed.). Cheltenham: Edward Elger.

Aurand, A. (2010). Density, housing types and mixed land use: Smart tools for affordable housing? *Urban Studies, 47*(5), 1015–1036.

Berke, P., & Kaiser, E. J. (2006). *Urban land-use planning* (5th ed.). University of Illinois Press.

Bolleter, J., & Ramalho, C. E. (2020). *Greenspace-oriented development: Reconciling urban density and nature in suburban cities*. Springer.

Boyko, C., & Cooper, R. (2009). Urban design decision-making: A new approach. In R. Cooper, G. Evans, & C. Boyko (Eds.), *Designing sustainable cities* (pp. 42–50). London: Wiley-Blackwell.

Bramley, G., Dunmore, K., Dunse, N., Gilbert, C., Thanos, S., & Watkins, D. (2010). *The implications of housing type/size mix and density for the affordability and viability of new housing supply*. Titchfield: National Housing and Planning Advice Unit.

Burgess, R., & Jenks, M. (2000). *Compact cities : Sustainable urban forms for developing countries*. Spon: Spon Press E. & F.N.

Burton, E. (2000). The compact city: Just or just compact? *A Preliminary Analysis. Urban Studies, 37*(11), 1969–2001.

Chan, Y.-K. (1998). Density, crowding, and factors intervening in their relationship: Evidence from a hyper-dense metropolis. *Social Indicators Research, 48*, 103–124.

Cheng, V. (2010). Understanding density and high density. In E. Ng (Ed.), *Designing high-density cities for social and environmental sustainability* (pp. 3–17). London: Earthscan.

Churchman, A. (1999). Disentangling the concept of density. *Journal of Planning Literature, 13*(4), 389–411.

S

Cohen, M., & Gutman, M. (2007). Density: An overview essay. *Built Environment, 33*(2), 141–144.
Cohen, S., & Sherrod, D. (1978). When density matters: Environmental control as a determinant of crowding effects in laboratory and residential settings. *Journal of Population, 1*, 189–202.
Dave, S. (2010). High urban densities in developing countries: A sustainable solution? *Built Environment, 36*(1), 9–27.
Dave, S. (2011). Neighbourhood density and social sustainability in cities of developing countries. *Sustainable Development, 19*, 189–205.
Doberti, R., & Giordano, L. (2007). Morphology of density. *Built Environment, 33*(2), 185–195.
Ellis, J. (2004). Explaining residential density (Case study for the proposed town of Coyote Valley, South of San Jose, California). *Places – A Forum of Environmental Design, 16*(2), 34–43.
Fabio Alberto Hernandez, P., Sabrina, S., & Yngve Karl, F. (2018). The value of urban density. *TeMA: Journal of Land Use, Mobility and Environment, 11*(2), 213–230. https://doi.org/10.6092/1970-9870/5484.
Feinstein, J. S. (2006). *The nature of creative development.* Stanford Business Books.
Forsyth, A. (2003). *Measuring density: Working definitions for residential density and building intensity* (Design brief 8). Minneapolis: Design Center for American Urban Landscape, University of Minnesota.
Friedman, A. (2007). *Sustainable residential development : Planning and design for green neighborhoods.* New York: McGraw-Hill.
Glass, D. C., & Singer, J. E. (1972). *Urban stress: Experiments on noise and social stressors.* New York: Academic.
Gleeson, B., & Low, N. P. (2000). *Australian urban planning : New challenges, new agendas.* Allen & Unwin.
Gordon, P., & Ikeda, S. (2011). In A. Andersson, D. E. Andersson, & C. Mellander (Eds.), *Does density matter?*
Grammenos, F. (2011). *European urbanism: Lessons from a city without suburbs.* Accessed from http://www.planetizen.com/node/48065.
Haarhoff, E., Beattie, L., & Dupuis, A. (2016). Does higher density housing enhance liveability? Case studies of housing intensification in Auckland. *Cogent Social Sciences, 2*(1). https://doi.org/10.1080/23311886.2016.1243289.
Hall, P. (1999). *Sustainable cities or town cramming?* London: Town and Country Planning Association.
Jenks, M., & Dempsey, N. (2005). The language and meaning of density. In *Future forms and design for sustainable cities* (pp. 287–309). Oxford: Architectural Press.
Jenson, R. (1966). *High density living.* Leonard Hill: The Book Service Ltd./Praeger.
McFarlane, C. (2016). The geographies of urban density: Topology, politics and the city. *Progress in Human Geography, 40*(5), 629–648. https://doi.org/10.1177/0309132515608694.
Mees, P. (2009). How dense are we? Another look at urban density and transport patterns in Australia, Canada and the USA. *Road and Transport Research, 18*(4), 58–67.
Ng, E. (2010). In E. Ng (Ed.), *Designing high-density cities for social and environmental sustainability.* London/ Sterling: Earthscan.
Pun, P. K. S. (1994). *Advantages and disadvantages of high-density urban development.* Consulate General of France in Hong Kong.
Raman, S. (2010). Designing a liveable compact city: Physical forms of city and social life in urban neighbourhoods. *Built Environment, 36*(1), 63–80.
Rao, V. (2007). Proximity distances: The phenomenology of density in Mumbai. *Built Environment, 33*(2), 227–248.
Rodin, J., Solomon, S., & Metcalf, J. (1978). Role of control in mediating perceptions of density. *Journal of Personality and Social Psychology, 36*, 989–999.
Rudlin, D., & Falk, N. (1999). *Building the 21st century home: The sustainable neighbourhood.* Oxford: Architectural Press.
Scoffham, E., & Vale, B. (1996). How compact is sustainable: How sustainable is compact? In M. Jenks, E. Burton, & K. Williams (Eds.), *The compact city: A sustainable urban form?* (pp. 66–73). London: E & FN Spon.
Schropfer, T., & Menz, S. (2019). *Dense and green building typologies: research, policy and practice perspectives.* Singapore: Springer.
Sherrod, D. R. (1974). Crowding, perceived control, and behavioral after effects. *Journal of Applied Social Psychology, 4*, 171–186.
Su, Q. (2011). The effect of population density, road network density, and congestion on household gasoline consumption in U.S. urban areas. *Energy Economics, 33*(3), 445–452. https://doi.org/10.1016/j.eneco.2010.11.005.
Tallon, A. (2010). *Urban Regeneration and Renewal* (Vol. IV). Routledge.
Vale, B., & Vale, R. (2010). Is the high-density city the only option? In E. Ng (Ed.), *Designing high density cities.* Earthscan.
Williams, D. C. (2000). *Urban sprawl : A reference handbook.* ABC-CLIO.

Smart Density

► Smart Densification

Smart Grids

► Smart Grids to Lower Energy Usage and Carbon Emissions: Case Study Examples from Colombia and Turkey

Smart Grids to Lower Energy Usage and Carbon Emissions: Case Study Examples from Colombia and Turkey

Hasan Volkan Oral[1], Hasan Saygın[2], Julián Andrés Mera-Paz[3] and Ramon Fernando Colmenares-Quintero[4]
[1]Faculty of Engineering, Department of Civil Engineering (English), Istanbul Aydın University, Istanbul, Turkey
[2]Application, and Research Center for Advanced Studies, Istanbul Aydın University, Istanbul, Turkey
[3]Faculty of Engineering, Universidad Cooperativa de Colombia, Popayán, Colombia
[4]Faculty of Engineering, Universidad Cooperativa de Colombia, Medellín, Colombia

Synonyms

Colombia energy transformation; Energy efficiency; Smart grids; Sustainable development goals; Turkey energy transformation

Definition

Currently, the United Nation's Sustainable Development Goals (SDGs) established in the Agenda 2030 have stimulated a series of actions by several countries, such as affordable and clean energy (SDG7) and sustainable cities and communities (SDG11). These objectives lead to thinking about the concept of smart grids (SG) so that the communities can become more sustainable in terms of energy and climate change. The paradigm is complex in a world where gas emissions, the greenhouse effect, and energy consumption from processes that include fossil fuels, damage the environment and undermine the conditions of well-being and good living. For this reason, project initiatives are being developed that seek the modernization of electricity generation and distribution systems in communities. One of the strategies is the creation and adaptation of micro-grid architectures that adapt to their operational context. The microgrid concept focuses on the controlled use of electrical energy, with a high degree of autonomy, monitoring, and control supported by information technology (IT), to optimize energy transfer while minimizing risks, and increasing quality, efficiency, and reliability of energy supply.

Introduction

The Sustainable Development Goals (SDGs), which are agreed by the United Nations (UN) (2017), aim to increase the welfare of societies, but at the same time to create strategies in combating important environmental problems such as climate change, global warming, biodiversity loss, and deforestation. The UN declared 17 SDGs, which are accepted by 193 countries. In 2018, these goals were updated and renewed. Among the SDGs, the purpose of referring to the concept of the smart city through the establishment of sustainable cities can be connected through SDG 11. The main aim of this goal is to make cities and human settlements inclusive, safe, resilient, and sustainable (Oral et al. 2020). According to UN Environment (2020), the concept of sustainable cities is sustainable consumption and production roadmap for cities covering all the sectors and upstream interventions through policy, technology, and financing to reduce and manage pollution and waste. The types of pollution are various, but one of the significant types, also known as air pollution, is based on Greenhouse Gas Emissions (GHGs) emissions from industrial plants in cities. The main sources of GHGs can be grouped under electricity generation, transportation, industrial processes, commercial and residential activities, agriculture, land use, and forestry. For instance, in the United States of America (USA) the contribution of these GHGs by economic sectors in 2018 were noted as Agriculture (10%), Transportation (28%), Electricity Generation (27%), Industry (22%), Commercial and residential (12%). Among them, electricity production generates the second-largest share of GHGs emissions. Approximately 63 % of the

S

country's electricity comes from burning fossil fuels, mostly coal and natural gas (EPA 2020).

A significant route to sustainable development (SD) is the convergence of smart grid technology (SGT), renewable energy infrastructure, and low-carbon emissions in power generation systems. In addition, SG strategies increase the quantity of variable renewable energy generation that can be used in power systems, thus increasing the ability of grid-connected clean energy systems such as solar, wind, and photovoltaic. Secondly, in the power sector, an SGT encourages energy saving. The key benefit of the SGT is that it will increase the efficiency of the usage of power grids and the efficiency of power consumption (Hu et al. 2014). The term "SGTs" is described by Milborrow (2016) as different technologies that might need to be introduced in the future to allow more efficient operation of electricity networks. Improving energy efficiency is one of the major advantages of intelligent grids. SGT optimizes demand and supply management for energy, minimizes the loss of electricity between power plants and customers thereby saving electricity. Related costs of constructing new power plants can also be avoided due to decreased peak demand. Via increased production and use of clean, renewable energy, SGTs also produce lower GHGs as a response to their contribution to climate change (Lee et al. 2012).

The research question of this study can be summarized as: "Which architecture model among SA applications is the most suitable for reducing energy use and carbon emission especially in off-grid communities?"

Consequently, the purpose of this book chapter is to present how lower energy usage and carbon emissions can be achieved based on SA architecture models in Colombia and Turkey to fill the gap in the literature.

The following is the book chapter's organizational structure:

- Section 4 provides a literature review on SGT architecture models, presenting the pros and cons of the systems
- Section 5 provides an overview of the current situation in Turkey and Colombia
- Section 6 presents the likely future applications in Turkey and Colombia
- Section 7 shows Results and Discussion
  - Section 7.1 proposes a data-structure architecture for smart grids
  - Section 7.2 summarises reference models for big data architecture
  - Section 7.3 outlines the challenges in designing big data architectures for smart girds
- Section 8 concluded the study

Smart Grids Architecture Models Available in the Literature

Below is an analysis of the advantages and disadvantages of different architecture models for SGTs, detailing the relevant components and characteristics. The architecture models reviewed in this section are SGAM (Smart grit architecture model), SGM (Smart micro-networks), AMI (Advanced metering infrastructure), ADA (Advanced distribution automation), DER (Distributed energy resources).

1. **SGAM - SA architecture model** It is an architecture model that is characterized by having a neutral technological position, adaptable to both traditional and unconventional energies. One of its main characteristics is having a set of interoperability layers that, in an articulated way, facilitate the proper functioning of the architecture. It is composed of: a commercial layer where the commercial goals, the regulatory, and policy approaches are outlined; a layer of functions where the use cases and functions for the micro-network are described; an information layer where the structure and modeling of the data are carried out; a communications layer where the protocols and rules for communication are established and finally, a layer of components where the domains are established: Process, field, station, operation, company and marketing and the other domains of Generation, transmission, distribution, distributed energy resources (DER), customer facilities (Trefke et al. 2013).

SGAM those involved in micro-network creation projects, by providing a comprehensive methodology to achieve a micro-network. In the case of management (Jacobson et al. 2016), it allows information to be shared between projects that implement similar use cases, with different technical solutions that lead to the so-called SGT.

2. **Smart micro-networks (SGM)** are a SA architecture, which is characterized by the interaction of interoperability functions. It is supported in three main layers that interconnect with each other. The first is a process layer in which all business and regulatory or normative processes are managed. The second is the station layer which includes activities, in addition to managing information, functional and non-functional requirements, procedures, records, storage, protocols, infrastructure, equipment, and communications. Finally, there is the operations control layer in which monitoring, follow-up, control, and supervision are carried out, with a special emphasis on the detection of alternative solutions to failures., This is a centralized control system generally accompanied by sensors, added to an automated control system supported on a distributed operating system for the management of SA resources and the management of the electrical network. Another important feature is the implementation of common methodologics that arc based on traditional and evolutionary processes. Currently, there is no documented evidence of the application of agile methodologies, which would surely be good practice due to the successful application in other architectures (CENELEC 2014; Zakariazadeh et al. 2014).
3. **Advanced metering infrastructure (AMI)** is an architecture proposed in a disruptive way, in it the paradigm of the layers or levels of its structure is broken and is replaced by the concept of various stages of communication, management, and engineering, where the AMI architecture becomes a system that combines smart meters, data management systems, technological communication networks in a bidirectional way, transforming the concept of architecture and allowing the effective interpretation of supply and demand management information. In some articles or scientific documents identified in the literature review, complex situations of security risk arising from alteration or modification of data are evidenced. However, it is mentioned that the solutions can be achieved through adequate public policies, regulations, and norms. Consequently, a suitable implementation of the AMI architecture is possible due to the strategic and rapid response to the needs of the demand, the monitoring of energy quality, the distribution process, efficiency of use, and the tendency to reduce the environmental impact. In documented cases, it mentions the application of agile methodologies such as Scrum, XP, lean, among others (Rua et al. 2010; Stellman and Greene 2014; Abdulla 2015).
4. **Advanced distribution automation (ADA)** In this architecture, bidirectional smart meters are consolidated and are expected to improve the quality of service, as the automation of energy distribution processes is increased, with an integrated, real-time monitoring system optimizing the efficiency in the delivery of energy, reducing failures and interruptions. The fundamental bases are the sensors, transducers, and intelligent electronic devices (IED) with a large amount of behavioral information being collected via the micro-network. Through the SCADA system (supervision, control, and data acquisition), the micro-network is managed with greater speed, reliability, and efficiency (Hanser 2010; Mohassel et al. 2014). In reports from the Center for Energy Advancement through Technological Innovation (CEATI) (Hanser 2010; Kolberg and Zühlke 2015), an attempt is made to define and show that for the network to behave intelligently, it must include automated monitoring of the network with sensors to improve reliability, automatic monitoring of equipment to improve maintenance, product monitoring based on supply and demand analysis to improve service quality. ADA architecture documentation mentions that experiments and projects are underway demonstrating its

implementation results in the reduction and control of the voltage force, or voltage, and the conservation and management of the flow of electric current or intensity is achieved. The architecture also gives greater accuracy in the detection of faults and their location plus analysis of the structure of the integrated monitoring and supervision system while providing results and reports on the operation of the micro-network.

5. **Distributed Energy resources (DER)** architecture aims to enhance energy resources in a way that optimizes planning the design of the network, for which a distributed electricity generation system is established (DG) (Beck et al. 2015; Bodek 2018). Key in planning is the analysis of geographic areas, the social and economic impact of the region (Highsmith and Cockburn 2001). An interesting characteristic of this architecture is its adaptability to accommodate conventional electrical energy systems with sensors and a system of micro generation with connections and control from alternating current (AC) to direct current (DC) (McGranahan et al. 2005). There is monitoring, control, and supervision system for energy storage (ESS) allowing virtualization of the use of load fluctuations, adapting to variables that influence the balancing of supply and demand (Brown 2008) in micro-grid processes managing: losses in the system, real-time fault detection, average interruption duration index (SAIDI), giving better power distribution, and a favorable cost-benefit ratio.

According to Zavoda (2008), a case study is reported where the DER architecture is used to reduce greenhouse gas emissions and while optimizing the use of electrical energy in a shopping center in Sydney Australia. It included integration between renewable energy (solar and thermal) with conventional energy. After implementation, a comparison is made taking into account: (1) cost and emissions solution compensation, (2) the cost of reduced emissions achieved in each investment scenario, and (3) benefits of investment concerning the business in a typical scenario. The results show that energy costs are reduced by 8.5%, carbon dioxide emissions are reduced by 29.6% and it is predicted that by making a greater investment in the construction of a micro-grid with a network architecture of 90% renewable energy can offer a 72% reduction in carbon dioxide emissions and a 47% reduction in energy costs. In the words of the authors: "the study demonstrates effectiveness, efficiency and flexibility of DER architecture for micro-networks in changing market conditions".

According to a literature survey by Kantarci and Mouftah (2011) of wireless sensor networks (WSNs), they will play a key role in the extension of the smart grid technology toward residential premises and enable various demand and energy management applications. The authors concluded that the packet delivery ratio, delay, and jitter of the wireless sensor home area network (WSHAN) improve as the packet size of the monitoring applications, that also utilize the WSHAN, decreases. Kanchev et al. (2011) proposed a model that aims for a deterministic energy management system for a microgrid, including advanced PV generators with embedded storage units and a gas microturbine. In conclusion, the authors suggested that the hypothetical business cases associated with smart grids and distributed resource integration provide more value to micro-grid management.

Pawar and Panduranga (2019) conducted a study about the design and development of a flexible Smart Energy Management System (SEMS) for optimal power negotiation and concluded that the design of a smart energy management system aims to replace the scenario of a complete power outage in a region with partial load shedding in a controlled manner as per the consumer's preference.

Rathor and Saxena (2020) summarized the studies relevant to the energy management system for a SA, that presents an overview and key issues. Hence, this review paper gives a critical analysis of the distributed energy resources behavior and different programs such as demand response, demand-side management, and power quality management implemented in the energy management system. Added to that Xu et al. (2016), Prakesh and Sherine (2017), and Etesami et al.

(2018a, b) proposed different approaches and novel models related to deploying efficient, smart energy grid management techniques.

Trefke et al. (2013) investigated the smart grid architecture model (SGAM) used for case management in a European Smart Grid Project. The authors highlighted the importance of determining the Key Performance Indicators (KPIs), which are closely related to having the highest efficient energy production.

Gottschalck et al. (2017) investigated SGAM methodology in a broad sense. In the study, the methodologies used under this model were examined in detail, and the usage of these methodologies was exemplified with case study examples. Uslar et al. (2019a, b) reviewed the studies in the literature under the title of "Applying the SA Architecture Model for Designing and Validating System-of-Systems in the Power and Energy Domain: A European Perspective." The authors provide a comprehensive overview of the state-of-the-art and related work for the theory, distribution, and use of the aforementioned architectural concept in Europe.

A limited number of studies (Pratt et al. 2010; Fu et al. 2012; Darby et al. 2013) have been found on the SA architecture on low carbon emission and energy use. Pratt et al. (2010) reported the reduced CO_2 benefits as the result of deploying SGAM, which are based on a survey of published results and simple analyses. Fu et al. (2012) investigated the potential use of SGAM in China, and Darby et al. (2013) examined the potential carbon impacts of smart grid development in some European countries. Moreover, as for Colombia, Rey et al. (2013) and Roldan et al. (2013) investigated the potential use of SGAM and for Turkey, Colak et al. (2014) presented the SA opportunities and applications relevant to the SGAM approach.

Current Situation for Smart Energy in Turkey and Colombia: Turkey

In Turkey, smart energy investments in the public utility sector are maintained by the government and stakeholders. Smart directives requiring the use of energy are another element that has led to the increase of investments in Turkey's energy sector. For instance, Northeast Group conducted a study in 2016 in Turkey with the eastern and central European countries (Keskin 2021) that focused on SA applications. Approximately $ 25.2 million investment has been made focused on eliminating power outages and reducing the power transmission and distribution losses. According to this report, the government's "intelligent distribution automation systems" in the distributed renewable energy resources will play an enabling role in promoting the use of rechargeable electric vehicles.

Smart Meter Systems

According to the 2009/72 / EC Electricity Directive, smart meter usage of countries in the European Union (EU) region requires a transition to 80% by 2020. In line with this directive, Turkey has committed to increasing investment in smart metering infrastructure to greatly reduce energy losses. A 2017 Frost & Sullivan report forecast that Turkey would need to install 3.6 million smart meters per annual installation of smart meters to meet the 2020 EC directive. According to this report, Turkey already surpassed this target in 2016 installing 4.6 million smart meter units (Nhede 2017).

However, in Turkey, the lack of standardization and public awareness of smart meter technologies impedes their successful use. As a result, in the first phase, intelligent systems are introduced in Turkey through pilot projects. A research analyst at Frost & Sullivan says: "The legislation provides the necessary infrastructure for the transition from old meters to electronic meters. Since the Turkish authorities do not have sufficient technical expertise in smart electricity, international companies are setting an example in this regard."

An Example

Enerjisa (Electricity Distribution Company of Sabanci Holding) serving 20 million consumers is Turkey's largest energy distribution service company and is deemed to provide a "public service." The European Bank for Reconstruction and Development (EBRD) forecast that $100 million needed to be invested in Turkey for intelligent-energy infrastructure, and Enerjisa has invested $ 28.6 million. According to The Financial Times, this investment by the private energy sector in Turkey aims to provide the advanced work of the three leading energy companies in the country. According to Power grid International, since 2016 to achieve clean energy in Turkey there has been a significant increase in investment in digital technologies, and this has led to Turkey's first digital power plant being established.

Energy Diversity

Today, 33% of Turkey's energy comes from natural gas. According to (Daily Sabah 2020) Energy, the Turkish government aims to generate 30% of its energy from renewable sources, including solar (solar) and wind, by 2023. Although the government does not fully plan to use solar and wind resources as the main source of energy production, it aims to meet a significant proportion of the country's energy demand from natural gas for a long time.

In April 2017, Wärtsilä, independent energy producer Yeşilyurt Enerji Elektrik Üretim A.Ş. signed an agreement to modernize a 73 MW natural gas facility. Within the scope of developments in the Solar PV (photovoltaic) industry, ET Energy is carrying out its 19 MW solar project in Kahramanmaraş, located in the south of Turkey. In October 2017, it was announced that the project will have an annual power generation capacity of 37,000MWh. According to PV Tech, this project will act as a supplement to the base-load energy generation to meet the country's peak demand.

A Turkish roadmap report, 'Turkey Smart Grid 2023 Vision and Strategy Determining Project, Short and Mid-Term', prepared for Electricity Distribution Services Association (Elder 2020) by AF MERCADOS EMI, collated information in 2014 from projects focused on research and development aimed at electrical energy infrastructures with advanced measurement, implementation of pilot plans with smart meters and communication controllers. The report developed a plan based on this information to focus technical efforts on the following topics:

- Advanced network monitoring, control, and management systems
- IT infrastructure and data analytics
- Enterprise application integration
- Distributed energy integration and storage
- Smart grid company vision and strategy
- Geographical Information Systems (GIS) and asset management
- Electric vehicles
- Customers and smart meter infrastructure
- Communication infrastructure
- Cyber security

To concentrate this work and research effort, it has been proposed to group these into 3 categories: (Elder 2020)

1. Smart Network Management: This category is aimed at the activities of distribution companies including among others administration, management, model design, and business processes, that are subject to advanced supervision of the network through an automated SCADA (Supervision, Control, and Data Acquisition) control system. This allows the activities to be both coordinated, examined, and evaluated in an automated way for the different operating scenarios and situations that may arise with the proper functioning between the power generation station(s) and the distribution feeders.
2. Smart Embedded System: This category is defined for components that are built or designed for the integration of low or medium voltage of electric current, and has a fundamental objective to achieve decreased carbon emissions. In the document, they are discussed at different scales and focus on electrical energy already in the distribution phase, i.e., associated with customer energy access, for example, energy storage in batteries, power plants, electric vehicles, etc.
3. Smart Markets and Customers: This category relates to elements that have to do directly with the consumption of electricity, for example, smart meters, management indicators such as balances in supply and demand, energy trading, and consumption functions, communications, the use of information and

communication technologies, the physical and logical security of devices. The highest investment costs that are incurred in this category are usually in infrastructure and devices and this is where the most innovation and development is required.

Running across these categories, it is proposed that there should be a component of training and strengthening in technical and systems competencies from academia, entrepreneurship, technology-based companies supported by industry to create technological prototypes that improve and allow a harmonious balance between energy consumption and environmental conservation. For Turkey as a nation, it is essential to reduce its carbon footprint and that distribution company reduces the levels of loss by technical and non-technical part of the energy flows while increasing network sustainability and energy quality. In parallel, competence development needs to support increased responsiveness and resilience of the network, and awareness of the situation of the network by companies and consumers.

Aligned to this competence development is the promotion and support of research and innovation in academic and industrial sectors to create devices and systems that support generation processes, measurement, and distribution with sophisticated technologies, that are equitable with the environment and real-time operation. These technologies need to be able to support the increase in capacity through the interconnection of renewable energies. The soft issues that these technologies must support include strengthening of strategies of prosumer communities (simultaneous consumers and generators of electrical energy), all in a constructive approach to a future where environmental sustainability and energy supply are economically viable for industrial and domestic consumers. These characteristics need to be in place to support continued growth in industrial and social prosperity for the participating countries.

Colombia

In Colombia the formulation of the document "Smar Grid Colombia Vision 2030", for the Colombian context, an analysis of the international context is carried out, where developments of countries such as Australia, Canada, the United States, Japan, China, South Korea are observed. Where it is noted that in the aforementioned nations there are elements in common, oriented to political-social and organizational elements, in particular, the ensuring the provision of a quality electricity supply, allowing economic growth, solutions aligned to the challenges facing Colombia in the face of the fulfillment of actions for the sustainable development goals.

The document "Study: Smart Grid Colombia Vision 2030 - Roadmap for the implementation of smart grids in Colombia" Created with financial funds from the Korean Fund for Technology and Innovation. Created within the framework of technical cooperation ATN-KK- 14254-CO (CO-T1337) with the Inter-American Development Bank - IDB, the Ministry of Mines and Energy, and the Ministry of Information and Communication Technologies. It is consolidated in 4 large parts:

1. General view and a summary of the analysis
2. Road map of the implementation of smart grids in Colombia
3. Analysis of policies and regulation of the electric sectors
4. 9 annexes that support the study.

With the consolidation of this document, true information is obtained on the current situation in Colombia and the real possibilities of the implementation of smart grids.

Smart Meter Systems

In Colombia, a limited number of distribution and commercialization companies have carried out pilots that include smart meter systems at different levels of development in cities such as Bogotá, Medellín, and Cali, however, they are isolated projects that are not framed within the guidelines of national order. There is also a roadmap that defines gradual implementation phases for the systems in Colombia by 2030. However, according to the (OECD 2014) in its study on telecommunications policies and regulation, it shows Colombia is the 3 country that advances the fastest in terms of interconnectivity, especially in mobile technology.

Several motivating elements stand out to guide a clear path to the 2030 vision, the elements are:

1. Improving the efficiency of electrical systems
2. Using and/or applying renewable energy standards and objectives
3. Improving the reliability of electricity systems
4. Promoting research and innovation with new products, services, markets
5. Generating environments for communities to be proactive
6. Optimize the use of network assets

On the issue of infrastructure, the development and strengthening of an AMI is a global priority, followed by the integration of the SGAM - SA architecture model, where most studies highlight the importance of countries having an energy policy composed of elements such as electrical energy assurance, environmental impact, conditions for electricity generation and independence, becoming prosumers, achieving a stable, flexible, and a sustainable energy supply.

The most relevant projects that are being developed with this approach are for instance, in Bogota, the Smart Metering Pilot of the Codensa company, as a result, the measurement systems have been normalized and the operation costs reduced, it is in the feasibility stage. In Antioquìa, the multiservice smart metering pilot project of the company Empresas públicas de Medellìn -EPM has impacted 1000 clients, improving the quality of service and identifying commercial energy losses and costs. In the company Internexa, the SA telecommunications project has obtained a business model and an interconnection regulatory framework, improving the telecommunications backbone (UPME 2016).

An Example

Currently, the world has been transforming the processes of the energy sector, this is how the International Renewable Energy Agency (IRENA 2016) stated that the costs of photovoltaic, solar, and wind electricity, among others, will continue to fall accordingly to the combination of increasing economies of scale, supply chains, and competitive environments. According to "Smart Grids Colombia Vision 2030," wind power is 29.5 gigawatts, solar energy is around 17 gigawatts, hydroelectric is 5 gigawatts, and biomass is 4 gigawatts. The Caribbean region and La Guajira have attracted academic and investment projects due to their political-demographic characteristics that facilitate the application of techniques, procedures, and processes to achieve project objectives.

Energy Diversity

In these regions the trend of use of architecture is DER to optimally manage operations, also to monetize the contribution of the system to customers, below is Fig. 1 that presents the roles of the integrators, optimizers, and aggregators.

To come to this view Colombia is implementing smart network projects in different parts of the country, where the implementation of AMI is illustrated in Fig. 2.

Projects will include Pilot smart metering, energy measurement technology, intelligent supervision and advanced control (iSAAC) Phase III, Multiservice Intelligent Measurement Pilot Project, IPV6 Protocol Study in the Distribution Domain Data Model in SA, Interconnection in Backbone Access Segments in Telecommunications Network, Implementation of the Measurement Management Center, software architecture for energy consumption management and measurement.

Future of the Applications Both in Colombia and Turkey

Two countries must have a clear roadmap where the sustainable development objectives are articulated, where strengthening is sought through the same education as that established in the 2030 agenda to provide and strengthen people in their knowledge, to reach their full potential within a framework of equity and dignity. Turkey had just signed the Paris Agreement when this work was written. As a result, Table 1 contains information that was current at the time the agreement was not signed. Therefore, Table 1 presents the trend about energy consumption is on the rise ranking among the top 20 countries with the highest consumption

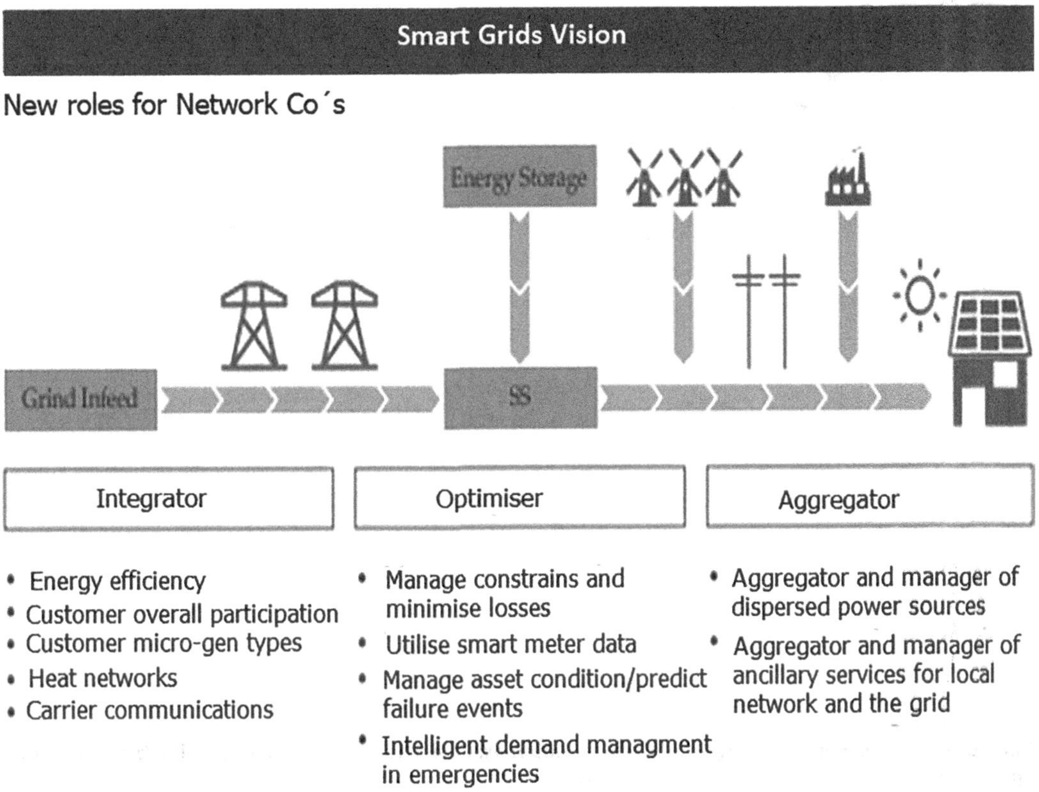

Smart Grids to Lower Energy Usage and Carbon Emissions: Case Study Examples from Colombia and Turkey, Fig. 1 SA Vision Source: Belmans (2010) (UPME 2016)

energy, emission levels increased 425 million metric tons of carbon dioxide in 2019, it is necessary and urgent to create an ecosystem that involves the government, companies, universities, organizations, among others, for strategic decision-making in the assurance of the economy, environment, regulations, education, policies.

In this framework, the future of applications focused on sustainable development and the possible opportunities for improving energy consumption. Oğuzhan (2014) stated that the Turkish government should focus its regulatory efforts on an energy policy that is linked to sustainable development, having three fundamental pillars: social equality, economic efficiency, and guaranteeing the ecological carrying capacity. Similarly, Erdin and Ozkaya (2019) also stated that the pressure on the environment of greenhouse gases released because of the use of fossil fuels complicates the guarantee of economic development in Turkey and puts in clear danger the existence of future generations. As a result, electrical energy must be considered a consumer commodity that is required for the day-to-day operations of business, community, and various areas of globalization. Consequently, it is concerning that the amount of energy consumed continues to rise, posing a threat to human well-being (Connoly and Prothero 2008).

This is where a big potential to construct a smart grid prototype that allows an appropriate design to take advantage of renewable energies and produce electric power prosumers arises because renewable energies play a significant role in ensuring sustainability. As mentioned, Lund (2010) explained that renewable energy can provide sustainable energy and energy security, generating an intensive global energy future that is technically possible. The government of Turkey is currently seeking to increase the

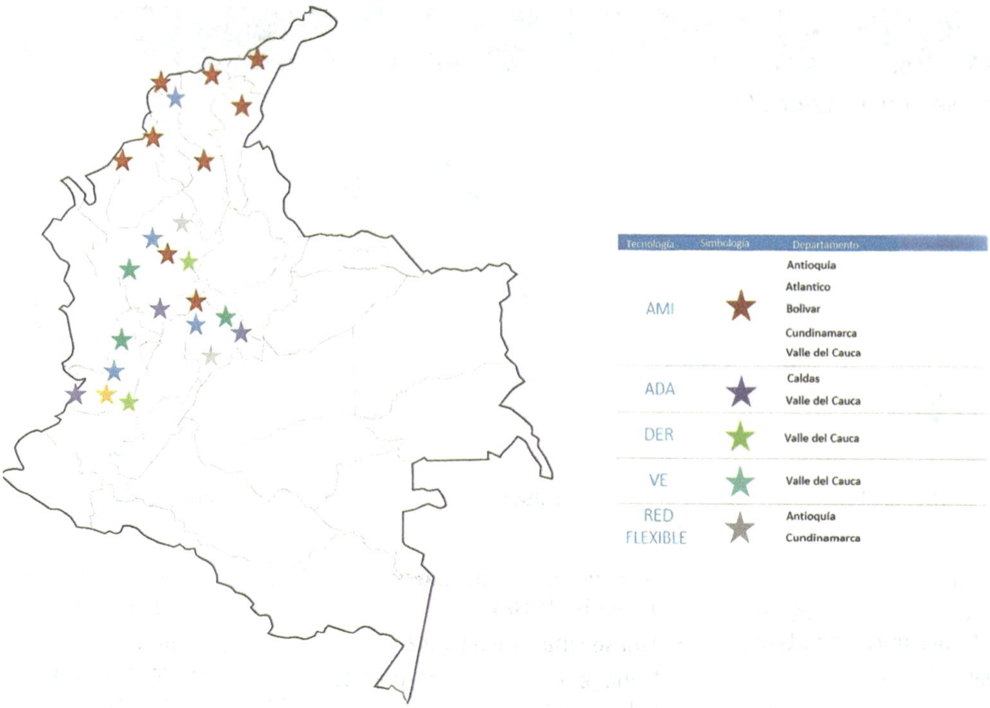

Smart Grids to Lower Energy Usage and Carbon Emissions: Case Study Examples from Colombia and Turkey, Fig. 2 AMI infrastructure in Colombia (UPME 2016)

Smart Grids to Lower Energy Usage and Carbon Emissions: Case Study Examples from Colombia and Turkey, Table 1 Turkey – Electricity consumption

Date	Generation GWh	Consumption GWh	Consumption per capita kWh
2019	290.442	251.376	3.023,0
2018	290.386	258.116	3.147,6
2017	283.086	247.851	3.067,1
2016	261.850	231.117	2.895,7
2015	249.245	216.658	2.751,5

Source: (IEA 2020)

proportion of renewable energy resources (RES), having an increase in an installed capacity greater than 30% of what is currently registered, this indicator has as a goal of compliance with the year 2023, it is contemplated in the process of investing approximately USD 110 billion (Erdin and Ozkaya (2019). Turkey represents one of the fastest-growing countries for the energy market (OECD 2019). There are a variety of options available in Turkey to create an alternative energy system that takes advantage of renewable energies while meeting energy demand, but this requires a long-term energy policy (Oğuzhan 2014). To do so, the country must prepare human talent with technical skills, soft skills, and transversal training that allows them to understand energy problems in the context of communities, from which assertive solutions for generation, monitoring, and control can be developed.

Colombia, for its part, is not far from the search for technological solutions appropriate to its context, to achieve the country's energy efficiency. According to Acosta Pérez (2019), Colombia is one of the biggest challenges is the incorporation of the consumer as an active role that changes the paradigm from only consumption to the generation of electrical energy, and its responsible use to this new role has been called "prosumer" (Castro et al. 2017). For this, an opportunity arises in the training and culture processes hand in hand with the implementation of a smart grid with robust characteristics in its architecture of self-management, cybersecurity, active consumer, asset optimization. To achieve effective proposals that contribute to a properly functioning electrical network, you must have an ICT infrastructure that facilitates and supports applications, for this, you must take into account prioritization elements:

1. Strengthen education and culture in community contexts against the energetic ecosystem.
2. Achieve the inclusion and implementation of renewable energy systems.
3. Systematization of energy consumption, through smart grid architectures.
4. Security, integrity, and privacy of information.

Consistent with the prioritization that was obtained from the review of documents from the Ministry of Mines and Energy, UPME, IPSE, and other entities directly related to the Colombian energy system, it is recognized that education, training, and awareness toward communities and/or end-users. It will be the most important exercise since the impact of the projects will bring greater benefits.

Smart Grids to Lower Energy Usage and Carbon Emissions: Case Study Examples from Colombia and Turkey, Table 2 Benefits and risks of implementing smart grids

Benefits	Risks
Creation of new jobs, economic improvement of the region.	It is necessary to strengthen policies or regulations against the modernization of conditions and other elements for the implementation of the smart grid.
Investment in infrastructure and more efficient technology.	Initially, investments in infrastructure, technology, and others have high costs.
Reduction of greenhouse gas emissions and protection of the environment.	It is observed the need for the smart grid to coexist with traditional networks, for a long time, until they are fully implemented.
Generation, distribution, and energy consumption with adequate monitoring and control.	Lack of qualified human talent to participate in the execution of smart grid implementation projects.
Updated information on supply and demand, additionally control of the quality of the service.	Insufficient parts, spare parts, or components in Colombia and/or Turkey to be able to massif the implementation of smart grid.
Energy management and administration through intelligent equipment that is reflected in energy efficiency and economic benefits.	Communities that inhabit the countries with distrust and doubt in the face of the maturity of the technology and its proper use in the territory.

Results and Discussion

Colombia and Turkey are collaborating to shape a new future for each of their countries, with smart grids being addressed in studies and research. There are also ongoing projects that are having a good influence. Smart grid deployment provides scalability, dependability, security, sustainability, and competitiveness, to name a few benefits. Table 2 shows the advantages and risks that smart grid adoption has brought to Colombia and Turkey.

The literature review presents that, Colombia and Turkey the path toward the energy transition is on the way, the planned projects, documents, norms, and laws allow migrating to the implementation of smart grids to reduce the use of conventional energy and thereby decrease carbon emissions. Among these, the appropriation of the AMI architecture stands out, as an effective transition strategy, which has been generating good results and tangible benefits such as those mentioned in this document.

For the two nations, there are future challenges such as the consolidation of public policies and a robust legal structure that allows a series of benefits to the end-user community of the energy service and to the industry that implements new technologies and architectures such as those mentioned in this document. Another great challenge is the intelligent control of energy storage that occurs in the regions, for which the technologies of Industry 4.0 play a fundamental role in terms of the optimization and quality of the devices that are involved in the smart grids. An additional challenge is the formulation of strategies for energy diversity. To this extent, countries must provide conditions for disruptive processes that allow the teaching-learning of local human talent that can face the challenges that arise for the energy transition and industry 4.0 learning from a perspective and worldview of the context. Therefore, the characterization of communities should be the starting point before planning any type of project. Colombia and Turkey have a wide horizon and great possibilities to take advantage of their natural resources and thus reduce carbon emissions and greenhouse gases, but a strong commitment is necessary at all levels, that is, from the government to the most remote community of the territory, and this process will be achieved in a phased manner, applying the correct technologies, understanding the context, and having legal support for it.

Proposing a Data Structure Architecture for Smart Grids

This section presents the necessary data structure architecture for smart grids. Therefore, the topics about big data, data structure are discussed, respectively.

Smart Grids and Big Data

Smart grid is the next-generation energy system capable of managing electricity demand, supply, and efficiency by using advanced digital information and communication technologies to generate the entire system's higher performance and efficiency (Diamantoulakis et al. 2015). The advance in big data trends and data processing makes its relationship with networks increasingly close and, in the same sense Wilcox et al. (2019), presents that the operation of smart grids and future energy management will be increasingly data-driven.

While Munshi and Mohamed (2017) argues that, the massive amount of data evolving from smart grids must be sufficiently managed, respectively. This is the great challenge from data science that requires advanced IT infrastructures, techniques, challenges associated with interoperability, storage, security, and integrity to deal with the amounts of data and their analysis generated from intelligent energy management systems. Big Data technology encompassing the analysis and transformation of data into useful knowledge with the help of different ICT technologies has become a priority development area for companies in general.

For organizations concerned with the generation and supply of electricity, this process is of great relevance (Colmenares-Quintero et al. 2021). Data analysis allows organizations to be proactive, forward-looking, anticipating outcomes, and behaviors based on data rather than hunches or assumptions (Kannan et al. 2018). It consists of theory-based mathematical models, such as regression models, decision trees, and neural networks or data-driven models, such as clustering or segmentation, that support the development of data mining and machine learning (Cattaneo et al. 2018) for descriptive and predictive analysis that are useful for decision-making (Hastie et al. 2008).

Data Architecture

Organizations today face both challenges and opportunities arising from the exponential generation and availability of data such as web application data, historical transaction data, and Internet sources.

According to Cai and Zhu (2015), Zhang et al. (2018), the characteristics of big data in smart grids are also following the universal modus operandi of big data of 5 Vs: Volume, variety, speed, veracity, and value.

Architecture describes the basic structure with its elements, the relationships between these elements, and the system's relationships with the environment (ISO 2000). It describes the principles for the design, development, and use of the system (IEEE 2000). Data architecture represents the data structure of an information system where entity types, modeling, storage, and processing of data and their relationships with each other are shown.

Data architecture is "useful for analyzing large existing data systems, providing the basis for the classification of data analysis processes and technologies. The classification of processes, technologies, and services into groups (components) further facilitates decision-making regarding the implementation of system processes and functionalities" (Sang et al. 2017).

Similarly, big data architectures may refer to the necessary data structures of an entire enterprise as the enterprise data architecture or refer to an application system's data architecture in a section of the enterprise (Mohammad et al. 2014). These architectures comprise an abstract view of the systems and the role of the various system components, their behavior, and how they interact with each other (Calheiros 2018).

Big data analysis requires all forms of different information and communication technologies to be related in an integrated analysis environment. Large-scale data architecture provides the framework for reasoning with all forms of data, consequently contemplates models, abstracts, and rules that direct how data should be stored, ordered, coupled, and implemented in data systems in an application domain.

Reference Models for Big Data Architecture

Table 3 presents the reference models. These models are also mostly presented in layers such as: (1) Data Source; (2) Storage, Processing, and Loading; (3) Data Analysis and Visualization, regardless of the names assigned to them.

Challenges in Designing Big Data Architectures for Smart Grids

there are various challenges related to big data and smart grids, such as ranging from the integration of different hardware, software, and communications technologies with different energy technologies such as wind, solar, biomass, or water that are acquiring a higher level of maturity and development. The interoperability of the different systems deployed on the smart grid makes it complicated and difficult to obtain data for their real application (Zhang et al. 2018).

Recently, software-defined cloud computing and networking technologies have proven to be very useful for efficiently implementing big data solutions: further work is needed to ensure that computing and networking facilities are scaled up to the ever-increasing scalc of the data (Ali et al. 2018). Another important aspect is the need for

Smart Grids to Lower Energy Usage and Carbon Emissions: Case Study Examples from Colombia and Turkey, Table 3 Reference Models Big Data Architectures

Reference model	Data source	Data storage, processing, and loading	Data analysis and visualization	
Microsoft	Data sources	Data transformation Data infrastructure	Data usage	
Big data architecture framework (BDAF)	Data models, structures, and types	Big data management infrastructure and services	Big data analytics and tools	
IBM reference architecture	Data sources	Streaming computing Data integration	Analytical sources	Actionable insight
ORACLE	Data sources	Infrastructure services Big data processing and discovery	Information analysis	
PIVOTAL	Capture infrastructure	Data storage and analytics Big data applications	Big data applications	
SAP big data architecture	Ingest	Store and process	Consume	
NBDRA	Collection	Preparation, curation Storage Data organization and distribution Computing and analytic	Analytics, visualization, access	

Source: (NIST 2015)

the human capital skills involved in these processes (technicians, analysts, and end-users) to generate value for the information analyzed and obtain relevant decision-making impacts. More emphasis on aspects such as business models from the user's perspective, consumer-oriented applications, best practices and process models, and the commercialization of large data analyses are possible trends.

The literature review about challenges in designing big data addressed that legal and ethical aspects such as regulation, privacy, and data security have emerged as remarkable issues in recent years. Such scenarios include cyber-attacks, metadata falsification, packaging, and phishing attacks are evident. These scenarios must be supported by legal issues that are a challenge for countries where the development of these technologies advances faster than the evolution of laws and regulations.

Conclusion

The study discusses how decreased energy usage and carbon emissions might be accomplished in Colombia and Turkey utilizing smart grid architectural models relevant to SDGs to fill a gap in the literature. Along with the smart grid models, Education 4.0, Energy 4.0, Artificial Intelligent Systems, and POWERBI (processing analysis) are all needed to attain this goal. On the other hand, the acceleration of technology adaptation in various communities has increased due to the pandemic generated by COVID-19. This pandemic period also created great opportunities to face the challenges of sustainable development. It is therefore essential to think about strengthening the pedagogical models in the processes of teaching defined in plans or curricular models that train children, adolescents, and young people.

In Colombia, through different institutions, organizations, among others, the option to improve the conditions and quality of life of its inhabitants is promoted by putting the SDGs into practice, based on the premise that the environment, society, and the economy, must interact together for optimal development of coexistence and respect for our ecosystem. Meanwhile in Turkey, the shift to digital infrastructure in energy production and consumption has accelerated, smart grid technologies have begun to be deployed, and the use of innovative technologies such as digital games for future generations' education has begun to expand in educational institutions.

Cross-References

▶ Big data for smart cities and inclusive growth
▶ Smart city: A universal approach in particular contexts

References

Abdulla. (2015). The deployment of advanced metering infrastructure. In *First workshop on Smart Grid and Renewable Energy (SGRE), 2015,* (pp. 1–3). https://doi.org/10.1109/sgre.2015.7208738

Acosta Pérez. (2019). Identificación de retos tic de los consumidores como actores activos en el marco de SA y propuesta de estrategia para afrontarlos en el contexto colombiano.

Ali, M. B., Wood-Harper, T., & Mohamad, M. (2018). Benefits and challenges of cloud computing adoption and usage in higher education: A systematic literature review. *International Journal of Enterprise Information Systems (IJEIS), 14*(4), 64–77. https://doi.org/10.4018/IJEIS.2018100105.

Beck, M. T., Fischer, A., Botero, J. F., Linnhoff-Popien, C., & de Meer, H. (2015). Distributed and scalable embedding of virtual networks. *Journal of Network and Computer Applications, 56*, 124–136.

Belmans, R., Buijs, P., & Bekaert, D. (2010). Seams issues in European transmission investments. *The Electricity Journal, 23*(10), 18–26.

Bodek, K. (2018). *Just-in time at Toyota*. New York: CRC Press/Routledge. https://doi.org/10.1201/9780203749715.

Brown, R. E (2008). Impact of smart grids on distribution system design. In *IEEE power and energy society general meeting-conversion and delivery of electrical energy in the 21st century, 2008*, (pp. 1–4). https://doi.org/10.1109/pes.2008.4596843

Cai, L., & Zhu, Y. (2015). The challenges of data quality and data quality assessment in the big data era. *Data Science Journal, 14*.

Calheiros, R. N. (2018). Big data architectures. *Encyclopedia of Big Data Technologies, 1–7*. https://doi.org/10.1007/978-3-319-63962-8_39-1.

Cattaneo, L., Fumagalli, L., Macchi, M., & Negri, E. (2018). Clarifying data analytics concepts for industrial engineering. *IFAC-PapersOnLine, 51*(11), 820–825.

Castro, N., Alves, J., Dantas, G., & Ferreira, D. (2017). Estado da arte da difusão de recursos energéticos distribuídos em quatro estados norte-americanos. Texto de discussão do setor elétrico, (72).

CENELEC. (2014). *The basic standard for the in-situ measurement of electromagnetic field strength is related to human exposure in the vicinity of base stations. European standard EN 50492:2008/A1: 2014.* European Committee for Electrotechnical Standardization: Brussels.

Colak, I., Bayindir, R., Fulli, G., Tekin, I., Demirtas, K., & Covrig, C.-F. (2014). SA opportunities and applications in Turkey. *Renewable and Sustainable Energy Reviews., 33*, 344–352. https://doi.org/10.1016/j.rser.2014.02.009.

Colmenares-Quintero, R. F., Valderrama-Riveros, O. C., Macho-Hernantes, F., Stansfield, K. E., & Colmenares-Quintero, J. C. (2021). Renewable energy-smart sensing system monitoring for an off-grid vulnerable community in Colombia. *Cogent Engineering, 8*(1), 1936372.

Connolly, J., & Prothero, A. (2008). Green consumption: Life-politics, risk and contradictions. *Journal of consumer culture, 8*(1), 117–145.

Daily Sabah. (2020, February 13). *Renewables account for almost half of turkeys installed power.* https://www.dailysabah.com/energy/2020/02/13/renewables-account-for-almost-half-of-turkeys-installed-power

Darby, S., Strömbäck, J., & Wilks. (2013). M. Potential carbon impacts of SA development in six European countries. *Energy Efficiency, 6*, 725–739. https://doi.org/10.1007/s12053-013-9208-8.

Diamantoulakis, P. D., Kapinas, V. M., & Karagiannidis, G. K. (2015). Big data analytics for dynamic energy management in SAs. *Big Data Research, 2*(3), 94–101.

EEE. (2000). Intelligent network workshop. https://doi.org/10.1109/inw.2000.868159.

Elder. (2020). *Turkey SA 2023 vision and strategy roadmap summary report.* http://www.elder.org.tr/Content/yayinlar/TAS%20EN.pdf, www.smartgridturkey.org

EPA. (2020). Retrieved from. https://www.epa.gov/ghgemissions/sources-greenhouse-gas-emissions#:~:text=Electricity%20production%20(26.9%20percent%20of,mostly%20coal%20and%20natural%20gas

Erdin, C. Y., & Ozkaya, G. (2019). Estrategias energéticas de Turquía para 2023 y oportunidades de inversión para fuentes de energía renovable: selección del sitio basado en electre. *Sostenibilidad, 11*(7), 2136.

Etesami, S. R., Saad, W., Mandayam, N. B., & Poor, H. V. (2018a). Stochastic games for the SA energy management with prospect prosumers. *IEEE Transactions on Automatic Control, 63*(8), 2327–2342.

Etesami, Saad, W., Mandayam, N. B., & Poor, H. V. (2018b). Stochastic games for the SA energy management with Prospect prosumers. *IEEE Transactions on Automatic Control, 63*(8), 2327–2342. https://doi.org/10.1109/TAC.2018.2797217.

Fu, L., et al. (2012). An analysis on the low-carbon benefits of SA of China. *Physics Procedia, 24*, 328–336.

Gottschalck, M., et al. (2017). *The use case and SA architecture model approach.* Cham: Springer.

Hanser. (2010). *Agile Prozesse: Von XP über Scrum bis MAP.* Berlin\Heidelberg: Springer. https://doi.org/10.1007/978-3-642-12313-9.

Hastie, T., Tibshirani, R., & Friedman, J. (2008). *The elements of statistical learning. Data mining, inference, and prediction* (2nd ed.). Springer.

Highsmith, & Cockburn, A. (2001). Agile software development: The business of innovation. *Computer (Long. Beach. Calif), 34*(9), 120–127. https://doi.org/10.1109/2.947100.

Hu, W., Zhou, H., Lai, J. & Deng, Q. (2014). Demand-side energy management: FTTH-based mode for smart homes. In *2014 American Control Conference* (pp. 1704–1709). IEEE.

IEA. (2020). *World energy balance 2020.* https://www.iea.org/countries/turkey

IRENA. (2016). *International renewable energy agency* https://www.irena.org/

Iso 20000. (2008). Process Improvement with CMMI® V1.2 and ISO Standards, 359–395. https://doi.org/10.1201/9781420052848.axg.

Jacobson, I., Spence, I., & Kerr, B. (2016). Use-case 2.0. *Communications of the ACM, 59*(5), 61–69.

Kanchev, H., Lu, D., Colas, F., Lazarov, V., & Francois, B. (2011). Energy management and operational planning of a microgrid with a PV-based active generator for SA applications. *IEEE Transactions on Industrial Electronics, 58*(10), 4583–4592.

Kannan, N., Sivasubramanian, S., Kaliappan, M., Vimal, S., & Suresh, A. (2018). Predictive big data analytic on demonetization data using support vector machine. *Cluster Computing.* https://doi.org/10.1007/s10586-018-2384-8.

Kantarci, E., & Mouftah, H. T. (2011). Wireless sensor networks for cost-efficient residential energy management in the SA. *IEEE Transactions on SA, 2*(2), 314–325. https://doi.org/10.1109/TSG.2011.2114678.

Keskin, M. (2021). *SAS and Turkey: An overview of the current power system and SA development.*

Kolberg, & Zühlke, D. (2015). Lean Automation enabled by Industry 4.0 Technologies. *IFAC-PapersOnLine, 48*(3), 1870–1875. https://doi.org/10.1016/j.ifacol.2015.06.359.

Lee, Y. et al. (2012). *SA and its application in sustainable cities*, Technical note 446, Inter-American Development Bank.

Lund, H., Connolly, D., Mathiesen, B. V., & Leahy, M. (2010). A review of computer tools for analysing the integration of renewable energy into various energy systems. *Applied energy, 87*(4), 1059–1082.

McGranahan, G., Marcotullio, P., Bai, X., Balk, D., Braga, T., Douglas, I., Elmquist, T., Rees, W., Satterwaite, D., Songsore, J., & Zlotnik, H. (2005). Urban systems. In

S

R. Hassan, R. Scholes, & N. Ash (Eds.), *Ecosystems and human well-being. volume 1: Current state and trends*. Washington: Island Press & Millennium Ecosystem Assessment.

Milborrow, S. (2016). *Multiview active shape models with SIFT descriptors*.

Mohammad, A., Mcheick, H., & Grant, E. (2014). Big data architecture evolution: 2014 and beyond. In *Proceedings of the fourth ACM international symposium on Development and analysis of intelligent vehicular networks and applications* (pp. 139–144).

Mohassel, A., Fung, F. M., & Raahemifar, K. (2014). Application of advanced metering infrastructure in SAs. In *22nd Mediterranean conference on control and automation 2014*, (pp. 822–828). https://doi.org/10.1109/med.2014.6961475.

Munshi, A. A., & Mohamed, Y. A. R. I. (2017). Extracting and defining flexibility of residential electrical vehicle charging loads. *IEEE Transactions on Industrial Informatics, 14*(2), 448–461.

Nhede. (2017). *Analysis: Smart energy investments in Turkey Turkey*. https://www.smart-energy.com/features-analysis/analysis-turkey-energy-market/

OCDE. (2014). *Estudio de la OCDE sobre politicas y regulación de telecomunicaciones en Colombia*. Paris OECD Publishing.

OECD. (2019). OECD Environmental performance reviews: Turkey 2019. https://www.oecd.org/turkey/oecd-environmental-performance-reviews-turkey-2019-9789264309753-en.htm

Oğuzhan. (2014). B. A. T. I. Türkiye'de Yenilenebilir Enerji Kaynaklarinin Sürdürülebilir Kalkinmaya Etkisi Konusunda Bir Alan Araştirmasi. *Trakya Üniversitesi Sosyal Bilimler Dergisi, 16*(2), 27–38.

Oral, H. V., Carvalho, P., Gajewska, M., Ursino, N., Masi, F., Hullebusch, E. D. V., . . . Zimmermann, M. (2020). A review of nature-based solutions for urban water management in European circular cities: A critical assessment based on case studies and literature. *Blue-Green Systems, 2*(1), 112–136.

Pawar, P., & Panduranga, V. (2019). *Performance analysis of a smart meter node for congestion avoidance and Los coverage*.

Prakesh, S., & Sherine, S. (2017). Forecasting methodologies of solar resource and Pv power for Sa energy management. *International Journal of Pure and Applied Mathematics, 116*(18), 313–318.

Pratt, R. G., Balducci, P. J., Gerkensmeyer, C., Katipamula, S., Kintner-Meyer, M. C., Sanquist, T. F., . . . Secrest, T. J. (2010). *The SA: an estimation of the energy and CO2 benefits* (No. PNNL-19112 Rev 1). Richland: Pacific Northwest National Lab. (PNNL).

Rathor, S., & Saxena, D. (2020). Energy management system for SA: An overview and key issues. *International Journal of Energy Research., 44*. https://doi.org/10.1002/er.4883.

Rey, J., Vergara, P. P., Osma-Pinto, G., Ordonez, G.. (2013). *Analysis for inclusion of SAs technology in Colombian electric power system*. https://doi.org/10.13140/RG.2.1.2863.7920.

Roldan, G. M. C. et al. (2013). Characterization model of SA in Colombia, VII International Symposium on Water Quality, 7.

Rua, D. Issicaba, D., Soares, F. J., Almeida, P. M. R., Rei, R. J., & Lopes, J. A. P. (2010). Advanced metering infrastructure functionalities for electric mobility. In *2010 IEEE PES innovative smart grid technologies conference Europe (ISGT Europe), 2010*, (pp. 1–7). https://doi.org/10.1109/isgteurope.2010.5638854

Sang, G. M., Xu, L., & De Vrieze, P. (2017). Simplifying big data analytics systems with a reference architecture. In *Working conference on virtual enterprises* (pp. 242–249). Cham: Springer.

Stellman, A., & Greene, J. (2014). *Learning agile: Understanding Scrum, XP, lean, and kanban*. O'Reilly Media, Inc.

The NIST Big Data Public Workinig Group (NBD-PWG). (2015). https://bigdatawg.nist.gov/home.php. Accessed 11 July 2021.

Trefke, S., Rohjans, M., Uslar, S., Lehnhoff, L. N., & Saleem, A. (2013). *SA architecture model use case management in a large European SA project* (pp. 1–5). Lyngby: IEEE PES ISGT Europe. https://doi.org/10.1109/ISGTEurope.2013.6695266.

UN Environment. (2020). Retrieved from https://www.unenvironment.org/regions/asia-and-pacific/regional-initiatives/supporting-resource-efficiency/sustainable-cities. Accessed 19 Nov 2020.

UPME. (2016). *SAs Colombia vision 2030*. Colombia, Unidad de planeación minero energetica, https://www1.upme.gov.co/Paginas/Smart-Grids-Colombia-Visi%C3%B3n-2030.aspx

Uslar, M., et al. (2019a). Applying the SA architecture model for designing and validating system-of-Systems in the Power and Energy Domain. *A European Perspective, Energies, 12*, 258. https://doi.org/10.3390/en12020258.

Uslar, M., Rohjans, S., Neureiter, C., Pröstl Andrén, F., Velasquez, J., Steinbrink, C., . . . Strasser, T. I. (2019b). Applying the SA architecture model for designing and validating system-of-systems in the power and energy domain: A European perspective. *Energies, 12*(2), 258.

Wilcox, T., Jin, N., Flach, P., & Thumim, J. (2019). A big data platform for smart meter data analytics. *Computers in Industry, 105*, 250–259.

Xu, H., Huang, R., Khalid, S., & Yu, H. (2016). Distributed machine learning-based smart-grid energy management with occupant cognition. In *IEEE international conference on SA communications (SmartGridComm)* (pp. 491–496). Sydney. https://doi.org/10.1109/SmartGridComm.2016.7778809.

Zakariazadeh, A., Jadid, S., & Siano, P. (2014). Smart microgrid energy and reserve scheduling with demand response using stochastic optimization. *International Journal of Electrical Power & Energy Systems, 63*, 523–533.

Zavoda. (2008). The key role of intelligent electronic devices (IED) in Advanced Distribution Automation (ADA). In *2008 China International Conference on Electricity Distribution, 2008*, (pp. 1–7). https://doi.org/10.1109/ciced.2008.5211637

Zhang, Y., Huang, T., & Bompard, E. F. (2018). Big data analytics in SAs: A review. *1*(1). https://doi.org/10.1186/s42162-018-0007-5.

Smart Sustainable City

▶ Green and Smart Cities in the Developing World

Smart(er) Cities

▶ The Governance of Smart Cities

Smart(er) Cities in the Time of Change

Vlada Kenniff[1] and David Mainenti[2]
[1]Long Island University, Brookville, NY, USA
[2]Palmer iSchool of Library and Information Studies, Long Island University, Brookville, NY, USA

Introduction

> The city is a fact in nature, like a cave, a run of mackerel or an ant-heap. But it is also a conscious work of art, and it holds within its communal framework many simpler and more personal forms of art. Mind takes form in the city; and in turn, urban forms condition mind.
> Today our world faces a crisis: a crisis which, if its consequences are as grave as now seems, may not fully be resolved for another century. If the destructive forces in civilization gain ascendancy, our new urban culture will be stricken in every part. Our cities, blasted and deserted, will be cemeteries for the dead: cold lairs given over to less destructive beasts than man. But we may avert that fate: perhaps only in facing such a desperate challenge can the necessary creative forces be effectually welded together. (Mumford 1938)

Urban centers house over 55% of the planet's population, a number that is poised to grow according to both past trends and future estimates (United Nations 2018a). Many urban centers strive to be better, smarter, provide greater services, protect its citizens from natural disasters and emergencies, provide education and health services, and stable infrastructures for water, sewer, transportation and recreation. Governments in many cities have become competitive in attempting to bring meaningful changes that govern the built environment through zoning and/or building codes, and in many ways that may be punitive or simply through the use of transparency and information to prepare such environments for transformative changes.

In many ways, a city is like a living, breathing organism. Until recently, the focus of cities on building out infrastructure were focused on conveyance systems to deliver goods and services: e.g., build or improve roads and transportation systems, deliver water, or take away sewage and garbage. These systems have governing authorities and organizations that raise significant capital, execute projects, provide maintenance, and coordinate repairs and rebuilding. They also establish a cost of service that is equitable and acceptable to all users, calibrating futures based on decades of established programming. While having done this for hundreds of years, in or around 2008, cities became home to more people in urban places than ever before. At the same time, citizens became untethered from wires, resulting in more Internet users preferring broadband over wired ways to access information and communicate. Finally, such users became "sentient" with networked objects in the reaches of cyberspace, consisting of billions of sensors generating data from building walls, city sidewalks, and cars reporting on minutiae of every kind: locations of vehicles, room temperatures, seismic tremors, health information, and so much more (Townsend 2013).

The pursuit of the "smart city" began at the start of the twenty-first century in many urban centers. The concept has wrestled with emerging transformations over the past two decades and rightfully evolved to recognize that the simple presence of information and communication technologies (ICT) is not enough to be "smart." Latest definitions suggest that smart cities "...raise new incarnations of recurring, fundamental questions about city management: who governs, how and what aspects of the city are governable?" These questions have been wrestled with by humanity in every urban center as long as they have existed, but in modern times and today comprise of "technical aids by government entities, citizens, community-based organizations and businesses" (Halegoua 2020).

Many cities have quickly become "smarter" in building out their sentient infrastructure, but governments are still catching up to make sure that information generated is either operationalized or useful for planning. While some regions have been using data generated by both people and sensors to both provide better services and plan for future scenarios that include more volatile weather and climate change, others are still struggling to provide basic services like clean water, housing, roads, education, and health facilities.

Smart(er) Cities in the Time of Change

> The contemporary debate commences with the collective recognition of the fact that digitalization-induced creative destruction is with us, while trying to articulate what creative construction may look like beyond what has materialized in the here and now. Taxi drivers may protest against Uber. Banks may try to survive in a form most familiar to them. Yet such forces can generally be regarded as atavist in nature. That is not to deny the real pain, fears and upheaval these changes continue to induce in people's lives – all these are, of course real, and in many segments of society they have only just started to be felt. But the paradigm shift has already taken place. We have moved beyond a critical crossroads, our world has changed forever, and many of us are trying to comprehend, envision and prepare for what lies ahead. And that is a formidable challenge because, as is the case with all paradigm shifts, the past has stopped to be a source of guidance to predict the future. Many forthcoming shifts may fundamentally change our world and challenge our beliefs and practices in a way a paradigm shift does. The next wave of technological innovations, including artificial intelligence, virtual reality, the harvesting of data and the introduction of "platform" principles are already here.

Boorsma argues that we are witnessing a significant paradigm shift, akin to the waves of the industrial revolutions that began in the late 1700s. If the past is any indication of the future, he may not be wrong. Looking at the industrial revolutions that took place approximately 100 years apart – the first wave was marked by the creation of steam-powered machines. This changed how railways were built as well as bridges and buildings. In the late 1800s, a second wave arrived with electricity, telecommunications, mass production, and urbanization. A third wave took place in the late 1900s introducing electronics, the Internet, automated production, and megacities. Highly populated cities are a relatively new concept in the history of *homo sapiens*. Until the 1800s, most cities remained small, ranging from 5000–15,000 people (Reichental 2020).

For decades, cities and their governing bodies were judged and measured by their ability to deliver hard infrastructure which comprises "tangible (physical) assets" where investment returns are easily measurable (Wataya and Shaw 2019). However, in today's world so intertwined with digital experience, such hard assets alone cannot deliver optimal "people-centered" services. While quality physical infrastructure is critical for a sustainable city, with accelerating population growth and overcrowding, urban communities are requiring delivery of less visible or intangible benefits, broadly "soft assets". The key need in making cities smarter is that the projected future of 68% of humanity will rely on such urban centers by 2050 (United Nations 2018).

Depending on which indicators are used to measure Smart City success, they may fall within a wide range of outcomes. Professor of Architecture Andrea Caragliu notes that the 1990s technology-centered definitions of Smart Cities should be viewed as inadequate in today's urban environment. Instead, modern definitions should focus less on ICTs and more on investment in human and social capital to produce sustainable economic growth, management of natural resources, and participatory governance (Caragliu et al. 2011). Urban centers remain the biggest producers of greenhouse gas emissions and drivers of climate change. While urban disadvantaged communities bear the brunt of extreme weather and flooding conditions, our greatest hope for a solution in using resources most efficiently (Townsend 2013). Recognizing these trends is key to resolving how sentient aspects of urban centers will shape the future of humanity as it wrestles with climate change. Some cities like New York, Vienna, and Amsterdam are using utility and sentient information to set their cities' long-term climate goals and enforce them through the administration of fines. Besides the long-term view, smart cities need to be not only sustainable but also resilient in order to mitigate the impacts of natural hazards, making cities more livable and resilient and, hence, able to respond more rapidly to new challenges (Kunzmann 2014).

In December 2019, a novel and contagious virus broke out in Wuhan, China. That virus has now been identified as a zoonotic coronavirus, similar to SARS coronavirus and Middle East respiratory syndrome (MERS) coronavirus, named COVID-19 (Liu et al. 2020). This latest threat to global health provides a reminder of emerging infectious pathogens and the need for medical surveillance, prompt and accurate diagnosis, and continuous research to understand our susceptibilities of humans and ability to devise effective countermeasures (Fauci et al. 2020). Cities and population health are interlinked (Alirol et al. 2011). Continued urbanization, industrialization, an aging population, climate change, and growing medical costs have placed ever-increasing demands and quality expectations on an already strained healthcare system. Technology can play a key role in improving the way we track and share knowledge regarding positive lifestyle habits, diagnostic trends, treatment recommendations, and prescription efficacy by increasing awareness and prioritizing health and wellness goals. However, along with these exemplary benefits, there is a need to ensure the safety and privacy of each patient's personal and medical data exchanged across connected devices and systems. To truly create a comprehensive smart city (e.g., an urban area that uses different types of electronic methods and sensors to collect data, gather insights, manage resources, and ultimately improve operations), it is essential to have an accessible, effective, and secure healthcare system for its inhabitants. While the COVID-19 pandemic has caught many governments, cities, regions, and medical professionals on the wrong foot, new digital health solutions continue to be developed and applied at a rapid pace in ways that are broadly transforming everyday life and accelerating trends in almost every industry, including medicine. Historically, cities have evolved in response to threats posed to health and other kinds of societal security. The events of the COVID-19 pandemic are expected to fundamentally change the way people and cities will function in the future. In fact, for this complex envisioned scenario, Smart Cities can be leveraged as one of the best resources to face this and future pandemics (Trencher and Karovonen 2019).

Sensing was one tool that was quickly deployed by large urban centers during the COVID-19 pandemic to determine social distancing and contact tracing. In the vast panorama of research and approaches in the context of smart cities, the need for sensing situations varied from traffic to energy consumption, pollution, and even viruses (Hancke and Hancke 2013). There is also the possibility of predicting the future evolution of such situations to prevent problems and to perform more informed decision-making and planning, so long as this is done without violating privacy concerns (Cecaj et al. 2021). The capacity to monitor and predict the behavior of crowds, in particular, is a fundamental enabling driver for Smart Cities (Hughe 2003). Vulnerable groups were compounded in having to juggle a major life-or-death issue during COVID-19 pandemic: heat. Extreme heat causes significant mortality around the globe every summer, yet the health risks of heat are systematically underestimated by the general public and even by those most vulnerable to them (Howe et al. 2019). As the planet continues to warm, urban centers will need to prepare to address another issue, in a smart way – cooling as a "right."

New York City was an early epicenter of the COVID-19 pandemic in the United States, due to the large numbers of international visitors coming through the city, its high density, and relatively large size. The COVID-19 pandemic suddenly and abruptly forced schools and education indeed to engage in major transformations (Iivari et al. 2020). Much of this change deeply depended on participants' access to broadband. As an example, while a majority of New Yorkers do have access to broadband, over 1.5 million residents do not have a connection at home or on a mobile device (The Mayor's Office of the Chief Technology Officer 2020). To help elders stay connected during the 2020 lockdown, the city partnered with LG and T-Mobile to freely distribute 10,000 tablets to seniors residing alone in public housing (The Eden Strategy Institute 2021).

While the response to the COVID-19 pandemic spread depended on digital tools for many urban centers, the crisis also exposed many societal flaws even in cities that were considered to be desirable and equitable pre-pandemic. On top of many of the hardships tied to social inequities, and sometimes a shortage of adequate housing due to a lack of heating or cooling, many of the same low-income households that were homeschooling children during the COVID-19 pandemic also experienced material hardships (i.e., difficulty paying for basic needs such as food, housing, and utilities). These hardships increased 46% since the beginning of the COVID-19 pandemic (Center for Translational Neuroscience 2020). Although projections point to the fact that as much as 68% of humanity will reside in urban centers by 2050, not all of these cities are layering sentient infrastructure into their built assets. In fact, many urban centers, particularly in the global south, lack infrastructure such as water treatment and distribution, transportation, and housing. Also, by 2050, three billion people, mostly in the global south, will be living in neighborhoods that have no mainstream governance, on land that is not zoned for development, and in places that are exposed to climate-related hazards such as floods. Models and analytical tools tailored for such communities need to be developed because the approaches used in cities in the global north cannot be transplanted. Unlike sentient cities where information is imbedded into its infrastructure, many of the informal urban centers lack data, and, as a result, informal socio-economic processes and limited local capacities must be considered (Bai et al. 2018).

Conclusion

The trend of urbanization is a relatively recent phenomenon in the modern existence of humanity, but key in how future threats, whether sudden or predicted, will be addressed. As scholars and governments continue to adapt the rapid changes in our built and sentient environments, the Smart Cities of the future are no longer simply defined in availability of ICTs but will instead be examined through various dimensions including digitalization, governance, vulnerability, and disaster and recovery preparedness.

Urban centers in the global south, as they transform, can benefit from the lessons learned of cities that are currently undergoing digital transformations as sentient infrastructure continues to be layered into such built environments. This layering of the urban nervous system should create necessary transparency, allowing governing bodies to set targets in collective pursuit of limiting the effects of climate change, particularly as they influence vulnerable populations.

Cross-References

▶ Big Data for Smart Cities and Inclusive Growth
▶ Smart City: A Universal Approach in Particular Contexts
▶ The Governance of Smart Cities
▶ The Sustainable and the Smart City: Distinguishing Two Contemporary Urban Visions
▶ Understanding Smart Cities Through a Critical Lens

References

Alirol, E., Getaz, L., Stoll, B., Chappuis, F., & Loutan, L. (2011). Urbanisation and infectious diseases in a globalised world. *Lancet Infectious Disease, 11*, 131–141.

Bai, X., Dawson, R., Ürge-Vorsatz, D., Delgado, G., Dhakal, S., Dodman, D., ... Schultz, S. (2018). Six research priorities for cities and climate change. *Nature, 555*, 23–25.

Caragliu, A., Del Bo, C., & Nijkamp, P. (2011). Smart cities in Europe. *Journal of Urban Technology, 18*, 65–82.

Cecaj, A., Mamei, M., & Zambonelli, F. (2021). Sensing and forecasting crowd distribution in smart cities: Potentials and approaches. *IoT, 2*, 33–49.

Center for Translational Neuroscience. (2020, September 28). No shelter from the storm: Higher income isn't protecting black and Latinx families from financial and material hardship during the pandemic. Retrieved from Medium: https://medium.com/rapid-ec-project/no-shelter-from-the-storm-88e290dad8e6

Fauci, A. S., Lane, H. C., & Redfield, R. R. (2020). Covid-19 – Navigating the uncharted. *The New England Journal of Medicine, 382*(13), 1268–1269.

Halegoua, G. R. (2020). *Smart cities*. Cambridge: The MIT Press Essential Knowledge.

Hancke, G., & Hancke, G. (2013). The role of advancing sensing in smart cities. *Sensors, 13*, 393–425.

Howe, P. D., Marlon, J. R., Wang, X., & Leiserowitz, A. (2019). A public perception of the health risks of extreme heat across US states, counties and neighborhoods. *Proceedings of the National Academy of Sciences of the United States of America, 116*(14), 6743–6748.

Hughes, R. (2003). The flow of human crowds. *Annual Review of Fluid Mechanics, 35*, 169–182.

Iivari, N., Sharma, S., & Venta-Olkkonen, L. (2020). Digital transformation of everyday life – how COVID-19 pandemic transformed the basic education of the young generation and why information management research should care. *International Journal of Information Management, 55*, 102183.

Kunzmann, K. (2014). Smart cities: A new paradigm of urban development. *Crios, 4*, 9–20.

Liu, Y., Gayle, A. A., Wilder-Smith, A., & Rocklöv, J. (2020). The reproductive number of COVID-19 is higher compared to SARS coronavirus. *Journal of Travel Medicine, 27*(4), 1–4.

Mumford, L. (1938). *The culture of cities*. New York: Integrated Media.

Reichental, J. (2020). *Smart cities for dummies*. Hoboken: Wiley.

The Eden Strategy Institute. (2021). *Top 50 smart cities governments*. Singapore: The Eden Strategy Institute.

The Mayor's Office of the Chief Technology Officer, M (2020). *The NYC Internet Master Plan*.

Townsend, A. M. (2013). *Smart cities: Big data, civic hackers, and the QUest for a New Utopia*. New York: W. W. Norton.

Trencher, G., & Karovonen, A. (2019). Stretching "smart": Advancing health and wellbeing through the smart city agenda. *Local Environment, 24*, 610–627.

United Nations. (2018a, May 16). *Department of Economic and Social Affairs*. Retrieved from 2018 Revision of World Urbanization Prospects: https://www.un.org/development/desa/publications/2018-revision-of-world-urbanization-prospects.html

United Nations. (2018b, May). *News*. Retrieved from Department of Economic and Social Affairs: https://www.un.org/development/desa/en/news/population/2018-revision-of-world-urbanization-prospects.html

Wataya, E., & Shaw, R. (2019). Measuring the value and the role of soft assets in smart city development. *Cities, 94*, 106–115.

Smartness

▶ The Sustainable and the Smart City: Distinguishing Two Contemporary Urban Visions

The Social and Solidarity Economy

Brendan Murtagh, Maria Pafi and Wesley Flannery
Urban Planning, School of Natural and Built Environment, David Keir Building, Queen's University Belfast, Belfast, UK

S

Synonyms

Community, Neighborhood, Social economy, Social enterprise, Social entrepreneurship, Social finance

Definition

This chapter will explore the scope, structure and potential of the Social and Solidarity Economy (SSE) in the future of urban regions. The SSE is, like the private economy, an assemblage of firms, entrepreneurs, intermediaries, and bespoke forms of finance that trade in goods and services for profit. What makes it different is that: surpluses are used for redistributive social purposes rather than personal gain; ownership is in the hands of stakeholders (such as communities) rather than individual shareholders; and economic activities (providing goods and services) reflect the needs of people rather than profit maximization. The exploitation of resources, capital accumulation, unsustainable forms of financialization and reductive labor have characterized market, and increasingly neoliberal forms of growth. Drawing on advanced social economies (such as the Basque Country), progressive financial environments (Québec in Canada) and networked cooperative economies (Emelia Romagna, Italy), the chapter evaluates the potential of the SSE as a site of political struggle as well as a practical agenda for the urban region. Arguing that a more sustainable planetary future is unlikely to be achieved without a different form of economics, it concludes by outlining the priorities for embedding, replicating and scaling the SSE as a long-term development goal.

Introduction

The COVID-19 pandemic highlighted the vulnerability of cities with a determination to "build back better," create more inclusive and prosperous communities and make places more resilient to health, environmental, and economic shocks. Brears (see chapter ▶ "Circular Economy Cities") argues that we need to reorientate the urban economy around a more sustainable "circular" relationship between production and consumption. Others advocate a planned approach to *Community Wealth Building* aimed at keeping investment moving within neighborhood economies, strengthening collective ownership of land and property, maximizing the impact of public spending on local supply chains and protecting the employment rights and income levels of the poorest paid (CLES 2018). At the heart of the project is a different type of *social* economy built not on profit maximization, but on how work, services, and facilities meet the needs of communities and how they want to live. The chapter describes urban regions where the social economy plays a central role in the redistribution of surplus from trading in goods and services, local ownership, and control of production and consumption and an underlying ethic of *solidarity* in how the city is organized. In short, the *Social and Solidarity Economy (SSE)* has the fundamental features of a trading economy, but it works for explicit social purposes and is embedded in solidaristic values in how it is developed, managed and its impact evaluated.

There is nothing particularly new or utopian about such a project and many places including Bilbao in the Basque Country, Bologna in Emelia Romagna and Montreal in Québec are built on a strong cooperative economic model. Here, the state, civic society and social enterprises have created a strong regulatory environment to grow the social finance market, preference social enterprises in procurement and introduced new legal forms (such as categories of cooperatives and Community Interest Companies) to grow the social economy ecosystem (Murtagh 2019).

These regions and cultures, regulatory environments and relations with local government show the variation in the SSE but this chapter is primarily concerned with conditions in the global North. Here, urban and state restructuring have repositioned the SSE as an actor with a range of overlapping and often conflicting functions by being close to the state, complicit in welfare reform and as a site of alternative economics and radical community politics. The chapter describes the scope and structure of the sector before looking at its distinct role in urban development, using examples drawn from the three cases. The analysis also reflects on criticisms of its ethical base, independence, economic impact and scale, especially in the context of urban deprivation and the increasing spatialization of poverty. The chapter concludes by highlighting some of the key components of the SSE as an arena of reform and potential in the future of the urban region more broadly.

The Social and Solidarity Economy

RIPESS is a global network of five continental social and solidarity economy networks which in turn brings together national and sectoral practitioners, academics and community enterprises. RIPESS (2015, p. 10) define the SSE as "an alternative to capitalism and other authoritarian, state dominated economic systems. In SSE ordinary people play an active role in shaping all of the dimensions of human life: economic, social, cultural, political, and environmental. SSE exists in all sectors of the economy—production, finance, distribution, exchange, consumption and governance." The SSE varies considerably across the globe. In the United Kingdom (UK) and the United States (USA) the emphasis is on a more entrepreneurial approach to enterprise development and profitability including operating in quasi-state markets, often with the private sector. In Southern Europe, where welfare systems are weaker, the sector plays a more prominent role in mainstream service delivery, while in northern Europe a more corporatist approach emphasizes a close partnership between the state (especially municipalities and local government) and social enterprises in supporting hard-to-reach groups (Defourny and Nyssens 2014). Amin (2009) also points out that the SSE survives and operates at scale independently from the state by responding to the needs of poor and marginal communities, particularly in the global South.

These relationships are at the heart of the tension in the social economy in general but social enterprises and cooperatives as core business models in particular. It is not a tension that can be easily resolved but needs to be managed to legitimize social enterprises to the communities they serve and their need to be financially sustainable and in particular, profitable. This 'steering' challenge between the social and the economic places considerable emphasis on the social entrepreneur (Nicholls 2010). For critics, too much emphasis is placed on the "heroic" individual rather than on the skills mix, services (such as legal and human resources), organizational structures and intermediaries, especially providing bespoke forms of social finance that support an alternative form of economics (Steyaert and Dey 2018).

The SSE therefore is a complex assemblage of businesses, finance, intermediaries, services and communities that sits between the public and private economy and while it is of course related to both, has a distinct logic, structure and support system. It includes a wide variety of business forms and legal structures including cooperatives, mutuals, associations, and foundations (CMAFs), distinct financial instruments (including credit unions, ethical banks and social investment products) instruments and intermediaries that provide research and technical support. However, it also includes innovative approaches to labor market integration by providing pathways to viable employment; encourages local exchange trading where people and neighborhood communities offer services to each other in units of time rather than currency; and advocacy networks, especially across the social enterprise sector. These networks are also important in creating the enabling environment for the social economy and have in some countries and city regions enacted new laws that support social value in public procurement; asset transfer that allows social enterprises to gain ownership over government land and buildings; and new legal forms such as Community Interest Companies, that are more hybrid private-social business entities capable of attracting (but also rewarding) commercial investors (Ridley-Duff and Bull 2016).

A key issue in the structure of the social economy is the enabling environment and how an urban ecosystem can scale the sector from the neighborhood level to offer a meaningful alternative to state or private markets. This highlights the central importance of tailored financial and fiscal support for social enterprises, primarily because they deliver blended social, economic and environmental value. Because of its distinct charitable model and the constraints this places on the creation and use of profits, the need to respond to communities of benefit, guarantee stakeholder governance and deliver social impact, bespoke forms of investment have grown up over the last thirty years. The social finance market remains weak and underdeveloped but is producing forms of *patient capital* that is better suited to social enterprise models. Here, loan finance, often matched by grant aid (from governments, charities, and donors) is lent at low interest rates on long terms and with repayment holidays as well as technical support to ensure the business operates profitably. Ethical funders and larger charitable banks, such as Tridos in the Netherlands, also offer loans and equity investment, especially where risk capital (such as in new ventures) is essential to business growth. For many smaller social enterprises who lack reserves or collateral to borrow at competitive interest rates, community shares and bonds have emerged, that also provide beneficiaries with a material (as well as a social) stake in local development (Nicholls 2010). In community shares, local people can buy a redeemable share within a set period of time to help get a project off the ground, leverage other grant and debt investment and develop financial solidarity across beneficiary communities.

But this raises a related concern for the sector, in that many social enterprises are not investment-ready, even if they could access more complex mixed funding options. The skills set, financial systems, regulatory control and understanding of risk are new areas for most community groups and neighborhood projects. Expertise and knowledge around financial literacy, cost accounting, balance sheet planning and profitability are critical to the long-term performance of social enterprises, especially if they are to scale as alternatives to mainstream models of urban development.

Social Enterprises

The urban economies of major Northern cities have shifted from labor-intensive heavy engineering and manufacturing to a more specialist service economy in which advanced education, new skills, knowledge (in information technology, finance, legal services and so on) and global business networks have taken precedence over 'local' factors of production, assets and resources (Harvey 2012). Harvey (2012) also argues this volume show that urban regeneration, especially in the global North, is being led by a distinct neoliberal ideology that involves a more direct role for the private sector in mega property projects, financial and fiscal incentives for developers and less emphasis on the provision of collective goods and services (housing, education, social care and so on) by the state. This has created deeply uneven social, economic and spatial effects with people and places able to connect to the knowledge economy separated from communities left behind by economic change.

Bridge et al. (2014) argue that social enterprises can also be viewed as a spectrum of activity but share a common concern for trading goods and services, making a financial surplus and redistributing profit for a defined social purpose. Their characteristics include the following attributes:

- First, social enterprises are businesses with explicit economic functions and characterized by a continuous activity producing goods and services *for profit*. This involves a significant level of economic risk, including, for example, using debt finance as well as donor investment and state grants to capitalize the business.
- Second, they have an explicit social benefit which, in the context of the city means working in disadvantaged places, under-invested communities, people at the edge of the labor market or those who face specific obstacles getting employment (including those with disabilities, older people or migrants).
- Third, to do this, the governance of social enterprises center on community ownership rather than on the financial interest of individual shareholders. This means that organizations are based on strong participatory processes, a high degree of autonomy and where parties (such as neighborhood communities) who are impacted by the activity, make decisions about the organization and how it works.
- Fourth, social enterprises maintain a strong set of ethical principles in working conditions, living wages, environmental principles and gender inclusion that are balanced with its commercial aims and collective ownership model.

What Do Social Enterprises Do for the City?

Clearly, social economies can emerge across regions, national scales and in different sectors (education, recycling, child services, and so on). What is of particular interest here is its performance and potential in the context of urban restructuring, uneven development and the increasing polarization of some people and places in the contemporary city. Laville (2014) argues that the SSE is a configuration capable of multiple (simultaneous) roles, but that it can work as an alternative to market-led urban development when its value base stays close to its social purpose, the needs of local communities, and to collective forms of ownership. Moreover, these are not incompatible with achieving scale and Table 1 describes three strong SEEs in Emelia Romagna, the Basque Country and Québec and while they are atypical, demonstrate how they function within the urban region. In short, the SSE has an impact on the city and its development in a number of interconnected ways:

- It creates employment, strengthen skills and creates pathways to work especially for people who lack education and experience to access high growth sectors of the economy. Moreover, it is good quality jobs, with fair pay, decent working conditions and inclusive employment practices that social enterprises protect at a time of increasingly precarious forms of work.
- The SSE also aims to create services that people need and not what makes most profit for private businesses including, for example, community transport, childcare services and affordable housing.

The Social and Solidarity Economy, Table 1 Urban social and solidarity economies

In **Québec, Canada** the government made the development of the social economy an explicit political and policy priority. They established *Chantier de l'économie sociale* (Chantier) in 1996 in response to a prolonged recission that had disproportionate impacts on the poorest communities, young people, women and first nations. The organisation supports social enterprises via training and skills development; research and intelligence; and advocacy networks. Chantier also provide integrated social finance and technical assistance and its loan funds are capitalised with a mix of public and private sources backed by tax incentives to attract investment into social enterprises and cooperatives. A distinctive feature of the approach is the 22 *social economy poles* distributed regionally but which also supports the development and coordination of community businesses within first nations. Québec now has more than 11,200 social enterprises, with 220,000 employees and with overall sales of CA$47.8bn. These poles are anchor institutions that support the incubation and development of social enterprises, build spatial clusters to create economies of scale and cooperate through consortia networks into higher value and more profitable supply chains.
The Basque Country has developed one of the most advanced SSEs globally and has legislation, policies and investment programmes that promote social enterprises and cooperatives that serve the needs of local people. The **Mondragon corporation** emerged from threatened closure of a factory in the small Basque town and employs 81,800 people across 150 countries. The cooperative is structured into four business areas including: Finance, industry, retail and knowledge. It has its own cooperative bank with 1.2 m customers, a revenue of around €14.8bn and assets valued at €33bn. The average salary ratio between highest and lowest paid employees/members in the co-operative is 1:6 and by contrast, the average equivalent salary ratio for a FTSE100 company is 1:130. Profits are reinvested via a solidarity fund, which in 2014, allocated €40 m to inter-cooperative linkages, training and research and development. The Mondragon university not only trains workers for the core businesses but invests heavily in innovation, especially to create the next generation of cooperatives in sectors such as nanotechnology, advanced fabrication and artificial intelligence.
Centred on **Bologna, the sector in Emelia Romagna** has its roots in the radical politics of the city, a strong trades unions movement and a cooperative agricultural tradition across the small family farm economy. Legal structures are constantly evolving and there is now a distinction between type A cooperatives that deliver a range of health, education and social services and type B coops that promote labour market integration although some businesses combine both types. The dense network of cooperatives, institutional support and technical assistance is a product of, and reproduces, a solidarity culture across businesses, government, trades unions and civil society. Thirty-five per cent of Italy's cooperatives are located in Emilia Romagna making it one of Europe's most concentrated sectors. In Bologna, two out of three citizens are members of a cooperative and cooperatives directly account for over 40% of the region's GDP. Moreover, cooperative networks have enabled small manufacturing firms to integrate and achieve scale, especially in supply chains and to pool resources to share marketing, Research and Development, training and distribution costs.

Source: Based on Murtagh (2019), Young Foundation (2017)

- Social enterprises help to build diversified economies by creating new services and alternative business models that survive (and thrive) in areas where private and public markets have failed. At a time when urban infrastructure is increasingly being privatized (such as in transport), the SSE has replaced services in financially sustainable ways.
- Linked to this, social enterprises nurture "local consumption" circuits by providing the types of goods and facilities that create local demand, support the profitability of community businesses, and keep money in the local economy. In short, spending, even in the most disadvantaged areas, can be recycled within the neighborhood, producing local multiplier effects and sustaining jobs, services and facilities in under-invested communities.
- Distinct forms of social savings, including Credit Unions, micro-finance, and community shares provide credit for people often denied access to bank accounts or who rely on predatory short-term finance or money lenders.

Social enterprises have a particular focus on asset-based forms of development. This means mobilizing social *capital* – social, financial, and physical assets, but especially land and buildings, rather than emphasizing need, poverty, and dependency on outsiders to fix local problems. Community Asset Transfer legislation in England has enabled neighborhood groups to get access to state-owned land and property at nil or nominal value and in Scotland land transfers are supported with capacity building, grant aid and soft loans. Being able to access underused or redundant assets, build collateral and an opportunity to leverage other investments have been critical to scaling regeneration projects where resources might appear limited (Murtagh and Boland 2019).

Limitations of the SSE

McMurtry (2015) questioned the ethical base of the SSE and the extent of its radicalism; while Nicholls (2010) sees its irreconcilable social and financial logics as barriers to growth; and Sunley and Pinch (2014) argue that social enterprises tend to concentrate on highly localized and weak markets. There are therefore a number of limitations to the SEE and social enterprises in particular and especially how they relate to private and state markets. These include:

- First, critics argue that it is effectively an arm of the state, responsibilizing (poor and marginal) people to look after their own affairs in employment, welfare provision, and coping with the uneven effects of economic change. For some, it is a neoliberal flanking measure, there to discipline communities in new governance regimes that simply enable the state to withdraw from housing, social care, education and so on.
- Second, and linked to this, labor market integration has been particularly criticized as "workfare," whose primary purpose is to get people off welfare support and benefit income, including those who are sick are disabled and into waged work.
- Third, social enterprises tend to operate in marginal sectors of the economy, low value labor markets (such as childcare, recycling, and community retailing) that are often characterized by poor pay, insecure conditions, and limited chances of progression.
- Fourth, the sector, certainly in the USA and the UK, tends to be dominated by a larger number of small, undercapitalized, and inexperienced enterprises that make growth and innovation difficult to achieve.
- Fifth, and related to the issue of scale, is the "local trap". Here, social enterprises are criticized for their concentration in underserved neighborhoods where prospects for growth are limited by the constraints of their market and weak integration with the wider social (or mainstream) economy. Here, enterprises, lack the finance, scale economies and business networks to grow out of their community to become a radical and sustainable alternative to the private sector.

- Sixth, many are criticized specifically because they are *social* and are operated by leaders with an interest in an ethical purpose rather than developing a resilient business model. The tension between the social and the economic is ever present making it difficult to scale and replicate successful companies, not least because of the lack of entrepreneurial knowledge and competence across the core staff.
- Finally, the sector is highly variegated but, in many countries and city regions, it lacks the bespoke, finance, necessary regulatory support or technical assistance to be anything other than a fringe, quasi-charitable activity. To grow the SSE, as more successful models in the Basque Country, Bologna and Québec have shown, requires an infrastructure that is missing in most city regions or is ideologically opposed in order to block competition with private and public sector providers.

Conclusions

The Basque, Italian and Canadian examples show that an alternative urban economy is possible and an effective response to unstable forms of private-led urban regeneration. It responds to fundamental questions about who the city is for and how economic activity (the production and consumption of goods and services) responds to socially defined needs and capabilities rather than endlessly de-risking the city for limited forms of property speculation. Solidarity values are embedded in the way in which services are defined; how businesses are structured and development is financed; and how *value* is constructed socially as well as economically. Creating space for the SSE, scaling its effects and replicating successful practice is an important (and urgent) project for the city, its sociospatial stability and to articulate what a right to the city could mean in practice.

This involves supporting the SSE through an integrated approach involving legal frameworks, finance and developing the skills set of social entrepreneurs and managers. Legislation that recognizes the distinct trading challenges and effects of social enterprises, preferences such entities in procurement and enables access to assets are proven actions successfully taken by a range of governments and municipalities. But the SSE also needs its own advocacy networks; R&D facility to strengthen innovation and diversification into high growth sectors; and the skills to manage and scale alternative area-based social economies as constitutive of the modern city. Finance is critical to social enterprises as it is to private businesses but so too is evidence that demonstrates community value impacts, manages the tension between the social and economic logics of the sector and how to build solidarity as a meaningful ethic for the future of the city region.

Cross-References

▶ Circular Economy Cities

References

Amin, A. (2009). *The social economy: International perspectives on economic solidarity.* London: Zed Books.

Bridge, S., Murtagh, B., & O'Neill, K. (2014). Understanding the social economy and the third sector. London: Palgrave.

Centre for Local Economic Strategies (CLES). (2018). *How we built community wealth in Preston: Achievements and lessons.* Manchester: CLES. Retrieved from https://cles.org.uk/publications/how-we-built-community-wealth-in-preston-achievements-and-lessons/

Defourny, J., & Nyssens, M. (2014). The EMES approach of social enterprise in a comparative perspective. In J. Defourny, L. Hulgård & V. Pestoff (Eds.), *Social enterprise and the third sector: Changing European landscapes in a comparative perspective* (pp. 42–65). London: Routledge.

Harvey, D. (2012). *Rebel cities: From the right to the city to the urban revolution.* London: Verso.

Laville, J.-L. (2014). The social and solidarity economy: A theoretical and plural framework. In J. Defourny, L. Hulgård & V. Pestoff (Eds.), *Social enterprise and the third sector: Changing European landscapes in a comparative perspective* (pp. 102–113). London: Routledge.

McMurtry, J.-J. (2015). Prometheus Trojan horse or Frankenstein? Appraising the social and solidarity

economy. In P. Utting (Ed.), *Social solidarity economy: Beyond the fringe* (pp. 57–78). London: Zed Books.
Murtagh, B. (2019). *Social economics and the solidarity city*. London: Routledge.
Murtagh, B., & Boland, P. (2019). Community asset transfer and strategies of local accumulation. *Social & Cultural Geography, 20*(1), 4–23. https://doi.org/10.1080/14649365.2017.1347270.
Nicholls, A. (2010). The legitimacy of social entrepreneurship: Reflexive isomorphism in a pre-paradigmatic field. *Entrepreneurship Theory & Practice, 34*(4), 611–633. https://doi.org/10.1111/j.1540-6520.2010.00397.x.
Ridley-Duff, R., & Bull, M. (2016). *Understanding social enterprise: Theory and practice*. Thousand Oaks: Sage.
RIPESS (Intercontinental Network for the Promotion of Social Solidarity Economy). (2015). *Global vision for a social solidarity economy: Convergences and differences in concepts, definitions and frameworks*. Madrid: RIPESS. Retrieved from http://www.ripess.org/wp-content/uploads/2017/08/RIPESS_Vision-Global_EN.pdf
Steyaert, C., & Dey, P. (2018). The book on social entrepreneurship we edit, critique and imagine. In P. Dey & C. Steyaert (Eds.), *Social entrepreneurship: An affirmative critique*. Cheltenham: Edward Elgar.
Sunley, P., & Pinch, S. (2014). The local construction of social enterprise markets: An evaluation of Jens Beckert's field approach. *Environment & Planning A, 46*(4), 788–802. https://doi.org/10.1068/a45605.
Young Foundation. (2017). *Humanity at work: Mondragon, a social innovation ecosystem case study*. London: Young Foundation. Retrieved from https://www.youngfoundation.org/publications/humanity-work-mondragon-social-innovation-ecosystem-case-study/

Social Economy

► The Social and Solidarity Economy

Social Enterprise

► The Social and Solidarity Economy

Social Entrepreneurship

► The Social and Solidarity Economy

Social Finance

► The Social and Solidarity Economy

Social Health

► Health and the Role of Nature in Enhancing Mental Health

Social Impact

► Community Engagement for Urban and Regional Futures

Social Justice

► Social Urbanism: Transforming the Built and Social Environment

Social Services

► Challenges of Delivering Regional and Remote Human Services and Supports

Social Sustainability

► Sustainable Development Goals

Social Urbanism

► New Forms of Shared Governance and Local Action Plan in Socially Vulnerable Settlements

Social Urbanism: Transforming the Built and Social Environment

Alejandra Rivera Vinueza
4CITIES Erasmus Mundus Joint Master Degree (EMJMD) in Urban Studies, Vrije Universitet Brussel (VUB), Brussels, Belgium
Institute for Human Rights and Business (IHRB), Built Environment Global Programme Manager, London, UK

Synonyms

Co-creation processes; Community-led interventions; Human Rights; Multidimensional urban development; Social Justice; Sustainable urban development; Urban upgrading

Definition of Social Urbanism

Social Urbanism (SU) is an urban governance model driving positive transformation through simultaneous actions in the built, natural, and social spheres of the city. This process aims to improve the quality of life of citizens by ameliorating the urban context in which they live. "Urban context" in a broad definition refers not only to physical infrastructure but also to environmental resources such as access to green belts, parks, clean and healthy water streams, etc., as well as the social context: the array of opportunities that support human development: easiness of mobility, quality of air, soundscape, and landscape, as well as the stock of social infrastructure, e.g., schools, hospitals, libraries, parks, cultural centers, the transport network, community housing, etc. Hence, SU is a comprehensive urban development model to transform multiple dimensions of the city, a process that is naturally carried out with the collaboration of multiple stakeholders. Five key and intertwined elements of SU are: (1) a physical enhancement of the urban fabric, (2) by the hand of social interventions, (3) driven by community-co-created processes, (4) facilitated by effective public institutional management, and (5) an economic model that is sustainable and redistributive.

Introduction

The purpose of this chapter is to provide a quick, brief, and general introduction to the concept of Social Urbanism and its basic characteristics. This chapter is divided into six main sections: (1) To provide background information about the concept, the first section explains the ideological foundations from where the idea of SU derives; then section (2) outlines four ambits of conditions of possibility that should be present in a territory, to serve as fertile ground for the SU model; section (3) describes the multitude of stakeholders that are involved in the process, the relationships they may adopt, and the necessary tools for their collaboration; then section (4) explains some considerations of the SU process to facilitate understanding of how the various characteristics of SU assemble to create a successful project; section (5) provides a brief mention of three case studies; and section (6) closes with a summarizing conclusion.

Ideological Foundations of Social Urbanism: Where the Idea Comes From

The term "social urbanism" was recently coined as such by Medellín's local government during the decade of the 2000s. The term designated a series of urban upgrading strategies to alleviate social inequalities, violence, and improve the city's standard of living in general. Despite the novelty of the term itself and the myriad of literature emerging on the subject in recent years, it is important to recognize the underlying foundations of the concept dating back to much older ideologies.

At its core, SU is based on the idea to respect, provide, and/or revindicate *human rights* and an environment of *dignity* for all city inhabitants. It seeks the full physical and social integration to urban life of underprivileged communities/districts (slums or areas characterized by the highest rates of insecurity and poverty). The premise is that those inhabitants are also worthy and deserving of being part of their city and accessing its resources. This *rights-based approach* is extended to include natural habitats and the right to live in healthy, clean, and beautiful environments. Thus, SU interventions also aim to revert and heal environmental damage previously caused, and protect and preserve nature.

SU also has ideological foundations in "*the right to the city*." The term was coined by Henri Lefebvre (1968) and retaken by other scholars like David Harvey (2003) and by social movements around the world. It is the idea that citizens hold the right and the power to shape their cities and the processes that define them. The idea claims that the city is a cohabited space; therefore, it should be cocreated by the inhabitants and users themselves. This idea grounds many of the urban development approaches that integrate participatory processes and citizen engagement strategies, to various degrees. The SU model embraces the right of citizens to decide and shape city-making processes by upholding participation as one of its main pillars and success factors.

SU's main objective is to satisfy basic needs and improve the quality of life of citizens. Hence it is a concept with large parallels to the processes of *social innovation*. These are socio-political processes in a territory that seek (1) the satisfaction of basic needs to the most vulnerable population, (2) to create the conditions for collective empowerment, and (3) to positively reconfigure social (power) relations (Moulaert et al. 2010). In fact, for socially innovative initiatives to fulfill their transformative potential, it is necessary to first have "positive conditions of possibility in the built environment (including social infrastructure stock) and in the social environment (individual and collective empowerment)" (Rivera Vinueza 2021), conditions that are exactly what SU seeks to improve.

Finally, the SU model of governance can also be seen as an institutional response to the calls for social and environmental justice that have emerged in a multitude of cities around the world. Such institutional response manifests in a working integral urban development model that redistributes the economic and non-economic benefits of planned urbanization. SU "simultaneously considers socio-economic and ecological components of space, with a greater focus on social sustainability" (Bellalta 2020). Therefore, it is a social and environmental justice-oriented form of urbanism.

Conditions of Possibility for Social Urbanism: Creating Fertile Ground

Political

City Mayors have a relatively high degree of *autonomy* to create redistributive mechanisms, to access or create various sources of funding, create civic and communication campaigns, and have a strong position to partner with multiple stakeholders to plan, design, and implement physical and social interventions. Having that said, the next most important political condition is *political will* and true interest from local political leaders to implement meaningful and impactful actions for citizens, with a focus on those marginalized and underprivileged, usually in peripheral areas. Such a clear objective at all government levels and between all public institutions facilitates communication and coordination.

This mission alignment allows for the construction of an urban policy that improves the quality of life of people in the city. It is reached with a balance between public policy and community activism (a harmonious *middle point* between guidance from top-down directives and receptiveness to bottom-up initiatives). It is also essential to count on the political commitment to *transparency and efficient management of public financial and non-financial resources*. This guarantees an environment of governability and trust within the local administration, along sectors, and with the citizens.

Lastly, the physical and social processes that produce a tangible urban transformation can be slow processes that take several years. Hence, SU is facilitated by the *continuation of an administrative ideology and alignment* of public policies throughout various consecutive public administrations. Therefore, political leaders think of themselves as a link in a chain of positive actions for citizens, rather than thinking in isolated terms of office for personal gain.

Social

SU is by definition a way of transforming and developing the city in a socially sustainable way. Thus, SU strives to address existing social inequality problems, but in receptivity and in conjunction with the implicated population. Such partnership is facilitated by a *strong and active civil society that is nonconformant* with the status quo of inequality and its associated urban problems, a civil society with a spirit of resilience that forms social movements, claims for social and environmental justice, and believes there are better ways of urban existence.

It is optimal when, on the one hand, the citizenship cares for urban, social, and environmental issues, and is motivated to propose potential solutions; on the other hand, city administration enables communication channels to embrace such possible solutions. Therefore, a culture of *citizen participation* is fertile ground for SU to be successful.

A strong and active civil society is not only a precondition of SU, but it is also its goal. This form of doing urbanism focuses on the provision of public spaces for the promotion of coexistence and respect for life. It seeks to strengthen the social tissue, with the construction or renovation of *social infrastructure*, such as schools, hospitals, libraries, parks, and with activities and programs that support human development, public health, safety, and welfare.

Cultural

Citizens' culture plays a key role in the transformation of a city under the SU model. SU needs (and constructs) *platforms* to facilitate, promote, and embrace *citizens' imaginaries* of the city they want to live in. As already mentioned in the social ambit, it is key that citizens count on communication channels to manifest what they desire and imagine for their city. Such ability makes civil society play an active role in informing, if not driving, urban and social processes. The cultural conditions of possibility also refer to the *attitude of citizens toward their city*. A strong city identity and culture call residents and visitors alike to love and care for the city. When that is the case, "social (positive) pressure" driven by locals, and usually followed by visitors, calls to care, preserve, protect, and respect public spaces, public infrastructure, and the city in general.

While culture is a vehicle for urban transformation, it is also a goal of the SU project to strengthen, consolidate, and disseminate the city's culture. This is reflected in the decisive physical change of the urban fabric by building or improving sports centers, recreational spaces, schools, libraries, and parks, along with citizen-driven activities to fill those spaces with autochthonous and unique art, music, food, and talent. By providing these social infrastructures, citizens receive the message that *they are worthy* and deserving of a place in the city and access to those spaces. In turn, that stimulates a sense of belonging and a feeling of civic ownership of the local community.

Economic

Another important condition of possibility for SU is *financial instruments* (or at least the potential to develop them) that can make the urbanistic model self-sustainable. The main source of revenue for city administrations is taxation. While this source is partly and initially used to finance urban interventions, the SU model is characterized by proposing innovative financial streams in the function of the common good. An example is a strong institutional base of *public companies* in key areas such as housing, transport, urban (re)-development, or utilities (water, electricity, and Internet). A well-managed public company is an important source of revenue. Such wealth produced by the urban activity is redistributed where it is most needed, e.g., in the form of social programs, slum upgrading, financing social infrastructure, transport network, etc. It is common, and desirable, that a city has two or more public companies of this kind so that higher revenue streams complement lower ones to meet different urban demands.

Contrary to the social and cultural ambits, economic profitability is not a goal in itself (for the sole purpose of revenue generation), but it is rather the condition that allows urban, social, and environmental interventions to be financed. Then, the SU model needs economic conditions of possibility and alternative economic strategies for its financial sustainability. These can include, for example, *city marketing and branding* to highlight the image of the city and increase its competitiveness as a destination for investors and tourists (Montoya Restrepo 2014).

A successful SU project may utilize *city-branding strategies*, partnerships with the private sector, and a narrative of technology and innovation to *attract foreign investment*, and thus open various sources of revenue. However, such one-off localized investments, usually going to already developed rich areas of the city, are accompanied with *local investments of larger scope and benefit distribution* (at neighborhood or district level). Thus, the SU model redistributes part of the generated revenue to areas of the city that are not recipients of foreign direct investments. This strategy balances the development of the territory and strives for socio-territorial equality.

Stakeholders: A Joint Effort

Due to being such an integral urban development model, SU processes involve a myriad of actors. SU is by definition a joint effort in which stakeholder diversity is praised and strived for since collaboration is trusted to facilitate consensus and produce better results. The following paragraphs describe the multitude of stakeholders that are involved in SU processes, the relationships they may adopt, and the necessary tools for their collaboration.

There are two axes in urban governance. *The horizontal axis* refers to the interaction *between sectors* (public, private, third sector, academia, etc.) at a specific geographic scale. The *vertical axis* refers to the interaction between various *geographic levels* or scales of decision-making, i.e., this refers to local, city, national scales, etc. SU processes require collaboration between a diversity of stakeholders in both the horizontal axis (between sectors) and the vertical axis (between levels of urban governance).

Multisectorial Collaboration

SU requires collaboration between the different sectors in the city as a condition to emerge, and once implemented, the urban model itself helps sustain such collaboration. SU is characterized by adopting a *Penta-helix approach* to multisectorial collaboration that is between the five main sectors of the city: (1) public administration, (2) private corporations, (3) NGOs or the third sector, (4) academia, and (5) civil society. This is important because other multisector collaboration models include only two or three sectors, e.g., public-private partnerships, public-academic, etc. However, the Penta-helix approach includes all five sectors, with an emphasis on the third sector, academic institutions, and civil society.

The relationships that may emerge between the five sectors are varied. In SU, processes are usually led by one sector (public, academia, or the community) and involve active participation and contributions from the other sectors (e.g., financial or in-kind support from the private sector and/or knowledge and capacity building from NGOs or foundations). The SU approach then seeks to give a greater protagonist role to sectors historically neglected in urban decision-making.

Multilevel Collaboration

The inclusion of multiple stakeholders in the decision-making processes of SU also extends along the *vertical axis to involve various levels of governance*. This includes actors from the micro to macro scale: household, neighborhood, district, city metropolitan, regional, national, and international levels. This type of multilevel collaboration requires in the first place an *alignment of purpose and vision*.

While it is a challenge to coordinate such a diverse and complex multitude of levels of action, exemplary cases of SU have displayed *leadership from one level to coordinate the rest*, for example, a city council or metropolitan area drafting a comprehensive sustainable development plan that delineates mechanisms to include and coordinate with smaller scales in the city (district councils, neighborhood associations, and community leaders), and also establishing good working relationships with higher levels of government and international actors by clearly communicating its goals, needs, and potential contributions.

In SU processes, multilevel collaboration usually originates from the public administration through innovative alliances that demonstrate openness to work with other sectors and other levels, thus serving as an example of scalar coordination. Tools for such multilevel collaboration can include physical or digital dialogue platforms, e.g., committees, councils, forums, etc. that allow constant communication between actors. Also, multidisciplinary and inter-institutional working groups assemble actors from various levels of governance and sectors at the same table; this can include, for example, representatives from a neighborhood association, a local NGO, the city's urban development company, an institute of regional development, and/or a national university. In this way, multiple stakeholders align objectives, define priorities, and work together toward unified social, environmental, and economic development goals.

The Concept of "Shared Governance"

In the SU contrivance, each stakeholder contributes with ideas, resources, administration, and/or execution of the city's urban development actions. When multisectorial and multilevel collaboration is achieved, at least to some degree, it is possible to state that the direction in which the city is transforming, evolving, and improving is being driven not by one actor, sector, or level, but by multiple ones; thus, the governance itself is shared between the contributing parties.

Going back to the ideological foundations of SU, one of its most basic principles is the respect and promotion of the *right of citizens to participate in this "shared governance,"* that is, allowing citizens to exercise their right to have a loud voice in decisions about interventions impacting their urban, social, or environmental habitat. In SU processes, the active participation of civil society is seen as a symbiotic partnership with the community to take advantage of local knowledge, rather than seen as an obstacle to unilateral goals.

The concept of "shared governance" implies an element of *trust* between working teams: between different government departments and institutions in the city, and between sectors. Specific *responsibilities and functions are entrusted* to each sector according to its specialty. City governance is the responsibility not only of the city council, but jointly also with other public and private entities, academia, and civil society organizations, *in coordination* with entities at other levels.

Considerations of the Social Urbanism Model

The *goal of SU* is to attend socio-spatial inequalities by using urbanism as a tool to bring social benefits through the ameliorating of the hard (built environment), soft (culture, arts, sports, civics, and economics), and natural environments (Rivera Vinueza 2021).

The SU process is complex due to the multitude of urban dimensions that the transformation involves and the multitude of stakeholders that should be aligned. Its *complexity* is also due to the necessary conditions of possibility that should be in place to serve as fertile ground for SU processes. Section Two has described some of the political, social, cultural, and economic conditions that are necessary but, at the same time, these are reinforced and strengthened by the implementation of the SU model. For example, it is a necessary condition to have a strong and vocal civil society that generates fresh ideas and solutions to their urban challenges, and at the same time, by carrying out participatory processes or supporting community-led initiatives, the social ambit is strengthened.

One key consideration of the Social Urbanism model is the *optimization of the purpose of interventions*. In other words, every intervention should meet territorial, social, and environmental goals. For example, greenways or linear parks fulfill an environmental function and reach the transformation of urban space, and when done in a participatory way, they yield to meaningful cocreation processes that build trust and partnership with the community. Other examples of the application of this consideration are the recovery of unused, abandoned, public infrastructure to make public spaces and ecological corridors that meet local needs.

Another key consideration of the SU model is its *integrality and wide scope.* Such ambition is manifested in *integral urban projects* that incorporate multiple subprojects simultaneously, from transport to landscaping, from street lighting to cultural centers (Malandrino 2017). Such projects are multidisciplinary and highly collaborative to understand different types of needs of a territory and its inhabitants and to reconcile interests, opinions, and methodologies. For example, transportation needs might conflict with the environmental preservation of an area, or a new-built library might need activities and programs to ignite its use, etc. Therefore, the strength of the integral urban projects is their ability to complement physical and social interventions to attend to diverse needs simultaneously.

Furthermore, the *order of priorities* of the SU project is worthy of close consideration. The primary objective is to improve the quality of life of citizens, along with accompanying social and environmental goals. It is also important to reach a certain level of economic sustainability to finance the physical interventions; however, revenue generation is not an end in itself, but rather a necessary condition for the primary goals. The SU model offers a "redistribution mechanism of the wealth generated by the urban development process itself" (Leite et al. 2020).

Case Studies

Medellín

The "Medellín model," thus named in 2008 by the Organization of American States at its xxxviii General Assembly, is the flagship case of social urbanism. The spectacular transformation of the city that occurred since the 2000s exhibited all the key elements described in this text: SU's foundational ideology, all its conditions of possibility, and multistakeholder collaboration. The "Medelln model" has defined and consolidated the concept, theory, and practice of SU itself. In addition, given the context of violence in this particular case, SU in Medellín has also built physical and social spaces, with structures and functions, to normalize and regain control of the territory (Montoya Restrepo 2014). The role of citizenship through education, civics, culture, arts, and sports has been fundamental in the city's transformation.

Sao Paulo

Another Latin American example comes from the socially inclusive land policies in Sao Paulo achieved in the 2013–2016 period, including land use regulation, city-wide programs, locally focused projects, and integrated plans and actions, aimed to promote social-territorial inclusion. The main achievement was the construction of a new urban regulatory framework for the city: a process driven by the local government with high citizen participation. It favored mechanisms to enhance the use of public land, recover public spaces, and in general, regain collective access to quality land (Leite et al. 2020). This epitomizes the ideological foundation of SU about the citizens' collective right to the city and to participate actively in the processes that shape it.

Cape Town

Kosovo is one of the largest informal settlements in Cape Town. It is characterized by low-quality single units from postapartheid subsidized housing, and problems of social exclusion, urban sprawl, and lack of social infrastructure. The upgrade process in 2006–2012 was funded by redirecting underspent government budget and CSR investments. It addressed several needs: (1) provided access to basic needs: water, sanitation, waste management, and public green spaces, (2) redesigned building typologies to be more adequate to current demographics, (3) incorporated sustainable infrastructure design and greening, e.g., mixed-use spaces and urban agriculture, and (4) promoted education, training, and skill development to improve access to jobs in surrounding central business districts (Goven 2005).

This multidimensional upgrade process fulfills various characteristics of SU: an integral urban project, addressing local needs, providing social infrastructure, and promoting more sustainable methods of living. However, this case lacks a strong financially sustainable model, and the incorporation of participatory and/or community-led approaches, which are key for the citizens' appropriation and endurance of the project.

Conclusion

This chapter has introduced the concept of SU and its basic characteristics. SU is an integral urban development model to improve the quality of life of citizens through simultaneous actions in the built, natural, and social environments of the city. SU finds ideological foundations in human rights, the right to the city, social innovation, and social and environmental justice. The following are four ambits of conditions of possibility that are considered ideal for SU processes to be able to emerge: (1) political will, autonomy, transparency, efficient management of resources, and consecutive progressive city administrations; (2) a strong and active civil society and effective channels for citizens participation and community-led initiatives; (3) embracement and promotion of the unique identity and culture of the city and its citizens; and (4) economic sustainability.

Due to being such an integral urban development model, to practice SU requires a myriad of actors. The process requires multisectorial and multilevel collaboration to form an environment of "shared governance" to plan and implement urban and social interventions. Other considerations of the Social Urbanism model include the optimization of the purpose of interventions, the ambition of integral urban projects attending various needs with the same project or related subprojects, and an order of priorities in which social and environmental goals prevail economic ones.

SU can also be seen as a strategy for redistribution of resources and revenue from the urbanization process itself. The model focuses on providing social infrastructure and attending to the (multidimensional) basic needs of the most deprived neighborhoods and communities. Hence, the success of this social and environmental justice-oriented form of urbanism is highly dependent on the common understanding that improvements done in the most marginalized and impoverished districts will ultimately benefit the entire city, its image, and national and international position, and improve quality of life for all its inhabitants.

Cross-References

► City Financing and Social Urbanism in Latin America: The Importance of Good Fiscal Management
► Community Engagement and Climate Change: The Value of Social Networks and Community-Based Organizations
► Improving Social Equity and Community Health and Well-Being in Low-Income Suburbs and Regions
► New Forms of Shared Governance and Local Action Plan in Socially Vulnerable Settlements
► The Social and Solidarity Economy
► Urban and Regional Leadership
► Urban Commons as a Bridge Between the Spatial and the Social

References

Bellalta, M. (2020). *Social urbanism: Reframing spatial design - Discourses from Latin America*. Applied Research & Design. ISBN 13: 978-1943532681.

Goven, G. (2005). *Green urbanism - Kosovo informal settlement upgrade case study*. ARG Design.

Harvey, D. (2003). The right to the city. *International Journal of Urban and Regional Research, 27*, 939–941. https://doi.org/10.1111/j.0309-1317.2003.00492.x.

Lefebvre, H. (1968). *Le droit à la ville*. Paris: Anthropos.

Leite, C., Acosta, C., Militelli, F., Jajamovich, G., Wilderom, M., Bonduki, N., Somekh, N., & Herling, T. B. (2020). *Social urbanism in Latin America. Cases and instruments of planning, land policy and financing the City transformation with social inclusion*. Springer Nature, Future City Series #13. https://www.springer.

S

com/gp/book/9783030160111. https://doi.org/10.1007/978-3-030-16012-8
Malandrino, C. (2017). *On the 'Medellin miracle' and the 'social urbanism' model*. The Urban Media Lab. https://labgov.city/theurbanmedialab/on-the-medellin-miracle-and-the-social-urbanism-model/
Montoya Restrepo, N. (2014). Urbanismo social en Medellín: una aproximación desde la utilización estratégica de los derechos. *Estudios Políticos, 45*, Instituto de Estudios Políticos, Universidad de Antioquia, pp. 205–222.
Moulaert, F., Martinelli, F., Swyngedouw, E., & González, S. (2010). *Can Neighborhoods save the city? Community development and social innovation*. Routledge.
Rivera Vinueza, A. (2021). Urban context's role in the emergence and development of social innovation. In Allegri et al. (Eds.), *Research tracks in urbanism: Dynamics, planning and design in contemporary urban territories* (1st ed., pp. 42–50). Taylor & Francis. https://www.taylorfrancis.com/chapters/edit/10.1201/9781003220855-6/urban-context-role-emergence-development-social-innovation-rivera-vinueza?context=ubx&refId=c78c34dc-80de-4f05-9103-b223cc6dd68b.

Further Reading

Green, J. (2018). *Medellín is healing itself with social urbanism*. The Dirt Uniting the Built and Natural Environments. https://dirt.asla.org/2018/11/28/medellin-is-healing-itself-with-social-urbanism/

Social-Ecological Systems

► Faith Communities as Hubs for Climate Resilience

Socially Vulnerable Neighborhoods

► Walkable Access and Walking Quality of Built Environment

Socio-ecological Resilience

► Concepts, Approaches, and Methodologies for Ecological Flood Resilience Assessment: A Review

Socio-ecological Urbanism

► Sustainable Community Masterplan

Socio-environmental Impact

► Community Engagement for Urban and Regional Futures

Soft-Governance

► Behavioral Science Informed Governance for Urban and Regional Futures

Soft-Law

► Behavioral Science Informed Governance for Urban and Regional Futures

Soft-Regulation

► Behavioral Science Informed Governance for Urban and Regional Futures

Soil Water Erosion Assessment for Conservation Planning in a Data-Pour Contest

Jean Nacishali Nteranya
Department of Geology, Faculty of Sciences, Université Officielle de Bukavu (UOB), Bukavu, Democratic Republic of Congo

Synonyms

Soil water erosion monitoring and management

Definition

- Soil: A terrestrial ecosystem that is at the interface between the hydrosphere, the biosphere, and the geosphere. It's the product of rock alteration through exogenous processes.
- Water erosion: A set of complex and interrelated processes that cause the detachment and transport of soil particles. It is defined as the loss of soil due to the water which tears off and transports the soil toward a place of deposit.
- Sheet erosion: A form of water erosion that is related to the detachment of soil particles caused by the impact of raindrops (splash effect) and to runoff when the intensity becomes greater than the infiltration rate.
- Linear erosion (micro-channel or rill erosion): A form of water erosion that manifests itself as micro-channels or rills. This form of erosion follows sheet erosion by concentration of runoff in hollows. At this stage, the gullies do not converge but form parallel streams.
- Gully erosion: A form of water erosion that manifests itself by the establishment of gullies (deepened channels where water streams are concentrated). Gully erosion is an advanced stage of erosion. This form of erosion can transform the landscape into "badlands" and also explains the undermining of structures (bridges, riffles, filtering dikes, etc.).

Introduction

Water erosion problems are increasingly felt in both developing and developed countries and have enormous economic as well as environmental consequences. Soil erosion is considered the most dangerous form of soil degradation on a global scale (Alexandridis et al. 2015). It accounts for 85% of land degradation (Singh and Panda 2017) and leads to the loss of 24 billion tons of soil (UNCCD 2017). It has become a global environmental and social problem and is currently one of the major environmental issues of water resource degradation. It endangers the sustainability of ecosystem services provided by soil at the global scale, and particularly in the humid tropics where the potential is great due to heavy rainfall (Labrière et al. 2015). Associated with chemical and physical degradation, they have been reported as the main types of soil degradation in the world (Avakoudjo et al. 2015). In addition, soil erosion by water has impacts on ecology through the destruction of natural environments (Issa et al. 2016) and on social and economic life at local, regional, and global scales (Elbouqdaoui et al. 2005). It is then important to better assess the risk of water erosion in order to develop preventive measures and adequate techniques to counter them. Therefore, a good planning of soil conservation is necessary. However, in developing countries, there is a scarcity or unavailability of relevant field data on the quantification of erosive processes and their causal factors. This makes it difficult to adopt appropriate conservation and management practices in erosion-prone areas. This lack of data limits the application of appropriate conservation measures. Tools to assess water erosion patterns can be of great importance in planning soil conservation measures. In order to address erosion problems, it is necessary to identify priority areas for the implementation of soil and water conservation measures that are based on erosion risk assessment. Since extensive erosion measurements are costly and time-consuming, the use of remote sensing and open access secondary data integrated with Geographic Information System (GIS) appears as an alternative for water erosion modeling. This chapter explores methods and concepts that are commonly used for water erosion assessment for soil conservation planning in areas where field data are scarce.

S

Background

Water erosion consists of the detachment and transport of soil particles under the combined mechanical action of rain and runoff (Dumas 2010). This erosion can take the form of sheet erosion, linear erosion (micro-channel or rill erosion), and gully erosion according to its increasing intensity. It is controlled mainly by natural factors (climate, topography, soil, vegetation) and anthropogenic factors (tillage systems, soil conservation measures, overgrazing, and deforestation) (Fig. 1). Soil loss through water erosion is influenced by a synergy of natural and anthropogenic factors such as soil type, topography, land use management, soil cover, and climate. Therefore, erosion is related to the synergy between human interactions and the environment through agricultural practices, urbanization, deforestation, construction, etc., which reduce the vegetation cover of the soil that contributes to soil stability. The erosion process is modified by the biophysical environment which includes soil, climate, terrain, land cover, and the interactions between them (Ganasri and Ramesh 2016). In addition, anthropogenic impact causes changes in the landscape, mainly on soil vegetation cover and soil physical properties (Thiemann et al. 2005). Each soil type will react differently to rain attack and runoff shear depending on its texture, structure, porosity, and organic matter content (Dumas 2010). Rain erosion occurs when there is no vegetative cover to dissipate the kinetic energy of the raindrop. In addition, plant roots also resist particle detachment by runoff, thus reducing erosion (Espiritu 1997). On the other hand, a loss of vegetation cover increases the risk of erosion.

Natural erosion known as geological erosion is one of the important processes that has shaped the earth's surface (Olson et al. 2016) and has produced some of the most spectacular geomorphic landscapes in the world. This continuous, slow, and constructive process can be significantly altered and exacerbated by anthropogenic activities and the effect of climate change. This creates a natural imbalance between soil loss and soil formation. The consequences of this imbalance are unfavorable in several cases for human society. Indeed, soil erosion in a watershed leads to increased sedimentation in streams and reservoirs, thus reducing their storage capacity and living space (Sinha and Regulwar 2015). In addition,

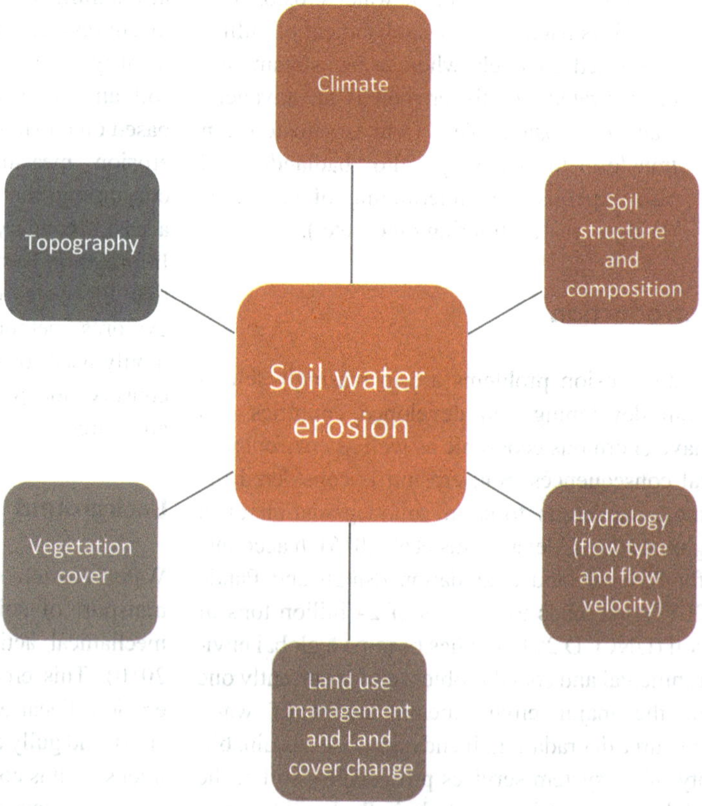

Soil Water Erosion Assessment for Conservation Planning in a Data-Pour Contest, Fig. 1 Factors affecting soil erosion

soil erosion leads to the removal and redistribution of the soil surface fraction including organic matter and clay content (Wijitkosum 2012). Acceleration of this process by anthropogenic activities can affect soil organic carbon stock, water resources, and negatively impact primary land productivity (Olson et al. 2016). Because soil organic matter content is concentrated in the surface portion of the soil, accelerated erosion leads to their decrease. Furthermore, soil erosion is a destructive process that alters and changes soil organic carbon causing the loss of a significant amount of organic carbon that has been retained in the soil system for millennia (Olson et al. 2016). This has implications for atmospheric CO_2 emissions (Chappell et al. 2014) and leads to loss of productivity (Wijitkosum 2012) and soil fertility by reducing areas suitable for agriculture (Farhan et al. 2015). Thus, it is very important to know the effects of erosion on the physicochemical characteristics of the soil in order to understand the dangers caused on productivity (Avakoudjo et al. 2015) and on the dynamics of soil organic carbon (Shukla and Lal 2005). Overall, soil erosion has negative impacts on soil physicochemical properties (Manyiwa and Dikinya 2013; Avakoudjo et al. 2015). The effects of water erosion are expressed on the ground by a decrease in vegetation cover density, soil loss, loss of organic matter and soil retention capacity, increased mineralization of the landscape, the appearance of active erosion notches, and progressive salinization of irrigated land. In addition, in urban areas, the sealing of surfaces following the construction of roads, houses, drainage networks, etc. contributes to the modification of surface runoff that becomes concentrated in some places. This concentration of surface runoff also contributes to the acceleration of erosive processes. The acceleration of water erosion in urban areas is not without consequences. It causes problems such as the destruction of urban infrastructures (bridges, roads, houses, etc.) and the reduction of the capacity of reservoirs and river beds due to the deposition of sediments, thus contributing to the increase in floods, the clogging of drainage networks, an increase in the cost of water treatment, etc. (Shikangalah et al. 2016).

Overview of Approaches Used in Water Erosion Assessment

Soil and water conservation requires the consideration of environmental, economic, and socio-political aspects in the development of management policies. To do so, a preliminary diagnosis of erosion problems and their causal factors is necessary. Furthermore, the selection of best management practices is important for soil conservation. Thus, the choice of management techniques must be adapted to the local context and take into account the characteristics of the natural vegetation and its uses, the characteristics of the farms, the financial and labor costs related to management, and the cropping systems practiced and their possible improvements.

Erosion risk reduction planning can be carried out once the magnitude, spatial distribution, and level of susceptibility are well assessed. The assessment of water erosion can be carried out on the basis of different approaches.

- The inventory and mapping approach based on field data collection is essential since the information is detailed and represents direct observations in the area of interest. In this perspective, permanent monitoring stations of soil loss by water erosion are installed for monitoring the dynamics of erosion. Field data also allow for an inventory of erosion patterns. The use of these conventional methods to assess erosion is costly in terms of time and money (Ganasri and Ramesh 2016).
- The second approach is by remote sensing. It consists of the use of satellite images and aerial photographs in the inventory of erosion forms.
- The third approach consists of the integration of data from remote sensing and field data in the GIS. GIS is a tool for the management, visualization, and analysis of georeferenced data.

The selection of an approach for erosion assessment is based on the available resources and the size of the study area. In developing countries that are characterized by limited data, remote sensing is a technology that can be used to assess erosion; combined with GIS and modeling, a reasonable estimate of soil erosion can be made (Dondofema 2007).

Remote Sensing as a Tool to Identify Erosion Prone Area

The descriptive water erosion assessment approach gives a true picture of the different forms of erosion located in the study area and their degree of exposure to degradation. The application of this approach requires inventory data of the different forms of erosion. Therefore, field campaigns are essential for monitoring of different forms of erosion and to assess soil losses due to water erosion with a high degree of accuracy. However, inventories based on field data alone are not sufficient for assessing erosion risk over a large area at the scale of a country, a region, or at the global scale, in remote areas or for long-term assessment. This is due to the fact that inventory work based only on field data requires high costs in terms of time and money. Therefore, in areas with low data coverage, remote sensing data are used for monitoring of water erosion patterns at different spatial and temporal scales. Aerial photographs and satellite images have the advantage of long-term monitoring of erosion dynamics and variations in surface conditions. Therefore, an inventory of erosion patterns can be made from satellite images and field work (Thiemann et al. 2005).

In the mapping of erosion patterns, the choice of images to be used depends on their qualities, which are a function of spectral, spatial, radiometric, and temporal resolution. Depending on the desired accuracy and the size of the study area, IKONOS, SPOT panchromatic, Landsat TM, and Orthophoto images can be used for erosion mapping in areas with low data coverage (Dondofema 2007; Wouters and Wolff 2010). The quality of the results therefore depends on the quality of the images chosen, but this choice is limited by the cost of image acquisition. Table 1 presents some of the satellite images used in the mapping and monitoring of water erosion according to the scale of observation.

In regions characterized by a severe lack of data, the use of open-source satellite data available on Google Earth can be a very effective tool for monitoring and mapping areas affected by erosion (McInnes et al. 2011; Ozer 2014). For this purpose a 3D visual inventory allowing the identification of erosion forms and a digitization on the digital globe are performed (McInnes et al. 2011). The observations made on Google Earth images can be transferred to a GIS using the Kml (Keyhole Markup Language) format (Frankl et al. 2013). Google Earth images are then used in gully mapping (McInnes et al. 2011), coastal erosion (Li 2016), and landslides (Maki Mateso 2014). Frankl et al. (2013) show that high-resolution Google Earth imagery yields more accurate results compared to field inventory done with a conventional GPS (Garmin GPSMap 60). However, McInnes et al. (2011) show that the use of Google earth images in gully erosion mapping gives satisfactory results in agricultural or uncovered areas and not in forested areas due to canopy

Soil Water Erosion Assessment for Conservation Planning in a Data-Pour Contest, Table 1 Satellite images used in the mapping of water erosion patterns

Image	Spatial resolution (m)	Spatial extension	Acquisition cost per km^2
Aerial photo	0–1	Local	High (more than 5$ Us)
Quickbird	0.65–2.4	Local	High (more than 5$ Us)
IKONOS	1–4	Local	High (more than 5$ Us)
Sentinnel	10–15	Local and regional	Fable (less than 0.5$ Us)
SPOT	1.5–20	Local and regional	High (more than 5$ Us) and Median (between 5$ and 0.5$)
LANDSAT	15–30	Local and regional	Low (less than 0.5$ Us)
ASTER	15–30	Local and regional	Low(less than 0.5$ Us)

coverage. The identification of gullies under the canopy cover can only be done through field work.

Quantitative Analysis of Soil Erosion Causative Factors for Susceptibility Assessment

In soil conservation planning, it is essential to understand the mechanism of erosion and the interactions between causative factors. Susceptibility analysis and erosion mapping are prerequisites for sustainable soil management and water erosion prevention. Therefore, the selection of erosion causal factors is important in the development of erosion occurrence models. Several factors are used in the modeling of erosive processes. Among these factors are soil factors (soil texture, soil permeability, soil moisture, soil organic matter content, etc.), lithological factors (degree of rock alteration, fracture density, etc.), climatic factors (rainfall intensity), hydrological and topographical factors (slope, profile curvature, stream power index, drainage density, sediment transport index and topographic wetness index, lineament density), and type of land use and land cover. These biophysical factors are not identical in all regions. Hence, erosion management requires site-specific studies.

To highlight potential erosion areas, statistical modeling based on logistic regression, analytical hierarchy process technique; sensitivity analysis approaches; stochastic gradient tree boost; weighting overlay; soft computing method; fuzzy and artificial neural-network evaluation methods; and bivariate method is used. The implementation of this approach requires taking into account the causal factors of erosion at the scale of the area under examination. These factors can be described as static or dynamic depending on whether or not they vary with the rainfall cycle. For example, rainfall erosivity, land surface temperature, and soil moisture are considered to be dynamic factors, while topographic factors are considered to be static. It should be noted that the quality of the model developed depends on the factors taken into account and the redundancy of the factors can lead to an overestimation of the model. Hence, a careful choice of variables and a reduction of variables is necessary before any application of the model. This reduction of variables can be done on the basis of correlation analysis and principal component analysis in order to avoid variables that are redundant. For example, hydrological and hydrogeomorphological factors such as length-slope (LS), topographic wetness index (TWI), plan curvature, and stream power index (SPI) can influence erosive processes. However, these factors are all related to slope angle. Thus, when these factors are associated with slope angle, they can be considered redundant (Abdulkadir et al. 2019). The redundancy of the factors may lead to the invalidity of the developed model, hence the need to do variable reduction. The accuracy of the validity of the erosion susceptibility map model can be improved by increasing the causal factors considered in the erosion occurrence models provided that these factors are not redundant. Before applying the models developed for soil conservation planning, it is essential to consider their performance.

In a data pour contest, the erosion susceptibility model can be realized by considering dependent variables which are freely available. Such variable can be extracted from remote sensing data or national and regional database. For instance, some topographic and hydrogeomorphological variables used in susceptibility models are extracted from the Digital Elevation Model derived from SRTM data, and the vegetation cover density is represented by the NDVI (Normalized Difference Vegetation Index) extracted from available remote sensing data (Landsat, Sentinnel, etc.). Soil and climatic variables can be extracted from national or regional database. However, the occurrence data of erosion which are used as independent variable of the model of susceptibility require a field or a remote sensing–based inventory of erosion prone area. Soil loss data from permanent monitoring stations of erosion are also considered when such data are available. Another approach uses the result of the quantitative assessment of soil loss by means of the USLE model as an independent variable and by considering the watershed as the unit of measure. This approach was used by Abdulkadir et al. (2019) in order to perform a quantitative analysis of soil erosion causative factors for susceptibility assessment in a complex watershed. Despite these considerations, field data are necessary for the validation of the developed models.

Quantitative Assessment of Soil Loss by Water Erosion for Conservation Planning

Soil erosion models are commonly used to investigate the physical processes and mechanisms that govern the rate of erosion and to identify areas at high risk of soil loss to aid conservation planning. A distinction is made between empirical models that estimate erosion based on the relationship of soil loss with the combination of predefined physical parameters and following standard coefficients or procedures, conceptual (partially empirical) models, and physical (or mechanistic) models that simulate the physical erosion process by solving the fundamental equations that describe runoff flow and sediment generation (Espiritu 1997; Ganasri and Ramesh 2016). These models include the Universal Soil Loss Equation (USLE), its modified (MUSLE) and revised (RUSLE) versions, Erosion Productivity Impact Calculator (EPIC), European Soil Erosion Model (EUROSEM), and Water Erosion Prediction Project (WEPP). These models are often integrated into GIS to facilitate spatial analysis of erosion risk. Due to the availability of low-cost remote sensing data with good spatial resolution and regional databases, these techniques are commonly used to assess water erosion processes at the watershed or larger-scale areas where data are limited.

The WEPP and R(USLE) models are the most commonly used models for predicting erosion at the landscape scale in the tropics. The WEPP model is a predictive (physical) process-based model: detachment, transport, and deposition of eroded soil. This model predicts soil erosion as a result of a rainfall event that can be projected to a year using detailed rainfall data to generate the annual soil loss prediction. This model takes into account soil movement and deposition processes. However, the application of this model requires detailed daily climate data, physiographic characteristics of crops at different stages of development, shape channel, substrate, and depth. Hence the use of this model in tropical areas where data are limited is difficult.

The Universal Soil Loss Equation (USLE), an empirical model developed by Wischmeier and Smith (1978), and GIS are the most popular because of their simple computational requirement and factors that are available on site (Espiritu 1997; Biswas and Pani 2015). Indeed, the (R)USLE model predicts average annual soil loss at the scale of a plot, watershed, or larger area based on six factors: rainfall erosivity (R), soil erodibility (K), slope length (L), slope steepness (S), soil cover (C), and supporting practices (P). These parameters can be estimated based on remote sensing and open access secondary data. This makes it an ideal candidate for determining erosion potential in areas with limited data. Another advantage of (R)USLE is that the parameters of this model can be easily integrated into a GIS for better analysis (Ganasri and Ramesh 2016). They are most widely used in soil erosion modeling with a considerable degree of certainty (Sinha and Regulwar 2015). The application of this model gives results that can provide valuable assistance, at a very low cost, to decision makers and land planners in order to simulate evolution scenarios and subsequently target priority areas that require conservation and erosion control actions. Indeed, thanks to the analysis of spatial and temporal variability of erosion risk, this model provides a good understanding of the relationships between erosion and land management (Hunink et al. 2015; Gashaw et al. 2018). Surveys of people's perceptions on land degradation from erosion can also be combined with (R)USLE modeling to gain a good understanding of the socioeconomic factors determining erosion and land conservation in a given setting (Farhan et al. 2015). This model also assesses the effect of seasonality on soil loss through water erosion. This effect depends on climate and spatial distribution of vegetation (Alexandridis et al. 2015) (Table 2).

Soil Water Erosion Assessment for Conservation Planning in a Data-Pour Contest, Table 2 Example of secondary data used for estimation of (R)USLE model factors

Factor	Associated data	Resolution
R	CRU TS 2.1 Climate Database	0,5 degree
	WorldClim v 1.4	1 km
	WorldClim v 2.0	1 km
	1979 to 1993 global daily meteorological data	2-degree
	CHIRPS database	0.05-degree
	CCI-LC	300 m
K	SOTERCAF 1.0	1/2000000
	SOTERSAF	–
	FAO DSMW	5-min
	AfSIS data	250 m
	HWSD V1.2	30 arc second
L & S	SRTM digital elevation data version 3	90 m
	SRTM digital elevation data version 4	3 arc second
	GTOPO-30 data set	1 km
	ASTER global digital elevation model (GDEM) version 2	30 m
C	Monthly Leaf Area Index (LAI) data layers were derived from the GlobCarbon project	1 km
	NDVI derived from Sentinnel	15 m
	NDVI derived from Landsat	30 m
	USGS global land cover map	1 km
	ESA global land cover	
	MODIS NDVI	250 m

CHIRPS: Climate Hazards Group InfraRed Precipitation; AfSIS: Africa Soil Information Service; SRTM: Shuttle Radar Topographical Mission Version; MODIS: Moderate Resolution Imaging Spectroradiometer; NDVI: Normalized Difference Vegetation Index; CCI-LC: Climate Change Initiative Land Cover project; HWSD V1.2: Harmonized World Soil Database version 1; CRU TS: Climatic Research Unit Time series; SOTERCAF: Soil and Terrain Database of Central Africa; SOTERSAF: Soil and Terrain Database for Southern Africa; DSMW: Digital Soil Map of the World; USGS: United States Geological Survey; ESA: European Space Agency

In areas where there is a severe lack of data on soil protection (P factor) and soil cover (C factor) alternative approaches have been developed. This is the case of the Land Erodibility Assessment Methodology (LEAM) model, developed by Manrique (1988), which was based on the Wischmeier equation chosen for its ease of use. It is a simple model designed for developing countries lacking complete and correct soil and climate data. It is also designed for regional and watershed studies and allows for a simple and quick diagnosis of potential erosion. The model includes among its factors soil erodibility, slope steepness, and rainfall erosivity. This model does not take into consideration the distribution of vegetation cover, which is a major deficiency since it can be a determining factor in assessing the erosion risk in the study area (Elbouqdaoui et al. 2005).

The (R)USLE model was developed on the basis of field parameters taken from agricultural lands in the United States to estimate sheet or rill erosion resulting from surface runoff/overland flow. These models do not account for gullying, sedimentation, and wind erosion. The results of (R)USLE models are considered valid only within the limits of the experimental conditions under which they were developed. However, this model is currently applied in contexts different from the original context for which it was developed. For instance, models developed in temperate countries are often applied in tropical areas, whereas the erosive processes are different in these two contexts. This may affect the validity of the results obtained from the (R)USLE model (Nacishali Nteranya 2021). It is therefore necessary to refer to field data (permanent soil loss monitoring station) in order to parameterize and validate the developed model while taking into account the observation scale. The model results can also be validated by referring to literature data while taking into account areas with similar climatic and geomorphological contexts. In areas where such soil loss monitoring stations are rare or non-existent, instead of relying on quantitative annual soil loss values, which should be used with caution since there is substantial uncertainty in the magnitude of the estimated values due to the lack of model calibration, the qualitative approach should be used. It is then customary to generate categorical data by delineating the study area into different erosion intensity zones in order to identify areas of low, moderate, high, and extreme erosion. This can be used to determine relatively high erosion intensity areas that merit special water and soil conservation efforts. Thus, on a broad scale, the results of the (R)USLE models reflect the relative pattern of potential erosion. Fig. 2 present a conceptual framework for soil erosion assessment using quantitative approach by integrating the (R)USLE model for sheet erosion assessment and Google Earth image analysis for gully erosion assessment.

Soil Loss Tolerance Limits Map as a Tool for Conservation Planning

Erosion risk is often estimated by integrating spatial data into GIS to estimate potential erosion rates and tolerable soil loss limits for conservation planning at the watershed or larger-scale areas. Erosion risk classes are prioritized based on differences between the prevailing erosion rate and allowable erosion limits. Criteria for judging whether or not a soil is susceptible to erosion are required for the adoption of appropriate erosion control measures. The tolerable soil loss limit is one of the criteria for identifying and prioritizing areas at risk of water erosion. This criterion is essential for the rational adoption of appropriate soil conservation measures in cultivated areas. It constitutes the foundation of soil conservation (Shi et al. 2004).

In the field of erosion, tolerance was first defined as the tolerable soil loss as it is balanced with the formation of the soil by the weathering of rocks. Tolerable soil loss represents the maximum rate of erosion that can occur while allowing economically sustainable agricultural productivity (Renard 1997). It varies from 1 to 12 t/ha/yr. (Roose 1994) depending on climate, rock type, soil thickness and soil types, and humus horizons. However, this approach denies the importance of selective erosion of nutrients and colloids that make up soil fertility. This same author then attempted to define tolerance as "erosion that would not cause a significant decrease in the productivity of the land." He then came up against certain major obstacles, such as the loss of productivity of different types of soil as a function of erosion.

The evaluation of the limit of tolerable soil loss requires taking into account the rate of soil formation and the time steps during which the land use would be sustainable. This tolerable soil loss limit is currently used in soil conservation planning (Sharda et al. 2013). Experimental data show that soils have an inherent capacity to allow some loss through water erosion without compromising the productivity and sustainability of land use in the long term, but this depends on the location. The tolerable or allowable rate of soil loss is a function of the balance between the forces causing soil loss and soil formation. Therefore, it is essential to consider the tolerable soil loss limit and the rate of erosion when planning soil conservation and erosion risk reduction in a given area. Consideration of the tolerable soil loss limit and quantitative measures of potential soil loss allows prioritization of areas for soil conservation. Thus, agricultural areas with potential soil loss above the tolerable limit are considered priorities for conservation, while areas with tolerable soil loss below the tolerable limit are considered to require no conservation action.

Soil Erosion Potential Map as a Tool for Modeling Best Management Practice

The soil erosion potential map is an essential tool in the fight against erosion. It provides an overview of the threatened areas and locates the

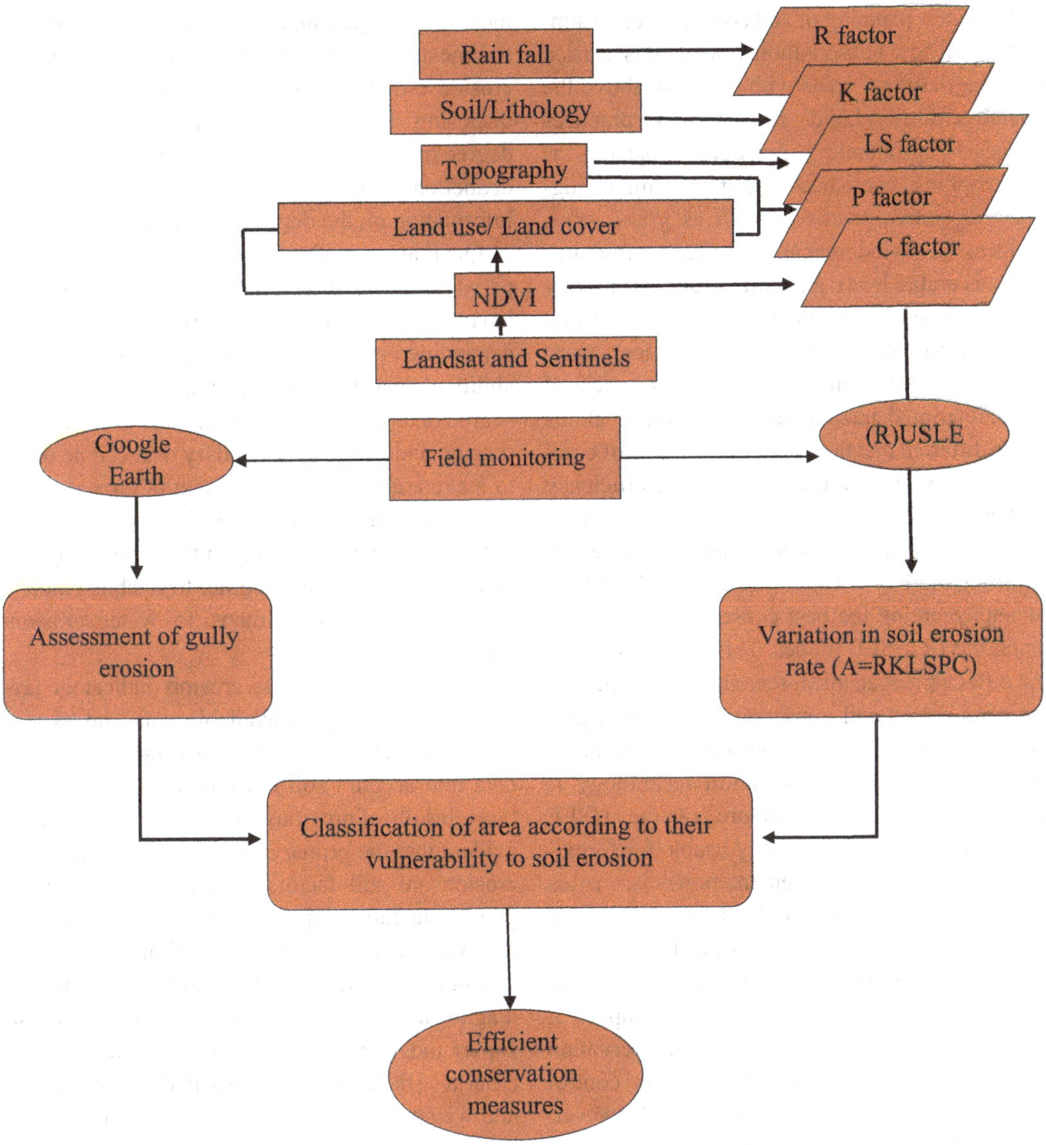

Soil Water Erosion Assessment for Conservation Planning in a Data-Pour Contest, Fig. 2 Conceptual framework for soil erosion assessment using quantitative approach

sectors requiring priority intervention in a sustainable soil management perspective. The spatial nature of water erosion is highlighted by integrating multi-source data into a geographic information system (GIS).

The potential erosion map obtained from the spatialization of the (R)USLE model results can be considered as a conservation planning tool. When coupled with the population density map, it is a tool that can be used to identify areas where the population is more vulnerable to erosion problems. It can also be combined with information on land use practices and climate predictions to identify areas where land use practices or future land use change is likely to increase erosion problems. It can be used to highlight areas most vulnerable to erosion and identify locations where erosion problems are likely to affect agricultural productivity and human well-being (Claessens et al. 2008).

For better planning of soil conservation, a simulation of best conservation practices is carried out in agricultural areas with soil loss above the tolerable limit. Land conservation practices include agronomic practices (such as intercropping, contour cropping, mulching, ridging), vegetative practices (such as grass strips, hedgerows, windbreaks), physical structures (such as arable land), and soil conservation practices), physical structures (such as terraces, benches, bunds, constructions, palisades), management modes (such as land use change, fenced areas, rotational grazing), as well as combinations of the different practices under conditions of complementarity and mutually reinforcing efficiency such as agroforestry. The establishment of different management scenarios can be done by adjusting the P or C values to simulate the effects of implementing the best conservation practices using the (R)USLE model. This allows to show the different management scenarios and to predict their impact on soil conservation. For this purpose, a value of P=1 is considered to simulate the case where no conservation technology is adopted. The simple structure of the USLE model equation makes it easily usable for formulating transparent management policy scenarios by changing the types of land uses (C and P factors) under different ecological conditions (R, K, L, and S factors). The USLE model offers the opportunity to easily evaluate the impact of different land uses (C factor) and soil conservation techniques (P factor) on water erosion control using data available in the literature. However, these parameters must be applied with caution since the ecological context in which they were previously calibrated (East of the Rocky Mountains, USA) may be different from the context of the study area.

Qualitative Assessment of Soil Water Erosion for Conservation Planning

Qualitative analysis methods of water erosion allow to prioritize the surface of the watershed or of a larger territory into distinct units according to the vulnerability to erosion and to determine the most fragile zones and thus potential providers of sediments. It is a predictive approach to water erosion based on the thematic mapping of predisposing factors (slope, lithology, land use, degree of vegetation cover, etc.). It ends with the deduction of the erosive state map which provides information on the degree of erosion.

Qualitative methods require little data compared to quantitative methods. These methods have the flexibility to modify the parameters depending on the scale of analysis and the availability of data and expertise of the operator. Multisource open-access biophysical data from remote sensing and available secondary data can be used to make a qualitative assessment of erosion risk. The methodology adopted in this approach consists of crossing parameters in the form of logical combinations in a GIS in order to establish multifactorial vulnerability maps to water erosion (Fig. 3).

In a first step, water erosion indicators are selected according to available data and expert judgments. The criteria for assessing erosion risk take into account soil erosion type, local conditions, results of previous work, and expertise. It is evident that the primary factors controlling water erosion are soil factors (soil erodibility), topographic factors (slope), climatic factors (rainfall erosivity), and land use (vegetation cover). The attributes of these different factors are selected as diagnostic criteria for erosion on the basis of expert judgment. In a second step, the selected criteria (factors) are classified, normalized, and weighted according to expert judgment. The classification of the different erosion factors according to their involvement in the occurrence of the phenomenon is established by taking into account field observations and bibliographic knowledge. The standardization of the values of the criteria is essential insofar as we have qualitative (categorical) variables such as the slope and quantitative (numerical) variables such as the type of land use which are not comparable between them. In this context, all variables are classified and normalized according to a standard scale taking into account the expert's knowledge and experience. This multi-criteria analysis is carried out in the GIS environment in order to determine the

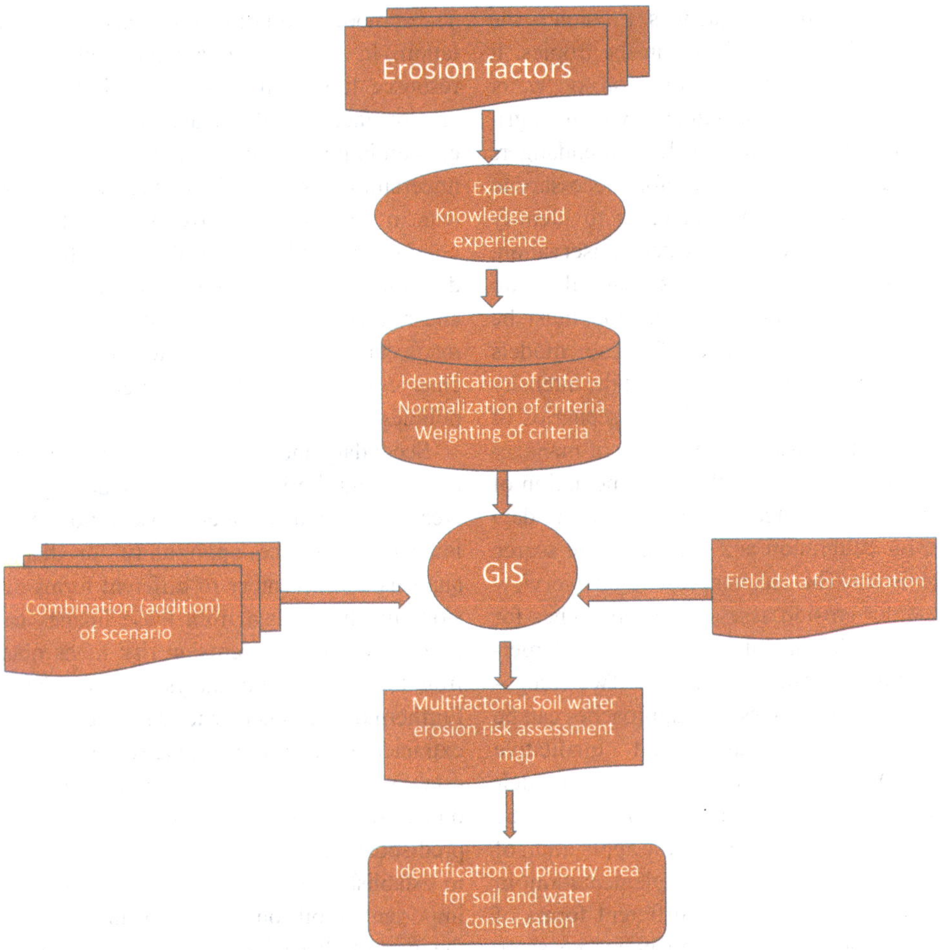

Soil Water Erosion Assessment for Conservation Planning in a Data-Pour Contest, Fig. 3 Conceptual framework for soil erosion assessment using qualitative approach

relative erosion risk by delimiting low, moderate, and very high risk areas. This prioritization of erosion risk is used to determine priority areas for soil and water conservation. The validation of this model requires inventory data of the different forms of erosion. To do so, an integration approach is adopted. The integration approach consists in superimposing the maps of erosive states obtained by thematic mapping (combination and prioritization of predisposing factors) and the map of erosion forms obtained by direct descriptive mapping of erosion forms on the ground or by satellite images.

Qualitative estimation based on multi-criteria analysis and expert knowledge can provide satisfactory results to planners in erosion conservation planning when data are limited or when large areas of analysis are considered. However, the results of this approach are a function of the expert's knowledge, and the results may vary depending on the expert's judgment.

Fig. 3 shows an example of a conceptual framework for the application of a qualitative water erosion assessment for soil conservation.

S

Summary/Conclusion

Soils provide many ecosystem services to humanity. However, soils are threatened by water erosion in both developed and developing countries where erosion problems have enormous economic and environmental consequences. Erosion is the main form of soil degradation in the tropics. It also contributes to the degradation of water resources, reduces the sustainability of the agricultural sector in the rural world, and endangers urban infrastructures. It is therefore necessary to assess the risk of erosion in order to identify priority areas for soil and water conservation. To do this, predictive models that take into account the causal factors of erosion must be applied. The development of these models requires quantitative data to estimate the intensity of erosion and to analyze different scenarios of soil loss due to erosion under different management practices. However, the implementation of these models is limited in areas where the data needed for calibration and validation are scarce or nonexistent. This chapter presents an overview of soil water erosion assessment approaches for conservation planning at the watershed or larger-scale area in a data-limited context. Two categories of erosion risk assessment approaches can be distinguished: quantitative and qualitative methods. Quantitative methods focus on modeling soil loss to assess erosion risk. However, the availability of data required for the application of these methods limits their implementation in data-poor areas. In addition, different forms of erosion can occur simultaneously in the same geographical area. As a result of this constraint and the limitations of available data, the application of a single quantitative method is not appropriate in the context of limited data. Thus, qualitative approaches are often developed to assess erosion risk and identify areas at high risk of erosion.

The use of open access secondary data and remote sensing data in the quantification of water erosion has limitations due to the heterogeneity of the data which are of different spatial and temporal resolution. Furthermore, in the context of poor data, the models commonly used to quantify soil losses due to water erosion, namely the USLE model and its modified versions (R)USLE, are used in contexts different from the original contexts for which the models were developed. These models do not take into account sedimentation. In addition, at a large scale, data and resource limitations and regional variability in erosion factors make quantitative assessment of erosion impossible in some cases. Because of the uncertainties associated with the use of secondary data in estimating soil loss, the soil loss values obtained should be considered qualitatively to determine relatively high erosion potential areas and not as actual soil loss values. The resulting maps can be used to simulate different land use change scenarios in order to develop conservation policies.

Secondary and remote sensing data can also be used for multi-criteria analysis to identify priority areas for soil and water conservation. Satellite imagery offers the opportunity for diachronic analysis and inventory of different forms of erosion. Inventory data from these images can be used to validate the erosion risk maps obtained from the multi-criteria analysis of secondary data. Furthermore, the occurrence data of erosion forms extracted from satellite images can be used to establish water erosion susceptibility maps based on statistical modeling according to the different predisposition factors. These models can be used to establish erosion susceptibility maps that are important in soil conservation planning. However, the use of these images is constrained by the variable quality of the images, the acquisition cost, and by the fact that free Google Earth images do not allow for an inventory of erosion under the canopy.

References

Abdulkadir, T. S., Muhammad, R. U. M., Yusof, K. W., Ahmad, M. H., Aremu, S. A., Gohari, A., & Abdurrasheed, A. S. (2019). Quantitative analysis of soil erosion causative factors for susceptibility assessment in a complex watershed. *Cogent Engineering, 6* (1), 1594506.

Alexandridis, T. K., Sotiropoulou, A. M., Bilas, G., Karapetsas, N., & Silleos, N. G. (2015). The effects of seasonality in estimating the C-factor of soil erosion studies. *Land Degradation & Development, 26*(6), 596–603.

Avakoudjo, J., Kouelo, A. F., Kindomihou, V., Ambouta, K., & Sinsin, B. (2015). Effet de l'érosion hydrique sur les caractéristiques physicochimiques du sol des zones d'érosion (dongas) dans la Commune de Karimama au Bénin. *Agronomie Africaine, 27*(2), 127–143.

Biswas, S. S., & Pani, P. (2015). Estimation of soil erosion using RUSLE and GIS techniques: A case study of Barakar River basin, Jharkhand, India. *Modeling Earth Systems and Environment, 1*(4), 1–13.

Chappell, A., Webb, N. P., Viscarra Rossel, R. A., & Bui, E. (2014). Australian net (1950s–1990) soil organic carbon erosion: Implications for CO_2 emission and land–atmosphere modelling. *Biogeosciences, 11*(18), 5235–5244.

Claessens, L., Van Breugel, P., Notenbaert, A., Herrero, M., & Van De Steeg, J. (2008). Mapping potential soil erosion in East Africa using the Universal Soil Loss Equation and secondary data. *IAHS Publication, 325*, 398.

Dondofema, F. (2007). *Relationships between gully characteristics and environmental factors in the Zhulube Meso-catchment: Implications for water resources management*. Faculty of engeneering, University of Zimbabwe, Zimbabwe.

Dumas, P. (2010). Méthodologie de cartographie de la sensibilité des sols à l'érosion appliquée à la région de Dumbéa à Païta-Bouloupari (Nouvelle-Calédonie). *Les Cahiers d'Outre-Mer. Revue de géographie de Bordeaux, 63*(252), 567–584.

Elbouqdaoui, K., Ezzine, H., Badrahoui, M., Rouchdi, M., Zahraoui, M., & Ozer, A. (2005). Approche méthodologique par télédétection et SIG de l'évaluation du risque potentiel d'érosion hydrique dans le bassin versant de l'Oued Srou (Moyen Atlas, Maroc). *Geo-Eco-Trop, 29*(1–2), 25–36.

Espiritu, K. (1997). *Development of a computerized version of the universal soil loss equation and the USGS pollutant loading functions*. University Libraries, Virginia Polytechnic Institute and State University, Blacksburg, Virginia.

Farhan, Y., Zregat, D., & Anbar, A. (2015). Assessing farmers' perception of soil erosion risk in Northern Jordan. *Journal of Environmental Protection, 6*(08), 867.

Frankl, A., Poesen, J., Haile, M., Deckers, J., & Nyssen, J. (2013). Quantifying long-term changes in gully networks and volumes in dryland environments: The case of Northern Ethiopia. *Geomorphology, 201*, 254–263.

Ganasri, B. P., & Ramesh, H. (2016). Assessment of soil erosion by RUSLE model using remote sensing and GIS-A case study of Nethravathi Basin. *Geoscience Frontiers, 7*(6), 953–961.

Gashaw, T., Tulu, T., & Argaw, M. (2018). Erosion risk assessment for prioritization of conservation measures in Geleda watershed, Blue Nile basin, Ethiopia. *Environmental Systems Research, 6*(1), 1–14.

Hunink, J. E., Terink, W., Contreras, S., & Droogers, P. (2015). Scoping assessment of erosion levels for the Mahale region, Lake Tanganyika, Tanzania. *Unpubl FutureW Report, 148*, 1–47.

Issa, L. K., Lech-Hab, K. B. H., Raissouni, A., & El Arrim, A. (2016). Cartographie quantitative du risque d'erosion des sols par approche SIG/USLE au niveau du bassin versant Kalaya (Maroc Nord Occidental). *Journal of Materials and Environmental Sciences, 7* (8), 2778–2795.

Labrière, N., Locatelli, B., Laumonier, Y., Freycon, V., & Bernoux, M. (2015). Soil erosion in the humid tropics: A systematic quantitative review. *Agriculture, Ecosystems & Environment, 203*, 127–139.

Li, J. (2016, June). Using Google Earth in the study of shoreline erosion process. In *ASEE's 123rd annual conference & exposition*. American Society for Engineering Education, New Orleans, Louisiana.

Maki Mateso, J. C. (2014). *Inventaire de glissements de terrain et étude des éléments à risque dans le rift ouest du bassin du lac Kivu*. Travail de fin d'études de Master complémentaire en gestion des risques naturels, Faculté des Sciences.

Manrique, L. A. (1988). *Land erodibility assessment methodology (LEAM): Using soil survey data based on soil taxonomy*. Editorial and Publication Shop, Honolulu, Hawaii, USA.

Manyiwa, T., & Dikinya, O. (2013). Using universal soil loss equation and soil erodibility factor to assess soil erosion in Tshesebe village, north East Botswana. *African Journal of Agricultural Research, 8*(30), 4170–4178.

McInnes, J., Vigiak, O., & Roberts, A. M. (2011). Using Google earth to map gully extent in the West Gippsland region (Victoria, Australia). In *19th international congress on modelling and simulation, Perth, Australia* (pp. 12–16).

Nacishali Nteranya, J. (2021). Cartographie de l'érosion hydrique des sols et priorisation des mesures de conservation dans le territoire d'Uvira (République démocratique du Congo). *VertigO-la revue électronique en sciences de l'environnement, 20*(3), 1–34.

Olson, K. R., Al-Kaisi, M., Lal, R., & Cihacek, L. (2016). Impact of soil erosion on soil organic carbon stocks. *Journal of Soil and Water Conservation, 71*(3), 61A–67A.

Ozer, P. (2014). Catastrophes naturelles et aménagement du territoire: de l'intérêt des images Google Earth dans les pays en développement. *Geo-Eco-Trop: Revue Internationale de Géologie, de Géographie et d'Écologie Tropicales, 38*(1), 209–220.

Renard, K. G. (1997). *Predicting soil erosion by water: A guide to conservation planning with the Revised Universal Soil Loss Equation (RUSLE)*. United States Government Printing, Washington, DC.

Roose, É. (1994). Introduction à la gestion conservatoire de l'eau, de la biomasse et de la fertilité des sols

S

(GCES). Bulletin pédologique de la FAO, vol. 70, Rome (Italie), 420 p.
Sharda, V. N., Mandai, D., & Ojasvi, P. R. (2013). Identification of soil erosion risk areas for conservation planning in different states of India. *Journal of Environmental Biology, 34*(2), 219.
Shi, Z. H., Cai, C. F., Ding, S. W., Wang, T. W., & Chow, T. L. (2004). Soil conservation planning at the small watershed level using RUSLE with GIS: A case study in the Three Gorge Area of China. *Catena, 55*(1), 33–48.
Shikangalah, R. N., Jeltsch, F., Blaum, N., & Mueller, E. N. (2016). A review on urban soil water erosion. *Journal for Studies in Humanities and Social Sciences*, 163–178.
Shukla, M. K., & Lal, R. (2005). Erosional effects on soil organic carbon stock in an on-farm study on Alfisols in west Central Ohio. *Soil and Tillage Research, 81*(2), 173–181.
Singh, G., & Panda, R. K. (2017). Grid-cell based assessment of soil erosion potential for identification of critical erosion prone areas using USLE, GIS and remote sensing: A case study in the Kapgari watershed, India. *International Soil and Water Conservation Research, 5* (3), 202–211.
Sinha, A. P., & Regulwar, D. G. (2015). Soil erosion estimation of watershed using Quantum Geographic Information System (QGIS) and Universal Soil Loss Equation (USLE). In *International Journal of Science and Research (IJSR) Proc. of National Conference on Knowledge, Innovation in Technology and Engineering (NCKITE)* (pp. 10–11).
Thiemann, S., Schütt, B., & Förch, G. (2005). Assessment of erosion and soil erosion processes–a case study from the Northern Ethiopian Highland. *FWU Water Resource Publication, 3*, 173–185.
UNCCD. (2017), Perspectives territoriales mondiales, Première Edition, 340 p.
Wijitkosum, S. (2012). Impacts of land use changes on soil erosion in Pa Deng sub-district, adjacent area of Kaeng Krachan National Park, Thailand. *Soil and Water Research, 7*(1), 10–17.
Wischmeier, W., & Smith, D. (1978). *Predicting rainfall erosion losses. Agriculture handbook*. USDA & Agricultural Research Service, (282).
Wouters, T., & Wolff, E. (2010). Contribution à l'analyse de l'érosion intra-urbaine à Kinshasa (RDC). *Belgeo. Revue belge de géographie*, (3), 293–314.

Soil Water Erosion Monitoring and Management

► Soil Water Erosion Assessment for Conservation Planning in a Data-Pour Contest

Solar Energy Communities in the Urban Environment

Danielle Griego[1], Prakhar Mehta[2] and Alejandro Nuñez-Jimenez[3,4]
[1]Center for Augmented Computational Design in Architecture, Engineering and Construction, D-BAUG, ETH Zurich, Zurich, Switzerland
[2]Digital Transformation: Bits to Energy Lab Nuremberg, School of Business, Economics and Society, Friedrich-Alexander University Erlangen-Nürnberg (FAU), Nuremberg, Germany
[3]Sustainability and Technology Group, D-MTEC, ETH Zurich, Zurich, Switzerland
[4]Belfer Center for Science and International Affairs, Harvard University, Cambridge, MA, USA

Introduction

Decarbonizing energy consumption in urban environments could contribute decisively to mitigating climate change and reducing local air pollution (de Coninck et al. 2018, 313–443). Over two-thirds of global primary energy consumption, and an even greater share of energy-related CO_2 emissions, are derived from energy use in cities (IRENA 2009). Within this context, increasing the adoption of solar photovoltaics (PV) is a promising route to decarbonize urban energy systems (Lee et al. 2018; Margolis et al. 2017). However, installing PV encounters barriers such as competing uses in already limited available space, which partially explains PV's modest diffusion in urban centers. Community solar is described as a group of owners and/or electricity consumers who jointly develop and/or use a PV system according to an agreed upon means of governance. This is a relatively novel model of PV installations and has the potential to lower some of PV's adoption barriers.

In recent years, there is a growing trend towards community solar in cities throughout Europe, the United States and many other regions (REN21 2021). Economies of scale from larger installations make community solar more affordable than individual installations for PV adopters (Bauer et al. 2019), and, by aggregating the demand of multiple users, community solar tends to use more of the electricity it generates (Awad and Gül 2018; Viti et al. 2020). The latter fact not only benefits adopters who save more by purchasing less electricity from the utility company, but it also benefits grid operators, who also have to accommodate less intermittent renewable energy production into the grid. Community solar also delivers social benefits by enabling individuals who would otherwise lack the possibility to own an individual PV system to become PV adopters. In addition, community solar promotes local innovation, sustainable social norms, joint-investments, and self-governance (Roelich and Knoeri 2015). For these reasons, policymakers, corporations, and electricity consumers increasingly see community solar as an attractive way to increase renewable energy in cities.

However, the highly idiosyncratic and varied ownership and user structures of solar energy communities in urban environments (SECUEs) could hinder their diffusion. Each SECUE faces a unique set of constraints that includes space availability, property ownership configurations, electricity and building regulations, pre-existing contractual arrangements, and social structures, among many others. For example, throughout Germany and Switzerland, housing cooperatives are a common means to jointly own apartment complexes in city centers (Seidl 2018). Diversity of SECUE configurations, while useful for adapting to their specific contexts, makes studying their characteristics and assessing their benefits, challenges, and opportunities difficult. This complicates the work of researchers in the growing body of literature studying SECUEs. More importantly, it makes it harder to draft effective policies and regulations and to develop successful business strategies.

This chapter tackles this challenge by presenting a taxonomy of PV installations based on ownership and user structures in urban environments. The taxonomy makes it easier to identify community solar installations and map economic, environmental and social incentives for participants. Before presenting the new taxonomy, the chapter provides a brief overview of existing classifications of renewable energy communities. In the later sections, this chapter includes examples, opportunities and challenges of each SECUE category from social, economic, and environmental perspectives.

Theoretical Background

Consensus around the definition of energy communities and useful classifications have yet to emerge in literature (Gui and MacGill 2018; Moroni et al. 2019). The difficulty originates, to a large extent, in how diversely community energy is implemented (Klein and Coffey 2016).

Classifications of renewable energy communities vary according to what dimensions they focus on. However, the most common criterion/dimension highlighted in the literature is the connection to a specific location. An early article by Walker (2011) presented six interconnected meanings of the term *community* in the context of the environment. Of these, community as place and community as network have come to be widely used in literature to describe existing energy communities (Bauwens and Devine-Wright 2018). Communities of place or place-based communities refer to local communities bound by geographical limits within which existing social relationships are leveraged to develop collective initiatives. In contrast, a community of network, community of interest or non-place-based community, is a collection of individuals or entities with shared goals, irrespective of geographical constraints. One example of a community of interest is the sonnenCommunity (Sonnen 2021) operating in Germany, Austria, Switzerland and Italy.

These two classifications are complemented by considering the purpose of the community (Moroni et al. 2019). Single-purpose communities are "set up for the sole purpose of producing, managing of purchasing energy in accordance with shared rules" (Moroni et al. 2019). Alternatively, multi-purpose communities aim to produce

S

and manage other goods and services along with energy, such as a water treatment plant or a marketplace.

Furthermore, Gui and MacGill (2018) categorize energy communities with reference to current energy systems and markets as centralized, distributed and decentralized. Centralized energy communities are described as "cohesive networks of household and businesses that collectively own or participate in energy-related projects." In distributed energy communities, distributed generation is owned individually in a network of households and businesses that share the same rules for energy consumption and supply. Typical examples included peer-to-peer (P2P) communities, for example, QuartierStrom in Switzerland (Wörner et al. 2019). Finally, decentralized energy communities are characterized as a group of households, businesses, and/or a municipality owned power generation that installs and consumes energy locally.

Although most literature acknowledges owners and users within renewable energy communities, only a few include ownership and user structures as core elements of the taxonomy. A notable exception is literature by Walker and colleagues (Walker 2008; Walker and Devine-Wright 2008), who classified energy communities by who developed (owners) and who was served (users) by the project. More generally, however, the literature tends to assume that, in renewable energy communities, the owners of distributed energy resource are also the users, even if it recognizes that this is not always the case. This assumption is in line with common views of energy communities, for example (IRENA Coalition 2018), as organizations which include at least two of the three characteristics: "a) local stakeholders own the majority or all of the renewable energy project; b) voting control rests with a community-based organization; c) the majority of social and economic benefits are distributed locally." Although these sources include all types of energy communities (wind, hydropower, combined-heat and power, etc.), they recognize that the owners may not be the end-users of the energy system. While owners of energy communities in rural areas are often the end-users, this is not always true in urban environments. Urbanites often lack space for installing renewable energy on their property, which makes owning distributed energy resources challenging. This chapter extends previous work by opening up assumptions around ownership and users in community solar to develop a new taxonomy for PV installations.

A Taxonomy for PV Systems in Urban Environments

Ownership and user structures of PV installations help reveal economic, environmental, and social benefits of different configurations and which ones are classified as solar energy communities (Fig. 1). The **ownership structure** of PV installations is analyzed by inquiring about who owns the system and if owners use the generated electricity. Ownership can be in the hands of one individual or firm (*single owner*) or shared among several entities (*multiple owners*). In addition, owners can consume electricity from the PV installation (*owner-user*) or instead supply it to other consumers (*owner-seller*). To characterize the **user structure** of PV installations further, a distinction is made between users who are physically connected to the PV installation (*local users*) and those who are not (*dispersed users*). Dispersed users form contractual agreements with owners to received electricity produced from a remote PV system.

According to their ownership and user structures, six PV installation models are presented in Fig 1. These include the tenant community, individual self-consumer, joint-ownership community, individual power plant, joint investment power plant, and virtual power plant. Each model is evaluated according to the add 'main' to communicate that they can generate other types of benefits but that is the major one type of benefits it generates; economic, energy/environmental, and social. Economic beneficiaries receive an economic incentive to utilize electricity generated by the PV system at a lower cost than electricity from the grid. Energy/environmental beneficiaries utilize the electricity generated by the PV system

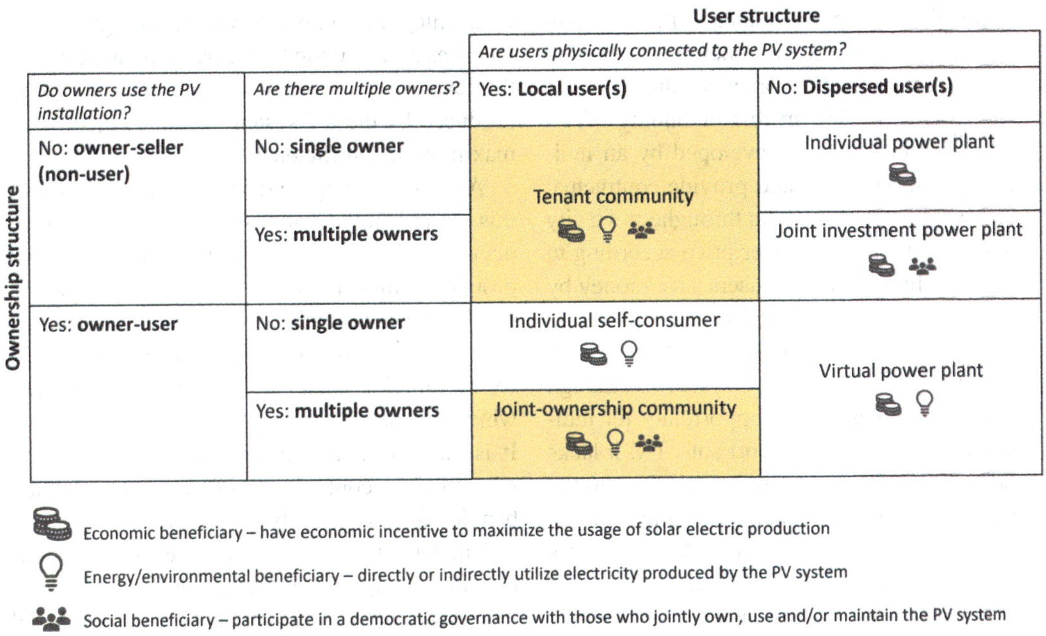

User structure
Are users physically connected to the PV system?
Do owners use the PV installation?
Are there multiple owners?
Yes: Local user(s)
No: Dispersed user(s)
Ownership structure
No: owner-seller (non-user)
No: single owner
Yes: multiple owners
Tenant community
Individual power plant
Joint investment power plant
Yes: owner-user
No: single owner
Yes: multiple owners
Individual self-consumer
Joint-ownership community
Virtual power plant
Economic beneficiary – have economic incentive to maximize the usage of solar electric production
Energy/environmental beneficiary – directly or indirectly utilize electricity produced by the PV system
Social beneficiary – participate in a democratic governance with those who jointly own, use and/or maintain the PV system

Solar Energy Communities in the Urban Environment, Fig. 1 Taxonomy for solar energy systems in urban environments characterized by ownership and user structures, and type of beneficiary

to reduce consumption of grid electricity. Lastly, social beneficiaries have a democratic governance for those who jointly own, use and/or maintain the PV system. The key distinction between SECUEs and other PV system models is the additional social benefits which include community empowerment, stronger community cohesion and increased social capital, among many others (Berka and Creamer 2018; Brummer 2018). Using this taxonomical approach, SECUEs are clearly identified as creating full beneficiaries who receive all three types of benefits from the PV system. This approach also reveals that there are two main models of SECUEs: joint-ownership and tenant communities, as highlighted in Fig. 1.

The joint-ownership community model represents SECUEs that have multiple individuals who jointly invest in a PV system and/or the associated infrastructure. Economies of scale can make joint-ownership more affordable than individually installing and operating smaller PV systems. The associated infrastructure includes a common electric network, a metering and monitoring system and batteries. This option is particularly well suited for individuals who live in housing cooperatives or condominium complexes which have an existing form of governance for shared building infrastructure. For independent building owners who come together to jointly invest in a new solar PV system and/or the associated infrastructure, they will need to establish organizational bylaws. These bylaws consist of agreements to share the PV system ownership and equitably distribute the energy and economic benefits. Hence, the joint-ownership community model has great potential in cities where the sharing-economy is already well established.

The tenant community model represents SECUEs which typically occur in rented apartments or rented building complexes. In this model, the PV system is installed, owned and maintained by a landlord or third-party company on or near the building. In this model, the local users are tenants who do not have partial ownership of the solar PV system. They do however pay for the electricity generated from the solar PV system that is consumed on-site at price agreed with the system owner, which is typically lower than the cost of electricity supplied by the electric utility.

Although virtual power plants (VPP) and joint investment power plants are sometimes referred to as 'community solar' in practice, they are not classified as a SECUE in this taxonomy. VPPs are solar PV installations developed by an individual or joint-investors and provide contractual agreements to dispersed users throughout the city who buy electricity at a lower price according to time of use. In other words, users save money by consuming more during the time the PV system generates electricity but there is no physical connection between user and PV system. Although this model provides a great opportunity for individuals to invest in and support solar PV, it lacks the social benefits that a community brings to the users. For example, the solarzüri program offered by the local electric utility (ewz 2021), provides an opportunity for anyone in Zurich, Switzerland, to own part of a PV system installed on a public building within the city. In this case, public buildings lease out their space for installing PV systems and the electricity is sent directly to the grid. The individual investors simply receive monetary credits for jointly investing in a PV system; however, there is no tangible community developed around the system.

Opportunities and Challenges for SECUE's

Each solar energy community model offers unique opportunities and faces distinct challenges related to their economic, environmental, and social dimensions. The opportunities and challenges are exemplified via existing case studies for the SECUE models identified in the taxonomy: the joint-ownership and the tenant community.

The joint-ownership model is the classic model for solar energy communities. One clear example is the Hvidorebo Housing Association near Copenhagen, Denmark (Roberts et al. 2014). In this SECUE, tenants jointly purchased a solar PV system which was installed on the rooftops of ten buildings within the housing estate and operated on a single electric network. This case study exemplifies joint-ownership since the tenants collectively own the system. Not only are the system costs integrated into the rent or mortgage payments as agreed upon by the participating tenants, the tenants also directly consume the electricity produced by the PV system with the objective to maximize self-sufficiency.

A clear advantage and opportunity presented in this case study is the self-proclaimed 'tenant democracy' even though tenants may not own property within the estate, they are all included in the decision-making processes for facility maintenance and renovations. This is common for Denmark's social housing association laws which are based on exercising self-governance. It is clear here that the tenants are active users who receive economic, social and environmental benefits from the jointly owned system.

The tenant democracy model enables individuals to participate who cannot or do not want to own their homes. However, this is a special model developed in Denmark. The limited influence that tenants have to make infrastructure investment decisions, remains a common challenge for such housing estates internationally. There are additional barriers for tenants who live in highly regulated buildings, such as historic buildings, who might have to deal with historic preservation laws which might make the implementation more challenging and/or expensive. This leads to another challenge for such SECUE model, tenants must create formal bylaws to govern the planning, installation, and maintenance of the system. Here the benefits of the shared infrastructure should outweigh the challenges of implementation.

The community tenant model for SECUE provides an alternative for individual tenants to have direct access to electricity generated by a PV system integrated into their community. A reference for this model is a regulation developed in Switzerland to form self-consumption communities directly translated as "merger for self-consumption" (Zusammenschluss zum Eigenverbrauch (ZEV) in German). One ZEV example includes the Ecoviva settlement, Niederlenz Aargau, Switzerland (Toggweiler et al. 2019). Here, the owner of the PV system is Energie 360°, a single third-party company, and the users are the residents of two apartment buildings with a total of 13 individual apartment units.

Opportunities presented by this example include benefit for tenants who do not have the direct responsibility to own, operate, and maintain the PV system. For example, the third-party owner integrated a smart home system to help tenants operate the system optimally and couple it with electromobility. Like the joint-ownership model, the single-owner of the tenant community profits from economies of scale where the capital investment is less for the entire system compared to installing individual systems of the same total capacity. This project has also led to additional innovations, for example, the installation of a geothermal heat pump and an intelligent control system which optimizes the building energy systems, electricity demands, and solar electricity generation. According to the ZEV model, the tenants receive cheaper electricity than from the local electric utility.

Although the ZEV stands as a leading model for tenant communities, some challenges include a relatively low return on investment for the single owner (Probst et al. 2019). While this varies depending on the prices charged by each local utility, low return on investments could slow down adoption. Furthermore, compared to the joint-owner model, the tenant community model might be less democratic, even though there is potential to allow non-owners to participate in the decision-making process. Here, the third-party owner should establish inclusive decision-making processes to consider the needs and concerns of the tenants.

Tenant communities can also have multiple joint investors, for example, the Lambeth Community Solar in London (Repowering 2021). In Lambeth, residents have an opportunity to purchase a share of community solar PV systems installed on the rooftops of local schools. The tenant communities include the users of the school building who integrate workshops about how to use solar electricity. The local schools also profit from the system because they have direct access to cheaper electricity produced by the PV system.

One key opportunity presented by this model is the ability for community members associated with the school to financially support pro-environmental infrastructure for their schools. In fact, it is possible for all households to become members of the "Community Benefit Society" with minimal investments and all joint investors have the same voting rights. This approach makes it easier for low-income households to also participate.

Similar to the single-ownership tenant community model, the joint investors face uncertainty about whether there will be a net positive return on investment. This risk is present even though Lambeth Community Solar shares are not regulated by the 'Financial Conduct Authority' because the organization is seen as a charity. It therefore focuses on social returns rather than financial returns, and does not have additional costs associated with financial regulations.

Lastly, a hybrid example is presented which incorporates aspects of both the joint-ownership and tenant community models. The Brooklyn Microgrid in New York City formed an energy marketplace in 2016 to connect individual PV system owners (prosumers) together with other local tenants (consumers) through a single microgrid (Brooklyn Microgrid 2019). Such a system presents various opportunities, for example prosumers and consumers have the flexibility to use, sell, and/or purchase locally generated solar electricity. Individual members of the community can display a certificate in their windows that provides social recognition for community members. This is also a dynamic system where new prosumers and consumers can join the community after the system was initiated.

The main challenge for such a community is the need for a mechanism to manage the daily energy auction through an external managing partner. Prosumers and/or consumers might eventually lose interest in the potentially time-consuming task of reviewing, assessing, and purchasing electricity from the most suitable source within the microgrid. These individual decisions might also be made to increase self-interest rather than to contribute to the collective benefit of the entire community. These community objectives therefore need to be highlighted and included transparently in the energy marketplace platform and adapted into the community governance.

Summary and Conclusions

This chapter draws on literature and case studies about energy communities to develop an updated taxonomy specific to PV systems in urban environments. This work aims to support stakeholders, practitioners, and scholars address the unique challenges and build upon the opportunities for SECUEs. By analyzing owners hip and user structures separately, the taxonomy presented in this chapter facilitates the process of assessing the benefits that each model generates and helps distinguish what actually constitutes a SECUE. The taxonomy is therefore based on ownership and user structures along with a mapping of the benefits specific to each model: energy, economic, and social benefits.

Cross-References

- ▶ Building Resilient Communities Over Time
- ▶ City Visions: Toward Smart and Sustainable Urban Futures
- ▶ Collaborative Climate Action
- ▶ Community Engagement and Climate Change: The Value of Social Networks and Community-Based Organizations
- ▶ Community Engagement for Urban and Regional Futures
- ▶ Community in a Changing Climate: Shaping Urban and Regional Futures
- ▶ Networking Collaborative Communities for Climate-Resilient Cities
- ▶ Smart grids
- ▶ Urban Densification and Its Social Sustainability

References

Awad, H., & Gül, M. (2018). Optimisation of community shared solar application in energy efficient communities. *Sustainable Cities and Society, 43*(March), 221–237. https://doi.org/10.1016/j.scs.2018.08.029

Bauer, C., Cox, B., Heck, T., & Zhang, X. (2019). Potenziale, Kosten und Umweltauswirkungen von Stromproduktionsanlagen: Aufdatierung des Hauptberichts (2017), 1–68. Retrieved from https://www.psi.ch/en/media/53333/download

Bauwens, T., & Devine-Wright, P. (2018). Positive energies? An empirical study of community energy participation and attitudes to renewable energy. *Energy Policy, 118*, 612–625.

Berka, A. L., & Creamer, E. (2018). Taking stock of the local impacts of community owned renewable energy: A review and research agenda. *Renewable and Sustainable Energy Reviews, 82*(October 2017), 3400–3419. https://doi.org/10.1016/j.rser.2017.10.050

Brooklyn Microgrid. (2019). www.brooklyn.energy. Retrieved June 30, 2021, from https://www.brooklyn.energy/

Brummer, V. (2018). Community energy – Benefits and barriers: A comparative literature review of Community Energy in the UK, Germany and the USA, the benefits it provides for society and the barriers it faces. *Renewable and Sustainable Energy Reviews, 94*-(November 2017), 187–196. https://doi.org/10.1016/j.rser.2018.06.013

de Coninck, H., Revi, A., Babiker, M., Bertoldi, P., Buckeridge, M., Cartwright, A., ... Sugiyama, T. (2018). IPCC special report on the impacts of global warming of 1.5 °C above pre-industrial levels and related global greenhouse gas emissions. *Global Warming of 1.5°C. An IPCC Special Report on the Impacts of Global Warming of 1.5°C above Pre-Industrial Levels and Related Global Greenhouse Gas Emission Pathways, in the Context of Strengthening the Global Response to the Threat of Climate Change.* Retrieved from https://www.ipcc.ch/site/assets/uploads/sites/2/2019/02/SR15_Chapter4_Low_Res.pdf

ewz. (2021). ewz-solarzueri. Retrieved June 30, 2021, from https://www.ewz.ch/de/private/solaranlagen/solarstrom-fuer-eigentuemer/ewz-solarzueri.html

Gui, E. M., & MacGill, I. (2018). Typology of future clean energy communities: An exploratory structure, opportunities, and challenges. *Energy Research and Social Science, 35*, 94–107. https://doi.org/10.1016/j.erss.2017.10.019

IRENA. (2009). Cities, towns & renewable energy: Yes in my front yard. *Cities, Towns and Renewable Energy: Yes in My Front Yard, 9789264076*(October 2016), 1–186. https://doi.org/10.1787/9789264076884-en

IRENA Coalition. (2018). Community energy: Broadening the ownership. *PhD Paper Citation (0)*. Retrieved from http://irena.org/-/media/Files/IRENA/Agency/Articles/2018/Jan/Coalition-for-Action_Community-Energy_2018.pdf?la=en&hash=CAD4BB4B39A381CC6F712D3A45E56E68CDD63BCD&hash=CAD4BB4B39A381CC6F712D3A45E56E68CDD63BCD

Klein, S. J. W., & Coffey, S. (2016). Building a sustainable energy future, one community at a time. *Renewable and Sustainable Energy Reviews, 60*, 867–880. https://doi.org/10.1016/j.rser.2016.01.129

Lee, M., Hong, T., Jeong, K., & Kim, J. (2018). A bottom-up approach for estimating the economic potential of the rooftop solar photovoltaic system considering the spatial and temporal diversity. *Applied Energy,*

232(October), 640–656. https://doi.org/10.1016/j.apenergy.2018.09.176

Margolis, R., Gagnon, P., Melius, J., Phillips, C., & Elmore, R. (2017). Using GIS-based methods and lidar data to estimate rooftop solar technical potential in US cities. *Environmental Research Letters, 12*(7). https://doi.org/10.1088/1748-9326/aa7225

Moroni, S., Alberti, V., Antoniucci, V., & Bisello, A. (2019). Energy communities in the transition to a low-carbon future: A taxonomical approach and some policy dilemmas. *Journal of Environmental Management, 236*, 45–53.

Probst, S., Kern, L., & Konersmann, L. (2019). Zusammenschluss zum Eigenverbrauch von Solarstrom auf Arealen Herausforderungen und Erfolgsfaktoren, (September), 1–40.

REN21. (2021). Renewables in Cities 2019 Status Report, 202. Retrieved from https://wedocs.unep.org/bitstream/handle/20.500.11822/28496/REN2019.pdf?sequence=1&isAllowed=y%0Ahttp://www.ren21.net/cities/wp-content/uploads/2019/05/REC-GSR-Low-Res.pdf

Repowering. (2021). www.repowering.org.uk. Retrieved June 30, 2021, from https://www.repowering.org.uk/lambeth-community-solar/

Roberts, J., Bodman, F., & Rybski, R. (2014). *Community Power: Model Legal Frameworks for Citizen-owned Renewable Energy. ClientEarth*. London. https://doi.org/10.4324/9780367821494-21.

Roelich, K., & Knoeri, C. (2015). *Governing the infrastructure commons: lessons for community energy from common pool resource management*. Leeds.

Seidl, I. (2018). *Collective financing of renewable energy projects in Switzerland and Germany*. Retrieved from https://www.wsl.ch/en/projects/collective-financing-of-renewable-energy-projects.html

Sonnen. (2021). sonnengroup.com. Retrieved June 30, 2021, from https://sonnengroup.com/sonnencommunity/

Toggweiler, P., Stickelberger, D., Krebs, A., Ammann, T., Spring, I., Tongi, M., … Hintz, W. (2019). *Leitfaden Eigenverbrauch, Version 2.0*. Bern.

Viti, S., Lanzini, A., Minuto, F. D., Caldera, M., & Borchiellini, R. (2020). Techno-economic comparison of buildings acting as single-self consumers or as energy community through multiple economic scenarios. *Sustainable Cities and Society, 61*(June), 102342. https://doi.org/10.1016/j.scs.2020.102342

Walker, G. (2008). What are the barriers and incentives for community-owned means of energy production and use? *Energy Policy, 36*(12), 4401–4405. https://doi.org/10.1016/j.enpol.2008.09.032

Walker, G. (2011). The role for "community" in carbon governance. *Wiley Interdisciplinary Reviews: Climate Change, 2*(5), 777–782. https://doi.org/10.1002/wcc.137

Walker, G., & Devine-Wright, P. (2008). Community renewable energy: What should it mean? *Energy Policy, 36*(2), 497–500. https://doi.org/10.1016/j.enpol.2007.10.019

Wörner, A., Meeuw, A., Ableitner, L., Wortmann, F., Schopfer, S., & Tiefenbeck, V. (2019). Trading solar energy within the neighborhood: Field implementation of a blockchain-based electricity market. *Energy Informatics, 2*, 1–12. https://doi.org/10.1186/s42162-019-0092-0

Solar Pumps

▶ Smart Agriculture and ICT

Solar-Powered Irrigation Systems

▶ Smart Agriculture and ICT

The Source Waters of Tanga

F. I. Kihara[1], M. K. Thomas[2], J. O. Kawira[3] and S. Thomas[2]
[1]The Nature Conservancy, Nairobi, Kenya
[2]Rural Focus Limited (RFL), Nanyuki, Kenya
[3]County Government of Laikipia, Laikipia, Kenya

Synonyms

Spices refers to cloves, candamon cinnamon and black pepper; Gemstones refers to alluvial gold found in the streams; Wetlands refers to bogs and other riparian vegetation; Forest refers to standing tree area used for no other purpose

Abbreviations

CSR	Cooperate Social Responsibility
EAMCEF	Eastern Arc Mountains Conservation Endowment Fund

S

EWURA	Energy and Water Utilities Regulatory Authority
GHG	Greenhouse gas
IUCN	International Union for Conservation of Nature and Natural Resources
MoW	Ministry of Water
Norad	Norwegian Agency for Development Corporation
RFL	Rural Focus Limited
TNC	The Nature Conservancy
TWF	Tanga Water Fund
UWAMAKIZI	Umoja wa wakulima wa hifadhi mazingira Kihuhwi Zigi
UWASA	Urban Water Supply and Sanitation Authority

Definition

Biodiversity is the variability among living organisms from all sources, including terrestrial, marine, and other aquatic ecosystems and the ecological complexes of which they are part; this includes diversity within species, between species, and of ecosystems' above ground as well as below ground carbon stock

Eastern Arc Mountains is a chain of mountains found in Kenya and Tanzania. The chain runs from northeast to southwest, with the Taita Hills being in Kenya and the other ranges being in Tanzania. They are delimited on the southwest by the fault complex represented by the Makambako Gap that separates them from the Kipengere Range.

Endemism is the ecological state of a species being native to a single defined geographic location, such as an island, nation, country or other defined zone, or habitat type; organisms that are indigenous to a place are not endemic to it if they are also found elsewhere.

Public-private partnership is a cooperative arrangement between two or more public and private sectors, typically of a long-term nature source waters.

Source waters refers to water sources (such as rivers, streams, lakes, reservoirs, springs, and groundwater) that provide water to public drinking water supplies and private wells as well as their replenishment sources.

Water funds refers to public-private partnerships that bring stakeholders from both sectors to pool financial resources to support conservation of their source waters. They are founded on the principle that it's cheaper to invest and keep surface water clean while still in the watershed than further downstream like in city water treatment plants.

Water security is the reliable availability of an acceptable quantity and quality of water for health, livelihoods, and production, coupled with an acceptable level of water-related risks

Part 1. Applying a Water Fund Approach to Conserve Tanga City's Source Waters

Introduction

Tanga is a port city located in the northeast region of Tanzania on the shores of the Indian Ocean. The city is the administrative headquarters of the Tanga Region and has an estimated 2019 population of 306,500, which is expected to rise to 406,135 by 2030. The main economic activities in the Tanga Region are agriculture, fishing, tourism, mining (cement, limestone), commercial sisal growing and processing, small-scale business, and commercial port activities.

The Tanga Region is characterized by a bimodal rainfall distribution peaking between March and May and between September and December. Annual rainfall in the region varies with altitude, ranging between 1,100 mm and 1,400 mm on the coast to above 2,000 mm in the Usambara Mountains. Much of the region experiences a warm and wet climate, while some parts of the lowlands are semiarid.

As with other African cities, the call to develop a water fund for Tanga arose from the realization that there was a need to respond to water security concerns – particularly rising demand that outstrips supply combined with declining water quality. The water fund model provides a solution to these challenges by offering a mechanism for stakeholders to work together to promote equitable access to water resources from the source to the points of consumption. At the same time, water funds promote sustainable water use and catchment conservation among stakeholders. It is a science-based approach that launched in Nairobi in 2012 and led to the establishment of the Upper Tana-Nairobi Water Fund. The Nature Conservancy (TNC) has replicated the water fund model in other African cities, including Cape Town, and has responded to the Tanzania Ministry of Water's (MoW) request during the 2016 Africa Water Week to implement the model. Since then, TNC has researched cities in which to establish water funds, resulting in the completion of a water fund in Tanzania (2018) and a feasibility study for Tanga (2019–2020).

This report summarizes the findings from the studies conducted on the Tanga Water Fund (TWF) and builds the case for its establishment by highlighting the benefits for Tanga City, the environment, beneficiary communities, and other stakeholders.

Tanga City relies on the Zigi River in Fig. 1, which feeds the Mabayani Dam for its domestic, commercial, and industrial water needs. Groundwater supplements about 1% of the supply to the city. The region's main economic activities all depend on water. As such, it is in the best interest of water users, catchment dwellers, and all stakeholders to ensure a sustainable and reliable water supply. Tanga Water Supply and Sanitation Authority (Tanga UWASA) handles the water distribution from the Mabayani Dam (28,000m^3/d) and the Mwakileo Borehole (3,800 m^3/d) to about 98% of the urban population. Tanga UWASA also covers the Pangani township and Muheza District following the MoW directive to cluster the three water supply utilities and expand Tanga UWASA's jurisdiction. This makes all of the region's water sources, including Muheza's intakes on the Mkulumuzi River, vital to the utility.

Tanga's landscape in Fig. 2, is characterized by diverse land cover and slopes. The Zigi catchment is dominated by forests (61.77%), with 50% of the land on a slope greater than 15%. The upper reaches of the Zigi River forested catchment are steep and mountainous and adjoin tea plantations. The lower reaches are hilly and gently sloping, with more open woodland, bushland, and farms.

The Source Waters of Tanga, Fig. 1 Zigi river flows out of the Amani Natural Reserve ahead of Mabayani dam. Photo by John Esther

S

The Source Waters of Tanga, Fig. 2 Water comes from Nature. Eastern Usambara mountain forest quenches Tanga. Photo by Roshni Lodhia

The Amani Nature Reserve (ANR), one of the most important protected forests in Tanzania due to its rich biodiversity and species endemism, covers over 7,000 ha of the upper Zigi catchment and is the source of the Kihuhwi tributary of the Zigi River. There are 15 forest reserves within the catchment, including the Kilanga (Nilo) Forest Reserve on the northeast end of the catchment area. Nilo occupies 1,415 ha and is the source of the Muzi tributary.

The Mkulumuzi River catchment is characterized by gentler slopes and vegetation, with 80.7% of the landscape consisting of low to moderate slopes (<5 to 15%). Grassland and bushland cover 50% of the catchment, while forests cover 24.6%. The Mkulumuzi catchment also hosts the Bassi, Magoroto, and mangrove forest reserves, with a total area of 1,160 ha.

The catchments spread across the districts of Korogwe, Muheza, Tanga, and Mkinga. These districts are predominantly rural and peri-urban, with agriculture being the main livelihood activity.

Other than its importance as a source of water to Tanga City, the Zigi watershed plays a critical role in providing ecosystem services and preserving biodiversity. The headwaters of the Zigi River emanate from the East Usambara Mountains, which are globally recognized for their exceptional biodiversity and species endemism. The richness in flora and fauna in this region has been compared to that of the Galápagos Islands.

> The East Usambara bloc boasts a variety of endemic and a near-endemic plant, vertebrate, and invertebrate species such as the African violet (*Saintpaulia*), the Sanje Mangabey (*Cercocebus sanjei*), Iringa red colobus (*Procolobus gordonorum*), the mountain dwarf galago (*Galagoides orinus*), and the Kipunji monkey (*Rungwecebus kipunji*). The Amani Nature Reserve alone has extensive floristic diversity (3,450 vascular plant taxa), 74 genera of animals, and more than 340 bird species (Fig. 3).

The health of the watershed contributes to the quality of water flowing through the Zigi River, with the watershed providing such critical ecosystem functions as flow regulation and sediment

The Source Waters of Tanga, Fig. 3 Silvery cheeked hornbill bird in the Amani nature reserve. Photo by Roshni Lodhia

retention. Land-use change and the loss of forest cover in both the Zigi and Mkulumuzi catchments, as well as across the East Usambara Mountains as a whole, have led to the deterioration of water quality and reliability to the detriment of water users in Tanga City and within the catchments. The Mkulumuzi River, a perennial coastal river, has recently experienced low to zero flows. The Zigi River suffers largely from soil erosion and sedimentation, with the region most prone to erosion being Bosha Ward in Mkinga District, where poor farming practices and alluvial gold mining are common. Gold mining in the Tanga watershed has increased the degradation of wetland ecosystems and natural habitats. The main threats to water resource sustainability are mostly due to changes in land use:

- Further degradation of the habitat, leading to loss of biodiversity
- Soil erosion and sediment yield, leading to loss of capacity at the Mabayani Dam
- Water quality deterioration on Zigi River, leading to high water treatment costs
- Climate change, which may result in increased intensity and frequency of storms, floods, and droughts

Concerns regarding the land-use practices within the East Usambara Mountains have been widely recognized and documented, and various projects focusing on forest conservation to protect biodiversity have been implemented. Examples include the Integrated Usambara Rain Forest project (1983), the Amani Forest Inventory and Management Plan project (1986–1987), and the IUCN/Forest Division/Norad Project in the East Usambara Mountains (1986–1987). More recent projects include the Equitable Payment for Watershed Services project (2009–2015); the Securing Watershed Services Through Sustainable Land Management in the Ruvu and Zigi catchments, Eastern Arc Region, Tanzania project (2016–2020); and the TWF feasibility study (2020). Continued efforts from interested parties in the conservation of the East Usambaras shows not only the importance of the mountains as a natural resource but also the potential for more significant impact from conservation activities through enhanced synergy among stakeholders, an opportunity that the water fund model offers.

Various policy and legislative directives across different sectors have also been put in place to support and prioritize catchment conservation, including protections for forest and wildlife reserves and the establishment of riparian areas. The MoW is responsible for the sustainable development and management of water resources, including overseeing urban and rural water and providing sanitation services. The Ministry of Natural Resources and Tourism; the Division of Environment in the vice president's office; the Ministry of Agriculture, Livestock, and Fisheries; the Tanzania Forestry Service; the office of the Regional Administrative Secretary; and the district councils all serve important roles related to the conservation and development of the people and environment within the two catchments. A key feature of Tanzania's policy and legislative framework is that it provides an enabling environment for multi-stakeholder engagement in catchment conservation and further encourages public-private partnership for enhanced resource mobilization, engagement, and implementation of activities.

The feasibility study (Rural Focus Ltd. 2020) identifies the following interventions as suitable for the landscape. These interventions could improve rural livelihoods through a systematic and targeted effort in which sustainable land management practices are adopted to improve water and food security while also enhancing catchment and forest conservation to protect biodiversity:

- Agroforestry and woodlots
- River riparian conservation
- Integrated watershed management
- Livelihood support
- Marketing and value chain support
- Improved rural roads, tracks, and pathways

The abovementioned interventions offer a variety of benefits to both the catchment and water users, including improved water quality and reliability; improved crop production, leading to better livelihoods (Fig. 4); improved access to markets for produce; and forgone costs of dam desilting and reduced water treatment costs.

This report analyses the benefits that could be accrued from the proposed sustainable land management interventions specifically for the **preservation of biodiversity, climate change mitigation impacts, clean water provision,** and **community well-being** and **livelihood benefits**.

Part 2. High Endemism and Biodiversity Significance of the Tanga Watershed and East Usambara Mountains, Tanzania

Introduction

Biodiversity is broadly understood to describe the variety of life-forms in a given community or habitat, but there are several aspects to this generic term. The first is the distinction between diversity and richness, both of which are amalgamated under the term "biodiversity." Richness refers to the total number of forms (e.g., species), while diversity incorporates the relative abundance of these different forms. The difference between these measures is significant for those threatened and endemic species that may have smaller populations but are of great conservation importance.

Biodiversity may be further separated to include ecological diversity, species diversity, and genetic diversity. Another element of biodiversity worth noting is endemism. This describes a species that is unique to a defined geographic region. This is important to areas such as the East Usambara Mountains of Tanzania, which possess

The Source Waters of Tanga, Fig. 4 Clean water in Tanga grows healthier families. Photo by Roshni Lodhia

an unusually high level of endemism, as well as various rare habitats and threatened species.

The East Usambara Mountains are one of 13 mountain blocks comprising the Eastern Arc Mountains, which stretch from southern Kenya through eastern Tanzania. This mountain range is renowned for its biodiversity and high endemism rates, with the East Usambaras having some of the highest (Burgess et al. 2007).

The natural primary forest of the East Usambara Mountains covers approximately 56,636 ha, 38,047 ha of which have experienced some disturbance since 1986 (Fig. 5). Just over 30% (30.6%) of this is classified as submontane rainforest; 63% is classified as lowland forest, predominantly defined by altitude with submontane forest generally occurring above 850 m. Low-altitude forest extending to the coastal plains bear compositional similarity to other coastal forests of East Africa (Hamilton and Bensted-Smith 1989) (Fig. 5). The high-altitude closed-canopy submontane forests carry the greatest biodiversity and are interspersed in the valleys and gentler slopes with swampy "wetland forest," itself a very important habitat for many species (Senzota and Mbago 2009).

Importance of Biodiversity

The most tangible benefits of biodiversity come in the form of ecosystem services and natural capital, for example, pollination. In a modified mosaic ecosystem where agroforestry is so prevalent, the biodiversity of pollinators (insects and birds) translates directly to higher yields and improved revenue, giving biodiversity monetary value too.

Healthy ecosystems also protect soils from eroding and leaching, resulting in improved water quality in the rivers that drain near them (Van Bierlviet et al. 2009). This is important in the Zigi catchment, as 96% of the water for Tanga is drawn from the Mabayani Dam on the Zigi River. The river is experiencing a sedimentation rate of 168,000 m^3 per annum due to runoff from the upper catchment, reducing reservoir capacity (Rural Focus Ltd. 2020).

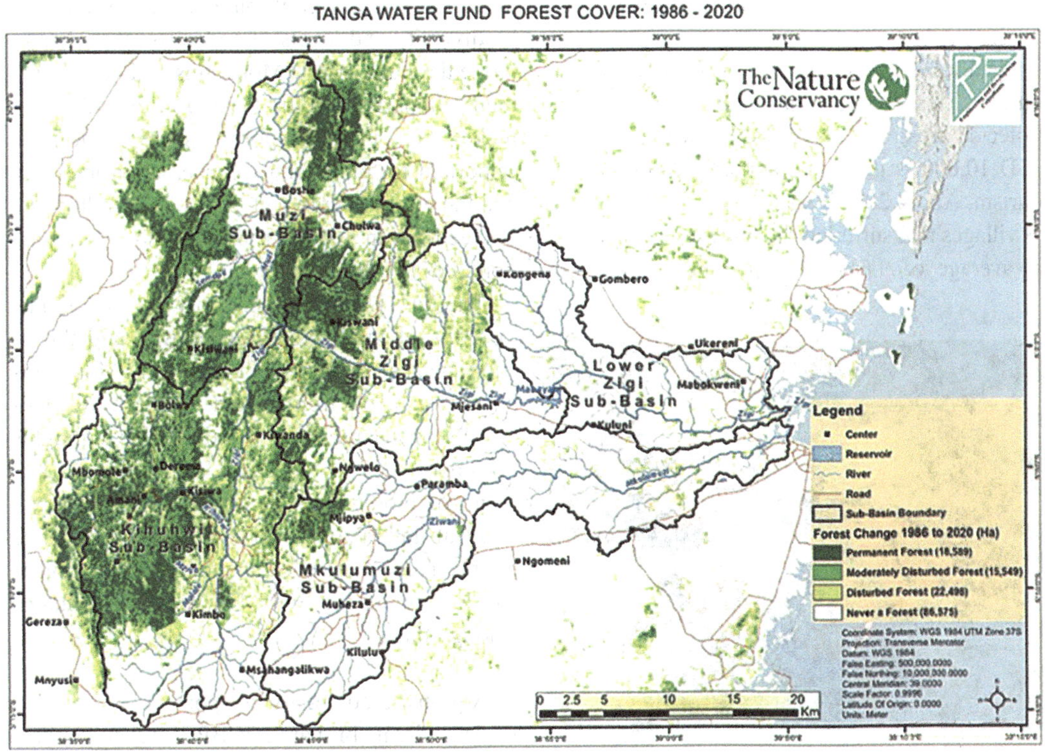

The Source Waters of Tanga, Fig. 5 Map of disturbed and undisturbed forests in the East Usambara mountains

High biodiversity in the Zigi watershed, especially within the protected areas, also directly benefits neighboring communities and Tanzania's economy through non-timber forest products (NTFPs) such as plants and food sources, for example, fruit and seeds (Kessy 1998). These benefits are exploited commercially in agroforestry enterprises like tree spice crops, mainly cardamom, cloves, and cinnamon, as well as black pepper in the understory (Bullock et al. 2013; Reyes 2008; Reyes et al. 2009). These production systems also heavily depend on the biodiversity of pollinators to maintain yields.

Figure 6, shows another biodiversity linked enterprise which is the legal trade in rare species, such as butterflies. The Amani Butterfly Project, for example, rears and sells over USD 60,000 worth of butterfly pupas annually, with 65% of profits going directly to local farmers in the East Usambara Mountains and another 7% going into financing development projects for villages involved in butterfly farming. This equates to an average 20% increase in household revenue for participating households (Morgan-Brown 2003, 2007).

Finally, biodiversity generates revenue through tourism. This is done in gazetted and demarcated protected areas such as the ANR, where roughly USD 10,000 is generated annually through ecotourism. About 20% is shared equally among the 18 villages that surround the reserve, contributing an average of 9.6% of total annual household income (Shoo and Songorwa 2013). However, there are challenges with ensuring equitable distribution of revenue within the community. Tourism is also highly dependent on global and national political and economic events and other market volatilities (Fig. 7).

The Source Waters of Tanga, Fig. 6 Unique butterflies of the Tanga watershed

Endemism and Biodiversity in the East Usambara Mountains

The Eastern Arc Mountains are known for their high biodiversity and endemism rates. This is thought to be due to their age, relative climatic stability, geographic character as isolated forest "islands," and their role in condensing moisture from the Indian Ocean (Hamilton and Bensted-Smith 1989). Many species have been separated from their closest relatives for long periods, leading to diversification and speciation. Additionally, the mountains provide refuge for species that were previously widespread but are now limited to these isolated patches of forest habitat. Together, these two factors contribute to the exceptional biodiversity and levels of endemism that characterize the Eastern Arc Mountains and have led to their comparison to the Galapagos Islands. This makes these mountain forests some of the most important for biodiversity conservation in Africa and the world (Burgess et al. 2007; Hamilton and Bensted-Smith 1989; Tropical Biology Association 2007) (Table 1).

The East Usambaras not only are exceptional in terms of biodiversity, endemism, and threatened species as part of the Eastern Arc Mountains but also stand out from the other Eastern Arc blocs on these measures (Figs. 7 and 8). This highlights the importance of the East Usambaras as a key target of biodiversity conservation at the local, national, and international levels. This importance has been recognized, leading to the designation of the forests of the East Usambara Mountains, among others:

- A global biodiversity hotspot
- An endemic bird area (ICBP 1992) and important bird area (BirdLife International, year 2001)

- A center of high plant diversity (WWF & IUCN, (1994))
- A globally important ecoregion (Worldwide Fund for Nature, 2019)
- Part of the UNESCO Man and the Biosphere Reserve network, 2000

Flora

Of global significance, the Eastern Arc Mountains have a 31% plant endemism rate (Senzota and Mbago 2009). This includes at least 800 vascular plant species, of which about 10% are trees, as well as 15 wild relatives of coffee and most species of African violet (*Saintpaulia*). An additional 32 species of bryophytes are also endemic (Burgess et al. 2007; Tropical Biology Association 2007).

The East Usambaras alone boast 3,450 species of vascular plants (Tropical Biology Association 2007). Of these vascular plant species, approximately 27% depend on primary forest habitats, and of the primary-forest-dependent species, 36–50% are endemic or near-endemic, depending on the forest area examined (Johansson et al.

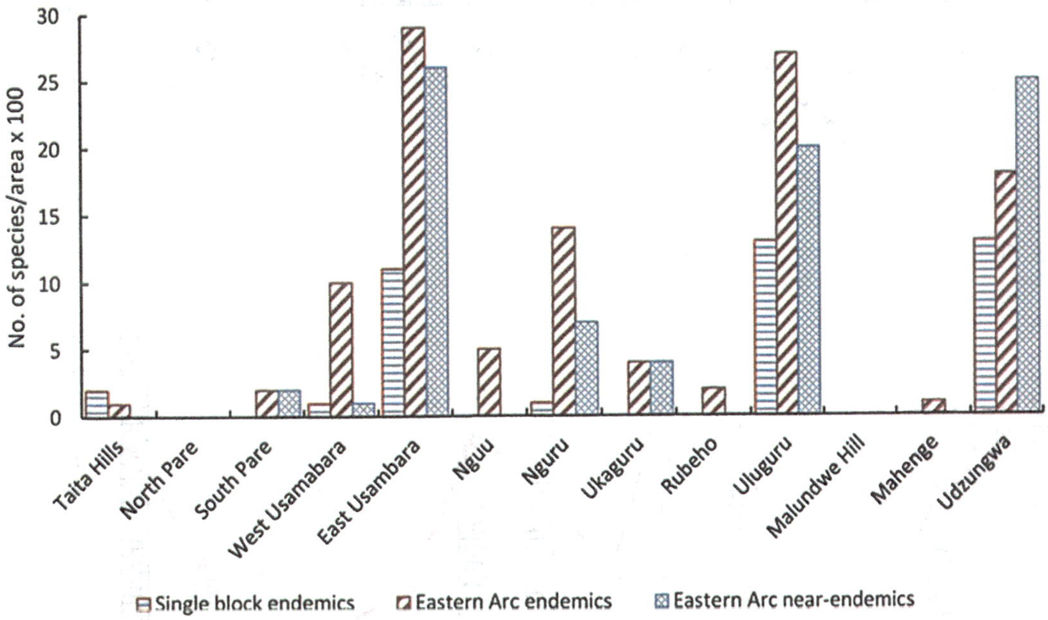

The Source Waters of Tanga, Fig. 7 Endemic vertebrate species relative to the remaining forest area of the Eastern Arc Mountain blocs. (Adopted from A globally important ecoregion Burgess et al. 2007)

The Source Waters of Tanga, Table 1 Summary of biodiversity in the Amani Nature Reserve. This is not exhaustive but is indicative of the species present. (Source: Tropical Biology Association 2017)

Taxon	Total no. of species	% forest-dependent	Non-forest species	Endemic species	Near-endemic species	Forest-dependent endemics and near-endemics
Mammals	59	15.3	6	0	3	2
Birds	65	33.8	15	2	3	3
Reptiles	49	46.7	6	3	15	17
Amphibians	27	66.6	0	2	14	16
Butterflies	112	20.5	4	1	10	9
Trees and shrubs	639	43	22	19	49	53
Total	**951**	–	**53**	**27**	**94**	**100**

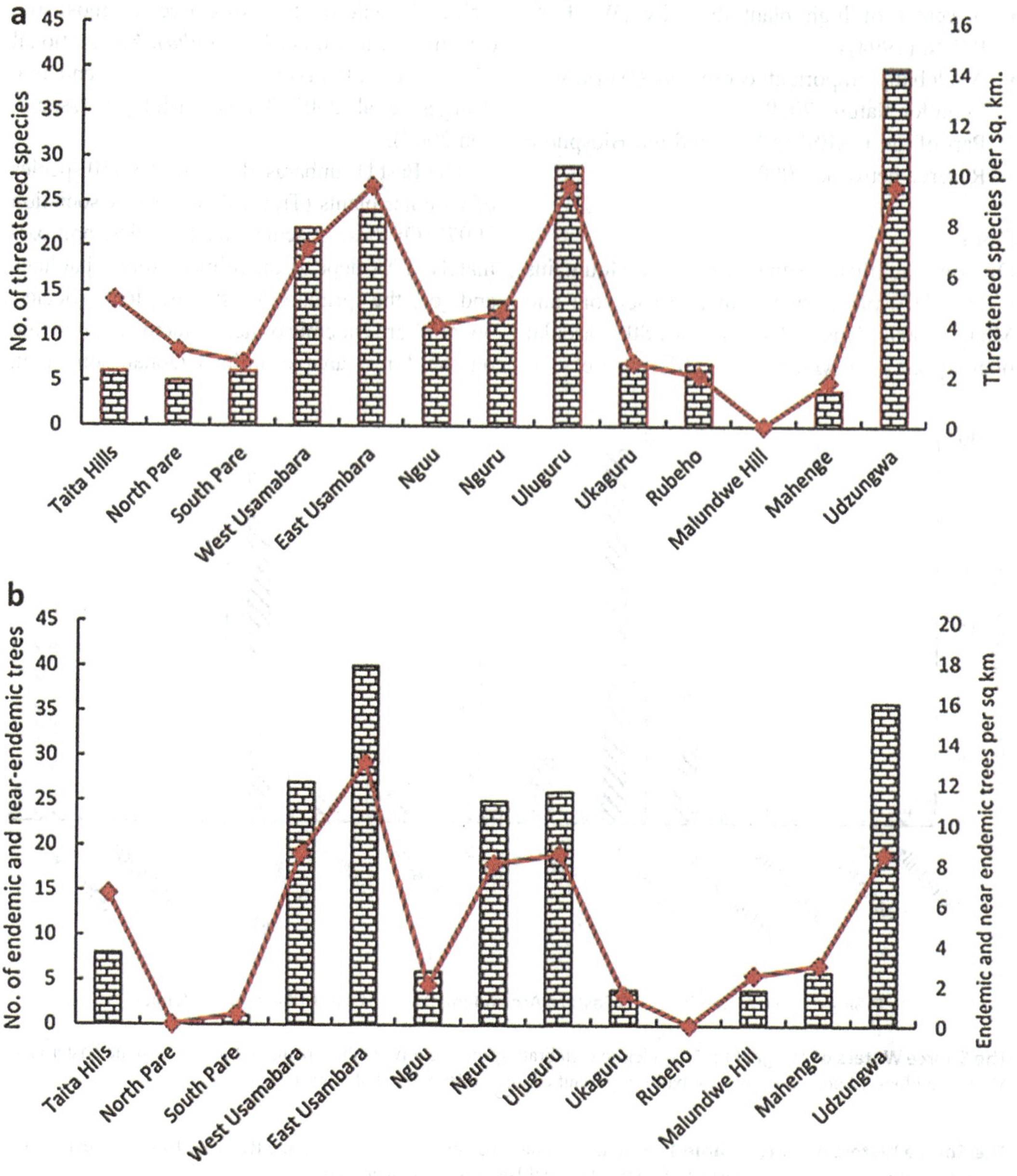

The Source Waters of Tanga, Fig. 8 (**a**) Threatened species trends across the Eastern Arc Mountains; (**b**) Endemic and near-endemic tree species trends across the Eastern Arc Mountains. (Adopted from Burgess et al. 2007)

1998). The East Usambaras also hold the highest number of endemic and near-endemic tree species of all the Eastern Arc Mountain blocs (Burgess et al. 2007).

Vertebrate Fauna

Across the Eastern Arc Mountains, there are at least 96 endemic vertebrate species, including 10 mammal, 19 bird, 29 reptile, and 38 amphibian species. A further 71 vertebrate species are near-endemic (Burgess et al. 2007).

The Eastern Arc is also home to four endemic or near-endemic species of primates – the Sanje Mangabey (*Cercocebus sanjei*), Iringa red colobus (*Procolobus gordonorum*), the Mountain Galago (*Galagoides orinus*), and the Kipunji

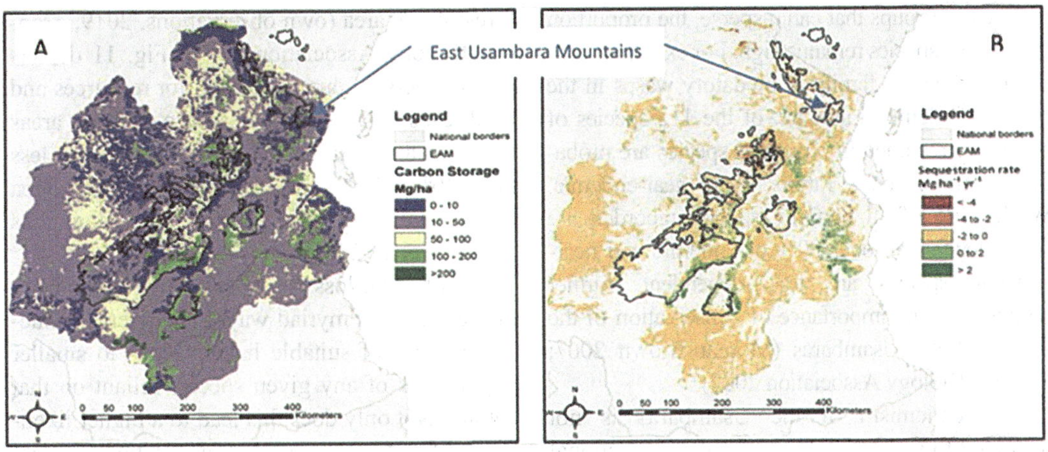

The Source Waters of Tanga, Fig. 9 Aboveground live carbon storage (**a**) and sequestration (**b**) for the Eastern Arc Mountains (Willcock et al. 2014)

monkey (*Rungwecebus kipunji*), which is the only species of that genus (Burgess et al. 2007).

Conservation priority assessments have ranked the Eastern Arc Mountains as the second or third most important area in Africa for the conservation of restricted-range bird species. All 19 endemic bird species depend on forest habitats, and at least seven of those species are limited to dense primary forest (Burgess et al. 1998, 2007). The East Usambara Mountains alone host 110 forest bird species (Tropical Biology Association 2007).

In the East Usambaras, there are 30 species of amphibians and reptiles that can be regarded as dependent on the natural forests of the Usambaras. Of these, 15 are endemic. In the East Usambaras alone, there are seven endemic taxa: two microhylid frogs, one of which is placed in an endemic genus; a rare chameleon (*Chamaeleo spinosus*) such the one in Fig. 10; a very rare lizard (*Bedriagaia moreaui*); and a colubrid snake (*Dipsadoboa werneri*) (Hamilton and Bensted-Smith 1989).

The Source Waters of Tanga, Fig. 10 Rare chameleon species of Tanga watersheds. Photo by Roshni Lodhia

Invertebrate Fauna

The few invertebrate groups that have been studied in the East Usambara Mountains show high rates of endemism, especially taxa with limited dispersal ability, such as millipedes (*Diplopoda*), of which 30 out of the 41 species around the ANR are believed to be endemic to the East Usambara Mountains. Of the terrestrial mollusks (*Gastropoda*), the East Usambaras have 55 endemic species (Tropical Biology Association 2007).

Even in groups that can disperse, the proportion of endemic species remains high. For example, 21% of the Sphecidae family of predatory wasps in the ANR is endemic. Likewise, of the 112 species of butterflies recorded in Amani, 2 species are probably endemic, and a further 9 are near-endemic. While 20% of all butterfly species recorded are forest-dependent, over 80% of endemic and near-endemic species are forest-dependent, further highlighting the importance of conservation of the forests of the Usambaras (Morgan-Brown 2007; Tropical Biology Association 2007).

The endemism of the Usambaras is not restricted to species; it also extends to functional traits and life histories. For example, the East Usambara tree-hole crab has adapted to living in the water-filled boles (tree holes) of various tree species at altitudes between 150 and 900 m above sea level (m.a.s.l.), with the *Myrianthus holstii* tree being the most common host (Bayliss 2002). This uniqueness of traits is likely due to the isolated island-like biogeography, which provides opportunities for species to adapt to niches that would commonly be occupied by species from other taxa.

Aquatic Biodiversity: Fish and Macroinvertebrates

Four types of fish are common in the waters of the Zigi River. These are tilapia, sardines, carps, and catfish. 68 taxa of macroinvertebrates found in the river include worms, crabs, water mites, mayflies, dragon- and damselflies, bugs, caddis flies, beetles, snails, shrimps, and flies (MOWI 2016).

Threats to Biodiversity

The forests of the East Usambaras and upper Zigi catchment have experienced exploitative human use for at least 2,000 years, but until the last century, this pressure was sustainable. However, human population expansion (influenced by economic opportunities such as employment in tea plantations, which require up to 4,000 people) is increasing the pressure on the remaining natural forest in the area (own observations, 2019; Tropical Biology Association 2007). Fig. 11 depicts land conversion due to demand for resources and food leads to significant clearance of forest areas for crops and housing, which has resulted in less than 30% of the original primary forest area remaining across the Eastern Arc Mountains (Burgess et al. 2007).

This habitat loss has affected the biodiversity of the region in myriad ways. The overall reduction of area of suitable habitat leads to smaller populations of any given species reliant on that habitat. Not only does this lead to a higher likelihood of extinction of that population (which, given the endemism rates of the area, may be the only population of a given species), but it reduces the genetic diversity, which limits a species' ability to adapt to shocks, including habitat loss, which may result in species extinction.

The second aspect of deforestation that negatively affects the biodiversity of the East Usambaras, with some estimates placing this as "the major threat to biological diversity," is habitat fragmentation (Johansson et al. 1998). This is crucial in the East Usambaras, where the forest has been reduced to fragmented protected areas in a matrix of agricultural land, with only the Derema Forest Corridor linking Amani Nature Reserve to other forested areas (Tropical Biology Association 2007). This fragmentation limits the movement of pollinator and dispersal species, including birds and insects, resulting in lower fertilization and dispersal and thus lower recruitment in forest plant species, ultimately leading to population extinction (Cordeiro and Howe 2003).

Land-use change has other more subtle negative effects, such as degrading soil or water quality. For example, tea cultivation in catchment areas has been found to result in significantly lower dissolved oxygen in tributary streams. This is associated with reduced species richness of more than 25% relative to streams with forest dominated catchments (Van Biervliet et al. 2009). Similarly, the unsustainable farming practices of many small-scale agriculturalists leads to leaching and reduced soil fertility, as well as reduced ground cover, leaving the soils exposed and prone to erosion. This reduces downstream water

The Source Waters of Tanga, Fig. 11 Peasant farming in the Tanga watershed. Photo by John Esther

quality and biodiversity, as well as degrades the terrestrial ecosystems in the catchment (Bullock et al. 2013).

Since 2003, small-scale alluvial gold mining operations have also become a significant threat to biodiversity due to their degrading impacts on aquatic and wetland forest ecosystems. This activity involves the excavation of open pits in riverine and wetland forest areas like the one in Fig. 12 and results in water-saturated soils with high organic matter, low pH, and low microbial activity. Nutrient recycling, especially nitrogen and phosphorus, is also negatively affected, with carbon-to-nitrogen ratios significantly higher in undisturbed forests, suggesting greater carbon sequestration capacity (Kweyunga and Senzota 2007). This degradation of water quality, soils, and wetland forest habitat significantly affects the ability of vegetation to reestablish itself, threatens amphibian and invertebrate aquatic fauna dependent on these habitats, and creates water quality concerns for water users in the lower catchment areas (Rural Focus Ltd. 2020; Senzota and Mbago 2009).

A further significant threat to the biodiversity of the East Usambaras is introduced species, especially *Maesopsis eminii* (Rhamnaceae) which has naturalized and begun to dominate areas of otherwise virgin forest. This tree naturally occurs throughout central Africa, including western Tanzania, and was first introduced to the East Usambaras in 1913 (Sheil 1994; Tropical Biology Association 2007). Its introduction has resulted in reduced regeneration of primary forest trees, thinner leaf litter, increased soil erosion, decreased organic matter in the topsoil, and reduced diversity in the soil microfauna. It is generally recognized that the tree's spread threatens the endemic and near-endemic tree species of the East Usambaras' primary forests (Binggeli and Hamilton 1993; Hamilton and Bensted-Smith 1989).

The final, and perhaps most significant, long-term threat to biodiversity is climate change. Projected changes for Tanzania by 2050 include (USAID 2018):

- Increased average annual temperature of 1.4–2.3 °C
- Increased duration of heat waves (by 7–22 days) and dry spells (up to 7 days)

S

The Source Waters of Tanga, Fig. 12 Wetland trees nicknamed Mdhahabu [gold tree] thanks to its fibrous roots that trap alluvial gold specks for miners. Photo by Fred Kihara

- Likely increase in average annual rainfall, with the greatest increase in the northeast
- Increased heavy rainfall event frequency (7–40%) and intensity (2–11%)
- Rise in sea levels of 16–42 cm
- Loss of Kilimanjaro glaciers

These effects cumulatively make Tanzania the 26th most vulnerable country to climate risks (USAID 2018). Studies on greenhouse gas (GHG) emissions show that the main contributing sectors are land-use change and forestry (87.33%), energy (6.39%), and agriculture (5.68%), with deforestation being the main contributing activity (Rural Focus Ltd. 2020).

These changes are expected to cause range shifts and ecosystem composition changes that threaten biodiversity. This is particularly true in island-like mountain ecosystems such as the East Usambaras because the potential range for expansion is limited due to altitude. This means that whereas species could ordinarily shift their ranges to compensate for environmental changes due to climate change, in a mountain-restricted species, a shift up the mountain in response to global warming results in a reduction in range area, ultimately leading to extinction. Additionally, changes to rainfall patterns and other climatic variables can result in changes in species' phenology. This can lead to dissociation of predator and prey population booms, leading to crop pest plagues and disease outbreaks, as well as reduced recruitment to populations which exacerbates the likelihood of extinction for threatened and endemic species (Post et al. 2001).

Potential Conservation Interventions

Soil and Water Conservation

Researchers identified 114 km^2 of agricultural land with especially high sediment loss (>10 t/ha/yr). Most of that land has slopes greater than 15%. It is recommended that terracing and agroforestry be implemented in these areas to protect against soil erosion (which helps with soil fertility and soil physical properties) and to provide ground cover. This will be crucial to protecting fragile aquatic ecosystems that harbor endemic and threatened amphibian species, as well as plant species that depend on wetland forest habitats.

Agroforestry and Woodlots for Carbon Sequestration

The Tanga Water Fund targets agroforestry and woodlots establishment on 10% of the 10,182 ha of farmland with slopes >5% within the Zigi catchment and where terracing is not targeted. Woodlots release the pressure on primary forests for firewood and construction materials, thereby helping to conserve the remaining primary forest that most species endemic to the East Usambaras depend on. Planting the targeted one million trees as part of agroforestry across these areas, as well as protecting soils from erosion, could create a sufficiently diverse habitat to provide dispersal corridors for birds and insects instrumental in pollination and the distribution of plant seeds. Agroforestry activities like the one shown in Fig. 13 are crucial for maintaining endemic and threatened species with small populations in a fragmented habitat matrix.

Riparian Conservation

Approximately 853 km and 269 km of river length within the Zigi and Mkulumuzi catchments, respectively, should be targeted for riparian conservation efforts. Tanzanian legislation stipulates that 5 m on either side of the river is not to be cultivated and a further 55 m is to remain under perennial cover. This implies that perennial crops such as fodder grass or spice trees are permitted within the riparian area. These practices are to be encouraged, as riparian conservation protects against soil erosion and leaching. This would thereby protect water quality, with ensuing biodiversity conservation implications, as outlined above. Additionally, continuous riparian habitats

The Source Waters of Tanga, Fig. 13 Integration of clove trees in agroforestry farming system in the Tanga watershed. Photo by John Esther

S

are ideal for species dispersal, as they transect all parts of an area and provide a diverse habitat with water availability. Riparian habitat conservation is therefore critical to biodiversity.

Integrated Watershed Management Plans

Integrated watershed management plans aim to address issues within a smaller (5–15 km^2) watersheds that are causing erosion or water pollution. These require a communal approach and provide an opportunity to carefully manage practices, such as mining, that damage unique habitats, including wetland forests. These plans would be integrated with catchment and sub-catchment plans, as well as other conservation initiatives, and are important for formalizing and implementing the necessary measures for limiting ecologically damaging behavior in the East Usambaras.

Livelihood Support

Various options to support rural livelihoods are proposed to address poverty and the unsustainable use of resources. These include high-value enterprises such as poultry, beekeeping, milk production from stall-fed livestock, and more efficient use of fuelwood through the adoption of energy-saving stoves. These practices will relieve pressure on the primary forest for food and fuel, thereby protecting these habitats for the endemic and threatened species that are wholly dependent on them.

Marketing and Value Chain Support

It is recommended that tree crops – for example, cloves – be processed, stored, and marketed after harvesting. Targeting efforts to crops that integrate agroforestry practices encourages the cultivation of those that can contribute to soil and water conservation measures in the East Usambaras. Fig. 14 shows some of the ways to increase revenue for the farmers, helping them shift from their reliance on forest products such as firewood and thus further contributing to habitat conservation.

Rural Roads, Tracks, and Pathways

Training on manual road and footpath maintenance and drainage is proposed in order to help keep roads passable and protect against soil loss. These tracks are a dominant source of erosion and turbidity in the watercourses. Managing this problem will improve water quality, which is crucial for amphibian, fish, and aquatic invertebrate species of the East Usambaras.

The Source Waters of Tanga, Fig. 14 Spreading fresh clove spices to dry. Photo by Roshni Lodhia

Conclusion

Fig. 15 depicts the forests of the East Usambaras unique biodiversity, with multitudes of endemic, near-endemic, and threatened species from across the tree of life. Indeed, these forests have some of the highest levels of endemism not only in Africa but across the world. However, these ecosystems are under threat from various natural and anthropogenic factors, and currently have less than 30% of the original primary forest remaining. With a growing human population, resolving these challenges requires collaboration between local communities and governing and technical authorities. TWF is one such initiative, and the interventions proposed here have the potential to expand biodiversity conservation in the East Usambara Mountains. The TWF stands as the first line of defense and helps foster the survival of forest-, wetland-, and river-dependent taxa.

The Source Waters of Tanga, Fig. 15 If tree and tree can live in harmony so can man and nature. By Roshni Lodhia

Part 3. Carbon Sequestration and High Climate Impacts of Conserving the East Usambara Mountains, Tanga

Introduction

Approximately 45,137 ha of the East Usambara Mountains remain as a natural forest, with 30.6% classified as submontane rainforest and 63% as lowland forest. These amounts vary by altitude, with submontane forest generally occurring above 850 m and low-altitude forest extending to the coastal plains. This composition is similar to other East African coastal forests (Hamilton and Bensted-Smith 1989).

The soils of the East Usambaras are predominantly clay and clay loams between 1 and 5 m in depth. Generally red and well-drained, most soils at higher altitudes exhibit an acidic and leached condition. Soil conditions under primary forest cover are better (Kashindye et al. 2018).

The potential for carbon sequestration and storage in the East Usambaras is significant, but mechanisms for quantifying and harnessing this to access global carbon markets are not in place. This would impact climate change mitigation, as well as improve local livelihoods, generate revenue, and conserve nature.

The Zigi watershed is dominated by forest, both natural primary forest and commercial teak and eucalyptus plantations. This is complemented by commercial agroforestry, primarily cardamom, cloves, and cinnamon, with an understory of black pepper and table sugarcane. Extensive tea plantations and small-scale subsistence agriculture are also prevalent, and, together, agriculture accounts for 69% of economic activities in the Zigi River catchment area (Rural Focus Ltd. 2020). These activities have devastated forests. Currently, only 18,589 ha of undisturbed primary forest remains (see Fig. 5).

S

Carbon: Natural Compound and Valuable Commodity

Climate Change

Climate change has become one of the leading concerns of the twenty-first century, with global warming, sea level rise, and increased variability and extremity in weather patterns all predicted as a result of anthropogenic activities since the industrial revolution. These changes are largely attributed to the release of CO_2 and other GHGs into the atmosphere. Activities contributing to this include the global transportation, agriculture, and energy sectors, as well as deforestation and forest degradation. Deforestation and forest degradation account for approximately 11% of global carbon emissions – more than the entire global transportation sector and second only to the energy sector (UNEP 2020). This makes deforestation the second leading cause of global warming (Forest Carbon Partnership Facility 2020), with tropical deforestation estimated to contribute 7–14% of global carbon dioxide emissions (Green et al. 2013).

Potential Solution: Conserving Nature, Generating Carbon Credits

With deforestation contributing so significantly to the climate change problem, afforestation and forest conservation efforts are an adaptable, effective, and cost-efficient solution that can provide up to a third of the mitigation required to keep global warming well below 2 °C; the benchmark set in 2015 (UNREDD 2020; UNFCCC 2015). To achieve this, it was recognized that creating mechanisms that provide financial incentives for afforestation and forest conservation initiatives were needed. To this end, carbon credits were formalized as a standardized means of measuring and valuing the contribution of natural habitats and areas to climate change mitigation in terms of "carbon stocks" comprising of carbon removed from the atmosphere by planted trees like in Fig. 16 (carbon sequestration) and stored (carbon storage). Trading in carbon credits allows revenue to be generated for the conservation of natural habitats and ecosystems that sequester carbon from the atmosphere, as well as providing states, companies, and businesses an opportunity to mitigate the carbon footprint of their activities. Figure 9 above illustrates the status of carbon storage in the Eastern Arc mountains.

The international market for carbon credits has been developed through initiatives such as REDD+ (reducing emissions from deforestation and forest degradation). This "creates a financial value for the carbon stored in forests by offering incentives for developing countries to reduce emissions from forested lands and invest in low-carbon paths to sustainable development. Developing countries would receive results-based payments for results-based actions. REDD+ goes beyond simply deforestation and forest degradation and includes the role of conservation, sustainable management of forests, and enhancement of forest carbon stocks" (UNREDD 2020). Fig. 17 shows tea planted at the edge of a natural forest serving as a permanent protective buffer and deterrent to encroachment.

The Source Waters of Tanga, Fig. 16 Growing black pepper spices as climbers on planted trees. Photo by Roshni Lodhia

There are many nuances and challenges in creating a globally recognized standard for calculating the carbon stocks of a given habitat or ecosystem and then converting these stocks to tradable carbon credits. Furthermore, ensuring that the sale of carbon credits is effectively reinvested in the people, institutions, and projects

The Source Waters of Tanga, Fig. 17 Tea growing at the edge of the cloud forest in Tanga watershed. Photo by Roshni Lodhia

working to conserve and regenerate natural forest habitats for carbon sequestration and storage remains complicated. However, through systems such as REDD+ – reducing emissions from deforestation and forest degradation in developing countries and the role of conservation, sustainable management of forests, and enhancement of forest carbon stocks in developing countries – these challenges have been addressed. Carbon credits are now a viable opportunity for funding conservation and climate change mitigation activities.

Saving Carbon in the East Usambara Mountains

Vegetation

Human population expansion (influenced by economic opportunities such as employment on tea plantations, which can require up to 4,000 people) is increasing pressure on the remaining natural forest in the area (Tropical Biology Association 2007). Higher demand for resources and food leads to forests being cleared to make way for crops and housing. Alternative revenue-generating activities such as alluvial mining for gold and other gemstones along rivers have further exacerbated the destruction of trees and wetland ecosystems in the last decade.

The rate of forest loss varies by forest type, with closed to nearly closed-canopy deciduous vegetation, known as "miombo woodlands," experiencing the greatest recent decline: approximately 43% between 1975 and 2000. This was partly due to this habitat existing mainly outside the network of protected areas across the Eastern Arc Mountains, which are dominated by primary closed-canopy forest as shown on Fig. 18 (Green et al. 2013).

Estimates of the carbon stocks (both vegetative and pedological) for each mountain block in the Eastern Arc Mountains are presented in Table 2, with estimated total carbon storage for the entire watershed of the Eastern Arc Mountains at 1.3 Mtonnes C (1.3×10^6 tonnes C).

Figure 9, below, spatially represents the carbon storage and sequestration across the Eastern Arc Mountains watershed.

Carbon storage and sequestration capacity also vary by habitat type. These estimates are given in Table 3.

S

The Source Waters of Tanga, Fig. 18 Tanga water fund protects millions of tonnes of carbon stocks from destruction. By Roshni Lodhia

The Source Waters of Tanga, Table 2 Carbon storage and sequestration estimates for the individual mountain blocks of the Eastern Arc Mountains. (Adapted from Willcock et al. 2014)

Eastern Arc Mountain Bloc	Area (km^2)	Aboveground live carbon storage (million tonnes)	Mean carbon sequestration (t/ha/yr)
North Pare	510	1.93	2.6
South Pare	2,327	8.96	2.41
West Usambara	2,945	13.52	3.64
East Usambara	1,145	5.91	2.79
Nguu	1,562	9.34	1.89
Nguru	2,565	15.11	1.79
Ukaguru	3,243	13.39	1.42
Uluguru	3,057	15.92	1.35
Rubeho	7,984	36.84	1.06
Malundwe	33	0.29	1.8
Udzungwa	22,788	101.73	1.01
Mahenge	2,606	23.58	0.19
Total	**50,765**	**246.52**	**21.95**

Soil

The organic matter in healthy soils contains carbon, which is stored in the soil and referred to as soil organic carbon (SOC). SOC is crucial to the global carbon cycle, with soils representing the world's largest terrestrial carbon stock. Tropical forests are estimated to account for about 32% of the SOC stored in the world's soils (Kirsten et al. 2016).

Carbon stocks in the soil are dynamic, capable of long-term persistence or loss through various processes. These processes include release as CO_2

The Source Waters of Tanga, Table 3 Mean carbon storage and sequestration by land cover type (Willcock et al. 2014)

Land cover category	Carbon storage (t/ha)	Range	Carbon sequestration (t/ha/yr)	Range
Lowland forest (<1,000 m)	182	152–360	−0.91	−7.08–4.29
Submontane forest (1,000–1,500 m)	189	95–588	−2.02	−11.06–1.29
Montane forest (1,500–2,000 m)	130	62–702	−2.03	−11.85–1.07
Upper-montane forest (>2,000 m)	166	69–533	−2.08	−10.49–1.23
Forest mosaic	121	55–485	−1.18	−6.69–2.92
Closed woodland	100	70–331	−1.24	−7.91–2.63
Open woodland	51	38–165	−1.49	−7.53–2.05

The Source Waters of Tanga, Table 4 Soil organic carbon estimate for the East Usambaras based on a per-area estimate from Lal (2004) and areas from Fig. 1, using a 35% depletion rate for disturbed forest soils

Type	Area (ha)	Estimate (t C/ha)	Discount factor (%)	Discounted Estimate (t C /ha)	Total (t C)
Undisturbed	18,589	197	100	197	3,662,033
Disturbed	38,047	197	65	128	4,871,918
Total	**56,636**				**8,533,951**

or CH_4 back into the atmosphere, loss in eroded soil material, or dissolution of organic carbon washed into rivers and oceans (FAO 2017).

According to data extrapolated from Amani Nature Reserve, in the East Usambaras, SOC stocks down to 1 m depth were 16.9–22.4 kg C m^{-2} (mean 19.7 kg C m^{-2}) (Kirsten et al. 2016), though some conversion of forest habitat to agricultural land use results in the depletion of SOC stock by 20–50% (Lal 2004, 2005). However, the SOC stock in secondary forest has also been found comparable to that in primary forest soils (20.2 kg C m^{-2}) (Kirsten et al. 2016). A time-series analysis of forest cover done in 2020 distinguishes undisturbed from disturbed forest but does not distinguish agricultural and secondary forest within the disturbed forest areas. Therefore, using the areal degradation since 1986 shown in Fig. 5 above and a 35% observed cover depletion rate for disturbed land, the net SOC stock of the East Usambaras is 8.5 million tonnes C, as shown in Table 4.

The SOC stock estimate for secondary forest (20.2 kg C m^{-2}) given above makes the case for plantation and agroforestry as appealing options for carbon stocks and for generating credits in ways that also provide livelihood opportunities for local community members.

The estimates of SOC stock above are fixed time measurements and do not give any indication of SOC sequestration or accumulation over time. Some estimates suggest the rate of SOC sequestration in tropical forest soils is 0.1–1 tonnes C ha^{-1} $yr.^{-1}$ (Lal 2004). Again, using area estimates given in Fig. 1, with no discounting for disturbed land and a midpoint estimate of 0.55 t C ha^{-1} $yr.^{-1}$, this would equate to carbon sequestration for the East Usambaras amounting to 31,150 t C $yr.^{-1}$.

S

Financial Analysis

The price of carbon credits constantly fluctuates on the international market, and there are myriad variables that are considered in their valuation. Generally, carbon quantity estimates are first converted to CO_2 equivalents (CO_2e) (This is achieved by multiplying by 44/12 due to the molecular weight ratio of CO_2 to carbon.). Even then, the value of carbon credits varies widely depending on the nature of the project and the standard used (Gold Standard 2020). Some national price estimates are given in Table 5. For the Tanga Water Fund, the average estimate of USD 11.4 per tCO_2e will be used (Fig. 19).

The valuation of the carbon stock in the East Usambaras is given in Table 6, and the per annum valuation of the carbon sequestration in the East Usambaras is given in Table 7.

The calculations in Tables 6 and 7 suggest the East Usambara carbon stocks are worth over USD 324 million, while the sequestration potential is worth a further USD 1.3 million per year. To realize this financial asset, elaborate valuation and verification by accreditation schemes are required. The final value will be arrived at based on the verification method used and may vary slightly in final quantities, more so if only considering the geographic coverage of the subbasins that supply water to Tanga in Fig. 19. However, this estimate makes a strong case for implementing TWF as a beneficial climate mitigation project. Figure 22 below shows forest cover trends over a 35-year span. This shows the effectiveness of conservation interventions in reducing forest degradation and improving forest cover.

The Source Waters of Tanga, Table 5 Price of carbon per unit tonne of CO_2 for selected countries (Kossoy et al. 2014; CDP 2013; Environmental and Energy Study Institute 2012)

Country	Price (USD tCO_2e^{-1})
Australia	21.1
Canada, British Columbia	29.1
European Union	5.9
India	0.9
Japan	2.9
South Africa	12.2
The United Kingdom	7.6
Average	**11.4**

Enhancing Carbon Sequestration through Sustainable Land Management

Soil and Water Conservation

Of the 114 km^2 of agricultural land with especially high sediment loss (>10 t ha^{-1} $yr.^{-1}$) identified,

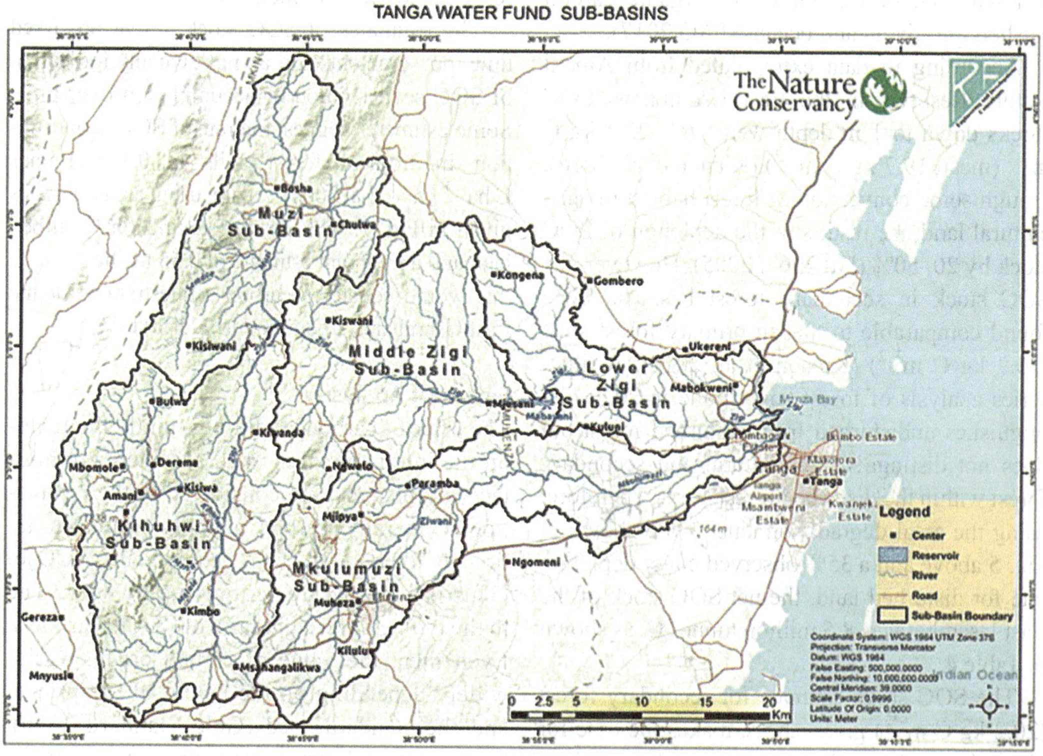

The Source Waters of Tanga, Fig. 19 Subbasins of the Tanga watershed

The Source Waters of Tanga, Table 6 Calculations of the estimated financial value of carbon stocks in the target areas of the Tanga Water Fund

Area	Altitude (m.a.s.l.)	Forested area (ha)	Aboveground carbon per unit area (t/ha)	Soil carbon per unit area (t/ha)	Combined total (t)		CO2 Equivalent (t CO2e)		Value (USD)	
Stage 1 – Kihuhwi	<1,000	25,161	182	197	9,536,106	11,484,704	34,965,722	42,110,580	398,609,236	480,060,617
	1,000–1,500	5,048	189	197	1,948,598		7,144,858		81,451,381	
Stage 2 – Zigi	<1,000	34,983	182	197	13,258,663	13,753,934	48,615,096	50,431,092	554,212,100	574,914,445
	1,000–1,500	1,283	189	197	495,271		1,815,995		20,702,346	
Stage 3 – Mkulumuzi	<1,000	8,480	182	197	3,214,019	3,214,019	11,784,737	11,784,737	134,345,999	134,345,999
	1,000–1,500	–	189	197	–		–		–	
Total		**74,956**				**28,452,657**		**104,326,409**		**324,360,289**

The Source Waters of Tanga, Table 7 Calculations of the estimated financial value of the annual carbon sequestration in the target areas of the Tanga Water Fund

Area	Altitude (m.a.s.l.)	Forested area (ha)	Aboveground carbon sequestration per unit area (t/ha/yr)	Soil carbon sequestration per unit area (t/ha/yr)	Combined Total (t/yr)		CO2 Equivalent (t CO_2e/yr)		Value (USD/yr)	
Stage 1 – Kihuhwi	<1,000	25,161	0.91	−0.55	−36,735	−49,709	−134,696	−182,267	418,784	566,685
	1,000–1,500	5,048	−2.02	−0.55	−12,974		−47,571		147,902	
Stage 2 – Zigi	<1,000	34,983	−0.91	−0.55	−51,076	−54,373	−187,277	−199,368	582,262	619,854
	1,000–1,500	1,283	−2.02	−0.55	−3,298		−12,091		37,592	
Stage 3 – Mkulumuzi	<1,000	8,480	−0.91	−0.55	−12,381	−12,381	−45,398	−45,398	141,145	141,145
	1,000–1,500	–	−2.02	−0.55	–		–		–	
Total		**74,956**				**−116,464**		**−427,033**		**1,327,684**

The Source Waters of Tanga, Fig. 20 Teak [*Tectona Grandis*] seedlings nurseries in the Tanga watershed. Photo by John Esther

The Source Waters of Tanga, Fig. 21 Fish farming is a major source of proteins for the community in the zigi river watershed. Photo by John Esther

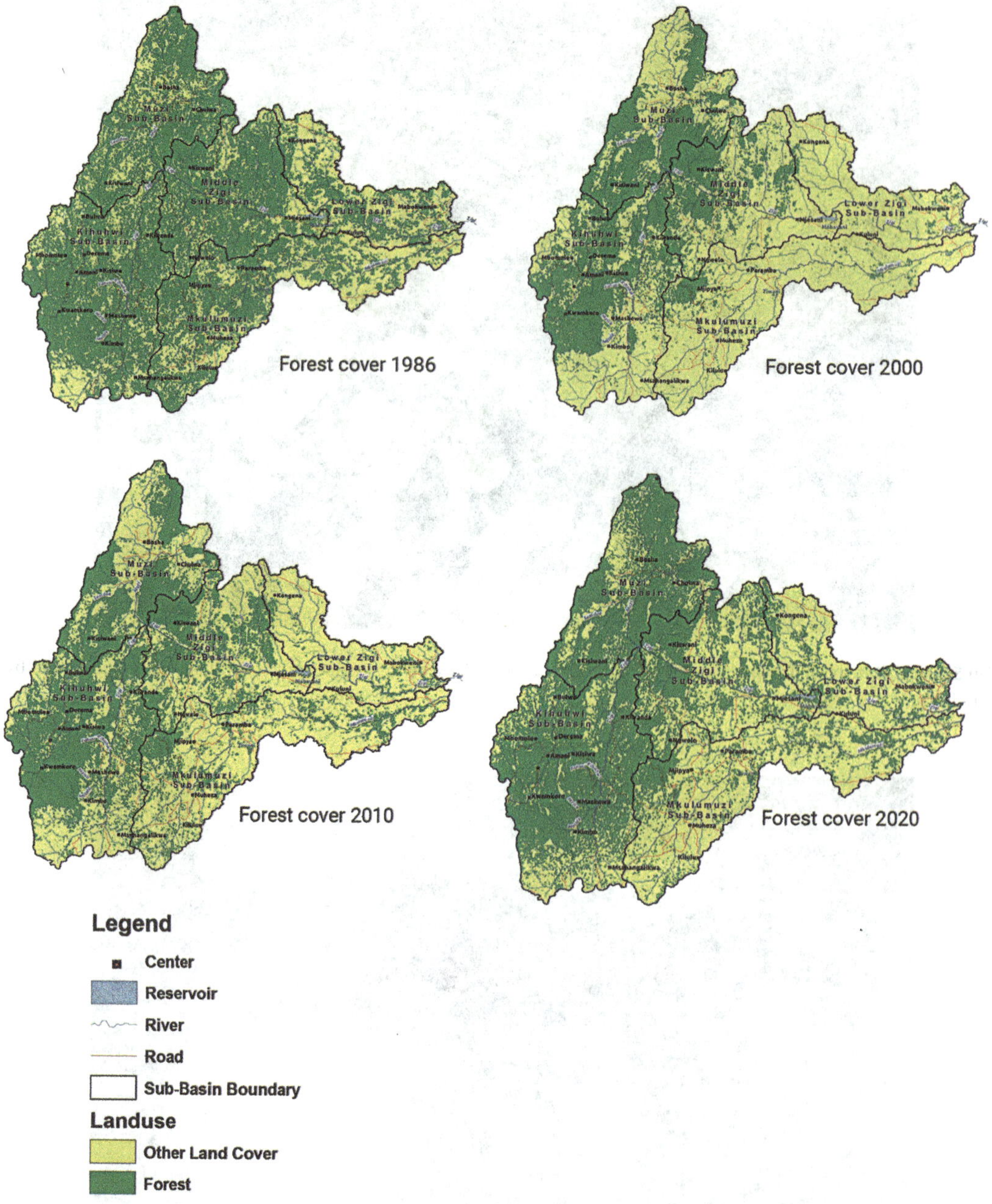

The Source Waters of Tanga, Fig. 22 Forest cover changes in the Tanga watershed between 1986 and 2020. Own analysis. Source: Landsat imagery

most have slopes greater than 15%. It was recommended that terracing and agroforestry be implemented in these areas to protect against soil erosion and to provide ground cover. This will result in improved soil properties, including nutrient cycling. This will conserve and increase the carbon sequestration and storage capacity of the soil, as well as protect against carbon loss through erosion.

Agroforestry and Woodlots for Carbon Sequestration

Tanzania is renowned for its participatory forest management efforts. It is proposed that agroforestry and woodlots be targeted on 10% of the 10,182 ha of farmland with slopes of >5% within the Zigi catchment and where terracing is not targeted. Woodlots can lessen the pressure on primary forest for firewood and construction materials, thereby helping to conserve the remaining primary forest. Implementing agroforestry across these areas, as well as protecting soils from erosion, can contribute to climate change mitigation by improving carbon sequestration and storage.

These agroforestry schemes will likely take the form of teak (*Tectona grandis*) plantations as shown on Fig. 20, which have been a highly successful commercial endeavor in many parts of Tanzania, including the East Usambaras (Bekker et al. 2004; Rance and Monteuuis 2004). Other agroforestry approaches with potential for success include growing sugarcane and cultivating spice tree crops, such as cardamom and cloves, with an understory of black pepper. These crops protect soil structures and create a mixed forest habitat with carbon sequestration and storage capacity. They have strong local and international market value, offering additional financial benefits. In the case of teak in the Tanga watershed, which drains the East Usambaras, the proposed agroforestry interventions are estimated to be worth about USD 3,170 ha^{-1} $yr.^{-1}$ (Rural Focus Ltd. 2020). This equates to a total of USD 48,415,410 once the interventions have been carried out over the full area for 15 years, or approximately USD 3.2 million $yr.^{-1}$.

The Tanzania Forest Conservation Group has successfully piloted REDD+ projects in the Tanzania districts of Kilosa and Lindi. A similar project could be extended to the Zigi catchment, providing an opportunity for additional benefits to the communities tasked with implementation and maintenance (Tanzania Forest Conservation Group 2020).

Complementary Activities

1. *Riparian protection:* 853 km of the Zigi River's riparian area will be conserved over the first 10 years, with another 269 km along the Mkulumuzi River planned to be done by the 15th year. The targeted area, which will consist mainly of cut-and-carry pasture and agroforestry, covers the 5 m from the riverbank that, by Tanzania state law, cannot be cultivated, and an additional 55 m buffer either side of the river that must be under perennial. This will result in increased carbon stocks and sequestration along the length of these rivers.
2. *Livelihood support activities:* As part of the conservation interventions, additional pasture will be established on the farmlands. Further, cookstoves, which require less biomass for household energy needs, will be adopted as well as fish farming as shown in Fig. 21. These changes should coincide with the anticipated rise in livestock numbers (which provide household protein in the form of milk and meat and revenue from the sale of these products) to ensure that any additional GHG emission from the livestock is offset by the interventions.
3. *Marketing of farm produce and increased motorized activities:* Increased crop production related to population growth in the catchment areas is likely to result in marginally increased carbon emissions from produce transportation, mechanized processing, and farm operations. These emissions, though marginal, should be easily offset by the anticipated sequestration gains generated by the proposed interventions.
4. *Rural roads, tracks, and pathway development:* Where possible it is recommended that manual road maintenance, drainage, and footpath maintenance be adopted to keep them passable and to protect against soil loss. Currently, the tracks are a prominent source of erosion and degradation of soil carbon stocks. The reduction of these processes will offset additional carbon from increased traffic.

Conclusion

The ecosystems of the East Usambaras are highly valuable for biodiversity and ecosystem services, including storing large stocks of aboveground soil carbon, as well as sequestering significant amounts of carbon from the atmosphere.

The Tanga Water Fund and its collaborating institutions should document and claim credit for the value of (i) carbon stocks being safeguarded by the project, (ii) the mitigation of carbon stock degradation resulting from activities of the project, and (iii) the increase in carbon stocks and sequestration capacity resulting from investments made by the fund.

The carbon value in the East Usambaras is estimated at over USD 324 million, plus an estimated per annum value of USD 1.3 million for carbon sequestration. There is a strong case for offering some of the CO_2 equivalent credits associated with the Zigi River watershed to TWF investors for their offset schemes or for directly trading the credits on the international carbon market to financially support the fund.

Part 4. Robust Governance and Stakeholders Engagement in Establishing a Lasting Water Fund in Tanga

The water fund model draws on the global payment for ecosystem services (PES) concept, in which downstream water users pay upstream watershed keepers to implement conservation activities that promote water security for the downstream users. Additionally, the water fund model incorporates a strong governance mechanism and is strengthened by the collaboration of both public and private sector stakeholders in its operation, financing, and governance. Thus, the engagement of stakeholders and their willingness and ability to collaborate is critical to the water fund's success.

The Tanga Water Fund feasibility study identified the following key stakeholder categories:

(a) **Water Service Providers in Tanga City and Its Environs** – Tanga UWASA
(b) **Major Water Users** – Tea plantations, sisal plantations, lime and cement manufacturers, Tanga fresh dairy processors, Tanga Port Authority, and hotels
(c) **Catchment Managers and Users** – Pangani Basin Water Board; district councils of Korogwe, Mkinga, and Muheza; UWAMAKIZI; and Zigi Water Users Association
(d) **Transport Infrastructure Developers** – Tanzania National Roads Agency, Tanga Port Authority, East African Crude Oil Pipeline and Tanzania Rural Roads Authority
(e) **Regulatory Agencies** – Energy and Water Utilities Regulatory Authority
(f) **Agencies Responsible for Protected Areas** – ANR
(g) **Conservation Entities in the Catchment Areas** – Tanzania Forest Conservation Group, International Union for Conservation of Nature (IUCN), TNC, Worldwide Fund for Nature, CARE Tanzania, ONGAWA, and Eastern Arc Mountains Conservation Endowment Fund (EAMCEF)
(h) **Public Funding Agencies** – United Nations Development Programme, Global Environment Facility, (Ministry of Water (MoW), USAID- RTI and GIZ
(i) **Research Institutions** – World Agroforestry and Tanzania Agricultural Research Institute

An analysis conducted in 2020 on how these stakeholders ranked in terms of interest and influence to the water fund found that the stakeholder landscape is complex, with institutions having different mandates and interests (Fig. 23). This is to be expected, as integrated water resource management cuts across many sectors. TWF aims to work with catchment users and managers, as well as water consumers, and to engage interested private sector leaders.

The water fund will establish appropriate structures to allow for district-level coordination, as there are development programs and relevant mandates across numerous institutions. The private sector's role within TWF is a critical ingredient to the fund's sustainability in regard to local funding, engagement, and governance. The private sector will also ensure financial efficiency and implementation accountability. In this regard, TWF will need to strengthen collaboration between public, private, and civil society sectors within the fund.

A review of past development projects within the Zigi watershed and East Usambara Mountain areas indicates that future initiatives such as the

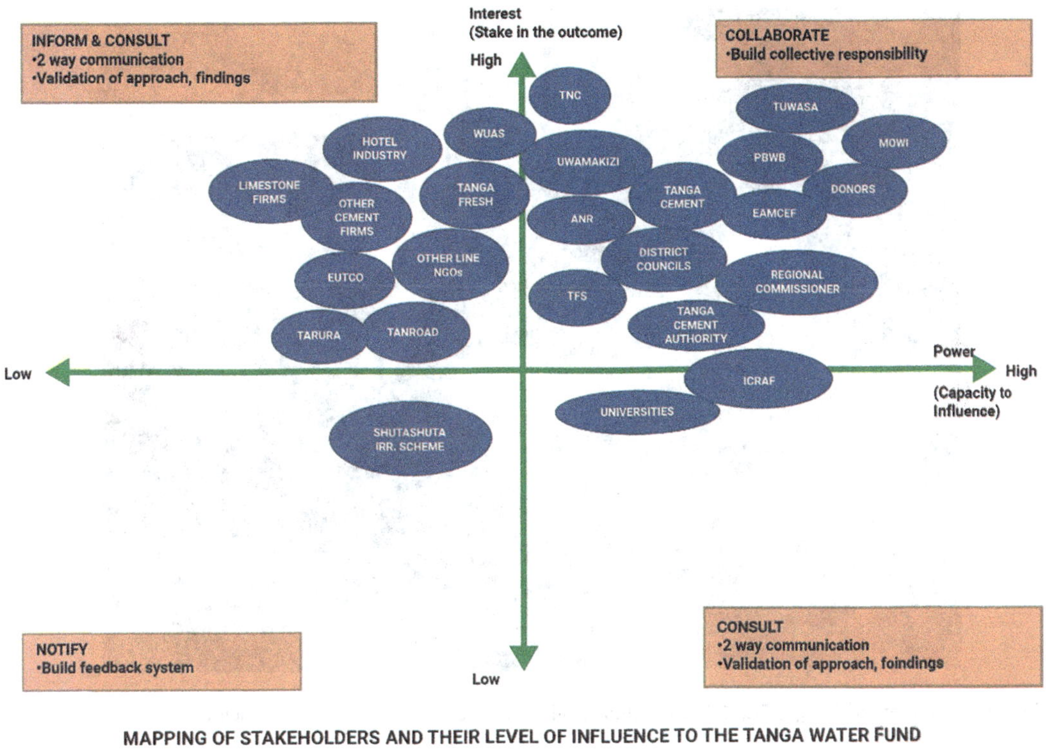

The Source Waters of Tanga, Fig. 23 Participatory mapping of stakeholders in the Tanga Water Fund (RFL 2020)

Tanga Water Fund will need to incorporate the following aspects into its stakeholder engagement strategy:

- Full support by the national government and national government agencies (e.g., Ministry of Water and Irrigation, the Pangani Basin Water Board, Tanzania Forestry Service)
- Strong and close links to regional and local government administrative and technical departments (Muheza, Tanga, Korogwe, and Mkinga districts)
- Provision of a structure that represents private sector interests and operates on principles of financial efficiency and performance accountability
- Engagement with other stakeholders for effective collaboration, governance, and implementation

Figure 24 shows stakeholder participation and commitment, TWF must present a strong business case. This can be accomplished by ensuring that potential stakeholders understand the value that can be accrued from their investment. As explained in the previous sections, the interventions to be implemented through TWF could generate numerous environmental and financial benefits to the various stakeholders. The feasibility study commissioned by TNC concluded that from a total investment of USD 17.8 million in sustainable land management interventions, stakeholders in the Zigi and Mkulumuzi watersheds can accrue annual net benefits amounting to USD 11.2 million from the 15th year. The financial analysis of carbon potential in the watershed estimates a value of carbon stocks and sequestration potential for the targeted areas of USD 324.4 million and USD 1.3 million, respectively.

However, given that the scale of the proposed interventions is resource-intensive and to maximize on available resources, a progressive three-phase approach is proposed based on the level of risk identified and urgency for intervention.

S

The Source Waters of Tanga, Fig. 24 Consulting Zigi river watershed communities on their conservation priorities. Photo by John Esther

Phase One would implement activities in the Kihuhwi River – the southern tributary of the Zigi River – and the area up to the confluence with the Muzi tributary from the north. That subbasin is 383.5 km^2. The area is largely covered by steeply sloping land with higher sediment yields than the other subbasins. In addition, it has extensive on-the-ground coverage by community-based operation UWAMAKIZI, a collaborator that can mobilize the implementation of proposed activities.

Phase Two covers the Muzi tributary and the mid-Zigi subbasin up to the Mabayani Dam and down to the delta, covering 1,088 km^2. The additional area of Muzi and the middle catchment will require the establishment and capacity development of local partners to help implement activities. The nongovernmental association ONGAWA has had experience working in this area, as has the Eastern Arc Mountains Conservation Endowment Fund (EAMCEF).

Phase Three adds Mkulumuzi River to the other two, increasing the area to 1,432 km^2. The Mkulumuzi area will benefit from a growing interest in the soon expanded Tanga port. The port suffers sediment deposition, which requires periodic desilting. As with Phase Two, Phase Three will require the establishment and capacity development of local partners to help implement activities. Phase Three is expected to be implemented after 10 years of work on the other two phases.

Successful execution of this strategy will require adequate financing, proper organization and management, and stakeholders' collaboration and participation. The EAMCEF and TNC, the leading organizations, share a common vision and baseline resources to steer the initial development of the Tanga Water Fund, as well as to coordinate stakeholder engagement and build on Tanga Urban Water Supply and Sanitation Authority's (UWASA) catchment conservation efforts. Two options for the water fund's establishment emerge: (1) embedding the TWF within EAMCEF and operating it as a project within EAMCEF with technical and fundraising support from TNC and peers and (2) registering TWF as a separate autonomous trust with an office in Tanga City. Regarding the first option, TNC needs to have further discussions with EAMCEF to chart the way forward and determine the TWF's governance structure.

An independent TWF would benefit from a three-tier governance structure in which (1) a board of trustees (BOT) draws in strategic government, private sector, and funding partners; (2) a board of management (BOM) provides technical and management oversight and ensures coordination across different agencies and districts; and (3) a management unit provides implementation capacity.

Management unit functions can be outsourced to an organization with an established office, capacity, or relevant experience that can be scaled up appropriately for the TWF. TNC and EAMCEF, and to some extent Tanga UWASA, have broad organizational capability to guide TWF's establishment. In due course and with sufficient capacity, TWF could attain more autonomy. TWF would require a lean administrative and technical staff that could draw technical capacity from the district departments.

Setting up and operating costs are estimated as:

- USD 96,000 (transport, furniture, equipment)
- USD 150,000 for the monitoring system (water quality and river flow monitoring equipment)
- USD 500,000 annual operational and maintenance costs, inclusive of a 20% communication, monitoring and evaluation budget

Long-term funding options may be explored as the water fund progresses. The public sector could co-fund the proposed interventions by aligning budgets and making funding commitments to achieve the identified outputs. This lessens the reliance on public funding beyond requiring commitments to co-fund specific activities.

One promising funding option is a catchment conservation levy applied to the Tanga UWASA water tariff. This method would leverage Tanga UWASA's good standing and existing revenue collection systems to target large consumers and industrial/commercial customers that have indicated their willingness to support a TWF. This could provide a transparent source of predictable funds earmarked for the TWF. A maximum of 5% levy could generate approximately USD 27,000 annually. To grow this revenue, the water utility would need to expand its service base and demand.

Grants and donations from development partners or commercial enterprises are recommended particularly for the initial stage. This requires strong cooperation and commitments from local private and public sector institutions to exhort external partners to invest or provide other leverage. Developing funding proposals requires continuous effort and outreach to potential donors. The MoW can play a significant role in securing high-level, national commitments to partner with development partners. Figure 25 shows a clear case why these investments are important for people.

As discussed previously, carbon markets are another option for long-term financing. Finally, loans and endowment and blended finance could also be considered.

Conclusions

The Tanga Water Fund can deliver a numerous benefits beyond its main objective of improved water security. Through the water fund, biodiversity preservation, climate change mitigation, and improved livelihoods could also be achieved.

There exists a healthy mix of stakeholders with the capacity to accomplish the water fund's objectives. The Tanga Water Fund will continuously structure its engagement with the different stakeholders to leverage their respective interests while ensuring appropriate role and levels of involvement in the fund's decisions and activities.

The Tanga Water Fund will maintain a sound, transparent, and accountable governance and implementation structure that well represents stakeholders' interests and capitalizes on their strengths and capacities. Options for this include (1) embedding the TWF within EAMCEF with technical support from TNC, and (2) registering TWF as a separate autonomous trust in the long run.

Implementation of the water fund's activities will be best done through a phased approach. The priority being the Kihuhwi sub-catchment, the area at highest risk. A phased approach will provide more realistic resource mobilization and help

The Source Waters of Tanga, Fig. 25 As Tanga city grows so does the clean water demand. Photo by Roshni Lodhia

concentrate efforts to result in visible, tangible impacts, to the benefit of stakeholders.

The Tanga Water Fund may adopt a blended structure for resource mobilization and funding that may incorporate catchment conservation levies, grants and donations, loans, endowment funds, and carbon offsets markets. While local enthusiasm is high, the reality is that with a small urban economy, the domestic contributions initially are likely to be low and insufficient to finance upscaling the project. Tanzania leadership has seen significant changes and so does external donor relationships. This may need careful navigation to identify sufficient donors or corporations willing to fully finance this public-focused work and outcomes.

References

Bayliss, J. (2002). The East Usambara tree-hole crab (Brachyura Potamoidea: Potamonautidae) — A striking example of crustacean adaptation in closed-canopy forest, Tanzania. *African Journal of Ecology, 40*(1), 26–34. https://doi.org/10.1046/j.0141-6707.2001.00333.x.

Bekker, C., Rance, W., & Monteuuis, O. (2004). Teak in Tanzania: II. The Kilombero Valley Teak company. *Bois et Forêts Des Tropiques, 279*(279), 11–22.

Binggeli, P., & Hamilton, A. (1993). Biological invasion by *Maesopsis eminii* in the East Usambara forests, Tanzania. *Opera Bot, 121*, 229–235.

Bullock, R., Mithofer, D., & Vihemaki, H. (2013). Sustainable agricultural intensification: The role of cardamom agroforestry in the East Usambaras, Tanzania. *International Journal of Agricultural Sustainability*, 1–21. https://doi.org/10.1080/14735903.2013.840436.

Burgess, N. D., Fjeldsa, J., & Botterweg, R. (1998). Faunal importance of the Eastern Arc Mountains of Kenya and Tanzania. *Journal of East African Natural History, 87*, 37–58.

Burgess, N. D., Butynski, T. M., Cordeiro, N. J., Doggart, N. H., Fjeldsa, J., Howell, K. M., Kilahama, F. B., Loader, S. P., Lovett, J. C., Mbilinyi, B., Menegon, M., Moyer, D. C., Nashanda, E., Perkin, A., Rovero, F., Stanley, W. T., & Stuart, S. N. (2007). The biological importance of the Eastern Arc Mountains of Tanzania and Kenya. *Biological Conservation, 134*, 209–231. https://doi.org/10.1016/j.biocon.2006.08.015.

CDP. (2013). Internal carbon Pricing for Low Carbon Finance. A briefing paper on linking climate-related opportunities and risks to financing decisions for investors and banks. A Carbon Pricing Unlocked White Paper.

Cordeiro, N. J., & Howe, H. F. (2003). Forest fragmentation severs mutualism between seed dispersers and an endemic African tree. *PNAS, 100*(24), 14052–14056.

EESI. (2012). Carbon pricing around the World Factsheet released. October 2012. Environment and Energy Study Institute.

FAO. (2017). *Soil organic carbon: The hidden potential.* https://doi.org/10.1038/nrg2350.

Forest Carbon Partnership Facility. (2020). *What is REDD+? | Forest carbon partnership facility.* https://www.forestcarbonpartnership.org/what-redd

Gold Standard. (2020). *CARBON PRICING: What is a carbon credit worth? | The Gold Standard.* https://www.goldstandard.org/blog-item/carbon-pricing-what-carbon-credit-worth

Green, J. M. H., Larrosa, C., Burgess, N. D., Balmford, A., Johnston, A., Mbilinyi, B. P., Platts, P. J., & Coad, L. (2013). Deforestation in an African biodiversity hotspot: Extent, variation, and the effectiveness of protected areas. *Biological Conservation, 164*, 62–72.

Hamilton, A., & Bensted-Smith, R. (eds) (1989). *Forest conservation in the East Usambara Mountains in Tanzania*. https://portals.iucn.org/library/efiles/documents/FR-TF-016.pdf

ICBP. (1992). *Putting biodiversity on the map: Priority areas for global conservation*. Cambridge: International Council for Bird Preservation (ICBP).

Johansson, S. G., Cunneyworth, P., Doggart, N., & Botterweg, R. (1998). Biodiversity surveys in the East Usambara Mountains: Preliminary findings and management implications. *Journal of East African Natural History, 87*(1), 139–157. https://doi.org/10.2982/0012-8317(1998)87[139:bsiteu]2.0.co;2.

Kashindye, A., Giliba, R., Sereka, M., Masologo, D., Lyatuu, G., & Mpanda, M. (2018). Balancing land management under livestock keeping regimes: A case study of Ruvu and Zigi catchments in Tanzania. *African Journal of Environmental Science and Technology, 13* (7), 281–290. https://doi.org/10.5897/AJEST2018.2648.

Kessy, J. F. (1998). *Conservation and utilization of natural resources in the East Usambara Forest Reserve: Conventional views and local perspectives*. Wageningen: Wageningen Agricultural University.

Kirsten, M., Kaaya, A., Klinger, T., & Feger, K. H. (2016). Stocks of soil organic carbon in forest ecosystems of the Eastern Usambara Mountains, Tanzania. *Catena, 137*, 651–659. https://doi.org/10.1016/j.catena.2014.12.027.

Kossoy, A., Oppermann, K., Platonova-Oquab, A., Suphachalasai, S., Höhne, N., Klein, N., Gilbert, A., Lam, L., Toop, G., Wu, Q., Hagemann, M., Casanova-Allende, C., Li, L., Borkent, B., Warnecke, C., Wong, L. (2014). State and trends of carbon pricing 2014. Washington, DC: World Bank. © World Bank. https://openknowledge.worldbank.org/handle/10986/18415 License: CC BY 3.0 IGO.

Kweyunga, C., & Senzota, R. (2007). Impact of small scale gold mining on soils of the wetland forests in East Usambara, Tanzania. *Tanzania Journal of Science, 33*. Pp 67- 78

Lal, R. (2004). Soil carbon sequestration in natural and managed tropical forest ecosystems. *Journal of Sustainable Forestry, 21*, 1–30. https://doi.org/10.1300/J091v21n01.

Lal, R. (2005). Forest soils and carbon sequestration. *Forest Ecology and Management, 220*(1–3), 242–258. https://doi.org/10.1016/j.foreco.2005.08.015.

Morgan-Brown, T. (2003). *Butterfly farming in the East Usambara Mountains*. Research report. Costech publishing

Morgan-Brown, T. (2007). *Butterfly farming and conservation behaviour in the East Usambara Mountains of Tanzania*. University of Florida, Florida.

MOWI. (2016). Securing watershed services through sustainable land management in the Ruvu and Zigi (Eastern Arc Mountains) Catchments. Quarter One Project Report. Ministry of Water & Irrigation, Government of Tanzania.

Post, E., Forchhammer, M. C., Stenseth, N. C., & Callaghan, T. V. (2001). The timing of life-history events in a changing climate. *Proceedings of the Royal Society of London, Series B: Biological Sciences, 268*, 15–23. https://doi.org/10.1098/rspb.2000.1324.

Rance, W., & Monteuuis, O. (2004). Teak in Tanzania: I. Overview of the context. *Bois et Forêts Des Tropiques, 279*(279), 5–10.

Reyes, T. (2008). *Agroforestry systems for sustainable livelihoods and improved land management in the East Usambara Mountains, Tanzania* (Issue March).

Reyes, T., Quiroz, R., Luukkanen, O., & De Mendiburu, F. (2009). Spice crops agroforestry systems in the East Usambara Mountains, Tanzania: Growth analysis. *Agroforestry Systems, 76*(3), 513–523. https://doi.org/10.1007/s10457-009-9210-5.

Rural Focus Ltd. (2020). *Tanga water fund feasibility study report*. A study commissioned by TNC.

Senzota, R., & Mbago, F. (2009). Impact of habitat disturbance in the wetland forests of East Usambara, Tanzania. *African Journal of Ecology, 48*, 321–328.

Sheil, D. (1994). Naturalized and invasive plant species in the evergreen forests of the East Usambara Mountains, Tanzania. *African Journal of Ecology, 32*(1), 66–71. https://doi.org/10.1111/j.1365-2028.1994.tb00556.x.

Shoo, R. A., & Songorwa, A. N. (2013). Contribution of eco-tourism to nature conservation and improvement of livelihoods around the Amani nature reserve, Tanzania. *Journal of Ecotourism, 12*(2), 75–89. https://doi.org/10.1080/14724049.2013.818679.

Tanzania Forest Conservation Group. (2020). *Making REDD Work – Tanzania Forest Conservation Group*. http://www.tfcg.org/what-we-do/redd/making-redd-work/

Tropical Biology Association. (2007). *Amani nature reserve: An introduction*. Field Guide. Department of Zoology. Downing Street, Cambridge. United Kingdom.

UNEP. (2020). *REDD+ | UNEP – UN Environment Programme*. https://www.unenvironment.org/explore-topics/climate-change/what-we-do/redd

UNFCCC. (2015). *The Paris Agreement*.

UN-REDD. (2020). *What is REDD+? - UN-REDD programme collaborative online workspace*. https://www.unredd.net/about/what-is-redd-plus.html

USAID. (2018). *Climate change in Tanzania: Country risk profile* (pp. 1–5). Factsheet prepared under the Climate Change Adaptation, Thought Leadership and Assessments (ATLAS) Task Order No.AID-OAA-I-14-00013

Van Biervliet, O., Wiśniewski, K., Daniels, J., & Vonesh, J. R. (2009). Effects of tea plantations on stream invertebrates in a global biodiversity hotspot in Africa. *Biotropica, 41*(4), 469–475. https://doi.org/10.1111/j.1744-7429.2009.00504.x.

Willcock, S., Phillips, O. L., Platts, P. J., Balmford, A., Burgess, N. D., Lovett, J. C., Ahrends, A., Bayliss, J.,

S

Doggart, N., Doody, K., Fanning, E., Green, J. M. H., Hall, J., Howell, K. L., Marchant, R., Marshall, A. R., Mbilinyi, B., & Munishi, P. K. T. (2014). Quantifying and understanding carbon storage and sequestration within the Eastern Arc Mountains of Tanzania, a tropical biodiversity hotspot. *Carbon Balance and Management, 9*(2), 1–17.

Space

► New Localism: New Regionalism

Spatial

► Spatial Demography as the Shaper of Urban and Regional Planning Under the Impact of Rapid Urbanization

Spatial – Geographic

► Spatial Justice and the Design of Future Cities in the Developing World

Spatial Demography as the Shaper of Urban and Regional Planning Under the Impact of Rapid Urbanization

Reconnoitering the Future

Halleluah Chirisa[1] and Maléne Campbell[2]
[1]Population Services International Zimbabwe, Harare, Zimbabwe
[2]Department of Urban and Regional Planning, University of the Free State, Bloemfontein, South Africa

Synonyms

Demography; Direction; Driver policy; Geographic; Population studies; Shaper; Spatial

Definition

Urban – a space with modern facilities such as electricity, running water, and developed infrastructure.

Demography – the study of the dynamics in population aspects; fertility, mortality, and migration.

Spatial data – information about a physical object that can be represented by numbers in a geographical coordinate.

Urbanization – the continuous increase in urban environs, populations becoming more urbanized.

Urban planning –a technical and political process that is focused on the development and design of land use and the built environment, including air, water, and the infrastructure passing into and out of urban areas, such as transportation, communications, and distribution networks and their accessibility.

Policy –a deliberate system of principles to guide decisions and achieve rational outcomes.

Management – all the activities and tasks undertaken for achieving goals by continuous activities like planning, organizing, leading, and controlling.

Introduction

There are attempts to improve traditional regression models of demographic processes operating in space through adoption of formal tools of spatial econometrics. This is a result of a re-emerging interest in spatial demography (Voss et al. 2006). During the 1990s in the US Great Plains, the concept of spatial autocorrelation and means to specify correctly multiple regression models in the company of spatial autocorrelation are made more concrete through an illustration of spatial modeling of county-level growth. As statistical models become more complex, as spatial processes are brought into empirical demographic studies to correct for potential misspecification, and as work begins to add in significant ways to the larger literature on spatial data analysis, spatial demography would have been moved forward in very exciting ways (Fischer and Wang 2011; Logan 2016). As a result of the increasing attention in the field of spatial econometrics among numerous disciplines in the social sciences, whereby the re-emergence of attention in spatial demography is a part, proposes a bright future for quantitative demographers.

Demography is fundamentally a spatial science; however, the application of spatial data as well as methods to demographic studies has a tendency to lag that of other disciplines. Recently, there has been a surge in interest in adding a spatial viewpoint to demography. Rapid advances in geospatial data, new technologies, and methods of analysis have been driven by a sharp rise in spatial demography interest. This chapter seeks to unpack importance of urban spatial demography in the face of increasing urban challenges as a result of natural population increase and rural to urban migration which is occurring in weak institutional policies.

Urban Spatial Demography: A Review

As more techniques for conducting spatial analysis are developed and more spatially referenced datasets become available, it becomes more vital to be cautious in terms of spatial thinking. It is crucial to identify key concepts that differentiate spatial social science (Logan 2016). Distinguishing is in relation to similarities among them and especially about where they are in relation to others. Considerable spatial analysis includes certain idea of distance. However, distance is multidimensional, as it refers to a diversity of phenomena such as proximity, connection, exposure, access, or even time (Wang and Kockelman 2013). Spatial dependence can easily be observed and measured, but its interpretation is highly dependent on theory.

Voss (2007, p. 457) defines spatial demography as "the formal demographic study of areal aggregates, i.e., of demographic attributes aggregated to some level within a geographic hierarchy." Robert Woods a British geographer who used first the term spatial demography in the title of a 1984 chapter appearing in an edited volume devoted to the topic of population geography (Woods 1984). Woods' desire was to give the study of migration "equal importance with fertility and mortality" (1984, p. 43; see also Woods and Rees 1986). The relationship between spatial demography and migration is also given prominence by demographer Kenneth Wachter whereby "Migration is the normal tag for the spatial subfield of demography, and movement is its preoccupation" (Wachter 2005, p. 15299).

Place, density, and movement are critical elements in population studies. Demographers' prime coordinates are the two time-like coordinates of time and age. The three space-like coordinates of physical location and the many-dimensional coordinates of social location have tended to play supporting roles (Wachter 2005). However, this has recently changed as there is an emergence of a mass of scientific inquiries that combine geography with demography and with the entire range of the social sciences that are coming to the front. This is explained by the expansion of spatial demography, as a result of the recent availability of fine-grained spatial data. Spatial data links geographic coordinates and categories to demographic, social, and economic variables, suited for analysis with the new computing tools of geographical information systems (GIS) (Jones 2014; Taylor et al. 2000). Political decisions concerning distributional equity across jurisdictions, involving representation, public housing, discrimination, and civil rights, are propelled into the courts, creating and funding a demand for expertise in spatial analysis, leading to a shift in spatial thinking. Another reason is the emotional acknowledgment that a once-rich variety of local particularities in customs, accents, values, legends, architecture, instincts, foods, and memories is vanishing or retreating into less visible forms.

Rationale for Spatial Demography

The central focus of demographic analysis is the spatial distribution of populations. Structured patterns in spatial distribution are evident from the highest levels of macro-spatial scale, for instance, global, national, and regional urban systems, to "fine-grained" patterns in metropolitan areas such as central cities, suburbs, neighborhoods, and blocks and nonmetropolitan hinterlands such as towns, villages, and hamlets (Fossett 2005, p. 479). Documenting and explaining spatial patterns have received much attention in social sciences. This has led to the establishment of a body of knowledge that is remarkable for its increasing nature, rigorous theoretical underpinnings, as well as extensive evidentiary base.

Despite its application in several fields, spatial statistics has drawn demographers' attention only recently. There is an absence of spatial perspectives in several current demographic studies despite demography having a rich body of methodologies. Several prevailing models of sociological demographic relate a geographic unit, like a census tract, a small town, or a county, as an independent isolated entity rather than as an object bounded by other geographic units with which it might interrelate (Chi and Zhu 2008). Numerous disciplines of social sciences such as geography and regional science have theorized spatial effects in population dynamics, comprising the spatial diffusion theory, central place theory, growth pole theory, as well as new economic geography theory (Chi and Ventura 2011). In disciplines such as human ecology, urban sociology, and rural demography, demographic and sociological theories, as well as empirical studies, spatial effects in demographic dynamics, alternatively, have been indirectly considered.

Spatial data is crucial for planning for elections, calculating per capita gross domestic product (GDP) and the denominator in disease incidence rates, assessing natural disaster impacts, and measuring demand for services (Tatem 2017). These activities require ongoing subnational scale data on population sizes and characteristics. In several developed countries having well-documented censuses, comprehensive civil and vital registration systems, and a wealth of other ongoing surveys and registers, it is often taken for granted that fine-grained, robust, consistent, and recent data on populations are readily available (Altman et al. 2018; Tatem 2017). However, in developing countries despite the growing capacity, obtaining consistent, comparable, and spatially detailed demographic data can be a challenge.

Space is a vital component for demographic studies. Migratory movements are existent merely as a result of people who perceive certain places to be more attractive than others. Fertility decline has been a result of the dissemination of ideas, that is, movement of ideas from people to people as well as from place to place (de Castro 2007). As a result of the lack of certain risk factors varying by location, mortality levels are far from being spatially homogeneous. However, many demography studies have not engaged population studies that formally address a spatial component. Moreover, spatial demography is practically not taken into account as part of the regular training of future professionals in the field. Nevertheless, the number of applications in spatial demography has been increasing lately. This is because of a number of reasons (Champion 2001). Spatial data is largely available, comprising all the data released by the US Census Bureau.

Utilization of spatial analysis has been enabled by various computer programs. Computer capabilities to store and examine large datasets have upgraded dramatically. There have been main initiatives to develop spatial thinking among the social sciences, and indeed the recognition that space is significant has been growing among social scientists (Logan 2012). The significance of spatially targeted policies has been acknowledged in diverse areas. The rising interest in evaluating the social context for a diversity of demographic outcomes contributed to an increase in the number of studies that examine the significance of community and neighborhood effects (Kwan 2018). Another reason for the increase in spatial demography applications is adversarial legalism, or resolving legal arguments involving the fair resource distribution across administrative units.

Approaches and Tools in Spatial Demography

Demography is both spatial and by nature interdisciplinary. Demographic transition, which provides the organizing framework for most demographic research, is certainly a complex set of transitions, each of which draws upon expertise in differing social science and health-related disciplines (Weeks 2004). The demographic transition typically initiates with the epidemiological transition, which is the shift over time from high death rates with deaths clustered at the younger ages and caused largely by communicable diseases to low death rates with deaths clustered at the older ages and caused largely by degenerative diseases (Islam and Tahir 2002).

A train of other transitions is therefore set in motion. An alteration from high fertility levels over which people have comparatively slight direct control to low fertility over which individuals have significant control represents the fertility transition. Population growth in rural areas and the desire by rural population to search for opportunities in other localities particularly urban areas lead to migration (Hare 1999). Rural to urban migration unleashes the urban transition, in which a population changes from being mainly rural to being mainly urban. Changes in mortality as well as fertility whereby high mortality and high fertility produce a very young age structure that is pyramid-shaped make the age structure transition foreseeable (De Silva 2012). Also, the declines in both mortality and fertility produce bulges in the young adult ages, leading ultimately to a barrel-shaped age structure.

The family and household transition signifies the modification from complex forms of family and household structure when mortality and fertility are both high, to less variability in the middle of the transition, to new forms of complexity when both fertility and mortality are low (Low et al. 2002). There is also the overall transition in population size that is evident when mortality declines sooner than fertility (the usual pattern in the demographic transition), from which massive changes follow with respect to resource use as well as allocation. All of these interrelated features of demographic change have a spatiotemporal element which contributes to knowledge of how and why these transitions occur. As a result of the rapid growth in computing power, availability of satellite imagery, and expansion of geospatial analysis tools over the past decade, there is provision of new opportunities for data integration to improve demographic mapping (Tatem 2017).

Development of new approaches as well datasets is continuously quickened by the requirement for reliable and timely subnational demographic data. Absence of up-to-date and reliable census data in some low-income countries is stimulating a move away from census disaggregation approaches to "bottom-up" mapping methods (Tatem 2017). Such mapping methods have very high spatial resolution mapping of buildings from satellite imagery which is integrated with small area micro-census surveys to forecast population distributions and demographics (Wardrop et al. 2018; Tatem et al. 2012). Mobile phone call data records are allowing timelier mapping of fluctuating population densities and migration patterns, and national household survey data provides valuable input to large area demographic mapping, whereas satellite data are assisting the development of urban growth mapping. In all of these efforts, the measurement as well as communication of uncertainty in outputs is a significant continuing research activity, providing associated uncertainty datasets a goal for all outputs.

Geo-statistical approaches aim at understanding the spatial distribution of values of an attribute of interest across an entire region under study, specified values at fixed sampling points, and are not only limited to focusing on observations as the location of events (Matthews and Parker 2013). Geo-statistics is grounded on the supposition that random processes with spatial autocorrelation can model at least some of the spatial variation in an attribute. They can also be utilized for provision of accurate as well as consistent approximations of attribute values at locations in which there is absence of measurements. The adoption, chiefly in spatial epidemiology, of advanced techniques for spatial pattern analysis, spatiotemporal analysis, and Bayesian mapping and modeling has been aided by developments in the materials and methods for geospatial data (Bivand et al. 2008). Contemporary applications of spatial pattern analysis methods in demography comprise the utilization of kriging and of local statistics of spatial association such as the G* statistic.

Spatial demography advances in two viewpoints. Firstly, there can be proposal of clear spatial demographic theories. Spatial effects in population dynamics are being suggested by certain demographic and sociological theories as well as empirical studies (Chi and Zhu 2008). However, existing demographic theories have not clearly specified spatial effects. It is clear that developments in spatial techniques as well as availability of spatial data are permitting for new demographic questions and development of new demographic theories. Geography and regional science theories can offer strong spatial components to spatial demographic studies. Secondly, in addition to spatial regression models, other spatial analysis techniques may be appropriate to demographic studies (Griffith et al. 1999). Spatial point data analysis, which has been extensively utilized in various disciplines, for instance, epidemiology as well as forestry, may possibly turn out to be a potentially valuable method for formal demographic studies, particularly with the development of geocoding techniques as well as different demographic survey database (Haining and Haining 2003; Chainey 2014). Geo-statistics, which has been regularly applied in physical as well as biological sciences, can be utilized potentially as an interpolation technique for demographic estimation. Another spatial data analysis technique, that is, spatial interaction modeling, can be very valuable for studying migration and demographic network. Hence, several spatial data analysis methods are available in other fields and can be well engaged in demographic studies.

Cases and Experiences

Spatial and demographic aspects related with cumulative air-toxic health risks at multiple geographic scales have been used in environmental quality studies in the United States (Liévanos 2015). A severe spatial cluster analysis of census tract-level projected lifetime cancer risk (LCR) in 2005 of ambient air-toxic emissions from stationary (e.g., facility) as well as mobile (e.g., vehicular) sources to locate spatial clusters of air-toxic LCR risk in the continental United States was employed. Assessments of intersectional environmental disparity hypotheses were carried on the predictors of tract presence in air-toxic LCR clusters with tract-level principal component factor measures of economic deprivation by race and immigrant status (Grineski et al. 2019). Logistic regression analyses demonstrate that net of controls, isolated Latino immigrant-economic deprivation is the strongest positive demographic predictor of tract presence in air-toxic LCR clusters. This is followed by black-economic deficiency as well as isolated Asian/Pacific Islander immigrant-economic deprivation.

In Ghana, the national population growth rate of approximately 2.7% has been outstripped by the 4.3% urban population growth rate (Osei and Duker 2008, p. 44). In 2000, the proportion of the population living in cities rose from 32% in 1984 to 43.8%. Such rapid urbanization stresses the existing resources meant for providing better service delivery such as provision of water and better sanitation. Urban populations are at risk of diseases such as cholera and typhoid as a result of inadequate sanitation systems coupled with intermittent supply of clean and sufficient water. Surface water pollution is mainly found to be worse where rivers pass through urban and overcrowded cities, and the commonest contamination is from human excreta and sewage (Osei 2010). Urban populations no longer adhere to traditional laws which are used to protect water bodies from pollution. In Ghana, it has been a well-known tradition that it is forbidden to eliminate waste or dispose waste in water bodies. It has become a common practice in urban suburbs to defecate and dump waste in and at the banks of surface water bodies.

Despite polluting water bodies, urban residents rely on polluted water bodies for various household activities during periods of water shortages. High rate of rural to urban migration, population growth, and redistribution have led to an increase in informality in urban suburbs. Informality in housing has led to development of slums. Slum conditions are characterized by poor service delivery. In relation to access and affordability to safe drinking water and sanitation, those living in urban slums and squatters are worse off than their rural counterparts. In many cases, public utility providers lawfully fail to provide sufficient services for slum dwellers as a result of technical and service regulations, land tenure system, and city development plans. Most slums and/or squatter settlements are also located at low-lying areas susceptible to flooding. Unfavorable topography, soil, and hydrogeological conditions make it difficult to achieve and maintain high sanitation standards among populations living in these territories. Given these spatial-related problems, there is evidence that spatial analysis and GIS have the capacity to analyze geographically referenced health data in Ghana (Krauss 2012). In Ghana, it has been proven that demographic risk factors of cholera differ from any region and therefore cannot be universally applied.

Urban spatial demography has been used in vulnerability assessment of Phoenix residents in 1990 and 2000 to extreme heat based on a composite index of vulnerability based on normalized indexes of physical exposure to heat and several socioeconomic adaptive capacity measures of equal weight (Chow et al. 2012). This study demonstrated that vulnerability varied significantly over space and time and that it is unequal across different demographic segments in Phoenix, with Hispanic populations having a disproportionate exposure to extreme heat versus other ethnic groups (Leal Filho et al. 2018). This marked difference is particularly apparent in both the increasing total Hispanic population and the ratio of Hispanics to total Phoenix population that were residing in more vulnerable areas from 1990 to 2000. In contrast, the proportion of non-Hispanic whites exposed to extreme heat decreased, despite a large increase in total elderly migrants in urban-fringe retirement communities. The need for climate adaptation in Phoenix is particularly acute; the research suggested that several Hispanics in inner-city neighborhoods, as well as some elderly in retirement communities, live at loci of heat vulnerability. This offered the need to identify areas with high vulnerability; city officials and policymakers could design more effective urban adaptation strategies, such as policies to improve social cohesion and integration within neighborhoods via widespread dissemination of heat-stress mitigation information in different languages.

Another study in Chicago aimed at assessing the role of demographic context in present-day models of neighborhood crime rates while discovering spatial heterogeneity of the state's census tracts (Arnio and Baumer 2012). From the study, there are conditions of modern-day models of neighborhood crime rates which showed that geographically weighted estimations offered a better fit to the data over "global" Ordinary Least Squares (OLS) models. They provided evidence of important spatial heterogeneity among several variables of interest across Chicago. From studies in Chicago, there is a noteworthy difference in the local parameter estimates for both burglary and robbery for logged percent black and immigrant concentration. There was also important local variation in the effects of socioeconomic disadvantage on robbery rates and residential stability on burglary rates (Ousey and Kubrin 2009).

Positive as well as negative approximations across the city were proposed by local variability, suggesting that the "global" patterns highlighted in the existing literature for these measures do not completely tackle the empirical existing complexity. Noteworthy variation in the measure of foreclosure (the change in logged REO foreclosures from 2007 to 2009) on both robbery and burglary, providing significant setting for current research findings that have produced disparate conclusions concerning the relation between foreclosure and crime in the city of Chicago. The study is crucial in spatial demography as it challenges the conventional approach to neighborhood studies of crime by proposing that methods accounting for spatial heterogeneity can augment capacities to enlighten neighborhood variation in crime rates and better inform the complex theoretical underpinnings of how demographic context is related with aggregate crime patterns.

In KwaZulu-Natal, South Africa, there was an investigation of spatial and demographic variations in HIV infection in small communities, utilizing a cohort of women engaged for several trials by means of population-based clinics (Ramjee et al. 2019). KwaZulu-Natal has always had the highest prevalence of HIV than in any other province in South Africa, and this trend has occurred since the early 1990s. The province of KwaZulu-Natal has shown significant variation, which cannot be identified in an aggregated data (Wand and Ramjee 2010). Explanation concerning possible proximal as well as distal contributors to the HIV/AIDS epidemic can be through knowledge of geographical variation and determination of the core areas of the disease.

It very crucial to determine as well as target the exact societies that need education, prevention, and treatment activities the most. This research offers attempts to visually and quantitatively describe the geographical characteristics of HIV infections in a region where the disease is considered to be widespread. Results from the study are crucial in informing development of prevention programs to address the HIV epidemic while considering those groups most affected differentially by geographical area (Wand and Ramjee 2010). It is difficult and almost impossible to examine the geographical structure of the HIV epidemic in sparsely populated, large geographical areas. Public demand for monitoring at localized level is an urgent need, allocating the resources cautiously to those communities where the infection is clustered. The study provided evidence of clusters of mainly susceptible women through research on the occurrence and frequency of HIV in the study setting. Local and authorities were urged to make provisions for a rapid response by scaling up HIV prevention, treatment, and care efforts in such societies.

Emerging Issues and Synthesis

The study shows the past decades have experienced increased computer capacity and software, analytical techniques, and availability of spatial data, intensely improving the ability to perform spatial demographic studies (de Castro 2007). Distinctive opportunities to advise policymaking are offered to researchers who include a spatial approach in their analyses (Jelokhani-Niaraki 2020). However, there is still a lack of strong commitment from population centers to include spatial analysis as part of their core training. The commitment can be attained through awareness of the potential of spatial demography that becomes widespread.

With the rapid advances in spatial analysis techniques as well as the increasing availability of geographically referenced data, spatial demography is moving to a new and exciting stage (Chi and Zhu 2008). This seems to be the time for demographers to explore as well as enrich the field of spatial demography. The demographic transformation of human societies into the urban era has pushed the monitoring of urban areas to the forefront of environmental and developmental agendas. As a result of this transformation, a higher percentage of the world's population currently resides in urban areas than ever before, and growth in urban areas is occurring at an unprecedented rate. Rapid urban growth has clearly moved from the Global North to the Global South. Urban growth has been associated with massive congestion, poor public transportation, informality, and poor service delivery. It is therefore imperative for planning authorities to mainstream development toward the future of urban spatial demography.

Demographic questions can be answered by spatial thinking and spatial analytical perspectives. This is because of the relationship between the spatial analysis of demographic processes and outcomes, which have typically drawn on macro-level or ecological data (Matthews and Parker 2013). Spatial demography becomes spatial analysis of demographic processes. The presence of graduate-level teaching in GIS, in general, and of detailed courses in advanced spatial data analysis with noteworthy social science or demographic content was inadequate.

Conclusion and Future Direction

The chapter concludes that as parts of the world are becoming more urban, it is crucial to understand urban growth and the implications of changes in an urban population. The growth of cities has attracted considerable scholarly attention during the last decade as it is becoming evident that influential forces of accumulation are strengthening the role of urban areas as the drivers of economic growth. As urbanization continues, no parts of the world will be protected from comprehensive changes in demographics as a result of aging population and increased migration. These changes will be amplified in urban areas. Developments in the application of spatial statistics, GIS, and remote sensing methods to sociological and demographic studies to the increase in the accessibility of geographically referenced data, the growth of user-friendly spatial data analysis software packages, and the increased computing power combined with affordable computers offer crucial projections in spatial demography. Spatial dynamics of population change should be formally incorporated into demographic models and empirical studies.

Cross-Referenes

▶ Hidden Enemy for Healthy Urban Life
▶ Improving Social Equity and Community Health and Well-Being in Low-Income Suburbs and Regions
▶ Spatial Justice and the Design of Future Cities in the Developing World

References

Altman, M., Wood, A. B., O'Brien, D., & Gasser, U. (2018). Practical approaches to big data privacy over time. *International Data Privacy Law, 8*, 29.

Arnio, A. N., & Baumer, E. P. (2012). Demography, foreclosure, and crime: Assessing spatial heterogeneity in contemporary models of neighborhood crime rates. *Demographic Research, 26*, 449–486.

Bivand, R. S., Pebesma, E. J., Gómez-Rubio, V., & Pebesma, E. J. (2008). *Applied spatial data analysis with R* (Vol. 747248717, pp. 237–268). New York: Springer.

Chainey, S. (2014). *Examining the extent to which hotspot analysis can support spatial predictions of crime*. Doctoral dissertation, UCL (University College London).

Champion, T. (2001) Urbanization, suburbanization, counterurbanization and reurbanization. In *Handbook of urban studies* (Vol. 160, p. 1). SAGE Landon.

Chi, G., & Ventura, S. J. (2011). An integrated framework of population change: Influential factors, spatial dynamics, and temporal variation. *Growth and Change, 42*(4), 549–570.

Chi, G., & Zhu, J. (2008). Spatial regression models for demographic analysis. *Population Research and Policy Review, 27*(1), 17–42.

Chow, W. T., Chuang, W. C., & Gober, P. (2012). Vulnerability to extreme heat in metropolitan Phoenix: Spatial, temporal, and demographic dimensions. *The Professional Geographer, 64*(2), 286–302.

de Castro, M. C. (2007). Spatial demography: An opportunity to improve policy making at diverse decision levels. *Population Research and Policy Review, 26*(5–6), 477–509.

De Silva, W. I. (2012). Sri Lankan population change and demographic bonus challenges and opportunities in the new millennium. Available online: http://crossasia-repository.ub.uni-heidelberg.de/3525/1/Sri%20Lankan%20Population%20Change.pdf. Accessed on 12 Sept 2020.

Fischer, M. M., & Wang, J. (2011). *Spatial data analysis: Models, methods and techniques*. New York: Springer Science & Business Media. Springer.

Fossett, M. (2005). Urban and spatial demography. In *Handbook of population* (pp. 479–524). Boston: Springer.

Griffith, D. A., Layne, L. J., Layne, L. J., Ord, J. K., & Sone, A. (1999). *A casebook for spatial statistical data analysis: A compilation of analyses of different thematic data sets*. New York: Oxford University Press on Demand.

Grineski, S., Morales, D. X., Collins, T., Hernandez, E., & Fuentes, A. (2019). The burden of carcinogenic air toxics among Asian Americans in four US metro areas. *Population and Environment, 40*(3), 257–282.

Haining, R. P., & Haining, R. (2003). *Spatial data analysis: Theory and practice*. Cambridge: Cambridge University Press.

Hare, D. (1999). 'Push' versus 'pull' factors in migration outflows and returns: Determinants of migration status and spell duration among China's rural population. *The Journal of Development Studies, 35*(3), 45–72.

Islam, A., & Tahir, M. Z. (2002). Health sector reform in South Asia: New challenges and constraints. *Health Policy, 60*(2), 151–169.

Jelokhani-Niaraki, M. (2020). Collaborative spatial multicriteria evaluation: A review and directions for future research. *International Journal of Geographical Information Science, 35*, 1–34.

Jones, C. B. (2014). *Geographical information systems and computer cartography*. London: Routledge.

Krauss, A. (2012). *External influences and the educational landscape: Analysis of political, economic, geographic, health and demographic factors in Ghana* (Vol. 49). New York: Springer Science & Business Media.

S

Kwan, M. P. (2018). The limits of the neighborhood effect: Contextual uncertainties in geographic, environmental health, and social science research. *Annals of the American Association of Geographers, 108*(6), 1482–1490.
Leal Filho, W., Icaza, L. E., Neht, A., Klavins, M., & Morgan, E. A. (2018). Coping with the impacts of urban heat islands. A literature-based study on understanding urban heat vulnerability and the need for resilience in cities in a global climate change context. *Journal of Cleaner Production, 171*, 1140–1149.
Liévanos, R. S. (2015). Race, deprivation, and immigrant isolation: The spatial demography of air-toxic clusters in the continental United States. *Social Science Research, 54*, 50–67.
Logan, J. R. (2012). Making a place for space: Spatial thinking in social science. *Annual Review of Sociology, 38*, 507–524.
Logan, J. R. (2016). Challenges of spatial thinking. In F. M. Howell, J. R. Porter, & S. A. Matthews (Eds.), *Recapturing space: New middle-range theory in spatial demography*. Cham: Springer.
Low, B. S., Simon, C. P., & Anderson, K. G. (2002). An evolutionary ecological perspective on demographic transitions: Modeling multiple currencies. *American Journal of Human Biology, 14*(2), 149–167.
Matthews, S. A., & Parker, D. M. (2013). Progress in spatial demography. *Demographic Research, 28*, 271–312.
Osei, F. B. (2010). *Spatial statistics of epidemic data: The case of cholera epidemiology in Ghana*. University of Twente, Faculty of Geo-Information Science and Earth Observation. Springer: New York.
Osei, F. B., & Duker, A. A. (2008). Spatial and demographic patterns of cholera in Ashanti region-Ghana. *International Journal of Health Geographics, 7*(1), 44–56.
Ousey, G. C., & Kubrin, C. E. (2009). Exploring the connection between immigration and violent crime rates in US cities, 1980–2000. *Social Problems, 56*(3), 447–473.
Ramjee, G., Sartorius, B., Morris, N., Wand, H., Reddy, T., Yssel, J. D., & Tanser, F. (2019). A decade of sustained geographic spread of HIV infections among women in Durban, South Africa. *BMC Infectious Diseases, 19*(1), 500.
Tatem, A. J. (2017). WorldPop, open data for spatial demography. *Scientific Data, 4*(1), 1–4.
Tatem, A. J., Adamo, S., Bharti, N., Burgert, C. R., Castro, M., Dorelien, A., Fink, G., Linard, C., John, M., Montana, L., & Montgomery, M. R. (2012). Mapping populations at risk: Improving spatial demographic data for infectious disease modeling and metric derivation. *Population Health Metrics, 10*(1), 8.
Taylor, M. A., Woolley, J. E., & Zito, R. (2000). Integration of the global positioning system and geographical information systems for traffic congestion studies. *Transportation Research Part C: Emerging Technologies, 8*(1–6), 257–285.
Voss, P. R. (2007). Demography as a spatial social science. *Population Research and Policy Review, 26*(5–6), 457–476.
Voss, P. R., White, K. J. C., & Hammer, R. B. (2006). Explorations in spatial demography. In *Population change and rural society* (pp. 407–429). Dordrecht: Springer.
Wachter, K. W. (2005). Spatial demography. *Proceedings of the National Academy of Sciences, 102*(43), 15299–15300.
Wand, H., & Ramjee, G. (2010). Targeting the hotspots: Investigating spatial and demographic variations in HIV infection in small communities in South Africa. *Journal of the International AIDS Society, 13*(1), 41–56.
Wang, Y., & Kockelman, K. M. (2013). A Poisson-lognormal conditional-autoregressive model for multivariate spatial analysis of pedestrian crash counts across neighborhoods. *Accident Analysis & Prevention, 60*, 71–84.
Wardrop, N. A., Jochem, W. C., Bird, T. J., Chamberlain, H. R., Clarke, D., Kerr, D., Bengtsson, L., Juran, S., Seaman, V., & Tatem, A. J. (2018). Spatially disaggregated population estimates in the absence of national population and housing census data. *Proceedings of the National Academy of Sciences, 115*(14), 3529–3537.
Weeks, J. R. (2004). The role of spatial analysis in demographic research. In *Spatially integrated social science* (pp. 381–399). Oxford University Press.
Woods, R., & Woodward, J. (Eds.). (1984). Urban disease and mortality in nineteenth-century England. BT Batsford Limited.
Woods, R., & Rees, P. H. (Eds.). (1986). *Population structures and models: Developments in spatial demography*. Allen & Unwin Australia.

Spatial Digital Twin

▶ Digital Twin and Cities

Spatial Distribution

▶ Multiple Benefits of Green Infrastructure

Spatial Equity

▶ Walkable Access and Walking Quality of Built Environment

Spatial Justice and the Design of Future Cities in the Developing World

Elton Manjeya[1], Innocent Chirisa[2] and Charles M. Chavunduka[1]
[1]Department of Architecture and Real Estate, University of Zimbabwe, Harare, Zimbabwe
[2]Department of Demography Settlement and Development, Social & Behavioural Sciences, University of Zimbabwe, Harare, Zimbabwe

Synonyms

Spatial – geographic; Future – the prospective; Justice – fairness, equity

Definition

Future city – a developed urban area that is sustainable, resilient, inclusive, and intelligent where the environmental, social, political, economic, and technological aims are met and resources are equitably distributed.

Street vending – the practice of selling goods and/or services to the general public in permanent designated spaces or temporary undesignated spaces in public spaces of cities by stationary or mobile operators. Street vending may be legal if it is regulated and complies with regulations governing the use of public space or illegal if it is does not comply with the regulations governing the use of public space.

Spatial justice – the fair and equitable allocation of social resources among all people. It is based on the notion that justice and injustice are visible in space as a result of the interaction between society and space.

Developing world – countries that have an underdeveloped industrial base, high rates of poverty and unemployment, low per capita income, low living standards, and low Human Development Index.

Urban design – the process of design and arranging buildings, public spaces, transport systems, amenities, and services. It is an approach to designing and managing public space to give form, shape, and character to groups of buildings, neighborhoods, and the entire city.

Introduction

The world is rapidly urbanizing. The year 2008 marked a milestone in the history of urbanization as, for the first time, half of the world's population lived in cities. More than half of the urban dwellers are found in cities in developing countries. Cities in developing countries are recognized for the concentration of economic and social opportunities which has led to their growth. The growth of cities in developing countries has seen a lot of challenges that emerge. The challenges include but are not limited to rapid population growth, urban poverty, and lack of access to shelter, infrastructure, and services by predominantly poor populations, weak economies, and unemployment. The challenges affect the social and spatial structure of cities. Further, they expose the growing inequality in terms of access to services and economic opportunities especially among the urban poor in cities of the developing world.

The growth of cities in developing countries is inevitable, yet in the process of growth are produced injustices that mainly affect the urban poor. There is consensus among researchers in various disciplines on the need to interrogate the question of spatial justice in cities. Urban design and urban planning deal with spatial justice issues in the land-use planning, allocation of resources and services, urban management, and urban governance on a day-to-day basis. The outcomes of urban design actions and planning have short-term and long-term impacts on spatial justice. Against this background, the chapter examines spatial justice issues in the urban design of cities of the developing world, paying particular attention to street vending in the urban informal sector. The chapter is organized into five sections. The next section defines spatial justice and gives a theoretical background of the concept and defines urban design and establishes the

S

relationship between spatial justice and urban design. It also offers the conceptual framework of the chapter. The other section examines some experiences of spatial justice and urban design in cities of developing countries. This is followed by a discussion on emerging issues and synthesis and finally the conclusion and future directions.

Conceptualizing Spatial Justice and Urban Design

All social problems in cities happen in space and time. Similarly, justice has a spatial dimension (Soja 2009). Spatial justice provides a lens to examine the processes that produce space and the implications of the spaces produced on social, economic, and political relations. Urban space is socially produced through a set of relations (political and economic) and processes (urban design and urban planning and investment decisions) (Dikeç 2001; Williams 2013). Usually, these processes and sets of relations further the interests of the elite groups of society at the expense of the disadvantaged groups. The interaction of these sets of relations sets in motion the process of producing urban space. Urban space which is the final product usually reflects the interest of the people who control the process and the exclusion, domination, and marginalization of disadvantaged groups such as street vendors. The dimensions of spatial justice include exclusion, marginalization, tribalism, racism, sexism, religious extremism, and spatial domination. The dimension is produced and reproduced in urban public spaces of cities in developing countries by relations of power established by the capitalist system (Dikeç 2001). Under these conditions, urban design trends to produce unjust outcomes, for instance, the marginalization and exclusion of "undesired" activities such as street vending through design and legal frameworks.

Urban space as a product of the process of the urban design reflects the vested interests of the designers, real estate investors, and politicians. The desire for clean modern urban spaces that attract investment is often prioritized over the needs of the street vendors. As a result, the informal economy and its variants and people with special needs are excluded and marginalized through the design of public space in cities. Lack of participation in the design processes by the urban poor and lack of transparency and accountability in the design processes further entrench spatial injustices. To design cities of the future, there is a need to revisit the processes of urban design that produce unjust outcomes by ensuring that the processes are inclusive and participatory. The stakeholders in the production of public space who include the local authority, business owners, real estate developers, and politicians need to collaborate to formulate public spaces where each group's interests are represented through design and policy (Fig. 1).

Spatial justice is a deliberate and directed attention to the spatial and geographical aspects of justice (Soja 2009). All social issues related to justice and injustice occur in space and time, meaning justice has a spatial and temporal dimension. Spatial justice is defined as "…the fair and equitable distribution in space of socially valued resources and the opportunities to utilize them"(Soja 2009, p. 3). The relationship between justice and space is intricately linked. All social issues have a justice component and all happen in space (Philopopoulos-Mihalopoulos 2015).

In coming up with the spatial justice theory, Soja (2009) uses the ideas from Lefebvre's (1974) *The Right to the City*, Rawls' (1974) *A Theory of Justice*, and Harvey's (1973) *Social Justice and the City*. Lefebvre (1974) argues that space is not a "thing" but rather it is a set of relations between things, that is, objects and products. Space is therefore a set of relations among things (Harvey 1973; Lefebvre 1974). Justice and injustice are created in the process of producing space and managing space – such a production determines the relations of power among stakeholders.

The process of producing space is determined by social relations of power in the capitalist system which results in marginalization, exclusion, and spatial domination of a certain population group, for instance, street vendors (Dikeç 2001; Dikeç and Gilbert 2002). This is the case with public spaces produced in cities of the developing countries where elites, through the process of

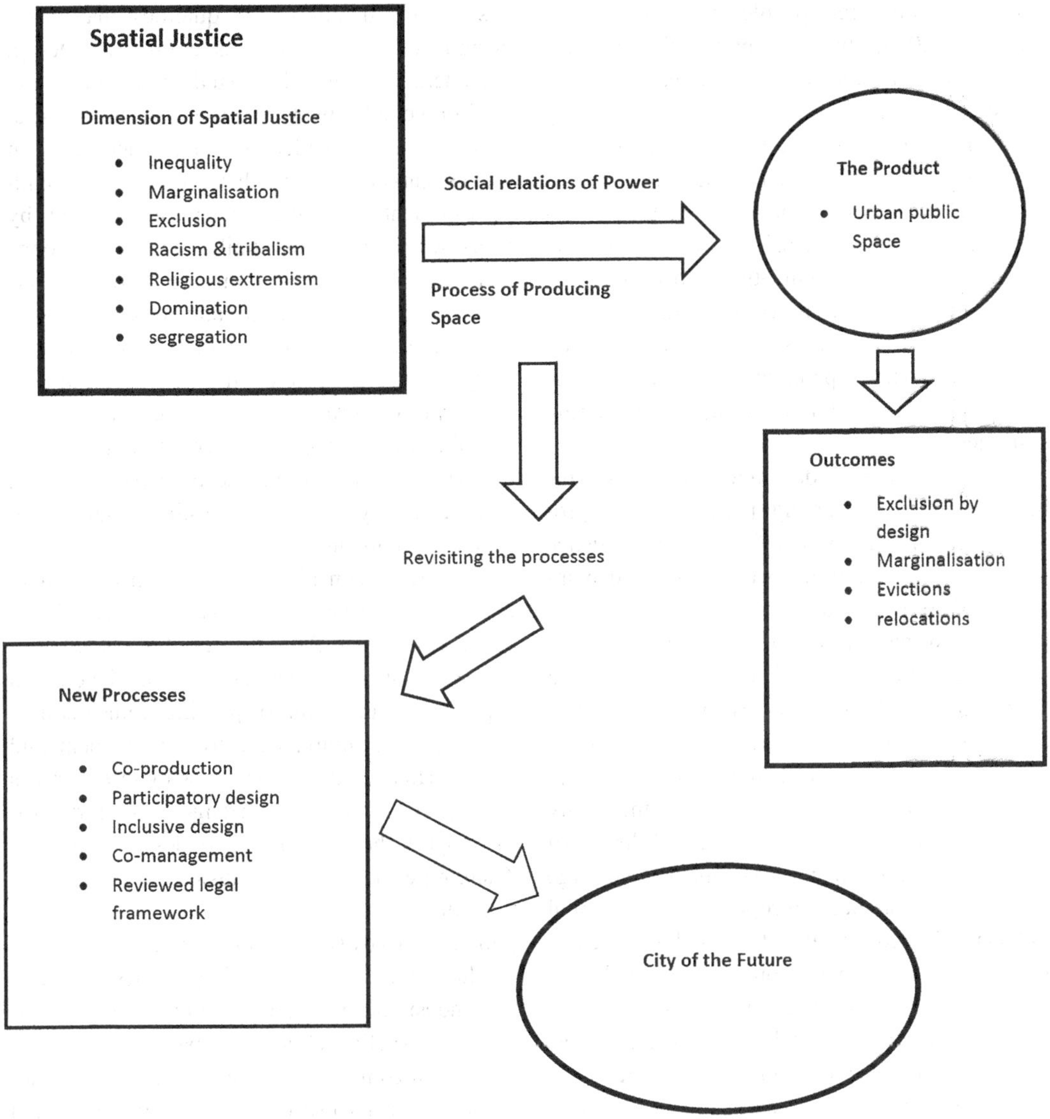

Spatial Justice and the Design of Future Cities in the Developing World, Fig. 1 The spatial justice urban design conceptual framework. (Authors 2020)

urban design and planning, produce public spaces that reflect their interest and exclude disadvantaged groups.

To understand what constitutes just spaces, Soja (2009) uses Rawls' (1974) concepts of "distributive justice" and the "difference principle." The distributive justice principle asserts that, for all people, there should be equality of liberties and fair distribution of social goods according to the greatest benefit to the least advantaged group (Rawls 2001). In the context of this chapter, it means urban space must be designed in a way that makes it accessible and usable by the people with the least advantage such as street vendors, women, children, the elderly, and people with special needs. In cities across the world, access to public spaces may be unrestricted but may not entail the freedom to use that space.

The principle of difference on the other hand asserts that a service should exist in such a way that it serves different people in the community (Williams 2013). In the context of this study, the infrastructure and social amenities to be used by street vendors are quite limited, poorly maintained, and only accessible to licensed vendors operating in designated locations.

Spatial injustices in space manifest themselves through systematic exclusion from design and planning processes, domination and oppression, racism and segregation, commodification, and touristification of public space and gentrification (Soja 2009). These spatial injustices are perpetuated by the process of urban design in developing countries.

The way cities in developed countries have been designed, regulating permitted and prohibited land uses in space, reflects spatial injustice (Williams 2013). Public space design and management are anchored on and controlled by the capitalist system which "excludes and sweeps away the poor in the planned city" (Watson 2009). One of the weaknesses of spatial justice as a theory and analytical framework is its neo-Marxist background that views anything to do with capitalism as unjust. However, the reality on the ground is that the planning and design of cities require investment. Investment is so large that governments need to rope in private capital to finance the projects. Private capital is interested in return on investment more than the inclusion of the disadvantaged groups in society.

Rowley (1994, p. 194) defines urban design as "an approach to how the public realm is designed and managed which is rooted in some commonly shared ideas about its value and important attributes." The public realm on the other hand is defined as "the public face of buildings, the spaces between the frontages, street, pathways, and parks. . ." (Rowley 1994, p. 194). Urban design is one other discipline responsible for the production and shaping of the built environment of cities. Although there is dearth of literature linking spatial justice and urban design, urban design is the process that is influenced by social relations in the production of space and whose outcomes can be spatially just or unjust.

In the production of public space, urban design uses several considerations and principles. Urban design considerations are qualities that urban design seeks to achieve as a process. Design considerations include visual, functional, and environmental considerations and the urban experience. Visual considerations are concerned with the appearance and aesthetic qualities of built environments and the perceptions created by users, which are almost similar to visual appropriateness. Visual considerations are useful in determining objects, artifacts, and ornaments that enhance the aesthetics of built environments. They are also used to identify activities and uses that bring discord to the aesthetic image of the city, for instance, the presence of street vendors in undesignated spaces and the litter they generate that negatively affects the aesthetic outlook of built environments.

Functional considerations are concerned with how built environments are used, the diversity of users, and their different needs. Functional considerations include issues such as road layout and capacity, the provision of parking, refuse collection, street furniture, pedestrian movement and routes, safety, and comfort. Rowley (1994) notes that spaces need to be arranged and furnished to support the most possible design activities. Therein lies the root of exclusionary design practices that create challenges of use and access to public spaces by street vendors. There is no clarity on the criteria for determining the desirable and the undesirable uses and functions. The desired uses are usually defined in urban design frameworks and zoning ordinances. In the same frameworks, informal use of space is prohibited and regarded as affecting the normal function of designed spaces in the built environment.

Responsive environments according to Bentley et al. (1985, p. 9) are "built environments that provide users with an essentially democratic setting, enriching their opportunities by maximizing the degree of choice available to them." Responsive environment principles include legibility, robustness, visual appropriateness, richness, and personalization of interest. Legibility, variety, permeability, and robustness are related to functional considerations. A design that achieves these principles is considered to be functional and responsive. However, public spaces in cities of developing countries are far from being responsive because urban design as a process and practice has remained static and unresponsive where new uses of public spaces have emerged. Urban design has failed to produce responsive environments because the social relations involved in the production of space do not recognize informal uses of space as a design need that warrants design action and attention. This rigidity creates spatial injustices, where certain uses such as street vending are excluded from the design process and design outcome.

The rigidities of urban design as far as accommodating informal uses in design create tension and contestations in the use of public space between street vendors and other users of space. The rigid nature of urban design is against the principles of diversity, variety, and robustness as the designed spaces are only used for certain uses prescribed in the design master and local plans. This also has a social and economic impact. The evictions, arrests, and relocation that street vendors are subjected negatively affect their livelihoods.

Spatial Justice and the Design of Cities in the Developing World: A Review

The economic landscape of cities in developing countries comprises of both the formal and the informal sectors operating side by side. The term informal sector was first used by the International Labour Organization (ILO) in its Kenya Mission Report of 1972. Since then the term informal sector has gained currency and generated a lot of academic research and debates concerning the formation, significance, and viability of the sector. The ILO (1972) defines the informal sector as a way of doing things characterized by (a) ease of entry, (b) reliance on indigenous resources, (c) family ownership, (d) small-scale operations, (e) labor-intensive and adaptive technology skills acquired outside of the formal sector, and (f) unregulated and competitive markets. Castells and Portes (1989, p. 12) define the informal economy as "process of income-generation. . .unregulated by the institutions of society, in a legal and social environment in which similar activities are regulated." While most definitions focus on formal vis-à-vis informal dichotomy, Castells and Portes (1989 argue that there are linkages between the formal and informal sectors. De Soto (1989, p. 14) notes that the "informal economy is the people's spontaneous and creative response to the state's incapacity to satisfy the basic needs of impoverished masses."

Due to political and economic shifts in the global markets where developing countries have not been spared, the informal economy has grown in developing countries. The informal sector has become a major source for the provision of essential services such as housing and employment. Informal sectors have moved into sectors previously operated by the public and formal private sectors, for instance, housing and construction services. The urban informal sector is quite visible and a permanent feature of cities of developing countries. Street vending is the most visible form of urban informality because it operates in public spaces of cities. Street vending employs millions of people in cities of the developing world. In - India, an estimated ten million people are employed in street vending (Bhowmik 2003). An estimated 1.8 million households depend on street vending for their primary source of income and livelihoods (Jha et al. 2018). In South America, street vendors represent between 5% and 10% of the total employed population (Roever 2010 in Linares 2018). In Colombia, 47.2% of the total working population in the 23 largest cities are informally employed, while 15–25% of all the people are employed in street vending (Bernal-Torres et al. 2020).

The informal economy in the Global South has become a constituent part of the growth dynamics of most societies in Africa which supports the livelihoods of millions of people (Rogerson 2016). The ILO estimates that more than 66% of total employment in sub-Saharan Africa is in the informal sector (Kathage 2018). In South Africa, street vending employs 22% of the working population and contributes 7% to the national gross domestic product (Gamieldien et al. 2017). In Zimbabwe, 25% of the people are employed in street vending. The Labour Force Survey of 2014 found that informal employment accounts for 94.5% of total employment (Moyo 2020). Street vending plays a very important role in cities of developing countries. Due to ease of entry and exit, low start-up capital, and limited skills requirement, street vending employs thousands of women and youths. Street vending acts as a safety net for those retrenched in the formal sector. Street enterprises support formal businesses and fresh produce farmers by buying in bulk and selling in smaller quantities to middle- and low-income citizens. In the process, street vending performs welfare distribution by providing the poor with affordable products and services while supporting formal businesses. The next section discusses street vending and urban design within the context of spatial justice.

Street Vending, Urban Design, and Spatial Justice

Despite its contribution to the urban economy, street vending is considered an illegal activity in public spaces of cities in the developing countries. Public space is a critical resource for the operation of street vending and the livelihoods of the urban poor, yet it is often not designed to accommodate the street enterprise (Brown 2001). The exclusion of street vending from the design of public space represents a form of spatial injustice where some uses are deemed legal and are designed for, while street vending is deemed illegal and excluded from design and by design. Because of economic recessions, weak and poor national and local governance, and corruption in the public and private sector, employment in the informal sector is not an essential preference of the urban poor but an alternative to the constraints of the formal sector (Pieterse 2008).

Where the design of urban public space is not responsive to the evolving needs and uses of the urban citizens such as the need for employment outside the formal sector, it creates "favorable" conditions for the poor to take matters in their own hands and occupy public spaces for street vending purposes. The conditions in public spaces in cities such as the high concentration of people with needs for goods and services create conducive settings that produce and reproduce urban informality particularly street vending (Banks et al. 2020). This is in line with Roy's (2005) assertion that informality is an "organizing logic" on its own. Street vending capitalizes on the inefficiencies of the urban design process and system by creating and "allocating" uses to spaces that are overlooked and ignored by urban design practitioners and policymakers. In the process, street vending transforms urban public space to accommodate the needs of street vendors. However, vendors seldom have control of the negative externalities generated by their transformation process. The negative externalities created include but are not limited to congestion as vendors occupy sidewalks and dirt caused by improper disposal of refuse due to unavailability of refuse receptacles.

Watson (2009) notes that urban spaces in developed and developing countries continue to change in terms of economy, society, spatial structure, and environment. Despite these changes, urban design and planning have changed very slowly with some design and planning systems reflecting ideas from developed countries. Urban design solutions end up being reactive to emerging and evolving uses of public spaces, which results in solutions and outcomes that rarely address the challenges at hand but create more problems. The theoretical foundations of urban design and the philosophical inclinations of colonialism which introduced urban planning and design to most developing countries are based on creating order, aesthetics, promoting public health and amenity, spatial domination, and segregation. Against this theoretical and philosophical background, urban design tends to discriminate and restrict uses like street vending or informal activities which are perceived as disturbing the visual and spatial order of public spaces. This can be related to the colonial ethos of trying to control native populations through segregation, exclusion, and marginalization and accessing and using public space in cities.

Control of access and restriction of certain uses through legal frameworks becomes a tool for exclusion. These restrictive legal frameworks are used in the process of producing and managing urban space in developing countries. The use of these legal frameworks produces unjust spaces where street vendors are excluded by design, by law, and in the planning process. In cities of the developing countries, urban design tends to increase social exclusion through anti-poor design strategies. The enforcement of regulations to maintain order and a semblance of modernity often results in arrests, evictions, and relocations of street vendors. For instance, in 2013, the city of Johannesburg in South Africa conducted "Operation Clean Sweep" which resulted in the eviction and relocation of all unlicensed street vendors (Bénit-Gbaffou 2016). In 2006, Malawi launched Operation "*Dongosolo*" (meaning order in the native *Chewa* language), wherein thousands of street vendors were evicted from the streets and lost their livelihoods (Kayuni and Tambulasi 2009). Using the Regional, Town and Country Planning Act (29:12), Zimbabwe launched Operation *Murambatsvina*/Restore Order where over 700,000 people employed in the informal sector were evicted and had their livelihoods destroyed. The motive behind these operations was to restore order in cities and towns of developing countries; however, this has spatial justice implications for the urban poor particularly street vendors; as such, a disruption to their livelihoods took time to recover.

Urban design has been criticized for serving the interests of the elites, for example, real estate developers and investors, business owners, and politicians, in the process of producing urban spaces at the expense of less powerful groups (Madanipour 2006; Bell and Loukaitou-Sideris 2014). Visions of modernity and image improvement strategies such as urban renewal or revitalization seldom consider the interests of the poor in the design and implementation. In Johannesburg, South Africa, urban revitalization policies to improve the city's image and competitiveness and attract foreign direct investment (FDI) resulted in street vendors being excluded in the design of public spaces. Street vendors had to continue their trade under extremely tense conditions (Huchzermeyer and Misselwitz 2016).

Finn (2018) notes that street vending in Kigali, Rwanda, is prohibited as city authorities want to maintain the image of a modern city that is attractive to investors. Attempts to beautify and revitalize public spaces in Mexico City to portray them as good places to live and invest under the *Recuperacion de Espacios Publicos (Recovery of the Public Space)* program saw the eviction of street vendors from plazas and streets as they were deemed unattractive elements. These image improvement and investment attraction initiatives further perpetuate spatial injustice in cities of the developing countries. Huchzermeyer (2014) observes that there is evidence of deep-rooted exclusion in cities across the African continent which can be attributed to the design processes and philosophy used in the design of African cities.

Against the background of growing spatial injustices and inequality in cities of developing countries, scholars argue that the urban poor is devising ways of claiming their "right to the city" (Williams 2013; Huchzermeyer and Misselwitz 2016; Huchzermeyer 2017). The urban poor resort to insurgent practices to continue trading in urban space. Insurgent practices are ways by the urban poor to operate below the radar of the regulatory framework governing the use of public space or symbolic prescriptions by the state (Pieterse 2008). Here street vendors live by their wits and depend on powerful support, social capital, and group solidarity to access and use public spaces without being arrested. Insurgent practices by the street vendors not only reflect the inherent structural rigidity of urban design but also the intentions of the power relations involved in the production of public spaces.

While the functional considerations of urban design focus on the provision of street furniture and urban amenities, street vendors are often deprived of infrastructure to support their operations. The limited designated trading site in unviable locations often has limited amenities such as ablution facilities, portable water, and shelter for storage and protection from the elements. In Accra, Ghana, street vendors spend a significant amount of their daily earnings paying for water, ablution, and storage facilities, yet this infrastructure could be provided by the local authority with responsive design and planning. The contest for public space in developing countries often affects street vendors. Strategies to raise income by city authorities result in the privatization and touristification of public spaces (Harvey 2018). This reduces the space available for street vendors to operate in. Vendors have to compete for space left on the sidewalks with pedestrians which usually results in congestion.

Conclusion and Future Direction

In cities of developing countries, the urban design creates spatial injustices in the production and management of public spaces through exclusionary policies and practices against street vendors. Urban design serves the interests of elite groups which entrench inequalities and bring to the fore the dualism such as the division between the rich and the poor, private and public, and formal and informal spaces (Loukaitou-Sideris and Mukhija 2016). To counter the exclusion by design, street vendors resort to insurgent and survivalist practices as coping mechanisms. Street vendors also resort to the production and provision of services they need without the assistance and supervision of local planning authorities (Balbo 2014). This reflects the failure of urban authorities to recognize street vending as an enterprise that uses space and in need of services that may otherwise be provided by the local authority.

Cities in developing countries often resort to lose-lose public space management strategies. These entail zero-sum approaches to evicting street vendors with little or no dialogue or understanding of the realities on the ground. Evictions, arrests, and confiscations are a lose-lose solution in the sense that they negatively affect the income of street vendors. Consumers who depend on street vendors for affordable goods and services are also affected. Also, businesses that supply street vendors with products for resale lose potential revenue by the disruption caused by the eviction and relocation of street vendors. The local and national governments may also lose tax revenue from the loss of business by shops that supply street vendors with products for resale (Harvey 2018).

Urban designers at times try to formulate solutions for street vendors without consulting the street vendors. The relocation and designation of vending sites are usually the preferred solution. Evidence from cities in developing countries shows that this solution has had limited success and at most failed. The failure is attributed to the solutions being implemented without the knowledge that street vendors have about the movement and concentration of customers, ideal spots for making sales, and knowledge of customer behavior (Harvey 2018). Relocations and designation of market stalls have proved to be ineffective in Accra, Ghana, where vendors were relocated from their trading site to make way for the construction of the *Makola* shopping mall. Although street vendors were assured that they would be accommodated in the shopping mall, it turned out that street vendors could not afford the high rentals that were required. Street vendors could not access customers because they were allocated upper floor stall, yet for their enterprise to be viable, it relies on direct access to customers on the ground floors that overlook the streets.

Urban design practice has remained static and unresponsive to the changing uses and needs of users in public spaces of cities of developing countries (Watson 2009; Njo 2003). Laws, regulations, and bylaws guiding urban design and development control still resemble those used by colonialists some 40–50 years ago with minor or no revisions at all. The Regional, Town and Country Planning Act (29:12) and Model Building By-Laws of Zimbabwe are examples that fit this category. Despite being outdated and rigid, these laws and regulations are continuously being used and as a result continue creating spatial injustices for street vendors.

The foregoing chapter has shown that urban design as a process and means of producing urban space creates unjust spaces through excluding street vendors from the design process. Outdated legal frameworks used in the design and management of public spaces in cities of developing countries also perpetuate spatial injustice by being unresponsive to the changing uses and spatial structure of urban space.

The power relations involved in the process of producing urban space play a key role in creating unjust public spaces in cities by ensuring their interests are expressed in the design at the expense of street vendors. Powerful pressure groups that represent the interests of the elites can influence urban design objectives and outcomes to further their interests. Moreover, street vendors' interests are seldom involved or represented in the production of public space.

The city of the future is one that supports the pillars of sustainable development and implements Sustainable Development Goals. It recognizes the need for inclusion of all groups and appreciates the diversity and conflicting interests. The city of the future ensures that there is distributional equity and supports the full development of each and all individuals (Marcuse 2009).

In future cities of the developing world, urban design systems and approaches need to work with street vendors to become pro-poor and inclusive. An inclusive agenda between street vendors and city authorities is important in creating vibrant and equitable cities. This implies working with street vendors to finding and implementing lasting solutions. This entails co-production which is participatory to involve all stakeholders in the design and governance of public spaces. This has the potential to create trust, buy-in, and accountability and may result in sustainable win-win solutions (Harvey 2018). Competing claims to public space need to be mediated (Marcuse et al. 2009). Street vendors need to organize themselves into social movements that will represent their interests in the process of design and also lobbying city authorities against unfair design outcomes. The legal and regulatory framework used in the production and management of public spaces needs to be reviewed to be responsive to the changing uses and needs of users in public space. This review should be a collaborative process among all stakeholders.

Cross-References

▶ Spatial Demography as the Shaper of Urban and Regional Planning Under the Impact of Rapid Urbanization

References

Balbo, M. (2014). Beyond the city of developing countries. The new urban order of the "emerging city". *Planning Theory, 13*(3), 269–287. https://doi.org/10.1177/1473095213496098.

Banks, N., et al. (2020). Urban informality as a site of critical analysis urban informality as a site of critical analysis. *The Journal of Development Studies, 56*(2), 223–238. https://doi.org/10.1080/00220388.2019.1577384. Routledge.

Bell, J. S., & Loukaitou-Sideris, A. (2014). Sidewalk informality: An examination of street vending regulation in China. *International Planning Studies, 19*(3–4), 221–243. https://doi.org/10.1080/13563475.2014.880333. Taylor & Francis.

Bénit-Gbaffou, C. (2016). Do street traders have the "right to the city"? The politics of street trader organizations in inner-city Johannesburg, post-Operation Clean Sweep. *Third World Quarterly, 37*(6), 1102–1129. https://doi.org/10.1080/01436597.2016.1141660.

Bentley, I., McGlynn, S., Smith, G., Alcock, A., & Murrain, P. (1985). *Responsive environment: A manual for designers*. Oxford: Architectural Press.

Bernal-Torres, C. A., Peralta-Gómez, M. C., & Thoene, U. (2020). Street vendors in Bogotá, Colombia, and their meanings of informal work. *Cogent Psychology, 7*(1). https://doi.org/10.1080/23311908.2020.1726095. Cogent.

Bhowmik, S. K. (2003). *Urban responses to street trading: India WEIGO*. Washington, DC. http://www.wiego.org/publications/urban-responses-street-trading-india. Accessed 2 May 2019.

Brown, A. (2001). Cities for the urban poor in Zimbabwe: Urban space as a resource for sustainable development. *Development in Practice, 11*(2–3), 319–331. https://doi.org/10.1080/09614520120056432. Taylor & Francis Group.

Castells, M., & Portes A. (1989). World underneath: The origins, dynamics, and effects of the informal economy. In Portes, A., Castells, M., Benton, L. A. (Eds.), *The informal economy: studies in advanced and less developed countries* (pp. 11–41). Baltimore and London: The Johns Hopkins University Press.

De Soto, H. (1989). *The other path. The invisible revolution in the third world.* New York: Harper and Row.

Dikeç, M. (2001). Justice and the spatial imagination. *Environment and Planning A: Economy and Space, 33*(10), 1785–1805. https://doi.org/10.1068/a3467.

Dikeç, M., & Gilbert, L. (2002). Right to the city: Homage or new societal ethics? *Capitalism, Nature, Socialism, 13*(2), 58–74. https://doi.org/10.1080/10455750208565479.

Finn, B. (2018). Quietly chasing Kigali: Young men and the intolerance of informality in Rwanda's capital city. *Urban Forum, 29*(2), 205–218. https://doi.org/10.1007/s12132-017-9327-y. Urban Forum.

Gamieldien, F., et al. (2017). Street vending in South Africa: An entrepreneurial occupation.

S

South African Journal of Occupational Therapy, 47(1), 24–29.
Harvey, D. (1973). *Social justice and the city*. Baltimore, MD: Johns Hopkins University Press.
Harvey, D. (2018). *Limits to capital*. London: Verso.
Harvey, J. (2018). How cities can achieve public space for all in Women in Informal Employment Globalizing and Organizing (WIEGO). Street Vendors and Public Space: Essential insights on Key Trends and Solutions. WIEGO. www.wiego.org. Accessed 14 August 2020.
Huchzermeyer, M. (2014). Invoking Lefebvre's 'right to the city' in South Africa today: A response to Walsh. *City, 18*(1), 41–49. https://doi.org/10.1080/13604813.2014.868166.
Huchzermeyer, M. (2017). Humanism, creativity, and rights: Invoking Henri Lefebvre's right to the city in the tension presented by informal settlements in South Africa today. In *Dialogues in urban and regional planning 6: The right to the city* (Vo. 85, No. 2014, pp. 83–104). London: Routledge. https://doi.org/10.4324/9781315628127.
Huchzermeyer, M., & Misselwitz, P. (2016). Coproducing inclusive cities? Addressing knowledge gaps and conflicting rationalities between self-provisioned housing and state-led housing programs. *Current Opinion in Environmental Sustainability*. https://doi.org/10.1016/j.cosust.2016.07.003.
International Labour Organisation (ILO). (1972). *Employment, incomes and equality: A strategy for increasing productive employment in Kenya*. Geneva: ILO.
Jha, A. K., et al. (2018). *Building urban resilience principles, tools, and practice environment and sustainable development*. Available at: http://documents.worldbank.org/curated/en/320741468036883799/pdf/Building-urban-resilience-principles-tools-and-practice.pdf. Accessed 27 June 2019.
Kathage, A. M. (2018). *Understanding the informal economy in African cities: Recent evidence from Greater Kampala*. Africa Can End Poverty, World Bank Blogs. https://blogs.worldbank.org/africacan/understanding-the-informal-economy-in-african-cities-recent-evidence-from-greater-kampala. Accessed 14 August 2020.
Kayuni, H. M., & Tambulasi, R. I. C. (2009). Political transitions and vulnerability of street vending in Malawi. *Theoretical and Empirical Researches in Urban Management, 3*(12), 79–96.
Lefebvre, H. (1974). *La production de l'Espace [The production of space]*. Paris, France: Éditions Anthropos.
Linares, L. A. (2018). The paradoxes of informalizing street trade in the Latin American city. *International Journal of Sociology and Social Policy, 38*(7), 651–672. https://doi.org/10.1108/IJSSP-09-2017-0119.
Loukaitou-Sideris, A., & Mukhija, V. (2016). Responding to informality through urban design studio pedagogy. *Journal of Urban Design, 21*(5), 577–595. https://doi.org/10.1080/13574809.2015.1071650. Routledge.
Madanipour, A. (2006). Roles and challenges of urban design. *Journal of Urban Design, 11*(2), 173–193. https://doi.org/10.1080/13574800600644035. Taylor & Francis Group.
Marcuse, P. (2009). From critical urban theory to the right to the city. *City, 13*(2–3), 185–197. https://doi.org/10.1080/13604810902982177.
Marcuse, P., et al. (2009). *Searching for the just city: Debates in urban theory and practice*. London: Routledge. https://doi.org/10.4324/9780203878835.
Moyo, S. (2020). *Navigating informality: Patterns of categorization in street vending*. UNDP. https://www.zw.undp.org/content/zimbabwe/en/home/blog/navigating-informality–patterns-of-categorization-in-street-ven.html. Accessed 13 August 2020.
Njo, A. J. (2003). Urbanization and development in Sub-Saharan Africa. *Cities, 20*(3), 167–174.
Philopopoulos-Mihalopoulos, A. (2015). *Spatial justice body, lawscape, atmosphere* (1st edn). New York: Routeledge.
Pieterse, E. (2008). *City futures: Confronting the crises of urban development*. London: Zed Books.
Rogerson, C. M. (2016). Responding to informality in urban Africa: Street trading in Harare, Zimbabwe. *Urban Forum, 27*(2), 229–251. https://doi.org/10.1007/s12132-016-9273-0. Urban Forum.
Rowley, A. (1994). Definitions of urban design: the nature and concerns of urban design. *Planning Practice and Research, 9*(3), 179–197.
Rawls, J. (1974). 1974, 'The Independence of Moral Theory', Proceedings and Addresses of the American Philosophical Association, 47, 5–22, in Collected Papers, 1999, pp. 286–302.
Rawls, J. (2001). *Justice as fairness: A restatement*. Cambridge MA: Harvard University Press.
Roy, A. (2005). Urban informality: Toward an epistemology of planning. *Journal of the American Planning Association, 71*(2), 147–158. https://doi.org/10.1080/01944360508976689.
Soja, E. W. (2009). The city and spatial justice. In *Justice et injustices spatial* (pp. 56–72). New York: Routledge. https://doi.org/10.4000/books.pupo.415.
Watson, V. (2009). "The planned city sweeps the poor away…": Urban planning and 21st-century urbanization. *Progress in Planning, 72*(3), 151–193. https://doi.org/10.1016/j.progress.2009.06.002.
Williams, J. (2013). Toward a theory of spatial justice. Paper presented at the annual meeting of the Western Political Science Association Los Angeles, CA, "Theorizing Green Urban Communities" Panel, March 28, 2013.

Spatial Planning

► Land Use Planning Systems in OECD Countries

Spatial Planning – Land-Use Planning

► Spatial Planning Under the Impact of Urbanization and Rural Transformation in Zimbabwe: A Review of Theories, Philosophies, and Practices

Spatial Planning Under the Impact of Urbanization and Rural Transformation in Zimbabwe: A Review of Theories, Philosophies, and Practices

Jeofrey Matai[1] and Innocent Chirisa[2]
[1]Department of Architecture and Real Estate, University of Zimbabwe, Harare, Zimbabwe
[2]Department of Demography Settlement and Development, Social & Behavioural Sciences, University of Zimbabwe, Harare, Zimbabwe

Synonyms

Spatial planning – land-use planning; Spatial transformation – urbanization and rural transformation; Rural transformation – rural change

Definition

Spatial Planning – It refers to the process of shaping to economic, cultural, physical, and ecological dimensions of space (Allmendinger and Haughton 2010). It is a state activity of influencing the distribution of activities in space, balancing the social, economic, and environmental demands for development through the vertical and horizontal integration of sectoral policies and programs.

Urbanization – It is described as the increase in the number of people living in urban areas that are small and medium towns and large cities. Shukla and Jain (2019) define it as the forceful socioeconomic process that transforms rural landscapes into urban landscapes. Thus, urbanization can be seen as a socioeconomic, political as well as physical process that changes the distribution of population, economic activities, way of life, and the landscape of places.

Rural Transformation – It refers to changes in rural places that are characterized by a decline in agriculture as an economic activity with a corresponding increase in non-farm economic activities (Matai et al. 2021). Like urbanization, rural transformation is also characterized by landscape changes where access to economic and social infrastructure increases and land use and land cover changes occur.

Philosophy – It is the pursuit of wisdom, reality, and knowledge (Metcalf 2020). It is that activity undertaken by people to understand essential truths about themselves, the world in which they live, and how they relate to the world and each other.

Spatial Transformation – It is described as the change in form and dimension of spaces as a result of social, economic, and political dynamics that are present at a given space in time. Urbanization and rural transformation are forms of spatial transformation.

Introduction

This entry examines the theoretical and philosophical underpinnings that inform the practice of spatial planning in Zimbabwe. This is against the background that urban and rural places are transforming, through the dual processes of urbanization and rural transformation, raising questions about the adequacy and capacity of the current spatial planning systems and practices. Spatial planning systems and practices in the country remain confined in the old system of doing things (Chigudu 2021; Chigudu and Chirisa 2020; Chirisa and Dumba 2012) despite the changing planning environment in both rural and urban spaces. The rate of urbanization has increased and is expected to increase (McGranahan and Satterthwaite 2014; Zhang 2016). The United Nations (2018) report on World Urbanisation Prospects shows that by

2018, 55% of the world's population resided in urban and is expected to reach 68% by 2050. On the other hand, there is increasing evidence pointing to the fact that rural places are transforming (see Belton and Filipski 2019; Berdegué et al. 2013; Long et al. 2011; Scoones and Murimbarimba 2021; Shukla and Jain 2019). While these changes in rural and urban spaces create opportunities for the betterment of people's wellbeing (Belton and Filipski 2019; UN-Habitat 2020), several negative externalities can be experienced in the absence of proper planning that adapts to the changes (Acheampong 2019b). The questions that arise are: what is adaptive spatial planning; what should inform spatial planning to respond to the changes in rural and urban places to capitalize on the opportunities and minimize the negative externalities therewith? An understanding of theories and philosophies that examine urbanization and rural transformation as well as questioning, explaining spatial planning practices amid transforming urban and rural places, is necessary to come up with spatial planning systems that are responsive, hence, the thrust of this entry.

Philosophies of Transformation

As elaborated earlier, philosophy is concerned with the need to understand fundamental truths about the world in which people live and how they relate with the world as well as how people relate with each other (Metcalf 2020). Lynch (2016) identifies four branches of philosophy: metaphysics, epistemology, axiology, and logic. Epistemology concerns the nature and origin of knowledge and truth. Devine, revelation, reason logic, and intuition are the main basis of epistemology. Axiology is about the study of principles and values and questions morals and values. Logic seeks to organize reasoning. Metaphysics studies the fundamental nature of reality such as wholes and their parts, processes, events, and cause and effect among others (Metcalf 2020). Cause and effect (also referred to as causation or casualty) explains the stimulus that one event, process, or object is responsible for the effect and the effect dependent on the cause. Lynch (2016) outlines that processes may have several causes which may be causal factors for the processes.

Based on the metaphysical branch of philosophy, the relationship between urbanization and rural transformation can be seen as a cause-effect one in which urbanization causes the transformation of rural places (see Berdegué et al. 2013; Matai et al. 2021; Shukla and Jain 2019). Rural transformation also contributes to urbanization as the economic and livelihood activities change from largely agricultural to non-farm activities (Belton and Filipski 2019; Berdegué et al. 2013; Long et al. 2011). The cause-effect relationship between urban and rural places through urbanization and rural transformation, and the subsequent rural-urban linkages influence spatial planning practice and the system therewith. Thus, the relationship between urbanization, rural transformation, and spatial planning practice can be best explained by the metaphysical branch of philosophy.

Karl Max's philosophy of revolution, involution philosophy by Alexander Goldenweiser, and Charles Darwin's evolution philosophy can be seen through the metaphysical branch of philosophy in that they investigate the existence and properties of rural and urban spaces over time as well as the causation relationships. The aforementioned philosophies are looked at in detail in the sections that follow.

Revolution Philosophy

A revolution is an unexpected radical change (White 1983). It is a form of transformation that is abrupt and is generally associated with changes in political economies and governments usually through violent protests. The revolution philosophy is linked to Karl Max based on the antagonism between labor and capital that is a resultant effect of the exploitative relationship between capital and labor (Holton 1981). The argument put forward by Max is that the exploitation of labor by capital would eventually lead to a revolution as labor will be fighting for equality (Wolff and Leopold 2021). Governments, their institutions, and laws are usually put in place to regulate how labor and capital relate but Max argues that eventually, labor will overthrow the capitalist class, seizing control of the economy (Allman 2007). While the philosophy is about labor and capital and the change resulting from the relationship between the two, it can also explain changes that occur in space particularly the motivation behind the "space revolution." Capitalism encourages the development of certain territories, usually the core by exploiting other places (the peripheral areas) (Baldwin 2001) as explained in the core-periphery model. Urban centers (towns and cities) represent the core while the rural areas are the peripheries. In this case, the rural areas lose natural resources, labor, and in some instances, economic surplus to the urban areas (Belton and Filipski 2019), a situation that perpetuates the underdevelopment of rural areas and furthers the development of urban areas. This situation acts as a driver to the transformation of rural places as people in the rural areas seeks to improve their quality of life. For example, engaging in better paying non-farm activities (Belton and Filipski 2019; Berdegué et al. 2013) can, however, raise food security issues in urban areas (Tacoli and Vorley 2015).

Involution Philosophy

The philosophy describes the adaptation by peasant society to the system that extracts local resources such as land, labor, and money resulting in a dual economic pattern that has a labor-intensive sector and a capital-intensive sector. The system promotes the capital-intensive sector and ensures that the sector grows rapidly while the labor-intensive remains behind (Geertz 2020). The penetration of information to the peasants is also prevented to ensure that agricultural modernization does not happen and that the life of the peasants does not change. The peasants respond to this situation by working harder amid the challenges of increasing population and declining land sizes as a result of land expropriation by the colonialists (McGee 2008). Through working harder on the small pieces of land, outputs increase. The scenario described above of the capital-intensive and the labor-intensive shows that through involution, transformation is not a result of applying new approaches or new ways of doing things as is the case with revolution, but rather a result of the rigidified pattern. In the field of spatial planning and development, involution can be used to explain changes that take place without the direct influence of exogenous factors. The unpleasant circumstances push people to make use of available resources to transform their lives. These have significant impacts on space. For example, intensified farming cause soil degradation with consequential effects on agricultural production. This can cause people to shift to non-farm activities or migrate to cities in search of sustainable livelihood sources.

Evolution Philosophy

Evolution describes a gradual change from one state to another. The change is however connected to the old one, hence White (1983) calls it the commencement of a new form based on the old one. Evolution is associated with continuity and discontinuity where noted trends stay in the same path in an unchanged direction or where things change with certain breaks being observed (Holton 1981). Evolution is associated with the survival of the fittest. Evolution can be described as laissez faire economics. Based on these views, evolution is linked to the concept of survival of the fittest which is based on the ability of an individual to outshine others. Charles Darwin is the major influence of the evolution philosophy based on the theory of natural selection processes in biological organisms. Natural selection produces a struggle for survival between species that have heritable traits and those with injurious heritable variations (Oppenheimer 1969). In this scenario, species with injurious heritable traits are destroyed while those with distinctive heritable characteristics survive. While this philosophy explains the transformation from a purely biological perspective that explains how other species survive and reproduce (Ruse 1975), it can also explain transformation in space, rural and urban areas in this case. Urban areas, due to their high production capacity and the influence of agglomeration of scale grow while rural areas remain behind or get worse off. As a result, people migrate to the cities, further strengthening urban areas and demanding more land for urban land uses that sees peri-urban land being succeeded and converted into urban land uses (see Matai et al. 2021; Ritchie and Roser 2018; Tacoli and Vorley 2015; Zhang 2016).

Evolution shows that transformation is an intrinsic process that is driven from within where an organism is an active participant in its adaptation. The role of external forces is not present in the evolution of organisms; this aspect of transformation from within is similar to involution (Geertz 2020), although, under involution, transformation is triggered by pressure exerted from external forces of colonialism. Evolution also emphasizes the importance of cooperation where organisms mutually benefit each other through social evolution. Cooperation explains spatial transformation through the integration of rural and urban spaces. Thus, as much as the two places compete for resources, rural-urban linkages explain the cooperation between rural and urban areas, which according to Shukla and Jain (2019) can have positive benefits for both rural and urban areas if properly managed. However, competition between rural and urban areas can negatively impact rural areas (Akkoyunlu 2015; Bah et al. 2003). Geddes and Newman (1999) recognize that the relationship between the organism and the surrounding environment is important in the evolution process. The organism shapes the surrounding environment and the organism also adapts to the environment. The relationship between territories (urban and rural) significantly impacts each other. For example, through the ecological footprint concept, cities influence what happens in rural areas. Rural areas, on the other hand, adapt by transforming.

Theories of Spatial Planning

The understanding of theories, ideas, philosophies, and thinking that informs what spatial planning is and how it is to be practiced is of critical importance in spatial planning. Planning theories are an important component of spatial planning as shown by the co-evolution of prevailing realities, practices, and the ideas that shaped planning (Acheampong 2019). This is seen by the increase in literature in planning theory (see Bertolini 2016; Davidoff 1965; Healey 2003, 2020). The entry is not an attempt to entirely cover the panning theories, but to highlight major thinking and main beliefs that continuously shape planning practice and scholarly work.

Most scholarly work acknowledges the role of normative planning theories in shaping spatial planning (Acheampong 2019; Healey 2003, 2020; Watson 2002). Some of the theories that fall under the normative planning theories include the systems theory, rational comprehensive theory, advocacy, communicative planning, and collaborative planning theory. The systems theory looks at cities and regions as spatial entities that are complex and interconnected to several parts (Chadwick 2013). As such Acheampong (2019) argues that the systems theory in spatial planning emphasizes the need to address the complex and dynamic interrelationships as a basis for addressing an existing problem. The rational comprehensive model partly draws from the systems theory and looks at planning as a technocratic exercise that is scientific and objective. The proponents of the rational comprehensive theory argue that planners should separate the means from the ends; this implies that planning and political processes should be separated, providing space for planning issues to be approached from a purely technical and rational process (Faludi 1973). Based on the overview of the above discussion on systems theory and the rational comprehensive theory, the two theories focus on a comprehensive evaluation of all possible courses of action, the consequences, and the selection of the best cause of action (Acheampong 2019a; Chadwick 2013; Faludi 1973). The approach to planning based on a purely technical approach to problem-solving and the value-free nature as well as the unitary approach to public interest exposed the theories to criticism based on lack of democracy and failure to engage with political processes in making decisions (Healey 2003; Rothblatt 1971). This gave birth to theories that recognize political processes.

Pragmatism recognizes the political nature of spatial planning (Acheampong 2019a). The theory approaches problems in an incremental manner that prioritizes getting things done by doing what works best in a given situation rather than theorizing (Healey 2009). Advocacy planning, another politically sensitive approach to spatial planning, takes a pluralistic and inclusive approach to the activity of planning. The theory is hinged on the need to engage and represent various groups in society whose views, preferences, and aspirations vary. The theory demystifies planning as a technical process. The communicative and collaborative approaches have been borne of the increasing need for discourse in determining planning outcomes (Acheampong 2019a; Morphet 2010). These approaches to spatial planning are centered on inclusiveness and dialogue that is intentionally made to pool resources together for problem-solving by stakeholders. These approaches to spatial planning are argued to allow for the pursuance of social justice through communication and negotiation as a mechanism of bringing divergent interests together (Acheampong 2019a; Healey 2009). They also put co-production as the major approach to solving problems of inequalities and other problems that affect communities.

The theories discussed above provide guiding principles for addressing dual processes of urbanization and rural transformation as spatial transformation drivers and outcomes. One of the key issues that spatial planning deals with is to cope with uncertainty (Belton and Filipski 2019). The systems and rational comprehensive theories reduce uncertainties in planning by finding a workable fit between local conditions and the motivation of spatial planning and by considering societal problems in their totality and the possible intervention measures and the likely consequences. Pragmatism, on the other hand, approaches the problems created by and between urbanization and rural transformation more practically and incrementally. As much as this can be regarded as a reactive approach to problems, which is less effective in dealing with long-range problems, it informs spatial planning practice by bringing in a more practical approach to problem-solving as opposed to philosophies and theorizing the problems (Acheampong 2019a). The pragmatic, advocacy, collaborative, and communicative approaches to spatial planning also play a critical role in shaping spatial planning practice. The approaches have strength in engaging communities, allowing discourses, dialogues, and inclusiveness in approaching community problems. Since urbanization and rural transformation involve individuals, groups, communities, and businesses with varying interests, preferences, views, and aspirations (Bertolini 2016), the theories are critical in informing spatial planning practice, providing a basis for the consideration of political processes to the technical and bureaucratic processes.

Summary/Conclusion

The entry concludes that spatial planning practice can be informed by the philosophies and theories as discussed herein. However, the philosophies and theories, when looked at independently cannot be effective in providing a strong and comprehensive basis for spatial planning practice. As such, a hybrid of the theories should be used to identify and examine planning problems and then develop spatial planning systems that are responsive to the changing environment that is a resultant effect of urbanization and rural transformation. Thus, legal frameworks, guiding principles, and institutional frameworks that can effectively deal with problems emanating from urbanization and rural transformation and capitalize on the opportunities created can be put in place.

Cross-References

- ▶ Digitalization, Urbanization, and Urban-Rural Divide
- ▶ Spatial Demography as the Shaper of Urban and Regional Planning Under the Impact of Rapid Urbanization

S

References

Acheampong, R. A. (2019a). The concept of spatial planning and the planning system. In R. A. Acheampong (Ed.), *Spatial planning in Ghana* (pp. 11–27). Springer International Publishing. https://doi.org/10.1007/978-3-030-02011-8_2.

Acheampong, R. A. (2019b). Urbanization and settlement growth management. In R. A. Acheampong (Ed.), *Spatial planning in Ghana: Origins, contemporary reforms and practices, and new perspectives* (pp. 171–203). Springer International Publishing. https://doi.org/10.1007/978-3-030-02011-8_9.

Akkoyunlu, Ş. (2015). The potential of rural-urban linkages for sustainable development and trade. *International Journal of Sustainable Development & World Policy, 4*(2), 20–40.

Allman, P. (2007). *On Marx: An introduction to the revolutionary intellect of Karl Marx*. BRILL.

Allmendinger, P., & Haughton, G. (2010). Spatial planning, devolution, and new planning spaces. *Environment and Planning C: Government and Policy, 28*(5), 803–818. https://doi.org/10.1068/c09163.

Bah, M., Cissé, S., Diyamett, B., Diallo, G., Lerise, F., Okali, D., Okpara, E., Olawoye, J., & Tacoli, C. (2003). Changing rural-urban linkages in Mali, Nigeria, and Tanzania. *Environment and Urbanization, 15*(1), 13–24.

Baldwin, R. E. (2001). Core-periphery model with forward-looking expectations. *Regional Science and Urban Economics, 31*(1), 21–49. https://doi.org/10.1016/S0166-0462(00)00068-5.

Belton, B., & Filipski, M. (2019). Rural transformation in central Myanmar: By how much, and for whom? *Journal of Rural Studies, 67*, 166–176. https://doi.org/10.1016/j.jrurstud.2019.02.012.

Berdegué, J. A., Rosada, T., & Bebbington, A. J. (2013). Rural transformation. https://idl-bnc-idrc.dspacedirect.org/handle/10625/51569

Bertolini, L. (2016). "Complex systems, evolutionary planning?" In *A planner's encounter with Complexity* (pp 81–98). Routledge.

Chadwick, G. (2013). *A systems view of planning: Towards a theory of the urban and regional planning process*. Elsevier.

Chigudu, A. (2021). The changing institutional and legislative planning framework of Zambia and Zimbabwe: Nuances for urban development. *Land Use Policy, 100*, 104941. https://doi.org/10.1016/j.landusepol.2020.104941.

Chigudu, A., & Chirisa, I. (2020). The quest for a sustainable spatial planning framework in Zimbabwe and Zambia. *Land Use Policy, 92*, 104442. https://doi.org/10.1016/j.landusepol.2019.104442.

Chirisa, I., & Dumba, S. (2012). Spatial planning, legislation and the historical and contemporary challenges in Zimbabwe: A conjectural approach. *Journal of African Studies and Development, 4*(1), 1–13. https://doi.org/10.5897/JASD11.027.

Davidoff, P. (1965). Advocacy and pluralism in planning. *Journal of the American Institute of Planners, 31*(4), 331–338. https://doi.org/10.1080/01944366508978187.

Faludi, A. (1973). *A reader in planning theory* (Vol. 5). Elsevier.

Geddes, M., & Newman, I. (1999). Evolution and conflict in local economic development. *Local Economy, 14*(1), 12–26. https://doi.org/10.1080/02690949908726472.

Geertz, C. (2020). Agricultural involution: The processes of ecological change in Indonesia. In *Agricultural involution*. University of California Press. https://doi.org/10.1525/9780520341821.

Healey, P. (2003). Collaborative planning in perspective. *Planning Theory, 2*(2), 101–123. https://doi.org/10.1177/14730952030022002.

Healey, P. (2009). The pragmatic tradition in planning thought. *Journal of Planning Education and Research, 28*(3), 277–292.

Healey, P. (2020). *Collaborative planning: Shaping places in fragmented societies*. Bloomsbury Publishing.

Holton, R. J. (1981). Marxist theories of social change and the transition from feudalism to capitalism. *Theory and Society, 10*(6), 833–867.

Long, H., Zou, J., Pykett, J., & Li, Y. (2011). Analysis of rural transformation development in China since the turn of the new millennium. *Applied Geography, 31*(3), 1094–1105. https://doi.org/10.1016/j.apgeog.2011.02.006.

Lynch, M. (2016, August 5). What you need to know as an educator: Understanding the 4 main branches of philosophy. *The Edvocate*. https://www.theedadvocate.org/need-know-education-understanding-4-main-branches-philosophy/

Matai, J., Musakwa, W., & Chirisa, I. (2021). Growth, expansion, and future of small rural towns. In *The Palgrave encyclopedia of urban and regional futures* (pp. 1–9). Springer Nature Switzerland. https://doi.org/10.1007/978-3-030-51812-7_72-1.

McGee, T. G. (2008). Managing the rural-urban transformation in East Asia in the 21st century. *Sustainability Science, 3*(1), 155–167. https://doi.org/10.1007/s11625-007-0040-y.

McGranahan, G., & Satterthwaite, D. (2014). *Urbanization concepts and trends*. International Institute for Environment and Development. https://www.jstor.org/stable/resrep01297

Metcalf, T. (2020, October 10). What is philosophy? 1000-word philosophy: An introductory anthology. https://1000wordphilosophy.com/2020/10/10/philosophy/

Morphet, J. (2010). *Effective practice in spatial planning*. Routledge.

Oppenheimer, H. (1969). III. Christian ethics: Helen Oppenheimer. *Religious Studies, 5*(2).

Ritchie, H., & Roser, M. (2018). Urbanization. Our World in Data. https://ourworldindata.org/urbanization

Rothblatt, D. N. (1971). Rational planning reexamined. *Journal of the American Institute of Planners, 37*(1), 26–37.

Ruse, M. (1975). Charles Darwin's theory of evolution: An analysis. *Journal of the History of Biology*, 219–241.

Scoones, I., & Murimbarimba, F. (2021). Small towns and land reform in Zimbabwe. *The European Journal of Development Research, 33*(6), 2040–2062. https://doi.org/10.1057/s41287-020-00343-3.

Shukla, A., & Jain, K. (2019). Critical analysis of rural-urban transitions and transformations in Lucknow city, India. *Remote Sensing Applications: Society and Environment, 13*, 445–456. https://doi.org/10.1016/j.rsase.2019.01.001.

Tacoli, C., & Vorley, B. (2015). Reframing the debate on urbanization, rural transformation, and food security. *IIED briefing paper – International Institute for Environment and Development*, No. 17281. https://www.cabdirect.org/cabdirect/abstract/20153155349

UN-Habitat. (2020). *World cities report 2020, the value of sustainable urbanization*. UN-Habitat.

United Nations. (2018). *World urbanisation prospects 2018: The 2018 revision* (Economic and Social Affairs) [Key facts]. United Nations.
Watson, V. (2002). The usefulness of normative planning theories in the context of sub-Saharan Africa. *Planning Theory, 1*(1), 27–52. https://doi.org/10.1177/147309520200100103.
White, B. (1983). "Agricultural involution" and its critics: Twenty years after. *Bulletin of Concerned Asian Scholars, 15*(2), 18–31. https://doi.org/10.1080/14672715.1983.10404871.
Wolff, J., & Leopold, D. (2021). Karl Marx. In E. N. Zalta (Ed.), *The Stanford encyclopedia of philosophy* (Spring 2021). Metaphysics Research Lab, Stanford University. https://plato.stanford.edu/archives/spr2021/entries/marx/
Zhang, X. Q. (2016). The trends, promises, and challenges of urbanization in the world. *Habitat International, 54*, 241–252. https://doi.org/10.1016/j.habitatint.2015.11.018.

Spatial Resilience

▶ Urban Resilience

Spatial Transformation – Urbanization and Rural Transformation

▶ Spatial Planning Under the Impact of Urbanization and Rural Transformation in Zimbabwe: A Review of Theories, Philosophies, and Practices

Spices

▶ The Source Waters of Tanga

Spiritual Services

▶ Faith Communities as Hubs for Climate Resilience

Sprawl

▶ Master Planned Estates and the Promises of Suburbia

Squatter Settlement – Slum, Informal Settlement

▶ Transnational Migrants on the Margin

Squatter Settlements

▶ Disaster Risk in Informal Settlements and Opportunities for Resilience

Stakeholder Consultation

▶ Community Engagement for Urban and Regional Futures

Stakeholder Engagement

▶ Community Engagement for Urban and Regional Futures

State

▶ Internationalization of Cities

The State of Extreme Events in India

Abinash Mohanty and Shreya Wadhawan
Council on Energy, Environment and Water (CEEW), New Delhi, India

Synonyms

Climate change; Climate risks; Climate vulnerability; Climate Risk Atlas (CRA); El Nino South Oscillation (ENSO); Geographic information system (GIS); Greenhouse Gas emissions (GHG);

Gross Domestic Product (GDP); Gross State Domestic Product (GSDP); India Meteorological Department (IMD); Indian summer monsoons (ISM); Land Use Land Cover (LULC); Micro, Small and Medium Enterprises (MSMEs); National Disaster Management Authority of India (NDMA); Press Information Bureau (PIB); Public Private Partnership (PPP); The Grid Analysis and Display System (GrADS); The Intergovernmental Panel on Climate Change (IPCC); World Meteorological Organization (WMO)

Definitions

1. ENSO – ENSO is a naturally occurring phenomenon that involves fluctuating ocean temperatures in the central and eastern equatorial Pacific, coupled with changes in the atmosphere (WMO).
2. Loss and damage – It is a general term used in UN climate negotiations to refer to the consequences of climate change that go beyond what people can adapt to, or when options exist but a community doesn't have the resources to access them (WRI, 2022).
3. Hydromet disasters – These disasters are a set of extreme climate events formed by the interaction of the weather with hydrological and climatic processes (GFDRR 2018).
4. Nature-based solutions (NbS) – Actions to protect, sustainably manage, and restore natural or modified ecosystems that address societal challenges effectively and adaptively, simultaneously providing human well-being and biodiversity benefits (IUCN).
5. Climate proofing – The explicit consideration and internalization of the risks and opportunities that alternative climate change scenarios are likely to imply for the design, operation, and maintenance of infrastructure (UNDP 2011).
6. ISM (Indian summer monsoons) – It typically lasts from June-September, with large areas of western and central India receiving more than 90% of their total annual precipitation during the period, and southern and northwestern India receiving 50%-75% of their total annual rainfall (IMD).
7. Climate variability – Variations in climatic conditions on time and space scales beyond that of individual weather events, but not persisting for extended periods of, typically, decades or longer (i.e., shorter term) (ADB).

Introduction

India is the third-largest economy in Asia (Silver 2021) and is projected to be one of the top three economic powers globally over the next 10–15 years. India seeks to invest more than USD 1.5 trillion to build its infrastructure over the next 10 years. India is also urbanizing fast, and hence its urban infrastructure will be pivotal to the economic manifesto. The Government of India has already invested more than USD 160 billion in creating and upgrading urban infrastructure over the last two decades; meanwhile, investments worth USD 28 billion are already in the pipeline for projects like Smart Cities Mission (ANI, NewsVoir 2021). The numbers are quite promising, but the flip side of the narrative is India's vulnerability to extreme events. India is the seventh most vulnerable country globally (Germanwatch 2020) and is already witnessing a glimpse of large-scale disasters and the resulting socioeconomic upheaval. India suffered an annual loss of USD 87 billion due to extreme climate events (WMO 2020). According to a report by IMD, 24 of the 36 meteorological subdivisions in India witnessed extreme events such as floods, landslides, and heatwaves during the southwest monsoon season (June to September) (IMD 2020), which significantly impacted the economies, lives, and livelihoods. The Intergovernmental Panel on Climate Change (IPCC) has projected that the frequency and intensity of extreme weather events will further increase in the twenty-first century (AR4 IPCC 2007). Climate change led to extreme weather events that resulted in more than four lakh human deaths worldwide in 1999–2018 (Mohanty 2020). Clearly, climate extremities are taking a toll on India's developmental pathways.

The sixth assessment report by IPCC confirms the human-induced changes in climate due to excess greenhouse gas emissions (GHGs) and unsustainable land use-land cover practices, which is all set to contravene the 1.5° mark. In a 1.5 °C climate scenario, climate change poses great challenges and threats to a significant portion of the emerging economies around the world as a country's economic growth is intricately linked with climate risks. The entry attempts to investigate some of these drivers through a robust coarse-grain spatial analysis and enumerate the State of Extreme Events across various climatic zones and sectors. The entry undermines the State of Extreme Events by identifying the ten most exposed (hotspot) states and the likelihood of sectors and populations at risk. The entry also identifies the landscape-based drivers of climate change. Further, it delves upon some tailor-made recommendations that can climate-proof India's sectors.

Methodological Approach

Any risk assessment study calls for a comprehensive hazard assessment (IPCC 2014). The principles of risk assessment (King et al. 2015) are used as the guiding norm for the development of this methodological approach for the hazard assessment, which is adopted from the study *Preparing India for Extreme Climate Events: Mapping Hotspots and Response Mechanisms* (Mohanty 2020). The focus of this subnational microlevel hazard assessment is to provide fundamental insights on the impacts of climate change through the lens of climate extremes. As stated by IPCC, any climate extremes study depends on both the quality and quantity of data. The study coupled information from the globally validated data sheets and data from other sources like IMD, WMO, and PIB, making this data collation quite comprehensive. The methodological approach has two key components: i) Development of gridded exposure sheet of climate events ii) Geospatial analysis of the extreme climate events using coarse-grained resolution temporal maps. Figure 1 illustrates the stepwise methodological approach adopted.

Though the above mentioned study presents a district-level profiling of India's extreme climate events, this study attempts to understand these extremes' complexity, nonlinear trends, and patterns across various zones at a subnational level. The prima facie extreme events catalogue was developed for a period of 50 years (1970–2019) using spatial and climatological modeling. As highlighted, the gridded exposure sheet provides a historical pentad decadal chronology of the extreme climate events between 1970 and 2019 to understand their pattern, frequency, and the areas affected in terms of districts across each zone and state. To maintain uniformity in the data set, the climate events that were supplemented to the base exposure sheet were based on the criteria of EM-DAT. (For a disaster to be entered into the Em-dat database, at least one of the following criteria must be fulfilled: i) Ten (10) or more people reported killed; ii) hundred (100) or more people reported affected;

S

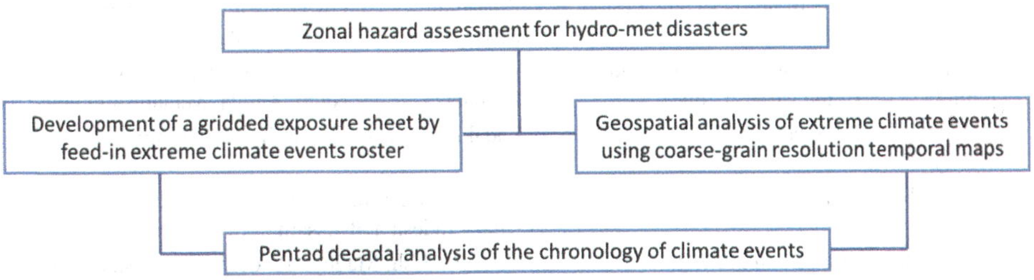

The State of Extreme Events in India, Fig. 1 Schematic representation of the approach and methodology. (Source: Authors' compilation based on Mohanty 2020)

iii) declaration of a state of emergency; and iv) call for international assistance.) The gridded exposure data set was developed using a downscaling approach. The downscaling methodology focused on the occurrence of the specific primary events (floods, drought, and cyclones) and their associated events. The study found a change in the trends and patterns of extreme climate events. Further, the geospatial analysis included the Land Use Land Cover (LULC) classification maps for each decade interval which were retrieved at 100-m resolution. According to the Koppen classification, India has six climatic zones: tropical humid climate, dry climate, warm temperate, cold climate, cold snow forest climate, and highlands. The climate zone classification base maps were derived from global reanalysis shape files modeled by (Beck et al. 2018) and were further clipped using Q-GIS 3.10. In order to develop a climate zone-gridded exposure data set, GrADS was used for the microlevel analysis. GrADS allows data from different data sheets to be graphically laid, with correct spatial and frequency of the extreme events for a given location. A rigorous pentad decadal geospatial analysis of the gridded exposure sheet based on its attribution, which includes frequency, districts affected, intensity, and associated events, was conducted. A temporal scale analyses of the maps was also conducted to generate empirical evidence on the changing extreme event profile across climate zones to derive at state level hotspot ranking across different zones. This methodological approach was adopted to investigate the intertemporal distribution of the events. Q-GIS 3.10 was used for carrying out the geospatial analysis of the extreme events at a temporal scale.

Mapping India's Exposure to Extreme Events

Mapping India's Exposure to Extreme Floods

India also has a unique geography, topology, and climatic regime. Floods have been devastating India's growth trajectory in recent decades. Floods are defined as "a general term for the overflow of water from a stream channel onto normally dry land in the floodplain (riverine flooding), higher-than-normal levels along the coast and in lakes or reservoirs (coastal flooding) as well as ponding of water at or near the point where the rain fell (flash floods)" (Em-dat Gloassary). India is divided into six climate zones: hot and dry, warm and humid, moderate, cold and cloudy, cold and sunny, and composite (when 6 months or more do not fall within any of the above categories) (Bansal and Minke 1988). Table 1 enumerates the zonal geographical classification vis-a-vis climate zones along with state listing.

The analysis suggests that the eastern zone of India is the most exposed to extreme flood events, closely followed by the northeastern and southern

The State of Extreme Events in India, Table 1 Geographical and climatic zones of Indian states

S. No.	Climatic zone	Geographical zone	States
1.	Hot and dry	Parts of northern and western zones	Rajasthan, Gujarat, Maharashtra, and some parts of Madhya Pradesh and Karnataka
2.	Warm and humid	Northeast, east, south, and parts of the western zone	Assam, Manipur, West Bengal, Odisha, Andhra Pradesh, Karnataka, Tamil Nadu, Kerala, Maharashtra, and Gujarat
3.	Cold and cloudy	Northern zone	Ladakh, Himachal Pradesh, and Jammu and Kashmir
4.	Cold and sunny	Northern zone	Jammu and Kashmir, Himachal Pradesh, and some parts of Uttar Pradesh and Assam
5.	Composite	The central zone, parts of northern, eastern, and western zones	Uttar Pradesh, Madhya Pradesh, Chhattisgarh, Bihar, Jharkhand, and parts of Rajasthan and Karnataka
6.	Moderate	No zone in India	Parts of Karnataka and Maharashtra

Source: Author's compilation based on Bhatnagar et al. (2019)

zones of India. The northeastern zone of India is becoming the flood capital, and EM-DAT further substantiates these findings; it suggests in the last two decades alone (2000–2020), South Asia has experienced 11% of the world's natural disasters and 12% of floods and droughts, exposing over 700 million people and 190 million ha of agricultural land (EM-DAT 2022). Figure 2 enumerates the state-wise exposure ranking of Indian states to extreme flood events. It is estimated that more than 97.51 million people are exposed to extreme flood events with reference to baseline data from Census 2011.

The analysis further finds that India's northern and western zones also face significant exposure to extreme flood events. Some of the nonflood hotspots are also witnessing an increased instance of urban flooding. The urban flooding spurt is intensified by microclimate zone shifts, urban-heat island effect (UHI), and land-surface temperature increase across the central, western, and southern regions. The study finds that changes in landscape indicators are some of the major drivers for the spurt in the extreme flood events; more than 40% change is found across the top ten exposed states. Table 2 enumerates the percentage-wise change in landscape indicators and the districts exposed.

An analysis by CEEW suggests that more than 40% of Indian districts are showcasing a swapping trend, i.e., traditional flood-prone areas are becoming drought-prone and vice versa (Mohanty 2020). Similarly, this study finds that more than 52% of the districts in the top ten flood hot spot states are showcasing a swapping trend. This calls for a more robust granular risk assessment. There has been a two-fold surge in extreme flood events across the top ten states in the last five decades. The State of Kerala has witnessed a three-fold increase, while the rest of the hot spot states have registered a two-fold surge. The analysis also infers that more than 46% of the districts in these hot spot states have undergone rapid landscape disruption triggering an exponential surge in these extremities. India witnesses a mixed bag of floods, i.e., riverine, coastal, and flash floods. There has been a substantial shift in the rainfall patterns, and primarily in the earlier decades,

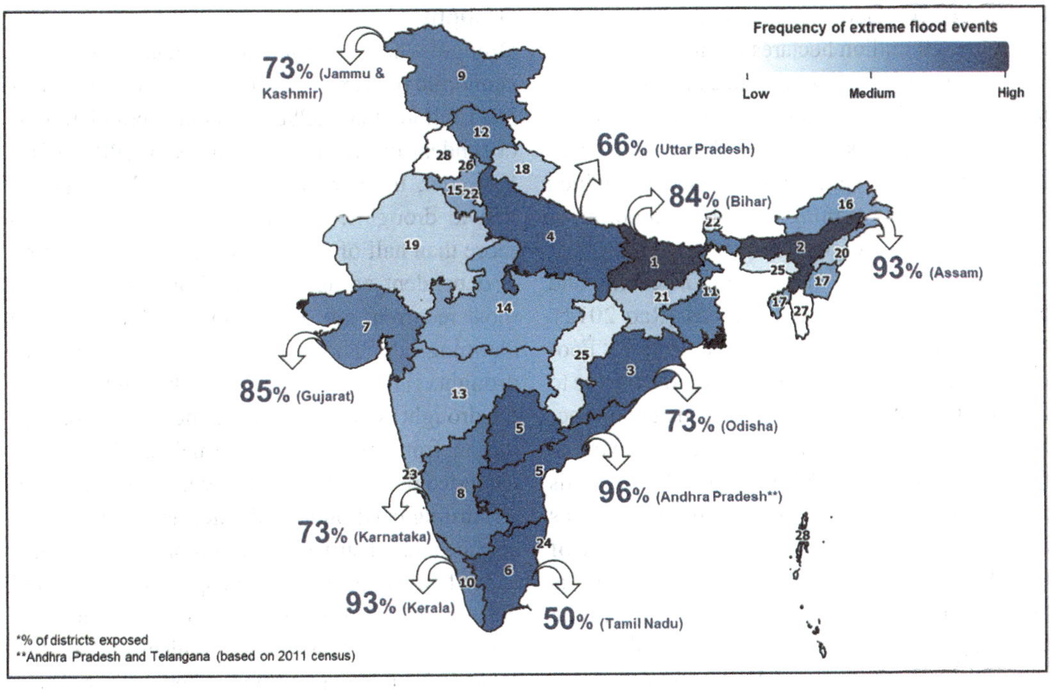

The State of Extreme Events in India, Fig. 2 Bihar and Assam are the most exposed to extreme flood events. (Source: Authors' analysis)

The State of Extreme Events in India, Table 2 Flood hotspot states at a glance

S. No.	State exposed	Percentage of districts exposed to extreme flood events	Percentage of area undergoing a change in LULC
1	Bihar	84	29
2	Assam	93	63
3	Odisha	73	53
4	Uttar Pradesh	66	22.5
5	Andhra Pradesh	96	43.4
6	Tamil Nadu	50	40
7	Gujarat	85	42.3
8	Karnataka	73	25
9	Jammu and Kashmir	73	82
10	Kerala	93	64

Source: Authors' analysis based on Mohanty and Wadhawan 2021.

floods were accompanied by the onset of Indian summer monsoons (ISM) through its inter-annual and intraseasonal phases of ISM (Ray et al. 2019). Favorable monsoonal conditions trigger heavy to very heavy extreme monsoon rainfall; low-pressure systems and high precipitation levels associate themselves to high downpour triggering disasters at an unprecedented scale (Dhar 2003). Floods have exposed ecosystems, lives, and livelihoods, and estimates suggest that more than 40 million hectares of land are exposed to floods and nearly 8 million hectares of land are impacted by floods annually (Ray et al. 2019). The analysis suggests that there has been a surge in extreme flood events in recent decades, which is adjunct to the warming of the Indian ocean. A warming Indian Ocean supplements to a surge in moisture leading to extreme rainfall events and hence floods in the recent decades (Rao 2012). Given India's exposure to floods, section "Economic Sectoral Exposure and Loss and Damage to Extreme Climate Events" provides insight on sectoral economic exposure of states to extreme floods. According to NDMA, the 2018 floods alone have affected more than 5.1 million hectares of land – which is equal to the size of the State of Punjab. While the surge in flood events is two-fold, there has been a six-fold increase in associated flood events (extreme rainfall, landslides, hailstorms, and thunderstorms) (Mohanty 2020). The warming climate coupled with rapid, unsustainable urbanization and indiscriminate encroachment of natural drainage systems have triggered some of the major floods across India in the last half a decade. Globally, India is becoming the world's flood capital, and clearly, it needs to revamp its flood mitigation strategy to insulate its communities, lives, livelihoods, and infrastructures.

Mapping India's Exposure to Extreme Droughts

Droughts have a political, climatological, and economic purview in India. According to Census 2011, more than 52% of India's population is dependent on the agricultural sector (PIB 2020). However, the agriculture sector is the worst hit due to droughts. This itself explains that when more than half of the county's earning population is dependent on agriculture, droughts which are most recurrent are devastating livelihoods at an alarming rate. There are primarily three kinds of droughts i) meteorological drought (Meteorological drought is defined as the deficiency of precipitation from expected or normal levels over an extended period of time), ii) hydrological drought (Hydrological drought is defined as deficiencies in surface and subsurface water supplies, leading to a lack of water for normal and specific needs), and iii) agricultural droughts (Agricultural drought is usually triggered by meteorological and hydrological drought and occurs when soil moisture and rainfall are inadequate during the crop-growing season, causing extreme crop stress and wilting). Droughts are defined as "an extended period of unusually low precipitation that produces a shortage of water, and operationally, it is defined as the degree of precipitation reduction that constitutes a drought, that varies by locality, climate and environmental sector" (Em-dat Gloassary). Change in inter-annual variability of monsoons linked to El Nino South Oscillation (ENSO), warming the Indian Ocean resulting in sea-surface temperature increase, is triggering the extreme drought and drought-like conditions across India (Kumar et al. 2013).

The pentad decadal analysis suggests that the southern and western zones are highly exposed to extreme drought events. The northern, eastern, and central zones also have greater exposure to extreme drought events. The northeastern zone has the least exposure to extreme drought events. The state-wise ranking of Indian States is derived through this pentad-decadal analysis inferred through frequency and intensity of drought extremes. There has been a two-fold increase in extreme drought events across the Indian states. In contrast, Karnataka has registered a four-fold increase in extreme drought events and Maharashtra, Andhra Pradesh, and Tamil Nadu have witnessed a three-fold increase in drought extremities, respectively (Fig. 3).

Further, the analysis suggests that 92.8% of the districts in the top ten hot spot states that are prone to extreme droughts have undergone significant land-use-surface changes. The land-surface changes have triggered extreme droughts in recent decades. The analysis also infers that 40% of the districts in these hot spot states have undergone rapid landscape disruption triggering an exponential surge in these extremities. Moreover, it is to be noted that more than 42% of the districts in hot spot states are showcasing a swapping trend, i.e., drought-prone areas are becoming flood-prone or vice versa. India cannot afford to operate in business as usual in drought mitigation, and investing in an effective drought public early warning system is imperative. India's national weather/

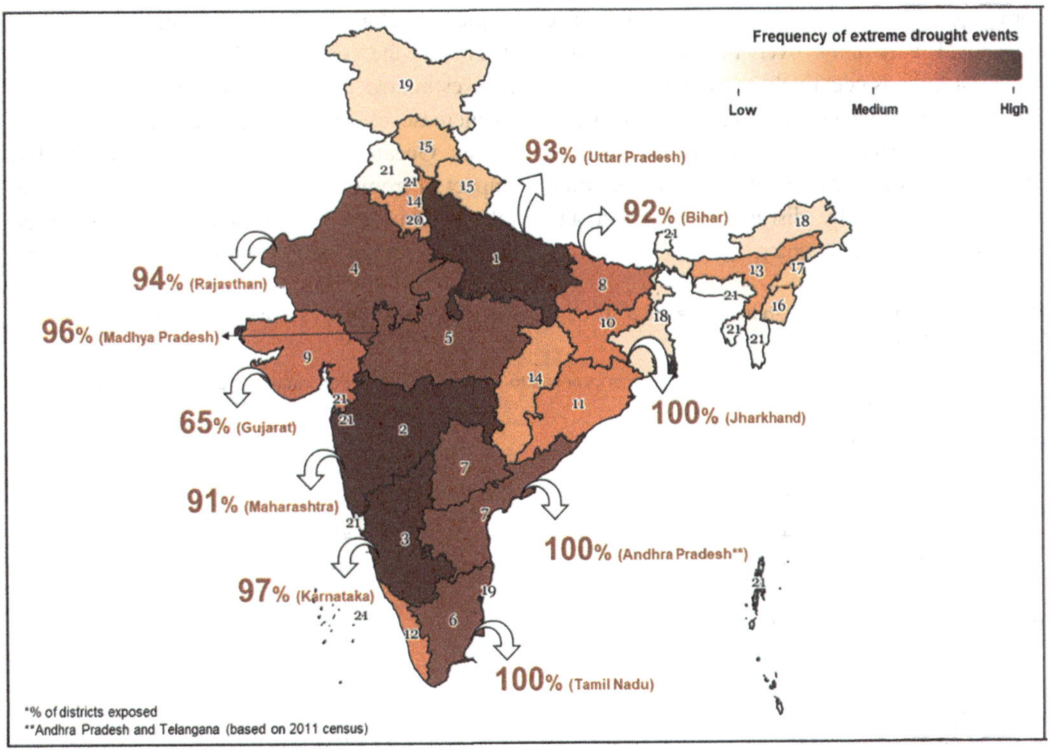

The State of Extreme Events in India, Fig. 3 Uttar Pradesh, Maharashtra, and Karnataka are the most exposed to extreme drought events in India. (Source: Authors' analysis)

S

meteorological monitoring agency-IMD uses only precipitation-based drought indexing, which has its own set of limitations (Table 3).

Post-2002, there has been substantial improvement in drought monitoring, but gaps still exist in predicting monsoonal drought on seasonal to decadal timescales (Rajeevan et al. 2012). The recent spurt in drought events is triggered by land-use change, urban heat island effect, and changes in precipitation levels. The pentad decadal analysis suggests that the annual average rainfall empirically links the climatological and meteorological linear relationship with drought events until 1990–1999. These findings are also confirmed in Mohanty 2020, at a district level. IMD also suggests that more than 25% of the country in 2001 and 29% of the country in 2002 was affected by moderate drought. The annual seasonal rainfall during the summer monsoons of 2002 was below 19%, thereby causing a country-level meteorological drought since 1987 in the recent decades (IMD 2005). This further prompted IMD to course correct its drought monitoring, especially in short-term forecasting. The monsoonal variability over the northeast and southeast India is negatively correlated compared to central and northwest India (Kumar et al. 2013). The drought extremities potentially threaten the economic sectors like agriculture, manufacturing, and MSMEs, thereby disrupting the food and water security of the country.

Mapping India's Exposure to Cyclones (Tropical Cyclones)

India's eastern coasts are most exposed to tropical cyclones; however, in recent decades, there has been a surge in extreme cyclonic events across western coasts as well. The northwestern Indian oceans register lesser cyclonic disturbances compared to the northeastern Indian oceans (Mohapatra 2012), but the intensity of the tropical cyclones is quite high. Cyclones are classified under the tropical storm and are defined as "a tropical storm originating over tropical or subtropical waters." "Cyclones characterized by a warm-core, non-frontal synoptic-scale disturbance with a low-pressure centre, spiral rain bands, and strong winds. Depending on their location, tropical cyclones are referred to as hurricanes (Atlantic, Northeast Pacific), typhoons (Northwest Pacific), or cyclones (South Pacific and the Indian Ocean)" (Em-dat Gloassary). India has witnessed more than three tropical cyclones yearly in recent decades (Mohanty 2020). IPCC estimates that maximum wind speed in case of a tropical cyclone will increase in the range of 2–11% by the year 2100 (IPCC 2013a, b), while it is estimated that the wind speed is projected to increase by the same range across the north Indian Ocean as well (Mohapatra 2012). The impact of tropical cyclone is sustained for a specific window postlandfall, but its scale of devastation is compounded by the associated events

The State of Extreme Events in India, Table 3 Drought hotspot states at a glance (Source: Authors' analysis based on Mohanty and Wadhawan 2021).

S. No.	State exposed	Percentage of districts exposed to extreme flood events	Percentage of area undergoing a change in LULC
1	Uttar Pradesh	93	22.5
2	Maharashtra	91	20
3	Karnataka	97	25
4	Rajasthan	94	48.4
5	Madhya Pradesh	96	46
6	Tamil Nadu	100	40
7	Andhra Pradesh	100	43.4
8	Bihar	92	29
9	Gujarat	65	42.3
10	Jharkhand	100	83.3

(heavy rainfall, floods, hailstorms, cold waves, and tornadoes).

The pentad decadal analysis suggests that the eastern zone of India is highly exposed to extreme cyclone events followed by southern and western zones. India's northern and north-eastern zones face less exposure; however, the central zone faces no exposure (Fig. 4). The state-wise ranking of Indian States is derived from the pentad-decadal analysis inferred through frequency and intensity of cyclones extremes. There has been a four-fold increase in extreme cyclone events across the Indian states, whereas Andhra Pradesh, Maharashtra, and Bihar have registered a two-fold increase in extreme cyclone events, and Odisha alone has registered a four-fold surge in extreme cyclone events.

Further, the analysis suggests that 34% of the districts in the top ten hot spot states are prone to extreme cyclones that have undergone significant land-use-surface changes. The land-surface changes have triggered extreme cyclones in recent decades. The analysis also infers that 42% of the districts in these hot spot states have undergone rapid landscape disruption, triggering cyclonic disturbances and an exponential surge in these extremities. An increase in sea surface temperature tips to an upsurge in the diameter of the cyclone, which further intensifies the cyclonic disturbances. The number of storms with more than 100 mm of rainfall in a day is reported to have increased by 10% per decade (UNEP 2009). The analysis also suggests that there has been a 12-fold surge in associated cyclone events in the recent decades alone, thereby compounding the impact (Table 4).

The surge in associated cyclone events can be mapped to a higher level of inundation; further, an

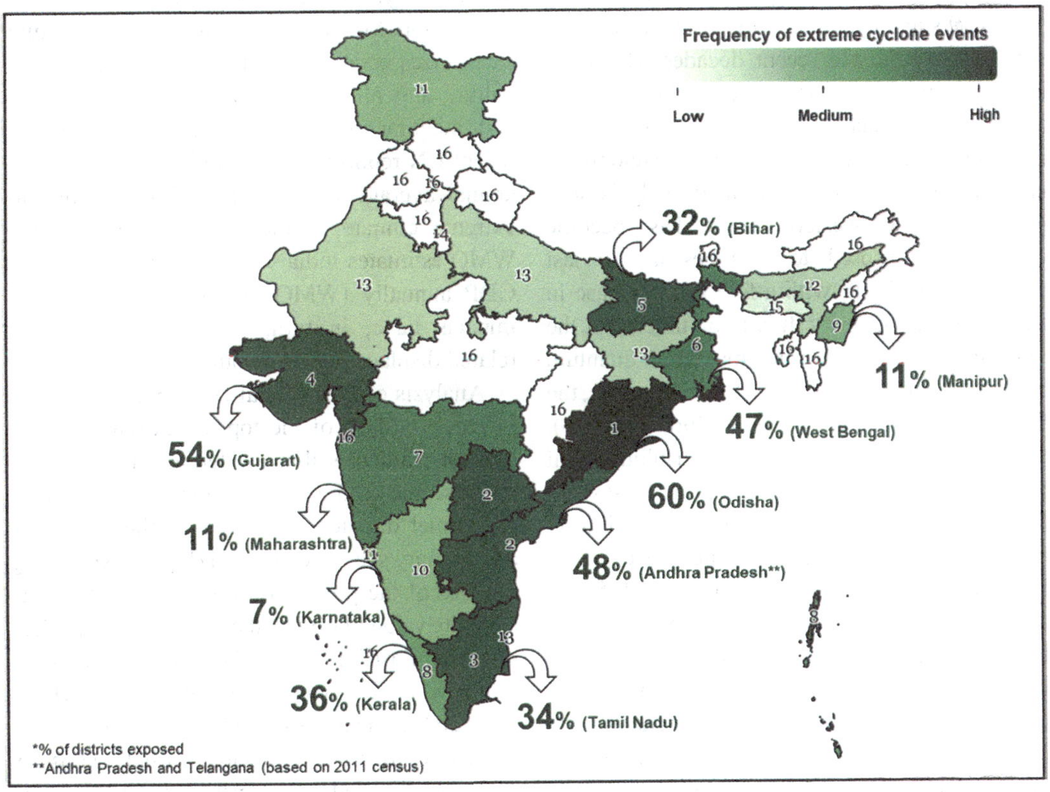

The State of Extreme Events in India, Fig. 4 Odisha is the most exposed to extreme cyclone events in India. (Source: Authors' analysis)

The State of Extreme Events in India, Table 4 Cyclone hotspot states at a glance

S. No.	State exposed	Percentage of districts exposed to extreme flood events	Percentage of area undergoing a change in LULC
1	Odisha	60	53
2	Andhra Pradesh	48	43.4
3	Tamil Nadu	34	40
4	Gujarat	54	42.3
5	Bihar	32	29
6	West Bengal	47	47.3
7	Maharashtra	11	20
8	Kerala	36	64
9	Manipur	11	55.5
10	Karnataka	7	25

Source: Authors' analysis based on Mohanty and Wadhawan (2021).

analysis by (Mohapatra 2012) confirms that Krishna, Godavari, Mahanadi, and Hooghly delta regions are prone to higher inundation levels under a climate change scenario. India has substantially improved the cyclone early warning post-1999 super cyclones, which has saved thousands of lives in the recent decades. However, insulating livelihoods and infrastructure is still a challenge that India is grabbling with as the spurt in cyclone extremities is ravaging its developmental trajectory. The pentad decadal analysis suggests that the western coast has become increasingly exposed to cyclones in the last decade (2010–2019) with a five-fold increase in frequency and intensity. It is to be noted that the State of Odisha witnesses the maximum quantum of loss and damage. Post-1999 Super Cyclone, the State witnessed Phailin (2013), Hudhud (2014), Titli, Phethai, and Daye (2018), Bulbul, Fani (2019), and Amphan (2020) incurring losses and damages worth more than USD 45 billion (TheHindu 2021). Island states and UTs of India, i.e., North and Middle Andaman, and South Andaman, are also exposed to extreme cyclone events. Changes in the forest management practices, increase in deforestation, reduction in forest cover, and agricultural practices aggravate the negative impacts of cyclones and even cause the onset of associated hazardous events such as inland flooding and landslides. Granular cyclonic risk assessment comprising of inundation levels across the eastern and western coast can help in strategizing and deriving hyperlocal developmental action plans for preparedness and mitigation.

Economic Sectoral Exposure and Loss and Damage to Extreme Climate Events

The Union Ministry of Statistics and Planning categorizes economic sectors into primary (agriculture and allied sectors), secondary (industry and manufacturing), and tertiary (services). A recent UN report notes that India has suffered an annual average loss of USD 87 billion due to extreme climate in the last two decades; the WMO estimates India's losses to exceed 0.5% of GDP annually (WMO 2020). These economic impacts have, in turn, triggered the disaster-related displacement of about 4.5 million people.

Analysis of the sectoral Gross State Domestic Product (GSDP) of the top ten extreme climate hotspots indicates that the tertiary, i.e., the services sector, is the most adversely impacted by hydro-met disasters, followed by the secondary and primary sectors combined (Fig. 5). Regarding the size of the population involved and the total absolute value of GSDP, the agriculture and allied sector is most impacted. These findings are consistent with those from various other studies (Kotz et al. 2022); the analysis additionally provides an indicative inference on sectoral GSDP exposure. A more detailed econometric analysis based on absolute numbers is needed to comment on India's overall GDP exposure to extreme climate events.

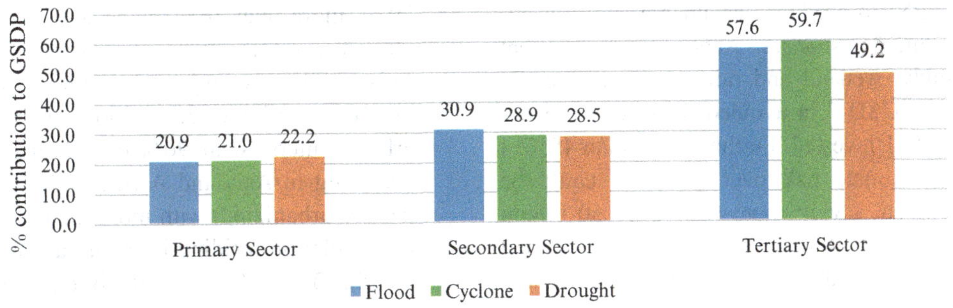

The State of Extreme Events in India, Fig. 5 Tertiary sector is the most adversely impacted by hydro-met disasters. (Source: Authors' analysis)

Building India's Resilience

The manuscript explicitly enumerates and generates empirical evidence on India's changing climate risk landscape. Given India's vulnerability, the following recommendations can mitigate the impacts of hydro-met disasters. As the chapter pans out, the thrust on climate-responsive risk financing, investing nature-based solutions, climate-proofed urban planning, and improved focus on impact-based public warning systems are some of the primary calls to action which can sustain the trinity of jobs, growth, and sustainability for a five trillion aspiring economic hub like India by climate-proofing its communities, economic sectors, and infrastructures. Applying the principles of risk assessment remains the core of any risk assessment process, and it should be applied to the broadest possible extent (CEEW 2015) to ensure further that the combination of developmental pathways is based on the pillar of jobs, growth, and sustainability.

(i) **Building India's resilience through improved public early warnings**: India has improved its early warning systems and public warning systems, but still, gaps need to be addressed through science and an evidence-based approach. As highlighted earlier, the granular risk assessments would be required to map, plan, and adapt better in a changing climate scenario. India needs to develop a high-resolution Climate Risk Atlas (CRA) that can map critical vulnerabilities (coasts, urban heat stress, water stress, crop loss, vector-borne disease, and biodiversity collapse) (Mohanty and Wadhawan 2021); this would help in having improved public awareness and information dissemination. A CRA further needs to be integrated to improve early warning systems that can project and predict these extremities at a short-term timescale. A comprehensive Climate Risk Atlas is needed with geo-tagged interfaces of critical infrastructure such as police and fire stations, hospitals, relief help desks, shelter houses, and warehouses. Integrating a public early warning system interfaced on a Climate Risk Atlas could have saved USD 89.7 billion with better disaster and emergency preparedness over the past two decades (Ghosh and Raha 2020). An effective impact-based early warning system can prepare lives and livelihoods and design infrastructures that can sustain climate extremities.

(ii) **Investing in nature-based solutions**: India needs to include nature-based infrastructure like wetlands, mangroves, and forest ecosystems, among others, under the ambit of critical infrastructure. Built-in infrastructure like buildings, roadways network systems, electric systems, dams, and bridges are currently considered under critical infrastructure's standard definition and practice. Broadening the definition of infrastructure to include natural ecosystems offers an opportunity to deploy and enhance nature-based solutions

S

(NBS) to produce sustainable and climate-resilient responses. According to a recent study, wetland and ridge restoration could save USD 7 in avoided damages for every USD 1 invested. Further, more than 45% of the climate risk over 20 years could be averted, saving more than USD 50 billion worth of damages against extreme flood events (Luedke 2019). Restoring, rebuilding, and investing in nature-based solutions can make our cities and villages more climate-resilient and alter the adverse impacts of climate change.

(iii) **Mainstreaming climate-response risk insurance**: Mainstreaming and promoting climate-responsive risk will fast-track the implementation of climate-resilient infrastructures and mobilize both public and private investments. Weather-related insurance losses have increased by five times to USD 55 billion since the 1980s, globally (WMO 2021). Insurance premiums are also rising steeply, especially in the most climate-vulnerable regions. Climate-responsive risk insurance calls for financial innovations that integrate physical climate impacts into investment decision-making. Climate-responsive risk insurance can reduce the cost of financing large-scale infrastructural projects and offer effective risk transfer mechanisms such as climate risk insurance, resilience bonds, and global risk investment pools. An efficient risk transfer mechanism will insulate both the public and private sectors from losses incurred due to the impacts of climate change. It can enhance Public-Private Partnership (PPP) models to finance climate-proofing projects based on effective risk transfer mechanisms. Further, the G20 countries, together with large private financial institutions, should incorporate climate-responsive risk insurance pools that can also provide a financial buffer against severe climate shocks and ensure building back better and forward.

(iv) **Climate-proofed urban planning**: According to the World Cities Report 2020, more than 56% of the world's population currently lives in urban regions. This number is projected to rise to 68% by 2050, tallying 2.5 billion people to urban centers, with 90% of this increase occurring in Asia and Africa. India too is rapidly urbanizing, with more than 35% of its population residing in the urban centers (World Bank 2022). In its Budget 2022, the government of India has also substantially increased the fiscal allocation to support urbanization-based development. Given the quantum of urban investment, India needs to make cities resilient by enhancing the urban green space that can act as a natural shock absorber against climate extremities and building climate action plans at a municipal level that integrates, promotes, and implements granular risk assessments, improved critical infrastructures, and urban policies to promote a better quality of life against both extreme and slow-onset events.

References

ANI, NewsVoir. 2021. India eyes USD 700 billion investments to boost urban infrastructure. 01 November. https://www.aninews.in/news/business/business/india-eyes-usd-700-billion-investments-to-boost-urban-infrastructure20211101121556/

Bansal, N. K., & Minke, G. (1988). *Climatic zones and rural housing in India. Part 1 of the Indo-German project on passive space conditioning*. Germany: N. p., Web. https://www.osti.gov/etdeweb/biblio/7784799.

Beck, H., Zimmermann, N., McVicar, T. et al. Present and future Köppen-Geiger climate classification maps at 1-km resolution. *Sci Data 5*, 180214 (2018). https://doi.org/10.1038/sdata.2018.214

Bhatnagar, M., Mathur, J., & Garg, V. (2019). *Climate zone classification of India using new base temperature*. IBPSA. Retrieved from http://www.ibpsa.org/proceedings/BS2019/BS2019_211159.pdf

Dhar, O. N. (2003). Hydrometeorological aspects of floods in India. *Natural Hazards, 28*, 1–33. https://doi.org/10.1023/A:1021199714487.

Eckstein, D., Künzel, V., & Schäfer, L. (2020). Global Climate Risk Index. Retrieved from https://www.germanwatch.org/sites/default/files/Global%20Climate%20Risk%20Index%202021_2.pdf

EM-DAT. (2022, January). The CRED/OFDA International Disaster Database. Retrieved from Available online: http://www.emdat.be/

EM-DAT Gloassary. The CRED/OFDA International Disaster Database. Retrieved from Available online: https://www.emdat.be/Glossary#letter_f

Ghosh, A., & Raha, S. (2020). *Jobs, growth and sustainability: A new social contract for India's recovery*. New Delhi: Council on Energy, Environment and Water: CEEW and NIPFP. Retrieved from https://www.ceew.in/publications/jobs-growth-and-sustainability

IMD. (2005). Climatological features of drought incidences in India. https://imdpune.gov.in/hydrology/Drought/drought.pdf

IMD. (2020). *End of the Season – Southwest Monsoon 2020*. New Delhi: PIB, GoI, Ministry of Earth Science. Retrieved from https://static.pib.gov.in/WriteReadData/userfiles/End%20of%20Season%20Report_2020.pdf

IPCC. (2007). Climate Change 2007: The physical science basis. Contribution of working group I to the fourth assessment report of the intergovernmental panel on climate change. https://www.ipcc.ch/site/assets/uploads/2018/03/ar4_wg2_full_report.pdf

IPCC. (2013a). Climate phenomena and their relevance for future regional climate change, Sec 14.6. Chapter 14: 1248–1251.

IPCC. (2013b). Climate phenomena and their relevance for future regional climate change – supplementary material, Sec 14.SM.4, pp 14SM-8–14SM-9.

IPCC. (2014). Climate Change 2014: Synthesis Report. Contribution of Working Groups I, II and III to the Fifth Assessment Report of the Intergovernmental Panel on Climate Change. https://www.ipcc.ch/site/assets/uploads/2018/05/SYR_AR5_FINAL_full_wcover.pdf

Kamaljit Ray, P. P. (2019). On the recent floods in India. *Current Science, 117*. https://doi.org/10.18520/cs/v117/i2/204-218.

King, D., Schrag, D., Dadi, Z., Ye, Q., & Ghosh, A. (2015). *Climate change: A risk assessment*. New Delhi: Council on Energy, Environment and Water. https://www.ceew.in/publications/climate-change

Kotz, M., Levermann, A. & Wenz, L. (2022). The effect of rainfall changes on economic production. *Nature 601*, 223–227. https://doi.org/10.1038/s41586-021-04283-8

Kumar, K., Rajeevan, M., Pai, D., Srivastava, A., & Preethi, B. (2013). On the observed variability of monsoon droughts over India. *Weather and Climate Extremes, 1*, 42–50. https://doi.org/10.1016/j.wace.2013.07.006.

Luedke, H. 2019. *Nature as resilient infrastructure – An overview of nature-based solutions*. October. https://www.eesi.org/papers/view/fact-sheet-nature-as-resilient-infrastructure-an-overview-of-nature-based-solutions

Mohanty, A. (2020). *Preparing India for extreme climate events: Mapping hotspots and response mechanisms*. New Delhi: Council on Energy, Environment and Water. Retrieved from https://www.ceew.in/publications/preparing-india-for-extreme-climate-weather-events

Mohanty, A., & Wadhawan, S. (2021). *Mapping India's climate vulnerability – A district-level assessment*.

Mohapatra, M. (2012). Classification of cyclone hazard prone districts of India. *Natural Hazards*. https://doi.org/10.1007/s11069-011-9891-8.

Rajeevan, M., Unnikrishnan, C. K., & Preethi, B. (2012). Evaluation of the ENSEMBLES multi-model seasonal forecasts of Indian summer monsoon variability. *Climate Dynamics, 38*(11–12), 2257–2274. https://doi.org/10.1007/s00382-011-1061-x.

Rao, S. D. (2012). Why is Indian Ocean warming consistently? *Climatic Change, 110*, 709–719. https://doi.org/10.1007/s10584-011-0121-x.

Ray, K., Prabha, P., Pandey, C., Dimri, A. P., Kishore, K. (2019). On the Recent Floods in India. Current science. 117. https://doi.org/10.18520/cs/v117/i2/204-218.

Silver, C. (2021). The top 25 economies in the world. *Investopedia*. 22 December. https://www.investopedia.com/insights/worlds-top-economies/

The Hindu. (2021, March). *Odisha suffered losses worth ₹31,945 cr. in eight cyclones*. Retrieved from https://www.thehindu.com/news/national/other-states/odisha-suffered-losses-worth-31945-cr-in-eight-cyclones/article34185730.ece

UNEP. (2009). Recent trends in melting glaciers, tropospheric temperatures over the Himalayas and summer monsoon rainfall over India. https://na.unep.net/siouxfalls/publications/himalayas.pdf

WMO. (2020). The State of the Climate in Asia 2020. Retrieved from https://library.wmo.int/doc_num.php?explnum_id=10867

WMO. (2021). Weather-related disasters increase over past 50 years, causing more damage but fewer deaths. https://public.wmo.int/en/media/press-release/weather-related-disasters-increase-over-past-50-years-causing-more-damage-fewer

World Bank. (2022). Urban population (% of total population) – India. https://data.worldbank.org/indicator/SP.URB.TOTL.IN.ZS?end=2020&locations=IN&start=1960&view=chart

S

Statute

► Regulation of Urban and Regional Futures

Stepping-Stone

► Habitat Provisioning

Stewarding Street Trees for a Global Urban Future

Paris, Taipei, and Washington, DC

Theodore S. Eisenman[1], Shenglin E. Chang[2] and Lucie Laurian[3]
[1]Department of Landscape Architecture and Regional Planning, University of Massachusetts-Amherst, Amherst, MA, USA
[2]National Taiwan University, Graduate Institute of Building and Planning, Taipei, Taiwan
[3]School of Planning and Public Affairs, The University of Iowa, Iowa City, IA, USA

Synonyms

Climate change adaptation; Green cities; Green infrastructure; Public realm; Stewardship; Streets; Urban commons; Urban design; Urban forests; Urban greening; Urban nature

Introduction

Street trees are one of the most prominent types of plants in the urban public realm. They define the street corridor, humanize the scale of cities, calm traffic, separate walkers from vehicles, and filter sunlight all while softening the urban fabric and introducing beauty in the form of flora. Importantly, trees can transform streets from utilitarian transportation corridors into places in which people want to be (Massengale and Dover 2014). This is especially important as human beings become an increasingly urban species; 2008 marked the first time that more people worldwide lived in urban than rural areas, and by the end of this century some three-quarters of humanity is projected to live in cities (Angel 2012), leading the contemporary era to be described as the "first urban century" (Hall and Pfeiffer 2000, p. 5).

In this dawning age of cities (Young and Lieberknecht 2019), people spend the vast majority of their time indoors (Brasche and Bischof 2005; Klepeis et al. 2001). Streets are by extension one of our most common experiences of outdoor settings, and these "travelscapes" represent an excellent opportunity to provide urban populations with the health and well-being benefits of nature contact, as evidenced by a robust body of literature (Frumkin et al. 2017; Hartig et al. 2014; Kuo 2015). This dovetails with increasing interest in urban greening, defined as a social practice of organized or semi-organized efforts to introduce, conserve, or maintain outdoor vegetation in urban areas (Eisenman 2016b; Roman et al. 2020). Greening includes a range of initiatives, policies, and incentives to vegetate the landscape of cities (Beatley 2016; Tan and Jim 2017), and it often includes ambitious tree planting initiatives (Eisenman et al. 2021; Nguyen et al. 2017; Young 2011). Of note, the systematic citywide planting of trees along streets was not common in most European and North American cities until the late nineteenth and early twentieth centuries (Campanella 2003; Dümpelmann 2019; Laurian 2019), but it has since become commonplace around the world (Lawrence 2006).

Yet, the actors and norms that guide street tree planting and management can vary in different cultural contexts. In North America, for example, urban forestry has traditionally focused on street trees, whereas European definitions of urban forestry relate more to forest ecosystems such as woodlands in or near cities (Konijnendijk et al. 2006). One study found substantial differences in why and how municipal leaders in North America and Scandinavia conduct inventories of urban trees. In both places, street trees figured prominently in urban forest inventories, and study participants mentioned operational planning and arboricultural maintenance as important rationales for this work. However, in North America citizen volunteers were important actors in conducting urban tree inventories, and this volunteer work may have spurred subsequent citizen engagement in local urban forestry activity. North American cities also emphasized a range of economic, environmental, and social benefits of urban trees as rationales for conducting inventories. In Scandinavian cities, by contrast, these benefits were not mentioned or recognized as important rationales for conducting urban tree inventories, nor did citizen volunteers participate in this work (Keller and Konjijnendijk 2012).

International dimensions are also important considerations when accounting for street tree planting and stewardship. While research suggests a basis for universal landscape preferences predicated on a shared evolutionary past (Appleton 1975; Ulrich et al. 1991), and studies consistently show reductions in stress when people have contact with vegetated landscapes (Frumkin et al. 2017; Hartig et al. 2014; Kaplan and Kaplan 1989)—including local trees (Suppakittpaisarn et al. 2019)—people have different perceptions of, and preferences for, urban trees (Konijnendijk 2008; Zhao et al. 2017). The same holds true for street-level vegetation. In Sapporo, Japan, for example, researchers found that people preferred sidewalk planting beds of flowers *without* trees over similar planting beds *with* trees (Todorova et al. 2004). By contrast, a study spanning four cities in the Netherlands found a strong preference for large trees along streets (Van Dongen and Timmermans 2019), while a study in Australia found that homes on streets with more than six different street tree species had reduced sale prices, suggesting a threshold beyond which people in this place will accept a diversity of street tree types (Plant and Kendal 2019). In Hong Kong, 94% of survey respondents supported street tree planting, but the most preferable streetscape attribute was high visual permeability (the openness of the street), suggesting that street trees should not be too large or too densely spaced.

International differences extend beyond landscape vegetation preference. For example, a comparative analysis of five capital cities in countries spanning three continents found substantial differences in street tree density and distribution; moreover, differences between cities in the same climate zone suggest that place-specific cultural dimensions such as urban form, aesthetic norms, and governance regimes are important factors in the density and distribution of urban street trees (Smart et al. 2020). People within a given city can also hold different perceptions of—and receptivity to—tree planting campaigns. In Detroit, Michigan, many neighborhoods targeted for street tree planting resisted such efforts, and this was explained by a lack of "procedural justice" and differing "heritage narratives" (perceptions of local history) between local residents and tree planting advocates (Carmichael and McDonough 2019).

The aforementioned distinctions illustrate the importance of comparative research on street tree planting and management, especially as greening (and associated constructs such as green infrastructure, ecosystem services, and nature-based solutions) becomes a common approach to planning for twenty-first-century cities worldwide. Unlike noncomparative research, comparative scholarship seeks to illuminate differences and similarities between the objects of analysis—in this case street trees—and their contextual conditions, such as culture and nationality. Comparative research can also illuminate the embedded customs and assumptions of a given place, which is especially important if they are taken to be universal (Esser and Vliegenthart 2017; Lewis-Beck et al. 2004). This is noteworthy in a globalizing world characterized by the widespread diffusion of information, values, and norms (Castells 1996). Vernacular distinctions are also important in an urban environmental discourse that is significantly influenced by Anglo-American and European tradition (Anguelovski and Martínez Alier 2014; Eisenman 2016a; Ernstson and Sörlin 2019).

This chapter seeks to enrich this conversation by offering brief case studies and comparative analysis of the typical actors and practices related to stewardship of urban street trees in three cities on different continents: Paris, France; Taipei, Taiwan; and Washington, DC in the USA. Each of these cities is the capital of their respective countries, so each subsection opens with a brief narrative addressing national and historic context. Each of these cases addresses both mature and newly planted street trees; and the respective cases draw upon a combination of academic literature, professional documentation, and select interviews with local experts.

Paris

National and Historical Context

The French tree-lined street and boulevard model was diffused throughout Europe and the Americas in the eighteenth and nineteenth centuries, and it is highly influential to this day. Within France, tree planting and management practices diffused from Paris to the provinces, reinforcing the special emphasis on Paris in this section. Until the nineteenth century, street tree planting decisions were made by kings and nobility. In Paris, Kings Charles V (fourteenth century), Henri IV's queen, Marie de Medici (seventeenth century), and Louis XIV and Louis XV (seventeenth and eighteenth centuries) had rows of elm, plane, linden, and mulberry trees planted at regular intervals along select streets, canals, ramparts, and boulevards (Dorion 2014; Lavedan 1993; Lawrence 1993, 2008). The first tree-lined promenades and boulevards include Henri IV's tree-lined mails and Marie de Medici's Cours-la-Reine. The latter, planted in 1628 in continuation of the Tuileries Garden, comprised three long allées lined with 1600 elms planted four meters apart. This has been described as creating the first urban tree canopy over pedestrians and vehicles (Bergeron 1989; Forrest 2002; Forrest and Konijnendijk 2005).

The French Revolution of 1789 shifted power over urban trees. Beyond the 60,000 Liberty Trees (mainly oaks and poplars) planted throughout the country as a political symbol, the Revolution laid the foundation for municipal governance. Since the nineteenth century, French urban tree planting and management has been under the purview of municipal agencies. An exception, however, is Paris, which remained under national control until 1977. Prior to this, the capital city was managed by prefects appointed by kings, emperors, and presidents, including Claude-Philibert Barthelot, Count de Rambuteau, and George-Eugène Haussmann who expanded the tree-lined boulevard model throughout the city (Jones 2006; Laurian 2019; Lawrence 2008).

Contemporary Paris

Today, Paris' street trees are managed by the Service of Trees and Woodlands (Service de L'Arbre et des Bois) of the Municipal Direction of Green Spaces and the Environment (Direction des Espaces Verts et de l'Environnement, DEVE). The DEVE answers directly to the mayor and city council (Ville de Paris 2019a). The agency's 3100 employees manage trees, including street trees and trees in more than 500 green spaces, two woodlands, a municipal nursery, 20 cemeteries, sports centers, and primary schools, with a €33 million ($38.9 million) operating budget in 2018 (Ville de Paris 2018a). Its staff includes planners, public outreach specialists, landscape designers (aménagement paysagers), and arborists (arborists-élagueurs), many of whom are certified arborists trained at the Paris School of Horticulture and Arboriculture (Ecole du Breuil des Arts et Techniques du Paysage). The DEVE partners with other municipal agencies, e.g., on the Paris Climate Plan, and with national agencies, e.g., the National Agency for Biodiversity.

While Paris' street tree planting is solely undertaken by municipal DEVE staff and funded through the municipal budget, the city also implements participatory programs. The Green Hand program (Main Verte), launched in 2003, supports 134 resident-led community gardens. Through the Greening Near My Home program (Du Vert Près de Chez Moi), launched in 2014, residents can suggest greening interventions for specific sites in their neighborhoods (Ville de Paris 2020a). Of 1500 proposals, 209 have been selected for implementation thus far and these include green walls, potted plants, and additional tree plantings conducted by municipal services. Residents can also apply for innovative Greening Permits (Permis de végétaliser) which allow them to garden in public spaces on sidewalks. Residents can install potted plants or grow micro-gardens, typically flowers and herbs, in street trees' planting beds (see Fig. 1). Permit holders are responsible for planting, watering, and maintenance, and they must publicly post their permit. A dedicated online interactive map provides the list, location, and photos of these resident-led projects (Ville de Paris 2020b).

Tree inventories and numerical tree planting goals drive urban forestry practices in French municipalities, including Paris, which aimed for 20,000 additional trees along streets and in parks

Stewarding Street Trees for a Global Urban Future, Fig. 1 Resident-led greening of street tree planting beds. (Sources from left to right: *1* Ville de Paris, H. Jarry, https://www.paris.fr/pages/un-permis-pour-vegetaliser-paris-2689; *2* Ville de Paris, Victor Connan, https://www.paris.fr/pages/un-permis-pour-vegetaliser-paris-2689; *3* Lucie Laurian; *4* Lucie Laurian)

and gardens between 2014 and 2020 (15,000 were added as of 2019). Of note, the newly reelected mayor ran on an ambitious platform of 170,000 more trees between 2021 and 2027, many of which will presumably be planted on streets. Currently, the city has on average 4.9 trees for every 100 m of street, but trees are not evenly distributed across street types: Collector streets have nearly three times as many trees as local streets (Smart et al. 2020). This is due to the narrow width of many streets in Paris, whose underlying settlement dates back some two millennia (Bournon 1888).

Programs to increase Paris' tree counts date back to Haussmann's projects and have steadily increased since then. The city had 38,000 trees in its first inventory in 1855, 88,000 by the end of the nineteenth century (Landau 1992), 96,000 in 1993, and 106,000 in 2020. In total, Paris is home to 504,000 trees: 106,000 street trees, 48,000 trees in 490 parks and gardens, 32,000 in cemeteries, 6000 along the périphérique highway, 7000 trees in municipal schools and day care centers, 4000 in sports complexes, and 300,000 in two woodlands. Since 2014, Paris' award-winning tree inventory—Paris Arbres Opendata, available online—includes for each tree the species and genus, planting date/age, size, health conditions, watering, pruning, and removal schedule. The database tracks data in real time, and is used for planning, analysis, and public information. The city also maintains a separate inventory of trees of special significance (*arbres remarquables*), noteworthy for their historical significance or morphology.

Paris' street trees are grown in a municipal nursery, the 44 ha municipal Horticulture Center, which provides about 80% of the city's plants and trees, meaning that Paris controls its tree source and supply. Street trees are planted when they are 5–10 years old in about $12 m^3$ of soil. Once planted, trees are staked, watered, and regularly pruned for 3 years. After this, trees are pruned to clear traffic signals, and remove low branches and dead limbs. All Paris' trees are inspected annually and one-fifth of trees receive a detailed diagnostic, the results of which are noted in the Arbres Opendata inventory.

Best management practices (BMPs) in urban tree management are implemented under the guidance of several charters. Paris signed the Regional Charter on Biodiversity and Natural Milieus in 2004, which commits the DEVE to supporting regional flora, fauna, and natural habitats, reducing mowing, introducing ponds and wetlands, planting native species, and limiting herbicides and pesticide use. The charter also commits the DEVE to considering street trees as living species rather than formal elements of urban design—which represent an important shift in ontological framing—leading to guidelines for planting diversified and native species, reducing pruning, adapting planting and maintenance to each species, and tracking tree maintenance. Paris' parks and gardens can also qualify for the Ecological Green Spaces label (Espace Verts Ecologique).

This designation implements the 1994 Aalborg Charter for European Sustainable Cities, the 2004 Regional Charter on biodiversity and natural milieus, the 2017 Paris Climate Plan (Plan Climat, Air, Energie), and the 2018 Paris Rain Plan (Plan Paris Pluie). For street trees, this translates into reduced pruning and chemical applications, and providing larger naturalized tree planting beds with native grasses and wildflowers where possible (Laurian 2012).

Beyond inventories and BMPs, the century-old practice of creating linear monocultures and regularly spaced street tree alignments (*arbres d'alignement*) has a long-lasting legacy. Original tree alignments generally relied upon one species to ensure formal regularity, and trees were planted all at once to ensure similar sizes. The most common species were elm (*Ulmus minor*, *campestris*, *pumila*, and, to a lesser extent, *sapporo gold* and *americana*), planetree (*Platanus acerifolia*), and linden/lime (*Tilia*) trees selected for their fast growth, wide canopies, and resistance to urban constraints and heavy pruning (Ville de Paris 2019b). Today, 58% of all street segments in Paris remain single-species (Ville de Paris 2019a). Alexandre Jouanet, head of the Service des Arbres et Plantations under Haussmann, led early diversification efforts: Half of the street trees his agency planted were planes and elms, but he diversified the mix with horse chestnuts (*Aesculus hippocastanum* and *Aesculus hippocastanum baumaii*), American walnuts (*Juglans nigra*), tree of heaven (*Ailanthus altissima*), and pagoda trees (*Sophora japonica*). Today, 37% of Paris' street trees are planes, 15% horse chestnuts, 10% linden, 10% pagoda trees, 3% maples, and 3% ash (Atelier Parisien d'Urbanisme 2010).

Diversification occurs with the tree replacement cycle. The Paris DEVE replaces 1500 street and 1500 park trees annually. Newly planted trees include 190 species, including regionally native species and Mediterranean species adapted to climate change. The Paris 2018–2024 Biodiversity Plan guides tree selection and management (Ville de Paris 2018b). It also highlights ecosystem functions provided by urban trees and associated goals related to climate change mitigation and adaptation (especially heat waves and urban heat island effects); air quality; stormwater runoff management; support for pollinators and wildlife; and biodiversity goals set in the 2016 National Law on Biodiversity, Nature, and Landscapes (*Loi pour la reconquête de la biodiversité, de la nature et des paysages*) and the 2009 Regional Ecological Plan (*Schéma Régional de Cohérence Écologique*). This is consistent with the National Environment Agency (*Agence de l'Environnement et de la Maitrise de l'Energie*) which refers to urban trees as "climate actors" (ADEME 2018).

The 2018–2024 Biodiversity Plan has also set a goal to assess Paris' canopy cover, and then to increase it by 1% by 2024 and 2% by 2030. This falls short of setting an actual numerical canopy cover goal (other French cities, in contrast, have adopted canopy cover goals, e.g., Lyon at 30% by 2030). Paris's canopy cover provided by street trees (excluding all parks and gardens) varies across district, from 0.5–3.5% when dividing the street tree canopy cover by each district's total land area, and from 2–11% when dividing the street tree canopy cover by the district's street area, i.e., excluding buildings' footprint (Atelier Parisien d'Urbanisme 2010). The MIT Green View Index (GVI), on the other hand, assesses the pedestrian perspective based on Google Street View panoramas. Among the 27 large cities investigated using this method, Paris has the lowest GVI at 8.8% (MIT Senseable City Lab 2020). Given Paris' very high density—over 20,000 residents/km^2 compared to London (4500/km^2), Amsterdam (4900/km^2), Berlin (3800/km^2), and New York City (10,200/km^2)—increasing canopy cover will require creative solutions.

In 2019, Mayor Anne Hidalgo announced the creation of new "urban forests" with 2000 trees set to be planted at key landmark locations: in front of the Hotel de Ville, behind the Opera Garnier, at Gare de Lyon, and along the Seine. This is predicated on goals to reduce urban temperature and to decrease the amount of impervious cover (O'Sullivan 2019a). Similar projects such as the 1993 Coulée Verte, which transformed a 4.7 km of railroad tracks into a linear garden, and the 1994 Jardin Atlantique with 150 trees planted above railroad tracks, suggest that the new tree planting campaign can be successful. The city is also removing asphalt (12.5 ha removed by 2020) to increase permeability, in concert with the Paris Rain Plan. These stormwater infiltration projects, often in schools and street medians, create new tree-planting opportunities (Ville de Paris 2019b).

Planting and managing street trees in Paris presents distinct challenges beyond urban density and underground utility and subway infrastructure. Some urban spaces were designed with the explicit exclusion of trees to preserve uninterrupted views of certain monuments and Beaux Arts facades, e.g., Place des Victoires, Place Vendôme, Rue de Rivoli, and Avenue de l'Opéra. Popular pressure could change this. For instance, Place des Vosges was designed without trees in 1605 and its first trees were planted 200 years later at residents' requests. In addition, tree pruning and shaping practices have strong cultural roots and values (see Fig. 2). In France, as in other European countries, linden and plane trees are heavily pollarded, a practice of removing the upper branches of a tree (Pacini 2007). This reduces trees' height and crown size, and can give form to outdoor spaces, e.g., linear edges delineating allées with "walls," creating "rooms" and "curtain" effects, and dense canopies that create outdoor "ceilings." Pollarded linden and plane trees are also a staple of French gardens and squares, e.g., at the Palais Royal and Jardin des Plantes, but extreme pruning practices are increasingly challenged today (Toussaint et al. 2002). This highlights the extent to which urban street trees in Paris and France are increasingly understood today as living organisms that serve a range of goals including biodiversity, sustainability, urban design, and cultural heritage.

Taipei

National and Historical Context

Historically, Taiwanese society has a long tradition of stewarding trees in public places such as temple squares. Long-lived trees and those associated with local legends have even been revered as holy or god-like. In some cases, villagers built small temples to worship tree spirits and pray for more prosperous lives for individuals, families, or the community. However, for contemporary Taiwan, trees became commonplace elements of the urban streetscape during Japanese colonial rule between 1895 and 1945.

During this colonial period, Taiwanese culture and urban form were heavily influenced by Japan, and the entire island (395 km long and 145 km across at its widest point) essentially served as a design laboratory for Japanese architects and urban designers trained in the West. Street trees became important urban design elements during this early twentieth-century period of Japanese

Stewarding Street Trees for a Global Urban Future, Fig. 2 Pruning and pollarded street trees in Paris. Left image source: Ville de Paris, Pierre Viguié: https://www.paris.fr/pages/chancre-colore-du-platane-paris-sous-surveillance-7476. Right: Pollarded lane trees https://pixabay.com/fr/photos/paris-france-trottoir-arbres-hiver-90938/. Licence: Pixabay (Free for commercial use, no attribution required)

rule (Tashir 1920). Initially, four types of trees gained special prominence for street planting: *Salix glandulosa var. warburgii, Alnus formosana, Pandanus otdoratissimus,* and *Bambusa stenostachya* (Ao 2000). But by the 1920s, more than 50 types of trees were commonly planted along Taiwanese streets. These plantings were noteworthy elements—symbolically and in practice—of a broad movement by the Japanese colonial government to modernize Taiwanese cities.

Nationwide today, trees along major highways are managed by the Federal Ministry of Transportation and Communication (MOTC). However, urban street tree planting and stewardship in Taiwan is managed at the municipal level, and each city (often in collaboration with county administrators) prepares management plans that are endorsed by local elected councils. The island spans humid subtropical and tropical climate zones, and has moist, hot summers from May to October, with rainstorms and occasional typhoons, and average high temperatures in July of 34 °C. Taiwan also has a strong cultural tradition of socializing outdoors. In light of these combined factors, street trees and shaded parks and plazas are highly valued.

Contemporary Taipei

In Taiwan's capital, Taipei, street trees are the sole responsibility of the Horticultural Engineering Team (HET) of the Park and Street Lights Office (PSLO) in the Public Works Department (PWD). Of the 196,000 trees on public land in Taipei City, roughly 89,000 are street trees; due to recent plantings, this is an increase from 88,000 street trees in 2017 (Taipei City 2019). Of these trees, most are individually tagged and registered in a central database (see Fig. 3). The HET is responsible for the daily management of these trees, but this office also subcontracts urgent tree pruning activities to private contractors during the typhoon season from July to September. In 2020, the PSLO allocated 18.1 million USD for all matters related to horticultural management of street trees, parks, and open spaces (R. Mo, personal communication, August 20, 2020).

The HET currently includes 250–260 staff members who manage the city's street trees based on area quadrants (east, west, north, and south), each of which is managed by a section leader. This includes some 150 trained arborists who do most of the hands-on work including pruning, weeding, fertilizing, and disease control (Taipei City 2017). Taiwan has three different systems for training and certifying arborists: municipal level, federal level, and through the Taiwan Arboriculture Society which is based on standards developed by the International Society of Arboriculture (ISA). In the case of Taipei, the city recruits entry-level applicants through written and physical tests, after which they proceed through two levels of training and certification. In addition to certifying HET staff, subcontractors

Stewarding Street Trees for a Global Urban Future, Fig. 3 Street trees in Taipei tagged and recorded by the city. (Source: Theodore S. Eisenman)

from private companies can also enroll in HET-led certification classes, which allows contractors to work on the city's tree management projects.

Both the city of Taipei and the Federal Forest Bureau publish tree trimming and maintenance guidelines for arborists to follow. Historically, this has been especially important prior to and during the annual typhoon season from roughly June to October. However, climate change is altering seasonal patterns, and typhoon-scale storms are becoming increasingly common throughout the year. This is creating maintenance challenges for the municipality, leading the city to initiate efforts to broaden the network of actors who steward trees. As of 2015, for example, Taipei allows schools, neighborhood leaders, nonprofit organizations, private companies, and individuals to adopt trees along streets and in parks and other public spaces; but most adoptees are private companies and local leaders (*lizhang*) of neighborhood groups called *li*. A distinct aspect of Taiwanese society is the establishment of formal neighborhood groups at the subdistrict level called "li," each of which has an elected leader called a "lizhang." In Taipei, there are 12 districts and 456 lizhang. The aforementioned tree adoption program consists of watering, weeding, fertilizing, monitoring tree health, and reporting triannually to the HET. If qualified adoptees do not meet certain management criteria, the HET can remove them. In 2016, 455 agents adopted trees in 440 locations across the city including parks, open spaces, and streets. The municipality estimates that this saved the city about 1.7 million USD (Xiao 2016).

In addition to the aforementioned voluntary stewardship, the city's efforts to maintain street trees can create disputes among citizens, city officials, and other stakeholders. As tree canopies grow, they can block street lights, requiring pruning to maintain sightlines and associated traffic and pedestrian safety. Many shop owners also believe that trees in front of their stores do not align with the spatial design principle of *feng-shui,* one of which holds that doors and passages should remain open, as this brings prosperity. This often requires tree planting teams to compromise with shop owner requests to move tree planting holes from directly in front of store entrances, even when the trees are located across several lanes of traffic in planted medians. To facilitate response to citizen complaints, the city provides a reporting system by phone and Internet. In 2015, the PSLO also launched a web-based mapping program and public tree database that allows people to monitor street trees, street lights, and related street furniture (Taipei City 2015).

Taipei has formal tree management guidelines based on biological characteristics and site context (Hsu 2010). This is important in a city with such a diversity of streetscape types (see Fig. 4). In the downtown area, for example, sidewalks are often up to 8 m (24 ft) wide and accommodate a range of uses including dedicated bike and pedestrian lanes, parking for mopeds, benches for sitting, bus stops, and single/double rows of trees and/or planting beds. Arterial streets often include landscaped medians planted with ground cover and trees while many of the sidewalks along local streets throughout the city are quite narrow, making tree planting difficult.

Of note, these guidelines stipulate that any street wider than 8 m should be planted with trees; but for sidewalks narrower than 2 m, no new trees should be planted or replaced. Small planting beds are installed in sidewalks between 2.5 and 3 m wide, and larger planting beds are installed in sidewalks wider than 3 m. The guidelines also identify 39 species as the top choices for street tree planting. This list is based on ten criteria: capacity to withstand air pollution; survival rate; air filtration capacity; attracting birds, butterflies, and other species; avoiding fallen fruits and leaves; avoiding pollen allergy; avoiding shallow and far-spreading root systems; strong and resilient branches to survive typhoons and severe winds; high pest tolerance and low risk for illness; and providing shade. Some of the more common street trees in Taipei include chinaberry or Indian bead tree (*Melia azedarach*), Toog tree or bishop wood (*Bischofia javanica),* camphor tree *(Cinnamomum camphora*), Japanese bay tree *(Machilus thunbergia),* and orchid tree (*Bauhinia variegate*).

This planting list was updated in 2014, and it also includes six trees to be avoided for new

Stewarding Street Trees for a Global Urban Future, Fig. 4 Range of streetscape types in Taipei. (Source: Top left, top right, and bottom left, Theodore S. Eisenman. Bottom right: Shenglin Chang)

planting and replacement due to a range of factors including shallow root systems, pollen allergenicity, dropping fruit, and fast-growing weak limbs. Of note, these trees were commonly planted during Japanese colonial rule and in the late twentieth century thereafter, and include nanyan or Indian laurel (*Ficus macrocarpa*); sacred fig or bodhi tree (*Ficus religiosa*); rubber tree (*Hevea brasiliensis*); yellow poinciana (*Peltophorum pterocarpum*); cotton tree (*Bombax ceiba*); and coral tree or tiger's claw (*Erythrina variegatea*). Inclusion of *Ficus macrocarpa* is particularly noteworthy, as this fast-growing tree is ubiquitous across Taipei due to widespread planting in the 1980s.

The aforementioned voluntary stewardship of street trees in Taipei also reflects a cultural affection for flora as well as a blurry line between the public and private realm. Trees in neighborhood parks, for example, are routinely adorned with orchids by local residents. Likewise, shop owners and residents commonly install containers with plants of various sizes in the adjacent sidewalk. These do-it-yourself sidewalk plantings can, however, create tension with neighbors, as well as the PSLO when street work needs to be conducted.

Citizen engagement in tree stewardship has also been advanced through a new nationwide tree planting proposal. Launched in 2019, the Patch by Planting (PBP) nonprofit group has identified places that can purportedly accommodate some 2.3 million new trees (PBP 2020). This includes highway medians and circles, corporate and industrial campuses, and government-owned lands (e.g., landscapes dedicated to power lines, and idle land formerly dedicated to sugar cane production). In August 2020, the PBP was listed among five finalists in a national "hackathon" for sustainable development. With this finalist status, the Taiwanese central government is likely to promote the PBP project and support the public-private partnership.

Washington, DC

National Context

According to a nationwide survey spanning 667 municipalities in the USA, nearly two-thirds (64%) of cities assume legal responsibility for trees in the right-of-way (street trees between the sidewalk and curb or ally), with nearly one-third managed jointly (16%) or solely (16%) by adjacent property owners. But this differs by region: sole municipal responsibility is highest in the northeast (79%) and midwest (74%), while abutting property owners have greater responsibility for street trees in the south and west. In the west, for example, 46% of municipalities have sole responsibility for street trees while adjoining property owners have sole (28%) or joint (21%) responsibility (Hauer and Peterson 2016). The US regions also have different histories related to municipal management of urban trees. In the northeast, where some communities have had formally designated "tree wardens" since the early 1900s (Ricard 2005), municipalities have had a person responsible for public trees for some 50 years on average, while this has been the case for shorter periods in the Midwest (34 years), west (28 years), and south (22 years) (Hauer and Peterson 2016).

Depending on location, municipal administration of public trees (of which streets and parks are principal sites) in the USA can be spread across several departments including public works, parks and recreation, streets/transportation, planning and community development, and urban forestry. However, parks and recreation, and public works departments were most common in 74% and 69%, respectively, of communities responding to the aforementioned survey. Of note, a designated forestry department is more common as population increases: 5% of municipalities with populations 2500–4999 have a forestry department whereas 46% of places with $\geq$ 50,000 people have such a department. The size of municipalities was also found to be an important consideration in who manages urban trees and how administrative departments interact. In small towns, public administrators and public works directors commonly lead public tree management in addition to other activities, while people identified as arborists/foresters become more common public tree managers as the size of the municipality increases. In small communities between 2500 and 9999, 12% have a certified arborist whereas 83% of municipalities with at least 50,000 people have a certified arborist on staff.

Yet, the disciplinary identity and expertise of US urban tree managers varies. According to two nationwide surveys, under half (45–46%) of the people who manage urban trees identified themselves as arborists or urban foresters (Hauer and Peterson 2016; O'Herrin et al. 2020). The others include a range of professionals commonly found in municipal government: public administrators (21%), horticulturalists (7%), outdoor recreationalists (6%), landscape architects (5%), urban planners (4%), foresters (3%), and civil engineers (2%). Of these, 80% are male and 90% are White (O'Herrin et al. 2020).

In addition to the distinctions noted above, there are differences in who manages mature trees versus new tree plantings in many US communities. Nationwide, some two-thirds of municipalities involve volunteers in tree activity, and tree planting is by far the most common volunteer activity (85% of communities) followed by watering (40%), awareness/education programs (39%), tree pruning (28%), and fundraising (20%) (Hauer et al. 2018). This is especially true for tree planting campaigns which have become quite common in the USA (Campbell 2017; Young 2011), including the successful planting of one million trees in New York City between 2007 and 2016. These campaigns rely on a hybrid network of public, private, and nonprofit actors for financing, administration, and on-the-ground planting and stewardship. Importantly, non-technical volunteers are essential for planting and watering, and ensuring the survival of trees installed during such campaigns (Roman et al. 2015; Vogt et al. 2015). These greening initiatives can also trigger reorganization of urban forestry governance (Campbell 2014). The State of Massachusetts, for example, launched a campaign in 2014 to plant tens of thousands of trees in 26 municipalities with below-average household incomes and educational attainment. But this can create tension and lack of clear management authority between municipal and state administrators, resulting in many newly planted trees not surviving (Breger et al. 2019).

As of 2014, US municipalities had on average 76 trees per street mile, and municipal tree activities had a mean annual budget of $801,595 per municipality, which works out to an average $8.76 per capita and 0.52% of the total municipal budget. Of this, US cities spent on average $42.60 per street tree, although this was roughly double ($82/street tree) in the south; cities also had on average 4821 street trees per full-time employee (FTE), although the number of street trees per FTE increased in tandem with city population. The municipal general fund accounted for 72% of urban forestry financing, and over half (53%) of respondents thought this was adequate. Two-thirds of financing went to tree planting (14%), tree pruning (23%), tree removal (25%), or stump removal (4%). Importantly, expenditures on street trees accounted for the largest portion of municipal tree management budgets: 62% for street trees versus 23% for park trees (Hauer and Peterson 2016).

Washington, DC

The US capital city is affectionately referred to as the City of Trees (Choukas-Bradley and Alexander 2008) (see Fig. 5). So essential to the character of Washington, DC are trees, that they were an integral part of the city's original design. In Pierre L'Enfant's 1791 Plan, space in the public right-of-way was exclusively reserved for trees. The city's sylvan moniker is also a legacy of an 1870 Parking Act that characterized public right-of-ways as linear parks (Government of the District of Columbia 2019), and an 1872 planting campaign that yielded 60,000 new street trees while pushing the city to the brink of bankruptcy (DDOT 2020c). Part of a large-scale modernization effort to build sewage infrastructure and paved streets, this has been described as the first city-wide tree planting of such magnitude in the USA, establishing a precedent where some 280 miles of streets would be lined with trees by 1912 (Dümpelmann 2019). This reflects a nation-wide movement in the late nineteenth and early twentieth century to green US cities through street tree planting and creation of large public parks (Eisenman 2016b). Focusing on the emergence of elm tree planting along streets first in New England and increasingly across the nation, landscape historian Thomas Campanella has described this turn-of-the-century greening as a democratic project and uniquely American aspiration to create the "pastoral city" (2003). This is echoed by another historian Eric Rutkow, who describes "trees as one of the great drivers of national development that helped to forge American identity" (2012, p. 314).

Stewarding Street Trees for a Global Urban Future, Fig. 5 Sylvan streetscape in Washington, DC. (Source: Government of the District of Columbia 2019)

Building upon this tradition and situating trees as important elements in a new sustainability plan, the city established in 2011 a goal to plant 8600 trees per year and achieve 40% urban tree canopy (UTC) cover by 2032 (District of Columbia 2011). At the time, this goal represented an ambitious 5% increase in UTC, and the city is making substantial progress, with a 2020 UTC of 38% (DDOT 2020c). Street tree planting has played an important role in the drive towards this 40% goal, and today the streets of the nation's capital are nearing 100% stocking level. In other words, spaces adjacent to a street that can accommodate a tree have a tree (Sanders, personal communication, August 17, 2020). The city averages 7.3 trees per 100 m of street, and these trees are evenly distributed across local, collector, and arterial streets, which is not the case in some other capital cities (Smart et al. 2020). Of the district's more than 200,000 trees on publicly managed land today, some 157,000 are street trees (Sanders, personal communication, August 17, 2020).

The following are some of the most common of these street trees: red maple (*Acer rubrum*), willow oak (*Quercus phellos*), pin oak (*Quercus palustris*), American elm (*Ulmus americana*), and red oak (*Quercus rubra*). But in recent years, Urban Forestry Division (UFD) has diversified its street trees to some 125 species, including many that are half to a third the size of large shade trees such as maples and oaks, e.g., serviceberry (*Amelanchier*), sweetbay magnolia (*Magnolia virginiana*), Japanese apricot (*Prunus mume*), American hornbeam (*Carpinus caroliniana*), and Persian parrotia (*Parrotia persica*). This has been characterized as a fundamental shift in what constitutes an appropriate street tree in the twenty-first century (Higgins 2020). In addition to diversifying the species pool and reducing pest risks associated with shade tree monocultures, small statured trees are less likely to damage electrical lines and property.

The municipality's Department of Transportation (DDOT) UFD has sole responsibility for street trees, as well as trees in other public landscapes such as parks and schools. It is worth noting that the 1870 Parking Act which characterized public right-of-ways as linear parks is still largely in effect today. This requires property owners to maintain the "public parking" directly abutting their property while giving the property owner the exclusive right to enter that public space (see Fig. 6).

DDOT UFD has over 20 full-time certified arborists on staff who do hands-on arboriculture but spend much of their time managing private contractors who do most of the technical work. This includes planting and stewarding some 8000–8500 street trees per year (DDOT 2020b). The city does not rely upon volunteers to plant trees, however, when new street trees are planted in front of homes, the UFD notifies the homeowner and provides recommendations for watering the tree, if they so choose. The department has even created a web-based software application that allows homeowners to report when they have watered a tree, and to record information about the health of the tree (DDOT 2020a).

To support tree planting, the city created a Tree Fund in 2002 (amended in 2016) that draws upon several sources beyond traditional financing from the municipal general fund. The city levies a fee for removing nonhazardous trees (usually due to building construction and development); and starting in 2011 these tree loss mitigation funds have directly supported street tree planting. Other financing includes grants from the city's Department of Energy and Environment, as well as the

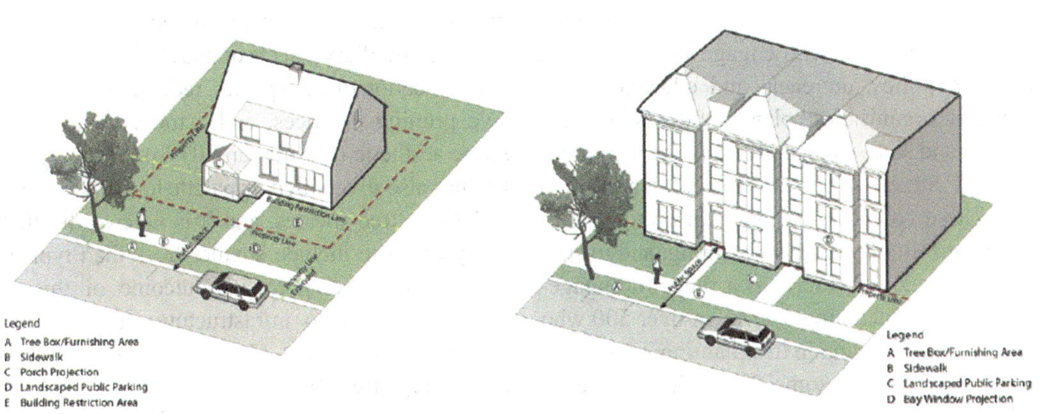

Stewarding Street Trees for a Global Urban Future, Fig. 6 Diagram depicting the landscaped "public parking" area adjacent to streets in Washington, DC. (Source: Government of the District of Columbia 2019)

Federal Clean Water Revolving Fund, both of which support the conservation or creation of vegetated green infrastructure systems to manage stormwater and protect the quality of local surface waters. All of these funds have allowed DDOT-UFD to increase street planting from 4000 locations annually to some 8000 over the past few years (DDOT 2020d).

However, this has not occurred without challenges. In low-income, underserved communities there can be resistance to tree planting initiatives. Local residents have communicated concerns that tree pollen aggravates allergies, shade draws drug dealers, leaves clutter the landscape and are difficult to rake, and that it is unclear who will manage the trees. Some have also complained that greening attracts affluent gentrifiers and higher taxes, pushing out older residents (Gowen and Mellnik 2013).

In addition to substantial public sector investment in street trees through the city's DDOT UFD, Washington, DC also has other private and nonprofit partners that engage in tree planting and stewardship in parks and on private lands that are not managed by DDOT. Most prominent is nonprofit organization Casey Trees, established in 2002 through a charitable donation by Betty Brown Casey who inherited $50 to $100 million upon the death of her husband, Eugene B. Casey, who accrued a fortune through real estate development across the district's metropolitan region (Jennings 1994). Casey Trees has grown into a major urban forestry actor in the city, with roughly 55 full-time staff including 10 certified arborists and others who engage in fundraising, planning, policy, outreach, and education. The nonprofit organization plants 3000–5000 trees per year and since its inception has engaged thousands of citizen volunteers in tree planting and care. These volunteers account for up to 50% of the organization's historical planting, and to support this work the group has a range of engagement models, including a corps of over 500 who have received training in tree planting, inventorying, and advocacy; some 2500 citizen science volunteers; 55 certified tree advocates; and about 100 volunteers who routinely engage in tree planting.

Discussion

Several noteworthy themes emerge from the aforementioned cases. Each of the cities, for example, has a unique history. The settlement of Paris extends back some two millennia, and early examples of tree-lined streets include seventeenth-century allées planted on behalf of kings, emperors, and the aristocracy. The underlying urban form of Washington, DC, by contrast, is heavily informed by the L'Enfant plan of 1791 (Kostof 1991); and citywide street tree planting a century later can be seen as a democratic project guided by an aesthetic aspiration for pastoral urbanism. Street tree planting in Taiwan, by extension, was heavily influenced by Japanese colonial rule 1895–1945. Such divergent histories reinforce the need to understand the historical legacies that undergird the structure and composition of contemporary urban forests (Roman et al. 2018).

Trees are some of the most potent and visible symbols of social process and collective identity (Rival 1998), and this is especially poignant when considering trees along streets, which are the most commonly used public spaces of cities (Jacobs 1993). In Taipei, *feng shui* design principles—dating back 3000 years (Marafa 2003; Xu 1997)—still hold cultural significance today and can inhibit the siting of trees in front of doorways. In Paris, the role of trees as place-making elements in urban design seems to be shifting to a more ecological orientation that foregrounds biodiversity and ecosystem functions such as cooling. Ancillary effects of this shift may include diversification of tree species, reduction in aggressive pruning practices, and a more rustic landscape aesthetic (O'Sullivan 2019b). Washington, DC has also diversified its palette to include some 125 street tree species, and it is noteworthy that many of these include smaller trees than typical shade trees. One potential outcome of this is reduced damage to infrastructure and property, which is an important—but often downplayed—risk of large trees (Roman et al. 2020).

Both Paris and Washington, DC have formal goals to increase canopy cover by 2030, and these cities are also pursuing efforts to diversify the types of trees planted along streets and in urban landscapes. This is a laudable goal that may reduce the likelihood of pests wiping out populations of tree monocultures while also supporting more diverse wildlife. But efforts to diversify street tree species would do well to consider insights derived from landscape preference research, which shows among other things that people desire a certain degree of visual order and "cues to care" (Nassauer 1995). This can, in turn, affect people's stewardship practices, perceptions of safety, and social cohesion (Nassauer 2011; Nassauer and Raskin 2014) as well as the coherence and legibility of streetscapes (Jacobs 1993; Massengale and Dover 2014). As greening efforts expand the quantity and diversity of plant material along urban streets, the work of Peter Trowbridge and Nina Bassuk offers valuable guidance. In *Trees in the Urban Landscape*, the coauthors provide 16 groups of biologically diverse yet visually compatible trees (Trowbridge and Bassuk 2004).

Another noteworthy theme that emerges from the cases described above is that all three cities have unique approaches to the governance and stewardship of streetscape vegetation. Paris has several initiatives that allow or actively encourage citizens to plant low-growing plant material in sidewalks; a less formal but culturally accepted norm also exists in Taipei, where it is common for shop owners and residents to install numerous planters on adjacent sidewalks. This type of resident-led streetscape greening does not seem to be as prevalent in Washington, DC, yet the city's municipal regulation requires property owners to maintain the "public parking" directly abutting their property. The district also has a formal process, including a robust website, that encourages residents to water newly planted trees and to record this activity. Paris also has a well-developed website that publicly tracks the health and management of the city's street trees. Reflecting yet another form of decentralized governance, streetscape stewardship in Taipei draws upon a network of 456 neighborhood groups called *li,* who often adopt trees and become de facto stewards. The Taiwanese capital also tags its street trees, which provides not only a formal recording mechanism for municipal staff, it also communicates to the public that the trees are actively cared for.

Of the three cities, Washington, DC may have the most diverse funding approach dedicated to new tree planting. In addition to traditional financing from the municipal general fund, the city levies fees for removing nonhazardous trees and it has grants via municipal departments focused on energy, environment, and water that support street tree planting. The district also has a major nongovernmental partner that focuses on parks and private lands that are not managed by the city, and while this group does not plant street trees in Washington, DC, it nevertheless reflects the prominent role of local nonprofit actors in urban tree planting nationwide, much of which focuses on streets (Eisenman et al. 2021).

An overarching theme that emerges from this study is that street tree planting and management figures prominently in each of the respective cities. All three municipalities show a net increase in the number of street trees in recent years, and there is substantial interest in the role that street trees can play in creating more livable and sustainable cities. This is good news, as streets represent one of—if not the most—common types of outdoor space that people engage on a regular basis. By extension, vegetated "travelscapes" represent an excellent opportunity to provide an increasingly urbanized human population with the benefits of nature contact. This will, however, require ongoing investment in the social infrastructure that stewards green infrastructure, and a commitment to the experiential dimension of street trees, as streets are the backbone of the urban public realm.

S

Cross-References

▶ Cities in Nature
▶ Emerging Concepts Exploring the Role of Nature for Health and Well-Being
▶ Green Cities
▶ Green Cities in Theory and Practice
▶ Green Cities: Nature-Based Solutions, Renaturing and Rewilding Cities
▶ Green Infrastructure

- ► Health and the Role of Nature in Enhancing Mental Health
- ► Innovation To Bring Nature-based Solutions to Life: Tales of Two Cities
- ► Nature-Based Solutions for River Restoration in Metropolitan Areas
- ► Need for Greenspace in an Urban Setting for Child Development
- ► Need for Nature Connectedness in Urban Youth for Environmental Sustainability
- ► Policy and Practices of Nature-based Solutions To Build Resilience in Seoul, Korea
- ► Role of Nature for Ageing Populations
- ► Strategies for Liveable and Sustainable Cities: The Singapore Experience
- ► Sustainable Cities via Urban Ecosystem Restoration
- ► Transport Resilience in Urban Regions
- ► Transportation and Mobility
- ► Urban Commons as a Bridge Between the Spatial and the Social
- ► Urban Ecosystem Services and Sustainable Human Well-being
- ► Urban Greening and Green Gentrification
- ► Urban Nature
- ► Wildlife Corridors

Acknowledgments Theodore Eisenman's contribution to this manuscript was supported by an Andrew W. Mellon fellowship through the Humanities Institute at The New York Botanical Garden, as well as the National Science Foundation award CNH–1924288.

References

ADEME, (Agence de l'Environnement et de la Maitrise de l'Energie). (2018). *L'arbre en milieu urbain, acteur du climat en Région Hauts-de-France*. www.arbre-en-ville.fr/wp-content/uploads/2019/05/Guide20l27arbre20acteur20du20climat20en20milieu20urbain2020 18.pdf

Angel, S. (2012). *Planet of cities*. Lincoln Institute of Land Policy.

Anguelovski, I., & Martínez Alier, J. (2014). The 'environmentalism of the poor' revisited: Territory and place in disconnected glocal struggles. *Ecological Economics, 102*, 167–176.

Ao, H. (2000). Taiwanese street trees. *Guangdong Yuanlin, 2*, 34–41.

Appleton, J. (1975). *The experience of landscape*. Wiley.

Atelier Parisien d'Urbanisme. (2010). Essai de bilan sur les arbres d'alignement dans. *Analyse statistique*. https://www.apur.org/sites/default/files/documents/APBROAPU506.pdf

Beatley, T. (2016). *Handbook of biophilic city planning & design*. Island Press.

Bergeron, L. (Ed.). (1989). *Paris. Genèse d'un paysage*. Picard.

Bournon, F. (1888). *Petite Histoire de Paris: Histoire–Monuments–Administration Environs de Paris*. Librarie Classique Armand Colin et Cie.

Brasche, S., & Bischof, W. (2005). Daily time spent indoors in German homes—Baseline data for the assessment of indoor exposure of German occupants. *International Journal of Hygiene and Environmental Health, 208*(4), 247–253.

Breger, B. S., Eisenman, T. S., Kremer, M. E., Roman, L. A., Martin, D. G., & Rogan, J. (2019). Urban tree survival and stewardship in a state-managed planting initiative: A case study in Holyoke, Massachusetts. *Urban Forestry & Urban Greening, 43*, 126382.

Campanella, T. J. (2003). *Republic of shade: New England and the American elm*. Yale University Press.

Campbell, L. K. (2014). Constructing New York City's urban forest: The politics and governance of the MillionTreesNYC campaign. In A. L. Sandberg, A. Bardekjian, & S. Butt (Eds.), *Urban forests, trees, and greenspace: A political ecology perspective* (pp. 242–260). Routledge.

Campbell, L. K. (2017). *City of forests, city of farms: Sustainability planning for New York city's nature*. Cornell University Press.

Carmichael, C. E., & McDonough, M. H. (2019). Community stories: Explaining resistance to street tree-planting programs in Detroit, Michigan, USA. *Society & Natural Resources, 32*(5), 588–605.

Castells, M. (1996). *The rise of the network society* (Vol. 1). Wiley-Blackwell.

Choukas-Bradley, M., & Alexander, P. (2008). *City of trees: The complete field guide to the trees of Washington, D.C* (3rd ed.). University Press of Virginia Press. https://www.upress.virginia.edu/title/31.

DDOT, (District Department of Transportation, Urban Forestry Administration). (2020a). *DDOT tree tool*. https://treewatering.ddot.dc.gov/treewatering/

DDOT, (District Department of Transportation, Urban Forestry Administration). (2020b). *Urban forestry division services*. https://ddot-urban-forestry-dcgis.hub.arcgis.com/pages/tree-services

DDOT, (District Department of Transportation, Urban Forestry Administration). (2020c). *Urban forestry in Washington, D.C*. https://urban-forestry-dcgis.opendata.arcgis.com/

DDOT, (District Department of Transportation, Urban Forestry Administration). (2020d). *Urban tree canopy in the nation's capital*. https://dcgis.maps.arcgis.com/apps/MapJournal/index.html?appid=0336fad670cb42ba8b894d57a827ecc3&webmap=c1b41d656894147a35e059a9d0774ff

District of Columbia. (2011). *Sustainable DC Plan* (p. 121). http://www.sustainabledc.org/about/

Dorion, N. (2014). *Petite histoire des alignements à Paris du xvie siècle au xixe siècle, Les dossiers Jardins de*

France (n°2). Société Nationale d'Horticulture de France.

Dümpelmann, S. (2019). *Seeing trees: A history of street trees in New York City and Berlin*. Yale University Press.

Eisenman, T. S. (2016a). Book review: The ecological design and planning reader [Review of *Book review: The ecological design and planning reader*, by F. O. Ndubisi]. *Journal of Planning Education and Research, 37*(3), 374–376.

Eisenman, T. S. (2016b). Greening cities in an urbanizing age: The human health bases in the nineteenth and early twenti-first centuries. *Change Over Time, 6*(2), 216–246.

Eisenman, T. S., Flanders, T., Harper, R. W., Hauer, R. J., & Lieberknecht, K. (2021). Traits of a bloom in urban greening: A nationwide survey of U.S. urban tree planting initiatives (TPIs). *Urban Forestry & Urban Greening*. in press, 127006.

Ernstson, H., & Sörlin, S. (Eds.). (2019). *Grounding urban natures: Histories and futures of urban ecologies*. The MIT Press.

Esser, F., & Vliegenthart, R. (2017). Comparative research methods. In J. Matthes, C. S. Davis, & R. F. Potter (Eds.), *The international encyclopedia of communication research methods* (pp. 1–21). Wiley.

Forrest, M. (2002). Trees in European cities—A historical review. In L. Dunne (Ed.), *Biodiversity in the city* (pp. 15–20). Environmental Institute, University College Dublin.

Forrest, M., & Konijnendijk, C. (2005). A history of urban forests and trees in Europe. In C. Konijnendijk, K. Nilsson, T. Randrup, & J. Schiperrijn (Eds.), *Urban forests and trees*. Springer.

Frumkin, H., Bratman, G. N., Breslow, S. J., Cochran, B., Kahn, P. H., Lawler, J. J., Levin, P. S., Tandon, P. S., Varanasi, U., Wolf, K. L., & Wood, S. A. (2017). Nature contact and human health: A research agenda. *Environmental Health Perspectives, 125*(7), 075001 (1–18).

Government of the District of Columbia. (2019). *Public realm design manual: A summary of Distric of Columbia Regulations and Specifications for the design of public space elements* (Version 2.1; p. 81). https://ddot.dc.gov/PublicRealmDesignManual

Gowen, A., & Mellnik, T. (2013). Environmentalists face challenges trying to plant in less-green neighborhoods. *Washington Post*. https://www.washingtonpost.com/local/environmentalists-face-challenges-trying-to-plant-in-less-green-neighborhoods/2013/04/25/21294968-ad27-11e2-a198-99893f10d6dd_story.html

Hall, P. G., & Pfeiffer, U. (2000). *Urban future 21: A global agenda for 21st century cities*. E & FN Spon.

Hartig, T., Mitchell, R., de Vries, S., & Frumkin, H. (2014). Nature and health. *Annual Review of Public Health, 35*(1), 207–228.

Hauer, R. J., & Peterson, W. D. (2016). *Municipal tree care and management in the United States: A 2014 urban & community forestry census of tree activities* (p. 71) [Special Publication 16–1]. College of Natural Resources, University of Wisconsin–Stevens Point. https://www.uwsp.edu/cnr/Documents/MTCUS%20-%20Forestry/Municipal%202014%20Report%20Executive%20Summary.pdf

Hauer, R. J., Timilsina, N., Vogt, J., Fischer, B. C., Wirtz, Z., & Peterson, W. (2018). A volunteer and partnership baseline for municipal forestry in the United States. *Arboriculture & Urban Forestry, 44*(2), 87–100.

Higgins, A. (2020). D.C. has become a leader in a movement to plant more diverse city trees. *The Washington Post*. https://www.washingtonpost.com/lifestyle/home/the-street-tree-gets-a-smart-makeover/2020/07/07/a7676f54-b71b-11ea-aca5-ebb63d27e1ff_story.html

Hsu, S. T. (2010). *Trimming and pruning guidelines and illustrations for trees in Taipei City*. Taipei City: Parks and Street Lights Office.

Jacobs, A. B. (1993). *Great streets*. MIT Press.

Jennings, V. T. (1994). A family fueds over millions. *Washington Post*. https://www.washingtonpost.com/archive/local/1994/01/11/a-family-feuds-over-millions/e7570d5c-5808-4c24-8ec3-55b00c74e677/

Jones, C. (2006). *Paris: Biography of a city*. Penguin Books.

Kaplan, R., & Kaplan, S. (1989). *The experience of nature: A psychological perspective*. Cambridge University Press.

Keller, J. K.-K., & Konjijnendijk, C. C. (2012). Short communication: A comparative analysis of municipal urban tree inventories of selected major cities in North America and Europe. *Arboriculture & Urban Forestry, 38*(1), 24–30.

Klepeis, N. E., Nelson, W. C., Ott, W. R., Robinson, J. P., Tsang, A. M., Switzer, P., Behar, J. V., Hern, S. C., & Engelmann, W. H. (2001). The National Human Activity Pattern Survey (NHAPS): A resource for assessing exposure to environmental pollutants. *Journal of Exposure Analysis and Environmental Epidemiology, 11*(3), 231–252.

Konijnendijk, C. C. (2008). *The forest and the city: The cultural landscape of urban woodland*. Springer.

Konijnendijk, C. C., Ricard, R. M., Kenney, A., & Randrup, T. B. (2006). Defining urban forestry – A comparative perspective of North America and Europe. *Urban Forestry & Urban Greening, 4*(3–4), 93–103.

Kostof, S. (1991). *The city shaped: Urban patterns and meanings through history*. Thames and Hudson.

Kuo, M. (2015). How might contact with nature promote human health? Promising mechanisms and a possible central pathway. *Frontiers in Psychology, 1093*.

Landau, B. (1992). La fabrication des rues de Paris au XIXe siècle: Un territoire d'innovation technique et politique. *Les Annales de la recherche urbaine, 57–58*, 24–45.

Laurian, L. (2012). Paris: An ecocity in the 21st century. In T. Beatley (Ed.), *Green cities of Europe: Global lessons on green urbanism*. Island Press.

Laurian, L. (2019). Planning for street trees and human-nature relations: Lessons from 600 years of street tree planting in Paris. *Journal of Planning History, 18*(4), 282–231.

Lavedan, P. (1993). *Histoire de l'urbanisme à*. Hachette.

S

Lawrence, H. W. (1993). The neoclassical origins of modern urban forests. *Forest & Conservation History, 37*(1), 26–36.

Lawrence, H. W. (2006). *City trees: A historical geography from the renaissance through the nineteenth century.* University of Virginia Press.

Lawrence, H. W. (2008). *City trees: A historical geography from the renaissance through the nineteenth century.* University of Virginia Press.

Lewis-Beck, M. S., Bryman, A., & Liao, T. F. (2004). *The Sage encyclopedia of social science research methods.* Sage.

Marafa, L. (2003). Integrating natural and cultural heritage: The advantage of feng shui landscape resources. *International Journal of Heritage Studies, 9*(4), 307–323.

Massengale, J., & Dover, V. (2014). *Street design: The secret to great cities and towns.* Wiley.

MIT Senseable City Lab. (2020). *Treepedia: Paris.* http://senseable.mit.edu/treepedia/cities/paris

Mo, R. (2020, August 20). *Pers. Comm.* (S. E. Chang, Interviewer) [Personal communication].

Nassauer, J. I. (1995). Messy ecosystems, orderly frames. *Landscape Journal, 14*(2), 161–170.

Nassauer, J. I. (2011). Care and stewardship: From home to planet. *Landscape and Urban Planning, 100*(4), 321–323.

Nassauer, J. I., & Raskin, J. (2014). Urban vacancy and land use legacies: A frontier for urban ecological research, design, and planning. *Landscape and Urban Planning, 125*, 245–253.

Nguyen, V. D., Roman, L. A., Locke, D. H., Mincey, S. K., Sanders, J. R., Smith Fichman, E., Duran-Mitchell, M., & Tobing, S. L. (2017). Branching out to residential lands: Missions and strategies of five tree distribution programs in the U.S. *Urban Forestry & Urban Greening, 22*, 24–35.

O'Herrin, K., Wiseman, P. E., Day, S. D., & Hauer, R. J. (2020). Professional identity of urban foresters in the United States. *Urban Forestry & Urban Greening, 54*, 126741.

O'Sullivan, F. (2019a). *Paris wants to grow 'urban forests' at famous landmarks.* Bloomberg CityLab.

O'Sullivan, F. (2019b). *Paris wants to grow 'urban forests' at famous landmarks.* CityLab. https://www.citylab.com/environment/2019/06/paris-trees-famous-landmarks-garden-park-urban-forest-design/591835/.

Pacini, G. (2007). A culture of trees: The politics of pruning and felling in the late 18th century France. *Eighteenth Century Studies, 41*(1), 1–15.

PBP. (2020). *Patch by planting.* https://sites.google.com/view/tree-taiwan/

Plant, L., & Kendal, D. (2019). Toward urban forest diversity: Resident tolerance for mixtures of tree species within streets. *Arboriculture & Urban Forestry, 45*(2), 41–53.

Ricard, R. M. (2005). Shade trees and tree wardens: Revising the history of urban forestry. *Journal of Forestry, 103*(5), 230–233.

Rival, L. M. (1998). Trees, from symbols of life and regeneration to political artefacts. In *The social life of trees: Anthropological perspectives on tree symbolism* (pp. 1–36). Berg.

Roman, L. A., Walker, L. A., Martineau, C. M., Muffly, D. J., MacQueen, S. A., & Harris, W. (2015). Stewardship matters: Case studies in establishment success of urban trees. *Urban Forestry & Urban Greening, 14*, 1174–1182.

Roman, L. A., Pearsall, H., Eisenman, T. S., Conway, T. M., Fahey, R. T., Landry, S., Vogt, J. M., van Doorn, N. S., Grove, J. M., Locke, D. H., Bardekjian, A. C., Battles, J. J., Cadenasso, M. L., van den Bosch, C. C. K., Avolio, M., Berland, A., Jenerette, G. D., Mincey, S. K., Pataki, D. E., & Staudhammer, C. (2018). Human and biophysical legacies shape contemporary urban forests: A literature synthesis. *Urban Forestry & Urban Greening, 31*, 157–168.

Roman, L. A., Conway, T. M., Eisenman, T. S., Koeser, A. K., Ordóñez Barona, C., Locke, D. H., Jenerette, G. D., Östberg, J., & Vogt, J. (2020). Beyond 'trees are good': Disservices, management costs, and tradeoffs in urban forestry. *Ambio*.

Rutkow, E. (2012). *American canopy: Trees, forests, and the making of a nation.* Scribner, Kindle edition.

Sanders, J. (2020, August 17). *Pers. Comm.* [Zoom].

Smart, N. A., Eisenman, T. S., & Karvonen, A. (2020). Street tree density and distribution: An international analysis of five capital cities. *Frontiers in Ecology and Evolution, 8*, 562646.

Suppakittpaisarn, P., Larsen, L., & Sullivan, W. C. (2019). Preferences for green infrastructure and green stormwater infrastructure in urban landscapes: Differences between designers and laypeople. *Urban Forestry & Urban Greening, 43*, 126378.

Taipei City. (2015). *Street trees and street lights informational network.* Parks and Street Lights Office.

Taipei City. (2017). *Parks and street light office records.* https://pkl.gov.taipei/News.aspx?n=67678EEB557696E7&sms=40B3A22009350941

Taipei City. (2019). *Parks, open spaces and street tree adaptation.* Parks and Street Lights Office. https://pkl.gov.taipei/cp.aspx?n=77AAFA25AB5C29AC.

Tan, P. Y., & Jim, C. Y. (Eds.). (2017). *Greening cities: Forms and functions.* Springer.

Tashir, Y. (1920). *Taiwan street trees and plantation guidelines.* Forest Department of Taiwan Soutokufu.

Todorova, A., Asakawa, S., & Aikoh, T. (2004). Preferences for and attitudes towards street flowers and trees in Sapporo, Japan. *Landscape and Urban Planning, 69*(4), 403–416.

Toussaint, A., Meerendre, V., Delcroix, B., & Baudoin, J. P. (2002). Analyse de l'impact physiologique et économique de l'élagage des arbres d'alignement en port libre. *Biotechnologie, Agronomie, Société et Environnement, 6*(2), 99–107.

Trowbridge, P. J., & Bassuk, N. L. (2004). *Trees in the urban landscape: Site assessment, design, and installation.* Wiley.

Ulrich, R. S., Simons, R. F., Losito, B. D., Fiorito, E., Miles, M. A., & Zelson, M. (1991). Stress recovery during exposure to natural and urban environments. *Journal of Environmental Psychology, 11*(3), 201–230.

Van Dongen, R. P., & Timmermans, H. J. P. (2019). Preference for different urban greenscape designs: A choice experiment using virtual environments. *Urban Forestry & Urban Greening, 44*, 126435.
Ville de Paris. (2018a). *Rapport d'activités 2018*. mairie-de-paris.agencezebra.net/ra-2018.
Ville de Paris. (2018b). *Le Plan Biodiversité de Paris 2018–2024*. paris.fr/pages/biodiversite-66#le-plan-biodiversite-2018-2024
Ville de Paris. (2019a). *L'arbre à Paris*. fr/pages/l-arbre-a-paris-199.
Ville de Paris. (2019b). *Des forêts urbaines bientôt sur quatre sites emblématiques*. https://www.paris.fr/pages/des-forets-urbaines-bientot-sur-quatre-sites-emblematiques-6899
Ville de Paris. (2020a). *Du Vert Pres de Chez Moi Program*. https://opendata.paris.fr/explore/dataset/du-vert-pres-de-chez-moi/table/
Ville de Paris. (2020b). *Le Permis de Végétaliser*. https://vegetalisons.paris.fr/vegetalisons
Vogt, J. M., Watkins, S. L., Mincey, S. K., Patterson, M. S., & Fischer, B. C. (2015). Explaining planted-tree survival and growth in urban neighborhoods: A social–ecological approach to studying recently-planted trees in Indianapolis. *Landscape and Urban Planning, 136*, 130–143.
Xiao, T. X. (2016). Adopting street trees, saving Taipei City 50 million TWD per year. *Epoch Times*. https://kairos.news/35809
Xu, P. (1997). Feng-shui as clue: Identifying prehistoric landscape setting patterns in the American southwest. *Landscape Journal, 16*(2), 174–190.
Young, R. F. (2011). Planting the living city: Best practices in planning green infrastructure – Results from major U.S. cities. *Journal of the American Planning Association, 77*, 368–381.
Young, R. F., & Lieberknecht, K. (2019). From smart cities to wise cities: Ecological wisdom as a basis for sustainable urban development. *Journal of Environmental Planning and Management, 62*(10), 1675–1692.
Zhao, J., Xu, W., & Li, R. (2017). Visual preference of trees: The effects of tree attributes and seasons. *Urban Forestry & Urban Greening, 25*, 19–25.

Stewardship

► Stewarding Street Trees for a Global Urban Future

Stock

► Closing the Loop on Local Food Access Through Disaster Management

Strategies for Liveable and Sustainable Cities: The Singapore Experience

Michael Koh and Alison Lee
Centre for Liveable Cities, Ministry of National Development, Singapore, Singapore

Introduction

Singapore faces the perennial challenge of balancing the essential needs of a nation within a small city-state of 728 km^2. Land has to be allocated for a spectrum of uses such as housing, infrastructure, defense, and greenery. Given its geographical constraints, Singapore has had to adopt innovative solutions to achieve a high standard of living for its residents.

In the 1960s, Singapore was a developing country with high unemployment, slums and squatter settlements, severe traffic congestion, and poor sanitation (Fig. 1). It has since transformed into a thriving and modern metropolis today. Despite its population density of 7,810 persons per km^2, the city ranked the top in Asia in Mercer's 2019 Quality of Living survey (Mercer 2019). Singapore is thus one of the few cities that has managed to attain high liveability in a high-density urban context.

The Centre for Liveable Cities (CLC) has studied the factors that led to Singapore's rapid urban development and distilled the learnings into the Singapore Liveability Framework (SLF) (Fig. 2). The SLF defines liveability as having a competitive economy, sustainable environment, and a high quality of life. To realize these desired outcomes, two systems were established: Integrated Master Planning and Development, and Dynamic Urban Governance. Under each pillar, five principles were identified to have contributed to Singapore's growth, and they continue to play key roles in guiding how planning and development is carried out today.

This chapter will describe some of the strategies Singapore has taken to become a liveable and sustainable city, as well as three underlying enablers that ensure successful implementation.

Strategies for Liveable and Sustainable Cities: The Singapore Experience, Fig. 1 In the 1960s, a large proportion of Singapore's population lived in overcrowded slums and shophouses. Images courtesy of the Housing & Development Board (top) and the Urban Redevelopment Authority (bottom)

Integrated Master Planning and Development

Singapore takes an integrated and long-term approach to urban planning to balance the various competing needs of a city-state. The Long-Term Plan acts as a strategic land use and infrastructure plan to guide development over a horizon of about 50 years. It undergoes regular reviews and is updated every 10 years by the national urban planning authority in Singapore, the Urban Redevelopment Authority (URA), together with the relevant government agencies.

The broad ideas introduced in the Long-Term Plan are then translated into the Master Plan at a more detailed level by delineating the permissible land use and density of each land parcel. Similarly, the Master Plan is reviewed every 5 years to ensure that it is kept up to date. Various planning instruments are employed to make sure that there is effective execution to fulfill what was envisioned. One key planning tool is the Government Land Sales (GLS) program where land is tendered out by the state to the private sector. Since most undeveloped land in Singapore is owned by the state, new developments often depend on the release of land through the GLS program. Detailed sale conditions are also prescribed such that developments that materialize are aligned with planning intentions.

Strategy 1: Provide Housing for All

The Housing & Development Board (HDB) was established in 1960 to address Singapore's acute

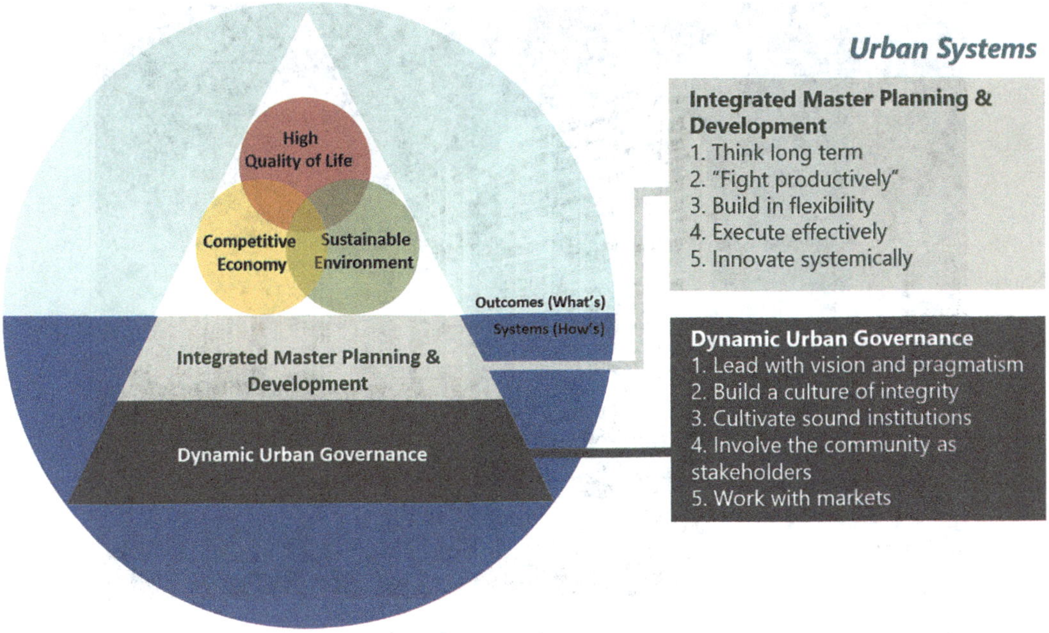

Strategies for Liveable and Sustainable Cities: The Singapore Experience, Fig. 2 The Singapore Liveability Framework. Image courtesy of the Centre for Liveable Cities

housing shortage at that time. It made great strides in the early years by building basic blocks of one-, two-, and three-room flats, completing more than 50,000 flats in 5 years (Housing & Development Board 1965). Once the housing shortage was solved, the HDB then turned toward providing better housing to meet rising aspirations for an improved quality of life. Catering to the low- and middle-income population in Singapore, larger flats in a variety of layouts, creative designs, and lush greenery were introduced.

More than 80% of Singaporeans live in HDB public housing today. Within this group, the rate of home ownership is over 90% which helps to cultivate a sense of belonging by giving residents a stake in the country among other benefits. The government sells HDB flats at highly-subsidized prices or provides housing grants for the purchase of flats on the open resale market. To qualify to buy a subsidized HDB flat, there are eligibility requirements in place such as those with regard to citizenship, income, and non-ownership of other private properties. Eligible flat buyers may also qualify for additional housing grants such as the Enhanced CPF Housing Grant and the Proximity Housing Grant. For lower-income households that find buying a flat out of their reach, there is a Public Rental Scheme to rent HDB units at heavily subsidized rental. The HDB also has several schemes set up to support these households in transitioning to owning a flat.

Over the years, the HDB has introduced new housing concepts such as waterfront housing (Fig. 3) and housing with skyrise greenery, as well as new flat types which allow multi-generational living where large families can stay together under one roof and 2-Room Flexi Flats for retirees and singles. Smart homes can also be found at Punggol Northshore where flats are equipped with smart distribution boards for residents to better manage energy consumption within their homes and facilitate the adoption of smart home solutions such as motion sensors and door contacts, and smart features to help in the management of the estate. Therefore, there are a multitude of lifestyle and residential options to suit various income levels.

On the opposite end of the spectrum, the private sector develops private property targeted at higher-income groups. This includes

Strategies for Liveable and Sustainable Cities: The Singapore Experience, Fig. 3 Waterfront public housing at Punggol town. Image courtesy of the Housing & Development Board

condominiums that often offer facilities such as swimming pools and gyms. In the 2010s, the government directed a large supply of condominium developments in non-prime suburban areas. This not only helped to meet the demand for more affordable private housing, but also gave homebuyers more choices in terms of location. While condominiums make up the majority of private properties, some land is also reserved for low-rise landed housing, usually in areas that cannot support high-density developments.

However, there is also a "sandwiched" class whose household incomes exceed the ceiling for public housing, but are insufficient to afford typical private housing. Hence, a new housing typology, the Executive Condominium (EC), was created in 1996 to plug this gap. ECs are developed and sold by private developers with initial eligibility conditions similar to the purchase of subsidized HDB flats. Requirements relating to income ensures that developers set a price that is reasonable for the "sandwiched" class. With standards comparable to private condominiums, ECs are an attractive option for many young families.

A range of housing types and schemes are thus available to meet the different needs and preferences of the population, making sure that all Singaporeans have quality homes and a pleasant living environment to live in.

Strategy 2: Planning for Mixed-Use

Singapore plans for mixed-use districts for people to live, work, and play in. This is reflected in the Master Plan which has zoned land parcels across Singapore with a variety of uses (Fig. 4). For example, the URA has injected a live-in population in the Central Business District (CBD) which has traditionally been primarily for office use. Despite being a prime location, there are also HDB public housing developments available.

For instance, the Tanjong Pagar area of the CBD (Fig. 4) has comprehensive provision of housing. This can be seen in the Master Plan where sites zoned Residential (orange), Residential with Commercial at 1st Storey (pink), and Commercial and Residential (light blue) can be found interspersed across the district. These residential developments comprise both private and public housing. In addition, there are large clusters of hotel developments (purple) that cater to tourists.

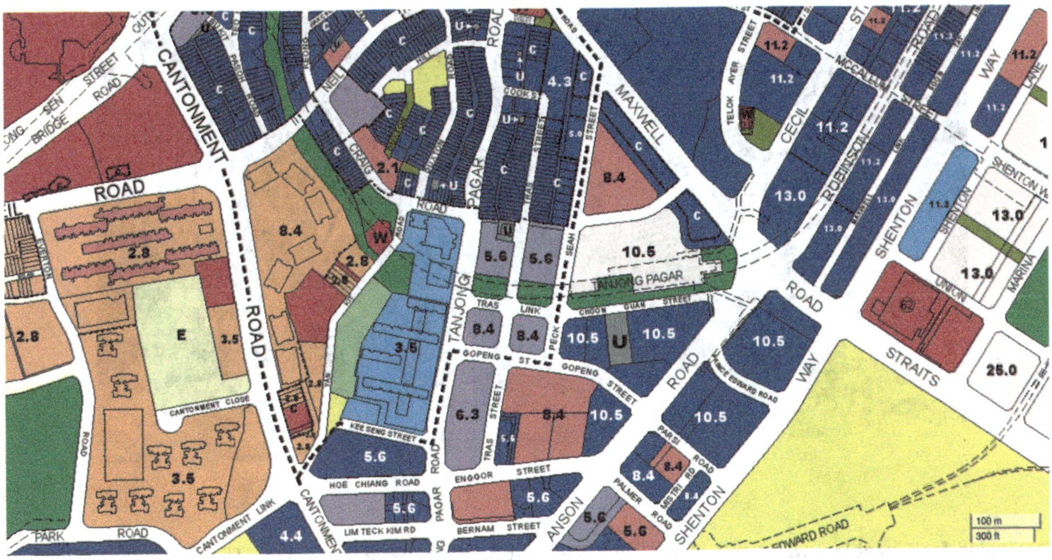

Strategies for Liveable and Sustainable Cities: The Singapore Experience, Fig. 4 Master Plan zoning in the Tanjong Pagar area. Image courtesy of the Urban Redevelopment Authority

Completed in 1977, Tanjong Pagar Plaza combines public housing units with retail shops and amenities such as a hawker center. The Pinnacle@Duxton is another notable project completed in recent years which added 1,848 units of public housing to the Central Area (Lee 2011). On the other hand, private residential developments and hotels were built by the private sector through the sale of sites under the GLS program.

Besides tall skyscrapers, Tanjong Pagar is also sensitively integrated with conservation areas. Labelled with "C" in Fig. 4, the conservation areas feature low-rise heritage shophouses which have been conserved, restored, and adaptively reused for purposes such as retail shops. The varying building heights as well as the blend of traditional and modern structures help to make for a visually striking landscape. In addition, ample open spaces provide respite among the dense cluster of buildings for residents, office workers, and members of the public.

To optimize the use of land, Singapore is moving toward having more mixed-use developments that offer greater convenience by meeting multiple needs of users. One way this is done is through a new zoning type – the white site. White sites are where a range of uses are allowed, often with minimum requirements for a certain use type. Developers are thus able to decide on a mix of uses for the project based on their analysis of the market and are motivated to update their buildings in line with changing real estate trends.

An example of this is the Guoco Tower and Sofitel Singapore City Centre hotel which are connected directly to the Tanjong Pagar MRT station at the basement level. It not only meets the required provision of at least 60% and 10% of Gross Floor Area (GFA) for office use and hotel use, respectively; it also includes a large public space of 30,000 sq. ft., also known as the City Room (Chow 2015). The City Room flows seamlessly into the public open space system in the area. It has since become a key destination for residents and office workers to unwind and participate in events (Fig. 5).

In 2019, the CBD Incentive Scheme was introduced to encourage building owners to convert existing, older, office towers in Singapore's financial hub into mixed-use developments with residential and hotel uses, in exchange for an increase in the development's allowable intensity (Urban Redevelopment Authority 2019c). This would rejuvenate and enhance the vibrancy of the CBD, especially after office hours and on weekends.

Strategies for Liveable and Sustainable Cities: The Singapore Experience, Fig. 5 The City Room at Guoco Tower used for an exercise session. Image courtesy of Ken Lee

Mixed-use developments can also be found across Singapore in the form of Integrated Transport Hubs (ITHs). Most ITHs are developed by private developers when the government sells premium sites located near public transport nodes such as Mass Rapid Transit (MRT) stations. Providing seamless connectivity, they are usually well connected to the transport nodes through sheltered or air-conditioned linkages. At the same time, a mix of uses in the development increases ridership and thus encourages greater utilization of the public transport infrastructure. In Singapore, ITHs are commonly mixed with retail and residential uses. For example, the Clementi Integrated Transport Hub has an MRT station and bus interchange with an adjoining mall and public housing blocks. The mall also includes amenities such as a public library and a post office.

The integration of mixed-use districts and developments starts at the planning stage through strategic zoning and the creation of innovative development concepts. Further downstream, building owners are motivated to include more uses into older developments through incentive schemes. Residents can thus benefit from the ease of meeting their daily needs within close proximity, while Singapore's limited land is better utilized.

Strategy 3: Planning for Polycentricity

In the 1960s, Singapore started urban renewal by facilitating private redevelopment of its city center. This helped to build up a modern CBD along Shenton Way where many commercial buildings for financial and business institutions could be found. By the 1990s, strong economic growth in Singapore had attracted many more businesses to locate their premises in the CBD. This resulted in a high volume of traffic, particularly during peak hours. To avoid further congestion in the core of the city, the 1991 Concept Plan thus put forth a polycentric strategy to decentralize commercial activities (Centre for Liveable Cities 2018a).

To bring jobs closer to homes, a hierarchy of suburban commercial centers, known as the Constellation Concept, was proposed. Outside of the Central Area, Singapore was split into regions, and each was to be served by a regional center – Jurong East in the west, Woodlands in the north, and Tampines in the east (Centre for Liveable Cities 2018a; Fig. 6).

Each regional center would also be complemented with industrial estates, business parks, and Institutes of Higher Learning, strengthening their role in redistributing commercial activities across Singapore. For example, the Woodlands regional center will be clustered with the upcoming Punggol Digital District (PDD),

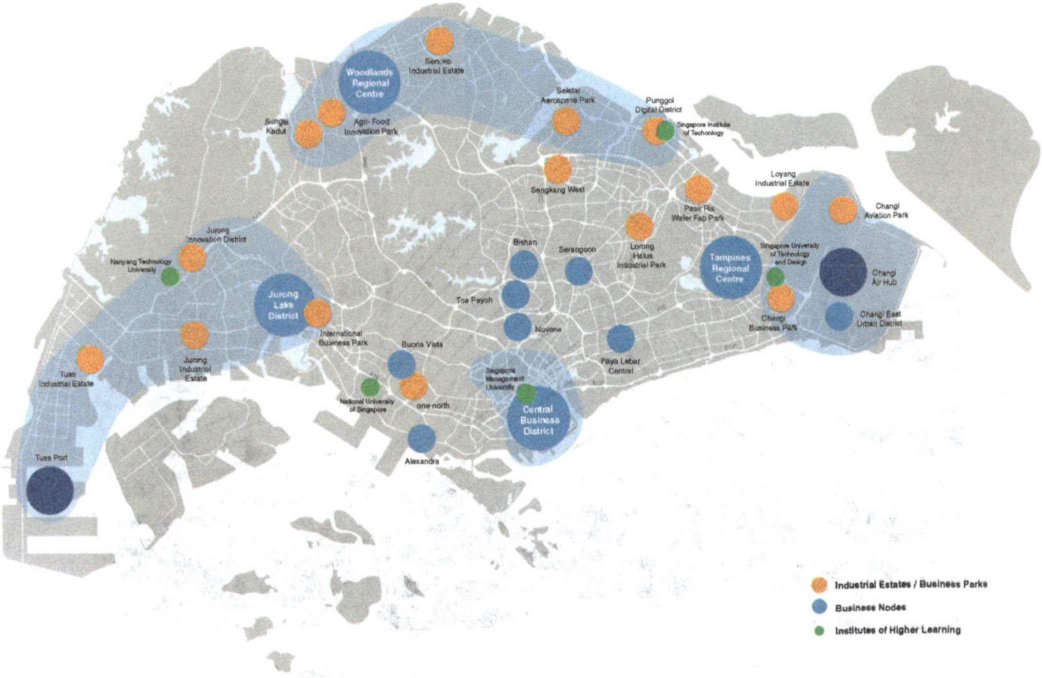

Strategies for Liveable and Sustainable Cities: The Singapore Experience, Fig. 6 Clusters of commercial nodes can be found outside the CBD in the west, north and east of Singapore. Image courtesy of the Urban Redevelopment Authority

which will become Singapore's first Enterprise District. At PDD, business park spaces for digital enterprises and the Singapore Institute of Technology will be co-located to facilitate greater collaboration.

In addition, regional centers are supported by subregional centers such as Buona Vista, Bishan, Serangoon, and Paya Lebar. Several fringe centers were also identified along the city fringe, including Newton, Novena, Outram, and Lavender. As a result, there is now a wider range of locations available for businesses. For instance, in the 1990s, several companies moved their backend operations to the Tampines regional center to save on rental costs (Urban Redevelopment Authority 2000). With comprehensive plans to build up these commercial nodes, some businesses have shifted entirely from the city center to newer areas like Paya Lebar Quarter (Centre for Liveable Cities 2021; Fig. 7).

Polycentricity has also benefitted residents as public housing towns were conceptualized to be self-sufficient. Hence, a majority of residents' needs can be fulfilled without having to travel out of the town. HDB towns are provided with basic facilities like schools as well as a town center. Acting as the main social and commercial hub serving the whole town, the town center is usually integrated with transport nodes like an MRT station and/or a bus interchange, and has a large concentration of amenities including commercial facilities such as shopping malls, retail shops, restaurants, food outlets, offices and banks, entertainment facilities such as cinemas, and social facilities such as libraries, polyclinics, and town plazas.

Each town comprises several neighborhoods, each of about 4,000 to 6,000 residential units, and served by a neighborhood center. Residents are able to visit their neighborhood centers which has amenities like large supermarkets and a variety of shops for day-to-day needs. Neighborhoods are further split into different precincts of about 400 to 800 residential units. These precincts usually have a small number of shops and facilities such as

S

Strategies for Liveable and Sustainable Cities: The Singapore Experience, Fig. 7 Paya Lebar Quarter is a development integrated with office, retail, and residential uses, as well as public spaces. Images courtesy of Alison Lee

fitness corners and playgrounds (Housing & Development Board n.d.).

Our Tampines Hub, which was opened in 2017, is an example of Singapore's first integrated community and lifestyle hub, located just outside the town center (Fig. 8). Besides retail shops, it combines public amenities including sports facilities, a library, a theatre, and a Public Service

Strategies for Liveable and Sustainable Cities: The Singapore Experience, Fig. 8 Our Tampines Hub is a key destination in Tampines town for many residents. Image courtesy of Alison Lee

Centre which brings together the services of multiple government agencies at one central location (Yong 2017). It has since become a popular activity hub for Tampines town. Following its success, more of such community hubs are coming up in other towns, built by both the public and private sector.

The HDB has also developed a new typology of neighborhood centers – the new generation neighborhood center – which places emphasis on facilitating social bonding by providing abundant community spaces. Completed in 2018 in Punggol town, Oasis Terraces is the first new generation neighborhood center (Fig. 9). Besides services provided by typical shopping malls, Oasis Terraces which has won many accolades and international awards, also features a community plaza, rooftop gardens, and a polyclinic (Seow 2019). More new generation neighborhood centers are in the pipeline to better meet the needs of today's residents.

The town center and neighborhood centers, as well as transportation networks are carefully planned and located to achieve the vision of highly-accessible towns, where such facilities are easily reached by walking, cycling, and public transport. Under the Land Transport Master Plan 2040, the Land Transport Authority (LTA) aims for residents to be able to reach their nearest neighborhood center within 20 min by walking or cycling.

During the COVID-19 pandemic, Singapore's polycentric structure also came in handy when many started working from home. Residents were able to carry out their daily activities within their towns and neighborhoods, minimizing the need to travel long distances.

Strategy 4: Ensuring Connectivity and Walkability

Currently, there are more than 9,000 km of roads and expressways in Singapore, taking up about 12% of land area (Centre for Liveable Cities 2018b). In view of the country's limited space, the government has been trying to reduce the demand for cars and private transport so as to allocate land to other essential uses. Hence, Singapore is working toward its goal of becoming a car-lite society by prioritizing public transport, walking, and cycling.

Strategies for Liveable and Sustainable Cities: The Singapore Experience, Fig. 9 Oasis Terraces (right) features extensive landscaping. Image courtesy of the Housing & Development Board

Promoting the use of public transport, Singapore will expand its rail network to about 360 km in 2030. This allows more people to have easy access to public transport — 8 in 10 households in Singapore will be within a 10-min walk from a train station by 2030 (Land Transport Authority 2021b). This is instrumental for the Land Transport Authority (LTA)'s eventual goal of creating a 45-minute city, where people can get from homes to their places of work more quickly.

Singapore is also working toward better first-and-last-mile connectivity. In light of Singapore's tropical climate, 200 km of covered linkways were installed from 2015 to 2018. By 2040, another 150 km will be added, improving the integration of MRT stations with nearby residential areas and amenities (Land Transport Authority 2020). This would further encourage active mobility and establish public transit as a main mode of travel.

Commuters also have the option of cycling to complete the last stretch of their travel. To enhance the cycling experience, it is important that the relevant amenities and infrastructure are in place (Fig. 10). The LTA has committed to building more than 1,300 km of cycling paths islandwide by 2030 (Land Transport Authority 2021a). Improvements are also in the works for wayfinding signs as guidelines are developed for consistent wayfinding and communication of information to land transport users. As private developers are required to submit a Walking and Cycling Plan as part of development proposals, this ensures that surrounding developments and key public transport nodes are well-connected for people who walk or use active mobility devices.

With these measures in place, there are also opportunities to start repurposing more road space for pedestrians. For instance, the 2019 Master Plan has proposed the transformation of one of the major roads in the Downtown area, Robinson Road, into a Transit Priority Corridor. By converting car lanes into bus lanes, cycling paths, or pedestrian walkways, this can help to improve the connectivity and walkability within the Central Area. At the same time, there is the potential to design vibrant walkways through an array of activity-generating uses on the first floor like alfresco dining (Urban Redevelopment Authority 2019a).

In the heartlands, regular upgrading programs for HDB towns like the Neighbourhood Renewal Programme (NRP) and Remaking Our Heartland (ROH) program are implemented to ensure that

Strategies for Liveable and Sustainable Cities: The Singapore Experience, Fig. 10 A curved cycling bridge at Braddell Road was completed in 2020. Image courtesy of Alison Lee

universal design is considered in the built environment, and that homes and precincts are well maintained and refreshed with new facilities and amenities.

Under the ROH program, the Bedok Town Centre was revitalized in 2015 with an enhanced pedestrian mall. By linking the MRT station with the town square, retail shops, shopping mall, and integrated community hub, the upgraded walkway boosts accessibility for residents travelling between the different facilities and commercial nodes. Inclusive design in the form of wheelchair-friendly ramps were also included as part of the revamp (Fig. 11).

These efforts reflect the government's investment in Singapore's internal connectivity, allowing people to travel efficiently and conveniently both within their residential areas and across the island.

Strategy 5: Convenient Access to Green and Blue

Far from being a concrete jungle, Singapore has incorporated plentiful green and blue spaces into the city. Given the high density in Singapore, it is essential that nature is easily accessible for the well-being of its residents. The idea of creating a Garden City started in 1967 to transform Singapore into a clean and green haven to attract tourism and investments while also improving liveability for residents. The vision has since evolved over the years to a City in a Garden, City of Gardens and Water and now, a City in Nature (see: Chap. X, Cities in Nature).

At a strategic level, Singapore has conserved four large biodiversity core areas despite intensive urbanization in other regions. These areas include the nation's nature reserves and surrounding these are networks of nature parks established by the National Parks Board (NParks). These nature parks help to buffer the reserves from the impacts of nearby development and also provide complementary habitats for biodiversity. In the heart of the city, there are national gardens such as Gardens by the Bay and the Singapore Botanic Gardens, the latter also being an institution for botanical research in the region. They provide

Strategies for Liveable and Sustainable Cities: The Singapore Experience, Fig. 11 The Bedok Town Center before (top) and after (bottom) the ROH program. Ramps have been added to ease mobility of wheelchair users. Images courtesy of the Housing & Development Board

green spaces for residents to enjoy and have also become popular tourist destinations. Another example is the Marina Barrage, a downtown freshwater reservoir and flood control tool. In contrast to conventional grey infrastructure, it features a large open space for the community to enjoy and is open for water sport activities within the Marina Reservoir.

Outside of the Central Area, there is a hierarchy of parks around Singapore. On the regional scale, there are larger parks such as East Coast Park and West Coast Park. Closer to homes, there are smaller neighborhood and community parks that residents can visit for leisure.

While the early greening efforts focused on building parks and tree planting initiatives, the 1991 Concept Plan put forth a Green and Blue Plan where green spaces and waterways would link up, enabling people to take part in recreational activities and have better access to nature. Using space innovatively, NParks developed a Park Connector Network (PCN) of shared paths, footpaths, and cycling lanes built along roads, storm water canals, and below viaducts. Some of these were strips of land that would have otherwise been left unused aside from occasional servicing of canals. To serve as a green mobility network, the current network of 340 km of park connectors will be further extended by 160 km to 500 km by 2030. As a result, all households would be within a 10-minute walking distance to a park (Heng 2020).

Singapore is also pushing for the integration of green spaces in private developments (Fig. 12). As more buildings come up, it is crucial that greenery is still preserved. Hence, the URA introduced the Landscaping for Urban Spaces and High-Rises (LUSH) program in 2009, where developers are required to replace greenery lost due to development within the new premises. There are also incentives available for the provision of green communal spaces like sky terraces and communal pavilions (Urban Redevelopment Authority 2021b). Since 2017, more sustainability-related amenities such as communal roof gardens and urban farms are also eligible to qualify for Gross Floor Area (GFA) exemptions (Urban Redevelopment Authority 2017). Since 2009, 176 hectares of vertical greenery have been incorporated into developments through the scheme, providing a backdrop for community engagement (Urban Redevelopment Authority 2020b). NParks also encourages the integration of greenery into private developments through its Skyrise Greenery Incentive Scheme. Introduced in 2009, it provides financial incentives for developers and building owners to incorporate rooftop and vertical greenery into their projects across Singapore.

In 2006, PUB, Singapore's National Water Agency, launched the Active, Beautiful, Clean Waters (ABC Waters) program. The ABC Waters program aims to transform drains, canals, and reservoirs beyond their utilitarian functions into multi-functional spaces along waterways that enhance the overall liveability of the surroundings. Through the use of these public spaces, the public would be able to better appreciate water as a precious resource and in turn, help keep waterways clean and litter-free.

A prime example is the rejuvenation of Kallang River at Bishan-Ang Mo Kio Park which not only increased the capacity of the waterway but also created vibrant community spaces for the park visitors and an additional habitat for biodiversity. Adopting the concept of integrating Kallang River with the adjacent park, the existing concrete canal was naturalized into a river using soil bio-engineering techniques (Fig. 13). The canal-turned river has become an accessible flood plain where people can enjoy recreational activities and observe wildlife during dry weather (Fig. 14). When it rains, water levels in the river will rise and the park space adjacent to the river will be used as a flood plain to channel storm water to the downstream Marina Reservoir. For public safety, there are warning systems in place to alert park users about rising water levels.

Kallang River at Bishan-Ang Mo Kio Park has become one of the flagship projects under the ABC Waters program. Since then, PUB has completed 48 projects island-wide and over 80 public and private developments have attained ABC Waters certification for their developments.

As a result of these measures, the total green cover has risen from about a third in the 1980s to around 40% today, even as Singapore continues to become more built-up. The Singapore Green Plan 2030 announced in February 2021 demonstrates Singapore's resolve to be even more sustainable. Its targets include planting one million more trees from 2020 to 2030 and introducing over 130 hectares of new parks by end-2026 (Singapore Green Plan 2021).

Strategies for Liveable and Sustainable Cities: The Singapore Experience, Fig. 12 The Khoo Teck Puat Hospital is an example of a development generously integrated with greenery. Images courtesy of Alison Lee

Strategy 6: Building a City for All Ages

Alongside Singapore's enhanced liveability, the life expectancy of its citizens has risen in parallel. Compared to 1960 when the average life expectancy was 62.9 years, life expectancy at birth in 2020 is 83.9 (Singstat 2020). Coupled with a low replacement rate, the country is now facing a greying society. By 2030, one in four Singaporeans will be aged 65 and above (Strategy Group 2021). It is thus important that there is long-term planning to cater to the future makeup of the population.

While there are nursing homes available for elderly who require long-term care and greater support as well as private retirement resorts for wealthier retirees, there was a gap in the market in terms of independent living options for middle-income seniors. The government thus sought to address this by introducing new public housing concepts.

Strategies for Liveable and Sustainable Cities: The Singapore Experience, Fig. 13 Kallang River before (top) and after (bottom) naturalization. Images courtesy of PUB, Singapore's National Water Agency

In 2017, the HDB completed a public housing development, Kampung Admiralty, which combined elderly housing with healthcare facilities, dining and retail areas, and communal facilities like community gardens (Fig. 15). Akin to a vertically connected one-minute city, the co-located facilities provide convenience to residents and reduce the need for travel outside of the integrated development. The residential units come in the form of Studio Apartments

Strategies for Liveable and Sustainable Cities: The Singapore Experience, Fig. 14 Visitors interacting with the water at Bishan-Ang Mo Kio Park. Image courtesy of PUB, Singapore's National Water Agency

sold with shorter 30-year leases, although some units (i.e., the balance/returned units) have since been offered to eligible seniors as short-lease 2-room Flexi flats. For the safety of residents, elderly friendly features like grab bars and slip-proof flooring are also built into each unit.

To keep the seniors at Kampung Admiralty engaged, the development has an Active Ageing Hub which plans activities and programs for the residents. Benches are also placed at unit entrances to facilitate interactions between neighbors. To promote inter-generational bonding, a childcare center is also housed in Kampung Admiralty so that residents can interact with the children through storytelling and craft workshops. With the success that Kampong Admiralty has seen, the HDB will build more of such integrated developments where there are suitable sites (Centre for Liveable Cities 2020).

A new assisted living public housing concept was also jointly developed and piloted by the Ministry of National Development (MND), Ministry of Health (MOH), and HDB in early 2021. Known as Community Care Apartments, the flats allow residents to live independently with the reassurance that help is readily available through regular checks and a 24-h emergency response system. Residents can also opt for additional services such as housekeeping and meal deliveries depending on their care needs. In addition, Community Care Apartments are designed with communal spaces on each level and an activity center within the block to encourage social interactions. There will also be a community manager appointed to organize activities and assist with arranging care services (Housing & Development Board 2020).

In addition, therapeutic gardens are being set up by NParks in parks and gardens across the island. Designed to facilitate interactions with nature and improve mental and physical well-being, they benefit a range of users including the elderly and those with dementia.

Strategies for Liveable and Sustainable Cities: The Singapore Experience, Fig. 15 Kampung Admiralty houses a multitude of uses for the convenience of its residents. Images courtesy of the Housing & Development Board

Beyond providing hardware, communities also need to adapt to become more senior- and dementia-friendly. Based on the Well-being of the Singapore Elderly (WiSE) study in 2013 spearheaded by the Institute of Mental Health, dementia affects 1 in 10 people aged 60 and above (Institute of Mental Health 2015). The Dementia-Friendly Singapore initiative by the Agency for Integrated Care (AIC) thus hopes to build a more inclusive society through the creation of Dementia-Friendly Communities (DFCs). A DFC aims to encourage persons living with dementia to continue living at home and go about their usual routines in their community. In a DFC, AIC works with various community stakeholders and partners to raise awareness of dementia and train frontline staff such as business and service staff to recognize, respect, and assist persons living with dementia. AIC also develops community resources to support persons living with dementia and their caregivers, as well as enhance the physical environment to make it safer and easier for them to move around. For instance, members of the public can bring those who may appear lost to Go-To Points to get help from trained staff to connect them with their caregivers. As of April 2020, there are 14 DFCs islandwide.

Strategy 7: Promoting Healthy Living

Besides creating a City in Nature which brings about physical and mental health benefits, there are other efforts in place that help to support a healthy population and environment.

Following a reorganization in 2017, healthcare facilities are now organized into three clusters taking care of the central, eastern, and western regions, respectively. This is similar to the distributed approach Singapore adopts for commercial, residential, and recreational spaces. Each healthcare cluster can better optimize resources while ensuring that residents have easy access to a range of facilities such as acute hospitals, community hospitals, and polyclinics close to their homes (Ministry of Health 2017).

To encourage active living, Sport Singapore launched ActiveSG in 2014 as a national movement to inspire and enable residents to lead healthier lives through sport and physical activity regardless of their age and ability. All Singaporeans and Permanent Residents were provided complimentary ActiveSG memberships with credits to access sporting facilities and programs around Singapore. There are also free programs available for anyone to join in at the 26 ActiveSG Sport Centers.

Catering to the older generation, Sport Singapore announced in 2020 that Singaporeans above the age of 65 will enjoy free entry to ActiveSG swimming pools and gyms. Gyms will also be made senior-friendly and all ActiveSG gyms will be made inclusive by 2026 (Chia 2020). Senior-friendly programs such as morning pool walks at the ActiveSG swimming pools get seniors involved in low impact physical activity and help them build water confidence before they eventually progress to activities such as aqua-aerobics. The easy access to sporting facilities and programs allows seniors to sustain their health and mobility longer, and continue to be engaged in the community.

The Sport Facilities Master Plan guides the development of quality, affordable common sport spaces at the national, regional, town, and neighborhood levels. Regardless of the scale and scope of a given facility, they all share the same underlying purpose: to encourage people to live actively through sport, anytime, anywhere. The Singapore Sports Hub, a Tier 1 national facility hosts community programs as well as many international sport events that enable people to come together to watch, cheer, and bond over sport. The Regional Sport Centres (RSCs) are focal points for sport programs, events, and activities for the five main regions in Singapore (Central, East, Northeast, North, and West). Opened in August 2017, Our Tampines Hub (OTH) was the first RSC and serves the Eastern region.

Residents in neighborhoods such as Jurong Spring, Taman Jurong, and Bukit Batok are also able to access sporting spaces under the Sport-in-Precinct initiative. These spaces make it convenient for children, adults, and seniors to enjoy sport and incorporate all-weather spaces, with sheltered and shaded areas, as well as hard courts designed for multi-purpose use that can be easily reconfigured for basketball, street soccer, badminton, sepak takraw, and other team sports. The exercise equipment provided has also been carefully selected for use by different ages and abilities.

Sport Singapore seeks to strengthen the sporting landscape and is committed to providing the majority of residents with access to a sporting facility within a 10-min walk from their homes by 2030.

On a daily basis, it is also imperative that the city is kept clean with proper waste collection systems in place. An example of how Singapore has innovated in this aspect is the Pneumatic Waste Conveyance System (PWCS). Spearheaded by the HDB, the PWCS is an automated system that uses air suction to transport solid waste from public housing blocks to a centralized bin center through an underground pipe network. The PWCS is especially useful in Singapore's high-density context as it reduces the manpower required to collect waste from individual buildings. It also makes for an overall cleaner and more hygienic living environment.

After piloting the system in public housing, the government has also mandated the use of the PWCS in condominium developments with more than 500 dwelling units (Boh 2017). The HDB has also scaled up the use of PWCS at the district level for newer HDB estates such as Bidadari and Tengah. The private sector is also expected to follow suit for a large master developer site currently up for sale at Kampong Bugis.

Strategy 8: Preserving Culture and Heritage

Although Singapore developed rapidly, planners made conscious efforts during the urban renewal process to ensure that there was conservation of built heritage. The Preservation of Monuments Board was established in 1971 to safeguard important monuments and landmarks. Detailed conservation plans were also made in the 1980s. Released in 1986, the Conservation Master Plan demarcated seven areas for conservation – Chinatown, Kampong Glam, Little India, Boat Quay, Clarke Quay, Emerald Hill, and Heritage Link (Centre for Liveable Cities 2019).

Singapore's built heritage mainly consisted of traditional shophouses of two- to three-storeys constructed prior to World War 2. By the 1980s, many of these buildings were rundown and commercially unattractive to the private sector. The URA thus took the initiative to restore 32 shophouses in Tanjong Pagar in 1987, showing through this pilot that restoration was indeed possible. As part of this effort, the government built back lanes for basic utilities such as water and electricity, while ensuring that the original character of the shophouses was respected.

However, a government-led restoration initiative would be unsustainable in the long run and the URA tapped on the land sales program in the late 1980s to sell unrestored shophouses to the private sector for adaptive reuse. Planners drew up detailed conservation guidelines and requirements to guarantee that restoration was properly executed. One of the shophouses restored by the government at 9 Neil Road set the restoration benchmark for the private sector (Fig. 16). Despite a poor economic outlook at the time, the land sale was a success and spurred private sector restoration of shophouses in cultural districts like Chinatown, Little India, and Kampong Glam (Centre for Liveable Cities 2021).

The year 1989 was another turning point in Singapore's conservation journey when the URA became the national conservation authority. It was also the year when 3,200 shophouses identified in the Conservation Master Plan were officially gazetted for conservation (Urban Redevelopment Authority 2021a). Since then, the URA has fostered a strong partnership between the public and private sectors in conservation.

At times, the government also sells sites with conserved buildings to the private sector through a Concept & Price Revenue tender under the GLS program. This mode of sale is typically used for strategic sites that require a high-quality design and development concept. A Concept Evaluation Committee would first evaluate all submitted concept proposals. Subsequently, only satisfactory concept proposals will compete based on price and the tenderer offering the highest bid price would win the tender, given that it is above the Reserve Price for the site (Centre for Liveable Cities 2021). Hence, this ensures that the conserved structure is sensitively incorporated into the development. Some examples include Capitol Singapore which houses the conserved Capitol Theatre, Capitol Building and Stamford House, and South Beach which includes four conserved buildings – the former Non-Commissioned Officers' Club and three blocks of the former Beach Road Camp. In each case, the conserved buildings have been meticulously restored and adaptively reused by the property developer.

With the majority of historically significant buildings having been conserved, more attention was put on enhancing the soft aspects of culture

Strategies for Liveable and Sustainable Cities: The Singapore Experience, Fig. 16 The shophouse at 9 Neil Road before (top) and after (bottom) restoration. Images courtesy of the Urban Redevelopment Authority

and heritage. This has been done through placemaking by introducing heritage trails and markers which help to highlight stories about the historical districts and create a more immersive experience for tourists.

Local stakeholders also took the initiative to revitalize their districts. For instance, street murals in Kampong Glam made the area an Instagrammable spot for locals and tourists alike. In addition, businesses and communities in cultural districts came together to form precinct associations to carry out place management. In collaboration with the government, these precinct associations often hold events that better activate heritage areas and boost vibrancy. To illustrate, the precinct association at Kampong Glam, One Kampong Gelam (OKG), was formed in 2014 to represent stakeholders in the area. OKG has worked with the government to implement road closures to open up space for activities and hold unique outdoor festivals and events, including a fashion show, an interactive gallery, and music competitions (One Kampong Gelam 2021; Fig. 17).

In fast-paced Singapore, these efforts are important to give Singaporeans a sense of identity. The conservation of architecture also helps to showcase the rich multi-racial and multi-religious heritage of the country.

Underlying Enablers

With these strategies in place, Singapore has also strengthened key enablers to ensure that plans can be implemented successfully.

Technology as an Enabler

Singapore is moving toward becoming a Smart Nation where digital technology is harnessed to

Strategies for Liveable and Sustainable Cities: The Singapore Experience, Fig. 17 Cultural performances were held at Kampong Glam during road closures. Image courtesy of the Urban Redevelopment Authority

transform the way we live, work, and play. For urban planners, technology is deeply rooted in the way plans are drawn up. At the URA, data analytics and geospatial technologies are used to study population and mobility trends. Insights are then channeled toward informing planning decisions.

In 2013, the URA also started the Digital Planning Lab which uses digital tools to enhance and improve planning processes. A suite of digital planning tools such as ePlanner was developed to allow for better visualization and analysis of data. For example, planners can easily study a town's demographics to produce a plan for renewal that is tailored to the residents' needs, or analyze the accessibility of amenities. Spatial data analytics has also been used to study how the COVID-19 pandemic has affected activity patterns. Going forward, the URA will also tap on artificial intelligence for anticipatory urban planning to better react to a dynamic environment (Urban Redevelopment Authority 2020a).

To tackle climate change, the Cooling Singapore project was started to better understand the Urban Heat Island (UHI) effect. It aims to build a Digital Urban Climate Twin (DUCT) for Singapore through the integration of various models and provide insights for climate-responsive design guidelines (Cooling Singapore 2021). The research project thus contributes toward a liveable and sustainable environment by ensuring thermal comfort for residents.

Innovations in technology have also enabled the city to become more resilient to disruptions. In response to new safe distancing measures enforced during the pandemic, government agencies swiftly released platforms such as the URA's Space Out and NParks' Safe Distance @ Parks which use real-time data to help people make informed decisions on their outdoor visits. To make contact tracing easier, the Government Technology Agency has also developed apps such as SafeEntry for visitors to digitally check-in at premises and TraceTogether to identify persons who were in close contact with infected individuals (Fig. 18).

Resilient Infrastructure as an Enabler

For a liveable environment, comprehensive urban infrastructure has to be in place to ensure that the everyday needs of residents can be met. At the same time, the infrastructure has to be resilient to shocks and stressors so that all aspects of life can continue to run smoothly.

S

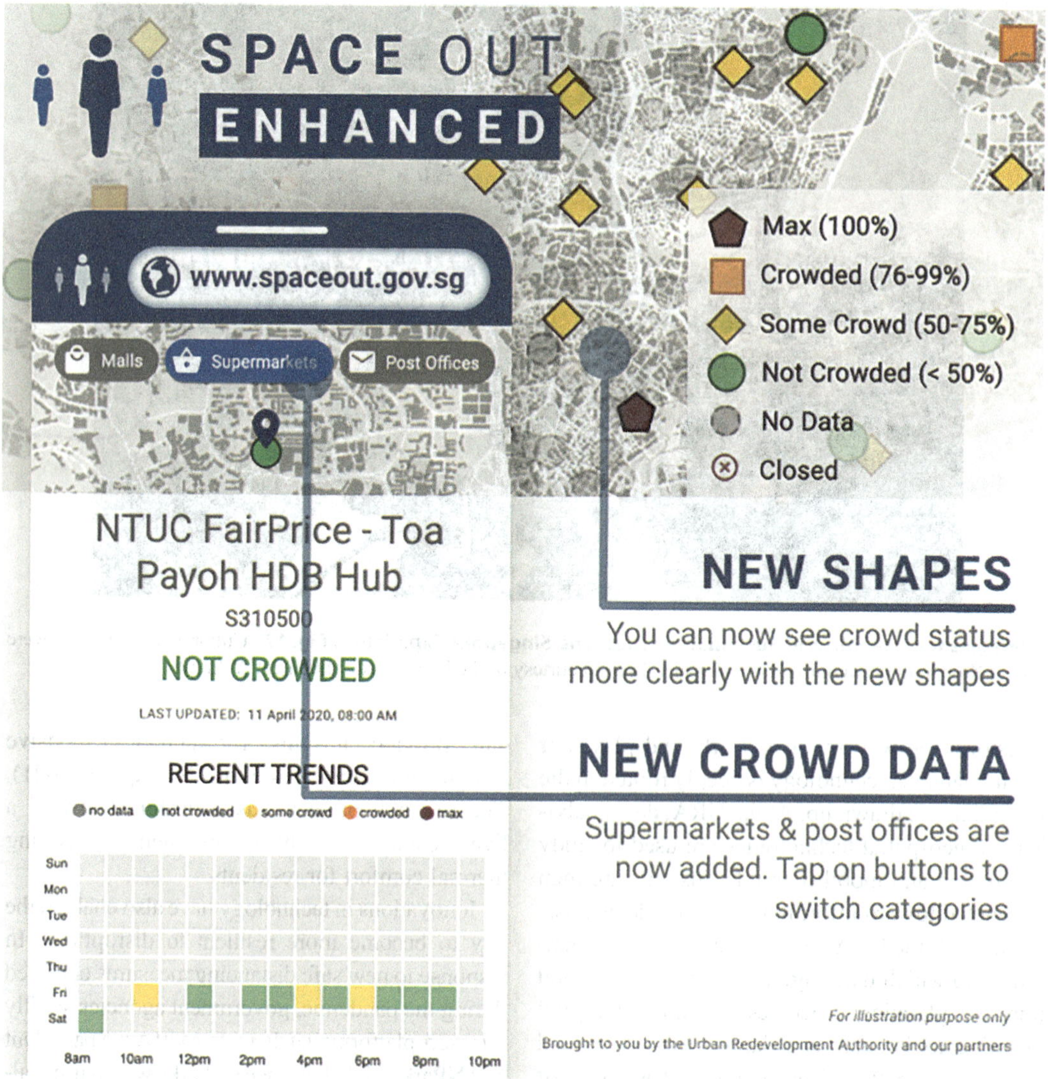

Strategies for Liveable and Sustainable Cities: The Singapore Experience, Fig. 18 Users of Space Out are able to check on crowd levels in locations such as supermarkets. Image courtesy of the Urban Redevelopment Authority

As a country that has limited natural resources and lacks the space to collect adequate rainwater to meet demand, Singapore has had to come up with innovative solutions to provide sustainable water supply. The government first studied the possibility of reclaiming used water as early as in the 1970s, and the first NEWater plants – where used water is treated and further purified into ultra-clean, high-grade recycled water – were eventually set up in 2003. NEWater is primarily channeled to industries for non-potable use (PUB, Singapore's National Water Agency 2021a). NEWater, together with local catchment, imported water, and desalinated water, form the Four National Taps that provide a diversified and sustainable water supply for Singapore.

Singapore's land size also imposes constraints on waste disposal facilities. Singapore's only operational landfill, Semakau Landfill, which commenced operations in 1999, occupies sea space between Pulau Semakau and former Pulau Sakeng (Fig. 19). However, it is projected to run

Strategies for Liveable and Sustainable Cities: The Singapore Experience, Fig. 19 Aerial view of Semakau Landfill. Image courtesy of the National Environment Agency

out of space by 2035 (Chong 2020) and it would not be sustainable to allocate more of Singapore's scarce land and sea space to accommodate another such landfill, even in the long run. To extend the lifespan of Semakau Landfill, one of the targets set under the Zero Waste Masterplan is to reduce the amount of waste sent to the landfill per capita per day by 30% by 2030 as Singapore shifts away from a linear economy and toward a circular economy for waste and resource management.

The future Integrated Waste Management Facility (IWMF), which will adopt an integrated approach to process multiple waste streams, will be a key initiative that will contribute to the Zero Waste Masterplan targets. Other initiatives to achieve these targets include developing Extended Producer Responsibility (EPR) policies such as the e-waste Producer Responsibility scheme implemented in July 2021, with plans to also implement EPR for packaging waste. Singapore is also developing local recycling facilities to support these policies and nurture the capability of the local recycling industry. Several e-waste recycling facilities have been set up, and Singapore is exploring mechanical and chemical recycling methods for plastics. Efforts are also underway to trial the use of incineration bottom ash (IBA) in constructing roads and non-structural concrete through project NEWSand (National Environment Agency 2019).

Singapore is also striving to increase the use of renewable energy such as solar energy and reduce its carbon footprint. A mega 60 megawatt-peak (MWp) floating solar farm of about 45 hectares located at Tengeh Reservoir was launched in July 2021. The solar energy produced is sufficient to power five local water treatment plants. It is thus a smart solution that not only addresses the energy issue but also cleverly makes use of space (PUB, Singapore's National Water Agency 2021b; Fig. 20).

S

Involve the Community as Stakeholders

As part of the planning process, the government employs a consultative mechanism by inviting the public to play a part in coming up with its plans. In June 2019, the Singapore Together movement was also launched to encourage the public to work with the government and also one another to build future Singapore. This approach

Strategies for Liveable and Sustainable Cities: The Singapore Experience, Fig. 20 The floating solar farm at Tengeh Reservoir. Image courtesy of PUB, Singapore's National Water Agency

is one of the principles of Dynamic Urban Governance in the CLC's Singapore Liveability Framework.

During regular reviews of the Master Plan, the URA gets feedback from the public by holding small scale exhibitions of proposed land use strategies and a larger-scale exhibition of the entire Draft Master Plan. This process takes place over a few years to allow planners sufficient time to consider the public's comments and revise the plans. Feedback is also obtained through other channels such as focus group sessions, community workshops and stakeholder meetings. The URA also works closely with stakeholders who would be affected by upcoming Master Plans. In such cases, the URA organizes engagement sessions and workshops to showcase proposed plans and understand any potential concerns. Through these in-depth sessions, planners are also able to tap on users on the ground for ideas to improve their plans (Urban Redevelopment Authority 2019b).

Another example is the development along the Rail Corridor, a former railway line and green belt. The URA held a long consultation process that spanned across 5 years. Various stakeholders, including interest groups, academics, and students were involved in the visioning for the Rail Corridor. This was done through a series of online and offline engagement sessions, as well as an Ideas Competition to solicit creative proposals. In 2015, the URA launched a Request for Proposal (RFP) for the private sector to come up with an overall plan to revitalize the Rail Corridor and the public's inputs were reflected in the design brief and criteria. The winning Concept Master Plan was later displayed in exhibitions around Singapore to obtain the community's views and further refine the plan. Concurrently, targeted workshops were organized for specific segments of the population, such as young families, seniors, and people with disabilities (Tan 2018).

Public engagement is also deeply embedded in the CLC's work. This is especially visible in projects such as Reimagining Tampines where CLC collaborated with government agencies, civic groups, students, and the private sector to come up with a comprehensive framework for the rejuvenation of ageing HDB towns. This helps to ensure that Singapore's polycentric approach remains relevant and towns continue to meet the needs of their residents. CLC also engaged residents at Cambridge Road, a flood prone area, to build up greater community resilience to the adverse effects of climate change. Even when participants were unable to meet up in person during the pandemic, discussions to co-create ideas to enhance their neighborhood continued online through Zoom (Fig. 21).

Ensuring Liveability and Sustainability in a Disrupted World

Singapore is one example of a city that has achieved a liveable and sustainable environment in a high-density context (Fig. 22). This can be attributed to the effective strategies formulated by

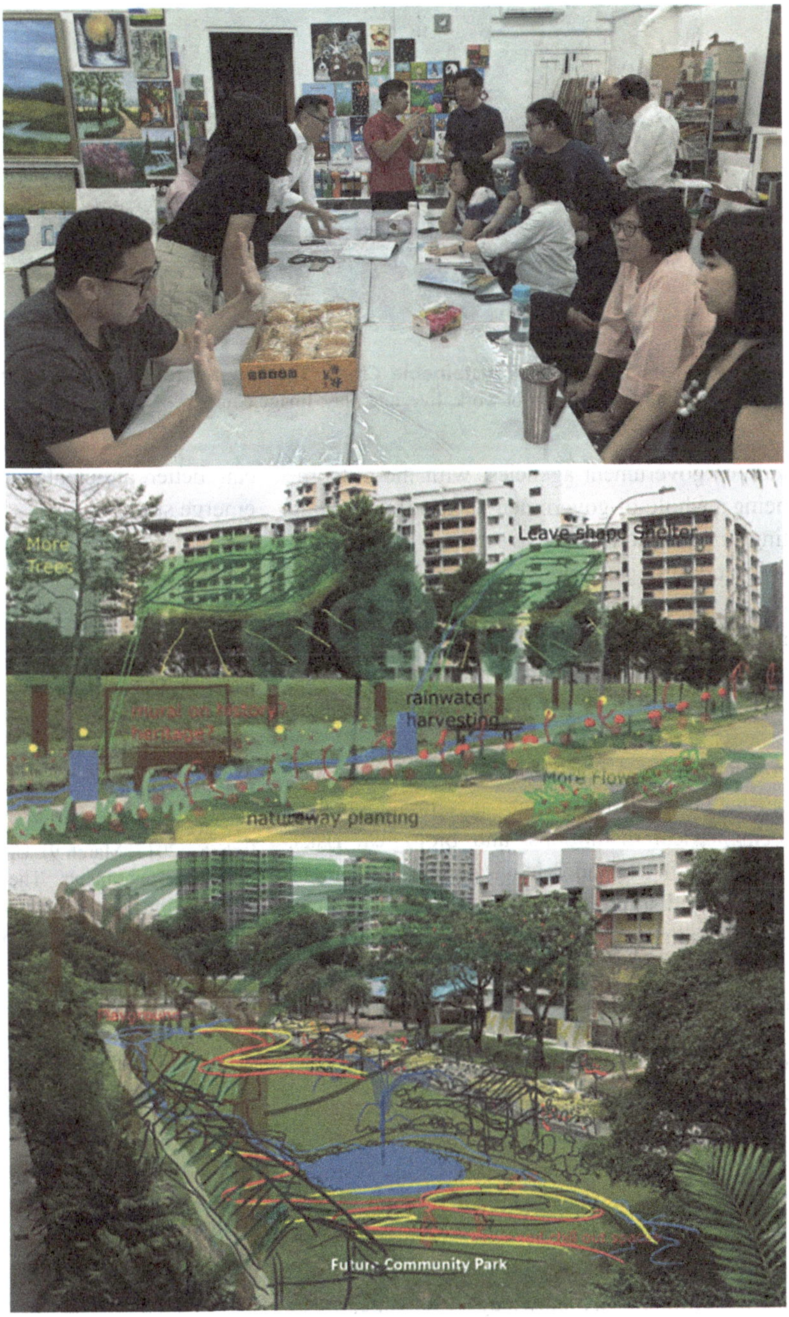

Strategies for Liveable and Sustainable Cities: The Singapore Experience, Fig. 21 The CLC facilitated meetings with residents to discuss climate resilience both in person and through virtual platforms. Image courtesy of the Centre for Liveable Cities

Strategies for Liveable and Sustainable Cities: The Singapore Experience, Fig. 22 Today, Singapore has developed into a thriving city for work, live, and play. Image courtesy of Yuchen Sun, Unsplash

various government agencies, with the outcome being a whole-of-government effort. Beyond putting in place comprehensive plans, good leadership and strong urban governance is critical to realize these plans. Urban planners should also make use of the right enablers by striving to innovate with technology, build resilient infrastructure, and partner the community in the planning process.

Cities around the world will inevitably face existential challenges such as climate change, technological disruptions, and biological risks. The COVID-19 pandemic has shown that shocks can easily disrupt essential facets of life such as supply chains and food resilience. Beyond liveability and sustainability, the next step would be to build resilience.

Building resilience requires better knowledge. With the wealth of knowledge gained in urban development, cities need to share experiences and lessons through constant dialogue. One such platform is the biennial World Cities Summit organized by the CLC and the URA which brings together government leaders and industry experts to discuss urban solutions and strengthen networks. Through this knowledge exchange, cities can better anticipate and react to shocks, and emerge stronger.

Cross-References

▶ Cities in Nature

References

Boh, S. (2017, March 9). Upgraded waste system for new private apartments. The Straits Times. https://www.straitstimes.com/singapore/housing/upgraded-waste-system-for-new-private-apartments.

Centre for Liveable Cities. (2018a). *Integrating land use & mobility: Supporting sustainable growth. Singapore urban systems studies series*. Singapore: Ministry of National Development.

Centre for Liveable Cities. (2018b). Streets for all: Designing multimodal streets for a car-lite Singapore. http://www.clc.gov.sg/docs/default-source/commentaries/bc-2018-05-multimodal-streets-for-all.pdf.

Centre for Liveable Cities. (2019). *Past, present and future: Conserving the nation's built heritage. Singapore urban systems studies series*. Singapore: Ministry of National Development.

Centre for Liveable Cities. (2020). Ageing well together. https://www.clc.gov.sg/docs/default-source/urban-

solutions/urb-sol-iss-16-pdfs/13_case_study-singapore-ageing-together.pdf.
Centre for Liveable Cities. (2021). *The government land sales programme: Turning plans into reality. Singapore urban systems studies series*. Singapore: Ministry of National Development.
Chia, N. (2020, March 6). Parliament: Free entry to all public gyms and swimming pools for Singaporeans aged 65 and above. The Straits Times https://www.straitstimes.com/politics/parliament-free-entry-to-all-public-gyms-and-swimming-pools-for-singaporeans-aged-65-and
Chong, C. (2020, September 26). Turning trash into treasure. The Straits Times. https://www.straitstimes.com/singapore/turning-trash-to-treasure-nea-seeks-to-reuse-landfill-materials.
Chow, C. (2015, March 2). Tanjong Pagar centre fills a void. The Edge Markets. https://www.theedgemarkets.com/article/tanjong-pagar-centre-fills-void.
Cooling Singapore. (2021). The project. https://www.coolingsingapore.sg/the-project.
Heng, M. (2020, March 5). S'pore's 2030 goal: More gardens, park connectors. The Straits Times. https://www.straitstimes.com/singapore/spores-2030-goal-more-gardens-park-connectors
Housing & Development Board. (1965). Annual report 1965. Housing & Development Board. p. 10.
Housing & Development Board. (2020). Singapore's first assisted living flats to be launched in February 2021. https://www.hdb.gov.sg/about-us/news-and-publications/press-releases/10122020-Singapores-First-Assisted-Living-Flats-to-be-Launched-in-February-2021.
Housing & Development Board. (n.d.). Town planning. https://www.hdb.gov.sg/about-us/history/town-planning
Institute of Mental Health. (2015). IMH link. https://www.imh.com.sg/uploadedFiles/Publications/IMH_Link/IMH%20Link%20Apr%20-%20Jun%202015.pdf.
Land Transport Authority. (2020). Land Transport Master Plan 2040. https://www.lta.gov.sg/content/dam/ltagov/who_we_are/our_work/land_transport_master_plan_2040/pdf/LTA%20LTMP%202040%20eReport.pdf.
Land Transport Authority. (2021a, April 9). Cycling. https://www.lta.gov.sg/content/ltagov/en/getting_around/active_mobility/walking_cycling_infrastructure/cycling.html.
Land Transport Authority. (2021b). Upcoming projects. https://lta.gov.sg/content/ltagov/en/upcoming_projects.html.
Lee, M. (2011). Pinnacle@Duxton. Singapore Infopedia. https://eresources.nlb.gov.sg/infopedia/articles/SIP_1779_2011-02-18.html.
Mercer. (2019, March 13). Quality of living city ranking. https://www.mercer.com/newsroom/2019-quality-of-living-survey.html
Ministry of Health. (2017, January 18). Reorganisation of healthcare system into three integrated clusters to better meet future healthcare needs. https://www.moh.gov.sg/news-highlights/details/reorganisation-of-healthcare-system-into-three-integrated-clusters-to-better-meet-future-healthcare-needs.
National Environment Agency. (2019, November 25). NEWSand: A key to closing Singapore's waste loop. https://www.nea.gov.sg/media/news/news/index/newsand-a-key-to-closing-singapore-s-waste-loop.
One Kampong Gelam. (2021). One Kampong Gelam. https://visitkamponggelam.com.sg/about/.
PUB, Singapore's National Water Agency. (2021a). NEWater. https://www.pub.gov.sg/watersupply/fournationaltaps/newater
PUB, Singapore's National Water Agency. (2021b). Floating solar systems. https://www.pub.gov.sg/solar/floatingsystems
Seow, J. (2019, February 17). First new-generation neighbourhood centre opens in Punggol. The Straits Times. https://www.straitstimes.com/singapore/housing/first-new-generation-neighbourhood-centre-opens-in-punggol.
Singapore Green Plan. (2021). Our targets. https://www.greenplan.gov.sg/key-focus-areas/our-targets/.
Singstat. (2020). Death and life expectancy. https://www.singstat.gov.sg/find-data/search-by-theme/population/death-and-life-expectancy/latest-data
Strategy Group. (2021). Population in brief 2021. https://www.strategygroup.gov.sg/files/media-centre/publications/Population-in-brief-2021.pdf
Tan, S. N. (2018, July 6). Co-creating the rail corridor's future. Civil Service College. https://www.csc.gov.sg/articles/co-creating-the-rail-corridor's-future.
Urban Redevelopment Authority. (2000). Annual report 1999/2000. Urban Redevelopment Authority. p. 25.
Urban Redevelopment Authority. (2017). Updates to the Landscaping for Urban Spaces and High-Rises (LUSH) Programme: LUSH 3.0. https://www.ura.gov.sg/Corporate/Guidelines/Circulars/dc17-06
Urban Redevelopment Authority. (2019a). Our downtown. https://www.ura.gov.sg/Corporate/Planning/Master-Plan/Regional-Highlights/Central-Area/Downtown.
Urban Redevelopment Authority. (2019b). Master plan. https://www.ura.gov.sg/Corporate/Planning/Master-Plan/Introduction.
Urban Redevelopment Authority. (2019c, March 27). Rejuvenation incentives for strategic areas: Central Business District (CBD) incentive scheme. https://www.ura.gov.sg/Corporate/Guidelines/Circulars/dc19-04
Urban Redevelopment Authority. (2020a). Digitalisation. https://www.ura.gov.sg/Corporate/Planning/Our-Planning-Process/Digitalisation.
Urban Redevelopment Authority. (2020b, August 25). Seeing the city in a new light. https://www.ura.gov.sg/Corporate/Resources/Ideas-and-Trends/seeing-the-city-in-new-light.
Urban Redevelopment Authority. (2021a). Brief history of conservation. https://www.ura.gov.sg/Corporate/Get-Involved/Conserve-Built-Heritage/Explore-Our-Built-Heritage/brief-history.

Urban Redevelopment Authority. (2021b). Commercial. https://www.ura.gov.sg/Corporate/Guidelines/Development-Control/Non-Residential/Commercial/Greenery.

Yong, C. (2017, August 7). One-stop lifestyle centre launched in Tampines. The Straits Times. https://www.straitstimes.com/singapore/one-stop-lifestyle-centre-launched-in-tampines.

Strategies for Taming the City

Debates in Managing Peri-Urbanization

Ndarova Audrey Kwangwama[1], Kadmiel H. Wekwete[2] and Innocent Chirisa[3]
[1]Department of Architecture and Real Estate, University of Zimbabwe, Harare, Zimbabwe
[2]Midlands State University, Gweru, Zimbabwe
[3]Department of Demography Settlement and Development, Social & Behavioural Sciences, University of Zimbabwe, Harare, Zimbabwe

Synonyms

Peri-urban – city-edge; Policy – direction; Strategy – action planned; Taming – containing

Definitions

Greening refers to "the process of making somewhere greener by planting grass, trees and plants there" or "the process of beginning to pay attention to the protection of the natural environment." Both definitions are relevant in the article. https://dictionary.cambridge.org/us/dictionary/english/greening.

Policy refers to "law, regulation, procedure, administrative action, incentive or voluntary practice of governments and other institutions" (Centres for Disease Control and Prevention).

Resilience can be traced back to the Latin word *resilio* meaning to bounce back/spring back (Dohwe and Kwangwama 2019). Resilience is also defined as the capacity of a system or community to survive, cope, recover, and reorganize efficiently and timeously after experiencing a physical hazard, shock, or disturbance (The World Bank 2012: 3; www.resalliance).

System originates from the Greek word "systems" that comes from "syn" defined as "together" and "histemi" defined as "to set" (Jenkins 1969: 2). A system refers to "a set of connected things or devices that operate together." https://dictionary.cambridge.org/us/dictionary/english/system

Sprawl when defined in the context of urbanization refers to "excessive spatial growth of cities taking up much open space and encroaching excessively on scarce agricultural land" (Brueckner 2000: 2, 3).

Introduction

Urbanization has increasingly become a common phenomenon globally. This has mainly been due to the increased flocking of people into urban centers as a way of looking for better employment opportunities, better health, and better educational facilities among other causes. The increase in numbers of urban dwellers has ultimately resulted in the increase in housing demand and, hence, the outward expansion of cities as edge-of-city areas form (Pradoto 2011). This outward expansion of cities have, in some instances, been associated with a positive economic advantage due to the spread of investment activities in the periphery, a feature which sometimes arises from foreign direct investment (Webster and Muller 2009). The spread of investment opportunities would then bring, in some instances, positive trickling down effects to the native lower-middle class communities through employment creation, which further improves the living standards of the locals (Pradoto 2011). However, the increased eating away of the periphery has been associated with more harm than good in many instances. This is because the increased eating away of the city's peripheral areas has been associated with land conflicts and water woes, as land for agricultural purposes is changed to other uses, in turn posing a threat to the urban region's food security. Studies have also revealed that peri-urbanization sometimes results in the appropriation of water and land, which may cause conflicts among the peri-urban water users, the peri-urban water users and the state, and/or between urban users and peri-urban users, as the scramble for land and water begins (Narain 2016). The mixing of cultures between the new residents and the local people of the area has also been received with much negativity (Pradoto 2011), on the fears that the local culture of the area will be diluted, resulting in the area losing its originality. This chapter reflects on the issue of peri-urbanization in the developing world, how it has affected sustainable development, and how various nations have engaged in various activities aimed at controlling the increased sprawling of their cities.

Conceptual Framework

The peri-urbanization concept is not new to the urban planning discourse. Due to the continued increase of people in the urban landscape, as they search for new opportunities and better living conditions, peri-urban landscapes have been formed. The types of outward developments have in some instances been regarded as chaotic (Ravetz et al. 2013), and in many instances, the eating away of agricultural land has been witnessed (Webster and Muller 2009), hence the disturbance in the peri-urban people's way of life. The increased shrinkage of urban agriculture would in a way pose a threat to the food security of the region. In this regard, many governments have engaged in various ways to control or manage the process of peri-urbanization, for it not to negatively affect the environment and the local people's livelihoods as well. This has resulted in adoption of policies such as greening the environment, through the adoption of green technologies (Mohazzam et al. 2020) and containment measures, aimed at containing the continued outward growth of cities through more compact and dense development processes where more people are accommodated within a smaller urban space. Other strategies involve the engagement of the local people in development for them to be on guard and stop peri-urbanization (Hudalah et al. 2007). Also, policy-makers must view the region as a system in order to stop the uncontrolled growth of cities through sprawl.

Managing Peri-Urbanization: A Review

Peri-urbanization refers to process where rural areas, which are located at the city edges of well-established cities, become more urban-like in character, with major transformations being recorded in the physical and socioeconomic spheres, usually in a piecemeal approach (Webster and Muller 2009). Peri-urban developments in Europe have been characterized by a swath of discontinuous low density development, and these areas represent a mixed rural-urban character, which is highly dynamic in nature, and studies have also showed that, besides socioeconomic changes, peri-urbanization is also characterized by land cover and land-use changes, planning process changes, and environmental and land management changes (Shaw et al. 2020). This is mainly as a result of the land-use transformations which sometimes exist from the zone of much agricultural activity to a zone with more residential and commercial or industrial uses. This is further supported by Webster and Muller's (2009) argument which stresses that the peri-urban people's way of life is quickly transformed from subsistence agricultural economies to the industrial or urban way of life within a short space of time, which usually results in infrastructural backlogs and the rapid deterioration of the environment.

Whether planned or unplanned, most peri-urban developments have been associated with negative environmental health risks. The nature of problems or challenges experienced in each peri-urban setup differs according to the peri-urban development's trajectory which differs according to the level of a country's development, from the more developed South Korea to the developing countries such as Thailand and the Philippines, which helps in anticipating future challenges for some regions (Webster and Muller 2002). A study carried out in a growing city in Mexico on 98 households in a planned community and 202 households in unplanned communities showed that both settlements were vulnerable to poor environmental health conditions, with other study results showing fewer cases of respiratory problems ($P = 0.039$) and dizziness ($P = 0.009$) in households belonging to planned settlements (Graham et al. 2004). This indicates that planned settlements are less likely to suffer from negative environmental health implications compared to unplanned peri-urban settlements. Other scholars have termed the peri-urban/the urban fringe in newer industrializing nations and most of the developing world, a zone of chaotic urbanization that usually leads to sprawl (Ravetz et al. 2013). The issue of chaos and/haphazardness usually results from the unplanned nature of settlements; as such, they may lack proper sewerage and proper water reticulation, which may in turn pose a health hazard to the residents. Peri-urban zones have also been associated with relatively low population densities, compared to urban areas, high dependence on transport for commuting, scattered and fragmented settlements, and the lack of spatial governance (Ravetz et al. 2013), which may mean haphazard uncontrolled developments in some cases.

One of the most obvious causes of peri-urbanization has been the rapid growth in population numbers and the increased demand for land for development. In Indonesia's Bandung Metropolitan region, three types of peri-urban developments have emerged, and these have been identified as urban, semi-urban, and potential urban areas, where the urban area is located in the Bandung boarder and has significant physical and socioeconomic characteristics, with the semi-urban and potential urban areas having mixture of both the rural and urban characteristics (Budiyantini and Pratiwi 2016). Peri-urbanization has therefore been associated with the widening of the gap as regards socioeconomic inequalities, which may promote resistance to change by the locals as people try to locate in the city boundaries as they may fear culture dilution (Zhao 2013).

While manufacturing remains a key driver to rapid economic and social regional transformations, the spread of higher economic activities has resulted in the creation of diversified and complex "second-generation" peri-urban landscapes due to improvement in the road network, and the changes in people's lifestyles in East Asia, for instance, have resulted in expansion of tourism and leisure activities and the building of second homes by people in the urban peripheries (Webster et al. 2014). In Ghana's Bosomtwe District, it has been observed that peri-urbanization has taken place as a result of the increased spread of residential, recreational (guest houses and hotels), and commercial land uses at the expense of agro-forest land uses, which in turn have a negative effect on food security and the local climate of the area (Appiah et al. 2014). Through increased expansion of more commercial uses, land for agriculture is eaten up, which in turn threatens the ability of local agricultural food producers to produce what can adequately sustain the country's population. The clearing of the greenbelt and areas covered with vegetation would have a bearing on the local climatic conditions, as it may result in climatic shifts. It has been observed that most of the Chinese cities that grew up around highly fertile land have lost cultivated land in the inner peri-urban area between 1991 and 2001, with the rate of the loss being 4.78 times faster than the population growth rate (Webster et al. 2003).

It is also argued that one of the most dominant forces that is shaping contemporary peri-urban political and spatial changes in Asia has been the rapid escalation of land values, leading state actors to come up with new strategies of land management practices that seek to extend state power through the exploitation of the urbanization process (Shatkin 2016). In this regard, it is evident that some of the peri-urbanization processes are sometimes politically motivated, as politicians thrive to aim for political gains instead of looking out for the general populace's well-being. Some of the Asian governments have therefore monetized land through the use of government's powers of land management to increase land revenues, through the extraction of revenues from land development, or through distribution of profits to from land development to powerful corporate backers of the state (Shatkin 2016). Marshall and Dolley (2019) further stress that peri-urbanization is at times characterized by the neoliberal reordering of space and co-option of environmental agendas by the powerful urban elites.

Due to increased foreign direct investment, some rural areas have been transformed creating transitional regions. Typical examples include the Yangtze and Pearl River Deltas of China, which are now characterized by a mixture of rural and urban land uses and a rapid socioeconomic structural change (Liu et al. 2004) This transformation sometimes benefits the locals through employment creation, which in turn impacts the people's standards of living through improved quality of life. The ability of the Chinese government to differentiate between peri-urban developments and of policy-makers to see the need for regional integration in order to guide development is a positive step toward improvement of the quality of developments that take place within the city fringes (Webster and Muller 2002). This reveals that a lot of factors (both natural and man-made) are all behind the formation of peri-urban areas in most urban setups. This shows the complexity of the issue of peri-urbanization as it may result from many factors and, hence, the need for policy-makers to be tactful when it comes to coming up with strategies that help in dealing with peri-urbanization or its management. The chapter now examines the various issues which have emerged on the platforms of developing countries, as well as the various ways which have been engaged by different countries regarding the taming of the city to prevent the unsustainable encroachment of the peri-urban zone by adjoining urban areas.

Results

The chapter appreciates that most of the developing world is now walking a path which has already been walked by the developing world. There is a need however to appreciate that peri-urbanization occurs differently depending on the level of development, and it varies from country to country. In Europe and China's Yangtze River Delta, peri-urbanization has been associated with features of a mix of rural-urban characteristics (Liu et al. 2004; Shaw et al. 2020), which in turn indicates a zone of not only infrastructural transformations but socioeconomic transformation as well. Despite positive gains through more employment opportunities for the locals, peri-urbanization has been linked to more environmental hazards and threats (Webster and Muller 2009). A study in Mexico revealed that planned settlements were associated with less respiratory problems and illnesses related to dizziness (Graham et al. 2004). Peri-urbanization especially in the developing world has also been described as chaotic or haphazard, which call for the need for policy-makers and urban planners to push for a more controlled and compact development for the urbanites, as it makes the urban space a healthier planet to live in. The controlled growth of the cities would help prevent conflicts and promote sustainable livable environments, which are well planned. The chapter also observes that sometimes the selfish practices by politicians to gain/at times abuse power has often resulted in the expansion of the peri-urban landscape (Shatkin 2016). There is also need to guard against corruption and selfish practices by politicians as they may end up fueling unsustainable peri-urbanization practices in the country.

Emerging Issues: Strategies for Taming the City

The issue of peri-urbanization is one that is no longer new to the developing world, as it has now become a common phenomenon to the development of most urban and peri-urban landscapes. The city of Leiria in central Portugal has experienced a lot of land-use changes between the period of 1958 and 2011, as witnessed by an increase in artificial areas, in a continuous and discontinuous fashion, resulting in a continual decrease of agricultural land use (Barros et al. 2018). This indicates a continuous transformation of land uses and the shrinking of land for agricultural purposes as substituted by other uses and would in turn pose food insecurities in the region. It is also observed that urban expansion has been linked to the expansion of the residential and industrial uses, which is then followed by tertiary growth, which brings one to the importance of strategic regional planning when managing urban sprawl and the artificialization process of medium-sized cities (Barros et al. 2018). It becomes necessary to engage in regional planning to effectively zone land uses accordingly, in order for an area to have well-balanced uses, which may enable it to effectively function on its own as a sustainable part of the urban system. Zoning helps in preventing having too many uses of something (say, residential) without having adequate uses to support the populations such as the commercial and industrial use.

Taking the case of China's Beijing after the year 2000, the number of temporary migrants (thousands of young and educated) continues to flood the peri-urban region, which has in turn resulted in the formation of slums and the widening the social inequality gap between the local people and the migrants (Zhao 2013). The formation of these slums has resulted in the unsustainable sprawling of the city. The introduction of new urban policies aimed at encouraging rural-urban integration has played a huge role in improving the rural people's living conditions and reducing the socioeconomic gaps between the rural and urban divide (Zhao 2013), which is a positive step toward ensuring positive integrated development of the country, thus reducing regional gaps. However, planning in the peri-urban region still faces challenges as a result of the ever-growing market forces and social discrimination which still exist due to the remnants of the Hukou system. This system makes it difficult to achieve rural-urban integration and a harmonized society. It is argued that more actions need to be taken to improve the political capacity of Beijing's planning system through institutional innovation arrangements in the form of power, rights, public resources, legitimacy, and accountability in the planning system (Zhao 2013).

Studies from India and China have revealed that the lack of understanding and policy engagement with peri-urbanization in its present form has led to increased exclusion and unrealized potential when it comes to the support of multiple sustainable urban development goals. Complex feedbacks across the rural-urban continuum have been created as a result of the change in land uses, resource extraction, and pollution, which in turn affects the socio-technical and socio-ecological systems (Marshall and Dolley 2019). Also, as a result of jurisdictional ambiguities, governance challenges are experienced, and this mainly comes as a result of increased growth of the informal market-based arrangements with little incentives for environmental management. It should however be highlighted that although the unique features of peri-urbanization may lack inclusion, they however present transformative innovation (Marshall and Dolley 2019). This calls for clarity on the responsible planning bodies and the Councils to clearly demarcate and specify areas under their jurisdiction, in order to prevent confusion regarding who owns the peri-urban areas and on whose account should these areas be developed.

In regulating peri-urban developments, the need to control the type and magnitude of developments that take place in peri-urban areas becomes necessary. Taking into cognizance that, one day, the urban area may need to expand into the greenbelt, or the peri-urban area in the near future, the need to control the size of developments that take place in the peri-urban area becomes necessary. The size of specialized uses such as parks within the inner peri-urban areas must be small in size, and not be too big, since it may end up being part of contiguous city fabric through evolution (Webster et al. 2003). As such, most of the land is not simply dedicated to one use as provisions are also left for the possibility of the change in uses, when it comes to future developments and expansion of the city boundaries. It is also observed that, in order to improve functionality of the peri-urban areas, amenities matter the most as they may play a part in increasingly attracting clients who may visit the extended industrial and commercial land-use functions. Furthermore, through infrastructural developments, a competitive advantage of the area is created as compared to the other surrounding rural or less developed areas (Webster et al. 2003). Positive infrastructural developments of the peri-urban zone would entail the development of a more efficient transport network which connects the inner peri-urban area with the core, and this allows the smooth flowing of business linkages between the core and the peripheral areas. Also, the ability to control developments, and not over invest in specialized land uses such as wild park reserves in the inner peri-urban area, would help preserve some of the land for possible land-use expansions which may branch from the core region. This chapter now looks at approaches which countries may engage when it comes to managing peri-urbanization.

Approaches to Managing Peri-Urbanization

The management of peri-urbanization is important as it helps in the balancing of uses. It has been observed that, in many instances, a region may end up being disadvantaged when it comes to agricultural production through the continued eating away of agricultural, peripheral land for other purposes. This may make it look as if some of the land uses are less important as compared to others, which may not be true. Disadvantaging the people by changing agricultural land to other uses may have long-term negative implications such as the beginning of the overreliance of the country on imports to boost food availability. Some of the measures for managing peri-urbanization which shall be discussed here include the use of the systems approach to addressing peri-urbanization, greening, and urban containment practices. These shall now be discussed in the following section.

The Systems Approach

The urban system is characterized by a system that is characterized by the flow of services, goods, and people, as well as the interdependence of infrastructure and socioeconomic networks. Systems thinking allows for a more holistic approach to viewing cities as complex living systems which are always undergoing numerous dynamic changes at any given point in time as they try to respond to internal and external factors and in order to address urban challenges. There is thus a need to look at the urban system as a whole and not its elements in isolation (Da Silva et al. 2012). This helps policy-makers to understand that the urban system as a unit with components have possibilities to affect the various facets of the environment. Countries such as China have started understanding the importance of the systems approach to solving the challenges related to the aspect of peri-urbanization.

Previously recognized as large- to small-sized settlements that are interspersed on rural land, peri-urban settlements have recently gained much recognition as a planning and policy matter by Chinese policy-makers (Webster and Muller 2002). Cities with a population of a million, for instance, were expected to solve own problems in relation to resources accorded to them and not worry about problems outside its city boundaries, yet in reality, that city was supposed to be viewed as a component of the broader urban system (Webster and Muller 2002). The changing in thinking from seeing city challenges as isolated to understanding that cities do belong to a bigger system with interlinkages helps policy-makers in coming up with strategies for promoting sustainable urban development in a holistic manner.

Greening

It is argued that through the practice of community gardens, concrete gardens are not merely created (Chirisa 2014). The process of peri-urbanization has, however, been observed to cause negative environmental impacts through the clearing of the land and the possibility for causing water conflicts. By clearing the land and eating up most of the greenbelt, the climatic trend of the area may be disturbed. This calls for the need to employ greening mechanisms when developing in order to ensure that the hard and soft landscapes are effectively blended and that the newly built-up environment would still maintain harmony with nature.

Studies have also revealed that, in virtually all of Southeast Asian states and China, industrial cleaner technologies have been adopted as compared to free-standing factories since the early 1980s. These green technologies are being driven by the rise of green-oriented estate developers and the existence of the environmental rating systems such as LEED and ISO 14001

(Webster et al. 2014). This shows increased awareness and positive developments regarding green issues in some of the Asian States.

While most of the developing countries' adoption of green technologies has been slow, the adoption of green renewable energy in Pakistan has already started creating new markets and job opportunities (Mohazzam et al. 2020). The adoption of green renewable technologies in respect of electricity generation and use would have a positive impact not only on the environment but on the people's incomes as well. This is largely attributed to the observation that it is less costly to use the readily available sun/solar energy compared to electricity.

Studies from Southeast Asia have shown that multinational companies in the region have been pressured by the domestic and international non-governmental organizations to improve on environmental performance especially in the developing world. Furthermore, the residents in the peri-urban areas have increasingly become environmentally conscious and have increasingly become less tolerant of pollution, and also the establishment of national government policies such as the current Chinese 12th Five Year Plan requires industrial zones to exhibit higher energy and land efficiency (Webster et al. 2014). Such policies help industries to become more aware of the environment, as they may suffer heavy penalties from government if they fail to honor the country's policies.

Containment

The aspect of containment focuses on containment of urban development through the promotion of sense cities which heavily rely on dense infrastructure to promote sustainable development (Nilsson et al. 2013). Compaction also deals with the reduction of vehicle travel distance and more investment in public transport with the main aim to lower carbon dioxide emissions and higher density developments with more utilization of the brownfields (Echenique et al. 2012). Containment can therefore be used by policymakers as a way to stop the sprawling of cities and encourage more compact developments as they also aim to reduce the levels of pollution as people travel longer distances to the city. In this regard, mixed-use developments may also come in handy as people get encouraged to stay where they work as the journey of commuting to and from work is cut.

Urban containment policies have been introduced in Beijing in the 1990s, and it is acknowledged that the overall urban compactness was enhanced, with studies showing that objectives of the municipal urban containment policies were partly achieved through local developments (Zhao et al. 2009). This shows positive efforts by the Chinese metropolitan to contain its urban space and prevent uncontrolled sprawl from happening. Ghana has also experienced sprawl in Accra and has earned the name Greater Accra as a result. Efforts to contain Accra have not been successful with major reasons being pointed out as weak public control over the land, as well as poor spatial planning. It becomes difficult to plan considering that sometimes land is held privately under customary institutions, while planning remains a function of the local government (Owusu 2013), hence the need to clearly differentiate powers and define in whose jurisdiction the land is in. Zhao et al. (2009) also advocate for the need for future developments to be controlled by urban containment policies.

The concept of functionalism has also been regarded as a way to manage peri-urbanization in the developing world. In managing peri-urban areas, the current planning system must be adjusted to respond to the dynamic changes of the peri-urban areas and move away from rigid standards to more flexible rules and the utilization of collaborative approaches through the involvement of the community in development (Hudalah et al. 2007). In this way, the community will not resist change and engaging the community would make them feel important, and it becomes easy for them to understand the importance of the environment as well as the negatives associated with unplanned peripheral developments in the urban fringes.

Discussion

It is quite evident that the concept of peri-urbanization is not new to the planning discourse.

Peri-urbanization has largely been experienced in Asian countries such as China and some African countries such as Ghana. Despite some of the positive effects of peri-urbanization such as the expansion of industrial activities into the periphery and the widening of employment opportunities for the residents of the periphery, the process of peri-urbanization, if not controlled, has caused more harm than good to the peri-urban residents. Some of the negative effects that have been associated with peri-urbanization include the uprising of land and water conflicts, as well as the negative environmental consequences such as increased pollution and the clearing of the greenbelt or land for subsistence farming, which in turn threatens the source of livelihoods for many.

Beijing has witnessed the emergence of slums in its urban fringes (Zhao 2013), while Portugal's city of Leiria has experienced massive land-use changes between 1958 and 2011, resulting in the shrinking of agricultural land (Barros et al. 2018), which indicates the aspect of uncontrolled and unsustainable urban development practices. The study also highlights the lack of jurisdictional ambiguities and governance challenges as one of the causes of peri-urbanization (Marshall and Dolley 2019), hence the need for the effective clarification and demarcation of the land on who owns it and who have the powers to control development in those areas. Also, in order to manage peri-urbanization, there is need for planners and policy makers to view the region as dependent on each other and to understand that settlements do not exist in isolation but are a part of the broader urban system (Da Silva et al. 2012), and this has seen the Chinese system appreciating that urban systems do not exist in isolation but form a system with interlinkages (Webster and Muller 2002), which forms a positive step to managing the spread peri-urbanization its urban fringes.

To manage peri-urbanization, community involvement is also critical (Hudalah et al. 2007), as this helps in teaching the locals about the negative effects of urban sprawl as well as get local views toward enhancing development. Involving the local people would promote a sense of oneness and prevent people from resisting change, in instances where local authorities may decide on expanding the city boundaries for the benefit of the locals as well. Greening the environment through the use of green technologies has been adopted (Mohazzam et al. 2020), and this helps in spreading developments that are environmentally sensitive even on the outskirts and preventing much environmental damage to the environment even in the case of peri-urbanization. The adoption of the environmental rating systems in development is also a plus to development as it creates a benchmark and a measure upon which industries may operate from.

The chapter also acknowledges the issue of urban containment where mixed-use developments and densely populated suburbs are created as a way to contain the urban landscape and prevent sprawl. In this case, more investments may be needed to develop high-rise flats to accommodate more people within the urban boarders and in turn cut the travelling distances of people to work. The cutting of the travelling distance would in turn reduce the rate of carbon dioxide emissions (Echenique et al. 2012), which would in turn have a positive bearing on the environment as environmental pollution is reduced.

Conclusion and Future Direction

In conclusion, though it is quite evident that the developed world has tried putting in place various measures to control peri-urbanization and the sprawling of cities, the developing world still has a long way to go as regards controlling the outward spread of cities into the peripheral regional. There is therefore a need for strict control measures by the government in terms of enforcement of policies, rules, and regulations and the paying of heavy penalties by illegal developments which may be found in the periphery. The control of the spread of peri-urbanization would help in creating more sustainable developments. In order to enhance the existing peri-urban areas and prevent the continued spread that may cause the sprawling of cities, there is need for policy-makers and city designers to:

- Encourage compact developments and mixed uses which help as urban containment measures to contain and prevent the unsustainable outward growth of cities.
- Promote the adoption of green technologies in city development.
- Involve the community in development to reduce resistance to change.
- View any urban or peripheral urban developments as part of the broader urban system in order to promote sustainable development.

Cross-References

▶ Green and Smart Cities in the Developing World

References

Appiah, D. O., Bugri, J. T., Forkuo, E. K., & Boateng, P. K. (2014). Determinants of peri-urbanization and land use change patterns in peri-urban Ghana. Available online: http://ir.knust.edu.gh/handle/123456789/10630

Barros, J. L., Tavares, A. O., Monteiro, M., & Santos, P. P. (2018). Peri-urbanization and rurbanization in Leiria city: The importance of a planning framework. *Sustainability, 10*(7), 2501.

Brueckner, J. K. (2000). Urban sprawl: diagnosis and remedies. *International Regional Science Review, 23*(2), 160–171.

Budiyantini, Y., & Pratiwi, V. (2016). Peri-urban typology of Bandung metropolitan area. *Procedia-Social and Behavioral Sciences, 227*, 833–837.

Cambridge English Dictionary. *Greening*. https://dictionary.cambridge.org/us/dictionary/english/system. Accessed 21 Jan 2021.

Cambridge English Dictionary. *System*. https://dictionary.cambridge.org/us/dictionary/english/system. Accessed 21 Jan 2021.

Centres for Disease Control and Prevention. *Policy*. Office of the Associate Director for Policy. https://www.cdc.gov/policy/analysis/process/docs/policydefinition.pdf. Accessed 21 Jan 2021.

Chirisa, I. (2014). Building and urban planning in Zimbabwe with special reference to Harare: Putting needs, costs and sustainability in focus. *Consilience, 11*(1), 1–26.

Da Silva, J., Kernaghan, S., & Luque, A. (2012). A systems approach to meeting the challenges of urban climate change. *International Journal of Urban Sustainable Development, 4*(2), 125–145.

Dohwe, P., & Kwangwama, N.A. (2019). Urban resilience in cities of the developing world with reference to Harare, Zimbabwe. In I. Chirisa & C. Mabeza (Eds.), *Community resilience under the impact of urbanisation and climate change: cases and experiences from Zimbabwe*, (pp. 53–94). Bamenda, Research and Publishing CIG.

Echenique, M. H., Hargreaves, A. J., Mitchell, G., & Namdeo, A. (2012). Growing cities sustainably: Does urban form really matter? *Journal of the American Planning Association, 78*(2), 121–137.

Graham, J., Gurian, P., Corella-Barud, V., & Avitia-Diaz, R. (2004). Peri-urbanization and in-home environmental health risks: The side effects of planned and unplanned growth. *International Journal of Hygiene and Environmental Health, 207*(5), 447–454.

Hudalah, D., Winarso, H., & Woltjer, J. (2007). Peri-urbanization in East Asia: A new challenge for planning? *International Development Planning Review, 29* (4), 503.

Jenkins, G. M. (1969). *The systems approach*. University of Lancaster, Department of Systems Engineering.

Liu, S. H., Chen, T., & Cai, J. M. (2004). Peri-urbanization in China and its major research issues. *Acta Geographica Sinica, 59*(Suppl 1), 101–108.

Marshall, F., & Dolley, J. (2019). Transformative innovation in peri-urban Asia. *Research Policy, 48*(4), 983–992.

Mohazzam, S., Ali, A., & Ali, S. H. (2020). Greening energy provision in urban Pakistan. In *Urban studies and entrepreneurship* (pp. 227–247). Cham: Springer.

Narain, V. (2016). Peri-urbanization, land use change and water security: A new trigger for water conflicts? *IIM Kozhikode Society & Management Review, 5*(1), 5–7.

Nilsson, K., Pauleit, S., Bell, S., Aalbers, C., & Nielsen, T. A. S. (Eds.). (2013). *Peri-urban futures: Scenarios and models for land use change in Europe*. Berlin/Heidelberg: Springer Science & Business Media.

Owusu, G. (2013). Coping with urban sprawl: A critical discussion of the urban containment strategy in a developing country city, Accra. *Planum, The Journal of Urbanism, 1*, 1–17.

Pradoto, W. (2011). Dynamics of peri-urbanization and socioeconomic transformation: Case of Metropolitan Yogyakarta, Indonesia. *International Journal of Arts & Sciences, 4*(27), 19.

Ravetz, J., Fertner, C., & Nielsen, T. S. (2013). The dynamics of peri-urbanization. In *Peri-urban futures: Scenarios and models for land use change in Europe* (pp. 13–44). Berlin/Heidelberg: Springer.

Shatkin, G. (2016). The real estate turn in policy and planning: Land monetization and the political economy of peri-urbanization in Asia. *Cities, 53*, 141–149.

Shaw, B. J., van Vliet, J., & Verburg, P. H. (2020). The peri-urbanization of Europe: A systematic review of a multifaceted process. *Landscape and Urban Planning, 196*, 103733.

Webster, D., & Muller, L. (2002). *Challenges of peri-urbanization in the Lower Yangtze Region: The case of the Hangzhou-Ningbo Corridor*. Stanford, CA: Asia/Pacific Research Center.

Webster, D., & Muller, L. (2009). Peri-urbanization: Zones of rural-urban transition. *Human Settlement Development, 1*, 280–309.

Webster, D., Cai, J., Muller, L., & Luo, B. (2003). *Emerging third stage peri-urbanization: Functional specialization in the Hangzhou peri-urban region (Research monograp)*. Stanford: Asia-Pacific Research Center, Stanford University.
Webster, D., Cai, J., & Muller, L. (2014). The new face of peri-urbanization in East Asia: Modern production zones, middle-class lifestyles, and rising expectations. *Journal of Urban Affairs, 36*(Supp 1), 315–333.
World Bank (2012) *Managing the risks of disasters in East Asia and the Pacific. Building urban resilience: Principles, tools and practice*. The World Group. USA. Washington DC.
Zhao, P. (2013). Too complex to be managed? New trends in peri-urbanization and its planning in Beijing. *Cities, 30*, 68–76.
Zhao, P., Lu, B., & Woltjer, J. (2009). Growth management and decentralisation: An assessment of urban containment policies in Beijing in the 1990s. *International Development Planning Review, 31*(1), 55–79.

Strategy

▶ Public Policies to Increase Urban Green Spaces

Strategy – Action Planned

▶ Strategies for Taming the City

Street-Engaged

▶ Global Homelessness: Neoliberalism, Violence, and Precarious Urban Futures

Streets

▶ Stewarding Street Trees for a Global Urban Future

Structural Racism

▶ Hidden Enemy for Healthy Urban Life

Sub-centers

▶ Urban Structure and Its Impact on Mobility Patterns: Reducing Automobile Dependence Through Polycentrism

Suburban

▶ Master Planned Estates and the Promises of Suburbia

Suburbanization

▶ Master Planned Estates and the Promises of Suburbia

Suburbia

▶ Master Planned Estates and the Promises of Suburbia

Sustainability

▶ An Overview of the Relationship of the Sustainable Development Goals and Urban and Regional Development
▶ Financing: Fiscal Tools to Enhance Regional Sustainable Development
▶ Master Planned Estates and the Promises of Suburbia
▶ Moving Towards Sustainable, Liveable, and Care-Full Urban Environments: Pre-schoolers' Rights and Visions for Planning Just, Socially, and Ecologically Integrated Cities
▶ Sustainable Development Goals and Urban Policy Innovation
▶ The Sustainable and the Smart City: Distinguishing Two Contemporary Urban Visions
▶ Urban Management in Bangladesh

Sustainability - Continuity

▶ Zooming Regions into Perspective

Sustainability Challenges

▶ Small Towns in Asia and Urban Sustainability

Sustainability Competencies in Higher Education

Jessica Ostrow Michel[1], Katja Brundiers[2], Matthias Barth[3], Daniel Fischer[2], Yoko Mochizuki[4], Aaron Redman[2] and Michaela Zint[1]
[1]School for Environment and Sustainability, University of Michigan, Ann Arbor, MI, USA
[2]School of Sustainability, Arizona State University, Tempe, AZ, USA
[3]Leuphana University, Lüneburg, Germany
[4]UNESCO, Paris, France

Synonyms

Attributes; Capabilities; Capacities; Learning outcomes

Introduction

The number of undergraduate and graduate programs focused on sustainability continues to grow in higher education. The emerging interdisciplinary field of study can be strengthened through a shared set of competencies that students develop in structured academic programs and curricula to successfully master real-world sustainability challenges in personal, civic, and professional domains. This entry summarizes half a decade of progress on defining, using, and assessing competencies as a guiding framework for the design of higher education sustainability programs.

Policymakers have deemed higher education a critical avenue for cultivating a more sustainable future and, in recent years, for contributing to the achievement of the Sustainable Development Goals (SDGs), adopted by the UN in 2015. Landmark scholarship, as far back as environmental educator David Orr's (2004) *Earth in Mind*, has contended that we are in urgent need of fundamentally different educational systems, given that our current ones have contributed to the unsustainability of today's world. Orr (2004) suggested that "education for sustainability" contains several aspects, including the need to accept the probability of survival of our species; an attitude of care and stewardship – particularly an "uncompromising commitment to life and its preservation" (p. 133); the knowledge necessary to comprehend interrelatedness of "disciplines and of the disparate parts of personality: intellect, hands, heart" (p. 137); and the practical competence required to act on these sets of values.

While the necessary radical shift in education has not occurred, and change to current education systems will most likely occur incrementally and at a slow pace, new sustainability degree programs have been introduced in institutions of higher education throughout the world. The potential of these programs to produce leaders and agents of change in the field of sustainability can be improved, however, by the development of an agreed-upon set of competencies that identify what students ought to be learning from their inter- and transdisciplinary studies in this field. To date, higher education is generally not producing students equipped to manage environmental and sustainability crises, for example. In addition, stakeholders – including students, employers, educators, and administrators – continue to question what it is that students ought to know and be able and willing to do to solve sustainability problems. In response to this question, the notions of *competence* or *key competencies* have emerged as a holistic concept to describe desired learning outcomes in the field.

Stimulated by the intensive scholarship, the concept of key competencies has also found its way into educational policy debates. In the United States, the National Council for Science and the

Environment (NCSE) launched a process in 2018 to complete a census of program objectives for sustainability and sustainability-related fields. Additionally, NCSE is currently developing a consensus statement for core competencies to be endorsed by academic deans and directors of environmental programs across the country.

History of Emerging Competencies for Sustainability

Agreement on a set of key competencies for sustainability has occurred over time as several strands of research and practice have come together. The renewed interest in competence as the framing for learning outcomes is closely tied to educational reform in general, in particular the shift to outcomes-based education. The discussion of competence as a learning outcome has a rich history that goes back decades and reflects different interdisciplinary understandings of the term. Popularized by the scientist and philosopher Noam Chomsky's work on linguistic development as mastery of a system of rules in the 1960s, and later adopted by the sociologist Jürgen Habermas in his work on the formation of communication (1970s and 1980s), the concept of competence was used to mark a substantive distinction between a generative system (competence) and a behavioral outcome (performance) (Klieme et al. 2008). Today's understanding of competence is largely shaped by vocational and occupational psychologists' work on understanding employees' responses to particular work-related challenges (Hyland 1993).

While educational psychologists called for a stronger recognition of learning outcomes that are relevant across different domains in life (vocational, personal, and social) in the early 1970s, it took more than a quarter of a century for this approach to key competencies to gain broader international traction when the Organisation for Economic Co-operation and Development (OECD) launched the Definition and Selection of Competencies (DeSeCo) initiative (Rychen and Salganik 2000). While the use of competencies has spread far beyond the auspices of the OECD, the organization's continued emphasis on them through activities such as the Programme for International Student Assessment (PISA), has propelled adoption internationally. It is worthy of mention that the OECD's work was in the context of primary and secondary education.

Defining Competencies for Sustainability in Higher Education

While competencies have emerged as a predominant approach within sustainability education research and practice, there has also been a "sea of labels" (Sterling et al. 2017, 153). The term competencies itself is understood dissimilarly in different cultural and disciplinary contexts. While sustainable or environmental literacy is utilized in some places, and attributes or learning outcomes in others, these approaches have not gained global adoption. The difference between these approaches and that of competencies is likely to be mostly semantic, though research exploring this is mostly lacking.

A competence has come to be understood as "a complex combination of knowledge, skills, understanding, values, attitudes and desire which lead to effective, embodied human action in the world, in a particular domain" (Crick 2008, 313). In application to the interdisciplinary field of sustainability, *sustainability competencies* are the specified cluster of the desired and related knowledge, skills, and attitudes that students graduating from a sustainability program ought to possess. Mastery of sustainability competencies would mean that graduating students are able to successfully solve real-world sustainability problems.

Building upon competencies, *key competencies* are a cluster of related competencies that are relevant across different domains and are essential for achieving successful performance in diverse contexts. Therefore, while graduates of higher education sustainability programs will need many general competencies (e.g., communication skills), the emphasis on key competencies highlights those competencies most critical to sustainability. Moreover, *key competencies for sustainability* emphasize specific competencies considered to be critical for sustainability and build on those competencies (like critical thinking), which are shared across academic programs. Some of the competencies are present in other traditional programs (e.g., values thinking competency is prevalent in the humanities, and systems thinking competency is shared across many programs). What distinguishes the key competencies in sustainability is that they cannot stand alone; they come as "package-deal" (i.e., as a framework of related competencies) in which knowledge, skills, and attitudes are interrelated and used together.

Scholarship on Key Competencies for Sustainability

Although the need to reorient the purpose of education systems to embrace sustainability was endorsed by global leaders in the United Nations' 1992 action plan for development, called Agenda 21, a coherent approach to translate these ambitions into learning outcomes was lacking. A critical development was the adoption of DeSeCo's competency approach in sustainability education. This was pioneered by Gerhard de Haan (2006), a professor at the Free University in Berlin, which was soon followed by its tentative specification for higher education in 2007 with an inaugural qualitative study by the scholars Matthias Barth, Jasmin Godemann, Marco Rieckmann, and Ute Stoltenberg (2007). Results of the work of Barth and his colleagues reinforced the definitional elements of competencies being based on cognitive (such as interdisciplinary knowledge acquisition) and the so-called noncognitive dispositions (such as motivating oneself and motivating others). This study convincingly described how a key competency could meet the goals of education for sustainable development. Along this line, scholars began to attempt to define key competencies (though not always using that particular terminology). Some focused on sustainability programs, whereas others developed competencies for sustainability as applied to other disciplines, in particular engineering, teaching, and business. This trend of parallel development within both disciplinary and interdisciplinary settings continues to this day.

In 2011, sustainability educators Arnim Wiek, Lauren Withycombe, and Charles Redman conducted a comprehensive literature review of scholarship on key competencies for sustainability. To review the proposed synthesis of competencies for sustainability, they administered a survey and conducted a full-day workshop at the American Association for the Advancement of Science (AAAS) Forum in 2010. The resulting article, *Key Competencies in Sustainability: A Reference Framework for Academic Program Development* is often recognized as the most frequently cited synthesis on sustainability competencies in higher education and as one of the most influential in the field of sustainability education research overall. Wiek and his colleagues define key competencies for postsecondary sustainability programs and provide a framework that explains how these competencies relate to sustainability outcomes. They also reinforce the notion that the goal of key competencies are "problem solving with respect to real-world sustainability problems, challenges, and opportunities" (2011, 204).

This work contributes to the literature a coherent sustainability problem-solving framework, including the five competencies derived from their synthesis: systems-thinking, anticipatory/futures-thinking, normative/values-thinking, strategic-thinking, and interpersonal/collaboration thinking. These competencies are outlined below.

1. Systems-thinking: The "ability to collectively analyze complex systems across different domains (society, environment, economy, etc.) and across different scales (local to global), thereby considering cascading effects, inertia, feedback loops and other systemic features related to sustainability issues and sustainability problem-solving frameworks" (Wiek et al. 2011, 207).
2. Anticipatory/Futures-thinking: The "ability to collectively analyze, evaluate, and craft rich 'pictures' of the future related to sustainability issues and sustainability problem-solving frameworks" (Wiek et al. 2011, 208–209).

3. Normative/Values-thinking: The "ability to collectively map, specify, apply, reconcile, and negotiate sustainability values, principles, goals, and targets" (Wiek et al. 2011, 209).
4. Strategic-thinking: The "ability to collectively design and implement interventions, transitions, and transformative governance strategies toward sustainability" (Wiek et al. 2011, 210).
5. Interpersonal/Collaborative: The "ability to motivate, enable, and facilitate collaborative and participatory sustainability research and problem solving" (Wiek et al. 2011, 211).

Since publication, Wiek and his colleagues have added an additional competence, entitled meta-competence, which consists of "meaningfully using and integrating the [other] five key competencies" to "[solve] sustainability problems and [foster] sustainable development" (2016, 243).

Since this study, additional research on sustainability competencies have been published. For example, scholars have developed additional studies focusing on intercultural perspectives on key competencies, the link between pedagogies and key competencies, and applications of key competencies to sustainability challenges like consumption.

Future Scholarship on Sustainability Competencies

To advance scholarship in the field of sustainability competencies in higher education, the research community needs to take into consideration the missing elements of past research, the importance of key competencies, the need for accurate assessment in the classroom, and the question of applicability across cultures. New research also needs to look into how findings are integrated into the curriculum, as well as how graduates utilize their education in the real world.

Consolidating Scholarship on Key Competencies for Sustainability

The political nature and origin of sustainability education (such as political declarations committing higher education institutions to sustainability as a policy imperative) have resulted in a research landscape that is characterized by experimentation, resulting in a large number of (often descriptive) case studies as well as conceptual (or even just programmatic) proposals. This situation has been intensively discussed as a challenge in the field and it is reflected in the competence literature. In a critical perspective it can be acknowledged that what is needed is consolidation and empirically rigorous research that goes beyond descriptive single-case studies to explore how these competencies can be assessed and developed in teaching and learning settings.

Advancing Key Competencies for Sustainability

While work on key competencies for sustainability has continued to advance, much of the current scholarship lacks either rigorous conceptual development or empirical data. Despite this large body of work, the key competencies framework described by Wiek and his colleagues is yet to be displaced as the standard within the field. Rather than emphasize continued restating of the key competencies, we see interesting developments aiming at addressing other limitations related to sustainability competencies.

Assessing Key Competencies for Sustainability

Assessing competencies has proven a challenge for educators in general and has received insufficient attention within sustainability education specifically. In light of this, future research ought to measure students' mastery of the knowledge, skills, and attitudes to address the real-world sustainability challenges. Such work, may in turn, yield substantial new insights into how to advance the current, mostly conceptual body of knowledge on key competencies for sustainability.

Diversifying Perspectives in Key Competencies for Sustainability

The dominance of North American and European scholarship has raised questions about the universality of competencies being proposed, as they may be culturally biased. We recognize this limitation in the scholarship completed to date, and call for further research involving more diverse scholars from additional countries and cultures (such as Africa, Asia, and South America), in particular from the Global South.

Supporting Pedagogies to Aid in Competency Development

Developing competencies for sustainability education may also imply the need for new pedagogies, and thus the research community has responded with a growing body of literature on the topic. So far, though, much of this work is either conceptual or consists of descriptive case studies of pedagogical enactments. Empirical studies investigating the links between specific pedagogical practices and the successful (or not) mastery of key sustainability competencies, have begun, but many more are needed.

Transitioning from Theory to Practice

Several higher education institutions have already begun to adopt competency frameworks for their sustainability programs, primarily drawing on those developed by Wiek and his colleagues. This use of competencies by universities presents a key frontier of future research. Moreover, what happens once students leave campus is almost completely unknown, leaving us with little insight about how students translate sustainability competencies into real-world action after graduation.

Conclusion

Through a diversity of methods (literature reviews, empirical research such as action research, case studies, and Delphi studies), and locations (Canada, Europe, Latin America, and the United States), scholars have proposed sustainability competencies through large-scale literature reviews and to a lesser extent empirical research. Consistent key competencies have emerged as a result (e.g., systems thinking and interdisciplinary knowledge), suggesting an emerging consensus. Yet little evidence exists that these competencies are being used to guide academic learning outcomes.

For the field to move toward establishing the competencies that will ultimately anchor it, several developments are needed. First, more research and praxis (as empirically informed practical solutions) will be needed to assess in how far the development of competencies takes place. Second, more tailored teaching and learning settings can and must be developed on the microlevel of courses and the macrolevel of programs that support explicitly the key competencies. And third, an agreed upon set of key competencies can and should be the foundation for accreditation measures and to ensure the comparability of programs across universities and regions, (i.e., to support the mobility of students). Overall, much work remains to be done on this front to ensure that competency-driven sustainability education will fulfill its promise in higher education, and in turn, to cultivate a more sustainable future.

Cross-References

- ► Education for Inclusive and Transformative Urban Development
- ► Environmental Education and Non-governmental Organizations

References

Barth, M., Godemann, J., Rieckmann, M., & Stoltenberg, U. (2007). Developing key competencies for sustainable development in higher education. *International Journal of Sustainability in Higher Education, 8*(4), 416–430.

Crick, R. D. (2008). Key competencies for education in a European context: Narratives of accountability or care. *European Educational Research Journal, 7*(3), 311–318.

De Haan, G. (2006). The BLK "21" programme in Germany: A "Gestaltungskompetenz"-based model for education for sustainable development. *Environmental Education Research, 12*(1), 19–32.

Hyland, T. (1993). Competence, knowledge and education. *Journal of Philosophy of Education, 27*(1), 57–68.

Klieme, Eckhard, Johannes Hartig, and Dominique Rauch (2008). "The concept of competence in educational contexts." Assessment of competencies in educational contexts 3 (22), Hogrefe & Huber Publishers Cambridge, MA.

Orr, D. W. (2004). *Earth in mind: On education, environment, and the human prospect*. Washington, DC: Island Press.

Rychen, D. S., & Salganik, L. H. (2000). Definition and selection of key competencies. In *The INES Compendium (Fourth General Assembly of the OCDE*

Education Indicators programmme) (pp. 61–73). París: OCDE.
Sterling, S., Glasser, H., Rieckmann, M., & Warwick, P. (2017). "More than scaling up": A critical and practical inquiry into operationalizing sustainability competencies. In *Envisioning futures for environmental and sustainability education* (pp. 681–700). Wageningen: Wageningen Academic Publishers.
Wiek, A., Withycombe, L., & Redman, C. L. (2011). Key competencies in sustainability: A reference framework for academic program development. *Sustainability Science, 6*(2), 203–218.
Wiek, A., Bernstein, M. J., Foley, R. W., Cohen, M., Forrest, N., Kuzdas, C., Kay, B., & Keeler, L. W. (2016). Operationalizing competencies in higher education for sustainable development. In M. Barth, G. Michelsen, M. Rieckmann, & I. Thomas (Eds.), *Routledge handbook of higher education for sustainable development* (pp. 241–260). London: Routledge.

Sustainability Transformation

► Sustainability Transition and Climate Change Adaption of Logistics

Sustainability Transition and Climate Change Adaption of Logistics

Martin Franz[1], Felix Bücken[1], Kim Philip Schumacher[1] and Kai Michael Griese[2]
[1]Institute of Geography, Osnabrück University, Osnabrück, Germany
[2]Hochschule Osnabrück University of Applied Sciences, Osnabrück, Germany

Synonyms

Climate change adjustment; Climate change mitigation; Resource-efficient logistics; Sustainability transformation; Sustainable logistics; Transport industry

Definitions

Logistics includes all transport, handling, and storage processes and aims to ensure the availability of goods and services (Tempelmeier 2018). Logistics processes can be carried out by manufacturing or trading companies as well as by specialized logistics companies. The implementation of logistics processes has economic as well as social and environmental implications that need to be considered if more sustainable logistics are to be achieved. Sustainable logistics serves the needs of the current generation without endangering the ability of future generations to meet their needs (WCED 1987). Therefore, a transition of the industry is necessary. Transition includes the "system-wide interaction and co-evolution of new technologies, changes in markets, user practices, policy and cultural discourses, and governing institutions" (Coenen and López 2010: 1150). A particular challenge, influenced by the environmental impact of logistics, is climate change. At the same time, climate change also has extensive impacts to which the logistics sector must adapt. The process of adjusting to the effects of current or expected climate change is called climate change adaptation (IPCC 2014).

Introduction

The demand for logistics services is constantly growing. An increase in the global flows of resources and commodities as well as a growing share of retail-related e-commerce has contributed to this development. The growth of the logistics industry and the number of logistics processes associated with it, however, have negative environmental and social impacts. There are, essentially, three sustainability issues related to logistics: (1) The impacts resulting from moving goods across the globe (which includes the use of energy and other resources) as well as the impacts and pollution associated with this use (e.g., noise, air pollution, CO_2, etc.), and the resources used for digital operations (e.g., energy consumption). (2) The land-take needed for the buildings and sealed surfaces of logistical companies as well as the land-take and resources used for the necessary (public) infrastructure such as roads, ports, airports, railway lines, and data cables (Behrends et al. 2008; McKinnon 2019; Hesse 2020). (3) Aspects of social sustainability, whereby competition and sometimes even predatory pricing forces companies to minimize transport and distribution costs, which results in the externalization of social costs – such as poor working conditions, labor relations, and remuneration (Jaffee and Bensman 2016).

Different consequences have emerged from these sustainability issues:

- The public image of the logistics industry has deteriorated. Lohre et al. (2015) attribute this to society's increasing awareness of sustainability issues as well as the transparency created through media coverage. Intricately linked to the resulting negative public image of logistics companies are increased protests in many locations by residents and environmental organizations against the presence of logistics companies in their areas. These protests are often not only against the environmental impact of the companies but also against low-paying jobs and low tax payments at their sites, or the feared destruction of other jobs (Neiberger 2015).
- Increasingly strict environmental regulations are being enacted in many countries. Murphy and Poist (2003) argue that environmental regulation is the most important reason for logistics companies to develop a corporate environmental policy.
- Rising pressure exists from customers to develop and implement more sustainable logistics concepts. Given the growing importance of green supply chain initiatives, it is expected that sustainability standards will play an increasingly important role in the selection of logistics service providers (Evangelista 2014).

In addition to these direct and indirect impacts, the logistics industry is also directly affected by the impacts of climate change – through heavy rain, storms, and heat events, which, in turn, can limit the usability of transport infrastructures and company premises (Bücken and Kanning 2021). A dedicated consideration of sustainability issues and climate change impacts can, therefore, help logistics companies to comply with environmental regulations (Lehmacher 2015), improve their image (Onischka 2009), avoid protests (Garbe and Hempel 2015), and prevent the physical impairment of company property due to environmental influences.

The focus of this chapter is on issues of climate change adaptation and environmental sustainability, in the sense of using resources as sparingly as possible.

Risk Management and Strategic Transition

Logistic companies facing sustainability and climate change challenges may consider different strategies to deal with these questions. Although both risk management and strategic transition are used to achieve sustainable solutions, each exhibit very different characteristics. In order to apply a risk-management approach to the challenges mentioned, it is first necessary to consider them as risks. Individual risk perception is determined by psychological, sociocultural, institutional, and historical influencing factors (Renn 2008; Lupton 2013). Not only individual experiences but also collectively shaped thought patterns, as well as the form of information generation and its processing, are decisive for risk perception (Vertzberger 1998). Christopher and Peck (2004) assign risk-sources to five categories based on issues of supply chain risk management. Unlike supply, process, demand, and control risks, environmental risks encompass factors that are external to the direct sphere of influence of the company. For example, as a result of the poor image of the logistics sector, public authorities are less willing to make land available for logistics uses (Garbe and Hempel 2015). In addition, the potential consequences of increased regulations expected for the future related to reducing greenhouse gases create further risks that require corporate management strategies (Onischka 2009), e.g. a possible ban on diesel-powered trucks. As these risks increase with the growing presence of logistic functions, capabilities that enable companies to provide adequate responses also become more important (Ponomarov and Holcomb 2009).

In the light of the increasingly unpredictable negative effects of climate change, conventional risk management suffers from a lack of reliable information and quantifiable cost-benefit considerations regarding the positive outcomes of measures that target climate impacts (Kauppi et al. 2016). The discrepancy between clearly identifiable costs of action and unclear returns from that action (Todaro et al. 2020) is particularly problematic for the logistics sector, where companies can hardly compete through quality features, are highly dependent on customers, and have low expected profit margins, which inhibit investments (Coe 2021). Firms commonly focus on short-term problem solving, existing customer satisfaction and, because they are often caught up in day-to-day business, they lack farsightedness on the management level (Linnenluecke and Griffiths 2015). Merely integrating the physical and regulative impacts of climate change and increasing sustainability demands into a regular risk-management framework, therefore, does not lead to sufficient anticipation of these changing circumstances.

The changes required to cope with long-term and gradual trends have to exceed the pragmatic problem-solving attitude and routines of small and medium logistic companies (Sullivan-Taylor and Branicki 2011). As Linnenluecke and Griffiths (2010) argue, gradual change creates the need for long-term adaptation, which means a strategic rethinking of organizational structures and processes. To enable effective adaptation, companies have to be aware of the actual risks they are confronted with – now and in the future (Weinhofer and Busch 2013). The ability to realize harmful developments "requires a culture that allows 'maverick' information to be heard, understood and acted upon" (Sheffi and Rice 2005: 47). The corporate attitude toward possible organizational transition, consequently, defines whether a company enters a sustainable development path.

Adaptation to climate change and sustainability demands are, however, not just difficult tasks, they also offer possibilities for innovation. When innovation is based on sustainable business models, a company transcends the phase of adaptation and enters the level of transition. Relating this to the logistics industry, firms need to more actively engage with adopting sustainable technologies, proactively shape markets through developing services that distinguish them from their competitors, e.g. due to a reduced carbon footprint, and they also need to reconsider certain practices, for example, focusing on leasing ecologically advantageous vehicles equipped with future-proof technologies instead of tying up their capital in purchasing trucks. Conversely, ownership can also facilitate transition. For instance, when firms use owned properties, they increase their opportunities to undertake climate adaptation measures or introduce space-efficient inventory management systems on these properties. This form of transition is essential, especially because innovation has long replaced productivity as the decisive factor for competitiveness (Porter and van der Linde 1995; Truffer 2008). This is particularly true for companies operating in global markets (Coe 2014).

Location Requirements, Site Selection, and Site Design

Location selection and site design are key issues to address when creating sustainable logistics that are better adapted to climate change and minimized resource use. In current business models, the ideal location of a logistics enterprise has to be large (ideally >5 ha), flat, rectangular, in very close proximity to major roads and transport axis, have a fast internet connection, allow for 24/7 operations, and be on the outskirts of a major conurbation (Langhagen-Rohrbach 2012). This model has not only led to a recent logistics sprawl, which stretches from the urban cores to the suburban and exurban areas, but also to diverging developments due to the relocation of city logistics to the urban core (Heitz et al. 2019). Distribution logistics (e.g., food, spare parts, medicine, and retail) is located in urban areas and in close proximity to its customers. The parcel, courier, and express services (items >31.5 kg) and forwarding companies dealing with heavier single dispatch items organize their operations in a hub-and-spoke system with the smaller distribution

S

centers in urban areas. Similarly, production logistics is located close to the customer, often in large logistic parks, that organize the supply of a large plant as well as the intra-logistic between plants (Langhagen-Rohrbach 2012: 219). Other sectors of the logistics industry have a wider choice of suitable and often cheaper locations. These are the central hubs of distribution and parcel logistics with 24/7 operations and large cross-docking facilities for swap bodies located at the crossroads of the national highway system (Hesse 2020).

Companies can choose their location in ways that are optimized from a resource point of view. The use of brownfield sites, for instance, is a great sustainability advantage because no new land needs to be sealed (Franz et al. 2006). However, although there are multiple possibilities for sustainable and climate-adapted site selection, there are no simple solutions because land prices, land availability, proximity to customers, labor demands, good infrastructure, and environmental impact (including the length of transport distances) are often in conflict with one another. For example, the density of logistics in urban areas has triggered opposition to real estate development related to logistics, but low traffic emissions due to the proximity to customers, with a possible further reduction by using electric vans or bikes in city-logistic concepts, are sustainability opportunities. Furthermore, the clustering of logistics properties allows for efficient collaborative use of facilities and means of transportation, based on digital platform solutions.

The design of the spaces and buildings themselves can also be optimized for efficient resource use. This includes multistory constructions, the flexible division of spaces into areas of different sizes, the potential for modular extensions or superstructures (Kujath 2003; Förster et al. 2017), and, more generally, third-party usability can be embedded into the planning. Further sustainability gains can be made through climate-adapted construction methods. For example, the installation of green roofs and facades can avert the development of heat islands, which, in turn, can have a direct impact on the performance of employees (Marx 2017). Moreover, the intensive greening of operational areas and the consideration of cold air supply can help to create a healthy microclimate (Sieber 2019). An additional factor to consider is how to protect properties from damage caused by extreme weather events. Investment in climate adaptation of land and buildings, like the installation of sufficient rainwater retention systems and construction of halls adapted to storms, hail, and snow, can protect properties from resulting damages (Weller et al. 2016).

A special role must be played by the municipalities and regional administration, because of their ability to influence the design of real estate through specifications of land use plans, development plans, and design statutes. Zoning should where possible avoid flood zones and other environmentally sensitive areas, and the use of valuable soils. In addition, zoning should reduce through traffic, specify green building solutions and soil sealing, and improve the landscape quality.

Communication and Cooperation for Sustainable Logistics Processes

In order to further develop logistics companies in ways that consistently pursue the demands of sustainable and climate change-adapted development, communication and cooperation processes are necessary at various levels. This applies both to exchanges via formal institutions, such as industry associations or contractual business collaborations, and to informal exchanges via network contacts. The latter aspect might be considered more important than formal institutions when implementing sustainability goals. Networks can also promote collaborative governance processes by (a) supporting the generation, appropriation, and diffusion of different types of knowledge and information about the systems under consideration, (b) enabling the mobilization and allocation of key resources for effective governance, (c) promoting greater commitment among stakeholders to common goals; and (d) providing a platform for conflict resolution (Bodin and Crona 2009).

In the context of adaptation to climate change and sustainability needs, two levels are especially significant:

1. Communications with residents, non-governmental organizations, and local and regional administrations help to prevent conflicts regarding the location of logistics companies in different areas and address the requirements for a design by different stakeholder groups at an early stage. The "acceptance of a project is generally higher in a [...] community that is able to participate actively in the decision-making process" and the "creation of a more cohesive community requires a certain degree of participation representing the minimum range of stakeholders from different fields (authorities, NGOs, associations) to make the process compatible with local and regional expectations, values and norms" (Thornton et al. 2007: 51). The development of trust between the actors based on effective communication can contribute to minimizing the risks for the logistics companies (Murphy 2006).
2. Cooperation with neighboring logistics companies, on the one hand, and companies in the same value chains, on the other hand, allows for the exploitation of synergies around land use and opens other potentials for effective resource utilization. Cooperation can also contribute to the diffusion of organizational and technological sustainability innovations through access to external knowledge. In this way, an expanded knowledge base and access to innovations can help minimize risks for the companies and support a sustainability transition.

The establishment of cooperative, inter-company relationships makes it possible to increase the effectiveness of logistics cooperation through the sharing of information and resources. Costs can be reduced when environmental goals come into focus, for example, by sharing warehouses or parking spaces (Soosay et al. 2008) or by sharing information in ways that reduce lead times (Pérez-Bernabeu et al. 2015) or safety stock (Soosay et al. 2008). A common information base can be built by establishing electronic links between the various actors located throughout the supply chain (Fulconis et al. 2007). Ideally, the potential synergies between different companies should already be considered by several companies when they plan locations in industrial parks.

Cooperative execution of logistics processes is also possible in last mile logistics, the transport of goods to the customer's doorstep. In this process, relevant data is harmonized and exchanged between cooperating companies to create a common plan for the shared use of resources (vehicles, infrastructures), which are accessible to all partners. For example, the joint use of vehicles can improve capacity utilization and lead to a reduction in the number of vehicles (de Souza et al. 2014), which lessens the required parking space. A similar optimization of the vehicle fleet can also be achieved through a cooperating hub-and-spoke system. In this system, deliveries are collected from different depots to be sorted at a hub and then transported in bundles to the destination region (Zäpfel and Wasner 2002). In order to exploit the sustainability benefits of cooperation, it is, therefore, worth analyzing the entire value chain in the context of supply chain management. This opens up more opportunities for sustainable optimization of flows of goods and information (Marshall et al. 2015).

Conclusion

Worldwide, the logistics industry is growing and, as a result, the environmental and social impacts of logistics processes are increasing. This includes both the use of energy and material resources with their associated impacts and pollution (e.g., noise, air pollution, CO_2, etc.) and the land taken for the buildings and sealed surfaces of logistics companies (for storage, packaging, and other services) as well as public infrastructures. In addition, labor relations in logistics often do not comply with the principles of social sustainability. These environmental and social impacts are also perceived by the public, politicians, and customers. This exposes the companies to various risks including reputational risk, increasing environmental and social regulation, and a competitive disadvantage in the allocation of space compared to manufacturing industries. At the same time, there are also direct risks from environmental hazards, especially from climate change. Concepts for more sustainable logistics, which are better adapted to climate risks, have therefore emerged in recent years. In practice, however, such approaches are still niche phenomena. Companies often respond to risks with short-term risk management rather than with a permanent transition strategy toward a more sustainable way of doing business and climate-adapted structures.

Approaches to more sustainable logistics should include strategies for more efficient use of resources within the company (e.g., less space required and lower CO_2 emissions) as well as the joint use of resources by several companies. Collaboration between companies could make an important contribution to these endeavors. Moreover, in order to achieve greater acceptance of logistics settlements and a better image of the industry, more intensive communication is deemed necessary with residents, non-governmental organizations, and local and regional administrations. In this way, design requirements can be addressed by various stakeholder groups at an early stage and, ultimately, later conflicts can be avoided when logistics companies settle in the area.

Cross-References

▶ Adapting Cities to Climate Change
▶ Future of Urban Governance and Citizen Participation

References

Behrends, S., Lindholm, M., & Woxenius, J. (2008). The impact of urban freight transport: A definition of sustainability from an actor's perspective. *Transportation Planning and Technology, 31*(6), 693–713.

Bodin, Ö., & Crona, B. I. (2009). The role of social networks in natural resource governance: What relational patterns make a difference? *Global Environmental Change, 19*(3), 366–374.

Bücken, F., & Kanning, H. (2021). Klimaanpassung von Logistikstandorten – eine Szenarienanalyse. Standort - Zeitschrift für angewandte Geographie https://doi.org/10.1007/s00548-021-00717-7

Christopher, M., & Peck, H. (2004). Building the resilient supply chain. *The International Journal of Logistics Management, 15*(2), 1–13.

Coe, N. M. (2014). Missing links: Logistics, governance and upgrading in a shifting global economy. *Review of International Political Economy, 21*(1), 224–256.

Coe, N. M. (2021). Coping with commoditization: The third-party logistics industry in the Asia-Pacific. *Competition & Change*. https://doi.org/10.1177/1024529420985240.

Coenen, L., & López, F. J. D. (2010). Comparing systems approaches to innovation and technological change for sustainable and competitive economies: An explorative study into conceptual commonalities, differences and complementarities. *Journal of Cleaner Production, 18* (12), 1149–1160.

de Souza, R., Goh, M., Lau, H.-C., Ng, W.-S., & Tan, P.-S. (2014). Collaborative urban logistics – Synchronizing the last mile: A Singapore research perspective. *Procedia – Social and Behavioral Sciences, 125*, 422–431. https://doi.org/10.1016/J.SBSPRO.2014.01.1485.

Evangelista, P. (2014). Environmental sustainability practices in the transport and logistics service industry: An exploratory case study investigation. *Research in Transportation Business & Management, 12*, 63–72.

Förster A., Wenzel S., Thierstein A., Gilliard L., Scholze L., Unland L., & Brunner, B. (2017). *Gewerbe & Stadt: Gemeinsam Zukunft gestalten*. Technische Universität München. https://mediatum.ub.tum.de/doc/1398132/file.pdf

Franz, M., Pahlen, G., Nathanail, P., Okuniek, N., & Koj, A. (2006). Sustainable development and brownfield regeneration. What defines the quality of derelict land recycling? *Environmental Sciences, 3*(2), 135–151.

Fulconis, F., Saglietto, L., & Paché, G. (2007). Strategy dynamics in the logistics industry: A transactional center perspective. *Management Decision, 45*(1), 104–117.

Garbe, C., & Hempel, J. D. (2015). Die Zukunft der Logistikimmobilie. In P. H. Voß (Ed.), *Logistik – eine Industrie, die (sich) bewegt: Strategien und Lösungen entlang der Supply Chain 4.0* (pp. 153–166). Springer Gabler.

Heitz, A., Launay, P., & Beziat, A. (2019). Heterogeneity of logistics facilities: An issue for a better understanding and planning of the location of logistics facilities. *European Transport Research Review, 11*(5). https://doi.org/10.1186/s12544-018-0341-5.

Hesse, M. (2020). Logistics: Situating flows in a spatial context. *Geography Compass, 14*(7). https://doi.org/10.1111/gec3.12492.

IPCC – Intergovernmental Panel on Climate Change. (2014). *Climate change 2014: Impacts, adaptation, and vulnerability. Annex II.* Contribution of Working Group II to the Fifth Assessment Report of the Intergovernmental Panel on Climate Change. Cambridge University Press. https://www.ipcc.ch/site/assets/uploads/2018/02/WGIIAR5-AnnexII_FINAL.pdf

Jaffee, D., & Bensman, D. (2016). Draying and picking: Precarious work and labor action in the logistics sector. *WorkingUSA, 19*(1), 57–79.

Kauppi, K., Longoni, A., Caniato, F., & Kuula, M. (2016). Managing country disruption risks and improving operational performance: Risk management along integrated supply chains. *International Journal of Production Economics, 182*, 484–495.

Kujath, H. J. (2003). *Logistik und Raum: Neue regionale Netzwerke der Güterverteilung und Logistik. Working Paper / Leibniz-Institut für Regionalentwicklung und Strukturplanung: Bd. 18.* Leibniz-Institut für Regionalentwicklung und Strukturplanung e.V. (IRS).

Langhagen-Rohrbach, C. (2012). Moderne Logistik – Anforderungen an Standorte und Raumentwicklung. *Raumforschung und Raumordnung, 70*(3), 217–227.

Lehmacher, W. (2015). Wirtschaft, Gesellschaft und Logistik 2050. In P. H. Voß (Ed.), *Logistik – eine Industrie, die (sich) bewegt: Strategien und Lösungen entlang der Supply Chain 4.0* (pp. 1–17). Springer Gabler.

Linnenluecke, M., & Griffiths, A. (2010). Beyond adaptation: Resilience for business in light of climate change and weather extremes. *Business & Society, 49*(3), 477–511.

Linnenluecke, M. K., & Griffiths, A. (2015). *The climate resilient organization: Adaption and resilience to climate change and weather extremes*. Edward Elgar Publishing.

Lohre, D., Pfennig, R., Poerschke, V., & Gotthardt, R. (2015). Schlüsselthema 4: Das Ansehen der Logistik in der Öffentlichkeit. In *Nachhaltigkeitsmanagement für Logistikdienstleister* (pp. 59–64). Springer Gabler.

Lupton, D. (2013). *Risk* (2nd ed.). Routledge.

Marshall, D., McCarthy, L., Heavey, C., & McGrath, P. (2015). Environmental and social supply chain management sustainability practices: Construct development and measurement. *Production Planning & Control, 26*(8), 673–690. https://doi.org/10.1080/09537287.2014.963726.

Marx, A. (Ed.). (2017). *Klimaanpassung in Forschung und Politik*. Springer Spektrum.

McKinnon, A. (2019). Freight transport and logistics. In J. Stanley & D. Hensher (Eds.), *A research agenda for transport policy* (pp. 99–107). Edward Elgar Publishing. https://doi.org/10.4337/9781788970204.00020.

Murphy, J. T. (2006). Building Trust in Economic Space. *Progress in Human Geography, 30*(4), 427–450.

Murphy, P. R., & Poist, R. F. (2003). Green perspectives and practices: A "comparative logistics" study. *Supply Chain Management: An International Journal, 8*(2), 122–131.

Neiberger, C. (2015). Leitbild Nachhaltigkeit – radikaler Wandel in Güterverkehr und Logistik? *Zeitschrift für Wirtschaftsgeographie, 59*(1), 77–90.

Onischka, M. (2009). Definition von Klimarisiken und Systematisierung in Risikokaskaden. *Diskussionspaper: Mainstreaming von Klimarisiken und -chancen im Finanzsektor.* Wuppertal. Institut für Klima, Umwelt, Energie. Wuppertal.

Pérez-Bernabeu, E., Juan, A. A., Faulin, J., & Barrios, B. B. (2015). Horizontal cooperation in road transportation: A case illustrating savings in distances and greenhouse gas emissions. *International Transactions in Operational Research, 22*(3), 585–606. https://doi.org/10.1111/itor.12130.

Ponomarov, S. Y., & Holcomb, M. C. (2009). Understanding the concept of supply chain resilience. *The International Journal of Logistics Management, 20*(1), 124–143.

Porter, M. E., & van der Linde, C. M. (1995). Toward a new conception of the environment-competitiveness relationship. *The Journal of Economic Perspectives, 9*(4), 97–118.

Renn, O. (2008). *Risk governance: Coping with uncertainty in a complex world.* Routledge.

Sheffi, Y., & Rice, J. B., Jr. (2005). A supply chain view of the resilient enterprise. *MIT Sloan Management Review, 47*(1), 41–49.

Sieber, S. (2019). Gewerbegebiete im Wandel – Wie Gewerbegebiete in Marl, Remscheid und Frankfurt Biodiversität und Klimaschutz verbinden. *Transforming Cities, 3*, 70–75.

Soosay, C. A., Hyland, P. W., & Ferrer, M. (2008). Supply chain collaboration: Capabilities for continuous innovation. *Supply Chain Management: An International Journal, 13*(2), 160–169. https://doi.org/10.1108/13598540810860994.

Sullivan-Taylor, B., & Branicki, L. (2011). Creating resilient SMEs: Why one size might not fit all. *International Journal of Production Research, 49*(18), 5565–5579.

Tempelmeier, H. (Ed.). (2018). *Fachwissen Logistik. Begriff der Logistik, logistische Systeme und Prozesse.* Springer Vieweg.

Thornton, G. J. P., Nathanail, C. P., Franz, M., & Pahlen, G. (2007). The development of a brownfield-specific sustainability and indicator framework for regenerating sites: Proposing a new definition of 'sustainable brownfield regeneration'. *Land Contamination & Reclamation, 15*(1), 41–54.

Todaro, N. M., Testa, F., Daddi, T., & Iraldo, F. (2020). The influence of managers' awareness of climate change, perceived climate risk exposure and risk tolerance on the adoption of corporate responses to climate change. *Business Strategy and the Environment, 30*(2), 1232–1248.

Truffer, B. (2008). Society, technology, and region: Contributions from the social study of technology to economic geography. *Environment and Planning A: Economy and Space, 40*(4), 966–985.

Vertzberger, Y. Y. I. (1998). *Risk taking and decisionmaking: Foreign military intervension decisions.* Stanford University Press.

WCED – World Commision on Environment and Development. (1987). *Our common future.* Oxford.

Weinhofer, G., & Busch, T. (2013). Corporate strategies for managing climate risks. *Business Strategy and the Environment, 22*(2), 121–144.

Weller, B., Fahrion, M.-S., Horn, S., Naumann, T., & Nikolowski, J. N. (2016). *Baukonstruktion im Klimawandel*. Springer Vieweg.

Zäpfel, G., & Wasner, M. (2002). Planning and optimization of hub-and-spoke transportation networks of cooperative third-party logistics providers. *International Journal of Production Economics, 78*(2), 207–220. https://doi.org/10.1016/S0925-5273(00)00152-3.

The Sustainable and the Smart City: Distinguishing Two Contemporary Urban Visions

Wolfgang Haupt
Leibniz-Insitute for Research on Society and Space, Erkner, Germany

Synonyms

Digitalization; Smart city; Smartness; Sustainability; Sustainable city; Urban sustainable development; Urban visions; Urbanization

Definition

Both the smart and the sustainable cities represent a positive vision of future urban development. This chapter analyzes and discusses definitions and notions of both concepts, as well as their emergence, establishment, and development. Both concepts share similarities, such as the far-reaching and integrative claim to bring positive change to cities or the notion to operationalize the terms sustainable and smart by defining subcategories. However, the differences between both concepts far *outweigh* the *similarities*. Indeed, sustainability and the sustainable city, on the one hand, and the smart city, on the other hand, are distinctly different in terms of their development history, the main driving forces behind, and the theoretical scope. The understanding of (urban) sustainability was strongly influenced and brought forward by the ideas and commitments of intellectuals, leading politicians, and environmental movements, whereas multinational technology companies mainly coined the development of the smart city concept. Besides, the significantly more comprehensively defined term sustainability had a consistent underlying idea from the beginning onward, whereas the smart city concept still lacks terminological and content-related clarity as well as a generally accepted definition.

Introduction

Cities worldwide are faced with a multitude of often interrelated challenges (Acuto 2013; Barber 2013; Haupt 2019). These are, for instance, dealing with growing inequalities, the impacts of climate change, infrastructure congestion, or increasing air pollution, just to name a few. At the same time, record numbers of people are living and continuously moving to cities and urban areas. Even under the current conditions, many cities fail to deal with several or all of the listed challenges. It is therefore evident that cities need to develop strategies, solutions, and models to tackle nowadays issues and to equip themselves for the future. In this context, *future city visions such as the sustainable city and the smart city are increasingly considered as potential answers to the challenges arising from progressing urbanization.*

In the past years, sustainable cities and smart cities h*ave generated widespread interest among politicians, policy-makers, practitioners, and scholars of various disciplines* (see Giffinger et al. 2007; Hiremath et al. 2013; Keivani 2010; Karvonen et al. 2020; Pufé 2017; Vanolo 2014). *Sustainable cities aim at achieving* a balance between the needs of social, economic, and environmental urban development in order to ensure that no *burdens are left to future generations* (Hiremath et al. 2013; Squires 2013). *Smart cities rely on innovations in* information and communication technology (ICT) and their potentials to optimize citizen's lives through the intelligent management and linking of urban infrastructures of various kinds (Angelidou 2015; Townsend 2013). Both concepts represent a positive vision of an urban future; however, the (possible) relations between them need to be better understood (Ahvenniemi et al. 2017; Bibri and Krogstie 2017; Bifulco et al. 2016; Höjer and Wangel 2014; Kunzmann 2014).

Drawing on a broad range of literature on both concepts this chapter reviews, structures, and compares the discourses on two key urban visions of the contemporary area. Both concepts are presented and discussed focusing on the following three thematic areas:

1. Definitions, notions, and key components
2. Origins and further development
3. Main points of criticism

Due to the intense and steadily growing body of literature on (urban) sustainability, and smart cities, and because of the various academic disciplines involved, this chapter has to remain on generic level. The literature survey makes no claim to completeness but aims to cover the main debates and discussions that are conducted in a variety of disciplines, mainly urban studies, geography, political sciences, and computer science.

Content-wise, the chapter is structured as follows: Within the first two sections, both concepts are presented separately, orienting along the presented three thematic areas. The following section discusses the main differences between the sustainable city and the smart city. The chapter closes with a conclusion summarizing the key observations.

Sustainability and Sustainable Cities

Definitions, Notions, and Key Components

The fundamental idea that has led to the contemporary understanding of sustainability was the observation of two mutually reinforcing dynamics: population growth on the one hand and an increasing scarcity of natural resources on the other hand (Pufé 2017). Being aware that these dynamics increasingly affect human livelihood, in 1987 the World Commission on Environment and Development (WCED) – also known as the Brundtland Commission – released the report *Our Common Future*, also known as Brundtland Report (see WCED 1987). The *Brundtland Report* contains the most recognized and most cited sustainability definition (Hilty et al. 2011; Pufé 2017): "Sustainable development is the development that meets the needs of the present without compromising the ability of future generations to meet their own needs" (WCED 1987: 41). To reach the ambitious goal of sustainable development, economic growth needs to be decoupled from environmental impacts and from the use of natural resources (Hilty et al. 2011). The *Brundtland Report* splits the very comprehensive sustainability term into three still quite comprehensive main dimensions: the economic, the social, and the environmental dimension. These dimensions of equal importance are considered to be key for meeting the goal of sustainable development (see WCED 1987).

Common ways to illustrate the sustainability concept are shown in Fig. 1. Often, sustainable development is pictured with three overlapping circles, whereby each circle stands for one of the three dimensions. The overlapping area of all circles symbolizes the concordance between the three dimensions and stands for the desired sustainable development. Another popular illustration shows the three dimensions as pillars carrying the sustainability rooftop. The probably most widespread illustration is the triangle of sustainability: In this equilateral triangle, the three dimensions encompass sustainability and in form of a joint interplay enable sustainable development.

As mentioned, there is quite a broad idea of what sustainability should be about. But what does this imply for a city? The *Brundtland Report* pointed to the challenges arising from a rapid global process of urbanization and the necessity to work toward urban sustainable development (WCED 1987). In the meantime, urban sustainability has become a distinct and field of practical action and research (see Allen 2001; Hiremath et al. 2013; Keivani 2010; Williams 2010). An often-used description of urban

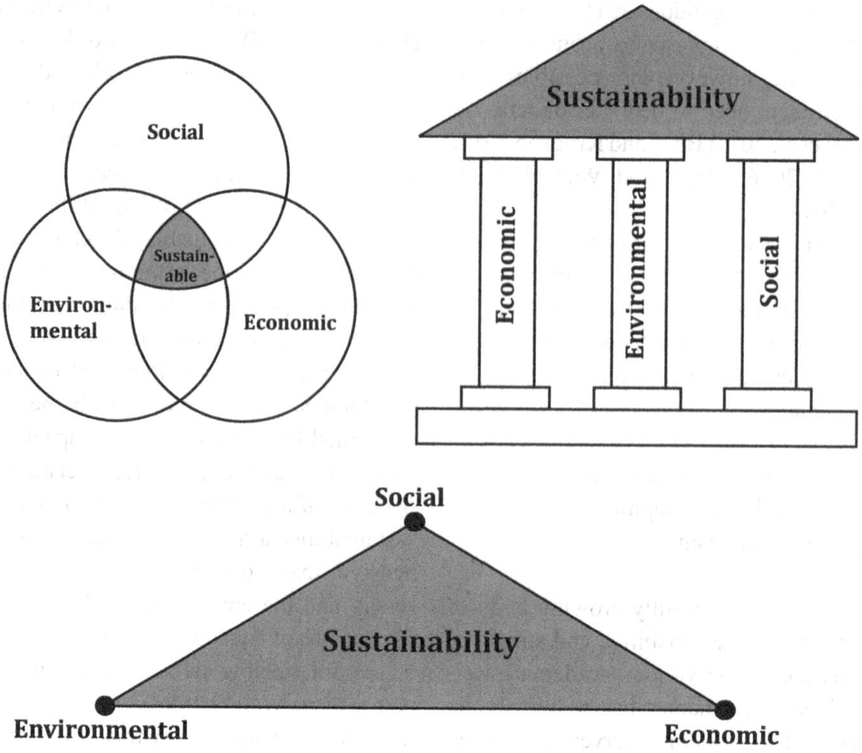

The Sustainable and the Smart City: Distinguishing Two Contemporary Urban Visions, Fig. 1 Different forms of visualizing the sustainability idea (1, Venn diagram of sustainable development; 2, three pillars of sustainability; 3, triangle of sustainability). (Source: own diagram)

sustainable development dates back to the Urban 21 Conference, which took place in Berlin in the year 2000:

> "Improving the quality of life in a city, including ecological, cultural, political, institutional, social and economic components without leaving a burden on the future generations. A burden, which is the result of a reduced natural capital and an excessive local debt. Our aim is that the flow principle, that is based on an equilibrium of material and energy and also financial input/output, plays a crucial role in all future decisions upon the development of urban areas." (Squires 2013: 198)

In a similar vein, Hiremath et al. (2013: 556) defined the goal of urban sustainable development as "achieving a balance between the development of the urban areas and protection of the environment with an eye to equity in income, employment, shelter, basic services, social infrastructure and transportation in the urban areas."

Working toward an overarching understanding of urban sustainability, several institutions and scholars added further dimensions in addition to the three original sustainability dimensions. For instance, the United Nations Department of Economic and Social Affairs (UN-DESA) introduced the urban governance dimension (see Fig. 2). Besides, Allen (2001) added a fourth and a fifth dimension: political sustainability and physical sustainability (see Fig. 3). What Allen defines as political sustainability shares several similarities with UN-DESA's understanding of the urban governance dimension (see Fig. 2). Indeed, Allen (2009: 3) describes political sustainability as "the quality of governance systems guiding the relationship and actions of different actors among the previous four dimensions. Thereby, it implies the democratization and participation of local civil society in all areas of decision-making" (Allen 2009: 3). Another dimension that, in Allen's view, deserves special attention is the built environment (mainly buildings and urban infrastructure), which she also refers to as "second nature"

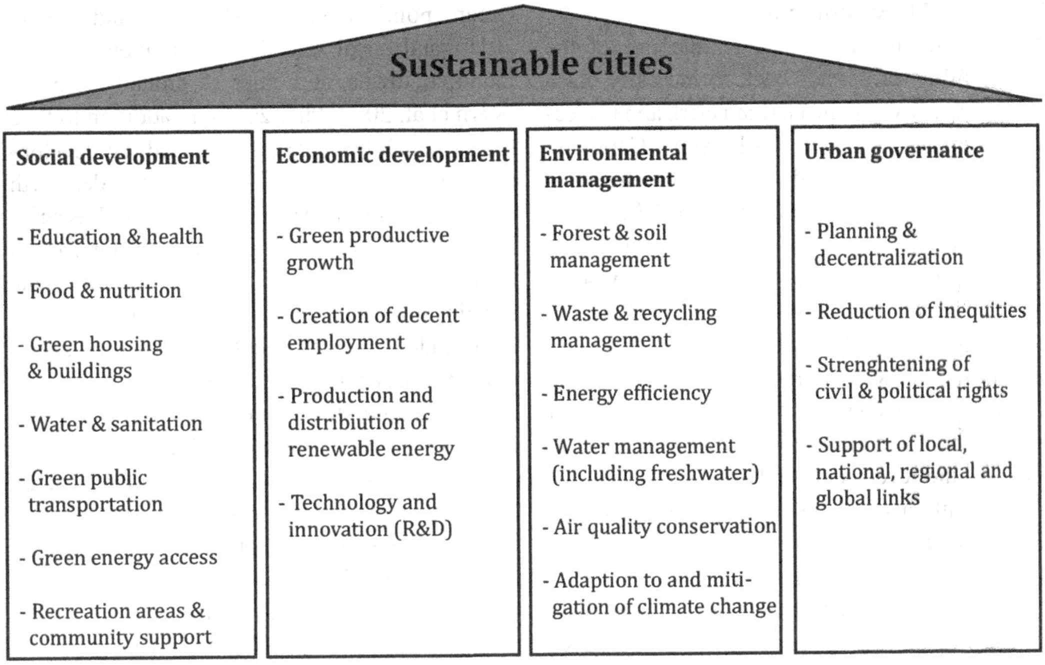
Sustainable cities
Social development
- Education & health
- Food & nutrition
- Green housing & buildings
- Water & sanitation
- Green public transportation
- Green energy access
- Recreation areas & community support
Economic development
- Green productive growth
- Creation of decent employment
- Production and distribiution of renewable energy
- Technology and innovation (R&D)
Environmental management
- Forest & soil management
- Waste & recycling management
- Energy efficiency
- Water management (including freshwater)
- Air quality conservation
- Adaption to and miti-gation of climate change
Urban governance
- Planning & decentralization
- Reduction of inequities
- Strenghtening of civil & political rights
- Support of local, national, regional and global links

The Sustainable and the Smart City: Distinguishing Two Contemporary Urban Visions, Fig. 2 Pillars for achieving sustainability of cities. (Source: UN-DESA 2013)

The Sustainable and the Smart City: Distinguishing Two Contemporary Urban Visions, Fig. 3 The five dimensions of urban sustainability as defined by Adriana Allen (2001). (Source: Allen 2001)

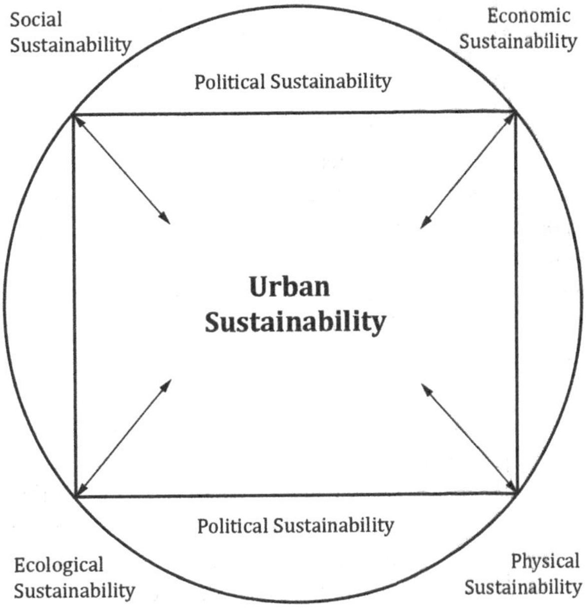

or "physical sustainability" (Allen 2009: 2). What these post-Brundtland urban sustainability definitions all share in common is that the original Brundtland spirit (social, economic, and environmental sustainability) is still put center stage. What is new is that these three dimensions were specified or – to put it in another way – were filled with life and complemented by additional dimensions that aim to meet the needs of cities and urban areas (Allen 2001, 2009).

Origins and Development

Aside from ordinary language, the use of the sustainability term dates back to the early eighteenth century when the German clerk and inspector of mines Hans Carl von Carlowitz (1645–1714) formulated his recommendations for desirable wild tree cultivation in *Sylvicultura oeconomica* (1713) (Pufé 2017). In the late eighteenth century, the Prussian scientist and explorer Alexander von Humboldt heavily criticized the rampant logging of construction and firewood and suggested several measures to reduce the demand for wood (Wulf 2016). In this context, it needs to be noted that wood could be regarded as the oil of the seventeenth and eighteenth centuries (ibid.). With this in mind, it does not surprise that the first reference to sustainability focused on this resource.

In the twentieth century, when wood had been long ago replaced by oil as the key driver of economic growth, the sustainability idea gained momentum again. This was particularly due to the oil crises in the 1970s (Mogens 2014) and the gradually growing awareness that burning fossil fuels leads to global climate change (Peñuelas and Carnicer 2010). Against this background, it is also understandable that the *Limits to Growth Report* published by the Club of Rome in 1972 attracted such a high level of interest and attention (see Meadows et al. 1972). The key message of the report was the warning that business as usual would lead to a disastrous decrease in population and living standards within the next 50 or 100 years. According to Pufé (2017), the key milestones that further coined and helped to establish the sustainability concept were the earlier mentioned *Brundtland Report* (1987), the *Rio Earth Summit and the Agenda 21* (1992), the *Millennium Development goals* (2000), and the *Durban Climate Summit* (2011).

From the beginning on, NGOs, local or global groups and movements, political parties, and local initiatives strongly referred to the sustainability idea (Pufé 2017). Examples include the antinuclear movement (Pufé 2017), environmental groups and social initiatives (Scott and Vare 2020; Pufé 2017; van Wehrden et al. 2019), Green political parties (Bossuyt and Savini 2018), and Local Agenda 21 resolutions in communities, towns, and cities all around the globe (Kern et al. 2007; Pufé 2017). In addition to that, the sustainability idea also inspired the development of further concepts such as the degrowth concept (Robra and Heikkurinen 2019). It is striking that over 300 years after von Carlowitz and over 30 years after the *Brundtland Report*, the sustainability idea appears to be still very much alive. Indeed, in 2016 the UN released its development goals for the year 2030 defining overall 17 Sustainable Development Goals (SDGs), one of them focusing explicitly on cities (SDG 11 Sustainable Cities and Communities) (Kroll et al. 2019; Robra and Heikkurinen 2019). Moreover, also contemporary climate justice movements such as Fridays for Future or Extinction Rebellion repeatedly bring the sustainability vision to the fore (Scott and Vare 2020; von Wehrden et al. 2019). In fact, sustainability has entered almost all aspects of human life. There are, for instance, sustainability standards and certification (e.g., for fair trade, fishing, farming, fashion), degree courses in sustainability, or sustainability strategies passed by cities, regions, or countries.

Working toward the vision of a sustainable city was underpinned by various funding programs in different countries. With regard to the financial budget, the programs of the EU do stand out. Between 2014 and 2020, the funding preferences of the European regional policy, one of the most important instruments of EU policy and one of the largest integrated development policies in the Western world (McCann and Varga 2015), were on urban areas (Fioretti et al. 2020). For their programs, the European Commission coined the overall goal of integrated sustainable urban development (Gargiulo et al. 2013). The European Commission emphasized the need to involve citizens and the civil society, the economy, and all relevant government levels (particularly on the local and regional level) (Fioretti et al. 2020). In this context, around 17 billion Euros were made available to be managed directly by European urban authorities (Fioretti et al. 2020).

Criticism

Several scholars have labeled sustainability as a very arbitrary concept. Indeed, Swyngedouw (2010: 190) noticed a "whole array of 'sustainabilities'" – such as sustainable environments, sustainable growth, sustainable companies, sustainable processes, sustainable technologies, sustainable markets, and many, many more. Apparently, the word can be used for everything, and everybody seems to associate it with something positive:

> "I have not been able to find a single source that is against sustainability. Greenpeace is in favour, George Bush Jr. and Sr. are, the World Bank and its chairman (a prime war monger in Iraq) are, the pope is, my son Arno is, the rubber tappers in the Brazilian Amazon are, Bill Gates is, the labour unions are." Swyngedouw (2010: 190)

Furthermore, Smythe (2014) criticizes that the concept hasn't changed (and improved) after the *Brundtland Report* and thus regards it as an inutile and eventually unsuccessful approach to tackle contemporary issues. The central point of her criticism is that economics has constantly been separated from the environment and society which resulted, in her view, in unsatisfactory human welfare. Smythe is convinced that just bringing these three components together in a polygon cannot generate sustainability. Instead, she suggests rethinking sustainability by starting with a more holistic approach that puts human welfare, in connection with the natural and the social world, into the center (Smythe 2014). Also, Hove (2004) expressed her disappointment with the results of sustainable development in practice: "Unfortunately, sustainable development has become more of a catch-phrase than a revolution of thought, and employing its use has simply fuelled the interests of advocates of exponential economic growth, undermining environmental reforms." She argues that applied sustainable development seems to be like trying to square the circle. Besides, until today attempts to interlock the opposing aims, environmental protection, on the one hand, and economic expansion, on the other hand, have failed (Hove 2004).

Smart Cities

Definitions, Notions, and Key Components

Defining what smartness means and what constitutes a smart city is more difficult than defining sustainability and the sustainable city. Indeed, a commonly accepted definition does (still) not exist for a smart city (Ahvenniemi et al. 2017; Angelidou 2015; Bibri and Krogstie 2017; Hollands 2008). Vanolo (2016: 27) summarized that, currently, the most common understanding of a smart city "relies on the implicit assumption that urban infrastructures and everyday life are optimized and 'greened' through technologies provided by ICT-companies." In a similar vein, Townsend (2013) understands a smart city as a city that can be monitored, managed, and regulated by using ubiquitous computing and ICT-infrastructure. Despite the lack of a definition, what is certain about the smart city is that it represents "a multidisciplinary field, constantly shaped by advancements in technology and urban development" (Angelidou 2015: 95).

As a reaction to "the large number and dispersion of smart city definitions," smart city assessment frameworks were developed. The purpose of these frameworks is to show policy-makers if their city's development "is proceeding towards the wanted direction" (Ahvenniemi et al. 2017: 241). For instance, there is the classification from a group of European researchers (Giffinger et al. 2007), who are considered to provide us with the most reliable definition so far (Vanolo 2014). The team developed an approach that should help to rank medium-sized European cities regarding their "smartness." The researchers specified that the smart city term is understood as a certain ability of a city and not as a sum of single aspects. Giffinger et al. (2007) identified six dimensions that are considered to be essential for the evaluation of a city and for assessing to what extent it deserves the adjective "smart." A look at these dimensions and the further specified keywords listed below them (see Fig. 4) demonstrates that a smart city – as understood by Giffinger et al. (2007) – basically covers all aspects of urban life. Indeed, this conglomeration of a definition shows

S

SMART ECONOMY
(Competitiveness)

- Innovative spirit
- Entrepreneurship
- Economic image & trademarks
- Productivity
- Flexibility of labour market
- International embededness
- Ability to transform

SMART PEOPLE
(Social & Human Capital)

- Level of qualification
- Affinity to life long learning
- Social & ethnic plurality
- Flexibility
- Creativity
- Cosmopolitanism/Open-mindedness
- Participation in public life

SMART GOVERNANCE
(Participation)

- Participaton in decision-making
- Public & social services
- Transparent governance
- Political strategies & perspectives

SMART MOBILITY
(Transport & ICT)

- Local accessibility
- (Inter-) national accessibility
- Availability of ICT-infrastructure
- Sustainable, innovative and safe transport system

SMART ENVIRONMENT
(Natural Resources)

- Attractivity of natural conditions
- Pollution
- Environmental protection
- Sustainable resource management

SMART LIVING
(Quality of Life)

- Cultural facilities
- Health conditions
- Individual safety
- Housing quality
- Education facilities
- Social cohesion

The Sustainable and the Smart City: Distinguishing Two Contemporary Urban Visions, Fig. 4 Dimensions of a smart city. (Source: Giffinger et al. 2007: 12)

that the smart city concept cannot be, other than the idea of sustainability, formulated in one catchy sentence.

Origins and Development History

The roots of the smart city concept are mainly to be found in the intelligent city and the smart growth concepts (see Fig. 5). The intelligent city represents a planning paradigm that aims at "bridging local resources, innovation institutions and broadband networks" (Komninos 2009: 338). In contrast, the smart growth concept is an urban planning theory that promotes a form of economic growth that avoids dissipation and damage to the environment (Gillham 2002). The main overlaps between both concepts can be seen in "learning and innovation" and in "infrastructure for communication and knowledge management" (Komninos 2006: 13). The smart city term first gained currency in the 1990s, when the focus was mostly on cities ICT-infrastructures (Albino et al. 2015). Over time, in many cities worldwide, the discourse about smart cities was driven by powerful and hegemonic economic actors and public-private coalitions (Rossi 2016; Söderström et al. 2014; Vanolo 2014; Wiig 2016). Over the last years, IBM has taken the lead over the smart city discourse. Indeed, in 2011, the term "smarter cities" was even registered as an IBM trademark (Söderström et al. 2014).

In addition to private actors, various funding programs on different levels (e.g., national state, EU) played a crucial role in promoting the spread of smart city projects (Gargiulo et al. 2013; Lange and Knieling 2020). It was also the absence of comparable funding programs in many EU-member states that helped to establish the EU's "driving role in setting priorities for the conception of smart city initiatives" (Lange and Knieling 2020: 109). Through the Seventh Framework Programme for Research and Technological Development (FP7), the term smart city was introduced to the EU (research) funding system, the main program to subsidize research on the

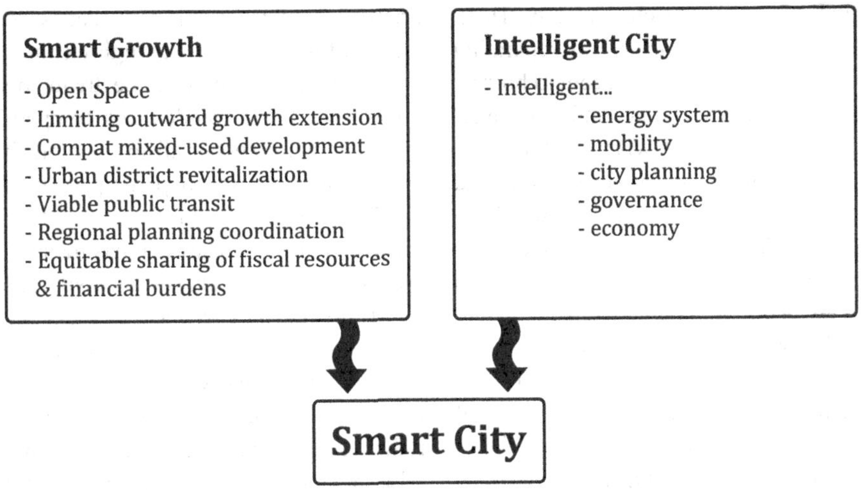

The Sustainable and the Smart City: Distinguishing Two Contemporary Urban Visions, Fig. 5 The evolution of the smart city concept. (Source: own diagram based on Gillham 2002 and BAUM 2013)

European level (Vanolo 2014). The program enabled the participation of local, regional, or national administrations and innovative businesses of different sizes (European Commission 2007). In the years that followed, a large number of European smart city projects were funded through the research and innovation program Horizon 2020 (Gargiulo et al. 2013; Lange and Knieling 2020). An integral part of the Horizon 2020 strategy was the Innovation Partnership on Smart Cities and Communities (EIP-SCC), launched in 2012 (European Commission n.d.). The program is divided into three categories, infrastructures and urban development, cooperation and capacity building and research, and innovation and competitiveness, and mainly targets different government and administration levels, innovative businesses of different sizes, banks, and research institutions (European Commission n.d.).

Despite the existence of numerous funding opportunities to support the development of smart cities in Europe, a general definition from EU-bodies is still lacking. Leastwise, there is a working definition of smart cities formulated under the leadership of the Danish think tank research and development (RAND) in a report requested by the European Parliament's Committee on Industry, Research and Energy (Manville et al. 2014). The suggestion of the report is to orient on the following working definition: "A city seeking to address public issues via ICT-based solutions based on a multi-stakeholder, municipally based partnership" (Manville et al. 2014: 24).

Criticism

Just as sustainability also the smart city and smartness term, in general, provoked criticism for its supposed lack of clarity. Indeed, the lack of definitional precision has led to some very different, sometimes unspoken, assumptions on the term (see Hollands 2008). The smart city, as it is understood presently, represents a relatively generic and quite optimistic concept (Vanolo 2014). The term has more the character of an evocative slogan than of a well-defined concept. This lack of clarity allows stakeholders to use the smart city term however it suits their agendas (ibid.). In this context, scholars have criticized its strong catchword character. Hollands (2008 and 2014) noted that terms like smart usually imply a positive and more likely uncritical view on the respective urban development. Which city does not want to be smart? Furthermore, he criticizes that people are "constantly bombarded with a wide range of new city discourses" (Hollands 2008: 305), be that the discourse about a smart, intelligent, innovative,

S

wired, digital, cultural, or a creative city. Here, Holland sees a big challenge in sharply separating all these different terms that often seem to borrow the assumptions from one another and also sometimes appear to fuse content-wise. Additionally, he distinguishes between an actual smart city and a city just wearing the smart label (ibid.). Ultimately, Hollands does not even acknowledge the smart city as a legit theoretical concept and instead refers to it as "concept/label" (Hollands 2008: 304). Also for Wiig (2016: 335), the smart city term is pure rhetoric. He claims coming up with descriptions like intelligent, transformative digital change usually serves quite a different purpose than improving urban inequalities. Instead, it is rather used to "sell a city in the global economy" (Wiig 2016: 335).

Besides the dissatisfaction with the inconclusiveness and vagueness of the smart city concept, the strong influence of business interests on the conceptual and semantic development of the smart city concept has been a main source of criticism among many scholars. Söderström et al. (2014) analyzed the "smarter city" campaign of IBM and came to the conclusion that the storyline of smart cities is at first glance about a sustainable and efficient city. However, at second sight, it is about occupying a dominant position in a huge market. Then again, Rossi (2016) warns to reduce smart urbanism to the economic contribution of influential and powerful actors, such as multinational companies. Indeed, Rossi's case study on Turin as a smart city has shown that economic city revitalization, for instance, in a postrecession capitalist environment, often also includes (and requires) less powerful actors, such as start-up innovative firms or the social economy.

In recent years, a growing number of scholars have started to ask for the role of the citizen within the smart city concept (see Harvey 2012; Hollands 2014; Kitchin 2015; Vanolo 2016; Colding et al. 2019). Indeed, many smart city projects "are characterized by the absence of citizen's voices," whereas there are also some projects that "are populated by active citizens operating as urban sensors" (Vanolo 2016: 34). In this context, Hollands (2014: 1) criticizes that the current orientation of the smart city that is "driven by the profit motive of global high-technology companies, in collusion with the trend toward city governance being wedded to a competitive form of 'urban entrepreneurialism' has left little room for ordinary people to participate in the smart city." In a similar vein, Harvey (2012) laments that the smart city discourse has mostly left aside the question of how cities work sociologically and politically as well as the complex and diverse conflicts and dynamics that come along with it. This is particularly problematic since most of the major urban issues are not of technological but rather of social nature (e.g., inequality or poverty). Also, there is no evidence for the assumptions that taking full advantage of existing innovations in ICT per se supports citizen involvement and more inclusive planning processes (Colding et al. 2019).

Comparing Both Concepts

It was shown that both concepts are divided into a number of subcategories. Sustainability in its original understanding has three dimensions (economic, ecologic, and social) (see WCED 1987). Moreover, depending on the focus and preferences, the components of urban governance (see UN-DESA 2013) and physical sustainability (see Allen 2009) can be added. Besides, the smart city consists of six subcategories (smart environment, smart economy, smart people, smart living, and smart governance) (see Giffinger et al. 2007). With regard to the areas of activity, these described dimensions or subcategories of the sustainable city, on the one hand, and the smart city, on the other hand, share many similarities and overlaps. In fact, one might argue that the differences are mostly of linguistic nature, wherefore some argue that both concepts might be integrated through so-called smart-sustainable cities. Nevertheless, when looking at the comprehensiveness and acceptance of their definitions as well as the development history and main driving forces behind both concepts, there are several substantial differences that are discussed in the following sections.

Urban Sustainability Through Smart Solutions?

Also because of the described similarities and overlaps with regard to the areas of activity, scholars have increasingly discussed the stronger integration of both concepts through smart sustainable cities (see Ahvenniemi et al. 2017; Bibri and Krogstie 2017; Bifulco et al. 2016; Höjer and Wangel 2014; Kunzmann 2014). However, several scholars see significant shortcomings – particularly of the smart city concept – that might speak against prospects for smart sustainable cities (Ahvenniemi et al. 2017; Cavada et al. 2016; Evans et al. 2019; Parks and Rohracher 2019). In this context, Ahvenniemi et al. (2017) found that while environmental sustainability is proclaimed to be a key target of smart cities, environmental indicators (surprisingly) only play a minor role in many smart city frameworks. Moreover, little is known about how smart cities can substantially contribute to meeting further sustainability goals like social well-being or the reduction of carbon emissions (Cavada et al. 2016; Evans et al. 2019; Parks and Rohracher 2019). Indeed, as Höjer and Wangel (2014: 333) have emphasized, the smart-sustainable city concept lacks clarity, and if it aspires "to have any useful meaning, it needs to be more strictly defined than it has previously been." Another suggestion for the integration of both concepts was outlined by McLaren and Agyeman (2015) who suggested to further develop the concept of the sharing city. They see the new digital technologies as a chance for cities to expand and connect smart technologies to fields like solidarity, justice, or sustainability (ibid.). In a similar vein, Ahvenniemi et al. (2017) and Bifulco et al. (2016) argue that smart city technologies might enable sustainable development. However, before this, the clear contribution of ICT to urban sustainable development needs to be specified and tested. Indeed, assuming that the integration of ICT within a city's service functions is per se beneficial for meeting sustainability goals is rather an uncritical observation and by no means represents a definite empirical finding (see Colding et al. 2019).

Different Comprehensiveness and Acceptance of Both Concepts

Both concepts have been criticized for their vagueness and lack of precision as well as for their catchword character (see Hollands 2008, 2014; Swyngedouw 2010). Indeed, the sustainable and the smart cities aim at no less than a "better" life in the city of tomorrow. However, beyond this understandable criticism, it needs to be emphasized that sustainability can be explained in one or two sentences or with the help of a simple illustration such as a triangle, while there is still no commonly accepted definition of a smart city (Ahvenniemi et al. 2017; Angelidou 2015; Bibri and Krogstie 2017; Hollands 2008). The classic sustainability definition formulated in the *Brundtland Report* in 1987 (WCED 1987: 42) – "Sustainable development is development that meets the needs of the present without compromising the ability of future generations to meet their own needs" – can basically be translated into all aspects of human life, the system of a city included. That the sustainability concept is not just a flash in the pan or an outdated idea from another time was – among others – demonstrated in 2016 when the UN announced to put sustainability into the center of the 2030 development goals (SDGs) (see Kroll et al. 2019; Robra and Heikkurinen 2019). On the contrary, it is widely acknowledged that the only uniting element of the different notions of a smart city is the central role of ICT (Angelidou 2015; Townsend 2013). However, by no means does this imply that future smart city development and research should be left solely to data analysts, computer scientists, or engineers. Quite the contrary, it cannot reasonably be expected from only a few professions or academic disciplines to deal with the diverse and complex implications that come along with the development and implementation of a concept that heavily impacts the economic, environmental, and social functioning of a city.

The described differences with regard to consistency and acceptance of definitions are also – in a way – reflected in the EU-funding programs that support the development of sustainable and smart cities. Through the European regional fund, well-endowed programs have supported integrated approaches to move toward sustainable urban development (see Fioretti et al. 2020). In contrast, the funding landscape for smart cities (mainly Horizon 2020) is significantly more scattered. Moreover, in this context, the question arises how an urban vision of this potential significance can be successfully steered, developed, and implemented in the total absence of a commonly accepted definition.

Different Development History and Different Driving Forces Behind Both Concepts

Furthermore, the two concepts differ considerably regarding their development history as well as the main driving forces behind it. The sustainability concept is over 300 years old, and urban sustainability has entered the arena with the *Rio Earth Summit* and the *Agenda 21* in 1992 at the very latest. In contrast, the smart city is mainly a product of the twenty-first century, and one might ask if we will still talk about smart cities in 20 or 30 years. Also, the fundamental beliefs that helped to establish the concepts and made them attractive for various actors could not be more different. Essential elements of sustainability have always been the management of scarcening resources (e.g., natural resources such as wood or oil), a reduction of resource consumption, or the adaption or expansion of existing infrastructures (e.g., car-free cities, bike lanes, social cohesion policies). Otherwise, smartness or the smart city focuses on more efficient use of resources. Behavioral change – e.g., consuming fewer resources – does not appear to be an integral part of the smart city concept. The notion is rather that smart and efficient technologies and digitalization will bring positive change and thus restrictions are not necessary.

Looking at the main driving forces behind both concepts, several differences could be observed. For the sustainability concept, the most notable driving forces were intellectuals (*Club of Rome:* 1972), politicians, and diplomats (*Brundtland Report*: 1987, and *Rio Earth Summit, Agenda 21*: 1992) (see Manville et al. 2014; Pufé 2017; WCED 1987). In comparison, the development of the smart city concept was strongly shaped by the interests and ideas of multinational technology companies that, until this day, influence the development of the concept to a certain degree (Hollands 2014; Rossi 2016; Söderström et al. 2014; Vanolo 2014; Wiig 2016). In the meantime, as part of various funding programs, both concepts were further influenced and developed by different EU-bodies (Fioretti et al. 2020; Gargiulo et al. 2013; Lange and Knieling 2020; Vanolo 2014). However, the probably most notable difference between both ideas can be seen in the role of society. From the beginning on, the sustainability idea was taken up and filled with life by social movements, political parties, or local initiatives (Bossuyt and Savini 2018; Kern et al. 2007; Pufé 2017) and inspires many of them until this day, including those that have emerged in recent times (Scott and Vare 2020; von Wehrden et al. 2019). Then again, civil society is often widely underrepresented within smart city projects (see Harvey 2012; Hollands 2014; Kitchin 2015; Vanolo 2016; Colding et al. 2019). Instead, it is rather representatives of businesses or ICT specialists that are mainly setting the tone (Albino et al. 2015; Angelidou 2015; Hollands 2014; Rossi 2016; Söderström et al. 2014; Vanolo 2014; Wiig 2016). In this context, it is striking that there also seems to be rather little interest of citizens or civil society to help co-shaping the smart city concept. And indeed it is legitimate to ask if there are any groups or movements out there that are asking for a smart city.

Summary

This chapter has elaborated on the discourses about two prominent, contemporary urban visions: the sustainable city and the smart city. The presentation and discussion of literature from various academic fields focused on the definitions, notions, and key components; on the origins and further development; and on main points of criticism on both concepts.

It was shown that the sustainable city and the smart city both represent visions that can be understood as very integrative and far-reaching with regard to the future development of cities and urban areas. There is, for example, the shared strong tendency to widely operationalize the concepts by defining subthemes: the three dimensions of sustainability (social, economic, and environmental sustainability), the two additional dimensions of urban sustainability (sustainable governance/political sustainability and physical sustainability), or the six key aspects of a smart city (smart environment, smart economy, smart people, smart living, and smart governance). Moreover, both concepts are subject to several programs aiming at various urban stakeholders, whereby especially the EU offers a broad range of funding opportunities.

Despite these similarities with regard to the areas of activity and the aspiration to provide a vision for the "better" city of tomorrow, both concepts vary to a great extent. Therefore, it does not surprise that attempts to integrate both concepts through smart-sustainable cities have not achieved wide acceptance yet. Indeed, until today it could not be demonstrated that smart city technologies such as ICT can substantially contribute to urban sustainability goals, particularly social sustainability.

Moreover, elaborating on the development history of both concepts, several further substantial differences become apparent. Not only can sustainability rely on a more coherent theoretical basis, but it also has – unlike the smart city a generally accepted definition. Indeed, the smart city never had much more than the ICT component as uniting element. Also, the main driving forces that mainly influenced the establishment and development of both concepts could not be more different: while intellectuals, leading politicians, and environmental movements had a major influence on the current understanding of sustainability, the smart city was and still is influenced by the ideas and interests of multinational technology companies. With all these differences in mind, people should be hesitant to put the sustainable city and the smart city on an equal footing just because they both represent a positive vision for the future and share several areas of activity.

References

Acuto M. (2013). *Global Cities, Governance and Diplomacy: The Urban Link*. London and New York: Routledge.

Ahvenniemi, H., Huovila, A., Pinto-Seppä, I., & Airaksinen, M. (2017). What are the differences between sustainable and smart cities? *Cities, 60*, 234–245. https://doi.org/10.1016/j.cities.2016.09.009.

Albino, V., Berardi, U., & Dangelico, R. M. (2015). Smart cities: Definitions, dimensions, performance, and initiatives. *Journal of Urban Technology, 22*(1), 3–21. https://doi.org/10.1080/10630732.2014.942092.

Allen, A. (2001). Urban sustainability under threat: The restructuring of the fishing industry in Mar del Plata, Argentina. *Development in Practice, 11*(2–3), 152–173. https://doi.org/10.1080/09614520120056324.

Allen, A. (2009). Sustainable cities or sustainable urbanisation? Retrieved from www.ucl.ac.uk/sustainable-cities

Angelidou, M. (2015). Smart cities: A conjuncture of four forces. *Cities, 47*, 95–106. https://doi.org/10.1016/j.cities.2015.05.004.

Barber, B. R. (2013). *If Mayors Ruled the World: Dysfunctional Nations, Rising Cities*. New Haven:Yale University Press.

BAUM. (2013). Intelligent cities – Routes to a sustainable, efficient and livable city. Retrieved from http://www.creatingurbantech.com/urban-tech-analysen/analyse/2015/03/intelligent-cities-routes-to-a-sustainable-efficient-and-livable-city/

Bibri, S. E., & Krogstie, J. (2017). Smart sustainable cities of the future: An extensive interdisciplinary literature review. *Sustainable Cities and Society, 31*, 183–212. https://doi.org/10.1016/j.scs.2017.02.016.

Bifulco, F., Tregua, M., Amitrano, C. C., & Anna, D. A. (2016). ICT and sustainability in smart cities management. *International Journal of Public Sector Management, 29*(2), 132–147. https://doi.org/10.1108/IJPSM-07-2015-0132.

Bossuyt, M., & Savını, F. (2018). Urban sustainability and political parties: Eco-development in Stockholm and Amsterdam. *Environment and Planning C: Politics and Space, 36*(6), 1006–1026. https://doi.org/10.1177/2399654417746172.

Cavada, M., Hunt, D. V., & Rogers, C. D. (2016). Do smart cities realise their potential for lower carbon dioxide emissions? *Proceedings of the Institution of Civil Engineers-Engineering Sustainability, 169*(6), 243–252. https://doi.org/10.1680/jensu.15.00032.

Colding, J., Barthel, S., & Sörqvist, P. (2019). Wicked problems of smart cities. *Smart Cities, 2*(4), 512–521. https://doi.org/10.3390/smartcities2040031.

European Commission. (2007). FP7 in brief. In *How to get involved in the EU 7th framework Programme for research*. Luxembourg City. https://ec.europa.eu/research/fp7/pdf/fp7-inbrief_en.pdf.

European Commission. (n.d.). Smart cities marketplace. https://smart-cities-marketplace.ec.europa.eu/

S

Evans, J., Karvonen, A., Luque Ayala, A., Martin, C., McCormick, K., Raven, R., & Voytenko Palgan, Y. (2019). Smart and sustainable cities? Pipedreams, practicalities and possibilities. *Local Environment, 24*(7), 557–564. https://doi.org/10.1080/13549839.2019.1624701.

Fioretti, C., Pertoldi, M., Busti, M., & van Heerden, S. (2020). *Handbook of sustainable urban development strategies*. Luxembourg: Publications Office of the European. https://doi.org/10.2760/32842, JRC118841.

Gargiulo, C., Pinto, V., & Zucaro, F. (2013). EU smart city governance. *TeMA – Journal of Land Use, Mobility and Environment*. https://doi.org/10.6092/1970-9870/1980.

Giffinger, R., Fertner, C., Kramar, H., Kalasek, R., Pichler-Milanović, N. A., & Meijers, E. (2007). Smart cities. Ranking of European medium-sized cities. Retrieved from http://www.smart-cities.eu/download/smart_cities_final_report.pdf

Gillham, O. (2002). *The limitless city: A primer on the urban sprawl debate*. Washington, DC: Island Press.

Harvey, D. (2012). *Rebel cities: From the right to the City to the urban revolution*. London: Verso.

Haupt, W. (2019). *City-to-city learning in transnational municipal climate networks: an exploratory study*. L'Aquila, Italy, Gran Sasso Science Institute. https://iris.gssi.it/handle/20.500.12571/9733#.YPFOsedCQ2w

Hilty, L., Lohmann, W., & Huang, E. (2011). Sustainability and ICT – An overview of the field. *Notizie di Politeia*, 13–28. https://doi.org/10.5167/uzh-55640.

Hiremath, R. B., Balachandra, P., Kumar, B., Bansode, S. S., & Murali, J. (2013). Indicator-based urban sustainability – A review. *Energy for Sustainable Development, 17*(6), 555–563. https://doi.org/10.1016/j.esd.2013.08.004.

Höjer, M., & Wangel, J. (2014). Smart sustainable cities definition and challenges. In L. Hilty & B. Aebischer (Eds.), *ICT innovations for sustainability* (pp. 333–349). Springer.

Hollands, R. G. (2008). Will the real Smart City please stand up? Intelligent, progressive or entrepreneurial? *City, 12*(3), 303–320. https://doi.org/10.1080/13604810802479126.

Hollands, R. G. (2014). Critical interventions into the corporate smart city. *Cambridge Journal of Regions, Economy and Society, 8*, 61–77. https://doi.org/10.1093/cjres/rsu011.

Hove, H. (2004). Critiquing sustainable development: A meaningful way of mediating the development impasse? *Undercurrent, 1*(1), 48–54.

Karvonen, A., Cook, M., & Haarstad, H. (2020). Urban planning and the smart city: Projects. *Practices and Politics, 5*(1). https://doi.org/10.17645/up.v5i1.2936.

Keivani, R. (2010). A review of the main challenges to urban sustainability. *International Journal of Urban Sustainable Development, 1*(1–2), 5–16. https://doi.org/10.1080/19463131003704213.

Kern, K., Koll, C., & Schophaus, M. (2007). The diffusion of local agenda 21 in Germany: Comparing the German federal states. *Environmental Politics, 16*(4), 604–624. https://doi.org/10.1080/09644010701419139.

Kitchin, R. (2015). Making sense of smart cities: Addressing present shortcomings. *Cambridge Journal of Regions, Economy and Society, 8*, 131–136. https://doi.org/10.1093/cjres/rsu027.

Komninos, N. (2006). The architecture of intelligent cities. Paper presented at the 2nd international conference on intelligent environments, Athens.

Komninos, N. (2009). Intelligent cities: Towards interactive and global innovation environments. *International Journal of Innovation and Regional Development, 1*(4). https://doi.org/10.1504/IJIRD.2009.022726.

Kroll, C., Warchold, A., & Pradhan, P. (2019). Sustainable development goals (SDGs): Are we successful in turning trade-offs into synergies? *Palgrave Commun, 5*(140). https://doi.org/10.1057/s41599-019-0335-5.

Kunzmann, K. R. (2014). Smart cities: A new paradigm of urban development. *CRIOS, 1*, 9–20. https://doi.org/10.7373/77140.

Lange, K., & Knieling, J. (2020). EU smart city lighthouse projects between top-down strategies and local legitimation: The case of Hamburg. *Urban Planning, 5*(1), 107–115. https://doi.org/10.17645/up.v5i1.2531.

Manville, C., Cochrane, G., Cave, J., Millard, J., Pederson, J. K., Thaarup, R., Liebe, A., Wissner, M., Massink, R., & Kotterink, B. (2014). *Mapping smart cities in the EU*. Brussels. http://www.europarl.europa.eu/RegData/etudes/etudes/join/2014/507480/IPOL-ITRE_ET(2014)507480_EN.pdf.

McCann, P., & Varga, A. (2015). Editorial: The reforms to the regional and urban policy of the European Union: EU cohesion policy. *Regional Studies, 49*(8), 1255–1257. https://doi.org/10.1080/00343404.2015.1048976.

McLaren, D., & Agyeman, J. (2015). *Sharing cities: A case for truly smart and sustainable cities*. Cambridge: The MIT Press.

Meadows, D. H., Meadows, D. L., & Randers, J. (1972). *The limits to growth*. New York: Universe Books.

Mogens, R. (2014). The 1973 oil crisis and the designing of a Danish energy policy. *Historical Social Research, 39*(4), 94–112. https://doi.org/10.12759/hsr.39.2014.4.94-112.

Parks, D., & Rohracher, H. (2019). From sustainable to smart: Re-branding or re-assembling urban energy infrastructure? *Geoforum, 100*, 51–59. https://doi.org/10.1016/j.geoforum.2019.02.012.

Peñuelas, J., & Carnicer, J. (2010). Climate change and peak oil: The urgent need for a transition to a non-carbon-emitting society. *Ambio, 39*(1), 85–90. https://doi.org/10.1007/s13280-009-0011-x.

Pufé, I. (2017). *Nachhaltigkeit* (Vol. 2). Stuttgart: UTB.

Robra, B., & Email Heikkurinen, P. (2019). Degrowth and the sustainable development goals. In W. L. Filho,

A. Azul, L. Brandli, P. Özuyar, & T. Wall (Eds.), *Decent work and economic growth. Encyclopedia of the UN sustainable development goals*. Cham: Springer. https://doi.org/10.1007/978-3-319-95867-5_37.
Rossi, U. (2016). The variegated economics and the potential politics of the Smart City. *Territory, Politics, Governance, 4*(3), 337–353. https://doi.org/10.1080/21622671.2015.1036913.
Scott, W., & Vare, P. (2020). Extinction? Rebellion? In W. Scott & P. Vare (Eds.), *Learning, environment and sustainable development*. London: Routledge.
Smythe, K. R. (2014). An Historian's critique of sustainability. *Journal of Current Cultural Research, 6*, 914–929. https://doi.org/10.3384/cu.2000.1525.146913.
Söderström, O., Paasche, T., & Klauser, F. (2014). Smart cities as corporate storytelling. *City, 18*(3), 307–320. https://doi.org/10.1080/13604813.2014.906716.
Squires, G. (2013). *Urban and environmental economics. An introduction* (Vol. 1). London/New York: Routledge.
Swyngedouw, E. (2010). Impossible sustainability and the post-political condition. In M. Cerreta, G. Concilio, & V. Monno (Eds.), *Making strategies in spatial planning. Knowledge and values* (Vol. 1, pp. 185–205). Dordrecht: Springer.
Townsend, A. M. (2013). *Smart cities: Big data, civic hackers, and the quest for a new Utopi*. New York: Ww Norton and Co.
UN DESA. (2013). An integrated strategy for sustainable cities. Retrieved from New York City. http://www.un.org/en/development/desa/policy/publications/policy_briefs/policybrief40.pdf
Vanolo, A. (2014). Smartmentality: The Smart City as disciplinary strategy. *Urban Studies, 51*(3), 883–898. https://doi.org/10.1177/0042098013494427.
Vanolo, A. (2016). Is there anybody out there? The place and role of citizens in tomorrow's smart cities. *Futures, 82*, 26–36. https://doi.org/10.1016/j.futures.2016.05.010.
Von Wehrden, H., Kater-Wettstädt, L., & Schneidewind, U. (2019). Fridays for future seen from a perspective of sustainability science. *GAIA, 28*(3), 307–309. https://doi.org/10.14512/gaia.28.3.12.
WCED. (1987). *Our common future*. Oxford: Oxford University Press.
Wiig, A. (2016). The empty rhetoric of the smart city: From digital inclusion to economic promotion in Philadelphia. *Urban Geography, 37*(4), 535–553. https://doi.org/10.1080/02723638.2015.1065686.
Williams, K. (2010). Sustainable cities: Research and practice challenges. *International Journal of Urban Sustainable Development, 1*(1–2), 128–132. https://doi.org/10.1080/19463131003654863.
Wulf, A. (2016). *The invention of nature: The adventures of Alexander von Humboldt, the lost Hero of science*. New York: Knopf.

Sustainable Cities

▶ Systemic Innovation for Thrivable Cities

Sustainable Cities via Urban Ecosystem Restoration

Arvind Kumar
India Water Foundation, New Delhi, India

In the pace of rapid urbanization and industrialization, sustainable development of cities has become one of the most important issues, especially for Asia-Pacific, which is home to more than 2.1 billion urban residents or 60 percent of the world's urban population. The spectacular demographic composition and economic momentum of the past two decades have turned Asia into one of the main engines of global prosperity and Asian cities into prominent symbols of this success. Furthermore, there is a growing interest in restoring urban ecosystems with commitments made by the parties to the Convention on Biological Diversity to restore at least 15% of degraded ecosystems by 2020.

We are entering a new urban era in which the ecology of the planet is increasingly influenced by human activities, with cities as crucial centers. Over the next decade, two-thirds of the demographic expansion in the world's cities will take place in Asia, which is already host to 50 https://www.unescap.org/our-work/environment-development/urban-development percent of the global urban population. Indeed, by 2020, of the 4.2 billion https://www.worldbank.org/en/topic/urbandevelopment/overview urban population of the world, 2.2 billion will be in Asia. In other words, it is estimated that between 2010 and 2020, a total 411 million people will be added to Asian cities or 60 percent of the growth in the world's urban population. Moreover, over one-half the world's urban population now lives in Asian-Pacific cities but so do most of the slum dwellers.

S

Challenges for Asian Cities

Over the next decade, two-thirds of the demographic expansion in the world's cities will take place in Asia, which is already host to 50 percent of the global urban population.

Climate change is one of the main driving forces, and if we talk about South Asia, it is particularly prone to weather-related disasters like floods, storm, landslide, heat waves, drought, and wildfire, all extreme weather conditions with urban and rural habitats at stake. All this portrays a trend towards a grim situation. The City of Hyderabad is under water with unleashed rains, Delhi is reeling under air pollution, and recently the northeastern region and states like West Bengal faced severe flood situation.

But urban habitats in Asian cities usually entail frenetic construction activities to meet the growing demand for homes, offices, and shops. These activities bring the habitat under pressure to share strong linkage with energy use, resource depletion, especially water, natural habitat destruction, and climate impacts. According to the 2018 Composite Water Management Index published by the NITI Aayog in India, it noted that 6% of economic GDP will be lost by 2050, while water demand will exceed the available supply by 2030.

Food supply is also at risk as areas for wheat cultivation and rice cultivation face extreme water scarcity. In the absence of timely regulatory interventions, the urban habitat can become vulnerable to enormous resource depletion, wastage, and exposure to the vagaries of climate change. As per the UN office for Disaster Risk Reduction (UNDRR) 2020, India ranks third in recording the highest number of natural disasters over the last 20 years after China and the United States of America (USA). Also, the world economy has lost an estimated $3 trillion in two decades due to the same reasons. Disasters are growing and hurting the urban landscape and opportunities. Whether it is a natural hazard or a man-made disaster, the level of preparedness of disaster management bodies to save lives and livelihoods is a challenge.

Cities sans Resilience

Migration in Asia is propelled by various "push" and "pull" factors. The "push" factors behind cross-border emigration include, inter alia, underemployment, protracted natural disasters, wars, and internal conflicts. Better economic opportunities, regional economic integration, changes in labor markets, and technical progress constitute some of the "pull" factors behind international migration. In many Asian countries, circular migration appears to be emerging as a dominant trend where trips vary from daily commutes to those lasting several months and where urban migrants retain strong links to rural areas.

Islamabad, Pakistan's capital, is a "planned city" that came into being in the 1960s. Over the years, however, many informal settlements and ad hoc developments have worsened housing and traffic woes for the city's residents. According to the United Nations Development Program, Pakistan has the highest rate of urbanization in South Asia, with 36.4% of the population living in urban areas, and estimates that by 2025 nearly half of the country's inhabitants will be living in cities https://www.unescap.org/publications/future-asian-and-pacific-cities-2019-transformative-pathways-towards-sustainable-urban.

Links between environmental commons and climate change have for a long time been ignored in international climate summits. But now, it has started seeking a central stage in regional platforms as well.

The rural poor and marginalized communities are directly dependent on natural resources, for food, fuel, medicine, shelter, and livelihoods, and are especially affected by resource depletion and environmental degradation. Any diminution or reduction in stocks of natural capital and flows of ecosystem services is prone to adversely impact the wellbeing and resilience of the poor. For instance, around 50 percent population of Karachi https://www.dw.com/en/is-rapid-urbanization-making-pakistans-cities-less-livable/a-55162735 is living informal buildup areas and has led to many problems including urban flooding, a new normal. According to the 2020 IQAir Air Visual report, Bangladesh topped the list of the world's most polluted countries in 2019 for PM2.5 exposure and Dhaka as one the most polluted city with Pakistan's Lahore and India's Delhi occupied the second and third spots in the list with scores of 178 and 176, respectively.

Besides weather uncertainties, several other factors interact in human-natural systems, such as unplanned urbanization, high rates of population growth, persistent poverty, loss of critical environmental services, and land degradation. This is a real challenge, which we all agree upon. Neither blame game nor Band-Aid approach to such menace will not help. It is only a temporary respite. Links between environmental commons and climate change have for a long time been ignored in international climate summits. But now, it has started seeking a central stage in regional platforms as well.

New Urban Paradigm Towards Ease of Living

Transversal' shift interlinking vertical linkages between water, energy, and environment with horizontal linkages like heath, education, agriculture, and entrepreneurship is an urgent opportunity for renewing urban ecosystems.

Urban Restoration Management is a new buzzword. Urban ecosystem is an ecological system located within a city or other densely settled area or, in a broader sense, the greater ecological system that makes up an entire metropolitan area. The largest urban ecosystems are currently concentrated in Europe, India, Japan, eastern China, South America, and the United States, primarily on coasts with harbors, along rivers, and at intersections of transportation routes https://www.britannica.com/science/urban-ecosystem. Today, however, the greatest urban growth occurs in Africa, South East Asia, and Latin America, and most megacities will be found there by 2030. There is growing recognition that the role ecosystems can play in helping urban people adapt to climate change, while working in tandem with nature and building with nature and green https://www.sciencedirect.com/science/article/pii/S1877343515000433#bib0445 infrastructure to envisage healthy ecosystems to ensure resilience to natural disasters such as floods, landslides, or storm surges. Only with healthy ecosystems can we enhance people's livelihoods, counteract climate change, and stop the collapse of biodiversity. It will only succeed if everyone plays a part. A "transversal" shift interlinking vertical linkages between water, energy, and environment with horizontal linkages like heath, education, agriculture, and entrepreneurship is an urgent opportunity for renewing urban ecosystems.

With more than 80% of global GDP generated in cities, urbanization can contribute to sustainable growth if managed well by increasing productivity, allowing innovation and new ideas to emerge. Asian cities are diversifying away from serving as the factories of the world to turn into innovative service providers and enjoy the unique status of "factory of the world, knowledge economies," financial centers, human capital, and Asian cities. Urban restoration can improve ecosystem biodiversity and enhance natural heritage, environmental aesthetics, educational opportunities, and people's connection with nature. Certain approaches like adaptive management, biodiversity, carbon offset, circular economy, decoupling, green economy, planetary boundaries, polluter pays principle, rewilding, etc. are a few organizing principles to understand, deal, and respond to contemporary environmental problems under different categories to create resilient ecosystem for inclusive green economy.

Deconstructing SDG11 in the light of urban ecosystem restoration involves housing burden, social equity, heritage protection, production and consumption

S

mode, social security, etc. that are the common short boards of urban sustainable development goals.

Water is the key to sustainable green habitat. Water, as a connector, offers holistic solutions to climate change resilience and the ability of water to strengthen multistakeholder collaboration across all sectors. This makes water as an integral part of initiatives to mitigate and adapt to climate change, especially in many efforts to reduce greenhouse gas emissions that depend on reliable access to water resources. It is known that water and ocean resources are trans-boundary in nature. There's need for promising blueprint towards climate change adaptation and building climate resilience. Climate resilience will be strengthened through healthy ecosystem services that rely on well-functioning river basins and effective water management in reducing vulnerability. Integrated basin-wise management like the Barack-Meghna river basin model though has encouraged knowledge sharing; community mobilization still needs to be taken to another level by focusing on adaptive management. Similarly, ecosystem-based approach can also be adopted for trans-boundary wetlands by harnessing circular water economy and green economy between various countries of South Asia as they serve as large carbon sinks.

This year World Calamity Control Day on 13 October 2020 assumes greater significance amid the corona virus and its growing pandemic, and managing rapidly growing cities and their urban regions is one of the most critical challenges facing Asia and the Pacific, especially regarding the relationship between urban development and natural resource management. The focus this year is "all about governance," and in the midst of climate emergency, we need clear vision, plans, and competent, empowered institutions acting on scientific evidence for the public good. During COVID-19, smart cities in India like Pune, Surat, Bengaluru, etc. are tracking the status of the disease spread and identifying localized hotspots through integrated data analytics monitored at their respective integrated command and control centers (ICCCs) http://bwsmartcities.businessworld.in/article/How-Tech-Enabled-Smart-Cities-Are-Helping-Fight-Covid19/23-09-2020-323716/. Vietnam's Ho Chi Minh City implemented an aggressive contact tracing, mass mobilization of citizens, and an unprecedented level of transparency emerging as an exemplary model for COVID-19 response but also an effective model of smart cities through its e-governance models.

Sustainable Opportunities

A sustainable green habitat can endow a city with food and shelter for people and, without resource depletion and green economy, shall lead to low carbon, resource efficient, and socially inclusive cities. It aims to halt the degradation of ecosystems and restore them to achieve global goals. The notion of green economy does not replace sustainable development but creates a new focus on the economy, investment, capital and infrastructure, employment and skills, and positive social and environmental outcomes across Asia and the Pacific. The UN Decade on Ecosystem Restoration https://www.decadeonrestoration.org/ (2021–2030) has given a clarion call towards conserving natural resources and preserving ecological integrity, restoration of degraded areas with impetus on coastal and marine ecosystem, capacity building of local communities and building partnerships, etc. Deconstructing SDG11 in the light of urban ecosystem restoration involves housing burden, social equity, heritage protection, production and consumption mode, social security, etc. that are the common short boards of urban sustainable development goals.

Implementation of SDGs require USD 90 trillion investment till 2030, with India alone requiring USD 2.5 trillion to achieve its committed climate targets by 2030.

We are entering a new urban era in which the ecology of the planet is increasingly influenced by human activities, with cities as crucial centers of development. Investing in urban green and blue infrastructure constitutes a tangible contribution that cities can make to the country's agenda on a green economy for the twenty-first century https://www.sciencedirect.com/science/article/pii/S1877343515000433. Building cities that "work," viz., inclusive, healthy, resilient, and sustainable, requires intensive policy coordination and investment choices. It is imperative for policymakers in every country to reflect on how cities and societies can be made more resilient while confronting the challenges, because transitioning towards more sustainable and resilient societies also requires an integrated approach that recognizes that these repercussions. Urban investments need to be enabled and supported through targeted public expenditure, policy reforms, and changes in taxation and regulation. It is said that implementation of SDGs require USD 90 trillion investment till 2030, with India alone requiring USD 2.5 trillion to achieve its committed climate targets by 2030, so definitely there is a need for adequate finances as well to strengthen the statistical capacities at the regional level. Green finance must be prominently stressed to support strategies for SDGs necessitated to raise money for low-carbon societies and influence to adopt more sustainable business practices.

Conclusion

Urban resilience is strengthened through healthy ecosystem services and must focus on "preparedness" and moved closer to "urban resilience" in both ambition and action. The tradition of indigenous knowledge and nature-based solutions needs to be recognized and coupled with modern science-based approach for appropriate technology interventions. In light of global climate change and other impacts on urban ecosystems, a truly sustainable restoration must look forward and build upon the nature that already exists within cities rather than backward in attempting to recreate from scratch what once existed https://www.ncrs.fs.fed.us/pubs/jrnl/2010/nrs_2010_gobster_003.pdf.

Leading cities not only bring the world into the city but also change the city world. Let us not forget that it is essential to adopt a broad, transdisciplinary perspective to tackle the present-day disaster risks and build climate resilience, "putting people first" approach. What is needed now is convergence and integration of such good initiatives as it is agreed that "South Asia is a family."

References

http://bwsmartcities.businessworld.in/article/How-Tech-Enabled-Smart-Cities-Are-Helping-Fight-Covid19/23-09-2020-323716/

HTTPS://WWW.BRITANNICA.COM/SCIENCE/URBAN-ECOSYSTEM

https://www.decadeonrestoration.org/

https://www.dw.com/en/is-rapid-urbanization-making-pakistans-cities-less-livable/a-55162735

https://www.ncrs.fs.fed.us/pubs/jrnl/2010/nrs_2010_gobster_003.pdf

https://www.sciencedirect.com/science/article/pii/S1877343515000433

https://www.sciencedirect.com/science/article/pii/S1877343515000433#bib0445

https://www.unescap.org/our-work/environment-development/urban-development

https://www.unescap.org/publications/future-asian-and-pacific-cities-2019-transformative-pathways-towards-sustainable-urban

https://www.worldbank.org/en/topic/urbandevelopment/overview

S

Sustainable City

► Green and Smart Cities in the Developing World
► Green Cities
► Green Cities in Theory and Practice
► The Sustainable and the Smart City: Distinguishing Two Contemporary Urban Visions
► Toward a Sustainable City
► Walkable Access and Walking Quality of Built Environment

Sustainable Community Masterplan

From Mitigation to Regeneration

Marina Matashova
Andorra-LAB, Forward Consulting Group, Andorra la Vella, Andorra

Synonyms

Green Building; Sustainable Development; Integrative Design Process; Socio-Ecological Urbanism; Regenerative Design

Definition

Sustainable Community Masterplan methodology represented in the chapter with a cross-discourse, illustrated by examples from the author's international practical experience, reveals how the socio-ecological approach and regenerative thinking challenge the urban planning definition and the professional role, expanding the list of competencies and design challenges by eliminating rigid divisions between disciplines.

Introduction

Green building strategies, performance goals, and associated assessment methods currently emphasize the ways and extent that the planned cities should mitigate global and local resource depletion and environmental degradation. By contrast, the emerging notion of "regenerative" design and development emphasizes a co-evolutionary, partnered relationship between humans and the natural environment, rather than a managerial one that builds, rather than diminishes, social and natural capitals (Mang and Reed 2012).

According to Bill Reed, the author of the "Trajectory of ecological design" diagram (Fig. 1): "Instead of doing less damage to the environment, it is necessary to learn how one can participate with the environment by using the health of ecological systems as a basis for design" (Mang and Reed 2012). In the past, attempts have been made to link natural and built environments in a holistic vision by the following authors: Lyle (1994), McHarg (1999), Yeang (1994), and Van der Rin and Cowen (1996).

The health emergency related to the COVID-19 pandemic has led to renewed interest in the topic, repositioning parks, gardens, and other "green infrastructure" resources as "essential infrastructure" supporting well-being. A series of national lockdowns limited the number of places individuals can use to support their mental and physical health. However, the quality, functionality, and location of green infrastructure in urban areas illustrated a disparity in distribution that meant that in many cases, communities with lower income and more significant health inequality suffered from insufficient access (Mell and Whitten 2021).

The development of green infrastructure within the framework of the socio-ecological urbanism paradigm makes it one of the leading regeneration actions of the current city aimed at addressing not only environmental problems but at the same time the problem of socio-spatial justice since it deals not only with carbon emission mitigation designs but also with adaptation measures to enhance adaptive capacities by integrating ecosystems and their services in planning and design (Mang and Reed 2012; Barthel and Colding 2013).

While merging reflective practice and applied theory, this entry reveals a series of socio-ecological approach strategies used in the work of the international architectural bureau Ricardo Bofill with the author's participation. This experience laid the basis for the author's Sustainable Community Masterplan methodology, which combines two approaches:

Green building – the transition from linear to circular patterns of resource consumption, including water, energy, matter, and the assessment of the effectiveness of measures to reduce environmental damage based on measurable parameters.

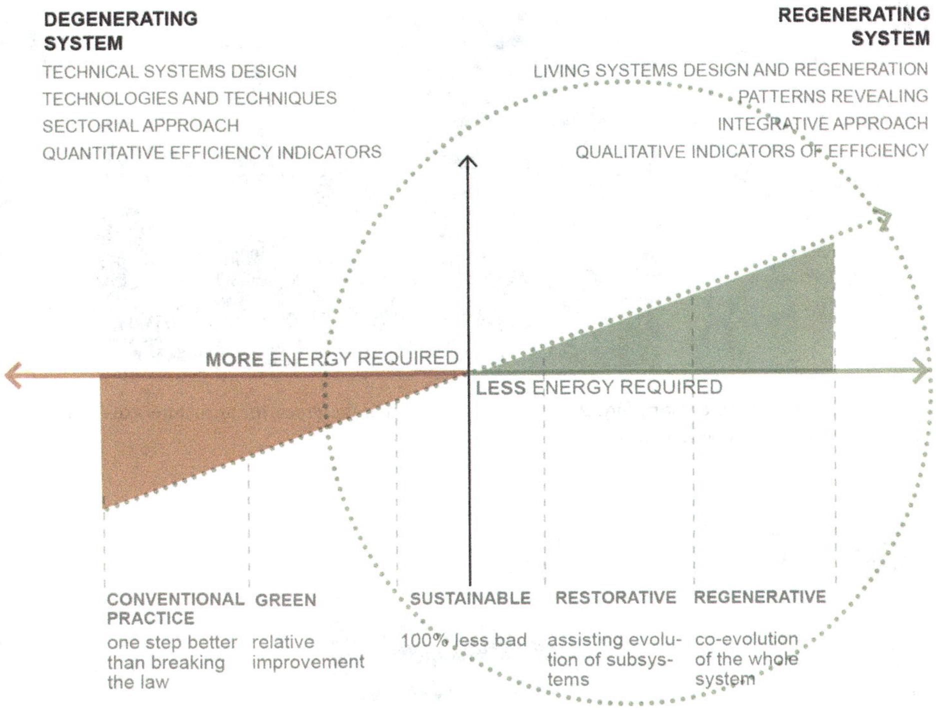

Sustainable Community Masterplan, Fig. 1 Trajectory of ecological design. (Source: © All rights reserved Regenesis – Bill Reed, bill@regenesisgroup.com. Translation into Russian by Marina Matashova)

Regenerative design is based on the ecosystem vision to urban habitat (Rueda 2018) and aspires to create the mutually beneficial relationship between nature and society through an understanding of the functioning of systems to restore society's health through the health of living natural systems.

Bringing gap between theory and practice, this entry has two parts: the first part analyzes the design strategies of socio-ecological urbanism used in international design practice; the second reveals the Sustainable Community Masterplan methodology as an integrative process.

Reflexive Practice: Examples of Socio-Environmental Approach Strategies

1. ***Urban Growth Through the Green Infrastructure Axes*** (Fig. 2) – The Greater Moscow project envisions a strategy for the Moscow Metropolitan Area development while changing the traditional principles of urban growth: from the growth through the transport route tracing to growth through identifying the green infrastructure axis based on the analysis of the geomorphological matrix. This principle ordinates new development around the central green public space as an axis of soft mobility, which is connecting the open spaces outside the city with those in the center, creating a slow (pedestrian, bicycle) radial axis as an alternative to the existing fast movement transport axis (Ricardo Bofill Taller de Arquitectura).
2. ***Nature-Based Solutions for Social Housing*** (Fig. 3) – An affordable housing project in Kolkata, a city of hundreds of lakes, overcomes the expansion/preservation dichotomy concerning natural areas and integrates natural lakes into the green structure of the residential complex. The proposal supposes regenerating the natural lakeshore using native plant

Sustainable Community Masterplan, Fig. 2 Urban growth through the green infrastructure axes ("Big Moscow," Russia). (Source: https://ricardobofill.com/)

Sustainable Community Masterplan, Fig. 3 Nature-based solutions for social housing (Konnagar). (Source: https://ricardobofill.com/)

species, creating a public access infrastructure based on soft impact design solutions with above-ground terraces and piers for recreation areas. Combined with natural ventilation solutions in buildings, this approach allows a high-quality environment with improved hygienic conditions to accessible housing by applying a natural-oriented approach (Ricardo Bofill Taller de Arquitectura).

3. ***Institutional Open Spaces as a Part of Urban Green Infrastructure*** (Fig. 4) – Aiming to resolve the conflict between social/environmental connectivity and make accessible Central Park within the borders of the Mohamed VI Polytechnic University in Morocco, the project proposal moves the park out of the campus area. This ensures continuity of the territorial green corridor and the soft mobility infrastructure passing through the central park as an integrated part of the metropolitan "Green City Project." Along with the Student Center, Congress Hall, and exhibition areas, the park includes areas for use by the adjacent urban residential area and the infrastructure of soft mobility of the territorial green corridor (Ricardo Bofill Taller de Arquitectura).
4. ***Green Infrastructure As a Flood Protection Strategy*** (Fig. 5) – The Xiongan Master Plan, China, proposes creating two new river corridors, wetlands at the interface with Lake Baiyang, as a flood management strategy and integration water cycle within the landscape. Artificial intelligence enables real-time hydrological data and weather monitoring to manage storage capacity and decide to keep or release water. Combining with a water supply and

Sustainable Community Masterplan, Fig. 4 Institutional open spaces as part of green infrastructure (Morocco). (Source: https://ricardobofill.com/, https://sadbenkirane.com/)

Sustainable Community Masterplan, Fig. 5 Green infrastructure as a flood protection strategy (Xiongan, China). (Source: https://ricardobofill.com/, https://expedition.uk.com/)

water reuse, this strategy reduces water use in the masterplan by 75%, along with the social effect of integrating the flood protection system with the system of public spaces (Ricardo Bofill Taller de Arquitectura; Expedition Engineering).

5. ***From Infrastructure Corridor to Green Infrastructure Axis*** (Fig. 6) – The strategic program for constructing a new transport highway, Y4 in Abidjan, one of the fastest-growing cities in the world with a low level of urban and social infrastructure, becomes an opportunity that goes beyond the implantation of engineering and transport infrastructure. An in-depth study of local natural and social conditions allows the implantation of a green corridor,

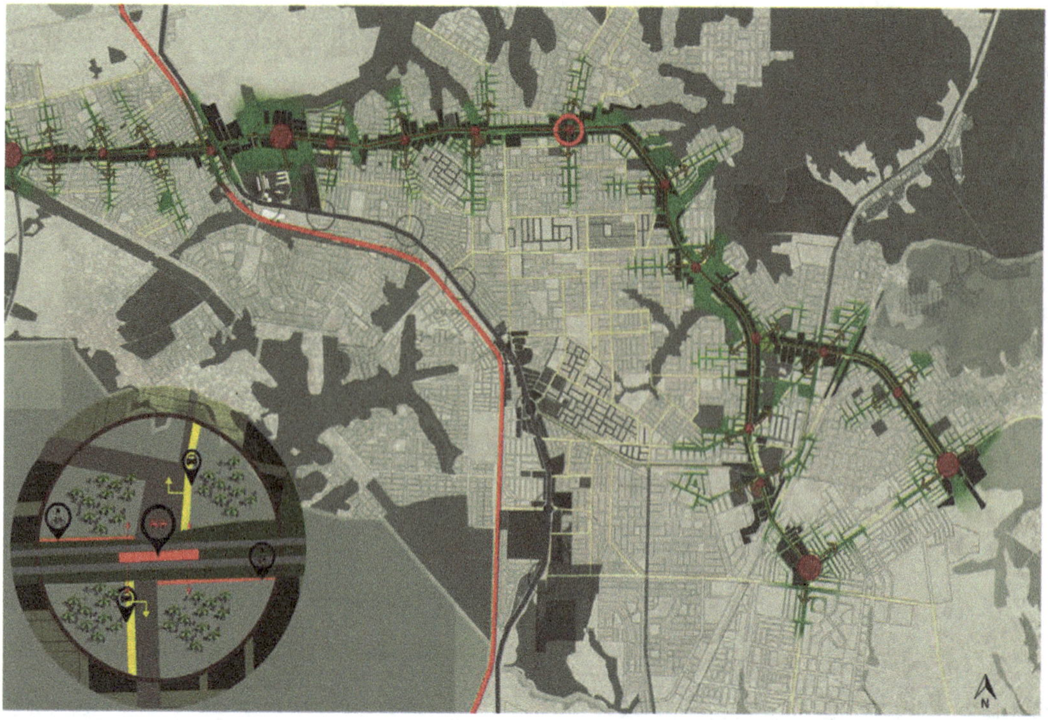

Sustainable Community Masterplan, Fig. 6 Transport axis as social and green infrastructure (Abobo, Abidjan, Ivory Coast). (Source: https://ricardobofill.com/)

which regenerates the natural water cycle, impulses the circular processes of urban metabolism, and increases energy consumption autonomy. It integrates the necessary new social infrastructure and economic activity and promotes the revitalizing of traditional activities, providing jobs for the local population within the framework of new economic models.

The street profile combines solutions for natural drainage, landscaping, implantation of slow, fast mobility and electric public transport lanes, garbage collection system, and intelligent lighting. Furthermore, the building parameters of the street facade are determined, considering the social structure and energy efficiency based on options testing, social infrastructure integration. New multimodal interchange hubs integrate the public electric transport stops as a carbon mitigation strategy and the adapted version of traditional activities: the markets historically located on the roads and interchange zones for existing district taxis. The project's objective is to form an action program for a step-by-step long-term strategy that allows integrating the existing way of life into a renewed urban environment, creating conditions for employment of the local population and competent management at the local level of the new intelligent, sustainable urban environment.

Integrative Design Process

> A critical capacity for practitioners seeking to work regeneratively will be consciousness of how they think, not just what they think about and what they do.
> Bill Rid

The review of design strategies reveals trends toward the transition from expansion to regeneration, an increase in the multimodality of urban planning tasks, including ecological, hydrological, and biological specializations; on the other hand, the strengthening of the role of engineering in the development of effective strategies for the use of resources. These trends challenge the definition of urban design and the designer's role, expanding the list of design competencies and challenges by eliminating rigid divisions between disciplines.

The Sustainable Community Masterplan Methodology was developed on the basis of international practice intending to move from linear models of the design process to the search of a holistic vision to natural and social living systems, along with the possibilities of necessary technologies integrating for the project life cycle modeling. Let us consider the main components of this methodology.

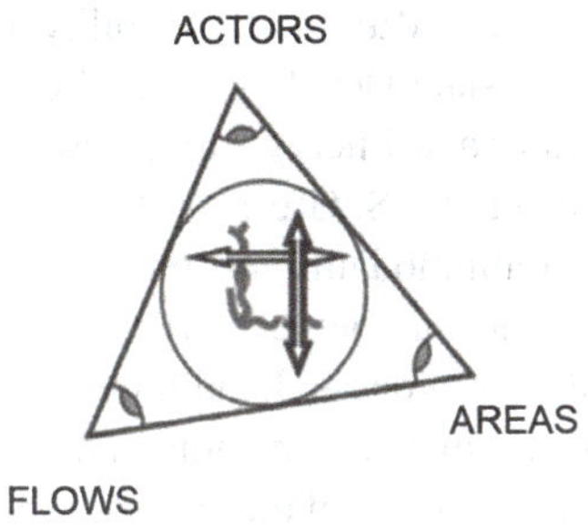

Sustainable Community Masterplan, Fig. 7 Three-axis approach, Tjallingii, S. (Source: Tjallingii (2015))

1. ***Three-Axis Approach to the Project Goals Setting*** (Fig. 7) – The transition from an expansive to a regenerative approach shifts the focus from an approach focused on territorial zoning to a more dynamic vision, which can be characterized in three directions identified in S. Tjallingi (2015):

   "Area" – Territory – A programmatic variety of spatial elements that provide the effect of a combination of social, environmental, and economic objectives. *How can spatial elements be combined?*

   "Flow" – Flow – Structural vision considering fast flows of engineering and transport infrastructure and slow water flows of recreational activities and soft mobility.

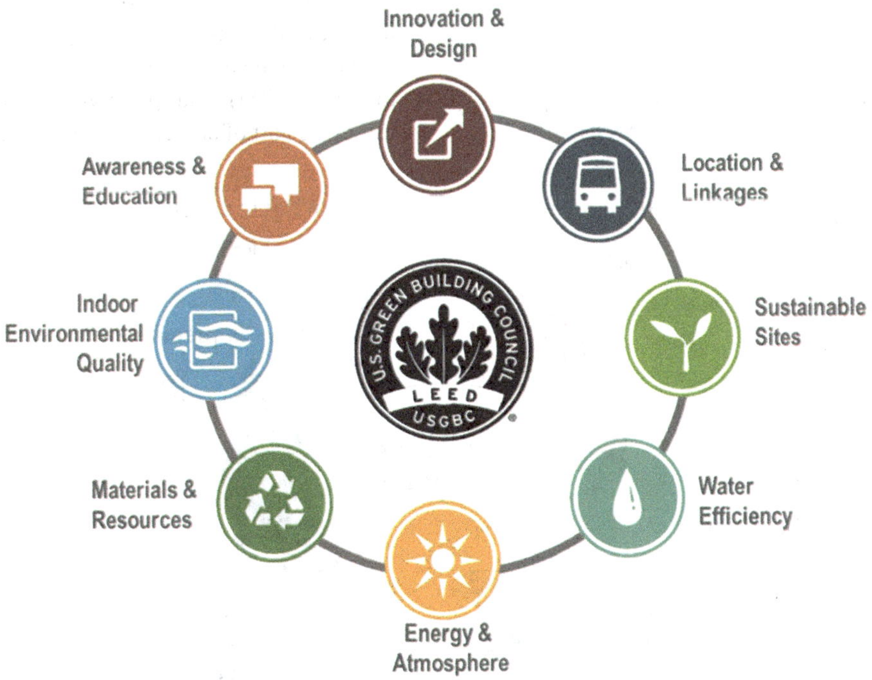

Sustainable Community Masterplan, Fig. 8 LEED categories. (Source: https://www.usgbc.org/)

S

Consideration of flows in a different plane: circular flows of energy, water, and matter. *How can flows be adapted to local conditions and each other?*

"Actor" – Human/User – The form of perception of spaces through the scenarios, the transition from normative assessment to identifying behavioral patterns. On the other hand, the regeneration of existing structures required expertise in everyday life – participatory design involving local stakeholders. *How could the framework create conditions to avoid conflicts and promote synergy?*

These three axes form the basis for the Sustainable Community Masterplan project goals definition (Fig. 11).

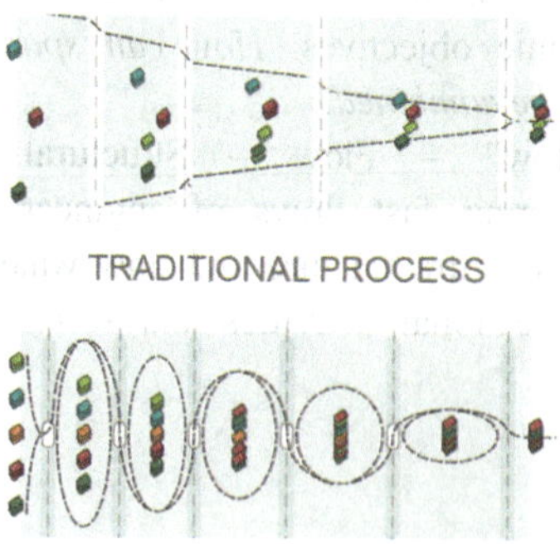

Sustainable Community Masterplan, Fig. 9 Integrative process. (Source: ANSI (2012))

2. ***Integrative Process*** (Fig. 8) – The use of Green Building efficiency indicators within the framework of the project goals set based on a socio-ecological approach ensures the consistency of techniques and technologies application and allows a synergistic effect between the categories. Green Certification Systems represent an attempt at a coordinated approach to assessing the environmental performance of a project. They can serve as a guideline in determining the scope of the deliverables from Feasibility Study to Construction Documentation, including the following categories: Green Infrastructure, Multimodal Mobility, Planning, and Real Estate Development, Circular Economy, Water and Energy Efficiency, and Waste Management (U.S. Green Building Council).

The multimodality of the project scope expands the number of participants required in the design process. In addition, the integrative process involves including all participants from the first stages of the project, in contrast to the traditional linear scheme of sequential inclusion of contractors (Fig. 9) (ANSI 2012).

The innovative nature of regenerative design expands the scope of works and involves the use of "Design Thinking" methods, which is expressed in early prototyping (CIM) and prototype performance testing as part of an iterative process (Fig. 10).

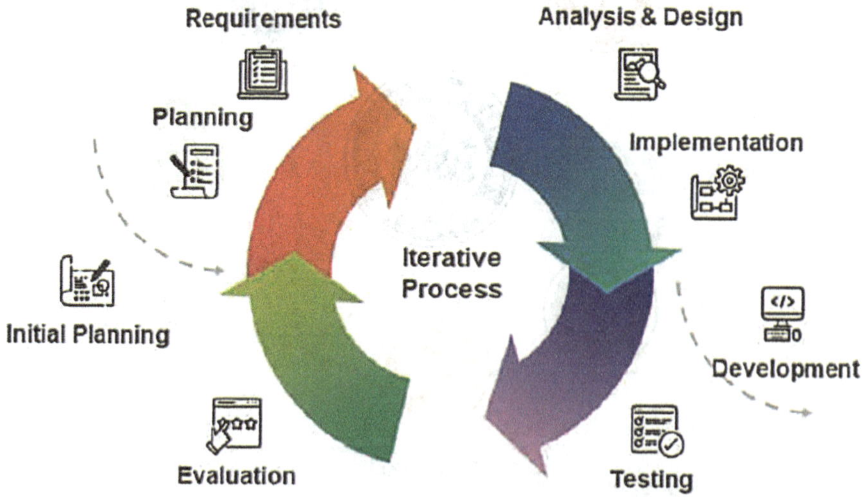

Sustainable Community Masterplan, Fig. 10 Iterative process. (Source: https://www.usgbc.org/)

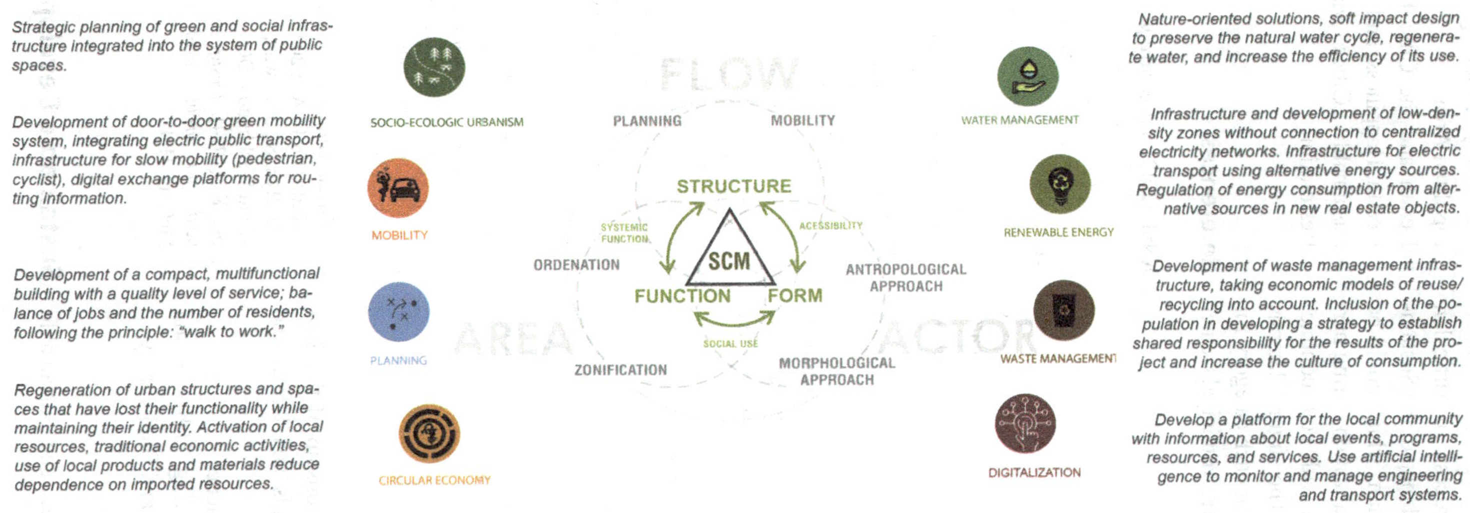

Sustainable Community Masterplan, Fig. 11 Sustainable urban masterplan: a specific case study. (Source: Own elaboration)

S

3. ***Design Process Mental Map*** – Visual communication tools allow to include a broader audience in the co-creation process. In particular, Fig. 11 represents a mental map of the Sustainable Community Masterplan, which could be used for holistic conceptualization during participatory design sessions with both contractors and stakeholders. The figure laterals include the tasks defined for each category for Sant Julia de Loria, Andorra, Municipal Center revitalization project case.

Conclusion

The analysis carried out in this entry allows us to see the synergistic interaction between the categories of sustainable planning due to the application of an integrative process that avoids the fragmented use of technologies based solely on numerical indicators.

The regenerative design allows to maximize the use of passive design tools and activate nature-oriented solutions while promoting the sustainability of natural components.

On the other hand, the development of green infrastructure makes it possible to restructure the city and overcome the disparity in the quality of the environment between the existing urban fabrics, thereby contributing to the improvement of the physical and mental health of the community.

Cross-References

- ▶ Adapting Cities to Climate Change
- ▶ An Overview of the Relationship of the Sustainable Development Goals and Urban and Regional Development
- ▶ Building Community Resilience
- ▶ Blue-Green Cities: Achieving Urban Flood Resilience, Water Security, and Biodiversity
- ▶ Circular Economy Cities
- ▶ Circular Water Economy
- ▶ City Visions: Toward Smart and Sustainable Urban Futures
- ▶ Green Cities: Nature-Based Solutions, Renaturing and Rewilding Cities
- ▶ Green Infrastructure in Metropolis Dimension: Case Study of Llobregat River, Barcelona Metropolitan Area
- ▶ Health and the City: How Cities Impact on Health, Happiness, and Well-Being
- ▶ Sustainable Cities via Urban Ecosystem Restoration

References

ANSI. (2012). *Integrative process (IP), ANSI consensus standard guide 2.0 for design and construction of sustainable buildings and communities*. New York: American National Standards Institute.

Barthel, S., & Colding, J. (2013). *Principles of social ecological design: Case study Albano Campus, Stockholm* (TRITA-ARK-Forskningspublikationer, Vol. 3). School of Architecture and the Built Environment.

Expedition Engineering. Electronic access: https://expedition.uk.com/. Acceded on 8 Aug 2021.

Lyle, J. T. (1994). *Regenerative design for sustainable development*. New York, NY: Wiley.

Mang, P., & Reed, B. (2012). Designing from place: A regenerative framework and methodology. *Building Research and Information, 40*(1), 23–38.

McHarg, I. (1999). *Design with nature*. Garden City, NY: Natural History Press.

Mell, I., & Whitten, M. (2021). Access to nature in a post Covid-19 world: Opportunities for green infrastructure financing, distribution and equitability in urban planning. *International Journal of Environmental Research and Public Health, 18*, 1527. https://doi.org/10.3390/ijerph18041527.

Ricardo Bofill Taller de Arquitectura. https://ricardobofill.com/. Acceded on 10 Aug 2021.

Rueda, S. (2018). *Carta para la planificación ecosistémica de ciudades y metrópolis*. Barcelona: Agencia d'Ecología Urbana de Barcelona.

Tjallingii, S. (2015). Planning with water and traffic networks: Carrying structures of the urban landscape. *Research in Urbanism Series, 3*, 57–80.

U.S. Green Building Council. https://www.usgbc.org/. Acceded on 10 Aug 2021.

Van der Ryn, S., & Cowan, S. (1996). *Ecological design*. Island Washington, DC.

Yeang, K. (1994). *Design with nature: The ecological basis for architectural design*. New York, NY: McGraw-Hill.

Sustainable Development

▶ European Green Deal and Development Perspectives for the Mediterranean Region

▸ Multiple Benefits of Green Infrastructure
▸ Philanthropy in Sustainable Urban Development: A Systems Perspective
▸ Sustainable Community Masterplan
▸ Sustainable Development Goals
▸ Wash (Water, Sanitation, and Hygiene): Infrastructure as a Measure of Sustainable Development

Sustainable Development and Responsible Tourism: The Grijalva-Usumacinta Lower River Basin

Luzma Fabiola Nava[1,2] and Celeste Nava Jiménez[3]
[1]CONACyT – Centro del Cambio Global y la Sustentabilidad, A.C. (CCGS), Villahermosa, Mexico
[2]International Institute for Applied Systems Analysis (IIASA), Laxenburg, Austria
[3]División de Ciencias Económico Administrativas, Campus Guanajuato, Universidad de Guanajuato, Guanajuato, Mexico

Synonyms

Ecotourism is *responsible tourism*; Responsible and equitable; *Sustainable development* is fair

Definition

Sustainable development and *responsible tourism* translate as a set of actions, behaviors, interventions, and practices aimed at maintaining balance between society and nature in a particular territory. Specifically, *sustainable development* suggests the satisfaction of human needs through the realization of economic and industrial activities without compromising natural resources. On the other hand, *responsible tourism* consists of temporal and territorial movement of people committed to the social, economic, ecological, cultural, and religious milieux. Broadly, both notions involve a long-term objective, which underlines (1) the responsibility of society toward conservation and preservation of the environment that surrounds us; and, (2) the need of an ad hoc decision process to meet the challenges and opportunities arising from activities aimed at sustainability and responsibility.

Introduction

During the 1970s began awareness of the proliferation of serious environmental issues. The rapid deterioration of the environment and its consequences for socioeconomic development urged the UN General Assembly (UNGA) to approve, in 1982, the Earth Charter and to create in 1983, the World Commission on Environment and Development (WCED). The works of the WCED were directed by Ms. Gró Harlem Brundtland who, after organizing numerous participating meetings in the whole planet, presented in 1987 at the UNGA, the report Our Common Future, better known as the Brundtland Report.

According to this document, the concept of sustainable development is a development that satisfies the needs of the present generations without compromising the ability of future generations to satisfy their own needs. According to Bermejo (2014), this interpretation is three-dimensional, because it brings together the social and economic dimensions, and sustainability, into the development. For Bermejo, since 1987, and until today, the notion of sustainable development has turned into a widely accepted expression, used by international organizations, governments, businesses, and civil society. Nevertheless, it is important to mention that the notion of sustainable development was used "[. . .] during the 18th century in biology to mean the evolution of young individuals towards the adult stage" (Bermejo 2014). Since then, it has been used in several contexts, such as during World War II, during the industrialization and capitalism processes, and for the environmental challenges and sudden loss of a basis of natural resources. This is how it has become widely acknowledged by the international community during the United Nations

Conference on Environment and Development (UNCED), also known as the Earth Summit, which took place in Rio de Janeiro in 1992, 3–14 June.

After several meetings, the Earth Summit resulted in the Rio Declaration on Environment and Development (often shortened to Rio Declaration); the program Agenda 21; the Convention on Biological Diversity (CBD); the United Nations Framework Convention on Climate Change (UNFCCC); and the Global Environment Facility (GEF). In particular, Agenda 21 is considered as the main outcome of this conference, as it represents a general strategy of sustainable development worldwide. Agenda 21 is a comprehensive plan of action addressing the pressing problems of today and aims at preparing the world for the challenges of the next century. Agenda 21 advocates that sustainable development should become a priority item on the agenda of the international community in order to meet the challenges of environment and development. Within this framework, particular attention should be directed toward comprehensive and sustainable water policies to ensure safe drinking water and sanitation to preclude both microbial and chemical contamination. Moreover, states are invited to promote research into the contribution of forests to sustainable water resources development (UN 1992a, b).

Agenda 21 has been closely monitored, leading to the development of adjustments and revisions, in particular the adoption of a supplementary agenda called Millennium Development Goals (MDGs). The United Nations Millennium Declaration (UNMD) aims to confirm the support of the signatories to the principles of sustainable development, including those stated in Agenda 21. Concerning sustainable development, one of the main values conveyed through this declaration consists in acting with caution when managing and planning the access and the use of all natural resources, in agreement with the principles of sustainable development.

Besides that, with regard to water, the signatories of this declaration decided to reduce by half, by year 2015, the percentage of population lacking access to potable water or unable to afford it. With this goal in mind, the UNMD, signed in 2000 and aiming to reduce poverty and improve the living conditions of the poorest people in the world, was organized under the guidelines of eight MDGs: (1) Eradicate extreme poverty and hunger, (2) Achieve universal primary education, (3) Promote gender equality and empower women, (4) Reduce child mortality, (5) Improve maternal health, (6) Combat HIV/AIDS, malaria, and other diseases, (7) Ensure environmental sustainability, and (8) Develop a global partnership for development (UN 2000). Particularly, goal 7 had two important targets related to water: first, to integrate the principles of sustainable development into country policies and programs and reverse the loss of environmental resources; and second, to halve the proportion of the population without sustainable access to safe drinking water and basic sanitation. Moreover, goal number 7 recognizes a fourfold problematic: (a) renewable water resources are becoming more and more scarce; (b) access to an improved drinking water source is still an issue; (c) many people still rely on unsafe water sources; and, (d) people in rural areas, the poor, and minorities have less access to both improved water and sanitation (UN 2014). As the year of 2015 was the deadline to achieve the MDGs, in 2014 the advances made have been assessed, which resulted in a more extensive list of objectives, namely, the Sustainable Development Goals (SDGs) (PNUD 2018).

In 2015, the UNGA agreed on the SDGs. More than 180 member countries signed the agreement, a new action plan to tackle the biggest worldwide issues. The 17 goals integrate as a whole a new agenda for development, Agenda 2030, and represent an universal call to action to end poverty, protect the planet, and ensure that all people enjoy peace and prosperity by 2030 (UNDP 2020; Ávila Akerberg et al. 2019). Agenda 2030, in place since 2016, aims to benefit the people, planet, and prosperity via sustainable development in its three dimensions: economic, social, and environmental (UN 2015). Related to the purposes of the MDGs, the SDGs include new scopes such as climate change, economic inequality, innovation, sustainable consumption, and peace and justice, among other priorities: (1) No poverty; (2) Zero hunger; (3) Good health and well-being; (4) Quality education; (5) Gender equality; (6) Clean water and sanitation; (7) Affordable and clean energy; (8) Decent work and economic growth; (9) Industry, innovation and infrastructure; (10) Reduced inequalities; (11) Sustainable cities and communities; (12) Responsible consumption and production; (13) Climate action; (14) Life Below Water; (15) Life on land; (16) Peace, justice, and strong institutions; and, (17) Partnerships for the goals (UNDP 2020). It is important to add that the SDGs cannot be viewed in isolation (Ávila Akerberg et al. 2019); that is, they recognize that action in one area will affect outcomes in others, and that development must balance social, economic, and environmental sustainability (UNDP 2020). With this is mind, and by means of a case study approach, the purpose of these definitions consists in exploring the relations between sustainable development and responsible tourism.

The Grijalva-Usumacinta Lower River Basin

The region of rivers Grijalva and Usumacinta, in the Southeast of Mexico, is one of the most diverse, biologically and culturally in Mexico (INECC 2007). The Grijalva River arises in Huehuetenango, Guatemala, and then flows in between the Chiapas highland ecosystem in Mexico. On the right bank its tributaries are Guatemalan, whereas on the left, they are completely Mexican. Between canyons, such as El Sumidero and La Angostura, the Grijalva River crosses the territory of Chiapas, traverses the state of Tabasco before emptying into the Gulf of Mexico. The Usumacinta River arises from the Chixoy and Lacantun rivers. The upper Usumacinta marks the border between Guatemala and Mexico. It wanders 200 km from the union of the Salinas and Pasion rivers, both main tributaries in the Gran Petén region in Guatemala. The Usumacinta River makes its way in between gorges and impressive cliffs more than 300 m high, until Boca del Cerro in Tabasco, Mexico. The Grijalva and the Usumacinta rivers converge at Tres Brazos in Tabasco; together for 20 km they flow into the Gulf of Mexico. Both rivers merge into the Grijalva-Usumacinta River Basin (GURB). Located in the southeast of Mexico, the GURB includes, for purposes of water administration in Mexico, the states of Tabasco, Chiapas, and small sections of Campeche. Worth to mention that, in 1957 the GURB was demarcated as a unique body of water and since then it has been recognized as a unique hydrographic space, subject to planning and management of water resources in the country (INECC 2007; Gallardo-Cruz and Carabias 2017; Gallardo-Cruz et al. 2017; Toledo 2003).

The GURB represents a kaleidoscope of ecosystems. Its ecosystems are home to 64% of the national biodiversity (INECC 2007). As a whole, coral reefs, seagrass beds, lagoons and coastal plain ecological complex, extensions of mangrove forests, forest areas, and freshwater reserves join up one of the richest regions among those located in the known intertropical strip as the Genetic bank of the Earth (INECC 2007; Toledo 2003). The GURB offers a magnificent series of ecosystem services consolidated into a single Decalogue of ecological functions: (1) climate regulation; (2) resilience in the face of disturbances caused by extreme weather events; (3) erosion and flood control; (4) freshwater supply; (5) soil formation sediment retention and accumulation of organic matter; (6) nutrient recycling; (7) treatment and biological control of waste; (8) creation of refuge areas for wildlife; (9) food production across a wide range of areas and conservation of genetic and biotic resources; and, (10) creation of living spaces for its human population. Thus, the GURB represents a vital process generation engine for the preservation of the Earth system and the current and future sustainability of societies living in this region (INECC 2007; Toledo 2003).

As mentioned by Yáñez-Arancibia, Day, and Currie-Alder (2009), the GURB represents the second highest river discharge in the Gulf of Mexico after the Mississippi River. Particularly, the Grijalva-Usumacinta lower river basin (GULRB) is very important ecologically, socially, and economically, but urban development, agriculture, coastal development, and oil and gas activities have led to environmental impacts on the largest lowland tidal wetland ecosystem in Mesoamerica. Nevertheless, projects of responsible tourism in the region aim at preserving the habitat and rationally using the natural and cultural goods, so that in the future, the new generations can appreciate this heritage (Maldonado-Alcudia et al. 2015). Therefore, sustainable development of this basin invites to highlight that water is the unifying element (García-García and Kauffer 2011). With its movements, water organizes life and connects the river and coastal and marine ecosystems (INECC 2007; Toledo 2003; Gallardo-Cruz and Carabias 2017). Water resources represent a regional attraction for sustainable development and responsible tourism.

Protected Areas and New Forms of Tourism

Traditional tourism has jeopardized the integrity of ecosystems and the services that they provide, as well as the traditional ways of using natural resources. Consequently, new forms of tourism have emerged as alternative sources of income and as strategies to create awareness of the importance of preserving natural resources and health of the ecosystems (Ávila Akerberg et al. 2019). Responsible tourism or ecotourism, as well as nature tourism, rural tourism, and adventure tourism, emerge as a responsible and concerned expression of an actor, – individual, group, public, or private –, who seek the sustainable development of a space through socio-environmental management of tourist attractions represented by natural and cultural resources (Maldonado-Alcudia et al. 2015). In Mexico, the creation of Protected Natural Areas (PNA) enables the advent of responsible tourism, thus setting a standard for the commitment of individuals toward the natural wealth (Martínez-Cervantes and Vargas-Martínez 2015).

A PNA is an initiative for the realization of sustainable development, since it represents an area meant for the conservation and protection of ecosystems and hydrographic basins (López-Hernández 2006). In other words, PNAs foster a pro-environment behavior. In Mexico, the protected spaces represent about 9.6% of the national territory. However, regionalization studies carried out by the Mexico's National Commission for the Knowledge and Use of Biodiversity (CONABIO) show that the terrestrial areas standing out for their ecosystem richness cover one quarter of the national territory (Martínez-Cervantes and Vargas-Martínez 2015).

In the GULRB, the state of Tabasco totalizes 17 conservation instruments. Twelve state-run PNAs occupying together 45,192.73 hectares and managed by the Minister of Well-Being, Sustainability and Climate Change. Two federal-run PNAs taking up 348,834.49 hectares of the state territory, managed by the National Commission on Natural Protected Areas (CONANP). In addition, three Areas Voluntarily Destined to Conservation (AVDC) established with a certificate, two of which created by the state government and one by the federation; in total they take up 446.22 hectares (Figs. 1 and 2). In percentages, the total of PNAs in Tabasco represents 16% of the state territory (Sistema Estatal de Áreas Naturales Protegidas 2021).

Protected areas in Tabasco provide to the citizens some space for recreation and contact with the natural wealth of the state, generate sources of employment and support an environmental culture at the local, regional, and national levels. As such, protected areas turn into the main reason why tourists decide to travel to a specific place (López-Hernández 2006). In the GULRB, the PNAs promise to encourage the development and the strengthening of an environmental culture based on integral enjoyment of natural resources. Nevertheless, one must consider the value society ascribes to them for their preservation and conservation (Martínez-Cervantes and Vargas-Martínez 2015), since sustainable development in the region depends on the good usage and exploitation of the natural environment by the citizens.

Let us remember that in a basin, water resource supports the supply of a wide range of ecosystemic goods and services to the population. Used in a fair and sustainable way for the tourist activity, these goods and services translate into a natural richness and a common heritage that all of us need to preserve and protect (Alpuche Álvarez et al. 2021). Thus, responsible tourism represents means to this objective and an activity of

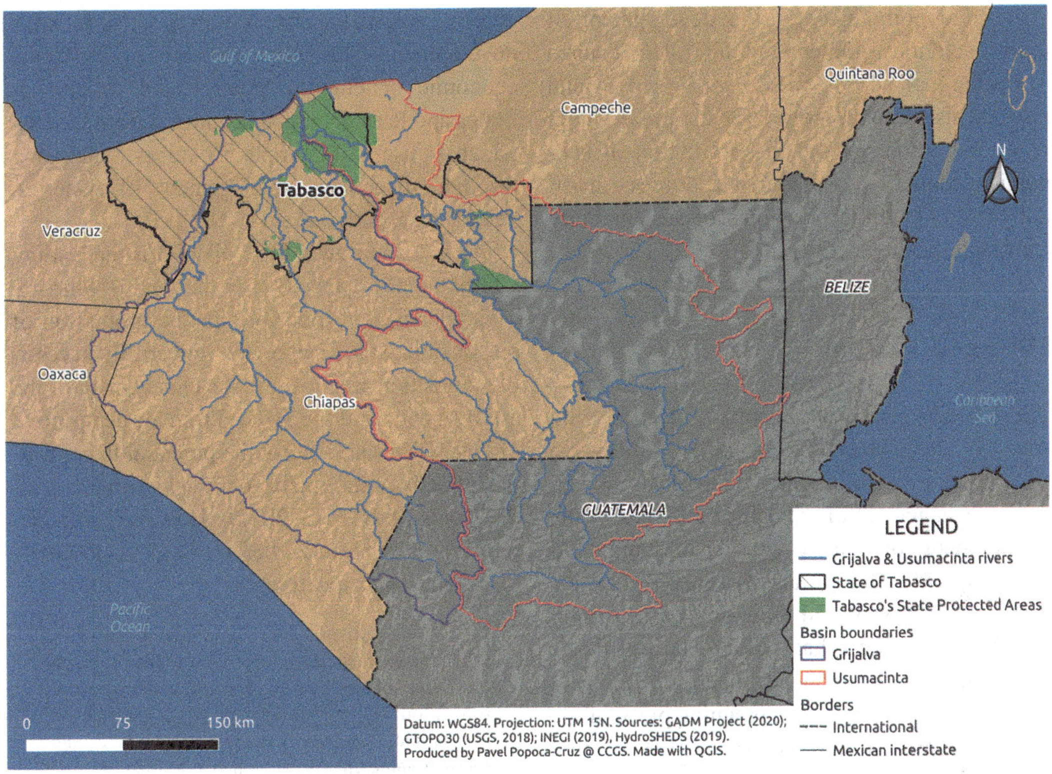

Sustainable Development and Responsible Tourism: The Grijalva-Usumacinta Lower River Basin, Fig. 1 The Grijalva-Usumacinta lower river basin and Tabasco's protected areas

No.	Management Category	Protected Natural Area, denomination	Type	Municipality	Surface (ha)
1	State Park	DE AGUA BLANCA	State	Macuspana	1,462.40
2	Ecological Reserve	CENTRO DE INTERPRETACIÓN Y CONVIVENCIA CON LA NATURALEZA YUMKA'	State	Centro	1,713.79
3	State Park	DE LA SIERRA DE TABASCO	State	Tacotalpa-Teapa	15,113.20
4	Natural Monument	GRUTA DEL CERRO COCONÁ	State	Teapa	442.00
5	State Park	LAGUNA DEL CAMARÓN	State	Centro	83.00
6	Ecological Reserve	LAGUNA DE LAS ILUSIONES	State	Centro	259.27
7	Ecological Reserve	DE LA CHONTALPA	State	Cárdenas	277.00
8	Ecological Reserve	LAGUNA LA LIMA	State	Nacajuca	36.00
9	Ecological Reserve	YU-BALCAH	State	Tacotalpa	572.00
10	Ecological Reserve	CASCADAS DE REFORMA	State	Balancán	5,748.35
11	Ecological Reserve	RÍO PLAYA	State	Comalcalco	711.00
12	Area Voluntarily Destined for Conservation	GUARITEC	State	Centla	7.00
13	State Park	LAGUNA MECOACÁN	State	Jalpa de Mendez, Parais	18,774.72
14	Area Voluntarily Destined for Conservation	TIERRA Y LIBERTAD	State	Macuspana	107.00
15	Biosphere Reserve	PANTANOS DE CENTLA	Federal	Centla, Jonuta y Macusp	302,706.00
16	Flora and Fauna Protection Area	CAÑÓN DEL USUMACINTA	Federal	Tenosique	46,128.49
17	Area Voluntarily Destined for Conservation	LOS MANGOS	Federal	Balancán	332.22
				Total	394,473.44

Sustainable Development and Responsible Tourism: The Grijalva-Usumacinta Lower River Basin, Fig. 2 List of protected areas in Tabasco

co-responsibility between the states, the economic and social stakeholders, and of course the tourists.

Therefore, the river basin, besides representing the territorial unity for water resources management, is bound to be a laboratory for public policies not only aimed at integrated water resources management, but also at sustained development of the natural resources located in this hydrographic space for the purposes of responsible tourism. Of course, each river basin has its own socioeconomic and environmental personality (Nava 2018), so each territorial unit provides different opportunities for the development of tourist activities. Hence, the GULRB requires assembling of ad hoc measures, programs and instruments of public policy for the preservation of natural resources and the promotion of economic, productive, and tourist activities carried out according to the principles of sustainability and responsibility.

SDGS and Tourism

The 17 SDGs have an important relation with tourism. They foster the creation of reflective thinking about sustainability of tourism, as well as the management of available resources in the development of tourist activity (Liburd and Edwards 2010; Murphy and Price 2005). In this respect, the UN World Tourism Organization (UNWTO) considers that tourism is a key factor for sustainable development, when it becomes a shared responsibility and an essential piece of the core of decision-making (OMT 2018), so that through the SDGs the purposes of the 2030 Agenda can be achieved in this domain.

In this regard, and in the framework of the 2030 Agenda, it is recognized that social and economic development depends on the sustainable management of natural resources. To promote sustainable tourism and to tackle water scarcity and water pollution are, among others, some of the means to achieve the Sustainable Development Goals and targets. Three are the SDGs related to tourism. (1) Goal 8 aims at promoting sustained, inclusive, and sustainable economic growth, full and productive employment and decent work for all. Within, by 2030 is expected to devise and implement policies to promote sustainable tourism enabling the creation of jobs and the promotion of local culture and products. (2) Goal 12, designed to ensure sustainable consumption and production patterns, seeks to contribute to the development and implementation of tools to monitor sustainable development impacts for sustainable tourism that creates jobs and promotes local culture and products. (3) Goal 14, intended to conserve and sustainably use the oceans, seas, and marine resources for sustainable development, projects to increase, by 2030, the economic benefits to Small Island developing States and least developed countries from the sustainable use of marine resources, including through sustainable management of fisheries, aquaculture, and tourism (UN 2015).

As is well known, the UNWTO is the leading international organization in the field of tourism. This agency is responsible for the promotion of responsible, sustainable, and universally accessible tourism geared toward the achievement of the 2030 Agenda and the SDGs. From the perspective of UNWTO, tourism has the potential to contribute, directly or indirectly, to the achievement of all the goals (UNWTO 2020). However, especial priority, as mentioned before, is given to Goals 8, 12, and 14, respectively.

> First at all, tourism, as services trade, is one of the top four export earners globally, currently providing one in ten jobs worldwide. Decent work opportunities in tourism, particularly for youth and women, and policies that favor better diversification through tourism value chains can enhance tourism positive socio-economic impacts. Secondly, the tourism sector needs to adopt sustainable consumption and production modes, accelerating the shift towards sustainability. Tools to monitor sustainable development impacts for tourism including for energy, water, waste, biodiversity and job creation will result in enhanced economic, social and environmental outcomes. Finally, coastal and maritime tourism rely on healthy marine ecosystems. Tourism development must be a part of Integrated Coastal Zone Management in order to help conserve and preserve fragile marine ecosystems and serve as a vehicle to promote a blue economy, contributing to the sustainable use of marine resources. (UNWTO 2020)

Tourism could be all of the above if we achieve tourism responsiveness. Thus, achieving responsible tourism is a continuous process and it requires constant monitoring of impacts, introducing the necessary preventive and/or corrective measures whenever necessary. In this regard, governments should consider whether they are paying sufficient attention to tourism within the field of sustainable development, and whether the tourism policies and actions adequately embrace concerns about sustainability (UNEP and UNWTO 2005).

Summary/Conclusion

The binomial made up of *sustainable development* and *responsible tourism* comes to life when both paradigms emerge into the World Heritage Convention Concerning the Protection of the World Cultural and Natural Heritage (1972), and the Rio Declaration on Environment and Development (1992). Both declarations, and respective events, represent pillars of environmental international law and enable to establish relations between sustainable development and responsible tourism.

The 1972 Convention highlights that (UNESCO 2020)

> [. . .] the natural heritage [is] increasingly threatened with destruction [. . .] and also by changing social and economic conditions which aggravate the situation with even more formidable phenomena of damage or destruction [. . .] Deterioration or disappearance of any item of the [. . .] natural heritage constitutes a harmful impoverishment of the heritage of all the nations of the world [Hence] natural heritage [is] of outstanding interest and therefore need to be preserved as part of the world heritage of mankind as a whole [and] it is incumbent on the international community [. . .] to participate in the protection of the [. . .] natural heritage of outstanding universal value [. . .]

It also recognizes that "natural heritage [could be] threatened by serious and specific dangers, such as the threat of disappearance caused by accelerated deterioration, large-scale public or private projects or rapid urban or tourist development projects" (Art. 11).

For its part, the 1992 Rio Declaration reaffirms the Declaration of 1972 and seeks to build upon it by working toward international agreements, which [. . .] protect the integrity of the global environment. Out of 27 principles (UN 1992a), the fourth states

> [. . .] in order to achieve sustainable development, environmental protection shall constitute an integral part of the development process and cannot be considered in isolation from it [. . .];

For its part, principle number eight stipulates that

> [. . .] to achieve sustainable development and a higher quality of life for all people, States should reduce and eliminate unsustainable patterns of production and consumption and promote appropriate demographic policies [. . .];

Furthermore, principle number 20 highlights the important role of women participation in environmental management in order to achieve sustainable development. Finally, principle 22 mentions that

> [. . .] indigenous people and their communities and other local communities have a vital role in environmental management and development because of their knowledge and traditional practices. States should recognize and duly support their identity, culture and interests and enable their effective participation in the achievement of sustainable development [. . .].

In this regard, it is then possible to elaborate a corollary based on the strategic means represented by water resources for the achievement of sustainable development and the responsible tourism. Water resources are a natural and cultural heritage (García 2004). Social and economic developments are inherent to the access, use, and exploitation of these resources. In the same way, the conservation and sustainability of the ecosystems and of the goods and services providing well-being to society depend on water resources (Alpuche Álvarez et al. 2021). Therefore, responsible tourism is an activity developed, in regions where it is possible, based on the recognition of water as a tourist attraction, whose access and uses provide benefits to the tourist users. That is how responsible tourism, organized around water resources, translates as an activity performed around the use of a natural resource humanity's heritage that must be preserved and protected by the states and the societies. In this case, the state is called upon to play the role of guardian of the world heritage. Henceforth, sustainable development and responsible tourism are projected goals for fair and equitable use and benefit of water resources and access to the ecosystemic goods and services provided by this resource.

S

Responsible tourism in Mexico must redirect toward the creation of a sustainable and responsible public policy stressing the importance of the supply and conservations measures of natural resources. Sustainable development in the

GULRB results from the economic viability of the project, possibilities of local prosperity, formation of human resources, social equity and community well-being; and from the governance and administration of the environment and the natural resources. Public policies about sustainable development and responsible tourism rely on the protection of natural resources for the purpose of achieving balance between socioeconomic development and environmental protection in accordance with the country context and regional and local conditions (Hunter 2002).

Cross-References

- ▶ Adapting Cities to Climate Change
- ▶ At the Intersection and Looking Ahead
- ▶ Blue-Green Cities: Achieving Urban Flood Resilience, Water Security, and Biodiversity
- ▶ Building Community Resilience
- ▶ Cities in Nature
- ▶ City Visions: Toward Smart and Sustainable Urban Futures

References

Alpuche Álvarez, Y. A., Nava, L. F., Carpio Candelero, M. A., & Contreras Chablé, D. I. (2021). Vinculando ciencia y política pública: La Ley de Aguas Nacionales desde las perspectivas sistémica y de servicios ecosistémicos. *Gestión y Política Pública, XXX*(2), 133–170. https://doi.org/10.29265/gypp.v30i2.881.

Ávila Akerberg, V., Martínez, T., Rodríguez-Soto, C., & Korosi, D. (2019). Biodiversidad, servicios ecosistémicos ylos objetivos del desarrollo sostenible en México. Universidad Autónoma del Estado de México, ISBN: 978 1 53239166 8.

Bermejo, R. (2014). *Del desarrollo sostenible según Brundtland a la sostenibilidad como biomimesis*. Bilbao: Universidad del País Vasco/Euskal Herriko Unibertsitatea. ISBN: 978-84-89916-92-0 [online]. Available at: https://www.upv.es/contenidos/CAMUNISO/info/U0686956.pdf. Accessed Feb 2021.

Gallardo-Cruz, J. A., & Carabias, J. (2017). *Presentación del Informe Técnico del Proyecto FORDECyT 273646 – Primera etapa [Cambio global y sustentabilidad en la cuenca del río Usumacinta y zona marina de influencia. Bases para la adaptación al cambio climático desde la ciencia y la gestión del territorio]* (p. 26). Tabasco: Centro del Cambio Global y la Sustentabilidad (CCGS).

Gallardo-Cruz, J. A., Charruau, P., & Rives, C. (2017). *Documento síntesis de la información disponible de los medios físico y biológico de la cuenca del río Usumacinta. Proyecto FORDECyT 273646 [Cambio global y sustentabilidad en la cuenca del río Usumacinta y zona marina de influencia. Bases para la adaptación al cambio climático desde la ciencia y la gestión del territorio]* (p. 43). Tabasco: Centro del Cambio Global y la Sustentabilidad (CCGS).

García, L. (2004). Agua y Turismo. Nuevos usos del recurso hídrico en la Península Ibérica, un enfoque integral. *Boletín de la Asociación de Geógrafos Españoles, 37*, 239–255.

García-García, A., & Kauffer, E. (2011). Cuencas compartidas entre México, Guatemala y Belice: Un acercamiento a su delimitación y problemática general. *Revista Frontera Norte, 23*(45), 131–162.

Hunter, C. (2002). Aspects of the sustainable tourism debate from a natural resources perspective. In R. Harris, T. Griffin, & P. Williams (Eds.), *Sustainable tourism: A global perspective* (pp. 3–23). Burlington: Elsevier Science.

INECC. (2007). La cuenca de los ríos Grijalva y Usumacinta [online]. Available at: http://www2.inecc.gob.mx/publicaciones2/libros/402/cuencas.html. Accessed Feb 2021.

Liburd, J. J., & Edwards, D. (Eds.). (2010). *Understanding the sustainable development of tourism*. Oxford, UK: Goodfellow.

López-Hernández, E. S. (2006). Áreas protegidas y ecoturismo. Una evaluación para su desarrollo sostenible en Tabasco. Colección José N. Rovirosa. Biodiversidad Desarrollo Sustentable y Trópico Húmedo. 1ª Edición. Universidad Juárez Autónoma de Tabasco, Villahermosa, p. 149.

Maldonado-Alcudia, M. C., Barragán-López, J. F., & Maldonado-Alcudia, C. M. (2015). Modelos teóricos aplicados al turismo. In *El turismo y el desarrollo comunitario. Investigaciones y propuestas* (pp. 9–42). Querétaro: Editorial Universitaria, Universidad de Querétaro.

Martínez-Cervantes, R. S., & Vargas-Martínez, E. E. (2015). Factores de comportamiento proambiental y uso turístico en parques nacionales. In *El turismo y el desarrollo comunitario. Investigaciones y propuestas* (pp. 175–199). Querétaro: Editorial Universitaria, Universidad de Querétaro.

Murphy, P. E., & Price, G. G. (2005). Tourism and sustainable development. *Global Tourism, 3*, 167–193.

Nava, L. F. (2018). La desafiante gestión integrada de los recursos hídricos en México: Elaboración de recomendaciones políticas. In J. Rojas Ramírez, A. Torres Rodríguez, & O. González Santana (Eds.), *Las ciencias en los estudios del agua. Viejos desafíos sociales y nuevos retos tecnológicos* (1st ed., pp. 26–42). Guadalajara: Editorial Universidad de Guadalajara.

OMT (Organización Mundial del Turismo). (2018). La contribución del turismo al desarrollo sostenible en Iberoamérica [online]. Available at: https://www.e-unwto.org/doi/pdf/10.18111/9789284420018. Accessed Feb 2021.

PNUD. (2018). Objetivos de Desarrollo Sostenible [online]. Available at: http://www.undp.org/content/undp/es/home/sustainable-development-goals.html. Accessed Feb 2021.
Sistema Estatal de Áreas Naturales Protegidas. (2021). Gobierno del Estado de Tabasco, listado de ANP [online]. Available at: https://tabasco.gob.mx/anps-tabasco-listado. Accessed Nov 2020.
Toledo, A. (2003). *Ríos, costas, mares, hacia un análisis integrado de las regiones hidrológicas de México* (p. 117). México: Secretaría de Medio Ambiente y Recursos Naturales, Instituto Nacional de Ecología y El Colegio de Michoacán.
UN. (1992a). Report of the United Nations conference on environment and development, Rio Declaration on Environment and Development, A/Conf.151/26 (Vol. I), Rio de Janeiro, 3–14 June 1992.
UN. (1992b). United Nations. Economic and Social Development, Division for Sustainable Development (DDS) conference on environment & development, Rio de Janerio, 3–14 June 1992, Agenda 21 [online]. Available at: https://sustainabledevelopment.un.org/content/documents/Agenda21.pdf. Accessed Feb 2021.
UN. (2000). *Declaración del Milenio*. Resolución aprobada por la Asamblea General, A/RES/55/2 [online]. United Nations. Economic and Social Development, Department of Economic and Social Affairs. Available at: https://documents-dds-ny.un.org/doc/UNDOC/GEN/N00/559/54/PDF/N0055954.pdf?OpenElement. Accessed Feb 2021.
UN. (2014). *The millennium development goals report 2014*. United Nations, Economic and Social Development, Department of Economic and Social Affairs, p. 59. ISBN 978-92-1-101308-5 [online]. Available at: http://mdgs.un.org/unsd/mdg/Resources/Static/Products/Progress2014/English2014.pdf. Accessed Feb 2021.
UN. (2015). *Transforming our world: The 2030 agenda for sustainable development*. Resolution adopted by the UN General Assembly on 25 September 2015 [online]. Available at: https://sustainabledevelopment.un.org/content/documents/21252030%20Agenda%20for%20Sustainable%20Development%20web.pdf. Accessed Feb 2021.
UNDP. (2020). What are the sustainable development goals? [online]. Available at: https://www.undp.org/sustainable-development-goals. Accessed Feb 2021.
UNEP and UNWTO (2005). Making tourism more sustainable: a guide for policy maker. Paris: UNEP, Division of Technology, Industry and Economics; Capitán Haya, Spain: World Tourism Organization. Available at: https://digitallibrary.un.org/record/561577. Accessed Feb 2021
UNESCO. (2020). *Convention concerning the protection of the world cultural and natural heritage*. UN Educational, Scientific and Cultural Organisation, 16 November 1972 [online]. Available at: https://www.refworld.org/docid/4042287a4.html. Accessed May 2020.
UNWTO. (2020). TOURISM 4 SDGs [online]. Available at: https://www.unwto.org/tourism4sdgs. Accessed Feb 2021.
Yáñez-Arancibia, A., Day, J. W., & Currie-Alder, B. (2009). The Grijalva-Usumacinta river delta functioning: Challenge for coastal management. *Ocean Yearbook, 23*, 473–501.

Further Reading

Boniface, B., & Cooper, C. (2005). *Worldwide destinations: The geography of travel and tourism*. London: Elsevier Science.
Denman, R. (2007). Objectives, policies and tools for sustainable tourism. In *Policies, strategies and tools for the sustainable development of tourism*. Madrid: World Tourism Organisation.
Escalona, F. (2019). Necesidad de reorientar la investigación y la enseñanza en el turismo. *Turismo Estudos y Práticas, 8*(1), 135.
Griffin, T., & DeLacey, T. (2002). Green globe: Sustainability accreditation for tourism. In R. Harris, T. Griffin, & P. Williams (Eds.), *Sustainable tourism: A global perspective*. Burlington: Elsevier Science.
Hardy, A., Beeton, R., & Pearson, L. (2002). Sustainable tourism: An overview of the concept and its position in relation to conceptualisations of tourism. *Journal of Sustainable Tourism, 10*(6), 475–496. https://doi.org/10.1080/09669580208667183.
Sharpley, R. (2000). Tourism and sustainable development: Exploring the theoretical divide. *Journal of Sustainable Tourism, 8*(1), 1–19. https://doi.org/10.1080/09669580008667346.
Zhenhua, L. (2003). Sustainable tourism development: A critique. *Journal of Sustainable Tourism, 11*(6), 459–475.

Sustainable Development Goals

An Overview of the Interconnected Relationship of SDGs and Urban and Regional Development

Ritesh Ranjan[1], Mohammed Firoz C[1] and Lakshmi Priya Rajendran[2]
[1]Department of Architecture & Planning, National Institute of Technology Calicut, Kozhikode, Kerala, India
[2]Bartlett School of Architecture, University College London, London, UK

S

Synonyms

Economic sustainability; Environmental sustainability; Social sustainability; Sustainable development

Definition

"Sustainable Development Goals (SDGs) are 17 goals that have been envisioned over the past decade with 169 targets under them that aim to eliminate poverty, hunger, and inequality within 15 years while monitoring and controlling climate change."

While safeguarding critical biological processes and ecosystems, the Sustainable Development Goals (SDGs) promote dignity and prosperity for humans. Apart from fighting poverty and inequality, the significance of supporting sustainable economic growth, peace, and justice; addressing fundamental social needs such as health and education; and taking steps to curb climate change and protect the environment is mainly emphasized.

Introduction

The world is constantly striving for development and prosperity. In pursuit of development, the indigenous and local features of a region are often overlooked. Historically, post Industrial Revolution period, the Western concept of development has dominated the minds of the whole world due to the trading and flow of information from the West to East. A result there has been socio-economic, (poverty, hunger, illiteracy, unhealthy life, and poor health) and environmental (spoiled aquatic life, pollution, and climate change) problems. There were substantial changes in societal structures and a change in the mindset of people following the end of World War II due to rapid industrialization. Initiatives and efforts have been made toward sustainable development to counterbalance the imbalance in the societal beliefs and infrastructure and establish equilibrium. The UN Sustainable Development Goals (SDGs) are one such initiation for appropriate policy interventions among member nations. SDG 11, one of the 17 goals, addresses the sustainability issues from the perspectives of urban and regional planning. This chapter therefore discusses a detailed timeline of interventions of the Sustainable Development Goals as a concept, its evolution, the latest interventions in the domain, and the SDGs and their relationship between Goal 11 related to urban and regional development along with its interconnection with all the 17 goals of the UN SDGs.

Timeline of Interventions in Sustainable Development

Sustainable development as a concept was existent since historical time. The way civilization's survived and their care for nature and society were all examples of sustainable development practices. Referring to the buildings and settlements from the ancient times, it is well observed that the use of vernacular materials in context to the local climate and relating to the major occupation of the people was in practice, whereas with the growth of industries and population, the expectations from the buildings increased, and market-driven construction techniques were in practice which hardly cared for the sustainability aspects. The word sustainability or sustainable development started appearing in literature post World War II. One such literature is the book named the *Silent Spring* in the year 1963 written by the American biologist Rachel Carson. This book warned the world about the potential environmental effects of chemical pesticides which inspired several global environmental movements. Following which in the year 1963, an international biological program was initiated by various nations around the world. The study that lasted for a decade, investigated environmental damage through biological and ecological mechanisms to lay the foundation for science-based environmentalism. In 1972, the first convention on the human environment of the United Nations (UN) member states was held. Club of Rome and the Massachusetts Institute of Technology (MIT) published *The Limits to Growth* in the same year, and the results were shocking to the world. A joint conference by the "United Nations Environment Program (UNEP)" and the UNESCO convened in 1977 in Tbilisi, Georgia, marked the first intergovernmental environmental education conference in the world. A global strategy for conservation was released by the "International Union for Conservation of Nature (IUCN)" in the year 1980. In 1983, in response to unsustainable economic growth, the UN established the "Brundtland Commission" which presented the report "Our Common Future" in 1987. This report defined the word "Sustainable development," and probably, even today, this is the most prominent definition for sustainable development among the several ones used in different literatures.

In 1988, the importance of climate change mitigation was understood, and therefore, a panel of intergovernmental experts on climate change was founded. In 1990, the World Conference on "Education for All" was held in Jomtien, Thailand, by the UN. Followed by the year 1992, The Earth Summit popularly known as the Rio Earth Summit that was focused on sustainable development based on which a shared global concept of sustainable development was adopted in that document.

In the year 1993, the world human rights conference took place which discussed the social dimensions of sustainability. In 1995, World Trade Organization (WTO) was established, and in 1999 the first Global Sustainability Index was launched to measure the level of sustainability universally. In September 2000, UN Millennium Summit was held where the global leaders committed their nation to a new global partnership to reduce extreme poverty and set out a series of time bound targets, with a period of 15 years. Further, in the year 2001, Qatar hosted WTO's 4th Ministerial Conference, in which the Doha Development Agenda was launched by all WTO members. A sustainable development summit was held in 2002 by the UN, it was focused to make political declaration and implementation plan in order to achieve development that respects environment. The Kyoto Protocol which was adopted in 1997, entered into force in 2005. The Kyoto Protocol operationalized the "United Nations Framework Convention on Climate Change (UNFCCC)" by committing for limiting and reducing greenhouse gas emission. In 2006, stern report on climate change was presented along with the SAARC development goals followed by the Nobel Peace Prize and the 4th International Conference on Environmental Education in 2007.

Later from the year 2009, the UNCCC conducted the 15th Conference of the Parties Series COP15, in Copenhagen, Denmark, with a goal to establish global climate agreement after the completion of the first commitment period under the Kyoto Protocol. COP16 was held in Cancun, Mexico, in 2010. The US$100 billion per annum "Green Climate Fund" and a "Climate Technology Centre" were proposed. This proposal was declined, concluding with the decision that the global warming potentials shall be those provided by the "Intergovernmental Panel on Climate Change (IPCC)." In 2011, COP17 was held in Durban, South Africa; this further progressed on the creation of the "Green Climate Fund (GCF)" for which a management framework was adopted.

At the time of the "Rio + 20 Conference," the UN states were working with the Millennium Development Goals (MDGs). Further, the Sustainable Development Goals (SDGs) were established in the year 2015. Globally, 10 million people participated in the consultation process to help shape the 2030 Agenda. Following the UN Summit in Paris, another international agreement was forged in 2015 to stop global warming and anthropogenic climate change. The Paris Climate Agreement is inextricably linked with the 2030 Agenda. A combination of these two key agreements provides the framework for global cooperation on sustainable development (Fig. 1).

Sustainable Development Goals by the United Nations

These SDGs were modeled after the Millennium Development Goals, their advancement, and structure. SDGs were created by the UN through the largest participatory process in world history, in which 10 million people expressed their views on shaping 2030, with 17 goals and 169 targets. Besides the MDGs, the SDGs include several new goals, such as ensuring resilient infrastructures, fostering industrialization that is sustainable and integral, and encouraging inventiveness (SDG 9);

Sustainable Development Goals, Fig. 1 "Sustainable Development Initiatives Timeline," an interpretation by authors based on open-source literature from the United Nations

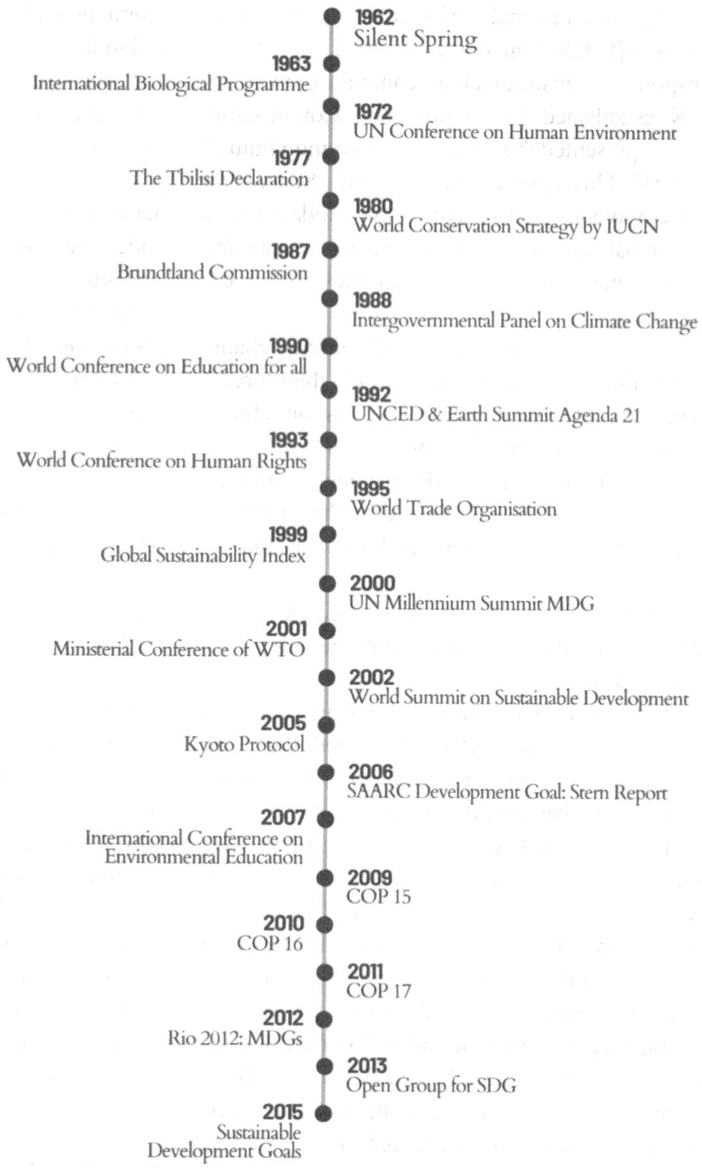

equalizing society and reducing inequality among nations (SDG 10); making safe, sustainable and resilient cities and human settlements (SDG 11); and promoting the sustainable and judicious consumption as well as production (SDG 12). MDG vs SDG targets are shown in Table 1. The MDGs were a symbol of solidarity with the poorest and the helpless, whereas the SDGs were holistic and broad (Cano et al. 2020).

A central theme of the 2030 development agenda is sustainability, as well as a system-wide approach to developmental paradigms. Ecological health and social justice must be balanced to attain sustainability. In other words, the 17 SDGs are not intended to settle a trade-off with multiple ideological perspectives about progress. Despite this, they are multidisciplinary and benefit each other in different ways (Barbier and Burgess 2017). Furthermore, (SDG 4) increasing access to good-quality education (SDG 6) and clean water in marginalized areas will enable those cities and communities to progress towards sustainability

Sustainable Development Goals, Table 1 The MDGs and SDGs

MDGs	GOAL	SDGs
"Eradicate extreme poverty and hunger"	1	"End Poverty in all its forms everywhere"
"Achieve universal primary education"	2	"End hunger, achieve food security and improved nutrition, and promote sustainable agriculture"
"Promote gender equality and empower women"	3	"Ensure healthy lives and promote well-being for all at all ages"
"Reduce child mortality"	4	"Ensure inclusive and equitable quality education and promote lifelong learning opportunities for all"
"Improve maternal health"	5	"Achieve gender equality and empower all women and girls"
"Combat HIV/AIDS and other diseases"	6	"Ensure availability and sustainable management of water and sanitation for all"
"Ensure environmental sustainability"	7	"Ensure access to affordable reliable, sustainable and modern energy for all"
"Develop a global partnership for development"	8	"Promote sustained, inclusive and sustainable economic growth, full and productive employment and decent work for all"
	9	"Build resilient infrastructure, promote inclusive and sustainable industrialization and foster innovation"
	10	"Reduce inequality within and among countries"
	11	"Make cities and human settlements inclusive, safe, resilient and sustainable"
	12	"Ensure sustainable consumption and production patterns"
	13	"Take urgent action to combat climate change and its impacts"
	14	"Conserve and sustainably use the oceans, seas and marine resources for sustainable development"
	15	"Protect, restore and promote sustainable use of terrestrial ecosystems, sustainably manage forests, combat desertification, and halt and reserve land degradation and halt biodiversity loss"
	16	"Promote peaceful and inclusive societies for sustainable development, provide access to justice for all and build effective, accountable and inclusive institutions at all levels"
	17	"Strengthen the means of implementation and revitalize the global partnership for sustainable development"

Source: Prepared by authors based on open-source data available under UN

(SDG 11) while improving the health of their citizens (SDG 3). This is also a measure of poverty reduction (SDG 1) since it strengthens community capacity, thereby reducing livelihood vulnerabilities. Hence, both the SDGs and their targets are multifaceted and interconnected (Cano et al. 2020).

Urban and Regional Development and SDGs

Multidisciplinary in nature, urban and regional development attracts scholars from a broad range of social and natural sciences such as planning, sociology, geography, political science, anthropology, and health and environmental science. SDG 11 acknowledges that cities are a key component of Agenda 2030. It is a fact that sustainable urban development is not a standalone goal, but is a way to achieve many other objectives, including addressing climate change, reducing poverty, increasing electricity access, and ensuring social inclusion. In addition, the New Urban Agenda of 2016 closely links SDG 11 to it. In order to foster urban sustainability, signatory countries must develop national urban policies. To make sure that SDG 2030 can be realized through urban development priorities, the National Urban Policy is supposed to observe and guide it. Localizing and mainstreaming SDG 11 within urban planning frameworks and adequate availability of budget are keys to making it successful. An integrated approach is required to address SDG 11 because of its interrelationships.

In developing countries, the majority of the population is growing exponentially. Following the current trends of urbanization, it can be anticipated that urban population will increase by around 68% by 2050 (Nations United 2018). Global environmental change has been exacerbated by urbanization's adverse impact. The recent past has been plagued by many natural disasters. The necessity of balancing development with nature was realized long ago. Over 55% of the planet's inhabitants live in cities, which contribute 85% to global GDP, but emit 75% of greenhouse gases as well. At the launch of the SDGs, former UN Secretary-General Ban Ki-Moon said: "Cities are where the battle for sustainable development will be fought" (Global utmaning 2017). The UN General Assembly adopted SDG 11 under its Agenda for Sustainable Development (2030) as a separate goal, in recognition of the importance of cities in a modern world.

Developing cities and regions has been made easier by the SDGs. To achieve holistic development, it is necessary to address each target and each goal. By solving one, other goals may be achieved as well. For sustainable development to be possible, social, economic, and environmental development pillars must be interdependent. Urban and regional development is directly addressed by SDG 11 by making cities and other human settlements sustainable, safe, resilient, and inclusive. Other goals and their targets have also an impact on SDG 11. This is where the relationship between SDG 11 and other goals becomes apparent (Wall 2020).

Sustainable Development Goal 11 and Other SDGs' Interlinkage

According to the UNDP (2019), "the SDGs are interconnected; the key to achieving one may lie in addressing issues more commonly associated with another." "The SDGs and the three pillars of sustainable development are indivisible, integrated and interdependent," said the United Nations Economic and Social Council President Mr. Oh Joon, who presided over the forum (Purvis et al. 2019; United Nations 2015). In addition to the UN, other organizations whose interests overlap with the SDGs demonstrate an understanding of the notion of interconnection.

It was a landmark agreement with the SDGs on cities and human settlements and demonstrates how "urban" has been elevated to the top of development agendas globally. In addition to SDG 11, other 2030 Agenda goals address urban issues.

SDGs 1 (poverty and tenure security), 3 (wellness), 6 (hygiene and water), 7 (an energy with no pollution), and 12 (conscious consumption and production, etc.) deal with issues relating to urbanization and human settlements. In line with SDG 11, "cities and human settlements should be inclusive, safe, resilient, and sustainable" by eliminating slum conditions, improving transportation access, reducing urban sprawl, engaging citizens in governance, and promoting cultural participation (UN-Human-Settlement-Programme 2018). Based on this UN Human Settlements Programme report 2018, 10 targets and 15 related indicators comprising SDG 11 have been worked out. The majority of them being measured on a city level and progress tracked nationally. This is shown in Table 2.

Other than the SDGs 1, 3, 6, 7, and 17, almost all the other goals contain similar challenges related to sustainable urbanization, such as poverty, inequality, pollution, and environmental degradation (UN-Human-Settlement-Programme 2018), as shown in Fig. 2.

"In addition to the 2030 Agenda for Sustainable Development, the New Urban Agenda, the Paris Agreement, and the Addis Ababa Action Agenda on Financing for Development, all of which have been adopted by the UN Member States, indicated a political commitment to eradicate poverty, save the ecosystem, strengthen ties, wellness, knowledge sharing, equal opportunity, urban sustainability, consumption, production, etc. Coordinating these agendas and frameworks at local, national, and

Sustainable Development Goals, Table 2 SDG 11 targets and indicators

Target	Current indicators
"**SDG Target 11.1** By 2030, ensure access for all to adequate, safe and affordable housing and basic services and upgrade slums"	"**11.1.1** Proportion of urban population living in slums, informal settlements or inadequate housing"
"**SDG Target 11.2** By 2030, provide access to safe, affordable, accessible and sustainable transport systems for all, improving road safety, notably by expanding public transport, with special attention to the needs of those in vulnerable situations, women, children, persons with disabilities and older persons"	"**11.2.1** Proportion of population that has convenient access to public transport, by sex, age and persons with disabilities"
"**SDG Target 11.3** By 2030, enhance inclusive and sustainable urbanization and capacity for participatory, integrated and sustainable human settlement planning and management in all countries"	"**11.3.1** Ratio of land consumption rate to population growth rate" "**11.3.2** Proportion of cities with a direct participation structure of civil society in urban planning and management that operate regularly and democratically"
"**SDG Target 11.4** Strengthen efforts to protect and safeguard the world's cultural and natural heritage"	"**11.4.1** Total expenditure (public and private) per capita spent on the preservation, protection and conservation of all cultural and natural heritage, by type of heritage (cultural, natural, mixed and World Heritage Centre designation), level of government (national, regional and local/municipal), type of expenditure (operating expenditure/investment) and type of private funding (donations in kind, private non-profit sector and sponsorship)"
"**SDG Target 11.5** By 2030, significantly reduce the number of deaths and the number of people affected and substantially decrease the direct economic losses relative to global gross domestic product caused by disasters, including water-related disasters, with a focus on protecting the poor and people in vulnerable situations"	"**11.5.1** Number of deaths, missing persons and directly affected persons attributed to disasters per 100,000 population" "**11.5.2** Direct disaster economic loss in relation to global GDP, damage to critical infrastructure and number of disruptions to basic services, attributed to disasters"
"**SDG Target 11.6** By 2030, reduce the adverse per capita environmental impact of cities, including by paying special attention to air quality and municipal and other waste management"	"**11.6.1** Proportion of urban solid waste regularly collected and with adequate final discharge out of total urban solid waste generated, by cities" "**11.6.2** Annual mean levels of fine particulate matter (e.g. PM2.5 and PM10) in cities (population weighted)"
"**SDG Target 11.7** By 2030, provide universal access to safe, inclusive and accessible, green and public spaces, in particular for women and children, older persons and persons with disabilities"	"**11.7.1** Average share of the built-up area of cities that is open space for public use for all, by sex, age and persons with disabilities" "**11.7.2** Proportion of persons' victim of physical or sexual harassment, by sex, age, disability status and place of occurrence, in the previous 12 months"
SDG Target 11.a Support positive economic, social and environmental links between urban, peri-urban and rural areas by strengthening national and regional development planning	"**11.a.1** Proportion of population living in cities that implement urban and regional development plans integrating population projections and resource needs, by size of city"
"**SDG Target 11.b** By 2020, substantially increase the number of cities and human settlements adopting and implementing integrated policies and plans towards inclusion, resource efficiency, mitigation and adaptation to climate change, resilience to disasters, and develop and implement, in line with the Sendai Framework for Disaster Risk Reduction 2015–2030, holistic disaster risk management at all levels"	"**11.b.1** Number of countries that adopt and implement national disaster in line with the Sendai Framework for Disaster Risk Reduction 2015–2030" "**11.b.2** Proportion of local governments that adopt and implement local disaster risk reduction strategies in line with national disaster risk reduction strategies"

(continued)

Sustainable Development Goals, Table 2 (continued)

Target	Current indicators
"**SDG Target 11.c** Support least developed countries, including through financial and technical assistance, in building sustainable and resilient buildings utilizing local materials"	"**11.c.1** Proportion of financial support to the least developed countries that is allocated to the construction and retrofitting of sustainable, resilient and resource-efficient buildings utilizing local materials"

Source: UN-Human-Settlement-Programme (2018)

Sustainable Development Goals, Fig. 2 Focal SDG 11 and other SDGs. (Source: UN Habitat; Cordio East Africa 2021)

international levels will also be essential" (UN-Human-Settlement-Programme 2018).

"2030 Agenda for Sustainable Development and its related goals are notable for addressing many global problems. By including a standalone goal addressing cities and human settlements (Goal 11), the SDGs acknowledge this complementary, mutually reinforcing relationship. Urbanization that is poorly managed threatens the SDGs, either directly or indirectly, through its effects on ecosystems, energy security, waste management, housing, urban mobility, etc. There are many targets of goal 11 that are linked to climate change, financing, sustainable production and consumption, gender-based violence, migration, food security and nutrition. It is important to integrate and reinforce relevant policies to support successful connections when these are established (Fig. 3). Existing policies must be able to facilitate lower negative externalities or trade-offs in places where links generate negative externalities" (UN-Human-Settlement-Programme 2018).

In order to develop sustainably, cities must address a number of challenges. Poverty, insufficient infrastructure, poor healthcare, increased slum dwellings, environmental degradation, and

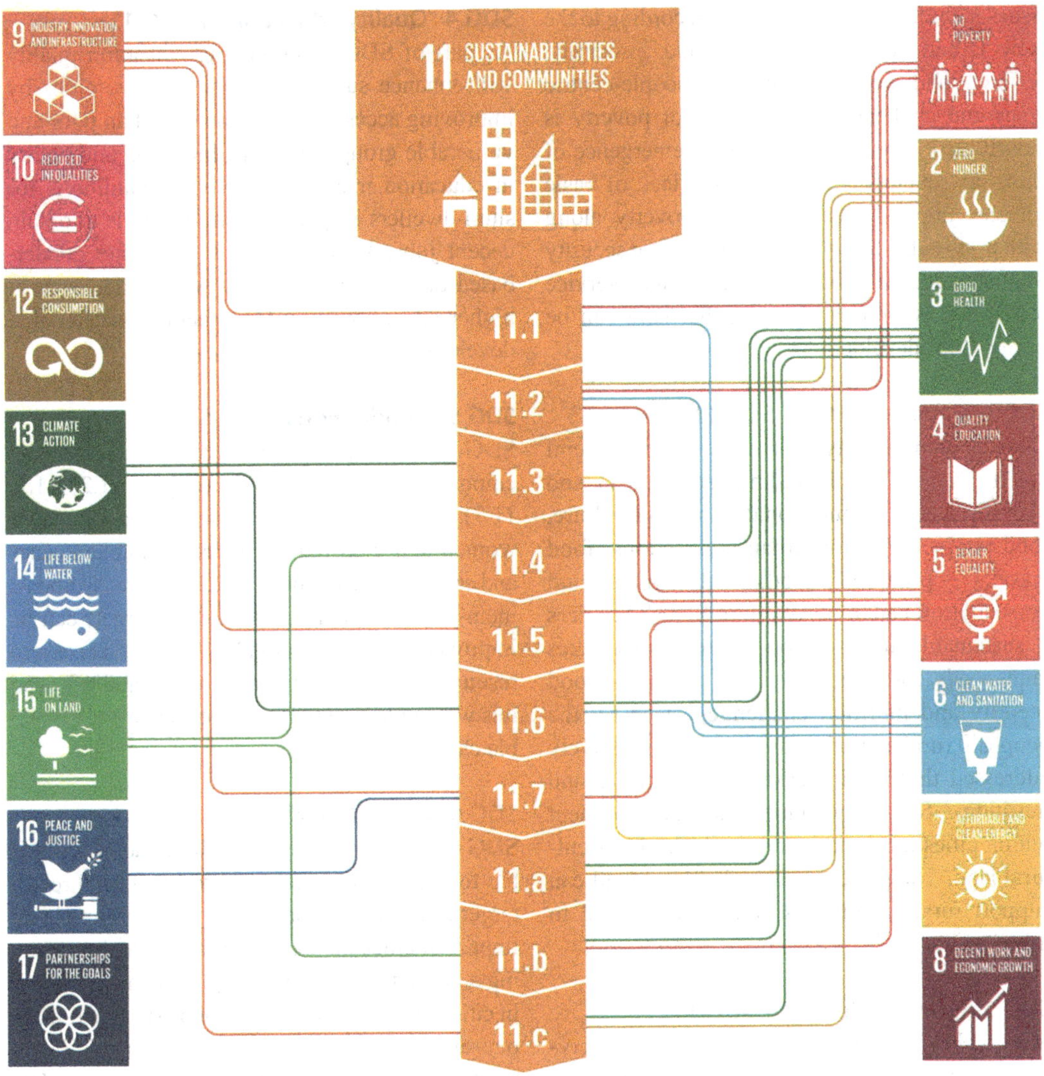

Sustainable Development Goals, Fig. 3 The interconnectivity between SDG 11 targets and the other SDGs and their respective targets. (Source: UN Habitat, Ohlander et al. 2019)

poor access to clean water are among these challenges. In order to combat these challenges, multiple agendas within the Sustainable Development Goals must be addressed simultaneously.

At the local, national, and global levels, achieving sustainable development depends on synchronizing the implementation of most SDGs. In addition to SDG 11, at least 11 other SDGs have targets and indicators that directly relate to it (Fig. 1). The city is one of the most important units to evaluate progress in meeting the Sustainable Development Goals since around a third of the indicators included in the global monitoring framework can also be measured locally. This section outlines in detail how the urban targets under SDG 11 relate to the other goals.

SDG 1 "End Poverty" + SDG 11

As part of the 2030 Agenda, every human settlement on Earth is to improve their living conditions, see their prosperity increase, and address issues including climate change. According to the New Urban Agenda, tackling food insecurity, poverty, and health issues are people-centric goals and challenges. Getting rid of poverty is closely tied into SDG 11 since the emergence of slums in developing countries and lack of basic services imply increasing urban poverty along with it. It is also essential that land tenure security is guaranteed in urban areas to ensure service provision, as well as ensuring that land can be acquired for development.

SDG 2 "Food Security" + SDG 11
Goal 2 (food security) is closely tied to several targets in goal 11, including targets 11.3 and 11.5. It focuses on nutritional status, agriculture, food production, rural-urban connections, food waste, city pollution's productivity impacts, and consumption patterns. Attaining SDG 2 requires a sustainable urbanization strategy that takes land needs for agriculture into account. Food security and nutrition challenges faced by the people in rural and urban areas can be effectively addressed through increased food security and nutrition. Social welfare and development within cities are hampered by a lack of agricultural land. The presence of agricultural land can support rural livelihoods and provide food to urban areas.

SDG 3 "Promote Health" + SDG 11
SDG 3 aims to promote healthy living for everyone. Road safety, health, and cities are inextricably linked since places affect health in many ways. By integrating urban planning, affordable housing, and reducing environmental hazards (targets 11.1, 11.2, 11.3, 11.7, 11.6), inclusive cities contribute to reduced non-communicable diseases and safer roads. By sacrificing productivity and air quality, poorly designed cities contribute to health challenges that affect cities' stability and the well-being of their citizens. Cities with unplanned and rapid urbanization suffer from traffic accidents as well as other health and environmental hazards. Thus, cities and infrastructure contribute to public health and well-being and are closely linked to the environment, poverty, and health.

SDG 4 "Quality Education" + SDG 11
As part of SDG 4, improving education quality will enhance sustainable and inclusive cities by improving access to education for urban poor and vulnerable groups in slums. Inclusion and equity in education may improve living conditions for slum-dwellers by giving them the skills to obtain decent jobs. It is essential to eradicate gender-based discrimination against women and girls in higher education in order to achieve sustainable development in inclusive cities.

SDG 5 "Gender Equality" + SDG 11
SDG 5 aims to achieve gender equality and empower women and girls (targets 11.2, 11.3, 11.7). SDG 11 is closely related to these because women and girls require access to public spaces and safety, as well as basic infrastructure. Women's inclusion in the pursuit of SDG 11 targets can be improved (e.g., by providing greater access to resources, productivity, and sustainability), and this will contribute to more sustainable and equitable cities.

SDG 6 "Clean Water" + SDG 11
SDG 6 aims to make water and sanitation accessible to all. Hence, it is linked to SDG 11 through target 11.6 that calls for reducing the environmental impact of cities per capita, for example, by reducing air pollution and improving waste management in cities. Water resources cannot be sustained without efficient and effective urban planning and waste management systems. By achieving SDG 6, we will not only promote decent housing and slum upgrade (target 11.1) but will also reduce water pollution-related deaths (target 11.5). We must maintain our efforts in supporting SDG 6 and SDG 11 in order to create the cities and world we envision. Cities produce the bulk of sewage and solid waste within urban areas, which is directly associated with several of these goals.

SDG 7 "Sustainable Energy" + SDG 11
Providing a stable, reliable, affordable, and sustainable supply of energy is among the goals of SDG 7. It is closely tied to SDG 11, since energy is a determinant of the economic, social, and environmental dimensions of SDGs. A safe, resilient,

inclusive, and sustainable human settlement requires the availability of clean, efficient energy systems, which can support their growth and performance. By providing sustainable transportation, housing, urban planning, pollutant reduction, and climate change mitigation, SDG 11 allows SDG 7 to be realized. However, unsustainable consumption patterns in urban areas may involve direct and embedded energy consumption.

SDG 8 "Sustainable Economic Growth" + SDG 11

SDG 8 emphasizes full and productive employment, economic growth that is inclusive and sustainable, and decent work. A city's role in driving innovation, consumption, and investment plays a critical role in economic development and prosperity. In fact, 80% of the world's GDP comes from cities. Therefore, innovation, entrepreneurship, job creation, and higher productivity are vital to achieving SDG 8. Cities will benefit from inclusive and sustainable economic growth (more affordable housing, better city planning, easy access to basic services).

SDG 9 "Resilient Infrastructure, Sustainable Industrialization & Innovation" + SDG 11

Urban development relies on physical infrastructure investments as well as innovative technologies, such as Intelligent Transportation Systems (ITS). In this regard, SDG 9 – To build resilient infrastructure, to achieve inclusive and sustainable industrialization, and to foster innovation – is closely linked to SDG 11 (targets 11.2, 11.3, 11.7, and 11.6). For cities to be sustainable and safe, smart infrastructure, industrialization, and innovation must be invested. This includes clean energy infrastructure, public transit systems, and improved civic engagement and management. Innovative technologies, like ITS, can leapfrog the technology curve, innovate, and diversify industries to create more sustainable, resilient, and inclusive cities.

SDG 10 "Reduce Inequalities Within and Among Countries" + SDG 11

Despite the fact that cities are susceptible to inequalities and they manifest in a variety of forms, they are also better positioned to address them through more employment opportunities, affordable housing options, and more accessible transportation. Thus, efforts to achieve SDG11 need to closely align with other goals, such as SDG 10, in order to eliminate inequalities in urban areas. Migrants, refugees, and disabled people also experience marginalization and exclusion, which can be exacerbated or presented with viable solutions by poor urban planning, design, and governance. But focusing on cities will ensure that inequalities in rural areas are not hidden behind national averages, as is done with SDG 11 and other SDGs that focus on rural inequity.

SDG 12 "Sustainable Consumption & Production" + SDG 11

In the quest for sustainable patterns of production and consumption, cities are essential as producers and consumers in the world. More than 70% of GHG emissions and natural resource consumption occurs in cities. By improving the management of natural resources, safely disposing of toxic waste and treating pollutants (target 11.6 and target 11.b), SDG 11 contributes to achieving SDG 12. By reducing sprawl and increasing sustainable consumption patterns, cities will increase their productivity and reduce waste. Resource innovation will improve productivity and reduce cities' environmental impacts. The reduction of material footprints and embedded energy may be possible when standards are used for buildings, energy, and transportation. In cities, sustainable consumption and patterns ensure resilience, inclusiveness, and sustainability by reducing latent stresses. There are also instances in which urban areas contribute to environmental degradation when they follow unsustainable patterns of production or consumption.

SDG 13 "Combat Climate Change" + SDG 11

Natural disasters and climate change pose a particular threat to cities. In the contemporary world, many residents of urban areas face direct and indirect impacts of climate change, such as more frequent and severe storms, floods and heatwaves, as well as a variety of diseases related to water, food and vectors. Furthermore, cities contribute to climate change by consuming 78% of the world's energy and produce more than 60% of greenhouse gas emissions (Nations 2021). Thus, SDG 13 – "Take urgent action to mitigate the impacts of climate change" – is central to achieving SDG 11 sustainability components. To combat climate change, SDG 11 offers a wide range of mitigation and adaptation strategies, mainly by sustainable and resilient urban planning (targets 11.2, 11.5, 11.b, 11.c), and by promoting judicious urban development through target 11.a.

SDG 14 "Conserve & Sustainably Use the Marine" + SDG 11

Goal 14 focuses on preserving and utilizing the large marine resources. SDG 14 is directly impacted by SDG 11, by improving waste management in the cities that pollute oceans. Environmental pressure is often increased by coastal cities and human settlements. In fact, cities are often responsible for pollution in the oceans. In addition, achieving SDG 14 strengthens the sustainability of urban planning and the sustainability of settlements given the importance of coastal areas to urban development for economic benefits and opportunities. The interrelationship between SDG 11 and SDG 14 can be seen clearly through the need to protect our biodiversity. Perceptions of urbanization influence our perceptions of its potential impact on farming and nutrition in the future. As urban populations grow and as diets and demands change, urbanization changes the demand for agricultural products.

SDG 15 "Protect & Restore Terrestrial Ecosystems" + SDG 11

As part of SDG 15, natural ecosystems, such as forests, wetlands, and arid lands, will be protected and restored. As a result of ecosystems and biodiversity, humans have the ability to produce, consume, and live. The earth and the ocean provide the human race with sustenance and livelihood. SDG 11 supports SDG 15 by aiming to promote sustainable urban development (target 11.3), maximizing the efficiency of urban planning (targets 11.2, 11.b, 11.c), the creation of environmentally conscious infrastructure (target 11.7), waste management and disposal (target 11.6), and preserving natural environments (target 11.4). Furthermore, SDG 15 aims to encourage the adoption of nature-based solutions and reduce disaster risk, which in turn ultimately contributes to sustainable cities and human settlements. SDG 15 may be negatively impacted by uncontrolled sprawl. Land degradation, environmental degradation, and increased risk of extreme weather events have been attributed to sprawling, low-density urban development.

SDG 16 "Peace & Efficient Governance" + SDG 11

To achieve SDG 11, peace and efficient governance (SDG 16) will need to be present as well as the necessary resources for its implementation. Developing an urban society at the right time will greatly shape our progress towards 2030 as humanity becomes increasingly urban. In order to achieve SDG 11 and SDG 16, the urban dimensions of crime, violence, and insecurity must be considered. Increasing numbers of crimes involving illicit flows and corruption are being committed within cities and are connected to several urban development projects. In order to achieve peace, inclusion, and sustainability in our cities, we need to pay attention to how our cities are governed, as well as how we are integrating urbanization into our cities.

SDG 17 "Strengthen the means of implementation and revitalize the global partnership for sustainable development" + 11

Goal 11 can succeed only if all other goals work in concert. As part of partnerships for sustainable urban development, the United Nations, member states, international and regional cities' associations, NGOs, private corporations, agencies for specialized funds, and National Commissions and Category 2 Centers, such as "United Nations Educational, Scientific and Cultural Organization (UNESCO)", are involved. The goals of many UN agencies and intergovernmental organizations have been coordinated through collaborative partnerships, including "United Nations Office for Disaster Risk Reduction (UNDRR)" formerly UNISDR, World Health Organization (WHO),

United Nations Education Programme (UNEP), United Nations Human Settlements Programme (UN-Habitat), UNESCO, etc. UN-Habitat and UNESCO renewed their partnership in 2017, committing to work together to integrate culture into SDG 11 and the City Prosperity Index (CPI) of UN-Habitat. Both the United Nations World Tourism Organization (2013) and the World Bank (2011) signed Memoranda of Understanding, strengthening cooperation between heritage management programs. UN-Habitat and UNEP have collaborated on the Greener Cities Partnership for many years. The goal of both of these agencies is to integrate urban infrastructure and solutions that are grounded in nature. As part of SDG 11, the Food and Agriculture Organization (FAO), the World Bank, the Organisation for Economic Co-operation and Development (OECD), the European Commission, and UN-Habitat have formed collaborations to assist with global monitoring of urban (city) and rural areas. UNDRR/UNISDR, UNESCO, the World Bank, and United Nations Development Programme (UNDP) have developed strong partnerships around a culturally appropriate program for disaster risk reduction (DRR). Partnerships formed for DRR in the cultural sector offer technical, economic, and operational support for post-disaster reconstruction and recovery programs.

Conclusion

By including SDG 11 in the Sustainable Development Agenda 2030, cities are acknowledged as an important factor in global development. Besides urban sustainability, there are several other objectives that can be achieved through this process, including fighting climate change, eradicating poverty, providing safe water, managing energy usage, promoting social inclusion, and promoting spatial justice.

In this chapter, SDG 11 has been used for urban and regional development as it is understood that developing the cities, in turn, will create a base work for the development of the region around it. Further, the interconnections of all the SDGs are discussed keeping SDG 11 in focus. In the pursuit of urban services and development, it is important to use a decentralized and fiscally federal approach that maintains a close relationship between local administrators and users. Building trust between communities and local leaders requires transparency and accountability based on advocacy, data, and evidence. Leadership should be encouraged to leapfrog from traditional tools and technologies, despite the risks. Hence, empowered metropolitan, urban, and rural local governments can play a significant role in achieving the SDGs for all urban and regional development projects.

Cross-References

- ▶ Smart Grids to Lower Energy Usage and Carbon Emissions: Case Study Examples from Colombia and Turkey
- ▶ Sustainable Cities via Urban Ecosystem Restoration

References

Barbier, E. B., & Burgess, J. C. (2017). The sustainable development goals and the systems approach to sustainability. *Economics, 11*, 1–24. https://doi.org/10.5018/economics-ejournal.ja.2017-28.

Cano, N. A., Velasco, J. O., & Franco, I. B. (2020). *Actioning the global goals for local impact towards sustainability science, policy, education and practice* (A. I. for B. and E., I. B. Franco, T. Chatterji, J. Tracey, & E. Derbyshire (Eds.); Issue January). Springer Nature Singapore Pte Ltd. https://doi.org/10.1007/978-981-32-9927-6

Cordio East Africa. (2021). *Focus SDG11 in connection with other SDGs*. Cordio East Africa. https://cordioea.net/sdg-waste/

Global utmaning. (2017). *Local implementation of the SDGs & the new urbanism agenda towards a Swedish National Urban Policy* (pp. 5–31). Global Utmaning Association. info@globalutmaning.se%7C.

Nations, t. U. (2021, September 22). *United Nations climate action*. Retrieved from Cities and pollution: https://www.un.org/en/climatechange/climate-solutions/cities-pollution

Nations United. (2018). *2018 Revision of World urbanization prospects*. Department of Economic and Social Affairs; United Nations. https://www.un.org/development/desa/publications/2018-revision-of-world-urbanization-prospects.html#:~:text=Today%2Cthemosturbanizedregions,Asiaisnowapproximating50%25

S

Ohlander, E., Pedersen, M., Wejs, A., Bonde, M. A. S., & Lehmann, M. (2019). *Guidance on integrating the sustainable development goals in urban climate change adaptation projects*. NIRAS, Aalborg University (Issue January). https://doi.org/10.1007/978-3-030-73575-3.

Purvis, B., Mao, Y., & Robinson, D. (2019). Three pillars of sustainability: in search of conceptual origins. *Sustainability Science, 14*(3), 681–695. https://doi.org/10.1007/s11625-018-0627-5.

UN Habitat, Sustainable Urbanization and Sustainable Development Goals, http://csud.ei.columbia.edu/files/2018/11/SDG-Booklet.pdf.

UN-Human-Settlement-Programme. (2018). *SDG 11 synthesis report: Tracking progress towards inclusive, safe, resilient and sustainable cities and human settlements, High level political forum 2018*. United Nations- Habitat. https://doi.org/10.18356/36ff830e-en.

United Nations. (2015). *Integrating the three dimensions of sustainable development*. United Nations ESCAP, 33. https://www.unescap.org/sites/default/files/IntegratingthethreedimensionsofsustainabledevelopmentAframework.pdf

Wall, T. (2020). *Encyclopedia of the UN sustainable development goals*. Springer, 46(02), 46-0890-46–0890.

Sustainable Development Goals and Urban Policy Innovation

Wendy Steele[1] and Lauren Rickards[2]
[1]Centre for Urban Research, RMIT University, Melbourne, VIC, Australia
[2]Urban Futures Enabling Capability Platform, RMIT University, Melbourne, Australia

Synonyms

Sustainability; Development; Transformation; Cities; Regions; Innovation; Ethics

Introduction

Cities and urban regions have the opportunity and capacity to move into a leading position on SDG leadership and innovation. This includes harnessing latent potential to deliver and demonstrate commitment to ethical SDG innovation across sector and scale that is responsible, inclusive, disruptive, and engaging. The SDGs offer a platform for fostering the innovative cross-scale, cross-sectoral linkages and experimentation needed to successfully pursue genuine sustainable development. By embracing the SDG framework, urban settlements could lead critical thinking and action on the SDGs, co-developing advice and advocacy in key areas such as action on climate change. However to enable this vision to be realized, institutional commitment and leadership is needed to embed the SDGs into strategies, practices, and plans from the ground up.

This chapter makes the case for towns, cities, and urban regions to embrace a deep commitment to the SDGs combined with a bold innovation culture. As outlined in this book, urban settlements face a myriad of challenges that impact on both people and planet including the climate emergency, rapid urbanization, urban sprawl, urban poverty, high unemployment rates, housing affordability, lack of urban investment, rising inequality, a growing digital divide, reduced biodiversity, environmental degradation, and increasing vulnerability to the speed and scale of catastrophic natural disasters. Addressing the sustainability of urban and regional futures requires a systems approach to the "wicked" complexity of societal challenges that are being felt at both the local and global scales. This requires the *scaling up, out, and deep* (see Moore et al. 2015) of the SDG agenda – from a niche concept on the fringe of urban policy and planning to one that is embedded within all levels of urban governance, including leadership, strategies, participatory practices, and culture.

Addressing Unsustainability

Under the leadership of the United Nations, the global community – including governments, businesses, and others in Australia – has committed to pursuing 17 ambitious Sustainable Development Goals (SDGs) over the next 15 years (e.g., eradicating poverty, tackling climate change, creating safe, resilient, and sustainable cities). These goals are unprecedented in their focus and intent. As the SDG Declaration by the United Nations General Assembly (2015, p. 1) states:

> Never before have world leaders pledged common action and endeavour across such a broad and universal policy agenda. We are setting out together on the path towards sustainable development, devoting ourselves collectively to the pursuit of global development and of "win-win" cooperation which can bring huge gains to all countries and all parts of the world.

Evidence is mounting that the world's current trajectory is unsustainable – socially, environmentally, and economically (Brundtland 1987). No longer can the costs and risks of existing ways of doing things be "externalized" as irrelevant, insignificant, accidental, or acceptable. Socioeconomic inequities, pollution, biodiversity loss, and climate change are among the Earth-destabilizing, life-endangering, reprehensible problems that are getting worse not better. It is increasingly clear that these outcomes are not unexpected side effects of existing development processes but are their highly predictable and logical consequences. To not only ameliorate or neutralize but actually reverse these trajectories – and to create not just a tolerable but a vibrant future – transformational, fundamental change is needed.

In recognition of this urgent situation, the SDG 2030 Agenda is a "universal call to action to end poverty, protect the planet and ensure that all people enjoy peace and prosperity" (see https://www.un.org/sustainabledevelopment/sustainable-development-goals/). The 17 interconnected goals illustrated in Fig. 1 build on the Millennium Development Goals but extend attention to all nations and critical contemporary key issues relevant to urban and regional futures such as climate change, sustainable consumption, peace and justice, partnerships, and economic inequality.

The 2030 Agenda is a radical plan for humanity and a new way of "doing" development, including in those countries typically thought of as "developed" (Steele and Rickards 2021). The SDG agenda recognizes that interventions in one area will affect outcomes in others and must be considered as part of a global systems framework. Sustainable development is underlined as relevant

Sustainable Development Goals and Urban Policy Innovation, Fig. 1 The 17 UN Sustainable Development Goals (UN 2015)

to all nations because none exists in a vacuum and all face internal inequities and sustainable development challenges. To this end, the SDG agenda more explicitly addresses oceans, ecosystems, energy, climate change, and sustainable consumption and production. It also underlines that these are inseparable from the equally important issues of inequalities, decent jobs, urbanization, industrialization, peace, and justice.

It is widely recognized that the SDG agenda is not perfect (Hope 2021; Kaika 2017; Moseley 2018; Sultana 2018). Sustained work is required to address its gaps (e.g., indigenous sovereignty) and risks (e.g., an atomization and bureaucratization of the goals; a reduction to targets, indicators, and reporting; and self-serving engagement designed to enhance and protect brands, not induce necessary change). Despite these limitations, the SDG agenda holds incredible transformative potential. It invites critical engagement from all motivated by its overarching goal of creating a better future. It brings together issues usually discussed in vacuums, actors usually separated by silos, and the economic, social, and environmental pillars of sustainability that are usually pursued in isolation. How to implement this shared SDG agenda is now the collective engagement challenge for urban and regional futures.

Engaging with the SDGs

Despite the huge amount of work now going into implementing the SDG agenda, vast knowledge gaps remain internationally around the SDGs in terms of what they are, how to plan and implement them, how to monitor and evaluate progress, and how to develop the skills and capabilities needed across governments, business, and NGOs/civil society. The transformative change demanded by the goals requires innovative new knowledge, approaches, and partnerships that take into account new, more distributed ways of connecting and creating change. It requires applied, creative engagement by cities and urban regions, including leadership, institutional commitment, and innovation, but time is running out to reverse current trajectories and achieve the transformative ambitions of the SDG agenda by 2030.

Because of the bold, transformative ambition of the SDG agenda and the need for global-local action that "leaves no one behind," all stakeholders (i.e., governments, civil society, the private sector, and other key players such as universities) are expected to contribute. The focus is on working through multi-stakeholder partnerships and the principle of subsidiary (devolving action to the local scale). Emerging efforts to implement the SDG agenda include capacity building, financing, the generation of new technologies and data, institutional buy-in, and governance innovations (Freistein and Mahlert 2016; Sachs 2017).

But how urban settlements and their communities respond to the SDG agenda is a strategic question. Even if the option is not to engage, this needs to be recognized as an active and consequential decision. Strategic foresight methods are now used across sectors to address two crucial factors:

1. The depth and breadth of institutional commitment
2. The boldness of the innovation culture

Where a city or region positions itself with regard to these two factors will determine the approach that is taken. Each can be mapped as an axis or continuum. Combining them into a two-by-two matrix yields four plausible scenarios that provide the basis for informed discussion about the strategic direction to be adopted to support the future of urban and regional settlements: Tolerant, Disengaged, Paternalistic, and Transformative.

At one end of the institutional commitment spectrum is *Shallow Commitment* which takes the form of tolerance for, or occasional endorsement of, SDG-related initiatives. Efforts around the SDGs exist, but are largely the work of isolated individuals or groups. They are generally ad hoc, disconnected, invisible to most people, and quickly forgotten. They include one-off events, single assessment tasks or courses, occasional publications and short-lived research, operational projects, or web pages. The SDGs are treated as an interesting but specialist topic with limited relevance to the functioning of the institution's core business. More specifically, the SDGs are misunderstood by many people as a traditional international development issue and only relevant to low-income countries and development specialists.

At the other end of the spectrum is *Deep Commitment*. Here, SDG engagement is characterized by strong institutional leadership, strategic prioritization, cultural commitment, and brand transformation. The SDGs are recognized and represented as part of a new worldwide agenda. They are used as an integrative, long-term, systematic framework of engagement that encompasses all university functions, components, and stakeholders. From the strategic planning to resourcing and implementation, the SDGs are used as a visible, cohering, focusing framework, with far-reaching impacts as a result.

Cutting across the question of commitment is the question of innovation culture, which can be characterized by how routine or imaginative it is. *Conventional innovation* can involve the prolific production of innovation products, done in a routine way. One of the ironies of innovation is that as a concept it is far from novel. Indeed, it is now mainstream, often forced and largely habitual, driven by an apparent need to produce new products for the market. "Innovative" innovation moves beyond this robotic approach. Situated at the other end of the spectrum is a bolder, more radical approach to innovation that nurtures creative shifts and scales them out to progressively alter, not reinforce, the existing institutional environment. This approach involves innovating the ways of doing things, including innovation itself.

Helping drive this new, more critical approach to *bold, ethical innovation* are two key agendas. One is the Impact Agenda which refers to the broad push to more explicitly, directly, and effectively meet societal needs. This involves a shift from top-down, linear, knowledge-centric models of innovation to more systemic, inclusive, action-oriented ways of doing innovation (Rickards et al. 2020). The second factor helping drive a more critical approach to innovation is the growing realization that the conventional approach to innovation is a source of problems as much as solutions. As an engine of consumption, conventional innovation processes underlie many environmental harms and social injustices.

An ethics-based approach to innovation is courageous, imaginative, and intelligent enough to not just change product specifications, but systems, goals, and paradigms – including the innovation culture itself – so that societal needs and goals are more effectively met. Ethical innovation can be summed up as responsible (anticipatory and precautionary), inclusive (collaborative, systemic), disruptive (bold, impactful), and engaged (democratic, debatable) (Rickards and Steele 2019).

A Framework for Transformative Action

A key *certainty* about the future is that urban settlements will sooner or later have to more directly address the escalating need for and pressure around the SDGs including action on climate change. The two axes outlined above represent two key uncertainties or questions: "How deeply will the commitment to the SDGs be?" and "How bold will the innovation culture be?" Combining the possibilities gives four possible pathways or scenarios which are outlined in Fig. 2 which provides a strategic framework for thinking through different SDG engagement options and their implications.

The first pathway is one that is *Disengaged*. This is where any existing work on the SDGs stagnates and fades to become just one of a number of reporting requirements and past enthusiasms. In this possible future, some SDG work continues, but it is largely ad hoc and driven by external requirements such as demands from industry partners. Meanwhile, the innovation culture is focused ever more narrowly on accelerating and refining existing product development processes and serving certain market players while remaining disengaged from most of society and

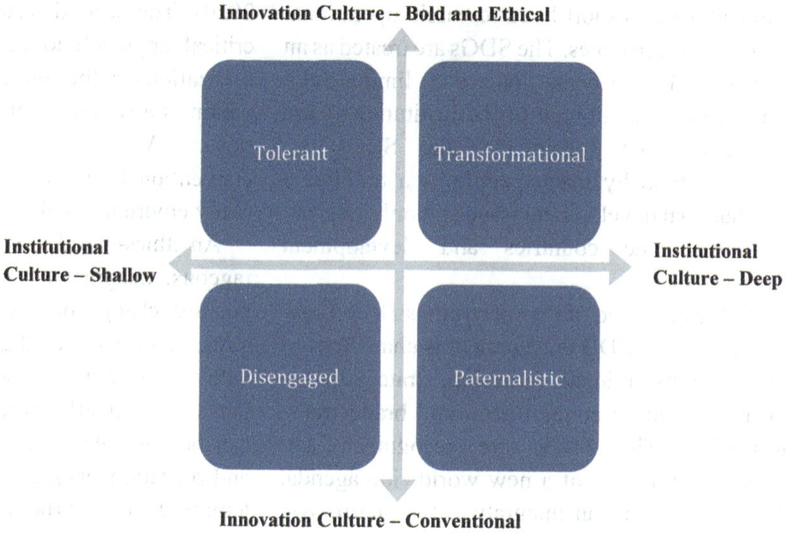

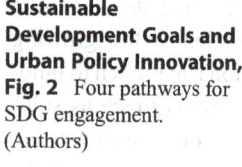

Sustainable Development Goals and Urban Policy Innovation, Fig. 2 Four pathways for SDG engagement. (Authors)

the processes' wider ramifications. Individuals striving to do things differently have been implicitly discouraged and have largely moved to other more open-minded institutions.

Second is the *Tolerance* pathway which frames the SDGs as a specialist or fringe topic. At the institutional level, the SDGs are resourced in a minor way, but are not recognized as a major societal challenge or guiding parameter or as relevant to the institution as a whole. Those actively working on the SDGs are largely left to their own devices, perhaps developing niches of radical innovation (e.g., bold experiments with partners in government, business, and community), but in an isolated manner.

The third *Paternalistic* pathway combines a conventional innovation culture with a deep commitment to the SDG agenda. Here, cities and communities take the need to address sustainable development seriously and work to embed the SDG agenda as a strategic priority from the top down. This includes engaging with the SDGs through a variety of compulsory and voluntary mechanisms including awareness raising and resourcing of some SDG research initiatives and the incorporation of the SDG agenda into strategic urban plans. However, despite these efforts at encouraging and directing SDG engagement, in practice, this does not substantially alter the status quo including existing approaches to innovation, its foci, its partnerships, or its image. This is SDG-flavored business as usual, using the language (if not transformative practice) of the SDGs at multiple scales.

The final pathway is a *Transformative* one. This involves a deep, ethical commitment to rapidly transitioning cities and urban regions into a better position from which to help transition the world onto a more sustainable, socially just pathway. This is a commitment to the principles and ethos of ethical innovation and to scale bold, ethical innovations for sustainable development up, out, and across places and spaces, institutional architecture, and community and stakeholder sectors. This institutional commitment to SDG innovation is deep, bold, and pioneering, showcasing and sharing different understandings and practices, underpinned by visionary leadership, resources, and support.

SDG Action at Scale

The SDG agenda is a powerful integrative framework that works at multiple scales. For example, the SDGs are being pursued via the new Urban Agenda led by UN Habitat, recognizing the foundational importance of urbanization issues to global outcomes. The United Nations Development Program (UNDP)'s 2018–2021 Strategic Plan similarly reinforces a strong commitment to working as a catalyst and facilitator for the SDG agenda in partnership with other UN initiatives, governments, civil society, and the private sector. In particular, it supports:

- Multi-stakeholder partnerships at national, regional, and global levels and mutual accountability for the SDGs in such partnerships
- Innovative platforms that strengthen collaboration with Governments as well as with civil society and the private sector
- National governments and partners to work toward common results and indicators and collectively report on them
- National progress in engaging citizens through volunteerism, empowerment, participation, and other means to strengthen national ownership and capacity and delivery of the sustainable development agenda
- Opportunities for civil society to effectively engage in sustainable development

The UN seeks to promote whole-of-government and whole-of-society responses in order to enable transformational change. Alongside government and community, the private sector is a critical partner in sustainable development, and many businesses now recognize this. NGOs, networks, and social movements are also essential in implementing the SDG global agenda. NGOs are mobilizing to address the SDGs in various ways, including communicating the SDGs to a broad audience, holding governments and industry to account, and implementing grassroots project.

Cities and urban regions have also risen to the SDG challenge. The City of Melbourne in Australia, for example, was an early adopter of the SDG framework, mapping each of its strategies and plans to assess delivery and interconnectedness, as well as identify gaps and opportunities. It deployed the SDGs both as a checklist, and to enable visioning, improve coordination of priorities, and provide a common language for the community, and as a future benchmarking opportunity (see https://www.melbourne.vic.gov.au/sitecollectiondocuments/sustainable-development-goals.pdf). Through involvement in transnational networks such as the C40 network and 100 Resilient Cities alliance, Melbourne and other global cities are further working collaboratively on SDG challenges.

Conclusion

As an integrative, transformational agenda, the SDG agenda demands approaches that not only work across boundaries but that also connect existing efforts on one issue with existing efforts on another issue to identify synergies and tensions. It is working at the nexus of issues such as water, food, carbon, climate, and health that is critical for sustainable urban and regional futures. For this reason, the SDG agenda is not just one of many strategic agendas; it is a framework that demands a new way of working.

First, the SDG agenda is less an internal, voluntary choice than a shift in the sociopolitical landscape that it needs to adapt to in the current context of growth-led development and the climate emergency. *Second*, the SDG agenda builds on and gives new meaning and intensity to existing drivers for change such as the need for renewable energy systems and low-carbon cities. The SDG agenda pushes in the same direction but with more positive purpose than simply justifying a financial investment. *Third*, the SDG agenda is more than social change, or environmental change. Every aspect of towns, cities, and regions is part of the SDG agenda as expressed across the 17 goals, whether recognized as such or not.

The SDG agenda is a cross-cutting agenda necessary for addressing the sustainability of cities and settlements. This includes a focus on pathways toward urban transformation that creates liveable future cities for rapidly changing communities. This is also a focus on better engaging communities to shape the transformations needed to ensure cities and regions are sustainable, inclusive in their governance, and equitable in the context of climate change. More than just a list of problems, the SDG agenda is increasingly recognized as a social force – an influential, galvanizing driver of change across sectors. Cities and urban regions have the capacity to be leaders in SDG innovation – the question is, will they?

References

Brundtland, G. (1987). Report of the World Commission on Environment and Development: Our common future. United Nations General Assembly document A/42/427.

Freistein, K., & Mahlert, B. (2016). The potential for tackling inequality in the Sustainable Development Goals. *Third World Quarterly, 37*, 2139–2155.

Hope, J. (2021). The anti-politics of sustainable development: Environmental critique from assemblage thinking in Bolivia. Transactions of the Institute of British Geographers. Accessed on https://rgs-ibg.onlinelibrary.wiley.com/doi/pdfdirect/10.1111/tran.12409

Kaika, M. (2017). "Don't call me resilient again!" The new urban order as immunology…or what happens when communities refuse to be vaccinated with smart cities and indicators. *Environment and Urbanization, 20*(1), 89–102.

Moore, M., Riddell, D., & Vocisano, D. (2015). Scaling out, scaling up, scaling deep: Strategies of non-profits in advancing systemic social innovation. *Journal of Corporate Citizenship, 58*, p67–p84.

Moseley, W. (2018). Geography and engagement with UN development goals: Rethinking development or perpetuating the status quo? *Dialogues in Human Geography, 8*(2), 201–205.

Rickards, L., & Steele, W. (2019). Towards a Sustainable Development Goals transformation platform. Accessed on https://www.rmit.edu.au/research/our-research/enabling-capability-platforms/urban-futures/sdg-transformation-platform

Rickards, L., Steele, W., Kokshagina, O., & Moraes, O. (2020). Research impact as ethos. Accessed on https://next.rmit.edu.au/wp-content/uploads/2020/09/Rickards-et-al-2020-Research-Impact-as-Ethos-1.pdf

Sachs, W. (2017). The Sustainable Development Goals and Laudato si: Varieties of post-development. *Third World Quarterly, 38*(12), 2573–2587.

Steele, W., & Rickards, L. (2021). *The Sustainable Development Goals and higher education: A transformative agenda?* London/New York: Palgrave.

Sultana, F. (2018). An(other) geographical critique of development and SDGs. *Dialogues in Human Geography, 8*(2), 186–190.

United Nations (UN) (2015). Transforming our world: The 2030 Agenda for Sustainable Development. Accessed on https://sustainabledevelopment.un.org/content/documents/21252030%20Agenda%20for%20Sustainable%20Development%20web.pdf

UN General Assembly (2015). Transforming our world : The 2030 Agenda for Sustainable Development, 21 October 2015, A/RES/70/1. Available at: https://www.refworld.org/docid/57b6e3e44.html. Accessed 28 Feb 2022.

Sustainable Development Goals from an Urban Perspective

María Carmen Sánchez-Carreira[1] and Bruno Blanco-Varela[2]

[1]Department of Applied Economics, Faculty of Economics, ICEDE Research Group, CRETUS Institute, Universidade de Santiago de Compostela, Santiago de Compostela, Galicia, Spain

[2]Department of Applied Economics, Faculty of Economics, Universidade de Santiago de Compostela, Santiago de Compostela, Galicia, Spain

Synonyms

2030 Agenda for Sustainable Development; Agenda post-2015; Global Goals; Global Partnership; Global Social Contract; Millennium Development Goals; Poverty Reduction; Prosperity; Urban Development; Urban Sustainable Development

Definition

The Sustainable Development Goals (SDGs) are an initiative of the United Nations launched in 2015 within the framework of the 2030 Agenda for Sustainable Development. This agenda set universal goals that commit United Nations Member States to transform the world. This initiative attempts to achieve a sustainable development in the three dimensions intrinsic to the term sustainable: economic, social, and environmental. Thus, it refers to economic growth, social inclusion, and environmental protection for the people and the planet. According to the well-known Brundtland report (World Commission on Environment and Development 1987), sustainable development means to meet the needs of the present without compromising the ability of future generations to meet their own needs. Thus, SDGs mean an international commitment to face the social, economic, and environmental challenges of globalization, focusing on people, planet, prosperity, peace, and partnerships, under the principle of leaving no one behind. This "no one" refers to any person or actor but also to any place or geographical area. The SDGs aim at eradicating poverty, protecting the planet, and ensuring prosperity for all people as part of a new sustainable development agenda. They can be considered as a global interconnection between conflicts, poverty, inequalities, and the environment.

The 2030 Agenda for Sustainable Development set 17 SDGs, which are specified in 169 targets and 234 indicators. This agenda frames the international, national, regional, and local development policies until 2030.

Introduction

The Sustainable Development Goals arise as a result of the United Nations (UN) General Assembly held in 2015, where 193 UN Member States adopted the 2030 Agenda for Sustainable Development. This agenda establishes a set of goals, targets, and indicators of universal application in order to achieve sustainable development in the three dimensions intrinsic to the term (social, economic, and environmental). Thus, it defines 17 Sustainable Development Goals (SDGs), 169 targets, and 234 indicators. It can be considered as an instrument of universal development but also as an international action agenda for development. It is articulated as a new global social contract that leaves no one behind (United Nations General Assembly 2015). This includes people, actors, and agents of society but also territories and geographical areas around the world.

Table 1 summarizes these goals. As it can be observed, the 17 goals show a global character, several of which are clearly transversal aspects. Moreover, they are interrelated that suggest a holistic approach to propose action that helps to achieve them.

The SDGs do not emerge as a completely new process but are part of a process started with the Millennium Development Goals (MDGs). This initiative was adopted in New York in September 2000 aimed at committing nations to a global partnership to reduce extreme poverty. It is based on the work of the United Nations during the former decade. It was in force between 2001 and 2015. It was composed of 8 objectives, specified in 28 targets and 48 indicators. Table 2 presents the eight MDGs.

Although the initiative failed, because it has not reached the intended objectives, it has achieved some progresses, mainly on reduction of poverty and undernourished people and improvement in education, health (concerning HIV/AIDS, malaria, and tuberculosis, reduction of child mortality and maternal mortality), as well as drinking water access. Moreover, it contributes to put the focus on world challenges and problems and the need of global alliances and multilateral cooperation for development. In this sense, the MDGs can be considered the precedent of the current SDGs while showing that there is still a long way to be done. The funding needed to achieve the goals is critical.

While the MDGs focused on reducing extreme poverty in all its forms, the SDGs encompass a broader agenda that includes the social, environmental, and economic aspects of sustainable development. Another difference is that SDGs are more global, including all countries, while MDGs were intended for developing countries, supported by developed countries. SDGs go a step further, involving all countries in the world, regardless of their development level. It does not only concern governments but also the private sector and society. In this sense, they can be considered universal, multilevel, and multistakeholder. The global nature of the objectives involves all the countries that ratify the agreement. Moreover, SDGs introduce prosperity and peace as key axes. Table 3 compares the addressed axes in the former MDGs and the current SDGs. The 17 SDGs can be classified into 5 axes based on the nature of the goals as presented in Table 3: people, planet, prosperity, peace, and partnerships. The MDGs were articulated into eight goals in three thematic areas. Six of the eight correspond to the most social area: eradicate extreme poverty and hunger, achieve universal

Sustainable Development Goals from an Urban Perspective, Table 1 Sustainable Development Goals

SDG	Description	Number of targets	Number of Indicators
1 NO POVERTY	End poverty in all its forms everywhere	7	14
2 ZERO HUNGER	End hunger, achieve food security and improved nutrition and promote sustainable agriculture	8	13
3 GOOD HEALTH AND WELL-BEING	Ensure healthy lives and promote well-being for all at all ages	13	25
4 QUALITY EDUCATION	Ensure inclusive and equitable quality education and promote lifelong learning opportunities for all	10	11
5 GENDER EQUALITY	Achieve gender equality and empower all women and girls	9	14
6 CLEAN WATER AND SANITATION	Ensure availability and sustainable management of water and sanitation for all	8	11
7 AFFORDABLE AND CLEAN ENERGY	Ensure access to affordable, reliable, sustainable and modern energy for all	5	6
8 DECENT WORK AND ECONOMIC GROWTH	Promote sustained, inclusive and sustainable economic growth, full and productive employment and decent work for all	12	17
9 INDUSTRY, INNOVATION AND INFRASTRUCTURE	Build resilient infrastructure, promote inclusive and sustainable industrialization and foster innovation	8	12

(continued)

Sustainable Development Goals from an Urban Perspective, Table 1 (continued)

SDG	Description	Number of targets	Number of Indicators
10 REDUCED INEQUALITIES	Reduce inequality within and among countries	10	11
11 SUSTAINABLE CITIES AND COMMUNITIES	Make cities and human settlements inclusive, safe, resilient and sustainable	10	15
12 RESPONSIBLE CONSUMPTION AND PRODUCTION	Ensure sustainable consumption and production patterns	11	13
13 CLIMATE ACTION	Take urgent action to combat climate change and its impacts	1	1
14 LIFE BELOW WATER	Conserve and sustainably use the oceans, sea and marine resources for sustainable development	10	10
15 LIFE ON LAND	Protect, restore and promote sustainable use of terrestrial ecosystems, sustainably manage forests, combat desertification, and halt and reverse land degradation and halt biodiversity loss	12	14
16 PEACE, JUSTICE AND STRONG INSTITUTIONS	Promote peaceful and inclusive societies for sustainable development, provide access to justice for all and build effective, accountable and inclusive institutions at all levels	12	23
17 PARTNERSHIPS FOR THE GOALS	Strengthen the means of implementation and revitalize the Global Partnership for Sustainable Development	19	25

Source: Own elaboration based on the United Nations General Assembly (2015) and United Nations (2020)

primary education, promote gender equality and empower women, reduce child mortality, improve maternal health, and combat HIV/AIDS, malaria, and other diseases. Thus, in the case of the MDGs, this area groups together goals 1–5. The seventh of the MDGs was focused on the second axis, the planet, to ensure environmental sustainability; in the case of the SDG, it is expanded to five goals

Sustainable Development Goals from an Urban Perspective, Table 2 Millennium Development Goals

Goal	Description	Number of targets	Number of indicators
	Eradicate extreme poverty and hunger	3	9
	Achieve universal primary education	1	3
	Promote gender equality and empower women	1	3
	Reduce child mortality	1	3
	Improve maternal health	2	6
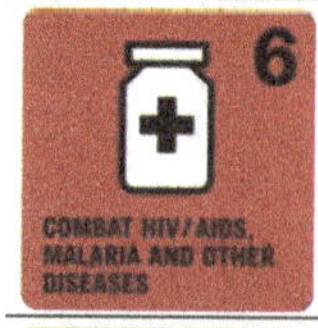	Combat HIV/AIDS, malaria and other diseases	3	10
	Ensure environmental sustainability	4	10
	Develop a global partnership for development	5	16

Source: Own elaboration based on the United Nations General Assembly (2000) and United Nations (2009)

(SDG6, SDG12, SDG13, SDG14, SDG15). The MDG8 focused on partnerships (develop a global partnership for development), as did goal 17 of the SDGs. In addition, the axis prosperity is expanded from SDG7 to SDG11 and focuses on economic sustainability.

In this context, SDGs arise as a real and global response to development and sustainability

Sustainable Development Goals from an Urban Perspective, Table 3 Comparison of addressed axes by MDG and SDG

Axis	MDG (2000–2015)	SDG (2015–2030)
People	MDG 1, MDG 2, MDG 3, MDG 4, MDG 5, MDG 6	SDG 1, SDG 2, SDG 3, SDG 4, SDG 5
Planet	MDG 7	SDG 6, SDG 12, SDG 13, SDG 14, SDG 15
Prosperity		SDG 7, SDG 8, SDG 9, SDG 10, SDG 11
Peace		SDG 16
Partnerships	MDG 8	SDG 17

Source: Own elaboration based on the United Nations and http://www.nu.org.bo/agenda-2030/transicion-de-los-odm-los-ods/

challenges of the planet, not only from reflection and previous experience of MDGs but also from the need for a more participatory dynamic. In this way, it is important to both know which actors participate (multi-stakeholder) and how they participate (multilevel). It also involves an unprecedented transformation starting the building of a new development model, given that it is a transformative, comprehensive, and inclusive agenda. In this multi-stakeholder and multilevel context, the criterion for decision-making is governance.

Funding is a critical issue to achieve the SDGs. It is estimated that annual investments of around $5–7 trillion are needed to fulfill the SDGs by 2030. The current investments are far from these values. It should be mentioned the creation of the UN SDG Fund in 2014, as an international mechanism to support sustainable development.

Relationship Between Cities and SDGs

Nowadays, cities are key players in economic development, acquiring relevance even in a context of globalization. Along the history, development has been accompanied by a continuous and increasing urbanization process. Thus, the majority of population lives in cities since 2007. The urban population accounts for 3.8 billion people worldwide in 2015, meaning a rate of urbanization of 54%, while only 37% (1.7 billion people) of the population lived in urban areas in 1970. The estimations rise urban population between 66% and 70% (around 6.5 billion people) in 2050 (United Nations 2018, 2019). It should be indicated that these estimations were carried out before the COVID-19 pandemic, which can alter the current configuration of urban and rural areas.

Population, employment, and economic activities tend to concentrate on certain areas, showing territorial disparities. Thus, different territorial agglomerations and spatial patterns arises from urban sprawl to megacities. Cities play a key role as economic drivers and places of services, culture, connectivity, and innovation. Besides being the main producers and consumer centers, cities are spaces where disparities emerge. Poverty, unemployment, slums, or socioeconomic inequalities emerge and increase in urban areas. Unsustainable consumption and production models coexist and have a decisive influence on climate change and environmental degradation. Moreover, cities accommodate most of the world enterprises, as well as informal businesses, provide markets for industry and employment, promote technological innovation, and support high housing density and efficient land use. Furthermore, unsustainable consumption and production models coexist and have a decisive influence on climate change and environmental degradation. Consequently, population, activities, culture, interactions, or environmental problems are mainly concentrated on urban environments. Sustainable development cannot be achieved without significantly transforming the way we build and manage our urban spaces (United Nations 2018, 2019; OECD 2020).

According to UN Habitat reports, cities currently represent around 80% of GDP, 60% of use of resources, 60% of energy consumption, 70% of waste, and 70% of global greenhouse gases (United Nations 2018, 2019). Cities occupy only 3% of the Earth's land, but the territory occupied by cities has increased 1.5 times more than the population over the last 20 years. Therefore, while the process of urbanization generates some of the greatest world development challenges, it also provides opportunities to advance sustainable development.

Population, productive activities, creative activities, employment, innovation, environmental problems, poverty, and inequalities are mainly concentrated on cities. Thus, cities are key areas to face the challenges of sustainability, and they also provide appropriate spaces for solutions and the promotion of inclusiveness, equity, and sustainable urban development.

Urban areas are the environments where problems arise and are also the most appropriate spaces for solutions. In this sense, Ban Ki-Moon, UN Secretary General during the 2007–2016 period, states that the battle for global sustainability will be won or lost in the cities. Thus, the importance of the local level in the SDG is highlighted, in particular the urban area.

In fact, cities and urban settlements are key players in the implementation and achievement of the SDGs. Within the SDG initiative, urbanization is so present that it is represented in one of the objectives, such as the SDG11 Sustainable Cities and Communities, which aims at more sustainable, inclusive, safe, and resilient cities. However, the actions of local level should not be limited to the SDG11. Figure 1 presents the links between the SDG11 and each of the SDGs. It shows the interaction among SDGs and the need and opportunity of local actions to implement and contribute to achieve any of the SDGs.

Apart from this specific goal (SDG 11), local level is crucial in the achievement of each of the SDGs. Thus, 92 targets and two-thirds of indicators are relevant for local governments. Table 4 classifies the number of these targets for each SDG. Thus, local level and specifically urban areas can act, implement, and contribute to the achievement of any of the SDGs.

SDGs in the Framework of the Urban Agenda

The previous section shows the relevance of urban areas for sustainable development and, in particular, for SDGs. A useful and guiding instrument for helping in the implementation of SDGs in urban areas is the framework of the New Urban Agenda. It is a non-compulsory international agreement, which generates a common view about the city adopted in the United Nations Conference on Housing and Sustainable Urban Development (Habitat III) held in Quito (Ecuador) in 2016. Thus, it is not a contract but a roadmap for public action. It is based on the previous work of the UN Habitat (the UN entity focused on urban development and future) and, in particular, the Action Plan (Habitat I) adopted in Vancouver 1976 and the Agenda Habitat (Habitat II) adopted in Istanbul in 1996. Thus, the declaration approved in Quito in 2016 is also known as Habitat III.

The global challenges, regardless of their social, economic, or environmental nature, are interrelated. Given that they mainly affect cities, urban development will only be possible with an integrated approach.

The Urban Agenda is a strategic process of reflection, definition, and transformation of the city. It can be related with the long trajectory and extensive work on planning made by cities. In this sense, the Urban Agenda provides a unified and comprehensive vision, unlike the fragmentation of territorial or sectorial plans. Apart from international urban agendas, each level of government can develop its urban agenda, whether local, regional, or national, creating a strategic planning waterfall as reference framework in order to fulfill the commitments of the 2030 Agenda. In the supranational area, it should be highlighted the Urban Agenda for the EU (also known as the Pact of Amsterdam), launched in 2016, as an integrated and coordinated approach to deal with the urban dimension of EU in legislation and policies. It focuses on the three pillars of better regulation, better funding, and better knowledge.

The term SDG localization refers to the process of subnational contexts in the achievement of the 2030 Agenda, from the setting of goals and targets to determining the means of implementation and using indicators to measure and monitor progress. Although the SDGs were agreed by national governments, regional and local levels can help to

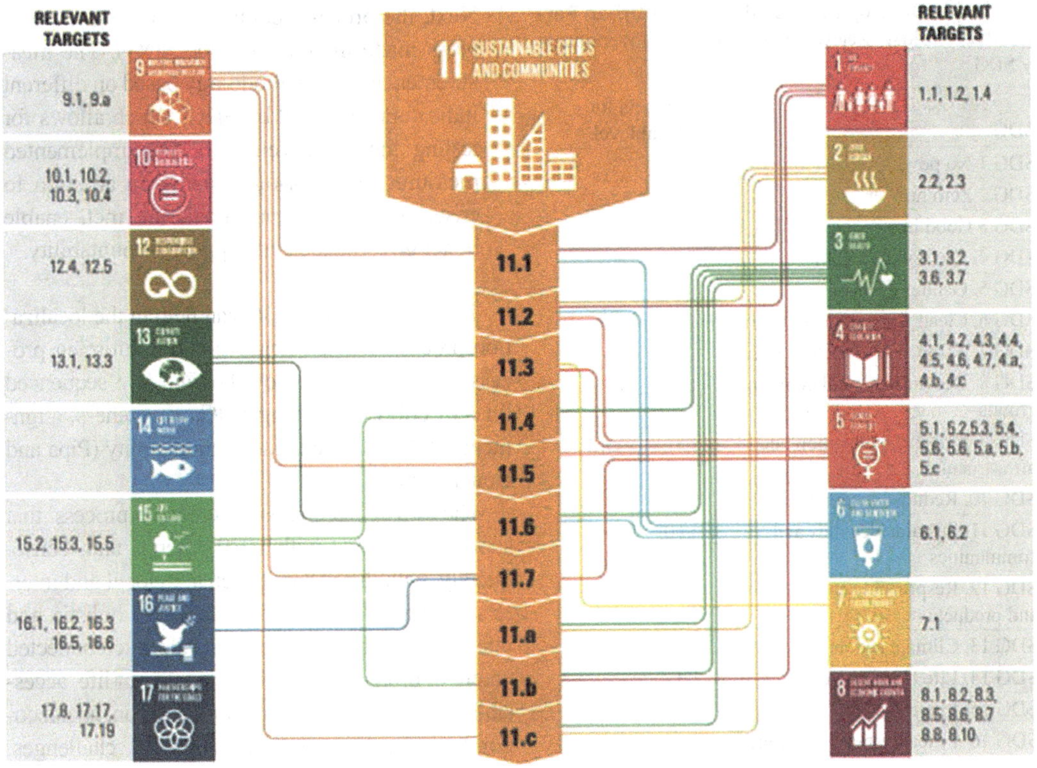

Sustainable Development Goals from an Urban Perspective, Fig. 1 Links between SDGs and SDG11. (Source: United Nations (2018, p. 10))

their achievement. Thus, local and regional governments can support the SDGs through action from a bottom-up and participatory approach, and at the same time, the SDGs can provide a framework for local development policy (Global Task Force of Local and Regional Governments 2016; United Cities and Local Governments 2019).

As it happens with the SDGs, the Urban Agenda does not emerge as a totally new initiative. It has a clear precedent in the Agenda 21, which was approved in the United Nations Conference on Environment and Development (UNCED) held in Rio de Janeiro (Brazil) in 1992. It underlines the role of local governments for achieving a sustainable development, giving rise to the implementation of action plans at the local level, considering environmental, economic, and social issues.

Another precedent of the Urban Agenda is the Integrated Sustainable Urban Development Strategy (ISUDS). This instrument is highlighted in the European Union, due to the implementation of the Europe 2020 Strategy in urban development. It should be indicated that the European Union has a long path of developing an urban model, since the end of the 1980s with the definition of an urban policy and specific founding for urban initiatives. This European tradition is based on the principles of integrity and sustainability in the urban planning and managing.

The United Nations-Habitat Urban Agenda is based on a collaborative and co-responsible method. In this way, the incorporation and planning of the SDGs in the local sphere requires a four-step path (Kanuri et al. 2016):

1. For the implementation of the SDGs in the urban context, an inclusive and participatory process of localization of SDGs should take place. This includes raising awareness of the SDGs at the local level, creating a framework for discussion and stakeholder involvement.

Sustainable Development Goals from an Urban Perspective, Table 4 Relevant targets for local governments by SDG

SDG	Targets for local level
SDG 1 No poverty	4
SDG 2 Zero hunger	4
SDG 3 Good health and well-being	6
SDG 4. Quality education	7
SDG 5. Gender equality	7
SDG 6. Clean water and sanitation	7
SDG 7. Affordable and clean energy	3
SDG 8. Decent work and economic growth	6
SDG 9. Industry, innovation and infrastructure	3
SDG 10. Reduced inequalities	5
SDG 11. Sustainable cities and communities	10
SDG 12. Responsible consumption and production	7
SDG 13. Climate action	3
SDG 14. Life below water	4
SDG 15. Life on land	5
SDG 16. Peace, justice and strong institutions	5
SDG 17. Partnerships for the goals	6

Source: Own elaboration based on the United Nations (2018), United Cities and Local Governments (2019), and OECD (2020)

Concerning decision-making, which should be oriented to the priority of sustainable development, political leadership and governance are key aspects.

2. Once the SDGs are localized within the urban environment, the second step consists of setting the local agenda for the SDGs. Locating SDGs allows for the inclusion of the diversity of people and places that make up the local environment. The purpose of the SDGs is to leave no one behind but through realistic and publicly supported decision-making.
3. The third step involves planning for implementation of the SDGs by 2030. To do this, the objectives need to take a long-term, multisectoral perspective, with adequate financial resources and partnerships among the stakeholders involved.
4. Next, the process requires monitoring the progress made in achieving the SDGs. The measurement uses a methodology based on different indicators classified by SDG, which allows for testing the effectiveness of the implemented measures. Monitoring systems, in addition to ensuring that designed goals are met, enable better governance and support accountability.

The cycle of local SDG can help in the localization process. It is an iterative and reinforcing process, which includes the following five sequenced lines of effort concerning SDGs: awareness, alignment, analysis, action, and accountability (Pipa and Bouchet 2020).

The urban agenda is a strategic process that should be aligned with 2030 Agenda and, consequently, with the SDGs, which are global and interrelated. Therefore, it should follow a holistic and integrated approach. Moreover, the interconnected nature of urban issues, such as quality of life, accessibility, mobility, economic revitalization, socioeconomic inequalities, or environmental challenges, suggests adopting a comprehensive approach to be more effective in addressing them. The integrated approach involves a multilevel, multisectoral, multistakeholder, collaborative, and participatory approach, where governance and leadership are crucial. Thus, some relevant issues for the urban agenda are the following ones:

- The multisectoral nature means that the urban issues and policies have different sectoral effects, as well as on the environmental, social, and economic dimensions. Thus, the interrelations and inter-sectoral effects need to be considered in the definition, implementation, and evaluation of urban agendas and policies. In this sense, policies should not be carried out on a summative basis but rather from a holistic perspective, focused on the interrelations.
- The multilevel character refers to the need of urban actions executed by the different levels of government: local, regional, national, and supranational. In this sense, it is relevant to take into consideration that urban policies are not limited to the municipal competences but also exceeds them. Thus, urban policy can be

defined as a whole set of measures at different levels that are addressed to cities (Van den Berg et al. 1998). This multilevel governance requires cooperation and vertical and horizontal coordination. The collaboration can generate synergies, optimize resources, and avoid overlaps and inefficient duplications. Moreover, it should be distinguished between urban policy, which means a policy that specifically addressed cities and policies that have urban impact, that is to say, that affect cities, such as transport or housing policies. The adoption of leverage policies can contribute to accelerate the implementation and achievement of the SDGs, based on the generation of synergies among SDGs and multiplier effects across cross-cutting instruments.

- The multi-stakeholder approach is related with the need of including and integrating all kinds of urban actors, whether they are public, private, or civil society actors. It requires a true participatory process, with high involvement and engagement of all kinds of stakeholders to co-create urban policies. Thus, it means a bottom-up approach, where the decisions are made in a collaborative way by the local community. The purpose is to build a consensus or widely shared view, strong partnerships, and networks to support and contribute to the implementation and achievement of the SDGs. In this sense, it should be highlighted the limited awareness and knowledge about SDGs by citizens, from the students to politicians. Different surveys find that between 28% and 45% of people have heard about the SDGs, although there are notable differences depending on the country. Young people present the higher awareness. It should be underlined that awareness is not the same as having a good knowledge about the SDGs (OECD Development Communication Network 2017). It indicates the need of including SDG training in different fields, from formal education to training for public officers and politicians, among other groups. Nussbaum (2002) underlines the relevance of having citizens trained and in global values. Concerning formal education, the authors of this chapter have implemented an initiative to learn about the SDGs in the subject of Regional and Local of the Economics Degree from the University of Santiago de Compostela (USC) (Galicia, Spain). It follows the learning-service methodology, which consists of learning by doing a community service. The main aim is to know the implementation of SDGs in the city of Santiago de Compostela, where the USC is located, and propose actions to a better implementation and achievement of all the SDGs.
- It also involves the reinforcement of territory, because any area cannot be left behind, regardless of its size, geographical position, rural or urban character, or specialization.

The complexity of the multidimensional approach of the urban agenda requires good governance and leadership. Thus, critical issues are the definition of priorities, the engagement with priorities, the alignment of local plans with SDGs, the mobilization of local resources, the building of institutional capabilities and leadership, and the promotion of ownership and co-responsibility for implementing strategic projects (Global Task Force of Local and Regional Governments 2016; United Nations 2017; United Cities and Local Governments 2019; Pipa and Bouchet 2020).

The urban agendas should lead to improvements at least in the areas of regulation and planning, financing, knowledge, governance and transparency, and participation. Other critical aspect in the urban agenda as well as in planning is the learning from the good practices and failures, as well as the building of capabilities for improving the design and implementation of local strategies and policies. Successful and previous experiences can serve as a basis for SDG localization, considering the specificities of each local context and one size does not fit all.

Although the New Urban Agenda is an instrumental tool to achieve the SDGs and, in particular, sustainable cities, it faces several challenges concerning the implementation and achievement of the results. In this sense, Saui (2017) identifies three main pitfalls in the implementation of sustainable cities: the idea of city as a business, the oversimplification of urban complexity, and the search for the ideal community.

Finally, different tools can be useful in the process of implementing the SDGs in the local level. Some of the most outstanding are the following (United Nations 2017; Global Task Force of Local and Regional Governments 2016; United Cities and Local Governments 2019; Pipa and Bouchet 2020; United Nations-Habitat 2020):

- Voluntary Local Review (VLR) is a tool for local governments created by the city of New York in 2018 to showcase achievements and challenges in reaching the Sustainable Development Goals. The Voluntary Local Review Declaration was presented in the United Nations General Assembly in 2019. VLRs form a framework to compare, exchange, evaluate, and learn what is already being done and how transformative impacts can be achieved. This tool can be used by cities of any size. Nowadays, 17 cities from different countries and continents signed it and committed to report on SDGs (United Nations 2017; Global Task Force of Local and Regional Governments 2016; United Cities and Local Governments 2019; Pipa and Bouchet 2020).
- Local 2030-Localizing the SDGs is a network that supports the delivery of the SDGs. The online platform provides tools, experiences, new solutions, and guides to support SDG localization. Special attention deserves the toolbox, which includes practical, concrete, and adaptable mechanisms and instruments to support the development, implementation, monitoring, and review of local SDGs actions.
- SDG Cities Flagship Program is a UN Habitat initiative that enables cities to accelerate SDG delivery, addresses key challenges, and sets aligned priorities. It follows a systematic methodology that links data, strategic planning, capacity strengthening, and impact investment to achieve large-scale impact. It includes constructive feedback from stakeholders, networking and exchange of global experiences and advice, citizen engagement, certification of SDG cities, and an investment facility. Moreover, the SDG City Wheel groups the SDG targets in three transformative commitments of the New Urban Agenda: sustainable urban development for social inclusion and ending poverty, sustainable and inclusive urban prosperity and opportunities for all, and environmentally sustainable and resilient urban development. It also provides support in building city capacities in key areas of policy, planning, governance, and local financing. Each city is encouraged to establish an urban lab to develop projects based on the urban strategies to accelerate SDG achievement and a multistakeholder process to identify challenges and priorities. The analysis of gaps will guide the interventions addressed to achieve the SDGs, which can be grouped into four types: enablers, capacity gaps, public infrastructure, and economic initiatives (United Nations-Habitat 2020).
- City Prosperity Initiative (CPI) is the global monitoring platform for SDGs through which UN Habitat supports cities over the world to monitor urban development. This framework is developed to formulate evidence-based decisions, policy recommendations, and urban data for cities. CPI helps identify, quantify, evaluate, monitor, and report on progress made toward the SDGs by cities and countries in a structured way. It has assisted more than 530 cities since its creation in 2012. It includes the CPI Index, which integrates and measures all indicators of SDG11 and a selection of other SDG indicators (United Nations 2018).
- New Urban Agenda Platform is a platform developed by UN Habitat for exchanging knowledge, best practices, data, and learning about the implementation of the New Urban Agenda and SDGs.
- Capital Advisory Platform is a partnership between UN Habitat and the Global Development Incubator that attempts to bridge the gap between sustainable urban development projects and investors. It aims at mobilizing private capital toward sustainable urban development through translating the urban planning needs of governments into a pipeline of investable projects for private asset allocators.
- The Global Competitiveness Report (2019–2020) introduces as a novelty the measurement

of the SDG implementation and progress from the urban perspective and, in particular, concerning SDG11 (Kamiya et al. 2020). It allows identifying the leader cities in the implementation of the SDGs. Thus, the top 20 world cities are New York, London, Tokyo, Paris, Singapore, San Francisco, Los Angeles, Boston, Dallas, Amsterdam, Atlanta, Sydney, Chicago, Seattle, Dublin, Philadelphia, Taipei, Houston, Copenhagen, and Melbourne. This shows an uneven geographical distribution, due to most of the cities concentrated on America, specifically in the USA (ten cities) and Europe (five cities). Among the top 200 cities, North America and Europe are highest with 68 and 67 cities, respectively. They are followed by Asia (56). There is only one city from Africa and six in Oceania (Kamiya et al. 2020).

In addition to these global tools, there are other specific to some regions, which are also relevant for the local level, such as the European multi-stakeholder platform on SDGs or the European Reference Framework for Sustainable Cities (RFSC).

Conclusions

The Sustainable Development Goals (SDGs) attempt to build a better world, based on sustainable development and the 2030 Agenda launched by the United Nations in 2015. It follows the Millennium Development Goals and it should not be end in 2030, due to the unfinished nature of development, understood as process to improve quality of life. The SDGs are multidimensional and embrace 17 goals. Although progress is being made on the achievement of SDGs, it does not seem enough to face the challenges.

Cities are key players to implement and achieve the SDGs. Apart from the SDG11 focused on cities, the other 16 goals can be accelerated in cities. In fact, urban areas face many challenges linked to SDGs but also provide an opportunity to overcome them and build a new and more sustainable development model. The current pandemic situation reinforces the challenges and opportunities for cities, because they are the hardest impacted areas. Thus, the achievement of SDGs depends mainly on cities, especially on their ability to adapt and be resilient to a changing global environment.

A good tool to drive the implementation of SDGs in cities is the New Urban Agenda. It follows an integrated approach, based on bottom-up, multilevel, multisector, and multi-stakeholder process to find and exploit opportunities and synergies. Planning, identification of priorities, alignment with existent strategies, leadership, governance, and mobilizing human and financial resources are critical issues. Monitoring is also a key aspect as tool to improve the implementation and source of learning, as well as leveraging synergies and cooperation. However, the focus should not be in gathering a lot of indicators but also on relevant ones that help to make decisions and improve. Learning and building of capabilities are relevant to improve the next urban planning processes, as well as the effective cooperation in a multidimensional approach. In this regard, work already done in the past is essential for the implementation of SDGs in cities. On the one hand, it is necessary to look at previous experiences, such as Agenda 21. On the other hand, it is necessary to focus on the analysis of good experiences and good practices in the territories that are aligned with the SDGs. In addition, the singularities of each territory and their resources and capabilities should be considered.

Cross-References

▶ Future of Urban Governance and Citizen Participation

References

Global Task Force of Local and Regional Governments. (2016). *Roadmap for localizing the SDGs: Implementation and monitoring at subnational level*. Global Taskforce of Local and Regional Governments, UN-Habitat and UNDP. Available at: https://www.

uclg.org/sites/default/files/roadmap_for_localizing_the_sdgs_0.pdf

Kamiya, M., Pengei, N., et al. (2020). *Global urban competitiveness report (2019–2020) the world: 300 years of transformation into city*. Nairobi/Beijing: United Nations Habitat, National Academy of Economic Strategy.

Kanuri, C., Revi, A., Espey, J., & Kuhle, H. (2016). *Getting started with the SDGs in cities*. Sustainable Development Solutions Network & German Cooperation. https://faud.unc.edu.ar/files/Cities-SDG-Guide.pdf

Nussbaum, M. (2002). Education for citizenship in an era of global connection. *Studies in Philosophy and Education, 21*, 289–303.

OECD Development Communication Network. (2017). *What people know and think about the Sustainable Development Goals*. Selected findings from public opinion surveys compiled by the OECD Development Communication Network (DevCom). Available at: https://www.oecd.org/development/pgd/International_Survey_Data_DevCom_June%202017.pdf

OECD. (2020). *A territorial approach to the sustainable development goals: Synthesis report* (OECD urban policy reviews). Paris: OECD.

Pipa, T., & Bouchet, M. (2020). *Next generation urban planning enabling sustainable development at the local level through voluntary local reviews (VLRs)*. Washington: Brookings.

Saui, V. (2017). The three pitalls of sustainable city: a conceptual framework for evaluating the theory-practice gap. *Sustainability, 9*(12), 2311.

United Cities and Local Governments. (2019). *Towards the localization of the SDGs. Local and regional government's report to the 2019 high level political forum*. Barcelona: United Cities and Local Governments.

United Nations. (2009). *The millennium development goals report 2009*. United Nations, Department of Economic and Social Affairs.

United Nations. (2017). *New urban agenda 2017*. Quito: United Nations.

United Nations. (2018). *SDG 11 synthesis report 2018: Tracking progress towards inclusive, safe, resilient and sustainable cities and human settlements*. Nairobi: United Nations.

United Nations. (2020). *The sustainable development goals report*. New York: United Nations.

United Nations. (2019). *World urbanization prospects: The 2018 revision*. New York: United Nations, Department of Economic and Social Affairs, Population Division.

United Nations General Assembly. (2000). United Nations Millennium Declaration, Resolution Adopted by the General Assembly, 18 September 2000, A/RES/55/2. Available at: https://www.refworld.org/docid/3b00f4ea3.html. Accessed 17 Oct 2020.

United Nations General Assembly. (2015). *Transforming our world: the 2030 Agenda for Sustainable Development*. A/RES/70/1. Available at: https://www.refworld.org/docid/57b6e3e44.html

United Nations-Habitat. (2020). *SDG cities flagship programme 5*. Nairobi: UN Habitat.

Van den Berg, L., Braun, E., & Van Der Meer, J. (1998). *National urban policies in the European Union*. London: Routledge.

World Commission on Environment and Development. (1987). *Our common future*. Oxford: Oxford University Press.

Further Reading

C40 Cities Climate Leadership Group: https://www.c40.org/

Cities Alliance: https://www.citiesalliance.org/

City Prosperity Initiative: https://unhabitat.org/programme/city-prosperity-initiative

Eixo Atlantico. (2018). Empowering cities: Urban Agenda of the Eixo Atlantico. Eixo Atlántico do Noroeste Peninsular. Available at: https://www.eixoatlantico.es/en/listado-publicaciones/3481-empowering-cities-urban-agenda-of-the-eixo-atlantico. This urban agenda is illustrative because it includes cities of different size of a cross-border Euro-region.

European Urban Knowledge Network (EUKN): https://www.eukn.eu/

Global Task Force of Local and Regional Governments: www.global-taskforce.org

Local 2030-Localizing the SDGs: https://www.local2030.org/

Local Governments for Sustainability (ICLEI): https://www.iclei.org/

Network of Regional Governments for Sustainable Development (NRG4SD): https://sustainabledevelopment.un.org/partnership/?p=1585

SDG Cities: https://unhabitat.org/programme/sustainable-development-goals-cities

SDG Knowledge Hub: https://sdg.iisd.org/

UCLG Forum on Intermediary Cities: https://intermediarycities.uclg.org/en/

UN Habitat New Urban Agenda: http://habitat3.org/the-new-urban-agenda/

United Cities and Local Governments. *The sustainable development goals: What local governments need to know*. Available at https://www.uclg.org/sites/default/files/the_sdgs_what_localgov_need_to_know_0.pdf

United Cities and Local Governments (UCLG): www.uclg.org

United Nations Habitat: https://unhabitat.org/

United Nations-Habitat. (2015). *International guidelines on urban and territorial planning*. Nairobi: ONU.

United Smart Cities: https://sustainabledevelopment.un.org/partnership/?p=10009

URBACT: https://urbact.eu/

Urban Agenda Platform: https://www.urbanagendaplatform.org/

Urban SDG: http://urbansdg.org/

Urban SDG Knowledge Sharing Platform: http://sdghelpdesk.unescap.org/node/301

Urban Sustainability Exchange (USE): https://use.metropolis.org/

Voluntary Local Reviews: https://sdgs.un.org/topics/voluntary-local-reviews

Sustainable Development Goals in Relation to Urban and Regional Development in Japan

Karine Tollari
Japan Local Government Centre, London, UK

Ever since the adoption of the Sustainable Development Goals (SDGs) in September 2015, member states of the UN have strived to implement the framework in an international effort to respond to cross-border challenges. Although the SDGs were negotiated and agreed upon by national governments, the Organization for Economic Co-operation and Development has estimated that 65% of the 169 targets which compose the 17 goals will require implementation by local authorities in order to be met.

Japan, an advocate of international cooperation for development since as early as the 1990s when it proposed "International Development Goals," has shown a strong commitment to those goals from national to regional level.

National Foundations for Regional SDG Strategies

Following the adoption of the SDGs, the Japanese government set up an "SDGs Promotion Headquarters" in May 2016. It is led by the Prime Minister and includes all ministers in order to ensure cross-ministerial cooperation and a comprehensive implementation of SDGs in national policies. In December of the same year, it finalized the "SDGs Implementation Guiding Principles" (revised in 2019), which were the fruit of the collaboration of a wide variety of stakeholders (government, private sector, academia, civil society...) and define the country's policy and priorities in regards to the implementation of the 2030 Agenda (Fig. 1).

Sustainable Development Goals in Relation to Urban and Regional Development in Japan, Fig. 1 The Japanese government's logo for SDGs Promotion. To be allowed to use it on their website or marketing material, companies or organizations must submit an application to the Ministry of Foreign Affairs describing their efforts toward realizing SDGs

SDGs Awards

In order to raise awareness of SDGs and foster initiatives furthering their advancement, in 2017 the SDG Promotion Headquarters launched the SDGs Awards. The program rewards local government bodies, companies, and organizations based in Japan which have made outstanding efforts toward the realization of SDGs either domestically or internationally, becoming agents of change in their communities.

Japan Food Ecology Center, Inc., which won the award in 2018, aims to achieve sustainable farming and create recycling networks by producing "eco-feed," a type of livestock feed created using recycled food waste. The waste is collected from local supermarkets and food companies and after being turned into eco-feed is delivered to 15 pig farms in the area. According to their estimates, the center receives approximately 33 tons of food waste per day, which after being processed and fed to livestock, enables the production of 40 tons of food per day.

Uomachi Shopping Street Association, located in Kitakyushu, won in 2019 for its initiative to become the first "SDG shopping street" which involved measures such as installing solar panelling; refurbishing empty spaces into restaurants, community areas or hotels; offering seminars to train its personnel to respect the environment; leading campaigns to eliminate food waste and promote local production; and organizing "wheelchair experiences" to raise awareness of the challenges faced by disabled people.

SDGs Future City Initiative

The SDGs Future City Initiative is another project developed by the Japanese government which supports the implementation of SDGs at an urban level by promoting cities which aim to be sustainable, inclusive, environmentally friendly, and resilient in the face of natural disasters.

The Future City Initiative was originally set up in response to increasing levels of urbanization around the world and other concomitant challenges, such as regional revitalization and an aging population. Although it started in 2010 and thus predates the adoption of the SDGs, since the concepts of sustainability in both policies coincided, the Future City Initiative was reformulated to include an SDG framework.

Since June 2018, the Japanese government has been annually calling for local authorities promoting SDGs to submit their proposals. Every year, approximately 30 are chosen to become "Future Cities," out of which 10 are designated "SDGs Model Cities." These are offered financial support to carry out their SDGs strategies and serve as inspiration for other local authorities.

In addition, cities are encouraged to set up local governance institutions conducive to the implementation of SDGs, such as a municipal "SDGs Headquarters" led by the mayor to coordinate the city's effort to carry out their program, which acts as a local counterpart to the Prime Minister's SDGs Promotion Headquarters.

An Example of an SDGs Model City: Shimokawa Town in Hokkaido

The town of Shimokawa was recognized in 2018 for its efforts to establish a sustainable forestry industry, promote energy self-sufficiency, and address the challenges linked to its aging population. From 2010 to 2016, Shimokawa's thermal energy self-sufficiency rate increased from 9% to 49% by using forest biomass, while cutting its CO_2 emissions by 18%. Among other measures, it raised awareness of SDGs via seminars and articles in its local magazine and it started a zero-emission timber process which attracted new businesses (for instance, aromatherapy production from fir needles). In terms of social improvements, the town launched a revitalization project and built a compact town with a residents' center and services nearby for the benefit of its elderly community, who had become isolated in an underpopulated village 10 km outside the town center.

The SDGs Future City Initiative also encourages actors from the private sector, academia, and the civil society to get involved in making the SDGs a reality. As seen in Model City Kitakyushu, this may be accomplished through the creation of a local "SDGs Council" whose members provide advice in regards to SDGs implementation, or through a municipal "SDGs Club" where all citizens can take part in activities aiming to raise awareness of the goals and foster collaboration.

Japan currently has several projects planned during the UN's "decade of action" leading to 2030, which will showcase its urban and regional implementation of SDGs: not only will the country be presenting its second Voluntary National Reviews at the High-Level Political Forum 2021, and it will also be holding the SDG Global Festival of Action (held online from Japan). In addition, the World Expo 2025 will be held in Osaka, which is also an SDG Model City. By choosing "Designing Future Society for Our Lives" as the theme for the expo, Japan hopes to bring attention to SDGs and inspire other countries to take more decisive steps toward their achievement.

Cross-References

► Sustainable Development Goals and Urban Policy Innovation
► Sustainable Development Goals from an Urban Perspective

Bibliography

Abe, S. (2015). *Statement at the United Nations sustainable development summit by the Prime Minister of Japan*. https://www.mofa.go.jp/policy/oda/sdgs/pdf/000198338.pdf. Last accessed 25 Dec 2020.

Future Cities Concept, Future City EcoModel City government website (in Japanese). https://future-city.go.jp/about/futurecity/. Last accessed 25 Dec 2020.

Japan Science and Technology Agency. (2018, May). *Book of Japan's practices for SDGs: Creating shared value by STI, business and social innovation* (p. 38). https://www.jst.go.jp/sdgs/pdf/sdgs_book_en_2018_2.pdf. Last accessed 25 Dec 2020.

Japan's SDGs Action Platform, Ministry of Foreign Affairs of Japan website (in Japanese). https://www.mofa.go.jp/mofaj/gaiko/oda/sdgs/award/index.html. Last accessed 25 Dec 2020.

Masuo, M. (2019, May). Kitakyushu city's SDGs indicators and ideas towards the common indicator framework of cities and regions under the OECD programme. At *OECD Working Parties on Urban Policy and Territorial Indicators – Joint Workshop* [Presentation]. https://www.oecd.org/regional/sdg-localised-indicator-framework.htm. Last accessed 25 Dec 2020.

Ministry of Foreign Affairs of Japan. (2019, January). *Japan's efforts for promoting the SDGs, January 2019*. https://sustainabledevelopment.un.org/content/documents/21603JAPANS_EFFORTS_FOR_PROMOTING_THE_SDGS.PDF. Last accessed 25 Dec 2020.

Ministry of Foreign Affairs of Japan, [Channel: 外務省 / MOFA]. (2020, July 17). *【外務省×SDGs】「行動の10年」できることから始めよう!*(“*‘Decade of Action’ Start from what we can*”) [Video]. YouTube. https://www.youtube.com/watch?v=DD4QtkZWkYw&feature=youtu.be. Last accessed 25 Dec 2020 .

Ministry of Foreign Affairs of Japan. (2020, November). *Japan's efforts for achieving the SDGs*. https://www.unescap.org/sites/default/files/1-3.%20Iwasaki%20Tetsuya.pdf. Last accessed 25 Dec 2020.

OECD Urban Policy Reviews. (2020, February). *A territorial approach to the sustainable development goals, synthesis report*. https://www.oecd.org/regional/a-territorial-approach-to-the-sustainable-development-goals-e86fa715-en.htm. Last accessed 25 Dec 2020.

Osaka Prefectural Government. (2020, March). *Osaka's vision for SDGs*. http://www.pref.osaka.lg.jp/attach/35381/00000000/Vision.pdf. Last accessed 25 Dec 2020.

SDGs for Regional Revitalization and the SDGs Future City Initiative, Japan's Office for the Promotion of Regional Revitalization, Cabinet Office website (in Japanese). https://www.kantei.go.jp/jp/singi/tiiki/kankyo/index.html. Last accessed 25 Dec 2020.

Shimokawa Town and the Institute for Global Environmental Strategies. (2018). *Shimokawa Town the sustainable development goals report – The Shimokawa challenge: Connecting people and nature with the future*. https://www.iges.or.jp/en/pub/shimokawa-town-sustainable-development-goals/en. Last accessed 25 Dec 2020.

Suginaka, A. (2012, October 15). *Future city initiative*. Sustainable Development Goals Partnership Platform. https://sustainabledevelopment.un.org/partnership/?p=1049. Last accessed 25 Dec 2020.

Sustainable Development Solutions Network Japan, UNSDSN website. https://www.unsdsn.org/japan. Last accessed 25 Dec 2020.

What is the Japanese Government's Logo for SDGs Promotion? Application process and advice, SDG Promotion Center for the Chubu region website (in Japanese). https://chubu-sdgs.com/sdgs-japanmark/. Last accessed 25 Dec 2020.

Sustainable Development is Fair

► Sustainable Development and Responsible Tourism: The Grijalva-Usumacinta Lower River Basin

Sustainable Economies

► Policies for a Just Transition

Sustainable Food Systems

► Formulating Sustainable Foodways for the Future: Tradition and Innovation

Sustainable Logistics

► Sustainability Transition and Climate Change Adaption of Logistics

Sustainable Urban Development

► Social Urbanism: Transforming the Built and Social Environment

S

Sustainable Urban Mobility

Interventions, Key Measures and Solutions, Actors, and Opportunities

Oliver Lah
Wuppertal Institute for Climate, Environment and Energy, Berlin, Germany
Urban Electric Mobility Initiative (UEMI) a UN-Habitat Action Platform, Berlin, Germany

The transport sector plays a major role in achieving the goals specified in the Paris Agreement, the Sustainable Development Goals, and the New Urban Agenda. Along with the provision of essential services to society, transport is also a significant contributor to the economy. However, it is also at the core of a number of major sustainability challenges, namely, climate change, air pollution, traffic safety, energy security, and resource efficiency. This section of the *Palgrave Encyclopedia of Urban and Regional Futures* will present a summary of recent literature on climate change mitigation actions relevant to the local and national level, opportunities for synergies between sustainable development and climate change objectives, and governance and institutional issues influencing measures' implementation.

The transport sector has an enormous and as yet largely untapped and cost-effective greenhouse gas emissions mitigation potential. There is a wide range of tested solutions that are readily available (see IPCC 2014; IEA 2020 for an overview), yet there is still a lack of implementation of measures, which is affected by technical and economic issues, but more importantly political and institutional barriers that affect countries' divergent progress in this area. Identifying institutional and governance barriers in the process of adopting low-carbon transport measures is relevant for all countries to create a path to a 1.5 °C stabilization scenario. The analysis presented here provides an overview on the feasibility, in political terms, of the implementation of a comprehensive strategy to decarbonize the transport sector, focusing on major industrialized and emerging economies, which will, in turn, provide broader insights on the prospects of decarbonizing the sector.

The analysis suggests that success for sustainable transport policies relies on two factors:

- An integrated approach to policies, combining a range of measures
- Continuity in policies and politics to engender a sense of stability for all actors

With its influence on both economics and the environment, the transport sector is a particularly interesting case for climate change policies. As such, transport-sector energy-efficiency polices will be used as an example of the differences in policy-making in different institutional frameworks.

This section of the *Palgrave Encyclopedia of Urban and Regional Futures* presents an overview of options for the decarbonization of the transport. The potential of climate change mitigation pathways is outlined, along with policy and institutional aspects related to these pathways' feasibility. This combined quantitative and qualitative analysis aims to synthesize recent papers on the subject and draw conclusions for future research.

Transport Sector Decarbonization: Policies, Co-benefits, and Coalitions

Transport is a complex and multifaceted activity, central to the economy and peoples' lives. Consequently, unintended consequences – both positive and negative – are common as policies seldom affect only one aspect of the sector (or other related sectors). For example, air quality measures can have adverse effects on fuel efficiency, or biofuels may affect land-use and food prices. Carefully selecting policies, and combining several in one package, is therefore vital to avoid unintended consequences, at least, and potentially even generate synergies and co-benefits. Moreover, such packages can provide the basis for coalitions of actors that can align different veto players.

This article provides an overview on the key elements of a low-carbon, sustainable development pathway for the transport sector:

- **Trends and drivers**: What are the transport sector's key trends and drivers? How large is the potential for GHG mitigation, and how can mitigation pathways contribute to sustainable development and broad coalitions?
- **Potential for co-benefits**: Which policies is a sustainable, low-carbon pathway for transport contingent upon, which barriers must be overcome to implement them, and what considerations are relevant to an integrated policy strategy's influence on political coalitions?
- **Coalitions and institutions**: What aspects of an institutional framework allow political stability and continuity, fostering the adoption of and durable support for sustainable transport strategies?

Trends and Drivers

Even according to very optimistic scenarios, transport sector GHG emissions are set to stay at current levels (Fulton et al. 2013; Harvey 2013), partly because increasing mobility demand outpaces efficiency gains. Even assuming a substantial adoption of higher-efficiency vehicles and some modal shift, transport CO_2 emissions in 2050 will remain at 2015 levels (≈7.5 gigatons of CO_2). If, however, current trends continue unabated, transport sector greenhouse gas emissions are forecast to double by 2050 (IPCC 2014). Setting the transport sector on a low-carbon development pathway is essential to stabilize global warming at well below 2 °C, the internationally agreed target under the United Nations Framework Conventions on Climate Change (UNFCCC). To meet this target, developed countries must rapidly decarbonize their transport sector over the coming decades (−80% by 2050), while developing and emerging countries must curb emissions growth (+70% by 2050), both of which will require substantial policy action. Further analysis is required to better understand the detailed differences between the pathways to 2 °C and 1.5 °C for specific sectors, including transport: this is the subject of an upcoming IPCC Special Report.

The scientific literature on decarbonization scenarios is not delivering a significantly changed message since the strengthened Paris Agreement targets were adopted (Creutzig 2016). However, according to several authors' analysis of Nationally Determined Contributions (NDCs), a substantial gap exists between the action needed and that proposed (Cooper 2016; Antimiani et al. 2016; Zhang and Pan 2016; Cassen and Gracceva 2016). Recent analysis shows that urban passenger transport and surface freight transport must play a major role in decarbonization, both in managing growth in emerging economies and drastically reducing industrialized economies' emissions, all the more so for a 1.5 °C pathway.

As previously mentioned, urban passenger transport is significant because it provides access to urban services, economic opportunities, and social participation (Admasu et al. 2016; Angel and Blei 2016; Bibas et al. 2015). Automobile and bus travel is increasing rapidly in developing and emerging economies and is forecast to continue doing so, driven by growing demand for travel in developing economies, itself both a vital component and consequence of economic development (Berry et al. 2016; Gschwender et al. 2016; Spyra and Salmhofer 2016).

Multinational analyses of the technological potential and effort required to decarbonize the transport sector (Dessens et al. 2016; Figueroa Meza et al. 2014; Fulton et al. 2013; IPCC 2014) have shown that moving transport to an emissions pathway consistent with global climate change targets will require substantial decarbonization – almost complete in industrialized countries – over the coming decades by the middle of this century. This development will unlock direct and indirect benefits outweighing the costs, with savings of between US$50t and $100t in fuel savings, reduced vehicle purchase costs, and necessary infrastructure (IEA 2014). The additional co-benefits and synergies generated by sustainable mobility – among them improved safety and air quality and reduced travel time – further strengthen the case for a shift to low-carbon transport.

Vehicle technology and fuel substitution provide the biggest GHG mitigation potential (Kahn Ribeiro and Figueroa 2012). However, these measures ignore the broader sustainable mobility perspective, including management of travel demand growth and modal shift, which have added benefits regarding air quality, traffic congestion, safety, and overall societal mobility, and thus may have substantial socioeconomic co-benefits and may also be more cost-effective (van Vuuren et al. 2015).

The mitigation potential of a number of measures in the transport sector has been well-detailed, for example, a shift to public and non-motorized transport, along with improvements to internal combustion engines' efficiency (Sims et al. 2014; Kok et al. 2011; Wright and Fulton 2005). However, a more integrated view, integrating technology shifts as only a part of a balanced perspective to the wider sustainable (urban) development approach, is underdeveloped (Creutzig 2016; Saujot and Lefèvre 2016).

Only a few high-level assessments of climate change mitigation potential have examined the relationship between fuel and technology elements, and the planning, and model shift aspects of decarbonization pathways for transport (Figueroa Meza et al. 2014; Fulton et al. 2013; Sims et al. 2014). However, there are a number of studies that provide indications of the costs and benefits of specific measures. These will be explored in the following section.

The main conclusion to be drawn from decarbonization scenarios is that light-duty vehicle (LDV) travel must change rapidly – foremost in industrialized countries – coupled with more efficient vehicle technologies and higher use of more efficient transport modes. Specifically, an emission reduction pathway for a 2 °C stabilization scenario (as suggested by the IPCC) will require a GHG reduction of 73–80% in industrialized economies, which will, in turn, require car travel to reduce by 4–37%, in combination with average vehicle fuel efficiency (reduction in energy/km) increasing by 45–56% (Fulton et al. 2013). In developing and emerging countries, per capita light-vehicle travel could grow by 130–350% (if accompanied by fuel efficiency and carbon intensity gains of 40–50%) even under a low-carbon development scenario (Fulton et al. 2013).

Decarbonizing Transport: Actions and Potential

Even under very optimistic circumstances, the transport sector is currently expected to continue to maintain current levels of GHG emissions (Fulton et al. 2013; Harvey 2013). Even taking into account the large-scale adoption of more efficient vehicle technologies, modest shifts to electric vehicles, and some modal changes, carbon dioxide emissions from transport will be nowhere near where they should be to be in line with the Paris Agreement as increase in demand for mobility outweighs the increase in efficiency. With around 8.5 Gt CO_2, transport accounted for nearly a quarter of energy-related global CO2 emissions in 2018 (IEA 2020), of which over 70% are land transport (6.2 Gt CO_2eq) and maritime and aviation account for 8% and 7%, respectively (0.7 and 0.6 Gt CO_2eq, 7%) (IEA 2020).

If current trends persist, GHG emissions from the transport sector will double by 2050 (IPCC 2014). Putting the transport sector on the path of low-carbon development is critical to global climate change mitigation efforts aimed at stabilizing global warming at well below 2 °C, and pursue efforts for the 1.5 °C limit, as agreed upon in the Paris Agreement. To achieve this goal, developed countries must quickly achieve decarbonization of the transport sector, while developing and emerging countries will have to manage travel demand and introduce low-carbon vehicle technologies (IEA 2020). Analysis of the NDCs show that there is a huge gap between the necessary mitigation actions and the policy actions proposed by countries (Cooper 2016; Antimiani et al. 2016; Zhang and Pan 2016; Cassen and Gracceva 2016).

Since the adoption of the enhanced targets of the Paris Agreement, the main information contained in the body of scientific literature on decarbonization scenarios for the transport sector has not changed substantially (Creutzig 2016). While the basic principles of decarbonizing the sector remain unchanged, the pace and intensity of the actions required to initiate the drastic shifts that are needed are changing substantially. The recent analysis of the International Energy Agency (IEA) for Net-Zero Roadmaps highlights that transport modes do not decarbonize at the same pace as the maturity in particular of technologies varies substantially (IEA 2021). Similarly, the levels of interventions and the ability to steer transitions in different subsectors differ remarkably. The technologies to decarbonize land transport, in particular urban mobility, are readily available and viable, and there are well-established policy measures that can be implemented by local, state, national, and supranational (where applicable) levels of government. For long-distance transport, such as heavy trucks, maritime, and aviation, there are fewer technological options readily available at the moment, and also the governance of these subsectors is more fragmented.

The IEA Net-Zero Emissions Scenario aims to illustrate a possible path toward near-decarbonization, i.e., 0.5 Gt CO_2, in 2050, with a drastic shift in CO_2 emissions from light-duty vehicles beginning already in 2021 and reductions of emissions from heavy trucks, aviation, and shipping beginning 2025. This also reflects on the assessment that many of the technologies for heavy-duty vehicles, ships, and aircrafts are still prototype and demonstration phases and will require more research and investment to reach a level of maturity where large-scale deployment becomes viable.

Moving transport to a decarbonization pathway will have different routes in industrialized and developing economies, taking into account the different capabilities and also the differing needs to enable mobility access. Taking this path will unlock direct and indirect benefits that outweigh the costs, with savings of between USD 50 trillion and 100 trillion in fuel savings, reduced vehicle purchases, needed infrastructure, and fuel costs (IEA 2014). The additional co-benefits and synergies generated by sustainable mobility, such as improved safety and air quality and reduced travel time, make an even stronger case for the shift toward low-carbon transport. The contribution of countries to the global decarbonization efforts of the (land-)transport sector is reflected in several studies that show travel demand, technology deployment, and policy interventions and their effect on different scenarios.

From a climate change perspective, vehicle technology and fuel conversion options offer the greatest mitigation potential (Kahn Ribeiro and Figueroa 2012; IEA 2020). However, this does not fully reflect the broader perspective of sustainable transport (Graph 1). A broader multimodal approach to managing travel demand growth and model splitting can generate significant benefits in air quality, traffic congestion, safety, and overall social mobility, can generate higher socioeconomic synergies, and is likely to be more profitable (Van Vuuren et al. 2015).

The mitigation potential of some mitigation measures in the transport sector has been well established. However, the ranges to which managing travel demand through urban planning and design, shifting to public and non-motorized transport, and reducing the carbon intensity of fuels and energy carriers vary greatly in each city and country (Sims et al. 2014; Kok et al. 2011). However, a more comprehensive view of that combines technology with a balanced perspective to transform the potential into a broader (urban) sustainable development mode of low-carbon travel options which still needs more research (Saujot and Lefèvre 2016; Creutzig 2016).

There are only a few high-level climate change mitigation potential assessments that have succeeded in showing the relationship between fuels, technology elements, and transport decarbonization route planning and model conversion (Sims et al. 2014; Figueroa Meza et al. 2014; Fulton et al. 2013). However, there are many case studies that provide evidence from unique experience regarding the costs and benefits of specific measures, which will be explored in the next section. The main message of the decarbonization scenario is that light-vehicle travel in industrialized countries must change rapidly, moving to more efficient vehicle technologies and more efficient modes of transport.

In industrialized economies, it is necessary to reduce car trips by between 4% and 37%, while

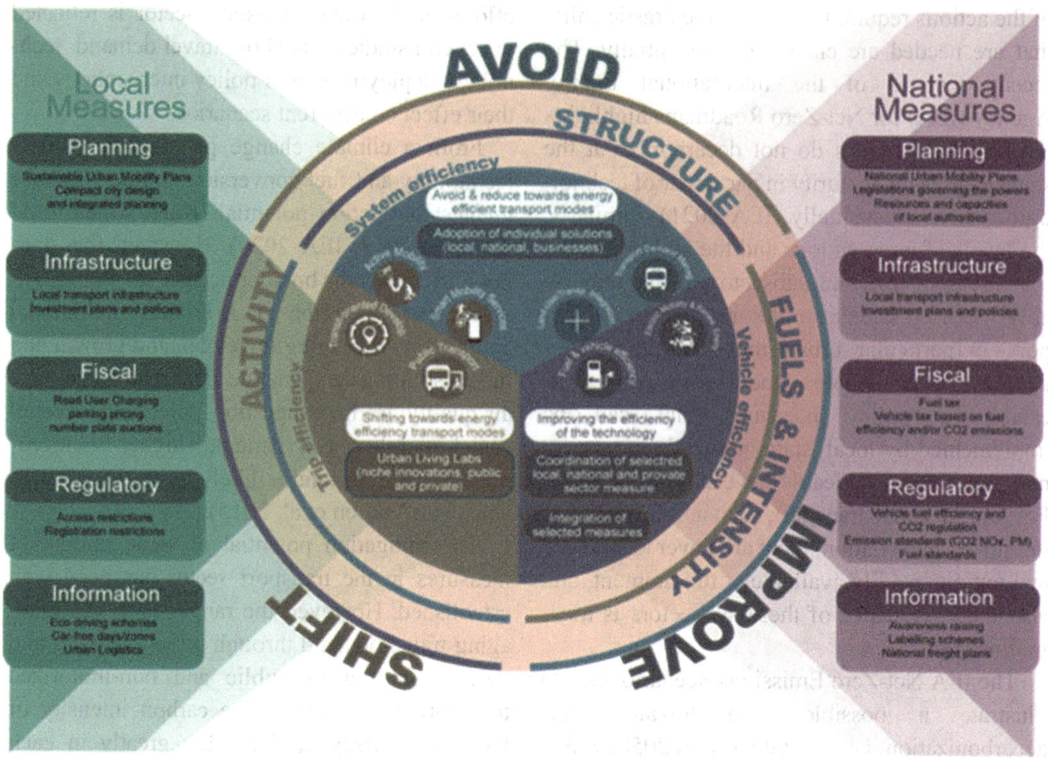

Sustainable Urban Mobility, Graph 1 Key measures in a comprehensive mobility transition strategy. (Based on Lah 2018)

the average fuel efficiency of vehicles (reduction in energy/km) is reduced by between 45% and 56% to achieve the expected reduction of 80%, which is roughly in line with the IPCC recommendation's emission reduction path of the 2 °C stabilization scenario (IEA 2012; Fulton et al. 2013). Considering the high level of saturation with regard to travel demand in industrialized economies and assuming that a transformative shift toward low-carbon modes and technologies will be pursued, one could argue that the shifts toward an enhanced net-zero target in OECD countries is a more feasible scenario. However, in developing and emerging economies, a net-zero scenario would be considerably more challenging. For a stabilization of global warming at 2 °C, fuel efficiency and carbon intensity will have to increase by 40–50%, but light-duty vehicle travel per capita still has the potential to increase by between 130% and 350% (IEA 2012; Fulton et al. 2013). Depending on the overall structure of the global shift toward net zero, such a target may require a far more rapid take-up of low-carbon vehicle technologies and fuels as well as an intensified shift toward low-carbon modes.

Transport decarbonization would entail urgent measures toward the transformation of the global vehicle fleet to attain drastic and quick improvements in overall efficiency and enable shifting away from oil, which, up to 2020, was estimated to account for 90% of the sector's energy consumption. The IEA's net-zero scenario assumes that 60% of the global car stock in 2040 would be electric (and almost fully electrified by 2050), compared to the current 1%, while the two- and three-wheeler fleet becomes fully electrified by 2050 (<20% currently). The scenario also emphasizes the continued tightening of fuel economy standards for passenger cars, leading up to a halting of sales of new cars with internal combustion engines by 2035. The transformation of the heavy-duty vehicle fleet toward being electric or powered by fuel cells is expected to happen at a slower pace (66% of global sales by 2050), and thus, continued improvements in fuel economy standards of internal combustion engine-based heavy-duty vehicles are assumed up to 2050 (47% improvement in 2050 compared to 2020). Continued improvements in efficiency in aviation and shipping are to be achieved through stock renewal (IEA 2021).

Decarbonization of the transport sector requires the transformation in vehicle fleets to go hand in hand with the transformation of the energy systems. Considering the scale of electrification required for future global fleets toward attaining the optimistic net-zero emission scenario, the importance of decarbonizing electricity generation is evident. Under IEA's net-zero emissions scenario, electricity would dominate the transport sector, accounting for more than 45% of total transport energy use in 2050. This would entail an eightfold increase in the generation of electricity from renewable sources in 2050 compared to 2020, which equates to a 90% share of total generation. The aggressive electrification of the vehicle fleet would require an estimated 200 million EV public charging units (from 1.3 million currently) (IEA 2021).

Furthermore, in the IEA's net-zero emissions scenario, hydrogen and hydrogen-based fuels are to penetrate long-haul trucking, while advanced biofuels and hydrogen-based fuels are to be increasingly used in shipping, and aviation is to be dependent on synthetic liquid fuels and advanced biofuels. Overall, the net-zero scenario assumes the need for hydrogen (and hydrogen-based fuels) to cater to 30% of transport energy consumption, while bioenergy would serve 15% (IEA 2021).

The transformation of the global energy system would require more than doubling of the annual investment in 2050 as compared to current levels (4.5 trillion USD versus 2 trillion) (IEA 2021). Investments in transport would need to increase to 1.1 trillion USD in 2050, compared to the current 150 billion USD. Such urgent shifts toward financing sustainable transport are critical. The UNEP (2020) estimates that in the top 50 economies in 2015, two-thirds of the transport infrastructure investments went to brown investments in road transport. It also suggests that many of the documented recovery packages that had been put up in response to the current COVID-19 pandemic have geared toward investing more into fossil-fuel-related entities than in clean energy. Purposive planning for mitigating other systemic impacts is also needed. For example, the electrification of transport is expected to contribute to the increasing demand for critical minerals such as lithium, which is expected to grow by a 100 times in 2050 as compared to 2020 (IEA 2021). This underlines the importance of supporting long-term planning of relevant supply chains.

While technology and operational improvements are crucial pillars for decarbonizing transport, holistic approaches that are geared toward synergistic effects are needed. UNEP (2020) suggests that while current climate commitments can significantly improve the overall efficiency of transport, the projected increase in transport demand – even considering the impacts of the current pandemic – will mean that decarbonization cannot be achieved. A meta-analysis of more than 50 life cycle assessment studies suggests that the estimated mitigation potential for measures that focus primarily on improving overall vehicle efficiencies vary the most and that combining wider consumption-focused measures with such technology-focused policies is essential for maximizing emission reduction. Such insights are also supported by IEA (2021), which estimates that without realizing the energy-related behavioral changes assumed in its net-zero scenario, emissions would be at 2.6 Gt CO_2eq in 2050.

Another key consideration refers to the distribution of the responsibility for bearing the global costs of decarbonization. The International Transport Forum (2021) argues that such responsibility needs to be linked with historical cumulative emissions. Regions which have historically benefited the most from fossil-based economies now have significantly better access to capacities, capital, and technologies needed for decarbonization. Thus, they can support efforts in emerging regions through capital investments and aiding technology transfer.

Potential for Co-benefits

The claim is frequently made that transport is one of the hardest sectors to decarbonize (Cai et al. 2015; Vale 2016; van Vuuren et al. 2015). However, a number of recent papers challenged this view, showing that an integrated approach to policy-making can create co-benefits with other key sectors, such as health, productivity, energy security, and safety, maximizing socioeconomic benefits (Bollen 2015; Dhar and Shukla 2015; Lah 2015; Schwanitz et al. 2015; Dhar et al. 2017). Additionally, by including benefits across sectors, policy packages have the potential to incorporate the positions of relevant veto players, thus aiding the forming of coalitions around the package and fostering its implementation (Lah 2017c).

All available mitigation options are required to bring the transport sector on a 2 °C stabilization pathway, including vehicle fuel efficiency, modal choice, and compact urban design among others, in line with the Avoid-Shift-Improve framework (Lah 2015). Efforts should cover all transport modes and require the involvement of national as well as local levels of government. As the uptake of low-carbon transport solutions has been lagging behind its potential, due to the existence of numerous hurdles, among which the initial-cost barrier, a vital element for transport decarbonization is to design an integrated policy package. This multimodal, multilevel sustainable transport package should tackle all elements of the Avoid-Shift-Improve paradigm and seek alignment and complementarity between national and local policies. Integrating national and local policies is critical to streamline decisions and ensure consistency in measures targeting sustainable mobility. The table below shows the critical interlinkages between systemic pillars and between local and national measures by highlighting some of the key interventions at the local and national level addressing system efficiency, trip efficiency, and vehicle efficiency (Table 1).

If applied in isolation, policy measures are unlikely to achieve goals without generating possible trade-offs and with them the risk of a veto player blocking the measure's implementation. Reducing fuel (and car) use by increasing fuel taxes, for example, in isolation would have a negative impact on transport affordability, and thus mobility (Greene et al. 2005; Sterner 2007), and which could then result in the initiative being blocked by veto players. In contrast, a balanced and integrated policy package including increased fuel taxes with measures like vehicle efficiency standards and additional differentiation of the vehicle taxes and with the provision of modal choices and compact city design would have the potential to address a wider range of policy objectives and thus ensure relevant veto player support (or at least allow) implementation.

Coalitions and Institutions

Energy- and climate change-related policies when applied in the transport sector generally require consensus regarding the need for policy intervention and a strategic, coherent, and stable operating environment. Given the transport sector's central economic and social position, measures in the sector, such as fuel and vehicle taxation, are highly visible and politically sensitive. They require strong political commitment to make it onto the policy agenda and to stay in the political process as they tend to be cost-effective only over the medium to long term (IEA 2010; IPCC 2014). Because transport politics is complex and multifaceted, and as policy interventions can have unintended consequences, developing and maintaining consensus can be difficult (Häussler et al. 2016; Klenk and Meehan 2015; Lijphard 1984). Linking and packaging policies can help maintain consensus by generating synergies and co-benefits between measures, including linking GHG reduction goals with other sustainable development goals, such as increasing energy security, road safety, and public health, increasing economic productivity and air pollution, and improving equity and access (Kanda et al. 2016; Wen et al. 2016). For this paper, local and national policy advisors in Europe, Asia, Africa, and the Americas were asked about the main barriers for the take-up of sustainable transport measures. Their responses were that lack of funds, lack of suitable technologies, and also public opposition

Sustainable Urban Mobility, Table 1 Elements of a multimodal, multilevel sustainable transport package and their complementarity (Lah 2018, 2019)

National measures	Complementarity of measures	Local measures
Planning • National Urban Mobility Plans • Legislations governing the powers, resources, and capacities of local authorities	• Objectives of national and local plans are in line and contribute to common goals • Compact and policy-centric planning enables short trips and increases access	• Sustainable Urban Mobility Plans • Compact city design and integrated planning
Infrastructure • National transport infrastructure investment plans and policies	• National and local transport infrastructure projects are following an integrated planning approach and contribute to the same objectives	• Local transport infrastructure investment plans and policies
Fiscal • Fuel tax • Vehicle tax based on fuel efficiency and/or CO_2 emissions	• Fuel tax encourages more efficient use of vehicles, which helps minimize rebound effects • Complementary measures at the local level help manage travel demand and can generate funds that can be re-distributed to fund low-carbon transport modes	• Road User Charging, parking pricing, number plate auctions • Provision of public transport, walking, and cycling infrastructure and services
Regulatory • Vehicle fuel efficiency and CO_2 regulation • Emission standards (CO, NOx, PM) • Fuel standards	• Vehicle standards and regulations ensure the supply of efficient vehicles together with local regulations, and access restrictions help in steering consumer behavior	• Access restrictions • Registration restrictions
Awareness and information • Awareness raising • Labelling schemes • National freight plans	• Consistent messaging can help in influencing choices and mobility behavior of individuals and businesses	• Eco-driving schemes • Car-fee days/zones • Urban logistics

are not the main barriers, as one might expect. Instead, insufficient knowledge of the various benefits of sustainable mobility, in particular among political decision-makers, and institutional barriers directly affect the implementation process. Additionally, (increased) knowledge of the potential co-benefits of sustainable transport policy can help align different policy actors and institutions.

An integrated policy approach, creating consensus and coalitions among diverse stakeholders and interests, can assist in overcoming implementation barriers, minimizing rebound effects, and motivating people, businesses, and communities (von Stechow et al. 2015). Such integrated approaches to policy-making are especially critical presently, because current GHG reduction measures alone are insufficient to achieve the reductions needed to shift to a 1.5 °C pathway (IPCC 2014).

Decision-making regarding transport (infrastructure) policies is as complex as the sector itself. A single measure will rarely have significant climate change impacts and generate economic, social, and environmental benefits (Creutzig 2016; Lah 2014). Many policy and planning decisions have synergistic effects: their impacts are larger if implemented together. Thus, if possible, integrated programs rather than individual strategies should be implemented and evaluated (Hüging et al. 2014). Improving public transport, for example, may have minimal impact on individual motorized travel and associated benefits such as reduced congestion, fuel costs, and air pollution. However, the same measure's effectiveness and benefit may be increased if implemented in conjunction with complementary incentives, such as efficient road and parking pricing (Cuenot et al. 2012; den Boer et al. 2011). In fact, the effectiveness of programs has been shown to be increased by including a combination of qualitative improvements to alternative modes (walking, cycling, and public transport services), incentives to discourage carbon-intensive modes (e.g., fuel pricing, vehicle fuel efficiency regulation, and taxation), and integrated transport and land use planning, to create more compact, mixed, and better-connected communities with less need to travel (Figueroa Meza et al. 2014; Sims et al. 2014).

An important driver for combining different fiscal, policy, regulatory and infrastructure measures is to create synergies and minimizing rebound effects. This effect refers to the tendency for the total demand for energy to decrease less than expected after the introduction of efficiency improvements, due to the resultant decrease in the cost of energy services (Lah 2014). Introducing fuel efficiency standards for light-duty vehicles, for example, may improve overall fleet efficiency, but may also induce additional travel through the reduced fuel costs for consumers (Yang et al. 2017). Ignoring or underestimating this effect when planning policies may lead to inaccurate forecasts and unrealistic expectations of the outcomes, and, in turn, the policies' payback periods (IPCC 2014). For household appliances such as fridges, washing machines, and lighting, the expected rebound effect is around 0–12%, while it can be up to 20% in industrial processes and 12–32% for road transport. The higher the potential rebound effect, the greater the uncertainty of a policy's cost-effectiveness and its effect upon energy efficiency (Ruzzenenti and Basosi 2008).

Current transport policy and infrastructure appraisals do not typically account for the wider socioeconomic benefits of sustainable mobility (Hüging et al. 2014). A number of studies posit that cost-effective reductions in transport sector GHG emissions rely on an integrated approach (IPCC 2014; Figueroa Meza et al. 2014). While these reductions can be achieved by various means, such as modal shift, efficiency gains, and reduced transport activity, a combination of measures is a key factor in maximizing synergies and reducing rebound effects. For example, much larger reductions in overall travel demand and larger modal shifts would be required if not accompanied by efficiency improvements in the vehicle fleet or vice versa (Figueroa Meza et al. 2014; Fulton et al. 2013).

Policy continuity and consensus – Political agenda setting and policy continuity are affected by political consensus, which is a result of political and institutional relationships (Fankhauser et al. 2015; Marquardt 2017). These relationships, including the interactions between different levels of government (e.g., local, state/regional, national, supra-national) and acknowledgment of scientific consensus on climate change policy, vary greatly between key political and societal actors (Never and Betz 2014). Political environments vary from country to country and change over time. This affects the rate of implementation of sustainable transport and other climate change mitigation measures and is one of the reasons for the significant differences between countries' progress reducing GHG emissions from the transport (and other) sectors. Instability in political environments also affects the relevant policy environments. As such, a shared set of methods and values – usually delivered through knowledge, or epistemic, communities – are generally considered vital for setting the policy agenda. Support from diverse political and public stakeholders is vital for the long-term success of policy and infrastructure measures. Often, this support will be a function of the level of trust between stakeholders and policy-makers and the role that agreed facts play in the decision-making process (Simmons 2016; Freitag and Ackermann 2016). Public perception and the influence of epistemic communities also influence political agenda setting and consensus on major policy issues such as climate change and energy efficiency (Cook and Rinfret 2015; Hagen et al. 2016).

Policy integration and coalition building – The policy environment, or setting in which decisions are made, is equally important to outcomes as the chosen combination of policies and decisions which make up a low-carbon transport strategy (Justen et al. 2014). This policy environment includes socioeconomic and political aspects of the institutional structures of countries, which affect coalition building, but can also increase the risk that a policy package fails because one part faces strong opposition (Sørensen et al. 2014). A core determinant of success is the involvement – at an early stage of the process – of potential veto players and the incorporation of their policy objectives in the agenda setting (Tsebelis and Garrett 1996).

Institutional context – The political and institutional setting in which policies are developed is influential in the success or failure of policies' implementation (Jänicke 1992). Institutional aspects, such as the presence or absence of a national environmental ministry, or a local environmental department, and their respective roles in the process are also likely to have an effect on the implementation of (primarily) climate-related transport measures (Fredriksson et al. 2016). The legal powers assigned to these agencies, and their budget and political influence, are equally important (Jänicke 2002).

The Role of Capacities in the Transformative Change Toward Net-Zero Mobility

For a truly transformative change, the approach to decarbonizing transport needs to go well beyond the vehicle and even the sector. All available options are required to bring the transport sector on a net-zero pathway and provide access to sustainable mobility for all. As the uptake of low-carbon mobility solutions has been lagging behind its potential, an integrated multimodal, multilevel sustainable transport package should tackle all aspects of the mobility system and seek alignment and complementarity between national and local policies as well as between public and private sector actions.

To achieve this, a move to a "safe system" approach for sustainable, decarbonized mobility may be required. This can build on years of experiences from the road safety realm, beginning from the first adoption of "vision zero" in Sweden in 1997 which has revolutionized the approach to improving road safety (Wegman 2017). There are now plenty of technological and operational options readily available which can drastically reduce CO_2 emissions and improve local air quality (Lah 2017c; Sims et al. 2014; IEA 2020). If we provide more sustainable choices to transport users and signal a clear preference – for example, through pricing or regulation – we can nudge consumers toward more sustainable choices.

Individual projects and technologies can contribute to the change, but only an integrated and systemic change across the whole sector and beyond – including the energy and resource dimensions – will enable a shift toward a net-zero transport system. For this, a societal perspective is needed to identify appropriate solutions. This is also vital to leverage on the potential for cost savings of a sustainable mobility system. A "safe system" approach for a transport sector that moves us toward net-zero emissions and that enables access to sustainable mobility for all needs to focus on four interconnected pillars: users, vehicles, services, and infrastructure.

Minimizing the carbon content in vehicle technologies is a key systemic change that is required in the mobility transition. Hence, the shift to electric mobility has a vital role to play in decarbonizing the sector. But the overall contribution of electric mobility to climate change mitigation and sustainable development depends critically on the integration with the other pillars of the system. Electric vehicles need to be fit-for-purpose – this means that they need to be resource and energy efficient, well integrated with other mobility services and infrastructure, and designed for mobility as a service, which provides access for all.

To adopt a "safe system" approach, improved capacities and a better understanding of the needs and opportunities for key players in the sector are important, including local and national authorities, vehicle manufacturers, and other technology developers, mobility service providers, and infrastructure developers. These are essential building blocks for the transition to sustainable mobility (Fig. 1).

This transition has the potential to unlock trillions of dollars in cost savings, at least from a whole-society perspective, by 2050, a low-carbon mobility system could cut transport related annual costs by over five trillion USD globally. But the resulting impacts can also shift value generation

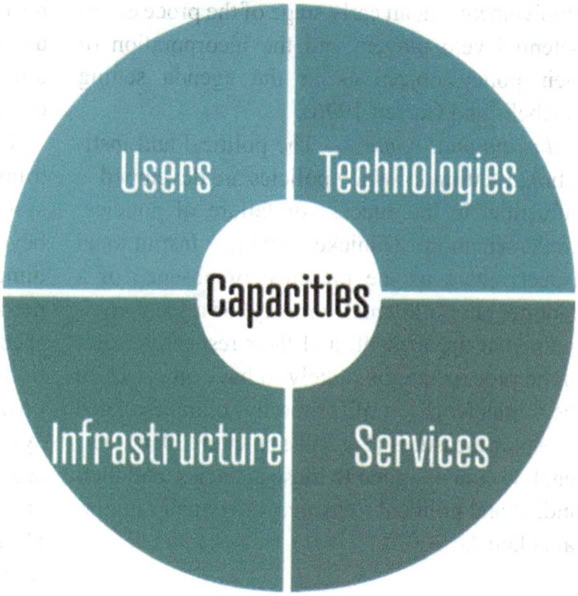

Sustainable Urban Mobility, Fig. 1 Safe system approach for net-zero mobility

and distribution. However, more sustainable travel patterns can generate substantial complementary benefits that could help in transforming cities into more liveable and economically efficient centers. Policy interventions to foster the electrification of the sector can help toward achieving a range of objectives, for example, air quality improvements and reductions of greenhouse gas emissions. But addressing all key objectives – such as access to safe roads and liveable cities for all – requires a much broader package of measures. Linking and packaging policies is also a key tool to generate synergies between different measures and to align different players. Capacity building on sustainable mobility needs to convey the need for an overarching approach that consists of several levels of intervention that shape not just vehicle technology but also mobility patterns and urban form. Such a capacity building approach includes:

- **Technologies:** For electric vehicles, there should be a clear focus on drastically downsizing vehicle size and power, fostering resource and energy efficiency, and boosting cost-effectiveness. This is countering the trend of the last few decades toward bigger, faster, and more powerful cars, which has eradicated almost all efficiency gains in powertrain technologies. Only then will the electrification of the entire vehicle fleet be viable and resource efficient as well as affordable. In addition, electric vehicle concepts should be designed for shared use cases, which will further foster access and affordability. Other technological innovations, such as automation, should focus on complementarity with public transport systems and should avoid competition with non-motorized transport. This is vital to the viability of public transport services and also encourages healthy and active mobility. Automation may play an important role in the provision of on-demand mobility services in rural areas where traditional public transport options are not viable.
- **Infrastructures**: Providing access for all to high-quality public transport services, walking, and cycling infrastructures is a vital part of a safe system for sustainable mobility. To enable this, compact city development can help with mixed-use, polycentric structures and short travel distances. To enable the shift to electric vehicles, a comprehensive network of charging solutions and reliable availability of charging points is a vital element for a systemic change.

- **Services:** Access to mobility services such as shared and ride hailing services is another critical element for a safe system for sustainable mobility. Services should be harmonized across available mobility services to encourage the use of the most efficient option.
- **Users:** Should have access to transparent information about the cost, time, safety, and other relevant aspects of different available mobility solutions and vehicles. To further guide consumer choices, differentiated pricing should favor efficient modes and vehicles. Regulation of vehicle standards and technologies further ensures that consumers have sufficient access to safe and efficient vehicles.

Discussion

Provided that GHG reductions technologies are widely available and policy mechanisms to support them have been proven to be effective, those factors that influence the policy environment for transport energy efficiency policies are the vital aspect in fostering their adoption and success.

In order to enable long-term investment decisions by industry and consumers, a stable political operating environment is crucial for the effectiveness of energy and climate change policies for the transport sector (Fais et al. 2016; Lakshmanan 2011; Spataru et al. 2015). Governance and institutional structures focused on consensus-making may foster such a strategic, coherent, and stable operating environment. Getting and maintaining transport sector energy use reduction policies on the policy agenda is crucial, as they are only cost-effective only over the medium to long term. Policy interventions in the transport sector, such as fuel and vehicle taxation, are highly visible and politically sensitive. To get a clearer picture of the feasibility of such interventions in advance, well-established political science theories that aim to identify key institutional characteristics influencing policy processes can be examined. For example, there are a number of studies examining the influence of institutions and actors within policy processes (e.g., Scruggs 1999, 2001; Jänicke 2002). Most of these studies focus on higher-level environmental performance indicators (e.g., Scruggs 1999; Jacob and Volkery 2004). Considering, though, the complexity of policy-making processes, it is challenging to draw direct conclusions relating institutional aspects with climate policy performance. However, several recent studies highlight the relevance of veto players in domestic and international environmental policy-making (Batalla 2012; Jahn 2014; Schulze 2014; Singh and Dunn 2013; Sotirov and Memmler 2012). Similarly, the role of corporatist structures in environmental policy is a subject of research (Benoit and Patsias 2014; Cairney 2014; Iguchi 2015; Jones 2014; Vink et al. 2015; Weiner 2014).

The relationship between institutional structures and the performance of climate policies is obvious when assessing the stability (or the lack thereof) of specific policies in different countries. After the 2016 US presidential elections, climate change policies in the USA have experienced a radical turnaround, highlighting the inherent political volatility of majoritarian democracies that rely almost entirely on minimal coalition and partisan support for climate change policies. In contrast, in the EU and Member States, there is a broad societal consensus for climate action. Establishing a relationship between institutional structures and capacities, and climate policy impacts, i.e., emission reductions, it is vital to first establish the ability of a policy package to deliver emission reduction impacts. A review of key policy measures (concerning light-duty vehicle (LDV) fleet emission reduction) in G20 countries showed that only *a combination* of vehicle fuel efficiency standards in conjunction with vehicle and fuel taxation delivers significant improvements in the efficiency of the fleet and in-use efficiency (Yang et al. 2017). These national-level policy measures are the backbone of a comprehensive policy package. When comparing these measures with institutional features of democracies, a positive relationship between the presence of institutions that enable a stable and consensus-oriented policy environment and the presence of key national-level policy measures can be observed (Lah 2017a). Relying on particular political parties to act may deliver swift and ambitious policy action, but this may be subsequently overturned with a change in government (Lah 2017b). The risk of political volatility is reflected in the ability or inability of political parties to find common ground on sustainable transport issues and that policies are often implemented in an isolated fashion. However, relevant authorities at local and national level are considered more likely to cooperate with counterparts on sustainable transport issues.

Institutional structures, policy continuity, and implementation are vital aspects in the delivery of global climate change goals in line with the Paris Agreement. The decarbonization pathways for various sectors are clearly outlined (IPCC 2014) and translated into action for the transport sector, highlighting that global climate change mitigation targets will not be reached without a substantial contribution by the transport sector (Fulton et al. 2013). The climate change mitigation potential of specific measures has been well established, showing that the technologies to reduce emissions from the transport sector are readily available (Figueroa Meza et al. 2014). An integrated policy approach that aims to generate synergies (or at least avoiding trade-offs) between various policy objectives can help maximize socioeconomic benefits and can help form coalitions, which are resilient in the face of political volatility.

Cross-References

- ▶ Disruptive Mobility: Sharing Electric Autonomous Vehicles (SEAVs) Reshape Our Future Cities
- ▶ Transportation and Mobility

References

Admasu, K., Balcha, T., & Getahun, H. (2016). Model villages: A platform for community-based primary health care. *The Lancet Global Health, 4*, e78–e79. https://doi.org/10.1016/S2214-109X(15)00301-0.

Angel, S., & Blei, A. M. (2016). The productivity of American cities: How densification, relocation, and greater mobility sustain the productive advantage of larger U.S. metropolitan labor markets. *Cities, 51*, 36–51. https://doi.org/10.1016/j.cities.2015.11.030.

Antimiani, A., Costantini, V., Kuik, O., & Paglialunga, E. (2016). Mitigation of adverse effects on competitiveness and leakage of unilateral EU climate policy: An assessment of policy instruments. *Ecological Economics, 128*, 246–259. https://doi.org/10.1016/j.ecolecon.2016.05.003.

Batalla, E. V. C. (2012). Veto players and state decisiveness: Negotiating bilateral economic partnership agreements between Japan and Southeast Asia. *Philippine Political Science Journal, 33*, 39–62. https://doi.org/10.1080/01154451.2012.684516.

Benoit, M., & Patsias, C. (2014). The implementation of territorialized agri-environmental measures in France: A broadening of democracy? The example of the Centre and Languedoc-Roussillon regions. *Innovation Journal, 19*, 45.

Berry, A., Jouffe, Y., Coulombel, N., & Guivarch, C. (2016). Investigating fuel poverty in the transport sector: Toward a composite indicator of vulnerability. *Energy Research and Social Science, 18*, 7–20. https://doi.org/10.1016/j.erss.2016.02.001.

Bibas, R., Méjean, A., & Hamdi-Cherif, M. (2015). Energy efficiency policies and the timing of action: An assessment of climate mitigation costs. *Technological Forecasting and Social Change, 90*(Part A), 137–152. https://doi.org/10.1016/j.techfore.2014.05.003.

Bollen, J. (2015). The value of air pollution co-benefits of climate policies: Analysis with a global sector-trade CGE model called WorldScan. *Technological Forecasting and Social Change, 90*(Part A), 178–191. https://doi.org/10.1016/j.techfore.2014.10.008.

Cai, Y., Newth, D., Finnigan, J., & Gunasekera, D. (2015). A hybrid energy-economy model for global integrated assessment of climate change, carbon mitigation and energy transformation. *Applied Energy, 148*, 381–395. https://doi.org/10.1016/j.apenergy.2015.03.106.

Cairney, P. (2014). The territorialisation of interest representation in Scotland: Did devolution produce a new form of group-government relations? *Territory, Politics, Governance, 2*, 303–321. https://doi.org/10.1080/21622671.2014.952326.

Cassen, C., & Gracceva, F. (2016). Chapter 7. Energy security in low-carbon pathways A2 – Lombardi, Patrizia. In M. Gruenig (Ed.), *Low-carbon energy security from a European perspective* (pp. 181–205). Oxford, UK: Academic Press.

Cook, J., & Rinfret, S. (2015). Are they really so different? Climate change rule development in the USA and UK. *Journal of Public Affairs, 15*, 79–90. https://doi.org/10.1002/pa.1512.

Cooper, M. (2016). Renewable and distributed resources in a post-Paris low carbon future: The key role and political economy of sustainable electricity. *Energy Research and Social Science, 19*, 66–93. https://doi.org/10.1016/j.erss.2016.05.008.

Creutzig, F. (2016). Evolving narratives of low-carbon futures in transportation. *Transport Reviews, 36*, 341–360. https://doi.org/10.1080/01441647.2015.1079277.

Cuenot, F., Fulton, L., & Staub, J. (2012). The prospect for modal shifts in passenger transport worldwide and impacts on energy use and CO_2. *Energy Policy, 41*, 98–106. https://doi.org/10.1016/j.enpol.2010.07.017.

den Boer, E., Van Essen, H., Brouwer, F., Pastori, E., & Moizo, A. (2011). *Potential of modal shift to rail transport*. Delft: CE Delft.

Dessens, O., Anandarajah, G., & Gambhir, A. (2016). Limiting global warming to 2 °C: What do the latest mitigation studies tell us about costs, technologies and other impacts? *Energy Strategy Reviews, 13–14*, 67–76. https://doi.org/10.1016/j.esr.2016.08.004.

Dhar, S., & Shukla, P. R. (2015). Low carbon scenarios for transport in India: Co-benefits analysis. *Energy Policy, 81*, 186–198. https://doi.org/10.1016/j.enpol.2014.11.026.

Dhar, S., Pathak, M., & Shukla, P. R. (2017). Electric vehicles and India's low carbon passenger transport: A long-term co-benefits assessment. *Journal of Cleaner Production, 146*, 139–148. https://doi.org/10.1016/j.jclepro.2016.05.111.

Fais, B., Sabio, N., & Strachan, N. (2016). The critical role of the industrial sector in reaching long-term emission reduction, energy efficiency and renewable targets. *Applied Energy, 162*, 699–712. https://doi.org/10.1016/j.apenergy.2015.10.112.

Fankhauser, S., Gennaioli, C., & Collins, M. (2015). The political economy of passing climate change legislation: Evidence from a survey. *Global Environmental Change, 35*, 52–61. https://doi.org/10.1016/j.gloenvcha.2015.08.008.

Figueroa Meza, M. J., Lah, O., Fulton, L. M., McKinnon, A. C., & Tiwari, G. (2014). Energy for transport. *Annual Review of Environment and Resources, 39*, 295–325.

Fredriksson, P. G., Sauquet, A., & Wollscheid, J. R. (2016). Democracy, political institutions, and environmental policy. In *Reference module in Earth systems and environmental sciences*. New York: Elsevier. https://doi.org/10.1016/B978-0-12-409548-9.09714-1.

Freitag, M., & Ackermann, K. (2016). Direct democracy and institutional trust: Relationships and differences across personality traits. *Political Psychology, 37*, 707–723. https://doi.org/10.1111/pops.12293.

Fulton, L., Lah, O., & Cuenot, F. (2013). Transport pathways for light duty vehicles: Towards a 2° scenario. *Sustainability, 5*, 1863–1874. https://doi.org/10.3390/su5051863.

Greene, D. L., Patterson, P. D., Singh, M., & Li, J. (2005). Feebates, rebates and gas-guzzler taxes: A study of incentives for increased fuel economy. *Energy Policy, 33*, 757–775.

Gschwender, A., Jara-Díaz, S., & Bravo, C. (2016). Feeder-trunk or direct lines? Economies of density, transfer costs and transit structure in an urban context. *Transportation Research Part A: Policy and Practice, 88*, 209–222. https://doi.org/10.1016/j.tra.2016.03.001.

Hagen, B., Middel, A., & Pijawka, D. (2016). European climate change perceptions: Public support for mitigation and adaptation policies. *Environmental Policy and Governance, 26*, 170–183. https://doi.org/10.1002/eet.1701.

Harvey, L. D. D. (2013). Global climate-oriented transportation scenarios. *Energy Policy, 54*, 87–103. https://doi.org/10.1016/j.enpol.2012.10.053.

Häussler, T., Schmid-Petri, H., Adam, S., Reber, U., & Arlt, D. (2016). The climate of debate: How institutional factors shape legislative discourses on climate change. A comparative framing perspective. *Studies in Communication Sciences, 16*, 94–102. https://doi.org/10.1016/j.scoms.2016.04.002.

Hüging, H., Glensor, K., & Lah, O. (2014). Need for a holistic assessment of urban mobility measures – Review of existing methods and design of a simplified approach. *Transportation Research Procedia, 4*, 3–13. https://doi.org/10.1016/j.trpro.2014.11.001.

IEA. (2010). *Cities, towns and renewable energy*. Paris: International Energy Agency (IEA)/OECD.

IEA. (2012). *Energy technology perspectives 2012*. Paris: International Energy Agency (IEA).

IEA. (2020). *Net zero by 2050*. Paris: International Energy Agency (IEA).

IEA. (2021). *Global energy review 2021*. Paris: International Energy Agency (IEA).

Iguchi, M. (2015). *Divergence and convergence of automobile fuel economy regulations: A comparative analysis of EU, Japan and the US*. Cham: Springer. https://doi.org/10.1007/978-3-319-17500-3.

IPCC. (2014). *Climate change 2014 – Mitigation of climate change, 5th assessment report*. Cambridge, UK: Cambridge University Press.

Jacob, K., & Volkery, A. (2004). Institutions and instruments for government self-regulation: Environmental policy integration in a cross-country perspective. *Journal of Comparative Policy Analysis: Research and Practice, 6*, 291–309.

Jahn, D. (2014). Changing of the guard: Trends in corporatist arrangements in 42 highly industrialized societies from 1960 to 2010. *Socio-Economic Review, 14*, mwu028.

Jänicke, M. (1992). Conditions for environmental policy success: An international comparison. *Environmentalist, 12*, 47–58. https://doi.org/10.1007/BF01267594.

Jänicke, M. (2002). The political system's capacity for environmental policy: The framework for comparison. In H. Weidner & M. Jänicke (Eds.), *Capacity building in national environmental policy* (pp. 1–18). Berlin/Heidelberg: Springer.

Jones, B. (2014). Pressure groups. In *Politics UK* (8th ed., pp. 178–201). London: Routledge. https://doi.org/10.4324/9781315740720.

Jordan, A., Wurzel, R. K. W., & Zito, A. R. (2013). Still the century of "new" environmental policy instruments? Exploring patterns of innovation and continuity. *Environmental Politics, 22*, 155–173.

Justen, A., Schippl, J., Lenz, B., & Fleischer, T. (2014). Assessment of policies and detection of unintended effects: Guiding principles for the consideration of methods and tools in policy-packaging. *Transportation*

Research Part A: Policy and Practice, 60, 19–30. https://doi.org/10.1016/j.tra.2013.10.015.

Kahn Ribeiro, S., & Figueroa, M. J. (2012). Energy end-use: Transportation. In *Global energy assessment – Toward a sustainable future* (pp. 575–648). Vienna/ Cambridge, UK/New York: International Institute for Applied Systems Analysis/Cambridge University Press.

Kanda, W., Sakao, T., & Hjelm, O. (2016). Components of business concepts for the diffusion of large scaled environmental technology systems. *Journal of Cleaner Production, 128*, 156–167. https://doi.org/10.1016/j.jclepro.2015.10.040.

Klenk, N., & Meehan, K. (2015). Climate change and transdisciplinary science: Problematizing the integration imperative. *Environmental Science & Policy, 54*, 160–167. https://doi.org/10.1016/j.envsci.2015.05.017.

Kok, R., Annema, J. A., & van Wee, B. (2011). Cost-effectiveness of greenhouse gas mitigation in transport: A review of methodological approaches and their impact. *Energy Policy, 39*, 7776–7793. https://doi.org/10.1016/j.enpol.2011.09.023.

Lah, O. (2014). The barriers to vehicle fuel efficiency and policies to overcome them. *European Transport Research Review, 25*, 5088–5098.

Lah, O. (2015). *Sustainable development benefits of low-carbon transport measures*. Eschborn: Deutsche Gesellschaft für Internationale Zusammenarbeit (GIZ) GmbH.

Lah, O. (2017a). Factors of change: The influence of policy environment factors on climate change mitigation strategies in the transport sector. *Transportation Research Procedia, 25*, 3495–3510.

Lah, O. (2017b). Continuity and change: Dealing with political volatility to advance climate change mitigation strategies – Examples from the transport sector. *Sustainability, 9*. https://doi.org/10.3390/su9060959.

Lah, O. (2017c). Decarbonizing the transportation sector: Policy options, synergies, and institutions to deliver on a low-carbon stabilization pathway. *WIREs Energy and Environment, 6*, e257. https://doi.org/10.1002/wene.257.

Lah, O. (2018). *Sustainable urban mobility pathways*. Amsterdam: Elsevier.

Lah, O. (2019). Cities as engines for mobility transitions: Co-benefits and coalitions as enablers for a low-carbon transport sector.

Lakshmanan, T. R. (2011). The broader economic consequences of transport infrastructure investments. *Journal of Transport Geography, 19*, 1–12. https://doi.org/10.1016/j.jtrangeo.2010.01.001.

Lijphard, A. (1984). *Democracies: Patterns of majoritarian and consensus governments in twenty-one countries*. London/New Haven: Yale University Press.

Lijphart, A. (2012). *Patterns of democracy: Government forms and performance in thirty-six countries*. London: Yale University Press.

Lundqvist, L. (1980). *The hare and the tortoise: Clean air policies in the United States and Sweden*. Ann Arbor: University of Michigan Press.

Marquardt, J. (2017). Conceptualizing power in multi-level climate governance. *Journal of Cleaner Production*. https://doi.org/10.1016/j.jclepro.2017.03.176.

Never, B., & Betz, J. (2014). Comparing the climate policy performance of emerging economies. *World Development, 59*, 1–15. https://doi.org/10.1016/j.worlddev.2014.01.016.

Ruzzenenti, F., & Basosi, R. (2008). The rebound effect: An evolutionary perspective. *Ecological Economics, 67*, 526–537. https://doi.org/10.1016/j.ecolecon.2008.08.001.

Saujot, M., & Lefèvre, B. (2016). The next generation of urban MACCs. Reassessing the cost-effectiveness of urban mitigation options by integrating a systemic approach and social costs. *Energy Policy, 92*, 124–138. https://doi.org/10.1016/j.enpol.2016.01.029.

Schulze, K. (2014). Do parties matter for international environmental cooperation? An analysis of environmental treaty participation by advanced industrialised democracies. *Environmental Politics, 23*, 115–139. https://doi.org/10.1080/09644016.2012.740938.

Schwanitz, V. J., Longden, T., Knopf, B., & Capros, P. (2015). The implications of initiating immediate climate change mitigation – A potential for co-benefits? *Technological Forecasting and Social Change, 90*(Part A), 166–177. https://doi.org/10.1016/j.techfore.2014.01.003.

Scruggs, L. A. (1999). Institutions and environmental performance in seventeen western democracies. *British Journal of Political Science, 29*, 1–31.

Simmons, R. (2016). Improvement and public service relationships: Cultural theory and institutional work. *Public Administration, 94*, 933–952. https://doi.org/10.1111/padm.12257.

Sims, R., Schaeffer, R., Creutzig, F., Cruz-Núñez, X., D'Agosto, M., Dimitriu, D., Figueroa Meza, M., Fulton, L., Kobayashi, S., & Lah, O. (2014). Transport. In O. Edenhofer et al. (Ed.), *Climate change 2014: Mitigation of climate change. Contribution of working group III to the Fifth Assessment Report of the Intergovernmental Panel on Climate Change*. New York: Cambridge University Press.

Singh, S. P., & Dunn, K. P. (2013). Veto players, the policy-making environment and the expression of authoritarian attitudes. *Political Studies, 61*, 119–141. https://doi.org/10.1111/j.1467-9248.2012.00959.x.

Sørensen, C. H., Isaksson, K., Macmillen, J., Åkerman, J., & Kressler, F. (2014). Strategies to manage barriers in policy formation and implementation of road pricing packages. *Transportation Research Part A: Policy and Practice, 60*, 40–52.

Sotirov, M., & Memmler, M. (2012). The advocacy coalition framework in natural resource policy studies – Recent experiences and further prospects. *Forest Policy*

and Economics, 16, 51–64. https://doi.org/10.1016/j.forpol.2011.06.007.
Spataru, C., Drummond, P., Zafeiratou, E., & Barrett, M. (2015). Long-term scenarios for reaching climate targets and energy security in UK. *Sustainable Cities and Society, 17*, 95–109. https://doi.org/10.1016/j.scs.2015.03.010.
Spyra, H., & Salmhofer, H.-J. (2016). The politics of decarbonisation – A case study. *Transportation Research Procedia, 14*, 4050–4059. https://doi.org/10.1016/j.trpro.2016.05.502.
Sterner, T. (2007). Fuel taxes: An important instrument for climate policy. *Energy Policy, 35*, 3194–3202.
Tsebelis, G. (2000). Veto players and institutional analysis. *Governance, 13*, 441–474.
Tsebelis, G., & Garrett, G. (1996). Agenda setting power, power indices, and decision making in the European Union. *International Review of Law and Economics, 16*, 345–361.
UNEP. (2020). Emissions gap report 2020. https://www.unep.org/emissions-gap-report-2020
Vale, P. M. (2016). The changing climate of climate change economics. *Ecological Economics, 121*, 12–19. https://doi.org/10.1016/j.ecolecon.2015.10.018.
van Vuuren, D. P., Kok, M., Lucas, P. L., Prins, A. G., Alkemade, R., van den Berg, M., Bouwman, L., van der Esch, S., Jeuken, M., Kram, T., & Stehfest, E. (2015). Pathways to achieve a set of ambitious global sustainability objectives by 2050: Explorations using the IMAGE integrated assessment model. *Technological Forecasting and Social Change, 98*, 303–323. https://doi.org/10.1016/j.techfore.2015.03.005.
Vink, M. J., Benson, D., Boezeman, D., Cook, H., Dewulf, A., & Termeer, C. (2015). Do state traditions matter? Comparing deliberative governance initiatives for climate change adaptation in Dutch corporatism and British pluralism. *Journal of Water and Climate Change, 6*, 71–88. https://doi.org/10.2166/wcc.2014.119.
von Stechow, C., McCollum, D., Riahi, K., Minx, J. C., Kriegler, E., van Vuuren, D. P., Jewell, J., Robledo-Abad, C., Hertwich, E., Tavoni, M., Mirasgedis, S., Lah, O., Roy, J., Mulugetta, Y., Dubash, N. K., Bollen, J., Ürge-Vorsatz, D., & Edenhofer, O. (2015). Integrating global climate change mitigation goals with other sustainability objectives: A synthesis. *Annual Review of Environment and Resources, 40*, 363–394. https://doi.org/10.1146/annurev-environ-021113-095626.
Wegman, F. (2017). The future of road safety: A worldwide perspective. *IATSS Research, 40*, 66–71. https://doi.org/10.1016/j.iatssr.2016.05.003.
Weiner, R. R. (2014). Les reciproqueteurs: Post-regulatory corporatism. *Journal of Environmental Policy and Planning, 20*(6), 775–796. https://doi.org/10.1080/1523908X.2014.947923.
Wen, J., Hao, Y., Feng, G.-F., & Chang, C.-P. (2016). Does government ideology influence environmental performance? Evidence based on a new dataset. *Economic Systems, 40*, 232–246. https://doi.org/10.1016/j.ecosys.2016.04.001.
Wright, L., & Fulton, L. (2005). Climate change mitigation and transport in developing nations. *Transport Reviews, 25*, 691–717.
Wurzel, R. K. W. (2010). Environmental, climate and energy policies: Path-dependent incrementalism or quantum leap? *German Politics, 19*, 460–478.
Yang, Z., Mock, P., German, J., Bandivadekar, A., & Lah, O. (2017). On a pathway to de-carbonization – A comparison of new passenger car CO_2 emission standards and taxation measures in the G20 countries. *Transportation Research Part D: Transport and Environment, 64*, 53–69.
Zhang, W., & Pan, X. (2016). Study on the demand of climate finance for developing countries based on submitted INDC. *Advances in Climate Change Research*. https://doi.org/10.1016/j.accre.2016.05.002.

Sustainable WaSH Infrastructure

▶ Wash (Water, Sanitation, and Hygiene): Infrastructure as a Measure of Sustainable Development

Sustainable Watershed Management

▶ Watershed Sustainability: An Integrated River Basin Perspective

Sustenance

▶ Closing the Loop on Local Food Access Through Disaster Management

Systemic Innovation

▶ Philanthropy in Sustainable Urban Development: A Systems Perspective

S

Systemic Innovation for Thrivable Cities

Alexander Laszlo[1], Karin Huber-Heim[2], Stefan Blachfellner[3] and Pavel Luksha[4]
[1]The Bertalanffy Center for the Study of Systems Science, Buenos Aires, Argentina
[2]Circular Economy Forum, Austria, Vienna, Austria
[3]The Bertalanffy Center for the Study of Systems Science, Vienna, Austria
[4]Global Education Futures, Moscow, Russia

Synonyms

Anticipatory cities; Circular economy; Circular lifestyle; Curated emergence; Design thinking; Disruptive innovation; Holistic well-being; Learning ecosystems; Lifewide learning; Sustainable cities

Definition

The topic area of *social equity and community health and well-being* comprises a broad range of quality-of-life issues in the context of urban and regional futures. From personal and interpersonal relations to the quest for psycho-emotional and physio-mental stability to the maintenance and renewal of economic and ecosystemic vitality, the issues are – by their very nature – complex, intertwined, and multifaceted. The emerging dynamics of disconnection in urban and regional environments raises concern for the types of systemic innovation approaches that foster countervailing dynamics of health and well-being for thrivability in urban communities. Learning how "to human well," at both individual and collective levels, is key to curating the emergence of urban communities and regional communities-of-communities as systemic nurturance spaces. Wisdom traditions that offer insight into the practice of what Thich Nhat Hanh called *Interbeing,* and what the Zulu and Xhosa know as *Ubuntu,* can help emerge an alternative to the dominant Western tradition of cut-throat competition, strident individualism, and self-righteousness independence. Looking to urban and regional forms of community and interdependence through design principles and practices that foster evolutionary learning communities and, in turn, communities of such communities will help set development attractors that move humanity from survival imperatives to thrival potentialities. Systemic innovations that encourage circular lifestyle engagements with nature, from principles of biomimicry to practices of circular economy, are proving to be valuable tools in the emergence of thrivable cities.

Introduction

Systemic innovation for thrivable cities is an area of praxis. As such, it combines living systems theory, social systems design methodology, and lifelong and transformative learning philosophy. This combination necessarily involves other sub-domains, including social psychology, environmentalism, urban planning, architecture, ethics and esthetics, governance, and the economics of value creation and exchange. By focusing on systemic innovation as a humanistic and environmentally attuned approach to urban and regional development, quality-of-life issues are prioritized over standard-of-living metrics. Attention is given to self-directed sustainable change dynamics that are inclusive, participatory, diversity rich, and life affirming. Frameworks that are leading the way in this area include those emerging from the study of healthy and authentic communities (HACs); youth involvement and inclusion in design practices; guidance structures based on elder councils and Indigenous wisdom circles; nature-inspired design practices; diversity training for anti-fragility; and the principles of circular lifestyles and practices of circular economics. Not all of these approaches are practiced or even recognized as valuable in every urban development project, but they are generally taken to be of potential value in the context of systemic innovation for thrivable cities.

Systemic innovation as a domain of design practice is defined, explored, and illustrated in this entry. Likewise, thrivability in the context of urban and regional life is also defined and distinguished from sustainability. The case for integral and holistic approaches for urban and regional development planning is made, and examples of successful and emerging initiatives in the application of systemic innovation for thrivable systems are presented.

Key to this area of praxis is the notion of collective intelligence (Hamilton 2014). Patterns of the growth and structure of cities based on centralized planning models have not yielded expected or desired sustainable development outcomes. The need for rhizomic "hub and spoke" type ecosystems of communities that span and integrate learning approaches and thrivability outcomes across a variety of urban centers promises a new model for the design and distribution of thrivable cities into thrivable regions. To develop the requisite competencies of collective intelligence for such processes to inhere in large-scale development initiatives, a focus on education and training is fundamental. Initial incursions involve the cultivation of *connective intelligence* such that an ability to focus on the resources (human, natural, informational, and economic) necessary to collective thrivability is developed and put into practice. With the consolidation of connective intelligence competencies, a learning community can begin to enact *collective intelligence* practices through which self-directed sustainable development pathways are collectively explored and practiced. This paves the way for the emergence of *collective creativity* in the community, liberating its participatory and inclusive potential through direct engagement in creative visioning and the translation of vision into action. By sharing both the successes and the frustrations among communities that engage in processes of systemic innovation, other communities can leapfrog their development potential, enriching the *collective wisdom* of the urban region of interlinked cities. Fundamentally, this approach helps strengthen the socio-ecosystemic fabric of society and creates the conditions for the emergence of a true learning society that fosters the basis for its own flourishing while ensuring that of future generations, as well (Milbrath 1989).

Terminology and Conceptual Frames

In order to consider the nature of systemic innovation in the context and service of thrivable cities, we must begin with a thorough consideration of key terms. Given that this topic deals with an area of real-world study and impact, the clearer we can be in the language of discourse used, the more practical will be the application of the key terms employed.

- Systemic Innovation

According to standard usage, an *innovation* is the concretization of a practical idea that augments human capability for action with societal impact, existing as an intermediate phase between the conceptual *invention* of an idea and its marketable *diffusion* in society. Clearly, advances in science and technology have created unprecedented opportunities for human development and well-being, many of which are showcased in the built landscapes of urban environments. And yet, as Jacques Ellul warned reprovingly over fifty years ago, "the machine tends not only to create a new human environment, but also to modify man's very essence" (Ellul 1964). As such, technological progress over the last 150 years has brought with it certain "side-effects" (*cf.* Meadows et al. 1972) that, although generally ignored for some time, have now become global issues that threaten the stability of societies and ecosystems the world over. The familiar litany of modern-day ills includes population growth, social inequities, hunger, armed conflicts, water shortages, health equity, pollution, climate change, and pandemic diseases – and these are but a few of the issues, many of which are often exacerbated in urban settings, and which together form a complex challenge for societal development (Merry 1995, 78). In ever more urgent and pressing ways, the finitude of resources on our planet calls for new forms of production, distribution, and consumption — and for new ways of

S

researching, developing, and innovating social and technological change in order to answer that call. As will subsequently be presented in greater detail, the United Nations Sustainable Development Goals (SDGs or Global Goals) provide a vision for 2030 and beyond, addressing the world's most pressing issues, like climate change, inequalities, overconsumption of resources, health and well-being, and justice, based on a multisolving, systemic perspective with the intention to "leave no one behind" (United Nations General Assembly 2015).

An important characteristic of systemic innovation is its fusion of scientific and ethical knowledge. Instead of only answering "know how" type of questions, such innovative advancement in the type of socio-technical ecosystems that comprise cities must also provide the means to answer questions of "know why" and "care why" in regard to the way in which we live, work, and learn together. Clearly, for innovation to be efficient, efficacious, and effective, as well as ethical, esthetic, empathetic, and humane, no single individual can be responsible for shaping it. This is another aspect of systemic innovation: It relies on collective intelligence. Systemic responses to the complexity of contemporary global and local challenges – personal, societal, and planetary – require an expanded perspective: a way of recognizing interconnections, of perceiving wholes and parts, of acknowledging processes and structures, and of blending apparent opposites. But most importantly, they require collaboration and an appreciation of reciprocity. Individual solutions and breakthrough ideas are necessary but not sufficient. Real opportunity to affect change arises from the systemic synergies that we create together. Those who are able to draw on contemporary insights from the sciences of complexity, computational and life sciences, and an embracive spirituality that reinstills a sense of integrity and ethical purpose in life are the leaders and designers of systemic innovation.

Systemic innovation calls for a clear and well-developed appreciation of how the dynamics of sociocultural change are linked to the dynamics of technological innovation. The set of interconnected and interdependent challenges that characterize global civilization in the first half of the twenty-first century directly impact, and are impacted by, the advancement of society at local and regional levels, and nowhere more so than in cities (Meadows et al. 2004).

The multilevel perspective (MLP) framework for the analysis of socio-technical systems change, originally developed by Frank Geels (2002a, 2002b), became a feasible systems analytic tool for understanding innovation transitions, including the phenomenon of systemic innovation for thrivable cities (OECD 2015).

The MLP framing distinguishes between three levels of the system to be innovated, their interdependencies, and interactions:

1. The landscape on the macrolevel
2. The dominant patchwork of the socio-technical regime on the meso-level
3. The niches on the microlevel

When applied to cities, the macrolevel landscape consists of the systems patterns, which

(a) Constitute and strengthen the meso-level patchwork of regimes, though when the landscape changes
(b) Open microlevel windows of opportunities for changes and innovations in the dominant regimes

Environmental and demographic changes, new social movements, changes in the prevailing political ideology, fundamental economic changes, new scientific paradigms, and sociocultural developments (such as increasing environmental awareness) are examples of landscape phenomena on the macrolevel. These phenomena can be observed and analyzed in terms of their interdependencies with the meso-level regimes. If several of these phenomena arise in codependency, the regimes will be pressured to adapt or innovate.

A socio-technical regime on the meso-level may include guiding principles and paradigms (such as growth), dominant technologies and infrastructures, industrial structure, market structures and the relationship between producers and consumers, politics and legal regulation, and the knowledge-base of the regime. The stability of existing socio-technical regimes occurs through interaction between the material aspects of the systems, embedded actors and organizational networks, and the rules and regimes which guide perceptions and actions. In general, the regimes tend to be resistant to change and rather reinforce their existing patterns, until the landscape changes and their environmental conditions ignite a necessary response.

Niches at the microlevel are most important for generating this type of response. Niches are the spaces for prototyping the experiments of innovative perceptions, actions and emerging patterns, and knowledge-bases. In essence, they manifest the spirit of innovation that Buckminster Fuller called for in his often-cited quote: "You never change things by fighting the existing reality. To change something, build a new model that makes the existing model obsolete." These niches can arise "accidentally," for example, through local problems within the regime, for which the niche innovations provide solutions. But niches can also be consciously developed. In a niche, innovations can develop without being exposed to the regime's selection pressure from the outset, as is typically the case with innovations that assert themselves in the existing socio-technical environment and therefore have to orient themselves from the outset to the existing regime structures. It should be noted that niches are not individual projects or experiments. They are initiatives in which alternative rules and practices are developed by a network of actors. Thus, these niches become vital for subsequent innovation leaps of the regime under pressure from the landscape. Niches are the systemic nurturance spaces of societal and technological innovation and, in this sense, are metaphorically akin to niches in a biological ecosystem. The innovation from a given niche can enter its regime during propitious windows of opportunity. In many cases, the regime will be adjusted by the perturbation stemming from the niche but will nonetheless assimilate the innovation. While radical innovation may take place in niches, the most common innovation pattern in the meso-level regimes involves incremental change.

As mentioned previously, a simple linear extrapolation of the trends that characterize the current set of challenges for humanity points toward ecological catastrophe and societal disintegration. However, with a solid grounding in systems thinking and the sciences of complexity, it is possible to explain why this is happening – and to develop policies and strategies to innovate the means of emerging a future that is not only sustainable, but also desirable and even thrivable, as well. There is an urgent societal need for research, development, and innovation based on the systems sciences, and in particular on the sciences of complexity and the study of socio-technical systems (*Cf.* Goerner 1994; Capra 1998; Pasmore & Sherwood 1978; Pasmore 1988).

A critical aspect of systemic innovation is that it is particularly applicable in VUCA and RUPT environments. VUCA stands for volatile, uncertain, complex, and ambiguous and is typically used in the sciences of complexity to characterize situations or conditions that express this combination of characteristics. RUPT stands for rapid, unpredictable, paradoxical, and tangled and is typically used in humanistic and sociological studies to characterize the lived experience of people who find themselves in VUCA situations or dynamics. A serious problem in contemporary issues of the growth and structure of cities lies in the fact that in the face of increasingly VUCA futures, many leaders, institutions, and conventional societal structures cling to yesterday's world as the referent for the behavior of dominant meso-level regimes until the pressures of macro-level landscapes and the perturbation of the niches open up their receptivity to change even in the face of uncertainty. If we are to respond appropriately to the demands of increasing complexity and to move across an ever higher "complexity barrier," new approaches for handling these challenges of change will be needed. The potential to foster a positive VUCA world – one based on vision, understanding, clarity, and agility (Johansen 2012) rather than on the negative reactive frame of the acronym – can best be advanced through systemic innovation that seeks to curate conditions of life and living that favor the dynamics of thrivability. As Janine Benyus notes, life creates conditions conducive to life (Benyus 2002a). Systemic innovation provides a path to connect life with life and to reimbue our relations at five levels of thrivability: with ourselves, with each other, with our more-than-human world, with past and future generations of all beings, and with the deeper patterns of being and becoming that inform all evolutionary processes in this universe (Laszlo 2020).

- Systems Thinking

Systems Thinking is an internalized manifestation (in the thinking of individuals or social systems) of systems concepts, systems principles, and systems models. It represents a paradigmatic effort at scientific integration and theory formulation on the transdisciplinary plane (*cf.* Scholz 2012; Scholz & Steiner 2015; Rousseau, Wilby, Billingham & Blachfellner 2018; Scholz 2020). Prior to the efforts of Ludwig von Bertalanffy and his colleagues in the 1950s and 1960s, no such effort derived from the natural sciences had been previously attempted. Nevertheless, many other expressions of holistic, integral, and indeed systemic thinking have been explored and developed in the history of Western thought. This is evidenced in the work of Thales of Miletus (on the unity of matter and form) as early as the beginning of the 500s BCE, and since then through many others, including in works of Nicholas of Cusa (on the coincidence of opposites) in the fifteenth century, of Alfred North Whitehead (process philosophy), Jan Smuts (holism), and notably of Alexander Bogdanov (tektology) – all in the 1920s, as well as of Norbert Wiener (cybernetics) in the 1940s, and of many others. While all of these expressions advanced various forms of systemic thinking, it is generally recognized that modern systems thinking came into being as the Systems Movement began with the founding of the Society for General Systems Research at the hands of Ludwig von Bertalanffy, Ralph Gerard, Anatol Rapoport, Kenneth Boulding, James Grier Miller, and Margaret Mead, among others.

Given the focus on systemic innovation for thrivable cities, it is worth distinguishing among different stances for how to engage with and apply a systems view of the city, its embedding context, and its nested component elements. To that end, it is important to differentiate between the transdisciplinary field of systems thinking, the loosely defined set of approaches that constitute systemic thinking, and the methodical approach to problem solving known as systematic thinking.

Systems thinking	An internalized manifestation (in the thinking of individuals or social systems) of systems concepts, systems principles, systems methods, and systems models
Systemic thinking	A tendency or natural predisposition to think in terms of systemic relationships without necessarily drawing upon systems concepts, systems principles, or systems models
Systematic thinking	Any methodical step-by-step approach that is carried out according to a predetermined algorithm or a fixed plan

To be a systems thinker, one needs both a systemic perspective and the type of orientation and intention of those involved in naturally holistic areas of inquiry (such as pattern language, ecological and living ecosystem dynamics, qualitative sociology, etc.), but even this is not sufficient to be a systems thinker. What is needed in addition is command of the conceptual tools required to bring those intentions into strategic models and, if possible, the practical methodological competencies to implement those models through real-world designs. So while one can be a systemic thinker, it does not necessarily follow that one is also a systems thinker since the latter implies an understanding of systems concepts not appreciated by the former. Systemic innovation for thrivable cities requires all three forms of thinking, with an emphasis on the mastery of systems thinking.

- Systems Design

Systems Design is a decision-oriented disciplined inquiry that aims at the construction of a model that serves as an abstract representation of a future system to be created. Such a framework, based on systems thinking, is relevant for the leadership of socio-technical systems innovation in VUCA contexts such as cities. As a relatively recent contribution to the area of systems thinking that comprises the field of the social systems sciences, Evolutionary Systems Design (ESD) responds to a need for a future-creating design praxis by embracing not only human interests and life spans but also those on ecosystemic and evolutionary planes as well (*cf.* Laszlo 2003, 29-46). The split between macro- and microscale conceptual frameworks in contemplation of human developmental concerns continues to provide a difference of perspective within systems science, tending to inspire either homo-centric change efforts or evolutionary interpretive frameworks for them. Little by way of evolutionary strategies for the design of healthy and sustainable modes of being and becoming in urban and regional development is contemplated in partnership with the life support systems of planet Earth. This is what ESD is all about.

As a species, our actions and interventions on this planet have been largely driven by chance and, at best, "20/20 hindsight." However, as Margaret Mead noted, we are at a point where, for the first time in human history, we are able to explain what is happening while it is happening (in Montuori 1989, 27). ESD builds on this relatively new metareflective competence by serving as an instrument for the evolution of consciousness and for conscious evolution. It suggests that with the new understanding of evolutionary dynamics and effective approaches to the participatory design of socio-technical systems, our species can stop drifting upon the currents of change and begin to adjust its sails in view of sustainable and even thrivable evolutionary futures. "As evolution becomes history, it can become conscious. As Jonas Salk put it: conscious evolution can emerge from the evolution of consciousness — and from the consciousness of evolution" (E. Laszlo 1996, 139). This is the understanding upon which ESD has been conceived.

The ESD orientation to future creation is essentially possibilistic. It assumes that human beings have the choice consciously to participate in the cocreation of the future. And yet, it seeks neither to predict nor to "socially engineer" the future. Rather, it seeks to create the conditions for the emergence of sustainable and evolutionary futures.

> "In systems such as contemporary society, evolution is always a promise and devolution always a threat. No system comes with a guarantee of ongoing evolution. The challenge is real. To ignore it is to play dice with all we have. To accept it is not to play God — it is to become an instrument of whatever divine purpose infuses the universe." (E. Laszlo 1996, 139)

The aphorism that captures the spirit of ESD is one of flow: *We cannot direct the wind, but we can adjust the sails.* Learning to sail the currents of evolution – not just to "go with the flow" but to become active participants in the journey – is at the heart of the ESD journey.

- Viability, Sustainability, and Thrivability

In the context of systemic innovation for thrivable cities, linear, reductionist, monodisciplinary, mechanistic thinking is not only hopelessly out of date, but also it is increasingly irrelevant – even dangerous. By their very nature, cities are contexts of social and socio-environmental interaction. Classical Darwinian development approaches focus on the survival needs of a given population and, in consequence, consider the parameters of systemic viability as the baseline reference upon which to explore developmental potentials. In this framework, sustainability is not always even incorporated into the development plans, but it is becoming more and more expected, and even required – due both to changing societal pressures and expectations as well as national and international standards such as the SDGs.

Beyond planning for viability, sustainability and then sustainable development became terms of reference for international, national, regional, urban, and local systemic change initiatives. Classical definitions of sustainability stipulate that any large-scale change, both in terms of process and outcome, must not exceed the carrying capacity of its contextual environment (be it an ecosystem, a bioregion, a biome, or the entire terrome of the planet itself). This is the bedrock of the UN SDGs, though clearly they reach well beyond this minimal objective. To the extent that the change process encourages, incorporates, and expresses manifestations of awe, joy, love, a sense of the sacred, celebrations of life, gratitude for any aspect of existence, or an appreciation for the magical/mystical quality of deep connections with the world around us, it begins to move beyond notions of sustainability and sustainable development. This is where notions of flourishing and thrivability take the discourse a step further. However, those who find themselves caring only for their own thrivability cannot participate in the cocreation of thrivable environments with and for others and their urban ecosystems. The need to focus also on the thrivability of others and ways to provide convivial contexts for them to engage with life is increasingly a requisite in urban contexts.

One of the widely used (though perhaps narrower) frameworks to comprehend thrivability is that of *holistic wellness,* understood by the World Health Organization (WHO) as "the optimal state of health of individuals and groups" – with *health* defined as "the state of complete physical, mental and social well-being" (WHO 2008). This definition extends the concepts of health and wellness far beyond the biomedical context: Holistic wellness has individual as well as communal, territorial, and even planetary dimensions and is contingent upon behavioral, cultural, institutional, relational, and environmental factors (Miller & Foster 2010). The convivial aspects of holistic wellness are clear: Conditions in urban settings (cultural backgrounds, economic opportunities, architectural styles, pollution levels, etc.) serve to contextualize and shape individual and collective wellness, and yet they themselves are determined by the choices and contributions of individuals and communities.

The frame of thrivability evokes exploration of, and engagement with, the way in which learning, playing, talking, working, and all aspects of life-as-art connect us to ourselves, to each other, to our more-than-human world, to past and future generations of all beings, and to the deeper dimension of the cosmos that patterns the evolutionary dynamics of emergence and coherence. These are the five intertwingled dimensions of systemic thrivability. Systemic innovation engages evolutionary systems designers across all five dimensions through dynamics that encourage consciously connecting, intertwingling, and cultivating the bigger story of our individual and collective being and becoming. The quality and character of this story depend on the way in which leaders of systemic innovation seek to author their life interactively along these five dimensions. Since these dimensions have been articulated and explored in detail elsewhere in the literature (*cf.,* Laszlo 2018, 2020), it suffices to list them here:

1. The intrapersonal dimension of coherence; thrivability within oneself
2. The inter-personal dimension of coherence; thrivability with one's communities and social systems
3. The trans-species dimension of coherence; thrivability with the more than human world
4. The transgenerational dimension of coherence; thrivability with past and future generations of all beings
5. The pan-cosmic dimension of coherence; thrivability with the deep dimension of immanent consciousness in the cosmos

By consciously, purposefully, and intentionally curating each of these dimensions *in dynamic relation to the other four,* it is possible to foster empathy-based innovation in service of a greater collective thrivability. The sense-abilities of the five dimensions and the concomitant sense of response-ability to which they give rise provide a self-generated ethical compass by which to guide systemic innovation. Indeed, it is not truly possible to address the second dimension without addressing the first – or the third without the other three. And the fifth dimension requires a truly transcendent capacity to move one's empathy and practice across time and space. It is possible to start with any of the dimensions – each of them is a portal for entry into this conscious and conscientious design space. But the portal only opens to all five dimensions when engaged with and enacted through integral *intention* informed by the living quality of *love.* For this to happen beyond the domain of naïve good intentions, the competence of visionary leadership must infuse all efforts of systemic innovation. Otherwise, the rest is just one or another form of dreaming about it.

As with the multilevel perspective (MLP) framework, a similar stance proves valuable when taken with regard to thrivability. Indeed, multidimensional thrivability can serve as a North Star for initiating, nurturing, cultivating, and aligning systemic innovations. Not only does it establish a set of criteria that help define what innovations are needed, but it also sets forth the principles by which these innovations should operate. Technological and economic plans, as well as the governance and communication models that incorporate them, should embrace thrivability not only as an outcome of their implementation, but also as an essential quality of the process of change to which they give rise. In other words, any effort at systemic innovation for thrivable cities that is designed or planned to be scaled up by means of manipulation, coercion, or deception of any kind ought to be rejected before it can be implemented since it does not conform with this principle of multidimensional thrivability.

Mahatma Gandhi is often quoted as having said, "first, be the change you wish to see in the world." Thc first word is oftcn omitted, yet it is key to being the change. Systemic innovation for thrivable cities draws upon this understanding, acknowledging that, if you are to "be the change," first you must be in convivial relationship with yourself and with everyone and everything with which you interact. Ultimately, this is an expression of love in the Greek sense of *agape*: It is the joy of connecting and being connected with, of affirming and being affirmed, and of dancing your path into existence with every breath, every thought, and every act that is at once both you and the rest of the universe. When engaged with in this spirit of systems design, cities become living greenhouses for life-affirming, future-creating, and opportunity-increasing expressions of individual and collective life.

The idea of "being the change" is also frequently defined in terms of "system being" (Laszlo 2015). In the systemic innovation design context, this invokes the idea of creating *protopian* futures. Protopia – as contrasted with utopia, dystopia, and myopia – is "a possible future that is inspiring, optimistic and achievable, based upon individual and collective action . . . of creating the future we want to see in the here and now" (Luksha *et al.* 2018). The inspiring future is brought into being by prototyping a diversity of possible solutions and scaling up those that work. In the process, learning to mimic the force of evolution occurs within communities of change leaders. Yet most important of all is that systemic innovators also need to think of themselves *as prototypes of their own "future selves,"* and therefore learn to become the system they want to see in the world. This is a critical condition for instigating the paradigm shift of our global civilization: The status will change only when a critical mass of humanity begins acting and living in accordance with the behavioral patterns, worldviews, and values of a thrivable civilization – and these patterns need to be established and embodied by those who lead the change (Laszlo *et al.* 2017).

Famed Buddhist scholar Robert Thurman asks us to "imagine a culture in which everything is geared toward helping all individuals become the best human beings they can be; in which individuals are driven to devoting their lives to becoming enlightened by the natural flood of compassion for others that arises from their wisdom" (Thurman 2005). Systemic innovation for thrivable cities embraces this type of higher objective *simultaneously* across all of the five intertwingled dimensions of systemic thrivability mentioned previously. This is the potential of systemic innovation for thrivable cities when it draws upon a deeper and higher framework of human potential. To generate the living greenhouses that this approach is oriented toward, the systems thinker and designer must see themselves as a type of gardener human potential, creating living spaces for human interaction to both *flourish* and *thrive.*

The analogy of the garden for the design of integral cities serves to distinguish between these two terms. A well-tended and healthy garden is a flourishing domain, and the plants and interdependent subsystems within it are the thriving beings within that domain. A flourishing garden can bring forth thriving plants, but it is also true that thriving plants – when structurally coupled through both auto-catalytic and cross-catalytic feedback cycles – can bring forth a flourishing garden, hence both the upward causation of systemic leverage points among the component parts of the system that can lead to higher levels of synergy and the downward causation of systemic nurturance spaces which elicit higher levels of synergy among the component parts. *Flourishing* relates to the characteristics of the domain, and *thrivability* (as with sustainability) relates to the potential realized/actualized expression of the beings within the domain. As such, it is worth bearing in mind that the term *flourish* comes from the Latin *florere* meaning to flower or blossom. By contrast, *thrive* comes from the Norse term *thrifask* meaning to grasp or get hold of. The agency involved in thriving connotes taking an active stand in creating conditions of well-being for oneself and for others in order that everyone, and everything, may flourish.

- The Ecosystemic Perspective

The incorporation of the term *ecosystem* in contemporary social science discourse has become a dominate feature in the domains of business and economics as well as in many public policy fora, from considerations of innovation and digitalization in all sectors of society to issues of sustainable development. Unfortunately, the way it is used often leads to misunderstandings.

The ecosystem view, which originated in the epistemological domains of systems science and human ecology, is based on what is known as Nested Systems Theory (NST). Nested systems are complex systems that exhibit emergent properties because they are open to information, material, and/or energy exchanges with their environments. By being open in this way, they have the capability to process resource inputs and outputs needed to adapt and grow. They are complex adaptive living systems because they show properties of structural resilience, including redundancy, modularity, and requisite variety (Walloth 2016).

The behavior, and thus the resulting patterns and observed rules of conduct, of each nested system is interdependent on its embedding and embedded systems in both directions, as inputs and as outputs. The interdependencies are the bases for further emergence. The technological systems are nested in the economic systems, which are nested in the social systems that are nested in the ecological systems. Thus, the thrivability of the ecological systems becomes the underlying imperative for all of the other systems to flourish. The outdated view of technology being the main driver of innovation is strongly challenged in the contemporary approaches of systemic innovation for thrivability.

Johan Rockström & Pavan (2016) popularized the ecosystemic framing of societal development in considering the economic, social, and ecological aspects of the UN Sustainable Development Goals (SDGs) as nested parts of the biosphere. This view has been empirically tested with a data simulation model called Earth3 (Randers *et al.* 2019) which demonstrated that of the 17 SDGs, conventional efforts to achieve the 14 socioeconomic goals will only serve to increase pressure on the boundaries of planetary sustainability (and even its life-supporting viability) in the long run, thereby moving the world away from the remaining 3 life-affirming environmental SDGs.

The epistemology of complex adaptive ecosystems and ways of modeling them challenges our paradigm for development in areas such as the growth and structure of cities (Bryn Mawr College n.d.) and urban development. It does so by replacing the current sectoral approach with an understanding of the essential interdependencies of all the systemic components involved, thereby transcending the siloed emphasized of the Brundtland Report (World Commission on Environment and Development 1987; Barbier & Burgess 2017) where social, economic, and ecological development are seen as separate, although connected parts. The NST model highlights the necessity of a transition from linear, nonsystemic thinking about city planning and development initiatives to an ecosystemic framework where the economy serves to help society evolve safely within the carrying capacity of our planet. This epistemology has also been further developed and strengthened through the work of Kate Raworth on the concept of what she has popularized as *doughnut economics* (2017).

Sustainable Cities in the Global Agenda 2030

In 2015, a gathering of 193 member states of the United Nations agreed on 17 goals to transform our world. These goals comprise 169 targets which nation-states can use to align their national policies with global sustainability objectives and to measure their progress in addressing them. Human rights were established as a cross-cutting concern of all these Sustainable Development Goals (SDGs). In contrast to the Millennium Development Goals (MDGs) that were developed fifteen years earlier by the UN, the new goals aim for comprehensive changes that also require industrialized countries to take their share of responsibility. This applies to such things as the careful use of resources, responsibility for social standards, and the emission of climate-damaging gases. The focus of the SDGs is squarely on sustainability. In addition, there are sociopolitical goals such as gender equality, a fair tax policy, the reduction of inequality between and within states, access to legal aid, and the operation of diversity inclusive institutions.

The first target of the first SDG proposed by the Open Working Group (OWG) of UN Member States is to "eradicate extreme poverty for all people everywhere" by 2030. As mentioned earlier, "Leave no one behind" (LNOB) is the central, transformative ideal of the 2030 Agenda for Sustainable Development, and it is widely recognized that attaining the goals can only be achieved through a radical transformation of our current economic, social, and political systems.

SDG 11 is titled Sustainable Cities and Communities and counts with a set of targets to "make cities inclusive, safe, resilient and sustainable." As all of the SDGs are of equal importance, this one applies to the half of the global population – almost 3.5 billion people – who currently live in cities and urban communities.

It is estimated that 70 per cent of the world population will live in urban settlements by the year 2050 (United Nations Human Settlements Programme (2019). The impact of the COVID-19 pandemic will be felt most acutely in poor and densely populated urban areas, especially for the one billion people living in informal settlements and slums worldwide, where the enormous density of population makes it difficult to follow measures such as social distancing and self-isolation, or to comply with public hygiene measures.

The SDGs promote positive economic, social, and environmental links between urban, peri-urban, and rural areas by strengthening national and regional development planning. In particular, they place attention on less developed countries (LDCs) by encouraging financial and technical assistance in the construction of sustainable and resilient buildings utilizing local materials.

All but one of the SDG targets are framed in a time horizon that reaches to the year 2030, and all place specific emphasis on the needs of people in vulnerable situations, including women, children, persons with disabilities, and older persons. A strong emphasis is also placed on the need for participatory, integrated, and sustainable human settlement planning and management.

By 2020, the number of cities and human settlements adopting and implementing integrated policies and plans that incorporate inclusion, resource efficiency, mitigation of and adaptation to climate change, and resilience to disasters has substantially increased, as well as the development and implementation of holistic disaster risk management at all levels.

To a great extent, these benefits are being achieved by changes in the way urban systems are planned, designed, and financed, and how they are made, used, and repurposed. Buildings, mobility, products and services, and food systems are urban systems that play an important role in people's lives. This is where systemic innovation is at its best. Within each of these systems, the opportunities offered by innovations in design, business models, and digital technology are manifold.

The City as a Living System

- The Circular Economy

The concept of the *circular economy* emerges from a variety of international schools of thought that date back to the Middle Ages and reach to times beyond the industrialization that emerged after the Second World War and the introduction of mass consumption with its incentive structures for perpetual economic growth. As such, the far-reaching origins of the social construct of circularity cannot be traced back to a specific date of origin or a single creator.

The modern narrative of a circular economy was developed and formulated by the Ellen MacArthur Foundation, a very influential charity dedicated to the establishment and acceleration of economic transition, strongly influencing the European perspective. It is based on schools of thought such as the cradle-to-cradle concept developed by Michael Braungart and William McDonough (2008) and Walter Stahel's performance economy (2006). It also integrates approaches from the research of Janine Benyus on biomimicry (2002b), as well as studies of material and energy flows through industrial systems, an area known as industrial ecology. It is further inspired by the work of Paul Hawken and Amory and Hunter Lovins on natural capitalism (2010), as well as studies on regenerative design by John Lyle (1994). As a field, it includes insights from the domains of bioeconomics, restorative economics, resource-based economics, and applied methodologies such as materials flow analysis and product life cycle analysis, all of which had significant influence on its formation (Ellen MacArthur Foundation 2019).

The work and studies of the MacArthur Foundation are based on the idea of a circular economy as a construct that mimics living systems. From this perspective, cities are compared with the living system of the human body, made up of many subsystems – such as the circulatory, respiratory, and digestive systems, for example – with each of these systems needing to work effectively in conjunction with the others to ensure the health and well-being of the whole. Value creation and well-being within systems are understood to be the result of independent dynamic systems, constantly responding and adapting to change. Just as our bodies constantly use material flows such as blood and oxygen to thrive, so too do city systems constantly have materials flowing through them in the form of capital, knowledge, data, energy, and materials, for example. At present, however, materials tend to flow in a linear way through cities. We source, produce, use, and within a short amount of time, dispose of the materials in flow. Over 65 years ago, economist and retail analyst Victor Lebov proclaimed:

> Our enormously productive economy . . . demands that we make consumption our way of life, that we convert the buying and use of goods into rituals, that we seek our spiritual satisfaction, our ego satisfaction, in consumption. . . . We need things consumed, burned, worn out, replaced, and discarded at an ever increasing rate. (1955)

It is a travesty of human progress to note that his characterization of *homo economicus* is still being held valid in today's day and age. But there is hope. A helpful reframing, consistent with a circular economy approach, is to think of cities as living systems that rely on the healthy circulation of resources.

The circular economy is based on three principles:

1. Preventing waste and pollution
2. Keeping products and materials in use for extended periods of time
3. Regenerating natural systems

The ideal objective of an industrial circular economy is therefore to maintain the value of manufactured goods at the highest level for the longest possible period of time. As such, managing existing resources by reusing, repairing, and remanufacturing them at the highest level of quality is paramount.

In urban and regional environments, the principles and practices of circular economy imply a system change in which services are provided using natural resources efficiently and in ways that optimize their reuse or regeneration. Furthermore, economic activities are planned and carried out in a way that closes, slows, and tightens information flows while maintaining responsiveness in feedback loops across value chains. As such, while the feedback loops are agile and efficacious, the flows they involve are slow. For example, the dynamics of mindful living form the basis of the slow food movement, and of slow money and slow cities. There is no rushing from one cycle to the next. People walk from place to place, they are involved in food growth and distribution cycles, and in short, they pay it forward. Such dynamics are the antidote to the rush/stress of modern Western fast flow living.

- Circular Cities

Cities are collectors of materials and nutrients, accounting for 75% of natural resource consumption. Cities also produce 50% of global waste and 60-80% of greenhouse gas emissions. These characteristics are all symptoms of the current global take->make->waste linear economy.

With their high concentration of resources, capital, data, and talent spread over a relatively small geographic area, cities are uniquely positioned to drive a global transition toward a circular economy. A Circular City is first and foremost the applied dynamics of a global circular economy – a system based on three principles:

1. Avoiding waste and pollution
2. Maintaining an unlimited circulation of material flows
3. Safeguarding the regeneration of embedded and embedding ecosystems

The collectively researched *Circular Cities 2030* initiative (2020) draws the conclusion that daily life in a circular city must be organized differently from the way we live today: Essentially, it must be inclusive and creative. Cities make perfect hubs for innovation. Lifewide education is at the heart of the collective experiential learning opportunity offered in urban environments, and community-based social entrepreneurship holds the potential to be a major financial pivot of the needed type of urban change (Hamilton 2019). The networks built by urban communities can be leveraged to provide access to meaningful work, housing, health care, and social mobility.

At the same time, Circular Cities are high-tech operations designed for economic and urban renewal. For instance, they encourage the formation of interconnected energy grids based on carbon-zero microgrids. These are decentralized, community-based renewable energy systems based on citizen's energy cooperatives. This adds a dynamic dimension to city life and redefines the concept of work, providing a means for the generation of both prosperity and community.

Systemic innovations in combined mobility offer comprehensive networks of public transport options such as buses, trains, bicycles, scooters, and a variety of other on-demand pickup and delivery services. Mobility as a service gives urban planners and engineers the freedom to rethink the movement needs of urban denizens. This results in prioritizing the design and implementation of important infrastructure services such as citywide water supply systems. In the course of such urban planning processes, wastewater treatment, stormwater management, and the remediation of contaminated soils can also be improved.

Urban production in a Circular City is focused on localized, bioregional inputs and circular materials use through the incorporation of regenerative supply chains and processes of industrial symbioses. Agriculture is integrated into the fabric of urban communities, combining the high-tech efficiencies of vertical farming, aquaponics, and photovoltaics, while the transformation of the built environment serves to create more space for regenerative urban farming practices and animal husbandry. In urban agriculture, the city seeks to provide some of its own food, reuse food waste, and keep wastewater in closed and local cycles to produce vegetables, fruits, and fish (Sukhdev et al. 2017, p.2).

- Anticipatory Learning Cities

To curate the city as a living system based on the principles of circular economics, the dynamics of evolutionary learning communities, and the practices of flourishing protopias, systemic innovation will increasingly need to focus on cultivating the rise of *the anticipatory city* (Inayatullah 2021). This conception of the city holds the potential of leading to a true flourishing of human potential: "Traditionally cities have been focused on short-term planning, expanding roads, and collecting garbage. In several recent projects, what has emerged are discussions on the rise of the anticipatory city. In this new image of the city, using big data, cities become sites where we can anticipate tomorrow's problems: flooding, psychological depression, pandemic spread and on the positive side, areas of well-being, longevity, and indeed even bliss" (Inayatullah *op. cit.*)

While the anticipatory city is as yet an anticipation of the future, it is also a city in making, now. It is being prototyped through systemic innovations for thrivability by systems thinkers and designers who are learning to use the SDGs and other transformative change scaffolding to bring the practice of circular lifestyles into manifest being. And while the full range of qualities of this protopian city still wait to be discerned, the anticipatory approach already makes possible the listing of several of its traits that are increasingly evident, and which stand in stark contrast to the industrial cities of the past (*cf.* Graham et al. 2020). The protopian anticipatory city is one that is as follows:

- Participatory and cocreated (vs. divided, alienated, and imposed)
- Complex, networked, and feedback based (vs. standardized, centralized, and linear)
- Living, and organically evolving (vs. mechanistic, and centrally designed)
- Regenerative and reviving (vs. extractive and exploitative)
- Postanthropocentric yet humanized (vs. human-system-centered yet dehumanized)
- Poly-temporal and poly-spatial (vs. synchronized and localized)

With the establishment of a Global Network of Learning Cities (GNLC) under the auspices of UNESCO, the figurative ideal of complementing the familiar label of "manufacture in..." with one that reads "mentefactured in..." is starting to gain traction. With a special focus on SDG 4 ("Ensure inclusive and equitable quality education and promote lifelong learning opportunities for all") and SDG 11 ("Make cities and human settlements inclusive, safe, resilient and sustainable"), the GNLC initiative promotes a form of peer-to-peer learning where, in this case, the peers are entire cities. The idea is that cities can learn from each other's successes and failures in meeting their respective development objectives such that they do not need to repeat the same mistakes or miss valuable opportunities if they can learn with and from each other directly.

UNESCO defines a learning city (UNESCO Institute for Lifelong Learning n.d.) as a one that:

- Effectively mobilizes its resources in every sector to promote inclusive learning from basic to higher education
- Revitalizes learning in families and communities
- Facilitates learning for and in the workplace
- Extends the use of modern learning technologies
- Enhances quality and excellence in learning
- Fosters a culture of learning throughout life

By promoting peer learning among member cities, cultural prosperity, social inclusion, economic development, and sustainability are all encouraged and enhanced. In the process, the ideal of lifelong learning is complemented and enriched by the equally powerful ideal of *lifewide* learning. The notion of lifelong learning is fairly straightforward and easy to understand. It denotes a learning path that does not end with the obtaining of an academic certificate or degree, or with the completion of studies at any level of formal schooling. Instead, lifelong learning encourages an approach to learning for life: learning all the time and throughout one's entire life.

Lifewide learning adds to the objective of seeking to learn throughout one's life the goal of learning in all aspects of one's life. When framed in the context of the anticipatory city, the learning city becomes a place that fosters full-spectrum learning in lived environments. The more diverse and heterogenous the learning environment, the richer the potential breadth of lifewide learning opportunities. As the article on "Learning Cities: ecosystems for lifewide-lifelong learning" proclaims (Lifewide Education n.d.), the very idea of a learning city "encourages us to think and act beyond the walls of established educational institutions and infrastructures and think about the **city as an ecosystem** containing vast opportunities, resources and potential for enabling people to learn and develop themselves in ways that meet their needs, interests and ambitions" [emphasis in original].

Use Cases

A variety of tested and applied systemic innovation approaches and tool sets are available, each incorporating various concepts previously highlighted. Most of these mainly address city administrations, as they have been identified to be the main initiators and funding sources of public sector innovation. Examples of such innovation include the local dimension of innovation for sustainable development (Tapia & Menger 2019), nature-based innovation systems (van der Jagt *et al.* 2020), emergence fostered by systemic analysis coupled with purposefully structured creative processes (Lanhoso & Coelho 2021), and cocreating sustainable urban metabolism toward healthier cities with strong stakeholder participation and "bottom-up" approaches (Froes & Lasthein 2020).

- Case One

One of the strongest use-case-based approaches was developed (and is continually being cocreated) with public administrations around the world by the OECD Observatory of Public Sector Innovation. This body also identifies cities as the key players in public sector innovation. Their so-called public sector innovation facets (OPSI 2018) and *de facto* tool sets comprise:

- *Enhancement-oriented innovation* to improve existing community development practice
- *Mission-oriented innovation* to identify pathways for reaching ambitious multilevel goals, with specific emphasis on systems thinking approaches
- *Adaptive innovation* to ease the incorporation of evolutionary changes in the sociotechnical ecosystems of cities
- *Anticipatory innovation* to accept and engage with the dynamics of significant uncertainty
    - and the four different types of change that occur at the intersection of these different facets:
        i. sustaining change
        ii. transformative change
        iii. disruptive change
        iv. optimizing change
- Case Two

The visual toolbox for system innovation (De Vicente Lopez & Matti 2016) is a collection of ready-to-implement tools to structure and manage the challenges, and exploit opportunities, of sustainability innovations and transitions. These are designed to help practitioners map, analyze, and facilitate sustainability transitions. The toolbox is built upon four modules:

1. Stakeholder management
2. Multilevel perspective
3. Visioning and backcasting
4. Niche management

Taken together, these modules facilitate the problem-solving process by setting out a pathway in the fuzzy and uncertain process of systemic innovation. The tools generate a multidisciplinary and multicultural setting, including transition management, innovation management, systemic thinking, design thinking, and project management. As such, the approach is based on a feasible systemic understanding of the problems and challenges at hand instead of on a linear process of reasoning. It incorporates a strong emphasis on bottom-up multistakeholder engagement. Particular attention is given to the niche management utilizing the multilevel perspective (MLP) described earlier. The microlevel is metaphorically conceptualized as the soil, the meso-level as the flowers, and the macrolevel as the sun of an organic ecosystem. In this sense, experimental prototyping of change initiatives that hold the potential of affecting the dominant system is encouraged through the shielding of existing fragile relations, the incorporation of nurturing networks, the nurturing of healthy and life-affirming visions and expectations, the nurturing of learning dynamics, and the empowering stretching and transforming of niches. This applied and tested approach has been developed in the framework of the Climate-KIC transition hubs (op. cit. De Vicente Lopez & Matti 2016). Climate-KIC is one of the largest public-private innovation partnerships and Knowledge and Innovation Communities (KIC) in the world, focusing on innovation to mitigate and adapt to climate change.

- Case Three

The most comprehensive use case for the realization of the conditions necessary for systemic innovation for thrivable cities is the methodological guide from the Thriving City Initiative (Doughnut Economics Action Lab and Biomimicry 3.8 and C40 Cities and Circle Economy 2020). The pilot project of this case involved an array of multidisciplinary qualitative and quantitative researchers, city staff representing a wide range of city departments, and city-based civic organizations and community networks. It was implemented and tested simultaneously in Philadelphia, Portland, and Amsterdam in 2019. The first version of the published guide stems from these experiences in the Global North. Future iterations are planned with a focus on the context of cities in the Global South, as well as on different scales, from neighborhoods to nations. The Thriving Cities Initiative takes the Doughnut Economics framework for meeting people's needs within the carrying capacity of our planet and downscales it to the city level. The freely available transformative development tool is based on a conceptual collaboration between Kate Raworth of Doughnut Economics Action Lab and Janine Benyus of Biomimicry 3.86, emphasizing the local-social, local-ecological, global-social, and global-ecological perspectives from an ecosystemic perspective, based on the principles entailed in bringing Doughnut Economics into practice. These are listed as:

- Aim to meet the needs of all people within the means of the living planet
- See the big picture
- Nurture human nature
- Think in systems
- Be distributive
- Be regenerative
- Aim to thrive rather than to grow

The so-called City Portrait methodology of the Thriving City Initiative invites the creation and pursuit of thought-provoking holistic visions of what it means to thrive through an iterative process of change. This vision is used to find collective answers to how a city can be a home to thriving people, in a flourishing place, while respecting the well-being of all people and the health of the whole planet in such a way that leads to shared inspiration for collective aspiration and transformative actions.

Summary and Conclusion

Systemic innovation for thrivable cities is fast becoming an indispensable focus area of urban and regional planning initiatives, especially in light of UN SDG objectives such as SDG 11: to make cities and human settlements inclusive, safe, resilient, and sustainable (see section on Sustainable Development Goals). This is because cities are complex adaptive systems (see section on Unpacking Cities as Complex Adaptive Systems) which cannot be approached through monodisciplinary, mechanistic, reductionistic frameworks. Indeed, cities are living, learning, evolving ecosystems with socio-technical, psycho-cultural, politico-economic, and biophysical dimensions that incorporate not only human needs, desires, hopes, and dreams, but also an array of nonhuman actors and their behavioral patterns, operational requirements, and systemic interdependencies. Systemic innovation takes into account this full spectrum of dynamic, evolving, interacting components and the fact that no city is an island unto itself (to paraphrase John Donne), but is interconnected with other cities and interdependent on other ecosystems. In short, cities are communities of beings in coevolutionary interaction with their embedding contexts.

In its most fundamental conception, a community can be considered "a *group of two or more individuals with a shared identity and a common purpose committed to the joint creation of meaning"* (Laszlo et al. 2008). Authentic communities are able to enhance their own development while at the same time enhancing that of each individual in the community, thereby promoting both freedom of personal choice and a sense of responsibility for the whole. In such communities, the operating principle is that of unity in diversity, on the understanding that what is sought for is full-spectrum unity. That is to say, unity without uniformity and diversity without fragmentation. In our current interconnected world, such orientations are increasingly indispensable. As Ruth Richards astutely points out, "our survival has become strongly dependent upon our commonalities as human beings, not on the differences between each other as individuals and as members of narrow reference groups" (Richards 1993, p. 168).

The notion of the city as a Learning Community (LC) that functions as an open dynamic system in which individuals collectively learn to adapt to their environment is becoming more and more compelling. Simple learning communities can be stepping stones toward an evolutionary learning community (ELC). An ELC can be considered an emergent (self-designing) learning system demonstrating dynamic stability by adapting *with* its environment and generating developmental pathways that are sustainable in the context of broader evolutionary flows. ELC is a human activity system that strives toward sustainable pathways for evolutionary development in synergistic interaction with its milieu. It does so through individual and collective processes of empowerment and learning how to learn and through an ongoing commitment to evolutionary learning (Laszlo 2015). "ELCs do not adapt their environment to their needs, nor do they simply adapt to their environment. Rather, they adapt *with* their environment in a dynamic of mutually sustaining evolutionary co-creation." (Laszlo & Krippner 1998). ELC is not to be considered separate from the environment, but a part of it. Just as the concept of "system" is more a pattern than a thing, ELC is more an ideal image of community than a particular social arrangement. Loye and Eisler (1987, p. 57) indicate the need for new visions of the future; a need for "a clearer sense of system goal states or prohuman images of the future." ELC is an image that can serve as a beacon for learning cities dedicated to forwarding a new culture of design appropriate to the challenges of life in an era radical interdependence, sustainability and thrivability.

Marilyn Hamilton (2014, 2019), these are meshworks of individual and collective evolutionary learning communities for thrivable futures, situated in localized spaces – urban, rural, and mixed, as well as global spaces – online, distributed, and virtual (Spencer-Keyse *et al.* 2020). ELEs become landscapes for fostering the type of future-creative competencies that fuse scientific and ethical knowledge as mentioned earlier: through responsive mindsets (know-what and know-why), skillsets (know-how and know-how-to), and heartsets (care-how and care-why). They begin to support and increase evolutionary coherence in our society by fostering learning that connects with our biosphere and even other planetary systems, cultivating consonant relations with life-affirming systems that celebrate "learning as a way of being." Finally, they start to operate as evolving, open-ended, prototype-and-scale-up systems that constantly generate new learning practices, curricula, and methods of thrivability, embracing an inclusive pathway for the generation of flourishing protopias (Smitsman *et al.* 2020).

In Jean Houston's book, *A Mythic Life,* she quotes Margaret Mead on her deathbed. "Forget everything I've been telling you about working with governments and bureaucracies! I've been lying here being an anthropologist in my own dying — fascinating experience, by the way; there is no hierarchy to it — and I've had an important insight into the future. The world is going to change so fast that people and governments will not be prepared to be stewards of change. What will save them is teaching-learning communities. They come together in churches or businesses or even in families. They could meet weekly and do your kind of exercises, especially ones that develop their capacities. There must be humor, laughter, games and good food as well. That will keep the participants coming back. Then, when they feel ready, they will choose projects to work on to help their *communities*. The *only way to have a possible society*, Jean, is to develop the *possible human* at the same time." Through systemic innovation for thrivable cities, we are beginning to find anticipatory ways of developing both the possible human and the possible society interdependently.

Cross-References

- ► An Overview of the Relationship of the Sustainable Development Goals and Urban and Regional Development
- ► Circular Cities
- ► Circular Economy Cities
- ► Connecting Urban and Regional Innovation Ecosystems to Enhance Competitiveness
- ► Innovation to Bring Nature-Based Solutions to Life: Tales of Two Cities
- ► Networking Collaborative Communities for Climate-Resilient Cities
- ► Sustainable Development Goals
- ► Sustainable Development Goals from an Urban Perspective
- ► Unpacking Cities as Complex Adaptive Systems

References

Publications

Barbier, E., & Burgess, J. (2017). The Sustainable Development Goals and the systems approach to sustainability. *Economics: The Open-Access, Open-Assessment E-Journal, 11*(2017-28), 1–22.

Benyus, J. (2002a). "Innovations inspired by nature" in *Doors of perception 7: Flow* (Conference in Amsterdam -14, 15, 16 November).

Benyus, J. (2002b). *Biomimicry: Innovation inspired by nature*. New York: Harper Perennial.

Braungart, M., & McDonough, W. (2008). *Cradle to cradle*. New York: Random House.

Capra, F. (1998). Evolution: The old view and the new view. In D. Loye (Ed.), *The evolutionary outrider: The impact of the human agent on evolution*. Adamantine: London.

Ellen MacArthur Foundation. (2019). *New plastics economy: Global commitments — progress report*. Cowes, UK: EMF pubs.

Ellul, J. (1964). *The technological society*. New York: Vintage Books.

Froes, I., & Lasthein, M. K. (2020). Co-creating sustainable urban metabolism towards healthier cities. *Urban Transform, 2*, 5.

Geels, F. (2002a). Technological transitions as evolutionary reconfiguration processes: A multi-level perspective and a case-study. *Research Policy, 31*, 1257–1274.

Geels, F. (2002b). From sectoral systems of innovation to socio-technical systems: Insights about dynamics and change from sociology and institutional theory. *Research Policy, 33*, 897–920.

Goerner, S. (1994). *Chaos and the evolving ecological universe*. Langhorne: Gordon and Breach.

Graham, Nick, et al. (2020). *Future Skills for the 2020s: A New Hope*. Global Education Futures & WorldSkills report. Moscow: GEF Press.

Hamilton, M. (2014). Meta-framework for security in the Human Hive: integrally aligning sustainability responses to trajectory of evolutionary threats. In ISSS Yearbook Special Issue of *Systems research & behavioral science*, A. Laszlo (Guest Ed.). Vol. 31, No. 5.

Hamilton, M. (2019). *Integral city 3.7: Reframing complex challenges for Gaia's Human Hives*. Minneapolis: Amaranth Press.

Hawken, P., Lovins, A., & Lovins, H. (2010). *Natural capitalism: The next industrial revolution* (2nd ed.). Oxfordshire: Routledge.

Houston, J. (1996). *A mythic life: Learning to live our greater story*. San Francisco: Harper.

Inayatullah, S. (2021). Creating a new renaissance: Can responses to Covid-19 Pivot us to a Transformed World? The *Journal of Futures Studies*, 13 August, online at https://jfsdigital.org/

Johansen, R. (2012). *Leaders make the future: Ten new leadership skills for an uncertain world*. Oakland: Berrett-Koehler Pubs.

Lanhoso, F., & Coelho, D. A. (2021). Emergence fostered by systemic analysis—Seeding innovation for sustainable development. *Sustainable Development, 29*, 768–779.

Laszlo, E. (1996). *Evolution: The general theory*. Cresskill: Hampton Press.

Laszlo, A. (2003). Evolutionary systems design: A praxis for sustainable development. In *Organisational Transformation & Social Change*, 1(1).

Laszlo, A. (2015). Living systems, seeing systems, Being systems: Learning to be the systems we wish to see in

the world. In *Spanda Journal,* 6(1), June issue on Systemic Change. Pp. 165–173.
Laszlo, A. (2018). Leadership and systemic innovation: Socio-technical systems, ecological systems, and evolutionary systems design. In M. Ruzzeddu and E. Ferone (Ed.), *International review of sociology,* Special Issue on "Governance of Innovation Processes: A Sociological Approach", 28(3), 1–12.
Laszlo, A. (2020). Practices that Ensoul the cosmos: expressions of connectedness, on Medium.com, 27 May.
Laszlo, A., & Krippner, S. (1998). Systems theories: Their origins, foundations, and development. In J. S. Jordan (Ed.), *Systems theories and a priori aspects of perception* (pp. 47–74). Amsterdam: Elsevier Science. Ch. 3.
Laszlo, A., Luksha, P., & Karabeg, D. (2017). Systemic innovation, education and the social impact of the systems sciences. In J. Kineman (Guest Ed.), The ISSS Yearbook Special Issue of *Systems Research & Behavioral Science,* 35(5), 601–608.
Laszlo, A., et al. (2008). The making of a new culture: Learning conversations and design conversations in social evolution. In P. M. Jenlink & B. H. Banathy (Eds.), *Dialogue as a collective means of design conversation* (pp. 169–186). Heidelberg: Springer Science +Business. Chapter 12.
Lebov, V. (1955). The real meaning of consumer demand. In the *Journal of Retailing.*
Loye, D., & Eisler, R. (1987). Chaos and transformation: Implications of non equilibrium theory for social science and society. *In Behavioral Science, 32,* 53–65.
Luksha, P., et al. (2018). *Educational ecosystems for societal transformation* (Global Education Futures report). Moscow: GEF Press.
Lyle, J. (1994). *Regenerative design for sustainable development.* Hoboken: Wiley Pubs.
Meadows, D., et al. (1972). *The limits to growth.* New York: Potomac Books.
Meadows, D., et al. (2004). *Limits to growth: The 30-year update.* White River Junction VT: Chelsea Green Publishing Company.
Merry, U. (1995). *Coping with Uncertainty: Insights from the new sciences of chaos, self-organization, and complexity.* Westport: Praeger.
Milbrath, L. (1989). *Envisioning a sustainable society: Learning our way out* (SUNY series in Environmental Public Policy). New York: SUNY Press.
Miller, G., & Foster, L. (2010). A brief summary of holistic wellness literature. *Journal of Holistic Healthcare,* 7(1), 4–8.
Montuori, A. (1989). *Evolutionary competence: Creating the future.* Amsterdam: J.C. Gieben.
Pasmore, W. A. (1988). *Designing effective organizations: The sociotechnical systems perspective.* New York: Wiley.
Pasmore, W. A., & Sherwood, J. J. (1978). *Sociotechnical systems: A sourcebook.* La Jolla: University Associates.
Randers, J., et al. (2019). Achieving the 17 Sustainable Development Goals within 9 planetary boundaries. *Global Sustainability, 2*(e24), 1–11.
Raworth, K. (2017). *Doughnut economics: Seven ways to think like a 21st-century economist.* London: Random House.
Richards, R. (1993). Seeing beyond: Issues of creative awareness and social responsibility. *Creativity Research Journal., 6*(1&2), 165–183.
Rousseau, D., Wilby, J., Billingham, J., & Blachfellner, S. (2018). *General systemology: Transdisciplinarity for discovery, insight and innovation.* New York: Springer.
Scholz, R. (2012). Transdisciplinarity. In H. Mieg & K. Töpfer (Eds.), *Institutional and social innovation for sustainable urban development.* London: Routledge.
Scholz, R. (2020). Transdisciplinarity: science for and with society in light of the university's roles and functions. *Sustainability Science., 15,* 1–17.
Scholz, R., & Steiner, G. (2015). Transdisciplinarity at the crossroads. *Sustainability Science., 10,* 521–526.
Smitsman, A., Laszlo, A., & Luksha, P. (2020). Evolutionary learning ecosystems for thrivable futures: Crafting and curating the conditions for future-fit education. *World Futures, 76*(1), 1–26.
Spencer-Keyse, J., Luksha, P., & Cubista, J. (2020). *Learning ecosystems: An emerging praxis for the future of education* (Joint report of the Global Education Futures initiative & the SKOLKOVO School of Management). Moscow: GEF Press.
Stahel, W. (2006). *The performance economy.* London: Palgrave Macmillan.
Stahel, W. (2019). Innovation in the circular and the performance economy. In F. Boons & A. McMeekin (Eds.), *The handbook of sustainable innovation* (pp. 38–58). Cheltenham: Edward Elgar Publishing.
Sukhdev, A., Vol, J., Brandt, K., & Yeoman, R. (2017). *Cities in the circular economy: The role of digital technology.* Cowes: Ellen MacArthur Foundation.
Thurman, R. A. F. (2005). *The jewel tree of tibet: The enlightenment engine of Tibetan buddhism.* New York: Free press.
van der Jagt, A., et al. (2020). Nature-based innovation systems. *Environmental Innovation and Societal Transitions, 35,* 202–216.
Walloth, C. (2016). *Emergent nested systems. A theory of understanding and influencing complex systems as well as case studies in urban systems.* Cham: Springer.

Weblinks

Bryn Mawr College. (n.d.). *Graduate program in the growth and structure of cities.* https://www.brynmawr.edu/cities/. Accessed 28 Sept 2021.
Circular Cities 2030. (2020). https://circularcities.com/home. Accessed 14 Sept 2021.
De Vicente Lopez, J. & Matti, C. (2016). *Visual toolbox for system innovation. A resource book for practitioners to map, analyse and facilitate sustainability transitions.*

S

Transitions Hub Series. Climate-KIC, Brussels. https://transitionshub.climate-kic.org/publications/visual-toolbox-for-system-innovation/. Accessed 29 Sept 2021.

Doughnut Economics Action Lab and Biomimicry 3.8 and C40 Cities and Circle Economy. (2020). *Creating city portraits. A methodological guide from the Thriving City Initiative.* Oxford. https://doughnuteconomics.org/Creating-City-Portraits-Methodology.pdf. Accessed 29 Sept 2021

Ellen MacArthur Foundation. (2015). *Growth within: A circular economy vision for a competitive Europe,* SUN and McKinsey Center for Business and Environment. https://ellenmacarthurfoundation.org/growth-within-a-circular-economy-vision-for-a-competitive-europe. Accessed 28 Sept 2021.

Ellen MacArthur Foundation. (n.d.). https://ellenmacarthurfoundation.org/. Accessed 14 Sept 2021.

Global Education Futures. (2020). *Future skills for the 2020s: A new hope.* GEF & WorldSkills https://futureskills2020s.com/. Accessed 14 Sept 2021.

Lifewide Education. (n.d.). *Learning cities: Ecosystems for lifewide-lifelong learning.* https://www.lifewideeducation.uk/learning-cities.html. Accessed 29 Sept 2021.

OECD. (2015). *System innovation: Synthesis report.* https://www.innovationpolicyplatform.org/www.innovationpolicyplatform.org/sites/default/files/general/SYSTEMINNOVATION_FINALREPORT_0/index.pdf. Accessed 29 Sept 2021.

OPSI. (2018). *Innovation facets.* https://oecd-opsi.org/proiects/innovation-facets/ Accessed 29 Sept 2021.

Rockström, J. & Pavan, S. (2016). *How food connects all the SDGs.* https://www.stockholmresilience.org/research/research-news/2016-06-14-how-food-connects-all-the-sdgs.html. Accessed 29 Sept 2021.

Tapia, C. & Menger, P. (2019). Innovation for sustainable development at local level: Instruments and examples to get started. *Brussels, Belgium.* https://www.inno4sd.net/uploads/originals/1/inno4sd-pub-mgd-03-2019-fnl-local-policy-guidelines.pdf. Accessed 29 Sept 2021

UNESCO Institute for Lifelong Learning. (n.d.). *UNESCO global network of learning cities.* https://uil.unesco.org/lifelong-learning/learning-cities/. Accessed 29 Sept 2021.

United Nations. (2015). *Sustainable development goals,* SDG 11: Sustainable Cities and Communities. https://www.un.org/sustainabledevelopment/cities/. Accessed 14 Sept 2021.

United Nations Department of Economic and Social Affairs. (2018). *Population dynamics, world urbanization prospects.* https://population.un.org/wup/. Accessed 14 Sept 2021.

United Nations Environment Program. (2011). *Cities investing in energy and resource efficiency.* https://wedocs.unep.org/bitstream/handle/20.500.11822/7979/GER_12_Cities.pdf?sequence=3&isAllowed=y. Accessed 14 Sept 2021

United Nations Environment Program. (2019). *Emissions gap report.* https://www.unep.org/resources/emissions-gap-report-2019. Accessed 14 Sept 2021.

United Nations General Assembly. (2015). *Transforming our world: The 2030 Agenda for sustainable development.* https://www.un.org/ga/search/view_doc.asp?symbol=A/RES/70/1&Lang=E. Accessed 14 Sept 2021.

United Nations Human Settlements Programme. (2019). *The Strategic Plan 2020–2030.* https://unhabitat.org/sites/default/files/documents/2019-09/strategic_plan_2020-2023.pdf. Accessed 14 Sept 2021.

World Commission on Environment and Development. (1987). *Our common future.* Oxford: Oxford University Press. http://www.un-documents.net/our-common-future.pdf. Accessed 29 Sept 2021.

World Health Organization. (2008). Commission on the Social Determinants of Health. *Closing the gap in a generation-health equity through action on the social determinants of health.* https://www.who.int/social_determinants/final_report/csdh_finalreport_2008.pdf. Accessed 29 Sept 2021.

Systems Thinking

▶ Philanthropy in Sustainable Urban Development: A Systems Perspective

T

Taming – Containing

► Strategies for Taming the City

Technology

► Theme Cities Networks

Territorial Competitiveness

► Connecting Urban and Regional Innovation Ecosystems to Enhance Competitiveness

Territorial Development

► An Overview of the Relationship of the Sustainable Development Goals and Urban and Regional Development
► Connecting Urban and Regional Innovation Ecosystems to Enhance Competitiveness

Territory

► New Localism: New Regionalism

The Grid Analysis and Display System (GrADS)

► The State of Extreme Events in India

The Intergovernmental Panel on Climate Change (IPCC)

► The State of Extreme Events in India

Theme Cities Networks

A Typology

Julien Forbat[1], Evelyne de Leeuw[2] and Jean Simos[1]
[1]University of Geneva, Institute of Global Health, Geneva, Switzerland
[2]Centre for Health Equity Training, Research and Evaluation, and Healthy Urban Environments (HUE) Collaboratory; University of New South Wales; Ingham Institute for Applied Medical Research, Liverpool, NSW, Australia

Synonyms

Health promotion; Healthy cities; Quality of life; Technology; Theme city network; Typology

R. C. Brears (ed.), *The Palgrave Encyclopedia of Urban and Regional Futures*,
https://doi.org/10.1007/978-3-030-87745-3

Definition

Over the past decades, following the creation of Healthy Cities by the WHO in the 1980s, several theme cities networks (TCNs) have been developed in order to tackle various issues related to quality of life in urban settings. This entry presents a typology of existing TCNs. Based on two key dimensions, namely the scope of health determinants considered and the role of technology, TCNs are classified in four types illustrating their considerable heterogeneity.

Introduction

Focusing on TCNs dealing with health promotion, this entry starts by presenting the selection of TCNs and relevant criteria used to build a typology. It then describes the four resulting categories of TCNs before concluding on the relevance of this typology in light of the recent coronavirus pandemic.

Why TCNs?

The term "theme cities network" was coined by Davies (2015) to describe novel entities appearing first in 1986 with the creation of Healthy Cities by the WHO. Since then, numerous institutions have developed programs of actions based on the city level in an attempt to bridge the gap between ambitious international objectives aiming at solving the health-sustainability nexus (Lawrence and Fudge 2009), such as those expressed in the Agenda 21, the Millennium Development Goals (MDGs) or the Sustainable Development Goals (SDGs), and States often devoid of the necessary expertise to implement relevant measures at the local level. Whereas the Agenda 21 was developed for local governments in general, including regions for instance, the specific role that needs to be played by cities has been formally acknowledged by the UN (2015) in the Agenda 2030 with the SDG 11, which aims at "making cities inclusive, safe, resilient, and sustainable." What is more, most of SDGs touch upon transversal health issues (inequality, governance, etc.) that can be addressed at the city level. Since TCNs are by essence transnational and their urban practices seem to scale outward (toward other cities from a specific TCN) rather than upward (toward regions and national states) (Davidson et al. 2019), this could lead to similar outcomes in cities across countries instead of within countries.

What Are TCNs?

TCNs aim by essence at tackling more or less specific sets of issues related to health promotion in urban settings. Some networks adopt a relatively narrow approach, for instance, "Smart Cities" focusing on the physical environment, while others have an all-encompassing approach, such as Cittàslow whose objectives range from the natural to the social and cultural environments. However, they all share the willingness to improve urban quality of life for *all* city dwellers. In that sense, they differ from gated communities, which target specific segments of population (Atkinson and Blandy 2006). TCNs, as well as local governments and communities they represent, also intend to actually implement policies to achieve better living conditions. This implies that advocacy networks and organizations, for instance ICLEI (local governments for sustainability) or Slumdweller International, are not considered here as TCNs.

Which TCNs?

The list of TCNs used to build this typology is not exhaustive. Every year new networks are being established and this trend is likely to continue in the foreseeable future. The selection of TCNs is based on two main criteria: first, the number of citations TCNs have received in the peer-reviewed scientific literature, and, second, their lifespan, implying that networks with little to no recorded history were excluded from the analysis. Table 1 offers a synoptic view of key elements related to selected TCNs.

The vast majority of TCNs share a similar governance mode, according to which

Theme Cities Networks, Table 1 Selection of TCNs. (Number of citations according to Google Scholar, July 5, 2021)

TCN	Date of origin	Main reference (number of citations)
Healthy Cities	1986	Duhl 1986 (173)
Sustainable Cities	1994	Haughton and Hunter 2004 (1355)
Child-Friendly Cities	1996	Riggio 2002 (211)
Cittàslow	1999	Pink 2008 (220)
Healthy Cities Bloomberg	2001	Twiss et al. 2003 (552)
Creative Cities	2004	Landry 2012 (6610)
Transition Towns	2006	Connors and McDonald 2010 (166)
Age-Friendly Cities	2006	WHO 2007 (1629)
Smart Cities	2013	Hollands 2008 (3181)
Resilient Cities	2013	Godschalk 2003 (1606)
Happy Cities	2013	Montgomery 2013 (610)
Inclusive Cities	2000s	Gerometta et al. 2005(461)

municipalities run a top-down policy process, supported by the institution having initiated the TCN, usually a UN organization, for instance the UNESCO (Creative Cities), or a private organization, such as the Rockefeller Foundation (Resilient Cities). Transition Towns and Inclusive Cities are the only TCNs selected here and following a bottom-up approach. It is also noticeable that TCNs are usually expanding at the global level and often originate in Europe or North America.

Typology

As stated before, TCNs aim at developing policies promoting better quality of life but they differ in their conception of health promotion objectives and the way they should be achieved. This considerable diversity has already been highlighted as a difficulty to build a typology (de Leeuw and Simos 2017). However, the concept of environmental health precisely emphasizes the complexity of the interrelation between numerous health determinants within urban settings. In that sense, city dwellers' well-being is not only affected by the quality of urban infrastructure (physical environment) or natural factors (for instance air quality) but also by a wider spectrum of determinants including social, economic, and cultural environments. What is more, factors promoting healthier lifestyles (physical activity, healthier diets, etc.), whether those are directly labeled as environmental factors or rather as interacting with them, can also be considered. Therefore, the type of health approach acknowledged by TCNs is reflected in the number of health determinants they refer to, more systemic approaches including more determinants. In the same vein, the complexity of environmental health is better taken into account when TCNs policies explicitly account for their health and environmental impact.

TCNs also have very different stances regarding the role of technology in implementing policies. Some networks, for instance Smart Cities, heavily rely on high-tech solutions, notably the Internet of Things (IoT), to deliver better living conditions in cities. Other networks adopt a more cautious, if not critical, perspective on technology, which is, for example, reflected in the concept of "local resilience" used by Transition Towns. It should be noted however that a strict divide between TCNs on technology issues could be misleading. Indeed, even proponents of "smart" solutions acknowledge the fact they are complex to implement and go beyond the provision of mere software and hardware products to encompass several aspects, including management, policy, and community issues to name a few (Garcia and Lippez-De 2016).

Figure 1 presents a typology of the 12 TCNs according to the two key features presented above: the type of approach to health promotion, and the role conferred to technology in responding

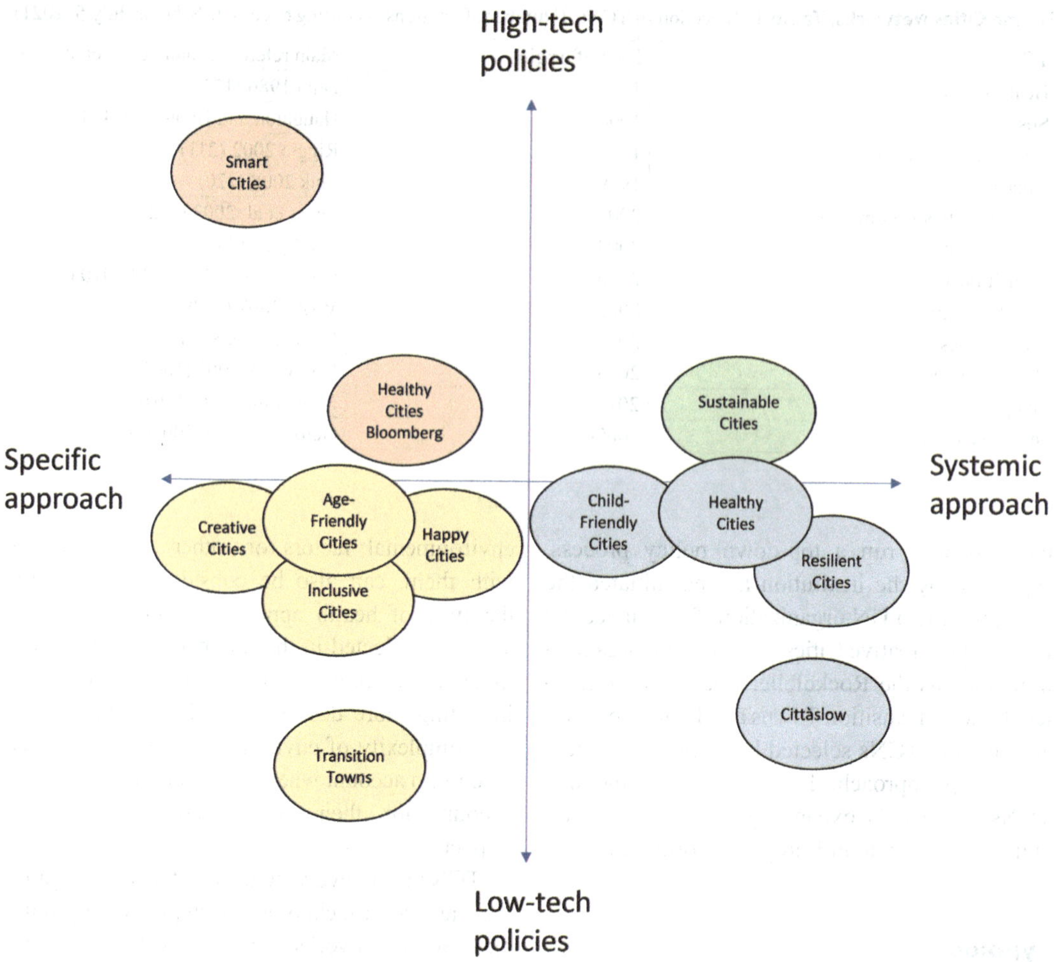

Theme Cities Networks, Fig. 1 Typology of TCNs

to the challenge of improved quality of life in cities. Elman (2005) distinguishes between three functions of typologies: descriptive, classificatory, and explanatory. The four clusters highlighted here have mostly a descriptive purpose, since proximity between several TCNs tends to overlap with such a clear-cut distinction and to undermine the significance of a classification that would be too rigid.

The typology calls three main observations. First, TCNs promoting high-tech policies ("technophile" networks) are relatively rare, especially in combination with a systemic approach of health determinants. This is likely due to the fact that technological policies are often devised for urban infrastructure, through IoT solutions for instance, and seem less relevant with social or cultural environments, whether it is for practical reasons (technology devices not readily available in a developing context for instance) or because of ethical considerations (the intrusive nature of some digital tools).

Second, the main divergence between TCNs relies on their approach toward health determinants. About 40% of TCNs lean toward systemic approaches, including an explicit accountability of health and environmental aspects of their policies, while the other 60% tend to adopt a narrower approach, extreme cases being represented by Smart Cities and Creative Cities.

Third, all clusters but one seems to integrate an "ideal type" (or pure type) emphasizing its features: Smart Cities combines narrow and technophile approaches, Transition Towns combines narrow and low-tech policies, and Cittàslow combines systemic and low-tech policies. Interestingly, the fourth cluster combining systemic and high-tech policies is the less represented among TCNs, highlighting the difficulty to develop technological tools beyond a certain point.

Conclusion

Cities are increasingly recognized by international institutions as key actors to respond to global challenges, such as climate change, biodiversity loss, and pandemics. In that sense, the distinction between the "city" – as a political community with a certain degree of collective agency – and the "urban," referring to the process of spatial transformation toward more built environment, appears particularly relevant (Acuto et al. 2021). Indeed, faced with similar trends, cities can decide or not to embrace their policymaking role at the local and international levels, by participating in TCNs, which can act as innovators and accelerators toward increased well-being for urban dwellers. Positive network externalities (Capello 2000) should be considered here, which is the fact that a city participating in a TCN is likely to receive positive outcomes (in the form of best practices translated into more efficient public administration for instance) largely superior to the "private" cost of belonging to a TCN. However, as the typology presented here shows, TCNs tend to follow different paths to achieve their objectives. Given the burgeoning number of TCNs, it is too early to determine whether one specific cluster of TCNs offers better prospects than the others. On the other hand, it is essential for cities and citizens to have a clear vision of the options currently available. As a matter of fact, the Covid crisis has increased the level of complexity faced by cities attempting to devise policies to build a more sustainable, resilient, and healthy future. During the pandemic, inequalities between citizens have been exacerbated in some cases, for instance in terms of private space at disposal per capita (people with more space being less impacted by lockdown measures) or in terms of job access (remote work opportunities have not been evenly distributed across communities). It is therefore crucial that best practices become readily available and shared within TCNs, and, who knows, between TCNs.

Cross-References

▶ Age-Friendly Future Cities
▶ The Governance of Smart Cities

References

Acuto, M., et al. (2021). The city as actor in UN frameworks: Formalizing 'urban agency' in the international system? *Territory, Politics, Governance,* 1–18.

Atkinson, R., & Blandy, S. (2006). *Gated communities.* Routledge: Abingdon.

Capello, R. (2000). The city network paradigm: Measuring urban network externalities. *Urban Studies, 37*(11), 1925–1945.

Connors, P., & McDonald, P. (2010). Transitioning communities: Community, participation and the transition town movement. *Community Development Journal, 46*(4), 558–572.

Davidson, K., Coenen, L., & Gleeson, B. (2019). A decade of C40: Research insights and agendas for city networks. *Global Policy, 10*, 697–708.

Davies, W. K. (Ed.). (2015). *Theme cities: Solutions for urban problems* (Vol. 112). New York: Springer.

de Leeuw, E., & Simos, J. (2017). *Healthy cities: The theory, policy, and practice of value-based urban planning.* New York: Springer Science.

Duhl, L. J. (1986). The healthy city: Its function and its future. *Health Promotion, 1*(1), 55–60.

Elman, C. (2005). Explanatory typologies in qualitative studies of international politics. *International Organization, 59*(2), 293–326.

Garcia, A. R., & Lippez-De, C. S. (2016). Technology helps, people make: A Smart City governance framework grounded in deliberative democracy. In J. Gil-Garcia, T. Pardo, & T. Nam (Eds.), *Smarter as the new urban agenda* (Public administration and information technology) (Vol. 11). Cham: Springer.

Gerometta, J., Häussermann, H., & Longo, G. (2005). Social innovation and civil society in urban governance: Strategies for an inclusive city. *Urban Studies, 42*(11), 2007–2021.

T

Godschalk, D. R. (2003). Urban hazard mitigation: Creating resilient cities. *Natural Hazards Review, 4*(3), 136–143.
Haughton, G., & Hunter, C. (2004). *Sustainable cities*. London: Routledge.
Hollands, R. G. (2008). Will the real smart city please stand up? Intelligent, progressive or entrepreneurial? *City, 12*(3), 303–320.
Landry, C. (2012). *The creative city: A toolkit for urban innovators*. London: Earthscan.
Lawrence, R., & Fudge, C. (2009). Healthy cities in a global and regional context. *Health Promotion International, 24*(S1), 11–18.
Montgomery, C. (2013). *Happy city: Transforming our lives through urban design*. New York: FSG.
Pink, S. (2008). Sense and sustainability: The case of the Slow City movement. *Local Environment, 13*(2), 95–106.
Riggio, E. (2002). Child friendly cities: Good governance in the best interests of the child. *Environment and Urbanization, 14*(2), 45–58.
Twiss, J., Dickinson, J., Duma, S., Kleinman, T., Paulsen, H., & Rilveria, L. (2003). Community gardens: Lessons learned from California healthy cities and communities. *American Journal of Public Health, 93*(9), 1435–1438.
United Nations (UN). (2015). *Transforming our world: The 2030 agenda for sustainable development*. New York: UN.
World Health Organization (WHO). (2007). *Global age-friendly cities: A guide*. Geneva: WHO.

Further Reading

Aalborg Charter. (1994). *Charter of European cities and towns towards sustainability*. European Conference on Sustainable Cities & Towns: Aalborg, Denmark, May 27, 1994.
ARUP. (2014). *City resilience framework*. London: The Rockefeller Foundation & ARUP.
Duhl, L. J. (1963). *The urban condition. People and policy in the metropolis*. New York: Basic Books.
Network, T. (2016). *The essential guide to doing transition: Getting transition started in your street, community, town or organisation*. Totnes: Transition Network.
Slow Food. (1989). *Slow Food manifesto: International movement for the defense of and the right to pleasure*. Paris: Slow Food.
World Bank. (2013). *Inclusion matters: The foundation for shared prosperity*. Washington, DC: World Bank.
Yencken, D. (1988). The creative city. *Meanjin, 47*(4), 597–608.

Theme City Network

▶ Theme Cities Networks

Threat of Lead Exposure

▶ Lead Exposure in US Cities

Threats

▶ Peri-urban Regions

Toward a Sustainable City

A Paradigm Shift in Children's Modern Recreational Opportunities

Raisa Sultana and Raisa Binte Huda
Department of Geography and Environment, University of Dhaka, Dhaka, Bangladesh

Synonyms

Built environment; Recreation; Recreational opportunities; Recreational space; Sustainable city

Definition

Recreation is an active or passive participation in any activity including relaxing, dancing, singing, and playing which can be enjoyed in a group or alone (Siddiqui 1990; Biswas 2002). **Urban recreational space** can be defined as the points in spatial fabric that are occupied by residents for recreation and which serve as an excellent platform for social, as well as environmental, interaction (Nilufar 1999; Ahmed 2010). **Built environment** is a man-made environment which mediates conflicting desires of the residents to ensure a healthy and sustainable place to live through infrastructures and other facilities. **Children** are the human beings below 18 years of age. **Sustainable city** refers to an urban area which is socially equitable, economically wealthy, environmentally balanced, fulfills the demands of residents and offers wider scopes for development.

Introduction

Cities are unique as it is the only environment built to cater specifically to the needs of residents. The built environment mediates conflicting desires to ensure a healthy and sustainable place to live. Rampant urbanization, extreme population pressure, and economic development in cities are creating an increase in property and land costs, which consequently have a terrible impact on the cities, mostly of developing countries. The mixed-zone strategy and compaction are slowly designing out recreational green spaces, such as parks and playgrounds, which are integral parts of the urban land coverage. However, the Sustainable Development Goal (SDG) section 11.7.1 states that by 2030, safe, inclusive, and accessible green and public spaces should be accessible to all the residents, especially for women, children, senior citizens, and the disabled population (Sustainable Development Goals: Bangladesh Progress Report 2018).

In the broadest terms, urban recreational space can be defined as the points in spatial fabric that are occupied by residents for recreation. These spaces play an important role in urbanites' well-being as it provides an excellent platform for recreation, social interaction, as well as interaction with a healthy environment (Nilufar 1999; Ahmed 2010). The benefits of recreation are multitude and the necessity of its presence on a daily basis is well documented in literature: ranging all the way from purely physiological to psycho-social well-being, including reduction in obesity (Sallis and Glanz 2006), better fitness (Marchione 2013), increased competence through play (Hart 2002), learning of critical thinking and prosocial behavior, establishment of a sense of self (Richardson et al. 2017), and cognitive development of a child. While private play is important for establishing autonomy over their own lives (Lester and Russell 2010), outdoor play is crucial for social navigation skills in unfamiliar environments, formation of friendships, and developing public trust (Hart 2002).

Unfortunately, in cities of developing countries with high level of poverty, there is a dearth of basic essential services and amenities like health, housing, and education, and recreational space is considered a low priority and ignored in the development agenda by the city development authorities (Bartlett 1999; Hart 2002; Krysiak 2017). This is often because research on the developmental importance of recreation for children is not easily available to decision-makers and because many view childhood as just a "temporary phase."

In Dhaka, planned neighborhoods are very few and availability of open space and recreational facilities for people are not found in most of the dense residential areas (Siddiqui 1990; Biswas 2002; Ahmed and Sohail 2008; Nilufar 1999). Parks and playgrounds have been encroached, lack good maintenance, display poor amenities, and are unsafe, unwelcoming, and unhealthy. Consequently, lifestyle is changing greatly and children are becoming more individualized, self-centered, and mechanical instead of being social and humanistic (Nilufar 1999). Most importantly, the repercussions are long term, as children are the future of the present society.

There is a gap in our knowledge of what adaptive coping mechanisms children are developing to fill up their need for recreation, what sort of new urban networks and spaces are being established in the process, and how we can integrate these new recreational opportunities into existing infrastructure without overhauling both established environments and lifestyle patterns. Appraising the remarkable benefits of recreation on children's overall well-being, the paradigm shift approach should begin with the question in a broad frame: How can we provide children with recreational opportunities to play freely, walk independently, live actively, and make them feel a sense of ownership and belonging to the community? The purpose of the article is to cross-examine three questions:

1. Why has urban children's opportunity of contact with recreational green space decreased?
2. How has this changed their recreational behavior?

3. How can recreational green space be reintroduced into their lives?

Method

This article is based on secondary data and empirical evidence from the authors' city of residence, Dhaka, to contextualize the issues of urban planning, which is a field that is highly spatially localized.

In defining search terms for the article, the central dimensions of the topic were first demarcated: urban recreational space, children's well-being and their sustainable future, and changing lifestyle and modern recreational opportunities (Fig. 1). Subtopics that fall under these topics were then listed in an iterative process with more relevant subtopics added as more papers were reviewed. Several scientific literatures, e.g., reports, research articles, dissertations, theses, and white papers published between 1980 and 2021 in the English Language were searched. This time period was selected for the study to explore changes in recreational behavior and to look for the modern approaches to recreational opportunities. In addition, using a snowball technique we hand-searched references cited in reviewed studies. More than 350 documents were reviewed of which 310 were finally included according to relevance.

Barriers to Urban Children's Accessibility to Recreational Green Space in Dhaka City

Existing literature is limited to either a single category of variable or covers limited geographical regions, particularly developing countries. The predominant emphasis is on barriers imposed by the built environment (physical availability and accessibility of parks and playgrounds, spatial proximity, traffic security, topographic attributes, weather condition, etc.) and their impacts on children's developmental needs (Abu-

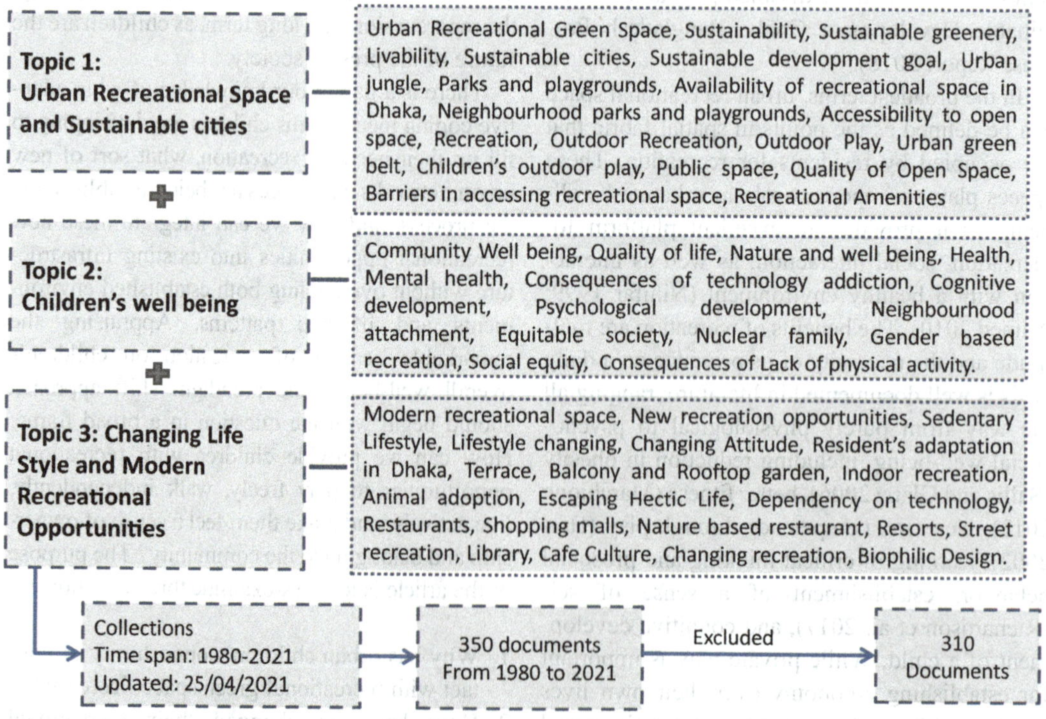

Toward a Sustainable City, Fig. 1 Concept operationalization and search strategy

Ghazzeh 1998; Bjorklid 1985; Ahmed 2010; Churchman 1980; Churchman 2003; Cooper and Sarkissian 1986; Freeman 1995; Gehl 1996; Moore et al. 1987; Verwer 1980; Watkins 1980). While it is all very well to focus on improving the quality of recreational space, children must be able to access these. Clearly several barriers exist within the very nature of cities that frustrate children's opportunities of contact with recreational green space, and to tackle these, one must first explore what these are.

Inadequate Provision

By 2030, it is estimated that more than 50% of the world's population will reside in urban areas (UN 2016). Rapid densification of cities results in an increasingly smaller portion of land being allocated for recreational open space in lieu of residential and other land usage. This creates the primary barrier to children's accessibility to recreational space: a lack of provision.

The standard amount of open space in an area is context specific and ultimately dependent on population density (Patel 2011). According to a report of RAJUK (2015), Dhaka has been suggested to have 0.065 hectares per 1000 people by DMDP or 0.388 hectares per 1000 people by DAP. However, currently, it offers a mere 0.028 hectares of recreational space per 1000 population, equal to only 1.51% of total land (462.32 hectares). This is significantly low. The recreational open space that does exist is highly spatially skewed (Sultana and Nazem 2016). Dhaka's Zone 4, for example, has no open space for its population density of 97,500 people/km^2. This is problematic as children's territorial extent (i.e., how far they are permitted to roam without parental supervision) is very limited: around 15 min or 300–500 m from their home (Freeman and Tranter 2011). Even for adults who accompany kids, the willingness to travel too far is limited. Research on urban parks suggests that on a regular basis most users want to travel there by foot and only if it takes 3–5 min of walking from the individual's residence/office (Kaplan and Kaplan 1989; Thompson 2002; Comedia/ Demos 1995).

Densification and pressure on cities to provide easily accessible service have caused schools to mushroom in every lane of the city, so much so that even school playgrounds have become rare, which used to previously resemble green oases in satellite images. Designating garages or the "left-over" space in front of the school building as playgrounds has become common practice. In fact, it is not uncommon to find schools where indoor corridors are the only "recreational space" available.

Lack of Modern Amenities

Even in areas where recreational facilities do exist, they fail to attract children anyway. In some designated parks, recreational equipments are out of order almost throughout the year. However, parents often aim to impart their kids at least one experience at these national theme parks, but people seldom return for a second turn. Recreational facilities must attract the child himself/herself (Abubakar and Aina 2006). However, what attracts a child evolves over time and differs by gender. While boys enjoy a more active, space-consuming, rough play style, and organized team sports (Børve and Børve 2017), girls seem to enjoy smoother interaction during play such as tag games, walking and relaxing, badminton, etc. (Hyndman and Chancellor 2015). Instead of a uniform open space-style playground, a design that offers variously structured subspaces of different sizes, among which at least one subspace offers amenity areas that girls prefer, can go a long way to attract girls to the recreational facilities.

Negative Perceptions

The hardest to overcome barrier that a city can impose on recreational space is "perception." Perceptions are tricky because they are often not based on objective truth but rather feed off assumptions: If a place "seems" to be dangerous, it will be avoided. In Dhaka city, lack of safety measures is the most important aspect which affects the children's enjoyment of recreation and outdoor play facilities (Ahmed and Sohail 2008; Biswas 2002), particularly for female teens and children dependent on parents' company. Young children are especially dependent on parental supervision and/or permission when accessing these facilities (Woolley 2006).

T

Unhealthy, unwelcome, and unsafe neighborhood environments snatch a wonderful childhood from children and their right to experience and explore spontaneously, safely, and on their own terms.

Teenagers have a more "sedentary" expectation from recreation and look forward to amenity areas where they can simply "hang-out." A cultural issue often arises in conservative societies about how welcome a group of teens are in the public sphere, i.e., whether a group of teens are readily labeled as delinquents by passersby or not (Owens 1988). Urban design plays a crucial role in this aspect: For example, locating the seats out in the open, from where there is a clear line of sight for passersby, but which are not so close to adjacent paths as to enable eavesdropping into the conversation, can create a sense of visual openness while also protecting privacy.

Poor Road Connectivity

As a city evolves, there is an increase in the number of urban primary- and secondary-level roads that cater to motorized vehicles at the complete disregard of the needs of pedestrians. This road connectivity impacts the play patterns of younger children by influencing parents' perceived level of safety. Parents are more willing to let their children roam about if there are few or no primary- and secondary-level roads, thus reducing threats from motorized vehicles and enabling street play. Distance is a major determinant of accessing recreational places for children, the more distant the playground from home, the lesser the frequency of visits to these places. Age is another factor, which decides how much distance a child can cover (Hand et al. 2016). Long-time residence in the same neighborhood and frequent social encounters facilitated by informal streets and local shops create a sense of kinship where neighbors look out for the kids. This way it gets easier to spot strangers that do not belong there.

Roads designed for pedestrians, for teenagers and for families to take a leisurely stroll on, are slowly becoming nonexistent. Streets offer more opportunities, than parks and plaza, to urbanites to engage with nature and the surrounding environment through shops, shady boulevards, and cafés. Not only are streets the true representative of public open space in which the users feel comfortable and welcome, but it also enables the maximum incidental interaction with nature and recreation (Jacobs 1961). However, poor condition of roads, vehicular threats, exhaust fumes, etc. are turning streets into a medium for getting from one indoor place to another instead of a space to be occupied by itself.

Urban Hectic Life

Urban workaholic life is one of the prominent and influential factors that determines accessibility to recreational time and space. Teenagers are granted more autonomy over their territorial range but urban teens with a hectic schedule (classes in the morning, tuition in the afternoon, and traffic congestion in between), paired with pubescent behavioral change and a different, more "sedentary" expectation from recreation, often do not find facilities within walking distance that cater to their needs.

Socioeconomic Inequity

In a study done by Islam et al. (2014), the cohort of children under 18 with the least exposure to outdoor recreational activity was found to be the upper-class kids. This can be explained by the fact that upper-class children are often from planned residential areas, where there is an increased presence of cars on streets and a greater desire of parents to control who their children mix with.

On the other hand, well-being of the less privileged children is often neglected by the city authority as well (Child 1983; Sharmin et al. 2020; Rissotto and Giuliani 2006). Encroachment and commercialization of open space has made free public "stay space" nonexistent, driving kids from socioeconomically poorer zones to participate in communal play in their local neighborhoods that are unfortunately usually both socially and physically unsafe. Low- and lower-middle-income families cannot afford the high entry fees to visit a city's amusement parks. Thus, the city is constantly failing to become inclusive and socially equitable regardless of age, gender, and class (Ahmed 2010; Siddiqui 1990; Islam et al. 2014). Nevertheless, cities need to ensure social cohesion, inclusion, solidarity with promoting equity, and avoiding discrepancy of opportunity (Rogers 1999).

A Paradigm Shift in Recreational Opportunities

In the rapidly urbanizing city, children barely see outdoor environments at all (Kyttä 2004). This marks the need to think creatively about transformation of built environments incorporating new recreational opportunities. Children have the capacity and motivation to search for and find recreational space no matter how small the scale. But the metamorphosing city puts an increasing number of barriers on their ability to do so. As the city grows and mutates, children cope. As children cope, there are large-scale cultural shifts in their lifestyle patterns. The contemporary changes are represented as cause-and-effect relationships in Fig. 2 briefly. It shows how the inaccessibility to recreational spaces is harming the well-being of residents (especially children) which in turn is triggering them to incorporate alternative yet modern recreational space into their lifestyle.

Evolving Café Culture and Commercialization of Recreational Space

One such shift is the modern "café culture" of urban teens: Recreational space to teenagers have started to exclusively mean eating out at cafés alone or with friends, the evidence of which can be found in the rapid proliferation of foreign food chain cafes and eateries all over Dhaka. Not only does this lead to gentrification and exclusion of peers who cannot afford to eat out regularly, but also encourages a consumption fetish among teens. The core reason behind café culture is the lack of any safe, free, quickly accessible recreational public space, even if small in scale.

Yet café culture has also proven to be one of the most accepted recreational opportunities for public sociability (Montgomery 1997). "Library café" is a new concept in urban area and attracts new users to libraries. Physical space that integrates amenities for both adults and children through a library and a café enhance the library's allure while encouraging a sense of community involvement (Woodward 2005). Similarly, restaurants offering kids' zone are attracting more users than their regular counterparts.

Growth of Nature Retreats and Resorts in Suburban Areas

There has been increased "destination activities" among the upper class that includes traveling to resorts and nature retreats. Albeit that such destination areas can be further modified to give children and their families an even better interaction with recreational space by incorporating varied natural activities, e.g., duck feeding, fishing, etc., such places are restricted economically only to the upper-rich and even then only to a few times a year.

Increased Dependence on Technology

The boom in technology and the diminishing outdoor space has resulted in children spending their leisure time more and more behind screens (Bratman et al. 2015; Mackerron and Mourato 2013; Ekkel and de Vries 2017). As accessible outdoor space becomes rarer over time, "extinction of experience" (Pyle 1978) will increasingly drive children toward the excessive usage of technology. The result of this would be dire: Unrestrained screen time has been linked to speech delay (Zimmerman et al. 2007), poor sleeping habits (Brambilla et al. 2017; Parent et al. 2016; Nathanson and Fries 2014; Magee et al. 2014), reduced mineral density of bones (Shao et al. 2015; Winther et al. 2015; Chastin et al. 2014), poor posture (Lui et al. 2011), increased depression, hypertension, aggression, and suicidal tendency among adolescents (Wood and Scott 2016; Maras et al. 2015) and much more (Lissak 2018).

What is not clear, however, is whether excessive use of screens caused reduced use of outdoor space or vice versa: Do children prefer screens over outdoor time or is it just where they can meet their friends than in parks? For example, a study done by Holloway and Valentine (2003) found that technology was less influential on children than expected and observed that children actually preferred to play outside if permitted.

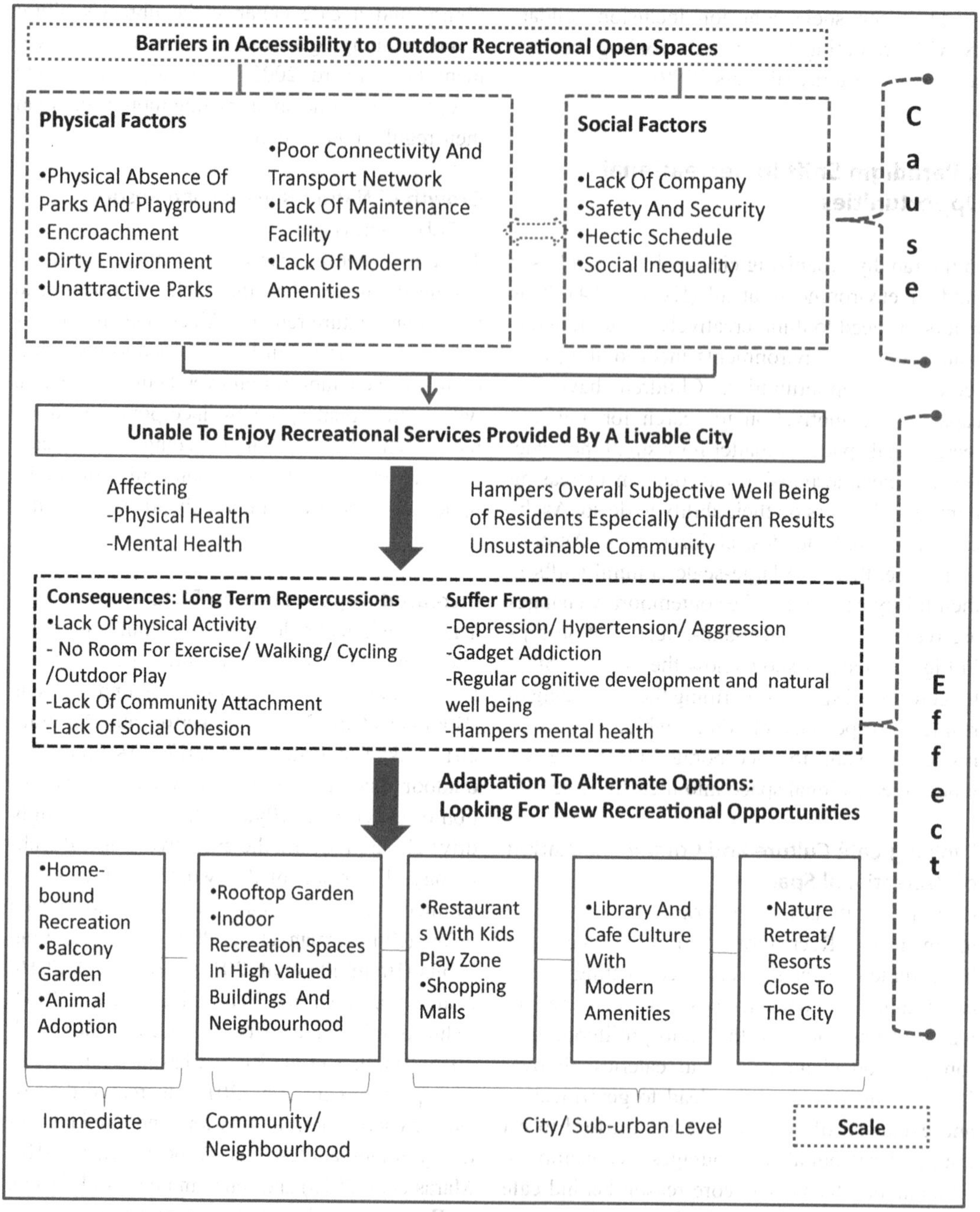

Toward a Sustainable City, Fig. 2 Shifting paradigm toward sustainable community well-being

Bringing Nature Indoors: Pets, Rooftop, and Balcony Gardens

There has been a movement toward bringing "nature" into homes in the form of pet adoption and house plants. Although adopting and taking care of a pet is not "true" exposure to nature or outdoors, encounters with pets can still serve as vital recreational opportunity. Dog walking, for example, has been found to be an effective way of increasing children's mobility and physical exercise, with parents being less worried about possible harm coming their children's way if a pet is with them (Christian et al. 2014). Moreover, animals have a long history of aiding emotionally or physically affected children: cerebral palsy, excessive weight, autism, hearing impairments, etc. (Timperio et al. 2008; Esposito et al. 2011).

On the other hand, plants inside the home, specifically terrace and balcony gardens, are getting popular these days for increasing creativity and providing psychological benefits to urbanites (Green 2004). Street trees, green walls, pot plants, bird box, and other domestic remnants of nature, despite their small scale, offer important opportunities of contact with nature for children.

What seems obvious is that as we move forward, newer technological advancements and newer patterns of open space networks will need to be conceptualized and utilized to satisfy children's old and unchanging requirement of recreational space.

Incorporate Modern Strategies into Built Environment

All children have a universal right to enjoy the benefits of recreational space regardless of whether they are in cities or not. The challenge, then, is how to achieve it. The urban residents now need a novel approach which will enable them to access recreational opportunities in a limited scale through an increasingly effective way. Instead of focusing on creating "pocket utopias" of parks and playgrounds, planning needs to go beyond functional thinking and move toward integrating recreation and nature into the city as a whole and into existing built infrastructures (malls, restaurants, hospitals, etc.). This is essential since urbanization and growing population means that recreational space will always remain in competition with other land usage.

A dense city like Dhaka has already filled up with concrete infrastructures with no room for breathing, hence ensuring a nearby, safe, well-maintained, attractive community park in every single neighborhood might be difficult. Therefore, indoor recreational space is greatly encouraged. Several research suggest that social interaction can be strengthened through strategic urban planning that encourages interactive spaces (Hadavi 2017; Jens and Gregg 2021). A well-balanced, enclosed, in-between informal space between built-in structures, corridors, and breakout spaces plays effective role in social activities (Beckers et al. 2016), promotes diversity and community bonding, as well as improves health and mental well-being (Jens and Gregg 2021). Traditionally, planners have put an increased emphasis on large-scale, idyllic versions of recreational experience while disregarding the more limited, incidental interactions with street trees or potted plants. This is unfortunate because children can interact with nature in meaningful ways even when nature is highly modified and at small scales. As new evidence appears, they provide newer insights into the relationships between the built environment, science, and nature (Ryan et al. 2014). The simple placement of a bench near a fountain in front of a shopping mall can act as an effective recreational environment for users (Goss 1993). Moreover, given the hectic urban schedule, incidental and regular exposure to hedges and shrubs on natural walkways, for example, might be more effective than rare visits to parks. Balcony gardens, potted plants, bird feeder, and a lone street tree are all limited natural experiences but still valuable.

There has also been a growing interest in transforming the building itself into green space for home-bound recreation by integrating rooftop gardens, green walls, or shared courtyard spaces with neighboring buildings, an effective response to growing densification. For example, terraces and playrooms in residential buildings have been considered beneficial and have precedents in many dense European cities (Ghosh et al. 2016). Urban roof garden has recently been promoted as an effective way of serving greenery in the cities and adding higher value to high-rise buildings by offering open space functions (Green 2004; Pouya et al. 2016).

It is also important to remember that nature is more than just "green": light and/or air, let in through clerestories, and water felt through the sound of fountains are all valid ways of incorporating nature in design. The concept of biophilic design is not new, rather it has been scientifically explored for decades (Ryan et al. 2014). Using natural elements, such as pebbles on pathways or wood benches, help to involve the patina of time that humans are particularly susceptible to. It promotes a sustainable life for the residents as the space offers stress relief, spiritual comfort, recreation, gardening, and commercial value.

Modern explorations in urbanism are challenging traditional ideas. For example, "urban acupuncture" is a school of thought that shuns massive urban renewal projects and overhaul of entire city blocks in favor of more microscale and community interventions which, in cities of limited resources and constrained budget, can cheaply provide children with that coveted green recreational space in the form of a series of "micro-parks" or "urban lounges" that they may enjoy while walking to school.

Conclusion

Faced with poverty and hunger, it is easy to trivialize children's access to recreation, but we must understand that recreation is a necessity for urbanites for proper functioning of everyday life amidst urban hectic schedule. Children's access to urban recreational space has decreased rapidly due to densification, increased road connectivity, and parents' growing sense of perceived danger and their desire to protect their children even at the detriment of cognitive development. However, this inaccessibility has also resulted in a movement toward bringing nature indoors, as well as home-bound recreation, biophilic design, and search for microscale recreational space to create a sustainable and resilient lifestyle. Moving forward, urban planning must move beyond the simplistic idea of "pocket utopias" of distant parks and playgrounds and rather integrate contact with recreational space into the city. To build a sustainable and livable city, modern strategies need to be incorporated into existing built environment for better recreational opportunities for children. Recreational space in building, rooftop, and balcony gardens as well as café and library with modern amenities are greatly encouraged. For further research, it is important to include children's opinions into recreational planning which should be very context specific (Hart 2002). There is also a need for a better understanding of what type of recreational spaces yield the best results and what kind of microlocations increase children's contact with recreational space the most.

Cross-References

- ► Built Environment
- ► Recreation
- ► Recreational Opportunities
- ► Recreational Space
- ► Sustainable City

References

Abubakar, I.., & Aina, A. (2006). GIS and space syntax: An analysis of accessibility to urban green areas in Doha District of Dammam Metropolitan Area, Saudi Arabia. Conference: Map Middle East 2006.

Abu-Ghazzeh, T. (1998). Children's use of the street as a playground in Abu-Nuseir, Jordan. *Environment and Behavior, 30*(6), 799–831.

Ahmed, A. (2010). Factors and issues related to children's play and their implications on play and recreation provision in Dhaka City. A doctoral thesis for Doctor of Philosophy of Loughborough University.

Ahmed, A., & Sohail, M. (2008). Child's play and recreation in Dhaka City. *Municipal Engineer, 161*(4), 263–270.

Bartlett, S. (1999). Children's experience of the physical environment in poor urban settlements and the implications for policy, planning and practice. *Environment and Urbanization, 11*(2), 63–73.

Beckers, R., van der Voordt, T., & Dewulf, G. (2016). Why do they study there? Diary research into students' learning space choices in higher education. *Higher Education Research and Development, 35*(1), 142–157.

Biswas, M., (2002). 'Pattern and trend of recreation activities in Dhaka city' facility unpublished MURP thesis, submitted to the urban and regional Planning, Bangladesh University of Engineering and Technology (BUET), Dhaka, Bangladesh.

Bjorklid, P. (1985). Children's outdoor environment from the perspectives of environmental and developmental psychology. In T. Garling & J. Valsiner (Eds.), *Children within environments*. New York: Plenum.

Børve, H. E., & Børve, E. (2017). Rooms with gender: Physical environment and play culture in kindergarten.

Early Child Development and Care, 187(5–6), 1069–1081.

Brambilla, P., Giussani, M., Pasinato, A., Venturelli, L., Privitera, F., del Giudice, E. M., & Chiappini, E. (2017). Sleep habits and pattern in 1-14 years old children and relationship with video devices use and evening and night child activities. *Italian Journal of Pediatrics, 43*(7).

Bratman, G. N., Hamilton, P., Hahn, K. S., Daily, G. C., & Gross, J. J. (2015). Nature experience reduces rumination and subgenual prefrontal cortex activation. *PNAS, 112*(28), 8567–8572.

Chastin, S. F. M., Mandrichenko, O., & Skelton, D. A. (2014). The frequency of osteogenic activities and the pattern of intermittence between periods of physical activity and sedentary behaviour affects bone mineral content: The cross-sectional NHANES study. *BMC Public Health, 14*(1), 1–12.

Child, E. (1983). Play and culture: A study of English and Asian children. *Leisure Studies, 2*, 169–186.

Christian, H., Trapp, G., Villanueva, K., Zubrick, S. R., Koekemoer, R., & Giles-Corti, B. (2014). Dog walking is associated with more outdoor play and independent mobility for children. *Preventive Medicine, 67*, 259–263.

Churchman, A. (1980). Children in urban environments: The Israeli experience. In W. Michelson & E. Michelson (Eds.), *Managing urban space in the interest of children*. Ottawa: Man and Biosphere Committee.

Churchman, A. (2003). Is there a place for children in the city? *Journal of Urban Design, 8*(2), 99–111.

Comedia/Demos. (1995). *Park life: Urban parks and social re-newal*. London: Comedia and Demos.

Cooper-Marcus, C., & Sarkissian, W. (1986). *Housing as if people mattered*. Berkeley: University of California Press.

Ekkel, E. D., & de Vries, S. (2017). Nearby green space and human health: Evaluating accessibility metrics. *Landscape and Urban Planning, 157*, 214–220.

Esposito, L., McCune, S., Griffin, J. A., & Maholmes, V. (2011). Directions in human-animal interaction research: Child development, health, and therapeutic interventions. *Child Development Perspectives, 5*(3), 205–211.

Freeman, C. (1995). Planning and play: Creating greener environments. *Children's Environments, 12*(3), 381–388.

Freeman, C., & Tranter, P. (2011). *Children and their urban environment. Changing worlds*. London: Earthscan.

Gehl, J. (1996). *Life between buildings*. Copenhagen: ArkitektensForlag.

Ghosh, S., Vanni, I., & Giovanangeli, A. (2016). Social aspects of institutional rooftop gardens. in Wilkinson, S. and Dixon, T. (eds.) *Green roof retrofit: Building urban resilience*. Chichester, UK: John Wiley & Sons. *27*, 189–215.

Goss, J. (1993). The "magic of the mall": An analysis of form, function, and meaning in the contemporary retail built environment. *Annals of the Association of American Geographers, 83*(1), 18–47.

Green, B. (2004). *A guide to using plants on roofs, walls and pavements*. London: Greater London Authority.

Hadavi, S. (2017). Direct and indirect effects of the physical aspects of the environment on mental well-being. *Environment and Behavior, 49*(10), 1071–1104.

Hand, K., Freeman, C., Seddon, P. J., Stein, A., & van Heezik, Y. (2016). A novel method for fine-scale biodiversity assessment and prediction across diverse urban landscapes. *Landscape and Urban Planning, 151*, 33–44.

Hart, R. (2002). Containing children: Some lessons on planning for play from New York City. *Environment and Urbanization, 14*(2), 135–148.

Holloway, S., & Valentine, G. (2003). *Cyberkids: Children in the information age*. London: Routledge Falmer.

Hyndman, B., & Chancellor, B. (2015). Engaging children in activities beyond the classroom walls: A social–ecological exploration of Australian primary school children's enjoyment of school play activities. *Journal of Playwork Practice, 2*, 117–141.

Islam, M. Z., Moore, R., & Cosco, N. (2014). Child-friendly, active, healthy neighbourhoods: Physical characteristics and children's time outdoors. *Environment and Behaviour, 48*(5), 711–736.

Jacobs, J. (1961). *The death and life of Great American cities*. New York: Random House.

Jens, K., & Gregg, J. S. (2021). How design shapes space choice behaviors in public urban and shared indoor spaces- a review. *Sustainable Cities and Society, 65*, 102592.

Kaplan, R., & Kaplan, S. (1989). *Experience of nature: A psycho-logical perspective*. New York: Cambridge University Press.

Krysiak, N. (2017). 'Designing child- friendly high density neighbourhoods: Transforming our cities for the health, wellbeing and happiness of children' Cities For Play. Churchill Trust.

Kyttä, M. (2004). The extent of children's independent mobility and the number of actualized affordances as criteria for child-friendly environments. *Journal of Environmental Psychology, 24*, 179–198.

Lester, S., & Russell, W. (2010). *Children's right to play: An examination of the importance of play in the lives of children worldwide* (Working paper no. 57). The Hague: Bernard van Leer Foundation.

Lissak, G. (2018). Adverse physiological and psychological effects of screen time on children and adolescents: Literature review and case study. *Environmental Research, 164*, 149–157.

Lui, D., Szeto, G., & Jones, A. (2011). The pattern of electronic game use and related bodily discomfort in Hong Kong primary school children. *Computers in Education, 57*, 1665–1674.

MacKerron, G., & Mourato, S. (2013). Happiness is greater in natural environments. *Global Environmental Change, 23*, 992–1000.

Magee, C. A., Lee, J. K., & Vella, S. A. (2014). Bidirectional relationships between sleep duration and screen

time in early childhood. *JAMA Pediatrics, 168*(5), 465–470.
Maras, D., Flament, M. F., Murray, M., Buchholz, A., Henderson, K. A., Obeid, N., & Goldfield, G. S. (2015). Screen time is associated with depression and anxiety in Canadian youth. *Prevention Medicine, 73*, 133–138.
Marchione, M. (2013). Study: Kids are less fit than their parents were. *The Denver Post.*
Montgomery, J. (1997). Café culture and the city: The role of pavement cafés in urban public social life. *Journal of Urban Design, 2*(1), 83–102.
Moore, R., Goltsman, S., & Iacofano, D. (1987). *Play for all guidelines*. Berkeley: MIG Communications.
Nathanson, A. I., & Fries, P. T. (2014). Television exposure, sleep time, and neuropsychological function among preschoolers. *Media Psychology, 17*(3).
Nilufar, F. (1999). Urban life and use of public space. Unpublished research report. Submitted to The Asiatic Society of Bangladesh.
Owens, P. (1988). Natural landscapes, gathering places, and prospect refuges: Characteristics of outdoor places valued by teens. *Children's Environments Quarterly, 5* (2), 17–24.
Parent, B. A., Weasley-Sanders, M. A., & Forehand, R. (2016). Youth screen time and behavioral health problems: The role of sleep duration and disturbances. *Journal of Developmental and Behavioral Pediatrics, 37*(4).
Patel, S. (2011). Analyzing urban layouts: Can high density be achieved with good living conditions? *Environment and Urbanization, 23*(2), 583–595.
Pouya, S., Bayramoğlu, E., & Demirel, Ö. (2016). Restorative garden as an useful way to relieve stress in megacities, a case study in Istanbul. *İnönü University Journal of Art and Design, 13*, 355–369.
Pyle, R. M. (1978). The extinction of experience. *Horticulture, 56*, 64–67.
RAJUK. (2015). *Draft Dhaka structure plan (2016–2035).* Rajdhani Unnayan Katripakkhya: Dhaka.
Richardson, E. A., Pearce, J., Shortt, N. K., & Mitchell, R. (2017). The role of public and private natural space in children's social, emotional and behavioral development in Scotland: A longitudinal study. *Environmental Research, 158*, 729–736.
Rissotto, A., & Giuliani, M. V. (2006). Learning neighbourhood environments: The loss of experience in a modern world. In *Children and their environments* Cambridge: Cambridge University Press, (pp. 75–90).
Rogers, R. (1999). *Towards an urban renaissance: Final report of the urban task force chaired by Lord Rogers of Riverside. Urban task force*. London: Department of the Environment, Transport and the Regions.
Ryan, C. O., Browning, W. D., Clancy, J. O., Andrews, S. L., & Kallianpurkar, N. B. (2014). Biophilic design patterns emerging nature-based parameters for health and well-being in the built environment. *International Journal of Architectural Research. Archnet-IJAR, 8*(2), 62–76.
Sallis, J. F., & Glanz, K. (2006). The role of built environments in physical activity, eating, and obesity in childhood. *The Future of Children, 16*(1), 89–108.
Shao, H., Xu, S., Zheng, J., Zheng, J., Zheng, J., Chen, J., & Huang, Y. (2015). Association between duration of playing video games and bone mineral density in Chinese adolescents. *Journal of the International Society for Clinical Densitometry, 18*(2), 198–202.
Sharmin, S., Kamruzzaman, M., & Haque, M. M. (2020). The impact of topological properties of built environment on children independent mobility: A comparative study between discretionary vs. nondiscretionary trips in Dhaka. *Journal of Transport Geography, 83*, p.102660.
Siddiqui, Md. M. R. (1990). Recreational facilities in Dhaka City: A study of existing parks and open space. MURP thesis, Submitted to the urban and regional planning, Bangladesh University of Engineering and Technology (BUET), Dhaka.
Sultana, R., & Nazem, N. I. (2016). Declining green space and livable environment: A study in Dhaka South City corporation. *CUS Bulletin on Urbanization and Development, 71*, 9–14.
Sustainable Development Goals: Bangladesh Progress Report 2018. (2019). Bangladesh planning commission, ministry of planning. Government Document.
Thompson, C. W. (2002). Urban open space in the 21st century. *Landscape and Urban Planning, 60*(2), 59–72.
Timperio, A., Salmon, J., Chu, B., & Andrianopoulos, N. (2008). Is dog ownership or dog walking associated with weight status in children and their parents? *Health Promotion Journal of Australia, 19*, 60–63.
UN. (2016). The world's cities in 2016. United Nation. Available at: https://www.un.org/en/development/desa/population/publications/pdf/urbanization/the_worlds_cities_in_2016_data_booklet.pdf. Retrieved 23 November 2019.
Verwer, D. (1980). Planning and designing residential environments with children in mind: A Dutch approach. In P. Wilkinson (Ed.), *Innovation in play environments*. New York: St. Martin's Press.
Watkins, W. (1980). Play environments in arid lands. In P. Wilkinson (Ed.), *Innovation in play environments* (pp. 152–170). New York: St. Martin's Press.
Winther, A., Ahmed, L. A., Furberg, A., Grimnes, G., Jordi, R., Nilsen, O. A., & Emau, E. (2015). Leisure time computer use and adolescent bone health – Findings from the Tromsø study, fit futures: A cross-sectional study. *BMJ Open, 5*(6).
Wood, H. C., & Scott, H. (2016). Sleepy teens: Social media use in adolescence is associated with poor sleep quality, anxiety, depression and low self-esteem. *Journal of Adolescence, 51*, 41–49.
Woodward, J. (2005). *Creating the customer-driven library*. Chicago: American Library Association.
Woolley, H. (2006). Freedom of the city: Contemporary issues and policy influences on children and young people's use of public open space in England. *Children's Geographies, 4*(1), 45–59.

Zimmerman, F. J., Christakis, D. A., & Meltzoff, A. N. (2007). Associations between media viewing and language development in children under age 2 years. *The Journal of Pediatrics, 151*(4), 364–368.

Toward Smart Public Lighting of Future Cities

Mohsen Mohammadzadeh[1] and Son Phung[2]
[1]School of Architecture and Planning, Auckland University, Auckland, New Zealand
[2]Department of Civil and Environmental Engineering, Auckland University, Auckland, New Zealand

Introduction

This entry explores smart public lighting-based digital twinning as a component of smart future cities. It first reviews the history and the different aspects of lighting in urban public spaces. It then investigates the new opportunities of rapidly developing lighting technologies, particularly digital twinning, toward improving energy efficiency as well as the quality of nightlife in future smart cities.

Public Lighting

Artificial lighting plays a significant role in people's everyday lives, including, but not limited to, their behaviors, perceptions of objects, and use of urban space and the built environment. In contemporary cities, nightlife activities are often structured by access to, and the use of, artificial lighting in private and public spaces. Artificial lighting is one of the prerequisite quality criterions of urban spaces in contemporary cities (Andrei et al. 2009) because it stands as a symbol of modernity, urbanity, security, technological advancement, and economic prosperity (Meier et al. 2014).

Historically, access to a reliable and affordable supply of fuel and lighting equipment has been a major constituent in providing public lighting in all parts of the city (Griffiths 2016). In ancient times, artificial lighting such as torches was extensively utilized in public urban spaces, often concentrated near the places of power (the castle or churches) for security and defense purposes. From 1820, marked as the beginning of light revolution, the invention of gas lanterns, incandescent lamps, and then discharge lamps made lighting feasible and affordable throughout the urban areas of industrial cities (Meier et al. 2014). Since the late nineteenth and early twentieth centuries, the distribution of electric power has reached all corners of the globe. Electricity is a reliable and affordable source of power for public lighting, and public lighting is one of the quality criteria and characteristics of the modern city (Narboni 2020).

Urban public lighting is a basic and essential public infrastructure in contemporary cities. It is a source of light on either a pole, a post, or a building at a public precinct, with the major functionality being to illuminate road accessibility for pedestrians, vehicles, or public space users. Urban nightlife and activities largely depend on the provision of public lighting.

Public lighting is a public service that all municipal authorities must provide in the areas they administrate in order to make the road, the street, and other public areas safer and more accessible for the city's inhabitants. Public lighting can be categorized in two main groups:

- Public authorities and local governments largely offer a street lighting service in different types of streets, roads, and vehicle-pedestrian mixed zones. The main purpose of the service is to facilitate the movement of vehicles safely during the night. Street lighting is often generic, impersonal, and static, and available in a number of forms, such as traditional lamp posts and wall lights, wind-powered streetlights, solar power streetlights, etc.
- Public authorities and local governments often provide public-space lighting as a required service on pathways and cycleways, and in sport grounds, parks, and public precincts. This service has at least two primary functions: (1) to ensure the accessibility, safety, and assembly

of pedestrians, and (2) to facilitate nightlife activities in urban areas. Public space lighting usually requires a higher esthetic standard than street lighting, targeting feature landscapes, landmarks, and public building.

A significant number of studies have investigated public lighting and its impacts on cities and people's everyday lives. In his study of street lighting, Bouman (1987) recognized that a demand for lights in public areas can arise for different reasons, including the lengthened work cycle of city dwellers resulting from the industrial revolution; the enhancement of nighttime activities ranging from traveling after work, to eating out and nighttime entertainment; the impetus of commercial and marketing activity; and the relationship between lighting and public safety. Many studies have shown a relationship between street lighting and road safety, pedestrian assurance, and crime rates (Boyce 2019; Levin et al. 2020; Steinbach et al. 2015; Tetri et al. 2018). Adequate public lighting can reduce 50% of crashes (Tetri et al. 2018), and Park and Garcia (2020) found that "proper street lighting is the main contributor to enhancing the feeling of safety on streets."

Boissevain (2018, p. 181) argues that "new developments in lighting systems are also enabling unexpected new ways of using light for communication, wayfinding, as well as cultural, artistic and civic expression and branding." Over the last three decades, under the flag of globalism, cities have competed in the global market to attract supplementary human and financial capital, which is crucial for their constant economic growth (Mohammadzadeh 2014). Place marketing has become the core of global city plans, policies, and development strategies. Arbab et al. (2020) coined the concept of "Lighting Branding" (LB) to explain a pervasive usage of lighting for place marketing that is aligned with other relevant concepts such as "place branding," "destination branding," and "city branding." Creating an attractive, unique, and identifiable city image that differentiates the city from other cities is one of the tools of place marketing. Arbab et al. (2020, p. 138) argue that lighting is "an element for the nocturnal branding of cities." Most branded cities such as New York, Tokyo, Dubai, and Singapore, among others, are recognized based on their night-light landscapes. Smart lighting technologies provide new capabilities in color control, levels of resolution, and an urban scale of lighting provision which generate new opportunities for city branding. Public lighting is becoming a crucial hallmark of civilization, safety, and a prosperous society.

Smart Lighting and New Opportunities

Smart lighting is "defined as a lighting system which gains real-time information of its environment and users through sensors and adapts its behaviour accordingly" (Pihlajaniemi et al. 2018, p. 60). The first attempts at smart lighting were initiated in Oslo, Norway, in the late 1990s to mitigate power consumption by 50%, minimize maintenance costs, and improve road safety (Suseendran et al. 2018). Smart lighting solutions in the intelligent transformation of cities have gained momentum in the last two decades. The confluence of multiple technological revolutions has generated new opportunities for using public lighting more efficiently and effectively in urban spaces.

Emerging smart lighting technologies help mitigate energy consumption and the cost of public lighting provision in cities. Approximately 2.4% of the annual global energy resources is used for lighting, resulting in 5 to 6% of the total greenhouse gas emissions in the atmosphere. Global electricity consumption is reaching considerably high figures and increasing by around 3% annually (Bachanek et al. 2021). According to the World Bank, the electricity consumed in public lighting globally is equivalent to Germany's total annual electricity consumption (Li and Makumbe 2017). The International Energy Agency (IEA) predicted that the overall demand for lighting will be 80% higher in 2030 than in 2005. Global lighting electricity demand will almost double the output of all nuclear power plants. IEA predicts that due to the lack of energy-efficiency policy and technologies, lighting-related annual CO_2 emissions will rise to almost 3 gigatons by 2030 (IEA 2006). The cost of public lighting may reach up to 65% of municipal electricity budgets. Outdoor lighting is responsible for 15 to 19% of global electricity consumption. According to estimates, cities consume almost 75% of global energy, and outdoor urban lighting alone can account for as much as 20 to 40% of budget expenditure related to power (Bachanek et al. 2021). Smart lighting solutions, as a component of smart city strategies and initiatives, can assist cities to mitigate the cost of public lighting by improving energy efficiency. However, existing street and other exterior lighting installations serve as a backbone of the lighting network. Today, most cities that install new smart lighting or retrofit existing fixtures choose a system that is already equipped with sensor technology or that can be upgraded easily to utilize the advantage of the Internet of Things (IoT) applications.

Over the last two decades, new emerging technologies and innovations have transformed the traditional lighting system into smart lighting. Researchers are constantly working on new technologies to improve the effectiveness and efficiency of the lighting system.

A lamp is a device that transforms electrical energy or gas into light. The amount of light produced by a lamp depends on the device's efficacy. Visible light output is usually measured in lumens. Light Emitting Diode (LED) is the most cost-effective lamp because of its low-energy consumption, long life, decreasing investment costs and maintenance costs, and lower environmental impact (Beccali et al. 2019). Pasolini et al. (2019, p. 2) argue that "LED fixtures can deliver electricity savings of up to 80% over classic lighting technologies." The deployment of LED will greatly reduce the cost of lighting in cities.

Sensor-based lighting is a popular way of controlling illumination. Beccali et al. (2017, p.287) argued that "the sensors that could be installed with the system are designed, i.e. to gather several types of data: presence and activities of people, vehicular traffic, air quality, weather data and acoustic levels." Sensors detect the movement of people and vehicles and translate it into a signal to denote that the space is occupied. The purpose is to save energy consumption by switching off or dimming lights to a lower level if no motion is detected for a predefined time.

Visible light communication (VLC) technology supports wireless network access as an alternative to traditional wireless technologies such as Wi-Fi. VLC converts the existing lighting systems into "Li-Fi" modems that assist in controlling the lighting system. VLC consists of a transmitter, a propagation channel, and receivers such as smart gadgets, including phones or tablets. The transmitter consists of Red-Green-Blue (RGB) LEDs or a single white LED. The LED is switched on and off quickly with the help of a modulator and thus sends light pulses. Since the communication between transmitter and receiver can only take place in the line of sight, the LEDs must be arranged accordingly (Füchtenhans et al. 2019).

IoT has generated new opportunities to utilize public light in urban spaces. LEDs can connect to the electrical control network that makes traditional lighting "smart." Smart lighting systems operate based on the intelligent interactions of light sources, sensors, and the communication network and react to external factors, including daylight, weather conditions, and people activities. Smart lighting is a component of IoT development. IoT connects everything around us through Information Communication Technologies (ICTs) that include, but are not limited to, sensors, smart devices, cars, and lighting infrastructure (Pihlajaniemi et al. 2018). IoT-based lighting can expand the provision of the Wi-Fi connection around the city. In addition, a fault detection system can be easily established on top of street light controllers. Pasolini et al. (2019, p. 2) point to "the remote monitoring of its functional parameters (power consumption, lamp temperature, electrical parameters)" and highlight that "this widespread network allows to collect data gathered by sensors (air pollution, vehicular traffic, water flooding, etc.) possibly mounted on streetlights, or transmitted by IoT wireless devices located in their surroundings." IoT helps to create an integrated lighting control system that features dynamic, sensor-based control, multiple luminaire states, and complex geometries. Other research studies on lighting control systems have focused on image processing, fuzzy systems, cooperative methods, wireless sensor network (WSN), simulation algorithms, and predictive control for energy optimization (Barra and Rahem 2014; De Paz et al. 2016; Zhang et al. 2013). A smart public lighting system connects to people's smart devices and locates the streetlight that requires intervention, issues actuation commands directly to the IoT node connected to the lamp, and signals the result of the intervention to a central system that can track every single lamppost and, hence, optimize the maintenance plan. Such a system can be successively extended to include other types of IoT nodes or clouds of IoT nodes, provided that each IoT peripheral system supports an HTTP-based interface, which makes it possible to interact with it in an open-, standard-, and technology-independent manner. Human Centric Lighting (HCL) is "evidence-based lighting solutions optimized for vision, performance, concentration, alertness, mood, and general human health and well-being. HCL balances visual, emotional, and biological benefits of lighting for humans, recognizing the role of light on human vision, psychology, and physiology" (Houser 2018, p. 213). The visual effect of smart lighting seeks to achieve the optimal lighting of an environment adapted to an activity, thus providing the basic requirements of lighting.

Smart lighting functions as an infrastructure for IoT and other smart solutions (Castro et al. 2013). Smart lighting generates new opportunities in monitoring the environment, improving urban quality standards, increasing public and traffic safety, expanding ICT as Wi-Fi hot spots, and delivering other location-based services like smart parking and smart navigation (Kokilavani and Malathi 2017). Smart lighting is increasingly recognized as a better illuminating methodology to save energy, lessen light pollution, and promote safe traffic. Vieira et al. (2019, p. 1) argue that "a wireless networked smart utility pole system based on IoT with centralized and remote-control technology that sports a photovoltaic panel, an on-board battery, a wireless communication and smart LED streetlighting is a smart city application with great potential to reduce energy cost and enhance public safety."

Artificial Intelligence (AI) is the next step in smart lighting. Using AI, machines can "learn from experience, adjust to new inputs, and perform human-like tasks" (Duan et al. 2019, p. 63). AI poses a great challenge to traditional control theories by managing systems without human interactions. An AI lighting control system is composed of system components, input components, and output components (Guo et al. 1993). AI compiles big data to meet the lighting requirements of different urban areas, ensuring the safe movements of vehicles and safe social activities, resulting in reduced lighting energy consumption and costs and an increase in residents' comfort levels and satisfaction (Lai et al. 2020). De Paz et al. (2016) discussed a combination of different statistics and Artificial Intelligence (AI) techniques such as artificial neural networks (ANN), the EM algorithm, methods based on ANOVA, and a Service Oriented Approach (SOA) to ensure the lighting system is innovative in terms of obtaining an intelligent prediction of consumption and cost. AI can automatically generate "smart lighting schedules based on input data such as flow of pedestrians or traffic, weather, or the cost of consumption associated with luminaires" (De Paz et al. 2016, p. 242).

A smart lighting system, empowered by IoT and AI, can collect, analyze, and predict people's activities, traffic movements, and even the characteristics of urban space and functions to provide the required illumination. Zanella et al. (2014, p. 29) argue that "streetlight is geographically localized on the city map and uniquely associated to the IoT node attached to it, so that IoT data can be enhanced with context information." AI constantly compiles collected urban big data in real time and offers the necessary light required in different parts of the city. "Public smart lighting becomes a dynamic platform serving as a backbone for 'smart city' developments. Cities can thus rethink the role of the public-light asset, which offers opportunities to deliver new services and generate revenues, instead of just being a cost factor" (Pasolini et al. 2019, p. 3).

This development trend will continue in upcoming years based on increased connectivity and industrial Internet of Things (IoT) solutions, with AI becoming a key element in most smart city strategies around the globe.

The Future of Digital Twin-Empowered Public Lighting

Michael Grieves in collaboration with John Vickers first coined the concept of digital twin in 2003 (Jones et al. 2020). A digital twin represents a real-time digital replica of a process, product, or service that is a cyber-representation capable of dynamically synchronizing a physical entity in actual time. A digital twin is a "mirror" made up of three components: (1) a physical product/environment; (2) a virtual representation of that product/environment; and (3) bidirectional data connections that feed data from the physical to the virtual representation, and information and processes from the virtual representation to the physical (Grieves and Vickers 2017; Jones et al. 2020).

Recent progress in ICTs, including, but not limited to, IoT, cloud computing, big data, AI, and digital twinning, generates new opportunities for the convergence of the physical and virtual worlds toward urban digitalization (Qi et al. 2019). A digital twin is a cyberphysical integration that improves the decision-making process of the physical model by using IoT to collect big data, visualizing real-time big data, and utilizing AI to learn, interpret, and simulate data. Digital twinning in combination with other ICT advancements offers a new capacity for continuous learning and updating the original physical model, and the data sources accumulated in the past and present, and has the capacity to simulate and predict possible faults and issues before they arise (Mohammadi and Taylor 2017). Large ICT corporations such as IBM, CISCO, ANSYS, GE Digital, Microsoft, and Siemens, among others, have developed digital twinning to assess physical product/environmental performance by examining "what-if" scenarios in the virtual product/environment. Cureton and Dunn (2021, p. 268) argue that "the prototype or enhancement to a physical asset can be designed in virtual space; thus a product or process can be tested for its future life cycle, performance, or maintenance without extensive costs." Digital twinning in combination with IoT and AI can adjust the physical product/environment if required based on the most efficient and effective desired outcomes.

Digital twinning in cities includes sensing systems enabled by IoT, cloud computing, big data, and AI (Cureton and Dunn 2021), and it is an effective tool for city governance in the design, build, and operation of a city. The birth of the digital twin marks a generation of combining symbiotic smart technologies. Mohammadi and Taylor (2017) coined the concept of the smart city digital twin to explain the emerging use of the digital twin as a component of smart city initiatives. A smart city "digital twin may be able to virtually capture interdependencies between infrastructure performance, technology interventions, and human dynamics" (Mohammadi and Taylor 2019, p. 1998). A smart city digital twin provides an ultrarealistic and real-time simulation of the city, from the microdwelling level to the macrometropolitan level, enabling urban planning and infrastructure optimization. AI progressively updates the digital replica of the city based on the constant collection of big data from the physical environment and social and economic activities to optimize the operation of the city in the present and provide predictive insights that answer the "what-if" scenarios of the future. It also provides a solution to the challenge of how to most effectively use the massive and constantly changing data collected by urban IoT (van der Heide et al. 2017) and create an infrastructure platform that enables AI to detect normal and abnormal patterns in the data on the operational processes of the city without requiring human intervention. Smart city digital twinning has proved it is flexible and accessible based on the variety of spatial and temporal scales (Mohammadi and Taylor 2017). A smart city digital twin offers a sophisticated, reliable, and data-driven city governance decision-making process that is a significant step toward shaping smart sustainable cities (Petrova-Antonova and Ilieva

2019; Wakil et al. 2019). Cureton and Dunn (2021, p. 270) state that "current city digital twin data selection has focused on transport planning and digital planning services and energy sectors." For example, Francisco et al. (2020) argue that smart city digital twinning can facilitate real-time urban building energy management. Planners and policymakers have increasingly utilized smart city digital twins to address urban complex issues. Cities such as Singapore (2018), Helsinki (2019), Munich (2018), Rotterdam (2019), and Wellington (2021), among others, use smart city digital twinning to address urban issues. In their paper "The digital twin of the city of Zurich for urban planning," Schrotter and Hürzeler (2020) address various urban issues in the city such as preparing environmental strategies, the 3D history city model, high-rise planning, and modeling urban climate. According to Research and Markets (2020), the digital twin market will be worth approximately $30 billion by 2025 with an increased CAGR of 38% between 2019 and 2025. City digital twinning is predicted as a future trajectory for local governments and urban management.

Digital twinning empowers public lighting systems through real-time operations that monitor and control faulty diagnostics, resulting in the optimization of energy efficiency and performance (Kaewunruen et al. 2019; Lu et al. 2019). Virtual replica technology can simulate big data in the city-level decision-making process, where most modeling and simulation tools struggle with scale and connectivity. Digital twins for public lighting can also be very flexible and accessible, based on the variety of spatial and temporal scales provided (Mohammadi and Taylor 2017). Public lighting can be digitalized on the scale of one light bulb to a single building, neighborhood, central business district, and national scale, using a variety of modeling approaches, including geometric and geospatial modeling, computation/numerical modeling, or AI/machine learning modeling (Grieves and Vickers 2017). Digital twinning empowers a smart lighting system that represents different points in the lighting life cycle, including feasibility studies, and asset management. Digital twinning in combination with other emerging technologies including, but not limited to, LEDs, AI, and IoT, generates an "autonomous urban lighting system." Autonomous urban public lighting can mitigate energy consumption and significantly reduce operation and maintenance costs by collecting and compiling real-time urban big data, predicating various scenarios, and providing the required illumination. Meleti and Delitheou (2020) define the autonomous urban lighting system as a component of "Smart Energy-Autonomous Cities."

An autonomous lighting system can give new life to modern urban centers by providing dynamic lighting and multiple benefits to humans and modern cities (Meleti and Delitheou 2020). It can generate unique identities for urban spaces and cities by constantly changing the level and color of lighting. Automating urban lighting has a positive impact on mitigating emissions from electricity and heat generation. The automation of public lighting should be considered a component of the urban automation process that aims to shape a smart sustainable city in the future.

Acknowledgments I want to acknowledge the New Zealand Ministry of Business, Innovation & Employment's supports of this work through National Science Challenge 11.

References

Andrei, H., Cepisca, C., Dogaru-Ulieru, V., Ivanovici, T., Stancu, L., & Andrei, P. C. (2009, June). Measurement analysis of an advanced control system for reducing the energy consumption of public street lighting systems. In *2009 IEEE Bucharest PowerTech* (pp. 1–6). IEEE.

Arbab, M., Mahdavinejad, M., Bemanian, M., & Arbab, M. (2020). Lighting branding: Lighting architecture and building nocturnal city identity. *International Review for Spatial Planning and Sustainable Development, 8*(1), 137–159.

Bachanek, K. H., Tundys, B., Wiśniewski, T., Puzio, E., & Maroušková, A. (2021). Intelligent street lighting in a smart city concepts – A direction to energy saving in cities: An overview and case study. *Energies, 14*(11), 3018.

Barra, K., & Rahem, D. (2014). Predictive direct power control for photovoltaic grid connected system: An approach based on multilevel converters. *Energy Conversion and Management, 78*(February), 825–834.

Beccali, M., Bonomolo, M., Galatioto, A., & Pulvirenti, E. (2017). Smart lighting in a historic context: A case study. *Management of Environmental Quality: An International Journal, 28*(2), 282–298.

Beccali, M., Bonomolo, M., Brano, V. L., Ciulla, G., Di Dio, V., Massaro, F., & Favuzza, S. (2019). Energy saving and user satisfaction for a new advanced public lighting system. *Energy Conversion and Management, 195*(September), 943–957.

Boissevain, C. (2018). Smart city lighting. In S. McClellan, J. A. Jimenez, & G. Koutitas (Eds.), *Smart cities* (pp. 181–195). Cham: Springer.

Bouman, M. J. (1987). Luxury and control: The urbanity of street lighting in nineteenth-century cities. *Journal of Urban History, 14*(1), 7–37.

Boyce, P. R. (2019). The benefits of light at night. *Building and Environment, 151*(March), 356–367.

Castro, M., Jara, A. J., & Skarmeta, A. F. (2013, March). Smart lighting solutions for smart cities (Paper presentation). In *27th international conference on advanced information networking and applications workshops* (pp. 1374–1379). IEEE.

Cureton, P., & Dunn, N. (2021). Digital twins of cities and evasive futures. In A. Aurigi & N. Odendaal (Eds.), *Shaping smart for better cities* (pp. 267–282). Academic.

De Paz, J. F., Bajo, J., Rodríguez, S., Villarrubia, G., & Corchado, J. M. (2016). Intelligent system for lighting control in smart cities. *Information Sciences, 372*, 241–255.

Duan, Y., Edwards, J. S., & Dwivedi, Y. K. (2019). Artificial intelligence for decision making in the era of Big Data–evolution, challenges and research agenda. *International Journal of Information Management, 48*, 63–71.

Francisco, A., Mohammadi, N., & Taylor, J. E. (2020). Smart city digital twin–enabled energy management: Toward real-time urban building energy benchmarking. *Journal of Management in Engineering, 36*(2), 04019045.

Füchtenhans, M., Grosse, E. H., & Glock, C. H. (2019). Literature review on smart lighting systems and their application in industrial settings. In 2019 6th International Conference on Control, Decision and Information Technologies (CoDIT) (pp. 1811–1816). IEEE.

Grieves, M., & Vickers, J. (2017). Digital twin: mitigating unpredictable, undesirable emergent behavior in complex systems. In F. J. Kahlen, S. Flumerfelt, & A. Alves (Eds.), *Transdisciplinary perspectives on complex systems: New findings and approaches* (pp. 85–113). Cham: Springer.

Griffiths, D. (2016). *The social and economic impact of artificial light at Pompeii*. Thesis, University of Leicester. https://hdl.handle.net/2381/39016

Guo, B., Belcher, C., & Roddis, W. K. (1993). RetroLite: An artificial intelligence tool for lighting energy-efficiency upgrade. *Energy and Buildings, 20*(2), 115–120.

Helsinki. (2019). *The Kalasatama digital twins project, the final report of the KIRA-digi pilot project*. https://www.hel.fi/stati c/liitteet-2019/Kaupu ngink ansli a/Helsi nki3D Kalas atama Digital_Twins.pdf. Accessed 10 June 2021.

Houser, K. (2018). Human centric lighting and semantic drift. *LEUKOS, 14*(4), 213–214.

IEA. (2006). *Light's labour's lost: Policies for energy-efficient lighting*. OECD Publishing.

Jones, D., Snider, C., Nassehi, A., Yon, J., & Hicks, B. (2020). Characterising the Digital Twin: A systematic literature review. *Journal of Manufacturing Science and Technology, 29*, 36–52.

Kaewunruen, S., Rungskunroch, P., & Welsh, J. (2019). A digital-twin evaluation of net zero energy building for existing buildings. *Sustainability, 11*(1), 159.

Kokilavani, M., & Malathi, A. (2017). Smart street lighting system using IoT. *International Journal of Advanced Research in Applied Science and Technology, 3*(11), 08–11.

Lai, X., Dai, M., & Rameezdeen, R. (2020). Energy saving based lighting system optimization and smart control solutions for rail transportation: Evidence from China. *Results in Engineering, 5*, 100096.

Levin, N., Kyba, C. C., Zhang, Q., de Miguel, A. S., Román, M. O., Li, X., & Elvidge, C. D. (2020). Remote sensing of night lights: A review and an outlook for the future. *Remote Sensing of Environment, 237*, 111443.

Li, J., & Makumbe, P. (2017, August 7). *LED street lighting: Unburdening our cities*. Retrieved on July 7, 2021, from. https://blogs.worldbank.org/energy/led-street-lighting-unburdening-our-cities

Lu, Y., Peng, T., & Xu, X. (2019). Energy-efficient cyber-physical production network: Architecture and technologies. *Computers & Industrial Engineering, 129*, 56–66.

Meier, J., Hasenöhrl, U., Krause, K., & Pottharst, M. (Eds.). (2014). *Urban lighting, light pollution and society*. Routledge.

Meleti, V., & Delitheou, V. (2020). Smart cities and the challenge of cities' energy autonomy. In J. C. Augusto (Ed.), *Handbook of smart cities* (pp. 563–592). Springer.

Mohammadi, N., & Taylor, J. E. (2017). Smart city digital twins (Paper presented). In *Proceedings of IEEE symposium series on Computational Intelligence (SSCI)* (Vol. 9 (1), pp. 1–5). IEEE

Mohammadi, N., & Taylor, J. (2019). Devising a game theoretic approach to enable smart city digital twin analytics (Paper presented). In *Proceedings of the 52nd Hawaii international conference on System Sciences* (pp. 1995–2002). IEEE

Mohammadzadeh, M. (2014). *The neoliberalised city fantasy: The place of desire and discontent*. Doctoral dissertation, ResearchSpace@ Auckland.

München. (2018). *Digitaler Zwilling: Bessere Luft durch intelligente Mobilität*. https://ru.muenchen.de/2018/194/Digitaler-Zwilling-Bessere-Luftdurch-intelligente-Mobilitaet-80933. Accessed 10 July 2021.

T

Narboni, R. (2020). Lighting public spaces: New trends and future evolutions. *Light & Engineering, 28*(2), 4–16.

Park, Y., & Garcia, M. (2020). Pedestrian safety perception and urban street settings. *International Journal of Sustainable Transportation, 14*(11), 860–871.

Pasolini, G., Toppan, P., Zabini, F., De Castro, C., & Andrisano, O. (2019). Design, deployment and evolution of heterogeneous smart public lighting systems. *Applied Sciences, 9*(16), 3281.

Petrova-Antonova, D., & Ilieva, S. (2019, June). Methodological framework for digital transition and performance assessment of smart cities (Paper presented). In *4th international conference on Smart and Sustainable Technologies (SpliTech)* (pp. 1–6). IEEE.

Pihlajaniemi, H., Juntunen, E., Luusua, A., Tarkka-Salin, M., & Juntunen, J. (2018). SenCity-piloting intelligent lighting and user-oriented services in complex smart city environments. *Light & Engineering, 26*(2), 60–67.

Qi, Q., Tao, F., Hu, T., Anwer, N., Liu, A., Wei, Y., . . . Nee, A. Y. C. (2019). Enabling technologies and tools for digital twin. *Journal of Manufacturing Systems, 58*(Part B), 3–21.

Research and Markets. (2020, March). *Digital Twin Market by Type (Asset, Process, System), Technology (Internet of Thing, Artificial Intelligence, Big Data Analytics, AR VR), End User (Automotive, Transport, Healthcare, Construction, Manufacturing, Retail) – Global Forecast to 2025.* https://www.researchandmarkets.com/reports/5004569/digital-twin-market-by-type-asset-process#src-pos-3. Accessed 10 Aug 2021.

Rotterdam. (2019). *Für IoT- und Smart City-Anwendungen: virtualcity SYSTEMS erstellt Digitalen Zwilling von Rotterdam.* https://www.business-geomatics.com/2019/09/09/fuer-iot-und-smart-city-anwendungen-virtualcitysystems-erstellt-digitalen-zwilling-von-rotterdam/. Accessed 10 Aug 2021

Schrotter, G., & Hürzeler, C. (2020). The digital twin of the city of Zurich for urban planning. *Journal of Photogrammetry, Remote Sensing and Geoinformation Science, 88*(1), 99–112.

Steinbach, R., Perkins, C., Tompson, L., Johnson, S., Armstrong, B., Green, J., & Edwards, P. (2015). The effect of reduced street lighting on road casualties and crime in England and Wales: controlled interrupted time series analysis. *Journal of Epidemiology and Community Health, 69*(11), 1118–1124.

Suseendran, S. C., Nanda, K. B., Andrew, J., & Praba, M. B. (2018, October). *Smart street lighting system* (Paper presented). In *3rd international conference on Communication and Electronics Systems (ICCES)* (pp. 630–633). IEEE.

Tetri, E., Chenani, S. B., & Rasanen, R. S. (2018). Advancement in road lighting. *Light & Engineering, 26*(2), 99–109.

van der Heide, J. J., Grus, M. M., & Nouwens, J. C. A. J. (2017). Making sense for society. *International Archives of the Photogrammetry, Remote Sensing and Spatial Information Sciences, 42*, 105–108.

Vieira, T. F., Brito, D. B., Ribeiro, M., & Araújo, Í. (2019, November). An IoT based smart utility pole and street lighting system (Paper presented). In *2019 IEEE Chilean conference on Electrical, Electronics Engineering, Information and Communication Technologies (CHILECON)* (pp. 1–5). IEEE.

VS. (2018). *Virtual Singapore 2019. Government of Singapore.* https://www.nrf.gov.sg/programmes/virtual-singapore. Accessed 12 Oct 2019.

Wakil, K., Nazif, H., Panahi, S., Abnoosian, K., & Sheikhi, S. (2019). Method for replica selection in the Internet of Things using a hybrid optimisation algorithm. *IET Communications, 13*(17), 2820–2826.

Wellington (2021). Wellington Digital Twin. https://buildmedia.com/work/wellington-digital-twin

Zanella, A., Bui, N., Castellani, A., Vangelista, L., & Zorzi, M. (2014). Internet of things for smart cities. *IEEE Internet of Things Journal, 1*(1), 22–32.

Zhang, J., Qiao, G., Song, G., Sun, H., & Ge, J. (2013). Group decision making based autonomous control system for street lighting. *Measurement, 46*(1), 108–116.

Towards a Social Capital Resilience Model in Coping with Floods and Droughts: The Case of Muzarabani, Zimbabwe

Rosemary Kasimba[1], Solomon Muqayi[2] and Innocent Chirisa[3]

[1]Department of Demography Settlement and Development, University of Zimbabwe, Harare, Zimbabwe

[2]Department of Governance and Public Management, University of Zimbabwe, Harare, Zimbabwe

[3]Department of Demography Settlement and Development, Social & Behavioural Sciences, University of Zimbabwe, Harare, Zimbabwe

Definition of Terms

Adaptive capacity: The "ability of the system to adjust to climate change (including climate variability and extremes), to moderate potential damages, to take advantage of opportunities or to cope with consequences" (Panel on Climate Change 2001, p. 6).

Vulnerability: There is no universal definition of the term vulnerability. Paul (2013, p. 1) defines it as a system to withstand against the perturbations of external stressors.

Community: This is an elusive concept. In this study, it is generally described "as a group of people with diverse characteristics who are linked by social ties, share common perspectives, and engage in joint action in geographical locations or settings" (MacQueen et al. 2001, p. 1).

Disaster: An event occurring that causes widespread material, human, environmental, and economic loss that normally exceeds community ability to cope with its own resources.

Climate change: Long-term shift in temperature and weather patterns.

Food security: Food security is an elusive concept. It exists when all people, at all times, have physical and economic access to sufficient safe and nutritious food to meet their dietary needs and food preferences for a healthy and active life (Pinstrup-Andersen 2009, p. 5). Kasimba (2018) conceptualized it "as a concept that puts more emphasis on the ability of people to have easy access to adequate nutritious and healthy food all the times and that their bodies should be able to ingest and metabolize such food."

Introduction

The Muzarabani community has suffered from floods and droughts for many years due to the climatic and geographical dynamics of the area (Chanza and Wit 2014). Residents' livelihood activities such as agriculture are heavily impacted and the most vulnerable people, such as the aged, women, and children, are severely affected. According to the IPCC (2007a), floods and droughts are likely to get worse as it is predicted that the magnitude and frequency of these will increase during the twenty-first century due to changes associated with climate variability (IPCC 2007b).

In that context, understanding the nature of the social capital that exists in the area is very critical. One way to do this is through increased community cooperation to coordinate community-based efforts (Aldrich and Meyer 2015). The connectedness of such efforts promotes a smooth implementation of disaster risk reduction strategies which will translate to a robust resilience to floods and drought effects. This process of building social networks is part of the development of social capital. This study focuses on how social capital and social networks have and could impact on residents' resilience to disasters, specifically floods and droughts in the lower Muzarabani. Kasimba (2018) looked at the role of social capital in disaster resilience focusing on Chadereka and Kapembere. However, it is the only study that looked at social capital and this study focused on Chadereka. Most of the studies examined survival strategies and socioeconomic impacts of floods and drought on the local people.

The Community and Regional Resilience Institute (CARRI) (2013, p. 10), defines community resilience as the capability of the community "to anticipate risk, limit impact and bounce back rapidly through survival, adaptability, evolution and growth in the face of turbulent change." Hoegl and Hartmann (2021) and Parker (2020) define it as "the ability of an entity or entities to positively adapt, emerge stronger or come back stronger, better positioned for the future, to bounce forward/bounce beyond." The study was guided by definition given by Hoegl and Hartmann (2021) and Parker (2020). The goal of resilience is "to reduce the conditions of vulnerability that a community faces" (Nyamwanza 2012, p. 3). As a result, the study sought to examine the applicability of social capital in enhancing the ability of the community to deal with floods and drought impacts. The study is significant because it adds knowledge to the prevailing academic literature in the sense that it informs local level responses to disasters, particularly, by the vulnerable groups in the society.

The Role of Social Capital in Disaster Resilience

This study was informed by the social capital community resilience model. The model is used to explain how the community overcomes stress, trauma, and other life challenges posed by disasters, by drawing from social networks and cultural resources embedded within communities (Fraser 2021). Social capital has several defining features and some of them are: participation, networks, provision of support, reciprocity, and trustworthiness. These defining features are theorized as being critical to promoting community resilience prior to and aftermath of a disaster (Aldrich and Meyer 2014). In most cases, it has been observed that tight bonds between relatives and neighbors led to collective action on the part of the community and the efficient allocation of resources, catalyzing communication to access assistance after disasters. Social capital community resilience model has two main dominant features that make it more crucial in disaster resilience: networks and norms, trust and reciprocity. Norms, trust, and reciprocity smoothen the functioning of networks that are indispensable in disasters. Networks provide the resources that are needed to "solve collective problems and pursue specific goals in the larger society" (Patton 1999, p. 6). Disaster researchers have built up a strong body of evidence about the role of social cohesion and networks during and after catastrophe (Aldrich and Meyer 2014, p. 6). This shows that social networks can provide financial and nonfinancial resources that can help individuals to successfully respond and recover from a disaster.

In addition, networks are fashioned out of a group of trustworthy individuals who are enthusiastic to cooperate toward collective goals, and thus, establish networks among. This is fundamental in managing disasters as a group. Societies are understood to have in-built aptitudes to adapt to disasters such as droughts and floods and these are promoted through communal behavior. Several scholars note that a close-knit community that has social capital responds to and recover more effectively than fragmented and isolated ones when it comes to disaster resilience.

Incorporating social capital in disaster resilience is controversial. It has been criticized for "focus[ing] on individual gains without regards to the broader community" (Delgado and Delgado-Humme 2013, p. 54). According to Delgado and Delgado-Humme (2013, p. 109), social capital is occasionally constructed as a way of improving people's lives without considering the influence of social structures and the negative side of social interactions. Ganapati (2012, p. 78) notes that "some networks in social capital are designed exclusively for and or by women whereas others were more inclusive in terms of their target group and membership." The study by Aldrich and Crook (2008) emphasized how social capital hindered important projects that were controversial and yet perhaps necessary for speedy recovery. For example, these communities did not allow trailer parks to temporarily house disaster victims. To some extent this shows that disaster victims cannot solely rely on social capital.

In addition, the word "community" does not have a single definition. Homan (2008) posits that a community is a number of people with something in common with one another that connects them in one way and distinguishes them from others. Fellin (1995) further states that a key feature of a community is the fact that participants share same mutual characteristics, such as location, interest, culture, and/or activities. In this study, community is conceptualized as a group of people that are staying in the same area, under one chief and one ward.

A number of empirical studies demonstrated the role social capital plays in enhancing the adaptive capacity of the community to disasters (Meyer 2012; Dynes 2005). As disasters tend to occur in certain geographical areas, it is crucial to comprehend how endogenous strategies lead to greater adaptive capacity (Aldrich 2010). Some scholars observed that information sharing is important in allowing victims to ascertain where support is being provided, and it can provide an important means for governments and nongovernmental organizations (NGOs) to reach vulnerable people (e.g., the elderly and disabled) in disaster-affected areas (Chamlee-Wright and Storr 2011).

Paton and Johnston (2001, p. 45) purport that social capital promotes community resilience to disasters as they argue that "social ties and networks enable communities to respond to adversity whilst retaining their core functions." However, that study was not conducted in Zimbabwe. This study sought to understand how social ties (informal ties) enhance people's resilience to floods and droughts in a specific location in Zimbabwe. Prior researches showed that social capital and social networks enabled people in disaster communities to withstand harsh conditions during and after a disaster than the less well-connected communities (Chamlee-Wright and Storr 2011; Meyer 2012). However, the link between the two, in Muzarabani, has not been studied yet. "Concepts like social resilience are related to theories of social capital which stress the importance of social networks, reciprocity and interpersonal trust" (Patterson et al. 2010, p. 127).

These allow disaster victims to successfully recover and respond than they can do in isolated efforts. From local to international levels, formal social capital improves disaster response and recovery (Varda et al. 2009). According to Chamlee-Wright and Storr (2011, p. 266), social capital facilitates community level planning for disaster mitigation, preparedness, evacuation, and provision of shelter before a disaster. For instance, residents in St Bernard utilized social capital to coordinate emergency management and community return to provide material resources (in the form of portable water, food, shelter, and clothing) to the vulnerable and to rebuild damaged houses, business, and other special spaces in their communities (Chamlee-Wright and Storr 2011, p. 266). This is in tandem with Magis (2010, p. 402) who notes that members of resilient communities intentionally develop personal and collective capacity that they engage to respond to and influence changes, to sustain and renew the community, and develop new trajectory for the community's future.

Elliott et al. (2010) conducted a study where they compared disaster outcomes for residents of two communities in New Orleans, the Lower Ninth Ward, a poor, majority African American community, and Lakeview, an affluent, majority white community. They found out that while Ninth Ward residents relied on bonding social capital for informal support during hurricane Katrina, they received less overall support in the year following the event. They concluded that a lack of bridging social capital to people outside the affected area and ties with individuals with more resources, resulted in reduced resilience for Ninth Ward residents compared with those in Lakeview (Aldrich and Meyer 2014, p. 8). Social relationships that exist among victims sometimes do not really bring fruitful results because everyone will be crying for help where the resources are limited, even where people would normally be willing to help each other, as shown in Elliott et al.'s (2010) study. However, this does not refute the role that social capital generally plays in enhancing community resilience to disasters. Thus, combining bridging and social capital is more likely to increase community's resilience to disasters (Aldrich and Meyer 2015). The study by Elliott et al. (2010) is different from this study in the sense that their study compared the use of social capital to bridging and bonding two different communities, while this study seeks to understand the extent of social capital that individuals have and how they make use of it in enhancing their resilience to droughts and floods. In addition, this researcher seeks to understand the forms of social capital that exists particularly in the Muzarabani area.

Barker's (2011) study in Brima, Australia, showed that the collective unit enhances community resilience as strangers and local residents helped each other. Furthermore, Kien (2011) conducted a study in the Vietnamese Mekong River Delta and found out that social capital (relationships with neighbors) is crucial in enhancing household resilience to floods. However, Barker (2011) and Kien (2011) did not examine how community relationships improve resilience, which is one of the aims of this study. As droughts and floods continue to get more prevalent, understanding residents' resilience to floods and droughts has become a necessity in the contemporary world. Studies on community resilience to disasters have become an issue in the contemporary world. Development agencies such as the Department of International Development (DFID) are "committed to build[ing] disaster resilience into all its programmes by 2015" (DFID 2011, p. 14) and "increasing attention paid to the capacity of disaster affected communities to recover with little or no external assistance following a disaster" (Manyena 2009, p. I).

Several studies outside Zimbabwe have shown that social capital plays a significant role in disaster resilience. Mogues' (2006) study revealed that social networks play an important role in asset recovery and growth after environmental shocks in Ethiopia. Manyena (2009) conducted a study where he sought to understand the extent to which development and humanitarian intervention programs promote resilience in disaster prone areas. His study focused on three case studies from different countries (Zimbabwe, Ethiopia, and East Timor). It established that local resilience to disasters is about agency albeit in a political and economic context. The study discerned that determining the extent to which development interventions enhance resilience is a substantive challenge.

Manyena et al. (2013) conducted a study on the connection between indigenous knowledge systems and disaster resilience in Muzarabani. Their findings revealed that indigenous knowledge systems played a significant role in reducing the impact of floods because they enhanced the resilience of the community. However, that study was centered on floods in Chadereka and Dambakurima Wards while this study will focuses on both floods and droughts and documenting the basis of people's resilience to these disasters. The study by Manyena et al. (2013) did not look at coping strategies employed by the vulnerable groups such as the elderly women and child-headed families as this study does.

Chingombe et al. (2014) conducted a study in Chadereka of Muzarabani on the usefulness of Participatory Geographic Information Systems (GIS) for the purposes of flood risk assessment in Zimbabwe. They found that PGIS can be a useful tool for collecting data on flood events. Nonetheless, they did not look at how floods affected vulnerable residents' livelihood security. Literature reviewed from the works of Gwimbi (2004), Murwira et al. (2012), and Mavhura et al. (2013) exhibited that enhancing community resilience is crucial in Muzarabani as floods and droughts have continued to threaten the lives of many. However, there are no studies on the role of social capital in coping with disasters in Muzarabani, yet according to Yandong (2007, p. 1), "early disaster researchers had noticed the role of social capital in disasters."

Below is a social capital community resilience model which is informing the study.

The social capital resilience model consists of the following: social capacity (people's ability to work together), social support, village to village support, sense of community and attachment to place, citizen/community participation, community competence, informal ties, and organizational linkages. These features enhanced the community resilience to floods and droughts. Carpenter (2014) defines community competence as organized action of people, communities, and institutions to prevent, manage, and learn from crisis.

There are two main types of social capital that are "bonding social capital characterized by strong, tight-knit relationships within peer networks and bridging social capital characterized by relationships between members of different social networks" (Murray et al. 2020, p. 2213). According to Coleman (1990), bonding social capital is formed among individuals in close social proximity, such as in families, churches, or neighborhoods. It relies on solidarity, reciprocity, familiarity, and trust. Bridging social capital results when people connect across social distances to share resources, knowledge, and spaces. Bridging social capital facilitates innovation and the diffusion of ideas and information across diverse stakeholders and groups (Murray et al. 2020, p. 2213). Figure 1 above has both bonding and bridging elements of social capital.

Methods

Study Area

The study area is the Lower Muzarabani District of Zimbabwe. Chadereka area of Muzarabani was purposively selected because it frequently experience floods and droughts. The area lies between two rivers, namely, Hoya to the east and Nzoumvunda to the west. The Hoya River often

Towards a Social Capital Resilience Model in Coping with Floods and Droughts: The Case of Muzarabani, Zimbabwe,

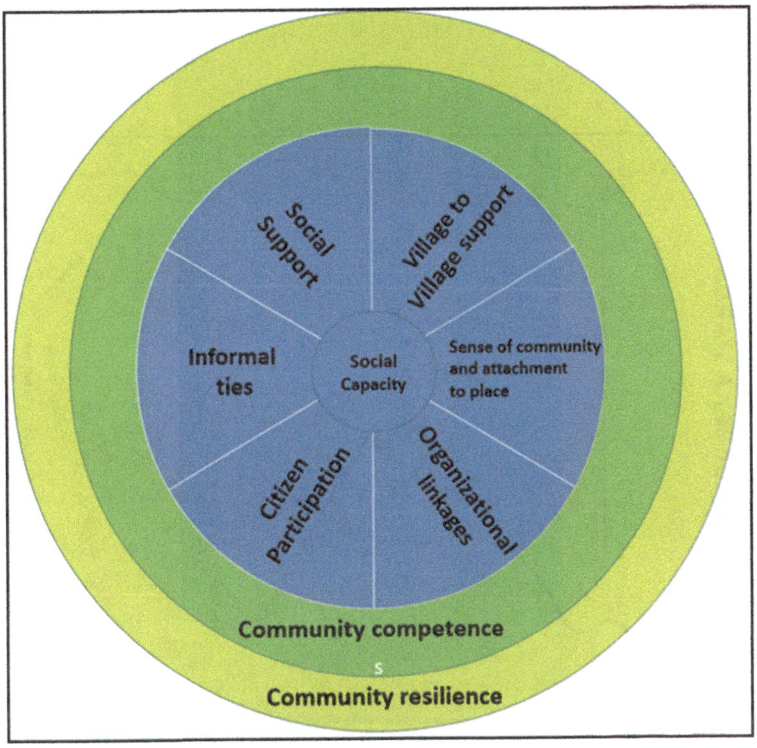

Fig. 1 Muzarabani Community Resilience Model. (Source: Rosemary Kasimba (2017))

discharges its water into the low lying Chadereka area, thereby exacerbating the flood problem.

Researchers were entirely convinced that the ward would enable them to obtain relevant information to fulfill the objectives of the study. The ward provided a general picture of how floods and droughts are affecting people and opportunity to show the role of social capital in enhancing community resilience. It would help show how the most vulnerable groups are responding to floods and droughts, among other disasters.

Whenever Zimbabwe is hit by floods and droughts, Muzarabani is usually one of the most affected districts (Mavhura 2017). The area experiences severe dry-spells during the rainy season (Ruswa 2017). Muzarabani District was chosen because we understood the local peoples' language and culture. Therefore, it became easier for us to interpret meanings and symbols. This increased the reliability of the research findings. Chadereka was preferred because it is frequently affected by floods and droughts (Mavhura 2017). a, many people lose their property (Chanza and Wit 2014).

Figure 2 above shows the location of study sites within Zimbabwe. It demonstrates the reason why the area is prone to floods as it is surrounded by big rivers such as Hoya to the East and Nzoumvunda to the West. In addition, the map illustrates that the district is part of the remote areas of Zimbabwe which are now densely populated. The majority of the people rely on farming and infrastructural development is lagging behind as residents still put up with bad roads and run down bridges which have not been repaired since Cyclone Eline in 2000 (Jack 2017). Some residents depend on cross-border trade into Mozambique, gold panning, and farming. The majority of the people who used to rely on cotton farming no longer grow that crop because of the incessant droughts and the fall in pricing on the market (Chanza and Wit 2014). Growing of sesame seed, groundnuts, and drought-resistant crops is becoming very common in the area. The people in the area highly depend on rain-fed and winter farming agriculture. Land presents more than an economic asset in Muzarabani but customary norms require women to access land through

T

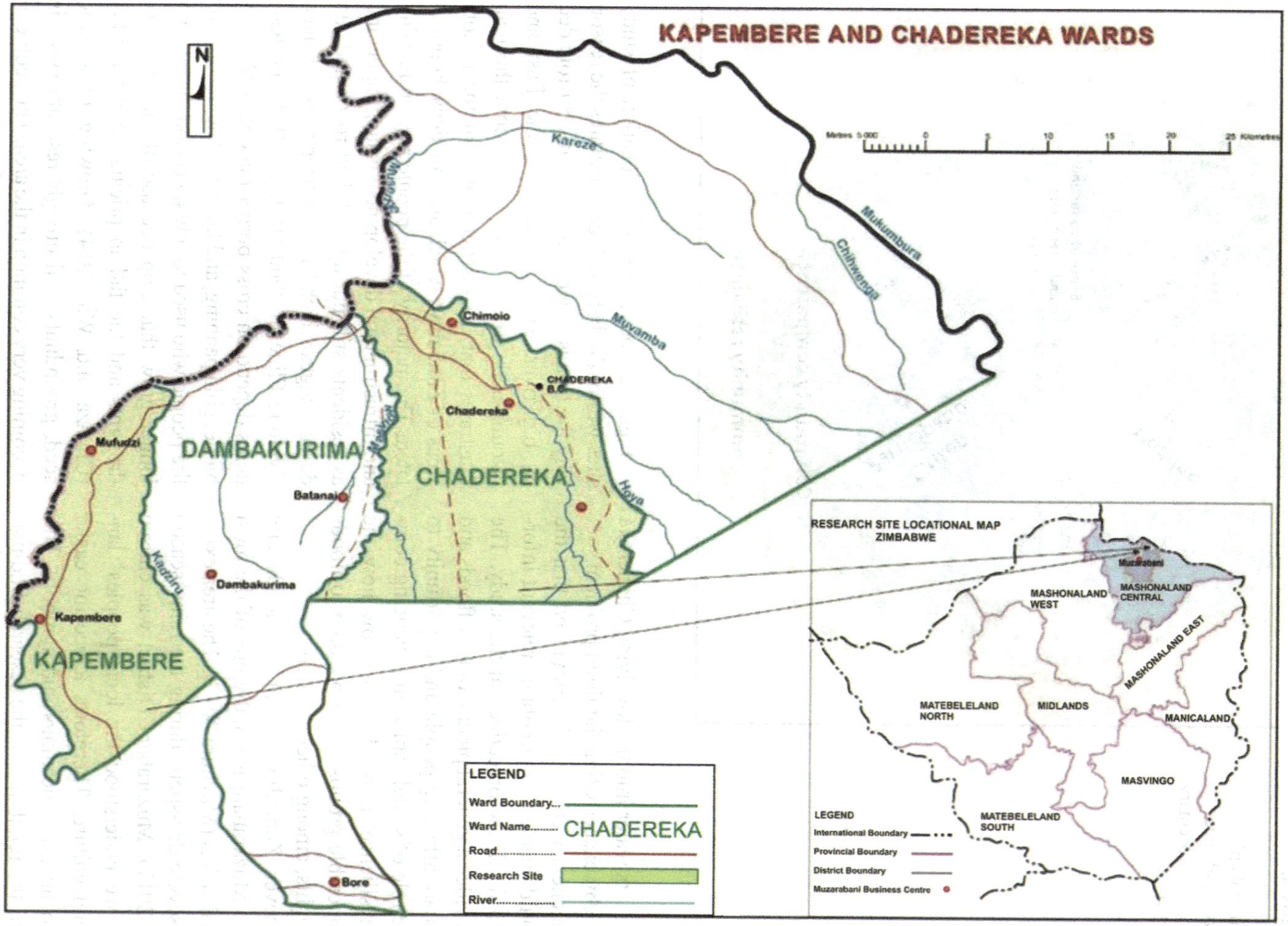

Towards a Social Capital Resilience Model in Coping with Floods and Droughts: The Case of Muzarabani, Zimbabwe, Fig. 2 The map of Zimbabwe showing the location of Muzarabani. (Source: Department of Surveyor General, Zimbabwe (2015))

their husbands or a male relative. Unmarried and divorced women often face difficulties in accessing land.

Several studies by Mudavanhu (2014), Louis et al. (2021), Madzana (2013), and Ruswa (2017) documented the effects of floods and droughts in Muzarabani. Food insecurity deepened poverty, early marriages, unemployment, stress and trauma, family disintegration, and poor enrolment in schools.

Research Approach and Design

This study employed a qualitative approach which provides a better understanding of the research problem. It involves methods that enable rural people to share, enhance, and analyze their knowledge of life and conditions, to plan and act, and to monitor and evaluate and reflect (Chambers 1994, p. 953). The qualitative method enabled the study to have a better understanding of system dynamics and appreciate the interlinked social capital factors' influence on the local people's coping capacities. The approach involved people in the processes that affected their livelihoods and empowered them in dealing with disasters. It is premised on the assumptions that local people possess adequate knowledge of their environment and situation. Understanding social capital required an in-depth participation of the research participants and that compelled the researcher to employ a qualitative methodology. The research employed a case study research design because it allowed an exploration and understanding of complex issues (Zainal 2007). It has enabled the intensive study of a single unit for the purpose of understanding a larger class of similar unit (Gering 2004). Although the results from a case study cannot be generalized to the larger population, "the results are useful for informing theory and providing novel insights which may warrant further research and policy consideration" (Darke et al. 1998). As the research attempted to understand the extent to which individuals used social capital in increasing their resilience to floods and droughts, the case study method provided us with the most appropriate data collection tools. We used key informant interviews and focus group discussion to collect data. Researchers collected information concerning how individuals were using social capital to increase their resilience to disasters. Key informant interviews were held with those who held strategic positions in the community such as village heads, councilor, Chief, and the District administrator. Each key informant interview took 45 minutes to 1 hour. Researchers sought to understand the role that social capital played in increasing people's resilience to disasters. Eight key informant interviews were conducted. Furthermore, researchers conducted six focus group discussions with heads of households (both males and females). These were done in cohorts. Respondents were conveniently selected.

We entered the field in January during the 2019/2020 agricultural season and finished the data collection in October 2019, in order to understand in detail, the problem at hand. The area experiences dry spells in mid-January to February and we were able to observe how the people employed social capital and networks to deal with such conditions and ensure food security. In addition, stretching the field work to October allowed us to have an in-depth-understanding of the role of social capital and networks in enhancing community resilience to floods and droughts. During the given period, we learned cultural aspects that were alien to us but helped in achieving the objectives of the research. The building up of social capital and networks is a process that demanded spending considerable time in understanding whether individuals had sufficient social capital and how they utilized it to ensure food security and emotional stability.

People may not be able to retrieve their past experiences in detail. So, we concentrated on the period 2010–2020. Within this period, there were several occasions in which floods and droughts devastated livelihoods and food security in Muzarabani. People who have stayed in this area for more than 20 years were able to reflect on their experiences with floods and droughts. We understood their perceptions of floods and droughts and the coping strategies they employ to reduce the effects of these disasters. Furthermore, we appreciated the changes in strategy that aimed at reducing the effects of floods and droughts. The

2010–2020/1 period is a contemporary period. This fact makes this study current and relevant given that the information collected, if continually updated, can be of use to people affected by floods and droughts in other areas. They can use it to bolster their resilience.

We took measures to ensure reliability of the research findings. Two research assistants who were proficient in the dialect of the Shona spoken in the area were recruited and trained to help us conduct focus group discussions. This ensured that they could communicate easily with the respondents. We sought and were granted permission to use voice recorders (from our phones) prior to conducting focus group discussions. Furthermore, audit trail was done which is defined as the process that encompasses a "thorough collection of documents regarding all aspects of research" (Carcaray 2009, p. 12). Thus, we were expected to "keep all records of study processes" (ibid, p. 12). We have kept all the documents that were used at each and every stage of our research. The videos and recordings are stored in our external hard drive and we have made sure that no one has easy access to them. All the materials with field notes and dates, and daily reports are kept in a safe place. These will be destroyed after 10 years. Research tools were pre-tested before the data gathering started in earnest. This enabled us to identify problems regarding our interview guides and facilitated necessary revisions.

In addition, researchers were considered as locals in the study. They understood the dialects used by the people in the community. One of the researchers grew up in an area 200 km from the study area. So, communication with the respondents was made easy. There was no need for an interpreter, and the respondents felt comfortable to share their experiences with the researchers.

Researchers are under the obligation to conduct a research in a way that does not raise reservations or suspicions of impartiality (ibid, p. 104). According to Brinkman and Kvale (2008, p. 263) "human interactions in qualitative enquiries affect researchers and participants; and knowledge produced through qualitative research affects our understanding of the human condition." From this perspective, we had to observe ethics from the beginning up to the end of the research. We sought permission for our study in accordance with all expected protocols. This was done in a descending hierarchical order: from the Provincial Administrator (PA) down to the community leaders (Councilor, Chief, Village Heads, and traditional leaders). To facilitate acceptance, one of the researchers carried all documents she had used to gain access in her previous study. She explained to all gatekeepers that the study had culminated from her previous research and the findings of the current study would influence policy making in line with disaster management. In addition, she was taken as a child from the area (*mwana wemumusha* as she hails from an area near the study area).

According to Muzvidziwa (2004, p. 302), "central to research process is the need for participation, integrity and responsibility on the part of researchers." We, therefore, ensured that no physical harm befell the researchers during the study and after the results were published. Furthermore, confidentiality and anonymity were ensured to the respondents. The respondents were informed that the information they provided would be treated confidentially and when published, would not be identified as theirs. This made them feel comfortable and be able to provide detailed information on their perceptions of floods and droughts and how they were coping. Pseudo names were used to ensure anonymity. In addition, we made it clear that there would be no direct benefits for participating in the research, and participants were explicitly informed that their involvement was voluntary. Establishing caring relationships and a concern for the well-being of the respondents was made paramount throughout the process. According to Adjibolosoo (1995, p. 3), "good and effective researchers should possess termed positive human factor attributes such as loyalty, accountability, responsibility, dedication, vision, honesty, motivation, wisdom, skills, knowledge, understanding and trustworthiness." Therefore, we made sure that these attributes were adhered to avoid complicating the research.

Data Analysis

We analyzed the data using social capital and social network analysis techniques in discussing the role of community relations, associations, and connections in enhancing the resilience of the community to floods and droughts. We used thematic analysis to analyze data. Content analysis is defined "as a research method for subjective interpretation of the content of data through the systematic classification process of coding and identifying themes or patterns" (Hsieh and Shannon 2005, p. 278). Under thematic analysis, we developed themes that concurred with my research objectives or research goals. After that, we used content analysis to develop themes under each objective. The method was time-consuming and challenging but it helped us to understand how social capital was being utilized in the community to enable residents withstand the harsh conditions such as acute food shortages imposed by floods and droughts. At the end of each data collection day, we produced a report on the key points. These reports were based on the notes we made. We did this so that we would not lose and forget the information since the phone that was used for recording sometimes acted up. Writing while we still had information avoided distortion of information. The content and thematic analysis was made possible by the manner in which we designed our data collection tools. We had organized questions in relation to objectives and this enabled us to check for inconsistences and consistencies. Thematic moves beyond counting explicit words or phrases and focus on identifying and describing both implicit and explicit ideas.

Results

Food Provision Through Social Capital

Community members were suffering from hunger. Some would withdraw their children from school. The research established that social capital assisted in the provision of food in schools. There are about two dominant groups in the area that have helped people to develop resilience in the community. These are Women in Action and Mother Support Group. In Chadereka, women formed groups to support pupils in schools. There has been a mushrooming trend where pupils were withdrawing from school due to hunger. The Mother Support group was formed in response to this problem. The group moves around in primary schools providing food to pupils thrice a week. Women in Action was formed in response to poverty which was exacerbated by floods and droughts. They formed this group to economically empower themselves. These women engage in a number of activities such as merry-go round, pool resources together for sharing, do gardening as well engaging in community based savings. Women in Action indicated that they normally gather every Wednesday afternoon to discuss the challenges that they are facing and share ideas on how they are tackling them. The group is very big and consists of more than hundred members. Women share experiences and new comers are welcome anytime. However, new members do not benefit from certain activities as soon as they join. They have to wait. For example, a new comer would not take a small loan from the savings before paying the required subscription fee. The group has helped people develop resilience. Sharing experiences strengthened their emotional capacity to withstand the harsh conditions imposed by floods and droughts. These women do sports on Fridays (a Chisi day) where community members are not allowed to work in their fields (This is the day people are not allowed to work in their fields. They have to stay at home or if they go to the field they are not allowed to work).

Mother Support Group consists of 50 women in Chadereka. This group conducts a number of economic activities such as providing food in primary schools (they move around primary schools, providing food to the pupils). The group works hand in hand with women in the communities. A number of women voluntarily contribute grain for mealie-meal to the group so that it can have food to provide the children in schools. Mother Support Group and the Women in Action emerged from the community. They all have structures where there is a Chairperson and Vice Chairperson, Secretary, Treasurer, and committee members. They are constituted by quite a number of people and membership is open to everyone. They are not formally registered but community leaders are aware of their existence and their purpose. These projects have been assisting several pupils in the community. Pupils used to faint due to hunger while at school. Cases of people who fainted at school are decreasing and the number of primary school going age children who are enrolling is slightly increasing because of the food scheme program that has been introduced by the Mother Support Group. One of the key informant interviewees in Chadereka said:

> We are very grateful to the women for the program that has been introduced by women in this community. Our children are no longer fainting in class and those who had dropped out of school are re-enrolling now. There are various groups that were formed by women in our communities and some of these groups help the most vulnerable like the elderly and child headed families with food.

The other respondent in focus group discussions in Chadereka said:

> There were several forms of activities that are carried out by women to increase community resilience to floods and droughts. These include exchanging labor for food stuffs or money and devising informal savings systems. Apart from these systems, there are those that involve women in a working system where there are psychological group supports, such as burial and church societies or prayer groups. We are very happy for their activities.

Women groups are working with the people and they help with food in schools and the community. They provide counseling services and help NGOs in their day to day execution of duties. They have been in existence for more than 7 years.

Women said that the local charity organizations play a pivotal role in primary schools. They indicated that they used to have higher cases of pupils who used to faint in the classrooms during lessons. The pupils' performance was below the standard. Some parents had opted to take their children out of school due to the shortage of food. It is viewed as a very good program as their children have food to eat whilst they are at school. A large number of the pupils who were taken out of school returned because there is now food in schools. They have their breakfast and lunch three times a week. These women are into income generating activities to promote the survival of their activities. They make shoe polish and uniforms which they sell at affordable prices.

Many of the respondents said that droughts have caused hunger in their community to such an extent that they ended up selling most of their cattle. Some single mothers said that they could not help their children on their own. They expressed gratitude toward their colleagues from the Mother Support Group.

Mother Support Group in Chadereka has helped quite a number of children. Social networking is being promoted as women in the group have the opportunity to link up with others who assist them when they are in need of help, especially food.

> I am very grateful to the club that was formed by women as a way of dealing with the hunger and trauma posed by the floods and droughts. Through their activities, my sons managed to go back to school two years ago and they did well in their ordinary levels. They are currently pursuing a teaching course in Harare under World Vision scholarship. What a blessing! Had it not been for the group that assisted with food at school, where would my sons be…?

One of the respondents said:

> Local organizations which were formed by women have helped us a lot in this area. They sometimes organize get together activities, especially on Fridays, when everyone is not allowed to work in the fields. We cook under a tree and the elderly women are invited to teach us how to preserve food for future use, cook and identify indigenous vegetables found in the forest. This has saved us a lot. We learn how to prepare traditional soap. Due to poverty, sometimes we cannot afford modern soap. So, traditional soap (made from trees) is really good for us especially when washing clothes that we normally wear at home.

Another responded said:

> Through local organizations which were formed by women in this area, we are able to tap knowledge from elderly women and this knowledge has come to our rescue. We have learnt how to prepare seeds for planting. We do not necessarily have to buy tested seed from the shops during the farming season. We learnt how to preserve vegetables in a way that makes them retain their nutritious value.

Many of the respondents reported that the elderly women are playing a pivotal role in enhancing community resilience to disasters. Some respondents indicated that people, particularly women, share survival strategies during meetings.

Social Capital Enhancing Community Harmony When Selling Assets

The study's findings showed that the residents of Muzarabani, mainly the most vulnerable, rely on their neighbors, relatives, and friends. Many of the most vulnerable do not have cattle. They have smaller pieces of land and fewer assets to sell in times of droughts in order to survive. Even if they are to sell assets such as cattle, they do not bargain successfully and may end up selling them at an unreasonably low price. During data collection in Muzarabani, people would come from Harare to buy cattle. Prices were ranging from US$250 to US$450. Most vulnerable people would sell their cattle at US$75 because they could not bargain successfully and did so in desperation. A lot of the respondents explained that they were now being assisted by their neighbors. They sold at a low price because what they wanted was money to buy food. When the residents in Muzarabani realized that there were some community members who had been selling their cattle at low prices, village meetings were held to educate each other on how best they could sell their assets like cattle and earn a living. They all agreed that no one was going to sell a cow for less than US$150. In addition, community leaders made it mandatory that every member of the community invites at least three other community members when selling a cow so that they could help in bargaining.

From there, we began to see residents selling their cattle or exchanging their cattle for food, without making losses, as previously happened. This shows how the most vulnerable people rely on social capital for their resilience to floods and droughts. In addition, Muzarabani community members of Chadereka support each other in different ways when it comes to food provision from the government. One of the authors of this article would go to the food rationing programs where people received maize grain from the government. Whenever the grains were inadequate, members sat according to their villages and selected the most vulnerable people who would then be given first preference. From these activities, it was discovered that social capital is a fundamental tool in disaster resilience. Community members interact and learn more about each other's social and economic status.

A Network of Voluntary Community Care-Givers

There are individuals who voluntarily visit the vulnerable households, taking note of how they are surviving and the challenges that they are facing due to floods and droughts. They normally visit the sick, the elderly, child headed families, and the households where there are people living with disabilities. These unpaid volunteers work with the Department of Social Welfare and Non-governmental Organizations operating in the area. These volunteers make sure that the most vulnerable people are given priority whenever there are food donations. One of the respondents who is living with disability said:

> Floods and droughts further complicate the situation of us who are living with disabilities. We cannot go find piece jobs. I cannot walk like the others. But I am glad that I am surrounded by people who are concerned about our condition. We are getting help through volunteers who visit us and relay information concerning our condition to the Department of Social Welfare and other Nongovernmental Organizations. We are the first to get food donations. Sometimes we get help from members of the community.

Another community member narrated that:

> We have community volunteers who always visit the elderly. It is through these people that I get help from community members. I get food from people. Sometime my neighbors assist me with food. We support each other.

Many of the vulnerable respondents said that they get help from other community members and a network of voluntary committed people who are in the community.

Enhancing the Interaction Among Various Stakeholders in Dealing with Floods and Droughts

Community and stakeholders such as Non-governmental Organizations, churches,

T

educational institutions, councilors, District Administrator, health officers, Environmental Management Agency officials, the Meteorological Department, village heads, and the chief ward work together to reduce the impacts of floods and droughts in Chadereka. There is the Ministry of Local Government, Public Works, Rural and Urban Development which is the entry point into the Districts of operation. There is the District Administrator (DA) who is the chairperson of the District Civil Protection Committee responsible for coordination during emergencies. Muzarabani Rural District Council (MRDC) is the local authority with jurisdiction over the area of operation.

The Ministry of Agriculture offers technical advice through AGRITEX officers and conducts trainings, monitoring and evaluation on the sustainable agriculture component of the project. The Ministry of Women's Affairs, Gender and Community Development is the task force member of the District Water Supply and Sanitation Committee (DWSSC) spearheading water and sanitation issues in the district of operation. The Ministry of Primary and Secondary Education is the entry point of schools in undertaking the Participatory Health and Hygiene Education (PHHE) component as well as the Zimbabwe Red Cross Society clubs. The Ministry of Health and Child Care, through the Environmental Health Department, provides data on the prevalence of diseases disaggregated per district, ward and village levels, through the Environmental Health Technicians. This data is used as comparisons against the initial contact document to check on progress achieved. District Development Fund (DDF) assists in the training of Water Point Committees, rehabilitation of defunct boreholes and pump minders in areas of operation (Water Point Committees are the people who are responsible for the maintenance of the boreholes.). It is part of the steering committee on water and sanitation which conducts meetings with communities during latrine construction and provides the technical expertise in the construction of bridges and culverts. In addition, there is the Department of Social Services which provides information on social services and distributes food hand-outs to community members in Chadereka. There are other Nongovernmental Organizations (NGOs). Examples include World Vision, SAT, Help from Germany, and MeDRA that complement Caritas' efforts in eradicating poverty in areas of operation. Caritas and the NGOs share reports in full council meetings and NGO forums, and there are traditional leaders and local leaders (councilor, village heads, and chiefs) who monitor environmental, child and women issues in their structures of their society as well as resolving conflicts within their jurisdiction among others. All these organizations are working hand-in-hand to increase the resilience of the community. Key informants highlighted that various stakeholders are working with the community on daily basis. They highlighted that the cash transfer program in 2019 saved a number of people although every household was not a beneficiary.

Neighbor to Neighbor Interaction

Neighbors in Muzarabani assist one another with food. When someone slaughters a goat or a cow, he or she cuts one leg of the beast into pieces and distributes it to neighbors and relatives. This reflects how communal the people are. Giving each other (neighbors, strangers coming from Muzarabani area, friends and relatives) part time jobs/casual jobs indicates that people are relying on social capital. Social capital helps people to be food secure as well as resilient to floods and droughts. The traditional social fabric in Muzarabani rural area is strong as neighbors and community members are guided by the belief that "*one finger cannot squash a louse or one man cannot surround the mountain.*" This is a common aphorism in Zimbabwean society. Thus, it is common and normal for people to assist each other in times of floods and droughts.

To What Extent Do Individuals Rely on Social Capital?

Respondents reportedly depended on social capital. Many of the respondents highlighted that being active in the community strengthens an individual's social capital. They highlighted that participation in community programs and activities, churches, paying visits to the sick, contributing to neighborhood fundraising and attending community meetings help people in building their network for information transmission as well as social standing. One woman said that participating in the Mother Support Group helped her a lot as her kids who had withdrawn from school re-enrolled. Many of the respondents indicated that building trust and reciprocity with neighbors, friends, family, and relatives helped them during floods and droughts. Community networks such as family, friends, and relatives were providing some kind of counseling to family members that would have lost their loved ones during the floods. The study established that societal relations are strategic to one's resilience to the impacts of floods and droughts. In times of emergency, for example, when one is bitten by a snake, members of the community assist with transport (they provide a scotch cart to take the victim of a snake bite to the hospital). In times of water borne diseases (like cholera), especially during flooding, there are volunteers who move around households, providing care and education during the recovery process. These people work with health experts like nurses. Some villages are far away from clinics and support from village and neighbors is very vital.

On respondent said:

> We are very poor. We are far away from hospitals. We have poor roads and we stay in remote parts of the country where leadership might not be aware of our genuine concerns. Floods and droughts pose a lot of challenges. We are mainly relying on the local resources which are community networks and Non-Governmental Organizations like MeDRA and Red Cross Society of Zimbabwe. Almost everyone in this community relies on social capital.

One key informant interviewee said:

> Our floods destroyed infrastructure. Roads and bridges were swept away. This resulted in some women giving birth in homes as roads were impassable. Some ended up dying while giving birth in the absence of nurses. Our children stooped going to school because there were no roads. The government stopped rehabilitating roads especially this side. We had to work together as a community to make foot bridges as well as repairing roads so that people can pass and travel to places where they can get social services like education and health care.

Respondents highlighted that they work hand in hand with NGOs. For example, in 2019, the Zimbabwe Red Cross Society provided building materials to Chadereka community to build foot bridges and toilets. Each household (without a Blair toilet) would get ten bags of cement for the toilet. Foot Bridges were built in strategic places that linked the public with social services (Figs. 3 and 4). Futhermore, households were able to build toilets with the help of other community members.

However, the study discovered that not all people can rely on social capital. Some

Towards a Social Capital Resilience Model in Coping with Floods and Droughts: The Case of Muzarabani, Zimbabwe, Fig. 3 A footbridge built by the community in Chadereka. (Source: Researchers)

Towards a Social Capital Resilience Model in Coping with Floods and Droughts: The Case of Muzarabani, Zimbabwe, Fig. 4 One of the culverts constructed in Chadereka by community members. (Source: Researchers)

respondents said that there has been discrimination and nepotism. One of the respondents said:

> I rely on my cattle and pension. I do not get help from the community because people think I am very rich because I have a car which I bought when I was still working in Harare (Capital city of Zimbabwe). I remember last year (2017), we had no food because our crops were washed away by the rains. We were not included on the register because community members said we had the resources to buy food. Since then I leaned that I should work hard for my family and I do not depend on social networks, or support from the community or neighbors. We know their true characters so I do not bother asking them for help.

Another respondent indicated that:

> I do not entirely depend on people from the community because our Village Head is too selective. We came from Masvingo and they think we are not part of the original people. We are always discriminated against because of our background. We stand on our own as a family. We are used to all sorts of problems caused by floods and droughts.

Results of the study showed that some families generally developed resilience on their own as they do not depend on the community.

Discussion

The study found out that there were high levels of community social resources (social capital) that were employed by the locals to increase their resilience to floods and droughts. Respondents believed in the ideals of community togetherness and clubs to reduce the problem of food insecurity. In line with the community resilience model, the study revealed that social support is very critical in enhancing the resilience of the community to disasters. Women in the Mother Support groups were commended. They were reliable and dedicated to whatever they do in order to ensure food security among the most vulnerable people such as school pupils. In addition, community leaders such as councilors, village heads, and church leaders supported the activities that were carried out by women. Community leaders in Chadereka are ready and willing to help the local people. They have been helping them in different ways. They allowed Nongovernmental Organizations to operate in the area. Thus, the findings have demonstrated that social capital plays a greater role in promoting the resilience of the community to disasters such as floods and droughts. The study coincides with Brennan et al. (2014, p. 1) who note that "when disasters occur, citizen groups and coordinated efforts of local volunteers can respond to lessen the impacts and "build back better." In addition to that, "in recent years, considerably emphasis has been placed on the role of community in disaster recovery and on the importance of local knowledge, action, participation and control in determining the nature of disaster response" (Brennan et al. 2014, p. 2). Thus, local people's participation is paramount in enhancing the resilience of the community to disasters. Mayunga (2007, p. 7) purported that "community ties and networks are beneficial because they allow individuals to draw on the social resources in their communities and increase the likelihood that such communities will be able to adequately address their collective concerns." Community members in Chadereka (mostly women) identified floods and drought-related problems and came up with projects that could lessen the impact of these problems. The study revealed that social capital facilitated people's ability to adapt to floods and droughts. The women in Chadereka pool their resources "or store foodstuffs [together] as mechanisms for handling risks to their livelihoods caused by climate change" (Agrawal 2008, p. 2).

Women in Africa have a long history of organizing themselves, formally or informally, in order to overcome the problems that they face (Raimundo 2009, p. 7). A study by Raimundo (2009) in Mozambique found out that "there were several forms of activities that are done by women to increase their resilience to floods and droughts". These ranged from exchanging of their labour for the labour of others, exchanging labour for foodstuffs or money and devising informal savings systems. The study demonstrated the importance of informal ties in enhancing community resilience to floods and droughts. Apart from these systems, there are those that involve women in a working system where there are psychological group supports, such as Burial and Church Societies or Prayer Groups Therefore, credit associations and women's clubs demonstrate that women play a pivotal role in disaster management (Raimundo 2009). They provide care to the community. This illustrates that social capital is significant in disaster management.

In line with the above, community participation is regarded by most scholars as an important tool to increase the resilience of community members to floods and droughts. Giddens's Agency concept is relevant in disaster resilience. According to Giddens (1984, p. 26), human agents know how that is in practical consciousness to "go on" in a wide variety of contexts. The study unleashed how community members in disaster-prone areas operate and survive in a world that may be at variance with their expectations. Local people's initiatives like the formation of local organizations as well as the will by voluntary individuals to work with the vulnerable population such as the elderly and the sick people, among others, demonstrated that people are not passive recipients. Rather, they improvise and come up with strategies that enable them to survive during and after disasters. Furthermore, local people's participation in activities that help them deal with floods and droughts shows the importance of a sense of citizen participation in disaster risk reduction. The study showed that people were not forced but because of the sense of community belonging, they had to act to serve themselves and other members of the community. Thus, citizen participation and the sense of belonging are critical aspects of social capital that help communities in dealing with disasters.

The findings of the study exhibit that neighbors support each other in times of crisis. They create jobs and provide food for each other. This means that people need each other in times of crises. In this case, norms of generalized reciprocity are being used (According to Voelkl (2015, p. 17), generalized reciprocity is described as a simple rule to "help somebody if you receive help from someone."). Although respondents stated some of the challenges they faced, the study observed that this type of social capital (generalized reciprocity) enables communities to survive. The research discovered that the most vulnerable members of the community are getting help from groups that were formed in the community. Continuous droughts in Chadereka have made people understand the need to pool their resources together, be innovative and creative for their survival. Furthermore, generalized reciprocity facilitates the exchange and borrowing of food among neighbors in the village. "While some researchers have begun to embrace social capital in their research, much work is needed to fully understand how social capital interacts with other forms of capital [and] how different forms of social capital contribute to food security" (Aldrich and Meyer 2014, p. 5). There are so many forms of social capital that exist in Chadereka that help the most vulnerable to ensure food availability. The researchers' interaction with community members, government official as well as NGOs, helped in understanding the role that social capital plays in enhancing the resilience of the community to floods and droughts.

However, these relationships bring successful results when there is the involvement of civic engagements that nurture good governance which supports civic engagement as espoused by Putnam's (1993) discussion. Respondents' narratives revealed that the government intervention influences the ability of community members to be able to deal with disasters. Hence, organizational linkages are very crucial in disaster resilience. The study demonstrated the significance of having different types of organizations in enhancing the resilience of the community to disasters.

The involvement of nongovernmental and government organizations/departments increased the local people's resilience to floods and droughts.

The research found that women and the elderly have adequate indigenous knowledge systems (IKS) that ensure food security. Indigenous knowledge systems enable the local people to predict the weather and the occurrence of floods and droughts so that they preserve and store food as well as grow drought tolerant crops. IKS is spread among community members through socialization. This confirms the idea that social capital is a crucial resource in the enhancement of disaster resilience and adaptive capacity. Thus, the study concurs with the findings of Agea et al. (2008, p. 70) in Masaka District of Central Uganda. They found out that IKS play an important role in enhancing food security. Kamwendo and Kamwendo (2014, p. 100) established that the IKS embedded in women, ensured food security in households in Kenya. Matsa and Manuku (2013) and Akpabio and Akankpo (2003, p. 50) denoted that the IKS embedded in local communities ensured food security.

The study revealed that people rely on bridging capital which "describes acquaintances or individuals loosely connected that span social groups, such as class or race" (Aldrich and Meyer 2014, p. 5). They are just helping each other regardless of not being blood related. "Individuals and groups can strengthen, increase and diversify their positive relationships with others (that is increasing social capital), [thereby expanding] their access to opportunities" (Marin et al. 2015, p. 455). These opportunities result in improving people's resilience to disasters such as floods and droughts. In Chadereka, women go to the Mozambican Border (that is close to Muzarabani) to look for jobs and this is only successful through links. When one does not have a link, it is difficult to get a job in order to get food. In the same tenor, Schramski's (2013, p. 76) study on adaptive capacity in rural South Africa, established that "at the community-wide scale non-parametric correlations indicate that there is a statistically significant relationship between adaptive capacity scores and exchanges of food, wood, water, money, and disease information". This means that these networks are significant predictors of adaptive capacity.

Recommendations

As a result of the foregoing conclusions, authors of this study recommend the following:

- A sophisticated social capital resilience model needs to be incorporated into the policy document for disasters by all stakeholders (NGOs and community leaders). Thus, bottom up measures should be taken to enhance bridging, bonding, and linking social capital among the community members.
- The community leaders are encouraged to enforce norms and values that strengthen community relationships and the spirit of togetherness.
- The establishment of a participatory group building framework may help members to establish a direct interaction with external stakeholders to ensure that local people who are not in leadership positions are not taken advantage of.
- Further researches need to be conducted to assess community capability to effectively deal with floods and droughts.
- There is a need to consider intra-community gendered dynamics in clubs to ensure a continued community participation in enhancing food security.
- Community leaders must be encouraged to put in place mechanisms and programs that sensitize men on their need to support women who are participating local organizations that are serving the community.
- Nongovernmental Organizations must be encouraged to rope in social workers so that they can provide awareness programs on discrimination and its negative impacts.

Conclusion

By and large, this entry shows the relationship between social capital and community resilience.

The overall findings from the study reflect that different types of social capital and collective efficacy enhance the ability of the Muzarabani community to respond and adjust to the impact of changing climatic conditions and related disasters (floods and droughts). The types of social capital include community interactions and initiatives, informal organizations, generalized reciprocity, NGOs, and social networks. One of the goals of the research was to understand if individuals had sufficient social capital in order to rebound in the aftermath of a disaster. Findings exposed several forms of social capital that were playing different and complementary roles in enhancing resilience of the community to climate change and floods and droughts. The study established that the bonding social capital in Muzarabani enabled individuals to receive immediate aid and warnings which allowed them to undertake disaster preparation. This finding is similar to the one by Tse et al. (2013) in China where households with larger Spring Festival networks (a social network that meets for yearly celebrations) "increased the likelihood that the household would rebuild their home after the 2008 earthquake" (Sadeka et al. 2015, p. 44). Thus, social capital and collective efficacy facilitated the exchange of ideas, provided social support and relief aid, and prompted the residents to be cooperative and work together as one. It positively influenced reciprocal help among the community members. However, in some few cases, discrimination based on sociocultural background, wealth status, and other considerations, discussed above, is perceived as an impediment towards the ability of the community to reduce its vulnerability to floods and droughts.

Cross-References

- ▶ Age-Friendly Future Cities
- ▶ Amsterdam's Pathway to Climate Neutrality: Creating an Enabling Environment
- ▶ An Overview of the Relationship of the Sustainable Development Goals and Urban and Regional Development

Acknowledgments We are grateful for the time and expertise provided by the respondents and enumerators, respectively. We would like to acknowledge two anonymous reviewers for their constructive comments on the first manuscript.

References

Adjibolosoo, S. (1995). *The human factor in developing Africa*. Westport: Praeger.

Agrawal, A. (2008). *The role of local institutions in adaptation to climate change*. Paper prepared for the Social Dimensions of Climate Change, Social Development Department, World Bank, Washington, DC.

Akpabio, I. A., & Akankpo, G. O. (2003). Indigenous knowledge practices and the role of gender in rice production in Ini, Nigeria. *Indilinga African Journal of Indigenous Knowledge Systems, 2*(1), 45–52.

Aldrich, D. P., & Crook, K. (2008). Strong civil Society as a double – Edged swords siting trailers in post Katrina New Orleans. *A Political Research Quarterly, 6*(1), 379–389.

Aldrich, D. P., & Meyer, M. A. (2014). Social capital and community resilience. *American behavioral scientist, 59*(2), 254–269.

Aldrich, D. P., & Meyer, A. M. (2015). Social capital and community resilience. *American Behavioural Scientists, 592*, 254–269.

Aldrich, D. P. (2010). Fixing recovery: Social capital in post-crisis resilience. *Journal of Homeland Security*.

Barker. (2011). *Flooding in, flooding out: How does post disaster volunteering build community resilience?* Master's thesis, Lund University International, Lunds University, Lund.

Brennan, M., Cantrell, R., Spranger, M., & Kumaran, M. (2014). *Effective community response to a disaster: A community approach to disaster preparedness and response. UF/IFAS Extension*. University of Florida.

Carpenter, A. C. (2014). *Community resilience to sectarian violence in Baghdad*. New York: Springer.

Chambers, R. (1994). The origins and practice of participatory rural appraisal. *World Development, 22*(7), 953–969.

Chanza, N., & de Wit, A. (2014). Harnessing indigenous knowledge for enhancing CBA and resilience in Muzarabani dryland, Zimbabwe. In *East and Southern Africa CBA & resilience learning conference* (pp. 1–4). Addis Ababa: ILRI Campus.

Chamlee-Wright, E. L., & Storr, V. H. (2011). Social capital as collective narratives and post-disaster community recovery. *The Sociological Review, 59*(2), 149–167.

Chingombe, W., Manatsa, D., Mukwada, G., Pedzisai, E., & Taru, P. (2014). A participatory approach in GIS data collection for flood risk management, Muzarabani district, Zimbabwe. *Saudi Society for Geosciences*. https://doi.org/10.1007/s12517-014-1265-6.

Coleman, J. (1990). Foundations of Social Theory. Cambridge: Cambridge University Press.

Community and Regional Resilience Institute. (2013). *Definitions of community resilience: An analysis. A CARRI report*. Meridian Institute.

Darke, P., Shanks, G., & Broadbent, M. (1998). Successfully completing case study research: Combining rigour, relevance and pragmatism. *Information Systems Journal, 8*(4), 273–289.

Delgado, M., & Delgado-Humme, D. (2013). *Assets assessments and community social work practice*. Oxford: Oxford University Press.

DFID. (2011). *Defining resilience: A DFID approach*. London: Department for International Development.

Dynes, R. R. (2005). *Community Social capital as the primary basis of resilience: Preliminary paper number 344*. Newark: University of Delaware, Disaster Research Centre.

Elliott, J. R., Haney, T. J., & Sams-Abiodun, P. (2010). Limits to social capital: Comparing network assistance in two New Orleans neighborhoods devastated by hurricane Katrina. *The Sociological Quarterly, 51*(4), 624–648.

Fellin, P. (1995). *The community and the social worker*. Itasca: Peacock.

Finucane, M. L., Blum, M. J., Ramchand, R., Parker, A. M., Nataraj, S., Clancy, N., Cecchine, G., Chandra, A., Slack, T., Hobor, G., & Ferreira, R. J. (2020). Advancing community resilience research and practice: moving from "me" to "we" to "3D". *Journal of Risk Research, 23*(1), 1–10.

Fraser, T. (2021). Japanese social capital and social vulnerability indices: Measuring drivers of community resilience 2000–2017. *International Journal of Disaster Risk Reduction, 52*, 101965.

Ganapati, E. N. (2012). Downsides of social capital for women during disaster recovery: Toward a more critical approach. *Administration and Society, 45*(1), 72–96. https://doi.org/10.1177/0095399712471491l.

Gering, J. (2004). What is a case study and what is it good for? *American Political Science Review, 98*, 341–354.

Giddens, A. (1984). *The constitution of society: An outline of the theory of structuration*. Berkeley: University of California Press.

Gwimbi, P. (2004). *Flood hazard impact and mitigation strategies in disaster prone areas of Muzarabani District (Zimbabwe): Exploring the missing link*. Ossrea: Addis Ababa.

Hoegl, M., & Hartmann, S. (2021). Bouncing back, if not beyond: Challenges for research on resilience. *Asian business & management, 20*(4), 456–464.

Homan, M. S. (2008). *Promoting community change: Making it happen in the real world*. Belmont: Brooks/Cole.

Intergovernmental Panel on Climate Change (IPCC). (2007a). *Impacts, adaptation and vulnerability. Working group 11 contribution to the fourth assessment report of the intergovernmental on climate change*. Cambridge, UK: Cambridge University Press.

Intergovernmental Panel on Climate Change (IPCC). (2007b). *Climate change 2007: Synthesis report*. Available at: http://www.IPCCch/publications_and_data/publications_ipcc_fourth_assessment_report_synthesis_report.Html

Jack, P. P. (2017). *An assessment of the impacts of drought on rural livelihoods: Ward 23, Muzarabani District*. Doctoral dissertation, Bindura University of Science Education.

Kamwendo, G., & Kamwmp;endo, J. (2014). Indigenious Knowledge-Systems and Food Security: Some Examples from Malawi. *Journal of Human Ecology, 48*(1) 97–101.

Kasimba, R. (2017). *Understanding the role of social capital in enhancing community resilience to natural disasters: A case study of Muzarabani District, Zimbabwe*. PhD thesis, Rhodes University, Grahamstown.

Kasimba, R. (2018). *Understanding the role of social capital in enhancing community resilience to natural disasters: a case study of Muzarabani District*, Zimbabwe.

Kien, N. V. (2011). Social capital, livelihood diversification and household resilience to annual flood events in the Vietnamese Mekong River Delta. Research Report number 2011-RR10. *Economy and Environment*. Programme for South East Asia EEPSEA.

Louis, N., Mathew, T. H., & Shyleen, C. (2021). Migration as a Determinant for Climate Change Adaptation: Implications on Rural Women in Muzarabani communities, Zimbabwe, Geneva. Springer.

Madzana, E. (2013). *The impact of floods on rural livelihoods: The case of Chadereka ward, Muzarabani*. Doctoral dissertation, Bindura University of Science Education.

Magis, K. (2010). Community resilience: An indicator of social sustainability. *Society and Natural Resources, 23*(5), 401–416.

Manyena, B. (2009). *Disaster resilience in development and humanitarian interventions*. Doctoral dissertation, Northumbria University.

Manyena, B. S., Collins, A. E., Mavhura, E., & Manatsa, D. (2013). Indigenous knowledge, coping strategies and resilience to floods in Muzarabani, Zimbabwe. *International Journal of Disaster Risk Reduction, 5*, 38–48.

Marín, L., Cuestas, P. J., & Román, S. (2015). Determinants of consumer attributions of corporate social responsibility. *Journal of Business Ethics, 138*, 247–260.

Matsa, W., & Manuku, M. (2013). Traditional Science of seed and crop yield preservation: Exploring the contribution of women to indigenous knowledge Systems in Zimbabwe. *International Journal of Social science and Humanities, 3*(4) 234–245.

Mavhura, E. (2017). Applying a systems-thinking approach to community resilience analysis using rural livelihoods: The case of Muzarabani district, Zimbabwe. *International Journal of Disaster Risk Reduction, 25*, 248–258.

Mayunga, J. S. (2007). Understanding and Applying the Concept of Community Resilience: A capital Based Approach. Department of Land Scape, Architecture and Urban Planning, Hazard Reduction and Recovery Centre, Texas, Texas A. M University.
Meyer, M. A. (2012). Social Capital and collective efficacy for disaster resilience: Connecting individuals with communities and vulnerability with resilience in hurricane-prone communities in Florida. PhD Thesis. Colorado: Colorado State University.
Mogue, T. (2006). *Shocks, livestock asset dynamics and DSGD. Discussion paper 38*. International Food Policy Research Institute (IFPRI).
Mudavanhu, C. (2014). The impact of flood disasters on child education in Muzarabani District, Zimbabwe. *Jàmbá: Journal of Disaster Risk Studies, 61*, Art. #138.
Murray, B., Domina, T., Petts, A., Renzulli, L., & Boylan, R. (2020). "We're in this together": Bridging and bonding social capital in elementary school PTOs. *American Educational Research Journal, 57*(5), 2210–2244.
Murwira, A., Masocha, M., Gwitira, I., Shekede, M. D., Manatsa, D., & Mugandani, R. (2012). *Vulnerability and adaptation assessment* (pp. 35–73). Harare: Zimbabwe Second National Communication to the United Nations Framework Convention on Climate Change.
Nyamwanza, A. M. (2012). Livelihood resilience and adaptive capacity: A critical conceptual review. *Jàmbá: Journal of Disaster Risk Studies, 41*, Art. #55.
Parker, D. J. (2020). Disaster resilience–a challenged science. *Environmental Hazards, 19*(1), 1–9.
Patton, M. Q. (1999). Enhancing the quality and credibility of qualitative analysis. *Health services research, 34*(5 Pt 2), 1189.
Paton, D., & Johnston, D. (2001). Disasters and communities: Vulnerability, resilience, and preparedness. *Disaster Prevention and Management, 10*(4), 270–277.
Patterson, O., Weil, F., & Patel, K. (2010). The role of community in disaster response: Conceptual models. *Population Research and Policy Review, 29*(2), 127–141.
Putnam, R. D. (1993). The prosperous community: Social Capital and public life. *The American Prospect, 41*(3), 11–18.
Raimundo, I. M. (2009). International migration management and development in Mozambique: What strategies? *International Migration, 47*(3), 93–122.
Ruswa, N. (2017). *An assessment of socio-economic impacts of floods in Chadereka ward, Muzarabani district, Zimbabwe*. Doctoral dissertation, Bindura University of Science Education.
Sadeka, S., Mohamad, M. S., Reza, M. I. H., Manap, J., & Sarkar, M. S. K. (2015). Social capital and disaster preparedness: conceptual framework and linkage. *Journal of Social Science Research, 3*, 38–48.
Schramski, C. S. (2013). *Adaptive capacity in rural South Africa: Temporal, network and agrarian dimensions*. PhD thesis, University of Florida.
Tse, C. W., Wei, J. & Wang, Y. (2013). Social Capital and Disaster Recovery: Evidence from Sichuan Earthquake in 2008. Washington, DC: Center for Global Development.
Varda, D. M., Forgette, R., Banks, D., & Contactor, N. (2009). Social network methodology in the study of disasters: Issues and insights prompted by post Katrina research. *Population Research and Policy Review, 2*(8), 11–29.
Voelkl, B. (2015). The evolution of generalized reciprocity in social interaction networks. *Theoretical population biology, 104*, 17–25.
Yandong, Z. (2007). *Social networks and reduction of risk in disasters: An example of Wenchuan earthquake*. Institute of Science, Technology and Society.
Zainal, Z. (2007). Case study as a research method. *Jurnal Kemanusiaan, 5*(1), 56–71.

Town and Regional Planning Real Estate

▶ The Urban Planning-Real Estate Development Nexus

Transformation

▶ Sustainable Development Goals and Urban Policy Innovation

Transformation - Change

▶ Growth, Expansion, and Future of Small Rural Towns

Transnational - International

▶ Transnational Migrants on the Margin

Transnational Crimes: Global Impact and Responses

Jade Lindley
Law School and Oceans Institute, The University of Western Australia, Crawley, Australia

Synonyms

Illegal International Movement; Organized Crime; International Law

Introduction

The opening of trade and travel borders due to globalization enabled transnational organized crime. As is the case with legitimate business, transnational criminal organizations transcend national borders to facilitate illegal operations. It finds sympathetic legal loopholes and gray areas to evade law enforcement and operate alongside legitimate business.

Not all criminal operations are transnational. Nor are they necessarily organized, however, the successful business model that necessitates criminal operations across borders for significant profit involves a high level of sophisticated business savvy. Stark variances between legal systems; tolerances for certain crimes; access to geographic locations; and lack of relevant training among law enforcement and the judiciary, increase vulnerability to certain transnational organized crimes.

Continuation of unfettered transnational organized crime in some locations may increase the potential vulnerability of its effects worldwide. This contribution explores transnational organized crime within the context of strengthening cross-border resilience. In this context, *resilience* refers to the ability to establish and maintain transparent; adaptable and secure supply chains; human security; financial systems; and political relationships. Specifically, it provides an overview of transnational organized crime, including relevant definitions, it then provides examples of how it may present. Further, it discusses the impact of transnational organized crime on global resilience. Finally, this contribution considers the role of international law in addressing transnational organized crime domestically, providing opportunities to strengthen responses.

Definition: What Is Transnational Organized Crime?

The United Nations (UN) Convention against Organized Crime (CTOC) is the guiding international law on transnational organized crime. Several elements collectively define transnational organized crime. CTOC provides clarity around the activity of transnational criminal groups, the serious crimes they may conduct, as well as the peripheral crimes that enable those serious crimes, such as money laundering and corruption (United Nations, 2000d: Articles 2, 7, and 8). In broad terms, Article 2(a) of CTOC explains organized crime as a serious crime that attracts a domestic penalty of 4 years imprisonment or longer, committed by a structured group of three or more people (United Nations, 2000d). Further explained in Article 3, crime involves transnational activity by crossing international borders, in some way (United Nations, 2000d).

Signed in Palermo, Italy in 2000, CTOC came into force in 2003 after achieving the requisite 40-member state signatures. By 2022, 190 UN member states agreed to cooperate with CTOC (United Nations Treaty Collection, 2022), indicating almost universal buy-in to prevent and prosecute transnational organized crime. Despite near-universal uptake of CTOC, not all member states meet obligations by adopting and implementing all required measures. The lack of harmonized responses, therefore, limits the ability for international law to be an effective vehicle for building resilience.

Types of Transnational Organized Crime

Transnational organized crime presents in a variety of ways. The web of transnational organized crime syndicates are facilitated by opportunity for criminal profit in locations where certain types of crimes may be viable. For example, while certain crime types may be viable in one location, they may not be in another due to varying tolerances for that crime based on cultural or religious beliefs; sympathetic or lack of modernized laws with low penalties; under-resourced, lax or untrained law enforcement and judiciary; low standards of human rights; and permissive financial systems. Access to supply and demand markets also combines to provide an opportunity for transnational organized crime.

Transnational organized crime is an umbrella term for a range of crimes that can be sorted into several categories: physical movement across borders; water-based crimes; environmental crime; virtual and financial crime; and enabling crimes. These categories are non-exhaustive, and crimes emerge and evolve constantly, though the crimes outlined below provide an overview of those most prevalent.

Physical Contraband Movement Across Borders

Illicit movement across international borders – trafficking or smuggling – is the containment and movement of contraband, whether the contraband agrees willingly or is moved unknowingly. This form of transnational organized crime can involve movement of any form of contraband including commonly known commodities such as drugs, weapons, and people, to less common commodities such as wildlife; foods and alcohol; priceless art and antiquities; cigarettes; charcoal; and precious gems and metals. Linked to CTOC are three related UN Protocols on trafficking in persons; smuggling of migrants; and trafficking of weapons (United Nations, 2000a, 2000b, 2000c, 2000d).

Trafficking in persons reached international concern initially due to the concern relating to women and children trafficked for the purpose of sexual, domestic, and military servitude, but later expanded to consider trafficking of men for all forms of labor, on land and at sea. Since the adoption of the *Trafficking in Persons Protocol*, extensive research has unraveled the nature and extent of trafficking in persons within various supply chains and many global corporations have committed to combat this modern form of slavery from within their business (United Nations, 2000c; US Department of State, 2021; Walk Free Foundation, 2018).

Drugs' trafficking is often a central focus of law enforcement and border officials. The UN established three guiding international laws within its toolkits to support control and resilience to drug-related transnational organized crime (United Nations Office on Drugs and Crime, 2022a, 2022b). The UN's World Drug Report provides the most expansive overview of the international drug trade across the various drug types; supply and demand; and movement between supply, transit, and destination locations (United Nations Office on Drugs and Crime, 2019).

While drugs trafficking causes extensive harm and generates substantial illegal profits, other internationally trafficked commodities, for example, fraudulently labeled food may generate similar profits, with fewer risks, but impact a much broader cross-section of the population (Lindley, 2021b; Mueller, 2007). Some of the less common transnational organized crimes are prevalent in particular locations due to supply and demand, for example, charcoal trafficking is limited to countries in and around the Middle East, often from Eastern Africa (White, 2019); while other commodities have global demand, such as precious gems and metals; and endangered and high demand wildlife species (United Nations Interregional Crime and Justice Research Institute, 2016; United Nations Office on Drugs and Crime, 2022d).

Water-Based Crimes

Over- and illegal fishing is decimating fish stocks reducing natural ocean habitats, food security, and ocean-based tourism. These effects are further worsened by climate change and pollution (Food and Agriculture Organization of the United Nations, 2016). Organized criminal fishing syndicates fuel further irreversible harm in the already devastating state of the world's oceans (de Coning, 2011). The Food and Agriculture Organization of the United Nations (FAO) assessments reveal 34.2% of fish stocks are classified as overfished with increasing trends in overfished stocks and declining trends in sustainably fished stocks, now resulting in virtually no underfished stocks remaining (Food and Agriculture Organization of the United Nations, 2020).

Fisheries lends itself to transnational organized crime due to the limited oversight available onboard vessels at sea, especially beyond territorial exclusive use boundaries (United Nations, 1982: Articles 3, 57 and 87). Already low profits are maximized by trafficking vessel workers and forced labor; trafficking contraband; and smuggling migrants among other opportunistic transnational organized crimes (de Coning, 2011; Lindley et al., 2018). Combined with enabling crimes to support illegal catch, through issuance of fake fishing licenses or entry of illegal catches into the legitimate market by corrupt fisheries officers at ports, transnational organized crime damages resilience within the fishing industry.

Despite the involvement of often transnational organized criminals, victims of crimes such as illegal fishing may be challenging to identify and therefore easier to justify reduced or absent penalties. "Organized criminal networks operating in the fisheries sector engage in illicit activities ranging from criminal fishing to tax crimes, money laundering, corruption, document fraud, and trafficking in persons, drugs and arms" (United Nations Office on Drugs and Crime, 2015). As such, it requires systemic, often industry-wide participation by highly organized criminal syndicates across jurisdictions at ports, markets, vessel ownership, and flag States (Lindley & Techera, 2020). Given the potential number of offenders, recipients, and conspirers involved in enabling illegal fishing, is unlikely that those involved are unaware of the intention to carry out illegal fishing (Lindley & Techera, 2020).

Often linked to transnational organized illegal fishing is maritime piracy. Targeting vessels at port or idle at a chokepoint for minor on-board thefts, to hijacking a steaming vessel along a major transit route, along with its crew and cargo, involves sophisticated, equipped, and organized criminals (Lindley, 2018). The potential harm from maritime piracy can be extreme: from hijacking cruise vessels full of tourists; oil tankers to commit terrorist attacks at port; or vessels containing food aid, the result has a broader global impact, in addition to disrupting supply chains increasing financial, environmental, and human costs. Often complicated by lack of clear law enforcement jurisdiction between vessel ownership, flag state, cargo source and destination, and seafarer homeland, international cooperation to effectively respond is essential.

Environmental Crime

Environmental crimes are often considered victimless, however, the potential victims are in fact endless (Lambrechts & Hector, 2016). Crimes against the environment may be against biodiversity, wildlife, land, air, and water, for example, deforestation; hazardous contamination of waterways; natural resource depletion and unsafe mining; and animal exploitation. The International Union for Conservation of Nature (IUCN) prioritizes environmental crime as a barrier to achieving environmental rule of law, specifically Part III(l) notes, addressing environmental crimes in the context of other types of crime such as money laundering, corruption, and organized crime (International Union for Conservation of Nature, 2016).

Environmental crime cannot exist in a vacuum as it relies on a various clandestine supply chains to evade authorities and yield profit. Often, environmental harms are linked to irresponsible corporate activity; however, much environmental crime is firmly embedded within organized criminal syndicate structures (Lindley, 2021a). For example, lawful disposal of waste can be expensive and inconvenient, and therefore acceptance of illegal and harmful waste, particularly by criminals in developing countries for profit may occur – though at the detriment of locals (Lindley, 2016). Internationally, there is a strong focus on preventing further harm to the environment and a plethora of relevant internationally and regionally-focused conventions exist to build environmental resilience (United Nations Environment Programme, 2022). Without regional cooperation to establish and police irresponsible practices, particularly between land and sea bordering nations, effective responses to environmental crimes cannot be possible.

Virtual and Financial Crime

Virtual and financial crimes are broad and often closely interlinked with other forms of transnational organized crime, particularly given the motivation for profit. The internet facilitates crime to an incomprehensible extent. Geographic locations are no barrier to online transnational organized criminal groups, or their victims. Cybercrimes such as identity theft and financial crimes can involve credit card or bank account phishing; romance and dating scams; investment and pyramid schemes, as well as using online portals for crimes such as child sex grooming and terror recruiting at the extreme (Australian Competition and Consumer Commision, 2022). Given the potentially large online transnational organized criminal syndicates involved in cybercrime, the UN's Global Program on Cybercrime's mandate is to assist member states in building resilient online communities to prevent and protect against online crime (United Nations, 2011).

Enabling Crimes

Linking all these transnational organized crimes together are the enabling crimes. Commonly, enabling crimes involve money laundering and corruption allows criminal groups to operate seamlessly across borders.

Money laundering is the process of legitimizing illegitimately obtained funds. The process may involve a range of transactions, often repeated, but ultimately resurfaces as legal tender (United Nations Office on Drugs and Crime, 2022c). The UN estimates that around 5% of global GDP, or up to $US2 trillion is laundered annually (United Nations Office on Drugs and Crime, 2022c). CTOC outlines measures to combat money laundering and provide training and technical assistance, especially for developing countries and countries with economies in transition (United Nations, 2000d: Articles 7 and 29).

Corruption is an enabling crime that can be described as involving a range of activities such as bribing government officials to act or fail to act in a way supportive of criminal activities. The UN adopted the UN Convention against Corruption (UNCAC) linked to CTOC. These conventions recognize the common threat to peace and security, sustainable development and justice to global resilience of transnational organized crime and corruption (Global Initiative Against Transnational Organized Crime, 2021a).

The Impact of Transnational Organized Crime on Global Resilience

Every form of transnational organized crime varies in prevalence, depending on a range of factors, and as such, each transnational organized crime must be considered singularly, as well as in conjunction with other related transnational organized crimes. Often, some forms of lower risk, high-profit crimes are conducted to build profit to fund riskier criminal activity that attracts greater payoffs (United Nations Office on Drugs and Crime, 2011). Unless adequately addressed transnational organized crime, challenges resilience relating to the environment, in supply chains, human security, financial systems, and political relationships. Building and maintaining crime-resilient systems is therefore necessary.

Despite extensive international support for CTOC, understanding transnational organized crime flows and strengthening important regional relationships is needed to address transnational organized crime in its various forms (Lindley, 2020). Extensive modeling research on the global flows of transnational organized crime is relevant to understand locations and industries in which greater resilience is needed (see, e.g., Global Initiative Against Transnational Organized Crime, 2021b; United Nations Office on Drugs and Crime, 2011). Understanding how transnational organized crime operates, impacts on global resilience and is, therefore, necessary to prevent it.

Building Law Enforcement Resilience

Responding to crime is a domestic law enforcement task, however, when it crosses borders, cooperative regional and international responses can improve outcomes. Interpol is the international leader in law enforcement with a focus on transnational organized crimes (Interpol, 2017). To operationalize, international law is useful to enable responses between borders. Due to the potential harm of transnational organized crime, there is a need to establish and maintain resilient futures, and international law is a vehicle to this.

International law can effectively bridge borders and streamline responses to transnational organized crime within a region. While domestic responses may differ due to political, cultural, social, and religious beliefs, among others, international law can unify a region and enable more effective responses to cross-border crimes. Tools within international law can reduce barriers for member states cooperating on investigations and prosecutions between borders. CTOC among other international laws provides a toolkit that member states may adopt to harmonize their relevant domestic laws with overarching international laws. CTOC tools such as extradition (Article 16), mutual legal assistance (Article 18), and joint investigations (Article 19) may be appropriate for investigating and responding to transnational organized crime (United Nations, 2000d). Cooperation using these tools is possible only between signatory member states, though due to near-universal adoption of CTOC, this may not present a barrier.

Domestic law enforcement task forces focused on specific or general transnational organized crime can be strengthened by working alongside relevant international and regional stakeholders, particularly for developing countries where human and financial law enforcement resources may be limited. Multi-language toolkits exist to develop and strengthen aspects of the investigative and prosecutorial responses. In 2020, UNODC introduced a toolkit to enable stakeholder engagement for transnational organized crime responses (United Nations Office on Drugs and Crime, 2020). Given opportunistic transnational organized criminals may target developing countries due to regulatory gaps and limited law enforcement resources to effectively respond to transnational organized crime, the opportunity to train and build capacity within law enforcement will optimize the ability to respond.

Conclusion

Transnational organized crime is an international problem and requires an international solution to increase global resilience. Near universal support for CTOC should indicate the issue is well controlled; however, highly sophisticated organized criminal groups cross international borders and conduct their business where sympathetic laws and societies exist. Further challenging authorities is the many and varied ways in which transnational organized crime presents, though common among all types are the enabling crimes such as money laundering and corruption. These enabling crimes provide an avenue in which global resilience to transnational organized crime can be focused.

Cross-References

▶ Fisheries Crime and Ocean Resilience
▶ Internationalization of Cities

References

Australian Competition and Consumer Commision. (2022). *Scamwatch: Types of scams*. https://www.scamwatch.gov.au/types-of-scams

de Coning, E. (2011). *Transnational organized crime in the fishing industry, focus on: Trafficking in persons, smuggling of migrants, illicit drugs trafficking*. Vienna: UNODC. Retrieved from http://www.unodc.org/documents/human-trafficking/Issue_Paper_-_TOC_in_the_Fishing_Industry.pdf.

Food and Agriculture Organization of the United Nations. (2016). *The state of world fisheries and aquaculture: Contributing to food security and nutrition for all*. Rome: FAO. Retrieved from http://www.fao.org/3/a-i5555e.pdf.

Food and Agriculture Organization of the United Nations. (2020). *The state of world fisheries and aquaculture 2020. Sustainability in action*. FAO: https://www.fao.org/3/ca9229en/ca9229en.pdf.

Global Initiative Against Transnational Organized Crime. (2021a, 13 December). *Crime and corruption: A tale of two conventions*. https://globalinitiative.net/analysis/crime-and-corruption-uncac/

Global Initiative Against Transnational Organized Crime. (2021b). *Global organized crime index 2021*. https://globalinitiative.net/wp-content/uploads/2021/09/GITOC-Global-Organized-Crime-Index-2021.pdf

International Union for Conservation of Nature. (2016, 26–29 April). *World declaration on the environmental rule of law*. https://www.iucn.org/commissions/world-commission-environmental-law/wcel-resources/environmental-rule-law

Interpol. (2017). *Global strategy on organised and emerging crime*. Interpol. https://www.interpol.int/content/download/5582/file/Global%20Strategy%20on%20Organized%20and%20Emerging%20Crime.pdf.

Lambrechts, D., & Hector, M. (2016). Environmental organised crime: The dirty business of hazardous waste disposal and limited state capacity in Africa. *Politikon, 43*(2), 251.

Lindley, J. (2016). *Somali piracy: A criminological perspective*. Routledge Publishing.

Lindley, J. (2018). Using regulatory pluralism to achieve effective control of Somali piracy: A model for other piracy-prone regions. In L. Y.-C. Chang & R. Brewer (Eds.), *Criminal justice and regulation revisited: Essays in honour of Peter Grabosky*. Routledge.

Lindley, J. (2020). Criminal threats undermining indo-pacific maritime security: Can international law build resilience? *Journal of Asian Economic Integration, 2*(2), 206–220. https://doi.org/10.1177/2631684620940477.

Lindley, J. (2021a). Crime and the environment. In E. Techera, J. Lindley, K. Scott, & A. Telesetsky (Eds.), *Routledge handbook of international environmental law* (2nd ed.). Routledge.

Lindley, J. (2021b). Food fraud: An international snapshot and lessons for Australia. *Journal of Financial Crime, 28*(2), 480. https://doi.org/10.1108/JFC-09-2020-0179.

Lindley, J., & Techera, E. (2020). Using routine activity theory to explain illegal fishing in the Indo-Pacific. In S. Hufnagel & A. Moiseienko (Eds.), *Criminal networks and law enforcement: Global perspectives on illegal enterprise*. Routledge.

Lindley, J., Percy, S., & Techera, E. (2018). Illegal fishing and Australian security. *Australian Journal of International Affairs, 73*(1). https://doi.org/10.1080/10357718.2018.1548561.

Mueller, T. (2007, 6 August). Slippery business: The trade in adulterated olive oil. *The New Yorker*. https://doi.org/https://www.newyorker.com/magazine/2007/08/13/slippery-business

United Nations. (1982). *Convention on the law of the sea*. UN.

United Nations. (2000a). *Protocol against the Illicit manufacturing of and trafficking in firearms, their parts and components and ammunition, supplementing the United Nations convention against transnational organized crime*. New York: UN. Retrieved from https://www.unodc.org/documents/middleeastandnorthafrica/organised-crime/UNITED_NATIONS_CONVENTION_AGAINST_TRANSNATIONAL_ORGANIZED_CRIME_AND_THE_PROTOCOLS_THERETO.pdf.

United Nations. (2000b). *Protocol against the smuggling of migrants by land, sea and air, supplementing the United Nations convention against transnational organized crime*. New York: UN. Retrieved from https://www.unodc.org/documents/middleeastandnorthafrica/organised-crime/UNITED_NATIONS_CONVENTION_AGAINST_TRANSNATIONAL_ORGANIZED_CRIME_AND_THE_PROTOCOLS_THERETO.pdf.

United Nations. (2000c). *Protocol to prevent, suppress and punish trafficking in persons, especially women and children, supplementing the United Nations convention against transnational organized crime*. New York: UN. Retrieved from https://www.unodc.org/documents/middleeastandnorthafrica/organised-crime/UNITED_NATIONS_CONVENTION_AGAINST_TRANSNATIONAL_ORGANIZED_CRIME_AND_THE_PROTOCOLS_THERETO.pdf.

United Nations. (2000d). *United Nations convention against transnational organized crime and the protocols thereto*. New York. Retrieved from https://www.unodc.org/documents/middleeastandnorthafrica/organised-crime/UNITED_NATIONS_CONVENTION_AGAINST_TRANSNATIONAL_ORGANIZED_CRIME_AND_THE_PROTOCOLS_THERETO.pdf

United Nations. (2011). *Resolution adopted by the General Assembly on 21 December 2010 (A/RES/65/230)*. https://www.un.org/ga/search/view_doc.asp?symbol=A/RES/65/230&Lang=E

United Nations Environment Programme. (2022). *Secretariats and conventions*. https://www.unep.org/about-un-environment/why-does-un-environment-matter/secretariats-and-conventions

United Nations Interregional Crime and Justice Research Institute. (2016). *Strengthening the security and integrity of the precious metals supply chain: Technical report*. UNICRI. http://www.unicri.it/sites/default/files/2019-11/Strengthening%20the%20Security%20and%20Integrity%20of%20the%20Precious%20Metals%20Supply%20Chain_0.pdf.

United Nations Office on Drugs and Crime. (2011). *Estimating illicit financial flows resulting from drug trafficking and other transnational organized crimes: Research Report*. UNODC. https://www.unodc.org/documents/data-and-analysis/Studies/Illicit_financial_flows_2011_web.pdf.

United Nations Office on Drugs and Crime. (2015). *Fact sheet: Fisheries crime*. http://www.unodc.org/documents/about-unodc/Campaigns/Fisheries/focus_sheet_PRINT.pdf

United Nations Office on Drugs and Crime. (2019). *World drug report*. New York: UNODC. Retrieved from https://wdr.unodc.org/wdr2019/en/prevalence_map.html.

United Nations Office on Drugs and Crime. (2020). *Toolkit on stakeholder engagement: Implementing the United Nations Convention against Transnational Organised Crime (UNTOC)*. UNODC. https://www.unodc.org/documents/NGO/SE4U/UNODC-SE4U-Toolkit-Interactive-WEB.pdf.

T

United Nations Office on Drugs and Crime. (2022a). *International drug control conventions: The single convention on narcotic drugs of 1961 as amended by the 1972 protocol, the convention on psychotropic substances of 1971 and the United Nations Convention against Illicit traffic in narcotic drugs and psychotropic substances of 1988*. UNODC. https://www.unodc.org/unodc/en/commissions/CND/conventions.html
United Nations Office on Drugs and Crime. (2022b). *Legal framework for drug trafficking*. UN. https://www.unodc.org/unodc/en/drug-trafficking/legal-framework.html.
United Nations Office on Drugs and Crime. (2022c). *Money laundering*. https://www.unodc.org/unodc/en/money-laundering/overview.html
United Nations Office on Drugs and Crime. (2022d). *Wildlife and forest crime: Target groups*. https://www.unodc.org/unodc/en/wildlife-and-forest-crime/target-groups.html
United Nations Treaty Collection. (2022). *Status of treaty: 12. United Nations Convention against Transnational Organized Crime*. UN. https://treaties.un.org/pages/ViewDetails.aspx?src=TREATY&mtdsg_no=XVIII-12&chapter=18&clang=_en.
US Department of State. (2021). *2021 Trafficking in persons report*. https://www.state.gov/reports/2021-trafficking-in-persons-report/
Walk Free Foundation. (2018). *Global slavery index 2018*. Perth, Austalia. Retrieved from https://www.globalslaveryindex.org/resources/downloads/
White, E. (2019, 29 April). *The unintended consequences of charcoal*. The Australian Strategic Policy Institute. https://www.aspistrategist.org.au/the-unintended-consequences-of-charcoal/

Further Reading

United Nations Convention on Transnational Organized Crime https://www.unodc.org/documents/middleeastandnorthafrica/organised-crime/UNITED_NATIONS_CONVENTION_AGAINST_TRANSNATIONAL_ORGANIZED_CRIME_AND_THE_PROTOCOLS_THERETO.pdf
United Nations Office on Drugs and Crime Organised Crime https://www.unodc.org/unodc/en/organized-crime/intro.html
United Nations Office on Drugs and Crime Tools and Publications https://www.unodc.org/unodc/en/organized-crime/tools-and-publications.html

Transnational Migrants on the Margin

Agency, Aspirations, and Perceptions for the Future Among Malawian Migrants in Zimbabwe's Norton Peri-urban

Bhanye Johannes[1], Vupenyu Dzingirai[1] and Innocent Chirisa[3]
[1]Department of Community and Social Development, University of Zimbabwe, Harare, Zimbabwe
[2]Department of Demography Settlement and Development, Social & Behavioural Sciences, University of Zimbabwe, Harare, Zimbabwe
[3]Department of Demography Settlement and Development, Social & Behavioural Sciences, University of Zimbabwe, Harare, Zimbabwe

Synonyms

Migration – movement; Squatter settlement – slum, informal settlement; Transnational – international

Definition

Migration – The movement of a person or a group of persons, either across an international border or within a State. It includes movement of refugees, displaced persons, economic migrants, and persons moving for other purposes, including family reunification.

Peri-urbanity – life and processes in urban fringe or transitional areas where people, resources, and goods connect and move between rural and urban areas.

Squatter Settlements – residential areas that develop without legal claims to the land and/or permission from the concerned authorities to build; as a result of their illegal or semi-legal status, infrastructure and services are usually inadequate.

Migrant – a person who moves away from his or her place of usual residence, whether within a country or across an international border, temporarily or permanently, and for a variety of reasons.

Agency – the capacity of individuals to act independently and to make their own free choices to achieve specific goals or their own interests.

Futurity – conditions or events of the time to come.

Introduction

The purpose of this chapter is to examine the agency, aspirations, and perceptions of migrants regarding their future in peri-urban squatter settlements of destination countries. The chapter specifically focuses on how migrants of international descent define themselves in space and time in peri-urban spaces. The chapter is based on ethnographic fieldwork carried between the period of May 2018 and December 2019, among Malawian migrants in Zimbabwe's Norton peri-urban area. The chapter identified tripartite perceptions among migrants regarding their future in peri-urban squatter settlements. Some, mostly elderly and long-term migrants perceive the settlement as a home of sorts, while the infirm and invalids and the poverty-stricken regard the settlement as a home of last resort. Recent and younger migrants on the other hand have a radical position against the peri-urban squatter settlement – regarding it as a launchpad from which to transit to better places particularly the metropolises. Yet other migrants have mixed perceptions regarding place, on one hand yoked to place while also eying for opportunities outside the settlement.

This chapter is divided into six major sections. The second section, which immediately follows after the introduction, presents the chapter background on transnational migrants and the peri-urban. The third section presents the migration infrastructure conceptual framework guiding the chapter. The fourth section presents the case study, upon which general points about how Malawian migrants (Lydiatians) make a living at Lydiate. The fourth section presents the partial ethnographic research methodology adopted for the study informing this chapter. The succeeding section is the core of the chapter, presenting the agency, aspirations, and perceptions of migrants regarding their future in the peri-urban. The last section concludes the chapter, providing some modest policy recommendations.

We now live in a world increasingly defined by population mobility, attracting much interest, but also growing concern in both policy and academic arenas (Forje 2019; Satterthwaite 2016). Transnational mobility has been driven by a number of factors, among them global liberalization, poverty, violent conflict, and environmental stress (Carling and Collins 2018; Van Hear et al. 2018). Scholars have had different views on the patterns of migration. Some scholars emphasize how there has been an upsurge of Africans migrating to the developed world particularly Europe and the Western countries that are widely perceived as centers of "development" (Castles et al. 2014; Flahaux and De Haas 2016). The underlying assumption by these scholars is that African migration is high and increasing, mainly directed toward the developed nations, and driven by poverty and violence (Castles et al. 2014; Flahaux and De Haas 2016). However, in recent years, scholars have begun documenting that far from being externally oriented, African migration is also internal (Nyamnjoh 2013; Segatti and Landau 2011). That is to say that Africans move into each other's countries as they do to the Global North (Flahaux and De Haas 2016). African scholars, too, have also begun documenting the nature of life at destination (Mushonga and Dzingirai 2020; Nyamwanza and Dzingirai 2020). Conclusions are varied, but there seems to be an agreement among themselves that migrants often settle in the peri-urban. These peri-urban destinations are often what Nyamwanza and Dzingirai (2020) term "rough neighborhoods." They are places of danger and risk. The settlements are no more than squatters, often characterized by closely packed, decrepit housing units and nonexistent social services (Schon 2019). In part, these conditions arise because of the unresolved nature of citizenship (Bhanye and Dzingirai 2020).

But how migrants define themselves in place and time in destination peri-urban settlements has not been extensively researched. Preliminary

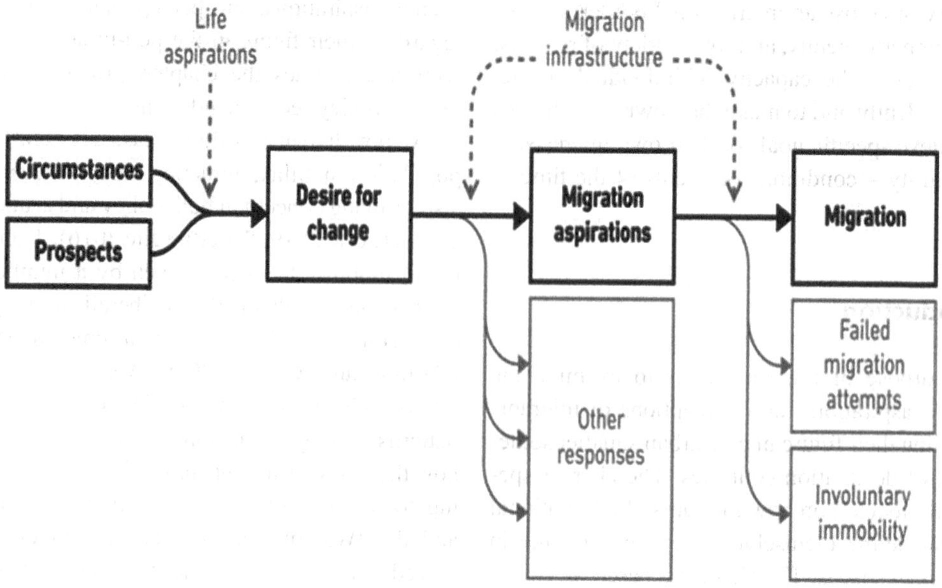

Transnational Migrants on the Margin, Fig. 1 Model of the causal chain leading to migration outcomes (Carling 2017)

evidence seems to suggest that migrants perceive destination settlements as permanent spaces or their final homes (Huchzermeyer and Karam 2006). These scholars assert that, despite the obvious difficulties and roughness of the settlements, migrants have come to terms in accepting them as the final home (Bulger 2008). But some scholars indicate that other migrants regard peri-urban squatter settlements as a springboard for them to transit to more affluent metropolises (Saunders 2011). This chapter therefore wades into this theorization, seeking to establish the truth about migrants and their sense of place.

Conceptual Framework

Migration is a product of complex physical and organizational infrastructures. Xiang and Lindquist (2014) explain "migration infrastructure" as "the systematically interlinked technologies, institutions and actors that facilitate and condition mobility." These infrastructures consist of the commercial migration industry, regulatory frameworks, migrants' social networks, digital transnational connectivity, and support systems. Importantly, migration infrastructure plays two distinct roles. First, it affects how people perceive the possibility of migration and whether or not they develop migration aspirations. Second, migration infrastructure affects whether or not such aspirations are realized (Carling 2017). Figure 1 shows a model of the causal chain leading to migration outcomes based on Carling (2017).

The central argument presented by the infrastructural turn in migration studies is around the significance and direction given to migrant mobilities through infrastructures (Khan 2018). However, such framing discounts the role of migrants as active agents with social and material resources that they use to interact with the infrastructure (Khan 2018). In this chapter, we argue that migrant agency – as manifest by radical young migrants eying for better opportunities in the city – is as much a part of the infrastructure as are its other social, commercial, regulatory, humanitarian, and technological dimensions. Such agency involves critical human interventions such as decision-making, strategy, transgression, and aspiration. In the next section, we present the case study area.

Research Methodology

This chapter is based on ethnographic fieldwork among Malawian migrants in Zimbabwe's Norton peri-urban. The first author immersed himself in the flow and rhythm of Lydiate squatter settlement, enabling him to venture into the lives of Malawian migrants, for a period of more than 1.5 years, between May 2018 and December 2019. Far away from traditional full immersion ethnography, the first author adopted "partial immersion ethnography" where he captured dynamics of both the greater sphere of Lydiate and also the traditional closed community. The researchers also spent time in the community, participating in various social spaces like community meetings and youth informal gatherings observing practices related to migrants' visions for the future. This was supplemented by more than 50 in-depth interviews with the youths, the women, the middle-aged, and the elderly. Questions that were asked aimed at understanding family histories, agency, perceptions, and aspirations of migrants regarding their future. This research required a consideration of ethical issues as it comprised of migrants with unresolved citizenship status, who are often the target of criminalization and abuse. Hence, to ensure their security, the chapter anonymized the participants who volunteered in the research by using pseudonyms.

The study informing this chapter was carried out in Lydiate, a peri-urban area in Norton Town located about 50 km away from Harare, the capital city of Zimbabwe. Lydiate settlement is located in Mashonaland West Province of Zimbabwe, in ward 14 under Chegutu Rural District Council. The first author ethnographically explored both the dynamics in the greater Lydiate area and the core Lydiate area where Malawian descendants (herein referred to as Lydiatians) reside. The core Lydiate area is a squatter settlement that houses Malawian migrants who were engaged as laborers on colonial white-owned farms. The migrants comprise of the first-, second-, third-, and fourth-generation Malawian descendants. There are also other recent migrants commonly referred to as "*vauyi vazvino.*" Lydiate squatter settlement is also punctuated by a frenzy of mobility and transitions; migrants continuously move in and out of the settlement in response to better opportunities elsewhere. The greater Lydiate area is made up of plots and agro-residential plots exclusively owned by indigenous Zimbabweans, which occasionally engage Lydiatians for labor. Livelihoods in Lydiate are diverse but also poor with subsistence farming, buying and selling, and casual labor being key ones. The majority of housing among Lydiatians constitutes temporary to semipermanent shacks made of pole and dagga, plastics, metal scrapes, and zinc roofing sheets. Basic infrastructure such as running water, indoor plumbing, and paved roads is virtually nonexistent, while electricity is only available to privileged few households.

Agency, Aspirations, and Perceptions for the Future Among Malawian Migrants in Lydiate

The study found out three major aspirations and perceptions among migrants in Lydiate peri-urban squatter settlement. These perceptions are perceptions of squatter settlements as home, perceptions of squatter settlements as a launchpad to transit to better places, and perception of settlements as dual functional places. These three broad categories of perceptions are discussed in the forthcoming sections.

The first broad category migrants who share the view of Lydiate as home can be divided into two groupings: "home of sorts" and "home of last resort." The first group of migrants consists of those who regard Lydiate settlement as a home of sorts. These people are often old and have been in the place for a very long time. These old timers are mostly first-generation Malawians who migrated directly from Malawi to Zimbabwe during the famous colonial labor migration period. Lydiate has become the only comfortable place for settlement and sustenance for them. A good case in point is of a 77-year-old Ephraim Malewezi who had this to say:

T

> I started staying at Lydiate in the year 1975. I was born in Malawi and I came here with my parents. We came to Zimbabwe, looking for job opportunities. My parents came directly at this place and we have not stayed at any other place besides Lydiate. I do not see myself going anywhere else beyond Lydiate, because this is the only place I know very well in this country. Making a living in this compound is no longer easy, but we have become comfortable with the life that we are living here. I do not have close relatives nor friends that can take me out of this community. Those who are around me are also poor and in need of support. My two sons Paul and Charles, for example, are also still staying in this compound. They have gotten their own space here and started their own families. Lydiate has become our home and we will die and get buried here (Interview with Ephraim Malewezi at Lydiate Farm, 05 April 2019).

While many old timers and long-term settlers see Lydiate as their final home, some of them have the opposite view regarding their children, encouraging them to migrate out of Lydiate. The reason why they want their children to make a living outside the settlement is that they can be looked after comfortably at the settlement. This was well demonstrated through the case of a 67-year-old Stella Howard who narrated:

> My hope is to see my children pursue a better life outside of this compound. Currently, my first-born child, Tatenda, works in Kariba. I have worked hard to raise money for him to start a better life, and he is now doing very well there. My second born, Talent is currently working at a butcher at the nearby Lucky store for more than 2 years now. If all goes well, he is set to leave for South Africa to get a better life. The main reason why I want my sons to leave this compound is because, this place is not ours *(pano apa hapasi pedu)*. We are foreigners here (Interview with Stella Howard at Lydiate Farm, 23 June 2018).

To be sure then, these migrants may desire a second future for their children elsewhere, but for them, Lydiate is now a final home.

The second group of migrants regards Lydiate as a home of last resort. This category of migrants falls into the following subcategories: the infirm and invalids and the poverty-stricken. These are described below, starting with the infirm and invalids.

The infirm and invalids subcategory consists of Lydiatians who are sick and bedridden and now regard the settlement as home of last resort. Some of them are old timers, while others are recent migrants who came to find a place to be looked after by sympathetic relatives, making Lydiate their final home in the process. We shall provide two cases to demonstrate this. The first case involves an 87-year-old Fahid Banda and his wife Hildah Banda (85). When the researchers were approaching the couple's place, they were welcomed by an unpleasant environment. The infrastructure for the two was a small round thatched hut and a kitchen shade made of pole and dagga. The wife who was so bedridden and helpless with swollen legs was seated in the small kitchen shade. She called the husband who came to put on a fire to warm her up as the interview proceeded. Fahid narrated:

> I stay together with my wife, Hildah. We came here a long time ago; I have even forgotten the year we came here. We do not have any children; thus, we have been staying on our own ever since we came here from Malawi. As you can see, we are now very old and my wife is sick and we cannot think of going anywhere. We are just waiting to die here (*tangomirirakufapano*).

The second case involves a 78-year-old widow, Lynda Kamanga, who narrated how her sickly condition has yoked her to Lydiate. Linda suffered from diabetes, swelling of legs, and several other ailments. She came at Lydiate compound in the year 2016 to be looked after by her sister. Because of that, she perceives Lydiate as her home of last resort.

The second subcategory includes those who are poor. For such migrants, poverty is a glue that binds them to their place of settlement. Those who are bound to place by poverty and its roughness include both the elderly, middle-aged, and youths. These are presented below.

There are old poor who are tied to Lydiate because of poverty. A good example is of an 82-year-old Margret Chirwa. When the researchers opened the small gate of the family's place, they were welcomed by two small huts made out of pole and dagga. The old widow was helpless upon receiving the researchers at her place, failing to find a proper stool for the researchers to sit on. The researchers however sat on the floor, assuring the old widow not to trouble herself. “Do not trouble yourself mother, I am comfortable here,” the researchers asserted. The old lady got settled and went on to converse with the researchers. She narrated her fate as follows:

> We survive from hand to mouth here my son, and we do not see ourselves going anywhere else. If we get piece jobs that give us $3 then we know we survive for that day. (Tikangowanawo maricho anotipawo ka $3 totoraramawo). It is poverty after poverty because of our old age (inhamo yoga yoga kuchembera kwataita kudai). We also survive on cutting and selling thatching grass (kutengesa huswa). After the rain season that is when we start to cut the grass and drying it. We then sell for $1 or $1.50 per bundle. We do not have land for farming here, we only cultivate small portions of land if we are given by indigenous people who own the land close by our stands. We do not have any other place to go to, this has now become our home (pano ndipo pava pamusha pedu) (Interview with Margret Chirwa at Lydiate Farm, 21 June 2018).

There are also middle-aged Lydiatians who are yoked to the place because of poverty. A middle-aged widow, Pricilla Ngozo, aged 43, explained how she is caught up in the vicious cycle of poverty at Lydiate. She narrated:

> I am currently staying with my three children. I started staying at Lydiate farm in 1986. We came with parents who used to work at a mine. My parents came from Malawi and we were born in Mutorashanga together with my siblings. My husband, James, died in 2016 and was buried here at Lydiate Farm. My first son Precious is now living in Epworth. He is just doing piece works. There is nothing in Epworth. The other three children are at home doing piece works. Usually we work in plots (maricho). We now consider ourselves people of this community as it is difficult to dream of a better life beyond this current poverty that we are in (Interview with Pricilla Ngozo at Lydiate Farm, 20 January 2019).

Just as the above cases illustrate how both old timers and the middle-aged regard the settlement as home because of poverty, the youths are also not exempt. A 27-year-old Focus Kwayera who is stuck at Lydiate had this to say:

> Although I desire to have a better life outside of this community, I have given up on that dream. There are several challenges in this community that have forced me to remain here. Firstly, my parents did not afford to provide me with adequate education, I only ended in grade 7. Secondly, in this compound, one is expected to start supporting the family in doing menial jobs at a very young age. I was born here and literally; this place has been my home from birth. Things are difficult here, there are no proper jobs, and there is nowhere I can get money to make savings to move out of this community. I have tried several piece jobs, but nothing is materializing. People are struggling in this compound (Interview with Focus Kwayera at Lydiate Farm, 13 January 2018).

Whatever the case, these groups find Lydiate as a home of last resort. The reason why migrants make peri-urban squatter settlements home of last resort is not because they do not have aspiration to migrate but rather the prevailing migration infrastructures do not permit them to leave the rough neighborhood. In the end, peri-urban squatter settlement becomes home of last resort.

The second broad category of migrants regards Lydiate squatter settlement as a “launchpad” or “gateway” from which to transit to better places. These are radical young people who see their future beyond their current place of settlement. The youths aspire to leave Lydiate because of various reasons as presented below.

Some youths are quite categorical about the rough living conditions in the community as the reason for their radical perception against being yoked to the settlement. This group of migrants sees their current place as a setback full of despair and misery. The case of a 28-year-old Kennedy Chimwemwe who worked on an agro residential plot near the researchers’ camp demonstrates this. Kennedy is a second-generation migrant, born in Lydiate. His parents who are now late were migrants from Malawi. As the researchers were interacting with Kennedy, he made a statement representative of many interactions between the radical youths and the investigator. He said, “Right now you and I are the same; we are poor. But later you will be rich, go back to your university and live in a luxury house.” The researchers explained that they do not live a luxurious life back home, but rather a simple life. Thomas clarified, “But you will have a proper meal every day, right, and you will be in the city full of opportunities *(kutaundi, kune*

T

mikana yakawanga)?" Kennedy further narrated the rough living conditions in the squatter settlement saying:

> In this community, there is no more life for the youths. There's nothing here. There's no work. When there is no work and no money, you don't know what to do with your future as a youth. Can you imagine that I can't even afford to buy proper clothing and only have this pair of shoes [pointing to his tennis shoes]. I had to work tirelessly for a month to buy this pair of shoes…Currently I am surviving on piece jobs in the plots. Per day, they're paying me an average of 4 dollars. I am tired of this kind of life and I have to migrate out of this community to survive (Interview with Kennedy Chimwemwe at Lydiate Farm, 28 June 2018).

Thus, rough living conditions in peri-urban squatter settlements instill agency and aspiration among young people not to remain yoked to the place but to migrate to the city where there are better opportunities.

Young Lydiatians are not only radical about the place because of the rough living conditions in the settlement. They are also persuaded by the desire to want to earn a better life in the towns and cities where they can get better jobs which are not found in the peri-urban squatter settlement. A good example is of a 29-year-old Dyman Dziko who was eying as far as the city of Harare as a place that the skills he has gained in bricklaying and professional thatching can be recognized. Dyman narrated:

> I want to move to Harare because there are certainly better opportunities there than in this community. You can easily get a job in town, which is much better than the menial jobs here. I have learnt several skills among them bricklaying, professional thatching and landscaping. I know with these skills I will be more marketable in the town. Further, I am still young and I have a lot of energy to hustle in the town. If my skills fail to yield something in town, I can still do anything that gives me money from buying and selling and changing money. What I know is that, my life in the city, will be far better than staying here. I will also be able remit money home for my parents to send my siblings to school. I am also motivated to move to town by the better lifestyle, lifestyle and electrcity in towns and cities (Interview with Dyman Dziko at Lydiate Farm, 04 July 2018).

The second case involves Jairos Mavuto, a 30-year-old young man, who always cheered up and chatted with the researchers about opportunities in the city. One afternoon, Jairos approached the researchers and he started explaining that every time he saw the researchers, he is reminded of his dream not to be a permanent resident of Lydiate. He told the researchers that life in the community was tough and he had plans to leave the community for a better life in the town. In an interview with the researchers, Jairos explained:

> My dream is to start a better life in the city. There are no meaningful jobs here. Thus, myself along with many other youths, want to move out of this community to get meaningful employment. I do not see myself in this impoverished community in the next two years. I am working very hard to have some savings to start a small business that will be operated by my wife in the town – preferably a small tuckshop. Currently, I am working as a fishmonger but I am hoping to get something better. I am quite confident that I will make it out of this community. If I get a good job after migrating out of this community, I will be able to take care of my parents (Interview with Jairos Mavuto at Lydiate Farm, 03 July 2018).

Thus, most young migrants in peri-urban squatter settlements have a desire for change. This desire is usually driven not only by their current rough living conditions but also their perception of prospects for a better future including personal security, improved living conditions, and professional development. In the next section, we discuss the migrants' targeted destinations.

The radical youths who want to move outside of the community target different places. Majority of them see their future in nearby towns like Norton, Chegutu, and the city Harare. During the study, the researchers came across a number of cases where young migrants had plans to move to nearby towns. In one case, a 35-year-old Thomas Kanyika had already secured a stand in Norton and was currently mobilizing for materials to build his house where he could move together with his family (Interview with Thomas Kanyika at Lydiate Farm, 02 April 2019). The second case involves a 29-year-old Mike Zikomo who indicated that his life in the compound was temporal as he was mobilizing money to cushion him as he moves to Norton town for a better life (Interview with Mike Zikomo at Lydiate Farm, 03 April 2019).

But, other young Lydiatians see their future outside Zimbabwe. Majority of these move to South Africa. A number of youths in the community have migrated out of Zimbabwe to South Africa for some years. Moving to faraway places like South Africa usually happens in stages with sending households sponsoring one member, who will later facilitate migration of the remaining siblings. A good case in mind is of Mrs. Chikumbutso whose three children have successfully moved outside Lydiate, migrating to South Africa. Mrs. Chikumbutso narrated:

> I have three children who now stay in South Africa – one girl and two boys. They did not go to South Africa, all at once; Ivy was the first one to leave in 2016. She was 26 then. I helped her to raise money for transport. She secured a job in South Africa working at a hotel. In 2017, she sent money for her brother Tina (24) who has just completed his Ordinary levels to also follow her for better opportunities in the diaspora. In the same year, Ivy also sends some money for More (22) to move to South Africa as well. Both Tina and More have also managed to secure jobs in South Africa and they occasionally send us money and groceries. They also visit on Christmas holidays (Interview with Mrs. Chikumbutso at Lydiate Farm, 05 August 2018).

Mobility out of Lydiate is not only characterized by chain migration as in the case of Mrs. Chikumbutso's children presented above. Mobility by young people is also characterized by staged migration with people moving to nearby places until they reach their preferred destination. Mrs. Kilembe narrated about his 32-year-old son:

My son Dickson is now based in South Africa for 2 years now. He works as a mechanic. He has always wanted to move outside of this community. From his teenage days, he was working towards moving out of Lydiate. He started doing menial jobs like thatching, brick molding and fish mongering. He later on secured a job as a care taker of a nearby agro-residential plot. After 2 years, he moved to work on another plot near Norton town. He later on secured a job as a shop keeper in Norton town, until he eventually moved out of this country to South Africa in 2017 (Interview with Mrs. Kilembe at Lydiate Farm, 06 August 2018).

Thus, destinations of radical youths vary. They target nearby towns to as far as metro-cities of neighboring countries. The journey to the final preferred destination may be very long, meandering, and difficult, but the radical youths often fight to endure it. This is because to them cities are the hubs of production that can enable them to realize their potential. However, most governments have failed to collaborate to realize the full potential of migrants in cities, as well as address the risks and ensure that their cities are a welcome destination for migrants of international descent – and one that ensures their full economic participation.

The outcomes of migration agency and aspirations differ. A wish to migrate could be converted into actual migration, depending on opportunities and resources. But it could also result in an unsuccessful migration attempt in the form of death, being trapped en route, or having to return against one's will (Carling 2017). As they seek new opportunities and dreams, young migrants from Lydiate often face various challenges. To begin with, for majority of the youths, Lydiate is the only place they have known. They do not possess a mental map of the realities of the places they aspire to make new homes outside Lydiate. Young Lydiatians have also stayed for years in precariousness, and they have remained socially and economically marginalized. Further, migration requires one to have savings that will sustain them in the new destination. This is quite not the case with young Lydiatians, who hardly have the capacity to raise such money. Additionally, mobility to other places becomes easier when migrants have established social networks. The networks comprise of the people that will smoothen the process of mobility through providing financial support, information, and temporal settlement in the destination area. Majority of Lydiatians do not have such a privilege.

Thus, it is not surprising that, while the youths make attempts to unyoke themselves out of Lydiate, they often found themselves coming back in the community, a place for their refuge *(gutareutiziro)*. One of the old residents, a 69-year-old Mathias Kamowa, narrated:

> Some of our children try to move out of this community but they will comeback. They have the desire to be part city dwellers *(vagary vemutaundi)*, however their poor education, poor social networks and general backwardness limit their success to become part of the city. Those who succeed usually have Ordinary level education, others have established networks, while others are just lucky. However, there is no harm in trying and failing, in the future, they will make it. Lydiate compound always has open arms for them in the event of them coming back. It is a place of refuge, when life gets tougher outside the community (Interview with Mathias Kamowa at Lydiate Farm, 02 August 2018).

In another case, a 28-year-old Hope Bante who had recently come back in the community after facing tough times in Norton town narrated:

> Life in the town was not as easy as I had expected. I left this community two years ago to seek for a new life in Norton. I had a friend who was already staying there. He wanted someone to help him operate his flea market. Initially, we were doing fairly well, however, in no time, the business crushed. I then started selling airtime and small accessories for survival, but the proceeds were not enough to sustain me in the town. Last month, I decided to come back home and I has happily welcome here. When the economy gets better, I will move out of this community again. At least I now have better knowledge of town life. I will be much wiser (Interview with Hope Bante at Lydiate Farm, 15 December 2018).

So clearly, in their attempt to unhook themselves from settlements of despair, young migrants often face a number of challenges, some so harsh enough to make them return back from their newly found destinations. Regardless of the number of failed attempts, majority of them still maintain a radical view and still see their future elsewhere. Their return to Lydiate is not because they want to settle there or regard it as home. Rather it is to cushion them as they make further plans for a future elsewhere.

The third broad category of migrants is those who have a dual sense of the place with regard to their destination settlement. Often middle-aged, these migrants want to benefit from opportunities that are presented by both the settlement and the nearby urban areas. From the data, it is evident that these middle-aged men and women are people engaged in petty commodity trading or people who cannot afford to pay a monthly rental in the nearby town of Norton. For example, Adrian Mutendere (43) and his wife Sylvia (38) survive on buying and selling in- and outside Lydiate compound. To them, Lydiate is a ready market owing to its growing and concentrated population. During the chapter, the researchers observed a number of selling points on different households. Some sell vegetables, while others are into selling basic food items. When the researchers approached the abovementioned Adrian and Sylvia's vegetable stall, the wife indicated that her husband had gone to Norton town to purchase more items for sale in the tuck shop (Interview with Sylvia Mutendere at Lydiate Farm, 05 April 2019). Just like the above couple Adrian and Sylvia, who survive on buying and selling, oscillating between Lydiate and the surrounding places, a 44-year-old Jonas Mbewe also oscillate between Lydiate and Mutare town, where he orders fruits for sell in the busy squatter settlement and in the nearby Norton town (Interview with Jonas Mbewe at Lydiate Farm, 29 March 2019).

In contrast to the above middle-aged traders who see the opportunity of a ready market, other middle-aged Lydiatians regard Lydiate as a strategic place of accommodation as they pursue work in the towns. A 37-year-old Irad Kumbukani who had no house in Norton nor can afford to rent one is a clear example from the remark below:

> I work in the nearby town of Norton as a security guard. It is much better to work in the town as you are guaranteed of a permanent job and better salary. It is now my fifth year working for the security company. However, housing in town is not cheap. I cannot afford to rent a house in town; therefore, I travel to work every day from Lydiate. Although located in the transitional zone of the town, Lydiate is just like other suburbs in Norton in terms of travel distance to work. I usually go to work by my bicycle and in tough times, I can even walk to and from work. I have built these two roomed structures and a kitchen after inheriting a piece of land from my deceased parents. I am planning to make the structures well plastered, put new floors and new iron sheet for roof to make it more appealing. Otherwise it will be unwise of me to rent in Norton. It is very expensive and some of the landlords even demand payment of rentals in US dollars (Interview with Irad Kumbukani at Lydiate Farm, 30 March 2019).

Thus, while some migrants respond to a desire for change by seeking a future elsewhere, other migrants prefer pursuing both local and city opportunities – dual sense of place. These migrants are often entrepreneurs and those who regard Lydiate as a strategic place of accommodation as they pursue work in the towns. What can be established then is that these middle-aged migrants are strategic and rational. They step their feet on both ends: staying yoked to the settlement while at the same time monkey branching for opportunities in the nearby towns.

Conclusion and Future Direction

In conclusion, migrants in peri-urban squatter settlements have diverse perceptions and aspirations regarding their future. This chapter identified tripartite perceptions among migrants. Some, mostly elderly and long-term migrants perceive squatter settlements as a home of sorts, while the infirm and invalids and the poverty-stricken regard settlements as a home of last resort. Recent and younger migrants on the other hand have a radical position against peri-urban squatter settlements – regarding them as a launchpad from which to transit to better places particularly the metropolis. Yet other migrants have mixed perceptions regarding the place, on one hand yoked to place while also eying for opportunities outside the settlement. What does this mean for Africa's emerging peri-urban? It means that migrant peri-urban squatter settlements are dynamic formations, capable of changing depending on the migrants' agency, perceptions, and aspirations as mediated by the prevailing nature of migration infrastructures.

Policy planning in managing migration in the peri-urban therefore needs to acknowledge different categories of migrants and their diverse aspirations and perceptions in space and time. This acknowledgment has a critical impact in successful integration of migrants settled in peri-urban spaces in line with Sustainable Development Goals including a call for "orderly, safe, and responsible migration and mobility of people" (target 10.7) as a pathway to reducing global inequalities. For young migrants who have a strong agency and radical aspirations to migrate to metropolises, host nations should come up with policies that ensure that the young migrants play an active role in the economy. This can be achieved through applying an equal opportunity policy that provides young migrants with access to education, training, and employment. Governments should also create migration infrastructures that facilitate mobility of young migrants from the peri-urban to the city where they perceive their lives can be transformed. For the old, infirm, and poor migrants who regard peri-urban squatter settlements as final home, host governments should come up with mechanisms that ensure that these migrants also enjoy quality of life. This can be achieved through giving them the same rights enjoyed by local citizens including granting them welfare benefits such as education facilities, housing, and healthcare services. Governments also need to collaborate with both rural and urban local authorities and other stakeholders, including the private and nongovernmental sectors, in delivering services in terms of housing, water, and sanitation for migrants who now regard peri-urban squatter settlements as their permanent home.

References

Bhanye, J., & Dzingirai, V. (2020). Plural strategies of accessing land among peri-urban squatters. *African and Black Diaspora: An International Journal, 13*(1), 98–113.

Bulger, P. (2008). Disposable people at the Peri-urban fringe. *Taming the disorderly city: The spatial landscape of Johannesburg after apartheid, 90.*

Carling, J. (2017). How does migration Arise? *Ideas to Inform International Cooperation on Safe, Orderly and Regular Migration*, 19–26.

Carling, J., & Collins, F. (2018). Aspiration, desire and drivers of migration. *Journal of Ethnic and Migration Studies, 44*(6), 909–926.

Castles, S., De Haas, H., & Miller, MJ. (2014). The age of migration: International population movements in the modern world. Houndmills/Basingstoke/Hampshire: Palgrave Macmillan Higher Education.

Flahaux, M. L., & De Haas, H. (2016). African migration: Trends, patterns, drivers. *Comparative Migration Studies, 4*(1), 1.
Forje, J. W. (2019). *In the heat of Africa's underdevelopment: Africa at the crossroads-time to deliver.* Langaa RPCIG.
Huchzermeyer, M., & Karam, A. (Eds.). (2006). *Informal Settlements: A perpetual challenge?* Juta and Company Ltd.
Khan, M. (2018). Conceptualising Migrant Agency: The Infrastructural Turn in Migration Studies. Unpacking Migration Blog. https://unpacking-migration.eu/countries_of_origin/. Accessed 30 June, 2020.
Mushonga, R. H., & Dzingirai, V. (2020). Marriage of convenience as a strategy of integration and accumulation among Nigerian migrant entrepreneurs in Harare, Zimbabwe. *African Identities*, 1–17. https://doi.org/10.1080/14725843.2020.1796588
Nyamnjoh, F. B. (2013). Fiction and reality of mobility in Africa. *Citizenship Studies, 17*(6–7), 653–680.
Nyamwanza, O., & Dzingirai, V. (2020). Big-men, allies, and saviours: Mechanisms for surviving rough neighbourhoods in Pretoria's plastic view informal settlement. *African and Black Diaspora: An International Journal*, 1–13.
Satterthwaite, D. (2016). Missing the Millennium Development Goal targets for water and sanitation in urban areas. *Environment and Urbanization, 28* (1), 99–118.
Saunders, D. (2011). *Arrival city: How the largest migration in history is reshaping our world.* Vintage.
Schon, D. A. (2019). Framing and reframing the problems of cities. In *Making cities work: The dynamics of urban innovation* (pp. 31–65). Routledge.
Segatti, A., & Landau, L. (Eds.). (2011). *Contemporary migration to South Africa: A regional development issue.* The World Bank.
Van Hear, N., Bakewell, O., & Long, K. (2018). Push-pull plus: Reconsidering the drivers of migration. *Journal of Ethnic and Migration Studies, 44*(6), 927–944.
Xiang, B., & Lindquist, J. (2014). Migration infrastructure. *International Migration Review, 48*(1_suppl), 122–148.

Transport Industry

► Sustainability Transition and Climate Change Adaption of Logistics

Transport Modes

► Pre-schoolers and Sustainable Urban Transport

Transport Resilience in Urban Regions

Yuerong Zhang and Stephen Marshall
Bartlett School of Planning, University College London, London, UK

Synonyms

Connectivity; Rapidity; Readiness; Recovery; Redundancy; Responsiveness; Robustness

Introduction

Resilience, as a property of a system, has recently gained significant interest among both practitioners and researchers from various disciplines because modern society is increasingly dependent on the stability of the transport system, particularly within systems that are increasingly complex and interdependent. This is seen both in the case of short-term disruptions like accidents, breakdowns, or terrorist attacks and the disruptions caused by climate change and extreme weather-induced events, which pose new challenges to transportation resilience. Therefore, it is important to study transport network resilience and its response to planned or unplanned disruptions, which could help engineers design a more robust system, planners to produce interventions, and scientists to understand complex network phenomena. This chapter presents a brief review of transport resilience studies, compares their approaches applied in measuring resilience, and provides several future research directions.

Transport Resilience in Urban Regions, Table 1 Resilience concepts. (Abstracted from Folke 2006; Carpenter et al. 2001; Gunderson and Holling 2001)

Resilience	Characteristics	Focus on	Diagram
Ecological resilience (Holling 1973)	Buffer capacity, withstand shock, maintain function, multiple locally stable equilibria	Existence of function	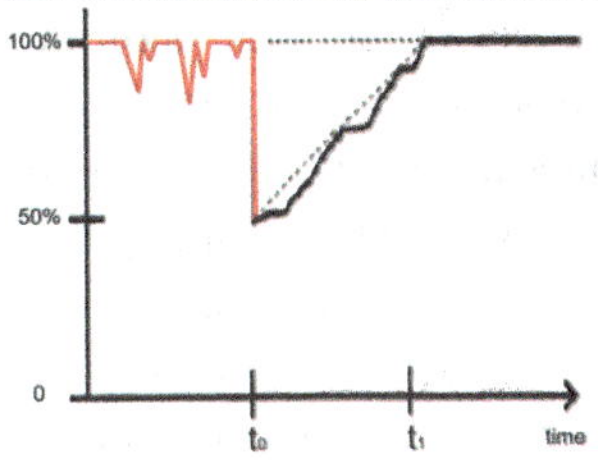
Engineering resilience (Pimm 1991)	Return time, efficiency, single equilibrium, constancy	Efficiency of function	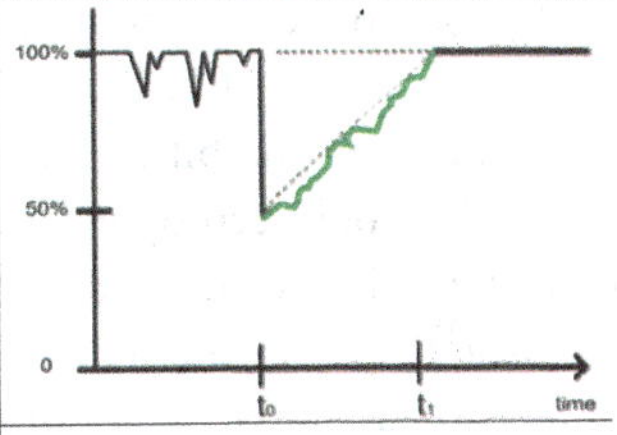
Socio-ecological resilience (Carpenter et al. 2001; **Gunderson and Holling** 2001)	Interplay disturbance, reorganization, sustaining and developing, multiple equilibria	Adaptive capacity, transformability, learning, innovation	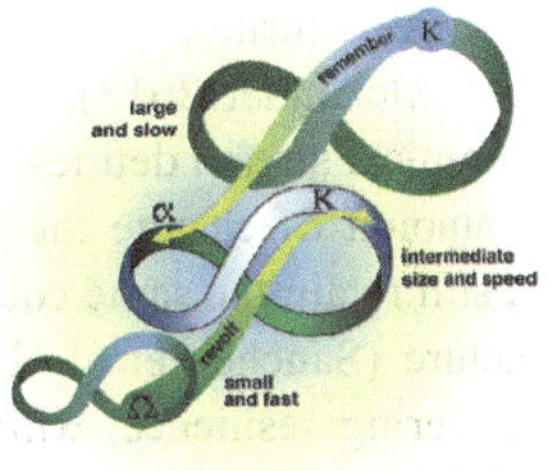

Origins and Concepts of Resilience

Resilience, stemming from the Latin *resilire*, indicates the ability to rebound, to spring back, or to resist (Rose 2007). Since Holling (1973) published his seminal work on resilience, a proliferation of studies has followed to develop a diverse understanding of resilience with different focuses. A commonly accepted taxonomy of resilience concepts is proposed by Folke (2006), who divided resilience concepts into two categories (see Table 1): (1) engineering resilience and (2) ecological/ecosystem. Engineering resilience is defined as the length of time that a system takes to return to equilibrium following a disturbance, focusing on maintaining the efficiency of function (Pimm 1991; Davoudi et al. 2012). In contrast to looking at the recovery speed ability, ecological resilience puts more emphasis on the ability of a system to withstand or cope with disturbance, either internal or external. The most classical definition of ecological resilience is from Holling (1973), who argued that resilience is a measure of the persistence of systems and of their ability to absorb change and disturbance and still maintain the same relationships between populations or state variables. As highlighted in red line in Table 1, ecological resilience concerns the ability to absorb shocks and retain performance; engineering resilience (highlighted in green line) emphases the capability for renewal, reorganization, and redevelopment (Folke 2006). As these two resilience interpretations focus on the system's ability at two periods, before and after disturbances, Bešinović (2020) named ecological and engineering resilience as proactive resilience and reactive resilience and commented that both resilience aspects are equally critical in interpreting resilience.

However, resilience is not only about persisting in the face of disturbance. It is also about the opportunities that disturbance opens

T

up. The third category, socio-ecological resilience (SER), is a transition from the ecological interpretations to one that at least includes the human or cultural ecology (Alexander 2013). Socio-ecological resilience raises the weighty question of whether a disruption to the system is always deemed a negative event. The disturbances can be regarded as a recombination of evolved structures and processes, renewal of the system, and emergence of new trajectories (Folke 2006; Gunderson and Holling 2001; Carpenter et al. 2001; Pendall et al. 2010). Perhaps the most ambitious conceptual structure under the socio-ecological resilience stream is Panarchy (Gunderson and Holling 2001), where the natural system and human systems are linked in nonstop adaptive cycles of growth, accumulation, restructuring, and renewal (Cutter et al. 2008; Alexander 2013). Therefore, the social-economical stream defines resilience in terms of the amount of change the system can undergo and still retain the same control on function and structure (Sanchez et al. 2018). In contrast to engineering resilience, which looks at a single balance, socio-ecological resilience embeds multiple equilibria across human and city systems. Meanwhile, it treats imbalance as a window of opportunity for innovation, while ecological resilience regards imbalance as a negative event or malfunction of the system.

The various schools' (ecological, engineering, and socio-ecological) resilience concepts seem to be conflicted or different and address different attributes. However, in effect, they are helping us map and interrogate a bigger picture of resilience at different time scales. That is, time scales have fundamentally shaped how resilience is defined, interpreted, and measured. Previous studies have already identified that time scales can be confounding factors in understanding resilience. For example, Meerow and his colleagues (2016) proposed the question of "resilience for when."

Transport Resilience in Urban Regions, Table 2 The multidimensional attributes of resilience

Study	Multidimensional attributes	Field
Bruneau et al. (2003)	1) Robustness 2) Rapidity 3) Resourcefulness 4) Redundancy	Seismic resilience of a community
Murray-Tuite and Mahmassani (2004)	1) Adaptability 2) Mobility 3) Safety 4) Recovery	Transport network
Cox et al. (2011)	1) Robustness (vulnerability) 2) Redundancy 3) Flexibility (responsiveness)	Transport security
Henry and Ramirez-Marquez (2012)	1) Reliability 2) Vulnerability 3) Recoverability	Transport network
Mattsson and Jenelius (2015)	1) Reduction (risk) 2) Readiness 3) Response 4) Recovery	Transport network
Bešinović (2020)	1) Robustness 2) Response 3) Survivability 4) Recovery 5) Mitigation 6) Preparedness	Railway network

Resilience as a Collection of Multi-attribute Properties

Although resilience can be general or specific, and can contain multiple levels of meanings, such as intervention policy (Sanchez et al. 2018; Linkov et al. 2013), the way of thinking (Meerow and Newell 2016; Cox et al. 2011; Linkov et al. 2014), metaphor (Carpenter et al. 2001; Pendall et al. 2010), and measurement (Östh et al. 2015, 2018), an agreement reached by existing studies is to interpret resilience through a collection of multidimensional attributes. As shown in Table 2, for example, Bruneau et al. (2003) characterized resilience into "3Rs": the system should (1) reduce failure possibilities before internal or external disruption; (2) reduce the consequences from failures, such as lives lost, damage, and negative economic and social consequences; and (3) reduce time recovery in order to minimize the loss. Accordingly, four R dimensions are proposed: robustness, redundancy, resourcefulness, and rapidity. Through investigating Londoners' travel patterns after the 2005 London tube and bus bombings, Cox and his colleagues (2011) highlighted the importance of vulnerability, flexibility, and availability to ensure resilience. Likewise, some studies (e.g., Henry and Ramirez-Marquez 2012; Mattsson and Jenelius 2015; Murray-Tuite and Mahmassani 2004) have proposed to interpret resilience in terms of a cluster of dimensions, such as reliability, vulnerability, recovery, response, readiness, and safety.

The finding that resilience is a multi-attribute property is in accordance with some prior studies. For example, the detailed review of transport resilience developed by Wan et al. (2018) summarized 12 resilience-related characteristics, namely, vulnerability, adaptability, robustness, flexibility, reliability, recoverability, redundancy, survivability, preparedness, resourcefulness, responsiveness, and rapidity. But one point that needs to be highlighted here is that these resilience-related characteristics are highly related to the interpretation of resilience. For example, robustness, flexibility, and redundancy are more closely related to Holling's interpretation of resilience, which emphasizes the system's ability to prevent disruption, whereas responsiveness and rapidity are more important in the case of Pimm's engineering resilience. Of course, there are some characteristics that are emphasized in both cases.

Transport Resilience-Related Studies

Current studies regarding transport resilience mainly follow the idea of ecological resilience and engineering resilience; few studies examine transport resilience from the perspective of socio-ecological resilience. Based on research methods, Rodriguez-Nunez and Garcia-Palomares (2014) further classify the existing studies into two main categories, which had limited interactions (see Table 3). The first one, structural resilience, is rooted in graph theory and complex network science. A real-world transport network is represented in the form of an abstract network, with a particular focus on the structure of the network, the positions of stations, and their relationship with resilience properties. Most transport resilience studies in this line are based on network science theories, on the basis of which we propose that the studies can be further subclassified into static network analysis (N1) and dynamic network analysis (N2). With regard to the static network analysis, a wide range of indicators have been suggested, such as connectivity, reciprocal distance, or efficiency (Zhang et al. 2015; Miller-Hooks et al. 2012; Latora and Marchiori 2005), modularity (Ash and Newth 2007; Derrible and Kennedy 2010a), assortativity (Derrible and Kennedy 2010b), cyclomatic number (Derrible and Kennedy 2009), and degree distribution. In addition to the conventional static network analysis, an emerging direction from dynamic network perspective investigates transport resilience through the percolation approach (Shante and Kirkpatrick 1971). This process mimics a disruption scenario by removing some fraction of nodes (together with the edges connected to the vertices) gradually and examining its influences on the overall network (Barabási 2016; Newman 2010). Compared to the static network analysis, the dynamic network analysis allows researchers to model various ways of disruption scenarios of transport

Transport Resilience in Urban Regions, Table 3 Summary of transport network resilience studies (F refers to studies having a full scan of the whole network, L refers to studies focusing on the local disruptions)

Study		Definition of resilience	Indicator	F	L	Subject	Cost
N1	Zhang et al. (2015)	Expected fraction of demand that can be satisfied post-disaster	Connectivity, average reciprocal distance, cyclicity	x		Theoretical model	
	Derrible and Kennedy (2010b)	Ability to offer an alternative route	Network assortativity, cycloramic number	x		Transit	
	Ash and Newth (2007)	Ability to withstand cascade failures	Assortativity, modularity, clustering	x		Theoretical model	
	Wong et al. (2020)	Ability to reduce negative impacts caused by disruptions	Average path length, cluster size, algebraic connectivity	x		Airline	
N2	Berche et al. (2009)	The efficiency of a network when the largest remaining cluster contains one-half of the original nodes of the network	Efficiency	x	x	Underground	
	Williams and Musolesi (2016) and Iyer et al. (2013)	The vulnerability of network, robustness threshold	Giant component size, efficiency	x	x	Underground	
T	Jenelius and Mattsson (2012)	The vulnerability of network and criticality (or importance) of nodes	Total travel delay	x	x	Transit	x
	Jin et al. (2014)	The fraction of travel demand that can be satisfied by the degraded transit network after disruptions;	The change of travel time		x	Transit	x
	Cats and Jenelius (2015)	The network's capability to resist disruption	Excess transfer and travel time		x	Underground	x
	Rodriguez-Nunez and Garcia-Palomares (2014)	The network's capability to resist disruption, which depends on (1) the availability of alternative route and (2) average excess travel time	Average excess travel time, number of trips that could not be completed	x	x	Underground	x
	Martín et al. (2021)	The capability to maintain the regional accessibility	The cumulative reduction of territorial/regional accessibility			Road	x
	Snelder et al. (2012)	Prevention, redundancy, robustness, compartmentalization, flexibility	Excess travel time, travel speed, redundancy, flexibility, et al.	x	x	Road	x
	D'Lima and Medda (2015)	The speed with which the passenger counts return to normal	Speed (mean reversion rate in stochastic model)		x	Underground	

network using measurements like the change of efficiency and giant component size. Some studies proposed to use the critical tolerance depending on the change of giant component size as the proxy of resilience. For instance, Berche et al. (2009) compared the resilience of 14 international city transit networks by calculating their percolation critical values where the largest remaining cluster (component size) contains one-half (i.e., 50%) of the original nodes of the network: that is, the fault-tolerance ability in network science.

The second vein is system-based transport resilience (also known as functional resilience, labelled as T in Table 3), which puts focus on estimating and predicting the impacts of disruption of several stations or routes based on demand and supply. For instance, there are two possible consequences after the closure of stations or lines: one is that travel demand cannot be satisfied where there are no alternative travel methods and the other is the decrease of transport serviceability, such as increasing the travel cost (time, money, or generalized travel cost). The commonly used indicators are additional travel time (Cats and Jenelius 2014; Rodriguez-Nunez and Garcia-Palomares 2014; Snelder et al. 2012) caused by disruptions. In addition to using the negative consequences caused by disruption to indirectly reflect the resilience, D'Lima and Medda (2015) developed a new measure depending on the speed with which a system returns to normal. This approach is an exemplar of indicating resilience based on the theory of engineering resilience (Pimm 1991). More details can be found in a detailed review by Reggiani and her colleagues (2015). Overall, most studies examined the importance of each transport component to maintaining the resilience, and little attention is paid to developing a complete evaluation of the whole transport network with few exceptions, such as fault tolerance (Berche et al. 2009; Wang et al. 2018).

Conclusion

This chapter briefly reviewed the current literature on the definition and understanding of resilience studies from multiple fields, including transport engineering, transport planning, network science, and urban planning. Throughout this chapter, particular focus has been placed upon how transport resilience is defined and measured and what gaps exist in current research.

With regard to the existing studies on resilience in general, there are three important findings. First, while considerable definitions have been given, the core content of resilience still revolves around its original definitions, either from Holling's (1973) ecological resilience, Pimm's (1991) engineering resilience, or Carpenter's (2001) socio-ecological resilience. The plausible distinct interpretations of resilience are under different time scales, which help us interrogate a synthetic picture of resilience: socio-ecological resilience emphasizes that disruption can sometimes be regarded as an opportunity for system upgrade for the long run, while ecological and engineering resilience focuses on the ability to prevent disruption before the event and the ability to reduce the impacts of disruption after the event. The second finding is that in line with previous studies (Wan et al. 2018), resilience embodies multi-attribute properties, and these properties also reflect time-varying characteristics: for example, redundancy is more critical before the event occurs, while responsiveness is more critical after the event. Third, it can be found that although there is proliferation of theoretical discussion on transport resilience, a comprehensive practical approach to reflect the multiple attributes of transport resilience is still lacking. Most existing transport studies remain at identifying critical transport infrastructure on a short-term basis, although it is significantly associated with resilience measurement and enhancement (Reggiani 2013). In future studies, it would be worthwhile to develop a comprehensive theoretical framework to interpret transport resilience at different time scales and their corresponding analytical schemes reflecting time-scale-sensitive attributes. This framework will also help open up avenues to incorporate ecological, engineer, and socio-ecological resilience into one unified scope, thus forming a bigger picture for understanding transport resilience. Fourth, current transport resilience studies remain at examining a single transport mode, which provides limited policy implications for building a more resilient transport system. Further research is needed to take into account multiple modes of transport in the interpretation and assessment of resilience.

References

Alexander, D. E. (2013). Resilience and disaster risk reduction: An etymological journey. *Natural Hazards and Earth System Sciences, 13*(11), 2707–2716.

Ash, J., & Newth, D. (2007). Optimizing complex networks for resilience against cascading failure. *Physica A: Statistical Mechanics and Its Applications, 380*, 673–683.

Barabási, A. L. (2016). *Network science*. Cambridge, UK: Cambridge University Press.

Berche, B., Von Ferber, C., Holovatch, T., & Holovatch, Y. (2009). Resilience of public transport networks against attacks. *European Physical Journal B: Condensed Matter and Complex Systems, 71*(1), 125–137.

Bešinović, N. (2020). Resilience in railway transport systems: A literature review and research agenda. *Transport Reviews, 40*, 1–22.

Bruneau, M., Chang, S. E., Eguchi, R. T., Lee, G. C., O'Rourke, T. D., Reinhorn, A. M., . . . Von Winterfeldt, D. (2003). A framework to quantitatively assess and enhance the seismic resilience of communities. *Earthquake Spectra, 19*(4), 733–752.

Carpenter, S., Walker, B., Anderies, J. M., & Abel, N. (2001). From metaphor to measurement: Resilience of what to what? *Ecosystems, 4*(8), 765–781.

Cats, O., & Jenelius, E. (2014). Dynamic vulnerability analysis of public transport networks: Mitigation effects of real-time information. *Networks and Spatial Economics, 14*(3–4), 435–463.

Cats, O., & Jenelius, E. (2015). Planning for the unexpected: The value of reserve capacity for public transport network robustness. *Transportation Research Part A-Policy and Practice, 81*, 47–61.

Cox, A., Prager, F., & Rose, A. (2011). Transportation security and the role of resilience: A foundation for operational metrics. *Transport Policy, 18*(2), 307–317. https://doi.org/10.1016/j.tranpol.2010.09.004.

Cutter, S. L., Barnes, L., Berry, M., Burton, C., Evans, E., Tate, E., & Webb, J. (2008). A place-based model for understanding community resilience to natural disasters. *Global Environmental Change, 18*(4), 598–606.

D'Lima, M., & Medda, F. (2015). A new measure of resilience: An application to the London Underground. *Transportation Research Part A: Policy and Practice, 81*, 35–46.

Davoudi, S., Shaw, K., Haider, L. J., Quinlan, A. E., Peterson, G. D., Wilkinson, C., . . . Porter, L. (2012). Resilience: A bridging concept or a dead end? "Reframing" resilience: Challenges for planning theory and practice interacting traps: Resilience assessment of a pasture management system in Northern Afghanistan urban resilience: What does it mean in planning practice? Resilience as a useful concept for climate change adaptation? The politics of resilience for planning: A cautionary note: Edited by Simin Davoudi and Libby Porter. *Planning Theory and Practice, 13*(2), 299–333.

Derrible, S., & Kennedy, C. (2009). Network analysis of world subway systems using updated graph theory. *Transportation Research Record: Journal of The Transportation Research Board*, 17–25.

Derrible, S., & Kennedy, C. (2010a). Characterizing metro networks: State, form, and structure. *Transportation, 37*(2), 275–297.

Derrible, S., & Kennedy, C. (2010b). The complexity and robustness of metro networks. *Physica A: Statistical Mechanics and Its Applications, 389*(17), 3678–3691.

Folke, C. (2006). Resilience: The emergence of a perspective for social–ecological systems analyses. *Global Environmental Change, 16*(3), 253–267.

Gunderson, L. H., & Holling, C. S. (2001). *Panarchy: Understanding transformations in human and natural systems*. Washington, DC: Island Press.

Henry, D., & Ramirez-Marquez, J. E. (2012). Generic metrics and quantitative approaches for system resilience as a function of time. *Reliability Engineering and System Safety, 99*, 114–122.

Holling, C. S. (1973). Resilience and stability of ecological systems. *Annual Review of Ecology and Systematics, 4*(1), 1–23.

Iyer, S., Killingback, T., Sundaram, B., & Wang, Z. (2013). Attack robustness and centrality of complex networks. *Plos One, 8*, E59613.

Jenelius, E., & Mattsson, L.-G. (2012). Road network vulnerability analysis of area-covering disruptions: A grid-based approach with case study. *Transportation Research Part A: Policy and Practice, 46*(5), 746–760.

Jin, J. G., Tang, L. C., Sun, L., & Lee, D.-H. (2014). Enhancing metro network resilience via localized integration with bus services. *Transportation Research Part E: Logistics and Transportation Review, 63*, 17–30.

Latora, V., & Marchiori, M. (2005). Vulnerability and protection of infrastructure networks. *Physical Review E, 71*(1), 015103.

Linkov, I., Bridges, T., Creutzig, F., Decker, J., Fox-Lent, C., Kröger, W., Lambert, J. H., Levermann, A., Montreuil, B., & NATHWANI, J. (2014). Changing the resilience paradigm. *Nature Climate Change, 4*, 407.

Linkov, I., Eisenberg, D. A., Bates, M. E., Chang, D., Convertino, M., Allen, J. H., . . . Seager, T. P. (2013). Measurable resilience for actionable policy. *Environmental Science & Technology, 47*(18), 10108–10110.

Martín, B., Ortega, E., Cuevas-Wizner, R., Ledda, A., & De Montis, A. (2021). Assessing road network resilience: An accessibility comparative analysis. *Transportation Research Part D: Transport and Environment, 95*, 102851.

Mattsson, L.-G., & Jenelius, E. (2015). Vulnerability and resilience of transport systems – A discussion of recent research. *Transportation Research Part A: Policy and Practice, 81*, 16–34.

Meerow, S., & Newell, J. P. (2016). Urban resilience for whom, what, when, where, and why? *Urban Geography, 40*, 1–21.

Miller-Hooks, E., Zhang, X. D., & Faturechi, R. (2012). Measuring and maximizing resilience of freight transportation networks. *Computers and Operations Research, 39*(7), 1633–1643. https://doi.org/10.1016/j.cor.2011.09.017.

Murray-Tuite, P., & Mahmassani, H. (2004). Methodology for determining vulnerable links in a transportation network. *Transportation Research Record, 1882*, 88–96.

Newman, M. (2010). *Networks: An introduction* (pp. 1–2). New York: Oxford University Press.

Östh, J., Reggiani, A., & Galiazzo, G. (2015). Spatial economic resilience and accessibility: A joint perspective. *Computers, Environment and Urban Systems, 49*, 148–159.

Östh, J., Reggiani, A., & Nijkamp, P. (2018). Resilience and accessibility of Swedish and Dutch municipalities. *Transportation, 45*(4), 1051–1073.

Pendall, R., Foster, K. A., & Cowell, M. (2010). Resilience and regions: Building understanding of the metaphor. *Cambridge Journal of Regions, Economy and Society, 3*(1), 71–84.

Pimm, S. L. (1991). *The balance of nature?: Ecological issues in the conservation of species and communities*. Chicago: University of Chicago Press.

Reggiani, A. (2013). Network resilience for transport security: Some methodological considerations. *Transport Policy, 28*, 63–68.

Reggiani, A., Nijkamp, P., & Lanzi, D. (2015). Transport resilience and vulnerability: The role of connectivity. *Transportation Research Part A: Policy and Practice, 81*, 4–15.

Rodriguez-Nunez, E., & Garcia-Palomares, J. C. (2014). Measuring the vulnerability of public transport networks. *Journal of Transport Geography, 35*, 50–63. https://doi.org/10.1016/j.jtrangeo.2014.01.008.

Rose, A. (2007). Economic resilience to natural and man-made disasters: Multidisciplinary origins and contextual dimensions. *Environmental Hazards, 7*(4), 383–398.

Sanchez, A., Heijden, J., & Osmond, P. (2018). The city politics of an urban age: Urban resilience conceptualisations and policies. *Palgrave Communications, 4*(1), 25.

Shante, V. K., & Kirkpatrick, S. (1971). An introduction to percolation theory. *Advances in Physics, 20*(85), 325–357.

Snelder, M., Van Zuylen, H., & Immers, L. (2012). A framework for robustness analysis of road networks for short term variations in supply. *Transportation Research Part A: Policy and Practice, 46*(5), 828–842.

Wan, C., Yang, Z., Zhang, D., Yan, X., & Fan, S. (2018). Resilience in transportation systems: A systematic review and future directions. *Transport Reviews, 38*(4), 479–498.

Wang, Y., Szeto, W. Y., Han, K., & Friesz, T. L. (2018). Dynamic traffic assignment: A review of the methodological advances for environmentally sustainable road transportation applications. *Transportation Research Part B: Methodological, 111*, 370–394.

Williams, M. J., & Musolesi, M. (2016). Spatio-temporal networks: Reachability, centrality and robustness. *Royal Society Open Science, 3*(6), 160196.

Wong, A., Tan, S., Chandramouleeswaran, K. R., & Tran, H. T. (2020). Data-driven analysis of resilience in airline networks. *Transportation Research Part E: Logistics and Transportation Review, 143*, 102068.

Zhang, X., Miller-Hooks, E., & Denny, K. (2015). Assessing the role of network topology in transportation network resilience. *Journal of Transport Geography, 46*, 35–45.

Transportation – Circulation

▶ Transportation and Mobility

Transportation and Land Use Integration: Shaping Transportation Demand and Delivering Transport Supply

Transportation Infrastructure to Reduce Carbon Intense Lifestyles, Improve Healthy Activity, and Increase Access to Jobs, Education and Services

Cole Hendrigan
University of Wollongong and Wollongong City Council, Wollongong, NSW, Australia

Introduction

In building cities, we must deliver services. Complimentary services make the spaces of apartment towers and houses into places for human lives and ambitions. Services make the hard concrete and stone of the buildings into a human habitat where civic life can flourish (Bartholomew and Ewing 2008). Whether built by government direct action, as in Singapore (Housing and Development Board), or by private actors such as large-scale residential tower developers, or the homeowner as builder (as seen in many informal settlements worldwide), the needs of all humans for sanitation as written in Sustainable Development Goal (SDG) #6, for education (SDG 4), jobs and transport (SDG 8,10) are the same: we all want them and need them to make the city work (SDG 11) (UN 2021). At the most basic without services people cannot thrive, fulfilling their professional lives and raising healthy families. Lastly, people will flow to quality services elsewhere, devoting their working lives there (Glaeser 2011).

Key among these services to be delivered is a coherent network of efficient transportation. In pervious iterations of human habitat, before the coming of the commuter rail services from cities and most certainly before the private motor car, walking was the primary mode of transport. Horses remained an expensive privilege without a self-sufficient grassed area for both feeding and a shelter for lodging.

The walking services as we know them today, being safe intersections and wide enough sidewalks, were mostly absent in the not-so-distant past. The streets were then what we today call "shared spaces". Then, all movement of all types mixed across the breadth of the street. If we look back to remnant Roman settlements such as Pompeii, at best we find street crossing stepping stones used to keep feet out of the wet street for "pedestrian" infrastructure. In later places such as colonial Paraty, Brazil there was no sidewalks, or in Old Quebec City the sidewalks appear to be later day additions, and even in modern Tokyo there are many streets with no sidewalks and only a painted striping to indicate where a pedestrian should walk when there is potential conflict with another mode. Walking service depends on the level of comfort and safety required given the transport mix of speed and volume context.

Distances in these walking cities were relatively short with a half hour being about the extent one walked to a farm or mill or processing site. Marchetti's Constant finds a half hour travel in each direction as a historical optimum distance (Marchetti 1994), with the corollary being we should strive for this time through delivered transport supply. Distances over an hour walking time would have led to a relocation of residence, at least temporarily or seasonally, thereby creating a nucleus around which a new settlement sprung. In such a concept, readily we can see the emergence of the walking region being a composition of small walkable "hamlets" linked on paths through fields to "towns" and clusters of towns feeding larger villages all based on a series of walkable day long travels to work, market or court and back. Such is the very real experience of the settling of the Canadian Prairie provinces with the larger "villages" being centers on the railway mainline, or as we can see today clearly in satellite imagery from rural India, and of course as per Christaller's "central place" theory viewable in much of Europe (King 1985).

Walking, with its limitation of distance, has long been the foundation of most of urban life. With many options to choose from today, all of them certainly more time efficient but also resource intensive than walking. What are they, what can they offer, or Supply, to meet the needs of our daily travel Demand will be described below.

Supply

Supply of transport services must be broken down by mode and the physical elements or design actions that support each mode. Below will be a list of supply, incomplete and expanding. To begin with the smallest and most urban (as they work best in space limited dense places, which are by definition "urban") in transport modes and to work onward to the largest land consumptive and least urban modes: Pedestrians to Cars.

Pedestrian Supply

Almost all travel in a city begins and ends with a walking trip. Walking somewhere, even from a short distance, in public space will be a part of a day for most people. When the supply – the physical elements in the environment – are designed and delivered appropriately in their urban habitat people *may* walk more, but when done poorly people will chose *not* to walk if they have an option.

As Pedestrian Supply is a prerequisite for all the other modes described below, aside from cars, if the Sustainable Development Goals and a progressive urban agenda is to be approached, there will be a concomitant long list. All other modes must have these considerations if the other modes are to work as well as intended, including private automobiles.

- Sidewalks/ Footpaths/ Pavements (or other names) being the primary surface upon which people will walk. They need to be clear of obstacles or trip hazards, smooth as possible with small construction joints between pavers to accommodate rolling prams or wheelchairs, with a 1 to 2% cross slope to shed water, and with sufficient widths to accommodate daily volumes at peak hours.
- Stairs are often an extension of the sidewalk, needing all the same characteristics, but with a change in vertical position. Stairs require a common riser and tread ratio of 150 mm up and 300 mm wide to be traversed with the least amount of tripping and greatest comfort. Stairs can be quite elaborate in their design, offering more than mere vertical access and become destinations in their as a place to see and be seen. Examples of such places are the Spanish Steps in Rome or at Federation Square in Melbourne. Stairs, beautiful or not, can also be a barrier to mobility impaired, necessitating ramps.
- Ramps, general, are a compliment to the stairs and the sidewalk, offering a smooth approach for those in wheelchairs, with walking canes or pushing prams. These should be designed with a "10 and 2 principle" of **10** meters at 6% slope to allow easy self-propulsion along with a **2**-meter landing at 1% to allow a person with mobility issues (of any sort) a place to pause and catch breath without concern (United Nations 2003). There are variations on the % slopes, depending on the specific distance to cover. One way to avoid expensive structures is to plan on having all surfaces rise at between 1 and 5% over longer distances.
- Dual curb (kerb) ramps are an outgrowth of single curb ramps (99% Invisible 2019), itself a major effort for equity of access. Curb ramps have evolved from their first iterations in California to a very specific design to allow plentiful room to wait, to descend at an expected slope, and to abut the asphalt of the road at an anticipated gradient. There should be tactile elements above the ramps so the wheeled mobility can "feel" the edge of the ramps and so that blind persons can also sense the change in elevation before it drops.
- Curb extensions are an extension of the sidewalk into the car-space of the street typically used for the quick turning of the car (NACTO 2012). By extending the curb four complimentary events happen: (1) the crossing distance between curbs is shortened, (2) the pedestrian is more visible to the car driver, (3) the car driver must slow down to close to a full stop to navigate the sharper turn of the vehicle thereby reacting to any pedestrian quicker, and at less speed (4) the driver's view is also forced to look to where the pedestrian is crossing on the perpendicular street. When combined with dual-curb ramps (see above), a curb plus judiciously placed bollards (see below), and zebra striping (see below), this can make for an almost effortless and safe condition for not only mobility challenged but all people no matter their ability.
- Mid-block crossings are an extension of the sidewalk, often with vertical deflection (speed table) and signals at best or mere zebra striping at least, in the middle of the urban block (NACTO 2012). These have the design significance of (1) increasing driver awareness that the street may have pedestrians on it (2) a safe place to cross for pedestrians without having to mount a bridge or descend through a tunnel (3) which incentivizes walking to destinations such as services and retail present on a high street. These are not required everywhere, but where they are used they are effective at linking two side of the street into a more cohesive "place" or destination and at slowing cars

to a pace conducive to a pedestrian environment.

- Bollards are primarily fixed structures arranged along a curb to limit the passage of a motor vehicle into the pedestrian realm, while also posing no barriers to pedestrians or wheeled prams or wheelchairs. Measuring between 1 and 1.3 meters in height and 0.1 to 0.3 meters in diameter they are often fixed into the ground via a dug foundation or with shear bolts. Some do flex for emergency vehicles and some do "knock-down" by gravity or mechanical means to permit certain vehicles through (i.e., delivery or emergency). They are typically not noticeable unless looked for in avoidance, yet they are remarkable in filtering the car from the pedestrian environments.
- Zebra Striping is controversial. Traffic Engineers view these as meddlesome false-promises of safety to pedestrians which achieve a lower Level of Service (discussed below) for the automobile drivers. Traffic engineers also often see these as items needing yearly budget and a project manager to repaint, draining resources. However, to a pedestrian looking for the slightest sense of a safe place to cross, when designed with curb extension and signage or flashing signals these can be all that gives them any hope to safely cross somewhere in the right of way.
- Street trees can be liked and disliked in equal measure. Those who dislike street trees – worth mentioning to be clear eyed about why they are not everywhere – are:
  - City staff accountants who see the cost to plant and maintain but not the benefits.
  - Civil engineers do not like trees near curbs or drainage or water utilities as the roots will penetrate any crack to seek water thereby cracking the infrastructure.
  - Traffic engineers feel they close the view cones and cause more accidents as a barrier to safe stopping to the car.
  - Architects feel they "block the view" to their buildings.
  - Home owners often find their fruit and leaves to be a chore to clean up after, and some tree species roots do emerge at the surface causing trip hazards to pedestrians and mobility impaired.
- However, almost everyone else likes them. They shade the street cooling it, they house birds which sing and raise young in their branches, they absorb rainwater in their trunk and branches via their roots while also keeping the surrounding soil porous to absorb the water through the maintenance of a vibrant biota, and they calm the senses of the urban dweller with the sounds of the leaves rustling and the dappled light.
- However, from a sustainable transportation perspective street trees cool the sidewalk and transit stop, reduce the speed of cars by decreasing visual certainty of the driver, and do indeed act as an effective buffer (like bollards) filtering the car environment from the pedestrian realm. Trees do cost a city to plant and maintain, and in this they have a value to replace. This value grows every year: indeed, money can grow on trees.
- Weather protection on buildings is catch all phrase to say awnings, arcades, and porches. While this may at first seem like an architectural device and having little to do with transportation, for pedestrians such frontages can make a hot day seem comfortably away from the sun, a wet day tolerably drier, and on all days they create a scaled-down and inviting human sized space in which to feel safe. Weather protection creates safe places connecting inside with outside of buildings, creating a semi-public/semi-private place into which greetings and messages of welcome are passed verbally or through symbols. In this such spaces are like the sinew of a great street tying active muscle with structural bone. While a pedestrian in a hurry may not have time to appreciate weather protection, it does make a difference to the level of comfort for the dozen types of trips made in the city. We miss it when it is gone.
- Wayfinding can be subtle, overt, or both. It can be appropriately sized street signs that can be read from 100 meters away (not 2), maps and signs with color coding for destinations and distances clearly marked, or it can also be a

surface application showing the appropriate use of the space and direction of travel (i.e., darker colored granite setts showing two "lanes"), and of course digital phone navigation or QR code type scanning to discover local services, retail, history, or transportation connections from a place.

- "Separation by speed and volume" is not a new concept, as it is used to separate pedestrians safely from high volume and high-speed cars. However, it is also being used to design out conflicts between pedestrians and cyclists and cyclists from cars, as well. It is a concept that will be more important in the Cycling Supply section below, but it is important to offset from Shared Space concepts.
- Shared Space is the idea, born or reborn by Hans Monerman of Netherlands (Hamilton-Baillie 2008, 2010), to remove the signage and marking and even curbs of the street. The idea was to remove the visual clutter which has served to signal priority only for cars in the right-of-way. By adding uncertainty to the streets, drivers slow down. By adding in a sense of ubiquitous opportunity to traverse the street, cyclists and pedestrians become a part of the everyday street scene, not surprising the driver, thereby increasing expectation and awareness of their rights to the street. The benefit of Shared Space is that it lends itself as an urban design strategy along high streets or adjacent to public plazas such as in front of a major transit station in specific small urban areas. The disbenefit is that it may not be an appropriate elsewhere, needing to default to Separation by speed and volume.
- Townscaping (iconic buildings and structures in the viewshed) (Cullen 1971) can be easy to overlook as a strategy to crafting walking places. Some will balk at the cost of trying to impose new iconic building just because it may create an "aesthetic," but that is not the request. Rather, townscaping is (1) to enhance existing buildings of note (or at least not knock them down) and by referencing them elsewhere (i.e., as an emblem on benches in the precinct), and (2) when a new building is proposed look to define a cornice or architrave or parapet wall or even add a cupola if possible, to create a building of recognizable (but not too garish) stature. This adds to the identity of a street or precinct, giving the place an image around which it might build a new series of memorable on-foot (not in car) experiences.
- Signaled intersections and signal equity are colleagues: they may work together but are not always friends. There are many intersections that do not have a signal phase for pedestrians, some which even restrict pedestrians by law, and the remainder that do have signals for pedestrians give very short amount of time to cross. Quite commonly 6 seconds may be all that is given to cross 6–9 meters of street, or about 1 second per meter. This is a rapid pace of walking for anyone able-bodied, but for anyone with a child in hand or any with mobility issues, this is quite a short amount of time to potentially cross the street before the red-hand caution flashes. The short reason for this is to limit the amount of time given to the pedestrian so that more time can be given to the motor car to make left or right turns at a signal phase. It is not so much to be cruel, for in the traffic engineers' mind they are making sure the pedestrian is not present during turning movements thereby saving lives. Yet, to the pedestrian it makes for a harried and stressful crossing. Into this policy realm has entered the notion of

Transportation and Land Use Integration: Shaping Transportation Demand and Delivering Transport Supply, Table a: Pedestrian capacity per day per urban 3.5 meter wide 'lane'

Mode	Pedestrian/walking/strolling/ambulating				
Capacity per 3.5 meters/hour	Daily throughput – 10 hours	Frequency	Low cost example	High cost example	City shaping capability
11,000–15,000	150,000	Ubiquitous	Any residential street	Prado, Barcelona	Very high, prerequisite for any urban area over medium density

signal equity, not equality. That is, there should be an equity in consideration of the pedestrian as a beneficial productive mode as the motor car, giving them *equitable* consideration. However, to be sure, not equal time – *equality* – (circa 45–90 seconds) as that would (a) create professional pushback and (b) lose some of the benefits (LOS, see below) of the street as a corridor for cars. Often doubling the pedestrian crossing time to 10 or 20 seconds makes little difference to a car, but an immense increase in satisfaction and comfort to the pedestrian.

- Flashing signals come in many varieties. Among the most popular are the RRFBs (regular rapid flashing beacons). Their effectiveness at alerting drivers to the crossing of a pedestrian, quick deployment, low cost to purchase and install, and ability to generate and store their own electricity needs for the low voltage light-emitting diodes (LEDs) is a combination used primarily on low to moderate volume streets with long spaces between regular intersections, such as at a school or playground.

Cycling and Micro-Mobility Supply

From a sustainability perspective, cycling is the most efficient use of kilojoules of energy by a wide measure. It makes us happier, fitter, more social, and uses almost no carbon but for bicycle production. Today most cities are attempting to design a network of separated and safe bicycle routes to lure everyday commuting residents back onto bikes. Even a slight increase in bicycle use can begin a cultural shift toward non-car modes as it becomes more visible, thereby making is acceptable. It also serves other modes when more trips are taken by bicycle on designed separated facilities as congestion *should* be better managed on all routes as a part of the overall network upgrades.

Yet, cycling, for many cities, will be a minor player in moving mass volumes of people. Many cities have, in the twentieth century, expanded beyond the scale for which most can reach destinations within their time budget by bicycle. Shanghai, China, in the 1970s had no metro, no high-speed trains to Suzhou or Nanjing, and almost no highways, existed only one side of the Huangphu River, and had *millions of cycling commuters* traveling to relatively close-by work. This city has spread far beyond those boundaries of 50 years ago. So too have any city with almost any measurable population growth. The growth in distances between origins and multiple destinations becomes overly complicated, lengthy, and potentially dangerous due to prioritization of cars.

For brevity, micro-mobility including Bicycles, e-Bike, Scooter, Skateboard, and others will be considered inside this category for this reason alone: they are as vulnerable as pedestrians on the street but with added speed leading to less reaction time, with graver consequences in a crash.

Sources for more information are found in the NACTO guides (NACTO 2012, 2014, 2016) as well as in several new key publications including the British Columbia Active Transportation Guide (BC MoT 2019), the recent Transport for New South Wales Cycle way Design Toolkit (TfNSW 2020), and the Crow Manual for Bicycle Traffic (CROW 2016).

What a cycling and micro-mobility network of corridors and nodes requires:

- A coherent network to expected destinations (jobs, schools, shops, recreation, malls) of consistent facilities including a quality surfaces, markings, wayfinding, separation by speed and volume, and special intersection consideration. Such a network is required for the cyclist to feel safe, connected, and with convenience. When this is well designed, there should be only faint decrease in urban travel times by other modes, including pedestrians, bus, or cars.
- Separation from other modes by speed and volume makes for fewer conflicts between pedestrians and motorized cars or buses. This will ensure the cyclist can maintain the attractive rolling forward momentum inherent with cycling in a safe manner between intersections. The separation can take a variety of forms from low pin-in-place curbs plus signage, to planter boxes with seasonal flowers, to poured-in-place curbs with elevated cycle track. There are several design guides to follow on the

design options, but in general these are either one directional or bi-directional with dimensions from 1.5 to 3.5 meters in width.

- Protected intersections are a further extrapolation of the dual-curb ramp and curb-extension for pedestrians from above, but with an extra lozenge or "banana" to create a safe zone for the queuing of bicycles and to create a safer curb-return radius for the slow turning of the cars. Indeed, for the driver to navigate the intersection they will perforce need to slow to look around the corner, seeing pedestrians and cyclist alike. One key feature is that the path of the cyclist must be maintained in as direct a line as possible, without deviation around the concrete curbing of the lozenge or banana. Deviations will require the cyclist to evasively navigate creating yet another conflict between curb and cyclists and cyclist on cyclists.
- Boxes are another means by which a cyclist, or especially a group of cyclists, gain a slight safety advantage – at lower road-redesign costs – over no changes to the street markings or curbing. The bike box is a green box of green paint at the placed at the end of a painted cycling lane, before the stop bar for cars and before the pedestrian painted cross walk. The cyclist can queue here during a red-light phase, but also be ahead of cars when the phase turns green, giving them an advantage of clearing the intersection before any turning or through traffic might overtake them. That is the theory, practice, and safety results are mixed especially on high volume corridors. Overall, this is not as advantageous to the cyclist as a fully protected intersection, as per above.
- Wayfinding serves to let the cyclist know directions and distances to known destinations and other unknown services nearby. Having a level of certainty only encourages more cycling activity, of benefit to the overall transportation network in reducing car congestion, increasing health and as a low cost means to potentially move large numbers of people, equivalent to a highway lane of trips per day or another all-day bus-per-hour. Wayfinding also helps pedestrians to know the same information and to make them aware that cyclists are frequenting the route. The same holds for drivers in cars, cycling wayfinding alters that cyclist are present and advertises that it is a mode made more available for consideration.
- Consistent surfaces and shared use design aspects. Consistent surfaces are important in having a surface which is not bumpy with needless sharp potholes, dips, or heaves in the surface. Likewise, when a corridor is designated as a cycle way, there should be two different types of surfaces for pedestrian zone and cycling zone. Such is the common treatment in Copenhagen where the pedestrian zone is paver setts and wider pavers and the cycle

Transportation and Land Use Integration: Shaping Transportation Demand and Delivering Transport Supply, Note 1 Note on lane capacity between modes, with an emphasis on bicycles

One bike lane of 1.5 meters width, per direction/lane, per hour = 1500 to 2000 bikes
One bus lane, 90 passengers seating and standing, 5-minute frequency, per hour = 2700
Cars per lane per hour at 50 km/h for LOS C (a very fast flow in urban setting) = 1000
Cars per lane per hour at 100 km/h for LOS A (a typical suburban or rural pace) = 2000
Proviso: the 2000 bikes per hour will be at an upper limit of both mode share and in dense urban areas and only when cycling is prioritized with separated bicycle tracks with protected intersections. Studies in dense urban setting of Chinese cities have found 2000 to be a practicable limit before excess congestion creates disincentives including crashes

Transportation and Land Use Integration: Shaping Transportation Demand and Delivering Transport Supply, Table b: Cycling and micro-mobility capacity per lane

Mode	Cycling/Bicycle/e-Bike/Scooter/Skateboard/				
Capacity per 3.5 meters/hour	Daily throughput	Frequency	Low cost example	High cost example	City shaping capability
15,000	150,000	Ubiquitous	5th st SW Calgary	Wilson Street, Sydney	Very high, offers

track is elevated with a half curb (~500 mm) and asphalt surfaced for smooth rolling of the wheeled vehicles.

- Secure and ubiquitous parking is important if cyclists are to feel respected and incentivized to safely park their – potentially expensive and valued – bicycle in full view of potential thieves. Other end of trip facilities such as bicycle maintenance stands, vending machines with typical bicycle spare parts, showers, change rooms, and water sources for refilling bottles all go toward making things easy for people if we want them to do it.
- Signal equity, like the needs for pedestrians, offers an explicit awareness of the needs for cyclists to cross at intersections. For cyclists, however, the needs are not so much about timing as they are for advance turning and to make drivers aware that cyclists are present in case they were not aware already.

Bus Supply

Buses are taken for granted, overlooked, in the funding of mass transit (public transportation). Often fast new underground metros or stylish LRTs elicit passionate debate due to costs or benefit. Yet buses – in many cities – do most work in moving people about. A few sources worth noting include Walker (2012), Higashide (2019), and Vuchic (2005, 2007), among many other esteemed authors.

The supply of buses is first and foremost about reliability (arriving when expected), legibility of routes, being fast (perhaps as fast as a private car) and direct (traveling in a straight line). To achieve such ridership service, several other factors go into providing bus supply.

- Hours of service need to be extensive, from early morning to late evening, with high frequency of at least 20 minutes but 10 or 5 minute headways (times between buses) being optimal. Together this adds up to a Service Hours figure. For example: 1 bus per hour over 12 hours = 12 hours. Whereas 6 buses per hour over 12 hours = 72 hours. With such leaps in service two things happen:
  - There are steps changes in the number of bus trips a rider can choose from.
  - The bus begins to be seen – having a higher visual presence – as a viable option in comparison to the car.
- Frequency, headway, needs to be as often as possible. The limitation on this is both the cost to purchase more buses to operate this headway as well as the considerable labor costs to drive, maintain, and clean the buses.
  - Headway is related to frequency, and is a term that is used interchangeably to mean: wait times between buses or trains.
- Bus capacity – the number if people a single bus can carry – changes as seating configurations may change, wheel-well design, rear door access, and internal steps alter the layout. Regulations regarding standing and crowding may reduce some capacity also. However, the true change in capacity comes with articulated and bi-articulated buses as discussed in the BRT section below. There must be sufficient capacity on a line in peak hour to make the service desirable, to achieve this agencies often make longer buses, run more buses or both.
- Reliability is paramount. Riders want to know that the bus will arrive when it is scheduled and that there will be room on board. When one or two buses are delayed, the backlog of riders at stops can overwhelm the arriving buses creating a bottleneck of service and yet longer dwell times – further delaying the bus – as riders embark or disembark from the bus. It is important to maintain the buses.
- Platform extensions reduce dwell time at each stop by eliminating the need for the driver to move out of the lane, into a bus bay, and then merge back into the lane of traffic. Instead, the driver keeps the bus in the lane of traffic and the pedestrian curb is extended through the space often designated for parallel parking to the traffic lane to the waiting door of the bus. This serves the riders very well, placing their safety and comfort above that of the car divers in the traffic lane. Yet, it also serves the bus to maintain a throughput, maintaining a schedule, without queue jump lanes or exclusive right of ways. In this way is makes for a more pleasant

Transportation and Land Use Integration: Shaping Transportation Demand and Delivering Transport Supply, Table c: Standard bus capacity per lane

Mode	Standard bus including electric battery or electric trolley types[a]				
Capacity per 3.5 meters/hour	Daily throughput	Frequency	Low cost example	High cost example	City shaping capability
5 minute frequency = 12 × 90 = 1080 10 minute = 540	10 minute frequency × 12 hours = 6480	Ideally every 5 minutes, per route, to avoid bus bunching but often as paltry as 20 or 30 minutes in low density residential areas	A highly efficient new system is in Banff, Canada, with a combination of high ridership per service hour and lower than average management costs	Transit systems that prove to be necessary for many but also expensive per rider km.	Serving cities where are and need to go, the land use response to be close to a bus route is limited to captive transit users

[a]Much of the capacity related information comes from: Transit Cooperative Research Program (2013) "Transit Capacity and Quality of Service Manual TCRP Report 165." Sponsored by the Federal Transit Administration

streetscape as more room is created on the sidewalk and waiting areas, and a faster more reliable bus service.

- Wayfinding is more than just the sign-post announcing the bus stop position on the curb. It can begin further back in the neighborhoods with signs, markers, icons placed on street signs showing the nearest bus stop and the walking distance. It can then also become a fully designed bus stop with a clear path to it, a unique roof line, plus information as per below. Wayfinding is more than a sign, its making the door to door journey easy and convenient.
- Information is important on the curb to know which bus comes when and where it is going as well as when it should depart. Yet, in recent years apps and "next bus" LED signage has worked together to provide this valuable information to the rider. These boards and apps can also be a place where the transit agency delivers other announcements related to their service and even broader civic announcements.
- Sidewalks to the bus stop itself must not be overlooked. There are many places, globally, where there is a bus service, with a sign-post designating a bus stop, but no apparent legal or safe way to cross the roads to get there. A coherent program of linking sidewalks and crossings to bus stops should seem a self-evident budget item on every town council's list of projects, but there are often conflicting purposes related to private motor vehicle movements (see LOS below).
- Shelters at the bus stop are ideal for waiting out of the sun, rain or snow but can need to be maintained with labor devoted to their care. Shelters also serve to advertise the bus service, giving the service a presence, but also a place to advertise third party services as a revenue stream.

Bus Rapid Transit (BRT) Supply

Bus rapid transit (BRT) supply needs are well documented and debated. What constitutes a true BRT is on a spectrum from bi-articulated buses with high-cost stations/high frequency/high capacity/separated lanes (almost a metro-type service) to conventional buses running at high frequency with limited stops (as an advanced bus service). In between these two points on the spectrum can be any variation of articulated buses, station and platform upgrades, lane exclusivity,

Transportation and Land Use Integration: Shaping Transportation Demand and Delivering Transport Supply, Table d: BRT line capacity per lane

Mode	BRT and BRT light				
Capacity per 3.5 meters/ hour	Daily throughput	Frequency	Low cost example	High cost example	City shaping capability
150 × 12 per hour = 1800	1800 × 12 hours = 21,600	Ideally every 5 minutes, but often 10 minutes	Curitiba and Bogota both have pioneering and low-cost BRT systems	Brisbane, Australia, has a high cost system which attracts circa 350,000 patrons per day	The land use response is often weak except for at significant destinations due to the noise and added congestion of the bi- or triarticulated buses

Transportation and Land Use Integration: Shaping Transportation Demand and Delivering Transport Supply, Table e: Tram line capacity per lane

Mode	Tram – a steel wheel on steel rail with one carriage and low speed restrictions				
Capacity per 3.5 meters/hour	Daily throughput	Frequency	Low cost example	High cost example	City shaping capability
Typically run as one carriage, but can be made into joined consists, or sets, of carriages	(200 capacity × 6 trips per hour) × 12 hours of service = 14,440 or around 15,000 per day per route.	5–20 minutes	Tallin, Estonia	Portland, Oregon; Toronto, Ontario	Modest, but memorable high streets. Example: Queen St West, Toronto

signal priority, and frequency scheduling. A well-known advocate of BRT is the Institute for Transportation & Policy Development, with a publication titled "The BRT Standard" (ITDP 2018). Robert Cervero (Cervero and Kang 2011; Cervero 1998) has also written widely on BRT.

For the most part, BRT have the same supply needs as regular bus, above, but with following added considerations:

- Advanced schedule information is often a key element of BRT services, with "next bus" electronic LED boards indicating the arrival times of buses and destinations. The Internet of Things (IoT) technologies and satellite tracking have aided the transit service agencies to both know where their buses are and to let the riding public know as well. This information reduces concerns about travel time and expected arrival to destination times, adding certainty and convenience to all.
- Platform size increase in width, length and height to accommodate more riders on a platform, often allowing "all door boarding." Such platform upgrades eliminate height differences so the mobility impaired or parents with children can more rapidly enter the bus which dwell time and increasing comfort.
- Lighting of the stops and platforms becomes a more prominent role as the volume of riders necessitates clear and apparent platforms for the driver, riders and surrounding interactions with other modes (trains, cars).
- Advanced pedestrian access with grade separation bridges and tunnels, as can be seen in Bogota Colombia, to keep the high volume of riders safely away from the motor cars in the same transport corridor.
- Corridor separation is often a prerequisite of a successful BRT, meaning the BRT will have limited interaction with other modes including pedestrians, cyclists, cars, trucks, or any other.

This ensures a clear running way so that speed can be attained and maintained, keeping to schedule by avoiding accidents.

- Signal priority is required at points where the BRT does intersect other modes, especially private cars and freight trucks so that the riders on the bus are given travel time priority over the others.
- Upgrade of the station from asphalt to concrete possibly including a complete upgrade of the subsurface gravels and asphalt layers to support the weight and volume of heavy bi or tri articulated buses. This is especially important at stations as the weight pressure placed on the running surface is magnified through the braking and acceleration at these sites.

Tram

Often trams are a relic of years gone by. They exist notably in some European cities such as Amsterdam and Hong Kong as double decker types, in Toronto and Melbourne. They typically span the breadth of the historic 1940s area of the city as any significant increase in area coverage were required then often a LRT or Metro is called in for service. However, there has been a slight resurgence in some American cities more as "pedestrian accelerators" or as development "trigger" in broader infill (re)urbanizing efforts over certain land areas. There have also been new entrants into the LRT space but which due to capacity and speed could be more accurately described as operating in a tram-like fashion rather than a LRT.

Though often mixed together, and not entirely different from each other, a clear separation of Tram from LRT is presented as:

- A tram often has but one carriage, not in consists, with a moderate capacity of 200 per carriage, can stop frequently and accelerate moderately fast, but is limited to speeds of less than 60 km/h due to shared operating conditions in urban corridors, stopping patterns being closely spaced, track conditions including sharp turning corners, lightness of body construction, and that most spaces in which a fast train-like carriage may be required have long ago been served by bus, BRT, LRT, Metro, or by private motor vehicle. It can go slow or moderately fast, with frequent stops.
- An LRT by contrast is more robustly built, running on ballasted or slab tracks, can go faster when needed in separated corridors, but can also stop and accelerate quickly (quicker than metro or commuter) a true advantage in the inner-city shared-corridors conditions. It can go slow or fast and stop frequently or travel longer distances on separated corridors.
- Variations on LRT are the autonomous elevated type in Vancouver, or the various Monorails of Japan and Malaysia.
- The difference is not mere semantics, or word choice, there are very real hourly throughput differences, speed and travel time considerations, road lane separation requirements and politically sensitive costs and timelines to be considered. All of these are important to accurately describe when asking for a format of high-capacity transport to support **SDG 9 on Innovation and Infrastructure, or SDG 11 Sustainable Cities and Communities, SDG 13 Climate Action.** To achieve SDG goals and target, dedicated funding will need to be secured, requiring a wide array of knowledge on the differences between the options.

To help with more clearly observing the differences, a comparison of the various higher capacity transit modes is presented below. Note, it is not considering costs or right of way allocations, but **in general** a bus is the most rapidly deployed on the street to fill a need, and inexpensive per unit to purchase, but with a higher labor cost per service hour or per rider/km than LRT or Metro. It scales up to High Speed Rail as being the most expensive, but with the most metro-region city-shaping effects. All of these are transformational on the labor and housing and employment zone land use changes, but on the personal level mean that one less or no car needs to be used for transportation to achieve a high-quality, low-carbon, living standard.

Continuing with what constitutes Tram service supply:

T

Transportation and Land Use Integration: Shaping Transportation Demand and Delivering Transport Supply, Table f: Light Rail line capacity

Mode	LRT – a moderate capacity and moderate cost "train" with high theoretical capacity				
Capacity per 3.5 meters/ hour	Daily throughput	Frequency	Low cost example	High cost example	City shaping capability
4000 per hour. The Skytrain in Vancouver – running at high frequency through purposefully planned density – carryies 7000 passengers or more per hour	12,000 to 30,000 per route depending on the frequency and onboard capacity	2 minutes, but 5 minutes to 20 minutes per route are more common	Calgary, Canada, had low construction costs per km on the original routes	Sydney, Australia LRT 2 to Randwick	Moderate, depending on land use rezoning in corridor, land use policies for consolidation and car parking policies

- Platforms for trams have been – historically – perfunctory suggestions of a stop.
  - This has been a function of the many stops expected in a dense urban environment potentially requiring many stops, an expensive proposition.
  - Second, this reflects the expectation of the relatively slow trams to be able to stop almost anywhere and for the able-bodied passengers (before universal access considerations).
  - However, modern variations do have offset from the sidewalk ramp-up platforms which raise the pedestrian or wheeled mobility person to the grade at which they can enter the tram directly or the trams themselves are "low" or "very low" arriving at sidewalk grade. Improvements to the wayfinding, safety, access, and overall presence of the stations have been greatly improved over the last decades.
- Signal priority at intersections is rare for trams, but where present the priority helps tram and its riders to travel at a reliable speed. This also sends a subtle signal to the car drivers that the larger capacity modes, carrying more people per meter square, deserves a subtle help to get the travellers to their destination before the private car, and perhaps they should consider the tram (or bus, or LRT) for their subsequent trips.

LRT Supply

Much of what an LRT supply – as an overall service – requires is what is also required for the Tram, above, but with a few added considerations.

- More robust tracks and bedding of sleepers than trams as they are often heavier rollingstock, with a higher crash rating, and have higher expectations to proceed at higher speeds with smoother ride.
- Often along with the heavier rollingstock and slab tracks comes a requirement to realign or to "sleeve" in the underground utilities below the street level in tube "conduit." This both protects the utilities (water, power, sewer, franchised cables) from the weight of the LRT but also makes any long-term utilities upgrades easier to undertake without disrupting the LRT (or tram) service.
- Doors should be a priority over seating. For quick embarking and disembarking, to lower the dwell time in a stop, it is important that there are several doors per carriage and that these are wide enough to accommodate the flow of pedestrians on and off simultaneously.
- Overhead on catenary or third rail electricity supply, depending on the street conditions for being a market high street favoring in-ground (covered) third rail, or a more general corridor where ease of repair is more a priority for the operations of the service.
- Separate corridors are often a prerequisite for most LRT technologies. If not, they are operating in mixed traffic like the

Transportation and Land Use Integration: Shaping Transportation Demand and Delivering Transport Supply, Note 2 Note 2, a transit mode comparison chart

Mode[a]	Surface	Capacity – range	Speed, Average	Years of service, rollingstock	Stopping pattern – meters	Optimal Urban environment	"City shaping" ability to change land use and induce travel patterns
Bus	Extra lifts (layers) of asphalt and subgrade and/or concrete at stops	90	30	15–20	200–300	Collectors, Arterials	Serving cities where are and need to go, the land use response to be close to a bus route is limited to captive transit users
BRT	Reinforced asphalt, extra lifts with improved subgrade and/or concrete	150	50	15–20	800	Arterials – Shared or Separated	The land use response is often weak except for at significant destinations due to the noise and added congestion of the bi or tri articulated buses
Tram	Rail with minimal ballast, often cast into a concrete bed	200	30	50	400	Arterials – Shared or Separated	Historically, most high streets across the world had tram and the residential streets facing the tram had smaller lots to maximize the return to developers
LRT[b]	Rail with ballast, often cast into a concrete bed	200–600	50–60	40–50	800–1200	Separated ROW	There is often a modest land use response from LRT, unless very frequent and with strong policies to densify, see Vancouver
Metro[b]	Elevated or tunneled rail beds with significant service conduits	600–1500	60–100	40–50	1200–2000	Separated ROW	Often a Metro station is indistinguishable from the surrounding dense urban environment it serves and has helped develop
Commuter	Rail with minimal ballast, often on concrete ties	600–800	60–200	50	4000–10,000	Separated ROW	Historically, small towns on the line grew in response to the service, though contemporary small towns resist dense co-location of land use and transit service
High Speed	Rail or Maglev often on trenched or elevated viaducts; slab track or ballastless on rubber shoes	600+	200–300	40[c]	100 km to 300 km	Separated ROW	China and Japan have relied heavily on land redevelopment schemes to fund the service through (a) higher ridership coming from (b) land development schemes both of which are recognized revenue streams

[a]Options for driverless (autonomous) operation exist
[b]Electric propulsion options are available; perforce Bus and BRT and some commuter rail are still carbon combustion
[c]0 Series Shinkansen models remained in use from 1964 until 2008, a span of 44 years though most are retired after twenty years
Note: Much of the content for this chart is from Vuchic (2005), Bertolini and Dijst (2003), Cervero (2007), Cervero et al. (2017), Condon (2010), Ewing and Cervero (2001), Frank et al. (2011), Frank and Pivo (1994), Handy et al. (2002), Kelbaugh (1989), Kenworthy and Laube (1999), Shoup (2011), Urbecon (2013), Vuchic (2007), Rodrigue (2020), MacKechnie (2019), TCRP (2013)

T

Transportation and Land Use Integration: Shaping Transportation Demand and Delivering Transport Supply, Table g: Modern Metro or Subway line capacity

Mode	Modern Metro				
Capacity per 3.5 meters/ hour	Daily throughput	Frequency	Low cost example	High cost example	City shaping capability
30,000 to 50,000	1,000,000 per day possible, such as Shanghai Metro carrying over 10 million per day on all its routes	5 to 10 minutes	Recent Spanish or Chinese examples	Sydney Metro	Possibly very high, with revenue streams to governments from the land use changes when combined with rezoning

Transportation and Land Use Integration: Shaping Transportation Demand and Delivering Transport Supply, Table h: Commuter rail line capacity per line

Mode	Commuter and Interurban Rail				
Capacity per 3.5 meters/ hour	Daily throughput	Frequency	Low cost example	High cost example	City shaping capability
Up to 25, 000	2 trips per hour, over 12 hours = 600,000	Varies widely from 1 or 2 trips per hour to ubiquitous in Tokyo or other megacities			

trams thereby negating much of their advantage.

- Right of ways to permit the optimal travel times (speed) without decreasing the liveability of the street and with a conscious mind to costs. Vukan Vuchic (see resources) speaks clearly of three distinct types of corridors including up to grade separation.
- Signal priority when at grade intersection with other modes.
- Physical separation from pedestrians, cyclists, buses, and cars.
- Prepayment options are often found at LRT stations, perhaps through turnstiles, though sometimes the card reader (Opal, Oyster, other) for the ride payment is onboard and exceedingly rare are options to pay a roaming ticket agent on board.
- Integration with other modes including walking sidewalks of sufficient width and safe crossings (Table 2)

Note: Regular buses are included on the chart as a high-capacity mode as, theoretically, high frequency of buses in a corridor can cross into the ranges of BRT or LRT for hourly throughput. However, in practice regular buses are not run at such frequency due to capital (# of buses), operational (labor) costs, and the negative effects of bus bunching on congested inner-city streets.

Metro Supply

Grade separation is a prerequisite to delivering a fast and reliable service to the riding public. There cannot be conflicts of signals or with vehicles, pedestrians, or buses blocking the corridor.

Furthering grade separation, there are instances where a high-capacity Metro service may have legacy level crossings. A program to either tunnel or bridge the train over the cars, or vice versa, where conflicts exist is a good capital expense to create reliable and safe services. An example of such a program is in Melbourne Australia.

- Stop spacing is contentious. In many older cities with a dense urban grain of much higher density and many mixed-use commercial and retail and residential areas, with older metro systems, the stop spacing is often within one kilometer. However, in many new world cities such as Sydney or Vancouver this stop spacing for the metro (Vancouver's is a light metro, or

Transportation and Land Use Integration: Shaping Transportation Demand and Delivering Transport Supply, Table i: High Speed Rail capacity

Mode	High Speed Rail				
Capacity per 3.5 meters/hour	Daily throughput	Frequency	Low cost example	High cost example	City shaping capability
Dependent on the hourly throughput and carriage configuration. As an example, an HSR train departs every 5 minutes on the Beijing to Shanghai line	420,000 passenger per day on the Shinkansen network or 220,000 on the Beijing to Shanghai line	Every 5 minutes on lines in China.	China	USA, California	Links cities together, increasing cultural and economic ties

an advanced LRT?) is often between 2 to 3 or more kilometers apart due to (a) cost to construct stations, (b) lack of potential ridership on the surface due to a lack of density in the corridor, and (c) travel time delays from many stops with the concomitant dwell time at each station. Modern metros can slow and accelerate much more efficiently, achieving the desired operational speed quicker, and the many wide doors can speed the onloading and offloading of passengers minimizing the dwell time.

- Doors on Metro systems need to be wide and frequent, to permit the free flow of passengers in and out of the carriages at stations to maintain the scheduled dwell time. Modern metros have on-platform doors that align with the doors and remove the possibility of a rider or an object finding itself before the train, which would force a delay on the entire network.
- Prepayment options are standard across most metro systems globally, either with a token, a ticket, or a card such as Oyster or Opal. However, increasingly apps and other online "scanning" payment systems are becoming more common.
- Integration with other modes is a prerequisite, and while often explicitly planned for there are instances, such as in Mexico City, where a wide array of para-transit options exist to move riders further into the neighborhoods where other forms of public transport options do not (yet) exist.

Commuter and Interurban Rail Supply

Commuter and Interurban rail lines used to be very common across the industrialized west, but now are rare. Yet, in Tokyo and many other major cities they are as important to the labor-market as the metro is in others. Often these operate at grade on older track, with some level crossings though of course better if these are removed, often a half hour or hour service at peak hours. These services often share track with freight, limiting travel paths, or time slots, from certain hours of the mid-day and evening, but certainly not always. In many cities the Metro and Commuter Services are only distinguishable by the width and seating versus standing space.

High Speed Rail Supply

Not all places are conducive to high-speed rail, attractive as it is as a high-capacity high speed and futuristic mode despite a corridor apparent and seeming need. While China's need is justified, with such a national population and rapidly growing cities forming a national network of trade and knowledge HSR was the best option. Yet, as we can see in California's experience, without a clear specification of construction and planning and without a set aside right of way, or ability to rapidly acquire the land, the High Speed Rail is remarkably slow in construction.

Japan, without the HSR, might physically "look" quite different had not been the HSR from the 1960s onwards to link the labor force together. It focused on moving people across the country to save on time, reduce the potential need for highways, and prove the technological gap thereby

T

placing itself at an advantage over peer countries. This was expensive, but much less costly than the alternatives considering both the wide circuitous highways that would have been required and/or the vast array of airports plus connecting modes to get residents from city to city as efficiently. The population numbers, ambitions to be technologically advanced, and the corrugated landscape all contributed to the rise of HSR as a prescient choice.

Typically, HSR requires the following as an addition to most of the requirements for the above rail modes:

- Complete grade separation from all other modes, including from other freight or passenger rail as well as all road-based modes and certainly humans on foot or bike. The travel speeds, the acceleration, deceleration, and voltage and signals are all highly specific. Likewise, where there is a temptation to mix freight or other passenger rail with HSR there will have to be such necessary caution of the "cants," or angle of elevation change (superelevation), on the curves to maintain speed that to avoid accidents the ultimate "design" will eventuate in a separated and special HSR track.
- All curves on a HSR alignment will need to be very gradual, or flattened out, so that the train may maintain its high speed. Most often this requires either entirely new alignment, tunnels, bridges, or viaducts away from any existing rail network.
- The ruling grades (slopes) must be consistent, as much as is possible, to maintain the speed and reduce the need for braking.
- Frequency is important, the Shinkansen trains.
- Direct service to the central business district, or close by, is an important consideration as edge of city stations are much less likely to have the Business to Business connections, create less land value increase, with less demand for adjacent services and overall be less attractive as an integrated part of the urban fabric.
- Last, to plan a HSR route it needs to pass through high activity zones or major cities on the corridor. It can deliver excellent travel times savings to many, but only if it serves the population – or potential future population – along its corridor with stop spacing at acceptable average distances with an array of stopping patterns.
  – For example, the Tokaido Shinkansen has a three types of services from express to slow with more over 515 km it has 17 stations.

Private Motor Vehicle Supply

In the past, for most of history, the majority of us walked. It was long and exhausting and limited the amount of productive time we could spend as economic units. We walked everywhere. This is so currently in many places, with walking to work or at least part way to a high-capacity public transportation line being very common. The affordable private automobile liberated us, or so it was imagined. The private Car is fast and frequent but has low individual capacity, unless upscaled to the size of a bus (see above) and occupies valuable space for relatively low-capacity lanes. It also requires at least a factor of three for its parking needs, for without these extra parking spaces the utility of the mode is lost. Donald Shoup has done extensive research into parking area, policy, and options for future consideration (Shoup 1999, 2011).

Automobiles are the exception to the hierarchy of transport needs above. They are the outlier. They do not build one up on the other to a crescendo of a walkable, useful, efficient use of energy or human talent. This is not to say they are entirely negative, they suit a role.

A good series of sources for further reading are IRC (1990), RMS NSW (2017), RTA (2002).

To commence a list of what a car and its car driver requires to operate as a mode in urban settings:

- View lines, at corners or along the corridor, are an important feature of motor vehicle throughput. They create a sense of security and surety about what else may be in the corridor and for which they may have to alter speed or direction, thereby altering the speed and throughput of their lane or lanes. This is deleterious to maintaining lane or corridor capacity, and

Transportation and Land Use Integration: Shaping Transportation Demand and Delivering Transport Supply, Table J: Private Motor Vehicle capacity per lane

Mode	Modern private motor vehicle				
Capacity per 3.5 meters/hour	Daily throughput 12 hours of activity	Frequency	Low cost example	High cost example	City shaping capability
Residential: 300–1000 though the theoretical capacity is high, too many cars often result in capacity upgrades to "collector" status. Collector: 1200 Arterial: 1600 normal to 2000 being at the high end often due to intersection spacing and lane widths. Low end is more typically 1200 due to land adjacent "abutters" High speed limited access "Highway": 1600 to 2400/h with 2000 vehicle per hour being a easily calculable – and average number	Residential: 12,000 Collector: 19,200 Arterial: 24,000 High speed limited access "Highway": 28,800	Ubiquitous	Lougheed Highway (#7) to #3 Highway British Columbia maintains connections between small towns while serving as the low speed "Main Street" in each small city it passes through. The difference between "highway" and "street" is lane numbers, curb and gutters and traffic signals	Katy Freeway, which cost 2.8 billion USD and resulted in no net time savings gains as during the construction, car dependent land uses continued to be built in anticipation of the increased access, generating more trips, leading to a failure to improve the situation	Requires tiers of roads including motorway, arterials, collectors and residential streets including tunnels and bridges. Also requires parking spaces of at least 3 parking spaces to 1 vehicle, generates "drive-to" large lot efficiencies which are contrary to urban life. One could say this mode negates "city shaping"

open view cones are often designed into the street, with a corresponding loss of building edges, street trees, or pedestrian safety. Yet, wide view cones are not absolutely necessary as there are many instances in urban areas where view cones around a corner at an intersection are occluded making the visibility of the driver less than what it could be. In Tokyo or in other historic parts of cities, it is understood that every corner must be approached with caution as pedestrians or other cars may be just outside of view.

- Lane width has an impact on speed and therefore capacity of the lane or corridor. With increase width the lane appears "faster" as an apparent speed limit despite regulatory signage to the contrary. When coupled with a straight view line (see above) this creates an environment in which an automobile may be driven faster than the speed limit, which in an urban setting is dangerous for anyone travelling by any other modes.
- Curb return radii are often needed to be "wide," that is with many meters between the two "flat" parts of the curb. The curved part of the curb being the part that faces – almost always – into the center of the intersection. A narrow curb return radii will have fewer meters. These wide curb return radii are necessary for private automobiles to operate at higher speeds as they turn the corner, thereby quickening the intersection exit time and allowing the lanes to flow at a

higher Level of Service (see below) than otherwise. This has a deleterious effect on pedestrians as the wider curbs returns as the speed of the car is higher, meaning and impact is more serious, and the driver is often looking to the opposite perpendicular view for oncoming cars – rather than the same view as they are turning – and not see the pedestrian making contact more likely. Also, wider curb returns further widens the street, which extends the distance and time required to walk between sides for pedestrians, thereby exposing them to greater potential conflict with private motor vehicles.

- Slip lanes are an extension of wider curb return radii (above), as the road itself is brought into the erstwhile urban area of the city block, occupying urban space, so that the private car may more readily maintain speed and clear the intersection soon with a more time efficient turning operation. Though rarely seen in dense urban areas, it is used extensively in suburban and in peri-urban growth areas; because of its ubiquity there is can be seen as an option for build back Level of Service car throughput in urban areas. For a bus this may be a benefit, but for a pedestrian these are often very high conflicts in such zones.
- "Shy distances" are often on the edges of the corridor in each direction. These are the spaces at the outside or inside edges of a direction of travel. These are the spaces – which further widen a street – for the automobile to travel with speed and within a tolerance of "mistakes" for driver weaving. This increases speed, as there is more apparent space in which to maneuver the automobile. Every extra width along a corridor given to maintain car speed is less space available to pedestrians, public transport operations, street trees, street front retail, or street furniture being those quantities that create quality urban environments.
- Signals are often, when well-designed, prioritized for the private automobile. The detection loops and cameras will detect "platoons" of cars and leave open a signal at "green" to move this platoon along the corridor. This can be frustrating to pedestrians or other cars who wish to cross this primary corridor from the perpendicular cross street, and also dangerous to cyclists in the corridor, as high vehicle speeds are encouraged. Rather than expecting pedestrians (being the majority in an urban setting), cars are thereby encouraged to travel at the speed limit toward their destinations to the detriment of other modes. Contrarily, signal priority for buses or at-grade operating trams/LRT (see above) is too often seen as beneficial to the more efficient operation of transportation corridors for the most potential travellers.
- Level of Service (LOS) is a calculation with outputs ranging from A to F, similar to a grading scheme. These outputs represent the easy with which a driver will be able to travel through a road corridor to with little to plenty of travel time delay. In short, the LOS is a calculation of volume of the cars on the lane or corridor divided by capacity (of the lane or corridor). It can also be the calculation of the time actual divided by time expected to traverse an intersection. Output "A" being fast, free flowing, arrival time is as expected, to "F" being congested with no appreciable expectation of accurate travel time by car. This can of course be temporal, being that at peak travel times in the morning it will be congested and D or F, say, but this is not how the measure is used. Rather, to try to ensure faster travel times for most car drivers, lanes are added, turning slip lanes are widened, access roads are removed and signal priority is given for the dominant travel paths if there is budget and space to do so for peak times.
- Traffic warrants are likewise a very transparent – once one sees on in action – calculation based on volume of users on a corridor by mode, distances between intersections, access roads, and adjacent speed limits to determine where to adjust an aspect of the physical design elements of the street to serve the street users. Unfortunate, most such warrants are designed around the free flow of private automobiles rather that the potential users of the street who might arrive by bus and then proceed to walk to the nearby destinations.

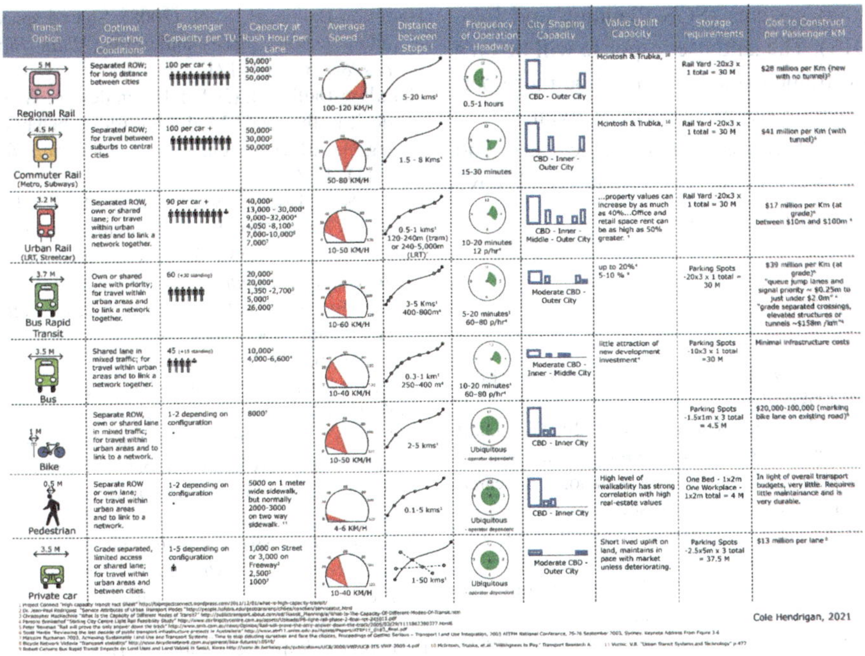

Transportation and Land Use Integration: Shaping Transportation Demand and Delivering Transport Supply, Fig. 1 Modes Comparison, by author, 2021

- Micro and Macro Simulation models have been around for many decades but have developed significantly in recent years. There are several providers and a few open-source software packages today offering this type of street operations analysis. Typically, these models identify opportunities to expand the width of roads, to widen intersections and to expand the high capacity road network further in anticipation of population growth and economic activity on "greenfield" sites. Yet, such modeling is rarely used to anticipate what public transport, walking or cycling might require under the same "growth" scenarios, creating a very large loss of opportunity to knit the urban fabric tighter, at lower cost, with increased revenues to governments, with less emissions and higher health outcomes over the "verified" car-oriented outcomes of such models.

Conclusion to Supply

Without getting into the details of the cost/km or cost /km/trip ratios that may emerge from a suite of mathematical models, there are productivity gains to be had by choosing the correct "mode" for the task. A few questions to ask when promoting, advocating, engineering, or financing a transport mode infrastructure is "This be enough supply to move the people needing the service today, but what about 5 or 20 years from now?"

What is the correct mode to invest in for a city depends largely on several factors, none of which are cultural, but are dependent on the cascade of concerns and criteria as outlined above. Figure 1 will summarize many of the critical points of the above, noting the capacity, frequency, and city shaping ability of each mode.

Primarily, there must be a need, as in riders to ride the transit or people to drive private cars. This, however, can be created through land supply on the urban edge or as infill dense and wonderfully humane urban habitat near high capacity transportation systems (SGS Planning and Economics 2013). The land use response will be on a longer time frame (Mees 2010) but be more pervasive and responsive to the supply given.

Note that Trackless Trams, Cable Cars, and Para-transit are all categories that could have been added to this array of work.

References

99% Invisible. (2019). 99% Invisible. In R. Mars (Ed.), *Curb cuts*. Oakland.

Bartholomew, K., & Ewing, R. (2008). *Land use-transportation scenario planning in an era of global climate change*. Transportation Research Board 87th Annual Meeting, Washington, DC.

BC MoT. (2019). *Active transportation design guide*. Victoria: Ministry of Transportation and Infrastrcuture.

Bertolini, L., & Dijst, M. (2003). Mobility environments and network cities. *Journal of Urban Design, 8*, 27–43.

Cervero, R. (1998). *The transit metropolis – A global inquiry*. Washington, DC: Island Press.

Cervero, R. (2007). Transit-oriented development's ridership bonus: A product of self-selection and public policy. *Environment and Planning A, 39*, 2068–2085.

Cervero, R., & Kang, C. D. (2011). Bus rapid transit impacts on land uses and land values in Seoul, Korea. *Transport Policy, 18*, 102–116.

Cervero, R., Guerra, E., & Al, S. (2017). *Beyond mobility: Planning cities for people and places*. Washington, DC: Island Press/Center for Resource Economics.

Condon, P. (2010). *Seven rules for sustainable communities. Design strategies for the post-carbon world*. Washington, DC: Island Press.

CROW. (2016). *CROW design manual for bicycle traffic*. Utrecht: CROW.

Cullen, G. (1971). *The concise townscape*. Architectural Press.

Ewing, R., & Cervero, R. (2001). Travel and the built environment: A synthesis. *Transportation Research Record: Journal of the Transportation Research Board, 1780*, 87–114.

Frank, L. D., & Pivo, G. (1994). Impacts of mixed use and density on utilization of three modes of travel: Single-occupant vehicle, transit, and walking. *Transportation Reseach Record, 1466*, 44–52.

Frank, L. D., Greenwald, M. J., Kavage, S., & Devlin, A. (2011). *An assesment of urban form and pedestrian and transit improvements as an integrated GHG reduction strategy*. Washington State Department of Transport.

Glaeser, E. (2011). *Triumph of the city: How our greatest invention makes us richer, smarter, greener, healthier, and happier*. Penguin Group US.

Hamilton-Baillie, B. E. N. (2008). Shared space: Reconciling people, places and traffic. *Built Environment (1978-), 34*, 161–181.

Hamilton-Baillie, B. (2010). Urban design: Why don't we do it in the road? Modifying traffic behavior through legible urban design. *Journal of Urban Technology, 11*(1), 43–62.

Handy, S. L., Boarnet, M. G., Ewing, R., & Killingsworth, R. E. (2002). How the built environment affects physical activity: Views from urban planning. *American Journal of Preventive Medicine, 23*, 64–73.

Higashide, S. (2019). *Better buses, better cities*. Washington, DC: Island Press.

IRC. (1990). *Guidelines for capacity of urban road in plains area*. New Delhi: Indian Roads Congress.

ITDP. (2018). *The bus rapid transit standard* [Online]. New York: Institute for Transportation and Development Policy. Available: https://www.itdp.org/library/standards-and-guides/the-bus-rapid-transit-standard/. Accessed June 2021.

Kelbaugh, D. (1989). *The pedestrian pocket book: A new suburban design strategy*. Princeton Architectural Press in Association with the University of Washington.

Kenworthy, J. R., & Laube, F. B. (1999). Patterns of automobile dependence in cities: An international overview of key physical and economic dimensions with some implications for urban policy. *Transportation Research Part A: Policy and Practice, 33*, 691–723.

King, L. J. (1985). *Central place theory*. SAGE Publications.

Mackechnie, C. (2019). *What is the passenger capacity of different modes of transit?* [Online]. Dotdash. Available: https://www.liveabout.com/passenger-capacity-of-transit-2798765. Accessed June 2021.

Marchetti, C. (1994). Anthropological invariants in travel behavior. *Technological Forecasting and Social Change, 47*, 75–88.

Mees, P. (2010). *Transport for suburbia: Beyond the automobile age*. London: Earthscan.

NACTO. (2012). *NACTO urban street design guide*. Washington, DC: Island Press.

NACTO. (2014). *NACTO urban bikeway design guide*. Washington, DC: Island Press.

NACTO. (2016). *NACTO transit street design guide*. Washington, DC: Island Press.

RMS NSW. (2017). Motorway design guide: Capacity and flow analysis. In R. A. M. S. Transport (Ed.). Sydney: NSW Government.

Rodrigue, J.-P. (2020). *The geography of transport systems*. Oxforshire: Routledge.

RTA. (2002). Guide to traffic generating developments. In Roads and Traffic Authority NSW (Ed.). Roads and Traffic Authority NSW.

SGS Planning and Economics. (2013). *Expanding land supply' through transport improvements* [Online]. Canbeerra: SGS Planning and Economics. Available: http://www.sgsep.com.au/insights/urbecon/expanding-land-supply-through-transport-improvements/. Accessed February 2014.

Shoup, D. (1999). The trouble with minimum parking requirements. *Transportation Research Part A, 33*, 549–574.
Shoup, D. (2011). *The high cost of free parking*. Planners Press, American Planning Association.
TCRP. (2013). Transit capacity and quality of service manual. In TCRP (Ed.). Sponsored by the Federal Transit Administration.
TFNSW. (2020). Cycleway designing toolkit. In TFNSW (Ed.). New South Wales Government.
UN. (2021). *The 17 sustainable development goals* [Online]. New York: United Nations. Available: https://sdgs.un.org/goals. Accessed Janauary 2020.
United Nations. (2003). *Accessibility for the disabled – A design manual for a barrier free environment* [Online]. United Nations. Available: https://www.un.org/esa/socdev/enable/designm/AD2-01.htm#Notes. Accessed June 2021.
Urbecon. (2013). 'Expanding land supply' through transport improvements. In SGS Planning and Economics (Ed.). Canberra: SGS Planning and Economics.
Vuchic, V. R. (2005). *Urban transit. Operations, planning and economics*. Hoboken: Wiley.
Vuchic, V. R. (2007). *Urban transit. Systems and technology*. Hoboken: Wiley.
Walker, J. (2012). *Human transit: How clearer thinking about public transit can enrich our communities and our lives*. Island Press.

Transportation and Mobility

Strides Toward Decarbonizing the Urban System

Tinashe Natasha Kanonhuhwa[1], Smart Dumba[1] and Innocent Chirisa[1]
[1]Department of Demography Settlement and Development, Social & Behavioral Sciences, University Zimbabwe, Harare, Zimbabwe
[2]Department of Urban & Regional Planning, University of the Free State, Bloemfontein, South Africa

Synonyms

Transportation – circulation; Mobility – movement

Definition

Decarbonization: the process of removing carbon in an environment.

Mobility: movement of goods, service, and people in defined pathways and vehicles.

Transport system: comprises the rolling stock, the way, termini, and institutions that guide movement.

Policy: guiding framework for decision-making and future shaping.

Management: the ordering of a system for functionality.

Introduction

Being at the heart of the urban activities, the mobility system is responsible for the movement of people, goods, and services from one point to the other and in the process, attract investment into cities (Venkat 2016). It connects people to different facilities and land uses in different areas, and through the transport system, it becomes possible for people to reach their areas of work, education, as well as areas of social interaction, which are fundamental to human development (Philp 2015). As a consequence to the increasing urbanization trends in cities, 60% of the world's population is projected to live in cities by 2030 (Bouton et al. 2015), and with more than two billion urbanites likely to enter the middle class, more people are expected to buy cars, with projected estimates likely to increase from 70 million a year in 2010 to 125 million by 2025 (Bouton et al. 2015), a development which is likely to cause a strain on existing urban transport infrastructure. It is further observed that cities consume over 80% of world's energy and are thus, responsible for 75% of total greenhouse gas emissions (Venkat 2016), a condition which further threatens the sustainable livability of cities, and this has prompted joint efforts by the developed world to decarbonize the mobility sector.

The study employs a desktop review of literature. Information gathered was mainly from Internet search engines such as Google scholar. Longo

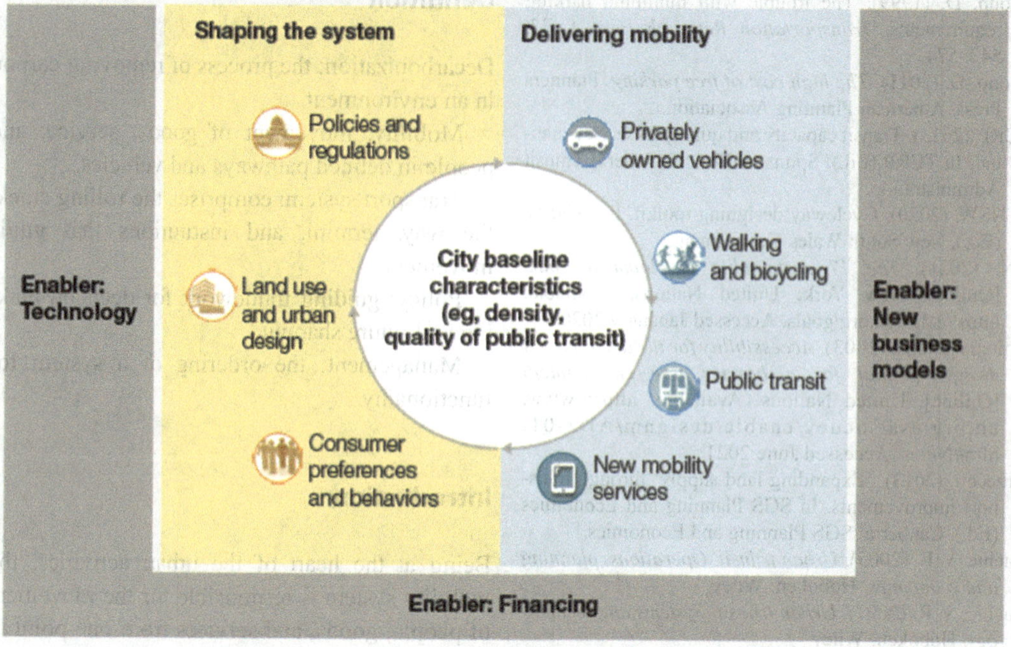

Transportation and Mobility, Fig. 1 A Framework for understanding urban mobility (Bouton et al. 2015)

et al. (2017) argues that the use of the Internet has now made research easy, due to availability of vast amounts of information on the Internet. The chapter makes more use of current scholarly literature on the transport mobility system in the urban sphere with particular focus on sub-Saharan Africa. This research, however, lacks the primary gathering of the relevant data through questionnaires and surveys, and this was also partly contributed by the resource and time constraints. The researchers, however, relied on the most current review of literature, as a way to ensure reliability and validity of the information gathered. This study can therefore be used to make conclusive remarks on the situation of urban transportation mobility in Africa.

This chapter embraces the positive role played by the transport sector in facilitating the smooth functioning of urban centers, as well as explores developments made globally by various nations in order to decarbonize the urban transport system, as a way to improve on the sustainable functioning of city centers. First priority shall now be given to an understanding of the conceptual framework guiding the transport and mobility in the urban sphere discourse.

Conceptual Framework (Fig. 1)

As a result of the different missions that guide people's lives, people travel to various destinations for different reasons. Various scholars have observed that some of the most dominant routes travelled by urbanites are centered round reaching areas of work and schooling (Philp 2015), and in so doing, various modes of transport may come in to play. These various models include the use of private motor vehicles or public transport in the form of buses and commuter buses, as well as through walking and cycling among other modes.

In shaping the urban transportation and mobility system, government can do this through establishment and enforcement of laws, policies, and regulations as well as through urban designing aimed at promoting smart and compact city designs which promote mixed uses, for instance. In this regard, people would work where they stay, thereby, cutting the number of traffic that enters the city center, and this would result in reduced levels of congestion on the city's major roads. Through improvement in technology and improved financing, a transport system which responds to the growing needs of society becomes possible. This chapter now examines global literature and experiences on transport and mobility practices.

Transport and Mobility: A Review

Evidence has shown that the urban transportation system provides link to the joining of various economic activities, goods, and people through a well-defined road network. Urban planning and design play a crucial role in determining the way in which urban traffic moves and circulates around the city's urban areas. It is often argued that land consumption depends on the compactness of a human settlement and residential density, while energy consumption is also dependent on the same variables, via their linkage with mobility patterns in the form of trip length, and modal choice which usually comes as choice between public and private means (Camagni et al. 2002). Various factors come into play in determining the effective mobility of urban transport, and these include street layouts, block sizes, and traffic control mechanisms among others.

Street layouts set the tone for the way vehicles move within and around urban centers. Streets can either be one-way, single-lane, or multilane and can allow either one-way or two-way movement of traffic (Mahajan et al. 2006). A multilane traffic lane with three or more traffic lanes would allow more traffic to pass through a particular point in less time when comparing to a one-way traffic lane, which may equally be busy in terms of traffic volumes being served. Multilane street designs help filter more traffic compared to single or dual lanes and, in many cases, help in reduce traffic congesting on the streets. Therefore, street lay out designs must suit the traffic demand of the area, and this must continuously be adjusted in line with the changing demands and the projected traffic demand of the area after a long period of time, say 10 years' time. Continuous upgrading of the transport system would not only save people's time on the road but also help improve the business networks in the country at local, as well as regional levels, as the road network becomes more reliable.

A block size is considered the smallest area surrounded by streets, and it is the block size which determines the number of intersections in an area, which in turn determines the frequency with which a vehicle stops (Mahajan et al. 2006). The more intersections, the more stops a vehicle makes, especially when controlled by road signs such as a stop or giveaway sign, and this has a bearing on the delays which traffic makes in the area. It is further argued that larger block sizes make the network more sensitive to clustering and this degrades performance (Mahajan et al. 2006).

Traffic control mechanisms in an area also have a bearing on the efficient and effective movement of traffic in the area. Some of the common traffic control mechanisms utilized by urban designers in street designs to promote functionality include traffic lights and stop and give way signs, and these also aim to reduce road accidents (Hussein 2013). These mechanisms sometimes form clusters and queues of vehicles at intersections as well as a subsequent reduction in the average speed of movement of vehicles (Mahajan et al. 2006). Reduced mobility would imply more static nodes and slower rates of route changes in the road network (Mahajan et al. 2006), and this would delay people from reaching their various destinations, so people spend more time on the road.

Interdependent vehicular motion also contributes to the rate of traffic movement or mobility in an area. The movement of every vehicle is influenced by the movement pattern of surrounding vehicles, and a regulated safe distance must be maintained between two vehicles, and this affects traffic speeds at specified times. The speed limit on the roads also directly affects the average speed of vehicles, as well as the frequency of how existing routes are broken and new routes established (Mahajan et al. 2006). Therefore, a map with fewer road intersections may encourage high acceleration speeds of vehicles, compared to maps with many intersections and smaller block

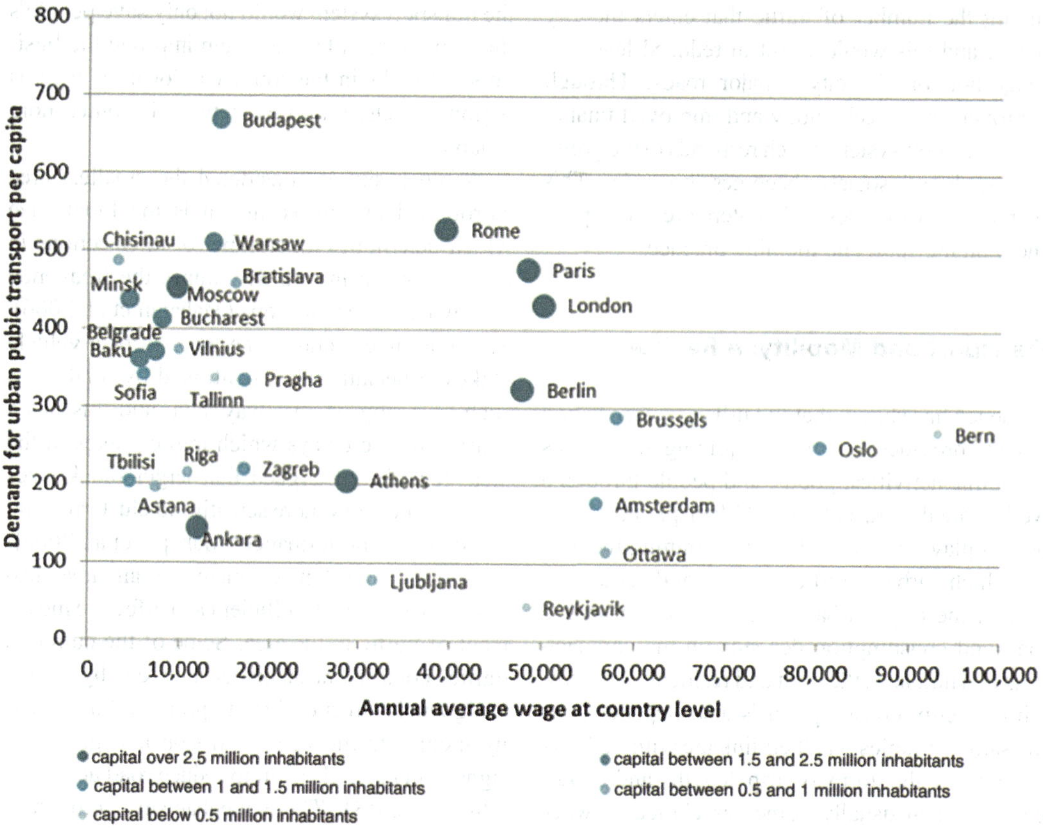

Transportation and Mobility, Fig. 2 Demand for urban public transport in UNECE capitals in relation to average annual wage at country level, 2011 (UNECE 2015)

sizes (Mahajan et al. 2006). A good design allows the easier movement of traffic to prevent traffic congestion.

Due to the increased urbanization trends which now characterize most urban landscapes globally (Drescher and Iaquinta 2002), people's demand for transport has increased. The choice of the mode of transport to use would depend also on the people's lifestyles (Philp 2015), with high-income earners opting for private motor vehicles and low-income earners, preferring subsidized public transport by government. The increase in the volume of traffic on the road has now made the challenge of traffic congestion a common phenomenon on the urban roads especially during the peak hours of the day. Traffic congestion comes with a myriad of challenges such as the loss in people's productive time, fatigue, and frustration of travelers as people spend more time on the road (Vencataya et al. 2018). The release of exhaust fumes and greenhouse gasses from diesel car engines also contributes to environmental pollution (Lu 2011; Reşitoğlu et al. 2015). Increased polluting levels in the atmosphere would have a negative effect on environmental cleanliness as well as the health of most urbanites.

The inevitable increased urbanization trends, which now face most urban areas in the developed world, many governments in cities such as Rome, Budapest, and London (Fig. 2), have all adopted the use of public transport systems (UNCECE 2015).

The diagram shows that, by the year 2011, many cities in Europe had already adopted the use of the public transport system. Cities with lower annual wage rates relied more on public transport such as rail and buses. Public transport becomes cheaper to use as it is usually subsidized by the government. It is often argued that public transport provides environmental benefits to

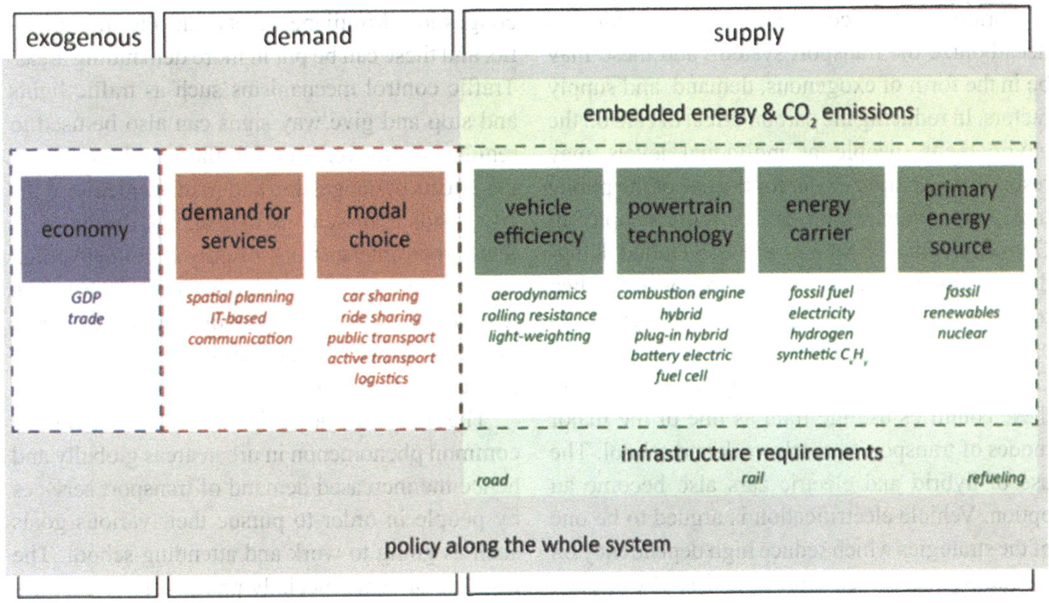

Transportation and Mobility, Fig. 3 A framework for decarbonization options (Boulouchos et al. 2019)

society (Boncenne 2012), and this is mainly due to lower levels of pollution as compared to the increased use of smaller cars on the road, which may end up being locked up in congestion. It is however arguing that public transport in the form of public buses accounts for 50–60% of public transport, of which 90% of this share is propelled to diesel engine technology (Ozener et al. 2018) and, hence, the continued release of greenhouse gas emissions into the atmosphere. Research have also shown that freight transport accounts for about 8% of CO2 emissions worldwide and due to its heavy reliance on fossil fuels, it is regarded as one of the hardest economic activities to decarbonize (McKinnon 2009). Demand for freight transport is, however, predicted to steeply rise by threefold by the year 2050 (McKinnon 2009), which in turn presents a challenge to most urban managers, as they are expected to come up with innovative ways aimed at managing the GHG emission phenomenon.

The growth of freight movement in the developed world has been driven by the need to transport freight over greater distances (McKinnon 2010), a development which may promote the interregional exchange of goods. In order to promote a sustainable future, where cities can become more competitive and where people, businesses, and culture can successfully thrive, low-energy request mobility becomes essential and, hence, the development of the energy efficient transportation vision by the International Association for Public Transport by 2025 which reduces the dependency of fossil fuels (Ozener et al. 2018).

In efforts to decarbonize its transport system in Europe and Asia, the need to decarbonize the transport sector becomes of paramount importance. Low-carbon transport system involves the comprehensive development which embraces dense but green and mixed use developments that enable shopping, leisure facilities, residential buildings, and workplaces to be closer together (Jennings 2011). In this regard, people travel less to their areas of interest and, hence, the reduction of congestion and associated challenges of time wastages, poor productivity, and the release of GHGs in to the atmosphere. Low-carbon transport involves the substitution of private motor vehicles with an efficient public transport system; the use of smart urban logistics, in the form of clean vehicles; and the use of electric hybrid engines on cars and motorbikes becomes essential (Jennings 2011). Figure 3 shows the various decarbonization options which the urban system can utilize.

Various factors come into play in efforts to decarbonize the transport system, and these may be in the form of exogenous, demand, and supply factors. In reducing the carbon effect of cars on the environment, people at individual levels may resort to car sharing or opt for the use of the public transport system which carries more people. In Germany, France, Spain, and the United Kingdom, many people use the metro system when traveling, which involves the use of high-speed electric trains (European Rail Research Advisory Council 2009); therefore the majority of people in these countries use the train as one of the major modes of transport to reach work and school. The use of hybrid and electric cars also become an option. Vehicle electrification is argued to be one of the strategies which reduce high dependency on dwindling fossil fuel sources and is also argued that it allows for the meeting of stringent emission targets by policy makers in metropolitan regions (Avci and Özener 2019). In this regard, massive air pollution in cities is reduced, and a healthier environment is created for the inhabitants.

The European Union, under the Paris Agreement, has committed to cut the GHG emissions to 40% below the 1990 levels by 2030, and between 1990 and 2015, emissions in the transport sector have actually increased by 20% and, hence, the need for more stricter and aggressive policies in order to decarbonize the sector (Tagliapietra and Zachmann 2018). This would entail investing more in clean vehicles, as well as follow examples of countries such as the United Kingdom and France by adopting plans to ban petrol and diesel vehicles by 2040, or even by 2030 (Tagliapietra and Zachmann 2018), and these would play a positive role in decarbonizing the European urban systems for many countries.

Key Determinants in Urban Transport and Mobility

Various factors play a critical role in promoting the easier movement of traffic on the urban roads. Urban designing is key to enabling the efficient movement of traffic in urban areas. The way in which streets are laid can prevent or encourage congestion. Multilane streets can filter more traffic, and these can be put in more demanding areas. Traffic control mechanisms such as traffic lights and stop and give way signs can also be used to regulate the movement of traffic in order to lessen the effects of congestion and road accidents on the city roads (Hussein 2013), and a design with fewer road intersections usually encourages high acceleration speeds (Mahajan et al. 2006). The way the city and the neighborhood are designed therefore speak volumes about the movement of traffic in the area of interest.

The study also notes that urbanization is now a common phenomenon in urban areas globally and hence the increased demand of transport services by people in order to pursue their various goals such as going to work and attending school. The increase of cars especially private motor vehicles has been argued to cause a lot of congestion problems in the city's main roads especially during the peak hours of the day. Congestion has therefore been recorded as one of the major causes of the emission of GHGs in to the atmosphere, with diesel cars cited as the major contributors (Lu 2011; Reşitoğlu et al. 2015). The increased emission of GHGs into the atmosphere would in turn have a negative impact on global temperatures as it is believed to contribute to global warming.

In efforts to reduce the problem of congestion and the emission of GHGs into the atmosphere, many cities such as Rome, Budapest, and London have adopted the use of the public transport system, as a major transporter of people to their various destinations. The carrying of more people in large buses and trains would reduce the number of cars which may flock to the city centers, thus, lessening the problem of congestion.

The adoption of the low-carbon transport system, which involves comprehensive development through adoption of the compact city approach and the mixed use development (Jennings 2011), not only allows people to work where they stay but also reduces the number of cars which travel long distances into the city center. Low-carbon mechanisms also encourage the use of the electrical car in place of the diesel and petrol fuelled car. Decarbonization of the transport system should therefore be regarded as a system which can start from the individual choices where people will choose public transport in place of private cars, and it would go on to the supplier, where governments may subsidize local suppliers to produce hybrid or electric cars. In enforcing the use of carbon cars, countries ought to be strict, and policies must speak an aggressive language of banning cars which cause environmental hazards such as those which are diesel- or petrol-powered (Tagliapietra and Zachmann 2018) and promote the use of renewable energies.

Evidence from Africa

The issue or GHG emissions in Africa is not new and is a cause for concern as it has been linked to unfavorable climatic changes. It is believed that the greatest contributor of GHGs is the developed world, with concerns however that a rise in greater percentage of GHG emissions shall come from the developing world (Jennings 2011). This comes at a time when the developing world, especially Africa, is heavily dependent on used imported cars from Asian countries such as Japan, and these have raised concerns that the continent is being used as a dumping ground for used cars by other countries (Roychowdhury 2018), a condition which would in turn culminate in an unclean environment through emission of exhaust fumes from the diesel and petrol-powered cars.

African countries rely on road as a major transporter or people, goods, and services, which makes private motor vehicles one of the major constituents of road transport. Jennings (2011) argues that 74% of all transport emissions are attributed to road transport (with private cars accounting for a greater proportion), while aviation is said to account for 12%, Shipping 10%, and rail transport having the least percentage of 2% global transport emissions. It is also argued that most of the rail systems especially in the developed world are now operating based on electricity, and they are also argued to be cheaper with low maintenance costs (Baxter 2015; Serrano-Jiménez et al. 2017), which greatly reduces the percentage of contribution of rail transport to GHGs. Most of Southern Africa is therefore heavily reliant on road which makes it difficult to control the rate of GHG emissions in Africa.

The use of low-carbon transport in Africa is quintessential to the development of the African continent as it is said to harbor a lot of advantages. A sustainable low-carbon transportation infrastructure does not only mitigate climate change but also increase energy security as less oil is imported and reduces traffic congestion (and, consequently, air and noise pollution), and by encouraging semi-dense, mixed use developments, urban sprawl is reduced (Jennings 2011). Low-carbon transport is, therefore, more economic and environmentally friendly. In efforts to cut on carbon emissions, some of the Southern African countries have also made huge strides in efforts to reduce the rate of CO2 emissions as caused by moving vehicles in their countries.

Announced in February, 2000, and launched in June, 2010, the Gautrain is considered a positive mark in easing congestion on Gauteng's highways (Jennings 2011). The Gautrain is an 80 km route between Pretoria and Sandton (Johannesburg), with nine stations which include OR Tambo International Airport, Rosebank, Marlboro, Midrand, and Centurion (Jennings 2011). The Gautrain is therefore a link which connects various areas of activity and, in the process, lessens the burden of congestion in Gauteng and, at the same time, ensures that areas are connected through a smooth traffic network and, at the same time, ensures convenience of many people in the area.

Egypt engaged in a vehicle scraping and recycling program in 2010, where a voucher of up to US$ 480 is offered to owners who voluntarily surrender their taxis, trailer trucks, microbuses, or buses for managed scraping and recycling (Hogarth et al. 2015). In this way, road unworthy vehicles are eliminated off the streets. The vouchers can then be used as down payment for loans in order to acquire newer and more efficient vehicles from participating dealers, and the Scrapping and Recycling Program is part of the Urban Transport Development Program which has seen the establishment of a law which renders owners of mass transport vehicles aged over 20 years ineligible for renewal of their operating licenses

(Hogarth et al. 2015). This development would make people more aware that they need to continuously upgrade their vehicles over time and, hence, the reduction of CO2 emissions from old vehicles which will be lagging behind in terms of technological improvements.

With an estimated population of 25 million people, Lagos is one of the fastest-growing megacities in the world, and people used to rely on an unreliable public transport system of 75,000 unreliable expensive and polluting minibuses, which would make 16 million trips daily (World Bank 2015). This indicates that the transport system would remain active for the greater part of the day and the city would suffer more from environmental pollution from the numerous public minibuses.

The World Bank has, however, financed the Lagos Urban Transport Project, and in March 2008, a 22 kilometer project connecting Lagos mainland with the island, became the first in sub-Saharan Africa (World Bank 2015). The Bus Rapid system is argued to provide a reliable way to move around the city and has been noted to save people's traveling time (Mobereola 2009). Apart from running 16 h a day with 220 buses to move 200,000 passengers daily, the system has managed to reduce CO2 emissions by 13%, and GHG emissions have been reduced by 20% (World Bank 2015). 2000 jobs for drivers, conductors, mechanics, and ticket sellers and an additional 10,000 indirect jobs to run informal park-and-ride facilities and mini fast-food services have also been generated (World Bank 2015).

The Lagos light rail network which is under construction is expected to carry 400,000 passengers daily and 700,000 once fully operational, and the rail network will be powered by electricity from natural gas sources (Hogarth et al. 2015). The project, which was initially funded by the Lagos State government and kick-started in 2009 and was expected to end in 2011, was still under construction in the year 2015, and the delays were said to be caused by funding challenges (Hogarth et al. 2015). This shows that, despite efforts by the Lagos State government to ease the unsustainable release of CO2 into the atmosphere through the use of the electric train, financial challenges remain a stumbling block for most African countries to implement positive change where it's needed.

Emerging Issues

The chapter notes that the African continent has also not been spared from the vices of overpopulation and the increased demand for transport in order to ferry people to their areas of work and schooling. The majority of countries in Southern Africa have therefore resorted to the importation of used cars from Asian countries, a scenario which further contributes to environmental pollution, due to the release of exhaust fumes and gases from used cars. This in turn have a bearing on the climatic trends of the continent as it may also contribute to global warming. In order to control the rate at which used cars are imported to Africa, African governments ought to be strict and may impose a ban in order to control the rate at which these cars are imported.

The chapter also notes the various efforts made by some African countries in making their operational environments cleaner through the adoption of greener technologies. The Gautrain, successfully launched in 2010, shows a positive step toward improvements in the transport sector. The Gautrain connects various areas including Sandton, Pretoria, and the OR Tambo International airport. It therefore serves the purpose of transporting more people in one goal, and by so doing, a large volume of traffic is cut in these areas as people rely more on the Gautrain. As such, the levels of pollution are cut, and transport efficiency and effectiveness are improved by the operation of the Gautrain.

Egypt has also engaged in a vehicle scrapping and recycling exercise which started off in 2010 (Hogarth et al. 2015). This exercise has played a significant role in eliminating road – unworthy cars on the road. Lagos has also promoted the use of public transpor.t and the construction of the light railway in Lagos would help lessen the burden of transport in the city. In order for projects to positively come into fruition, government funding becomes essential, and at times, the finishing of essential public transport projects becomes difficult due to shortages of funds.

Conclusion and Future Direction

In conclusion, as a result of increased urbanization and the increased demand for transport by the population, the urban sphere has suffered massively at the negative effects of congestion which have prompted the increased emission of GHG emissions as car spends more time on the road. Many hours of productivity have been lost, and urban heats have been created, and this has prompted many governments to consider decarbonizing their transport systems. Some governments have therefore engaged in decarbonization exercises, and this has seen the increased intensification of the public transport system in the African landscape and the introduction of the Gautrain in South Africa. There is, however, a need to also draw lessons from the international platform and the need for increased government support and funding in order to implement workable decarbonization projects in Africa's transport system as it comes along with a lot of advantages such as a cleaner operating environment and improved efficiency on the road. In order to improve the transport and mobility flow of vehicles in city centers and to lessen the challenge of congestion on the African landscape, there is need for:

- Increased commitment and funding by the government to sponsor the development of public transport systems as a way to cut the number of mini-commuter buses and private cars from entering the city and causing congestion
- Encouragement of local industries to manufacture electric cars, as well as facilitate for the importation of electrified cars
- Governments to ban the importation of diesel and petrol fueled cars, as well as impose high taxes for the importers and users of petrol- and diesel-powered cars
- Governments to invest more in electric trains as they allow people to move in large numbers
- Capacity building of the locals by educating them on the importance of using electric cars and renewable energies on the environment

Cross-References

▶ Amsterdam's Pathway to Climate Neutrality: Creating an Enabling Environment

References

Avci, G. Z., & Özener, O. (2019). An electric public transportation vehicle modelling and comparison with conventional diesel vehicle. *International Journal of Engineering Research and Advanced Technology, 5* (4), 59–66.

Baxter, A. (2015). *Network rail a guide to overhead electrification. no. February.* London: Network Rail.

Boncenne, A. (2012). Assessing the impact of the Canadian tax credit for public transit passes. Available online: https://papyrus.bib.umontreal.ca/xmlui/bitstream/handle/1866/8832/rapport_recherche_ABoncenne.pdf

Boulouchos, K., Sturm, P. J., Kretzschmar, J., Duic, N., Laurikko, J., Bradshaw, A., Harmacher, T., Bettzüge, M. O., Giannopoulos, G., La Poutré, H., & Blok, K. (2019). Decarbonisation of transport: Options and challenges. Available online: https://www.sccer-mobility.ch/export/sites/sccermobility/p_supporting_measures/Annual-Conferences/AC2019/dwn_AC19/PDFPraesentationen/02_SCCER_Annual_Conference_2019_KB.pdf

Bouton, S., Knupfer, S. M., Mihov, I., & Swartz, S. (2015). Urban mobility at a tipping point. *McKinsey and Company.* Available online: http://worldmobilityleadershipforum.com/wp-content/uploads/2016/04/urban_mobility_at_tipping_point_final.pdf

Camagni, R., Gibelli, M. C., & Rigamonti, P. (2002). Urban mobility and urban form: The social and environmental costs of different patterns of urban expansion. *Ecological Economics, 40*(2), 199–216.

Drescher, A. W., & Iaquinta, D. L. (2002). Urbanization – Linking Development across the Changing Landscape. Final Draft. State of Food and Agriculture (SOFA) – Special Paper on Urbanization.

European Rail Research Advisory Council. (2009). Metro, light rail and tram systems in Europe. https://www.uitp.org/sites/default/files/cck-focus-papersfiles/errac_metrolr_tramsystemsineurope.pdf

Hogarth, J. R., Haywood, C., & Whitley, S. (2015). *Low-carbon development in sub-Saharan Africa: 20 cross-*

T

sector transitions. London: Overseas Development Institute (ODI).

Hussein, H. A. (2013). The role of street traffic signs in reducing road accidents. In *First international symposium on urban development: Koya as a case study* (Vol. 303). WIT Press.

Jennings, G. (2011). A guide to low carbon transport. Transportation, climate change and the UN Framework Convention on Climate Change 17th conference of the Parties (COP17) in Durban.

Longo, J., Kuras, E., Smith, H., Hondula, D. M., & Johnston, E. (2017). Technology use, exposure to natural hazards, and being digitally invisible: Implications for policy analytics. *Policy & Internet, 9*(1), 76–108.

Lu, J. (2011). Environmental effects of vehicle exhausts, global and local effects: A comparison between gasoline and diesel. *Journal of Energy in Southern Africa, 6*, 171.

Mahajan, A., Potnis, N., Gopalan, K., & Wang, A. (2006). Urban mobility models for vanets. In: *2nd IEEE International Workshop on Next Generation Wireless Networks* (Vol. 33, p. 38).

McKinnon, A. (2009). Decarbonizing freight transport. In *European responsible care conference*. Available online: http://www.climatecouncil.ie/media/Transpor%20Transition%20Paper%202%20McKinnon%20Decarbonizing%20Freight.pdf

McKinnon, A. (2010). Green logistics: The carbon agenda. *Electronic Scientific Journal of Logistics, 6*(1), 1–9.

Mobereola, D. (2009). *Africa's first bus rapid transit scheme: The Lagos BRT-Lite system* (No. 53497, pp. 1–54). Washington DC: The World Bank.

Ozener, O., Ozkan, M., Orak, E., & Acarbulut, G. (2018). A fuel consumption model for public transportation with 3-D road geometry approach. *Thermal Science, 22*(3), 1505–1514.

Philp, M. (2015). RP-2015 Carbon Reductions and Co-benefits: Final Report–Part I, Literature. Available online: https://apo.org.au/sites/default/files/resource-files/2015-05/apo-nid61816.pdf

Reşitoğlu, İ. A., Altinişik, K., & Keskin, A. (2015). The pollutant emissions from diesel-engine vehicles and exhaust after treatment systems. *Clean Technologies and Environmental Policy, 17*(1), 15–27.

Roychowdhury, A. (2018). *Clunkered: Combating dumping of used vehicles – A roadmap for Africa and South Asia*. New Delhi: Center for Science and Environment.

Serrano-Jiménez, D., Abrahamsson, L., Castaño-Solís, S., & Sanz-Feito, J. (2017). Electrical railway power supply systems: Current situation and future trends. *International Journal of Electrical Power & Energy Systems, 92*, 181–192.

Tagliapietra, S., & Zachmann, G. (2018). *Addressing Europe's failure to clean up the transport sector. Policy Brief 2018/02*. Bruegel.

United Nations Economic Commission for Europe (UNECE). (2015). In United Nations (Ed.), *Sustainable urban mobility and public transport in UNECE capitals*.

Vencataya, L., Pudaruth, S., Dirpal, G., & Narain, V. (2018). Assessing the Causes & Impacts of traffic congestion on the society, economy and individual: A case of Mauritius as an emerging economy. *Studies in Business and Economics, 13*(3), 230–242.

Venkat, K. (2016). Indicator model for benchmarking the transition to a low carbon urban mobility system: Application results from three Scandinavian cities. *IIIEE Master's Thesis*.

World Bank. (2015). Shifting gears: Toward resilient and low-carbon transport how the World Bank integrates climate in its transport lending and policy work. Available online: http://pubdocs.worldbank.org/en/171991478841153943/4pagerShiftingGearRev6-web.pdf

Treatment Urban Wetlands

▶ Artificial Urban Wetlands

Tree-Based Farming

▶ *Wadi* Sustainable Agriculture Model, The

Turkey Energy Transformation

▶ Smart Grids to Lower Energy Usage and Carbon Emissions: Case Study Examples from Colombia and Turkey

Typology

▶ Theme Cities Networks

U

Understanding Smart Cities Through a Critical Lens

Claudia Fonseca Alfaro[1], Lorena Melgaço[2] and Guy Baeten[3]
[1]Institute for Urban Research, Malmö University, Malmö, Sweden
[2]Department of Human Geography, Lund University, Lund, Sweden
[3]Urban Studies, Malmö University, Malmö, Sweden

Introduction

Defined broadly, smart cities are policies or urban development projects that harness digital technologies to respond to the challenges and risks generated by urban growth. Smart city agendas are being promoted by private and public entities at different levels (from transnational to local) and mobilized by local governments to achieve social inclusiveness, economic growth, and sustainability. The proposed interventions – often designed in close collaboration with the private sector – range from strategies to mitigate natural disasters and the management of urban services to achieve efficiency, to the design of digital interfaces between citizens and government and between residents and urban services providers. Judging by the number of projects across the globe being advertised and developed, it is fair to say that smart cities are becoming a favored strategy and policy framework for international organizations, local and national governments, and the private sector. Recent examples of smart city interventions include the following: (a) IBM's "Smarter Cities Challenge" which, with the exception of Antarctica, boasts projects in all continents (IBM n.d.); (b) India's plan to build "100 smart cities" (Datta 2015a); (c) 200 smart pilot project s in China (Yin et al. 2015); (d) New Clark City in the Philippines (Mouton 2020); (e) Google's now defunct Canadian project to redevelop Toronto's Eastern waterfront through its sister company Sidewalk Labs (SidewalkLabs 2018; Baeten 2020); and (f) what Kitchin (2015) refers to as the "canonical examples" – Songdo in South Korea, Masdar in the United Arab Emirates, PlanIT Valley in Portugal, and Rio de Janeiro's Operations Centre in Brazil. Less prominent projects have also been launched in Latin American countries such as Argentina, Chile, and Mexico (Vivas et al. 2013). Intergovernmental institutions are engaging, as well, with smart cities, and their activities range from concrete strategies and initiatives to roundtables, briefs, and publications. For example, these activities include the United Smart Cities program from the United Nations Economic Commission for Europe (UNECE); the Roundtable on Smart Cities and Inclusive Growth launched by the Organisation for Economic Co-operation and Development (OECD); the EU-financed Smart Cities Marketplace that brings together cities, industries, small and medium-sized enterprises, investors, researchers,

R. C. Brears (ed.), *The Palgrave Encyclopedia of Urban and Regional Futures*,
https://doi.org/10.1007/978-3-030-87745-3

and other smart city actors; the World Bank's brief on smart cities; and the Inter-American Development Bank (IDB) publication *The Road Toward Smart Cities: Migrating from Traditional City Management to the Smart City.*

A growing adhesion to the smart city agenda may be partially attributed, at least at the local level, to the effects of austerity policies and neoliberalization strategies. Yet, the need for efficiency felt by local governments goes hand in hand with pressure from the private sector in its search for new markets (Kitchin et al. 2019). A second factor fueling the popularity of these types of interventions – against a complex background of measurement and ranking systems (Mora et al. 2019) – is a "globalizing narrative" that positions smart cities as best practices and, thus, creates worldwide competition among what are perceived to be "lighthouse" and "follower" cities (Joss et al. 2019). This creates a third driving force. Faced with these circumstances, local governments may also strategically operationalize the smart city agenda to support existing priorities at the city level (Haarstad and Wathne 2019; Wiig 2015).

This entry seeks to provide an overview of smart city critiques that have been formulated in recent years by mainly social scientists. We will use this overview to distil potential paths of smart cities in the near future and make suggestions to adapt the smart city research agenda to these potential future developments.

A Critique of Smart Cities

Critical scholarship has increasingly dedicated attention to the phenomenon of the smart city. Scholars have (a) highlighted the dangers when corporate power influences the design of cities, (b) questioned the smart city's one-size-fits-all policies, (c) called attention to how smart projects might exacerbate urban inequalities, and (d) problematized the use of big data to steer urban development (Evans et al. 2019; Hollands 2008; Kitchin 2015; Marvin et al. 2016; Shelton and Clark 2016; Shelton and Lodato 2019; Söderström et al. 2014; Söderström and Mermet 2020; Vanolo 2016). A commonality in this critique is a concern with the neoliberal underpinning embedded in the proposed technological solutions to urban inequalities. Smart city frameworks, visions, and projects have trouble addressing social challenges that are gendered, racialized, uneven (locally and globally), or determined by class and age. A lack of commitment to social justice creates a context where visions of *smart* future cities are mainly understood in economic terms. Feminist interventions raise concerns the smart city can have an effect in the reproduction and deepening of gender inequalities (cf. Listerborn and Neergaard 2021). From a postcolonial perspective, scholars question the role of the smart city paradigm in reproducing an urbanism road map that is entrenched in the contradictions of capitalist modernity and developmentalist understandings of the global South, and as such, reproduces power, knowledge, and technological hierarchies and inequalities (Datta 2018; Datta and Odendaal 2019). At the center of a sociopolitical discussion is the concern that the economic attractiveness and popularity of the smart city discourse might depoliticize the urban arena through universal models of modernization and development (Vanolo 2014). The city, thus, becomes a blank canvas for the trial and operationalization of technological interventions (Evans et al. 2019; McFarlane and Söderström 2017). In other words, smart city frameworks, if unchecked and not debated, might not only create sociopolitical challenges by exacerbating already existing patterns of unequal development and socio-spatial inequalities, but also have profound implications on democracy, governance, and citizen's freedom. The following sections expand on the points raised by critical scholars.

Urban Inequalities and Power

There are concerns that the smart city is actually not capable of addressing social inequalities despite stated aims to solve urban challenges. Analyzing the case of Digital On-ramps – a project to promote inclusion in Philadelphia, USA – Wiig (2016) argues that smart city projects have a tendency to simplify the complexity of urban challenges (i.e., marginalization) and come up with what seem to be straightforward solutions that are, in reality, problematic. Examining the workings of Uber in Washington DC and exploring the labor that is needed to collect enough data for digital solutions to function, Attoh et al. (2019) conclude that the "idiocy of the smart city" (i.e., its need for data) reproduces "the exploitation and asymmetrical power relations" of the gig economy. Looking at the European context, Martin et al. (2019) make a case that flagship projects largely neglect social equity while, for example, pursuing environmental agendas that aim to green and decarbonize urban economies. Applying their lens to other parts of the world, scholars that have researched cases in emerging or developing economies have criticized how *smart* projects ignore the realities of poverty and inequality. Datta (2015a), looking at a case in India, compares the immense gaps in infrastructure between existing settlements and envisioned smart cities, concluding that the projects are business-oriented "utopian imaginings." Following a similar line of argument, Watson (2015) warns that projects in Africa could potentially redirect investment from the basic infrastructure that is needed toward a mirage of *smart* that she labels "fantasy cities." Providing a case from Nairobi, Guma (2019) describes how information and communications technology (ICT) solutions for water and electricity supply provide a "differentiated" and "splintered" service to poor communities. Looking at Brazil, Luque-Ayala and Neves Maia (2019) challenge the idea suggested that mapping favelas in Rio de Janeiro using digital technology is a "mechanism for socio-economic inclusion."

Gender and Intersectionality

Feminist contributions foreground the implications of the smart city in the reproduction and deepening of gender and other intersectional inequalities in the urban context. Listerborn (2022) – thinking about the everyday – argues that digitalization does not act in a "no-(wo-) mans-land." Instead, these processes and discourses emerge from norms and practices embedded in the sexism found within the technology industry, existing gender digital bias, and data and gender gaps. Similarly, but turning their eyes to the impacts of the platform economy, Bauriedl and Strüver (2020) claim gender norms are "reproduced by the platformisation of everyday life." Rose (2017) highlights how the smart city reproduces a white-male-centered model that enables already mobile people to be even more mobile, affecting other groups and hindering local forms of agency and governance. Bringing in an intersectional perspective, Wigley and Rose (2020, p. 195) highlight that, when spatialized, the smart city not only impacts women, but also gendered *and* racialized populations. Classism and ageism are also a problem. For example, offering a class perspective on the matter, Aurigi and Odendaal (2020, p. 4) argue the idea of *smart* reinforces a "middle-class oriented drive toward urban development and management trajectories," leaving little room for reinventing urban space along other lines. Suopajärvi (2018, p. 79), on the other hand, brings light to the marginalization senior citizens suffer when they are stripped from the right to "make a voluntary decision to become smart citizens."

One-Size-Fits-All? Corporate Interests and Urban Entrepreneurialism

The influence of the private sector in defining the smart city agenda is highly debated and questioned by scholars. Aurigi and Odendaal (2020, p. 1) argue smart city projects are often designed by large corporations according to "delineated norms and standards" that are technological and business-oriented in nature and where the outcome is a "pre-packaged, product-like version of the smart city." Such a model of city-making contrasts with the actual messiness or complexity of the urban fabric (i.e., different socio-spatial and political contexts) of both the "high-tech metropolis" (Aurigi and Odendaal 2020), which serves as reference for the smart city, and less prominent cities that face, for example, problems with urban inequalities. While this is particularly true for the Global South, such concerns should not be overlooked in the Global North either (Aurigi and Odendaal 2020; see also Söderström et al. 2014; Viitanen and Kingston 2014). Looking at the Global South, Jirón et al. (2021) describe how policymakers and private companies attempted to emulate a "global narrative and aesthetic" through a smart city proposal in Santiago do Chile that neglected context. Unrooted from the local, it became a type of "placebo urban intervention" – a performative action that fails to address inequality but is successful in *camouflaging* neoliberal urbanization. From a perspective in the Global North, Viitanen and Kingston (2014) argue that the geographical choices to invest in smart city initiatives and infrastructure, such as in the "Manchester Oxford Corridor" – developed to host knowledge-intensive industries – create "premium networked-spaces" that deepen existing social inequality in the city. A factor that explains the influence of corporate actors in cities is a climate of urban entrepreneurialism that trusts tech companies to successfully steer urban development through a vision of technological intervention (Barns and Pollio 2018; Hollands 2008; Wiig 2015). According to Söderström et al. (2014), this climate produces a form of "corporate smart storytelling" that promotes a "technocratic fiction" where data and software trump knowledge and specific expertise and where public investment is channeled to technology. Through a "rhetorics of urgency" (Datta 2015a), corporations claim challenges posed by urbanization, resources depletion, and climate change can only be addressed through a paradigmatic change centered on technological fixes (Aurigi and Odendaal 2020; Kaika 2017; Martin et al. 2019).

Technocratic Governance and Neoliberal Governmentality

There are also challenges posed by technocratic governance and neoliberal governmentality. The existence of the smart city often relies on the close cooperation between large corporations and local governments. When this is the case, McGuirk et al. (2021) argue the state (central to smart city governance), embarks on an experimental exercise through which it adopts a probusiness and entrepreneurial model of urban development that provides little room for citizen input and diminishes the possibilities for democratic decision-making (cf. Hollands 2015; Shelton and Lodato 2019). Kitchin et al. (2019) see this process as an example of neoliberal governmentality and citizenship – a context where citizens participate in the urban through consumption and not through democratic processes. Blinded by what Levenda (2019, p. 566) calls a "spectacle" of smartness, individual consumption trumps the strive for systemic change that addresses urban inequalities. Vanolo (2014) describes the discourse fueling this logic of depoliticization as "smartmentality" or practices that help produce "docile subjects and mechanisms of political legitimisation." As the devolution principle burdens cities, smart cities expand "urban austerity" and justify the popularity of public-private partnerships, according to Hollands (2015). This form of governance seems to prevail in the West, and Curran and Smart (2021) highlight that differences in governance worldwide need to be considered. In China, for example, state entrepreneurialism may rely on state-owned enterprises, not on private actors. Looking at India, Datta (2015b), Datta et al. (2022) argues the smart city is energized by a young population seeking modernity and driven by the state for the possibilities that smart strategies offer to centralized authoritarian ruling. While the specific interaction between private and state actors may vary considerably across geographies and needs to be further studied – thinking through Braun (2014) – it seems like the state is increasingly working *through* technological objects to govern urbanites in the smart city.

Big Data, Algorithmic Violence, and Cybersecurity

The possible repercussions that big data (vis-à-vis the smart city) can have on privacy, integrity, and cybersecurity are also topics that draw academic interest. There is critical scholarship on platform capitalism (Srnicek 2016), surveillance capitalism (Zuboff 2019), data capitalism (West 2019), and algorithm and data-driven governance (Curran and Smart 2021; Just and Latzer 2017) attempting to explore how societies are being shaped and impacted by big data. According to Firmino et al. (2019), the embeddedness of digital technology in the everyday not only influences the urban form but also leaves a digital trace that is increasingly mobilized for profit and control. For example, in authoritarian countries with little data privacy or protection, big data is being accumulated and used for social classification. Curran and Smart (2021) highlight how the Chinese government relies on the algorithmic classifications of people to analyze the country's security and social credit system. Data collected allows the state to expose specific groups (for example, people that work in the informal economy or individuals that borrow money) who are subsequently categorized as part of a "risk-class." The surveillance of citizens through the use of big data opens up questions of social justice (cf. Heeks and Shekhar 2019). Looking at the case of Blue CRUSH – a policing program developed by IBM and Memphis municipality in the United States – Tulumello and Iapaolo (2021) unveil the tensions surrounding "technical neutrality." The authors argue programs that are believed to be "neutral" often ignore the political nature, and possible bias, of how "social problems" are defined. Other research shows how decisions based on algorithms may negatively impact historically disadvantaged groups either because of bias or lack of data (cf. Barocas and Selbst 2016). According to Benjamin (2019 in Tulumello and Iapaolo 2021), digital data is "produced through histories of exclusion and discrimination." For example, little data regarding women's unpaid care labor results in less investment in social and welfare infrastructure (Perez 2019). Another example is how digital surveillance systems disproportionally target racialized groups (Noble 2018, see also Chang 2019; Eubanks 2018 for further discussion on how digital data furthers existing biases and discrimination). Safransky (2020) urges us to "investigate [the] geographies of algorithmic violence" thinking through the operationalization of Market Value Analysis, a data-driven tool used to evaluate and guide spatial investment and disinvestment in US cities. There is a risk these types of systems reproduce "racial, infrastructural, and epistemological violence" (ibid). In short, automated algorithm selection can shape reality, "increas[ing] individualization, commercialization, inequalities, and deterritorization and decreas[ing] transparency, controllability and predictability," in the words of Just and Latzer (2017, p. 238). A final important issue to highlight is cybersecurity. Kitchin and Dodge (2019) argue that, despite the goal of smart city technologies to address urban challenges and risks, these types of interventions can "paradoxically [...] create new vulnerabilities and threats, including making city infrastructure and services insecure, brittle, and open to extended forms of criminal activity."

The Global South: Colonialism, Imperialism, and Development

Another strand of research questions if smart cities are colonial and imperial modes of development that will simply bring homogenization and a specific type of modernity to the Global South. Jirón et al. (2021) see the smart city agenda as a form of "monist core theory." Watson (2014) agrees arguing that the universalization of the *smart* framework attempts to homogenize cities around the globe when, instead, specific historical conditions and socio-spatial realities should be considered. There is not only homogenization but also echoes of colonialism and imperialism. The smart city framework advances the idea of modernization and development through a modernity/coloniality tension, according to Melgaço and Milagres (forthcoming). Along the same lines, Datta (2015b) concludes smart cities reflect a type of utopian urbanism that dictates a "strong postcolonial model of modernity, rationality in development." Moser (2015, p. 33), drawing on empirical material in India, asserts smart city projects still reproduce "the language of the colonial 'civilizing mission'" whose modernization is dependent upon "technologically savvy Western-sanctioned global 'experts.'" Looking at South Africa and its relationship with surveillance capitalism, Kwet (2019) describes smart cities as a form of "digital colonialism" that is enabled and promoted by US-based tech corporations. In Kwet's view, this is a new type of imperialism that nowadays haunts the Global South. With the reproduction of colonial and imperial ideals comes symbolic violence in the sense that the smart city bypasses "citizenship, democracy, privacy and data justice" while reproducing geopolitical relations of inequality (Datta and Odendaal 2019). Yet, the smart city should not be reduced to a "Western import" since, at least in the case of India, the framework was a global model brought in but *provincialized* at various scales of the state (Datta et al. 2022). In short, local power structures in place and operationalizations of the concept need to be taken into account.

Resisting: The Emergence of Antismart City Grassroots

Since the Facebook-Cambridge Analytica scandal in 2018 and a range of other debates centered around privacy, antitrust, and misinformation (Paul 2019), citizens and organizations have begun to adopt a more skeptical attitude toward tech giants and their ways of collecting and processing data. This so-called "techlash" has found its way into growing critiques of and active resistance toward smart city projects and other urban initiatives by tech giants. Arguably the most high-profile case of smart city resistance has taken place in Toronto, Canada, where Google's sister company Sidewalk Labs had agreed with Waterfront Toronto, a three-government-level planning authority, to develop plans for parts of the brownfields located in the shadow of the city's downtown. A loose but sizeable collection of activists, academics, intellectuals, union leaders, and entrepreneurs have voiced concerns over Sidewalk Labs' plans ever since the announcement of the project in 2017. The protesters eventually organized themselves in 2019 behind the banner "Blocksidewalk" – arguably the first-ever form of organized resistance against smart cities. Looking at an example from the global South, in Guadalajara, the central Mexican government and the city government lined up with MIT and the Arup Group in the 2010s to build a Creative Digital City. The proposal aimed to house a variety of ICT industries by replacing an inner-city residential and commercial neighborhood. The plan triggered protest from a range of local groups, including local residents and businesses who fear displacement (de la Peña 2017). In another case from Latin America, the project for a model city in Honduras – the Zone for Economic Development and Employment (ZEDE) in the Gulf of Fonseca – also met local opposition. Peasant activists, a local environmental NGO, and representatives of the Catholic Church created an awareness-raising campaign to confront the elite and the transnational actors involved in the project, who in their view threatened their livelihoods through, for example, the expropriation of land and environmental degradation (Lynch 2019). Moving to Asia, as in the Gulf de Fonseca ZEDE, local movements challenged the fast-track project development of Dholera smart city in India. Fearing eviction and dispossession, the Land Rights Movement Gujarat organized a series of protests in the region, causing bottle-necks to the city's construction (Datta 2015a). These protests have in common the emergence of an anti-smart city grassroots movement – across different constellations and contexts – in the wake of planned proposals by big global corporations or powerful state actors. This supports the observation made by Leitner et al. (2007, p. 8) that resistance against neoliberal urbanism is not merely a "reaction" but a contestation that can "rework neoliberal practices and imaginaries" or, in this case, smart city practices and narratives. These varied examples point to the existence of the "proliferation of practices of dissent, dissatisfaction and disagreement across the world" (Kaika 2017, p. 98) that question the undergirding tropes of sustainable and smart cities.

In addition to grassroots resistance, other actors, such as local administrations, are furthering the use of digital technology to, perhaps paradoxically, resist the corporate smart city. Barcelona has become a reference for radical local experiments, in which the municipality is repurposing digital infrastructure to foster access to information and participation in decision-making (Charnock et al. 2021). However, technologies are not the only tools being repurposed. There are examples of local governments reworking the very conceptualization of *smart* in order to access funding and carry out what they consider are more important infrastructural interventions (Datta et al. 2022). Turning to everyday resistance against the corporate smart city, this can be found in more localized, nonconfrontational attempts to rework the idea of smart through, for example, "small-scale sociotechnical interventions" (Hollands 2015). Leontidou (2015)), looking at Mediterranean Europe, highlights how the use of digital technology not only facilitates resistance to political and economic autocracies but may also support new cultural forms of countermainstream grassroots. Following this line of thought, Odendaal (2021, p. 651) suggests engaging with smart urbanism "as a continuous process of emergence and remaking." Analyzing two grassroots projects that use digital technologies and social media to expose social inequalities (MapKibera in Nairobi, Kenya, and Reclaim the City in Cape Town, South Africa), the author argues that emancipatory smart practices must consider how "technology appropriation is deeply attached to livelihood strategies and place." In a similar vein, Hollands (2015) gives the example of Brickstarter – an urban crowdsourced platform – to highlight how technology can help citizens engage with local city planning. Luque-Ayala et al. (2020) also discuss how platforms may be mobilized toward aims that exist outside the realm of capital accumulation. For the authors, *data_labe* – a local initiative based in the North of Rio de Janeiro – is an example of how "citizen-generated data" and digital technologies can be used by favela residents to support or carry out "political activism on issues of gender, race and the environment."

Engaging with the Messy Futures of Smart Cities: Possible Research Agendas for the Future

As the nascent field of critical smart urbanism (cf. Luque-Ayala and Marvin 2015) moves forward, possible research agendas for the future include four priorities: (a) investigate the actual implementation of smart city projects; (b) include more cases from across the Global South and rural contexts to, in turn, develop comparative approaches; (c) rethink the smart citizen; and (d) study with more nuance the realities of resistance.

The Implementation of Smart City Projects

Luque-Ayala and Marvin (2015) have argued that despite their now almost ubiquitous presence, there are gaps in our understanding of how smart city projects are being implemented. In their view, there is a need to not only reflect what the social and political implications for the urban might be (e.g., on democracy, governance, and visions of "the good city"), but also to investigate how the concept *smart* is actually being operationalized by local actors (e.g., in partnerships between municipalities and ICT companies). Reaching a similar conclusion, Shelton and colleagues (2015) propose a shift between the analysis of the smart city as a "paradigm" to the study of "actually existing smart cities." Answering the call, scholars have begun to analyze empirical examples of smart city projects and initiatives. The edited volume *Inside Smart Cities: Place, Politics and Urban Innovation* (Karvonen et al. 2019) – which includes cases from the Global North and South – stands out for its attention to the scale of the neighborhood and the city. Complementing this piece, a Special Issue in Urban Planning, edited by Karvonen et al. (2020), looks at concrete examples from Canada, Germany, Scandinavia, and China to explore how urban planners navigate and put to practice smart city projects.

Smart Across Different Geographies

A second point raised by Luque-Ayala and Marvin (2015) is the need for smart urbanism to develop an "understanding of smart across contrasting geographies." Given the previous examples, it is clear that our knowledge of how *smart city* projects, agendas, or policies are being implemented is expanding little by little. However, it is important to point out that, up to now, critical smart urbanism has mostly focused on investigating cases from the Global North, with only a few studies from the Global East or South: the African continent in general, Brazil, Chile, China, India, Kenya, Philippines, South Africa, and the United Arab Emirates (cf. Datta, 2015b; Guma 2019; Jirón et al. 2021; Karvonen et al. 2019; Luque-Ayala and Neves Maia 2019; Mouton 2020; Odendaal 2021; Tironi and Valderrama 2018; Watson 2015). As explained before, the uneven relevance given to the Global South is compounded by approaches to the study of smart cities that reproduce Eurocentrism and coloniality.

U

In response to this – and following a general trend in urban studies toward postcolonizing and decolonizing the field and moving toward comparative tactics (see Robinson 2016; Roy 2009) – the Special Issue "Smart cities between worlding and provincializing" in the journal *Urban Studies* (Burns et al. 2021) dislocates the center of smart urbanism. The Special Issue also brings to light another important factor: The critical field of smart urbanism has ignored the realities of rural contexts. Looking at two small and remote communities in Canada, Spicer et al. (2021) argue that the implementation of digital technologies has allowed local government to collaborate and provide services that, in the end, have improved the quality of life of residents. For the authors, this is an example of citizen-first, horizontal governance from which larger municipalities can learn.

Rethinking the Smart Citizen

Reflecting on Rossi's (2016) argument that it is not productive to assume that all smart cities are designed by corporations in a top-down movement that ignores the political, Caprotti (2019, p. 2476) asks: "where can the non-corporate, non-governmental smart city be found in the ordinary cities of today?" While the *smart citizen* is stereotypically seen as a neoliberal subject, "the 'actually existing smart citizen' plays a much messier and more ambivalent role in practice" (Shelton and Lodato 2019, p. 36). Zandbergen and Uitermark (2020) provide an example from Amsterdam where residents have used Smart Citizen Kits provided by public-private actors to measure air quality. In the authors' view, the project enabled messy articulations of citizen participation as the users used the collected data to negotiate their relationship with the government and participate in political debate. The edited volume *The Right to the Smart City* (Cardullo et al. 2019) seeks to reflect on similar questions of citizen participation, providing a few examples of civic engagement through hackathons and big data appropriation. In short, for these authors, smart urbanism can also offer potential for a project of urban social justice where technology is reappropriated. This an important aspect of *smart* that remains underrepresented in the literature.

Sovereignty in the City

Building on the work of Sadowski (2021) and Pasquale (2017), among others, there is also a need to investigate *if* and *how* future smart city projects will entail contestation over authority in the city. What is at stake is not only sovereignty over certain functions such as transit, but also control over entire city districts (e.g., the case of Sidewalk Labs in Toronto). In smart cities, urban planning as has traditionally been implemented can become redundant if data and algorithmic come to continuously define and redefine optimal uses of the built environment (e.g., from park benches that measure their own user frequency to the occupancy of modular buildings that can be continuously altered). Future research will need to carefully scrutinize the relation between smart city products and their actual impact on levels of private and public control over the design of city districts and their services. Broadening our understandings toward difference and beyond universalist views of the smart city can benefit from grounded discussions of resistance against corporate-led smart cities (see Irazábal and Jirón 2021; Perng and Maalsen 2020, among others). Nevertheless, there will still be a need to consider how, swimming against the trends of oligopoly and standardization of platforms, these initiatives are always vulnerable and may not thrive (Curran and Smart 2021). While small-scale approaches may be less effective political strategies in changing the structures that sustain the corporate smart city, investigating them may give researchers pointers on how communities can engage in challenging the mainstream (Mouton and Burns 2021).

Conclusion

As a general assessment, the three occurrences of smart city interventions (i.e., the provision of digital governance tools to municipalities, digital service provision, and smart cities on property development) will continue to develop side-by-side, and together they will form the constituting elements of ever-expanding attempts to reduce the urban and its inhabitants to a data mine upon which profit can be made. The range of critiques formulated by academics and activists alike – together with the everyday practices of resistance – will continue to be reiterated in different forms, and they will provide intellectual input to acts of resistance and protests against smart city initiatives and projects across the globe. The discrepancy between marketing promises and material realities of actually existing smart cities will continue to manifest itself, as well. The smart city concept will remain a constitutive element of urban planning for years to come, but its various implementations in various parts of the world will unavoidably be patchy, fragmented, and often contested. A closer look at the actual unfolding of smart visions, projects, developments, implementation, and repurposing at all levels will need further attention.

References

Attoh, K., Wells, K., & Cullen, D. (2019). "We're building their data": Labor, alienation, and idiocy in the smart city. *Environment and Planning D: Society and Space, 37*(6), 1007–1024. https://doi.org/10.1177/0263775819856626.

Aurigi, A., & Odendaal, N. (2020). From "Smart in the Box" to "Smart in the City": Rethinking the socially sustainable smart city in context. *Journal of Urban Technology, 0*(0), 1–16. https://doi.org/10.1080/10630732.2019.1704203.

Baeten, G. (2020). Sidewalk Labs' plans for Toronto shake the foundations of planning as we know it. *Plan Canada, 2020*, 26–28.

Barns, S., & Pollio, A. (2018). Parramatta smart city and the quest to build Australia's next great city. In A. Karvonen, F. Cugurullo, & F. Caprotti (Eds.), *Inside smart cities* (1st ed., pp. 197–210). Routledge. https://doi.org/10.4324/9781351166201-13.

Barocas, S., & Selbst, A. D. (2016). Big data's disparate impact. *California Law Review, 104*, 671.

Bauriedl, S., & Strüver, A. (2020). Platform urbanism: Technocapitalist production of private and public spaces. *Urban Planning, 5*(4), 267–276. https://doi.org/10.17645/up.v5i4.3414.

Braun, B. P. (2014). A new urban dispositif? governing life in an age of climate change. *Environment and Planning D: Society and Space, 32*(1), 49–64. https://doi.org/10.1068/d4313.

Burns, R., Fast, V., Levenda, A., & Miller, B. (2021). Smart cities: Smart cities between worlding and provincializing [Special Issue]. *Urban Studies, 58*(3), 461–673. https://doi.org/10.1177/0042098020975982.

Caprotti, F. (2019). Spaces of visibility in the smart city: Flagship urban spaces and the smart urban imaginary. *Urban Studies, 56*(12), 2465–2479. https://doi.org/10.1177/0042098018798597.

Cardullo, P., Di Felicaiantonio, C., & Kitchin, R. (Eds.). (2019). *The right to the smart city* (1st ed.). Emerald Publishing.

Chang, E. (2019). *Brotopia: Breaking up the boys' club of silicon valley*. Portfolio.

Charnock, G., March, H., & Ribera-Fumaz, R. (2021). From smart to rebel city? Worlding, provincialising and the Barcelona model. *Urban Studies, 58*(3), 581–600. https://doi.org/10.1177/0042098019872119.

Curran, D., & Smart, A. (2021). Data-driven governance, smart urbanism and risk-class inequalities: Security and social credit in China. *Urban Studies, 58*(3), 487–506. https://doi.org/10.1177/0042098020927855.

Datta, A. (2015a). New urban utopias of postcolonial India: 'Entrepreneurial urbanization' in Dholera smart city, Gujarat. *Dialogues in Human Geography, 5*(1), 3–22. https://doi.org/10.1177/2043820614565748.

Datta, A. (2015b). A 100 smart cities, a 100 utopias. *Dialogues in Human Geography, 5*(1), 49–53. https://doi.org/10.1177/2043820614565750.

Datta, A. (2018). The digital turn in postcolonial urbanism: Smart citizenship in the making of India's 100 smart cities. *Transactions of the Institute of British Geographers, 43*, 405–419.

Datta, A., & Odendaal, N. (2019). Smart cities and the banality of power. *Environment and Planning D: Society and Space, 37*(3), 387–392. https://doi.org/10.1177/0263775819841765.

Datta, A., Mackinnon, D., Fast, V., & Burns, R. (2022). Dialogue with Ayona Datta. In D. Mackinnon, V. Fast, & R. Burns (Eds.), *Digital (in)justice in the smart city*. University of Toronto Press view.

de la Peña, A. (2017). *Contesting the smart city: A case study from the Global South*. CRUSH working paper.

Eubanks, V. (2018). *Automating inequality: How high-tech tools profile, police, and punish the poor*. St. Martin's Press.

Evans, J., Karvonen, A., Luque-Ayala, A., Martin, C., McCormick, K., Raven, R., & Palgan, Y. V. (2019). Smart and sustainable cities? Pipedreams, practicalities and possibilities. *Local Environment, 24*(7), 557–564. https://doi.org/10.1080/13549839.2019.1624701.

Firmino, R., Cardoso, B., & Evangelista, R. (2019). Hyper-connectivity and (Im) mobility: Uber and surveillance capitalism by the global south. *Surveillance and Society, 17*(1/2), 205–212.

Guma, P. K. (2019). Smart urbanism? ICTs for water and electricity supply in Nairobi. *Urban Studies*. https://doi.org/10.1177/0042098018813041.

Haarstad, H., & Wathne, M. W. (2019). Are smart city projects catalyzing urban energy sustainability? *Energy Policy, 129*, 918–925. https://doi.org/10.1016/j.enpol.2019.03.001.

Heeks, R., & Shekhar, S. (2019). Datafication, development and marginalised urban communities: An applied

U

data justice framework. *Information, Communication & Society, 22*(7), 992–1011. https://doi.org/10.1080/1369118X.2019.1599039.

Hollands, R. G. (2008). Will the real smart city please stand up? Intelligent, progressive or entrepreneurial? *City, 12*(3), 303–320. https://doi.org/10.1080/13604810802479126.

Hollands, R. G. (2015). Critical interventions into the corporate smart city. *Cambridge Journal of Regions, Economy and Society, 8*(1), 61–77. https://doi.org/10.1093/cjres/rsu011.

IBM. (n.d.). *IBM smarter cities challenge*. Retrieved April 7, 2018, from https://www.smartercitieschallenge.org

Irazábal, C., & Jirón, P. (2021). Latin American smart cities: Between worlding infatuation and crawling provincialising. *Urban Studies, 58*(3), 507–534. https://doi.org/10.1177/0042098020945201.

Jirón, P., Imilán, W. A., Lange, C., & Mansilla, P. (2021). Placebo urban interventions: Observing Smart City narratives in Santiago de Chile. *Urban Studies, 58*(3), 601–620. https://doi.org/10.1177/0042098020943426.

Joss, S., Sengers, F., Schraven, D., Caprotti, F., & Dayot, Y. (2019). The smart city as global discourse: Storylines and critical junctures across 27 cities. *Journal of Urban Technology, 26*(1), 3–34. https://doi.org/10.1080/10630732.2018.1558387.

Just, N., & Latzer, M. (2017). Governance by algorithms: Reality construction by algorithmic selection on the Internet. *Media, Culture & Society, 39*(2), 238–258. https://doi.org/10.1177/0163443716643157.

Kaika, M. (2017). 'Don't call me resilient again!': The New Urban Agenda as immunology . . . or . . . what happens when communities refuse to be vaccinated with 'smart cities' and indicators. *Environment and Urbanization, 29*(1), 89–102. https://doi.org/10.1177/0956247816684763.

Karvonen, A., Cugurullo, F., & Caprotti, F. (Eds.). (2019). Inside smart cities: Place, politics and urban innovation. Routledge Taylor and Francis Group.

Karvonen, A., Cook, M., & Haarstad, H. (2020). Urban planning and the smart city: Projects, practices and politics. *Urban Planning* (2020th-03–13th ed.)*, 5* (1), 4.

Kitchin, R. (2015). Making sense of smart cities: Addressing present shortcomings. *Cambridge Journal of Regions, Economy and Society, 8*(1), 131–136. https://doi.org/10.1093/cjres/rsu027.

Kitchin, R., & Dodge, M. (2019). The (in) security of smart cities: Vulnerabilities, risks, mitigation, and prevention. *Journal of Urban Technology, 26*(2), 47–65.

Kitchin, R., Cardullo, P., & Di Feliciantonio, C. (2019). Citizenship, justice, and the right to the smart city. In *The right to the smart city* (pp. 1–24). Emerald Publishing.

Kwet, M. (2019). Digital colonialism: US empire and the new imperialism in the Global South. *Race & Class, 60*(4), 3–26. https://doi.org/10.1177/0306396818823172.

Leitner, H., Peck, J., & Sheppard, E. S. (2007). *Contesting neoliberalism: Urban frontiers*. Guilford Press.

Leontidou, L. (2015). «SMART CITIES» OF THE DEBT CRISIS: GRASSROOTS CREATIVITY IN MEDITERRANEAN EUROPE. *Επιθεώρηση Κοινωνικών Ερευνών, 144*(144). https://doi.org/10.12681/grsr.8626.

Levenda, A. M. (2019). Thinking critically about smart city experimentation: Entrepreneurialism and responsibilization in urban living labs. *Local Environment, 24*(7), 565–579. https://doi.org/10.1080/13549839.2019.1598957.

Listerborn, C. (2022). Who is telling the smart city story? Feminist diffractions of smart cities, Digital (in)justice in the smart city. In D. Mackinnon, V. Fast, & R. Burns (Eds.). University of Toronto Press view.

Listerborn, C., & Neergaard, M. (2021). Uncovering the 'cracks'?: Bringing feminist urban research into smart city research. *ACME, 20*(3), 294–311.

Luque-Ayala, A., & Marvin, S. (2015). Developing a critical understanding of smart urbanism? *Urban Studies, 52*(12), 2105–2116. https://doi.org/10.1177/0042098015577319.

Luque-Ayala, A., & Neves Maia, F. (2019). Digital territories: Google maps as a political technique in the re-making of urban informality. *Environment and Planning D: Society and Space, 37*(3), 449–467. https://doi.org/10.1177/0263775818766069.

Luque-Ayala, A., Firmino, R. J., Fariniuk, T. M. D., Vieira, G., & Marques, J. (2020). Platforms in the making. In M. Hodson, J. Kasmire, A. McMeekin, J. G. Stehlin, & K. Ward (Eds.), *Urban platforms and the future city* (1st ed., pp. 248–261). Routledge. https://doi.org/10.4324/9780429319754-21.

Lynch, C. R. (2019). Representations of utopian urbanism and the feminist geopolitics of "new city" development. *Urban Geography, 40*(8), 1148–1167. https://doi.org/10.1080/02723638.2018.1561110.

Martin, C., Evans, J., Karvonen, A., Paskaleva, K., Yang, D., & Linjordet, T. (2019). Smart-sustainability: A new urban fix? *Sustainable Cities and Society, 45*, 640–648. https://doi.org/10.1016/j.scs.2018.11.028.

Marvin, S., Luque-Ayala, A., & McFarlane, C. (2016). *Smart urbanism. Utopian vision or false dawn?* Routledge.

McFarlane, C., & Söderström, O. (2017). On alternative smart cities: From a technology-intensive to a knowledge-intensive smart urbanism. *City, 21*(3–4), 312–328. https://doi.org/10.1080/13604813.2017.1327166.

McGuirk, P., Dowling, R., & Chatterjee, P. (2021). Municipal statecraft for the smart city: Retooling the smart entrepreneurial city? *Environment and Planning A: Economy and Space*, 0308518X211027905. https://doi.org/10.1177/0308518X211027905.

Melgaço, L., & Milagres, L. (forthcoming). On the contradictions of the (climate) smart city in the context of socio- environmental crisis. In D. Mackinnon, V. Fast, & R. Burns (Eds.), *Digital (in)justice in the smart city.* University of Toronto Press view.

Mora, L., Deakin, M., & Reid, A. (2019). Strategic principles for smart city development: A multiple case study

analysis of European best practices. *Technological Forecasting and Social Change, 142*, 70–97. https://doi.org/10.1016/j.techfore.2018.07.035.
Moser, S. (2015). New cities: Old wine in new bottles? *Dialogues in Human Geography, 5*(1), 31–35. https://doi.org/10.1177/2043820614565867.
Mouton, M. (2020). Worlding infrastructure in the global South: Philippine experiments and the art of being 'smart'. *Urban Studies*. https://doi.org/10.1177/0042098019891011.
Mouton, M., & Burns, R. (2021). (Digital) neo-colonialism in the smart city. *Regional Studies*, 1–12. https://doi.org/10.1080/00343404.2021.1915974.
Noble, S. U. (2018). *Algorithms of oppression: How search engines reinforce racism*. nyu Press.
Odendaal, N. (2021). Everyday urbanisms and the importance of place: Exploring the elements of the emancipatory smart city. *Urban Studies, 58*(3), 639–654. https://doi.org/10.1177/0042098020970970.
Pasquale, F (2017) From territorial to functional sovereignty: The case of Amazon. *Law and Political Economy*, 6 December. lpeblog.org/2017/12/06/from-territorial-to-functional-sovereignty-the-case-of-amazon/
Paul, K. (2019, December 28). A brutal year: How the "techlash" caught up with Facebook, Google and Amazon. *The Guardian*. https://www.theguardian.com/technology/2019/dec/28/tech-industry-year-in-review-facebook-google-amazon
Perez, C. C. (2019). *Invisible women: Exposing data bias in a world designed for men*. Random House. https://books.google.se/books?id=MKZYDwAAQBAJ
Perng, S.-Y., & Maalsen, S. (2020). Civic infrastructure and the appropriation of the corporate smart city. *Annals of the American Association of Geographers, 110*(2), 507–515. https://doi.org/10.1080/24694452.2019.1674629.
Robinson, J. (2016). Comparative urbanism: New geographies and cultures of theorizing the urban. *International Journal of Urban and Regional Research, 40*(1), 187–199. https://doi.org/10.1111/1468-2427.12273.
Rose, G. (2017). Posthuman agency in the digitally mediated city: Exteriorization, individuation, reinvention. *Annals of the American Association of Geographers, 107*(4), 779–793. https://doi.org/10.1080/24694452.2016.1270195.
Rossi, U. (2016). The variegated economics and the potential politics of the smart city. *Territory, Politics, Governance, 4*(3), 337–353. https://doi.org/10.1080/21622671.2015.1036913.
Roy, A. (2009). The 21st-century metropolis: New geographies of theory. *Regional Studies, 43*(6), 819–830.
Sadowski, J. (2021). Who owns the future city? Phases in technological urbanism and shifts in technology. *Urban Studies, 58*(8), 1732–1744.
Safransky, S. (2020). Geographies of algorithmic violence: Redlining the smart city. *International Journal of Urban and Regional Research, 44*(2), 200–218. https://doi.org/10.1111/1468-2427.12833.
Shelton, T., & Clark, J. (2016). Technocratic values and uneven development in the "smart city". *Metropolitics*, online.
Shelton, T., & Lodato, T. (2019). Actually existing smart citizens: Expertise and (non)participation in the making of the smart city. *City, 23*(1), 35–52. https://doi.org/10.1080/13604813.2019.1575115.
Shelton, T., Zook, M., & Wiig, A. (2015). The 'actually existing smart city'. *Cambridge Journal of Regions, Economy and Society, 8*(1), 13–25. https://doi.org/10.1093/cjres/rsu026.
SidewalkLabs. (2018). *Sidewalk Labs*. https://www.sidewalklabs.com. Accessed 27 May 2020.
Söderström, O., & Mermet, A.-C. (2020). When Airbnb sits in the control room: Platform urbanism as actually existing smart urbanism in Reykjavík. *Frontiers in Sustainable Cities, 2*, 15. https://doi.org/10.3389/frsc.2020.00015.
Söderström, O., Paasche, T., & Klauser, F. (2014). Smart cities as corporate storytelling. *City, 18*(3), 307–320. https://doi.org/10.1080/13604813.2014.906716.
Spicer, Z., Goodman, N., & Olmstead, N. (2021). The frontier of digital opportunity: Smart city implementation in small, rural and remote communities in Canada. *Urban Studies, 58*(3), 535–558. https://doi.org/10.1177/0042098019863666.
Srnicek, N. (2016). *Platform capitalism*. Wiley.
Suopajärvi, T. (2018). From tar city to smart city. *Ethnologia Fennica, 45*, 79–102. https://doi.org/10.23991/ef.v45i0.68961.
Tironi, M., & Valderrama, M. (2018). Acknowledging the idiot in the smart city: Experimentation and citizenship in the making of a low-carbon district in Santiago de Chile. In *Inside Smart Cities* (pp. 163–181). Routledge.
Tulumello, S., & Iapaolo, F. (2021). Policing the future, disrupting urban policy today. Predictive policing, smart city, and urban policy in Memphis (TN). Urban Geography, 0(0), 1–22. https://doi.org/10.1080/02723638.2021.1887634.
Vanolo, A. (2014). Smartmentality: The smart city as disciplinary strategy. *Urban Studies, 51*(5), 883–898. https://doi.org/10.1177/0042098013494427.
Vanolo, A. (2016). Is there anybody out there? The place and role of citizens in tomorrow's smart cities. *Futures, 82*, 26–36. https://doi.org/10.1016/j.futures.2016.05.010.
Viitanen, J., & Kingston, R. (2014). Smart cities and green growth: Outsourcing democratic and environmental resilience to the global technology sector. *Environment and Planning A: Economy and Space, 46*(4), 803–819. https://doi.org/10.1068/a46242.
Vivas, H. L., Britos, P. V., García-Martinez, N., & Cambarieri, M. (2013). Investigación en Progreso: Estudio y Evaluación de Tecnologías de la Información y la Comunicación para el Desarrollo de Ciudades Inteligentes. *Revista Latinoamericana de Ingeniería de Software, 1*(1), 146–151.
Watson, V. (2014). African urban fantasies: Dreams or nightmares? *Environment and Urbanization, 26*(1), 215–231.

Watson, V. (2015). The allure of 'smart city' rhetoric: India and Africa. *Dialogues in Human Geography, 5*(1), 36–39. https://doi.org/10.1177/2043820614565868.
West, S. M. (2019). Data capitalism: Redefining the logics of surveillance and privacy. *Business & Society, 58*(1), 20–41.
Wigley, E., & Rose, G. (2020). Will the real smart city please make itself visible? In K. Willis & A. Aurigi (Eds.), *The Routledge companion to smart cities* (pp. 301–311). Routledge.
Wiig, A. (2015). IBM's smart city as techno-utopian policy mobility. City, 19(2–3), 258–273. https://doi.org/10.1080/13604813.2015.1016275.
Wiig, A. (2016). The empty rhetoric of the smart city: From digital inclusion to economic promotion in Philadelphia. *Urban Geography, 37*(4), 535–553. https://doi.org/10.1080/02723638.2015.1065686.
Yin, C., Xiong, Z., Chen, H., Wang, J., Cooper, D., & David, B. (2015). A literature survey on smart cities. *SCIENCE CHINA Information Sciences, 58*(10), 1–18. https://doi.org/10.1007/s11432-015-5397-4.
Zandbergen, D., & Uitermark, J. (2020). In search of the smart citizen: Republican and cybernetic citizenship in the smart city. *Urban Studies, 57*(8), 1733–1748.
Zuboff, S. (2019). *The age of surveillance capitalism: The fight for a human future at the new frontier of power*. Public Affairs.

Understanding Urban Engineering

M. Cavada
School of Architecture, Imagination Lancaster, Lancaster University, Lancaster, UK

Synonyms

Smart cities; Urban futures; Urban policy; Urban transformation

Definition

This chapter explains the meaning of the term "Urban Engineering" by describing the relationship of the two terms – urban and engineering (Cavada et al. 2022). It discusses how the two terms converges and the ways urban engineering can provide beneficial solutions to design future urban transformations. This chapter also relates urban engineering to other urban concepts to indicate similarities. Urban engineering can relate, for example, to smartness, as it has been explored in recent research to offer an assessment for decision-making and establish communication between those involved and those affected by the proposed urban transformations.

Introduction

An urban transformation can significantly impact the way people live in the urban context. A wide range of professionals is responsible for making decisions that positively impact the urban context, and therefore, peoples' lives. Often, urban transformation is the outcome of professional views merged under a shared goal that brings positive change. The urban agenda is multifaceted; matters involve the built environment, planning and development, transportation, infrastructure, economy, services are some of them. This urban agenda requires considerations from every professional angle involved; often, one professional angle addresses one subject and can overlook the impact that can cause to another area. For example, rapid urban development might be good for the short-term economy and produce an adverse effect on the future well-being of people. However, positive change happens when different approaches and those who might contradict combine their knowledge and expertise into the shared goals. These different approaches need to amalgamate into solutions that are fit for purpose seamlessly and bridge any professional boundaries for the shared benefits for people and the urban environment. It is often the case that urban professionals fall into the silo of their profession, focusing on the objectives within their professional boundaries. Urban matters, however, require professionals to deal with complex issues, with different goals, systems, and overarching solutions at any stage.

For example, experts need to consider the applicability of their proposed solution in overarching matters that are not firmly close to their profession and add positive impact into overarching matters. Those are pressing issues; for example, sustainability and prosperity are the overarching issues that can offer shared benefits to people and environment. Understandably, these are not entirely linked to a single profession and are equally shared across an interdisciplinary urban agenda that can be followed by many professionals. Therefore, an urban solution includes several overarching approaches that cannot be distinguished, or they are either an abstract idea or a concrete strategy. This chapter discribes urban engineering as the term to bridge these professional gaps; whether an urban transformation is a product of urban quality or engineering ingenuity. Based on earlier research, the chapter considers some contradictions between the different approaches in the urban and engineering professions. There is still much discussion about whether the urban transformation is more of a product of urban quality or engineering ingenuity and this chapter establishes a discussion on this subject.

Nevertheless, this chapter does not attempt to separate the two but rather underlines the shared goal of positively impacting peoples' lives in the urban context. The combined urban and engineering approaches share a goal that brings them together, aiming at sustainable living today and allowing for prosperous living in the future. Yet the term urban engineering has been discussed briefly and needs to be explored further to get a comprehensive view of the complexities involved in urban transformations.

Why Do We Need Urban Engineering, and What Is It?

Urban transformation is a complex issue in its own right and requires a wide range of solutions simultaneously at any size or scale. For example, considering the timeframe to design and implement an urban solution, this can take place on a different timescale for different urban areas, depending on the size and the population dimensions. For example, an urban intervention can impact mobility transformation and can affect a whole metropolitan area, increase walkability on a neighborhood scale, or create congestion issues in other areas. Such transformations can impact whole urban systems; for example, electric cars require a charging system; this needs an entire infrastructure system that might adversely affect other areas. Several decisions need to be made to achieve the desired impact on the solution's efficacy and, most crucially, the quality of these during implementation and the future impact of these solutions. In an attempt to define urban engineering, the term has been introduced to describe a systematic approach responding to complex urban issues (Cavada et al. 2021). This term consists of two established professional approaches; there is a contrast in the meaning between the "urban" and the "engineering" which each carries. Urban refers to a highly populated spatial area, where professionals are primarily concerned with the planning and design of the urban areas where planning sets the parameters for problem-solving. Urban design can be considered a creative approach to these parameters, for example, how to best plan for the streets, parks, and building environment into workable and enjoyable spaces. According to Rogers (2018, p5), engineering, in the broader sense, can embrace overarching solutions "*sustainability, resilience, adaptability, liveability and smartness*" into the engineering agenda and test the efficacy, eliminating the risk of the design decision. Engineering professionals consider these parameters in the civil, infrastructure, and systems' agendas. Consider urban engineering as the theoretical unrepining of urban interventions, where urban and engineering converge toward an interdisciplinary urban agenda.

Is Urban Engineering Smart?

Urban engineering adheres to several theoretical approaches to solve complex issues. For example, sustainability has been at the forefront of guiding urban transformation in the recent past. Since then, sustainable solutions have been implemented into the urban agenda to support decision making. Not just local but international agendas promote sustainability, as the Goal 11 of the Sustainable Development Goal helps make "*cities and human settlements inclusive, safe, resilient, and sustainable*" (UN 2015). Recently, other agendas can offer alternatives in urban engineering, using digitalization and other tools to bring significant change. The smart agenda is widely known for embedding digital technology to provide efficiency into city systems. Cutting-edge technologies may promise efficient solutions; however, the delivery may lack quality during their efficacy. That is happening because the smart agenda is different for each city and needs solutions to fit these different needs. Recent research relates urban engineering to smartness to evaluate urban transformation for delivering the societal, environmental, economic, and governance lenses (Cavada et al. 2021). A smart assessment offers an analysis of the possible impacts and can prioritize the decision-making toward the benefits and opportunities for any urban transformation proposal in a systemic way (Cavada 2019).

For example, taking Moreno's 15–20-minute city concept through an urban smart engineering approach, researchers explored the benefits of the n-minute city (Cavada et al. 2022). The narrative of this research suggests that smart evaluation can unveil the societal, environmental, economic, and governance impacts of that change on the city and its people (Cavada et al. 2019). An example is using an urban case study to explore the smart system approach in urban engineering (Cavada et al. 2022). This research uses a transport intervention to describe the 15–20-minute city – a case study aiming to implement a car-free and more walkable city in a realistic scenario. The Smart Model Assessment Resilient Tool (SMART) was used to evaluate the impact of this scenario across the four smart lenses assessed the societal, environmental, economic, and governance impact of the n-minute solution (Cavada 2019). Although the n-minute solution provided ecological and societal benefits, the assessment showed that further action and smart initiatives need to be implemented to provide additional economic and governance benefits for a holistic smart urban engineering approach. Overall, the research underlines the need for understanding urban matters in a holistic and overarching approach, establishing the cooperation between the higher idea of the solution and engineering precision in the various stages of decision making.

A Complex Urban System

This chapter has discussed urban engineering as a system over the complexity of interdisciplinary action. This system aims to bring engineering precision together designing urban quality for solving urban problems. The author suggests that rapid urban development in many underdeveloped areas globally led to separate benefits, such as economic-focused outcomes. This is often seen as a single-focus engineering exercise. This disconnected urban transformation has failed to provide overall benefits for these areas' environmental, social, and living quality. Thus, urban engineering provides a systemic foundation for urban transformation for the quality and quantity of an urban solution. For example, this chapter discussed how concepts (sustainability and smartness) align with urban engineering. Based on academic research, urban solutions can adopt a smart approach to evaluate the shared benefits and base decision-making on the SMART assessment's environmental, societal, economic, and governance lenses. This process will allow smartness as a new concept to systematically embed overarching solutions for urban engineering. As an exercise done before, decision-makers can evaluate, discuss, discuss, and prioritize the smart lenses to provide a balanced approach to the overall benefits.

Conclusion

Urban engineering has been discussed aimed to respond to the urban transformation complexity. The term is used to describe a broad thinking about the urban transformation and bridge the professional silos that exist when proposing urban solutions. Although Urban Engineering was not established through formal education, this chapter used existing research to show that a new urban theoretical underpinning is possible and used examples to provide evidence for development. Explaining Urban Engineering as a concept can also explain how overarching ideas of sustainability and smartness can be used and explore the benefits to improve urban transformation design. Additionally, on a policy level, Urban Engineering can provide better communication with policymakers. Meaning both as a concept and providing the measurable impacts can explain the opportunities and pitfalls of the urban solutions and make appropriate decisions with all actors. Urban engineering is conceived to be open and understood by many professionals. Furthermore, urban engineering is about between all involved actors, not just urban professionals, strengthening the shared understanding between those who make decisions and those impacted by these decisions. Overarching concepts, specifically those in urban engineering, should understand their involvement and everyone's shared benefit.

Cross-References

▶ City Visions: Toward Smart and Sustainable Urban Futures
▶ Smart City: A Universal Approach in Particular Contexts
▶ The Governance of Smart Cities

Acknowledgements The author gratefully acknowledges the financial support of the Expanding Excellence in England Fund (E3) provided by Research England to Imagination Lancaster.

References

Cavada, M. (2019). *Smart Model Assessment Resilient Tool (SMART): A tool for assessing truly smart cities* (PhD Thesis). Department of Civil Engineering. Birmingham University. Etheses.

Cavada, M., Tight, M., & Rogers, C. (2019). Smart Singapore case study: Is Singapore truly smart? Chapter 16. In *Smart city emergence: Cases from around the world*. Elsevier. Edition 1. ISBN: 9780128161692.

Cavada, M., Bouch, C., Rogers, C., Grace, M., & Robertson, A. (2021). A soft systems methodology for business creation: The lost world at Tyseley, Birmingham. *Urban Planning, 6*(1), 32–48. https://doi.org/10.17645/up.v6i1.3499.

Cavada, M., Rogers, C. D. F., & Dalton, R. C. (2022). A smart system approach for urban engineering. In *Urban assemblance AMPS conference, Hatfield the city architecture, media, ai and big data*. Chapter 21. Hatfield London. Proceedings Journal Series Publication ISSN 2398-9467. http://architecturemps.com/wp-content/uploads/2022/02/Amps-Proceedings-Series-25.pdf

Rogers, C. D. F. (2018). Engineering future liveable, resilient, sustainable cities using foresight. *Proceedings of the Institution of Civil Engineers – Civil Engineering, 171*(6), 3–9.

UN Goal 11: Make cities and human settlements inclusive, safe, resilient and sustainable. 2015. Found at: https://sdgs.un.org/goals/goal11. Accessed 27 Nov 2021.

Understanding Women's Perspective of Quality of Life in Cities

Fathima Zehba M. P.[1] and Mohammed Firoz C.[2]
[1]Department of Architecture and Planning, Calicut, National Institute of Technology, Calicut, Kerala, India
[2]Urban and Regional Planning, National Institute of Technology, Calicut, Kozhikkode, India

Synonyms

Gender Inclusive Planning; Livability

Definition

"Quality of Life," in urban planning terms, is a multi-dimensional concept used to evaluate people's actual living conditions apart from their economic living standards. Quality of Life (hereafter used as QOL) should measure a person's wellbeing, satisfaction, and happiness in all the elements of conditions that influence their lives such as housing, education, health, infrastructure, environment, governance, etc. (Ballas 2013; Diener and Suh 1997). The concept often overlaps with "Livability," which has the same factors as that of QOL, but measures from a geographical aspect. QOL of women is the

U

degree to which women are healthy, comfortable, and able to participate in life events. Studies on women's QOL have similar influencing elements of "Gender Equality". It is the equal opportunities, rights, and responsibilities for all genders.

Introduction

There are immense researches on QOL in different domains and directions since its conceptualization. QOL of women is often understood and read from a health-related perspective. Many pieces of such researches are mostly conducted in the field of medicine and women's health in this regard because of the obvious understanding of biological differences between men and women. However, the well-being of women, though studied, is not a progressing or updated area of research. Though urban women enjoy more opportunities and access to services when compared to their rural counterparts, they are not as equally benefitted as urban men (Pozarny 2016). While studying QOL in a city, a variety of domains are considered under various sectors of urban planning. The services and opportunities provided for its citizens may be evaluated in this process by studying the citizen's needs. But it is also necessary to understand the gender-based differences of needs and priorities in the QOL elements so that sustainable development can be achieved. Unfortunately, this aspect is least considered in most such cases. Women perceive life and its values differently from men. Their issues need to be addressed for the betterment of their QOL in an urban context. In this chapter, the QOL concerns of women (and not other gender minorities) in cities are discussed in detail.

Cities can be a tool for addressing multiple global challenges such as unemployment, poverty, climate change, inequality, and environmental degradation if used wisely. Any such city is a product of urbanization which brings economic development. When urbanization happens, like any other gender, women are benefitted largely. "Urban women" avail more opportunities to flourish than women in a rural setting, which can contribute to economic sustainability. In developing countries, this geographical difference is so evident and cities have a considerable impact on the QOL of women (Menon and Sukumarazn 2015). Urban women and female-headed households are increasing in their number in cities. Their experiences vary according to their age, wealth, education, household profiles, and care responsibilities. Urban women have comparatively more access to infrastructure, services, and opportunities for paid employment than women in rural settings. The homogeneous characteristic of a rural area limits exposure and awareness in alternating gender roles. But in an urban setting, a heterogeneous community fosters tolerance of differences and encourages equality. Favorable conditions for education also improve urban women's social status (Chant 2013; Pozarny 2016).

However, there are many challenges, inequities, and insecurities faced by urban women. They encounter gender inequalities in many contexts such as access to decent work opportunities, services, housing security and financial assets, fair tenure rights, engaging in public governance structures, and personal security. Issues of safety are associated with poor infrastructure and transport designs. Though a city offers a heterogeneous space to develop, women face increased workloads due to a double-burden from job and care work. There are issues raised by the unorganized informal sector like street vending which are poorly paid, unregistered, and without any social protection. In poor urban households, girls have unfavorable conditions for attaining education (Chant 2013; Pozarny 2016).

Women's Perspective of Urban QOL

Before analyzing the gender differences, the different sectors of QOL need to be identified. Most of the recognized QOL indicators consider subdomains such as EUROSTAT (2015), Michalos et al. (2011), and OECD (2011).

1. **Income, Wealth, and Employment**
2. **Housing and Services**
3. **Health**
4. **Education**
5. **Safety and Security**

6. **Transport and Mobility**
7. **Work-life balance, Leisure**
8. **Environment**

Although gender role differences in cities and the challenges of urban women are studied, the QOL perspective of urban women is not identified much. Men and women have different roles, needs, and access to resources. QOL measures the available and accessible social and physical infrastructure and the satisfaction of people affected by it. As human beings perceive problems and find solutions according to their social roles, it is clear that the QOL perspective of women has more layers and differences from that of men. The concept of "gender" cuts across all social classifications, spheres of activities, and institutional structures (Jiron 1999). Issues of QOL of women in a city are deeply associated with the gender inclusiveness of cities. Gender-inclusive urban planning is an approach that considers people of all genders and sexualities, to ensure that their needs are addressed and used in project design, evaluation, and delivery, to promote gender equity. This is a basic need rather than considering as a special approach, as cities around the world are mostly "gender-blind" – more suitable for heterosexual able-bodied cis-gender men (The World Bank 2020). The World Bank identifies six issue areas where gendered challenges occur in an urban environment. These are (The World Bank 2020):

1. Access – Use of public services and spaces without any barriers or constraints
2. Mobility – Moving in the city using safe, affordable, and easy transport services
3. Safety and freedom from violence – Safety and freedom from real and perceived danger, both in private and public spheres
4. Health and hygiene – Keeping a healthy and active lifestyle, free from any risks caused by the built environment
5. Climate resilience – Preparing for, responding to, and coping with the effects of a disaster
6. Security of tenure – Owning or accessing land or housing to live or work

How these areas of challenges overlap with various domains of QOL from women's perspectives is discussed in the following section.

Income, Wealth, and Employment

This domain accounts for most of the income and wealth-generating factors and the employment-related factors to be taken while studying QOL. It includes income and wealth status, poverty rate, employment/unemployment rate, low-wage earners, material deprivation, life expenses, etc. Although urban women earn more and have higher wealth status than rural women, there exists inequality in the same from men. The issues are closely associated with the "security of tenure" aspect. Women around the globe do not have equal rights on land as men. This restricts them from attaining wealth and financial stability because, the land is considered as an income generator, a key input for accessing many financial resources and services, a space for housing, and a major asset that enables a citizen to participate in a city and its planning processes. Tenure mostly relies on men and their relations, which makes women exposed for exploitation and eviction. This mostly affects the women in informal settlements of a city. They are constrained in generating income and accumulating wealth because of the inability to access services and opportunities entitled to land ownership (The World Bank 2020). Among the earning population in a city, women contribute to most of the low-wage workers and unorganized employment sector. In addition, the gender pay gaps at workplaces, unemployment among women, workplace violence are other important issues of concern.

U

Housing and Services

This domain accounts for all the factors under housing including house ownership, housing conditions, and the services associated with housing like water supply, sanitation, electricity, and drainage. Since the universal gender role ascertained to women is to invest most of her life in caregiving and home-making, the quality of housing and services affects her life considerably. The spatial divisions as "public" and "private" in a city are often following the traditional gender division of labor such that "productive" income-generating work is for men and unpaid "reproductive" work is for women. A city's zoning of spaces into industrial/commercial (productive, income-generating, commerce) and residential (private, caregiving, reproductive) are thus found to have inequality in the use of spaces between men and women. In this context, lesser access and availability to basic services like water supply and sanitation creates an additional burden to women's time and energy, which is prominent in informal settlements. The distance between basic resources and infrastructure from housing location affect women's time spent on caregiving which in turn affect their productive time for seeking employment or education. House ownership, similar to land ownership, is held with the men in the family in most cases (The World Bank 2020).

Health

Health is considered one of the most important factors affecting QOL. This domain accounts for all the variables which define the health status such as life expectancy, maternal mortality rate, lifestyle diseases, mental health, the practice of physical activity, fertility rate, etc. Globally, women are found insufficiently physically active than men which in turn results in chronic diseases affecting their well-being and lifespan. A reason for this in cities is the lesser access, mobility, and comfort to use public spaces such as parks. Studies show that urban women feel lesser mental well-being than those women in rural areas due to the stress and additional burden asserted by an urban setting on their shoulders by the uncomfortable living conditions. Hygiene-related health problems are mostly seen in informal settlements where women lack access to basic sanitation facilities. Inadequate sanitation infrastructure results in serious health conditions like urinary and reproductive tract diseases among women and girls in such settings.

Education

The education domain accounts for variables like female literacy rate, educational attainment, number of early leavers, school and college enrollment, etc. An urban setting gives women more access and opportunities for attaining education. The number of early leavers is lesser and female school and college enrollments are greater in cities when compared to villages in any part of the world. However, girls from economically weaker settings face difficulties in attaining education even in cities since they are given responsibilities of caregiving at an early age itself.

Safety and Security

This domain can be considered the most important in women's QOL, which accounts for safety and crime against women. Although crime affects both men and women in a city, women are more affected and express higher levels of fear than men. The crime rate per year including crime against women reported in a city reduces the feeling of safety, especially in its public spaces. This results in women's behaviors such as staying home at night, avoiding the use of public spaces, etc. which eventually reduce their opportunities (Gordon et al. 1980). The high crime rate in a city increases the fear level of women commuters and hence negatively affects their everyday life. Underprivileged women in low-income neighborhoods and the unorganized employment sector feel less safe. Safety has a considerable impact on the use of urban open spaces such as parks, squares, plazas, streets, etc. Women do not prefer their routes to work or commerce to be altered due to the same reason (Sham et al. 2013; Tandogan and Simsek 2018). The social and spatial divisions in a city not only restrict women from using public spaces but also put them at high risk of violent "policing" and hate crimes. In addition, the rate of non-intimate partner violence is more in cities than in rural areas especially in informal settlements (The World Bank 2020).

Transport and Mobility

This domain accounts for the factors under transportation infrastructure and mobility such as coverage and availability of public transport, availability of parking, footpaths, pedestrian crosses, and bicycle paths, etc. Urban transport facilities are often designed according to the patterns of able-bodied men and ignore the complex needs of women commuters. Women travel in a city for several purposes using many modes of transportation. Their trips as caregivers are mostly short, localized, multi-purpose, and multi-stop. The transportation infrastructure offered by a city thus is an important influential factor of QOL of women. High crime rates and availability of street lights are catalytic variables here as they affect the fear level of women travelers in both public and private modes of transportation. Insecurity and fear of harassment in public transport limits women's everyday movement. Women spend more time and money traveling, due to safety concerns and their complex travel needs. This in turn reduces their opportunity to access employment and education, thus directly linked with economic stability. Many studies suggest a compact city for gender inclusiveness where travel fewer distances for fulfilling their social and economic roles, where their everyday needs are addressed, which in turn promotes sustainable mobility – accessible public transport and short walking distances (Gauvin et al. 2020). In addition, women are more likely to walk longer distances in a city than men to accomplish many activities like shopping, leisure, escorting a child, or elderly. In such cases, their mobility is hindered due to poor infrastructure such as lack of paving or pedestrian crossing.

Work-Life Balance and Leisure

This domain accounts for the factors such as work ing hours, time and opportunity for leisure, commuting time, the flexibility of work schedule, etc. The major concern here is the use of public spaces like parks and streetscapes. Gender inclusiveness of public spaces is as important as its safety. Public spaces enable women, the elderly, and other marginalized groups to engage in public life. But women tend to use parks and streets lesser than men due to reasons of fear and lack of other inclusive factors. Many studies on the same reveal that women usually prefer public spaces with safer perimeter, cleanliness, and safety and find lack of proper lighting, deserted roads, absence of street vendors and shops as unsafe situations. Lack of proper usable public toilets makes a public space unfriendly for women. They face special challenges in accessing and discomfort in using open spaces for recreational purposes. The use of public spaces also varies according to the age and economic status of women. Another concern is the nonflexibility of work schedules which makes it difficult for many women to accomplish their daily tasks. This in turn affects the overall work-life balance for working women in cities (Scraton and Watson 2010).

Environment

This domain accounts for all the factors under environmental conditions such as quality of air and water, pollution, green and open spaces, climate risks, etc. As a part of the gender minority, women and girls are more vulnerable to poor environmental conditions and climatic risks, especially in informal settlements. They are more at risk of death when a disaster occurs due to their household responsibilities (The World Bank 2020).

Working for Better QOL of Urban Women

As cities are incubators of social change, they can make the QOL of women better. A women-inclusive city design and planning is necessary to achieve this. According to the World Bank, a women-inclusive city should be (The World Bank 2020),

- Accessible – Public spaces and services should be accessible easily and used comfortably by everyone
- Connected – Moving around to accomplish daily needs should be easy, affordable, and safe for everyone
- Safe – Private and public spaces should be safe from real and perceived danger for everyone
- Healthy – There should be an opportunity for leading a healthy and active lifestyle for everyone
- Climate Resilient – Tools and social networks required for preparing for and cope with a disaster should be available for everyone
- Secure – Access and ownership to land and housing should be available for everyone

Accordingly, how the QOL of urban women can be made better through the various QOL domains discussed before is explained in the following section.

To Improve Income, Wealth, and Employment

- Build a gender-disaggregated land ownership database to identify the gap in existing wealth accumulation
- Implement mechanisms to achieve security of tenure such as long term rental contracts, collective land titles
- Identify alternative, accessible, and affordable finance mechanisms for startup loans for income-generating activities
- Reinforce equal pay legislation
- Implement women skill development programs, women education programs
- Enhance income-generating opportunities within dwelling units for economic activities to take place inside a neighborhood, mixed-use.zoning
- Institute child-care services in office spaces

To Improve Housing Conditions and Services

- Offer small loan programs for housing improvements, eliminate discrimination among women in housing finances
- Implement affordable housing in areas accessible by rapid transit
- Design housing layouts having practical solutions catering to women's requirements
- Locate public housing within 500 meters walking distance from public transport and basic amenities
- Introduce WASH (Water and Sanitation Hygiene) services 500 meters of households in informal settlements
- Introduce sewerage and sanitation structures in urban upgrading projects
- Stipulate a minimum of 50% women representation in any urban service provision projects

To Improve Health

- Design and install clean, secure, accessible public WASH facilities with a focus on the menstrual hygiene needs of women and girls in informal settlements.
- Provide toilet cubicles in public spaces with storage space to keep menstrual products dry and provide disposal bins with lids.
- Create accessible open and recreational/ sports spaces for women and girls

To Improve Education

- Improve access to schooling and create awareness about education in informal settlements
- Promote the and use of digital devices and broadband infrastructure, and make it accessible and affordable
- Provide fee concessions for students in public transportation increase the accessibility of educational opportunities
- Implement adult literacy programs for women and girl child literacy schemes for girls
- Encourage girls to come to school by providing subsidy programs and schemes such as free food and nutrition, especially in developing and under-developed countries

To Improve Safety and Security

- Provide adequate lighting in public spaces to increase visibility at night
- Integrate reduction of gender-based violence into development policies and plans
- Provide signage buildings and streets for wayfinding and improving access
- Install clear sightlines in parks where it is interrupted by trees
- Provide adequate lighting in streets and bus shelters to increase women's mobility without having the risk of gender-based violence

To Improve Transport and Mobility

- Provide multi-modal transportation options that are affordable, safe, convenient, and accessible for women
- Place transport nodes at a distance of 500 meters between each other and from housing, workplaces, and amenities
- Promote the concept of a polycentric "city of short distances"
- Provide wide and properly paved sidewalks and pedestrian infrastructure
- Increase the frequency of reliable public transit services at nights and on weekends

- Create a walking-friendly street network and a citywide cycling network
- Provide spaces reserved for women, such as women-only buses or metro cars in any public transportation service

To Improve Work-Life Balance and Leisure

- Provide public spaces and parks within walking distance from the housing.
- Segregate sports fields properly for multiple sports suitable for both men and women
- Public spaces should be designed in a women inclusive manner by Considering factors (United Cities and Local Governments 2016):
  - Proper lighting and visibility
  - Women friendly landscaping
  - Clean and safe toilets with proper lighting and diaper changing spaces
  - Motorized and Pedestrian traffic
  - Signages and Security personnel
  - Proximity to other public spaces and emergency services
  - Equitable number of amenities and mixed uses, flexible spaces
  - Breastfeeding rooms/private chambers

To Improve Environment

- Integrate gender equity and empowerment of women in disaster risk reduction
- Create gender-disaggregated data in damage and loss assessments in a disaster, especially for thc informal sector

Participatory Urban Planning and Governance for Improving Women QOL

Women's participation in urban planning, decision-making, and their presence in urban governance is essential for understanding their QOL perceptions and implementing appropriate solutions. Women are considered the experts of everyday life. Studies have developed planning tools and techniques for including women both as objects and subjects of urban planning. This approach not only benefits the QOL of women but also that of children, youth, elderly, and people of different races, ethnicity, income, and sexual identities (Ortiz Escalante and Gutiérrez Valdivia 2015).

Bottom-up participatory processes are required to collect gender-disaggregated data and studying women's priorities. Women should be included in consultation during the assessment of requirements and needs using tools such as interviews and focused group discussions. For example, while planning for a neighborhood, the experience of the community who live in and uses the spaces should be taken, which is mostly women. A variety of methods can be adopted for accommodating women's needs through their participation such as awareness workshops, assessment of everyday networks, community mapping, safety audits, etc. Another important factor is the women's representation in the city administration. In India, 33% women representation is stipulated in Local Self Government bodies as part of a decentralized planning system. It is seen from many such cases that women's representation in the administration has a considerable positive impact on pro-women urban governance (Dellenbaugh 2020; Ortiz Escalante and Gutiérrez Valdivia 2015).

Case Study – Vienna, Austria

Vienna, the capital city of Austria has been working toward gender mainstreaming since 1990. It has been ranked as the world's most livable city consistently as a result of the attempts to address the gaps between men and women in its public policy and urban planning. Within 11 years, 60 pilot projects have been executed in various sectors by three feminist urban planning experts, which have resulted in a high QOL of women in the city. Vienna serves as one of the most successful examples of women's inclusiveness around the world. Some of the efforts adopted by the city are,

- In one of its inner districts, alongside a metro line, in a pilot project, wider sidewalks, improved lighting, and ramps to improve access for people with strollers and wheelchairs were implemented.
- Recognized equal importance of pedestrian walkways as that of public transport lanes in every new transport project.

- Revamped six public parks to make them women inclusive – new recreation facilities like badminton and volleyball courts, cozy seatings, extra footpaths, streetlights, benches, and trees.
- New housing projects with practical layouts of flexible spaces, spacious stairwells and entrances, large and naturally lit kitchens and located in close distance to a supermarket, tram stop, and a primary school.
- Any new public building has gender-inclusive quality criteria to be satisfied (e.g., schools with a strong focus on open spaces and playgrounds).

Conclusion

Although QOL is a much-researched topic in urban planning, its gender perspective is less explored. How women perceive their QOL while living in a city is different from that of men and other genders. Hence what a city offers them is highly important. Inclusion of a women's perspective in urban planning is necessary as it allows different aspects of women's everyday life to be prioritized and planned accordingly. Such practices of gender-inclusive designs have been implemented in public spaces in many cities (e.g., Gender-sensitive public parks and squares in Vienna). Safety is the most prioritized concern of QOL while considering women in a city. However, it shouldn't be the only factor in achieving better QOL. Cities in the twenty-first century are far ahead in progressive development and should provide better amenities, skills, and employment opportunities for women. Concepts like "Feminist Urbanism" are innovative tools to focus on building gender-equal cities by allocating resources and services fairly to all social groups, especially to women and sexual minorities. Gender-based analysis of spaces, gender-sensitive infrastructure planning, and gender-responsive decision-making have become some of the necessary concepts to be studied in this regard. In addition, the perspective QOL of women shouldn't be confined to safety factors in public spaces, but also social factors such as awareness and capacity building especially in women among low-income neighborhoods. This shall improve women's QOL and enable sustainable development.

Cross-References

► Feminist Planning and Urbanism: Understanding the Past for an Inclusive Future
► Gender Inequalities in Cities: Inclusive Cities
► Health and the City: How Cities Impact on Health, Happiness, and Well-Being
► Urban Well-Being

References

Ballas, D. (2013). What makes a 'happy city'? *32*. https://doi.org/10.1016/j.cities.2013.04.009

Chant, S. (2013). Cities through a "gender lens": A golden "urban age" for women in the global South? *Environment & Urbanization, 25*(1), 9–29. https://doi.org/10.1177/0956247813477809

Dellenbaugh, M. (2020). *Cities through a "gender lens"*. Retrieved June 30, 2021, from https://urbact.eu/cities-through-gender-lens

Diener, E. D., & Suh, E. (1997). Measuring quality of life: Economic, social, and subjective indicators. *Social Indicators Research, 40*. https://doi.org/10.1023/A:1006859511756

EUROSTAT. (2015). *Quality of life facts and views*.

Gauvin, L., Tizzoni, M., Piaggesi, S., Young, A., Adler, N., Verhulst, S., … Cattuto, C. (2020). Gender gaps in urban mobility. *Humanities and Social Sciences Communications, 7*(1), 1–13. https://doi.org/10.1057/s41599-020-0500-x

Gordon, M. T., Riger, S., Robert, K., & Heath, L. (1980). Crime, women, and the quality of urban life. *Journal of Women in Culture and Society, 5*(3).

Jiron, P. (1999). Quality of life and gender: A methodology for urban research. *Environment and Urbanization, 11* (261). https://doi.org/10.1630/095624799101285011

Menon, R., & Sukumarazn, V. P. (2015). Urbanization and women empowerment an overview. *International Journal of Human Resource Management and Research (IJHRMR), 5*(2), 35–40.

Michalos, A. C., Smale, B., Labonté, R., Muharjarine, N., Scott, K., Moore, K., … Hyman, I. (2011). *Technical paper: Canadian index of wellbeing*.

OECD. (2011). *Compendium of OECD well-being indicators*.

Ortiz Escalante, S., & Gutiérrez Valdivia, B. (2015). Planning from below: Using feminist participatory methods to increase women's participation in urban planning.

Gender and Development, 23(1), 113–126. https://doi.org/10.1080/13552074.2015.1014206
Pozarny, P. F. (2016). *Gender roles and opportunities for women in urban environments* (GSDRC Helpdesk Research Report 1337). Birmingham.
Scraton, S., & Watson, B. (2010). Gendered cities: Women and public leisure space in the 'postmodern city'. *Leisure Studies, 17*(2), 123–137. https://doi.org/10.1080/026143698375196
Sham, R., Shah Muhammed, M. Z., & Ismail, H. N. (2013). A review of social structure, crime and quality of life as women travelers in Malaysian cities. In *Procedia – Social and behavioral sciences* (Vol. 101, pp. 307–317). Elsevier B.V. https://doi.org/10.1016/j.sbspro.2013.07.205
Tandogan, O., & Simsek, B. (2018). Fear of crime in public spaces: From the view of women living in cities. In *Procedia engineering* (Vol. 161, pp. 2011–2018). The Author(s). https://doi.org/10.1016/j.proeng.2016.08.795
The World Bank. (2020). *Handbook for gender-inclusive urban planning and design.*
United Cities and Local Governments. (2016). *UCLG public space policy framework by and for local governments.* United Cities and Local Governments. Retrieved from https://www.uclg.org/sites/default/files/public_space_policy_framework.pdf

UNDRR – United Nations Office for Disaster Risk Reduction

▶ Adapting to a Changing Climate Through Nature-Based Solutions

UNEP – United Nations Environment Programme

▶ Adapting to a Changing Climate Through Nature-Based Solutions

UNFCCC – United Nations Framework Convention on Climate Change

▶ Adapting to a Changing Climate Through Nature-Based Solutions

Universities Expansion

▶ Impact of Universities on Urban and Regional Economies

Unpacking Cities as Complex Adaptive Systems

Ditjon Baboci
Independent Researcher, Tirana, Albania

Introduction

Global Urban Challenges

Starting with James Watt's invention of the steam engine in 1784, the impact of humanity on the global environment has avalanched to the point where in 2002 Paul Crutzen formalized mankind's omnipresence on the planet by terming it the beginning of the "anthropocene." The Anthropocene is not simply a marker of history but rather a monocultural mindset that must be questioned.

The current urban challenges feel like an impending tsunami because civilization is at a crucial point in its exponential growth. Both world population and GDP growth have dramatically multiplied since the industrial revolution. It is in the very nature of exponential expansion that the future accelerates – change happens more quickly. The urban challenge the world is facing; and just as importantly the environmental and energy burden that will have to be confronted in the next 30 years, will likely never be equaled again in the future (West 2017, loc. 229–254).

From the standpoint of the economy, cities account for 85% of global GDP generation. The environmental impact of urban areas cannot be understated: cities represent approximately one half of the world population but account for 75% of natural resource consumption, and between 60% and 80% of greenhouse gas emissions (Ellen Macarthur Foundation 2017, 4). Urban areas are the most important consumers of materials and nutrients, while at the same time being the places where future growth and construction will happen.

The current single largest global resource footprint is urban infrastructure. This includes not only buildings, housing, and offices but also roads and other kinds of mobility infrastructure. More than 42 billion tonnes of resources are consumed by this sector every year (Verstraeten-Jochemsen et al. 2018, 6). These challenges are especially relevant to the developing world, which encounters the dual problem of high urban growth coupled with the need for cheap construction. The present planning and construction technologies will have a long-lasting impact on the future of cities, especially when considering that 60% of the building volume required by 2050 has not been built yet; in emerging economies that figure reaches as high as 70% (Ellen Macarthur Foundation 2017, 11).

Context of Urban Planning

In the twenty-first century of tech and information, urban planning often boils down to an obsolete set of unsubstantiated top-down verdicts with almost no correlation between decisions and desired outcomes. Planning per se tries to simplify and formalize inherently informal components of complex processes (Scott 1998). It is a modernist ideological effort to quantify and understand the human condition. Often however actions do not correlate to consequences – there is more resilience in localism and complexity, a kind of resilience which is often difficult to achieve through simplification (Scott 1998). Furthermore – like most bureaucratic enterprises – urban planning is highly path-dependent. Diamond explains how "minor cultural features may arise for trivial, temporary local reasons, become fixed, and then predispose society toward more important cultural choices." (Diamond 1997, loc. 6759) The choices of generations-removed planners have become ingrained not only in the patterns exhibited by our cities, but more crucially in the structures, processes, and procedures of urban plans.

Professionals have argued in favor of a more sophisticated form of planning for almost a century. The concept of *"eyes on the street"* is most often perceived as one of Jane Jacobs' anecdotal findings from observing life in Greenwich Village. We relate Jacobs to social "bottom-up processes" but not to problems of "organized complexity" – despite her dealing with both topics in her *oeuvre* (Jacobs 2016). Her work seems qualitative in nature, but she is coarse-graining her experience in patterns, which we struggle to quantitatively describe more than 60 years after her writing.

Cities are incremental: successful cities are able to informally digest and metabolize change, furthermore they often invite and welcome it (Lindblom 1959). Christopher Alexander affirms that neither towns, nor neighborhoods "can be created by centralized authority, or by laws, or by master plans. We believe instead that they can emerge gradually and organically, almost of their own accord." (Alexander 1977, 3). He argues that traditionally there was little need for modern urban planning because the slow pace of change allowed sufficient time for adaptation.

More recently Bertaud argued that a planner's role should be akin to a midwife for cities. They should keep their fingers on the urban pulse and heed what people need, rather than implement arbitrary formal schemes which – as seen in examples from Brasilia to Chandigarh – often do not even reflect the needs of citizens, let alone fulfill them (Bertaud 2018).

Urban Complex Systems

Cities as Complex Adaptive Systems

In this uncertain context it is crucial to look at cities through the lens of systems thinking in an attempt to improve the understanding and the outcomes of plans in the crucial upcoming decades. Cities can be interpreted as the interaction of diverse networks; because of this they are often well adapted to survive sudden growth or shrinking (West 2017). The time dimension is just as important as the physical dimension to the planning of cities. Urban plans should be reimagined in terms of circular processes, which adapt to changing contexts rather than time-finite deliverables. Important references from similar complex systems can be helpful in better leveraging adaptive tools. Ant colonies grow more stable with time for example – with a clear cycle of infancy, adolescence, and maturity, despite the limited lifespan of individual ants (Johnson 2001). Similarly different cities might need diverse strategies depending on the pattern of their growth. Complex adaptive systems exhibit some signature concepts that seem to have direct analogies to urban areas: these systems are typically emergent, resilient, and self-similar.

Signature Concepts of Complex Systems

Emergent

Cities exhibit emergence: this means that the behavior of the city as a whole is more complex than the sum of all individual actions and transactions. In this context the third law of British science fiction author Arthur C. Clarke might be reworded as: "*any sufficiently complex entity is indistinguishable from magic.*" Technology is helping in understanding and generating behaviors that 100 years ago might have been equated to magic: the blinking coordination of certain species of fireflies, recommendation algorithms, or urban mobility systems. In each of these processes, behavior emerges bottom-up through simple rules leading to higher-order abstractions (Johnson 2001).

More pertinent to planning, this kind of nonlinear complex behavior makes quantitative prediction extremely difficult; often mathematically unfeasible (Taleb 2018). Although having a detailed snapshot of all relevant data, it would still be impossible to calculate potential interactions: the computational requirements would be orders of magnitude more complex than the already immense amount of data required (Thurner et al. 2018, 21). This is evident in much simpler systems: because of advances in sequencing DNA we can trace monogenic diseases, but as soon as the number of genes involved in a disease goes up, our ability to understand and unpack the disease falls apart (Taleb 2018).

Power law behavior emerges from cities and is omnipresent in complex systems (Thurner et al. 2018, 18). Cities make parsimonious use of infrastructure. Furthermore services are provided more efficiently. Urban infrastructure and service provision exhibits a clear trend scaling sublinearly at a slope of 0.85 related to demographics: for every doubling of the population, there is only a corresponding 85% increase in these infrastructural needs. On the other hand other socioeconomic variables such as crime or wages scale superlinearly at a slope of 1.15 (West 2017, loc. 585). A simple but powerful illustration of these power laws is the proportional increase of walking speed in larger cities (West 2017, loc. 5519). Until recently this hidden sense of order lurked at the boundary of understanding – but it is slowly coming to the surface.

Resilient

By their very *nomenclature* adaptive systems are resilient. Different from most biological complex systems, cities never seem to die – the few cities that have disappeared in history are outliers whose fall can be attributed to conflict or abuse of their immediate environment (West 2017). The secret behind resiliency is a city's ability to adapt to change: this typically happens on long timescales. From our limited perspective these transformations might seem invisible. In comparing urban development with other processes of long and organic accretion (i.e., evolution), one notices that they are often characterized as operating in *punctuated equilibrium*: long periods of stability interrupted by sharp intervals of sudden environmental change and fast adaptability (Gould and Eldredge 1993).

Redundancy is inherent to complex systems – the more redundant a system is, the less it is prone to fail completely (Cook 2000). Furthermore, the individual agents of cities routinely engage in low-risk high-reward behavior: they can explore potentially successful niches with little costs to the system as a whole (Holland 2014; Krakauer

2018). For example a citizen could invest her lifesavings in starting a new business, which could grow into an important enterprise for the city. The city stands to gain, while only the citizen stands to lose if the said business fails. This is akin to scout bee behavior where individual bees will shift from group behavior to research for new sources of food (Seeley and Buhrman 1999). Cities do not have resting points and do not strive toward any simple kind of equilibrium. They are living entities always changing and adapting to environmental changes.

Self-similar

At a high enough level of abstraction, all complex adaptive systems are self-similar: modular, recursive, and incremental. The degree of abstraction is lower for hymenopteran insects and higher for cities, but ultimately the terminal unit behavior of each agent of the ant or human system is similar in scope. Self-similarity in this sense concerns the invariant terminal units that make up each city: individual citizens (West 2017, loc. 1928).

Fractal self-similarity is evident in the natural world – from coastlines and tree branching, to mammalian circulatory systems (Holland 2014). Most evidently the power of self-similar units is demonstrated by programming: even the most complex computer program can be broken down to bits of 0 s and 1 s. This is similar to the recursive, object-oriented abstraction of life from atoms, to molecules, organs, the body, social groups, and ultimately cities: "particular combinations of agents at one level become agents at the next level up." (Holland 2014).

The multitude of small-scale, competing actors bring about more interactions and hence create more innovation and resilience. This is similar to the differences in linguistics and political geography when comparing the historical development of European countries to China: the former has a history of smaller city-states warring and innovating against one another, while the latter a millennia hegemonic system. Because of its hierarchical top-down decision making, innovation in historical China was often halted by decree; while the series of self-similar city-states needed to fight for their viability – making sure only the fittest survived. That's part of the reason why at the beginning of the twentieth century these regions had colossal technological and political disparities (Diamond 1997).

Practical Approaches and Applications to Planning

Typically wired to simplify and formalize systems, solutionist-oriented human thinking is unequipped to outsized results stemming from small actions. It naturally follows that some of the approaches to resolving conflict in complex adaptive systems are counterintuitive and achieve results in a nonlinear way. What superficially seems clear often isn't. On the contrary, approaches to these problems seem the opposite of what one should do. This concept is very relevant to cities, particularly to issues of urban mobility: Braess' paradox and induced demand being two prime examples. Braess determined that at times adding an extra road segment to an existing mobility network might impede its flow, and equivalently removing a road segment might decrease travel times (Braess et al. 2005). Similarly, induced demand illustrates how increasing the capacity of infrastructure can have a negative long-term impact on traffic and waiting times (Hymel et al. 2010). Bearing in mind the characterization of complex systems as emergent, resilient, and self-similar, there might be some practical approaches that can capture the nature of these characteristics in the form of feedback loops, lever points, and statistical trends.

Feedback Loops

Feedback loops are the key regulating component of complex adaptive systems. They are functional elements in all decentralized systems that enable consistency, growth, and self-regulation (Johnson 2001, loc. 1673). Negative feedback loops attempt to keep the system in stasis: a good example is a thermostat – no matter the environment, a thermostat will always attempt toward equilibrium at a certain preset temperature (Holland

2014, loc. 1041). On the other hand, positive feedback loops expedite the adoption of consistently beneficial elements. They are autocatalytic processes that exponentially speed up the rate of acceleration – just like innovation after the industrial revolution sped up the development of further technologies (Diamond 1997, loc. 4151).

One related and important element of feedback loops are response thresholds: the amount of stimuli that provokes a response or activates a loop (Krakauer 2018, loc. 1133). The variance of thresholds can impart program-like behavior to systems, allowing them to self-regulate through autonomous subroutines (Holland 2014, loc. 1041). It is easy to imagine similar behaviors already surreptitiously taking place in cities: infrastructure expansion, various mobility investments depending on the volume of traffic, foot-traffic generating new small businesses, etc.

Lever Points

In complex adaptive systems small directed actions can cause large outsized predictable changes in aggregate behavior. A good example is how vaccines produce long-term changes to immune systems (Holland 2014, loc. 861). To date, however, lever points can only be identified bottom-up through trial and error and do not have universal application; on the contrary they are highly context-dependent. In the urban sphere, transformation through urban acupuncture in the past decade seems to have had a much higher signal-to-noise ratio and quality of life impact than any comprehensive planning effort.

The lack of resilience in a system that is overly interdependent can also make these lever points drivers of systemic risk. They can have outsized unpredicted negative effects. Such interdependence in financial markets, for example, might be the cause of financial crises, with small, seemingly innocuous events avalanching into full blown disasters (Billio et al. 2012, 555). Similarly air travel volume – and its corollary interconnectedness – is a key factor in increasing the spread of epidemics because of the outsized risk posed by an infected traveler (Grais et al. 2004).

Statistical Trends

At the current state of technology it is next to impossible to begin to map, let alone predict and plan urban regions. At the same time it is an accepted fact that formal approaches to shaping cities have rarely been successful. These arbitrary approaches have only succeeded at the conception stage of a city, or – with limited scope – through sheer authority and capital injection (Scott 1998). There are alternatives to improve understanding however. Statistical mechanics have shifted the zeitgeist of what it means to know, toward a less deterministic, more macroscopic regime (Thurner et al. 2018, 5). For example, coarse-graining has aided comprehension in biomolecular protein modeling – a more limited, but similarly difficult enterprise to urban planning (Ingólfsson et al. 2014).

Concluding Thoughts

> *The totally planned city is . . . a myth. Therein lies the historic error of urban planners and designers and of architects: they fail to see, let alone analyze or capitalize upon, the informal aspects of urban life, because they lack a professional vocabulary for describing them. – Brillembourg and Klumpner.* (McGuirk 2015, loc. 326)

Urban plans and city development more generally cannot be interpreted as finite processes: ones that begin with a vision go through a regulated process and conclude with a physical deliverable leaving a mark on the urban fabric. Plans should not have an expiry date – they should be circular, continuous, and responsive. Planners need to think of a city plan more like urban DNA: not a formal top-down blueprint of the human body, but rather a multitude of bottom-up instructions and switches that express themselves when exposed to local signals and conditions (Pistoi 2020). In this sense DNA is less a formula to make a human body, and more a field of opportunities which expresses specific genes depending on the context. Flawless organisms emerge out of the multitude of singular decisions taken at the cell level.

A complex systems approach would transform an urban plan from a static document to a distributed, blockchain-based, tended set of conditional rules, which could express themselves through feedback loops dependent on a series of response thresholds at the local level. It would warrant true citizen-driven planning. Furthermore it would enable sustainable and responsible planning by ensuring that consideration of neighboring units is systematically weighed and taken into account. The local environmental context would drive the collaboration between citizens and planners further enhancing decision making by making the neighboring urban *cells* an active part of the planning dialogue. Learning from fields with a more advanced understanding of complex systems can be helpful in shaping a new paradigm of planning.

References

Alexander, C. (1977). *A pattern language: Towns, buildings, construction*. Oxford University Press.

Bertaud, A. (2018). *Order without design*. Cambridge, MA: MIT press.

Billio, M., Getmansky, M., Lo, A. W., & Pelizzon, L. (2012). Econometric measures of connectedness and systemic risk in the finance and insurance sectors. *Journal of Financial Economics, 104*(3), 535–559. Science Direct. https://doi.org/10.1016/j.jfineco.2011.12.010.

Braess, D., Nagurney, A., & Wakolbinger, T. (2005). On a paradox of traffic planning. *Transportation Science*, (39), 446–450. Researchgate. https://doi.org/10.1287/trsc.1050.0127.

Cook, R. I. (2000). How complex systems fail. Cognitive Technologies Laboratory. University of Chicago, Chicago IL.

Diamond, J. (1997). *Guns, germs, and steel: The fates of human societies*. W. W. Norton.

Ellen Macarthur Foundation. (2017). *Cities in the circular economy: An initial exploration*. https://www.ellenmacarthurfoundation.org/assets/downloads/publications/Cities-in-the-CE_An-Initial-Exploration.pdf

Gould, S. J., & Eldredge, N. (1993). Punctuated equilibrium comes of age. *Nature, 366*, 223–227. https://doi.org/10.1038/366223a0.

Grais, R., Ellis, J., & Kress, A. (2004). Modeling the spread of annual influenza epidemics in the U.S.: The potential role of air travel. *Health Care Management Science*, (7), 127–134. SpringerLink. https://doi.org/10.1023/B:HCMS.0000020652.38181.da.

Holland, J. H. (2014). *Complexity: a very short introduction*. Oxford University Press.

Hymel, K. M., Small, K. A., & Van Dender, K. (2010). Induced demand and rebound effects in road transport. *Transportation Research Part B: Methodological, 44* (10), 1220–1241. Elsevier. https://doi.org/10.1016/j.trb.2010.02.007.

Ingólfsson, H. I., Lopez, C. A., Uusitalo, J. J., De Jong, D. H., Gopal, S. M., Periole, X., & Marrink, S. J. (2014). The power of coarse graining in biomolecular simulations. *Wiley Interdisciplinary Reviews: Computational Molecular Science, 4*(3), 225–248. US National Library of Medicine. https://dx.doi.org/10.1002%2Fwcms.1169.

Jacobs, J. (2016). *The death and life of great American cities*. Vintage.

Johnson, S. (2001) Emergence: the connected lives of ants, brains, cities, and software. New York: Scribner

Krakauer, David C. ed. (2019). Worlds Hidden in Plain Sight: The Evolving Idea of Complexity at the Santa Fe Institute, 1984–2019. Santa Fe Institute Press.

Lindblom, C. E. (1959). The science of "muddling through". *Public Administration Review, 19*(2), 79–88. JSTOR. https://doi.org/10.2307/973677.

McGuirk, J. (2015). *Radical cities: Across Latin America in search of a new architecture*. Verso.

Pistoi, S. (2020, February 6). *DNA is not a blueprint*. https://blogs.scientificamerican.com/observations/dna-is-not-a-blueprint/. Retrieved September 29, 2020, from https://blogs.scientificamerican.com/observations/dna-is-not-a-blueprint/

Scott, J. C. (1998). *Seeing like a state: How certain schemes to improve the human condition have failed* (Paperback ed.). Veritas Paperbacks.

Seeley, T. D., & Buhrman, S. C. (1999). Group decision making in swarms of honey bees. *Behavioral Ecology and Sociobiology*, (45), 19–31. Springer. https://doi.org/10.1007/s002650050536.

Taleb, N. N. (2018). *Skin in the Game: Hidden Asymmetries in Daily Life*. New York: Random House.

Thurner, S., Hanel, R., & Klimek, P. (2018). *Introduction to the theory of complex systems*. Oxford University Press. https://doi.org/10.1093/oso/9780198821939.001.0001.

Verstraeten-Jochemsen, J., Kouloumpi, I., Russell, M., de Wit, M., Douma, A., & Friedl, H. (2018). *City-as-a-circle*. Retrieved September 29, 2020, from https://www.circle-economy.com/resources/city-as-a-service

West, G. (2017). Scale: The Universal Laws of Growth, Innovation, Sustainability, and the Pace of Life in Organisms, Cities, Economies, and Companies. New York: Penguin Press.

Urban

▶ Moving Towards Sustainable, Liveable, and Care-Full Urban Environments: Pre-schoolers' Rights and Visions for Planning Just, Socially, and Ecologically Integrated Cities

▶ Shrinking Towns and Cities

Urban – Non-rural

▶ Urban Policy and the Future of Urban and Regional Planning in Africa

Urban Agenda

▶ Housing and Development

Urban Agriculture

▶ Future Foods for Urban Food Production
▶ Urban Food Gardens

Urban and Regional Economies

▶ Impact of Universities on Urban and Regional Economies

Urban and Regional Leadership

Fatime Barbara Hegyi
Joint Research Centre – European Commission, Seville, Spain

Definition

Leadership stands as one of the key success factors in achieving growth at urban and regional levels (OECD 2010), while strong and focused leadership contributes to the success of places (Beer and Clower 2014). Furthermore, dedicated leadership contributes to an efficient governance system and impacts the motivation and commitment of stakeholders, who are indispensable in transforming places, organizations, and capabilities (Trickett and Lee 2010). These transformations lead cities and regions to become resilient in adapting to social, economic, and environmental changes (Hegyi 2020). This publication presents results of a 2020 survey on the impact and interdependence of leadership on governance systems, stakeholders' involvement, and monitoring and evaluation linked to research and innovation strategies of Smart Specialization.

Introduction

Adapting from the longstanding tradition of measuring the impact of leadership in the business sphere allows leaders of cities and regions to meet the demands of their stakeholders. Such demands include creating and nurturing cities and regions that are safe and resilient while offering livable and healthy environment built together with their own communities (Hegyi 2020). Integral leadership requires new approaches of leadership with diverse qualities and skills that promote urban and regional agendas addressing current and future societal, economic, environmental, and health challenges (see chapter on ▶ "Regulation of Urban and Regional Futures"). The commitment of diverse actors throughout policy cycles across institutional boundaries leading to organizational renewal requires a new understanding of the impact of leadership (Sotarauta 2018). Consequently, when considering urban and regional systems, the dynamics of change as regards the value network structure and the types of actors involved the need to be closely observed, as innovative local communities contribute greatly in redefining spaces and functions, thereby reinventing places from bottom-up (Hegyi et al. 2019 and Vandecasteele et al. 2019).

Accordingly, along with assessing leadership, organizational systems and stakeholder involvement need to be analyzed, which have been two of the main objectives of the 2020 process evaluation exercise of the Smart Specialization Platform of the Joint Research Center (Hegyi et al. 2021). The evaluation focused on discovering the interdependencies of governance systems, stakeholders' involvement, and monitoring and evaluation linked to research and innovation strategies of Smart Specialization (S3). S3 is a place-based policy experiment offering a methodology along which regions and the Member States of the European Union have been developing their research and innovation strategies leading to a limited number of clearly defined research and innovation priorities. Given that experimentation is at the heart of S3 and that it is place-based strategy (Hegyi and Rakhmatullin 2020), the results of the evaluation exercise provide a solid insight into how leadership impacts regional and urban policy implementation in the field of research and innovation (see chapter on ▶ "Connecting Urban and Regional Innovation Ecosystems to Enhance Competitiveness").

Place Leadership in Urban and Regional Innovation Policy Agendas

Believing that place leadership improves the capacity of generating and implementing a future-oriented vision through the interaction of stakeholders supported by adequate governance structure, the before mentioned process evaluation exercise zeroed in at qualities of leadership and the impact of such qualities on various aspects of the S3 policy experience (see chapter on ▶ "Participatory Governance for Adaptable Communities"). Correspondingly, given that places with visible leadership prove to better adapt to changing social, economic, and environmental challenges, governance structures tend to adapt more to place-based approaches when it comes to development strategies (Sotarauta and Suvinen 2019; Stimson et al. 2009; Beer and Clower 2014). Therefore, it is crucial to measure the impact of leadership in ensuring efficiency, sustainability, and success of urban and regional objectives (Hegyi et al. 2020a).

The 2020 S3 survey resulted in 79 responses covering 8 national level and 71 regional level research and innovation strategies of S3 (Hegyi et al. 2021). Sixty-two percent of the S3 implementing authorities have indicated that S3 process is relying on dedicated political leadership (according to 18% it strongly contributes, while according to 44%, it somewhat contributes). While 87% have indicated that the S3 process is relying on dedicated management leadership (44% have indicated that it strongly contributes, while 43% that it somewhat contributes), presented in Fig. 1.

In the case of political leadership, 8% of the S3 implementing authorities have indicated that the political leadership is strongly holding back the

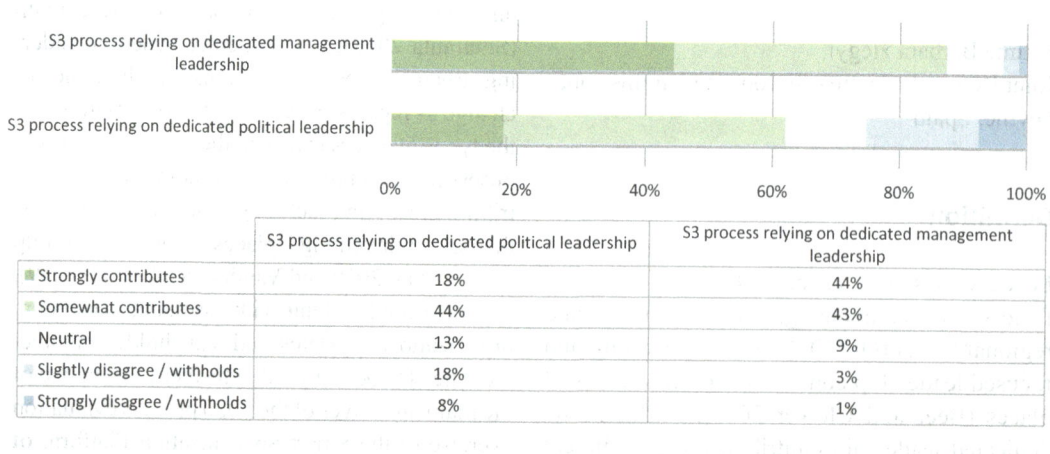

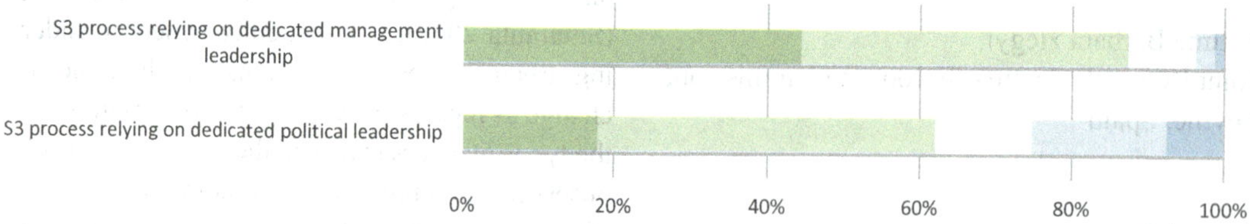

	S3 process relying on dedicated political leadership	S3 process relying on dedicated management leadership
Strongly contributes	18%	44%
Somewhat contributes	44%	43%
Neutral	13%	9%
Slightly disagree / withholds	18%	3%
Strongly disagree / withholds	8%	1%

Strongly contributes Somewhat contributes Neutral Slightly disagree / withholds Strongly disagree / withholds

Urban and Regional Leadership, Fig. 1 Place leadership in urban and regional innovation policy agendas_S3 process' reliance on dedicated political and management leadership

S3 process, while 18% indicated that it is slightly holding it back. However, 1% has indicated that the management leadership is holding back the S3 process and only 3% that this effect is present slightly.

Fifty-seven percent of S3 implementing authorities have responded that leadership has a strong role and further 33% that it has a role in the effective implementation of S3, while 5% have indicated that leadership slightly holds it back. The role of leadership in enhancing the commitment of stakeholders towards S3-related objectives is seen positive by 90% of respondents, out of whom 37% signal that leadership has a strong role in it. When observing the role of leadership in S3 governance structure in becoming a learning organization, 66% of respondents feel that leadership has a positive role, while 8% think that S3 leadership holds back such process. When it comes to promoting and diffusing new ideas and narratives on innovation strategies and promoting trust among stakeholders, S3 leadership again is viewed as a positively contributing factor. Figure 2 presents the perceived role of leadership in various aspects of S3 implementation.

S3 implementing authorities have also expressed their views on the type of contribution provided by groups of stakeholders in the S3 implementation process. The types of contributions varied from specialized knowledge and expertise, legitimacy, resources, and capacity to develop initiatives and leadership. Leadership has been indicated as a contribution from each group of stakeholders, as shown in Fig. 3. S3 responsible bodies, intermediary organizations, regional government, and administration were indicated most often as contributors of leadership in the S3 process. Vocational educational institutions, training institutions, and civil society have been indicated at least contributing to the S3 process with leadership (see chapters on ▶ "Education for Inclusive and Transformative Urban Development").

Impact of Leadership on Motivation and Commitment

As the results of the survey show, leadership impacts various aspects of policy implementation from learning through trust and from higher

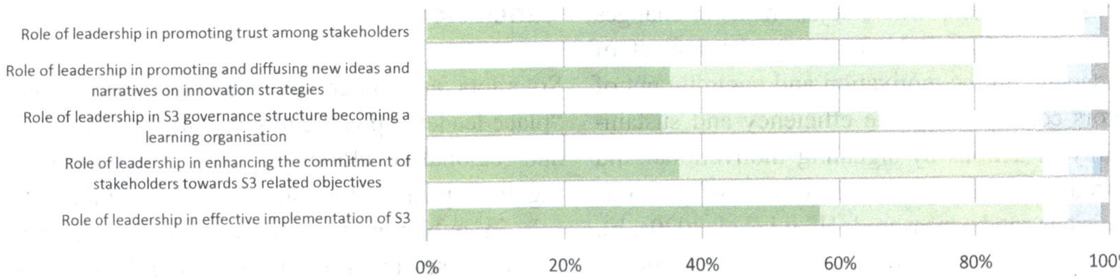

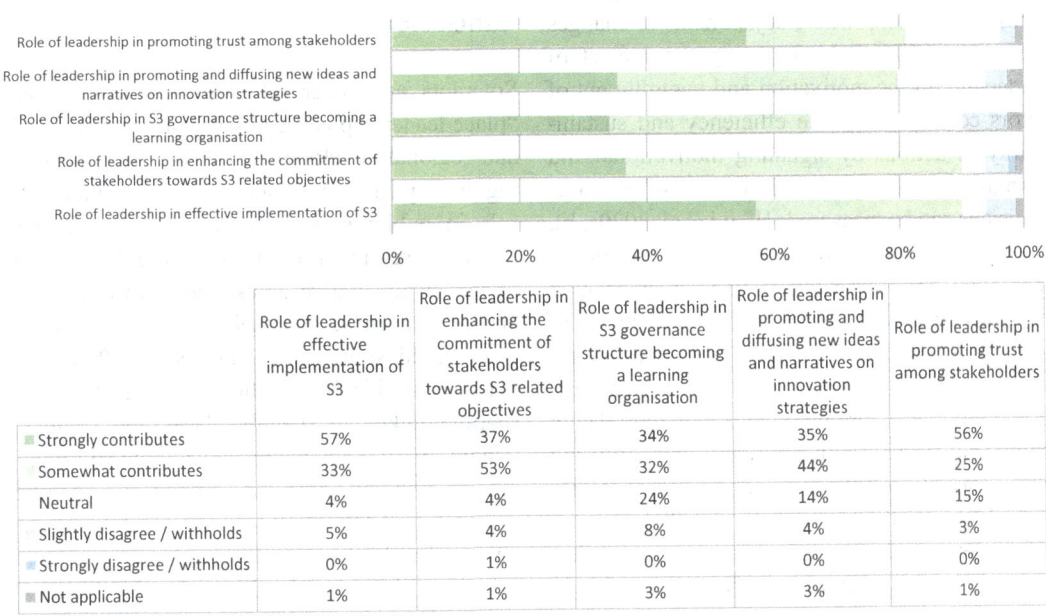

	Role of leadership in effective implementation of S3	Role of leadership in enhancing the commitment of stakeholders towards S3 related objectives	Role of leadership in S3 governance structure becoming a learning organisation	Role of leadership in promoting and diffusing new ideas and narratives on innovation strategies	Role of leadership in promoting trust among stakeholders
Strongly contributes	57%	37%	34%	35%	56%
Somewhat contributes	33%	53%	32%	44%	25%
Neutral	4%	4%	24%	14%	15%
Slightly disagree / withholds	5%	4%	8%	4%	3%
Strongly disagree / withholds	0%	1%	0%	0%	0%
Not applicable	1%	1%	3%	3%	1%

Urban and Regional Leadership, Fig. 2 Place leadership in urban and regional innovation policy agendas_Role of leadership in S3 implementation

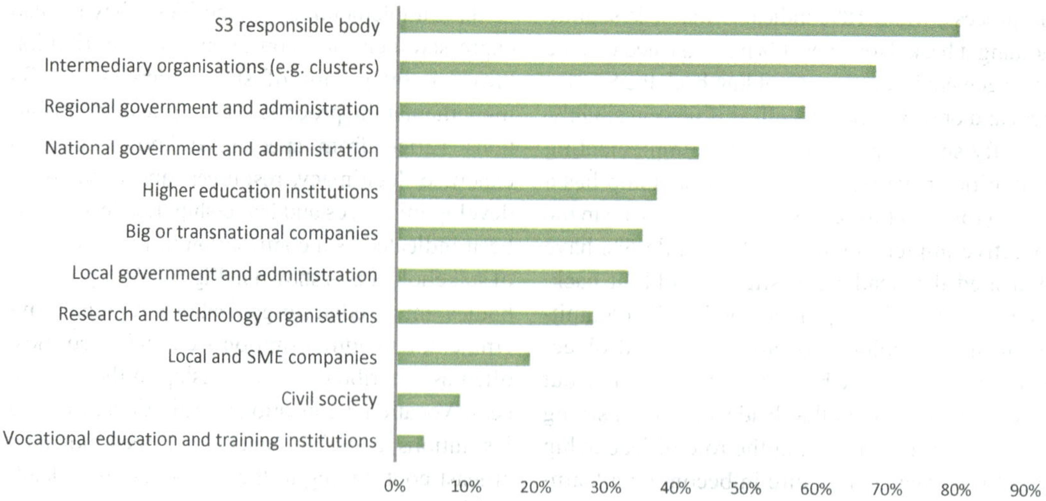

Urban and Regional Leadership, Fig. 3 Place leadership in urban and regional innovation policy agendas_Leadership as a contribution from the stakeholder group in the S3 policy making and implementation process

educational institutions to clusters. Leadership affects the motivation and commitment of stakeholders, which in return have an impact on the efficiency, sustainability, and success of projects, policies, and agendas (Hegyi et al. 2020a, b). Therefore, the impact of leadership in the public sphere is to be assessed to be able to highlight the areas of leadership where adjustments or changes are needed. Regularly assessing the impact of leadership on the motivation and commitment of actors contributes to the efficiency and sustainability of actions by signaling motivational and attitudinal challenges, while effects of previous decisions can be measured (Hegyi et al. 2020a, b).

The same survey results show that the role of leadership with respect to effective implementation of S3 correlates with enhancing the commitment of stakeholders towards S3-related objectives (correlation coefficient 0.917), with S3 governance structure becoming a learning organization (correlation coefficient 0.683), with promoting and diffusing new ideas and narratives on innovation strategies (correlation coefficient 0.685) and with thickening relationships and promoting trust among stakeholders (0.892), whereas the role of leadership in S3 governance structure becoming a learning organization shows a high degree of correlation with enhancing the commitment of stakeholders towards S3-related objectives (correlation coefficient 0.683). Interestingly, the data from the S3 2020 survey does not show a correlation between the level of development of regions and any measured factors of leadership.

Conclusion

Sotarauta and other scholars have agreed that "place leadership is the missing piece in the local and regional development puzzle," thus measuring the impact of leadership can play an important role in public policy setting (Sotarauta 2016). Piloted assessment frameworks highlight areas of leadership, where adjustments are needed to ensure efficiency, sustainability, and success of place-based strategies (Hegyi et al. 2020a), as place leadership is a "collective form of agenda shaping linked to specific places and context" (Hegyi 2020). Leadership shows to be contributing to winning agendas, proved by specific responses to COVID-19 from world leading institutions. Leadership has never been so important than in a global pandemic, as the C40 mayors' agenda for a green and just recovery puts it: [city] leaders ensuring to turn the tragedies of a pandemic into a better tomorrow (C40 2020). As the

2020 Strategic Foresight Report of the European Commission states, "after an uneasy start, the EU and its Member States pulled together to deal with the crisis [COVID-19] and agility and leadership at all levels of government played a key role in the response" (European Commission 2020), just as San Francisco's "hard-won immune system" has been dedicated to community leaders joining hands with experts to fight the spring COVID-19 epidemic, highlighting the historically built and maintained trust of the diverse population of the city, and the health department having direct access to the mayor and both working with top-line researchers at the University of California San Francisco (Duane 2020). Leadership, legitimacy, resources, and trust are named as indispensable for institutions, cities, and nations to reach resilience and to be able to respond well to crisis situations linked to health, environment, economy, or society (Goldin and Muggah 2020). All these examples illustrate the importance of being aware and measuring and facilitating the impact of leadership at the urban and regional level in actions addressing specific agendas of shared interests of local communities.

Cross-References

- ▶ Connecting Urban and Regional Innovation Ecosystems to Enhance Competitiveness
- ▶ Education for Inclusive and Transformative Urban Development
- ▶ Regulation of Urban and Regional Futures

References

Beer, A., & Clower, T. (2014). Mobilising leadership in cities and regions. *Regional Studies, Regional Science, 1*(1), 10–34.

C40. (2020). *C40 Mayors' agenda for a green and just recovery*. Available at: https://c40-production-images.s3.amazonaws.com/other_uploads/images/2093_C40_Cities_%282020%29_Mayors_Agenda_for_a_Green_and_Just_Recovery.original.pdf?1594824518

Duane, D. (2020). San Francisco was uniquely prepared for Covid-19. *Wired Magazine*, September.

European Commission. (2020). *2020 strategic foresight report. Charting the course towards a more Resilient Europe*. Available at: https://ec.europa.eu/info/sites/info/files/strategic_foresight_report_2020_1.pdf

Goldin, I., & Muggah, R. (2020). *Terra incognita: 100 maps to survive the next 100 years*. Random House UK.

Hegyi, F.B., Guzzo F., Perianez-Forte I. & Gianelle C. (2021). The JRC Survey on Smart Smart Specialisation: 2020 edition", Joint Research Centre Technical Reports, Seville: European Commission.

Hegyi, F. B. (2020). Leadership to address urban environmental challenges. *Mark and Focus, 3*(1), 16.

Hegyi, F. B., & Rakhmatullin, R. (2020). *Developing an evaluation framework integrating results of the thematic approach to smart specialization*. Luxembourg: Publications Office of the European Union, isbn 978-92-76-17904-7, https://doi.org/10.2760/636484.

Hegyi, F. B., Morgan, H., & Vendel, M. (2019). Urban mobility in transformation: Demands on education to close the predicted knowledge gap. In H. B. Jørgensen, K. Andersen, & O. Anker Nielsen (Eds.), *DTU international energy report 2019: Transforming urban mobility*. Copenhagen: Technical University of Denmark.

Hegyi, F. B., Borbely, L., & Bekesi, G. (2020a). *Factors of leadership attitude enhancing interregional collaboration. Dynamic interregional strategic partnerships' leadership impact on motivation and commitment*. Seville: European Union.

Hegyi, F. B., Borbely, L., & Bekesi, G. (2020b). *Leadership attitude impact on motivation and commitment in interregional collaboration: Pilot cases of thematic smart specialization partnerships*. Seville: European Union.

Organisation for Economic Co-operation and Development. (2010). *Regions matter*. Paris: OECD.

Sotarauta, M. (2016). Place leadership, governance and power. *Administration, 64*(3/4), 45–58. https://doi.org/10.1515/ad-min-2016-0024.

Sotarauta, M. (2018). Smart specialization and place leadership: Dreaming about shared visions, falling into policy traps? *Regional Studies, Regional Science, 5*(1), 190–203. https://doi.org/10.1080/21681376.2018.1480902.

Sotarauta, M., & Suvinen, N. (2019). Place leadership and the challenge of transformation: Policy platforms and innovation ecosystems in promotion of green growth. *European Planning Studies*. https://doi.org/10.1080/09654313.2019.1634006.

Stimson, R., Stough, R. R., & Salazar, M. (2009). *Leadership and institutions in regional endogenous development*. Northampton: Edwar Elgar.

Trickett, L., & Lee, P. (2010). Leadership of 'subregional' places in the context of growth. *Policy Studies, 31*(4), 429–440.

Vandecasteele, I., Baranzelli, C., Siragusa, A., & Aurambout, J. P. (Eds.). (2019). *The future of cities – Opportunities, challenges and the way forward* (pp. 8–9 and 113–122). Luxembourg: Publications Office.

U

Urban Atmospheric Microbiome

Justin D. Stewart[1] and Peleg Kremer[2]
[1]Department of Ecological Science, Vrije Universiteit Amsterdam, Amsterdam, Netherlands
[2]Department of Geography and the Environment, Villanova University, Villanova, PA, USA

Definitions

Microbe: Bacteria, fungi, and archaea are largely microscopic organisms that are globally dispersed and influence nutrient cycling, animal health, and human well-being.

Urban structure: The collection of built and natural elements in an urban area that give rise to the arrangement of a city in three-dimensional space.

Diversity: The number of species or functional traits with their abundance and distribution.

Niche: A position for a species or function in an environment.

Community composition: The distribution and abundances of potentially interacting microbial species with an environment.

Microbial biogeography: Patterns in diversity when viewed across spatial scales such as a city or a room.

Introduction

Assemblages of microbes into communities (often termed microbiomes) are ubiquities throughout the atmosphere and influence biotic members and abiotic components within the urban landscape. The majority of microbes cannot be grown in labs and thus there are limited studies of urban microbiomes and the atmosphere. Recent advances in biological, spatial, and computational tools have allowed for high-throughput and efficient analysis of microbiome membership and reveal a wide spread of microbial diversity; however, current knowledge is skewed toward bacterial studies due to difficulty with analyzing archaea and fungi. These methods have been applied to identify that urban microbiomes host unique communities that interact with the natural and built elements of cities and may influence ecosystem services, and human well-being.

Microbes in the Air

Outdoor Microbial Diversity

Urban air hosts diverse communities of microbes (Bowers et al. 2011; Brodie et al. 2007; Chaudhary et al. 2020; Pan et al. 2019; Stewart et al. 2020, 2021) with up to 10^6 cells per cubic meter of air (Zhen et al. 2017). Often these microbes are lifted from the Earth's surface by wind and air movement that then settles down (Fig. 1) onto the same environment or undergoes transport to a new location. Of these organisms, outdoor urban air is composed of diverse bacteria (phylum: Proteobacteria, Firmicutes, and Actinobacteria (Dueker et al. 2018; G. Á. Mhuireach et al. 2019; Stewart et al. 2020, 2021), archaea (phylum: Thaumarchaeota and Euryarchaeota (Fröhlich-Nowoisky et al. 2014), and fungi (phylum: Ascomycota and Basidiomycota (Chaudhary et al. 2020; Yamamoto et al. 2012). Microbial diversity (the number of species and the relative composition of their assemblages) is also vertically stratified with higher diversity closer to the Earth's surface (Robinson et al. 2020). Potential sources for microbes that are resuspended into the atmosphere including soil (Stewart et al. 2020, 2021), animals (Stewart et al. 2021), water (Genitsaris et al. 2017), and vegetation. Vegetation in particular is a large source for atmospheric microorganisms (Lindemann and Upper 1985; G. Mhuireach et al. 2016; G. Á. Mhuireach et al. 2019; C E Morris and Kinkel 2002) that are released from plant leaves through wind, precipitation, and other processes.

Furthermore, urban landscapes are spatially complex and composed of numerous features in various configuration and thus host numerous niches for microbes to occupy, leading to patchy atmospheric biodiversity across cities (G. Á. Mhuireach et al. 2019; Stewart et al. 2021)

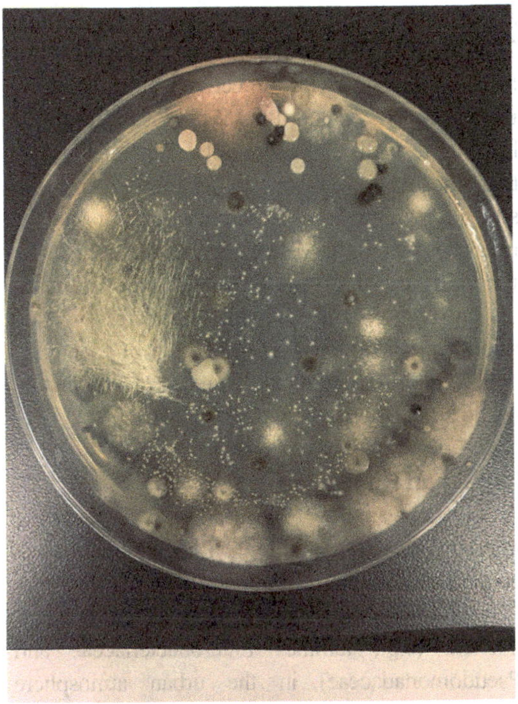

Urban Atmospheric Microbiome, Fig. 1 Example of passively collected bacteria and fungi from dry deposition of atmospheric particles in Philadelphia, PA, USA. (Photo used with permission by Justin D. Stewart)

(Fig. 2). High turnover rate in community composition has demonstrated rapid (daily) temporal variability of microbial diversity in the atmosphere (Bowers et al. 2013; G. Á. Mhuireach et al. 2019; Stewart et al. 2020) that is dependent on meteorology (Uetake et al. 2019; Zhen et al. 2017) and atmospheric chemistry (Pan et al. 2019).

Indoor Microbial Diversity

Atmospheric microbiomes also exist inside buildings where individuals live, play, and work. The orientation and structure of the indoor built environment influences microbial biogeography. Patterns in how many species are found, how related they are, and their original source environment have been observed in subways (Stewart et al. 2021), commercial buildings (Amend et al. 2010; Hewitt et al. 2012; Meadow et al. 2014; Tringe et al. 2008), hospitals (Jaffal et al. 1997; C. S. Li and Hou 2003), and residential homes (Gandara et al. 2006; Moon et al. 2014). As humans spend approximately 90% of their lives indoors (defined as inside a relatively closed space), indoor microbiomes are often shaped by the microorganisms humans bring with them from

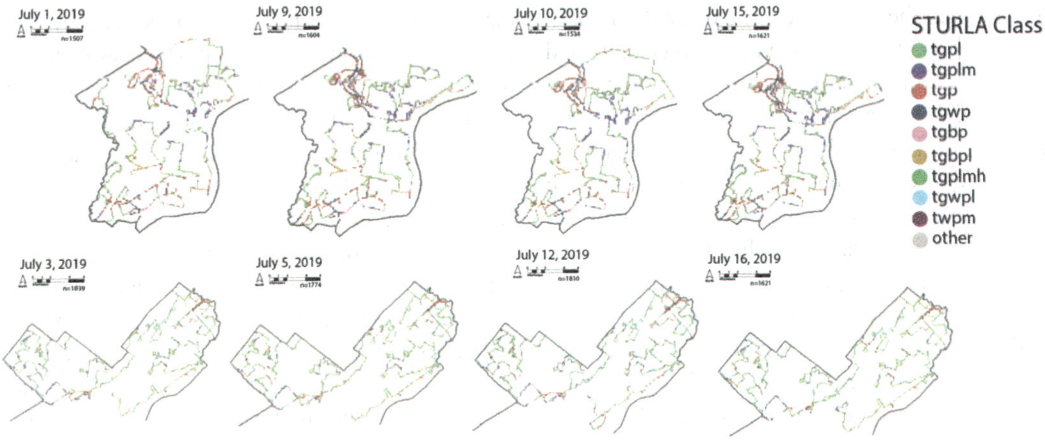

Urban Atmospheric Microbiome, Fig. 2 Example of atmospheric microbiome sampling routes colored by units of urban structures from the STURLA model. In this model 120 m^2 blocks of land are coded based on if they contain specific landscape elements (t=trees, g=grass, b=bare, w=water, p=paved, l=lowrise building, m=midrise building, h=highrise building) that are summed up into a single unit (e.g., class tgpl has trees, grass, pavement, and lowrise buildings). (Image is reproduced with permission of the authors of the study (https://doi.org/10.3389/fevo.2021.620461))

outside (Hospodsky et al. 2012; James F Meadow et al. 2015; Tringe et al. 2008) such as the Gammaproteobacteria.

Likewise, building design (Peters 2017) has long been understood to influence human well-being (Peavey 2011) as demonstrated by nurse Florence Nightingale's early observation that open windows for improved ventilation made for a healthy hospital (Nightingale 1860). Since then we have learned that some diseases are caused by microbes (Grimes 2006) and more recently that certain microorganisms are beneficial to human health. Within the indoor environment, building architecture also influences microbiome diversity and composition. Ventilation, in particular, structures microbial communities (Y. Li et al. 2007; J. F. Meadow et al. 2014) as it allows for active transport of microbes by wind from outside the built environment into a new niche. For example, microbial composition of air outside seems to influence (J. F. Meadow et al. 2014; Stewart et al. 2021) which microbes are inside as a function of wind dispersal with regard to bacteria but may not with fungi (Adams et al. 2013). Due to the outdoor sources of many indoor microbes, indoor microbiomes also experience seasonality that is also observed outdoors (Adams et al. 2013).

Microbial Functions and Traits

Microbes in the atmosphere are difficult to quantify functionally due to sampling constraints and technical limitations of measuring functions outside the laboratory. It is often easier to measure byproducts and metabolites that these organisms produce or even quantify a process they are performing such as nitrogen fixation. Functional characterizations are largely limited to the culturable fraction (often collected from cloud water or atmospheric deposition), which severally underrepresents total biodiversity. Functional diversity has also been extrapolated by predicting genomes (Douglas et al. 2019) of closely related and well-characterized relatives.

In particular, microbes high in the atmosphere (e.g., above average building heights) are associated with ice nucleation and cloud formation (Burrows et al. 2009; Hassett et al. 2015; Cindy E. Morris et al. 2013; Warren and Wolber 1991) and absorb/reflect solar radiation (Spänkuch et al. 2000). Likewise, airborne microbial communities host functional profiles reflective of their underlying ecosystem (Tignat-Perrier et al. 2020) despite hosting less diverse/fewer total functional traits than those below them. For example, air above soil hosts microbes with similar functions to the microbes in soil; however, a total fewer number of functions are observed. Genome prediction based on the known suite of genes of closely related microbes has inferred a diverse set of metabolic genes, antibiotic production, and virulence factors in the atmosphere (Stewart et al. 2020, 2021).

Impact on Human Health

Associations are now being made between airborne microbial exposure and human health in an urban context especially given the generally high proportion of medically relevant class Gammaproteobacteria (e.g., families Enterobacteriaceae and Pseudomonadaceae) in the urban atmosphere (Bowers et al. 2013; Cáliz et al. 2018; Stewart et al. 2021). Exposure to increased microbial diversity has been linked to allergy desensitization and both lessen (Illi et al. 2012; Lynch et al. 2014; Riedler et al. 2001; Ruokolainen et al. 2015; Schaub et al. 2009; Steele and Streit 2005) and worsen (Jie et al. 2011) cases of asthma depending on the type of microbe to which an individual is exposed. Given the high diversity of microbial species (Locey and Lennon 2016) and almost limitless niches they can fill, the atmosphere often contain pathogenic microbes (Aziz et al. 2018; Fan et al. 2019; Liu et al. 2018); however the ecological and functional implications of their concentrations and inhalation are still being uncovered (Willis et al. 2020). Likewise, the postmortem human microbiome (a proxy for microbial exposure when an individual is alive) has been associated with urban greening efforts as well as exposure to harmful bacteria and decreased microbial diversity where vegetation and greenspace is deficient (Pearson et al. 2019, 2020).

Conclusion

It is becoming increasingly clear that humans live in complex states of symbiosis with equally complex microbial communities in urban atmospheres. Just as humans have altered natural landscapes to fit their needs by designing cities and other built environments, microbial communities have adapted and become structured toward these alterations. We already know that once a microbe is present in a specific ecosystem such as the soil or the human body it can be beneficial, deleterious, or cause no response depending on its identity. Research is still ongoing as to how the role of dispersal of microbes in the atmosphere occurs in urban environments and what the potential implications are. With continuing technological advancements and integration of ecological theory, relationships between humanity, urban landscapes, and microbiomes will be further elucidated.

References

Adams, R. I., Miletto, M., Taylor, J. W., & Bruns, T. D. (2013). Dispersal in microbes: Fungi in indoor air are dominated by outdoor air and show dispersal limitation at short distances. *The ISME Journal, 7*(7), 1262–1273. https://doi.org/10.1038/ismej.2013.28.

Amend, A. S., Seifert, K. A., Samson, R., & Bruns, T. D. (2010). Indoor fungal composition is geographically patterned and more diverse in temperate zones than in the tropics. *Proceedings of the National Academy of Sciences of the United States of America*. https://doi.org/10.1073/pnas.1000454107.

Aziz, A. A., Lee, K., Park, B., Park, H., Park, K., Choi, I.-G., & Chang, I. S. (2018). Comparative study of the airborne microbial communities and their functional composition in fine particulate matter (PM2. 5) under non-extreme and extreme PM2. 5 conditions. *Atmospheric Environment, 194*, 82–92. https://www.sciencedirect.com/science/article/pii/S1352231018306228.

Bowers, R. M., Sullivan, A. P., Costello, E. K., Collett, J. L., Jr., Knight, R., & Fierer, N. (2011). Sources of bacteria in outdoor air across cities in the midwestern United States. *Applied and Environmental Microbiology, 77*(18), 6350–6356. https://doi.org/10.1128/AEM.05498-11.

Bowers, R. M., Clements, N., Emerson, J. B., Wiedinmyer, C., Hannigan, M. P., & Fierer, N. (2013). Seasonal variability in bacterial and fungal diversity of the near-surface atmosphere. *Environmental Science & Technology, 47*(21), 12097–12106. https://doi.org/10.1021/es402970s.

Brodie, E. L., DeSantis, T. Z., Parker, J. P. M., Zubietta, I. X., Piceno, Y. M., & Andersen, G. L. (2007). Urban aerosols harbor diverse and dynamic bacterial populations. *Proceedings of the National Academy of Sciences of the United States of America, 104*(1), 299–304. https://doi.org/10.1073/pnas.0608255104.

Burrows, S. M., Elbert, W., Lawrence, M. G., & Pöschl, U. (2009). Bacteria in the global atmosphere – Part 1: Review and synthesis of literature data for different ecosystems. *Atmospheric Chemistry and Physics, 9*(23), 9263–9280. https://doi.org/10.5194/acp-9-9263-2009.

Cáliz, J., Triadó-Margarit, X., Camarero, L., & Casamayor, E. O. (2018). A long-term survey unveils strong seasonal patterns in the airborne microbiome coupled to general and regional atmospheric circulations. *Proceedings of the National Academy of Sciences of the United States of America, 115*(48), 12229–12234. https://doi.org/10.1073/pnas.1812826115.

Chaudhary, V. B., Nolimal, S., Sosa-Hernández, M. A., Egan, C., & Kastens, J. (2020). Trait-based aerial dispersal of arbuscular mycorrhizal fungi. *New Phytologist*. https://doi.org/10.1111/nph.16667.

Douglas, G. M., Maffei, V. J., Zaneveld, J., Yurgel, S. N., Brown, J. R., Taylor, C. M., Huttenhower, C., & Langille, M. G. I. (2019). PICRUSt2: An improved and extensible approach for metagenome inference. *bioRxiv*. https://doi.org/10.1101/672295.

Dueker, M. E., French, S., & O'Mullan, G. D. (2018). Comparison of bacterial diversity in air and water of a major urban center. *Frontiers in Microbiology*. https://doi.org/10.3389/fmicb.2018.02868.

Fan, C., Li, Y., Liu, P., Mu, F., Xie, Z., Lu, R., Qi, Y., Wang, B., & Jin, C. (2019). Characteristics of airborne opportunistic pathogenic bacteria during autumn and winter in Xi'an, China. *Science of the Total Environment*. https://doi.org/10.1016/j.scitotenv.2019.03.412.

Fröhlich-Nowoisky, J., Ruzene Nespoli, C., Pickersgil, D. A., Galand, P. E., Müller-Germann, I., Nunes, T., Gomes Cardoso, J., Almeida, S. M., Pio, C., Andreae, M. O., Conrad, R., Pöschl, U., & Després, V. R. (2014). Diversity and seasonal dynamics of airborne archaea. *Biogeosciences*. https://doi.org/10.5194/bg-11-6067-2014.

Gandara, A., Mota, L. C., Flores, C., Perez, H. R., Green, C. F., & Gibbs, S. G. (2006). Isolation of Staphylococcus aureus and antibiotic-resistant Staphylococcus aureus from residential indoor bioaerosols. *Environmental Health Perspectives*. https://doi.org/10.1289/ehp.9585.

Genitsaris, S., Stefanidou, N., Katsiapi, M., Kormas, K. A., Sommer, U., & Moustaka-Gouni, M. (2017). Variability of airborne bacteria in an urban Mediterranean area (Thessaloniki, Greece). *Atmospheric Environment*. https://doi.org/10.1016/j.atmosenv.2017.03.018.

Grimes, D. J. (2006). Koch's postulates – Then and now. *Microbe*. https://doi.org/10.1128/microbe.1.223.1.

Hassett, M. O., Fischer, M. W. F., & Money, N. P. (2015). Mushrooms as rainmakers: How spores act as nuclei for raindrops. *PLoS One*. https://doi.org/10.1371/journal.pone.0140407.

Hewitt, K. M., Gerba, C. P., Maxwell, S. L., & Kelley, S. T. (2012). Office space bacterial abundance and diversity in three metropolitan areas. *PLoS One*. https://doi.org/10.1371/journal.pone.0037849.

Hospodsky, D., Qian, J., Nazaroff, W. W., Yamamoto, N., Bibby, K., Rismani-Yazdi, H., & Peccia, J. (2012). Human occupancy as a source of indoor airborne bacteria. *PLoS One, 7*(4), e34867. https://doi.org/10.1371/journal.pone.0034867.

Illi, S., Depner, M., Genuneit, J., Horak, E., Loss, G., Strunz-Lehner, C., Büchele, G., Boznanski, A., Danielewicz, H., Cullinan, P., Heederik, D., Braun-Fahrländer, C., & von Mutius, E. (2012). Protection from childhood asthma and allergy in Alpine farm environments—The GABRIEL advanced studies. *Journal of Allergy and Clinical Immunology, 129*(6), 1470–1477.e6. https://doi.org/10.1016/j.jaci.2012.03.013.

Jaffal, A. A., Nsanze, H., Bener, A., Ameen, A. S., Banat, I. M., & El Mogheth, A. A. (1997). Hospital airborne microbial pollution in a desert country. *Environment International*. https://doi.org/10.1016/S0160-4120(97)00003-2.

Jie, Y., Houjin, H., Feng, J., & Jie, X. (2011). The role of airborne microbes in school and its impact on asthma, allergy, and respiratory symptoms among school children. *Reviews in Medical Microbiology, 22*(4), 84. https://doi.org/10.1097/MRM.0b013e32834a449c.

Li, C. S., & Hou, P. A. (2003). Bioaerosol characteristics in hospital clean rooms. *Science of the Total Environment*. https://doi.org/10.1016/S0048-9697(02)00500-4.

Li, Y., Leung, G. M., Tang, J. W., Yang, X., Chao, C. Y. H., Lin, J. Z., Lu, J. W., Nielsen, P. V., Niu, J., Qian, H., Sleigh, A. C., Su, H. J. J., Sundell, J., Wong, T. W., & Yuen, P. L. (2007). Role of ventilation in airborne transmission of infectious agents in the built environment – A multidisciplinary systematic review. *Indoor Air*. https://doi.org/10.1111/j.1600-0668.2006.00445.x.

Lindemann, J., & Upper, C. D. (1985). Aerial dispersal of epiphytic bacteria over bean plants. *Applied and Environmental Microbiology*. https://doi.org/10.1128/aem.50.5.1229-1232.1985.

Liu, H., Zhang, X., Zhang, H., Yao, X., Zhou, M., Wang, J., He, Z., Zhang, H., Lou, L., Mao, W., Zheng, P., & Hu, B. (2018). Effect of air pollution on the total bacteria and pathogenic bacteria in different sizes of particulate matter. *Environmental Pollution, 233*. https://doi.org/10.1016/j.envpol.2017.10.070.

Locey, K. J., & Lennon, J. T. (2016). Scaling laws predict global microbial diversity. *Proceedings of the National Academy of Sciences of the United States of America, 113*(21), 5970–5975. https://doi.org/10.1073/pnas.1521291113.

Lynch, S. V., Wood, R. A., Boushey, H., Bacharier, L. B., Bloomberg, G. R., Kattan, M., O'Connor, G. T., Sandel, M. T., Calatroni, A., Matsui, E., Johnson, C. C., Lynn, H., Visness, C. M., Jaffee, K. F., Gergen, P. J., Gold, D. R., Wright, R. J., Fujimura, K., Rauch, M., . . . Gern, J. E. (2014). Effects of early-life exposure to allergens and bacteria on recurrent wheeze and atopy in urban children. *Journal of Allergy and Clinical Immunology, 134*(3), 593–601.e12. https://doi.org/10.1016/j.jaci.2014.04.018.

Meadow, J. F., Altrichter, A. E., Kembel, S. W., Kline, J., Mhuireach, G., Moriyama, M., Northcutt, D., O'Connor, T. K., Womack, A. M., Brown, G. Z., Green, J. L., & Bohannan, B. J. M. (2014). Indoor airborne bacterial communities are influenced by ventilation, occupancy, and outdoor air source. *Indoor Air, 24*(1), 41–48. https://doi.org/10.1111/ina.12047.

Meadow, J. F., Altrichter, A. E., Bateman, A. C., Stenson, J., Brown, G. Z., Green, J. L., & Bohannan, B. J. M. (2015). Humans differ in their personal microbial cloud. *PeerJ, 3*, e1258. https://doi.org/10.7717/peerj.1258.

Mhuireach, G., Johnson, B. R., Altrichter, A. E., Ladau, J., Meadow, J. F., Pollard, K. S., & Green, J. L. (2016). Urban greenness influences airborne bacterial community composition. *Science of the Total Environment, 571*, 680–687. https://doi.org/10.1016/j.scitotenv.2016.07.037.

Mhuireach, G. Á., Betancourt-Román, C. M., Green, J. L., & Johnson, B. R. (2019). Spatiotemporal controls on the urban aerobiome. *Frontiers in Ecology and Evolution, 7*, 43. https://doi.org/10.3389/fevo.2019.00043.

Moon, K. W., Huh, E. H., & Jeong, H. C. (2014). Seasonal evaluation of bioaerosols from indoor air of residential apartments within the metropolitan area in South Korea. *Environmental Monitoring and Assessment*. https://doi.org/10.1007/s10661-013-3521-8.

Morris, C. E., & Kinkel, L. L. (2002). Fifty years of phyllosphere microbiology: Significant contributions to research in related fields. *Phyllosphere Microbiology*.

Morris, C. E., Monteil, C. L., & Berge, O. (2013). The life history of pseudomonas syringae: Linking agriculture to earth system processes. *Annual Review of Phytopathology*. https://doi.org/10.1146/annurev-phyto-082712-102402.

Nightingale, F. (1860). *Notes on nursing*. New York: D. Appleton and Company. https://doi.org/10.2105/ajph.37.4.470-b.

Pan, Y., Pan, X., Xiao, H., & Xiao, H. (2019). Structural characteristics and functional implications of PM2.5 bacterial communities during fall in Beijing and Shanghai, China. *Frontiers in Microbiology, 10*. https://doi.org/10.3389/fmicb.2019.02369.

Pearson, A. L., Rzotkiewicz, A., Pechal, J. L., Schmidt, C. J., Jordan, H. R., Zwickle, A., & Benbow, M. E. (2019). Initial evidence of the relationships between the human postmortem microbiome and neighborhood blight and greening efforts. *Annals of the American Association of Geographers*. https://doi.org/10.1080/24694452.2018.1519407.

Pearson, A. L., Pechal, J., Lin, Z., Benbow, M. E., Schmidt, C., & Mavoa, S. (2020). Associations detected between measures of neighborhood environmental conditions and human microbiome diversity. *Science of the Total Environment*. https://doi.org/10.1016/j.scitotenv.2020.141029.

Peavey, E. (2011). Book review: Healing spaces: The science of place and well-being. *HERD: Health Environments Research & Design Journal*. https://doi.org/10.1177/193758671100400213.
Peters, T. (2017). Super architecture: Building for better health. *Architectural Design, 87*, 24–31.
Riedler, J., Braun-Fahrländer, C., Eder, W., Schreuer, M., Waser, M., Maisch, S., Carr, D., Schierl, R., Nowak, D., von Mutius, E., & Team, A. S. (2001). Exposure to farming in early life and development of asthma and allergy: A cross-sectional survey. *Lancet, 358*(9288), 1129–1133. https://doi.org/10.1016/S0140-6736(01)06252-3.
Robinson, J. M., Cando-Dumancela, C., Liddicoat, C., Weinstein, P., Cameron, R., & Breed, M. F. (2020). Vertical stratification in urban green space aerobiomes. *Environmental Health Perspectives*. https://doi.org/10.1289/EHP7807.
Ruokolainen, L., von Hertzen, L., Fyhrquist, N., Laatikainen, T., Lehtomäki, J., Auvinen, P., Karvonen, A. M., Hyvärinen, A., Tillmann, V., Niemelä, O., Knip, M., Haahtela, T., Pekkanen, J., & Hanski, I. (2015). Green areas around homes reduce atopic sensitization in children. *Allergy, 70*(2), 195–202. https://doi.org/10.1111/all.12545.
Schaub, B., Liu, J., Höppler, S., Schleich, I., Huehn, J., Olek, S., Wieczorek, G., Illi, S., & von Mutius, E. (2009). Maternal farm exposure modulates neonatal immune mechanisms through regulatory T cells. *Journal of Allergy and Clinical Immunology*. https://doi.org/10.1016/j.jaci.2009.01.056.
Spänkuch, D., Döhler, W., & Güldner, J. (2000). Effect of coarse biogenic aerosol on downwelling infrared flux at the surface. *Journal of Geophysical Research Atmospheres*. https://doi.org/10.1029/2000JD900173.
Steele, H. L., & Streit, W. R. (2005). Metagenomics: Advances in ecology and biotechnology. *FEMS Microbiology Letters*. https://doi.org/10.1016/j.femsle.2005.05.011.
Stewart, J. D., Shakya, K. M., Bilinski, T., Wilson, J., Ravi, S., & Choi, C. S. (2020). Variation of near surface atmosphere microbial communities at an urban and a suburban site in Philadelphia, PA, USA. *Science of the Total Environment, 138353*. https://doi.org/10.1016/j.scitotenv.2020.138353.
Stewart, J. D., Kremer, P., Shakya, K. M., Conway, M., & Saad, A. (2021). Outdoor atmospheric microbial diversity is associated with urban landscape structure and differs from indoor-transit systems as revealed by mobile monitoring and three-dimensional spatial analysis. *Frontiers in Ecology and Evolution, 9*, 9. https://www.frontiersin.org/article/10.3389/fevo.2021.620461.
Tignat-Perrier, R., Thollot, A., Magand, O., Vogel, M., & T., & Larose, C. (2020). Microbial functional signature in the atmospheric boundary layer. *Biogeosciences*. https://doi.org/10.5194/bg-17-6081-2020.
Tringe, S. G., Zhang, T., Liu, X., Yu, Y., Lee, W. H., Yap, J., Yao, F., Suan, S. T., Ing, S. K., Haynes, M., Rohwer, F., Wei, C. L., Tan, P., Bristow, J., Rubin, E. M., & Ruan, Y. (2008). The airbone metagenome in an indoor urban environment. *PLoS One*. https://doi.org/10.1371/journal.pone.0001862.
Uetake, J., Tobo, Y., Uji, Y., Hill, T. C. J., DeMott, P. J., Kreidenweis, S. M., & Misumi, R. (2019). Seasonal changes of airborne bacterial communities over Tokyo and influence of local meteorology. *Frontiers in Microbiology*. https://doi.org/10.3389/fmicb.2019.01572.
Warren, G., & Wolber, P. (1991). Molecular aspects of microbial ice nucleation. *Molecular Microbiology*. https://doi.org/10.1111/j.1365-2958.1991.tb02104.x.
Willis, K. A., Stewart, J. D., & Ambalavanan, N. (2020). Recent advances in understanding the ecology of the lung microbiota and deciphering the gut-lung axis. https://doi.org/10.1152/Ajplung.00360.2020, https://doi.org/10.1152/ajplung.00360.2020.
Yamamoto, N., Bibby, K., Qian, J., Hospodsky, D., Rismani-Yazdi, H., Nazaroff, W. W., & Peccia, J. (2012). Particle-size distributions and seasonal diversity of allergenic and pathogenic fungi in outdoor air. *ISME Journal*. https://doi.org/10.1038/ismej.2012.30.
Zhen, Q., Deng, Y., Wang, Y., Wang, X., Zhang, H., Sun, X., & Ouyang, Z. (2017). Meteorological factors had more impact on airborne bacterial communities than air pollutants. *Science of the Total Environment*. https://doi.org/10.1016/j.scitotenv.2017.05.049.

Urban Biodiversity

► Habitat Provisioning

Urban Capacity

► Urban Resilience

Urban Centers

► Multi-stakeholder Partnerships to Support Climate Migrants in Fragile Cities

Urban Climate Governance

► Carbon Neutral Adelaide

U

Urban Climate Resilience

João Cortesão[1] and Samantha Copeland[2]
[1]Landscape Architecture and Spatial Planning, Wageningen University, Wageningen, The Netherlands
[2]Ethics and Philosophy of Technology, Delft University of Technology, Delft, The Netherlands

Synonyms

Climate adaptation; Climate-adaptive urban areas; Climate-resilient city; Climate-responsive urban areas; Resilient cities; Resilient urban climates

Definition

The capacity of an urban area to maintain or quickly return to desired functions in the face of climate-related chronic stresses and/or acute shocks, to adapt to changing and uncertain climatic conditions, or to rapidly transform in ways that build climate-adaptive capacity across temporal and spatial scales. The result of infrastructural measures contributing to urban resilience that include climate-responsive design principles targeted at absorbing, reacting, recovering, adapting, and reorganizing in response to climate phenomena, and with the ultimate purposes of preserving infrastructure and increasing the capacity of urban populations to cope with the impacts of climate change in ways that maintain a shared urban identity. The outcome of integrating urban socio-technical-environmental systems to increase the climate resilience of an urban area that comprises both infrastructural and ethical implications.

Introduction

Urban climate resilience is an approach to mitigating the impacts of climate change in the urban built environment that fits within the scope of resilience and urban resilience. The urban built environment describes all land developed within known physical boundaries that contains buildings and associated infrastructure (Tunstall 2006), and high population density. "Resilience" can be defined as "the ability of a socio-technical-environmental system to sustain, improve and innovate its key functions – through absorbing, reacting to, recovering from, adapting to or reorganizing – in response to chronic stresses, abrupt shocks, and disruptions" (DeSIRE 2020). Resilience, thus, comprises the development of solutions for reducing the severity of disruptions to human living environments caused by issues that are complex and hard to fully predict, such as climate change, ecological restoration, equity, or economic opportunity. Resilience engineering has become the primary method for addressing the pressing challenges that current and future societies face by embracing the "deep uncertainty" intrinsic to any future projection and by adopting a "complex systems" approach. Deep uncertainty can be defined as "a situation in which analysts do not know or cannot agree on (1) models that relate key forces that shape the future, (2) probability distributions of key variables and parameters in these models, and/or (3) the value of alternative outcomes" (Hallegatte et al. 2012: 2). Complex systems involve "a large number and diversity of components and relationships between the components, as well as a high degree of uncertainty regarding the system's current state and its path of future development" (Pols et al. 2015: 50). Dealing with deep uncertainty and systemic reasoning calls for "transdisciplinarity," which is about "research collaborations among scientists from different disciplines and non-academic stakeholders from business, government, and the civil society" targeted at developing solutions (Lang et al. 2012: 26) under a multi-perspective approach, leading to more holistic and, therefore, robust answers to a problem.

"Urban resilience" fits within the scope of resilience engineering and it relates to "the ability of an urban system-and all its constituent socio-ecological and socio-technical networks across temporal and spatial scales-to maintain or rapidly return to desired functions in the face of a disturbance, to adapt to change, and to quickly transform systems that limit current or future adaptive capacity" (Meerow et al. 2016: 39). Urban areas tend to be particularly vulnerable to cascading system failures following natural disasters due to the concentration of people and services, as well as to the concentration and interconnectivity of infrastructure (Elmqvist et al. 2019). On top of this, due to the predicted increase in urban populations, the coming decades are expected to bring substantial urban expansion and so increase the demand for infrastructure, food, and energy (McPhearson et al. 2016). Accelerating action on urban resilience is, thus, of utmost importance and can no longer be postponed.

Among the several challenges resilience engineering and urban resilience address, there is climate change. "Climate adaptation" is the overarching answer to this challenge, together with climate mitigation. Adaptation is "the adjustment in natural or human systems in response to actual or expected climatic stimuli or their effects, which moderates harm or exploits beneficial opportunities" (Parry et al. 2007: 6). In line with this definition, climate adaptation is about making changes to practices, processes, and structures to reduce vulnerability to extreme weather events and strengthen community resilience (Henstra 2012: 189). "Vulnerability" in this context is "the degree to which a system is susceptible to, and unable to cope with, adverse effects of climate change, including climate variability and extremes" (Parry et al. 2007: 6).

"Urban climate resilience" (also referred to as "resilient urban climates" or "climate-resilient cities") is a term currently used to describe the goal/outcome of addressing climate adaptation in urban areas. It refers to the capacity of an urban area "to withstand climate change stresses, to respond effectively to climate-related hazards, and to recover quickly from residual negative impacts" (Henstra 2012: 178). In more detail, urban climate resilience describes the capacity of an urban area to maintain or quickly return to desired functions in the face of climate-related chronic stresses (e.g., urban heat islands) and/or acute shocks (e.g., heatwaves or intense rainfall), to adapt to changing and uncertain climatic conditions, or to rapidly transform/improve the urban climate system, while ensuring the safety, health, and well-being of its community. Hence, urban climate resilience comprises both an infrastructural (i.e., urban areas as physical assets; spatial design and engineering; the material dimension of urban areas) and a social (i.e., urban areas as socio-cultural assets; inclusivity and ethics; the immaterial dimensions of urban areas) dimension. On this premise, action toward urban climate resilience calls for bringing closer together two core components: climate-adaptive design and ethics.

A Matter of Climate-Adaptive Design

This dimension of urban climate resilience refers to "climate-adaptive spatial design" (also referred to as "climate-responsive," "climate-conscious," or "bioclimatic" design) solutions (see Lenzholzer 2015) developed by spatial design disciplines such as architecture, urban design/planning, or landscape architecture. Climate-adaptive spatial design is about site-specific design solutions broadly targeted at "controlling the amount of solar radiation, controlling exposure to wind, finding the right balance between air temperature and relative humidity, enabling evaporative cooling and enabling heat losses from surfaces" (Cortesão 2020: 43). This is achieved by choosing and combining spatial design elements (e.g., trees, paving, or cladding materials) and parameters (e.g., orientation of buildings or height-to-width ratio) in synergetic ways. The expected outcome is an urban area, or parts of it, that is resilient to extreme weather phenomena, whose outdoor spaces provide protection from negative effects and exposure to positive aspects of climate throughout the year, and that comprises energy-conscious and climate-neutral buildings (Brears 2018; Lenzholzer 2015; Nikolopoulou 2004; Santamouris 2001; Stremke et al. 2011). The choices made regarding the combination of spatial design elements and parameters determine these attributes and, thus, urban climate resilience. For example, creating an unpaved, green area or a hard-paved, stony area results in rather

Urban Climate Resilience, Fig. 1 The climate-adaptive design component of urban climate resilience deals with planning and implementing infrastructural interventions in the built environment, i.e., spatial design and engineering solutions shaping or reshaping the physical layout of an urban area and, thereby, determining its climate resiliency

contrasting micro-scale effects, such as mean radiant temperature or water infiltration rate, which might be positive or negative to urban climate resiliency and outdoor human thermal comfort (Fig. 1).

Climate-adaptive design in the context herewith provided also involves design solutions created across engineering fields such as hydraulic engineering, for instance, flood defenses (e.g., Jonkman et al. 2018), or civil engineering, for example, critical infrastructures (e.g., Bozza et al. 2017). Climate-adaptive design and engineering concur to addressing both infrastructural and social components of urban climate resilience.

The design solutions developed toward urban climate resilience lead to building climate-adaptive capacity across temporal and spatial scales. This means solutions both immediately or gradually responding to disruption/change, i.e., "speed of recovery" (Meerow et al. 2016), and covering intervention scales from the microclimate to the local climate scale. In the urban context, the "microclimate scale" refers to confined and well-defined spaces within the air layer near the ground surface, such as a street or a square, where the climatic variables are greatly conditioned by the immediate surroundings, and "local climates" refer to areas broadly between 100 and 10,000 m, one or more city districts, largely determined by local conditions (Cuadrat and Pita 2009).

The public and the private sector, together, shape (physically and symbolically) urban areas. The relationship(s) between one and the other are therefore paramount in addressing urban climate resilience. Broadly put, academia creates evidence-based design knowledge on climate-adaptive design; spatial designers introduce site-specific design solutions into a design schema; contractors and builders materialize the design schema, which makes their mindset and the qualities of their building practices a cornerstone of implementing an urban climate resilience vision. Citizens can be engaged in implementing climate-adaptive solutions, for example, through streetscape gardening or depaving (Tamminga et al. 2020) and, last but not least, civic authorities determine new urban (re)developments and, depending on how the political agenda includes climate action, regional and local governments can create or hinder the chances for the implementation of climate-adaptive solutions (Pot 2020). On the one hand, the relationships between these parties determine the extent to which, and the quality with which, climate-adaptive design schemas are actually delivered. On the other hand, with the conscientious inclusion of considerations of justice and normativity, it is possible to come to a greater understanding of these relationships while attending to opportunities for improving them, by appropriately framing urban climate resilience as a continuous process of adaptation.

A Matter of Ethics

This dimension of urban climate resilience, ethics, attends to the impact that climate-adaptive solutions inevitably have on the populations of people and animals that live within and who move through and interact with the urban environment. Large numbers of people within a small area mean that ethical concerns, that is, concerns about how one should behave toward one another, including concerns of distributive and social justice, are of particular importance in the context of urban climate resilience. Migration into and out of urban areas, as well as between areas within the boundaries of a city, mean that social resilience, community resilience, personal resilience, and urban resilience are interrelated scales at which urban climate resilience can be measured and in respect to which design plays a vital role. This component acknowledges that the end beneficiaries of urban resilience are the people that populate an urban area (Fig. 2). While comprising nature-based solutions fostering urban biodiversity and ecosystems that hold the potential to improve the living conditions of animals, in urban areas it is the conditions that climate-responsive design delivers for people's safety, health, and well-being (including the social dimensions of health) that are its end goal.

Social resilience and vulnerability are regularly addressed in discussions related to urban climate resilience, although their importance demands still further elucidation on a case-by-case basis. Social resilience has to do with the capacities of individuals and groups to create and to utilize institutional structures and social relationships in order to adapt, cope, or transform in response to a crisis and in preparation for future crises (e.g., Keck and Sakdapolrak 2013). Vulnerabilities are sometimes the result of historical inequalities and institutions that constrain where people live within or commute to or through an urban environment; it is well known there is a correlation between economic and environmental vulnerability. Inequalities and vulnerability, both in respect to the effects of climate change and in socioeconomic terms, can be exacerbated by efforts toward urban climate resilience; they can also be thereby improved.

Urban Climate Resilience, Fig. 2 The ethics component of urban climate resilience acknowledges that the end beneficiaries of urban resilience are the people that populate an urban area; it deals with social resilience, attending to the potential impact of design on populations, the environment, and over time

Social resilience is generally measured indirectly via indicators, but this raises practical issues such as how to aggregate diverse factors in an overall assessment, or how to draw boundaries around particular social groups and communities within the urban environment in order to evaluate and compare their resilience (Copeland et al. 2020). Further, categorizing particular groups as vulnerable, according to their current status, can hinder efforts to make changes required for urban climate resilience. This happens when the effects of measures to increase resilience inadequately address chronic or emergent vulnerabilities and inequalities (Shokry et al. 2021). Designing for urban climate resilience requires the active and ongoing involvement of the city's private citizens, both because they own property and make use of public property, and because social resilience and attention to questions of justice are necessary for urban climate resilience. Any transformative design within the urban built environment comprising a paradigm shift, such as the shift needed toward urban climate resilience, requires the agreement of such citizens. However, the urban population is not only made up of property-owning private citizens, but includes also migrant populations and those who use the interconnecting public spaces. Whether urban dwellers cooperate, for instance, with depaving operations, will determine not only the pace at which these operations will take place, but also their preservation and functionality after completion. When such changes to shared space are not accepted, people may take alternative routes or even stop using an outdoor space.

Strategies to increase urban climate resilience may also require small- or large-scale cooperation from citizens who need to be relocated, for example, to enhance the resilience of a coastline by reclaiming land. Greening efforts in neighborhoods that contribute to reducing heat stress or that improve transportation access will also raise property values, resulting in the migration of local residents. Without considering the social impact and ethical implications of plans for enhancing urban climate resilience, the ultimate result may be to have no overall effect – for instance, if citizens forced to relocate do so in a way that reduces their and the city's overall resilience in relation to climate change.

Summary/Conclusion

This chapter provided a review of the fundamental concepts of urban climate resilience while drawing attention to the need for operationalizing these concepts in a way that reinforces and deepens the association of urban climate resilience with climate-adaptive design and with ethics. This perspective reflects the view that climate adaptation can no longer be postponed and that public-private relationships are key in urban climate resilience.

Increasing urban climate resilience requires infrastructural design solutions developed with the ultimate purpose of increasing the capacity of urban populations to cope with climate change through mitigation, adaptation, and recovery from crises, in ways that maintain a shared urban identity. Transformative approaches to resilience are particularly appropriate for urban socio-technical-environmental systems, insofar as changing populations and movement of inhabitants within the urban area call for technical solutions at multiple scales and across time. Further, measures to increase urban resilience to climate change, insofar as they are likely to require radical changes to ways of life and behavior in urban environments, require an ethics-based approach.

In sum, the concepts of climate-responsive design and ethics are ought to be seen as inextricably connected while conceiving and implementing climate adaptation measures within the built environment. Reinforcing and deepening the association between climate-adaptive design and ethics enables acceleration toward climate adaptation, as it entails site- and culture-specific ways of interweaving the infrastructural and social components of urban resilience. Together, climate-responsive design and ethics are the key components of effective measures to increase urban climate resilience.

Cross-References

► From Vulnerability to Urban Resilience to Climate Change
► How Cities Can Be Resilient
► Urban Resilience: Moving from Idealism to Systems Thinking

References

Bozza, A., Asprone, D., & Fabbrocino, F. (2017). Urban resilience: A civil engineering perspective. *Sustainability, 9*(103), 1–17. https://doi.org/10.3390/su9010103.

Brears, R. C. (2018). *Blue and green cities. The role of blue-green infrastructure in managing urban water resources. Blue and green cities*. London: Palgrave Macmillan. https://doi.org/10.1057/978-1-137-59258-3_10.

Copeland, S., Comes, T., Bach, S., Nagenborg, M., Schulte, Y., & Doorn, N. (2020). Measuring social resilience: Trade-offs, challenges and opportunities for indicator models in transforming societies. *International Journal of Disaster Risk Reduction, 51*, 101799. https://doi.org/10.1016/j.ijdrr.2020.101799.

Cortesão, J. (2020). Bioclimatic urban design. Goals and methods. In P. Burlando & F. Mazzino (Eds.), *New challenges for contemporary landscape architecture. Toward nature?* (pp. 40–49). Florence: Altralinea Edizioni.

Cuadrat, J. M., & Pita, M. F. (2009). *Climatología*. Madrid: Ediciones Cátedra.

DeSIRE. (2020). Mission statement. Website: https://www.4tu.nl/resilience/research/desire-mission-statement/

Elmqvist, T., Andersson, E., Frantzeskaki, N., McPhearson, T., Olsson, P., Gaffney, O., Takeuchi, K., & Folke, C. (2019). Sustainability and resilience for transformation in the urban century. *Nature Sustainability, 2*(4), 267–273. https://doi.org/10.1038/s41893-019-0250-1.

Hallegatte, S., Shah, A., Lempert, R., Brown, C., & Gill, S. (2012). Investment decision making under deep

uncertainty: Application to climate change. *Policy research working paper, 6193*. https://doi.org/10.1596/1813-9450-6193.
Henstra, D. (2012). Toward the climate-resilient city: Extreme weather and urban climate adaptation policies in two Canadian provinces. *Journal of Comparative Policy Analysis: Research and Practice, 14*(2), 175–194. https://doi.org/10.1080/13876988.2012.665215.
Jonkman, S. N., Voortman, H. G., Klerk, W. J., & van Vuren, S. (2018). Developments in the management of flood defences and hydraulic infrastructure in the Netherlands. *Structure and Infrastructure Engineering, 14*(7), 895–910. https://doi.org/10.1080/15732479.2018.1441317.
Keck, M., & Sakdapolrak, P. (2013). What is social resilience? Lessons learned and ways forward. *Erdkunde, 67*(1), 5–19. https://doi.org/10.3112/erdkunde.2013.01.02.
Lang, D. J., Wiek, A., Bergmann, M., Stauffacher, M., Martens, P., Moll, P., Swilling, M., & Thomas, C. J. (2012). Transdisciplinary research in sustainability science: Practice, principles, and challenges. *Sustainability Science, 7*, 25–43. https://doi.org/10.1007/s11625-011-0149-x.
Lenzholzer, S. (2015). *Weather in the city*. Rotterdam: nai010 Publishers.
McPhearson, T., Parnell, S., Simon, D., Gaffney, O., Elmqvist, T., Bai, X., Debra, R., & Revi, A. (2016). Scientists must have a say in the future of cities. *Nature, 538*(7624), 165–166. https://doi.org/10.1038/538165a.
Meerow, S., Newell, J. P., & Stults, M. (2016). Landscape and urban planning defining urban resilience: A review. *Landscape and Urban Planning, 147*, 38–49. https://doi.org/10.1016/j.landurbplan.2015.11.011.
Nikolopoulou, M. (Ed.). (2004). *Designing open spaces in the urban environment: A bioclimatic approach. RUROS: Rediscovering the Urban Realm and Open Spaces*. Pikermi: Centre for Renewable Energy Sources.
Parry, M., Canziani, O., Palutikof, J., van der Linden, P., & Hanson, C. (Eds.) (2007). Climate change 2007 – Impacts, adaptation and vulnerability contribution of working group II to the Fourth Assessment Report of the IPCC. Cambridge University Press, Cambridge, UK. https://doi.org/10.1016/B978-008044910-4.00250-9.
Pols, L., Edelenbos, J., Pel, B., & Dammers, E. (2015). Urbanized deltas as complex adaptive systems. In J. E. Han Meyer, A. Bregt, & E. Dammers (Eds.), *New perspectives on urbanizing deltas. A complex adaptive systems approach to planning and design* (pp. 47–60). Amsterdam: MUST Publishers.
Pot, W. (2020). *Deciding for tomorrow, today. What makes governmental decisions about water infrastructure forward looking?* (PhD thesis). Wageningen University.
Santamouris, M. (2001). *Energy and climate in the urban built environment*. London: James & James.
Shokry, G., Anguelovski, I., Connolly, J. J. T., Maroko, A., & Pearsall, H. (2021). "They didn't see it coming": Green resilience planning and vulnerability to future climate gentrification. *Housing Policy Debate*. https://doi.org/10.1080/10511482.2021.1944269.
Stremke, S., Van Den Dobbelsteen, A., & Koh, J. (2011). Exergy landscapes: Exploration of second-law thinking towards sustainable landscape design. *International Journal of Exergy, 8*(2), 148–174. https://doi.org/10.1504/IJEX.2011.038516.
Tamminga, K., Cortesão, J., & Bakx, M. (2020). Convivial greenstreets: A concept for climate-responsive urban design. *Sustainability, 12*, 3790. https://doi.org/10.3390/su12093790.
Tunstall, G. (2006). *Managing the building design process*. Oxford, UK: Butterworth-Heinemann.

Urban Commons

► Stewarding Street Trees for a Global Urban Future

Urban Commons as a Bridge Between the Spatial and the Social

Marianna Charitonidou
Department of Art Theory and History, Athens School of Fine Arts, Athens, Greece
School of Architecture, National Technocal University of Athens, Athens, Greece
Department of Architecture, ETH Zurich, Zurich, Switzerland

Synonyms

Community resources; Shared urban codes; Shared urban conventions; Shared urban practices; Shared urban resources

Definition

The notion of "urban commons" refers to the shared codes and conventions characterizing the production of urban space. Conceiving urban space as 'urban commons" goes hand in hand with shaping strategies of urban planning that go beyond the distinction between public and private space. Shaping urban planning strategies that aim to promote "urban commons" goes hand in hand with designing urban spaces aiming to accommodate the daily activities of citizens, cultivating their tendency to enhance commoning practices, on the one hand, and contributing to their sense of sharing the "urban commons," on the other. "Urban commons" should be understood beyond their reduction to natural resources and aim to shed light on the concept of "urban commons". Useful for understanding the role of "urban commons" in re-inventing urban planning practices are the debates around the benefits of the so-called negotiated planning approach and co-production" approach. These approaches are at the center of the current debates in the field of spatial planning. Particular emphasis is placed on the interaction of the technological, economic, and cultural aspects of "urban commons," on the one hand, and on the capacity of commoning practices to promote a sense of community.

Introduction

To better grasp the implications behind the concept of urban commons, we can bring in mind David Harvey's critique of Garrett Hardin's approval of privatization. Harvey, in "The Future of the Commons" (Harvey 2011), apart from criticizing Hardin's approach, in "The Tragedy of the Commons" (Hardin 1968), underscores the fact that Elinor Ostrom, in *Governing the Commons,* focused her understanding of commons mainly on natural resources (Ostrom 2015). More specifically, Harvey argues that all resources are socially defined in the sense that they are always related to technology, economy, and culture. John Bingham-Hall, in "Future of Cities: Commoning and Collective Approaches to Urban Space," argues that the notion of the common "suggests a community of commoners that actively utilise and upkeep whatever it is that is being commoned" (Bingham-Hall 2016).

As David Bollier remarks, in "A new politics of the commons," "[r]einventing the commons is still a fledgling vision, but its spontaneous embrace by so many different constituencies suggests a deep human yearning to explore new modes of social connection and collaboration" (Bollier 2007; Bollier and Helfrich 2012). Ida Susser and Stéphane Tonnelat, in "Transformative Cities: The Three Urban Commons," draw upon Henri Lefebvre's theory and distinguish three ways of understanding urban commons. More specifically, they draw a distinction between the following three types of urban commons: (a) a first type of urban commons concerning "issues of production, consumption, and use of public services and public goods reframed as a common means for a decent everyday life"; (b) a second type of urban commons referring to "public spaces of mobility and encounter collectively used and claimed by citizens, such as streets, subways, cafés, public gardens, and even the World Wide Web"; and (c) "a third type of urban commons under the form of collective visions within which each individual may find a place" (Susser and Tonnelat 2013, p. 108).

Commons Versus Commoning: Understanding Urban Commons as Public Space or as Community?

A dilemma that dominates the debates around urban commons is the question whether understanding commons as public space or as community is preferable. Useful for responding to this question is Jeremy Németh's definition of the commons, in "Controlling the Commons: How Public is Public Space?" as "any collectively owned resource held in joint use or possession to which anyone has access without obtaining permission of anyone else" (Németh 2012, p. 815). Understanding commons as community implies that community is conceived as a homogeneous group of people, while comprehending the commons as public space is based on the intention to take into consideration the relation between heterogeneous communities (Charitonidou 2021a).

Another issue that is also present in the scholarship around commons is the interrogation concerning replacement of the notion of commons by that of commoning. In relation to this, Patrick Bresnihan, in "The More-than-human Commons:

From Commons to Commoning," highlights that "[t]he noun 'commons' has been expanded into the continuous verb 'commoning', to denote the continuous making and remaking of the commons through shared practice" (Bresnihan 2016, p. 96). Another remark that could help us better grasp the impact of commoning practices on the relations between citizens is Stavros Stavrides's claim, in *Common Space: The City as Commons*, that "[c] ommoning practices importantly produce new relations between people" (Stravides 2016, p. 2; 2015; Borch and Kornberger 2015). David Harvey remarked, in *Rebel cities: From the right to the city to the urban revolution*, regarding the concept of "commoning": "At the heart of the practice of commoning lies the principle that the relation between the social group and that aspect of the environment being treated as a common shall be both collective and non-commodified-off limits to the logic of market exchange and market valuations" (Harvey 2012, p. 73).

Michael Hardt and Antonio Negri remark, in *Commonwealth,* regarding the role of commons within the context of capitalism and the expansion of commons: "Contemporary forms of capitalist production and accumulation in fact, despite their continuing drive to privatize resources and wealth, paradoxically make possible and even require expansions of the common" (Hardt and Negri 2009, ix; 2001; 2004). Hardt and Negri use the notion of the common to refer not only to the natural resources such as "the earth we share but also the languages we create, the social practices we establish, the modes of sociality that define our relationships, and so forth" (Hardt and Negri 2009, p. 350). Hardt and Negri's theory is based on the intention to reflect upon the concept of the commons in relation to the concept of urbanity. This is evidenced in their claim that "the metropolis [. . .] [is] a factory for the production of the common" (Hardt and Negri 2009, p. 350).

Karl Marx, in his *Critique of Hegel's Philosophy of Right,* notes that "at its highest point the political constitution is the constitution of private property" (Marx 1970). Hardt and Negri, starting out from Marx's understanding of property, remark that "[p]rivate property in its capitalist form [. . .] produces a relation of exploitation in its fullest sense" (Hardt and Negri 2009, p. 23). They also underline that the relations of exploitation in capitalism are based on "the production of the human as commodity," on the one hand, and the exclusion "from view the materiality of human needs and poverty," on the other hand (Hardt and Negri 2009, p. 23). Hardt and Negri, in *Commonwealth*, redefine the poor as follows: "The poor [. . .] refers not to those who have nothing but to the wide multiplicity of all those who are inserted in the mechanisms of social production regardless of social order or property" (Hardt and Negri 2009, p. 40). They also highlight that "the multitude of the poor is a real and effective menace for the republic of property" (Hardt and Negri 2009, p. 40). They also mention that among the practical maneuvers aiming to divide the poor and to deprive "them of the means of action and expression, and so forth" are the "efforts to tame, undermine, and nullify [. . .] [their] power" (Hardt and Negri 2009, p. 46).

Conclusion: "Collaborative Approaches," "Co-production," or "Negotiated Planning?"?

A distinction that is helpful for realizing the implication of the implementation of participation-oriented strategies is that between the so-called collaborative approaches and the co-production approaches (Charitonidou 2021a, b, c, 2022a).

Of pivotal importance for the history of Advocacy planning movement in the United States of America are the debates concerning the critiques of urban renewal in Philadelphia during the late 1950s. Denise Scott Brown has commented on advocacy planners' critique of urban renewal program, highlighting that it "derived from the problem that urban renewal had become 'human removal'" (Scott Brown 2009, p. 32; Charitonidou 2021c; Lung-Amam et al. 2015; Charitonidou 2022b). As I have analyzed, in "Denise Scott Brown's active socioplastics and urban sociology: From Learning from West End to Learning from Levittown," Scott Brown has "underscored that the main argument of advocacy planners was that architects and urban planners'

U

leadership had diverted urban renewal from a community support to a socially coercive boondoggle" (Charitonidou 2022b; Scott Brown 2009, p. 33; Pacchi 2018). Among the most important advocacy planners was Paul Davidoff, who taught at the City Planning Department of the University of Pennsylvania between 1958 and 1965. The pedagogical approaches at the Department of City Planning at the University of Pennsylvania when Scott Brown resettled there was influenced by social sciences and New Left critiques (Charitonidou 2022b, d, e).

The commonalities and differences of the collaborative approaches and the co-production approaches are examined by Vanessa Watson, in "Co-production and collaboration in planning: The difference," where she remarks that, "co-production" and "collaborative or communicative planning," despite their shared concern "with how state and society can engage in order to improve the quality of life of populations [. . .] with an emphasis on the poor and marginalized," differ in the sense that co-production "works outside (and sometimes against) established rules and procedures of governance in terms of engagement with the state, while this is much less usual (although not impossible) in collaborative and communicative planning processes" (Watson 2014, p. 71). The reflections developed here regarding the role of urban commons in shaping urban planning strategies aimed to render explicit the relevance and importance of establishing methods intending to examine the actual practices of citizens while making decisions related to spatial planning. Within the framework if this endeavour to better grasp the role of urban commons in shaping urban planning strategies a concept that is of great importance is that of negotiated planning, which "focuses less on normative expressions of how planning should be (i.e., informed by evidence and participation) and more on the actual practices evident in cities" (Cirolia and Berrisford 2017, p. 73). As Vanessa Watson has highlighted, "negotiated planning" strategies should be based on a close analysis of "the difficulties of [. . .] [the] processes as well as to the range of contexts and conditions within which participation takes place" (Watson 2014, p. 63), which would save them (the "negotiated planning" strategies) from the traps of an idealized image of collaborative planning based on the "Habermasian" model (Tewdwr-Jones and Allmendinger 1998). The shift from "collaborative" toward "negotiated planning" is related to the intensification of the interest not only in the commoning practices, but, most importantly, in the actual "actors and power dynamics [. . .] involved," as well as in "the 'virtuous cycle' of planning, infrastructure, and land" (Cirolia and Berrisford 2017, p. 77).

A characteristic of "negotiated planning" that is note-worthy is the attention it pays to the actual "power-laden compromises, contests [. . .] among various arms of the state, civil society, and the local and international private sector" (Cirolia and Berrisford 2017, p. 71). More specifically, "negotiated planning" approaches place particular emphasis on "the actions and agendas of a whole range of stakeholders who together work to configure a fragile system which is constituted through and co-constitutive of each urban context" (Cirolia and Berrisford 2017, p. 71). The implementation of "negotiated planning" strategies in urban planning would imply that the architecture and urban planning strategies concerning the design of pro-poor housing should be shaped in close dialogue with the actual commoning practices. Within such a perspective, architecture and urban planning should act as actors connecting planning, infrastructure, and land.

Another aspect concerning the role of urban commons in spatial planning decision-making that is at the core of the current debates is the intention to combine the use of advanced digital tools such as digital twins and civic-oriented participatory design methods. However, despite the aspirations of urban scale digital twins to enhance the participation of citizens in the decision-making processes relayed to urban planning strategies, the fact that they are based on a limited set of variables and processes makes them problematic (Charitonidou 2022c). During the coming years, given the galloping development of urban scale digital twins applications globally, it will be of pivotal importance to shape methodological tools offering the possibility to develop new forms of social advocacy around big data, bringing together the reflection on smart cities and the debates around urban commons.

Cross-References

- ▶ Collaborative Climate Action
- ▶ Collective Emotions and Resilient Regional Communities
- ▶ Community Engagement and Climate Change: The Value of Social Networks and Community-Based Organizations
- ▶ Growth, Expansion, and Future of Small Rural Towns

References

Bingham-Hall, J. (2016). *Future of cities: Commoning and collective approaches to urban space*. Government Office for Science.

Bollier, D. (2007). A new politics of the commons. *Renewal: A Journal of Labour Politics*, 17 December 2007. https://dlc.dlib.indiana.edu/dlc/bitstream/handle/10535/3379/Renewal_Essay_Dec_2007%5B1%5D.pdf?sequence=1&isAllowed=y. Accessed 10 Jan 2022.

Bollier, D., & S. Helfrich. (Eds.). (2012). *Wealth of the commons a world beyond market and state*. The commons strategies group [in cooperation with]. Levellers Press.

Borch, C., & Kornberger, M. (Eds.). (2015). *Urban commons: Rethinking the city*. Routledge.

Bresnihan, P. (2016). The more-than-human commons: From commons to commoning. In S. Kirwan, L. Dawney, & J. Brigstocke (Eds.), *Space, power and the commons: The struggle for alternative futures*. Routledge.

Charitonidou, M. (2021a). Housing programs for the poor in Addis Ababa: Urban commons as a bridge between spatial and social. *Journal of Urban History*. https://doi.org/10.1177/0096144221989975.

Charitonidou, M. (2021b). Revisiting Giancarlo De Carlo's participatory design approach: From the representation of designers to the representation of users. *Heritage, 4*(2), 985–1004. https://doi.org/10.3390/heritage4020054.

Charitonidou, M. (2021c). The 1968 effects and civic responsibility in architecture and urban planning in the USA and Italy: Challenging 'nuova dimensione' and 'urban renewal'. *Urban, Planning and Transport Research, 9*(1), 549–578, https://doi.org/10.1080/21650020.2021.2001365.

Charitonidou, M. (2022a). Revisiting civic architecture and advocacy planning in the US & Italy: Urban planning as commoning and new theoretical frameworks. In *The proceedings of the ACSA 110th annual meeting "EMPOWER"*, Los Angeles, May 18–20, https://doi.org/10.3929/ethz-b-000528991.

Charitonidou, M. (2022b). Denise Scott Brown's active socioplastics and urban sociology: From Learning from West End to Learning from Levittown. *Urban, Planning and Transport Research, 10*(1), 131–158. https://doi.org/10.1080/21650020.2022.2063939.

Charitonidou, M. (2022c). Urban scale digital twins in data-driven society: Challenging digital universalism in urban planning decision-making. *International Journal of Architectural Computing*. https://doi.org/10.1177/14780771211070005

Charitonidou, M. (2022d). *Drawing and experiencing architecture: The evolving significance of city's inhabitants in the 20th century*. Transcript Verlag. https://doi.org/10.14361/9783839464885

Charitonidou, M. (2022e). Denise Scott Brown's nonjudgmental perspective: Cross-fertilization between urban sociology and architecture. In F. Grahn (Ed.), *Denise Scott Brown in other eyes: Portraits of an architect*. (98–106) Birkhäuser. https://doi.org/10.1515/9783035626254-008

Cirolia, L. R., & Berrisford, S. (2017). Negotiated planning': Diverse trajectories of implementation in Nairobi, Addis Ababa, and Harare. *Habitat International, 59*, 71–79. https://doi.org/10.1016/j.habitatint.2016.11.005.

Hardin, G. (1968). The tragedy of the commons. *Science, 162*, 1243–1248.

Hardt, M., & Negri, A. (2001). *Empire: The new world order*. Harvard University Press.

Hardt, M., & Negri, A. (2004). *Multitude: War and democracy in the age of empire*. The Penguin Press.

Hardt, M., & Negri, A. (2009). *Commonwealth*. Harvard University Press.

Harvey, D. (2011). The future 8ft he commons. *Radical History Review, 109*, 101–102. https://doi.org/10.1215/01636545-2010-017.

Harvey, D. (2012). *Rebel cities: From the right to the city to the urban revolution*. Verso Books.

Marx, K. (1970). *Critique of Hegel's philosophy of right* (1843) (O'Malley J, Trans.). Oxford University Press.

Lung-Amam, W. W., Anne Harwood, S., Francisco Sandoval, G., & Sen, S. (2015). Teaching equity and advocacy planning in a multicultural 'post-racial'. *Journal of Planning Education and Research, 35*(3), 337–342. https://doi.org/10.1177/0739456X15580025

Németh, J. (2012). Controlling the commons: How public is public space? *Urban Affairs Review, 48*(6), 811–835. https://doi.org/10.1177/1078087412446445.

Ostrom, E. (2015). *Governing the commons: The evolution of institutions for collective action*. Cambridge University Press.

Pacchi, C. (2018). Epistemological critiques to the technocratic planning model: The role of Jane Jacobs, Paul Davidoff, Reyner Banham and Giancarlo De Carlo in the 1960s. *City, Territory and Architecture, 5*(17), 1–8. https://doi.org/10.1186/s40410-018-0095-3.

U

Scott Brown, D. (2009). Towards an 'Active Socioplastics'. In *idem., Architecture words 4: Having words.* (pp. 22–54). Architectural Association.

Stavrides, S. (2015). Common space as threshold space: Urban Commoning in struggles to re-appropriate public space. *The Footprint, 16*, 9–20. https://doi.org/10.7480/footprint.9.1.896.

Stavrides, S. (2016). *Common space: The City as commons*. Zed Books.

Susser, I., & Tonnelat, S. (2013). Transformative cities: The three urban commons. *Focaal- Journal of Global and Historical Anthropology, 66*(2013), 105–132. https://doi.org/10.3167/fcl.2013.660110.

Tewdwr-Jones, M., & Allmendinger, P. (1998). Deconstructing communicative rationality: A critique of Habermasian collaborative Planning. *Environment and Planning A: Economy and Space, 30*(11), 1975–1989. https://doi.org/10.1068/a301975.

Watson, V. (2014). Co-production and collaboration in planning: The difference. *Planning Theory & Practice, 15*(1), 62–76. https://doi.org/10.1080/14649357.2013.866266.

Urban Decarbonization

► Carbon Neutral Adelaide

Urban Densification and Its Social Sustainability

Rebecca Cavicchia[1] and Roberta Cucca[2]
[1]Department of Urban and Regional Planning, BYREG – Norwegian University of Life Science, Ås, Norway
[2]BYREG – Norwegian University of Life Science, Ås, Norway

Introduction

Since the Brundtland Report (WCED 1987), urban densification has been almost universally considered one of the most sustainable ways to develop contemporary cities. Generally speaking, urban densification refers to a set of measures that "constrain the expansion of urban areas, restrain development in rural areas and maintain the separation of settlements, thereby preventing urban sprawl and focusing resources on the re/development of existing towns and cities" (Bibby et al. 2020, p. 2). Such measures aim to (i) increase proximity among people and urban functions; (ii) create an efficient transportation system, which encourages walkability and the use of bikes and public transport (reducing car-related travels and associated emissions); (iii) create mixed-use urban areas; and (iv) save farmland and natural areas (OECD 2012). Land take (i.e., the loss of undeveloped land) is one of the most worrying trends for climate change (European Commission 2016), and countries and cities often react by adopting pro-densification policies. In Germany and Austria, maximum limits have been fixed for land take; in Britain, 60% of new construction has to be on brownfield; in Denmark, there is a "station proximity principle" for urban development since 1989; and in Norway and Switzerland, urban densification is the national urban development strategy. In some contexts (e.g., Bulgaria, Poland, and some Italian counties), the quality of the soil is a criterion in the decision for development, and development fees are imposed on the basis of the soil quality. Tax abatements and reduction of development fees are often used as incentives for developers to concentrate on brownfield development, thus limiting the exploitation of undeveloped land (Teller 2021). Although densification and urban density are today subject to wide consensus, they have been – and to some extent still are – contested ground for research and practice (Holman et al. 2015). They have been accompanied by various degrees of acceptance and rejection (McFarlane 2016) in different historical moments and geographical contexts. Today – despite opposition and disagreement – densification satisfies multiple and diverging interests: politicians who want to attract new activities and new populations (Idt and Pellegrino 2021), environmental activists who aim to preserve natural areas and biodiversity (Tretter 2013), and real estate developers who want to maximize their profits by intensification of land-use (Kotsila et al. 2020). This apparent consensus around densification is perceived by some scholars as a manifestation of post-politics: to them, it reflects the emergence of de-politicized and consensual technocratic forms of governance (Charmes and Keil 2015).

Although studies on the sustainability dimensions of urban densification are extensive, theoretical and empirical investigations of its socio-spatial implications are still a relatively new avenue for the social sciences. Focusing on three major topics – urban housing affordability, gentrification, and socio-spatial segregation – this chapter sheds light on the multiple and complex relationships between urban densification policies and their socio-spatial implications, here considered as fundamental indicators of the social sustainability of urban densification. To reach this goal, the chapter proceeds as follows. In section "Densification as Urban Sustainability" Historical Overview and Ideals, after an historical overview of urban densification, the rhetoric of "densification as urban sustainability" is presented. In section "Urban Densification and Social Inclusion Opportunities and Challenges," the complex relationships between urban densification and its socio-spatial implications are explored. Finally, the chapter draws conclusions on the opportunities and limitations of urban densification with respect to the social pillar of sustainability.

"Densification as Urban Sustainability" Historical Overview and Ideals

With the Industrial Revolution, cities started to become denser, more crowded, and more polluted. Engels' description of industrial Manchester, where he lived between 1842 and 1844, was of a place of disease, pollution, and incredible poverty (Engels 1844). These observations became particularly influential in urban thinking, stimulating a discourse about bad density as associated with slums, poverty, and unhealthy living conditions (McFarlane 2016). Cities were the places where these conditions could be found, and density was the representation of the deprivation for working classes of all the possible positive aspects associated with everyday life (green space, clean air, and so on), which were transferred to the suburbs and enjoyed by the aristocracy and bourgeoisie.

As a reaction to the bad and degraded densities that characterized industrial cities, in the earlier part of the twentieth century, Ebenezer Howard and other scholars proposed new city models that could offer a green and healthy environment. Howard's vision of the Garden City was an "amalgam of the best features of city and countryside, was a constellation of inter-connected, self-contained new towns, surrounded by a greenbelt and placed around a large main city" (Sharifi 2016, p. 4). Howard proposed a polycentric city model, envisaging Garden Cities as the satellites of the main city where industries were located. In the Garden City, it is possible to discern the first signs of urban sprawl, based on lower densities and a separation of functions. The rise of the automobile in the first half of the twentieth century was the main trigger in translating such ideas into practice.

Urban sprawl has been defined as "the physical pattern of low-density expansion of large urban areas, under market conditions, mainly into the surrounding agricultural areas" (Ludlow 2006, p. 20), and it has produced different suburban geographies in Europe and the USA. As noted by McFarlane (2016), while the UK reaction to inner city density was the Garden City, the US reaction to it was an uncontrolled suburbanization based on car-oriented living, private property, and individual living. This kind of urban expansion has resulted in spatial city models (see, for instance, the Burgess monocentric city model inspired from Chicago; (Burgess 1925), which reflected specific social compositions. Compared to the European context, the social hierarchy of space in US cities was more segregated as a result of the suburbanization of the social elite and middle classes and the racist regulation of residential space that drove discrimination, white flight, and the formation of ghettos, leading to socially and racially homogeneous neighborhoods (Arbaci 2019). In European cities, suburbanization was not as intense and divisive. Social elites did not abandon the central areas, and residential mobility was traditionally low (Inverse Burgess model; see Leontidou 1990).

If the Garden City expressed a search for an alternative solution to the bad density produced

U

during the Industrial Revolution, the 1960s saw an inversion of orientation due to an increasing awareness of the limits and the costs of uncontrolled urban sprawl (Neuman 2005). Jane Jacobs was probably the most eminent figure advocating for urban density at that time. She advocated for cities against what she defined as the "orthodox urbanism" of the Garden Cities and of other modernist city models. Jacobs emphasized density as an urban quality, something that cities should pursue to create vital spaces (Jacobs 1961). Density is a fundamental aspect of *planning for vitality*, as it allows the creation of social and physical diversity (Moroni 2016), against the tendency towards demolition that is typical of the modernism. Jacobs' search for social and physical urban diversity, however, was not free from critique. Susan Fainstein (2005), for instance, criticized Jacobs for having influenced planning approaches – such as Transit-Oriented Development, New Urbanism, and Smart Growth (Sharifi 2016) – which were able to create only the appearance of diversity – the diversity of elites and of the creative class (Florida 2005) – while obscuring the rise of socio-spatial inequalities.

Once the way towards the possibility of good urban densities was paved, urban densification started to be implemented in multiple contexts. Among the various planning approaches that have urban densification as a central tool, the Compact City has probably been the most influential in the European context (Jenks et al. 2000), on which this review is mainly focused. Cities, regions, and nations have been eager to adopt policies and tools to avoid sprawl and to ensure the aforementioned urban qualities. From being blamed as the cause of slums, pollution, and unhealthy conditions, density – as a basic characteristic of the compact city – is now furthered through a particularly powerful narrative: that of "sustainability as density" (Quastel et al. 2012). In the 1990s, the European Commission recommended the "compact city" as the most sustainable and livable model for urban development (Commission of European Communities 1990). Density is a key dimension of urban sustainability, and densification is widely considered a key tool to make the three dimensions of urban sustainability – environmental, economic, and social – achievable in practice.

From an environmental sustainability perspective, urban densification has to do with how intensely land is used and, consequently, with how natural resources can be used sustainably. Densification may make it possible to contain population within already built-up areas with positive effects in terms of saving natural land and farmland (Hofstad 2012). Additionally, densification is generally associated with sustainable transport choices, reduction of car traffic, and consequently, reduced greenhouse gases (GHG) emissions (OECD 2012). To achieve these goals, two main strategies are generally adopted: (1) areas for densification are generally organized in the proximity of transport hubs, through transit-oriented approaches, and (2) the distance between buildings and functions is reduced – through mixed land-use – with the result that people are more incentivized to use pedestrian paths, bicycle lanes, and public transport. Additionally, when cars, buses, and other motorized means of transport travel shorter distances, their energy use and emissions are reduced (Hofstad 2012).

Although primarily an environmental policy tool, urban densification is also supported for its potential economic benefits. It is deemed to require fewer urban costs, especially in terms of infrastructure, compared to low-density developments (Neuman 2005). Additionally, by increasing economic density – that is, the concentration of people and work – urban densification is considered to have positive outcomes in terms of productivity, innovation, and service accessibility (Ahlfeldt et al. 2018; Jenks and Jones 2009). Thus, urban densification is increasingly adopted in entrepreneurial policy approaches to boost cities' economic growth. In this respect, the main claim to support the adoption of densification strategies is deeply rooted in the Ecological Modernization ideology, which has become dominant in urban planning and policies (Næss and Saglie 2019). The core of Ecological Modernization is that the solution to environmental degradation can be found within the capitalist system, provided that capitalism undergoes a process of transformation. This transformation would consist in decoupling economic growth from environmental degradation, and urban densification, thanks to the mentioned environmental and economic benefits, is supposed to be able to achieve this decoupling. However, evidence suggests that densification does not guarantee the alleged benefits (Næss et al. 2020).

With respect to social sustainability – the main concern of this contribution – density is claimed to be related to social equity and diversity (Fig. 1), although the direction of this relationship is far from clear, and the claims are not always supported by empirical evidence. Increasing urban density is deemed to favor access to social infrastructure and cheaper transport and to encourage a more diverse, inclusive, and livable urban environment by facilitating opportunities for social interaction (Ståhle 2017).

However, density alone is not enough to reach these goals. Indeed, studies have shown that a compact urban form – as a combination of density with other neighborhood characteristics such as proximity to transport, mixed land use, lively public spaces, and so on – may enhance subjective well-being and neighborhood satisfaction (Mouratidis 2018), reduce social segregation, and favor social mix because people can live in closer proximity (Boyko and Cooper 2011; Burton 2000). In a recent policy brief from the OECD (Monroy et al. 2020), and increasingly in several European contexts and elsewhere, compact urban development is depicted as the best way to make urban growth not just more environmentally sustainable, but also more socially inclusive. Such inclusivity mainly refers to the capacity, through densification, of accommodating a fast-growing urban population (UN 2019), enhancing good accessibility to cheaper transport solutions, urban opportunities, and housing. Other things being equal, in terms of growth boundaries and developable land available, densification can provide a wider housing supply compared to low-density development (Aurand 2010). Additionally, thanks to the wider variety of housing types

Urban Densification and Its Social Sustainability, Fig. 1 Urban densification and its alleged social benefits. (Authors' elaboration)

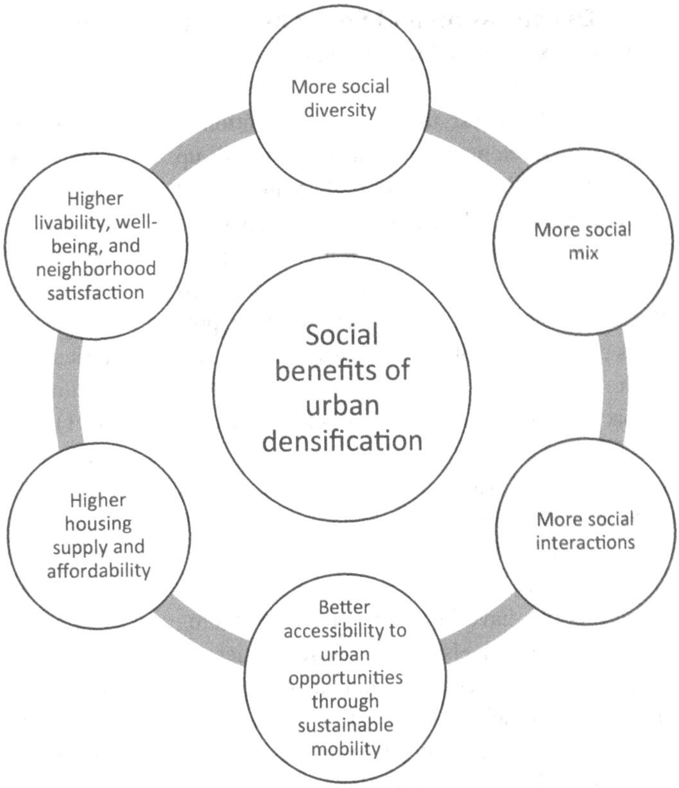

and sizes, it may offer more affordable solutions for lower income people (Aurand 2010).

Despite all these potential benefits, examples of densification in several contexts show the complexities of reaching urban sustainability goals in practice. The topic is so controversial that a debate around the so-called compact city paradox (Neuman 2005) emerged almost 20 years ago. Such debate highlights the several trade-offs entangled with urban densification, which often triggers gains in one dimension of sustainability and losses in others (Westerink et al. 2013). This shows that simply equating densification to urban sustainability is too limited. The next section provides further illumination on this topic. It focuses on the limitations of urban densification from a social sustainability perspective and specifically explores social and spatial inequalities related to housing (un)affordability, gentrification, and segregation as possible outcomes of the widespread market-oriented approaches to urban densification.

Urban Densification and Social Inclusion Opportunities and Challenges

The renewed interest to urban densification – channeling development within already built-up areas and funneling the attention of urban renewal programs in inner urban areas – has largely coincided with the passage from a Fordist to a post-Fordist society. The large legacy of dismissed urban areas left by the deindustrialization processes during the 1980s and the neoliberal wave, which arose from the crisis of Keynesian approaches to urban and social policies, have profoundly affected urbanization frameworks and their outcomes (Cucca and Ranci 2016; Rousseau 2015). Welfare retrenchment, market-based urban policies, housing financialization, and the rise of private actors as the de facto directors of urban transformations have been widespread phenomena in Europe and elsewhere. Growth-oriented approaches have become dominant in planning (OECD 2012), with global cities increasingly competing for a more attractive and sustainable image (Anderberg and Clark 2013).

Against this background, the pursuit of sustainability goals through densification have often overlapped with economic growth interests, profit-oriented development, and market logics, in many cases triggering or exacerbating socio-spatially exclusionary dynamics.

Urban Densification and Housing (Un) Affordability

The challenges put forward by urban population growth (UN 2019), as well as by climate change and the global urban housing affordability crisis (Wetzstein 2017), have contributed to increase the consensus about the potential for urban densification to accommodate more people in less space (Monroy et al. 2020). In some contexts, increasing housing affordability is an explicit goal linked to the implementation of densification strategies (Churchman 1999; Debrunner and Hartmann 2020; Giddings and Rogerson 2021; Westerink et al. 2013). However, and despite the aforementioned potential, the link between densification and housing affordability is far from straightforward. This section explores three aspects that exemplify this complex relationship.

1. The first concerns the availability and value of urban land. Typically, efficient land-use is the main goal of urban densification. In some cases, densification is actively pursued, for example, through transit-oriented approaches, while in others, it is rather the result of sectoral land-use policies and urban growth restrictions (Addison et al. 2013). When natural areas and agricultural land (Westerink et al. 2013) or cultural heritage areas (Idt and Pellegrino 2021) are protected, development must concentrate in specific areas of the city, often brownfields and already built-up areas, resulting in densification. Because of land-use restrictions, developable land becomes scarcer, and this may generate important consequences for housing affordability. Indeed, as its availability declines, urban land becomes also more valuable (Addison et al. 2013). This is likely to affect housing affordability, as land forms the highest share of the final price of housing (Knoll et al. 2017). Additionally, the need to

densify and – as a consequence of bringing more people into an area – to intensify land-use means that various functions compete for increasingly scarce urban land, which can generate further pressure. The results are often heated and exclusionary housing markets that increasingly push vulnerable groups – mostly renters and first-time buyers, among whom young, low-income, and immigrant groups are overrepresented (Haffner and Hulse 2021) – out from the greenest, "sustainable," and well-located urban areas. A now wide field of literature on green/eco-gentrification shows the importance of combining urban densification with specific measures to make new developments more inclusive (see affordability programs in Nantes, Vienna and Zurich; (Cucca 2012; Debrunner and Hartmann 2020; Garcia-Lamarca et al. 2019), thus providing an opening reflections on the limitations – with respect to housing affordability – of urban densification per se, especially in market-driven planning contexts. In this respect, whether land is privately or publicly owned represents a key point for the inclusion of housing affordability programs in new compact developments. Indeed, land ownership can give municipalities more power to negotiate with private developers (often the de facto protagonists of urban transformation) for the provision of fixed amounts of affordable housing units and to make a more active use of land use policies (Debrunner and Hartmann 2020; Nordahl 2014).

The "land factor" is particularly relevant for disentangling the relationships between urban densification and housing affordability when considering urban geographies of densification. Indeed, in several contexts (Cavicchia 2021; Giddings and Rogerson 2021), densification is primarily developed in central and semi-central locations and in close connection to public transport (OECD 2012), where land is supposedly more costly. Considering the strong link between urban housing affordability and land rent (Peverini 2021), this might generate an intrinsic trade-off for which, under urban densification policies, the high land rent makes housing affordability very hard to accomplish.

2. The second aspect concerns housing supply, which is often used as an argument in favor of adopting urban densification to increase housing affordability. Developers often argue, for example, that the more units they can fit into a lot, the cheaper the sale or rental prices for those units (Dalton 2016). However, the fact that many compact cities have significant affordability issues (Cavicchia 2021; Dalton 2016; Tretter 2013) indicates that the relationship of densification (as increased housing supply) with better affordability is not straightforward. Two main points might be stressed here. The first concerns the relationship between housing supply and housing affordability. Housing supply is price inelastic. Housing takes time to build (plots of land to be regulated, planning processes to be developed) and before an effect on prices is visible, there could be a significant time lag (Hinton 2003). It is difficult to estimate the extent to which housing supply can be substantially beneficial for housing affordability (Rodríguez-Pose and Storper 2020). The second concerns the characteristics of the housing supply provided through urban densification. If, for instance, the housing supplied through densification is limited to well-located and attractive areas of the city and to market-rate housing, or if it privileges specific social group (Kern 2007), then vulnerable households might continue to have difficulty in finding affordable solutions within the compact city.
3. The third aspect concerns brownfield redevelopment. Channeling development into brownfield areas – commonly priority areas for development in densification policies (Rousseau 2015) – often implies high up-front costs, such as site cleaning and new infrastructure. The Concerted Action on Brownfield and Economic Regeneration Network (CABERNET) has identified three categories of brownfields in Europe, according to the level of on-site contamination. (CABERNET is a multidisciplinary network. It aims to enhance the rehabilitation of

brownfield sites, according to the context of sustainable development of European cities.) The more contaminated the site is, the higher the expenses – often needing concerted public-private funding – needed to make the areas ready for development. When these expenses – impact fees – are regulated to be sustained by private developers, they have the potential to be passed on to housing consumers, thus decreasing affordability (Addison et al. 2013). It should also be noted that brownfield sites are often located within urban boundaries with good connections to local infrastructure. This makes such sites a competitive alternative to greenfield as the potential rent gap and value extraction from their development might be very profitable for developers and landowners, triggering a "business of densification" with significantly exclusionary implications (Debrunner et al. 2020; European Commission 2016).

All three of these factors are deeply entangled with spatial choices of (or for) densification and thus invite a critical spatial reading of densification in relation to housing affordability and, as a consequence, urban justice. In their study on suburban densification in the Frankfurt region, Jehling et al. (2020) discuss the spatiality of densification from different perspective of urban justice (libertarian, egalitarian, and social). Increasing density in central and semi-central locations and around transportation nodes means densifying in the most accessible areas of the city. While this may be positive in terms of higher opportunities for accessibility for everyone (in addition to all the environmental advantages), housing affordability may be negatively affected. A critical look at this clearly shows an intrinsic trade-off. Dense housing settlements located in the most attractive and accessible locations have the environmental benefits of requiring less land consumption and of reducing car dependency. However, they are also likely to be more expensive and thus exclusionary for vulnerable groups. More studies are needed to explore whether and to what extent the increased transport affordability, linked to a central location, compensates for the higher housing expenses. Various socio-spatial implications may be considered among the results of increasingly unaffordable and inaccessible areas in central and semi central locations of the city. The first is a continuous suburbanization of poverty, which reverses the traditional mono-centric Burgess model mentioned above. Poverty suburbanization is the result of variegated processes of direct and indirect displacement and is likely to deepen existing patterns of inequalities and socio-spatial polarization. The second is a "lock in" effect, which means that vulnerable groups may have very limited residential choice as a result of a shrinking affordable housing pool (Slater 2009), which can result in stronger segregation patterns.

Oslo and Austin are two examples of densification strategies implemented with the aim of saving natural areas and boosting the local economy but without attention to housing affordability (Cavicchia 2021; Tretter 2013). In both cities, exclusionary phenomena related to densification have been documented, specifically in the form of low housing accessibility for first-time buyers and vulnerable groups in Oslo and displacement of Latino communities in Austin. Conversely, with its democratic housing policies and programs (and despite the limitations also present in this context), Vienna represents a good example of reverting the poverty suburbanization trend. Here, compact and affordable housing solutions are often combined in attractive and central locations of the city, and the trade-off between densification and social inclusion appears less marked compared to other contexts.

Densification and Gentrification

The relationships between urban densification and gentrification dynamics have been mainly explored in the fields of urban political ecology and urban geography and in connection to three main factors: brownfield redevelopment, low-carbon lifestyle, and social mixing.

Brownfield redevelopment. As mentioned above, the redevelopment of brownfield is an important tool for developing compact city strategies, as it makes it possible to channel development in already built-up areas. However, as the pioneer work of Davidson and Lees (2005) has explored, the redevelopment of brownfield can trigger gentrification dynamics. Following the theory of "new-build gentrification," they extended the notion of gentrification to new constructions and addressed issues of both direct and indirect displacement (Davidson and Lees 2005, 2010; Rerat et al. 2009, 2010; Rérat 2012). In this theoretical perspective, densification-related gentrification acquires three main characteristics. The first is that private developers and public authorities are the actors triggering the process of rent gap and consequent gentrification through area investment and redevelopment of wasteland (Rerat et al. 2009). The second concerns landscape change in the form of brownfield redevelopment, infill, and the addition of new floors (Davidson and Lees 2005; Franz 2011). If Victorian townhouses and brownstones have been the peculiar traits of gentrification aesthetics in London and New York in the 1960s–1970s, central, compact, and well-connected neighborhoods have often become the new aesthetic of gentrification, answering the needs of a new middle class that is increasingly mobile, technological, and "green." The third aspect concerns displacement and the nuanced ways through which it might take place. For example, Rerat et al. (2009) have shown how the conversion of urban wastelands through densification in various Swiss cities has not only produced highly exclusionary contexts from which vulnerable groups are indirectly displaced (being excluded or pressured into being displaced because spaces and services in their neighborhood have been gentrified), but that such exclusionary effects even spill over into their surroundings, triggering and spreading new gentrification processes.

Low-carbon lifestyle. Gentrification is, among other aspects, a cultural phenomenon (Annunziata 2007). As such, lifestyle and consumption choices can be contributing drivers of gentrification dynamics (Hjorthol and Bjørnskau 2005). For example, the promotion of "climate friendliness" by local governments and the increasing desirability of "green living" might affect the choice of middle- and upper-income urban residents to live near public transit and in higher-density mixed-use areas (Rice et al. 2020). Car-free areas and walkability are particularly valued factors driving residential choice, as well as powerful elements for branding densification areas as "sustainable" (Quastel et al. 2012). In their study about densification in existing neighborhoods in Vancouver, British Columbia, Quastel et al. (2012) used a municipal walkability index, building density, and proximity to transport hubs as independent variables describing densification. Using principal component analysis, they found a statistically significant correlation between densification and socio-economic neighborhood changes associated with gentrification (they controlled for occupation and education level, age, household composition, and ethnic background). Among the causes of the found gentrification patterns, they emphasized the lack of affordable housing, developers' profit-oriented behaviors, and the cultural and life-style value represented by walkability, the most emphasized quality of densification areas in Vancouver. Their explanations give a clear idea of densification-driven gentrification as a socio-economic but also profoundly cultural phenomenon. Densification, indeed, answers the cultural and life-style necessities of an increasingly eco-conscious middle-upper class. Additionally, there is a key political economic shift that appears to accelerate this trend: the increased presence of creative and high-tech industries and their workers (with their lifestyle and consumption choices; (Rice et al. 2020). This shift might be supported by densification. Indeed, given its widely documented positive effects on local economies (Ahlfeldt et al. 2018), densification is often implemented with the aim of attracting new creative industries and investments and of boosting city competitiveness (Rousseau 2015; Tretter 2013). As recently discussed (Freemark 2020), the presence of attractive job markets may be one of the main factors affecting exclusionary dynamics related to the housing market. Attractive and accessible urban locations, opportunities, and job markets are characteristics often associated with urban densification and that might be considered as enabling exclusionary dynamics.

Social mixing. In addition to favoring environmental and economic advantages, urban densification is often adopted with the aim of increasing social mix, either at the urban or at the regional level (Rosol 2015; Rousseau 2015). There are two main rationales behind this. One is that densification may entail social mix because it allows a higher proximity among people and a higher variety of dwelling types compared to low density developments. The other is that new buildings are usually costlier than existing ones, so including new densification developments in previously not so well-off areas can make it more mixed, through the influx of middle-income groups.

Socially mixed neighborhoods are a major and very popular planning and policy goal of state-led interventions in many countries in Western Europe and North America. This is closely related to re-urbanization trends – that is, the movement of middle- and high-income groups back to the inner city – and may lead to conflicts with long-term residents of usually a lower social status. In consequence, the re-urbanization of the middle-classes often results in forcing poorer households from the center of cities to the periphery. The idea of social mix has evolved from a progressive urban policy, aiming at contrasting segregation and up-lifting degraded areas in the 1960s and 1970s, towards a policy leading to gentrification. As Bridge and Butler (2011) contend, gentrification – understood as the movement of middle-income people into low-income neighborhoods causing the displacement of the low income residents – is now often "rhetorically and discursively disguised" as social mixing. It is, moreover, criticized as a one-sided strategy, as it is rarely advocated in more affluent neighborhoods, but channeled into low-income ones through urban renewal programs and the addition of middle-income housing. The case of the greater Lyon region, investigated by Rousseau (2015), is a clear example of these dynamics and also shows that there could be a significant gap between the content and the goals of densification policies and their translation into practice. Among the motivations to implement densification in this context, there was the aim of balancing the polarization between the wealthy municipalities in the western part of the region and the lower income municipalities in the east, which were the result of a suburbanization wave in the 1940–1970s, characterized by large state-led estate projects. In the western municipalities, the power coalitions among incumbent residents objected to the new developments (a well-known practice called NIMBYism – or, "Not In My Back Yard"), and conservative municipal administrations managed to significantly reduce the scope of the densification interventions in this part of the region. Only small limited-density interventions were allowed. On the east side, conversely, where densification was positively accepted as an opportunity to increase services and connections, the new interventions were mainly organized around transport hubs. The pattern aimed at attracting middle income people into poorer areas and not the opposite. The result has been that densification, at a regional scale, has not achieved the goal of better mix but has rather helped in consolidating the high status of the west and triggering new-build gentrification in the east. This case exemplifies the thin border between densification being a benefit and a source of profit for some, and a burden for others – usually less powerful and more vulnerable social groups.

Densification and Segregation

This section explores the possible relationships between urban densification and patterns of socio-spatial concentration of specific social groups in cities. Urban densification, when implemented as a central part of urban sustainability agendas, may be associated with different processes of residential segregation including (1) a lack of inclusion of diverse social groups, such as the densification of central areas or high-rise development through high energy-efficiency buildings, which may translate into exclusionary enclaves, as in the case of waterfront and eco-districts (Andersen and Røe 2017); and (2) with micro-segregation that, at the same time mitigates but also complements neighborhood segregation, such as processes of vertical segregation (Maloutas 2020).

Densification may be associated with the concentration of high-income groups in privileged areas with high environmental standards. This is the case for the eco-cities or eco-districts (Bardaka et al. 2018; Flurin 2017), which can be seen as a continuation of planning, architectural and design trends aiming to reconcile nature and the city from the nineteenth century onwards, while at the same time limiting urban sprawl, and fostering environmental innovations in the building sector (Friedman 2014). Eco-city projects are being built across the globe (Holden et al. 2015), and several are already in advanced stages of construction. However, "eco" has been defined as the discursive construction of an environmentally friendly city *for its inhabitants*, filtering and protecting through a highly technological envelope; these are places within which urban life can be made clean, healthy, and comfortable, but they become areas of self-segregation for a green elite, who are able to afford the cost of the high building standards (Caprotti 2014). In other contexts, eco-projects have mainly targeted the middle class with social-mix expectations. As part of its sustainable urban planning agenda, the municipality of Paris has established the target of increasing the availability of housing that is both low energy and affordable. This is the case of the eco-quartier scheme, an environmental program that includes social aspects such as population diversity and affordable housing. All of the Parisian eco-quartiers are located in low-income areas; however, most of the green subsidized housing in those eco-quartiers is intended to accommodate the middle class. Given the distribution of the different categories of subsidized housing, few low-income families will be able to stay in the eco-quartiers or move into them (Machline et al. 2018).

A pattern of segregation also characterizes the social composition of the inhabitants of densified areas built in privileged locations. A special focus has been given to the renewal of obsolete urban industrial harbor locations – showing again how brownfield areas are central in the understanding of densification-related socio-spatial inequalities – to be replaced by very exclusive new urban developments along the waterfront. Around the globe, cities are trying to redefine their relationship with the water and develop now defunct harbor sites and other brownfields into new, compact, high-quality urban districts. In Europe, this process is particularly evident in cities such as Copenhagen and Oslo (Grønning 2011), which have been strongly affected by processes of self-segregation by affluent groups in very luxurious, brand-new neighborhoods characterized by direct access to the water. Alongside creative hubs and cultural quarters, the waterfront is a site for intensified planning attention (Keil 2007), including not only expensive apartments, but also creative, cultural, and technological industries and commodified leisure and entertainment spaces. Mega structures and signature buildings also epitomize the performativity of neoliberal competitiveness on the waterfront (Doucet et al. 2011). According to Oakley (2011, p. 234) "waterfront renewal is being driven by a neoliberal competitive city paradigm" fostering an appropriation of the economic values of an environmental resource by one class from another.

Densification can also be associated with micro-segregation patterns. One example of this is the vertical dimension of inequality associated with urban densification and the increasingly widespread high-rise urban living in Europe. In new densified areas, whenever market rules shape vertical patterns of housing prices, social stratification may be found among apartments on upper versus lower floors (Maloutas 2020) as a result of individual economic status, but also as a result of housing choices and strategies (Flint 2017). Advocates of densification claim that the co-existence of various population groups in the same buildings or building blocks may solve the emerging problems caused by the city's own growth, while simultaneously preventing the formation of slums (Flint 2017). However, although vertical differentiation often results in interclass sharing of living space, there is not a necessary relationship between spatial contiguity and more positive interactions (Bridge and Butler 2011); social conflicts and exclusive use of social and commercial infrastructures by specific groups can be possible outcomes of increased social mix.

Concluding Remarks

Urban densification has the potential to generate benefits on several levels and in connection to all three urban sustainability dimensions. However, especially in light of global challenges related to climate change, uncontrolled economic growth, and urban affordability crisis, it is important to shed light on its limitations. This chapter has offered a comprehensive, if not exhaustive – as geographically limited to the global north – review of the possible socio-spatial implications of densification in urban contexts. Three main phenomena, strongly connected to each other, have been investigated: housing affordability, gentrification, and socio-spatial segregation. Some aspects have emerged as central in both the understanding of these phenomena (as indicators of the social sustainability of urban densification) and in delineating new research pathways.

First, a critical approach towards the socio-spatial implications of urban densification cannot just focus on densification as such but has to critically engage with the wider policy context in which densification is implemented and with the dynamics between the actors directly or indirectly involved. This review has illustrated that, if not complemented with policy tools to enhance affordability and limit displacement, urban densification might become a tool for capital accumulation and rent value extraction from powerful actors such as developers and land and property owners.

Second, it is central to address the spatiality of urban densification for a better understanding of the changing social geographies it can produce. More than 10 years ago, Soja (2009) argued the importance of exploring much more explicitly and actively the spatial dimension of justice, not as an alternative to other conceptual forms of justice but as a way to look at it from a critical spatial perspective. Extending the same critical spatial approach to urban densification can open up a better understanding of the mechanisms behind densification's socio-spatial implications in urban contexts, with potential enrichments for research, both empirical and theoretical (see, for instance, the discussion on land and land rent developed above).

Finally, the socio-spatial implications of urban densification (specifically, new-build gentrification and vertical differentiation) might not be fully understood with the more traditional theoretical approaches on gentrification and urban segregation but rather require a nuanced perspective on the possible exclusionary consequences that urban densification might generate (different and nuanced forms of displacement and micro segregation patterns in connection to high-rise densification). When exploring urban densification and social sustainability, existing research has tended to focus on aspects such as well-being, neighborhood satisfaction, and livability. In many contexts, these goals are achieved. However, critical research on urban densification should always pay attention to who wins and who loses, to where densification brings social and environmental qualities and where it brings burdens, and such research must acknowledge that densification, while not standing alone as a potential driver of inequalities, is also not sufficient – alone – to enhance social inclusion.

References

Addison, C., Zhang, S., & Coomes, B. (2013). Smart growth and housing affordability: A review of regulatory mechanisms and planning practices. *Journal of Planning Literature, 28*(3), 215–257.

Ahlfeldt, G., Pietrostefani, E., Schumann, A., & Matsumoto, T. (2018). Demystifying compact urban growth. *Coalition for Urban Transitions*.

Anderberg, S., & Clark, E. (2013). Green and sustainable Øresund region: Eco-branding Copenhagen and Malmö. In I. Vojnovic (Ed.), *Urban sustainability: A global perspective* (pp. 591–610). Lansing: Michigan State University Press.

Andersen, B., & Røe, P. G. (2017). The social context and politics of large scale urban architecture: Investigating the design of Barcode, Oslo. *European Urban and Regional Studies, 24*(3), 304–317.

Annunziata, S. (2007). Se tutto fosse gentrification: possibilità e limiti di una categoria descrittiva. In A. Balducci & V. Fedeli (Eds.), *I territori della città in trasformazione: tattiche e percorsi di ricerca*. Milano: Franco Angeli.

Arbaci, S. (2019). *Paradoxes of segregation: Housing systems, welfare regimes and ethnic residential change in Southern European cities*. Wiley.

Aurand, A. (2010). Density, housing types and mixed land use: Smart tools for affordable housing? *Urban Studies, 47*(5), 1015–1036.

Bardaka, E., Delgado, M. S., & Florax, R. J. G. M. (2018). Causal identification of transit-induced gentrification and spatial spillover effects: The case of the Denver light rail. *Journal of Transport Geography, 71*, 15–31.

Bibby, P., Henneberry, J., & Halleux, J.-M. (2020). Incremental residential densification and urban spatial justice: The case of England between 2001 and 2011. *Urban Studies*. https://doi.org/10.1177/0042098020936967.

Boyko, C. T., & Cooper, R. J. P. I. P. (2011). Clarifying and re-conceptualising density. *Prog Plann, 76*(1), 1–61.

Bridge, G., & Butler, T. (2011). *Mixed communities: Gentrification by stealth?* Bristol: Policy Press.

Burgess, E. (1925). The growth of the city. In R. Parks, E. W. Burgess, & R. D. McKenzie (Eds.), *The city*. Chicago: University of Chicago Press.

Burton, E. (2000). The compact city: Just or just compact? A preliminary analysis. *Urban Studies, 37*(11), 1969–2006.

Caprotti, F. (2014). Eco-urbanism and the eco-city, or, denying the right to the city? *Antipode, 46*(5), 1285–1303.

Cavicchia, R. (2021). Are green, dense cities more inclusive? Densification and housing accessibility in Oslo. *Local Environment, 26*(10), 1250–1266.

Charmes, E., & Keil, R. (2015). The politics of post-suburban densification in Canada and France. *International Journal of Urban Regional Research, 39*(3), 581–602.

Churchman, A. (1999). Disentangling the concept of density. *Journal of Planning Literature, 13*(4), 389–411.

Commission of European Communities. (1990). *Green paper on the urban environment*. Luxembourg: Commission of the European Communities.

Cucca, R. (2012). The unexpected consequences of sustainability. Green cities between innovation and ecogentrification. *Sociologica, 6*(2), 1–21.

Cucca, R., & Ranci, C. (2016). *Unequal cities: The challenge of post-industrial transition in times of austerity*. London: Taylor & Francis.

Dalton, M. (2016). Does planning acknowledge the cost of redevelopment on housing affordability? Nova Scotia: School of Planning, Dalhousie University.

Davidson, M., & Lees, L. (2005). New-build 'gentrification' and London's riverside renaissance. *Environment and Planning A, 37*(7), 1165–1190.

Davidson, M., & Lees, L. (2010). New-build gentrification: Its histories, trajectories, and critical geographies. *Population, Space and Place, 16*(5), 395–411.

Debrunner, G., & Hartmann, T. (2020). Strategic use of land policy instruments for affordable housing – Coping with social challenges under scarce land conditions in Swiss cities. *Land Use Policy, 99*, 104993.

Debrunner, G., Hengstermann, A., & Gerber, J.-D. (2020). The business of densification: Distribution of power, wealth and inequality in Swiss Policy Making. *The Town Planning Review, 91*(3), 259–281.

Doucet, B., Van Kempen, R., & Van Weesep, J. (2011). Resident perceptions of flagship waterfront regeneration: The case of the Kop Van Zuid in Rotterdam. *Tijdschrift voor Economische en Sociale Geografie, 102*(2), 125–145.

Engels, F. (1844). *The condition of the English working class in England*. St. Albans: Granada Publishing.

European Commission. (2016). FUTURE BRIEF: No net land take by 2050?

Fainstein, S. S. (2005). Cities and diversity: Should we want it? Can we plan for it? *Urban Affairs Review, 41*(1), 3–19.

Flint, S. (2017). *Residential choices as a driving force to vertical segregation in Whitechapel* (pp. 39–57). London: Centre for Advanced Spatial Analysis.

Florida, R. (2005). *Cities and the creative class*. New York: Routledge.

Flurin, C. (2017). Eco-districts: Development and evaluation. A European Case Study. *Procedia Environmental Sciences, 37*, 34–45.

Franz, Y. (2011). Gentrification trends in Vienna. In V. Szirmai (Ed.), *Urban sprawl in Europe*. Budapest.

Freemark, Y. (2020). Upzoning Chicago: Impacts of a zoning reform on property values and housing construction. *Urban Affairs Review, 56*(3), 758–789.

Friedman, A. (2014). *Fundamentals of sustainable neighbourhoods*. Basel: Springer.

Garcia-Lamarca, M., Anguelovski, I., Cole, H., Connolly, J. J., Argüelles, L., Baró, F., Loveless, S., Pérez del Pulgar Frowein, C., & Shokry, G. (2019). Urban green boosterism and city affordability: For whom is the 'branded' green city? *Urban Studies, 58*(2), 004209801988533.

Giddings, B., & Rogerson, R. (2021). Compacting the city centre: Densification in two Newcastles. *Buildings & Cities, 2*(1), 185.

Grønning, M. (2011). What is the Fjord City? *Territorio, 56*(1), 141–150. http://digital.casalini.it/3096706.

Haffner, M. E., & Hulse, K. (2021). A fresh look at contemporary perspectives on urban housing affordability. *International Journal of Urban Sciences, 25*(supp 1), 59–79.

Hinton, I. (2003). *Market failure and the London housing market*. London: London (United Kingdom): Greater London Authority.

Hjorthol, R. J., & Bjørnskau, T. (2005). Gentrification in Norway: Capital, culture or convenience? *European Urban and Regional Studies, 12*(4), 353–371.

Hofstad, H. (2012). Compact city development: High ideals and emerging practices. *European Journal of Spatial Development, 1*, 1–23.

Holden, M., Li, C., & Molina, A. (2015). The emergence and spread of ecourban neighbourhoods around the world. *Sustainability, 7*(9), 11418–11437.

Holman, N., Mace, A., Paccoud, A., & Sundaresan, J. (2015). Coordinating density; working through

U

conviction, suspicion and pragmatism. *Progress in Planning, 101*, 1–38.

Idt, J., & Pellegrino, M. (2021). From the ostensible objectives of public policies to the reality of changes: Local orders of densification in the urban regions of Paris and Rome. *Land Use Policy, 107*, 105470.

Jacobs, J. (1961). *The death and life of great American cities*. New York: Random House.

Jehling, M., Schorcht, M., & Hartmann, T. (2020). Densification in suburban Germany: Approaching policy and space through concepts of justice. *Town Planning Review, 91*, 217.

Jenks, M., & Jones, C. (2009). *Dimensions of the sustainable city* (Vol. 2). Springer Science & Business Media.

Jenks, M. J., Burgess, M. J. R., Acioly, C., Allen, A., Barter, P. A., & Brand, P. (2000). *Compact cities: Sustainable urban forms for developing countries*. London/New York: Taylor & Francis.

Keil, R. (2007). Sustaining modernity, modernizing nature. In R. Krueger & D. Gibbs (Eds.), *The sustainable development paradox: Urban political ecology in the US and Europe* (pp. 41–65). New York: Guilford Press.

Kern, L. (2007). Reshaping the boundaries of public and private life: Gender, condominium development, and the neoliberalization of urban living. *Urban Geography, 28*(7), 657–681.

Knoll, K., Schularick, M., & Steger, T. (2017). No price like home: Global house prices, 1870–2012. *American Economic Review, 107*(2), 331–353.

Kotsila, P., Connolly J., Anguelovski, I. & Dommerholt, T. (2020). Drivers of injustice in the context of urban sustainability.

Leontidou, L. (1990). *The Mediterranean city in transition: Social change and urban development*. Cambridge University Press.

Ludlow, D. (2006). Urban sprawl in Europe: The ignored challenge: European Environment Agency.

Machline, E., Pearlmutter, D., & Schwartz, M. (2018). Parisian eco-districts: Low energy and affordable housing? *Building Research & Information, 46*(6), 636–652.

Maloutas, T. (2020). Vertical social differentiation as segregation in spatial proximity. In S. Musterd (Ed.), *Handbook of urban segregation*. Edward Elgar Publishing.

McFarlane, C. (2016). The geographies of urban density: Topology, politics and the city. *Progress in Human Geography, 40*(5), 629–648.

Monroy, A. M., Gars, J., Matsumoto, T., Crook, J., Ahrend, R., & Schumann, A. (2020). Housing policies for sustainable and inclusive cities: How national governments can deliver affordable housing and compact urban development. In O. Publishing (Ed.), *OECD regional development working papers*. Paris: OECD.

Moroni, S. (2016). Urban density after Jane Jacobs: The crucial role of diversity and emergence. *City, Territory and Architecture, 3*(1), 13.

Mouratidis, K. (2018). Is compact city livable? The impact of compact versus sprawled neighbourhoods on neighbourhood satisfaction. *Urban Studies, 55*(11), 2408–2430.

Næss, P., & Saglie, I.-L. (2019). Ecological modernisation: Achievements and limitations of densification. In D. Simin, C. Richard, W. Iain, & B. Hilda (Eds.), *The Routledge companion to environmental planning* (pp. 63–72). London: Routledge.

Næss, P., Saglie, I.-L., & Richardson, T. (2020). Urban sustainability: Is densification sufficient? *European Planning Studies, 28*(1), 146–165.

Neuman, M. (2005). The compact city fallacy. *Journal of Planning Education and Research, 25*(1), 11–26.

Nordahl, B. (2014). Convergences and discrepancies between the policy of inclusionary housing and Norway's liberal housing and planning policy: An institutional perspective. *Journal of Housing and the Built Environment, 29*(3), 489–506.

Oakley, S. (2011). Re-imagining city waterfronts: A comparative analysis of governing renewal in Adelaide, Darwin and Melbourne. *Urban Policy and Research, 29*(3), 221–238.

OECD. (2012). *Compact city policies: A comparative assessment. OECD green growth studies*. Paris: OECD Publishing.

Peverini, M. (2021). Grounding urban governance on housing affordability: A conceptual framework for policy analysis. Insights from Vienna *Partecipazione e Conflitto, 14*(2), 848–869.

Quastel, N., Moos, M., & Lynch, N. (2012). Sustainability-as-density and the return of the social: The case of Vancouver, British Columbia. *Urban Geography, 33*(7), 1055–1084.

Rérat, P. (2012). Housing, the compact city and sustainable development: Some insights from recent urban trends in Switzerland. *International Journal of Housing Policy, 12*(2), 115–136.

Rerat, P., Söderström, O., Piguet, E., & Besson, R. (2009). From urban wastelands to new-build gentrification: The case of Swiss cities. *Population, Space and Place, 16*, 429–442.

Rérat, P., Söderström, O., & Piguet, E. (2010). New forms of gentrification: Issues and debates. *Population, Space Places, 16*(5), 335–343.

Rice, J. L., Cohen, D. A., Long, J., & Jurjevich, J. R. (2020). Contradictions of the climate-friendly city: New perspectives on eco-gentrification and housing justice. *International Journal of Urban and Regional Research, 44*(1), 145–165.

Rodríguez-Pose, A., & Storper, M. (2020). Housing, urban growth and inequalities: The limits to deregulation and upzoning in reducing economic and spatial inequality. *Urban Studies, 57*(2), 223–248.

Rosol, M. (2015). Social mixing through densification? The struggle over the Little Mountain public housing complex in Vancouver. *Journal of the Geographical Society of Berlin, 146*(2-3), 151–164.

Rousseau, M. (2015). 'Many rivers to cross': Suburban densification and the social status quo in greater Lyon.

International Journal of Urban and Regional Research, 39(3), 622–632.
Sharifi, A. (2016). From Garden City to eco-urbanism: The quest for sustainable neighborhood development. *Sustainable Cities and Society, 20*, 1–16.
Slater, T. (2009). Missing Marcuse: On gentrification and displacement. *City & Society, 13*(2-3), 292–311.
Soja, E. (2009). The city and spatial justice. *Justice Spatiale/Spatial Justice, 1*(1), 1–5.
Ståhle, A. (2017). *Closer together: This is the future of cities*. Årsta: SCB Distributors.
Teller, J. (2021). Regulating urban densification: What factors should be used? *Buildings & Cities, 2*(1), 302–317.
Tretter, E. (2013). Sustainability and neoliberal urban development: The environment, crime and the remaking of Austin's downtown. *Urban Studies, 50*(11), 2222–2237.
UN. (2019). *2018 revision of world urbanization prospects*. New York: United Nations. Department of Economic and Social Affairs.
WCED. (1987). *Our common future*. Oxford.
Westerink, J., Haase, D., Bauer, A., Ravetz, J., Jarrige, F., & Aalbers, C. B. (2013). Dealing with sustainability trade-offs of the compact city in peri-urban planning across European city regions. *European Planning Studies, 21*(4), 473–497.
Wetzstein, S. (2017). The global urban housing affordability crisis. *Urban Studies, 54*(14), 3159–3177.

Urban Design

▶ Computational Urban Planning
▶ Stewarding Street Trees for a Global Urban Future
▶ Walkable Access and Walking Quality of Built Environment
▶ Women in Urbanism, Perpetuating the Bias?

Urban Development

▶ Housing and Development
▶ Peri-urban Regions
▶ Philanthropy in Sustainable Urban Development: A Systems Perspective
▶ Sustainable Development Goals from an Urban Perspective

Urban Digital Twin

▶ Digital Twin and Cities

Urban Ecology

▶ Habitat Provisioning

Urban Ecosystem Services and Sustainable Human Well-Being

M. Balasubramanian
Centre for Ecological Economics and Natural Resources, Institute for Social and Economic Change, Bangalore, Karnataka, India

Introduction

Urban ecosystem services (UES) provide a number of benefits to human well-being. The benefits include provisioning ecosystem services (food, water, and energy); regulating services (carbon sequestration, soil and water regulation); and cultural ecosystem services (tourism and recreational services) (Mea 2005; TEEB 2010; Remme et al. 2021). Urban ecosystem services are also important to human health in the context of maintaining physical and mental well-being (Bratman et al. 2019). Nearly, 55 percent of global population lives in the urban areas (United Nations 2019). An increasing population growth and unsustainable consumption of material, goods, and services are the major challenges to sustainable urban ecosystem services in the twenty-first century. There are a number of studies investigated in the context of urban ecosystem services for example, recreational ecosystem services, Fischer et al. (2018) studied an urban ecosystem services especially park and green space are vital contribution to the sociocultural aspects of five European cities. UES play a vital role of economic aspects, for example, in most of the urban park and green space generating income sources for urban municipalities as well as employment opportunities for thousands of people in many developing and developed countries. Schirpke et al. (2018) examined the economic value of urban ecosystem services estimated at 48.56 euro (mean willingness to pay) per visit to the recreational sites in Italy. Many developing countries have assessed the value of urban recreational ecosystems services. Tibesigwa et al. (2020) estimated the average willingness to pay at US$ 0.40 to 0.79 per month to visit the urban park in Tanzania. China's urban park and green spaces are a vital role to urban dwellers and other tourist visitors (Ranhao Sun et al. 2019). UES are also important for maintaining sustainable ecological conditions, for example, multifunctional green infrastructure provides a number of benefits such as ecological functions, production functions, and cultural functions to urban dwellers (Lovell and Taylor 2013); carbon storage of urban forest, park, and other urban ecosystem areas (Klein et al. 2021); and urban air pollution removal (Nowak et al. 2008). Provisioning urban water services are sources for millions of urban dwellers for drinking and urban agricultural productivity. Further, urban ecosystem services also play important role in urban sustainability in the context of maintaining green park and water bodies which are vital part of human well-being. However, little research has been done in the context of urban ecosystem services and sustainable human well-being, and therefore urban planners are not able to design better urban planning. This article is discuss on the opportunities and challenges of urban ecosystem services and integrating with the urban planning for achieving urban sustainable development goals especially cities should favourable for human settlements.

Why Urban Ecosystem Services Are Important?

Urban areas are important for economic growth or development, for example, more than 80 percent of global Gross domestic products (GDP) generated from urban areas (World Bank 2020). Urban areas are the major sources for labor market (formal and informal employment) of millions of people around the world. For example, urban labor force participation rate is 47.8 percent in India (MoSPI 2020). At the same time, urban areas are facing a number of environmental issues such as air pollution, solid waste, climate-related events including flood, and other extreme events. The sixth IPCC assessment report mentioned increased heat, flood, and heavy precipitation in urban areas, especially in coastal cities (IPCC 2021). Therefore, urban green space or urban ecosystem is important for maintaining social, economic, and environment benefits of thousands of cities around the world (Natural Economy North West 2010a).

Economic Benefits

There are a number of studies that have examined how urban ecosystem services contribute to economic growth or development. For instance, a recent study by Kwon et al. (2021) examined the positive association between urban green space and economic growth of 30 rich counties around the world. In addition this study has also found urban green space as the major tool for improving quality of life or happiness and wealth (GDP) of urban dwellers of these countries. Millward and Sabir (2010) study showed that urban green park has the main role of economic benefits for city residents in Toronto, Canada. Sim (2020) analyzed the urban green space has improved the economic benefits of local business in cities of South Korea. Urban green space or ecosystem has the major role for improving urban property value of cities in the USA, Singapore, and Malaysia (Sadeghian et al. 2013). Economic benefits of urban ecosystem services had been estimated for various components such as environmental and property value at US$ 1.2 million and the value of tourism and employment generation at US$ 111 million in the North West America (Molla 2015). Urban ecosystem services played a very important role for economic development in the context of tourism, employment, and hotel industries in the Lingqiu,

China (Dai et al. 2021). In addition, an urban ecosystem service has also improved property value (McPhearson et al. 2013). The economic value of central park in the New York City has been estimated at US$ 500 billion through the developable urban land (Sutton and Anderson 2016). However, there are a number of valuation studies that have been estimated, for example, economic value of climate regulation at Beijing (Leng and Badarulzaman 2014); New York (Peper et al. 2007, Chau et al.2010); Hong Kong and the value of air quality regulation, for example, Lanzhou (Zhang et al. 2006); Hong Kong (Peng and Jim 2015); Rome (Capotorti et al. 2019). Therefore, economic benefits of urban ecosystem services have been well studied but there is lack of implementation at the policy level.

Social Benefits

Social value of urban ecosystem services has been ignored or still not much recognized. Urban ecosystem services provide a number of social benefits such as recreation, sidewalks, cycling in the urban parks, research and education activities (Sutton and Anderson 2016), and reduced social inequality through the open access of urban green spaces (Wolch et al. 2014). Furthermore, urban green space are important relationship between human and nature (Zhou et al. 2018). Cultural ecosystems services is a vital role in the urban dwellers such as good quality of life including physical and mental health well-being (TEEB 2011). A recent study has estimated the social benefits of urban ecosystems services such as aesthetic value, culture, recreation, future, and history (Chen et al. 2020). In addition, good social relation will make better quality of life; therefore urban ecosystem services have been one of the important tools for creating good human well-being (Lapointe et al. 2021; Stiglitz et al. 2009). However, attention should be paid to social aspects of urban ecosystem services for better urban sustainability.

Health Benefits

UES are very important for human health and well-being. For example, most of the urban dwellers receive physical and mental health benefits through various ecosystem services. Ecosystem services provide positive benefits to human health (James et al. 2019; Kondo et al. 2020; Hunter et al. 2019). Study by Remme et al. (2021) examined urban ecosystem services are key role of maintaining physical activities especially for type 2 diabetes and cardiovascular diseases of Amsterdam, the Netherlands. In addition, this study also found that within 1000 meter 1.8 million physical activities happened per park and that these physical activities avoided treatment costs for present and future. Urban recreational ecosystems services have reduced health stress and improved mental well-being (Keeler et al. 2019). Positive physical and mental health has created good social relation and well-being (Hartig et al. 2014).

Urban Ecosystem Services and Human Well-Being

As mentioned in the previous section urban ecosystem services are important for human well-being. Provisioning ecosystem services has been provide food materials for urban dwellers through urban garden and vegetable cultivated areas. However, urban areas have less land for provisioning services, and therefore a few vegetables and fruits are the main sources from urban lands (Anderson et al. 2013). Second, regulating ecosystems services is very crucial for maintaining the climate and livelihood. For example, water erosion prevention, climate regulation, and air quality regulation all the ecosystem services are important for human well-being (Russo et al. 2017). Third, cultural ecosystem services are very critical to human well-being through visiting recreational sties, walking in the urban green space; and playing in the urban parks (Jennings and Bamkole 2019).A number of studies have investigated the perspective of urban ecosystem services, and more studies are needed on human well-being aspects because recent studies have mentioned that loss of biodiversity may affect human health and well-being. Further, increasing air pollution and decreasing of quality and quantity of urban water bodies, improper solid waste management are the major environmental problem on the one hand, an increasing health concern of urban dwellers due to changes of life styles are on the another hand is the major attention to an improve the urban green cover area is the main goal of many urban municipalities and policy makers in many developed and developing countries. However, the rapid urbanization has been created more demand for urban lands and other open space for industrial and commercial purpose. Over the period urban ecosystem services have degraded in many developing cities, for example expansion of urban land use areas are deteriorating environment in China (Yuan et al. 2018; Li et al. 2016). Therefore, urban ecosystem services such as urban green space, parks, urban common, water bodies, and wetlands are vital for human health and well-being.

Challenges and Opportunities of Urban Ecosystem Services

Urban ecosystem services play the key role in human well-being. Further, there is a lack of integration into the policy is another problem most of the urban planners. Further, there are a number of challenges in the urban ecosystem services research: first the concept of urban ecosystem services needs to be clear in the context of indicators. Second, the data or statistics is another issue for urban ecosystem services research in many developing countries. Third, valuation of ecosystem services or accounting for urban ecosystem services is very helpful to designing better policy for changes in stock and flow of urban ecosystem services in the rapid urbanization and deterioration of ecosystem services in the urban areas. In addition, identifying major indicators of urban ecosystem services helps to conserve for socio-economic and environmental well-being. However, creating data sets or statistics in the field of urban ecosystem services is very essential for designing urban planning and management at the sustainability aspects. Further, valuation of ecosystems services in the urban areas is very crucial for investing in the urban ecosystem services. The abovementioned points are the big challenges to sustainable urban human in relation with ecosystem services.

Conclusion

The ongoing consumption of urban lands is unsustainable in the many parts of the world. Also, increasing urban CO_2 emission and air pollution are other challenges to most of the urban dwellers especially because of the health problems they cause to generation population and vulnerable population such as urban informal workers, those staying at urban slums. Therefore, urban ecosystem services are vital provisioning services (sources of fruits and vegetables), regulating ecosystem services (maintaining climate and weather), and cultural services (recreational services) to millions of urban population. In addition, urban ecosystem services plays a major role in the context of sustainable human well-being for the present and future generation.

References

Anderson, P. M. L., Okereke, C., Rudd, A., & Parnell, S. (2013). *Regional assessment of Africa urbanization, biodiversity and ecosystem services: Challenges and opportunities: A global assessment* (pp. 453–459). Dordrecht: Springer. https://doi.org/10.1007/978-94-007-7088-1.

Bratman, G. N., Anderson, C. B., Berman, M. G., Cochran, B., De Vries, S., Flanders, J., Folke, C., Frumkin, H., Gross, J. J., Hartig, T, & Kahn, P.H. (2019). Nature and mental health: An ecosystem service perspective. *Science Advances, 5*(7), aax0903.

Capotorti, G., Alós Ortí, M. M., Copiz, R., Fusaro, L., Mollo, B., Salvatori, E., & Zavattero, L. (2019) Biodiversity and ecosystem services in urban green infrastructure planning: A case study from the metropolitan area of Rome (Italy). *Urban For. Urban Green, 37*, 87–96.

Chau, C. K., Tse, M. S., & Chung, K. Y. (2010). A choice experiment to estimate the effect of green experience on preferences and willingness-to-pay for green building attributes. *Building and Environment, 45*, 2553–2561

Chen, S., Wang, Y., Ni, Z., Zhang, X., & Xia, B. (2020). Benefits of the ecosystem services provided by urban green infrastructures: Differences between perception and measurements. *Urban Forestry & Urban Greening, 54*, 126774.

Dai, X., Johnson, B. A., Luo, P., Yang, K., Dong, L., Wang, Q., Liu, C., Li, N., Lu, H., Ma, L., & Yang, Z. (2021). Estimation of urban ecosystem services value: a case study of chengdu, Southwestern China. *Remote Sensing, 13*(2), 207.

Fischer, L. K., Honold, J., Botzat, A., Brinkmeyer, D., Cvejić, R., Delshammar, T., Elands, B., Haase, D., Kabisch, N., Karle, S. J., & Lafortezza, R. (2018). Recreational ecosystem services in European cities: Sociocultural and geographical contexts matter for park use. *Ecosystem Services, 31*, 455–467.

Hartig, R. M., de Vries, S., & Frumkin, H. (2014). Nature and health. *Annual Review of Public Health, 35*, 207–228.

Hunter, et al. (2019). Environmental, health, wellbeing, social and equity effects of urban green space interventions: A meta-narrative evidence synthesis. *Environment International, 130*, 104923.

IPCC. (2021). Climate Change 2021: The physical science basis. contribution of working Group I to the sixth assessment report of the intergovernmental panel on climate change [Masson-Delmotte, V., P. Zhai, A. Pirani, S. L. Connors, C. Péan, S. Berger, N. Caud, Y. Chen, L. Goldfarb, M. I. Gomis, M. Huang, K. Leitzell, E. Lonnoy, J. B. R. Matthews, T. K. Maycock, T. Waterfield, O. Yelekçi, R. Yu, and B. Zhou (eds.)]. Cambridge University Press.

James, J. J., Christiana, R. W., & Battista, R. A. (2019). A historical and critical analysis of park prescriptions. *Journal of Leisure Research, 50*, 311–329.

Jennings, V., & Bamkole, O. (2019). The relationship between social cohesion and urban green space: An avenue for health promotion. *International Journal of Environmental Research and Public Health, 16*(3). https://doi.org/10.3390/ijerph16030452.

Keeler, B. L., Hamel, P., McPhearson, T., Hamann, M. H., Donahue, M. L., Prado, K. A. M., Arkema, K. K., Bratman, G. N., Brauman, K. A., Finlay, J. C., & Guerry, A. D. (2019). Social-ecological and technological factors moderate the value of urban nature. *Nature Sustainability, 2*(1), 29–38.

Klein, C., Kuempel, C., Watson, R., Coll, M., Teneva, L., & Mora, C. (2021). Global fishing and seafood trade burdens places with ineffective fisheries management.

Kondo, C., et al. (2020). Nature prescriptions for health: A review of evidence and research opportunities. *International Journal of Environmental Research and Public Health, 17*, 4213.

Kwon, O. H., Hong, I., Yang, J., Wohn, D. Y., Jung, W. S., & Cha, M. (2021). Urban green space and happiness in developed countries. *EPJ data science, 10*(1), p. 28.

Lapointe, M., Gurney, G. G., Coulthard, S., & Cumming, G. S. (2021). Ecosystem services, well-being benefits and urbanization associations in a Small Island Developing State. *People and Nature, 3*(2), 391–404.

Leng, K.S., & Badarulzaman, N. (2014). Branding George Town world heritage site as city of gastronomy: prospects of creative cities strategy in Penang. *International Journal of Culture, Tourism and Hospitality Research.*

Li, B., Chen, D., Wu, S., Zhou, S., Wan, T., & Chen, H. (2016). Spatio-temporal assessment of urbanization impacts on ecosystem services: Case study of Nanjing City, China. *Ecological Indicators, 71*, 416–427.

Lovell, S. T., & Taylor, J. R. (2013). Supplying urban ecosystem services through multifunctional green infrastructure in the United States. *Landscape ecology, 28*(8), 1447–1463.

McPhearson, T., Kremer, P. & Hamstead, Z. A. (2013). Mapping ecosystem services in New York City: Applying a social–ecological approach in urban vacant land. *Ecosystem Services, 5*, 11–26.

Mea, M. E. A. (2005). Ecosystems and human well-being: Wetlands and water synthesis.

Millward, A. A., & Sabir, S. (2010). Structure of a forested urban park: Implications for strategic management. *Journal of environmental management, 91*(11), 2215–2224.

Molla, M. B. (2015). The value of urban green infrastructure and its environmental response in urban ecosystem: A literature review. *International Journal of Environmental Sciences, 4*(2), 89–101.

MoSPI (2020) Annual Report 2018–19, Periodic Labour Force Survey, Ministry of Statistics and Programme Implementation, Government of India.

Natural Economy North West. (2008) Developing an outline strategy for linking grey and green infrastructure. Kendal, NENW

Nowak, D. J., Crane, D. E., Stevens, J. C., Hoehn, R. E., Walton, J. T., & Bond, J. (2008). A ground-based method of assessing urban forest structure and ecosystem services. *Aboriculture & Urban Forestry, 34*(6), 347–358.

Pascual Unai et al., (2010). The Economics of Ecosystem and Biodiversity

Peng, L. L., & Jim, C. Y. (2015). Economic evaluation of green-roof environmental benefits in the context of climate change: The case of Hong Kong. *Urban Forestry & Urban Greening, 14*(3), 554–561.

Peper, P. J., McPherson, E. G., Simpson, J. R., Gardner, S. L., Vargas, K. E., & Xiao, Q (2007) New York City, New York Municipal Forest Resource Analysis; U.S. Department of Agriculture, Forest Service, Pacific Southwest Research Station, Center for Urban Forest Research: Newtown Square, PA, USA, 2007

Remme, R. P., Frumkin, H., Guerry, A. D., King, A. C., Mandle, L., Sarabu, C., Bratman, G. N., Giles-Corti, B., Hamel, P., Han, B., & Hicks, J. L. (2021). An ecosystem service perspective on urban nature, physical activity, and health. *Proceedings of the National Academy of Sciences, 118*(22).

Russo, A., Escobedo, F. J., Cirella, G. T., & Zerbe, S. (2017). Edible green infrastructure: An approach and review of provisioning ecosystem services and disservices in urban environments. *Agriculture,*

Ecosystems and Environment, 242(4), 53–66. https://doi.org/10.1016/j.agee.2017.03.026.
Sadeghian, M. M., & Vardanyan, Z. (2013). The benefits of urban parks, a review of urban research. *Journal of Novel Applied Sciences, 2*(8), 231–237.
Schirpke, U., Scolozzi, R., Da Re, R., Masiero, M., Pellegrino, D., & Marino, D. (2018). Recreational ecosystem services in protected areas: A survey of visitors to Natura 2000 sites in Italy. *Journal of outdoor recreation and tourism, 21*, 39–50.
Sim, J. (2020). Seeing impacts of park design strategies on local economy through big data: A case study of Gyeongui Line Forest Park in Seoul. *Sustainability, 12*(17), p. 6722.
Stiglitz, J. E., Sen, A., & Fitoussi, J. P. (2009). Measurement of economic performance and social progress. Online document http://bit.ly/JTwmG. Accessed June, 26, p. 2012.
Sun, R., Li, F., & Chen, L. (2019). A demand index for recreational ecosystem services associated with urban parks in Beijing, China. *Journal of environmental management, 251*, 109612.
Sutton, P. C., & Anderson, S. J. (2016). Holistic valuation of urban ecosystem services in New York City's Central Park. *Ecosystem Services, 19*, 87–91.
TEEB. (2011). *TEEB manual for cities: Ecosystem services in urban management.* Geneva: The Economics of Ecosystems and Biodiversity (TEEB).
Tibesigwa, B., Ntuli, H., & Lokina, R. (2020). Valuing recreational ecosystem services in developing cities: The case of urban parks in Dar es Salaam, Tanzania. *Cities, 106*, 102853.
United Nations, Department of Economic and Social Affairs, Population Division (2019). World Urbanization Prospects: The 2018 Revision (ST/ESA/SER.A/420). New York: United Nations.
World Bank. (2020). Urban Development. https://www.worldbank.org/en/topic/urbandevelopment/overview#1.
Yuan, Z., Chen, L., & Li, D. (2018). Urban stormwater management based on an analysis of climate change: A case study of the Hebei and Guangdong provinces. *Landscape and Urban Planning, 177*, 217–226.
Zhang, W. (2006). Initial analysis on the ecological service value of the greening land in Lanzhou city. *Prataculturál Science, 23*, 98–102.
Zhou, D., Tian, Y., & Jiang, G., (2018). Spatio-temporal investigation of the interactive relationship between urbanization and ecosystem services: Case study of the Jingjinji urban agglomeration, China. *Ecological indicators, 95*, 152–164.

Urban Eco-systems of innovation

► Connecting Urban and Regional Innovation Ecosystems to Enhance Competitiveness

Urban Edge

► Peri-urbanization

Urban Food Gardens

Kristina Ulm
Faculty of Arts, Design and Architecture, University of New South Wales, Sydney, NSW, Australia

Synonyms

Allotment; Community garden; Guerilla gardening; Home garden; Urban agriculture

Definition

Urban food gardens like community gardens, home gardens, and allotments are a form of urban agriculture that describes spaces in towns and cities where plants for human consumption are grown with a main purpose other than economic profit. Thus, urban food gardens are noncommercial forms of urban agriculture (Hou, 2020). While some urban food gardens may sell their produce, it is only a complementary function in addition to recreation, socializing, exercise, and growing plants for self-consumption. Legislation on sale of edible products from urban food gardens varies from one country to another, and one city to another (Pires, 2011). Urban food gardeners grow plants to be used for consumption, medical purposes, and aesthetics. They also participate in other gardening activities such as composting, weeding, and harvesting. Animals like bees, worms, or chickens can be part of urban food gardens.

Introduction

International research and practice use a diverse terminology to describe urban food gardens, ranging from allotments and balcony gardens to school and verge gardens. This entry provides an overview of the diverse types of urban food gardens and the different terminologies that are used to describe them. Terminologies for urban food gardens along their social organization are most common in research and widely used in policy, practice, and everyday language. These three main types of urban food gardens are home (or household) gardens, allotments, and community gardens. These urban food gardens can further be classified along different spatial, political, and legal dimensions. Economic and environmental dimensions are rarely used to name specific types of noncommercial urban food gardens in academic literature, and thus are not part of this entry. After first introducing home, allotment, and community gardens (including their respective subtypes), this entry elaborates on the spatial, political, and legal dimensions of urban food gardens as shown in Table 1.

Within the spatial dimension the primary question on naming these types of gardens is: Where does it occur? On what kind of land or structure? Rooftop gardens and food gardens in the street space, such as verge, nature strip, or footpath gardens, are described in that section. Because some spatial dimensions are inextricably linked to their social organization, like home or school gardens, there is an overlap with the social organization dimension. Within the political dimension the primary questions on naming these types of gardens are: How do the gardens relate to actions of governmental authorities? Who initiated them? To what extent are gardeners politically motivated? The described concepts include citizen- and community-led gardens, bottom-up and grassroots gardening, political gardens, radical gardens, and guerilla gardens. Within the legal dimension, the questions are: How does the garden relate to the law and land ownership? That section discusses public gardens and the continuum between legal/formal and illegal/informal gardens.

An urban food garden can be described along any one or multiple of these dimensions. This means that there are connected meanings and uses of these terminologies. While some people use the same name to mean distinct types of urban food gardens (e.g., community gardens), others use different names for the same type of urban food garden. For example, collective garden, neighborhood garden, and community garden can describe the same type of urban food garden. An urban food garden can be categorized along any of the dimensions shown in Table 1. For example, an informal rooftop community garden could be called a rooftop garden, a community garden, or an informal garden depending on what a research study focusses on.

This entry starts with home and allotment gardens, where only a single household or family is involved in urban food gardening, before elaborating on community gardens with their different subtypes.

Urban Food Gardens, Table 1 Terminologies of urban food gardens after political, legal, and spatial dimensions

Main types (Social organization)	Political	Legal	Spatial
Household	Citizen-led	Public	Rooftop
Allotment	Bottom-up	Informal	Street
Community	Grassroots	Illegal	
	Radical		
	Guerilla		

Home and Household Gardens

Considered the most "private" form of urban food gardens, home gardens are cultivated by only one household or family. Home gardens are also referred to as residential or private gardens as they are defined by being managed by an individual household on land that is part of or adjacent to the main residence of that household (Kirkpatrick & Davison, 2018; Taylor & Lovell, 2014). A home garden is exclusively accessible to that household through formal agreement like ownership or lease (Taylor & Lovell, 2014). Home gardens are sometimes further specified by referring to spatial dimensions, such as balcony, rooftop, or backyard gardens. They are globally widespread and one of the oldest historically documented form of urban food gardens, for example, reaching millennia back to Maya and Aztec cities (Isendahl & Smith, 2013). Home gardens are found to primarily serve recreation and further assist with food production to the household (Kirkpatrick & Davison, 2018). They are thus identified as a livelihood strategy (Chambers, 1995). Home gardens are also called household gardens or kitchen gardens and were encouraged under the name "victory gardens" in the USA (United States of America) during World War II (Horst et al., 2017).

Allotment Gardens

Allotment gardens are like household gardens that are not part of the physical "home." They are like home gardens in the way that access, and gardening activities are limited to the household that has the right to manage the cultivated land, either leased or owned, through a formal agreement. The difference is that, while home gardens are seen as directly adjacent and accessible from the gardening household's primary residence, allotment gardens can be up to several kilometers away.

Allotment gardens are common across Europe, ranging from the UK (United Kingdom) to Eastern Europe. They are known as "Kleingärten" ("little gardens") or "Familiengärten" ("family gardens") in German speaking countries (Göttl & Penker, 2020; Rosol, 2010), "jardins familiaux" ("family gardens") in France (Pourias et al., 2016), "volkstuin" ("gardens for people") in the Netherlands (Keshavarz et al., 2016), or "dachas" in states of the former Soviet Union (Zavisca, 2003). There are also informal allotment gardens, set up "non-legally" in underutilized space surrounding large-scale apartment buildings or in "nature protection zones" (Djokić et al., 2018). Allotment sizes can range from 100 to 500 sqm (Pourias et al., 2016). Allotment gardens are usually spatially clustered in groups of individually leased allotments, sharing access paths to the individual gardens. Consequently, they require a certain degree of collaboration and rule-setting regarding shared access points. Compared to home gardens, gardeners can meet other allotment gardeners on shared spaces or through "conversations over the fence." Nonetheless, it is also possible to manage one's own allotment garden without any necessity for communication with neighboring gardeners, depending on the organization of the allotment garden group. As a result, even though some spaces and facilities may be shared in allotment gardens, they are typically considered as "individual gardening" and clearly separated from more collective forms of gardening, like community gardening (Göttl & Penker, 2020). The boundaries get blurred nonetheless where explicit communally shared spaces are provided. In some countries, groups of spatially clustered allotment gardens can be structured into a formal organization that manages the rules of gardening and sometimes provides communally used spaces, such as seating, playgrounds, or community buildings (Costa et al., 2016). Allotments are commonly located on publicly owned land (Tappert et al., 2018) and usually leased for several years, if not decades, by individual private households. Municipalities usually manage the allocation of gardens. Due to long-term leases, limited allotments and increasing demand, waiting lists are commonly reported, for example in France (Pourias et al., 2016). In North America, the historical lack of allotment gardens is assumed to have led to long waiting list for community gardens (Bach & McClintock, 2020; Göttl & Penker, 2020). Although allotment gardens are on publicly owned land, they are traditionally seen as private space (Rosol, 2010). Thus, compared to community gardens that are hailed for their multiple benefits, European allotment gardens are sometimes perceived more critically because of the "private" management which leads to only selected parts of the population benefiting from the social benefits (Tappert et al., 2018). They are commonly (and sometimes controversially) regarded as "land in waiting" for development (Tappert et al., 2018).

Historically, allotment gardens thrived in economic crises as their primary function was to increase food security by offering spaces for self-sufficient food production (Lawson, 2005; Pourias et al., 2016). They evolved from the nineteenth century "worker's gardens" in France (Pourias et al., 2016) or World War Two "victory gardens" in the UK and the USA (Keshavarz et al., 2016; Lawson, 2005). Since the nineteenth century, "relief gardens," often organized in allotment form, were introduced in the USA to enable people in challenging economic condition to take care of themselves through food production in urban areas instead of applying for government aid (Kurtz, 2001).

Community Gardens

Community gardens involve members from different households acting collaboratively around a garden space, which is the distinguishing factor from home and allotment gardens (Göttl & Penker, 2020; Guitart et al., 2012). The term community garden has a wide range of meanings. The definition of community gardens varies from one cultural background to another, differing between geographic regions, as well as individual researchers (Pourias et al., 2016). Thus, Rosol (2010, p. 552) sums it up accurately: "Community gardens have no widely acknowledged common definition." Some research studies use the term without explicitly defining it (e.g., van Holstein, 2020). An example of a wide definition is "open spaces which are managed and operated by members of the local community in which food or flowers are cultivated." (Guitart et al., 2012, p. 364). In the USA, community gardens are not necessarily growing edible plants, which is however commonly part of their definition in Australia (Nettle, 2016). It was found that across Australia "councils tend to define community gardens differently, but generally community gardens are understood to be public open spaces operated by the community for personal food production, and to serve as sites for environmental activities and community education." (Pires, 2011, p. 5).

An important aspect of community gardens, which explains the choice of naming it after its social organization, is that part of their main function is to "grow a sense of community" (Rogge & Theesfeld, 2018).

Different Levels of Collaboration

A core characteristic of community gardens is a degree of collaboration between individual gardeners from different households (Christensen et al., 2018; Draper & Freedman, 2010; Okvat & Zautra, 2011). However, there are different perceptions on what is shared and to what degree. For example, collaboration is necessary for agreeing on shared rules, the use of common tools, or shared spaces. The categorization of community gardens in this entry uses the following levels of degree of collaboration. The first level is shared agreement on the management and rules of the garden. The second level is shared use of physical tools and resources, and the third level is shared cultivated space, which equates more time spent together gardening. Community gardens vary in their social organization between these levels of collective gardening and be constituted by mixed forms by combining the organization styles in different ways.

The relationships between these levels of collaboration and the different terminologies for community gardens as well as home and allotment gardens are illustrated in Fig. 1. While some names can be clearly differentiated from others, like home gardens and its synonyms, community gardens have multiple meanings. A clear distinction is more feasible when it is easy to perceive if an urban food garden is "private" or "public." Home gardens are seen as a dominantly private activity, overlapping with concepts of private home and activities that occur within one's own residence. Community gardens, on the other hand, are predominantly seen as public, involving multiple households (Draper & Freedman, 2010; Rosol, 2010). However, in between these urban food gardens seen as "private" home gardens or "public" community gardens are a variety of gardens were there are various views on how private or public they are. A range of overlapping names are used to describe these undefined (in terms of public-private) spaces, like allotment gardens, community-allotment gardens, or neighborhood gardens.

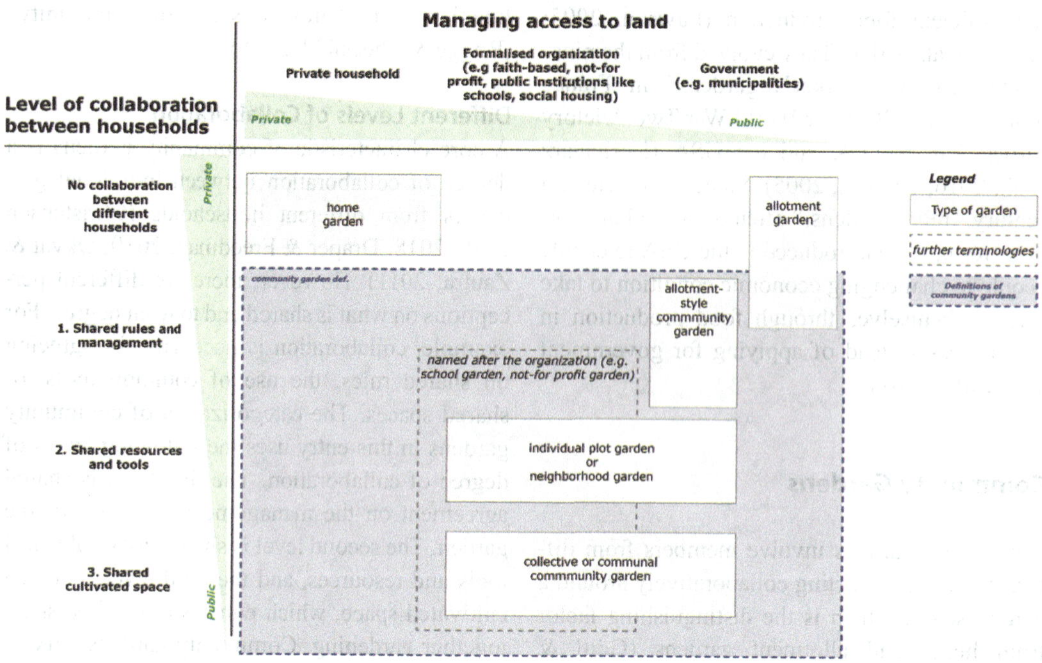

Urban Food Gardens, Fig. 1 Types of urban food gardens along social organization

First Level Allotment-Style Community Gardens

The first level of collaboration is shared agreement on the management and rules of the garden. This applies to allotment gardens as well. Consequently, depending on the dominance of individually assigned lots, community gardens can resemble and overlap significantly with the concept of allotment gardens. This applies especially to North American community gardens as there are no traditional concept of the term allotment gardens (Lawson, 2005; Lovell, 2010). Lawson (2005) views these individual allotment style gardens not as "community" gardens and argues to call them urban gardens instead. Other researchers classify allotment gardens as a subtype of community gardens (Tracey et al., 2020), which underscores the problem of no widely accepted definition of what a community garden means. Allotment-style community gardens can be contested in their social benefits. While Lovell (2010) criticizes "allotment style community gardens" in Canada for benefiting selected individuals rather than a larger public, Gaynor (2006) advocates for "community allotment gardens" in Australia to give gardeners a certain level of independence in collective gardening forms.

Second Level Community Gardens with Individual Plots

The second level of collaboration is shared use of physical tools and resources. Different to the allotment style gardens, where individual gardening spaces are separated, these community gardens offer individual gardening plots of smaller sizes that are located within a larger shared garden space, (Bach & McClintock, 2020; Göttl & Penker, 2020; Pourias et al., 2016). Bach and McClintock (2020) distinguish these understandings of traditional community gardens in North America from "collective gardens," to which they attribute the additional third level of sharing spaces that are cultivated.

While allotment-style community gardens are debated if they are part of community gardening or not, individual plots community gardens are widely seen as a form of community gardening. With a lack of allotment gardens in North America, community gardens took that role, which means that compared to community gardens with individual lots in Europe, Canadian community gardens ("jardins communautaires" in Montreal) have larger individual lots (15–20 m^2) compared to individual lots in Parisian shared gardens ("jardins partages") that can range from 2 to 20 m^2 (Pourias et al., 2016).

Community gardens with individual plots are also referred to as neighborhood gardens (Pourias et al., 2016). Subsequently, within research in the USA, as well as in other regions, the term neighborhood garden, while less common, is used interchangeably with the term community garden (McClintock & Cooper, 2010; Pothukuchi, 2015; Purcell & Tyman, 2015; Spijker & Parra, 2018).

Sometimes, neighborhood gardens are seen as a subtype of community gardens (Rogge & Theesfeld, 2018; Tracey et al., 2020). Research in the USA often refers to the definitions of community and neighborhood gardens of the American Community Garden Association (Lawson & Drake, 2013; Mok et al., 2014).

Third Level Communal and Collective Gardening

The third level of collaboration is shared cultivated space in community gardens, which is also referred to as communal or collective gardening. The difference in sharing space that is cultivated is the degree to which time spent together gardening is shared. If more spaces are collectively gardened, there are consequently more opportunities for sharing time together gardening (van Holstein, 2020). This is rarer in community gardens that are organized in individual lots with time spent gardening independently.

Communal gardens in Australia are seen as a subtype of community gardens where there are no individual plots assigned (Mintz & McManus, 2014). The term "communal garden" itself is less widely used than community garden. It is more commonly referred to as describing the way a community garden is organized (Eizenberg, 2012; Göttl & Penker, 2020; Palau-Salvador et al., 2019; Rogge & Theesfeld, 2018; Rosol, 2010) and used in this way interchangeably with "collective gardening" (Bach & McClintock, 2020; Göttl & Penker, 2020; Kingsley & Townsend, 2006).

Many community gardens across Europe, North America, and Australia are found to have a combination of individually assigned gardening lots, which vary in size, and communally cultivated spaces (ACFCGN, 2019; Göttl & Penker, 2020; Palau-Salvador et al., 2019; Pourias et al., 2016).

Subtypes of Community Gardens

The term community garden is likewise used as an umbrella term that encompasses various kinds of community gardens that are affiliated with a formal organization, like nonprofit or school gardens. Urban food gardens managed by a nonprofit organization can be called nonprofit gardens. Some researchers consider nonprofit and institutional gardens as types of urban agriculture different from community gardens (McClintock, 2014; Meenar et al., 2012). More often however, gardens managed by not-for-profit organizations are included within the concept of community gardens (Beavers et al., 2021; Lawson & Drake, 2013). Community gardens can be further subclassified as part of different public organizations like schools, prisons, hospitals, where members of that organization are involved in gardening (McClintock, 2014). The community gardening organization survey of the American Community Gardening Association adds further "community gardens types" such as public housing gardens, church gardens, and senior housing gardens (Lawson & Drake, 2013). Even though these are located on land associated with an institution, many of them are open to the wider community (Lawson & Drake, 2013) and seen as "public" (Draper & Freedman, 2010; Tracey et al., 2020). It is further recognized that the variety of involved organizations increased in the last decade (Lawson & Drake, 2013; Pothukuchi, 2015).

School gardens are sometimes seen as a subtype of community gardens (Draper & Freedman, 2010; Lawson & Drake, 2013) but at other times as different to community gardens (Audate et al., 2019; Beavers et al., 2021; McClintock, 2014; Zeunert et al., 2018) with a much stronger educational purpose (Mok et al., 2014). The difference might be the extent to which the garden is accessible to the public beyond the school community. In Australia, for example, around a tenth of community gardens are affiliated with a school or church (ACFCGN, 2019) and thus considered a form of community garden (Mintz & McManus, 2014).

Community gardens can also be differentiated according to special functions such as educational, mental health, or intercultural gardens (Rogge & Theesfeld, 2018). In Germany, intercultural gardens were established in Berlin and other cities as community gardens with a special focus on supporting migrants (Rogge & Theesfeld, 2018; Rosol, 2010).

Community gardens are often seen as public spaces, compared to private home or allotment gardens, as they commonly allow for "members of the public" to participate or physically access the space (Draper & Freedman, 2010; Rosol, 2010). Regarding the relationship with the government, community gardens are often seen as grassroots organizations (Tornaghi, 2014; van Holstein, 2020) and sometimes as part of guerilla gardening (Apostolopoulou & Kotsila, 2021; Zhu et al., 2020). Further political and legal dimensions of community gardens are discussed in the following sections.

Political Dimensions of Urban Food Gardens

Studying the relationship between the state and its citizens, political dimensions are a growing research area in the urban food gardening literature. Terminologies used to describe political features of an urban food garden range widely on a scale of how politically motivated the initiation and implementation of the gardens are. Some urban food gardens start out from citizens taking initiative, which *can* but does not have to be either politically motivated, or outside the legal realm. These urban food gardens are called citizen-led or community-led primarily to emphasize the efforts of the people starting a garden without any support of a formal organization. Bottom-up and grassroots gardening are overlapping concepts. Political, radical, and guerilla gardening are, however, more commonly seen as part of explicit political activism.

Citizen-Led and Community-Led Gardening

Research that emphasizes the actors involved in initiating urban food gardens uses terminologies like citizen-led, community-led, or resident-led to make a distinction from gardens that were initiated by organizations such as, for example, businesses, schools, or local governments (Bach & McClintock, 2020). Citizen means in this case "the civic identity of an urban inhabitant, rather than to nationality or citizenship status" (Bach & McClintock, 2020, p. 16).

On the one hand, citizen-led does not necessarily imply a political or legal status of the urban food gardens studied and government support is explicitly suggested (Marshall et al., 2019). On the other hand, citizen-led and community-led are sometimes seen as political activism and a countermovement against government-led urban planning (Bach & McClintock, 2020; Certomà & Tornaghi, 2015). A stronger focus on these political motivations of urban food growing activities is implied by terms such as grassroots-led or activist-led (Certomà & Tornaghi, 2015; Coles & Costa, 2018), which overlaps with terms like bottom-up and grassroots.

Bottom-Up and Grassroots Gardening

The terms bottom-up and grassroots describe urban food gardens that are primarily initiated or managed by people or organizations outside government, like individual citizens or other formal or informal nongovernmental organizations (Fox-Kämper et al., 2018; Lovell, 2010). Their meaning, thus, overlaps with the concepts of citizen-led, community-led, participant-initiated, or community-driven (Lovell, 2010; Nettle, 2016). Due to no universal definition of these terms, a wide range of urban food gardens, from home gardens to community gardens, can be classified as bottom-up gardening (Casazza & Pianigiani, 2016). First, when describing the initiation phase of an urban food garden, grassroots and bottom-up can mean that a garden was started by the gardeners themselves or a "grassroots organization." A widespread view of "grassroots organizations" is that they can be any form of social organization excluding projects initiated by public authorities (Pothukuchi, 2015). Drake and Lawson (2015) used these admittedly ill-defined terms in a survey on North American community gardens, leading to most community gardens identifying themselves as grassroots-initiatives. Second, when describing the management arrangement of an urban food gardens, it is commonly distinguished between the initiation phase of a garden and the later stages (Fox-Kämper et al., 2018; Rogge & Theesfeld, 2018). A wide range of approaches between "bottom-up" and "top-down" (meaning government-initiated and managed) can be found in urban food gardens (Fox-Kämper et al., 2018). It is argued that most community gardens are managed bottom-up and top-down at the same time as these are a complex interplay ensuring long-term success (Fox-Kämper et al., 2018; Rosol, 2010). Lastly, while the term "grassroots" is often used synonymously to bottom-up, it is sometimes more strongly associated with political activism (Pothukuchi, 2015). Especially critical geographers use "grassroots" in conjunction with activism and citizen-led social movements (Rosol, 2010), opposing government and associated with food sovereignty (Tornaghi, 2014).

Bottom-up and grassroots gardening translate inconsistently to the legality of urban food gardens. For example, bottom-up as a form of management primarily relying on citizens or nongovernmental organizations can be part of legally fully acknowledged urban food gardens (Fox-Kämper et al., 2018).

Political and Radical Gardening

Political and radical gardening explicitly refers to a strong activist role of urban food gardens. "Political gardeners" see themselves as activists demanding new ways of democratic participation in city planning (Certomà & Tornaghi, 2015; Follmann & Viehoff, 2015). Some scholars regard political forms of urban food gardens as a grassroots response to the shortcomings of the neoliberal economy (Certomà & Tornaghi, 2015), addressing what is perceived as "abandoned by government" (Kurtz, 2001; Pothukuchi, 2015). Political gardens are seen as a form of "bottom-up" gardening (Fox-Kämper et al., 2018) and associated with radical acts like guerilla gardening (Adams et al., 2015; Bach & McClintock, 2020). Others dispute these critical positions and refer to the wide range of what is considered a community garden, arguing that the emphasis on self-reliance fuels neoliberal ideals (Pudup, 2008; Rosol, 2010). Thus, it is cautioned that even traditionally radical forms of political gardening like guerilla gardening can be a practice that aligns with dominant politics and can be practiced in "non-participatory" ways (Certomà & Tornaghi, 2015). Lastly, some forms of political gardens are considered less radical but nonetheless can have a subtle political impact on changing their immediate neighborhoods (Certomà & Tornaghi, 2015).

Guerilla Gardening

Guerilla gardening describes typically nonlegal forms of urban food gardening on land for which the gardeners do not have sufficient permission like vacant lots or footpaths and is often considered a political action (Certomà & Tornaghi, 2015). This is emphasized by its association with warfare "guerilla" terminology. However, it can also be used describing the illegal, informal, or unclear legal types of urban food gardens without necessarily assuming a political motivation on behalf of the gardeners, including any gardening without formal approval from the landowner (Crane et al., 2013; Tornaghi, 2014). In research on urban food gardens, the term guerilla gardening is geographically predominantly used in industrialized countries such as the UK (Hardman et al., 2018; Thompson, 2015; Tornaghi, 2014), Greece (Apostolopoulou & Kotsila, 2021), the USA and Canada (Bach & McClintock, 2020; Crane et al., 2013; Morrow & Martin, 2019), Australia (Marshall et al., 2019), and more recently China (Hung, 2017; Zhu et al., 2020). Research on similar urban food gardens labels guerilla gardens as "non-legal practices" and informal, for example, in Serbia (Djokić et al., 2018).

Despite usually viewed as radical forms of gardening, some guerilla gardens are less activist, even supported by public agencies (Hardman et al., 2018), which can be seen as an oxymoron. Bach and McClintock (2020) argue that the lines between guerilla gardening and institutional initiatives can be blurred, taking the radicality out of the political ambitions of these forms of gardening.

Legal Dimensions of Urban Food Gardens

Legal features of urban food gardens can, first, describe the legality of land ownership, for example, separated into private and public land, and second, the formal approval of the gardening activity. There is a wide range between what is explicitly legal and illegal as some "illegal" activities may be tolerated or even encouraged by authorities at later stages. Thus, instead of clear distinctions on what is legal, illegal, formal, or informal, it is a continuum between formality and informality.

Informal and Illegal Gardens

Perceiving formal urban food gardens as within the legal realm (Blomley, 2005), informal food gardens are generally seen with occurring outside the legal realm (Djokić et al., 2018).

A more nuanced view is defining informal urban food gardens as not being associated with local authorities in the initiation and maintenance of the gardens as well as no formal record to authorities of what is happening (Hardman et al., 2018). This definition is established within the context of a municipality in the UK, where "guerilla gardening" is officially legalized, causing activists to prefer the term informal gardening as they perceive the crucial political aspect of "guerilla gardening" being lost (Hardman et al., 2018). Similarly, other researchers clearly distance their use of the term informal gardening from politically connotated terms like guerilla gardening. For example, the gardeners of "self-claimed" gardens, occurring informally in shared green spaces between apartment buildings in China report a wide variety of motivations – other than political – for gardening (Zhu et al., 2020). Nonetheless, informal gardening is frequently used interchangeably with more political terms like radical, grassroots, and guerilla gardening (Palau-Salvador et al., 2019; Tornaghi, 2014). Overall, informal food growing is widely spread in research on urban food gardens in Africa, South America, and Asia (Freeman, 1991; Thornton, 2020). The term illegal is a lot less used than informal and only sometimes mentioned as a descriptive feature, but rarely as the main term for naming an urban food garden (Tornaghi, 2014; Zhu et al., 2020).

Public Garden

The term public garden is rarely used in the scholarly literature. Public gardens can refer to municipal gardens, owned and operated by local authorities (Palau-Salvador et al., 2019).

As the term public garden is commonly used as an additional descriptor, Blomley's (2005) research in which public gardens is the preferred dominant terminology constitutes an exception. A lens of legal geography is applied to Canadian "public gardens," urban food garden that are challenging traditional notions of the law on what is public and private (Blomley, 2005). These are gardens that are a "private encroachment on public land" (Blomley, 2005, p. 281), that are privately managed spaces on publicly owned land along the street space and thus have commonly a precarious legal standing as informal initiatives. Blomley's (2005) research found that what is perceived as public and private by neighborhood residents is much more complex than what current government regulation bases itself on.

Spatial Dimensions of Urban Food Gardens

Many urban food gardens have a certain spatial dimension inherently implied from a socio-spatial condition, like home or school gardens. Spatial dimensions are rarely used to label a type of urban food garden unless the spatial characteristic significantly defines the garden, which is the case with rooftop and various names for street gardens.

Rooftop Food Gardens

A rooftop food garden describes growing edible plants on the roof, the topmost structural part of a building, and can be located on privately as well as publicly owned buildings. Compared to soil-based gardens on the ground, a rooftop garden comes with structural challenges as the building must be designed to hold heavy weights on its roof (Sanyé-Mengual et al., 2016). Furthermore, due to challenging accessibility and higher investment costs, rooftop gardens occur more commonly as commercial rooftop farming and are not as widespread as noncommercial urban food gardens (Bach & McClintock, 2020; James, 2016; Mok et al., 2014; Pinheiro & Govind, 2020). Most commonly, literature on urban food gardens mentions rooftops as a space where a food garden could occur as part of a spatial categorization (Lovell, 2010; Mougeot, 2000). The primary focus in this research is however on social dimensions of urban food gardens (Coles & Costa, 2018; Taylor & Lovell, 2014). For example, Hou (2020) describes rooftop food gardens that can be on private dwellings, community-centers, or schools, as a subset of home or community gardens.

Street Food Gardens, Footpath, Sidewalk, and Easement Gardens

Urban food gardens that occur within or along streets are often explicitly named after their unique spatial characteristics. Street food gardens are further differentiated into more detailed terminologies depending on where in the street they occur, for example, footpath garden or verge garden. Also, different names exist in different countries, for example, an easement garden in the USA is called a verge garden in the United Kingdom, Australia, and New Zealand (Hunter & Brown, 2012). An easement garden is defined by Hunter and Brown (2012, p. 407) as "a privately installed garden within the property's easement area, the street-side space that is owned and regulated by local government." Bach and McClintock (2020) call it a sidewalk garden in Canada. If municipalities are involved in planting edible plants in the streets, terms like edible green infrastructure, edible landscapes, edible urban greening, or edible urbanism are used (Hardman et al., 2018; Russo et al., 2017).

Even though research on street food gardens is limited, the practice is estimated to occur widely across the world (Marshall et al., 2019). Due to their unique location within the street space, street food gardens have additional benefits as well as limitations. For example, a critical limitation are the numerous services that occur in the street space. Service providers for electricity, gas, and water have an interest in how urban food gardens in the street are executed and maintained.

Street food gardens are characterized by a multiplicity of involved actors, adding significant complexity to social interactions. For example, ownership of the street space is often different from who maintains the urban food garden. The street, including the food gardens, is usually owned by local, state, or national authorities. Maintenance and management of the garden, however, is commonly performed by organizations, groups, and individuals other than government like not-for-profits, community organizations, or informal neighborhood groups (Hunter & Brown, 2012). Street food gardens, compared to other urban food gardens, are usually publicly accessible due to their location (Hunter & Brown, 2012). Consequently, passers-by and neighbors who pick parts of plants are a common, and at times desired, feature of street food gardens (Lopes & Shumack, 2012). On the one hand, this demonstrates the high educational value of street food gardens, easily reaching a variety of people incidentally who otherwise might not be able to observe how food is grown (Hsu, 2018). On the other hand, because of their public accessibility, theft of whole plants is more likely to occur in street food gardens. Compared to other urban food gardens, street food gardens are more severely shaped by local environmental factors which explains why a separate terminology is used to discern these gardens from other gardens.

Conclusion

Even though researchers may use the same term for describing an urban food garden, it can mean something different from study to study, depending on the cultural context of the urban food garden and the researchers. Especially the term community garden is contentious because it is widely used and at the same time shows the widest range of different meanings assigned to it. Terminologies used to describe urban food gardens depend on the research focus and can draw from the social organization of a garden, like home gardens or community gardens, as well as from spatial, political, or legal dimensions. There are overlaps between these concepts as an urban food garden can be named after any of these dimensions. For example, an urban food garden started and maintained by residents on the rooftop of a residential building without formal approval could be named as following: rooftop garden (spatial dimension), community garden (social organization dimension), citizen-led garden (political dimension), and informal garden (legal dimension). Furthermore, a combination of these is possible, for example, a grassroots rooftop community garden. Researchers choose to employ the terminology best suited to their purposes, focus of study, and cultural context.

Cross-References

▶ Circular Cities

References

ACFCGN. (2019). *Community Garden Survey 2019*. Australian City Farms and Community Garden Network

Adams, D., Hardman, M., & Larkham, P. (2015). Exploring guerrilla gardening: Gauging public views on the grassroots activity. *Local Environment, 20*(10), 1231–1246. https://doi.org/10.1080/13549839.2014.980227

Apostolopoulou, E., & Kotsila, P. (2021). Community gardening in Hellinikon as a resistance struggle against neoliberal urbanism: Spatial autogestion and the right to the city in post-crisis Athens, Greece. *Urban Geography*. https://doi.org/10.1080/02723638.2020.1863621. Scopus.

Audate, P. P., Fernandez, M. A., Cloutier, G., & Lebel, A. (2019). Scoping review of the impacts of urban agriculture on the determinants of health. *BMC Public Health, 19*(1), 672. https://doi.org/10.1186/s12889-019-6885-z

Bach, C. E., & McClintock, N. (2020). Reclaiming the city one plot at a time? DIY garden projects, radical democracy, and the politics of spatial appropriation. *Environment and Planning C: Politics and Space*. https://doi.org/10.1177/2399654420974023. Scopus.

Beavers, A. W., Atkinson, A., Ma, W., & Alaimo, K. (2021). Garden characteristics and types of program involvement associated with sustained garden membership in an urban gardening support program. *Urban Forestry & Urban Greening, 59*, 127026. https://doi.org/10.1016/j.ufug.2021.127026

Blomley, N. (2005). Flowers in the bathtub: Boundary crossings at the public–private divide. *Geoforum, 36*(3), 281–296. https://doi.org/10.1016/j.geoforum.2004.08.005

Casazza, C., & Pianigiani, S. (2016). Bottom-up and top down approaches for urban agriculture. *Civil Engineering and Urban Planning: An International Journal (CiVEJ), 3*(2), 49–31. https://doi.org/10.5121/civej.2016.3204

Certomà, C., & Tornaghi, C. (2015). Political gardening. Transforming cities and political agency. *Local Environment, 20*(10), 1123–1131. https://doi.org/10.1080/13549839.2015.1053724

Chambers, R. (1995). Poverty and livelihoods: Whose reality counts? *Environment and Urbanization, 7*(1), 32.

Christensen, S., Dyg, P. M., & Allenberg, K. (2018). Urban community gardening, social capital, and "integration" – A mixed method exploration of urban "integration-gardening" in Copenhagen, Denmark. *Local Environment*. https://www-tandfonline-com.wwwproxy1.library.unsw.edu.au/doi/abs/10.1080/13549839.2018.1561655

Coles, R., & Costa, S. (2018). Food growing in the city: Exploring the productive urban landscape as a new paradigm for inclusive approaches to the design and planning of future urban open spaces. *Landscape and Urban Planning, 170*, 1–5. https://doi.org/10.1016/j.landurbplan.2017.10.003

Costa, S., Fox-Kämper, R., Good, R., Sentić, I., Treija, S., Atanasovska, J. R., & Bonnavaud, H. (2016). The position of urban allotment gardens within the urban fabric. In *Urban allotment gardens in Europe*. Routledge, London.

Crane, A., Viswanathan, L., & Whitelaw, G. (2013). Sustainability through intervention: A case study of guerrilla gardening in Kingston, Ontario. *Local*

Environment, 18(1), 71–90. https://doi.org/10.1080/13549839.2012.716413

Djokić, V., Ristić Trajković, J., Furundžić, D., Krstić, V., & Stojiljković, D. (2018). Urban garden as lived space: Informal gardening practices and dwelling culture in socialist and post-socialist Belgrade. *Urban Forestry and Urban Greening, 30*, 247–259. https://doi.org/10.1016/j.ufug.2017.05.014

Drake, L., & Lawson, L. J. (2015). Results of a US and Canada community garden survey: Shared challenges in garden management amid diverse geographical and organizational contexts. *Agriculture and Human Values, 32*(2), 241–254. https://doi.org/10.1007/s10460-014-9558-7

Draper, C., & Freedman, D. (2010). Review and analysis of the benefits, purposes, and motivations associated with community gardening in the United States. *Journal of Community Practice, 18*(4), 458–492. https://doi.org/10.1080/10705422.2010.519682

Eizenberg, E. (2012). Actually existing commons: Three moments of space of community gardens in New York City. *Antipode, 44*(3), 764–782. https://doi.org/10.1111/j.1467-8330.2011.00892.x

Follmann, A., & Viehoff, V. (2015). A green garden on red clay: Creating a new urban common as a form of political gardening in Cologne, Germany. *Local Environment, 20*(10), 1148–1174. https://doi.org/10.1080/13549839.2014.894966

Fox-Kämper, R., Wesener, A., Münderlein, D., Sondermann, M., McWilliam, W., & Kirk, N. (2018). Urban community gardens: An evaluation of governance approaches and related enablers and barriers at different development stages. *Landscape and Urban Planning, 170*, 59–68. https://doi.org/10.1016/j.landurbplan.2017.06.023

Freeman, D. B. (1991). *City of farmers: Informal urban agriculture in the open spaces of Nairobi, Kenya*. MQUP. http://ebookcentral.proquest.com/lib/usyd/detail.action?docID=3330857

Gaynor, A. (2006). *Harvest of the suburbs: An environmental history of growing food in Australian cities*. University of Western Australia Press. https://trove.nla.gov.au/version/45001315

Göttl, I., & Penker, M. (2020). Institutions for collective gardening: A comparative analysis of 51 urban community gardens in anglophone and German-speaking countries. *International Journal of the Commons, 14*(1), 30–43. https://doi.org/10.5334/ijc.961

Guitart, D., Pickering, C., & Byrne, J. (2012). Past results and future directions in urban community gardens research. *Urban Forestry & Urban Greening, 11*(4), 364–373. https://doi.org/10.1016/j.ufug.2012.06.007

Hardman, M., Chipungu, L., Magidimisha, H., Larkham, P. J., Scott, A. J., & Armitage, R. P. (2018). Guerrilla gardening and green activism: Rethinking the informal urban growing movement. *Landscape and Urban Planning, 170*, 6–14. https://doi.org/10.1016/j.landurbplan.2017.08.015

Horst, M., McClintock, N., & Hoey, L. (2017). The intersection of planning, urban agriculture, and food justice: A review of the literature. *Journal of the American Planning Association, 83*(3), 277–295. https://doi.org/10.1080/01944363.2017.1322914

Hou, J. (2020). Governing urban gardens for resilient cities: Examining the 'Garden City Initiative' in Taipei. *Urban Studies, 57*(7), 1398–1416. https://doi.org/10.1177/0042098018778671

Hsu, J. P. (2018). The pedagogical life of edible verge gardens in Sydney: Urban agriculture for the urban food imaginary, Ph.D., University of Hawai'i at Manoa. In *ProQuest Dissertations and Theses* (2115845768). ProQuest dissertations & theses global. https://login.wwwproxy1.library.unsw.edu.au/login?qurl=https%3A%2F%2Fwww.proquest.com%2Fdocview%2F2115845768%3Faccountid%3D12763

Hung, H. (2017). Formation of new property rights on government land through informal co-management: Case studies on countryside guerilla gardening. *Land Use Policy, 63*, 381–393. https://doi.org/10.1016/j.landusepol.2017.01.024

Hunter, M. C. R., & Brown, D. G. (2012). Spatial contagion: Gardening along the street in residential neighborhoods. *Landscape and Urban Planning, 105*(4), 407–416. https://doi.org/10.1016/j.landurbplan.2012.01.013

Isendahl, C., & Smith, M. E. (2013). Sustainable agrarian urbanism: The low-density cities of the Mayas and Aztecs. *Cities, 31*, 132–143. https://doi.org/10.1016/j.cities.2012.07.012

James, S. (2016). *Farming on the Fringe: Peri-urban agriculture, cultural diversity and sustainability in Sydney*. Springer International Publishing. https://doi.org/10.1007/978-3-319-32235-3

Keshavarz, N., Bell, S., Zilans, A., Hursthouse, A., Voigt, A., Hobbelink, A., Zammit, A., Jokinen, A., Mikkelsen, B. E., Notteboom, B., Ioannou, B., Certomà, C., Schwab, E., Sentić, I., Barstad, J., Willman, K., Calvet-Mir, L., Baležentienė, L., Weirich, M., … Gogová, Z. (2016). A history of urban gardens in Europe. In *Urban allotment gardens in Europe*. Routledge, London.

Kingsley, J., & Townsend, M. (2006). 'Dig In' to social capital: Community gardens as mechanisms for growing urban social connectedness. *Urban Policy and Research, 24*(4), 525–537. https://doi.org/10.1080/08111140601035200

Kirkpatrick, J. B., & Davison, A. (2018). Home-grown: Gardens, practices and motivations in urban domestic vegetable production. *Landscape and Urban Planning, 170*, 24–33. https://doi.org/10.1016/j.landurbplan.2017.09.023

Kurtz, H. (2001). Differentiating Multiple Meanings of Garden and Community. *Urban Geography, 22*(7), 656–670. https://doi.org/10.2747/0272-3638.22.7.656

Lawson, L. (2005). *City bountiful: A century of community gardening in America*. University of California Press. http://ebookcentral.proquest.com/lib/usyd/detail.action?docID=227286

Lawson, L., & Drake, L. (2013). Community gardening organization survey 2011–2012. *Community Greening Review, 18*, 20–47.

Lopes, A. M., & Shumack, K. (2012). "Please Ask Us" conversation mapping as design research: Social learning in a verge garden site. *Design Philosophy Papers, 10*(2), 119–132. https://doi.org/10.2752/089279312X13968781797832

Lovell, S. T. (2010). Multifunctional urban agriculture for sustainable land use planning in the United States. *Sustainability, 2*(8), 2499–2522. https://doi.org/10.3390/su2082499

Marshall, A. J., Grose, M. J., & Williams, N. S. G. (2019). Footpaths, tree cut-outs and social contagion drive citizen greening in the road verge. *Urban Forestry & Urban Greening, 44*, 126427. https://doi.org/10.1016/j.ufug.2019.126427

McClintock, N. (2014). Radical, reformist, and garden-variety neoliberal: Coming to terms with urban agriculture's contradictions. *Local Environment, 19*(2), 147–171. https://doi.org/10.1080/13549839.2012.752797

McClintock, N., & Cooper, J. (2010). *Cultivating the commons an assessment of the potential for urban agriculture on Oakland's public land*. 75.

Meenar, M. R., Featherstone, J. P., Cahn, A. L., & McCabe, J. (2012). *Urban agriculture in post-industrial landscape: A case for community-generated urban design*. 14.

Mintz, G., & McManus, P. (2014). Seeds for change? Attaining the benefits of community gardens through council policies in Sydney, Australia. *Australian Geographer, 45*(4), 541–558. https://doi.org/10.1080/00049182.2014.953721

Mok, H.-F., Williamson, V. G., Grove, J. R., Burry, K., Barker, S. F., & Hamilton, A. J. (2014). Strawberry fields forever? Urban agriculture in developed countries: a review. *Agronomy for Sustainable Development, 34*(1), 21–43. https://doi.org/10.1007/s13593-013-0156-7

Morrow, O., & Martin, D. G. (2019). Unbundling property in Boston's urban food commons. *Urban Geography, 40*(10), 1485–1505. https://doi.org/10.1080/02723638.2019.1615819

Mougeot, L. J. A. (2000). *Urban agriculture: definition, presence, potentials and risks, and policy challenges*. 62.

Nettle, C. (2016). *Community gardening as social action*. Routledge. https://doi.org/10.4324/9781315572970.

Okvat, H. A., & Zautra, A. J. (2011). Community gardening: A Parsimonious path to individual, community, and environmental resilience. *American Journal of Community Psychology, 47*(3), 374–387. https://doi.org/10.1007/s10464-010-9404-z

Palau-Salvador, G., de Luis, A., Pérez, J. J., & Sanchis-Ibor, C. (2019). Greening the post crisis. Collectivity in private and public community gardens in València (Spain). *Cities, 92*, 292–302. https://doi.org/10.1016/j.cities.2019.04.005

Pinheiro, A., & Govind, M. (2020). Emerging global trends in urban agriculture research: A scientometric analysis of peer-reviewed journals. *Journal of Scientometric Research, 9*(2), 163–173. https://doi.org/10.5530/JSCIRES.9.2.20. Scopus.

Pires, V. (2011). *Planning for urban agriculture planning in Australian cities*. State of Australian Cities.

Pothukuchi, K. (2015). Five decades of community food planning in Detroit: City and grassroots, growth and equity. *Journal of Planning Education and Research*. https://doi.org/10.1177/0739456X15586630

Pourias, J., Aubry, C., & Duchemin, E. (2016). Is food a motivation for urban gardeners? Multifunctionality and the relative importance of the food function in urban collective gardens of Paris and Montreal. *Agriculture and Human Values; Dordrecht, 33*(2), 257–273. http://dx.doi.org.wwwproxy1.library.unsw.edu.au/10.1007/s10460-015-9606-y

Pudup, M. B. (2008). It takes a garden: Cultivating citizen-subjects in organized garden projects. *Geoforum, 39* (3), 1228–1240. https://doi.org/10.1016/j.geoforum.2007.06.012

Purcell, M., & Tyman, S. K. (2015). Cultivating food as a right to the city. *Local Environment, 20*(10), 1132–1147. https://doi.org/10.1080/13549839.2014.903236

Rogge, N., & Theesfeld, I. (2018). Categorizing urban commons: Community gardens in the Rhine-Ruhr agglomeration, Germany. *International Journal of the Commons, 12*(2), 251–274. https://doi.org/10.18352/ijc.854

Rosol, M. (2010). Public participation in post-fordist urban green space governance: The case of community gardens in Berlin. *International Journal of Urban & Regional Research, 34*(3), 548–563. https://doi.org/10.1111/j.1468-2427.2010.00968.x

Russo, A., Escobedo, F. J., Cirella, G. T., & Zerbe, S. (2017). Edible green infrastructure: An approach and review of provisioning ecosystem services and disservices in urban environments. *Agriculture, Ecosystems & Environment, 242*, 53–66. https://doi.org/10.1016/j.agee.2017.03.026

Sanyé-Mengual, E., Anguelovski, I., Oliver-Solà, J., Montero, J. I., & Rieradevall, J. (2016). Resolving differing stakeholder perceptions of urban rooftop farming in Mediterranean cities: Promoting food production as a driver for innovative forms of urban agriculture. *Agriculture and Human Values, 33*(1), 101–120. https://doi.org/10.1007/s10460-015-9594-y

Spijker, S. N., & Parra, C. (2018). Knitting green spaces with the threads of social innovation in Groningen and London. *Journal of Environmental Planning and Management, 61*(5–6), 1011–1032. https://doi.org/10.1080/09640568.2017.1382338. Scopus.

Tappert, S., Klöti, T., & Drilling, M. (2018). Contested urban green spaces in the compact city: The (re-) negotiation of urban gardening in Swiss cities. *Landscape and Urban Planning, 170*, 69–78. https://doi.org/10.1016/j.landurbplan.2017.08.016

Taylor, J. R., & Lovell, S. T. (2014). Urban home food gardens in the Global North: Research traditions and future directions. *Agriculture and Human Values; Dordrecht, 31*(2), 285–305. http://dx.doi.org.wwwproxy1.library.unsw.edu.au/10.1007/s10460-013-9475-1
Thompson, M. (2015). Between boundaries: From Commoning and Guerrilla gardening to community land trust development in Liverpool. *Antipode, 47*(4), 1021–1042. https://doi.org/10.1111/anti.12154
Thornton, A. (2020). *Urban food democracy and governance in North and South*. Springer International Publishing. https://doi.org/10.1007/978-3-030-17187-2.
Tornaghi, C. (2014). Critical geography of urban agriculture. *Progress in Human Geography, 38*, 551–567. https://doi.org/10.1177/0309132513512542
Tracey, D., Gray, T., Sweeting, J., Kingsley, J., Bailey, A., & Pettitt, P. (2020). A systematic review protocol to identify the key benefits and associated program characteristics of community gardening for vulnerable populations. *International Journal of Environmental Research and Public Health, 17*(6), 2029. https://doi.org/10.3390/ijerph17062029
van Holstein, E. (2020). Strategies of self-organising communities in a gentrifying city. *Urban Studies, 57*(6), 1284–1300. https://doi.org/10.1177/0042098019832468
Zavisca, J. (2003). Contesting capitalism at the Post-Soviet Dacha: The meaning of food cultivation for urban Russians. *Slavic Review, 62*(4), 786–810. https://doi.org/10.2307/3185655
Zeunert, J., Waterman, T., & Waterman, T. (2018). *Routledge handbook of landscape and food*. Routledge. https://doi.org/10.4324/9781315647692.
Zhu, J., He, B.-J., Tang, W., & Thompson, S. (2020). Community blemish or new dawn for the public realm? Governance challenges for self-claimed gardens in urban China. *Cities, 102*, 102750. https://doi.org/10.1016/j.cities.2020.102750

Urban Food Production

► Role of Urban Agriculture Policy in Promoting Food Security in Bulawayo, Zimbabwe

Urban Forestry

► Urban Forestry in Sidewalks of Bogota, Colombia

Urban Forestry in Sidewalks of Bogota, Colombia

Laura Patricia Otero-Durán
Urban Development Institute, Bogotá, Colombia

Synonyms

Green infrastructure; Landscaping; Nature-based solutions; Public space; Urban forestry; Urban planning

Definition

The urban infrastructure is a system made up of the natural network (green infrastructure) and the built infrastructure (grey infrastructure). Hence, the European Strategy on Green Infrastructure urges member states to "ensure that the protection, restoration, creation and enhancement of green infrastructure becomes an integral part of spatial planning and territorial development whenever it offers a better alternative, or is complementary, to standard grey choices" (European Commission 2019).

According to the Food and Agriculture Organization of the United Nations, the green infrastructure backbone are the urban forests comprising all woodlands, groups of trees, and individual trees located in urban and peri-urban areas (FAO 2016). Within the categories of urban forests are peri-urban forests, urban parks greater than 0.5 Ha, pocket parks and gardens with trees less than 0.5 Ha, urban road trees, and public spaces and other green spaces with trees such as fields, open and botanical gardens (FAO 2016).

Considering that, street corridors have a high potential to host green infrastructure, this chapter will address the case of urban trees on the Bogota sidewalks due to the multiple environmental, economic, and social benefits they provided.

Introduction

The Sustainable Development Goal number 11 "Sustainable Cities and Communities" highlights the importance of improving urban management so that cities are more resilient and sustainable. Similarly, the New Urban Agenda recognizes infrastructure as a driver of cost efficiency and resource use by promoting environmental protection and highlights the commitment in land use planning to improve urban ecosystems and their environmental services.

Likewise, the Nature based Solutions (NbS) are actions to protect, manage sustainably, restore natural ecosystems, or modified and thus effectively resolve social challenges, while adaptive and simultaneously provide human well-being and benefits derived from biodiversity (IUCN 2016).

In the framework of the NbS, the infrastructure approach makes it possible to address urban problems such as the hardening of public space that affects the urban water balance by increasing runoff (Figueroa Arango 2020). This is done through green infrastructure (GI) solutions such as vegetated surfaces, green roofs, public parks, green walls, urban forests, green alleys and streets, community gardens, and urban wetlands and Sustainable Drainage systems (SuDS) (Van Oijstaeijen et al. 2020).

In that sense, it is important to consider that green infrastructure is a multiscale territorial planning tool that serves ecological, productive, and cultural functions while contributing to the resilience of the territories (Calaza Martínez 2019). If we considered that street corridors configure a network to connect citizens through different modes of transport, they are a perfect opportunity to promote the connectivity of urban biodiversity through the forestry in sidewalks.

Urban Forestry in Bogotá

Local Context

Latinoamerica is the most urbanized development region in the world, with the 80% of the population in urban zones (United Nations 2016) in Colombia by 2018 more than 77% of the population were in the municipal capitals (DANE 2018). Bogotá is the capital city of Colombia, located at 2650 m, with a rainfall multiyear averages spatial variation in the urban area, ranging from 600 mm to 1430 mm per year, with two peaks of rain and two dry (IDIGER 2021) Similarly, the average monthly temperature varies between 12 °C and 14 °C, which denotes the existence of different climatic zones throughout the city (Domínguez et al. 2010).

In the last decades, a strong demographic explosion has taken place, and the consequent urbanization of rural areas, hence the green areas have been reduced considerably, with loss of spaces for recreation and conservation and the increase of the atmospheric pollution. Temperature evaluation studies explain the existence of a heat island in the center of the city of Bogotá, whose formation dates to the 1970s, which expands from the center to the north of the city and exceeds in three degrees the climate of the periphery (Domínguez et al. 2010).

Bogotá is a highly vulnerable city to the effects of climate change, due to the high levels of vulnerability of the city and its population and the low adaptation capacity of its institutions. In the capital region, an increase of between 6% and 8% in rainfall is expected for the period 2011–2100 and of more than 2 °C in the average temperature (IDEAM 2017).

Bogotá has 163.635 Ha and those are distributed in 20 administrative locations and three types regarding land use planning: 23% is urban, 75% rural, and 2% expansion land (SDP 2020). According to the national population census carried out in 2018 and population projections generated by the National Statistics Department, by 2020 Bogotá has 7.743.955 inhabitants (SDP 2020).

Since 1997 the national government issued the 388 law, which established the territorial order plans looking for development and sustainable. In 1998 the decree 1504 defined that the public space is made up of: (1) constitutive elements, which include natural elements (such as hills, wetlands, and rivers), and built elements such as sidewalks and road separators and (2) complementary elements such as the vegetation used in gardens and trees.

In the year 2000, conservation biologists in Colombia started to recognize the conservation value of urban green spaces through the first Territorial Ordering Plan (POT) and incorporated the concept of Ecological Main Structure (EMS) in the sense of emphasize biological connectivity over other environmental and social functions (Andrade et al. 2013).

In the Bogota context, the urban street trees (ST) are considered part of the natural elements of the public space, that's why in the first POT approved in 2000 the built public space system is made up of district parks and pedestrian public spaces for staying, meeting, or moving. Regarding the location of the urban trees in the system of built public space, there are regulations for the spatial distribution in the cross sections of the roads. Hence, the urban forestry has been legally recognized as an essential part of urbanism in the city.

Nevertheless, the provision of effective public space is below the desirable standards, reaching 4.53 m^2 per inhabitant, and there is no functional articulation of this with the areas of environmental importance that enhances its values in the city (SDA 2019). Also, the low urban arborization, on average 0.17 trees per inhabitant, causes that in some parts of the city, there are islands of heat and atmospheric conditions and poor urban quality (SDA 2019). It is important to note that the trees distributed along the city are concentrated in the high socioeconomic sectors, so the greenery in the city can also be seen as an inequality indicator.

Urban Forestry Management

It is worth mentioning that during the planning and design process, there is a local authority, District Planning Secretariat (SDP), in charge of defining the models for the public space elements, in the places of permanence and the circulation corridors for all modes of transport. While the definition of the design of urban forestry projects corresponds to the Botanical Garden of Bogotá (JBB).

Furthermore, the management of public space and urban trees is distributed among several public entities that oversee the planning, implementation, and maintenance as shown in Fig. 1. For this reason, inter-institutional coordination is essential to achieve the objectives of tree planting and contribute to the quality of urban life.

In addition, the District Environment Secretary (SDA) local environmental authority, adopted the Plan for Urban Forestry, Green Zones and Gardening for Bogotá D.C. (2019–2030), with the main purpose of consolidate trees, green areas, and gardening as integrating and structuring elements of the urban and environmental design of the city. That document relies on the establishment for the green infrastructure in the city where the forestry is in the center of some of the main politics in the Territorial Ordering Plan, as the Fig. 2 illustrates.

Meanwhile, the responsibility to manage the inventory of the urban trees in the city lies with the Botanical Garden of Bogota (JBB); hence, they created an interactive tool than can be consulted in the web, which is possible because every single tree in the city has an identification code. That is the Information System for the Management of Urban Trees (SIGAU) and thanks to this, the environmental authority took the SIGAU data to propose actions related to forestry, green areas, and gardening (Fig. 3).

The tree cover occupies 2.47% of the urban area of Bogotá and the distribution of arboreal individuals according to their size indicates that 82.26% are arboreal with an average height of 4.7 m., while the proportion of shrubs is 13.23% and that of palms is 3.94% (SDA 2019). In the SIGAU, there is no information about the about the herbaceous stratum due to its difficulty in keeping the inventory and location updated due to the dynamics of urbanization. Further analysis can be done to establish the multistratum index in the city.

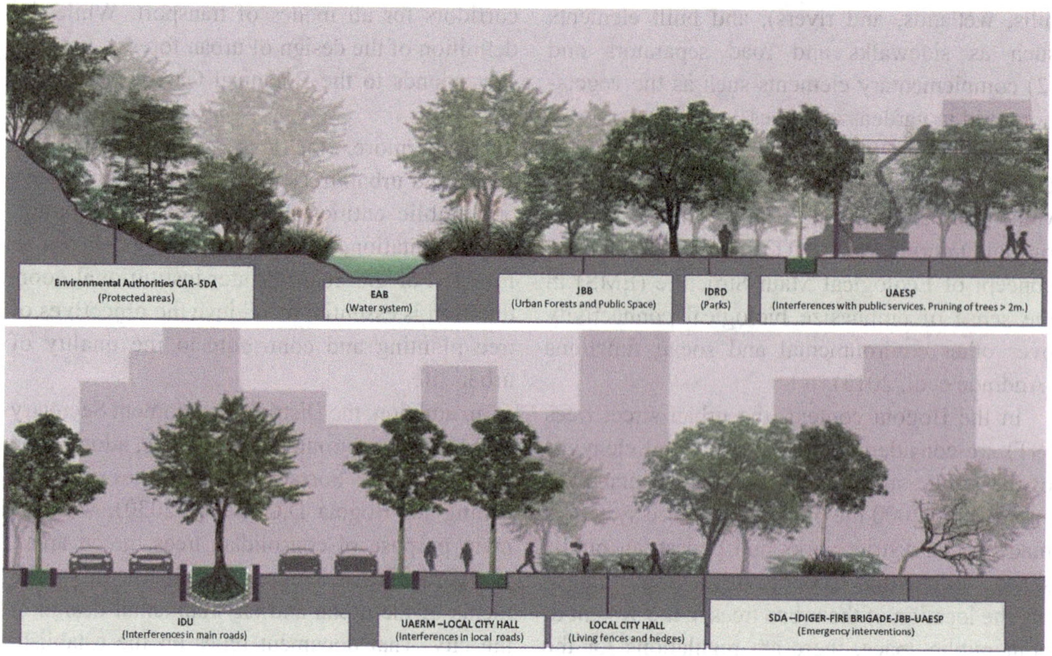

Urban Forestry in Sidewalks of Bogota, Colombia, Fig. 1 Urban forestry competences in Bogotá according to Decree 531–2010 and 383–2018. (Source: Adapted from "Technical guidelines for the vegetal cover in the green corridor of Carrera Septima". JBB 2021)

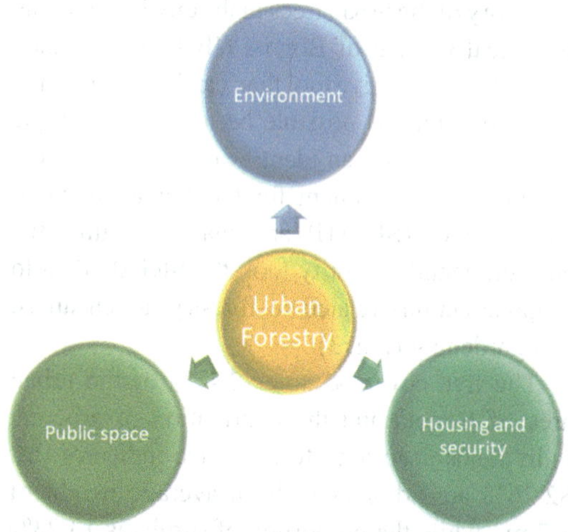

Urban Forestry in Sidewalks of Bogota, Colombia, Fig. 2 Urban Forestry in the politics of the Territorial Urban Plan. (Source: Author from District Plan for Urban Forestry, green zones, and gardening for Bogota (2019–2030))

Regarding the distribution of the 1.340.145 trees, according to the SIGAU data by August 2021, 32.21% are in the urban circulation system, 27.6% are in the recreational system, 22.54% in the canals and water rounds, 17.23% in the protection system, and 0.42% in right-of-way strips and degraded areas (https://sigau.jbb.gov.co/, August 2021).

The Interference Approach: Trees in the Sidewalks and Silvicultural Treatments

The trees incorporated into the built public space configures an hybrid system in which natural and built elements interact (Alberti 2008); thus, ecosystem services of provision, regulation, and culture are generated, as well as problems or dis-services associated with their characteristics and location. The main dis-service is the interference with the grey infrastructure because that is the principal cause of cutting trees in the public space.

From this approach, it is useful to study the characterization of the existing trees on the

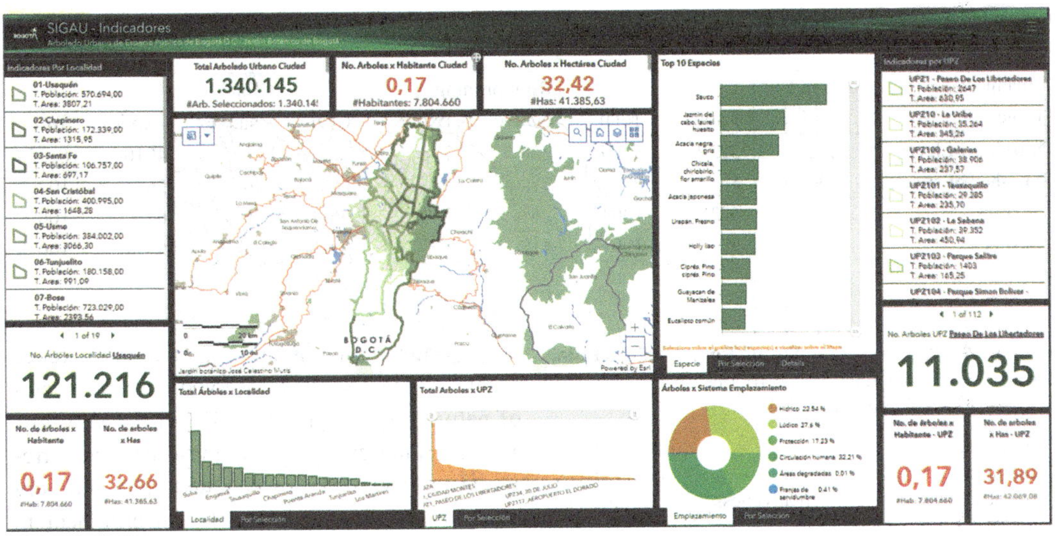

Urban Forestry in Sidewalks of Bogota, Colombia, Fig. 3 Urban forestry indicators in Bogotá. (Source: Information System for the Management of Urban Trees (SIGAU) JBB. https://jbb.maps.arcgis.com/apps/MapSeries/index.html?appid=ae3ab3570dcb4a8ab2b2acfbb9607e00, August 2021)

Urban Forestry in Sidewalks of Bogota, Colombia, Table 1 Forestry Characterization in sidewalks of Bogota. (Source: Otero-Durán 2021)

Origin / Topic	Native	Exotic	Total
Number of trees	54.967	154.924	209.891
Number of species	112	122	234
% trees	26%	74%	100%
% species	48%	52%	100%

sidewalks and of the silvicultural permits for root pruning and cutting for each species regarding the roads interference. The results of a study carried out in April 2020 based on data provided by the entity in charge of silvicultural permits and the inventory of trees located on platforms are discussed below.

First, the characterization obtained from the 209,891 arboreal individuals present on the sidewalks is shown in Table 1. It is observed that although the diversity of exotic species is similar to the native ones, there is a predominance of arboreal individuals belonging to exotic species that represent 74%, compared to the natives that reach only 26% of the total.

On the other hand, regarding the silvicultural treatments, it is important to note that the records corresponding to road reserve areas for new road projects were eliminated, as they were not considered relevant for the interferences approach.

As a result of the analysis of the information provided by the SDA by April 2020, for silvicultural treatments applied to arboreal individuals on platforms, the following characterization was obtained. It is noted that some species were subject only to cutting, others only to root pruning, and others to both types of treatments; the total number of interventions was 6235 and corresponds to 102 species.

Second, for each species subject to silvicultural treatment of cutting or root pruning in Table 2, the number of interventions was quantified, then the species were ordered from highest to lowest according to their percentage of participation in the total of interventions carried out between 2016 and April 2020. Afterwards, the data were separated by quartiles, to give the level of interference to the species as follows: low first quartile, middle second and third quartile, and high last quartile.

On the other hand, to determine the representativeness of the 102 species subject to

Urban Forestry in Sidewalks of Bogota, Colombia, Table 2 Silvicultural permissions 2016–2021 characterization according with interference with main roads. (Source: Otero-Durán 2021)

Treatment	Cut			Root pruning			Total		
Topic	Native	Exotic	Total	Native	Exotic	Total	Native	Exotic	Total
Number of trees	1957	3252	5209	268	758	1026	2225	4010	6235
Number of species	43	56	99	17	24	41	46	56	102
% trees	38%	62%	100%	26%	74%	100%	36%	64%	100%
% species	43%	57%	100%	41%	59%	100%	45%	55%	100%

Urban Forestry in Sidewalks of Bogota, Colombia, Table 3 Potential Interference of tree species in sidewalks. (Source: Otero-Durán 2021)

Interference according to treatments by species (1)	Representativeness (2)	Potential interference according to (1) y (2)
Low	Low	Very low
	Medium	Low
	High	Medium
Medium	Low	Low
	Medium	Medium
	High	High
High	Low	Medium
	Medium	High
	High	Very high

silvicultural treatment and located in the site under study, the information regarding the number of arboreal individuals of each species present on the sidewalks was ordered highest to lowest according to their percentage of participation in the total number of individuals on sidewalks. Later, the data were separated by quartiles, to grant the representativeness to the species as follows: low first quartile, middle second and third quartile, and high last quartile.

Then, the two criteria reviewed so far were combined, and it was assumed that the number of individuals of the species affects the potential level of interference, and a new qualitative classification was proposed with interference categories ranging from very low to very high, as shown in Table 3.

So, after applying the potential interference criteria to the data, it is observed that the species of native origin represents 61% of the very low and low interference, while those exotics correspond to 75% of the species with high and very high interference. This can be an argument to recommend the selection of native species as a protection measure for the public space infrastructure (Table 4).

Conclusions

Arborization must balance ecological, landscape, social, economic, functional, and technical aspects. In addition, the selection of species must consider the adaptability to each humidity zone, the type of space to be planted, and the possible interferences with elements present such as public utility networks, the objective pursued with arborization, the physiological characteristics of the species, and its vulnerability to urban pollution (JBB 2010).

Adopting and interference approach in urban forestry planning can improve the associate ecosystem services and minimize the disservices, for example, the simple decision of the origin of the species can affect the protection of the infrastructure. In Bogota, 64% of the trees with silvicultural treatment belong to exotic species as seen in Table 2, which may indicate that this type of flora generates greater interference with the grey infrastructure. (see chapter ▶ "Integrated Urban Green and Grey Infrastructure")

However, to verify this statement, it is necessary to analyze the growth rates of these species in

Urban Forestry in Sidewalks of Bogota, Colombia, Table 4 Species distribution according to potential interference with infrastructure. (Source: Otero-Durán 2021)

Origin Potential interference	Native	%	Exotic	%	Total
Very low and low	25	61	16	39	41
Medium	13	45	16	55	29
High and very high	8	25	24	75	32

the conditions of Bogotá, which differ from their environments of origin, and consider that the development of trees can be faster in the tropics than in places with seasonal changes. (see chapter ▶ "Green Cities: Nature-Based Solutions, Renaturing and Rewilding Cities")

In addition, it is important to note that only 3% of the arboreal individuals existing on sidewalks were subjects of silvicultural treatments in the study period. Then, to deepen the interference approach, it is necessary to review the state of the road corridors with damage to the infrastructure caused by the trees and cross it with the data of the species present there. (see chapter ▶ "Multiple Benefits of Green Infrastructure")

It is also important to consider that the diversity of native species in urban trees is decisive to prevent the homogenization of urban fauna (Alvey 2006). However, the data analyzed shows in Bogotá the privilege of exotic species, which can affect urban biodiversity, given that these sidewalk corridors are key in functional connectivity between peri-urban areas and patches of vegetation in urban areas (Angold et al. 2006).

References

Alberti, M. (2008). Advances in urban ecology: integrating humans and ecological processes in urban ecosystems (No. 574.5268 A4). New York: Springer.

Alvey, A. A. (2006). Promoting and preserving biodiversity in the urban forest. *Urban Forestry and Urban Greening, 15*, 153–164.

Andrade, G. I., Remolina, F., Wiesner, D. (2013). Assembling the Pieces. A framework for the integration of multifunctional ecological main structure in the emerging urban region of Bogotá, Colombia. Urban Ecosystems (Springer). https://doi.org/10.1007/s11252-013-0292-5

Angold, P. G., Sadler, J. P., Hill, M. O., Pullin, A., Rushton, S., Austin, K., et al. (2006). Biodiversity in urban habitat patches. *Science of the Total Environment, 360*, 196–204.

Calaza Martínez, P. (2019). Guía de la Infraestructura verde municipal. https://www.aepjp.es/wpcontent/uploads/2019/07/AEPJP-Guia-Biodiversidad.pdf

Departamento Administrativo Nacional de Estadística DANE. Censo Nacional de Población y Vivienda 2018. https://www.dane.gov.co/index.php/estadisticas-por-tema/demografia-y-poblacion/censo-nacional-de-poblacion-y-vivenda-2018

Domínguez, E., Ángel, L., Ramírez, A. (2010). Isla de calor y cambios espacio-temporales de la temperatura en la ciudad de Bogotá. *Revista de la Real Academia de Ciencias Exactas, Físicas y Naturales – Serie A: Matemáticas, 34*(131), 173–183. ISSN 0370-3908. https://www.accefyn.com/revista/Vol_34/131/173-183.pdf

European Commission. (2019). Ecosystem services and green infrastructure. Retrieved from https://ec.europa.eu/environment/nature/ecosystems/index_en.htm

FAO. (2016). Directrices para la silvicultura urbana y periurbana, por Salbitano, F., Borelli, S., Conigliaro, M. y Chen, Y. 2017. Directrices para la silvicultura urbana y periurbana, Estudio FAO: Montes N° 178, Roma, FAO. http://www.fao.org/3/b-i6210s.pdf

Figueroa Arango, C. (2020). *Guía para la integración de las Soluciones Basadas en la Naturaleza en la planificación urbana. Primera aproximación para Colombia*. http://hdl.handle.net/20.500.11761/35585

Instituto de Hidrología, Meteorología y Estudios Ambientales IDEAM. Caracterización Climática de Bogotá y Cuenca Alta del Río Tunjuelo. 2017. http://www.ideam.gov.co/documents/21021/21135/CARACTERIZACION+CLIMATICA+BOGOTA.pdf/d7e42ed8-a6ef-4a62-b38f-f36f58db29aa

Instituto Distrital de Gestión de Riesgos y Cambio Climático IDIGER. (2021). Caracterización Climatológica De Bogotá, Como Un Aporte Al Fortalecimiento De La Red Hidrometeorológica De Bogotá (RHB). https://www.idiger.gov.co/documents/20182/558631/Caract+Climatol%C3%B3gica+-+Bogot%C3%A1+%281%29.pdf/b5dbcea1-d291-40a0-8ee8-71ca322edcab

U

IUCN. (2016). CEM work on nature-based solutions. https://www.iucn.org/commissions/commission-ecosystem-management/our-work/nature-based-solutions

Jardín Botánico de Bogotá José Celestino Mutis JBB. (2010). Manual de Silvicultura Urbana de Bogotá. Decreto 531 de 2010.

Jardín Botánico de Bogotá José Celestino Mutis JBB, Universidad de los Andes. (2021). Manual de Coberturas Vegetales de Bogotá D.C.

Otero-Durán, L. P. (2021). *El arbolado urbano protector de la infraestructura de espacio público en Bogotá*. Recuperado de: http://hdl.handle.net/10554/53173

Secretaría Distrital de Ambiente SDA. (2019). Plan Distrital de Silvicultura Urbana, Zonas Verdes y Jardinería para Bogotá (2019–2030).

Secretaría Distrital de Planeación SDP. (2020). Documentos técnicos de soporte del Plan de Ordenamiento Territorial de Bogotá. http://www.sdp.gov.co/micrositios/pot/documentos

United Nations Habitat III Secretariat. (2016). Regional Report for Latin America and the Caribbean. https://uploads.habitat3.org/hb3/HabitatIII-Regional-Report-LAC.pdf

Van Oijstaeijen, W., Van Passel, S., & Cools, J. (2020). Urban green infrastructure: A review on valuation toolkits from an urban planning perspective. *Journal of Environmental Management, 267*, 110603. https://doi.org/10.1016/j.jenvman.2020.110603

Further Reading

Otero-Durán, L. P. (2021). *El arbolado urbano protector de la infraestructura de espacio público en Bogotá*. In: http://hdl.handle.net/10554/53173

Secretaría Distrital de Ambiente. (2019). Plan Distrital de Silvicultura Urbana, Zonas Verdes y Jardinería para Bogotá (2019–2030). https://oab.ambientebogota.gov.co/?post_type=dlm_download&p=18398

Urban Forests

► Stewarding Street Trees for a Global Urban Future

Urban Futures

► Green and Smart Cities in the Developing World
► The Governance of Smart Cities
► Understanding Urban Engineering

Urban Futures: Pathways to Tomorrow

Nick Dunn
Lancaster University, Lancaster, UK

Synonyms

Future cities

Definitions

Urban futures have two principal components. The first responds to the significant transformation of the world in becoming an increasingly urban place. Over the last century, cities and suburbs worldwide have seen gradual, and in some instances rapid, population growth. The second refers to the unfolding or unknown conditions of how such areas, and those yet to still emerge, will be in the future. This latter aspect is often alloyed to technological developments due to the way futures are more widely perceived as being synonymous with technology. This is because these innovations have fundamentally changed how we live in the past, as well as where and why, and this often continues to be the case. However, profound changes in either of these two components present major challenges for urban areas, including cities. Urban futures, therefore, is a field that explores ways to tackle challenges in relation to climate change, social inequalities, pollution, lack of access to transportation and utility infrastructures, shortage of affordable housing, ageing, etc. Given the complexity of these challenges, many of which are interrelated, it is perhaps no surprise that there are a variety of ways through which urban futures are conceived, designed, and implemented. These may share similar goals yet different drivers behind types of urban future has led to their characterization including: equitable, healthy, resilient, smart, sustainable, etc. The creation of urban areas for unpredicted and uncertain futures, therefore, requires and involves a broad range of disciplines.

Introduction

Urbanization has resulted in the significant transformation of our planet. Indeed, the scale, scope, and speed of urbanization over the last few decades are unprecedented. The publication of the UNFPA's 2007 report identified a watershed moment since it recognized that, for the first time in history, more people around the world lived in cities and towns than in rural areas. This population shift coupled with the concentration of interrelated systems that urban areas often represent is having profound effects upon our environment, our health, and our societies. Urban landscapes are complex. Their combination of people, economy, and technology is frequently positioned as positive, providing as it does the opportunities for us to flourish and change our lives in positive ways. However, it is also the dynamic interactions between these elements that result in many of the problems in urban landscapes.

To understand the scale of this transition, the 2018 report by the UN DESA Population Division (UN DESA 2018) projects that by 2050, 68% of the world's population will be urban. Across the next few decades of the twenty-first century, numerous nations of Africa, Asia, and Latin America will continue in their efforts to manifest their urban futures. Crucially though, many of these countries do not seek to reproduce the old urbanism of 50 years ago but are instead investigating urban futures across a range of measures. Sustainability, in its many forms, is typically a key goal of these urban futures yet it is challenged by rapid growth. This is particularly true of those contexts where basic infrastructures cannot be assembled to support it. This has led to the vast number of urban poor, the hundreds of millions of people living in substandard conditions with huge health implications (WHO and UN-Habitat 2016). Urban futures is a multi-disciplinary field that investigates different ways to address such challenges. These issues may relate to environmental, social, technological, or political dimensions, though they are frequently a combination of more than one of these. Given the difficulty and scale of these problems, some are global challenges whereas others are more specific and local, there are a number of approaches for how urban futures are conceived and implemented. What is increasingly apparent is that there is no one-size-fits-all approach and urban futures will be complex, diverse, and heterogeneous.

Working with the Future

From the beginning of time, we have longed to know what is ahead of us. For thousands of years we have sought to predict, control, manage, and understand the future. Whether through astrology, philosophy, creative visions, and scientific analysis of data to identify patterns and trends, we have attempted to know the future (Gidley 2017). Indeed, the rapid growth in the use of the term "urban futures" since the mid-1960s is reflective of the accelerated rate of change that we have faced in the interim period. Previously accepted certainties such as ecological stability have become increasingly complex, ambiguous, uncertain, and, in some instances, volatile. Due to the major advances in technology and changes in societies and environment throughout the twentieth century, it also resulted in the latter being characterized as the first century to be arguably be more preoccupied with the future than the past (Samuel 2009). This has meant that efforts to comprehend the future and respond to its challenges are increasingly difficult.

There are various methods for thinking about the future. The majority of them are based on the four-stage methodological approach developed by Richard Slaughter (1997). The benefits of this approach lie in its flexibility and choice within each major step, and that the methods are integrated into a generic foresight process that can be applied to many different situations. Given its generous scope, the four stages – input methods, analytic methods, paradigmatic methods, and iterative and exploratory methods – feature in many of the futures methods found elsewhere. Of particular relevance to urban futures are iterative and exploratory methods which include backcasts, scenarios, and visioning. Joseph Voros (2003) has referred to these methods as being forms of prospective methods that seek to produce future images. Many futurists use visioning as a method to encourage potential futures by creating and describing alternative futures from business-as-usual ones. This is the principal value of iterative and exploratory methods, to reveal new ideas and challenge existing, sometimes deeply entrenched, assumptions about the future using imagination and creativity. The role of visualization as part of these techniques is often vital to enabling those exploring potential futures to articulate their ideas and share them. In their expression of the "not-yet," Fred Polak (1973) has suggested that such imagery shapes our ideas of, and intentions toward, futures. This resonates with what Montfort (2017) has termed "future-making," the act of imagining a specific future and deliberately seeking to contribute to it.

Plurality of Futures

The idea that we are all headed toward one common point has been questioned, especially since the mid-twentieth century when the scientific positivist concept of a single future could be usefully challenged. Many futurists realized that attempting to predict the future in this manner was not the most valuable approach for thinking about the complexity of the world. As the field of future studies developed, Galtung (1982) was an early advocate of different types of futures, referring to probable futures, possible futures, and preferred futures. Probable futures usually relate to extrapolation of trend data and toward more negative forecasts. Possible futures utilize imagination in the production of alternative visions. Preferred futures concern critical and normative values. Bjerstedt (1982) identified a fourth type, prospective futures, which related to activism when faced with probable futures. Important in the context of this chapter is the overall shift from positivism to pluralism. The former is rooted in the hard sciences and predicates an empiricist approach to "the future". The latter originates from the social sciences and promotes a diversity of approaches to "multiple futures". This is particularly relevant to urban futures.

Perspectives on urban futures are diverse, but a common thread connecting them is the recognition that cities are complex systems comprising of many interconnected parts that together can result in unintended consequences. In practice this requires thinking about how strategies and policies applied to shape urban communities might have ripple effects throughout that system. Because of the way cities are systems of systems in themselves yet are also typically integrated within wider national and global systems and networks, these effects are not necessarily limited to one place. Urban futures will not be distributed homogeneously. Multiple futures will coexist, as there are currently multiple presents across different geographies and societal contexts, experienced differently by those in them (Sardar 2010; Savransky et al. 2017). There is considerable work being done to find the most appropriate way to engage as many stakeholders as possible in their respective futures.

Pathways to Tomorrow

Urban futures, therefore, is a field that explore ways to tackle challenges in relation to climate change, social inequalities, pollution, lack of access to transportation and utility infrastructures, shortage of affordable housing, ageing, etc. Given the complexity of these challenges, many of which are interrelated, it is perhaps no surprise that there are a variety of ways the urban futures are conceived, designed, and implemented. These may share similar goals but different drivers behind types of urban future has led to their characterization including: equitable, healthy, resilient, smart, sustainable, etc. The creation of urban areas for unpredicted and uncertain futures, therefore, requires and involves a broad range of disciplines. With increased pressures on urban settlements, both internal (inequalities, population increases, density, etc.) and external (climate change, extreme weather, resource security, etc.), it is becoming more and more urgent for us to think, plan and envision for long-term futures (Karuri-Sebina et al. 2016; Boyko et al. 2020). It is important to highlight that urban futures work is rarely about predictions but instead draws upon analysis from the "science of cities" (UK Government Office for Science 2016; Batty 2017) to inform its approaches.

Perhaps the most compelling way urban futures are conceptualized and communicated is through visions. Throughout history many architects, artists, designers, planners and writers have imagined how urban futures might be (Goodman 2008; Dunn et al. 2014). It is through the midwifing of ideas about potential future places where "urban futures" as a term overlaps considerably with "future cities" (Moir et al. 2014). The conceptual origins of an "ideal future city" are commonly traced back to the imaginary cities of Thomas More's (1516) *Utopia* and Francis Bacon's (1627) *New Atlantis*, though the idea is evident in the writings of Plato. Numerous works of fiction have explored notions of urban futures, such as H.G. Wells (1905), Aldous Huxley (1932), and J.G. Ballard (1975). The impact of such fictional work should not be underestimated since it can often explore urban futures that not only speculates about how we might live, where and why but also can provide critiques of contemporary societies (Abbott 2016). Parallel to this literary strand has been a fertile seam of architects, planners, and urban designers investigating ways in which urban futures might form (Hall 2014). These visions for urban futures range from incremental developments to existing cities to radical ideas for urban life (Dahinden 1972; Banham 1976; Fishman 1982).

Whether as visions for real, constructed places, un-built ideas or fiction, the production of these urban futures is crucial for being able to learn about the past and better orientate ourselves for what comes next (Dunn and Cureton 2020). Far from being emancipatory exercises or flights of fantasy, visions for urban futures can enable us to examine what our cities could and should be like. In addition, they can provide us with a method through which we can explore how such futures will operate, including what infrastructural and governance systems might be needed to deliver them. This is particularly valuable if we want to investigate alternative urban futures that lie beyond existing path dependencies which are often driven by technology alone (Urry 2016; Dunn 2018). It is increasingly recognized that participatory and inclusive approaches to the formation of visions for urban futures are needed. In particular, three core principles of participation, co-production, and engagement are vital to this process (Pollastri et al. 2018; Dixon and Tewdwr-Jones 2021) bringing together government, industry, academia, civil society, and others.

Visions for urban futures can, therefore, represent shared expectations about how the places where we will live, work, and play in the long term will be. They contribute toward objective and strategic methods by being both expressive and instrumental to the way futures are articulated. There is no illusion that visions alone will bring about the urgent and important transitions needed if we are to secure a sustain urban areas that cause minimal or, ideally, zero degradation to our planet and the other species we share it with. Yet, the practices that the visioning process for urban futures entails – conceptualization, envisioning, and performing – are also vital to our capability to responding to their growing complexity (Neuman and Hull 2009). Visioning has the scope to provide urban futures that have integrity and are able to capture and articulate the complexity of dynamic situations and relevant contextual characteristics.

In order to formulate robust pathways to tomorrow, suitable delivery mechanisms for the translation of vision into actions to achieve these futures are needed (Rogers 2018). Making urban futures visible requires us to examine the history of the future, question its dominant voices and better understand those that have been marginalized, underrepresented, silenced or unable to speak (Sand 2019). The recent coronavirus pandemic has profoundly disrupted what we thought we knew about cities which makes the field of urban futures even more significant for how we navigate our way through the further challenges that lie ahead. Although the impacts of the pandemic may prove to be ongoing in the medium-term, the effects of climate change will be a long-term crisis for urban areas. It is imperative we find ways to tackle such 'wicked problems' in urban settlements (Rittel and Webber 1973). Urban futures thus provide an opportunity to develop a practical framework to conceive, design and deliver the equity, safety, security, and sustainability that is crucial for tomorrow's world and the future generations that will inhabit it.

Summary

The global pattern of increased urbanization over the last few decades in particular has led numerous commentators to reasonably assume the future will be urban. Urban futures is a multi-disciplinary field that investigates different ways to address the challenges we face in the medium- and long-term. Such challenges may relate to environmental, social, technological or political dimensions, though they are often a combination of more than one of these. Urban futures work draws from the wider field of future studies. Core to its approaches are iterative and exploratory methods including visioning as an effective way to bring different stakeholders together to share their expectations. There is a long history of thinking about urban futures. However, the accelerated rate of change around the world since the mid-1960s has led to greater efforts to investigate urban futures. Visions are perhaps the most compelling way through which urban futures are conceptualized and communicated. These can take a variety of forms from radical proposals to incremental developments and are expressed through a range of media. It is increasingly acknowledged that core principles of participation, co-production, and engagement are essential for visioning process to be effective. Visions fit within a wider framework to ensure the delivery of an equitable, safe, secure, and sustainable world for future generations. Urban futures, therefore, is a burgeoning and increasingly significant field of study for how pathways to tomorrow are conceived, designed, and implemented.

Cross-References

► Age-Friendly Future Cities
► Augmented Reality: Robotics, Urbanism, and the Digital Turn
► City Visions: Toward Smart and Sustainable Urban Futures
► Community Engagement for Urban and Regional Futures
► Feminist Planning and Urbanism: Understanding the Past for an Inclusive Future
► Spatial Justice and the Design of Future Cities in the Developing World

References

Abbott, C. (2016). *Imagining urban futures: Cities in science fiction and what we might learn from them*. Middletown: Wesleyan University Press.

Bacon, F. ([1627] 2010). *The New Atlantis*. Ocean: Watchmaker Publishing.

Ballard, J. G. (1975). *High rise*. London: Jonathan Cape.

Banham, R. (1976). *Megastructure: Urban futures of the recent past*. London: Thames and Hudson.

Batty, M. (2017). *The new science of cities*. Cambridge, MA: The MIT Press.

Bjerstedt, A. (1982). *Future consciousness and the school*. Malmo: School of Education, University of Lund.

Boyko, C., Cooper, R., & Dunn, N. (Eds.). (2020). *Designing future cities for wellbeing* (1st ed.). London: Routledge.

Dahinden, J. (1972). *Urban structures for the future*. London: Pall Mall Press.

Dixon, T. J., & Tewdwr-Jones, M. (2021). *Urban futures: Planning for foresight and city visions*. Bristol: Policy Press.

Dunn, N. (2018). Urban imaginaries and the palimpsest of the future. In C. Lindner & M. Meissner (Eds.), *The Routledge companion to urban imaginaries* (1st ed., pp. 375–386). London: Routledge.

Dunn, N., & Cureton, P. (2020). *Future cities: A visual guide*. London: Bloomsbury.

Dunn, N., Cureton, P., & Pollastri, S. (2014). *A visual history of the future*. London: Foresight Government Office for Science, Department of Business Innovation and Skills, HMSO.

Fishman, R. (1982). *Urban utopias in the twentieth century*. Cambridge, MA: The MIT Press.

Galtung, J. (1982). *Schooling, education and the future* (Vol. 61). Malmo: Department of Education and Psychology Research, Lund University.

Gidley, J. M. (2017). *The future: A very short introduction*. Oxford: Oxford University Press.

Goodman, D. (2008). *A history of the future*. New York: Monacelli Press.
Hall, P. (2014). *Cities of tomorrow: An intellectual history of urban planning and design since 1880* (4th ed.). Hoboken: Wiley-Blackwell.
Huxley, A. ([1932] 2007). *Brave new world*. London: Vintage Books.
Karuri-Sebina, G., Haegeman, K., & Ratanawaraha, A. (2016). Urban futures: Anticipating a world of cities. *Foresight, 18*(5), 449–453. https://doi.org/10.1108/FS-07-2016-0037.
Moir, E., Moonen, T., & Clark, G. (2014). *What are future cities? Origins, meanings and uses*. London: Foresight Future of Cities Project and the Future Cities Catapult.
Montfort, N. (2017). *The future*. Cambridge, MA: The MIT Press.
More, T. ([1516] 2012). *Utopia*. London: Penguin.
Neuman, M., & Hull, A. (2009). The futures of the city region. *Regional Studies, 43*(6), 777–787. https://doi.org/10.1080/00343400903037511.
Polak, F. (1973). *The image of the future* (E. Boulding, Trans.). San Francisco: Jossey-Bass. (Original work published 1961).
Pollastri, S., Dunn, N., Rogers, C., Bokyo, C., Cooper, R., & Tyler, N. (2018). Envisioning urban futures as conversations to inform design and research. *Proceedings of the Institution of Civil Engineers – Urban Design and Planning, 171*(4), 146–156. https://doi.org/10.1680/jurdp.18.00006.
Rittel, H. W. J., & Webber, M. M. (1973). Dilemmas in a general theory of planning. *Policy Sciences, 4*(2), 155–169.
Rogers, C. (2018). Engineering future liveable, resilient, sustainable cities using foresight. *Proceedings of the Institution of Civil Engineers – Civil Engineering, 171*(6), 3–9. https://doi.org/10.1680/jcien.17.00031.
Samuel, L. R. (2009). *Future: A recent history*. Austin: University of Texas Press.
Sand, M. (2019). On "not having a future". *Futures, 107*, 98–106. https://doi.org/10.1016/j.futures.2019.01.002.
Sardar, Z. (2010). Welcome to postnormal times. *Futures, 42*(5), 435–444. https://doi.org/10.1016/j.futures.2009.11.028.
Savransky, M., Wilkie, A., & Rosengarten, M. (2017). The lure of possible futures. On speculative research. In A. Wilkie, M. Savransky, & M. Rosengarten (Eds.), *Speculative research: The lure of possible futures* (1st ed., pp. 1–17). New York: Routledge.
Slaughter, R. A. (1997). Developing and applying strategic foresight. *ABN Report, 5*(10), 13–27.
UK Government Office for Science. (2016). *Future of cities: The science of cities and future research priorities*. London: Foresight Government Office for Science, Department of Business Innovation and Skills, HMSO.
UN DESA Population Division. (2018). *World urbanization prospects: The 2018 revision, key facts*. New York: United Nations.
UNFPA. (2007). *State of world population 2007: Unleashing the potential of urban growth*. New York: United Nations Population Fund.
Urry, J. (2016). *What is the future?* Cambridge: Polity Press.
Voros, J. (2003). A generic foresight process framework. *Foresight, 5*(3), 10–21. https://doi.org/10.1108/14636680310698379.
Wells, H. G. ([1905] 2005). *A modern utopia*. London: Penguin.
WHO, & UN-Habitat. (2016). *Global report on urban health: Equitable healthier cities for sustainable development*. Geneva: World Health Organization.

Urban Governance

► New Forms of Shared Governance and Local Action Plan in Socially Vulnerable Settlements
► Participatory Governance for Adaptable Communities

Urban Green

► Innovation to Bring Nature-Based Solutions to Life: Tales of Two Cities

Urban Green Infrastructure

► Health and the Role of Nature in Enhancing Mental Health

Urban Greenery

► Innovation to Bring Nature-Based Solutions to Life: Tales of Two Cities

Urban Greening

► Stewarding Street Trees for a Global Urban Future

Urban Greening and Green Gentrification

Alessandro Busà
School of Geography, Geology and the Environment, University of Leicester, Leicester, UK

Introduction

The many environmental, social, and economic benefits associated with urban "green" development initiatives are broadly recognized: the creation of green space in cities contributes to a better physical and mental health for residents and can lead to improved social interactions and stronger community cohesion; investments in public transportation systems, energy and waste management, or energy-efficient retrofits strongly reduce pollution and can translate into significant cost savings and economic efficiency improvements for municipal administrations; remediation of contaminated industrial areas creates new developable land, bringing back investment and opening up space for revitalized neighborhoods; and resiliency planning can protect risk-prone areas from natural hazard events or reduce future disaster-related response and recovery costs.

Greening interventions however have also been associated with substantial transformations in the socioeconomic fabric of urban neighborhoods and urban regions. Over the last 15 years, a significant body of research in geography, urban planning, sociology, and environmental studies has highlighted a close correlation between the implementation of green-friendly, sustainable, or climate-resilient planning initiatives and inequitable patterns of demographic and socioeconomic change, particularly in disadvantaged or impoverished regions or neighborhoods with below-market property values, and where local governments and the development community see a potential for high-end physical upgrading derived from environmental cleanup or greening efforts. This growing field of study has raised issues around the uneven distribution of green space among different socioeconomic groups, and has brought to the fore the occurrence of "green gentrification" (Gould and Lewis 2012), a process through which rising housing and living costs resulting from ecological improvements can lead to the exclusion or displacement of vulnerable longtime residents and businesses. Over time, this process can ultimately cause a wholesale social reconfiguration of neighborhoods, with more affluent groups replacing less affluent ones who are forced to relocate to less desirable urban areas.

Urban Growth and Urban Greening

Since the early 2000s, urban greening initiatives have become a pivotal ingredient of economic development agendas in cities in advanced economies, where broader processes of deindustrialization have been compounded at the local level by specific land use policies and regulations aimed at advancing the vision of postindustrial, green, livable, and resilient cities. Entrepreneurial, competition-oriented urban development agendas have increasingly incorporated sustainable guidelines (energy-efficient public transit, green retrofitting of buildings, remediation of Brownfield sites and contaminated waterways, among others) not only as a way to protect the environment and create more livable cities, but also as a pathway to economic development, and as a strategy to enhance their attractiveness to high-value industries and residents, which translates into increased tax revenues to support municipal services and higher average incomes to keep the economy vibrant (Busà 2013). The development community has also acknowledged that the elimination of environmental burdens and the provision of green amenities in urban areas can be a powerful tool to promote real estate growth by facilitating the extraction of value from underperforming urban land. When ecological burdens are removed and new environmental amenities are created in their place, property values experience a rebound, attracting a higher-income demographic and generating demand for high quality housing and services. While local homeowners may stand to benefit from the increased equity of their homes, heavier tax burdens may convince those of fewer means to sell their properties and move elsewhere; in communities characterized by high shares of renters and low-income households, residents may found themselves unable to bear soaring rental costs or to find alternative living arrangements in an increasingly tight and competitive housing market. Over time, long-standing residents without the means to stay can be left with no choice but to relocate to less expensive, less attractive, and often more polluted areas, thus creating new geographies of social exclusion and segregation within urban regions (Lees et al. 2016; Gould and Lewis 2017). The potential ambivalent repercussions of urban greening initiatives on vulnerable groups highlight a "green space paradox" (Wolch et al. 2014), whereby greening efforts aimed at creating healthier and more desirable neighborhoods can ultimately lead to the exclusion of those residents and users who supposedly stood to benefit from environmental improvements. The occurrence of greening initiatives whose outcomes have proven socially exclusive raises the question of whether greening actions may imply conscious trade-offs between social and ecological sustainability (Haase et al. 2017).

Research in green gentrification has identified a range of green development initiatives that can act as local drivers of gentrification pressures on existing communities. Among those are the design or upgrading of public parks, environmental remediation plans on former industrial land, the construction of new green-certified buildings, and climate-resilient development initiatives.

Parks and Gardens

Several studies from real estate economics have used hedonic pricing models to link the presence of green space and urban nature with increases in residential property values in adjacent areas (Conway et al. 2008; Bockarjova et al. 2020). Much research on green gentrification has focused closely on the impact of park creation and restoration on housing market dynamics and other indicators of demographic and socio-economic change: an early study in New York City used census data to analyze the impact of the restoration of Prospect Park in Brooklyn, and found strong evidence of high residential turnover in the park's neighboring areas (Gould and Lewis 2012). In Taipei, the creation of the monumental Da-an Forest Park, created in 1994 on an area previously occupied by informal shack settlements, led to a sharp decrease in affordable housing options in neighboring lots, including in the adjacent Da-an public housing estate (Huang 2015). The clash between the demands of environmental justice movements striving to improve existing park space and local officials' efforts to boost property values through greening improvements has been explored in the analysis of another New York City park in a section of Harlem primed for gentrification after the launch of the city's sustainability Plan NYC2030 (Checker 2011). More studies have exposed extensive patterns of residential and commercial displacement that have followed the development of internationally renowned, design-intensive parks: research in the US has shown how centrally located parks that serve as green corridors are most likely to contribute to the gentrification of their surrounding neighborhoods (Rigolon and Németh 2019). In New York City, for instance, researchers have documented staggering increases in housing prices and evidence of intense commercial and residential gentrification following the creation of the Highline, a globally acclaimed elevated park designed by international architects on a once-derelict freight railway line that runs near the Hudson River in a transitioning area of Manhattan (Black and Mallory 2020). Similarly in Chicago, the 2013 groundbreaking of the 606 Trail, resulted from the conversion of a dismissed railway track into a multi-use 2.7 mile linear greenway, has led to dramatic changes in property values particularly in its Western portion, which had previously been characterized by a relatively affordable housing stock (Smith et al 2016). Another study in the US has examined the effect on the housing market of a new 22-mile green beltline connecting 45 Atlanta neighborhoods through a loop of parks, trails, and a streetcar, and found rapid and significant increases in property values for homes within a half-mile of the Beltline (Immergluck and Balan 2017); similar was the impact of the creation of the Gyeongui Line Forest Park in Seoul, another case of a disused railway that was converted to a linear urban park (Kwon et al., 2017). Evidence of a direct causal correlation between the creation or upgrading of park space and gentrification patterns, however, is not always demonstrable nor easily defined. Following the history of a German park created in the early 2000s on the area of a dismissed railway station, a recent study in Leipzig suggests that the park did not contribute significantly to social change through the years in which the city suffered high housing vacancy rates, but became a strong catalyst for gentrification only during the most recent phase of relentless expansion of the local housing market (Ali et al 2020). In Europe, another longitudinal study of the impact of municipal parks across the city of Barcelona has documented that gentrification trends are mostly to be observed in desirable neighborhoods where parks have been refurbished or newly created, but not as much in underserved areas characterized by a less desirable housing stock (Anguelovski et al. 2017). Finally, while green gentrification research has hitherto suggested a causal correlation between park space development and ensuing gentrification patterns, recent studies are questioning whether an opposite trajectory might be equally valid: research on public park spending in the city of Los Angeles (Riebel et al., 2021) has found that recent municipal investments in park creation or upgrading are being disproportionally targeted at LA districts in the early-stages of gentrification—this could be a result of strategic calculations by public officials and development advocates eager to direct investment to areas that are ripe with development potential, or because the presence of early gentrifiers may bring the necessary leverage and political clout to successfully advocate for green investments in up-and-coming neighborhoods.

Remediation

In areas that have long suffered from environmental degradation, profit-seeking regeneration efforts incorporating greening strategies are generally aimed at capturing the untapped environmental value of underperforming urban land, and are usually associated with high-profile development schemes to ensure their financial viability. Environmental gentrification scholarship has used the notion of "green gap"—the hiatus between the ground rent in a disinvested area and the potential ground rent that could be accumulated by maximum exploitation of the land with the most profitable uses, to explain this process of value extraction: the realization of a "green gap" is possible "when land deemed vacant, underused, or contaminated is identified by developers as a possible area to be 'greened,' generating amenities that may allow for higher economic value and profit accumulation" (Anguelovski et al. 2018). Hazardous waste sites resulting from dismissed industrial uses are often located in disadvantaged urban areas whose communities are predominantly low income and suffer the negative externalities of air, land, or water contamination. Several studies of remediation in North America have shown that remediation has significant ripple effects on property values (De Sousa et al. 2009). In addition, large-scale remediation projects in these areas require massive public and private funding, and the high costs of cleanup can make the incorporation of affordable housing provisions in these developments risky and challenging (Squires and Hutchison 2021). A 2011 systematic study of environmental remediation efforts on former waste sites across the USA highlighted a pattern connecting such improvements and an increase in mean household income in the proximity of remediated areas (Gamper-Rabindran and Timmins 2011), although a study in Portland published in the same year challenged the notion of a direct correlation between remediation and gentrification (Eckerd 2010). Several studies on populations that are vulnerable to displacement as land values escalate in remediated areas have been conducted in the USA (Pearsall 2010; Curran and Hamilton 2012; Miller 2016) and Canada (Dale and Newman 2009). In Nanjing, China, researchers have shown how the environmental restoration of the polluted Qinhuai River has led to the mass resettlement of thousands of low-income residents, leaving room for the proliferation of high-end gated communities along the regenerated waterfront (Yu et al. 2020). More research has illustrated the displacement of small manufacturing activities in former industrial areas of New York City (Curran 2007), also as a result of specific rezoning guidelines contained in New York City's sustainability plan (see Busà 2017). In South Korea, the restoration of the heavily polluted (and once-covered by an elevated multi-lane highway) Cheonggye Stream in the heart of Seoul's central business district and its transformation into a 5.8 km long linear park has contributed to staggering increases in land prices and accelerated land use change along the newly created waterfront, leading to the displacement of clusters of small industrial manufacturers and the expansion of high-end commercial and office uses (Lim et al. 2013).

Green Buildings

Green buildings can reduce waste, pollution, and environmental degradation through building technologies aimed at cutting emissions, using energy efficiently and promoting the health of their inhabitants or users. As the global green buiding market expanded relentlessly over the last 15 years, researchers have questioned the ripple effect of green-certified buildings on their surrounding neighborhoods. In London, research has examined the impact of green office buildings on nearby commercial property values, concluding that newly built green properties had a direct spillover effect on commercial rents and prices in neighboring areas (Chegut et al. 2013). A study in North America has focused on regional policies mandating or tightening green building regulations and certification systems, concluding that these may result in greater social segregation, as rising building costs may hamper the ability of low-income communities to build new homes (Mehdizadeh and Fischer 2013). More research in the USA also analyzed the impact of the LEED-ND (LEED for Neighborhood Development) rating system introduced in 2007 to promote mixed-income neighborhood development, but found that only 40% of certified projects reviewed included a component of affordable housing, concluding that "planners should not assume that equity or affordable housing goals will be met just because projects are LEED-ND certified" (Szibbo 2015). Another study conducted in the USA has examined the impact of over 200 proposed, completed, and/or in progress LEED-ND sites on their immediate and adjacent neighborhoods, and found statistically significant changes to educational attainment, median rent, and other indicators of neighborhood gentrification (Benson and Bereitschaft 2019).

Climate-Resilient Infrastructure

Climate-resilient infrastructure and adaptation planning are increasingly employed as a way to mitigate the challenges of weather disruption while reducing future disaster-related response and recovery costs. While no action in high-hazard regions or areas can result in increased vulnerability for the populations inhabiting them, recent scholarship is starting to assess the uneven impacts of adaptation and resiliency planning (Long and Rice 2018). A comprehensive study investigating the effects of mitigation planning on vulnerable communities conducted in eight cities across the globe has found evidence of uneven enforcement of land use regulations, uneven development which has left poor communities vulnerable to environmental risks, and mass resettlement of informal dwellers living in high-hazard areas (Anguelovski et al. 2016). In the USA, Gould and Lewis (2018) claim that new large-scale developments on flood-prone areas along the coastline of New York City are incorporating costly flood protection infrastructures (elevated berms above flood levels and physical barriers such as dams and floodwalls) that make them unaffordable but to the wealthy, potentially leading to new forms of "resilience gentrification." Other studies connecting elevation above the flood plain with gentrification have been conducted in Florida (Keenan et al. 2018). Another study in New York City has documented the complexity

U

of public, financial, and real estate interests involved in the planning of a multibillion dollar design for resiliency planning post-Hurricane Sandy along the East River in the Lower East Side of Manhattan, in a low-income tract dominated by a large number of distressed public housing projects: to be financially viable to investors, the weatherization plan called for the development of new market-rate housing and semi-privatized green space geared towards higher-income occupants. While the plan didn't materialize, it resulted in a wave of high-end investment to parcels of land in the immediate proximity of the area, fueling a proliferation of luxury residential towers along the East River flood plain (Dupuis and Greenberg 2019).

Conclusions

Over the last 20 years, green gentrification research has provided important and thought-provoking insights and informed a broader interdisciplinary debate around the social dimension and implications of urban sustainability planning. The occurrence of green gentrification sheds light on a paradox within discourses around sustainability, and on the way these discourses can be recurrently promoted to conceal "business-as-usual" practices of urban development, while failing to address considerations of their social costs. While sustainable development initiatives should aim for a balance between the environmental, economic, and social dimensions of development, "actually existing sustainabilities" (Krueger and Agyeman 2005) show that hitherto adopted solutions to environmental challenges may at times exacerbate, rather than resolve social challenges, as greening blueprints can be incorporated into development schemes whose results are neither inclusive nor socially sustainable. From a policymaking perspective, the incorporation of stronger concerns for the social impacts of urban greening policies and practices is required. Sustainability appraisals designed to assess a fuller spectrum of social repercussions of urban greening strategies are being gradually introduced in a growing number of countries to support decision-making and planning strategies; further progresses in this direction can be made by promoting a stronger inclusion of local stakeholders and affected residents in the design, implementation, and managements of greening projects—a demanding task, but necessary if the aim is to realize the promise of green, livable, and inclusive communities. At present, researchers in the US are measuring successes and limitations of "Park-related Anti-displacement Strategies" (PRADS) implemented through joint efforts of private developers, park agencies, policymakers, environmental groups, community organizations and housing advocates, and aimed at preventing displacement in areas targeted for new park development (Rigolon and Christensen 2019). Project-specific measures include tax abatements or loans to promote local homeownership, inclusionary zoning mandates to support creation or preservation of affordable housing, tenant protection regulations, low interest loans to preserve or expand neighborhood businesses, local hiring ordinances and job training opportunities, and community land trusts for permanently affordable housing, among others. Out of 27 parks analyzed in 19 cities across the US, the study found several of these strategies being actively employed in about half of the case studies, indicating a strong shift in public perception and collective understanding of the potential threats of green gentrification, and a growing awareness of the value of community-led organizing efforts to curb its most disrupting consequences.

Cross-References

► Green Cities
► Green Infrastructure
► Housing and Development
► Sustainable Development Goals
► Transportation and Mobility

References

Ali, L., Haase, A., & Heiland, S. (2020). Gentrification through Green Regeneration? Analyzing the Interaction between Inner-City Green Space Development and Neighborhood Change in the Context of Regrowth: The Case of Lene-Voigt-Park in Leipzig, Eastern Germany. *Land, 9*(1), 24. https://doi.org/10.3390/land9010024.

Anguelovski, I., Shi, L., Chu, E., Gallagher, D., Goh, K., Lamb, Z., et al. (2016). Equity impacts of urban land use planning for climate adaptation. *Journal of Planning Education and Research, 36*(3), 333–348. https://doi.org/10.1177/0739456x16645166.

Anguelovski, I., Connolly, J. J., Masip, L., & Pearsall, H. (2017). Assessing green gentrification in historically disenfranchised neighborhoods: A longitudinal and spatial analysis of Barcelona. *Urban Geography, 39* (3), 458–491. https://doi.org/10.1080/02723638.2017.1349987.

Anguelovski, I., Connolly, J., & Brand, A. L. (2018). From landscapes of utopia to the margins of the green urban life. *City, 22*(3), 417–436. https://doi.org/10.1080/13604813.2018.1473126.

Benson, E. M., & Bereitschaft, B. (2019). Are LEED-ND developments catalysts of neighborhood gentrification? *International Journal of Urban Sustainable Development, 12*(1), 73–88. https://doi.org/10.1080/19463138.2019.1658588.

Black, K. J., & Richards, M. (2020). Eco-gentrification and who benefits from urban green amenities: NYC's high line. *Landscape and Urban Planning, 204*, 103900. https://doi.org/10.1016/j.landurbplan.2020.103900.

Bockarjova, M., Botzen, W., Schie, M. V., & Koetse, M. (2020). Property price effects of green interventions in cities: A meta-analysis and implications for gentrification. *Environmental Science & Policy, 112*, 293–304. https://doi.org/10.1016/j.envsci.2020.06.024.

Busà, A. (2013). The unsustainable cost of sustainability: PlaNYC 2030 and the future of New York City. *Critical Planning Journal, 20*, 193–217.

Busà, A. (2017). *The creative destruction of New York City engineering the city for the elite*. Oxford: Oxford University Press.

Checker, M. (2011). Wiped out by the "Greenwave": Environmental gentrification and the paradoxical politics of urban sustainability. *City & Society, 23*(2), 210–229. https://doi.org/10.1111/j.1548-744x.2011.01063.x.

Chegut, A., Eichholtz, P., & Kok, N. (2013). Supply, demand and the value of green buildings. *Urban Studies, 51*(1), 22–43. https://doi.org/10.1177/0042098013484526.

Conway, D., Li, C. Q., Wolch, J., Kahle, C., & Jerrett, M. (2008). A spatial autocorrelation approach for examining the effects of urban Greenspace on residential property values. *The Journal of Real Estate Finance and Economics, 41*(2), 150–169. https://doi.org/10.1007/s11146-008-9159-6.

Curran, W. (2007). From the frying pan to the oven: Gentrification and the experience of industrial displacement in Williamsburg, Brooklyn. *Urban Studies, 44*(8), 1427–1440. https://doi.org/10.1080/00420980701373438.

Curran, W., & Hamilton, T. (2012). Just green enough: Contesting environmental gentrification in Greenpoint, Brooklyn. *Just Green Enough*, 15–31. https://doi.org/10.9774/gleaf.9781315229515_3.

Dale, A., & Newman, L. L. (2009). Sustainable development for some: Green urban development and affordability. *Local Environment, 14*(7), 669–681. https://doi.org/10.1080/13549830903089283.

De Sousa, C. A., Wu, C., & Westphal, L. M. (2009). Assessing the effect of publicly assisted brownfield redevelopment on surrounding property values. *Economic Development Quarterly, 23*(2), 95–110. https://doi.org/10.1177/0891242408328379.

Dupuis, E. M., & Greenberg, M. (2019). The right to the resilient city: Progressive politics and the green growth machine in New York City. *Journal of Environmental Studies and Sciences, 9*(3), 352–363. https://doi.org/10.1007/s13412-019-0538-5.

Eckerd, A. (2010). Cleaning Up Without Clearing Out? A Spatial Assessment of Environmental Gentrification. *Urban Affairs Review, 47*(1), 31–59. https://doi.org/10.1177/1078087410379720.

Gamper-Rabindran, S., & Timmins, C. (2011). Hazardous waste cleanup, neighborhood gentrification, and environmental justice: Evidence from restricted access census block data. *American Economic Review, 101*(3), 620–624. https://doi.org/10.1257/aer.101.3.620.

Gould, K. A., & Lewis, T. L. (2012). The environmental injustice of green gentrification. In J. DeSena & T. Shortell (Eds.), *The world in Brooklyn: Gentrification, immigration, and ethnic politics in a global city* (pp. 113–146). Plymouth: Lexington Books.

Gould, K. A., & Lewis, T. L. (2017). *Green gentrification: Urban sustainability and the struggle for environmental justice*. London: Routledge Taylor & Francis Group.

Gould, K. A., & Lewis, T. L. (2018). From green gentrification to resilience gentrification: An example from Brooklyn. *City & Community, 17*(1), 12–15. https://doi.org/10.1111/cico.12283.

Haase, D., Kabisch, S., Haase, A., Andersson, E., Banzhaf, E., Baró, F., et al. (2017). Greening cities – To be socially inclusive? About the alleged paradox of society and ecology in cities. *Habitat International, 64*, 41–48. https://doi.org/10.1016/j.habitatint.2017.04.005.

Huang, L. (2015). Promoting private interest by public hands? The gentrification of public lands by housing policies in Taipei City. In L. Lees, H. B. Shin, & E. Lopez-Morales (Eds.), *Global gentrifications: Uneven development and displacement* (pp. 223–244). Bristol: Policy Press.

Immergluck, D., & Balan, T. (2017). Sustainable for whom? Green urban development, environmental

U

gentrification, and the Atlanta beltline. *Urban Geography, 39*(4), 546–562. https://doi.org/10.1080/02723638.2017.1360041.

Keenan, J. M., Hill, T., & Gumber, A. (2018). Climate gentrification: From theory to empiricism in Miami-Dade County, Florida. *Environmental Research Letters, 13*(5), 054001. https://doi.org/10.1088/1748-9326/aabb32.

Kwon, Y., Joo, S., Han, S., & Park, C. (2017). Mapping the Distribution Pattern of Gentrification near Urban Parks in the Case of Gyeongui Line Forest Park, Seoul, Korea. *Sustainability, 9*(2), 231. https://doi.org/10.3390/su9020231.

Krueger, R., & Agyeman, J. (2005). Sustainability schizophrenia or "actually existing sustainabilities?" toward a broader understanding of the politics and promise of local sustainability in the US. *Geoforum, 36*(4), 410–417. https://doi.org/10.1016/j.geoforum.2004.07.005.

Lees, L., Shin, H. B., & Lopez-Morales, E. (2016). *Planetary gentrification*. Cambridge: Polity Press.

Lim, H., Kim, J., Potter, C., & Bae, W. (2013). Urban regeneration and gentrification: Land use impacts of the Cheonggye stream restoration project on the Seouls central business district. *Habitat International, 39*, 192–200. https://doi.org/10.1016/j.habitatint.2012.12.004.

Long, J., & Rice, J. L. (2018). From sustainable urbanism to climate urbanism. *Urban Studies, 56*(5), 992–1008. https://doi.org/10.1177/0042098018770846.

Mehdizadeh, R., & Fischer, M. (2013). The unintended consequences of greening America: An examination of how implementing green building policy may impact the dynamic between local, state, and federal regulatory systems and the possible exacerbation of class segregation. *Energy, Sustainability and Society, 3*(1). https://doi.org/10.1186/2192-0567-3-12.

Miller, J. T. (2016). Is urban greening for everyone? Social inclusion and exclusion along the Gowanus Canal. *Urban Forestry & Urban Greening, 19*, 285–294. https://doi.org/10.1016/j.ufug.2016.03.004.

Pearsall, H. (2010). From brown to green? Assessing social vulnerability to environmental gentrification in New York City. *Environment and Planning C: Government and Policy, 28*(5), 872–886. https://doi.org/10.1068/c08126.

Reibel, M., Rigolon, A., & Rocha, A. (2021). Follow the money: Do gentrifying and at-risk neighborhoods attract more park spending? *Journal of Urban Affairs*, 1-19. https://doi.org/10.1080/07352166.2021.1886857.

Rigolon, A., & Christensen, J. (2019). *Greening without Gentrification: Learning from Parks-Related Anti-Displacement Strategies Nationwide (Rep.)*. Retrieved October 22, 2020, from Institute of the Environment and Sustainability at UCLA website: https://www.ioes.ucla.edu/wp-content/uploads/Greening-without-Gentrification-report-2019.pdf

Rigolon, A., & Németh, J. (2019). Green gentrification or 'just green enough': Do park location, size and function affect whether a place gentrifies or not? *Urban Studies, 57*(2), 402–420. https://doi.org/10.1177/0042098019849380.

Smith, G., Duda, S., Lee, J.M., & Thompson, M. (2016). *Measuring the Impact of the 606: Understanding How a Large Public Investment Impacted the Surrounding Housing Market (Rep.)*. Retrieved July 13, 2020, from Institute for Housing Studies at DePaul University website: https://www.housingstudies.org/media/filer_public/2016/10/31/ihs_measuring_the_impact_of_the_606.pdf

Squires, G., & Hutchison, N. (2021). Barriers to affordable housing on brownfield sites. *Land Use Policy, 102*, 105276. https://doi.org/10.1016/j.landusepol.2020.105276.

Szibbo, N. (2015). Lessons for LEED® for neighborhood development, social equity, and affordable housing. *Journal of the American Planning Association, 82*(1), 37–49. https://doi.org/10.1080/01944363.2015.1110709.

Wolch, J. R., Byrne, J., & Newell, J. P. (2014). Urban green space, public health, and environmental justice: The challenge of making cities 'just green enough'. *Landscape and Urban Planning, 125*, 234–244. https://doi.org/10.1016/j.landurbplan.2014.01.017.

Yu, S., Zhu, X., & He, Q. (2020). An assessment of Urban Park access using house-level data in urban China: Through the lens of social equity. *International Journal of Environmental Research and Public Health, 17*(7), 2349. https://doi.org/10.3390/ijerph17072349.

Urban Greenspace

► Innovation to Bring Nature-Based Solutions to Life: Tales of Two Cities

Urban Health

► Urban Well-Being

Urban Health and Well-Being

► Health and the City: How Cities Impact on Health, Happiness, and Well-Being

Urban Health Epistemic Communities

▶ Urban Health Paradigms

Urban Health Ideas

▶ Urban Health Paradigms

Urban Health Paradigms

Jinhee Kim[1], Evelyne de Leeuw[1,2],
Ben Harris-Roxas[3] and Peter Sainsbury[4]
[1]Centre for Health Equity Training, Research and Evaluation (CHETRE), UNSW Australia Research Centre for Primary Health Care & Equity, South Western Sydney Local Health District, Ingham Institute, Sydney, NSW, Australia
[2]Healthy Urban Environments (HUE) Collaboratory, Maridulu Budyari Gumal Sydney Partnership for Health, Education, Research and Enterprise SPHERE, Sydney, NSW, Australia
[3]School of Population Health, University of New South Wales, Sydney, NSW, Australia
[4]School of Medicine, University of Notre Dame, Sydney, NSW, Australia

Synonyms

Urban health epistemic communities; Urban health ideas; Urban health policy ontologies

Definition

A ***paradigm*** is a coherent body of work that shares a standard set of concepts, theories, methods, and instruments that researchers and scientists within the paradigm take for granted (Kuhn 1962). It is also commonly used as an epistemological term that stands for the disciplinary frames, beliefs, ideas, norms, ontological positions, or "thought patterns" in disciplines and sectors (Seel 2012).

Urban health is a field of research, practice, and policy that addresses the health of people that is impacted by the physical and social environments of the urban setting (Wuerzer 2014).

Urban health paradigms are distinct approaches to understanding and addressing urban health issues that manifest from the underlying ideations and beliefs on how the problem is conceptualized, the best methodologies to create knowledge on the issue and the optimal solutions to address them.

Transdisciplinary research is a type of cross-disciplinary research that aims for a greater degree of integration than multidisciplinary or interdisciplinary research. Two or more disciplines transfer and transcend knowledge and research methods and create coherent knowledge without any boundaries between disciplines (Lawrence 2021).

Policy ideas are causal beliefs about the relationships between things and people in the world that guide people's decisions and actions (Béland and Cox 2010).

Introduction

Urban health is a complex field of research and action that requires intersectoral collaboration and multidisciplinary methodologies. Therefore, researchers and policy actors that study and address urban health issues come from a wide range of disciplines, sectors, and trades. This diversity of backgrounds invites multiple conceptual, theoretical, methodological, and instrumental views on urban health. These views, or paradigms, are reflected in the language and terminologies researchers and actors use that are frequently regarded as barriers to intersectoral action. More importantly, differing paradigms shape the underlying beliefs about which urban health problems are important and worth addressing and their appropriate and preferred solutions. The four urban health paradigms that are presented in this entry illustrate the different foundational positions of urban health researchers and policy actors. Each of these urban health paradigms supports a distinct set of research traditions and policy ideas.

The Need to Understand Urban Health Paradigms

The conceptual arena around cities and health is occupied by many different concepts, approaches, ideas, and theories. This is inherent to the nature of the field. But not all of these freely pervade all others. There are a few hermetic paradigms that are conceptually and practically exclusive. Understanding urban health paradigms entails more than the identification of which paradigms exist. It is more important to articulate how the paradigms differ in their views. In doing so, participants can acknowledge the views and begin to search for coherence in understanding knowledge produced from differing urban health paradigms. As a result, scientists and policy actors can interpret knowledge produced from one paradigm, make sense of the paradoxes between disciplinary knowledge, and ultimately seek strategies to combine or connect with the coexisting paradigms.

Recognizing urban health paradigms is essential in two key areas to promote urban health – (a) facilitating transdisciplinary knowledge in research and practice and (b) influencing the policy process. Transdisciplinary approaches help establish better urban health research and policy. Urban health is a complex field, and each disciplinary interpretation focuses on specific characteristics of the object of study. The increasing availability and diversity of empirical knowledge on urban health ironically creates larger uncertainty on the issue (Lawrence and Gatzweiler 2017). The conflicting meanings and values of data reduce the capability of humans to act on the information that they do not necessarily understand. The objective of transdisciplinarity is to preserve the different realities and confront them, and thus the goal is not to search for consensus but to search for coherence (Ramadier 2004). Understanding the grounding and views of different urban health paradigms not only allows us to reinterpret the information, but furthermore leads to opportunities to develop transdisciplinary research programs.

Furthermore, paradigms are the foundations for ideas and beliefs that are central in policymaking. Policymaking is a process of conflict and negotiation over policy ideas (Leftwich 2015). In fact, the rules and norms of the decision-making structures that influence urban health processes and outcomes are ideas that have been institutionalized (Cairney 2012). To influence policy change, policy actors search for other policy actors who share their beliefs and aspire to form like-minded coalitions (Jenkins-Smith et al. 2018). Because deep core beliefs are foundational and less subject to change, policy actors generally target incremental change in lower-level belief systems that define the preferred policy instruments and settings while maintaining the overall policy paradigm. To achieve this, policy actors frame their policy proposals within the conceptual understanding of others to gain support (Kingdon 1984). Having a clear understanding of existing urban health paradigms can assist policy actors to frame their ideas within the language and worldviews of different paradigms.

Four Types of Urban Health Paradigms

Four distinctive paradigms currently coexist in urban health scholarship and policy – the "*medical-industrial city*," "*urban health science*," "*healthy built environments,*" and "*health social movement*." While some overlaps and similarities can be found, each one of these is characterized by their prominent beliefs on:

- Which urban health issues are more important (a *conceptual* gaze)
- What causes these urban health issues (*theoretical* frameworks)
- Which data collection or analytical method would best measure and seek information (*methodologies*)
- Which solutions effectively resolve the prioritized issues (*instrumental dimensions*)

Medical-Industrial City

The "*medical-industrial city*" paradigm is driven by business, industry, and the government generally in the healthcare, construction, and technology sectors where healthcare infrastructure and services are treated as commodities to be invested in for urban change. Views of this paradigm align with neoliberal principles and the urban growth machine theory (Molotch 1976). The concept of health is viewed in a biomedical, individualistic, pathogenic model and coupled with the image of economic prosperity in the form of livability and healthy lifestyles. Actors in this paradigm include place entrepreneurs, businesses, and industries that profit from such development, politicians, universities, media, and corporate capitalists. This view is less active in scientific research, but often is the dominant position in current urban development practices such as large-scale healthcare industry- and infrastructure-centered urban development or smart city (for health) initiatives.

Urban Health Science

The "*urban health science*" paradigm emphasizes epidemiological and classic Cartesian methodologies to empirically analyze the complex causal associations between the urban environment and health. The urban environment is viewed as a critical layer in the multilevel framework of the social determinants of health and the urban health system is a complex network of multidirectional causal relationships, feedback loops, and unintended consequences (Ettman et al. 2019; Rydin et al. 2012). Evidence is generally produced through the measurement and quantification of the urban condition and its health impacts and is critical when developing solutions to improve health. Therefore, recommended solutions are generally technocratic and usually propose lists of "evidence-based" interventions or best buys.

Healthy Built Environment

The "*healthy built environment*" paradigm sets the (re-)integration of health as a main objective for spatial planning. Ideas in this paradigm focus on transforming the sets of procedures, institutions, and regulations of the urban planning system (Barton et al. 2015; Kent et al. 2018). Researchers and policy actors following this paradigm develop and propose codes or guidelines as criteria or benchmarks for urban planning projects to ensure health is considered in the projects. Rigorous quantitative methods are regarded as high-quality evidence, but proponents of this paradigm also advocate for more participatory and comprehensive methods to understand the complexity of the issues. Healthy urban development checklists and healthy urban planning guidelines are examples following this paradigm (UN-HABITAT and World Health Organization 2020).

Health Social Movement

The "*health social movement*" paradigm supports a "value-based" approach to health promotion that values principles such as health equity and empowerment in identifying and solving of urban health issues (Brown and Fee 2014; de Leeuw and Simos 2017). Views of this paradigm align with the definition and principles of health promotion as outlined in the Ottawa Charter for Health Promotion (World Health Organisation 1986). Ideally driven by the empowered community, solutions focus both within (in community-driven action) and outward (in mobilizing for policy and systems change). According to the perspectives of this paradigm, a healthy city is not defined by a set of standards, but rather a healthy city is one that commits to healthy city values and continuously creating supportive environments for health. The WHO Healthy Cities movement is an example that follows the principles of this paradigm (Table 1).

Conclusion

The growing global challenges we face today influence a myriad of complex urban and regional issues that call for multitude of responses across all disciplines, sectors, regions, and scales. Criticisms that the responses have been inadequate or fragmented cannot be reconciled unless we identify where and how the

Urban Health Paradigms, Table 1 Four urban health paradigms (summary)

	Four urban health paradigms			
	Medical-industrial city	Urban health science	Healthy built environment	Health social movement
View on urban health	Biomedical and individualistic approach to health and illness. Healthcare infrastructure and health-related technologies influence health outcomes	Focus on risk factors that lead to illness or disease outcomes. The urban built environment is a critical layer in the multi-determinant model	Focus on health-promoting lifestyles, wellbeing, quality of life, or flourishing. Elements of urban planning process and governance structure impact these health outcomes	Socio-ecological view on health and an explicit focus on health equity. The sociopolitical factors underlying the urban governance systems are the main drivers of urban health
Urban health solutions	Investment in healthcare infrastructure as drivers of economic growth	Expert-led empirical evidence-based interventions and technological solutions	Influencing the planning system and urban planning regulations and processes	Value-driven community empowerment approach to transform the urban environment
Examples	Health and innovation precincts models for urban development	Urban health indicators, Partnership for Healthy Cities (Bloomberg Foundation)	Healthy urban planning principles and guidelines	WHO Healthy Cities movement

similarities and differences occur. Without the recognition of the underlying foundational beliefs that form different positions in addressing urban issues, responses will remain fragmented, or collaboration efforts will be at best superficial. Articulating the paradigms to address conflicts or paradoxes between different worldviews is the first step in reconstructing for transdisciplinary evidence and decision-making that result in better and more equitable economic, environmental, and social consequences.

Understanding urban health paradigms provides opportunities to facilitate transdisciplinary urban, peri-urban, and regional health research and to influence land use, transport, and other planning policies that impact health. The different urban health paradigms that exist are differentiated by the conceptual, theoretical, methodological, and instrumental beliefs on how cities create and enhance health. Researchers, practitioners, and policy actors can use the urban health paradigms as a framework to interpret and apply knowledge produced across diverse belief systems. Instead of searching for unity or consensus, approaches to improve urban health should seek coherence in the application of knowledge, methodologies, and solutions that transcend urban health paradigms.

Cross-References

- ▶ City Visions: Toward Smart and Sustainable Urban Futures
- ▶ Health and the City: How Cities Impact on Health, Happiness, and Well-Being
- ▶ The Sustainable and the Smart City: Distinguishing Two Contemporary Urban Visions

References

Barton, H., Thompson, S., Burgess, S., & Grant, M. (2015). *The Routledge handbook of planning for health and well-being: Shaping a sustainable and healthy future*. Taylor & Francis Group.

Béland, D., & Cox, R. H. (2010). *Ideas and politics in social science research*. Oxford University Press.

Brown, T. M., & Fee, E. (2014). Social movements in health. *Annual Review of Public Health, 35*(1), 385–398. https://doi.org/10.1146/annurev-publhealth-031912-114356.

Cairney, P. (2012). The role of ideas. In *Understanding public policy: Theories and issues* (pp. 220–243). Palagrave Macmillan.
de Leeuw, E., & Simos, J. (2017). *Healthy cities: The theory, policy, and practice of value-based urban planning*. Springer.
Ettman, C., Vlahov, D., & Galea, S. (2019). Why cities and health?: Cities as determinants of health. In S. Galea, C. Ettman, & D. Vlahov (Eds.), *Urban health* (pp. 15–24). Oxford University Press. https://doi.org/10.1093/oso/9780190915858.003.0002.
Jenkins-Smith, H., Nohrstedt, D., Weible, C., & Ingold, K. (2018). The advocacy coalition framework: An overview of the research program. In C. Weible & P. Sabatier (Eds.), *Theories of the policy process* (pp. 135–171). Routledge. https://doi.org/10.4324/9780429494284-5.
Kent, J., Harris, P., Sainsbury, P., Baum, F., McCue, P., & Thompson, S. (2018). Influencing urban planning policy: An exploration from the perspective of public health. *Urban Policy and Research, 36*(1), 20–34. https://doi.org/10.1080/08111146.2017.1299704.
Kingdon, J. W. (1984). *Agendas, alternatives, and public policies*. Little, Brown.
Kuhn, T. (1962). *The structure of scientific revolutions*. University of Chicago Press.
Lawrence, R. (2021). *Creating built environments: Bridging knowledge and practice divides*. Routledge. https://www.routledge.com/Creating-Built-Environments-Bridging-Knowledge-and-Practice-Divides/Lawrence/p/book/9780815385394.
Lawrence, R., & Gatzweiler, F. (2017). Wanted: A transdisciplinary knowledge domain for urban health. *Journal of Urban Health, 94*, 592–596. https://doi.org/10.1007/s11524-017-0182-x.
Leftwich, A. (2015). *What is politics?: The activity and its study*. Wiley.
Molotch, H. (1976). The city as a growth machine: Toward a political economy of place. *American Journal of Sociology, 82*(2), 309–332. JSTOR. https://www.jstor.org/stable/2777096.
Ramadier, T. (2004). Transdisciplinarity and its challenges: The case of urban studies. *Futures, 36*(4), 423–439. https://doi.org/10.1016/j.futures.2003.10.009.
Rydin, Y., Bleahu, A., Davies, M., Dávila, J. D., Friel, S., De Grandis, G., Groce, N., Hallal, P. C., Hamilton, I., Howden-Chapman, P., Lai, K.-M., Lim, C., Martins, J., Osrin, D., Ridley, I., Scott, I., Taylor, M., Wilkinson, P., & Wilson, J. (2012). Shaping cities for health: Complexity and the planning of urban environments in the 21st century. *The Lancet, 379*(9831), 2079–2108. https://doi.org/10.1016/S0140-6736(12)60435-8.
Seel, N. M. (2012). Paradigm. In *Encyclopedia of the sciences of learning* (pp. 2552–2553). Springer. https://doi.org/10.1007/978-1-4419-1428-6_2314.
UN-HABITAT and World Health Organization. (2020). *Integrating health in urban and territorial planning: A sourcebook* (Licence: CC BY-NC-SA 3.0 IGO) (p. 108). UN-HABITAT and World Health Organization.
World Health Organisation. (1986). *The Ottawa charter for health promotion*. World Health Organisation (WHO). http://www.who.int/healthpromotion/conferences/previous/ottawa/en/.
Wuerzer, T. (2014). Urban health. In A. C. Michalos (Ed.), *Encyclopedia of quality of life and well-being research* (pp. 6835–6837). Dordrecht: Springer. https://doi.org/10.1007/978-94-007-0753-5_3127.

Urban Health Policy Ontologies

► Urban Health Paradigms

Urban Heat Islands

Humera Afaq[1], Deborah Nabubwaya Chambers[2] and Tara Rava Zolnikov[3]
[1]National University, San Diego, CA, USA
[2]Community Health, National University, San Diego, CA, USA
[3]Behavioral Sciences, California Southern University, Costa Mesa, CA, USA

Introduction

The urban heat island (UHI) effect occurs when the outside air temperature is significantly higher in urban areas than surrounding suburban areas (Da-Lin Zhang et al. 2011); this is a common environmental phenomenon occurring in most cosmopolitan areas. This difference occurs because the area's temperature is affected by how well the surfaces in each environment absorb and retain heat (Rupard 2019). The UHI effect can contribute to smog and signifies one of the most substantial human-made alterations to surface climate (Rupard 2019; Emmanuel 2004). It can be envisioned as a dome of stationary warm air over densely populated urban areas (Emmanuel 2004). Additionally, it can significantly impact populations and human health through high air pollution, reduced nighttime cooling, and increased temperatures, which can have adverse health effects.

This situation is a causative factor for the development of global warming, heat-related casualties, and erratic climatic changes (Deilami et al. 2018). Moreover, the effects of UHI in cities of low latitude could cause the increase of energy consumption for cooling and peak demand in summer temperatures (Radhi et al. 2015; Santamouris 2014a). Overall, these effects can contribute to a modern endemic threat to thermal luxury (e.g., air conditioning) and human health. In addition, it has also been demonstrated that weather conditions of extremely high temperatures would cause an increase in mortality and hospital admissions (Chan et al. 2012); this has occurred historically in regions such as the Chicago, United Kingdom, and China (Campbell et al. 2018; Arbuthnott and Hajat 2017; Luo et al. 2018).

In the warm season, the inner-city heat and temperature are directly related to land use, for example, industrial areas have the warmest average temperatures, whereas the beaches are the coolest (Roth et al. 1989). The city's infrastructure, such as roads, buildings, and other groundwork, retains the ability to capture and reemit the sun's heat more than the natural sites, like woodlands or water bodies. Greenery or greenspace within an urban setting is quite limited while construction is highly concentrated, so "islands" of higher temperatures develop within this area rather than remote areas. These compartments of heat can consistently form in any season, condition, or climate, though the intensity of an urban heat island can increase during the day; this can cause even more significant effects during warmer seasons.

An urban heat island can be responsible for a daytime temperature rise of 1–7 °F and an average rise of 2–5 °F (Hibbard et al. 2017). According to Hibbard (1971), the heat island effect will gradually and progressively increase as urban areas' infrastructure and population density grow; this has been further proven valid over time as increased exposure to extreme heat from both climate change and the UHI effect threatens the sustainability of rapidly growing urban settlements worldwide. Regions that are exposed to severely high temperatures endanger urban health and development, increase reductions in labor productivity and economic output, and contribute to increased morbidity and mortality (Tuholske et al. 2021).

Causes

There are several different factors responsible for the development of urban heat islands. These can include loss of natural environment and materials or the built environment of cities. Loss of natural land and its utilization for building metropolitan areas contribute to UHI, for example, plants, trees, and water bodies are natural resources to keep the environment cooler. By providing shade, the perspiration of water from the trees and the plants contributes by cleaning the air. Likewise, the replacement of natural soil or vegetation by materials, such as concrete and asphalt, construction of hard surfaces such as roads, buildings, and parking areas, does not provide shade or moisture through evaporation and plant transpiration, which leads to high temperatures in the environment. Aspects like physical properties regarding structures in cities, like artificial materials such as buildings, roads, parking areas, and walking paths, absorb most of the heat during the day and release it slowly after the sunset. Natural surfaces – soil or plants – do not readily absorb heat (Simmons 2008). City structures, and how cities and inner cities are vertically built-up, can hinder airflow. Heat and hot air can become trapped between buildings, blocking natural airflow that can bring in cooler effects. Additional factors can come from heat production generated from machines, motor vehicles, trains, air-conditioning and heating systems, factories, and industries. Another cause can be geographic landscapes, as neighborhood mountains do not allow the free flow of air and block the wind from reaching the inner-city, so warm air is not redistributed to outside areas. Finally, urban constructions, roads, pavements, and parking areas block the pollutants from escaping (Hibbard et al. 2017).

Outside natural influences can also be contributing factors. The emission of infrared rays from buildings and street surfaces can contribute to UHIs. Warm surfaces with elevated amounts of infrared radiation emitted from the ground can then be discharged to a higher amount of radiation back into the environment (Shahmohamadi et al. 2011). The urban greenhouse effect increases when the radiation reaches the polluted urban climate (Simmons 2008; Ramanathan and Feng 2009; Kweku et al. 2017). Consequently, it keeps the Earth warm because some of the planet's heat that may escape from the atmosphere out to space is retained. Studies have reported that the Earth's average temperature would be significantly colder without the greenhouse effect, and life on Earth would be difficult (Kweku et al. 2017).

Features of Heat Islands

Temperatures may vary city by city; heat islands can be measured by the temperature difference among towns and surrounding areas. These areas have a higher temperature secondary to heat-absorbing buildings, roads, and pavements, whereas other sites have more grass and trees that keep the atmosphere cooler. These temperature differences create intraurban heat islands. Urban parks, ponds, and residential areas are typically cooler than downtown. The heat island phenomenon can occur in any part of the day or night; maximum elevation of temperature occurs during clear and still-air nights. Under these conditions, a 3–5 °F temperature increase can occur; however, trends can vary, and 8–10 °F peaks have also been observed (Shahmohamadi 2010b).

Temperature differs at the earth's surface and in the atmospheric air because heat islands are surface heat islands and atmospheric heat islands. Consequently, there will be varied formation, techniques used to identify and measure UHIs, associated impacts, and the strategies available to cool them (Simmons et al. 2008).

Surface Heat Islands

This type of heat island forms because urban surfaces, such as roadways and rooftops, absorb and release heat to a greater extent than most natural surfaces. On hot days, the temperature of roof material may become warmer than the air temperature (Simmons et al. 2008). Surface heat islands tend to be most intense during the day when the sun is shining.

Atmospheric Heat Islands

Heat islands can form due to warmer air in urban areas compared to the cooler air in outlying areas. Atmospheric heat islands vary much less in intensity than surface heat islands. For example, summer mortality rate analysis in Shanghai revealed a high rate of heat-related mortality in urban regions. It has been established that UHI is responsible for the health hazards in Shanghai that are secondary to life-threatening thermal conditions (Tan et al. 2010).

Strategies to Reduce the Urban Heat Island Effect

To curb urban heat island effects, alterations in urban environment infrastructure should be made to add greenspace, for example, green belts should be introduced along streets or parks, gardens, and other green zone plantations could be included throughout the city. That said, sufficient space might not be available to add trees or other greenery; while these challenges may exist, small green infrastructure practices can be incorporated into grassy or unfertile areas, empty lands, and street rights-of-way. Transformation of the city with local plants, trees, shrubs, grasses, and vegetation can encourage dissipation or decrease temperatures in urban heat islands. Green belts and green roofs can provide direct and indirect cooling. Increasing trees and vegetation in surrounding areas can absorb pollution and improve the environment and air quality. Using light-colored materials or paints on buildings can help, as light colors reflect more sunlight and trap less heat. These strategies can be promoted through governmental support, by supporting positive activities through incentives or a tax exemption (Imam and Banerjee 2016; Gunawardena et al. 2017; Velázquez et al. 2019; Li and Yeung 2014).

Reducing the Effect of Urban Heat Island Energy Consumption

As described, the urban heat island effect is caused by less rigorous wind, high density, shading, and the higher temperature of the environment. Some suitable implementations must be made to achieve a better climatic condition to overcome the effect of excessive heating. High environmental temperature contributes to the overutilization of air-conditioning systems, overconsumption of electric energy, and increased energy demand; this situation then creates a harmful cycle between energy supply and heat production. Strategies should be made to decrease excessive heat production and energy preservation while improving systems for buildings.

Properly developed landscape plans for the built urban environment can help in reducing energy consumption. For example, suitable landscaping around a building could be positive by reducing energy consumption. To improve the outdoor climate condition, landscaping neighborhoods is an essential requirement. Greeneries, plantations, and vegetation in urban areas can positively affect thermal temperatures toward a cooler ambient atmosphere (EPA 2021). Natural greeneries such as trees, vegetation, and greenery on roof tops may help reduce urban heat island effects by sheltering and embracing buildings surfaces, artificial constructions, and reflecting radiations coming from the sun. In addition, these plants release moisture into the atmosphere, contributing to the cooling effect (EPA 2021). Installation of cool or reflective roofs helps reflect sunlight and heat away from the building structures, reduce roof temperatures, and minimize the need for air conditioning. One study proved that cool roofs could save annual energy consumption as much as 50 cents per square foot, contributing to better air quality and less emission of greenhouse gases into the atmosphere (EPA 2021).

Application of some cooling effects through natural means, such as plantation, vegetation, and promotion of greeneries, is advised because plantation reduces UHI effects by shading building surfaces, deflecting radiation from the sun, and releasing moisture into the atmosphere. Trees increase the shade coverage which helps reduce UHI, improve quality of the air by increasing evapotranspiration, which finally cools the air. Additionally, trees provide ecosystem services including habitat, erosion control, water quality, human health, and esthetics. The green roof protects the building structures from direct heat from sun. In winter, the green roof reduces heat loss through added roof insulation (BCIT 2021).

Moreover, water surfaces and wind channeling can subsidize through natural or artificial fences and lessen the effect of solar radiation in summer and shelter buildings in winter. Plantation and greeneries must be encouraged and promoted in cities, as it is environmentally friendly and can positively impact neighborhoods. Trees filter water supply; they absorb stormwater run-off before releasing it back into the atmosphere. Roots allow soil to retain water, filter out toxins, and impact the health of its citizens. Green foliage decreases stress levels and increases attention spans (Seitz and Escobedo 2008).

Summary

Because of urbanization, the urban heat island (UHI) effect has become a modern threat worldwide; UHIs can affect human health through increased average temperatures. Individuals with respiratory illnesses have difficulty breathing because sunlight reacts with smog from vehicles to produce polluted air. Long-term exposure to polluted air causes cancer, emphysema, bronchitis, and accelerated aging of the lungs (NIH 2019).

Pocket parks or other areas of greenspace could be a key contributor to altering these negative effects. Lin et al. (2017) provided evidence that pocket parks were cooler than their surrounding urban streets for daytime and nighttime, which means that pocket parks could help other metropolitan cities alleviate UHI intensity at the micro-level. Planting trees is also an effective method to reduce the UHI intensity inside the parks for high-rise, high-density urban environments (Lin et al. 2017). The cooling effect provided by the trees during the daytime may have an extended impact after sunset. Some researchers found that trees have more influence on reducing air temperature than grass since their shelter can provide shading in addition to the evapotranspiration effect (Giridharan et al. 2008; Yang et al. 2011). It has been suggested that the implementation of grass planting should be combined with trees (Srivanit and Hokao 2013).

Another study showed that smaller parks are better at lowering temperatures than larger ones. More extensive parks can contribute to the thermal environment in opposing ways, especially in the subtropical area. At one end, larger parks explain a more significant portion of land exposed to solar radiation and a higher temperature during the daytime. On the other end, an enormous surface area means more vegetation that could alleviate UHI by evaporation and transpiration (Chang et al. 2007). The intensity of UHI in urban gardens is subject to the balance between these two opposing effects. The larger parks in overpopulated areas do not always lead to lower outdoor air temperature, though avoiding direct radiation from the sun could significantly lower daytime UHI intensity (Lin et al. 2017). During the night, the intensity of UHI is controlled by heat dissipation from the park ground and neighboring buildings. However, during early nighttime, the temperature of the air benefits from less heat received during the daytime.

The urban environment involves many factors, such as air temperature, ventilation, air pollution, perceptive experience, etc. These factors are usually negatively influenced by the high density of adversely contributing factors to the local environment. However, vast masses of green natural areas can carry immense value from both social and environmental perspectives; therefore, to abandon the effect of UHI, comprehensive environmental policies must be developed, and practical strategies should be formulated.

Conclusion

The average temperature of the earth's surface is rising due to climate change. Thus, contributing factors, such as energy utilization, must be reduced, while green energies and climate-reducing strategies must be implemented to protect the environment. These strategies are useful in urban areas. UHI effect are one negative outcome that affects human health. Individuals may be susceptible to heatwaves or suffer from ailments such as cancer and asthma. Heat affects life and safety. To combat UHI, painting a building white reduces the internal temperature by about 10–30%. Consequently, energy efficiency is promoted, and cities can combat climate change effectively. Encouraging and promoting greenspace can reduce UHI intensity. Planting trees in the parks and maintaining brighter buildings and roofs are an effective method to diminish the UHI intensity in the urban environment. These types of solutions could enhance microclimate regulation, drive more sustainable urban development, address the environmental health threat, and mitigate UHI.

Cross-References

- ▶ Building Community Resilience
- ▶ Children, Urban Vulnerability, and Resilience
- ▶ Circular Economy and the Water-Food Nexus
- ▶ City Visions: Toward Smart and Sustainable Urban Futures
- ▶ Connecting Urban and Regional Innovation Ecosystems to Enhance Competitiveness
- ▶ Future of Urban Governance and Citizen Participation
- ▶ Smart Cities
- ▶ Urban and Regional Leadership
- ▶ Urban Atmospheric Microbiome
- ▶ Urban Climate Resilience
- ▶ Urban Commons as a Bridge between the Spatial and the Social
- ▶ Urban Ecosystem Services and Sustainable Human Well-being
- ▶ Urban Food Gardens
- ▶ Urban Forestry in Sidewalks of Bogota, Colombia
- ▶ Urban Futures
- ▶ Urban Greening and Green Gentrification
- ▶ Urban Health Paradigms
- ▶ Urban Management in Bangladesh

- ▶ Urban Nature
- ▶ Urban Policy and the Future of Urban and Regional Planning in Africa
- ▶ Urban Resilience
- ▶ Urban Resilience: Moving from Idealism to Systems Thinking
- ▶ Urban Structure and Its Impact on Mobility Patterns: Reducing Automobile Dependence Through Polycentrism
- ▶ Urbanization, Planning Law, and the Future of Developing World Cities
- ▶ Urban Well-Being
- ▶ US Urban and Suburban Yardscaping

References

Arbuthnott, K. G., & Hajat, S. (2017). The health effects of hotter summers and heat waves in the population of the United Kingdom: A review of the evidence. *Environmental Health, 16*(1), 1–13.

BCIT. (2021). *Why green roofs? Benefits?* Centre for Architectural Ecology. BCIT. Retrieved from https://commons.bcit.ca/greenroof/faq/why-green-roofs-benefits/

Campbell, D., Shotter, J., & Draper, R. (2018). The socially constructed organization. In D. J. Wuebbles, D. W. Fahey, K. A. Hibbard, D. J. Dokken, B. C. Stewart, & T. K. Maycock (Eds.), *Climate Science Special Report: A Sustained Assessment Activity of the U.S. Global Change Research Program* (pp. 405–442). Washington, DC: Global Change Research Program.

Chan, E. Y. Y., Goggins, W. B., Kim, J. J., & Griffiths, S. M. (2012). A study of intracity variation of temperature-related mortality and socioeconomic status among the Chinese population in Hong Kong. *Journal of Epidemiology Community Health, 66*(4), 322–327.

Chang, C. R., Li, M. H., & Chang, S. D. (2007). A preliminary study on the local cool island intensity of Taipei city parks. *Landscape and Urban Planning, 80*(4), 386–395.

Da-Lin Zhang, Yi-Xuan, S., Russell, R. D., & Fei, C. (2011). Impact of Upstream Urbanization on the Urban Heat Island Effects along the Washington–Baltimore Corridor, 2012–202. https://doi.org/10.1175/JAMC-D-10-05008.1

Deilami, K., Kamruzzaman, M., & Liu, Y. (2018). Urban heat island effect: A systematic review of spatio-temporal factors, data, methods, and mitigation measures. *International Journal of Applied Earth Observation and Geoinformation, 67*, 30–42.

Emmanuel, M. R. (2004). An Urban Approach To Climate Sensitive Design: Strategies for the Tropics.

EPA. (2021). United States Environmental Protection Agency. Heat Island Compendium. Retrieved from https://www.epa.gov/heatislands/heat-island-compendium

Giridharan, R., Lau, S. S. Y., Ganesan, S., & Givoni, B. (2008). Lowering the outdoor temperature in high-rise high-density residential developments of coastal Hong Kong: The vegetation influence. *Building and Environment, 43*(10), 1583–1595.

Gunawardena, K. R., Wells, M. J., & Kershaw, T. (2017). Utilizing green and bluespace to mitigate urban heat island intensity. *Science of the Total Environment, 584*, 1040–1055.

Hibbard, K. A., Hoffman, F. M., Huntzinger, D., & West, T. O. (2017). Changes in land cover and terrestrial biogeochemistry. In D. J. Wuebbles, D. W. Fahey, K. A. Hibbard, D. J. Dokken, B. C. Stewart, & T. K. Maycock (Eds.), *Climate science special report: Fourth national climate assessment* (Vol. I, pp. 277–302). Washington, DC: U.S. Global Change Research Program. https://doi.org/10.7930/J0416V6X.

Imam, A. U., & Banerjee, U. K. (2016). Urbanization and greening of Indian cities: Problems practices, and policies. *Ambio, 45*(4), 442–457. https://doi.org/10.1007/s13280-015-0763-4.

Kweku, D. W., Bismark, O., Maxwell, A., Desmond, K. A., Danso, K. B., Oti-Mensah, E. A., & Adormaa, B. B. (2017). Greenhouse effect: Greenhouse gases and their impact on global warming. *Journal of Scientific Research and Reports*, 1–9.

Lin, P., Lau, S. S. Y., Qin, H., & Gou, Z. (2017). Effects of urban planning indicators on urban heat island: A case study of pocket parks in high-rise high-density environment. *Landscape and Urban Planning, 168*, 48–60.

Luo, Y., Li, H., Huang, F., Van Halm-Lutterodt, N., Xu, Q., Wang, A., & Guo, X. (2018). The cold effect of ambient temperature on ischemic and hemorrhagic stroke hospital admissions: A large database study in Beijing, China between years 2013 and 2014–utilizing a distributed lag non-linear analysis. *Environmental Pollution, 232*, 90–96.

NIH. (2019). *Climate and human health. Health impacts*. National Institute of Environmental Health Sciences. Retrieved from https://www.niehs.nih.gov/research/programs/climatechange/health_impacts/index.cfm

Radhi, H., Sharples, S., & Assem, E. (2015). Impact of urban heat islands on the thermal comfort and cooling energy demand of artificial islands–a case study of AMWAJ Islands in Bahrain. *Sustainable Cities and Society, 19*, 310–318.

Roth, M., Oke, T. R., & Emery, W. J. (1989). Satellite-derived urban heat islands from three coastal cities and the utilization of such data in urban climatology. *International Journal of Remote Sensing, 10*(11), 1699–1720.

Rupard, M. (2019). Urban Heat Islands: Causes, Impacts, & Mitigation. B.S. City & Regional Planning, California Polytechnic State Univeristy San Luis Obispo.

Santamouris, M. (2014a). On the energy impact of urban heat island and global warming on buildings. *Energy and Buildings, 82*, 100–113.

Santamouris, M. (2014b). Cooling the cities–a review of reflective and green roof mitigation technologies to fight heat island and improve comfort in urban environments. *Solar Energy, 103*, 682–703.

Seitz, J., & Escobedo, F. (2008). Urban forests in Florida: Trees control stormwater runoff and improve water quality. *EDIS, 2008*(5).

Shahmohamadi, P., Che-Ani, A. I., Ramly, A., Maulud, K. N. A., & Mohd-Nor, M. F. I. (2010a). Reducing urban heat island effects: A systematic review to achieve energy consumption balance. *International Journal of Physical Sciences, 5*(6), 626–636.

Shahmohamadi, P., Che-Ani, Maulud, K. N. A., Tawil, N. M., & Abdullah, N. A. G. (2010b). Mohd-Nor Faculty of Engineering and Built Environment, University Kebangsaan, Malaysia. Faculty of the Built Environment, University of Malaya, Malaysia. *International Journal of Physical Sciences, 5*(6), 626–636.

Shahmohamadi, P., Che-Ani, A. I., Maulud, K. N. A., Tawil, N. M., & Abdullah, N. A. G. (2011). The impact of anthropogenic heat on formation of urban heat island and energy consumption balance. *Urban Studies Research, 2011*.

Simmons, M. T., Gardiner, B., Windhager, S., & Tinsley, J. (2008). Green roofs are not createdequal: The hydrologic and thermal performance of six different extensive green roofs and reflective and non-reflective roofs in a sub-tropical climate. *Urban Ecosystem, 11*(4), 339–348.

Srivanit, M., & Hokao, K. (2013). Evaluating the cooling effects of greening for improving the outdoor thermal environment at an institutional campus in the summer. *Building and Environment, 66*, 158–172.

Tan, J., Zheng, Y., Tang, X., Guo, C., Li, L., Song, G., & Chen, H. (2010). The urban heat island and its impact on heat waves and human health in Shanghai. *International Journal of Biometeorology, 54*(1), 75–84.

Tuholske, C., Caylor, K., Funk, C., Verdin, A., Sweeney, S., Grace, K., & Evans, T. (2021). Global urban population exposure to extreme heat. *Proceedings of the National Academy of Sciences, 118*(41).

Velázquez, J., Anza, P., Gutiérrez, J., Sánchez, B., Hernando, A., & García-Abril, A. (2019). Planning and selection of green roofs in large urban areas. Application to Madrid metropolitan area. *Urban Forestry & Urban Greening, 40*, 323–334.

Yang, F., Lau, S. S., & Qian, F. (2011). Urban design to lower summertime outdoor temperatures: An empirical study on high-rise housing in Shanghai. *Building and Environment, 46*(3), 769–785.

Urban Hinterland

▶ Planning for Peri-urban Futures

Urban Management in Bangladesh

Rules and Regulations

Prabal Barua and Syed Hafizur Rahman
Department of Environmental Sciences, Faculty of Physical and Mathematical Sciences, Jahangirnagar University, Dhaka, Bangladesh

Synonyms

Climate change; Disaster management; Sustainability; Urbanization

Definition

The urban environment comprises many fundamental elements: housing, water supply, trades, transportation, electricity supply, gas supply, waste disposal, telecommunications, different service facilities, sludge and drainage systems, open space, and park and field facilities. Some autonomous bodies and private or local administrations manage the fundamental facilities of the urban environment for urban dwellers (Devas 2001).

Urban management is a wide-ranging expression covering a huge set of functions. The core purpose of an urban city is to give communities a livable atmosphere, including shelter, employment, livelihoods, entertainment, food, and other necessary things for healthy, sound living conditions. Urban management is a mixture of all elements converging jointly to provide communities with all kinds of basic amenities as citizens in the urban city. When inhabitants find income-generating opportunities, they shift to an urban center from marginalized rural areas. A city with job opportunities can attract people only when it has basic infrastructure available. Cities with all infrastructure in place but no economic opportunities are of no use to people, who will move to other cities even if their infrastructure is not as good as that of the city they are living in. With the growth of urbanization, it is becoming quite hard to ensure that city environments provide good quality of life for urban populations. Nowadays, most urban cities have the disadvantages of high air pollution, insufficient pure water for drinking and other uses, and inadequate waste disposal and management processes. Moreover, rapid population growth and migration from rural to urban areas – as well as lack of city planning and design, lack of implementation of environmental laws, etc. – further worsen the city environment (Jabeen et al. 2010).

For development of sustainable urban management, city planners and researchers have taken three different approaches to ensure that sound environmental facilities are available for citizens, development activities take place for citizens that are community friendly and provide accessible communication and other basic amenities, and there is freedom from conflict during distribution of resources and services. The conflicts among the communities is within the city itself, among various social groups where easily accessible social programs, cultural activities, recreation, and transportation facilities are available and all amenities are environmentally friendly and sustainable for long-term capacity of effective urban management (Fig. 1).

So, it can be concluded that urban management appears to have the following objectives: to design, provide, and sustain a city's basic services and different infrastructures; and to ensure that the city's governance is in a robust state, institutionally and economically, to ensure adequate provision and safeguarding for its inhabitants.

Urbanization of Bangladesh and Requirement for Protection

Bangladesh is a unitary state system with constitutional obligations of local government to provide every kind of facility for the population. For urban citizens, local government bodies, such as city corporations for large cities and municipalities for secondary cities and small towns, are created. There are 12 city corporations and 328 municipalities in Bangladesh at present. The classification of urban areas in Bangladesh was established in 1977 during the time of the declaration of the Pourashava Ordinance of 1977, and after the Local Government (Municipality) Ordinance of 2009 was declared in the constitution of the country, a modified concept of urban areas was introduced (Banks et al. 2011).

According to the existing laws in Bangladesh, an urban area can be declared under four conditions:

1. Three-fourths of the mature male population is involved in nonagricultural professions.
2. The use of almost 33% of the land is nonagricultural in nature.
3. The community population is at least 50,000.

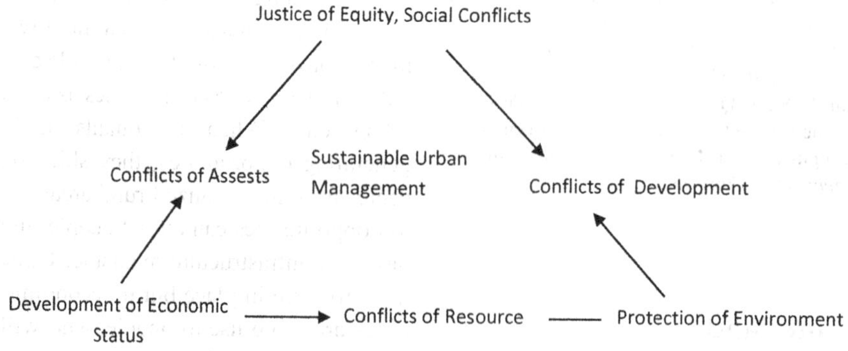

Urban Management in Bangladesh, Fig. 1 Principles of sustainable urban management

4. The population density is not less than 1500 people per square mile.

Bangladesh – a developing country in South Asia – had an urban population of 60 million people in 2018, although its rate of urbanization (1.40) was lower than that of lower-middle-income countries. It is predicted that by 2050, 56% of the nation will live in urban communities. The urban communities exhibit very high human population density, with inadequate space for living and little road infrastructure for movement. The rapid urban population growth in metropolitan cities and their outskirts is increasing environmental pollution, posing problems for human health, and threatening the general quality of life. Unmanageable traffic troubles city inhabitants. Currently, urban growth is extremely unplanned, showing high degrees of congestion, overcrowding, and malfunctioning of the traffic system (Begum and Hasan 2017; Barun and Abheuer 2011). High-rise residential buildings are springing up in large cities, and many such buildings tend to serve commercial purposes. Business activities are found to greater or lesser degrees all along the roads and pavements but are intensified at crossroads. The traffic system has cost millions of dollars and has allowed thousands of defective, outmoded buses and other commercial vehicles to travel everywhere and freely take over the roads. Life has become miserable with load shedding, low gas pressure, and shortages of pure drinking water. The changing face of what was once called a green city is now characterized by apartment blocks sprouting like mushrooms. It has thus become an urban jungle. Nowadays, hardly any houses with gardens and orchards are available in the city. Multistory apartment blocks are close to each other with little space and without parking and playgrounds (Swapan et al. 2017; Masheque and Kabir 2016).

Environmental degradation stems from rural-to-urban migration from villages and small market towns with rural characteristics to metropolitan urban centers full of development activities. All such development activities centering on urbanization conflict with ecological factors, leading to gradual degradation of life support systems, including the air, water, and land. Enormous population pressure in the core of the city, even downtown, aggravates this situation. Development trends at the urban level are not environmentally friendly, contributing much to the havoc created by climate change.

In the wake of misdirected urban development, the human–environment interface in Dhaka, Chittagong, and other major urban areas seriously impairs their ecology. This has become a matter of concern to urban planners. The city environment has worsened in the wake of unplanned urbanization. There has been hardly any well planned development over the years. The authorities are spending lavishly, mostly in unproductive sectors, without producing any commensurate results. Rampant corruption and the blame game account for much fuzzy urban governance.

A desirable state of urbanization with the overriding considerations of communications, the traffic system, and housing and residential patterns has been stressed recently by urban planners. What city dwellers expect from the city development authority is a well-planned city with systematic clustering of houses in the residential area, well connected to the city's main focus points – its administrative and commercial districts – through linking roads that are accessible to transport (Roy 2009; Panday 2011). Distressingly, the tremendous growth of urban populations in Dhaka and Chittagong have aggravated the crisis of unplanned urbanization in these cities, with concrete jungles growing thickly all around the cities and growing numbers of vehicles thronging limited numbers of roads. Dhaka and Chittagong are becoming urban jungles, and everywhere people have found apartment blocks sprouting like mushrooms. There are fewer houses with gardens. Apartments are so close to each other that there is a lack of privacy. Children are growing up in apartments that do not have any outside access to recreation – no garden, no playground. Over a couple of decades, urbanization has assumed greater significance in influencing the growth, distribution, density, and structure of the population.

Even then, there is no proper system for waste disposal. There is generation of "organic pollution hazards" because garbage spreads everywhere in the form of heaps all over filthy city points. Now there are dumping grounds in outer, loosely built-up communities. A well-maintained garbage disposal system is lacking in the urban cities of Bangladesh. The city corporations in statistical metropolitan areas collect garbage through use of scanty resources and manpower. The garbage collected from residents and offices by vans is disposed of at dumping grounds. At many city points, amounts of solid waste are discharged into drains or canals, causing a great deal of water pollution (Panday 2017; Khan et al. 2012).

In urban areas of Bangladesh, the drainage systems are poor; most drainage systems remain choked, and flooding and waterlogging during the rainy season cause much trouble for pedestrians. Waste accumulates in the rainwater. As a result, roads, lanes, and by-lanes become pools of water. Wastewater goes out through open drains, which run along the roads. Waterlogging during the rainy season – which is due to filled-up canals, defective drainage systems and sewage – is another menace.

This is common in all urban cities of Bangladesh, which are growing vertically because of lack of space. Planned urbanization with proper schemes for housing and settlement of low- and medium-income groups is woefully lacking. The large cities of the country, such as Dhaka and Chittagong, are dotted with small-nucleated settlements on the fringes of each residential locus. The poor, mostly working class – consisting of factory laborers, rickshaw pullers, vendors, and maidservants – live in slum areas in squatter settlements. In Chittagong, slum areas are situated on hillsides or on hilltops, which are frequently prone to natural hazards (Ahmed et al. 2018).

Housing for all is an aim of the Bangladesh Vision 2021 political manifesto. The government intends to provide accommodation for rural residents through rural housing schemes around growth centers in each union and *upazila* [sub-district], with modern facilities in urban areas. The government's plan to provide residential accommodation for all also includes insolvent freedom fighters. An initiative has been taken to build shelter homes for the floating population in urban areas in order to create employment opportunities. For the poor and for fixed-income groups, possession of an apartment in metropolitan urban areas has become very difficult to achieve and is like a golden deer. Everybody is hankering after money to chase the golden deer.

Admittedly, the housing problem is now an immense one all over the country. Residential accommodation is a basic need of the disadvantaged population that is supposed to benefit from the social safety net. For this reason, housing for all should be brought into the social safety net and extended with a much enhanced allocation.

Good governance in the main cities will not be a distant dream if urban citizens are provided with accommodation. Apartment culture in Dhaka, Chittagong, and other urban cities has developed, with high-rise building mushrooming. Because of the escalating cost per square foot, it has become difficult for most citizens doing white-collar jobs to purchase an apartment. The affluent and some workers sending remittances from abroad can afford to possess one or sometimes even more apartments. Management of tenders for contractorship and approval of blueprints for residential building and commerce by offering kickbacks has become proverbial. Under the Building Construction Act, all types of buildings (including dwelling units and commercial establishments) require permits from the relevant government department. A few persons do comply with building codes. Conversely, it is easy to arrange permits by exerting undue influence and by exploiting any kinship or more fictive connections (Ahmed et al. 2013).

Congestion on roads brings traffic to a standstill for hours, causing tremendous hardship for pedestrians. It is paralyzing normal day-to-day business and public and social services, making it difficult for traders, merchants, bankers, and utility service personnel to accomplish their respective tasks in time. The beneficiaries of services rendered by the public and private sectors are unable to get necessary things done without delays, as the locations of institutional buildings and commercial establishments are hard to reach. Recently, traffic management in urban cities such as Dhaka and Chittagong has become the worst in the world, and the traffic system has come to the verge of collapse (Ahmed et al. 2014). There has been hardly any marked improvement in the management of traffic movement at the main city points despite a plethora of meetings and seminars. Nothing tangible is being done to prevent the traffic system from stumbling into a dark hole. Urban cities are becoming crowded with huge numbers of city dwellers, and there have been marked increases in the floating populations moving to and from the busy commercial and administrative districts in cities. As a result of tremendous population pressures in Dhaka and Chittagong, the numbers of private cars, family pick-up vans, utility vehicles, and sport utility vehicles on city roads have increased enormously.

According to statistics on the transportation systems of urban cities in Bangladesh, the numbers of private vehicles on city roads increase daily by 150 in Dhaka and by 70 in Chittagong, whereas the increases in the numbers of new public transport vehicles are negligible (6 per day in Dhaka and 3 per day in Chittagong). In Chittagong, which is the commercial capital of Bangladesh, only 8% of the land area is occupied by the road network, whereas internationally, a road network occupying 25% of urban land area is considered to be the minimum standard. Dhaka, too, requires a vast extension of its road network to make it proportionate to the size of the city. There are only a few alternative roads. In comparison, the Indian city of Kolkata is better off in this respect. Now Dhaka and Chittagong have become urban jungles with a mass of cars gridlocking the city streets. At this critical juncture, it is important that some remarkable steps are taken so that vehicles can move smoothly and reach their destinations in time. Plans for opening of four-lane VIP roads will do little to rid us of troublesome road jams if no serious action is taken to limit the increases in the numbers of cars on the roads (Ahmed et al. 2012).

Despite repeated warnings and responses from the relevant administrations, tragic incidents occur in Chittagong almost every year during the monsoon. During a heavy downpour on July 1, 2011, devastating landslides hit hill slums in Chittagong, killing 120 people living in houses on the hill slopes. Some families have continued to take risks by remaining there despite that catastrophe, whereas others have shifted to safer places. It should be noted that in 2011, the Chittagong City Corporation initiated construction of a protective wall to prevent landslides. It was, of course, intended to be a protective measure against natural devastation, but its construction was faulty, as different experts noted. Following the disaster, a hill management committee (headed by the Chittagong divisional commissioner) was formed to compensate and rehabilitate the affected people (Ahmed et al. 2015).

The causes of landslides are unstable geographical formations, reckless construction, heavy rainfall, lack of proper sewerage systems, and lack of knowledge about nature's ability to inflict huge numbers of human fatalities. Yet, the most remarkable factor seems to be lack of governance in urban Chittagong, where growth of squatter settlements on risky hill slopes has mushroomed. There have been reckless earthworks on those hills, while the relevant authorities in Chittagong have turned a deaf ear to allegations about illegal earthworks on public land, leveling of hills, and movement of hill rocks. It can be concluded that the original residential pattern, which was full of well-planned accommodation built around open spaces, is gradually fading. Among the high-rise buildings in a city with a seemingly fast-growing skyline, there are numerous dilapidated huts and metal-clad houses built mostly by people in the lower deciles in terms of income and calorie consumption. Bangladesh is now at risk of devastating earthquakes, which would be a great catastrophe (Brammer 2018).

Urban Management in Bangladesh

Bangladesh's speedy urbanization has occurred faster than the nation's social and economic advancements, making the country one of the earth's fastest-rising economies. Worldwide, urbanization has spurred economic development and has added considerably to national economies. Urban Bangladesh contributes more than 65% of the national gross domestic product (GDP), but its urbanization process is still uncoordinated, is somewhat unplanned, and, most importantly, lacks a favorable policy framework. We still have not agreed on what kind of cities we want. While economic progress is helping Bangladesh to graduate out of the least-developed country (LDC) category, urban poverty and development require significant attention.

The total population living in urban areas of Bangladesh was only 6.27 million in 1974 but increased to over 39 million by 2011. In 2020, it was approximately 60 million. The level of urbanization leaped from 8.78% to 40% during the same period. The long-run exponential growth rate in the urban population over the period of 1974–2020 was 7.50%, versus only a 1.36% growth rate in the rural population. When this is seen in conjunction with the fact that in the period of 1974–2011, the long-run trend of growth in the total population was only 2.18%, it gives an idea of the rapidity of the pace of urbanization. The effect of speedy urbanization on the urban environment in Bangladesh is alarming. The overall reduction in urban quality of life is due to increasing mass poverty, unpleasant social inequality, rising unemployment, high population density in small areas, insufficiency of clean water, unclean sanitation, increasing outbreaks of diseases and infections, high crime rates, unpleasant social disturbances, and so on (Mia et al. 2015). In addition, unplanned usage of urban land (e.g., destruction of hills and increasing utilization of hills for housing settlements) impact the normal flows of drainage schemes, rivers, and canals. In summary, the rapid urban population increase is rapidly degrading the urban environment, and such environmental degradation is apparent in the worsening of the physical and community environments in urban places and obliteration and degradation of various ecosystems connected with urban cities (Kawsar 2012).

The government of Bangladesh has constantly given priority to planned urbanization in its national development plans, beginning with the first Five Year Plan (FYP). Almost all succeeding FYPs, including the seventh FYP (2016–2020), have highlighted its significance. The current plan is focused on adopting an appropriate vision for a sustainable urban future through development of appropriate policies. According to the seventh FYP, Bangladesh is still lacking a comprehensive national corporeal plan or plan of land use, which was a focus of the first FYP, although the Planning Commission has considered the significance of the urban sector. Over time, the population of Bangladesh will become more urban. Nearly half of all Bangladeshis are anticipated to live in prestigious urban areas by 2035. At present, more than 60% of the urban communities are concentrated mainly in the five metropolitan cities of Dhaka, Chittagong, Khulna, Sylhet, and Rajshahi. By the year 2030, this tendency toward migration from rural to urban places will be more definite.

Pattern of Urban Climate Change

Living in cities, particularly in high-density slum settlements, can mean exposure to a number of disaster risks, health risks, and environmental risks, which particularly affect the poor. As the world becomes more urbanized, urbanization is increasingly viewed as the heart of the climate agenda. The impacts of modern urbanization and climate change are converging in dangerous ways that seriously threaten the world's environmental, economic, and social stability.

Human vulnerability is about the prospect of eroding human development achievement and sustainability. Climate change vulnerability is a new form of marginalization and intensifies urban poverty. Cities are important entities in the climate change arena, both as contributors to greenhouse gas (GHG) emissions and as centers of activity for reducing emissions. Climate change presents unique challenges for urban areas and their growing populations.

Bangladesh symbolizes an area where a complex set of climate-driven outcomes have already been demonstrated as land is submerged and communities are forced to migrate in large numbers. Progression of climate change drives migration of extremely poor people living in marginalized rural areas, particularly those in coastal areas, to urban areas. Rapid urbanization and consumption of fossil fuels, which cause large proportions of greenhouse gas emissions, are the main causes of climate change. According to a population census conducted by the Bangladesh Bureau of Statistics in 2011, approximately 28% of the total population of Bangladesh (which was 144 million at that time) lived in urban areas of the country, and different statistics previously indicated that by 2020, about 40% of communities would be living in urban areas. Meanwhile, the annual growth rate of urban areas increased from 25% in 2000 to 40% by 2020. This is very alarming for local government, which needs to ensure availability of basic amenities as urban facilities for the population (Rashid 2018: Saleh 2018).

In contrast, the rural growth rates in Bangladesh were only 1.7% in 2000 and 0.2% in 2020 – much lower than the growth rates in urban areas. In addition, the nation is a land-limited country; nearly half of the country's rural marginalized communities are effectively landless. The main forces causing landlessness and thereby driving rural-to-urban migration are the economic pull, recurrent natural disasters, land scarcity, environmental effects such as soil salinity and rising sea levels, and more persistent joblessness and underemployment.

In addition to social and economic matters, other environmental matters that have direct effects on migration are river erosion, flooding, cyclones, storm surges, and other natural disasters. Sociopolitical influences also affect the internal migration process. Extensive poverty combined with elevated population growth, landlessness, and inequality of income opportunities force rural poor people to shift to locations that are extremely exposed to natural hazards, such as low-lying coastal areas or *char* land [accretion land]. Various natural hazards such as river erosion and flooding have direct effects on the livelihoods of vulnerable coastal communities, forcing them to relocate to pursue alternative livelihoods in newly settled urban areas. In most cases, the destinations of homeless and landless climate migrants are urban slums (Barua et al. 2017; Barua and Rahman 2018, 2019, 2020).

Urbanization relentlessly increases the numbers of urban poor, and this is recognized as an important challenge for Bangladesh. All of the big urban cities in the country have already experienced large influxes of disaster victims, with consequent pressures on housing, urban service amenities, and employment opportunities, and with increasing urban poverty. In fact, urban areas of Bangladesh are distinguished by rapid increases in redundancy and underemployment, and linked with this are rapid increases in poverty and slum communities (Barua et al. 2017). Roy et al. (2012) stated that the chief challenges suffered by the urban poor and migrated communities of Bangladesh are:

- Insecure and inadequate incomes
- Imperfect or nonexistent access to public service facilities (water supply, drainage, hygiene, sanitation, and roads)
- Insufficient provision of basic facilities such as health care and education
- Insecure access to property and lack of housing
- Lack of access to natural capital such as forestry and fisheries

Islam and Ahmed (2011) found that within a decade, urban built-up areas of Bangladesh almost doubled in dimension, whereas all other types of location decreased. They assessed three issues related to these changes:

- Rapid urbanization
- Extensive housing requirements in urban areas
- Absence of appropriate urban land use policies and controls

Climate change has induced forced displacement or migration from disaster hot spot rural areas to developed urban areas, putting great pressures on urban land use, housing conditions, water resources, and sanitation, increasing the number of vulnerable urban communities, and thereby impacting the whole urban environment (Barua et al. 2017; Ahsan et al. 2014). In different studies, it was found that almost 60% of the populations of Dhaka and Chittagong are living in different slums with high population densities, and all of them are climate-displaced persons. Crime and drug addiction are common problems in these slums (Ghosh 2014).

Urban Management and Status of Rules and Regulations

Rapid urbanization, weak city governance, and the macroeconomic background all expose urban communities to multiple challenges. Article 15 of the Constitution of Bangladesh declares an obligation "to improve the living conditions of Bangladesh through planned development." To fulfill this obligation, the Planning Commission develops various long-, mid-, and short-term development and economic plans, such as the Perspective Plan (PP) of 2010–2021, Five Year Plans (FYPs), and Annual Development Programs (ADPs). In addition, different international development agendas and goals – such as the Millennium Development Goals (MDGs) of 2000, Sustainable Development Goals (SDGs) of 2016–2030, and New Urban Agendas of 2016 – all inform national development policies and strategies, including the National Strategy for Economic Growth, Poverty Reduction and Social Development (I-PRSP) of 2002–2005; Steps Toward Change: National Strategy for Accelerated Poverty Reduction II of 2007–2011; National Sustainable Development Strategy (NSDS) of 2010–2021; and Perspective Plan of 2010–2021. A handful of acts, rules, regulations, and sector-specific policy documents are the fundamental bases for physical planning and development in urban areas. Among these, the dominant statutes include the Town Improvement Act of 1953 (Ahmed et al. 2012); East Bengal Building Construction Act of 1952 (Ahmed et al. 2012; Choguill 2012), along with its six amendments up until 2006 (GOB 2015b); Bangladesh National Building Code (BNBC) of 2006; and Building Construction Rules of 2008. The primary urban sector–specific policies include the National Urban Sector Policy (NUSP) of 2010 and National Housing Policy (NHP).

The Bangladesh government developed the National Urban Sector Policy of 2010, which envisions strengthening the beneficial aspects of urbanization and, at the same time, effectively dealing with its negative consequences so as to achieve sustainable urbanization, keeping in view the multidimensional nature of the urbanization process. The policy is designed to be gender sensitive and sensitive to the needs of children, the aged, and the disadvantaged.

The key components of the policy are (1) patterns and processes of urbanization; (2) local urban planning; (3) local economic development and employment; (4) urban local finance and resource mobilization; (5) urban land management; (6) urban housing; (7) urban poverty and slum improvement; (8) urban environmental management; (9) infrastructure and services; (10) social structure; (11) rural-to-urban linkage; (12) urban governance; and (13) research, training and information on urban cities.

The policy recognizes that urban areas will form a network of distribution, where each center will fall into a hierarchy. The policy also recognizes that rural-to-urban migration plays a key role in urbanization and has both positive and negative consequences. To achieve balanced urbanization, rural-to-urban migration must be properly guided to avoid overconcentration of populations in one or a few cities. The policy has a special focus on urban land management (paragraph 5.5). The policy emphasizes that the government must exert some degree of control over use and development of urban land–based policies and regulations. A range of urban planning tools – including land use planning, transportation planning and management, site planning, subdivision regulations, and building regulations – can be applied to minimize environmental impacts of urban development activities.

In the urban land management section (paragraph 5.5.7), the policy emphasizes (1) reforming land transfer laws to counter trends toward land accumulation; (2) implementation of land-banking and land-pooling programs that allow the government to increase its pool of land, which can be exchanged for low-cost housing sites in the city; (3) undertaking land readjustment projects that include low-cost land and housing sites; (4) allocating *khas* land (Government state land) for housing the poor; and (5) allocating a reasonable proportion of land in urban places for housing the poor.

The policy also highlights issues related to land development (paragraph 5.5.10) and states that the government can intervene in the land market either by developing land itself or by enabling the private sector to carry out land development activities and to take up special schemes to develop land for housing low-income groups and the poor. The National Urban Sector Policy includes a provision for in situ upgrading and improvement of slums, and for resettlement of slum dwellers, and it seeks to ensure tenure security for the urban poor. The policy states that there should not be any eviction of slum dwellers without proper rehabilitation. The policy also states that master plans should designate areas for slum rehabilitation and that the government should provide the urban poor with access to infrastructure and services for all inhabitants of slums/informal settlements.

There are also other policies and strategies in Bangladesh for water conservation, environmental management, and human settlement of urban areas: the National Industry Policy of 2010; National Drinking Water Policy of 1999; National Policy for Safe Water Supply and Sanitation of 1998; National Environmental Management Plan of 1995; National Environment Policy of 1992; Water Act of 2013; National River Protection Commission Act of 2013; Real Estate Development and Control Act of 2010; Local Government (Municipal) Act of 2009; Local Government (City Corporation) Act of 2009; Local Government (Union Council) Act of 2009; Rivers, Flood Plains and Water Bodies Conservation, Development and Reclamation and Filling Restriction Act of 2005; *Jaladhar Songrokkhon Ain* [Reservoir Conservation Act] of 2000; All Playgrounds of Municipal Area, Open Area, Park and Natural Water Reservoir Including Municipal Area of City, Divisional Town & District Town, Conservation Act of 2000; Water Development Board Act of 2000; *Poribesh Songrokkhon Ain* [Environmental Protection Act] of 2000; Environment Conservation Act of 1995 and its amendment passed in 2002; Environmental Court Act of 2000; Building Construction Acts of 1952 and 1996; Water Supply and Drainage Authority Act of 1996; Environment Conservation Act of 1995; Water Resource Planning Act of 1992; Environmental Pollution Control Ordinance of 1977; Water Pollution Control Ordinance of 1970; Ground Water Management Ordinance of 1985; Environment Pollution Control Ordinance of 1977; Environmental Court Rules of 2000; and Environmental Conservation Rules of 1997 (Young Power in Social Action and Displacement Solutions 2014).

The Bangladesh Environmental Conservation Act of 1995 and the accompanying 1997 rules are arguably the most important legislative documents addressing industrial water pollution. The act is dedicated to "conservation, improvement of quality standards, and control through mitigation of pollution of the environment" (Environment Conservation Act of 1995). The 1997 Environment Conservation Rules, made in accordance with the 1995 act, provide additional guidance for specific components of the act. The act is, in theory, enforced by the Department of Environment (DoE), which has responsibility for:

- Coordinating with other authorities or agencies that have relevance to the objectives of the act
- Adopting safety measures and determining abatement measures to prevent accidents that may cause environmental degradation
- Advising persons on environmentally sound use, storage, transportation, importation, and

exportation of hazardous materials or their components

- Conducting research and assisting other authorities and agencies in conservation and improvement of the environment
- Investigating locations, equipment, manufacturing or other processes, ingredients, or materials, to ensure improvement of the environment, and to control and mitigate pollution
- Collecting, publishing, and disseminating information regarding environmental pollution
- Advising the government on manufacturing processes and materials that may cause pollution
- Ensuring potable water quality

The Environmental Court Act of 2000 supports the Environmental Conservation Act of 1995 and the Environmental Conservation Rules of 1997 by providing for establishment of environmental courts to conduct trials for offenses relating to environmental pollution. It includes protocols for establishment of the court and defines the court's jurisdiction, appropriate penalties, powers of search and entry, and procedures for investigations, trials, and appeals. The Environmental Conservation Act of 1995 and the Environmental Court Act of 2000 were amended in 2002, and the Environmental Conservation Rules of 1997 were extended to incorporate ambient air quality and vehicle emissions.

The Environmental Court Rules of 2000 mention the All Playgrounds of Municipal Area, Open Area, Park and Natural Water Reservoir Including Municipal Area of City, Divisional Town & District Town, Conservation Act of 2000. These rules are imposed for metropolitan areas, municipal areas, divisional towns, and district towns. The following point is also highlighted in the rules: "Natural water body means any river, creek, lake or pond, or a water body indicated as such in a master plan, or any area declared as a flood flow zone by the government, local government or an organization through government gazette notification and any land that contains flowing water and rainwater shall also be included." If any spaces or portions of these areas are owned by any individual and they change those places, they must submit an application to the relevant government authority within 60 days and adhere to that authority's recommendations. The relevant aspects are:

- Any loss to the master plan
- The magnitude of any such loss
- Any loss in the surrounding environment
- Any possibility that the change will affect local inhabitants

If the application is rejected, the applicant may reapply within 30 days, and the relevant authority will reconsider it. If the application is rejected again, the landowner must abide by the government authority's decision. If any person violates the relevant 2000 law, that person faces a prison term of 5 years or a financial penalty of BDT 50,000. If any person changes their land or portions of their land without the necessary consent from the relevant officials, the authority can halt the landowner's actions by issuing a formal notice and is authorized to destroy any structure built on the land without paying any kind of compensation to the landowner. In addition, if anybody violates the 2000 law and establishes any structure on the land, the relevant authority can demolish it. According to the Code of Criminal Procedure of 1898, a first-class magistrate can impose a financial penalty on any person who violates this law.

The problems facing Bangladesh include the following:

- Lack of coordination between different government autonomous bodies and government departments.
- Lack of proper implementation of laws and policies of existing lawful agencies. Some of these laws apply to preservation of playgrounds (e.g., the outer stadium of Kazir Dawri), open spaces (everywhere), and water reservoirs (e.g., Fay's Lake), but preservation of water bodies involves micromorphological

phenomena and, as a result, it is more difficult for officials to monitor.

- Corruption of government bodies or departments involved in approval of plans, who may turn a blind eye to filling-in of water bodies.
- Political influence over appointed officials.
- Inadequate budgets for operation of the relevant authorities; thus, their personnel do not perform thorough monitoring or visit the relevant areas.
- Lack of dissemination of relevant information (e.g., via leaflet distribution programs) in the Chittagong Metropolitan City; as a result, city dwellers are not aware of the relevant laws and the importance of the land.
- Political influence enabling lawbreakers to evade financial penalties.
- Legal violations by real estate companies and property developers who fill in water bodies and thereby dramatically increase land prices; sometimes, even high officials are involved in these activities directly or indirectly.
- Failure by the relevant authorities to detect filling-in of water bodies situated inside landowners' boundaries; this approach has been used to fill in water bodies in Dhaka, Chittagong, Khulna, and other metropolitan cities.

Islam et al. (2014) mentioned that urbanization brings economic and social benefits, but it also poses some serious challenges, especially when it takes place at a pace as rapid as that now in evidence in Bangladesh. Rapid and unplanned urbanization in Bangladesh is creating the following challenges to sustainable development:

- The challenge of providing for an enormous human population
- The challenge of addressing vast inequities in urbanization
- The challenges of increasing urban economic efficiency and employment opportunities
- The challenge of eradicating urban poverty
- The challenges of equitable distribution and accessibility of land and housing
- The challenges of supplying urban material resources and services
- The challenge of facilitating standard education
- The challenge of easy accessibility of standard health facilities
- The challenge of easy accessibility of modern transportation facilities
- The challenges of climate change–induced disaster management, environmental pollution, and sustainability
- The challenges of violent extremism and different social, cultural, and political conflicts
- The challenge of facilitating appropriate utilization of natural resources
- The challenge of good governance approaches

Conclusions

Urbanization and development go hand in hand. An important dimension of the urbanization–growth interface in Bangladesh is that the country's development has not followed the models of enclave-type development (which is an exclusive export sector–driven process of growth and urbanization) or narrow local development unconnected to the global economy. Rapid urbanization in Bangladesh began in the 1980s, a process initially driven by rural poverty and hardship rather than by industrialized labor market opportunities. The level of urbanization has reached 35% and is expected to continue growing rapidly. Consequently, urbanization is changing the dynamics of poverty in Bangladesh, and the tipping point at which Bangladesh's poor population will become predominantly urban is expected to occur within this generation. Urban cities must be an abode where citizens can lead fulfilling lives in complete safety and equanimity. Distressingly, years of neglect in addressing contemporary problems of urban governance have made Bangladesh's cities landscapes of environmental terrorism.

This book chapter has focused on sustainable urban management strategies in Bangladesh and highlighted the different rules and strategies existing in the country for resource conservation and environmental management in urban Bangladesh. However, the implementation of the existing acts and rules is not satisfactory. Contemporary trends in the urbanization of Bangladesh, with rapid growth in urban populations, pose a formidable threat to the management sector. Demands to establish development agencies, as well as public utilities for services and supplies, are increasing. In the absence of a suitable policy framework, the coordination of the promotional activities of all public utilities and departments has assumed an ominous shape, because many urban policy-makers are devoid of skills in policy analysis and policy-making. They shape urban policy with little knowledge about the objective conditions of complicated urban life. Increasing disenchantment with erratic urban management has generated institutional concern about suitable policy planning for urban development to combat problems such as congestion, pollution, deteriorating amenities, population concentration in metropolitan cities, subhuman living conditions in slum areas, and increasing violence. Thus, a recent concern on the part of policy-makers is to formulate appropriate policies for urban governance. Planned city development with modern amenities for life and desirable population levels requires suitable policy strategies to make human settlements, housing, health, social services, and planned development of residential enclaves – with the amenities of schooling, hospital facilities, communications, and market facilities – easily accessible.

Cross-References

▶ Climate Change

References

Ahmed, S. J., Bramley, G., & Dewan, A. M. (2012). Exploratory growth analysis of a megacity through different spatial metrics: A case study on Dhaka, Bangladesh (1960–2005). *Journal of the Urban & Regional Information Systems Association, 24*(1), 45–65.

Ahmed, B., Kamruzzaman, M., Zhu, X., Rahman, M. S., & Choi, K. (2013). Simulating land cover changes and their impacts on land surface temperature in Dhaka, Bangladesh. *Remote Sensing, 5*(3), 5969–5998.

Ahmed, B., Hasan, R., & Maniruzzaman, K. M. (2014). Urban morphological change analysis of Dhaka City, Bangladesh, using space syntax. *ISPRS International Journal of Geo-Information, 3*(4), 1412–1444.

Ahmed, N., Boex, J., Momen, M., & Panday, P. (2015). The local government system in Bangladesh: A comparative analysis of perspectives and practices. *Union Parishad Governance Project and Upazila Parishad Governance Project. Social Science, 30*(3), 45–65.

Ahmed, S., Nahiduzzaman, K. M., & Hasan, M. M. U. (2018). Dhaka, Bangladesh: Unpacking challenges and reflecting on unjust transitions. *Cities, 77* (3), 142–157.

Ahsan, R., Kellett, J., & Karuppannan, S. (2014). Climate induced migration: Lessons from Bangladesh. *Journal of Climate Change Impacts and Responses, 5*(2), 1–14.

Banks, N., Roy, M., & Hulme, D. (2011). Neglecting the urban poor in Bangladesh: Research, policy and action in the context of climate change. *Environment and Urbanization, 23*(2), 487–502.

Barua, P., & Rahman, S. H. (2018). Community-based rehabilitation attempt for the solution of climate displacement crisis in the coastal area of Bangladesh. *International Journal of Migration and Residential Mobility, 1*(4), 358–378.

Barua, P., & Rahman, S. H. (2019). Impact of river erosion on livelihood and coping strategies of displaced people in south-eastern Bangladesh. *International Journal of Migration and Residential Mobility, 2*(1), 34–55.

Barua, P., & Rahman, S. H. (2020). Relationship between climate change adaptation and vulnerability aspects for the coastal Bangladesh. *Asian Profile, 48*(1), 60–80.

Barua, P., Rahman, S. H., & Morshed, M. H. (2017). Sustainable adaptation for resolving climate displacement issues of south eastern islands in Bangladesh. *International Journal of Climate Change Strategies and Management, 15*(2), 440–465.

Barun, B., & Abheuer, T. (2011). Floods in megacity environments: Vulnerability and coping strategies of slum dwellers in Dhaka, Bangladesh. *Natural Hazards, 58*(3), 45–65.

Begum, A., & Hasan, S. (2017). Urban local governance in Bangladesh: Practice and problems of personal administration. *Asian Profile, 45*(5), 400–420.

Brammer, H. (2018). After the Bangladesh flood action plan: Looking to the future. *Environmental Hazard, 9* (1), 118–130.

Choguill, C. L. (2012). New Communities for Urban Squatters: Lessons from the Plan That Failed in Dhaka, Bangladesh. *Urbanization, 45*(3), 70–85.

Devas, N. (2001). Does city governance matter for the urban poor? *International Planning Studies, 6*(4), 393–408.

Ghosh, J. (2014). The Challenges of Urbanisation may be even Greater in Smaller Towns. *City, 45*(5), 50–65.
GOB. (2015b). New Communities for Urban Squatters: Lessons from the Plan That Failed in Dhaka, Bangladesh. Ministry of Disaster Management and Relief, Government of Bangladesh, 80p.
Islam, M. S., & Ahmed, R. (2011). Land use change prediction in Dhaka City using GIS aided Markov chain model. *Journal of Life and Earth Science, 6*(1), 81–89.
Islam, M. S., Rana, M. M. P., & Ahmed, R. (2014). Environmental perception during rapid population growth and urbanization: a case study of Dhaka city. *Environment, development and sustainability, 16*(2), 443–453.
Jabeen, H., Johnson, C., & Allen, A. (2010). Built in resilience: Learning from grassroots coping strategies for climate variability. *Environnment and Urbanization, 22*(2), 415–431.
Kawsar, M. A. (2012). Urbanization, economic development, and inequality. *Bangladesh Research Publications Journal, 6*(4), 440–448.
Khan, M. H., Grubner, O., & Kramer, A. (2012). Frequently used healthcare services in urban slums of Dhaka and adjacent rural areas and their determinants. *Journal of Public Health, 34*(2), 261–271.
Majumder, A. K., Hossain, M. K., Islam, M. N., & Sarwar, N. I. (2007). Urban environmental quality mapping: A perception study on Chittagong Metropolitan City. *Kathmandu University Journal of Science, Engineering, and Technology, 3*(2), 35–48.
Masheque, S., & Kabir, N. (2016). Urban affairs in Bangladesh: A study of contemporary policy issues. *Indian Institute of Public Administration, 16*(3), 45–60.
Mia, M. A., Nasrin, N., Zhang, M., & Rasiah, R. (2015). Urban development and problem in Chittagong, Bangladesh. *Cities, 48*(3), 31–41.
Panday, P. K. (2011). Local government system in Bangladesh: How far is it decentralised? *Journal of Local Self Government, 9*(3), 2005–2030.
Panday, P. (2017). Decentralization without decentralization: Bangladesh's failed attempt to transfer power from the central government to local governments. *Asia Pacific Journal of Public Administration, 39*(3), 177–188.
Rashid, S. F. (2018). Strategies to reduce exclusion among populations living in urban slum settlements in Bangladesh. *Journal of Health and Population Nutrition, 27*(5), 574–586.
Roy, M. (2009). Planning for sustainable urbanization in fast growing cities: Mitigation and adaptation issues addressed in Dhaka, Bangladesh. *Habitat International, 33*(3), 276–286.
Roy, M., Jahan, F., & Hulme, D. (2012). *Community and institutional responses to the challenges facing poor urban in Khulna, Bangladesh in an era of climate change* (Global Development Institute Working Paper Series 16312). Brooks World Poverty Institute, University of Manchester, Manchester.
Saleh, A. Z. M. (2018). Growth, poverty and employment sectors of Bangladesh. *Urban Studies, 30*(3), 60–75.
Swapan, M. S. H., Zaman, A. U., Ahsan, T., & Ahmed, F. (2017). Transforming urban dichotomies and challenges of South Asian megacities: Rethinking sustainable growth of Dhaka, Bangladesh. *Urban Science, 1* (4), 31–40.
Young Power in Social Action and Displacement Solutions. (2014). *Mapping study of climate displacement in Bangladesh*. Chittagong: Young Power in Social Action and Displacement Solutions.

Urban Nature

▶ Stewarding Street Trees for a Global Urban Future

Urban Open Space

▶ Public Space

Urban Parks

▶ Butterfly Gardening in Colombo, Sri Lanka: Approach to Biodiversity Conservation, Monitoring, Education, and Awareness in Urbanizing Habitats

Urban Planning

▶ Augmented Reality: Robotics, Urbanism, and the Digital Turn
▶ Multiple Benefits of Green Infrastructure
▶ The Urban Planning-Real Estate Development Nexus
▶ Urban Forestry in Sidewalks of Bogota, Colombia
▶ Women in Urbanism, Perpetuating the Bias?

U

The Urban Planning-Real Estate Development Nexus

Mapping the Future for Developing Countries

Yvonne Munanga[1], Walter Musakwa[2] and Innocent Chirisa[3]
[1]Department of Architeture and Real Estate, University of Zimbabwe, Harare, Zimbabwe
[2]Future Earth and Ecosystem Services Research Group, Department of Urban and Regional Planning, University of Johannesburg, Johannesburg, South Africa
[3]Department of Demography Settlement and Development, Social & Behavioural Sciences, University of Zimbabwe, Harare, Zimbabwe

Synonyms

Urban planning; Town and regional planning real estate; Land and buildings development; Progression nexus; Link/relationship

Definition

Urban planning is a technical and political process that is focused on the development and design of land use and the built environment. It involves development of master plans on the social and economic activities of urban areas.

Real estate development is the continual reconfiguration of the built environment to meet society's needs.

Land use refers to the purpose a particular piece of land serves; it indicates how people are using the land, for example, residential land use, commercial, industrial.

Sustainability refers to the ability to maintain an object at a certain rate or level, avoiding its depletion or avoiding depletion of natural resources in order to maintain an ecological balance.

Urbanization refers to the population shift from rural to urban areas or the process by which large numbers of people become permanently concentrated in certain areas thereby forming cities.

Introduction

Sustainable urban planning and development is not possible without growth in real estate that provides facilities for all human activities while improving the public spaces in our cities and towns (Kaklauskas et al. 2015). Sustainable development is also a political goal of many nations. Urban planning plays a crucial role in this respect, as a majority of people live in cities where most of the economic value is added. In addition, urban areas are also consuming most resources, which suggests that an efficient organization of urban areas is a key element of achieving this goal. Real estate development plays a central role in the transformation of built space (Healey 1991).

Sustainable real estate development was called upon as a result of assessment claims that pointed on chaotic urban development, depletion of nonrenewable resources, and climate change. Currently, the process of real estate development is dominated by applying the sustainability principles more and more widely in designing, assessing, constructing, using, and demolishing of buildings. The sustainable development basically means that priorities are given to mixed use of buildings, social diversity of people, high-quality projects, and sustainable buildings at the same time observing the urban planning laws of the respective towns and cities (Kaklauskas et al. 2015).

Real estate developments and urban landscape transformation experienced by most nations globally are results of urbanization. Urbanization is a transformational phenomenon that can help improve the lives of hundreds of millions of people during the coming decades. Urban policymakers and planners have an important role to play in ensuring that urban expansion, and the economic growth it brings, is efficient and inclusive, because once cities are built, their urban form and land-use patterns are locked in for generations, making it critical for cities to get their urban form right today, or spend decades and large sums of money trying to undo their mistakes. Although the growth of urban areas provides opportunities for the poor, urban expansion, if not well planned, can also exacerbate inequality in access to services, employment, and housing (Urban Development Series 2015).

Urban planners and policy-makers at the national and municipal levels have important roles to play in ensuring that real estate development in response to urbanization proceeds in an economically efficient, sustainable, and inclusive manner. If market forces are left to themselves, they may not appropriately manage the positive and negative externalities of urbanization or provide public goods necessary to support the transformations brought about by urbanization. There is also need for government support to avoid uncoordinated and inefficient development which leads to irreversible spatial patterns that result in urban ills such as pollution, inequalities, and congestion. Governments, particularly in lower- and middle-income countries with rapid urban population growth, can prepare for future spatial expansion by facilitating the supply of urban land. Policy-makers should come up with policies that govern governments on future urban expansions and transformations, development of economically efficient cities, and growth of cities in environmentally sustainable ways (World Bank and Development Research Center of the State Council, P.R China 2014; Urban Development Series Report 2015).

Defining Urban Planning-Real Estate Development Nexus

Great cities are born of and give rise to great infrastructure. In history, city planning has been intensely engrained in real estate development and physical planning. One of its ontological bases has been to create urban places out of space through the intermediary of infrastructure. At present, the links between real estate development and urban planning may be described as numerous but non-strategic and non-comprehensive, even as the bond between infrastructure and cities remains tight (Neuman and Smith 2010). Real estate development and urban planning are intertwined in a way. While urban planning influences the future spatial distribution of activities with the aim of creating a more rational territorial organization of land uses and the linkages between them and also to balance demands for development with the need to protect the environment and to achieve social and economic objectives, real estate development helps deliver some of these objectives as it ensures physical infrastructure development (Sartorio 2005; Miles et al. 2007).

Real estate development trends have however changed over the years, and with the call for developing compact cities and new urbanism concept by urban planning professionals around the globe, real estate sector has come up with the concept of mixed-use developments which were once prevalent before the industrialization era and introduction of the automobile to ensure compact settlements (Sartorio 2005; Barros 2005; Herndon and Drummond 2011; Wyatt 2013). This calls for adjustment of planning laws that guide urban planning so that they accommodate the current changes and trends in real estate so as to avoid conflicts in urban development as these two fields are greatly intertwined. Both urban planning and real estate development deliver their objectives through land use, hence the need to integrate these two. Urban planning endeavors to create a more rational territorial organization of land uses and the linkages between them, to balance demands for development with the need to protect the environment, as well as to achieve social and economic objectives (Albrechts 2004). It tries to coordinate and improve the impacts of other sectorial policies on land use in order to achieve a more even distribution of economic development within a given territory that would otherwise be created by market forces. It is therefore an important lever for promoting sustainable development and improving the quality of life in urban landscapes (Berkes et al. 2003; Zigraj 2009).

Healey (1994) acknowledges the nexus between urban planning and real estate and notes that there is need for professionals in the development process to understand the complexity of interactions within and around the processes of urban planning and property development. She argues that understanding these linkages is a necessary task, to assist in both the practical activities of urban management and property and economic developments of cities (Metzeger 2015). In her work on land and property development, Healey has displayed a particularly keen interest in the relationship between urban planning and the development industry, investigating the potential role of methods and devices for partially guiding or directing private investments in a fundamentally capitalist society marked by strong property rights, a high level of private land ownership, and socio-spatially highly uneven concentrations of capital, as well as a strong political will toward promoting market-led spatial development in urban areas. Healey (2010) notes that the public sector through the urban planning system plays a key role in capacitating a locality to supply land and property to meet economic demand and to contribute to the health of the local economy through responsive and sustainable real estate developments.

One of the key influences on national public policy toward property development in Britain in the 1980s was a view that urban planning policies were holding back the ability of the real estate development industry to respond to emerging demand (Healey 2010; Metzeger 2015). This however is not synonymous to Britain alone, but even urban planners in Zimbabwe share the same sentiments as they recognize the bottlenecks presented by the current planning regulations to the building industry which is part of the real estate development sector (Chirisa 2014). To address this challenge with respect to planning regulations, patterns of subsidy, and development activity in Britain, policy initiatives introduced were aimed at breaking this inertia, to allow the development industry to respond unfettered to market demand (Healey 2010). Belaieff et al. (2007) asserts that planners have long known that urban and real estate developments are intertwined and correlate with land use, particularly in urban areas. However, urban planning and property development issues are too often viewed and addressed separately. As a result, opportunities to collectively address individual concerns in each sector are lost. Planning approaches are often described as adversarial or lacking in proper representation, and this needs to be addressed as sustainable urban development hinges on proper planning.

For example, urban planning standards, espoused in Zimbabwean statutes, are said to have inhibited the majority of players in the building industry or real estate development sector (Chirisa 2014). Specifically, the separation of land uses and bureaucratic procedures in planning are the cause of inefficiency in the general urban system which greatly inhibits sustainable real estate developments, and this is common in many developing nations as they just inherited planning standards from their colonizers and very little has been done up to date in terms of updating these planning regulations (Belaieff et al. 2007). The relationship between urban planning and real estate development really has to be recognized by professionals in the development process as failure to appreciate this negatively affects physical development of settlements.

Scholarship and Experiences in Urban Planning and Real Estate Development

International experience shows that urbanization issues are driving city transformation, planning, and management (Spatial Transformation of Cities Conference Report 2014). As cities urbanize, real estate development is inevitable since citizens require accommodation with supporting services. Since 2006, more than half of the world's population has been living in urban environments. In Asia, India to be specific, rapid population growth has outpaced the planning interventions in most of its large cities resulting in unplanned, leapfrogged physical development and substantial transport challenges for intra-city and regional traffic (Mittal and Kashyap 2015). This is also the case in Jakarta, Indonesia, Gangnam District in Seoul, Hanoi and Ho Chi Minh City in Vietnam, and Singapore where densities are growing each year in urban areas (World Bank and Development Research Center of the State Council PR China 2014; Urban Development Series Report 2015). However, these concerns are being addressed by formulation of spatial development strategies to guide hurried urban expansion creating planned real estate development opportunities and to ease regional traffic movements in Asia and Europe mostly (Urban Development Series Report 2014; Mittal and Kashyap 2015).

More so, China is facing fast and constant changes occurring in its built environment sector due to urbanization. This is a challenging situation for real estate because their character does not inherently support rapid changes and constant development (Toivonen and Viitanen 2015). This challenge is not only unique to China and Asia, but it is a global challenge affecting mostly the developing nations because they do not have the capacity to support fast changes in real estate trends and demands. Their capacity is hindered by a variety of factors which include both financial and institutional challenges. The responsible authorities which include mainly local authorities (Councils) and government ministries and departments are limited mainly by lack of finances and existing planning regulations which are not flexible enough to accommodate the ever-changing real estate needs of citizens (UN 2018).

Real estate is defined as the economy's stock of buildings, the land on which they are built, and all vacant land. Real estate makes up the largest single component of a nation's tangible assets which along with its financial assets make up the gross assets of national wealth and hence has a significant bearing on a country's economy and its development. Real estate development has the potential to advance sustainability in terms of meeting economic, physical, and social development goals of any nation. Real estate development has an immense impact on the spatial transformation of cities, and nations' development can be profound. If efficiently mobilized, the real estate industry can contribute to meeting economic- and social sustainability-related criteria such as providing local employment and creating economic growth through efficient utilization of resources, both human and material to construct new or renovate existing buildings and services (Mouzughi et al. 2014). In view of this, real estate development cannot be divorced from urban planning and spatial transformation of urban landscapes.

Emerging Issues and Synthesis

Since 2006, more than half of the population is now living in urban areas, and the rate is even higher in Europe and is expected to increase to above 70% by 2030 (Bran et al. 2012). Therefore, urban and spatial planning is now expected to consider these trends of suburban expansion that create pressure on infrastructure such as transport, utilities, and waste disposal (Loan and Radulescu 2010). This pressure also calls for a change in real estate development since the same factors influence property development patterns (Pauleit et al. 2010; Kotze et al. 2014). It is argued that transformation of landscapes is driven by natural and societal processes in the context of global change. Main drivers are social and demographic changes, economic changes, technological changes, and environmental changes (Anderson 2006). This therefore calls for integration of urban plans and real estate development plans so that urban landscape transformations are done in an orderly manner, with the involvement of professionals from the fields of planning and real estate development.

Focusing on Africa, its urban landscape has been changing in the past years. Most nations are experiencing a massive rural to urban migration which calls for improved spatial planning and a new type of real estate. Some scholars (Carmona 2010; Chirisa 2013; Chirisa et al. 2016; Chenal 2016) call for an integrated approach toward the creation of sustainable urban landscapes. Urbanization and rapid demographic changes can be strong drivers of growth (either planned or organic), but they also have the potential to exacerbate already pressing problems related to urban planning (Elshater 2012; Chenal 2016). In a study carried out in 2016 by Chenal on the importance of planning infrastructure and finance for Africa's growing cities, Burkina Faso, Ethiopia,

Kenya, Nigeria, Senegal, South Africa, Tanzania, and Uganda were found to be experiencing changes in the form, nature, and appearance of their cities and urban areas, and there is a growing demand of a new type of real estate. These changes are palpable to not only the mentioned countries alone but the whole of Africa (Chenal 2016). Therefore, planning professionals are encouraged to design more sustainable compact cities that are less sprawling yet catering for the economic, social, and physical needs of the urban populace (Tibaijuka 2006; Chenal 2016). However, planning of such cities cannot be done in isolation; realtors have to be involved in the process so that the urban plans also embrace the new or current trends of real estate developments.

Looking at Zimbabwe, for example, one of the challenges facing urban planning is that it relies heavily on outdated standards and master and local plans (Chirisa 2004). The Harare Combination Master Plan, for instance, is yet to be updated since it was crafted; therefore, it does not reflect the current socioeconomic realities prevalent, hence failing to cater for the current urban transformations (Toriro 2007; Kamete 2009). The social, economic, and physical environment in which real estate development and construction take place has changed, yet most town councils are holding on to outdated irrelevant master plans. Apart from these master plans which guide development, both spatial and real estate, planning still relies heavily on the outdated building standards set by the British (Ndlovu and Umenne 2008; Chirisa 2014), and this imposes a great challenge to real estate development as this impedes on the ability to cope with the current trends in the sector. Issues of enforcement worldwide have also proved to be another challenge as there is heavy political interference, thereby hindering local authorities from addressing some of the challenges associated with organic growth facing the urban centers (Chenal 2016).

The reigning changes in the economy, technologies, and environment, across the globe, call for a new type of real estate to cater for the emerging needs of the urban populace (Smersh et al. 2003; Elshater 2012; Chenal 2016). Due to static spatial plans, there is considerable growth of the informal sector ranging to include informal enterprising and housing. The challenges faced now are to do with accommodating the informality, for example, informal traders, in urban areas (Nhongo and Mafusire-Kamanga 2016; Mutondoro et al. 2017). Consequently, to promote sustainable transformation of urban landscape that brings robust, functional, and livable urban areas, there is need for an integration of spatial planning and real estate development (Nhongo and Mafusire-Kamanga 2016). This can only be done when the practitioners are aware of the effects of one urban sub-system on the other, in this case spatial planning effects on real estate.

Zimbabwe's urban landscape encompasses the large metropolitan areas of Harare and Bulawayo, large cities and towns, and many other small urban centers (Muronda 2008). Physical and spatial planning for these urban areas are administered by the Department of Physical Planning (DPP), a technical arm of the government that is in charge of managing the spatial planning system and providing technical advice for the implementation of the development planning systems (Toriro 2007; IBRD/WB 2012). On one hand, most urban centers in the past 20 years have experienced a high rate of urbanization and a growth of the informal sector (Tibaijuka 2005) which has led to a transformation in the urban landscape which is organic to a greater extent. On the other hand, real estate development in Zimbabwe has been lagging behind in terms of responding to the obtaining needs, especially that of mixed-use developments that promote sustainable settlements in the face of the current social, economic, and technological changes (Nhongo and Mafusire-Kamanga 2016).

In recent years, Harare urban landscape has transformed due to the changes in the economy. The informal sector has become a source of income for the greater population as the economy continues to plunge (Muronda 2008; Nhongo and Mafusire-Kamanga 2016). This has seen most buildings in and around the city being transformed to shopping "stalls" to accommodate the sprouting small businesses. The central business district has become an eye sore due to the unplanned organic transformations. The master plan and local development plans guiding spatial planning in the city of Harare are outdated (Kamete 2009). Failure to guide current physical development needed for the city in response to the changing socioeconomic trends is now inevitable (Nhongo and Mafusire-Kamanga 2016).

Real estate needs are changing and this cannot be divorced from spatial planning. The carrying capacity of Harare has been exceeded (Toriro 2007; Muronda 2008; Nhongo and Mafusire-Kamanga 2016). The transport network has become stressed, and the urban infrastructure and basic amenities supply have been overloaded, causing maintenance problems for the city planners (Muronda 2008). This is also a challenge in most African cities where the cities are also being affected by the massive rural to urban migration, demographic changes, an increase in informal activities, and changes in their economies, thus leading to urban landscape transformations which are mostly unplanned (Nhongo and Mafusire-Kamanga 2016). There is therefore a need to come up with development plans for these cities and towns that embrace the current trends in real estate before the situation worsens like that of Harare.

In Harare, efforts are currently underway, with the Harare City Council working on a review of the Local Development Plan No. 22 (LDP22). The Local Development Plan 22, according to the City Officials, is going to incorporate mixed-use development (buildings and land) and will also address informal sector infrastructure needs. This is in response to the organic urban landscape transformation facing the city (Interview by HCC Town Planner, 8 February 2018). LDP22 was last updated in 2000, and a lot of changes have since happened in Harare which have promoted organic transformation (Mabaso 2015; Nhongo and Mafusire-Kamanga 2016). Real estate development trends are changing around the globe (Bhatti 2014), and Harare as the capital city needs to move with the times as well. Despite these changes to the LDP, there is no tangible evidence on how the spatial planning and real estate development can be integrated and the real effect of the integration on real estate development. Much of the developments in the city have been ill-advised and ill-timed with the impact, worsening the predicament of the residents and business people in Harare (Mabaso 2015). The growth of informality in housing or businesses in African cities is on the rise, and it is an indication of the gaps being left out by the formal systems (Muronda 2008; Kamusoko 2013; Mabaso 2015).

Conclusion and Future Direction

Sustainable development is not possible without growth in real estate that provides facilities for all human activities while improving the public spaces in our cities and towns (Kaklauskas et al. 2015). Urban planning and development play a crucial role in this respect, as a majority of people live in cities where most of the economic value is added. In addition, urban areas are also consuming most resources, which suggest that an efficient organization of urban areas is a key element of achieving this goal. At the same time, real estate development plays a central role in the transformation of built space (Healey 1991). It is quite evident that urban plans are delivered through real estate development projects; hence, separating urban planning and real estate development makes no sense as they are intertwined in a way. Development plans for any settlement have to adhere to the planning laws of that land; at the same time, embracing current real estate trends as urban development has an impact on the economy of any nation. Integrating these two sectors helps in shaping the urban landscape into livable and sustainable settlements.

In addition, integrating real estate development and urban planning is becoming inescapable by the day as spatial patterns are closely related to the capacity of an economy to grow and produce jobs.

As real estate development expands, the economy also grows and employment increases. Moreover, as urban centers grow, there is concentration of labor, market skills, goods, and network of social and business relationships; as a result, there is a benefit of agglomeration economies. However, for urban centers to grow, there is a need for proper spatial planning and proper real estate development; hence, these two cannot be considered in isolation. Urban settlements should consider mixing commercial space, offices, and residential areas to reduce the distance residents have to travel for any services. In turn, the increased densities will allow more efficient and cleaner transport modes to become viable and affordable, and these transport modes include biking, walking, and public transit systems.

Amsterdam is one city which has implemented such modes successfully; hence, lessons can be drawn from there and many other cities which have managed to have viable mixed-use developments. Moreover, the mixing of land uses is now being practiced not only in Asia but in Europe, America, the Middle East, and even Africa (Dumaine 2012). Mixed-use development suggests real estate development that combines more than one land use. It is an ambiguous multifaceted concept (Rowley 1996; Herndon and Drummond 2011). Mixed-use developments consist of three or more significant revenue-producing uses that in well-planned projects are mutually supporting. They also consist of significant physical and functional integration of project components which ensure a relatively close knit and intensive use of land (ULI 1976). However, the integrated uses in a mixed use must be substantial enough to attract a significant market in their individual capacities, and these uses exclude uses that simply serve as amenities for a primary use; hence, the planning and real estate development professionals should keep this in mind.

Planning in most developing nations, Zimbabwe included, is hinged on the Euclidian zoning; therefore, embracing mixed-use developments will help in altering the current spatial patterns as the primary goals for mixed use revolve around the desire to alter the current patterns of urban growth and rectify the detrimental effects that Euclidian zoning and sprawl have had on urban areas. It also forms part of a strategy for sustainable development and a theory of good urban structure with the objectives of economic vitality, social equity, and environmental quality (Grant 2002; Schwanke and Phillips 2003; Herndon and Drummond 2011). The emerging consensus is that development is more sustainable if it produces a mixture of uses. Segregation of land uses previously encouraged is becoming irrelevant (City of Atlanta 2002). Mixed-use development ensures vitality through activity diversity. It makes areas safer, reduces the need to travel, and also adds to the liveliness and significance of town centers. Different but complementary uses reinforce each other, thus making urban centers more attractive to residents, businesses, shoppers, and visitors (Department of Environment 1995; Coupland 1997).

However, for this to be possible in the Zimbabwean context, there is a need for a shift in the planning process of local development plans which guide physical development in urban areas. The current pieces of legislation guiding urban planning also need to be updated as a matter of urgency. The planning process has to be all inclusive during the initial phases so that professionals in the real estate development sector are also included and not let the planners come up with the concept plans on their own. A good starting point to implement this would be in the planning and development of the proposed new city – Mount Hampden. This new city provides the best platform to address the current ills faced in most Zimbabwean cities and towns which resulted from not appreciating the nexus between urban planning and real estate development.

References

Albrechts, L. (2004). *Strategic (spatial) planning re-examined*. Leuven: Catholic University of Leuven.

Anderson, E. (2006). *Urban landscapes and sustainable cities*. Oxford: Oxford University.

Barros, J. (2005). *Simulating urban dynamics in Latin American cities*. London: University College of London.

Belaieff, A., Moy, G., & Rosebro, J. (2007). *Planning for a sustainable nexus urban land use, transport and energy*. Karlskrona: Blekinge Institute of Technology.

Berkes, F., Colding, J., & Folke, C. (2003). *Navigating social-ecological systems: building resilience for complexity and change*. Cambridge: Cambridge University Press.

Bhatti, M. (2014). *Real estate 2020 building the future*. Available online: https://www.pwc.com/gx/en/asset-management/publications/pdfs/real-estate-2020-pwc.pdf. Accessed on 23 October 2020.

Bran, F., Popa, D., & Popa, C. (2012). Spatial planning: global challenges and European approaches. In *Proceedings of the 6th International Management Conference: approaches in organisational management. Management Acad. Soc. Romania (SAMRO)* (pp. 462–467). Available online: conference.management.ase.ro/archives/2012/pdf/58.pdf. Accessed on 24 October 2020.

Carmona, M. (2010). Contemporary public space: critique and classification, part one: critique. *Journal of Urban Design, 15*(1), 123–148.

Chenal, J. (2016). *Capitalizing on urbanization: The importance of planning infrastructure and finance for Africa's growing cities*. Maputo: Ecole Polytechnique.

Chirisa, I. (2013). Increased squalor in urban and peri-urban Zimbabwe under the GPA? assessing the rhetoric, practices and contradictions. *Southern Peace Review Journal, 2*(1), 118–135.

Chirisa, I. (2014). Building and urban planning in Zimbabwe with special reference to Harare: Putting needs, costs and sustainability in focus. *Consilience: The Journal of Sustainable Development, 11*(1), 1–26.

Chirisa, I., Mazhindu, E., & Bandauko, E. (2016). *Peri-urban developments and processes in Africa with special reference to Zimbabwe*. New York: Springer International.

Coupland, A. (1997). *Reclaiming the city: Mixed-use development*. London: E & FN.

Dumaine, F. (2012). When one must go: the Canadian experience with Strategic Review and judging program value. *New Directions for Evaluation, 2012*(133), 65–75.

Elshater, A. (2012). *New urbanism principles versus urban design: Dimensions towards behaviour performance efficiency in Egyptian neighbourhood unit*. Cairo: Ain Shams University.

Grant, D. (2002). *Rethinking organizational change*. New York: John Wiley & Sons.

Healey, P. (1991). Models of the development process: a review. *Journal of Property Research, 8*(3), 219–238.

Healey, P. (1994). *Urban policy and property development: The institutional relations of real estate development in an old industrial region*. Newcastle: University of Newcastle.

Healey, P. (2010). *Re-thinking the relations between planning, state and market in unstable times*. Porto: FEUP edicoes.

Herndon, V., & Drummond, S. B. (2011). *Holy clarity: the practice of planning and evaluation*. New York: Alban Institute.

Kaklauskas, A., Zavadskas, E. K., Dargis, R., & Bardauskienė, D. (2015). *Sustainable development of real estate*. Vilnus: Vilnus Gediminas Technical University.

Kamete, A. Y. (2009). In the service of tyranny: debating the role of planning in Zimbabwe's urban 'Clean Up' operation. *Urban Studies, 46*(4), 897–922.

Kamusoko, T. (2013). *Monitoring urban spatial growth in Harare Metropolitan Province, Zimbabwe*. Tokyo: Seikei University.

Kotze, N., Donaldson, R., & Visser, G. (2014). *Life in a changing urban landscape*. Johannesburg: University of Johannesburg.

Mabaso, T. (2015). *Urban physical development and master planning in Zimbabwe. An assessment of conformance in the City*. Harare: ICED.

Metzeger, J. (2015). *The planning/development nexus: How places are produced and changed*. London: Royal Institute of Technology.

Miles, M. E., Netherton, L. M., & Schmitz, A. (2007). *Real estate development principles and processes*. Washington DC: Urban Land Institute.

Mittal, J., & Kashyap, A. (2015). Real estate market led land development strategies for regional economic corridors–a tale of two mega projects. *Habitat International, 47*, 205–217.

Mouzughi, Y., Bryde, D., & Al-Shaer, M. (2014). The role of real estate in sustainable development in developing countries: the case of the Kingdom of Bahrain. *Sustainability, 6*(4), 1709–1728.

Muronda, T. (2008). *Evolution of Harare as Zimbabwe's capital city and a major central place in Southern Africa in the context of by Byland's model of settlement evolution*. New Delhi: Nehru University.

Mutondoro, F., Chiweshe, M., Mlilo, M., Ncube, M. J., & Rehbock, N. (2017). *Land governance in the context of the new urban agenda: experiences from Harare (Zimbabwe) and Johannesburg (South Africa)*. Johannesburg: Africa Research Institute.

Ndhlovu, L. B., & Umenne, S. I. (2008). Trends in earthen construction for rural shelter in Zimbabwe: the case of tsholotsho in matebeleland north province. Available online: https://www.semanticscholar.org/paper/Trends-in-Earthen-Construction-for-Rural-Shelter-in-Ndlovu-Umenne/f39ef74c70c36d048eacfee5fb7d9ba893dae2b9. Accesed on 29 October 2020.

Neuman, M., & Smith, S. (2010). *City planning and infrastructure: Once and future partners*. Texas: SAGE.

Nhongo, J., & Mafusire-Kamanga, B. (2016). *Foreign real estate investment opportunities and investment protection in Zimbabwe*. Harare: D.L.A.

Pauleit, S., Breuste, J., Qureshi, S., & Sauerwein, M. (2010). Transformation of rural-urban cultural landscape in Europe: integrating approaches from ecological, socio-economic and planning perspectives. *Landscape Online, 20*, 1–10.

Rowley, A. (1996). *Mixed-use development: Ambiguous concept, simplistic analysis and wishful thinking, planning, practice and research*. Atlanta: University of Atlanta.

Sartorio, F. S. (2005). Strategic spatial planning: a historical review of approaches, its recent rival and an overview of the state of the art in Italy. *disP-The Planning Review, 41*(162), 26–40.

Schwanke, D., & Phillips, P. (2003). *Mixed-use development handbook* (2nd ed.). Washington, DC: The Urban Land Institute.

Smersh, G., Smith, M., & Schwartz, Jr., A. (2003). Factors affecting residential property development patterns. *Journal of Real Estate Research, 25*(1), 61–76.

Tibaijuka, A. K. (2005). *Report of the fact finding mission to Zimbabwe to assess the scope and impact of operation Murambatsvina*. Nairobi: UN Habitat.

Tibaijuka, A. K. (2006). *Africa on the move: An urban crisis in the making*. Nairobi: UN-Habitat.

Toivonen, S., & Viitanen, K. (2015). Forces of change shaping the future commercial real estate market in the Helsinki Metropolitan Area in Finland. *Land Use Policy, 42*, 471–478.

Toriro, P. (2007). Town planning in Zimbabwe: history, challenges and the urban renewal operation murambatsvina, Chapter 11 in Maphosa, F., Kujinga, K., & Chingarande, S. D., (eds). *Zimbabwe's development experiences since 1980: challenges and prospects for the future*. Addis Ababa: OSSREA.

The World Bank Development Research Center of the State Council, The People's Republic of China. (2014). Urban china toward efficient, inclusive and sustainable urbanization. Available online: https://elibrary.worldbank.org/doi/abs/10.1596/978-1-4648-0206-5. Accessed on 21 October 2020.

UN. (2018). *Urbanization and national development planning in Africa*. Addis Ababa: UN Economic Commision for Africa.

Urban Land Institute. (1976). *Mixed-use developments: new ways of land use*. Washington DC: Urban Land Institute.

Wyatt, P. (2013). *Property valuation*. New Jersey: John Wiley & Sons.

Urban Policy

► Understanding Urban Engineering

Urban Policy and the Future of Urban and Regional Planning in Africa

Willoughby Zimunya[1,2] and Innocent Chirisa[3]
[1]Department of Demography Settlement and Development, University of Zimbabwe, Harare, Zimbabwe
[2]Department of Urban and Regional Planning, University of the Free State, Bloemfontein, South Africa
[3]Department of Demography Settlement and Development, Social & Behavioural Sciences, University of Zimbabwe, Harare, Zimbabwe

Synonyms

Planning – Organization; Policy – Direction; Urban – Non-rural

Definition

Urban policy is a high-level coordinated guide that is prepared and implemented by governments in with other stakeholders to address urban issues with the aim to build productive, socially and environmentally sustainable cities.

Planning is a tool for making informed decisions that enhances the quality of life through land use, development control, and provision of services, facilities, and infrastructure in an economic and sustainable manner.

Urban development refers to the creation and sustenance of cities through the building and restructuring processes in order to meet their functional and livability requirements.

Urban sustainability refers to a condition or improved levels of social and economic development and ecological balance and is supportive of continued existence of cities.

Spatiality refers to the impacts of urban policy and planning with the distribution and manifestation of effect across a given space.

Introduction

The building of sustainable cities is now an imperative developmental issue in the new millennium as the world is rapidly urbanizing and also confronted crises like the COVID-19 pandemic have been experienced globally. Cities as the largest and specialized socioeconomic settlements are important places where the condition of sustainability has to be achieved so that they continue to exist and function efficiently. The poor management of urban growth that is undermining the future of cities in some parts of the world including Africa is a cause for concern. Several instruments have been developed and used including urban policy and urban regional and planning (called planning hereafter) to effectively manage urban growth. There is long historical role and inextricable connection between urban policy and planning as public tools for guiding urban growth spanning over a century (Cochrane 2007; Galès 2015; Teitz 2015), and no city has not been influenced by planning policy (Pillay 2008). It is important to continually reflect on the effectiveness of these two instruments with a view to improve their effectiveness in tackling urban issues and build better cities particularly in Africa where urbanization is fastest.

Planning as one of the policy tools for managing urban growth has played a vital role in providing guidance to urban development in some parts of the world. It has aided the sustainable urban growth of sustainable cities through coordinating development efforts, directing infrastructure development, shaping urban spatial structure and form, and controlling urban expansion. However, while urban policy and planning have led to sustainable urbanization in some regions of the world, it is failing to properly guide the connected development of cities in others due to certain deficiencies. This failure is disturbing in the context of rapid urbanization particularly Africa and Asia where it is occurring faster and exerting pressure on cities. The deficiencies in urban policy and planning are manifested by dysfunctional cities that are growing in a disorderly manner undermining urban sustainability and good quality of life. Notwithstanding the shortcomings of urban policy and planning elsewhere, the utility of these two instruments in facilitating planned and sustainable development has been rediscovered. There is rejuvenated interest on the two instruments based on their positive impacts as some of the most viable mechanisms of dealing with the upward urbanization trends and concomitant challenges that are being encountered globally (Dede 2016; Watson 2016; World Bank 2016). The essence of these tools is that they provide both proactive and reactive strategic solutions to critical urban issues that are affecting the sustainable growth cities.

In light of this background, it is necessary to interrogate the current urban policy and planning systems with a view to reflect on issues surrounding their effectiveness in tackling urban issues in relation to the obtaining and desirable future urbanization trends in Africa. The two major questions to be addressed by this chapter are as follows: What are the existing planning policy frameworks and their outcomes? What is the nature of planning policy that is appropriate for urban Africa? To address these research questions, this study was guided by a qualitative approach and used literature review and content analysis to tape into ideas from scholarly work on experiences concerning the application of urban policy and planning. These ideas were used to suggest an appropriate planning policy framework for Africa.

Based on the research questions, the chapter is organized into five sections which include an introduction providing a contextual back of the issues under study, a conceptual framework of the role of urban policy and planning in guiding urban development, literature review of the scholarly work and experiences in the implementation of urban planning and policy, and a synthesis of emerging issues. The chapter ends with a conclusion and recommendations about the future direction of planning policy for Africa.

Defining Urban Policy and Planning

This section conceptualizes and puts into perspective the importance of using appropriate urban policies and planning strategies to support sustainable urban development in the context of rapid urbanization. It also presents an analysis of factors that affect the effectiveness of twin instruments in directing urban development and suggest how these issues can be addressed.

Cities are now the major settlements globally, and as such, they exert influence on the socioeconomic development of humanity (Cochrane 2011; Glaeser 2012; Watson 2016). To this end properly planned and well-managed cities are vital for anchoring a sustainable urban future and human development, and connected policy and planning can assist in achieving that goal (Glaeser 2012; Dede 2016; Watson 2016). Proper policy and planning are mutually supporting public tools among several other mechanisms that are applied to guide the development and management of cities (Parker and Doak 2012). As complementary tools, their goal is to make cities sustainable and resilient hubs of socioeconomic development for all.

Urban policy is a high-level coordinated guide that is prepared and implemented by governments in which other stakeholders address urban issues with the aim to build productive, socially, and environmentally sustainable cities (Glaeser 2012; Teitz 2015; OECD 2019). It comprises vision, values, and principles that represent the desired outcomes about urban land management, urban governance, service delivery, and environmental sustainability (Glaeser 2012; USAID 2013; Teitz 2015). It also assists governments to coordinate its efforts with other actors in urban development (USAID 2013) since building cities is a collective process. Overall, the importance of urban policy is to provide a framework about the envisioned nature and quality of urban development and management as well as the governance structure that forms the basis of planning.

Urban policy guidelines and proposals are implemented to support the building and running of better cities through planning (Rydin 1993; Cullingworth and Nadin 2006; Parker and Doak 2012; Berg 2019). Planning is defined differently, but in this chapter, it refers to a statutory, strategic, and ongoing public-driven collaborative proactive and reactive process of formulating and designing strategies for guiding the development of sustainable cities through promotion of connected spatial patterns. In this case, planning generates plans and strategies that are used to translate policy proposals into actuality (Cullingworth and Nadin 2006; Parker and Doak 2012; Berg 2019). These planning strategies are prepared in line with the urban policy which precedes their formulation (Parker and Doak 2012; Berg 2019). This implies that sound policy leads to proper planning which in turn promotes urban sustainability. Corollary deficient policy affects the planning system with detrimental impacts on urban development. Thus, where urban policies and plans are not linked, this has implications on the quality of urban development. In realization of this fact, it is emphasized that urban policies and planning are integrated at various levels (Parker and Doak 2012).

Planning is applied in the implementation of policy to promote connected and sustainable development by organizing and restructuring urban land uses, controlling development activities, and guiding infrastructure investments (UN-HABITAT 2008; Rydin 2011; Hyman and Pieterse 2017; Hall and Tewdwr-Jones 2020). It is also used to perform the governance role of coordinating the policies and actors that are involved in urban development and management processes (Gurran 2011). In addition, planning also operationalizes urban policy through enhancing ecological balance by fostering conservation and preservation of key resources. The essence of adopting planning as a mechanism for facilitating urban development is that it minimizes negative effects of actions and improves outcomes than without planning (Ngah 1998; Hall and Tewdwr-Jones 2020). It aids in creating better cities by proactively appraising urban development actions and determining their possible impacts (Gurran 2011; Couch 2016). This approach ensures that urban development action taken satisfies both the needs of the current generations and posterity. However, planning can be applied reactively to formalize unplanned developments (Gurran 2011), but this approach is not commendable in the context of sustainable development paradigm.

It is noteworthy that the application of urban policy and planning in guiding urban development results in spatial changes. Urban development refers to the creation of cities through the building and restructuring processes in order to meet their functional and livability requirements (Quadeer 2012). It is manifested by physical developments in the form of infrastructure facilities and land uses and also by non-physical outcomes of urban change emanating from planning policy implementation (Rydin 2010; Parker and Doak 2012; Quadeer 2012). These developments and outcomes can improve or affect the livability and functionality of cities, but from the urban sustainability perspective, society expects positive outcomes on planning policy interventions in the form of better urban conditions, but this is not always the case (Blackman 1995; Parker and Doak 2012). The cumulative outcomes of urban development are reflected by the livability, productivity, and sustainability of cities that are impacted by the planning policy decisions.

In light of the above, the outcomes of urban policy and planning have an element of spatiality. The spatiality of urban policy and planning outcomes concerns the placed-based impacts of urban development (Rydin 1993; Parker and Doak 2012). These effects of planning policy efforts in both the planned and unplanned and the positive and negative outcomes are usually spatially distributed and manifest across the urban space at various spatial levels (Galès 2015). As such, the notion of spatiality suggests that policies and plans need to facilitate balanced development in all urban spaces in order to foster sustainability. Further, planning and policy on their own cannot influence the sustainability and spatiality of urban development. Their effectiveness, that is, the extent to which they aid in the achievement of development outcomes, is determined by institutional factors, the availability of resources to support planning policy implementation, and political priorities which may support or undermine planning objectives (Parker and Doak 2012; Berg 2019). The effectiveness of planning and policy on directing urban development is also influenced by their responsiveness to the changing circumstances and emerging urban issues. In addition, the availability of planning skills, urban land management, legislative framework, and economic and environmental and social factors (Fall and Coulibaly 2016; Schindler et al. 2018; Berg 2019) affect the efficacy of planning policy in an interconnected way.

In order to enhance the effectiveness of urban planning and policy in directing urban development, these factors need to addressed in a holistic and coordinated manner (Acheampong and Ibrahim 2016; OECD 2016, 2018; Schindler et al. 2018). It is important to consider the effect of these factors in the formulation and implementation of urban policy and planning strategies. Additionally, it is also necessary to periodically assess the efficacy of urban policies and planning systems in terms of the extent to which they will be supporting sustainable urban development. Thus, failing to address these factors will undermine the utility of urban policy and planning as viable tools for guiding urban development.

A Survey of Scholarship and Experiences in Urban Policy and Planning

Urban policy emerged as tool to resolve urban problems in the western world. It has evolved over time in response to changes in the urban landscape resulting in traditional anti- or counter-urbanization "traditional" policies and pro-urbanization "new" policies. The traditional urban policy aims at addressing urban problems by restraining growth of cities through various strategies. It is still being applied in some countries, although it was abandoned by the developed countries in the 1970s. There are several shortcomings that prompted its abandonment.

The traditional urban policy abandoned for its counter-urbanization thrust was introduced through anti-rural urban migration policies and in the decentralization and growth point strategies. This policy concentrated on restraining growth of cities at the expense of expanding infrastructure and service provision to accommodate the natural increase of population and future growth of cities. South Africa and Brazil once relied on these retrogressive policies before they adopted proactive policies that support urban growth (UN-HABITAT 2014; Turok 2015; COGTA 2016). The adoption of the former policies consequently led to infrastructure deficits which are now difficult to clear.

These policies are also sector-biased and fragmented, and this weakened their effectiveness in guiding urban development. Additionally, the traditional urban policies failed to integrate the operations of multiple urban development actors (Hölzl and Nuissl 2014; Korah et al. 2017). This lack of coordinative policy approach is worsened by the stop-go and intermittent responses to urban problems which lead to the unending urban challenges (Todes et al. 2010; Hölzl and Nuissl 2014). Bangladesh, Zimbabwe, and Zambia (Panday and Jamil 2010; Muzondi 2014; Government of Zambia 2015) are such countries that are experiencing persistent urban challenges because they rely on unconnected urban policy. Thus, in the absence of coherent policy strategies in place, most urban areas continue to suffer from urbanization challenges even with adequate resources.

Previous studies indicate that the traditional policies also disregarded some facets that are essential for the sustenance of cities, and this led to the poor service provision of services that affected the life of quality. The most ignored facets were urban transport and socioeconomic and environmental concerns (Söderholm and Wihlborg 2016). The effect of ignoring these facets was to undermine the efficiency of cities and urban sustainability, and this policy deficiency is manifested in the dysfunctional urban environments, mostly in Africa, Latin America, and Asia. The challenges of urban sustainability being experience in Bangladesh demonstrates the problem of relying traditional polices to guide urban development (Islam 2006). Besides that, these policies are spatially biased as they concentrated on segregated distribution of services which results in socio-spatial inequality and polarization that undermines the stability, inclusivity, and safety of urban areas. South Africa experienced this problem during the apartheid era before it transformed and shifted to broad-based urban development policies (COGTA 2016).

The literature also shows that traditional polices do not focus on institutional structures as important aspects of the development and management of cities (Korah et al. 2017). As such, municipalities remain focused on the traditional role of administration, and they are unable to formulate policy strategies to deal with emerging challenges affecting cities (Jordan and Magnoli 2002). Thus, this policy deficiency weakens the urban management system which is essential in promoting sustainability in cities. Without proper urban management system, there are bound to be serious consequences on the citizens. For instance, urban development in Ghana and Bangladesh is being affected by institutional challenges (Sabbi and Mensah 2016; Korah et al. 2017). Previous research also shows that the traditional policies were adopted blindly, and they failed to match demands of the local situation (Watson 2009; Korah et al. 2017). This unsuitability of the policies has become a constraint to the development and management of cities that are affected by them. The consequences of this policy deficiency include a host of challenges in most urban challenges in the Global South.

Some studies indicate that the traditional urban policies did not provide for the delivery of all the collective utilities that are essential for good living conditions (United Nations 2008; Korah et al. 2017). The emphasis of these traditional urban policies is on welfare and people-centered development. The policies concentrated on the treatment of symptoms of the challenges rather than focusing on place-based interventions that ensure strategic development by providing adequate services and facilities for the betterment of living conditions in cities. This policy deficiency partly explains the existence of squalid living conditions in Ghana (Korah et al. 2017).

The disenchantment with traditional urban policy is leading to the adoption of a hybrid of new policies that are supportive of urban growth. These policies are being adopted in response to the changing circumstances and complexities in cities that are associated with the phenomenal urban growth that is being experienced globally. The new set of urban polices has been attractive because they realize the positive role of urbanization in socioeconomic development. South Africa and Brazil have discarded the anti-urbanization traditional urban policies and adopted proactive policies that are supportive of urban growth, thereby enhancing urban sustainability (UN-HABITAT 2014; Turok 2015; Berg 2019). Additionally, these policies also focus on increased multilevel coherence and integration. Besides that, they also give attention to several aspects which include land management, housing, infrastructure development, service provision, economic development, and social equity among others that affect the urban sustainability, thereby leading to better quality of life. To this end, countries like Germany, South Korea, and Australia have benefitted immensely, applying integrated policies to guide urban development (Australian Government 2011; Park et al. 2011; UN-HABITAT 2014). Further, the new policies also aim at empowering institutions involved in urban development at various levels. For instance, Ethiopia guided by its policies has invested heavily in institutional capacity building in urban development (UN-HABITAT 2014). Generally, the new set of urban policies is effective because it is integrative, holistic, long term, home grown, and responsive to the local situation.

It is noted that the existence of sound urban planning frameworks subverts anarchy and makes development in cities controllable (UN-HABITAT 2009, 2016). Sound and proper urban planning result in orderly compact, serviced, connected, and environmentally sustainable urban settlements. Lack of a proper planning leads to uncontrolled development that harms the social, economic, and ecological fabric of urban areas (Rydin 1993). Overall, sound planning frameworks are necessary to prevent or ameliorate negative externalities and for producing positive impacts of urban growth.

Planning is one of the mechanisms that are employed by governments to guide urban development in response to the urbanization process. Previous studies have observed that planning is dynamic as it has been evolving in response to changing urban conditions (Rydin 1993; Ward 2004). These changes are partly influenced by the societal values and higher aspirations for better life in these ever-changing conditions. Past studies have indicated that these changes are necessary as they ensure that planning remains relevant, effective, and responsive to prevailing situation (Ward 2004; Watson 2009; UN-HABITAT 2009). Initially, planning was adopted in practice and has incrementally changed from blue print, structure, development control, and master planning approaches that comprise (modernist) traditional planning system. It has also progressed from these early approaches through radical changes resulting in the adoption of contemporary planning systems which comprise new approaches (post-modernist) in some countries (UN-HABITAT 2009). The new approaches are wider in scope and include other non-spatial aspects. These new approaches include offshoots of collaborative, governance, urban design, and spatial planning approaches. Although both approaches are still in use, many countries are abandoning the traditional approach because of its weaknesses.

The traditional planning system has been criticized for being narrow in its scope as it emphasizes on the land-use planning only. Some scholars have observed that the traditional planning framework only focuses on physical and numerical outputs without relating these to outcomes and impacts on the society (Nadin 2007; Morphet 2011). Further, these scholars argue that because of its narrow approach, this planning system disregards the socioeconomic, environmental, and institutional aspects which are part of planning goals that underpin sustainable urban development (Nadin 2007; Morphet 2011). The experience of Zimbabwe and some countries in East Africa shows that they are failing to tackle urban challenges because they rely on this planning approach with spatial issues (Chirisa and Dumba 2012; Lwasa and Kinuthia-Njenga 2012).

Further, the traditional planning system is bureaucratic and static and does not match the demands of the current environment. These challenges are characterized by the delays in the preparation and approval of new plans and the review of existing plans that are usually outdated (Watson 2006; Morphet 2011; UN-HABITAT 2009). By failing to be responsive to situations, this planning system hinders the development that it is required to guide (Ward 2004; Watson 2009; Morphet 2011). This shortcoming creates a lacuna in the planning system that affects proper city development and management. In the absence of a planning framework for decision-making, urban development proceeds through conjectural approach and crisis management with consequences on the cities affected by such a planning system. The urban challenges being experienced in Zimbabwe, Turkey, and Tobago are explained by relying on this flawed traditional planning system characterized by rigidity (Chirisa and Dumba 2012; Dede 2016; Mycoo 2017). The negative effect of this unresponsiveness is that development ends up in undesignated areas, thereby undermining the functionality cities. The shortcoming is worsened by the non-existence of a proper reviewing mechanism for the planning system.

The traditional planning approach has also been critiqued for being overly regulatory. It emphasizes on the conformance of development to the plans without considering the cumulative impacts of the individual decisions on the localities. The expense of facilitating development with the aim to manage urban growth challenges also becomes a major problem (Morphet 2011). While this regulatory aspect is a necessary component of any sound planning system (Ward 2004), failure to balance it with the promotion of development undermines the urban sustainability (Morphet 2011). This shortcoming is affecting the effectiveness of planning in guiding urban development in East Africa countries that are still applying this system (Lwasa and Kinuthia-Njenga 2012).

The traditional planning has been associated with lack of adequate public participation in its approaches (Morphet 2011; Korah et al. 2017). This system relies on a technocratic and top-down approach to preparation of planning strategies which is not comprehensible to most citizens. The system lacks proper mechanisms to engage with the various stakeholders in cities (Morphet 2011). The framework for preparing plans does not explicitly provide for community participation, and such a framework imposes plans on the communities. This is the case with the Anglophone Caribbean countries and Turkey which are experiencing urban challenges because of relying on the inappropriate top-down master planning systems (Dede 2016; Mycoo 2017). This lack of proper community participation results in lack of ownership and lack of cooperation in the implementation of plans prepared under this system. This causes anarchy and leads to developments that violate planning decisions (UN-HABITAT 2009; Korah et al. 2017). Further, the lack of participation undermines sustainability by failing to generate sound decisions that can be sustained and supported by local actors.

Previous research has observed that one major weakness of the planning system is its failure to coordinate various sectoral activities (Nadin 2007; Hölzl and Nuissl 2014; Abou-Korin and Al-Shihri 2015). Nadin (2007) posits that while coordination is a key function of planning, this integrative role is not practiced in the traditional planning style. The Turkish experience in urban development being guided by the master planning approach which is failing to coordinate various actors gives credence to this observation (World Bank 2015; Dede 2016). Basically planning has just remained an administrative tool for controlling and managing development without coordinating multi-sectoral development planning activities that affect the cities. This shortcoming is further worsened by lack of an explicit statutory framework that provides for such coordinative role and has resulted in negative outcomes in cities.

Nadin (2007) posits that traditional planning has not been very ambitious on evidence-based approach to decision-making. Planning decisions have proceeded, guided by outdated plans or conjecture where such plans are not in existence (UN-HABITAT 2009; Hölzl and Nuissl 2014). The lack of analytic rigor in traditional planning sometimes contributes to flawed decisions and conclusions that are not backed by evidence. The flawed decisions affect the efficacy of plans in addressing the problems that bedevil urban environments. This problem is being experienced in the Anglophone Caribbean countries that are still relying on this archaic planning system (Mycoo 2017). Overall, it has failed to promote proper development of cities. This, therefore, prompts the adoption of alternative planning approaches which address these shortcomings.

The aim of adopting new planning approaches was that they "achieve greater ambition and more" (Nadin 2007), for the urban areas. These new approaches include offshoots of collaborative, governance, urban design, and spatial planning approaches (Nadin 2006, 2007; UN-HABITAT 2009).

The alternative planning approaches are being embraced for their ability to facilitate the integration of various development efforts. One dimension of this integration aspect includes facilitating vertical integration of sectors at various levels through spatial planning. Spatial planning fosters coordination of the action at hierarchical levels, and this improves the efficacy of planning system. The aim is to deliver desirable outcomes as the various structures mutually support each around a place or a common goal (Vigar 2009; Todes et al. 2010). To this end, Vigar (2009) has posited that the alternative planning approaches are assisting in coordinating the fragmented institutional arrangements with a view of improving the results of urban development. For instance, Singapore is using this planning approach to integrate institutions at various hierarchical levels (Choe 2016; Liu 2016), resulting in coordinated urban development.

The other integrative dimension of planning is its ability to facilitate horizontal coordination of actors in urban affairs. An assessment of literature reveals that the spatial planning and collaborative planning approaches have the strength of "soft planning" as they realize the shortcoming of depending on "hard planning" only (Healey 2006; Watson 2008; Vigar 2009). These new approaches are framed on the principles of deliberative democracy in planning which emphasize on inclusivity, co-governance, the engagement of actors, and the joining up of stakeholders in tackling public problems. The collaboration among the stakeholders assists in consensus building, sharing responsibilities, mobilizing resources, ownership of programs, and building trust among the players. In light of this, one scholar has noted that the bottom-up approach to planning espoused by the new planning paradigm has served to generate a win-win situation for both the authorities and the public (Vigar 2009). Singapore has benefitted greatly by using this planning approach in engaging with stakeholders in urban development, resulting in better outcomes (Yuen 2009; Liu 2016). The win-win situation comes through the commitment of various actors and the all-inclusive solutions, and this leads to better conditions in cities.

Prior research also reveals that the alternative approaches to urban planning are less directive and bureaucratic, and they are responsive to the prevailing situation (Ward 2004; UN-HABITAT 2009; Mäntysalo et al. 2015). This implies that the planning framework under these approaches is dynamic and effective as it responds to problems with speed. Also implied in this observation is that the planning system is less rigid and provides room for discretion in making decisions. The system and its instruments are reviewed regularly with the aim of addressing current problems. It is these two aspects of responsiveness and dynamism that have made the new planning approaches favorable tools which facilitate the management of current and future urban problems.

The alternative planning approaches are being adopted because they are broad in scope, and this enables them to address a wide range of urban issues in comparison to the traditional planning approaches. Past studies indicate that the urban environment is fast changing, thereby generating a wide range of issues which require a robust urban planning approach to address them (Vigar 2009; UN-HABITAT 2009; Mäntysalo et al. 2015). In light of this realization, progressive urban planning is responding to these changes by broadening its scope to become holistic and flexible. Progressive urban planning aims to address the urban problems that are emerging in response to the changing contexts (Rydin 1993; UN-HABITAT 2009; Todes et al. 2010). The new issues that are being tackled by the new planning paradigm include sustainability and resilience issues, poverty alleviation, and informality in all its forms and many other spatial issues (UN-HABITAT 2009; Chirisa and Dumba

2012). South Africa and Ethiopia are some of the countries that have modernized their planning systems by adopting the spatial planning style in order to resolve a wide range of emerging urban issues (UN-HABITAT 2014; Turok 2015; Berg 2019).

The growing complexity in cities and the incidence of new urban problems are also presenting difficulties from planning, making perspective such that new methods of panning are required to solve these problems. An assessment of literature shows that new planning approaches are considering these technical requirements by becoming evidence-based and rigorous and by providing monitoring and review mechanisms (Albrechts 2006; Liu 2016). This implies that with this modernization of the planning system, there will be an improvement in the making of planning decisions and the quality of planning outcomes through proper analysis and monitoring of plan performance. The successful Singaporean experience in urban development shows the benefits of modernizing the planning system in line with these ideas (Yuen 2009; Lang 2016; Liu 2016).

It is argued that the effectiveness of a planning system depends among other factors on the nature of the planning instruments of that system (Rydin 1993; Nadin 2007). In support of this view, Albrechts (2006) and Mäntysalo et al. (2015) noted that new urban planning approaches are being adopted on the strength of using innovative sets of planning instruments to address urban problems. The planning instruments include the use of strategic plans, strategic projects, new techniques, policies, institutions, and processes. While these new planning instruments have played a significant role in the performance of the new planning system, one of the problems is that some of them are informal and non-statutory, and they have come under criticism where the legitimacy of certain planning decisions is questioned. In light of this, Mäntysalo et al. (2015) opines that it is important to see how the informal instruments can be fused with the statutory planning instruments in order to avoid problems, thereby strengthening effectiveness of the planning system in urban areas.

Conclusion and Future Direction

This chapter has identified sound urban policy instruments and planning frameworks as fundamental tools for dealing with the urbanization issues. This conclusion is informed by a review of the various urban policy tools and planning approaches at various scales which was informed by the concept of urban sustainability. The review was supported by an analysis of the experiences of various countries in order to draw lessons and insights on the utility of various urban policy and planning systems. Based on the overall results of the review of this chapter, several recommendations can be made. All actors in urban affairs should acknowledge rapid urbanization as an inevitable and positive phenomenon which needs to be embraced rather than resenting it. Secondly, there is need to initiate strategic reform processes that result in embracing responsive planning policy systems that are sensitive to the dynamics in the urban environment. Thirdly, these planning policy systems should be guided by the principles of sustainable urban development. Fourthly, this reform process should also address the institutional aspects, issues of sectoral coordination, and capacity building which are key determinants of the efficacy of the policy instruments and planning systems.

In addition, there is also need to deal with aspects of urban finance, urban economy, participation and inclusivity, poverty reduction, housing, land, and environmental and transport management that affect urban sustainability. Fifthly, the planning policy should address the issue of spatiality by integrating all facets of development in place. Sixthly, the central government should lead reform process because of its strategic position and intimate linkages to other urban development institutions.

Finally, Africa has an opportunity to manage the rapid urban growth if it develops sound planning policy that is guided by lessons from practices of the other regions. The battle for urban sustainability is won through strategic, effective, state-supported, and comprehensive planning policy.

Cross-References

▶ Spatial Demography as the Shaper of Urban and Regional Planning Under the Impact of Rapid Urbanization

References

Abou-Korin, A. A., & Al-Shihri, F. S. (2015). Rapid urbanization and sustainability in Saudi Arabia: The case of dammam metropolitan area. *Journal of Sustainable Development, 8*(9), 52–65. https://doi.org/10.5539/jsd.v8n9p52.

Acheampong, R. A., & Ibrahim, A. (2016). One nation, two planning systems? Spatial planning and multi-level policy integration in Ghana: Mechanisms, challenges and the way forward. *Urban Forum*, (27), 1–18. https://doi.org/10.1007/s12132-015-9269-1

Albrechts, L. (2006). Bridge the gap: From spatial planning to strategic projects. *European Planning Studies, 14*(10), 1487–1500.

Australian Government. (2011). *Our cities, our future a national urban policy for a productive, sustainable and livable future*. Department of Infrastructure and Transport. Melbourne, Australia: Australian Government.

Blackman, T. (1995). *Urban policy in practice*. London: Routledge.

Chirisa, I., & Dumba, S. (2012). Spatial planning, legislation and the historical and contemporary challenges in Zimbabwe: A conjectural approach. *Journal of African Studies and Development, 4*(1), 1–13.

Choe, A. F. C. (2016). The early years of nation-building: Reflections on Singapore's urban history. In C. K. Heng (Ed.), *50 years of urban planning in Singapore* (World scientific series on Singapore's 50 years of nation-building) (pp. 3–22). Singapore: World Scientific Publishing.

Cochrane, A. (2007). *Understanding urban policy. A critical approach*. Malden: Blackwell.

Cochrane, A. (2011). Making up global urban policies. In G. Bridge & S. Watson (Eds.), *The new Blackwell companion to the city* (1st ed., pp. 738–746). Chichester: Blackwell.

COGTA. (2016). *Integrated urban development framework. A new deal for South African towns and cities*. Pretoria: Department of Cooperative Governance and Traditional Affairs.

Couch, C. (2016). *Urban planning. An introduction* (1st ed.). London, UK: Palgrave.

Cullingworth, B., & Nadin, V. (2006). *Town and country planning in the UK* (14th ed.). New York: Routledge.

Dede, O. M. (2016). The analysis of Turkish urban planning process regarding sustainable urban Development. In M. Ergen (Ed.), *Sustainable urbanization* (pp. 270–290). London, UK: InTech.

Fall, M., & Coulibaly, S. (eds) (2016). *Diversified urbanization: The case of côte d'ivoire*. The World Bank. https://doi.org/10.1596/978-1-4648-0808-1.

Galès, P. L. (2015). Urban policy in Europe. In *International encyclopaedia of the social & behavioural sciences* (pp. 900–907). Cambridge, USA: Elsevier.

Glaeser, E. L. (2012). The challenge of urban policy. *Journal of Policy Analysis and Management, 31*(1), 111–122.

Government of Zambia. (2015). *The case for national urban policy for Zambia*. Lusaka, Zambia: Government of Zambia.

Gurran, N. (2011). *Australian urban land use planning: Principles, systems and practice* (2nd ed.). Sydney, Australia: Sydney University Press.

Hall, P., & Tewdwr-Jones, M. (2020). *Urban and regional planning* (6th ed.). New York: Routledge.

Hölzl, C., & Nuissl, H. (2014). Urban policy and spatial planning in a globalized city – A stakeholder view of Santiago de Chile. *Planning Practice & Research, 29* (1), 21–40.

Hyman, K., & Pieterse, E. (2017). Infrastructure deficits and potential in African cities. In S. Hall & R. Burdett (Eds.), *Sage book of the 21st century city* (pp. 429–451). London: SAGE Publications.

Islam, N. (2006). Bangladesh. In B. Roberts & T. Kanaley (Eds.), *Urbanization and sustainability in Asia* (pp. 43–70). Manila, Philippines: Asian Development Bank.

Jordan, J., & Magnoli, K. (2002). *Towards more sustainable cities: The Urban Management and Sustainable Development Project in Latin America and the Caribbean*. WIT Press.

Korah, P. I., Cobbinah, P. B., & Nunbogu, A. M. (2017). Spatial planning in Ghana: Exploring the contradictions. *Planning Practice & Research, 32*(4), 361–384.

Lang, N. (2016). Planning to overcome the constraints of scarcity. In C. K. Heng (Ed.), *50 years of urban planning in Singapore* (World scientific series on Singapore's 50 years of nation-building) (pp. 71–80). Singapore: World Scientific Publishing.

Liu, T. K. (2016). Planning & urbanization in Singapore: A 50-year journey. In C. K. Heng (Ed.), *50 years of urban planning in Singapore* (pp. 23–44). Singapore: World Scientific Publishing.

Lwasa, S., & Kinuthia-Njenga, C. (2012). Reappraising urban planning and urban sustainability in East Africa. In S. Polyzos (Ed.), *Urban development* (pp. 3–22). London, UK: InTech.

Mäntysalo, R., Kangasoja, J. K., & Kanninen, V. (2015). The paradox of strategic spatial planning: A theoretical outline with a view on Finland. *Planning Theory & Practice, 16*(2), 169–183.

Morphet, J. (2011). *Effective practice in spatial planning*. New York: Routledge.

Muzondi, L. (2014). Urbanization and service delivery planning: Analysis of water and sanitation management

U

systems in the city of Harare, Zimbabwe. *Mediterranean Journal of Social Sciences, 5*(20), 2905–2915.
Mycoo, M. (2017). Reforming spatial planning in anglophone Caribbean countries. *Planning Theory & Practice, 18*(1), 89–108.
Nadin, V. (2007). The emergence of the spatial planning approach in England. *Planning Practice & Research, 22*(1), 43–62.
Ngah, I. (1998). Urban planning. A conceptual framework. *Jurnal Alam Bina Jilid, 1*(1), 1–9. https://www.academia.edu/387598/UrbanPlanningaConceptualFramework.
OECD. (2016). *Better policies for sustainable development 2016: A new framework for policy coherence.* Paris: OECD Publishing.
OECD. (2018). *Policy coherence for sustainable development 2018: Towards sustainable and resilient societies.* Paris: OECD Publishing.
OECD. (2019). *OECD principles on urban policy.* Paris, France: OECD Centre for Entrepreneurship, SMEs, Regions and Cities.
Panday, P. K., & Jamil, I. (2010). Challenges of coordination in implementing urban policy: The Bangladesh experience. *Public Organization Review, 11*, 155–176. 2011.
Park, J., et al. (2011). *Urbanization and urban policies in Korea.* Gyeonggido, South Korea: Korea Research Institute for Human Settlements (KRIHS)
Parker, G., & Doak, J. (2012). *Key concepts in planning.* London: SAGE Publications (key Concepts in Human Geography).
Pillay, U. (2008). Urban policy in post-apartheid South Africa: Context, evolution and future directions. *Urban Forum, 2008*(19), 109–132.
Quadeer, M. A. (2012). Urban Development. In B. Sanyal, L. J. Vale, & C. D. Rosan (Eds.), *Planning ideas that matter. Livability, territoriality, governance and practice* (pp. 206–232). Cambridge: The MIT Press.
Rydin, Y. (1993). *The British planning system. An introduction.* London: Macmillan Press.
Rydin, Y. (2010). *Governing for sustainable urban development* (1st ed.). New York: Earthscan.
Rydin, Y. (2011). *The purpose of planning: Creating sustainable towns and cities* (1st ed.). Bristol: Policy Press, University of Bristol.
Sabbi, M., & Mensah, C. A. (2016). Juggling administrative institutions: Local state actors and the management of urban space in Kumasi, Ghana. *Urban Forum, 27*(1), 59–78.
Schindler, S., Mitlin, D., & Marvin, S. (2018). National urban policy making and its potential for sustainable urbanism. *Current Opinion in Environmental Sustainability, 34*, 48–53.
Söderholm, K., & Wihlborg, E. (2016). Striving for sustainable development and the coordinating role of the central government: Lessons from Swedish housing policy. *Journal of Sustainability, 8*(827), 1–13.
Teitz, M. B. (2015). Urban policy in North America. In *International encyclopedia of the social & behavioural sciences* (pp. 908–914). Cambridge, USA: Elsevier.
Todes, A., et al. (2010). Beyond master planning? New approaches to spatial planning in Ekurhuleni South Africa. *Habitat International, 34*(4), 414–420.
Turok, I. (2015). Turning the tide? The emergence of national urban policies in Africa. *Journal of Contemporary African Studies, 33*(33), 348–369. https://doi.org/10.1080/02589001.2015.1107288
UN-Habitat. (2008). *Urban planning best practices on creating harmonious cities. City experiences.* UN-HABITAT: Nairobi, Kenya.
UN-HABITAT. (2009). *Planning sustainable cities: Global report on human settlements.* Nairobi: UN-HABITAT.
UN-HABITAT. (2014). *The evolution of national urban policies. A global overview.* Nairobi: UN-HABITAT.
UN-HABITAT. (2016). *Urbanization and development. Emerging futures. World cities report.* Nairobi: UN-HABITAT.
United Nations. (2008). *Spatial planning key instrument for development and effective governance with special reference to countries in transition.* Geneva, Switzerland: United Nations.
USAID. (2013). Sustainable service delivery in an increasingly urbanizing world. USAID policy. USAID.
van der Berg, A. (2019). Can South African planning law and policy promote urban sustainability in the Anthropocene? In M. Lim (Ed.), *Charting environmental law futures in the Anthropocene* (pp. 203–220). Singapore: Springer Nature Singapore.
Vigar, G. (2009). Towards an integrated spatial planning? *European Planning Studies, 17*(11), 1571–1590.
Ward, S. V. (2004). *Planning and urban change* (2nd ed.). London: SAGE Publications.
Watson, V. (2006). Deep difference: Diversity, planning and ethics. *Planning Theory, 5*(1), 31–50.
Watson, V. (2009). "The planned city sweeps the poor away..." urban planning and 21st century urbanization. *Progress in Planning, 72*(3), 151–193.
Watson, V. (2016). Locating planning in the new urban agenda of the urban sustainable development goal. *Planning Theory, 15*(4), 435–448.
World Bank. (2015). *Rise of the Anatolian tigers. Turkish urbanization review.* 87180 (pp. 1–125). Washington, DC: World Bank.
World Bank. (2016). *Republic of Kenya urbanization review. Review Report AUS8099* (pp. 1–178). Washington DC, USA: The World Bank.
Yuen, B. (2009). Guiding spatial changes: Singapore urban planning. In S. V. Lall et al. (Eds.), *Urban land markets improving land management for successful urbanization* (pp. 363–384). London: Springer.

Urban Poor

▶ Participatory Governance for Adaptable Communities

Urban Resilience

Michael Henderson
Ramboll Ltd and Oxford Brookes University, London, UK

Synonyms

Disaster preparedness; Resilient cities; Spatial resilience; Urban capacity

Definition

Resilience is the ability to recover quickly to a precrisis state, to bounce back. Urban resilience then relates to the ability of human settlements to respond in the face of adversity. It denotes a shift in the focus of disaster risk management (DRM) from disaster preparedness, primarily hazards emergency response measures, to actively reducing vulnerability of urban systems in the first place. Pivotal in this shift has been the work of 100 Resilient Cities (100RC), pioneered by the Rockefeller Foundation. 100RC defines urban resilience as *the capacity of individuals, communities, institutions, businesses, and systems within a city to survive, adapt, and grow no matter what kinds of chronic stresses and acute shocks they experience.* The Organization for Economic Cooperation and Development (OECD) also provide a definition for resilient cities as *cities are cities that have the ability to absorb, recover, and prepare for future shocks (economic, environmental, social, and institutional). Resilient cities promote sustainable development, well-being, and inclusive growth.* (OEDC https://www.oecd.org/cfe/regionaldevelopment/resilient-cities.htm)

Rapid urbanization across the globe has the effect of concentrating loved ones, expensive infrastructure assets, and economic/social systems into increasingly dense, centralized locations. This agglomeration means that increased value is vulnerable to hazard exposure. Furthermore, fueled by a changing climate, we are seeing an increase in the number, severity, and unpredictability of hydrometeorological hazards, resulting in catastrophic loss of life and unsustainable economic losses.

The shift toward building "resilience" represents in a maturing of the DRM sector, which was traditionally focused on containing the impact of hazards through effective emergency response to disaster risk reduction (DRR), which uses more comprehensive understanding of risk to both reduce vulnerability and limit hazard exposure. This has happened with the backdrop of looming climate change which mobilized intense research into "future" projections, vastly increasing our understanding of future risks and provided the intent to do something about it ahead of time. As climate change has steadily transitioned from future scenario to present-day hazard, so to the DRM and climate change communities have converged bringing together an appreciation empirical analysis to reduce impacts of disasters in the first place, while ensuring post-disaster responses "build back better."

There have been several conceptualizations of urban resilience. The UNDRR (formally UNISDR) has developed a "disaster resilience scorecard for cities" structured around the "Ten Essentials for Making Cities Resilient," first developed as part of the Hyogo Framework for Action in 2005, and then updated to support implementation of the Sendai Framework for Disaster Risk Reduction: 2015–2030. (UNDRR Disaster Resilience Scorecard for Cities (updated 2017) https://www.unisdr.org/campaign/resilientcities/toolkit/article/disaster-resilience-scorecard-for-cities) This provides a structured approach for cities and their stakeholder to ascertain how well resilience thinking is embedded in their city and to help provide steps for improvement. Similarly, in preparation for 100RC, the Rockefeller Foundation with Arup developed the City Resilience Framework which "*provides a lens through which the complexity of cities and the numerous factors that contribute to a city's resilience can be understood.*" (The Rockefeller Foundation and Arup (updated 2015) https://www.rockefellerfoundation.org/wp-content/uploads/City-Resilience-Framework-2015.pdf) It introduces seven qualities that make resilient

U

cities as having flexibility, build in redundancy, build robustness, are resourceful, reflective, inclusive, and integrated.

While these qualities have several parallels with sustainability, and resilience and sustainability are often spoken about in the same context – especially in relation to building climate resilience – they are distinct. It is highly possible to be more resilient without being more sustainable. This is particularly apparent in the qualities of redundancy and robustness which can lead to increased specification and resource use in infrastructure, such as rarely use backup systems or increased volume of concrete/steel reinforcements. It is important therefore that in building resilience we do not undermine our ability to move toward more sustainable futures.

Critically, the 100RC definition includes both the shocks typically associated with disaster risk management, such as earthquakes, floods, and explosions, with the underlying stresses that can exasperate risk, including social inequality, environmental degradation, and aging infrastructure. By addressing the stresses, vulnerability is reduced limiting the impact of hazard exposure and empowering coping mechanisms to enable a faster recovery or even to "build back better."

Urban areas, and cities in particular, are complex. They are a manifestation of physical infrastructure and built environment elements with contrived social, economic, and governance systems. Intervention in one aspect of urban management can have serious knock-on consequences for others. As such, concept of urban resilience embraces urban planning as a key part of DRM; taking a "systems thinking" approach to urban development that not only assesses any intervention on its own merit, but considers its impact holistically across the urban ecosystem.

Urban resilience comes in many forms, can blur what we think of as infrastructure, and does not necessarily need to be in urban areas. For instance, traditional responses to managing flood risk has been through the development of hard, engineered flood defences. They are positioned to "defend" against the hazard. Increasingly however, we are learning that for some instances it is better to "absorb" the hazard. Nature-based solutions (NbS) or green infrastructure, utilizing and manipulating natural processes to deliver prescribed objectives, are often effective mechanisms to help absorb impacts. Within the flood risk example, this could be by using vegetated sustainable drainage systems or even using forestation or rewilding techniques to manage upstream water flows to reduce the downstream impacts on cities. NbS often has added secondary benefits, known as ecosystem services, such as supporting biodiversity and carbon sequestration which help to further reduce externalities and build resilience.

Urban resilience is therefore about creating robust physical, institutional, social, economic, and environmental systems within settlements that are capable to understand, mitigate, and respond to acute shock events by ameliorating underlying stresses.

Cross-References

▶ Policy and Practices of Nature-Based Solutions to Build Resilience in Seoul, Korea

Urban Resilience: Moving from Idealism to Systems Thinking

Michael Robbins
HIPR, New York City, NY, USA

Definition

Ecosystem- an ecosystem consists of all the organisms and the physical environment they interact with.

Value- relative worth.

Green Space- places in urban environments that incorporate different species of plants and grasses to sequester carbon and also to create a semblance of nature in an otherwise industrial or commercial area.

Resilience- the capacity for people to recover from difficulty.

Globalization- the process of interaction and integration among people, companies, and governments worldwide.

The primary focus of this chapter is to discuss how young people, which will incorporate mostly the millennial class, can become more environmentally aware or at least begin to understand environmentalism in the modern urban setting.

The approach will differ from the more conventional science and topical arrangements of environmental awareness. More than simply the knowledge of climate change, or the carbon footprint attached to eating certain food groups on a regular basis (which often passes as environmental advocacy), we will instead discuss the concept of resiliency in urban settings, its approach and applicability to the modern urban space, as well as ecosystems and consumer behaviors. Resilience is at the heart of so many issues in modern cities.

Uncovering this topic will provide a more all-encompassing picture of how young people – the consumers who still have the most life span, occupy the majority of the workforce, and have the most human capital to share – can understand the true impact of living in resilient urban spaces and the fundamentality of resiliency.

When discussing whether or not people are living in resilient urban spaces, we will also uncover when people are not, and the consequences this poses to human capability and to society more generally. As such, many research questions guide our thought process as we start. For example, is it consumerism and living comfortably with no concern for the outside world the major problem of urban cities? Is it an awareness of the environment around oneself but too much societal/social pressure to conform to sustainability measures? And finally, are young people aware that their inhabitance in urban areas also plays a role in how such an ecosystem is transformed by their presence? At its best, the systems thinking approach employed in this section is about normalizing the fact that urban spaces are ecosystems, resilient or not, and understanding how structural arrangements impede on cities will help us discover their true nature.

One of the major issues that will serve as a general underlying theme within this chapter is a mainstream problem that environmental knowledge is passed on (in schools, editorials, word of mouth, etc.) but in effect that knowledge becomes lost within a hyper-connected globalized world where decision-making processes become skewed and people with widely different interests and lifestyles undermine true environmental behaviors and attitudes (Hukkinen et al. 2022).

Understanding the Urban Ecosystem

If you live in a city in America, you have probably rarely considered it an ecosystem as well. Most of the newer, younger classes of millennials and Gen X'ers who live in cities simply divide up the city into their respective parts – the Mission District in San Francisco, the Lower East Side of Manhattan, Denver's LoDo District, and the list goes on. Doing this helps people make sense of their cities, particularly where the attractions are, where the financial district is, where the best places to eat are, and of course where the parks are to relax and unwind. Dividing a city up into its respective parts, for what it's worth, is a way to understand how to make efficient use of each part of town and how to maximize value at certain parts of the day depending on congestion or traffic.

Spatially and scientifically, however, we don't usually divide up city districts by the amount of water flowing through each respective region, and we also don't think about cities with regard to how many trees are in each neighborhood, but indeed we do think about which areas have the nicest parks, unconsciously, without making the distinction that the urban canopies and places with the most biodiversity when it comes to trees, shrubs, plants, etc. are highly important to our mental and physical well-being (https://www.weforum.org/agenda/2018/05/when-societies-feel-good-forests-get-bigger-says-study/. Adam Jezard).

In fact, the British Ecological Society and the University of Vermont conducted research on the effects of nature therapy on urbanites in 2019. For

3 months, a team from the University of Vermont studied hundreds of tweets every day and ran them through its sentiment analysis tool, hedonometer. The tool rates words on their representative happiness, awarding each word it scrutinizes a score. That enables it to estimate how happy Twitter users are at any given time. In this case, it looked at how people were tweeting from 160 parks in San Francisco and concluded that people used more upbeat words when surrounded by trees and greenery (https://www.weforum.org/agenda/2019/08/nature-mood-happiness-health-outdoors/).

For many of us living in urban centers, going to the park to unwind and heal in a sense is self-evident – one only needs to sit in a field for a few hours, disconnected from our smartphones, staring out at the horizon, to understand the healing power of nature. For our purposes, this *value* of green space is important because having access to such spaces promotes resilience. If you are able and have the means to travel to a nice park several times a week and unwind, your work performance might also increase. This is our first point with respect to resiliency. For humans, resilient behaviors must be learned, understood, and repeated.

This happens in a systemic fashion. We have to leave our apartments and travel to a green space. If the drive is too long or there's too much traffic, we have to settle for perhaps a park that is a little more polluted but easier to get to. Then we actually have to disconnect once we get to the seminatural destination and focus on our health. If we can manage all of these elements and make it part of our daily routine, then we have a chance of becoming more resilient in our daily lives. While many scholars have focused on the resiliency of nature, we're focusing on the resilience of humans in urban ecosystems.

True resilience has a cycle. The worker who leaves their apartment to go running almost nightly might fare better during flu season than someone who has a compromised immune system because they rarely exercise. That's why becoming resilient, in terms of the urban individual example, can vary from person to person if one citizen has the best access to parks and organic foods and another citizen has neither of these things.

Economists actually analyze how well whole economies "bounce back" after a recession, and the reasoning for why some do better than others has to do with a cycle of affairs that were already pre-existing in that economy, similar to the cycle of behaviors of an individual before flu season. Sure, the analogy is a stretch, but the point is that resilient behavior doesn't just spring up from anywhere. It's intertwined with so many daily aspects of life. For the purpose of looking at urban ecosystems, we're in a great place to dig a little deeper.

Globalization Makes Resiliency Harder

One of the problems with millennial culture, a generation born into an already globalized and interconnected economy, is that it is incredibly hard to become resilient as previously mentioned when there are so many distractions in modern life, and therefore modern cities. With so many consumer decisions to make today, so many brands and choice overload, how can one develop a good cycle that leads to resiliency? That's a more difficult question to answer.

This is accompanied by corporate structures that are defined for us, difficult to change, and easy to become accustomed to. If we think about areas that have been newly gentrified in cities, we know that these huge new developments are biodiversity deserts, mostly with poor use of green spacing, and where consumer decisions are basically picked for us by urban planners seeking to maximize profits. But if millennials want to live somewhere relatively affordable, in housing that is secure and modern, and with people of their own age, they don't really have too many choices in the matter – moving to a gentrified neighborhood becomes part of the demand housing developers wanted in the first place.

While gentrification helps millennials move into cities and start to understand the urban spheres they are living in from a safe starting point, these new areas in cities also destroy local ecosystems, are biodiversity deserts, and unfairly attract investments and unequal distribution to whole urban ecosystems.

This is one of the main contradictions of globalization, actually, that more young people, namely, millennials, have been awakened by vast social and environmental movements but are still not able to make any headway in their own communities in terms of urban ecosystems because corporate interests are the ones putting them in newly gentrified parts of towns that promote inequality. As of July 1, 2019, millennials, whom we define as ages 23 to 38, numbered 72.1 million (https://www.weforum.org/agenda/2019/08/nature-mood-happiness-health-outdoors/).

The generation has been praised as being the most aligned to support global causes such as climate change. But in fact this says little about how to navigate urban life in a sustainable fashion, or how to advocate for the biodiversity of city landscapes, or how to advocate for more green space in cities, or how to understand that celebrations in parks often lead to excess waste because parks in most municipalities in the USA pay for one dumpster and zero recycling units as a means to keep costs down.

The point is this – simply having a vegan lifestyle and buying eco-brands accessible in a newly gentrified part of town is a start to responsible consumerism, but sustainable consumerism must not be confused with resilient cities and green spacing. Resilience is about cycles, and consumer decisions are often erratic and nonsensical.

The Uneven Distribution of Green Space

If millennials are unable to do anything about the highly corporate world we are living in and the gentrified neighborhoods that act as biodiversity deserts and the distracted mass of citizens that can't see inequities in their own community, then what does that say about the state of US cities and this whole ecosystems approach we are talking about? For one, it signals that the online world and consumer-centered world of buying and supporting sustainable brands is way different than the reality of uneven distribution of human wellness in urban areas and of the wellness of the actual ecosystems we supposedly support.

Let's consider a classic example of how badly uneven distribution has become even with so many ideologically aligned millennials to support the environment.

In the city of Denver, Colorado, there is a downtown skate park where skateboarders and bikers frequent to test out their skills on this concrete obstacle course. The park is actually located next to a bigger, more sprawling park in a recently gentrified part of Denver – the Highlands. Confluence Park seems to have the majority of tree cover in the area, the most grass, along with a river cutting through with a bridge for runners. It's one of the more affluent neighborhoods in the city, and it's also quite nice to walk through because the trees and grass absorb some of the pollution during peak hours of congestion in the city. In general, Confluence Park represents an area where a sort of urban greenery makes visits comfortable and relaxing.

What is incredibly interesting for the purpose of this topic, and for dealing with urban resilience, is that the skate park, an older construction, but just a 2-minute walking distance from Confluence Park, has very little tree cover and is much closer to the highway than Confluence Park. There is a noticeable difference in the air quality between the two spots, and on hotter days, you can't escape the sun at the skate park. Thus, athletes who skate there frequently are breathing more polluted air with less natural covering. After frequenting the park, there is so much pollution on some days around rush hour that you can leave with a thin film of dirt and debris on your skin.

There is clearly more money and investment in maintaining the urban greenery of Confluence Park, but shouldn't athletes everywhere be entitled to clear air?

The interesting and paradoxical part of the Denver situation is that most people who live in the Confluence neighborhood are young millennials, environmental enthusiasts who, upon being asked, will state how bad climate change is and how they support eco-clothing brands. But structurally many are simply unaware of how investments in Confluence Park, and almost net zero investments at the skate park across the street, have manifested in different levels of breathed pollution – which might lend itself to chronic health conditions later in life. This structural and perhaps systems thinking aspect of the interdependent parts in play (the trees to absorb pollution, the investments to plant new trees, etc.) shows how injustices due to globalization, gentrification, and larger corporate interests in city planning do in fact reduce resilience for major urban areas.

It is up to the current and future generations to put down their smartphones, engage with the community they are in, and realize the small interconnected parts, such as the need for social bonding, that end up being opportunity costs to a wider population because of poor city planning.

Examples of Resilient Planning in the European Union

Distributional injustices as mentioned above seem to be commonplace in US cities, and one could continue to list the grievances across Latin communities in South Los Angeles to Afro-Caribbean communities in Brooklyn, New York. Urban planning and green spacing is not easy. And one point we've arrived at so far is that millennials in large part support a new wave of ideological sustainability, but the planning and interests of those building the gentrified neighborhoods they live in the urban sphere sometimes make problems worse for existing communities. While more strategic green space is desperately needed in America, especially in some lower socioeconomic neighborhoods with higher childhood rates of asthma, some European states have already realized what a resilient and thriving urban space looks like. Let's take a look at how urban resilience is built, and thought of, in the European examples.

Amsterdam comes to mind first, which has more bicycles than people and in which citizens see the benefit of exercising on the way to work in high numbers. The ability for a critical mass of citizens to see the opportunity in biking as a collective good for national pollution levels and well-being in general is a true milestone on the road to a resilient urban sphere.

In Copenhagen too, the meaning of resilience goes beyond affordable housing and clean air. It also means creating physical spaces in the city where people can meet, gather, play, and engage as active citizens. One example of this approach in Copenhagen is the Superkilen Park in Norrebro. The park is divided into three zones dedicated to sports, games, or outdoor activities and is a symbol of the "living together" approach. Another example is the swimming possibilities in the harbor of Copenhagen, which are the result of the climate prevention infrastructure that has been put in place (https://journals.openedition.org/factsreports/4750).

In both of these examples, citizens are benefitting from measures put in place by local and national authorities who understand that when you give people more room and green space to play, it in turn increases their standard of living, their health, and their ability to return to work each day satisfied with their quality of their life. That in turn supports a healthy and thriving economy where people want to go to work. That is resilience.

Resilience, the Economy, and Productivity

The Amsterdam example above also alludes to another factor we haven't discussed yet, productivity and economic gains. When states do a great job of promoting whole economies and systems that work for their citizenry, productivity goes up, sometimes exponentially. If it's easier to get to work and you're on time and it's not a hassle, that might amount to higher levels of productivity when you're in the office. One of the reasons productivity has gone up in the COVID economy is because knowledge laptop workers have adjusted to home working routines and have cut the hassle of inefficient US transportation systems out of their weekly schedules. It makes a difference.

In fact in America, we typically see economic losses due to traffic and hours lost sitting in traffic.

The total cost of lost productivity caused by congestion amounted to $87 billion in 2019. The average Bostonian driver lost 164 working hours last year thanks to traffic, which cost the local economy $4.1 billion (https://www.cnbc.com/2019/02/11/americas-87-billion-traffic-jam-ranks-boston-and-dc-as-worst-in-us.html).

Thus the economy, the politics of good city planning, and resilience are interwoven. Urban spheres that promote resilience, either through the use of green space for restorative purposes or through better bike routes, for example, will see a boost or "growth" in economic output, whereas cities that are more unevenly distributed are suffering from daily losses in output.

Can You Teach Resilience?

The problem with resilience is one of timing and foresight. It's not easy to examine the consequences of eating fast food today for ourselves in 20 years' time, and for individuals who are governed by their feelings more than their direct observations, what feels good here and now usually wins. That's how a consumer economy largely works. We buy and consume to activate our neurotransmitters to the point where it becomes addicting, and then we rationalize to ourselves that our behaviors have no long-term consequences because thinking about the future is, one, fear inducing and, two, difficult.

From this perspective, resilience can be taught when other similar topics are taught – such as consumer behaviors and lifestyle choices. One of the points of bringing up some of the European examples is that in more collectivist societies with choices made for people, resilience can be understood simply by participating in such a society. But it can't be taught in the American example if what we value is erratic consumerism that doesn't follow a cycle and only follows profit.

References

Hukkinen, J. I., Eronen, J. T., Janasik, N., Järvensivu, P., & Kaaronen, R. O. (2022). Coping with policy errors in an era of chronic socio-environmental crises. *Ecological Economics, 199*, 107489.

Urban Sprawl

▶ Urban Structure and Its Impact on Mobility Patterns: Reducing Automobile Dependence Through Polycentrism

Urban Structure

▶ Urban Structure and Its Impact on Mobility Patterns: Reducing Automobile Dependence Through Polycentrism

Urban Structure and Its Impact on Mobility Patterns: Reducing Automobile Dependence Through Polycentrism

Jeffrey Kenworthy
Curtin University Sustainability Policy Institute, Curtin University, Perth, W.A., Australia
Frankfurt University of Applied Sciences, Frankfurt am Main, Germany

Synonyms

Automobile dependence; Centralization; Decentralization; Decentralized concentration; Density; Mixed land use; Polycentrism; Sub-centers; Urban sprawl; Urban structure

Definition

Reducing automobile dependence has become critical in all cities today. Local, regional, and global problems are demanding solutions, ranging from local traffic congestion, noise and other traffic impacts, regional air pollution, and global greenhouse gas emissions and the linked problem of how to reduce consumption of fossil fuels in cities, especially in transport. A key factor in

U

tackling automobile dependence is the way that different land uses are arranged in urban regions. At opposite ends of the spectrum, this arrangement can be through low-density, heavily zoned urban sprawl with a lack of major focal points or centers (dispersal and decentralization) or through higher-density, mixed land uses, often with special attention to the development of strong sub-centers (polycentrism or decentralized concentration). This article describes these matters of urban structure with examples and explains how higher-density, mixed land uses arranged into strong centers within urban regions can be used to help reduce automobile dependence.

Introduction

An important aspect of any metropolitan region is its internal structure. In other words, how are the main physical elements of the region, such as residences, places of work, shops, businesses, government departments, schools, hospitals, and so on, laid out across the landscape? Are they sprinkled across the urban region like "salt and pepper" with few significant or easily recognized concentrations, or are there easily identified places where such uses are concentrated and mixed together? In other words, are there clear, dense, mixed use centers within the metropolitan region?

Transport demand and the modes that are used to meet this demand arise from land use, and the arrangement of land uses is critically important in determining how much people need to rely on cars. This article explains, with examples, how minimizing urban sprawl and creating cities with multiple strong centers can be effective in helping to reduce automobile dependence and strengthen the use and convenience of public transport, walking, and cycling.

Urban Structure and Its Impact on Mobility Patterns: Reducing Automobile Dependence Through Polycentrism, Fig. 1 The city center of Freiburg im Breisgau showing its pedestrian orientation. (Source: Jeffrey Kenworthy)

Types of Urban Centers and Their Exemplars

Cities with clearly recognizable centers, most often termed the central business district (CBD) and sub-centers, are distinguished by the number and distribution of such centers within their urbanized territory. The character, size, and number of these centers, how they are connected to each other by different transport modes, as well as the dominant modes of circulation within them, are very important in determining how they function. Such urban structure is also very important in how the overall metropolitan region functions; is it a highly car-dependent region with unsustainable transport, or is it possible to move around easily on public transport or even by bike or on foot?

Centers within cities can be significantly rail-based or walking and cycling-based at one extreme, or at the other end of the spectrum, they can be primarily car-based, with some combinations in between. For example, virtually all the centers in the Tokyo and Stockholm metropolitan regions are strongly rail-based, as well as many of the centers across the Greater Toronto Area and increasingly in Metro Vancouver. Significant numbers of cities in Europe have centers where cycling and walking provide a high proportion of their accessibility needs (e.g., Freiburg im Breisgau, (Germany); Amsterdam, Groningen, and Delft (the Netherlands); Copenhagen (Denmark), and many others (Figs. 1, 2, 3, and 4). Conversely, the Los Angeles, Detroit, Atlanta, and Houston metropolitan areas (USA), indeed most other American cities, have primarily car-based access in whatever "centers" exist. Cities such as Los Angeles and Detroit are classed as "extremely decentralized," while Atlanta and Houston are cast as "decentralized," indicating the weak nature of any form of centering (Romein et al. 2009). In between these extremes, there are cities that have centers with both significant car- and bus-based access

Urban Structure and Its Impact on Mobility Patterns: Reducing Automobile Dependence Through Polycentrism, Fig. 2 The center of Amsterdam and its strong orientation to non-motorized modes. (Source: Jeffrey Kenworthy)

Urban Structure and Its Impact on Mobility Patterns: Reducing Automobile Dependence Through Polycentrism, Fig. 3 The center of Delft and its focus primarily on bike access and mobility. (Source: Jeffrey Kenworthy)

(Boulder, USA – see Schiller and Kenworthy 2018) and sometimes a sizeable light rail transit (LRT) system connecting important centers (e.g., Edmonton and Calgary, Canada).

It is also possible to see such polycentrism (see later discussion) at work in whole countries. In some sense, small countries such as Switzerland, which has a wide variety of cities of different sizes within its confined boundaries (Zurich, Bern, Lucerne, Lausanne, Montreux, Locarno, and many more), perform somewhat akin to one large polycentric megalopolis, with each city acting as a sub-center of the whole. Binding all these cities together into a seamless whole is the Swiss Federal Railway system (SBB/CFF/FFS), which for the most part offers 30-min train services between all the cities, regardless of time of day. SBB offers attractive fare "abonnements," including an annual pass that allows passengers' free and unlimited travel across the whole country on most trains, as well as much of the public transport systems within cities (see https://www.sbb.ch/en/travelcards-and-tickets/railpasses/ga.html accessed November 10, 2020). Such attractive public transport services and fare offers encourage free and easy movement within this effectively large, national, polycentric urban region.

The existence and expansion of significant non-auto-based sub-centers within cities also play a role in determining how livable the urban region is overall and how much it may be moving toward or away from automobile orientation in its transport system. That is, if population and jobs, indeed most of the critical needs that people have in any city (mixed land use), are focused into discernable, relatively high-density, compact centers, serviced by quality public transport systems (transit for short) and having an excellent, safe, and human-quality public realm (e.g., pedestrian and bike-only or traffic calmed streets), people will most likely find it convenient to walk or cycle for local needs and use transit for longer

Urban Structure and Its Impact on Mobility Patterns: Reducing Automobile Dependence Through Polycentrism, Fig. 4 Copenhagen's city center is pedestrian and bicycle-friendly. (Source: Jeffrey Kenworthy)

trips. This is an effective way to reduce automobile dependence (Wen et al. 2019, 2020; Newman and Kenworthy 1999, 2006, 2015; Schiller and Kenworthy 2018). Often there is also greater potential for community and the development of social capital and economic prosperity in such places (Wen et al. 2020). This is because public spaces are attractive, encouraging people to meet in them, and businesses are attracted to locate in such centers because of the high quality of life, opportunities for networking, and ease with which their employees can access the centers, very often by transit, foot, or bike without enduring endless daily traffic congestion (Florida, 2005, 2010, 2012a, b, c).

The Vancouver region in British Columbia, Canada, has been moving in the direction of strong sub-centers for decades due to the implementation of an elevated, driverless train system called SkyTrain, around which high-density, mixed land uses have been developed at key stations such as Main Street/Science World, Joyce/Collingwood, Metrotown, Edmonds, and New Westminster (Schiller and Kenworthy 2018). Figures 5, 6, and 7 partly demonstrate this. Punter (2003) has additionally shown how first-rate urban design in the public spaces of such centers has been instrumental in changing the nature of the Vancouver region into a much more livable place (e.g., it regularly makes it into the top five cities of the world on livability indices (see https://en.wikipedia.org/wiki/Most_livable_cities accessed October 6, 2020 and http://web.archive.org/web/20121128200352/ http://www.mercer.com/press-releases/quality-of-living-report-2011 accessed October 6, 2020).

U

Urban Design and Preservation of Natural Landscapes in Cities

Indeed, urban design in the centers dotting metropolitan regions is critically important, as demonstrated repeatedly over many years in the prolific

Urban Structure and Its Impact on Mobility Patterns: Reducing Automobile Dependence Through Polycentrism, Fig. 5 Transit-oriented sub-centers at stations on Vancouver's SkyTrain (Joyce/Collingwood in the foreground and Metrotown in the distance). (Source: Jeffrey Kenworthy)

works on sustainable urbanism by Jan Gehl and his associates in Copenhagen (e.g., Gehl 2010; Gehl et al. 2011; Matan and Newman 2016). Schiller and Kenworthy (2018) devote a chapter of their book on sustainable transportation to "Urban Design for Sustainable and Active Transportation and Healthy Communities."

If cities are concentrating new development into compact sub-centers, it is also less likely that there will be widespread destruction of natural areas and agricultural land, simply because the demand for development space, including roads and parking, will be minimized. Green space for enhancing air quality, improving water catchment, protecting natural habitats, and providing important recreational opportunities will be maximized. Such urban regions, developed significantly around strong centers, are thus characterized by generous amounts of green space, forests, and near-city agriculture, community allotments and gardens to provide for local food needs, as well as the more sustainable mobility patterns already mentioned (based on transit, walking, and cycling and less need to use a car) – see Fig. 8. The city's "ecological footprint" can be much smaller (Wackernagel and Rees 1996).

The Stockholm region has been portrayed in this way as a "transit metropolis" (Cervero 1998; Kenworthy 2019). The city has generous green areas between and around major satellite centers on its tunnelbana (metro) system (e.g., Kista, Vällingby, etc.) – see Figs. 9, 10, and 11. The Rhein-Main Region centered on Stadt Frankfurt (City of Frankfurt) is also known as a "green region" with large areas of "Stadtwald" (city forest) and other green space, agricultural land, and garden association areas interspersed between often well-defined urban settlements of various sizes and density. Taken at the scale of a conurbation, sub-centers within the Rhein-Main region can actually be viewed as whole cities such as Darmstadt or Mainz or at a finer scale as distinct, well-defined medium-sized "Kommune" within

Urban Structure and Its Impact on Mobility Patterns: Reducing Automobile Dependence Through Polycentrism, Fig. 6 New Westminster sub-center within 400 m easy walk of its SkyTrain station. (Source: Jeffrey Kenworthy)

the region, (see https://www.region-frankfurt.de/Services/Statistiken-Prognosen/Statistik-Viewer accessed October 6, 2020).

The Rhein-Main Region has very large areas of agricultural land and forest separating the major centers that comprise its polycentric structure. It is recognized as a major "polycentric" urban region along with others such as the Ruhrgebiet in Germany (Essen, Duisburg, Bochum, Düsseldorf, etc.) and Randstad in the Netherlands (e.g., Amsterdam, Rotterdam, Leiden, the Hague, etc.) each similarly displaying a landscape that for the most part clearly separates urban and non-urban areas and provides many opportunities for both human recreation as well as plant animal habitats. All these large polycentric conurbations are also well-serviced with extensive networks of S-Bahn (fast suburban rail), regional rail, and regional bus services, which assist in minimizing automobile dependence.

The notion of sub-centers works at a much smaller scale too, such that there are various-sized centers to be observed within Stadt Frankfurt alone, such as Nordweststadt, Riedberg, the new Europaviertel, and many others (e.g., see Gesellschaft für Markt and Absatzforschung mbH, 2010).

Monocentric, Polycentric, and "Full-Motorization" Models of Urban Development and Their Implications for More Sustainable Transport

Individual cities and whole urban regions therefore come in various forms. Some cities are more "monocentric," meaning they still have one primary, strong city center compared to other smaller centers, and in the context of this article, this center is, for the most part, still the primary force in shaping the way the city works in transport terms. The concept of monocentric cities is discussed extensively in the literature and emerged chiefly in early economic explanations

Urban Structure and Its Impact on Mobility Patterns: Reducing Automobile Dependence Through Polycentrism, Fig. 7 Transit-oriented, high-density residential development within 400 m of Edmonds SkyTrain station in the Vancouver region. (Source: Jeffrey Kenworthy)

of the way cities worked (Alonso 1964; Mills 1969; Muth 1969).

In the USA, Arribas-Bel and Sanz-Gracia (2014) demonstrated that 57.7% of metropolitan areas in the USA in 2010 still conformed to a definition of the monocentric model. One can visualize this monocentric form as a bike wheel with all the spokes leading to the central hub. New York, London, Paris, and Tokyo historically were monocentric cities, and most of their extensive rail systems were focused on their city centers. While their massive rail systems (subways and longer-distance suburban rail), which were mostly developed many decades ago, are still concentrated on their original city centers, such megalopolises have grown so big that they do now have numerous other significant sub-centers in their regions (very often linked to key railway stations), as well as some possibilities for circular car and public transport travel around their regions, increasingly by rail (e.g., Fig. 12 – the LRT lines that operate around the inner part of the Paris region, chiefly the Ville de Paris, forming cross-city travel possibilities through key interchange points).

The New York region, which extends across the three US states of New York, New Jersey, and Connecticut with over 23 million people, has a part of Manhattan as its "Central Business District" (the area south of 60th Street, sometimes referred to as the Hub), with over 2 million jobs located in a relatively small area of around 2330 ha. New York's CBD is also a very strong neighborhood, housing between 500,000 and 600,000 people (Kenworthy and Laube 1999).

New York's vast subway and commuter rail system and many of its freeways are focused on this one major center like spokes on a wheel. Indeed, the New York region is the most transit-oriented region in the USA with the highest per

Urban Structure and Its Impact on Mobility Patterns: Reducing Automobile Dependence Through Polycentrism, Fig. 8 Strong centers such as in Frankfurt, Germany (Frankfurt Central Business District some 2 km away in background), leave ample room for green space and productive land. (Source: Jeffrey Kenworthy)

capita use of public transport of all US metropolitan areas (Kenworthy and Laube 2001; Newman and Kenworthy 2015). This is primarily because of the very dense and large single center of lower Manhattan (the CBD) and, more widely, the high-density City of New York as a whole, with a massive rail infrastructure leading into it and binding it together, supplemented by an extensive bus system (Kenworthy and Laube 1999).

There are enormous tidal flows every day of people into and out of the New York central area and the whole City of New York that forms the *inner area* of the New York Tri-State Metropolitan Region. The City of New York, with about 8.2 million people, is very dense and transit-oriented. It is impossible to move so many people (especially into and around the crowded Manhattan Island) in such a short, peak period timeframe without a high-capacity rail system. It would be physically impossible to achieve this task mainly with cars because of the road and parking capacity constraints. Indeed, the Manhattan CBD has only about 67 parking spaces for every 1000 jobs located there, so most people (about 93%) either use transit to get to work there or they walk, or increasingly today, ride bikes, including through new shared bike schemes (Newman and Kenworthy 1989 and 2015).

Monocentric cities are generally becoming less common because of the need to maintain adequate accessibility in increasingly complex and congested cities. A high proportion of cities today demonstrate some form of polycentrism (Ahlfeldt and Wendland 2010; Arribas-Bel and Sanz-Gracia 2014; Romein et al. 2009). The form of that polycentrism is, however, critical in reducing dependence on the automobile.

A primarily monocentric pattern is evident in Perth, Western Australia, a city of around 2.1 million people. Perth's CBD is still the largest and most influential of centers in the metropolitan

U

Urban Structure and Its Impact on Mobility Patterns: Reducing Automobile Dependence Through Polycentrism, Fig. 9 Stockholm rail-based sub-center showing dense clustering around a tunnelbana station, with surrounding green space. (Source: Jeffrey Kenworthy)

region, with all rail lines and a high proportion of bus routes currently still concentrated on the CBD. In such monocentric cities, it may not be easy to move circumferentially around the city on the public transport system, although Perth does have one large circular bus route connecting universities, hospitals, and shopping centers over a wide area. However, it takes several hours to travel the whole route. If there are no significant sub-centers within the urban landscape, then high-capacity public transport modes (rail or bus rapid transit (BRT)) have no areas of "critical mass" on which to focus and justify strong and frequent circumferential services with possibly significant back-loading of passengers between them during peak times (back-loading refers to reverse flows of passengers between centers at either end of a public transport line, which improves the utilization of public transport services). It is difficult to service low-density, heavily zoned, dispersed, and scattered urban development or "urban sprawl" with anything more than somewhat inferior and infrequent bus services. Demand is simply too low. Such cities, like Perth, become rapidly automobile-dependent.

Public transport needs concentrations of activities (population, employment, businesses, schools, etc.) to work properly. Where a city is monocentric and sprawling, without any significant sub-centers, captive public transport users (those with no choice) may find they need to travel on public transport toward the city center and then travel

Urban Structure and Its Impact on Mobility Patterns: Reducing Automobile Dependence Through Polycentrism, Fig. 10 The recent Hammarby Sjöstad sub-center in Stockholm focused on a light rail line which provides excellent public transport access and non-motorized mode facilities. (Source: Jeffrey Kenworthy)

out again from the city center to get where they want to go, instead of a more direct trip across the city. This can be very inefficient and time-consuming and tends to push people toward the faster but more expensive car option, even if they would struggle to afford it.

On the other hand, if a city with a very strong central area or CBD, such as in Singapore, also develops major sub-centers at railway stations on its radial rail system at significant density in more distant parts of the urban region, then it becomes possible to join these sub-centers up directly with high-quality, direct, cross-city transit lines, including rail lines (Fig. 13). The transit services can also be frequent because the sub-centers generate sufficient demand throughout the day, due to their density and their mixture of land uses, ensuring people have different reasons to go to the centers at various times of the day. At ground level the station precincts in Singapore have intense mixed land uses that are pedestrian-oriented (Fig. 14). Singapore has built such non-radial mass rapid transit (MRT) lines using fully automated, driverless light rail transit (LRT). This complements the major MRT radial train lines leading from sub-centers into the central area of Singapore. Singapore has rapidly evolved and continues to evolve into a green, livable, and strongly polycentric urban form with very low dependence on the car.

Toronto was able to build the cross-city Sheppard subway line because it developed two major sub-centers dating back as far as the 1950s (North York Center and Scarborough Center) which the subway now connects. Such a city structure is more like a spider web than a bike

Urban Structure and Its Impact on Mobility Patterns: Reducing Automobile Dependence Through Polycentrism, Fig. 11 Hammarby Sjöstad sub-center with its excellent pedestrian and cycling orientation. (Source: Jeffrey Kenworthy)

Urban Structure and Its Impact on Mobility Patterns: Reducing Automobile Dependence Through Polycentrism, Fig. 12 Part of the circular LRT around the inner area of the Paris region. (Source: Jeffrey Kenworthy)

Urban Structure and Its Impact on Mobility Patterns: Reducing Automobile Dependence Through Polycentrism, Fig. 13 LRT line and station in Singapore showing high-density clustering at the station. (Source: Jeffrey Kenworthy)

wheel, with good connections that cut across or join the radial transport routes (see Thomson (1977), Figs. 10, 14, and 16). This type of urban form, which sees major centers grow up outside the CBD, is known as "polycentrism" or "decentralized concentration." Li (2020) discusses this whole issue of polycentrism, concentration, and decentralization in relation to changes in urban spatial structure between 2001 and 2016 within 286 Chinese cities. From the perspective of reducing automobile dependence, polycentrism only works effectively when centers are transit-based and fundamentally pedestrian- and bicycle-oriented in their internal movements.

Thomson (1977) referred to city structures involving polycentrism as having "weak-center," "strong-center," "traffic-limitation," or "low-cost" strategies of metropolitan structure, depending on the extent and strength (size and density) of the sub-centers around the city and the type of transport infrastructure provided to connect them all. He contrasted these strategies to the ultimate form of decentralization or "no centers" as a "full-motorization" strategy. Figures 15 and 16 show Thomson's "full-motorization" and "weak-center" strategies, while Fig. 17 shows the Los Angeles urban landscape, a prime example of full motorization.

The extreme opposite to both monocentric and polycentric urban forms is the "dispersed city" or "decentralized city," or as already mentioned according to Thomson (1977), the "full-motorization" strategy. The type of urban development that epitomizes this kind of city has sometimes been referred to derogatorily as "urban splatter." Such urban development is somewhat like the effect one gets if one shakes salt and pepper over a plate. Some of the prime examples of cities that have developed along such lines are found in the USA (e.g., Los Angeles, California, Denver, Colorado, and Phoenix, Arizona).

Urban Structure and Its Impact on Mobility Patterns: Reducing Automobile Dependence Through Polycentrism, Fig. 14 LRT station in Singapore showing pedestrian-oriented mixed uses at ground level. (Source: Jeffrey Kenworthy)

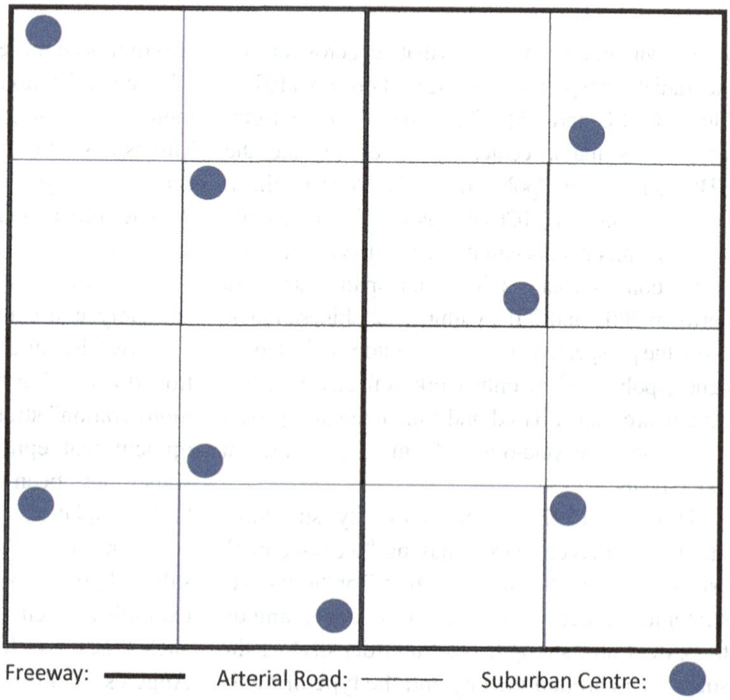

Urban Structure and Its Impact on Mobility Patterns: Reducing Automobile Dependence Through Polycentrism, Fig. 15 Thomson's full-motorization strategy. (Source: Author's drawing, based on Thomson (1977))

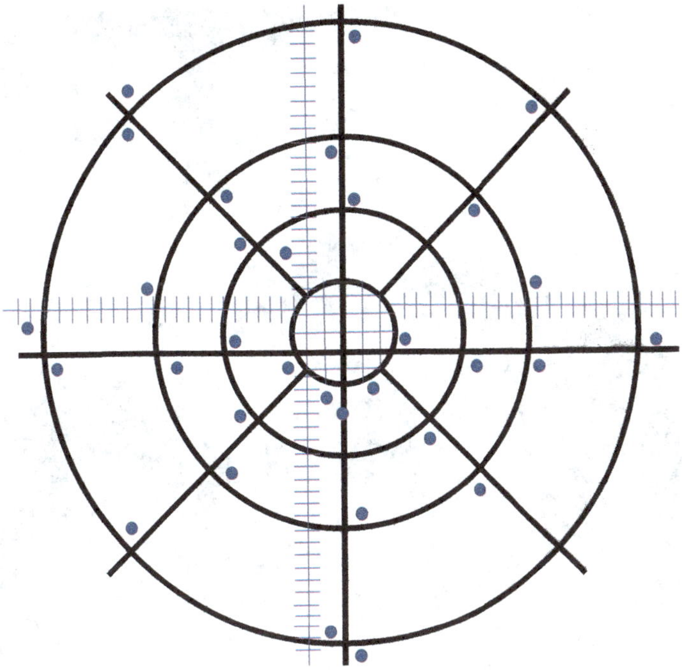

Urban Structure and Its Impact on Mobility Patterns: Reducing Automobile Dependence Through Polycentrism, Fig. 16 Thomson's "weak-center strategy. (Source: Author's drawing, based on Thomson (1977))

Unlike most cities in the world of comparable size (around 5 million), Phoenix hardly has any real city center or CBD. There is a central area or CBD of sorts, and today it is getting a little larger and more significant within the region because the city has installed its first light rail line leading into it and through it, which is making it more attractive for investment and denser urban development. However, this "central city area" has traditionally been little more than the equivalent of a suburban car-based center with some offices and shops, a little residential development, and lots of parking.

Partly for these reasons, the Phoenix metropolitan area is highly dependent on the automobile, has a very poor, predominantly bus-based public transport system with exceptionally low patronage, and its "sub-centers" are mainly car-based shopping centers in the suburbs surrounded by a sea of parking or freeway-based business parks with similar vast areas of car parks. All its centers are highly car-dependent and pedestrian-unfriendly. Indeed, any such urban form can only be serviced effectively by the automobile because distances are too great for walking and cycling, and public transport is ineffective and infrequent. The public realm is also hostile and dangerous due to roads and parking taking up so much space, air pollution, and noise, and few pedestrian and cyclist facilities make walking and cycling dangerous and unattractive.

Conclusions

To help to reduce automobile dependence in cities, a polycentric structure is needed based on significant dense centers containing mixed land use, serviced by a backbone of high-quality urban rail transit. As well, such centers need to have a first-class urban public realm with urban design qualities that are inviting and safe for pedestrians and cyclists. Urban dispersion and decentralization are recipes for locking in increased dependence on the automobile and fostering increasingly unsustainable transportation systems.

U

Urban Structure and Its Impact on Mobility Patterns: Reducing Automobile Dependence Through Polycentrism, Fig. 17 Part of Los Angeles showing a "full-motorization" strategy which is lacking in centers. (Source: Jeffrey Kenworthy)

Cross-References

- ▶ Green Cities
- ▶ Low-Carbon Transport
- ▶ Transportation and Mobility

References

Ahlfeldt, G. M., & Wendland, N. (2010). *How polycentric is a monocentric city? The role of agglomeration economies*. Munich: MRPA paper no. 24078. https://mpra.ub.uni-muenchen.de/24078/. Accessed 10 Nov 2020.

Alonso, W. (1964). *Location and land use: Toward a general theory of land rent*. Cambridge, MA: Harvard University Press.

Arribas-Bel, D., & Sanz-Gracia, F. (2014). The validity of the monocentric model in a polycentric age: US Metropolitan Areas in 1990, 2000 and 2010. *Urban Geography, 35*(7), 980–997.

Cervero, R. (1998). *The transit metropolis: A global inquiry*. Washington, DC: Island Press.

Florida, R. (2005). *Cities and the creative class*. New York: Routledge.

Florida, R. (2010). *The great reset: How new ways of living and working drive post-crash prosperity*. New York: Harper Collins.

Florida, R. (2012a). Cities with denser cores do better. *The Atlantic*. [Online]. Available from: www.theatlanticcities.com/jobs-and-economy/2012/11/cities-denser-cores-do-better/3911/. Accessed 9 Sept 2016.

Florida, R. (2012b). *The rise of the creative class: Revisited*. New York: Basic Books.

Florida, R. (2012c). How and why American cities are coming back. *The Atlantic*. [Online]. Available from: theatlanticcities.com/jobs-and-economy/2012/05/how-and-why-american-cities-are-coming-back/2015/. Accessed 9 Sept 2016.

Gehl, J. (2010). *Cities for people*. Washington, DC: Island Press.

Gehl, J., Svarre, B. B., & Risom, J. (2011). Cities for people. *Planning News, 37*, 6–8.

Gesellschaft für Markt and Absatzforschung mbH. (2010). *Baustein 2/10: Fortschreibung Einzelhandels- und Zentrenstruktur, Frankfurt am Main 2010*. Frankfurt am Main: Stadtplanungsamt, Stadt Frankfurt.

Kenworthy, J. R. (2019). Urban transport and eco-urbanism: A global comparative study of cities with a special focus on five larger Swedish urban regions. *Urban Science, 3*(25), 1–44. https://doi.org/10.3390/urbansci3010025.

Kenworthy, J. R., & Laube, F. B. (1999). *An international sourcebook of automobile dependence in cities, 1960–1990*. Boulder: University Press of Colorado.

Kenworthy, J., & Laube, F. (2001). *The millennium cities database for sustainable transport*. Brussels: International Union of Public Transport (UITP), and Perth: Institute for Sustainability and Technology Policy (ISTP), Murdoch University – CD ROM database.

Li, Y. (2020). Towards concentration and decentralization: The evolution of urban spatial structure of Chinese cities, 2001–2016. *Computers, Environment and Urban Systems, 80*, 101425. https://doi.org/10.1016/j.compenvurbsys.2019.101425.

Matan, A., & Newman, P. (2016). *People cities: The life and legacy of Jan Gehl*. Washington, DC: Island Press.

Mills, E. S. (1969). The value of urban land. In H. S. Perloff (Ed.), *The quality of the urban environment*. Baltimore: The Johns Hopkins Press for Resources for the Future.

Muth, R. F. (1969). *Cities and housing: The spatial pattern of urban residential land use*. Chicago: University of Chicago Press.

Newman, P. W. G., & Kenworthy, J. R. (1989). *Cities and automobile dependence: An international sourcebook*. Aldershot: Gower.

Newman, P. W. G., & Kenworthy, J. R. (1999). *Sustainability and cities: Overcoming automobile dependence*. Washington, DC: Island Press.

Newman, P., & Kenworthy, J. (2006). Urban design to reduce automobile dependence. *Opolis, 2*(1), 35–52.

Newman, P., & Kenworthy, J. (2015). *The end of automobile dependence: How cities are moving beyond car-based planning*. Washington, DC: Island Press.

Punter, J. (2003). *The Vancouver achievement: Urban planning and design*. Vancouver: UBC Press.

Romein, A., Verkoren, O., & Fernandez-Maldonado, A. M. (2009). Polycentric metropolitan form: Application of a 'northern' concept in Latin America. *Footprint, Metropolitan Forum, 2009*, I27–I45.

Schiller, P. L., & Kenworthy, J. R. (2018). *An introduction to sustainable transportation: Policy, planning and implementation*. London: Earthscan (Taylor and Francis/Routledge).

Thomson, J. M. (1977). *Great cities and their traffic*. Middlesex: Penguin Books.

Wackernagel, M., & Rees, W. (1996). *Our ecological footprint: Reducing human impact on the earth*. Gabriola Island: New Society Publishers.

Wen, L., Kenworthy, J., Guo, X., & Marinova, D. (2019). Solving traffic congestion through street renaissance: A perspective from dense Asian cities. *Urban Science, 3* (18), 1–21. https://doi.org/10.3390/urbansci3010018.

Wen, L., Kenworthy, J., & Marinova, D. (2020). Higher density environments and the critical role of city streets as public open spaces. *Sustainability, 12*, 8896. https://doi.org/10.3390/su12218896.

Urban Sustainability: Multifunctional and Multipurpose Planning of Urban Space

Aynaz Lotfata
Geography Department, Chicago State University, Chicago, IL, USA

Introduction

As the urban population increases, cities face a dilemma. On the one hand, horizontal growth causes devastating effects on the environment, and on the other hand, vertical growth leads to a decrease in residents' quality of life (Ghafouri and Weber 2020). For instance, growing urbanism faced problems providing equal access to public facilities due to the lack of urban lands to accommodate new facilities for public use. The focus of this study is to explore the notion of multi-functionality in urban planning and design research in terms of improving equal access to public spaces impacting urban well-being. That is to say that allocations of multiple uses to an urban land promote equity in access, improving the quality of life.

The entry raises the question of "What multi-functionality of urban spaces offers equal access to public spaces in densely built-up urban areas and consequently contributes to urban wellbeing of cities?" The multifunctional organization of urban space optimizes urban land use planning efficiency by repurposing urban areas towards a new function(s) in different circumstances and times. For instance, half-day public school can be repurposed for other activities after school, extending schools' life cycle and life duration. The multifunctionality of urban spaces aids in

U

responding quickly to new circumstances, which prevents delays in reusing and repurposing of the metropolitan area and urban demolishing by keeping urban spaces in the life cycle of the built environment (Foster 2021; Parris and Kiroff 2017).

Multifunctionality enables the urban elements to adopt different functions in different circumstances (Jampani et al. 2020). For instance, combining compatible functions increases the diversity of uses in place. Likewise, the multifunctionality of urban elements at the micro-scale increases urban network flexibility at the macro-scale (Sharifi 2019). Urban networks are multiple systems characterized by the interplay of ecological, social systems, political systems, and spatial/physical systems (e.g., Pickett et al. 2011). In contrast, the monofunctional urban space consumes more urban land that accelerating urban sprawl. This trend of urban development threatens the livelihood of cities due to the disrupted balance between built-up and non-built-up areas.

Despite the abundant research on the multifunctionality of urban elements (e.g., Meerow 2020; Alves et al. 2021), studies on their linkage with urban networks that enhance their multifunctionality are missing. This study explored the multifunctionality concept within a nested hierarchy of urban scales and dynamics among them. In this regard, we interrogate the links between the multifunctionality of urban elements and the multi-scale dynamics of land uses associated with urban space network characteristics. Does relationship between urban networks and urban elements support alterations in the usage of lands at different urban scales in different circumstances and times?

Multifunctional Urban Space

Most literature debated the multifunctionality of green areas and agricultural lands (e.g., Alves et al. 2021; Hansen et al. 2019; Breuste and Qureshi 2011; Chiesura 2004). These studies (e.g., Breuste and Qureshi 2011; Chiesura 2004) highlight the different types of natural, economic, and social services that green areas can provide without reflecting on urban networks' role on the multifunctional performance of the green regions (Haines-Young and Potschin 2010; Peschardt et al. 2012). Multifunctionality is diverse uses of space that can change over time, stated as multi-use spaces (e.g., Tartaglia et al. 2021). Parallel to this definition, it is highlighted as the capacity of urban areas to provide multiple urban services (Rolf et al. 2019; Roe and Mell 2013; Belmeziti et al., 2018) and "an integration and interaction between services" (Roe and Mell 2013). Multifunctionality aims to intertwine different functions and thus use limited space more effectively (Ahern 2011). Understanding cities as complex urban systems requires capturing them with their nested hierarchy of scales and cross-scale dynamics among them (Sharifi 2019) so that micro to macro cross-scale dynamics gain importance. The research on the impacts of these dynamics on multifunctionality is absent.

In network-oriented planning, interactive system thinking supports and facilitates the multifunctional operation of urban elements. So, multifunctional land use (MLU) integrates various land use functions in a determined area and time (Van Broekhoven and Vernay 2018). Interesting aspects of the MLU complement other concepts that address the idea of mixing urban functions and flows, such as mixed land use, compact city, and low carbon city. MLU is especially interesting to focus on studying the specificities and challenges of integrating functions. MLU's core ambition is the integration of (physical) functions. In contrast, concepts such as compact city and low carbon city are broader; they, respectively, are about designing cities to have high density, besides integrating material flows, the promotion of active modes (walking and biking) of transportation, and efficient building. MLU seeks to create synergies between previously separated urban functions (Vreeker et al. 2004).

In this regard, multifunctionality needs to be assessed within the context of a multi-scalar urban network. Urban elements need to be integrated into the urban structure as one system for multifunctional capacity to arise. This mutual relation between urban elements and their surrounding environment addresses the links between cities' form, function, and structure where goods, sources, information, and people flow. Hence, multifunctionality occurs when systems (social, economic, environmental, political, and physical systems) interact and inclusively engage with each other. As such, the effects on one system will influence another or result in the inability to operate it positively because the urban system is dynamic, unpredictable, and multidimensional, including a collection of interconnected relationships and parts (e.g., Yang et al. 2017; Tobey et al. 2019; Yang and Yamagata 2019).

Connectivity can be taken as an essential core of urban elements' multi-scalar network characteristics. It indicates the spatial distribution and relations of elements and, consequently, the benefits they provide. Connectivity is often referred to as ecological connectivity (e.g., Chang et al. 2012). Ecological connectivity is not only meant in a physical sense but also functionally. For instance, the random and equal distribution of public facilities in the urban network can impact urban functions on a larger scale, like changing walking habits, walking for recreational use, and controlling urban traffic. That is to say that urban elements with stabilized identity generally provide service and receive benefits from the surrounding environment. The balance between service provisioning and receiving benefits make urban elements potential to take multifunctional roles. Fisher et al. (2009) distinguished between "service production areas" and "service benefit areas" in an urban context. For instance, while giving service to the neighborhood, public schools should take advantage of integrated street networks connecting them to other parts of the city. By assumption of urban elements' diversity, they classified urban elements into three categories:

- "In situ" is when services are provided in the exact location as the benefits received.
- "Omni-directional" when services benefit the surrounding landscape without a specific directional bias.
- "Directional" if services provided by one area benefit another location.

Likewise, Syrbe and Walz (2012) suggested distinguishing urban elements between "service providing areas," "service benefiting areas," and "service connecting areas." Such frameworks can explore the spatial interactions of urban elements and place the foundation for the multifunctional oriented development of urban context. The multifunctional-oriented development needs thorough investigations as each urban element are identified with specific urban agents, urban systems, and urban network characteristics. In addition, the non-integrated/non-interrelated urban elements are vulnerable to the decay of urban livability. Wilson (2008) debates this issue as "weak multifunctionality" that has been used to describe "negative" processes that are often increasing the vulnerability of urban regions. Additionally, multifunctionality is associated with spatio-temporal characteristics of land (see Fig. 1): (1) a spatial combination of separate land units with different functions; (2) different functions devoted to the same land unit but separated in time, and (3) the integration of functions on the same unit of land at the same time.

The "multifunctional spaces" concept is used besides the mixed-use, multi-use, and multipurpose definitions, often without any distinction. **Mono-functional space**: covers just one activity and one group of users. **Periodic-functional space**: the new activity begins immediately after the previous, and the area is empty for a short

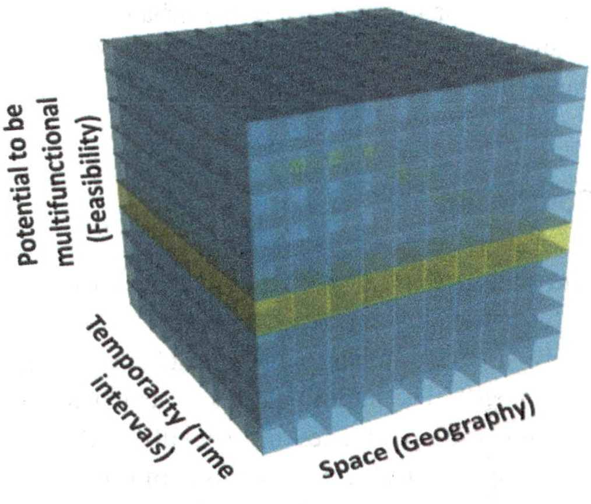

Urban Sustainability: Multifunctional and Multipurpose Planning of Urban Space, Fig. 1 The conceptual model of space-time multifunctionality

U

period such as art galleries. **Multi-purpose space**: The space can be adapted to the new function and circumstances with a few changes or modifications, for example, sports salons. **Mixed-use space**: there is always more than one activity going on in the area. **Complex space**: These spaces usually have different parts and a sequence of use which permits the new group of users to enter when the previous group has not entirely vacated the complex, such as Cinema Complex. **Shared space**: Space has the primary function, but in some parts of the activity cycle, it can be replaced with another activity or have the two activities simultaneously, for example, weekly markets at parks.

Multifunctional land-use planning is an integrated approach aimed at increasing how land is used in different areas and at different scales. This capacity helps planners control urban sprawl and growth by reviving the existing mono-functional spaces with a function other than their mains' and optimizing the operating rate of multi-use spaces toward a multi-purpose project. This approach can enhance the quality of life in the city and thus create valuable spaces. However, there is not a holistic schema for assessing the multifunctional capacity of urban elements. At the same time, academic studies often emphasized the multi-use of urban spaces without giving a comprehensive guideline of its implementation in the urban context. In the practical term, multifunctionality is used as a temporary approach to solve urban problems, for instance, using the urban sidewalk as an outdoor dining place during the COVID-19 Pandemic.

Conclusion

The multifunctionality term has been used in architecture and landscape architecture and is rarely used in urban planning debates (Lang et al. 2020). The multifunctional-oriented urban development coordinates the socially and economically sustainable multiple uses of urban elements. Here, the multifunctionality of existing urban spaces is proposed for service provision in densely built-up urban environments by considering the characteristics of the multi-scale networks of urban blocks, its mobility network, the composition of land use, and distribution of density support spontaneous alterations in land use over time.

The multifunctionality approach supports the bottom-up re-discovery of social places for collective appropriations, meeting places, and shared experiences by residents and local organizations. The bottom-up spontaneous spatial activities can be considered a conscious planning strategy to examine the multifunctional capacity of urban space. This approach should be integrated into the planning and design agenda of cities. The multifunctionality adds practical insights to support urban elements' flexibility to timely respond to changing circumstances. This study invites the practitioners and policymakers to design and plan different functional scenarios for individual urban blocks while training and preparing urban networks to accommodate new functions.

We suggest a phenomenology approach in testing the multifunctionality of urban spaces. Those participating in the research share their experiences, and thus they could reach some design recommendations. Also, in future studies, we recommend focusing on the regulative aspects of multifunctionality. Regular planning should include different stakeholders, including community groups, civic organizations, and residents, in decision making, policy design, and planning process to ensure the efficient function of the place.

Cross-References

- ▶ Adapting Cities to Climate Change
- ▶ City Visions: Toward Smart and Sustainable Urban Futures
- ▶ Toward a Sustainable City

References

Ahern, J. (2011). From fail-safe to safe-to-fail. Sustainability and resilience in the new urban world. *Landscape and Urban Planning, 100*, 341–343.

Alves, A., Sanchez, A., Gersonius, B., & Vojinovic, Z. (2021). Selecting multi-functional green infrastructure to enhance resilience against urban floods. In M. Babel, A. Haarstrick, L. Ribbe, V. R. Shinde, & N. Dichtl (Eds.), *Water security in Asia. Springer water*. Cham: Springer.

Belmeziti, A., Cherqui, F., & Kaufmann, B. (2018). Improving the multi-functionality of urban green spaces: Relations between components of green spaces and urban services. *Sustainable Cities and Society, 43*, 1–10.

Breuste, J. H., & Qureshi, S. (2011). Urban sustainability, urban ecology, and Society for Urban Ecology (SURE). *Urban Ecosyst (Springer, USA), 14*(3), 313–317.

Chang, Q., Li, X., Huang, X., & Wu, J. (2012). A GIS-based green infrastructure planning for sustainable urban land use and spatial development. *Procedia Environmental Sciences, 12*, 491–498.

Chiesura, A. (2004). The role of urban parks for the sustainable city. *Landscape and Urban Planning, 68*, 129–138.

Fisher, B., Turner, R. K., & Morling, P. (2009). Defining and classifying ecosystem services for decision making. *Ecological Economics, 68*, 643–653.

Foster, G. (2021). Circular economy strategies for adaptive reuse of cultural heritage buildings to reduce environmental impacts. *Resources, Conservation and Recycling, 152*, 104507. ISSN 0921-3449.

Ghafouri, A., & Weber, C. (2020). Multifunctional urban spaces a solution to increase the quality of urban life in dense cities. *Manzar, The Iranian Academic Open Access Journal of Landscape, Nazar Research Center for Art, Architecture and Urbanism, 12*(51), 34–45. https://doi.org/10.22034/manzar.2020.214183.2023. hal-02914038.

Haines-Young, R., & Potschin, M. (2010). The links between biodiversity, ecosystem services and human well-being. In D. Raffaelli & C. Frid (Eds.), *Ecosystem ecology: A new synthesis* (pp. 110–139). Cambridge University Press, BES.

Hansen, R., Olafsson, A. S., van der Jagt, A. P. N., Rall, E., & Pauleit, S. (2019). Planning multifunctional green infrastructure for compact cities: What is the state of practice? *Ecological Indicators, 96*, 99–110.

Jampani, M., Amerasinghe, P., Liedl, R., Locher-Krause, K., & Hülsmann, S. (2020). Multi-functionality and land use dynamics in a peri-urban environment influenced by wastewater irrigation. *Sustainable Cities and Society, 62*, 102305. https://doi.org/10.1016/j.scs.2020.102305

Lang, W., Hui, E. C. M., Tingting, C., & Xun, L. (2020). Understanding livable dense urban form for social activities in transit-oriented development through human-scale measurements. *Habitat International, 104*, 102238. ISSN 0197-3975.

Meerow, S. (2020). The politics of multifunctional green infrastructure planning in New York City. *Cities, 100*, 102621. ISSN 0264-2751.

Parris, S., & Kiroff, L. (2017). *Adaptive reuse and repurposing of industrial buildings to residential dwellings in Auckland City*. University of Adelaide (Ed.), State of Australian Cities (SOAC 2017) conference, p. 113.

Peschardt, K. K., Schipperijn, J., & Stigsdotter, U. K. (2012). Use of small public urban green spaces (SPUGS). *Urban Forestry & Urban Greening, 11*(3), 235–244.

Pickett, S. T. A., Cadenasso, M. L., Grove, J. M., Boone, C. G., Groffman, P. M., Irwin, E., Kaushal, S. S., Marshall, V., McGrath, B. P., Nilon, C. H., Pouyat, R. V., Szlavecz, K., Troy, A., & Warren, P. (2011). Urban ecological systems: Scientific foundations and a decade of progress. *Journal of Environmental Management, 92*, 331–362.

Roe, M., & Mell, I. (2013). Negotiating value and priorities: Evaluating the demands of green infrastructure development. *Journal of Environmental Planning and Management, 56*, 650–673.

Rolf, W., Pauleit, S., & Wiggering, H. (2019). A stakeholder approach, door opener for farmland and multifunctionality in urban green infrastructure. *Urban Forestry & Urban Greening, 40*, 73–83. ISSN 1618-8667.

Sharifi, H. (2019). Resilient urban forms: A macro-scale analysis. *Cities, 85*, 1–14.

Syrbe, R.-U., & Walz, U. (2012). Spatial indicators for the assessment of ecosystem services: Providing, benefiting and connecting areas and landscape metrics. *Ecological Indicators, 21*, 80–88.

Tartaglia, A., Terenzi, B., & Castaldo, G. (2021). Landscape as strategy for environmental multi-functionality. In F. Bianconi & M. Filippucci (Eds.), *Digital draw connections* (Lecture notes in civil engineering) (Vol. 107). Cham: Springer.

Tobey, M. B., Binder, R. B., Chang, S., Yoshida, T., Yamagata, Y., & Yang, P. P. J., (2019). Urban systems design: A conceptual framework for planning smart communities. *Smart Cities, 2*(4), 522–537. https://doi.org/10.3390/smartcities2040032

Van Broekhoven, S., & Vernay, A. L. (2018). Integrating functions for a sustainable urban system: A review of multifunctional land use and circular urban metabolism. *Sustainability, 10*, 1875.

Vreeker, R., Groot, H. L., & Verhoef, E. T. (2004). Urban multifunctional land use: Theoretical and empirical insights on economies of scale, scope and diversity. *Built Environment, 30*, 289–307.

Wilson, G. A. (2008). From weak to strong multifunctionality: Conceptualizing farm-level multifunctional pathways. *Journal of Rural Studies, 24*, 367–383.

Yang, P. P., & Yamagata, Y. (2019). Urban Systems design: From "science for design" to "design in science". *Environment and Planning B: Urban Analytics and City Science, 46*, 1381–1386.

Yang, P. P.-J., Wiedenback, A., Tobey, M., Wu, Y., Quan, S. J., & Chauhan, Y. (2017). Material based urban modeling: An approach to integrate smart materials in a near-zero community design. *Energy Procedia, 105*, 3765–3771.

Urban Sustainable Development

► Sustainable Development Goals from an Urban Perspective
► The Sustainable and the Smart City: Distinguishing Two Contemporary Urban Visions

Urban Systems

► Resource Effectiveness in and Across Urban Systems

Urban Transformation

► Understanding Urban Engineering

Urban Upgrading

► Social Urbanism: Transforming the Built and Social Environment

Urban Visions

► The Sustainable and the Smart City: Distinguishing Two Contemporary Urban Visions

Urban Vulnerability – Poverty

► Children, Urban Vulnerability, and Resilience

Urban Well-Being

Rachel Cooper and Christopher T. Boyko
Lancaster University, Lancaster, UK

Synonyms

Healthy cities; Urban health

Definition

People's capacity to balance healthy, fulfilling lives within and beyond the context of a dense, geographical location that recognizes city-based morphologies and stressors alongside processes of urbanization.

Introduction: Defining Urban Well-Being

"Urban" and "well-being" are both very broad terms that encompass many issues and topics and relate to both human and planetary prosperity.

Urban. Urban may be described as a physically bounded unit or geographical location, an "agglomeration." This description is rather static and discrete and does not consider the historical and geographic reality associated with urbanization, that is, the global processes of networks and flows that bring about circuits of capital, whether production and trade, or the built environment (Pavoni and Tulumello 2020). Thus, "urban" is more than a place; in Lefebvre's (1989/2014) words, it is "conceived and lived as a social practice" (p. 204), and "the sum of productive practices and historical experiences" (p. 203).

City also is an important synonym for "urban." Friedman (1986) defined the concept as a spatially integrated, social, and economic system at a particular location. In effect, cities are a product of the socio-organizational process of urbanization (Harvey 1996), of which capital, labor (Holston and Appadurai 1996), and sharing (Boyko et al. 2017) are a vital part. Thus, there is a moral and a physical configuration to cities – though not always agreed upon or claimed by its citizens and decision-makers (Sassen 1996) – that mutually interact and shape one another, and which may be expressed territorially, ecologically, economically, politically, and socially. Crucially, cities are more than their artifacts (e.g., buildings, roads, and policies) (Park 1915); they are dense agglomerations of heterogeneous people in lived space that continuously form and reform in the development of citizenship (Holston and Appadurai 1996).

Well-being. Finding an all-encompassing definition is difficult, mainly because there are many ways of achieving well-being (see Fuller et al. 2012, for a strong critique of existing definitions). How the concept is defined is contingent on what is valued (Dodge et al. 2012; Kahneman 2011), and who has the power to create and perpetuate those values. The concept of well-being also has a long history, involving new insights and nuances and often is synonymous with other terms, such as "health" – "a state of complete physical, mental and social well-being and not merely the absence of disease or infirmity" (World Health Organization 2021) – and "happiness" – a combination of feeling good (a person's psychological state) and being good (living a good life) (Phillips et al. 2014).

Beginning in Ancient Greece, well-being was considered as either the experience of pleasure or enjoyment (*hedonia*), or the pursuit of a virtuous, "good life" through worthwhile endeavors (*eudaimonia*) (Henderson and Knight 2012). In more recent years, scholars have developed approaches that recognize the multidimensional nature of well-being (e.g., Diener 2009; Michaelson et al. 2009; Stiglitz et al. 2009), including the idea that well-being can be explored from subjective and/or objective perspectives. One of the most prominent of these approaches is "flourishing," which acknowledges the experience of life going well, of feeling good, and of functioning effectively (Huppert and So 2013). Flourishing also may involve the upkeep of strong, mental well-being, bringing together both *hedonia* and *eudaimonia* (Huppert 2009a, b; Huppert and So 2013; Keyes 2002; Ryff and Singer 1998). Furthermore, Seligman's (2011) PERMA model adopts the metaphor of building blocks for a flourishing life. Through an exploration of positive emotion, engagement, relationships, meaning, and accomplishment, Seligman suggests that well-being is multidimensional, and that it embodies both subjectivity and objectivity.

An additional focus of some additional well-being approaches has centered around reaching balance, equilibrium, or homeostasis. For example, Sen's (1985) capability approach suggests that well-being should be concerned with what individuals are actually able to achieve, and seen as a kind of balance between people's capabilities (what people feel they are able to do and be if they so choose) and functionings (the various things that people succeed in doing or being). Dodge et al. (2012) also believe that individuals are driven to achieve a balance between the psychological, social, or physical resources they may have or not have, and the psychological, social, or physical challenges they may face. That is, when individuals have the resources they need to meet a specific challenge, they attain stable well-being.

Taking inspiration from the notion of flourishing as a dynamic state, the multidimensional nature of well-being, the inclusion of subjective and objective dimensions, and the need for balance, White's (2010) description of well-being feels appropriate: doing well, feeling good, doing good, and feeling well (see also Biloria et al. 2019). "Doing well" refers to a certain, material standard of living and/or prosperity. "Feeling good" involves personal perceptions of levels of satisfaction. "Doing good" involves a shared understanding of how the world is and should be. "Feeling well" emphasizes the significance of health to well-being.

Putting *urban* and *well-being* together, *urban well-being* may be considered as people's capacity to balance healthy, fulfilling lives within and beyond the context of a dense, geographical location that recognizes city-based morphologies and stressors alongside processes of urbanization.

U

Contributions to the Field

The field of urban well-being is truly interdisciplinary, spanning multiple disciplines with unique perspectives. These include, among others, economics, engineering, environmental sciences, geography, health, neuroscience, psychology, social sciences, sociology, and urban design. The following paragraphs contain examples of academic research that have been undertaken, and which demonstrate the strong relationship between urban and well-being (see also a more practice-based project about young people, food security, and well-being that directly involved communities and social enterprises in the research, Dombroski et al. 2019).

Geographers have discovered a negative relationship between well-being at the intraurban scale and urbanization levels, measured as a combination of population density and opportunities for participation in activities in different neighborhoods (Schwanen and Wang 2014; see also Brereton et al. 2008; Morrison 2011). In a different study, geographers explored how rural areas attract urbanites to visit for well-being purposes and how it is important from an urban well-being and nature conservation perspective to make these urban-rural ties more visible (Bijker et al. 2014). *Environmental engineering* scholars also have promoted the idea that nature conservation is key: Particularly in cities, it can contribute to biodiversity and well-being by converting urban pollution into ecosystem services (Mitsch and Jorgensen 2003). One example of nature conservation is the creation and maintenance of urban forests in megacities, which can provide multiple benefits to human well-being (Endreny et al. 2017).

Urban designers and environmental psychologists have studied the impact of urban density and deprivation on walkability and well-being, noting that residents living in low-density, high-deprivation urban neighborhoods had the poorest well-being, based on well-being questionnaires. In contrast, those living in low-density, low-deprivation areas had the best well-being (Boyko et al. 2020a). Furthermore, *environmental scientists* have examined the relationship between characteristics of urban parks, such as quantity, quality and accessibility, and well-being. They noted that park quantity was one of the strongest predictors of well-being and suggested that the creation and sustainability of park networks is key to facilitating social capital in cities (Larsen et al. 2016).

Finally, *multidisciplinary researchers,* involving *neuroscientists, psychologists,* and *geographers,* have posited that biometric data on urban well-being, such as electrodermal activity and heart rate, can be used to address some of the pressing, often negative issues associated with contemporary working life in cities (Pykett et al. 2020a). Such interdisciplinary research may be valuable for informing policies around urban well-being, including healthier commuting and reducing urban stress (Pykett et al. 2020b).

In most of these studies, urban well-being is found to be affected by multiple trade-offs that center around physical/environmental, social, economic, and governance issues. For example, Endreny et al. (2017) suggest that urban decision-makers (e.g., local authority planners, architects) and other stakeholders (e.g., residents, business owners) need to have frank discussions about the financial cost of creating urban forests. These conversations need to happen alongside a better understanding of the limits to tree growth in some contexts, the potential increase in air pollution if tree boundary layers become very low, and so forth, before committing to forestry programs in cities.

Principle Issues and Discourses

Traditionally, much of the research on urban well-being has been on the relationship between well-being and the environment, often considered to be the physical environment (Cooper et al. 2014). There are other factors that contribute to the well-being of people in urban environments, such as social factors and economic performance, as well as governance and decision-making. The following section will attempt to illustrate the interdependencies between these and the urban environment.

People and places are the central dimensions of well-being and the urban environment. To understand this relationship, it is important to uncover: (1) how humans engage at a multisensory level with their environment; (2) the type and quality of environment; and (3) its impact on people throughout their life-course. This is a very complex relationship that has developed over the evolution of humans and their inhabitation of the planet, and results in a very simple cause and effect relationship. That is, people are affected by their environment through their senses, which results in both psychological and physical impacts, and they are also the agents that create, modify, and maintain much of the material world in which citizens live and work (Cooper 2014).

In the first instance, human well-being is influenced by our sensory stimulation (i.e., what is seen, smelled, touched, tasted, and felt). Both the physical and the ambient (i.e., relating to immediate surroundings) environment have an effect on our senses. Furthermore, our well-being is influenced by the capacity of the environment to cause physical and/or mental danger. These can be categorized as factors in the environment that affect our health and well-being as physical, ambient, and psychological, and these have an effect on both our physical and mental health (see Table 1):

Physical Environment

The fabric of the physical environment includes the design and construction of buildings, the spaces between buildings (e.g., parks, roads, and paths), and associated infrastructure (e.g., bridges, canals). This, therefore, concerns not just the design and construction of the built form, but also the use, management, maintenance, and regeneration of these spaces and places (Cooper et al. 2009). Furthermore, the materials and impacts of those material can cause harm and danger, for example, asbestos in house walls, or inflammable materials.

Looking at the constructs involved in the quality of the physical environment, these can be viewed at different scales, from the entire urban scale to individual buildings via the neighborhood. Consequently, the influence of these constructs on human well-being is wide-ranging. If residential location is examined, for instance, findings demonstrate that poorer housing quality can lead to poorer mental health (Cooper et al. 2010). Hunt (1990) found that, among children and adults, psychological distress was positively correlated with a number of housing problems, as was feelings of isolation, depression, and excessive worrying (Evans et al. 2003; Payne 1997).

The type of home has an impact, too, for example, individuals living in high-rise buildings suffer

Urban Well-Being, Table 1 Factors affecting urban well-being. (Source: Cooper et al. 2010a)

Main factors	Constructs	Examples
Quality of the fabric of the physical environment	Design, construction and maintenance of buildings, the spaces between buildings and associated infrastructure	Houses requiring major repair, broken structural elements, graffiti, rubbish, vandalism, lack of recreation space, public drinking, public drug use, abandoned buildings
Quality of the ambient environment	Lighting Noise/acoustics Thermal quality Access to nature	Access to sunlight and windows Living near airports, neighbour noise Mould and damp Excessive built form with no green space or no views (i.e., concrete jungle)
Psychological impacts of the physical and ambient environment	Density Accessibility Safety and fear Way finding	Numbers of residents, crowded settings and lack of privacy or spaces for play inside dwellings High-rise and numbers of floors Crime and fear of crime as a result of urban form and lighting Poor layout, pavements, access

U

significantly higher levels of mental health problems compared with those in low-rise developments (Moore 1975); they also feel less sense of control and safety, less social support and social relations, and less attachment to the community (McCarthy and Saegert 1979). Furthermore, elderly residents living in high-rise dwellings were more depressed, had higher rates of psychiatric disorder, and were more socially isolated compared with those living in detached homes in the community (Husaini et al. 1991).

The other principal, built environment that has a significant impact on human well-being is the workplace. Over the course of the last century, the harm that work buildings do to people has moved significantly from the physical harm of the factories and toxic workplaces of the industrial revolution to the mental harm resulting from stress in the office (although this balance varies globally). Nevertheless, the built environment is a major culprit in making people sick at work (Myerson 2014). For example, sick building syndrome causes flu-like symptoms, lethargy, and headaches and has been recognized as a major issue in relation to office settings: the development of office blocks "was ruthlessly streamlined to meet business objectives: high-speed lifts, computer-controlled lighting, and centralized air conditioning removed environmental control from the individual" (Myerson 2014, p. 378), resulting in both physical and mental stress.

At the wider urban scale, it is recognized that the quality of the fabric of the neighborhood affects those living there. For instance, physical decay in US and UK neighborhoods was found by Maxfield (1987, as cited in Perkins et al. 1992), which was related to higher levels of fear among residents. Furthermore, the lack of sidewalks and the increase in traffic in some cities may contribute to low walkability and lack of exercise among residents. This, in concert with urban planners enabling fast food outlets to proliferate in areas, has been proven to increase obesity levels and noncommunicable diseases (Cooper et al. 2010). These are just an indication of the negative impact that a poorly designed and constructed physical urban environment can have on our well-being.

Ambient Environment

The quality of the ambient environment is related to the physical design of the environment and includes elements such as acoustics, lighting, and air quality, as well as temperature, color, ventilation, humidity, access to nature, having views of nature and natural sunlight, and having plants in offices and homes. All of these ambient qualities also have specific impacts on our health and well-being. For example, noise from neighboring apartments is seen as a major cause of annoyance and produces reductions in individuals' quality of life (Evans and Cohen 1987; Goldstein 1990). Also, residents living in damp dwellings have been known to experience more emotional distress (Martin et al. 1987), and poorer mental health in general (Hopton and Hunt 1996, as cited in Krieger and Higgins 2002). Indeed, dampness, mold, and cold indoor conditions are significantly associated with anxiety and depression (Hyndman 1990), especially among children. High levels of air pollution predict levels of psychological distress (Evans et al. 1987), but the physical harm is noted more specifically with increases in lung-related diseases and poor performance at school among children (Brunekreef et al. 1997).

While there are negative aspects of environmental feature, it is well established that there are positive impacts too; often, the sound of birds, running water, and certain types of music are used as soothing effects, to reduce stress (Ulrich 2000) and to buffer general noise (Winter et al. 1994). Also, what has been identified in terms of esthetics and beauty contributes much to restorative benefits and a higher sense of well-being. For instance, exposure and access to views of nature from a variety of physical environments can improve individuals' health and well-being by providing restoration from stress and mental fatigue (Kaplan 2001; van den Berg et al. 2003). Moreover, inside dwellings, direct contact with natural elements alongside views of nature have an enhanced, restorative effect on individuals (Evans and McCoy 1998; Hartig and Evans 1993; Kaplan and Kaplan 1989). Finally, residents living in greener settings demonstrate reliably better performance on measures of attention and effective management of major life issues compared with those who reside in less green environments (Kuo 2001).

Psychological Environment

As discussed above, both the physical and ambient environment have a significant psychological effect on people living in cities. These psychological impacts are especially important in terms of our perceptions of density and crowding, sense of safety and fear, and wayfinding. Crowding is believed to have substantial negative impacts on social relations (Baum and Paulus 1987), and psychological health (Edwards et al. 1996; Evans et al. 1989; Gove and Hughes 1983; Lepore et al. 1991). Density is a significant psychological issue, especially with people who are not part of the same family. Also, for residents living in high-density dwellings, they tend to develop higher levels of emotional illnesses, hostility (Mitchell 1971), and neuroticism (Bagley 1974), compared with individuals living in lower-density dwellings.

Psychological factors will affect us differently across our life-course (Cooper 2014). For example, in childhood, high noise levels have been associated with increased, psychological stress, poor reading skills, and general learning. With light that is too high or too low, this will have an impact on children's ability to learn and their cognitive performance. Similarly, in adulthood, high noise levels will negatively impact on blood pressure, long-term memory, stress, exhaustion, and anxiety. Additionally, not having any views, especially of nature, can result in job stress, depression, claustrophobia, and tension among adults. Later in life, noise levels, light levels, and having no views can affect sleeping patters and levels of general agitation. At all stages, unsafe and poor infrastructure has an effect of quality of life and safety.

Additional Factors: Social and Economic Factors

The design and construction of the urban environment has an influence beyond individuals' health and well-being; cities also provide the infrastructure to enable its citizens to access services, to deliver utilities, and to travel to employment, education, and health. In short, cities enable citizens to operate as a community and take pleasure in supporting each other (see Jacobs 1961).

There are aspects that make cities work well for different generations, genders, and ethnicities. Mobility supports social interaction, and it is important to provide safe streets, safe not only in terms of physical infrastructure, but also in terms crime and fear of crime. Martin and Wood (2014) recommend streets that are well connected, with traffic calming, and if streets are designed for children, they should work for everyone. Brown and Lombard (2014) suggest that mixed building uses, shorter blocks, and enhanced street connectivity offer higher levels of walking and greater social interaction, and there is evidence that children in walkable neighborhoods are most likely to walk to school and are less likely to be obese. Furthermore, for older adults with accessibility issues, public transport especially is important to enable independent lives and social connectivity.

Socialization often is increased by the provision of formal leisure and recreation facilities, such as parks and other green spaces. However, they are not always designed for everyone, and it is important to consider a variety of demographics, such as gender and ethnicity (Ellaway 2014). For example, Martin and Wood (2014) recommend that child-friendly neighborhoods have play areas with less landscaping, that are less manicured, and have natural bases on playgrounds with trees, rocks, and water. While playgrounds should foster a sense of risk and adventure, they also need balance, safety, and natural surveillance, with the ability for parental observation and general eyes on the street from the community (Jacobs 1961).

Finally, urban economies have a significant influence on the residents and their ability to thrive. Low-income communities often have less access to good-quality food, education, infrastructure, and health and social services. Without investment, the infrastructure, and the neighborhoods in which they sit, becomes less attractive to employers. This, indeed, is a huge problem: There is a global need to challenge the economic, employment, and social inequalities that exist in urban environments. Too often, the debate is about where to intervene in the cycle of renewal or decay: Invest either in people, via skills and education, or in infrastructure. It should not be a question of either/or, but of both.

Additional Factors: Climate Change and Net Zero

There are many other factors to consider that go beyond human well-being, including the well-being of land, water, atmosphere, and other organisms (i.e., a more-than-human approach to well-being). Climate change has illustrated that humans and the planet are symbiotic systems; for urban well-being to flourish, the underlying causes of climate change must be addressed. The ambition for net zero cities and urban environments means rethinking all aspects of the way the urban environment is designed, developed, constructed, managed, and maintained. This includes redesigning the use of roads and pavements, reducing patterns of movement, the use of cars, changing the materials and the construction of buildings, introducing more nature, and minimizing energy use through the entire lifecycle of the urban environment.

Conclusion: The Future of Urban Well-Being

"Urban well-being" is characterized by people's capacity to balance not only healthy, fulfilling lives within and beyond the context of a dense, geographical location, but also the nature of urbanization and the decisions made to ensure that the place enables both the planet and the person to flourish. Therefore, there is an urgency in ensuring that urban governance, urban design, and decision-making go beyond designing our cities to cope with political and economic drivers. Cities also need to consider carefully the evolution of their spaces and places, the stresses on the planet's resources and systems, and the pressures on individuals living and working in urban environments, not just now, but for the next century.

In their current form, government departments do not work together to make the above happen, as they are generally siloed. For example, Public Health England in the UK sits among departments of transport, housing, social services, and business and industry, each having their own agendas for the future. These departments and portfolios are complex and complicated structures, but in order to deliver urban well-being, they need to be porous, more transparent in their decision-making, and focused on the long term as well as the short term. One of the biggest issues is many people have difficulty imagining alternative futures and also the transition from the current situation to alternative futures.

New processes and structures are necessary: First, researchers, practitioners, and decision-makers must establish "what is," including understanding what is happening in an urban environment and the levels of well-being (Boyko et al. 2020b). To do this, digital systems and sensors are needed to provide significant amounts of baseline data alongside further behavioral and social, qualitive data. Assumptions of well-being must be challenged, for instance, examining links between inequality, deprivation, social support, built environment, and well-being, much of which has been illuminated during the 2020/2021 COVID-19 pandemic. Second, values and aspirations of the population need to be uncovered to understand "what could be." Third and most important, governance, leadership, and decision-making must develop a new vision of transitioning what exists now to achieve a long-term goal of urban well-being. Here, the full panoply of professions is needed to envision the end goal, and design places, products, services, and systems that transition over a specific timescale to create flourishing places where both people and planet coexist in harmony.

Cross-References

► Age-Friendly Future Cities
► Cities in Nature

- ▶ Getting Our Built Environments Ready for an Aging Population
- ▶ Green Cities
- ▶ Health and the City: How Cities Impact on Health, Happiness, and Well-Being
- ▶ Healthy Cities
- ▶ Planning Healthy and Livable Cities
- ▶ Role of Nature for Ageing Populations
- ▶ Urban Densification and Its Social Sustainability
- ▶ Urban Health Paradigms
- ▶ Urban Nature

References

Bagley, C. (1974). The built environment as an influence on personality and social behaviour: A spatial study. In D. Canter & T. Lee (Eds.), *Psychology and the built environment* (pp. 156–162). London: Wiley.

Baum, A., & Paulus, P. B. (1987). Crowding. In D. Stokols & I. Altman (Eds.), *Handbook of environmental psychology* (pp. 530–570). New York: Academic.

Bijker, R. A., Mehnen, N., Sijtsma, F. J., & Daams, M. N. (2014). Managing urban wellbeing in rural areas: The potential role of online communities to improve the financing and governance of highly valued nature areas. *Land, 3*, 437–459.

Biloria, N., Reddy, P., Fatimah, Y. A., & Mehta, D. (2019). Urban wellbeing in the contemporary city. In N. Biloria (Ed.), *Data-driven multivalence in the built environment*. New York: Springer.

Boyko, C. T., Clune, S. J., Cooper, R. F. D., Coulton, C. J., Dunn, N. S., Pollastri, S., Leach, J. M., Bouch, C. J., Cavada, M., de Laurentiis, V., Goodfellow-Smith, M., Hale, J. D., Hunt, D. K. G., Lee, S. E., Locret-Collet, M., Sadler, J. P., Ward, J., Rogers, C. D. F., Popan, C., Psarikidou, K., Urry, J., Blunden, L. S., Bourikas, L., Büchs, M., Falkingham, J., Harper, M., James, P. A. B., Kamanda, M., Sanches, T., Turner, P., Wu, P. Y., Bahaj, A. S., Ortegon, A., Barnes, K., Cosgrave, E., Honeybone, P., Joffe, H., Kwami, C., Zeeb, V., Collins, B., & Tyler, N. (2017). How sharing can contribute to more sustainable cities. *Sustainability, 9*(5), 701.

Boyko, C. T., Cooper, C., Coulton, C., & Hale, J. D. (2020a). Health, wellbeing and urban design. In C. T. Boyko, R. Cooper, & N. Dunn (Eds.), *Designing future cities for wellbeing* (pp. 158–170). New York: Routledge.

Boyko, C. T., Cooper, C., & Dunn, N. (Eds.). (2020b). *Designing future cities for wellbeing*. New York: Routledge.

Brereton, F., Clinch, P., & Ferreira, S. (2008). Happiness, geography and the environment. *Ecological Economics, 65*(2), 386–396.

Brown, S. C., & Lombard, J. (2014). Neighborhoods and social interaction. In R. Cooper, E. Burton, & C. L. Cooper (Eds.), *Wellbeing and the environment* (pp. 91–117). Chichester: Wiley Blackwell.

Brunekreef, B., Janssen, N. A. H., de Hartog, J., Harssema, H., Knape, M., & van Vliet, P. (1997). Air pollution from truck traffic and lung function in children living near motorways. *Epidemiology, 8*(3), 298–303.

Cooper, R. (2014). Wellbeing and the environment: An overview. In R. Cooper, E. Burton, & C. L. Cooper (Eds.), *Wellbeing and the environment* (pp. 1–19). Chichester: Wiley Blackwell.

Cooper, R., Boyko, C., Pemberton-Billing, N., & Cadman, D. (2009). The urban design decision-making process: Definitions and issues. In R. Cooper, G. Evans, & C. Boyko (Eds.), *Designing sustainable cities* (pp. 3–16). Chichester: Wiley Blackwell.

Cooper, R., Boyko, C., & Codinhoto, R. (2010a). *The effect of the physical environment on mental wellbeing. State of science review: SR-DR2*. London: Government Office for Science.

Cooper, R., Boyko, C. T., & Cooper, C. L. (2010b). Design for health: The relationship between design and non-communicable diseases. *Journal of Health Communication, 16*(Suppl. 2), 134–157.

Cooper, R., Burton, E., & Cooper, C. L. (Eds.). (2014). *Wellbeing and the environment*. Chichester: Wiley Blackwell.

Diener, E. (2009). Subjective well-being. In *The science of well-being* (pp. 11–58). New York: Springer.

Dodge, R., Daly, A. P., Huyton, J., & Sanders, L. D. (2012). The challenge of defining wellbeing. *International Journal of Wellbeing, 2*(3), 222–235.

Dombroski, K., Diprose, G., Conradson, D., Healy, S., & Watkins, A. (2019). *Delivering urban wellbeing through transformative community enterprise*. Building Better Homes Towns and Cities National Science Challenge.

Edwards, J. N., Fuller, T. D., Sermsri, S., & Vorakitphokatorn, S. (1996). Chronic stress and psychological well-being: Evidence from Thailand on household crowding. *Social Science & Medicine, 42*(2), 265–280.

Ellaway, A. (2014). The impact of local social and physical local environment on wellbeing. In R. Cooper, E. Burton, & C. L. Cooper (Eds.), *Wellbeing and the environment* (pp. 51–68). Chichester: Wiley Blackwell.

Endreny, T., Santagata, R., Perna, A., de Stefano, C., Rallo, R. F., & Ulgiati, S. (2017). Implementing and managing urban forests: A much needed conservation strategy to increase ecosystem services and urban wellbeing. *Ecological Modelling, 360*, 328–335.

Evans, G. W., & Cohen, S. (1987). Environmental stress. In D. Stokols & I. Altman (Eds.), *Handbook of environmental psychology* (pp. 571–610). New York: Academic.

Evans, G. W., & McCoy, J. M. (1998). When buildings don't work: The role of architecture in human

health. *Journal of Environmental Psychology, 18*(1), 85–94.
Evans, G. W., Jacobs, S. V., Dooley, D., & Catalano, R. (1987). The interaction of stressful life events and chronic strains on community mental health. *American Journal of Community Psychology, 15*(1), 23–33.
Evans, G. W., Palsane, M. N., Lepore, S. J., & Martin, J. (1989). Residential density and psychological health: The mediating effects of social support. *Journal of Personality and Social Psychology, 57*(6), 994–999.
Evans, G. W., Wells, N. M., & Moch, A. (2003). Housing and mental health: A review of the evidence and a methodological and conceptual critique. *Journal of Social Issues, 59*(3), 475–500.
Friedman, J. (1986). The world city hypothesis. *Development and Change, 17*(1), 69–83.
Fuller, S., Atkinson, S., & Painter, J. (2012). *Wellbeing and place*. Farnham: Ashgate.
Goldstein, G. (1990). Urbanization, health and mental wellbeing: A global perspective. *The Statistician, 39*(2), 121–133.
Gove, W. R., & Hughes, M. (1983). *Overcrowding in the household*. New York: Academic.
Hartig, T., & Evans, G. W. (1993). Psychological foundations of nature experience. In T. Gärling & R. G. Golledge (Eds.), *Behavior and environment: Geographic and psychological approaches* (pp. 427–457). North-Holland: Elsevier.
Harvey, D. (1996). Cities or urbanization? *City, 1*(1–2), 38–61.
Henderson, L. W., & Knight, T. (2012). Integrating the hedonic and eudaimonic perspectives to more comprehensively understand wellbeing and pathways to wellbeing. *International Journal of Wellbeing, 2*(3), 196–221.
Holston, J., & Appadurai, A. (1996). Cities and citizenship. *Public Culture, 8*(2), 187–204.
Hunt, S. (1990). Emotional distress and bad housing. *Health and Hygiene, 11*, 72–79.
Huppert, F. A. (2009a). A new approach to reducing disorder and improving well-being. *Perspectives on Psychological Science, 4*(1), 108–111.
Huppert, F. A. (2009b). Psychological well-being: Evidence regarding its causes and consequences. *Applied Psychology: Health and Well-Being, 1*(2), 137–164.
Huppert & So. (2013). Flourishing across Europe: Application of a new conceptual framework for defining well-being. *Social Indicators Research, 110*(3), 837–861.
Husaini, B., Moore, S., & Castor, R. (1991). Social and psychological well-being of black elderly living in high- rises for the elderly. *Journal of Gerontological Social Work, 16*(3-4), 57–78.
Hyndman, S. J. (1990). Housing dampness and health amongst British Bengalis in East London. *Social Science and Medicine, 30*(1), 131–141.
Jacobs, J. (1961). *The death and life of great American cities*. New York: Vintage Books.
Kahneman, D. (2011). *Thinking, fast and slow.* Toronto: Doubleday Canada.
Kaplan, R. (2001). The nature of the view from home: Psychological benefits. *Environment and Behavior, 33*(4), 507–542.
Kaplan, R., & Kaplan, S. (1989). *The experience of nature: A psychological perspective*. Cambridge, UK: Cambridge University Press.
Keyes, C. L. M. (2002). The mental health continuum: From languishing to flourishing in life. *Journal of Health and Social Behavior, 43*(June), 207–222.
Krieger, J., & Higgins, D. L. (2002). Housing and health: Time again for public health action. *American Journal of Public Health, 92*(5), 758–768.
Kuo, F. E. (2001). Coping with poverty: Impacts of environment and attention in the inner city. *Environment and Behavior, 33*(1), 5–34.
Larsen, L. R., Jennings, V., & Cloutier, S. A. (2016). Public parks and wellbeing in urban areas of the United States. *PLoS One, 11*(4), e0153211.
Lefebvre, H. (1989/2014). Dissolving city, planetary metamorphosis. *Environment & Planning D, 32*, 203–205.
Lepore, S. J., Evans, G. W., & Palsane, M. N. (1991). Social hassles and psychological health in the context of chronic crowding. *Journal of Health and Social Behavior, 32*(4), 357–367.
Martin, K. E., & Wood, L. J. (2014). 'We live here too'... what makes a child friendly neighbourhood? In R. Cooper, E. Burton, & C. L. Cooper (Eds.), *Wellbeing and the environment* (pp. 147–184). Chichester: Wiley Blackwell.
Martin, C. J., Platt, S. D., & Hunt, S. M. (1987). Housing conditions and ill health. *British Medical Journal, 294*(6580), 1125–1127.
McCarthy, D. P., & Saegert, S. (1979). Residential density, social overload, and social withdrawal. In J. R. Aiello & A. Baum (Eds.), *Residential crowding and density* (pp. 55–75). New York: Plenum Press.
Michaelson, J., Abdallah, S., Steuer, N., Thompson, S., & Marks, N. (2009). *National accounts of wellbeing: Bringing real wealth onto the balance sheet*. London: New Economics Foundation.
Mitchell, R. E. (1971). Some social implications of high density housing. *British Journal of Psychiatry, 36*(1), 18–29.
Mitsch, W. J., & Jorgensen, S. E. (2003). Ecological engineering: A field whose time has come. *Ecological Engineering, 20*(5), 363–377.
Moore, N. C. (1975). Social aspects of flat dwellings. *Public Health, 89*(3), 109–115.
Morrison, P. S. (2011). Local expressions of subjective well-being: The New Zealand experience. *Regional Studies, 45*(8), 1039–1058.
Myerson, J. (2014). Workplace and wellbeing. In R. Cooper, E. Burton, & C. L. Cooper (Eds.), *Wellbeing and the environment* (pp. 373–389). Chichester: Wiley Blackwell.
Park, R. E. (1915). The city: Suggestions for the investigation of human behavior in the city environment. *American Journal of Sociology, 20*(5), 577–612.

Pavoni, A., & Tulumello, S. (2020). What is urban violence? *Progress in Human Geography, 44*(1), 49–76.
Payne, S. (1997). Poverty and mental health. In D. Gordon & C. Pantazis (Eds.), *Breadline Britain in the 1990s* (pp. 159–177). Bristol: Summersleaze House Books.
Perkins, D. D., Meeks, J. W., & Taylor, R. B. (1992). The physical environment of street blocks and resident perceptions of crime and disorder: Implications for theory and measurement. *Journal of Environmental Psychology, 12*(1), 21–34.
Phillips, J., Nyholm, S., & Liao, S. (2014). The good in happiness. In T. Lambrozo, J. Knobe, & S. Nichols (Eds.), *Oxford studies in experimental philosophy* (Vol. 1, pp. 252–293). Oxford, UK: Oxford University Press.
Pykett, J., Osborne, T., & Resch, B. (2020a). From urban stress to neurourbanism: How should we research city well-being? *Annals of the American Association of Geographers, 110*(6), 1936–1951.
Pykett, J., Chrisinger, B. W., Kyriakou, K., Osborne, T., Resch, B., Stathi, A., & Whittaker, A. C. (2020b). Urban emotion sensing beyond 'affective capture': Advancing critical interdisciplinary methods. *International Journal of Environmental Research and Public Health, 17*(23), 9003.
Ryff, C. D., & Singer, B. (1998). The contours of positive human health. *Psychological Inquiry, 9*(1), 1–28.
Sassen, S. (1996). Whose city is it? Globalization and the formation of new claims. *Public Culture, 8*(2), 205–223.
Schwanen, T., & Wang, D. (2014). Well-being, context, and everyday activities in space and time. *Annals of the Association of American Geographers, 104*(4), 833–851.
Seligman, M. E. P. (2011). *Flourish: A new understanding of happiness and well-being- and how to achieve them*. London: Nicholas Brealey Publishing.
Sen, A. (1985). *Commodities and capabilities*. Amsterdam: North Holland.
Stiglitz, J., Sen, A., & Fitoussi, J. P. (2009). *Report by the commission on the measurement of economic performance and social progress*. Paris: Gouvernement de la République francaise.
Ulrich, R. S. (2000). Evidence based environmental design for improving medical outcomes. In *Proceedings of the Healing by design: Building for health care in the 21st century conference*, Montreal.
van den Berg, A. E., Koole, S. L., & van der Wulp, N. Y. (2003). Environmental preference and restoration: (How) are they related? *Journal of Environmental Psychology, 23*(2), 135–146.
White, S. C. (2010). Analysing wellbeing: A framework for development practice. *Development in Practice, 20*(2), 158–172.
Winter, M. J., Paskin, S., & Baker, T. (1994). Music reduces stress and anxiety of patients in the surgical holding area. *Journal of Post Anesthesia Nursing, 9*(6), 340–343.
World Health Organization. (2021, April 30). Constitution. https://www.who.int/about/who-we-are/constitution

Urbanism

▶ Feminist Planning and Urbanism: Understanding the Past for an Inclusive Future
▶ Women in Urbanism, Perpetuating the Bias?

Urbanization

▶ Butterfly Gardening in Colombo, Sri Lanka: Approach to Biodiversity Conservation, Monitoring, Education, and Awareness in Urbanizing Habitats
▶ The Sustainable and the Smart City: Distinguishing Two Contemporary Urban Visions
▶ Urban Management in Bangladesh

Urbanization – Spatial Transformation

▶ Urbanization, Planning Law, and the Future of Developing World Cities

Urbanization, Planning Law, and the Future of Developing World Cities

Andrew Chigudu[1], Charles M. Chavunduka[1], Innocent Maja[2] and Innocent Chirisa[1]
[1]Department of Demography Settlement and Development, Social & Behavioural Sciences, University of Zimbabwe, Harare, Zimbabwe
[2]Faculty of Law, University of Zimbabwe, Harare, Zimbabwe

Synonyms

Developing world cities – cities of the global south; Future – prospects; Planning Law – legislative instruments guiding planning; Urbanization – spatial transformation

U

Definition

Planning Law – Statutory guidelines and the process of adhering to them in the making and management of places through land-use planning

Urbanization – Demographic and territorial process tranforms rural to urban areas, in terms of population and areal expansion, respectively

Introduction

Urban planning was introduced in European cities during the nineteenth century industrialization period (Scott 1971; Chigudu 2021). Prior to the industrialization, the development of cities used to be chaotic. In cities, houses were built near available and endowed resources such as water sources (Woltjer and Al 2007). The development and growth of cities took place with the evolution of new technology and materials. However, it was affected by the population increase in major cities. The change in modes of urban transport led to advanced growth of cities. Modes of urban transportation changed from horses and horse-drawn carriages to automobile (Mabogunje 2015; Chigudu 2021).

The industrialization era stirred the need to develop standard housing, industries, and road infrastructure in expectation of increase in population, speculation of profit, and growth of unrestricted business enterprises in cities (Kraftchik 1990; Chigudu and Chirisa 2020; Chigudu 2021). This resulted in the swift population growth in major cities, coupled with sprawling cities and several forms of physical damages to the urban environment. The need for planned cities was realized, due to undue growth of slums, high level of traffic congestion, disorder, unpleasant environment, and high threat of diseases in cities (Woltjer and Al 2007; Chigudu 2019). This resulted in the introduction of planning laws to regulate development and growth of cities.

This chapter focuses on the realities in growth and development of cities, drawing lessons from some African major cities. Most African nations adopted colonial spatial planning legislation that shaped growth and development of their major cities. The adopted planning legislative and institutional frameworks influenced city growth and continue to do so in most African cities. Differences in the growth and development of cities are notable in especially terms of the standard and state of urban infrastructure. The chapter provides a reflection and experiences in the growth and development of cities and urban planning law. The emerging issues were discussed and synthesized. Conclusion and future directions were drawn from the study.

Defining Cities and Urban Planning

The visual image of cities gives special pleasures. The city is a construction in space of huge scale and architectural impression (Lynch 1960). Cities can be recognized as large inhabitable settlements which serve as foci of population, economic, and culture diversity. A city can be defined as a highly organized community, a theatre for social drama, prominent center with huge population, and a place larger than a village or town (Mumford 1973; UN 2014). However, maintaining the clear lines or boundaries between the city and rural areas can understate the continued expansion of areas with urban characteristics (UN 2014). The UN-HABITAT (2018) affirms that in cities is where the issue of sustainable development is won or lost. Urban planning refers to the creation of maps that show which activity is undertaken in the quest to control the city's development (Dear and Scott 2018). Planning came into use as a way to reduce health risks in cities after the Industrial Revolution in Britain. Cities were overcrowded, with poor sanitation and sources of drinking water, which led to the spread of diseases such as cholera, typhoid, and tuberculosis. Urban planning provided knowledge about where human waste goes and the sources from which to get potable drinking water. When urban planning was adopted in African cities, it was used to demarcate the residential areas from the commercial and industrial areas (Rathore et al. 2016).

City planning is also defined as a type of government intervention in developmental processes.

It can take the form of constructing roads or resurfacing them, subdividing, or consolidating urban land. Basically, planning is important as it provides solutions to problems of the city relating to housing, health and education facilities, and water and sewer systems (Cooke 2016; Kunzmann 2017; Chigudu and Chirisa 2020). Urban planning seeks to control use of land and other physical resources for the general good of the public and has potential to give significant quality of life and well-being for the people living in cities. Planning separates land uses which cannot be mixed, as in the case of housing developments within an industrial site (Rathore et al. 2016). Such a mix may lead to noise pollution and health hazards because of noxious industrial emissions. Planning, therefore, provides a way of making land uses compatible and complement each other for the best possible outcome.

In planning terms, development refers to the alteration of or improvement on or under a piece of land (Kunzmann 2017; Chigudu 2021). Development in cities involves the construction of new infrastructure and replacement of old infrastructure on a piece of urban land. Such infrastructure includes roads and highways, dams, bridges, water and sewer treatment plants, schools, hospitals, and buildings in general. Urban development refers to the improvement of cities or settlements to adopt more modern features (Hilson and Potter 2005; Chigudu and Chirisa 2020). Largely, the growth and development of cities involve the alteration of the physical attributes of urban land either on the surface or beneath it (Cooke 2016; Chigudu 2021). This can either be in the form of construction of a new building, a cluster of buildings, or the renovation of dilapidated buildings.

However, the growth and development of cities need to be controlled because if unchecked, other urban challenges can emerge. If growth of a city is left to run by the invisible hand, there is bound to be informal settlements and slums, which are difficult to manage and restore (Louw et al. 2003; Devine-Wright et al. 2015). For instance, this includes dilapidation of parts of the city center, developments on road reserves, construction on water sources or wetlands and encroachment onto farmlands (Chigudu 2021). This drives the need for urban planning laws that give effect to enforcement of standards and principles in growth and development of cities.

Planning law refers to the legal documents put in place by the government and used by local authorities for the control of urban development, planning processes, preparation, consent, and alteration of development plans (Waterhout et al. 2005; Chigudu 2019). It governs how spatial planning and urban development take place. Planning law creates the process and institutions that guide urban growth and ensures that development follows the set principles and procedures (Booth 2003; Chigudu 2021). It focuses on ensuring that growth of cities occurs in an orderly and approved way. There are several procedures to be observed to control and guide growth of cities. Planning law includes processes such as application for development permits as well as enforcement and prohibition orders (Alexander 2001; Chigudu and Chirisa 2020). The city council authorities are concerned with planning law since it ensures that development takes place on proper land and does not infringe upon the environment or adversely affect other uses.

Reflections and Experiences in Cities and Urban Planning Law

In 2018, an estimated 55, 3% of the world's population lived in urban settlements (UN-HABITAT 2018). It has been projected that 60% of the global population will reside in urban areas and one in every three people will live in cities with at least half a million of population by 2030 (ibid.). Cities contribute more than 90% of GDP in Malaysia and Thailand and close to 100% in Singapore, Hong Kong, and China, with their strong, highly competitive and resilient economies (World Bank 2015). Even in countries with low urbanization rates like Sri Lanka and Bangladesh, more than 65% of GDP is produced in the urban areas. By 2030, more than 55% of the population of Asia will be urban (World Bank 2015; UN-HABITAT 2018). Towns and cities are growing outward, engulfing semi-urban rural areas and forming new business and entrepreneurial agglomerations.

The rapid growth of most African cities has in turn led to boom in urban population, which local authorities are not prepared to handle (Chigudu 2019). A lot of people are migrating to major cities, leading to unemployment, since the labor available exceeds the workforce needed. This results in a reduction in the Gross Domestic Product (GDP) per capita, leaving most families without sufficient income to cater for their basic needs (Muzzini et al. 2016). There is pressure on resources, social amenities, and the environment in major cities in Africa, since local authorities failed to develop and maintain urban infrastructure to keep in pace with the rapid urbanization and peri-urban development (UN-HABITAT 2008). Inadequate employment and wages force the urban poor citizens to resort to desperate livelihood measures. Despite that African cities contributing about 55% to the continent's total GDP, 43% of the urban population live below the poverty line and urban slums population continues to grow (UN-HABITAT 2008, 2018).

Cities around the world are faced with urbanization challenges, owing largely to the rapid urban population growth caused by rural-urban migration. Within Africa, Sub-Saharan Africa has the highest annual population growth rate of 4% (Schram et al. 2013), which is higher than Northern Africa. Growth is expected to include an annual urban population growth rate of 2.5% (Chigudu 2021). Sub-Saharan Africa is therefore predicted to experience rapid urbanization in the next three decades. This, against a background in which 55% (±289 million) of the urban population in sub-Saharan Africa lives in slums, 70% (±367 million) has no access to safe water and sanitation (Schram et al. 2013; Chigudu 2019).

Most urban citizens in Harare, Johannesburg, Nairobi, Lusaka, Kampala, and Lagos cannot afford to buy houses or pay rent, and end up constructing shacks to live in (UN-HABITAT 2008). Due to the lack of potable water, people resort to just whatever water they can find. This creates favorable conditions for the outbreak of waterborne diseases such as cholera and typhoid, as occurred in Harare and Lusaka in 2008 and 2018, respectively (Wood et al. 2017; Chigudu 2019). Another consequence of the rapid urbanization is land degradation, since shacks and squatter settlements are not planned for and are constructed in a haphazard manner. They cause the ground to become bare, which reduces infiltration, and accelerates soil erosion. Since people also cut down trees for shelter construction, deforestation occurs resulting in the destruction of vegetation and increases the risk of land degradation and climate change (da Silva et al. 2017).

Within a city, planning is centered on making decisions about land allocation and the distribution of land-use activities (Nadin 2007; Aribigbola 2008). Therefore, it involves the demarcation of commercial, industrial, residential, and green spaces (Albrechts 2017). In Harare and Lusaka, urban planning is carried out using several plans such as master plans, local development plans, integrated plans, local area plans, schemes, and subject plans (Chigudu 2021). Statutes and bylaws exist to govern how urban planning is carried out in different parts within a city. Urban planning helps to create order by bringing together land uses that are compatible and separating those that are not.

The main purpose of urban planning is to ensure that the utilization of land resources is intentional and executed in an organized manner to meet the realities and needs of present and future generations (Chigudu and Chirisa 2020). Planning seeks to enhance links between housing, transport, energy, and industrial sectors to improve national and local development. This encompasses environmental considerations with a view to promoting sustainable urban development. By so doing, planning delivers economic, social, and environmental benefits, creating a more stable and predictable environment for investment and development and contributing to the achievement of national development goals (Carter 2007; Barton 2009; Chigudu and Chirisa 2020).

Most urbanization occurs as people search for favorable economic opportunities and a better life (Chigudu 2021). The growth of cities and urbanization are a defining phenomenon of the twenty-first century and is the locus of demographic and economic transformation (Cohen 2004). Global trends in human population growth and economic development point to the excessive growth of major cities (Chigudu 2021). This creates an even greater imperative for strategies that make cities competitive and relevant in the twenty-first century. Among the major strategies intended to unlock spatial economic growth in cities and spatial planning toward economic development nodes and corridors and techno-centers or techno districts (Chigudu 2019; Chigudu and Chirisa 2020). The rapid growth of cities induces a high demand for housing, urban transport, water, and sewerage (Wang 2014; Nhapi 2015; Siziba 2017; Chigudu and Chirisa 2020). On the other hand, the swift growth of cities has positive effects in the form of high productivity and cheap labor due to excess competition.

Excessive urbanization in cities in developing countries greatly surpasses their ability to satisfactorily deliver basic amenities to the urban populace (Cohen 2004). For decades, scholars have been worried about the pace at which cities are growing within Africa and the associated consequences for urban development and management (Rakodi 1990; Davis 2006; Silva 2012). It has become increasingly urgent to address the problem of how to plan for urban growth and regulate the use of scant, valuable, and environmentally overused urban land (Chigudu 2019). Without some level of state control through urban planning legislative framework, the prospects of addressing the problems faced by disadvantaged urban dwellers are slim (Davis 2006). The pressure to provide municipal services and responsive urban planning is particularly evident in major cities (Chigudu 2021).

The issue of urbanization is a type of response to a puzzling set of economic, social, demographic, cultural and environmental developments (Vernon 2005; UN-HABITAT 2008; Reinhard and Yasin 2011). Possible causes of worldwide urbanization include natural population increase, migration to urban areas and the extension of boundaries of urban centers (Knox 2009; Chigudu 2019). Individuals relocate to urban areas in search of social and financial opportunities in different cities. This tendency is more notable in African major cities such as Kigali, Nairobi, Addis Ababa, Maputo, Johannesburg, Harare, Lusaka, Kampala and Lagos (Cohen 2004; Reinhard and Yasin 2011).

The provision of amenities has not been keeping up with the level of growth of major cities in post-independent Zimbabwe and Zambia (Chigudu and Chirisa 2020). Furthermore, mismanagement, corruption, and conflicts in local authorities are such that cities fail to provide amenities such as water and sanitation, housing, public transport, and refuse collection (Muchadenyika and Williams 2016; Chigudu 2019). The wave of growth of cities like Harare, Lusaka, Lagos, and Nairobi is intensified by increased informality, high levels of joblessness, shortage of housing, scarcity of resources, industrial development, and environmental challenges (Agbola 2004; UN-HABITAT 2008; Chigudu 2019).

City authorities in sub-Saharan Africa are failing to provide adequate basic urban infrastructure to cope with the increasing rate of urban development and ensure sustainable urban growth (Chigudu 2019). This may be attributed to the failure by urban planning institutions, practitioners, and policy-makers to project possible growth trends of major cities. The obsolete and insufficient urban infrastructure has become a source of great concern to spatial planning practitioners, researchers, and government in major cities (UN-HABITAT 2009). The absence of requisite infrastructure in major cities result in problems such as urban sprawl, slum development, traffic congestion, outbreak of communicable diseases, and sprouting of informal activities (UN-HABITAT 2009; Nyambo 2010; Chigudu and Chirisa 2020). Such problems require strong institutions and sustainable statutory frameworks to guide urban planning and growth of cities.

Urban planning experts indicated that there are weekly pipe bursts in Misisi and Kanyama in Lusaka, Kibera in Nairobi and Mbare, and Epworth in Harare. The incapacity of cities to handle pressure has led to the excavation of pit latrines in Kanyama and Epworth as a measure of resilience (Chigudu 2021). These latrines give off a very unpleasant smell; are sources of diseases such as diarrhea, typhoid, and cholera since flies can access them easily; and also pollute the groundwater taken up by people through boreholes and wells. The sinking of boreholes and digging of wells were measures taken by citizens to provide water to their own families (Muchingami et al. 2019; Chigudu 2021). Several new residential areas in Harare and Lusaka lack running and potable water. Not everyone can afford to sink boreholes on their stand or plot, and this has led to the development of long water queues, which places physical and mental strain on people. Some wells have collapsed, and they have also been a source of disease like typhoid, diarrhea, and cholera since the groundwater is being contaminated and therefore unsafe to drink (Chigudu and Chirisa 2020).

In South Africa, the legislation on urban planning has been borrowed from British colonialists (Berrisford 2011). The 1995 Development Facilitation Act was enacted to make the spatial planning system unitary and ensure uniform development of cities. Before South Africa attained independence, urban planning was done to perpetuate the apartheid system (Landman 2004; Chigudu 2019). After independence, urban planning became a means of eliminating the development gap between areas inhabited by the blacks and those inhabited by the whites. The leading statutory framework was the constitution that gives urban planning power to local planning authorities. The Spatial Planning and Land Use Management Act (SPLUMA) of 2013 provides for the statutory framework that guides urban planning and growth of cities (Nel 2016; Chigudu and Chirisa 2020). It gives city, municipal, and town councils the power to plan for the improvement of their municipality or city region. The institutions that guide urban planning and development include the Ministry of Land Affairs and Housing and the Department of Rural Development and Land Reform (Joscelyne 2015; Chigudu 2021).

On the other hand, Zimbabwe adopted the British colonial the urban planning system. Infrastructure growth and development were mainly concentrated in major cities, while small urban centers and communal areas remained neglected. The urban planning institutions are guided by the national constitution of 2013 (Ncube 2013). The legislative framework consists of the Regional Town and Country Planning Act (RTCPA) (Chapter 29:12) of 1976, which is the pillar of urban planning in the country (Wekwete 1989; Chigudu and Chirisa 2020). Urban planning is also guided by the Urban Councils Act (Chapter 29:15) and the model building bylaw of 1977. Other statutes that guide urban planning in Zimbabwe include the Environmental Management Act (Chapter 20:27), Mines and Minerals Act, and the Forestry Act (Chigara et al. 2013; Chigudu 2021). They provide much-needed information that guides urban planning, growth, and development of cities and protection of the environment.

The Zimbabwe RTCPA maintained the British legacy in urban planning issues, which includes characteristics of rigid standards and approaches (Ncube 2013). This explains gaps between what is in the legislative frameworks and bylaws and the unfolding reality on growth and development of major cities. The issue of development control that is highlighted in part five of the RTCPA (Chapter 29:12) demands for permits to be secured for development before any improvements or alterations are made to the land (Chirisa 2010; Chigudu 2021). However, some developers ignore such planning laws and go ahead with developments without permit. The act needs to be amended to become more flexible and accommodate slum upgrading and regularization rather than resorting to demolishing ("use of the axe"). Flexible urban planning legislation that accommodates the urban poor and low-income groups is critical.

The urban planning practice in Zambia was borrowed from pre-independence system. The inflexible the urban planning system has created several challenges in Zambia. The successive post-independence Zambian governments struggled to find a permanent solution in the provision of decent housing in rapidly growing Lusaka city (Mulenga 2003; Chigudu and Chirisa 2020). The slums in Lusaka owe their origins to the lack of provision of low-cost housing schemes and a rather biased colonial planning statutory framework and housing policy. Such slums include Misisi, Kuku, Chibolya, Bauleni, and Kanyama were inhabitants still use pit latrines for sanitation and urban crime, for instance, prostitution and theft is rampant (Mutale 2017; Zulu and Oyama 2017; Chigudu and Chirisa 2020). Although the Housing (Statutory and Improvement Areas) Act of 1976 has shown that critical housing shortage can be resolved through participation of the slum dwellers (Mwimba 2006), the government seems not have understood the essential lessons that should have been learnt from the first slum upgrading projects.

The form of urban planning fostered segregation with areas inhabited by whites being developed, while those inhabited by blacks were neglected (Rakodi 1986; Chigudu 2019). Zambia therefore required to revise its planning statutory framework after independence. The Town and Country Planning Act (TCPA) of 1962 was reviewed in 2015 to give the Urban and Regional Planning Act (URPA) (CAP 283) (GoZ 2015; Taylor et al. 2015; Chigudu 2021). Although there have been minor adjustments, the planning legislation's fundamental principles are still the same. The URPA provides for the flexible integrated development and local area planning. In 2018, the Lusaka City Council began the implementation of the Lusaka Decongestion Integrated Project that consisted of upgrading major highways and construction of ring roads and over- and underpasses along major arterial roads (Mutale 2017; Chigudu and Chirisa 2020; Chigudu 2021).

Emerging Issues and Synthesis

If properly managed, the growth of cities has the potential to advance the standards of the world's population. The shift into an urbanized world has huge implications for the world economy, social, and physical conditions and the environmental state of the world cities. An increasing number of cities have assumed significant roles in the globalization of the economies, particularly in the area of financial services, commerce, information technology, transport, and telecommunication, among others (Cohen 2004). When competently regulated, the growth of cities can give rise to civilized societies as a result of economic development and other multiplier effects. Therefore, there is need for a viable growth and management of cities through sustainable urban planning legislative and institutional framework.

The rapid rate at which growth of cities is occurring makes it imperative for local authorities in sub-Saharan Africa to rethink the current urbanization paradigm (Schram et al. 2013; Chigudu 2021). Local authorities are increasingly called upon to mainstream urban management and development interventions that promote sustainable growth of cities and efficient service delivery. The implementation and attainment of the 2030 Agenda for Sustainable Development and Sustainable Development Goal number 11, which seeks to make cities and human settlements inclusive, safe, resilient and sustainable, have been topical in Africa (Chigudu and Chirisa 2020).

Urban planning in Africa has remained influenced by master planning and spatial development frameworks (Kadiri and Oyalowo 2011). The master planning approach is based on the assumption that future land uses can be precisely predicted and planned for. This thinking contradicts with the realities of slow or lack of growth of industries and formal commercial activities and the rapid expansion of informal residential settlements and slums in most African cities (Chakwizira and Mashiri 2008; Kadiri and Oyalowo 2011; APA 2012). The growth and development of slum and informal settlements have become a reality in major cities in sub-Saharan Africa. There has been growth of residential areas in capital cities without industrialization widening unemployment levels.

The available urban planning laws and ordinances have failed to effectively address the challenges of urban development and management in most African cities (Oyesiku 2004; Chigudu 2019). The administrative performance of planning authorities in developing nations has been inhibited by their structure, which has proved ineffective since most cities in developing countries have expanded without any comprehensive or organized plan over the years (Mabogunje 1990; Oyesiku 2004). This highlights the need for the development of an all-inclusive and collective organizational structure in which local authorities are related to each other and through which information can be shared in all directions (Chigudu and Chirisa 2020).

Issues relating to cities' infrastructure expansion, establishment, and funding have not been given sufficient attention in urban planning institutional and legislative frameworks (Jiriko 2008; Chigudu 2021). The adopted urban planning practices have not been able to forecast land uses in major cities that include recreational, residential, industrial, and commercial use, among others. Any improvements such as road construction, new neighborhoods, and schools and adjustment to change are a challenge for most local authorities using the master plan (Agbola 2004; Jiriko 2008; Chigudu 2021). The adopted spatial planning system is inflexible (Chigudu and Chirisa 2020), time-consuming, and restrictive for growth development of cities. There is need for the adoption by cities of sustainable spatial planning as a strategic tool for the future. This ensures that infrastructure investment and urban sprawl in major cities within developing nations are properly managed.

Conclusion and Future Direction

The major cities in sub-Saharan Africa have experienced rapid urban growth and excessive pressure on amenities. Cities face difficulties with adopted planning legislation especially in adjusting their spatial plans to meet new urban demands. The countries in developing world experience urban planning failure to deal with challenges of rapid expansion of major cities. Most African governments lack a deliberate urban policy to address urbanization challenges in the capital city by developing compact and sustainable satellite towns. There is need for a deliberate effort to attract investment and industrial growth in other small towns and municipalities through spatial planning and urban development projects. The African governments should realize that growth of cities must be accompanied with industrialization for sustainable urban development.

The adopted planning law tends to promote segregatory land use and use of automobile, stimulating environmental and social problems in cities. Sustainable and futuristic cities require conscious urban planning that integrates economic, social, and environmental factors in the preparation and execution of spatial plans and collaboration of all local stakeholders. An integrated approach to urban planning is substantial since it views all related aspects such as environment, social, economic, and transport issues holistically. The planning legislation in place and the governing authorities influence the growth and development of cities (Kunzmann 2017; Chigudu and Chirisa 2020).

The existing planning legislation, which regulates urban development, is falling short for the future of African cities. The experienced challenges of growth of cities require dynamic approaches to urban planning realities. There is need for flexible, legally enforceable planning legislative frameworks to guide growth and development of cities. African cities and local authorities need to come up with planning structures that keep up with changing environments, to achieve SDG 11. Evaluation and review of urban planning legislation are critical for planning frameworks to remain relevant and sustainable. There is need to come up with spatial planning approaches that embrace partnerships, infrastructure financing, and maintenance options.

Spatial planners, together with the government, need to acknowledge planning realities, focusing on the realities and challenges of growth and development of cities and their implications on economic development. The other reality that should be taken into consideration in transforming planning practice is the fact that households and firms, not governments, shape the growth and development of cities. Urban planning should be part of the solution to urban challenges.

Cross-References

- ▶ Spatial Demography as the Shaper of Urban and Regional Planning Under the Impact of Rapid Urbanization

References

Agbola, T. (2004). *Readings in urban and regional planning*. Ibadan: Foludex Press.

Albrechts, L. (2017). From traditional land use planning to strategic spatial planning: The case of Flanders. In *Revival: The changing institutional landscape of planning* (pp. 95–120). London: Routledge.

Alexander, E. R. (2001). Governance and transaction costs in planning systems: A conceptual framework for institutional analysis of land-use planning and development control – The case of Israel. *Environment and Planning B: Planning and Design, 28*(5), 755–776.

APA (African Planning Association). (2012). *The state of African planning*. Nairobi: Shutterstock.

Aribigbola, A. (2008). Improving urban land use planning and management in Nigeria: The case of Akure. *Cercetări practice și teoretice în managementul urban, 3*(9), 1–14.

Barton, H. (2009). Land use planning and health and well-being. *Land Use Policy, 26*, 115–123.

Berrisford, S. (2011). Revising spatial planning legislation in Zambia: A case study. *Urban Forum, 22*(3), 229–245.

Booth, P. (2003). *Planning by consent: The origins and nature of British development control*. London: Routledge.

Carter, J. G. (2007). Spatial planning, water and the water framework directive: Insights from theory and practice. *Geographical Journal, 173*(4), 330–342.

Chakwizira, J., & Mashiri, M. (2008). *Some Insights into the Intersection of Physical Planning and Governance in Zimbabwe*. Planning Africa 2008: Shaping the Future, ISBN 978-0-620-40287-3, pp. 178–201.

Chigara, B., Magwaro-Ndiweni, L., Mudzengerere, F. H., & Ncube, A. B. (2013). An analysis of the effects of piecemeal planning on development of small urban Centres in Zimbabwe: Case study of Plumtree. *International Journal of Management and Social Sciences Research, 2*(4), 2319–4421.

Chigudu, A. (2019). Institutional frameworks for urban development in Zimbabwe and Zambia. *Journal for Public Administration and Development Alternatives, 4*(2), 1–16.

Chigudu, A. (2021). The institutional and legislative planning framework of Zambia and Zimbabwe: Nuances of urban transformation. *Land Use Policy, 100*(1), 1–10.

Chigudu, A., & Chirisa, I. (2020). The quest for a sustainable spatial planning framework in Zimbabwe and Zambia. *Land Use Policy, 29*(1), 1–7.

Chirisa, I. (2010). Inner-City revitalization and the poor in Harare: Experiences, instruments and lessons from Mbare. Chapter 6. In *Urbanising Africa: The City Centre Revisited Experiences with Inner-City Revitalisation from Johannesburg (South Africa), Mbabane (Swaziland), Lusaka (Zambia), Harare and Bulawayo (Zimbabwe)* (pp. 3–11). Rotterdam: Institute for Housing and Urban Development Studies.

Cohen, B. (2004). Urban growth in developing countries: A review of current trends and a caution regarding existing forecasts. *World Development, 32*, 23–51.

Cooke, P. (2016). *Routledge revivals: Theories of planning and spatial development (1983)*. New York: Routledge.

da Silva, J. M. C., Prasad, S., & Diniz-Filho, J. A. F. (2017). The impact of deforestation, urbanisation, public investments, and agriculture on human welfare in the Brazilian Amazonia. *Land Use Policy, 65*, 135–142.

Davis, D. (2006). Continue education guideline implementation and the emerging transdisciplinary field of knowledge translation. *Journal of Continue Education Health Profession, 26*, 5–12.

Dear, M., & Scott, A. J. (2018). Urbanisation and urban planning. In *Capitalist society* (Vol. 7). London: Routledge.

Devine-Wright, P., Price, J., & Leviston, Z. (2015). My country or my planet? Exploring the influence of multiple place attachments and ideological beliefs upon climate change attitudes and opinions. *Global Environmental Change, 30*, 68–79.

GoZ (Government of Zambia). (2015). *The urban and regional planning act*. Lusaka: Central Statistics Office.

Hilson, G., & Potter, C. (2005). Structural adjustment and subsistence industry: Artisanal gold Mining in Ghana. *Development and Change, 36*(1), 103–131.

Jiriko, K. (2008). *Urban master planning paradigm in Nigeria: What future?* Kaduna: Mba.

Joscelyne, K., 2015. *The nature, scope and purpose of spatial planning in South Africa: Towards a more coherent legal framework under SPLUMA* (Doctoral dissertation, University of Cape Town).

Kadiri, W. A., & Oyalowo, B. (2011). Land alienation and sustainability issues in the Peri-urban Interface of south-West Nigeria. *Development, 54*(1), 4–20.

Knox, P. (2009). Urbanisation international. Chapter 22. In *Encyclopaedia of human geography*. New Jersey: Wiley-Blackwell.

Kraftchik, W. A., 1990. Small-scale enterprises, Inward Industrialisation and Housing. Unpublished Masters Dissertation, School of Economics, University of Cape Town.

Kunzmann, K. (2017). Spatial development and territorial cohesion in Europe. In *Spatial planning and urban development in the new EU member states* (pp. 33–44). London: Routledge.

Landman, K. (2004). Gated communities in South Africa: The challenge for spatial planning and land use management. *Town Planning Review, 75*(2), 151–172.

Louw, E., van der Krabben, E., & Priemus, H. (2003). Spatial development policy: Changing roles for local and regional authorities in the Netherlands. *Land Use Policy, 20*(4), 357–366.

Lynch, K. (1960). *The image of the City*. Massachusetts/London: MIT Press.

Mabogunje, A. L. (1990). Urban planning and the post-colonial state in Africa: A research overview. *African Studies Review, 33*(2), 121–203.

Mabogunje, A. (2015). *The development process: A spatial perspective*. London: Routledge.

Muchadenyika, D., & Williams, J. (2016). Social change: Urban governance and urbanization in Zimbabwe. *Urban Forum, 27*, 253–274.

Muchingami, I., Chuma, C., Gumbo, M., Hlatswayo, D., & Mashingaidze, R. (2019). Approaches to groundwater exploration and resource evaluation in the crystalline basement aquifers of Zimbabwe. *Hydrogeology Journal, 27*(3), 915–928.

Mulenga, C. L. (2003). *Understanding slums: Case studies for the global report on human settlements*. Lusaka: University of Zambia.

Mumford, L. (1973). What is a city. In T. LeGates & F. Stout (Eds.), *The City reader* (5th ed.). Abingdon: Routledge, 2011.

Mutale, E. (2017). *The management of urban development in Zambia*. London: Routledge.

Muzzini, E., Eraso Puig, B., Anapolsky, S., Lonnberg, T. and Mora, V., 2016. Urbanisation and growth.

Mwimba, C. (2006). We will develop without urban planning: How Zambia's urban planning practice is removing value from the development process. *Paper prepared for the SAPI conference*, 21–24 March.

Nadin, V. (2007). The emergence of the spatial planning approach in England. *Planning, Practice and Research, 22*(1), 43–62.

Ncube, C. (2013). The 2013 elections in Zimbabwe: End of an era for human rights discourse? *Africa Spectrum, 48* (3), 99–110.

Nel, V. (2016). Spluma, zoning and effective land use management in South Africa. *Urban Forum, 27*(1), 79–92.

Nhapi, I. (2015). Challenges for water supply and sanitation in developing countries: Case studies from Zimbabwe. In *Understanding and managing urban water in transition* (pp. 91–119). Dordrecht: Springer.

Nyambo, E. M. (2010). Environmental consequences of rapid urbanisation: Bamenda City, Cameroon. *Journal of Environmental Protection, 1*, 15–23.

Oyesiku, O. K. (2004). Town and country planning law and Administration in Nigeria. In Agbola (Ed.), *Readings in urban and regional planning* (pp. 257–269). Ibadan: Macmillan.

Rakodi, C. (1986). Colonial urban policy and planning in northern Rhodesia and its legacy. *Third World Planning Review, 8*(3), 193.

Rakodi, C. (1990). *Policies and preoccupations in rural and regional development planning in Tanzania, Zambia and Zimbabwe* (pp. 129–153). London: Paul Chapman Publishing.

Rathore, M. M., Ahmad, A., Paul, A., & Rho, S. (2016). Urban planning and building smart cities based on the internet of things using big data analytics. *Computer Networks, 101*, 63–80.

Reinhard, M., & Yasin, S. (2011). Impacts of urbanisation on urban structures and energy demand: What can we learn for urban energy planning and urbanisation management? *Urban & Regional Planning. Sustainable Cities and Society, 1*(1), 45–53.

Schram, S., Flyvbjerg, B., & Landman, T. (2013). Political science: A Phronetic approach. *New Political Science, 35*(3), 359–379.

Scott, M. (1971). *American city planning since 1890: A history commemorating the fiftieth anniversary of the American Institute of Planners* (Vol. 3). University of California Press.

Silva, C. N. (2012). Urban planning in sub-Saharan Africa: A new role in the urban transition. *Cities, 29*(3), 155–157.

Siziba, N. (2017). Effects of damming on the ecological condition of urban wastewater polluted Rivers. *Ecological Engineering, 102*, 234–239.

Taylor, T. K., Banda-Thole, C., & Mwanangombe, S. (2015). Characteristics of house ownership and tenancy status in informal settlements in the City of Kitwe in Zambia. *American Journal of Sociological Research, 5* (2), 30–44.

UN (United Nations). (2014). *Cities of today, cities of tomorrow! Unit 1: What is a City?* New York: Cyberschoolbus.

UN-HABITAT. (2008). *The state of African cities report: A framework for addressing urban challenges in Africa*. Nairobi: UN-HABITAT.

UN-HABITAT. (2009). *Planning sustainable cities: Global report on human settlements*. London: Earthscan.

UN-HABITAT. (2018). *Regional training workshop on human settlement indicators: Global City definition*. Bangkok: UN-HABITAT.

Vernon, H. J. (2005). Urbanisation and growth. *Handbook of Economic Growth, 1*(3), 1543–1591.

Wang, Q. (2014). Effects of urbanisation on energy consumption in China. *Energy Policy, 65*, 332–339.

Waterhout, B., Zonneveld, W., & Meijers, E. (2005). Polycentric development policies in Europe: Overview and debate. *Built Environment (1978–)*, 163–173.

Wekwete, K. (1989). *Planning Laws for urban and regional planning in Zimbabwe – A review*. Harare: UZ (DRUP).

Woltjer, J., & Al, N. (2007). Integrating water management and spatial planning: Strategies based on the Dutch experience. *Journal of the American Planning Association, 73*(2), 211–222.

Wood, C. L., Mcinturff, A., Young, H. S., Kim, D., & Lafferty, K. D. (2017). Human infectious disease burdens decrease with urbanisation but not with biodiversity. *Philosophical Transactions of the Royal Society B: Biological Sciences, 372*(1722), 20160122.

World Bank. (2015). *Zambia living conditions monitoring survey 2015*. Lusaka: Central Statistics Office.

Zulu, R., & Oyama, S. (2017). Urbanisation, housing problems and residential land conflicts in Zambia. *Japanese Journal of Human Geography, 69*(1), 73–86.

Urban–Rural Fringe

▶ Peri-urbanization

Urban–Rural Interface

▶ Peri-urbanization

Urban-Rural Interfaces

▶ The Challenges for Wildland-Urban Interfaces (WUI) in Metropolitan Areas: Reducing Fire Risk, Providing Employment Opportunities, and Preserving Natural Habitat

US Urban and Suburban Yardscaping

From Conventional to Regenerative

Zdravka Tzankova and Christopher Vanags
Vanderbilt University, Nashville, TN, USA

Introduction: The US Urban and Suburban Yardscape

Climatically, hydrologically, and ecologically diverse, the many urban areas of the USA are unified by a single and prominent landscape feature – turfgrass lawns and the heavily manicured, regionally homogenous yardscapes that surround them. From San Diego, Los Angeles and Phoenix to St. Louis, Nashville, Boston, Washington, DC, and Miami, turfgrass monoculture dominates a majority of private as well as public outdoor spaces – front and back yards, parks and open spaces, university and corporate campuses (Groffman et al. 2014, 2016; P. Robbins and Birkenholtz 2003; Wheeler et al. 2017).

Urban and suburban turfgrass takes up over 40 million acres of US land, or nearly 2% of all land in the United States (Milesi et al. 2005). Another way to put this towering figure in perspective: In a context where America is the world's largest corn producer, US turfgrass acreage is 3 times that of US corn (*FAOSTAT* n.d.; Milesi et al. 2005). Estimated to account for 35–58% of the US turfgrass total, urban and suburban lawns are expected to continue expanding with the projected expansion of urban and suburban land (Milesi et al. 2005; P. Robbins and Birkenholtz 2003).

The lawn-dominated American residential yardscape is energy, water, and chemical inputs intensive, raising significant ecological, human health, and climate concerns – from air pollution and climate forcing to water pollution and depletion of scarce water resources; from wildlife disruption and biodiversity loss to the loss and disruption of ecosystem functions and services.

Irrigation of the lawn-dominated US yardscapes soaks up 40% to 70% of residential water use, depending on the climate, hydrology, and dominant yard management practices of a city – an estimated total of 9 billion gallons per day (st. Hilaire et al. 2008; *US Outdoor Water Use | WaterSense | US EPA* n.d.). In comparison, farming in California – a large and fairly dry agricultural state – calls for an estimated 27 billion gallons of water per day (*Water Use in California Public Policy Institute of California* n.d.).

Americans spend over $ 35 billion per year on lawn and garden products, including garden chemicals (Mintel 2020), with 90 million pounds of fertilizer and 60 million pounds of pesticide active ingredients put to residential uses in the USA each year (EPA et al. 2017; The Week Staff 2015). Further, US yard care is a projected growth market for the pesticide industry and its products (Grand View Research 2019).

While a majority of pesticides used in the USA – nearly 90% of all pesticides by volume – still go to agricultural uses, home and garden applications account for 23% of all insecticide use by volume and 50% of insecticide use by value. Home and garden use also makes up 24% of all US pesticide use by value (EPA et al. 2017). If measured in volume of chemical inputs per hectare, fertilizer and pesticide use on lawns is several times more intensive than agricultural land applications (Carrico et al. 2012).

In a context where 2000 acres of US farmland is converted to urban and suburban use daily (Freedgood et al. 2020), where lawns, not corn, are America's largest irrigated crop (Milesi et al. 2005) and 50–70% of US [sub]urban households are managing their yards with pesticides and fertilizers (EPA et al. 2017; Fissore et al. 2011; Groffman et al. 2016; Lofquist et al. 2012), the management of US [sub]urban yards is in pressing need of a sustainability transition.

The rest of this chapter discusses the barriers and opportunities for a sustainability transition in the management of US urban and suburban residential yard spaces. It starts with an overview of key environmental, health, and climate costs of dominant, input-intensive yardscaping practices (Part 2). This is followed by a summary of key factors perpetuating ecologically problematic status quo practices in urban and suburban yard management (Part 3). The chapter then shifts to discussing environmentally preferable alternatives, from incremental improvements on the environmentally problematic status quo practices to more fundamental changes in managing residential yard spaces, as pioneered or practiced in different US cities (Part 4). The chapter briefly concludes with a summary of key mechanisms of environmentally positive change – and outline of key actors, dynamics, and forces with the potential to advance sustainability transitions in the management of US urban yardscapes (Part 5).

The Environmental Impacts of US [sub] urban Yardscaping

Energy, water, fertilizer, and pesticide-intensive, US urban and suburban yardscaping is implicated in

- Exacerbating water scarcity in water-stressed regions (Pride-Brown and Hess 2017; Hogue and Pincetl 2015)
- Contributing to climate forcing through N_2O emissions from [over]fertilized lawns, as well as GHG emissions from mowers and other lawn and yard management equipment (Carrico et al. 2012; Townsend-Small and Czimczik 2010)
- Eroding water quality through nutrient and pesticide runoff and pollution (Cole et al. 2011; Janasie 2015; Nowell et al. 2021; USGS 2006, 2015; Whitney 2010)
- Threatening both human health and biodiversity (especially insect and pollinator diversity) through fertilizer and pesticide use, runoff, and pollution (Hladik et al. 2018; Meftaul et al. 2020; Nathan et al. 2020).

While even the most intensively managed [sub]urban turfgrass monoculture may be ecologically preferable to the impervious surfaces (asphalt, concrete, tar, and tile) common across US urban and suburban landscapes, the recreational, runoff management, and heat-reducing benefits provided by conventionally managed urban and suburban lawns come with a heavy ecological, water, and carbon footprint of their own (Larson et al. 2020b).

On the one hand, urban and suburban lawns can help mitigate the urban heat island effect, capture, filter, and reduce stormwater runoff and even play a role in carbon sequestration. Maximizing a lawn's carbon sequestration potential is water and fertilizer-intensive, however (Carrico et al. 2012; Townsend-Small and Czimczi 2010), meaning that increases in carbon sequestration by urban lawns come with steep biogeochemical and ecological tradeoffs (Larson et al. 2020b): Fertilizer overuse, loss, and pollution is a leading cause of global environmental change and ecological disruption across scales and throughout the USA (Galloway et al. 2008), while a growing number of US cities and regions are increasingly water-stressed and looking for ways to increase water efficiency and reduce water use (Berkowitz 2019; Brown and Hess 2017). Further, the carbon sequestered in intensively managed lawns tends to be more than offset by increased GHG emissions from fossil-fueled lawn equipment and fertilizer nutrients off-gassed as N_2O (a potent GHG with nearly 300 times the global warming potential of CO_2) (Carrico et al. 2012; Lerman and Contosta 2019; Townsend-Small and Czimczik 2010).

Impacts Stemming from Nutrient Loss and Pollution

Residential fertilizer use is the second largest source of household nitrogen in the USA – following dietary sources, which account for 40% of household N (compared to yard fertilizer's 26%), but ahead of travel-related N emissions from internal combustion (which account for 25% of US household N) (Fissore et al. 2011; Souto et al. 2019). Lawn fertilizers are also heavily implicated in phosphorus (P) overloading and pollution of coastal and fresh water throughout the USA. N and P overloading from the residential use of fertilizers is a well-known contributor to the disruption and decline of entire aquatic ecosystems in the Great Lakes, Chesapeake Bay, Gulf of Mexico, and beyond.

Like its agricultural counterpart, then, the urban residential use, overuse, and mismanagement of fertilizer is a major cause of ecologically consequential nutrient loss and pollution (Carrico et al. 2012; Gu et al. 2015). This comes in several forms, driving a range of negative environmental impacts and feedbacks, as follows:

- Nitrate contamination of groundwater and drinking water, which poses a range of health hazards and affects many communities throughout the USA (Nolan et al. 1998; Pennino et al. 2017).
- Eutrophication of lakes, rivers, and coastal areas, with excess nutrients driving algal blooms and the onset of anoxic conditions that harm the health and functioning of aquatic ecosystems (Janasie 2015; Tzankova 2013)
- Climate forcing through N_2O emissions, resulting from excess application of lawn fertilizers (Gu et al. 2015).

Impacts Stemming from Pesticide Use and Pollution

A number of herbicides, insecticides, and other general use chemicals that work, but are not explicitly produced and marketed as pesticides, are commonly used in US residential yardscaping, both by households and professional applicators, for the purpose of maintaining manicured, weed-free lawns and gardens, and controlling insect and vertebrate species perceived and experienced as pests (EPA et al. 2017; P. Robbins and Sharp 2008).

Lists of the lawn and yard pesticides most commonly used in the USA, along with data on application amounts and common environmental fates and impacts, are compiled by the US Environmental Protection Agency (EPA et al. 2017), as well as pesticide reform organizations and coalitions, such as Beyond Pesticides (Beyond Pesticides 2015). Commonly used yardscaping pesticides are further chronicled through their aquatic residue signatures, as captured and measured by USGS water quality monitoring (Nowell et al. 2021).

Herbicides are commonly mixed into lawn fertilizers, and different pesticides have different environmental fates following their use on US yards and gardens. Some sorb more readily to soil or organic matter, while others are more prone to runoff and leaching into ground water. Some degrade rapidly in response to UV exposure, while others are more persistent (Meftaul et al. 2020; MU Extension n.d.; National Pesticide Information Center (NPIC) n.d.).

Assessments by the USGS National Water-Quality Assessment (NAWQA) Program show that pesticides are present throughout most of the year in most of the streams that drain urban areas. A stunning 97% of urban stream samples collected by USGS detected the presence of one or more pesticides, with 6.7% of sampled urban streams registering annual mean pesticide concentrations that exceed human health benchmarks, and an alarming 83% of urban streams exceeding aquatic-life benchmarks for water and streambed sediment. Five herbicides (simazine, prometon, tebuthiuron, 2,4-D, and diuron) and three insecticides (diazinon, chlorpyrifos, and carbaryl) that were commonly used in urban yardscaping at the time of the USGS assessment were frequently detected in urban streams throughout the USA (USGS 2006). More recent USGS research further indicates that urban streams across different regions of the USA share a common, 16-cpompound "urban pesticide signature," comprised of seven herbicides (prometon, 2,4-D, diuron, tebuthiuron, sulfometuron-methyl, bromacil, triclopyr, and hexazinone), three insecticides (fipronil, carbaryl, and the neonicotinoid imidacloprid), and five fungicides (carbendazim, azoxystrobin, propiconazole, 4-hydroxychlorothalonil, and tebuconazole) (Nowell et al. 2021).

Yardscaping use of pesticides leads to a range of direct and indirect human exposures – during application, through contact with treated plants and lawn-pesticide exposed pets, and also through pesticide migration into the home, where pesticide-contaminated household dust has been found to result in child exposure of still undetermined consequences (Morgan et al. 2008; P. Robbins and Sharp 2003; Solomon et al. 2005; Zartarian et al. 2000).

Neonicotinoids, a popular new class of systemic pesticides, have been widely used by US nursery growers of garden ornamentals, despite being suspected as a key driver of pollinator declines and a leading cause of the colony collapse disorder decimating bee populations worldwide (Friends of the Earth 2016; Hladik et al. 2018; Simon-Delso et al. 2014; Woodcock et al. 2017).

In sum, climate mitigation, biodiversity protection, and the over-all health and resilience of terrestrial and aquatic ecosystems across the urban-rural continuum is just as dependent on understanding and reforming ecologically problematic patterns of urban and suburban lawnscaping as it is on advancing biodiversity-friendly and climate-smart agriculture.

Key Forces Driving Conventional Yardscaping Preferences and Practice

Attitudinal and demographic factors are a focus for much of the research seeking to explain US yardscaping preferences and practices (Carrico et al. 2012; Larson et al. 2016; Martini et al. 2015; Wheeler et al. 2020). It is social norms and expectations, however – along with community-level processes of enforcing and re-enforcing these norms and expectations – that have proven pivotal in perpetuating the input-intensive, lawn-centric, and environmentally problematic American yard aesthetic (Carrico et al. 2018; Larson and Brumand 2014; Sisser et al. 2016).

Structural and institutional factors (regional planning and development patterns, the public and private governance of US urban spaces), along with the power and influence of agro-chemical, turf, and real estate industries, have, in turn, helped to create or re-enforce such social norms and expectations (Brown and Hess 2017; Larson et al. 2017; Larson and Brumand 2014; P. Robbins and Sharp 2003; Souto et al. 2019; Wheeler et al. 2020). Political ecologists, for example, have clearly documented how the American manicured lawn esthetic has been -produced through the discursive power of a chemical industry that has succeeded in equating input-intensive yard management with neighborliness, civic pride, and respect for the environment and family values (P. Robbins and Sharp 2003).

Regarding the attitudinal and demographic influences on yardscaping preferences and practice, many studies empirically grounded within specific metro areas have examined how esthetic and recreational preferences, environmental views and concerns, and demographic factors like age, income, and education affect yard management decisions and practices at the residential parcel scale (Carrico et al. 2012, 2018; Larson et al. 2010). A number of these studies have found positive correlations between higher levels of income and education and more input-intensive yard management (i.e., more intensive fertilizer and/or pesticide use). They have also identified positive correlations between water- and chemical-intensive management practices and structural factors, such as lot size and property value. These studies have found that homeowners with larger, higher-value homes are more likely to use input-intensive management practices, compared to smaller-parcel, lower-property value counterparts within the same study area (Feagan and Ripmeester 1999; Larson et al. 2017;

P. Robbins et al. 2001; Souto et al. 2019; Wheeler et al. 2020).

Notable and initially puzzling, research has found no logical correlation between individuals' levels of environmental awareness and concern and household use of yard chemicals. In fact, a number of studies have found high levels of yard chemical use among individuals expressing high levels of environmental awareness and concern (Robbins and Sharp 2003).

This initially puzzling dynamic brings into focus the significance of broader social norms and community-level expectations in shaping household-level decisions and management practices (Carrico et al. 2012; Larson and Brumand 2014; Sisser et al. 2016). It also centers the role of municipal planning, policies and ordinances, as well as their private regulatory counterparts – Homeowner Associations' covenants, conditions, and restrictions – in encoding and underwriting the environmentally problematic status quo of managing for impeccable yard spaces (Carrico et al. 2012; J. C. Fraser et al. 2013; Wheeler et al. 2020).

Two-thirds of the 156 broadly representative, geographically diverse US municipalities examined in a comprehensive recent study, for example, have municipal goals emphasizing the esthetic maintenance of "neat," "healthy," and "attractive" residential landscapes – goals and rules that effectively, if sometimes tacitly, drive and re-enforce the input-intensive management practices of the American yard aesthetic (Larson et al. 2020b). What municipal regulations omit can be just as important as what they stipulate when it comes to the regulatory perpetuation of an input-intensive, manicured yard aesthetic and practice. Over half of the diverse US municipalities studied by Larson et al. (2020a), for example, have some regulatory restrictions on landscape irrigation, yet hardly any of them are regulating the presence of water-thirsty lawns and grass within residential yardscapes. Further, a majority of municipal water conservation rules come in the form of time-of-day, days-of-week, and/or seasonal watering restrictions, rather than stricter volume-based restrictions backed by tiered and steeply ascending water pricing or by climate-smart irrigation technology requirements. The formerly lawn-heavy desert city of Phoenix has recently as a notable and highly promising exception to this pattern, however: Its increased summer rates for water are credited with reducing per capita water usage by 30% over the last 20 years and an associated reduction in residential lawn presence from 80% of households in 2000 to an estimated 14% of households today (J. Robbins 2019).

The private governance of lawns and yardscapes, as encoded in the covenants, conditions, and restrictions of Homeowners Associations (HOAs), can be even more detailed and prescriptive than their public, municipal counterparts, and even more likely to underwrite and re-enforce input-intensive management for a manicured yard aesthetic (Carrico et al. 2012, 2018). Created to safeguard the economic and other interests of homeowners in common interest developments (CIDs) such as planned and gated communities and subdivisions, HOAs have the power to set and enforce rules for almost anything that affects property values and tangible or intangible amenities in the residential spaces they govern (Clarke and Freedman 2019; J. Fraser et al. 2015). Indeed, research of city-level yardscaping practices finds the yardscaping practices of HOA residents to be more input-intensive than those of other, non-HOA neighbors, and points to a number of formal and informal mechanisms through which the rules and norms of a pristine yard esthetic are being enforced in America's growing number of privately governed [sub] urban communities (J. Fraser et al. 2015; J. C. Fraser et al. 2013).

With 60% of recently build single-family homes and 80% of houses in new subdivisions being part of Homeowners Associations, a fifth of American households reside in neighborhoods privately governed by HOA covenants, conditions, and restrictions. Private municipal governance is therefore emerging as a critical force shaping the US yardscape – and its growing environmental and carbon footprint (Clarke and Freedman 2019; J. Fraser et al. 2015; J. C. Fraser et al. 2013; Souto et al. 2019).

In sum, community-level processes are critical in perpetuating and [re-]enforcing ecologically problematic norms of neighborliness and community belonging through the water-, energy-, and chemical-intensive management of residential outdoor spaces. The instruments of public and private municipal governance are also important in tacitly or explicitly encoding these norms and expectations, and underwriting their enforcement.

Emerging Alternatives to Conventional US Yardscaping

If landscape sustainability is defined as "the capacity of a landscape to consistently provide long-term, landscape-specific ecosystem services essential for maintaining & improving human wellbeing," then the sustainability of the urban and suburban landscapes of which residential yards are an important part can be defined by their capacity to consistently provide ecosystem services such as wildlife habitat, water quality protection, flood control, and climate and heat mitigation (Larson et al. 2020a).

While status quo practices of US urban yardscaping have not been historically conducive to ecosystem health and service provision, local and state governments, civil society groups, and even businesses actors have undertaken a range of actions and initiatives that promise to advance sustainability transitions and ecosystem service provision across the US urban landscape.

Below, we review four types of departure from conventional, input-intensive, and multiply ecologically problematic practices that still characterize much of US urban yardscaping. We underscore key features and sustainability advantages of each alternative practice, while also identifying key societal drivers and current levels of uptake.

Increased Mowing Height and Decreased Mowing Frequency

Even in a context of conventional turfgrass lawns and yards, small, incremental changes in yard management, such as increases in mowing height, decreases in mowing frequency, and recycling of lawn clippings, can generate tangible environmental and biodiversity benefits.

Higher mowing height (e.g., mowing to a height greater than 4 or 5 inches) protects against weeds and pest insects, and increases root depth and biomass, as well as the rooting capacity of lawn grasses. These changes, in turn, bring further positive spillovers: The weed and insect pest protection conferred by lower intensity management translates to lower [perceived] needs for pesticide treatment, while larger root systems translate to increased carbon and nutrient storage, greater drought tolerance, and lower irrigation requirements (Smitley n.d.; Watson et al. 2020). Similarly, recycling of grass clippings by leaving them on the lawn is a natural source of nutrients, precluding the need for fertilizer additions and helping protect water quality and aquatic ecosystems (Kopp and Guillard 2004).

Decreases in mowing frequency have proven beneficial to the significant urban population of pollinators, by allowing wildflowers and other forage species to flourish, thus enhancing the quality of both habitat and food supply for bees and other pollinators in the urban lawnscape (del Toro and Ribbons 2020; Hall et al. 2017; Xerces Society 2017). Lower mowing frequencies have been further associated with a higher abundance and species richness of bees through increasing the floral resources available to them (Lerman et al. 2018).

In light of the benefits from lower intensity lawn management, a number of US municipalities are taking part in "No Mow May" initiatives and campaigns. These encourage or incentivize citizens to forego lawn mowing during the month of May, in order to leave flowering plants like clover and dandelion undisturbed and available for bee foraging, that is, in order to provide better habitat and a more secure food supply for bees and other pollinators. As part of such "No Mow May" initiatives and campaigns, US cities suspend enforcement of mowing ordinances and commit to no-mow practices in city-managed spaces, though commitments by some cities are limited to nonenforcement in back yards only, and municipal no-mow practices extend to a few selected municipally managed green spaces only (Burrows 2020; Fort Atkinson n.d.). Expansion of "No

Mow May" and other municipal incentives or mandates for transition to lower-intensity lawn management practices can have significant and scaleable ecological benefits. Residential yards following "No Mow May" practices, for example, have been found to harbor greater plant diversity and abundance and a higher pollinator diversity and species richness compared to conventionally managed counterparts (del Toro and Ribbons 2020).

Improvements to Pollinator Habitats

The creation and improvement of pollinator habitat is underway in many US cities, as part of city-level sustainability initiatives or as part of larger conservation practice networks and partnerships, like the National Pollinator Garden Network, and its Million Pollinator Garden Challenge (The National Pollinator Garden Network n.d.; Xerces Society 2021).

The Pollinator Vision Plan commissioned by the City of Portland, Oregon, illustrates the type of rigorous, comprehensive, and long-term approach to designing and managing pollinator-friendly urban-green spaces long recommended by ecologists, inter-governmental organizations, and bee conservation organizations such as the Xerces Society (Derby Lewis et al. 2019; Hall et al. 2017; Wilk et al. 2019).

The Pollinator Vision Plan of Portland, OR

Building on numerous and diverse, but largely fragmented previous efforts at creating and enhancing urban habitat for insect and bee pollinators, Portland's Pollinator Vision Plan proposes the development of an ecologically sound, conservation-science-guided network of patches and corridors of pollinator-friendly habitat. Starting with comprehensive mapping to determine the pollinator habitat potential of the city's residential yard spaces, as well as its parks, schools, community gardens, libraries, highways, power line corridors, hospitals, and correctional facilities, Portland's planning efforts were explicitly focused on understanding the size, connectivity, and other attributes of sites with pollinator habitat potential, then analyzing how these sites can be integrated into a larger habitat network. This approach represents an important and promising departure from past pollinator protection and habitat efforts in Portland and other cities, where numerous pollinator gardens planted by backyard and community gardeners, schools, and institutional landowners remain isolated and subjected or adjacent to pesticide use and spraying, while much of the urban space beyond them remains hostile to pollinators.

In addition to its overarching approach of advancing pollinator health and diversity through creating an interconnected habitat network, Portland's pollinator plan drills down to a number of specific recommendations for the successful implementation of this networked habitat approach. It considers, for example, the particulars of engaging different types of institutional actors and landowners and those of accommodating pollinator habitat needs during planned redevelopments of various urban neighborhoods. It also includes specific examples and recommendations for the types of management practices – and management practice changes – required for creating and/or maintaining the proposed pollinator habitat network; changes such as the introduction of no-mow grass alternatives in residential yard spaces and implementation of selective mowing regimes in public parks and spaces and the restoration of native meadow habitats on golf courses.

By constructing a continuous and integrated network of pollinator habitat throughout the city of Portland, and laying out a series of detailed recommendations for the long-term pollinator friendly management of this habitat network and its urban surroundings, the City of Portland has made a major step towards the regenerative management of urban land and yard space.

A number of other US cities are undertaking significant and comprehensive efforts to create and maintain pollinator friendly urban habitats. The efforts of the 144 current city affiliates of the program, created by the Xerces Society for Invertebrate Conservation, for example, have major potential to advance pollinator conservation by creating millions of acres of suitable pollinator habitat throughout the USA, thus helping offset some of the habitat loss and multiple environmental stressors threatening the health and viability of many pollinator species and communities in the USA and globally (Xerces Society 2021).

Pesticide Bans and Restrictions

Even the best pollinator habitats and habitat networks may not, by themselves, overcome the pollinator hazards presented by the widespread urban use and presence of yardscaping pesticides. Eliminating the cosmetic yard and landscaping use of pesticides is another critical step to advancing sustainability of urban land management in residential yard spaces and beyond.

The *Safe Grow Act of 2013*

Passed by the municipality of Tacoma Park, Maryland, this municipal law offers an instructive case study in the drivers, dynamics, and value of municipal regulations banning the cosmetic use of pesticides in private yards and public outdoor spaces (City of Takoma Park 2013; City of Takoma Park MD 2013a). Motivated by concern for public health, water quality, pollinator health, and the health of aquatic and other ecosystems, the City of Takoma Park prohibited the landscaping use of a long list of carcinogenic, endocrine disrupting, or otherwise human-health-hazardous pesticides – pesticides that are mostly restricted or banned for cosmetic use in Canada and flagged as highly hazardous by the US Environmental Protection Agency, the Canadian Ministry of the Environment, and the European Commission, yet commonly applied in US lawn and yard care (City of Takoma Park MD 2013b).

Takoma Park municipal regulations restrict the cosmetic yardscaping use of pesticides by commercial applicators, as well as property owners and tenants. They also encourage the use of cultural, physical, biological, and mechanical methods of pest control and prioritize education on hazardous pesticide alternatives (City of Takoma Park MD 2013a).

Takoma Park's regulatory leadership on pesticides has helped spur similar action by other Maryland municipalities (Safe Grow Montgomery MD 2016). Though its behavioral, environmental, and health effects have not been rigorously studied, research on similar municipal pesticide restrictions and bans in Canada, where they are more common, indicates that clearly disseminated and well-enforced pesticide bans can produce substantial decreases in the proportion of households applying pesticides or hiring lawn care providers to apply such pesticides, along with increases in the use of natural lawn care methods (Cole et al. 2011).

Municipal restrictions on the yardscaping use of pesticides have been pursued by municipalities across several different states. A majority of US municipal governments are severely restricted or altogether precluded from regulating the yardscaping use of pesticides, however. This is the result of state laws that vest all authority for any regulation of pesticide use with state governments, explicitly or implicitly banning municipal governments from regulating the advertising, marketing, distribution, sale, and use of pesticides (among other aspects of pesticide regulation) (Beyond Pesticides 2013; Centner and Heric 2019).

Present in 43 of the 50 US states, these state preemption laws were passed under pressure from the chemical and agricultural industries, with significant involvement and direct facilitation by the American Legislative Exchange Council – a corporate funded association of conservative state legislators who are working to erode the regulatory power and capacity of government through designing and promoting bills that undermine environmental regulation, facilitate school privatization, undercut health care reform, mandate strict and discriminatory election laws, and limit legislature's ability to raise revenues through taxes (Centner and Heric 2019).

Alaska, Hawaii, Maine, Maryland, Nevada, Utah, and Vermont are the seven states where municipalities do have the capacity to regulate yardscaping and other pesticide uses within their jurisdictions. And even though 5 of the 43 states with pesticide preemption laws make it possible for local governments to petition state pesticide boards or commissioners for permission to regulate pesticide use within their jurisdictions, and another seven states make it possible for citizens to petition the state for authorizing local pesticide regulations, the current crop of state pesticide preemption laws is a significant obstacle that needs to be overcome as part of advancing meaningful and comprehensive sustainability transitions in US urban yardscaping (Beyond Pesticides 2013; Centner and Heric 2019). Environmental NGOs and civil society groups are actively working to overturn these state preemption laws. They are openly and explicitly seeking to return regulatory jurisdiction to local governments, who are, by virtue of their multiplicity alone, much less susceptible to industry influence and capture and thus more likely to succeed in regulating pesticide use in the public environmental and health interest (Beyond Pesticides 2013; Diane Sofranec 2021).

Another tactical, if partial way that municipal governments have been able to circumvent state-level pesticide ban preemptions is though municipal restrictions on the use of synthetic lawn fertilizers. Dane County, Wisconsin, for example, passed a 2004 county ban on the use of phosphorus-containing synthetic lawn fertilizers (a ban passed to protect local lakes). Although Wisconsin is one of the 29 US states which explicitly ban municipal pesticide regulation, the Dane County fertilizer ban, which applied to 61 municipalities overseen by the county, had the effect of limiting the use of lawn herbicides, by virtue of restricting the sale and use of popular "weed and feed" products that combine synthetic fertilizers with herbicide (Beyond Pesticides 2013).

Bans and Restrictions on the Urban Use of Fertilizers

Municipal and state regulation on residential and other urban uses of lawn fertilizers is another key step on the path to sustainable yardscaping and urban land management transitions in the USA.

A number of US counties and municipalities, and at least 11 US states, have regulatory restrictions on the residential use of lawn fertilizers, along with controls on the formulation and sale of such fertilizers (Christopher D. Ryan et al. 2019; Janasie 2015; S Lee and McCann 2018; Souto et al. 2019).

Motivated by concerns over water quality and aquatic health, particularly concerns over the ecological and health effects of excessive phosphorous (P) and nitrogen (N) loading of surface and ground waters, states, and municipalities have imposed various limits on the lawn application of fertilizers. These regulatory controls differ in scope, stringency, and focus. Inter-jurisdictional differences in fertilizer use regulation occur for a combination of reasons. The locally specific nature of ecological concerns and ambient conditions as well as regional variation in political opportunities and constraints are key among these.

One US state that stands out with the scope and detail of local government efforts to control the use and impacts of lawn fertilizers is Florida. The Florida Department of Environmental Protection has developed a *Model Ordinance For Florida-Friendly Use Of Fertilizer On Urban Landscapes*, and a majority of Florida counties, as well as a number of Florida municipalities, have urban fertilizer use ordinances that reflect key elements of the model ordinance, banning summer and rainy-weather applications of fertilizer, mandating water setbacks, stipulating N and P content limits, and calling for the use of slow release N fertilizers (UF IFAS Extension n.d.).

Table 1 provides an overview of key statutory provisions in five other US states, which have some of the longest histories of regulating the urban use of lawn fertilizers (based on data and initial analysis provided by Janasie 2015).

Conclusion: Mechanisms and Forces of Change

As with climate policy and governance, bottom-up approaches – approaches that start with local and regional initiatives, then scale up to broader, more coordinated governance and management regimes – hold considerable promise for advancing sustainability transitions in the management of US urban land and yard spaces.

City-level planning for climate mitigation and resilience can become one major avenue for advancing sustainability transitions in the management of residential (and nonresidential) yard space. Even as conventional yardscaping contributes to climate forcing and climate change, the rights kinds of green infrastructure are critically important for urban climate adaptation. The currently dynamic space of municipal climate planning thus represents an unparalleled opportunity

US Urban and Suburban Yardscaping, Table 1 State law restrictions on the urban use of lawn fertilizers

State	Reg. since	Fertilizer statute provisions		
		Use restrictions	Sale and display restrictions	Enforcement and penalties
Minnesota	2002	Prohibits the application of P (phosphorus) fertilizer to turf	No regulation of fertilizer sales	Minimal
Maine	2008	Does not regulate use – no limits or restrictions on fertilizer use	Requires store signs urging customers to abstain from using P-containing turfgrass fertilizers	None
Wisconsin	2010	Prohibits the application of P (phosphorus) fertilizer to turf	Prohibits sale of P fertilizers for nonapproved uses (but allows stores to stock and sell them for approved ones)	$ 50 fine for first-time violators $ 200–500 fine for repeat violators
New York	2010	Prohibits the application of P (phosphorus) fertilizer to turf Bans use of any lawn/nonagricultural fertilizer b-n December 1–April 1. Ban on the use of any fertilizer near waterways	Requires stores to separate display of P and non-P fertilizers Requires store signs stating the regulatory restrictions on residential use of P-containing fertilizers	Civil penalty up to $ 500 for the first violation and up to $ 1000 for any subsequent violations
New Jersey	2011	Prohibits the application of P (phosphorus) fertilizer to turf Regulates the use of N (nitrogen) fertilizer on turf Ban on winter application of turf fertilizer (and ban on applying to frozen ground at any time) Ban on turfgrass application of fertilizers before or during rainfall Ban in turfgrass applications of fertilizer when soils are water-saturated Ban on the use of N and P fertilizers near waterways	Labeling requirements similar to those in NY Ban on the sale of fertilizers exceeding the statutorily stipulated N content Restrictions on consumer sales of P containing fertilizer (only OK in certain limited circumstances)	Professional fertilizer applicators are subject to civil penalties of $ 500 for a first-time violation, and up to $ 1000 penalty for subsequent violations

Based on data from Janasie (2015)

for designing and implementing the right kinds of urban green infrastructure – green infrastructure that is tailored to the climate, hydrology, and landscape of each urban region and designed for management with regenerative, climate-friendly, and ecologically compatible land care practices.

Municipal climate planning also presents a clear opportunity for integrating the environmental and climate justice dimensions of urban land management and green infrastructure planning. Historically redlined neighborhoods, which are experiencing significantly higher flood hazards and heat island effects than otherwise similar neighborhoods within the same city, are in particularly urgent need of good green infrastructure (Hoffman et al. 2019; Messager et al. 2021) – the kinds of green infrastructure that mitigates climate risks without introducing additional landscaping chemicals and toxic exposures to their residents. City planning for both climate mitigation and climate adaptation thus represents a logical target for advocates of sustainable urban land management transitions. And it should arguably be a priority target for existing or new coalitions between environmental and climate justice activists and the more traditional antitoxics, conservation, and climate groups and movements.

Large corporate actors within the complex value chains for yard and garden products are another promising target of activism and a potentially promising lever of change. Stymied by the political power and influence of the chemical and agri-business industries, who have been notably successful in blocking federal regulatory controls of environmentally problematic yardscaping inputs like pesticides, environmental NGOs like Friends of the Earth have taken the fight for safe, ecologically sound agricultural and urban land management to the market. After long-standing and unsuccessful advocacy for reform in the federal regulation of pesticides and other toxics, these environmental NGOs are directly targeting large, consumer-facing, and reputationally sensitive corporate retail brands like Lowe's, Home Depot, Ace Hardware, asking big retailers to flex their buyer power, and the private authority that comes with such power, and use such buyer power and private authority to stop the nursery growers who supply their garden centers from using some of the most ecologically and health-problematic pesticides (like neonicotinoids and chlorpyrifos).

Even the partial success of such efforts at private, market-driven governance of ecologically problematic practices within horticultural value chains may, in turn, open new political opportunities for a public regulatory reform of pesticide licensing, use, and marketing. In particular, the many branded, consumer-facing, and reputationally sensitive garden supply retailers, who find themselves pressured into commitments for sourcing and selling horticultural products grown without the use of hazardous pesticides, may well find it advantageous to start advocating for the types of public policies and regulation that make the fulfillment of such sustainable sourcing commitments easier and more straightforward (Tzankova 2020a, 2020b). Politically influential corporate retailers of garden supplies and yardscaping products can be enrolled as members of pesticide reform and sustainable land management policy coalitions, thus helping tip the scale in favor of sustainability transitions in US urban yardscaping and beyond.

Finally, the work of culture change – of undoing the discursive feats of a chemical industry that cemented a manicured lawn and yard aesthetic – could spell considerable promise for a move towards more regenerative urban yard management in the USA. Given the role of current norms as a driver for input-intensive and socio-ecologically problematic status-quo practices, norm change, challenging as it is to attain in practice, may prove a critical accelerator for sustainability transitions in the management if US urban yardscapes.

U

Cross-References

- ▶ Butterfly Gardening in Colombo, Sri Lanka: Approach to Biodiversity Conservation, Monitoring, Education, and Awareness in Urbanizing Habitats
- ▶ Green Cities in Theory and Practice
- ▶ Water-Smart Cities

References

Berkowitz, B. (2019, August 6). Mapping the strain on our water supply – The Washington Post. *The Washington Post*. https://www.washingtonpost.com/climate-environment/2019/08/06/mapping-strain-our-water/

Beyond Pesticides. (2013). *State preemption law: The battle for local control*. https://www.beyondpesticides.org/assets/media/documents/lawn/activist/documents/StatePreemption.pdf

Beyond Pesticides. (2015). *Environmental effects of 30 commonly used yard pesticides*. https://doi.org/10.1111/jawr.12159/abstract

Brown, K. P., & Hess, D. J. (2017). The politics of water conservation: Identifying and overcoming barriers to successful policies. *Water Policy, 19*(2), 304–321. https://doi.org/10.2166/wp.2016.089

Burrows, S. (2020, May 18). "No Mow May" campaign asks us to leave the lawn alone to help save bees – PopularResistance.Org. *PopularResistance.Org*. https://popularresistance.org/no-mow-may-campaign/

Carrico, A. R., Fraser, J., & Bazuin, J. T. (2012). Green with envy: Psychological and social predictors of lawn fertilizer application. *Environment and Behavior, 45*(4), 427–454. https://doi.org/10.1177/0013916511434637

Carrico, A. R., Raja, U. S., Fraser, J., & Vandenbergh, M. P. (2018). *Household and block level influences on residential fertilizer use*. https://doi.org/10.1016/j.landurbplan.2018.05.008

Centner, T. J., & Heric, D. C. (2019). Anti-community state pesticide preemption laws prevent local governments from protecting people from harm. *International Journal of Agricultural Sustainability, 17*(2), 118–126. https://doi.org/10.1080/14735903.2019.1568814

City of Takoma Park. (2013). *Safe grow act*. https://takomaparkmd.gov/government/police/neighborhood-services/safegrow/

City of Takoma Park MD. (2013a). *Chapter 14.28 restricted lawn care pesticides*. https://www.codepublishing.com/MD/TakomaPark/#!/TakomaPark14/TakomaPark1428.html

City of Takoma Park MD. (2013b). *List of restricted pesticides*. City of Takoma Park. https://takomaparkmd.gov/government/police/neighborhood-services/safegrow/list-of-restricted-pesticides/

Clarke, W., & Freedman, M. (2019). The rise and effects of homeowners associations. *Journal of Urban Economics, 112*, 1–15. https://doi.org/10.1016/j.jue.2019.05.001

Cole, D. C., Vanderlinden, L., Leah, J., Whate, R., Mee, C., Bienefeld, M., Wanigaratne, S., & Campbell, M. (2011). Municipal bylaw to reduce cosmetic/non-essential pesticide use on household lawns – A policy implementation evaluation. *Environmental Health: A Global Access Science Source, 10*(1), 1–17. https://doi.org/10.1186/1476-069X-10-74

del Toro, I., & Ribbons, R. R. (2020). No Mow May lawns have higher pollinator richness and abundances: An engaged community provides floral resources for pollinators. *PeerJ, 8*, e10021. https://doi.org/10.7717/PEERJ.10021

Derby Lewis, A., Bouman, M. J., Winter, A. M., Hasle, E. A., Stotz, D. F., Johnston, M. K., Klinger, K. R., Rosenthal, A., & Czarnecki, C. A. (2019). Does nature need cities? Pollinators reveal a role for cities in wildlife conservation. *Frontiers in Ecology and Evolution, 7*(JUN), 220. https://doi.org/10.3389/FEVO.2019.00220

Diane Sofranec. (2021, March 26). *5 regulatory issues to watch in 2021 – Pest Management Professional: Pest Management Professional*. Pest Management Professional. https://www.mypmp.net/2021/03/26/5-regulatory-issues-to-watch-in-2021/

EPA, OCSPP, & OPP. (2017). *US EPA – Pesticides industry sales and usage 2008–2012*.

FAOSTAT. (n.d.). Retrieved June 27, 2021, from http://www.fao.org/faostat/en/#rankings/countries_by_commodity

Feagan, R. B., & Ripmeester, M. (1999). Contesting natural(ized) lawns: A geography of private green space in the Niagara region. *Urban Geography, 20*(7), 617–634. https://doi.org/10.2747/0272-3638.20.7.617

Fissore, C., Baker, L. A., Hobbie, S. E., King, J. Y., Mcfadden, J. P., Nelson, K. C., & Jakobsdottir, I. (2011). Carbon, nitrogen, and phosphorus fluxes in household ecosystems in the Minneapolis-Saint Paul, Minnesota, urban region. *Ecological Applications, 21*(3).

Fort Atkinson, W. (n.d.). *No Mow May initiative*. Retrieved July 8, 2021, from https://heartofthecity.us/no-mow-may-initiative/

Fraser, J. C., Bazuin, J. T., Band, L. E., & Grove, J. M. (2013). Covenants, cohesion, and community: The effects of neighborhood governance on lawn fertilization. *Landscape and Urban Planning, 115*, 30–38. https://doi.org/10.1016/j.landurbplan.2013.02.013

Fraser, J., Bazuin, J. T., & Hornberger, G. (2015). The privatization of neighborhood governance and the production of urban space. *Environment and Planning A, 48*(5), 844–870. https://doi.org/10.1177/0308518X15621656

Freedgood, J., Hunter, M., Dempsey, J., & Sorensen, A. (2020). *The state of the states*. www.csp-inc.org.

Friends of the Earth. (2016). *Gardeners beware: Bee toxic pesticides found in "bee-friendly" plants sold at garden centers across the US*. www.foe.org

Galloway, J. N., Townsend, A. R., Erisman, J. W., Bekunda, M., Cai, Z., Freney, J. R., Martinelli, L. A., Seitzinger, S. P., & Sutton, M. A. (2008). Transformation of the nitrogen cycle: Recent trends, questions, and potential solutions. *Science, 320*, 889–892. http://science.sciencemag.org/

Grand View Research. (2019). *Home & garden pesticides market size report, 2019–2025*. https://www.grandviewresearch.com/industry-analysis/home-garden-pesticides-market

Groffman, P. M., Cavender-Bares, J., Bettez, N. D., Grove, J. M., Hall, S. J., Heffernan, J. B., Hobbie, S. E.,

Larson, K. L., Morse, J. L., Neill, C., Nelson, K., O'Neil-Dunne, J., Ogden, L., Pataki, D. E., Polsky, C., Chowdhury, R. R., & Steele, M. K. (2014). Ecological homogenization of urban USA. *Frontiers in Ecology and the Environment, 12*(1), 74–81. https://doi.org/10.1890/120374

Groffman, P. M., Grove, J. M., Polsky, C., Bettez, N. D., Morse, J. L., Cavender-Bares, J., Hall, S. J., Heffernan, J. B., Hobbie, S. E., Larson, K. L., Neill, C., Nelson, K., Ogden, L., O'Neil-Dunne, J., Pataki, D., Chowdhury, R. R., & Locke, D. H. (2016). Satisfaction, water and fertilizer use in the American residential macrosystem. *Environmental Research Letters, 11*(3), 034004. https://doi.org/10.1088/1748-9326/11/3/034004

Gu, C., Crane, J., Hornberger, G., & Carrico, A. (2015). The effects of household management practices on the global warming potential of urban lawns. *Journal of Environmental Management, 151*, 233–242. https://doi.org/10.1016/j.jenvman.2015.01.008

Hall, D. M., Camilo, G. R., Tonietto, R. K., Ollerton, J., Ahrné, K., Arduser, M., Ascher, J. S., Baldock, K. C. R., Fowler, R., Frankie, G., Goulson, D., Gunnarsson, B., Hanley, M. E., Jackson, J. I., Langellotto, G., Lowenstein, D., Minor, E. S., Philpott, S. M., Potts, S. G., ... Threlfall, C. G. (2017). The city as a refuge for insect pollinators. *Conservation Biology, 31*(1), 24–29. https://doi.org/10.1111/COBI.12840

Hladik, M. L., Main, A. R., & Goulson, D. (2018). Environmental risks and challenges associated with neonicotinoid insecticides. *Environmental Science & Technology, 52*(6), 3329–3335. https://doi.org/10.1021/ACS.EST.7B06388

Hoffman, J. S., et al. (2019). The effects of historical housing policies on resident exposure to intra-urban heat: A study of 108 US urban areas. *Climate, 8*(1). https://scholar-google-com.proxy.library.vanderbilt.edu/scholar?hl=en&as_sdt=0%2C43&q=the+effects+of+historic+housing+policies+on+resident+exposure&btnG=

Hogue, T., & Pincetl, S. (2015). Are you watering your lawn: High resolution data may help devise effective water conservation strategies in urban areas around the world. *Science, 348*(6241), 1319–1320. https://science-sciencemag-org.proxy.library.vanderbilt.edu/content/sci/348/6241/1319.full.pdf

Janasie, C. (2015). State fertilizer bills: The greenest way to a more natural landscape? *Rutgers Journal of Law & Public Policy, 13*, 1. https://advance-lexis-com.proxy.library.vanderbilt.edu/document/?pdmfid=1516831&crid=7164f370-f71c-47de-9bbd-d02e86f1bbf2&pddocfullpath=%2Fshared%2Fdocument%2Fanalytical-materials%2Furn%3AcontentItem%3A5HVS-0340-0240-Y0NV-00000-00&pdcontentcomponentid=292204&pdteaserkey=sr5&pditab=allpods&ecomp=ybvnk&earg=sr5&prid=468aa4ab-e5cb-405e-95ab-8639f87f4172

Kopp, K. L., & Guillard, K. (2004). *Decomposition rates and nitrogen release of turf grass clippings*. https://opencommons.uconn.edu/plsc_confs

Larson, K. L., & Brumand, J. (2014). Paradoxes in landscape management and water conservation: Examining neighborhood norms and institutional forces. *Cities and the Environment, 7*(1).

Larson, K. L., Cook, E., Strawhacker, C., & Hall, S. J. (2010). The influence of diverse values, ecological structure, and geographic context on residents' multifaceted landscaping decisions. *Human Ecology, 38*, 747–761. https://doi.org/10.1007/s10745-010-9359-6

Larson, K. L., Nelson, K. C., Samples, S. R., Hall, S. J., Bettez, N., Cavender-Bares, J., Groffman, P. M., Grove, M., Heffernan, J. B., Hobbie, S. E., Learned, J., Morse, J. L., Neill, C., Ogden, L. A., O'neil-Dunne, J., Pataki, D. E., Polsky, C., Chowdhury, R. R., Steele, M., & Trammell, T. L. E. (2016). Ecosystem services in managing residential landscapes: Priorities, value dimensions, and cross-regional patterns, *19*, 95–113. https://doi.org/10.1007/s11252-015-0477-1

Larson, K. L., Hoffman, J., & Ripplinger, J. (2017). Legacy effects and landscape choices in a desert city. *Landscape and Urban Planning, 165*, 22–29. https://doi.org/10.1016/j.landurbplan.2017.04.014

Larson, K. L., Andrade, R., Nelson, K. C., Wheeler, M. M., Engebreston, J. M., Hall, S. J., Avolio, M. L., Groffman, P. M., Grove, M., Heffernan, J. B., Hobbie, S. E., Lerman, S. B., Locke, D. H., Neill, C., Chowdhury, R. R., Trammell, T. L. E., & Trammell, T. L. E. (2020a). Municipal regulation of residential landscapes across US cities: Patterns and implications for landscape sustainability. *Journal of Environmental Management, 275*, 111132. https://doi.org/10.1016/j.jenvman.2020.111132

Larson, K. L., et al. (2020b). Municipal regulation of residential landscapes across US cities: Patterns and implications for landscape sustainability. *Journal of Environmental Management, 275*, 111132. https://doi.org/10.1016/j.jenvman.2020.111132

Lee, S., & McCann, L. (2018). Passage of phosphorus-free lawn fertilizer laws by U.S. states on JSTOR. *Journal of Natural Resources Policy Research, 8*(1–2), 66–88. https://www-jstor-org.proxy.library.vanderbilt.edu/stable/10.5325/naturesopolirese.8.1-2.0066

Lerman, S. B., & Contosta, A. R. (2019). Lawn mowing frequency and its effects on biogenic and anthropogenic carbon dioxide emissions. *Landscape and Urban Planning, 182*, 114–123. https://doi.org/10.1016/j.landurbplan.2018.10.016

Lerman, S. B., Contosta, A. R., Milam, J., & Bang, C. (2018). To mow or to mow less: Lawn mowing frequency affects bee abundance and diversity in suburban yards. *Biological Conservation, 221*, 160–174. https://doi.org/10.1016/J.BIOCON.2018.01.025

Lofquist, D., Lugaila, T., O'connell, M., & Feliz, S. (2012). *Census briefs*. www.census.gov/population

Martini, N. F., Nelson, K. C., Hobbie, S. E., & Baker, L. A. (2015). Why "Feed the Lawn"? Exploring the influences on residential turf grass fertilization in the Minneapolis–Saint Paul Metropolitan Area. *Environment and Behavior, 47*(2), 158–183. https://doi.org/10.1177/0013916513492418

U

Meftaul, I. M., et al. (2020). Pesticides in the urban environment: A potential threat that knocks at the door. *Science of the Total Environment, 711*, 134612. Elsevier B.V. https://doi.org/10.1016/j.scitotenv.2019.134612

Messager, M. L., Ettinger, A. K., Murphy-Williams, M., & Levin, P. S. (2021). Fine-scale assessment of inequities in inland flood vulnerability. *Applied Geography, 133*, 102492. https://doi.org/10.1016/J.APGEOG.2021.102492

Milesi, C., Running, S. W., Elvidge, C. D., Dietz, J. B., Tuttle, B. T., & Nemani, R. R. (2005). Mapping and modeling the biogeochemical cycling of turf grasses in the United States. *Environmental Management, 36*(3), 426–438. https://doi.org/10.1007/s00267-004-0316-2

Mintel. (2020). *Lawn and garden products: Incl impact of COVID-19 – US – April 2020 – Market Research Report*. https://reports.mintel.com/display/986936/

Morgan, M. K., Stout, D. M., Jones, P. A., & Barr, D. B. (2008). An observational study of the potential for human exposures to pet-borne diazinon residues following lawn applications. *Environmental Research, 107*(3), 336–342. https://doi.org/10.1016/J.ENVRES.2008.03.004

MU Extension. (n.d.). *Pesticides and the environment*. Retrieved July 7, 2021, from https://extension.missouri.edu/publications/g7520

Nathan, V. K., Jasna, V., & Parvathi, A. (2020). Pesticide application inhibit the microbial carbonic anhydrase–mediated carbon sequestration in a soil microcosm. *Environmental Science and Pollution Research, 27*(4), 4468–4477. https://doi.org/10.1007/s11356-019-06503-1

National Pesticide Information Center (NPIC). (n.d.). *Fipronil technical fact sheet*. Retrieved July 7, 2021, from http://npic.orst.edu/factsheets/archive/fiptech.html

Nolan, B., et al. (1998). National look at nitrate contamination of Ground Water. *Water Conditioning and Purification, 39*(12), 76–79. https://water.usgs.gov/nawqa/nutrients/pubs/wcp_v39_no12/

Nowell, L. H., Moran, P. W., Bexfield, L. M., Mahler, B. J., van Metre, P. C., Bradley, P. M., Schmidt, T. S., Button, D. T., & Qi, S. L. (2021). Is there an urban pesticide signature? Urban streams in five U.S. regions share common dissolved-phase pesticides but differ in predicted aquatic toxicity. *Science of the Total Environment, 793*, 148453. https://doi.org/10.1016/J.SCITOTENV.2021.148453

Pennino, M. J., Compton, J. E., & Leibowitz, S. G. (2017). Trends in drinking water nitrate violations across the United States. *Environmental science & technology, 51* (22), 13450–13460.

Robbins, J. (2019, February 7). In era of drought, phoenix prepares for a future without Colorado river water. *Yale Environment, 360*. https://e360.yale.edu/features/how-phoenix-is-preparing-for-a-future-without-colorado-river-water

Robbins, P., & Birkenholtz, T. (2003). Turfgrass revolution: Measuring the expansion of the American lawn. *Land Use Policy, 20*, 181–194. https://doi.org/10.1016/S0264-8377(03)00006-1

Robbins, P., & Sharp, J. T. (2003). Producing and consuming chemicals: The moral economy of the American lawn. *Economic Geography, 79*(4), 425–451. https://doi.org/10.1111/j.1944-8287.2003.tb00222.x

Robbins, P., & Sharp, J. T. (2008). Producing and consuming chemicals: The moral economy of the American lawn. In *Urban ecology: An international perspective on the interaction between humans and nature* (pp. 181–205). Springer US. https://doi.org/10.1007/978-0-387-73412-5_11

Robbins, P., Polderman, A., & Birkenholtz, T. (2001). Lawns and toxins: An ecology of the city. *Cities, 18* (6), 369–380. https://doi.org/10.1016/S0264-2751(01)00029-4

Ryan, C. D., et al. (2019). Culture, science, and activism in Florida lawn and landscape fertilizer policy. *HortTechnology, 29*(6), 854–865. https://journals-ashs-org.proxy.library.vanderbilt.edu/horttech/view/journals/horttech/29/6/article-p854.xml

Safe Grow Montgomery MD. (2016). *MEDIA RELEASE: Health and environmental coalition urges vigorous Montgomery County defense of historic lawn care pesticide law against industry lobbyist's lawsuit. Safe Grow Montgomery*. Health and Environmental Coalition Urges Vigorous Montgomery County Defense of Historic Lawn Care Pesticide Law against Industry Lobbyist's Lawsuit. http://safegrowmontgomery.org/media-release-health-and-environmental-coalition-urges-vigorous-montgomery-county-defense-of-historic-lawn-care-pesticide-law-against-industry-lobbyists-lawsuit/

Simon-Delso, N., Amaral-Rogers, V., Belzunces, L. P., Bonmatin, J. M., Chagnon, M., Downs, C., Furlan, L., Gibbons, D. W., Giorio, C., Girolami, V., Goulson, D., Kreutzweiser, D. P., Krupke, C. H., Liess, M., Long, E., McField, M., Mineau, P., Mitchell, E. A. D., Morrissey, C. A., … Wiemers, M. (2014). Systemic insecticides (neonicotinoids and fipronil): Trends, uses, mode of action and metabolites. *Environmental Science and Pollution Research, 22*(1), 5–34. https://doi.org/10.1007/S11356-014-3470-Y

Sisser, J. M., Nelson, K. C., Larson, K. L., Ogden, L. A., Polsky, C., & Chowdhury, R. R. (2016). Lawn enforcement: How municipal policies and neighborhood norms influence homeowner residential landscape management. *Landscape and Urban Planning, 150*, 16–25. https://doi.org/10.1016/j.landurbplan.2016.02.011

Smitley, D. (n.d.). *Mow high for weed and grub control*. Retrieved July 8, 2021, from www.msue.msu.edu

Solomon, K., et al. (2005). Nonagricultural and residential exposures to pesticides. *Scandinavian Journal of Work Environmental and Health, 31*(Suppl 1), 74–81. https://www-jstor-org.proxy.library.vanderbilt.edu/stable/pdf/40967440.pdf

Souto, L. A., Listopad, C. M. C. S., & Bohlen, P. J. (2019). Forging linkages between social drivers and ecological processes in the residential landscape. *Landscape and*

Urban Planning, 185, 96–106. https://doi.org/10.1016/j.landurbplan.2019.01.002

st. Hilaire, R., Arnold, M., et al. (2008). Efficient water use in residential urban landscapes. *HortScience, 43*(7), 2081–2092. https://journals-ashs-org.proxy.library.vanderbilt.edu/hortsci/view/journals/hortsci/43/7/article-p2081.xml

The National Pollinator Garden Network. (n.d.). *Partners – Million pollinator garden challenge*. Retrieved July 8, 2021, from http://millionpollinatorgardens.org/partners/

The Week Staff. (2015, January 8). *Blades of glory: America's love affair with lawns | The Week*. https://theweek.com/articles/483762/blades-glory-americas-love-affair-lawns

Townsend-Small, A., & Czimczik, C. I. (2010). Carbon sequestration and greenhouse gas emissions in urban turf. *Geophysical Research Letters, 37*(2), 2707. https://doi.org/10.1029/2009GL041675

Tzankova, Z. (2013). The difficult problem of nonpoint nutrient pollution: Could the endangered species act offer some relief. *William & Mary Environmental Law and Policy Review, 37*, 709–757.

Tzankova, Z. (2020a). Can private governance boost public policy? Insights from public–private governance interactions in the fisheries and electricity sectors. *Regulation & Governance*. https://doi.org/10.1111/REGO.12317

Tzankova, Z. (2020b). Public policy spillovers from private energy governance: New opportunities for the political acceleration of renewable energy transitions. *Energy Research & Social Science, 67*, 101504. https://doi.org/10.1016/J.ERSS.2020.101504

UF IFAS Extension. (n.d.). *Florida fertilizer ordinances*. Retrieved July 7, 2021, from https://ffl.ifas.ufl.edu/fertilizer/#!

US Outdoor Water Use | WaterSense | US EPA. (n.d.). Retrieved June 17, 2021, from https://19january2017snapshot.epa.gov/www3/watersense/pubs/outdoor.html

USGS. (2006). *Pesticides in the nation's streams and ground water, 1992–2001 – A summary*. https://pubs.usgs.gov/fs/2006/3028/

USGS. (2015). *Irrigation water use*. https://www.usgs.gov/mission-areas/water-resources/science/irrigation-water-use?qt-science_center_objects=0#qt-science_center_objects

Water Use in California – Public Policy Institute of California. (n.d.). Retrieved July 1, 2021, from https://www.ppic.org/publication/water-use-in-california/

Watson, C. J., Carignan-Guillemette, L., Turcotte, C., Maire, V., & Proulx, R. (2020). Ecological and economic benefits of low-intensity urban lawn management. *Journal of Applied Ecology, 57*(2), 436–446. https://doi.org/10.1111/1365-2664.13542

Wheeler, M. M., Neill, C., Groffman, P. M., Avolio, M., Bettez, N., Cavender-Bares, J., Roy Chowdhury, R., Darling, L., Grove, J. M., Hall, S. J., Heffernan, J. B., Hobbie, S. E., Larson, K. L., Morse, J. L., Nelson, K. C., Ogden, L. A., O'Neil-Dunne, J., Pataki, D. E., Polsky, C., ... Trammell, T. L. E. (2017). Continental-scale homogenization of residential lawn plant communities. *Landscape and Urban Planning, 165*, 54–63. https://doi.org/10.1016/j.landurbplan.2017.05.004

Wheeler, M. M., Larson, K. L., & Andrade, R. (2020). Attitudinal and structural drivers of preferred versus actual residential landscapes in a desert city. *Urban Ecosystems, 23*(3), 659–673. https://doi.org/10.1007/s11252-020-00928-0

Whitney, C. (2010). Living lawns, dying waters: The suburban boom, nitrogenous fertilizers, and the nonpoint source pollution dilemma. *Technology and Culture, 52*(3), 652–674. https://scholar-google-com.proxy.library.vanderbilt.edu/scholar_lookup?title=Living%20lawns%2C%20dying%20waters&publication_year=2010&author=K.%20Whitney

Wilk, B., Rebollo, V., & Hanania, S. (2019). *A guide for pollinator-friendly cities: How can spatial planners and land-use managers create favourable urban environments for pollinators? Guidance prepared by ICLEI Europe for the European Commission*. https://eur-lex.europa.eu/legal-content/EN/TXT/?uri=CELEX:52018DC0395

Woodcock, B. A., Bullock, J. M., Shore, R. F., Heard, M. S., Pereira, M. G., Redhead, J., Ridding, L., Dean, H., Sleep, D., Henrys, P., Peyton, J., Hulmes, S., Hulmes, L., Sárospataki, M., Saure, C., Edwards, M., Genersch, E., Knäbe, S., & Pywell, R. F. (2017). Country-specific effects of neonicotinoid pesticides on honey bees and wild bees. *Science, 356*(6345), 1393–1395. https://doi.org/10.1126/SCIENCE.AAA1190

Xerces Society. (2017). *Model local resolution to protect pollinators*. https://xerces.org/pesticides/model-local-resolution

Xerces Society. (2021). *Bee City USA – Current Bee City USA affiliates*. https://beecityusa.org/current-bee-city-usa-affiliates/

Zartarian, V. G., Özkaynak, H., Burke, J. M., Zufall, M. J., Rigas, M. L., & Furtaw, E. J. (2000). A modeling framework for estimating children's residential exposure and dose to chlorpyrifos via dermal residue contact and nondietary ingestion. *Environmental Health Perspectives, 108*(6), 505–514. https://doi.org/10.1289/EHP.00108505

V

Vegan Food Justice

► The Vegan Food Justice Movement

The Vegan Food Justice Movement

Teagan Murphy[1] and Anne Mook[2]
[1]University of Maryland, College Park, MD, USA
[2]University of Georgia, Athens, GA, USA

Synonyms

Vegan food justice; Vegan social justice

Definition

The vegan food justice movement combines the aims of veganism and food justice, seeking to provide communities with nutritious, affordable, and culturally appropriate foods, free from exploitation of and cruelty to all human and nonhuman animals.

Introduction

Veganism is a growing trend in the United States, both in terms of the number of individuals who identify as vegan – with some sources reporting up to 6% of the US population as vegan, up from 1% in 2014 – and the availability and sales of vegan or "plant-based" foods in restaurants and grocery stores, particularly in liberal urban centers in the United States (Forgrieve 2018). Despite the growing availability and consumption of vegan options, dairy and meat consumption has also continued to grow in the past few years in the United States, achieving relatively little reduction in animal cruelty, social inequality, and environmental degradation (Kuck and Schnitkey 2021). Therefore, gaining support for veganism and other dietary choices that reduce or eliminate the consumption of animal products is an important step toward achieving animal and social justice. While veganism is often labeled simply as a diet, many vegan activists, scholars, and organizations are increasingly adopting a more ideological definition, referring to veganism as "*a philosophy and way of living which seeks to exclude – as far as is possible and practicable – all forms of exploitation of, and cruelty to, animals for food, clothing or any other purpose; [which] by extension, promotes the development and use of animal-free alternatives for the benefit of animals, humans and the environment*" (The Vegan Society n.d.; Greenebaum 2012; Kalte 2020). Under this definition, veganism is more than just a dietary preference: it is a conscious choice to participate in a movement that seeks to promote justice and equality, not just for farmed animals, but among all human and nonhuman species.

R. C. Brears (ed.), *The Palgrave Encyclopedia of Urban and Regional Futures*,
https://doi.org/10.1007/978-3-030-87745-3

How does veganism fit into the growing food justice movement? Food justice refers to "communities exercising their right to grow, sell, and eat [food that is] fresh, nutritious, affordable, culturally appropriate, and grown locally with care for the well-being of the land, workers, and animals" (Just Food 2010; Alkon and Agyeman 2011). The food justice movement aims specifically to provide this agency to low-income communities, communities of color, and Indigenous communities, which have been historically and institutionally denied access to food security (Alkon and Agyeman 2011). Food justice has also been referred to as a feminist or ecofeminist effort, drawing connections between women and the environment as exploited for profit in a capitalist and patriarchal society (Mallory 2013). The vegan and food justice movements are often discussed and studied separately and are sometimes even framed at odds with one another, particularly in critiques of veganism (Dickstein et al. 2020; Greenebaum 2018; Harper 2011, 2012; Polish 2016). However, several activists and scholars have worked to shed light on how veganism is a necessary part of food justice. This chapter aims to highlight the works of these scholars, as well as the inherent multi-issue nature of vegan food justice.

Speciesism

One of the most well-known goals of the vegan food justice movement is protecting animal lives and promoting animal wellbeing. Many vegans believe that discrimination between different types of animal species including humans, or '*speciesism*', is the root cause for animal abuse. Animal rights organizations such as PETA (2021), define speciesism as "a misguided belief that one species is more important than another" and argue that speciesism explains why non-vegans feel no remorse after eating beef, pork, or chicken, but typically do feel sorry for an abused dog or cat. This hierarchy of species and human superiority is taught in many cultures. Abrahamic societies, for example, teach their populations that it is acceptable, even virtuous, to eat animal meat or use animal products such as dairy and eggs (Waldau and Patton 2006). Therefore, given this believed human superiority over other nonhuman species, humans are free to use their milk, eggs, skin, and flesh for consumption. On the contrary, individuals who oppose this view of human superiority, or anti-speciesists, may choose a vegan diet as they believe it is unethical to eat meat or use animals (Rosenfeld 2019).

There are two ways veganism reduces speciesism. The first, and most obvious, is that not eating and using animal products avoids the exploitation and killing of farm animals for consumption (Francione and Garner 2010; Janssen et al. 2016; Perz et al. 2018). Peter Singer, a philosopher of ethics, argues that animals – like humans – are sentient beings with the capacity to suffer. Therefore, ethics must be applied to all human and nonhuman animal species (Singer 2009). However, the debate over whether and how we can achieve an anti-speciesist society is a complicated one. Some animal rights activists advocate for better treatment of farm animals, using labels such as free-range chickens, grass-fed cows, and rating systems for the quality of life of the animal (e.g., the *Beter Leven* label in the Netherlands), so that consumers can make more ethical decisions when consuming animal products (Girish and Barbuddhe 2020; Mook and Overdevest 2021). While a more humane treatment of farm animals might be a step in the right direction, from a utilitarian perspective this approach might be counterproductive as improved animal welfare could make the consumption of animals seem less unethical and therefore increase consumption (Muraille 2018). Ultimately, many advocates against speciesism seek a total abolition of animal agriculture, believing there is no ethical or humane way to exploit an animal for profit and consumption (Rosenfeld 2019).

The second way veganism promotes anti-speciesism is that an individual on a plant-based diet requires about 60 to 80% less agricultural land to sustain their diets than an omnivorous individual would require (Aleksandrowicz et al. 2016). Billions of animals are threatened and an increasingly growing list of species are endangered because of the rapid expansion of human development. Currently, about one-third of the planet's land is used for agriculture or livestock – and with a growing human consumer population, significant amounts of habitats for non-livestock animals are invaded or destroyed. Simultaneously, meat production causes significant environmental pollution and is a major contributor to greenhouse gas emissions (approximately responsible for 30% of greenhouse gasses in the western world) which causes climate change (Petrovic et al. 2015). Environmental pollution and climate change directly affect many human and non-human species alike. The high volumes of manure from hog farming in concentrated animal feeding operations for example contaminate the water, soil, and air quality. Furthermore, climate change threatens species to such an extent that they will either need to move to more suitable habitats or go extinct. For example due to rising temperatures, water temperatures increase which leads to lower levels of oxygen and acidification. Corals and many sea creatures cannot survive this, while other fish species move toward the poles or deeper waters where the temperatures are lower (Cashion et al. 2020). In other cases, species are threatened by desertification, wildfires, and floodings (Waller et al. 2017; Soultan et al. 2019). Therefore, taking an anti-speciesist stance and choosing a vegan diet both directly and indirectly impacts the wellbeing of animals (Wrenn 2011).

Race, Class, and Decolonization

Veganism is frequently perceived as an exclusive and privileged lifestyle choice limited to white and middle- to upper-class populations. The mainstream vegan movement has been criticized for perpetuating an elitist and inaccessible image of veganism (Greenebaum 2018; Harper 2012; Polish 2016), lacking representation of vegans of color in the media (Greenebaum 2018), pushing a "colorblind" single-issue focus on animal cruelty (Harper 2011, 2012), and emphasizing consumer-oriented activism (Dickstein et al. 2020; Polish 2016). Vegan scholars of color are working to shape perceptions and understandings of veganism to make it more widespread through two means: normalization and radicalization.

The association of veganism with privilege and whiteness reportedly serves as a barrier to adopting veganism for people of color (Greenebaum 2018). Several studies on veganism indicate that there is a strong stigma against vegans, as they are often labeled by non-vegans as killjoys, attention-seeking, pretentious, annoying, rude, overbearing, and maintaining an air of moral superiority (Bresnahan et al. 2016; Cole and Morgan 2011; Greenebaum 2012; MacInnis and Hodson 2017; Markowski and Roxburgh 2019; Twine 2014). Noting that these studies are primarily limited to white, middle-class participants, Greenebaum (2018) finds that the "vegan stigma" is compounded for people of color, who may face accusations of "acting white" or rejecting their own culture and ethnicity by adopting a vegan diet. In a similar vein, studies related to eating habits in Black adults have found that one of the barriers to plant-based or meat-reduced diets for Black Americans is the perception that cutting out certain foods means giving up a part of their cultural identity (James 2004).

Normalizing and destigmatizing veganism is a matter of improving representation, accessibility, and cultural compatibility. Amplifying vegan voices of color through social media, blog posts, published works, conference panels, and other mediums can create a diverse representation of veganism to juxtapose the monolithic stereotype of the white, elitist "Whole Foods" vegan (Greenebaum 2018; Harper 2012). Broader representation allows for the promotion of vegan foods using ingredients and techniques that are more familiar and culturally relevant to broader groups of people.

This push to make veganism more accessible has prompted several Black-owned vegan soul food restaurants to open up throughout the South, providing completely plant-based versions of traditional soul food dishes and creating a healthier way to maintain these culturally significant foods (Crimarco et al. 2020). Many of these restaurant owners see themselves as taking on roles as promoters of health and veganism in their communities by making vegan meals more accessible and educating their customers on both health-related and non-health-related reasons to go vegan (Crimarco et al. 2020). By emphasizing familiar ingredients and recipes, vegans of color such as these restaurant owners are making veganism seem more attainable (Crimarco et al. 2020; Greenebaum 2018).

For many vegan scholars of color, the liberatory potential of veganism goes beyond normalization. Using frameworks of intersectionality and decolonization, these scholars aim to radicalize veganism by highlighting what a race- and class-conscious vegan movement could look like, going as far as to show that a full commitment to veganism is simultaneously a commitment to dismantling white supremacy.

Many scholars highlight veganism as a mechanism to "decolonize one's diet," or reject dietary habits formed out of a history of European colonization and institutionalized food insecurity (Harper 2009; Polish 2016). Harper's (2009) anthology *Sistah Vegan*, for example, uplifts the narratives of Black women aiming to decolonize their bodies from heavily processed diets high in meats, sugars, and fats (Harper 2009; James 2004). Veganism allows scholars like Harper and those included in her anthology to resist a colonized food system that has historically perpetuated food insecurity within many low-income communities of color (Harper 2009, 2011, 2012).

Scholars have also highlighted that speciesism and racism are connected through how our understanding of "Human" has been shaped by white supremacy (Belcourt 2014; Polish 2016). Speciesism places nonhuman animals as the inferior other to humanity, whereas racism places people of color as the inferior other to whiteness. Black and Indigenous bodies, in particular, have been historically dehumanized and labeled as "animal" or animalistic, a process that refuses humanness to colonized subjects (Belcourt 2014). To Belcourt (2014), an Indigenous scholar from the Driftpile Cree Nation, these labels perform epistemic violence that both marginalizes racialized bodies and denies animals their subjectivity.

It is important to note that vegan scholars of color are highly critical of analogous comparisons between racism and speciesism, such as claims that factory farming is "the new slavery" (Harper 2012; Polish 2016). These comparisons perpetuate damaging associations of people of color with negative assumptions about animality, and they suggest that the lasting impacts of racial oppression are over (Harper 2012; Polish 2016). Instead, vegan scholars of color highlight how these forms of oppression are connected – as opposed to parallel – and all work to maintain the power of white supremacy. As such, veganism serves as a mechanism to reject both the speciesism and racism of colonized food systems.

Vegan Feminism

Vegan anti-speciesism seeks equality between human and nonhuman species. By extension, vegan feminism seeks to dismantle a patriarchal consumer system of animal products. Corey Wrenn (2017) argues that consuming animal productions upholds a patriarchal society with two aspects of capitalism, namely, subordination and consumption. In a capitalist system, animals are subordinate to humans, women are subordinate to men, and workers are subordinate to capitalists. The works of vegan feminist scholars like Wrenn have highlighted how nonhuman female animals suffer from subordination to enable capitalist consumption.

Female bodies of both humans and nonhuman animals are objectified and exploited to achieve pleasure and financial gain (Adams 2017). Given the reproductive capabilities of human and nonhuman female bodies, they are often used to achieve sexual gratification (sex objects) and/or pregnancy (for meat and milk). There is a link between human and nonhuman suffering as uninvited sexual advances such as assault and rape are prevalent in the human species, with many members of (sub) cultures continuing to believe that it is a woman's responsibility to satisfy men's sexual needs (Adams 2006, 2015; Gaard 2017). Social pressure on human women to produce progeny is also common practice. For animals – in particular livestock – reproduction is often desired to maintain and/or increase meat and/or dairy products and is often achieved through artificial insemination, which is the practice of inserting sperm into the female's reproductive tract (Cusack 2013). Embryo transfers of high-value animals such as Wagyu beef cattle are also increasingly common, where the cows are injected with hormones to "super-ovulate" and any embryos are flushed out and frozen to be inserted into a less valuable cow species (Gomes et al. 2017). In other words, the patriarchal capitalist system shows parallels between the maltreatment and exploitation of human and nonhuman females.

Human diets are gendered too. Eating a bloody steak, for example, is considered masculine while eating a salad is considered a more feminine diet option. Therefore, men are to a greater extent responsible for driving meat factories (Love and Sulikowski 2018). Vegan ecofeminists, therefore, argue that women should stand together and recognize that eating meat and consuming dairy products is dehumanizing, unethical, and exploitative, just like the underpaid and cruel work hours of sweatshop workers in Bangladesh (Gaard 2017). However, vegan ecofeminists do not reject meat consumption solely to oppose animal cruelty and environmental degradation by empathizing with nonhuman animals – they also seek to address larger systems of oppression including sexism. Not all feminists support this vegan ecofeminist view. George (1994) for example argues that it is unfair to put the burden of fighting animal cruelty and exploitation on women, especially given that men are less likely to develop nutritional deficiencies as they do not bear children or lactate. While more research is necessary to understand the long-term effects of a vegan diet, current case studies suggest that veganism has positive short-term <24 health outcomes for both men and women (Medawar et al. 2019).

Conclusion

The vegan food justice movement reveals an approach to veganism that goes beyond health fads and consumer culture, instead encompassing both ideology and praxis geared toward liberation. The works and scholars referenced in this chapter shed light on how veganism ties not only to anti-speciesism and animal rights, but to racial justice, decolonization, and feminism. To minimize harm and achieve liberation for all human and nonhuman animals, the vegan food justice movement asserts that vegans must understand these issues as interconnected, and must commit to dismantling all forms of oppression to achieve food justice. Conversely, the movement also asserts that food justice is incomplete without a commitment to veganism.

Continued research on veganism and food justice must not shy away from highlighting the intersections between these two seemingly disparate segments of food activism. Furthermore, vegan food justice advocates must continue to target broader social systems such as white supremacy, capitalism, and colonization as the core drivers of food injustice. Identifying these systems highlights the intersecting interests of vegan food justice more clearly, and will ultimately aid advocates in achieving a socially just relationship with food.

Cross-References

▶ Food Justice

References

Adams, C. J. (2006). An animal manifesto gender, identity, and vegan-feminism in the twenty-first century. *Parallax, 12*(1), 120–128. https://doi.org/10.1080/13534640500448791.

Adams, C. J. (2015). *The sexual politics of meat: A feminist-vegetarian critical theory*. Bloomsbury Publishing USA.

Adams, C. J. (2017). The poetics of Christian engagement: Living compassionately in a sexual politics of meat world. *Studies in Christian Ethics, 30*(1), 45–59. https://doi.org/10.1177/0953946816674148.

Aleksandrowicz, L., Green, R., Joy, E. J., Smith, P., & Haines, A. (2016). The impacts of dietary change on greenhouse gas emissions, land use, water use, and health: A systematic review. *PLoS One, 11*(11), e0165797. https://doi.org/10.1371/journal.pone.0165797.

Alkon, A. H., & Agyeman, J. (2011). *Cultivating food justice: Race, class, and sustainability*. Cambridge, MA: MIT Press.

Belcourt, B. R. (2014). Animal bodies, colonial subjects: (Re)Locating animality in decolonial thought. *Societies, 5*. https://doi.org/10.3390/soc5010001.

Bresnahan, M., Zhuang, J., & Zhu, X. (2016). Why is the vegan line in the dining hall always the shortest? Understanding vegan stigma. *Stigma and Health, 1*(1), 3–15. https://doi.org/10.1037/sah0000011.

Cashion, T., Nguyen, T., Ten Brink, T., Mook, A., Palacios-Abrantes, J., & Roberts, S. M. (2020). Shifting seas, shifting boundaries: Dynamic marine protected area designs for a changing climate. *PLoS One, 15*(11), e0241771. https://doi.org/10.1371/journal.pone.0241771.

Cole, M., & Morgan, K. (2011). Vegaphobia: Derogatory discourses of veganism and the reproduction of speciesism in UK national newspapers1. *The British Journal of Sociology, 62*(1), 134–153. https://doi.org/10.1111/j.1468-4446.2010.01348.x.

Crimarco, A., Turner-McGrievy, G. M., Botchway, M., Macauda, M., Adams, S. A., Blake, C. E., & Younginer, N. (2020). "We're not meat shamers. We're plant pushers.": How owners of local vegan soul food restaurants promote healthy eating in the African American Community. *Journal of Black Studies, 51*(2), 168–193. https://doi.org/10.1177/0021934719895575.

Cusack, C. M. (2013). Feminism and husbandry: Drawing the fine line between mine and bovine. *Journal for Critical Animal Studies, 11*(1), 24–45. https://doi.org/10.5422/fordham/9780823254156.003.0002.

Dickstein, J., Dutkiewicz, J., Guha-Majumdar, J., & Winter, D. R. (2020). Veganism as left praxis. *Capitalism Nature Socialism, 0*(0), 1–20. https://doi.org/10.1080/10455752.2020.1837895.

Forgrieve, J. (2018). The growing acceptance of veganism. *Forbes* Available at: https://www.forbes.com/sites/janetforgrieve/2018/11/02/picturing-a-kindler-gentler-world-vegan-month/?sh=78e1ebb22f2b

Francione, G. L., & Garner, R. (2010). *The animal rights debate: Abolition or regulation?* New York: Columbia University Press.

Gaard, G. (2017). *Critical ecofeminism*. Lexington Books.

George, K. (1994). Should feminists be vegetarians? *Signs, 19*(2), 405–434. https://doi.org/10.1086/494889.

Girish, P. S., & Barbuddhe, S. B. (2020). Meat traceability and certification in meat supply chains. In *Meat quality analysis: Advanced evaluation methods, techniques, and technologies* (Vol. 2(3), pp. 153–170). London: Academic. https://doi.org/10.1016/B978-0-12-819233-7.00010-0.

Gomes, G. C., Rae, D. O., Block, J., & Risco, C. A. (2017, September). Superovulation and embryo transfer in Wagyu donors and cross-bred recipients. In *American Association of Bovine Practitioners proceedings of the annual conference* (pp. 214–214).

Greenebaum, J. (2012). Veganism, identity and the quest for authenticity. *Undefined*. /paper/Veganism%2C-Identity-and-the-Quest-for-Authenticity-Greenebaum/b62c7a7c1c37ad62314d73c10cd5dce4e55554b8

Greenebaum, J. (2018). Vegans of color: Managing visible and invisible stigmas. *Food, Culture & Society, 21*(5), 680–697. https://doi.org/10.1080/15528014.2018.1512285.

Harper, A. B. (2009). *Sistah vegan: Black female vegans speak on food, identity, health, and society*. Brooklyn: Lantern Books. ISBN:1590561457.

Harper, A. B. (2011). Vegans of color, racialized embodiment, and problematics of the "exotic". In A. H. Alkon & J. Agyeman (Eds.), *Cultivating food justice: Race, class and sustainability* (pp. 221–238). Cambridge, MA: MIT Press. ISBN:9780262516327.

Harper, A. B. (2012). Going beyond the normative white "Post-racial" vegan epistemology. In P. W. Forson & C. Counihan (Eds.), *Taking food public: Redefining foodways in a changing world*. New York: Routledge. ISBN:9780415888554.

James, D. (2004). Factors influencing food choices, dietary intake, and nutrition-related attitudes among African Americans: Application of a culturally sensitive model. *Ethnicity & Health, 9*(4), 349–367. https://doi.org/10.1080/1355785042000285375.

Janssen, M., Busch, C., Rödiger, M., & Hamm, U. (2016). Motives of consumers following a vegan diet and their attitudes towards animal agriculture. *Appetite, 105*, 643–651. https://doi.org/10.1016/j.appet.2016.06.039.

JUST FOOD. (2010). *JUST FOOD*. Retrieved July 12, 2021, from https://www.justfood.org

Kalte, D. (2020). Political veganism: An empirical analysis of vegans' motives, aims, and political engagement. *Political Studies, 0032321720930179*. https://doi.org/10.1177/0032321720930179.

Kuck, G., & Schnitkey, G. (2021). An overview of meat consumption in the United States. *Farmdoc Daily (11):76*. Department of Agricultural and Consumer Economics, University of Illinois at Urbana-Champaign, May 12, 2021.

Love, H. J., & Sulikowski, D. (2018). Of meat and men: Sex differences in implicit and explicit attitudes toward meat. *Frontiers in Psychology, 9*, 559. https://doi.org/10.1007/s10806-011-9373-8.

MacInnis, C. C., & Hodson, G. (2017). It ain't easy eating greens: Evidence of bias toward vegetarians and vegans from both source and target. *Group Processes & Intergroup Relations, 20*(6), 721–744. https://doi.org/10.1177/1368430215618253.

Mallory, C. (2013). Locating ecofeminism in encounters with food and place. *Journal of Agriculture and Environmental Ethics, 26*, 141–189. https://doi.org/10.1007/s10806-011-9373-8.

Markowski, K. L., & Roxburgh, S. (2019). "If I became a vegan, my family and friends would hate me:" Anticipating vegan stigma as a barrier to plant-based diets. *Appetite, 135*, 1–9. https://doi.org/10.1016/j.appet.2018.12.040.

Medawar, E., Huhn, S., Villringer, A., & Witte, A. V. (2019). The effects of plant-based diets on the body and the brain: A systematic review. *Translational Psychiatry, 9*(1), 1–17. https://doi.org/10.1038/s41398-019-0552-0.

Mook, A., & Overdevest, C. (2021). What drives market construction for fair trade, organic, and GlobalGAP certification in the global citrus value chain? Evidence at the importer level in the Netherlands and the United

States. *Business Strategy and the Environment*. https://doi.org/10.1002/bse.2784.
Muraille, E. (2018). Debate: Could anti-speciesism and veganism form the basis for a rational society? *The Conversation* Available at https://theconversation.com/debate-could-anti-speciesism-and-veganism-form-the-basis-for-a-rational-society-100151.
Perz, S. G., Covington, H., Espin Moscoso, J., Griffin, L., Jacobson, G., Leite, F., Mook, A., Overdevest, C., Samuel-Jones, T., & Thomson, R. (2018). Future directions for applications of political economy in environmental and resource sociology: Selected research priorities going forward. *Environmental Sociology, 4*(4), 470–487. https://doi.org/10.1080/23251042.2018.1446678.
PETA. (2021, May 18). *What is speciesism and how you can overcome it*. PETA. https://www.peta.org/features/what-is-speciesism/
Petrovic, Z., Djordjevic, V., Milicevic, D., Nastasijevic, I., & Parunovic, N. (2015). Meat production and consumption: Environmental consequences. *Procedia Food Science, 5*, 235–238. https://doi.org/10.1016/j.profoo.2015.09.041.
Polish, J. (2016). Decolonizing veganism: On resisting vegan whiteness and racism. In J. Castricano & R. R. Simonsen (Eds.), *Critical perspectives on veganism* (pp. 373–391). Springer. https://doi.org/10.1007/978-3-319-33419-6_17.
Rosenfeld, D. L. (2019). Ethical motivation and vegetarian dieting: The underlying role of anti-speciesist attitudes. *Anthrozoös, 32*(6), 785–796. https://doi.org/10.1080/08927936.2019.1673048.
Singer, P. (2009). *Animal liberation: The definitive classic of the animal movement*. New York: Harper Perennia. ISBN:9780061711305.
Soultan, A., Wikelski, M., & Safi, K. (2019). Risk of biodiversity collapse under climate change in the Afro-Arabian region. *Scientific Reports, 9*(1), 1–12. https://doi.org/10.1038/s41598-018-37851-6.
Twine, R. (2014). Vegan killjoys at the table – Contesting happiness and negotiating relationships with food practices. *Societies, 4*(4), 623–639. https://doi.org/10.3390/soc4040623.
Waldau, P., & Patton, K. C. (Eds.). (2006). *A communion of subjects: Animals in religion, science, and ethics*. New York: Columbia University Press. ISBN:9780231509978.
Waller, N. L., Gynther, I. C., Freeman, A. B., Lavery, T. H., & Leung, L. K. P. (2017). The Bramble Cay melomys Melomys rubicola (Rodentia: Muridae): a first mammalian extinction caused by human-induced climate change? *Wildlife Research, 44*(1), 9–21. https://doi.org/10.1071/WR16157.
Wrenn, C. L. (2011). Resisting the globalization of speciesism: Vegan abolitionism as a site for consumer-based social change. *Journal for Critical Animal Studies, 9*(3), 9–27.
Wrenn, Corey (2017) Toward a vegan feminist theory of the state. In: Nibert, David, ed. *Animal oppression and capitalism*. Praeger Press: Santa Barbara.

Vegan Social Justice

▶ The Vegan Food Justice Movement

Vertical Farming

▶ Future Foods for Urban Food Production

Vertical Integration

▶ Collaborative Climate Action

Virtual City

▶ Digital Twin and Cities

Virtual Reality

▶ Emerging Concepts Exploring the Role of Nature for Health and Well-Being

Voluntary Agreements

▶ Voluntary Programs for Urban and Regional Futures

Voluntary Approaches

▶ Voluntary Programs for Urban and Regional Futures

Voluntary Environmental Governance Arrangements

▶ Voluntary Programs for Urban and Regional Futures

Voluntary Environmental Programs

▶ Voluntary Programs for Urban and Regional Futures

Voluntary Programs

▶ Voluntary Programs for Urban and Regional Futures

Voluntary Programs for Urban and Regional Futures

Jeroen van der Heijden
School of Government, Victoria University of Wellington, Wellington, New Zealand
School of Regulation and Global Governance, Australian National University, Canberra, Australia

Synonyms

Advanced environmental practices; Alternative compliance tools; Industry self-regulation; Non-state market-driven (NSDM) governance systems; Private regulation; Private standards; Self-regulation; Voluntary agreements; Voluntary approaches; Voluntary environmental governance arrangements; Voluntary environmental programs; Voluntary programs; Voluntary regulation; Voluntary regulatory regimes

Definition

Voluntary programs have rapidly gained prominence to accelerate urban climate action. Effectively, voluntary programs are rule systems that set requirements for their participants, but they do not have to force of law behind them. The challenge is to develop programs that combine requirements (rules) and incentives (rewards) in such a way that they achieve voluntary compliance. Lessons from voluntary programs for urban climate action may help to develop voluntary programs to tackle other urban challenges too.

Taking fossil-fuelled cars off city streets by banning trips that can relatively easily be done by other modes of transport considerably reduces global carbon emissions (Hickman & Banister, 2014). Thus, it seems logical if city governments would say:

> Because of the climate emergency, driving your children to school in a fossil fuel car is prohibited within our city's boundaries. This holds also for commuting to work, doing groceries, and visiting relatives and friends. Non-compliance will be severely penalised. Effective immediately.

There are many positive things to say about such a direct, government-led intervention: It applies to everyone, it is clear (at least at first glance), it is a strong response to an existential threat (climate change), and because governments can enforce the intervention with the full force of law it stands a good chance to achieve its effects.

Unfortunately, one of the major challenges to accelerate the transition towards sustainable, inclusive, and healthy urban and regional futures is the development and implementation of these kinds of mandatory requirements (laws, regulations, rules, codes, etc.). The political process of developing them involves lengthy debates about their need or lack thereof; it may be hijacked by interest groups; and, it may get stuck in discussions about the best way of incentivizing people to follow the mandate. Likewise, the process of implementing them may be hampered by all sorts of exemptions that need to be made to people who have a good reason to use their car for the sort of trips the mandate seeks to ban. Finally, governments may lack the capacity and expertise to enforce the mandate; and, it may take some (or even a lot of) time before the results of the mandate become visible, which may trigger a process of amending the mandate or terminating it altogether under new political leadership (Van der Heijden, 2014).

Luckily, there is an alternative to mandatory interventions: voluntary programs. Like mandatory interventions, voluntary programs are rule-systems that set rules that need to be met, and rewards for compliance with these rules or penalties for non-compliance. The main differences are (i) that voluntary programs cannot be enforced with the force of law, and (ii) that they do not apply to everyone (they require voluntary opting in) (Coglianese & Nash, 2009). Because of their promises, governments and others (including firms and NGOs) have embraced voluntary programs as an alternative to mandatory interventions.

Some Examples of Voluntary Programs for Urban and Regional Futures

When overlooking the broad literature on voluntary programs, three broad types stand out: certification and rating, innovative forms of financing, and action networks (Van der Heijden, 2017). Among the best-known examples of certification and rating are the international building assessment programs LEED (Leadership in Energy and Environmental Design) and BREEAM (BRE Environmental Assessment Methodology). These voluntary programs seek to improve the environmental sustainability of buildings by making visible their environmental performance. To do so, buildings are assessed against a set of sustainability requirements and their performance is certified in a particular class. In this way buildings can be compared according to their relative score. These programs are widely implemented around the globe and normally seek to push the performance of buildings beyond national construction codes. For building developers, building owners and building occupants, there are clear incentives to have their buildings certified under such voluntary programs: They can make visible their environmental credentials, and they may yield higher rents and higher sales prices than non-certified buildings (Eichholtz et al., 2010).

An example of innovative forms of finance is 1200 Buildings in Melbourne, Australia. The voluntary program seeks to help commercial property owners to access long-term loans for energy retrofits and upgrades. To this end, the government of Melbourne has entered into agreements with finance providers. For finance providers, property owners may be too risky clients. Therefore, the city government (as a very low-risk client) borrows funds and supplies these to property owners. Funds are paid back by property owners to the government through a newly introduced property tax. The program sets strict criteria for property owners who want to join the program. Among others, it requires them to commit to improving the energy efficiency of their buildings by 38 per cent by 2020 (compared to 2010). This requirement relates to the city's ambitions of reducing resource consumption and carbon emissions (Van der Heijden, 2015).

Examples of action networks are city networks such as ICLEI and C40s that seek to bring together cities from around the world to learn from each other and experiment with novel forms of climate action. Likewise, the Transition Town Network is a voluntary program that provides support to urban transition initiatives at the level of streets, neighborhoods, whole cities, and communities. These initiatives range from local energy production and urban agriculture to the improvement of health and education at city level. Yet another example is the Rockefeller Foundation's 100 Resilient Cities program that has merged into the Resilient Cities Network. This voluntary program provided city governments funds to hire a "resilience officer" to create and implement urban (climate) resilience actions and share lessons about local experiences with participating cities and others (Van der Heijden, 2017).

What Are the Main Promises of Voluntary Programs?

Voluntary programs appear ideally suited for the governance of urban and regional futures: They provide a vehicle for taking action in situations in which it is too costly or difficult to implement mandatory interventions, for instance because of political unwillingness (Darnall & Carmin, 2005). They also provide an opportunity for showcasing and marketing desired "beyond compliance" behavior (Saurwein, 2011). Further, voluntary programs open up opportunities for governments

V

to collaborate closely with citizens and firms, without forcing the latter to be involved in such collaborations or join the resulting programs (Hofman & De Bruijn, 2010).

Another promise of voluntary programs is that they move away from the traditional negative incentives that often come with mandatory interventions (i.e., fines or even imprisonment for non-compliance). Rather, voluntary programs often focus on the economic or reputational rewards that come to those that join the programs and meet their requirements. Reaping the positive advantages of a voluntary program may be more attractive to people and firms than seeking to avoid the negative consequences of not complying with mandatory interventions. For example, owners and tenants of LEED or BREEAM-certified buildings often use their buildings' environmental credentials to add to their firms' overall social corporate responsibility images (Van der Heijden, 2017).

A further promise addressed in the literature is that voluntary programs can be developed and implemented with or without government involvement. Government involvement brings the advantage of providing legitimacy to voluntary programs in the eyes of the wider public, and the government may carry (part of) the costs of developing and implementing voluntary programs (Delmas & Terlaak, 2001). Yet, the advantage of purely private programs is their explicit "by the industry, for the industry" branding. This may get on-board those individuals or firms that are critical about the role of governments in the transitioning towards regional and urban futures. Last, but certainly not least, voluntary programs may be used by local governments (solely or in collaboration with citizens and firms) to bypass higher levels of government and take urban climate action before a higher level of government introduces a mandatory requirement (Parnell, 2016).

What Are the Main Challenges for Voluntary Programs?

For voluntary programs to be effective, at least three challenges need to be overcome (Potoski & Prakash, 2009; Van der Heijden, 2017). First, the program needs to attract a meaningful number of participants. Obviously, without participants, a voluntary program will not be able to see its good intentions materialize – without participants, there is no one to 'walk the talk'. A central challenge for voluntary programs is to find a good balance between program criteria (i.e., the rules participants need to meet) and program rewards. If the rules are too demanding, or the rewards too few, the program will not be attractive to prospective participants.

Second, the requirements of the program should not be too low. If the program does not set strict enough requirements, it will not make a meaningful contribution to achieving a societal desirable end. More problematically, the program may give the illusion of something being done to achieve that societal desirable end, which may hamper or even stall the development of more meaningful programs (or mandatory requirements). Also, particularly for prospective participants who want to be seen as leaders in their cohort, or at least want to be seen as contributing to a meaningful goal, too low requirements will not allow them to use their participation in the program for reputational purposes.

Third, not only must programs attract participants and set meaningful requirements, they also must prevent that some participants join the program but not comply with its requirements. To this end, some form of monitoring and enforcement is required. Yet, too strict monitoring and enforcement may go against the "feel" of voluntariness of the program for some prospective participants. Also, monitoring and enforcement comes with additional costs, which will (likely) be borne by program participants (and their clients or stakeholders). Finally, with monitoring and enforcement the question arises who is in the best position to carry out these processes: program participants, the program administrators, a third party, or even the government?

Conclusion

Over the last three decades, or so, voluntary programs have often been suggested as an alternative to mandatory (government-led) interventions to accelerate the transition towards sustainable, inclusive, and healthy urban and regional futures. At first glance, voluntary programs offer many opportunities and advantages over mandatory interventions at city and regional levels. At second glance, however, voluntary programs come with their own design and implementation challenges. Particularly, the complex interaction of program participation criteria, program rewards, and program monitoring and enforcement prove a conundrum.

When overlooking the literature on voluntary programs, it becomes clear that they are no panacea for overcoming the shortfalls of mandatory interventions. Positive and negative experiences with voluntary programs for urban and regional futures have been reported, and systematic reviews of the academic literature indicate that the interaction of a voluntary program's design, implementation, and context matter for its performance (see among others, Coglianese & Nash, 2009; Delmas & Terlaak, 2001; Potoski & Prakash, 2009; Saurwein, 2011; Van der Heijden, 2017). For academics that is gentle way of saying: it remains to be seen whether voluntary programs will deliver on their promises in the long run.

Cross-References

- ▶ Collaborative Climate Action
- ▶ Circular Economy Cities
- ▶ City Visions: Toward Smart and Sustainable Urban Futures
- ▶ Green Cities in Theory and Practice
- ▶ Green Economy Policies to Achieve Water Security
- ▶ Urban and Regional Leadership
- ▶ Urban Policy and the Future of Urban and Regional Planning in Africa

References

Coglianese, C., & Nash, J. (2009). Government clubs: Theory and evidence from voluntary environmental programs. In M. Potoski & A. Prakash (Eds.), *Voluntary programs* (pp. 231–257). Cambridge, MA: MIT Press.

Darnall, N., & Carmin, J. (2005). Greener and cleaner? The signaling accuracy of U.S. voluntary environmental programs. *Policy Sciences, 38*(2–3), 71–90.

Delmas, M. A., & Terlaak, A. K. (2001). A framework for analyzing environmental voluntary agreements. *California Management Review, 43*(3), 44–62.

Eichholtz, P., Kok, N., & Quigley, J. (2010). Doing well by doing good? Green office buildings. *American Economic Review, 100*, 2492–2509.

Hickman, R., & Banister, D. (2014). *Transport, climate change and the City*. London: Taylor & Francis.

Hofman, P., & De Bruijn, T. (2010). The emergence of sustainable innovation: Key factors and regional support structures. In J. Sarking, J. Cordeiro, & D. Vasquez Bruzt (Eds.), *Facilitating sustainable innovation through collaboration* (pp. 115–133). Amsterdam: Springer.

Parnell, S. (2016). Defining a global urban development agenda. *World Development, 78*(February), 529–540.

Potoski, M., & Prakash, A. (2009). *Voluntary programs: A club theory perspective*. Cambridge: MIT Press.

Saurwein, F. (2011). Regulatory choice for alternative modes of regulation: How context matters. *Law & Policy, 33*(3), 334–366.

Van der Heijden, J. (2014). *Governance for urban sustainability and resilience: Responding to climate change and the relevance of the built environment*. Cheltenham: Edward Elgar.

Van der Heijden, J. (2015). Voluntary programmes for building retrofits: Opportunities, performance and challenges. *Building Research & Information, 43*(2), 170–184.

Van der Heijden, J. (2017). *Innovations in urban climate governance: Voluntary programs for low carbon buildings and cities*. Cambridge: Cambridge University Press.

Voluntary Regulation

▶ Voluntary Programs for Urban and Regional Futures

Voluntary Regulatory Regimes

▶ Voluntary Programs for Urban and Regional Futures

Vulnerability to Food Insecurity Among the Urban Poor in Sri Lanka: Implications for Policy and Practice

Thilini Navaratne[1] and Sylvia Szabo[2]
[1]Department of Business Economics, Faculty of Management Studies and Commerce, University of Sri Jayewardenepura, Nugegoda, Sri Lanka
[2]Department of Social Welfare and Counselling, University of Seoul, Seoul, South Korea

South Asia has added an estimated 130 million people to its towns and cities, since the turn of the century (Roberts 2018). Sri Lanka has been undergoing a major demographic transformation resulting in the movement from a predominantly rural to a mainly urban population (Weeraratne 2016). In 2020 it was recorded that there were approximately 4.1 million people living in urban areas in Sri Lanka. This accounts to around 19% of the overall population and the annual urban population continues to growth at a rate of 1.2% (The World Bank 2021). However, scholars highlight that these statistics do not reflect the true level of urbanization in Sri Lanka and that it is considerably understated (Weeraratne 2016). The manner in which urbanization is taking place in Sri Lanka has been classified as "messy urbanization" due to the pressures and congestion it has created in the main cities of the country. Colombo, Galle, and Kandy are the main cities, and these cities are considered as an urbanization belt (FAO 2018).

As the country faces rapid urban growth, a plethora of challenges arise in relation to households' food security. In fact, vulnerability to food security in the urban areas is considered to be more severe when compared to the rural areas (Kalansooriya et al. 2020). The ever-increasing cost of living in the urban areas, especially in relation to shelter, transport, and healthcare facilities create adverse impacts on the accessibility to food, which in turn results in higher prevalence of hunger, obesity, malnutrition, and unhealthy diets among the urban population (Kalansooriya et al. 2020). More importantly, the urban dwellers residing in informal settlements are considered a neglected community and their access to basic human needs and their fulfillment is at a stake. When families do not have sufficient access to affordable and nutritious food there is a risk of being trapped in poverty as food insecurity has negative consequences on many other aspects of well-being (Gunawardhana and Ginigaddara 2021).

A study conducted in urban slums in the Colombo municipality revealed that around 72% of the households were highly food insecure as per the Household Food Insecurity Access Score (HFIAS) (Gunawardhana and Ginigaddara 2021). The same study indicated that the use of low-quality food, purchasing food items on credit from grocery shops, and relying on less expensive foods were the most widely used coping strategies for food insecurity among urban slum dwellers. Another study which was carried out in Gampaha, the second highly populated district after Colombo, revealed that approximately one in three primary school children suffered from malnutrition. Furthermore, the results highlighted a double burden of malnutrition within households, with co-existence of child thinness and maternal overweight and obesity (Shinsugi et al. 2019). Another study conducted in the city of Kandy revealed that in many cases there existed intra-household food insecurity among the urban dwellers. Results of this study showed that there was a significant difference among calorie adequacy ratios of fathers, mothers, and children. Fathers had the highest mean calorie adequacy ratios and children had the lowest mean calorie adequacy ratios (Rathnayake and Weerahewa 2002).

The global pandemic has further exacerbated the existing vulnerabilities among the urban dwellers in Sri Lanka. Many scholars put forth that the rising food prices in the country has a significant influence on the food security status of the country. Colombo Consumer Price Index (CCPI) based Inflation increased to 6.0% in August 2021. This increase was mainly due to price increases observed in items of both Food and Non-food categories. Monthly changes of

Food and Non-food categories recorded at 0.21 %" and 0.06 %, respectively. Accordingly, within the Food category, prices of fresh fish, fresh fruits, sugar, and chicken increased, whereas prices of rice and coconut recorded a decrease (CBSL 2021). With lockdowns being imposed from time to time and with the economy slowly slipping into an economic crisis, the exchange rate has depreciated considerably. This unfavorable exchange rate movement has negatively affected the essential commodity exports of the country. As the urban communities are highly dependent on the markets for securing access to their food supplies, as opposed to the rural communities, the urban dwellers and especially the urban poor households are the worst hit by this threat (Development Asia 2021).

The entire food system of the country has been affected by the pandemic. Scholars highlight that the pandemic has created shocks on global as well as national food systems (Schmidhuber et al. 2020). The impacts of the pandemic can be felt on both supply and demand sides and will be felt differently across specific markets and communities. However, it is evident that the urban poor will be one of the negatively affected and thus, policymakers should pay special attention to this challenge, if they are to bring back the country into the economic growth path. In particular the urban poor, who are housed in informal settlements and those engaged in informal labor market needs extra attention. They may not be covered by any of the social protection schemes or COVID-19 relief schemes. Hence the first policy focus should be to promptly identify the most at risk urban dwellers and to enroll them into a special social protection scheme or program. Thereafter, medium- and long-term planning should be undertaken in the areas of price control, urban farming, and initiatives aiming at overall increase of domestic food production.

References

Central Bank of Sri Lanka [CBSL]. (2021). *CCPI based Inflation increased to 6.0 per cent in August 2021.* Retrieved September 10, 2021, from https://www.cbsl.gov.lk/en/news/inflation-in-august-2021-ccpi

Development Asia. (2021). *Building a pandemic-resilient food system in Sri Lanka.* Retrieved September 12, 2021, from https://development.asia/insight/building-pandemic-resilient-food-system-sri-lanka

FAO. (2018). *Food security and nutrition in City region food system planning.* Colombo: Policy brief. Food and Agriculture Organization, United Nations.

Gunawardhana, N. L., & Ginigaddara, G. A. S. (2021). Household food security of urban slum dwellers: A case study in Colombo municipality, Sri Lanka. *Journal of Food Chemistry and Nanotechnology, 7*(2), 34–40.

Kalansooriya, C. W., Gunasekara, W. G. V., & Jayarathne, P. G. S. A. (2020). Food security in urban households: The role of women in an Asian context. *Economy, 7*(1), 11–18.

The World Bank. (2021). *United Nations population division's world urbanization prospects.* Retrieved September 10, 2021, from https://data.worldbank.org/indicator/SP.URB.GROW?locations=LK

Rathnayake, I., & Weerahewa, J. (2002). An assessment of intra-household allocation of food: A case study of the urban poor in Kandy. *Sri Lankan Journal of Agricultural Economics, 4*(1381-2016-115737), 95–105.

Roberts, M. (2018). Urban growth in South Asia: A view from outer space. In *Alternative approaches in macroeconomics* (pp. 269–302). Cham: Palgrave Macmillan.

Shinsugi, C., Gunasekara, D., Gunawardena, N. K., Subasinghe, W., Miyoshi, M., Kaneko, S., & Takimoto, H. (2019). Double burden of maternal and child malnutrition and socioeconomic status in urban Sri Lanka. *PLoS One, 14*(10), e0224222.

Schmidhuber, J., Pound, J., & Qiao, B. (2020). *COVID-19: Channels of transmission to food and agriculture.* Covid.

Weeraratne, B. (2016). *Re-defining urban areas in Sri Lanka* (Working paper series no. 23). Institute of Policy Studies of Sri Lanka.

V

W

Wadi Sustainable Agriculture Model, The

Ganesh Keremane
Adelaide, South Australia

Synonyms/Related Terms

Agroforestry; Alternative agriculture; Community-supported agriculture; Holistic management; Integrated farming systems; Tree-based farming

Definition

Sustainable agriculture concept is very vague and ambiguous in its meaning probably because agriculture is practiced in so many climates and different cultural contexts. Sustainable agriculture is an umbrella term and includes different approaches (Hansen 1996); as such, there is no one correct, universal definition, rather it has to adapt to the respective context (Pretty 2008). In the context of this chapter which discusses the *wadi* model, "sustainable agriculture is one that, over the long term, enhances environmental quality and the resource base on which agriculture depends; provides for basic human food and fibre needs; is economically viable, and enhances the quality of life for farmers and society as a whole" (Kluson n.d.).

Wadi is an integrated farming system comprising agricultural, horticultural, and forestry components implemented on small plots of underutilized farmland. Since this model was first designed and implemented in the state of Gujarat, India, the model assumed the name *wadi* – a Gujarati term used to describe a small plot of land allotted for growing fruits, vegetables, and flowers (Shah 2005).

Introduction

World agriculture has undergone a major transformation during the past four decades. Production per unit area of land has increased as a result of the adoption of conventional agriculture model characterized by monocultures of crops, use of fertilizers for crop nutrition and agrochemicals for plant protection, and mechanization for farm operations (Knorr & Watkins, 1984 as cited in Hansen 1996). While these practices may yield short-term economic gains, the negative impacts of these practices on the natural environment and the long-term ecological sustainability are well-known (Plumecocq et al. 2018; German et al. 2016), thereby forcing farmers to think about alternatives to conventional agriculture. Alternative approaches to agriculture most commonly referred to as "alternative agriculture" are often likened with sustainable agriculture (Dahlberg 1991). Then what is sustainable agriculture?

R. C. Brears (ed.), *The Palgrave Encyclopedia of Urban and Regional Futures*,
https://doi.org/10.1007/978-3-030-87745-3

Sustainable Agriculture

Interest in the concept of sustainability in general and sustainable agriculture, in particular, has been growing over the years out of the concern for the fast-depleting natural resources. The concept generated a lot of international debate following the publication of the Brundtland Report by the World Commission on Environment and Development (1987) which widely broadcast the use of the term sustainable development (Dahlberg 1991). Since then, people have tried to define the concept in many different ways resulting in an array of definitions of sustainable agriculture. The ambiguity is probably because agriculture is practiced in so many climates and different cultural contexts across the world; also people across the world perceive "agriculture" and "sustainability" differently (Smith and McDonald 1998). There have been many attempts to try and find one, universally accepted definition, but due to the complex and contested nature of the concept, these efforts were doomed to failure (Velten et al. 2015). Therefore, the questions to be considered are: What constitutes sustainable agriculture? And, how can it be achieved?

All the discussions and implementation of the idea of sustainable agriculture rests on the Brundtland principle which states that we must meet the needs of the present without compromising the ability of future generations to meet their own needs (Brundtland report, 1987). The Food and Agriculture Organization (FAO) of the United Nations states that "sustainable agriculture must meet the needs of present and future generations for its products and services while ensuring profitability, environmental health, and social and economic equity. . .It must also strike a balance between protecting agro-ecosystems and meeting society's growing needs by offering decent and resilient livelihoods for rural populations" (FAO 2014, p. 12). In essence, sustainable agriculture means giving equal importance to long-term stewardship of both natural and human resources to attain economic gain (Brodt et al. 2011). According to Christen (1996), the important attributes of sustainable agriculture include ensuring intergenerational equity; preserving the resource base of agriculture and preventing adverse environmental externalities; protecting biological diversity; assuring the economic viability of agriculture; enhancing job opportunities in farming and preserving local rural communities; achieving food security; and finally contributing to global sustainable development.

Sustainable agriculture is not a single, well-defined end goal, and there is no need to seek the one universal definition; instead, we need to adapt the definition of sustainable agriculture to the respective context. In principle, sustainable agriculture is a system-describing concept incorporating the key elements of achieving yield stability, reducing input use, conserving natural resources, and enhancing the environment (Siebrecht 2020). Pretty (2008, p.447) opines that "Sustainability in agricultural systems incorporates concepts of both resilience (the capacity ofsystems to buffer shocks and stresses) and persistence (the capacity of systems to continue over long periods), and addresses many wider economic, social and environmental outcomes. Accordingly, Hansen (1996) classified the different interpretations of sustainable agriculture into two broad approaches (Table 1).

This chapter discusses one such adaptation called the *wadi* sustainable agriculture model. Adapting Hansen's classification illustrated in Table 1, the *wadi* model can be seen both as an approach to agriculture and as a property of agriculture because *wadi* by design is a set of strategies developed to enable small and marginal landholders to realize a set of goals.

***Wadi* Sustainable Agriculture Model, The, Table 1** Conceptual approaches to agricultural sustainability

Sustainability as an approach to agriculture	Sustainability as a property of agriculture
Sustainability as an ideology	Sustainability as the ability to fulfil a set of goals
Sustainability as a set of strategies	Sustainability as the ability to continue

Source: Hansen (1996)

Wadi Sustainable Agriculture Model

A *wadi* is a holistic approach to dryland farming that takes into account all aspects of rural life. It is a strategy that encourages efficient use of resources available within the immediate production environment and adopting technologies that are suited to the local environments (Francis and Hilderbrand 1989) to satisfy a set of goals including maintenance or enhancement of the natural environment, provision of human food needs, economic viability, and social welfare (Hansen 1996; Smith and McDonald 1998).

The Genesis of the *Wadi* Model

For centuries, farmers, especially in the developing world, have been practicing agriculture using traditional and indigenous methods. As such, these farmers have an intimate knowledge of their surroundings, which as Altieri (2004) argues needs to be gathered by having a dialogue with the traditional farmers because any strategy developed to make agriculture sustainable has to ensure that people with the local knowledge participate in development strategies. This is exactly how the *wadi* model was designed and implemented for the first time in southern Gujarat, India.

The idea of *wadi* evolved through discussions with the tribal communities in the hill tracts of southern Gujarat where BAIF Development Research Foundation (BAIF), a Pune-based organization, was working intensively to find a permanent solution to the typical problems affecting most tribal areas/communities, viz., lack of livelihood options due to lack of resources ultimately leading to migration for employment. As a first step, BAIF officials assessed the traditional knowledge framework and resource management practices used by rural communities and subsequently based on scientific evidence modified and translated it into holistic development approach – one that takes into account all aspects of rural life. The model was called *wadi* because it was first developed and implemented in Gujarat, and *wadi* in Gujarati implies a small orchard (approximately 1 acre) of fruit crops. Figure 1 illustrates the conceptualization process of the *wadi* model.

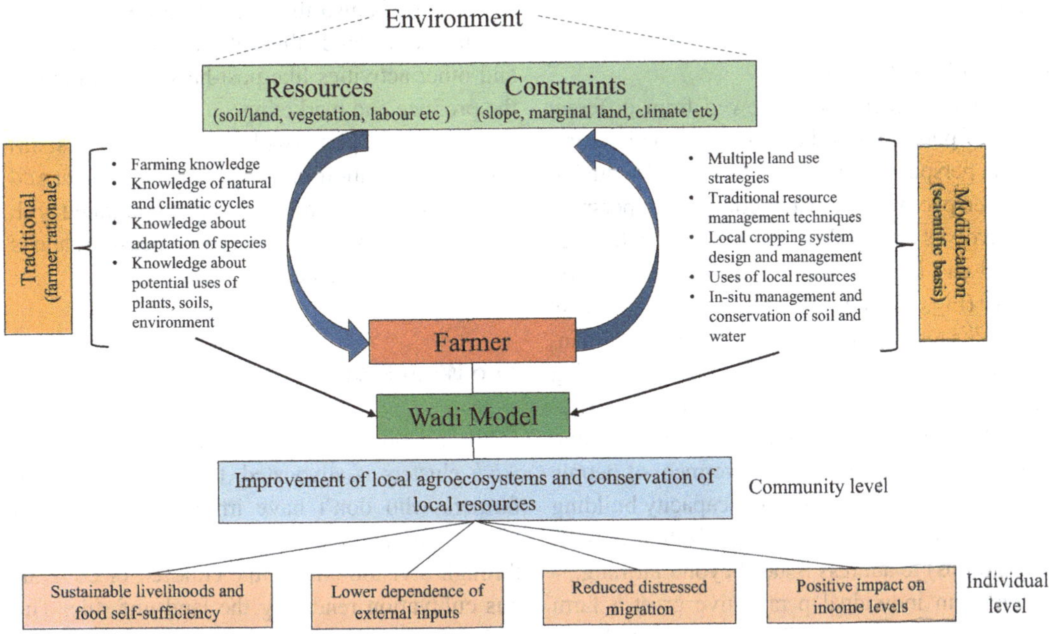

Wadi Sustainable Agriculture Model, The, Fig. 1 Conceptualizing *Wadi* model. (Source: Modified from Altieri 2004)

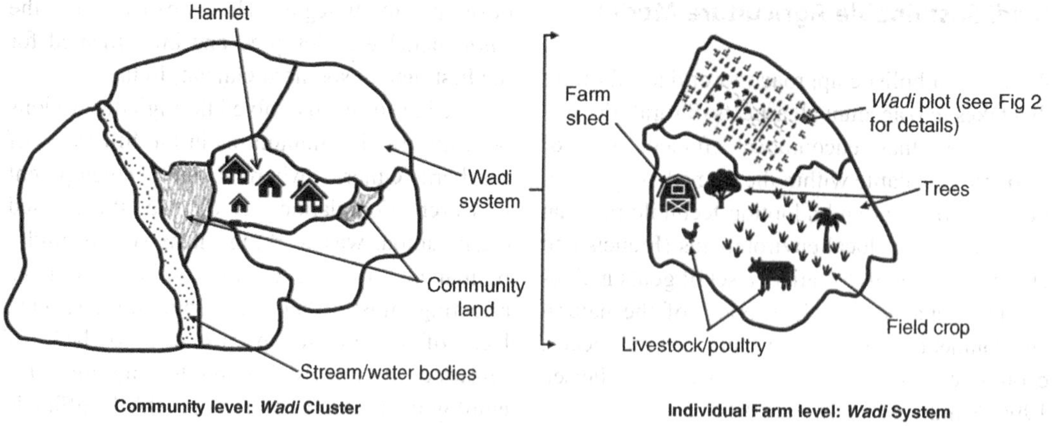

***Wadi* Sustainable Agriculture Model, The, Fig. 2** Schematic of a *wadi* cluster and *wadi* system. (Source: Sohani et al. 2011)

The first *wadi* was established in Vansda village, in Navsari district of southern Gujarat, in 1982 with the overall objective to rehabilitate the tribal communities in their natural environment by promoting sustainable livelihood programs (Tripathy 2018; Brockington et al. 2016). Since then, the model has been experimented and replicated in other parts of India dominantly inhabited by tribal population with a great degree of success (NABARD 2019).

Concept Description

The *wadi* concept can be viewed from different levels or perspectives, i.e., through a wider community perspective, and also from an individual or farm perspective. The wider community perspective involves the development of a designated area of land and its inhabitants in the form of a ***wadi* cluster** (Fig. 2). Activities at the community level (*wadi* cluster) include the natural resource management, the adoption of sustainable farming practices, and the overall socioeconomic elevation of communities. Another important feature of the *wadi* cluster is the empowerment of people through social mobilization and capacity building (e.g., formation of people's organization or self-help groups) to address issues beyond farming.

From an individual perspective or at a farm level, it is called as ***wadi* system** (a tree-based farming system) wherein the physical land unit (***wadi* plot**) interacts with other production components of the farm such as annual crop fields and livestock (Fig. 2). And the interaction (s) between the different components extends beyond just the tree-crop interface in a carryover form. For example, the fodder from the forestry tree species in the *wadi* system provides food for livestock, and the farmyard manure from the livestock is returned to the land in the form of organic manure to crops grown in interspaces. Furthermore, there is also the interaction between *wadi* and non-wadi lands through the sharing of labor and other activities like post-harvest handling of the produce and marketing.

The *wadi* plot (within the *wadi* system) involves plantation of a combination of fruit and forest trees on underutilized lands integrated with some soil and water conservation measures and is the emphasis of this chapter.

The *Wadi* Plot

While area-based treatment through developing *wadi* clusters is advocated, the focus here is on farmers who don't have irrigation facility and whose farms have either become fragmented through division among the children or continuous cultivation rendering the land unfertile. The added impact of climate change has resulted in agriculture on small farms being unviable and vulnerable to crop failure. Establishing a *wadi*

***Wadi* Sustainable Agriculture Model, The, Fig. 3** Layout of a *wadi* plot. (Source: Modified from Sawant and Ajwani (2011) and Shah 2005)

plot addresses these issues and makes farming profitable even on small farms, thereby providing a pathway out of poverty for small and marginal farmers.

Layout

A *wadi* plot, usually established on a portion of underutilized land, comprises of agricultural, horticultural, and forestry components integrated with soil and water conservation measures (Fig. 3). The *wadi* may be of any fruit crop (or a combination of tree-crops) suited to the region, planted in crop fields, with multipurpose tree species on the periphery and soil and water conservation structures (e.g., farm pond and trench-cum-bunds) built through the land (Balamatti 2013).

As mentioned earlier, a *wadi* plot is typically established on a portion (usually covering between 0.4 and 1.0 ha) of underutilized farmland with agricultural, horticultural, and forestry species. The arrangement of these species mostly centers around the horticultural component with intercrops grown in the interspaces of fruit trees and forestry species planted at relatively close spacing along the border of the plot (Fig. 3). Some key elements to be considered when establishing a *wadi* are:

- Planting grafted fruit trees in recommended spacing, usually 5 m–7 m from the plot boundary in wide-spaced rows, typically 8 m–10 m apart
- Multipurpose tree species (MPTS) planted 2 m from the boundary and all around the plot at approximately 1 m intervals
- Perimeter fencing of the plot by using thorny plants (e.g., agave, cacti) to protect the trees from livestock, particularly during the early years of establishment
- Using the interspaces between fruit trees to cultivate annual subsistence and/or cash crops
- Constructing physical soil and water conservation measures (e.g., trenches, bunding) and low-cost water harvesting structures (e.g., farm pond)

Components of a Wadi Plot

Each *wadi* looks at agricultural development that bears fruit in 5 to 7 years, this being the amount of time that it takes for saplings to bear fruit, for the *wadi*s to become self-sustained, and for the

Wadi Sustainable Agriculture Model, The, Fig. 4 Components of a *wadi*

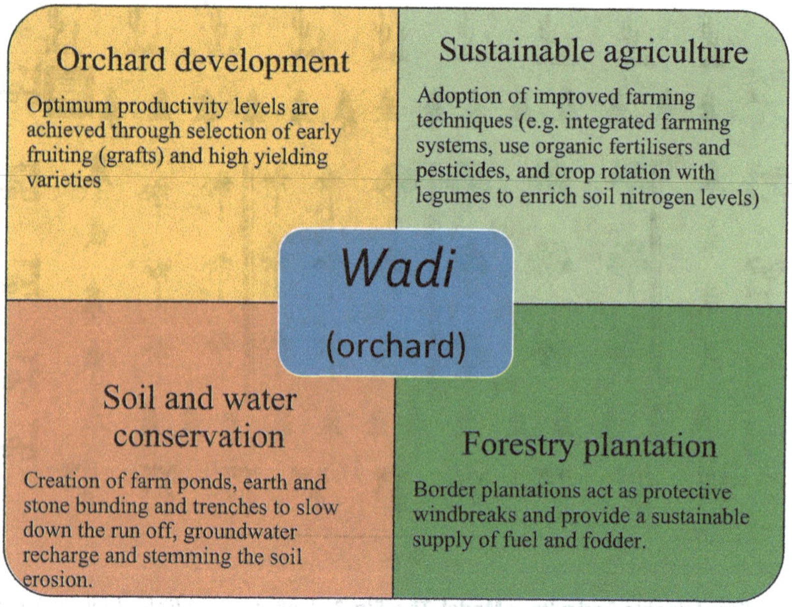

farmers to have acquired a wider basket of cultivation practices and water management. The key components of a *wadi* include (Fig. 4):

Orchard development involves planting a variety of suitable fruit trees to ensure income and livelihood security under varying climatic conditions. The fruit trees are the primary commercial crop on the plot, but they have a long gestation period. Use of grafts or other vegetative propagation practices reduces the gestation period of perennial crops to about 4 years. Also, fruit trees that are less vulnerable to climate irregularities (can grow on arid land) than conventional farming regimes should be planted. This protects the farmer against complete crop failure and provides an additional high-value income stream, thereby enhancing income and livelihood security.

Soil and water management includes the construction of farm ponds, farm bunds, and trenches to conserve water and reduce soil erosion. Considering that the *wadi* plots are established on underutilized drylands, water security is a prerequisite for ensuring the survival of the *wadi* through the harsh summers. The soil and water conservation measures/structures help minimize soil erosion and retain moisture in situ as well as support irrigation.

Sustainable agriculture component involves intercropping seasonal crops between trees for supplementary income and ensuring crop rotation with one leguminous crop to enrich soil nitrogen levels. Also, adopting sustainable practices such as composting, green manuring, and mulching improves and maintains soil health for sustained productivity levels over the years. Given that the main component of the *wadi* plot – fruit trees – has a long gestation period, raising intercrops with improved cultivation practices provides an alternative source of income.

Fencing and border plantation are basically for providing shelter to the *wadi* and helping farmers meet their needs for fuel, fodder, and small timbers, thereby reducing the pressure on existing forests. Border plantations serve multiple functions; besides keeping grazing animals at bay, use of MPTS along the plot boundaries acts as a windbreak and also acts as a source of fodder, firewood, green manure, poles, and timber. Additionally, if timber species such as teak and bamboo are planted, they can also become a major source of income.

A *wadi* is an individual physical unit within the *wadi* cluster, and therefore it is unlikely that changes in the management of a single farm or *wadi* plot will result in detectible changes in water quality or biodiversity. But adopting an area-based approach through simultaneous planning and implementation of *wadi* plots in a cluster can lead to system-level impacts.

Key Insights

Wadi may not be the best option for irrigated fertile land where two to three crops are raised successfully every year. It is an ideal system for the land of marginal fertility and limited water. The *wadi* is a versatile and eco-friendly model designed to address the problem of sustainable livelihood and nutritional security of the rural poor while also helping to regenerate the natural resource base. The diversified nature of the *wadi* model enhances multiple livelihood opportunities, thereby opening up multiple income streams. The advantages accrued from a *wadi* can be summarized as:

- The orchard provides sustainable income every year.
- Intercropping provides earlier returns during the gestation period of fruit trees.
- The fruit and crop produce ensures food security to the family.
- Assured irrigation because of the water harvesting structures means farmers can grow two to three crops per year.
- Reduced or no use of external inputs minimizes the cost of production.
- Additional income generation through alternative livelihood activities such as vermicomposting and nursery raising.

By mapping the *wadi* concept to Hansen's classification of sustainable agriculture approaches (see Table 1), it is evident that *wadi* satisfies most of the strategies and goals associated with sustainability (see Table 2). While this characterization of the *wadi* sustainable agriculture model is based on a literal interpretation of sustainability (Hansen 1996), a more comprehensive approach should incorporate various elements/factors, viz., biophysical, economic, social, and – at varying scales – field, farm, watershed, regional, and national (Smith and McDonald 1998).

Wadi Sustainable Agriculture Model, The, Table 2 Characterization of *wadi* sustainable agriculture model

Strategies	
Self-sufficiency through preferred use of on-farm or locally available resources to purchased resources	✓
Reduced use or elimination of soluble or synthetic fertilizers	✓
Reduced use or elimination of chemical pesticides	✓
Increased or improved use of crop rotations for diversification, soil fertility, and pest control	✓
Increased or improved use of manures and other organic materials as soil amendments	✓
Increased diversity of crop (and animal) species	✓
Maintenance of crop or residue cover on the soil	✓
Reduced stocking rates for animals	NA
Goals	
Enhances environmental quality and the resource base on which agriculture depends	✓
Provides for basic human food and fiber needs	✓
Is economically viable	✓
Enhances the quality of life for farmers and society as a whole	✓

Source: Adapted from Hansen (1996)
Note: A check (✓) indicates the *wadi* system incorporates the specified strategy and goal; NA – not applicable

The *wadi* model addresses critical requirements of resource-poor farmers, such as food and employment security, increase in income, etc., and at the same time also has the potential to revitalize the rural communities through *wadi* cluster. Therefore, as more and more farmers around the world are beginning to accept or support features of the "sustainable agriculture" agenda such as preserving small and medium family farms, protecting the natural environment, and uplifting rural communities (Beus and Dunlap 1993), the *wadi* which by design encompasses these features is a perfect model for farmers especially in the arid and semi-arid tracts of the world. In India, successful implementation of the *wadi* model in Gujarat paved the way for it to be extended to other states, and today the model is being implemented in 29 Indian states covering

553,000 families through a special Tribal Development Fund (NABARD 2019).

Conclusion

Majority of farmers in the developing world tend small plots in marginal environments; therefore, diverse agricultural systems that can resist changing circumstances are extremely valuable to poor farmers (Altieri 2004). Some of the key principles that underlie such agricultural systems include species diversity, organic matter accumulation through enhanced recycling of biomass and nutrients, and minimization of resource losses through soil and water conservation measures. For centuries, farmers have developed diverse and locally adapted agricultural systems, and research conducted on traditional agriculture has shown that most indigenous modes of production exhibit a strong ecological basis and lead to the regeneration and preservation of natural resources (Denevan 2001). But what is missing is a "dialogue of wisdom" (Altieri 2004) between scientists/researchers and farmers, thereby preventing the translation of these practices into a holistic approach or strategy that is based on scientific evidence. The *wadi* model evolved from such a process and is a testament to how having a dialogue between farmers and scientists could lead to participatory planning and development of a sustainable agriculture system aimed at transforming the lives of small and marginal farmers.

In a sustainable system, agriculture has to remain the dominant land use over time, and the resource base should continually support production levels needed for economic gains or survival, which exactly is the objective of a *wadi*. It essentially is a strategy to stabilizing yields, promoting dietary diversity, minimizing the risks, and maximizing returns using low levels of technology and limited resources. The *wadi* is a sustainable solution to farming on small, underutilized plots in dry areas, but addressing watershed- or landscape-scale problems is possible only when the model is adopted at the community level in the form of *wadi* clusters.

References

Altieri, M. A. (2004). Linking ecologists and traditional farmers in the search for sustainable agriculture. *Frontiers in Ecology and the Environment, 2*(1), 35–42. https://doi.org/10.1890/1540-9295(2004)002[0035:leatfi]2.0.co.

Balamatti, A. (2013). New approaches by NGOs in augmenting rural development. In P. Nath (Ed.), *The basics of human civilization-Food, agriculture and humanity*, (Volume-I: Present scenario, pp. 287–301). New India Publishing Agency.

Beus, C. E., & Dunlap, R. E. (1993). Agricultural policy debates: Examining the alternative and conventional perspectives. *American Journal of Alternative Agriculture, 8*(3), 98–106. https://doi.org/10.1017/S0889189300005129.

Brockington, J. D., Harris, I. M., & Brook, R. M. (2016). Beyond the project cycle: A medium-term evaluation of agroforestry adoption and diffusion in a south Indian village. *Agroforestry Systems, 90*(3), 489–508. https://doi.org/10.1007/s10457-015-9872-0.

Brodt, S., Six, J., Feenstra, G., Ingels, C., & Campbell, D. (2011). Sustainable agriculture. *Nature Education Knowledge, 3*(10), 1.

Dahlberg, K. A. (1991). Sustainable agriculture – fad or harbinger? *Bioscience, 41*(5), 337–340. https://doi.org/10.2307/1311588.

Denevan, W. M. (2001). *Cultivated landscapes of native Amazonia and the Andes*. New York: Oxford University Press.

Food and Agriculture Organisation of the United Nations. (2014). *Building a common vision for sustainable food and agriculture: Principles and approaches*. Rome: FAO.

Francis, C.A. & Hilderbrand, P.E. (1989). Farming systems research/extension and the concepts of sustainability. *Agronomy & Horticulture Faculty Publications*. 558. https://digitalcommons.unl.edu/agronomyfacpub/558

German, R. N., Thompson, C. E., & Benton, T. G. (2016). Relationships among multiple aspects of agriculture's environmental impact and productivity: A meta-analysis to guide sustainable agriculture. *Biological Reviews, 92*(2), 716–738. https://doi.org/10.1111/brv.12251.

Hansen, J. W. (1996). Is agricultural sustainability a useful concept? *Agricultural Systems, 50*(2), 117–143. https://doi.org/10.1016/0308-521X(95)00011-S.

Kluson, R. A. (n.d.). *Sustainable agriculture: Definitions and concepts [Fact sheet]*. Institute of Food and Agricultural Sciences, University of Florida. http://sfyl.ifas.ufl.edu/sarasota-docs/ag/SusAgFAQ.pdf.

National Bank for Agriculture and Rural Development. (2019). *2018–2019 Annual Report*. Mumbai: NABARD.

Plumecocq, G., Debril, T., Duru, M., Magrini, M. B., Sarthou, J., & Therond, O. (2018). The plurality of values in sustainable agriculture models: Diverse

lock-in and coevolution patterns. *Ecology and Society, 23*(1), 21. https://doi.org/10.5751/ES-09881-230121.
Pretty, J. (2008). Agricultural sustainability: concepts, principles and evidence. *Philosophical Transactions of The Royal Society B-Biological Sciences, 363* (1491), 447–465. https://doi.org/10.1098/rstb.2007.2163.
Sawant, Y. G., & Ajwani, S. (2011, November). BAIF's *wadi* programme: A case study on a comprehensive model for sustainable livelihood. In *Proceedings of the 2nd international conference on sustainability: People, plant and prosperity* (pp. 203–225). Shillong: Rajiv Gandhi Indian Institute of Management.
Shah, D. (2005). *Sustainable Tribal Development Model Case of Wadi*. Occasional paper 43. Department of Economic Analysis & Research, National Bank for Agriculture & Rural Development, Mumbai.
Siebrecht, N. (2020). Sustainable agriculture and its implementation gap – Overcoming obstacles to implementation. *Sustainability, 12*(9), 3853. https://doi.org/10.3390/su12093853.
Smith, C. S., & McDonald, G. T. (1998). Assessing the sustainability of agriculture at the planning stage. *Journal of Environmental Management, 52*(1), 15–37. https://doi.org/10.1006/jema.1997.0162.
Sohani, G., Desai, J., & Sawant, Y. (2011). *Wadi programme: A tree-based farming system user guide* (5th ed.). Pune: BAIF Development Research Foundation.
Tripathy, S. (2018). Tribal development through horticultural plantations under *WADI*. *Horticulture International Journal, 92*(3), 90–93. https://doi.org/10.15406/hij.2018.02.00033.
Velten, S., Leventon, J., Jager, N., & Newig, J. (2015). What is sustainable agriculture? A systematic review. *Sustainability, 7*(6), 7833–7865. https://doi.org/10.3390/su7067833.
World Commission on Environment and Development. (1987). *Our common future*. Oxford: Oxford University Press.

Walkability

▶ Hidden Enemy for Healthy Urban Life
▶ Walkable Access and Walking Quality of Built Environment

Walkable Access

▶ Walkable Access and Walking Quality of Built Environment

Walkable Access and Walking Quality of Built Environment

A Case Study of Englewood, Chicago City Metropolitan

Aynaz Lotfata
Department of Geography, Chicago State University, Chicago, IL, USA

Synonyms

Walkable access; Walkability; Walking; Pedestrian density flow; 15-Minute city; Socially vulnerable neighborhoods; Spatial equity; Urban design; Sustainable city; Health environment

Definitions

Enhancing walkability provides a holistic solution to a variety of urban problems. Walkability is a quantitative and qualitative measurement of how inviting or un-inviting an area is to pedestrians. Adding a quantitative approach to qualitative research can provide a more holistic (and quantifiable) view of the built environment on a micro-scale. New Urbanism practices have emphasized the principle of walkability and motivating walking behavior as crucial for environmental and social sustainability. Walkable access is the primary metric to enhance the neighborhood's livability while human activities are minimized locally during urban uncertainties. The walking accessibility enhances neighborhood diversity, participation, intergenerational accessibility, comfort, environmental integrity, and economic viability. This chapter highlights the co-existence of "walkability" and "walking access" measures in urban practice and research to create resilient and sustainable neighborhoods. The "walkable access" term used in this study addresses joint measures. Equity in walkable access to urban services should be studied as a critical element of social sustainability and an essential attribute of healthy and livable urban neighborhoods.

Introduction

Does walkability matter? When is the built environment walkable? What is the primary trigger of walking? This study aims to answer these questions by testing the walking quality of one of the most socioeconomically vulnerable neighborhoods of the Chicago Metropolitan Area. The primary concern is to create a livable and safe place for different racial groups (Stratford 2016; Giap et al. 2014). The walkable access gained momentum during the COVID-19 pandemic, while gentrification is a growing concern in the walkable environment that may contribute to socioeconomic inequity in walking access to urban facilities (Adkins et al. 2017). The walkable access is the primary metric to enhance the neighborhood's livability while human activities minimized to local scale during the COVID-19 pandemic, and existing inequity in walking access contributes to spatial injustice (Su et al. 2019; Wang and Yang 2019; Ruiz-Padillo et al. 2018; Saelens et al. 2003). The walking accessibility enhances neighborhood diversity, participation, intergenerational accessibility, comfort, environmental integrity, and economic viability (Stratford et al. 2019). Equity in walking access to urban services should be studied as a critical element of social sustainability and an essential attribute of healthy and livable urban neighborhoods. The urban services in the 400-m distance (0.25 miles) to residential lands contribute to an active and healthier lifestyle (Gutiérrez and García-Palomares 2008; Hsiao et al. 1997). Specifically, immediate access to urban services gets priority for the elderly, people with disabilities, and people without private vehicles.

The walkable access to urban services (grocery, restaurants, medical centers, etc.) is one primary indicator among frequently used indicators, as connectivity, suitability, serviceability, and perceptibility in assessing walkability. Some researchers identify the walkable accessibility based on walking as walking for groceries, walking for transportation, etc. Different walking goals are weighted in the walking indices by variable weights, with home and grocery weighting higher (e.g., Shashank and Schuurman 2019; Brownson et al. 2009). Meanwhile, to evaluate how walkable a place is, it is essential to test the quality of the built environment beyond the distance to urban facilities such as enclosure and imageability (Blečić et al. 2020). The walking access spatial distribution of urban amenities may not necessarily build a suitable environment for walking. Recently, walkability has been debated within spatial justice and environmental justice literature (e.g., Scott 2021; Telega et al. 2021; Kim et al. 2020). Kevin Lynch (1960) and Jane Jacobs (1961) discussed the relations among walkability, quality of life, and spatial justice.

Most studies in walkability are published in medical and health fields (e.g., Shashank and Schuurman 2020; Su et al. 2017; Zhao and Chung 2017; Coffee et al. 2013; Andrews et al. 2012). From a health perspective, urban services in walking access contribute to shaping a healthier lifestyle. Walking access to urban facilities encourages physical activities associated with lower cardiometabolic risk (Coffee et al. 2013), lowering body mass (Antonakos et al. 2020; Hoenher et al. 2011), enhancing mental health (Wang et al. 2019), congestion (NICE 2012), and increasing creativity (Oppezzo and Schwartz 2014). In addition, the walking access quality of the neighborhood reduces vehicle-oriented transportation and decreases the emission of air pollutants (James et al. 2017). Nevertheless, displacement of the elderly, minorities, and people with no high school diploma occurs in walkable neighborhoods. These groups of people are isolated in non-walkable environments. Some do not have access to private vehicles. While their mobility is restricted, walking access to public facilities gets critical importance (Bereitschaft 2017a).

Most quantitative and GIS-based studies examine the density of pedestrians, pedestrian network, and connectivity and their distribution in the structure of cities (e.g., Telega et al. 2021; Keyvanfar et al. 2018; Walkscore.com 2010). The limited number of studies take the qualitative and quantitative approaches in micro-scale to study the quality of a walkable environment (e.g., Bereitschaft 2017). In this paper, we used the quantitative-qualitative method to study the walkable environment. The walkable access was quantified using pedestrian-based network analysis, and the quality of walkability was interpreted using a walking survey. Five street-scale indicators are used to describe the quality of the walkable environment: façade continuity, transparency, visual richness/diversity, human scale (e.g., urban furniture), and those spatial/physical features (tidiness and safety) that might harm the streetscape environment (Porta and Renne 2005). Ewing and Handy (2009) used the imageability, enclosure, human scale, transparency, and complexity attributes to illustrate the quality of walking sidewalks. The limited studies are focused on the walkability of areas with low income and minorities in urban design and urban planning studies. For instance, the negative relationship between walkability and population under 18 years old was examined by Cutts et al. (2009). Su et al. (2019) developed the Integrated Walkability Index (IWI) by aggregating the pedestrian quality into four categories (connectivity, accessibility, suitability, serviceability, and perceptibility), a case of Hangzhou metropolitan area. Their results have shown socioeconomic inequality in accessing the walkable environment. Riggs (2016) highlighted that Black residents were more likely to live in less walkable neighborhoods in the San Francisco Bay area.

While the existing studies examined the walkable accessibility associated with specific land uses and urban forms (e.g., Majik and Pafka 2019; Balcctis ct al. 2020; Samantha et al. 2020), this study examined the walkable accessibility per residential units to urban services. In addition, this study used quantitative and qualitative approaches to highlight the mutual benefits of walkable access and walking environment quality. The urban service in walkable accesses is not a sufficient driver to encourage the population to walk. Drawing upon a quantitative measure of the walkable access and a qualitative walking survey of an Englewood neighborhood of the Chicago Metropolitan Area, this paper explores inequities in the walking access to multiple urban services of the Engelwood neighborhood with a Walk Score of 65 (WalkScore 2021). The primary concern is whether the areas with fair walking accessibility benefit the residents equally in terms of the quality of the walking environment (Middleton 2018). The incorporated walking quality of the streetscapes to walkable accessibility gives a comprehensive insight into the walking capacity of the Englewood neighborhood. This paper uses quantitative-qualitative and participant observation approaches, applying Ewing and Handy's (2009) and Tveit et al. (2006) conceptual framework for testing design elements and planning approaches in the micro-scale (streetscape) which may impact the walking behavior of residents. The photographic surveying was carried out similarly to Bereitschaft's (2017) analysis of urban design features.

Data and Methods

Study Area

Four streets within the Englewood neighborhood of the city of Chicago were selected (Table 1; Fig. 1). The Englewood neighborhood exhibited quantitative Walk Scores® of 65 (indicating "walkable" condition). A district predominately defines with "high" social vulnerability (SV). The social vulnerability was quantified at the Block Group level by normalizing and summing five socioeconomic and demographic variables from the US Census Bureau. SV values range from 0 (lowest vulnerability) to 1 (highest vulnerability). To get detailed information about the method used to quantify the Social Vulnerability Index (SVI), the Centers for Disease Control and Prevention (CDC) 2018 SVI (CDC 2018) measures social vulnerability to natural and human-made hazards. The SVI comprises four themes: (1) socioeconomic status, (2) household composition, (3) race/ethnicity/language, and (4) housing/transportation. The four main streets of S Halsted St, S Ashland St, W 63 St, and W 69 St are selected as they are the main hubs of socioeconomic activities. The block groups with the lowest SV indicate empty spaces predominantly occupied by vacant lands.

Data Collection

We obtained estimates of current walkable access among single-family and multifamily residential

Walkable Access and Walking Quality of Built Environment, Table 1 Social vulnerability index of the four streets

Study area	S Ashland St	S Halsted St	W 63 St	W 69 St
# Block groups	14	18	21	18
African American population to White population percentile	12.31 to 1.62	16.71 to 0.94	18.41 to 2.68	16.4 3to 0.71
SVI Theme 1 min-max range	0.01–0.78	0.01–0.84	0.01–0.83	0–0.65
SVI Theme 2 min-max range	0.02–0.91	0–0.96	0.02–0.88	0–0.85
SVI Theme 3 min-max range	0.05–0.73	0.01–0.73	0.01–0.90	0–0.71
SVI Theme 4 min-max range	0.09–0.89	0.10–0.99	0.06–0.89	0.02–0.98
All themes min-max range	0.01–0.87	0–0.96	0.01–0.91	0–0.76

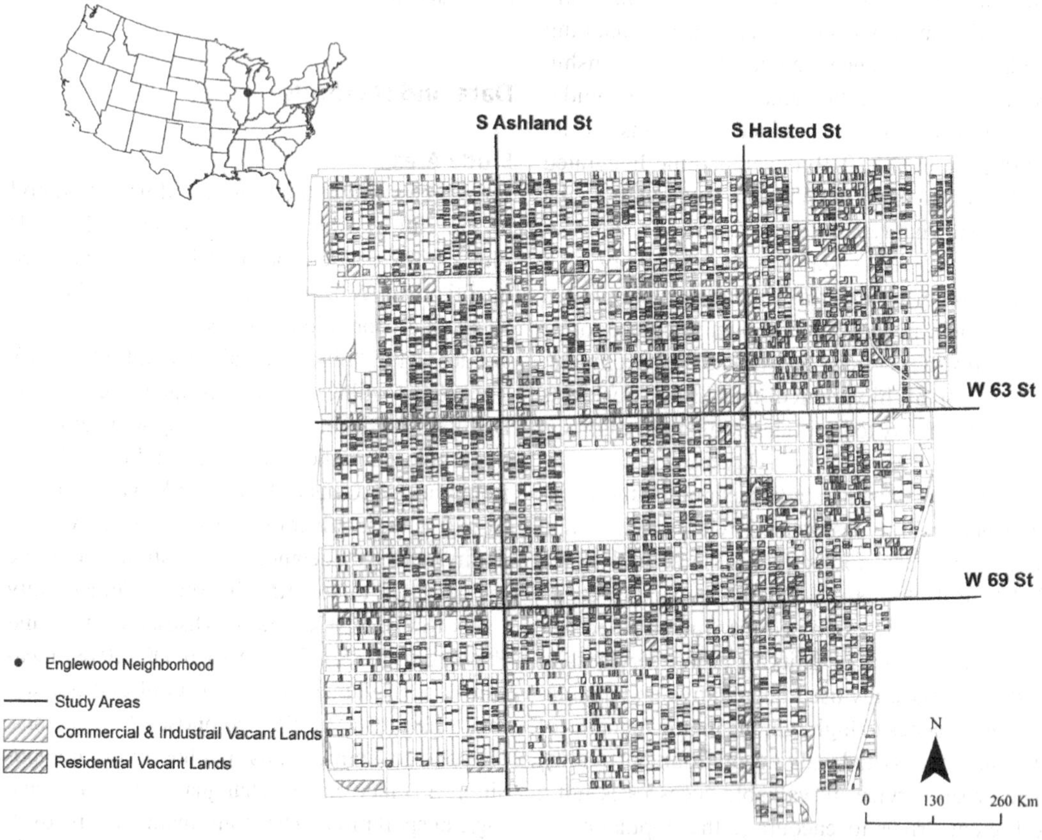

Walkable Access and Walking Quality of Built Environment, Fig. 1 The study area, Englewood neighborhood, and four main streets

units of a total of 6,500 in the Englewood neighborhood. The residential and cultural land use data was obtained from the 2015 Land Use Inventory of the Chicago Metropolitan Agency for Planning (CMAP 2015). The medical care services data was obtained from the Illinois Department of Public Health (IDPH 2020). Grocery, bike station, bus station, public schools, restaurants, and rail station data was provided by the Chicago Data Portal (CDP 2020). For the network analysis, Chicago Sidewalk Inventory Data used includes both and one-side street sidewalks from the Chicago Data Portal (CDP 2020).

Pedestrian-Based Network Analysis

Network analysis was used widely to investigate the spatial accessibility to numerous types of services, such as green space (e.g., Qiu et al. 2019; Czembrowski et al. 2019; Iojă et al. 2014), food (e.g., Kuai and Zhao 2017; Polzin et al. 2014), and COVID-19 testing sites (e.g., Tao et al. 2020). This study utilized network analysis to find the nearest routes from each single- and multiple-family house to nearby facilities, including schools, medical centers, bus stations, rail stations, groceries, restaurants, and parks. Accessibility results are divided by 400 m to measure the variation of walking distance around the 400 m access. The 400 m (0.25 miles) address the 5 min of access to urban facilities used as a measure to build healthy, active, and sustainable neighborhoods (e.g., Azmi and Karim 2012). Finally, the variations averaged to estimate the total walking access score based on accessibility to all services. In addition, we separate access to the urban services based on urban functions as access to the essential services including grocery and medical; recreational services encompassing restaurants and parks; and access to transportations including bike, bus, and rail stations. Fig. 2 legend is classified based on all fields' range, mean, and standard deviation. To the accuracy of the accessibility estimates, we have run the analysis while including urban facilities within 400-m distance to the neighborhood boundary. In addition, the Pedestrian Flow Densities are quantified by summing up nearest accessibility values of sidewalk segments per residential unit to urban services. Pedestrian Flow Densities depend on the degree of sidewalk connectivity (Peimani 2019).

Estimating Residential Building Heights

Light Detection and Ranging (Lidar) Point Cloud Data (a technology used to create high-resolution models of ground elevation) was obtained from Illinois Height Modernization (ILHMP 2020). The height of spatial features is quantified using LAS Height Metrics using a geographic information system or GIS (ArcGIS Pro, ESRI Inc.). The spatial resolution of height values was set to 9 * 9 m' resolution, the width of average building units in the Englewood neighborhood. The pixel values are classified into four classes – 2.28 m (basement), 4.26 (an average story height), 8.52 m (two stories), and more than two stories. The classifications align with the Chicago Legal Design Codes (CLDC 2020). In addition, using Moran's I, spatial autocorrelation is quantified to test the degree to which height features tend to be clustered spatially or dispersed. The test calculates the likelihood that clustering within a dataset appears due to random chance. An index value of +1 equals perfect spatial correlation or clustering, 0 equals a random spatial pattern, and −1 equals perfect dispersion (ESRI 2021). The Moran's I test is a global spatial autocorrelation test, where the variance of an individual point is measured against the entire dataset. At the same time, a local point pattern spatial autocorrelation test can delineate individual clusters or "hot spots" within a dataset. The Getis-Ord Gi* test statistic is measured by comparing the sum of a point and its nearest neighbors to the sum of all points in a given study area. The statistic, a z-value, indicates high or low values (e.g., values with high standard deviations from the overall mean) clustering spatially (Getis and Ord 1992). Such an approach has been widely used in applied geographic research to identify clustering of diseases (e.g., Şener and Türk 2021), crime incidence (e.g., Haleem et al. 2020), availability of medical care (e.g., de Moura and Procopiuck 2020), and food retailers (Yoshimura et al. 2020).

Walking Survey

The author conducted the walking survey of S Halsted St, S Ashland St, W 63 St, and W 69 St, the primary hubs of socioeconomic activities, in June 2021. The survey was conducted by walking distance of about 2.48 miles (each street) along one side and then crossing the street to complete a survey in the opposite direction. The study for each street was conducted once in the afternoon (about 5 pm) and during the weekday. Survey

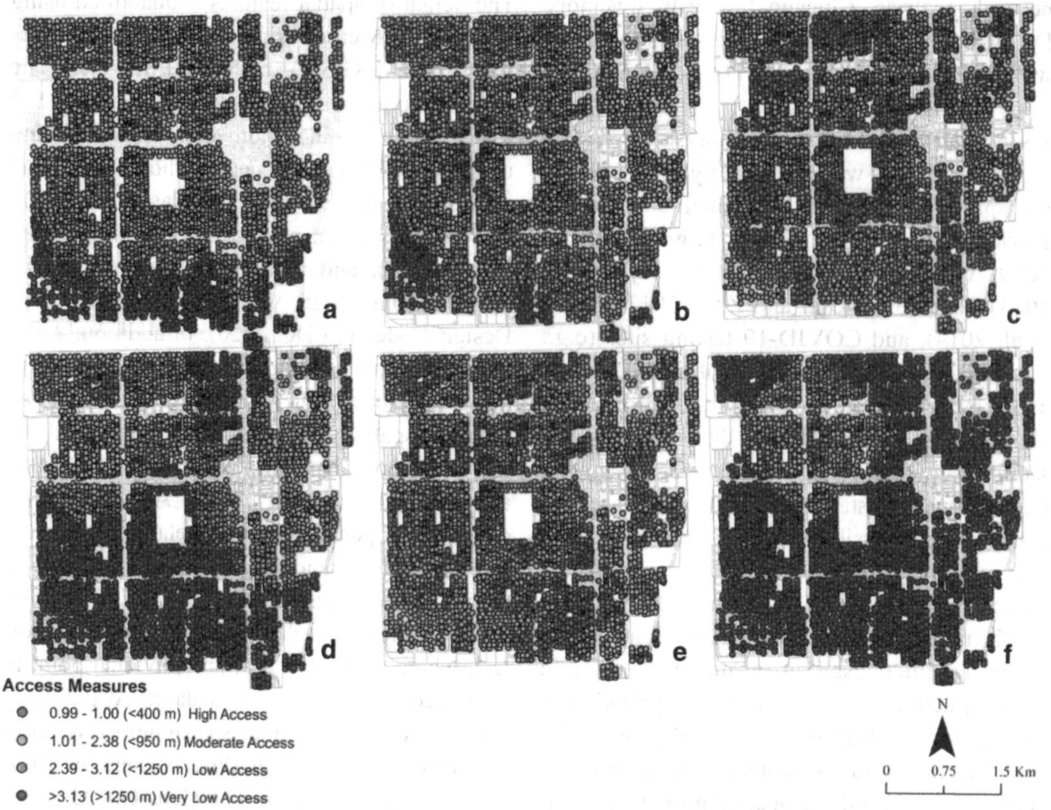

Walkable Access and Walking Quality of Built Environment, Fig. 2 Illustration of walking access to urban services from residential units: (**a**) access to multiple urban services; (**b**) access to public transportation (bus, bike, and rail stations); (**c**) access to recreational facilities (restaurant and parks); (**d**) access to essential services (grocery and medical); (**e**) access to schools; (**f**) access to cultural centers

took 8 h. Photographs were used to record the observation, and over 1000 digital photos were collected.

Data Interpretation

A detailed field survey was conducted to observe and record the street-level pedestrian environment in high and low pedestrian density flow street segments. Visuality, enclosure, human scale (e.g., the scale of urban furniture), tidiness, and safety attributes were highlighted as qualities of the street environment that might affect the use of pedestrians (Ewing and Handy 2009). These encompass average sidewalk width, the presence of pedestrian crossings, traffic lights, and trees. Information collected from the walking survey is categorized based on Ewing and Handy (2009), Tveit et al. (2006), and Porta and Renne (2005) streetscape elements, and those address the author's walking experience and generalize the walking environment patterns.

Walking Environment Indicators

Ewing and Handy's (2009), Tveit et al.'s (2006), and Porta and Renne's (2005) definitions on qualities of urban space were used to interpret the walking survey data. The indicators include imageability, enclosure, human scale, transparency, tidiness, and safety. Imageability addresses uniqueness, place identity, the mental image of the environment, historical richness, and sense of place; enclosure refers to the extent to which buildings, walls, trees, and other vertical items frame a street and public space; human scale is associated with both human-sized urban furniture and ratio of a facade to sidewalk/street width;

transparency accounts for the degree of connection between interior and exterior environments; tidiness relates to urban elements maintenance as green space maintenance; and safety addresses safety infrastructures such as crosswalk signals, traffic density, and speed.

Results

Figure 2 illustrates walking access to urban facilities for the residential units in Englewood neighborhood. Figure 3 shows pedestrian density flows and clustered patterns of building heights. Figures 4, 5, 6, 7, 8, and 9 illustrates the walking elements and designs of four streetscapes in the Englewood neighborhood to complement and extend the study on walkability.

Walking Access

Analyzing network walking distance in conjunction with land use data has shown that most urban facilities are not within walking distance to residential units. Figure 2a shows the access to the multiple urban facilities, while most residential lands are not within 400 m of walking access. While few residential units in the intersection of S Ashland St and W 63 St are within walking distance of public transportations, most residential ones are within 400–950 m (Fig. 2b). Figure 2c reveals the fair distribution of recreational facilities compared to public transports, while areas

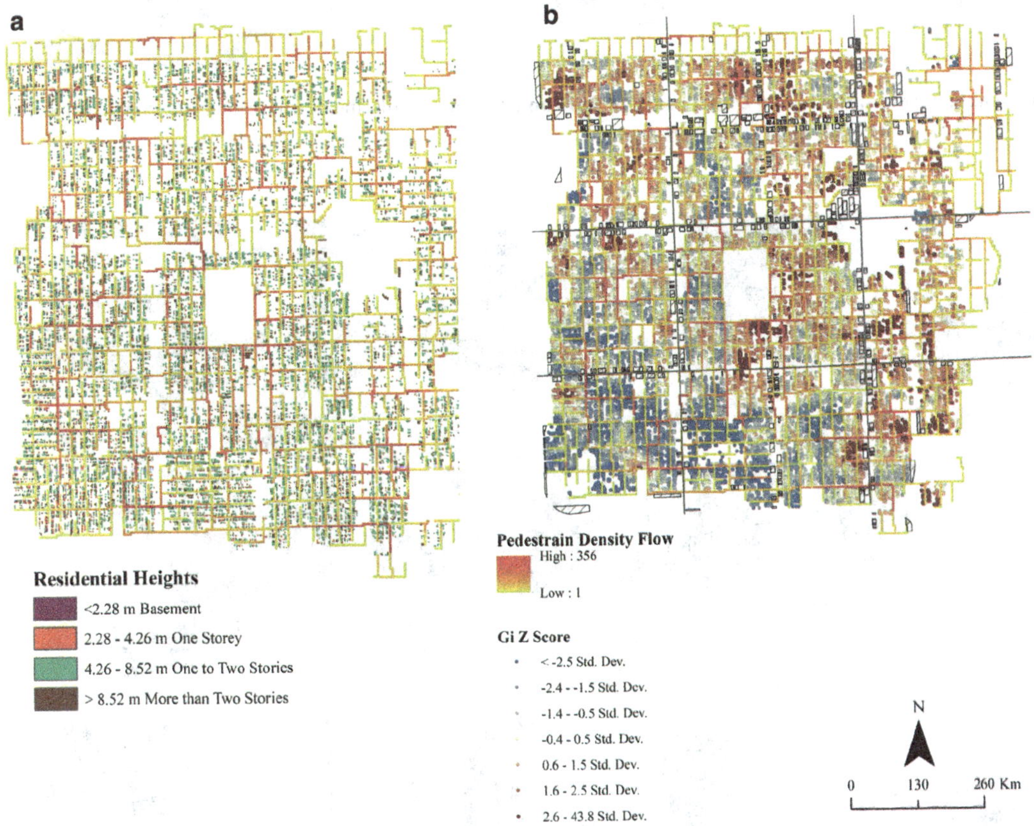

Walkable Access and Walking Quality of Built Environment, Fig. 3 (**a**) Pedestrian density flows and residential building heights. (**b**) A Getis-Ord Gi* statistical "hot spot" analysis of estimated total height values in Englewood. The Gi* score is also the standard deviation (SD) from the average value of a point's neighbors. A high Z-score (red) indicates clustering of high height values, while a low Z-score (blue) indicates spatial clustering of low height values. Median z-scores (yellow) indicate that there is no significant spatial relationship between a site's height value and that of neighboring points

W

Walkable Access and Walking Quality of Built Environment, Fig. 4 Illustrating imageability: (**a**) S Ashland St; (**b**) S Halsted St; (**c**, **d**) W 63 St; (**e**–**f**) W 69 St

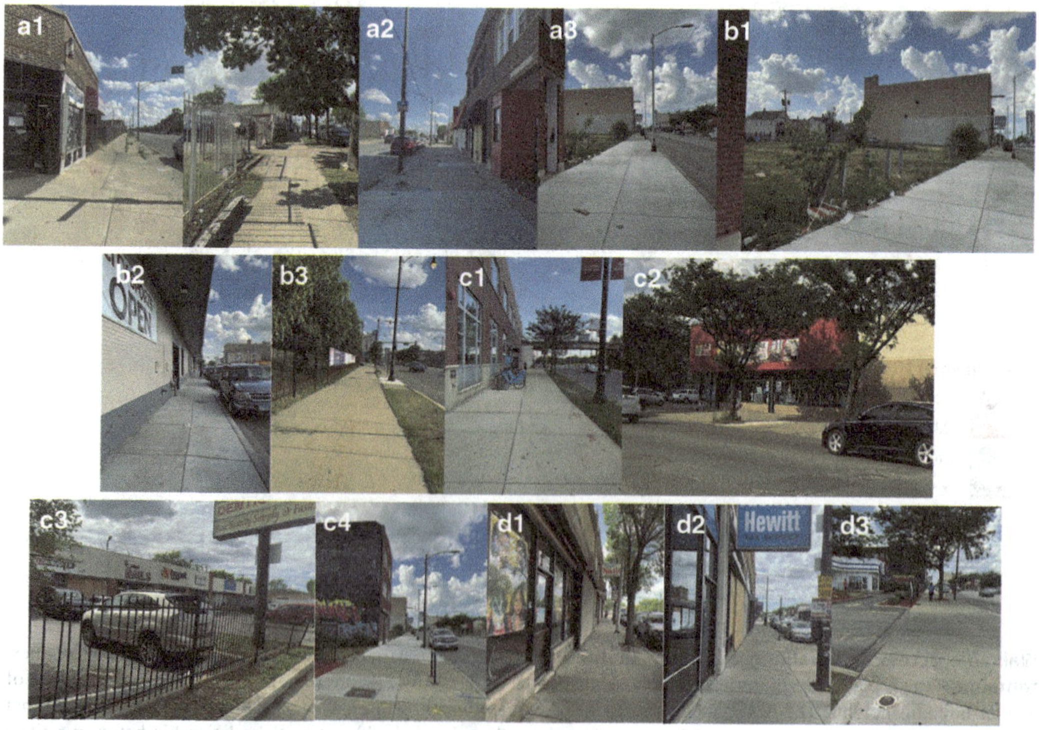

Walkable Access and Walking Quality of Built Environment, Fig. 5 Illustrating enclosure: (a1, a2, a3) S Ashland St; (b1, b2, b3) S Halsted St; (c1, c2, c3,c4) W 69 St; (d1, d2,d3,d4) W 63 St

Walkable Access and Walking Quality of Built Environment, Fig. 6 Illustrating human scale: (a1, a2, a3,a4) S Ashland St; (b1, b2, b3,b4) S Halsted St; (c1, c2, c3,c4); W 63 St; (d1, d2,d3,d4) W 69 St

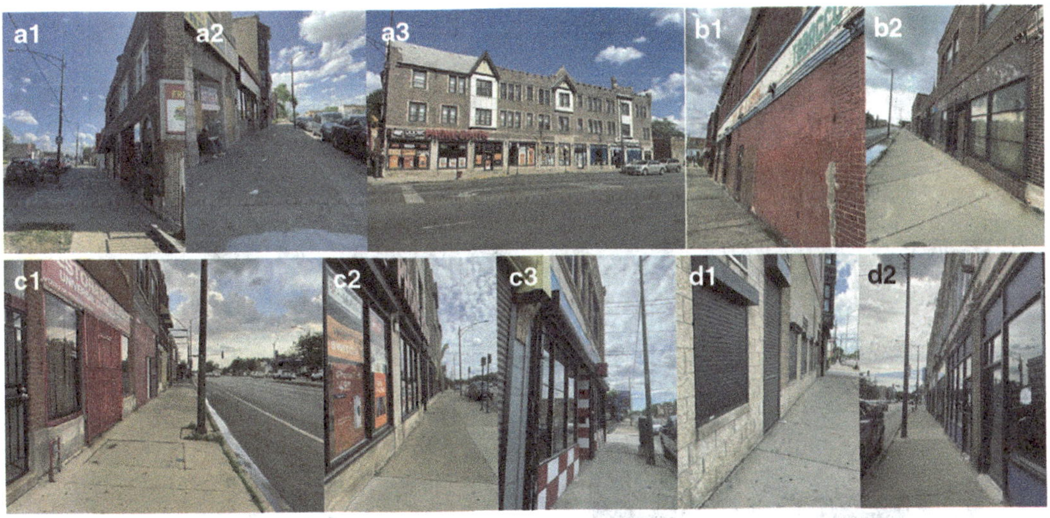

Walkable Access and Walking Quality of Built Environment, Fig. 7 Illustration of transparency: (a1, a2, a3) S Ashland St; (b1, b2) S Halsted St; (c1, c2,c3) W 63 St; (d1, d2) W69 St

between S Halsted St and S Ashland St do not have walking accessibility to recreational amenities. Figure 2d shows that all residential places are not within 400-m walking distance to urban services except a small number of residential units in the northwest of S Ashland St. Figure 2e

Walkable Access and Walking Quality of Built Environment, Fig. 8 Illustration of tidiness: (a1, a2, a3) S Ashland St; (b1, b2,b3,b4) S Halsted St; (c1,c2, c3,c4) W 63 St; (d1, d2,d3) W69 St

Walkable Access and Walking Quality of Built Environment, Fig. 9 Illustration of safety: (a1, a2, a3,a4) S Ashland St; (b1, b2,b3) S Halsted St; (c1, c2, c3) W 63 St; (d1, d2,d3) W69 St

illustrates fair access to public schools. Still, few areas on the north and south sides of S Ashland St. Nevertheless, Fig. 2f shows the deprived walking accessibility to cultural centers throughout the neighborhood.

Pedestrian Density Flow and Spatial Distributions of Residential Building Heights

The distribution of density flows is the number of residential uses to access urban amenities at the sidewalk segment level. The density flow varies from 1 (low density) to 356 (high density). While density flow is low and does not have a significant difference in the four main streets of the neighborhood, high-density flow is mostly in secondary streets in the north, south, and west of Ogden (William) Park (Fig. 3a and b). In addition, the results reveal that one- to two-story residential units, dominant residential land use patterns, are distributed evenly in high- and low-density sidewalks of the neighborhood (Fig. 3a). The Englewood neighborhood approximately includes 230,800 m^2 one-story buildings and 1,340, 900 m^2 two-story height buildings. The Global Moran's I results in a significant spatial clustered pattern with a Z-score of 6.2 ($p<0.01$). While Global Moran I ignores local variations, Local Moran I (Fig. 3b) shows clustering patterns of the height of residential units on the local level. That contributes to estimating the population distribution throughout the neighborhood. In Fig. 3b, a low negative Z-score with a small p-value ($p<0.01$) indicates a significant cold spot. The cold spot (blue marks) shows the clustering patterns of single-floor residential buildings. However, a high positive Z-score with a small p-value ($p<0.01$) reveals a significant cold spot. Hot spot (red marks) addresses the clustering patterns of multi-floor residential buildings. The results show hot spots along S Halsted St, north of S Ashland St, and residential units near Ogden Park (Fig. 3b).

Streetscape Walking Qualities

Imageability

Differences in imageability between streets were minor (Fig. 4). It is hard to say that each streetscape has unique landmarks. The most prevalent landmarks are wall art (Fig. 4a), educational places (Fig. 4b), churches (Fig. 4c), graffiti (Fig. 4d), buildings with different color compositions (Fig. 4d), and different architectural designs (Fig. 4e-f 4e–i). Additionally, the north and east sides of the streets tended to have more places for passing the time on the sidewalks. The south side of S Ashland St and S Halsted St showcased the graffiti, large parking lots, and empty lands. The western side of W 63 St and W 69 St exhibited the ignored spaces in terms of the non-maintained sidewalk, no street lighting, and vacant lands. In addition, the ratio of the buildings' height to street width is too low, the street will feel too wide, and the imageability of the street is lost. There were recessed buildings in disordered manners that negatively affected the imageability of the streetscapes, while the height of the building did not show the abrupt changes. However, the relatively compact spatial density of the eastern side of W 63 St and W 69 St helped create a sense of place and imageability of the streetscapes. Meanwhile, the concentration of spatial uses in the S Ashland St and S Halsted St gives the street prominence in the minds of the observers though it is not a prevalent spatial pattern along the street.

Enclosure

As for imageability, there was no significant difference in the degree of enclosure among streets. The non-continuity of the street wall where industrial and commercial vacant lands delineate the streetscapes is a concern of streetscapes (Fig. 1). All streetscapes contained more gaps between buildings, mainly vacant lands and parking lots (Fig. 5). Most of these spatial gaps are poorly maintained (Fig. 5 A2), contributing to a sense of discontinuity and vulnerability. Additionally, the presence of several suburban-like building setbacks reduces the sense of the enclosure (e.g., parking lots facing the streets rather than the building) (Fig. 5 D1). The street wall continuity happened after the sequential spatial gaps, where people experience the pedestrian. It looks like socializing spots along the street without providing the real pedestrian-use experience. Additionally, trees can define space and increase the degree of enclosure. However, there were not enough trees on either side of the streets.

Human Scale

Design features and elements are not primarily based on human size. Although buildings are between one and two stories tall, the proportion of the building height to street and sidewalk width is not designed properly. In general, streets are very wide and negatively affect the sense of the place. However, there were some differences between the north and south of the streetscapes and the west and east sides of the streets. Street width to building height proportion is better designed in the north and eastern sides. All streetscapes noticeably contained less and no street furniture in S Ashland St, as well as no defined bus station. There were no street cafes and dining out. The food graphics on the outdoor wall of the restaurant showcased the dining place. In general, there is no pedestrian-oriented infrastructure in streetscapes, particularly in S Ashland St and the Southern side of S Halsted St. There are no bike racks, bus stops, and planters. On the southern side of S Ashland St, retail shops showcased items for sale outside the store on the adjacent sidewalk adding to the sense of the enclosure, complexity, and human scale of the streetscape (Fig. 6)

Transparency

As with enclosure, there was significant concern about the overall transparency of the streetscapes. Transparency correlates with the continuity of street walls. Without a street wall, there can be no "human activity beyond the street." Thus, there was a lack of transparency due to the presence of spatial gaps, fewer windows at street level, and non-utilized spaces. Street wall continuity and windows at street level enhance the street permeability and add vitality to the street. Figure 7 illustrates the transparency after every sequential spatial gap though most of these spaces are non-utilized or provide the sociocultural and recreational opportunities to enrich the pedestrian experience.

Tidiness

There were no significant differences in the tidiness and overall maintenance of all streetscape environments. Empty lands, discontinuous street wall, and a mix of trash and vegetation are the clearest sign of neglect (Fig. 8a). In general, trash was more common among streetscapes as unmaintained vegetation. The vegetation throughout the north side of S Halsted St streetscape tended to be well manicured by comparison, often consisting of sculpted trees and flowering (Fig. 8b). Graffiti, however, was sustainability observed on either side of all streetscapes. Additionally, parking meters, bike racks, pedestrian safety signs, outdoor lighting, telephone booths, and trash are old and rusted in each of the streetscapes (Fig. 8). New and modern urban furniture was mostly found in the north side of S Halsted St.

Safety

The streets are mainly designed for motor vehicles. Human senses impact pedestrian experience. The smell of automotive exhaust (instead of the smell of cooking at a nearby cafe) was particularly noticeable and corresponded with heavy traffic (especially truck traffic). Motor vehicle sounds disturb people experiencing the pedestrian. In addition, the large parking lot and non-maintained vacant lands create unsafe and unpleasant conditions for pedestrian experience. Locomotive factories near pedestrians further make unpleasant visual experiences. Their environmental pollution makes the experience worse. More importantly, non-maintained and non-paved pedestrians decrease the quality of pedestrian experience. They are not usable by disabling people and children. Poor urban lighting, street crossing signs, and lack of bus stations further make conditions worse for pedestrians. Stations were observed at the junctions of S Ashland St and W 63 St and S Halsted St and W 69 St (Fig. 9).

Discussion and Conclusion

Our main findings are explained as follows. First, spatial equity in urban services has received much attention due to the importance of walking in health-related issues, controlling air pollutants emissions, and minimizing human activities to locals. Using GIS network analysis, walking access to urban services was studied to investigate the possible differences in pedestrian accessibility to urban services. The findings reveal that most regions have an undersupply of concentrations of multiple services and have insufficient walking access to urban services. The lack of good quality services in socially and economically disadvantaged areas can be interpreted as an effect and a cause of inequality. The distance between the living and the servicing levels and the quality of the networks create barriers to accessing employment for disadvantaged groups. Access, or lack of access, to these resources negatively impacts the exercise of a sense of community. In addition, we showcased the accessibility in terms of population density, using building heights as an estimator of population density. The results showed that the regions with the highest residential density have few walking accessibilities to urban facilities. Future studies might weigh the sidewalk segments based upon population density distribution.

Second, the walking survey of four streets in Englewood neighborhood revealed poor maintenance of the built environment. The low pedestrian activities have highlighted the desert of appropriate design and features of sidewalks though the Walk Score metric showcased the high score of the walkability. Walk Score fails to reveal significant inequities in walking access to urban services due to the limit of the walkability to access transportation. Nevertheless, the walkability score should be revised based on the quality of the pedestrian network and the spatial distribution of services.

On a micro-scale, the lack of walkability might be a common feature in large US cities. The photographic survey and observations have shown that that might affect walking behavior. The number of pedestrian users was very few in four streets except in areas with urban services. The intersection of S Halsted St and W 63 St is an area with relatively high pedestrian users. In general, the streets are designed for motor vehicle users with a wide width. The large and many gaps in the street wall and a few fair quality urban services are not attracting pedestrians for recreational and utilitarian purposes. The sidewalk should not be considered as a two-dimensional statics feature. In sidewalk designing, the first step is to understand how the human body experiences space. The multisensory development of the sidewalks is required to improve the spatial experience of pedestrians. Along the sidewalks, several spatial variables work together to enhance the multisensory characteristics of the space. One of the spatial elements is the building wall plane, which passes immediately beside pedestrians as they are walking.

The photos and detailed observations address neglected spaces that were not previously debated in the literature. More notably, the detailed descriptions of imageability, enclosure, and non-visual sensation at the micro-scale contribute to expanding knowledge of high SV neighborhoods. Gaps in building wall planes impact the pedestrian experience in the streets. There were few positive enclosures except for a few parcels. It is required to assess the revitalized high SV neighborhoods to raise sustainable solutions for the Englewood neighborhood. Though this study was limited to the preliminary quantitative and qualitative study, it may be used as the basis for additional studies with potentially spatial and temporal statistics analysis. Future studies need to investigate precisely how each of these elements affects pedestrian behavior.

The field survey of four streetscapes revealed problems in the quality and maintenance of the built environment and the level of pedestrian activity. There were no noticeable differences in enclosure, transparency, safety, human scale, and tidiness among streetscapes. There was no continuous walking environment because of several spatial gaps among buildings. While the quality of the walking environment does not invite the pedestrians to use it as a public space and for daily communication, the proximity of homes and destinations such as groceries cannot solve walkability issues of the neighborhood by itself. The existing quality of sidewalks addresses the conceptualized model of Carmona (2015). Urban space invaded by motor vehicles with commercial and industrial vacant lands and the main streets highlights the neglected area. In addition, the non-functioning and large parking lots create a sense of loneliness and fear that build the exclusionary space where it is isolated from the rest of the city. The solution might be increasing compactness bringing numerous activities closer together, thus increasing their accessibility from residential origins, and it contributes to improving a sense of enclosure. The wall plane and tree canopies along the sidewalk may create a contained space within the streets. The sense of enclosure provides the natural and resident-based surveillance on the urban environment debated by Newman (1996) in Defensible Space Theory. To illustrate, it increases informal social control. People tend to monitor suspicious behavior in this environment. That encourages them to know their neighbors, which contributes to neighboring, empowering, and networking. Broadly, it enhances a sense of security (Cozens et al. 2005).

Furthermore, as an enclosure, the psychological well-being and sense of safety depend on the mental images and transparency of the built environment. The edges, paths, nodes, and districts debated by Lynch (1960) should be revitalized to contribute to the wayfinding in the neighborhood. This study further invites attention to the green space quality at the micro-scale. While the aerial remote sensing data implicate the equal distribution of green spaces in multiple scales, the collected micro-scale data highlights non-usable green space mixed with litter (weedy and shrubs). Future studies might investigate the ecological, accessible, and functional benefits of green spaces in socially vulnerable neighborhoods.

The results further suggest that neglected neighborhoods exist on a micro-scale and that neglected spaces should be considered alongside the macro-scale elements density, connectivity, transit accessibility, and land use mix. It is required to encourage infill development along the main streets. That contributes to decreasing spatial gaps in the street wall continuity and also helps better mix of land uses and better neighborhood density. This approach brings social opportunities to neighborhoods as places for outdoor dining and cafes.

The design and planning base testing the sidewalks is an efficient and holistic approach to promote urban sustainability that addresses three pillars of sustainability: environmental, social, and economic (e.g., Lotfata and Ataov 2020). This study highlights walking mode of transportation to improve public health, increase community diversity, enhance a sense of community, and reduce environmental degradation (e.g., Zuniga-Teran et al. 2017; Grant et al. 2017).

Additionally, infill revitalization strategy and other strategies may also encourage gentrification, leading to the displacement of the residents. The infill development strategies in US cities lead to inequity (Kim and Larsen 2017). Community engagement design, planning, and preservation of local commercial activities can mitigate the displacement. Jacobs (1969) debates the conservation of the local market while adding new social and commercial activities to the neighborhood.

References

Arlie, A., Carrie, M., Michele, S., Maia, I., & Gretchen, L. (2017). Contextualizing walkability: Do relationships between built environments and walking vary by socioeconomic context? *Journal of the American Planning Association, 83*(3), 296–314.

Azmi, D. I., & Karim, H. A. (2012). Implications of walkability towards promoting sustainable urban neighbourhoods. *Procedia - Social and Behavioral Sciences, 50*, 204–213.

Balcetis, E., Cole, S., & Duncan, D. T. (2020). How walkable neighborhoods promote physical activity: Policy implications for development and renewal. *Policy Insights From the Behavioral and Brain Sciences, 7*(2), 173–180.

Bereitschaft, B. (2017a). Equity in neighbourhood walkability? A comparative analysis of three large U.-S. cities. *Local Environment, 22*(7), 859–879.

Bereitschaft, B. (2017b). Equity in microscale urban design and walkability: A photographic survey of six Pittsburgh streetscapes. *Sustainability, 9*(7), 1233.

Blečić, I., Congiu, T., Fancello, G., & Trunfio, G. A. (2020). Planning and design support tools for walkability: A guide for urban analysts. *Sustainability, 12*(11), 4405.

Brownson, R. C., Hoehner, C. M., Day, K., Forsyth, A., & Sallis, J. F. (2009). Measuring the built environment for physical activity: State of the science. *American Journal of Preventive Medicine, 36*, S99–S123. https://doi.org/10.1016/j.amepre.2009.01.005.(e12).

Carmona, M. (2015). Re-theorising contemporary public space: A new narrative and a new normative. *Journal of Urbanism: International Research on Placemaking and Urban Sustainability, 8*(4), 373–405.

Carmona, M., Heath, T., Oc, T., & Tiesdell, S. (2003). Public places-urban spaces. The Dimensions of Urban Desi.

Cathy, A., Ross, B., Tamara, D., Philippa, C., & Natalie, C. (2020). Associations between body mass index, physical activity and the built environment in disadvantaged, minority neighborhoods: Predictive validity of GigaPan® imagery. *Journal of Transport & Health, 17*, 100867. ISSN 2214-1405.

Christine, M., Hoehner, S. L., Handy, Y. Y., Steven, N., & Blair, D. B. (2011). Association between neighborhood walkability, cardiorespiratory fitness and body-mass index. *Social Science & Medicine, 73*(12), 1707–1716. ISSN 0277-9536.

Coffee, N. T., Howard, N., Paquet, C., Hugo, G., & Daniel, M. (2013). Is walkability associated with a lower cardiometabolic risk? *Health & Place, 21*, 163–169. ISSN 1353-8292.

Cozens, P. M., Saville, G., & Hillier, D. (2005). Crime prevention through environmental design (CPTED): A review and modern bibliography. *Property Management, 23*(5), 328–356.

Cutts, B. B., Darby, K. J., Boone, C. G., & Brewis, A. (2009). City structure, obesity; environmental justice: An integral analysis of physical and social barriers to walkable streets and park access. *Social Science & Medicine, 69*, 1314–1322.

Czembrowski, P., Łaszkiewicz, E., Kronenberg, J., Engström, G., & Andersson, E. (2019). Valuing individual characteristics and the multifunctionality of urban green spaces: The integration of sociotope mapping and hedonic pricing. *PLoS One, 14*(3).

de Moura, E. N., & Procopiuck, M. (2020). GIS-based spatial analysis: Basic sanitation services in Parana State, Southern Brazil. *Environmental Monitoring and Assessment, 192*, 96.

Ewing, R., & Handy, S. (2009). Measuring the unmeasurable: Urban design qualities related to walkability. *Journal of Urban Design, 14*, 65–84.

Gavin, J. A., Edward, H., Bethan, E., & Rachel, C. (2012). Moving beyond walkability: On the potential of health geography. *Social Science & Medicine, 75*(11), 1925–1932. ISSN 0277-9536.

Getis, A., & Ord, J. K. (1992). The analysis of spatial association by use of distance statistics. *Geographical Analysis, 24*(3), 189e206.

Giap, T. K., Thye, W. W., & Aw, G. (2014). A new approach to measuring the liveability of cities: The global liveable cities index. *World Review of Science, Technology and Sustainable Development, 11*, 176–196.

Grant, G., Machaczek, K., Pollard, N., & Allmark, P. (2017). Walking, sustainability and health: Findings from a study of a walking for health group. *Health & Social Care in the Community, 25*(3), 1218–1226.

Gutiérrez, J., & García-Palomares, J. C. (2008). Distance-measure impacts on the calculation of 27 transport service areas using GIS. *Environment and Planning. B, Planning & Design, 28*(35), 480–503.

Haleem, M. S., Do Lee, W., Ellison, M., et al. (2020). The 'exposed' population, violent crime in public space and the night-time economy in Manchester, UK. *European Journal on Criminal Policy and Research, 2020*. https://doi.org/10.1007/s10610-020-09452-5.

Hsiao, S., Lu, J., Sterling, J., & Weatherford, M. (1997). Use of geographic information system 2 for analysis of transit pedestrian access. *Transportation Research Record, 1604*, 50–59.

Iojă, C. I., Grădinaru, S. R., Onose, D. A., Vânău, G. O., & Tudor, A. C. (2014). The potential of school green areas to improve urban green connectivity and multifunctionality. *Urban Forestry & Urban Greening, 13*(4), 704–713.

Jacobs, J. (1961). The death and life of great American cities.

James, P., Kioumourtzoglou, M. A., Hart, J. E., Banay, R. F., Kloog, I., & Laden, F. (2017). Interrelationships between walkability, air pollution, greenness, and body mass index. *Epidemiology (Cambridge, MA), 28*(6), 780–788.

Keyvanfar, A., Ferwati, M. S., Shafaghat, A., & Lamit, H. (2018). A path walkability assessment index model for evaluating and facilitating retail walking using decision-tree-making (DTM) method. *Sustainability, 10*(4), 1035.

Kim, J., & Larsen, K. (2017). Can new urbanism infill development contribute to social sustainability? The case of Orlando, Florida. *Urban Studies, 54*(16), 3843–3862.

Kim, E. J., Kim, J., & Kim, H. (2020). Does environmental walkability matter? The role of walkable environment in active commuting. *International Journal of Environmental Research and Public Health, 17*(4), 1261.

Kuai, X., & Zhao, Q. (2017). Examining healthy food accessibility and disparity in Baton Rouge, Louisiana. *Annals of GIS, 23*(2), 103–116.

Lotfata, A., & Ataöv, A. (2020). Urban streets and urban social sustainability: A case study on Bagdat street in Kadikoy, Istanbul. *European Planning Studies, 28*, 1735–1755.

Lynch, K. (1960). *The image of the city*. Cambridge, MA: MIT Press.

Majic, I., & Pafka, E. A. (2019). An open-source GIS tool for measuring walkable access. *Urban Science, 3*(2), 48.

Matthew, C. (2015). Re-theorising contemporary public space: A new narrative and a new normative. *Journal of Urbanism: International Research on Placemaking and Urban Sustainability, 8*(4), 373–405.

Middleton, J. (2018). The socialities of everyday urban walking and the 'right to the city'. *Urban Studies, 55*, 296–315.

National Institute for Health and Care Excellence (NICE). Public health guideline: Physical activity: Walking and

W

cycling; National Institute for Health and Care Excellence (NICE): London, 2012.

Newman, O. (1996). *Creating defensible space*. Washington, DC: U.S. Department of Housing and Urban Development.

Oppezzo, M., & Schwartz, D. L. (2014). Give your ideas some legs: The positive effect of walking. *Journal of Experimental Psychology. Learning, Memory, and Cognition, 40*, 1142–1152.

Peimani, N. (2019). Transit-oriented morphologies and forms of urban life. Available online: http://contour.epfl.ch/?p=1137&lang=en.

Polzin, P., Borges, J., & Coelho, A. (2014). An extended kernel density two-step floating catchment area method to analyze access to health care. *Environment and Planning. B, Planning & Design, 41*(4), 717–735.

Porta, S., & Renne, J. L. (2005). Linking urban design to sustainability: Formal indicators of social urban sustainability field research in Perth, Western Australia. *Urban Design International, 10*(51–64), 34.

Qiu, J., Bai, Y., Hu, Y., Wang, T., Zhang, P., & Xu, C. (2019). Urban green space accessibility evaluation using age-based 2-step floating catchment area method. IGARSS 2019–2019 IEEE International Geoscience and Remote Sensing Symposium.

Riggs, W. (2016). Inclusively walkable: Exploring the equity of walkable housing in the San Francisco Bay Area. *Local Environment, 21*, 527–554.

Ruiz-Padillo, A., Minella Pasqual, F., Uriarte, A. M. L., & Cybis, H. B. B. (2018). Application of multi-criteria decision analysis methods for assessing walkability: A case study in Porto Alegre, Brazil. *Transportation Research Part D: Transport and Environment, 63*, 855–871.

Saelens, B. E., Sallis, J. F., & Frank, L. D. (2003). Environmental correlates of walking and cycling: Findings from the transportation, urban design and planning literature. *Annals of Behavioral Medicine, 25*, 80–91.

Samantha, L. P., KangJae, J. L., Nicholas, A. P., Alan, R. G., & Andrew, J. M. (2020). Understanding access and use of municipal parks and recreation through an intersectionality perspective. *Journal of Leisure Research, 51*(4), 377–396.

Scott, R. P. (2021). Shared streets, park closures and environmental justice during a pandemic emergency in Denver, Colorado. *Journal of Transport & Health, 21*, 101075.

Şener, R., & Türk, T. (2021). Spatiotemporal analysis of cardiovascular disease mortality with geographical information systems. *Applied Spatial Analysis, 14*, 929–945.

Shashank, A., & Schuurman, N. (2019). Unpacking walkability indices and their inherent assumptions. *Health & Place, 55*, 145–154.

Stratford, E. (2016). Mobilizing a spatial politics of street skating: Thinking about the geographies of generosity. *Annals of the American Association of Geographers, 106*, 350–357. https://doi.org/10.1080/00045608.2015.1100062.

Su, S., Zhou, H., Xu, M., Ru, H., Wang, W., & Weng, M. (2019). Auditing street walkability and associated social inequalities for planning implications. *Journal of Transport Geography, 74*, 62–76.

Tao, R., Downs, J., Beckie, T. M., Chen, Y., & McNelley, W. (2020). Examining spatial accessibility to COVID-19 testing sites in Florida. *Annals of GIS, 26*(4), 319–327.

Telega, A., Telega, I., & Bieda, A. (2021). Measuring walkability with GIS – Methods overview and new approach proposal. *Sustainability, 13*(4), 1883.

Tveit, M., Ode, Å., & Fry, G. (2006). Key concepts in a framework for Analysing visual landscape character. *Landscape Research, 31*(3), 229–255.

U.S Control Diseases Center (CDC). (2018). Social Vulnerability Index (SVI). Available from: https://www.atsdr.cdc.gov/placeandhealth/svi/index.html

U.S. Chicago Data Portal (CDP). (2020). Regional Sidewalk Inventory. Available from: https://datahub.cmap.illinois.gov/dataset/regional-sidewalk-inventory

U.S. Design Legal Regulation Codes. Available from: https://library.municode.com/il/chicago_heights/codes/zoning?nodeId=CH3GEPR)

U.S. Environmental Systems Research Institute (ESRI) (2021). https://pro.arcgis.com/en/pro-app/latest/tool-reference/spatial-statistics/h-how-spatial-autocorrelation-moran-s-i-spatial-st.htm

U.S. Illinois Department of Public Health (2020). Available from: https://www.dph.illinois.gov/

U.S. Illinois Height Modernization (ILHMP) Lidar Data (2020). Available from: https://clearinghouse.isgs.illinois.edu/data/elevation/illinois-height-modernization-ilhmp

U.S. WalkScore. Available from: https://www.walkscore.com/

Wang, H., & Yang, Y. (2019). Neighbourhood walkability: A review and bibliometric analysis. *Cities, 93*, 43–61.

Wang, R., Lu, Y., Zhang, J., Liu, P., Yao, Y., & Liu, Y. (2019). The relationship between visual enclosure for neighbourhood street walkability and elders' mental health in China: Using street view images. *Journal of Transport & Health, 13*, 90–102.

Yoshimura, Y., Santi, P., Arias, J. M., Zheng, S., & Ratti, C. (2020). Spatial clustering: Influence of urban street networks on retail sales volumes. *Environment and Planning B: Urban Analytics and City Science, 48*, 1926–1942.

Zhao, Y., & Chung, P.-K. (2017). Neighborhood environment walkability and health-related quality of life among older adults in Hong Kong. *Archives of Gerontology and Geriatrics, 73*, 182–186. ISSN 0167-4943.

Zuniga-Teran, A. A., Orr, B. J., Gimblett, R. H., Chalfoun, N. V., Guertin, D. P., & Marsh, S. E. (2017). Neighborhood design, physical activity, and wellbeing: Applying the walkability model. *International Journal of Environmental Research and Public Health, 14*(1), 76.

Walking

► Pre-schoolers and Sustainable Urban Transport
► Walkable Access and Walking Quality of Built Environment

Walking Behavior

► Hidden Enemy for Healthy Urban Life

Wash (Water, Sanitation, and Hygiene): Infrastructure as a Measure of Sustainable Development

Anuja Joy[1,2], Shyni Anilkumar[1] and C. Mohammed Firoz[1]
[1]National Institute of Technology Calicut, Kozhikode, Kerala, India
[2]Kozhikode, Kerala, India

Synonyms

Sustainable development; Sustainable WaSH infrastructure; Water, Sanitation and hygiene (WaSH)

Definition

Safe drinking water, proper sanitation, and hygiene practices are critical in developing a healthy community. Water Sanitation and Hygiene commonly referred to as (WaSH) has been widely recognized as an essential determinant for a sustainable community, development, and these are realized to be interrelated.

In 2010, United Nations General Assembly stated that proper water and sanitation facilities are human rights which that provide full enjoyment in life. Following this, United Nations International Children's Emergency Fund (UNICEF) established the need for WaSH practices for the better future of children (UNICEF 2016), which aimed to enhance sustainable future generations.

One of the major goals of Sustainable Development Goals (SDGs), initiated by UN is the- (SDG 6), which focus on clean water and sanitation facilities. Following this, UN and several other international agencies attempted to develop sustainable approaches for planning, implementation, and monitoring of WaSH infrastructure.

Introduction

Sustainable development fosters, development using natural resources and enables to the preserve them for future generations as well (UN 1987). It is an utmost requirement for enhancing the future of the developing world, hence requires continuous planning and management of actions as well as effective utilization of resources. Many studies have established that the vast majority of communities across the globe, especially in the developing and underdeveloped nations are experiencing major environmental, economic, and social consequences of unsustainable practices. These compounding issues humanity encounter globally, later necessitated formulation of universally acceptable guidelines for sustainable development. The need for proper sanitation and access to drinking water has been established through different treaties and summits since 1930. UN developed Millennium Development Goals in the year 2000 based on Millennium Summit to achieve eight goals related to sustainable development by 2015. It considered the sustainable access to safe drinking water and proper sanitation facilities in their target 7c under goal no 7- *Environmental Sustainability*. In 2010, UN stated that proper water and sanitation are human rights and abbreviated Water, Sanitation, Hygiene as WaSH and also developed a framework to manage WaSH systems and practices.

Later in 2015, the UN general assembly set up the SDGs for the sustainability of global community. Seventeen SDGs are introduced for the period of 2015–2030, each one with different targets, implementation projects, and policies which differ for different countries. The 6th goal of SDG targets to achieve the provision of clean water and sanitation facilities. It includes various targets which aim at water use and scarcity, water resources management, ecosystem, cooperation and participation, sanitation and hygiene, water quality and wastewater, water-related disasters, and drinking water. These targets further intend to achieve specific sub-targets such as water quality, proper sanitation facilities, and good hygiene practices. Apart from the 6th goal, SDGs further indicate a specific set of goals related to development of WaSH sector. For example, SDG1 focuses to eradicate poverty everywhere in the world. One of the targets to achieve goal no-1 is to improve the access to basic drinking water, sanitation, and proper hygiene. Similarly, SDG3 aims for good health and wellbeing. Additionally, SDG11 focuses on the sustainable and resilient development of safe and inclusive cities with proper access to basic infrastructure and services.

In 2016, UNICEF developed a strategic plan for WaSH practices for the period of 2016–2030 which focused on water and sanitation. The strategy emphasized that everyone in the world should get good drinking water and proper sanitation facilities. They also developed a framework for programming sustainability in WaSH sector which included the community, sector, and service level framework for the WaSH infrastructure. This framework helps to incorporate sustainability with the WaSH programs in a project cycle.

Spatio-temporal impact of global climate change has resulted into scarcity of good resources of drinking water. The alarming changes in the climate pattern also identified to be hindering the effective planning and management of healthy sanitation and hygiene practices. Large sections of population especially in the developing nations are still struggling to manage and overcome the issues and challenges of safe drinking water and sanitation especially during natural disasters. Safely managed drinking water is not available for one in four people in the world and also half of the world population is lacking proper sanitation facilities (WHO/UNICEF JMP 2021). As per the data by WHO/UNICEF JMP 2021around 3.6 billion people in the world lack safely managed sanitation facilities. The report also indicate that around 670 million people lack basic hand-washing facilities and almost 60% of the rural areas experience issues of availability and accessibility to safe drinking water compared to the urban areas

The Components of WaSH Infrastructure and Their Interlinkages

WaSH encompasses the three major sectors essential for community development, namely, water, sanitation, and hygiene. The functions and services rendered by these sectors are inter-related to each other. Their forward and backward linkages are key to achieve healthy community. For example, water is a basic component in the WaSH system. Every human being requires sufficient amount of water to facilitate their day-to-day activities and other basic requirements of life. Despite water being a basic necessity, a huge number of communities still lack proper access to safe drinking water sources and are unaware of proper management of these resources. Most of the people rely on the direct usage of water from the river, dam, lake, etc. The impurity present in these sources of water is likely to cause extreme health issues and slowly result in the impairment of children. Moreover women are forced to collect water by walking long distances for their family's daily needs.

Sanitation is another major component in the WaSH sector. Sanitation is the safe disposal or management of human waste, waste water, and garbage. Proper sanitation facilities are also important for sustainable development of humanity. People without appropriate sanitation facilities practice open defecation and other unhygienic practices which lead to the contamination of the water sources and environment. The issue of unscientific waste management and unhygienic practices of water collection, usage, sanitation systems, etc., reported to be causing major health issues, specifically for children. Though community latrine practices are encouraged in some communities, improper maintenance of these spaces have most often resulted in to a failed attempt in many contexts. Women often feel unsafe to use community latrine facilities. Studies also indicate that women face multiple issues with public sanitation facilities and are at times socially harassed by the public which discourages them from choosing these facilities.

Personal hygiene is the third component in the field of WaSH. Hygiene is a practice which helps the community to maintain healthy life. Access to safe water and appropriate sanitation facilities are key to maintain good practices of hygiene. Proper hygienic facilities protect people and children from major infections and diseases. It also helps women and adolescent girls to maintain appropriate menstrual hygiene. However, the scope for adopting good hygiene practices vary from community to community. Relevant data published by WHO/UNICEF JMP, 2021 revealed that around 2.3 billion people worldwide are still not adopting proper hygienic practices at their home. Lack of availability of facilities for good hygienic practices such as basic infrastructure ensuring supply of good quality water and customized sanitation systems are identified as major detriment to address this challenge. The data also indicate that some community also lack proper access to soap to follow basic hygienic practices at the household level. Moreover, some rural communities are still unaware of maintaining hygiene environment at household level and its importance for healthy life.

The lack of WaSH facilities causes multiple issues in the sustainable development of the population worldwide. Improper WaSH practices directly affect social welfare of women and children. In 2016, it is found that 3.3% of global deaths and 4.6% become disabled due to inadequate WaSH practices (WHO 2019). Diarrhea and other infectious diseases are increasing because of the negligence toward WaSH practices. It majorly affects the children under age five by adversely hampering their nutrition level. Globally among this age group, 13% of children are reported to be deceased and 12% became disabled (WHO 2019). Moreover, lack of provision of proper WaSH facilities at schools in many regions has resulted in to attendance shortage of children, especially in adolescent girls during their menstrual cycle. Usage of impure water for preparing food and cleaning utensils increases the risk of food contamination, leading to severe health issues in people, particularly in children. In many households, especially in the rural regions, women are forced to fetch water from long distances. This adversely affects their mental and physical health results in psychosocial stress (Mills and Cumming 2016). Due to the poor access to safe drinking water, rural households are forced to use surface water for their primary needs because the social inequity of water distribution forces them to do so. Studies indicate that rich people get easy access to water while vulnerable poor sections are neglected (Joy et al. 2014). Further the lack of proper sanitation facilities have compelled the community to practice open defecation.

In short lack of appropriate infrastructure facilities of various elements of WaSH sector result in gender specific and age-specific negative externalities, there by hindering sustainable development of the community at large. Hence future efforts in the provision of WaSH infrastructure calls for a service-oriented approach by adopting context-specific sustainable techniques and methods. The foreground literature underlines that sustainable approaches for planning and management of WaSH systems and their elements are the need of the hour to ensure a resilient and healthy community.

WaSH and Sustainable Community Development

Sustainable development demands the integration of various sectors of development such as social and economic infrastructure preserving the ecological environment. Such growth of any community calls for adopting sustainable interventions for planning, management, and implementation of infrastructure and services appropriate to the bio-physical, socio-economic, and cultural context. It also envisages development practices minimizing air, water, and soil pollution which are detrimental to the community at large. Deforestation, improper management of waste and depletion of biodiversity, water pollution, air pollution, etc. are the results of adopting unsustainable practices. Social sustainability helps to strengthen and develop gender equality in societies and community's welfare. This approach helps to develop the marginalized population and women by ensuring their participation in the social development process along with providing opportunities for growth. Social development with focus on psychological growth of women and children will help in building a sustainable society in the future. Economic sustainability is another pillar emphasizing sustainable practices for enhancing micro and macro economy of the community. It considers ways and methods for income generations, processes, and productions of goods and services that are economically viable, and consumption of goods for the long-term benefit of the community.

Water, sanitation, and hygiene practices play a catalytic role in leading to environmental, social, and environmental sustainability of a community. Conversely, inappropriate and inadequate provision of WaSH facilities would result in to major health risk at the household level especially for children. Table 1 encapsulates the inter linkages between sustainable development pillars and the major dimensions of WaSH infrastructure.

Sustainable WaSH Infrastructure

Sustainable infrastructures are the facilities inclusive of ecological, social, and economic aspects, which reduce the potential risk and provide numerous benefits to the community. Such infrastructures vary in different communities and depends on the local context, size of the community, etc. Availability and accessibility to these facilities have been recommended for the successful implication of various goals targeted in SDGs. Thus, WaSH infrastructure facilities could be considered as a measure to achieve sustainable development in a given region.

Numerous innovative and sustainable infrastructures have been either proposed or in practice developed by different NGOs and organizations in the field of WaSH sector. Such infrastructures include appropriate points for toilets and water collection specifically in schools and households, community-based water storage tanks, recycling and reusable methods for wastewater management, promoting water hygiene points for hygiene practices, infrastructure facilities for proper water resource management, etc. (Swiss Red Cross 2014). Urine diversion dehydration toilets in rural and urban areas, household pit latrines with urine diversion, community-level wastewater management facilities, grey water tower (Muench and Ingle 2012), economically sustainable ferro cement tanks for water storage, sand dam for retaining rainwater, etc., are some of the methods of such infrastructures. The following section presents a few sustainable WaSH infrastructure facilities being practiced in various communities and the positive impact on their community development.

Case 1: Water Access in Rwanda

Life in rural communities of Rwanda was miserable due to the poor accessibility to water. The "Water Access Rwanda" is a social enterprise promoting water accessibility in the country. They developed a mini-grid called "Inuma for the development of water kiosks for the public." "Inuma Nexus" is the main filtration point of water which can filter around 60-l water per minute. "Inuma Nodes" are around 800m away from Inuma Nexus, located in the community concentrated area from that point water is distributed to household also. Most people used to collect water from the Inuma nodes (Rwanda 2021). Other than Inuma, they also promote services such as water sales, borehole drilling, management of water points, pump and filters, fitting and sales.

For example, "Uhira" is a service that focuses on the farmers to get water for farming. It provides

Wash (Water, Sanitation, and Hygiene): Infrastructure as a Measure of Sustainable Development, Table 1 Pillars of sustainable development and WaSH components

	Water	Sanitation	Hygiene
Environmental sustainability	Availability of adequate quantity of water in rural areas helps to manage sustainable agriculture practices. Availability of good quality water ensures sustainable ecology and environment. Excess or shortage of water due to climate change hinders the sustainable infrastructure development Excess surface run-off and reduced groundwater percolation affects the water ecosystem.	Proper sanitation facilities lead to environmental sustainability by preventing open defecation and water pollution (Joanna Esteves Mills and Oliver Cumming, 2016) Reduction, Recycling and reuse of human waste minimize the need for centralized waste disposal systems which in turn reduces environmental contamination. A safely managed sanitation facility should have onsite management of human waste using septic tank or appropriate methods and proper connection to the secondary treatment plant for waste water management.	Household level hygienic practices ensures better housing environment Awareness of hygiene practices in public spaces minimizes spreading of contagious diseases.
Social sustainability	Community management of water supply systems strengthens the social tie-ups Participation of households, especially women and marginalized community in the decision-making process leads to responsible management of water supply facilities Adopting socially accepted tools and technologies minimizes conflicts with the socio-cultural behavior	Provision of socially accepted sanitation facilities improves human well-being, social and development. Household level sanitation facilities reduces anxiety especially among children and risk of sexual assault. Well managed sanitation systems at public spaces, especially in schools minimize lost educational and job opportunities.	Awareness regarding proper hygiene practices from the childhood level helps the communities to reduce the spread of diseases like Covid-19. Hygienic practices of adolescent girls in the menstrual cycle can achieve by positive care of social issues like gender inequalities, poverty, cultural taboos, social stigma etc., by communities.
Economical sustainability	Improved access to water services at the individual or household level, brings in savings in time and provides livelihood opportunities especially for the poor. Improved management of water resources contribute to all economic generating activities directly or indirectly especially in the agricultural and food sector. Effective management of water ecosystem services benefits urban economy Proper mitigation measures for water-related natural disasters can support economic sustainability.	Proper sanitation facilities help to reduce health-related cost In regions with water scarcity , recycling of wastewater, and sludge can support local economy Effective time saved due to reduced illness and direct access to sanitation facilities translates to better productivity	Proper hygiene practices can reduce burden of diseases and related health system cost Development of locally sustainable systems and techniques for practicing household or individual hygiene boost generation of local livelihood opportunities.

W

proper access to water at a reliable price. It also supports the community by digging boreholes for their farming practices. "Amazi" is another service that ensures proper rainwater harvesting. Rainwater harvesting methods help people to store rainwater and treat them for future use. By using this method, the excess unutilized water is saved for the dry season. Another system named "Voma" is a service that focuses on community development through improving water accessibility and quality. It develops water collection points in between the community (Rwanda 2021).

Case 2: WaSH in Ghana

In Ghana, people were suffering due to lack of safe drinking water and sanitation facilities. Various NGOs and organizations in Ghana implemented sustainable systems for ensuring access to safe drinking water and sanitation facilities for communities in different regions of the country.

People in the central Ghonja district was using dam water directly for drinking. In the dry season, many of their sources were drying up. Rainwater harvesting tanks introduced by WaSH Alliance helped them to store water in the rainy season for the dry season. This system helped to save water in nearby schools as well. People used 1000 l polyethylene tanks with 15 years of life for hand washing purposes in the institutional latrines. "Presby Water Project," an organization started in Northern Ghana by the Presbyterian Church of Ghana in 2005, introduced a Ferrocement tank with a 1500 l capacity and 50 years of life. This infrastructure facility enhanced the economic sustainability of the community (WaSH Alliance international Accelerating WaSH 2015).

Sanitation Market (SANIMART) is a method of marketing latrines to households developed by artisans from the same community. The household members have to option to select latrines affordable to them. URBANET is a local NGO in Ghana provided a drip system for selected farmers that treat large quantity of wastewater from toilets and kitchens (WaSH Alliance international Accelerating WaSH 2015).

Another major intervention was treatment of fecal sludge and organic solid waste. In Kumasi, a city in Ghana, 40% of people are using public toilet facilities. This fecal sludge is dried along with municipal solid waste and co-composted together, which is used for farming (Muench and Ingle 2012).

Conclusion

Sustainability is a necessary practice for the future existence of the humanity. Sustainable development goals help to accelerate this. The 6th goal of SDGs focus on clean water and sanitation facilities for all. However, proper mechanisms for the implementation and monitoring of WaSH infrastructure are necessary for achieving the expected outcome of SDG-6. Hence the various sectors of these infrastructure needs to be planned, designed, delivered, and managed effectively by minimizing the negative impacts and thereby achieving sustainability of the community. This calls for integration of project-oriented and service-oriented approach in the organization and implementation of infrastructure. This framework also must seek the ways and methods to integrate the major pillars of sustainability. Finally the sustainability of the already implemented of WaSH infrastructure system in practice for various community also should be assessed for better results. The sustainability assessment of WaSH infrastructures projects and their components along with systems for proper monitoring and maintenance can help the community to achieve the 6th goal of SDGs.

Cross-References

▶ Water, Sanitation, and Hygiene Question of Future Cities of the Developing World
▶ Water Security and Its Role in Achieving SDG 6
▶ Water Security, Sustainability, and SDG 6

References

von Muench, E., Ingle, R. (2012). Compilation of 25 case studies on sustainable sanitation projects from Africa. In *Sustainable Sanitation Alliance* (Issue February). http://www.susana.org/en/resources/case-studies/details/1623

Mills, J. E., & Cumming, O. (2016). The impact of water, sanitation and hygiene on key health and social outcomes: Review of evidence (Vol. 7, Issue 3), UNICEF.

Joy, K. J., Paranjape, S., & Bhagat, S. (2014). *Conflicts Around Domestic Water and Sanitation in India Cases, Issues and Prospects.*

Rwanda, W. A. (2021). *Water Access Rwanda annual report 2020–2021.*

Swiss Red Cross. (2014). *Water, Sanitation and Hygiene (WASH) Guidelines.*

UN. (1987). *Our common future.* World Commission on Environment and Development. https://doi.org/10.9774/gleaf.978-1-907643-44-6_12

UNICEF. (2016). Strategy for water, sanitation and hygiene 2016–2030. In *UNICEF Website.* https://www.unicef.org/wash/files/UNICEF_Strategy_for_WASH_2016_2030.PDF

WaSH Alliance international Accelerating WaSH. (2015). *Accelerating WASH in Ghana best practices from the 2011–2015 WaSH programme.*

WHO/UNICEF JMP. (2021). Progress on household drinking water, sanitation and hygiene 2000–2020 Five years into the SDGs. In *Launch version July 12 Main report Progress on Drinking Water, Sanitation and Hygiene.*

WHO. (2019). Safe water, better health. Geneva: World Health Organization; 2019. *Licence: CC BY-NC-SA 3.0 IGO.* https://apps.who.int/iris/bitstream/handle/10665/329905/9789241516891-eng.pdf

Wastewater Processing

▶ Water Pollution and Advanced Water Treatment Technologies

Wastewater Recycling

▶ Water Pollution and Advanced Water Treatment Technologies

Water – Hydrologic Resources

▶ Water, Sanitation, and Hygiene Question of Future Cities of the Developing World

Water Conservation

▶ Circular Economy and the Water-Food Nexus
▶ Circular Water Economy
▶ Water Security and the Green Economy
▶ Water-Smart Cities

Water Efficiency

▶ Circular Water Economy
▶ Water Security and the Green Economy
▶ Water-Smart Cities

Water for All

▶ Meeting SDG6: Ensuring Safe Drinking Water for All in Rural India

Water Governance

▶ Watershed Sustainability: An Integrated River Basin Perspective

Water Policy

▶ Water Policy in the State of Tabasco

Water Policy in the State of Tabasco

Luzma Fabiola Nava[1,2] and Ojilve Ramón Medrano Pérez[1]
[1]CONACYT-Centro del Cambio Global y la Sustentabilidad, A.C. (CCGS), Villahermosa, Tabasco, Mexico
[2]International Institute for Applied Systems Analysis (IIASA), Laxenburg, Austria

Synonyms

Adaptation; Hydrographic; Lack of political will; Water policy; Wealthy natural resources

Definition

A review of water policy of the state of Tabasco reveals a certain degree of continuity of longest-standing sociopolitical practices. Despite the constraints imposed by climate change and the rapid growth of the population, water policy in the state of Tabasco has not experienced a deep transformation. The water policy and management profile, under different political orientations from those of previous governments, remains the same. Tabasco is a wealthy water region and rich in natural resources. However, water resources management is poor and insufficient. A good dose of political will is required to foster an ad hoc adaptation of the normative framework in order to promote greater favorable institutional resilience and public awareness and participation.

Introduction

Mexico is a hydrological mosaic. As mentioned in Alpuche Álvarez et al. (2021) and Nava (2006), its main characteristics are plentiful water resources and aquatic ecosystems, heterogeneity in its distribution in space and time, and the visions linked to various water management policies, as well as differing models and programs of economic development.

In Tabasco, a complex ensemble of governmental and social interests swirls around water as a natural resource. Tabasco is very rich in natural resources crowned by a lack of socially sustainable and responsible policies on water resources. In the early twentieth century, Tabasco was a state isolated from the rest of the country. The focus of its management of water resources was conditioned by its ecological specificity which translated into a subsistence economy of, dependent on, the exploitation and exportation of tropical forest resources, plantation cultivation, and cow's hides, especially since the Porfiriato era. Nevertheless, it was during the nineteenth century when most of the tendencies, which characterize the relationship between the regional society and the environment, appeared. To this day, these tendencies have been characterized by the strategies of the use of natural resources without mechanisms of regulation of the regional ecosystems. In other words, this predatory attitude in relation to the biophysical environment has promoted the extraction of natural resources without any concern for their conservation or replacement (Tudela 1989). The transformation of nature in Tabasco into an arid zone has had irresponsible repercussions on the performance of water bodies (Sánchez Munguía 2005).

Water bodies in Tabasco have lost their extension to the growth of agriculture. Similarly, fires, the opening of networks of roads and highways, activity of the oil industry, overexploitation of resources, the expansion of cattle ranching, deforestation for extracting timber and intentional burning used to capture animals have all resulted in pollution. In addition, the displacement of water bodies and the lack control of the groundwater table and the lack of flood control are considered as a problem (Guerra Martínez and Ochoa Gaona 2006; Ramos and Palomeque 2017; Zamudio Chimal et al. 2016; Gracia Sánchez and Fuentes Mariles 2005). Consequently, all of this has led to the natural tendency of flooding brought on by runoff, the hydro-morphology of rivers, the lowland plains, the amount of sediment carried by the currents, as well as the integral problem of the drainage system (Gracia Sánchez and Fuentes Mariles 2005). Therefore, water security of Tabasco, with respect to its Achille's heel, lies in the design and ad hoc carrying out of public policies for managing water resources. In this relation, public policy based on a holistic vision is needed, one that goes beyond providing public services, direct utilization, and economic programs. What is needed is a long-term vision water policy, which takes into account the historic evolution of water resources management within the territory, but above all the specificities of this and the society living there.

Water Resources in Tabasco

Shared River Basins

Around the world, shared river basins are distributed unevenly on four continents. Europe has 69 watersheds, Africa 59, Asia 57, and the Americas 78 for a total of 263 river basins which account for 45% of the Earth's surface. Of the 78 shared river basins on the American continent, the most important are the Amazon, Colorado, Mississippi, Bravo, Orinoco, San Juan, and Usumacinta. In the case of Latin America, 60% of the territory drains into shared water systems (Contreras Chablé and Nava 2020).

According to data from the National Water Commission (Comisión Nacional del Agua, CONAGUA 2018), the dynamics of the hydrological cycle in Mexico is equipped to receive 1,449,471 million m^3 of rainwater. Of these, 72.1% returns to the atmosphere, 21.4% drains into rivers, and 6.4% is deposited in underground aquifers. Mexico has 757 superficial watersheds, 8 of which are shared with the United States of America, Guatemala, and Belize. Altogether, 51 main rivers flow into these shared basins. Of these eight basins, three are shared with the United States of America (Bravo, Colorado, and Tijuana), four with Guatemala (Grijalva-Usumacinta, Suchiate, Coatan, and Candelaria), and one with Belize and Guatemala (Hondo) (CONAGUA 2018) (Fig. 1). The Grijalva-Usumacinta river basin is made up of the two rivers of the same name and is the most important basin in Mexico in terms of natural surface runoff (CONAGUA 2018). Tabasco hosts this natural treasure.

The Grijalva-Usumacinta River Basin

The hydrological complex, which makes up the Grijalva-Usumacinta River basin, is shared by various territories. This region is one of the most biologically and culturally diverse in the Southeast of Mexico. The Grijalva River arises in Huehuetenango, Guatemala, and then flows in between the Chiapas highland ecosystem in Mexico. The Usumacinta River arises from the Chixoy and Lacantun rivers in Guatemala. The Grijalva and the Usumacinta rivers converge at Tres Brazos in Tabasco; together for 20 km they flow into the Gulf of Mexico. Both rivers merge into the Grijalva-Usumacinta River Basin (GURB). Located in the southeast of Mexico, the GURB includes, for purposes of national water administration, the states of Tabasco, Chiapas, and small sections of Campeche in Mexico (Nava and Nava Jiménez 2022).

Water from the states of Chiapas, Tabasco, and Campeche drain into this basin, which constitutes an area of 83,553 km^2. A total of hydrological sub-watersheds makes up this hydrological conglomerate. In all, these account for an average natural surface drainage of 60,269 hm^3/year (CONAGUA 2018). The heterogeneity which characterizes each of the main watersheds that constitute the hydrological complex made up of the Grijalva and Usumacinta river basin is notorious. The Grijalva River basin is not the same as the Usumacinta, and it is expected that the Usumacinta conserve the exuberant biodiversity that characterizes it (Contreras Chablé and Nava 2020). Furthermore, with a total territorial extension of 130,087 km^2, this complex body of water is shared by Mexico (63.17%), Guatemala (36.82%), and Belize (0.1%) (García García 2013).

The Grijalva-Usumacinta floods at the end of each rainy season, and in Tabasco, there is no area which escapes the flooding (Tudela 1989). According to data from INEC (2020), the precipitation in the region is the highest in the whole country and one of the highest in the world. The

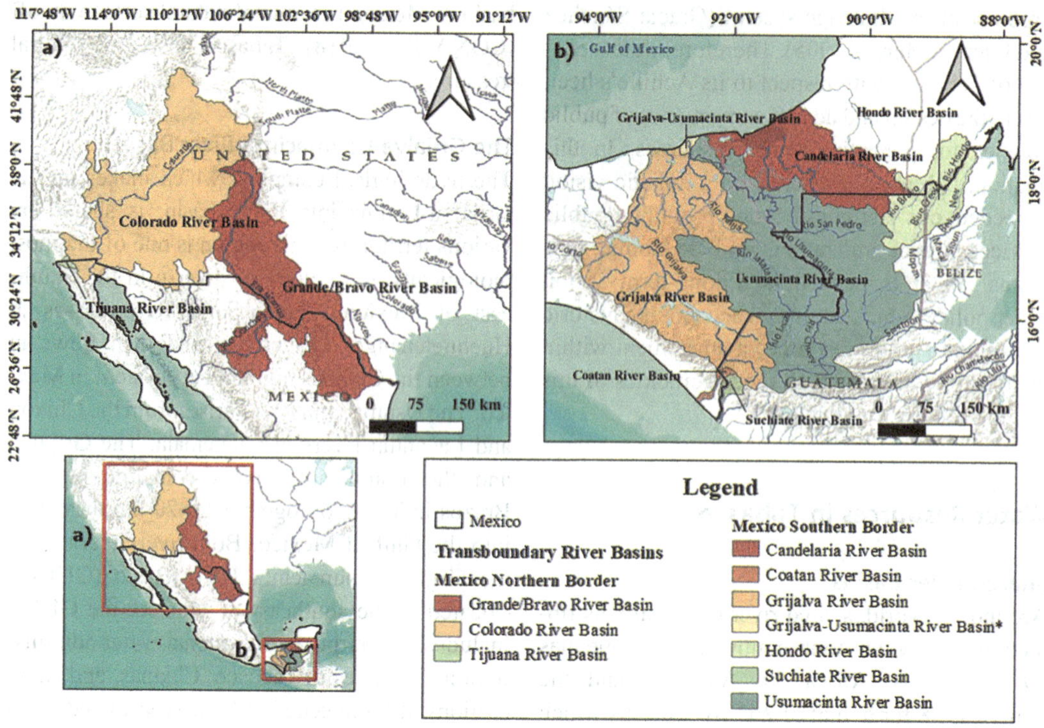

Water Policy in the State of Tabasco, Fig. 1 Mexico shared river basins. (Source: Proprietary with spatial information from TFDD et al. (2018) for frame (**a**); and from the Organismo de Cuenca Frontera Sur (Agency for the shared river basins of the Southern Border) for frame (**b**)) Organismo de Cuenca Frontera Sur, CONAGUA (2021)

annual average is 2143 mm, 2.6 times the national average. In the Sierra of Chiapas and in the Sierra Lacandonia, in some places, the rainfall exceeds 4000 mm annually and sometimes reaches 5000 per year, while on the northern coast, the average is 2093 mm per year and reaches 2750 mm. The masses of cold air from the north and the tropical humid air from both the Atlantic and Pacific cause most of the annual precipitation in the region. Rain falls practically year-round, 83% of the 365 days a year but especially during the summer, autumn, and winter months. Spring is relatively dry. During the summer, rains are intense. During autumn and winter, the northers blow brings prolonged, torrential rain. Rivers and lagoons reach their highest levels between September and November, turning the plains into water mirrors. The season of floods unleashes disasters in agriculture and for people living in the northern coastal plains. The lithological conformation of the riverbanks and the alluvial soil, the sinuosity of the channels with many meanders, and the sediment discharges of material of the upper watersheds make the plains easily erodible and subject to the great floods which occur periodically on the coastal plain, especially in the area known as the "Olla de la Chontalpa." This situation has made building special structures for permeable protection necessary. However, the problems of extreme flooding and phenomena such as those of 1995, 1999, 2007, 2008, and 2010 have not been solved.

Socioeconomic Features of the State of Tabasco

Tabasco is one of the 32 states in Mexico and is located in the southeastern region of the country. It borders on the north with the Gulf of Mexico, on the east with Campeche, on the south with Chiapas, on the southeast with Guatemala, and on the west with Veracruz. Tabasco covers 26,000 km^2 and has a humid tropical climate with an average

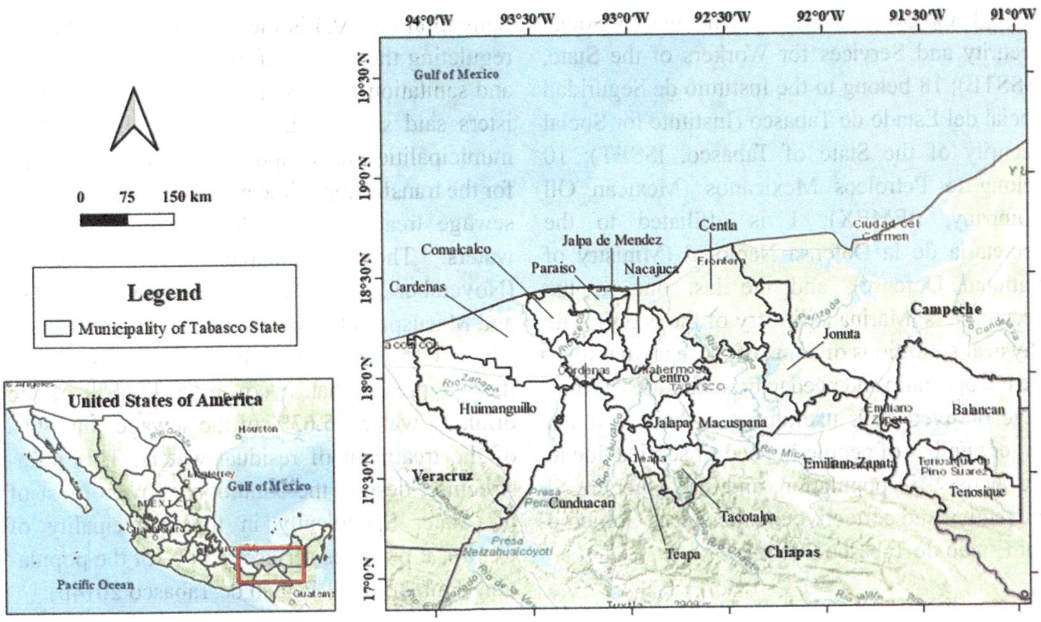

Water Policy in the State of Tabasco, Fig. 2 Tabasco and its municipalities. (Source: Proprietary with spatial information from INEGI)

temperature of 25 °C. It is divided into 2 large hydrological regions, the Grijalva and the Usumacinta regions; into 5 subregions: Centro, Chontalpa, Sierra, Rios, and Pantanos; and into 17 municipalities (Fig. 2).

Population

Tabasco's population in 2015 was 2,426,269 inhabitants of which 49.23% are males and 507% females. It makes up 2% of the national population of 121,347,800 inhabitants. According to projections of the National Demographic Council (CONAPO), the population will reach 2,822,526 with an annual growth rate of 0.8%. In 2050, there will be 3,163,647 inhabitants at a lower annual growth rate of 0.35% (CONAPO 2019).

Housing

According to the 2015 Intercensal Survey by INEGI, in 2015, in Tabasco, there were 646,059 private dwellings, of which: 73.1% had indoor running water, 99.3% had electricity, and 53.3% had drainage connected to public sewage networks.

Healthcare

According to the most recent information provided by the Sectorial Program of Health 2013–2018, out of the total population in 2013 (2,333,646 inhabitants), 1,624,583 inhabitants (69.62%) had no social security and relied on the Ministry of Health for healthcare, while 709,063 (30.38%) receive their healthcare in social security institutions. The infrastructure for attending this population consists of 725 medical units, made up of 33 hospitals, 689 out-patient facilities, and 3 support units. To attend the population without social security, the Ministry of Health has 635 units, of which 5 are hospitals for specialized medicine, 10 are general hospitals, 8 are community hospitals, 538 are out-patient fixed units, 70 are caravans for healthcare, 1 regional laboratory for public health, 1 state hemotherapy center, 1 odonatological unit, and 1 Mexican Red Cross medical unit.

Those residents who have social security are cared for in 90 medical units of which: 34 belong to the Instituto Mexicano del Seguro Social (Mexican Institute of Social Security, IMSS); 26 belong to the Instituto de Seguridad y Servicios Sociales de los Trabajadores del Estado (Institute for Social Security and Services for Workers of the State, ISSSTE); 18 belong to the Instituto de Seguridad Social del Estado de Tabasco (Institute for Social Security of the State of Tabasco, ISSET); 10 belong to Petroleos Mexicanos (Mexican Oil Company, PEMEX); 1 is affiliated to the Secretaría de la Defensa Nacional (Ministry of National Defense); and the last one to the Secretaría de Marina (Ministry of the Navy). The physical conditions of some of these medial units are not optimum and need to be substituted as they have outlived their usefulness or have a much larger number of people in need of service due to the increase in population, and still others need preventive and corrective maintenance (Gobierno del Estado de Tabasco 2014a).

Economy

In economic terms, the contribution of Tabasco to the national GDP is 2.3%, which places the state in the fourteenth position, with contributions similar to the states of Queretaro and Sinaloa. Mining is the sector with the greatest contribution to the state's GDP and consists of the extraction of sulfur, crude oil, and natural gas. Tabasco holds the first place at the national level in the extraction of sulfur and crude oil, and second in natural gas. With respect to agricultural production, the cultivation of yuca, cacao, bananas, pineapple, and squash are the highest in the country. The main fishery products are snood, flag fish, snapper, oysters, bass, and wahoo. The production of the first three lead the country at the national level (INEGI 2015).

Infrastructure

Tabasco has a network of 10,710 km of highways and roads. 596 km are federal highways, 5340 are state roads. 4631 km are rural roads, and 143 km are improved paths: the length of the railroad is 300 km of tracks. There is also an international airport and two ports (Frontera and Dos Bocas) (INEGI 2015).

Public Water and SANITATION

According to the Special Program of the State (2013–2018), the Commission of Water and Sanitation (CEAS) is the agency responsible for regulating the service of drinking water, sewage, and sanitation in the state. CEAS directly administers said services in 14 municipalities. Three municipalities have applied to the government for the transferring of the public service, drainage, sewage treatment, and distribution of residual waters. These municipalities are Balancan (November 29, 2005); Centro (April 23, 2003); and Macuspana (August 8, 2005). The infrastructure of CEAS along with that of these municipalities at present has a coverage 87.79% of the drinking water, 55.63% of the sewage, and 40% of the treatment of residual waters. This infrastructure, despite the conditions, covers most of the state. Specifically in the municipality of Centro, it provides service to 90% of the population (Gobierno del Estado de Tabasco 2014b).

Education

The average grade reached in Tabasco in the population 15 years and over is 9.3, which is equivalent to finishing middle school. Besides, for every 100 people 15 years and older, 4.8 have received no schooling; 52.7 have finished primary education; 23.5 have finished high school; and 18.8 have finished higher education. However, in Tabasco, 5 out of every 100 people 15 years and older do not know how to read or write (INEGI 2015).

Hydrologic Characterization of the State of Tabasco

Tabasco is the state with the most complex hydrological network in the country. It has 184 km of coasts, 1.58% of the national total (F-ODM 2012), and has the most voluminous rivers in Mexico. The Grijalva and the Usumacinta rivers account for 30% of the surface water drainage in the country (López Castañeda and Zavala Cruz 2019; Plascencia Vargas et al. 2014). Mexico has a hydrological system of 633,000 km in length. However, for administrative and management purposes, the Mexican territory has been organized into 13 hydrological administrative regions (RHA). The hydrological complex made up of the Grijalva and Usumacinta river basin constitutes the Southern border XI RHA with a length of

1521 km. The annual average natural drainage of water is 124,477 hm^3 per year to which is added 23 aquifers with filtration of 22,218 HM^3 per year. There are 8,000,000 inhabitants in the region averaging 18,776 m^3 per person per year (CONAGUA 2018).

In Tabasco, the total volume of extraction of surface water is 321,219,040 m^3/year (REPDA 2019). The largest volume of water, 42.33%, is given over to agricultural use, 34% is for urban public use, 15.42% for industrial use, 5.75% is for aquaculture, and 1.44% for services. The livestock sector and other multiple uses has less than 1% awarded to them. Villahermosa is the capital of the state and is located in the southeastern part of the Centro subregion at the coordinates 17°59′ north and 92°56′ west with an average altitude of 10 m above sea level. Villahermosa is nestled between the Grijalva and Carrizal Rivers. It has a surface area of 61 km^2 and a population of 684,847 inhabitants according to the II Population Census in 2015 (INEGI 2020).

Tabasco is a flood plain which drains into the Grijalva-Usumacinta system and a state characterized by the profuse presence of water (Salazar Ledesma 2013) and the movement of river water (García 2013), is a fragmented hydro-social territory due to its relations with power and the political-administrative instruments for managing its hydric resources. The hydric exuberance of Tabasco has led to the emergence of feuds over the ownership of the land, military confrontations, but most importantly, to the coexistence of administrative policy which has contributed to the historic and present-day conformation of the territory with respect to the river basins within it (Salazar Ledesma 2013; Tudela 1989).

The Flood Plain and Its Achille's Heel

Tabasco is situated on a large plain, a vast alluvial plain, barely above sea level. Water is omnipresent and takes on a main role (Tudela 1989). Approximately 60% of the territory of Tabasco is less than 20 meters above sea level (masl). The average altitude of 11 out of 17 municipal capitals is below this level: Cárdenas, 15 masl; Comalcalco, 9 masl; Cunduacán, 10 masl; Frontera, 2 masl; Huimanguillo, 20 masl; Jalpa, 8 masl; Jonuta, 15 masl; Macuspana, 17 masl; Nacajuca, 5 masl; Paraíso, 2 masl; and Villahermosa, 10 masl (Díaz Perera, 2014). Accordingly, the inhabitants of Tabasco have developed a culture of water defined as an experience acquired through millennial interaction with water in the context of two river basins and a coastal plain (Díaz Perera 2014). The Grijalva-Usumacinta floods at the end of each rainy season, flooding large sectors of the plain and determining the interconnection of a group of intermittent or abandoned river basins and lagoons, many of which are permanent. To this, we must add the fact that the soil is not very permeable, causing any precipitation over 50 mm over a period of 24 h to bring about some kind of flooding (Tudela 1989).

In this respect, Arreguín-Cortés et al. (2014) offer an analysis of the floods of the plains of Tabasco in the period of 1995–2010. These authors argue that the factors which influence the flooding are mainly a result of the lack of a proper territorial framework, the deforestation of the higher regions of the basins, the false concept of the diminishing hydrologic regime associated with the construction of large dams on the Grijalva River and climate change. The flood of 1995 was caused by the cyclones Opal and Roxanne which resulted in an average accumulated precipitation of 1792 mm for the period of June 1 to October 31. In 1999 and during the same period, the average precipitation was|1720 mm, with isohyets concentrated in Peñitas on the order of 2450 mm. The phenomenon was due to a combination of tropical fronts 26 to 30, tropical depression 11, and cold fronts 4 and 7 which brought on three surges in the rivers in the mountains, and with no control caused flooding near the city of Villahermosa. The flood of 2007 was characterized by abrupt and intense cooling from the Pacific Ocean, which caused a cold phase of El Niño -Southern Oscillation (ENSO). The rainy season came early with the impact of tropical storm Barbara on the upper Grijalva on June 2. The intense precipitation continued until the end of October with an accumulation of 1423 mm, with its nucleus in the Peñitas basin with about 2500 mm. This event was produced by the incidence of cold fronts 4 and 7 and reinforced four days later by tropical storm Noel in which in less than 72 h over 1000 mm of rain accumulated in the Peñitas basin. A year later, the heaviest rain was in the basin of the tributary of the Usumacinta River. The Grijalva River basin registered accumulated precipitation of 1510 mm, a volume far greater than the previous year. The level of precipitation at the El Muelle station surpassed the NAMO (Nivel de aguas máximas oridinaria) (maximum ordinary level of water) for September 23–29. The embankments built in 2000 did not reach their maximum ordinary level of water, although the use of gunny sacks with sand were anticipated for this season. Finally, in 2009, a *déjà-vu* event took place. The rainy season was similar to the one of 1999, subject to the presence in the intertropical area of the convergence of the upper Grijalva River basin for 40 days, along with tropical storms Karl and Mathew, with accumulated precipitation for the period of 1572 mm, greater than during the rainy seasons of 2007 and 2008. These floods caused very severe catastrophic loses, some of which brought about profound political effects by accelerating the government's plans to intervene the hydrological and hydraulically systems.

Priorities of Public Policy

In Tabasco, flooding is an expression of nature with which the plain must coexist by means of managing water resources and the ability of adaptation of society. In this respect, public policy talks a lot about mitigating the adverse effects of floods through infrastructure projects. While these projects are necessary due to the complex and dynamic hydric system, they must go hand in hand with measures of adaptation and plans of resilience which take into account local knowledge and allow for strengthening response capacities of the local population and institutions. However, the cost of inaction and inefficiency in scientific reports and the federal and state government to persuade the citizens of this reality represent an important challenge in facing the adverse effect of flooding in Tabasco.

Non-exhaustive Review of Public Regulations

Since the first report of flooding conditions and hydric wealth in the Spanish chronicles of 1579 (Rivera-Trejo 2020), Tabasco has recurrently been affected by floods (Table 1). This is also highlighted by Mundo-Molina (2020): Since 1782, Tabasco has been affected by at least 20 events of flooding.

Despite the long list, institutional political efforts in water resources matters aimed at solving the problems in Tabasco are relatively recent. In the first place is the creation, by the federal administration of Miguel Aleman Valdes (1946–1952), of the Commission of the Grijalva River (June 27, 1951), which was created as a technical agency in which experts would propose solutions to solve the problems of flooding in Tabasco. Said Commission is a dependency of the then Secretaría de Agricultura y Recursos Hidráulicos Ministry of Agriculture and Hydraulic Resources, SARH).

Later, during the federal administration of Adolfo Ruiz Cortines (1952–1958), the first project for controlling flooding on the plain of Tabasco was begun. It was declared that Tabasco would be "free of flooding," with the structures and measures to control the Grijalva River and the plans for intensive agriculture on the plain on no fewer that 500,00 hectares (Rivera-Trejo 2020). However, it is not until 1963 when the works of this plan were concreted with the completion of the Malpaso Dam. Along with this project, others such as the channel to alleviate the Samaria Gulf (1965–1978) and the Chontalpa Plan (1966) were promoted. Later, the Angostura Dam (1975), Chicosen Dam (1980), and the Peñitas Dam (1987) were finished, and with this, the system of dams on the upper and middle Grijalva was complete. The Commission of the Grijalva River concluded its activities in 1985. In 1989, CONAGUA was created as the entity which administered national waters.

Within the context of this new institutional scene, and despite the early efforts to control flooding, severe new floods occurred in Tabasco in 1989 and 1995, and along with these, a new plan for controlling flooding was announced. In 1995, the Integral Program for Controlling Flooding (PICI) was presented as a proposal for

Water Policy in the State of Tabasco, Table 1 Historic flooding and plans for controlling it in Tabasco[a]

Year	Reference of the flooding
1579	The Spanish chroniclers Vasco de Rodríguez and Melchor Alfaro in their memoirs talk about conditions of flooding, the richness of the rivers, and the everlasting summers, which characterize Tabasco.
1782	Record of the "Diluvio de Santa Rosa" (flood of Santa Rosa)
1820	Record of the "Diluvio grande" (great flood)
1868	Record of a flood which affected the neighborhoods of Santa Cruz, Mustal, Mayito, and Curahueso; flooding of arroyos and bodies of water like La Pólvora
1879	800 houses flooded
1886	Level: 13.71 masl
1888	Cyclone with torrential rain and strong wind; floods in Villahermosa
1889	Houses flooded (155), deaths, and missing ships
1909	Rain for a month from October to November 20 with 2953 people affected
1912	The Grijalba River overflows
1927	Record of swelling and flooding of the Grijalva
1932	The rain which was registered caused the digression ("breaking") of the basin towards the "Hoya de la Chotalpa," thus creating the present-day Samaria River
1936	Over 50% of the general spendings were used to attend to damages in agriculture and livestock caused by the floods of October and November
1944	Record of a cyclone affecting the municipalities of Nacajuca, Jalpa de Méndez, Cunduacán, and Huimanguillo, as well as the communication routes and banana industry
1951	Creation of the Commission of the Grijalva River
1952	Chotalpa affected by flooding with damage of over 100,000,000 pesos of that time
1953	Upon advancing projects to control the Grijalva River and plans for intensive agriculture in 500,000 hectares and beginning the first project for controlling flooding on the plain of Tabasco President Adolfo Rui Cortines declared that Tabasco would be "free from flooding"
1955	Cyclone Janet caused substantial damages and losses of human lives
1963	Overflowing of the Grijalva River
1964–1966	The Malpaso Dam (1964) is completed, a project for a channel to alleviate Samaria Gulf (1965–1978), and the Plan Chontalpa (1966) to develop 300,000 hectares for agriculture
1969	Flooding of the Grijalva River
1973	Record of intense rain with property damage in the city and public services paralyzed
1975	Unveiling of the angostura dam
1980	The historic precipitation of October 29 causes collapsing of bridges, highways washed out and damage to housing. The Chicosen dam is unveiled
1987	Unveiling of the Peñitas dam and along with it four dams of the Grijalva hydroelectric system
1988	Record of 641 mm with 7000 people affected, mainly in the municipality of Centro
1989	During 19 days in October, rain accumulated to the tune of 609 mm, causing flooding in Villahermose. The Comision Nacional del Agua (CONAGUA) (National Water Commission) was created
1995	Hurricanes Roxanne and Opal in the Gulf of Mexico bring 622 mm of rain causing flooding, 36,900 families affected, 13 deaths and damage to agriculture
1995	The Proyecto Integral Contra Inundaciones (PICI) (The Integral Project for Controlling Floods) (1995–2007) for Tabasco (1995–2007) was announced
1999	With cold front no. 7, a monthly accumulation of 696 mm of rain caused flooding of the tributaries, flooding in Villahermosa with 161,521 people affected, and 23,000 hectares of corn and bananas flooded
2007	Reports of flooding caused by two cold fronts and the releasing of water from the Peñitas dam left 500,000 people affected in 17 municipalities, 1 death, and 100% of crops lost
2008–2012	Proyecto Hídrico Integral de Tabasco (2008–2018) (Hydrological Project for Tabasco) was announced
2020	Floods caused by cold front no. 4 and releasing water from the Peñitas Dam (1750 m^3/sec) caused flooding and a Declaration of a State of Emergency for 17 municipalities (Agreement 3568 published in the Official State Newspaper)
2020	Announcing of a new project to prevent flooding in Tabasco

Source: Proprietary with information from Rivera-Trejo (2020) and Arreguín-Cortes et al. 2014. *Plans for control in Tabasco*[a] in gray added by the authors

W

solving the problem of the systemic flooding, with a duration of from 1995 to 2007 and a planned cost of two billion pesos. PICI planned to control the drainage from the upper part of the Grijalva basin with various hydraulic works such as embankments and underrun protection, drainage, crossing structures, dredging natural causeways, and drainage of the Mezcalapa-Samaria hydraulic systems, the Sierra and Carrizal-Medellin Rivers (Arreguín-Cortés et al. 2014). It is worth mentioning that in 1999, cold front no. 7 flooding occurred, leaving 161,521 people and 23,000 hectares affected. Later, in 2007, severe floods were produced by two cold fronts and the opening of the gates of the Peñitas Dam. This situation left 500,000 people affected in 17 municipalities, 1 death, and the loss of 100% of the crops (Rivera-Trejo 2020). As an institutional reaction to the event of the new flooding, the Integral Hydric Plan for Tabasco (PHIT, 2008–2012) was presented with budgetary costs of 9300 billion pesos. However, in 2013, this program was substituted by the Proyecto Hidrológico de Tabasco (Hydrological Project of Tabasco, PROHTAB), which with 185 different actions hoped to continue the programs for controlling flooding which had been presented earlier. In the beginning, the investment for PROHTAB was 16 billion pesos, but from the first disburse of 2 billion pesos for the program, the amount has decreased year by year. As a result, in 2018, the budget was reduced to zero, and in 2019 and 2020, the investment was minimum, with a budgetary deficit for PROHTAB of an estimated 13 billion pesos for works not carried out since its official creation in 2015 (Albert-Hernandez 2020). Therefore, PROHTAB has been undermined by the budgetary reductions and today is for all practical purposes, nonexistent.

Lastly, in 2020, floods were registered in extensive areas of the state caused by cold front no. 4 and the opening of the gates of the Peñitas Dam (1750 m^3/second), bringing about a Declaration of Emergency for 17 municipalities (Agreement 3568 published in the Official Newspaper of the State). Because of this event of severe flooding, the federal administration of Andres Manuel Lopez Obrador (AMLO) (2018–2024) announced a new project to control flooding in Tabasco. Although the details of this program are not known, several of the measures such as dredging the causeway, the creation of the Commission of the Grijalva River and a decree for an Integral Plan for preventing flooding in Chiapas, Tabasco, and Veracruz have been announced. To this end, the Integral Plan contemplates "an improved handling of dams, water purification plants, hydroelectric projects, flood control, cleaning up rivers, wellbeing and other matters" (Morales and Villa y Caña 2020). Along these same lines, in a visit to Tabasco, AMLO announced that "a Commission for the Grijalva River would be created, without bureaucratic devices or any of that stuff, with a person in charge who (I will) name, for managing the four hydroelectric plants on the Grijalva: Angostura, Chicoasen, Malpaso and Peñitas" (AMLLO 2020a). In addition, AMLO has expressed the need for civil protection in the face of events of flooding, expressing the need for the flood basins of the Grijalva River to be kept empty during the rainy season in August, September, October, and November (AMLO 2020b). In fact, recently a decree was published which establishes the measures of coordination which should be observed by the agencies and entities of the Federal Public Administration, for the management of dams and the reduction of disasters caused by flooding of the basin of the Grijalva River and its relation to the control and distribution of electric energy, with social responsibility and civic protection (DOF: 01/12/2020), which indicates coping capabilities shown by the institutions in the face of floods which occur.

Derived from this, and since the first plan or bill was proposed by Adolfo Ruiz Cortines (1952–1958) until the most recent Integral Plan announced by AMLO (2018–2024), there has been a total of five plans to solve the problem of flooding in Tabasco. Nevertheless, up until now, many of the plans have focused on mitigating the effects of flooding by way of infrastructure works, but not measures of adaptation and strengthening the local ability to face these events. This having been said, the problems of flooding continues to be an unfinished work in which blame has always been sought (for example, the dams) and a mistaken rhetoric has been used (for example, "avoid floods"), all of these continuously ignoring the nature of the flood plain in Tabasco, forgetting that it is a part of the largest delta flood plain in Mexico, into which the mightiest rivers of Mexico flow (Mundo-Molina 2020). Thus, in the face of this palpable reality of the territory, over and above preventing floods, the capacity to respond and the adaptation and resilience at the local level to confront these natural events must be adapted.

Civil Protection and Infrastructure

Floods are a serious problem in the Mexican southeast. However, it is a well-known fact that there is scarce responsive ability in the face of these phenomenon in this region in both institutional and socioeconomic terms (Andrade-Velázquez and Medrano-Pérez 2020). In this respect, the states of Chiapas and Tabasco report larger numbers of damages to houses, dwellings, and hospitals, as well as the number of people affected in these kinds of events (PNUD México-INECC 2018). In fact, the attention for those with damaged property includes cleaning, disinfecting, repairing, and fumigating housing, together with replacing household goods and distributing basic grocery supplies. These actions are promoted and headed by federal institutions such as the Ministry of the Navy, the Ministry of Defense, the Ministry of Wellbeing, CONAGUA, etc., and by state agencies such as the Institute for Civil Protection of the State of Tabasco (IPCET), in the various localities affected by the events. In terms of civil protection, actions have included alerts before the storm hits, setting up shelters, closing highways, evacuating vulnerable places and people, as well as preventive measures such as dredging the riverbeds. However, these actions tend to be responsive in the face of events of flooding, very often being the product of local immediatism and urgency. On the other hand, the occurrence of severe flooding in Tabasco goes hand in hand with the contemporary implementation of integral projects and plans for facing flooding in the state. With respect to PICI and PHIT, while the functioning of these plans cannot be evaluated globally since they must be considered incomplete (Rivera-Trejo 2020), both plans were interrupted by severe floods in Tabasco. In turn, for PHIT, there were reports of not only unfinished works but also omissions, apparent acts of corruption, and noncompliance with regulations (Badillo 2020).

Challenges and Opportunities

Since the first call to "free from or avoid flooding in Tabasco" by the federal administration under Adolfo Ruiz Cortines (1952–1958), there has been a parade of hydraulic projects and plans to prevent flooding. Since then, there has been repeated talk of mitigating the adverse effects of flooding with little attention paid to adapting to the floods. Nevertheless, within the context of climate change, and taking into account the geomorphological characteristics of Tabasco, adapting to these kinds of events is imperative. In this sense, the predominance of a focus on engineering, as well as political and institutional inefficiency, and the lack of any scientific narrative which reflects on and persuades citizens of this reality represent, all together, important challenges for confronting the problem. Therefore, for a territory that often faces the adverse effects of extreme hydrometeorological phenomenon, adapting not only deserves priority attention but also should be included as a guiding principle in public policies and plans and programs to be implemented in these territories.

From this perspective, next we will propose a *Decalogue* of the challenges and opportunities for facing the problems of flooding in Tabasco:

W

1. Adapt the institutional framework and regulating norms of the complex hydrology of the floodplain of Tabasco in order to promote

greater favorable institutional resilience for citizens' adaptation and participation.

2. Promote territorial regulation, which defines the guidelines for planning in the territory with a participative focus on development and sustainable management of human settlements, and one, which responds to the nature of the territory of Tabasco, thus contributing to the prevention and adaptation in the face of flooding.
3. Identify and conclude the priority hydraulic works and establish a maintenance program for the whole system for protection against flooding.
4. Establish a network of climatic and hydrometric measuring and monitoring with real-time time and spatial data and information using the appropriate technology in each strategic location in the state, effectively contributing to the management and decision-making processes in the face of events of flooding.
5. Coordinately design and implement actions of civil protection in the face of flooding events between the various entities and agencies of the federal, state, and local governments.
6. Implement actions and programs focused on strengthening the response capacity at the local level.
7. Strengthen technical capabilities with training programs at various levels of government, local organization, and for anyone interested.
8. Promote social participation and integrate local knowledge into the design of public policies and the implementation of programs and actions.
9. Establish educational plans and campaigns and social awareness on the problems of flooding on a permanent basis using various means of social communication.
10. Change the vision of public policies and narrative in the face of the problems of flooding, progressing from "there will be no more flooding" to "adapting to living in a floodplain."

Cross-References

- ► Adapting Cities to Climate Change
- ► At the Intersection and Looking Ahead
- ► Blue-Green Cities: Achieving Urban Flood Resilience, Water Security, and Biodiversity
- ► Building Community Resilience
- ► Cities in Nature
- ► City Visions: Toward Smart and Sustainable Urban Futures
- ► Water Security and Its Role in Achieving SDG 6

References

Albert-Hernández, M. (2020). Del PICI al PROHTAB, miles de millones de pesos invertidos en evitar inundaciones en Tabasco. En XEVA 91.7 FM. Available at: http://xeva.com.mx/nota.cfm?id=121081&t=del-pici-al-prohtab-miles-de-millones-de-pesos-invertidos-en-evitar-inundaciones-en-tabasco. Accessed Oct 2020.

Alpuche Álvarez, Y. A., Nava, L. F., Carpio Candelero, M. A., & Contreras Chablé, D. I. (2021). Vinculando ciencia y política pública. La Ley de Aguas Nacionales bajo la perspectiva sistémica y de servicios ecosistémicos. *Gestión y Política Pública, 30*(2). División de Administración Pública del Centro de Investigación y Docencia Económicas, A.C. https://doi.org/10.29265/gypp.v30i2.881.

Andrade-Velázquez, M., & Medrano-Pérez, O. R. (2020). Precipitation pattern in Usumacinta and Grijalva basins (southern Mexico) under a changing climate. *Revista Bio Ciencias, 7*, e905. https://doi.org/10.15741/revbio.07.e905.

Arreguín-Cortés, F. I., Rubio-Gutiérrez, H., Domínguez-Mora, R., & de Luna-Cruz, F. (2014). Análisis de las inundaciones en la planicie tabasqueña en el periodo 1995-2010. *Tecnología y Ciencias del Agua., V*(3), 5–32.

AMLO. (2020a). Conferencia de prensa. Avances en la atención a Tabasco por inundaciones. En Youtube: Andrés Manuel López Obrador, (minuto 38–39). Available at: https://www.youtube.com/watch?v=iG5pzzWihxc&feature=emb_title. Accessed Nov 2020.

AMLO. (2020b). Presidente firmará decreto para regular presas, prevenir inundaciones y proteger a la población. En boletines de presidente.gob.mx. Available at: https://presidente.gob.mx/presidente-firmara-decreto-para-regular-presas-prevenir-inundaciones-y-proteger-a-la-poblacion/. Accessed Nov 2020.

Badillo, D. (2020). ¿Por qué Tabasco es un desastre en prevención de inundaciones? En El Economista,

Estados. Available at: https://www.eleconomista.com.mx/estados/Por-que-Tabasco-es-un-desastre-en-prevencion-de-inundaciones-20201115-0001.html. Accessed Nov 2020.

Contreras Chablé, D. I., & Nava, L. F. (2020). La cooperación en cuencas transfronterizas: una oportunidad para la cuenca del río Usumacinta. *Kuxulkab,* Revista de divulgación de la División Académica de Ciencias Biológicas. *Universidad Juárez Autónoma de Tabasco (UJAT), 26*(56), 15–30. https://doi.org/10.19136/kuxulkab.a26n56.3342.

CONAGUA. Comisión Nacional del Agua. (2018). Estadísticas del agua en México. CONAGUA. México: Secretaría de Medio Ambiente y Recursos Naturales. P. 306. Available at: http://sina.conagua.gob.mx/publicaciones/EAM_2018.pdf. Accessed Aug 2019.

CONAPO. Consejo Nacional de Población. (2019). *Colección. Proyecciones de la población de México y las entidades federativas 2016-2050Tabasco*, Primera edición: julio 2019. ISBN: 978-607-427-320, Available at: https://www.gob.mx/cms/uploads/attachment/file/487371/27_TAB.pdf. Accessed Nov 2020.

Díaz Perera, M. Á. (2014). La construcción de las condiciones históricas de posibilidad de un desastre: el caso de dos colonias de Villahermosa, Tabasco. Mario González y Claudia Brunel (Coordinadores), *Montañas, pueblos y agua. Dimensiones y realidades de la cuenca Grijalva.* México DF, El Colegio de la Frontera Sur: Juan Pablos Editor, pág. 181–212.

DOF: 01/12/2020. Diario oficial de la federación, *DECRETO por el que se establecen las medidas de coordinación que deberán observar las dependencias y entidades de la Administración Pública Federal, para el manejo de presas y la reducción de desastres por inundaciones en la cuenca del Río Grijalva, y su relación en el control y despacho de generación eléctrica, con sentido social y de protección civil,* Available at: https://www.dof.gob.mx/nota_detalle.php?codigo=5606505&fecha=01/12/2020. Accessed Nov 2020.

F-ODM. Fondo para el Logro de los Objetivos para el Desarrollo del Milenio. (2012). Documento Estatal de Tabasco. Programa Conjunto para Fortalecer la Gestión Efectiva y Democrática del Agua y Saneamiento en México para el logro de los Objetivos de Desarrollo del Milenio (PCAyS), México, 183p.

García García, A. (2013). Las inundaciones fluviales históricas en la planicie tabasqueña: un acercamiento integral de largo aliento en la perspectiva de cuencas hidrográficas. En *Cuencas en Tabasco: una visión a contracorriente* (61–99). México, D. F.: Centro de Investigaciones y Estudios Superiores en Antropología Social (CIESAS).

Gracia Sánchez, J. & Fuentes Mariles, Ó. A. (2005). La problemática del agua en Tabasco: inundaciones y su control. (Eds). Blanca Jiménez y Luis Marín, *El agua en México vista desde la academia,* México: Academia Mexicana de Ciencias, p. 177-185.

Guerra Martínez, V., & Ochoa Gaona, S. (2006). Evaluación espacio-temporal de la vegetación y uso del suelo en la Reserva de la Biosfera Pantanos de Centla, Tabasco (1990–2000). Investigaciones geográficas, (59), 7–25. Available at: http://www.scielo.org.mx/scielo.php?script=sci_arttext&pid=S0188-46112006000100002&lng=es&tlng=es. Accessed Sep 2020.

Gobierno del Estado de Tabasco. (2014a). *Programa Sectorial de Salud 2013–2018,* Primera Edición, Septiembre de 2014. Impreso en Talleres Gráficos del Gobierno del Estado de Tabasco. Secretaría de Salud. 104 p. Available at: https://tabasco.gob.mx/sites/default/files/users/setabasco/06-PROGRAMASECTORIALDESALUD.pdf. Accessed Nov 2020.

Gobierno del Estado de Tabasco. (2014b). *Programa Especial de la Comisión Estatal de Agua y Saneamiento 2013–2018,* Primera Edición, Septiembre de 2014. Impreso en Talleres Gráficos del Gobierno del Estado de Tabasco. Comisión Estatal de Agua y Saneamiento (CEAS), 68 p. Available at: https://tabasco.gob.mx/sites/default/files/users/setabasco/05-PROGRAMAESPECIALDECEAS.pdf. Accessed Nov 2020.

INECC. Instituto Nacional de Ecología y Cambio Climático. (2020). La cuenca de los ríos Grijalva y Usumacinta. Available at: http://www2.inecc.gob.mx/publicaciones2/libros/402/cuencas.html. Accessed Nov 2020.

INEGI. Instituto Nacional de Estadística y Geografía. (2020). Available at: https://www.inegi.org.mx/datos/. Accessed Sept 2020.

INEGI. Instituto Nacional de Estadística y Geografía. (2015). Resumen por entidad estatal. *Encuesta Intercensal 2015.* Available at: http://cuentame.inegi.org.mx/monografias/informacion/tab/default.aspx?tema=me&e=27. Accessed Sept 2020.

López Castañeda, A., & Zavala Cruz, J. (2019). *Hidrología. En La biodiversidad en Tabasco. Estudio de Estado* (Vol. I, pp. 51–59). México: Conabio.

Morales, A. & Villa y Caña, P. (2020). AMLO contempla decreto para controlar presas en Tabasco, Chiapas y Veracruz. En El Universal, Nación. Available at: https://www.eluniversal.com.mx/nacion/amlo-contempla-decreto-para-controlar-presas-en-tabasco-chiapas-y-veracruz. Accessed Nov 2020.

Mundo-Molina, M. (2020). El edén estragado. En Ojarasca, La Jornada, 11/12/2020. Available at: https://ojarasca.jornada.com.mx/2020/12/11/el-eden-estragado-2576.html Accessed Dec 2020.

Nava, L. F. (2006). Cuando la gestión del agua se vuelve problemática: el caso de México. *En Observatoire des Amériques, 6*(38), 1–10.

Nava, L. F., & Nava, J. C. (2022). Sustainable development and responsible tourism: The Grijalva-Usumacinta lower River Basin. In R. Brears (Ed.), *The Palgrave Encyclopedia of urban and regional futures.* Cham:

Palgrave Macmillan. https://doi.org/10.1007/978-3-030-51812-7_148-1.
Organismo de Cuenca Frontera Sur, CONAGUA. Mapa de cuencas hidrológicas transfronterizas. Programa Hídrico Estatal 2014–2018 del Estado de Chiapas. Available at: http://www.igh.com.mx/programa_hidrico_chiapas/. Accessed Jan 2021.
Plascencia Vargas, H., González Espinosa, M., Ramírez Marcial, N., Álvarez Solis, D., & Musálem Castillejos, K. (2014). Características físico-bióticas de la cuenca del río Grijalva. En *Montañas, pueblos y agua. Dimensiones y realidades de la cuenca Grijalva.* Vol. I, (coords). Mario González Espinosa y Marie Claude Brunel Manse. El Colegio de la Frontera Sur (ECOSUR), México, pp. 29–79.
PNUD México-INECC. (2018). Vulnerabilidad actual y futura de los recursos hídricos ante el cambio climático en los estados del sureste de México, con enfoque en el desarrollo urbano sustentable. Proyecto 86487 "Plataforma de Colaboración sobre Cambio Climático y Crecimiento Verde entre Canadá y México". 206 pp. Centro del Cambio Global y la Sustentabilidad en el Sureste, A.C., México.
Ramos, R., & Palomeque, M. (2017). Modelación del cambio de uso del suelo en Comalcalco, Tabasco, México. *Revista de Urbanismo, 37*, 1–17. https://doi.org/10.5354/0717-5051.2017.47986.
Rivera-Trejo, F. (2020). Inundaciones en Tabasco Mitos y realidades. En Youtube: Gerencia AMH Oficial. Evento en línea organizado por la Asociación Mexicana de Hidráulica (AMH). Available at: https://www.youtube.com/watch?v=sffeg92r2zk Accessed Nov 2020.
Salazar Ledesma, F. (2013). Las cuencas fluviales de Tabasco como recurso teórico-metodológico en el estudio de la organización territorial prehispánica y española de los siglos XVI y XVII. En *Cuencas en Tabasco: una visión a contracorriente* (27–60). México, D. F.: Centro de Investigaciones y Estudios Superiores en Antropología Social (CIESAS).
Sánchez Munguía, A. (2005). Uso del suelo agropecuario y deforestación en Tabasco 1950–2000. Colección José N. Rovirosa. Biodiversidad, Desarrollo Sustentable y Trópico Húmedo. Villahermosa: Universidad Juárez Autónoma de Tabasco (UJAT), p.
Transboundary Freshwater Dispute Database (TFDD), College of Earth, Ocean, and Atmospheric Sciences, Oregon State University. (2018). Available at: http://transboundarywaters.science.oregonstate.edu Accessed Nov 2020.
Tudela, F. (1989) [1992]. *La modernización forzada del trópico: el caso de Tabasco.* Proyecto integrado del Golfo, México: El Colegio de México.
Zamudio Chimal, B. F.; Rosas Castro, J. & Mapén Franco, F. de J. (2016). Estudio de 1985 a 2010 del impacto ambiental y sus efectos en el desarrollo económico y social generado por el campo petrolero Puerto Ceiba-Paraíso, Tabasco México, *Análisis Organizacional,* Volumen 1, Número 8, Año 2016, Available at: http://remineo.org/repositorio/rao/aonc/raoncv1n8.pdf#page=33 Accessed Nov 2020.

Water Pollution and Advanced Water Treatment Technologies

Manasi R. Mulay[1,2] and Natalia Martsinovich[1]
[1]Department of Chemistry, University of Sheffield, Sheffield, UK
[2]Grantham Centre for Sustainable Futures, Sheffield, UK

Synonyms

Wastewater processing; Wastewater recycling; Water purification

Definitions

- Water treatment – Treating wastewater to make it suitable for reuse, using chemical, physical, or biological methods
- Advanced oxidation processes (AOPs) – Processes that involve generation of oxidizing species to chemically destroy pollutants
- Micropollutants – Water pollutants with concentrations of several ng/L to μg/L, usually pharmaceuticals, personal care products, pesticides, and herbicides
- Microplastics – Pieces of plastics with length less than 5 mm, these are emerging water pollutants in urban settings

Introduction

The demand for safe potable water is growing exponentially due to the rapid growth of the global population. The United Nations sustainable development goal (SDG) 6 has set the target to provide clean water access to everyone by 2030 (UN 2018). However, supplying clean water to the population that is currently deprived of access to safe potable water remains a global challenge. Moreover, existing freshwater resources are under increasing threat of pollution caused by industrialization and intensive agriculture, which expanded to satisfy the needs of the growing population (UN 2018). Additional issues are emerging as a result of climate change, e.g., climate change has triggered severe and intense drought conditions in some parts of the world (Huber 2018; WWF 2019). The rise in atmospheric temperatures caused by climate change is also resulting in accelerated demand for drinking water in urban settlements, even within the population that already has access to clean and safe drinking water (Brears 2020a). Simultaneously, frequent floods are occurring in some parts of the world, which disturb freshwater ecosystems (Talbot et al. 2018).

Water demand is mainly influenced by five key factors: population growth and migration, industrialization, agriculture, changing lifestyle, and climate change, as shown in Fig. 1. These factors are interrelated. In urban areas, high population density and industrial pollution present major challenges. In rural areas, the key driver of water demand is agriculture, but access to safe drinking water also remains a challenge (MacAllister et al. 2020). According to the United Nations Department of Economic and Social Affairs (UN-DESA), 55% of the world's population was urban in 2018, and the urban population is anticipated to grow to 68% by 2050, with almost 90% of this increase predicted to be in Asia and Africa (UN-DESA 2018). Urbanization accelerates water demand and contributes more to water pollution compared to rural regions (McGrane 2016).

The overall demand for clean water is already higher than usable clean water reserves across the globe (FAO 2021). This is compounded by the increase in drought conditions, which is causing the natural water sources to dry faster (MacAllister et al. 2020). This means that there is a situation of increased water stress at the global level (FAO 2021). Thus, to address the world water problem, it is crucial to develop the water infrastructure and establish sustainable water management.

There are two approaches to water management: the first is to increase water supply, while the second is to improve the effectiveness of existing water supply channels. Conventionally, the first approach is followed by building new infrastructure to deliver water to consumers, but now the focus is moving towards the second approach in the framework of developing water-smart cities (Brears 2020b). Responsible use of available water resources, e.g., by reducing water consumption, is one of the routes towards achieving water security. In addition, sustainable reuse of water is an essential step to lower the water stress. This means, for example, that "grey" water (domestic wastewater from all channels in households, except sewage) can be re-used for the irrigation (UN-Water 2015). Moreover, after suitable treatment to achieve high purity,

Water Pollution and Advanced Water Treatment Technologies, Fig. 1 Factors that affect water demand

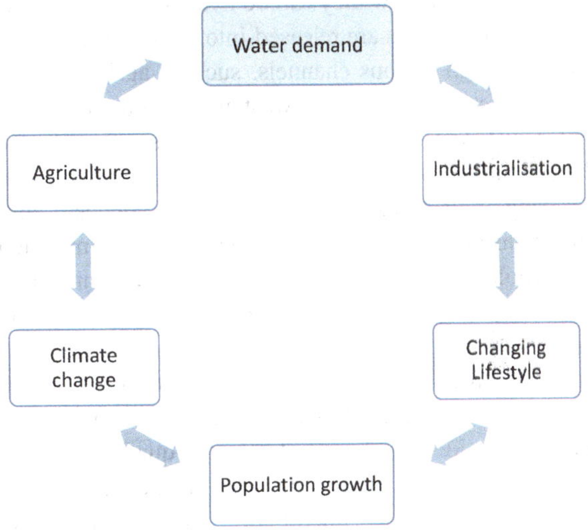

W

recycled water can be used for drinking purposes. Thus, water recycling and wastewater treatment become the inevitable solution to achieve the circular water economy (Brears 2020a).

Water Pollution

Eighty percent of the wastewater across the globe is discharged untreated into natural water resources, resulting in water pollution (UNESCO 2020). The water pollution is affecting both humans and the environment. The pollutants reach humans either via direct intake of polluted water or via consumption of food grown using the polluted water, e.g., vegetables grown with contaminated water supply were found to have the content of metal pollutants higher than the natural levels (Gebre and Van Rooijen 2009). The hazards associated with water pollution include, but are not limited to, diarrhea, typhoid, and cholera. Some of these adverse effects are immediate and may result in death. For example, diarrhea caused by water pollution due to the presence of bacteria, parasites and viruses in drinking water kills almost half a million people every year (WHO 2019). Moreover, acute and chronic toxicity associated with water micropollutants can result in hormonal disruptions, antimicrobial resistance, cancer, and difficult pregnancies (Stasinakis and Gatidou 2010). Thus, it is crucial to understand what pollutants are present in water, where they come from, and how they can be treated.

Water pollutants are released into the environment through various channels, such as agriculture, industry, and domestic wastewater, and then enter different water matrices, such as rivers, lakes, surface water, and groundwater. Rapidly growing industrialization, agricultural practices, direct discharge of wastewater, and waste from household and fuel stations are some of the main sources of water pollution. Notably, some of the factors creating water demand are also the factors or key sources of water pollution (UN 2018).

The World Health Organization (WHO) has provided guidelines for monitoring the quality of drinking water, which include three main aspects of water pollution – microbial, chemical, and radiological (WHO 2017). Wastewater quality is described by two important parameters: biochemical oxygen demand (BOD), which indicates the amount of biologically degradable matter, and chemical oxygen demand (COD), which shows the amount of chemically oxidizable matter (Drinan et al. 2000). Water contaminants can be broadly categorized into biological or chemical. Figure 2 lists some of the key persistent water pollutants: natural organic matter, silts, clays, pathogens, dyes, heavy metals, microplastics, pharmaceuticals, pesticides, and herbicides (Parsons and Jefferson 2006).

Biological pollutants or pathogens include bacteria, such as *E. coli*, which is mainly responsible for spread of infectious diseases through water, protozoa, e.g., cryptosporidium that causes gastroenteritis, as well as viruses and parasites (Parsons and Jefferson 2006). WHO has developed guidelines for drinking water quality regarding the concentration levels of pathogens in a variety of matrices (WHO 2017).

Heavy metal pollution includes metals such as arsenic, lead, nickel, tin, molybdenum, antimony, copper, and selenium, which are recognized as some of the high toxicity materials and are released into the environment by metallurgical industry effluents (Václavíková et al. 2009). Presence of metals in wastewater has a beneficial side-effect of killing pathogens. However, if metal impurities are present in excess, they cause a problem for sludge disposal, resulting in higher process cost and environmental impact (Drinan et al. 2000).

Another emerging type of pollutants is microplastics (Di and Wang 2018). Microplastics are plastic pollutants or plastic fragments with length less than 5 mm. These microplastics in the ocean are consumed by fish and thus may end up in human bodies, but they can also be consumed by humans with drinking water and breathing air (Gasperi et al. 2018; Lim 2021).

Perfluoroalkyl substances (PFAS), commonly known as deadly "forever chemicals," are another type of critical water pollutants, which have strong resistance to degradation (Franke et al. 2019; Meegoda et al. 2020). PFAS can be polymeric or non-polymeric and originate from

Water Pollution and Advanced Water Treatment Technologies, Fig. 2 Key water pollutants

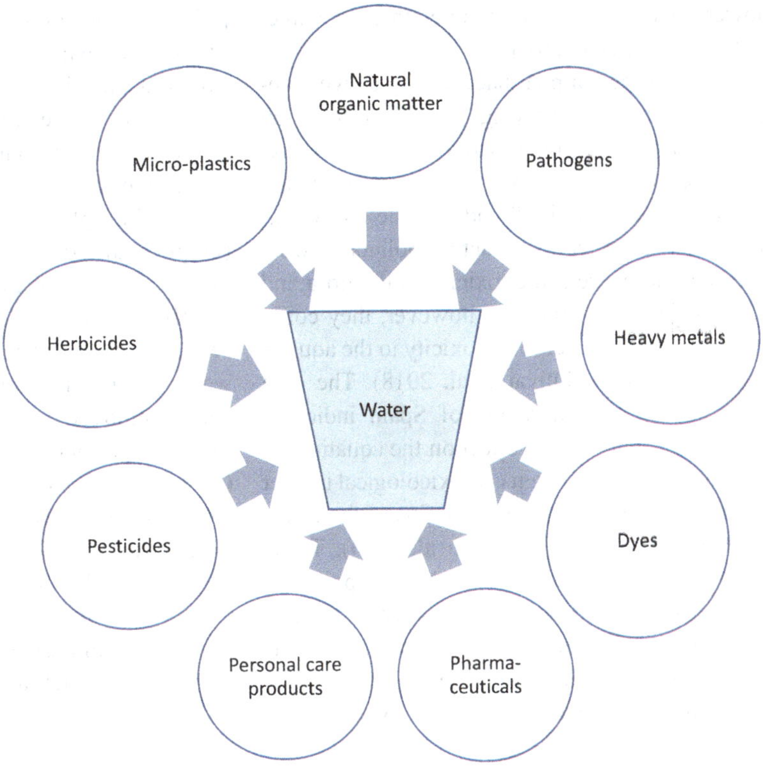

chemical industries or consumer products (Meegoda et al. 2020). The high electronegativity of fluorine results in a strong bond between carbon and fluorine, thus making degradation of these pollutants difficult even by some of the advanced oxidation processes (Franke et al. 2019).

Pollutants coming from the textile industry largely constitute of dye molecules which colorize water if they are discharged untreated (Vandevivere et al. 1998). Water contaminated with high concentrations of colorful dyes from textile industrial wastewater has received attention due the immediate hazards associated with them (Lellis et al. 2019). Water contaminated with dyes cannot be used for bathing, household, or irrigation applications due to the acute toxicity associated with it (Lellis et al. 2019).

Water polluted by colorless pollutants, including pharmaceuticals, hormones, personal care products, and pesticides, also has long-term hazards, such as bioaccumulation and endocrine disruption (Stasinakis and Gatidou 2010). Pharmaceutical pollutants are often released to the environment from hospitals, households, veterinary clinics, and farms (Pérez and Barceló 2008). These colorless pollutants occur in concentrations of several picograms per liter to micrograms per liter and thus are known as micropollutants (Schröder et al. 2016). The concentration of colorless micropollutants in river surface waters as well as sediments is reported to be proportional to the density of population and livestock units (Osorio et al. 2016).

The concentrations of pollutants vary at different collection points over the water course due to natural degradation methods such as biodegradation, sorption, and sunlight-driven transformation along the path, dependent on the season (Meierjohann et al. 2016). Degradation rates were found to vary for different compounds depending on their degradation mechanisms, e.g., carbamazepine was not degraded during any season, but for a photosensitive drug, such as ketoprofen, its sunlight- and temperature-dependent degradation resulted in significantly lower concentrations in summer than in winter (Meierjohann et al. 2016).

The presence of pharmaceutically active compounds can have varying adverse effects on humans as well as the environment, e.g., a case study performed at the river Yamuna, downstream of Wazirabad in India, found that the concentrations of pharmaceutically active pollutants were too small to cause acute toxicity to the flora and fauna of the river Yamuna; however, they could potentially result in chronic toxicity to the aquatic and human life (Mutiyar et al. 2018). The case study at the Iberian region of Spain indicated relatively small toxicity effect on the aquatic life, but potentially significant ecotoxicological impact on algae in the most polluted basins of the rivers Llobregat and Ebro (Osorio et al. 2016). These case studies showed that although water pollution is a global issue, occurrence and concentrations of pharmaceutically active compounds, such as micropollutants, are site-specific (Meierjohann et al. 2016; Mutiyar et al. 2018; Osorio et al. 2016).

Information on all pollutants posing potential hazards to human health, ecology, and environment is necessary to be used as a guideline for monitoring water quality. The UN SDG 6 report recommended monitoring water quality as one of the measures to provide sustainable water for all (UN 2018). The SDG 6 update report in 2021 further highlighted the importance of water quality monitoring, noting that more than three billion people are at risk due to lack of data about their water quality, as these data are not collected on a regular basis in a majority of countries across the globe (UN-Water 2021). Furthermore, guidelines on water quality are needed for control of wastewater discharge and for design of water treatment facilities.

Water Treatment Technologies

Water treatment can be classified into wastewater treatment and drinking water treatment, depending on the extent of treatment (Parsons and Jefferson 2006). Water treatment has three stages: primary, where the large portion of pollutants or solid biomass is separated from wastewater using processes such as sedimentation and filtration; secondary, where biological and organic matter are removed using biological processes; and tertiary, for further purification before the water can be used for purposes, such as drinking, pharmaceutical, and food industry (Gupta et al. 2012). In particular, treatments for odor and taste are essential for drinking water supply (Montiel 1983; Zoschke et al. 2012). A separate type of water treatment is desalination water treatment, where minerals from sea water can be removed to make the sea water usable for industrial, agricultural, or domestic applications (Lattemann and Höpner 2008). Oil-water separation is another type of water treatment, which is required for oil-spills, and much research is being carried out on developing these techniques (Bayat et al. 2005; Xue et al. 2014). Figure 3 summarizes the key water treatment techniques, which are discussed in the following sections.

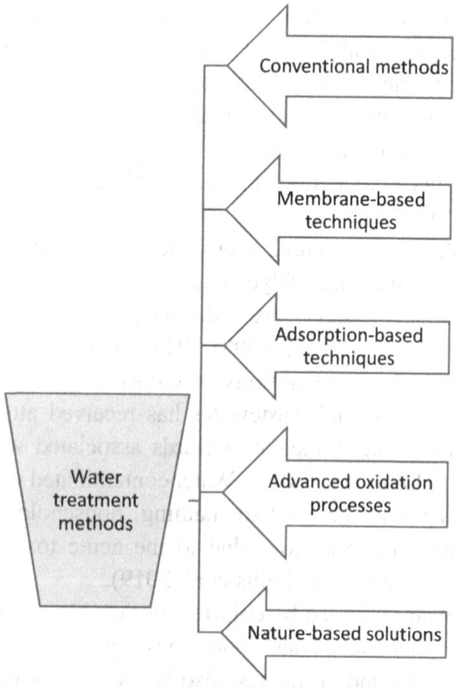

Water Pollution and Advanced Water Treatment Technologies, Fig. 3 Key water treatment methods

Conventional Water Treatment Techniques

Conventional water treatment techniques involve physical methods for pollutant removal, such as clarification, filtration, sedimentation; chemical methods, such as flocculation/coagulation, chemical disinfection, and distillation; and biological methods, such as microbial treatment. Some of the conventional treatments are effectively applied in primary treatment of wastewater, for separation of solid sludge from the wastewater and for separation of natural organic matter and soil contamination. For example, clarification, also known as sedimentation, is a method of purification which involves settling the impurities with the help of gravity (Parsons and Jefferson 2006). Filtration is a method that involves slowly passing contaminated water through porous media where the impurities remain on the mesh and clean water passes through it. The efficiency of filtration or sedimentation is greatly dependent on the particle size, dimensions, and solubility of the pollutants (Jefferson and Jarvis 2006).

Primary treatment using coagulation involves precipitating the impurities by adding metal salts, such as ferric chloride or potassium aluminum sulfate (alum) (Parsons and Jefferson 2006). Coagulation is usually followed by a process of agglomerating the precipitated impurities, which is known as flocculation (Chong 2012). Treatment and disposal of coagulated sludges are challenges in this process, as they involve highly concentrated waste (Bratby 2016). Water disinfection to kill pathogens is carried out by applying chemicals, such as chlorination and ozonation, or physical processes, such as UV disinfection (Parsons and Jefferson 2006). Although chlorination has been in wide use, the by-products of chlorine disinfection are known to be carcinogenic, therefore, requiring stricter regulations and development of alternative methods (Boorman 1999; Drinan et al. 2000). Ozonation and UV disinfection emerged as the two most common alternative disinfection methods (Drinan et al. 2000; Parsons and Jefferson 2006).

Pollutants that cannot be separated by conventional methods during the primary stage of treatment require advanced methods of treatment. These methods include adsorption technologies, membrane-based methods, advanced oxidation processes, and nature-based solutions.

Adsorption-Based Methods

Adsorption of water pollutants on sorbent materials can occur via two mechanisms: physisorption or chemisorption. A variety or carbon materials are commonly used as sorbents. For example, carbon nanotubes are excellent in sorption processes (Sharma et al. 2012). Granular activated carbons (GAC) have been widely researched and are effective for PFAS removal, except for short chain PFAS (Khaydarov and Gapurova 2009; Meegoda et al. 2020). The cheaper alternative to activated carbon is agro-industrial waste products, such as banana stalk waste, jackfruit peel, and pomelo peel, which have been used as treatment for water decolorization (Teng and Low 2012). The natural compounds that are mainly responsible for earthy taste and odor of drinking water, geosmin and methylisoborneol (MIB), can also be removed from water by sorption on activated carbon (Ridal et al. 2001). Other advanced materials are being researched for adsorption of pollutants, such as graphene, zeolites, and metal-organic frameworks (Cao and Li 2014; Joseph et al. 2019; Wang and Peng 2010). In addition to chemical adsorption, biosorption is a technique that is suitable for removing heavy metals and organic pollutants (Sun et al. 2012).

Membrane-Based Technologies

The conventional method of filtration has been in use since ancient times. Membrane-based techniques have been developed as an advanced type of filtration techniques for removal of key persistent pollutants (Parsons and Jefferson 2006). These methods are of particular interest due to the chemical-free nature of treatment, and they are very effective. Membrane-based methods are further classified based on the type and size of membrane (Singh and Hankins 2016). Membranes can be polymeric, glass, sintered metals, ceramic, charged, or ion exchange. Some of the commercially popular membrane-based techniques are reverse osmosis (pressure-based technique) and forward osmosis, nanofiltration (NF),

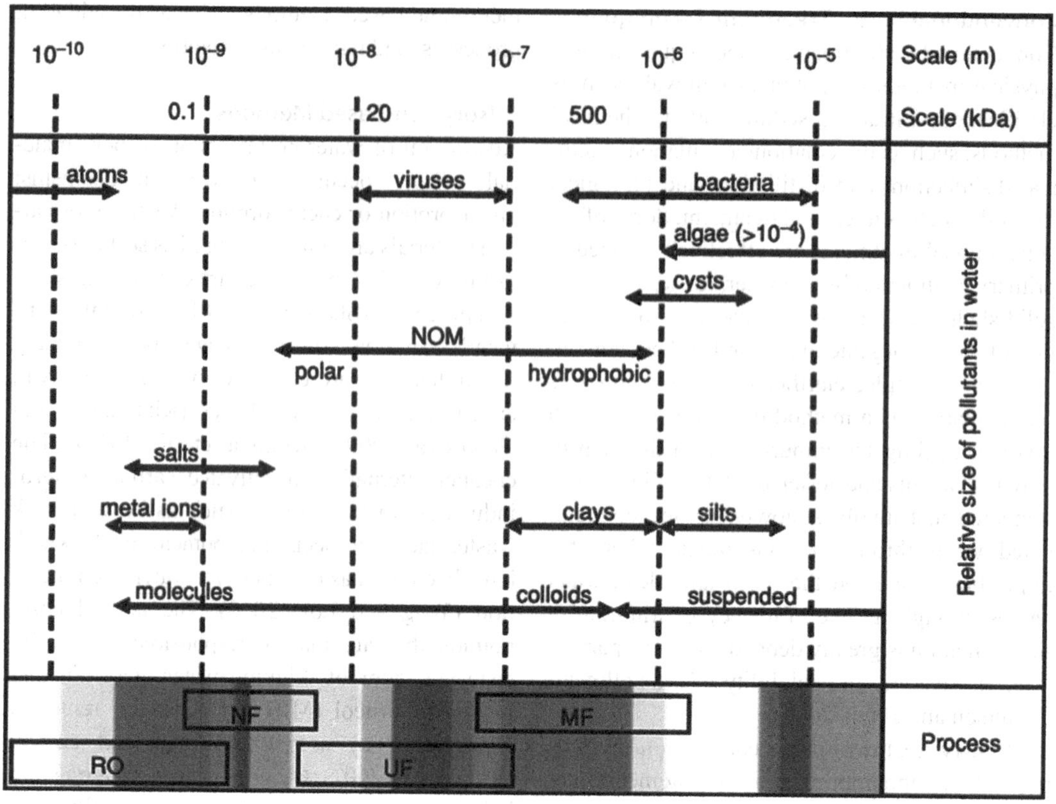

Water Pollution and Advanced Water Treatment Technologies, Fig. 4 Membrane-based separation techniques after Parsons et al. (Parsons and Jefferson 2006)

microfiltration (MF), and ultrafiltration (UF). The effectiveness of membrane-based methods is dependent on the size of the membrane pores and the size of the pollutants, chemical and thermal stability of membranes and pollutants, reusability of membranes, and their selectivity towards pollutant removal (Singh and Hankins 2016).

Figure 4 presents an overview of membrane-based techniques effective for the removal of pollutants of different sizes. For example, reverse osmosis (RO) is suitable for the removal of metal ions or salts and is useful in desalination treatment. Nanofiltration can remove some salts effectively; it has also been used for removal of color from the textile wastewater, whereas ultrafiltration is effective for removal of viruses and natural organic matter (NOM). Microfiltration can be useful for the removal of microplastics, clay particles, hydrophobic NOMs, and some bacteria (Parsons and Jefferson 2006). Membrane-based techniques are also effective for removal of all types of PFAS, but membrane-methods such as RO and NF are more expensive than GAC-based adsorption (Meegoda et al. 2020). Membrane-based techniques have their limitations: one of the main challenges associated with these techniques is the fouling of membrane over the period of several reuses (Singh and Hankins 2016).

There are two main problems associated with the techniques discussed so far. The first problem is that the conventional techniques are still not able to completely remove all pollutants and ensure safe drinking water. The second problem is that when the pollutants are separated from water, they still need to be disposed of, which can be hazardous to the environment. So, there is a need for methods that can destroy pollutants instead of just separating them from water. Therefore, new techniques are being researched. The most important of them are advanced oxidation processes, which instead of separating or removing the pollutants attempt to destroy those pollutants and convert them into chemically simpler, relatively less toxic or non-toxic form.

Advanced Oxidation Processes

Advanced oxidation processes (AOPs) use strong oxidizing species to destroy pollutants (Deng and Zhao 2015). Pollutants that have high resistance to degradation processes and are relatively stable can be destroyed to less complex and relatively less toxic or non-toxic molecular forms by interacting with reactive oxygen species (Deng and Zhao 2015). AOPs are further classified based on the method of generation of the oxidizing species into photochemical and non-photochemical methods, as shown in Fig. 5.

Photochemical processes are light-driven chemical treatment processes, such as photocatalysis where light-activated photocatalyst materials in the presence of light (either solar light or lamp) generate the oxidizing species to destroy the pollutants (Chong et al. 2010; Malato et al. 2009; Mulay and Martsinovich 2021). Photochemical techniques include solar disinfection – use of sunlight for water disinfection (Marugán et al. 2020).

Non-photochemical methods include electrochemical oxidation, Fenton oxidation, and short-lived radical-based methods. Electrochemical oxidation uses an electrochemical cell, where voltage applied to electrodes results in the formation of reactive chemicals such as hydrogen peroxide, ozone, and hydroxyl radicals, which chemically react with pollutants and destroy them (Martínez-Huitle and Panizza 2018). Fenton oxidation process uses ferrous salts together with hydrogen peroxide to produce reactive species that destroy water pollutants (Deng and Zhao 2015). Microwave radiation and ultrasound energies are also used for generation of oxidizing species (Mudhoo 2012). Advanced oxidation processes are suitable for destroying organic pollutants and micropollutants, such as pharmaceuticals and personal care products, and industrial pollutants, such as dyes, as well as pesticides and herbicides (Garcia-Segura et al. 2018; Mulay and Martsinovich 2021; Vandevivere et al. 1998). The drawback is that AOPs are not suitable for treatment of PFAS pollution as PFAS contain strong carbon-fluorine bonds which are difficult to break by AOPs. As an alternative, high temperature and pressure

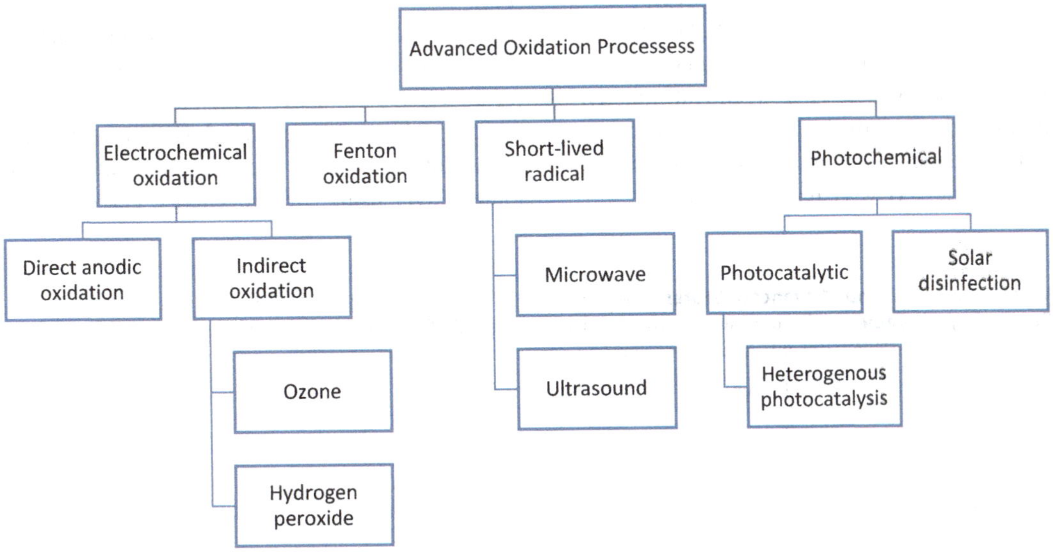

Water Pollution and Advanced Water Treatment Technologies, Fig. 5 Types of advanced oxidation processes

W

methods have been used to decompose PFAS, but their cost is prohibitive (Meegoda et al. 2020).

Nature-Based Methods

Nature-based methods have been used since distant past by indigenous communities and can be categorized as a sub-group of conventional methods. However, these methods have recently found increased attention due their effectiveness and sustainability aspects, and their relatively fewer or no side-effects. Nature-based methods involve removal of pollutants with the help of entities found in nature. For example, phytoremediation is one of the nature-based methods. "Phyto" means "plant" in Greek; therefore, phytoremediation literally means plant-based remediation (Sharma et al. 2013). Phytoextraction processes involve interaction with plant roots to remove heavy metal pollutants from the soil. A similar process used for wastewater treatment is known as rhizofiltration, where plants are used in either hydroponic settings (grown in wastewater) or in constructed wetland settings for heavy metal removal (Dushenkov et al. 1995). This method needs harvesting of plants over time, as the plants get saturated with the contaminants (Sharma et al. 2013). In another approach, living or dead microorganisms can interact with metal pollutants and absorb them in the process known as bio-sorption or bioaccumulation (Sun et al. 2012). Dye pollutants can be removed with algal treatment using species such as spirogyra. This treatment method is performed at relatively low energy; therefore, it not only saves costs but also reduces greenhouse gas emissions (Dwivedi and Vats 2013).

Plant biomass, such as dried leaves, flowers, and bark, have been extensively researched for heavy metal removal and found to be effective (Srinivasan 2013). For example, *Moringa oleifera* is a tree, which is also known as drumstick vegetable; its leaves and seed pods are often used in Asian cuisine. Extracts of *Moringa oleifera* seeds have been found to have properties of a coagulant that can remove dyes by adsorption (Sánchez-Martín and Beltrán-Heredia 2012). It was found to be effective for anionic dyes, such as azo, anthraquinone, and indigoid dyes, but not effective for cationic dyes, such as methylene blue (Sánchez-Martín and Beltrán-Heredia 2012). Some tannin-based compounds have also been found effective as coagulants in removal of commonly used surfactants, such as sodium laurel sulfate (Sánchez-Martín and Beltrán-Heredia 2012).

Thus, nature-based and nature-inspired methods promise cost-effective, low-energy, and low environmental impact processes for water remediation. Besides the process benefits, nature-based methods offer wider environmental and community benefits (Boano et al. 2020).

Hybrid Methods of Water Treatment

Efficiency and cost-effectiveness are important considerations in the choice of suitable techniques for water treatment facilities. Combinations of techniques were often found to be more effective than standalone techniques. Table 1 after Torres et al. (2007) compares some of the AOPs for decomposition of Bisphenol-A (BPA) when used standalone and when used in combination with other AOP, using electric energy per order of

Water Pollution and Advanced Water Treatment Technologies, Table 1 Electric energy cost estimates for Bisphenol A (BPA, 300 ml, initial concentration of 118 μmol L^{-1}) degradation by various AOPs, after Torres et al. (2007) with permission from ACS

Process	Power (W)	Time (min)	% TOC removed	EE/O (kWh/m^{-3})
UV	25	600	Less than 60%	–
US	80	600	Less than 60%	–
US/Fe(II)	80	600	64	6010
UV/US	105	300	66	3735
UV/US/Fe(II)	105	120	79	1033

pollutant removal (EE/O) as the figure of merit. For example, combination of the ultrasound (US) method with the Fenton oxidation method yielded 64% removal of total organic carbon (TOC) from water in 600 min, whereas only the ultrasonic method removed $<60\%$ TOC under the same conditions. Furthermore, this performance of the ultrasound/Fenton method was outperformed by a combination of ultraviolet (UV) light with US (UV/US) in just 300 min, although with higher power. The values of EE/O in the table clearly show that when three methods – UV, ultrasonic, and Fenton – were combined, this combination resulted in a drastic reduction in the energy cost. In another example, intermediates of photocatalytic degradation accumulated at the photocatalyst surface and degraded the performance, but a combination of photocatalysis with sonolysis improved the mass transfer by preventing adsorption of the intermediates at the surface, resulting in better performance (Neppolian et al. 2012). Other studies showed that combinations of biological treatments with advanced oxidation processes can deliver the advantages of advanced oxidation process at relatively lower costs (Oller et al. 2011).

Analysis and Comparison of Water Treatment Techniques

Table 2 presents a summary of persistent pollutants and the key treatment techniques that can be suitable for treating the water pollution, based on compilation of data from references (Meegoda et al. 2020; Mulay and Martsinovich 2021; Parsons and Jefferson 2006; Salimi et al. 2017; Singh and Hankins 2016; Sánchez-Martín and Beltrán-Heredia 2012). It is clear that there is no single method that works for all types of pollutants. Effectiveness of the methods strongly depends on the chemistry of the pollutants. For example, removal or separation methods are effective for chemically stable pollutants, such as PFAS. For microplastics, the commonly used methods are microfiltration, coagulation, and magnetic separation. However, coagulation-based methods may require high amounts of salts for removal of microplastics, which is a drawback. In contrast, advanced oxidation methods are able to destroy a variety of pollutants but are not very effective for microplastics (Shen et al. 2020). Although individual methods cannot achieve complete elimination or degradation of all pollutants, combinations of several methods are often more effective and can be used to eliminate multiple types of pollutants.

Based on the capabilities of these water treatment techniques discussed so far, analysis of strengths, weaknesses, opportunities, and threats (SWOT) of the key water treatment methods can be carried out as presented in Table 3. The weaknesses of the methods can be turned into opportunities with more investment into the research & development.

Challenges and Future Scope

The efficiency of water treatment depends on the type of the technique, operational conditions of treatment, as well as the properties of the pollutants. Figure 6 represents the challenges associated with existing water treatment technologies. One of the key challenges associated with the advanced techniques for water treatment is their high costs. Another challenge associated with some of the advanced oxidation processes, such as photocatalysis, is secondary pollution, i.e., the possibility of formation of equally or more toxic intermediates by destruction of organic micropollutants. Thus, sustainability of advanced oxidation processes requires further investigation. Some of the methods have proven to be efficient in destroying pollutants on the laboratory scale; however, their efficacy at the pilot level or larger commercial scale needs further investment into research and development.

Currently available water treatment technologies are excellent for carbon, heavy metal, nitrogen, and microbial elimination. However, complete elimination of some of the emerging pollutants, such as PFAS and critical PhAcs, is challenging with most of the existing technologies. This is because even if they are removed from water, they still need to be disposed of, so their complete elimination from the environment is yet not achieved. Existing methods are also very

Water Pollution and Advanced Water Treatment Technologies, Table 2 Water treatment methods used at primary, secondary, and tertiary stages, and commonly found pollutants which can be removed using these methods. A single tick indicates that the method is effective for removing this pollutant; a double tick indicates that the method is highly effective, while a cross indicates that this method is ineffective for this pollutant

Pollutants (row)/ methods (column)	Requirements	NOM	Pathogens	Heavy metals	Dyes	Micropollutants	PFAS	Microplastics	Taste and odor compounds
Origin	->	Animal and plant litter	Fecal waste, animal manure	Industrial effluents	Textile industrial wastewater	Pharmaceutical industry, domestic industrial wastewater, agricultural run-off	Consumer products, foams, chemical industry products	Consumer products, plastic waste	Metabolism of algae
Sedimentation/ clarification, chemical precipitation		✓							
Disinfection	Chlorine, ozone, UV, solar, etc.		✓						
Coagulation	Coagulating agent – salts	✓		✓	✓			✓	
Nature-based coagulation	Natural products with coagulating properties		✓	✓	✓				
Flocculation	Flocculating agent	✓		✓					
Magnetic separation				✓				✓	
Reverse osmosis	Membrane						✓		
Nanofiltration	Membrane	✓		✓			✓✓		
Ultrafiltration	Membrane		✓						
Microfiltration	Membrane	✓	✓					✓✓	
Adsorption	Carbonaceous materials, nanomaterials, zeolites, metal-organic frameworks			✓	✓		✓		✓

Biosorption, bioremediation	Microorganisms, fungi, algae, yeast			✓	✓				
Phytoremediation/ rhizofiltration	Plants			✓					
Advanced oxidation processes	Source to generate oxidizing species	✓	✓		✓	✓✓	X		✓

W

Water Pollution and Advanced Water Treatment Technologies, Table 3 Strength, weakness, opportunity, threat (SWOT) analysis of common water treatment methods

Methods	Strengths	Weaknesses	Opportunity (scope to improve)	Threat (hazards or alarming side-effects)
Primary separation methods	Easy to use, require little investment	Not effective to remove all pollutants	Can be used in combination with the advanced methods to have higher efficiency	Not all pollutants are removed
Adsorption-based methods	Good for removal of pollutants that cannot be chemically destroyed	Scaling up is a challenge; high cost of manufacturing of nanomaterials	Scope for selectivity – selective removal of hazardous chemicals, persistent pollutants	Possibility of chemical contamination from the treatment materials
Membrane-based techniques	Chemical-free treatment	Membrane fouling, high operational cost	Research to improve the recyclability and reusability of the membrane	Disposal of the separated pollutants is a problem
Advanced oxidation processes	Pollutants can be destroyed to less toxic form	Process cost and higher environmental impact	Complete destruction of pollutants	Secondary pollution by intermediates
Nature-based solutions	Potentially sustainable alternative with low environmental impact, wider community benefits	Scaling up can be challenging	More research is needed for improving the range of applicability and scale up	It should not result in excess burden on land, forest, or nature when scale up

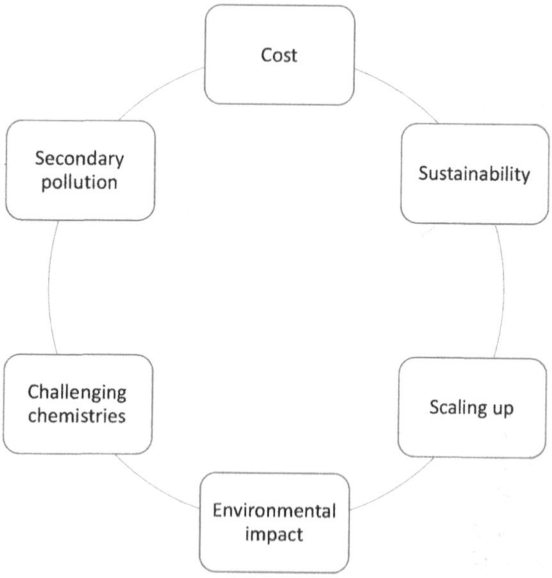

Water Pollution and Advanced Water Treatment Technologies, Fig. 6 Open challenges associated with advanced water treatment techniques

energy intensive (Schröder et al. 2016). Thus, existing methods are not sufficiently effective and more advanced methods are needed. Moreover, policy guidelines are needed for reducing water pollution and maximizing effectiveness of water treatment technologies. WHO has provided guidelines for drinking water quality; however, some of the emerging persistent organic pollutants need more attention and further study (WHO 2017).

Recommendations

Improving the effectiveness of treatment of polluted water is one of the key strategies for water-resilient future. Strategic planning and management of water resources and research in treatment technologies promise to provide clean water resources for sustainable and healthier life. At the same time, sustainable lifestyle and agriculture, such as reduction in the use of pesticides and pharmaceuticals, can result in decreased pollution. The following necessary steps for achieving clean water resources can be recommended:

- Investing in research and development for advanced water treatment technologies in academia and industry. This includes research and development of new methods and materials, as well as development of modeling techniques to model pollutant removal.
- Cross-sector collaboration and partnerships between research laboratories, industries, and policymakers.

Treatment of polluted water for re-use is essential for sustainable water supply. However, minimizing water pollution in the first place is equally important to protect the environment from irreversible damage. Minimizing water pollution can be achieved with collective efforts through public-private partnerships. The following steps can be recommended to minimize water pollution:

- Make the environmental impact assessment mandatory before approval of new industrial projects
- Conduct regular environmental audits, including water audits
- Set effluent discharge permits
- Sustainable use of pesticides and herbicides
- At domestic level: responsible disposal of pharmaceuticals
- Policymaking for industrial wastewater disposal
- Documentation and monitoring of potential health hazards associated with emerging pollutants

Additionally, treatments such as desalination offer vast potential to tap into sea and ocean-based water (Laffoley et al. 2021). Ninety-seven percent of all water on Earth is in the ocean. However, existing desalination treatments have very high environmental and marine impact, which may be reduced by using renewable energy sources (Lee and Jepson 2021). More research and development is needed to make desalination processes more sustainable to alleviate the stress on the freshwater resources.

Water, food, and energy are entangled with community lives. An integrated approach at the water-food-energy nexus is needed. Only with clean water resources can food production be truly sustainable, and only with sustainable use of water in agriculture and industry can water stress be reduced and eliminated. At the same time, policies on renewable energy generation affect water resources, and usage of clean energy can ensure lower environmental impact of water treatment techniques (Beck and Walker 2013; Endo et al. 2017).

Cross-References

▶ Circular Water Economy
▶ Meeting SDG6: Ensuring Safe Drinking Water for All in Rural India
▶ Water Security and the Green Economy
▶ Water Security, Sustainability, and SDG 6
▶ Water-Smart Cities

Acknowledgments The authors would like to acknowledge the Grantham Centre for Sustainable Futures for funding, training, and scholarship for MRM.

References

Bayat, A., Aghamiri, S. F., Moheb, A., & Vakili-Nezhaad, G. R. (2005). Oil spill cleanup from sea water by sorbent materials. *Chemical Engineering & Technology: Industrial Chemistry-Plant Equipment-Process Engineering-Biotechnology, 28*(12), 1525–1528.

Beck, M. B., & Walker, R. V. (2013). On water security, sustainability, and the water-food-energy-climate nexus. *Frontiers of Environmental Science & Engineering, 7*(5), 626–639.

W

Boano, F., Caruso, A., Costamagna, E., Ridolfi, L., Fiore, S., Demichelis, F., ... Masi, F. (2020). A review of nature-based solutions for greywater treatment: Applications, hydraulic design, and environmental benefits. *Science of the Total Environment, 711*, 134731.

Boorman, G. A. (1999). Drinking water disinfection byproducts: Review and approach to toxicity evaluation. *Environmental Health Perspectives, 107*(Suppl 1), 207–217.

Bratby, J. (2016). *Coagulation and flocculation in water and wastewater treatment*. IWA Publishing.

Brears, R. C. (2020a). Circular water economy. In *The Palgrave encyclopedia of urban and regional futures*. Springer International Publishing. https://doi.org/10.1007/978-3-030-51812-7_49-1.

Brears, R. C. (2020b). Water-smart cities. In *The Palgrave encyclopedia of urban and regional futures*. Springer International Publishing. https://doi.org/10.1007/978-3-030-51812-7_44-1.

Cao, Y., & Li, X. (2014). Adsorption of graphene for the removal of inorganic pollutants in water purification: A review. *Adsorption, 20*(5-6), 713–727.

Chong, M. F. (2012). Direct flocculation process for wastewater treatment. In *Advances in water treatment and pollution prevention* (pp. 201–230). Springer.

Chong, M. N., Jin, B., Chow, C. W., & Saint, C. (2010). Recent developments in photocatalytic water treatment technology: A review. *Water Research, 44*(10), 2997–3027.

Deng, Y., & Zhao, R. (2015). Advanced oxidation processes (AOPs) in wastewater treatment. *Current Pollution Reports, 1*(3), 167–176.

Di, M., & Wang, J. (2018). Microplastics in surface waters and sediments of the Three Gorges Reservoir, China. *Science of the Total Environment, 616*, 1620–1627.

Drinan, J., Drinan, J. E., & Spellman, F. (2000). *Water and wastewater treatment: A guide for the nonengineering professional*. CRC Press.

Dushenkov, V., Kumar, P. N., Motto, H., & Raskin, I. (1995). Rhizofiltration: The use of plants to remove heavy metals from aqueous streams. *Environmental Science & Technology, 29*(5), 1239–1245.

Dwivedi, S., & Vats, T. (2013). Remediation of dye containing wastewater using viable algal biomass. In *Green materials for sustainable water remediation and treatment* (pp. 212–228). Cambridge, UK: RSC Green Chemistry Series.

Endo, A., Tsurita, I., Burnett, K., & Orencio, P. M. (2017). A review of the current state of research on the water, energy, and food nexus. *Journal of Hydrology: Regional Studies, 11*, 20–30.

FAO. (2021). *Indicator 6.4.2- Level of water stress*. Food and Agricultural Organisation of United Nations. Retrieved 26 Oct 2021 from https://www.fao.org/sustainable-development-goals/indicators/642/en/

Franke, V., Schäfers, M. D., Lindberg, J. J., & Ahrens, L. (2019). Removal of per-and polyfluoroalkyl substances (PFASs) from tap water using heterogeneously catalyzed ozonation. *Environmental Science: Water Research & Technology, 5*(11), 1887–1896.

Garcia-Segura, S., Ocon, J. D., & Chong, M. N. (2018). Electrochemical oxidation remediation of real wastewater effluents – A review. *Process Safety and Environmental Protection, 113*, 48–67.

Gasperi, J., Wright, S. L., Dris, R., Collard, F., Mandin, C., Guerrouache, M., ... Tassin, B. (2018). Microplastics in air: Are we breathing it in? *Current Opinion in Environmental Science & Health, 1*, 1–5.

Gebre, G., & Van Rooijen, D. J. (2009). Urban water pollution and irrigated vegetable farming in Addis Ababa. In: Shaw, R.J. (ed). *Water, sanitation and hygiene - Sustainable development and multisectoral approaches: Proceedings of the 34th WEDC International Conference, Addis Ababa, Ethiopia, 18-22 May 2009*. https://repository.lboro.ac.uk/articles/conference_contribution/Urban_water_pollution_and_irrigated_vegetable_farming_in_Addis_Ababa_/9585773

Gupta, V. K., Ali, I., Saleh, T. A., Nayak, A., & Agarwal, S. (2012). Chemical treatment technologies for wastewater recycling – An overview. *RSC Advances, 2*(16), 6380–6388.

Huber, K. (2018). Resilience strategies for drought. https://www.c2es.org/document/resilience-strategies-for-drought/

Jefferson, B., & Jarvis, P. R. (2006). Practical application of fractal dimension: Theory and applications. In G. Newcombe & D. Dixon (Eds.), *Interface science in drinking water treatment: Theory and applications*. Academic.

Joseph, L., Jun, B.-M., Jang, M., Park, C. M., Muñoz-Senmache, J. C., Hernández-Maldonado, A. J., ... Yoon, Y. (2019). Removal of contaminants of emerging concern by metal-organic framework nanoadsorbents: A review. *Chemical Engineering Journal, 369*, 928–946.

Khaydarov, R., & Gapurova, O. (2009). Application of carbon nanoparticles for water treatment. In *Water treatment technologies for the removal of high-toxicity pollutants* (pp. 253–258). Springer.

Laffoley, D., Baxter, J. M., Amon, D. J., Claudet, J., Downs, C. A., Earle, S. A., ... Levin, L. A. (2021). The forgotten ocean: Why COP26 must call for vastly greater ambition and urgency to address ocean change. In *Aquatic conservation: Marine and freshwater ecosystems, 32* (1), 217–228. Wiley, https://doi.org/10.1002/aqc.3751

Lattemann, S., & Höpner, T. (2008). Environmental impact and impact assessment of seawater desalination. *Desalination, 220*(1–3), 1–15.

Lee, K., & Jepson, W. (2021). Environmental impact of desalination: A systematic review of life cycle assessment. *Desalination, 509*, 115066.

Lellis, B., Fávaro-Polonio, C. Z., Pamphile, J. A., & Polonio, J. C. (2019). Effects of textile dyes on health and the environment and bioremediation potential of living organisms. *Biotechnology Research and Innovation, 3*(2), 275–290. https://doi.org/10.1016/j.biori.2019.09.001.

Lim, X. (2021). Microplastics are everywhere – But are they harmful? *Nature, 593*, 22-25.

MacAllister, D. J., MacDonald, A., Kebede, S., Godfrey, S., & Calow, R. (2020). Comparative performance of rural water supplies during drought. *Nature Communications, 11*(1), 1–13.

Malato, S., Fernández-Ibáñez, P., Maldonado, M. I., Blanco, J., & Gernjak, W. (2009). Decontamination and disinfection of water by solar photocatalysis: Recent overview and trends. *Catalysis Today, 147*(1), 1–59.

Martínez-Huitle, C. A., & Panizza, M. (2018). Electrochemical oxidation of organic pollutants for wastewater treatment. *Current Opinion in Electrochemistry, 11*, 62–71.

Marugán, J., Giannakis, S., McGuigan, K. G., & Polo-López, I. (2020). Solar disinfection as a water treatment technology. In W. Leal Filho, A. M. Azul, L. Brandli, A. Lange Salvia, & T. Wall (Eds.), *Clean water and sanitation* (pp. 1–16). Springer International Publishing. https://doi.org/10.1007/978-3-319-70061-8_125-1.

McGrane, S. J. (2016). Impacts of urbanisation on hydrological and water quality dynamics, and urban water management: A review. *Hydrological Sciences Journal, 61*(13), 2295–2311.

Meegoda, J. N., Kewalramani, J. A., Li, B., & Marsh, R. W. (2020). A review of the applications, environmental release, and remediation technologies of per-and polyfluoroalkyl substances. *International Journal of Environmental Research and Public Health, 17*(21), 8117.

Meierjohann, A., Brozinski, J.-M., & Kronberg, L. (2016). Seasonal variation of pharmaceutical concentrations in a river/lake system in Eastern Finland. *Environmental Science: Processes & Impacts, 18*(3), 342–349.

Montiel, A. (1983). Municipal drinking water treatment procedures for taste and odour abatement – A review. *Water Science and Technology, 15*(6–7), 279–289.

Mudhoo, A. (2012). Microwave-assisted organic pollutants degradation. In *Advances in water treatment and pollution prevention* (pp. 177–200). Springer.

Mulay, M. R., & Martsinovich, N. (2021). TiO_2 photocatalysts for degradation of micropollutants in water. In W. Leal Filho, A. M. Azul, L. Brandli, A. Lange Salvia, & T. Wall (Eds.), *Clean water and sanitation* (pp. 1–19). Springer International Publishing. https://doi.org/10.1007/978-3-319-70061-8_194-1.

Mutiyar, P. K., Gupta, S. K., & Mittal, A. K. (2018). Fate of pharmaceutical active compounds (PhACs) from River Yamuna, India: An ecotoxicological risk assessment approach. *Ecotoxicology and Environmental Safety, 150*, 297–304. https://doi.org/10.1016/j.ecoenv.2017.12.041.

Neppolian, B., Ashokkumar, M., Tudela, I., & González-García, J. (2012). Hybrid Sonochemical treatment of contaminated wastewater: Sonophotochemical and sonoelectrochemical approaches. Part I: Description of the techniques. In *Advances in water treatment and pollution prevention* (pp. 267–302). Springer.

Oller, I., Malato, S., & Sánchez-Pérez, J. (2011). Combination of advanced oxidation processes and biological treatments for wastewater decontamination – A review. *Science of the Total Environment, 409*(20), 4141–4166.

Osorio, V., Larrañaga, A., Aceña, J., Pérez, S., & Barceló, D. (2016). Concentration and risk of pharmaceuticals in freshwater systems are related to the population density and the livestock units in Iberian Rivers. *Science of the Total Environment, 540*, 267–277.

Parsons, S. A., & Jefferson, B. (2006). *Introduction to potable water treatment processes.* Wiley Online Library.

Pérez, S., & Barceló, D. (2008). Advances in the analysis of pharmaceuticals in the aquatic environment. In D. S. Aga (Ed.), *Fate of pharmaceuticals in the environment and water treatment systems*. Boca Raton: CRC Press.

Ridal, J., Brownlee, B., McKenna, G., & Levac, N. (2001). Removal of taste and odour compounds by conventional granular activated carbon filtration. *Water Quality Research Journal, 36*(1), 43–54.

Salimi, M., Esrafili, A., Gholami, M., Jafari, A. J., Kalantary, R. R., Farzadkia, M., ... Sobhi, H. R. (2017). Contaminants of emerging concern: A review of new approach in AOP technologies. *Environmental Monitoring and Assessment, 189*(8), 414.

Sánchez-Martín, J., & Beltrán-Heredia, J. (2012). Nature is the answer: Water and wastewater treatment by new natural-based agents. In *Advances in water treatment and pollution prevention* (pp. 337–375). Springer.

Schröder, P., Helmreich, B., Škrbić, B., Carballa, M., Papa, M., Pastore, C., ... Molinos, M. (2016). Status of hormones and painkillers in wastewater effluents across several European states – Considerations for the EU watch list concerning estradiols and diclofenac. *Environmental Science and Pollution Research, 23*(13), 12835–12866.

Sharma, S. K., Sanghi, R., & Mudhoo, A. (2012). Green practices to save our precious “water resource”. In *Advances in water treatment and pollution prevention* (pp. 1–36). Springer.

Sharma, R. K., Alok, A., Manab, D., & Aditi, P. (2013). Green materials for sustainable remediation of metals in water. In A. Mishra & J. H. Clark (Eds.), *Green materials for sustainable water remediation and treatment* (p. 11). RSC Publishing.

Shen, M., Song, B., Zhu, Y., Zeng, G., Zhang, Y., Yang, Y., ... Yi, H. (2020). Removal of microplastics via drinking water treatment: Current knowledge and future directions. *Chemosphere, 251*, 126612.

Singh, R., & Hankins, N. P. (2016). Introduction to membrane processes for water treatment. In R. Singh & N. P. Hankins (Eds.), *Emerging membrane technology for sustainable water treatment* (pp. 15–52). Elsevier.

Srinivasan, R. (2013). Role of plant biomass in heavy metal treatment of contaminated water. In A. Mishra & J. H. Clark (Eds.), *Green materials for sustainable water remediation and treatment* (p. 30). RSC Publishing.

Stasinakis, A. S., & Gatidou, G. (2010). Micropollutants and aquatic environment. In J. Virkutyte, R. Varma, & V. Jegatheesan (Eds.), *Treatment of micropollutants in water and wastewater.* IWA Publishing.

Sun, J., Ji, Y., Cai, F., & Li, J. (2012). Heavy metal removal through biosorptive pathways. In *Advances in water treatment and pollution prevention* (pp. 95–145). Springer.

Talbot, C. J., Bennett, E. M., Cassell, K., Hanes, D. M., Minor, E. C., Paerl, H., . . . Xenopoulos, M. A. (2018). The impact of flooding on aquatic ecosystem services. *Biogeochemistry, 141*(3), 439–461.

Teng, T. T., & Low, L. W. (2012). Removal of dyes and pigments from industrial effluents. In *Advances in water treatment and pollution prevention* (pp. 65–93). Springer.

Torres, R. A., Pétrier, C., Combet, E., Moulet, F., & Pulgarin, C. (2007). Bisphenol a mineralization by integrated ultrasound-UV-iron (II) treatment. *Environmental Science & Technology, 41*(1), 297–302.

UN. (2018). *SDG 6 synthesis report 2018 on water and sanitation*. United Nations. https://doi.org/10.18356/e8fc060b-en.

UN-DESA. (2018). *68% of the world population projected to live in urban areas by 2050, says UN.* United Nations Department of Economic and Social Affairs. Retrieved 26 Oct 2021 from https://www.un.org/development/desa/en/news/population/2018-revision-of-world-urbanization-prospects.html

UNESCO, U.-W. (2020). *United Nations World Water development report 2020.* UNESCO.

UN-Water. (2015). *Compendium of water quality regulatory frameworks: Which water for which use?* UN-Water.

UN-Water. (2021). *Summary Progress Update 2021 – SDG 6 – water and sanitation for all.* UN-Water.

Václavíková, M., Vitale, K., Gallios, G. P., & Ivanicová, L. (2009). *Water treatment technologies for the removal of high-toxity pollutants*. Springer.

Vandevivere, P. C., Bianchi, R., & Verstraete, W. (1998). Treatment and reuse of wastewater from the textile wet-processing industry: Review of emerging technologies. *Journal of Chemical Technology & Biotechnology: International Research in Process, Environmental AND Clean Technology, 72*(4), 289–302.

Wang, S., & Peng, Y. (2010). Natural zeolites as effective adsorbents in water and wastewater treatment. *Chemical Engineering Journal, 156*(1), 11–24.

WHO. (2017). *Guidelines for drinking-water quality: Fourth edition incorporating the first addendum*. WHO.

WHO. (2019). *Drinking water.* World Health Organisation. Retrieved 26 Oct 2021 from https://www.who.int/news-room/fact-sheets/detail/drinking-water

WWF. (2019). *Drought risk: The global thirst for water in the era of climate crisis*. WWF Report. https://www.wwf.de/fileadmin/fm-wwf/Publikationen-PDF/WWF_DroughtRisk_EN_WEB.pdf

Xue, Z., Cao, Y., Liu, N., Feng, L., & Jiang, L. (2014). Special wettable materials for oil/water separation. *Journal of Materials Chemistry A, 2*(8), 2445–2460.

Zoschke, K., Dietrich, N., Börnick, H., & Worch, E. (2012). UV-based advanced oxidation processes for the treatment of odour compounds: Efficiency and by-product formation. *Water Research, 46*(16), 5365–5373.

Water Purification

► Water Pollution and Advanced Water Treatment Technologies

Water Quality

► Green Economy Policies to Achieve Water Security
► Meeting SDG6: Ensuring Safe Drinking Water for All in Rural India

Water Quantity

► Green Economy Policies to Achieve Water Security

Water Recycling

► Circular Economy and the Water-Food Nexus
► Circular Water Economy

Water Security

► Green Economy Policies to Achieve Water Security
► Responsibility to Prepare and Prevent (R2P2): Applying Unprecedented Foresight to Addressing Unprecedented Climate Risks
► Water Security and the Green Economy

Water Security and Its Role in Achieving SDG 6

Elif Çolakoğlu
Department of Security Sciences, Gendarmerie and Coast Guard Academy, Ankara, Turkey

Synonyms

Ecological security; Economical security; Human security; Peace–conflict security; Right to water; Water sustainability

Definition

Water security is quickly becoming one of the most pressing concerns facing humanity in the twenty-first century. People do not have enough water to meet their fundamental needs since they do not have access to it. The number of them who do not have access to clean water and do not have sufficient hygienic conditions on the planet today is in the billions. Due to lack of basic water and sanitation, these people, the majority of whom are children and the elderly, may die each year from water-related diseases. Individual safety may be jeopardized when proper conditions are not given for persons or their access to water is restricted, owing to poor health and livelihood conditions. Human health and well-being are dependent on safe drinking water, sanitation, and hygiene. In this context, water security can be described as ensuring that everyone has access to sufficient and safe water at an affordable price so that they can live a healthy, dignified, and productive life. Water security encompasses human security – everyone has access to safe drinking water; economic security – the ability of water resources to meet normal economic development needs; and ecological security – the ability of water supplies to meet ecosystems' lowest water demands without causing harm.

Introduction

In the twenty-first century, the number of people who do not have access to water and live-in unsanitary conditions has risen to billions. An estimated 2.2 billion people (26%) lack access to safe drinking water and 4.2 billion people (46%) lack access to safe sanitation, according to the data from UNICEF and WHO (2021: 9; UN Water, 2021b). Due to a lack of access to safe and clean water, these people, the most of whom are children and the elderly, may die each year from water-related diseases. For example, every year, 297.000 children under the age of five die from diarrhea caused by poor sanitation. Diseases like cholera, dysentery, hepatitis A, and typhoid have been linked to poor sanitation and contaminated water (WHO, 2019). If current trends continue, it is expected that a major section of the world's population would face acute water shortages, or scarcity.

Despite the fact that water is the most basic component of all living things, people cannot directly benefit from it or meet their needs, because the amount of fresh water available to humanity on the planet's surface is highly limited. The saline water in the seas and the frozen water at the poles are two factors that prohibit us from enjoying the benefits of water. At the same time, fast population growth, particularly in developing nations, and the resulting increase in water demand can put a strain on limited water resources. Water resources tend to diminish gradually as the annual amount of water that may be utilized per person each year declines as much as water demand, which is increasing year by year. Additionally, pollution of underground and surface water resources by agricultural, industrial, and residential pollutants can also deplete them. Human-caused water pollution and the overuse of water basins result in the ecological devastation of water resources on the one hand, and the unusability of these already precious resources in terms of quality on the other. In 2015, water pollution claimed the lives of 1.8 million people. Also, every year, nearly 1 billion people become ill as a result of contaminated water. Moreover, low-income groups are particularly vulnerable because their dwellings are frequently located near polluting enterprises (The Lancet, 2017). Warming of the earth's surface can also have a negative impact on water cycles and resources, both in terms of quantity and quality. Large and sudden changes in the water cycle can occur as a result of climate change, which is directly related to water quality, that is, extreme temperatures and colds. Droughts, floods, and reduced precipitation that feeds groundwater all have the potential to wipe out entire freshwater ecosystems. Finally, water management policies focused on water price and buying and selling as an economic good have the potential to deplete water resources and make them scarce. Increases in water prices that surpass purchasing power can lead to water shortages because access to sufficient water is intimately linked to the economy as well as the quality and amount of water (Çolakoğlu, 2008). Millions of regular Americans, for example, are said to be in risk of losing their houses because they are unable to pay their bills due to the rising cost of mains water. The overall cost of utility water and sewer service climbed by nearly 80% between 2010 and 2018, according to a comprehensive assessment of 12 different communities in the United States published in 2020. Water bills in one city climbed by 154% in less than ten years, and more than two-fifths of inhabitants in several cities suffer extremely high prices. Furthermore, while the COVID-19 pandemic has highlighted the necessity of access to clean water, this crisis in water affordability has only underlined the harm that bills have caused to the poorest (Colton, 2020). While the study focuses on the scope of the rising water issue in the United States, similar situations can be found all around the world.

The production and consuming activities that people engage in for their economic interests and accept themselves at the center of their perspective of the environment are surely the causes listed for the scarcity of water resources. Due to insufficient and polluted water supplies, people in many parts of the world currently lack access to adequate and clean drinking water. However, the protection of fresh water resources, which are limited in quantity and quality, is vital for the future supply of drinking and utility water. If water-related issues are to be solved, it is first required to adopt an ecologically sensible attitude toward water and realize that water has a superior and inherent value. Water is a priceless resource that is not only necessary for survival but also plays an important part in the life support system, economic development, community well-being, and cultural values. If water resources are to be managed in a sustainable manner, they must first be protected in terms of quality and quantity. Sustainability is characterized as an ethic that incorporates resource efficiency concepts that can help with water planning and management, as well as equity of distribution between current and future generations, ecosystem protection, and public engagement. Since morality and fairness originate in certain social and cultural contexts, any decision that can be made within the framework of an ethical understanding that may be created in this direction can penetrate individual and social actions and help decide the concepts of morality and justice (Çolakoğlu, 2008).

In brief, if individuals are denied access to water that is not provided under sufficient conditions in terms of quality and quantity, they may face substantial issues in terms of hygiene and livelihood. It is vital to provide enough clean and safe water, as well as make it sustainable, in order to meet their basic needs. The United Nations (UN) has been actively involved in building water security around the world since the 2000s, identifying and revealing clear targets for countries in this issue. This study looks at water security in the context of access in general in a theoretical setting and investigates it conceptually. Its main aim is to provide a better understanding of the concept of water security, which has recently emerged as a stronger concept for the protection and sustainable use of water resources on the planet and is increasingly appearing on the agendas of states, global discussions, academic and international studies. Its subgoals are herewith to contribute to the creation of the concept of water security, to inspire conversation on the subject, to take a step toward concretizing the notion, and to eventually lead policymakers. The major goal of ensuring clean water and sanitation for all, as well as accessible water and wastewater services for everyone, and sustainable water management, which is accepted as the 6th goal in a series of universal calls for action by the UN member states, and also recognizes other issues that surround achieving the human right to water and sanitation, is then discussed. Finally, the notion of the right to water, which is advocated as part of the UN's goals and programs and is meant to provide a lifeline for those forced to live under water scarcity, is included in this framework.

Water Security as a Concept

Ever since their existence on earth, humans have shaped and changed their environment in line with their own wishes, interests, and activities. Such external interventions to the environment cause the functioning of all cycles in nature, such as the water cycle, to be adversely affected, and many problems concerning the environment. The basis of this is that they accept themselves as the center in their perception of the environment and their production and consumption activities for their economic interests. One of the main effects of this is the pollution, destruction, and destruction of water, which is one of the most important sources of life. The water coming to the earth through the cycles can no longer meet the needs, both with population growth and with the pollution and mismanagement of water resources, and causes people to be deprived of adequate and safe water and sanitation in many parts of the world. This also poses serious threats to human health and ecosystems. In this respect, it is not difficult to say that water security is an integral part of the more comprehensive concept of human security. Water security is securing access to sufficient and safe water at an affordable price for every person to lead a healthy, dignified, and productive life. When individuals are not provided with appropriate conditions or their access to water is denied, human safety may face serious threats due to poor health and livelihood conditions (United Nations Development Programme, 2006: 2). Sufficient water available to meet their basic needs, which include sanitation and hygiene and safeguarding of health and well-being, should be provided for individuals. Water security is access to safe water in order to provide access to safe drinking water, improve conditions for hygiene and food safety, and enhance nutritional status.

The term water security encompasses the safety of the entire population, everyone's ability to obtain safe water for domestic use and economic security, the ability of water resources to meet the normal requirements of economic development, and the ability of ecosystems to meet their lowest water demands without causing damage to water resources. As an integral part of socio-economic and human development and growth, water security is the ability to provide water under appropriate conditions and at an affordable price, based on criteria set for households, industrial and commercial units, and agricultural areas. (Xia et al., 2007: 10).

Being insecure of water leads to a violation of some of the most basic principles of social justice. First of all, the civil, political, and social rights ensured to each individual may be endangered due to water insecurity. This adversely affects the principle of citizenship equality. For example, in some underdeveloped societies, the duty of carrying water is generally accepted as the duty of women. These women, who spend long hours on this issue and constantly suffer from water-related diseases, have less power to participate in their own societies, even if they can participate in elections in determining management and later processes. The principle of minimum social needs is that all citizens have access to sufficient water to meet their basic needs and lead a dignified life. This requirement is 20 liters per day. Gleick (2003) recommends that the basic needs including drinking, personal sanitation, washing of clothes, food preparation, and personal and household hygiene are total about 50 liters per person per day. The principles of equal opportunity which is a fundamental requirement for social justice, and of fair distribution, can also be affected by water insecurity (United Nations Development Programme, 2006: 2–3). If we accept as a basic human right to water on the other hand, individuals may be provided with access to safe and affordable water and thus will have confidence in the direction of the water.

Elements of Water Security

There are three basic elements of water security so that every person can lead a clean, healthy, and productive life, while ensuring that the natural environment is protected and enhanced (Global Water Partnership, 2000). Water access, safety, and affordability are core elements of water security.

Water access is the first element of water security. Accordingly, all the people in a community should have access to water of sufficient quantity and quality for the basic human needs, small-scale livelihoods, and local ecosystem services all the year round. Water security is reliable access. Unreliable water supplies expose communities to negative effects on physical health and livelihoods. Therefore, sufficient quantities of water for household use, including drinking, cooking, bathing, sanitation, and hygiene – that is, domestic water – must be ensured. But also, water services must be available within, or in the immediate vicinity of health and educational institutions, workplaces, public places, and workplaces, without discrimination. Securing access for all must be ensured in a sustainable manner (Siwar & Ahmed, 2014: 284).

As the second element of water security, *water safety* is related to its quality. There must be sufficient quantity of water to satisfy all personal and domestic needs of the households. The quality of water should be such that no significant health risk arises from its use and should be acceptable to users in appearance, taste, and odor. Contaminant levels should not exceed the usually accepted water quality standards that are generally defined by national and/or local levels (WaterAid, 2012). For this, the World Health Organization Guidelines (The World Health Organization issued a 2017 addendum to its fourth edition of the 2011 Guidelines for Drinking-Water Quality, and updated.; (World Health Organization, 2017).) for drinking-water quality provide a basis for the development of national regulations and standards (UNDESA, 2014) for water safety.

Water affordability is related to sufficient purchasing power for households to afford access to the water, and water facilities and services. The price of these services must be affordable for all, including those living in extreme poverty. According to UNDP, water costs should not exceed 3% of household income (UNDESA, 2014). National regulations have to set standards regarding pricing of water (Siwar & Ahmed, 2014: 284).

Briefly, achieving water security for people is possible with adequate access to safe water at an affordable price. General Comment No. 15 states that *"to ensure that water is affordable, States parties must adopt the necessary measures that may include, inter alia: (a) use of a range of appropriate low-cost techniques and technologies; (b) appropriate pricing policies such as free or low-cost water; and (c) income supplements. Any payment for water services has to be based on the principle of equity, ensuring that these services, whether privately or publicly provided, are affordable for all, including socially disadvantaged groups. Equity demands that poorer households should not be disproportionately burdened with water expenses as compared to richer households."* (art. 27) (Committee on Economic, Social and Cultural Rights, 2002). In other words, in November 2002, this General Comment of the Committee on Economic, Social, and Cultural Rights stresses the economic accessibility dimension of the right to water should be affordable for everyone. Water affordability means that all the people should access to safe and sufficient drinking water at an affordable cost in order to meet basic needs, and access to safe and affordable water is considered to be a human right.

Dimensions of Water Security

Some studies (See also for the studies; (FAO, 2000; Gleick & Iceland, 2018; Global Water Partnership, 2016; Greya & Sadoff, 2007; UNEP, 2009; UNESCO, 2021; UNICEF, 2021; United Nations, 2021c).) have highlighted mainly four fundamental dimensions of water security, namely human security, ecological security, peace–conflict security, and economical security, as a major key to reach the sustainable development. These dimensions, which are outlined below, measure a state's national water security.

Human Security

As noted earlier, water is the life for us all and is an integral part of the more comprehensive concept of human security. No life can be sustained without water. Water is a necessity for life and indispensable for human health and well-being. For the realization of the right to life, water must be safe and accessible. When people have not gain access to clean water, billions of people around the world die every year due to various diseases caused by contaminated water and are continuing to suffer from poor access to water, sanitation, and hygiene. Therefore, as part of human security, water security is essential to ensuring that every individual has access to sufficient and safe water at an affordable price to lead a healthy, dignified, and productive life, as well as continuity and timing of water service, and also, every individual has a fundamental right to access water.

Ecological Security

Water has also a vital importance for all other living and nonliving beings. However, the water that comes to the earth with the water cycles can no longer meet the needs due to population growth, pollution, and mismanagement of water resources and causes people to be deprived of adequate and safe water in many parts of the world. Hence, this consequence poses increasingly severe risks for ecosystems and the planet's life support systems beyond human health.

Illich stated that it is wrong to approach water from an engineering point of view and try to explain and describe it with a simple chemical formula and that it has turned into a common enemy that dries out the soils or that it has turned into a common enemy that kills people with floods and hurricanes and harms other living and nonliving things because it rains heavily (2007: 15). Because the necessity of explaining and recognizing the rights of all living and nonliving beings in nature, which do not claim rights but need rights in order to exist, on the basis of an ethical assumption is clear for today. As long as the anthropocentric view does not change, the destructions against all living and nonliving beings in nature will continue, together with humans, and this process will continue to progress towards an end that also eliminates the life opportunities of all living things. Living with such an ethic means using water in a more balanced and sustainable way as far as possible and sharing what is owned. The basis of this is basically to protect aquatic ecosystems and to adopt an integrated approach that is far from utilitarianism.

Peace–Conflict Security

In addition to being a basic element of life, water resources have become one of the most important components in terms of welfare, development, and meeting basic needs of societies today, and their importance has increased considerably. In this sense, water is also indispensable for states. Therefore, states facing a common problem in ensuring water security have to address the relationship between water and security. Because water insecurity has the potential to pose a threat to national security, it can be a source of conflicts within or between states. Today, water problems, which have reached the level to cover military actions, can now be handled in the context of environmental security, and therefore, water and security are seen and developed within the scope of "common security." Because in many societies, water as a natural resource is a power, and inequalities in the distribution of economic power and deep inequalities in access to water can be created. Water allocation can lead to conflict; the fact that all water management serves many purposes can trigger a competition based on power between competing interest groups. This can lead to intense political pressures and even cause instability and conflicts at the regional and international level. These conflicts can also occur on a political, ethnic, and religious basis to a large extent, as the reason for the disagreement may differ.

The history of water-related violence is actually very old. Examples of competition and conflict over the allocation or sharing of freshwater resources as a matter of national security are quite common. In 1503, Leonardo de Vinci conspired with Machiavelli to change the flow of the Arno River in the war between Pisa and Florence, and for 80 years against Spain in the sixteenth century, the Dutch created floods in their own lands to protect their towns from Spanish military forces ("the Dutch Water Line"), can be given as examples.

On the other hand, water resources can be used as military and political targets. If water powers a state as an economic or political resource, it provides a justification for going to war to secure access to water, and in this respect, water transmission systems can be a target of military occupation. But at the same time, these systems can be the target of any terrorist act, or the water resources can be deliberately poisoned or contaminated with disease-causing substances. As a result, individuals may be harmed, water may be rendered unusable, or water supply and treatment infrastructure may be destroyed (Gleick, 2006: 482). Water resources and its systems can be polluted and turned into an instrument of war or conflict, which can cause serious problems that can threaten public health and the environment, that is, the country's water security can be threatened by terrorists or other forces antagonistic to the interests of the states.

Economical Security

Considering water only as an economic good can create a water insecurity problem both in terms of environmental ethics and human rights. But at the same time, the recognition of water as an economic good as a scarce resource can lead to conflicting problems. Because the water management policy, which is based only on the pricing of water and its buying and selling as an economic good, can lead to the destruction of water resources and making them scarce. Economic structures that shape water policies, including multinational companies in the world, may cause water shortages through privatizations in the countries where they are located, increases in water prices that exceed purchasing power, and/or other activities, because having access to sufficient water is closely related to its economy, as well as its quality and quantity.

Perceiving water only as an "economic good" neither eliminates the water scarcity and its negative consequences, nor is it the only way to ensure a more effective and measured use of water. On the contrary, such a policy creates and increases socio-economic inequalities as well as asymmetric power relations in society. Those who provide water service are not able to run the system properly and reduce the quality of services. On the one hand, water is transported to the homes of the wealthy and influential layers of the society directly and cheaply through pipes, on the other hand, poor and vulnerable layers have to drink unsafe and unhealthy water or pay high amounts of money to water sellers. Therefore, accepting that water "belongs to the public," that is, it is a "common value" or "a public good," is necessary. Water is a public service that requires both central and local solutions and should be evaluated in line with the principle of "indivisibility of services" at the regional scale. In this respect, water should be accepted as a service that should be used by the state on behalf of the "public." It is surely beyond doubt that this may not be enough to eliminate existing problems. Therefore, if the aim is to ensure access to water for all people in the society, it can be said that water and access to water should be accepted as a human right first and then institutionalized in this direction in accordance with this aim (Çolakoğlu, 2008).

Water Security, Sustainable Development Goals, and SDG 6

Water security in society necessitates guaranteeing the sustainability of water services in all its dimensions. First and foremost, the quality and quantity of these services must be protected if they are to be managed in a sustainable manner. Sustainability includes an ethic that adopts the principles of resource efficiency that can provide benefits in the planning and management of water, equality of distribution between present and future generations, ecosystem protection, and public participation. Each decision to be taken within the framework of an ethical understanding that has been created in this direction can penetrate individual and social behaviors and help determine the concepts of morality and fairness since it emerges according to the social and cultural structure of that society. Therefore, this new water ethics or behavior provides benefits in guiding production and explaining the policies and laws arising from social institutions, as well as influencing water use decisions (Beveridge, 2006: 26). Briefly, water security should not be understood in this context as simply balancing natural water resources with the country's economy; rather, it should be recognized as a multidimensional concept that includes social, cultural, and political institutional processes, as well as a variety of indicators (Mengi & Algan, 2003: 3), that is, water withdrawal, renewable water resources, dependency ratio.

By defining water security as *"...the capacity of a population to safeguard sustainable access to adequate quantities of and acceptable quality water for sustaining livelihoods, human well-being, and socio-economic development, for ensuring protection against water-borne pollution and water-related disasters, and for preserving ecosystems in a climate of peace and political stability..."* (United Nations University, 2013: 1), the UN has clearly linked sustainability and water resources. In this respect, especially since the 2000s, water security has been accepted as one of the important criteria and principles in the realization of sustainable development goals. Water security is widely acknowledged to be included in development goals and to be at the heart of progress. In line with the main goal of "ensuring environmental sustainability," which is one of the eight goals adopted in the Millennium Declaration (2000), the subtarget of reducing the proportion of people who cannot access safe drinking water and basic health care by half by the year 2015 (principle 19) has been adopted. The declaration also underlined the need to prevent excessive use of water resources (principle 23). Accordingly, the year 2003 has been declared as the "International Year of Fresh Water" (Habitat, 2021). According to the results obtained that since 1990, 2.6 billion people have gained access to improved drinking water, with 1.9 billion receiving piped drinking water on premises. Over half of the world's population (58%) now has access to this higher level of service. Globally, 147 countries have met the target for drinking water, 95 countries have met the target for sanitation, and 77 countries have met both. Additionally, improved sanitation is now available to 2.1 billion people worldwide. Since 1990, the proportion of people who practice open defecation has decreased by nearly half (United Nations, 2015: 7). Nonetheless, some subtargets have not been met, particularly in the poorest regions, due to various challenges (e.g., the lack of synergies among the goals, the economic crisis, etc.) (WHO, 2018; Lomazzi et al., 2014). Today, while significant progress has been made in increasing access to safe drinking water and sanitation, billions of people, primarily in rural areas, continue to lack these essential services. One in every three people in the world does not have access to safe drinking water, two out of every five people do not have access to a basic hand-washing facility with soap and water, and more than 673 million people continue to practice open defecation (United Nations, 2021a).

However, the UN, which has been successful in drawing the attention of the international public to these issues and creating sensitivity, has also established new targets for the post-2015 year. The UN adopted the Sustainable Development Goals (SDGs), also known as the Global Goals, in 2015 as a universal call to action to end poverty, protect the environment, and ensure that by 2030, all people enjoy peace and prosperity. As a result, the 17 SDGs are integrated – they recognize that development must balance social, economic, and environmental sustainability. For the period 2015–2030, Sustainable Development Goal 6 on water and sanitation (SDG 6), adopted by UN Member States at the 2015 UN Summit as part of the 2030 Agenda for Sustainable Development (UN Department of Economic and Social Affairs, 2021), seeks to ensure the availability and sustainable management of drinking water and sanitation for all as one of these targets. However, these new targets, as shown below, are more comprehensive than the Millennium Development Goals (MDGs) and include issues such as water quality and wastewater, water use and efficiency, and integrated water resource management, in addition to drinking water supply and sanitation. According to the SDG 6 (UNDP, 2021):

(a) Ensure universal and equitable access to safe and affordable drinking water for all by 2030.
(b) By 2030, all people will have access to adequate and equitable sanitation and hygiene, and open defecation will be eliminated, with a special focus on the needs of women and girls, as well as those in vulnerable situations.
(c) Improve water quality globally by 2030 by reducing pollution, eliminating dumping, and minimizing the release of hazardous chemicals and materials, halving the proportion of untreated wastewater, and significantly increasing recycling and safe reuse.
(d) Significantly increase water-use efficiency across all sectors by 2030, while also ensuring sustainable withdrawals and supply of freshwater to address water scarcity and significantly reduce the number of people suffering from water scarcity.
(e) Implement integrated water resource management at all levels by 2030, including transboundary cooperation as needed.
(f) Protect and restore water-related ecosystems, such as mountains, forests, wetlands, rivers, aquifers, and lakes, by 2020.
(g) Extend international cooperation and capacity-building assistance to developing countries in water- and sanitation-related activities and programs, such as water harvesting, desalination, water efficiency, wastewater treatment, recycling, and reuse technologies, by 2030.
(h) Encourage and strengthen local community participation in improving water and sanitation management.

In the 2030 Agenda for Sustainable Development, countries have agreed to use a set of global indicators to track and analyze progress toward the goals (UN Water, 2021b). When the 2030 Agenda was adopted, all goals were designed to be integrated and indivisible, with the goal of balancing the social, economic, and environmental dimensions of sustainable development. As a result, the Agenda recognizes that social development and economic prosperity are dependent on the sustainable management of freshwater resources and ecosystems and emphasizes the SDGs' interconnected nature. From this aspect, the SDG 6 which includes eight universally applicable and aspirational global targets is, above all, a public investment program in core infrastructure (roads and power), digital, water and sanitation, human capital (health and education), and the environment. Also, depending on national realities, capacities, levels of development, and priorities, each government must decide how to incorporate them into national planning processes, policies, and strategies (United Nations, 2018). It does not appear possible to achieve the SDGs if it is not realized. For example, many indicators for 2020 are not yet available due to time lags in international statistics caused by the increased poverty rates and unemployment following the outbreak of the COVID-19 pandemic, so the global decline in SDG performance is due to this main reason. The highest priority of every government has been the fight against the pandemic which has impacted all dimensions of sustainable development (Sachs et al., 2021).

As a direct response to the Decade of Action and Delivery for Sustainable Development called for by Heads of State and Government at the SDG Summit in 2019, the UN system launched the SDG 6 Global Acceleration Framework in July 2020, with the goal of accelerating progress toward the SDGs and putting the world on track to meet their targets by 2030. Additionally, in support of SDG 6 and other water-related targets, the UN General Assembly unanimously adopted the resolution "International Decade for Action – Water for Sustainable Development" (2018–2028) in December 2016, and on 21 December 2020, the resolution on "the UN Conference on the Midterm Comprehensive Review of the Implementation of the Objectives of the International Decade for Action, 'Water for Sustainable Development', 2018–2028," the first UN Conference on water since 1977. Water is also central to landmark agreements like the Sendai Framework for Disaster Risk Reduction and the 2015 Paris Agreement (UN Department of Economic and Social Affairs, 2021).

Defining the Right to Water in the Context of Water Security: Challenges and New Developments

Defining and accepting water as a human right can help to build a trustworthy environment for millions of people who are currently water insecure. The right to water is a right that stems from the need for water security; it is the right of all people to be provided with water of adequate quality and quantity to meet their biological, natural, social, and human needs at the bare minimum.

People may face serious problems due to poor hygiene and livelihood conditions if they do not have access to water that is provided under suitable conditions in terms of quality and quantity. It is essential to provide an adequate and safe drinking water supply for people's basic needs. Today, if water is accepted as a human right in the context of human rights, and if strong measures are taken and necessary institutional and legal obligations can be created, the conditions in which individuals must live that are caused by water insecurity can be improved, in other words, people's access to potable (safe), usable, physically accessible, and affordable water can be strengthened. Because the right to water refers to a right that can provide people's access to water in the face of water insecurity that may arise from water scarcity or pressure. It has been observed that an intense effort is being made on an international rather than a national scale to make the right to water binding all over the world.

The right to water can provide an individual security demand for all people in the face of water being accepted solely as an economic good and the rights violations that this acceptance may result in. This right is also the right to benefit from water at a "reasonable price" (Smets, 2000: 55), which is determined by the minimum amount of water that a person can consume in order to meet his basic needs. The human need for water serves as the foundation for this right; that is, there is a principle that water should be economically available, particularly for the poorer sections of society, and that this situation does not create dependency, whether provided by the public or private sector. As a result, when developing a state's water management policy and defining water as a human right as part of that policy, the principles of equality and equity should be considered, and the "social justice dimension" should not be overlooked.

Furthermore, one of the primary reasons for asserting the right to water as a human right is the belief that water is one of the most essential elements of human life. The right to water can be recognized as a human right when water is recognized as an essential quality for sustaining human life. At the same time, this can provide a basic basis for the legitimacy of the right to water. Even if the subject of this right is "human," according to the ecocentric ethical approach, human beings have a duty to protect and observe the rights of other living and nonliving things, and as a result of this right, people do not have the right and authority to use water indefinitely. In this context, when evaluating water in terms of human rights, it is critical to take this approach. When people use water, they must fulfill their obligations to respect, protect, and benefit from the water that other living and nonliving creatures need. Water is significant because it is required for the continuation of life and because the living environment of other living and nonliving things is water-dependent. An ecocentric approach will allow us to define the value of water, which has inherent value in terms of sustainability for humans, all other living things, and nonliving things. In this regard, the right to water concerns not only to plants, animals, and microorganisms, but also to the ecosystems that surround river basins, lakes, aquifers, oceans, and waterways.

The essence of this approach is that people question their relationships with their surroundings and attempt to explain them using an ethical value. This is also supported by the concept of "environmental justice." Environmental justice is founded on the principles of fairness and justice. Given the intertwined nature of human rights and sustainable development, environmental justice becomes increasingly important in ensuring social justice at all stages, from equitable distribution of water resources to the foundation of public participation in the water management process. In this regard, the human right to water should be considered alongside the discourse of environmental justice and should be made a part of the right to water. Mecelle-i Ahkâm-Adliye, often known as Mecelle, is a codex of Islamic private law regulations that was used as a legal basis in religious courts in the Ottoman Empire's last half-century. It was compiled between 1868 and 1876 by a committee directed by Ahmet Cevdet Pasha. This right is extended to all living beings, not just people. Mecelle is the world's first legislative rule to acknowledge this privilege, which is known as "Hakk-ı Şefe." (See articles 1234, 1235, 1237, 1238, 1263 and 1264 of Mecelle; (Berki, 1978: 252, 257).) The objective here is to preserve people's rights against the right to water, particularly private property rights (Çolakoğlu, 2008).

Recognition of water as a human right and its international acceptance, on the other hand, will first and foremost provide an international legal framework within which standards for human access to water can be established. Of course, this is insufficient. Regulations and safeguards must be developed to ensure that everyone has this right. It will be able to provide safe, usable, accessible, and affordable water, especially to the most vulnerable and powerless members of society, with the recognition of this right. However, it is possible that acceptance of the right to water will create a chaotic environment in the future, both in terms of human rights and international relations. This can be caused by the fact that the content of this right is not well filled, allowing states' sovereign rights to become porous. At the same time, it evokes the idea that individuals' basic needs may not be adequately met in the long run and that the right to water may be inadequately met and ignored, because the protection and applicability of this right gives important powers to actors other than states, whereas in reality it is "a rights and freedoms from which the entire society can benefit" (Sav, 2007: 137–138). Because it operates within the price mechanism that underpins the neoliberal economy, it calls into question whether individuals' right to water can be protected as a human right. As a result, it is obvious that without effective state protection, it is impossible to realize water, which is a common value requiring public benefit and also a public good.

Furthermore, the adoption of the right to water may cause states to demand water from other states that are relatively well-off in terms of water resources in order to provide the right to water to their citizens in the long run. This has the potential to pose significant risks for countries and regions where water resources are scarce in the future, such as the Middle East, as well as to cause confusion by paving the way for some demands that will not be met in the future. On the other hand, it is clear that states will not support the right to water in a way that limits or abolishes their sovereignty or sovereign rights over water resources within their control (Sav, 2007: 6–13). This makes it difficult for all states to accept the concept of the right to water. Therefore, governments are primarily responsible for ensuring the fulfillment of this right and should do so as accurately and promptly as possible at the national and local levels. When establishing and shaping the right to water, this must be taken into account. It is obvious that water service, which requires public benefit and is both a common value and a public good, cannot be realized without the state's effective protection. This right, which is an indispensable right for every human being, is an integral part of other rights such as the right to life, health, food, education, peace, environment, and development and is a prerequisite for the realization of these rights (Çolakoğlu, 2011: 231).

On the other hand, the UN pursues a different strategy at the international level for the acknowledgment of the right to water, emphasizing the sanitation aspect in order to increase acceptance of this right. Initially, the Committee on Economic, Social, and Cultural Rights issued and adopted General Comment No. 15 on the right to water at its 29th session, held in Geneva from November 11 to November 29, 2002. The right to water was officially recognized in international law for the first time. As previously stated, the main point of General Comment No. 15 is stated in paragraph 2, which says; *"the human right to water entitles everyone to sufficient, safe, acceptable, physically accessible and affordable water for personal and domestic uses. An adequate amount of safe water is necessary to prevent death from dehydration, to reduce the risk of water-related disease and to provide for consumption, cooking, personal and domestic hygienic requirements."* However, in the years since, the right to water has been regulated as a right under a single title, alongside sanitation, due to the importance of sanitation and the fact that it should not be considered separately from right to water. For example, there was no target for access to sanitation in the UN Millennium Declaration, which was adopted in 2000. According to Khalfan and Kiefer (2008), this situation was corrected two years later at the 2002 World Sustainable Development Summit, when sanitation was included in the UN MDGs and states were obligated in this direction. With the increasing importance of sanitation services in the following years, it has been accepted that both services are by definition fundamental rights and should be evaluated together (Çolakoğlu, 2011: 223). Maybe, it is an effort to boost the chances of international recognition of the right to water. As a result of this effort, access to water and sanitation has been recognized as a human right essential for the full enjoyment of life and all human rights; on July 10, 2010, the UN General Assembly adopted Resolution 64/292, eight years after the Committee issued General Comment No. 15.

The right to water and SDG 6 are linked in numerous ways. We can talk about a mutual fostering process in this way, which promotes both the SDG 6 aim and the implementation of the right to water. Firstly, it was shaped in response to the requirements of many internationally binding and nonbinding legal documents such as the Convention on the Rights of the Child, the Convention on the Elimination of Discrimination Against Women, the General Comment No. 15 of the Committee on Economic, Social, and Cultural Rights, based on SDG 6's main targets, and the substance of SDG 6 reflects this. Secondly, in accordance with the UN definition, it is proposed that nonstate actors, such as the Corporate Water Stewardship promoted by the UN Water Compact, as well as various types of public-private partnerships, in addition to the state, be protected under this right. As a result, the right to water aligns with SDG 6's goals. Thirdly, in terms of effectiveness, the SDG 6 strategy helps to the development and hardening of international law. The SDGs have mobilization power that human rights may benefit from if they are well-run and executed. Finally, there are flaws in the SDG 6 implementation process in terms of ensuring water security. For example, although various follow-up and review mechanisms are mentioned in line with this main target, the relevant process cannot be carried out properly for a variety of reasons, including the fact that it is based on volunteerism, uncertainty in measuring accessibility, availability, and affordability, all of which are considered essential elements of the right to water. Additionally, the 2030 Agenda for Sustainable Development omitted a framework for monitoring and review to ensure that other stakeholders, such as private players, are aligning their operations and policies with the SDGs. It is a methodical technique that has been implemented. Checks and balances, as well as peer pressure, must find a place in order to adapt and remedy flaws (Boisson de Chazournes, 2021: 144–145). As a result, tracking progress toward SDG 6 is a way to ensure that the target is met on an ongoing basis. The UN-Water Integrated Monitoring Initiative for SDG 6 assists nations in tracking water and sanitation-related issues as part of the 2030 Agenda, as well as compiling country data to report on global progress (The progress can be monitored in the reports released in August 2021. In 2020, the most recent global data drive took place, yielding status updates on nine of SDG 6's global indicators. See for these reports; United Nations, 2021b).) toward SDG 6 (UN Water, 2021a).

Finally, the SDG 6 Global Acceleration Framework was launched in 2020 due to the limited amount of time left to accomplish the SDG 6 and the urgent need for coordinated action to address the complexity and scale of water and sanitation issues. According to the SDG 6 Synthesis Report 2018 on Water and Sanitation, the world will not meet the SDG 6 targets by 2030 if present trends continue, and have barely made an average of 1% annual progress on improving access to basic water supply and sanitation since 2000, according to progress rates (United Nations,

2018). Therefore, the Framework attempts to achieve rapid results at a larger scale as part of the Decade of Action to deliver the SDGs by 2030. At the Framework, the ability to do this depends on the four pillars of action that accelerate broad stakeholder action by significantly increasing the international community's support for countries to achieve SDG 6 on water and sanitation:

(a) *Engage:* The Regional Coordination Mechanism and the UN Resident Coordinator system will connect available knowledge and scale up support within the system and the international community by leveraging the UN's convening power. By uniting external support around government-led objectives and offering technical expertise and resources, this will lead to better engagement with countries.
(b) *Align:* Synergies across multiple SDGs can be utilized through improving ways of cooperation and collaboration within the UN system and the international community.
(c) *Accelerate:* The international community's assistance for country advancement will be greatly increased because to five accelerators: efficient financing, improved data and information, capacity development, innovation, and governance.
(d) *Account:* This entails encouraging all actors to share accountability by reviewing progress and learning together. To build a planning and delivery culture where stakeholders use the most up-to-date evidence on what works, learn swiftly from failure, and adjust to changing realities, nimble evidence-based implementation will be used. For this, the worldwide community will convene in New York in 2023 for the UN General Assembly (UN-Water, 2020).

Also, prioritizing the vulnerable, inclusivity, conflict sensitivity, releasing gender and youth potential, planning for resilience, and developing and implementing transformations based on scientific evidence are among the guiding principles of the Framework (UN-Water, 2020).

Conclusion

Today, water resources are one of the natural resources that have become limited due to a variety of factors including pollution, population increase, and mismanagement and are unable to meet rising demands. Therefore, millions of people around the world are at risk of losing their lives or being forced to fight epidemics due to a lack of safe water. While people in developed countries are only directly affected by these consequences due to a lack of purchasing power, people in developing countries particularly disadvantaged groups suffer the most severe repercussions of water scarcity. In this context, the UN's efforts to reduce water insecurity become increasingly important. The UN has been developing and coordinating a set of shared principles, policies, and strategies for its member countries to assure the quality and quantity of water sustainability for the previous two decades. The MDGs were created by the UN in the year 2000. Some progress has been made toward these goals over the years, and people all across the world now have access to safe water and sanitation. Despite this, the final product has fallen short of expectations, and problems have persisted due to factors such as climate change and population increase. As a result, 15 years later, the UN revised the Goals, making them more precise and comprehensive, resulting in the 2030 SDGs, also known as the Global Goals. As one of the key goals, the SDG 6, with six targets, includes all water and sanitation issues and ensures universal and equitable access to safe and affordable drinking water for all by 2030. But also, as a measure of maintaining sustainability, the human right to water overlaps importantly with these targets under the SDG 6 and is a basis that can allow people who are water insecure to secure their access to water and to get safe and sufficient water. As a result, the

UN member nations' policies and plans will decide the process's destiny and success.

References

Berki, A. H. (1978). *Açıklamalı Mecelle (Mecelle-i Ahkâm-ı Adliyye)*. Hikmet Yayınları.

Beveridge, M. (2006). *Proposing a water ethic: A comparative analysis of water for life: Alberta's strategy for sustainability*, A thesis presented to the University of Waterloo in fulfilment of the thesis requirement for the degree of master of environmental studies in environment and resource studies, Waterloo.

Boisson de Chazournes, L. (2021). The sustainable development goals (SDGs) and the rule of law: A propose SDG 6 on access to water and sanitation. *American Society of International Law. Proceedings of the ASIL Annual Meeting: The Promise of international law, sustainable development and international law* (pp. 143–147). Cambridge: Cambridge University Press.

Çolakoğlu, E. (2008). *Suya Erişim Bağlamında Su Güvenliği* (Unpublished Phd Thesis), T.C. Ankara Üniversitesi, Sosyal Bilimler Enstitüsü, Siyaset Bilimi ve Kamu Yönetimi Anabilim Dalı, Ankara.

Çolakoğlu, E. (2011). Emniyetli İçme Suyu ve Sanitasyon Hakkı. *Mülkiye, XXXV*(272), 217–238.

Colton, R. (2020). *The affordability of water and wastewater service in twelve U.S. Cities: A social, business and environmental concern*. Prepared for The Guardian, New York.

Committee on Economic, Social and Cultural Rights. (2002). *General Comment No. 15 (2002) The right to water (arts. 11 and 12 of the International Covenant on Economic, Social and Cultural Rights)*, Twenty-ninth session Geneva, 11–29 November 2002 Agenda item 3.

FAO (2000). New dimensions in water security: Water, society and ecosystem services in the 21st century, .

Gleick, P. H. (2003). Water Use. *Annual Review of Environment and Resources, 28*, 275–314.

Gleick, P. H. (2006). Water and terrorism. *Water Policy, 8*(2006), 481–503.

Gleick, P., & Iceland, C. (2018). Issue brief: Water, security, and conflict. World Resources Institute.

Global Water Partnership. (2000). *Towards water security: A framework for action*. Retrieved July 04, 2021, from https://www.gwp.org/globalassets/global/toolbox/references/towards-water-security.-a-framework-for-action.-executive-summary-gwp-2000.pdf

Global Water Partnership. (2016). *Linking ecosystem services and water security – SDGs offer a new opportunity for integration.*

Greya, D., & Sadoff, C. W. (2007). Sink or swim? Water security for growth and development. *Water Policy, 9*, 545–571.

Habitat. (2021). *BM Binyıl Bildirgesi*. Retrieved July 29, 2021, from http://www.habitat.org.tr/gundem21/40-gundem21/129-bm-binyil-bildirgesi.html

Illich, I. (2007). *H2O ve Unutmanın Suları* (Çev.: Lizi Behmoaras), Yeni İnsan Yayınevi. İstanbul.

Khalfan, A., & Kiefer, T. (2008). *The human right to water and sanitation: Legal basis, practical rationale and definition (26 March 2008)*. Retrieved August 10, 2021, from https://www.joinforwater.ngo/sites/default/files/library_assets/W_ALG_E36_human_right.pdf

Lomazzi, M., Borisch, B., & Laaser, U. (2014). The millennium development goals: Experiences, achievements and what's next. *Global Health Action, 7*(1), 1–9.

Mengi, A., & Algan, N. (2003). *Küreselleşme ve Yerelleşme Çağında Bölgesel Sürdürülebilir Gelişme: AB ve Türkiye Örneği*. Siyasal Kitabevi.

Sachs, J., Kroll, C., Lafortune, G., Fuller, G., & Woelm, F. (2021). *The decade of action for the sustainable development goals: Sustainable development report 2021*. Cambridge University Press.

Sav, Ö. (2007). Su Hakkı Olmalı mı? *Türkiye Barolar Birliği Dergisi, 72*, 134–151.

Siwar, C., & Ahmed, F. (2014). Concepts, dimensions and elements of water security. *Pakistan Journal of Nutrition, 13*(5), 281–286.

Smets, H. (2000). *L'eau au XXIe Siècle: De La Vision à L'Action*. Futuribles.

The Lancet. (2017, October 19). *The Lancet commission on pollution and health*. Retrieved August 22, 2021, from https://www.thelancet.com/commissions/pollution-and-health

UN Department of Economic and Social Affairs. (2021). *Water and sanitation*. Retrieved July 30, 2021, from https://sdgs.un.org/topics/water-and-sanitation.

UN Water. (2021a). *Monitoring water and sanitation in the 2030 agenda for sustainable development*. Retrieved August 27, 2021, from https://www.sdg6monitoring.org/

UN Water. (2021b). *Sustainable development goal 6 on water and sanitation (sdg 6)*. Retrieved August 27, 2021, from https://www.sdg6data.org/

UNDESA. (2014). *International decade for action 'Water for Life" 2005–2015*. Retrieved July 05, 2021, from https://www.un.org/waterforlifedecade/human_right_to_water.shtml

UNDP. (2021). *Goal 6: Clean water and sanitation*. Retrieved July 30, 2021, from https://www.undp.org/sustainable-development-goals#clean-water-and-sanitation

UNEP. (2009). Water security and ecosystem services: The critical connection.

UNESCO. (2021). The United Nations world water development report 2021: Valuing water.

UNICEF. (2021). *Reimagining WASH, water security for all.*

UNICEF and WHO. (2021). The measurement and monitoring of water supply, sanitation and hygiene (WASH) affordability: A missing element of monitoring of sustainable development goal (SDG) targets 6.1 and 6.2.
United Nations. (2015). The millennium development goals report 2015.
United Nations. (2018). Sustainable development goal 6 synthesis report 2018 on water and sanitation.
United Nations. (2021a). *Goal 6: Ensure access to water and sanitation for all*. Retrieved July 30, 2021, from https://www.un.org/sustainabledevelopment/water-and-sanitation
United Nations. (2021b). *SDG 6 progress reports*. Retrieved August 27, 2021. https://www.unwater.org/publication_categories/sdg6-progress-reports/?utm_content=buffer90ff6&utm_medium=social&utm_source=twitter.com&utm_campaign=buffer
United Nations. (2021c). The sustainable development goals report 2021.
United Nations Development Programme. (2006). Human development report (beyond scarcity: Power, poverty and the global water crisis).
United Nations University. (2013). *Water security & the global water agenda an UN-Water analytical brief*.
UN-Water. (2020). The sustainable development goal 6 global acceleration framework.
WaterAid. (2012). Water security framework.
WHO. (2018, February 19). *Millennium Development Goals (MDGs)*. Retrieved July 30, 2021, from https://www.gwp.org/globalassets/global/toolbox/references/towards-water-security.-a-framework-for-action.-executive-summary-gwp-2000.pdf
WHO. (2019, June 18). *1 in 3 people globally do not have access to safe drinking water – UNICEF, WHO*. Retrieved August 21, 2021, from https://www.who.int/news/item/18-06-2019-1-in-3-people-globally-do-not-have-access-to-safe-drinking-water-unicef-who
World Health Organization. (2017). *Guidelines for drinking-water quality: Fourth edition incorporating the first addendum*.
Xia, J., Lu, Z., Liu, C., & Yu, J. (2007). Towards better water security in North China. *Water Resources Management, 21*(1).

Water Security and the Green Economy

Robert C. Brears
Our Future Water, Christchurch, Canterbury, New Zealand

Synonyms

Green economy; Green growth; Water conservation; Water efficiency; Water security

Introduction

The traditional economic model of employing various types of capital, including human, technological, and natural, to produce goods and services has brought about many benefits, including higher living standards and improved human well-being. At the same time, economic growth has resulted in environmental degradation. Also, the global economic model is confronted by a wide array of trends including rapid population growth, urbanization, increasing poverty, rising demand for energy and food, as well as climate change, resulting in resource scarcity and social challenges. In response, many multi-lateral organizations have called for the development of a green economy that improves human well-being and social equity and reduces environmental degradation. A vital aspect of the green economy is that it achieves water security. This chapter defines the concept of the green economy and green growth before discussing water's role in the green economy. After which, the chapter provides a series of case studies of locations implementing technologies and practices to achieve water security and green growth.

The Green Economy

The green economy results in improved human well-being and social equity, while significantly reducing environmental risks and ecological scarcities. In its simplest form, a green economy is low carbon, resource-efficient, and socially inclusive. In this type of economy, growth in income and employment are driven by both public and private investments that reduce carbon emissions, enhance resource efficiency, and prevent the loss of biodiversity and ecosystem services. A vital component of this economy is that economic development views natural capital as a critical economic asset and as a source of public benefit. The overall aim of a transition toward a green economy is to enable economic growth and investment while increasing environmental quality and social inclusiveness.

Green Growth

In the green economy, green growth is about fostering economic growth and development while ensuring that natural assets continue to provide the resources and environmental services on which our well-being relies. There are a variety of characteristics attributed to green growth, including:

- More effective use of natural resources in economic growth
- Valuing ecosystems
- Inter-generational economic policies
- Increased use of renewable sources of energy
- Protection of vital assets from climate-related disasters
- Reduced waste of resources

The main overall objectives of the green economy and green growth include:

- *Improving resource-use efficiency*: a green economy is one that is efficient in its use of energy, water, and other material inputs.
- *Ensuring ecosystem resilience*: it also protects the natural environment, its ecosystems, and ecosystem flows.
- *Enhancing social equity*: it promotes human well-being and equality.

Water's Role in the Green Economy

Water, unlike any other natural resource, affects every aspect of society and the environment and is essential for human well-being. Specifically, water is embedded in all aspects of development, including food security, health, and poverty reduction, and in sustaining economic growth in agriculture, industry, and energy generation. As such, the transition toward the green economy requires not only the conservation of water resources but also finding new and innovative economic growth and social development opportunities that embrace the sustainable management of water resources. A key component of creating the green economy is ensuring water security for all users and uses, both human and natural, where water security is defined by the United Nations as "the capacity of a population to safeguard sustainable access to adequate quantities of and acceptable quality water for sustaining livelihoods, human well-being, and socio-economic development, for ensuring protection against water-borne pollution and water-related disasters, and for preserving ecosystems in a climate of peace and political stability."

Overall, ensuring water security in the green economy can be achieved by:

- Creating policy instruments that promote complementary benefits
- Developing fiscal instruments that give a price to environmental goods
- Strengthening institutional arrangements that enable the management of water across sectoral silos and even political/administrative boundaries
- Developing financial instruments that share risks between governments and investors and make new water technology affordable
- Developing skills that support the sustainable management of water in the green economy
- Establishing information and monitoring systems that set targets, define trajectories, and monitor progress on water efficiencies
- Developing innovative plans that increase water productivity, protect groundwater and surface water resources, and ensure adequate levels of water quality (Brears 2016, 2017, 2020a)

Case: Dallas Water's Demand Management Initiatives

Dallas Water Utilities (DWU) is a major retailer and wholesaler provider of water in North Texas, serving over 2.5 million people across a nearly 700 square mile service area. In 2016, DWU released its Water Conservation Plan, which aims to reduce Dallas' per capita water use by an average of one percent per year over the next 5 years. Raising public awareness is one of the key approaches DWU is taking to achieve the water conservation target, with the utility using a well-diversified approach that includes television and radio ads and social media advertising along with its updated, flagship website SaveDallasWater, which offers a range of demand management initiatives for domestic and non-domestic customers, including:

- *New Throne for Your Home Program*: The New Throne for Your Home Program offers customers free high-performance high-efficiency toilets. These WaterSense®-certified toilets ensure high performance and reduced water consumption, with customers able to apply to replace up to two toilets.
- *Free automatic irrigation system evaluation*: The free automatic irrigation system evaluation initiative involves a licensed landscape irrigation specialist assessing the irrigation system for programming errors, leaks in the system, broken or misaligned equipment, problems with pressure (too high or too low), and coverage problems.
- *Free water efficiency assessments*: The free water efficiency assessment initiative offers up to $100,000 in site-specific rebates to industrial, commercial, and institutional (ICI) customers to help them save water and money by identifying opportunities to increase water-use efficiency and reduce water, wastewater, and electricity costs.

DWU has two planned public education measures for FY 2020, in particular:

- *ICI training*: DWU plans to develop, lead, and manage ongoing water efficiency programs for ICI facility managers and irrigators, with topics including industrial cooling and process, food processing, irrigation management, and leakage control. DWU will also work with local businesses, green building organizations, and energy utilities for input on the curriculum development and certification process.
- *ICI Business Partnership Program*: DWU plans to establish an ongoing Business Partnership Task Force to engage the ICI community in DWU's water conservation program, particularly business leaders of large water using organizations, with the Task Force facilitating the discussion of water conservation practices, sharing of conservation success stories, and discussing DWU ICI water conservation programs (Brears 2019a).

Case: San Francisco's Water-Efficient Urban Farms

The San Francisco Public Utilities Commission (SFPUC) is aiming to build sustainable food systems that provide education, economic, and health benefits for residents. SFPUC has developed a range of programs that support the efficient use of water throughout local urban agricultural, demonstration, and community gardens. The Community Garden Irrigation Meter Grant Program is designed to help urban agriculture projects, and community and demonstration gardens better track and manage their irrigation water use. The grant offers a one-time waiver of up to $12,000 in SFPUC fees for the installation of a new dedicated irrigation water service and meter for projects that have a minimum 10-year span. Urban agriculture and community garden projects are required to stay within a calculated water budget. The water budget sets an annual amount of water to be used in the garden to maintain proper plant health. The Maximum Applied Water Allowance (MAWA) is calculated as follows:

$$MAWA = 35.1 \times 0.62 \times [(0.7 \times \mathrm{LA}) + (0.3 \times \mathrm{SLA})]$$

Where:

- 35.1 = San Francisco's reference evapotranspiration rate, in inch/year
- 0.62 = Conversion factor to convert inches/year to gallons/year
- 0.7 = Evapotranspiration adjustment factor
- LA = Size of the total landscape area, in square feet
- 0.3 = Additional water allowance determined by the state
- SLA = Size of the Special Landscape Area, in square feet (edible landscape areas are considered Special Landscape Area) (Brears 2020b)

Case: Santa Monica Using Green Infrastructure to Turn Stormwater into a Resource

In the United States, the Water Infrastructure Improvement Act amends the Federal Water Pollution Control Act to provide for the use of green infrastructure to reduce stormwater flows. One city that is demonstrating how green infrastructure can turn stormwater into a resource is Santa Monica. Santa Monica, in partnership with the Santa Monica-Malibu Unified School District and the Metropolitan Water District of Southern California, has constructed the Los Amigos Park Storm Water Harvesting and Direct Use Demonstration Project. The project involves capturing stormwater runoff from a storm drain near the park, pretreating flows with a hydrodynamic separator, storing flows in a subsurface storage system, and treating the water with ultraviolet light before use for indoor flushing and park irrigation, both of which currently use potable water. The project stores around 53,000 gallons of urban runoff and offsets up to 550,000 gallons of potable water per year, ensuring urban runoff can become a resource rather than a waste that carries pollution into Santa Monica Bay. In addition to reducing the amount of polluted runoff going into the ocean, the project demonstrates to the broader community the benefits of capturing and using urban runoff and stormwater for uses that do not require potable water. Overall, the project contributes toward the city's broader goal of reducing water use by 20% and being 100% water self-sufficient by 2020 (Brears 2019b; City of Santa Monica 2020).

Case: Singapore Seeking Energy-Efficient Desalination Solutions

Currently, Singapore uses around 430 million gallons of water per day. By 2060, the water demand is projected to double, with seawater desalination expected to become a significant source of the city-state's water by then. Recognizing the country's growing dependency on energy-intensive seawater desalination, Singapore's Public Utilities Board (PUB) has for a considerable amount of time focused on seeking energy efficiency solutions for desalination. The current state-of-the-art seawater desalination plant consumes around 3.5 kWh/m^3 at about 50% recovery. Through R&D, the short-term goal is to reduce energy consumption for seawater desalination to less than 2 kWh/m^3 at the system level. PUB states that there needs to be considerable research directed toward improving the main desalting step through innovative solutions as well as increasing energy recovery to achieve this goal. With pre- and posttreatment for seawater desalination typically consuming about 1 kWh/m^3, PUB has identified this as a focus area of R&D. To find innovative ways of reducing the water-energy nexus pressures of desalination, PUB has launched a request for proposal to invite researchers and industries to develop low-energy and low-chemical solutions for pretreatment processes and posttreatment processes for seawater desalination:

- *Pretreatment*: PUB is seeking solutions that can enhance the effectiveness and/or efficiency of pretreatment for seawater desalination. The technologies proposed shall show how it uses less energy and/or less chemicals as compared to the standard treatment train.
- *Posttreatment*: PUB is soliciting solutions that can enhance the effectiveness and/or efficiency of posttreatment (particularly boron removal) for seawater desalination. The technologies proposed shall show how it uses less energy and/or less chemicals compared to the standard treatment train (Our Future Water 2020; Public Utilities Board 2019).

Case: Testing Smart Sensing Network Technology in Sydney

The New South Wales Smart Sensing Network (NSSN) was established in 2016 with funding from the NSW State Government to position the state as a leader in sensing technology. As part of the initiative, a $3-million project, funded by Sydney Water, along with other water utilities, such as

Hunter Water, SA Water, Melbourne Water, Intelligent Water Networks (Victoria), Queensland Urban Utilities, and the NSW State Government, was launched in 2019 to solve the global challenge of water leakage and supply disruption. One of the initiatives of the project is an acoustic sensing pilot in Sydney. NSSN, Sydney Water, SA Water, and the University of Technology Sydney are collaborating on an acoustic sensing pilot in Sydney that aims to proactively reduce leaks and breaks in the water network using cutting-edge acoustic technology. As part of the project, SA Water has deployed its state-of-the-art acoustic sensors within Sydney Water's CBD water main network. Since their installation in 2017, SA Water's acoustic sensors have helped detect around half of all water main leaks and breaks in Adelaide's CBD, enabling them to be proactively repaired, minimizing interruption to customers and commuters while reducing operational costs. In Sydney, these sensors monitor a total of 13 kilometers of pipes across the CBD to predict leaks and enable preventative maintenance: necessary as in Sydney's current drought, the dry soil is exacerbating water main breaks. In return for contributing to leak detection in Sydney, SA Water will be able to increase the range of data it has to baseline acoustic patterns against and further fine-tune their algorithms used to monitor leakage in Adelaide's CBD network. Overall, Sydney Water estimates that in 3 years, this new technology will help reduce significant breaks by 50% in the CBD (Brears 2020c).

Conclusion

The green economy enables economic growth and investment while increasing environmental quality and social inclusiveness. Specifically, the main objectives of the green economy include improving resource-use efficiency, ensuring ecosystem resilience, and enhancing social equity. A vital aspect of the green economy is that it achieves water security for all users and uses, both human and natural. Overall, ensuring water security can be realized by utilizing a range of demand management initiatives including education and financial tools to save water indoors and outdoors and in commercial operations, providing financial tools to help urban agricultural producers become more water-efficient, utilizing green infrastructure to turn stormwater into a resource for non-potable uses, developing technologies that reduce water-energy pressures in desalination plants, and fostering partnerships to develop advanced technologies that reduce leakage in the water distribution system.

Cross-References

▶ Green and Smart Cities in the Developing World
▶ Green Economy Policies to Achieve Water Security

References

Brears, R. C. (2016). *Urban water security*. Chichester/Hoboken: Wiley.

Brears, R. C. (2017). *The green economy and the water-energy-food Nexus*. London: Palgrave Macmillan.

Brears, R. C. (2019a). Dallas water utilities is the star of Texas. Retrieved from https://medium.com/mark-and-focus/dallas-water-utilities-is-the-star-of-texas-b7f04c32df48

Brears, R. C. (2019b). Green Infrastructure Turning Stormwater Into a Resource. Retrieved from https://medium.com/mark-and-focus/green-infrastructure-turning-stormwater-into-a-resource-bcd5b81e3bbb

Brears, R. C. (2020a). *Developing the circular water economy*. Cham: Palgrave Macmillan.

Brears, R. C. (2020b). The rise of water and food-secure cities. Retrieved from https://medium.com/mark-and-focus/the-rise-of-water-and-food-secure-cities-58b13f6b8615

Brears, R. C. (2020c). Testing smart water tech down under. Retrieved from https://medium.com/mark-and-focus/testing-smart-water-tech-down-under-d08688f7609a

City of Santa Monica. (2020). Los Amigos Park stormwater demonstration project. Retrieved from https://www.smgov.net/bebp/project.aspx?id=53687095405

Our Future Water. (2020). Low-energy desalination on the horizon. Retrieved from https://www.ourfuturewater.com/2020/03/10/low-energy-desalination-on-the-horizon/

Public Utilities Board, S. (2019). Fact sheet on PUB's directed Request for Proposal (RFP) 19/01 on low-energy, low-chemical pre- and/or post-treatment for seawater desalination. Retrieved from https://www.pub.gov.sg/globalhydrohub/Documents/Annex%20A%20Directed%20RFP%201901%20Fact%20Sheet_Final.pdf

Water Security, Sustainability, and SDG 6

Alan Shapiro
British Columbia Institute of Technology, Vancouver, BC, Canada

> "As our global economy grows, so will its thirst... Water security is not an issue of rich or poor, North or South... And yet there is still enough water for all of us if we keep it clean, use it more wisely, and share it fairly." – Quote taken from United Nations Secretary General Ban Ki-Moon's speech at the World Economic Forum Annual Meeting in Davos-Kloster. January 24, 2008.

Synonyms

Water sustainability

Definition

"Water security is defined as the capacity of a population to safeguard sustainable access to adequate quantities of acceptable quality water for sustaining livelihoods, human well-being, and socio-economic development, for ensuring protection against water-borne pollution and water-related disasters, and for preserving ecosystems in a climate of peace and political stability" (UN-Water 2013).

Introduction

Over the past decade, water security has emerged as the dominant narrative for global water management and risks. This paradigm encompasses a variety of issues, from water scarcity and quality to flooding and drought, and offers a common language and framework for a wide range of water challenges and crises often felt and managed at a local or regional scale.

Unsurprisingly for such a broad concept, definitions of water security vary based on discipline, scale, and priorities. One of the first broad-reaching attempts to synthesize the concept into a concise definition was by the Global Water Partnership in their 2000 report, *Towards Water Security: A Framework for Action*:

> Water security at any level from the household to the global means that every person has access to enough safe water at affordable cost to lead a clean, healthy and productive life, while ensuring that the natural environment is protected and enhanced.

While this definition served to propel water security to global center stage, the concept has continued to evolve. Today, it can best be thought of through the lens of UN-Water's 2013 working definition, which explicitly integrates the dimensions of sustainability and water-related risks:

> Water security is defined as the capacity of a population to safeguard sustainable access to adequate quantities of acceptable quality water for sustaining livelihoods, human well-being, and socio-economic development, for ensuring protection against water-borne pollution and water-related disasters, and for preserving ecosystems in a climate of peace and political stability.

The growing popularity of the water security paradigm is not unexpected. Unlike past dominant narratives such as integrated water resources management, water security invites participation from a range of communities not typically involved in water management conversations, including global security, human welfare, and economic development, and consolidates water development and risk under a single umbrella. This is perhaps best articulated by the World Economic Forum in their 2009 report on global water security: "Water security is the gossamer that links together the web of food, energy, climate, economic growth, and human security challenges that the world economy faces over the next decades."

The Evolution of Water Security

Mentions of water security can be traced back to the 1990s in the context of broader conversations around food, military, and environmental security (Cook and Bakker 2012). The concept gained prominence in 2000 with the publication of two notable reports: *A Water Secure World* by the World Water Council (WWC 2000) and *Towards Water Security: A Framework for Action* by the Global Water Partnership (GWP 2000). Since then, it has evolved and gained popularity, garnering widespread and often disparate use in a range of research, practice, and policy communities.

Water security is by no means exclusive as an overarching water narrative. Water management paradigms are in a constant state of evolution, integrating new concepts and responding to global trends. The evolution of these narratives over the past four decades is summarized by Hoekstra et al. (2018). The authors describe the emergence of the integrated water resources management (IWRM) framework in the 1980s, sustainable water management in the 1990s, water security in the 2000s, and adaptive water management in the 2010s.

While all of these remain active in research and practice today, water security can be viewed as their synthesis, incorporating key elements of integration, sustainability, and adaptation. Aside from its broader focus and its integration of issues outside the traditional scope of water management, water security's unique appeal lies in its definition of an end goal for water resources (Cook and Bakker 2012). While this goal may vary from region to region, it is possible to envision a water secure future for a river basin and even for the globe as a whole (it is towards this future that Sustainable Development Goal 6 is targeted). An end goal, by extension, can be operationalized through concrete goals and indicators – a critical step to translating a broad paradigm into action at local and regional scales.

The water sustainability narrative, on the other hand, offers a lens through which to view water resources and issues, without delineating specific outcomes (Norman et al. 2010). Water security and water sustainability can thus be viewed as broadly synonymous concepts, with the former offering greater value to water managers and policymakers, as well as more explicit recognition of human well-being and water-related risks.

Water Security and the Sustainable Development Goals

In September 2015, the United Nations adopted the 2030 Agenda for Sustainable Development, outlining 17 Sustainable Development Goals (SDGs) for the global community. SDG 6 lays out a goal to "ensure availability and sustainable management of water and sanitation for all." SDG 6 defines targets around provision of drinking water; sanitation and hygiene services; water quality and wastewater reuse; water-use efficiency and water scarcity; integrated water resources management; water-related ecosystems; international cooperation and capacity-building; and local participation in water and sanitation management. These targets are supported by a range of indicators, with progress monitored and reported by UN-Water and a network of agencies at the global and regional levels (UN-Water 2018).

Water is by no means confined to SDG 6; many of the SDGs are rooted in or impacted by water security. A HLPW (2018) by the High Level Panel on Water, convened by the United Nations and the World Bank Group in response to the growing urgency of global water challenges, stresses that "water is the common currency which links nearly every SDG, and it will be a critical determinant of success in achieving most other SDGs – on energy, cities, health, the environment, disaster risk management, food security, poverty, and climate change among others."

SDG 6 has served the important role of galvanizing the global conversation around water management towards shared targets and indicators. In this way, SDG 6 can be seen as the leading edge of global water security implementation, supported by numerous public, private, and nonprofit sector initiatives at multiple scales. While the language of water security is not always used in the context of the SDGs (the language of IWRM is often more common), there is strong alignment between SDG targets and water security principles.

Elements of Water Security

A comprehensive review of academic and policy literature by Cook and Bakker in 2012 found that academic disciplines referencing water security ranged from engineering and natural sciences to social sciences and health, with each defining water security based on its distinct perspectives, approaches, and scales. Broad and narrow framings of water security are seen as complementary, enabling high-level priority setting and policymaking while supporting operationalization of water security at the management level (Varis et al. 2017).

Alongside its 2013 working definition, the UN-Water outlines a number of core elements to achieving and maintaining water security:

- Access to safe and sufficient drinking water at an affordable cost in order to meet basic needs;
- Protection of livelihoods, human rights, and cultural and recreational values
- Preservation and protection of ecosystems in water allocation and management systems
- Water supplies for socioeconomic development and activities
- Collection and treatment of used water
- Collaborative approaches to transboundary water resource management
- The ability to cope with uncertainties and risks of water-related hazards
- Good governance and accountability and due consideration of the interests of all stakeholders

Hoekstra et al. (2018) consolidate these elements into four distinct focus areas: increasing economic welfare, enhancing social equity, moving towards long-term sustainability, and reducing water-related risks. Collectively, the first three can be viewed as the developmental approach to water security perhaps best exemplified by SDG 6 (van Beek and Lincklaen Arriens 2014). This approach focuses on improving water security over time through the definition of goals and targets. The fourth element draws on a risk management approach to reduce hazards, exposure, and vulnerability related to climate and water-related stressors and disasters (Grey et al. 2013). Van Beek and Lincklaen Arriens (2014) argue that developmental and risk management approaches must be balanced and pursued concurrently.

More recently, the concept of urban water security has emerged as a distinct sub-theme, which applies to urban areas and introduces additional water management considerations specific to these contexts (Brears 2017, see ▶ "Water Security and the Green Economy"). Unique considerations for urban environments include high density of population and economic activities, which concentrate risks to water resources and systems. Most large urban areas are also unable to meet their resource needs from within their own territory, requiring the import of natural resources, including food and water supply. This reliance on external water resources, both directly and indirectly, such as through water embedded in imported food supplies, adds an additional level of water risk and exposure for urban areas (Hoekstra et al. 2018).

Conclusion

Water management is complex by its nature, and water management paradigms are constantly evolving. But the integrative power, broad appeal, and clear goals offered by the water security concept offer advantages over competing narratives. It seems likely that the global community will continue to view water issues through a water security lens for some time to come.

How then can water security be operationalized and translated into practice? Global goals and targets under the SDG umbrella are shaped and advanced on the ground by regional organizations and policies responding to local water challenges and crises. By extension, operational definitions of water security inevitably vary regionally based on the local context of water supply, quality, access, and risks.

In Australia, water security is being applied as a toolkit for managing water scarcity. In northern China, water security has a combined focus on water availability and pollution. And in Canada, water security has become a framing narrative for a wide range of regional water issues, including transboundary water management, indigenous drinking water supply, source water protection, and management of climate-driven water risks such as flooding and drought (Pomeroy et al. 2019).

While regional focuses vary, the UNESCO (2019) offers a range of concrete actions that can be taken to advance water security at multiple scales:

- Expanding wastewater management and reuse
- Encouraging investment and collaboration around new technology development
- Enhancing partnerships with the private sector
- Raising the profile of water security on national government agendas
- Involving local populations in the development process and running capacity-building programs at different levels
- Generating better policies through dialogue, knowledge exchange, and communication
- Improving data quality in order to generate better policies

Water management narratives will inevitably continue to change and evolve. But in the present day and for the foreseeable future, the water security toolkit and SDG agenda offer a powerful vision and opportunity to redefine water management, align the global community, and mobilize action around water issues.

Cross-References

► Water Security and the Green Economy

References

Brears, R. (2017). *Urban water security*. Wiley https://doi.org/10.1002/9781119131755.

Cook, C., & Bakker, K. (2012). Water security: Debating an emerging paradigm. *Global Environmental Change, 22*, 94–102. https://doi.org/10.1016/j.gloenvcha.2011.10.011.

Global Water Partnership. (2000). *Towards water security: A framework for action*. https://www.gwp.org/globalassets/global/toolbox/references/towards-water-security.-a-framework-for-action.-mobilising-political-will-to-act-gwp-2000.pdf

Global Water Partnership. (2014). *GWP strategy towards 2020: A water secure world*. https://www.gwp.org/globalassets/global/about-gwp/strategic-documents/gwp_strategy_towards_2020.pdf

Grey, D., Garrick, D., Blackmore, D., Kelman, J., Muller, M., & Sadoff, C. (2013). Water security in one blue planet: Twenty-first century policy challenges for science. *Philosophical Transactions of the Royal Society A: Mathematical, Physical and Engineering Sciences, 371*. https://doi.org/10.1098/rsta.2012.0406.

High Level Panel on Water. (2018). *Making every drop count: An agenda for water action*. https://sustainabledevelopment.un.org/content/documents/17825HLPW_Outcome.pdf

Hoekstra, A. Y., Buurman, J., & van Ginkel, K. C. H. (2018). Urban water security: A review. *Environmental Research Letters, 13*. https://doi.org/10.1088/1748-9326/aaba52.

Norman, E., Bakker, K., Dunn, G., Allen, D., Cook, C. (2010) Water Security: A Primer.

Pomeroy, J., Merrill, S., DeBeer, C., Adapa, P., Phare, M-A., Overduin, N., Miltenberger, M., Maas, T., Pentland, R., Brandes, O., & Sandford, R. W. (2019). *Water security for Canadians: Solutions for Canada's emerging water crisis*. Canadian Water Security Initiative. https://gwf.usask.ca/outputs-data/major-outcomes/water-security-4-canada.php

UNESCO. (2019). *Water security and the Sustainable Development Goals*. https://digitallibrary.un.org/record/3807832

UN-Water. (2013). *Water security and the global water agenda: A UN-Water analytical brief*. https://www.unwater.org/publications/water-security-global-water-agenda/

UN-Water. (2018). *Sustainable Development Goal 6: Synthesis report 2018 on water and sanitation*. https://www.unwater.org/publication_categories/sdg-6-synthesis-report-2018-on-water-and-sanitation/

van Beek, E., & Lincklaen Arriens, W. (2014). *Water security: Putting the concept into practice*. Global Water Partnership Technical Committee. https://aquadoc.typepad.com/files/gwp_tec20_web.pdf

Varis, O., Keskinen, M., & Kummu, M. (2017). Four dimensions of water security with a case of the indirect role of water in global food security. *Water Security, 1*, 36–45. https://doi.org/10.1016/j.wasec.2017.06.002.

World Economic Forum. (2009). *The bubble is close to bursting: A forecast of the main economic and*

geopolitical water issues likely to arise in the world during the next two decades. https://www.weforum.org/reports/bubble-close-bursting
World Water Council. (2000). *A water secure world: Vision for water, life, and the environment.* https://www.ircwash.org/resources/world-water-vision-commission-report-water-secure-world-vision-water-life-and-environment

Water Sustainability

▶ Water Security and Its Role in Achieving SDG 6
▶ Water Security, Sustainability, and SDG 6
▶ Watershed Sustainability: An Integrated River Basin Perspective

Water, Sanitation and Hygiene (WaSH)

▶ Wash (Water, Sanitation, and Hygiene): Infrastructure as a Measure of Sustainable Development

Water, Sanitation, and Hygiene Question of Future Cities of the Developing World

Shamiso Hazel Mafuku[1], Walter Musakwa[2] and Innocent Chirisa[1,3,4]
[1]Department of Architecture and Real Estate, University of Zimbabwe, Harare, Zimbabwe
[2]Department of Urban and Regional Planning, University of Johannesburg, Johannesburg, South Africa
[3]Department of Demography Settlement and Development, Social & Behavioural Sciences, University of Zimbabwe, Harare, Zimbabwe

Synonyms

Hygiene – Cleanliness; Sanitation – Facilities for Cleanliness; Water – Hydrologic Resources

Definition

Water, Sanitation, and Hygiene (WASH) refers to the provision of safe water for drinking, washing, and domestic activities, the safe removal of waste and health promotion activities to encourage healthy behavioral practices among populations.

Developing world comprises of a class of economically underdeveloped and developing nations which generally have low production rates and a struggling labor market usually paired with relatively low levels of education, poor infrastructure, improper sanitation, limited access to health care, and lower costs of living.

Future cities is a term used to imagine the ideal cities of the future, how they will operate, what systems will orchestrate them, and how they will relate to their stakeholders, that is, citizens, governments, businesses, investors, and others. Future cities can convey either environmental, social, economic, or governance aims or a hybrid of some or all of these elements.

Introduction

Water has always been an issue of consideration across cultures and regions in defining how cities are developed. As settlements evolve and grow or shrink, water, sanitation, and hygiene (WASH) have become more and more important in developing and governing human settlements. Providing clean water and sanitation to all is one of the biggest challenges of the twenty-first century, especially in developing countries, where 50% of the population is projected to live in cities by 2050. It is important to note that in most developing countries, where WASH services are considered to be sufficient and safe, the systems are often poorly maintained (WHO 2009).

In 2015, at the end of the Millennium Development Goals (MDGs), a third of the world still defecated in the open or used unsafe or shared toilets. The new Sustainable Development Goals (SDGs) are much more ambitious. For sanitation, everyone needs access not only to a toilet but also to a sanitation system that safely captures and manages waste (Mason and Le Seve 2017). Over

60% of the world, 4.5 billion people, currently lack that. With population growth, the target is to reach at least 5.2 billion more people by 2030. Meanwhile, at least 2.5 billion more people will need a reliable water supply service that is available at home and free from contamination. Historically, cities in poorer regions have struggled to extend formal, centralized water and sewerage networks to growing populations living on their edges. In sub-Saharan Africa, the share of the urban population receiving a piped water supply has fallen from 67% to 56% since 1990 (WHO and UNICEF 2017). This raises a major concern on how future cities of the developing world will thrive concerning the water, sanitation, and hygiene question. The aim of this chapter is to therefore bring to light the key challenges that cities of developing countries currently face and how these can be tackled in order to create sustainable WASH systems for future cities.

Conceptualizing Water, Sanitation, and Hygiene (WASH)

There are different definitions of developing countries; however, the broad definition is that this is a term often used to refer to countries with medium to low human development index (HDI) or sometimes the gross domestic product (GDP) is used. These countries are mainly found in Africa and Asia and some in South America and generally have low income per capita of population that are trying to improve their conditions through industrialization. Developing countries generally have lower living standards among their population, low industrialization levels, as well as low infrastructural developments (Kuepper 2017).

In addition, the United Nations (UN) describes a developing country as a country with a relatively low standard of living, an undeveloped industrial base, and moderate to low HDI, with a high employment share of 60–70% in agriculture. However, regardless of the low income in developing countries, there is high population growth in these countries. It is this high population growth in developing countries that is seriously outstripping the capacity of most countries to provide adequate services for their citizens (Cohen 2006). Currently, many developing countries already struggle to cope with consistent water shortages, and they lack adequate water infrastructure. High population growth and rapid urbanization cause more pressure to these limited water resources which in turn results in poor access to improved water sources and limits hygienic practices in the urban communities, as these hygienic practices highly depend on the availability of safe water and adequate sanitation services (Eid 2015).

Having defined developing countries, it is important to also note that cities are not homogeneous, especially in developing nations. Lüthi et al. (2011) identified four typical urban contexts, i.e., the planned areas, the squatter settlements, the middle- and high-income settlements in the inner part of the city, and the city edge settlements as illustrated in Fig. 1. Contrasts between these contexts can be striking, for instance, skyscrapers next to slum pockets in India or favelas next to villas with swimming pools in Brazil (Reymond et al. 2016). It is not an abnormal thing to find low-income squatter settlements which are not properly planned and have no basic services right next to modern and lavish gated medium- to low-density suburbs. Current urbanization patterns are highly characterized by segregation. Levels of WASH service provision vary greatly in all these areas, with the poorest neighborhoods having the most inadequate services.

Water, sanitation, and hygiene (WASH) is defined as the delivery of clean and safe water for domestic consumption and other household uses and a secure removal of wastewater and human excreta as well as health promotion activities to encourage healthy behavioral practices among populations (WHO 2009). WASH unites the three linked aspects of health- and water-related social services. This conveys the message that achieving health benefits depends on three mutually reinforcing aspects, i.e., clean water, safe sanitation, and changed hygiene behavior (Huston and Moriarty 2018). Water is supplied for various reasons which include domestic, agricultural, and industrial uses. Domestic water supply on the other hand means the source and

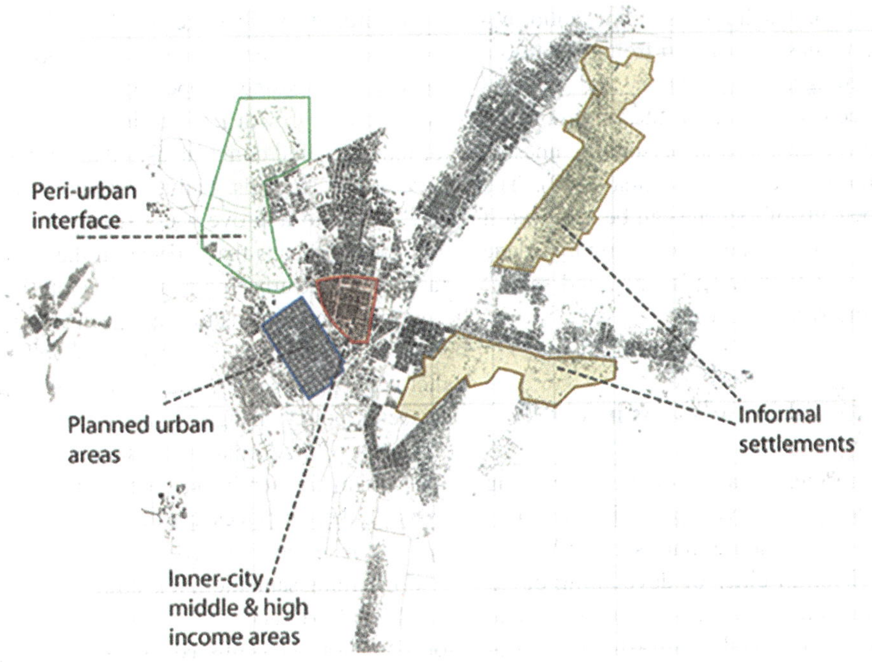

Water, Sanitation, and Hygiene Question of Future Cities of the Developing World, Fig. 1 Main settlement contexts that need to be considered in planning for WASH in developing countries (Lüthi et al. 2011, p. 79)

infrastructure that provides water to households. A domestic water supply can take different forms which include streams, springs, wells, boreholes, rainwater collection systems, piped water supply with taps, or water vendors (UNICEF 2015b).

Water supply and sanitation are closely related, and they both influence human health and development. According to WHO (2008), sanitation refers to the promotion of hygiene and prevention of disease by maintenance of sanitary conditions. Also based on WHO's definition, sanitation is defined as the collection, transportation, treatment, and disposal or reuse of human excreta or domestic waste water. This can be through collective systems or by installations serving a single household or undertaking.

The World Health Organization (WHO) (2017) defines hygiene as conditions or practices conducive to maintaining health and preventing disease, especially through cleanliness. Kumwenda (2019) points out that hygiene is however far more than just cleanliness because cleanliness mostly involves the removal of dirt, wastes, or unwanted things from the surface of objects using detergents and other necessary equipment. On the other hand, hygiene practices focus on the prevention of disease through the use of cleaning as one of the several inputs. Most disease control interventions to a greater extent rely on hygiene for them to achieve their goal.

Hygiene is applied in different areas with the aim of prevention of disease transmission and promoting health. The common types of hygiene include personal hygiene which has to do with taking care of one's body and clothes. There is also water hygiene which involves the collection, transportation, storage, and use of water without contaminating it and food hygiene which is the practical process of ensuring that food is fit to eat without any contamination. There is also waste handling hygiene which relates to how solid, liquid, and gas wastes are handled from generation, collection, storage, transportation, and disposal to prevent contamination of the environment. Hygiene can be practiced at personal, domestic, and community levels (Kumwenda 2019). The maximum benefits of hygiene can be achieved if improvements in hygiene are concurrently made with improvements in water supply, and sanitation, hence the WASH concept.

Scholarship and Experiences in WASH

Most urban residential areas of the developing world lack adequate service provision, and this section provides literature that assesses WASH service provision in cities of developing countries. This includes services such as water and sewer reticulation as well as infrastructural provision. The section gives a picture of the current WASH scenario in the developing world, particularly sub-Saharan Africa, Asia, and South America. Inadequate provision of water and sanitation is a major challenge in most urban residential settlements of the developing world, and this has detrimental effects on people's hygiene and health. Given the general inadequacy of service provision especially in the peri-urban context of cities, most households spend a significant percentage of their income to obtain water. As a result, these areas are greatly affected by a number of water- and sanitation-related diseases such as cholera and typhoid. The poorer citizens are often the most disadvantaged in this regard.

Planners and policymakers must bear in mind that investment of urban basic services in these areas is paramount in preventing health disasters. A comparative study conducted on the governance of water supply and sanitation in the peri-urban context of three metropolitan areas (Mexico, District Federal; Chennai, India; and Caracas, Venezuela) revealed a significant gap in access to water supply and sanitation between urban and peri-urban dwellers, with the latter suffering significant conditions of deprivation (McGregor et al. 2006).

According to McGregor et al. (2006), in and around cities, water is commonly in short supply and under increasing competition among different users. Urban growth leads to increasing demand for industrial and domestic uses that conflict with agricultural demands. Failure by public and private sectors to improve water supply and sanitation usually means that citizens in the periphery, particularly the urban poor, are often left to device their own mechanisms to access WASH services. To add on to that, since their pleas as well as their alternative practices are often ignored by the public sector, policy changes aimed at improving the efficiency of the formal water supply and sanitation system frequently do little to improve their access to these services and often even constitute an obstacle to them (ibid.).

Concerning sanitation, it is critical to point out that the global and sub-Saharan MDG targets for sanitation were missed in 2015. The global target for sanitation was for 77% of the world population to use improved sanitation facilities, while sub-Saharan Africa had a target of 62%. Unfortunately, the progress achieved with the water target could not be replicated as only 68% of the world population had access to improved sanitation facilities, living a gap of 9%. The situation in sub-Saharan Africa was worse than this global picture, where only 30% of the population had access to adequate sanitation facilities and 32% of the population missed the sanitation target by 2015 (Ohwo and Agusomu 2018).

Globally, 2.4 billion people use unimproved sanitation facilities, with sub-Saharan Africa accounting for 695 million (28.96%) of the population. Similarly, of the 638 million people sharing sanitation of an otherwise improved type, the region accounted for 194 million people (30.41%). In addition, 23% of the population in sub-Saharan Africa still practice open defecation as against the global average of 13% in 2015 (WHO and UNICEF 2017). Hence, it was not surprising that only 3 countries (Cape Verde, Réunion, and Seychelles) in sub-Saharan Africa met the sanitation target among 95 countries globally, while 43 countries missed the target, and 4 other countries (Sudan, South Sudan, Somalia, and Congo) were not considered due to lack of reliable data (WHO and UNICEF 2017). This shows that sub-Saharan Africa constitutes a significant drag on the attainment of the sanitation target at the global level (Ohwo and Agusomu 2018).

In the majority of developing countries, the poor citizens are the ones who are mostly affected by the health issues emanating from inadequate water, sanitation, and hygiene systems. The population of most urban centers in developing countries is ever increasing at a very rapid rate, and this has resulted in the formation of various informal settlements, most of which are located in the periphery of the cities. Most of those areas are characterized by a lack and/or inadequacy of WASH services, with some people resorting to open defecation as there is no proper disposal systems of human excreta. Hygiene in these areas is often poor as there is no access to clean water. Inadequate WASH services have health and socioeconomic consequences. For example, diarrheal diseases are the most common WASH-related diseases, which kill about 1.7 million people yearly, of which sub-Saharan Africa contributes a significant proportion. Diarrheal diseases are one of the major consequences of the conditions in these areas, and this has resulted in a lot of deaths and illnesses mainly among young children (World Health Organization 2017).

Various challenges are faced in developing nations in trying to implement adequate WASH systems. Poverty is a chief reason in developing countries, with most households having no adequate income to prioritize paying for water and sanitation services. Majority of the poor citizens prioritize their income to buy food, clothing, and other immediate needs placing sanitation and hygiene low in the priority list. There is also a lack of political commitment as most of the hygiene initiatives are implemented by nongovernmental organizations (NGOs) and rarely by the government. Despite advocating for water, sanitation, and hygiene (WASH), there are not enough initiatives introduced by the government through relevant ministries (Kumwenda 2019).

Lack of community participation during planning phase is another major WASH issue. As much as there might be a solution, if the people who are receiving the solution do not realize the need for that solution, then the solution becomes ineffective. That is why it is of utmost importance to involve the community in all WASH initiatives. This offers proper understanding of the whole project, and when the people are involved, they get a feeling of ownership of the project and also understand the benefit of the solution. Additionally, some of the sanitation and hygiene technologies that are introduced may contradict with some cultural beliefs, and this affects adoption and implementation of the projects in most developing nations. Most sanitation and hygiene technologies are considered by citizens to be not user-friendly which make acceptability a challenge for them (Tearfund 2007).

The WASH scape in developing countries is also characterized by a lack of information on hygiene infrastructure and practices. There is generally a lack of recent, reliable information on the condition of existing WASH infrastructure and practices, including whether or not the infrastructure is actually functioning or benefits of some hygiene practices. This makes needs and demands, particularly in informal settlements within cities, frequently unknown, making the task of setting implementation priorities more difficult.

Emerging Issues and Synthesis

It can be concluded that hundreds of millions of people do not have access to clean water in developing nations (UNICEF 2015b). Despite the high reported figures from most nongovernmental reports, the situation on the ground reveals that most people do not have access to improved water sources due to various previously outlined challenges. This poses great challenges for developing nations when it comes to sanitation and hygiene implementation as most of the hygiene practices require the use of clean water. Furthermore, insufficient water supply limits good hygiene practices such as bathing and hand washing. Children in developing countries sometimes clean only some parts of their bodies and not the whole body due to inadequate water, and this affects their health (Kumwenda 2019).

Concerning sanitation, many analysts around the globe and governments are increasingly recognizing that decentralized and small-scale systems are quite viable alternatives for WASH in cities of developing countries as opposed to centralized conventional systems. There are various reasons why conventional urban water and sewer management are not the only solution to WASH problems. The reasons include lack of capital as well as unstable and unreliable energy supplies. The institutional environments which are also fragmented are quite disabling in terms of managing centralized water and sewer systems. Therefore, it should be noted that the majority of urban dwellers now rely on individual onsite systems such as septic tanks and latrines. There is therefore a need to shift toward more resource-efficient and cost-effective systems that can help deliver the necessary WASH services for people of all income groups within cities (Reymond et al. 2016).

Also, it is important to note that one of the major challenges for urban service providers in the developing world is that they often do not have the political support, incentives, financial resources, or capacity to address the challenges of the poorest people. Despite the overarching responsibility to fulfil the human rights to water and sanitation, there is often no clearly defined agency to take this on, and this mainly points to institutional gaps in WASH management in developing countries. Even when there is a mandate to expand WASH services in urban contexts, service providers can be ill-equipped or unwilling to address the needs of low-income settlements (Tearfund 2007). Customers in these areas are perceived as an unviable market segment and not a priority. Informal, unmapped, and peripheral settlements may fall outside the jurisdiction of municipal providers, which creates practical challenges in terms of billing. Even where the will exists, limited disaggregated data prevents targeted efforts to reach the poorest urban communities, which are missed in planning and service delivery.

Poor WASH access makes urban populations more vulnerable to ill-health and climate risks. The substantial growth of the population in at-risk areas, particularly through unplanned urban development, is one of the most important drivers of disaster risk. High inequalities and fast disease propagation increase people's vulnerability to health and climate risks. This in turn has impacts on social cohesion and well-being. Conversely, providing climate-proof WASH services makes vulnerable groups more resilient (Mason and Le Seve 2017).

In the future, an increasing demand from growing populations and economies will have more widespread effects on water availability than climate change. Climate change will reduce water availability in some regions, costing as much as 6–14% of growth, but the most consistent effect will be increasing variability (Mason and Le Seve 2017). The growing population in cities will result in the expansion of peri-urban areas, and this presents various health and environmental issues that will not be solved by conventional approaches to WASH planning. Given these circumstances, it is important to address how the delivery of WASH services can be improved and expanded. The following section presents some approaches by which sustainable WASH can be accomplished in future cities.

Conclusions and Future Direction

There is a growing need for participatory planning to achieve sustainable WASH initiatives and systems. All relevant urban sanitation stakeholders should be consulted in the planning process. This helps to understand the diverse nature of cities and the different perspectives of various stakeholders from different fields (Reymond et al. 2016). Some of the stakeholders who should be consulted include the residents of the settlements, the service providers, responsible WASH institutions, as well as local and central government bodies reliable for water and sanitation. This can be done through public awareness of the participation procedures as well as interviews with experts and so on. Most importantly, the strategies to be implemented must be above all acceptable and compatible with the residents' cultural and religious beliefs. Services that the people want will often result in their cooperation in the maintenance and upgrading of the systems as well as their willingness to pay any tariffs which may be required from them.

One of the major issues that limit WASH service provision in the developing world is that of financing. Given the general scarcity of public revenues, the levels of external funding historically provided for drinking water service are no longer available. Water authorities realize that they must look to their customers for the funds to construct, operate, and maintain improved facilities. This is achievable because generally even the poor people are willing to pay a reasonable fee for reliable WASH services. The practice of cost recovery requires the technical and administrative ability to operate an effective pricing system (Huston and Moriarty 2018). A pricing system requires metering of the water supply to each customer. However, unless the public has confidence in the system, they will not connect, will not pay, and may break the meters. Thus, the initial installation of meters must be supported by an adequate system of maintenance, and the meters must also be read at regular intervals. Water accounts should be prepared and bills delivered with the least possible delay after meter reading. Also, payment of accounts within a reasonable time must be enforced, with genuine penalties for late payment (National Research Council 1996). Concerning partnerships, public-private partnerships can be very useful if there is cooperation from the local authorities in relation to terms and conditions.

Water conservation and reuse can also go a long way in improving WASH services for future cities. Both quality and quantity should be taken into consideration when within catchment areas. Reuse of municipal wastewater for nonportable uses can be a very cost-effective measure if proper treatment is carried out. Conservation of water should be a priority for all future cities, and construction of dams for water harvesting is very critical.

Although conventional and centralized water supply and sanitation systems seem to be the most ideal form of water and sanitation systems, many scholars and governments are increasingly recognizing that decentralized and small-scale systems are quite viable alternatives for WASH in cities of developing countries. This includes using alternatives such as onsite septic tanks and other fecal sludge management mechanisms such as EcoSan toilets and so on. There is still a need to educate more and more people on the advantages of these alternative systems as most stakeholders still have a rigid mindset and are not open to some of the new alternatives (Reymond et al. 2016).

There is also a need for better training of sanitation professionals. In developing countries, improvements in sanitation will require better training and involvement of sanitation professionals. This is due to a number of factors, including decreased emphasis on sanitation in public health programs, fewer specialized academic and continuing education programs, relatively low salaries for sanitation professionals, and a lack of knowledge and understanding by the public about sanitation services relative to other public services. The growing sanitation challenges of cities in developing countries require expanding the scientific and technical base of sanitation, developing systems to disseminate that information, and fielding trained workers to apply the knowledge and make it available to communities. Hygiene education should also be an integral part of future water and sanitation programs (Van Rooijen 2009).

Creating strong legislation, regulations, and policies will also go a long way in improving WASH in future cities of the developing world. Governments should develop national hygiene strategies and create necessary regulations and policies to advance the strategies. Additionally, the roles and responsibilities of different national institutions to implement the law must be defined properly. Stakeholders must be involved in all the stages of the process to ensure the acceptance of the legislation and policy by the public. In addition, there must be creation of the mechanisms for monitoring and enforcing implementation of the regulations. This will help those implementing hygiene programs to request specific hygiene regulations to make their programs successful. Lastly, officials that check for compliance of WASH requirements should be committed to reduce corruption as this will help to ensure quality production and healthy environments which will prevent the transmission of diseases and infections (Kumwenda 2019).

Cross-References

- ▶ Health and the City: How Cities Impact on Health, Happiness, and Well-Being
- ▶ Hidden Enemy for Healthy Urban Life
- ▶ Improving Social Equity and Community Health and Well-Being in Low-Income Suburbs and Regions

References

Cohen, B. (2006). Urbanization in developing countries: Current trends, future projections and key challenges for sustainability. *Technology in Society, 28*, 63–80. Science Direct.

Eid, E. (2015). *The importance of water, sanitation and hygiene as keys to national development*. Baltimore: John Hopkins Water Institute.

Huston, A., & Moriarty, P. (2018). *Understanding the WASH system and its building blocks. Working paper series building strong WASH systems for the SDGs*. The Hague: IRC.

Kuepper, J. (2018). What defines developing countries? The Balance; 2017. Available online: https://www.thebalance.com/what-is-a-developing-country-197898. Accessed 7 Aug 2020.

Kumwenda, S. (2019). Challenges to hygiene improvement in developing countries. The relevance of hygiene to health in developing countries, Natasha Potgieter and Afsatou Ndama Traore Hoffman. IntechOpen. https://doi.org/10.5772/intechopen.80355. Available: https://www.intechopen.com/books/the-relevance-of-hygiene-to-health-in-developing-countries/challenges-to-hygiene-improvement-in-developing-countries. Accessed Aug 10 2020.

Lüthi, C., Panesar, A., & Schütze, T. (Eds.) (2011). *Sustainable sanitation in cities – A framework for action* (1st ed.) Rijswijk: Papyroz Publishing House. 148p. 10.1.1.365.6104.

Mason, N., & Le Seve, M. D. (2017). *10 things to know about the future of water and sanitation*. London: Overseas Development Institute.

McGregor, D., Simon, D., & Thompson, D. (2006). *The - peri-urban interface: Approaches to sustainable natural and human resource use*. London: Earthscan.

National Research Council. (1996). *Meeting the challenges of megacities in the developing world: A collection of working papers*. Washington, DC: National Academy Press.

Ohwo, O., & Agusomu, T. (2018). Assessment of water, sanitation and hygiene services in Sub-Saharan Africa. European Scientific Journal December 2018 edition Vol. 14, No. 35. ISSN: 1857 – 7881 (Print) e – ISSN 1857-7431. Available online: https://www.researchgate.net/publication/330427841. Accessed 10 Aug 2020.

Reymond, P., Renggli, S., & Luthi, C. (2016). Towards sustainable sanitation in an urbanizing world. Sustainable Urbanization, Mustafa Ergen, IntechOpen. https://doi.org/10.5772/63726. Available online: https://www.intechopen.com/books/sustainable-urbanization/towards-sustainable-sanitation-in-an-urbanizing-world. Accessed 10 Aug 2020.

Tearfund. (2007). Sanitation and hygiene in developing countries: Identifying and responding to barriers. A case study from Burkina Faso. http://tilz.tearfund.org/Research/Water+and+Sanitation. Accessed 3 Aug 2020.

UNICEF. (2015a). *Programming for sustainability in water services: A framework*. New York: UNICEF.

UNICEF. (2015b). Water sanitation and hygiene. The case for support. Available online: https://www.unicef.org. Accessed 1 Aug 2020.

Van Rooijen, D. (2009). Urbanization, water demand and sanitation in large cities of the developing world: An introduction to studies carried out in Accra, Addis Ababa and Hyderabad. Available: https://www.researchgate.net/publication/228474925. Accessed 10 Aug 2020.

World Health Organization (WHO) (2008). *Water, sanitation and hygiene in health care facilities: Status in low and middle income countries and way forward*. Geneva: WHO.

WHO. (2009). *Global Health risks: Mortality and burden of disease attributable to selected major risks*. Geneva: World Health Organization.

World Health Organization (WHO) (2017). *Water sanitation and health*. Available on: http://www.who.int/water_sanitation_health/en/

World Health Organization (WHO) and United Nations Children's Fund (UNICEF) (2017). *Progress on drinking water, sanitation and hygiene: 2017 Update and SDG Baselines*. Geneva: WHO and UNICEF. Available on: https://www.communityledtotalsanitation.org/sites/communityledtotalsanitation.org/files/JMP_2017.pdf

Water-Food Nexus

▶ Circular Economy and the Water-Food Nexus

Watershed Management on Water Sustainability

▶ Watershed Sustainability: An Integrated River Basin Perspective

Watershed Sustainability: An Integrated River Basin Perspective

Felipe Armas Vargas[1], Luzma Fabiola Nava[2,3], Oscar Escolero[4], Eugenio Gómez Reyes[1] and Samuel Sandoval Solis[5]
[1]Departamento de Ingeniería de Procesos e Hidráulica, CBI, Universidad Autónoma Metropolitana-Iztapalapa, Ciudad de México, Mexico
[2]CONACYT-Centro del Cambio Global y la Sustentabilidad A.C. (CCGS), Villahermosa, Tabasco, Mexico
[3]International Institute for Applied Systems Analysis (IIASA), Laxenburg, Austria
[4]Departamento de Dinámica Terrestre y Superficial, Instituto de Geología, Universidad Nacional Autónoma de México, Ciudad de México, Mexico
[5]Department of Land, Air and Water Resources, University of California Davis, Davis, CA, USA

Synonyms

Water governance; Water sustainability; Sustainable watershed management; Watershed management on water sustainability

Definition

The current pressures exerted naturally and anthropogenically on water resources have repercussions on the sustainable development of ecosystem services provided by watersheds. The study of the theoretical framework of the Watershed Governance Prism (WGP), through the lens of the Duero River Basin (DRB) in Mexico, allows identifying which of the multiple perspectives of water governance present the greatest restriction among the rest of the perspectives related to adverse decision-making processes, critical sustainable development of water resources, and restraint environmental services. The association of the main problems of the DRB with the axes of the WGP demonstrates that the use of the WGP helps to identify the opportunities to promote the desired sustainable development. Consequently, these results offer a new assessment on how watershed issues can interact with the WGP thought the identification of perspectives to create and enable justice and equity in watershed sustainability.

Introduction

Large-scale driving forces of change (e.g., urbanization, climate change, floods, and droughts) require effective public policies that reinforce the resistance of both human and environmental water systems to preserve the ecosystem services provided by watersheds (Alpuche Álvarez et al. 2021). Addressing these challenges will require coordinated efforts at macro and local levels, such as strengthening resilience and implementing environmental flows and policies to combat water pollution (World Bank 2018). The search for solutions to these challenges demands inter- and transdisciplinary approaches (Arrojo et al. 2005).

The need to develop a holistic approach for integrated water management was boarded in 1992 (Arrojo et al. 2005), and this approach incorporates social, environmental, cultural,

W

institutional, and political aspects (Domínguez 2012). Additionally, achieving sustainable systems for resource management is not possible in the absence of effective governance (Arrojo et al. 2005). This integrated participation will allow better management of the resources of a watershed (Andrade-Pérez 2007; Nava 2013).

Watersheds are an accepted scale for water governance activities (Cohen and Davidson 2011), and currently, the evolution of water-environmental management approaches conducted by government agencies and scientific research groups nationally and internationally has been notable in literature. These theoretical frameworks include, from the highest to lowest order, Integrated Water Resources Management (IWRM; Savenije and Van der Zaag 2008), Ecosystem Approach to Health (Ecohealth; Lebel 2003), Water Framework Directive (WFD; Water Framework Directive 2000), Ecosystem Approach (EA; CBD 2014), Watershed Governance Prism (WGP; Parkes et al. 2010), and the Sustainability Wheel (SW; Schneider et al. 2015).

The objective of the present work was to analyze the theoretical framework of the Watershed Governance Prism (WGP) in the Duero River Basin (DRB) in order to identify which of the multiple perspectives of water governance present the greatest restriction (that is, is more problematic) among the rest of the perspectives, as a consequence of the deficiencies in the decision-making processes, reduced sustainable development of water resources, and environmental services of the watershed. The study also aims to identify which aspects of the prism (watersheds, ecosystems, social systems, and health and well-being) are neglected or overlooked.

WGP Framework and Study Area

The WGP is a contemporary conceptual framework that presents multiple facets of governance, characterizing water resources management linking social and environmental aspects with the social determinants of health in a watershed context. The WGP comprises four constitutive elements or vertices: watersheds, ecosystems, health/well-being, and social systems. The interaction between them forms six linear links: (1) ecosystems–health/well-being, (2) watersheds ecosystems, (3) watersheds–health/well-being, (4) watersheds–social systems, (5) social systems–health/well-being, and (6) ecosystems–social systems. The interaction between the axes forms four surfaces that represent the different perspectives of water governance.

Perspective A comprises water governance for sustainable development (links: watersheds, ecosystems, and social systems).

Perspective B comprises water governance for ecosystems and well-being (links: watersheds, ecosystems, and health/well-being).

Perspective C comprises water governance for social determinants of health (links: watersheds, social systems, and health/well-being).

Perspective D comprises water governance for the promotion of socio-ecological health (links: ecosystems, social systems, and health/well-being).

Finally, integrating the four perspectives (A, B, C, D) makes up the WGP, facilitating integrated watershed governance and understanding between the four perspectives (Parkes et al. 2010).

The Case of the Duero River Basin

The DRB is located northwest of the state of Michoacan, Mexico (Fig. 1). It possesses a wealth of natural and water resources, such as rivers, lakes, springs, aquifers, pine and oak forests, and geysers. The aquatic biological diversity is represented by numerous fish and macroinvertebrate species. Additionally, there are places where people can enjoy recreational and cultural activities (ecosystem services), such as Lake Camecuaro National Park, La Beata hill, the Geiser de Ixtlan Recreation Center, and various spas, all of which are located in the Duero region, providing the opportunity to interact with the environment closely. There are also storage dams, a hydroelectric power plant (El Platanal), agricultural areas, canals, extraction wells, treatment plants, and drinking water systems, making up the hydraulic infrastructure. Moreover, this region provides water to a population of close to 400,000 inhabitants (Comisión Nacional del

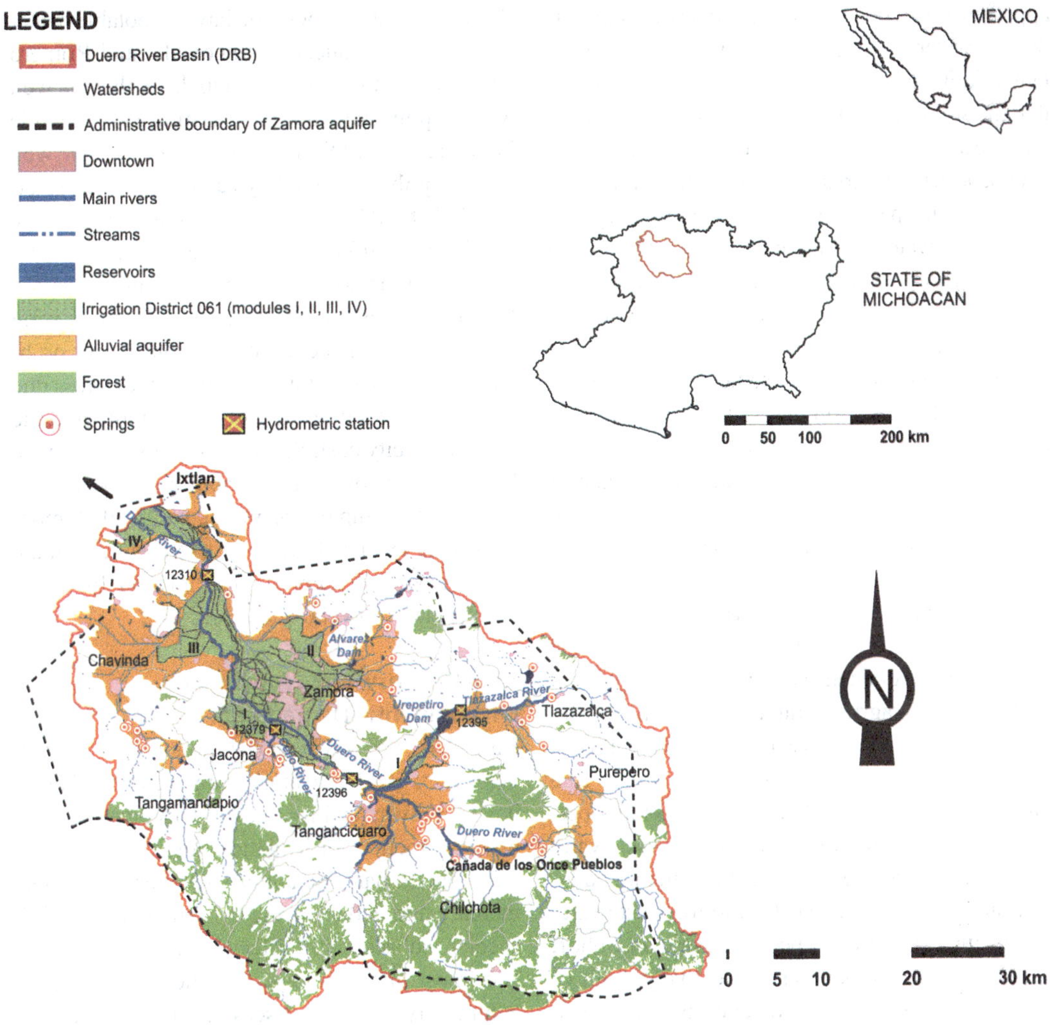

Watershed Sustainability: An Integrated River Basin Perspective, Fig. 1 The Duero River Basin (DRB)

Agua-Instituto Politécnico Nacional (Conagua-IPN) 2009; Velázquez 2005; Velázquez et al. 2011).

Despite its natural wealth, the DRB faces adversities that affect the river flow regimes, river corridors, aquatic habitats, deforestation, erosion, land-use change for agricultural activities, introductions of exotic species, wastewater discharge into rivers, and the lack of wastewater treatment. Furthermore, the lack of technicalized irrigation, eutrophication in irrigation channels, loss of biodiversity, and degradation of water quality due to pollutants exacerbate water availability and accentuate the fragility of ecosystems. Additionally, there is increasing urbanization of river stretches, lack of specific sites for solid waste storage, and constant social pressure for improved water resources management (Conagua-IPN 2009, Velázquez 2005; Velázquez et al. 2011). In this order of ideas, the main topics that limit effective water governance in the DRB are mentioned below.

Discharge of Wastewater into the River Network

The Duero River receives direct discharges of residual waste from three main cities: Tangancicuaro, Zamora, and Jacona. The null register or

W

appearance of collective enteric diseases suggests that the water dilutes pollutants, which is an important task, in addition to the retention or filter action that the soil exerts on pollutants. However, studies are needed on the degree of survival of enterobacteria in agricultural soils, the proportion of bacteria that pass into vegetables due to irrigation using wastewater, and the incidence of diseases in farmers who use this polluted water. Urban areas, such as Zamora and Jacona, add pollution derived from industrial, agro-industrial, and service activities. Moreover, the irregular settlements located on the banks of springs and rivers (Tangancicuaro and Chilchota) add to the pollution problem. The sewage discharge into rivers exists due to the lack of regulation on discharges and functional, operational infrastructure for its treatment (Velázquez 2005).

Lack of Wastewater Treatment

The sanitation system in the region is represented by sewage and drainage services, which generally discharge into agricultural irrigation infrastructure. Although they have low-efficiency levels, some populations have treatment plants (the municipality of Zamora being an exception, operating at 90% of the total volume of wastewater generated; Velázquez 2005). In Jamanducuaro (Tlazazalca) and Atacheo (Zamora), wastewater treatment is supposed to be done using an Imhoff tank and oxidation lagoon, respectively, both of which are not operating. However, two treatment plants – the Anaerobic Reactor of Carapan (BOD 30%) and Extended Aeration System (EAS) in the Zamora supply region (FC, 90%) – have variable efficiency in terms of removal of biological oxygen demand (BOD) and fecal coliforms (FC), respectively. The installed treatment capacity in the EAS is 6 l/s with a treated flow rate of approximately 4–6 l/s. Finally, Zamora has another treatment plant that operates through a stabilization lagoon system (FC, 99%; BOD, 72%) with a capacity of 330 l/s (Pimentel et al. 2011; Velázquez 2005).

Shortage of Water in the Watershed Localities

With regard to water supply and scarcity, Pimentel (2007) identified some localities that suffer from low water availability for primary consumption. While La Labor does not have a potable water service, in La Sauceda, Rinconada, and Romero de Torres (in the same municipality), there is well water supply every third day, lasting 3–10 h, while in Atacheo and Ojo de Agua, there is daily well water supply (managed by each community) lasting 1–3 h. All these localities are within the municipality of Zamora. In La Luz, El Valenciano, and El Limon localities (in the municipality of Ixtlan), there is a daily well water supply lasting 3–8 h, although the water presents boron problems. All these communities live under stress due to the short duration of the water supply and the high electricity cost. Some communities also have legal disputes over the distribution of water volumes: for example, the municipality of Zamora consumes 180 l/s of the Del Bosque spring (in the municipality of Jacona), but only 60 l/s arrive during the day (Velázquez, 2005).

Aquatic Ecosystems Affected by Environmental Degradation

Karr (1981) argued that fish quantity and biodiversity are good indicators of environmental quality. Moreover, anthropogenic activities (such as water extraction for irrigation, loss of river continuity due to barriers, and wastewater discharges) increase and endanger the existence of some fish species, subsequently affecting the environment. The most stressful time for the ichthyological community is the dry season (from January to May) due to the natural decrease of river flows. For example, from 2000 to 2012, some identified species such as *Menidia jordani* still preserve their "tolerant" status, while in the Cyprinidae family, the status of *Algansea tincella* changed from "tolerant" to "moderately tolerant," and the status of *Aztecula sallaei* changed from "moderately tolerant" to "sensitive"; *Goodea atripinnis* maintains a "highly tolerant" status, but *Alloophorus robustus* changed from "moderately tolerant" to "sensitive" (Ramírez-Herrejón et al. 2012; Lyons et al. 2000; Mercado-Silva et al. 2006).

Influence of Human Activity on Water Bodies

López-Hernández (1997) identified the central to lower part of the DRB as having the lowest biological index values, qualifying the waters as contaminated and with less capacity for self-purification. Moncayo-Estrada et al. (2011) have highlighted the necessity of maintaining good water quality in the rivers of the DRB. Subsequently, Moncayo-Estrada et al. (2015) evaluated the biotic integrity index to compare it with previous years (1986, 1991). They found that the Etucuaro region has retained its “regular” condition, while Lake Camecuaro changed from a “good” to “regular” status; additionally, the El Platanal watershed status was “poor” and that of Zamora, La Estanzuela, and San Cristóbal “A” changed from “regular” to “poor.” Bacterial contamination was also found from the Cañada de Los Once Pueblos to the limits of the Zamora valley, except for the Carapan and Camecuaro springs (Velázquez 2005).

The Use of the Irrigation District

The irrigation district (ID-061) presents various issues, such as the loss of cultivation areas due to urbanization, water scarcity, irregular settlements in canal maintenance areas, urbanization in river stretches, flooding, changes in land use, disabling canals by the presence of subdivisions, differences between irrigation modules, and lack of civic culture. Furthermore, as the hydraulic gradient descends along the Duero River, water quality decreases. For example, sewage limits the cultivation of higher-value fruits and vegetables (such as strawberries), as Irrigation Modules II and III users receive water from the river already contaminated with drainage and other discharges based on their location. However, when the river crosses Module IV (located in the lower zone, toward the DRB boundary), it has already received all the upstream discharges, necessitating higher expenses in the production processes (Velázquez 2005).

Limitation of Multilevel Governance

An analysis of a key actors’ map of the three levels of government (federal, state, and municipal), local organizations and associations, agricultural producers, and irrigation modules, among others, for the management of watershed resources, revealed no links between them despite being in the same region (DRB). However, some information flows are fed back to the rest of the actors in the watershed. Thus, it is necessary to identify and formulate mechanisms that promote social participation. Regarding the three government levels, it is also necessary to develop and strengthen the institutional image and revise cross-cutting public policies (Conagua-IPN 2009). These topics contribute to the loss of ecosystem capabilities and water quality degradation (Velázquez 2005), putting the socioeconomic development of the watershed at risk. Consequently, it is vitally necessary to halt these adverse conditions (Moncayo-Estrada et al. 2015).

Just as the prism axes can be interpreted as isolated links, the perspectives of the prism can be interpreted as a union or integration. Subsequently, we can take advantage of this quality of the prism to associate the perspectives with the proposals formulated for the improvement of the sustainable development of water resources and the watershed. These improvement actions were identified from technical studies made by Conagua-IPN (2009) and Velázquez (2005).

The DRB Under the WGP Framework (Axis Analysis: The Relationship Between the Vertices)

Figure 2 illustrates the connection between the representative problems of the DRB and the links between the six axes of the WGP. According to Bunch and Waltner-Toews (2015), not all the prism axes will be identified as important in each situation, but by identifying the line of problems and developing a description, the prism can provide information about the search for the problem.

The most common issues arising from failures in water governance, related through the WGP, are as follows:

1. Ecosystems and health/well-being (Parkes et al. 2010). As there is no site in the hydrographic network that can escape wastewater pollution, the wastewater problem has serious consequences for public health (Velázquez 2005). The chemical composition of groundwater in the watershed demonstrates high

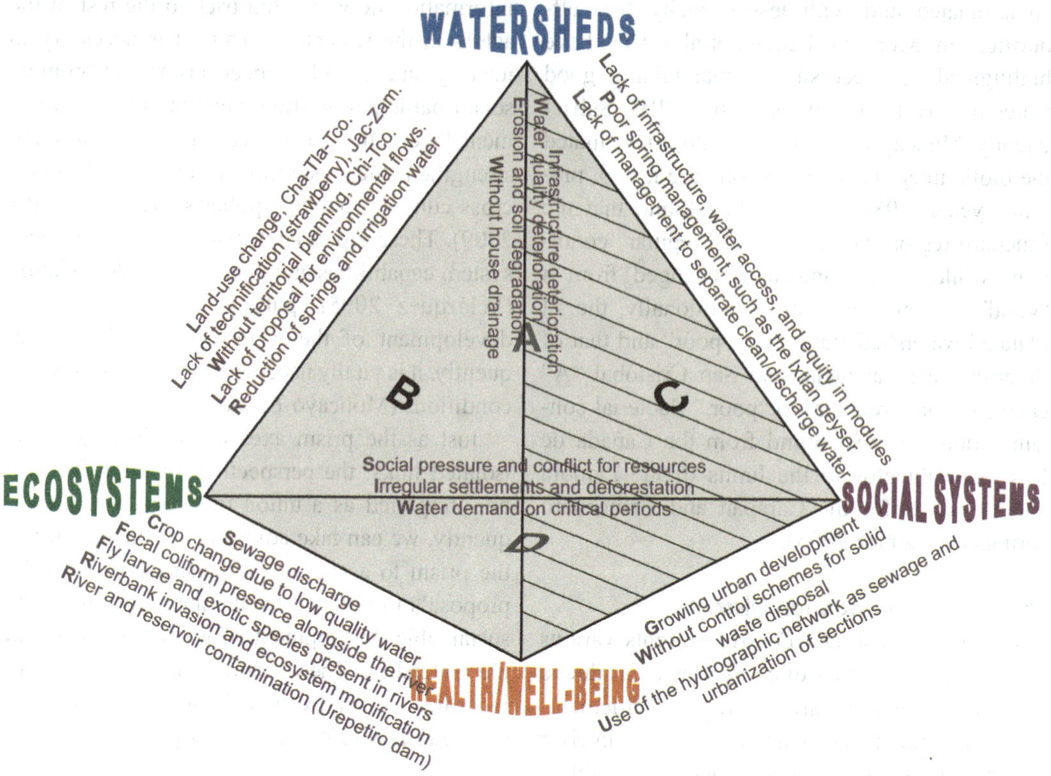

Watershed Sustainability: An Integrated River Basin Perspective, Fig. 2 Link between the axes of the WGP, and the issues of the DRB. Municipalities: Jacona (Jac), Zamora (Zam), Chilchota (Chi), Tangancicuaro (Tco), Tlazazalca (Tla), Tangamandapio (Tmp), Chavinda (Cha), and Ixtlan (Ixt). Prism adapted from Parkes et al. (2010) and modified with information from Conagua-IPN (2009) and Velázquez (2005)

concentrations of toxic elements such as boron (B) and lead (Pb) (Velázquez et al. 2011). Pb in groundwater is probably related to volcanic origin materials such as siliceous sands (Inocencio-Flores et al. 2013). As groundwater is meant for domestic and urban use, it is necessary to perform a detailed analysis of the chemical, physical, and microbiological composition in the sources detected to prevent potential health issues. For example, crop change, exotic species introduced in the river, and the loss of vegetation in banks are all issues caused by water quality degradation (Conagua-IPN 2009).

2. Watersheds and ecosystems (Parkes et al., 2010). Proposals are needed to implement environmental flow regimes. Because of poor infrastructure and lack of governmental financial support, the lack of irrigation technology causes great difficulties that prevent irrigation modules from achieving financial self-sufficiency (Velázquez, 2005). The lack of territorial planning caused by changes in land use (Conagua-IPN 2009) is evidenced in streambed regulation, degradation of slopes, flow rate, channeling, and alterations in river morphology, all of which are mainly caused by agricultural activity.
3. Watersheds and health/well-being (Parkes et al. 2010). Some populations have wastewater treatment plants, although most operate with low levels of efficiency. The sanitation system is represented by sewage and drainage services that regularly discharge waste into the river, which end up in the irrigation infrastructure. For example, the Dren Chavinda drain and the Ixtlan municipality drainage system are connected to the Duero River. Dams and

hydraulic infrastructure suffer from deterioration and flooding. In the wet season, Dren Chavinda risks overflowing, and there are mosquito outbreaks (Velázquez, 2005). Only 19 springs (that supply the population) benefit from conservation efforts, while 26 others do not have a similar program (Zavala-López 2011).

4. Watersheds and social systems (Parkes et al. 2010). Module I (ID-061) has a natural supply from the Orandino and Tamandaro springs, with adequate water quality and quantity. However, Module IV (downstream from the Duero River) receives wastewater and discharges from the Zamora-Jacona conurbation. The advantage in Module I is that many of its farmers can reach an insured market with agribusiness companies closing the production-marketing cycle. Regarding water access, 17% of the population (out of 402,698 inhabitants in 2000) did not have tap water access (Velázquez 2005); moreover, the financial situation of ID-061 has deteriorated, and there is no local investment in local projects to improve water management (Conagua-IPN 2009).
5. Social systems and health/well-being (Parkes et al. 2010). Landfills and leachate generation present an infiltration risk into the aquifer (Velázquez 2005). In the DRB, there are 13 final disposal sites without control schemes (unprotected landfills); one is found in the municipality of Zamora (Conagua-IPN 2009). In 2009, the DRB region had 447,324 inhabitants who generated solid waste of 0.72 kg/hab/day. Of the amount of urban solid waste generated, only 70% is collected for disposal in respective landfills, with the rest being dispersed in the environment. The accumulation of solid waste on riverbanks and tributaries is a common sighting (Conagua-IPN 2009), posing great risks to public health.
6. Ecosystems and social systems (Parkes et al. 2010). Pimentel et al. (2011) stated that conflicts between communities over water allocations have worsened due to the claim of some municipalities (mainly Zamora). They proposed that municipalities with resources (Tangancicuaro and Jacona) purchase water from their springs. However, agricultural users have prevented these requests from being fulfilled. The physical and chemical properties of the spring water in this area exceed the wells' water quality in Zamora; thus, it is more important (Conagua-IPN 2009); moreover, there is great demand for water in critical periods (dry season).

Figure 3 illustrates the improvement actions proposed for the desired sustainable development in the DRB.

Perspective A: Water governance for sustainable development (linking watersheds, ecosystems, and social systems; Parkes et al., 2010). Payment for ecological services (PES) in the upper watershed has been proposed to counter clandestine logging. Sites in forests, sections of rivers, and riverbanks that require reforestation and restoration and ecological improvement programs should be identified and evaluated to achieve beneficial impacts related to erosion and soil recovery, increase aquifer recharge, capture CO_2, and facilitate local biodiversity recovery. The modernization of agricultural activities will optimize water use (Conagua-IPN 2009; Velázquez 2005). Modernization in the hydro-agricultural infrastructure and technification of the agricultural surface can be realized in coordination with users and local authorities. This initiative comprises the exchange of 5 Mm3 of surface water for 5 Mm3 of groundwater. This being the highest quality water, it will increase produce exports to international markets (Jiménez 2011).

Perspective B: Water governance for ecosystems and well-being (linking watersheds, ecosystems, and health; Parkes et al. 2010). Proposing environmental flow regimes (EFR) in the DRB can serve as a mitigation measure against the pressure of climate change. These flows can continue to maintain and preserve the functionality and structure of ecosystems and the environment (habitats, banks, and aquifers), in addition to increasing resilience and reducing the loss of ecological integrity. Armas-Vargas et al. (2017) made an EFR proposal based on the physical habitat

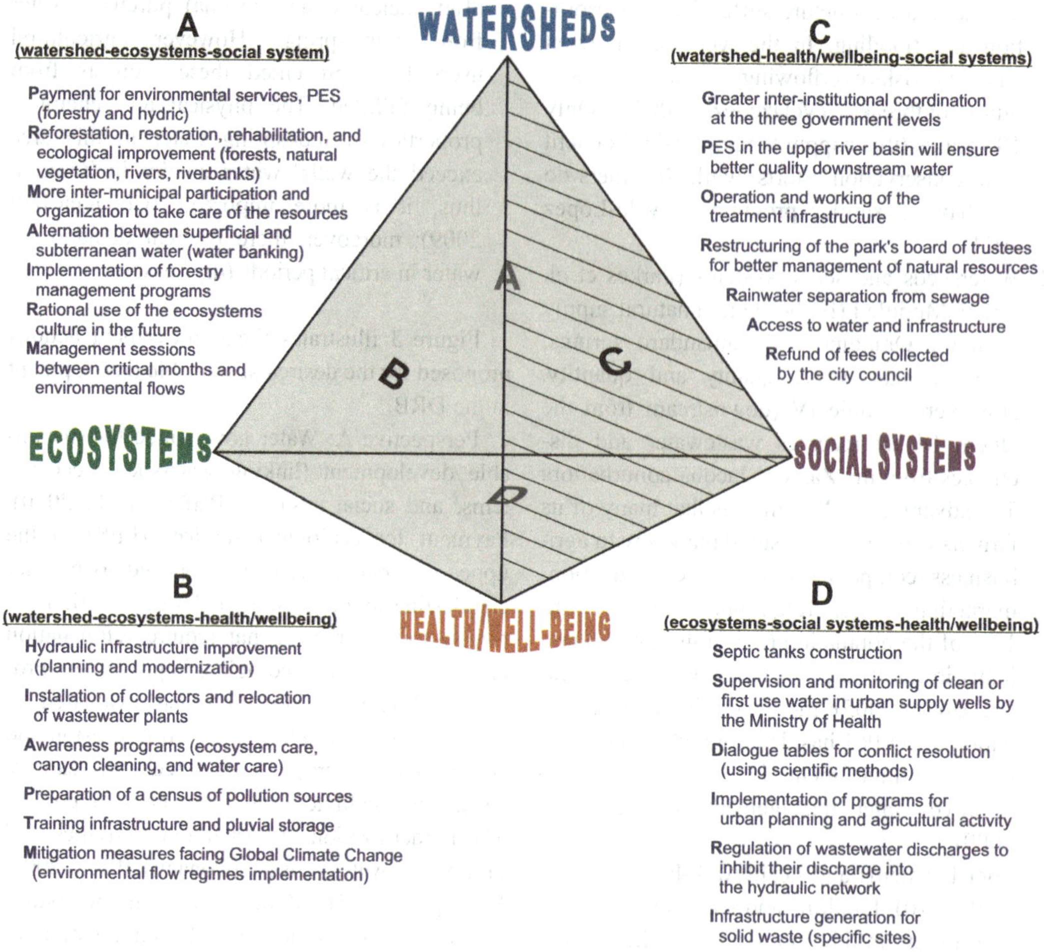

Watershed Sustainability: An Integrated River Basin Perspective, Fig. 3 Prospects of the WGP with proposals aimed at promoting sustainability in the DRB. (Prism adapted from Parkes et al. (2010) and modified with information from Conagua-IPN (2009) and Velázquez (2005))

simulation method, using the Physical Habitat Simulation System (PHABSIM) model. However, more planning is required, especially concerning developing a pollution source roll due to wastewater discharges. A decrease in the concentration of fecal coliforms has been observed in the upper and middle part of the DRB (between Chilchota and Tangancicuaro) due to the contributions of springs (Conagua-IPN 2009; Velázquez 2005).

Perspective C: Water governance for social determinants of health (linking watersheds, social systems, and health; Parkes et al. 2010). It is necessary to continue promoting social participation in the decision-making process because water plays a fundamental role in all aspects of public and private life. The Duero River Basin Commission (DRBC) can organize a joint coordination effort for sanitation actions (Pimentel et al. 2011). There are projects underway to reopen and operate abandoned treatment infrastructure, access to drinking water, and sewer coverage. As there are communities governed by consuetudinary law, it is necessary to continue motivating and encouraging them to use the drainage infrastructure. Payment through economic compensation is recommended to avoid the use and pouring detergents into the river (Conagua-IPN 2009; Velázquez 2005).

Perspective D: Governance for socio-ecological health promotion (linking ecosystems, social systems, and health; Parkes et al., 2010). The use of septic tanks in small towns should be promoted (Velázquez 2005). Inocencio-Flores et al. (2013) and Velázquez et al. (2011) confirmed the presence of contaminants in water sources (wells and springs). Additionally, it is convenient for a third party (Ministry of Health) to supervise and monitor water quality at the sources identified. Pimentel et al. (2011) suggested analyzing bacteriological and other contaminants, such as heavy and organic metals. To protect the environment and public health, direct discharge into the river watershed network should be avoided and treatment systems reactivated. It is also important to establish solid waste deposit sites (Conagua-IPN 2009) and canyon cleaning programs by involving citizens and encouraging state participation (Velázquez 2005).

A Threefold Discussion

Good and Bad Water Governance

Much more is known about bad governance than good governance (Rogers 2002). From our interpretation, Fig. 2 illustrates that the DRB is undergoing a crisis, which is reflected (or diagnosed) by the various issues facing watershed resources. According to Nava (2013), this crisis will persist until governance is renewed, innovated, and adapted at the watershed level. For De Carvalho and Angulo (2014) and the Water Governance Facility (2016), if there are insufficient human and financial resources, lack of investment funds, inadequate or deficient management of resources, bureaucracy, corruption, and water organizations that do not contribute to the formulation of public policies, there will be poor water governance, making it impossible to solve or even identify the cause of water problems. According to Pahl-Wostl (2009), most resource management problems originate from governance failures. The presence of these symptoms in the DRB makes sense when, in the local press, the municipalities of the watershed are regularly exhibited and urged to participate and join efforts to achieve the sanitation of resources by calling on local and federal authorities to integrate and stop pollution from wastewater discharge into the rivers and canals of the DRB. After a decade of deliberations and negotiations, in 2017, the work of the DRBC was resumed to resolve some of the present issues (lack of financial resources to build treatment plants and conduct sanitation actions), in addition to the creation of a trust with contributions from governance actors. Figure 2 illustrates four perspectives representing different types of water governance, where we observe relatively more issues in the axis composed of ecosystems and health/well-being, followed by the watersheds and ecosystems axis, and the watersheds and health/well-being axis. The prism indicates a priori that water governance for ecosystems and well-being (referred to in Perspective B) is currently identified as the most problematic compared to the other perspectives (D, A, and C) and, together, impact the resilience of the water-environmental resources of the DRB.

The results demonstrated that the governance of water for ecosystems and well-being (Perspective B) is the governance with the most issues in the DRB. The study carried out by Bunch et al. (2014) concluded that out of a sample of 100 articles reviewed, the study of Perspective B was dominant over the rest of the perspectives of the prism. This implies that the scientific community prefers to focus on issues related to eco-hydraulic and hydrological aspects of the watershed, as well as environmental and watershed management. According to Parkes et al. (2010), when attention is focused on Perspective B, the vertex that can be overlooked is that of social/equity issues. As Pimentel (2007) pointed out, there are still DRB communities where the drinking water service is intermittent or every third day, and high-energy costs lead to inequality. Additionally, there is an increasing urban development; lack of infrastructure, access to water, and household drainage; shortages; irregular settlements; poor management of separating rainwater from discharges further; and conflicts and social pressures for water resources.

Pimentel et al. (2011) highlighted conflicts between communities over water allocations. However, reducing wastewater discharges to the river network is more of a priority than reducing water shortages in the localities of the basin (not to mention the stress and low availability to which they are subjected). The problem of the discharge of wastewater without treatment in the hydraulic network has existed for more than 20 years, with implications for human health, as well as economic effects for irrigation modules downstream of the Duero River, in addition to the degradation of river ecosystems. For example, if farmers located downstream of the Duero River want to grow strawberries and use the river's water (as an ecosystem service), they would not market their products in the same way due to the degradation of the quality of the river water. Consequently, there is unequal competition among farmers who grow strawberries with high-quality water from a well or spring, presenting a greater economic advantage when marketing their products. Conagua-IPN (2009) and Velázquez (2005) observed that the main problem of ID-061 is that it is located within urban areas receiving municipal discharge. An attempt to solve this problem was made with the construction of treatment plants, most of which do not operate. However, the organizations' performances have been insufficient in sanitizing the water in the watershed, and there have been high levels of bacteriological and chemical contamination (Pimentel et al. 2011).

For Hurlbert and Diaz (2013), the limitations are related to the existence of neoliberalism, which is characterized by the limited role of the state in the economy and an active and enthusiastic role of the private sector as the main engine of economic development, with superfluous attention to environmental issues, and considering water as a market product or a privately owned good. For Rogers (2002), if water resources are excessively managed through private markets, only those with higher incomes will have access to water. However, if it is administered by public authorities, it is also not certain whether there will be equal water access. For this reason, we understand that bad practices in water governance depend on decision-making. For Batchelor (2007), water governance encompasses how allocation and regulation policies are exercised in water management. For example, in the first months of 2015 in Mexico, the emblematic case of privatizing water emerged in the national public press, making modifications to the National Water Law. The considered measures that were planned, such as mega aqueducts, intensive use of surface and underground resources, and fracking, caused social discontent, in addition to the scandal experienced by the highest authority of the water agency (Conagua), due to questions against it about direct allocations (without tenders), conflict of interest, and lack of experience in management, among others. For these reasons, water users distrust the terms privatization and neoliberalism, as corruption and unethical practices are still present in decision-making, with the repercussion on water governance.

To achieve good governance, negotiation; agreements for water use, clear and transparent regulations; the recognition of rights, obligations, roles, and responsibilities (De Carvalho and Angulo 2014); access to information; cooperation between stakeholders; accountability; recognition of water uses and customs; and collective decision-making (Domínguez 2012) are all necessary. Good governance is fundamental for improving access to water and sanitation services, ensuring the sustainable and equitable use of water resources, and expanding the provision of sanitation and drinking water supply services (Stockholm International Water Institute [SIWI] 2015). In general, without good water governance, it will not be possible to achieve long-term water security (Tortajada 2010). For the Water Governance Facility (2016), effective governance of water resources will be key to achieving the fair allocation of water resources and conflict resolution. In 2019, a wastewater discharge dispute arose surrounding Lake Camecuaro, one of the main ecotourism attractions of the DRB (municipality of Tangancicuaro). After a year of negotiations and management efforts between the municipal and state governments, a wastewater treatment plant and a sanitary sewer were installed to guarantee the sustainability of the water bodies and their interaction with adjacent ecosystems. Government authorities must ensure the correct application of their normative base not solely by raising fines, as according to Lebel (2003), the health of the ecosystem is equal to human health.

Figure 3 illustrates some of the proposed actions derived from good governance in the DRB. These include various developments and dialogue between multiple stakeholders. For example, Perspective A highlights the PES, as well as the alternation between surface and groundwater. Perspective B highlights the modernization of hydraulic infrastructure, as well as a pollution source registry. Perspective C highlights greater transverse and horizontal coordination, as well as access to water and infrastructure. Finally, Perspective D highlights the construction of septic tanks and the integral management of solid waste (at specific sites). According to Parkes et al. (2010), the WGP has the property of unifying different perspectives of water governance to facilitate integrated watershed governance. At the end of 2007, the DRBC was created to improve water management, develop hydraulic infrastructure, increase the environmental recovery of the watershed, and contribute to the conflict resolution associated with competition between water uses and water users. The DRBC integrates various governance actors, allowing sanitation action and coordination. These interactions promote and facilitate water governance in the watershed, whose purpose lies in managing water resources and providing water services.

Participation in Water Governance

The water crisis has motivated governance actors to participate and organize strategies to achieve sustainability and balance in the watershed. This has led to a consensus among the actors to assess and organize the watershed, and related ecosystems before environmental threats and adverse risks appear, with consequences for socioeconomic well-being. For Pimentel and Velázquez (2015), the participation of landowners and communities is essential in making decisions about water management due to their extensive knowledge and operations in the territory. The creation of organizations such as the Duero River Basin Commission (DRBC), according to Nava (2013), allows actors to reaffirm their willingness to work together for the integrity of ecosystems, preserve the traditional way of life, and promote societal participation in decision-making. For Pimentel et al. (2011), although the formation of the DRBC is an important step, it does not guarantee the sustainable management of water resources due to extensive bureaucracy and little social participation.

Society can increase its participation in protecting the environment in many ways. According to the Ottawa Charter (WHO 1986), this is possible through the empowerment of people by increasing their control over their health, thereby improving it and creating favorable environments for health promotion. This can also be done by developing public policies to strengthen the community's socio-ecological environment and requiring the designated authorities to fulfil their obligations and responsibilities. Similarly, financing is essential for the implementation of programs. According to Pimentel and Velázquez (2015), the DRBC does not have an operational budget for buildings and infrastructure projects, as its budget covers management, studies, and conciliation.

To Warner (2005), as the population size in most cities, towns, and municipalities prevents the direct participation of all stakeholders in decision-making at the watershed level, participation is not a simple task. Furthermore, the question of who represents the large groups of stakeholders is highly politicized. For example, in the DRB for more than a decade, the local press has been following the meetings, which have presented disappointing results, urging the participants to join forces and work decisively to achieve common long-term benefits, such as the sanitation of the Duero River, a constant concern for water actors. Peters and Pierre (2002) argued that all negotiation implies discussion and divergent interests among the participants. For Warner (2005), a multi-stakeholder dialogue is not just a conversation but also an interactive approach to meet and interact in a forum for conflict resolution, negotiation, and collective decision-making, seeking consensual solutions through mutual understanding. Legal, political, or bureaucratic concerns also limit the negotiations of multiple stakeholders. In the case of the DRB, various actors have been participating due to their concern for the environment. For example, strawberry producers in the Zamora Valley, faced with the need for high-quality water for agriculture and to comply with international market regulations, carried out a study coordinated by Velázquez (2005) that addresses the problem of wastewater in the Duero River.

Van Buuren et al. (2019) classify this participation as "Project-oriented initiatives," in which stakeholders/citizens mobilize to develop their own project proposal, challenging government decision-making. However, due to the requests and demands established by users of the DRB's water resource, a collaboration agreement was established between the National Water Commission and the National Polytechnic Institute, Michoacán Unit, to carry out the Comprehensive Sanitation project of the Duero River Basin (Conagua-IPN 2009). Van Buuren et al. (2019) refer to this participation as "Capacity-driven participation," where interested parties are invited to participate and strengthen the capacity of governance and empower stakeholders through collective action. Pimentel and Velázquez (2015) propose integrating four watershed level into the DRBC, where peoples and communities can participate, thus allowing direct and accurate information on resource issues, generating rapid diagnoses for decision-making.

The Prism as a New Opportunity for Integration

The issues and initiatives of the DRB, presented through the WGP, interact in the vertices of the prism. The simple structure of the WGP justifies linking the issues with the axes and the solution actions with the prism perspectives. The DRB-WGP interaction allows us to visualize the issues interacting with other prism environments (ecological, health, and socioeconomic) and identify which perspectives are limiting or restrictive for integrated watershed governance. In the same way, in the four perspectives of the prism, the actions that will be implemented in the watershed are visualized through the different types of water governance, which, applied together, can produce improvements in the DRB in the short and long term. The concentration of the issues identified in the WGP allows us to suggest the following governance restricting order, starting with the governance of water for (1) ecosystems and well-being, (2) promotion of socio-ecological health, (3) sustainable development, and (4) the determinants of social health (Perspectives B, D, A, and C, respectively). Among the vicissitudes in the DRB, polycentric governance is practised due to the verticality between organizations (Conagua-Basin Council-DRBC); participatory horizontality between municipalities; and transfer of operations, conservation, and management of ID-061 to the users of the irrigation modules. For Parkes et al. (2010), the WGP proposes a shift toward the integrated governance of watersheds as a basis for fostering health, sustainability, and social-ecological resilience.

It would be interesting to incorporate the direct participation of the governance structure into this study to identify the degree of support or obstruction towards various adaptation actions, using the Water Governance Assessment Tool (WGAT) (Bressers et al. 2013). The analysis of organizations – such as the Conagua, the Lerma-Chapala Basin Council, the Duero River Basin Commission (DRBC) – and agency bodies – such as municipalities, water operating agencies, water users' committees, indigenous communities, and irrigation modules – could be of help to better understand why problems such as wastewater discharge are caused or accentuated by the lack of treatment, water services, or environmental protection.

Conclusions

Through this study, the WGP has allowed us to link various watershed issues with the prism axes and propose various solutions for watershed sustainable development from the prism perspectives, moving from an isolated (traditional) vision to an integrated perspective. This led to identifying the most relevant issues (symptoms) of the DRB, which are mainly concentrated in the ecosystem–health/well-being axis, followed by the watersheds–ecosystems and watersheds–health/well-being axes, and to a lesser extent on the remaining axes. This demonstrated that Perspective B, the governance of water for ecosystems and well-being, is the main limiting perspective generating failures in water governance. Likewise, Perspectives D, A, and C are gradually contributing to the water crisis in the DRB, consequently hindering integrated watershed governance. In other words, Perspective B requires the highest priority or attention when initiating improvement actions to achieve the desired sustainable development in the DRB, requiring decision-making for proposals or alternative solutions indicated in each perspective of the WGP-DRB. With this analysis applied to the DRB, it was also possible through the theoretical framework of the WGP to identify that the social systems vertex overlooks social and equality problems, such as houses without drainage services, direct access to water, or electrical power.

This entry offers a different perspective on how the issues in the watershed can interact with the WGP, contributing to the identification of perspectives that generate ineffective water governance practices and proposals or improvement initiatives that generate effective practices in water governance aimed at mitigation, improving resilience, and the sustainable development of the watershed. This is analogous to a medic consultation, where there is a symptom analysis, followed by a diagnosis and then a treatment plan for the symptoms, and more importantly, the disease.

Cross-References

▶ Water Security and Its Role in Achieving SDG 6

Acknowledgments We express our gratitude to the *División de Ciencias Básicas e Ingeniería* (DCBI) of the *Universidad Autónoma Metropolitana-Iztapalapa (UAM-I)* for the support received.

References

Alpuche Álvarez, Y. A., Nava, L. F., Carpio Candelero, M. A., & Contreras Chablé, D. I. (2021). Vinculando ciencia y política pública. La Ley de Aguas Nacionales bajo la perspectiva sistémica y de servicios ecosistémicos. Gestión y Política Pública, Vol. 30, núm. 2, División de Administración Pública del Centro de Investigación y Docencia Económicas, A.C. https://doi.org/10.29265/gypp.v30i2.881

Andrade-Pérez, A. (2007). *Aplicación del Enfoque Ecosistémico en Latinoamérica*. Bogotá: IUCN.

Armas-Vargas, F., Escolero, O., Garcia de Jalon, D., Zambrano, L., González del Tánago, M., & Kralisch, S. (2017). Proposing environmental flows based on physical habitat simulation for five fish species in the lower Duero River basin, Mexico. *Hidrobiologica, 27*(2), 185–200. https://doi.org/10.24275/uam/izt/dcbs/hidro/2017v27n2

Arrojo, A. P., Assimacopoulos, D., Barraque, B., Bressers, J. T. A., & Esteban, C. (2005). *European declaration for a new water culture*. Madrid: Foundation for a New Water Culture. Retrieved from https://research.utwente.nl/en/publications/european-declaration-for-a-new-water-culture.

Batchelor, C. (2007). *Water governance literature assessment*. International Institute for Environment and Development. Retrieved from https://pubs.iied.org/G02523.

Bressers, H., Boer, C. D., Lordkipanidze, M., Özerol, G., Kruijf, J. V.-D., Farusho, C., & Browne, A. (2013). *Water governance assessment tool: With an elaboration for drought resilience. Report to the DROP project*. Enschede: INTERREG IVb DROP Project.

Bunch, M. J., Parkes, M., Zubrycki, K., Venema, H., Hallstrom, L., Neudorffer, C., . . ., & Morrison, K. (2014). Watershed management and public health: An exploration of the intersection of two fields as reported in the literature from 2000 to 2010. *Environmental Management, 54*(2), 240–254. Retrieved from https://doi.org/10.1007/s00267-014-0301-3

Bunch, M. J., & Waltner-Toews, D. (2015). Grappling with complexity: The context for one health and the ecohealth approach. In J. Zinsstag, E. Schelling, D. Waltner-Toews, M. Whittaker, & M. Tanner (Eds.), *One health: The theory and practice of integrated health approaches* (pp. 415–426). Boston, MA: CAB International.

CBD. (2014). *Convention on biological diversity*. Ecosystem Approach. Retrieved from http://www.cbd.int/ecosystem/.

Cohen, A., & Davidson, S. (2011). An examination of the watershed approach: Challenges, antecedents, and the transition from technical tool to governance unit. *Water Alternatives, 41*(1), 1–14.

Comisión Nacional del Agua-Instituto Politécnico Nacional (Conagua-IPN). (2009). Programa Detallado de Acciones Para el Proyecto Emblemático: Saneamiento Integral de la Cuenca Del Río Duero, 142.

De Carvalho, M. V., & Angulo, R. (2014). Gobernanza de los Sistemas Locales de Gestión del Agua en Bolivia. Asociación Internacional para la Gobernanza, la Ciudadanía y la Empresa. Retrieved from http://www.aigob.org/?s=Gobernanza+de+los+sistemas+locales+de+gestion+del+agua+en+bolivia.

Domínguez, S. J. (2012). *Hacia Una Buena Gobernanza para la Gestión Integrada de los Recursos Hídricos*. Proceso Regional de las Américas VI Foro Mundial del Agua. Retrieved from https://www.gwp.org/globalassets/global/gwp-cam_files/gobernanza-para-girh-2012.pdf.

Hurlbert, M. A., & Diaz, H. (2013). Water governance in Chile and Canada: A comparison of adaptive characteristics. *Ecology and Society, 18*(4), 61. Retrieved from https://doi.org/10.5751/ES-06148-180461

Inocencio-Flores, D., Velázquez-Machuca, M. A., Pimentel-Equihua, J. L., Montañez-Soto, J. L., & Venegas-González, J. (2013). Hidroquímica de las aguas subterráneas de la cuenca del río Duero y normatividad para uso doméstico. *Tecnología y Ciencias del Agua, 4*(5), 111–126.

Jiménez, S. R. (2011). La gestión social del agua: El programa K030 en el distrito de riego 061, Zamora Michoacán, México. *Agricultura, Sociedad y Desarrollo, 8*(3), 329–344.

Karr, J. R. (1981). Assessment of biotic integrity using fish communities. *Fisheries, 6*(6), 21–27. Retrieved from https://doi.org/10.1577/1548-8446(1981)006<0021:AOBIUF>2.0.CO;2

Lebel, J. (2003). *Health: An ecosystem approach*. Ottawa: International Development Research Centre.

López-Hernández, M. (1997). Caracterización Limnológica del Río Duero, Michoacán. Tesis Doctoral, Posgrado en Ciencias Biológicas, Universidad Nacional Autónoma de México, México. Retrieved from http://132.248.9.195/ppt1997/0252352/0252352.pdf

Lyons, J., Gutiérrez-Hernández, A., Díaz-Pardo, E., Soto-Galera, E., Medina-Nava, M., & Pineda-López, R. (2000). Development of a preliminary index of biotic integrity (IBI) based on fish assemblages to assess ecosystem condition in the lakes of Central Mexico. *Hydrobiologia, 418*(1), 57–72. Retrieved from https://doi.org/10.1023/A:1003888032756

Mercado-Silva, N., Lyons, J., Díaz-Pardo, E., Gutiérrez-Hernández, A., Ornelas-García, C. P., Pedraza-Lara, C., & Zanden, M. J. V. (2006). Long-term changes in the fish assemblage of the Laja River, Guanajuato, Central Mexico. *Aquatic Conservation: Marine and Freshwater Ecosystems, 16*(5), 533–546. Retrieved from https://doi.org/10.1002/aqc.737

Moncayo-Estrada, R., Lyons, J., Ramírez-Herrejón, J. P., Escalera-Gallardo, C., & Campos-Campos, O. (2015). Status and trends in biotic integrity in a sub-tropical river drainage: Analysis of the fish assemblage over a three decade period. *River Research and Applications, 31*(7), 808–824. Retrieved from https://doi.org/10.1002/rra.2774

Moncayo-Estrada, R., Silva-García, J. T., & Ochoa-Estrada, S. (2011). Identificación de Zonas de Mayor Riesgo (Focos Rojos). II Congreso Nacional de Manejo de Cuencas Hidrográficas, Villahermosa, México, 18–20 Mayo 2011. Retrieved from http://www.remexcu.org/documentos/cnmch/II-CNMCH-2011_memoria.pdf

Nava, L. F. (2013). Gobernanza global del agua. In M. López-Vallejo, A. B. Mungaray-Moctezuma, F. Quintana-Solórzano, & R. Velázquez-Flores (Eds.), *Gobernanza Global en un-Mundo Interconectado* (pp. 113–121). UAEBC-AMEI-UPAEP.

Pahl-Wostl, C. (2009). A conceptual framework for analysing adaptive capacity and multilevel learning processes in resource governance regimes. *Global Environmental Change, 19*(3), 354–365. Retrieved from https://doi.org/10.1016/j.gloenvcha.2009.06.001

Parkes, M. W., Morrison, K. E., Bunch, M. J., Hallström, L. K., Neudoerffer, R. C., Venema, H. D., & Waltner-Toews, D. (2010). Towards integrated governance for water, health and social–ecological systems: The watershed governance prism. *Global Environmental Change, 20*(4), 693–704. Retrieved from https://doi.org/10.1016/j.gloenvcha.2010.06.001

Peters, B. G., & Pierre, J. (2002). La gobernanza en niveles múltiples: ¿Un pacto fáustico? (Multilevel governance: A Faustian bargain?). *Foro Internacional, 42(3)*(169), 429–453.

Pimentel, E. J. L. (2007). Construyendo la Problemática de la Gestión de las Aguas Superficiales y Subterráneas de la Cuenca del Río Duero, Michoacán, México. In *XXVI Congreso de la Asociación Latinoamericana de Sociología* (pp. 1–14). Guadalajara: Asociación Latinoamericana de Sociología. Retrieved from https://www.aacademica.org/000-066/1093.

Pimentel, E. J. L., & Velázquez, M. M. A. (2015). Modelo organizativo para la gestión integral de la cuenca del Río Duero, Michoacán. In Burgos, A., Bocco, G. y Sosa-Ramírez, J. (Eds.), Dimensiones Sociales en el Manejo de Cuencas (pp. 69–85). Morelia: UNAM-SEMARNAT.

Pimentel, E. J. L., Velázquez, M., Sánchez, M., & Seefoó, L. (2011). Gestión y calidad del agua en la cuenca del río Duero, Michoacán. In U. Oswald (Ed.), *Retos de la Investigación del Agua en México* (pp. 521–530). Cuernavaca: UNAM, Red Temática.

Ramírez-Herrejón, J. P., Mercado-Silva, N., Medina-Nava, M., & Domínguez-Domínguez, O. (2012). Validación de dos índices biológicos de integridad (IBI) en la subcuenca del río Angulo en el centro de México. *Revista Biología Tropical, 60*(4), 1669–1685.

Rogers, P. (2002). *Water governance in Latin America and the Caribbean*. Washington, DC: Inter-American Development Bank.
Savenije, H. H. G., & Van der Zaag, P. (2008). Integrated water resources management: Concepts and issues. *Physics and Chemistry of the Earth, Parts A/B/C, 33*(5), 290–297. Retrieved from https://doi.org/10.1016/j.pce.2008.02.003
Schneider, F., Bonriposi, M., Graefe, O., Herweg, K., Homewood, C., Huss, M., . . ., & Weingartner, R. (2015). Assessing the sustainability of water governance systems: The sustainability wheel. *Journal of Environmental Planning and Management, 58*(9), 1577–1600. Retrieved from https://doi.org/10.1080/09640568.2014.938804
Stockholm International Water Institute (SIWI). (2015). Improved Water Governance. Retrieved from https://www.siwi.org/priority-area/water-governance/#:~:text='Good%20governance'%20is%20thus%20crucial,to%20water%20and%20sanitation%20services.&text=It%20determines%20the%20equity%20and,socio%2Deconomic%20activities%20and%20ecosystems
Tortajada, C. (2010). Water governance: Some critical issues. *International Journal of Water Resources Development, 26*(2), 297–307. Retrieved from https://doi.org/10.1080/07900621003683298
Van Buuren, A., van Meerkerk, I., & Tortajada, C. (2019). Understanding emergent participation practices in water governance. *International Journal of Water Resources Development, 35*(3), 367–382. Retrieved from https://doi.org/10.1080/07900627.2019.1585764
Velázquez, M. M. A. (2005). *Diagnóstico para el Saneamiento del Río Duero. Informe Técnico*. Michoacán: SAGARPA–COEFREM.
Velázquez, M. A., Pimentel, J. L., & Ortega, M. (2011). Estudio de la distribución de boro en fuentes de agua de la cuenca del río Duero, México, utilizando análisis estadístico multivariado. *Revista Internacional de Contaminación Ambiental, 27*(1), 19–30.
Warner, J. (2005). Multi-stakeholder platforms: Integrating society in water resource management? *Ambiente & Sociedade, 8*(2), 4–28. https://doi.org/10.1590/S1414-753X2005000200001.
Water Framework Directive. (WFD). (2000). Directive 2000/60/EC of the European Parliament, and of the Council of 23 October 2000 Establishing a Framework for Community Action in the Field of Water Policy. OJEC, L 327/1. Retrieved from https://eur-lex.europa.eu/eli/dir/2000/60/oj
Water Governance Facility. (2016). *Water governance*. Retrieved from http://www.watergovernance.org/water-governance/
WHO. (1986). *Ottawa charter for health promotion*. Geneva: World Health Organization.
World Bank. (2018). *Watershed: A new era of water governance in China*. Washington, DC: World Bank.
Zavala-López, L. J. (2011). *Inventario y Caracterización de los Manantiales en la Cuenca del Río Duero*. Michoacán: Tesis de Licenciatura, Universidad Autónoma Chapingo (UACh), Chapingo.

Water-Smart Cities

The Role of Demand Management

Robert C. Brears
Our Future Water, Christchurch, New Zealand

Synonyms

Water conservation; Water efficiency; Climate change; Demand management

Introduction

Traditionally, urban water managers, faced with increasing demand for water alongside varying levels of supplies, have relied on large-scale, supply-side infrastructural projects, such as dams and reservoirs, to meet increased demands for water. This supply-side approach, however, is under increasing pressure from rapid urbanization and climate change impacting the availability of good quality water of sufficient quantities.

Currently, 55% of the world's population lives in cities, and by 2050 this will rise to 68%. Urbanization, combined with the overall growth of the world's population, could add another 2.5 billion people to urban areas by 2050 (UN Department of Economic and Social Affairs 2018). Cities impact the hydrological cycle by extracting significant amounts of water from surface and groundwater sources as well as degrade surface and groundwater resources. Urbanization is increasing demand for water resources with large cities estimated to obtain around 78% of their water from surface sources, some of which are far away (McDonald et al. 2014). However, water for cities will become scarcer as the century progresses: reduced freshwater availability and competition from other uses including energy and agriculture could reduce water availability in cities by as much as two thirds by 2050, compared to 2015 levels (World Bank 2016).

Climate change is projected to lead to more frequent and intense droughts as well as increase the frequency and magnitude of flooding, diminishing the availability of water for cities as well as degrading water quality (Brears 2020). Climate change is projected to impact on urban water resources in the following ways. Storm events (flooding) wash pollutants from urban areas into surface water bodies as well as contaminate groundwater supplies. Air temperatures in urban areas compared to surrounding rural areas are 3.5 to 4 °C higher. Rising temperatures will increase the demand for water for cooling and drinking. During heat waves and droughts, the demand for water increases (drinking water and water for cooling). Also, higher temperatures mean algal levels increase, which degrades the quality of water resources and leads to increased treatment costs and energy use in the treatment process. Globally, cities are mainly concentrated in coastal zones, leaving a large portion of the world's urban population exposed to the risk of sea-level rise and intensifying storm surges, which contaminate groundwater supplies and damage water infrastructure (Brears 2014).

To become water-smart, urban water managers are transitioning towards demand management, which promotes the better use of existing water supplies before plans are made to increase supply further. Specifically, demand management promotes water conservation, during times of both normal conditions and uncertainty, through changes in practices, cultures, and attitudes of society towards water resources (Brears 2016).

Demand Management

Urban water managers can implement a range of demand management tools to promote water conservation and water efficiency, including water pricing, smart metering, and the use of subsidies and rebates to modify water users' behavior in a predictable, cost-effective way. Other demand management tools available to promote water conservation and water efficiency include regulations, water efficiency labelling, as well as education and public awareness.

Water Pricing

The main pricing structures used by water utilities to promote the conservation of water resources are volumetric rates, increasing block tariffs, and two-part tariff systems. A volumetric rate is a charge based on the volume used at a constant rate. Therefore, the amount users pay for water is strictly based on the amount of water consumed. An increasing block tariff contains different prices for two or more pre-specified quantities (blocks) of water, with the price increasing with each successive block. The pricing of water can also be done using a two-part tariff system: a fixed and a variable component. In the fixed component, water users pay one amount independently of consumption, and this covers infrastructural and administrative costs of supplying water. Meanwhile, the variable amount is based on the quantity of water consumed and covers the costs of providing water as well as encouraging conservation (Brears 2016).

Case 1: Scottish Water's Water and Wastewater Charges

Scottish Water aims to make charges reflect, as closely as possible, the cost of the service provided. Metered customers are billed directly for all water and wastewater services by Scottish Water. Water bills sent to customers generally consist of the following elements:

- *Annual fixed charges*: The annual fixed charge is based on the size of the meter serving of the house or property (Table 1).
- *Volumetric charges*: The water and wastewater charges are each based on the size of the meter and the volume of water recorded on the meter serving the property (Table 2).
- *Property and road drainage charges*: These charges are based on the council tax band for your home/property (Table 3) (Scottish Water 2020).

Water-Smart Cities, Table 1 Scottish Water's annual fixed charges

Fixed charges	Water (£/meter)	Waste water (£/meter)
Up to 20 mm	£155.64	£159.84
25/30 mm	£462.00	£475.00
40 mm	£1309.00	£1344.00
50 mm	£2910.00	£2989.00

Water-Smart Cities, Table 2 Scottish Water's volumetric water charges

Volumetric water charges	**£/m^3**
For the first 25 m^3 – up to 20 mm meters only	£2.4438
For volumes after the first 25 m^3 – up to 20 mm only	£0.8855
Volume charge for larger meters	£0.8855
Volumetric waste water charges	**£/m^3**
For the first 23.7 m^3 – up to 20 mm meters only	£3.1598
For volumes after the first 25 m^3 – up to 20 mm only	£1.4942
Volume charge for large meters	£1.4942

Water-Smart Cities, Table 3 Scottish Water's property and road drainage charges

Council tax band	Property drainage	Council tax band	Property drainage
A	£32.10	E	£58.85
B	£37.45	F	£69.55
C	£42.80	G	£80.25
D	£48.15	H	£96.30

Smart Meters

A crucial part of the ICT network is smart meters, which enable water utilities to conduct regular meter reads of customers throughout the day, provide customers with real-time water consumption data, as well as quickly detect water losses in the system. Currently, many cities are using automatic meter readers (AMR), which are "one-way" automated meter readers that send water usage data back to the utility. In contrast, advanced metering infrastructure (AMI), or "smart meters," is a two-way solution in which a network is created between the meters and the utility's Information system. In this network, smart meters not only allow for remote meter reading but also allow high-resolution consumption data to be sent to the customer. This data can be used to raise awareness of water consumption and allow customers to develop their own strategies to reduce water usage.

From the water utility's side, smart meters provide multiple benefits including leak detection, energy reduction, demand forecasting, enhanced awareness campaigns, promotion of efficient appliances, and performance indicators. From the customer's side, smart meters can provide information on when/where is water being used, comparisons of their water use against other customers, and quick leak detection. Smart apps can also be developed for customers, so they can, for example:

- Compare their water usage with neighbors in the same street or suburb.
- Compare water consumption with standard profiles (customers with the same socio-demographic factors).
- Compare their water consumption with the most efficient users in the city.
- Forecast their next water bill (Brears 2018).

Case 2: Yorkshire Water's Smart Meter Trial

Yorkshire Water is conducting its second smart meter trial in Sheffield as part of its leakage reduction strategy of ensuring the security of supply and saving up to 250,000 l of water per burst. Around 1000 m will be deployed on commercial properties and will remotely send 15-min water flow information back to the water

(continued)

W

Case 2: Yorkshire Water's Smart Meter Trial (continued)

company every 4 h. Using this data, Yorkshire Water will be able to detect increases in demand due to leakage and respond quickly. The meters will be deployed across 30 network zones – district metered areas – which monitor water flow into and out of the areas of the network. Sheffield was chosen for the trial for geographical reasons with the hilly terrain testing the capability of the wireless network for transferring flow data as well as its elevation causing leakage from winter freeze-thaw events (Yorkshire Water 2020).

Subsidies or Rebates

Subsidies or rebates are used to modify water users' behavior in a predictable, cost-effective way, that is, reduce wastage and lower water consumption. Subsidies (incentives) are commonly used to encourage the uptake of water-saving devices or water-efficient appliances as positive incentives are found to be more effective than disincentives in promoting water conservation. Also, incentives have been found to reduce the gap between the time the incentive is presented and behavioral change as compared to disincentives. Water utilities commonly offer rebates to customers who purchase water-efficient toilets, taps, and showerheads to accelerate the replacement of old water-using fixtures (Brears 2016).

Case 3: San Francisco Public Utilities Commission's Water-Efficient Equipment Rebates

The San Francisco Public Utilities Commission (SFPUC) offers water-efficient equipment rebates to nonresidential customers who can reduce their use of potable water through upgrade or replacement of existing onsite water-using equipment. Eligible projects must achieve a water savings of 200 ccf (149,000 gallons) or more per year to qualify, where1 ccf = 748 gallons. SFPUC provides qualifying projects rebate funding of $1.00 per ccf over a 10-year lifespan up to 50% of the project's equipment costs. A single customer may apply for more than one project. SFPUC provides two types of rebates:

- *Fixed water-saving retrofit projects*: These projects consist of standardized equipment that provide predictable water savings, as listed in Table 4.
- *Custom retrofit projects*: These consist of unique or site-specific equipment retrofits that result in project-specific water savings. They include any water-saving equipment retrofit not listed in the fixed water-saving equipment projects list and result in a minimum of 200 ccf of annual potable water savings. Custom retrofit projects are approved on a case-by-case

(continued)

Water-Smart Cities, Table 4 Fixed water-saving equipment project list

Equipment type	Estimated annual water savings	Rebate amount
Cooling tower pH controllers	1734 ccf	50% of the equipment cost, up to $8000
Medical equipment steam sterilizers	670 ccf	50% of the equipment cost, up to $2500
Water-efficient ice machines	364 ccf	50% of the equipment cost, up to $1800
Commercial laundry retrofits	344 ccf	50% of the equipment cost, up to $1700
Dry vacuum pumps	279 ccf	50% of the equipment cost, up to $1000
Connectionless food steamers	109 ccf per compartment	50% of the equipment cost, up to $545 per compartment

Case 3: San Francisco Public Utilities Commission's Water-Efficient Equipment Rebates (continued)
basis and require metering 60 days pre-equipment installation and 60-day post-equipment installation to verify equipment retrofit savings (San Francisco Public Utilities Commission 2020).

Regulations

Water management generally comes in the form of temporary and permanent regulations. Temporary regulations restrict certain types of water use during specified times and/or restrict the level of water use to a specified amount. These programs are usually enacted during times of severe water shortages and cease once the shortage has passed. Restrictions are placed on nonessential water uses. Meanwhile, permanent regulations include amendments to building codes or ordinances requiring the installation of water-saving devices and maximum water use standards for plumbing fixtures (Brears 2016).

Case 4: City of Guelph's Outdoor Water Use Restrictions
The City of Guelph, Canada, is the largest community in Canada that relies almost entirely on groundwater for its water supply. To conserve water resources, Guelph has established outdoor water restriction levels. Allowed at all levels are the watering of trees and vegetable gardens any day, any time, the use of sprinklers and splash pads, and the filling of swimming pools, hot tubs, garden ponds, or fountains (must recirculate water). Meanwhile, Table 5 details the outdoor water restriction levels that can be applied (City of Guelph 2020).

Water-Smart Cities, Table 5 City of Guelph's outdoor water use restrictions

Restrictions	Level 0 blue	Level 1 yellow	Level 2 red
Lawn watering	Permitted from 7 to 9 a.m. and 7 to 9 p.m. If your street address is an even number, you can water on even-numbered calendar days. Odd street numbers can water on odd calendar days	Permitted from 7 to 9 a.m. and 7 to 9 p.m. If your street address is an even number, you can water on even-numbered calendar days. Odd street numbers can water on odd calendar days	Not permitted
Car washing	Permitted at home with a shut-off nozzle	Permitted at home with a shut-off nozzle	Not permitted
Watering flowers and plants	Permitted any time	Permitted any time	Permitted 7–9 a.m./p.m. odd/even dates by address
Enforcement	$250 ticket or court summons for repeat violations	$350 ticket or court summons	$550 ticket or court summons
Water supply	No serious storage, rainfall, or stream flow issues	Less rain, low flow rates, or low water storage levels: Less than 80% of historical average precipitation over 1 and/or 3 months, or 2 weeks without rain Eramosa River flow (less than 70% of minimum low flow) Water storage level (less than 75% of average)	Serious rainfall, streamflow, or storage concerns: Less than 60% of historical average precipitation over 1 and/or 3 months, or 3 weeks without rain Eramosa River flow (less than 50% of minimum low flow) Water storage level (less than 65% of average)

Water Efficiency Labelling

The labelling of household appliances according to water efficiency is essential in reducing household water consumption by eliminating unsustainable products from the market; however, this is provided the labelling scheme is clear and understandable and identifies both private and public benefits of conserving water. Nonetheless, people are more likely to respond to eco-labels if the environmental benefits match closely personal benefits such as reduced water bills (Brears 2020).

Case 5: Mandatory Water Efficiency Labelling in Singapore

Initiated in 2009, Singapore's Mandatory Water Efficiency Labelling (Mandatory WELS) is a grading system with a 0/1/2/3-tick rating denoting the water efficiency level of a product. Mandatory WELS covers taps and mixers, dual-flush low-capacity flushing cisterns, urinal flush valves and waterless urinals (2/3-tick rating), and clothes washing machines intended for household use (2/3/4-tick rating). Since 2018, the Mandatory WELS extends to covering dishwashers for household use. It is mandatory for suppliers and retailers to obtain the relevant water efficiency labels for their products before advertising and displaying them for sale in Singapore. Also, all products must publicly showcase their water efficiency label at all times. The Mandatory WELS can be read as such:

1. Products with the most ticks are recommended.
2. The label shows a product's water consumption, wash program, type, brand, and model.
3. Each label carries a registration number for validation (Public Utilities Board 2020).

Education and Awareness

Education of the public is crucial in generating an understanding of water scarcity and creating acceptance of the need to implement water conservation programs. Water utilities can promote water conservation in schools to increase young people's knowledge of the water cycle and encourage the sustainable use of scarce water resources. Meanwhile, water utilities can raise public awareness of the need to conserve water resources through many formats, including:

- *Public information*: Printed literature distributed or available for the general public, public service announcements and advertisements on billboards and public transportation, television commercials, newspaper articles and advertisements, and Internet and social media campaigns.
- *Public events*: Customers can receive information on water conservation tips and receive water-saving devices at conservation workshops, expos, fairs, etc. as people frequently make poor choices concerning environmentally friendly products or services due to misinformation or lack of information.
- *Information in water bills*: Water bills should be understandable, enabling customers to identify volume of usage, rates, and charges quickly. Water bills should be informative enabling customers to compare their current bill with previous bills. Finally, water bills should contain water conservation tips to help customers make informed decisions on future water use (Brears 2020).

Case 6: Hamburg Wasser's Aqua Agents

Hamburg Wasser's Aqua Agents program for elementary students in the third and fourth grade provides a series of digital missions for students to understand the water cycle. For example, where does the water go when it rains? How does the water get into the tap, and how is the dirty water cleaned later? With simple household objects, the Aqua Agents conduct exciting experiments to learn about the precious resource of water in a playful way. The Aqua Agents project

(continued)

Case 6: Hamburg Wasser's Aqua Agents (continued)

is an educational offer from the Michael Otto Environmental Foundation, which is intended to draw the attention of children to the importance of sustainable water use. In cooperation with Hamburg Wasser, 20 school classes each year can take part in free adventure days in the waterworks or the sewage treatment plant (Hamburg Wasser 2020).

Conclusion

Traditionally, urban water managers have relied on large-scale, supply-side infrastructural projects to meet increased demands for water. This supply-side approach is under pressure from rapid urbanization and climate change impacting the availability of good quality water of sufficient quantities. To become water-smart, urban water managers are transitioning towards demand management, which promotes the better use of existing water supplies before plans are made to increase supply further. Urban water managers can implement a range of demand management tools to promote water conservation and water efficiency. Water pricing structures can be devised to promote water conservation. Scottish Water's metered customers are billed directly for all water and wastewater services with customers receiving a bill with a fixed and volumetric component for both water and wastewater. Smart meters provide customers with real-time water consumption data, as well as enable water utilities to detect water losses in the system quickly, for example, Yorkshire Water conducting its second smart meter trial as part of its leakage reduction strategy. Subsidies are used to encourage the uptake of water-saving devices or water-efficient appliances. Rebates are used to accelerate the replacement of old water-using fixtures, for example, SFPUC offers water-efficient equipment rebates to nonresidential customers who can reduce their use of potable water through upgrade or replacement of existing onsite water-using equipment. Regulations can be either temporary, restricting certain types of water use during certain times, or permanent, including amendments to building codes or ordinances requiring the installation of water-saving devices. The City of Guelph relies almost wholly on groundwater for its water supply and therefore, to conserve this resource, has established outdoor water restriction levels with enforcement action taken for noncompliance. Water efficiency labelling schemes help reduce household water consumption by eliminating unsustainable products from the market, with Singapore's mandatory water efficiency labelling scheme covering a range of domestic water-using devices as well as appliances used for household use. Finally, education and public awareness are crucial in generating an understanding of water scarcity and encouraging society to use scarce water resources wisely, an example of which is Hamburg Wasser's program for young children that teaches them about the water cycle.

Cross-References

▶ Circular Economy and the Water-food Nexus
▶ Circular Water Economy
▶ Hidden Potential of Wastewater
▶ Water Security and the Green Economy

References

Brears, R. C. (2014). Urban water security in Asia-Pacific: Promoting demand management strategies. Retrieved from https://refubium.fu-berlin.de/bitstream/handle/fub188/18349/pp414-urban-water-security-asiapacific.pdf?sequence=1&isAllowed=y

Brears, R. C. (2016). *Urban water security*. Chichester/Hoboken: Wiley.

Brears, R. C. (2018). Smart water, Smart metering. Retrieved from https://medium.com/mark-and-focus/smart-water-smart-metering-4eff05fca4e9

Brears, R. C. (2020). *Developing the circular water economy*. Cham: Palgrave Macmillan.

City of Guelph. (2020). Outdoor water use and restrictions in Guelph. Retrieved from https://guelph.ca/living/

W

house-and-home/lawn-and-garden/outdoor-water-use-and-restrictions-in-guelph/#about-guelph%e2%80%99s-outside-water-use-program-and-bylaw
Hamburg Wasser. (2020). Digital orders for small Aqua Agents. Retrieved from https://www.hamburgwasser.de/privatkunden/themen/digitales-ferienangebot-der-aqua-agenten/
McDonald, R. I., Weber, K., Padowski, J., Flörke, M., Schneider, C., Green, P. A., ... Montgomery, M. (2014). Water on an urban planet: Urbanization and the reach of urban water infrastructure. *Global Environmental Change, 27*, 96–105. https://doi.org/10.1016/j.gloenvcha.2014.04.022.
Public Utilities Board, S. (2020). About water efficiency labelling scheme. Retrieved from https://www.pub.gov.sg/wels/about
San Francisco Public Utilities Commission. (2020). Water efficient equipment rebates. Retrieved from https://sfwater.org/index.aspx?page=512
Scottish Water. (2020). Metered charges 2020 – 2021. Retrieved from https://www.scottishwater.co.uk/your-home/your-charges/your-charges-2020-2021/metered-charges-2020-2021
UN Department of Economic and Social Affairs, P. D. (2018). 2018 revision of world urbanization prospects. Retrieved from https://www.un.org/development/desa/publications/2018-revision-of-world-urbanization-prospects.html
World Bank. (2016). High and dry: Climate change, water, and the economy. Retrieved from https://openknowledge.worldbank.org/handle/10986/23665?utm_source=Global+Waters+%2B+Water+Currents&utm_campaign=9905bbdc1e-Water_Currents_Water+Utiliti_12_dec_2018&utm_medium=email&utm_term=0_fae9f9ae2b-9905bbdc1e-25803553
Yorkshire Water. (2020). Yorkshire water looking to save millions of litres of water with new smart meters. Retrieved from https://www.yorkshirewater.com/news-media/2020/yorkshire-water-arqiva-smart-meter-trial/

Weak Governance

► Multi-stakeholder Partnerships to Support Climate Migrants in Fragile Cities

Wealthy Natural Resources

► Water Policy in the State of Tabasco

Weather pattern turnover and surface water resources in Sri Lanka

► Climate Change and Surface Water Resources in Sri Lanka

Weathering Change

Lessons for Adaptive Cities in Latin America and the Caribbean

Tatiana Gallego Lizon
Washington, DC, USA

Introduction

Cities are the region's heartbeat. Home to 81% of the population, cities represent 65% of Latin America and the Caribbean's (LACs') gross domestic product (GDP) growth today, and this share is set to keep growing. This urbanization of the region has proved to be prosperous, and even though many LAC cities are less productive in average than cities in European or North American regions, still a 1% increase in urbanization is associated with a 3.8% increase in GDP per capita. This high concentration of people and assets makes them both vulnerable to climate-related shocks and large emitters.

Climate change effects are already taking place in the region. A 2 °C scenario could cost the region between 10.8% and 13% of the GDP by mid-century, an impact about four times that resulting from COVID-19 (Swiss Re 2021). LAC cities simply cannot afford to be unprepared: to avoid future losses, economic recovery in the region needs to specifically target the adaptation infrastructure gap, as the return for every dollar

The opinions expressed in this publication are those of the author and do not necessarily reflect the views of the Inter-American Development Bank, its Board of Directors, or the countries they represent.

spent in adaptation is expected to be 3.5 times in avoided costs.

While cities have a key role to play in the quest to carbon neutrality, and the localization of nationally determined contributions will be key to achieving it, this entry stresses the urgency of pursuing adaptation strategies in cities in the region, highlighting opportunities for leveraging the natural assets in what is probably the most biodiverse-rich continent in the planet. For this, section "Climate Change: Exposure and Vulnerability in Latin America and the Caribbean" summarizes the expected effects of climate change in LAC; section "Cost and Benefits of Adaptation" makes an argument for investing in adaptation on the basis of economic savings; sections "Cities and the Twenty-First Century: Changing Directions" and "BiodiverCities: Biodiversity, Adaptation, and Resilience" describe emerging paradigms in urban development, as we engage in biodiversity positive outcomes; section "Adaptive Territorial and City Planning" reviews opportunities to better integrate territorial and urban planning as a path to defining adaptation strategies; while section "Green Infrastructure" highlights how green infrastructure can contribute practically to the implementation of adaptation plans.

Climate Change: Exposure and Vulnerability in Latin America and the Caribbean

Climate patterns in LAC, much as in the rest of the world, are changing. Temperatures are increasing and some areas are experiencing variations in the frequency and intensity of precipitation. Some regions are seeing higher rainfall, while others are experiencing shortages. The sixth Intergovernmental Panel on Climate Change (IPCC) reports (IPCC 2021, 2022) highlight that (i) mean temperatures have very likely increased in LAC and will continue to do so at rates greater than the global average; (ii) mean precipitation is projected to change, increasing in northwest and southeast South America, while decreasing elsewhere; and (iii) relative sea level rise is extremely likely to continue in the oceans around Central and South America, contributing to increased coastal flooding in low-lying areas and shoreline retreat along most sandy coasts. Central America is expected to see greater aridity, droughts, and fires. The Andean region will also see glacier volume loss and permafrost thawing, decreases in snow and ice, and increases in pluvial or river flooding.

Globally, cities occupy an estimated 3% of the land, but they consume two-thirds of the world's energy resources and emit three-fourths of all carbon dioxide. A high concentration of people and assets also makes cities particularly vulnerable to climate-related shocks. Raising sea levels and the increasing frequency of storm surges are expected to disproportionally affect coastal cities, which include 60 of the 70 most populated cities in the region, but also those in the Caribbean basin countries, whose proportion of the population in low elevation zones (below 10 m above sea level) can be as high as 80% of the countries' total (e.g., Bahamas or Suriname). Similarly, riparian cities are expected to be affected by increasing rainfall and flashflood events, increasing the probability of flooding and the need for expanding riverbeds.

Urban centers also generally experience substantially higher temperatures than their surrounding areas (e.g., 8–9C differences have been reported in Santiago de Chile, closer to 20C under extreme conditions, in Buenos Aires). City geometry (height and orientation might cause reduced ventilation), city design and materials (with heat retaining properties, particularly concrete), and anthropogenic activities further trap heat in what is commonly known as urban heat island effect. Warmer temperatures have been associated to (i) changes in meteorological conditions that favor pollution and lower air quality levels, (ii) lacking vegetation to provide shade, and (iii) fewer water bodies, capable of absorbing heat and radiation at microscales.

Many of the components making our cities (roads, buildings, much of the infrastructure, and some public space) are also characterized by design and material choices with impervious properties, which have sealed surfaces, hampering water infiltration, water table replenishment and, in extreme cases of overextraction (such as in Mexico City [Poreh et al. 2021] or Jakarta), can lead to land subsidence, but also increases water runoff and flood risks.

Cost and Benefits of Adaptation

The effects of climate change translate into increasingly severe weather or health problems, which in turn can result in significant economic costs. Reports (Statista 2022) suggest that in 2021, the estimated economic loss of natural disasters worldwide was USD 343 billion, while the estimated insured loss amounted to USD 130 billion. Although these figures vary (e.g., German reinsurer Munich Re placed the value of overall losses at USD280 billion, of which USD120 billion insured), they are consistent and confirm an upward trend associated with economic costs of natural disasters. Recent reports (OCHA 2020) list LAC as the second most disaster-prone region in the world, with 1,205 disasters recorded between 2000 and 2019, of which floods are the most common calamity in the region (548 events, affecting 41 million people and damages estimated at USD26 billion), followed by storms (330, including 23 Category 5 hurricanes).

Business disruptions are expected to become increasingly important. CDP's Global Supply Chain Report (2021) reported a total financial impact of USD1.26 trillion by suppliers due to environmental risks (climate change, deforestation, and water insecurity) over a 5-year period, while Morgan Stanley estimated potential cost of damages in USD54 trillion by 2010 due to global warming.

Global low-emission resilient finance infrastructure investment needs have been estimated (CCFLA 2015) at around USD93 trillion, with 9–27% of the total associated with climate enhancement. Out of this total figure, 70% of the value is expected to be invested in urban areas. At a cost of around USD4.5 trillion to USD5.4 trillion per year, cities will need to dedicate an important part of their budgets to climate proofing, much above the USD 384billion estimated to have been globally invested annually in urban climate finance, on average, in 2017–2018 (CCFLA 2021). Moreover, finance flows to support decarbonization substantially outweigh adaptation finance in cities, as does the use of gray over green infrastructure and nature-based solutions.

Nature generates not only environmental value but also social and economic value. According to the World Economic Forum (WEF 2022), nature positive investments could, by 2030, generate more than 59 million jobs in cities, nearly half of which would be dedicated to restoring and protecting ecosystems, and generate over USD1.5 trillion in annual business value, and yet cities currently spend less than 0.3% of their infrastructure investment on nature-based solutions (equivalent to USD28 billion). Beyond cities, another recent report (UNEP 2021) finds that approximately USD 133 billion flows annually into nature-based solutions, 86% from the public sector and 14% from private sources.

Cities and the Twenty-First Century: Changing Directions

Cities are engines of economic and social development. Globally they house half of the people and contribute up to 80% of the GDP, but in regions like LAC, where they are home to 81% of population, this value can be much higher. This clustering of people and assets makes them especially vulnerable to climate change, and yet, it is also this concentration that can turn them into powerhouses of climate action. For example, as 76% of energy demand comes from cities, mostly associated with transport and buildings, achieving carbon neutrality is largely dependent on the action taken at urban level. On the other hand, inefficiencies are riddling urban growth, while developmental solutions – predominantly built on anthropocentric approaches – are proving asymmetric, particularly in cities.

Ineffective urban expansion, or sprawl, has characterized much of growth in cities in the region. The recent World Cities Report (UN-Habitat 2020) estimates the expansion of urban areas in relation to urban population growth increased by a ratio of 1.5 in 25 years (1990–2015). In LAC, urban densities have traditionally been above average; however, this has also changed. For example, Mexico doubled its country population between 1980 and 2010, period in which cities grew a sevenfold, often resulting from suburbanization policies and low-density peripheral growth (IADB 2018).

Beyond land use inefficiencies, cities consume three-fourths of the world's natural resources. As a result, barely one-fourth of the land surface on Earth is now qualified as natural environment, not urbanized or under intense agricultural extension. According to the World Wildlife Fund (WWF 2020), 58% of the earth's land surface is under intense human pressure. Driven by increased consumption demand and prevailing production (food, energy) practices, ecosystem degradation and habitat contraction are some of the main causes for a reduction in fauna of 68% since 1970, with LAC emerging as the region with largest losses. A cutback in the number of species can significantly affect natural lifecycles or give rise to the uncontrolled development of invasive species. Yet, habitat fragmentation and contraction are forcing many species to move into other habitats – including those occupied by humans, increasing environmental health risks. Nature itself provides with unique regulating instruments that can help counter many of the effects of climate change – including land erosion or changing water cycles – and minimize temperature rise – through carbon capture.

Thus, to truly enable sustainability, we need to urgently review urban planning, design and management practices, and project a more balanced socio-ecological path to development and growth in and in connection with cities. The sections that follow offer thoughts on how cities may choose ways to adapt to the imminent effects of climate change, while simultaneously searching for ecologically positive impacts.

BiodiverCities: Biodiversity, Adaptation, and Resilience

With over half of the world's biological diversity, 7 of the 25 global biodiversity hotspots, and 11 of the 14 terrestrial biomes, LAC is one of most biodiverse regions. Many cities in LAC sit and/or interact (receiving waters, emissions, etc.) with immediate or nearby ecosystems. Healthy natural ecosystems provide essential regulating, provisioning, and cultural services that can help cities positively adapt to adverse climate change scenarios. As cities expand into peripheral areas, measures are needed to conserve critical habitats, control and manage land conversion, and integrate these landscapes into the urban fabric. Conversely, cities need to restore degraded consolidated urban areas through local government and community action into biodiversity micro-hotspots. Many cities also encompass protected areas (e.g., Nairobi's National Park or Washington DC's Rock Creek Park) within or in its limits that can provide important contributions to biodiversity.

An emblematic reference case, Singapore has actively developed for over a decade the concept of city-garden, which simultaneously provides for essential water and air quality management, and biodiversity stewardship. Singapore has built action from self-assessment, using the City Biodiversity Index (Chan and others, 2021). The weighed indicators focus on three areas of work: (i) native biodiversity in the city, (ii) ecosystem services provided by biodiversity, and (iii) governance and management of biodiversity in the city. As a result, Singapore has not only promoted conservation of key habitats, habitat enhancement, and restoration and recovery of species (especially endemic) but has also fostered applied research in conservation biology and planning, and community oversight. In LAC, this index has also been adopted in Curitiba and Antigua.

In addition to the adoption of the City Biodiversity Index, other tools and city-biodiversity-based strategies have been developed for LAC (IADB 2020). Sao Paulo, located in Brazil's biodiversity-rich Atlantic Forest region, has conducted detailed diagnostics, identified priority areas for conservation and recovery, and drafted a plan with supporting actions to (i) protect the local rain forest, (ii) provide wildlife care, (iii) pay for provision of environmental services, (iv) establish systems of protected and green areas, and wildlife management, and (v) monitor vegetation cover and the built environment. Bogotá is in the process of building a 57-kilometer ecological corridor in the city's forested Eastern Hills – an ecological hotspot, serving as both the lungs of the city and a carbon sink for emissions, which had been degraded over time – prioritizing activities that focus on participatory reforestation with native species, conservation by local communities, habitat connectivity, and equitable access to all.

To ensure replicability and scalability, the support of national and supranational institutions will be essential. In 2019, the European Union, launched its "BiodiverCities" program, a pilot aimed at improving civil society participation in planning and decision-making in and around 13 European cities with a shared vision of the green city of tomorrow that accounts for nature and biodiversity. And in 2020, Colombia's government, through its Ministry of Environment and Sustainable Development, launched the national initiative on "Biodiverciudades," whose objective is to preserve and integrate local biodiversity and its benefits in sustainable urban development, connecting natural capital with growth strategies, as a catalyst for improving quality of life and enabling resilience and low carbon development. Achieving biodiversity positive cities requires a new planning and development paradigm.

Adaptive Territorial and City Planning

The climate commitments taken by individual nations under the Paris Agreement were primarily defined at national level, through a sectoral lens. The need for a territorial approach to "localize" the nationally determined contributions has become evident as implementation advances, but adaptation planning is lagging, and similarly to mitigation, needs to take into consideration the vast geographical, economic, and social differences that define areas within a country.

But even where adaptation plans are underway, challenges are emerging. Cities like Tokyo or Mexico City have developed specific climate adaptation plans, but administrative boundaries can limit the effectivity of adaptation actions, as these tend to focus on maximizing local benefits and might not take into consideration ecological connectivity, effects up- and downstream, and interdependences, thus highlighting the importance of regional dimensions and institutions.

Integrated Resource Management

The allocation and management of natural resources, such as water or land, require the coordination and commitment of multiple actors across jurisdictions, particularly in the face of scarcity or imbalance.

The variability of rainfall patterns and raising temperatures, associated with climate change, can contribute to longer and more intense periods of drought – as projected by the IPCC for Southwestern South America and Northeastern South America – and be aggravated by unsustainable levels of abstraction. Events such as the prolonged drought that affected Sao Paulo in 2015 saw the Cantareira Dam's capacity levels fall below 15%, endangering water supply to the city. Similarly, the intensity and frequency of extreme precipitation and seasonal variations, projected to increase in Northern and Northwestern South America, is likely to increase the risk of flooding and landslides, as seen in several provinces of Peru in 2021 and Quito in 2022. Integrated cross-sectoral territorial approaches focusing on watershed management are necessary to curb climate change impacts at basin scale. Simulation tools for the management and planning of water resources, like HydroBID (a public platform for modeling hydrology and climate change in LAC), can help project conditions under changing scenarios and evaluate quantity, quality, and infrastructure needs for adaptation. Complementing technological and modeling solutions, preventive risk management also needs to encourage different institutional approaches – from community-based self-help solutions, such as those selected in Lao PDR, to river management planning, practiced for the Rhine Delta in The Netherlands (Cap-Net 2020) – to guide water resource planning.

Coastal areas also need to be managed holistically, attending to both land-based and marine life aspects. For LAC, managing climate risks and accelerating adaptation to sea level rise, erosion, and coastal flooding, is key to the future wealth of the region. While some hard protection measures linked to gray infrastructure such as dikes and seawalls – widely implemented in northwestern Europe or East Asia, but also in LAC's coastal cities in Guyana, Surinam, or Panama – may at times be inevitable, they can exacerbate erosion and affect adjacent coastline and seabeds, reducing the natural ecosystem's response ability. Soft protection measures (such as dune rehabilitation or sand nourishment) can provide integrated (yet temporary) responses against sea level rise but may incur negative physical and biological changes on beach ecosystem services. Alternative responses, nested on coastal planning, focus on modifying or retrofitting (through elevation, floating, etc.) existing infrastructure and built environment, may constitute an intermediate pathway for many densely populated areas. Ecosystem-based adaptation – based on mangrove, salt marsh, coral reef or oyster bed restoration – might provide the most integrated approach to coastal management, reducing risks while rebuilding natural systems that can attenuate wave power or erosion. One such example is being piloted in the city of Barranquilla, Colombia, through public-private alliances and community participation, aims to recover and preserve local fauna and flora in surrounding wetlands and marshes and incorporate them to the development of the city. Most extreme measures (under evaluation in Jakarta or in California) consider the development and implementation of "managed retreat" strategies (Bongarts et al. 2021), as a piece of anticipatory planning that accepts the need for relocation at different scales to protect people and assets from coastal hazards.

Rural-Urban Connectivity

Ecological and economic connectivity also offer opportunities to plan and manage adaptation at a regional scale. Rural and urban areas are linked economically, socially, and environmentally. Thus establishing such connections can generate mutual benefits, including those associated with adaptation, through ecosystem services and food systems (the majority of which flow from rural to urban areas).

Urban ecosystem-based adaptation can encompass services provided both (i) within cities and (ii) from rural to urban areas. According to the Millennium Ecosystem Assessment, services may be associated with provisioning (water, food, fuel), regulating (climate, flood, heat), supporting (production, nutrients), or cultural services (recreational, aesthetic, etc.), but in most cases, specific actions can generate multiple benefits. For example, in the Aburrá Valley, surrounding Medellin, or in the State of Bahia, near its capital, Salvador, the preservation of upland forests was promoted to prevent soil degradation and landslides; while, in São Paulo, restoring 4,000 hectares in a nearby watershed forest was associated with improvements in water quality, including cuts in sediment pollution by a third, and an increase water supply during the dry season, all of which generated a 28% return on investment for the local water company (WRI 2018). Similarly, seasonal floodplains which may help mitigate, regulate, or prevent floods, avoiding costly property and infrastructure loss in nearby urban areas, can also provide critical habitat for wildlife. Although still at early stages of application, ecosystem services valuation methodologies and payment for ecosystem services are expected to grow in upcoming years (Geneletti et al. 2020).

Urban areas are also dependent on sustainable food systems produced in rural areas, which call for collaboration on the efficient use of shared land and water resources, and environmental practices (including soil enrichment and pest control management methods which balance intensive production goals with ecosystemic capacity). Similarly, market access and the existence of appropriate urban infrastructure can cut down food loss and waste, as can logistics and better integration among markets and local suppliers.

Urban Planning and Design

Urban planning decisions generally have long-term impacts on the morphology and function of a city, as decisions taken today will take physical form and be present for periods of 20 years or longer (Skelhorn et al. 2020). Urban and environmental planning frameworks and processes are progressively incorporating adaptation into design principles and enforcement instruments such as land use regulation, building codes, zoning schemes, or spatial plans, but these need to be more aggressively implemented in areas of urban expansion and urban regeneration, all while pursuing incentives for retrofit in consolidated areas. Two aspects are particularly urgent when it comes to city planning: temperature and socio-spatial differences within cities.

Tackling urban waste heat will need a multipronged approach. Firstly, by redefining the urban form (street layout, building orientation and heights) with respect to its immediate natural settings (topography of adjacent hills, coastal winds etc.), cities can allow for natural ventilation. Secondly, while seeking more efficient land use, cities also need to define per capita green space standards connecting dwellings to nearby habitat. Increasing the vegetative cover in a city can simultaneously lower outdoor temperatures, water runoff, and water and air pollution. Thirdly, principles of urban design (captured in building codes and other regulatory instruments) need to incentivize the transition towards low heat capacity construction materials, and the use of reflective coatings and solar capture (photovoltaic) surfaces. City design also needs to promote the capture/exchange of urban waste heat from buildings, transportation, and industry, in a way that accelerates circularity.

Socio-spatial segregation characterizes many cities in LAC. The effects of climate change (e.g., flash floods, heat waves) disproportionally affect informal settlements, where around one-fifth of LAC's urban population resides, calling for both citywide and localized urban adaptation interventions (IDB 2021). Vulnerabilities derive, among others, from both geographical conditions (in low lying lands, riverbeds, or deforested hills) and qualitative factors (design and construction of housing, but also habitat). In Chile, for example, where recent studies (building on household cadasters, climate risk mapping, and vulnerability profiling) indicate that a third of informal settlements are in areas subject to climate risks, digital instruments (e.g., geo-referenced data, artificial intelligence, and climate risk modeling) are being developed to anticipate, measure, and prioritize interventions in cities. In the city of Buenos Aires, targeted interventions in the emblematic neighborhood of Barrio 31 have enabled its socioeconomic integration to the rest of the city. As part of a comprehensive program which included both habitat and housing adaptation improvements, residents have seen comprehensive improvements to drainage and road networks, the development and greening of parks (which not only provided recreational grounds for families and community but also assisted water infiltration), and a thermal efficiency and safety housing program assisting technically self-managed retrofit and housing upgrades.

Green Infrastructure

Countries in LAC face a substantial gap in infrastructure. Nature-based solutions can build climate resilience, lessening the impact of infrastructure on their surrounding ecosystems, while leveraging social and economic benefits. In the urban realm, the region is advancing nature-based solutions in three distinct areas: the public space, buildings and housing, and the water sector.

Preservation and Greening of Public Space

Tree canopies bring a myriad of benefits to urban areas, not only supporting mitigation goals through the capture of carbon dioxide, but also adaptation objectives, as they help regulate water flows through their root system. May be most important of all, tree canopies can help reduce local temperatures, blocking shortwave radiation and increasing water evaporation. Although a few cities in LAC, like Curitiba, are blessed with over 50 square meters of green space per capita, many cities, like Buenos Aires or Mexico City, have less than 5 square meters in average per person. In addition to high urban densities and compactness (Buenos Aires has a population density of 13,680 persons per square kilometer, while in Quito and Sao Paulo, these amount to 7,200 and 7,913 persons per square kilometer), low canopy coverage is generally observed (the Green View Indexes for Buenos Aires, Quito, and Sao Paulo were measured as 14.5%, 10.8%, and 11.7%). (Source: Treepedia (http://senseable.mit.edu/treepedia/cities/buenos%

20aires), accessed on 30 January 2022.) Cities, like Barcelona or Paris, are in the process of converting large sections of their road networks, redesigning them to incorporate high volumes of vegetation, but may be more notable in the region (IDB 2021b) are the efforts of Medellin and Mexico City. Under its project "30 Green Corridors," Medellin provided ecological connectivity for 18 road axes in the city, 12 ravines, and 3 hills, which resulted in temperature reduction of 2–3C. Mexico City, under the city's "Green Plan," has pledged to plant 8 million new trees and shrubs, create 450 pollinator gardens, and boost the plant production of native species in its nurseries.

Complementary to the development of green corridors is the preservation, greening, and new development of urban parks, particularly metropolitan parks, which can be particularly effective at creating microclimate regulation, generating habitats for biodiversity, carbon sequestration, and in the case of submergible or floating parks, flood prevention. Examples in LAC include the Hill of Chapultepec in Mexico City, La Sabana in San José de Costa Rica, or Panama City's Metropolitan Natural Park. Despite the tremendous benefits brought about by these spaces, development pressures for conversion of natural land for residential purposes, maintenance needs, the prolific uptake of invasive species, or inequitable access remain a frequent challenge.

The Building Sector

In Latin America, buildings consume 21% of treated water and 42% of the electricity and produce 25% of emissions and 65% of waste (Cesano and Russell 2013). Green certification schemes across the region (based on both international standards, such as LEED or EDGE, and local climatic conditions) have advanced in the last decade, focusing primarily on efficiency goals (electricity, water and waste reduction) and emission reduction (use of renewable energy, and nontoxic materials, ethically and sustainably sourced), but the potential for greening construction can also support urban adaptation strategies.

For example, green roofs can significantly reduce both peak flow rates and total runoff volume of rainwater by storing it in plants and substrate and releasing it back to the atmosphere through evapotranspiration (Solecki and Marcotullio 2013). Similarly, the use of nature in green walls and hanging gardens can significantly assist cooling. Advocating nature use in public buildings, the Ministry of Education of the City of Buenos Aires "green roofs" pilot at "School 6: French and Beruti" delivered not only a thermal and water insulation solution, but also provided scarce green space and learning opportunities for children to explore the newly created habitat. The new green roof is estimated to absorb nearly 90% of all rainfall and to have led to energy savings of around 50% (IADB 2021b).

Blue-Green Infrastructure

Urban blue-green infrastructure can also contribute to solving changes in precipitation associated with both frequency and intensity, providing alternatives to standard approaches to storm water, but also strategic water resource management. As a result, urban blue-green infrastructure such as coastal wetlands or submergible parks (at larger scales), and bioswales, rain gardens, or planter boxes (at community and household level) can help reduce pluvial floods, mitigate the effects of droughts, and improve water quality. Blue-green infrastructure builds on physical and biological processes associated with topography and natural gravity flows (including aquifer recharge), filtration, and purification (Meney and Pantelic 2021).

To respond to increasing water shortages, poor drainage conditions and land subsidence (from groundwater overextraction), Iztapalapa, one of Mexico City's largest municipalities, is implementing a decentralized agenda for water supply, which in addition to household rainwater harvesting includes the hydrological park of "La Quebradora" over a surface of 4 hectares. La Quebradora seeks to (i) harvest rainwater and recharge the aquifer and (ii) redirect rainwater flows from surrounding avenues and treat these in an activated sludge plant and wetlands in the park. Similarly, to respond to increasingly damaging flooding events, Sao Paulo is constructing of 20,000 square meters of bioswales and rain gardens along main axes of the city to increase soil permeability and groundwater recharge.

Conclusions

Confronted by the accelerating effects of the climate change, cities in the region need to speed up the development and implementation of climate action plans capable of dynamically respond to varying conditions. High concentration of people and assets makes them particularly vulnerable to climate-related shocks, putting entire countries at risk. As a biodiversity-rich region, adaptive urban planning and development can leverage its natural capital to strengthen resilience and secure a balanced socio-ecological path to development. Such approach needs to build on better integrated territorial and urban planning, and infrastructure development which progressively offsets gray with green infrastructure. In a planet where resources will soon be scarce, integrated resource management will be the sole path to ensuring water and land efficiencies. Adaptation at a regional scale can build on ecological and economic connectivity between urban and rural areas, offering essential ecosystem services, while cities specifically tackle heat island effects and water challenges deriving from surface impermeabilities. Green infrastructure provides us with solutions based on nature which can address both temperature and water cycle problems at small and large scales, allowing for actors at all levels to participate. Issues remain in this road map, including metrics, monitoring, and financing.

Cross-References

- ▶ Adapting to a Changing Climate Through Nature-Based Solutions
- ▶ Climate Resilience in Informal Settlements: The Role of Natural Infrastructure

References

Bongarts Lebbe, T., Rey-Valette, H., Chaumillon, E., Camus Guigone, A. R., Cazenave, A., Claudet, J., Rocle, N., Meur-Férec, C., Viard, F., Mercier, D., Dupuy, C., Ménard, F., Rossel, B. A., Mullineaux, L., Sicre, M. A., Zivian, A., Gaill, F., & Euzen, A. (2021). Designing coastal adaptation strategies to tackle sea level rise. *Frontiers in Marine Science, 8*, 740602. https://doi.org/10.3389/fmars.2021.740602.

Cap-Net. (2020). Climate change adaptation and integrated water resources management training manual. https://cap-net.org/iwrm4climateresilience/

Cesano, D., & Russell, J. (2013). ELLA Policy Brief: Green Building in Latin America. ELLA, Practical Action Consulting, Lima

Chan, L., Hillel, O., Werner, P., Holman, N., Coetzee, I., Galt, R., & Elmqvist, T. (2021) *Handbook on the Singapore Index on Cities' Biodiversity (also known as the City Biodiversity Index)*. Montreal/Singapore: Secretariat of the Convention on Biological Diversity/National Parks Board, Singapore.

Cities Climate Finance Leadership Alliance (CCFLA). (2015). State of city climate finance 2015. New York. https://citiesclimatefinance.org/publications/the-state-of-city-climate-finance-2015-2/

Cities Climate Finance Leadership Alliance (CCFLA). (2021). The state of cities climate finance – Part1. Washington. https://citiesclimatefinance.org/publications/2021-state-of-cities-climate-finance/

Geneletti, D., Cortinovis, C., Zardo, L., & Adem, E. B. (2020). Reviewing ecosystem services in urban climate adaptation plans. In *Planning for ecosystem services in cities* (SpringerBriefs in Environmental Science). Cham: Springer. https://doi.org/10.1007/978-3-030-20024-4_3.

Inter-American Development Bank. (2018). *Program to strengthen urban development and land-use management reform: Loan proposal*. Washington, DC.

Inter-American Development Bank. (2020). *Ciudades biodiversas y resilientes en América Latina y el Caribe*. Intelligent Social Investment; Scott-Brown, M.; Rodríguez, E.L. Washington, DC: IADB. https://doi.org/10.18235/0002618

Inter-American Development Bank. (2021). In: Vera, F., & Sordi, J. (Eds.). *Ecological design: Strategies for the vulnerable city: Adapting precarious areas in Latin America and the Caribbean to Climate Change*. https://doi.org/10.18235/0003271.

Inter-American Development Bank. (2021b). Infraestructura Verde Urbana I: Retos, oportunidades y manual de buenas practices. Washington, DC. https://doi.org/10.18235/0003748

IPCC. (2021). Climate change 2021: The physical science basis. Contribution of working group I to the sixth assessment report of the Intergovernmental Panel on Climate Change [V. Masson-Delmotte, P. Zhai, A. Pirani, S. L. Connors, C. Péan, S. Berger, N. Caud, Y. Chen, L. Goldfarb, M. I. Gomis, M. Huang, K. Leitzell, E. Lonnoy, J. B. R. Matthews, T. K. Maycock, T. Waterfield, O. Yelekçi, R. Yu & B. Zhou (Eds.)]. Cambridge University Press. In Press. https://www.ipcc.ch/report/ar6/wg1/#FullReport

IPCC. (2022). Climate change 2022: Impacts, adaptation, and vulnerability. Contribution of Working Group II to the Sixth Assessment Report of the Intergovernmental Panel on Climate Change [H.-O. Pörtner, D. C. Roberts, M. Tignor, E. S. Poloczanska, K. Mintenbeck, A. Alegría, M. Craig, S. Langsdorf, S. Löschke,

V. Möller, A. Okem, & B. Rama (Eds.)]. Cambridge University Press. In Press. https://www.ipcc.ch/report/sixth-assessment-report-working-group-ii/
Meney, K. A., & Pantelic, L. (2021). Decentralized water and wastewater systems for resilient societies: A shift towards a green infrastructure-based alternate economy. In R. C. Brears (Ed.), *The Palgrave handbook of climate resilient societies*. Cham: Palgrave Macmillan. https://doi.org/10.1007/978-3-030-42462-6_32.
OCHA. (2020). Natural disasters in Latin America and the Caribbean, 2000–2019. Panama. https://www.humanitarianresponse.info/sites/www.humanitarianresponse.info/files/documents/files/20191203-ocha-desastres_naturales.pdf
Poreh, D., Pirasteh, S., & Cabral-Cano, E. (2021). Assessing subsidence of Mexico City from InSAR and LandSat ETM+ with CGPS and SVM. *Geoenvironmental Disasters, 8*, 7. https://doi.org/10.1186/s40677-021-00179-x.
Skelhorn, C., Ferwati, S., Shandas, V., & Makido, Y. (2020). Urban form and variation in temperatures. In *Urban adaptation to climate change* (Springer briefs in environmental science). Cham: Springer. https://doi.org/10.1007/978-3-030-26586-1_5.
Solecki, W., & Marcotullio, P. (2013). Climate change and urban biodiversity vulnerability. In *Cities and biodiversity outlook: Urbanization, biodiversity and ecosystem services: Challenges and opportunities* (pp. 85–504). https://doi.org/10.1007/978-94-007-7088-1_25.
Statista. (2022). https://www.statista.com/statistics/612561/natural-disaster-losses-cost-worldwide-by-type-of-loss/. Accessed on 20 Feb 22.
Swiss Re Institute. (2021). The economics of climate change: No action not an option. Swiss Re Institute. Switzerland https://www.swissre.com/dam/jcr:b257cfe9-68e8-4116-b232-a87949982f7c/nr20210421-ecc-publication-en.pdf
United Nations Environment Program. (2021). State of finance for nature. https://www.unep.org/resources/state-finance-nature
World Economic Forum. (2022). BiodiverCities by 2030: Transforming cities' relationship with nature. https://www3.weforum.org/docs/WEF_BiodiverCities_by_2030_2022.pdf
World Resources Institute. (2018). Natural infrastructure in São Paulo's Water System. https://www.wri.org/research/natural-infrastructure-sao-paulos-water-system
WWF. (2020). In: R. E. A. Almond, M. Grooten, & T. Petersen (Eds.), *Living planet report 2020 – Bending the curve of biodiversity – Bending the curve of biodiversity loss*. Gland: WWF. https://livingplanet.panda.org/en-us/

Welfare Services

► Challenges of Delivering Regional and Remote Human Services and Supports

Well-Being

► Emerging Concepts Exploring the Role of Nature for Health and Well-Being
► Health and the Role of Nature in Enhancing Mental Health

Wetlands

► The Source Waters of Tanga

Why Large Cities Won't Survive the Twenty-First Century

William E. Rees
School of Community and Regional Planning, University of British Columbia, Vancouver, BC, Canada

Introduction: Setting the Scene

The population of the world's largest city, metropolitan Tokyo was near its all-time peak at 37.3 million people in 2021. This is roughly equivalent to thc population of Canada, geographically the second largest country on Earth. This is truly an extraordinary population for a single city, but analysts expect it will soon be surpassed by several mega-cities in the developing world; by 2050, the largest megalopolis will be Mumbai at 42.4 million; by 2100 Lagos will take the population prize at 88.3 million – and 9 other developing country mega-cities are projected to top 50 million people by century's end (Hoornweg and Pope 2016).

Urbanization generally is accelerating. Half the world's people have been living in cities and towns since 2007; by 2018, over 4.1 billion people or 55% of the total population were urbanized and the UN projects this proportion to rise to 68% by 2050. Rural-urban migration combined with general population growth could add another

~2.5 billion people to the world's cities by 2050 (an increase of 61%). Ninety percent of this increase will take place in developing Asian and African cities (UN 2018) as the world population soars to 9.7 billion in 2050 on its way to perhaps 11 billion in 2100 (UN 2019).

Or maybe not.

Population projections are typically made by manipulating purely demographic data – present population, age distribution, age-specific fertility and mortality rates, net migration – *as if there were no "environment."* This approach can succeed only when nothing happens to affect key demographic variables. For example, during the past half century of continuous growth and increasing economic prosperity in much of the world, UN population forecasts have been fairly accurate and improving with increasing data reliability.

But coming decades will look nothing like the recent past. The human enterprise is now in a state of advanced ecological overshoot (Rees 2020). Eco-overshoot (hereafter, "EO") exists when the consumption of bioresources and the production of wastes exceed the regenerative and assimilative capacities respectively, of supportive ecosystems. When in EO, we can achieve further growth only by depleting essential natural capital and overtaxing the life-support functions of the ecosphere including the climate system, i.e., by destroying the biophysical basis of our own existence.

And that is precisely what we are doing. The global footprint network monitors the annual occurrence of "Earth Overshoot Day," the date in the year when humanity's demand for ecological resources exceeds nature's budget (supply) for that year (GFN 2021a). Each year, Overshoot Day occurs a little earlier as demand increases and eco-production declines with accelerating ecosystems degradation – in 2021, it fell on July 29. Remember, the difference between demand and supply can be made up *only* by depleting remaining natural capital stocks – fish stocks, forests, soil organic matter and nutrients, ground water, etc. – that took thousands of years to accumulate in nature, and by over-filling nature's waste sinks. (Even climate change is a waste-management issue – CO_2 is the greatest waste by weight of industrial economies.)

EO means that humanity is running an ecological deficit, an energy and material deficit far more important that the fiscal deficits that preoccupy politicians. Yet most politicians, like their constituents, have never heard of overshoot. Instead, popular interest swings with media attention among its various *symptoms* – climate change, plunging biodiversity, plastic pollution of the oceans, landscape and soil degradation, tropical deforestation, the SARS-CoV-2 pandemic – without connecting the dots. Even when commentators talk about the need for "multi-solving" they usually mean coordinated efforts to fight the diverse effects (intense heat-waves, extended drought, increasingly violent storms, unprecedented wildfires, rising sea-levels, accelerating forest die-back, etc.) of just one human-induced phenomenon, climate change.

Acknowledging EO is important because it is the *ultimate* meta-problem, the overlying proximate cause of all the other problems associated with humanity's ecological predicament. Biodiversity loss, air/land/water pollution, climate change, impending resource scarcity – pick your issue – all result from EO, too many people consuming excess energy/material and over-polluting their supportive ecosystems. We cannot "solve" any major symptom of EO, including climate change, in isolation from any other. Conversely, tackling EO directly would address all its symptoms simultaneously.

But here's the rub; by definition, the only way to "tackle" EO is by significant absolute reductions in energy/material consumption and human numbers.

Which brings us back to population projections, urbanization and the future of cities.

Most national governments and international organizations see the future as a technologically more advanced and socially more inclusive extension of the recent past. They acknowledge environmental problems, of course, but again the major focus respecting cities is on climate change. For example, the United Nation's Sustainable Development Goal #11, aims to "Make cities and human settlements inclusive, safe, resilient and sustainable" by, among other things, substantially "increasing the number of cities and human settlements adopting and implementing integrated policies and plans towards inclusion, resource efficiency, mitigation and adaptation to climate change [and], resilience to disasters" (UN n.d.). Similarly, the C40 cities network, an association of 100 of the world's major cities is "working to deliver the urgent action needed right now to confront the climate crisis and create a future where everyone, everywhere can thrive." C40 city mayors "are committed to using a science-based and people-focused approach to help the world limit global heating to 1.5 C° and build healthy, equitable, and resilient communities" (C40 Cities 2021).

In short, neither the UN, the C40 network nor similar organizations acknowledge overshoot – overconsumption and overpopulation – and the attendant possibilities of significant resources shortages and widespread systems collapse. Modern techno-industrial (MTI) society radiates confidence in the ability of human ingenuity and technology to power through the ecological crisis however it might currently be defined. Climate change an existential threat? Not to worry – humanity's future hangs on belief that the transition to wind, solar and hydrogen electricity will enable a smooth transition to a zero-carbon economy, paving the way to a bountifully safe and sustainable future.

Socially Constructed Shared Illusions

What ordinary citizens don't appreciate is that human beings characteristically "socially construct" their own realities, or rather their *perceptions* of reality. Virtually everything we think we "know" – political ideologies, religious beliefs, disciplinary paradigms, cultural norms and even scientific truths – are products of the human mind, conceived in words and massaged into received wisdom through social discourse among participants in the exercise, whether they be street-wise villagers, priests or scientists. The two most important things to keep in mind about any social construct is that: (1) it may or may not contain an accurate "map" of any part of biophysical reality it purports to represent and; (2) accurate or not, people live out of their constructed perceptions *as if they were real*.

Why raise this issue here? Because it flags the possibility that mainstream understanding of economic/population growth and its implications for urbanization/cities is fatally flawed. Consider that the most prominent development-related social construct in play in the world today is growth-oriented neoliberal economics. Virtually all senior and local governments are in thrall of the double-barreled myth of infinite economic growth enabled by continuous technological development. This socially constructed narrative has birthed the conviction that climate change (remember, overshoot isn't part of the discussion) can be solved through human ingenuity – new technologies – and is therefore no impediment to maintaining the *status quo*.

The problem is that neoliberal economic models contain no useful information whatever about the complex structure and behavioral dynamics of the ecosystems and even social systems with which the economy interacts in the real world. This is a crucial failing. The first law of cybernetics (systems regulation) states that if we hope to maintain control, the internal variety or complexity of our management system must at least match the variety/complexity of the system being managed. Given our vacuous economic models, it should be no surprise that the ecosphere is increasingly in turmoil and that the evening news frequently features stories of mounting socio-political unrest. Neoliberal theory is already floundering; mainstream political and economic constructs are wholly incompetent to guide the future development of the human enterprise.

In this light, the purpose of this paper is threefold: first I describe just two dimensions of urban biophysical reality that are missing from mainstream thinking but essential to understanding prospects for cities in the twenty-first century. Second, I explore what including these elements implies for further urbanization and the future of existing cities in the context of climate change and the renewable energy (non)transition and; third, I outline an alternative settlement pattern that conforms to biophysical reality. This analysis and perceptual "reset" are sufficient both to explain why large cities and mega-cities may not survive the twenty-first century and to stimulate the quest for alternatives consistent with one-planet living.

The Secret Life of Cities

Cities are many things; some conceive of cities as concentrations of people crowded into areas dominated by the "built-environment" (urban areas now comprise the principal habitat of *H. sapiens*). Others see cities as engines of economic growth; centers of commerce; seats of government; the loci of great universities and well-springs of arts and culture. Cities are, indeed, all of these things simultaneously – but something is missing. Not many urban dwellers or even urban scholars think first of cities as biophysical entities subject to the same natural laws and constraints as all other complex living systems.

This is a serious failing. The capacity of cities to function in their myriad ways depends utterly on the integrity of their biophysical relationships. Indeed, the state and fates of modern cities may well be determined as much by the operation of the law of conservation of matter and the laws of thermodynamics as by economic, social or political conditions (see Box 1).

Box 1 Fundamental Physical Laws and Concepts

- The law of conservation of mass dictates *that matter is neither created nor destroyed*, e.g., the mass (weight) of reacting substances at the beginning of a chemical reaction is precisely equal to the mass of new compounds at the end of the reaction. What goes in all comes out, albeit in altered form.
- The first law of thermodynamics is a restatement of the law of conservation of energy: energy is neither created nor destroyed. Thus, energy may be transformed from an "available" form (e.g., chemical energy in gasoline) into useful work (e.g., kinetic energy of the moving vehicle) plus a useless degraded form (e.g., heat radiating from the engine or dissipated in the exhaust), but the total quantity of energy remains constant.
- The second law of thermodynamics dictates that, with any change in an isolated system, the entropy (disorder, randomness) of the system always increases. In fact, *any real process increases the entropy of the universe*. The second law also implies that no energy/material transformation (e.g., chemical energy into useful work) can be 100% efficient. Some energy is always lost as low grade heat; matter "rusts," crumbles and disintegrates. Entropy increases.

Cities and the Basic Laws of Nature

Have you ever heard your home town referred to as a "dissipative structure"? Probably not – this is a term one would apply only if describing a city from the perspective of the second law of thermodynamics.

The second law states that any spontaneous change in the state of an isolated system (one unable to exchange energy or matter with its environment) increases the "entropy" of that system. This means that the system becomes more disordered – it loses structure and potential, concentrations disperse, available energy is degraded and dissipated as low-grade heat. In an isolated system, therefore, each successive event brings it closer to a state of maximum local entropy. This is a state of thermodynamic equilibrium in which *no further change is possible*. In the extreme case, all form and function is lost; matter would be randomly dispersed; no point in the "system" would be distinguishable from any other.

Of course, not all dynamic systems are isolated; exchanges with their surroundings are possible. Consider first the human body. Like other complex systems, our physical selves are subject to entropic decay; we are continuously wearing out under the dictates of the second law. However, living systems are characterized by metabolic processes that seem to defy the entropy law. The human body is an "autopoietic" or self-producing system of sub-systems exquisitely structured to perform numerous biological and social functions simultaneously without running down. This is because healthy bodies are *open* systems, able to produce and maintain themselves in a *far-from-equilibrium* steady-state by importing highly-structured energy-rich material (we call it "food") from their environments. We use a portion of this imported energy/matter for repair and growth (self-production) but "dissipate" most of it back into the environment as bodily waste and low-grade heat energy.

Now consider "the city." In many respects, every city is a complex, highly-structured multi-functional super-organism. Indeed, various urban sub-systems – water and sewage, solid waste disposal, electricity and communications, streets and roads, inter-city transportation, etc. – are directly analogous to functionally similar human organ systems. Cities are also subject to second law erosion but, like our bodies, are open systems maintained by a compound metabolic process that is even more complex than our own. In addition to the collective metabolic demands of its human inhabitants, cities have an *industrial* metabolism. To grow or simply maintain themselves in a smoothly functioning operational state far-from-equilibrium, cities must import large quantities of low-entropy energy/matter. This includes all the food and fiber to satisfy the biological needs of their inhabitants, plus all the fossil and electrical energy, and all the raw materials, manufactured goods and equipment required to construct and maintain the built environment and supportive infrastructure. To this we must add the energy and material resources embodied in the appliances, tools, electronic gadgets, toys, and other artifacts of modern consumer society. One result of producing themselves is that cities necessarily generate prodigious quantities of material waste and low-grade heat energy all of which is "exported" – dissipated – into their ambient environments. It should be no surprise that cities account for 60–70% of global material consumption, expected to grow from 40 billion tonnes in 2010 to 90 billion tonnes in 2050 (IRP 2018). They also log up to 80% of global energy consumption and 70% of greenhouse gas (GHG) emissions. Cities are indeed archetypal "dissipative structures".

Of course, cities are not all thermodynamically equal. High-quality energy and resources (negentropy) are expensive. Wealthy consumer cities can therefore afford the maintenance costs of keeping material entropy at bay locally while exporting their second law dregs to rural areas and the global commons. By contrast, the crumbling buildings, run-down infrastructure and general squalor of impoverished cities bear witness to the relentless corroding effect of the second law when it cannot constantly be papered over. Such evidence of the entropy law at work accurately reflects the egregious wealth/income gap between rich and poor cities (and, for that matter, between rich and poor neighborhoods within cities). Keep in mind, however, that for all their glittery splendor, high-income cities actually impose a much greater per capita entropic burden on the ecosphere than do low-income cities. Extremes of consumption imply extremes of entropic dissipation.

In summary, cities thrive and grow by extracting negentropy (high-grade energy and resources) from their environments and exporting entropy (low grade heat and useless waste) back into those same environments, i.e., the ecosphere. However, because no energy transformation is close to 100% efficient, the price of any increment of urban growth, or even simple maintenance, is a much greater increase in the entropic disordering of the ecosphere. Indeed, the law of conservation of mass and the first law of thermodynamics (conservation of energy) ensure that 100% of the energy/material inputs imported to maintain or expand the city eventually joins the entropic waste stream. Simply put, *a little order over here (the city) means much greater disorder over there (elsewhere in the ecosphere)*.

All of which suggests another analogy. We can define a parasite as any organism that gains its vitality at the expense of the vitality of its host. It should be obvious from the foregoing description of urban metabolism that cities, as presently conceived, exist in a potentially parasitic relationship with the rest of the ecosphere (Rees 2021).

Cities per se Are Inherently Unsustainable

Cities are generally perceived as productive wonders – economic powerhouses, founts of cultural creation, etc. – and in a strictly anthropocentric sense they are. However, we have shown that, from a biophysical perspective, all economic processes, cultural events, and other urban activities are mainly consumptive. This begs the question: if cities are mostly about consumption, who or what is doing the production? The short answer: the ecosystems that constitute the rest of the ecosphere.

Cities are themselves sometimes called ecosystems but are anything but. Complete ecosystems include: (1) producer organisms (mostly green plants); (2) macro-consumers (mostly multicellular animals, including humans); and (3) micro-consumers (bacteria and fungi). Green plants self-produce using extra-planetary solar energy to assemble biomass from carbon dioxide, water, and trace nutrients. Macro-consumers self-produce by consuming plants or other macro-consumers and micro-consumers self-produce by decomposing the bodies of both plants and animals and returning nutrients to the soil so the cycle can repeat continuously. In short, complete ecosystems are exquisitely complex quasi-independent systems that can maintain themselves and thrive "far-from-equilibrium" indefinitely by continuously transforming and recycling matter and assimilating and dissipating solar energy. The resultant waste heat is radiated off the planet which increases the entropy of the universe.

By contrast, cities are dominated by a single macro-consumer species, their human inhabitants. There are insufficient producers and decomposers to sustain the system, particularly if we factor in cities' industrial metabolism. Cities and their human populations can maintain themselves "far-from-equilibrium" *only* by assimilating and dissipating biomass (food and fiber), and fossil fuel and material resources imported from their extra-urban "environments." They "radiate" the resultant waste (pollution) back into those environments thereby increasing the entropy of the ecosphere. Clearly, cities per se are inherently unsustainable. Enclosed in an impermeable glass bell-jar (i.e., unable to exchange with its environment), any city would simultaneously starve and suffocate in its own entropic excreta in an inexorable descent toward thermodynamic equilibrium.

The Real Urban Human Ecosystem

We often hear that cities occupy only 2–3% of Earth's land surface. In ecological terms, such estimates are meaningless – they consider only the mostly lifeless built-up lands physically occupied by human settlements. A more relevant approach might be to ask "how large an area of productive ecosystems is necessary to support a given urban population at a specified material standard of living?" The answer to this question, combined with the built-up land, would constitute the area of that city population's de facto, functionally complete, ecosystem.

We can estimate this area using ecological footprint analysis (EFA). The ecological footprint (EF) of any study population – an individual to an entire nation – is defined as:

> the area of productive land and aquatic ecosystems required, on a continuous basis by that population, to produce the bio-resources that the population consumes and to assimilate its carbon wastes.

Numerous studies have shown that average human per capita eco-footprints range from over 10 global average hectares (gha) in rich countries to as little as half a gha in the poorest nations. Western European countries typically have EFs of 4–5 gha/capita; Canadians and Americans "enjoy" EF of ~8.1 gha; Japan's average per capita EF is 4.6 gha (see GFN 2021b).

We can readily show from these data that city eco-footprints are enormous. Indeed, the EFs of rich-country cities may be a 100 or more – even a 1000 – times the size of their geographic areas (Warren-Rhodes and Koenig 2001). Consider metropolitan Tokyo: with population of 37.3 million people (~30% of Japan's domestic population) and a per capita EF of 4.6 gha, the total eco-footprint of metro-Tokyo is

~171,580,000 gha. This is nominally 127 times larger than the city's metropolitan area of 1,350,000 ha. More telling, Japan's domestic bio-capacity is only ~75,600,000 gha, so *the EF of Tokyo alone is 2.3 times greater than that nation's entire productive area.* The residents of Tokyo are running a huge eco-deficit; they live, in large part, off the productive and assimilative capacities of ecosystems in distant countries and the global commons. So large is Tokyo's deficit that Japan could not support the population of just its national capital at current material standards if the country were cut off from the rest of the world.

It is worth noting in passing that humanity as a whole is running a massive ecological deficit but, unlike Toyko or Japan, cannot cover it through trade or natural material flows. Instead, we can (only temporarily) suspend the human enterprise far-from-equilibrium by depleting vital ecosystems, destroying non-human species and undermining global life support systems. Such are the consequences of overshoot. Also, while EF results may appear frighteningly extreme, they are typically underestimates for several technical reasons. For example, if data sets conflict, analysts typically use the more conservative numbers; not all waste streams are included in the EF; and the method estimates ecosystem areas-in-use but not whether such land/water use is sustainable (i.e., EFA does not account for ecosystem degradation or over-harvesting).

Bottom line: Cities may be where most people "live," but built-up areas constitute less than 1% of the functional human urban ecosystem. Each city is a compact node of intense consumption and energy/matter dissipation (i.e., pollution); the vastly larger and arguably more important productive and assimilative component of the urban ecosystem is the city's rural hinterland, the globally scattered aggregate eco-footprint of its human inhabitants. This reality remains largely out-of-mind – globalization and trade have isolated urbanites both spatially and psychologically from the ecosystems that support them. But the fact remains: no city could survive in the absence of distant supportive ecosystems. (By contrast the latter would thrive splendidly in the absence of cities.)

The Existential Threat to Cities

Modern cities and mega-cities exist because they can. No one sat down to plan a metro New York of 18 million people, a Shanghai of 28 million or a Tokyo of 37 million. These and like cities are truly "emergent phenomena" of the modern techno-industrial age, manifestations of humanity's explosive growth in the past two centuries. More than 300,000 years passed before the human population reached its first billion in the early nineteenth century. Then, in just 200 years, less than 1/1500th as much time, humanity expanded sevenfold and will top eight billion by 2023. This brief period of continuous growth and urbanization, a state that economists, politicians, and many ordinary citizens take to be the norm, is actually *the single most anomalous period in human history*!

It is a little known but crucial fact that this explosive anomaly was made possible by fossil fuels. Fossil fuels (FF) are a prodigious source of "negentropy," of potential and possibilities. Other factors, particularly, improving public health and longevity, contributed. However, it is fossil energy that made the modern world possible.

Modern cities and mega-cities in particular are the most spectacular products of FF. As already argued, abundant cheap energy was – and still is – necessary not only to "build out" our cities, but also to supply them with all the food, consumer goods, and low entropy materials needed to defend urban infrastructure against the corrosive workings of the second law. Consider that fossil fuels and petroleum-derived inputs (e.g., pesticides, fertilizers) inject ten times as much energy into agriculture and food processing as does photosynthesis and are thus crucial to food production. Stand on the sidewalk near a major construction site on a busy road in any city anywhere – the near-deafening din of excavators, cement mixers, dump trucks, and power tools of all kinds blending with the road noise generated by passing delivery and passenger vehicles is the sound of raw energy – mostly FF – at work (and being permanently dissipated).

Because cities are consumptive black holes, everything essential to cities' growth and maintenance – including all that that raw energy – has to be brought in from cities' global EF hinterlands. Cities are therefore dependent for survival on the global and national marine, air and highway transportation networks that represent almost 20% of final energy demand, the bulk of which is provided by fossil fuels (see Friedemann 2016). Passenger cars are the largest energy hogs using 59% of transportation energy. Road freight accounts for another 27%, much of it to service cities. In the USA, for example, more than 80% of towns and cities are provisioned *only* by trucks; heavy duty diesel-powered Class 8 trucks haul 70% of the nation's freight. Air, rail, and marine transportation also contribute significantly, accounting for 7%, 2%, and 2% of transportation energy respectively.

The Climate Change – Energy Conundrum

Cities' profound dependence on FFs raises several issues bearing on the future of urbanization and urban life. First, the fossil-fueled expansion of the human enterprise has taken us well into potentially fatal ecological overshoot (EO). Without abundant cheap energy, the overexploitation of both ecosystems and non-renewable resources (including FFs themselves) would not have been possible. Second, FFs are a major source of carbon dioxide (CO_2) emissions. CO_2 is an unavoidable entropic product of fossil fuel combustion and the principal driver of the most obvious symptom of overshoot, anthropogenic climate change.

Atmospheric CO_2 and other GHG concentrations are increasing. The current trajectory implies 3–4 C° mean global warming in this century, far above the existing 1 C°+ warming that is already causing unprecedented climate havoc around the world. In recognition that even 3 C° warming spells climate disaster, parties to the United Nations Framework Convention on Climate Change committed in the 2015 COP21 Paris Agreement to hold the increase in global average temperature to "well below 2 C° above pre-industrial levels and pursue efforts to limit the temperature increase to 1.5 C° above pre-industrial levels" (IPCC 2018).

To meet the 1.5 C° challenge, carbon emissions (basically fossil fuel use) would have to be reduced by ~50% by 2030 on the way to full decarbonization by 2050 (Rockström et al. 2017; IPCC 2018). Some authorities argue that complete decarbonization must be achieved by 2030 (Spratt et al. 2020). (Meanwhile, the voluntary emissions reduction commitments – nationally determined contributions – made in Paris constitute only a third of the reductions needed to limit warming to even 2 C°.)

The Paris targets obviously pose an unprecedented challenge to a world primarily powered by fossil fuels. So-called modern renewable energy (RE), mostly wind turbines and solar PV, has made significant inroads displacing FF (mainly coal) in electricity production. However, in 2020, a year in which FF use and emissions actually declined by over 6% due to the CoViD-19 pandemic, FF still provided 83% of primary energy while wind and solar (where most investment in renewables is going) the equivalent of only 4.4%. FF even accounted for 61% of electricity generation while wind and solar provided only 9.1%, or less than 2% of total final energy consumption (data from BP 2021).

Any political leader who moved aggressively to cut FF use by the minimal 50% in this decade without viable substitutes and a comprehensive socioeconomic restructuring plan would be courting economic and political disaster. Most countries would suffer the pain of strict rationing of energy to essential uses, serious energy shortages and shrinking economies. With reduced services and goods production and the collapse of tourism, we would see declining incomes, rampant unemployment and rising inequality. Reduced agricultural output, combined with broken international supply lines and failing intercity transportation, would lead to local famines and global food shortages. The expected 60%+ expansion of cities (by 2050) could not occur; it would likely even be impossible to maintain large cities and mega-cities. Whither their existing populations? Civil disorder and geopolitical tension would rise perhaps to the breaking point. All would be complicated by continuing climate change – even if atmospheric GHG concentrations stabilize, there is already an additional 0.5 C° warming "in the pipe" due to the thermal inertia of the oceans.

All of which explains why global MTI society has taken an alternative course. Most senior governments, urban administrations, international organizations, many academic analysts and even environmental organizations have bought into a new mythic construct, the so-called green renewable energy (RE) transition as reflected in such concepts as the Green New Deal, the circular economy and green growth. Numerous promotional brochures and formal studies argue that falling costs and increasing efficiency make 100% renewable energy – mainly wind turbines and solar photovoltaics, but now also hydrogen – possible, by no later than 2050. "Net zero by 2050" (meaning no new manmade additions to atmospheric CO_2) is part of the new energy mantra. It seems we can eat the climate challenge and have our energy cake too – what's not to like?

Plenty, as it turns out. Most of these ebullient assessments are incomplete analyses that ignore important technical issues, material supply problems, land shortages, ecological and social impacts, and the overall scale of the exercise. Seibert and Rees (2021) review the evidence showing that modern REs are actually not renewable (merely replaceable); that their production from mine-head to installation is itself fossil-energy-intensive; that they cannot deliver the same quantity and quality of energy as FFs (in much of the world, there are inadequate energy returns on energy invested); and that their life-cycles entail egregious social injustice and significant ecological degradation. Moreover, according to Michaux (2021), there are simply not enough key material resources or time to replace the existing fossil fuel powered system with renewable technologies on the schedule set by the IPCC. Some climate scientists refer to net zero by 2050 as an illusion or dangerous trap that, at best, unnecessarily extends the FF era (Dyke et al. 2021; Spratt and Dunlop 2021).

Consider just one dimension of the scaling-up problem. In 2020, fossil fuels supplied 462.9 exajoules (Ej) of primary energy to the world. To displace 50% of this quantitatively with wind and solar electricity by 2030 implies constructing new wind and solar capacity sufficient to displace 25.7 Ej of FF energy *each year* for the next 9 years (231.5 Ej/9 years). If we (generously) assume a conversion ratio of 2.47:1 for wind and solar energy (i.e., one unit of wind/solar electricity = 2.47 units of fossil energy when converted to electricity), we would need to construct 10.4 Ej of new wind and solar generation capacity *annually* through 2030. But this increment exceeds the entire 8.8 Ej of wind and solar generation in 2020. In short, to replace just half of fossil fuel usage with electricity by 2030 would require that the world construct *every year* for almost a decade, more than the entire global multi-decade cumulative physical stock of wind turbines and solar panels (energy data from BP 2021). We must also assume that many difficult or impossible to electrify uses of FF will be electrified, that there is no need for the high-heat and other special qualities of FF in multiple end-uses, that the demand for investment capital in an already stressed market doesn't collapse the economy, and that there will be no growth in demand for energy. (In fact, analysts expect demand to grow by 40%+ by 2050.) This last is an important consideration – in recent years, growth in electricity consumption *alone* has exceeded new renewable supply, a problem that is anticipated to resume in 2021 as demand rebounds from the pandemic slump. A smooth transition away from fossil fuels is an impossibility theorem.

Just what is going on here? Mainstream governments, major corporations and their allies are behaving as exemplary discounters: they prefer to accept the uncertain risk of future catastrophic climate change which (they hope) will mainly affect other people somewhere else, rather than the immediate certain risk of economic and social chaos at home. Moreover, they are bound to seek solutions self-referentially from within the neoliberal techno-expansionist paradigm. Assertive policies that would actually work to reduce carbon emissions but create energy supply shortages or other threats to economic growth are inadmissible; population or family planning is still taboo; significant lifestyle changes are not on the table. The only politically feasible "solutions" to climate change – high-tech wind turbines, solar photovoltaics, hydrogen fuels, electric vehicles, and as yet unproved (and totally impractical) carbon-capture and storage technologies all require major capital investment. These techno-fixes serve as stimulants for economic growth, provide well-paying jobs and generate opportunities for profit. However, far from addressing our eco-predicament, these technologies would extend the *status quo*. As Spash (2016) and others have observed, acceptable "climate action" makes capitalist growth economies appear to be the solution to, rather than the cause of, our ecological crisis. The mainstream is essentially promoting *business-as-usual-by-alternative-means*; this will not "solve" climate change and *does not even acknowledge overshoot*.

It also means, of course, that fossil fuel use will continue for years and decades to come (as long recognized by the International Energy Agency, the US Energy Information Administration, Canada's Energy Regulator, and similar national entities). Some argue that even the IPCC has long been politically motivated to underestimate the scale of the problem. Thus, contrary to the Paris Accord, there is no chance the world can avoid 1.5 C° mean global warming and we will likely see a potentially disastrous 2 C° increase by 2050. Indeed, a prudent course would assume *no remaining carbon budget* even for the 2 C° target (Spratt et al. 2020).

Why so? Because even 2 C° warming may well trigger irreversible runaway "hothouse Earth" conditions (Steffen et al. 2018). In coming years, we will see an ice-free Arctic Ocean, more rapidly melting permafrost, methane releases, an increase in wildfires, and other short-term positive feedbacks that could put climate change on steroids.

Even in the best case, the world can expect more and longer heat waves and droughts, more violent tropical storms, extended wild-fire seasons, accelerating desertification, water shortages, crippled agriculture, food shortages, rising sea levels, and broken supply lines. Coastal cities will be flooded and some may eventually be abandoned. Many other cities are likely to be cut off from food-lands, energy, and other essential resources with the breakdown of national highway and marine transportation networks; this alone would make urban life untenable. According to the recent *Environmental Risk Outlook 2021* (2021), at least 414 cities with a total 1.4 billion plus inhabitants, are at high or extreme risk from a combination of pollution, dwindling water supplies, extreme heat stress, and other dimensions of climate change.

From this perspective, it appears that the sun is setting on the era of urbanization – how can anyone think seriously that we can build out cities to accommodate an *additional* 2.5 billion people? (Using what source of energy?) Devoid of cheap energy and economically drained, existing large cities and megacities will succumb to the entropy law. No longer able to remain "far from equilibrium," or even feed their human inhabitants, they can only contract or be abandoned. Many will not survive the end of the century. In the more vulnerable parts of the world, severe heat and drought will render even rural regions uninhabitable. Various studies estimate there could be mass migrations involving one to two billion eco-refugees by mid to late century (see Baker 2021). Domestic chaos and widespread geopolitical conflict is inevitable.

In *Triumph of the City*, his paean to human achievement, urban economist Edward Glaeser (2011) posits that "If the future is to be greener, then it must be more urban. For the sake of humanity and our planet, cities are – and must be – the wave of the future" (p. 222). Ironically, the ecological catastrophe that Glaeser supposed cities could head off may, instead, stop urbanization in its tracks.

Missed Opportunity: Can We Still Achieve "One-Planet Living"?

The foregoing analysis shows that that neoliberal economics, the economic hand-maiden to expansionist capitalism, is a multi-flawed construct. It not only contains no "map" of biophysical reality, it is positively hostile to the ecosphere within which the real economy is embedded. Note, too, that neoliberal models of human economic behavior are also crude caricatures of the real thing. Regrettably, the global mainstream continues to live out of this destructively distorted constructed reality. The inevitable result is rampant consumerism, eco-overshoot (EO), egregious inequality, and the pan-cultural delusion that technology can dissolve any constraints on growth.

The aggregate symptoms of EO leave little doubt that the continuity of civilization – urban or otherwise – requires that the world community socially construct a new way of being on Earth, one that transcends MTI sensibilities. Confronting EO demands a conscious transformational paradigm shift, i.e., the abandonment of the foundational beliefs, values, and assumptions of neoliberal capitalism and their replacement with a framework that better reflects biophysical reality. This implies nothing less than a personal-to-civilizational metamorphosis from contemporary growth-obsessed juvenility to adult steady-state maturity. The goal would be a world in which fewer people can enjoy emotionally satisfying, materially sufficient lives in community without wrecking the planet. This is the essence of one-planet living – the balancing of population and material well-being within the regenerative and assimilative capacities (biocapacity) of Nature. Clearly, EO can be "solved" only through significant reductions in energy and material throughput (Fig. 1).

The quest for one planet living may be far too ambitious an undertaking for an over-crowded, competitive, increasingly fractious species in overshoot. Obviously, too, chances of success would be much greater had we begun the task a half century ago. Nevertheless, if humanity does not attempt such a pre-emptive correction, an overstressed ecosphere will impose its own painful solution.

Economy as Eco-Niche

Ecologists who study the material and social relationships of non-human species say they are mapping those species' ecological niches. An organism's "niche" describes its food, habitat and related resource demands and the role that the species plays in maintaining the function and structure of its ecosystem. Well-adapted niches are non-disruptive; they define the relevant species' economic relationships within, while contributing to the structural integrity of, relevant ecosystems. The time has come to redefine the human eco-niche – the material economy – so that *H. sapiens* becomes a harmoniously integrated component of the ecosystems that support our species on the only planet we have. (This and the immediately following sub-section are revised from Rees 2021.)

This vision suggests that one possible form of a new civilization might be a network of

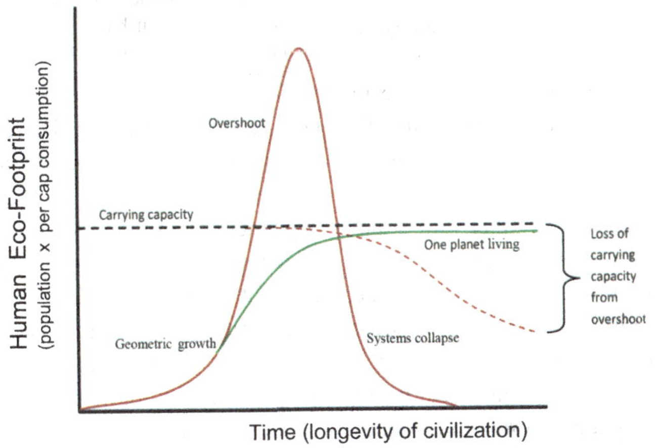

Why Large Cities Won't Survive the Twenty-First Century, Fig. 1 The human enterprise is in overshoot, nearing the peak of a one-off population outbreak (solid red line). The cost of overshoot is a reduction of long-term carrying capacity (biocapacity). A more sophisticated civilization would have self-regulated to achieve "one planet living" (solid green line). The best our MTI society can do now is a controlled contraction that comes off peak and stabilizes at or below Earth's remaining biocapacity (dotted red line)

W

cooperative regional eco-economies supporting many fewer people thriving more equitably within the regenerative capacity of local ecosystems. For so "radical" a transformation to succeed, the denizens of MTI cultures must abandon and evolve beyond the core paradigm that defines their present way of being in the world.

To begin the metamorphosis, major national governments would formally have to:

- Accept the conceptual limitations of neoliberal economic thinking. In particular, we must abandon the myth of perpetual growth and our overweening confidence in human technological prowess;
- Acknowledge the need to reduce the human ecological footprint. As long as humanity remains in EO, sustainable production/consumption means absolutely less production/consumption;
- Concede the theoretical and practical difficulties/impossibility of a green, quantitatively equivalent, energy transition;
- Recognize that equitable sustainability requires economic leveling; i.e., we need fiscal and other regulatory mechanisms to ensure income/wealth/opportunity redistribution between and within countries – greater equality is better for everyone (Wilkinson and Pickett 2009);
- Participate in a global population strategy to enable a managed non-coercive, economically secure and cooperative descent to the one to two billion that could live comfortably indefinitely within the biophysical means of nature.

We will also have to resurrect other values that have been sacrificed to market capitalism. The cult of competitive individualism must concede to the need for cooperative collective solutions. A sense of unity with – or at least respect for – nature, recognition of material limits, loyalty to place, greater social equality, community cohesion, regional self-reliance and local economic diversity are all prerequisites for, long-term economic security, social well-being, and ecological stability. Above all, the new human eco-niche must be regenerative, i.e., the economy should be re-embedded in *local* community and that this (re)union developed as a fully-integrated mutualistic component of its sustaining ecosystems.

Those dubious of the beneficial pull of relocalization should consider the push factor. Globalization and unfettered trade – i.e., dependence on distant "elsewheres" for food and many other resources – will no longer be possible in the emerging resource and climate-constrained world. This is not entirely a bad thing. Globalization is a driver of overshoot – so-called free trade, particularly in the past half-century, greatly accelerated resource (over)exploitation and pollution, and facilitated population growth. It follows that adaptive eco-economies be more eco-centric local economies. Agriculture and essential light manufacturing – e.g., food processing, textiles, clothing, furniture, tools – will all be relocalized providing ample meaningful employment. There will be a resurgence of personal skills and pride in workmanship. As an immediate additional benefit, when citizens become acutely aware of their dependence on *local* ecosystems they become more actively concerned about the health and integrity of those systems. A sense of conscious participation in one's eco-niche is not possible if the relevant ecosystems are half a planet away.

"One Planet" Eco-Cities and Bioregions

> Localisation stands, at best, at the limits of practical possibility, but it has the decisive argument in its favour that there will be no alternative. (Fleming 2021)

Denizens of urban MTI society are generally unconscious of their individual eco-footprints or the extent to which their home towns and cities are dependent on productive hinterlands dispersed all over the planet. However, as we have shown, it will not likely be possible in coming decades to feed and otherwise provision large cities and megacities. Many urban populations will have to be dispersed and redistributed.

Most urbanites also forget that industrial energy now does the work that people and animals use to perform. Various studies show that people in high-income countries have the energy equivalent of hundreds of human slaves per capita in continuous employment providing them with the goods and services they have come to take for granted. Some of this energy represents displaced draft animals. For example, the population of working horses and mules in the US peaked at 26 million in about 1915 – when the human population was about 100 million – only to be gradually replaced by fossil-powered farm and industrial equipment. The post-carbon US economy would once again need as many work-horses (and about 20 million ha – 50 million acres – of dedicated fodder-producing land) *even if the human population shrinks back from 331 to 100 million.*

Consistent with these data and the re-localization imperative, the following policies/objectives comprise just one example of how we might reconfigure present settlement patterns into more functionally self-contained urban-centered human econo-ecosystems. Senior governments should cooperate with regional and local officials to:

- Create national sub-systems of self-reliant bioregions or ecoregions *centered on existing smaller cities* with boundaries based on ecologically meaningful land-forms and biophysical features (e.g., watersheds, heights of land);
- Size each urban-centered eco-region initially to contain, where possible, a productive ecosystem area equivalent to its population's currently globally dispersed supportive hinterland; i.e., internalize their de facto eco-footprints. (As our example of Toyko revealed, there will be insufficient domestic land/water in many countries, forcing recognition of the need for much lower levels of material consumption and reduced populations.);
- Re-localize government services and decision-making authority, i.e., devolve sufficient governance and taxation powers to the new urban eco-regions to enable effective management of their internal resource- and ecosystems;
- Organize the regional economy and commerce to sustain the population as much as possible on domestic bio-resources and ecosystems, thus reducing reliance on trade. There will still be some trade but:
  - imports should be restricted to true necessities that cannot be produced locally;
  - exports should be limited to bio-resources in true eco-surplus, i.e., harvest rates must be less than regeneration rates to prevent natural capital depletion;
- Facilitate the organization of producer and consumer co-ops – every working person should have a genuine stake in the eco-economy. The ratio of highest paid management to average worker wages should be no greater than 5:1 (the average for Spain's well-known Mondragon cooperatives);
- Phase out trivial and non-essential uses of fossil energy (e.g., private automobiles, off-road vehicles, pleasure motor craft, jet skis, most snow-mobiles, and leaf-blowers);
- Allocate any remaining carbon budget (there may soon be none) to absolutely essential uses, for example, in agriculture and transportation;
- Invest in truly renewable energy sources: mechanical wind and water power; managed biomass, and in technologies that make efficient use of human and animal labor;
- Facilitate breeding programs to supply the draft horses and mules that will be needed to work the land, particularly in agriculture, as fossil-fueled equipment is phased out. Include approximately 0.8 ha (two acres) of forage-land per animal in the regional land bank (see second point above);
- Reintegrate animal husbandry with food-cropping in keeping with sound soils management and to reduce the need for artificial fertilizer with its associated ground- and surface-water pollution;
- Re-design urban waste management to convert settlements from resource-depleting throughput systems into self-sustaining circular-flow ecosystems. For example, collect, treat, and recycle animal and domestic nutrient-containing wastes onto the eco-region's farm and forest lands whence it came. (Circularity in nutrient flows is structurally and functionally necessary for any ecosystem's continuity.);
- Invest in natural capital restoration; regenerate depleted soils, degraded landscapes, wooded areas, and other wildlife habitats to promote

biodiversity, enhance regional productivity, increase carbon sink capacity, and mitigate climate change. (Human overuse has already dissipated half the world's topsoil but soil still contains several times as much carbon as the atmosphere.);

- Recognize that governance of regional ecosystems and landscapes for the common good will sometimes require stinting customary private property rights. Importantly, citizens who realize that their security depends on maintaining the integrity of local ecosystems have an incentive to support such measures.

Concluding Reflections

> Due to the power/interest structures of global capitalism and the juggernaut-like momentum of the global economy, it is most unlikely that any of the [proposed] radical changes to society and the economy. . . will be adopted in time [to avoid catastrophe]. (Dilworth 2010)

The adaptations to EO proposed in this paper run 180° from the capital-intensive growth-oriented "solutions" supported by governments, corporations, and international organizations anxious for the economy to come roaring back from the CoViD-19 pandemic. The mainstream vision is, however, fatally flawed. It is reductionist, narrowly focused on a solitary symptom of overshoot, ignorant of energy realities and devoid of biophysical insight. In particular, it acknowledges neither existing overshoot nor its roots in humanity's increasingly parasitic relationship with an increasingly turbulent ecosphere.

By contrast, the alternative above starts from EO and advances an adaptive approach to human ecological dysfunction that is wholly consistent with biophysical evidence and trends. The downsizing and re-localization of economic activities and their reintegration with communities and supportive ecosystems disaggregates the human enterprise into manageable spatial and eco-economic units consistent with the necessity of one-planet living. Most importantly, the transformation of modern *H. sapiens* from parasite on the ecosphere, to mutualistic participant in local ecosystems, represents humanity's ascent beyond even ecological literacy to lived experience.

Assuming that our best science is valid, the proposed approach clearly has the higher probability of successfully extending the longevity of civilization. However, there is as yet scant evidence that the world community or any individual nation is preparing voluntarily to embark on any form of deliberate long descent toward one-planet sustainability. Rather, it's full ahead on the RE transition and promotion of post-pandemic economic recovery. City administrations seem preoccupied with such marginally useful things as greener buildings, reduced carbon emissions from engineering operations, better public transit, enhanced green space, and "smarter traffic" control all of which, ironically, makes them more attractive to investment and growth. The circular economy (not wholly possible) and green growth (an oxymoron) are popular if somewhat delusional concepts for a society in overshoot. Certainly no city planning department has yet announced a scheme for the equitable contraction of the city's operations, the downsizing of its economy or the dispersal of its population. It is safe to say that no city or megacity on earth is remotely sustainable or even in managed control of its supportive ecosystems. On its present "developmental" path, global civilization is destined to have an interesting encounter with biophysical reality.

Is it too late to wake whole nations of sleepwalkers? Perhaps not. Increasing numbers of thoughtful citizens, activist organizations, and NGOs are taking to the streets. Politicians may yet be forced to take note; the kind of dramatic socioeconomic reset proposed herein may yet be within reach. As the human eco-predicament worsens, there is (shrinking) room for hope that there will yet be a grand popular awakening, one sufficient to catalyze the greater transformation needed to create a true global eco-civilization.

Dare we contemplate that *H. sapiens* can rise to full potential? Will our species rally and gift itself with the chance to take one more step up the evolutionary ladder?

References

Baker, L. (2021). More than 1 billion people face displacement by 2050 – Report. https://www.reuters.com/article/ecology-global-risks-idUSKBN2600K4

BP. 2021. *Statistical Review of World Energy 2021*. London: British Petroleum. https://www.bp.com/content/dam/bp/business-sites/en/global/corporate/pdfs/energy-economics/statistical-review/bp-stats-review-2021s-full-report.pdf

C40 Cities and McKinsey Sustainability. (2021). Focused adaptation – A strategic approach to climate adaptation in cities. https://www.c40.org/press_releases/c40-mckinsey-focused-adaptation

Dilworth, C. (2010). *Too smart for our own good. The ecological predicament of humankind*. Cambridge, UK/New York: Cambridge University Press.

Dyke, J., Watson, R., & Knorr, W. (2021). Climate scientists: Concept of net aero is a dangerous trap. *The Conversation*. https://theconversation.com/climate-scientists-concept-of-net-zero-is-a-dangerous-trap-157368

Environmental Risk Outlook 2021. (2021). London: Verisk Maplecroft. https://www.maplecroft.com/insights/analysis/asian-cities-in-eye-of-environmental-storm-global-ranking/#report_form_container

Fleming, D. 2021. David Fleming quotes. *Quotes.net*. https://www.quotes.net/authors/David+Fleming

Friedemann, A. J. (2016). *When trucks stop running – Energy and the future of transportation* (Springer briefs in Energy). Cham: Springer.

GFN. (2021a). *Earth Overshoot Day*. Global Footprint Network. https://www.footprintnetwork.org/our-work/earth-overshoot-day/

GFN. (2021b). *Open data platform*. Global Footprint Network. https://data.footprintnetwork.org/?_ga=2.229270748.1563653894.1629587653-885488540.1614640094#/

Glaeser, E. (2011). *Triumph of the city*. New York: Penguin Books.

Hoornweg, D., & Pope, K. (2016). Population predictions for the world's largest cities in the 21st century. *Environment and Urbanization, 29*(1), 195–216.

IPCC. (2018). Summary for policymakers. In *Global warming of 1.5 C°. An IPCC special report on the impacts of global warming of 1.5 C° above pre-industrial levels…* Geneva: World Meteorological Organization. https://www.ipcc.ch/sr15/chapter/spm/

IRP. (2018). *The weight of cities: Resource requirements of future urbanization. A report by the International Resource Panel*. Nairobi: United Nations Environment Programme.

Johan, Rockström Owen, Gaffney Joeri, Rogelj Malte, Meinshausen Nebojsa, Nakicenovic Hans Joachim, Schellnhuber (2017) A roadmap for rapid decarbonization. Science 355(6331) 1269–1271 https://doi.org/10.1126/science.aah3443

Michaux, S. P. (2021). *Assessment of the extra capacity required of alternative energy electrical power systems to completely replace fossil fuels*. Report 42/2021, Geological Survey of Finland. https://tupa.gtk.fi/raportti/arkisto/42_2021.pdf

Rees, W. E. (2020). Ecological economics for humanity's plague phase. *Ecological Economics, 169*, 106519. https://doi.org/10.1016/j.ecolecon.2019.106519.

Rees, W. E. (2021). Growth through contraction: Conceiving an eco-economy. *Real World Economics Review, 96*, 98–118. http://www.paecon.net/PAEReview/issue96/Rees96.pdf

Seibert, M. K., & Rees, W. E. (2021). Through the eye of a needle: An eco-heterodox perspective on the renewable energy transition. *Energies, 14*(15), 4508. https://doi.org/10.3390/en14154508.

Spash, C. L. (2016). This changes nothing: The Paris Agreement to ignore reality. *Globalizations, 13*(6), 928–933. https://doi.org/10.1080/14747731.2016.1161119.

Spratt, D., & Dunlop, I. (2021). *"Net zero 2050": A dangerous illusion*. Melbourne: Breakthrough National Centre for Climate Restoration. https://52a87f3e-7945-4bb1-abbf-9aa66cd4e93e.filesusr.com/ugd/148cb0_714730d82bb84659a56c7da03fdca496.pdf

Spratt, D., Dunlop, I., & Taylor, L. (2020). *Climate Reality Check 2020*. Melbourne: Breakthrough National Center for Climate Restoration.. https://www.climaterealitycheck.net/flipbook

Steffen, W., Rockström, J., Richardson, K., et al. (2018). Trajectories of the Earth System in the Anthropocene. *Proceedings of the National Academy of Sciences of the United States of America, 115*(33), 8252–8259.

UN. (2018). *68% of the world population projected to live in urban areas by 2050, says UN*. New York: United Nations. Retrieved from https://www.un.org/development/desa/en/news/population/2018-revision-of-world-urbanization-prospects.html

UN. (2019). World total population (graph). *World Population Prospects 2019*. Retrieved from https://population.un.org/wpp/Graphs/Probabilistic/POP/TOT/900

UN. (n.d.). *#Envision2030 goal 11: Sustainable cities and communities*. New York: United Nations. Retrieved from https://www.un.org/development/desa/disabilities/envision2030-goal11.html

Warren-Rhodes, K., & Koenig, A. (2001). Ecosystem appropriation by Hong Kong and its implications for sustainable development. *Ecological Economics, 39*, 347–359.

Wilkinson, R., & Pickett, K. (2009). *The spirit level: Why equality is better for everyone*. London: Penguin Books.

W

Wildfire

▶ Managing the Risk of Wildfire Where Urban Meets the Natural Environment

Wildlife Corridors

Urban Wildlife Corridors and Their Multiple Benefits

Tanya Clark[1], Tara Rava Zolnikov[1,2] and Frances Furio[1]
[1]California Southern University, School of Behavioral Sciences, Costa Mesa, CA, USA
[2]National University, Department of Community Health, San Diego, CA, USA

Introduction

Wildlife corridors are vital routes by which both flora and fauna engage in the activities necessary to carry out their life cycles (Liu et al. 2018b). Urban wildlife corridors provide a link between habitats that have been disconnected due to urbanization (Ogden 2015). These passages can be land-based or water-based, so small as to enable the spread of the smallest spore or vast enough to span an entire continent. Simply put, without the proper wildlife corridors, the planet's life-forms are severely compromised (Epps et al. 2007).

Each time the human species increases its urban footprint, it does so at the risk of endangering native wildlife dependent on the land for mating, foraging, and migration. When such intrusion disrupts natural habitat, species survival can be threatened to the point of extinction (Kremen and Merenlender 2018; Theobald et al. 2012). Connectivity refers to the extent to which a landscape helps or hinders movement between habitats, and maintaining a proper degree of connectivity for species is a major concern (Abrahms et al. 2017; Caro et al. 2009; Moilanen et al. 2005).

Urban growth inevitably causes environmental change which can interfere with biodiversity (Concepción et al. 2016; Keeley et al. 2018). Urbanization constitutes a key threat to biodiversity when the destruction of natural habitats restricts resource availability, colonization, and migration (Murray et al. 2019; Vimal et al. 2012). An urban wildlife corridor is a man-made solution to this problem.

Wildlife corridors are also referred to as linear habitats, wildlife movement corridors, biological corridors, green corridors, and dispersal corridors. A habitat-linking corridor is a recognized solution for enabling the migration and interbreeding of natural populations whose habitats are eroded by urbanization and climate-induced destruction (Ogden 2015). These landscape linkages have been shown to be efficacious in mediating the adverse effects of devastated natural wildlife corridors, but their use is not without controversy (Haddad et al. 2015; Liu et al. 2018a, b). Implementation of wildlife corridors in areas affected by habitat fragmentation have been met with resistance due to concerns over the loss of land use for economic gain and a relative lack of scientific studies on their use to date (Habib et al. 2016; Peng et al. 2017).

Wildlife Corridors

Wildlife corridors are built in rural, suburban, and urban settings. They are meant to reconcile habitat disruption due to the intrusion of man-made impediments, such as highways and housing tracts. The ultimate goal of wildlife corridors in all geographic settings is to preserve biodiversity compromised by habitat fragmentation.

Biodiversity

The quality of life on the planet depends on a robust state of biodiversity (Haahtela 2019). Biodiversity refers to the variability among all living organisms, both terrestrial and aquatic, and includes diversity within species as well as between species (Bell 1992). According to the United Nation's 2005 Millennium Ecosystem Assessment, ecosystems have changed faster and more extensively over the past 50 years versus any other historical period. Loss of biodiversity is too often the destructive outcome of this swift and sweeping change. Species diversity is a significant predictor of ecosystem stability and productivity, and as such the restoration, conservation, and preservation of biodiversity have been recognized as major global priorities (Tilman et al. 2014; Moilanen et al. 2005).

Habitat Fragmentation

Habitat loss represents a significant threat to biodiversity (Fletcher et al. 2018). Habitat fragmentation does not annihilate the environment, as does habitat loss; rather it impedes it (e.g., via freeway dissection) (Liu et al. 2016). Such habitat interruptions can affect a species ability to maintain life by isolating it from the environment in which it feeds, migrates, and breeds. Habitat fragmentation in urbanized areas can be mitigated by providing land linkages, which facilitate safe movement for affected species.

Examples of Wildlife Corridors

One of the most well-known wildlife corridors was constructed across the Trans-Canada Highway and includes overpasses and underpasses. It was built in Banff Park in order to address the problem of wildlife fatalities due to traffic collisions (Ford et al. 2020). The Eco-Link is the first ecological bridge in Asia (Wang 2019). It is a 203-foot-long man-made bridge that arches over a busy freeway in Singapore and links the Bukit Timah Nature Reserve with the Central Catchment Nature Reserve. On Christmas Island, Australia, a 5-meter crab bridge was built to re-establish connectivity for the annual migration of millions of crabs (Prestamburgo et al. 2016).

The Benefits of Urban Wildlife Corridors

The United Nations has projected that nearly all global population growth from 2017 to 2030 will be driven by urbanization with approximately 1.1 billion new city dwellers expected by 2025 (Cohen 2015). Wildlife corridors have been shown to mitigate the effects of habitat fragmentation due to urbanization (Panzacchi et al. 2016). One example of a wildlife corridor enabling safe species movement is a highway overpass which allows for grizzly bear passage between habitat areas split apart by urbanization (Ogden 2015).

Colonization, migration, and interbreeding are the three primary beneficial outcomes expected from the implementation of an urban wildlife corridor. Colonization refers to the ability of species to resettle in new areas when natural resources, such as food, are lacking in their original habitat (Collinge 2000). Migration is the practice of species relocation on a seasonal basis (Aziz and Rasidi 2014). Interbreeding refers to the ability of species to mate in connected regions thereby promoting genetic diversity (Aars and Ims 1999). For those species that are more sensitive to the negative effects of inbreeding, a wildlife corridor represents a literal life bridge by expanding opportunities for gene flow (Keller and Waller 2002).

Conclusion

Curtailing the loss of biodiversity that can coincide with habitat fragmentation is the primary purpose of implementing a wildlife corridor (Resasco 2019; Keeley et al. 2018). Habitat fragmentation can result in the loss of movement that is vital for species survival, gene flow, and colonization. Urban wildlife corridors represent an efficacious solution in regions where urbanization obstructs these activities (Panzacchi et al. 2016; Abrahms et al. 2017).

It is predicted that by 2030 an additional 1.2 billion people will dwell in urban areas with 290,000 square kilometers of natural habitat expected to be adapted to urban land use (Cohen 2015). Such massive takeovers will inevitably impact wildlife habitats and by extension threaten the earth's biodiversity (McDonald et al. 2019). Wildlife corridors should be considered as a viable option for protecting biodiversity and thereby safeguarding robust lifecycles for impacted species (Kremen and Merenlender 2018; Resasco 2019). Additionally, the evidence between biodiversity and human well-being is compelling and constitutes a convincing argument for the expenditure of financial capital on urban wildlife corridors (Reid et al. 2005; Raffaelli and Frid 2010; Sandifer et al. 2015).

References

Aars, J., & Ims, R. A. (1999). The effect of habitat corridors on rates of transfer and interbreeding between vole demes. *Ecology, 80*(5), 1648–1655.

Abrahms, B., Sawyer, S. C., Jordan, N. R., McNutt, J. W., Wilson, A. M., & Brashares, J. S. (2017). Does wildlife resource selection accurately inform corridor conservation? *Journal of Applied Ecology, 54*(2), 412–422.

Aziz, H. A., & Rasidi, M. H. (2014). The role of green corridors for wildlife conservation in urban landscape: A literature review. In IOP Conference Series: Earth and Environmental Science (Vol. 18, No. 1, p. 012093). Bristol, UK: IOP Publishing.

Bell, D. E. (1992). The 1992 convention on biological diversity: The continuing significance of US objections at the earth summit. *George Washington Journal of International Law and Economics, 26*, 479.

Caro, T., Jones, T., & Davenport, T. R. (2009). Realities of documenting wildlife corridors in tropical countries. *Biological Conservation, 142*(11), 2807–2811.

Cohen, B. (2015). Urbanization, City growth, and the new United Nations development agenda. *Cornerstone, 3*(2), 4–7.

Collinge, S. K. (2000). Effects of grassland fragmentation on insect species loss, colonization, and movement patterns. *Ecology, 81*(8), 2211–2226.

Concepción, E. D., Obrist, M. K., Moretti, M., Altermatt, F., Baur, B., & Nobis, M. P. (2016). Impacts of urban sprawl on species richness of plants, butterflies, gastropods and birds: Not only built-up area matters. *Urban Ecosystem, 19*(1), 225–242.

Epps, C. W., Wehausen, J. D., Bleich, V. C., Torres, S. G., & Brashares, J. S. (2007). Optimizing dispersal and corridor models using landscape genetics. *Journal of Applied Ecology, 44*(4), 714–724.

Fletcher, R. J., Jr., Didham, R. K., Banks-Leite, C., Barlow, J., Ewers, R. M., Rosindell, J., et al. (2018). Is habitat fragmentation good for biodiversity? *Biological Conservation, 226*, 9–15.

Ford, A. T., Sunter, E. J., Fauvelle, C., Bradshaw, J. L., Ford, B., Hutchen, J., et al. (2020). Effective corridor width: Linking the spatial ecology of wildlife with land use policy. *European Journal of Wildlife Research, 66*(4), 1–10.

Haahtela, T. (2019). A biodiversity hypothesis. *Allergy, 74*(8), 1445–1456.

Haddad, N. M., Brudvig, L. A., Clobert, J., Davies, K. F., Gonzalez, A., Holt, R. D., et al. (2015). Habitat fragmentation and its lasting impact on Earth's ecosystems. *Science Advances, 1*(2), e1500052.

Habib, B., Rajvanshi, A., Mathur, V. B., & Saxena, A. (2016). Corridors at crossroads: Linear development-induced ecological triage as a conservation opportunity. *Frontiers in Ecology and Evolution, 4*, 132.

Keeley, A. T., Ackerly, D. D., Cameron, D. R., Heller, N. E., Huber, P. R., Schloss, C. A., . . ., & Merenlender, A. M. (2018). New concepts, models, and assessments of climate-wise connectivity. *Environmental Research Letters, 13*(7).

Kremen, C., & Merenlender, A. M. (2018). Landscapes that work for biodiversity and people. *Science, 362*(6412), 1–11.

Keller, L. F., & Waller, D. M. (2002). Inbreeding effects in wild populations. *Trends in Ecology & Evolution, 17*(5), 230–241.

Liu, J., Wilson, M., Hu, G., Liu, J., Wu, J., & Yu, M. (2018a). How does habitat fragmentation affect the biodiversity and ecosystem functioning relationship? *Landscape Ecology, 33*(3), 341–352.

Liu, C., Newell, G., White, M., & Bennett, A. F. (2018b). Identifying wildlife corridors for the restoration of regional habitat connectivity: A multispecies approach and comparison of resistance surfaces. *PLoS One, 13*(11), e0206071.

Liu, Z., He, C., & Wu, J. (2016). The relationship between habitat loss and fragmentation during urbanization: An empirical evaluation from 16 world cities. *PLoS One, 11*(4), e0154613.

McDonald, R. I., Mansur, A. V., Ascensão, F., Crossman, K., Elmqvist, T., Gonzalez, A., et al. (2019). Research gaps in knowledge of the impact of urban growth on biodiversity. *Nature Sustainability, 3*, 1–9.

Moilanen, A., Franco, A. M. A., Eary, R. I., Fox, R., Wintle, B., & Thomas, C. D. (2005). Prioritizing multiple-use landscapes for conservation: Methods for large multi-species planning problems. *Proceedings of the Royal Society, Series B, Biological Sciences, 272*, 1885–1891.

Murray, M. H., Sánchez, C. A., Becker, D. J., Byers, K. A., Worsley-Tonks, K. E., & Craft, M. E. (2019). City sicker? A meta-analysis of wildlife health and urbanization. *Frontiers in Ecology and the Environment, 17*(10), 575–583.

Ogden, L. E. (2015). Do wildlife corridors have a downside? *Bioscience, 65*(4), 452–452.

Panzacchi, M., Van Moorter, B., Strand, O., Saerens, M., Kivimäki, I., Colleen, C. S. C., et al. (2016). Predicting the continuum between corridors and barriers to animal movements using step selection functions and randomized shortest paths. *Journal of Animal Ecology, 85*(1), 32–42.

Peng, J., Zhao, H., & Liu, Y. (2017). Urban ecological corridors construction: A review. *Acta Ecologica Sinica, 37*(1), 23–30.

Prestamburgo, S., Premrù, T., & Secondo, G. (2016). Urban environment and nature. A methodological proposal for spaces' reconnection in an ecosystem function. *Sustainability, 8*(4), 407.

Raffaelli, D., & Frid, C. (2010). The evolution of ecosystem ecology. In D. Raffaelli & C. Frid (Eds.), *Ecosystem Ecology: A New Synthesis (ecological reviews)* (pp. 1–18). Cambridge: Cambridge University Press. https://doi.org/10.1017/CBO9780511750458.002.

Reid, et al. (2005). *Millennium ecosystem Assessment. Ecosystems and human well-being: Synthesis*. Washington, DC: Island Press. Copyright © 2005 World Resources.

Resasco, J. (2019). Meta-analysis on a decade of testing corridor efficacy: What new have we learned? *Current Landscape Ecology Reports, 4*(3), 61–69.

Sandifer, P. A., Sutton-Grier, A. E., & Ward, B. P. (2015). Exploring connections among nature, biodiversity, ecosystem services, and human health and well-being:

Opportunities to enhance health and biodiversity conservation. *Ecosystem Services, 12*, 1–15.
Theobald, D. M., Reed, S. E., Fields, K., & Soulé, M. (2012). Connecting natural landscapes using a landscape permeability model to prioritize conservation activities in the United States. *Conservation Letters, 5*(2), 123–133.
Tilman, D., Isbell, F., & Cowles, J. M. (2014). Biodiversity and ecosystem functioning. *Annual Review of Ecology, Evolution, and Systematics, 45*, 471–493.
Vimal, R., Geniaux, G., Pluvinet, P., Napoleone, C., & Lepart, J. (2012). Detecting threatened biodiversity by urbanization at regional and local scales using an urban sprawl simulation approach: Application on the French Mediterranean region. *Landscape and Urban Planning, 104*(3–4), 343–355.
Wang, J. (2019). Re-imagining urban movement in Singapore: At the intersection between a nature reserve, an underground railway and an eco-bridge. *Cultural Studies Review, 25*(2), 8.

WMO – World Meteorological Organization

► Adapting to a Changing Climate Through Nature-Based Solutions

Women in Urbanism, Perpetuating the Bias?

The shortfalls of feminist reactions to traditional urban planning and design

Soumaya Majdoub
Research Group Interface Demography, Department of Sociology, VUB Free University of Brussels, Brussels, Belgium
Brussels Center for Urban Studies (BCUS), Brussels, Belgium
Brussels Interdisciplinary Research Centre for Migration and Minorities (BIRMM), Brussels, Belgium

Synonyms

Equal cities; Gender blind urban planning and design; Gender responsive planning and design; Gender-neutral cities; Inclusive cities; Urban design; Urban planning; Urbanism

Definition

A gender-neutral city derives from a feminist perspective on urbanism that seeks to avoid gendered and racially unequal outcomes of urban policies. It assumes that both sexes are included equally in the urban planning process, are not shaped by or in the interest of a particular sex and are also affected equally by various urban problems.

A gender-neutral city adopts an intersectional perspective to equity including reflections on gender identities and seeking to remediate exclusion of all marginalized people.

An intersectional approach links all types of oppression and reveals the interconnected nature of social categorizations. These categorizations such as race, class, and gender are seen as overlapping, interdependent mechanisms of oppression or discrimination.

Decolonial urban practice refers to the process of unlearning and contesting the dominance of Western knowledge on cities and urbanization.

Introduction

Urban resilience requires inclusive planning and design processes. Inclusive urban planning cannot be reduced to the obvious signs of gender balance or the topics that are most often linked with women in public spaces. Street lighting, gender balance of names of public spaces and the number of women architects working in design forms are the most obvious points of attention but do distract from the core issues. Representation is a matter of paramount importance since it directly affects the form of the build environment. The exclusion of women from urban planning and design means that women's daily lives and perspectives do not shape urban form and function but in turn that environment shapes how they live, move, and engage in and with public spaces. In other words, city planning overlooks the specific challenges and concerns that women and girls face. This underlines the fact that the city is not inclusive and equitable in its design, infrastructure, facilities, and services. Even when striving for gender-neutrality, the sheer definition of this concept has a male perspective and represents men's interest. This male gaze is encoded in cities across the world. Although women are affected differently and disproportionately by urban problems, their issues and perspectives are not taken into account by default. Women who bear the impact of the location of public transports stops since they spend more time traveling and prefer public transport to do so. Women who come from disadvantaged backgrounds living in insecure neighborhoods, commuting more likely in the late or early hours. Women who need ramps for wheelchairs or zigzagging through the city with buggies. Women who fear physical or sexual violence in public space for whatever reason they are out there. Women who complain about a lack of public bathrooms since statistically they have a higher chance of wanting to use one, are on their menstrual cycle or have children with them.

Moreover, urban planning and design is largely done through the lens of eurocentric values. And design history and theory inordinately overshadowed by male white authors. Something that traditional feminist organizations lack to recognize sufficiently. Which voices are given greater or less value by feminist organization reveals whether an intersectional approach is been adopted or if it is a shear perpetuation of the distortion in the field of architecture and urbanism. Sara Ahmed (2013) refers to this in her work "Making Feminist Points" when she states that it is often feminists themselves that end up framing their propre work in relation to a male intellectual tradition. This reinforces the power relations at work since referencing can be seen as "an ethical act", and citation as an academic correspondent to that act, and more broadly about how these operated as forms of "critical spatial practice" (Rendell 2006). In addition, women tend to cite themselves less than men do (Savonick and Davidson 2017). While it is important to be appreciative of this bias, it requires a different lens to rewrite the canon that has shaped and is still shaping different disciplines informing urban policies and shaping cities worldwide. This lens, intersectional, also feeds the prerequisite of a decolonial approach, rendering seen what is unseen. Here lies the challenge of women in urbanism movements. How to challenge the gatekeepers of the canon? How to decenter the whiteness of and within those organizations. How to render visible what history concealed, all while revealing the asymmetries and tackling unjust and inequitable structures of power along gender, class and racial lines. This requires an examination of the intertwined relationship of gender inequality, race, the built environment and urban planning and design.

Intertwined and Invisible Affects

If race is absent in architecture, it is because "the discourse of modernism gave rise to modern architecture, and... architecture has done very little to address how race, racial representation an racial thinking have shaped its own practices and discourse" (Cheng et al. 2020). The architectural scholar and cultural historian Mabel O. Wilson goes even further and claims that "Modern architecture builds the world for the white subject, maintaining the logics of racism while also imagining a future world in which nonwhite subjects remain exploitable and marginal"...."the power of the architecture and its archive is to produce 'whiteness' by design" (https://www.latimes.com/entertainment-arts/story/2021-03-19/racism-in-architecture-moca-show-on-race-by-mabel-wilson). A critical practice of architecture and urbanism draws attention to the use of language to confront the supposed dichotomy between white and non-white. An interesting approach can be found in the work of Marie-Louise Richards who addresses the effects of the unseen in architectural practice. She states that "the theme of invisibility also makes visible how whiteness remains unmarked when the racial is deployed to organize society and conceptualize spaces. Unmarked, whiteness becomes a 'blind spot', describing a certain unconscious viewing of oneself and others, which results in an unequal distribution of power and privilege associated with skin" (Richards 2018). Whiteness then is visible and invisible, referring to what is considered normal and neutral, even universal and mostly unquestioned. Just like the male dominance in the respective disciplines remained unchallenged up until recently. Richards (2018) states that since whiteness exists everywhere and nowhere, it becomes visible on the individual body. Not only bodies considered "white" because in order to survive, all bodies make whiteness visible. In public space this process is very palpable.

Race and Gender, Tensions Inscribed in Public Spaces

When referring to Wilson's definition of race, "as the controlling of space, controlling where people are able to go" it becomes clear how the interlaced relation between race and gender is most explicit in public spaces. Oftentimes when addressing gender equality in cities, the focus remains exclusively on the safety of public spaces which accentuates them as spaces of control and linking them to urban fears and relations of threat. The source of that threat not being white people, since the construction of "whiteness" grants them the privilege of not being feared in public spaces. Spaces that are "produced in part through gendered ideologies about cities, bodies, safety and fear" and a reflection of "the patriarchal power relations that continue to structure contemporary urban life" (Kern 2010). Because in those spaces "the matter of race is very much about embodied reality; seeing oneself or being seen as white or black or mixed does affect what one can do, even where one can go, which can be re-described in terms of 'what is and is not within reach'." (Ahmed 2006). Where space and race intersect, a norm submerges of what is being assessed as "acceptable" and legitimate to take up ownership in these public spaces. On top of the notion of "whiteness," "acceptability" needs to be added to understand the tensions that arose during and after the Black Lives Matter protests. A movement, incidentally, started by women. Many memorials honoring recognized colonialists or slave dealers in the United States, Britain and Europe, have been dismantled following the protests. But not without fierce public debate about the behavior of protesters, the legitimacy and acceptability of those protests, more than about the history of these contested leaders. Acceptable behavior in public spaces is determined by an unwritten norm impregnated by the notion of whiteness which not only defines what a public space looks like but also who is allowed to engage in and with these spaces. Examples worldwide show different ways of reacting to claims of space and what is perceived as a transgression. In New Zealand, the Māori party launched a petition (https://www.maoriparty.org.nz/nz_to_aotearoa) demanding to change New Zealand's official name to Aotearoa which is the indigenous language name for the country. In a reaction, Prime minister Jacinda Ardern stated that an official name-change was "not something we've explored". In Belgium, on the initiative of the Brussels State Secretary for Heritage and Urbanism, a working group on decolonization of public space was launched addressing the discussion on whether or not to remove colonial symbols from public spaces (https://www.themayor.eu/en/a/view/brussels-launches-working-group-on-decolonisation-of-public-space-6289). The open call instigated a reflection on whether alternatives for removal could be formulated, through contextualization or a completely different statue or name. Less is true on the other side of the Canal. In London, the Prime Minister described the Black Lives Matter protestors, who made the state of slave trader Edward Colton end up in the harbor, as "thugs" (https://www.bbc.com/news/uk-52960756). At the same time, no such comment was made to describe the far-right protestors that tried to violently block an authorized anti-racism march through London (https://www.theguardian.com/world/2018/oct/13/anti-fascists-block-route-of-democratic-football-lads-alliance-london-march). One group's presence and actions in public space is perceived as deviant and asking for policing.

While the other group's presence, peaceful or not, is being identified with "normality". Owing to the fact that black and white bodies are not coded in the same way. The way those bodies move through public space neither. The latter neutral and decent while black bodies carry an unerasable transgressiveness. Whiteness being the characteristic of public spaces, the sheer presence of black bodies is a transgression in itself.

The Lack of Self-Critique and Cultural Humility

In the quest of gender neutrality, feminist movements in urbanism need to integrate the demographic realities of and dynamics in cities that predetermine the position of different groups given that race needs to be considered as relational not binary. In cities characterized by super-diversity, a feminist critique to traditional urban planning, design and policies, needs to entail a self-critique starting with the acknowledgment that "imagining spaces also depends on who is doing the imagining and what traditions, history, languages and mythologies they have inherited" and who is telling them. Richards (2018) goes further and proposes to "identify how norms are reproduced in order to locate what values must be examined and challenged" in the field of architecture and urbanism. Beyond gender. The canons of architecture and urbanism asks for an approach based on cultural humility. While first coined in an article on training outcomes in multicultural education in the health sector (Tervalon and Murray-Garcia 1998), it was not until 2014 that professor of social and cultural analysis James Arvanitakis thought of applying the concept to academia as a whole. Cultural humility requires more than the competency of understanding different cultural backgrounds. It is indeed a "more productive approach to engaging with communities, especially communities of color'"(Sweet 2016). If embraced by the disciplines of urbanism and architecture and the feminist organizations that mobilize against the long history of gendered discrimination, a decolonial practice would unfold itself. However, only if acknowledged that this history also entails severe racial and cultural discrimination. Seeing that "cultural humility can illuminate the prevailing power position of experts in the field and their broader positionality vis-a-vis the communities with which they work" (Sweet 2016). Contemplating ones positionality is what feminist organizations lack to integrate in their urban activism.

Conclusion

Resilient and socially equitable cities cannot be achieved if our debates are still dominated by biased knowledge and narratives of whiteness. A decolonized perspective on urban transformations points out the areas of attention when striving for urban resilience. When tracing the neo-colonial patterns of inequalities, the historical colonial roots of power, identity, space, and belonging emerge to the surface. Therefore, a feminist city cannot be solely about claiming space in a man-made world, as formulated in Leslie Kern's book with the same title. Foremost, the underlying mechanisms need to be deconstructed. Illuminating the intertwined relationship between race, gender inequality, the built environment, and urban planning and design is not sufficient. A critical look on the use of language and the assertion that race as a concept is not stable but relational, is necessary but remains incomplete. An intersectional approach as well. Feminist urban activism seeks to remediate exclusion of all marginalized people by first acknowledging the presence of the trauma of colonization in public spaces repurposing those places of trauma for healing. Only then an inclusive urban future can be imagined.

Cross-References

▶ Age-Friendly Future Cities
▶ Feminist Planning and Urbanism: Understanding the Past for an Inclusive Future
▶ Understanding Women's Perspective of Quality of Life in Cities
▶ Urban Well-Being

References

Ahmed, S. (2013). *Making feminist points*. Feminist killjoys, 11. https://womantheory.wordpress.com/2014/02/06/making-feminist-points-by-sara-ahmed/

Ahmed, S. (2006). Orientations: Toward a queer phenomenology. *GLQ: A Journal of Lesbian and Gay Studies, 12*(4), 543–574.

Cheng, I., Davis, C. L., & Wilson, M. O. (Eds.). (2020). *Race and modern architecture: A critical history from the enlightenment to the present*. University of Pittsburgh Press.

Kern, L. (2010). Selling the 'scary city': Gendering freedom, fear and condominium development in the neoliberal city. *Social & Cultural Geography, 11*(3), 209–230. https://www.tandfonline.com/doi/abs/10.1080/14649361003637174.

Rendell, J. (2006). *Art and architecture: A place between* (pp. 1–240). London: IB Tauris.

Richards, M. L. (2018). *Hyper-visible invisibility: Tracing the politics, Poetics and.* Becoming a Feminist Architect, 39.

Savonick, D., & Davidson, C. (2017). *Gender bias in academe: An annotated bibliography of important recent studies*. https://academicworks.cuny.edu/cgi/viewcontent.cgi?article=1166&context=qc_pubs

Sweet, E. L. (2016). Cultural humility: An open door for planners to locate themselves and decolonize planning theory, education, and practice. *EJournal of Public Affairs, 7* (2). http://www.ejournalofpublicaffairs.org/cultural-humility/

Tervalon, M., & Murray-Garcia, J. (1998). Cultural humility versus cultural competence: A critical distinction in defining physician training outcomes in multicultural education. *Journal of Health Care for the Poor and Underserved, 9*(2), 117–125. https://muse.jhu.edu/article/268076/summary.

World Meteorological Organization (WMO)

► The State of Extreme Events in India

Y

Young Children

► Moving Towards Sustainable, Liveable, and Care-Full Urban Environments: Pre-schoolers' Rights and Visions for Planning Just, Socially, and Ecologically Integrated Cities

Young People

► Pre-schoolers and Sustainable Urban Transport

Youth and Public Transport

Technology and the Making of the "Responsible" Citizen in Dubai

Abdellatif Qamhaieh
American University in Dubai, Department of Architecture, Dubai, United Arab Emirates

Introduction

Cities of the Arabian Gulf grew rapidly during a period of abundant oil and "infatuation" with modernist city-planning ideas (Khalaf 2006). During the 1960s and 1970s, car-centric urban planning was dominant in the region and arguably continues until today. As a result, most of the cities are planned to accomodate the car (Elsheshtawy 2008). The dependence on automobiles is continuously reinforced by the excellent road infrastructure, hot and arid climatic conditions, and a car culture that has become an integral part of life for the residents of this region (Qamhaieh and Chakravarty 2020).

Nevertheless, the car's complete and undisputed dominance has been challenged lately by major investments in public transport systems (Almardood and Maghelal 2020; Elsheshtawy and Bastaki 2011; Rizzo 2014). While the cities of the region continue their plans of globalization and growth, the negative impacts of automobile dependence have become very apparent. Hence, the need to invest in more sustainable modes of transport has become a pressing issue. This is especially true as the region hosts major global events such as the Dubai Expo 2020 and Doha's World Cup 2022 – with the expected large influx of tourists into the area. The different lifestyles and needs of residents and tourists are forcing authorities to provide transport options beyond the personal motor vehicle.

Dubai, the largest city in the United Arab Emirates and arguably the most famous in the region, was one of the first to embark on the journey toward more sustainable public transport systems. Though the city had a basic and minimal bus system since the 1960s (Gokulan 2015), by 2005, the Road and Transport Authority (RTA) was created, and the bus system was then revamped and transferred into a modern system.

R. C. Brears (ed.), *The Palgrave Encyclopedia of Urban and Regional Futures*,
https://doi.org/10.1007/978-3-030-87745-3

The city was also the first in the region to establish a metro, which began operation in 2009 (RTA 2021). Since then, the metro has succeeded in changing the city, reducing some traffic, and contributing to Dubai's modern and global appeal. Most importantly, it made it easier for tourists and residents to move around the city without using a personal automobile or a taxi.

Despite Dubai's success in providing a range of sustainable transport options, and while the authorities recognize and aim toward more sustainable urban policies, the car remains the defacto mode of transport in the city, especially for the middle class and wealthier segments of society. As of 2018, there were 514 cars for every 1000 residents, which is a staggering ratio (Tesorero 2018). The country as a whole has a very strong car culture, and the ease with which one can buy, operate, and drive a car makes it the most appealing mode of transport for those who can afford it (Qamhaieh and Chakravarty 2020).

As of lately, in an attempt to improve public transport (PT) integration into the urban fabric and influence ridership numbers, the focus of the authorities has been directed toward the issues of the first/last mile (RTA 2020b). Shared mobility solutions, including e-scooters, e-bikes, and shared cars, have been provided in dense areas, in tourist-oriented areas, and close to primary public transport nodes. While these systems are still relatively new, they appear to be gaining popularity within specific demographics – especially the youth. Access to these shared mobility systems is made easier by the wide availability of mobility-related phone apps and e-payment systems, which are part of Dubai's smart-city initiative.

This chapter primarily takes a closer look at public transport and shared mobility systems in Dubai. While cars remain dominant, these shared mobility solutions appear to be growing in popularity, especially among the middle-class youth and young professionals. Meanwhile, context-specific social and physical barriers keep PT ridership limited to poorer segments of society and tourists. The chapter poses the following questions: Could these shared mobility solutions make PT more appealing to the more affluent demographics? And could they get youth more integrated into more sustainable, socially responsible lifestyles? The chapter overviews literature related to ridership in general and Dubai in particular. Interviews with college students relative to their use of public transport and shared mobility solutions are also provided.

Background: Choice of Transport Mode and Negative Attitudes Toward PT

When people choose travel options, several economic, social, and practical factors impact their transport mode choice (B D Taylor and Fink 2013). The one indisputable fact remains that the personal automobile remains a dominant and sought-after mode of transport (Jensen 1999). Despite the various advancements in public transport (PT), and in the various levels of success – depending on the context, no public transport system can offer the freedom, privacy, independence, and the set of other complex emotions evoked by the car (Anable 2005; Kent 2015; Sheller et al. 2000; Steg 2005).

Understanding why people choose to use or not use public transport remains a challenging question and highly dependent on context and a range of factors (Beirão and Sarsfield Cabral 2007; Burian et al. 2018; Manville et al. 2018). Taylor and Fink (2003, 2013) suggest that these factors could be grouped into elements intrinsic to the systems itself (such as quality and level of service) and those external to the transport system (such as attitudes and access to cars). In a study by Grengs (2005), the author argues that the provision of public transport systems should have two overarching goals. On the one hand, public transport systems should reduce congestion and dependency on cars, reverse urban sprawl, and address the environmental cost of car ownership. On the other hand, PT should provide transportation means for those without cars. The author further argues that these goals are sometimes conflicting due to socio-spatial inequalities within the city, especially in the North American context. They are also impacted sometimes by the type of PT system; for example, light-rail systems receive more attention as they appeal to the wealthier segments and are considered to be of higher quality. The bus system, meanwhile, appears to cater mainly to the poorer segments and receives less attention (Grengs 2005; Manville et al. 2018; Taylor and Morris 2015).

Krizek and El-Geneidy (2007) breakdown potential PT riders into two main categories: First, car owners who use public transportation and are therefore considered to be *choice riders*; these riders typically live in dense cities with mature public transport systems, are usually more affluent, and some have some concerns for the environment (Collins and Chambers 2005). These choice riders tend to be "pickier," relevant to transport system quality since they have the option to drive, and only choose PT over the car when they consider the former as superior (Beimborn et al. 2003). Second, *captive riders* are those who do not own/operate a car or cannot afford a car and usually fall within the lower-income demographics. Some also could have medical conditions or age restrictions that could prevent driving. Such populations usually have complete dependence on public transport systems regardless of their quality or provided level of service (LOS) (Beimborn et al. 2003; Krizek and El-Geneidy 2007). Public transport plays a critical role for the urban poor beyond mere movement. Without adequately planned and executed PT systems, these vulnerable populations have limited access to employment opportunities, essential needs, and services. This limited access prevents their meaningful participation and therefore, in some cases, excludes them from mainstream society (Ureta 2008).

Around the world, the complexity of public transport systems and annual ridership numbers vary considerably. In the US context, for example, and outside of the large, dense, and traffic-congested cities, most Americans do NOT use public transport and depend on the automobile for their movement (Manville et al. 2018). Those riding public transport are usually the urban poor, or residents of denser cities such as New York – where congestion and parking shortages are very prominent. In a study by Beirão and Sarsfield Cabral (2007) examining ridership in Porto, Portugal, the authors indicated that several factors influence how residents move around the city using public transport. In their study, respondents indicated that their perceived level of service for the transport system played a significant role in the ridership numbers. Elements such as comfort, travel time, and ease of navigation helped make transport systems more accessible. Whereas, car owners were usually the ones to view public transport with the most negative lens, particularly the bus. The authors suggest that for public transport systems to be successful, both the actual level of service and the negative attitudes toward these systems need to be addressed systematically through improvements to PT and informative advertisement campaigns (Beirão and Sarsfield Cabral 2007).

Similar trends could be seen in other studies relative to transit ridership. In a study by Murray et al. (2010) about attitudes toward public transport in New Zealand, prejudice toward the system was highest among non-users, whereas direct contact with PT seemed to reduce this prejudice. A study within the Australian context by Xia et al. (2017) indicated that the *perceived* safety concerns were significant in impeding the use of public transportation – even for other sustainable forms of travel such as walking and biking. The authors argued for a *push* and *pull* approach, which makes driving and parking more difficult, and improves awareness and quality of the PT systems and public realm infrastructure. The authors also argued that increasing awareness relative to automobile dependence's environmental and health costs could be critical in influencing public policy and improving PT ridership. In a study by Burian et al. (2018), the authors demonstrated how the city's spatial layout and the *speed* at which PT moves had the most significant impact on ridership. The study also indicated that walking distance to metro stops was not necessarily an obstacle to increased ridership, but fewer and smoother PT connections were. Similar thoughts were echoed by Ramos et al. (2019) in a study of transport ridership in Lisbon. In the study, the multiple connections needed to arrive at one's destination made using PT difficult. Other notable results of the study were the preference toward rail systems, the sensitivity toward timetables, and the overall negative attitudes toward overcrowding the public transport systems.

One of the elements to impact PT transport ridership and help maintain the car's dominance is the *first and last-mile* connections (Shaheen and Chan 2016). As transport routes are usually fixed and operate based on specific stops, the journey beyond the transport stop could make or break the transport system. This is especially true in lower-density suburban residential areas, where the type of density, sprawl, and land-use patterns do not encourage PT usage (Cervero and Landis 1994). Even if the PT element itself is of the highest quality, the connections beyond the transport stops are crucial to its success. Excessively long, dangerous, or unpleasant connections guarantee that those who have access to cars will prefer to use them. Therefore, lacking proper integration and seamless PT connections, these systems do not fare well beyond captive riders (Xia et al. 2017).

As of recently, and with smartphones and data available to the masses, the concept of *shared mobility* has gained importance and has been proposed as a solution for first and last-mile connections (Shaheen and Chan 2016). Ride-sharing apps, car-sharing, bike-sharing, and e-scooter sharing have helped bridge the gap between traditional PT systems such as rail systems, metro, or buses, and the riders' final destinations. Such ride-sharing technologies are gaining importance in globalizing cities worldwide and could be considered a meaningful solution to first and last-mile problems. They also help reduce car ownership and are deemed efficient and environmentally friendly since they help reduce greenhouse gas (GHG) emissions (Shaheen and Chan 2016). Though questions of equity remain significant relative to these shared mobility solutions, if properly implemented, they appear to reduce transport inequalities and help better connect potential riders to public transportation as the cost of using these systems is still much lesser than that of owning and operating the personal motor vehicle (Zuo et al. 2020).

Loyalty to Public Transportation and the Importance of Youth Engagement

Encouraging people to use public transportation (and active transport systems such as biking) should be a very important issue for cities and policymakers. The use of public transportation systems and making people loyal advocates for such systems have far-reaching environmental, social, and economic benefits – while making cities and communities more sustainable (Van Lierop et al. 2017). When the public buys into and uses PT, especially across the different socio-economic classes, the systems usually succeed and improve with time as they respond to users' demands. Furthermore, they become an integral part of city life and a point of pride. According to Van Lierop et al. (2017), improving passenger experience while riding PT (or waiting for it) and passenger perception of the system help induce loyalty. Age also tends to play a role in transit readership. Youth and younger adults (under 30 years old) tend to use public transport more, especially in big cities with well-established and mature systems. This is usually due to limited car access, lifestyle choices, and the highly mobile nature of these populations. As they grow older and move to suburban living or have families, they tend to shift to cars to accommodate their lifestyle changes (Grimsrud and El-Geneidy 2014).

Unfortunately, those with access to cars at an early age tend to bypass public transport altogether. Typical car-dependent users view society as drivers versus nondrivers. They view the city differently – as an observer through the car windows instead of an active participant (Sheller et al. 2000). They also view pedestrians and public transport riders as "others." Such notions could have implications for society as it keeps different classes segmented, with some even socially excluded, whereas using public transport involves the learning experience of riding with and dealing with others in the city (Currie and Stanley 2008). When riding public transportation, different groups will interact with others, whether purposefully or casually. This interaction strengthens community belonging and increases awareness toward others. In a study by Green et al. (2014), the authors documented the benefits of the free "*travel for all campaign*" in London, where youth between 16 and 18 were allowed access to buses free of charge. The study documents the benefits of this campaign beyond the apparent freedom of movement. Youth viewed the trips as an act of socializing – either with friends or strangers – in addition to their acquired freedoms. Youth who are regular users of PT understand the social benefits of riding public transport; they also view it as an opportunity to socialize with others physically and digitally (i.e., work/study while on the move) while being responsible toward the environment (Brown et al. 2016). Therefore, transport agencies must attract younger generations and keep them satisfied and loyal to PT. This exposure at earlier ages means that they are most likely to use PT for longer times and transition to cars later in life.

Public Transport in Dubai: Peculiarities of the System

As discussed in the introduction, Dubai, a pioneering city in the GCC region, has had a public transportation system – represented by the public bus, since the early 1960s (Gokulan 2015). Though mostly popular within the lower-income, captive segments (more on this issue below), the bus system serves different areas of the city. Since the Roads and Transport Authority's creation in 2005, the bus system has been greatly expanded. Today's fleet includes regular and double-decker buses, and some double-carriage (articulated) buses on some routes. It is a large fleet that includes around 1518 buses as of 2021. The bus and all PT systems in Dubai are based on a prepaid card system known as NOL card (RTA 2021), which assures a seamless transition between the different PT modes. Due to the scorching climate, some of the bus stops are air-conditioned and provide riders relief from the brutal conditions, especially in the hot summer months. Bus networks, in general, are common in the middle-east and the GCC region in particular, but the characteristics and ridership demographics differ considerably across countries (Belwal and Belwal 2010).

The claim for fame regarding PT in Dubai has been the metro system, which was the first in the region. It is a modern, mostly elevated metro system that commenced operation in 2009 (Farooqui 2019). It consisted of two lines initially: the *red-line,* which runs 52 kms between the north and south of Dubai. In contrast, the *green-line* travels 23kms east to west within the older, higher-density areas. As of recently, a new line was added to serve the Expo 2020 site. Known as *Route 2020*, it runs 15 kms between east and west, and with more expansion planned for the near future (Fig. 1). The metro system has proven to be popular with the city residents and tourists alike due to its comfortable, air-conditioned, and efficient operation. It is elevated on a track and provides pleasant views of the city. It also includes WIFI, but most importantly, riders get to bypass the suffocating traffic conditions on the street below.

According to RTA (2019), the metro carried 204 million riders in 2018, compared to 39 million riders in its first year of operation. In general, the metro and the overall improvements to the PT network increased public transport's share of total *journeys/year* from 6% in 2006 to 17.5% in 2018. The city also has a tram system, which connects to the metro system – though limited to Al-Safouh and Dubai marina areas. A mono-rail system also runs along the spine of the famous Dubai palm, and a water ferry system connects some areas along the city's coast and waterways. A fleet of 3854 taxis completes and complements the overall transportation network and connects the different areas of the otherwise fragmented city (RTA 2021).

Still, despite the availability of public transport and the improvements in ridership across all sectors, the Middle East region, in general, and Dubai, in particular, appear to perform poorly relative to the sustainability and performance of the public transport systems – at least in comparison to other similar cities (Almardood and Maghelal 2020; De Gruyter et al. 2016; Kaiser 2007). This is likely due to the dominance of automobiles, and the high per capita car ownership rates (Qamhaieh and Chakravarty 2020; Tesorero 2018), and the car-based urban form dominant in the region and in Dubai in particular (Alawadi and Benkraouda 2019; Elsheshtawy 2010; Khalaf 2006). The hot and arid climate and the typical lack of shading make walking and moving outside very uncomfortable (Al-Ali et al. 2020; Alawadi et al. 2021). Beyond these

Youth and Public Transport, Fig. 1 Dubai metro and transport map. Source OpenStreetMaps

rather apparent challenges to PT, other context-specific factors are at play in keeping ridership somewhat limited. In a study by Alkaabi (2014), the author examines *social attitudes* toward public transport in Dubai by surveying 340 respondents. The study highlighted nationality as an important predictor of transit ridership. The author found UAE nationals are less likely to use the Dubai metro versus other nationalities, with those from the Indian subcontinent representing the majority of users. The author concluded that dominant car culture, suburban living patterns, and the stigma of riding in PT were the causes of such findings, especially among the country's nationals. In the study of the bus system in Abu Dhabi, the UAE's capital, various authors highlighted similar trends in their findings (Almardood and Maghelal 2020; Qamhaieh and Chakravarty 2017). National origin, which happened to align with socioeconomic class, appeared to impact ridership numbers. The authors suggested that the PT ridership reflected divisions within the overall society, as the PT system was an extension of public spaces within the city.

Shared Mobility: "Upping" the Status of PT in Dubai

Like other globalizing cities, technologies and their use in city management (i.e., *smart cities)* have gained significant momentum in Dubai over the last few years (Khan et al. 2017). The Smart Dubai initiative has become one of the most important government projects lately, with the stated goal of rendering the government sector completely paperless by the end of 2021 (Smart Dubai, 2021). Innovation in city management is not new in Dubai, as the city is known for attracting the latest and greatest in technologies. As of late, the transportation sector has seen its share of innovations. From experimental flying taxis to driverless vehicles, in addition to some of the most cutting-edge digital traffic and PT

Youth and Public Transport, Fig. 2 Shared mobility solutions are emerging in different areas of the city. These solutions are provided by private corporations and require a credit card and a smartphone app to operate. Source (Author)

management tools – Dubai has become a hub for technology, and this direction is visible throughout the city (Badard 2016; RTA 2017, 2020a).

During the last 2 years, and in response to poor integration between PT and the urban environment, new ride-sharing apps and equipment started to appear throughout the city. These shared mobility and first–/last-mile initiatives have gained importance and have become quite popular among the population, at least for those who can afford to use them. Following international trends, the RTA seems to consider shared mobility the missing piece of the puzzle relative to PT ridership. Therefore, the city is investing heavily in these systems (RTA 2020b) – most notably, e-bikes, e-scooters, and shared cars. All these systems are smartphone app-based and require a credit card to use. They are provided by private operators and are not necessarily government-sponsored (Fig. 2). They appear to be gaining popularity and attract the younger "hip" urban crowd. Still, they are relatively new, with a limited spread, and are mostly located in the more business, tourist, and entertainment-oriented areas (Tesorero 2020). It is worth noting that poor public realm planning and conditions make riding such mobility solutions rather tricky, if not dangerous, in some cases.

Methods

Due to the emphasis on youth mobility, the selected participants in this study were college students at a university in Dubai. The university itself is unique due to its direct metro access, which made it easy for students to use the metro, if not other forms of PT. The students were primarily middle-class from a socioeconomic perspective, and they varied in national backgrounds, which is rather typical in the city. In order to understand youth engagement in public transport and subsequently shared mobility, a number of steps were taken.

First, an online survey was sent to the students (n = 35) titled *"How I Move."* The survey consisted of two groups of questions related to public transport: One set targeted those who use PT, looking at actual feedback about the system. The other group of questions targeted students who were primarily automobile dependent, gauging their perceived opinions about PT, with the intention of examining the prejudice toward the system. The distinction between the two groups was made based upon the question, *"do you use PT on a regular basis?"* For this survey, the sampling technique was *convenience sampling* (i.e., sampling those who were in the class at the time and willing to respond) (Etikan 2016). A larger and more purposeful sample was beyond the scope and limitations of this study. It is worth noting that the participants were asked to reflect on their perceptions and comments before Covid-19, as the ridership patterns during the pandemic are most likely very different and not the focus of this study.

For the second part of the research, the same students were asked to participate in an experiment. The willing students were asked to rent some of these commercially available shared mobility solutions and reflect on their experiences. After completing their experiment, students were given a further reflective survey with questions about their experience and were also asked to participate in a focus group for further discussion. The number of participants was 14 for both the second survey and the focus group.

Results from the First Survey

Out of the 35 surveys sent to the group, 24 responses were recorded (68% response rate). The demographic profile of the respondents reflected the typical college-aged student characteristics in the city of Dubai. The participants' age ranged between 20 and 26 years, with the majority between 20 and 22. Another important demographic was the national origin, as the majority were Arab expatriates, with Asian expatriates (mostly from India and Pakistan) representing the second most common group (Fig. 3).

75% of the respondents reported having a driving license, **whereas 62% owned a car**, which is rather significant for a youthful population, but at the same time expected in a highly automobile-dependent city such as Dubai within a relatively affluent population. Sixteen respondents also indicated using the family car to move around the city, and not necessarily their own cars. Some of them

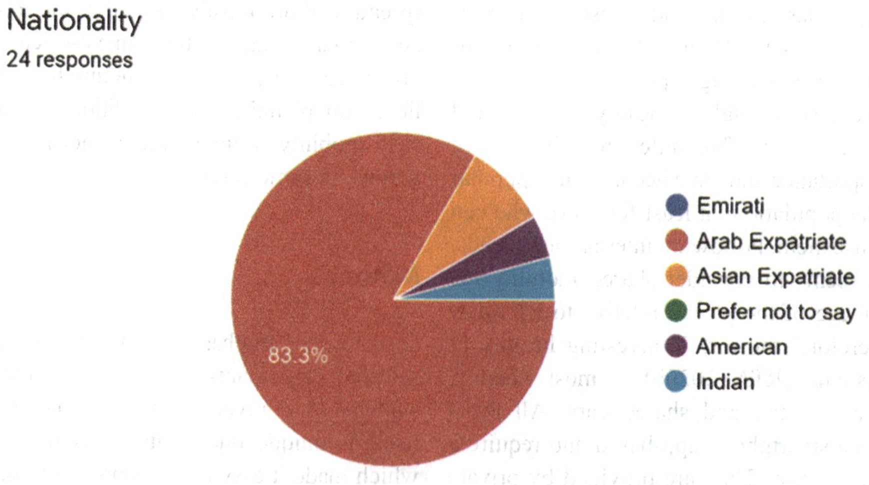

Youth and Public Transport, Fig. 3 Nationality of respondents

though own their vehicle, yet use the family car also, so they had multiple options relative to car usage.

The question about the *usage of public transport on a regular basis was* included in the survey. Upon responding with Yes, respondents (riders) were asked to evaluate the PT system based upon their actual experience. For those who answered No (nonriders), the survey automatically took them to a different section, where they were also asked about their opinions of PT in Dubai. Highlights of the results from both groups could be seen below.

Responses from Riders

The regular riders of PT (total ten respondents) mainly indicated that they use the metro system and the tram system only. Few indicated that they use the bus, though some suggested that some combination is used (Fig. 4). The sample size is very small to produce meaningful results, but it was evident from subsequent discussions with some of these students that the bus was not really something they dealt with, as it was not viewed as a good system. 70% reported that the metro/tram station was within walking distance from where they lived.

According to these regular users of PT, the strengths of the system included speed (80%), cleanliness (80%), and safety (70%) as the most dominant responses. When asked about the weakness of the systems, 90% of the responses indicated that being **crowded** was the biggest problem, then being **expensive** (70%) and poor **coverage of the city** (50%). All respondents to these questions could pick three options for system strengths and weaknesses out of a list of 11 items (each). When asked to comment about the areas for improvement, the **multiple stops** and the system's perceived **slow speed** were the most mentioned areas. **Cost and reduction of crowding** were some of the other comments made. Most of these comments were similar to the points discussed in the literature earlier, which are pretty typical for most public transport systems.

Responses from Drivers/Nonriders

Students who reported that they *do not use public transport* were asked about their opinions/impressions about the system. These nonusers were asked open-ended questions about their view of PT. Out of the 24 original survey respondents, 14 were nonusers of public transport except on rare occasions (Fig. 5). Taxis were excluded from the responses as they are not shared by multiple riders.

When checked against car ownership, 86% who owned cars did not use PT which is typical of such a group (Fig. 6).

As for respondents' opinions regarding public transport's strengths, different responses were recorded, but the most frequent response words included *pleasant looking (metro and tram), fast,* and *cheap.* Whereas, when asked about the system's perceived weaknesses, the most frequent

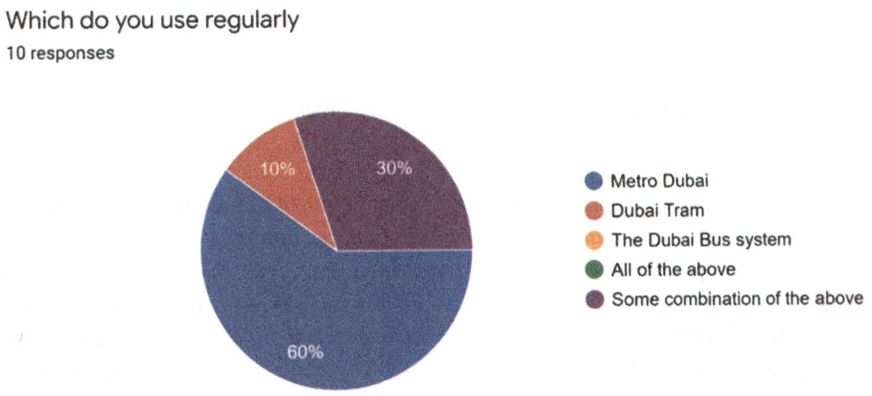

Youth and Public Transport, Fig. 4 Mode distribution

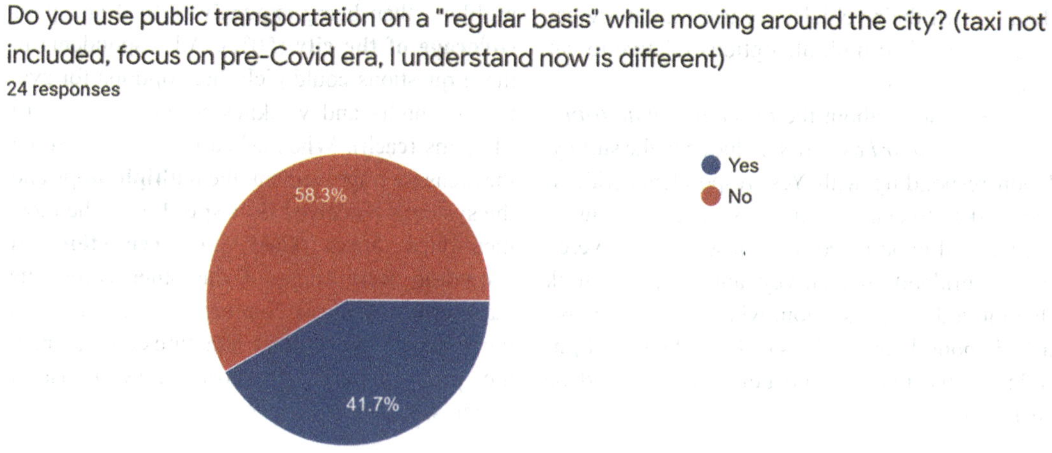

Youth and Public Transport, Fig. 5 Usage of PT in Dubai

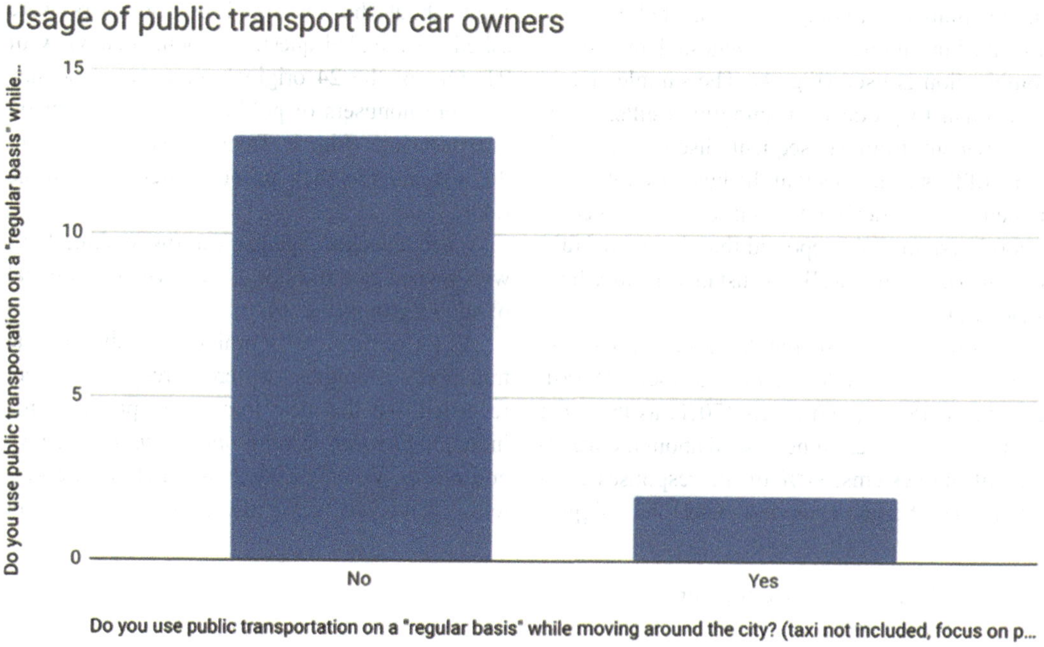

Youth and Public Transport, Fig. 6 Public transport usage for car owners

negative responses were *slow in comparison to the car, crowded, and does not reach the exact destination, and requires people to walk.*

Results from Shared-Mobility Test

Though the original 35 survey respondents were invited to take part in the shared-mobility experiment, only 14 students participated. According to the experiment requirements, the students were to select a location of their choice and provide pictures proving their participation. The time of day and duration to ride was left for them to decide to make it easier for them to participate. Afterward, they were required to participate in a small reflective survey, and subsequently a brief focus-group discussion.

Reflective Survey Results

The participants were asked which mode of transport they used (e-bike or e-scooter), and the results were split with seven users of e-bikes and seven users of e-scooters. Out of all respondents, 71% had previously tried at least one of these commercially available solutions, as indicated by the survey. When asked about the price to rent an e-bike or e-scooter, 77% were satisfied with the rental cost (though later in the focus group, the responses were slightly different, as students expressed concern regarding the long-term rental costs if it was to be used daily). Regarding the ease of riding, 57% thought these shared mobility solutions were easy to rent, whereas the other responses ranged between neutral or dissatisfied (Fig. 7).

The respondents were then asked to comment on the best and worst aspects of the system after their trial. The responses varied between the respondents, but on the positive side, **the new experience, the ease of use, and the convenience/fun** were among the words used to describe their experience. Regarding the negative aspects, the **distribution of the renting spots, price versus duration of the ride**, and the **lack of dedicated lanes** were words used in some of the comments.

The respondents were also asked, "what could encourage the wider usage of these systems?" and "what could be an obstacle to their wide usage in Dubai?" According to the responses received, adding **more renting spots was a top priority to increase usage**, as were **the placement in strategic locations** and **lower prices.** Regarding the obstacles, **the roads/cars and the lack of dedicated infrastructure** appeared to be important; the **weather** was also mentioned in the comments, and **the inability to go for long distances** or **cross busy streets.**

Some of the final open-ended comments in the survey included responses such as **fun, exciting, surprising, and enjoyable experiences,** which bode well for these shared mobility platforms. It was worth noting that all participants in this experiment were regular car drivers. They were also evenly split between males and females.

Focus Group Discussion

After completing the survey, the students were given the opportunity to reflect further on their experience riding the shared mobility solutions as part of a small focus group. Overall, the students were excited about the experience. They appeared to be enthusiastic and happy to have taken part in the trial. Although some indicated in the survey initially that the rental price was low, during the discussion, they appeared to backtrack as they considered the long-term rental costs, which would be expensive in such cases. Several comments were made about the lack of dedicated bike/scooter lanes and about having to navigate busy

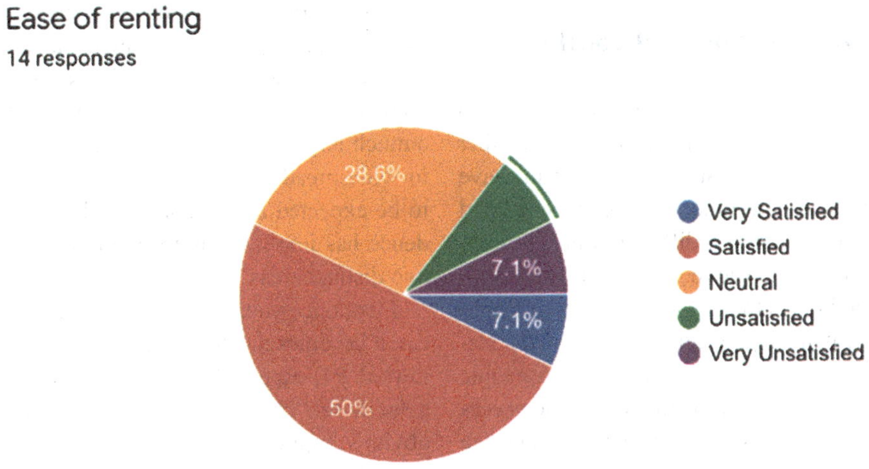

Youth and Public Transport, Fig. 7 Ease of renting

Y

sidewalks. Concerns about vehicle traffic were also made, especially when crossing streets. Some female participants made comments about the size of the e-bike as being large and not suitable for their heights. Some thought the e-bikes were heavy, whereas the e-scooters were too dangerous. One female participant indicated discomfort as she felt others in the street were staring at her while riding the e-bike, possibly pointing to some gender dimension.

The students were then asked if the solutions were practical for daily use instead of one "fun" trial. Some stated that it was hard to carry items such as backpacks or bags while riding; others commented that these solutions were for tourists and not necessarily for daily usage. Comments were also made about users' age and that these solutions will only work for the young (hip) crowd. Older generations and the elderly would not be willing or able to use these. Comments about types of clothes were also made, as a rider in a suit would be able to use an e-scooter, but not an e-bike.

Finally, as drivers of cars, the participants were asked if they would leave their cars and use PT and a combination of shared mobility instead. Few indicated that they would consider this if the systems were cheaper, more available, and easier/safer to ride. The majority, though, still stated that they would not give up their cars for such devices and preferred to continue driving.

Summary, Discussion, and Conclusions

In Dubai, a pioneering city in the region relative to transport, the provision of the metro and other supplemental systems such as the tram have been important milestones. These have helped expose the residents to a different kind of city – one beyond the car. The metro and the tram have become symbols of the city and are often featured in advertisements representing Dubai. Metro stations are now landmarks, and millions of people use the system to move around the city and access its multiple attractions. The system has also been a hit among tourists, who use it to shuttle from one side of the city to another in air-conditioned comfort, escaping the harsh desert heat. One could argue that the metro system has managed to "elevate" the public transport status in Dubai. It also managed to increase the ridership beyond the usual captive riders otherwise found in slow buses stuck in traffic.

Still, despite its success, the metro and other PT systems in Dubai suffer from lower ridership numbers in comparison to other similarly sized cities around the world. The *trifecta* of automobiles, car-centric city planning, and cheap fuel relative to income levels assures a complete dominance of cars among Dubai's well-to-do residents, nationals, and expatriates alike. While some do use PT, especially the metro, this use remains temperamental, and in some cases, seasonal depending on climate and conditions. Tourists do use the public transport systems, mainly the metro and tram, as they are well connected to the city's attractions. Meanwhile, the sizable lower-income segments are completely dependent on official public transport and other forms of informal transport to move around the city, and their daily trips are often complex, slow, and involve multiple poorly coordinated connections.

The status of youth relative to public transport is a critical issue in a city such as Dubai. The middle class in the city have "abundant" access to cars early in life, at least compared to other countries around the world. In the data collected for the chapter, most students owned a car or operated the family vehicle, and few used public transport. In the subsequent shared-mobility experiment, all participants owned a car. Though the sample size for this study is small, and a complete understanding of this issue will require a much larger and comprehensive study, it points to widespread automobile dependence, which is to be expected in this context. This early dependence has implications for the overall society in two distinct areas.

First, once younger generations are used to the car, it becomes much harder for them to transition toward PT, especially as they get older and life gets more "complicated." This has direct and obvious implications relating to sustainability at the urban scale. As the car remains dominant, society expects and demands more car-related infrastructure. Second, by being solely automobile dependent, the youth miss out on much-needed urban experiences nor develop PT loyalty. In the car, they are shielded from others and do not casually interact with other transport-captive groups. They also do not experience the city from the pedestrian perspective and remain outsiders in their own society. In the context of Dubai, this latter point is significant, as the majority of residents are expatriates who do not necessarily develop a sense of belonging, even after living in the country for a long time. It is also significant because the society appears to be divided along socioeconomic lines, which also happen to coincide with national origin. Riding public transport and navigating the city and society are learning experiences, and these car-dependent youth are missing out on these valuable experiences!

So how does shared mobility fit into this picture? Shared mobility, namely e-scooters and e-bikes, provides something new for the city residents. These technology-based solutions seemed to fit well with younger generations. In the study, the most noticeable effect of the riding experiments with the students was *the excitement factor*. The participants appeared to be genuinely excited to use these solutions, especially since it was the first time for some. They also managed to deal with the whole process with relative ease as a tech-savvy demographic. Renting the equipment was easy for them, and they did not see the cost as a major problem, at least not for occasional use. Also, riding an e-scooter or e-bike was considered a *fun experience*.

Beyond the study participants, it is becoming common to see these e-scooters/e-bikes zipping around some of the denser and newer urban areas in Dubai, while often operated by younger riders. Tourists also appear to use them in tourist-oriented areas. It is common to find e-bikes/e-scooters completely rented out during the weekend and busy times as a testament to their popularity. Despite the enthusiasm toward these solutions, questions remain relative to their integration within the overall PT system. Some of the participants' notes highlighted inconsistent coverage, the lack of pickup and drop-off points, and the absence of safe and dedicated lanes. More importantly, the privatization of these solutions and their associated costs remains another critical factor, as they are too expensive to use on a daily basis, and this automatically excludes the poorer segments of society.

In conclusion, this chapter advances the argument that these shared mobility solutions might be a good option for Dubai, considering the context and climate. They are relatively easy to deploy and could help compensate for the poor PT system integration into the urban fabric due to the city's original car-based and fragmented planning. More importantly, they help residents experience and view the city in a completely different light. They also might have benefits beyond the obvious ease of movement, as they could help get more people onto public transport, especially if they are planned correctly and are viewed as part of a more extensive accessibility planning approach. Shared mobility solutions seem to be popular as they make movement easier, even in the hotter climate. They do not need much physical effort to move, and they are based on smart technologies that younger generations find appealing and intuitive. For these systems to reach their full potential, they need to be better distributed in the city and taken seriously as a more sustainable transport solution that could help reduce automobile dependence. Transportation authorities should make these systems accessible to the masses by regulating and subsidizing the cost and preventing private operators from charging expensive fairs. This will broaden the user base for these systems and prevent additional transport inequalities and social divisions.

As more usage is adopted and better supporting infrastructure is provided, these shared mobility solutions could help elevate the perceived lower status of public transport and its users. Although some currently view them as fun/pass-time activities, as opposed to serious mobility solutions, the mere exposure to these options could eventually induce more serious and meaningful usage. Even with this little exposure, it could be that these younger generations will get used to driving less and consider using public transportation more. Maybe in a few years, college-aged students will skip car ownership altogether. Such a notion will help create more socially and environmentally *responsible* members of society, ones who appreciate their city, its other residents, and the unique urban experience Dubai provides.

Cross-References

- ► Disruptive Mobility: Sharing Electric Autonomous Vehicles (SEAVs) Reshape Our Future Cities
- ► Sustainable Urban Mobility

References

Al-Ali, A., Maghelal, P., & Alawadi, K. (2020). Assessing neighborhood satisfaction and social capital in a multicultural setting of an Abu Dhabi neighborhood. *Sustainability, 12*(8), 3200. https://doi.org/10.3390/su12083200.

Alawadi, K., & Benkraouda, O. (2019). The debate over neighborhood density in Dubai: Between theory and practicality. *Journal of Planning Education and Research, 39*(1), 18–34. https://doi.org/10.1177/0739456X17720490.

Alawadi, K., Hernandez Striedinger, V., Maghelal, P., & Khanal, A. (2021). Assessing walkability in hot arid regions: The case of downtown Abu Dhabi. *URBAN DESIGN International, February.* https://doi.org/10.1057/s41289-021-00150-0.

Alkaabi, K. A. (2014). Analyzing the travel behaviour and travel preferences of employees and students commuting via the Dubai Metro. In *The Arab World Geographer/Le Géographe du monde arabe* (Vol. 17, Issue 1).

Almardood, M. A., & Maghelal, P. (2020). Enhancing the use of transit in arid regions: Case of Abu Dhabi. *International Journal of Sustainable Transportation, 14*(5), 375–388. https://doi.org/10.1080/15568318.2018.1564405.

Anable, J. (2005). 'Complacent car addicts' or 'aspiring environmentalists'? Identifying travel behaviour segments using attitude theory. *Transport Policy, 12*(1), 65–78. http://eprints.uanl.mx/5481/1/1020149995.PDF.

Badard, O. (2016). Innovation drives more efficient public transport | the national. *Gulf News*. https://www.thenationalnews.com/opinion/innovation-drives-more-efficient-public-transport-1.198729

Beimborn, E. A., Greenwald, M. J., & Jin, X. (2003). Accessibility, connectivity, and captivity: Impacts on transit choice. *Transportation Research Record: Journal of the Transportation Research Board, 1835*(1), 1–9. https://doi.org/10.3141/1835-01.

Beirão, G., & Sarsfield Cabral, J. A. (2007). Understanding attitudes towards public transport and private car: A qualitative study. *Transport Policy, 14*(6), 478–489. https://doi.org/10.1016/j.tranpol.2007.04.009.

Belwal, R., & Belwal, S. (2010). Public transportation Services in Oman: A study of public perceptions. *Journal of Public Transportation, 13*(4), 1–21. https://doi.org/10.5038/2375-0901.13.4.1.

Brown, A. E., Blumenberg, E., Taylor, B. D., Ralph, K., Bloustein, E. J., Voulgaris, C. T., & Turley Voulgaris, C. (2016). A taste for transit? Analyzing public transit use trends among youth. *Journal of Public Transportation, 19*(1), 4.

Burian, J., Zajíčková, L., Ivan, I., & Macků, K. (2018). Attitudes and motivation to use public or individual transport: A case study of two middle-sized cities. *Social Sciences, 7*(6), 83. https://doi.org/10.3390/socsci7060083.

Cervero, R., & Landis, J. (1994). Making transit work in suburbs. *Transportation Research Record, 1451*, 3.

Collins, C. M., & Chambers, S. M. (2005). Psychological and situational influences on commuter-transport-mode choice. *Environment and Behavior, 37*(5), 640–661. https://doi.org/10.1177/0013916504265440.

Currie, G., & Stanley, J. (2008). *Transport reviews investigating links between social capital and public transport.* https://doi.org/10.1080/01441640701817197.

De Gruyter, C., Currie, G., & Rose, G. (2016). Sustainability measures of urban public transport in cities: A world review and focus on the Asia/Middle East region. *Sustainability, 9*(1), 43. https://doi.org/10.3390/su9010043.

Elsheshtawy, Y. (2008). The evolving Arab City. In Y. Elsheshtawy (Ed.), *The evolving Arab City: Tradition, modernity and Urban development.* Routledge. https://doi.org/10.4324/9780203696798.

Elsheshtawy, Y. (2010). *Dubai: Behind an urban spectacle.* Routledge.

Elsheshtawy, Y., & Bastaki, O. (2011). *The Dubai experience: Mass transit in the Arabian peninsula.* http://globalvisions2011.ifou.org/Index/Group4/FOUA00065-00072P2.pdf

Etikan, I. (2016). Comparison of convenience sampling and purposive sampling. *American Journal of Theoretical and Applied Statistics, 5*(1), 1. https://doi.org/10.11648/j.ajtas.20160501.11.

Farooqui, M. (2019). 10 glorious years of Dubai metro: A runaway success story | transport – Gulf News. *Gulf News*. https://gulfnews.com/uae/transport/10-glorious-years-of-dubai-metro-a-runaway-success-story-1.66226435

Gokulan, D. (2015). Must read: The life and times of the Dubai bus since 1968 – News | Khaleej Times. *Khaleej Times*. https://www.khaleejtimes.com/nation/general/must-read-the-life-and-times-of-the-dubai-bus-since-1968

Green, J., Steinbach, R., Jones, A., Edwards, P., Kelly, C., Nellthorp, J., Goodman, A., Roberts, H., Petticrew, M., & Wilkinson, P. (2014). On the buses: A mixed-method evaluation of the impact of free bus travel for young people on the public health. *Public Health Research, 2* (1), 1–206. https://doi.org/10.3310/phr02010.

Grengs, J. (2005). The abandoned social goals of public transit in the neoliberal city of the USA. *City, 9*(1), 51–66. https://doi.org/10.1080/13604810500050161.

Grimsrud, M., & El-Geneidy, A. (2014). Transit to eternal youth: Lifecycle and generational trends in greater Montreal public transport mode share. *Transportation, 41*(1), 1–19.

Jensen, M. (1999). Passion and heart in transport – A sociological analysis on transport behaviour. *Transport Policy, 6*(1), 19–33. https://doi.org/10.1016/S0967-070X(98)00029-8.

Kaiser, J. (2007). Dubai public transport bus master plan: a new era of public transport services in the world's fastest developing city. *International conference on competition and ownership in land passenger transport, 10Th, 2007, Hamilton Island*, 37–81 [WORKSHOP 3B]. https://trid.trb.org/view/855162

Kent, J. L. (2015). Still feeling the Car – The role of comfort in sustaining private Car use. *Mobilities, 10* (5), 726–747. https://doi.org/10.1080/17450101.2014.944400.

Khalaf, S. (2006). The evolution of the Gulf city type, oil, and globalization. In J. W. Fox, N. Mourtada-Sabbah, & M. Al Mutawa (Eds.), *Globalization and the Gulf* (pp. 254–275). Routledge.

Khan, M., Woo, M., Nam, K., & Chathoth, P. (2017). Smart City and smart tourism: A case of Dubai. *Sustainability, 9*(12), 2279. https://doi.org/10.3390/su9122279.

Krizek, K. J., & El-Geneidy, A. (2007). Segmenting preferences and habits of transit users and non-users. *Journal of Public Transportation, 10*(3), 5.

Manville, M., Taylor, B. D., & Blumenberg, E. (2018). Transit in the 2000s: Where does it stand and where is it headed? *Journal of Public Transportation, 21*(1), 104–118. https://doi.org/10.5038/2375-0901.21.1.11.

Murray, S. J., Walton, D., & Thomas, J. A. (2010). Attitudes towards public transport in New Zealand. *Transportation, 37*(6), 915–929. https://doi.org/10.1007/s11116-010-9303-z.

Qamhaieh, A., & Chakravarty, S. (2017). Global cities, public transportation, and social exclusion: A study of the bus system in Abu Dhabi. *Mobilities, 12*(3), 462–478. https://doi.org/10.1080/17450101.2016.1139805.

Qamhaieh, A., & Chakravarty, S. (2020). Drive-through cities: Cars, labor, and exaggerated automobilities in Abu Dhabi. *Mobilities, 15*(6), 792–809. https://doi.org/10.1080/17450101.2020.1822103.

Ramos, S., Vicente, P., Passos, A. M., Costa, P., & Reis, E. (2019). *Perceptions of the public transport service as a barrier to the adoption of public Transport: A Qualitative Study*. https://doi.org/10.3390/socsci8050150.

Rizzo, A. (2014). Rapid urban development and national master planning in Arab gulf countries. Qatar as a case study. *Cities, 39*, 50–57. https://doi.org/10.1016/j.cities.2014.02.005.

RTA. (2017). *Dubai self-driving transport strategy*. https://www.rta.ae/links/sdt/en/news2.html

RTA. (2019). News Details- Dubai Metro accomplishes a decade of happy journeys for 1.5 b riders. September 8, https://www.rta.ae/wps/portal/rta/ae/home/news-and-media/all-news/NewsDetails/dubai-metro-accomplishes-a-decade-of-happy-journeys-for-riders?lang=en

RTA. (2020a). *Premium initiatives rolled out to mark 'UAE Innovation Month.'* https://www.nol.ae/wps/portal/rta/ae/home/news-and-media/all-news/NewsDetails/premium-initiatives-rolled-out-to-mark-uae-innovation-month

RTA. (2020b). Endorsing the first and last-mile strategy to link with public transport network. https://www.rta.ae/wps/portal/rta/ae/home/news-and-media/all-news/NewsDetails/endorsing-the-first-and-last-mile-strategy-to-link-with-public-transport-network

RTA. (2021). *Roads & transport authority – Public transport*. https://www.rta.ae/wps/portal/rta/ae/public-transport

Shaheen, S., & Chan, N. (2016). Mobility and the sharing economy: Potential to facilitate the first- and last-mile public transit connections. *Built Environment, 42*(4), 573–588. https://doi.org/10.2148/benv.42.4.573.

Sheller, M., Urry, J., & Purcell, D. (2000). The city and the car. *International Journal of Urban and Regional Research, 24*(4), 737–757. https://doi.org/10.1111/1468-2427.00276.

Steg, L. (2005). Car use: Lust and must. Instrumental, symbolic and affective motives for car use. *Transportation Research Part A: Policy and Practice, 39*(2–3), 147–162. https://doi.org/10.1016/j.tra.2004.07.001.

Taylor, B. D, & Fink, C. N. Y. (2003). The factors influencing transit ridership: A review and analysis of the ridership literature publication date. In *UCLA department of urban planning*. https://escholarship.org/uc/item/3xk9j8m2

Taylor, B. D., & Fink, C. N. Y. (2013). Explaining transit ridership: What has the evidence shown? *Transportation Letters, 5*(1), 15–26. https://doi.org/10.1179/1942786712Z.0000000003.

Taylor, B. D., & Morris, E. A. (2015). Public transportation objectives and rider demographics: Are transit's priorities poor public policy? *Transportation, 42*(2), 347–367. https://doi.org/10.1007/s11116-014-9547-0.

Tesorero, A. (2018). Dubai has so many cars, but what about parking spaces? *Khaleej Times*. https://www.khaleejtimes.com/nation/dubai/dubai-has-so-many-cars-but-what-about-parking-spaces-free

Tesorero, A. (2020). How Dubai commuters can now use e-scooters to or from the metro | transport – Gulf news. *Gulf News*. https://gulfnews.com/uae/transport/how-dubai-commuters-can-now-use-e-scooters-to-or-from-the-metro-1.74857382

Ureta, S. (2008). To move or not to move? Social exclusion, accessibility and daily mobility among the low-income population in Santiago, Chile. *Mobilities, 3*(2), 269–289. https://doi.org/10.1080/17450100802095338.

Van Lierop, D., Badami, M. G., & El-Geneidy, A. M. (2017). *Transport reviews what influences satisfaction and loyalty in public transport? A review of the literature*. https://doi.org/10.1080/01441647.2017.1298683.

Xia, T., Zhang, Y., Braunack-Mayer, A., & Crabb, S. (2017). Public attitudes toward encouraging sustainable transportation: An Australian case study. *International Journal of Sustainable Transportation, 11*(8), 593–601. https://doi.org/10.1080/15568318.2017.1287316.

Zuo, T., Wei, H., Chen, N., & Zhang, C. (2020). First-and-last mile solution via bicycling to improving transit accessibility and advancing transportation equity. *Cities, 99*, 102614. https://doi.org/10.1016/j.cities.2020.102614.

Z

Zero Emissions

► Carbon Neutral Adelaide

Zombie Subdivisions

Victoria Kolankiewicz and David Nichols
Faculty of Architecture, Building and Planning, University of Melbourne, Melbourne, VIC, Australia

Synonyms

Ghost estates and new ruins (David McWilliams in Kitchin et al. 2014, p. 1070); Interstitial spaces (Phelps and Silva 2018, p. 1206)

Definition

Zombie subdivisions are subdivided tracts of land where development has failed to proceed, either in part or full. The term "zombie" is lent by Holway et al. (2014) based on their work at the Lincoln Institute of Land Policy. They describe these sites as the "living dead of the real estate market" as they are planned, even attempted, but not completed (Holway et al. 2014, p. 5).

Introduction

Three typologies of the zombie subdivision exist. The first is where the developer or landowner is actively managing or completing a previously stalled development. A second typology finds development temporarily stalled, awaiting more favorable circumstances. The final typology sees these sites totally abandoned or vacated, with no intention to recommence or finalize development (Kolankiewicz et al. 2019). These subdivisions typically exist on paper. They can often be found in cartographic materials such as street directories, where proposed street layouts and names are shown. Zombie subdivisions are typically under private ownership, whether in the hands of developer interests, or sold to private landholders. These sites are closely tied to speculative development, and their ultimate fulfillment is precluded by site conditions or economic circumstance which renders their completion or intended use unviable.

Economic Zombies

The creation of zombie subdivisions is closely tied to institutional or market forces, as well as broader site conditions affecting the viability of development. Economic conditions are formative in the "zombification" of subdivisions, drawing

R. C. Brears (ed.), *The Palgrave Encyclopedia of Urban and Regional Futures*,
https://doi.org/10.1007/978-3-030-87745-3

upon a long history of real estate speculation in industrializing nations as an endeavor of corporate profiteering (Vance 1971). The zombie subdivision problem is exacerbated by the speculative nature of real estate development, where buyers are often enticed to purchase land with developer assurances of significant increases in value. Speculative estates can be quickly drawn up and sold with few buyer protections. Local governments are also enticed into cooperation with subdivision schemes by the prospect of solidifying their financial position through local population growth. Speculation can often lead to the subdivision of more lots than what is viable, particularly where developer cash flow precludes the provision of key infrastructure and services. Ireland's "ghost estates" arose in a period of rapid economic growth preceding the 2008 Global Financial Crisis, facilitated by neoliberal economic policies collectively known as the "Celtic Tiger." This was further facilitated by lax planning regulations and favorable tax incentives to encourage development. Today, there are over 2500 abandoned or unfinished estates across Ireland (Kitchin et al. 2014). They exemplify the precarious conditions of capitalism where the provision of housing was driven by corporate interests rather than being responsive to the needs of the public.

Teton Valley, a largely agricultural locality in Idaho, is another example of this, which experienced haphazard and speculation-driven subdivision throughout the 2000s. Its own planning processes facilitated the subdivision of 20-acre lots into smaller 2.5-acre parcels, with no broader provisions for how the County would accommodate urban development and population growth (Trentadue 2012). Over 10,000 lots were rapidly approved and sold at inflated prices, serviced only by incomplete roads and sidewalks (Peterson et al. 2013). Providing infrastructure to these lots would prove costly as they were often in unincorporated and poorly sited areas. Foreclosures and revised land values in the wake of the 2008 GFC saw a significant reduction in Teton County's property tax base. This, alongside the high capital costs of lot servicing, would have resulted in a budget deficit of approximately $17 million dollars if development were to recommence. Landholders and developers were left to shoulder the financial burdens of debt and foreclosure. Today, the County continues to host over 8000 undeveloped allotments along culs-de-sac and looped streets, now gradually being revegetated and reincorporated back into the rural environment (Dunham-Jones and Brown 2015).

Environmental Zombies

Zombie subdivisions can also develop where development is financially viable but ultimately stymied by environmental or site conditions. Political or regulatory obstacles, such as incompatible zoning or the realization of local environmental significance, can partially or wholly limit development activities in spite of their profitability (Taylor et al. 2015). These tracts of land can further "zombify" where municipalities lack the planning tools or financial resources to acquire and remedy the problems affecting these difficult sites. Such developments are ultimately not matched to local conditions. An Australian neologism sees these sites described as "old and inappropriate subdivisions" (Spragg 1976). These subdivisions are typically suburban-sized lots in non-urban areas, where their development would have implications for environmental health and bushfire safety. Such subdivisions are subjected to restructure plans which, with landholder cooperation, result in their consolidation into larger lots under municipal planning powers (Shire of Yarra Ranges, Restructure Plan for Old and Inappropriate Subdivisions, 2015). Another example is Ninety Mile Beach, a rural beachside locality of Gippsland, Victoria. Here, sensitive environmental conditions became apparent following the subdivision and sale of land in the 1950s (Victorian Ombudsman 2019). Small allotments were sold along a sliver of land between Lake Reeve and the Bass Strait, which not only contained delicate sand dunes but was also susceptible to erosion and flooding (Wellington Shire Council 2018). The local municipality has been supported by the Victorian State Government to engage landholders in a Voluntary Assistance Scheme, which as of 2011 has facilitated the transfer of approximately 1500 lots to municipal ownership for remediation.

Subdivisions are also "zombified" by their unintended co-location with hazardous industry. Another Australian example is the Burns Road Estate, located in the Melbourne suburb of Altona. Although the Estate was subdivided in the early twentieth century, the growth of a local petrochemical industry has stymied any capacity to develop the land as a result of planning regulations that limit the number of people on any given lot (Hobsons Bay City Council 2014, 2019). This could also be resolved through the restructuring and consolidation of allotments. However, this problematized by the sheer number of landholders – 505 blocks are held between 170 owners – as well as the value of the land itself, highlighting the importance of funding and government support in "reanimating" the zombie (HBCC 2014, 2019).

Summary/Conclusion

Zombie subdivisions reveal oversights in the development process. They can mark a failure to anticipate changes in economic conditions. They can also illustrate an oversight of broader environmental or site factors which impact the viability of a development. They illustrate the need for strong governance and planning regulations to guide outcomes that are in the interests of the environment and the community, rather than financial and developer interests. In an era of so much uncertainty – notably climate change and the COVID-19 pandemic – there may be more zombie subdivisions to come. At time of writing, a debt crisis for the China-based Evergrande Group is mounting with significant implications for its numerous real estate projects still under construction (Shepherd 2021). Its Evergrande Technology and Tourism City has investors calling for clarity on the project's future: "What are we thousands of owners to do?". Understanding the genesis of "zombified" sites may not only provide some lessons for housing developments of the future but also illuminate how such problematic sites can be redressed and resolved.

Cross-References

▶ Master Planned Estates and the Promises of Suburbia

References

Dunham-Jones, E., & Brown, W. (2015). The public sector steps up – And retrofits a zombie subdivision. In E. Talen (Ed.), *Retrofitting sprawl: Addressing seventy years of failed urban form* (pp. 139–156). Athens, GA: University of Georgia Press.

Hobsons Bay City Council. (2014). *Schedule 4 to the Special Use Zone, Hobsons Bay Planning Scheme.* http://planning-schemes.delwp.vic.gov.au/schemes/hobsonsbay/ordinance/37_01s04_hbay.pdf

Hobsons Bay City Council. (2019). *Burns Road Industrial Estate,* https://www.hobsonsbay.vic.gov.au/Services/Planning-Building/Planning-Scheme-Amendments-and-Current-Projects/Current-Projects/Burns-Road-Industrial-Estate

Holway, J., Elliott, D. L., & Trentadue, A. (2014). *Arrested developments: Combating zombie subdivisions and other excess entitlements.* Cambridge MA: Lincoln Institute of Land Policy.

Kitchin, R., O'Callaghan, C., & Gleeson, J. (2014). The new ruins of Ireland? Unfinished Estates in the Post-Celtic Tiger era. *International Journal of Urban and Regional Research, 38*(3), 1069–1080.

Kolankiewicz, V., Phelps, N., Nichols, D., & Taylor, E. (2019). Tracing the 'Zombification' of Undeveloped Estates in Greater Melbourne and its Outlying Regions. *9th State of Australian Cities National Conference.* https://doi.org/10.25916/5f0ceb1737686

Peterson, M. N., Peterson, T. R., & Liu, J. (2013). "Housaholism" in the greater Yellowstone ecosystem. In M. N. Peterson, T. R. Peterson, & J. Liu (Eds.), *The housing bomb: Why our addiction to houses is destroying the environment and threatening our society* (pp. 55–81). Baltimore, MA: John Hopkins University Press.

Phelps, N. A., & Silva, C. (2018). Mind the gaps! A research agenda for urban interstices. *Urban Studies, 55*(6), 1203–1222.

Shepherd, C. (2021, September 22). China's Evergrande veers toward default — and a $300 billion global shock. *Washington Post.* https://www.washingtonpost.com/world/asia_pacific/china-evergrande-debt-markets/2021/09/22/eeb80fd4-19cc-11ec-bea8-308ea134594f_story.html

Spragg, R. C. (1976). *Inappropriate rural subdivision: a review.* Melbourne, Australia: Town and Country Planning Board.

Taylor, E., Nichols, D., & Kolankiewicz, V. (2015). Solomon Heights – A Zombie Subdivision?. In P. Burton & H. Shearer (Eds.), *7th State of Australian Cities Conference* (pp. 1–11).

Trentadue, A. (2012). *Addressing excess development entitlements: Lessons learned in Teton County, Idaho*. Cambridge, MA: Lincoln Institute of Land Policy.

Vance, J. E. (1971). Land assignment in the precapitalist, capitalist, and postcapitalist city. *Economic Geography, 47*(2), 101–120.

Victorian Ombudsman. (2019). *Investigation into Wellington Shire Council's handling of Ninety Mile Beach subdivisions*. Melbourne: Victorian Government Printer.

Wellington Shire Council. (2018). *Ninety Mile Beach Plan*. www.wellington.vic.gov.au/Developing-Wellington/Planning-Projects/Ninety-Mile-Beach-Plan

Zoning

► Participatory Planning: A Useful Tool for the Development of Sustainable Mega-City Regions

Zoning Regulation

► Land Use Planning Systems in OECD Countries

Zooming Regions into Perspective

Climate Change, Resilience, and Settlement Planning Systems

Innocent Chirisa
Department of Demography Settlement and Development, Social & Behavioural Sciences, University of Zimbabwe, Harare, Zimbabwe

Synonyms

Land use – land organization; Region – precinct; Sustainability – continuity

Definition

Urban planning – technical and political process resulting in the organisation and re-organisation of space.

Land use – organization of space for defined functions and activities.

Sustainability – continuity with a balance.

Urbanization – a demographic and spatial transformation of the social and landscape from rural to urban state.

Introduction

The gale force winds of the tropical cyclones that have inundated Southern Africa since 2000 continue to be a serious cause of concern among practitioners in the built environment. Clearly the built environment comprising infrastructure and buildings are under threat of destruction. The destruction of infrastructure carries with its huge costs for the governments relating to the evacuation of victims at mission stations the reconstruction and maintenance of infrastructure and housing facilities in the affected areas. Most of the mainstream studies have generally centered on informal and slum settlements considering them to be more prone to cyclonic disasters (e.g., Muller et al. 2003; Huchzermeyer and Karam 2006; Oldewage-Theron et al. 2006; Pelling and Wisner 2012) than rural settlements. The most recent tropical cyclones code-named "Idai" has ravaged Malawi, Mozambique, and Zimbabwe, while "Kenneth" targeted Mozambique and Tanzania. The massive devastation wrought by these cyclones has not only attracted international attention but has shifted the radars of research interest to rural areas as the most affected areas.

The study in support of this chapter is largely informed by a document review. The study employs literature, reports, and newspaper articles on rural poverty, vulnerability, and resilience building strategies in Zimbabwe. The data collected during the document reviews was then presented and interpreted using thematic content analysis. The process entails grouping data in themes, like rural housing, rural construction, legislation for rural housing, and resilience in rural housing in Zimbabwe. These themes will help in responding to the aim of the study.

Background and Overview

Disaster planning and disaster resilience have increasingly occupied the center stage of global concern. The combined interest in disaster planning and resilience has mobilized research underpinning the creation of adaptive designs and policy frameworks to address the growing socio-economic and spatial problems. In some of the coastal areas, such as Maputo and the low-lying regions of Tete in Mozambique, flooding has become rampant (Kay and Alder 2017). In areas prone to active earthquakes, innovative strategies regarding earthquake adaptation have become popular, while in the countries prone to mud slides, similar threats have been addressed accordingly (Thomas and Griffiths 2017). Cyclones have become detrimental globally, and the loss to cyclonic disasters of properties worth trillion of dollars has been common (Payo Garcia et al. 2017). Consequently, the systematic crafting of resilient responses to cyclonic disasters in the affected rural areas has become one of the global Sustainable Development Goals (SDGs).

The need for cyclonic disaster planning targeting most rural areas in Africa seems to be long overdue as developments in these areas have generally not been guided by any form of systematic planning and are piecemeal. This is so since spatial and economic development of most rural Africa is largely independent of the development planning for urban areas (McManus et al. 2012). The former colonial development policy ideals and principles traditionally segregated against the development of the African rural areas to match the development of the urban areas where the majority White population lived. This white supremist approach to development created a planning system that neglected rural planning (Muller et al. 2003). As such, development in rural areas has been routinely incremental and haphazard.

However, since independence, in many Southern African countries, rural development planning is largely managed by central governments through decentralized local governments although most of these lower level local authorities have invariably failed to address local needs (Sørensen and Torfing 2012). The major causes of this failure have been attributed to the lack of sufficient technical and financial to support local development initiatives as well as political interference in local governance by central authorities. As such, rural development planning has not been productive, while the rural constituencies have tended to survive on handouts from politicians in exchange for their votes come election time. This kind of survival has left rural Africa depending on whatever they can get their hands on. In most cases, the rural dwellers survive on land-based economic and social activities thereby leaving them prone to climate variabilities. A report by UN (2012) indicated that most (93%) of the African rural residents survive on subsistence agriculture thereby exposing them to climate change related food security threats. Wary of possible loss of agricultural produce due to hostile weather conditions, most rural residents have settled in flood plains that are fertile and productive albeit the proneness to wash away through flooding.

In considering the demands of affordability and cultural norms, rural inhabitants in Africa are adept in deploying vernacular architecture and the readily available building material including stones, mud, wood, and grass (Sørensen and Torfing 2012) for the construction of their homes. The use of these easily accessible building materials has dominated African housing construction since the eclipse of the Stone Age when the indigenous population relocated from the cave into other forms of shelter (Pelling and Wisner 2012). While the housing is normally cheap to build in African rural areas, however, the durability of the structures in the wet weather is highly questionable. The buildings are designed to withstand the generally hot climates in most rural Africa; however, under prolonged periods of extremely wet weather and flooding conditions, the poorly constructed mud huts can easily disintegrate leading to deaths and loss of livelihoods. In efforts to avoid and ameliorate such calamities, the governments of Zimbabwe, South Africa, Botswana, and Nigeria have mobilized funding for rural housing construction programs (Pelling and Wisner 2012; Scott 2013; Murata et al. 2019). However, the funding for such massive investment programs has generally been paltry.

By and large, regional planning for rural development in most Southern African countries has traditionally focused on economic development (Murata et al. 2019). The importance of planning for disaster resilience on the African subcontinent has increased relative to the damages wrought by cyclones, volcanic eruptions, and earthquakes among other natural disasters (Devi 2019). In most rural areas in Africa, the systematic planning of settlements common in the global North is absent. A look at rural Mozambique can explain this. Since Samora Machel's declaration of free land for all, Mozambique has followed an anarchist path of unplanned rural development (Greco 2016). This path has exposed the country to most natural disasters, with the most recent disaster, Cyclone Idai, causing the death of more than one thousand people and leaving a hundred thousand victims in need of assistance.

The ravaged country now requires $2 billion dollars to recover from the damage caused by cyclones Idai and Kenneth (Devi 2019). The tradition of "villagization" rural development planning in Tanzania dates back to the 1960s when Julius Nyerere conceived *ujamaa* (villagization) policy. Nyerere's *ujamaa* approach was based on the concept of modelling collectivized villages sharing commonage facilities, such as cropping fields, irrigation schemes, and paddocks that would generate agglomeration economies for rural development. The main focus on economic development was embedded in settlement planning.

Rural settlement planning in Zimbabwe has been closely associated with piecemeal planning. Settlement planning was segregatory in nature as it prioritized white settled areas as compared to areas where the blacks were residing. On the one hand, there exist government policies and legislation that attempt to respond to the needs of housing settlers. On the other hand, there is a wave of uncoordinated development, as the government does not have enough capacity to successfully carry out systematic rural settlement planning. Precolonial Zimbabwe had a divided rural planning into two, the part that was developed for Agricultural Development Authority that was well planned and the unplanned Tribal Trust Lands where the majority African population was forced to live by the Rhodesian colonial and subsequent settlers' governments (Stoneman 2017).

By 1978, the Ministry of Lands, Natural Resources and Rural Development had responsibility for ARDA, TILCOR, Intensive Conservation Areas (ICAs), and the Water Authorities. This Ministry was solely responsible for rural development. During colonial settlerist era, the name of ARDA changed to Agriculture and Rural Development Authority. The new name justified the need for an integrated plan for rural development under the then Ministry of Finance. At the attainment of independence in1980, Zimbabwe saw a shift into a unitary planning system when the new Mugabe-led government embarked on the Growth with Equity Policy approach (Makumbe 1996). A new initiative to stimulate district level planning was needed. Accordingly, a new approach – bottom-up planning – was inserted in the development planning system and enshrined in legislation. The idea was to institutionalize a planning system that allowed development plans to commence from the village and influence each level upward to ward, district, and province level (Hoddinott 2006). The 1990s were a decade of major spatial planning upheaval in the country. The failure of the existing development planning modus operandi to provide an integrated and participatory planning system that could respond to people's local needs and priorities, led to the growth of a number of new planning initiatives. Surprisingly, albeit the relevance of disaster and flood mitigation measures to physical planning, such weather calamities ever threatening the country's space economy engaged the imagination of the day's land-use practitioners.

Undoubtedly, Zimbabwe was caught in a disaster emergency planning nap when tropical Cyclone Idai drowned Maputo before sweeping through the country's eastern highlands in early May 2019. Cyclone Idai left painful memories with stakeholders in government, the research and academic fraternity, civil society, and the affected communities in the devastated parts of the country. Despite forewarnings about the approaching cyclone, the following questions were asked: why were the weather forecasts concerning the cyclone ignored? What were the long-, medium-, and short-term measures that should have been put in place? Who should have done what and why or why not? It is very important that any stakeholder does not end up in paralysis by analysis of why they did not do what was at their disposal to help avert a disaster. Although the early warning shots were fired, no action, in terms of disaster awareness and preventative evacuations, was taken resulting in communities being affected in situ.

It is important that one also understands the dynamics surrounding the genesis and evolution of the cyclone. It began in the northeastern coast of Madagascar in the Indian Ocean. A vivid description of the life of the cyclone has been presented as follows:

> "The cyclone started off the eastern coast of Mozambique in early March and hit the country's coast for the first time before heading back out into the Mozambique Channel. It intensified, weakened and intensified again before hitting the Mozambique coast for the second time on 14 March. Its winds reached up to 177 km/h (106 mph) and heavy rainfall caused disastrous flooding across a number of countries in its path."

Figure 1 below is portrayal of its path depicting the wind speeds and the affected areas (Rodgers et al. 2019).

Overall, the intensity of the winds of Cyclone Idai grew in force leaving a trail of destruction along its path. In Zimbabwe, the areas that were most affected included Chimanimani, Chipinge, some parts of Masvingo Province, and Chikomba Rural District in Mashonaland East Province. Some houses were destroyed, people lost lives (so were livestock and poultry), and social infrastructure including schools as well as roads and bridges were destroyed. The evidence of rocks and boulders coming to the surface in Chimanimani suggested that an earthquake might have happened before the cyclone hit.

Literature Review

To understand the processes underlying planning for rural resilience and sustainable rural communities, the term "rural" should be defined albeit there is no consensus regarding its definition (Jordan and Hargrove 1987). This is largely because

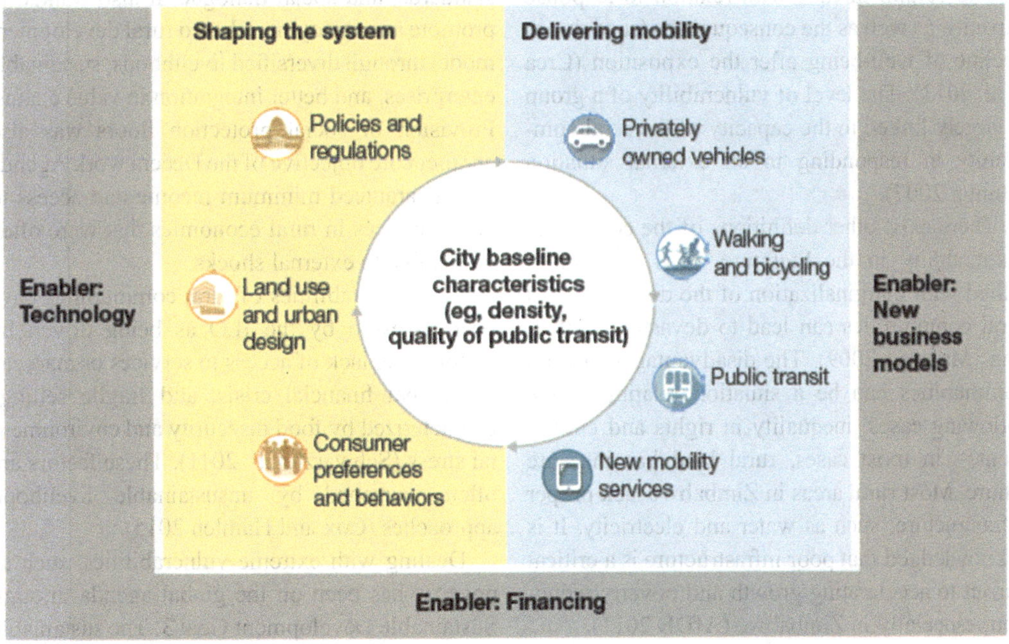

Zooming Regions into Perspective, Fig. 1 The path of Cyclone Idai (Cai et al. 2019)

rural areas differ from place to place and over time. The definition of "rural" has been based on population sizes, activities in the area, and economic development – among other conceptualizations (Flora 2018). The OECD defines "rural" as a community with a threshold population of 150 or less people per square kilometer. A further look at literature suggests that the definitions of "rural" are rather complex leading to (Cobbett 2016) defining "rural" simply as "what is not urban." Clearly, defining rural makes it easy to understand what the rural development plans are meant to address and how they are designed (Shadish et al. 2002, p. 65).

Rural areas are largely associated with widespread poverty, climate change, natural resource degradation, conflict, weak institutions, and poor agricultural conditions (Marsden 2009). Rural poverty is a driver of a host of social problems, including hunger and malnutrition, poor working conditions, and exploitation of children (Mayer et al. 2005). In addition, rural communities are particularly vulnerable to climate change due to their dependence on rainfed agriculture (Asian Development Bank 2014). Vulnerability as a concept is largely associated with the level of exposure to certain threats, with relation to response capacity, as well as the consequences faced due to decline of well-being after the exposition (Crea et al. 2013). The level of vulnerability of a group is largely linked to the capacity of the rural community in responding to an external situation (Hunter 2007).

There exist other definitions of the concept of vulnerability in the literature. This situation if tallied with marginalization of the disadvantaged rural communities can lead to devastating situation (Marsden 2009). The disadvantages of rural communities can be a situation relating to the following cases, inequality in rights and entitlement – in most cases, rural land has insecure tenure. Most rural areas in Zimbabwe lack proper infrastructure, such as water and electricity. It is acknowledged that poor infrastructure is a critical barrier to accelerating growth and poverty reduction, especially in Zimbabwe (AfDB 2011).

Having understood the risk and the vulnerabilities associated with rural communities, it becomes apparent that some communities are more vulnerable than others (McManus et al. 2012). In the context of this study, risk is defined as to uncertainty about and severity of the events and consequences (or outcomes) of an activity with respect to something valued by humans (Aven and Renn 2009). This imply that they lack the resilience capacity to cope, react, resist, anticipate, and recover from various external factors that communities are exposed to (Franklin et al. 2011). The living conditions of these vulnerable groups can be enhanced through government interventions and policies (Glover 2012). Efforts to build resilience can be done through formal and informal ways. Some of these ways include adaptation, reduction, sharing or transfer mechanism (insurances), avoidance, retention, and preparedness among others.

The International Labor Organization (ILO) has looked at increasing rural resilience to deal with rural poverty. Rural resilience has been done in the context of providing decent employment in what is called the Decent Work Agenda (Scott 2013). The Decent Work Agenda was aimed at increasing the voice of rural people through organization of communities and promotion of rights, standards, and social dialogue. It also wanted to promote an employment-based rural development model through diversified livelihoods, sustainable enterprises, and better integration in value chains. Provision of social protection floors was also another core objective of the Decent work Agenda as it guaranteed minimum income and access to basic services in rural economies that were often vulnerable to external shocks.

The vulnerabilities of rural communities have been assessed by the ILO as being driven by factors like, lack of access to services or markets, the global financial crisis, and fragile settings characterized by food insecurity and environmental stress (Schwarz et al. 2011). These factors are often worsened by unsustainable livelihood approaches (Cox and Hamlen 2015).

Dealing with extreme vulnerabilities, such as poverty, has been on the global agenda through Sustainable Development Goal 3. The sustainable development goal is aimed at eradicating extreme poverty that is largely a result of environmental stress and various vulnerabilities in most rural and urban areas. The literature reveals the gap where the settlement-based rural development approach is left out. Most settlement-based planning approaches are inclined toward urban settlement planning, unlike the rural counterparts.

Results and Analysis

Rural housing is one of the major components that affect rural resilience. From the study it was noted that rural housing in Zimbabwe has insecurity of tenure and the houses are being constructed using poor-quality building materials. This section looks at the current situation in Zimbabwe with regard to rural settlement planning as well as rural housing construction and management.

It has emerged from the study that most rural houses have not been built to standard to allow them to withstand wind and related pressures. Most rural houses are sited haphazardly and in environmentally sensitive areas. Building on environmentally sensitive areas has become rampant to the extent that wetland invasion has become a common practice. In rural areas, settling in grazing lands has become common. Although, the fast-track land resettlement program (FTLRP) was conducted in the spirit of decongesting the communal rural areas, fresh problems have been observed in both the sending and receiving areas. Some vacated areas as a result of fast-track land resettlement program (FTLRP) have become so "deserted" that the vegetation has brought about new problems including threats to wildlife. In the areas where people were settled to (receiving areas), infrastructure deficits accompanied by threshold capacities are rampant. Apart from that, in A1 farms, that are usually on average six hectares (6 ha) in size, for example, the increasing household formations pose serious threats to homesteads' creation. The farm, in no time, becomes littered with so many built structures compromising space for farming. The farming area is overcrowding and requires attention to reduce land degradation.

Rural settlement planning in Zimbabwe is in a piecemeal state. This means that there is no orthodox way of developing plans or rural expansion, rather than a series of incremental land allocations (Dent et al. 2013). Inadequate resources, both financial and technical, have attributed to poor resettlement planning. Currently, the local chiefs and local headmen do most rural planning. These people allocate land to the villagers to construct their homes, based on tradition and the availability of developable land. Rural communities grow incrementally. The absence of concrete planning framework for rural settlements has left some rural dwellers be located in dangerous locations, like waterways and mountain edges, among other dangerous sites for human settlements.

Lack of planning in Zimbabwe can also be understood by looking at the existing tenure systems in managing rural settlements in Zimbabwe. Rural land in Zimbabwe is largely classified into seven main tenure systems that include large-scale commercial farming land, small-scale commercial farming land, communal areas, resettlement areas, government estates, forest areas, and wildlife areas (Wekwete 1989). The most common tenure system in most of Zimbabwe's rural areas is the communal tenure (86% of rural dwellers). The land, that is under the communal land, is governed by the Communal Lands Act (Chapter 20:04). This act indicates that the only kind of land rights that the occupier has are "use rights" and cannot be sold, but it can be hereditarily transferred. Since planning is affected by legislation, it indicates that most rural spaces in Zimbabwe are just land banks that are waiting for the arrival of urban areas and be converted into council land or state land for urban development. This can be used to explain why rural settlement planning is still in Zimbabwe.

Rural Housing Construction in Zimbabwe

Rural housing construction in Zimbabwe is normally done by the owners (Zami 2015). The housing construction process is done without any proper design of the house; the builder just uses his/her experience from the past housing projects done. The housing is also constructed based on the resources of the owner. As such, in rural Zimbabwe, there exists a vast range of housing types constructed using material ranging from pole and mud to modern houses (Dube et al. 2018).

The diverse types of rural houses in Zimbabwe manifest the absence of any form of regulations in housing construction. This places all the poor rural dwellers at most vulnerable position in case of disaster outbreak.

Under the central government framework, there is no single or leading coordinating agency for rural development and planning. There is a Cabinet Committee on Rural Development, but it has a very broad remit and seems to act to sort out problems between potentially competing ministries (Dube et al. 2018). By the nature of their remits, the Ministry of Local Government Public Works and National Housing (MLGPNH) and the Ministry of Rural Resources and Water Development (MRRWD) have overarching roles. The MLGPNH coordinates and guides local government and supports capacity building for the rural district councils (RDCs).

The Ministry of Local Government Public Works and National Housing (MLGPNH) houses the Department of Physical Planning (DPP). The Ministry of Rural Resources and Water Development (MRRWD) was created in 1997 and has important implementation powers in rural areas through the District Development Fund (DDF) (Mashizha and Mapuva 2018). The Rural District Councils play a pivotal role in Rural Housing and Legislation in Zimbabwe. These are supposed to manage rural development and constructions. However, there is no harmony among the acts that regulates rural development. For example, Roads and Traffic Act (13:11) provides a guideline on how rural dwellers can develop their dwellings from major roads, but the stipulations of building codes for rural dwellings are silent.

A Traditional Leaders Act (TLA) was passed in 1998 that sought to make the old Ward and Village Development Committees (WADCOs and VIDCOs) elected committees of new structures – Ward Village Assemblies. The functions of the VIDCO remained as described in the Rural District Councils Act, and those of the WADCO, previously undefined, were set out in the new LTA. However, the provisions of the LTA had not been implemented by November 2000 due to a lack of resources. The traditional leaders continue to operate in the context of a village court without much influence in rural housing planning and the rural settlement development. The traditional leaders also allocate land without any orthodox guideline but based on the opinion or request of the leader and the villager, respectively.

Having looked at the housing design, construction, and settlement planning and tenure systems in rural Zimbabwe, the question is how resilient are rural housing communities in Zimbabwe (Siwawa 2018)? A lack of rural planning systems that are efficient has reduced rural resilience in Zimbabwean settlements. A recent example is the case of Cyclone Idai. The cyclone hit Zimbabwe's Masvingo and Manicaland Province leaving most rural dwellers vulnerable.

Evidence from the images shows some huts that were destroyed in Manicaland and Masvingo Provinces of Zimbabwe. Most of these rural housing structures are made of pole, bricks, and mud that are not durable in times of disaster occurrence. As depicted from the images, rural dwellings are vulnerable to disasters due to poor structures that cannot succumb the ravages of Cyclone Idai.

Discussion

There is no clear, coordinated national rural settlement planning policy for Zimbabwe. While there are several different policy strands that can be said to influence rural development, these have not resulted in an overall integrated and holistic rural development strategy. This has resulted in a sector-led approach to rural development that has led to gaps and overlaps in activities. The lack of a coordinated rural development strategy has also led to an ad hoc gender policy in rural areas. The lack of a long-term vision for Zimbabwe led to the launch of the Vision 2020 process in 1996. Following a number of consultative exercises, a broad national vision was agreed. This placed a heavy emphasis on industrializing the economy of Zimbabwe along the route of endogenous development. But the vision has still not been realized albeit its far-reaching implications for a long-term development investment in the country's rural areas. Admittedly, the current planning approaches to rural development are disjointed. As a case in point, the existing planning legislation is silent on the design of rural settlements except for the services in rural areas as provided for in the Town and Regional Planning Act of 1996. The present situation implies that rural housing development is largely determined by one's own income. Given the dominant tenure systems in the communal lands of Zimbabwe, it is unprofitable for rich people to develop houses on land with no entitlement to ownership. Thus, most housing construction materials in rural areas are not durable and they are easily destroyed in case of disasters occurrence.

Conclusions and Policy Options

In conclusion, clearly the planning for resilient rural settlements in Zimbabwe is still at a nascent stage if not poorly understood. The existing planning legislation does guide proper housing development in rural communities saving for the controlled rural service centers where standardized developments have long taken shape. The anti-urban policy bias practices that tradition in most independent Africa seem to have conspired against the charting of planning legislation for housing and settlement planning in rural Zimbabwe. The policy bias continues to expose rural settlement planning to unforeseeable weather calamities as evidenced by the widespread damage that followed Cyclone Idai in recent times. The massive loss and damage of the ill-sited social infrastructure and housing in the affected areas of Chipinge, Chimanimani, and parts of Masvingo Province serve as wake-up call for the introduction and the need to comply with rural housing construction and related infrastructure planning standards. Compliance with these planning standards and land-use regulations will help with guaranteeing the sustainable layout, amenity, and safety regarding the place-specific characteristics of the different rural settlements over time and space. By extension, these planning standards will determine the kind of designs and materials appropriate for region-wide housing construction programs.

Cross-References

- ▶ Financing: Fiscal Tools to Enhance Regional Sustainable Development
- ▶ Improving Social Equity and Community Health and Well-Being in Low-Income Suburbs and Regions
- ▶ Local and Regional Development Strategy
- ▶ New Localism: New Regionalism

References

AFDB (African Development Bank Group). (2011). Climate change action plan 2011–2015. Available online: https://www.afdb.org/fileadmin/uploads/afdb/Documents/Policy-Documents/ClimateChangeActionPlan%28CCAP%292011-2015.pdf. Accessed on 3 December 2020.

Asian Development Bank. (2014). Climate change and rural communities in the greater mekong subregion a framework for assessing vulnerability and adaptation options CEP knowledge series # 1. Available online: https://www.adb.org/sites/default/files/publication/42089/climate-change-rural-communitiesgms.pdf. Accessed on 3 September 2020.

Aven, T., & Renn, O. (2009). On risk defined as an event where the outcome is uncertain. *Journal of Risk Research, 12*(1), 1–11.

Cai, W., Mccann, A., Patel J. K. (2019). Mozambique's cyclone: Mapping the destruction of Idai. *New York Times*. Available online: https://www.nytimes.com/interactive/2019/03/19/world/africa/mozambique-cyclone-idai-maps.html

Cobbett, W. (2016). *Rural rides*. Read Books Ltd..

Cox, R. S., & Hamlen, M. (2015). Community disaster resilience and the rural resilience index. *American Behavioral Scientist, 59*(2), 220–237.

Crea, T. M., Lombe, M., Robertson, L. A., Dumba, L., Mushati, P., Makoni, J. C., Mavise, G., Eaton, J. W., Munatsi, B., Nyamukapa, C. A., & Gregson, S. (2013). Asset ownership among households caring for orphans and vulnerable children in rural Zimbabwe: The influence of ownership on children's health and social vulnerabilities. *Aids Care, 25*(1), 126–132.

Dent, D., Dubois, O., & Dalal-Clayton, B. (2013). *Rural planning in developing countries: Supporting natural resource management and sustainable livelihoods*. Routledge.

Z

Devi, S. (2019). Cyclone Idai: 1 month later, devastation persists. *The Lancet, 393*(10181), 1585.

Dube, E., Mtapuri, O., & Matunhu, J. (2018). Flooding and poverty: Two interrelated social problems impacting rural development in Tsholotsho district of Matabeleland North province in Zimbabwe. *Jàmbá: Journal of Disaster Risk Studies, 10*(1), 1–7.

Flora, C. B. (2018). *Rural communities: Legacy+ change*. Routledge.

Franklin, A., Newton, J., & McEntee, J. C. (2011). Moving beyond the alternative: Sustainable communities, rural resilience and the mainstreaming of local food. *Local Environment, 16*(8), 771–788.

Glover, J. (2012). Rural resilience through continued learning and innovation. *Local Economy, 27*(4), 355–372.

Greco, E. (2016). Village land politics and the legacy of Ujamaa. *Review of African Political Economy, 43* (sup1), 22–40.

Hoddinott, J. (2006). Shocks and their consequences across and within households in rural Zimbabwe. *The Journal of Development Studies, 42*(2), 301–321.

Huchzermeyer, M., & Karam, A. (Eds.). (2006). *Informal settlements: A perpetual challenge?* Cite the city of publication: Juta and Company Ltd..

Hunter, L. M. (2007). *Climate change, rural vulnerabilities, and migration*. Population Reference Bureau.

Jordan, S. A., & Hargrove, D. S. (1987). Implications of an empirical application of categorical definitions of rural. *Journal of Rural Community Psychology, 8*(2), 14–29.

Kay, R., & Alder, J. (2017). *Coastal planning and management*. CRC Press.

Makumbe, J. M. (1996). *Participatory development: The case of Zimbabwe*. University of Zimbabwe.

Marsden, T. (2009). Mobilities, vulnerabilities and sustainabilities: Exploring pathways from denial to sustainable rural development. *Sociologia Ruralis, 49*(2), 113–131.

Mashizha, T.M. and Mapuva, J., 2018. The colonial legislation, current state of rural areas in Zimbabwe and remedial measures taken to promote rural development. Journal of Asian and African Social Science and Humanities. (ISSN 2413-2748), 4(3). 22-35.

Mayer, M. L., Slifkin, R. T., & Skinner, A. C. (2005). The effects of rural residence and other social vulnerabilities on subjective measures of unmet need. *Medical Care Research and Review, 62*(5), 617–628.

McManus, P., Walmsley, J., Argent, N., Baum, S., Bourke, L., Martin, J., Pritchard, B., & Sorensen, T. (2012). Rural community and rural resilience: What is important to farmers in keeping their country towns alive? *Journal of Rural Studies, 28*(1), 20–29.

Muller, E., Diab, R. D., Binedell, M., & Hounsome, R. (2003). Health risk assessment of kerosene usage in an informal settlement in Durban, South Africa. *Atmospheric Environment, 37*(15), 2015–2022.

Murata, C., Mantel, S., De Wet, C., & Palmer, A. R. (2019). Lay knowledge of ecosystem Services in Rural Eastern Cape Province, South Africa: Implications for intervention program planning. *Water Economics and Policy (WEP), 5*(02), 1–29.

Oldewage-Theron, W. H., Dicks, E. G., & Napier, C. E. (2006). Poverty, household food insecurity and nutrition: Coping strategies in an informal settlement in the Vaal triangle, South Africa. *Public Health, 120*(9), 795–804.

Payo Garcia, A., Wood, B., Kessler, H., Lee, J., Burke, H., Barkwith, A., Pennington, C., Hobbs, P., Rees, J., Ellis, M. and Gatliff, R., 2017. Trimingham Erosion project: Can next decades coastal erosion rates be more accurately assessed.

Pelling, M., & Wisner, B. (2012). *Disaster risk reduction: Cases from urban Africa*. Routledge.

Rodgers, L., Fletcher, G., & Bryson, M. (2019). Cyclone Idai: How the storm tore into southern Africa, *BBC*. March 22, 2019. Available online: https://www.bbc.com/news/world-africa-47638696.

Schwarz, A. M., Béné, C., Bennett, G., Boso, D., Hilly, Z., Paul, C., Posala, R., Sibiti, S., & Andrew, N. (2011). Vulnerability and resilience of remote rural communities to shocks and global changes: Empirical analysis from Solomon Islands. *Global Environmental Change, 21*(3), 1128–1140.

Scott, M. (2013). Resilience: A conceptual lens for rural studies. *Geography Compass, 7*(9), 597–610.

Siwawa, V. (2018). The implementation of site and service schemes in a depressed economy. The case study of Beitbridge, Zimbabwe (Doctoral dissertation).

Sørensen, E., & Torfing, J. (2012). Introduction: Collaborative innovation in the public sector. *The Innovation Journal, 17*(1), 1–14.

Stoneman, C. (2017). *Land reform in Zimbabwe: Constraints and prospects*. Routledge.

Thomas, R. G., & Griffiths, J. S. (2017). An overview of the main landslide field sites in the 9th ICFL. In *Landslides* (pp. 3–13). Routledge.

Wekwete, K. H. (1989). Physical planning in Zimbabwe: A review of the legislative, administrative and operational framework. *Third World Planning Review, 11*(1), 49.

Zami, M. S. (2015). Drivers and their relationship with inhibitors influencing the adoption of stabilized earth construction to alleviate urban housing crisis in Zimbabwe. *Key Engineering Materials, 632*, 119–144.

9783030877446VOL03